中国探月工程科学成果（第二辑）
航天科技图书出版基金资助出版

嫦娥四号科学研究创新成果

（上册）

刘继忠　严　俊
邹永廖　唐玉华　等　编著

中国宇航出版社
·北京·

图书在版编目（ＣＩＰ）数据

嫦娥四号科学研究创新成果 : 汉文、英文 / 刘继忠等编著. -- 北京 : 中国宇航出版社, 2024.7
ISBN 978-7-5159-2185-3

Ⅰ. ①嫦… Ⅱ. ①刘… Ⅲ. ①月球探测器－科学研究－中国－汉、英 Ⅳ. ①V476.3

中国国家版本馆CIP数据核字(2023)第022637号

责任编辑　张丹丹　　　**封面设计**　宁星文化

出版发行　中国宇航出版社

社　址	北京市阜成路8号　**邮　编**　100830 (010) 68768548	**版　次**	2024年7月第1版 2024年7月第1次印刷
网　址	www.caphbook.com	**规　格**	787×1092
经　销	新华书店	**开　本**	1/16
发行部	(010) 68767386　(010) 68371900 (010) 68767382　(010) 88100613（传真）	**印　张**	167.75
		字　数	4076千字
零售店	读者服务部　(010) 68371105	**书　号**	ISBN 978-7-5159-2185-3
承　印	北京厚诚则铭印刷科技有限公司	**定　价**	898.00元（全三册）

本书如有印装质量问题，可与发行部联系调换

编审委员会

编审办公室

前　　言

月球承载了中华民族许多美好的传说和向往。实施探月工程是党中央高瞻远瞩，着眼现代化建设全局，根据世界科技发展大势，为推动航天事业发展、促进科技进步和创新、提高综合国力做出的重大战略决策。2004 年，我国启动了探月工程，并制定了“绕、落、回”三步走的发展路线。2007 年 10 月，嫦娥一号任务成功实施，实现了探测器环绕月球飞行，标志着第一步“绕”的成功。2010 年 10 月，嫦娥二号任务成功实施，获得了 7m 分辨率全球影像图，并对小行星 Toutatis 进行了飞越探测。2013 年 12 月，嫦娥三号任务实现我国第一次着陆月球，开展“巡天、观地、测月”探测，圆满完成第二步“落”的战略目标。

2018 年 12 月 8 日，嫦娥四号任务在西昌卫星发射中心，由长征三号乙运载火箭发射升空，进入奔月轨道，于 2019 年 1 月 3 日成功着陆在月球背面南极-艾特肯盆地（SPA）预选着陆区并正常在轨运行至今。嫦娥四号实现了人类首次月球背面着陆，树立了人类探月史新的里程碑。月球南极-艾特肯盆地是月球上最大、最深、最古老的撞击坑，具有重要的科学研究价值。嫦娥四号探测器通过就位和巡视探测，获取了大量月球深部物质、月表浅层结构、月球空间环境及月面天文观测等数据。通过对数据的详细分析，科学家们取得了一系列创新成果，并在国内外权威和重要期刊上发表。

为系统总结嫦娥四号科学成果，方便资料保存和后续相关科研人员查阅参考，探月与航天工程中心、中国科学院月球与深空探测总体部、深空探测实验室联合组织编写了本书。本书由刘继忠、严俊、邹永廖、唐玉华担任编委会主任并对全书进行审定。参与编写的作者都是工作在探月工程和行星科学研究一线的人员。全书共 6 章，第 1 章介绍嫦娥四号任务概况，主要包括嫦娥四号的任务构成、工程和科学目标、有效载荷配置、数据发布规则等。第 2 章介绍嫦娥四号任务有效载荷指标及数据分析方法，包括着陆区概况、有效载荷功能和技术指标、科学数据分析方法等。第 3 章介绍月球背面着陆巡视区的深部物质成分研究方面的科学成果。第 4 章介绍在月球背面着陆巡视区形貌构造、区域地质演化历史等方面取得的科学成果。第 5 章介绍月球背面环境特征等方面的科学成果。第 6 章介绍月球存在的主要科学问题及未来展望。

全书收集了 2019 年成功降落月球背面以来至 2023 年正式刊出的主要科学论文，共 130 篇，在此特别感谢被收录文章作者的支持。

目　录

① 为了保持与原刊载媒体一致，本书沿用了每篇论文发表时对该词的表达方式，全书未做统一，far - side 等词同理。

第 1 章　嫦娥四号任务概述

1.1　工程背景实施和组成

2013 年 12 月，嫦娥三号成功降落月球正面虹湾地区，实现了我国在月球表面就位和巡视探测的目标，探月工程“绕、落、回”三步走规划的第二步“落”圆满完成。嫦娥四号作为嫦娥三号任务的备份星，担负着新的责任和使命。国家航天局组建了论证专家组，深入研究了国内外月球和深空探测发展趋势，聚焦月球探测科学价值，以“效益最优、承上启下、创新引领、体现开放”的原则，在嫦娥四号探测器框架基本不变的前提下，深入论证任务新的科学目标和工程可实现性，最终确定去人类从未到达过的月球背面，开展人类历史上首次月球背面的软着陆和巡视探测。

月球背面探测器无法直接对地球进行通信，对工程提出了新的需求和挑战。为顺利实现在月球背面的着陆和巡视探测，工程增加了一颗运行于地月 L2 点的“鹊桥”中继卫星。为此，嫦娥四号工程实施了两次发射任务，分别为中继星任务和探测器任务。2018 年 5 月，长征四号丙运载火箭将“鹊桥”中继星成功发射，最终进入地月 L2 点使命轨道并稳定运行；2018 年 12 月，采用长征三号乙运载火箭发射探测器，在地面测控支持下，完成地月转移和环月飞行，自主完成动力下降和软着陆，通过中继测控通信，开展月球背面就位和巡视探测。2019 年 1 月 11 日下午，嫦娥四号着陆器与玉兔二号巡视器工作正常，在“鹊桥”中继星支持下顺利完成两器互拍，地面接收图像清晰完好，中外科学有效载荷工作正常，探测数据有效下传，搭载的科学实验项目顺利开展，达到工程既定目标，这些标志着嫦娥四号任务圆满成功。截至 2023 年 1 月底，着陆器和玉兔二号已在月球背面崎岖的地形地貌上连续工作逾 51 个月昼，巡视器累计行驶近 1500m，成为在月球工作时间最长的探测器。

嫦娥四号任务有工程总体，以及探测器（含中继星）、运载火箭、发射场、测控、地面应用等五大系统，如图 1－1 所示。五大系统在工程总体组织下完成工程任务。

(1) 工程总体

工程总体是工程的组织实施单位，是工程总指挥和总设计师系统，即两总系统的支撑机构和具体办事机构。工程两总主要负责工程实施的总体方案、工程质量、重大技术问题和总体计划进度的研究、协调和决策，解决工程实施中的关键问题，保证研制建设工作的顺利开展。探月与航天工程中心负责探月工程的组织管理和总体技术工作，开展规划论证、总体设计、系统协调、组织实施、科学研究、国际合作等，通过总体系统设计，保证工程性能最优，计划、经费、质量最优，实现对系统风险的有效管控。

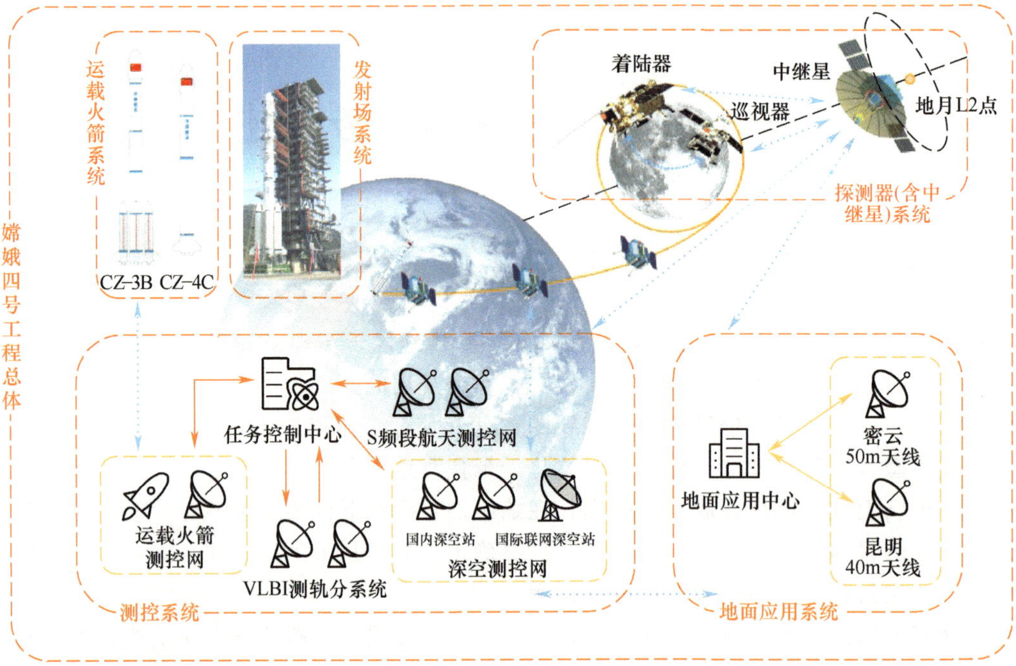

图 1－1　嫦娥四号任务工程组成

(2) 探测器系统

探测器系统包括嫦娥四号探测器和鹊桥号中继星。

嫦娥四号探测器包含着陆器和巡视器两个独立运行平台。探测器发射后，与运载火箭分离，进入地月转移轨道，在地面测控支持下，经中途修正，在近月点实施近月制动，实现月球捕获。在近月点实施软着陆制动，探测器完成主减速段、调姿下降段、悬停段、缓速下降段、缓冲着陆段等降落过程，在月球背面实现软着陆。着陆月面后，着陆器和巡视器分离，在中继星支持下，两器携带的有效载荷开展科学探测，在月夜期间通过同位素电源（RTG）的支持，开展月夜温度测量。

着陆器由 11 个分系统及工程参数测量设备组成，有效载荷分系统配置有地形地貌相机、降落相机、低频射电频谱仪和月表中子与辐射剂量探测仪（LND），并搭载生物科普试验载荷。巡视器由 8 个分系统组成，有效载荷分系统配置有全景相机、测月雷达、红外成像光谱仪、中性原子探测仪（ASAN）。

“鹊桥”中继星选用成熟的 CAST100 卫星平台研制。运载火箭将中继星送入地月转移轨道后，在地面测控支持下，经中途修正，在近月点实施近月制动和月球借力，进入环绕地月 L2 点的 Halo 使命轨道。在探测器任务实施发射和着陆月球背面后，开展中继通信，并择机开展科学探测。有效载荷配置中-荷低频射电探测仪（NCLE）。

(3) 运载火箭系统

中继星任务采用 CZ－4C 运载火箭发射，将中继星直接发射至近地点高度 200km、远地点高度约 400000km 的地月转移轨道。探测器任务采用 CZ－3B 运载火箭，将探测器直接发射至近地点高度 200km、远地点高度约 420000km 的地月转移轨道。

（4）发射场系统

选用西昌卫星发射中心发射，发射场系统具备探测器测试、加注、塔上环境和无线转发等保障条件；具备运载火箭测试、转运、吊装、对接、加注、发射等保障条件；可为发射任务提供通信、气象、计量、安控等保障。

（5）测控系统

测控系统承担运载火箭、探测器、中继星在各飞行阶段探测器的测控、轨道测量与确定，以及落月后通过中继星对着陆器和月面巡视器的遥操作控制。主要由运载火箭测控网、S 频段航天测控网、深空测控网、甚长基线干涉（VLBI）测轨 4 个分系统组成。

（6）地面应用系统

地面应用系统负责根据科学探测任务，制定科学探测计划和有效载荷运行计划，监视着陆器和巡视器有效载荷的运行状态，编制有效载荷控制指令和注入数据，完成有效载荷运行管理；使用北京密云 50m 接收天线和云南昆明 40m 数据接收天线，两个地面站并行工作，同时接收着陆器和巡视器的下行探测数据，进行本地存储和备份；对科学探测数据进行处理、解译，组织开展科学应用研究；管理科学探测数据并按规定分发数据产品。

1.2 工程和科学目标

（1）工程目标

1）研制月球中继通信卫星，突破高增益天线研制、地月 L2 点中继卫星轨道设计、轨道维持及天线高精度指向等技术，实现国际首次地月拉格朗日 L2 点的测控通信及中继通信。

2）研制月球着陆器和巡视器，突破月球背面软着陆、同位素核源研制及试验、低频射电频谱探测等技术，实现国际首次在月球背面软着陆和巡视探测。

（2）科学目标

1）月基低频射电天文观测与研究；

2）月球背面巡视区形貌和矿物组分探测与研究；

3）月球背面巡视区浅层结构探测与研究；

4）试验性开展月面中子辐射剂量、中性原子等月球环境探测研究。

1.3 着陆区地质背景

嫦娥四号月球探测器于 2019 年 1 月 3 日在月球背面的冯·卡门撞击坑内成功着陆。

通过利用月球侦察轨道器（LRO）影像底图、着陆器降落相机和着陆器监视相机影像数据，并借助影像匹配和视觉测量定位技术，确定了着陆点的精确坐标为177.588°E，45.457°S。另外，利用CE-4着陆相机、导航相机和CE-2地形数据重建了CE-4的动力下降轨迹，得到着陆点的精确位置为177.5991°E，45.4446°S，海拔为-5935m。两组坐标的差异反映了LRO数据和CE-2数据之间的偏差。

冯·卡门撞击坑直径约186km，深度约3km，位于月球上最大、最古老的南极艾特肯盆地（South Pole-Aitken，SPA）内部，SPA的直径约2600km，深度约13km。前人研究推测SPA形成过程中挖掘出了大量的月球深部物质，溅射物的原始深度可能在下月壳甚至上月幔。玉兔二号月球车的巡视探测对认识月球深部物质以及月球背面的地质演化过程具有重要的科学意义。

嫦娥四号探测器着陆之前，遥感观测已较好地预示了着陆区的地质背景，为解译玉兔二号获取的科学数据提供了地质框架。冯·卡门撞击坑形成在约39.2亿年以前的月球前酒海纪，其坑底在漫长的岁月中经历了撞击溅射物、火山活动、构造运动等多重地质过程的改造；冯·卡门撞击坑的底部大部分区域被雨海纪的玄武岩物质填充，地形整体平坦，在玄武岩平原之上，覆盖着大量由较年轻的撞击坑形成的二次坑。其中，来自冯·卡门撞击坑东侧的芬森撞击坑（Finsen，直径72km，42.0°S，177.9°W）形成的二次坑对着陆点和玉兔二号巡视路线的地形影响最突出，是着陆区域地貌的重要特征；芬森撞击坑的二次坑显著退化，在地形地貌上呈现为发散式的线性凹陷和凸起。

对于着陆区地形地貌的研究，可用的观测数据有以往轨道器任务获得的遥感影像和数字高程模型（DEM），以及着陆后更高分辨率的就位探测数据。嫦娥四号着陆后，玉兔二号月球车的全景相机和导航相机具有获取高分辨率立体图像的能力，这些立体图像应用于巡视探测区域的三维地形重建和月球车定位路径规划，所产生的毫米-厘米级分辨率的DEM和正射影像图为着陆区形貌的精细分析提供了基础数据。

对着陆器着陆位置的详细遮挡分析显示，着陆器对太阳和中继卫星具有良好的能见度。着陆器位于倾向西南、倾角约4.5°的斜坡上，着陆地点周围岩石丰度几乎为0。

1.4 科学探测价值

(1) 月球背面地质特点

根据Clementine多波段光谱数据和月球勘探者号的伽马谱数据，人们根据地球化学和地质演化的差异，将月壳划分为三个独立演化的地体：高地斜长岩地体（FHT），风暴洋克里普岩地体（PKT）和南极艾特肯地体（SPA）（图1-2）。高地斜长岩地体是分布区域和体积最广泛的月球地体，超过月球表面的60%，最为显著的表面特征是具有高的反照率、分布广泛的撞击坑、较高的地形，且有较高的长石含量；风暴洋克里普岩地体以玄武岩为主，由两种反照率相差比较大的物质组成，即比较亮的非玄武岩质的高地岩石和比较暗的火山熔岩，特点是富集Th元素。

南极艾特肯地体是一个特殊的构造单元，直径约为2600km，是太阳系中最大的撞击盆地，形成于月球历史的早期，也是最古老的撞击坑，深度约13km，是一个天然的深度

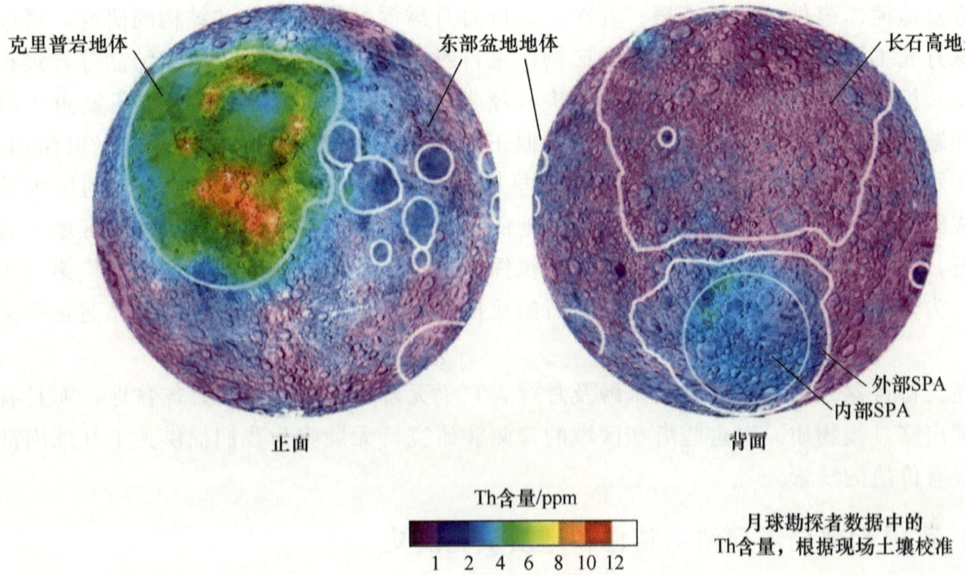

图 1-2　根据地球化学和地质演化的差异，将月壳划分为三个独立演化的地体

剖面，也许能揭示出月球深部（下月壳或上月幔）的物质组成与结构，从而对月球壳-幔的分异机制和过程给出最直接的制约。SPA 有比较独特的地质地球化学和演化特征，相对于典型的非月海玄武岩月壳具有较高的 FeO 丰度。与 PKT 相比，Th 含量较低，而 FeO 含量相近；与 FHT 相比，FeO 含量较低而 Th 含量相近。探测和研究 SPA，对于揭示月球最早期的撞击历史，特别是判定内太阳系在约 39 亿年前是否发生过密集轰击事件，具有决定性的作用，可以为月球的起源和演化提供新的证据。

SPA 这个月球上最大、最深和最古老的撞击盆地，其地形、反射率特征、物质成分与嫦娥三号任务的着陆区有较大差异，其成因是月球科学研究的重大难题之一。同时月球本身还遮挡屏蔽了来自地球的各类干扰，月球背面是一个难得的干净的射电探测环境，利用这个特点，可以在月球背面开展对太阳爆发以及太阳系行星的低频射电场特性的探测。另外，对月球背面着陆区上空的月球空间环境的探测和认知也具有开创性价值，具有重要的科学意义。

(2) 建立月背综合地质剖面

任务探测了嫦娥四号任务着陆区和巡视区的地形地貌、物质成分、浅层结构，在国际上首次建立了月球背面综合地质剖面。

月球物质组成和结构是行星科学最基本的问题之一，与月球起源、演化、岩浆岩结晶过程、月球圈层结构和月球发电机等重大科学问题密切相关。首先，目前普遍接受的月球成因假说是大碰撞理论，这个理论可以解释月球和地球的同源性，但大碰撞后的吸积时间、同位素异常等方面证据不足，特别需要月球深部物质组成和结构等新证据和约束。南极艾特肯盆地撞击数值模拟证明其挖掘深度可以到达月幔，但到底是来自下月壳还是上月幔，有待确定。艾特肯盆地形成末期在盆地聚集了大量的撞击熔融物，熔融物的厚度、结晶分异的程度、目前该位置的月壳厚度，以及与出露物质的关系等都是认知月球早期演化和艾特肯盆地撞击事件的重要证据。其次，通过对月球深部物质成分和结构的研究，可以获取月球月壳不同深度乃至深层月幔的矿物信息，对于细致研究月球岩浆洋结晶分异过程和月幔反转机制等均具有重要价值，是解决月球岩浆洋假说和月球壳-幔分异机制重大科学争议问题的关键。艾特肯盆地没有经历类似正面风暴洋区域那样巨大的月海充填作用，这可能暗示了月球内部成分的不均一性或者早期月壳结构和性质（如热状态）的巨大差异，其成因一直是一个未知领域。艾特肯盆地的盆底代表了月壳平均厚度最薄的区域，有大量辉石，其岩石组成未被探测过，着陆巡视探测将为该地区的物质成分研究提供第一手的资料，为全方位立体化研究月球深部岩石组成提供重要契机，提供月球岩浆洋演化等关键线索。

即使艾特肯盆地的挖掘深度尚未触及月幔，它仍无疑是挖掘最深的月球盆地，而且有可能局部出露月幔物质。对这些潜在区域的探测和研究，无疑将为我们提供关于月球内部属性的最有价值的线索。

（3）可揭示部分浅层结构特性和地质演化历史

月球壳-幔的形成也是月球关键科学问题之一，包括其形成的机制、过程和时间等。现在的主流观点是月球岩浆洋假说，即地-月系统形成之后，月球很可能处在一个高温、全球或深度达近 1000km 的熔融状态。由于热辐射，月球温度开始降低，从岩浆洋依次结晶出橄榄石、辉石、长石等不同矿物，由于比重不同，重的橄榄石和辉石下沉，较轻的长石上浮，分异形成了月幔和月壳。较晚结晶并且位于上部的富铁质、富钛等矿物，由于其密度较大，会因重力不稳而发生翻转，这些较重的物质下沉，底部富镁物质被往上推，从而改变了月幔原始的层状分布。月球岩浆洋假说也有人质疑，而其结晶分异和翻转过程人们更不清楚，月球物质组成在深度方向上的分布信息为解决这一重大科学问题提供了重要证据。因此，对南极艾特肯盆地的探测，将首次获得月球深部物质组成的分布信息，直接验证月球岩浆洋假说，提出月球壳-幔分异的过程和机制。探测巡视区域下的浅层结构，特别是玄武岩层的厚度、喷发期次等信息，并结合影像和光谱数据，有望揭示艾特肯盆地内的岩浆活动历史。

（4）进行低干扰状态的宇宙射电研究

低频射电频谱仪获取的太阳低频射电和月表射电环境探测数据，在太阳风激波、日冕物质抛射和高能电子束的产生机理等方面取得了原创性的成果。

宇宙天体起源演化是自然科学中的基本问题之一，利用射电望远镜对来自宇宙的射电辐射进行探测是研究这一问题的重要手段。甚低频频段，作为最后几个未被实质探测过的射电天文频谱窗口之一，对其进行探测将有助于解决一些重大的科学问题，如宇宙的暗黑时代以及黎明时期，全天 21cm 辐射的分布、暗物质与暗能量、高能宇宙射线起源等。在低频射电波段对来自宇宙空间背景、银河系、太阳、行星的信号进行测量，可为探究这些问题做出独特的贡献。

在地球上由于受到地球电离层的干扰，难以开展频率低于 10MHz 的射电天文观测，甚至在地球轨道以及月球正面开展的空间射电天文观测也受到地球电离层反射和人工无线电的干扰。而月球背面屏蔽了来自地球的无线电干扰、闪电和极光无线电发射，月夜期间还屏蔽了强烈的太阳射电辐射，因此月球背面被认为是开展低频射电天文学观测的绝佳地点。嫦娥四号任务可以实现人类历史上首次月球背面的甚低频频段射电天文探测。

任务首次在月球背面和月球轨道上连续、系统地监测到类地行星射电爆发，并进行了空间电磁波环境辐射变化特性监测，包括对地球千米波辐射观测，获取月球边缘电离层分布特性，将有利于揭示地-月空间天气的规律特性，以及地-月空间等离子体空间环境特征。利用低频射电频谱仪观测 10MHz 以下的电磁波谱频带，可以揭示不同尺度的宇宙天体的起源与演化规律；给出目前 HF 波段空间观测能力条件下，空间 21cm 中性氢辐射及其对宇宙黑暗时代和宇宙黎明的约束；基于观测数据绘制天空背景低频射电巡天图像；开展太阳和行星射电辐射变化的监测，研究行星际日冕物质抛射（ICME）和行星际激波的动力学过程和结构演化。

低频射电观测研究可以通过两个途径实现：一是利用着陆器搭载的低频射电频谱仪，在月昼期间（特别是黎明或傍晚）探测来自太阳的爆发，从爆发的低频截止频率获取月球上空的等离子体频率，进一步获得其最大密度信息；二是利用中继卫星上搭载的中-荷低频射电探测仪，在类似掩星探测的几何构型情况下，从地球 AKR 爆发信息、太阳爆发信息中获得被地-月空间等离子体调制了的信息，提取地-月空间等离子体空间环境特征。

考虑到嫦娥四号任务的约束条件（如夜间不工作、探测器寿命、没有专门设计低频射电探测的电磁兼容等），上述科学探测与研究不一定能完全实现，最为可能的探测目标只能是太阳低频射电和月表射电环境探测。但即使如此，也将填补 100kHz～1MHz 射电天文观测的空白，在太阳风激波、日冕物质抛射和高能电子束的产生机理等方面取得原创性的成果。

(5) 首次开展能量中性原子探测

本任务是国际上首次在月表开展能量中性原子探测，有望在太阳风-月表相互作用的微观物理机制、月表逃逸层的形成和维持机制等方面取得突出的研究成果。

月球的大气非常稀薄，太阳辐射、太阳风和宇宙线均可直接到达月面，与月壤相互作用，形成特有的近月面空间环境。月表空间环境包括中性粒子、等离子体、粒子辐射、尘埃等多个要素，其形成涉及许多微观的物理过程，如太阳风粒子在月面的溅射/反射、高能粒子与月壤的作用等。目前我们对月表环境的认知十分有限，通过研究太阳风、宇宙线与月面相互作用的微观过程，既可以进一步认识月表辐射环境，还可以为未来月面科考站的建设乃至保障航天员登月的安全，提供观测依据。

中性原子探测仪可以测量太阳风与月面相互作用产生的中性原子，包括太阳风溅射的月壤物质以及被月面散射的太阳风物质。研究太阳风与月壤相互作用的机制，包括太阳风对月表物质的溅射，月面对入射太阳风的反射和散射；确定月表溅射物质的成分、能量分布，推演月壤物质与入射太阳风的作用过程；确定太阳风在月面的散射率、散射粒子的能量和空间分布；考察太阳风与月壤相互作用与太阳风速度、密度、温度的关系，不同地形地貌条件下溅射和散射情况的变化。

月表中子与辐射剂量探测仪可以测量月面高能粒子辐射剂量，包括银河宇宙线、太阳高能粒子事件，以及地球辐射带的粒子，同时，还可以测量银河宇宙线与月表相互作用产生的次级粒子。研究可以确定包括质子、电子、阿尔法粒子等的月面高能粒子能谱、高能粒子辐射剂量及其随时间的变化，考察太阳高能粒子事件期间辐射剂量的变化；研究高能粒子与月面相互作用的机制，及由此产生的次级粒子尤其是月面中子的剂量；研究如月表电磁场、次级粒子效应、地形地貌等月面条件因素对月表辐射环境的影响；分析评估月面辐射剂量效应。

利用月表中子与辐射剂量探测仪、中性原子探测仪数据综合分析，研究揭示太阳风、磁尾等离子体与月表相互作用的过程和特性，确定月表等离子体环境的形成机制及其与太阳活动、月面地质条件的关系，研究太阳风、磁尾等离子体与月表相互作用的过程和机制，获得月表溅射粒子的特征，研究月球稀薄大气成分的逃逸机制。

1.5 有效载荷配置

为实现上述科学目标，嫦娥四号任务着陆器、巡视器、中继星都配置了有效载荷，具体配置和主要科学探测任务见表 1-1。

表 1-1 嫦娥四号任务有效载荷配置及探测任务

科学目标	主要科学探测任务	有效载荷配置		
		着陆器	巡视器	中继星
月基低频射电天文观测与研究	对太阳低频射电特征进行探测； 试验性探测来自太阳系内天体和银河系的低频射电辐射	1. 低频射电频谱仪		1. 低频射电探测仪
月球背面巡视区形貌和矿物组分探测与研究	获取着陆区和巡视区的光学图像； 巡视区月表红外光谱分析和成像探测	2. 降落相机 3. 地形地貌相机	1. 全景相机 2. 红外成像光谱仪	
月球背面巡视区浅层结构探测与研究	月壤厚度与月壳次表层物质结构探测		3. 测月雷达	
月面中子辐射剂量、中性原子等月球环境探测研究	测量月表的综合粒子辐射剂量和 LET 谱； 观测巡视探测点的能量中性原子及正离子	4. 月表中子与辐射剂量探测仪	4. 中性原子探测仪	

嫦娥四号任务的部分有效载荷继承于嫦娥三号任务，包括降落相机、全景相机、地形地貌相机、测月雷达、红外成像光谱仪，Astronomy and Astrophysics 2014 Vol 14，No12 专刊（https：//www. raa - journal. org/issues/all/2014/v14n12/）登载了它们的定标和数据处理方法。

嫦娥四号任务着陆于月球背面，其光谱特性与月球正面有很多不同，载荷研制和科学研究团队不断改进红外成像光谱仪、测月雷达、低频射电频谱仪等的算法和模型，提升光谱反演精度。

1.6　科学数据产品和发布

嫦娥四号任务的科学数据包括着陆器、巡视器、中继星所搭载的降落相机、地形地貌相机、低频射电频谱仪、月表中子与辐射剂量探测仪、全景相机、红外成像光谱仪、测月雷达、中性原子探测仪以及低频射电探测仪等有效载荷探测数据及其相关的工程参数数据，以及有效载荷性能、指标及其地面和在轨定标等相关数据。嫦娥四号任务科学探测数据发回地面后，经预处理后的科学探测数据分为0、1、2三级。

0级：Level0A，经地面调解、帧同步、解扰、信道译码、分路解帧等处理后形成的探测仪器源包数据。Level0B，在0A级基础上经过多个地面接收站数据合并（适用时）、排序去重复、去除包结构、解压缩（适用时）后形成的探测仪器采集数据。

1级：Level1，在0B级数据产品的基础上进行温度、电压、电流等仪器参数的数值转换，并按探测周期重新组织的数据。

2级：Level2，在1级数据产品的基础上利用定标结果进行校正，并提供了几何信息的数据。

0级数据产品格式为二进制文件；1级和2级数据产品格式一般采用PDS标准格式。

按照《月球与深空探测工程科学数据管理办法》工程任务的要求，地面应用系统对嫦娥四号任务探测数据进行接收、处理、解译和管理，制定科学数据接收、处理、解译、数据格式等技术操作细则，确定公开发布数据的种类及对应的应用规范说明，对科学数据使用申请进行技术评议，按规定分发数据产品。在中国探月与深空探测网（www. clep. org. cn）建立和维护科学数据信息平台，按月发布科学数据动态信息。

国际合作载荷科学探测数据按照中外共同签署的国际载荷政府间谅解备忘录，由相关各方共同签署了技术合作协议，规定了科学探测数据共享、联合发布的相关条款。探月与航天工程中心在《月球与深空探测工程科学数据管理办法》框架下制定了国际合作载荷数据的发布细则。2018年6月14日，国家航天局与沙特阿拉伯在北京举行沙特载荷图像联合发布仪式。2019年4月18日，国家航天局在北京举办了中-德、中-荷与中-瑞合作载荷科学探测数据交接仪式，国家航天局向瑞典交接了中性原子探测仪数据，向德国交接了月表中子与辐射剂量探测仪数据，向荷兰交接了低频射电频谱仪数据。

1.7　科学研究的组织管理

“快出成果、多出成果、出好成果、出大成果”是工程要努力实现的科学成果目标。为实现这个目标，嫦娥四号任务采用了“科学研究核心团队”的组织管理模式，根据科学目标和有效载荷科学探测任务，成立了“探月工程嫦娥四号任务科学研究核心团队”。设立了“科学数据处理方法和探测区地质背景研究”“月球背面巡视区形貌、浅层结构和矿物组分研究”“月球低频射电环境研究”“月表中性原子及粒子辐射环境研究”4个研究团组，共28家单位、120余学者参加。工程总体专门制定了嫦娥四号任务科学数据对核心团队申请和发布的“绿色通道”，使核心科学家团队能够快速获得数据，开展数据分析和研究，尽早产出科学成果。同时，为更好地取得新发现、新成果，工程总体组织核心科学家团队代表和工程队伍，包括载荷工程师、地面应用系统和飞控组成员，在每月夜着陆器和巡视器休整期间，召开专题研讨会，对前一月昼获得的数据及遇到的问题以及后续科学探测需求进行讨论，以便按照科学家需求更好地制定探测计划，获得更优质的科学数据。核心科学家团队在工程领导、首席科学家组成的科学家委员会指导下，构成了科学研究、载荷研制、在轨探测、数据处理一体化协同工作团队，有序并高效开展了探测数据的研究工作，形成了实现我国月球与深空探测重大工程任务科学目标的组织模式。

“科学研究核心团队”模式弥补了探月工程重大专项各次任务对组织科学应用研究的不足，构建了科学工程的互动模式。团队成为嫦娥四号任务科学成果研究的主力军，也成为行星科学知识创新的主导力量、专业人才的成长基地。

核心团队组织架构如图 1-3 所示。

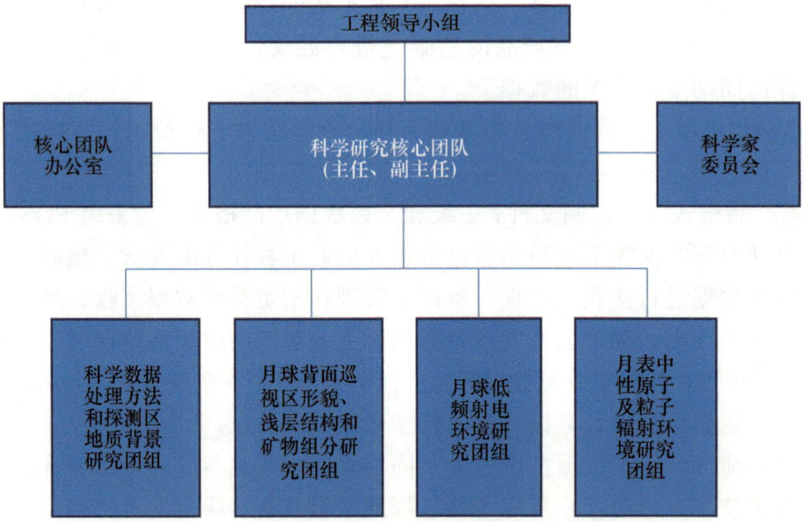

图 1-3　嫦娥四号任务科学研究核心团队构成

嫦娥四号任务月球科学与应用首席科学家担任核心团队主任，领导和组织科学研究，决策科学研究重大问题，对科学探测数据申请进行审批，审核和监督检查各团组的研究方案及其落实情况。团组组长对团组成员的研究分工、合作关系、竞争程度、研究目标、研究方向等进行把关，组织研究并部署本团组的研究目标、研究分工、研究任务、研究方案、研究计划，组织提出相关载荷科学探测规划建议。核心团队办公室负责核心团队日常工作的组织管理和支撑服务，负责制定和完善核心团队管理办法，召开工作会，协调相关问题；组织对各团组工作进行考核和评价；组织汇总统计研究成果并适度宣传。

第2章　嫦娥四号任务有效载荷指标及数据分析方法

2.1　嫦娥四号任务着陆区的选址

从大范围选址看，月球南极艾特肯盆地直径约2600km，深度达12km，是月球上乃至太阳系最大的撞击盆地，是提供不同深度物质组成分布的重要剖面，可以获得月球下月壳甚至上月幔的重要信息，且以往的探测任务都未曾到达，无论是从科学价值还是工程显示度方面都具有超出以往任务的重大意义，是嫦娥四号任务的选定探测区域。

具体着陆选址从科学价值和工程实现两个方面考虑。从科学价值方面考虑，拟选主着陆区为冯·卡门撞击坑（直径约190km，深度约5km，中央峰高度约1.55km），备选着陆区为克雷蒂安撞击坑（撞击坑直径约为140km，深度约为1.8km，没有中央峰）。两个着陆区都是月球南极艾特肯盆地的典型地貌类型，且高程很低，有利于开展月壳活动研究，这两个区域不仅在物质成分和地质年代上极具代表性，而且极可能含有月幔物质成分，尤其是Th、FeO、TiO_2等含量较高。同时这些区域物质组成的异常空间分布暗藏该地区的火山活动以及月壳活动线索。撞击坑表面覆盖着雨海纪物质，通过对相关物质成分和结构的深入探测，研究月球南极艾特肯盆地区域的物质来源和形成过程。

从工程实现方面考虑，南极艾特肯盆地以高地为主，遍布撞击坑，地形较为崎岖，无大面积平坦区域，地形可能对着陆区内探测器造成光照、测控遮挡等难题，对工程实现提出了约束。工程上采用实现性、可达性、安全性等多层次优选方法，在SPA内进一步优选着陆范围，分析中继星通信、热控适应能力、月面能源供给、航迹高程起伏变化、着陆区范围需求、动力下降测控弧段、着陆时太阳高度角约束、轨道可达性约束、具备推迟着陆需求等工程可实现影响因素，精选后确定的主着陆区位于冯·卡门撞击坑东南部，覆盖范围是：177.6±1.2°E，45.5±0.5°S，范围约为50km×30km；备选着陆区位于克雷蒂安环形坑内，覆盖范围是：163.1±1.2°E，46.1±0.5°S，如图2-1所示。

2.2　有效载荷功能和技术指标

着陆器有效载荷部分沿用嫦娥三号任务探测器载荷，并在此基础上根据科学目标及平台约束新增相应有效载荷，3个探测平台共配置9台有效载荷，其中有3台来自国际合作。具体配置为：着陆器配置降落相机、地形地貌相机、低频射电频谱仪、月表中子与辐射剂量探测仪4台有效载荷；巡视器配置全景相机、测月雷达、红外成像光谱仪、中性原子探测仪4台有效载荷；中继星配置中-荷低频射电探测仪1台有效载荷。

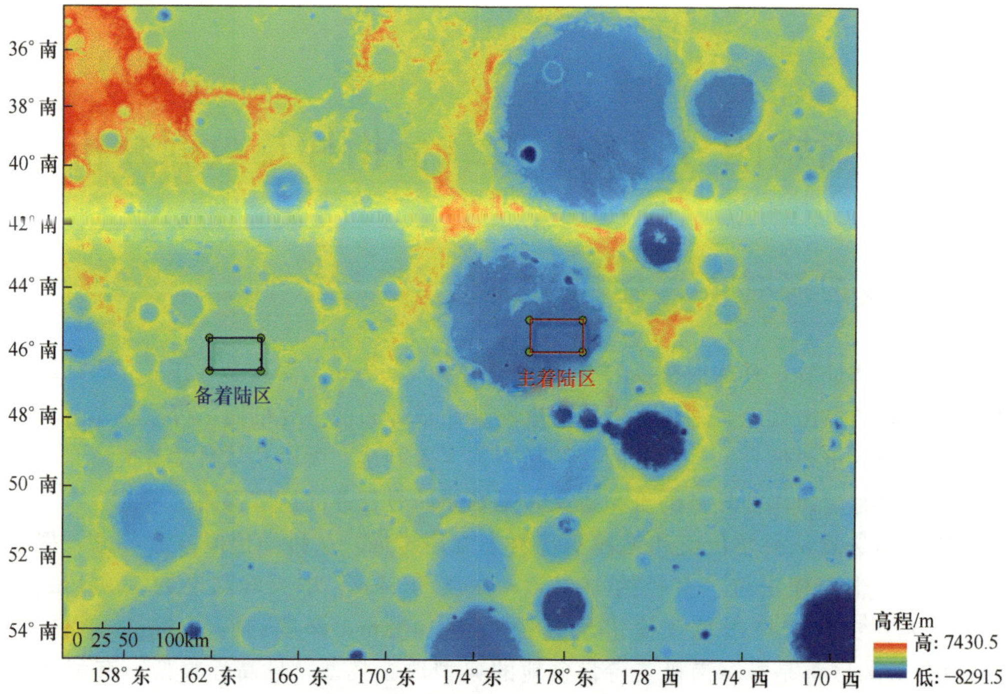

图 2-1　嫦娥四号预选着陆区位置分布示意图

2.2.1　降落相机

降落相机（图 2-2）安装在着陆器底部，在着陆器动力下降过程中开机，在2km～4m 各个高度获取降落区域的月貌特征图和地形地貌图，用于分析着陆区月表的地形地貌和区域地质情况。

对降落相机从不同高度获得的月表图像覆盖范围以及相应的空间分辨率进行分析，结果表明，在 2km～4m 高度范围内，相机可获取降落区域内成像幅宽为 3.31～1656.85m，空间分辨率在 3.23～1618.02mm 的图像。

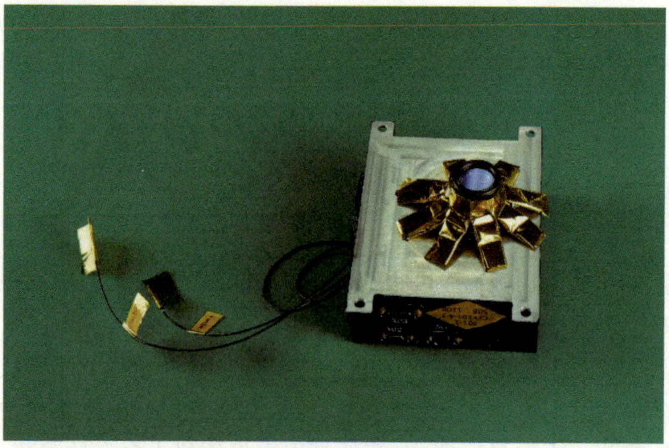

图 2-2　降落相机

降落相机的主要性能指标见表 2 - 1。

表 2 - 1　降落相机的主要性能指标

序号	技术指标	参数
1	波段范围	可见光
2	正常成像距离	4m 至无穷
3	有效像元数量	≥1024×1024
4	视场角/(°)	45×45（偏差不超过 5%）
5	系统静态传函	≥0.20（全视场）
6	信噪比/dB	≥40（最大信噪比） ≥30（反照率 0.09，太阳高度角 30°）

为了验证降落相机的性能，研制团队在地面进行了辐射定标和几何定标。辐射定标的试验目的是确定降落相机的辐射响应特性，测试相机自动曝光能力及其动态范围。试验结果表明，在试验辐亮度等级范围内，CMOS 响应线性度较好，相机在不同曝光情况下相对定标校正后图像质量较好。几何定标的试验目的是确定相机的内方位元素（即主点、主距）及畸变性能，试验结果表明，内方位元素和整个相面内畸变分布满足要求。

2.2.2　地形地貌相机

地形地貌相机（图 2 - 3）安装在着陆器顶部的相机支架上，成像距离约为 5m～∞，相机支架具有俯仰转动和方位转动功能，通过控制相机支架的二维转动，地形地貌相机可以获取着陆点光学图像，用于月表地形地貌的调查研究。

通过分析和计算，地形地貌相机在物距为 5～100m 时图像的分辨率为 0.85～16.99mm，幅宽为 2.00～39.97m。

图 2 - 3　地形地貌相机

地形地貌相机的主要性能指标见表 2－2。

表 2－2　地形地貌相机的主要性能指标

序号	名　　称	主要参数和性能
1	波段范围	可见光
2	颜色	彩色（R，G，B）
3	正常成像距离	5m 至无穷
4	有效像元数量	≥2352×1728
5	视场角/(°)	22.9×16.9（偏差不超过 5%）
6	信噪比/dB	≥40（最大信噪比） ≥30（反照率 0.09，太阳高度角 30°）
7	系统静态传函	≥0.20（全视场，静态成像时测）

为验证相机科学探测内容，安排了 4 项定标试验项目，包括平场定标、绝对辐射定标、内方位元素定标、彩色定标。试验结果表明，相机满足正样产品任务书与定标试验大纲要求。

2.2.3　全景相机

全景相机（图 2－4）安装在巡视器顶部的相机指向机构上，相机的正常成像距离为 3m 至无穷，相机指向机构可以进行二维转动。利用巡视器的移动和相机指向机构的转动，能够获取巡视区月表的三维光学成像，用于巡视区地形地貌研究、巡视区撞击坑调查与研究、巡视区月球地质构造解析和综合研究。

图 2－4　全景相机

全景相机空间覆盖长度最终取决于巡视器实际的行走距离，覆盖宽度取决于镜头的高度。在成像 1～100m 范围内，相机分辨率为 0.15～14.62mm，幅宽为 0.34～34.38m。借助桅杆的 360°方位运动能力和一定角度的俯仰运动能力，相机分时、分幅拍摄不同方向景物，通过拼图处理成为一幅全景图像。全景相机具有成图范围广的特点，有利于全面了解着陆区或探测区周围的景物和地貌。

全景相机主要性能指标见表 2-3。

表 2-3　全景相机主要性能指标

序号	技术指标	参　数
1	波段范围	可见光
2	成像模式	彩色成像、全色成像（可切换）
3	正常成像距离	3m 至无穷
4	视场角/(°)	19.7×14.5（偏差不超过 5%）
5	信噪比/dB	≥40（最大信噪比） ≥30（反照率 0.09，太阳高度角 30°）
6	系统静态传函	≥0.20（全视场）

为验证相机科学探测内容，开展了暗电流采集、整机平场、模式归一化定标和彩色定标等 4 项定标试验项目。试验结果表明各项定标结果均满足要求。

2.2.4　红外成像光谱仪

红外成像光谱仪（图 2-5）安装在巡视器前部，开展巡视区月表红外光谱分析和成像探测，可以获取探测目标 450～2400nm 全波段的光谱及 450～950nm 波段范围的高光谱图像数据，用于巡视区月表矿物组成和分布的分析，以及巡视区能源和矿产资源的综合研究。

月表矿物如辉石、橄榄石等在 450～2400nm 波段范围内有明显的吸收峰特征，根据红外成像光谱仪光谱范围和光谱分辨率的设置，可以很好地对月面这些矿物进行识别与研究。表 2-4 列出了从可见光到短波红外区间月表矿物的主要矿物学特征及可探测的主要矿物目标。

表 2-4　月表主要矿物的分布、元素组成及主要光谱特征

矿物名称	隶属	分布	主要元素	光谱特征
辉石 $(Ca, Mg, Fe)_2Si_2O_6$	硅酸盐	月海玄武岩 高地斜长岩	Mg，Fe，Ca，Na，Al，Si，O	0.95μm 附近和 1.8～2.4μm 的双吸收带，及 0.75μm 和 1.4μm 附近的双反射峰
斜长石 $Ca_2Al_2Si_2O_3$	硅酸盐	月海玄武岩 高地斜长岩	K，Ca，Na，Al，Si，O	1.25μm 处的吸收带及 0.8μm 附近的反射峰
橄榄石 $(Mg, Fe)_2SiO_4$	硅酸盐	月海玄武岩	Mg，Fe，Ca，Mn，Cr，Al，Si，O	1.0μm 处的吸收峰及 0.65μm 附近的反射峰
钛铁矿 $FeTiO_3$	氧化物	月海玄武岩	Ti，Fe，O	0.5μm 和 1.5μm 附近的吸收峰及 1.0μm 附近的反射峰

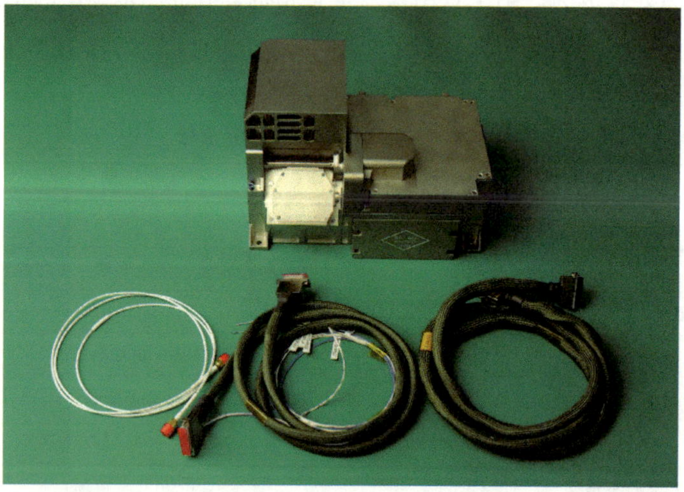

图 2-5　红外成像光谱仪

红外成像光谱仪的性能指标见表 2-5。

表 2-5　红外成像光谱仪的性能指标

序号	名称	可见近红外通道	近红外短波红外通道
1	光谱范围/nm	450～950	900～2400
2	光谱分辨率/nm	2～10	3～12
3	视场/(°)	≥6×6	≥2×2
4	系统静态传函 MTF	>0.1	—
5	信噪比/dB	≥40（最大信噪比） ≥30（反照率 0.09，太阳高度角 45°）	≥40（最大信噪比） ≥30（反照率 0.09，太阳高度角 15°）

为验证红外成像光谱仪科学探测内容，安排了光谱定标、辐射定标、信噪比定标等 3 项定标试验项目，定标试验表明各项结果均满足要求。

2.2.5　测月雷达

测月雷达（图 2-6）用于巡视路线上月壤厚度、结构以及月壳浅层结构探测。测月雷达天线为收发分置形式，第一通道发射天线和接收天线安装于巡视器后部的底板下侧，第二通道天线安装于巡视器底部的底板下侧。巡视器移动时，测月雷达开机工作，以天线共偏移点方式进行探测，可实现对巡视器路线下方的月壤厚度、结构以及月球浅层岩石结构的探测。

仿真分析计算表明，利用测月雷达第一通道中心频率 60MHz、工作带宽 40MHz、动态范围约为 170dB 的天线，可探测到深度不小于 100m 的月壳浅层结构数据，其层厚度分辨率为米级；利用测月雷达第二通道中心频率 500MHz、工作带宽 450MHz、动态范围约为 140dB 的天线，可探测深度不小于 30m 的月壤结构，其层厚度分辨率为分米级。

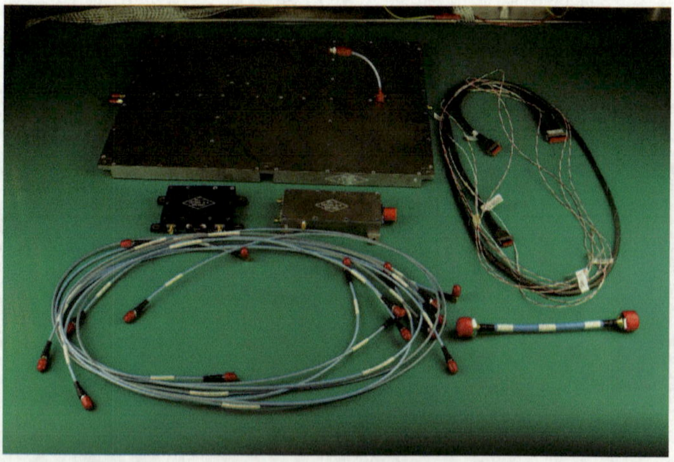

图 2-6　测月雷达

测月雷达的性能指标见表 2-6。

表 2-6　测月雷达的性能指标

<table>
<tr><th rowspan="2">序号</th><th rowspan="2" colspan="2">名　称</th><th colspan="2">主要参数和性能</th></tr>
<tr><th>第一通道</th><th>第二通道</th></tr>
<tr><td rowspan="3">1</td><td rowspan="3">发射机</td><td>脉冲电压幅度/V</td><td>1000（偏差不超过 5%）</td><td>≥400（偏差不超过 5%）</td></tr>
<tr><td>脉冲重复频率/kHz</td><td>0.5/1/2（可切换）</td><td>5/10/20（可切换）</td></tr>
<tr><td>脉冲上升时间/ns</td><td>≤5</td><td>≤1</td></tr>
<tr><td rowspan="2">2</td><td rowspan="2">接收机</td><td>工作频带/MHz</td><td>10～175</td><td>10～1000</td></tr>
<tr><td>输入动态范围/dB</td><td>优于 90</td><td>优于 90</td></tr>
<tr><td rowspan="3">3</td><td rowspan="3">发射和接收天线</td><td>中心频率/MHz</td><td>60</td><td>500</td></tr>
<tr><td>工作带宽/MHz</td><td>≥40</td><td>≥450</td></tr>
<tr><td>驻波系数</td><td>≤3</td><td>≤2.5</td></tr>
<tr><td>4</td><td colspan="2">探测深度/m</td><td>≥100</td><td>≥30</td></tr>
<tr><td>5</td><td colspan="2">厚度分辨率</td><td>米级</td><td>≤30cm</td></tr>
</table>

测月雷达标定试验的项目包括第一通道、第二通道的系统增益。测试内容包括测月雷达系统增益与发射机输出脉冲幅度、收发天线增益和反射系数（电压驻波比）、接收机最小可检测信号功率等参数，通过对雷达系统增益的标定实现对测月雷达探测深度指标的标定。定标试验表明各项结果均满足要求。

2.2.6　低频射电频谱仪

低频射电频谱仪（图 2-7）安装在着陆器上，在月昼期间工作，对太阳低频射电特征和月表低频辐射环境进行探测。在 0.1～80MHz 探测太阳爆发产生的低频电场，给出太阳爆发低频电场的幅度随频率变化的特性，通过观测月球电离层对太阳爆发电场的截止特性，研究着陆点上空月球电离层的特性，并反演着陆点上空电离层等离子体频率峰值和电子总含量，同时在月昼期间监测月表低频辐射电场的变化情况。

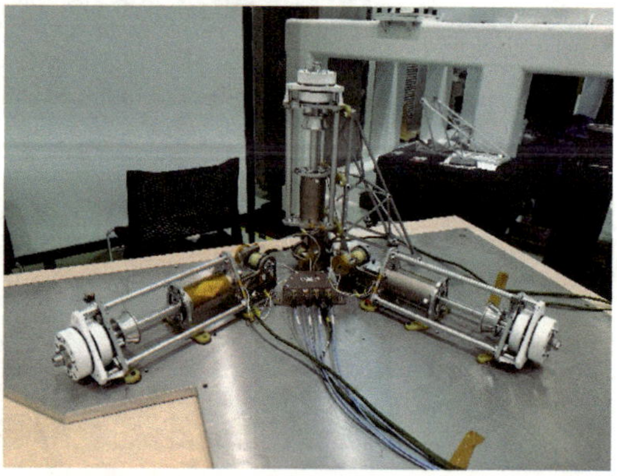

图 2-7　低频射电频谱仪

低频射电频谱仪的性能指标见表 2-7。

表 2-7　低频射电频谱仪的主要性能指标

序号	名　称	主要参数和性能
1	工作频段	0.1～40MHz
2	接收机灵敏度	≤10nV/$\sqrt{Hz}$
3	动态范围	≥75dB
4	频率分辨率	1～10kHz（0.1～1.0MHz） 100～200kHz（1.0～40MHz）

低频射电频谱仪定标试验的目的是测试系统增益及灵敏度指标，评估低频射电频谱仪对空间电场的探测能力。安排了低频射电频谱仪的 5 项定标试验。接收机定标，在不同温度条件下测试接收机的灵敏度、增益频率特性和动态范围；前置放大器定标，包括测试输入阻抗、不同温度条件下接收机灵敏度、增益频率特性以及动态范围；接收天线定标，包括分析天线的输入阻抗、最大接收方向的有效长度、归一化接收方向图、接收方向的单位矢量等；内定标测试；高频电缆定标。定标试验结果表明，低频射电频谱仪满足任务要求。

2.2.7　月表中子与辐射剂量探测仪（LND）

月表中子与辐射剂量探测仪（图 2-8）安装在着陆器有效载荷舱内，包括一个传感探头和一个电子学箱。用于测量月面高能粒子辐射剂量，包括银河宇宙线、太阳高能粒子事件，以及地球辐射带的粒子，同时，还可测量银河宇宙线与月表相互作用产生的次级粒子。研究月面的高能粒子（包括质子、电子、阿尔法粒子等）的能谱、高能粒子辐射剂量及其随时间的变化，考察太阳高能粒子事件期间辐射剂量的变化；研究高能粒子与月面相互作用的机制，及由此产生的次级粒子尤其是月面中子的剂量；研究月表电磁场、次级粒子效应、地形地貌等月面条件因素对月表辐射环境的影响；分析评估月面辐射剂量效应。

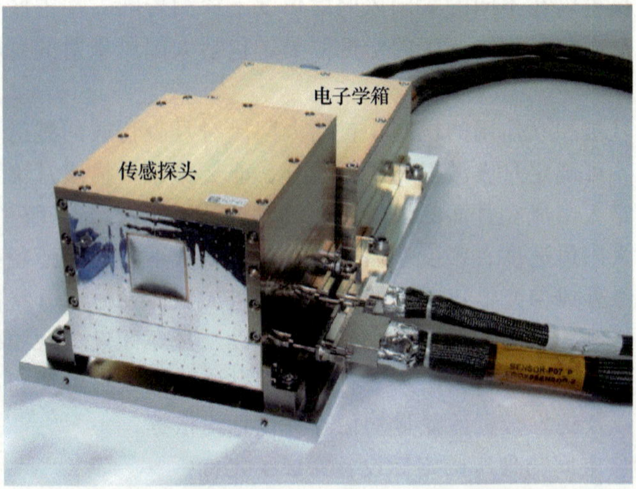

图 2-8　月表中子与辐射剂量探测仪

月表中子与辐射剂量探测仪的性能指标见表 2-8。

表 2-8　月表中子与辐射剂量探测仪的主要性能指标

序号	项　目	指　标
1	快中子能谱	2～20MeV，32 个能道
2	热中子通量	10～10^4 个/min
3	质子能谱	7～30MeV，32 个能道
4	电子能谱	60～500keV，32 个能道
5	α 粒子能谱	7～20MeV/n，32 个能道
6	重离子能谱	10～30MeV/n，32×32 个矩阵
7	LET 谱范围	0.1～430keV/μm，64 个能道
8	时间分辨率	剂量率 1min；质子 5min 电子 5min；重离子 30min

着陆器上有 4 个为仪器提供热量和动力的放射源，包括一个放射性同位素热电发生器（RTG）和三个放射性同位素加热器装置（RHU），这些放射源发出的辐射会对 LND 的测量形成不可忽略的背景干扰。为了消除这些放射源对 LND 测量数据的影响，在地面实验室里测量了 RTG 和 RHU 对 LND 的影响。考虑到地球环境与月球上不同，使用数值方法和蒙特卡罗模拟相结合的方法得到了这些放射源的综合影响，总剂量本底为 5.2μGy/h，中性粒子剂量本底为 1.69μGy/h。

2.2.8　中性原子探测仪（ASAN）

中性原子探测仪（图 2-9）安装在巡视器的+X 舱板上，仪器隔热安装在有效载荷舱内，传感器的入口系统通过+X 舱板的开孔伸出舱外。传感器入口采用一次性开启的保护盖，防止着陆时月尘进入。巡视器到达月表后，保护盖永久打开。结合巡视器的机动性，中性原子探测仪可以在不同的太阳风照射角度下探测着陆器冲击区以外未受扰动的月壤，测量太阳风与月面相互作用产生的中性原子，包括太阳风溅射的月壤物质以及被月面散射的太阳风物质。

研究太阳风与月壤相互作用的机制，包括太阳风对月表物质的溅射，月面对入射太阳风的反射和散射等，可确定月表溅射物质的成分、能量分布，推演月壤物质与入射太阳风的作用过程，可确定太阳风在月面的散射率、散射粒子的能量和空间分布；可考察太阳风与月壤相互作用与太阳风速度、密度、温度的关系，以及不同地形地貌条件下溅射和散射情况的变化。结合红外成像光谱仪获取的月壤矿物组成和挥发分组分特征，研究月球挥发分在月壤中的赋存机制、来源、早期挥发分组成，以及月球挥发分（如硫元素等）与月表物质之间的相互作用等，并分析太阳风稀有气体在月壤中的赋存特征和迁移规律，为月壤的太阳风改造过程以及挥发分资源储量的估算提供理论依据。

ASAN 使用瑞典空间物理所的定标设备完成了定标。

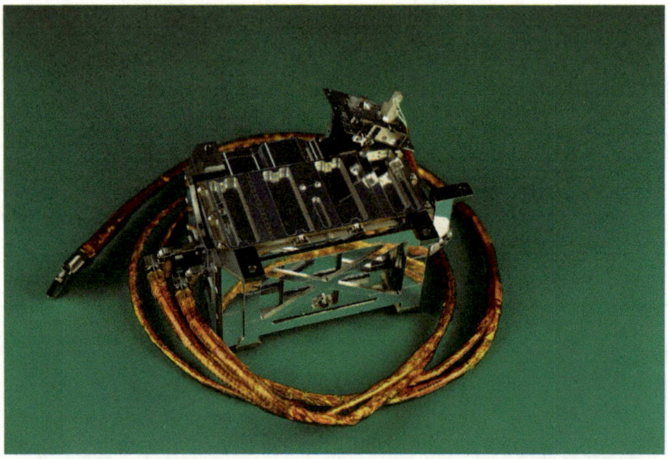

图 2-9　中性原子探测仪

中性原子探测仪的主要性能指标见表 2-9。

表 2-9　中性原子探测仪的主要性能指标

序号	项　目	指　标
1	观测要素	能量中性原子（ENA），正离子
2	能量范围	10eV～10keV
3	质量分辨率	能量中性原子：氢原子，其他成分原子组 正离子：质荷比 1，2，4，8，16，32
4	能量分辨率	7%（正离子），30%（能量中性原子）
5	时间分辨率	10s

2.2.9　中-荷低频射电探测仪（NCLE）

低频射电探测仪（图 2-10）安装在“鹊桥”中继星上，用于在低频波段获得天空背景射电巡天图像，开展对行星射电辐射变化的监测研究，检验行星磁层低频射电辐射随着行星尺度变化的幂率谱规律假说；开展灾害性空间天气事件和太阳射电爆发的监测；开展地-月空间低频射电天文空间干涉测量技术试验。

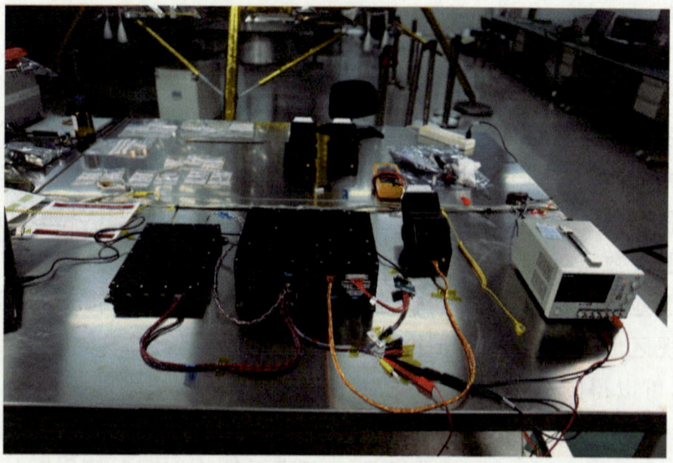

图 2-10　低频射电探测仪

低频射电探测仪的主要性能指标见表 2-10。

表 2-10　低频射电探测仪的主要性能指标

序号	项　目	指　标
1	工作频段/频谱分辨率	100kHz～1MHz/1kHz
		1～10MHz/10kHz
		10～80MHz/100kHz
2	时钟稳定度	$<5\times10^{-11}$（短稳）
3	接收机灵敏度	<−160dBm/Hz@10MHz
4	动态范围	≥84dB
5	空间干涉基线	>400000km
6	干涉角分辨率	≤1Mbps

2.2.10　生物科普试验载荷

生物科普试验载荷（图 2-11）安装于着陆器上，质量为 2.608kg，尺寸为 ϕ173mm×198.3mm，携带有棉花、马铃薯、油菜、拟南芥、酵母和果蝇共 6 种生物，开展在月球低重力、强辐射、自然光照条件下的生长等试验，观察微循环生态系统内动植物的一些重要生命活动特征。

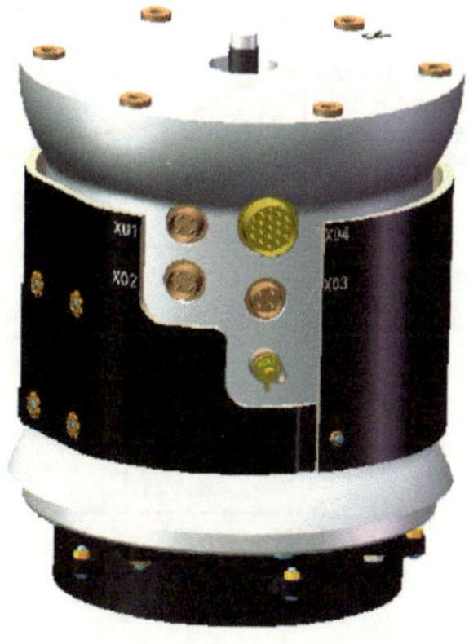

图 2-11　生物科普试验载荷

2.3　科学目标可实现性和可实现程度分析

科学团队和工程研制团队一起，经过对嫦娥四号任务科学目标的分析，将科学探测任务分配到各有效载荷，确定了有效载荷技术指标和技术方案，经过定标、科学研制试验等专项研究，对嫦娥四号任务各项科学目标的可实现性和可实现程度分析如下。

1）在 2km～4m 高度范围内，降落相机可获取降落区域内成像幅宽为 3.31～1656.85m、空间分辨率在 3.23～1618.02mm 范围内的图像；地形地貌相机的成像距离范围约为 5m 至无穷，可以实现对着陆区周围 360°范围内的全景成像；全景相机可获取分辨率为厘米级的图像剖面，在探测点上全景相机具备 360°获取数据的能力，通过数据处理可实现全景拼接和彩色复原；红外成像光谱仪可以获取 450～2400nm 月面探测目标全波段的光谱数据，通过数据处理可以实现对月面巡视区主要矿物成分的识别和反演分析。利用降落相机、地形地貌相机、全景相机、红外成像光谱仪和测月雷达的探测数据，可以实现“月球背面巡视区形貌和矿物组分探测与研究”的科学目标。

2）测月雷达可以得到巡视线路上 60MHz 和 500MHz 频率的雷达回波数据，通过后续数据处理可以计算行进线路上月表相对介电常数及获得浅表层的分层结构。利用测月雷达的回波探测结果，可以实现“月球背面巡视区浅层结构探测与研究”的科学目标。

3）当太阳流量密度小于 10^{-21} W・m^{-2}・Hz^{-1}，低于低频射电频谱仪灵敏度时，低频射电频谱仪无法探测该状态下的太阳低频射电辐射；当发生太阳爆发事件、爆发的流量密度大于 10^{-18} W・m^{-2}・Hz^{-1}时，采用短天线频域对消算法降噪后，低频射电频谱仪在 100kHz 到 8MHz 能够探测到该状态下的太阳低频射电辐射；当太阳爆发峰值流量密度大于10^{-16} W・m^{-2}・Hz^{-1} 时，采用短天线频域对消算法降噪后，低频射电频谱仪在 100kHz 到 40MHz 全频段能够探测到该状态下的太阳低频射电辐射，如图 2－12 所示。

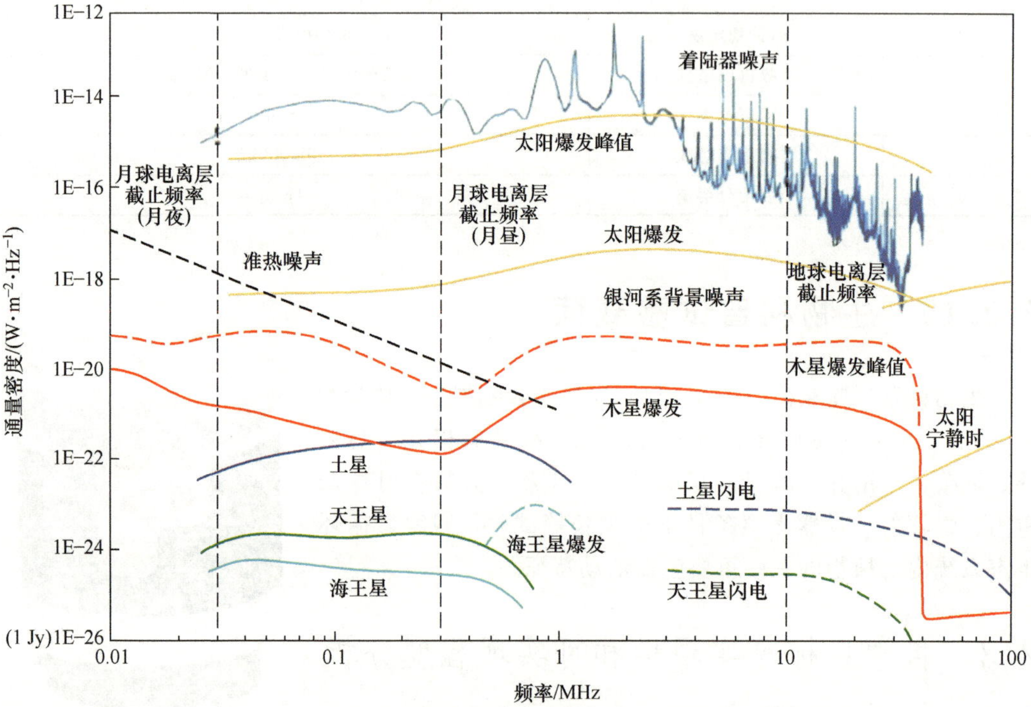

图 2－12　月球表面的电磁波通量密度

利用中继星低频射电探测仪探测数据和低频射电频谱仪在有较大太阳爆发或较强的低频宇宙射电现象时获得的有效科学探测数据，可以实现“月基低频射电天文观测与研究”的科学目标。

4）利用月表中子与辐射剂量探测仪和中性原子探测仪的科学探测数据，可实现“月面中子辐射剂量、中性原子等月球环境探测研究”的科学目标。

2.4　在轨科学探测

2.4.1　精确定位落月坐标

嫦娥四号在月球背面表面软着陆的动力下降是一个时间短、速度变化快的过程，难以依靠地面实时控制，只能采用探测器自身携带的敏感器实现自主测控。嫦娥四号着陆后，科学和工程团队对着陆点精确定位进行了确定。

利用嫦娥二号高分辨率地形数据，结合嫦娥四号下降过程和月面探测期间获得的多源影像数据，经精细重构嫦娥四号在月球背面粗避障、精避障和悬停避障等自主导航降落过程，对探测器的位置确定为东经177.5991°，南纬45.4446°，高程−5935m。基于嫦娥二号正射影像、LROC NAC正射影像、嫦娥四号降落相机影像、嫦娥四号监视相机影像等多源数据，利用影像特征匹配定位和单像视觉测量定位技术，确定的嫦娥四号着陆点的精确位置为177.588°E，45.457°S。利用嫦娥四号探测器动力下降过程中的降落相机序列图像以及着陆区高精度地形图，基于高精度图像匹配和几何变换的探测器着陆点高精度快速定位方法，解算的嫦娥四号探测器着陆点位置为177.5884°E，45.4565°S（图2-13）。

2.4.2　月球背面探测约束

冯·卡门撞击坑内地形高低起伏，撞击坑边缘或着陆点周围复杂的地形可能导致在月面工作的着陆器和巡视器存在光照和通信被遮挡的可能，如发生遮挡，对任务影响较大。任务组利用嫦娥二号、LRO等月球卫星探测得到的图像、激光测高等数据，对冯·卡门撞击坑着陆区的高程起伏、坡度、撞击坑和石块分布密度，以及着陆区内地形对光照和中继链路遮挡进行了详细分析，提出了月球车图像的地形遮挡角高精度计算方法。分析认为，在考虑月面曲率影响的情况下，当遮挡山体距离着陆点大于150km时，只要该山体高度不超过6.5km，该山体即位于着陆点的水平面以下，对着陆点不构成遮挡。

因为月球也进行自转运动，所以月球上也像地球一样有白天黑夜之分，嫦娥四号在轨科学探测按照一个月球公转周期（约28天）为一个工作周期，对于嫦娥四号着陆点，自太阳从东侧天际线升起到从西侧天际线落下的时间为月昼，其余见不到太阳的时间为月夜。由于月球表面没有大气，整体保温效果较差，昼夜温差很大，白天温度很高，能达到100℃以上，夜晚温度又会降到−190℃以下。对于嫦娥四号巡视器来说，中午高温不利于设备工作，因此定义当地太阳高度角大于33°之间的时段为月午，月午期间巡视器转向一个安全的航向静置，同时关闭巡视器上个别设备，以静止的姿态度过月午。

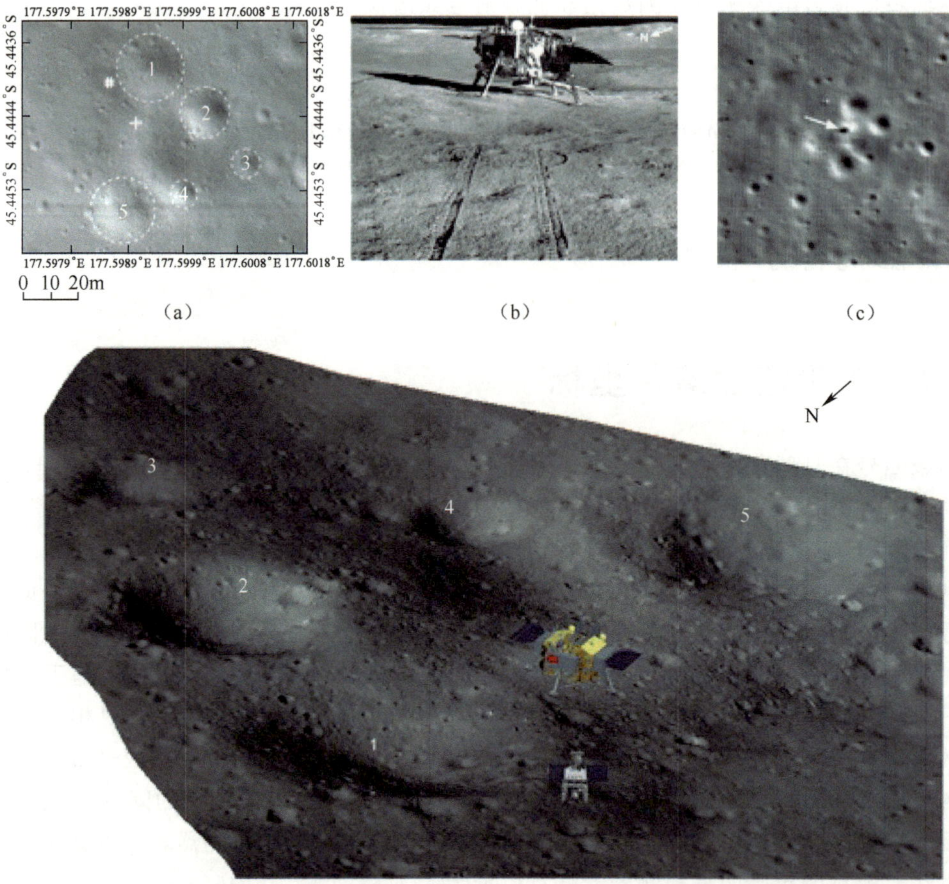

图 2-13　CE-4 着陆点在影像上的位置示意图

(a) 着陆点在 5cm 分辨率降落相机 DOM 上的位置（“+”位置）；(b) 巡视器全景相机获取的着陆器照片；
(c) 美国月球勘测轨道器（LRO）上的窄角相机（NAC）于 2019 年 1 月 12 日获取的 CE-4 着陆点照片
（箭头所指位置）；(d) 基于降落相机 DEM、DOM 数据制作的 CE-4 着陆点三维地形渲染图

嫦娥四号着陆器和巡视器的工作周期分为月夜休眠、月昼探测、巡视器月午休眠等工作阶段，嫦娥四号科学探测任务以月昼为周期进行。月昼工作期内嫦娥四号巡视器利用导航相机对月面地形进行感知拍照，处理导航相机图像数据，利用立体匹配等技术对月面地形进行三维还原，生成月面地形和三维场景等一系列地形产品，并利用视觉定位、外测数据和遥测数据等方法确定巡视器位置和姿态。

探测计划的实施主要由飞控组执行。每月昼前由地面应用系统根据上个月昼获取的图像，综合工程目标和前期探测需求，明确本月昼科学探测需求并提交飞控组。飞控组根据科学探测目标，完成探测周期规划与验证，对科学探测需求进行可行性分析，形成月昼工作实施方案。根据地形产品，结合科学探测需求和巡视器移动越障能力进行探测或移动行为的规划，指定探测周期内的导航点和探测点，生成移动指令和注入数据，通过地面测站实施上注。经嫦娥四号探测任务飞控组审议通过后开始实施。

每个工作周期内，在巡视器移动过程中，测月雷达开机工作；探测模式时，巡视器全

景相机、红外成像光谱仪、中性原子探测仪开机工作。期间监视巡视器月面工作过程，应急情况下及时发送移动停止指令。飞控组根据地面应用系统需求发送巡视器遥测数传数据、探测周期规划、导航单元规划、探测器位置姿态等，接收地面应用系统发送的最新科学探测需求信息、全景相机信息、地形地貌相机信息等。

2.4.3　有效载荷探测情况

(1) 着陆器载荷探测情况

降落相机在动力下降段开机，对着陆点进行拍照，相机共开机 1 次，开机时长 9min。探测器成功软着陆后，降落相机完成探测使命。

地形地貌相机在探测器着陆后开机工作。第一月昼共开机 46 次，时长 52h41min，完成了低频射电频谱仪天线解锁过程的成像、巡视器 X 点移动和原地转弯成像、巡视器 A 点拍照 [图 2-14 (a)]、$-60°\sim+45°$间俯仰角环拍、巡视器 D'点侧面拍照、巡视器 D'点原地转弯至两器互拍位置动态摄像、巡视器 D'点正面静态拍照等。着陆器进入第一月夜后，地形地貌相机完成探测使命。

(a)

(b)

图 2-14　两器互拍照片

(a) 地形地貌相机对在 A 点的巡视器拍照；(b) 全景相机在 A 点对着陆器拍照

因测月雷达在巡视器行走期间开机工作，其发射的信号可能对低频射电频谱仪有干扰，因此，低频射电频谱仪在巡视器行走期间保持关机，在月午巡视器休眠时开机工作。

月表中子与辐射剂量探测仪（LND）在着陆器月昼唤醒后开始工作，休眠设置前择机关机。

生物科普试验载荷于 2019 年 1 月 3 日随嫦娥四号着陆至月球背面，2019 年 1 月 5 日成功在月球表面培育出棉花嫩叶。

(2) 巡视器有效载荷探测情况

全景相机在月球车到达探测站点时开机成像。第一月昼，在 A 点对着陆器进行了拍照［图 2－14（b）］、D 点对着陆器拍照；第二月昼使用全景相机完成环拍彩色成像，如图 2－15 所示。后续月午前、休眠前进行全景相机两圈全色环拍；遇到感兴趣的探测对象时可定点规划多组全色或者彩色成像。

图 2－15　全景相机环拍，三圈镶嵌圆柱投影图

测月雷达在每个月昼巡视器移动时同步开机探测。

红外成像光谱仪在月午前后太阳高度角较高时或遇到感兴趣的探测对象时开机，开展成像或定标工作，对巡视器车辙，沿途月壤、石块、撞击坑进行成像探测。

中性原子探测仪在月午前后太阳高度角较高时或遇到感兴趣的探测对象时开机探测，每次开机高压模式 54min，离子模式 6min。

(3) 以典型月昼为例，嫦娥四号任务探测情况

以第 36 月昼为例，巡视器前 36 月昼行驶路线如图 2－16 所示。

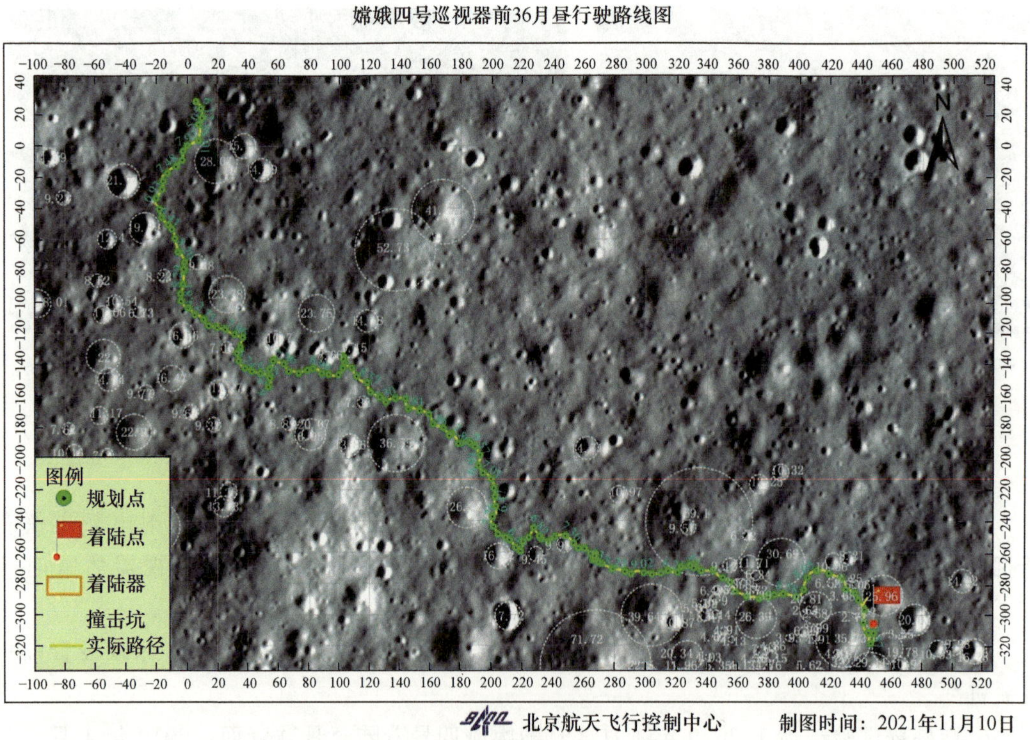

图 2－16　巡视器行驶路线图

嫦娥四号任务探测情况如下：

2021 年 10 月 30 日着陆器自主唤醒进入月昼，11 月 11 日着陆器完成月夜休眠。10

月 29 日巡视器自主唤醒进入月昼；11 月 1 日，完成进月午设置；11 月 8 日，进行出月午设置；11 月 11 日，巡视器完成休眠控制。

第 36 月昼工作期间，着陆器月表中子与辐射剂量探测仪累计工作 224h10min；低频射电频谱仪累计工作 126h8min。巡视器行走 5 步，行走约 35.42m，巡视器全景相机成像 160 对，测月雷达开机 5 次，红外光谱探测仪实施采集流程 2 次（定标 1 次），中性原子探测仪工作 1 次。

2.5　科学探测数据处理

2.5.1　月表物质的光谱特性

红外成像光谱仪是探测月球背面物质成分的关键科学载荷，可开展现场、实时及目标无损的精细光谱及成像，为达到对月球表面成分分析与资源勘查的科学探测需求，载荷研制克服了重量、体积、功耗等资源的严酷限制和原位场景中光照、温度等恶劣环境影响，成功获取了月球背面月表红外光谱及图像科学数据。

针对月球表面复杂环境下光谱精准探测的基础性科学问题，研究团队在月球表面光照、温度等复杂环境影响下的红外光谱特征生成、传输演化与光电效应机理等研究基础上，提出了凝视型时序扫描的新型声光光谱原位探测机制（图 2－17），构建了场景自适应光电参数调控探测理论模型。采用凝视目标、声光时序调控光谱、自适应探测的方法，建立了静态无运动声光电耦合高频调控模型，基于光极化及静态高频调制的光束操控，采用了信号光束的光正交极化、构型消光及陷光，静态高频调制，解决了原位应用场景中杂光抑制难题。提出了天体表面现场与地面实验室协同定标、溯源交叉验证等数据处理及在轨同步探测与定标的方法，构建了全信息链的高精度定量化数据反演模型，实现了全链路、宽谱段、多溯源的定标及定量化反演，实现了现场、实时及目标无损探测成像。

提出的温度效应测试及校正方法消除了在轨探测的月面环境影响，在较大程度上抑制了仪器内部温度敏感器件温度变化对光谱探测结果造成的影响，提高了声光红外成像光谱仪的定量化水平。为了定量评估不同红外成像光谱仪数据处理方法的可靠性，利用一块矿物组成与月球高地岩石相似的苏长-辉长岩（CR－1）进行光谱地面验证试验，CR－1 的 VNIS 光谱用 Hapke 模型计算出样品中矿物模式含量与其真实结果在误差范围内一致。由红外成像光谱仪鉴定件测得的 CR－1 的光谱在 971（±1）nm 和 1957（±8）nm 波段处表现出明显的吸收特征，这一吸收特征与玉兔二号巡视器上 VNIS 在第 3 月昼探测到的岩石的吸收特征相似。

月球车在月面工作时，在复杂光照的条件下，会存在一些外在干扰影响红外成像光谱仪获得的原位数据。统计前 24 月昼光照及车体角度信息，发现获得的光谱数据可能会受到阴影和杂光的影响，如图 2－18 所示。针对复杂光照影响，从照明角度与车身朝向的几何角度关系、比较相邻探测的可见光-近红外图像中亮斑的相对位置移动、异常区域的平均反射光谱与实验室测量的热控多层反射光谱的比较等三个方面来确定杂光影响，构建了阴影校正算法，提高了月球表面矿物成分解释的准确性。

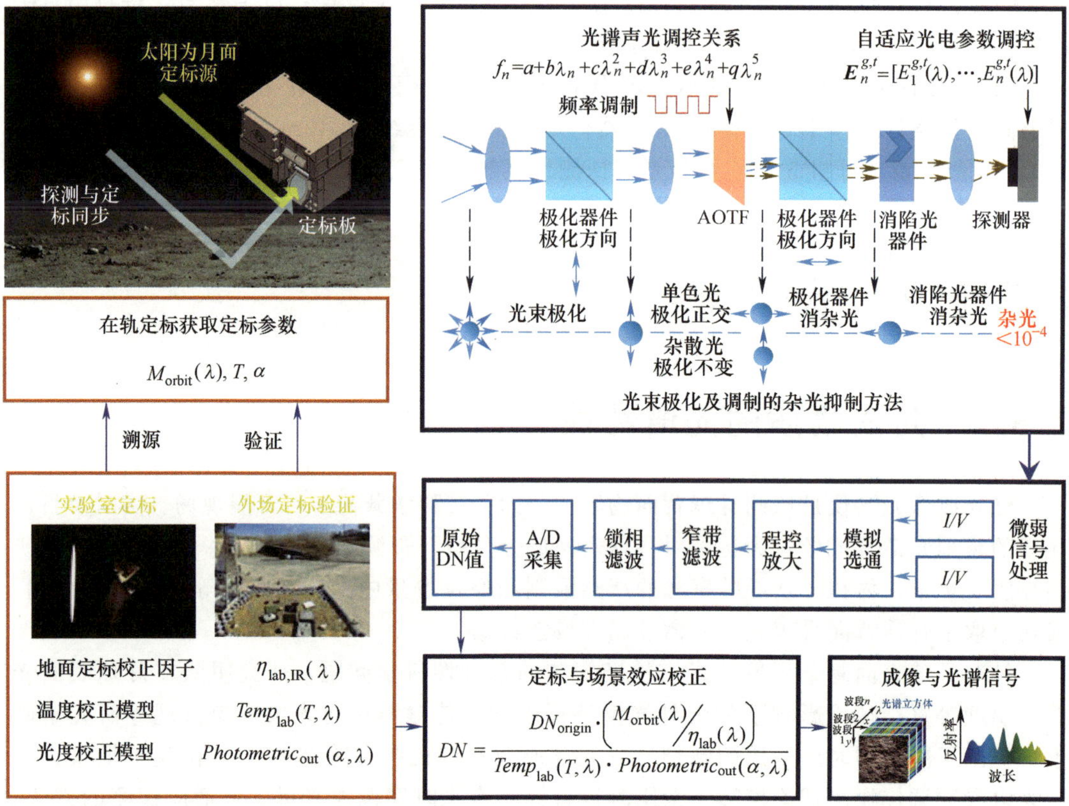

图 2-17 凝视型声光时序扫描的光谱原位探测机制

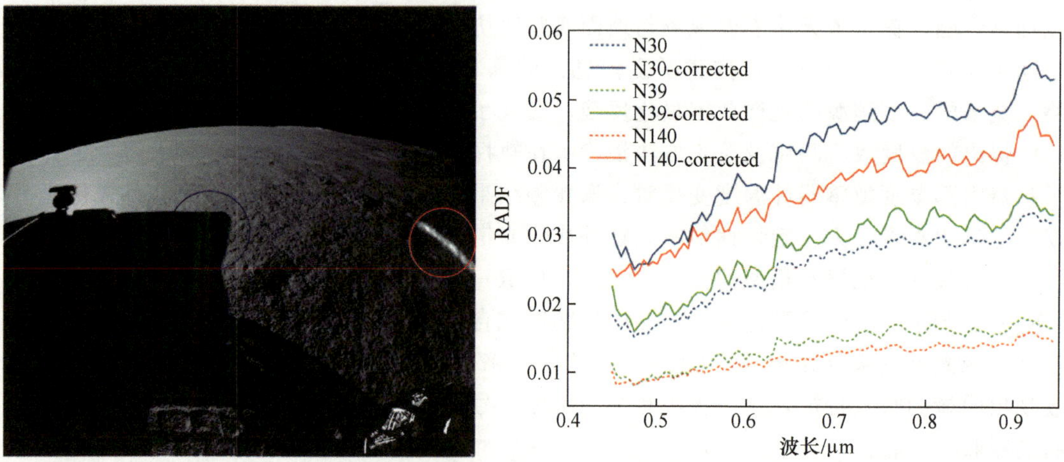

图 2-18 嫦娥四号第四月昼避障图像及阴影校正结果图

2.5.2 低频射电探测载荷的数据处理

在嫦娥四号任务中，着陆器搭载了低频射电频谱仪（LFRS）、中继星搭载了中-荷合作 NCLE 超长波探测仪、龙江 2 号搭载了短波波段接收设备。其重要的科学目标是探索月球背面和月球空间的电磁波辐射环境，在太阳活动峰年探测太阳 HF－VHF 频段爆发，特别是被电离层阻断、在地面看不见的太阳射电爆发。

LFRS 月面天线展开正常，探测工作正常，取得了理想的观测数据，LFRS 工作在 0.1～40MHz，其天线是由 3 个 5m 长的单极子构成的三极子正交天线，理论上 LFRS 完全可以利用 DOA 估计的方法来对太阳射电暴发进行定位。但由于嫦娥四号着陆器采用的是嫦娥三号的平台，并没有针对低频射电观测进行专门的电磁兼容设计，因此平台本身存在着大量的低频射电干扰，给射电观测带来了严重的影响。为减少和消除平台的干扰，载荷产品设计时加入了一个 1m 长的短天线，采用短天线频域对消算法降低平台噪声的影响。

结合 CE－4 测月雷达及全景相机的探测数据，构建了较为准确的着陆区地形、地质结构及电磁参数模型，对应观测数据的卫星位置、姿态信息、地面观测数据等，开展了观测数据初步定标和数据解译算法研究，并以此开展了低频射电频谱仪全系统仿真，获取了天线输入阻抗、方向图、增益等特性参数。LFRS 测试数据证实了地面验证差分流量扣除噪声的观测效果与方法策略的仿真效果基本一致。使用 LFRS 的实际测量数据，模拟仿真空间干涉测量的结果表明，载荷基本满足月-地之间的协同或干涉测量条件。使用 2020 年 4 月份获得的经过再次更正的部分低频射电频谱仪 2C 数据，通过生成二维图像遴选了典型定标正常的数据，约占 50%，生成了三天线频谱（图 2－19）。

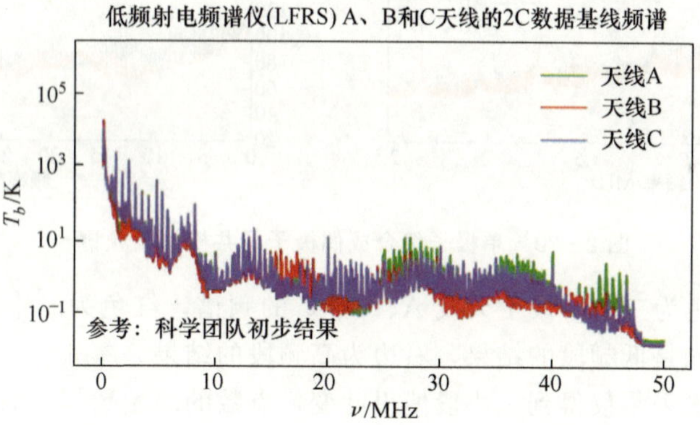

图 2－19　LFRS 载荷二级数据频谱特性

为了利用 1 级观测数据开展科学研究，必须要对数据进行去除干扰的处理。在实际的数据处理中，一种方式是采用短天线消除本底，另一种方式是采用将两个单极子天线组合成偶极子的方法，来消除两个单极子天线上的共模分量。针对第一种方式，整理了存在显著定标跳跃异常前后不一致的一天的频谱后，分析判断认为，从 1 级数据出发优化定标流程策略，可以对大部分定标失败的 2C 数据重新定标、获得有效数据。采用这种方法重新定标的 2C 数据，可以用于对类似太阳爆发这样的射电现象进行捕获。嫦娥四号低频射电频谱仪 2C 数据对宇宙黑暗时代的探测能力分析表明，其四个量级的本底被扣除，使得与宇宙中性氢射电辐射相关的信号也基本被扣除。这种定标方式可以用于对类似太阳爆发的现象进行捕获。

在第二种方式中，针对低频段天线有效长度计算误差较大的问题开展研究，提出了面电流积分法，即通过将天线进行三角面元剖分并进行面电流积分计算的方法，此方法相比于传统的线电流积分方法的计算准确度有一定的提升。考虑来自平台的干扰基本是近场辐射信号，在不同的天线上会产生一定的共模信号，因此这种方法可以在一定程度上减小来自平台的干扰，但这需要用到天线接收到的原始信号。以 LFRS 在 2019 年 2 月 8 日观测的数据为例，对其进行处理，如图 2-20 所示，可以看到经过偶极子组合后的频谱中干扰信号在整个频段均得到了一定程度的抑制，在某些频段甚至达到 20～30dB。

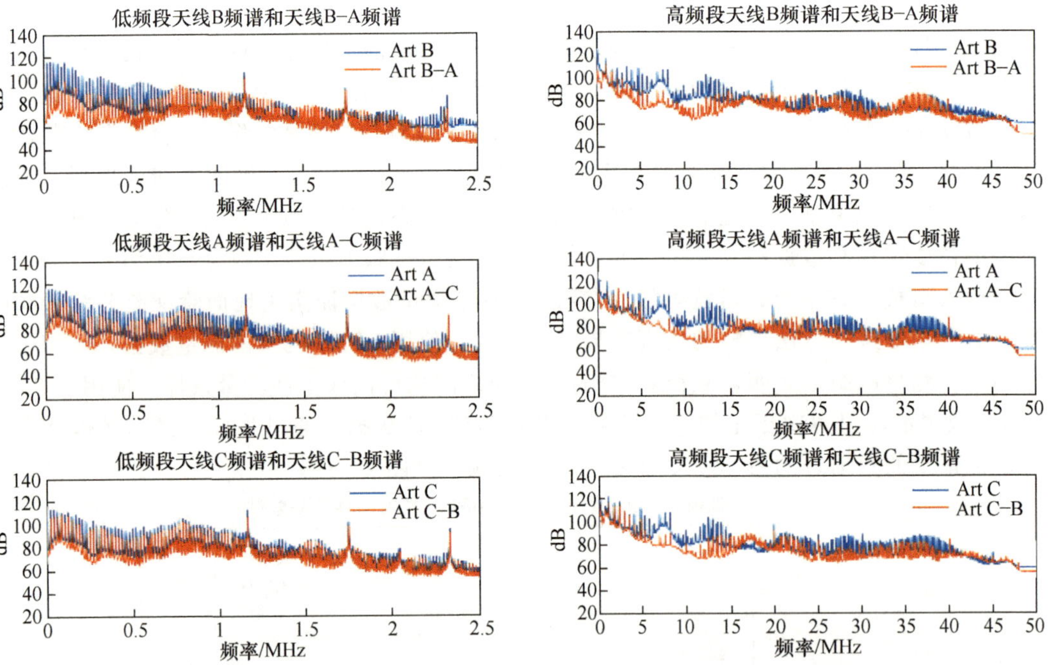

图 2-20　单极子组合成偶极子的共模抑制频谱

图 2-20 中蓝色为三个单极子天线 A、B、C 的频谱，红色为三个组合的偶极子的频谱。其中左边显示的是低频段的结果，右边为高频段的结果。

由于 LFRS 单次采集仅得到一小段傅里叶变换点数的原始数据，时间太短，导致数据的灵敏度不够，因此频谱中的噪声起伏很大。为此，对一天的数据进行积分，获得了相当于在低频段 2s 和高频段 100ms 积分时间的频谱，如图 2-21 所示，长时段积分后频谱的灵敏度得到了明显的提高，频谱曲线变得平滑，这种方法同时也损失了时间分辨率，对于研究宇宙背景辐射等这类稳定的谱信号非常有效。另外，在整个频谱中仍然有大量的干扰存在，这主要来源于平台的低频干扰产生的二次和三次谐波。

图 2-21 中蓝色为三个单极子天线 A、B、C 的频谱，红色为三个组合的偶极子的频谱。其中左边显示的是低频段的结果，右边为高频段的结果。

进一步地，任务组根据低频射电频谱仪获得的数据分析月表低频辐射特性和 NCLE 部分数据，结合其他射电空间和地面项目分析月背空间低频辐射特性，并对二者进行对比分析，探索出了利用绕月卫星研究月球电离层吸收辐射特性的方法。结合着陆器中性原子探测器，分析了月球电离层在频谱仪观测过程中的影响，仿真分析了月球电离层截止频率变化特性。

任务组开展了着陆器低频射电背景噪声抑制及弱信号提取技术研究，提出了通过短天

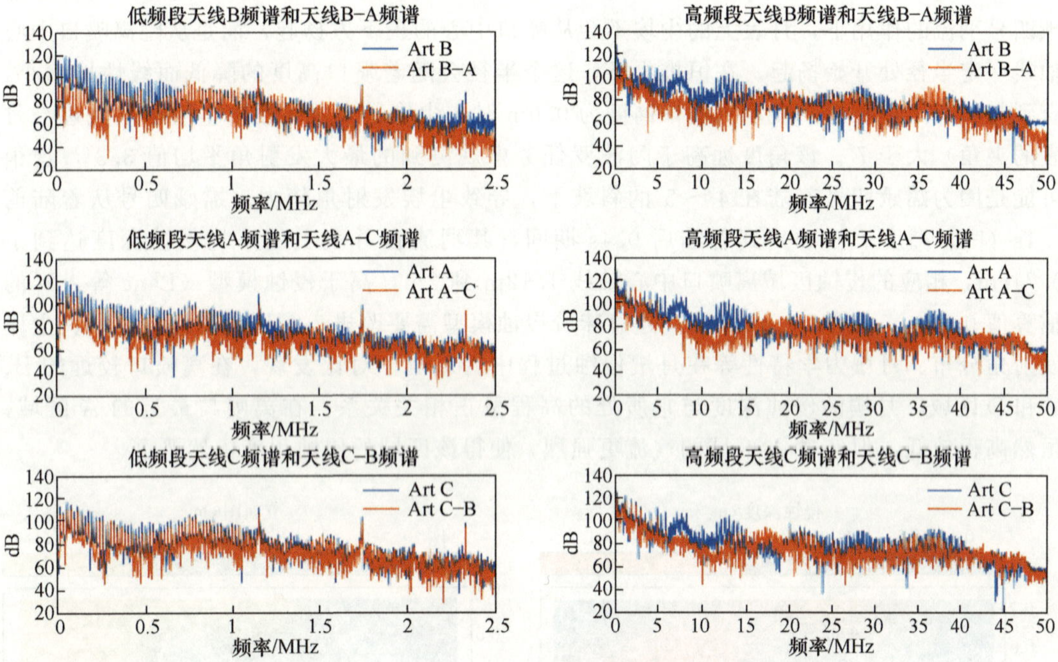

图 2-21　单极子组合成偶极子的积分频谱

线近场对消抑制着陆器噪声干扰的方法。经处理，在 100kHz～10MHz 频带噪声降低约 20～30dB，在 10～40MHz 频带噪声降低约 10～20dB。同时，提出了一种基于 Wiener 估计的噪声抑制技术，可以有效地提高低频射电频谱仪的探测能力。通过有效的噪声抑制，使得本来淹没于着陆器背景噪声中的弱信号能够被提取和检测，有效提高了低频射电频谱仪的信号质量和探测能力。

综上，嫦娥四号任务着陆器平台上的 LFRS 载荷和中继通信卫星平台上的 NCLE 载荷都是已有平台上的搭载载荷，两个平台都存在较大的本底固有电磁噪声，且在 EMC 或本底噪声考虑方面，都没有被专门采取严格适当的低频电磁波屏蔽措施。虽然 LFRS 增加了一个短天线用于线性消除本底巨大（40dB）的信号噪声，用于提取微弱的爆发信号；NCLE 采用表校和偶子对消方式抑制噪声，但均未取得预期的效果。嫦娥四号在月球背面开展低频射电探测的探索性工作，对未来继续开展类似的任务探索积累了宝贵经验，对平台的约束、月球背面自然环境有了新的认知。

2.5.3　发动机羽流对月表的风蚀作用

风蚀，即风的侵蚀作用，是在具有大气的类地行星表面发生的基本地质变化过程之一，它会引起地形、环境变化甚至经济损失。月球作为一个无大气天体，很难有条件研究风蚀作用对其表面形貌的影响。然而，在航天器着陆过程中，变推力发动机产生的羽流会对月球表面产生强烈的冲击，导致月面尘埃的快速输运和表面侵蚀。

对嫦娥四号降落相机及玉兔二号全景相机的数据进行分析，发现着陆器上变推力发动机喷口距离月面 13m 高时，羽流开始侵蚀月球表面，扬起尘埃，影响表面能见度。在嫦娥四号羽流的作用下，月表上的尘埃不会从喷口中心的正下方扬起，而是从距离喷口中心轴线一定半径处开始扬起。在恒推力下，这个半径会随着喷口高度的降低而线性地缩小，直到与喷口半径相近。发动机喷口高度为 0.6m 时，尘埃最大发射角（尘埃运动轨迹与月表的夹角）大于 7°。该角度远高于阿波罗任务中观测到的最大发射角平均值 3.3°，这很可能是因为嫦娥四号着陆在 4°～5°的斜坡上，导致尘埃发射角较大。嫦娥四号从着陆前 1.1s（喷口高度 2.67m）到着陆后 0.4s 期间，其羽流在月表上的平均侵蚀深度达到了 0.71cm，相应的侵蚀区域离喷口中心轴线 1.12m 到 2m，高于侵蚀模型（Lane 等人）的估算值 0.3cm，如图 2－22 所示。这意味着侵蚀模型需要改进，需要考虑局部地形、尘埃发射角分布、月壤力学特性等在月壤侵蚀过程中的影响。对比发现，在离喷口较远的 B、C 和 D 区域，月壤的侵蚀深度与它所处的高程呈正相关关系。在离喷口最近的 A 区域，虽然高程较低，但经过 A 区域的气流更强烈，使得该区域的侵蚀程度仍然严重。

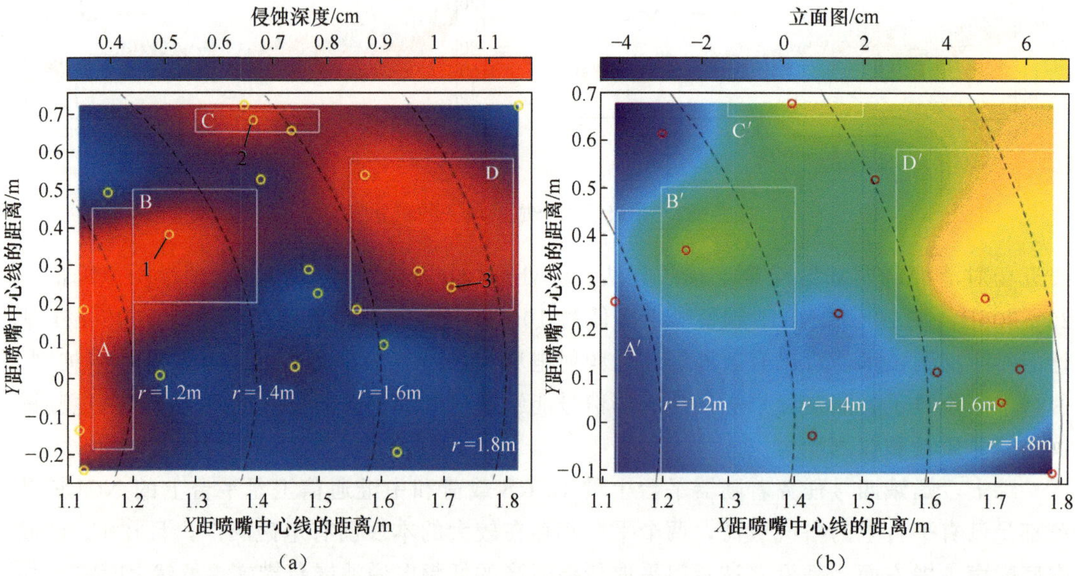

图 2－22　任务中测量的月表侵蚀深度和局部月表高程图

注：(a) 基于嫦娥四号降落图像测量的月表侵蚀深度，黄圈位置是测量点，相应的平均侵蚀深度是（0.71 ± 0.20）cm。其他区域的侵蚀深度是通过插值测量值获得的。(b) 基于嫦娥四号降落图像测量的局部月表高程图。红圈位置是测量点，其他区域的高程是通过插值测量值获得的。高程图的基平面是嫦娥四号着陆点的斜面。

嫦娥四号着陆时，喷出的气流在喷口正下方挖掘出许多小坑。基于玉兔二号拍摄的全景图，测得小坑的最大深度为 1.84cm，比阿波罗 11 号着陆时产生的侵蚀坑深度 2.54cm 小了约 1/3。因为嫦娥四号的推力小于阿波罗 11 号的推力，并且嫦娥四号以 4 倍于阿波罗 11 号的降落速度着陆月表，使得嫦娥四号的气流与月表的作用时间更短，产生的侵蚀坑也就更浅。

基于嫦娥四号降落图像和玉兔二号全景图的观测结果，为月表侵蚀过程提供了新的见解，有助于侵蚀模型的改进。同时，也对后续的登月计划提供了重要的参考信息。

主要成果附录

[1] Chen L J, Yan Y H, Fan Q X, et al. An agile very low frequency radio spectrum explorer [J]. Research in Astronomy and Astrophysics, 2021, 21 (4): 63 - 72.

[2] Chang R, Yang W, Lin H, et al. Lunar Terrestrial Analog Experiment on the Spectral Interpretations of Rocks Observed by the Yutu - 2 Rover [J]. Remote Sensing, 2022, 14 (10): 2323.

[3] Di K C, Liu Z Q, Liu B, et al. Chang'e - 4 lander localization based on multi - source data [J]. Journal of Remote Sensing, 2019, 23 (1): 177 - 184 [DOI: 10.11834/jrs.20199015].

[4] He Z, Li C, Xu R, et al. Spectrometers based on acousto - optic tunable filters for in - situ lunar surface measurement [J]. Journal of Applied Remote Sensing, 2019, 13.

[5] He Z P, Li J N, Li C L, et al. Measurement and Correction Model for Temperature Dependence of an Acousto - Optic Tunable Filter (AOTF) Infrared Spectrometer for Lunar Surface Detection [J]. Applied spectroscopy, 2020, 74 (1).

[6] Li C L, Xu R, Lv G, et al. Detection and calibration characteristics of the visible and near - infrared imaging spectrometer in the Chang'e - 4 [J]. Review of Scientific Instruments, 2019, 90 (10): 103106.

[7] Li C, Wang Z, Xu R, et al. The Scientific Information Model of Chang'e - 4 Visible and Near - IR Imaging Spectrometer (VNIS) and In - Flight Verification [J]. Sensors, 2019, 19 (12): 2806. https://doi.org/10.3390/s19122806.

[8] Liu C, Liu L, Chen J, et al. Mafic mineralogy assemblages at the Chang'e - 4 landing site: A combined laboratory and lunar in situ spectroscopic study. A&A, (2021).

[9] Liu D W, Zhang Y Z, Zhang G L, et al. An Empirical Model to Estimate Abundance of Nanophase Metallic Iron ($npFe^0$) in Lunar Soils [J]. Remote Sensing, 2020, 12 (6).

[10] Liu J J, Ren X, Yan W, et al. Descent trajectory reconstruction and landing site positioning of Chang'E - 4 on the lunar farside [J]. Nature Communications, 2019, 10.1038/s41467 - 019 - 12278 - 3.

[11] M J Bentum, M K Verma, R T Rajan. A Roadmap towards a Space - based Radio Telescope for Ultra - Low Frequency Radio Astronomy [J]. Advances in Space Research, 2020.

[12] Qi X, Ling Z, Zhang J, et al. Photometric Normalization of Chang'e - 4 Visible and Near -Infrared Imaging Spectrometer Datasets: A Combined Study of In - Situ and Laboratory Spectral Measurements [J]. Remote Sensing, 2020, 12 (19): 3211.

[13] Xu J F, Wang M Z, Lin H L, et al. Effect of Lunar Complex Illumination on In Situ Measurements Obtained Using Visible and Near - Infrared Imaging Spectrometer of

Chang'E - 4 [J]. Remote Sensing, 2021, 13 (12): 2359. https://doi.org/10.3390/rs13122359.

[14] You J L, Zhang X P, et al. Analysis of plume - lunar surface interaction and soil erosion during the Chang'E - 4 landing process [J]. Acta Astronautica, 2021, 185: 337 - 351.

[15] Zhu X Y, Su Y, Ji Y C, et al. Ground experiments and performance evaluation of the Low - Frequency Radio Spectrometer onboard the lander of Chang'e - 4 mission [J]. Research in Astronomy and Astrophysics, 2021, 21 (5): 140 - 147.

[16] 陈昊，刘世杰，童小华，等．嫦娥四号着陆点区域高分辨率数字高程模型构建 [J]. 上海航天，2019，36 (3)：36 - 40. DOI：10.19328/j.cnki.1006 - 1630.2019.03.005.

[17] 陈林杰，颜毅华，谭宝林．基于空间矢量天线的太阳低频射电爆发探测研究 [J]. 深空探测学报（中英文），2021，8 (1)：100 - 108.

[18] 何志平，李春来，吕刚，等．月球表面原位光谱探测技术研究与应用 [J]. 红外与激光工程，2020，49 (5)：62 - 69.

[19] 焦康，王明远，张同杰，等．嫦娥四号低频射电频谱仪 2C 数据对宇宙黑暗时代的探测能力分析 [J]. 天文研究与技术，2021，18 (4)：472 - 476.

[20] 李飞，吴波，杨眉，等．嫦娥四号冯·卡门撞击坑着陆区地形及月面遮挡分析 [J]. 中国科学：技术科学，2019，49 (12)：1385 - 1396.

[21] 刘晨迪，周建锋，苏彦．用 CLEAN 算法对嫦娥四号低频射电频谱仪信号进行干扰抑制 [J/OL]. 天文研究与技术，2021：1 - 16 [2021 - 11 - 01].

[22] 马友青，彭松，温博，等．基于嫦娥四号月球车图像的地形遮挡高精度预报试验研究 [J]. 中国科学：技术科学，2020，50 (1)：54 - 68.

[23] 司马兵，彭松，马友青，等．嫦娥四号月面巡视器全景相机制图方法 [J]. 中国科学：技术科学，2020，50 (2)：151 - 160.

[24] 王镓，万文辉，赵焕洲，等．基于视觉的嫦娥四号探测器着陆点定位方法 [J]. 载人航天，2019，25 (1)：12 - 18.

[25] 王镓，吴伟仁，李剑，等．基于视觉的嫦娥四号探测器着陆点定位 [J]. 中国科学：技术科学，2020，50 (1)：41 - 53.

[26] 武宇翔，温卫斌，平劲松，等．利用 Halo 轨道仿真开展嫦娥四号中继星与月面设备干涉基线的研究 [J]. 天文研究与技术，2021，18 (1)：69 - 76.

[27] 张韬，苏彦．嫦娥四号低频射电频谱仪降低背景噪声方法的研究 [J]. 天文研究与技术，2019，16 (3)：312 - 320.

An agile very low frequency radio spectrum explorer

Lin-Jie Chen[1], Yi-Hua Yan[1,2], Qiu-Xiang Fan[3], Li-Hong Geng[1] and Susanta Kumar Bisoi[1,4]

1. Key Laboratory of Solar Activity, National Astronomical Observatories, Chinese Academy of Sciences, Beijing 100101, China; ljchen@nao. cas. cn
2. School of Astronomy and Space Sciences, University of Chinese Academy of Sciences, Beijing 100049, China
3. Institute of Automation, Chinese Academy of Sciences, Beijing 100190, China
4. National Institute of Technology, Rourkela 769008, India

Abstract: The very low frequency (VLF) regime below 30MHz in the electromagnetic spectrum has presently been drawing global attention in radio astronomical research due to its potentially significant science outcomes exploring many unknown extragalactic sources, transients, and so on. However, the nontransparency of the Earth's ionosphere, ionospheric distortion and artificial radio frequency interference (RFI) have made it difficult to detect the VLF celestial radio emission with ground-based instruments. A straightforward solution to overcome these problems is a space-based VLF radio telescope, just like the VLF radio instruments onboard the Chang'E-4 spacecraft. But building such a space telescope would be inevitably costly and technically challenging. The alternative approach would be then a ground-based VLF radio telescope. Particularly, in the period of post 2020 when the solar and terrestrial ionospheric activities are expected to be in a 'calm' state, it will provide us a good chance to perform VLF ground-based radio observations. Anticipating such an opportunity, we built an agile VLF radio spectrum explorer co-located with the currently operational Mingantu Spectra Radio Heliograph (MUSER). The instrument includes four antennas operating in the VLF frequency range 1-70MHz. Along with them, we employ an eight-channel analog and digital receivers to amplify, digitize and process the radio signals received by the antennas. We present in the paper this VLF radio spectrum explorer and the instrument will be useful for celestial studies of VLF radio emissions.

Keywords: very low frequency-instrumentation; polarimeters-methods; data analysis

1 Introduction

The very low frequency (VLF) regime is the last possible unexplored regime to be studied in the whole electromagnetic (EM) spectrum in order to reveal several important astronomical sciences such as dark matter, galactic interstellar medium, transients and variable sources, etc (Jester & Falcke 2009; Boonstra et al. 2016). However, due to limitation of the Earth's ionosphere and strong artificial radio frequency interference (RFI), the celestial radio emissions below 30MHz are not studied in detail with ground-based radio telescopes. To study the low frequency universe several ground based radio telescopes have been built to date, such as the Low Frequency Array (LOFAR) in Europe (van Haarlem et al. 2013), the Long Wavelength Array (LWA) in New Mexico, USA (Ellingson et al. 2009), and the Giant Ukrainian Radio Telescope in Ukraine (Konovalenko et al. 2016). However, they mostly operate above 10MHz and are severely limited in radio observations between 10 to 30MHz.

In order to explore the VLF spectral regime, building radio telescopes locating beyond the Earth's ionosphere is the most viable solution. But the telescopes should be either located far way from the Earth or the observations from them should be shielded from the Earth-based RFI so that the RFI impact from the Earth can be reduced to the minimum level (Alexander et al. 1975; Kaiser et al. 1996). Keeping these in mind, several spacebased instruments have already been built to carry out the VLF observations (Herman et al. 1973; Kaiser et al. 1996; Gopalswamy et al. 2014) such as the very low frequency radio spectrometer (Ji et al. 2017) and the Netherlands-China Low Frequency Explorer (NCLE) onboard Chang'E-4 mission (Boonstra et al. 2017), which is the latest radio instrument operating in this field. Nevertheless, all of these instruments are mostly single element types, and it is also costly and technologically challenging to develop such space-based instruments.

The solar activity varies periodically every 11 years and it is known as a solar cycle. Currently, we are experiencing the minimum of solar cycle 24 and, similar to the previous minimum of solar cycle 23, it is expected that we will experience a sunspotless activity for around 2-3 years during the current solar cycle minimum. In order to access how the Earth's ionosphere would behave in the post-minimum of cycle 24, Janardhan et al. (2015) examined the correlation of the critical frequency in MHz of F-region ($foF2$) of Earth's ionosphere with the smoothed sunspot number (SSN) for the time period 1994-2014, and reported a strong correlation with a correlation coefficient, r of 0.96. From the estimation of the quantity $foF2$, we would know the value of the minimum reflection cut-off frequency of the Earth's ionosphere. The authors estimated that the $foF2^2$ for the period 2008-2009, covering the duration of the minimum of cycle 23 when there is no sunspot altogether, was 10 $(\text{MHz})^2$ implying that the maximum reflection cut-off frequency would be ≤3. 5MHz. Janardhan et al. (2015) therefore reported that the period after 2020, with a prolonged low levels of night - time maximum reflection cut - off frequency well below 10MHz, would help to obtain systematic ground based VLF radio observations. It is important to note that quiet - Sun conditions post 2020 would provide an excellent opportunity to perform VLF radio astronomy. Therefore, we propose and discuss a VLF radio spectrum explorer, which is located at Mingantu station.

The system composition and specification of the VLF radio explorer is briefly described in Section 2, while the design of the radio explorer is discussed in Section 3. In Section 4, scientific observations and analyses related to the radio spectrum explorer are presented, followed by the conclusion.

2 System Composition and Specification

The proposed VLF radio spectrum explorer is composed of four low - frequency active antennas, two analog receiving modules, a GPS receiver, and a digital receiver combined with a storage array. Each antenna receives radio signals with two linear polarizations. The eight - channel radio frequency (RF) signals from the four antennas are amplified and filtered by an amplifier module and a filter bank separately. Then the radio signals will be digitized by analog to digital converter (ADC) with suitable power levels and frequency bandwidths. In the digital receiver synchronized by a GPS receiver the sampled signals will be processed and stored for scientific studies. A systematical schematic of the radio explorer is shown in Figure 1.

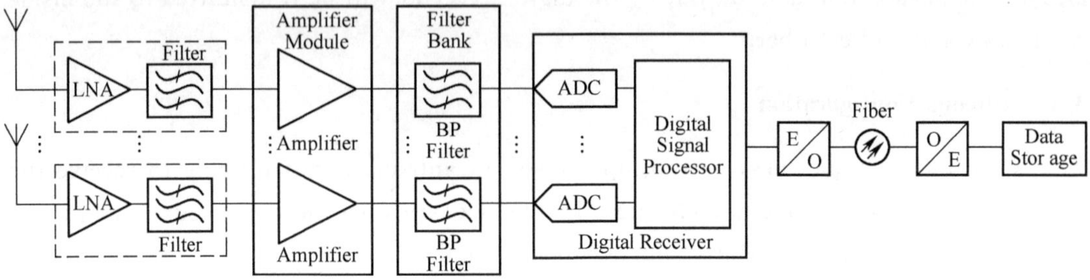

Fig. 1　System schematic of the VLF radio spectrum explorer.

The upper limit of operating frequency of the instrument is purposefully set to 70MHz. This would provide a common frequency window of 30 - 70MHz with other available low frequency radio telescopes so as to perform joint observations. The joint observations would, in particular, be used for instrumental calibration purposes and common science case studies. The setting of upper limit frequency to 70MHz would allow our radio observations free from the FM band (87 - 108MHz) signals. Thus, we could avoid the strong RFIs, and it also allows the band - pass filter to have a high stop - band rejection for the sampling bandwidth of 80MHz. The lower frequency limit is designed to be 1MHz so as to prevent the strong MW and LW RFIs. In the whole frequency band, 2048 frequency channels are implemented to get the high resolution dynamic spectrum with the minimum integration time of 10 milliseconds. The system dynamic range of the instrument is about 35dB. The main specifications are listed in Table 1.

Table 1　Basic Specifications of the VLF Radio Spectrum Explorer

Antenna	5m, 4 × 2dipoles
Polarization	Dual linear
RF channel	8
Frequency range	1～70MHz
Frequency channel	2048
Frequency resolution	～39kHz
ADC	12bits @ 160MHz
Time resolution	⩾10ms
Sky noise limited	5dB, 8～65MHz
System dynamic range	⩾35dB
Sensitivity	$\sim 10^4$ Jy @ 30MHz (10kHz, 100ms)

3　System Design

The whole system of the radio spectrum explorer mainly includes two parts, outdoor units and indoor units. The outdoor units include the active antennas, the analog receiving modules, the digital receiver, and the GPS receiver, which are hosted by an electronics cabinet. The indoor units include the monitoring and control computers, and the storage array. The observation data output by the digital receiver will be transmitted to the inside receivers via an optical fiber.

3.1　Antenna Configuration

At very low frequencies, setting up a resonant antenna is not feasible to receive the radio signal due to its long wavelength. In this scenario, we need an active antenna in order to obtain a good gain and a wide frequency band.

The resonant frequency of the active antenna should be set around 35MHz so as to achieve a frequency range from 1 to 70MHz. However, we set the resonant frequency of the antenna to be around 30MHz. Thus, we could avail more observations at the low end of the frequency band to study in detail and explore the science at a very low frequency regime. However, changing the resonant frequency to 30MHz we make a trade - off between the sensitivity and the frequency band. Therefore, the length of the antenna is set to be around 5meters. An inverted - V shaped dipole is designed as the receiving antenna. The antenna rests on a ground plane consisting of a metal mesh to avoid the reflected radio signals from the ground. The inverted - V shaped dipole antenna resting on a metal mesh is shown in Figure 2.

In order to have a wider field of view (FoV), the inverted - V dipole antenna is

Fig. 2 Radio spectrum explorer built at Mingantu Observing Station. It includes four inverted - V cross - dipole antennas standing above the ground metal meshes. The low noise amplifiers are at the top of the four supporting poles. For each antenna three cables inside the tubular pole are used to power the antenna and transmit its radio signals to the outdoor receiver, which is hosted by an RF - shielded cabinet hanging on the base block of the left parabolic antenna.

designed with an angle of 109° between the two monopoles. This, in principle, will broaden the radiation pattern of the antenna. Two perpendicular inverted - V dipoles are combined together to form a cross - dipole antenna as the receiving antenna element. The radio spectrum explorer consists of four inverted - V cross - dipole antennas. Three of them stand on the ground forming a regular triangle, the other one lies at the center of the triangle. All these three antennas have the same distance of 6meters from the central antenna. The antenna layout is shown in Figure 3. The four antennas of the radio spectrum explorer form a mini array. So we can get different radiation patterns for the mini - array by the method of beam - forming.

As shown in Figure 3, the single inverted - V antenna has an E - Plane beam widths varying from 153. 9 to 47. 6 degree between 1 to 70MHz. At lower frequencies, it can almost observe all the sky. Using detailed simulations, it is shown that the meshed ground plane improves the gain of the antenna at most of the frequencies. Although the distances between the four antennas are short - spaced, as shown in Figure 3, it is, however, still possible to obtain a rough pointing at higher frequencies via beam - forming.

In an active antenna, a low noise amplifier (LNA) connected with the antenna is used to receive and amplify the radio signals with low noise. In a 50 - ohm system the LNA of the active inverted - V antenna can achieve a low noise figure of ~2. 0dB at most of the operating frequency band. Also, the LNA allows a matching between the antenna and the

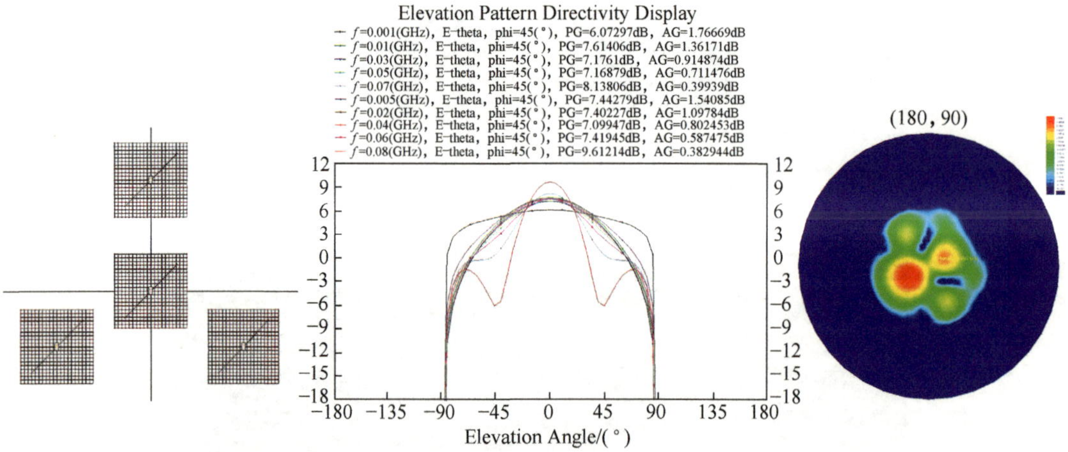

Fig. 3 Antenna layout and simulations for the radio spectrum explorer. *Left*: Layout of the four inverted - V dipole antennas. *Middle*: The simulated radiation patterns of the single antenna. *Right*: The simulated beamforming pattern at 30MHz for the four antennas.

transmission cable. In the design of this LNA, a two - stage amplifier (Chen et al. 2018) is employed to obtain a reasonable gain and low noise. While the use of the LNA reduces the noise contributions, the available gain of the active antenna is decreased as well due to mismatching between the active antenna and LNA. For one inverted - V cross - dipole antennas, two LNAs are used and placed on one printed circuit board (PCB) combined together. The PCB is kept inside a 3D printing box, which is fixed on the top of the antenna supported by a pole as shown in Figure 2.

At very low frequencies, the sky noise limited performance is a primary requirement for the antenna system to carry out observations with high spectral and temporal resolutions. This shows how large the sky noise is during the output of the antenna as compared to the noise from the antenna system itself. In order to meet this requirement, the LNA is designed to achieve a sky noiselimited performance within each operational frequency band. Using a sky temperature model (Jester & Falcke 2009), we simulated the sky noise limited performance of the active antenna. As shown in Figure 4, it is seen that an approximate of 5dB sky noise limitation is achieved between 8 and 65MHz. Using a method proposed in Chen et al. (2018), we calculated the sky noise - limited performance for the active inverted - V antenna inside the chamber and in the field as well. From Figure 4 it is clear that the measured results match well with the simulation results. However, it is noticed that above 42MHz the measured results differ from the simulation results. It is probably caused due to the difference in the noise level of LNA as estimated by the measurement and the simulation. In spite of this, it is evident that the measured results meet the required antenna noise level needed for our operational frequency range and clearly also illustrate the performance of the inverted - V dipole antenna designed for the radio explorer. It is to be noted that we did not measure the sky noise - limited performance below the frequency of 25MHz due to the RFI impact.

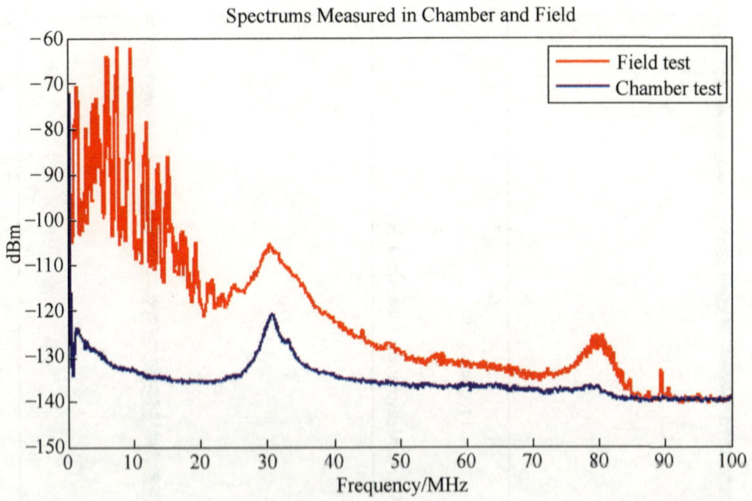

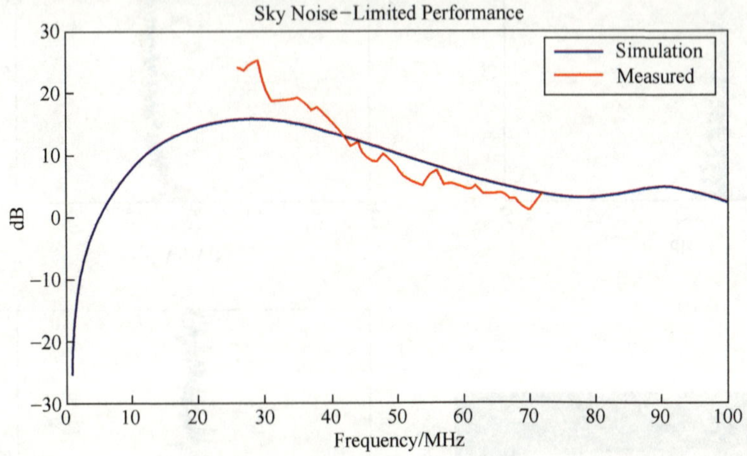

Fig. 4　Sky noise - limited values as measured for an inverted - V dipole antenna. *Left*: Calibrated noise power density as measured in the chamber and at Mingantu Observing Station. *Right*: The performance of the simulated and measured results of the sky noise - limited values.

3. 2　Analog Receiving Module

As mentioned earlier, an amplifier module following the antenna is employed to amplify the radio signal in order to make its power suitable for sampling. Due to the low available gain of the antenna as a result of mismatching with the amplifier, we consider to reduce the receiver noise after the antenna to improve the system sensitivity. The amplifier has a gain of around 22dB and a low noise level of 2dB. This, in turn, significantly reduces the noise contribution of the digital receiver. In order to avoid the signal aliasing in ADC, we use an anti - aliasing bandpass filter to achieve the operating frequency band of (1 - 70) MHz.

As shown in Figure 5, the gain of the amplifiers is about 22dB, and the VSWR are

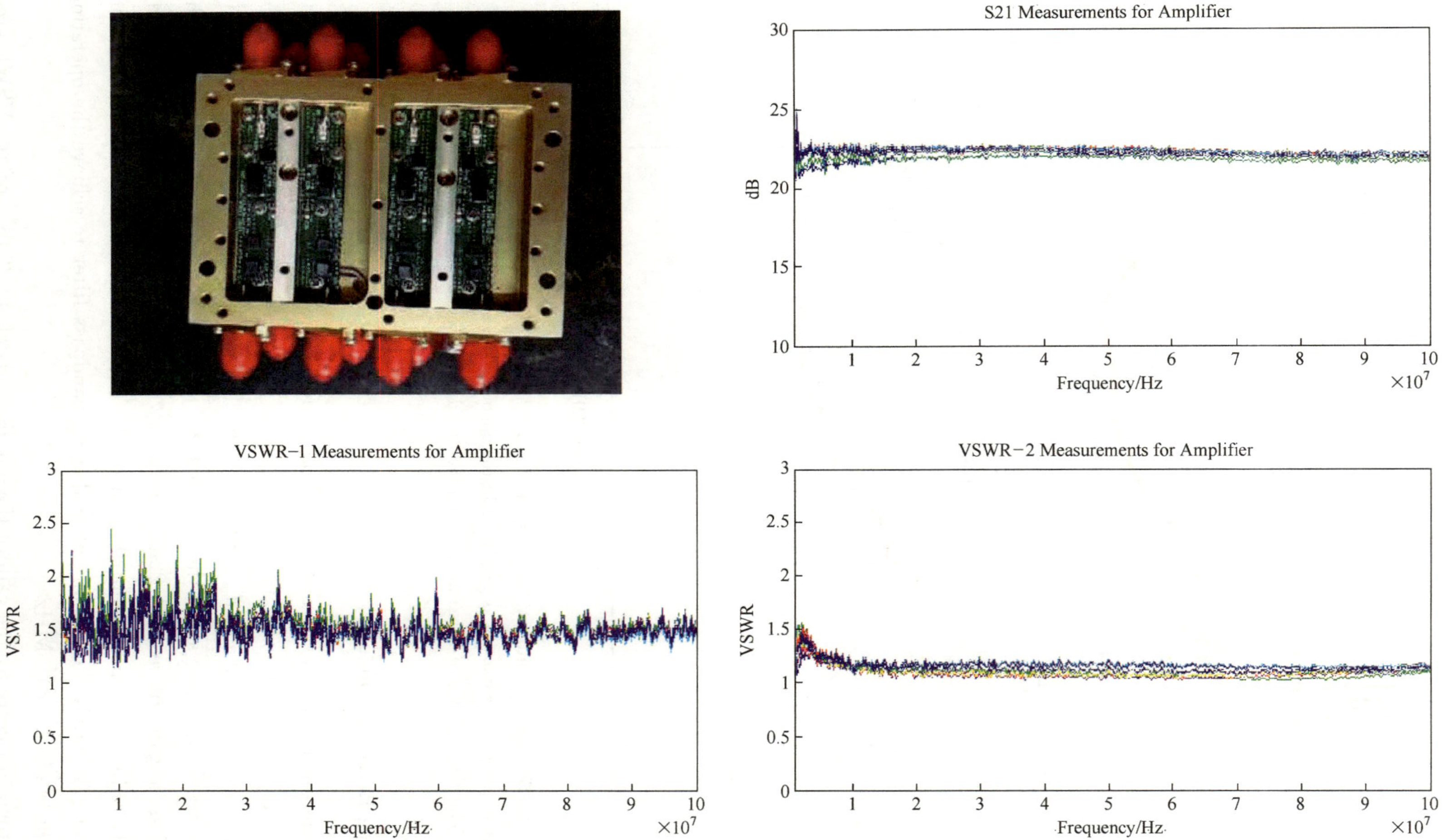

Fig. 5　Amplifier measurements, different colors represent different amplifiers. *Upper-Left*: The amplifier module. *Upperright*: The measured S-parameter S21, it shows the gain of the amplifier. *Lower-left*: The measured voltage standing wave ratio (VSWR) at the input port. *Lower-right*: The measured VSWR at output port.

less than 2 and 1.5, respectively, at input and output ports. We can also see that the measurement results coincide with each other among the amplifiers at most of the frequency range. Figure 6 plots the measurement results for the filter bank. As we can see that the 3dB bandwidths of the filters are from 1 to 68MHz, it is slightly narrower than the design. The measurements show the filters accord with each other well. From both Figures 5 and 6, it can be seen that there is a lesser difference for the same measurements of both the amplifier and filter. This is mainly attributed to the nonidentical characteristic of the electronic components.

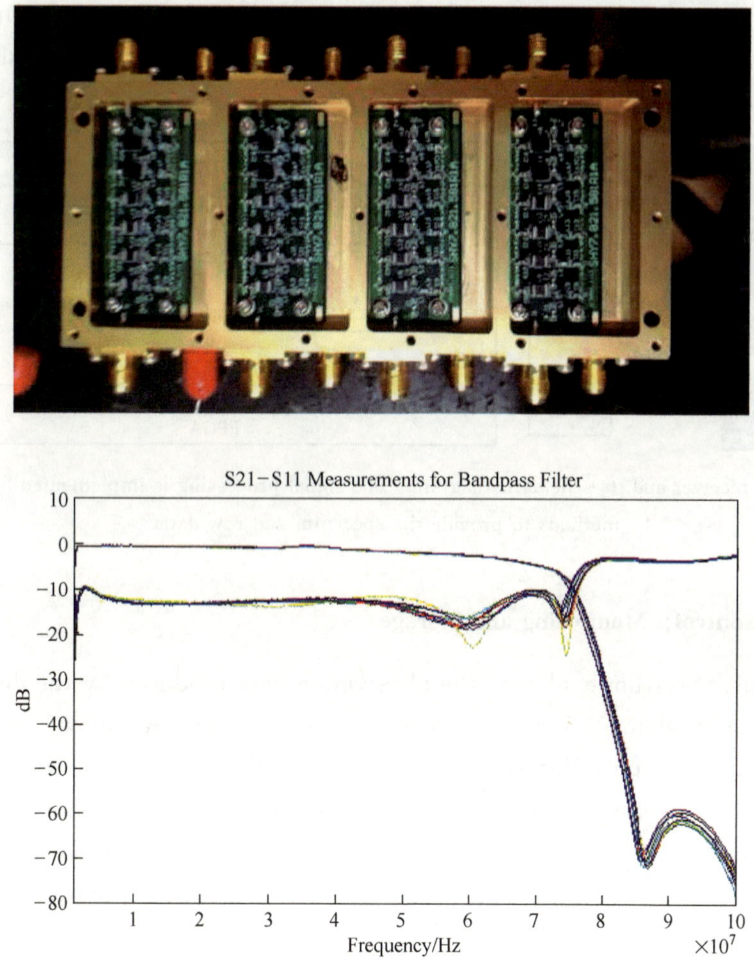

Fig. 6　Bandpass filter measurements, different colors represent different filters. *Left*: The filter bank. *Right*: The measured S-parameter S21 showing the the frequency response of the filter, and S11 showing the reflection coefficient at the input port. Since the symmetrical characteristic of the filter, only half of the parameters are plotted.

3.3　Digital Receiver

For a low-frequency radio array, the digital receiver is a core instrument. The digital receiver mainly performs all the required signal processing in the digital domain. In our proposed VLF radio spectrum explorer, the digital receiver used would sample the received signals by the active antenna with a 12 - bit ADCs operating at 160 million samples per second (MSPS). We then use a 4096 - point fast Fourier transform (FFT) to channelize the received signals from all the four antennas. In this way, we can obtain 2048 sub - band signals with a bandwidth of ~39kHz. Each sub - band signal would be integrated next for at least 10milliseconds, which can be changed as per requirement. Meanwhile, a piece of 4096 - point raw sampled data would be sent for output in a single integration time period for all the antennas that can further be used for more flexible offline processing. For all the four dual - polarized dipole antennas, the maximum output data rate obtained is around 12.5MB s^{-1}.

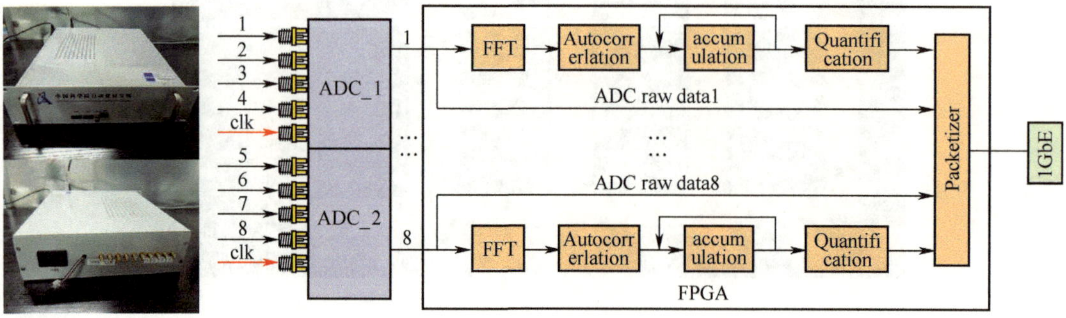

Fig. 7 Digital receiver and its schematic diagram. The signal processing is implemented in two different methods to provide the spectrum and raw data.

3.4 System Control, Monitoring and Storage

In the radio spectrum explorer, the observation data produced by the digital receiver would be transmitted to an indoor receiver via optical fiber. As shown in Figure 8, a high-performance computer located in a high - level EM shielding room is used to receive the data and store them in a disk array. Meanwhile, the computer also controls the digital receiver and monitors the observation results in real time. For the convenience of observing, the system control and monitoring can be achieved remotely on any other authorized computer with the internet connection.

3.5 System Analysis

In order to analyze noise distribution for the radio spectrum explorer, we define the antenna noise and receiver noise with a reference plane between the antenna and the LNA used in Chen & Yan (2019). The antenna noise includes the background sky noise and the thermal noise from the antenna, while the receiver noise includes the noise from the LNA, analog and digital receiving units. We use the temperature model in Jester & Falcke (2009) and simulate the different noise contributions from the whole system. The noise

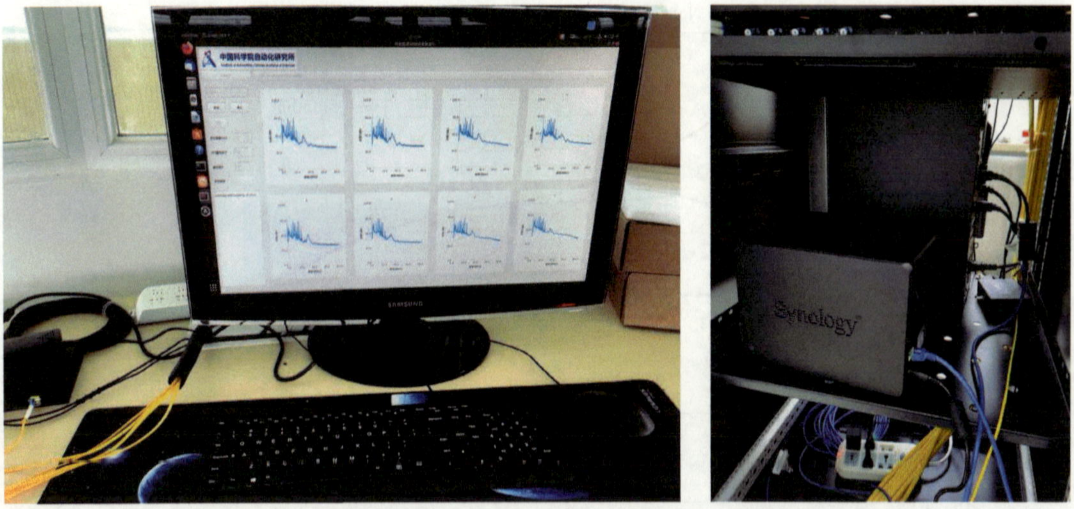

Fig. 8　System control, monitoring and storage for the radio spectrum explorer. *Left*: the control and monitoring interface displaying on a screen locating in an EM shielding observation room. *Right*: the control computer and disk array in a highlevel EM shielding room. The screen and computer are connected with each other via a fiber to avoid the RFI leaking.

figure of the amplifier after the antenna is around 1. 0dB. Thus, its noise temperature is around 75K. As the gain of the amplifier is 22dB and the available gain of the LNA is about 20dB, we can ignore the equivalent noise of the amplifier and digital receiver at the reference plane.

The simulated noise contributions of the system at the reference plane are plotted in Figure 9. It can be seen that the antenna noise exceeds the receiver noise above 7MHz. Generally, the sky noise is very high and the antenna thermal noise is too small. Hence, it can be concluded that the sky noise is always larger than the receiver noise. Further analysis shows that the sky noise is at least 5dB more than the receiver noise above 8MHz, which agrees well with the sky noise - limited result shown in Figure 4. Similar to the simulation of noise contribution, the sensitivity of the antenna as well as the mini array was also simulated and plotted in Figure 9. From Figure 9, it is clear that the antenna and mini array sensitivities lie well below the peak and average radio emission fluxes from Jupiter. We thus conclude that even the single antenna sensitivity is good enough to detect the radio emission from Jupiter. In the other hand, the solar radio bursts, which generally produce strong radio emission at very low frequency range, could be as well detected by the antenna and mini array.

The system dynamic range for the radio spectrum explorer depends on all the active devices such as the active antenna, amplifier and digital receiver. Simulations of the system show that, in absence of RFI, the total power of the radio signal received by the antenna is about −48dBm. This roughly agrees with the measurement carried out at the instrument site. Similarly, the gain of the amplifier is 22dB and its input power of 1dB

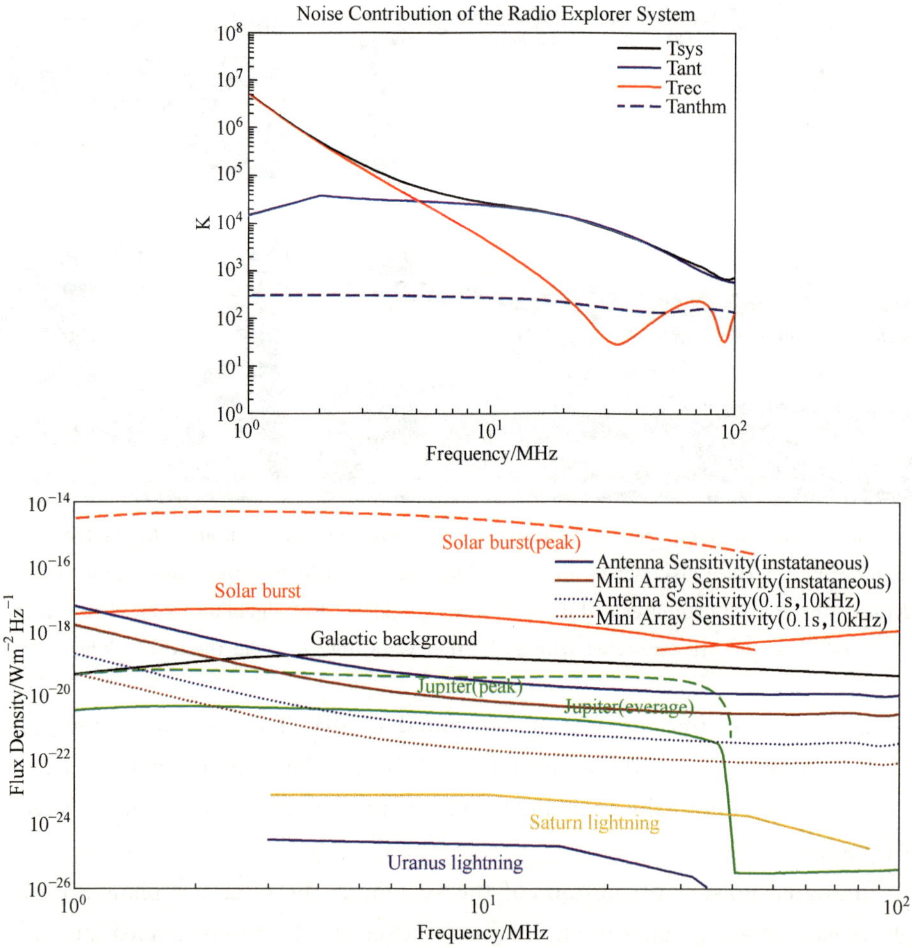

Fig. 9　*Left*: Simulated noise contributions of the receiving chain for the radio spectrum explorer. The black, blue, red, and dashed blue plots represent respectively the system noise, the antenna noise, the receiver noise and the antenna thermal noise. *Right*: Flux densities of radio emissions from different cosmic objects along with the antenna and array sensitivities are plotted. For plotting the different cosmic radio emissions, we used the data in Zarka et al. (2012), while for the sensitivities of the antenna and array system, we performed simulations with different integration time and bandwidth.

compression point is around −8dBm. Thus, we estimate that the net signal power at the ADC input would be −26dBm. If we use a 12-bit ADC with a full-scale of 10dBm, then the net signal power at the ADC input would fill only 6 bits. The remaining 6bits could be used for obtaining the high dynamic range signals. Thus, in absence of the RFIs, the radio spectrum explorer could have a system dynamic range of around 36dB.

4　Scientific Observations

The system design as aforementioned for the radio spectrum explorer would make it a competent instrument at VLF, which can be useful for carrying out studies of various cosmic objects and phenomena such as solar radio bursts, planetary radio emissions and joint low frequency observations with other available facilities.

4.1 Expected Scientific Studies

A round clock monitoring of solar activity is required to understand space weather in near - Earth space. In particular, changes pertaining to eruptions of solar flares and coronal mass ejections (CMEs) are crucial as such phenomena are not well studied in near -Earth space. One way to quantify the changes is the study of radio burst activity associated with them. There are several types of burst activity, but the two burst types are of particular interest: Type Ⅱ and Ⅲ. Type Ⅱ bursts are produced by fast driving shocks associated with the flares or CMEs, while Type Ⅲ bursts are a result of propagating electron beams in solar corona and interplanetary space. These bursts are produced near the local electron plasma frequency or/and its harmonics and as the plasma frequency depends on the electron number density, the source position of each burst corresponds to a certain height in the corona. Using our proposed VLF radio spectrum explorer, which can be used as a highly effective solar monitoring system, we aim to identify these Type Ⅱ and Ⅲ burst types from the solar dynamic spectra. In coordination with LOFAR and LWA solar imaging spectroscopy, we will study the plasma processes associated with these bursts to prepare a density diagnostics of the energetic electrons at different heights of the corona. This, in turn, will enable the study of density diagnostics of the different structure of CMEs as well as determining their velocity and acceleration. Also, through observation of absorption bursts from solar dynamic spectra of VLF radio spectrum explorer, we plan to locate the source region of CMEs and plasma characteristics associated with the emerging magnetic flux channels. Hence, this instrument can effectively be used to study the propagation and evolution of CMEs as well as the non - thermal electron beams associated with the flares in the interplanetary space through the monitoring of solar radio burst activity.

In addition to solar activity, the VLF radio instrument can also be used to study planetary radio emissions. It is known that the planet Jupiter emits strong radio emissions (~MJy), which are point - like, fully polarized and quasi - permanent in a 0.3—3MHz range. In particular, the Jupiter emissions can be well measured in the decameter (30MHz) range (Zarka 2004; Nigl et al. 2007). Thus, the Jupiter emissions can be useful for ground calibration for the radio spectrum explorer. While the planetary radio emissions from Saturn and Uranus can be used to study quasi - continuous radio emissions and their seasonal effects.

The ground VLF radio spectrum explorer can be used to perform joint observations with the Netherlands - China low frequency explorer (NCLE), which is a radio instrument onboard the Chinese Chang'E - 4 lunar mission (Boonstra et al. 2017) to perform radio observations in the frequency range 0.08 – 80MHz. The key science objectives of the NCLE, that operates in the similar frequency range like our VLF radio spectrum explorer, are overlapped with that of our instrument, such as measuring radio emission from different planets in our solar system, studying Earth's ionospheric fluctuations, and solar radio activity and space weather. We, therefore, target for the joint spectrum observations from the VLF radio spectrum explorer and the NCLE instrument in the frequency range 1 – 70MHz for the studies of planetary radio emission, solar radio bursts, and space weather in near – Earth space.

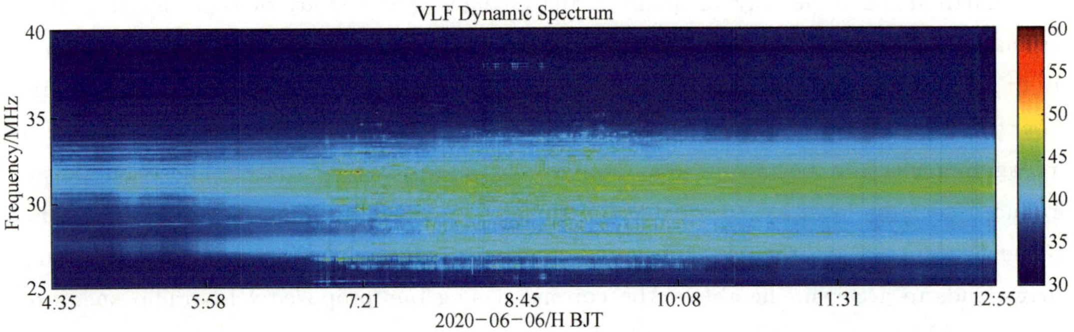

Fig. 10 Spectrum of ionospheric low frequency radio emission observed by the radio spectrum explorer on 2020 June 6.

4.2 Observing Modes

As mentioned in Section 3, this radio spectrum explorer contains four VLF radio antennas, and the digital receiver will produce two different data outputs including the spectrum data and raw sampling data, which make it capable to do the observations in different working modes. If the four antennas work independently, each of them can be considered as a spectrometer as shown in Figure 11. These four antennas can also work together as a mini array to improve the sensitivity. For the spectrum data, we can sum the same polarized signals of the four antennas together to get the combined signal with an incoherent beam, as shown in Figure 12. Suppose the four antennas are identical, the incoherent beam is the same with the beam of one single antenna. In this way, the sensitivity

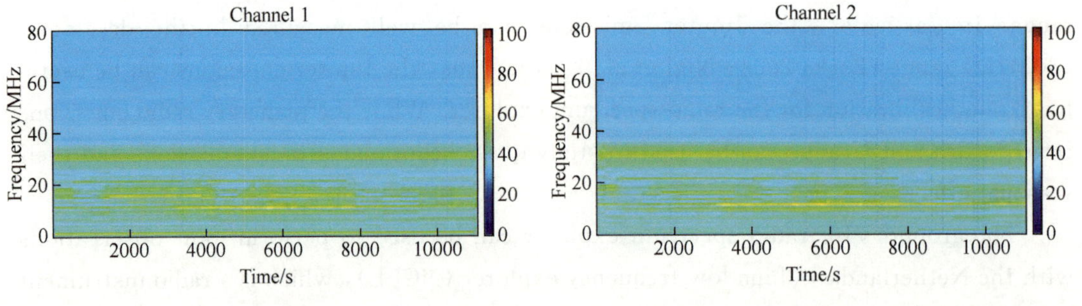

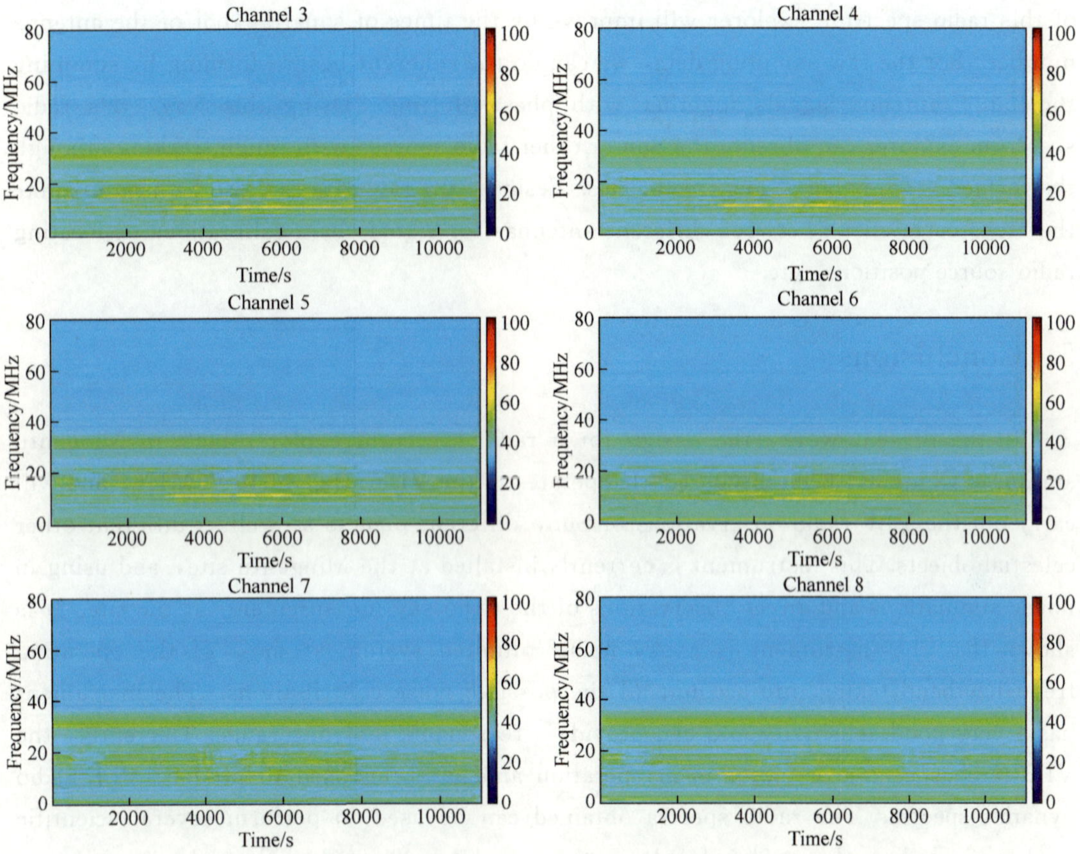

Fig. 11　Dynamic spectrum observed by the four dual - polarization antennas of the built radio spectrum explorer on 2020 May 31.

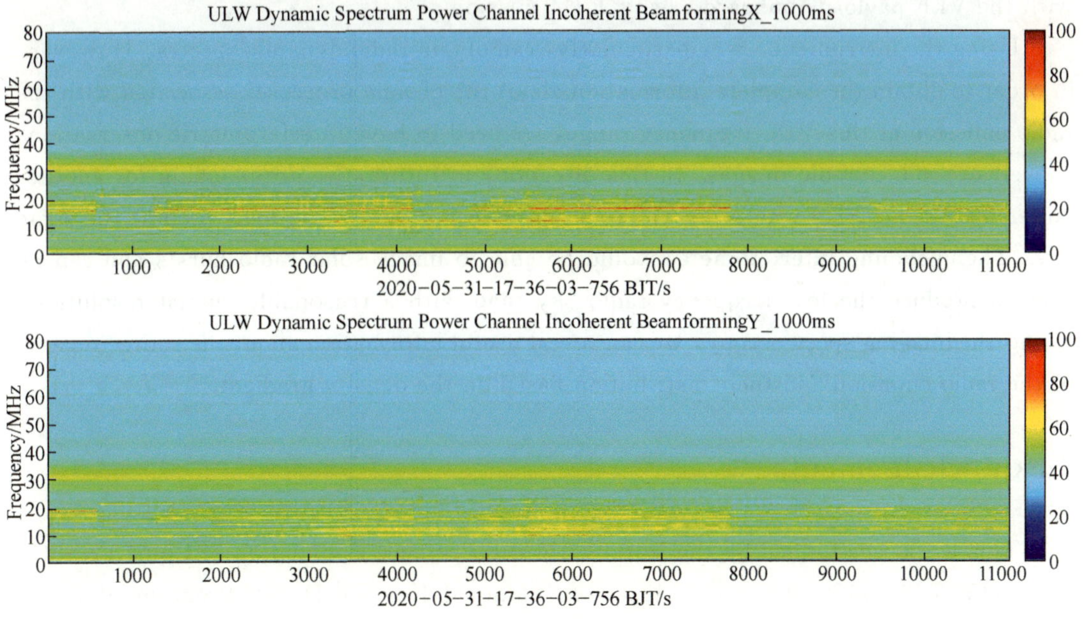

Fig. 12　Dynamic spectrum observed with the incoherent beam pointing at the Zenith on 2020 May 31.

of this radio spectrum explorer will improve by the times of square - root of the antenna number. For the raw sampling data, we can do the coherent beam - forming by summing the four - antenna signals together with phase shifting. Using this way, this radio spectrum explorer can observe the Sun or other radio sources with rough tracking, though the formed beam is big at low frequencies. Besides, the raw data can also be used to make the cross correlations between different antennas for signal chain calibrations or locating radio source positions, etc.

5 Conclusions

In the present work, the design for a radio spectrum explorer built in Mingantu station in China has been discussed. It operates in the VLF range of 1 - 70MHz, and will carry out the VLF radio observations of Sun, different planets as well as different other celestial objects. This instrument is currently installed at the Mingantu site, and using in house simulations and direct observations of the radio sky measurements at the site. It is shown that the instrument is a sky noise - limited system at most of the operating frequency band taking into account all the possible noise contributions. The instrument has a very good sensitivity and can provide a reasonable dynamic range. Currently, the VLF radio spectrum explorer is in operation and has been used to produce VLF radio dynamic spectra. The radio spectra obtained can be used to perform several scientific studies, including the study of solar activity and space weather through a continuous monitoring solar radio burst activity associated with the energetic electrons from the solar flares and CMEs, planetary radio emissions, and carrying out joint observations as well with the VLF payloads onboard Chang'E - 4 mission.

Using the instrument, we, particularly, can obtain radio dynamic spectra. However, in order to obtain the complete information about the plasma processes associated with the radio emission at this VLF frequency range, we need to have interferometric observations as well to produce radio images. In this direction, in future, we are planning for another larger VLF radio array consisting of 36 antennas within 3km to be built soon at the same site. The upcoming interferometer would be able to image solar radio bursts and can be used to produce the low frequency radio sky map with a reasonable spatial resolution. Using the imaging spectroscopy, we can reveal useful information about the source regions of the radio emission and their distribution based on the density gradient.

Acknowledgements

This work was funded by the National Natural Science Foundation of China (NSFC, Nos. 11573043, 11790305 and 11433006), National Key R&D 278 Program of China (2018YFA0404602), the CE - 4 mission of the Chinese Lunar Exploration Program: the

Netherlands - China Low Frequency Explorer (NCLE), and Chinese Academy of Sciences (CAS) Strategic Priority Research Program (XDA15020200) are acknowledged. The authors would like to thank Mr. Jianxi Ren and Ms. Zhijun Chen from NAOC for the help in building this radioexplorer.

References

[1] Alexander, J. K., Kaiser, M. L., Novaco, J. C., Grena, F. R., & Weber, R. R. 1975, A&A, 40, 365

[2] Boonstra, A., Garrett, M., Kruithof, G., et al. 2016, Discovering the Sky at the Longest Wavelengths (DSL), in 2016 IEEE Aerospace Conference, 1, Yellowstone Conference Center, Big Sky, Montana, Mar 5 - 12, 2016

[3] Boonstra, A. J., Wise, M., Marel, J. v. d., et al. 2017, The Netherlands - China Low Frequency Explorer, in Porceeding of 32nd URSI GASS, Montreal, 19 - 26 August 2017

[4] Chen, L., Aminaei, A., Gurvits, L. I., et al. 2018, Experimental Astronomy, 45, 231

[5] Chen, L., & Yan, Y. 2019, Radio Science, 54, 660

[6] Ellingson, S. W., Clarke, T. E., Cohen, A., et al. 2009, in IEEE Proceedings, 97, 1421

[7] Gopalswamy, N., Makela, P., & Yashiro, S. 2014, in 2014 United States National Committee of URSI National Radio Science Meeting (USNC - URSI NRSM), 1

[8] Herman, J. R., Caruso, J. A., & Stone, R. G. 1973, Planet. Space Sci., 21, 443

[9] Janardhan, P., Bisoi, S. K., Ananthakrishnan, S., Sridharan, R., & Jose, L. 2015, Sun and Geosphere, 10, 147

[10] Jester, S., & Falcke, H. 2009, New Astron. Rev., 53, 1

[11] Ji, Y., Zhao, B., Fang, G., et al. 2017, Journal of Deep Space Exploration, 4, 150

[12] Kaiser, M. L., Desch, M. D., Bougeret, J. L., Manning, R., & Meetre, C. A. 1996, Geophys. Res. Lett., 23, 1287

[13] Konovalenko, A., Sodin, L., Zakharenko, V., et al. 2016, Experimental Astronomy, 42, 11

[14] Nigl, A., Zarka, P., Kuijpers, J., et al. 2007, A&A, 471, 1099

[15] van Haarlem, M. P., Wise, M. W., Gunst, A. W., et al. 2013, A&A, 556, A2

[16] Zarka, P. 2004, Planet. Space Sci., 52, 1455

[17] Zarka, P., Bougeret, J. L., Briand, C., et al. 2012, Planet. Space Sci., 74, 156

Lunar Terrestrial Analog Experiment on the Spectral Interpretations of Rocks Observed by the Yutu－2 Rover

Rui Chang[1,2], Wei Yang[1], Honglei Lin[1], Rui Xu[3], Sheng Gou[4], Rong Wang[3] and Yangting Lin[1]

1. Key Laboratory of Earth and Planetary Physics, Institute of Geology and Geophysics, Chinese Academy of Sciences, Beijing 100029, China; changrui@mail.iggcas.ac.cn (R. C.); linhonglei@mail.iggcas.ac.cn (H. L.); linyt@mail.iggcas.ac.cn (Y. L.)
2. University of Chinese Academy of Sciences, Beijing 100049, China
3. Key Laboratory of Space Active Opto－Electronics Technology, Shanghai Institute of Technical Physics, Chinese Academy of Sciences, Shanghai 200083, China; xurui@mail.sitp.ac.cn (R. X.); wangrong@mail.sitp.ac.cn (R. W.)
4. State Key Laboratory of Remote Sensing Science, Aerospace Information Research Institute, Chinese Academy of Sciences, Beijing 100101, China; gousheng@aircas.ac.cn

Abstract: A visible and near－infrared imaging spectrometer (VNIS) loaded by the Chang'e－4 rover is the primary method for detecting the mineral composition of the lunar surface in the landing region. However, different data processing methods yield inconsistent mineral modes in measured lunar soil and rocks. To better constrain the mineral modes of the soil and rocks measured by Chang'e－4 VNIS, a noritic－gabbroic rock with a mineral composition similar to that of the lunar highland rocks is measured by scanning electron microscopy (SEM), the spare. ight model of Chang'e－4 VNIS and TerraSpec－4 of ASD. Backscattered electron and energy dispersive spectrometry show that olivine, pyroxene, and plagioclase modal mineral abundances are 12.9, 35.0, and 52.2%, respectively. The estimated results of the spectrum by the Hapke radiative transfer model are 7.5, 39.3, and 53.2% for olivine, pyroxene, and plagioclase, respectively, which is consistent with to those of SEM mapping within error. In contrast, the estimated results of the modified Gaussian model are 29 and 71% for olivine and pyroxene, respectively, indicating the absence of plagioclase. Based on our implemented Hapke model, we decode the data of the two rocks detected by the rover on the 3rd and 26th lunar days of mission operations. The obtained results suggest that both rocks are norite or gabbro with noticeable differences. The first rock, with more olivine and pyroxene, may have been excavated from the Finsen crater. The second rock, with more plagioclase, may have been ejected from the southwestern edge of the Von Kármán crater, indicating the initial lunar crust.

Keywords: visible and near - infrared spectra; Hapke radiative transfer model; mineral abundance; rock; Chang'e - 4

1 Introduction

The Chang'e - 4 (CE - 4) spacecraft of China successfully landed on 3 January 2019 at the Von Kármán crater of the South Pole Aitken (SPA) basin on the lunar farside. The Yutu - 2 rover of the Chang'e - 4 mission traveled >700m until July 2021 and performed a detailed investigation of the lunar surface. A panoramic camera, visible and near - infrared imaging spectrometer (VNIS), and ground - penetrating radar were equipped with the Yutu - 2 rover to explore the topography, composition, and subsurface structure of the landing area, respectively. The VNIS of the Yutu - 2 rover covers the spectral range from 450 to 2395nm with a wavelength interval of 5nm [1, 2], providing an unprecedented opportunity to study the compositions of deep lunar materials.

Based on Clementine 500m data, the composition of the lunar crust of the SPA is uniformly noritic [3]. It is mainly covered with well - developed mature and anorthositic regolith accompanied by the outcropping of norite, gabbro, and olivine - rich rocks [3]. The landing site is located in the Mg - pyroxene annulus of the SPA basin [4], which is the largest and deepest basin on the Moon, allowing us to touch deep lunar materials. The Von Kármán crater is filled with pre - Nectarian basalts and Imbrian and Eratosthenian ejecta from nearby craters [5 - 7]. The exact landing site is located on ejecta strips radiating from the Finsen crater. Previous studies have revealed that although the materials around the landing site are unlikely to be local basalts, they could represent the SPA basement [5 - 8].

The mineral abundances of lunar regolith decoded from the VNIS data in previous studies yielded significant differences [7, 9 - 12]. VNIS is the primary method used to detect the composition of soil and rocks on the lunar surface and trace their origins. However, complex factors, such as space weathering, particle size, multiple scattering of particles, spectral response of the instrument, and viewing geometries [3, 13 - 15] affect the spectral features, which can result in large uncertainties in the quantitative mineral compositions from the spectra of the lunar surface. For example, whether the estimated result contains plagioclase depends on the estimation method, the modified Gaussian model (MGM), or the Hapke radiative transfer model. MGM is primarily used to quantify mafic minerals, such as olivine and pyroxenes, because they have unique Fe^{2+} absorption features centered near 1 and 2μm. However, it is difficult to estimate the abundance of plagioclase, owing to its weak spectral absorptions in the near - infrared bands. Li et al. (2019) suggested lunar regolith with ~50% olivine, ~40% low - calcium pyroxene (LCP), and ~10% high - calcium pyroxene (HCP) using MGM [9]. The results indicated that the materials at the landing site could have originated from the original lunar mantle [9] without considering plagioclase. Other results using the Hapke radiative transfer model suggested that the source of the materials observed by the Yutu - 2 rover could be the lunar crust or impact melt [7, 10, 11]. A study with additional data on lunar regolith indicated ~60% plagioclase, ~30% pyroxene, and ~10% olivine at the landing site [10, 11]. In addition, fresh rock was observed by the Yutu - 2 rover on the 3rd lunar day of the mission. Its spectrum shows apparent absorption at wavelengths of 1 and 2μm. The results estimated by the Hapke radiative transfer model for the rock indicate ~50% plagioclase, ~40% pyroxene, and ~10% olivine [7, 11]. The results estimated by the Hapke radiative transfer model may indicate an impact - induced melt pool source from the lower crust and upper mantle [7, 11]. The gap between data and interpretation is that the ratio of LCP to HCP has considerable diversity, and the absolute errors of the results are larger than 10%. Therefore, standard calibration in data processing methods with lunar terrestrial analogs is necessary to better interpret the Yutu - 2 spectral data.

To quantitatively evaluate the reliability of different VNIS data processing methods, we selected a rock (Figure 1) with a reflectance spectrum similar to that of the rock

Figure 1　(a) The gray panoramic camera image of the rock. (b) The complementary metal - oxide semiconductor (CMOS) image of the rock observed by the visible and near - infrared imaging spec - trometer onboard the Yutu - 2 rover. The yellow circle in the CMOS image shows the viewed area of the short - wavelength infrared detector. (c) The rock used in the terrestrial analog experiments.

observed by the Yutu - 2 rover on the 3rd lunar day of the mission [7] . We performed terrestrial analog experiments using the spareflight model of the Chang'e - 4 VNIS (VNIS SF) . The "ground truth" composition of the selected rock was obtained by geochemical analysis in the laboratory. Thus, the present study aimed to optimize the processing method and estimate the model of VNIS data to obtain a more accurate modal mineral abundance of the detected objects.

2 Sample Description and Analytical Methods

2.1 Sample Description

The terrestrial igneous rock CR - 1 studied herein is a norite - gabbro. CR - 1 had a brown - black color and was 15cm × 8cm × 4cm in size (Figure 1c) . This sample had an ophitic texture and was mainly composed of pyroxene and plagioclase, with minor olivine crystals based on binocular observations. A fragment of CR - 1 was cut into slices with a thickness of approximately 2mm and polished to prepare for petrographic and mineral chemistry analyses. In order to avoid the potential influence of low - temperature alteration, the fresh inner part of the rock was crushed to approximately 60 mesh using an agate mill to separate olivine, orthopyroxene, clinopyroxene, and plagioclase grains. The separated minerals were then handpicked under a binocular microscope to achieve 100% purity.

2.2 Backscattered Electron Imaging

Petrographic observations of the comparison samples were performed by backscat - tered electron (BSE) imaging using a Thermo Fisher Apreo field emission scanning electron microscope with a Bruker XFlash 60 energy dispersive spectroscopy (EDS) detector at the Institute of Geology and Geophysics, Chinese Academy of Sciences (CAS), Beijing, China. The instrument was operated at accelerating voltages of 10 - 15 kV with 1.6 - 13 nA beam currents. The mineral distribution map was constructed using Maps and Nanomin software [16] . The calculation method of mineral composition involves collecting BSE and X - ray signals at each point and comparing them with recipes of the database to recognize the minerals automatically.

2.3 Ground Spectroscopic Verification Experiment

An experiment by TerraSpec - 4 of ASD was performed to test the end - member spectra of the rock - forming minerals to interpret the spectrum of CR - 1. The ground spectroscopic verification experiment consisted of two steps. The VNIS SF experiment was conducted to obtain data at the laboratory using the same instrument as that at the lunar surface.

The separated pure minerals, which were used as end – members to interpret the spec – trum, were detected using TerraSpec – 4. The process of testing the spectra by TerraSpec – 4 was performed in a darkroom, and the Hi – Bright contact probe was used. The incidence and emission angles were 0° and 30°, respectively. The instrument TerraSpec – 4 was operated at the setting of 136ms integration time of visible spectra testing and 16 gain with 2084 offset of near – infrared spectra testing.

The VNIS SF experiment was performed in an optical darkroom (Figure 2a) at the Key Laboratory of Space Active Opto – Electronics Technology, Shanghai Institute of Technical Physics (CAS). The VNIS SF experiment was based on a motionless platform that stimulated the Yutu – 2 rover and detected the lunar surface target in situ as the rover stopped. A collimating light source sphere composed of tungsten halogen lamps was used, and the radiation reflected by the surface of the rock entered the VNIS SF. It wasfirst collimated into a parallel beam by the instrument and modulated by an acousto – optic tunable filter to form quasi – monochromatic light of a specific wavelength (45°– 2400nm) that converged on the detector. Thus, the single – band spectrum and image of the observation target were obtained. The incidence angle was 45. during the experiment to ensure the intensity of light. The VNIS SF was similar to the VNIS onboard the rover, and the experimental conditions were the same as the detection conditions on the lunar surface (Figure 2b). The sample was measured from a height of 0.69m above the ground surface at an emission angle of 45°.

(a) (b)

Figure 2 (a) The picture of the VNIS SF experiment condition. (b) The model picture of the detecting condition of VNIS on the lunar surface [17].

2.4 Data Processing Methods

The common methods for quantifying mineral abundance on the lunar surface are the

Hapke radiative transfer model [14] and MGM [18]. We applied the Hapke model described by Lin et al. (2020) [7] to retrieve the mineral fractions from the rock spectrum acquired in the laboratory using the VNIS SF. The Hapke model describes the relationship between reflectance and single-scattering albedo, phase function, multiple scattering properties, and so on [14]. To avoid the thermal effects of the hot lunar surface [19], we only used spectra from 0.75 to 1.6μm, which covers all absorption features near 1μm, to perform spectral unmixing. The pure end-members selected from the rock sample and the spectra of the Apollo samples collected in the RELAB database were used. MGM, which does not rely on predetermined assumptions of the spectral constituents in a mixture, was also applied to the spectrum of the laboratory rock sample. The normalized band strength ratio [20] was calculated from the MGM-derived band depths to estimate the mineral fractions in the rock sample [21].

3 Results

3.1 Scanning Electron Microscopy Results

Scanning electron microscopy (SEM) mapping indicated that CR-1 comprised olivine, pyroxene, and plagioclase with a gabbro texture. The modal mineral abundances of olivine, pyroxene, and plagioclase were 12.9, 35.0, and 52.2%, respectively (Figure 3). The error bar for the obtained modal mineral abundance was approximately

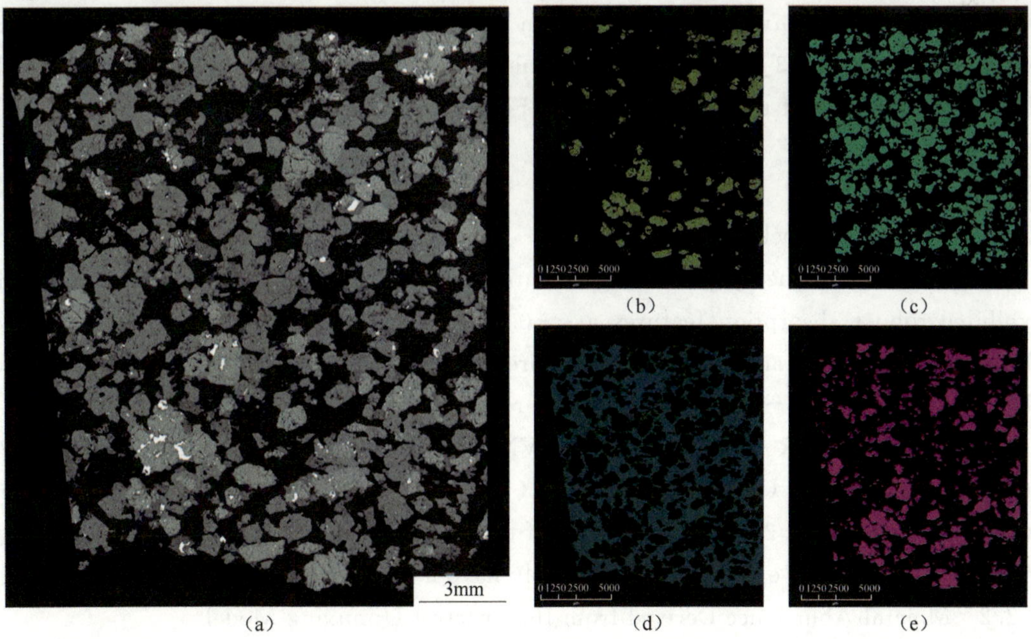

Figure 3 The mineralogy of CR-1. (a) Backscattered electron image of CR-1. (b-d) The distribution of olivine, pyroxene, and plagioclase. (e) The distribution of calcium of CR-1. Different grayscale shows different mineral phases.

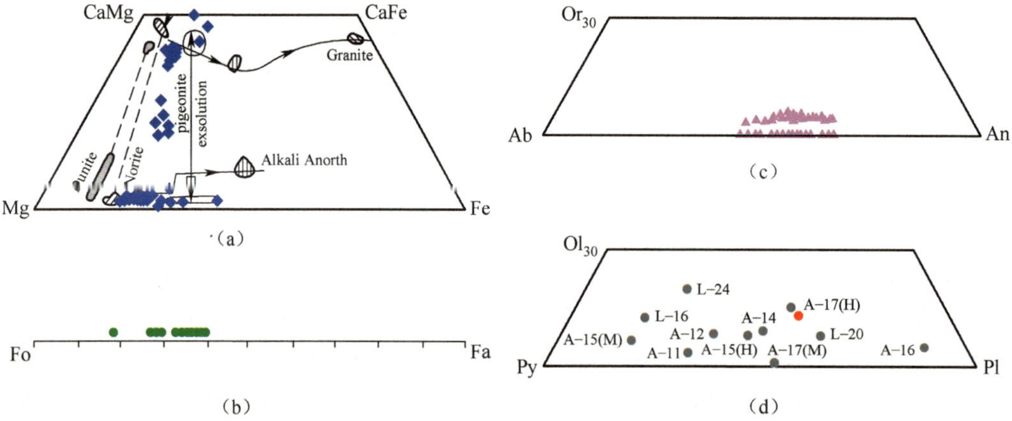

Figure 4　(a) Composition of pyroxene in the Ca - Mg - Fe quadrilateral [24]. (b) Range of compositional variation in olivine. (c) Composition of feldspars in the Or (30%) - Ab - An triangle. (d) Composition of CR - 1 in the olivine (30%) - pyroxene - plagioclase triangle [24]. The results of this study are marked as color points.

5%. The calcium content of the pyroxenes ranged from Wo_{01} to Wo_{50} (Figure 4a). A total of 22% of pyroxene was LCP with calcium contents of 0.36 to 0.67%, whereas the remaining 78% of pyroxene was HCP with calcium contents of 6.85 to 8.50%. The Mg# values of pyroxene and olivine ranged from 58 to 83 and 60 to 82, respectively. The An of plagioclase varied from 45 to 67. The grain size of the sample ranged from 60 to 2000μm, with an average of 400μm. Thus, CR - 1 can be classified as olivine noritic gabbro of lunar highland rocks [22, 23], the features of which are similar to Apollo 17 highland soil with the feature of norite of the lunar highland (Figure 4) [24].

3.2 VNIS Results

3.2.1 Laboratory Rock Sample (CR - 1)

Figure 5 shows the spectrum of the pure mineral separated by TerraSpec - 4. LCP exhibited robust absorption features at ~0.93 and ~1.90μm, whereas HCP showed absorptions at ~1.04 and ~2.25μm (Figure 5). Olivine and plagioclase demonstrated absorption features at ~1.07 and 1.25μm, respectively (Figure 5).

The spectra of CR - 1 detected by the VNIS SF exhibited evident absorption features at wavelengths of 971 (±1) nm and 1957 (±8) nm (Figure 6), which were longer at 1000nm and shorter at 2000nm than those of the Chang'e - 4 rock. The results are similar to those of the Chang'e - 4 rock but probably have more calcium and less iron [25].

3.2.2 Mineral Abundance Derived from the Spectral Unmixing Model

The absorption features of the mafic minerals (olivine and pyroxene) can be observed in the rock spectrum, which is consistent with the ground truth. Using the Hapke model with our laboratory end - members, the modeled mineral abundances were 7.5% olivine,

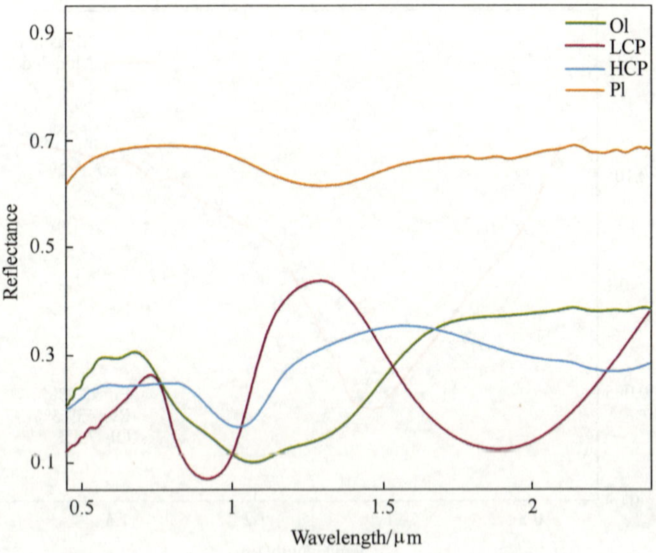

Figure 5　The spectra of end - members of CR - 1 detected by TerraSpec - 4. The green, pink, blue, and orange lines show the spectra of olivine, LCP, HCP, and plagioclase, respectively.

39.3%pyroxene, and 53.2% plagioclase (Figure 7a), which became 11% olivine, 20% pyroxene, and 69% plagioclase with Apollo end - members. The result estimated by MGM of the comparison sample detected by VNIS SF suggested 29% olivine, 30% orthopyroxene, and 41% clinopyroxene (Figure 7b).

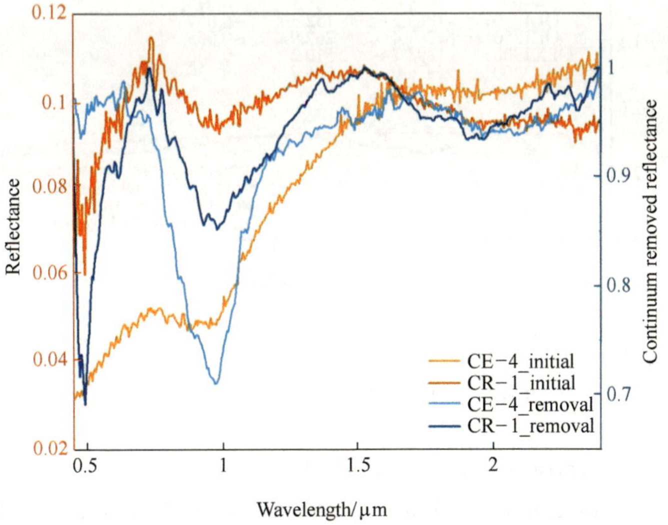

Figure 6　Spectra of the Chang'e - 4 rock and CR - 1 by VNIS. The light orange and blue lines are the initial and continuum removed spectra of the Chang'e - 4 rock, respectively; the dark orange and blue lines are the initial and continuum removed spectra of CR - 1, respectively.

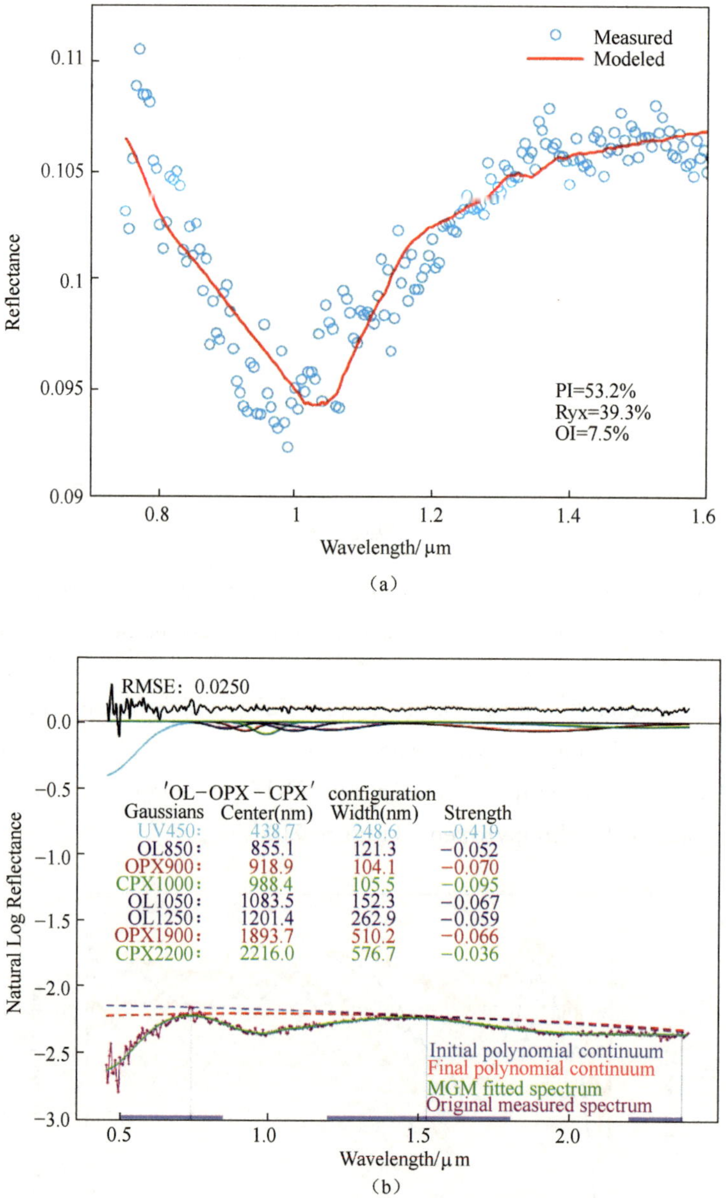

Figure 7　The estimated results of CR-1 by the Hapke model (a) and modified Gaussian model (b), whose spectrum was detected by VNIS SF.

3.2.3　Chang'e-4 Spectral Data (Lunar Rock)

The spectra are illustrated in Figure 6, where the rock was detected on the 3rd lunar day of the mission. Using the data provided by the Lunar and Deep Space Exploration Scientific Data and Sample Release System (China National Space Administration, CNSA) (Table S1) and the method described by Yang et al. [26], the absorption peaks around 1 and 2μm were determined to be 958 (± 12) nm and 2043 (± 12) nm, respectively. The Hapke model results indicated 11.7% olivine, 42.8% pyroxene, and

45.5% plagioclase, whereas MGM indicated 31% olivine, 30% orthopyroxene, and 40% clinopyroxene.

4 Discussion

Liu et al. described a laboratory experiment associated with the Chang'e-4 lunar in situ spectroscopic study [27]. Mafic minerals can be estimated accurately using MGM according to the mixing powder of mafic minerals. However, plagioclase cannot be analyzed in a plagioclase-rich sample. CR-1, as a plagioclase-rich rock sample, is a suitable object for evaluating the method that can estimate the VNIS data of the plagioclase-rich rock more accurately and can be applied to estimate the VNIS data of Chang'e-4. However, this study only contained noritic rock for detection and did not consider the influence of space weathering.

4.1 Error and Correction of Verification Experiment

Chemical analysis results indicate that CR-1 comprises 12.9% olivine, 35.0% pyroxene, and 52.2% plagioclase. The difference between this result and that estimated by the Hapke model (7.5% olivine, 39.3% pyroxene, and 53.2% plagioclase) was approximately 5.3%. Moreover, the precision of the estimated results using the Hapke model was 10%. The error probably arises from the different responses of the environment or the optical scattering characteristic diversity of the rock and powder. Another critical factor is the particle size of minerals in the rock. During repeated testing, the pyroxene content increased with an increase in the assumed particle size of the rock, and the olivine content decreased with an increase in the assumed particle size. The abundance of plagioclase varied widely, ranging from 47 to 73%. This is mainly because of the weaker absorption depth of plagioclase than those of pyroxene and olivine.

Because of the similar spectral absorption characteristics between CR-1 and lunar rock detected by Yutu-2, our implemented Hapke model can be applied to estimate the VNIS data of Chang'e-4.

4.2 Spectral Comparison of CR-1 and Chang'e-4 Rock

According to Section 3.2.1., CR-1 has spectral absorption features at the ~1 and ~2μm absorption bands, similar to those of the rock detected by Chang'e-4 on the 3rd lunar day of the mission (Figure 2). The pyroxene composition of CR-1 consisted of pyroxene from lunar norite (Figure 4a). However, there are two differences between them. First, the ~2μm absorption band of the Chang'e-4 rock is slightly longer than that of the norite. This may have been caused by more iron in the Chang'e-4 rock. Second, the space weathering resulted in the more uplifted shape of the Chang'e-4 rock spectrum.

The influence of multiple factors on the mineral content of the rock must be controlled using a model for calculating powder samples. The particle size of minerals in rocks is critical for estimating modeled mineral abundances using the Hapke model. The real value of the particle size of the minerals in the rock samples ranged from 60 to 2000μm. Through repeated tests using different particle sizes in the actual value range, 400μm was determined to be a suitable particle size to obtain a result consisting of a chemical composition test by BSE and EDS.

Based on the accurate chemical composition of the rock and the results estimated by the Hapke model and MGM, up to 50% of plagioclase in the rock could not be decoded by MGM. This illustrates that the MGM is unsuitable for estimating spectra with similar absorption features and shapes without other constrained conditions.

4.3 Application of Hapke Radiative Transfer Model to the Chang'e-4 Rock Type

Based on the method in this study, the more accurate modal mineral abundances of the rock detected on the 3rd lunar day of the mission are 11.7% olivine, 42.8% pyroxene, and 45.5% plagioclase, which use photometrically corrected data according to Yang et al. (2020) [26]. In this correction, the photometric properties of the lunar regolith were corrected such that the average values of $b=-0.17$ and $c=0.70$, which were derived from photometric modeling using the data from the VNIS on the Yutu-2 rover, were used as the typical values in the Legendre pronominal phase function instead of those from Lucey (1998) [26, 28]. This result is consistent with that of Lin et al. (2020) [7] within the error range. Another critical factor in the model, the particle size, was set to a unified value of 80μm. This indicates that the particle sizes of the lunar soil and the surface of the rock are affected by space weathering.

Additionally, another rock was detected on the 26th lunar day of the mission by the Yutu-2 rover. Its spectral absorption features were almost identical to those detected on the 3rd lunar day of the mission (Figure 8). Similar absorption positions, 0.98 and 2.12μm, illustrate the similar compositions of the two rocks. By estimating the spectra of detecting the second rock three times, the compositions of the landing area's surface materials were 15%, or even 27% plagioclase, more than that of the first rock (N15) (Table 1). The difference between data N167 and data N168 and N169 may have been caused by an area of the short-wavelength infrared (SWIR) detector being viewed differently. The N167 viewed area of the SWIR detector includes a small amount of lunar soil (8% area) and more rock shadow (20% area), resulting in a spectrum with lower reflectance and weaker absorption features than the others. The results suggest the presence of anorthositic layers around the Chang'e-4 landing area or the Von Kármán crater and that the rock sputtered at the detected site.

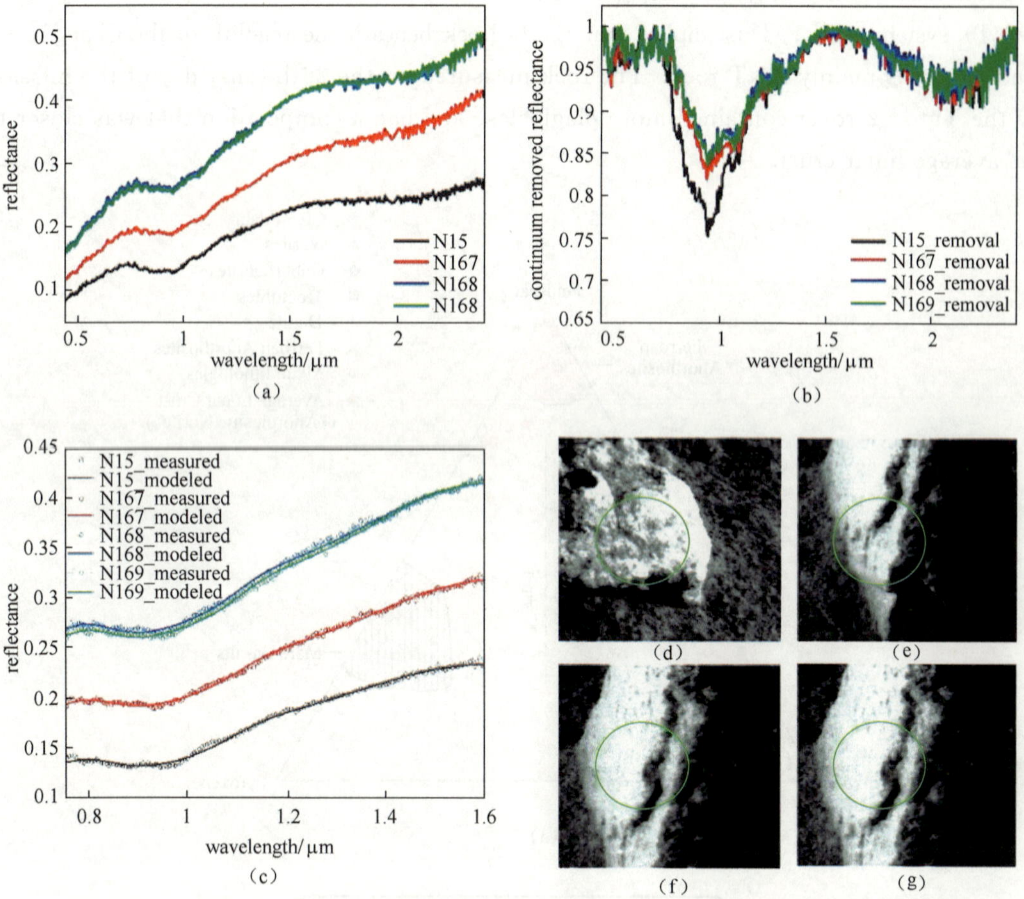

Figure 8　VNIS spectra of the lunar surface rocks measured on the 3rd and 26th lunar days of the mission (http: //moon. bao. ac. cn/searchOrder _ dataSearchData. search. accessed on 27 December 2021) . The spectra were photometrically corrected using the phase function derived by Yang et al. (2020) [26] . (a) The initial spectra of the Chang'e－4 rocks. (b) The continuum removed spectra of the Chang'e－4 rocks. (c) The estimated results of the Chang'e－4 rocks by the Hapke model. (d－g) the CMOS images of the Chang'e－4 rocks, which were N15, N167, N168, and N169, respectively.

Table 1　Estimation of the Chang'e－4 rocks' VNIS data

Number	Hapke Model				Modified Gaussian Model			
	Particle Size (μm)	Pyx (wt%)	Olv (wt%)	Plg (wt%)	SMFe	Olv (wt%)	Opx (wt%)	Cpx (wt%)
N15	80	42.8	11.7	45.5	0.07%	0.31	0.30	0.40
N167	80	30.6	9.3	60.1	0.05%	0.35	0.28	0.37
N168	80	24.2	3.8	72.0	0.04%	0.36	0.26	0.37
N169	80	25.1	2.6	72.3	0.04%	0.35	0.28	0.37

By plotting the data of the four detected sites of the Chang'e－4 rocks (Figure 9), one can observe that they all belong to the norites in the anorthosite - norite - troctolite (ANT) system [24]. This implies that the bedrock beneath the regolith of the Chang'e - 4 landing site is mainly ANT rock. The rock measured on the 26th lunar day of the mission by the Yutu - 2 rover contained more plagioclase and had a composition that was closer to the average lunar crust.

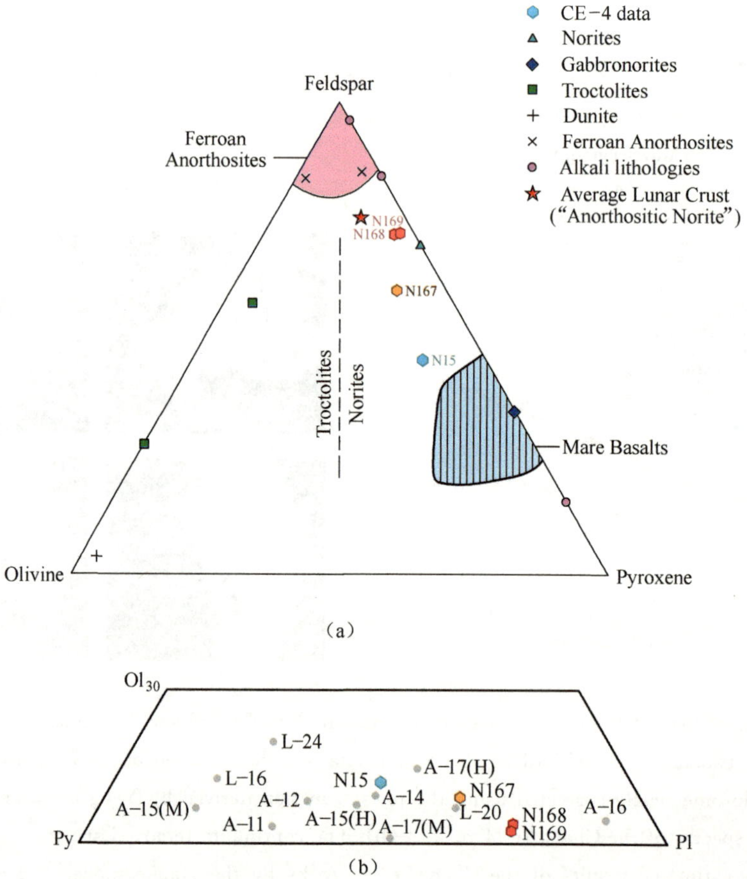

Figure 9　(a) Modal proportions of major minerals in the highland pristine igneous rocks and monomictic breccias [24]. (b) Composition of the Chang'e - 4 rocks in the olivine (30%) - pyroxene - plagioclase triangle compared with data of Apollo and Luna. Sampling sites are indicated; for example, A - 11 is Apollo 11 and L - 16 is Luna 16. (H) and (M) indicate highland soil and mare soil, respectively.

According to the distribution of mineral abundance in the area around the Von Kármán crater (Figure 10), most areas are covered by well - developed mature lunar soils, norites, and basalts. However, a few anorthositic components are exposed at the southwestern edge of the Von Kármán crater, which has an abundance of more than 70%. Connected with the NE - SW linear features at the landing area surface (Figure 10), the rock could have been ejected from the southwestern edge of the Von Kármán crater, which was detected by the Yutu - 2 rover on the 26th lunar day of the mission. This indicates that the materials of the early lunar crust can be conserved at the crater edges inside the SPA basin.

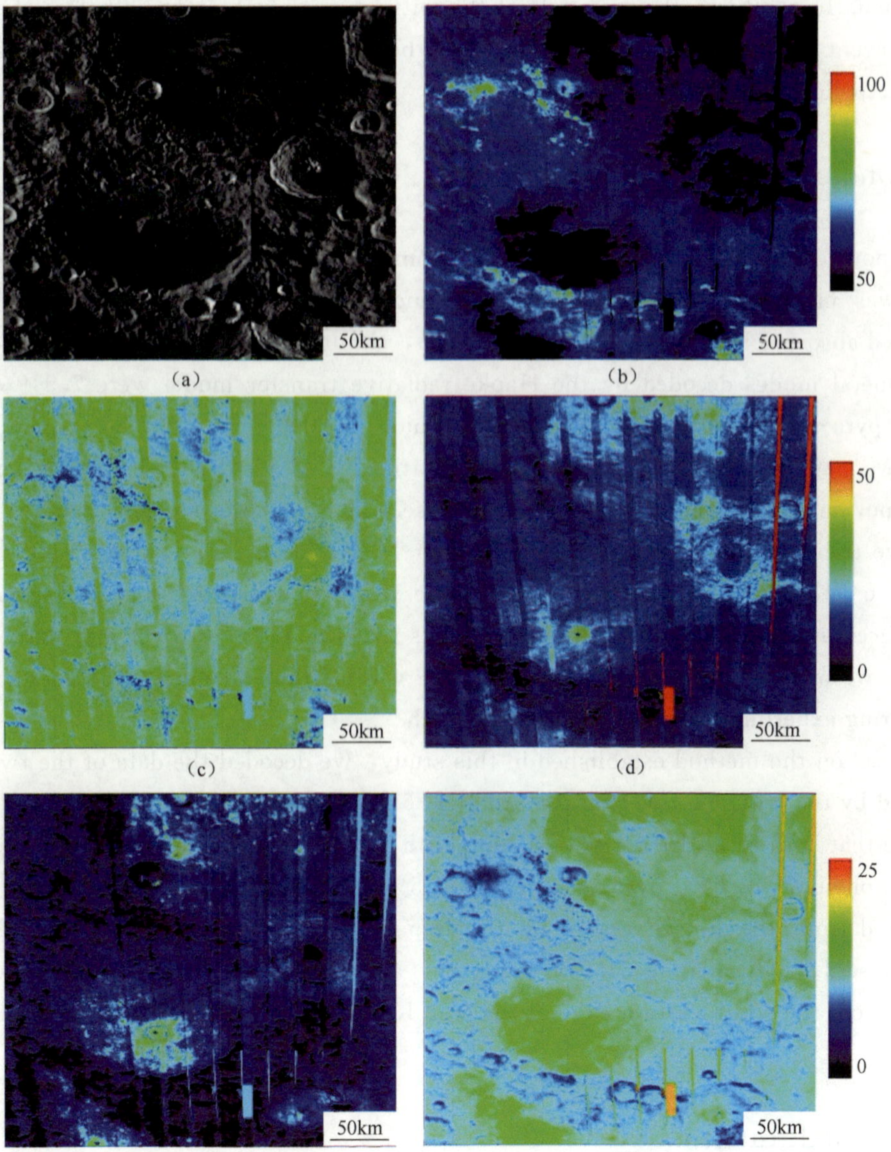

Figure 10　Distribution of mineral abundance of the area around the Von Kármán crater (https: //quickmap. lroc. asu. edu/) (accessed on 27 December 2021) . (a) WAC Mosaic and NACs of LROC. Mineral abundance (wt%) detected using Kaguya multi - band imagery data: (b) plagioclase, (c) orthopyroxene, (d) clinopyroxene, (e) olivine, and (f) FeO.

In summary, noritic and anorthositic rocks are present in the landing region, repre -

senting materials of fast crystallization from the impact melt pond and the average lunar crust. The materials excavated during the impact events are also present beneath the lunar regolith, indicating the characteristics of the plutonic rocks crystallized from the SPA melt pond. In addition, the materials of the initial lunar crust, formed before the SPA-impact event, can be conserved in the SPA, whose mineral composition is between Luna 20 and Apollo 16.

5 Conclusions

A noritic-gabbroic rock with a mineral composition similar to that of lunar highland rocks was measured by SEM, VNIS SF, and TerraSpec-4 of ASD. The spectra exhibited absorption features at ~1 and ~2μm, similar to that of the Chang'e-4 rock. The mineral modes decoded by the Hapke radiative transfer model were 7.5% olivine, 39.3% pyroxene, and 53.2% plagioclase, which is consistent with the estimates from the SEM mapping within error. In contrast, the MGM yielded 29% olivine and 71% pyroxene, indicating the absence of plagioclase. These results suggest that the Hapke radiative transfer model is more reliable than MGM for estimating the VNIS data of Chang'e-4. The particle size was set as the weighted mean value of the surface. To obtain accurate results for other lithologic rocks that the Yutu-2 rover may detect in the future, we will test more lithologic rocks with various mineral modes and conduct space weathering experiments to improve our research.

Based on the method established in this study, we decoded the data of the two rocks detected by the Yutu-2 rover on the 3rd and 26th lunar days of the mission. The results indicate that both rocks are norite or gabbro with noticeable differences. The first rock had 11.7% olivine, 42.8% pyroxene, and 45.5% plagioclase, which may have been excavated from the Finsen crater by the impact event. The second rock with more plagioclase (3.2% olivine, 24.6% pyroxene, and 72.2% plagioclase) may have been ejected from the southwestern edge of the Von Kármán crater, indicating the initial lunar crust.

Acknowledgments

The authors would like to thank Hengci Tian for providing the rock sample CR-1 and Renhao Ruan for help in performing the experiment. The Chang'e-4 data used in this work were processed and produced by "Ground Research and Application System (GRAS)" of China's Lunar and Planetary Exploration Program; they can be downloaded at http://moon.bao.ac.cn/searchOrder_dataSearchData.search (accessed on 15 March 2022).

References

[1] Li, C. L. ; Wang, Z. D. ; Xu, R. ; Lv, G. ; Yuan, L. Y. ; He, Z. P. ; Wang, J. Y. The Scientific Information Model of Chang'e - 4 Visible and Near - IR Imaging Spectrometer (VNIS) and In - Flight Verification. Sensors 2019, 19, 2806.

[2] Jia, Y. Z. ; Zou, Y. L. ; Ping, J. S. ; Xue, C. B. ; Yan, J. ; Ning, Y. M. The scientific objectives and payloads of Chang' E - 4 mission. Planet Space Sci. 2018, 162, 207 - 215.

[3] Pieters, C. M. ; Head, J. W. ; Gaddis, L. ; Jolliff, B. ; Duke, M. Rock types of South Pole - Aitken basin and extent of basaltic volcanism. J. Geophys. Res. Planet 2001, 106, 28001 - 28022.

[4] Moriarty, D. P. ; Pieters, C. M. The Character of South Pole - Aitken Basin: Patterns of Surface and Subsurface Composition. J. Geophys. Res. Planet 2018, 123, 729 - 747.

[5] Huang, J. ; Xiao, Z. Y. ; Flahaut, J. ; Martinot, M. ; Head, J. ; Xiao, X. ; Xie, M. G. ; Xiao, L. Geological Characteristics of Von Karman Crater, Northwestern South pole - Aitken Basin: Chang' E - 4 Landing Site Region. J. Geophys. Res. Planet 2018, 123, 1684 - 1700.

[6] Qiao, L. ; Ling, Z. H. ; Fu, X. H. ; Li, B. Geological characterization of the Chang'e - 4 landing area on the lunar farside. Icarus 2019, 333, 37 - 51.

[7] Lin, H. ; He, Z. ; Yang, W. ; Lin, Y. ; Xu, R. ; Zhang, C. ; Zhu, M. H. ; Chang, R. ; Zhang, J. ; Li, C. ; et al. Olivine - norite rock detected by the lunar rover Yutu - 2 likely crystallized from the SPA - impact melt pool. Natl. Sci. Rev. 2020, 7, 913 - 920.

[8] Ling, Z. C. ; Qiao, L. ; Liu, C. Q. ; Cao, H. J. ; Bi, X. Y. ; Lu, X. J. ; Zhang, J. ; Fu, X. H. ; Li, B. ; Liu, J. Z. Composition, mineralogy and chronology of mare basalts and non - mare materials in Von Karman crater: Landing site of the Chang' E - 4 mission. Planet Space Sci. 2019, 179, 4741.

[9] Li, C. L. ; Liu, D. W. ; Liu, B. ; Ren, X. ; Liu, J. J. ; He, Z. P. ; Zuo, W. ; Zeng, X. G. ; Xu, R. ; Tan, X. ; et al. Chang' E - 4 initial spectroscopic identification of lunar far - side mantle - derived materials. Nature 2019, 569, 378.

[10] Hu, X. Y. ; Ma, P. ; Yang, Y. Z. ; Zhu, M. H. ; Jiang, T. ; Lucey, P. G. ; Sun, L. Z. ; Zhang, H. ; Li, C. L. ; Xu, R. ; et al. Mineral Abundances Inferred from In Situ Reflectance Measurements of Chang' E - 4 Landing Site in South Pole - Aitken Basin. Geophys. Res. Lett. 2019, 46, 9439 - 9447.

[11] Gou, S. ; Yue, Z. Y. ; Di, K. C. ; Wang, J. ; Wan, W. H. ; Liu, Z. Q. ; Liu, B. ; Peng, M. ; Wang, Y. X. ; He, Z. P. ; et al. Impact melt breccia and surrounding regolith measured by Chang'e - 4 rover. Earth Planet Sci. Lett. 2020,

544, 6378.

[12] Zeng, Q.; Chen, S.; Zhang, Y.; Mu, Y.; Dai, R.; Yang, C.; Li, A.; Lu, P. Mineralogical and chemical properties inversed from 21 - lunar - day VNIS observations taken during the Chang' E - 4 mission. Sci. Rep. 2021, 11, s41598 - s42021.

[13] Clark, R. N.; Roush, T. L. Reflectance Spectroscopy - Quantitative - Analysis Techniques for Remote - Sensing Applications. J. Geophys. Res. 1984, 89, 6329 - 6340.

[14] Hapke, B. Bidirectional Reflectance Spectroscopy. Theory. J. Geophys. Res. 1981, 86, 3039 - 3054.

[15] Lin, H. L.; Xu, R.; Yang, W.; Lin, Y. T.; Wei, Y.; Hu, S.; He, Z. P.; Qiao, L.; Wan, W. X. In Situ Photometric Experiment of Lunar Regolith with Visible and Near - Infrared Imaging Spectrometer on Board the Yutu - 2 Lunar Rover. J. Geophys. Res. Planet 2020, 25, 76.

[16] Gu, L. X.; Hu, S.; Anand, M.; Tang, X.; Ji, J. L.; Zhang, B.; Wang, N.; Lin, Y. T. Occurrence of tuite and ahrensite in Zagami and their significance for shock - histories recorded in Martian meteorites. Am. Miner. 2022, 17, 564.

[17] Lin, H. L.; Yang, Y. Z.; Lin, Y. T.; Liu, Y.; Wei, Y.; Li, S.; Hu, S.; Yang, W.; Wan, W. H.; Xu, R.; et al. Photometric properties of lunar regolith revealed by the Yutu - 2 rover. Astron. Astrophys. 2020, 638, 7859.

[18] Sunshine, J. M.; Pieters, C. M.; Pratt, S. F.; McNaron - Brown, K. S. Absorption Band Modeling in Reflectance Spectra: Availability of the Modified Gaussian Model. In Proceedings of the 30th Lunar and Planetary Science Conference, Houston, TX, USA, 15 - 29 March 1999; p. 1306.

[19] Lin, H. L.; Li, S.; Lin, Y. T.; Liu, Y.; Wei, Y.; Yang, W.; Yang, Y. Z.; Hu, S.; Wu, X.; Xu, R.; et al. Thermal Modeling of the Lunar Regolith at the Chang' E - 4 Landing Site. Geophys. Res. Lett. 2021, 48, 1687.

[20] Kanner, L. C.; Mustard, J. F.; Gendrin, A. Assessing the limits of the Modified Gaussian Model for remote spectroscopic studies of pyroxenes on Mars. Icarus 2007, 187, 442 - 456.

[21] Gou, S.; Di, K. C.; Yue, Z. Y.; Liu, Z. Q.; He, Z. P.; Xu, R.; Lin, H. L.; Liu, B.; Peng, M.; Wan, W. H.; et al. Lunar deep materials observed by Chang'e - 4 rover. Earth Planet Sci. Lett. 2019, 528, 5829.

[22] Streckeisen, A. Classi. cation and Nomenclature of Plutonic Rocks. Rocks Geo. Times 1973, 18, 26 - 30.

[23] Stoeffler, D.; Knoell, H. D.; Marvin, U. B.; Simonds, C. H.; Warren, P. H. Recommended classification and nomenclature of lunar highland rocks - A committee report. In Proceedings of the Conference on the Lunar Highlands Crust, Houston, TX, USA, 14 - 16 November 1979; pp. 51 - 70.

[24] Heiken, G.; French, B. Lunar Sourcebook: A User's Guide to the Moon; Cambridge University Press: Cambridge, UK, 1991; pp. 357 - 474.

[25] Klima, R. L.; Pieters, C. M.; Boardman, J. W.; Green, R. O.; Head, J. W.; Isaacson, P. J.; Mustard, J. F.; Nettles, J. W.; Petro, N. E.; Staid, M. I.; et al. New insights into lunar petrology: Distribution and composition of prominent low - Ca pyroxene exposures as observed by the Moon Mineralogy Mapper (M - 3). J. Geophys. Res. Planet 2011, 116, 3719.

[26] Yang, Y. Z.; Lin, H. L.; Liu, Y.; Lin, Y. T.; Wei, Y.; Hu, S.; Yang, W.; Xu, R.; He, Z. P.; Zou, Y. L. The Effects of Viewing Geometry on the Spectral Analysis of Lunar Regolith as Inferred by in situ Spectrophotometric Measurements of Chang' E - 4. Geophys. Res. Lett. 2020, 47, 80.

[27] Liu, C.; Liu, L.; Chen, J.; Cao, H.; Qu, H.; Qiao, L.; Zhang, J.; Qi, X.; Lu, X.; Xu, R.; et al. Mafic mineralogy assemblages at the Chang'e - 4 landing site: A combined laboratory and lunar in situ spectroscopic study. Astron. Astrophys. 2022, 658, 1398.

[28] Lucey, P. G. Model near - infrared optical constants of olivine and pyroxene as a function of iron content. J. Geophys. Res. Planet 1998, 103, 1703 - 1713.

Chang'e – 4 lander localization based on multi – source data

DI Kaichang, LIU Zhaoqin, LIU Bin, WAN Wenhui, PENG Man, WANG Yexin, GOU Sheng, YUE Zongyu, XIN Xin, JIA Mengna, NIU Shengli

State Key Laboratory of Remote Sensing Science, Institute of Remote Sensing and Digital Earth, Chinese Academy of Sciences, Beijing 100101, China

Abstract: High precision localization of Chang'e – 4 lander is critical to support safe and efficient mission operations. This paper presents the lander localization techniques, including image feature matching and monoscopic image measurement, using digital orthophoto maps of Chang'e – 2 and LROC NAC images, descent camera images and monitoring camera image of Chang'e – 4 lander. The lander location is precisely determined to be (177. 588°E, 45. 457°S) using these techniques. Localization of the lander in multiple datasets is valuable for synergistic scientific investigation of the landing site.

Keywords: Chang'e – 4; lander localization; descent camera image; monitoring camera image; multi – source data

Chang'e – 4 (CE – 4) is one of the key missions in the Chinese Lunar Exploration Program (Jia et al., 2018). With the support of the Queqiao (Magpie Bridge) relay satellite, CE – 4 lunar probe has successfully landed on the far side of the moon in Von Kármán crater inside the South Pole – Aitken (SPA) basin at 10: 26 am on January 3, 2019. The rover was successfully released from the lander and touched on the lunar surface on the same day. The success of this task marked the first soft landing of human spacecraft on the far side of the moon, which opened a new era of lunar exploration.

CE – 4 lander localization is a fundamental work to support mission operations and science applications. High precision lander localization is of great significance for mission planning and safe and efficient implementation of the plans (Jia et al., 2014; Wan et al., 2014; Liu et al., 2015). In addition, determination of lander location in multiple datasets is particularly important for synergistic use of these datasets in scientific investigation of the landing site.

In the CE – 4 mission operations, the Planetary Remote Sensing team of State Key

Laboratory of Remote Sensing Science has been participating in the teleoperation tasks in Beijing Aerospace Control Center. Based on digital orthophoto maps (DOMs) generated from multiple orbital images, descending images from the lander's descent camera, and the ground image from the lander's monitoring camera, we have determined the lander location with high precision, using techniques of image feature matching, monoscopic image measurement, etc. The lander localization result has been used to support mission operations. In this short article, we briefly present the lander localization techniques and the results.

1 Data

The high-resolution DOMs generated from orbital images are used as the reference maps for CE-4 lander localization. At present, Chang'e-2 (CE-2) orbital image data is the highest-resolution stereo image data that covers the entire moon surface (Ye et al., 2013). The 7m resolution DOM and 20m resolution digital elevation model (DEM) products of the entire moon have been produced and released (Li et al., 2018). The Lunar Reconnaissance Orbiter Camera (LROC) Narrow Angle Camera (NAC) has acquired high resolution (up to 0.5m) images of nearly the entire moon. Because the LROC NAC is not a stereo imaging camera, it only acquired stereo images by side swing in a very limited number of regions (Robinson et al., 2010). The SLDEM2015 (Barker et al., 2016), generated by integration of a DEM generated by the Japanese Selenological and Engineering Explorer (SELENE) terrain camera and product of Lunar Reconnaissance Orbiter Laser Altimeter (LOLA), was also utilized for terrain analysis and sky visibility analysis.

A downward-looking descent camera identical to that on Chang'e-3 is mounted in the bottom of CE-4. The field of view (FOV) is 45.4° with the image size of 1024×1024 pixels (Liu et al., 2015; Wan et al., 2014; Liu et al., 2014). The camera took a series of images during the descending process. However, because of the limitation of the communication bandwidth, the strategies of selective sampling and image compression with a ratio of 1∶64 were used when transferring the descent images for quick view.

Three monitoring cameras, which have the image size of 1024×1024 pixels and 60° FOV, are mounted on the lander of CE-4. Monitoring cameras A and B are on the top of the lander, and monitor camera C is on the front panel of the lander with pitch angle of 15° and height of 1.42m. Monitoring camera C can surveille the front view of the lander, which provides important information for safe separation of the rover from the lander.

2 Lander Localization Techniques and Results

A rough location of the lander is first determined by matching a medium resolution (similar to that of the orbital DOM) descent camera image to the orbital DOM. Next, the lander location is transferred to other descent images by sequential image matching. Finally, the lander location is refined by combining the monoscopic measurements from the monitoring camera image (Wan et al., 2014; Liu et al., 2015).

Because the descent images have been heavily compressed and even have some mosaic effects, these images have been pre - processed with the parabolic model to enhance the image quality and consequently to improve the image matching accuracy. For comparison purpose, the descent images are matched to both CE - 2 and LROC NAC DOMs. Fig. 1 shows the image matching results, the middle image is a descent image taken at about 4km altitude, the left and right images are the CE - 2 and LROC NAC DOMs respectively. The labels A, B and C are the matched corresponding craters for lander localization.

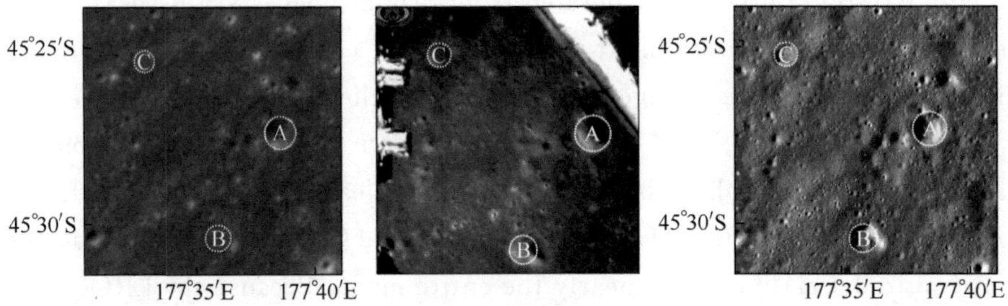

(a) CE-2 DOM　(b) Descent image　(c) LROC NAC DOM

Fig. 1　Feature matching results of multiple data

During the mission, monoscopic measurement technique was applied to monitoring camera C (referred as the monitoring camera) in order to accomplish a series of tasks, including distance measurement from the lander, obstacle size measurement, and calculations of the illumination and communication occlusion envelopes. These measurements supported the decision making for rover - lander separation and high precision lander localization. More specifically, under the assumption that the front of the landing point is a horizontal plane, with monitoring camera image coordinates, its object space coordinates in the lander local coordinate system can be calculated according to the lander position and orientation and the camera installation matrix. Then the sizes of the rocks and impact craters as well as their distances to the lander can be calculated (Zhao et al., 2014). Based on the same method, the illumination and communication occlusion envelopes can be calculated according to the azimuth and elevation angles of the sun and the relay satellite, in order to avoid that the rover enters the lander - induced shadow or blind zone of communication when the rover is separated from the lander. Fig. 2 shows the monoscopic measurement results, in which the white lines indicate the distances to the front panel of the lander, and the diameter of the front crater is about 29m.

Finally, the lander location was refined by combing the measurement results from the monitoring camera image and high resolution descent images. The lander location was

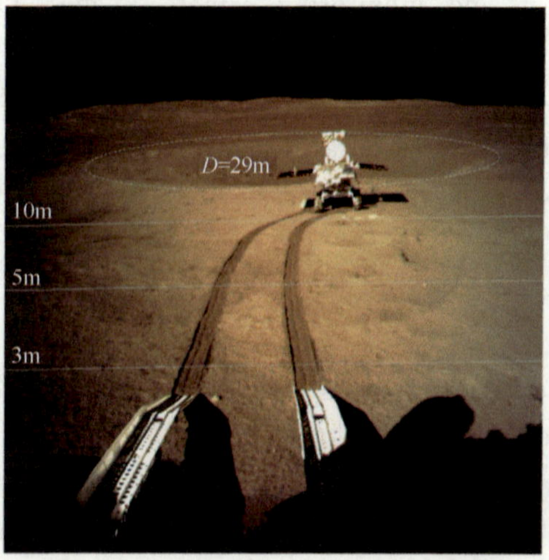

Fig. 2　Monoscopic measurement results of the monitoring camera image
(The white lines indicate the distances to the front panel of the lander,
and the diameter of the crater is about 29m)

precisely determined to be (177.588°E, 45.457°S) on the LROC NAC DOM. Meanwhile, the lander location was also determined on CE－2 DOM and SLDEM2015. Fig. 3 shows the CE－4 lander localization maps based on LROC NAC DOM, in which the right map is the zoom－in view of the rectangle in the left map.

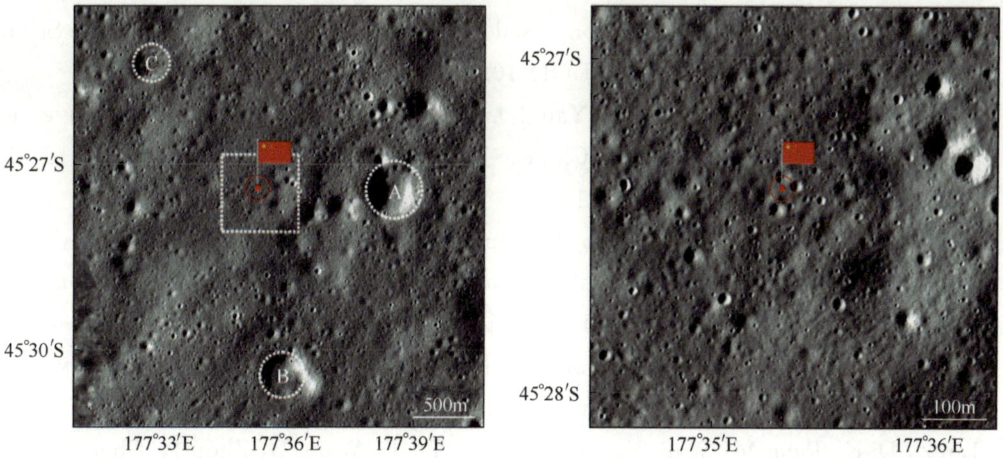

(a) Landing location　　(b) Local magnification

Fig. 3　CE－4 lander localization result on LROC NAC DOM

3　Conclusion

This paper presented the techniques of CE－4 lander localization using orbital basemaps, descent images and monitoring camera image. The lander location was determined to be (177.588°E, 45.457°S) on the LROC NAC DOM. Meanwhile, the lander location was also determined on CE - 2 DOM and SLDEM2015. High precision lander localization in multi - source data is valuable to support mission operations and synergistic scientific investigation of the landing site.

Acknowledgements

The authors gratefully acknowledge Beijing Aerospace Control Center for their support to this study. The authors also thank the institutions/organizations for making the data used in this paper available. The original images of Fig. 1 (b) and Fig. 2 were downloaded from the website of China's Lunar and Deep Space Exploration (http://www.clep.org.cn), thank you all here.

References

[1] Barker M K, Mazarico E, Neumann G A, Zuber M T, Haruyama J, Smith D E. 2016. A new lunar digital elevation model from the Lunar Orbiter Laser Altimeter and SELENE Terrain Camera. Icarus, 273: 346 - 355 [DOI: 10.1016/j.icarus.2015.07.039]

[2] Jia Y, Liu S, Li M, Li Q, Peng S, Wen B, Ma Y and Zhang S. 2014. Chang'E - 3 system pinpoint landing localization based on descent image sequence. Chinese Science Bulletin, 59 (19): 1838 - 1843 [DOI: 10.1360/N972014 - 00020]

[3] Jia Y, Zou Y, Xue C, Ping J, Yan J and Ning Y. 2018. Scientific objectives and payloads of Chang'E - 4 mission. Chinese Journal of Space Science, 38 (1): 118 - 130 [DOI: 10.11728/cjss2018.01.118]

[4] Li C, Liu J, Ren X, Yan W, Zuo W, Lingli M, Zhang H, Yan S, Wen W, Tan X, Xiaoxia Z, Wenrui W, Qiang F, Liang G, Guangliang Z, Baochang Z, Jianfeng Y and Ouyang Z. 2018. Lunar Global High - precision Terrain Reconstruction Based on Chang'e - 2 Stereo Images. Geomatics and Information Science of Wuhan University, 43 (4): 485 - 495 [DOI: 10.13203/j.whugis20170400]

[5] Liu Z, Di K, Peng M, Wan W, Liu B, Li L, Yu T, Wang B, Zhou J, Chen H. 2015. High precision landing site mapping and rover localization for Chang'e - 3 mission. Science China - physics Mechanics & Astronomy, 58 (1): 1 - 11

[6] Robinson M S, Brylow S M, Tschimmel M, Humm D C, Lawrence S J, Thomas P C, Denevi B W, Bowmancisneros E, Zerr J, Ravine M A. 2010. Lunar Reconnaissance Orbiter Camera (LROC) Instrument Overview. Space Science Reviews, 150: 81 - 124 [DOI: 10.1007/s11214 - 010 - 9634 - 2]

[7] Wan W , Liu Z, Liu Y, Liu B, Di K, Zhou J, Wang B, Liu C and Wang J. 2014.

Descent Image Matching Based Position Evaluation for Chang'e - 3 Landing Point. Spacecraft Engineering, 23 (4): 5 - 12 [DOI: 10.3969/j.isn.1673 - 8748.2014, 04.002]

[8] Ye P, Huang J, Zhang Y and Meng L. 2013. Technological achievements of Chang'E - 2 probe and prospect of deep space exploration in China. Science China: Technical Science, 7: 467 - 477 [DOI: 10. 1360/092013 - 229]

[9] Zhao Q, liu Z, Wan W, Li W, Sun Y, Di K, Zhou J, Xi L and Tie W. 2014. Decision support for seperation of Chang'e - 3 rover and lander based on monoscopic measurement using monitoring camera image. Journal of Remote Sensing, 18 (5): 981 -987 [DOI: 10. 11834/jrs. 20144070]

Spectrometers based on acousto – optic tunable filters for in – situ lunar surface measurement

Zhiping He, Chunlai Li, Rui Xu, Gang Lv, Liyin Yuan, and Jianyu Wang

Chinese Academy of Science, Shanghai Institute of Technical Physics, Key Laboratory of Space Active Opto – Electronics Technology, Shanghai, China

Abstract: The lunar surface consists of rocks of varying sizes and shapes, which are made of minerals, such as pyroxene, plagioclase, olivine, and ilmenite, that exhibit distinctive spectral characteristics in the visible and near – infrared (VIS – NIR) and short – wave infrared (SWIR) regions. To analyze the composition of the lunar surface minerals, several spectrometers based on acousto – optic tunable filters (AOTFs) have been developed to detect lunar surface objects and to obtain their reflectance spectra and geometric images. These spectrometers, including the VIS – NIR imaging spectrometer onboard China's Chang'e 3/4 unmanned lunar rovers and the Lunar Mineralogical Spectrometer onboard the Chang'e 5/6 lunar landers, use AOTFs as dispersive components. Both are equipped with a VIS/NIR imaging spectrometer, one or several SWIR spectrometers, and a calibration unit with dust – proofing functionality. They are capable of synchronously acquiring the full spectra of the lunar surface objects and performing in – situ calibrations. We introduce these instruments and present a brief description of their working principle, implementation, operation, and major specifications, in addition to the initial scientific achievement of lunar surface exploration.

Keywords: spectrometer; acousto – optic tunable filter; lunar surface measurement; in situ.

1 Introduction

Morphological and spectral measurements are the two major methods of analyzing rock structures and compositions. [1] An imaging spectrometer has the ability to simultaneously obtain both the images and the spectral signatures of targets and is widely used in terrestrial and space – based remote sensing applications. Mineralogical composition analysis is one of the major tasks of China's lunar exploration project. [2] Minerals such as pyroxene, plagioclase, olivine, and ilmenite, in different sizes and shapes, constitute most of the lunar surface rocks.[3] The minerals have distinctive spectral characteristics in the VIS/NIR and short - wave infrared (SWIR) wavebands that can be used for identification. Figure 1 shows the spectral reflectance curves of the primary rock - forming single minerals that can be used for mineral identification.[4]

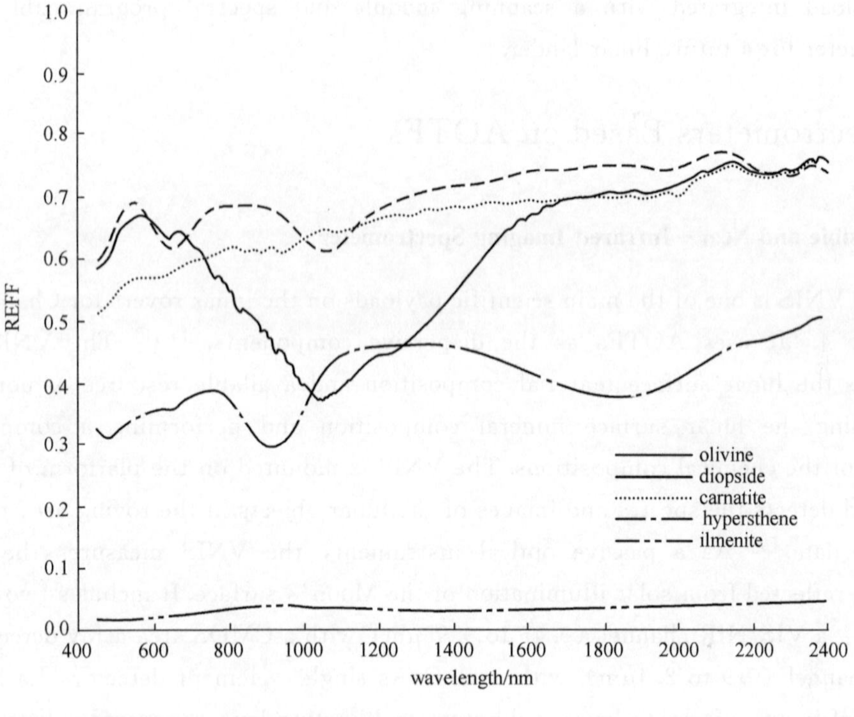

Fig. 1　Spectral reflectance curves for the main single minerals on the lunar surface

Imaging spectral remote sensing is one of the most important scientific achievements in the past 20 years of the 20th century. Imaging spectrometry combines the imaging technology of traditional remote sensing and spectrum analysis technology. Acousto - optic tunable filter (AOTF), which operates based on acousto - optic interaction, is an electronically tunable optical filter. AOTF has special features compared to other dispersive parts, such as solid state and programmable. An imaging spectrometer based on AOTF is especially suitable for deep space exploration applications (Moon, Mars, and asteroid detection), because of the characteristics such as electronically tunable spectral selectivity, environmental adaptability, rapid response, and simple structure.[5-7]

As early as 1987, a spaceborne spectrometer based on AOTF was developed in the former Soviet Union for ocean exploration. European Space Agency (ESA) successfully launched the Mars Express loaded with the SPICAM (Spectroscopy for Investigation of Characteristics of the Atmosphere of Mars) spectrometer in 2003.[8] In 2005, ESA

Venus Express satellite was also loaded with AOTF spectrometer for Venus detection.[9] Since 2006, the Shanghai Institute of Technical Physics began to study imaging spectrometers based on AOTF. Visible and nearinfrared imaging spectrometer (VNIS) is a payload of lunar rover for 0.45 – to 2.4 – μm spectral bands detection and will be able to inspect and probe minerals in the rover region. Lunar mineralogical spectrometer (LMS) is a payload integrated with a scanning module and spectral programmable imaging spectrometer for a future lunar lander.

2 Spectrometers Based on AOTFs

2.1 Visible and Near – Infrared Imaging Spectrometer

The VNIS is one of the main scientific payloads on the lunar rovers for Chang'e 3 and Chang'e 4. It uses AOTFs as the dispersive components.[10,11] The VNIS mostly addresses the lunar surface material composition and available resource exploration for determining the lunar surface mineral composition and performing a comprehensive analysis of the chemical compositions. The VNIS is mounted on the platform of the lunar rover and detects the spectra and images of the lunar objects in the roving area to provide scientific data.[9] As a passive optical instrument, the VNIS measures the radiance diffusely reflected from solar illumination of the Moon's surface. It includes two detection channels: a VIS/NIR channel (0.45 to 0.95μm) with a CMOS area array detector and a SWIR channel (0.9 to 2.4μm) with an InGaAs single – element detector. Each channel consists of image – forming lens, field stop, collimating lens, convergent lens, and the AOTF. Furthermore, a calibration unit with dust – proofing functionality, motor drive circuit, radio frequency (RF) drive circuits, and main control circuits has been integrated in this payload as the public component, as shown in Fig. 2. The structure, optical module, and design of the VNIS are detailed in Figs. 3 and 4.

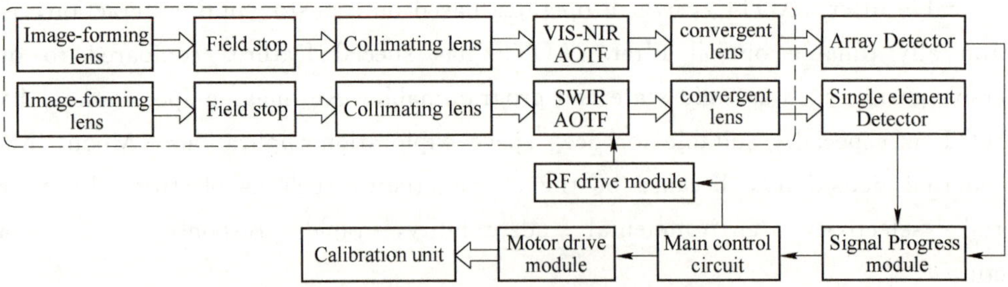

Fig. 2　Schematic diagram of the optical, electronic, and calibration modules of the VNIS

As shown in Fig. 4, two types of AOTFs have been used in the optical system. Each

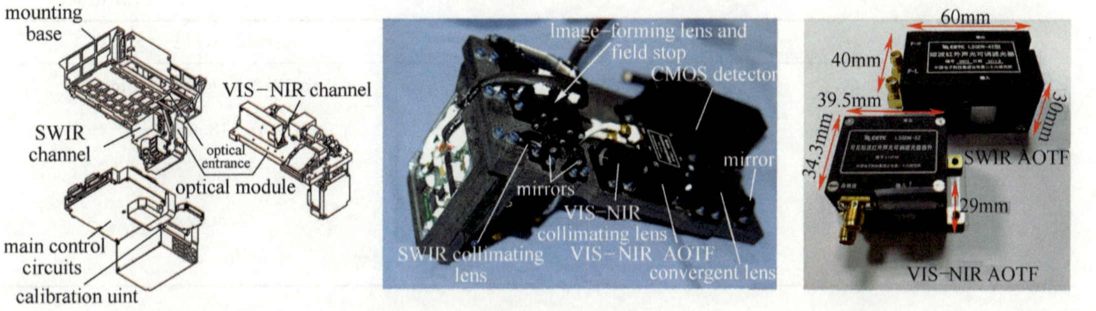

(a)　　　　(b)　　　　(c)

Fig. 3　(a) Assembly diagram of the VNIS, (b) the optical modules, (c) AOTFs used in VNIS; the aperture of visible and near-infrared (VIS-NIR) AOTF is 8 × 8mm², and the SWIR one is 12 × 12 mm²

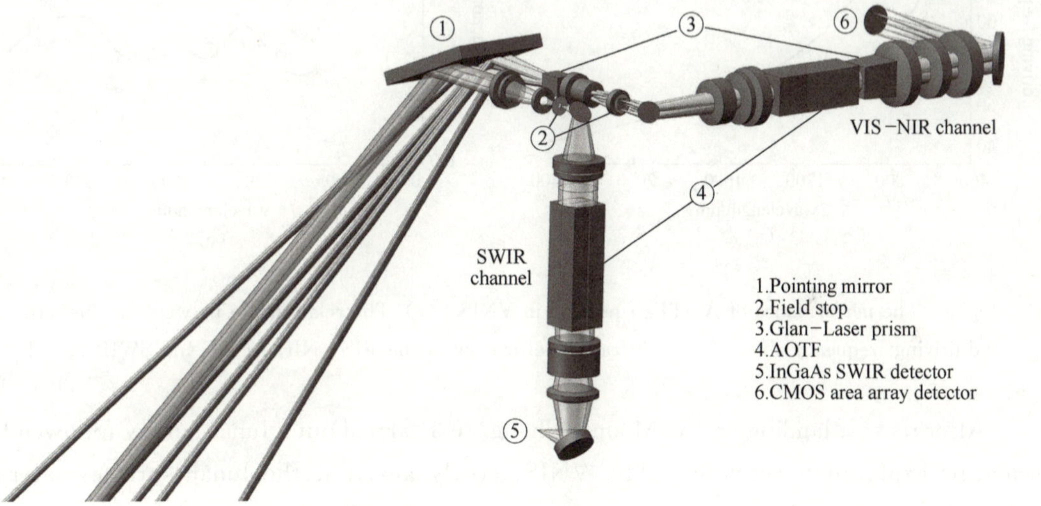

Fig. 4　Diagram of the optical design of the VNIS

AOTF has two transducers with separate ports to optimize the bandpass (FWHM) and diffraction efficiency. Furthermore, an optical wedge has been designed at the exit port to eliminate chromatic aberration in both AOTFs. The main characteristics of the VNIS AOTFs are shown in Table 1. The performance of AOTFs such as acoustic frequency range and diffraction efficiency is shown in Fig. 5.

Table 1　Main characteristics of the VNIS AOTFs

	VIS-NIR AOTF	SWIR AOTF
Material	TeO_2	
Spectral range	0.45 to 0.95μm	0.9 to 2.4μm
FWHM	2.0 to 6.0nm @ <633nm	3.0 to 6.0nm @ <1380nm

Continued

	VIS - NIR AOTF	SWIR AOTF
	2. 5 to 7. 0nm @ >633nm	4. 0 to 12. 0nm @ >1380nm
Acoustic frequency range	70. 7 to 178. 6MHz	41. 9 to 118. 9MHz
Angular aperture	>7deg	>8deg
Diffraction angle	>5. 6deg	>7. 5deg
Power	~2W	

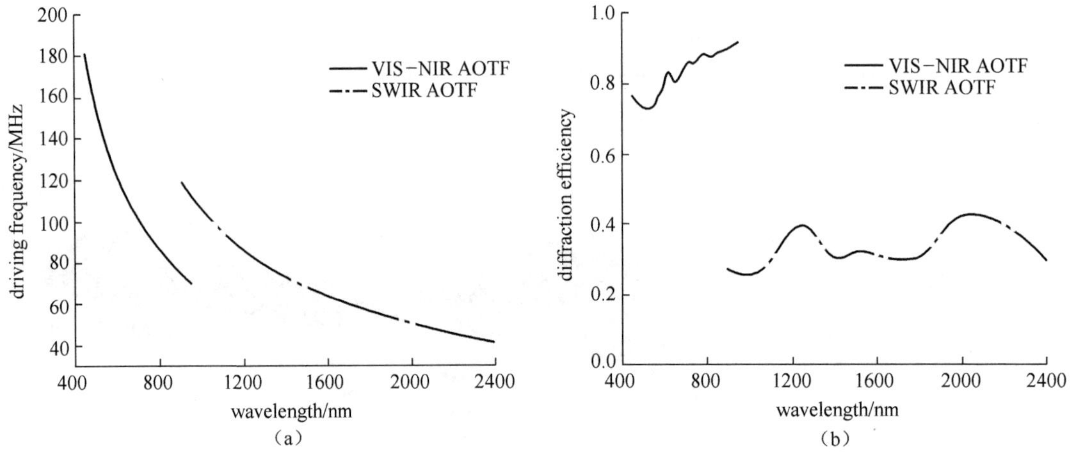

Fig. 5　The performance of AOTFs equipped in VNIS. (a) The relationship between wavelength and driving frequency and (b) the diffraction efficiency of the VIS - NIR AOTF and SWIR AOTF

After its soft landing on the Moon, Chang' e 3 carried out a lunar survey followed by scientific exploration activities. The VNIS mostly analyzed the lunar surface material composition and explored the available resources. The VNIS operated in two modes: detection and calibration. In the detection mode, the VNIS acquired scientific data from the lunar surface objects; the angle between the framework of the calibration unit and the horizontal mounting face of VNIS was about 55 deg. In the calibration mode, using solar radiation as the calibration source, the diffusing calibration plate of the calibration unit was set to a position parallel to the mounting plane to allow calibration of the instrument. The calibration unit of the VNIS at the light entrance consisted of an ultrasonic motor (USM), harmonic reducer, framework, and an internal diffuser panel. The operating modes and dust - proofing function were realized by the self - locking characteristics of the USM.[10] The structure and function of the calibration unit are shown in Fig. 6.

The VNIS was mounted in front of the lunar rover to detect lunar surface objects at a 45 - deg viewing angle and to obtain the spectra and geometric data at a height of 0. 695m, as shown in Fig. 7.

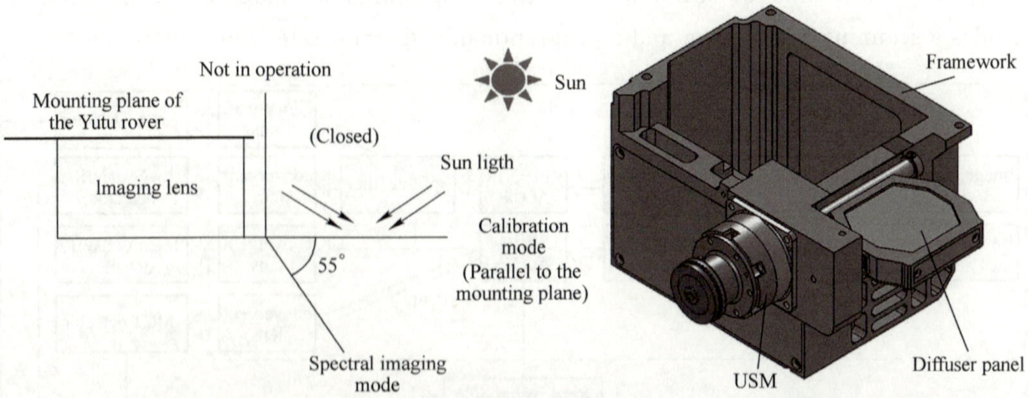

Fig. 6　Schematic view of the structure and functioning of the calibration unit

(a)　(b)

Fig. 7　(a) Structural dimensions and mounting position of the VNIS. (b) The photograph of the Yutu lunar rover taken by the terrain camera (TCAM) at point A of the soft landing site

2.2 Lunar Mineralogical Spectrometer

The LMS is one of the main scientific payloads onboard the lunar landers for Chang'e 5 and Chang'e 6. The LMS primarily follows the technology of the VNIS but expands its spectral range from 0.48 to 3.2μm. It has the following functions: (1) to obtain the spectral image data of the specified objects in the VIS - NIR and spectral data in the SWIR channels, (2) to perform in - situ calibration, and (3) spectral imaging and spectral detection with multiband in full scanning region.

The LMS includes four detection channels: one is a hyperspectral imager with a spectral range of 0.48 to 0.95μm, and the others are three spectrographs and their total spectral range being 0.9 to 3.2μm. The LMS includes a two - dimensional (2 - D) scanning mechanism, imageforming lens, collimating lens, AOTF, convergent lens, the detector components, motor drive, RF drive, and a main control circuit, as shown in Fig. 8. It is a kind of scanning, spectral programmable imaging spectrometer and includes a scanning mechanism and a programmable spectral selection spectrometer.

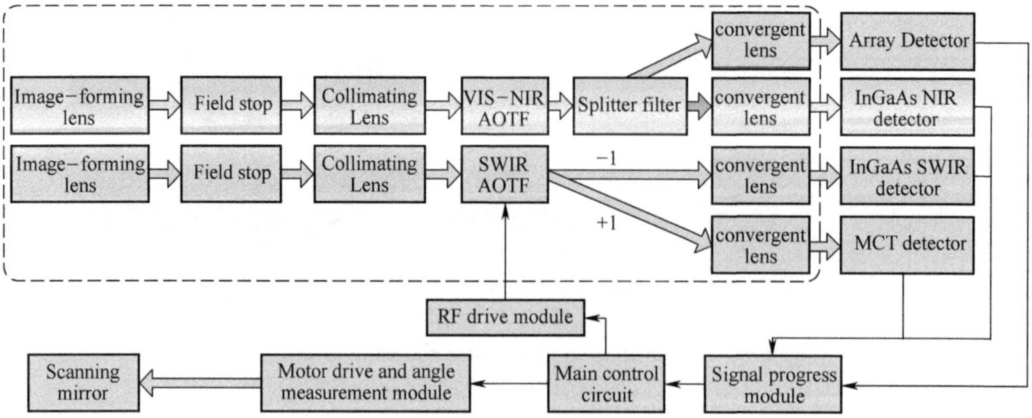

Fig. 8　System diagram of LMS

As shown in Fig. 9, two AOTFs have been used in the optical system. They also have ports to optimize the bandpass (FWHM) and diffraction efficiency. The -1 order diffraction light of the VIS-NIR AOTF was used for the VIS spectral imaging and NIR spectra acquirement, where an optical wedge was designed for the AOTF to eliminate

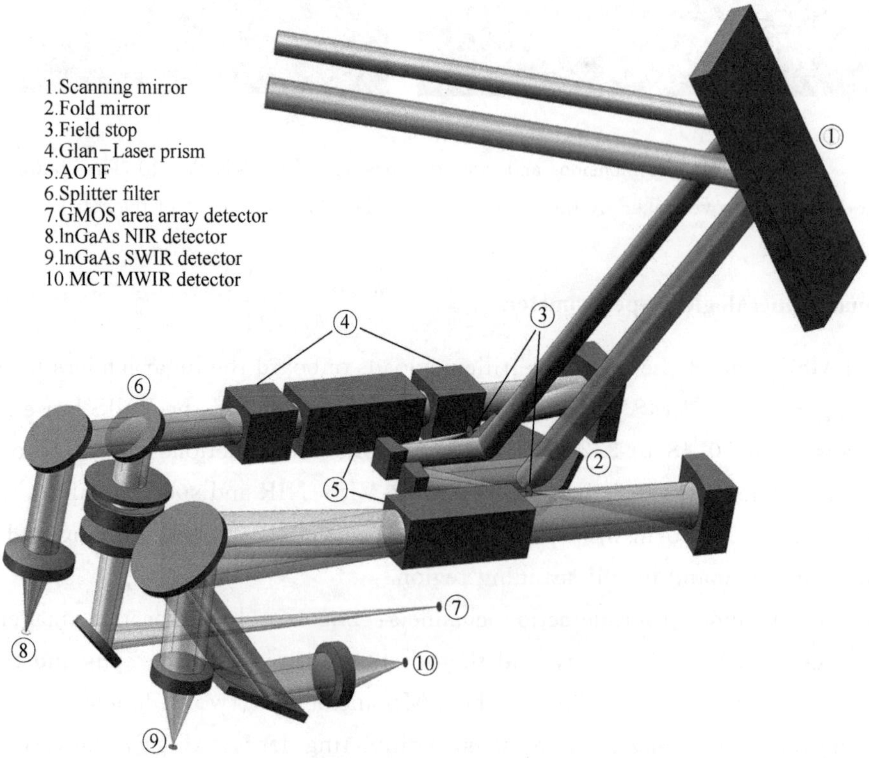

Fig. 9　Diagram of the optical design of LMS

chromatic aberration. The ±1 order diffraction lights of the IR AOTF were used for the SWIR and MWIR channels, respectively. The main characteristics of these AOTFs are shown in Table 2.

Table 2　Main characteristics of the VNIS AOTFs

	VIS - NIR AOTF	IR AOTF
Material	TeO_2	
Spectral range (μm)	0. 48 to 1. 45	1. 4 to 3. 2
FWHM	2. 6 to 9. 4 nm @ <780nm	7. 6 to 20. 8 nm @ 1. 4 - 2. 3μm
	2. 4 to 9. 0nm @ >780nm	11. 6 to 24. 9 nm @ 2. 2 to 3. 2μm
Acoustic frequency range (MHz)	45. 2 to 163. 6	27. 7 to 66. 2
Angular aperture (deg)	>7	>3
Diffraction angle (deg)	>5. 9	>6. 5
Power (W)	~2	

After spectral calibration, the spectral resolution was 2. 4 to 9. 4nm in the VIS - NIR and 7. 6 to 24. 9nm in the IR channels, as shown in Fig. 10.

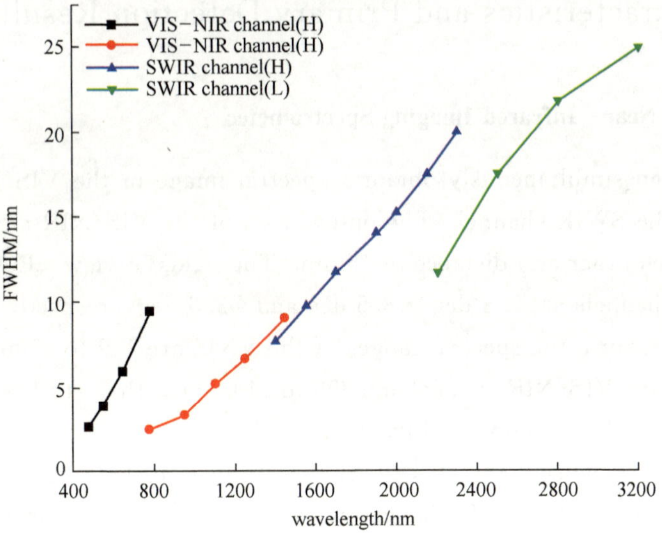

Fig. 10　Spectral resolution of LMS, separated into four channels by AOTF ports

Images of the lander and LMS are shown in Fig. 11. The LMS is mounted on the $+Z\ -Y$ panel of the lander to detect the lunar surface objects and obtain the spectra and geometric data at a height of about 1. 55m. Because the platform on which the LMS is mounted is unmovable, a pointing component is required for expanding the observable areas. The 2 - D scanning mechanism includes USMs, harmonic reducer, rotary potentiometers, and an aluminum - based scanning mirror, which ensure that the system is light and minimized and the system can adapt to a high surrounding temperature of up to 94℃.

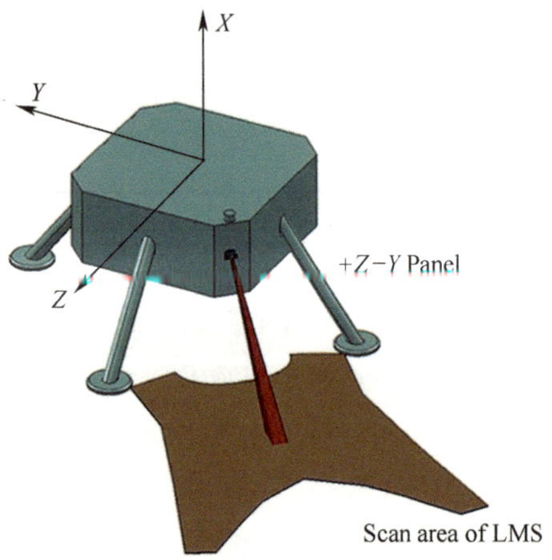

(a)

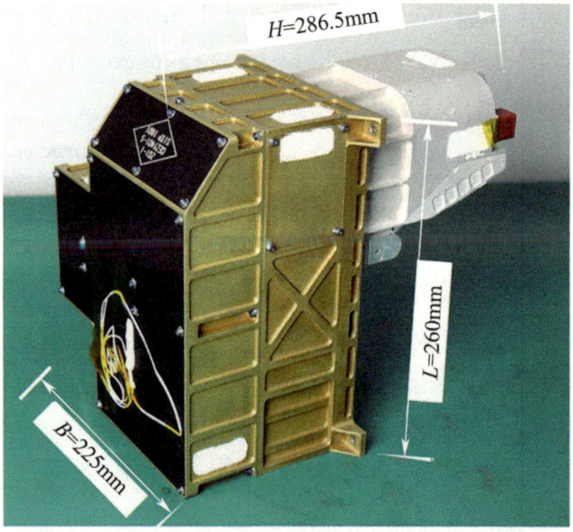

(b)

Fig. 11 Schematic diagram of Chang' e 5 mission: (a) mounting position of the LMS on the lander of Chang' e 5 probe and (b) the flight product of the LMS

3 Main Characteristics and Primary Detection Results

3.1 Visible and Near - Infrared Imaging Spectrometer

The VNIS can simultaneously obtain a spectral image in the VIS/NIR channel and spectral data in the SWIR channel. The optical axis of the VIS/NIR and SWIR channels are parallel to each other at a distance of 18mm. The fields of view (FOVs) in the VIS/NIR and SWIR channels are 8.5 deg ×8.5 deg and Φ3.6 deg, respectively. According to the spectral calibration, the spectral ranges of the VNIS are 449 to 950nm with 2 - to 7 - nm resolution in the VIS/NIR channel and 899 to 2402nm with 3 - to 12 - resolution in the SWIR channel, which are shown in Fig. 12.

The signal - to - noise ratio (SNR) of the VNIS is tested and analyzed by ground radiometric calibration. The SNR changes with the input signal, and it is an important indicator of the radiometric response characteristics of the VNIS. The SNR is >31 dB in the VIS - NIR band when the albedo is 0.09 and the solar incident angle is 45 deg; and the SNR is >32 dB in the SWIR channel when the albedo is 0.09 and the solar incident angle is 75 deg.[10] The main performance specifications of the VNIS are shown in Table 3.

After the Chang' e 3 mission began its scientific exploration stage, the VNIS instrument successfully completed the first lunar surface spectral acquisition at BTC 10:10 on December 23, 2013. During the first two lunar days, the VNIS performed measurements at four locations and obtained data in detection mode (four times) and calibration mode (three times). The total size of data was 350 MB.[11] The topographical images of N205 and the REFF of all detection nodes are shown in Fig. 13.

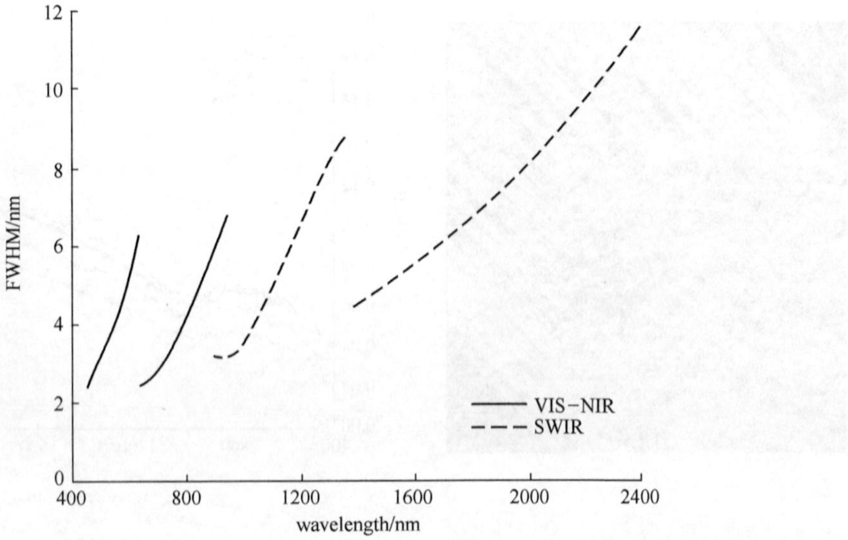

Fig. 12　Spectral resolution of VNIS/Yutu, separated into four channels by AOTF ports

Table 3　Main performance specifications of VNIS/Yutu[10]

Description	Chang' e 3's specification performances	
	VIS - NIR	SWIR
Spectral coverage (nm)	449 to 950	899 to 2402
Spectral resolution (nm)	2 to 7	3 to 12
FOV (deg)	8.5 × 8.5	Φ3.6 deg
Effective pixels	256 × 256	1
Quantization (bits)	10	16
SNR (dB)	≥31 (albedo is 0.09 and solar incident angle is 45 deg)	≥32 (albedo is 0.09 and solar incident angle is 75 deg)
Spectral sampling interval (nm)	5	
Power consumption (W)	19.8	
Weight (kg)	4.7 (spectrometer probe) 0.7 (logical control component in Rover payload Electric Control Box)	

During the Chang'e - 4 mission, VNIS was carried on Yutu - 2 lunar rover. Since the successful soft landing in Von Kármán crater on the far side of the Moon on January 3 2019,[12] VNIS had been working for five lunar days and performed 27 detections. The selected targets such as Yutu - 2's wheel track and rock, obtained in the detection node A and LE00303, and their corresponding REFF features are shown in the Fig. 14.

3.2　Lunar Mineralogical Spectrometer

In the detection mode, the LMS obtained spectral images in the VIS/NIR channel

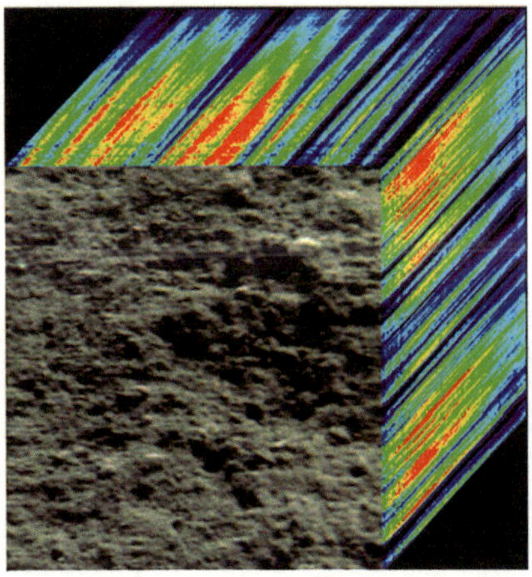

(a)

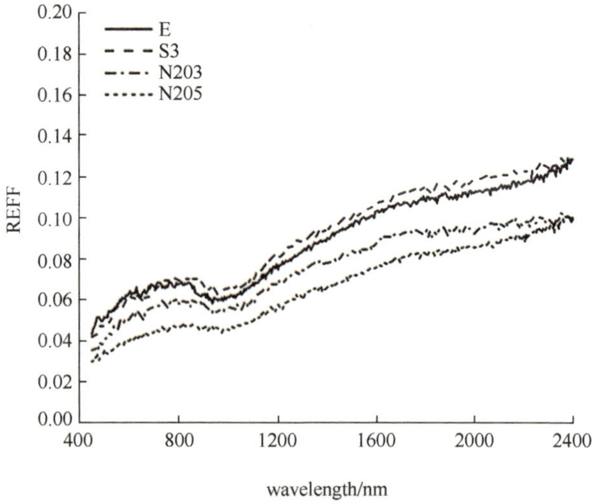

(b)

Fig. 13 (a) Data from VNIS/Yutu at N205 note, false color picture (500, 550, and 645nm, in the right) and (b) REFF curves of four lunar exploration nodes during two lunar days by VNIS/Yutu

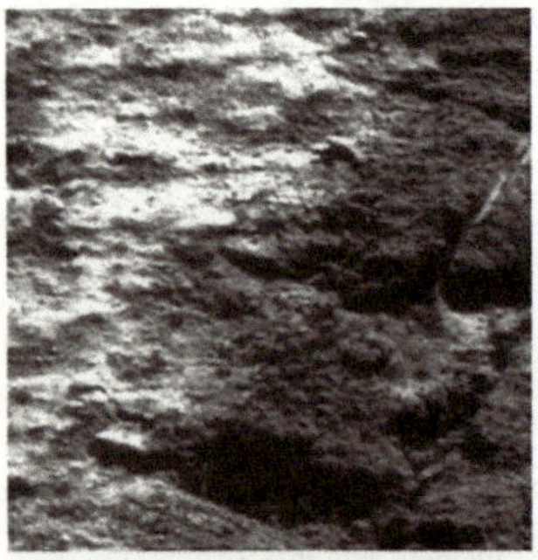

(a)

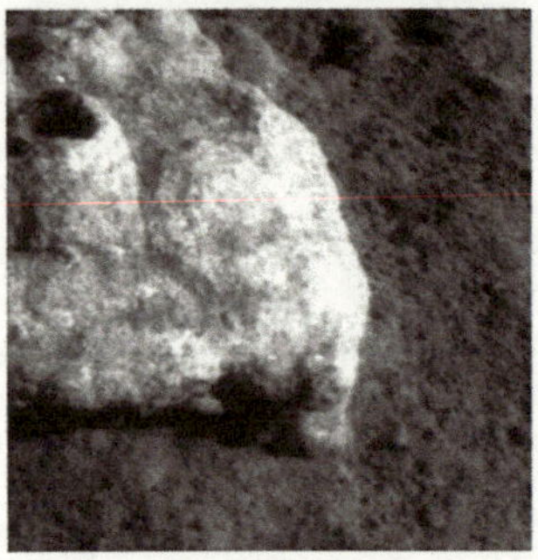

(b)

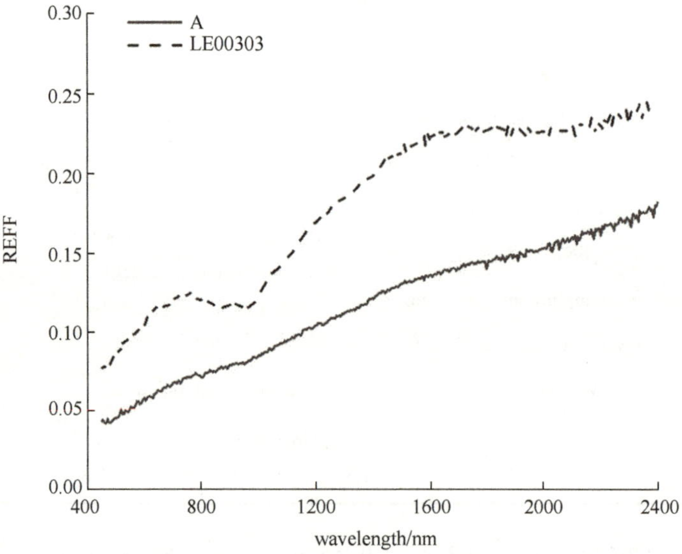

(c)

Fig. 14 (a) A 635 - nm image from VNIS/Yutu - 2 at node A, (b) 635 - nm image from VNIS/Yutu - 2 at node LE00303, and (c) REFF curves of nodes A and LE00303

and spectral data in three SWIR channels in sequence. The main performance specifications of the LMS are shown in Table 4.

Table 4　Main performance specifications of the LMS

Description	Specification requirements		Chang'e 5's specification performances	
	VIS-NIR	SWIR	VIS-NIR	SWIR
Spectral coverage (nm)	480 to 950	900 to 3200	479 to 955	896 to 3212
Spectral resolution (nm)	5.0 to 25.0	5.0 to 25.0	2.4 to 9.4	7.6 to 24.9
FOV (deg)	≥3.0 × 3.0	≥3.0 × 3.0	4.17 × 4.17	4.17 × 4.17
Effective pixels	256 × 256	1	256 × 256	1
Quantization (bits)	≥10	≥10	10	16
SNR (dB, albedo is 9% and solar incident angle is 45 deg)	≥30	≥30	≥34	≥39
Spectral sampling interval (nm)	5		5	
Power consumption (W)			16.0	
Weight (kg)			5.6	

Performance parameters such as FOV, spectral range, spectral resolution, radiance, and geometric imaging features were calibrated in the ground experiment. The FOV of the LMS was 4.17 deg ×4.17 deg in all channels.

The SNR is >34 dB in the VIS/NIR band, and >39 dB in the IR band, as shown in Fig. 15.

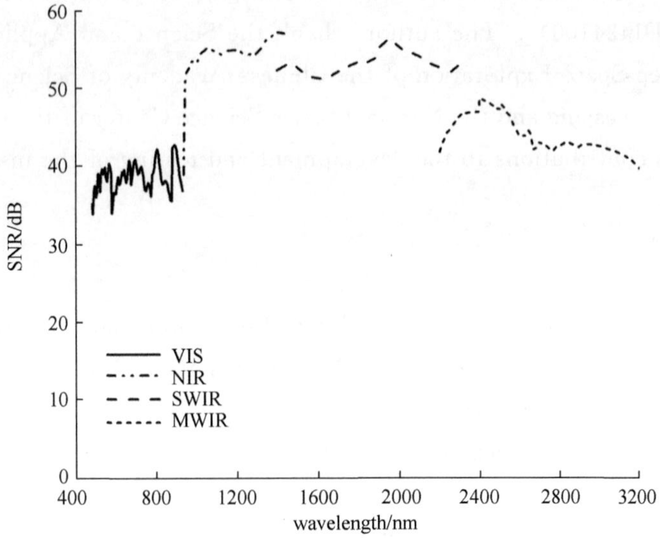

Fig. 15　SNR of LMS. The performance of LMS is calculated by the radiance calibration at the typical temperature (20℃)

4　Conclusions

To analyze the composition of the lunar surface minerals, several spectrometers based on AOTFs have been developed to detect lunar surface objects and to obtain their reflectance spectra and geometric images; these include the VNIS onboard China's Chang'e 3 and Chang'e 4 lunar rovers and the LMS onboard Chang'e 5 and Chang'e 6 lunar landers. These spectrometers are capable of synchronously acquiring the full spectra of the lunar surface objects and performing in-situ calibrations. The VNIS/Yutu has performed several explorations and calibrations, and obtained several spectral images and spectral reflectance curves of the lunar soil in the Imbrium region following its first successful operation at the landing site on December 23, 2013. The VNIS/Yutu-2, which is travelling in Von Kármán crater after the soft landing on January 3, 2019,[12] will provide valuable in-situ hyperspectral images and infrared spectra of the soil, rocks, and other materials on the far side of the Moon. The flight module of LMS has completed all the ground calibration and environmental verification tests and will observe the sampling area in the lunar surface sampling mission with the Chang'e 5 probe by the end of 2019.

Acknowledgments

This paper was supported by the special funds for the second and third phases of the Chinese lunar exploration program, the National Natural Science Foundation (Nos. 21105109 and 61605231), and the Program of Shanghai Academic/Technology Research Leader (No. 19XD1424100). The authors thank the Science and Application Center for the Moon and Deep Space Exploration of the Chinese Academy of Sciences for the ground test and data preprocessing and the National Space Science Center of the Chinese Academy of Sciences for its contributions to the development and testing of the instruments.

References

[1] Z. —P. He et al., "Operating principles and detection characteristics of visible and nearinfrared imaging spectrometer (VNIS) in Chang'e 3," Res. Astron. Astrophys. 14 (12), 1567-1577 (2014).

[2] P. J. Ye and J. Peng, "Deep space exploration and its prospects in China," Eng. Sci. 8 (10), 13-18 (2006).

[3] B. Liu et al., "Reflectance conversion methods for the VIS/NIR imaging spectrometer aboard the Chang' E-3 lunar rover: based on ground validation experiment data," Res. Astron. Astrophys. 13 (7), 862-874 (2013).

[4] Z. He et al., "Visible and near-infrared imaging spectrometer (VNIS) and its preliminary results from the Chang'E 3 Project," Rev. Sci. Instrum. 85, 083104 (2014).

[5] O. Korablev et al., "Development of a mast or robotic arm-mounted infrared AOTF spectrometer for surface Moon and Mars probes," Proc. SPIE 9608, 960807

(2015) .

[6] J. - P. Bibring et al. , "The MicrOmega investigation onboard Hayabusa2," Space Sci. Rev. 208, 401 - 412 (2017) .

[7] Zh. He et al. , "Visible and near - infrared imaging spectrometer (VNIS) for Chang' E - 3," Proc. SPIE 9263, 92630D (2014) .

[8] J. Wang et al. , "Visible and near - infrared imaging spectrometer aboard Chinese Chang'E - 3 spacecraft," in Optical Payloads for Space Missions, Ch. 5, S - E Qian, Ed. , John Wiley & Sons, Chichester (2015) .

[9] J. - L. Bertaux et al. , "SPICAV on Venus Express: three spectrometers to study the global structure and composition of the Venus atmosphere," Planet. Space Sci. 55 (12), 1673 - 1700 (2007) .

[10] Zh. He et al. , "Visible and near - infrared imaging spectrometer (VNIS) for in - situ lunar surface measurement," Proc. SPIE 9639, 96391S (2015) .

[11] Zh. He et al. , "The spectrometers based on AOTF for in - situ Lunar surface measurement," Proc. SPIE 10780, 107800N (2018) .

[12] C. Li et al. , "Chang'E - 4 initial spectroscopic identification of lunar far - side mantle - derived materials," Nature 569, 378 - 382 (2019) .

Measurement and Correction Model for Temperature Dependence of an Acousto - Optic Tunable Filter (AOTF) Infrared Spectrometer for Lunar Surface Detection

Zhiping He[1,2], Jinning Li[1,2], Chunlai Li[1,2], and Rui Xu[1,2]

1. Key Laboratory of Space Active Opto - Electronics Technology, Shanghai Institute of Technical Physics of the Chinese Academy of Science, Shanghai, China
2. University of Chinese Academy of Sciences, Beijing, China

Abstract: The photoelectric response characteristics of an infrared spectrometer based on an acousto - optic tunable flter (AOTF) are greatly affected by the temperature change of the radio frequency power amplifier and shortwave infrared detector. If calibration is not conducted, that will affect the quantitative level of the data. This paper puts forward a measurement and correction method for the temperature characteristics of an AOTF infrared spectrometer which is used in lunar surface detection and sets up a temperature characteristics correction model. This model was applied to an AOTF infrared spectrometer mounted on China's unmanned lunar rover. Laboratory tests show that the temperature causes an instrument variation of ~20% when the temperature is between −20℃ and +55℃, but this model reduces this variation to<6%. Evaluating data from the lunar surface also verifies the effectiveness of this method.

Keywords: Acousto - optical tunable filter; infrared spectrometer; lunar surface detection; temperature correction model

Introduction

An acousto - optic tunable filter (AOTF), a spectroscopic device based on the principle of the acousto - optical effect, can change the diffraction wavelength from a broadband or multiline laser source through a crystal by controlling the radio frequency (RF) power amplifier.[1-3] Some advantages of AOTF infrared spectrometers include their small size, light weight, no moving parts, flexible wavelength selection and strong environmental adaptability.[4,5] They are especially suitable for deep space exploration studies, which have strict requirements on weight and volume and must function in the relatively harsh space environment.[6-13] In some of wide variations temperature environment, such as the surfaces of the Moon and Mars, causing the spectral characteristics of the output and response curves to differ when provided with the same input spectral information. This directly affects the inversion accuracy of spectral data and the quantization level of the spectral information. In order to ensure consistent spectral instrument output in wide operating temperature range environments, it is necessary to define and correct for the temperature effects on AOTF infrared instruments.

At present, studies mainly focus on the effect of temperature changes on transfer functions of AOTF, and correct the transfer functions for different temperature ranges,[14-18] and less research on the temperature effect of the whole infrared imaging spectrometer. This paper explores the influence of temperature change on the response of an AOTF infrared spectrometer which is used in lunar surface detection. A general temperature effect test and correction method is proposed to achieve consistent spectral detection over a wide operating temperature range.

Principle and Measurement Methods

The photoelectric response of the AOTF infrared spectrometer is shown in Fig. 1. It is mainly composed of input optics, an AOTF crystal, an imaging system, an RF drive control circuit and a CPU. When spectrum detection begins, the reflected light from the target object passes through the input optics collimation and enters the AOTF crystal. The RF signal is loaded onto an AOTF crystal by the RF power amplifier, and the wavelength of the AOTF output is selected. The monochromatic light enters the detector and is converted into an electrical signal, which is processed by an A/D conversion circuit and then outputted to obtain the required spectral information.

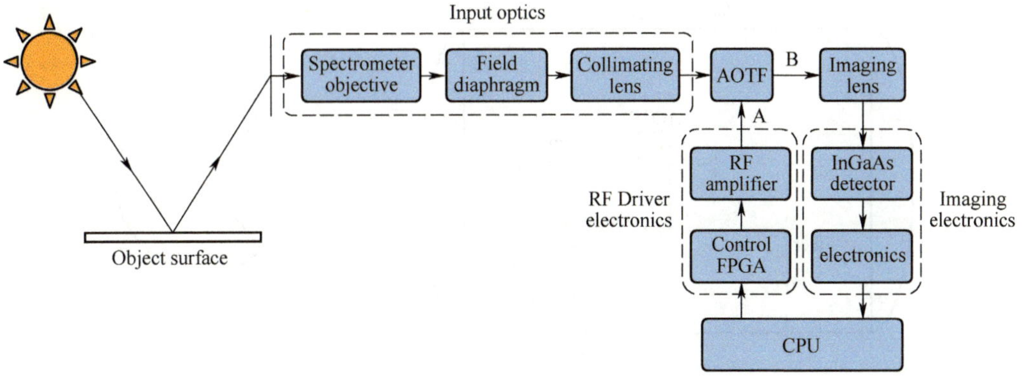

Fig. 1　Photoelectric response link of the AOTF infrared spectrometer

The RF power amplifier and infrared detector are the two main temperature sensitive devices in the AOTF infrared spectroscopy instrument. When working in wide temperature range environments, the output signal power (A point) of the RF power amplifier will be affected by the change in ambient temperature or by its own temperature. This will affect the diffraction efficiency of the crystal and lead to a change in the optical signal at point B.

When the temperature varies, the spectral response rate of the infrared detector will also change, causing the output analog signal (C point) to change. The comprehensive effect of temperature on the RF power amplifier and the infrared detector leads to variations in the data values collected by the AOTF infrared spectrometer. If not cor - rected, the accuracy of the data will be seriously affected. Therefore, a proposed measurement and correction method for this temperature effect is shown in Fig. 2.

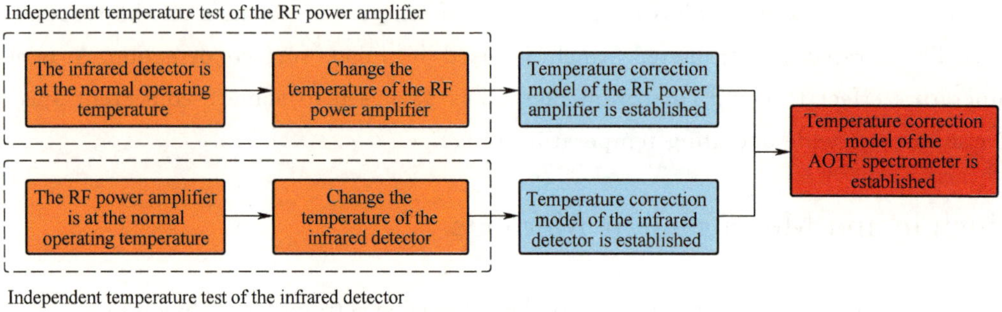

Fig. 2　Temperature effect measurement and calibration method

When testing the temperature effect on the RF power amplifier, the detector would be maintained at a stable temperature and the RF power amplifier is placed in a vacuum chamber. The temperature of the vacuum chamber is changed and the photoelectric response of the instrument as a function of temperature would be measured and a model can be established. Independent temperature experiment of RF power amplifier is shown in Fig. 3. Likewise, while testing the temperature effect on the infrared detector, the RF

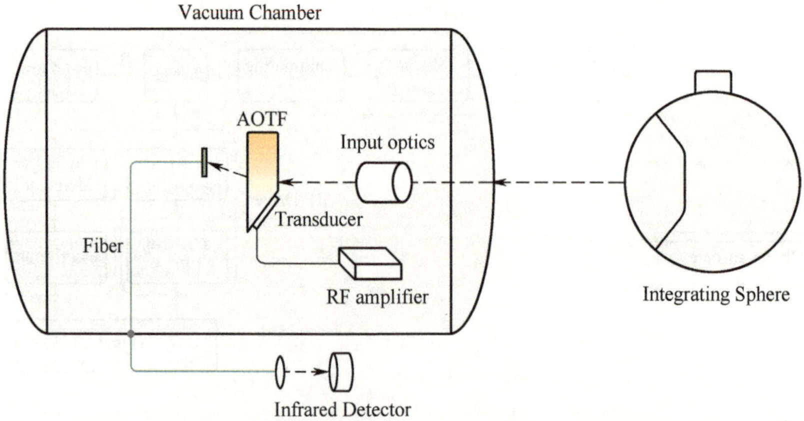

Fig. 3　Independent temperature experiment of RF power amplifier

power amplifier would be maintained at a stable temperature and photoelectric response can be measured and used to develop a model. By combining these two independent temperature tests, a temperature correction for the AOTF infrared spectrophotometer can be established.

Correct Model Building

Temperature Effect Test and Model Building of the RF Power Amplifier

To begin the process of developing the AOTF spectrometer correction model, the independent temperature test of the RF power amplifier was carried out using a LabSphere US-200 SF integrating sphere as the radiation light source. The RF power amplifier was placed in a vacuum chamber and the detector was in a room temperature environment. The temperature of the vacuum chamber ranged between −20℃ and 55℃ and the working wavelengths were 900nm to 2400nm. The digital number (DN) value vary with wave-length and various temperatures is shown in Fig. 4a. The temperature of RF power amplifier change with the wavelength and the temperature of the vacuum chamber as shown in Fig. 4b.

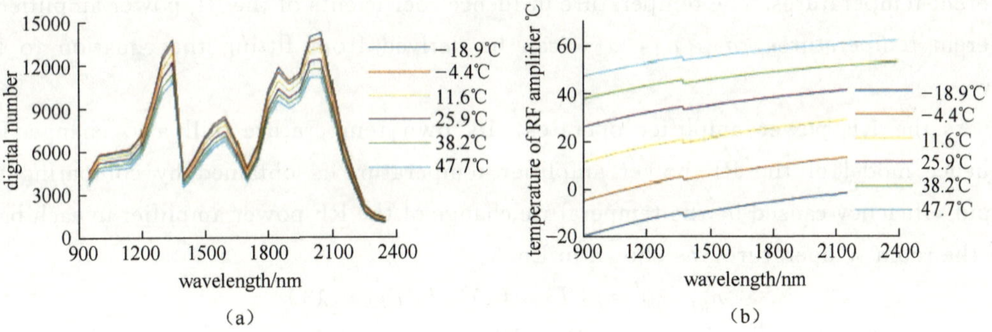

Fig. 4 (a) Spectral response curves at various temperatures.
(b) Temperature variation curve of RF power amplifier during measurement under different vacuum chamber temperatures

Referring to Fig. 4a, when the temperature of the RF power amplifier increased, the DN value of the spectrometer's response to the same light source gradually decreased. The spectral response curve of the RF power amplifier at room temperature (25.9℃) was interpolated, and this value was used as a benchmark to normalize the other DN value curves. In addition, Figs. 4a and 4b illustrate that the variation trend of RF power ampliier affected by temperature at different wavelengths is basically the same. These results indicate that the temperature effect from the RF power amplifier is only influenced by temperature and has nothing to do with wavelength. Therefore, a one-dimensional function model with temperature as the variable can be established to fit the relationship between the RF power amplifier temperature and the normalized spectral response DN value under the full band by using a quadratic polynomial. Figure 5 is the fitting curve of the relationship between the normalization factor and the temperature of RF power amplifier, and the corresponding relationship is shown in Eq. I.

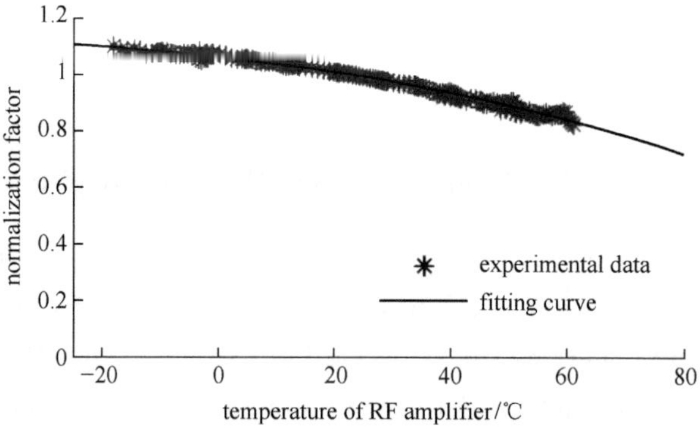

Fig. 5　Fitting results of temperature influence coefficient of RF power amplifier

$$f(T_{RF}) = a_2 T_{RF}^2 + a_1 T_{RF} + a_0 \tag{1}$$

where the input variable T_{RF} is the set temperature of the RF power amplifier in the test, $f(T_{RF})$ is the normalized spectral response DN value of the RF power amplifier at different temperatures. The temperature influence coefficients of the RF power amplifier at different temperatures, a_0, a_1, a_2, can be derived from fitting the equation to the curve.

As the RF power amplifier operates, its own temperature will also change. The influence model of the RF power amplifier temperature is obtained by comparing the output efficiency caused by the temperature change of the RF power amplifier in each band and the room temperature, as shown in Eq. 2:

$$\rho_{RF}(\lambda) = f(T_{RFC}(\lambda)) / f(T_{RFW}(\lambda)) \tag{2}$$

where $\rho_{RF}(\lambda)$ is the influence model of the RF power amplifier temperature changing with band $T_{RFW}(\lambda)$ and $T_{RFC}(\lambda)$ are the temperatures of RF power amplifier at band λ when the spectrometer is operating and the calibration standard at room temperature. For different types of RF power amplifiers, the temperature influence model can be obtained by measuring the fitting curve of the temperature influence coefficient of the RF power amplifier and the temperature change curve of the RF power amplifier itself at room temperature.

Temperature Effect Test and Model Building of the Detector

The independent temperature experiment of the detector also used LabSphere US - 200 SF integrating sphere as the radiation light source. The InGaAs infrared detector was placed in the vacuum chamber and the RF power amplifier was placed at room temperature. The temperature of the vacuum chamber ranged from −20℃ to 55℃ and the working wavelengths were 900 to 2400nm.

The digital number values of the spectral response vary with wavelength and vacuum chamber temperatures as shown in Fig. 6a. The temperature of InGaAs detector vary with wavelength at different ambient temperatures are shown in Fig. 6b. The spectral response curve of the InGaAs infrared detector at room temperature was interpolated, and this value was used as a benchmark to normalize the data shown in Fig. 6a.

Referring to Fig. 6a, the DN value variation at different wavelengths is not the same. At some wavelengths the DN value variation increases and in the other decreases. The reason is that the detection rate of InGaAs detector is related to the absorption coefficient, and its spectral response curve varies with wavelength. For semiconductor materials, the forbidden band width changes with temperature which affects the detector cut - off wavelength, and the response rate drifts with temperature. [19] So the detection rate is a function of temperature and wavelength.

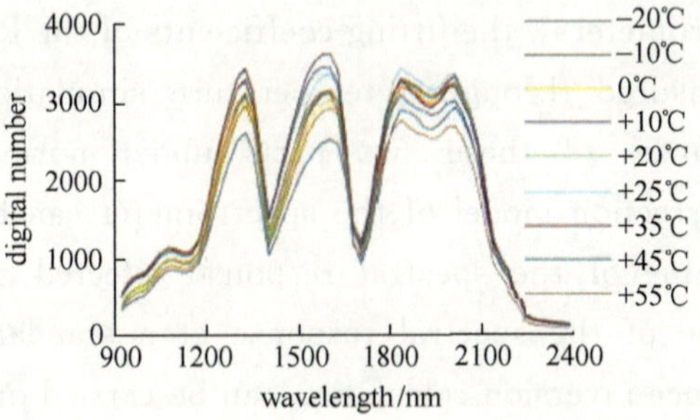

(a)

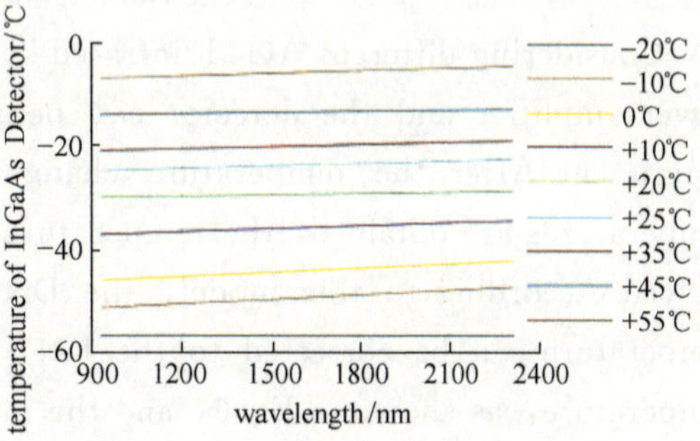

(b)

Fig. 6　(a) Spectral response DN value of detector temperature characteristic test.
(b) Temperature curves of InGaAs detector at various ambient temperatures

When studying the temperature correction model of the InGaAs detector, the detection rate variation of each wavelength at different temperatures should be considered. In this paper, binary polynomial fitting is used as the basis for establishing the temperature correction model for the InGaAs detector. The linear relationship is used to simulate the drift of the response rate with the detector's refrigeration temperature, and the cubic polynomial is used to simulate the change of response rate with wavelength.

Equation 3 shows the normalized response rate fitting function:

$$
\begin{aligned}
r(T_D(\lambda), \lambda) = & p_1 + p_2(\lambda + q_1 + q_2 T_D(\lambda)) \\
& + p_3(\lambda + q_1 + q_2 T_D(\lambda))^2 \\
& + p_4(\lambda + q_1 + q_2 T_D(\lambda))^3
\end{aligned} \tag{3}
$$

where $T_D(\lambda)$ is the temperature change value of the InGaAs detector in the simulation experiment. The temperature influence model of the InGaAs detector is described by Eq. 4:

$$g_D(\lambda) = r(T_{DC}(\lambda), \lambda) / r(T_{DW}(\lambda), \lambda) \tag{4}$$

where $T_{DW}(\lambda)$ is the temperature of the InGaAs detector at different ambient temperatures, $T_{DC}(\lambda)$ is the calibration temperature curve of the InGaAs detector at room temperature, and p_1, p_4 and q_1, q_2 are fitting coefficients. Generally, the temperature influence model of an InGaAs detector can be obtained by measuring the fitting coefficient and the temperature change curve of the InGaAs detector at room temperature.

Temperature Effect Correction Model for the AOTF Infrared Spectrometer

The temperature effect on an AOTF infrared spectrometer is affected by both the infrared detector temperature and the RF power amplifier temperature. Both the temperature effect of the RF power amplifier and the temperature effect of the detection exist at the same time, affecting the final spectrum detection results. According to the transmission characteristics of the circuit system, we can cascade the temperature effects of the two parts, and get the temperature correction model of the infrared imaging spectrometer. The mathematical model is shown in Eq. 5

$$g(\lambda) = g_{RF}(\lambda) \cdot g_D(\lambda) \tag{5}$$

Considering different AOTF infrared spectrometers, the fitting coefficients of the RF power amplifier and the detector can be measured through a temperature simulation experiment. After the temperature change curves of these two parts under normal temperatures are obtained, the temperature correction model of the spectrometer can be derived. According to this model, the DN value of the spectral response affected by temperature can be corrected to the DN value of the spectral response at a standard temperature, as shown in Eq. 6, and the radiance inversion calculation can be carried out according to the corrected spectral response curve.

$$DN_{\text{standard}}(\lambda) = DN(\lambda) \cdot g(\lambda) \tag{6}$$

Verification Tests

Ground Verification Test

The effect of the temperature correction model was verified by a ground validation test. The test object was the visible - near - infrared (Vis - NIR) imaging spectrometer (VNIS) which is a payload instrument for Chang'e - 3 and Chang'e - 4 lunar detection. The short - wave infrared (SWIR) channel of the spectrometer is a typical AOTF infrared spectrometer, which needs to operate reliably at -20℃ to $+50$℃. The structure, optical module, and design of the VNIS are detailed in Figure S1 (Supplemental Material). The main performance specifications of the VNIS - SWIR channel are shown in Table Ⅰ.[20-23]

Ground verification test used a LabSphere US - 200 SF integrating sphere as the light source, and the test is shown in Figure S2 (Supplemental Material), and the VNIS was on the vacuum chamber which varied in temperature from -20℃ to 55℃. We derived the

DN value curve of the short－wave infrared wavelength channel at different temperatures and the DN value curve after a temperature correction，as shown in Figs. 7a and 7b.

Table Ⅰ　Main performance specifications of the VNIS－SWIR channel

Description	Performance specifications
Spectral coverage (nm)	899 to 2402
Spectral resolution (nm)	3 to 12
FOV	Φ3.6°
Effective pixels	1
Quantization (bits)	16
SNR (dB)	≥32 (albedo is 0.09 and solar incident angle is 75°)

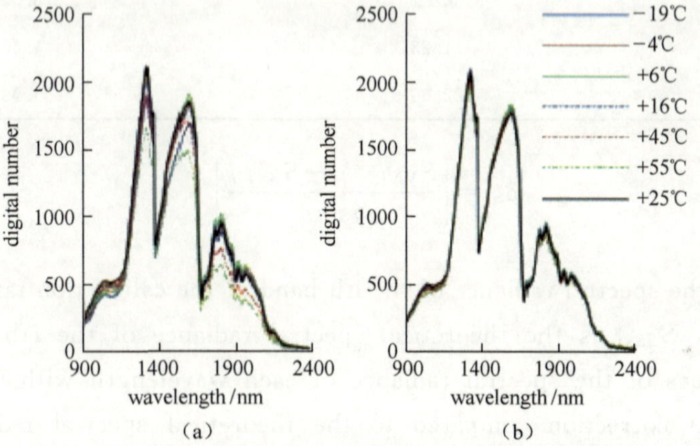

Fig. 7　Observation curve of the spectrometer on the same stable light source at different ambient temperatures

(a) Original spectral curves at different surrounding temperature，and (b) spectral curves after correction

Using the data at 25℃ as a benchmark，the average deviation of data with and without a temperature correction was calculated，and the results are shown in Table Ⅱ.

As shown in Table Ⅱ，the average deviation of the whole band between the temperature correction data and the standard data was less than 6%. Compared with the data before correction，the deviation of the data is significantly reduced. Therefore，the application effect of the temperature correction model in the VNIS spectrometer is better in the ground test.

Practical Application Test

In order to evaluate the practical application of the temperature correction model，we selected a group of calibration data from the VNIS used for research on the lunar surface. The spectral data of each band were corrected by using the temperature correction model, and we used the calibration matrix to invert the spectral radiance with and without temperature correction. The two groups of data were compared to the theoretical calibration spectral radiance data and we calculated the relative error of each band. The equation is as follows:

Table Ⅱ　The average deviation of data with and without temperature correction

Environment temperature (℃)	Without temperature correction (%)	With temperature correction (%)
−20	11.71	3.89
−4	7.43	3.95
6	6.31	4.15
16	5.13	3.95
45	11.23	4.60
55	22.66	5.99

$$\delta_S = \left| \frac{S_{VNIS,\ i} - S_{SS,\ i}}{S_{SS,\ i}} \right| \tag{7}$$

where $S_{VNIS,i}$ is the spectral radiance of the ith band of the calibration target obtained by the spectrometer, $S_{SS,i}$ is the theoretical spectral radiance of the ith band, and the relative error values of the spectral radiance of each wavelength with and without the temperature effect correction compared to the theoretical spectral radiance is shown in Fig. 8.

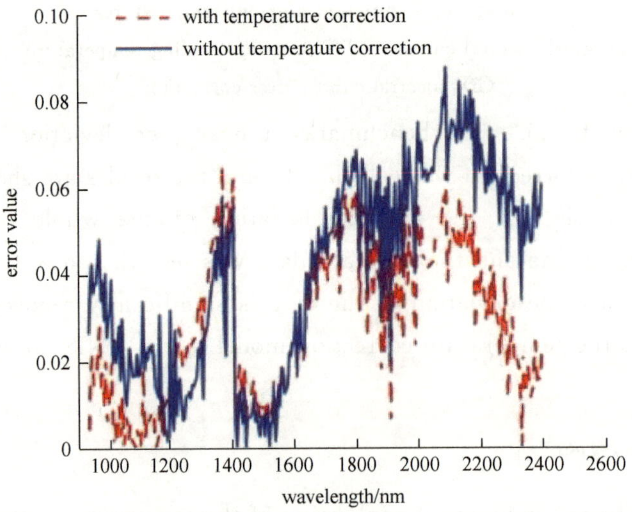

Fig. 8　The relative error of the spectral radiance of each band with and without the temperature effect correction compared to the theoretical spectral radiance

The results in Fig. 8 indicate that the error in the spectral radiance data was effectively suppressed when the temperature correction model was used. The mean error of the entire band is 2.9% and the maximum residual error was less than 6.8%. It conclusively shows that the temperature correction model can effectively suppress the temperature effect of the VNIS and improve the quantitative level of the collected data from the lunar surface.

Conclusion

This paper presents a method for testing and correcting AOTF infrared spectrometer data for the temperature effects that influence the RF generator circuit and the infrared detector. This correction method for temperature effects proved to be effective in improving the accuracy of the spectral response data by inhibiting the influence of tem - perature change on spectrum detection results and improving the quantization level of the AOTF infrared spectrometer. Our temperature effect correction model for the AOTF infrared spectrometer was successfully applied to the data processing of the infrared imaging spectrometer of Chang'e - 3 and eventually will be applied to the data from Chang'e - 4 and Chang'e - 5, providing significantly more accurate spectral radiance data and an important reference value.

Acknowledgments

The authors thank the Science and Application Center for the Moon and Deep Space Exploration of the Chinese Academy of Sciences for the ground test and data preprocessing; and the National Space Science Center of the Chinese Academy of Sciences for its contributions to the development and testing of the instruments.

References

[1] R. W. Wallace, S. E. Harris. "Acousto - Optic Tunable Filter". J. Opt. Soc Am. 1969. 59 (6): 96 - 98.

[2] G. Georgiev, D. A. Glenar, J. J. Hillman. "Spectral Characterization of Acousto - Optic Filters Used in Imaging Spectroscopy". Appl. Optics. 2002. 41 (1): 209 -217.

[3] M. Burmen, F. Pernus, B. Likar. "Spectral Characterization of Near - Infrared Acousto - Optic Tunable Filter (AOTF) Hyperspectral lmaging Systems Using Standard Calibration Materials". Appl. Spectrosc. 2011. 65 (4): 393 - 401.

[4] J. F. Turner, P. J. Treado. "Near - Infrared Acousto - Optic Tunable Filter Hadamard Transform Spectroscopy". Appl. Spectrosc. 1996. 50 (2): 277 - 284.

[5] N. Gupta. "Acousto - Optic Tunable Filters for Infrared Imaging". In: Proceedings

Volume 5953, Acousto - Optics and Photoacoustics. 2005. 59530O.
[6] O. I. Korablev, D. A. Belyaev, Y. S. Dobrolenskiy, et al. "Acousto - Optic Tunable Filter Spectrometers in Space Missions" . Appl. Optics. 2018. 57 (10): C103 - C119.
[7] J. L. Bertaux, D. Nevejans, O. I. Korablev, et al. "SPICAV on Venus Express: Three Spectrometers to Study the Global Structure and Composition of the Venus Atmosphere" . Planet. Space Sci. 2007. 55 (12): 1673—1700.
[8] D. Emmanuel, L. Nicolas, V. O. Bert, et al. "Tunable Acousto - Optic Spectral lmager for Atmospheric Composition Measurements in the Visible Spectral Domain". Appl. Optics. 2012. 51 (25): 6259—6267.
[9] D. A. Belyaev, K. B. Yushkov, S. P. Anikin, et al. "Compact Acousto - Optic lmaging Spectro - Polarimeter for Mineralogical Investigations in the Near Infrared". Opt. Express. 2017. 25 (21): 25980 - 25991.
[10] V. Y. Molchanov, V. M. Lyuty, V. F. Esipov, et al. "An Acousto - Optical lmaging Spectrophotometer for Astrophysical Observations" . Astron. Lett. 2002. 28 (10): 713 - 720.
[11] O. I. Korablev, J. L. Bertaux, E. Dimarellis, et al. " An AOTF - Based Spectrometer for Mars Atmosphere Sounding". Infrared Spaceborne Remote Sensing X. 2002. 4818: 261 - 271.
[12] V. Leroi, J. P Bibring, M. Berthe. "Micromega/IR: Design and Status of a Near - Infrared Spectral Microscope for In Situ Analysis of Mars Samples". Planet. Space. Sci. 2009. 57 (8 - 9): 1068 - 1075.
[13] O. I. Korablev, Y. Dobrolensky, N. Evdokimova, et al. "Infrared Spectrometer for Exomars: A Mast - Mounted Instrument for the Rover" . Astrobiol. 2017. 17 (6 - 7): 542 - 564.
[14] S. N. Mantsevich, O. I. Korablev, Y. K. Kalinnikov, et al. "Wide - Aperture TeO_2 AOTF at Low Temperatures: Operation and Survival" . Ultrasonics. 2015. 59: 50 -58.
[15] S. N. Mantsevich, E. I. Kostyleva. "Examination of the Temperature Influence on Phase Matching Frequency in Tunable Acousto - Optic Filters". Ultrasonics. 2019. 91: 45 - 51.
[16] A. C. Vandaele, A. Mahieux, S. Robert, et al. "Improved Calibration of SOIR/ Venus Express Spectra". Opt. Express. 2013. 21 (18): 21148 - 21161.
[17] A. Mahieux, S. Berkenbosch, R. Clairquin, et al. "In - Flight Performance and Calibration of SPICAV SOIR Onboard Venus Express" . Appl. Optics. 2008. 47 (13): 2252 - 2265.
[18] O. I. Korablev, J. L. Bertaux, A. Fedorova, et al. "SPICAM IR Acousto - Optic Spectrometer Experiment on Mars Express" . J. Geophys. Res - Panet. 2006. Ⅲ (E9) . doi: 10.1029/2006JE002696.
[19] R. W. M. Hoogeveen, J. V. D. A. Ronald, A. P. H. Goede. "Extended Wavelength InGaAs Infrared (1.0 - 2.4μm) Detector Arrays on SClAMACHY for Space -

Based Spectrometry of the Earth Atmosphere". Infrared Phy. Technol. 2001. 42 (1)：1-16.

[20] Z. P. He，C. L. Li，R. Xu，et al. "Spectrometers Based on Acousto - Optic Tunable Filters for In - Situ Lunar Surface Measurement". J. Appl. Remote Sens. 2019. 13 (2)：027502.

[21] Z. P. He，B. Y. Wang，G. Lu，et al. "Operating Principles and Detection Characteristics of the Visible and Near - Infrared Imaging Spectrometer in the Chang'e - 3". Res. Astron. Astrophys. 2014. 14 (12)：1567-1577.

[22] Z. P. He，B. Y. Wang，G. Lv，et al. "Visible and Near - Infrared lmaging Spectrometer and its Preliminary Results from the Change'3 Project". Rev. Sci. Instrum. 2014. 85 (8)：083104.

[23] C. L. Li，Z. D. Wang. R. Xu，et al. "The Scientific Information Model of Chang'e - 4 Visible and Near - IR Imaging Spectrometer (VNIS) and In - Flight Verification". Sensors. 2019. 19 (12)：E2806.

Detection and calibration characteristics of the visible and near – infrared imaging spectrometer in the Chang'e – 4

C. L. Li, R. Xu, G. Lv, L. Y. Yuan, Z. P. He,[a] and J. Y. Wang[b]

Key Laboratory of Space Active Opto – Electronics Technology, Shanghai Institute of Technical Physics of the Chinese Academy of Science, Shanghai 200083, China

a) Electronic mail: hzping@mail. sitp. ac. cn

b) Electronic mail: jywang@mail. sitp. ac. cn

Abstract: Chang'e 4 is a robotic lunar exploration mission operated by the China National Space Administration and was to pull off the first – ever soft landing of space hardware on the mysterious lunar far side. The Visible and Near – IR Imaging Spectrometer (VNIS) is one of the main scientific instruments onboard the Chang'e 4 lunar rover. In order to analyze the lunar surface mineral composition, the VNIS was mounted on the front of the rover and detects lunar objects with a 45° visual angle to obtain spectra and geometry data to make mineralogical and compositional measurements. The VNIS is the same as the one in the Chang'e 3 Yutu rover and consists of a visible and near – infrared imaging spectrometer (0. 45 – 0. 95μm), a shortwave IR spectrometer (0. 9 – 2. 4μm), and a calibration unit with dust – proofing functionality. Here, we describe detection and calibration characteristics of the VNIS, which provides valuable information for scientific data processing and applications.

I Introduction

Chang'e 4 is a robotic lunar exploration mission and it was to pull off the first – ever soft landing of space hardware on the mysterious lunar far side. The mission is the fourth in the Chang'e program, and its main objectives were the first soft landing and exploration on the far side of lunar. The Chang'e 4 spacecraft consisted of the lander and the Yutu – 2 lunar rover, each of which carried appropriate scientific payloads. An imaging spectrometer based on acousto – optic tunable filter (AOTF) is especially suitable for deep space exploration applications (Moon, Mars, and asteroid detection).[1-6] The visible and near – IR imaging spectrometer (VNIS) is an AOTF –

based imaging spectrometer; after soft landing on the moon, the VNIS was involved mostly in the exploration of the lunar surface mineral composition and available resource.[5] As shown in Fig. 1, the VNIS is mounted on the front of the Yutu-2 rover with a 45°visual angle to obtain the spectra and the geometry images at a height of 0.69 m. The spectral range of VNIS is 0.45 - 2.40μm, with a spectral resolution of 2.4 - 9.6 nm. Minerals such as pyroxene, plagioclase, olivine, and ilmenite, in different sizes and shapes, constitute most of the lunar surface rocks. They have distinctive spectral characteristics in the visible and near-infrared (VIS/NIR) waveband (0.45 - 0.95μm) and the shortwave infrared (SWIR) waveband (0.90 - 2.40μm).[6,7] Chang'e 4 is a robotic lunar exploration mission and it was to pull off the first-ever soft landing of space hardware on the mysterious lunar far side. Compared with the one onboard Chang'e-3 spacecraft, the new VNIS is for the measurement on the far side of lunar and has been upgraded in the control software, optimized in the logic and efficiency of data acquisition, and added a step of stray light measurement for the 0-order beam produced by acousto-optic interaction for the purpose of higher accurate radiation quantification.

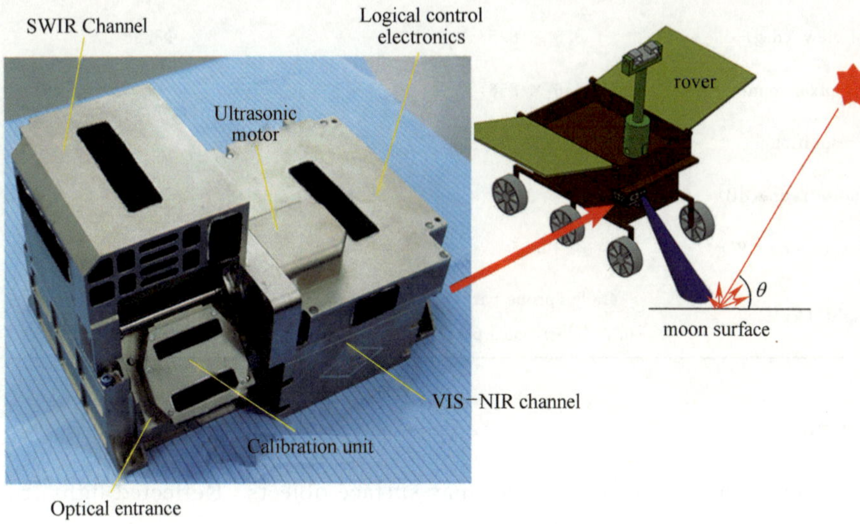

Fig. 1　Schematic of mineralogical detection by the Yutu-2 rover and visible and near-infrared imaging spectrometer (flight model)

This paper describes the detection and calibration characteristics of the VNIS, which provides valuable information for scientific data processing and applications.

Ⅱ　Detection Characteristics of VNIS

The VNIS has two detection channels: a VIS/NIR channel (0.45 - 0.95μm) with a CMOS area array detector and a SWIR channel (0.9 - 2.4μm) with an InGaAs single-element detector to obtain image spectra and mineralogical spectrum of the lunar surface, respectively. Calibration unit, which is situated outside the main optics, performs *in situ* calibration and provides dust - proofing and insulation. The main performance specification of the VNIS is shown in Table I. Compared with the one onboard Chang'e - 3 space - craft, the spectral resolution of the new VNIS is 2.4 - 6.5nm but not 2 - 7 nm in the VIS/NIR channel and 3.0 - 9.0nm but not 3 - 12 nm in the SWIR channel.

The VNIS has two operating modes: lunar surface detection and *in situ* calibration. The spectral radiance range of the detected objects determines the dynamic range of the detection instruments, and the VNIS was expected to explore objects at solar elevation angles of 15°- 45°and a lunar surface object reflectance of 5%- 98% (lunar surface objects and the diffuser panel of calibration unit) . [5]

Table Ⅰ　Major specifications of VNIS

Description	Specification	
	VIS/NIR	SWIR
Spectral range (nm)	450 - 950	900 - 2400
Spectral resolution (nm)	2.4 - 6.5	3.6 - 9.6
Field of view (deg)	8.5 × 8.5	Φ3.58
Effective pixel count	256 × 256	1
Bit resolution	10	16
Signal to noise ratio (dB)	>33	>31
Power consumption (W)	16.95	
Weight (kg)	4.65 (probe part) 0.7 (electronics part)	

A. Detection mode

The VNIS obtains scientific data on lunar surface objects. Reflected light from lunar objects arrives at the AOTFs, where only narrow wavelength band data or imaging information passes through. The quasi - monochromatic light converges on the detector, and the spectrometer can realize flexible and rapid choices of wave - length by altering the driving frequency exerted on the AOTF, so we can acquire spectra or image data at different wavelengths over the available spectral coverage.

B. Calibration mode

Using solar spectral irradiation as the traceability source, the VNIS's diffuser panel of the calibration unit is set to a horizontal position to calibrate the device. The diffuser panel is located at the light entrance and made of polytetrafluoroethylene (PTFE) material. In order to achieve *in situ* calibration, the diffuser panel inside the calibration unit will be located parallel to the mounting plane of the VNIS by the lightweight ultrasonic motor.

The VNIS is a passive optical instrument and measures the radiance diffusively reflected from solar illumination on the Moon's surface. With a spectral sampling interval of 5nm, the VNIS detects the lunar surface composition automatically by default, and it has sequentially sampled 100 frames of spectral images in the VIS/NIR band and 300 frames of spectral data in the SWIR band.[4-6] In addition, the VIS/NIR and SWIR channels both need to collect 20 frames of dark current information for data processing. Different from the one onboard CE-3, the VNIS takes extra 20 frames of bias light in the imaging channel for higher accuracy of radiometric calibration.

Ⅲ CALIBRATION CHARACTERISTICS OF VNIS

A. Spectral calibration

The AOTF and its performance have a direct impact on the spectral range and resolution of the VNIS. The parameters of the AOTFs, such as the relationship among diffraction wavelengths, RF driving frequency, diffraction efficiency, and spectral resolution, should be tested before system integration and calibration. An AOTF performance analysis system is designed to test the parameters of the AOTF,[8,16] and the detail of parameters of the AOTF equipped in the flight model is shown in Fig. 2.

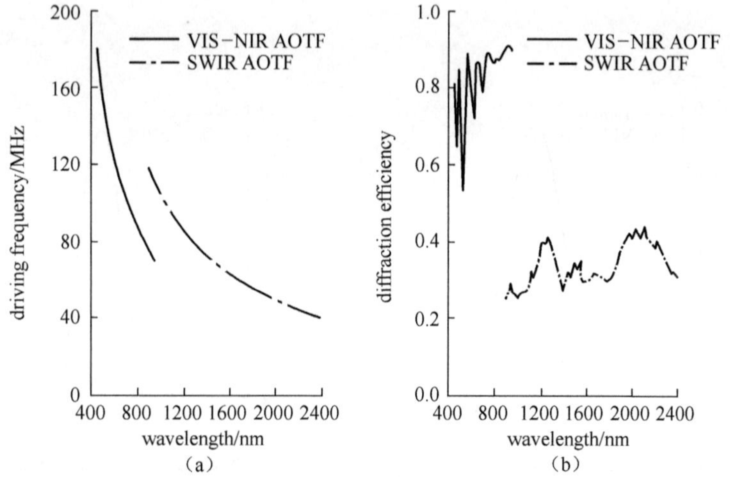

Fig. 2　AOTF performance:[10] (a) driving frequency and (b) diffraction efficiency

For laboratory spectral calibration, a monochromator (Horiba *i*HR320) calibrated with a Hg/Ne-Ar source is used. After collimation, monolight enters the optical system of the VNIS, as shown in Fig. 3.[9]

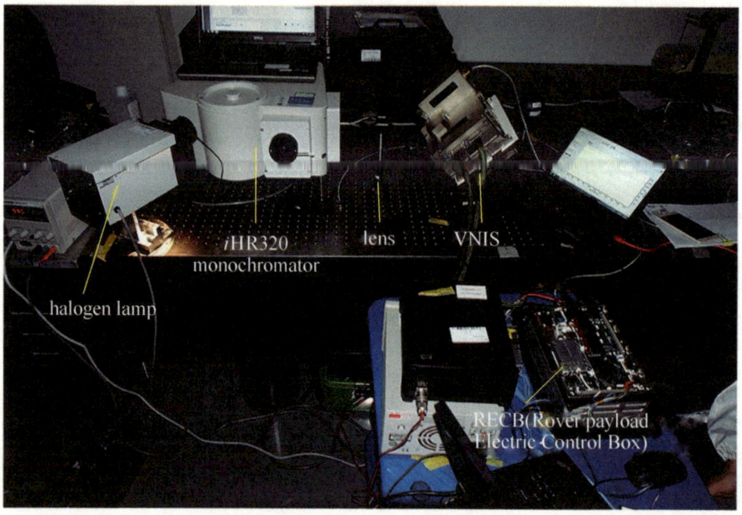

Fig. 3 Configuration of laboratory spectral calibration

After calibration and confirmation on the ground, the spectral range of the VNIS was found to be 450 - 2400nm, and the typical spectra response curves in two channels are plotted in Fig. 4. The spectral resolutions are 2.4 - 6.5nm in the VIS/NIR band and 3.6 - 9.6nm in the SWIR band, as shown in Fig. 5.

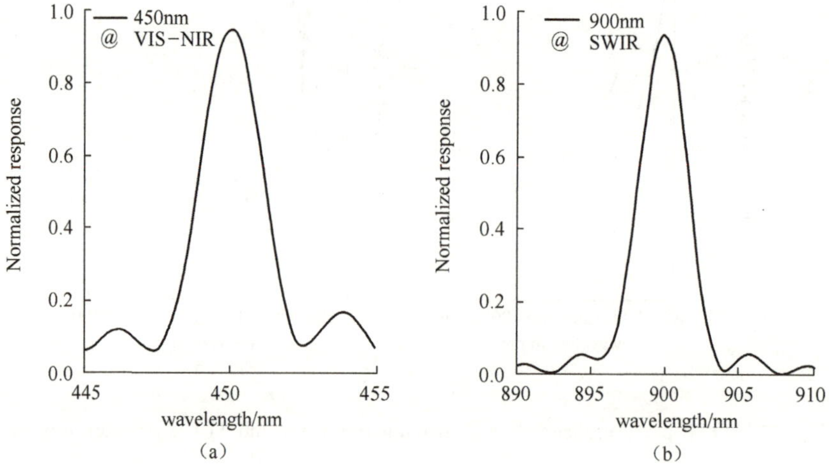

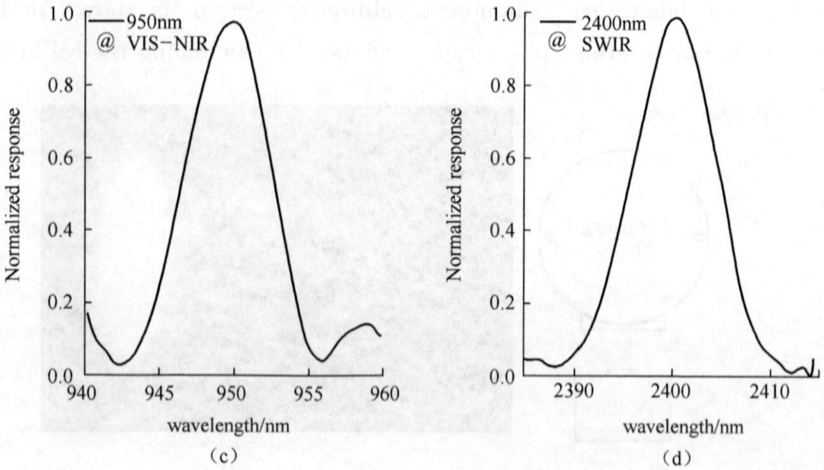

Fig. 4 The normalized response in spectral calibration. The light source was generated by the monochromator. (a) and (c) are characteristics at the both ends of the spectral coverage in the VIS－NIR channel, while (b) and (d) are characteristics at the both ends of the spectral coverage in the SWIR channel

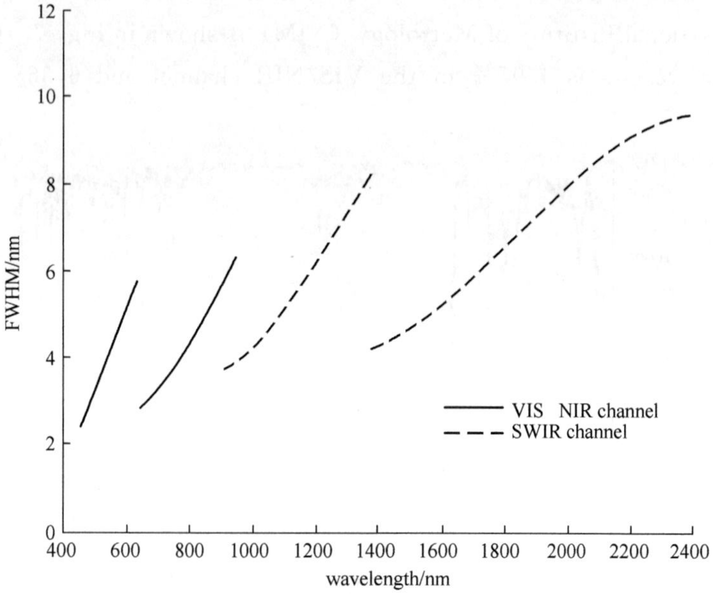

Fig. 5 Spectral resolution of the VNIS

B. Radiometric calibration

Radiometric calibration is a procedure that associates the value measured by the instrument [e. g. , the data number (DN) of a CMOS image pixel or single－element detector] with absolute physical quantities (such as the spectral radiance). The inversion model for the radiance of a target is established using a standard radiation source.[11] The configuration of the laboratory radiometric calibration system is shown in Fig. 6; the radiance level of the integrating sphere can be adjusted by switching the calibrated lamps.

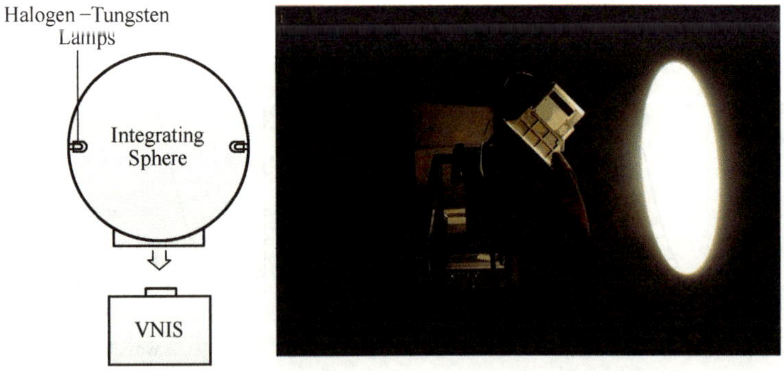

Fig. 6　Laboratory radiometric calibration system

In radiometric calibration, the spectral response data are acquired under different integrating sphere radiance levels. Therefore, the radiance inversion model is finally obtained by linear fitting to the DN of each pixel in each band with the trend of the variation in the light source. The linearity between the response and the spectral radiance traced by the National Institute of Metrology (NIM) is shown in Fig. 7. The uncertainty of the inversion results is 4.97% in the VIS/NIR channel and 6.36% in the SWIR channel.

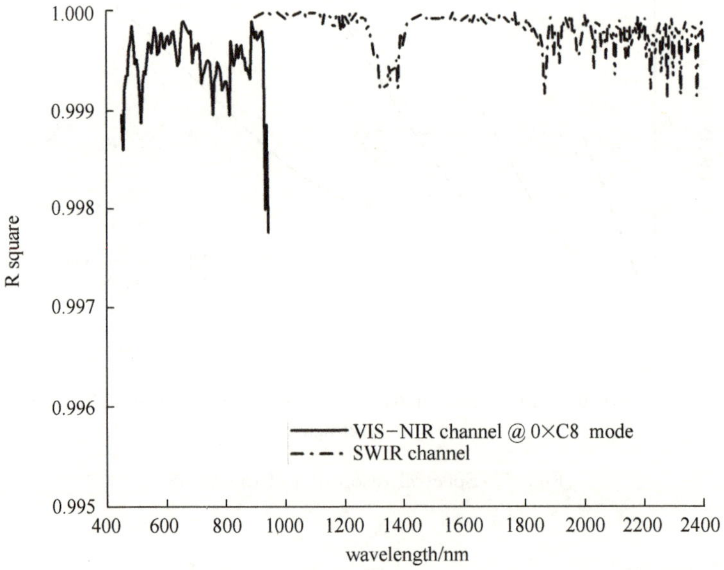

Fig. 7　The linearity of laboratory radiometric calibration between the response of VNIS and the U6500 integration sphere traceable to NIM, Chinese Academy of Science

The SNR of the VNIS was tested and analyzed by ground radiometric calibration, and the SNR was obtained by calculation as shown in Fig. 8. When the albedo is 0. 09 and the solar incident angle is 45° in the VIS/NIR band, and the albedo is 0. 09 and the solar inci - dent angle is 75° in the SWIR band, the SNR must be greater than 30dB.

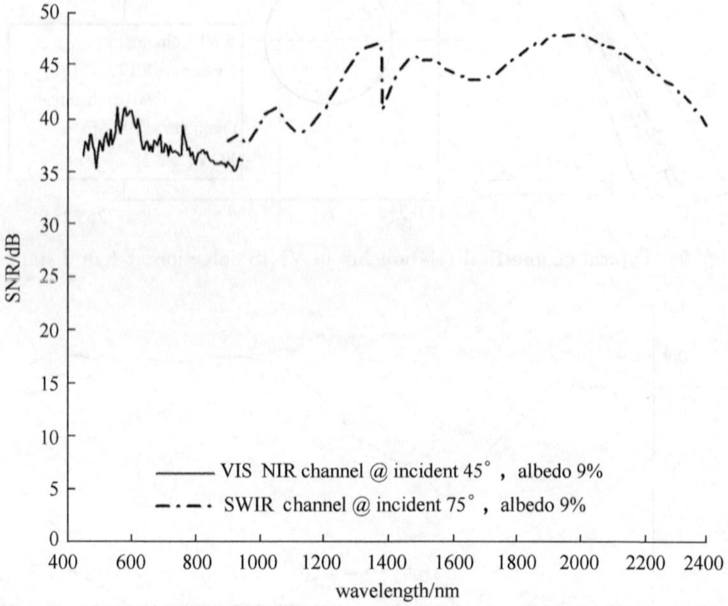

Fig. 8　Typical SNR of the VNIS calculated by in - flight calibration data

C. Geometric calibration

The VNIS is used to detect lunar surface objects, and the optical axis of the VIS/NIR and SWIR channels is paralleled each other at an 18mm distance. [12-14] The FOVs in the VIS/NIR and SWIR are 8. 5° × 8. 5° and Φ3. 58°, respectively. The geometrical characteristics of imaging in the VIS/NIR band and spectral detection in the SWIR, obtained by testing and calibration on the ground, are shown in Fig. 9. The circle represents the SWIR channel ' s FOV, which has a diameter of 107. 6 pixels and is centered at the coordinate (98, 127. 5) of the VIS/NIR image in 1m detection distance typically.

Ⅳ　OPERATION

The flight model of VNIS was developed in January 2018, and the ground verification experiment was set to verify the performance of VNIS; the powder samples were made of olivine, augite, hypersthene, and plagioclase, which are typical components used to simulate lunar soil. The spectra of these mineral samples are shown in Fig. 10.

VNIS was successfully booted up on January 3, 2019. In Fig. 11, the images and REFF of the first Scientific site (note A) and Intact rocks on the moon's surface (note

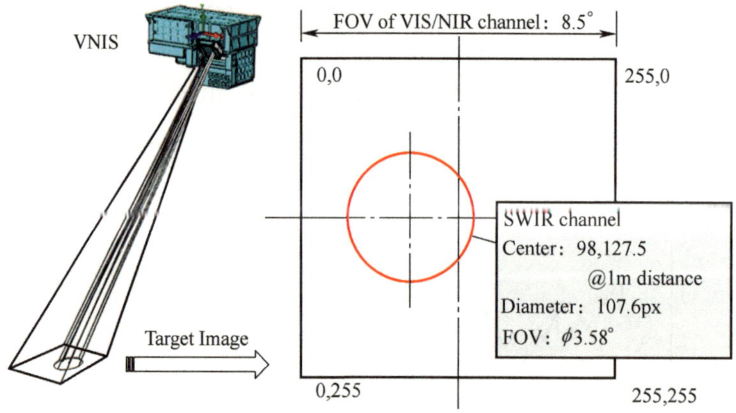

Fig. 9　Typical geometrical relationship in VNIS detection at 1 m distance

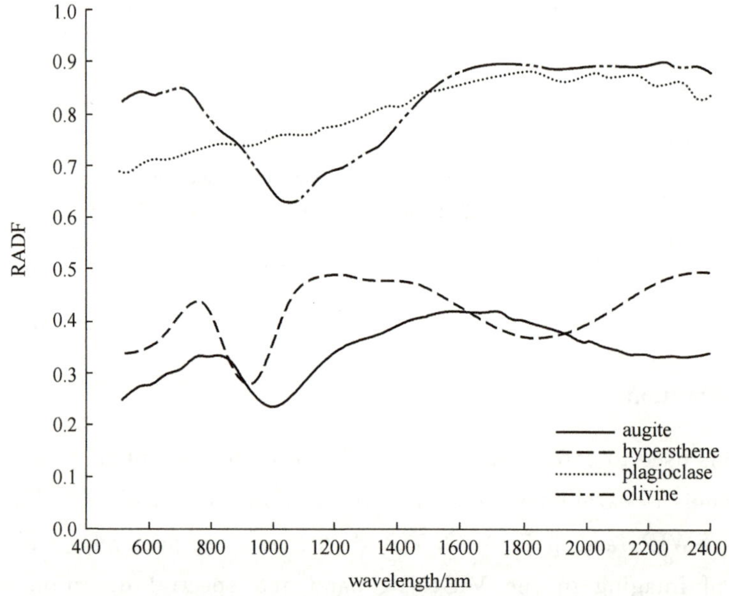

Fig. 10　Typical mineral spectra acquired by the VNIS smooth with the Savitzky – Golay filter

LE00303), which detected by VNIS/Yutu – 2, were shown. The VNIS had been working for five lunar days and performed 27 detections.

Ⅴ　Conclusion

As one of the main scientific payloads of the Chang'e 4 rover, the VNIS, which is the same as the one on the Chang'e 3 rover, was mounted on the front of the Yutu – 2 rover to acquire spectral and geometric data in order to achieve the scientific goals of obtaining data for the composition of lunar surface materials and its resources exploration. Before it booted up, the VNIS completed a ground test, calibration, and environmental simulation test. Since the successful boot on December 8, 2018, the VNIS has already completed several lunar detection and calibration runs, where in it has obtained spectral image data, providing reference material for scientific applications. This paper describes main detection and calibration characteristics of the VNIS, which provides valuable information for scientific data processing and applications.

Acknowledgments

This paper was supported by the Chinese lunar exploration program's special funds for the second phase and the National Natural Science Foundation (Grant Nos. 21105109 and 61605231). The authors thank the Science and Application Center for Moon and Deep Space Exploration of the Chinese Academy of Sciences for the ground test and data preprocessing, and the National Space Science Center of the Chinese Academy of Sciences for its contributions to the development and testing of the instruments.

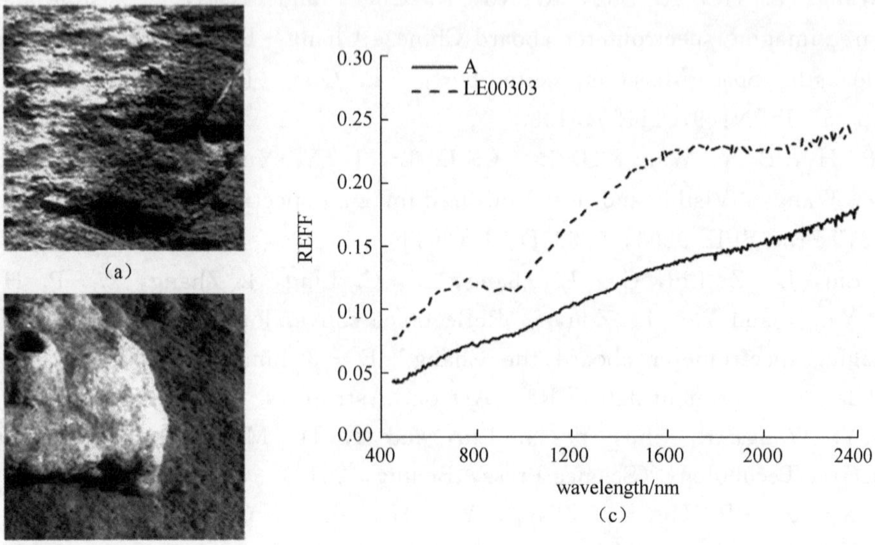

Fig. 11　(a) A 635 – nm image from VNIS/Yutu – 2 at node A, (b) 635 – nm image from VNIS/Yutu – 2 at node LE00303, and (c) REFF curves of nodes A and LE00303.[15]

References

[1] O. Korablev et al., "Development of a mast or robotic arm – mounted infrared AOTF spectrometer for surface Moon and Mars probes," Proc. SPIE 9608, 960807 (2015).

[2] J. – P. Bibring et al., "The MicrOmega investigation onboard Hayabusa2," Space Sci. Rev. 208, 401 – 412 (2017).

[3] D. A. Belyaev et al., "Compact acousto – optic imaging spectro – polarimeter for mineralogical investigations in the near infrared," Opt. Express 25 (21), 25980 – 25991 (2017).

[4] S. N. Mantsevich et al., "Examination of the temperature influence on phase

matching frequency in tunable acousto - optic filters," Ultrasonics 91, 45 - 51 (2019) .

[5] S. W. Dai, J. Wu, H. X. Sun, B. M. Zhang, J. F. Yang, G. Y. Fang, J. Y. Wang, H. Y. Wang, and J. S. An, "Chang' E - 3 Lunar Rover's scientific payloads," Chin. J. Space Sci. 34 (3), 332 - 340 (2014) .

[6] Z. P. He, B. Y. Wang, G. Lv, C. L. Li, L. Y. Yuan, R. Xu, K. Chen, and J. Y. Wang, "Visible and near - infrared imaging spectrometer and its preliminary results from the Chang' E 3 project," Rev. Sci. Instrum. 85 (8), 083104 (2014) .

[7] Z. - P. He, B. - Y. Wang, G. Lv, C. - L. Li, L. - Y. Yuan, R. Xu, B. Liu, K. Chen, and J. - Y. Wang, "Operating principles and detection characteristics of visible and near - infrared imaging spectrometer (VNIS) in Chang'e 3," Res. Astron. Astrophys. 14 (12), 1567 (2014) .

[8] J. Wang, Z. He, R. Shu, R. Xu, K. Chen, and C. Li, "Visible and near - infrared imaging spectrometer aboard Chinese Chang' E - 3 spacecraft," in Optical Payloads for Space Missions, edited by S. - E. Qian (John Wiley & Sons, 2015), Chap. 5, ISBN: 9781118945148.

[9] Z. P. He, B. Y. Wang, G. Lv, C. L. Li, L. Y. Yuan, R. Xu, K. Chen, and J. Y. Wang, "Visible and near - infrared imaging spectrometer (VNIS) for Chang' E - 3," Proc. SPIE 9263, 92630D - 1 (2014) .

[10] B. Liu, J. - Z. Liu, G. - L. Zhang, Z. - C. Ling, J. Zhang, Z. - P. He, B. - Y. Yang, and Y. - L. Zou, "Reflectance conversion methods for the VIS/NIR imaging spectrometer aboard the Chang' E - 3 lunar rover: Based on ground validation experiment data," Res. Astron. Astrophys. 13 (7), 862 (2013) .

[11] J. Y. Wang, R. Shu, Y. N. Liu, and Y. H. Ma, Introduction of Imaging Spectral Technology (Science Press, Beijing, 2011) .

[12] R. Xu, Z. - P. He, H. Zhang, Y. - H. Ma, Z. Q. Fu, and J. - Y. Wang, "Calibration of imaging spectrometer based on acousto - optic tunable filter (AOTF)," Proc. SPIE 8527, 85270S (2012) .

[13] Z. P. He, R. Xu, C. L. Li, G. Lv, L. Y. Yuan, B. Y. Wang, R. Shu, and J. Y. Wang, "Visible and near - infrared imaging spectrometer (VNIS) for in - situ lunar surface measurement," Proc. SPIE 9639, 96391S (2015) .

[14] Z. P. He, C. L. Li, R. Xu, G. Lv, L. Y. Yuan, and J. Y. Wang, "Spectrometers based on acousto - optic tunable filters for in - situ lunar surface measurement," J. Appl. Remote Sens. 13 (2), 027502 (2019) .

[15] C. Li et al., "Chang' E - 4 initial spectroscopic identification of lunar far - side mantle - derived materials," Nature 569, 378 - 382 (2019) .

[16] R. Xu, Z. - P. He, Z. H. Zhang et al., "The research on characteristics of acousto - optic tunable filter by space radiation experiment," in International Symposium on Photoelectronic Detection and Imaging 2011: Space Exploration Technologies and Applications (International Society for Optics and Photonics, 2011), pp. 819627 - 1 - 819627 - 8.

The Scientific Information Model of Chang'e - 4 Visible and Near - IR Imaging Spectrometer(VNIS) and In - Flight Verification

Chunlai Li, Zhendong Wang, Rui Xu, Gang Lv, Liyin Yuan, Zhiping He and Jianyu Wang

Key Laboratory of Space Active Opto—Electronics Technology, Shanghai Institute of Technical Physics, Chinese Academy of Sciences, Shanghai 200083, China; lichunlai @ mail. sitp. ac. cn (C. L.); wangzhendong@ mail. sitp. ac. cn (Z. W.); xurui @ mail. sitp. ac. cn (R. X.); lvgang @ mail. sitp. ac. cn (G. L.); yuanliyin@mail. sitp. ac. cn (L. Y.)

Abstract: The Chang'e - 4 (CE - 4) lunar rover, equipped with a visible and near - IR imaging spectrometer (VNIS) based on acousto - optic tunable filter spectroscopy, was launched to the far side of the moon on December 8,2018. The detection band of the VNIS ranges from 0. 45 to 2. 4μm. Because of the weak reflection of infrared radiation from the lunar surface, a static electronic phase - locked acquisition method is adopted in the infrared channel for signal amplification. In this paper, full - link simulations and modeling are conducted on the infrared channel information flow of the instrument. The signal characteristics of the VNIS are analyzed in depth, and the signal to noise ratio (SNR) prediction and laboratory verification are presented. On 4 January 2019, the VNIS started working successfully and acquired high - resolution spectrum data of the far side of the moon for the first time. Through analysis we have found that the SNR ratio is in line with our predictions, and the data obtained by VNIS in orbit are consistent with the information model proposed in this paper.

Keywords: the Chang'e - 4 lunar rover; phase - locked; signal - to - noise ratio; spectral resolution; infrared focal plane components

1 Introduction

The Chang'e - 4 (CE - 4) lunar rover is the first man - made spacecraft launched to the far side of the moon, and its rover (Yutu - 2) is equipped with a visible and near - IR imaging spectrometer (VNIS), which is used to analyze the composition of lunar surface minerals [1, 2]. The VNIS inherits information from the Chang'E-3 Lunar Rover's scientific payloads, and the software is optimized. The VNIS is a spectrum detector based on acousto-optic tunable filter (AOTF) spectroscopy [3, 4], and it has two detection channels - visible near-infrared (450-950nm) and short-wave infrared (900-2400nm). It performs spectral analysis and imaging detection of minerals on the lunar surface under appropriate solar illumination and assists in the comprehensive detection of minerals and chemical compositions in the patrol areas.

The AOTF spectrometer has been used in deep space exploration for a long time. The SPICAM [5], which was loaded on the ESA Mars Express mission (2003), was one of the first spectroscopic applications of the AOTF on a civilian spacecraft. It is a single-pixel spectrometer which can sequentially measure the spectrum of reflected solar radiation from Mars between 1.1 and 1.7μm with spectral resolution of 3.5cm^{-1}, and spatial resolution of ~6km from 250km of the Mars express orbit. Similar instruments, such as the SPICAV [6], were loaded on the ESA Venus express mission (2005). NASA has also been working on spectrometers based on the AOTF. AIMS [7] is a compact, two-channel AOTF imaging spectrometer developed by NASA/GSFC, and it is one of the alternative payloads of the NASA Mars rover. Since 2006, the Shanghai Institute of Technical Physics began to study imaging spectrometers based on AOTF. The VNIS is a payload of the lunar rover for 0.45-2.4μm spectral bands detection and will be able to inspect and probe minerals for the rover region [2] (China's Chang'e 3 and Chang'e 4 unmanned lunar rover).

The CE-4 spacecraft was successfully launched on 8 December 2018, and it landed on the moon on 3 January 2019. Then the VNIS was powered on and acquired infrared spectral data of the lunar surface. This paper focuses mainly on simulation analyses and laboratory tests based on the infrared channel design features and information flow model of the instrument and carries out verification according to the acquired in-orbit data.

2 Instrument Description

2.1 Basic Principle of The Visible and Near-IR Imaging Spectrometer (VNIS)

The VNIS uses an acousto-optic tunable filter (AOTF) for the spectral filtering of light. When a beam of broadband light passes through an optically elastic crystal that vibrates at high frequencies, monochromatic light of a certain wavelength will be diracted inside the crystal and transmitted from it at an angle, and the non-diffracted light travels through the crystal directly in the original direction, thereby achieving the goal of light filtering [8, 9].

The AOTF is a spectroscopic device based on the principle of acousto-optic diraction and is composed of a crystal and a transducer bonded on it [10, 11]. The transducer converts the electric signal into ultrasonic vibrations in the crystal, which generate spatially periodic modulations. When incoming light is incident on the AOTF, the wavelength of the diffracted light is related to the frequency of the driving electric signal, and it can be changed by varying the frequency of the electric signal [12].

According to this principle, by changing the driving frequency of the AOTF crystal through rapid scanning, the wavelength of the first - order diffraction light passing through the AOTF changes sequentially. In this way, the spectral information of the target can be obtained by VNIS [10]. The VNIS can obtain a spectral image in the VIS/NIR channel (visible and near - infrared waveband) and spectral data in the SWIR channel (short waveband) simultaneously. Figure 1 shows the components and basic principle of the VNIS, and Table 1 presents the main specification of the VNIS.

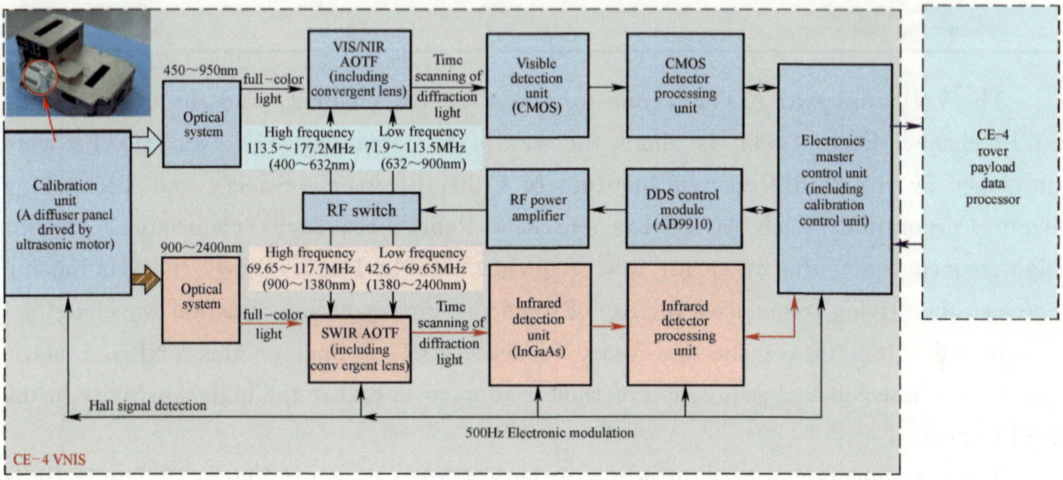

Figure 1　The components and basic principle of the VNIS

Table 1　The characteristics of the Chang'e-4 visible and near-IR imaging spectrometer (VNIS)

Description	Specification	
	VIS/NIR	SWIR
Spectral range (nm)	450-950	900-2400
Spectral resolution (nm)	2-10	3-12
Number of bands	100	300
Field of view (°)	8.5×8.5	ϕ3.58
Number of valid pixels	≥256×256	1
Quantized value (bit)	10	16

Continued

Description	Specification	
	VIS/NIR	SWIR
S/N ratio (dB)	≥43 (maximum SNR) ≥33 (albedo 0.09, solar elevation angle 15°)	≥46 (maximum SNR) ≥31 (albedo 0.09, solar elevation angle 15°)
RF ranges (MHz)	High frequency (F - H): 113.5 - 177.2 (400 - 632nm) Low frequency (F - L): 71.9 - 113.5 (632 - 900nm)	High frequency (F - H): 69.65 - 117.7 (900 - 1380nm) Low frequency (F - L): 42.6 - 69.65 (1380 - 2400nm)
Modulation frequency (Hz)	—	500
Integration time (ms)	18.2 - 256 (adjustable)	—
Detection range (m)	0.7 - 1.3	
Measurement time (min)	2	

The VNIS has two AOTFs, one for the VIS/NIR channel, and the other for the SWIR channel. Each AOTF is about the size of a matchbox. The two AOTFs were developed by the 26th Research Institute of China Electronic Science and Technology Group Corporation. Each AOTF has two RF (Radio Frequency) connectors, one for high frequency and the other for low frequency. Table 1 also shows the relationship between the driving frequencies for AOTFs and the corresponding diraction wavelength.

In order to improve the sensitivity of the SWIR channel of the VNIS, a static electronic phase - locked acquisition method is adopted to realize the high sensitivity in the SWIR channel.

The components and basic principle of the VNIS are shown in Figure 1. The VNIS is equipped with a calibration unit at the light entrance. The calibration unit consists of an ultrasonic motor, framework, and diffuser panel inside. The calibration unit could be located at a different position by the ultrasonic motor for lunar surface detection, in - orbit calibration and dust - proof functions. While the Chang’ 4 spacecraft was operating in - orbit flight and the soft - landing phase, the calibration unit was locked to thermal insulation and it was dust - proof. After the Yutu - 2 rover separated from the lander on the moon, the VNIS was remote controlled to collect soil images and spectrum and the calibration unit was set to be fully open in order not to block the reflectance light into the imaging lens. The angle between the framework of calibration unit and horizontal mounting face of the VNIS was about 55°. When the VNIS was required to perform an in - orbit calibration, the calibration unit would be located at a horizontal position by the hall sensor and the sun light could be diused into the VNIS from diuser panel inside. The structure and function of the calibration unit is shown in Figure 2.

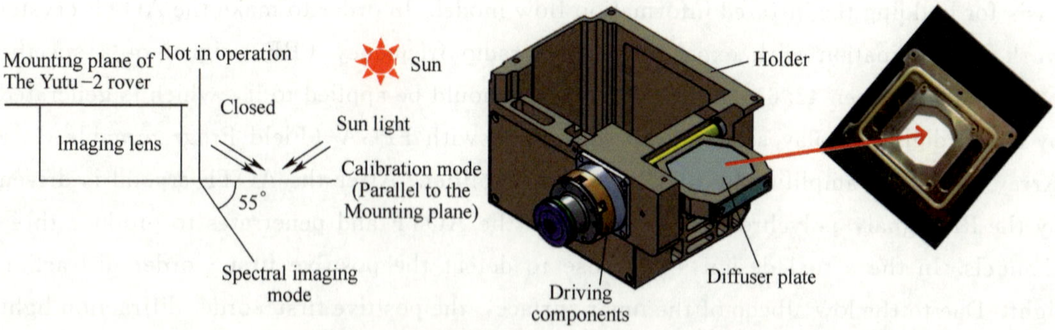

Figure 2　Structural and functional schematic view of the calibration unit

2.2　Optical Design of the SWIR Channel

The optical system of the VNIS in the infrared channel is shown in Figure 3. The target input rays enter into the instrument through an imaging lens, after being collimated into a parallel beam, which drives the AOTF to work, so that the emitted light passing through the AOTF forms an infrared monochromatic light of a specific wavelength and then converges to the detector through a convergent lens.

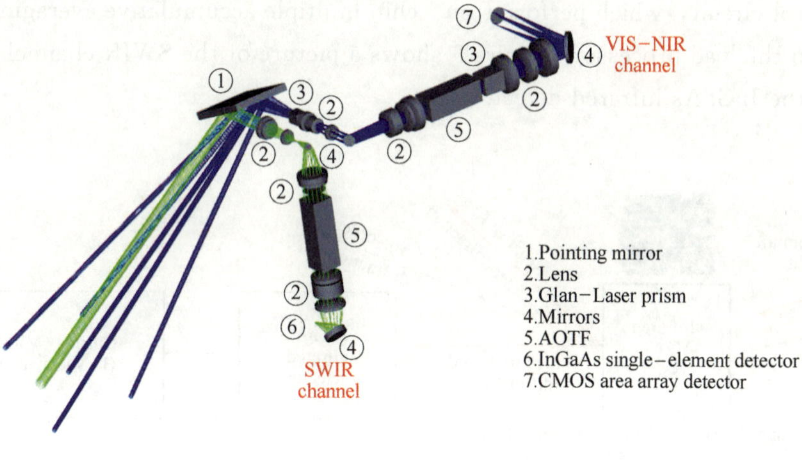

Figure 3　The optical design diagram of the VNIS

According to the basic principle of AOTF spectrophotometry, after passing through the AOTF, three light rays are formed through the convergent lens, which are positive first-order diraction light, negative first-order diraction light and zero-order diffraction light [10, 11]. An infared (the material is InGaAs) detector with a diameter of 1 mm is placed at the convergence of the positive first-order diraction light, and a light filter is designed for the zero-order diraction light, thereby suppressing the stray light.

2.3　The Information Link of Infrared Channel

Figure 4 shows the block diagram of the SWIR channel of the VNIS, which is the basis for building the infrared information flow model. In order to make the AOTF crystal work in combination with expectations, an radio frequency (RF) signal of a specific frequency (between 42. 6MHz to 117. 7MHz) should be applied to it, which is generated by DDS (digital display scope) chip configured with FPGA (Field Programmable Gate Array) and then amplified by the RF power amplifier. After the AOTF crystal is driven by the RF signal, polychromatic light enters the AOTF and penetrates to produce three channels. In the actual design, we chose to detect the positive first - order diffraction light. Due to the low albedo of the lunar surface, the positive first - order diffraction light after AOTF is very weak. In order to improve the ability to detect such a weak infrared signal, a static electronic phase - locked acquisition method is adopted.

As shown in Figure 4, by controlling the output RF signal amplified by the power amplifier, the periodic modulation of the output two channels of monochromatic infrared light at the AOTF outlet can be controlled. The modulation frequency is 500Hz. The positive first - order diffraction light after modulation is received by an InGaAs infrared detector, and the corresponding voltage signal is obtained through current - voltage conversion. The signal is processed by a phase - locked amplifier and after low - pass filtering, sent to the ADC (analog - digital conversion) chip. It is processed by the FPGA control circuit, which performs on - chip multiple accumulative averaging and then uploads it to the load processor. Figure 5 shows a picture of the SWIR channel processing circuit and the InGaAs infrared detector.

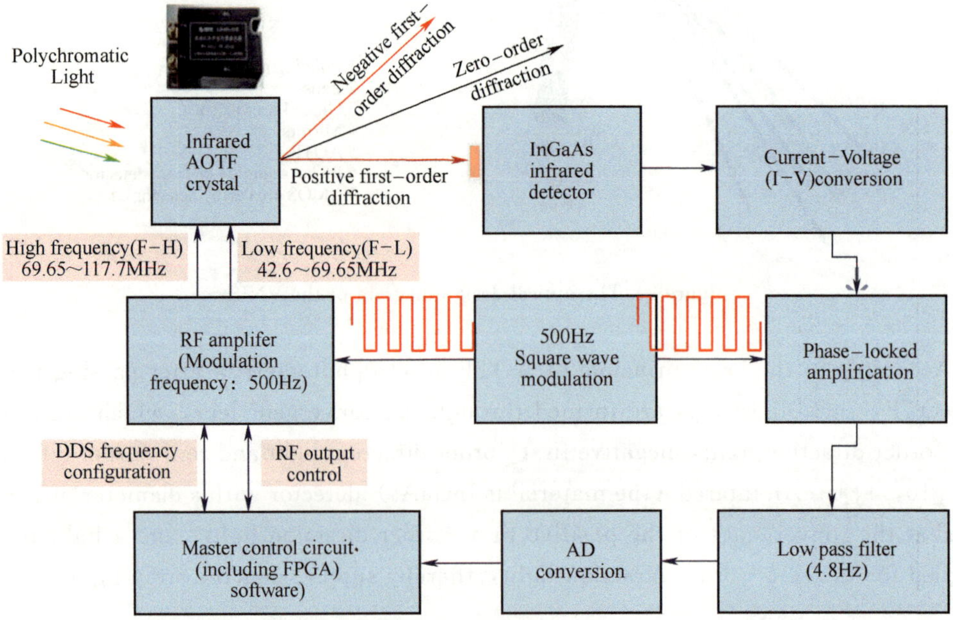

Figure 4 Schematic diagram of the phase - locked processing circuit of infrared channel

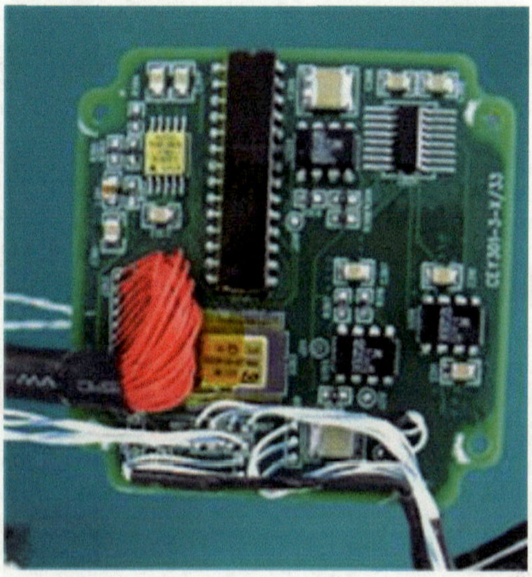

Figure 5　The picture of the SWIR channel processing circuit and the detector. The infrared detector type is J23TE2 - 66C - R01M - 2. 6, and is manufactured by Judson, with a peak current response rate of 1. 2A/W

3　Signal Flow Model Simulation and Testing

3. 1　The Signal Acquisition Model of Infrared Spectral

The phase - locked amplification method of infrared channel electronics is introduced in Section 2. 3. In this section, we further discuss the infrared channel information flow model and signal characteristics. The infrared spectral signal acquisition model is established based on the AOTF spectroscopic system, as shown in Figure 6.

For weak infrared signal processing technology with phase - locked amplification, moving parts (such as the modulating reticle and chopper) are generally used for modulation between the signal light and background, such as the spatial target infrared spectroscopy system [6]. In the VNIS, a phase - locked amplification model without moving parts is designed according to the characteristics of AOTF, as shown in Figure 6. By controlling the output RF signal amplified by the power amplifier, the periodic modulation of the output positive first - order diffraction light at the AOTF outlet can be controlled. The modulation frequency is designed to be 500Hz. This 500Hz frequency is generated by the FPGA chip, not by the moving parts (such as motors or galvanometers), it has a great advantage in reliability for space applications.

The positive first - order diraction light after passing through the AOTF crystal is received by the InGaAs detector. After current - voltage conversion by a preamplifier, it is sent to the subsequent stage for signal conditioning. As shown in Figure 6, for two

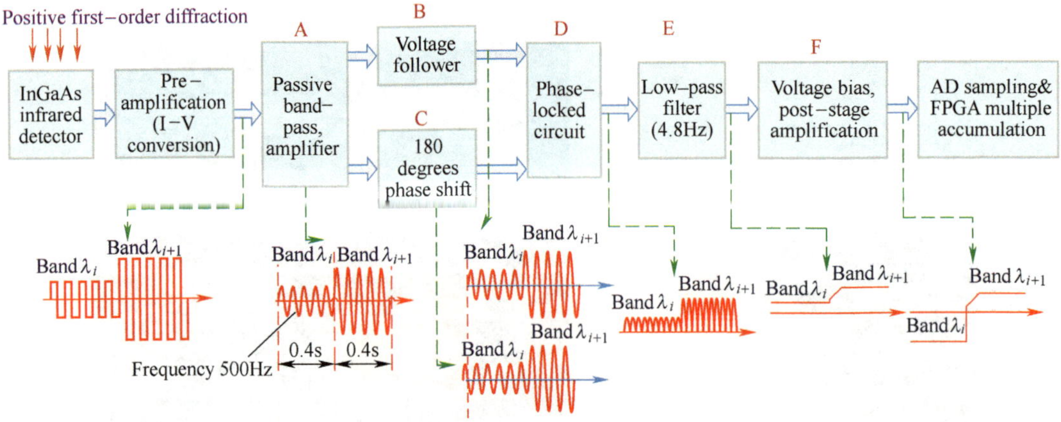

Figure 6 The information flow model of infrared spectrum acquisition based on acousto-optic tunable filter (AOTF) modulation

consecutive wavelength signals (band λ_i and band λ_{i+1}), after 500Hz electronic modulation, what emerges is a periodic (500Hz) infrared analog signal similar to a square wave after passing through the preamplifier. After passive band-pass and pre-amplification, the square wave appears to be an approximate sinusoidal signal, and the peak of the waveform reflects the intensity of the infrared monochromatic light of the wavelength. Then, it converts the signal into two signals of the same frequency and the same amplitude but with an heir phase differing by 180 degrees.

Specifically, one chain is designed as a voltage follower circuit and the other is designed as a reverse circuit, thus realizing a phase shift of 180 degrees. The phase-locked circuit in Figure 5 realizes the chip selection of two signals. It uses a high-speed analog switch (ADG409) to realize the phase-locked output so that the signal becomes half the sinusoid shown in the figure. After low pass filtering (4.8Hz), the signal is converted into a corresponding DC level signal, which is fed for AD sampling after voltage bias and post-stage amplification.

In order to further improve the detection sensitivity, the signal of the same wavelength is collected eight times by ADC chip (AD976), and then the sampled digital number (DN) values from the eight times are averaged and conducted within the FPGA. For infrared full-spectrum detection of lunar surface targets, the time-division detection of different wavelengths is achieved by varying the frequency of the RF signal applied to the AOTF crystal. The VNIS has 300 infrared sampling bands. The acquisition time of a single band is 0.4s, so the acquisition time of the full spectrum channel is approximately 2min.

The above model describes the mechanism of the infrared channel signal generation of the VNIS. At the forefront of the model, the infrared detector produces a photo-generated signal current I (λ) due to receiving infrared light energy P (λ). The model aims to enhance the sensitivity of the infrared channel. Based on the above information flow model, the source of the photo - generated current I (λ) at the input end can be further analyzed so as to obtain the signal - to - noise ratio of the system.

For the lunar surface spectral detection model, the surface of the moon can be approximated as a Lambert body. The target energy P (Δλ) received by the VNIS can be expressed as [13, 14]:

$$P(\Delta\lambda)=\frac{1}{4}\cdot E(\lambda)\cdot\frac{A}{(F\#)^2}\cdot\tau_0(\lambda)\cdot\rho(\lambda)\cdot\sin\theta\cdot\Delta\lambda \tag{1}$$

where E (λ) represents the spectral irradiance of the sun near the lunar surface. A is the pixel area of detector, F# is the optical aperture, τ_0 (λ) is the optical system efficiency (include the efficiency of AOTF), θ is the solar elevation angle, Δλ is the spectral resolution, and ρ (λ) is the reflectance factor. Here is also a concept that the actual energy received by the instrument is the spectral radiance $B(\lambda)$ forced by the target. The relationship between the radiance $B(\lambda)$ and irradiance $E(\lambda)$ is [1, 2]:

$$B(\lambda)=\frac{E(\lambda)\cdot\rho(\lambda)\cdot\sin\theta}{\pi} \tag{2}$$

Generally, the target energy P (λ) is determined after the instrument system parameters are determined. That is, given the target radiant power received by the InGaAs detector in the information flow model in Figure 6, the signal current I (λ) of the infrared detector response can be expressed as:

$$I(\Delta\lambda)=P(\Delta\lambda)\times R_\lambda \tag{3}$$

where R_λ represents the current response of detector. The VNIS uses the J23TE2 - 66C - R01M - 2. 6 infrared detector manufactured by Judson, and the peak current response rate is about 1. 2A/W. After the signal current of the detector response is determined, the noise current $I_{n\lambda}$ can be calculated by the following equation.

$$I_{n\lambda}=\sqrt{2\times q\times(I_{s\lambda}+I_{dark}+I_{black})\Delta f} \tag{4}$$

where q represents the electron charge, I_{dark} denotes the dark current of the detector, I_{black} denotes the current caused by the thermal background radiation, and Δf is the bandwidth of the information processing circuit. Due to the rapid development of the detector technology, the dark current of the InGaAs infrared detector at a low temperature of around −40℃ is basically zero. The SNR can be simplified as:

$$\mathrm{SNR}(\lambda)=\frac{I_{s\lambda}}{I_{n\lambda}}=\frac{I_{s\lambda}}{\sqrt{2\times q\times I_{s\lambda}\Delta f}}=\sqrt{\frac{I_{s\lambda}}{2\times q}}\frac{1}{\sqrt{\Delta f}} \tag{5}$$

As can be seen, given the system design parameters and the selected infrared detector, the system SNR is directly related to the information processing bandwidth Δf of the circuit. The system adopts the information flow model shown in Figure 6. Here, Δf is actually the low - pass filtering 4. 8Hz mentioned in the Figure 6, and the system noise beyond((500 - 2. 4) Hz - (500+2. 4) Hz) is filtered out.

In contrast, if the direct signal acquisition method (without phase lock - in method) is adopted, the bandwidth Δf is generally around 2000Hz, which is the mechanism of increasing the SNR ratio. The Figure 7 shows the estimated SNR curve based on the static electronic phase - locked acquisition method (bandwidth is about 4.8Hz), which is compared with the direct signal acquisition method (bandwidth is about 2000Hz).

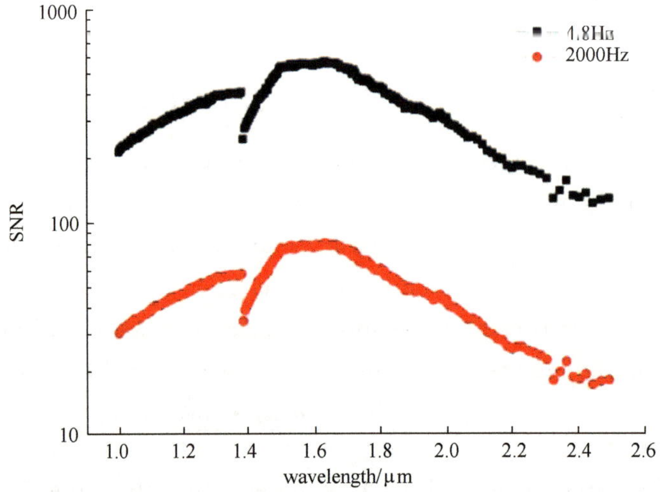

Figure 7　The signal - to - noise ratio (SNR) curve comparison between the static electronic phase - locked acquisition method (black) and the direct signal acquisition method (red)

As shown in Figure 7, there exists a step at 1380nm. The reason is that the SWIR AOTF crystal has two driving frequency range (F - H and F - L), and the 1380nm is exactly the frequency switching point of the SWIR AOTF crystal. It can be seen from the Figure 7 that the SNR is about 600 at 1.7μm.

The parameters used in the calculations are:

➢ $E=217W/m^2/\mu m$@1.7μm;

➢ $A=1mm^2$;

➢ F#: 2.8;

➢ $\tau_0=0.35$@1.7μm;

➢ $\Delta\lambda=8nm$@1.7μm;

➢ $R_\lambda=1.2A/W$@1.7μm;

➢ $\theta=15°$;

➢ $\rho=0.09$@1.7μm;

➢ $B=1.61W/m^2/\mu m/Sr$@1.7μm.

3.2 Laboratory Testing and Evaluation

After the information flow model is established, the signal characteristic and the SNR of VNIS are tested in the laboratory. Figure 8 is a photo of the laboratory testing, where the light source is a halogen lamp and the target is a diuse reflection plate.

Adjusting the intensity of the Halogen lamp can simulate the change of the spectral radiance forced by the diuse reflection plate, and synchronous measurement is carried out by a commercial spectrometer (FieldSpec 4) .

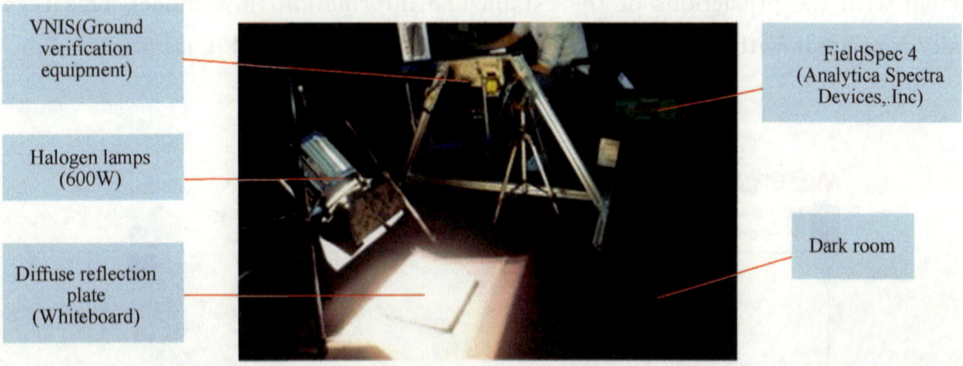

Figure 8　The experimental test was carried out in a dark room, using a halogen lamps to illuminate the diuse reflection plate (reflectance >90%), and then at the same angle, using both VNIS (ground verification equipment) and a commercial spectrometer ((FieldSpec 4, Analytica Spectra Devices), Inc.) measure the spectral radiance forced by the diuse reflection plate [15]

When the spectral radiance at 1.7μm measured by the FieldSpec 4 is about 1.61W/m^2/μm/Sr, the intensity of the halogen lamp is stabilized, and the test is started. Figure 9 shows the original digital number (DN) values of 300 infrared spectral bands, and the SNR is about 500@1.7μm (the signal DN is 1600, and the noise DN is 3.2) . This is slightly different from the value (SNR≈600@1.7μm) given in Section 3.1, because the light source here is not the sunlight with an equivalent elevation angle of 15 degrees and some transmission errors caused by the commercial spectrometer.

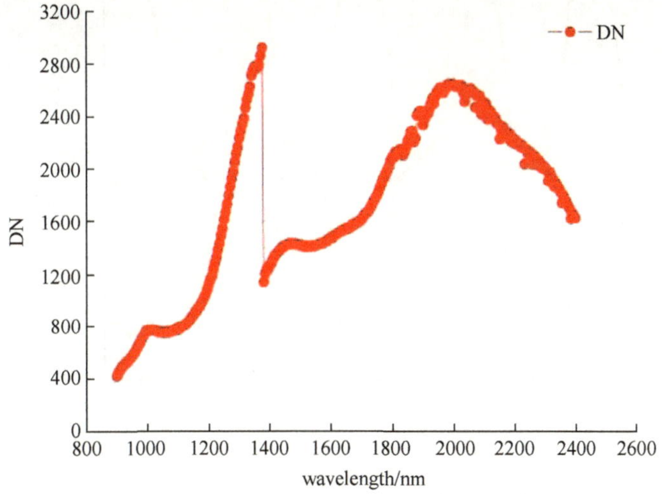

Figure 9　The original infrared full spectrum signal measured in the laboratory

In order to further validate the information flow model, the analog signals (at 1.7μm) were captured by an oscilloscope. Figure 10 shows the signal waveforms of A, B, C and D in the model in Figure 6, and the characteristics of those waveforms are consistent with the predictions of the established information flow model. This result is basically consistent with the information flow model described in Figure 6.

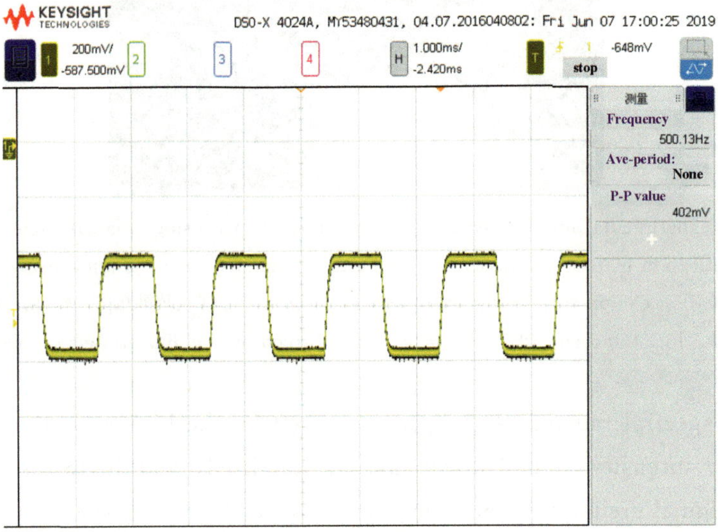

(a)

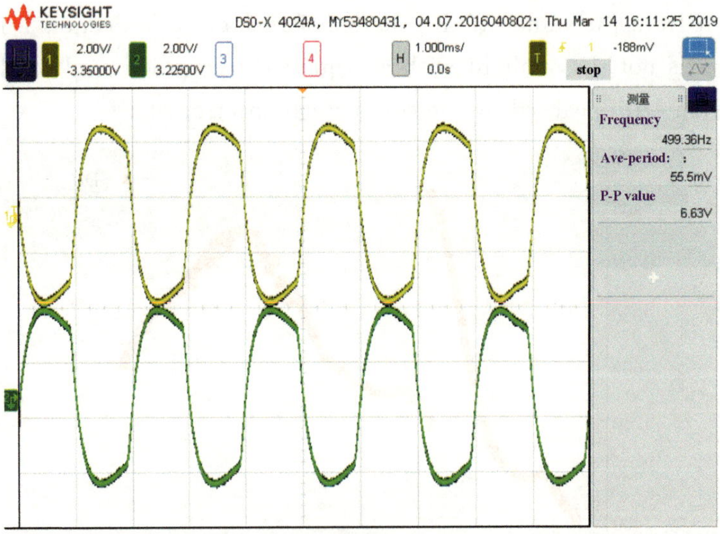

(b)

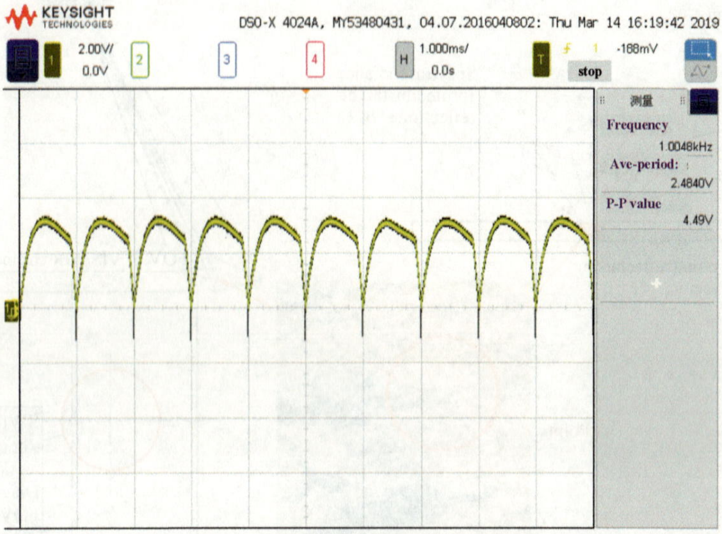

(c)

Figure 10　The analog signals of VNIS at 1.7μm captured by oscilloscope (corresponding to A, B, C and D in Figure 6). (a) The analog signal after current to voltage (I-V) conversion captured by oscilloscope (point A in Figure 6); (b) The analog signal before phase-locked circuit captured by oscilloscope (the yellow is for point B in Figure 6, and the green is for point C in Figure 6); (c) The analog signal after the phase-locked circuit captured by oscilloscope (point D in Figure 6)

4　In-Flight Test

On 4 January 2019, the instrument obtained the infrared spectrum data of the first scene on the far side of the moon. Figure 11 shows the scene of the first infrared spectrum on the far side of the moon. Through absolute lab and in-flight radiometric calibration, we can get the relationship between DN values and spectral radiance at the entrance pupil, which is consistent with B (λ) in Figure 11. The spectral radiance B (λ) divided by the solar irradiance E (λ), multiplied by π, and further divided by the cosine of the incident angle θ can obtain the reflectance factor ρ (λ) [1]. The DN values of the original infrared spectrum data of A Point and its full-spectrum SNR are show in Figures 12 and 13, and Figure 14 gives the infrared spectral reflectance curve (900-2400nm) at A point. In this test, the solar elevation angle is about 15degrees.

According to the obtained raw data, the signal-to-noise ratio at 1.7 m is calculated to be around 470, which is basically consistent with the prediction in Section 3.1 (Figure

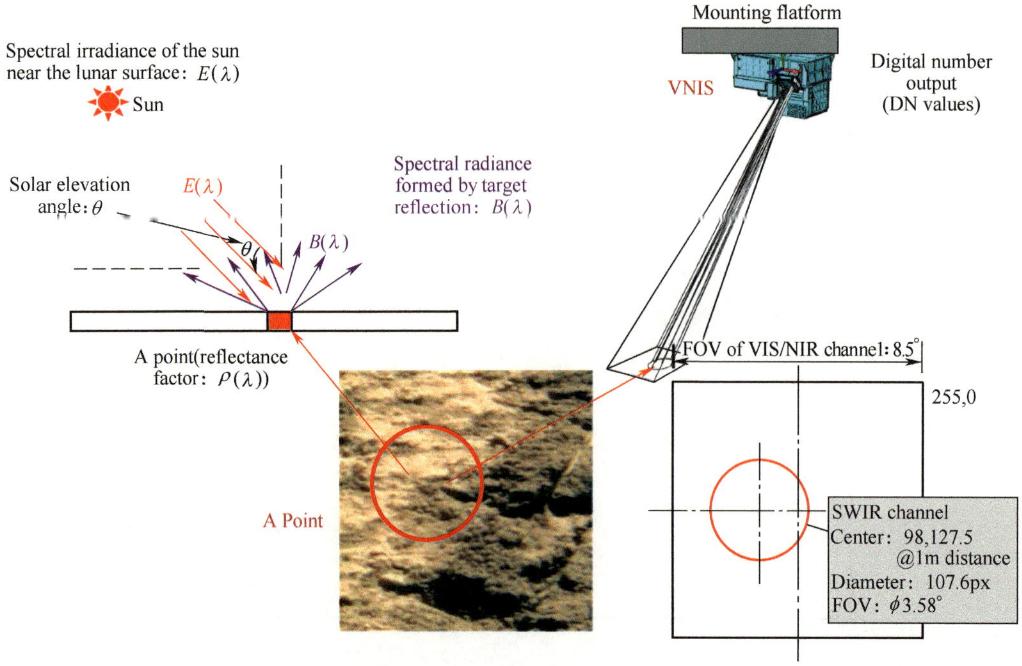

Figure 11　The first scene (which is defined as point A) obtained by VNIS on the far side of the moon. The VNIS is used to detect lunar surface objects and the optical axis of the VIS/NIR channel and SWIR channel are parallel to one another at an 18mmdistance [15, 16]. The FOV (Field of view) in the VIS/NIR and SWIR are 8.5°×8.5° and Φ 3.58°, respectively. The circle represents the SWIR channel's FOV, which has a diameter of 107.6 pixels and is centered at the coordinate (98, 127.5) of the VIS/NIR image in 1 m detection distance typically

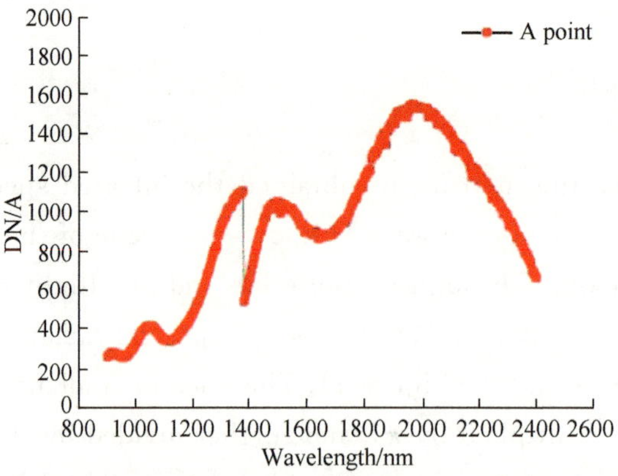

Figure 12　The original digital number (DN) values of the far side of the moon surface at the A point

9). This test validates the proposed AOTF system infrared spectral information processing model based on phase - locked amplification technology and also verifies the high sensitivity of the of the lunar surface weak infrared signal detector.

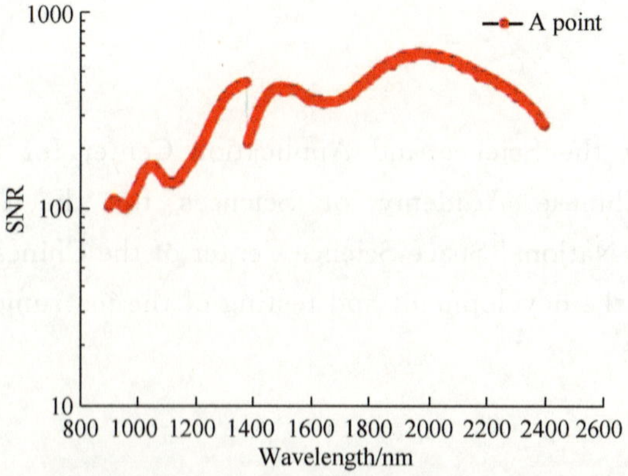

Figure 13　The full－spectrum signal to noise ratio（SNR）curve of the far side of the moon surface at point A

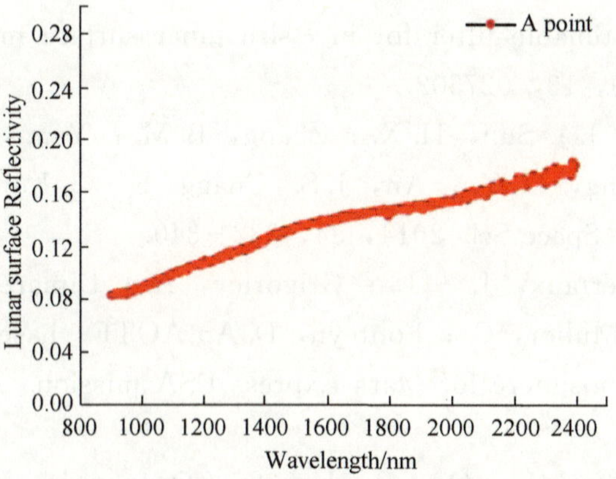

Figure 14　The spectral reflectance curve of the far side of the moon surface at point A

5　Conclusions

In view of the extremely weak infrared reflectance spectral signals on the lunar surface，we proposed an information flow model，on the basis of which the signal－to－noise ratio has been predicted. As of the end of February 2019，the VNIS has operated six times on the lunar surface and has acquired 12 sets of infrared spectral data. It is found that an average SNR of 300 can still be obtained when the lunar albedo is around 9% and the solar elevation angle is 15 degrees，which further verifies the effectiveness of the proposed method for weak infrared signal. The SNR values of in－flight testing are basically consistent with the predictions given in this paper. In China's future deep space exploration programs，the method proposed in this paper will be helpful to further study the acquisition of weak infrared spectral information on the surface of planets.

Acknowledgments

The authors thank the Science and Application Center for Moon and Deep Space Exploration of the Chinese Academy of Sciences for the ground test and data preprocessing, and the National Space Science Center of the Chinese Academy of Sciences for its contributions to the development and testing of the instruments.

References

[1] Li, C.; Liu, D.; Liu, B.; Ren, X.; Liu, J.; He, Z.; Zuo, W.; Zeng, X.; Xu, R.; Tan, X.; et al. Chang'E-4 initial spectroscopic identification of lunar far-side mantle-derived materials. Nature 2019, 569, 378-382.

[2] He, Z.; Li, C.; Xu, R.; Lv, G.; Yuan, L.; Wang, J. Spectrometers based on acousto-optic tunable filter for in-situ lunar surface measurement. J. Appl. Remote Sens. 2019, 13, 027502.

[3] Dai, S. W.; Wu, J.; Sun, H. X.; Zhang, B. M.; Yang, J. F.; Fang, G. Y.; Wang, J. Y.; Wang, H. Y.; An, J. S. Chang'E-3 Lunar Rover's Scientific Payloads. Chin. J. Space Sci. 2014, 34, 332-340.

[4] Korablev, O.; Bertaux, J.-L.; Grigoriev, A.; Dimarellis, E.; Kalinnikov, Y.; Rodin, A.; Muller, C.; Fonteyn, D. An AOTF-based spectrometer for the studies of Mars atmosphere for Mars Express ESA mission. Adv. Space Res. 2002, 29, 143-150.

[5] Bertaux, J. L.; Korablev, O.; Perrier, S.; Quemerais, E.; Montmessin, F.; Leblanc, F.; Lebonnois, S.; Rannou, P.; Lefèvre, F.; Forget, F.; et al. SPICAM on Mars Express: Observing Modes and Overview of UV Spectrometer Data and Scientific Results. J. Geophys. Res. 2006, 111, E10.

[6] Bertaux, J. L.; Nevejans, D.; Korablev, O.; Villard, E.; Quémerais, E.; Neefs, E.; Montmessin, F.; Leblanc, F.; Dubois, J. P.; Dimarellis, E.; et al. SPICAV on Venus Express: Three Spectrometers to Study the Global Structure and Composition of the Venus Atmosphere. Planet. Space Sci. 2007, 55, 1673-1700.

[7] Glenar, D. A.; Blaney, D. L.; Hillman, J. J. AIMS: Acousto-optic Imaging Spectrometer for Spectral Mapping of Solid Surfaces. Acta Astronaut. 2003, 52, 389-396.

[8] He, Z. P.; Wang, B. Y.; Lv, G.; Li, C. L.; Yuan, L. Y.; Xu, R.; Chen, K.; Wang, J. Y. Visible and Near-Infrared Imaging Spectrometer (VNIS) and Its Preliminary Results from the Chang'E 3 Project. Rev. Sci. Instrum. 2014, 86, 8.

[9] He, Z. P.; Wang, B. Y.; Lü, G.; Li, C. L.; Yuan, L. Y.; Xu, R.; Liu, B.;

Chen, K.; Wang, J. Y. Operating principles and detection characteristics of Visible and Near - Infrared Imaging Spectrometer (VNIS) in Chang'e 3. Res. Astron. Astrophys. 2014, 14, 1567.

[10] Wang, J.; He, Z.; Shu, R.; Xu, R.; Chen, K.; Li, C. Visible and Near - infrared Imaging Spectrometer aboard Chinese Chang'E - 3 Spacecraft. Chapter 5; In Optical Payloads for Space Missions; Qian, S. - E., Ed.; John Wiley & Sons: Chichester, UK, 2016.

[11] Yuan, L.; He, Z.; Lv, G.; Wang, Y.; Li, C.; Wang, J. Optical design, laboratory test, and calibration of airborne long wave infrared imaging Spectrometer. Opt. Express 2017, 25, 22440 - 22454.

[12] He, Z.; Wang, B.; Lv, G.; Li, C.; Yuan, L.; Xu, R.; Chen, K.; Wang, J. Visible and Near - infrared Imaging Spectrometer (VNIS) for Chang'E - 3. Proc. SPIE 2014, 9263, 92630D - 1.

[13] Liu, B.; Liu, J. Z.; Zhang, G. L.; Ling, Z. C.; Zhang, J.; He, Z. P.; Yang, B. Y.; Zou, Y. L. Reflectance conversion methods for the VIS/NIR imaging spectrometer aboard the Chang'E - 3 lunar rover: Based on ground validation experiment data. Res. Astron. Astrophys. 2013, 13, 862.

[14] Wang, J. Y.; Shu, R.; Liu, Y. N.; Ma, Y. H. Introduction of Imaging Spectral Technology; Science Press: Beijing, China, 2011.

[15] Xu, R.; He, Z. P.; Zhang, H.; Ma, Y. H.; Fu, Z. Q.; Wang, J. Y. Calibration of imaging spectrometer based on acousto - optic tunable filter (AOTF). Proc. SPIE 2012, 8527, 85270S.

[16] He, Z.; Wang, B.; Lv, G.; Li, C.; Yuan, L.; Xu, R.; Chen, K.; Wang, J. Visible and Near - infrared Imaging Spectrometer (VNIS) for In - situ Lunar Surface Measurement. Proc. SPIE 2015, 9639, 96391S.

Mafic mineralogy assemblages at the Chang'e - 4 landing site: A combined laboratory and lunar in situ spectroscopic study

C. Liu[1], L. Liu[1], J. Chen[1], H. Cao[1], H. Qu[1], L. Qiao[1], J. Zhang[1], X. Qi[1], X. Lu[1], R. Xu[2], Z. Wang[2], Z. He[2], and Z. Ling[1,3]

1. Shandong Key Laboratory of Optical Astronomy and Solar - Terrestrial Environment, School of Space Science and Physics, Institute of Space Sciences, Shandong University, Weihai 264209, PR China
e - mail: zcling@sdu. edu. cn
2. Shanghai Institute of Technical Physics, Chinese Academy of Science, Shanghai 200083, PR China
e - mail: hzping@mail. sitp. ac. cn
3. CAS Center for Excellence in Comparative Planetology, Chinese Academy of Sciences, Hefei 230026, PR China

Abstract: *Context*. Chang'e - 4 (CE - 4) provides unprecedented information about lunar materials exposed by the South Pole - Aitken (SPA) basin. Diverse results have been obtained from previous interpretations of CE - 4 visible and near - infrared (VNIR) spectra. Some studies suggest that materials at the CE - 4 landing site are dominated by olivine and orthopyroxene, but others argue that only a small amount of olivine should be exposed at the CE - 4 landing site.

Aims. Laboratory spectroscopy studies using the Engineering Model of CE - 4 Visible and Near - infrared Imaging Spectrometer (VNIS) are critical in constraining the accurate mineral proportions and composition of soils and boulders at the CE - 4 landing site.

Methods. VNIR spectra of nine lunar analogs - prepared by mixing orthopyroxene (OPX), clinopyroxene (CPX), olivine (OL), and plagioclase - were acquired using the CE - 4 VNIS Engineering Model. The spectral indices model and modified Gaussian model were developed to estimate CPX/(CPX+OPX) and OL/(OL+CPX+OPX) and are applicable to the in situ spectra acquired by the Yutu - 2 VNIS spectrometer.

Results. The lunar rocks and regolith at the CE - 4 landing site excavated by the Finsen impact are CPX - rich with limited OL (CPX : OPX : OL = 56 : 29 : 17). The mineral chemistries of the four lunar rocks show Mid - Ca, Fe pyroxene, and Mid - Mg OL (Fo_{60-79}), providing critical constraints for mineral compositions in the SPA compositional anomaly. These rocks exhibit high M1 intensity ratios, indicating that they were crystallized at a high temperature (980 – 1300℃) and a rapid – cooling magmatic system produced by impact melt differentiation or volcanic resurfacing events.

Keywords: Moon – planets and satellites; composition – techniques; spectroscopic – planets and satellites; surfaces – infrared; planetary systems – techniques; imaging spectroscopy

1 Introduction

The Chang'e – 4 (CE – 4) spacecraft with the Yutu – 2 rover successfully landed on the floor of the Von Kármán crater (44.4°S, 176.2 E) within the South Pole – Aitken (SPA) basin (Wilhelms et al. 1979; Moriarty & Pieters 2018; Di et al. 2019; Liu et al. 2019; Wu et al. 2019), facilitating the first in situ investigations of materials on the far side of the Moon. The Von Kármán crater floor has previously been flooded by several lava flows and later extensively covered by ejecta from the Finsen crater (Huang et al. 2018, 2020; Ling et al. 2019; Qiao et al. 2019; Fu et al. 2020; Guo et al. 2021). The Finsen crater is located at the boundary of the SPA compositional anomaly (SPACA) with high Ca – and Fe – pyroxene compositions compared with the surrounding regions, which are rich in Mg – pyroxene, and are therefore named the Mg – pyroxene annulus in the SPA basin (Moriarty & Pieters 2018; Moriarty et al. 2020). The materials at the CE – 4 landing site excavated and ejected by the Finsen crater may therefore provide clues regarding mineral assemblages and the composition of the SPACA.

The Yutu – 2 rover carried a visible and near – infrared imaging spectrometer (VNIS) composed of two detection channels: a visible and near – infrared (VNIR) imaging spectrometer (450 – 950nm) and a short – wave infrared spectrometer (SWIR, 900 – 2400nm), enabling an in situ mineralogy survey of materials exposed by SPA impact events (He et al. 2019; Li et al. 2019a, b). However, the reported mineralogy from VNIS spectral investigations appears to contain anomalies and has been disputed (Moriarty et al. 2020). A modified Gaussian model (MGM) was recently used to show that materials at the CE – 4 landing site are likely dominated by olivine (OL) and orthopyroxene (OPX) (Li et al. 2019c; Gou et al. 2019). Hu et al. (2019) and Lin et al. (2020), based on Hapke's radiative transfer model, argued that materials at the CE – 4 landing site are rich in plagioclase (PLAG) and OPX only with a small amount of OL. Through a preliminary spectral check of CE – 4 spectra, Chen et al. (2020) suggested that the mafic mineralogy at the CE – 4 landing site may be mainly comprised of CPX – bearing minerals with intermediate composition and probably dominated by pigeonite. According to the remote observations of the Kaguya Multi – band Imager, the materials in the Von Kármán crater are dominated by PLAG (~54wt.%) and OPX (~24wt.%) with only a low abundance of OL (~9wt.%) (Lemelin et al. 2016, 2019; Qiao et al. 2019). The abundance ratio of clinopyroxene (CPX) and OPX is ~0.7 at the CE – 4 landing site based on the VNIR spectra of the Chandrayaan – 1 Moon Mineralogy Mapper (Ling et al. 2019).

In addition to direct observation and interpretation of in situ VNIS spectra, laboratory spectroscopic studies on lunar soil analogs provide another means to establish more reliable models and better constrain the mineralogy at the CE - 4 landing site. In this study, we prepared a series of lunar soil analogs by mixing four terrestrial minerals: CPX, OPX, OL, and PLAG. The VNIR spectra of these analogs were characterized using a backup instrument of the CE - 4 VNIS payload (Engineering Model, EM) to build improved models and better constrain the mineralogy and composition of lunar rocks and regolith at the CE - 4 landing site.

2 Experimental and analytical methods

Four natural terrestrial minerals including augite (CPX, $\rho = 3.34\text{g} \cdot \text{cm}^{-3}$), hypersthene (OPX, $\rho = 3.54\text{g} \cdot \text{cm}^{-3}$), OL ($\rho = 3.42\text{g} \cdot \text{cm}^{-3}$), and PLAG ($\rho = 2.88\ \text{g} \cdot \text{cm}^{-3}$) purchased from Shuiyuan Shanchang Mineral Specimen Co. LTD were used as endmembers. Minerals were first ground and sieved to a grain size range of 20—100μm. Subsequently, mineral powders were mixed in different volume fractions (Tab. B. 1) to obtain lunar soil analogs (~170g for each mixture) with variable CPX/(CPX+OPX) values (19. 4—65. 6. %) and plagioclase volume fractions (34. 1—59. 5vol.%) according to remote sensing detections (Ling et al. 2019; Qiao et al. 2019). It should be noted that the spectral features of OL were not observed in the orbital VNIR spectra at the CE - 4 landing site; however, OL below the detection limit could not be excluded (Ling et al. 2019). Therefore, OL was added to the mixtures up to 27. 5vol.%. Furthermore, ilmenite and nanophase - iron as well as other space weathering products were not considered in this study.

Elemental compositions (i. e., Si, Ti, Al, Cr, Fe, Mn, Mg, Ca, Na, K, and P shown in Tab. B. 2) and mineralogical characterization (pyroxene and OL) of the samples were confirmed by X - ray fluorescence (XRF) and Raman spectroscopy. VNIR reflectance spectra of mixtures were obtained in the laboratory using the CE - 4 VNIS EM, which is a copy of the VNIS payload aboard the Yutu - 2 rover. In addition, VNIR spectra of these samples were also acquired by another VNIR spectrometer (FieldSpec 4 Hi - Res VNIR spectrometer, Analytical Spectral Devices Inc., i. e., ASD) to evaluate the instrumental effects and spectral properties of the VNIS payload.

3 VNIR spectral production

3.1 CE - 4 spectral data and preprocessing

Here, we analyzed a level 2B VNIS radiance dataset collected from the first to the twenty - sixth lunar day of the CE - 4 mission carried out as part of the Ground Research and Application System of Chinese Lunar Exploration Program. As part of this work, we used 10 spectra of four lunar rocks and 89 spectra of lunar regolith, including the spectra of the same lunar soils obtained at different optical geometry (i. e., Photometric Experiments 1 and 2). The images and spectra of the four lunar rocks are shown in Fig. 1 and 2. The data were corrected using a compensation matrix owing to the change in the working environment of the CE-4 VNIS payload. The data were then divided by the distance-corrected solar irradiance data (Gueymard 2004; Chen et al. 2020) and converted into reflectance factors (REFFs) (Hapke 2012). Subsequently, REFF spectra were photometrically corrected to the same illumination-viewing geometry as the laboratory spectra ($i = 60°$, $e = 45°$, $g = 90°$) using the Lommel-Seeliger model built based on the data acquired on the tenth lunar day (Qi et al. 2020). The VNIR bands (450—945nm) were connected to the SWIR bands (900—2395nm) based on the average value in the overlapping spectral region (900—945nm). The reflectance spectra were then smoothed using the Savitzky-Golay algorithm (Savitzky & Golay 1964).

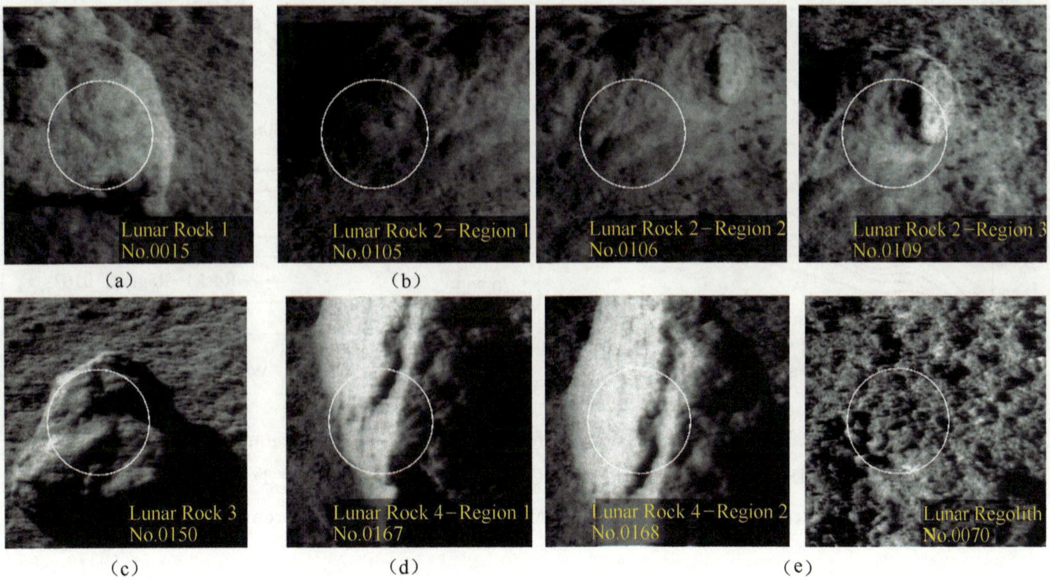

Fig. 1　CMOS images of lunar rocks and regolith observed by the Yutu-2 rover

3.2　Continuum removal

Continuum removal was conducted before characterizing the spectral absorption features. A linear spectral continuum in wavenumber space (appear curved in wavelength space) with free slope and offset (Fig. 2a) was defined by two tie points in the range of 475—815nm and 815—1985nm for ASD and VNIS EM data (Sunshine et al. 1990; Klima et al. 2008). However, this method did not work well for the spectra acquired by the CE-4 VNIS payload, likely because of the complex space weathering effects on the lunar surface. Therefore, a two - part linear continuum technique (Fig. 2b) was applied to the in -flight VNIS spectra automatically determined by three tie points (743±24nm, 1438±69nm, 1629±13nm) (Martinot et al. 2018; Moriarty & Pieters 2018; Chen et al. 2020).

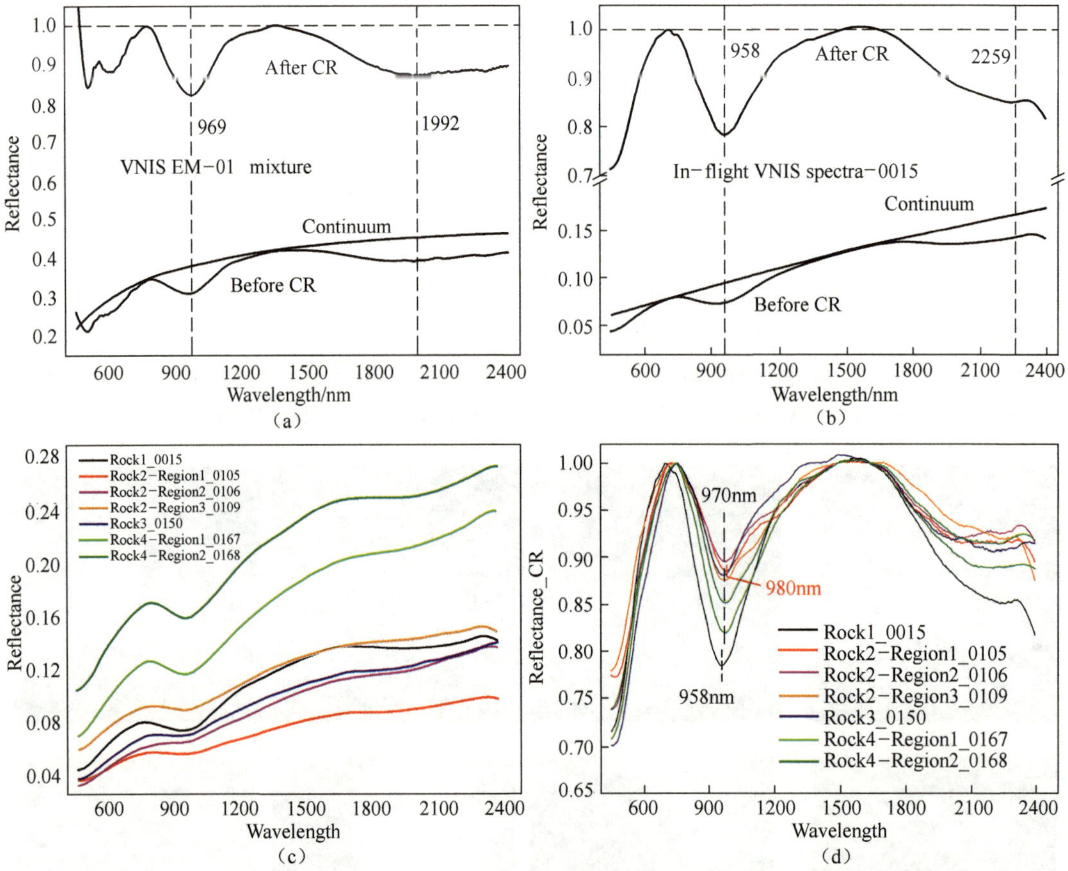

Fig. 2 (a—b) Reflectance spectra before and after continuum removal acquired by VNIS Engineering Model and CE - 4 VNIS payload. (c - d) VNIR spectra of CE - 4 lunar rocks before and after continuum removal

3.3 Spectral indices model

The CPX, OPX, and OL contents can be characterized using the Spectral Indices Model (SIM) including OPX _ INDEX (Eq. (1)), CPX _ INDEX (Eq. (2)), and the integrated band depth ratio (IBDR) (Eq. (3)) (see Cloutis et al. 1986, Mustard et al. 2011, Staid et al. 2011 and Horgan et al. 2014). OPX _ INDEX was designed to detect OPX absorption features at 930nm and 1815nm relative to the spectral shoulder of OPX at ~1330nm, and CPX _ INDEX was defined as an indicator of two CPX absorption features at 1000 and 2200nm and the reflectance shoulder at ~1500nm. The IBDR is the relative fraction of integrated band depth at 1000nm (IBD1 in 750—1300nm) of the total IBD at 1000 and 2000nm (IBD2 in 1800 — 2310nm). The spectral indices were calculated as follows:

$$OPX\_INDEX = 100 \times \frac{R(1350) - R(930)}{R(1350) + R(930)} \times \frac{R(1350) - R(1950)}{R(1350) + R(1950)}, \quad (1)$$

$$CPX\_INDEX = 100 \times \frac{R(1500) - R(1000)}{R(1500) + R(1000)} \times \frac{R(1500) - R(2200)}{R(1500) + R(2200)}, \quad (2)$$

$$IBDR = 100 \times \left(\frac{\sum_{n=750}^{n=1300} (1 - R_n / RC_n)}{\sum_{n=750}^{n=1300} (1 - R_n / RC_n) + \sum_{n=1800}^{n=2310} (1 - R_n / RC_n)} \right), \quad (3)$$

where R_n represents the reflectance, and RC_n is the value of a linear continuum between 750 and 1300nm.

3.4 Modified Gaussian Model

The MGM is an improved approach for VNIR spectral deconvolution from composite absorption features to individual absorption bands produced by Fe^{2+} electronic transitions in pyroxene and OL (Sunshine et al. 1990; Sunshine & Pieters1998). The MGM was used to estimate the modal abundance of mineral end members in our lunar soil analogs. The reflectance spectra of nine mixtures were analyzed with MGM to study the relationship between the relative band strengths and mineral abundance. It should be noted that OL was considered during the MGM analysis despite the low abundance and may be masked by the spectral features of pyroxene. Two additional spectral bands were added in the MGM deconvolution which were attributed to the spectral feature of pyroxene at ~450nm and a falldown spectral feature at ~2500nm for CE - 4 in - flight spectra (Fig. 2d).

4 Results and discussion

4.1 Comparison of ASD and VNIS EM spectra

Comparison of ASD and VNIS EM spectra led to a better understanding of spectral properties due to instrument effects. The spectra of mixed sample No. 01 acquired by ASD and VNIS EM were different (Fig. B. 1). Generally, the spectral absorption features agree well with each other in terms of the peak shapes and positions. The reflectance values of the ASD spectra before continuum removal were higher than those of the VNIS EM in all channels. For the continuum removed spectrum, an obvious difference was observed below ~730nm. A doublet feature at ~505nm and ~600nm was observed in the VNIS EM spectra. In addition, the absorption features at ~1μm of the ASD spectra are significantly wider than those in the VNIS EM spectra. These differences could be due to the different detection modes of ASD and VNIS EM. An acousto - optic tunable filter was used to collect the reflected light at a certain energy and record the reflectance, channel by channel

(He et al. 2019; Li et al. 2019a, b). For ASD, the reflectance values of all channels were acquired once using a spectrometer to split light based on different wavelengths. Despite this, the Str _CPX/ (Str _CPX+Str _OPX) and Str _OL/ (Str _OL+Str _CPX1+Str _OPX1) acquired using MGM with VNIS EM spectra and ASD spectra, respectively, are comparable, except for the OL spectra feature at ~850nm (Fig. B. 2a, b). Subsequently, we built two polynomial models using the Str ratios obtained from the ASD spectra, and the models were used to predict CPX/(CPX+OPX) and OL/(OL+CPX+OPX) using the spectra of VNIS EM. The CPX/(CPX+OPX) ratios are slightly underestimated but acceptable (RMSE = 4.38) (Fig. B. 2c), but the OL/(OL+CPX+OPX) ratios are clearly underestimated, which may be caused by the different Str ratios of the OL spectral feature at ~850nm and pyroxene features at ~1000nm derived from the spectra of VNIS EM and ASD (blue triangle in Fig. B. 2b).

The absorption feature at ~1000nm is composed of the spectra collected by two VNIS AOTFs in VNIR bands (450—945nm, including the 850nm feature of OL) and SWIR bands (900—2395nm, including spectral features of CPX and OPX at ~1000nm) (He et al. 2019; Li et al. 2019a, b). The different spectral detection modes of the two AOTFs (VNIR: spectra and imaging; SWIR: spectra) may be the major factor driving the inconsistency between the Str _OL/(Str _OL+Str _CPX1+Str _OPX1) ratio and the values acquired from ASD spectra, which affect the predicted CPX/(CPX+OPX) ratios. It is therefore import to build suitable models for the interpretation of the spectra acquired by the CE-4 VNIS payload using the dataset obtained by the VNIS EM.

4.2 SIM results of VNIS EM spectra

The spectral indices of the VNIS EM are listed in Table B. 3. The CPX/(CPX+OPX) values linearly increase with the corresponding spectral index ratios (Fig. 3a), with an R-square of 0.92, which can be expressed as follows:

$$\frac{CPX}{CPX+OPX}\% = 0.96 \times \frac{CPX\_INDEX}{CPX\_INDEX + OPX\_INDEX}\% + 19.27. \quad (4)$$

The predicted CPX/(CPX+OPX) values (Tab. B. 3) using the leave-one-out cross-validation (LOOCV) method are similar to the actual values (Fig. 3b). All scattered plots lie along the 1∶1 line within two dashed lines of ±3%, and the root mean square error (RMSE) is 4.20, indicating a good correlation between the predicted and actual values. Equation (4) can be applied to constrain the relative abundance of CPX and OPX at the CE-4 landing site based on the spectra acquired by the CE-4 VNIS payload.

The IBDR values tended to increase with increasing OL volume fractions in mafic minerals (OL/(OL+CPX+OPX)), as shown in Fig. 3c, which can be expressed as follows:

$$\frac{OL}{OL+CPX+OPX}\% = 3.20 \times IBDR - 246.63, \quad (5)$$

where the IBDR is calculated using Eq. (3). The predicted OL/(OL+CPX+OPX) values of the samples were derived using the LOOCV algorithm. Scattered plots of predicted and actual OL/(OL+CPX+OPX) values shown in Fig. 3d lie along the 1 ∶ 1 line with an RMSE of 18. 65. The performance of this model was better (RMSE = 16. 33) for samples with an IBDR of greater than 80% or predicted values of greater than 10 vol.%. Hence, this method cannot obtain accurate OL volume fractions in mafic minerals but can be applied to qualitatively constrain OL content at the CE - 4 landing site.

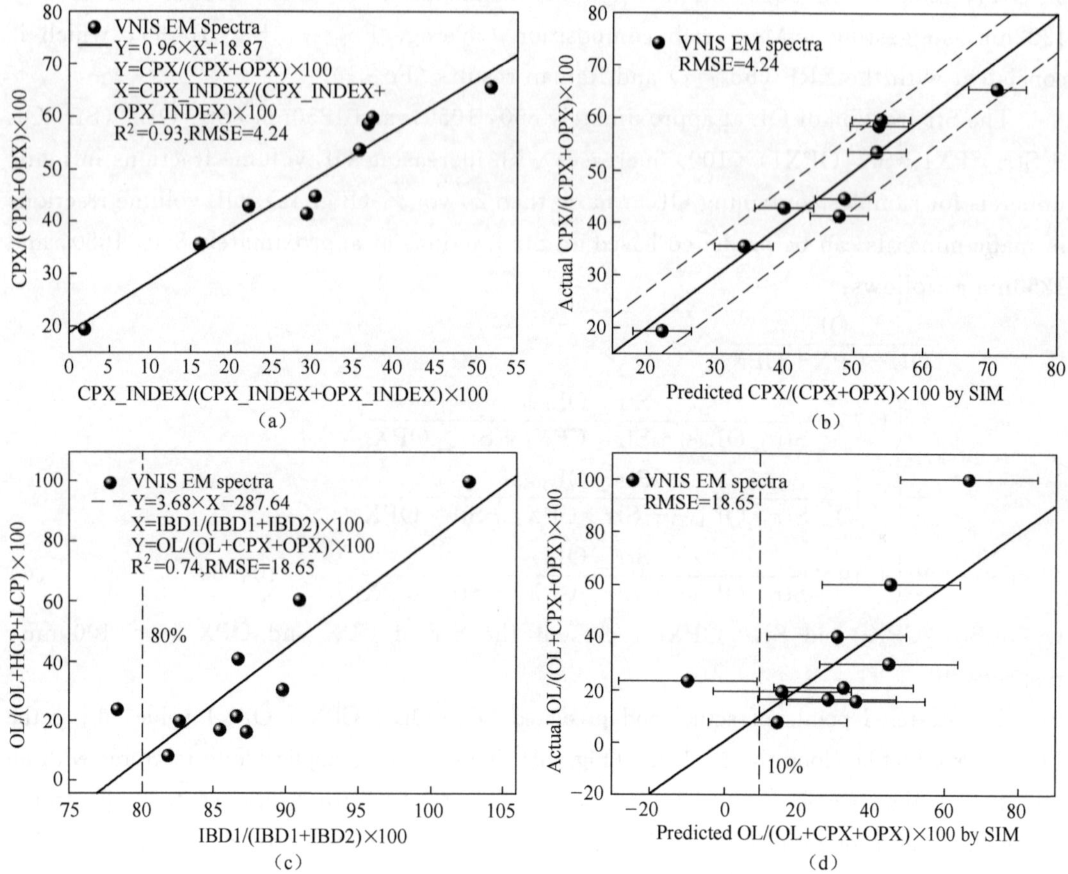

Fig. 3　Results of SIM models using VNIS EM spectra

(a) Scattered plots of the actual CPX/(CPX+OPX) values and the corresponding index values obtained using the SIM. (b) Comparison of the actual and predicted CPX/(CPX+OPX) using the LOOCV method. (c) Scatter plots of the OL/(OL+CPX+OPX) ratios and IBD1/(IBD1+IBD2) acquired using the SIM. (d) Comparison of actual and predicted OL/(OL+CPX+OPX) values using the LOOCV method

4. 3　MGM results of VNIS EM spectra

The absorption parameters including band center (BC), strength (Str), and full width at half maximum (FWHM) of spectral features of CPX, OPX, and OL were derived from their continuumremoved spectra acquired by the VNIS EM using MGM deconvolution. For lunar rocks with an IBDR of greater than 10%, the MGM results with the OPX - CPX - OL configuration are given in Tab. B. 4.

The BCs of OL derived with MGM linearly move to longer wavelengths with lower forsterite numbers (Fo = Mg/(Mg + Fe) molar ratio) due to substitution of Mg^{2+} (radius of 0.72Å) with larger Fe^{2+} (radius of 0.77Å) (Sunshine & Pieters 1998). The fitted M2 band (~1050nm) of OL is difficult to determine by MGM deconvolution because of the strong absorption features of pyroxene (Sunshine & Pieters 1998; Varatharajan et al. 2014; Ling et al. 2015). The fitted M1 bands occur at 831 — 852nm and 1185 — 1222nm, suggesting a Mg - rich composition (average $Fo_{89.17}$, Fig. B. 3), which is consistent with the XRF ($Fo_{89.24}$) and Raman results ($Fo_{87.8}$).

The Str fraction of OL at approximately 850, 1050, and 1250nm (Str _OL/ (Str _OL +Str _CPX1+Str _OPX1)×100) increases with increasing OL volume fractions in mafic minerals for samples containing OL at more than 20 vol.% (Fig. 4a). OL volume fractions in mafic minerals can be predicted based on Str fractions at approximately 850, 1050, and 1250nm as follows:

$$\frac{OL}{OL+CPX+OPX}\% = \left(1.75\times\frac{Str\_OL_{850}}{Str\_OL_{850}+Str\_CPX_1+Str\_OPX_1}\right.$$
$$+\left(1.15\times\frac{Str\_OL_{1050}}{Str\_OL_{1050}+Str\_CPX_1+Str\_OPX_1}\right.$$
$$+\left.\left(4.19\times\frac{Str\_OL_{1250}}{Str\_OL_{1250}+Str\_CPX_1+Str\_OPX_1}\right)\%-104.69 \qquad (6)$$

where Str $\_CPX_1$ and Str $\_OPX_1$ represent the Str of CPX and OPX at ~1000nm, respectively.

The scattered points of actual and predicted OL/(OL + CPX + OPX) values using the LOOCV method lie along the 1 : 1 line (Fig. 4d). This model is applicable to mixtures with an OL of more than 20vol.% (No. 1, 5, 7, 9 in Tab. B. 4) with RMSE of 4.89. For other samples, the absorption features of OL were significantly affected by pyroxene, which diminishes the accuracy of the model (RMSE = 8.07). Therefore, OL should not be considered in samples with an OL of lower than 20 vol.% (No. 2, 3, 4, 6, 8), and results derived from MGM deconvolution with OPX - CPX configuration are given in Tab. B. 5.

The Str fractions of CPX in pyroxene appear to vary linearly with the volume percentage of CPX (Tab. B. 4 and B. 5, Fig. 4c), which could be quantified by the Str fractions of CPX as follows:

$$\frac{CPX}{CPX+OPX}\% = 0.91\times\sum_{n=1}^{n=2}\frac{Str\_CPX_n}{Str\_CPX_n+Str\_OPX_n}\% - 38.26, \qquad (7)$$

where Str $\_CPX_1$ and Str $\_OPX_1$ represent the Str of CPX at ~1000nm and ~2000nm, respectively. The predicted CPX/(CPX+OPX) derived using the LOOCV algorithm and the actual values are shown in Fig. 4d. All scatter plots lie along the 1 : 1 line within the two dashed lines of ±10%, indicating a good correlation ($R^2 = 0.96$, RMSE = 3.16). The equation provides a method for estimating the CPX volume fractions in pyroxene based on the MGM deconvolution of spectra acquired by the CE-4 VNIS payload.

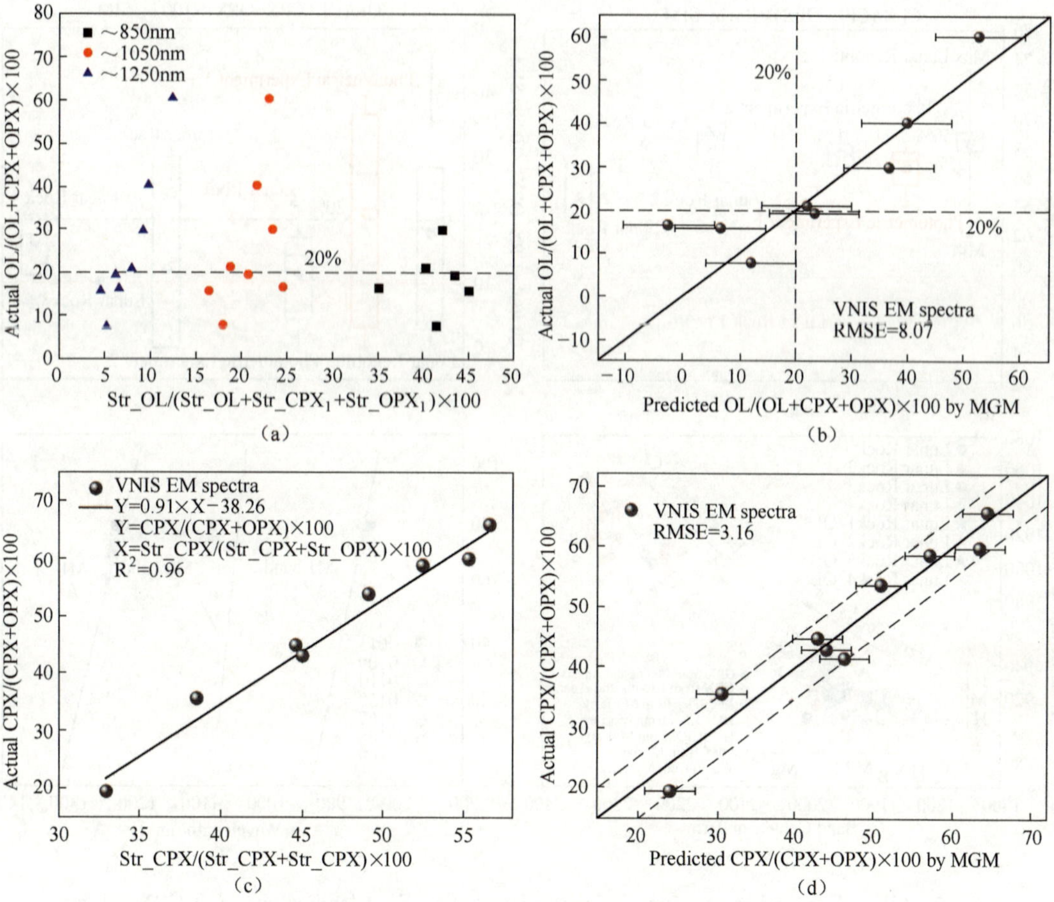

Fig. 4　Results of MGM models using VNIS EM spectra

(a) Scatter plots of actual OL/(OL+CPX+OPX) values and Str_OL/ (Str_OL+Str_CPX1+Str_OPX1) obtained using MGM with VNIS EM spectra. (b) Comparison of actual and predicted OL/(OL+CPX+OPX) using the LOOCV algorithm. (c) Scatter plots of CPX/(CPX+OPX) and corresponding Str from MGM with VNIS EM spectra. (d) Comparison of actual and predicted CPX/(CPX+OPX) values using the LOOCV algorithm

5　Applications on CE-4 spectral analysis

5.1　Mineralogy of lunar rocks and regolith

The volume fractions of CPX and OL derived using the SIM algorithm (Eqs. (4) and (5)) using the spectra acquired by the VNIS payload on the Moon are given in Tab. B. 6 and Fig. 5a-b. The predicted OL volume fractions lower than zero indicate a low OL content, and these values were replaced by zero. CPX/(OL+CPX+OPX) values shown in Tab. B. 6 were obtained based on the predicted results of CPX/(CPX+OPX) and OL/(OL+CPX+OPX).

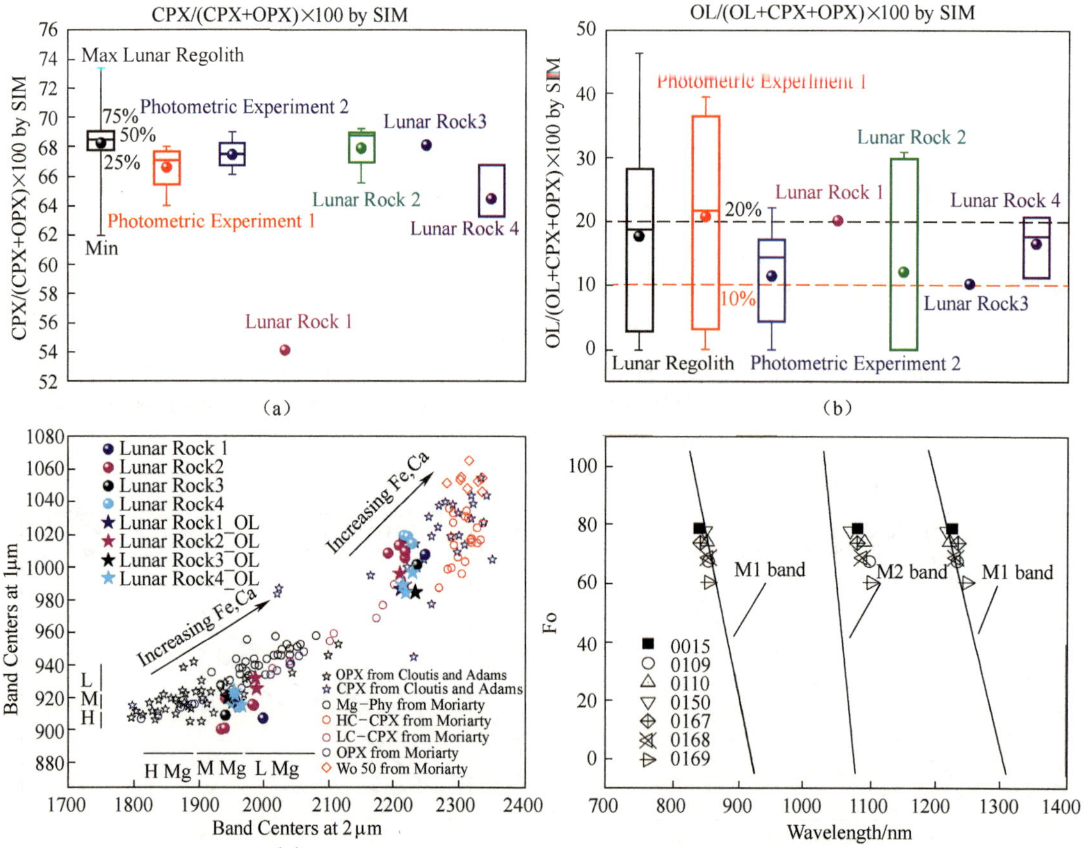

Fig. 5 Mineralogy and compositions of materials at the CE - 4 landing site. (a) CPX volume fractions in pyroxene and (b) OL volume fractions in mafic minerals using the SIM algorithm. (c) Band centers of the pyroxene and (d) Fo values of OL derived using the MGM algorithm

The CPX/(CPX+OPX) and OL/(OL+CPX+OPX) values in the same region of lunar soil and rocks where several VNIR spectra have been obtained (i. e., regions of two photometric experiments, Region 2 and Region 3 of Lunar Rock 2, as well as Region 2 of Lunar Rock 4) remain constant in the error range of Eq. (4) (RMSE=4. 24) and Eq. (5) (RMSE=16. 33). For example, the CPX/(CPX+OPX) and OL/(OL+CPX+OPX) values of lunar soil in Photometric Experiment 2 were 66−69 vol.% and 0−20 vol.% based on 12 VNIR spectra in different viewing geometries, indicating a steady evaluation of mafic mineral fractions in mixtures.

The lunar rocks and regolith at the CE - 4 landing site are CPX - rich and poor in OL (CPX : OPX : OL = 56 : 29 : 17). Lunar Rock 2 exhibits mineralogy inhomogeneity of

OL content in different regions (Region 1 and Region 2 : 0 vol.%; Region 3 : ～30 vol.%). The mineralogical features of Lunar Rocks 2－4 are similar to those of lunar regolith, indicating their similar origins. Nevertheless, Lunar Rock 1 had a lower CPX content (CPX : 55±4 vol.%; OPX : 45±4 vol.%) compared with other rocks and surrounding lunar regolith, which may provide evidence that Rock 1 has a different geological origin.

5.2 Mineral compositions of lunar rocks at the CE-4 landing site

MGM deconvolution was conducted for the CE-4 spectra of the four lunar rocks. For Region 1 and Region 2 of Lunar Rock 2, the OL content is not considered because of the extremely low OL according to the results of the SIM algorithm (shown in Tab. B. 6). For other lunar rock spectra, two MGM configurations (OPX－CPX and OPX－CPX－OL) were carried out, and the results are given in Tables 1 and 2, respectively. The predicted OL volume fractions in Lunar Rocks 1, 2, and 4 were more than 20 vol.% based on the MGM algorithm (Fig. 6b), indicating acceptable prediction results of Eq. (6).

The BCs of OPX and CPX in lunar rocks acquired by the MGM deconvolution method are plotted in Fig. 5c and compared with those obtained by Adams (1974), Cloutis & Gaffey (1991), and Moriarty & Pieters (2018). The results of MGM with OPXCPX－OL and OPX－CPX configurations exhibited similar BCs. The OPX of these lunar rocks contains more Fe than the OPX samples from Adams (1974) and Cloutis & Gaffey (1991), and more Mg than OPX samples from Moriarty & Pieters (2018). In addition, OPX exhibits mid-Mg characteristics based on the results of Klima et al. (2011). The CPX of four lunar rocks exhibits relatively mid-Ca and Fe contents compared with HCCPX and LC-CPX from Moriarty & Pieters (2018). For OL, the Fo values and BCs derived from MGM are given in Fig. 5d, indicating that OL in lunar rocks exhibits mid-Mg content (Fo_{60-79}).

5.3 Comparison of CE-4 mineralogy from different models

The mineral contents of the four lunar rocks obtained by the SIM algorithm were compared with the results of the MGM deconvolution algorithm (Fig. 6). The CPX/(CPX+OPX) derived using MGM with the OPX-CPX configuration (Tab. 1) was comparable to the results of the OPX-CPX-OL pattern (Tab. 2). All scattered points of CPX/(CPX+OPX) acquired using the SIM and MGM algorithms lie along the 1 : 1 line considering the RMSE values, and the OL contents predicted by the SIM algorithm increase linearly with the results of the MGM. However, the OL contents obtained by MGM were slightly higher than those obtained by the SIM algorithm. This may be caused by the fact that the absorption feature at ～1200nm is neglected for mid-to-high Fe-pyroxene (Klima et al. 2007, 2011), which is difficult to consider for the OPX-CPX-OL configuration of the MGM deconvolution algorithm. Thus, the results of CPX/(CPX+

OPX) and OL/(OL+CPX+OPX) predicted by the SIM algorithm are comparable with the MGM results, indicating a reliable prediction by models built by SIM, and that SIM can better constrain the mineralogy assemblages (as shown in Fig. 5 a–b) for the spectra acquired by the CE–4 VNIS payload.

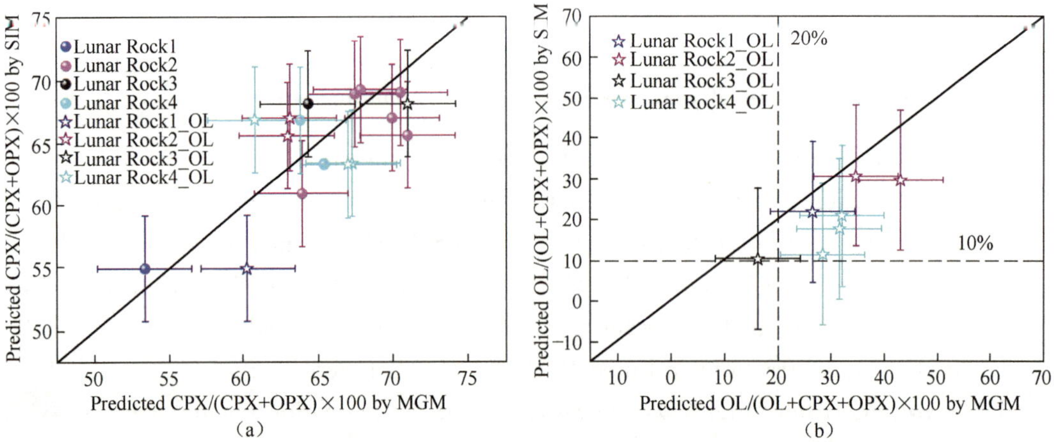

Fig. 6　Comparison of predicted CPX/(CPX+OPX) and OL/(OL+CPX+OPX) by SIM and MGM methods

Tab. 1　MGM results with OPX–CPX configuration of lunar rocks at the CE–4 landing site

Target		Rock1	Rock2—Region1	Rock2—Region2		Rock2—Region3		Rock3	fRock4—Region1	Rock4—Region2	
Spectral No.		0015	0105	0106	0107	0109	0110	0150	0167	0168	0169
Py-450	BC	447.9	461.6	440.5	451	440.2	438.1	426.2	428.6	431.3	434.7
	FWHM	205.3	236.8	246.9	232.5	230.6	242	295.2	261.8	264.8	267.7
	STR	−0.33	−0.26	−0.31	−0.32	−0.25	−0.27	−0.38	−0.36	−0.31	−0.31
OPX—1000	BC	906.9	919.1	900.1	900.8	914.8	915.4	908.8	913.7	918.4	917.1
	FWHM	182.4	147.6	141.4	139.8	148.1	143.5	140.5	147.2	139	129.3
	STR	−0.14	−0.06	−0.05	−0.05	−0.07	−0.07	−0.07	−0.11	−0.09	−0.08
CPX-1000	BC	1007.3	1012.8	1008	1005.2	1017	1009.1	1001.2	1019.2	1018.5	1014.4
	FWHM	198.7	183.5	178.1	170	187.7	180.8	160.3	181.9	178.9	168.1
	STR	−0.14	−0.09	−0.09	−0.08	−0.09	−0.1	−0.09	−0.13	−0.1	−0.1
Py-1200	BC	1191.5	1217	1214.6	1205.9	1180.1	1175.9	1171.3	1178.4	1154.7	1154.3
	FWHM	282	267.2	260.7	260.3	219.7	223.3	226	271.7	278.6	296
	STR	−0.05	−0.06	−0.05	−0.05	−0.05	−0.05	−0.04	−0.07	−0.06	−0.07
OPX-2000	BC	1998.8	1939.4	1933.1	1938.3	1984.3	1982.1	1939.4	1959.8	1954.3	1949.8
	FWHM	356.2	334.7	320.3	308.5	334.9	336.8	329.3	341.7	339.6	336.6
	STR	−0.11	−0.06	−0.05	−0.06	−0.04	−0.04	−0.07	−0.05	−0.07	−0.07
CPX-2000	BC	2247.7	2208.1	2190.9	2217.2	2215.8	2217.4	2235.1	2215.7	2221.2	2228.4
	FWHM	319.3	355.5	369.4	366.1	347.9	353.6	345.4	349.1	350.6	349.3
	STR	−0.11	−0.08	−0.07	−0.08	−0.07	−0.06	−0.09	−0.08	−0.1	−0.1

Continued

Target		Rock1	Rock2−Region1	Rock2−Region2	Rock2−Region3	Rock3	fRock4−Region1	Rock4−Region2			
Fall−down feature	BC	2500.2	2554	2519.7	2554.7	2558.7	2546.6	2548.7	2552	2548.2	2548
	FWHM	300	297.6	282.9	297.9	294.3	307.6	304.5	300.4	304.7	305.3
	STR	−0.18	−0.17	−0.1	−0.08	−0.23	−0.16	−0.09	−0.11	−0.14	−0.12
M1 intensity ratio		0.19	0.29	0.28	0.25	0.34	0.33	0.2	0.33	0.24	0.3

Tab. 2　MGM results with OPX−CPX−OL configuration of lunar rocks at the CE-4 landing site

Target		Rock1	Rock2−Region3		Rock3	Rock4−Region1	Rock4−Region2	
Spectral No.		0015	0109	0110	0150	0167	0168	0169
Pyroxene−450	BC	442.8	434.6	433.2	433.1	420.2	429.6	431.7
	FWHM	213.4	242.5	252.4	285.4	276.9	270.3	276.2
	STR	−0.34	−0.26	−0.28	−0.37	−0.37	−0.31	−0.32
OL−850	BC	842.6	854.2	852.3	845.7	841.7	854	854.8
	FWHM	109.6	110.8	104.2	82.3	95.1	97.8	95.3
	STR	−0.1	−0.04	−0.04	−0.04	−0.05	−0.04	−0.04
OPX−1000	BC	920.1	931.5	925.1	909.5	915	921.3	923.5
	FWHM	103	93.9	95.8	80.3	101.1	96.1	97.2
	STR	−0.15	−0.08	−0.07	−0.07	−0.11	−0.09	−0.09
CPX−1000	BC	989.3	995.5	988.7	978.1	984.2	988.8	996.6
	FWHM	110.1	111.2	111	98.7	111.5	104.4	112.1
	STR	−0.15	−0.08	−0.09	−0.12	−0.13	−0.12	−0.13
OL−1050	BC	1080.5	1100	1092.3	1075.4	1081.4	1089.1	1102.9
	FWHM	144.5	158.4	158.4	117	163.4	129.7	122.9
	STR	−0.11	−0.08	−0.07	−0.06	−0.11	−0.09	−0.09
OL−1250	BC	1226.9	1234.3	1222.4	1218.5	1235.4	1232	1249
	FWHM	271	224	243.3	235	260.2	256.8	262.7
	STR	−0.04	−0.03	−0.03	−0.03	−0.05	−0.04	−0.05
OPX−2000	BC	1967	1987	1988.5	1925.1	1965.1	1954.5	1952.9
	FWHM	343.1	334.3	343.8	327	345.2	351	349.6
	STR	−0.09	−0.04	−0.05	−0.07	−0.06	−0.07	−0.07
CPX−2000	BC	2214.9	2209.3	2216.4	2217.4	2218.2	2213.7	2228.7
	FWHM	319.2	326.2	340.9	354.8	335.8	338.5	338.4
	STR	−0.13	−0.06	−0.06	−0.09	−0.07	−0.1	−0.1
Fall−down feature	BC	2508.5	2515.3	2502.9	2502.5	2503.4	2502.2	2502.1
	FWHM	295	290.5	299.6	298.8	298.8	299.3	299.6
	STR	−0.22	−0.16	−0.12	−0.07	−0.07	−0.11	−0.09

5.4 Geologic interpretation

The pyroxenes in the CE-4 rocks and regolith are dominated by CPX (55—73 vol.% in Tab. B. 6), which is consistent with the higher CPX/OPX values (intensity ratios of 0.77—1.99 or CPX volume fractios of 41—83 vol.% based on Eq. (7)) exposed in the wall and ejecta blanket of the Finsen crater (Ling et al. 2019; Qiao et al. 2019; Fu et al. 2020; Guo et al. 2021). Lunar Rock 1 had a lower CPX (~55 vol.%) than other rocks and regolith (62—73 vol.%), which may be formed in a lower stratum with smaller CPX/OPX ratios exposed by the Finsen crater (Ling et al. 2019; Qiao et al. 2019; Guo et al. 2021). The lunar regolith around the CE-4 landing site exhibits mid-Ca-Fe pyroxene, and mid-Mg OL compositions, which is consistent with the mineral compositions at the Finsen crater wall in SPACA, indicating that these materials originate from the upper stratum of the SPA basin with intermediate mineral compositions (Moriarty & Pieters 2018; Moriarty et al. 2020).

The relative intensity of the absorption features at ~1200nm for pyroxene and ~2200nm for CPX (M1 intensity ratio) is a critical indicator of the crystallization temperature of pyroxene (Klima et al. 2008). The M1 intensity ratio values are defined as $Str_{1200nm}/Str_{1200nm}+Str\_OPX_2+Str\_OPX_2$, where Str values are derived from the MGM deconvolution algorithm with the OPX—CPX configuration (Tab. 1). The absorption features of both OPX and CPX were considered because of their similar energy of M1 absorption at ~1200nm. The M1 intensity ratio values are 0.25—0.29 for Region 1 and Region 2 of Lunar Rock 2, which have a low OL content and thus a negligible effect from the OL feature at ~1250nm.

The M1 intensity ratio values of CE-4 VNIS spectra for Lunar rock 2 are 0.25—0.29, which are slightly higher than the intensity ratio of HED meteorites (0.14—0.26) and synthetic OPX No. 004 crystallized from 980℃, but lower than that of synthetic CPX No. 011 (0.31) (Klima et al. 2007, 2008), indicating that the pyroxenes in lunar rocks are crystallized at a high temperature (980—1300℃) and are cooled quickly, which is consistent with the mid-Ca and mid-Fe content in pyroxenes of CE-4 rocks (Lindsley 1983). Hence, the lunar rocks at the CE-4 landing site may have crystallized in a high-temperature and rapid—cooling magmatic system caused by impact melt differentiation or volcanic resurfacing processes in the SPA basin (Moriarty & Pieters 2018; Moriarty et al. 2020) before being ejected by the Finsen impact event. The unique pyroxene mineral compositions and mafic mineral proportions are critical for constraining the mineral compositions in SPACA and provide important clues for evaluation of the geologic evolution of the inner SPA basin after the huge impact event.

6 Conclusions

In this study, the VNIR spectra of nine lunar analog samples were obtained using the

VNIS EM. Four models to predict the CPX/(CPX+OPX) and OL/(OL+CPX+OPX) of mafic minerals were proposed using SIM and MGM algorithms and were applied to VNIR spectra obtained by CE－4 VNIS payload on the lunar farside surface. In addition, we suggest the use of a threshold of 20 vol.% of OL content in mafic minerals when using the MGM deconvolution algorithm with the OPX−CPX−OL configuration.

The mineralogy of lunar rocks and regolith at the CE－4 landing site is rich in CPX with a low abundance of OL (CPX : OPX : OL = 56 : 29 : 17). The four lunar rocks are composed of mid－Ca－Fe pyroxene, and mid－Mg OL contents, providing critical constraints for mineral compositions in SPACA. The lower CPX fraction of Lunar Rock 1 suggests that it may come from a lower stratum exposed by the Finsen crater than other rocks and regolith. The high M1 intensity ratio values imply that the lunar rocks at the CE－4 landing site may have been crystallized in a high temperature (980−1300℃) and rapid－cooling magmatic system produced by impact melt differentiation or volcanic resurfacing events in the SPA basin before subsequent excavation by the Finsen impact.

Acknowledgements

The Chang'E－4 mission was carried out by the Chinese Lunar Exploration Program, and the scientific data are provided by China National Space Administration. This work is supported by National Natural Science Foundation of China (11941001, 41972322, U1931211), the Strategic Priority Research Program of Chinese Academy of Sciences (XDB 41000000), the National Key Research and Development Program of China (2020YFE0202100, 2019YFE0123300), the Pre－research project on Civil Aerospace Technologies (Nos. D020102 and D020201) funded by China National Space Administration (CNSA), the Natural Science Foundation of Shandong Province (ZR2019MD008), China Postdoctoral Science Foundation (2020M682164), and Physical－Chemical Materials Analytical and Testing Center of Shandong University at Weihai. The CE－4 spectral database reported in this work is available at https://moon.bao.ac.cn/ce5web/searchOrder_dataSearchData.search. The experimental and simulation data are available at https://pds.wh.sdu.edu.cn/English/Data_PDS_Service/Publication_Data_Share.htm. This is the SDU－CPS publication #88.

References

[1] Adams, J. B. 1974, J. Geophys. Res., 79, 4829

[2] Chen, J., Ling, Z., Qiao, L., et al. 2020, Sci. China Inf. Sci., 63, 1

[3] Cloutis, E. A., & Gaffey, M. J. 1991, J. Geophys. Res. Planets, 96, 22809

[4] Cloutis, E. A., Gaffey, M. J., Jackowski, T. L., & Reed, K. L. 1986, J. Geophys. Res. Solid Earth, 91, 11641

[5] Di, K., Liu, Z., Liu, B., et al. 2019, J. Remote. Sens., 23, 177

[6] Fu, X. H., Qiao, L., Zhang, J., Ling, Z. C., & Li, B. 2020, Res. Astron. Astrophys., 20, 008

[7] Gou, S., Di, K., Yue, Z., et al. 2020, Icarus, 345, 113776

[8] Guo, D., Fa, W., Zeng, X., Du, J., & Liu, J. 2021, Icarus, 359, 114327

[9] Gueymard, C. A. 2004, Sol. Energy, 76, 423

[10] Hapke, B., 2012, Theory of Reflectance and Emittance Spectroscopy (Cambridge: Cambridge university press)

[11] He, Z., Li, C., Xu, R., et al. 2019, J. Appl. Remote. Sens., 13, 027502

[12] Horgan, B. H., Cloutis, E. A., Mann, P., & Bell III, J. F. 2014, Icarus, 234, 132

[13] Hu, X., Ma, P., Yang, Y., et al. 2019, Geophys. Res. Lett., 46, 9439

[14] Huang, J., Xiao, Z., Flahaut, J., et al. 2018, J. Geophys. Res. Planets, 123, 1684

[15] Huang, J., Xiao, Z., Xiao, L., et al. 2020, Geology, 48, 723

[16] Klima, R. L., Pieters, C. M., & Dyar, M. D. 2007, Meteorit Planet Sci, 42, 235

[17] Klima, R. L., Pieters, C. M., & Dyar, M. D. 2008, Meteorit Planet Sci, 43, 1591

[18] Klima, R. L., Pieters, C. M., Boardman, J. W., et al. 2011, J. Geophys. Res. Planets, 116

[19] Kuebler, K. E., Jolliff, B. L., Wang, A., & Haskin, L. A. 2006, Geochim. Cosmochim. Acta, 70, 6201

[20] Lemelin, M. 2016, The Composition of the Lunar Crust: an in - depth Remote Sensing View (USA: University of Hawaii at Manoa)

[21] Lemelin, M., Lucey, P. G., Miljković, K., et al. 2019, Planet. Space Sci., 165, 230

[22] Li, C., Wang, Z., Xu, R., et al. 2019a, Sensors, 19, 2806

[23] Li, C., Xu, R., Lv, G., et al. 2019b, Rev. Sci. Instrum., 90, 103106

[24] Li, C., Liu, D., Liu, B., et al. 2019c, Nature, 569, 378

[25] Lin, H., He, Z., Yang, W., et al. 2020, Natl. Sci. Rev., 7, 913

[26] Lindsley, D. H. 1983, Am. Mineral., 68, 477

[27] Ling, Z., Jolliff, B. L., Wang, A., et al. 2015, Nat. Commun., 6, 1

[28] Ling, Z., Qiao, L., Liu, C., et al. 2019, Planet. Space Sci., 179, 104741

[29] Liu, J., Ren, X., Yan, W., et al. 2019, Nat. Commun., 10, 1

[30] Lucey, P. G. 2004, Geophys. Res. Lett., 31, L08701

[31] Martinot, M., Besse, S., & Flahaut, J. 2018, J. Geophys. Res. Planets, 123, 612

[32] Moriarty, D., & Pieters, C., 2018, J. Geophys. Res. Planets, 123, 729

[33] Moriarty, D., Watkins, R., Valencia, S., et al. 2020, J. Geophys. Res. Planets, e2020JE006589

[34] Mustard, J. F., Pieters, C. M., Isaacson, P. J., et al. 2011, J. Geophys. Res. Planets, 116, E00G12

[35] Qi, X., Ling, Z., Zhang, J., et al. 2020, Remote Sens., 12, 3211

[36] Qiao, L., Ling, Z., Fu, X., & Li, B. 2019, Icarus, 333, 37
[37] Savitzky, A., & Golay, M. J. 1964, Anal. Chem., 36, 1627
[38] Staid, M. I., Pieters, C. M., Besse, S., et al. 2011, J. Geophys. Res. Planets, 116, E00G10
[39] Sunshine, J. M., & Pieters, C. M. 1998, J. Geophys. Res. Planets, 103, 13675
[40] Sunshine, J. M., Pieters, C. M., & Pratt, S. F. 1990, J. Geophys. Res. Planets, 95, 6955
[41] Varatharajan, I., Srivastava, N., & Murty, S. V. S. 2014, Icarus, 236, 56
[42] Wang, A., Jolliff, B. L., Haskin, L. A., Kuebler, K. E., & Viskupic, K. M. 2001, Am. Mineral., 86, 790
[43] Wilhelms, D. E., Howard, K. A., & Wilshire, H. G. 1979, Geologic map of the south side of the Moon (Department of the Interior, US Geological Survey)
[44] Wu, W., Li, C., & Zuo, W. 2019, Nat. Geosci., 12, 222

Appendix A　VNIR spectra measurements

VNIR reflectance spectra in 450—2395nm were obtained using CE-4 VNIS EM, which is a copy of the VNIS payload aboard the Yutu-2 rover. Powder samples were laid on a disk with a volume of 12cm by ×16cm by ×5mm. An 800Wtungsten halogen lamp was selected as a light source, and the distance between the lamp and the sample dish was fixed at ~0.7m, which is similar to the parameters of the VNIS payload aboard Yutu-2. The incident and emission angles were fixed at 60° and 45°, respectively, according to the average solar incident angles and emission angles of the VNIS measurements on the lunar surface. A phase angle of 90° (azimuth angle of 55°) was selected in this study based on actual measurements by the VNIS payload. The spectrometer was calibrated by an approximately reflecting Lambertian surface (SRT-99-100) from Spectralon before each measurement. The relative reflectance was converted into absolute reflectance by multiplying the absolute reflectance of SRT-99-100.

To evaluate the instrumental effects and spectral properties of VNIS, we also acquired absolute reflectance spectra of the same lunar soil analogs at 350—2500nm using a FieldSpec 4 Hi-Res VNIR spectrometer (Analytical Spectral Devices, Inc.) coupled with an 8° field of view (Unit 18380) at the same viewing geometry as CE-4 VNIS EM (i = 60°, e = 45°, g = 90°). A 75 W ASD illuminator lamp was selected as the light source. Powder samples were filled in a ~55mm diameter culture dish to avoid additional inaccuracy caused by the varying distance between the ASD fiber and the sample surface. Spectra from three locations were acquired to account for the heterogeneity of the samples. In total, 100 scans were performed at each location to avoid possible instability of the instrument. Before each measurement, the spectrometer was calibrated using an approximately reflecting Lambertian surface (SRS-99-010) from Labsphere.

Appendix B　Chemical and mineralogical characterization

The elemental compositions of nine lunar soil analogs and end members (shown in Tab. B. 2) were measured using an ARL Perfom'X 4200 X-ray fluorescence (XRF) spectrometer at the State Key Laboratory of Ore Deposit Geochemistry, Institute of Geochemistry, Chinese Academy of Sciences. The mineralogical information of the four end members was characterized using Raman spectroscopy (100—1400cm^{-1}) using an inVia Raman System (Renishaw Company) at a spectral resolution better than 1cm^{-1}. The purity of the end members was ensured using a 20-point laser Raman point-counting technique. Raman peak positions of end members were acquired by the spectral deconvolution method using a mixed Gaussian-Lorentzian algorithm.

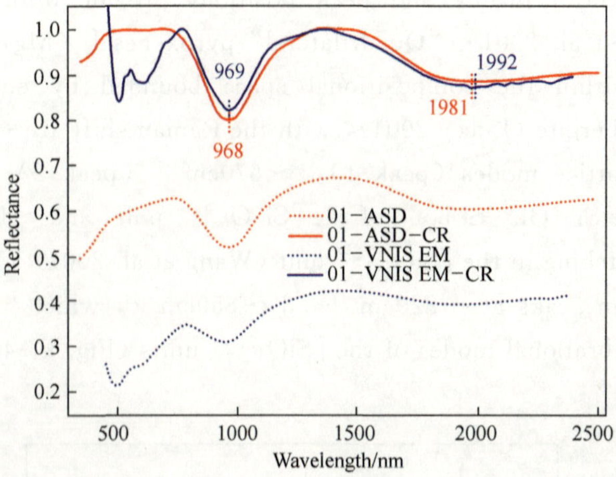

Fig. B. 1　Comparison of the spectra of No. 01 mixture acquired by ASD and VNIS EM

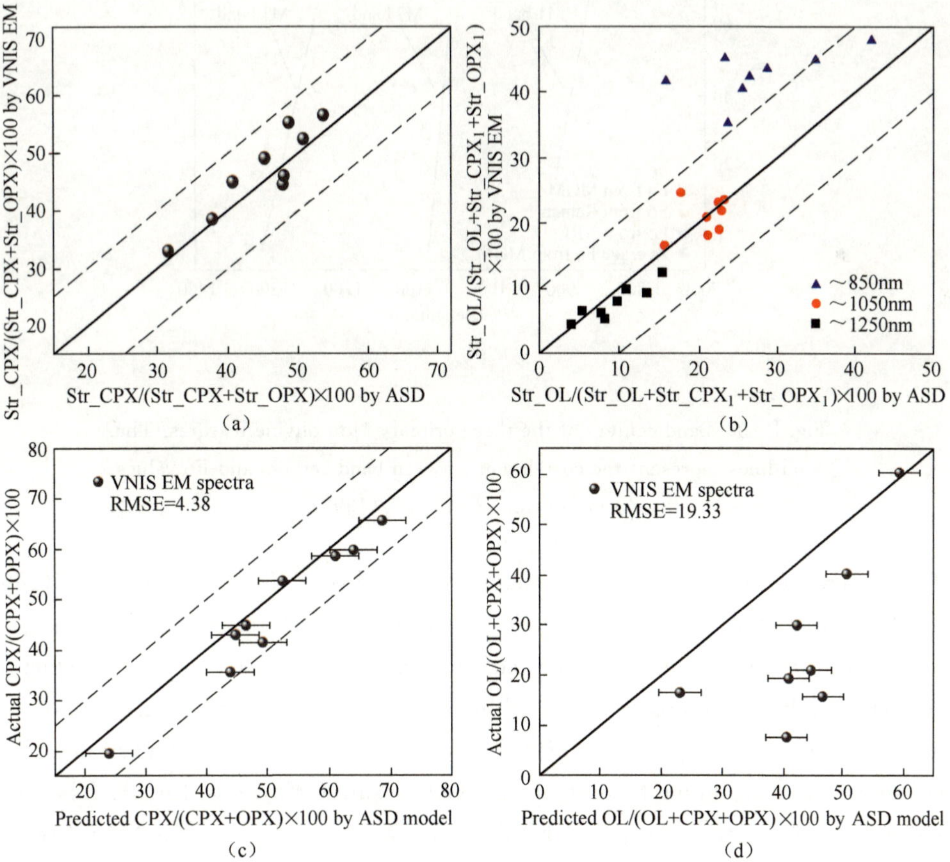

Fig. B. 2　Comparison of the Str _CPX/(Str _CPX+Str _OPX (a) and Str _OL/(Str _OL+Str _CPX1+Str _OPX1 (b) derived from MGM algorithm using spectra acquired by ASD and VNIS EM, and the predicted CPX/(CPX+OPX) (c) and OL/(OL+CPX+OPX) using the spectra of VNIS EM by the MGM model built using the ASD spectra

The Raman spectral profile and peak positions provide information on mineral composition (Wang et al. 2001). "Quadrilateral" pyroxenes [$(Mg, Fe, Ca)_2Si_2O_6$], primarily defined within the compositional space bounded by enstatite, ferrosilite, diopside, and hedenbergite (Lucey 2004), with the Raman shift at ~320cm^{-1} commonly referred as crystal lattice modes (peak 1), ~670cm^{-1} (peak 2) caused by bridging oxygen stretch in the Si－Ob－Si bond of the $[Si_2O_6]^4$ unit, and ~1000cm^{-1} attributed to Si－O_{nb} bond stretching in the $[SiO_4]^{4-}$ unit (Wang et al. 2001) (Fig. B. 4a). Olivine has two major Raman peaks at ~820cm^{-1} and ~850cm^{-1}, which are attributed to the internal stretching vibrational modes of the $[SiO_4]^{4-}$ units (Fig. B. 4c).

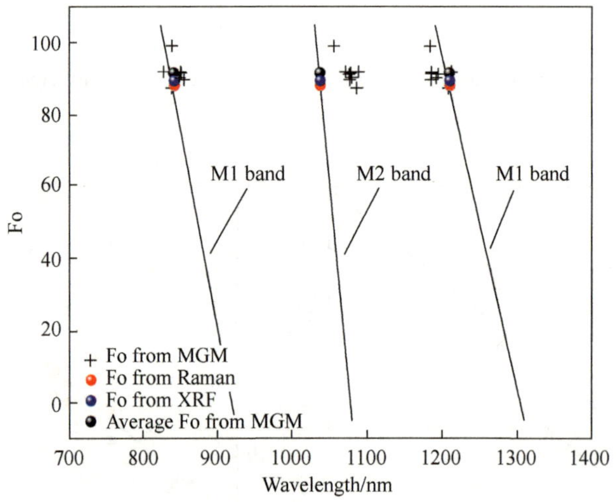

Fig. B. 3 Band centers of the three primary 1μm olivine features. The solid lines represent the correlation between band centers and Fo values from Sunshine & Pieters (1998)

The mineral compositions of pyroxene (Wo=Ca/(Mg+Fe+Ca), Fs=Fe/(Mg+Fe+Ca), En=Mg/(Mg+Fe+Ca) and Fo of OL were calculated using their Raman peak positions based on the equations described by Wang et al. (2001) and Kuebler et al. (2006). Hypersthene and augite exhibited high Mg and Fe contents (Fig. B. 4b), and an obvious boundary (~30 vol.%) of Ca content. The average composition of augite ($Wo_{38}Fs_{12}En_{49}$) was similar to that of the XRF result ($Wo_{38}Fs_{17}En_{45}$). The different compositions of hypersthene acquired by Raman ($Wo_{15}Fs_{18}En_{67}$) and XRF ($Wo_{26}Fs_{22}En_{51}$) could be attributed to different detection areas and a small amount of mineral for both techniques. OL is Mg－rich with an average Fo of 87.8 (Fig. B. 4d), which is consistent with the XRF results ($Fo_{89.2}$). Moreover, the peak positions (Peak 1 and Peak 2) of OL systematically shifted to higher wavenumbers with increasing Fo values.

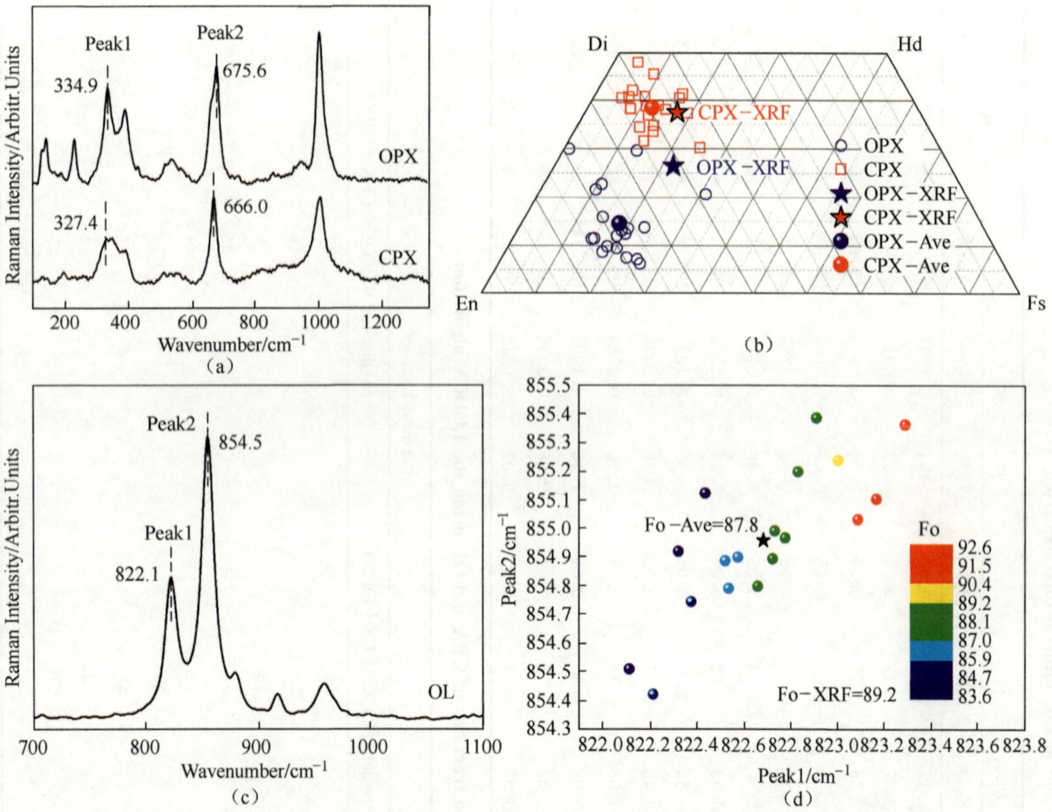

Fig. B. 4　(a) Raman spectra and peak positions of augite and hypersthene. (b) En - Di - Hd - Fs (Enstatite - Diopside - Hedenbergite - Ferrosilite) quadrilateral composition of hypersthene (red) and augite (blue). (c) Representative Raman spectra of olivine. (d) Variations of Peak1 and Peak2 position with Fo of olivine

Tab. B. 1　Mineral volume fractions in lunar soil analogs studied in this work

No.	Volume Proportion/vol.%				CPX/(CPX+OPX)	OL/(OL+CPX+OPX)
	OPX	CPX	PLAG	OL		
1	19.1	27.1	41.5	12.3	58.6	21.0
2	19.7	22.8	47.3	10.2	53.7	19.4
3	21.2	16.0	55.5	7.3	42.9	16.5
4	22.0	12.1	59.5	6.4	35.6	15.8
5	15.0	12.2	54.4	18.4	44.8	40.3
6	22.6	33.6	39.0	4.7	59.8	7.7
7	10.6	7.5	54.4	27.5	41.4	60.3
8	40.7	9.8	34.1	15.4	19.4	23.3
9	13.4	25.7	44.1	16.7	65.6	30.0

Tab. B. 2 Composition of lunar soil analog mineral mixtures from XRF

Sample	Compositions/wt.%															
No.	SiO_2	Al_2O_3	Fe_2O_3	MgO	CaO	Na_2O	K_2O	MnO	P_2O_5	TiO_2	Cr	Total	Mg #	Wo	Fs	En
1	53.52	11.99	8.83	13.87	6.58	3.76	0.41	0.13	0.04	0.54	0.1087	99.77	75	21	19	60
2	54.42	12.76	8.26	12.65	5.95	4.16	0.47	0.12	0.04	0.54	0.0917	99.46	75	20	20	60
3	55.92	14.13	7.88	10.81	5.1	4.71	0.56	0.11	0.05	0.46	0.0806	99.81	73	20	22	59
4	57.16	14.82	7.43	9.81	4.56	5	0.59	0.1	0.05	0.43	0.0809	100.03	72	19	22	58
5	52.39	10.14	11.63	17.1	4.09	3.08	0.37	0.17	0.08	0.59	0.1253	99.76	74	11	23	66
6	53.6	12.24	9.26	11.99	7.51	3.67	0.4	0.13	0.05	0.69	0.0712	99.61	72	24	21	54
7	54.44	12.62	6.74	16.35	3.7	4.44	0.53	0.09	0.03	0.24	0.1206	99.31	83	12	15	73
8	54.88	13.35	7.31	13.99	4.48	4.62	0.53	0.1	0.04	0.35	0.1053	99.76	79	15	18	67
9	53.51	12.11	7.93	14.42	6.32	3.9	0.43	0.11	0.03	0.49	0.1048	99.35	78	20	17	63
CPX	48.15	8.59	10.43	13.74	15.9	1.65	0.06	0.15	0.01	1.16	0.0584	99.90	[illegible]	38	17	45
OPX	48.76	7.56	13.74	16.1	11.53	1.38	0.08	0.19	0.06	1.13	0.1065	100.64	[illegible]	26	22	51
OL	42.56	1.4	10.23	42.81	1.69	0.35	0.02	0.14	0.01	0.08	0.3042	99.58	89	2	11	87
PLAG	63.9	22.06	0.82	0.07	3.12	8.37	0.92	0.01	0.02	0.03	0.0312	99.35	[illegible]	82	15	3

Tab. B. 3 Spectral indices and predicted volume fractions of CPX and OL using the LOOCV algorithm

ID	OPX_ INDEX	CPX_ INDEX	IBD1	IBD2	Predicted values	
					Predicted CPX/(CPX+OPX)	Predicted OL/(OL+CPX+OPX)
1	0.46	0.27	5.93	0.92	54	32
2	0.44	0.25	5.43	1.14	53	16
3	0.76	0.22	6.31	0.56	40	28
4	0.91	0.17	5.55	0.95	34	36
5	0.48	0.21	5.78	0.84	49	31
6	0.97	0.58	5.25	0.8	54	15
7	0.24	0.1	6.99	1.55	48	45
8	1.15	0.02	4.75	0.47	22	−10
9	0.29	0.31	5.2	1.44	71	45
OL	—	—	6.46	−0.16	—	66

Tab. B. 4　MGM results with OPX - CPX - OL configuration using continuum - removed spectra acquired by CE - 4 VNIS EM

No.	Additional Band			OL - 850			OPX - 1000			CPX - 1000			OL - 1050			OL - 1200			OPX - 2000			CPX - 2000		
	BC	FWHM	STR	BC	FWHM	STR	BC	FWHM	STR	BC	FWHM	STR	BC	FWHM	STR	BC	FWHM	STR	BC	FWHM	STR	BC	FWHM	STR
1	**574**	**206. 1**	**−0. 19**	**855. 6**	**97. 4**	**−0. 07**	**926. 3**	**102. 7**	**−0. 13**	**997. 1**	**107**	**−0. 15**	**1077. 1**	**106. 2**	**−0. 06**	**1184. 9**	**216. 4**	**−0. 02**	**1851. 8**	**404. 3**	**−0. 1**	**2245**	**462. 9**	**−0. 11**
2	627. 5	144. 2	−0. 2	849. 4	95. 7	−0. 08	928. 4	111. 2	−0. 13	993. 8	106. 6	−0. 13	1079. 8	99. 5	−0. 07	1190. 9	197. 9	−0. 02	1893. 9	541. 9	−0. 07	2202. 6	549. 3	−0. 06
3	604. 3	214. 9	−0. 16	835	86. 5	−0. 08	916. 4	98. 3	−0. 16	992. 4	101. 8	−0. 15	1079. 2	117. 4	−0. 08	1194. 6	193. 3	−0. 02	1844. 6	440. 1	−0. 16	2267. 6	392. 6	−0. 13
4	590. 6	198. 3	−0. 13	839. 8	87	−0. 06	918. 9	113. 1	−0. 14	982. 2	86. 8	−0. 1	1055. 7	87. 2	−0. 08	1183. 2	198. 4	−0. 02	1841	479. 1	−0. 16	2284. 2	417. 7	−0. 13
5	**518. 6**	**271. 1**	**−0. 22**	**851. 4**	**73. 3**	**−0. 04**	**922. 6**	**108. 5**	**−0. 15**	**989. 4**	**98. 3**	**−0. 12**	**1071. 3**	**82. 4**	**−0. 05**	**1185. 3**	**196. 6**	**−0. 01**	**1842. 4**	**489. 4**	**−0. 18**	**2301. 7**	**460. 6**	**−0. 12**
6	574. 3	226. 4	−0. 17	852. 4	76. 4	−0. 05	927. 6	114. 2	−0. 11	990. 7	117. 3	−0. 09	1078. 4	123. 3	−0. 06	1184. 9	184. 6	−0. 02	1875. 8	463. 3	−0. 12	2300. 5	398. 3	−0. 09
7	**555. 4**	**224. 6**	**−0. 23**	**840. 2**	**67**	**−0. 04**	**918. 8**	**108. 4**	**−0. 15**	**996. 5**	**110. 4**	**−0. 17**	**1086. 5**	**98. 2**	**−0. 07**	**1207. 9**	**231. 6**	**−0. 02**	**1838. 4**	**472. 1**	**−0. 15**	**2274. 7**	**513. 3**	**−0. 17**
9	**567. 5**	**224**	**−0. 26**	**846. 3**	**82. 2**	**−0. 11**	**921. 8**	**103. 4**	**−0. 21**	**987. 4**	**79. 9**	**−0. 11**	**1050. 8**	**102. 1**	**−0. 06**	**1151. 1**	**205. 2**	**0**	**1845. 5**	**500**	**−0. 21**	**2293. 8**	**457**	**−0. 1**

Tab. B. 5　The MGM results with OPX−CPX configuration for samples with OL less than 20 vol.% based on continuum - removed spectra acquired by the CE - 4 VNIS EM

No.	Additional Band			OPX−1000			CPX−1000			Py−1200			OPX−2000			CPX−2000		
	BC	FWHM	STR	BC	FWHM	STR	BC	FWHM	STR	BC	FWHM	STR	BC	FWHM	STR	BC	FWHM	STR
2	635	198. 8	−0. 14	923. 9	153	−0. 14	1004. 5	148. 7	−0. 12	1175. 3	249. 4	−0. 03	1812. 4	452	−0. 08	2201. 5	453. 4	−0. 09
3	592	197. 2	−0. 13	903. 5	142. 9	−0. 15	1009. 9	137. 2	−0. 12	1171. 2	249. 5	−0. 02	1826. 3	498. 4	−0. 15	2269. 3	502. 6	−0. 13
4	522. 3	256. 9	−0. 22	927. 1	144. 7	−0. 15	1008. 4	158. 1	−0. 08	1150. 7	196. 8	−0. 01	1839. 3	487. 7	−0. 18	2300. 1	476. 1	−0. 12
6	558. 1	206	−0. 22	919. 5	147. 6	−0. 13	1005. 1	155. 5	−0. 15	1162. 2	247. 1	−0. 03	1834. 6	547. 9	−0. 13	2247. 6	553. 3	−0. 17
8	569. 1	200. 6	−0. 23	914. 2	132. 2	−0. 22	1003. 5	120. 1	−0. 09	—	—	—	1816. 6	453. 1	−0. 19	2209. 8	452. 3	−0. 12

Tab B. 6 CPX, OPX and OL volume fractions using SIM algorithm based on continuum - removed spectra acquired by CE - 4 VNIS payload

Target		No	Predicted CPX content using Eq. 4				Predicted OL content using Eq. 5				Derive CPX and OPX content	
			OPX_ INDEX	CPX_ INDEX	CPX_ INDEX/(CPX_ INDEX+ OPX_ INDEX)×100	Predicted CPX/(CPX + OPX)×100	IBD1	IBD2	IBD1 /(IBD1 + IBD2) ×100	Predicted OL/(CPX + OPX + OL)×100	Predicted CPX/(CPX + OPX + OL)×100	Predicted OPX/(CPX + OPX + OL)×100
Rock1		0015	−1.72	−1.03	37.47	54.93	10.6	2.08	83.62	20.14	43.87	35.99
Rock2	Region 1	0105	−1.2	−1.29	51.94	68.86	5.13	1.51	77.21	0	68.86	31.14
	Region 2	0106	−2.11	−2.29	52.07	68.98	4.9	1.77	73.5	0	68.98	31.02
		0107	−2.09	−2.29	52.3	69.2	4.58	1.66	73.34	0	69.2	30.8
	Region 3	0109	−1.23	−1.15	48.5	65.54	6.51	1.04	86.23	29.75	46.04	24.21
		0110	−1.35	−1.35	49.96	66.95	6.51	1.01	86.53	30.86	46.29	22.85
Rock3		0150	−1.67	−1.75	51.18	68.13	5.73	1.35	80.94	10.29	61.12	28.59
Rock4	Region 1	0167	−1.91	−1.89	49.8	66.8	9.04	1.75	83.82	20.88	52.85	26.27
	Region 2	0168	−1.05	−0.9	46.21	63.34	7.44	1.72	81.18	11.18	56.26	32.56
		0169	−1.12	−0.96	46.12	63.25	7.32	1.51	82.93	17.63	52.1	30.27
Photometric Experiment1		0068	−1.53	−1.58	50.86	67.82	6.17	0.84	87.99	36.24	43.25	20.52
		0069	−1.57	−1.58	50.19	67.18	6.54	0.83	88.73	38.98	40.99	20.03
		0070	−1.55	−1.37	46.87	63.98	6.02	0.87	87.37	33.94	42.27	23.8
		0071	−1.69	−1.58	48.34	65.39	5.56	0.9	86.09	29.25	46.26	24.49
		0072	−1.76	−1.61	47.89	64.96	5.98	0.75	88.81	39.26	39.46	21.28
		0074	−1.51	−1.57	51.02	67.97	4.91	1.08	81.93	13.94	58.5	27.56
		0075	−1.81	−1.84	50.48	67.45	5.11	1.21	80.87	10.03	60.69	29.28
		0076	−1.69	−1.69	49.96	66.95	4.7	1.25	78.98	3.08	64.89	32.03
		0077	−1.71	−1.69	49.73	66.73	4.45	1.2	78.78	2.35	65.16	32.49
		0078	−1.9	−1.95	50.76	67.72	4.27	1.53	73.59	0	67.72	32.28

Continued

Target	No	Predicted CPX content using Eq. 4				Predicted OL content using Eq. 5				Derive CPX and OPX content	
		OPX_ INDEX	CPX_ INDEX	CPX_ INDEX/(CPX_ INDEX+ OPX_ INDEX)×100	Predicted CPX/(CPX + OPX)×100	IBD1	IBD2	IBD1 /(IBD1 + IBD2) ×100	Predicted OL/(CPX + OPX + OL)×100	Predicted CPX/(CPX + OPX + OL)×100	Predicted OPX/(CPX + OPX + OL)×100
Photometric Experiment 2	0079	−1.96	−2.02	50.76	67.72	4.88	1.19	80.36	8.15	62.2	29.65
	0080	−1.84	−1.94	51.31	68.25	4.73	1.22	79.44	4.77	65	30.23
	0081	−1.81	−1.91	51.31	68.25	4.05	1.18	77.5	0	68.25	31.75
	0082	−1.74	−1.83	51.33	68.27	3.74	1.29	74.38	0	68.27	31.73
	0083	−1.68	−1.83	52.1	69.01	4.52	1.19	79.14	3.67	66.48	29.86
	0084	−1.63	−1.63	49.97	66.96	3.85	0.83	82.35	15.47	56.6	27.92
	0085	−1.6	−1.63	50.42	67.39	3.8	0.81	82.46	15.88	56.69	27.43
	0086	−1.61	−1.59	49.72	66.72	3.73	0.79	82.5	16.02	56.04	27.95
	0087	−1.74	−1.69	49.37	66.38	3.8	0.71	84.16	22.15	51.68	26.17
	0088	−1.8	−1.78	49.8	66.79	3.94	0.77	83.58	20.02	53.42	26.56
	0089	−1.73	−1.67	49.13	66.15	3.63	0.81	81.72	13.17	57.44	29.39
	0090	−1.24	−1.29	50.95	67.91	3.92	0.8	83.11	18.26	55.51	26.23
Lunar Regolith	0001	−1.98	−2.28	53.49	70.34	6.55	1.1	85.6	27.43	51.05	21.52
	0003	−1.44	−1.52	51.35	68.28	4.18	0.94	81.58	12.65	59.64	27.7
	0004	−0.82	−0.95	53.78	70.62	5.17	0.93	84.7	24.11	53.6	22.29
	0005	−1.54	−1.63	51.34	68.28	4.88	0.7	87.48	34.37	44.81	20.82
	0008	−2.1	−2.24	51.63	68.56	3.88	1.05	78.66	1.89	67.26	30.85
	0009	−1.75	−1.9	52.06	68.97	3.56	0.72	83.21	18.63	56.12	25.25
	0012	−1.64	−1.61	49.5	66.51	5.45	0.83	86.73	31.61	45.48	22.9
	0013	−2.19	−2.31	51.36	68.3	7.44	1.61	82.23	15.04	58.02	26.93

Continued

Target	No	Predicted CPX content using Eq. 4				Predicted OL content using Eq. 5				Derive CPX and OPX content	
		OPX_ INDEX	CPX_ INDEX	CPX_ INDEX/(CPX_ INDEX+ OPX_ INDEX)×100	Predicted CPX/(CPX + OPX)×100	IBD1	IBD2	IBD1 /(IBD1 + IBD2) ×100	Predicted OL/(CPX + OPX + OL)×100	Predicted CPX/(CPX + OPX + OL)×100	Predicted OPX/(CPX + OPX + OL)×100
Lunar Regolith	0014	−1.99	−2.06	50.94	67.89	8.35	1.57	84.18	22.23	52.8	24.97
	0016	−1.17	−1.05	47.32	64.41	3.39	0.53	86.5	30.74	44.61	24.65
	0017	−1.89	−2.11	52.78	69.67	5.8	1.65	77.88	0	69.67	30.33
	0020	−1.68	−1.81	51.87	68.79	5.35	0.88	85.86	28.4	49.25	22.35
	0021	−1.08	−1.11	50.67	67.63	4.44	0.55	88.95	39.77	40.73	19.5
	0023	−1.88	−1.9	50.26	67.24	5.81	0.68	89.53	41.91	39.06	19.03
	0024	−1.57	−1.27	44.78	61.97	3.85	0.95	80.29	7.88	57.08	35.04
	0025	−1.66	−1.69	50.47	67.45	5	0.75	86.99	32.56	45.48	21.95
	0026	−1.21	−1.27	51.25	68.2	4.77	1.29	78.77	2.31	66.62	31.07
	0027	−1.63	−1.75	51.73	68.66	4.33	1.25	77.61	0	68.66	31.34
	0028	−1.72	−1.81	51.3	68.24	4.64	0.98	82.61	16.42	57.03	26.55
	0029	−1.53	−1.56	50.58	67.55	4.44	1.22	78.45	1.12	66.8	32.09
	0031	−1.27	−1.32	50.96	67.91	5.14	1.02	83.47	19.59	54.61	25.8
	0032	−1.48	−1.6	52.02	68.93	4.64	0.93	83.29	18.94	55.88	25.19
	0034	−1.46	−1.6	52.3	69.21	2.88	1.13	71.92	0	69.21	30.79
	0036	−1.43	−1.42	49.84	66.84	3.35	0.85	79.78	6.03	62.81	31.16
	0038	−1.22	−1.26	50.94	67.9	4.57	0.96	82.63	16.51	56.69	26.8
	0040	−1.37	−1.33	49.31	66.33	4.5	0.96	82.39	15.64	55.96	28.41
	0041	−1.38	−1.5	51.94	68.86	4.41	1.23	78.22	0.28	68.67	31.05
	0043	−1.84	−1.93	51.21	68.16	10.18	1.04	90.72	46.28	36.62	17.11

Continued

Target	No	Predicted CPX content using Eq. 4				Predicted OL content using Eq. 5				Derive CPX and OPX content	
		OPX_INDEX	CPX_INDEX	CPX_INDEX/(CPX_INDEX+OPX_INDEX)×100	Predicted CPX/(CPX+OPX)×100	IBD1	IBD2	IBD1/(IBD1+IBD2)×100	Predicted OL/(CPX+OPX+OL)×100	Predicted CPX/(CPX+OPX+OL)×100	Predicted OPX/(CPX+OPX+OL)×100
Lunar Regolith	0045	−1.27	−1.66	56.66	73.4	8.45	1.46	85.24	26.11	54.24	19.65
	0047	−0.93	−0.97	50.89	67.84	4.35	0.79	84.56	23.62	51.82	24.56
	0051	−1.68	−1.74	50.87	67.82	5.3	1.01	83.98	21.47	53.26	25.27
	0052	−1.32	−1.41	51.75	68.67	5.6	0.86	86.72	31.55	47	21.44
	0053	−1.25	−1.3	50.94	67.89	6.47	0.87	88.08	36.58	43.06	20.36
	0055	−1.53	−1.66	52.07	68.98	6.38	1.34	82.62	16.48	57.61	25.9
	0058	−1.09	−1.01	48.15	65.2	4.67	1.1	80.94	10.28	58.5	31.22
	0059	−0.75	−0.67	47.3	64.39	4.61	0.63	87.96	36.11	41.14	22.75
	0060	−1.57	−1.6	50.56	67.53	4.59	1.21	79.17	3.78	64.97	31.24
	0064	−1.47	−1.4	48.73	65.77	4.84	0.79	85.94	28.68	46.91	24.41
	0091	−1.33	−1.38	51	67.95	5.9	0.99	85.69	27.77	49.08	23.15
	0093	−1.68	−1.75	51.04	67.99	4.61	0.8	85.17	25.85	50.42	23.73
	0096	−1.59	−1.75	52.47	69.37	5.14	1.33	79.4	4.62	66.16	29.22
	0100	−1.73	−1.88	52.01	68.93	4.62	1.46	75.98	0	68.93	31.07
	0102	−1.93	−2.1	52.19	69.1	3.15	1.62	65.99	0	69.1	30.9
	0103	−1.42	−1.48	51.12	68.06	4.76	1.32	78.34	0.73	67.57	31.7
	0104	−1.16	−1.27	52.1	69.01	5.34	1.48	78.3	0.55	68.63	30.82
	0108	−1.4	−1.43	50.49	67.46	6.38	1.08	85.58	27.36	49.01	23.64
	0113	−1.7	−1.82	51.72	68.64	4.61	1.2	79.29	4.21	65.75	30.04
	0114	−1.72	−1.89	52.36	69.26	4.81	1.38	77.72	0	69.26	30.74
	0166	−1.57	−1.76	52.94	69.82	3.85	1.45	72.63	0	69.82	30.18

Continued

Target	No	Predicted CPX content using Eq. 4				Predicted OL content using Eq. 5				Derive CPX and OPX content	
		OPX_ INDEX	CPX_ INDEX	CPX_ INDEX/(CPX_ INDEX+ OPX_ INDEX)×100	Predicted CPX/(CPX + OPX)×100	IBD1	IBD2	IBD1 /(IBD1 + IBD2) ×100	Predicted OL/(CPX + OPX + OL)×100	Predicted CPX/(CPX + OPX + OL)×100	Predicted OPX/(CPX + OPX + OL)×100
Lunar Regolith	0120	−1.52	−1.6	51.34	68.28	5.33	0.9	85.53	27.19	49.72	23.09
	0124	−1.84	−2.07	52.99	69.86	4.24	1.53	73.51	0	69.36	30.14
	0125	−1.48	−1.58	51.67	68.59	6.45	0.98	86.75	31.69	46.36	21.45
	0128	−1.91	−2.04	51.59	68.52	8.47	1.02	89.25	40.88	40.51	18.61
	0129	−1.09	−1.09	50.03	67.02	7.52	1.17	86.51	30.79	46.38	22.82
	0131	−0.75	−0.78	51.01	67.96	7.75	1.43	84.43	23.12	52.25	24.63
	0132	−0.44	−0.49	52.52	69.41	4.74	1.59	74.88	0	69.41	30.59
	0133	−1.5	−1.61	51.9	68.82	5.26	1.59	76.84	0	68.82	31.18
	0136	−1.29	−1.39	51.93	68.85	6.66	1.34	83.23	18.7	55.97	25.33
	0138	−1.39	−1.52	52.22	69.12	6	1.11	84.34	22.8	53.33	23.84
	0142	−1.81	−1.94	51.67	68.59	4.97	1.32	79.03	3.24	66.37	30.39
	0144	−1.42	−1.66	53.81	70.66	7.13	1.33	84.25	22.46	54.79	22.75
	0146	−1.32	−1.47	52.68	69.57	5.61	1.07	84.02	21.61	54.54	23.86
	0148	−1.65	−1.66	50.07	67.05	7.78	1.09	87.76	35.39	43.33	21.29
	0153	−2.36	−2.66	52.96	69.84	8.54	1.75	82.97	17.75	57.44	24.81
	0154	−1.24	−1.38	52.68	69.57	4.77	1.13	80.91	10.18	62.49	27.33
	0161	−1.5	−1.64	52.3	69.21	5.45	1.46	78.92	2.84	67.24	29.92
	0164	−4.94	−5.29	51.74	68.66	11.99	1.6	88.2	37.02	43.24	19.74
	0166	−1.57	−1.76	52.94	69.82	3.85	1.45	72.63	0	69.82	30.18

An Empirical Model to Estimate Abundance of Nanophase Metallic Iron ($npFe^0$) in Lunar Soils

Dawei Liu[1], Yuanzhi Zhang[1,2], Guangliang Zhang[1], Bin Liu[1], Xin Ren[1], Rui Xu[3] and Chunlai Li[1]

1. Key Laboratory of Lunar and Deep Space Exploration, National Astronomical Observatories, Chinese Academy of Sciences, Beijing 100101, China; zhangyz@nao.cas.cn (Y. Z.); zhanggl@nao.cas.cn (G. Z.); liub@nao.cas.cn (B. L.); renx@nao.cas.cn (X. R.); licl@nao.cas.cn (C. L.)

2. School of Astronomy and Space Science, University of Chinese Academy of Sciences, Beijing 100049, China

3. Key Laboratory of Space Active Opto - Electronics Technology, Shanghai Institute of Technical Physics, Chinese Academy of Sciences, Shanghai 200083, China; xurui@sitp.ac.cn

Abstract: Lunar soils gradually become mature when they are exposed to a space environment, and nanophase metallic iron ($npFe^0$) generates within them. $npFe^0$ significantly changes the optical properties of lunar soils and affects the interpretation of the remotely sensed data of the lunar surface. In this study, a correlation analysis was conducted between $npFe^0$ abundance and reflectance spectra at short wavelengths for lunar soil samples in four size groups based on their spectral and compositional data, collected by the Lunar Soil Characterization Consortium (LSCC). Results show that 540nm single scattering albedo (SSA) of lunar soils correlates well with their corresponding $npFe^0$ abundance for each size group of lunar soil samples. However, it is poorly correlated with $npFe^0$ abundance when all size groups were considered because of the strong interference from grain size variation of lunar soils. To minimize the effect of grain size, the correlation of $npFe^0$ abundance with the spectral ratio of 540nm/810nm SSA of all size groups for LSCC samples was calculated and results show that a higher correlation existed between them ($R^2 = 0.91$). This ratio can serve as a simple empirical model for estimating $npFe^0$ abundance in lunar soils. However, bias could be introduced to the estimation result when lunar soils possess a high content of agglutinitic glass and ilmenite. Our future work will focus on improving the model's performance for these lunar soils.

Keywords: $npFe^0$; single scattering albedo; lunar soil characterization consortium

1 Introduction

Space weathering represents any natural processes pursuing an airless planetary body that largely modify its surface material's physical and compositional characteristics [1, 2]. It is mainly induced by solar wind ion implantation and micrometeoroid bombardment and has been generally applied to planetary observations, such as on the Moon or Mercury as well as asteroids [3 - 6]. Space weathering also influences the reflectance spectra of planetary surface materials and reduces our capability to remotely evaluate the planetary mineralogy and composition on their surface [1, 7].

Samples from the Moon and asteroids provide us an opportunity to research space weathering of an airless body. Actually, when the Apollo soil samples were brought back to the Earth-based laboratories, the measured results of lunar soils quickly showed that their reflectance is very different from that of pulverized lunar rocks [7 - 9]. In comparison with the optical properties of pulverized lunar rocks, the spectral reflectance of lunar soils shows a red-sloped continuum (namely reddening), a lower albedo (namely darkening), and attenuated absorption bands [1, 10]. Early literature attributed these optical changes to the presence of dark amorphous "glass" or "agglutinate" in lunar soils [11]. However, subsequent laboratory and remote sensing measurement have showed that these optical properties of lunar soils are caused by the accumulation of fine grained nanophase metallic iron ($npFe^0$) residing in amorphous rims of lunar soil grains after a long duration exposure of fresh lunar soils to a space weathering environment [12 - 16]. Two major mechanisms have been proposed to explain the accumulation of $npFe^0$ within lunar soils. One is solar wind ion implantation stating that surface material of lunar soil grains such as O atoms are preferentially sputtered from their lattice sites as a result of ion-irradiation, generating a reducing environment in which iron particles within residue surface material are reduced to form $npFe^0$ [10, 17 - 19]. The other mechanism proposed that because of continuous impact of micrometeoroids, lunar soil grains are vaporized so that some FeO molecules are dissociated into their constituent neutral atoms [10, 20]. The O atoms dissipate, and the iron atoms deposit and accumulate on the surface of lunar soil grains as $npFe^0$ [21]. A number of laboratory experiments have been carried out to simulate the process of solar wind bombardment and micrometeorite impact events to investigate space weathering and the formation of $npFe^0$ [21 - 28]. For example, Duke et al. [22] irradiated an olivine sample with 1.0keV H^+ and 4keV He^+ to simulate solar wind implantation and found that the surface of iron of olivine is dramatically reduced to the metallic form. The process of micrometeorite impacts was simulated by [25, 26] using pulse-laser irradiation on the olivine and pyroxene grains. Results show that lunar-like $npFe^0$ could form within the amorphous rims of mineral grains and the darkening and reddening effect on the spectra of these grains can be observed after pulse - laser irradiation. By performing slow - and rapid - heating experiments on mature lunar soils, Thompson et al. [21] simulated micrometeorite impact events and provided the first direct in - situ observation of $npFe^0$ formation within lunar soil grains. In addition to the generation of $npFe^0$, recent work also suggests that nanoparticles of Fe^{2+} and Fe^{3+} oxidation states could also form within sub - mature and mature lunar soils as a result of space weathering [29].

Estimation of $npFe^0$ abundance in lunar soils is a very interesting topic for many lunar scientists. It is helpful to accurately evaluate the lunar surface's mineralogy and composition using remotely sensed measurement. Currently, Hapke's radiative transfer model (RTM) has been generally applied to assess the mineralogy of the lunar surface owing to its ability to account for the effects of space weathering. Estimation of $npFe^0$ abundance in lunar soils is a necessary critical input for accurate Hapke's RTM mineral abundance inversion. The $npFe^0$ has been also associated with the presence of H_2O/OH^- due to reduction of Fe^{2+} to $npFe^0$ by impinging solar wind with the release of H_2O/OH^-. in lunar soils [30]. This brings about a highly possible usage of $npFe^0$ to determine lunar surface regions with H_2O/OH^- [31]. In addition, determination of the abundance of $npFe^0$ in lunar soils helps us to understand the formation of lunar swirls, which were once attributed to deficient $npFe^0$ as a result of retarded space weathering [32 - 34].

Despite the significant importance of $npFe^0$, its abundance in lunar soils has not been widely quantitatively evaluated [35 - 38]. Some previous studies attempted to establish models that can relate the measured reflectance spectra of lunar soils to their characteristic ferromagnetic resonance (I_s) induced by $npFe^0$. Is is highly linearly correlated with the abundance of $npFe^0$ and can be used as a measure of $npFe^0$ in lunar soils [39 - 41]. According to the measured reflectance spectra and I_s of 35 lunar soil samples, Hiroi et al. [35] found that I_s shows a good linear correlation with the spectral slope defined by a straight - line tangent to the shoulders of a 1μm absorption band of lunar soils. Mouélic et al. [36] expanded the number of investigated lunar samples to 50 and found that I_s can be also evaluated by the continuum slope derived from the ratio of 1500nm/750nm reflectance. Pieters et al. [37] developed several statistically optimized formulations that can link I_s of lunar soils with their corresponding Clementine - bands - based spectral parameter according to the compositional and spectral data acquired by the Lunar Soil Characterization Consortium (LSCC). Compared with the work of [35, 36], more samples of lunar soil were used in their investigation and the sample dataset shows a larger variation in composition and maturity which is well suited for characterizing the spectral properties of lunar soils. Recent work estimating the abundance of $npFe^0$ in lunar soils has been done by Trang and Lucey [38], who derived the $npFe^0$ abundance by using Hapke's RTM and constraining the input FeO and ilmenite content of lunar soils.

While these previous studies have provided possible ways to predict $npFe^0$ or I_s of lunar soils, the practical use of these methods will be impeded by several limitations. For example, most of these works require band information at longer wavelength (e. g., 1500nm) to define the spectral slope [35, 36], which is difficult in some cases. On the one hand, the spectral range of many remote sensing sensors stops at around 1μm and no longer - wavelength data are available (e. g., Clementine, Chang'E - 1 Interference Imaging Spectrometer). On the other hand, it is a challenge to accurately define a continuum slope when lunar soils become highly mature or lack apparent absorption features at around 1μm [35]. In addition, the model for $npFe^0$ estimation could be too complex to be applied. Many input parameters and pre - knowledge on the compositional information of lunar soils are required before the model can be used, which is difficult to obtain in advance [38].

Here, we developed a simple empirical model to estimate $npFe^0$ abundance in lunar soils from measured reflectance data. This model correlates the $npFe^0$ abundance with the ratio of single scattering albedo (SSA) of two wavelengths outside the absorption bands of lunar soils and requires no longer - wavelength spectral data. Grain size effects of lunar soils on $npFe^0$ abundance estimation are partially removed according to this model.

2 Dataset

The dataset of the Lunar Soil Characterization Consortium (LSCC) [7, 15] was applied in the study to develop the model. The dataset includes 10 highland and 9 mare soil samples. Each soil sample was grouped into four types of grain sizes (<10μm, 10 - 20μm, 20 - 45μm, and <45μm). The maturity index (I_s/FeO), composition (e. g. FeO content) and reflectance spectra of each size group (altogether 76 data samples) were measured by the LSCC (Table 1). The spectral data of these soil samples can be downloaded from Reflectance Experiment Laboratory (Relab) of Brown University (http: //www. planetary. brown. edu/relabdocs/LSCCsoil. html). The maturity index, composition and mineralogy information of corresponding soil samples can be acquired from the Planetary Geoscience Institute of the University of Tennessee (https: //pgi. utk. edu/lunar - soil - characterization - consortium - lscc - data/). Correlation was analyzed for all the samples except for sample 71061 in that it is abundant in black beads [42], with very different optical properties from those of other samples [15, 43].

Table 1　Lunar Soil Characterization Consortium (LSCC) dataset used in this study

Sample	FeO (%)	I_s/FeO	Ilmenite (%)	Agglutinitic Glass (%)	Grain Size	Highland/Mare
10084	12	145	5	62.6	<10μm	Mare
12001	12.5	115	1.6	61.9	<10μm	Mare
12030	14.3	32	3	55	<10μm	Mare

Continued Table 1

Sample	FeO (%)	I_s/FeO	Ilmenite (%)	Agglutinitic Glass (%)	Grain Size	Highland/Mare
15041	11	161	0.7	70.4	<10μm	Mare
15071	9.59	159	1.2	59.7	<10μm	Mare
70181	12.7	104	3.4	58.3	<10μm	Mare
71501	13.5	88	7.6	53.1	<10μm	Mare
79221	11.3	169	5.2	61.5	<10μm	Mare
14141	7.66	14.5	1.7	45.9	<10μm	Highland
14163	8.83	87	1.1	66.3	<10μm	Highland
14259	7.82	174.8	1.5	71.6	<10μm	Highland
14260	8.1	144.9	1.3	66.5	<10μm	Highland
61141	3.66	119.3	0.3	61.6	<10μm	Highland
61221	3.64	19.8	0.9	41.6	<10μm	Highland
62231	3.63	169	0.4	69.5	<10μm	Highland
64801	3.84	115.2	0.2	63.6	<10μm	Highland
67461	3.35	35.2	0.2	35.8	<10μm	Highland
67481	3.61	38.5	0.2	35.2	<10μm	Highland
10084	14.7	87	5.2	57	10 - 20μm	Mare
12001	15.9	67	1.8	56.8	10 - 20μm	Mare
12030	17.2	17	3.2	49.8	10 - 20μm	Mare
15041	14.4	92	0.8	56.7	10 - 20μm	Mare
15071	15.4	80	1.8	49.2	10 - 20μm	Mare
70181	15.5	63	6.7	51.7	10 - 20μm	Mare
71501	16.4	50	9.7	44.8	10 - 20μm	Mare
79221	15	78	6	54.3	10 - 20μm	Mare
14141	9.46	11.6	1.1	48.6	10 - 20μm	Highland
14163	10.1	64.8	0.9	58.5	10 - 20μm	Highland
14259	9.71	101.8	1.2	68.7	10 - 20μm	Highland
14260	9.84	98.9	1	65.2	10 - 20μm	Highland
61141	5.14	81.6	0.3	53.9	10 - 20μm	Highland
61221	4.4	13.89	0.3	32.6	10 - 20μm	Highland
62231	4.86	109.9	0.5	55	10 - 20μm	Highland
64801	4.78	84.9	0.2	61	10 - 20μm	Highland
67461	4.64	23.9	0.3	32.4	10 - 20μm	Highland
67481	4.04	33	0.2	28.6	10 - 20μm	Highland

Continued Table 1

Sample	FeO (%)	I_s/FeO	Ilmenite (%)	Agglutinitic Glass (%)	Grain Size	Highland/Mare
10084	15.5	67	6.4	53.9	20 – 45μm	Mare
12001	16.9	51	2.6	56.2	20 – 45μm	Mare
12030	17.6	12	2.6	39.4	20 – 45μm	Mare
15041	15.2	66	1.2	51.3	20 – 45μm	Mare
15071	15.6	49	1.9	47.6	20 – 45μm	Mare
70181	16	53	8.9	43.4	20 – 45μm	Mare
71501	17.8	28	12.3	38.3	20 – 45μm	Mare
79221	15.8	57	7.3	46.5	20 – 45μm	Mare
14141	11.6	5.8	1.9	41	20 – 45μm	Highland
14163	11.5	43.2	0.8	56.4	20 – 45μm	Highland
14259	11	77.2	1.3	60.5	20 – 45μm	Highland
14260	10.7	80.2	0.9	64	20 – 45μm	Highland
61141	5.15	75.5	0.3	50.1	20 – 45μm	Highland
61221	4.62	8.4	0.6	28.9	20 – 45μm	Highland
62231	5.31	80.7	0.3	50.6	20 – 45μm	Highland
64801	4.82	83.4	0.3	53.6	20 – 45μm	Highland
67461	4.93	22.3	0.3	25.4	20 – 45μm	Highland
67481	5.19	20.7	0.1	27.6	20 – 45μm	Highland
10084	14.8	88			<45μm	Mare
12001	16	62			<45μm	Mare
12030	16	20			<45μm	Mare
15041	14.2	93			<45μm	Mare
15071	14.9	71			<45μm	Mare
70181	15.3	61			<45μm	Mare
71501	16.5	44			<45μm	Mare
79221	14	91			<45μm	Mare
14141	9.81	9.7			<45μm	Highland
14163	9.94	66.5			<45μm	Highland
14259	9.54	108.6			<45μm	Highland
14260	9.65	93.3			<45μm	Highland
61141	4.8	94.5			<45μm	Highland
61221	4.47	13.6			<45μm	Highland
62231	4.87	116.7			<45μm	Highland
64801	4.68	92.2			<45μm	Highland
67461	4.24	29.8			<45μm	Highland
67481	4.38	33.5			<45μm	Highland

Note: agglutinitic glass and ilmenite contents for size group <45μm are not provided in LSCC dataset

3　Method

3.1　Strong Influence of $npFe^0$ at Short Wavelengths

Liu et al. and Liu. [44, 45] conducted sensitivity analyses of a suite of Hapke's RTM - simulated reflectance spectra of lunar soils with the aim of quantitatively determining the relative importance of $npFe^0$ abundance, grain size and mineral abundance in controlling the optical properties of lunar soils. Results show that $npFe^0$ has strong influences in the full spectral ranges as compared to mineral abundance. The simulated spectral reflectance of lunar soils is more sensitive to the variation of $npFe^0$ at shorter spectral regions than longer spectral ranges. The strong $npFe^0$ effect on short - wave bands is evident in the spectral reflectance of LSCC data, which tend to be convergent in short spectral regions for different sizes of grains in each sample of lunar soils (Figure 1) . This indicates that the spectral reflectance of lunar soils in short wavelengths is less sensitive to the variation of composing minerals of lunar soils, while mainly controlled by the variation of $npFe^0$ abundance [44, 45] . The strong influence of $npFe^0$ at short wavelengths can be also demonstrated by the work of [46], who prepared a suite of analog soils of varying abundance of $npFe^0$ to study the optical effects of $npFe^0$ on the measured reflectance spectra. Result shows that $npFe^0$ (< 10nm) greatly reddens short wavelengths spectra while leaving the longer wavelengths spectra mostly unaffected. Using lunar samples of similar composition but distinct exposure age, Fischer et al. and Fischer and Pieters [47, 48] attempted to develop a mathematical model that can describe the optical alteration of lunar soils due to space weathering. Their work suggests that the effect of $npFe^0$ on the optical properties of lunar soils can be evaluated as a function of

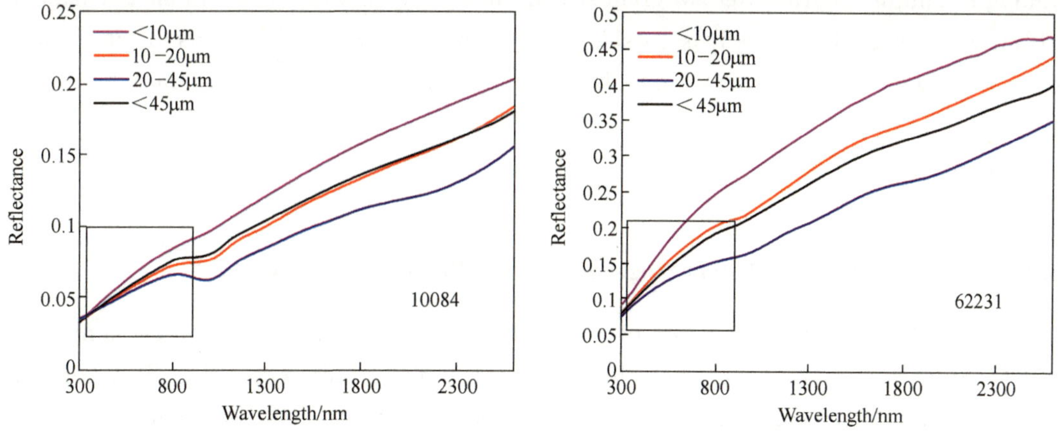

Figure 1　Reflectance spectra of four grain size groups of lunar soil samples 10084 and 62231. Black squares show the convergence of reflectance spectra at short wavelengths

wavelength with an increasing rate of darkening toward short wavelengths. This result is also consistent with Mie theory, which states that if the $npFe^0$ is smaller than the wavelength of incident light, the greatest amount of light extinction will be at the short wavelengths region [48]. Combing Hapke's RTM and Mie theory to model the scattering behavior of $npFe^0$, wohlfarth et al. [49] also found a strong influence of $npFe^0$ at short wavelengths of simulated reflectance spectra.

3.2 Estimation of $npFe^0$ Abundance in Lunar Soils

The strong effect of $npFe^0$ on short wavelengths spectra motivates us to explore the correlation between the $npFe^0$ abundance of lunar soils and their spectral of corresponding reflectance values only at short spectral ranges. Through conducting ferromagnetic resonance measurement on lunar soil samples, Morris [41] found a linear relation among $npFe^0$ abundance, FeO content and maturity index (I_s/FeO), reproduced below:

$$npFe^0 = (3.2 \times 10^{-4})(FeO)(I_s/FeO) \quad (1)$$

In this study, the abundance of $npFe^0$ for each lunar sample was first obtained via Equation (1) based on the measured maturity index (I_s/FeO) and FeO content (Table 1). Then, the LSCC measured spectral reflectance of each sample was re-sampled to the same bands as Moon Mineralogy Mapper (M^3) data and transformed to single scattering albedo (SSA) using a simplified Hapke's RTM (Appendix A and Equation (A2) of [50]). This transformation has potential to remove the effects of multiple scattering because $npFe^0$ residing in the rims of soil grains mainly influences SSA of lunar soils. Considering the influence of $npFe^0$ increases with decreasing wavelength [44, 45], 540nm SSA, which is the first available band of M^3 data of high quality, was chosen for the correlation analysis with the obtained $npFe^0$ abundance. Finally, the correlation of $npFe^0$ abundance with the spectral ratio of 540nm SSA/810nm SSA was also calculated, aiming at minimizing the effect of grain size of lunar soils on $npFe^0$ estimation. The band at 810nm was selected because it is outside the absorption wavelength regions of lunar soils, and the influence of composing minerals on $npFe^0$ estimation could be minimized.

4 Result

4.1 Correlation between 540nm Single Scattering albedo (SSA) and $npFe^0$

Figure 2 shows the correlation of $npFe^0$ abundance with 540nm SSA of all samples of LSCC soils. In general, as the abundance of $npFe^0$ increases, the SSA of the corresponding lunar soil decreases, which is consistent with the darkening effect of $npFe^0$, as expected. However, for all groups of investigated soil samples, the $npFe^0$ does not show a good correlation with the corresponding SSA at 540nm. This could be attributed to the influence of grain size variation of lunar soils at short wavelengths. In addition to the varying abundance of $npFe^0$, the grain size of lunar soils is also a significant factor driving the measured reflectance. For example, the spectral reflectance of a suite of transparent materials with different sizes of grains was measured by Adams and Filice [51] and Pieters [52], whereby it was noted that there is a general increase in the overall reflected spectra with decreasing grain size.

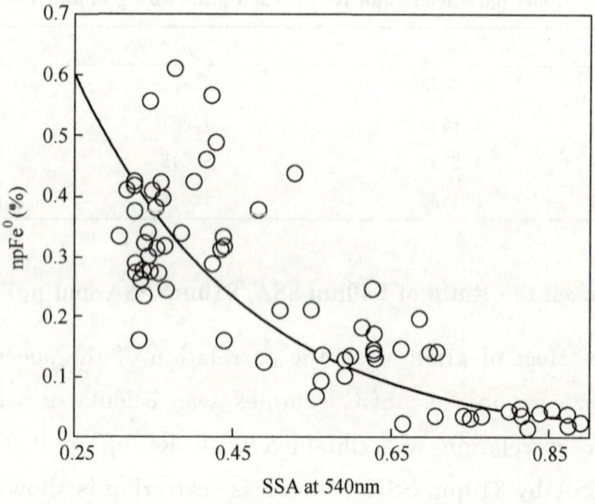

Figure 2　Correlation of SSA at 540nm with $npFe^0$ for all size groups of LSCC soils

To validate this speculation, correlation was analyzed for each group of sizes of LSCC samples seen in Figure 3. It is clear that the $npFe^0$ of each size group highly correlates to the SSA at 540nm, with the determination (R^2) close to or higher than 0.90 (Table 2). The abundance of $npFe^0$ in each size group of LSCC data can be well fitted in the form of

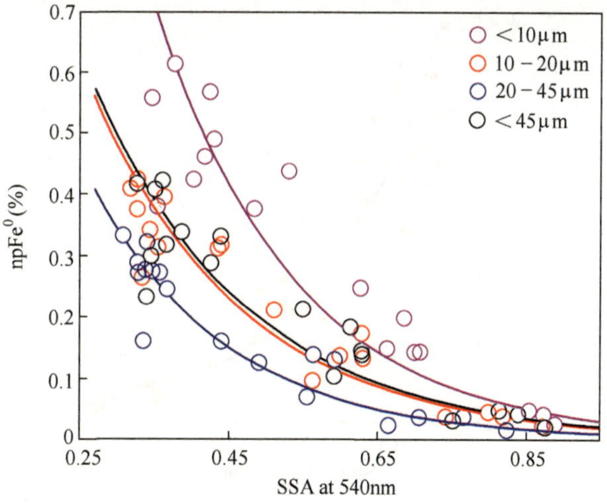

Figure 3　Correlation of SSA at 540nm and $npFe^0$ for each grain size group of LSCC samples. Solid color lines represent regression lines for different size groups of lunar samples

an exponential function as $Y = \alpha e^{(-\beta X)}$. Here, X is SSA at 540nm, Y is the abundance of npFe^0, and α and β are two regressed constants. Detailed values of α, β and R^2 for each size group samples are referred to Table 2. The regression lines for 10 – 20μm and the < 45μm size groups have similar α and β (Table 2) because the reflectance spectra of bulk soil (< 45μm) are most similar to that of a 10—20μm size fraction [1] (Figure 1).

Table 2　Model parameters and R^2 for each grain size group of LSCC data

Grain Size	α	β	R^2
<10μm	4.6478	5.375	0.90
10—20μm	2.1549	4.991	0.92
20—45μm	1.8737	5.648	0.89
<45μm	2.1155	4.836	0.89

4.2　Correlation between the Ratio of 540nm SSA/810nm SSA and npFe^0

To minimize the effect of grain size, the correlation of the spectral ratio of 540nm/810nm SSA of all size groups of LSCC samples was calculated with npFe^0. It clearly reveals that a higher correlation was obtained with R^2 up to 0.91 (Figure 4) after normalizing 540nm SSA by 810nm SSA. Much less scattering is shown in Figure 4 than in Figure 3, which demonstrates the success of using the spectral ratio of 540nm/810nm SSA in minimizing the grain size effect. This pronounced improvement in npFe^0 estimation can be accounted for by several reasons. First, grain size has a significant influence on

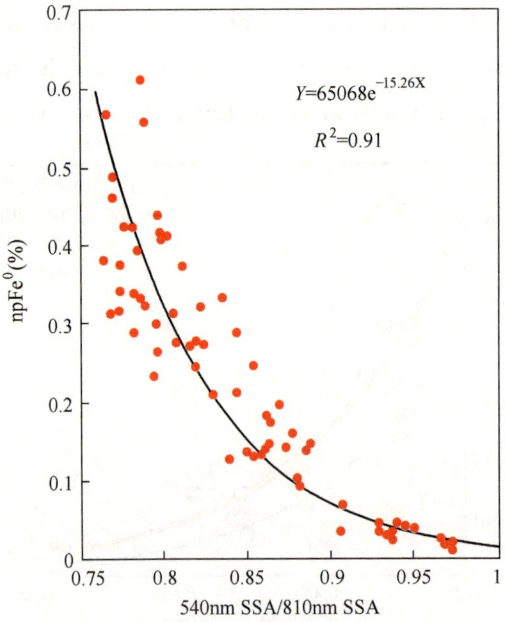

Figure 4　Correlation between the ratio of 540nm SSA/810nm SSA and npFe^0 for all size groups of LSCC soil samples

reflectance spectra at all wavelength regions [44, 45, 52]. By normalizing SSA at 540nm to that at 810nm, the effects of grain size could be significantly cancelled out. Second, both 540nm and 810nm are outside the absorption wavelength regions of lunar soils and cannot be largely affected by the Fe^{2+} absorption of lunar mafic minerals. Third, the ratio of 540nm SSA over 810nm SSA can be regarded as a slope measure for the reflected continuum at short spectral regions; the higher the $npFe^0$ included in a lunar soil is the redder the slope in the reflected continuum is, and vice versa.

Our result also suggests that this simple SSA ratio is suited for $npFe^0$ estimation of lunar soils showing a wide range of maturity (fresh, sub - mature, mature). Even for the mature lunar samples (I_s/FeO > 60), variation of $npFe^0$ can to some degree be reflected by the variation of the 540nm/810nm SSA ratio (Figure 5).

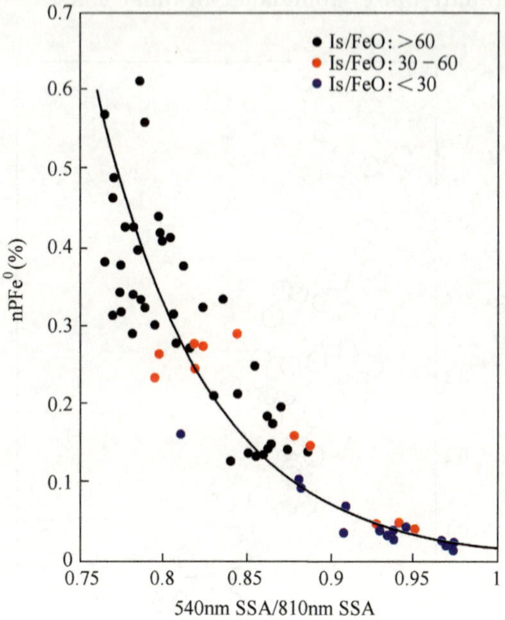

Figure 5　Correlation between the ratio of 540nm SSA/810nm SSA and $npFe^0$. Different solid color circles represent different maturity groups (I_s/FeO) of LSCC lunar samples. The black line is the regression line defined in Figure 4

Although this empirical relationship was developed on the basis of LSCC data samples with a grain size < 45m, it might be also suited for estimating the $npFe^0$ of lunar soils with a larger grain size. Pieters et al. [1] measured the spectra of lunar soils as a function of grain size and found that while these lunar soils are extremely poorly sorted (varying from several microns to several hundred microns), the spectra of the bulk of the lunar soils were most similar to that of the finest size fraction (< 45μm). The optical properties of lunar soils are dominated by this narrow size range because of their disproportionately large surface areas [1, 53].

We also investigated the correlation between the $npFe^0$ and the ratio of 540nm to 810nm reflectance rather than SSA for all groups of LSCC samples (Figure 6). It can be seen that abundance of $npFe^0$ is poorly correlated with the ratio of 540nm/810nm reflectance of lunar soils. This is not unexpected because SSA is used to describe the scattering behavior of a single particle of lunar soil. $npFe^0$ in the amorphous rims of soil grains significantly influences a single particle's capability in reflecting or absorbing the incident light [10] and causes the SSA of lunar soils to be highly sensitive to the variation of $npFe^0$. However, in addition to SSA, the measured reflectance spectra of lunar soils are also influenced by other factors, such as multiple scattering and opposition effect [54], resulting in the unsensitivity of lunar soil spectra to the variation of $npFe^0$. Therefore, it is necessary to covert remotely obtained reflectance spectra to their corresponding SSA to estimate $npFe^0$ abundance in lunar soils.

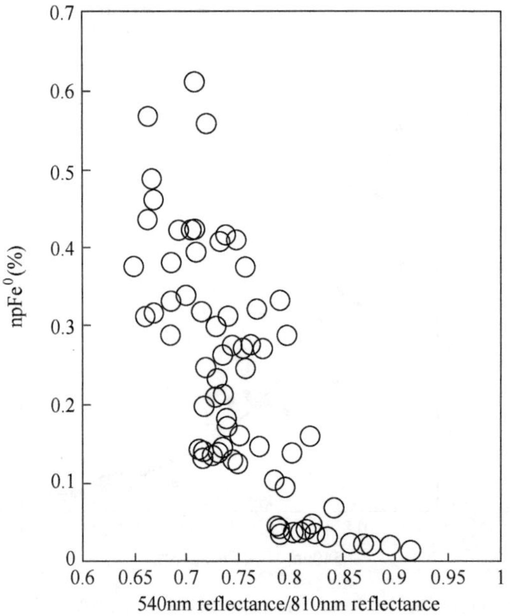

Figure 6 Correlation between the ratio of 540nm reflectance/810nm reflectance and $npFe^0$ for all size groups of LSCC soil samples

5 Discussion

5.1 Scattering Points Analysis

Although normalizing 540nm SSA by 810nm SSA can partially remove the grain size effect of lunar soils and improve the correlation between $npFe^0$ abundance and the 540nm/810nm SSA ratio, there are still many points not tightly close to the regression line

(Figure 4) . This might be explained by two reasons. First, the scattering could result from the high ilmenite content of lunar soils. Trang and Lucey [38] attempted to derive $npFe^0$ abundance using Hapke's RTM and found that the variation of ilmenite content of lunar soils has a strong impact on the estimation result. This is consistent with our finding that many scattering points occur for soil samples with ilmenite content higher than 5% (Figure 7) . The content of TiO_2, as the major composition of ilmenite, will increase in the soil samples with increasing ilmenite. Enhanced TiO_2 could result in a blue slope (in contrast to the reddening effect of $npFe^0$) of the reflectance spectra at short wavelengths [55, 56] . Our regressed relationship as a measure of the slope for the reflectance continuum at short wavelengths could be affected by the enhanced TiO_2 content in lunar soils. We analyzed the correlation between $npFe^0$ and the 540nm/810nm SSA ratio for lunar samples of varying TiO_2 content. The result shows that some scattering points are samples with TiO_2 content higher than 6%. The correlation between $npFe^0$ abundance and the SSA ratio for lunar soils samples with TiO_2 content $< 6\%$ is remarkably better than ($R^2=0.92$) (Figure 8a) that for samples with TiO_2 content $> 6\%$ ($R^2=0.11$) (Figure 8b), indicating higher TiO_2 content within lunar soils will to some extent interfere with the abundance prediction of $npFe^0$. Further correlation analysis on LSCC data suggests an obvious linear relationship between the ilmenite and TiO_2 content of lunar soils. 6% TiO_2 roughly corresponds to 5% ilmenite in lunar soils (Figure 9), which agrees well with our observation.

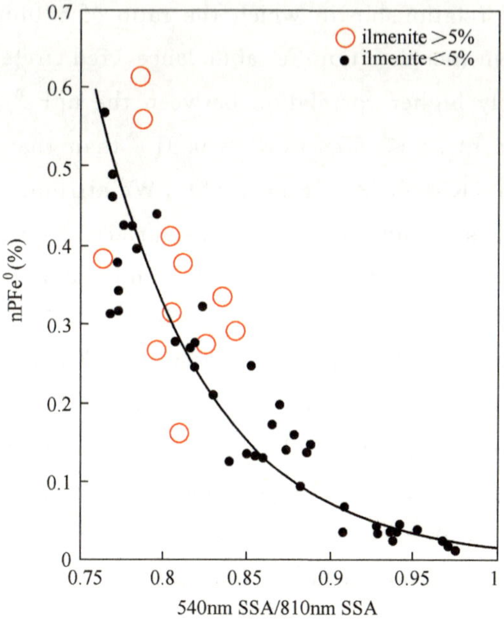

Figure 7 Correlation between the ratio of 540nm SSA/810nm SSA and $npFe^0$ denoted by different groups of LSCC lunar samples with varying ilmenite content. Solid black line is the regression line defined in Figure 4

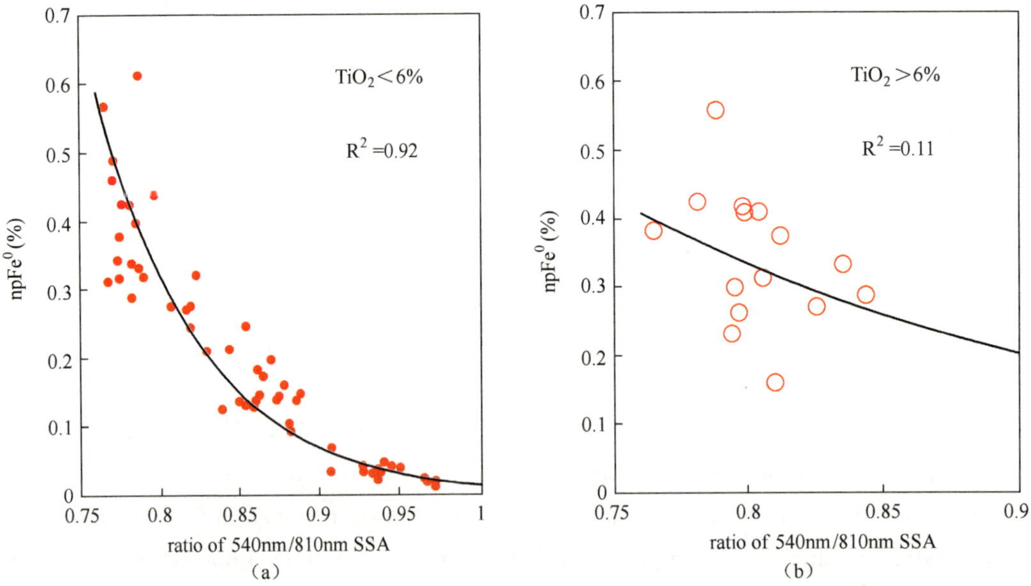

Figure 8　Correlation between the ratio of 540nm SSA/810nm SSA and $npFe^0$ for (a) lunar soil samples with TiO_2 content < 6% and for (b) lunar soil samples with TiO_2 content > 6%

In addition to the interference from the ilmenite/TiO_2 content of lunar soils, some scattering points may result from samples possessing abundant (> 55%) agglutinitic glass (Figure 10). Higher agglutinitic glass content in lunar soils could even cause the saturation of a regressed relationship in which the ratio of 540nm/810nm SSA changes very slightly with apparent increase in $npFe^0$ abundance (red circle in Figure 10). This is evidenced by the relatively higher correlation between the $npFe^0$ and SSA ratio for soil samples with agglutinitic glass < 55% ($R^2 = 0.91$) than that for soil samples with agglutinitic glass > 55% ($R^2=0.78$) (Figure 11). We attributed this result to the fact that in addition to the fine size $npFe^0$ coating on the surface of lunar soil grains, there are also large size metallic irons (>50nm), termed as microphase iron or Britt - Pieters particle, residing in the interior of agglutinitic glass [38, 57, 58]. These larger size microphase irons mainly darken the lunar soils without much reddening effect compared to that of smaller size $npFe^0$ [58, 59]. Higher abundance of large size microphase iron could accumulate in lunar soils as the agglutinitic glass content increases, resulting in a much flatter reflectance spectra of lunar soils at short wavelengths [46, 49]. Therefore, the 510nm/810nm SSA ratio as a slope measure (reddening) at short wavelengths becomes insensitive to the variation of $npFe^0$ abundance.

5.2　Non - linear Correlation Analysis

Figure 4 shows that the correlation between $npFe^0$ and the ratio of 540nm SSA / 810nm SSA can be regressed using an exponential function rather than a linear

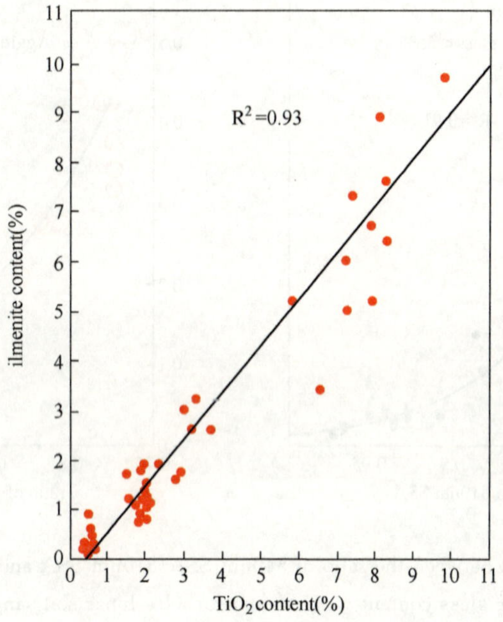

Figure 9　Linear correlation between ilmenite content and TiO_2 content of lunar soil samples

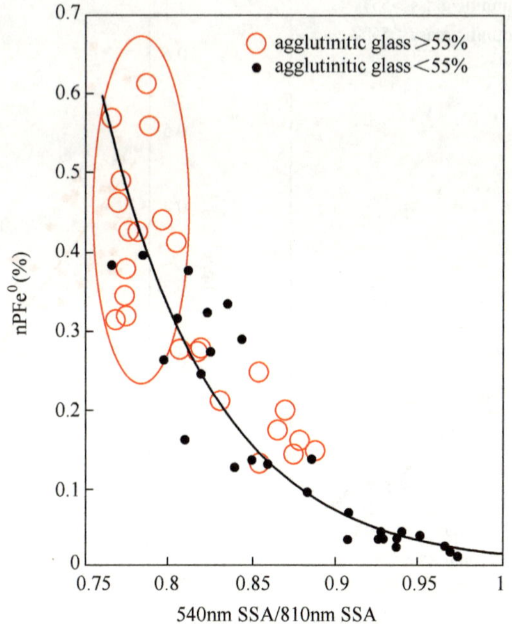

Figure 10　Correlation between the ratio of 540nm SSA/810nm SSA and $npFe^0$ denoted by different groups of LSCC lunar samples with varying agglutinitic glass content. Solid black line is the regression line defined in Figure 4

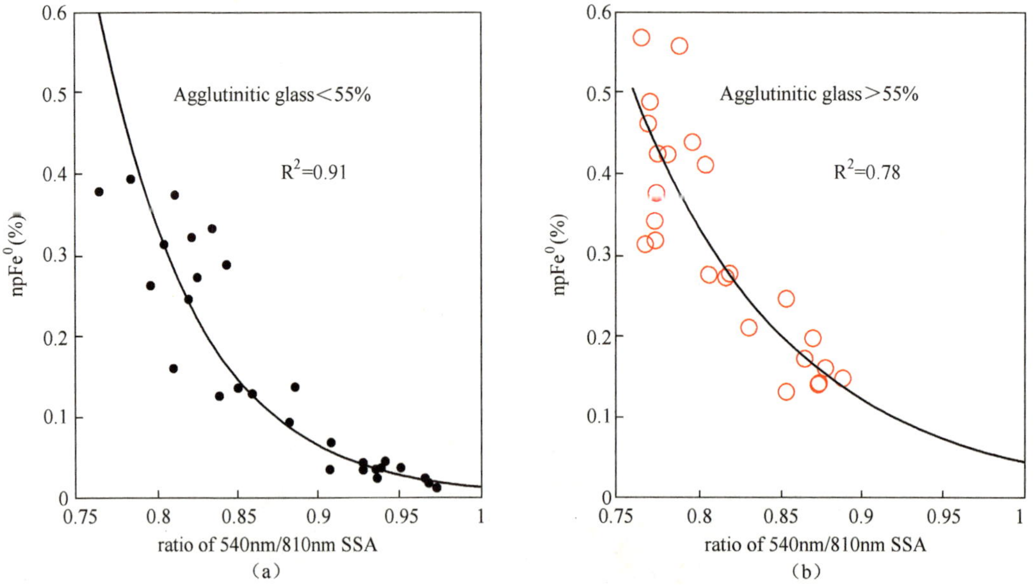

Figure 11　Correlation between the ratio of 540nm SSA/810nm SSA and $npFe^0$ for (a) lunar soil samples with agglutinitic glass content < 55% and for (b) lunar soil samples with agglutinitic glass content >55%

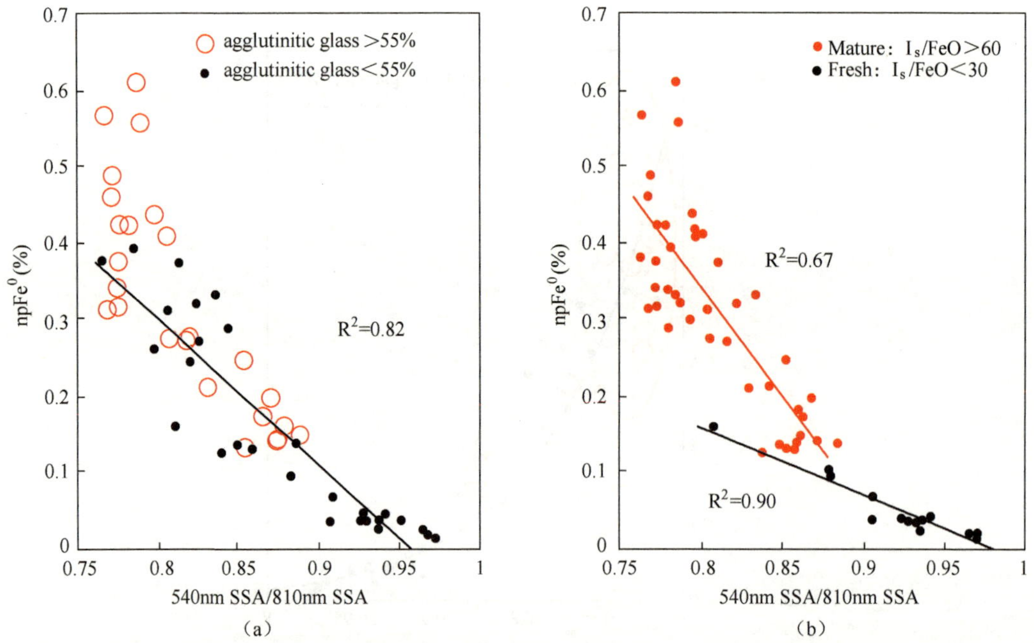

Figure 12　(a) Correlation between the ratio of 540nm SSA/810nm SSA and $npFe^0$ for lunar soil samples with agglutinitic glass content < 55%. Red circles are soil samples ignored in the regression process with agglutinitic glass content > 55%; (b) Correlation between the ratio of 540m SSA/810nm SSA and $npFe^0$ for mature (red dots and line) and fresh (black dots and line) lunar soil samples function. One reason might be that higher agglutinitic glass content within lunar soils, as discussed in Section 5. 1, could result in the unsensitivity of the 540nm/ 810nm SSA ratio to the variation of $npFe^0$ abundance. This can be demonstrated by the fact that a relatively acceptable linear relationship ($R^2 = 0.82$) exists between soil samples of lower agglutinitic glass content and $npFe^0$ abundance when soil samples of higher agglutinitic glass content were ignored (Figure 12a) . Alternatively, it was found that if sub - mature soil samples were not considered, the correlation between $npFe^0$ and the ratio of 540nm SSA/810nm SSA can be also fitted by two separate linear functions corresponding to mature and fresh samples, respectively (Figure 12b) . The exponential form seems to be the combination of these two linear forms. Mature soil samples show a steeper regressed slope and a lower R^2 compared to that of fresh soil samples. This is because the maturity of lunar soils roughly positively correlates with the their agglutinatuc glass content (Figure 13) . Mature lunar soils might have an increased agglutinitic glass content relative to that of fresh soils. As discussed in 5. 1, this could lead to the 540nm/810nm SSA changing very slightly with apparent variation in $npFe^0$ abundance. The two - segment linear fitting also implies that the maturity of lunar soils should be taken into account (e. g. , optical maturity parameter: OMAT) in future work to further refine the model, and a linear correlation between $npFe^0$ abundance and the refined spectral parameter could be expected.

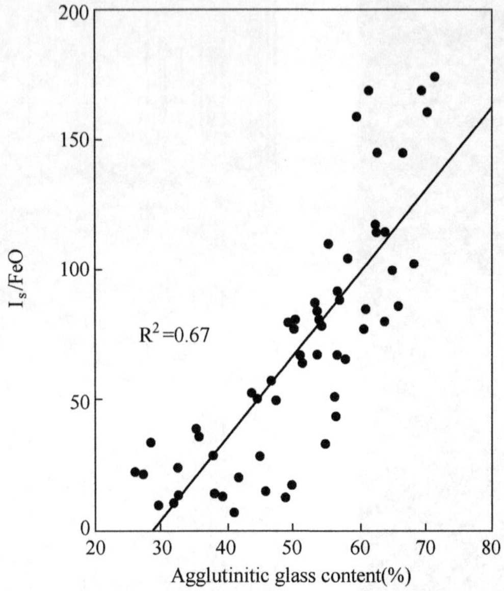

Figure 13　Correlation between the maturity (I_s/FeO) and agglutinitic glass content of LSCC lunar soil samples. Solid black line is the linear regression line

5. 3　Case Application and Comparison

Although accurate estimation of $npFe^0$ abundance in lunar soils based on the 540nm /

810nm SSA ratio could be impeded by the agglutinitic glass and ilmenite content, this ratio can still be used to investigate the variation trend of $npFe^0$ for lunar surface areas. In this study, this empirical model was applied to a swirl and a mare region on the lunar surface to evaluate their regional $npFe^0$ distribution. The estimation results for the two regions were also compared, aiming at testing whether this model is sensitive to the variation of $npFe^0$ abundance in lunar soils.

1) Lunar swirls

Lunar swirls are bright curvilinear markings on the lunar surface. They are optically immature compared to the surrounding areas and associated with regions having high magnetic field strength [60, 61]. One prevailing hypothesis for the origin of lunar swirls states that the magnetic field above swirl surfaces deflects the solar wind ions away from swirl regions and directs them to the off - swirl regions. Less $npFe^0$ is generated in swirl regions than surrounding off - swirl regions owing to reduced solar wind flux [62, 63]. This provides the best target area for our model evaluation.

We applied the model (exponential function in Figure 4) to one sub - image of M^3 level 2 reflected spectral data covering the central region of a famous lunar swirl 'Reiner Gamma' (Figure 14a) to assess whether the mapped $npFe^0$ distribution geographically agrees well with the bright swirl observed in the remote sensing image.

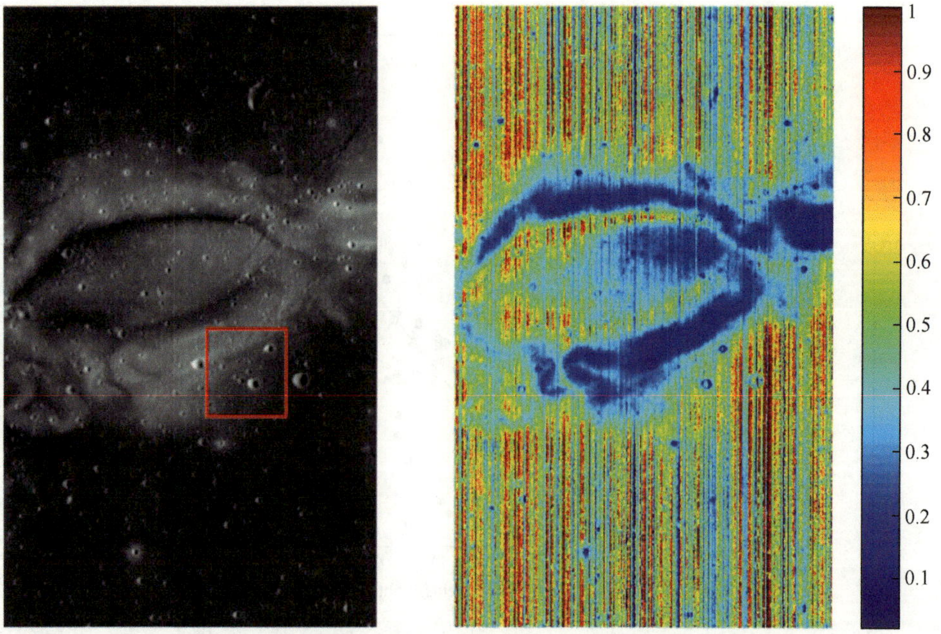

(a)　　(b)

Figure 14　(a) M^3 750nm reflectance image for lunar swirl 'Reiner Gamma';
(b) $npFe^0$ abundance distribution of 'Reiner Gamma'

Our results show that the swirl region evidently possesses a lower amount of $npFe^0$

relative to the off - swirl region (Figure 14b). The bright swirl in the image (Figure 14a) geographically correlates well with the lunar surface of decreased $npFe^0$, and the outline of swirl can be fully characterized by the areas deficient in $npFe^0$ abundance.

The geographical agreement between the 750nm reflectance image and $npFe^0$ distribution map is reasonable and not an artifact resulting from higher correlation between reflectance data and their corresponding SSA. Although reflectance at 750nm correlates with the single band SSA at 540nm or 810nm to some degree, it is not necessarily correlated with the ratio of 540nm SSA/810nm SSA. Shown in Figure 15a, b are the reflectance image at 750nm of one sub - area in Figure 14a (red square) and the ratio image of 540nm SSA/810nm SSA of that sub - area, respectively. It can be seen that although some areas are positively correlated (bright area indicated by red arrows), other surface areas are negatively correlated. Parts of crater walls or floors (yellow arrow) with higher reflectance at 750nm show a decreased ratio of 540nm SSA/810nm SSA, contrasting to the trend seen in the bright areas (red arrow). Some small craters in the 750nm reflectance image (Figure 15a) even disappear in the SSA ratio image (Figure 15b) (green arrow). These observations suggest that the geographical correlation between derived $npFe^0$ distribution and 750nm reflectance is not an artifact of the way the data reduced. Single band reflectance (e. g., 750nm reflectance) reflects the brightness variation of the lunar surface, and it can be influenced by many factors such as variation in grain size and composing minerals of lunar soils, shadow effects, topography variation

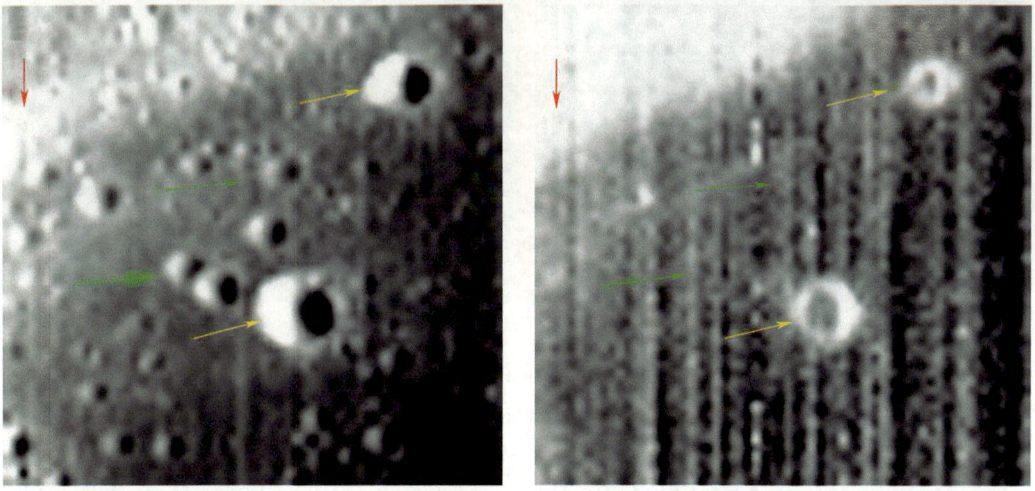

(a)　　　　(b)

Figure 15　(a) M^3 750nm reflectance image of one sub - area in Figure 14a (red box); (b) the ratio of 540nm/810nm SSA of this sub - area. Red arrow shows positively correlated bright swirl regions between the two images. Yellow arrows show some parts of crater wall or floor having higher brightness in the 750nm reflectance image while having low values in the 540nm/810nm SSA ratio image. Green arrows show the craters disappeared in the 540nm/810nm SSA ratio image and viewing geometry. Using one band reflectance or SSA alone cannot provide a reliable prediction of $npFe^0$ abundance (Figure 2) . However, the ratio of 540nm SSA/810nm SSA can partially remove these effects, amplifying the $npFe^0$ effect on regulating the short wavelength spectra (Figure 4) . It is capable of reflecting the regional $npFe^0$ variation trend when there is indeed $npFe^0$ abundance difference within lunar soils. The effectiveness of this ratio in $npFe^0$ estimation can also be validated by the fact that the derived abundance of $npFe^0$ within this region ranges from 0% to 1% close to the reported values of lunar soil samples [41] and the predicted values from remote sensing data [38] .

2) Lunar maria

To make a comparison with the swirl region, distribution of $npFe^0$ of a mare region to the north of 'Reiner Gamma' was also investigated. This mare region is relative homogenous without apparent albedo difference across its surface, except some distributed fresh impact craters (Figure 16a) . It is expected that the $npFe^0$ variation for this region should not be as strong as that of the lunar swirl (Figure 14b) .

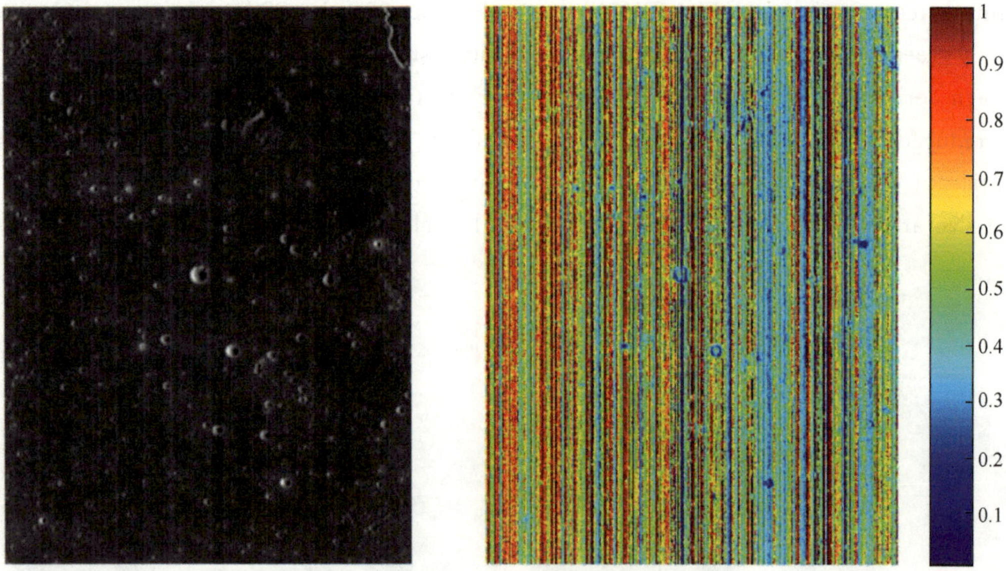

(a)　　(b)

Figure 16　(a) M^3 750nm reflectance image of the mare region to the north of Reiner Gamma; (b) $npFe^0$ abundance distribution

By applying this empirical model to this mare region, it was found that, in contrast to the large $npFe^0$ abundance difference between the swirl and off - swirl regions in Figure 14b, no surface area with obvious decreased or increased $npFe^0$ can be observed within this mare region (Figure 16b) . The $npFe^0$ distribution of the whole region only bears resemblance to that of the off - swirl regions in Figure 14b.

It is worth noting that even in such areas without significant $npFe^0$ variation, the fresh craters on the surface of the mare can still be clearly revealed by regional distribution of $npFe^0$ (blue round spots or rings in Figure 16b). These fresh craters show a decreased amount of $npFe^0$ consistent with what is expected on the lunar surface. Because of the short exposure time to a space weathering environment, a lower amount of $npFe^0$ could have accumulated in these craters relative to the surrounding lunar soils [43]. This also demonstrates the effectiveness of this empirical model in estimating $npFe^0$ abundance of the lunar surface. However, many strips can be seen in both investigated swirl and mare regions (Figures 14b and 16b). This can be ascribed to the noise originating in the original M^3 data (strips in Figures 14a and 16a) rather than a true variation of $npFe^0$ abundance.

6 Conclusions

The 540nm/810nm SSA ratio is highly correlated to $npFe^0$ abundance in lunar soils. This spectral ratio highlights the spectral importance of $npFe^0$, reduces the grain size effect of lunar soils on the short wavelengths reflectance spectra and, thus, can be considered as a useful spectral index for the prediction of $npFe^0$ abundance within lunar soils.

However, the performance of this simple empirical model does show some interference from the high content of agglutinitic glass and ilmenite in lunar soils. More work regarding removing the effects of high aglutitinic glass and ilmenite/TiO_2 content should be considered in future work to further refine the model's performance in estimating $npFe^0$ abundance of lunar soils.

Acknowledgments

We thank Kevin Cannon and an anonymous reviewer for their constructive reviews and comments, which greatly improved this manuscript. We also thank the Reflectance Experiment Laboratory (Relab) of Brown University and Planetary Geoscience Institute of the University of Tennessee for making the LSCC dataset publicly available.

Appendix A

An approximate solution to the radiative transfer equations was proposed by [10, 54, 64] to describe the light scattering behavior of lunar soils based on the assumption that lunar surface materials are intimately mixed and lunar soil grains are closely spaced with their size larger than spectral wavelengths. In his work, the reflectance spectra of lunar soils can be converted to its SSA by the following equation:

$$R=\frac{\omega}{4}\frac{1}{\mu_0+\mu}\{[1+B(g)]\times P(g)+H(\mu_0, \omega)\times H(\mu, \omega)-1\} \quad (A1)$$

Here, R is remotely obtained or lab - measured reflectance, ω is the SSA of lunar soils, μ_0 and μ are the cosine of the incidence light angle (i) and emittance light angle (e), respectively. g is the phase angle. B (g) and P (g) are the backscattering function and phase function of lunar soils, and H represents the multiple scattering effect between lunar soil grains.

It was found that the backscattering function B (g) defining the opposition effect of lunar soils decreases significantly with increasing phase angle g and can be ignored when g is greater than 15° [54]. Because all the reflectance spectra of the soil samples used in this study were measured at phase angle 30° and most of the remotely obtained reflectance spectra such as M^3 data had also been photometrically corrected to the phase angle 30°, B (g) was set to be zero to ignore the backscattering effect of lunar soils. To further simplify the model, the lunar soils were considered as isotropic scattering and the phase function P (g) was set to be unity in this study. This assumption was adopted and supported by the work [65], which found that it will not strongly affect the performance of Hapke's RTM. According to these assumptions, Equation (A1) can be simplified as follows:

$$R = (\omega \times H(\mu_0, \omega) \times H(\mu, \omega))/(4 \times (\mu + \mu_0)) \quad (A2)$$

Here, H (x, ω) is Chanderasekhar's function for isotropic scattering of lunar soils, and x represents μ_0 (incident light angle) or μ (emittance light angle) [50].

Setting the incidence angle and emission angle as 30° and 0°, the same as the measured reflectance data of lunar soil samples in this study, the SSA of lunar soils could be derived from Equation (A1) and Equation (A2) as follows [50]:

$$\omega = 1 - \left(\frac{-27.865R + \sqrt{4.029R^2 + 602.932R + 268.696}}{51.71R + 16.392}\right)^2 \quad (A3)$$

These steps are the same as the work of Yan et al. [50].

References

[1] Pieters, C. M.; Fischer, E. M.; Rode, O.; Basu, A. Optical effects of space weathering - the role of the finest fraction. J. Geophys. Res. 1993, 98, 20817 - 20824.

[2] Chapman, C. R. Space weathering of asteroid surfaces. Annu. Rev. Earth Planet. Sci. 2004, 32, 539 - 567.

[3] Hapke, B. Interpretations of optical observations of Mercury and the moon. Phys. Earth Planet. Interi. 1977, 15, 264 - 274.

[4] Hapke, B.; Cohen, A.; Cassidy, W.; Wells, E. Solar radiation effects on the optical properties of Apollo 11 lunar samples. In Proceedings of the Apollo 11 Lunar Science Conference, Houston, TX, USA, 5 - 8 January 1970; pp. 2199 - 2212.

[5] Noble, S. K.; Pieters, C. M. Space weathering on Mercury: Implications for remote sensing. Sol. Syst. Res. 2003, 37, 31 - 35.

[6] Brunetto, R.; Vernazza, P.; Marchi, S.; Birlan, M.; Fulchignoni, M.; Orofino, V.; Strazzulla, G. Modeling asteroid surfaces from observations and irradiation experiments: The case of 832 Karin. Icarus 2006, 184, 327 - 337.

[7] Pieters, C. M.; Taylor, L. A.; Noble, S. K.; Keller, L. P.; Hapke, B.; Morris, R. V.; Allen, C. C.; McKay, D. S.; Wentworth, S. Space weathering on airless bodies: Resolving a mystery with lunar samples. Meteorit. Planet. Sci. 2000, 35, 1101 -1107.

[8] McCord, T. B.; Johnson, T. V. Lunar spectral reflectivity (0.30 to 2.50 microns) and implications for remote mineralogical analysis. Science 1970, 169, 855 - 858.

[9] McCord, T. B.; Adams, J. B. Progress in remote optical analysis of lunar surface composition. Moon 1973, 7, 453 - 474.

[10] Hapke, B. Space weathering from Mercury to the asteroid belt. J. Geophys. Res. 2001, 106, 10039 - 10073.

[11] Conel, J.; Nash, D. Spectral reflectance and albedo of Apollo 11 lunar samples: Effects of irradiation and vitrification and comparison with telescopic observations. Geochim. Cosmochim. Acta 1970, 1, 2013 - 2024.

[12] Cassidy, W.; Hapke, B. Effects of darkening processes on surfaces of airless bodies. Icarus 1975, 25, 371 - 383.

[13] Keller, L. P.; Mckay, D. S. Discovery of vapor deposits in the Lunar Regolith. Science 1993, 261, 1305 - 1307.

[14] Keller, L. P.; McKay, D. S. The nature and origin of rims on lunar soil grains. Geochim. Cosmochim. Acta 1997, 61, 2331 - 2341.

[15] Taylor, L. A.; Pieters, C. M.; Keller, L. P.; Morris, R. V.; McKay, D. S. Lunar mare soils: Space weathering and the major effects of surface - correlated nanophase Fe. J. Geophys. Res. 2001, 106, 27985 - 27999.

[16] Taylor, L. A.; Pieters, C. M.; Patchen, A.; Taylor, D. S.; Morris, R. V.; Keller, L. P.; Mckay, D. S. Mineralogical and chemical characterization of lunar highland soils: Insights into the space weathering of soils on airless bodies. J. Geophys. Res. 2010, 115, E02002.

[17] Hapke, B. Effects of a simulated solar wind on the photometric properties of rocks and powders. Ann. N. Y. Acad. Sci. 1965, 123, 711 - 721.

[18] Stern, A. The lunar atmosphere: History, status, current problems, and context. Rev. Geophys. 1999, 37, 453 - 491.

[19] Crider, D. H.; Vondrak, R. R. Hydrogen migration to the lunar poles by solar wind bombardment of the moon. Adv. Space Res. 2002, 30, 1869 - 1874.

[20] Housley, R. M. A model for chemical and isotopic fractionation in the lunar regolith by impact vaporization. In Proceedings of the Lunar and Planetary Science Conference, Houston, TX, USA, 19 - 23 March 1979.

[21] Thompson, M. S.; Zega, T. J.; Howe, J. Y. In Situ experimental formation and

growth of Fe nanoparticles and vesicles in lunar soil. Meteoritics Planet. Sci. 2017, 52, 413－427.

[22] Dukes, C.; Baragiola, R.; McFadden, L. Surface modification of olivine by H^+ and He^+ bombardment. J. Geophys. Res. Planets 1999, 104, 1865－1872.

[23] Loeer, M.; Dukes, C.; Baragiola, R. Irradiation of olivine by 4keV He^+: Simulation of space weathering by the solar wind. J. Geophys. Res. Planets 2009, 114, E03003.

[24] Christoersen, R.; Rahman, Z.; Keller, L. Solar ion sputter deposition in the lunar regolith: Experimental simulation using focused－ion beam techniques. In Proceedings of the Lunar and Planetary Science Conference, The Woodland, TX, USA, 19－23 March 2012; p. 2614.

[25] Sasaki, S.; Nakamura, K.; Hamabe, Y.; Kurahashi, E.; Hiroi, T. Production of iron nanoparticles by laser irradiation in a simulation of lunar－like space weathering. Nature 2001, 410, 555－557. [PubMed]

[26] Sasaki, S.; Hiroi, T.; Nakamura, K.; Hamabe, Y.; Kurahashi, E.; Yamada, M. Simulation of space weathering by nanosecond pulse laser heating: Dependence on mineral composition, weathering trend of asteroids and discovery of nanophase iron particles. Adv. Sp. Res. 2002, 29, 783－788.

[27] Gillis－Davis, J.; Gasda, P.; Bradley, J.; Ishii, H.; Bussey, D. Laser space weathering of Allende (CV2) and Murchison (CM2) carbonaceous chondrites. In Proceedings of the Lunar and Planetary Science Conference, The Woodlands, TX, USA, 16－20 March 2015; p. 1607.

[28] Matsuoka, M.; Nakamura, T.; Kimura, Y.; Hiroi, T.; Nakamura, R.; Okumura, S.; Sasaki, S. Pulse－laser irradiation experiments of Murchison CM2 chondrite for reproducing space weathering on C－type asteroids. Icarus 2015, 254, 135－143.

[29] Thompson, M. S.; Zega, T. J.; Becerra, P.; Keane, J. T.; Byrne, S. The oxidation state of nanophase fe particles in lunar soil: Implications for space weathering. Meteorit. Planet. Sci. 2016, 51, 1082－1095.

[30] Housley, R. M.; Grant, R. W.; Paton, N. E. Origin and characteristics of excess Fe metal in lunar glass welded aggregates. In Proceedings of the Lunar Science Conference, Houston, TX, USA, 5－8 March 1973; pp. 2737－2749.

[31] Bandfield, J. L.; Poston, M. J.; Klima, R. L.; Edwards, C. S. Widespread distribution of OH/H_2O on the lunar surface inferred from spectral data. Nat. Geosci. 2018, 11, 173－177.

[32] Hood, L. L.; Schubert, G. Lunar magnetic anomalies and surface optical properties. Science 1980, 208, 49－51.

[33] Hood, L. L.; Williams, C. R. The lunar swirls: Distribution and possible origins. In Proceedings of the Lunar and Planetary Science Conference, Houston,

TX, USA, 14 - 18 March 1989; pp. 99 - 113.

[34] Blewett, D. T.; Coman, E. I.; Hawke, B. R.; Gillis - Davis, J. J.; Purucker, M. E.; Hughes, C. G. Lunar swirls: Examining crustal magnetic anomalies and space weathering trends. J. Geophys. Res. 2011, 116, E02002.

[35] Hiroi, T.; Pieters, C. M.; Morris, R. V. New Considerations for estimating lunar soil maturity from VIS - NIR reflectance spectroscopy. In Proceedings of the Lunar and Planetary Science Conference, Houston, TX, USA, 16 - 20 March 1998; p. 1152.

[36] Mouélic, S. L.; Langevin, Y.; Erard, S. Discrimination between maturity and composition of lunar soils from integrated Clementine UV - visible/near - infrared data: Application to the Aristarchus Plateau. J. Geophys. Res. 2000, 105, 9445 - 9455.

[37] Pieters, C.; Shkuratov, Y.; Kaydash, V.; Stankevich, D.; Taylor, L. Lunar soil characterization consortium analyses: Pyroxene and maturity estimates derived from Clementine image data. Icarus 2006, 184, 83 - 101.

[38] Trang, D.; Lucey, P. G. Improved space weathering maps of the lunar surface through radiative transfer modeling of Kaguya multiband imager data. Icarus 2019, 321, 307 - 323.

[39] Morris, R. V. Surface exposure indices of lunar rocks: A comparative FMR study. In Proceedings of the Lunar and Planetary Science Conference, Houston, TX, USA, 15 - 19 March 1976; pp. 315 - 335.

[40] Morris, R. V. The surface exposure (maturity) of lunar soils: Some concepts and Is/FeO compilation. In Proceedings of the Lunar and Planetary Science Conference, Houston, TX, USA, 13 - 17 March 1978; pp. 2278 - 2297.

[41] Morris, R. V. Origins and size distribution of metallic iron particles in the lunar regolith. In Proceedings of the Lunar and Planetary Science Conference, Houston, TX, USA, 17 - 21 March 1980; pp. 1697 - 1712.

[42] Heiken, G.; McKay, D. S. Petrography of Apollo 17 soils. In Proceedings of the Lunar and Planetary Science Conference, Houston, TX, USA, 18 - 22 March 1974; pp. 843 - 860.

[43] Noble, S. K.; Pieters, C. M.; Taylor, L. A.; Morris, R. V.; Allen, C. C.; McKay, D. S.; Keller, L. P. The optical properties of the finest fraction of lunar soil: Implications for space weathering. Meteorit. Planet. Sci. 2001, 36, 31 - 42.

[44] Liu, D.; Li, L.; Zhang, Y. Z. Sensitivity analysis for Hapke's radiative transfer model. In Proceedings of the Lunar and Planetary Science Conference, The Woodlands, TX, USA, 18 - 22 March 2013; p. 1290.

[45] Liu, D. Improvement and Use of Radiative Transfer Models to Assess Lunar Space Weathering and Mechanisms for Swirl Formation. Ph. D. Thesis, Indiana University, Indiana, IN, USA, 2015.

[46] Noble, S. K. ; Pieters, C. M. ; Keller, L. P. An experimental approach to understanding the optical effects of space weathering. Icarus 2007, 192, 629 - 642.

[47] Fischer, E. M. ; Pieters, C. M. ; Pratt, S. F. Modeling the space weathering - induced optical alteration of lunar soils: First results. In Proceedings of the Lunar and Planetary Science Conference, Houston, TX, USA, 15 - 19 March 1994; pp. 371 372.

[48] Fischer E. , M. ; Pieters, C. M. A Model for lunar soil optical alteration due to space weathering. In Proceedings of the Lunar and Planetary Science Conference, Houston, TX, USA, 14 - 18 March 1995; pp. 395 - 396.

[49] Wohlfarth, K. S. ; Wöhler, C. ; Grumpe, A. SpaceWeathering and Lunar OH/H_2O—Insights from Ab Initio Mie Modeling of Submicroscopic Iron. Astron. J. 2019, 158, 80.

[50] Yan, B. ; Wang, R. ; Gan, F. ; Wang, Z. Minerals mapping of the lunar surface with Clementine UVVIS/NIR data based on spectra unmixing method and Hapke model. Icarus 2010, 208, 11 - 19.

[51] Adams, J. B. ; Filice, A. L. Spectral reflectance 0. 4 to 2. 0 microns of silicate rock powders. J. Geophys. Res. 1967, 72, 5705 - 5715.

[52] Pieters, C. M. Strength of mineral absorption features in the transmitted component of near - infrared reflected light: First results from RELAB. J. Geophys. Res. 1983, 88, 9534 - 9544.

[53] Lucey, P. G. Radiative transfer modeling of the effect of mineralogy on some empirical methods for estimating iron concentration from multispectral imaging of the Moon. J. Geophys. Res. 2006, 111, E08003.

[54] Hapke, B. Bidirectional reflectance spectroscopy: 1. Theory. J. Geophys. Res. 1981, 86, 3039 - 3054.

[55] Lucey, P. G. ; Blewett, D. T. ; Hawke, B. R. Mapping the FeO and TiO_2 content of the lunar surface with multispectral inagery. J. Geophys. Res. 1998, 103, 3679 - 3699.

[56] Johnson, J. R. ; Larson, S. M. ; Singer, R. B. Remote sensing of potential lunar resources: 1. Near - Side compositional properties. J. Geophys. Res. 1991, 96, 18861 - 18882.

[57] Pieters, C. M. ; Noble, S. K. Space weathering on airless bodies. J. Geophys. Res. 2016, 121, 1865 - 1884. [PubMed]

[58] Lucey, P. G. ; Riner, M. A. The optical effects of small iron particles that darken but do not redden: Evidence of intense space weathering on Mercury. Icarus 2011, 212, 451 - 462.

[59] Liu, D. ; Li, L. ; Sun, Y. An improved radiative transfer model for estimating mineral abundance of immature and mature lunar soils. Icarus 2015, 253, 40 - 50.

[60] Schultz, P. H. ; Srnka, L. J. Cometary collisions on the Moon and Mercury. Nature

1980，284，22 - 26.

[61] Richmond，N. C.；Hood，L. L.；Mitchell，D. L.；Lin，R. P.；Acuna，M. H.；Binder，A. B. Correlations between magnetic anomalies and surface geology antipodal to lunar impact basins. J. Geophys. Res. 2005，110，E05011.

[62] Besse，S.；Dhingra，D.；Nettles，J.；Klima，R.；Garrick - Bethell，I.；Clark，R. N.；Combe，J. P.；Head，J. W.，III；Taylor，L. A.；Pieters，C. M.；et al. M^3 spectral analysis of lunar swirls and the link between optical maturation and surface hydroxyl formation at magnetic anomalies. J. Geophys. Res. 2011，116，E9.

[63] Kramer，G. Y.；Combe，J. P.；Harnett，E. M.；Hawke，B. R.；Noble，S. K.；Blewett，D. T.；McCord，T. B.；Giguere，T. A. Characterization of lunar swirls at Mare Ingenii：A model for space weathering at magnetic anomalies. J. Geophys. Res. Planet 2011，16，E04008.

[64] Hapke，B. Theory of Reflectance and Emittance Spectroscopy；Cambridge University Press：New York，NY，USA，2005.

[65] Mustard，J. F.；Pieters，C. M. Photometric phase functions of common geologic minerals and applications to quantitative analysis of mineral mixture reflectance spectra. J. Geophys. Res. Solid Earth 1989，94，13619 - 13634.

Descent trajectory reconstruction and landing site positioning of Chang'E - 4 on the lunar farside

Jianjun Liu[1,2], Xin Ren[1], Wei Yan[1], Chunlai Li[1,2], He Zhang[3], Yang Jia[3], Xingguo Zeng[1], Wangli Chen[1], Xingye Gao[1], Dawei Liu[1], Xu Tan[1], Xiaoxia Zhang[1], Tao Ni[1,2], Hongbo Zhang[1], Wei Zuo[1], Yan Su[1] & Weibin Wen[1]

1. Key Laboratory of Lunar and Deep Space Exploration, National Astronomical Observatories, Chinese Academy of Sciences, Beijing 100101, China

2. School of Astronomy and Space Science, University of Chinese Academy of Sciences, Beijing 100049, China

3. China Academy of Space Technology, Beijing 100094, China. Correspondence and requests for materials should be addressed to C. L

Abstract: Chang'E - 4 (CE - 4) was the first mission to accomplish the goal of a successful soft landing on the lunar farside. The landing trajectory and the location of the landing site can be effectively reconstructed and determined using series of images obtained during descent when there were no Earth—based radio tracking and the telemetry data. Here we reconstructed the powered descent trajectory of CE - 4 using photogrammetrically processed images of the CE - 4 landing camera, navigation camera, and terrain data of Chang'E - 2. We confirmed that the precise location of the landing site is 177.5991°E, 45.4446°S with an elevation of —5935m. The landing location was accurately identified with lunar imagery and terrain data with spatial resolutions of 7m/p, 5m/p, 1m/p, 10cm/p and 5cm/p. These results will provide geodetic data for the study of lunar control points, high - precision lunar mapping, and subsequent lunar exploration, such as by the Yutu - 2 rover.

The Chang'E - 4 (CE - 4) spacecraft successfully landed on the lunar farside on January 3, 2019. The planned durations of the scientific investigations for the lander and the Yutu - 2 rover are 6 and 3 months, respectively. After launch on December 8, 2018, CE - 4 conducted several stages, including transfer from the Earth to the Moon, orbiting around the Moon and a powered descent. The powered descent stage was the most important stage for soft landing and was implemented under complete autonomous control[1]. According to the orbit design, a 7500 Newton variable propulsion engine was used for deceleration during the CE - 4 powered descent stage, which involved phases of main braking, attitude adjusting, vertical descent, hover and hazard - avoidance, slow descent[1-5]. To better understand the autonomous control result and to service orbit control strategy analysis and detection mission planning for subsequent missions, it is of great significance to accurately reconstruct the landing trajectory and determine the location of the landing site after a safe landing.

Radio measurements[6-12], real - time telemetry data[2,13], and image - based matching[3,14-19] are commonly used for powered descent trajectory reconstruction and landing site positioning on a planetary surface. Because there was no radio measurement equipment between the CE - 4 and the relay satellite Queqiao, it was unable to perform direct or indirect radio measurements by the ground tracking network on the lunar farside. In addition, the telemetry data received by Queqiao (including the detector altitude, acceleration, and attitude) were not released and could not to be used. As a result, it is difficult to accurately reconstruct the spacecraft's landing trajectory and to confirm the precise landing site location. However, these problems can be effectively solved through localization technology based on landing images, which is not affected by factors such as the lunar gravity field and the dynamical model.

In this study, high - frequency landing sequence images transmitted by Queqiao after safe landing were used for descent trajectory reconstruction and landing site positioning of CE - 4, which completely recorded the entire process of the powered descent. We reconstructed the descent trajectory, showing even barely perceivable maneuvers of the spacecraft during the landing approach and that the landing site can also be precisely localized. Here, the digital orthophoto map (DOM) and digital elevation model (DEM) of Chang' E - 2 (CE2TMAP2015, see the data descriptions of the Methods section[20,21]) were used as the geographical reference data. The CE - 4 landing camera (LCAM, see the Methods section for details) sequence images were applied to reconstruct the CE - 4 powered descent trajectory by the photogrammetry bundle adjustment technology[22,23] (see the Methods for a detailed technical flow). Using combined binocular stereoscopic images obtained by the navigation camera (NCAM, see the Methods), which consists of two cameras, we confirmed that the precise location of the landing site is 177.5991°E, 45.4446°S with an elevation of −5935m.

Results

Descent trajectory from LCAM images. The sequence images obtained by the LCAM in vertical attitude after the attitude adjusting phase of CE - 4 were used for the reconstruction of the landing trajectory in this study.

According to the reconstructed trajectory, CE - 4 adopted a vertical descent approach to gradually approach the landing zone after the attitude adjusting phase with an altitude of 5635m and a velocity of ~85m/s. During the period of descending to an altitude of 4130m above the lunar surface, the trajectory moved 77m to the north and crossed a crater with a diameter of approximately 200m. CE - 4 reached over area A, as indicated in Fig. 1, with a velocity of ~78m/s. Over area A, CE - 4 decreased from 4130m from the lunar surface to 1495m, the velocity decreased to ~ 51m/s, and the trajectory was adjusted from the northeast; there are some craters with diameters of ~70—100m to the northwest, where is more flat (Fig. 1d, e). Then, the trajectory moved 244m to the northwest from the altitude of 1495m to 99m and the velocity decreased to 0m/s. CE - 4 reached over area B in Fig. 1 and entered the hover and hazard - avoidance phase.

At an altitude of 99m above the lunar surface (B in Fig. 1), CE - 4 hovered for~13s. Then, the trajectory moved 12m to the southwest and crossed the crater with a diameter of 25m. The altitude above the lunar surface decreased to 30m, and the velocity changed to ~1. 5m/s. Finally, the CE - 4 trajectory slowly descended vertically and, the spacecraft landed safely.

Landing site localization based on multisource image data. We generated LCAM topographic maps (including the DOM and DEM data) with different spatial resolutions by photogrammetry bundle adjustment technology[22,23]. These maps used a total of 180 images from the LCAM sequence images below an altitude of 5635m (one image per second), twelve horizontal and vertical control points (GCPs), and one vertical GCP selected on the CE2TMap2015 (Fig. 2a). Using the stereoscopic images obtained by the NCAM from the Yutu - 2 rover on the top of the lander and a total of 14 GCPs selected from the LCAM DOM and DEM (Fig. 2b), the DOM with a resolution of 2cm/p was produced by the close - range photogrammetry technology. In this way, the NCAM DOM was precisely registered to the LCAM DOM. According to the position relationship between the NCAM and the origin of the lander body coordinate system, the horizontal position and elevation of the lander in the LCAM terrain data were measured, which were determined as the final position of the CE - 4 landing site (see the Methods). Detailed information on the location of the landing site is shown in Tab. 1 and Fig. 3.

We also marked the location of the landing site on the LCAM images and CE - 2 DOM subdivision with a resolution of 7m/p (Tab. 1 and Fig. 4).

The images of the LCAM and Yutu - 2 panoramic camera (PCAM) (Fig. 5) show that CE - 4 landed on a gentle slope of a degraded crater, only 8. 35m from the rim of a 25m crater to the north and was surrounded by 5 craters with diameters of 10. 21~25. 00m and depths of 1. 21 ~ 3. 03m. CE - 4 landing site can be also located in the Lunar Reconnaissance Orbiter (LRO) narrow angle camera (NAC) images, whose relative position to the surrounding terrain is consistent with our results in Fig. 5.

Discussion

Similar to Chang'E - 3 (CE - 3), three types of sensors were employed by CE - 4, including a laser ranging sensor, a microwave ranging and velocimeter sensor and an optical imaging sensor and a laser imaging sensor[1] . The powered descent process was

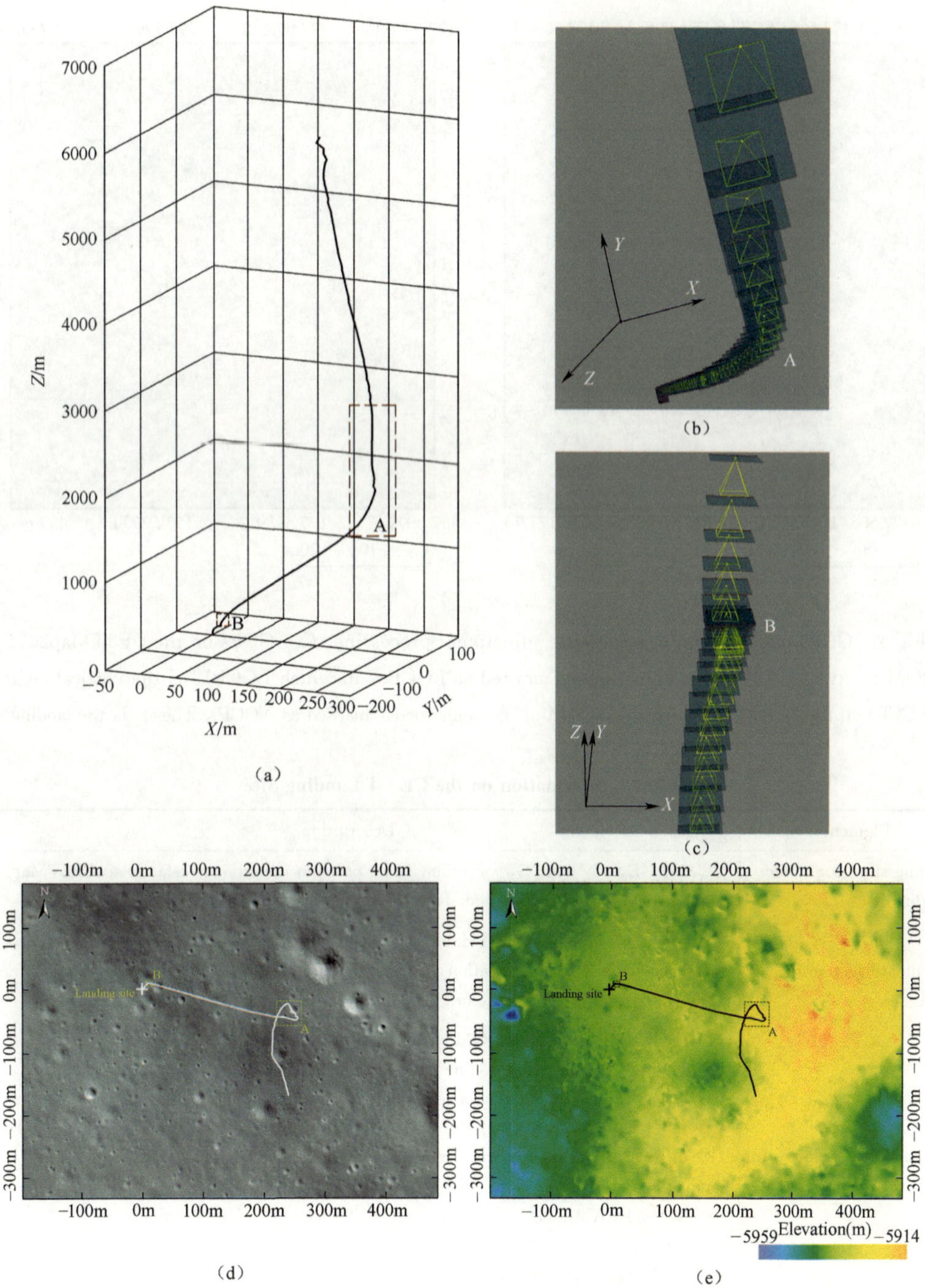

Fig. 1　CE－4 powered descent trajectory. The ＋ is the identified landing location. (a) CE－4 descent trajectory from the altitude of 6000m to the lunar surface; (b), (c) are zoomed images of the descent trajectories of A and B, respectively. The green box represents the position of the LCAM focal plane, and the yellow cone represents the field of view of the LCAM. (d) The solid white line is the projection of the descent trajectory on the LCAM DOM. (e) The solid black line is the projection of the descent trajectory on the LCAM DEM. The coordinate system in this figure is the tangent plane coordinate system of the landing site. This system uses the location of the landing site as the coordinate origin, with its X and Y axes pointing towards the geographical east and the north of the landing site, respectively. The Z－axis, X－axis, and Y－axis form a right－hand system, with the Z－axis pointing to the zenith direction of the landing site

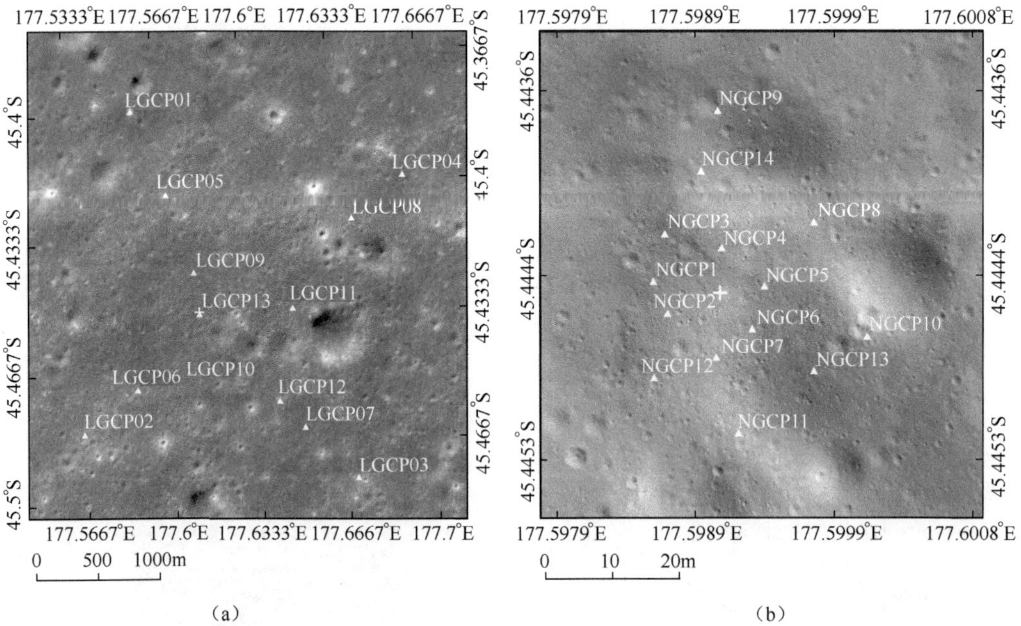

Fig. 2　GCPs map for photogrammetric adjustment processing. (a) GCPs on the CE2TMap2015 DOM for the LCAM image adjustment, marked as LGCPs, in which LGCP13 is the vertical GCP; (b) GCPs on the LCAM DOM for the NCAM image adjustment, marked as NGCPs. The + is the landing site

Tab. 1　Information on the CE－4 Landing Site

Element	Description
Landing site geographic coordinates	177.5991°E, 45.4446°S, −5935m. (The position uncertainty relative to CE2TMap2015 is meter level; See the Methods for details)
Coordinate system	The lunar coordinate is based on the mean Earth/polar axis coordinate system. The reference surface of the elevation is the surface of the Moon's spheroid with a radius of 1737.4km, and the reference origin is the mass center of the Moon.
Landing site location in the images[a]	(1) In the CE2Tmap2015 DOM with a resolution of 7m, landing site pixel coordinate (line 13602, column 21647) (2) In the LCAM image with a resolution of 5m, landing site pixel coordinate (line 412, column 472) (3) In the LCAM image with a resolution of 1m, landing site pixel coordinate (line 403, column 288) (4) In the LCAM image with a resolution of 10cm, landing site pixel coordinate (line 592, column 389) (5) In the LCAM image with a resolution of 5cm, landing site pixel coordinate (line 500, column 426)

[a] The subdivision number for the CE2TMap2015 DOM and DEM with a resolution of 7m is K136 in this study (Fig. 4a, b). The corresponding file names for LCAM 5m, 1m, 10cm, and 5cm resolution images are as follows:
CE4_LCAM_20190103022303_1740_08.tif (Fig. 5c),
CE4_LCAM_20190103022414_2444_08.tif (Fig. 5d),
CE4_LCAM_20190103022500_2906_08.tif (Fig. 5e),
and CE4_LCAM_20190103022525_3160_08.tif (Fig. 5f)

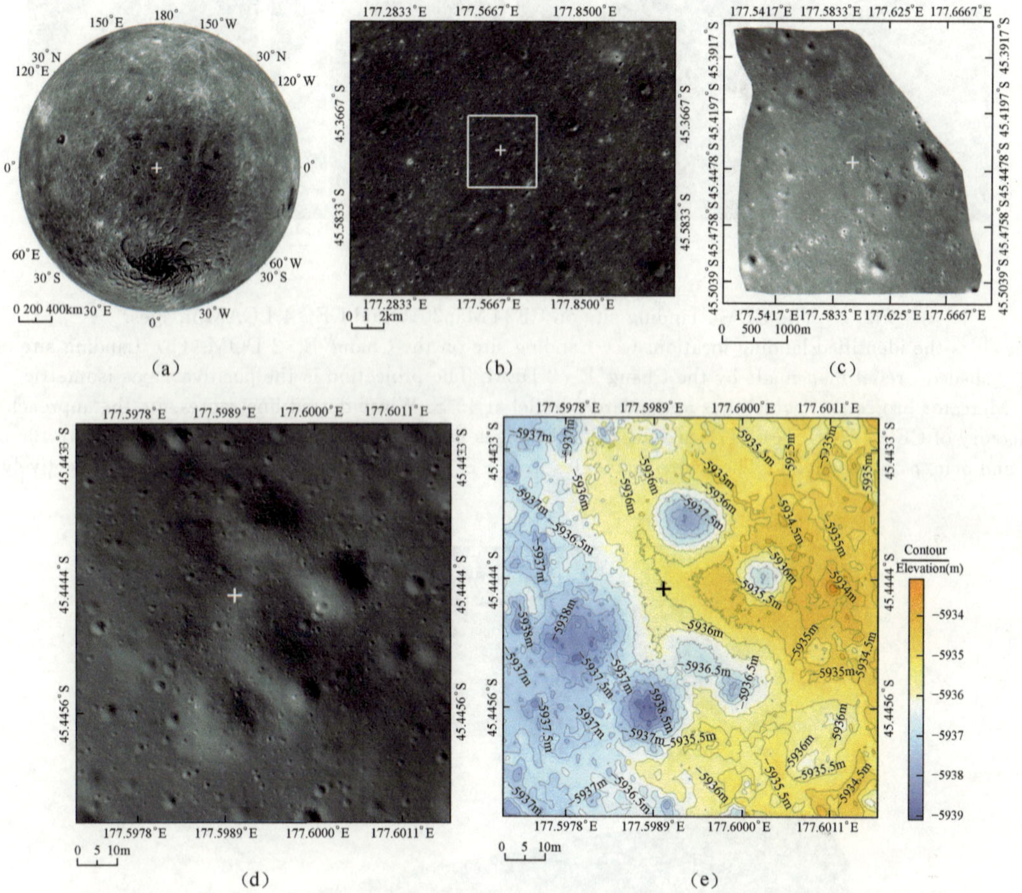

Fig. 3　Location of the CE - 4 landing site. The + is the identified landing location. (a), (b) are the CE - 2 DOMs, using the CE - 4 landing site as the projection center of the azimuth projection. (c) is the LCAM DOM with a resolution of 5m/p. (d) is the LCAM DOM, which is generated by the LCAM images with resolutions from 21. 4cm/p to 0. 2cm/p and is uniformly resampled at 5cm/p during mapping. (e) is a shaded - relief map with contour line produced by the LCAM DEM, which area and resolution are the same as in (d)

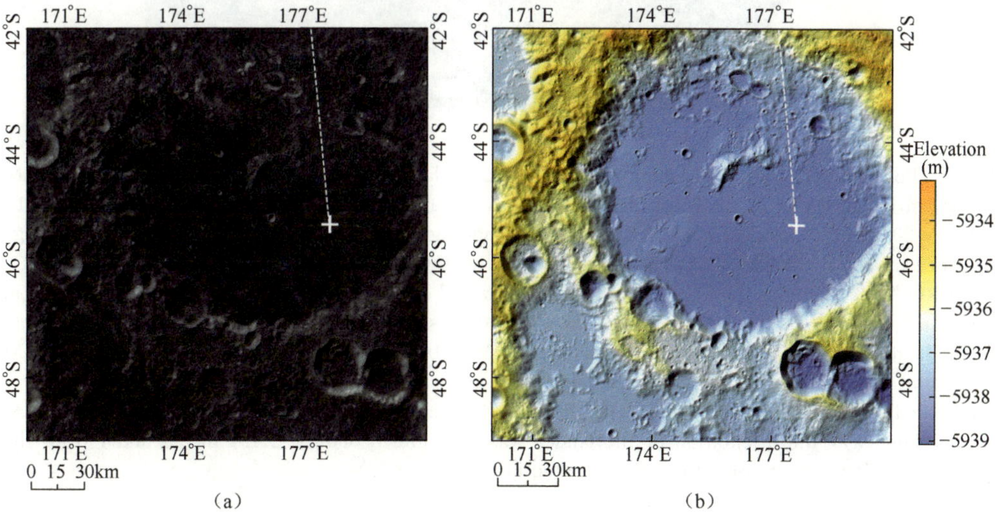

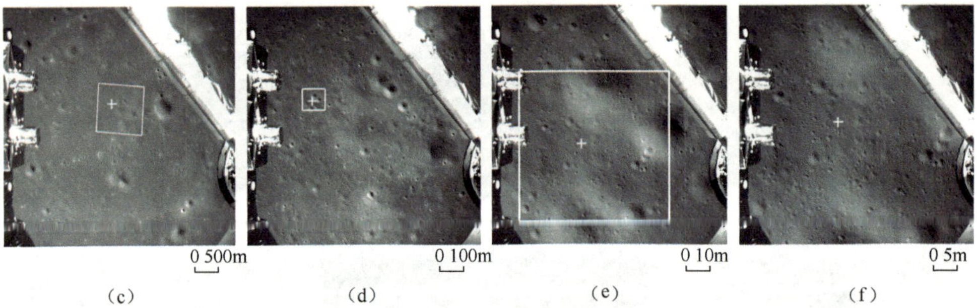

Fig. 4　The CE－4 landing site on CE2TMap2015 and CE－4 LCAM images.
The ＋ is the identified landing location. (a) Landing site on the Chang'E－2 DOM. (b) Landing site on a shaded－relief map made by the Chang'E－2 DEM. The projection is the positive－axis isometric Mercator projection, which has a standard parallel at 45°S. White dotted line represents the approach trajectory of CE－4; (c) (e), (f) are the LCAM images with spatial resolutions of 5m/p, 1m/p, 10cm/p, and 5cm/p, respectively. The white rectangles in (c－e) represent the borders of (d－f), respectively

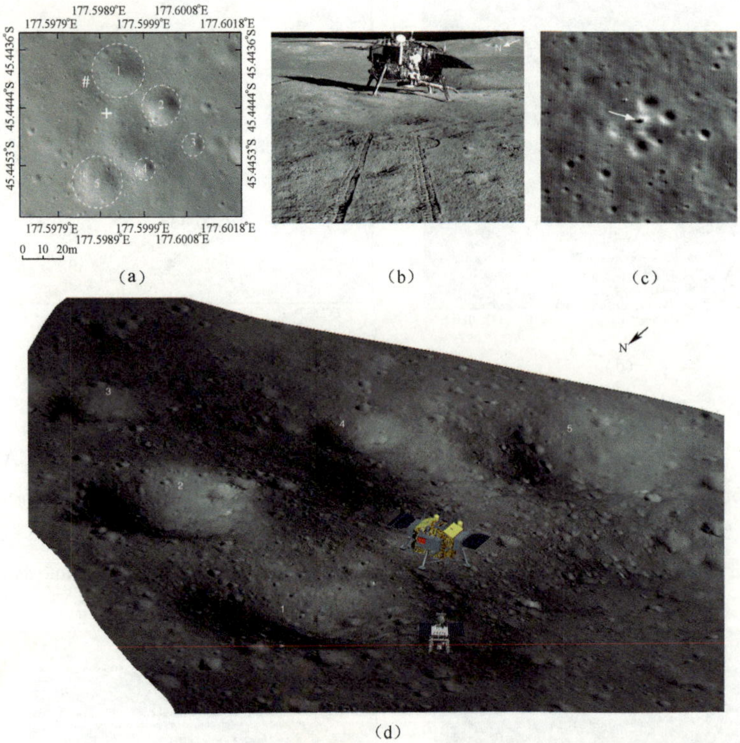

Fig. 5　Position of the CE－4 landing site. a The CE－4 landing site on the 5 cm/p resolution LCAM DOM, in which the ＋ is the identified landing location and the ＃ is the location of Yutu－2 on 12 January 2019. Five prominent craters around the landing site are marked. (b) CE－4 lander image obtained from the northwest of the landing site by the Yutu－2 PCAM at the location labeled in (a). (c) LRO NAC image of the CE－4 landing site collected on 1 February 2019 (M1303640934LR), in which the arrow indicates the location of CE－4 landing site. (d) Three－dimensional landscape map of the landing site generated by the DEM and DOM data through reconstruction of CE－4 LCAM sequential images acquired within 100m above the lunar surface, in which the location of CE－4 lander and Yutu－2 is the same as labeled in (a). While landing position was determined from LCAM orbital data (a), the orientation of the lander was determined using PCAM image taken from the ground (b). Prominent craters are also marked in (d), as in (a)

realized by autonomous control technology with adaptive powered explicit guidance, multibeam fault - tolerant navigation, and partition quaternion control. However, the trace control strategies for CE - 3 and CE - 4 are different, and this difference is also reflected in the trajectory recovered in this article.

The CE - 3 landing site is located on the northern Mare Imbrium of the lunar nearside. The length of the CE - 3 nadir flight trace is ~450km from south to north. The terrain along the nadir flight trace shows a flat, small variation, gradually decreasing from south to north with a maximum variation of ~800m (Fig. 6). In the main braking phase, ranging measurements were introduced to correct the bias and drift of the inertial measurement unit (IMU) in time and ensure the stability of the powered descent trajectory. At an altitude of 3~2.4km, the CE - 3 lander performed quick attitude adjustments to a vertical attitude.

In contrast, the landing site of CE - 4 is located southeast of the Von Kármán crater on the lunar farside. The length of the CE - 4 nadir flight trace is also ~450km from north to south. However, the terrain of the nadir flight trace shows a large variation with a maximum variation of 6000m (Fig. 6). Because the pointing position of the ranging beam is different from the final landing site and the terrain variation is obvious, if ranging measurements were introduced for a correction in advance, there would be a safety risk. Under its trace control strategy, CE - 4 performed rapid attitude adjustments at an altitude of 8~6km, and ranging measurements were introduced to correct the IMU bias and drift. This strategy ensured a safe landing of CE - 4. The recovered trajectory showed that the lander attitude was adjusted to enable thrusting for near—vertical descent at an altitude of 5635m. Thus, the powered descent trajectory of CE - 4 is obviously different from that of CE3 (Fig. 7).

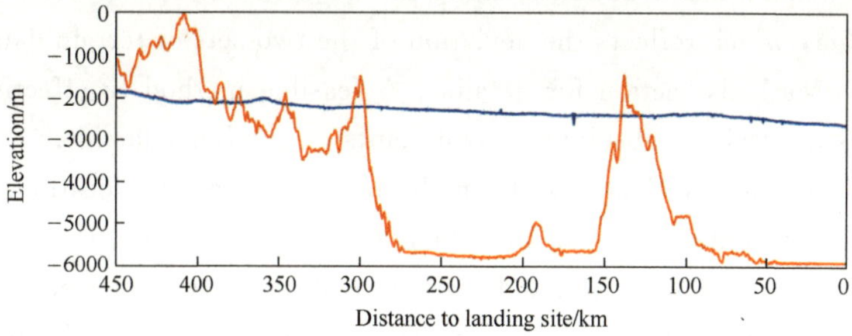

Fig. 6 Elevation variations of the lunar terrain along the CE - 3 and CE - 4 nadir flight traces. The blue and orange lines represent the elevation variation for the CE - 3 and CE - 4 nadir flight trace, respectively. Elevation is given with respect to the Moon's spheroid with a radius of 1737.4km

Based on the sequence images of the LCAM, we reconstructed the powered descent trajectory of the CE - 4 lander through photogrammetry. Combining the generated cm - resolution terrain data near the landing site with the NCAM stereoscopic images, the geographical location of the landing site was calculated in this study. The horizontal and

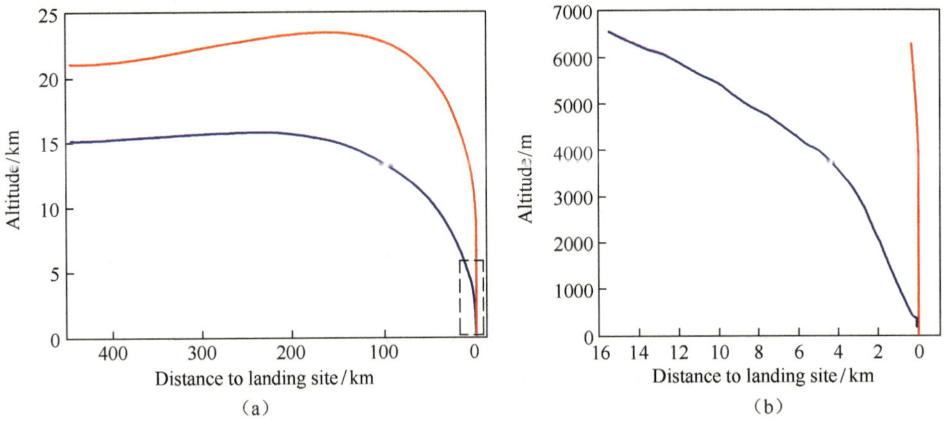

Fig. 7 Comparison of powered descent trajectory between the CE－3 and CE－4. The blue and orange lines represent trajectories of the CE－3 and CE－4, respectively. (a) The designed trajectories of the entire powered descent, in which the marked rectangle corresponds to designed trajectories below the altitude of ~6km; (b) Magnified portion of the reconstructed trajectory below the altitude of ~6km. Altitude is given with respect to final landing site level

the vertical RMS errors of the landing site are 0.7m (1δ) and 1.0m (1δ) respectively (see the Methods section for details).

Compared with the positioning results of the landing site based on LRO terrain data (177.5885°E, 45.4561°S, －5927m)[24], our results show a 226m deviation along the latitude direction and a 348m deviation along the longitude direction. The total positional deviation is 415m, which reflects the deviation of the two sets of terrain data on the lunar farside (see the Methods section for details). A feasible method to effectively eliminate this deviation is to establish absolute control points on the lunar farside.

As a permanent artificial landmark on the lunar farside, the location of the CE－4 lander was precisely confirmed using CE－2 and CE－4 images and can serve as a potential control point on the farside. The result will provide a worthy geodetic data point for studies on lunar control points, high－precision lunar mapping, and subsequent lunar exploration, such as for the Yutu－2 rover.

Methods

Instruments and dataset descriptions. The dataset used in this study includes landing camera (LCAM) images, navigation camera (NCAM) images and the Chang'E－2 (CE－2) global lunar terrain (CE2TMap2015)[20,21].

The LCAM is one of the scientific payloads installed on the bottom of the Chang'E－4

(CE-4) lander. According to the analysis of the characteristics of the LCAM sequence images, the CE-4 lander gradually approached the landing area using a vertical descent mode after attitude adjusting phase. The overlap of the LCAM sequence images exceeds 94%. Over 40% of the images cover the landing site. The rotation and translation between adjacent images are slow. Therefore, 180 images in 1s intervals during this phase were selected for the reconstruction of the powered descent trajectory and landing area topography. The main performance parameters of the LCAM are shown in Tab. 2.

Tab. 2 The performance parameters of the LCAM

No.	Name	Performance parameters
1.	Wavelength range/nm	419～777
2.	Field of view/°	45.3° × 45.3°
3.	Focal Length/mm	8.5
4.	Effective pixel numbers	1024 × 1024
5.	Pixel size on Focal plane/μm	6.7
6.	Frame rate/fps	10
7.	Quantized value/bit	8
8.	Data compression ratio	8∶1

The NCAM is an engineering payload onboard the CE-4 rover and is installed on the top of the rover mast. It consists of two optical systems. The NCAM obtained images surrounding the landing site at the top of the lander before the rover separated from the lander. Based on these images, topographic data of the landing site with cm-level accuracy can be reconstructed by close-range photogrammetry. Debris scattered near the landing site can be accurately identified. Because the relationship between the location of the NCAM and the CE-4 lander is known, the precise position of the landing site can be located by combining the images of the LCAM and NCAM[25]. The main performance parameters of the NCAM are shown in Tab. 3.

Tab. 3 The performance parameters of the NCAM

No.	Name	Performance parameters
1.	Focal Length/mm	Left camera: 17.69
		Right camera: 17.98
2.	Effective pixel numbers	1024 × 1024
3.	Pixel size on Focal plane/μm	15
4.	Baseline length/mm	269.69
5.	Quantized value/bit	8

The georeferenced data used in this study are CE2TMap2015, which include the DOM and DEM with resolutions of 7m/p and 20m/p, respectively. These data are produced using CE-2 stereo images with a resolution of 7m. The average horizontal and elevation relative position errors are 5m and 2m, respectively. Compared to the absolute positions of 5 laser reflectors located on the Moon, the horizontal positional deviation of CE2TMap2015 is 21～97m, and the elevation deviation is 2～19m[20,21].

LRO terrain data is used for CE－4 landing site position comparison. Orbit overlap analyses show that, LRO spacecraft ephemeris can be improved to less than 10m horizontally and 1m vertically by combining radiometric tracking data and LOLA data with the GRAIL gravity model[25-27]. As a result, the LRO terrain data uncertainty is about 20 meters[28] and the positional deviation from the 5 laser reflectors is <5.2m[29]. A pair of LRO NAC observations of the CE－4 landing site were collected on 1 February 2019 (M1303619844LR, M1303640934LR), and were used to create a digital terrain model (DTM) on the LOLA－plus－GRAIL coordinate framework. The coordinates of CE－4 lander were derived (177.5885°E, 45.4561°S, －5927m)[24] based on the DTM.

It can be seen that both CE2TMap2015 and LRO terrain data show good internal consistency. However, the positioning deviation of the two sets of terrain data on the lunar farside will both increase due to several reasons, such as orbital measurement error, the lumpy gravity field of the Moon on the lunar farside, small uncertainties in camera model (distortion, FL), etc. Compared to the LRO terrain data, the average global positional deviation of CE2TMap2015 is 354m with a standard deviation of 228m (1δ)[20,21,30], which mainly results from large deviation of two sets of terrain data on the farside. The 415m discrepancy between our coordinates and the LRO terrain data coordinates for the CE－4 lander is within the positional deviation range.

Technical flowchart. A total of 180 images (one image/per second) from the LCAM sequence images and 18 pairs of NCAM stereo images captured at the top of the CE4 lander were selected for this work. Fig. 8 shows the technical flowchart of the trajectory recovery and landing site position determination.

CE－4 lander descent trajectory reconstruction. First, evenly distributed points (or tie points) were automatically extracted using scale－invariant feature transform (SIFT) feature matching[31]. Then, least squares matching was applied to these tie points to achieve more precise matching results, and a random sample consensus (RANSAC)－based optimization algorithm was used to automatically eliminate mismatches[32,33]. A total of 30992 tie points, evenly distributed throughout the overlaps of the LCAM images, were extracted. The relative positional relationship of these images can be precisely restored according to these tie points.

Facilitated by one LRO NAC image (M178833263LC), 13 evenly distributed GCPs close to the landing site (including 12 horizontal and vertical GCPs and 1 vertical GCP) were extracted from the CE2TMap2015 map and LCAM images. The centers of some easily recognized small craters were mainly selected to ensure the selection accuracy of the GCPs.

According to photogrammetric bundle adjustment theory[22,23], the relationship

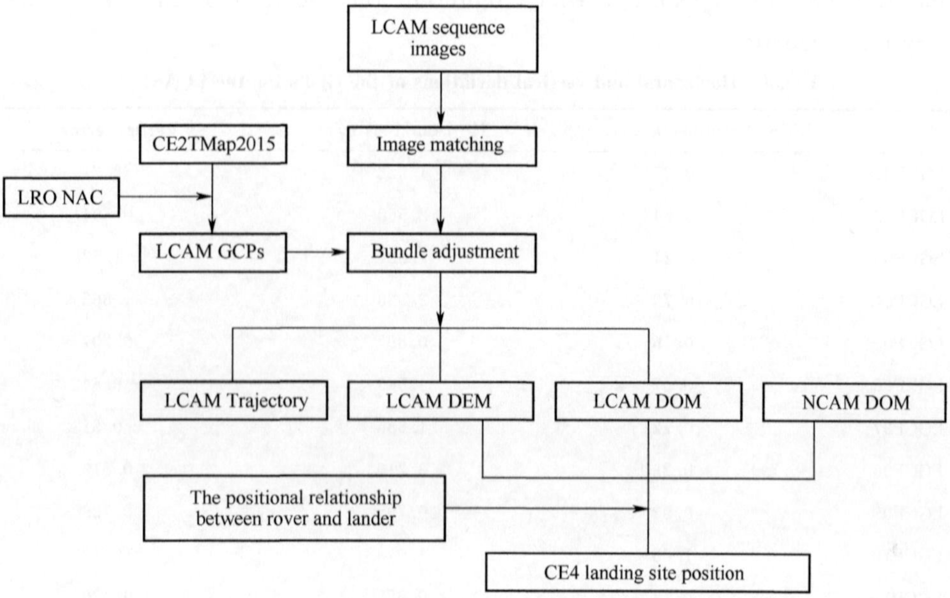

Fig. 8 Data processing flowchart

between the lunar surface points, the corresponding image points and the camera projection center can be expressed by the classical collinear equation, which is as follows.

$$x - x_0 + \Delta x = -f \frac{r_{11}(X - X_s) + r_{21}(Y - Y_s) + r_{31}(Z - Z_s)}{r_{13}(X - X_s) + r_{23}(Y - Y_s) + r_{33}(Z - Z_s)}$$
$$y - y_0 + \Delta y = -f \frac{r_{12}(X - X_s) + r_{22}(Y - Y_s) + r_{32}(Z - Z_s)}{r_{13}(X - X_s) + r_{23}(Y - Y_s) + r_{33}(Z - Z_s)} \quad (1)$$

where (x, y) are the image coordinates of the feature points, (x_0, y_0) are the image coordinates of the image principal point, $(\Delta x, \Delta y)$ are self - calibration correction terms of the camera, including lens distortion correction, (X, Y, Z) are the lunar space coordinates of the feature points, (X_s, Y_s, Z_s) are the lunar space coordinates of the image principal point, f is the focal length of the camera, and $r11-r33$ are the elements of the rotation matrix R $(\varphi, \omega, \kappa)$, which is formed by the Euler attitude angle $(\varphi, \omega, \kappa)$ of the LCAM and used to translate feature points from the lunar space coordinates to the image coordinates. $(X_s, Y_s, Z_s, \varphi, \omega, \kappa)$ forms the exterior orientation (EO) parameters of the LCAM. Using Taylor's formula, the linearized error equations of each pixel can be derived by formula (1). The EO parameters of each image and the lunar space coordinates of each feature point can be solved by the least squares method. We implemented the photogrammetric bundle adjustment theory by ContextCapture Master software in this study. Therefore, we can use the location of each image to reconstruct the descent trajectory of the CE4 lander.

After the adjustment, the RMS error of the reprojection error for all the tie points is 0.5 pixels. The horizontal and vertical RMS errors of the GCPs are, respectively, 0.715m

(1δ) and 1.040m (1δ) (Tab. 4), which represent the accuracy of the CE - 4 descent trajectory reconstruction.

Tab. 4 Horizontal and vertical deviations of the GCPs for the LCAM

Name	RMS of reprojection error/pixel	Horizontal error/m	Vertical error/m
LGCP01	0.35	1.522	−2.319
LGCP02	0.74	2.560	0.504
LGCP03	0.21	1.327	−1.528
LGCP04	0.72	2.933	1.365
LGCP05	0.55	0.897	0.807
LGCP06	0.67	1.565	−0.642
LGCP07	0.72	1.685	−0.513
LGCP08	0.23	0.710	0.610
LGCP09	0.93	0.795	0.799
LGCP10	1.08	1.048	−0.788
LGCP11	0.75	0.855	0.420
LGCP12	0.84	0.865	−0.483
LGCP13	0.11	/	0.065
Global RMS	0.61	0.715	1.040

Precise localization of the landing site. First, the position and attitude of the LCAM images were obtained by the landing trajectory reconstruction. Then, a pixel - level point cloud was generated by the image dense matching technique[34,35], and a DEM of the LCAM with a 5cm spatial resolution within 80m around the landing site was interpolated from the triangulated irregular network (TIN). Finally, the same resolution and coverage DOM of the LCAM was produced using the image reprojection technique with the DEM and EO parameters.

Similarly, we produced the DEM and DOM of the NCAM with a 2cm spatial resolution within 30m around the landing site using the NCAM images. The only difference for this process is that the GCPs were selected from the LCAM terrain data. Considering that the spatial resolution of the NCAM image is close to that of the descent camera terrain data, some small stones scattered around the landing site which can be easily identified were selected as GCPs to ensure the accuracy of the GCP selection (Fig. 9). A total of 14 GCPs were selected.

A total of 21, 091 tie points were extracted using the image matching method. The RMS error of the reprojection error for all tie points is better than 0.2 pixels after the adjustment. The GCP horizontal RMS error is 0.017m (1δ), and the vertical RMS error is 0.031m (1δ) (Tab. 5). According to the positional relationship between the NCAM and the origin of the lander's body coordinate system (LBCS), the horizontal position of the vertical projection of the LBCS origin in the LCAM terrain data is 177.5991° E,

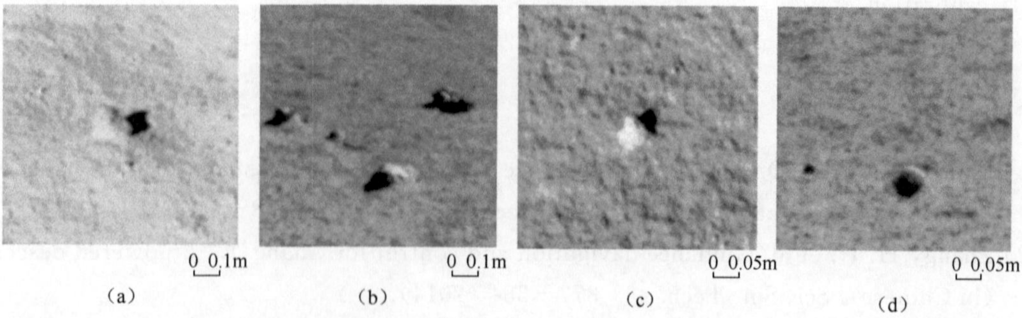

Fig. 9　Small rocks used as GCPs. (a—d) are stones in the NCAM image that can also be easily identified in the LCAM DOM

45.4446°S, and the elevation value is −5935m.

Tab. 5　Horizontal and vertical deviations of the GCPs for the NCAM

Name	RMS of the reprojection error/pixel	Horizontal error/m	Vertical error/m
NGCP01	0.077	0.053	−0.038
NGCP02	0.006	0.034	0.038
NGCP03	0.007	0.014	−0.037
NGCP04	0.010	0.032	0.048
NGCP05	0.005	0.021	0.017
NGCP06	0.007	0.028	−0.065
NGCP07	0.003	0.027	−0.004
NGCP08	0.007	0.040	−0.009
NGCP09	0.013	0.075	0.013
NGCP10	0.005	0.049	−0.027
NGCP11	0.005	0.031	0.014
NGCP12	0.005	0.020	0.001
NGCP13	0.005	0.015	0.040
NGCP14	0.017	0.014	0.015
Global RMS	0.012	0.017	0.031

Acknowledgements

This research was funded by the Chang' E-4 mission of Chinese Lunar Exploration Program (CLEP) and the National Natural Science Foundation of China (Grant No. 11941002). We thank the team of the CE-4 Project for their successful work, especially the GRAS of CLEP for their valuable and efficient assistance with providing the data and data calibration.

References

[1] Li, S., Jiang, X. Q. & Tao, T. Guidance Summary and Assessment of the Chang'e-3 Powered Descent and Landing. J. Spacecraft Rockets 53, 258—277 (2016).

[2] Zhang, H. H. et al. Guidance navigation and control for Chang'E-3 powered descent (in Chinese). Sci. Sin. Tech. 44, 377—384 (2014).

[3] Jia, Y. et al. Chang' E-3 system pinpoint landing localization based on descent image sequence (in Chinese). Chin. Sci. Bull. (Chin. Ver.). 59, 1838—1843 (2014).

[4] Klumpp, A. R. Apollo Lunar Descent Guidance, R-695. (Charles Stark Draper Laboratory of Massachusetts Institute of Technology, Cambridge, 1971).

[5] Li, F. et al. Influence analysis of terrain of the farside of the Moon on softlanding. J. Deep Space Exploration 4, 143—149 (2017).

[6] Cao, J. F. et al. Determination of Impact Site of CE-1 Using USB and VLBI Data. J. Astronautics 31, 1724-1729 (2010).

[7] Edwards, C. D. et al. Telecommunications relay support of the Mars Phoenix lander mission. IEEE Aerospace Conference. pp. 1-13 (2010).

[8] Folkner, W. M. et al. Interior structure and seasonal mass redistribution of Mars from radio tracking of Mars Pathfinder. Science 278, 1749-1751 (1997).

[9] Li, J. L. et al. Determination of the controlled landing trajectory of Chang'E-1 satellite and the coordinate analysis of the landing point on the Moon. Chin. Sci. Bull. 55, 752-757 (2010).

[10] Li, P. J. et al. Kinematic statistical method using in the orbit determination of CE-1. Prog. Astron. 28, 290-300 (2010).

[11] Li, R. et al. Initial results of rover localization and topographic mapping for the 2003 Mars Exploration Rover mission. Photogrammetric Eng. Remote Sens. 71, 1129-1142 (2005).

[12] Yoder, C. F. & Standish, E. M. Martian precession and rotation from Viking lander range data. J. Geophys. Res. 102 (E2), 4065-4080 (1997).

[13] Lee, A. Y. et al. Preliminary design of the guidance, navigation, and control system of the altair lunar lander. Conference: AIAA Guidance, Navigation, and Control Conference, https://arc. aiaa. org/doi/10. 2514/6. 2010-7717 (2010).

[14] Ansar, A. & Cheng, Y. An analysis of spacecraft localization from descent image data for pinpoint landing on Mars and other cratered bodies. Photogramm. Eng. Remote Sens. 71, 1197-1204 (2005).

[15] Liu, B. et al. Descending and landing trajectory recovery of Chang'e-3 lander using descent images. J. Remote Sens. 18, 981-987 (2014a).

[16] Liu, J. J. et al. Positioning of the CE-1' s impact site based on CCD image data

photographed during the controlled impact on the Moon. Sci. China Earth Sci. 41, 1714 - 1720 (2011).

[17] Liu, J. J. et al. Reconstructing the landing trajectory of the CE - 3 lunar probe by using images from the landing camera. Res. Astron. Astrophys. 14, 1530 - 1542 (2014).

[18] Matthies, L. et al. Computer vision on Mars. Int J. Comput Vis. 75, 67 - 92 (2007).

[19] Xiong, Y., Olson, C. F. & Matthies, H. Computing depth maps from descent images. Mach. Vis. Appl. 16, 139 - 147 (2005).

[20] Ren, X., et al. A global adjustment method for photogrammetric processing of Chang'E - 2 stereo images. IEEE Transactions on Geoscience and Remote Sensing. https: //doi. org/10. 1109/TGRS. 2019. 2908813 (2019).

[21] Li, C. L. et al. Lunar global high precision terrain reconstruction based on Chang'e - 2 stereo images. Geomat. Inf. Sci. Wuhan. Univ. 43, 486 - 495 (2018).

[22] Wang, Z. Z. Principles of Photogrammetry [with Remote Sensing]. Press of Wuhan Technical University of Surveying and Mapping. (Publishing House of Surveying and Mapping, Beijing, 1990).

[23] Brown, D. C. A solution to the general problem of multiple station analytical stereo - angulation. Air Force Missile Test Center Report No. 58 - 8, Patrick AFB, Florida. (1958).

[24] Topographic Map of the Chang'e 4 Site. http: //www. lroc. asu. edu/posts/1100. April 30 (2019).

[25] Lemoine, F. G. et al. GRGM900C: A degree 900 lunar gravity model from GRAIL primary and extended mission data. Geophys. Res. Lett. 41, 3382 - 3389 (2014).

[26] Mazarico, E. et al. Orbit determination of the lunar reconnaissance orbiter. J. Geod. 86, 193 - 207 (2012).

[27] Mazarico, E., et al. Improved orbit determination of lunar orbiters with lunar and gravity fields obtained by the GRAIL mission. In: Proceedings of the 44th Lunar and Planetary Science Conference. The Wood - lands, TX, Abstract # 2414 (2013).

[28] Speyerer, E. J. et al. Pre - flight and on - orbit geometric calibration of the Lunar Reconnaissance Orbiter Camera. Space Sci. Rev. 200, 357 - 392 (2016).

[29] Wagner, R. V. et al. Coordinates of anthropogenic features on the Moon. Icarus 283, 92 - 103 (2017).

[30] Yan, W. et al. Orbit optimization of Chang' E - 2 by global adjustment using images of the moon. Adv. Space Res. 56, 2389 - 2401 (2015).

[31] Lowe, D. G. Distinctive image features from scale - invariant keypoints. Int. J. computer Vis. 60, 91 - 110 (2004).

[32] Fischler, M. A. & Bolles, R. C. Random sample consensus: a paradigm for model fitting with applications to image analysis and automated cartography. Commun. ACM 24, 381 - 395 (1981).

[33] Zhang, L. Automatic digital surface model (DSM) generation from linear array images. Mitteilungen - Institut fur Geodasie und Photogrammetrie an der Eidgenossischen Technischen Hochschule Zurich. (2005)
[34] Hirschmüller, H. Semi - global matching - motivation, developments and applications. Photogrammetric Week 11, 173 - 184 (2011).
[35] Re, C., et al. Evaluation of area - based image matching applied to DTM generation with Hirise images. ISPRS Annals of the Photogrammetry, Remote Sensing and Spatial Information Sciences (2012).

A Roadmap towards a Space – based Radio Telescope for Ultra – Low Frequency Radio Astronomy

M. J. Bentum[1,4], M. K. Verma[2], R. T. Rajan[2], A. J. Boonstra[4], C. J. M. Verhoeven[2], E. K. A. Gill[3], A. J. van der Veen[2], H. Falcke[5,4], M. Klein Wolt[5], B. Monna[7], S. Engelen[7], J. Rotteveel[8], L. I. Gurvits[6,3]

1. Eindhoven University of Technology, Faculty of Electrical Engineering, 5600 MB Eindhoven, The Netherlands
2. Delft University of Technology, Faculty EEMCS, Mekelweg 4, 2628 CD Delft, The Netherlands
3. Delft University of Technology, Faculty of Aerospace Engineering, Kluyverweg 1, 2629 HS Delft, The Netherlands
4. ASTRON, Netherlands Institute for Radio Astronomy, Oude Hoogeveensedijk 4, 7991 PD Dwingeloo, The Netherlands
5. Radboud University Nijimegen, P. O. Box 9010, 6500 GL Nijmegen, The Netherlands
6. Joint Institute for VLBI ERIC, Oude Hoogeveensedijk 4, 7991 PD Dwingeloo, The Netherlands
7. Hyperion Technologies, Vlinderweg 2, 2623 AX Delft, The Netherlands
8. Innovative Solutions in Space, Motorenweg 23, 2623 CR Delft, Netherlands

Abstract: The past two decades have witnessed a renewed interest in low frequency radio astronomy, with a particular focus on frequencies above 30MHz e. g. , LOFAR (LOw Frequency ARray) in the Netherlands and its European extension ILT, the International LOFAR Telescope. However, at frequencies below 30MHz, Earth – based observations are limited due to a combination of severe ionospheric distortions, almost full reflection of radio waves below 10MHz, solar eruptions and the radio frequency interference (RFI) of human – made signals. Moreover, there are interesting scientific processes which naturally occur at these low frequencies. A space or Lunar – based ultra – low – frequency (also referred to as ultra – long – wavelength, ULW) radio array would suffer significantly less from these limitations and hence would open up the last, virtually unexplored frequency domain in the electromagnetic spectrum.

A roadmap has been initiated by astronomers and researchers in the Netherlands to explore the opportunity of building a swarm of satellites to observe at the frequency band below 30MHz. This roadmap dubbed Orbiting Low Frequency Antennas for Radio Astronomy (OLFAR), a space – based ultra – low frequency radio telescope that will explore the Universe's so-called dark ages, map the interstellar medium, and study planetary and solar bursts in the solar system and search them in other planetary systems. Such a radio astronomy system will comprise of a swarm of hundreds to thousands of satellites, working together as a single aperture synthesis instrument deployed sufficiently far away from Earth to avoid terrestrial RFI. The OLFAR telescope is a novel and complex system, requiring yet to be proven engineering solutions. Therefore, a number of key technologies are still required to be developed and proven. The first step in this roadmap is the NCLE (Nether-lands China Low Frequency Explorer) experiment, which was launched in May 2018 on the Chinese Chang'e 4 mission. The NCLE payload consists of a three monopole antenna system for low frequency observations, from which the first data stream is expected in the second half of 2019, which will provide important feedback for future science and technology opportunities.

In this paper, the roadmap towards OLFAR, a brief overview of the science opportunities, and the technological and programmatic challenges of the mission are presented.

Keywords: OLFAR; Nanosatellite; Satellite Swarms; Ultra-Long Wavelength Astronomy; Ultra-Low Frequency Radio Astronomy

1 Introduction

One of the last unexplored domains of the electromagnetic spectrum unexplored by astronomers is the ultra-low frequency band. This frequency domain, which covers frequencies from 0.1 to 30MHz, contains footprints of the early Universe (between 0.38 and 400 million years after the Big Bang) Jester and Falcke. 2009 Raian et al., 2016), magnetospheric radio emissions from planets in the Solar System and perhaps from exoplanets (Zarka, 1998, 2007; Bentum, 2017, 2018), and imprints of a broad variety of astrophysical processes (Jester and Falcke, 2009).

For a variety of reasons, astronomical observations in this frequency domain have been rare. One of the key reasons is a major impact of ionosphere on radio propagation. The ionosphere reflects and refracts radio waves at low frequencies, and completely blocks signals from space below the ionospheric cut-off frequency. The cutoff frequency depends on the time of the day and the activity of the Sun, but can be as high as 10MHz. In addition, human-made long range radio communications use this reflection property to broadcast signals across continents, as it keeps the radio waves inside the atmosphere (so called 'ducting effect'). This makes the situation with RFI for radio astronomy problematic, as it is almost impossible to find or create a radio quiet zone. Both these two factors make it impossible to make observations at the surface of the Earth in the ultra-low frequency domain.

A space or Lunar - based radio telescope will overcome these problems and enable observations in the ultra - low frequency domain. There have been four space missions dedicated to low - frequency radio astronomical observations. The first radio astronomy explorer RAE - 1 was deployed in an Earth orbit and made the first measurements of the Galaxy's spectrum, between 0.4 and 6.5MHz (Alexander et al., 1969). In 1971, the Interplanetary Monitoring Platform (IMP - 6) conducted measurements of the Galactic radiation at 22 frequencies between 130 and 2600kHz (Brown, 1973). The third mission was the radio astronomy explorer RAE - 2 placed in a lunar orbit and observed between 25kHz and 13MHz (Alexander et al., 1975). The fourth, and the most recent experiment on long - wavelength observations is the Netherlands Chinese Low Frequency Explorer (NCLE), which was launched as a part of the Chinese Chang'E 4 Lunar mission on May 21, 2018. Another synergistic development of the past decades is the implementation of the first generation Space Very Long Baseline Missions, the Japan - led VSOP/HALCA and Russia - led RadioAstron (Gurvits (2019) and references therein).

The NCLE consists of three monopole antennas of 5 meter each, low noise amplifiers, a receiver system and a calibration unit, for monitoring the sky at low frequencies between 80kHz to 80MHz (Prinsloo et al., 2018). In addition to these four implemented missions, there have been numerous concept studies for space - based radio arrays in the past (French et al., 1967; Weiler et al., 1988; Basart et al., 1997; Jones et al., 2000). The more recent initiatives will be discussed in the next section.

In order to match in angular resolution and sensitivity modern Earth - based low - frequency radio astronomy systems as LOFAR - ILT at ultra - low frequencies, a system with a synthesized aperture of at least ~100km is needed. Such a system can be composed of multiple satellites forming an interferometer, similar in principle to Earth - based low - frequency instruments e.g., LOFAR (van Haarlem et al., 2013). A space - based radio instrument as OLFAR will use small (nano - cube) satellites instead of conventional micro to large satellites. Several small satellites can be built with off - the - shelf components, making them affordable and simple to assembly. Hundreds to thousands of satellites can be deployed in a swarm formation, offering a good aperture (u, v) coverage and also baseline redundancy.

Satellite swarms in Earth orbits can in principle be formed while meeting the required aperture (u, v)- geometry and sufficient orbit stability. However, such a system will not be protected against human - made RFI. In addition, the Auroral Kilometre Radiation (AKR) (Mutel et al., 2003) which is the strongest near - Earth source between a few tens of kHz and approximately 700kHz, will contribute to the interference sustained by OLFAR. On Lunar orbits a swarm formation could be shielded from the terrestrial RFI (Bentum and Boonstra, 2016). However, recent studies have shown that on the Lunar orbits suitable for effective shielding, the relative motion of the satellites are unacceptably fast (Bentum et al., 2019).

A solution for reducing RFI might be deploying OLFAR at larger distances from Earth where the intensity of terrestrial RFI is lower. lower RFI allows for a smaller dynamic range of receiver systems on the satellites and therefore requires smaller data storage capacity. However, a larger distance over which the data need to be transmitted back to Earth increases the data transport challenge - a clear subject of a trade - off consideration. Deploying OLFAR closer to Earth requires more data to achieve the same measurement fidelity compared with a higher altitude. On the other hand, the shorter distance to Earth makes it possible to transfer the data a a higher speed.

In this article the roadmap for implementing OLFAR is presented. Several intermediate steps are envisioned. The article is organized as follows. In Section 2, a brief overview of recent initiatives for an ultra - low frequency space - borne radio astronomy telescope is given. In Section 3, the OLFAR science cases and high level requirements are presented after which, in Section 4, the roadmap towards a full OLFAR is discussed. Conclusions are presented in Section 5.

2 Overview of recent initiatives

During the last decade, several concept studies and projects have been drafted with the aim of building an ultra - low - frequency radio interferome - ter in space, and thus allows sky imaging down to a few kHz. This has long been impossible to implement due to the required aperture size. But it should be possible with current technology. The space - based studies have significantly improved the technology readiness levels (TRLs) of various subsystems, such as navigation, communications, small - spacecraft technology and swarm analysis. At this moment several teams in the world are working towards an implementation of such a radio telescope, including:

• In the Netherlands, a team consisting of several universities (Eindhoven University of Technology, Radboud University, Delft University of Technology), institutes (ASTRON, Joint Institute for VLBI ERIC, SRON), and companies (of which: Innovative Solutions in Space, Hyperion, NLR, TNO) started several initiatives: the ESA - funded DARIS Distributed Aperture Array for Radio Astronomy In Space) project, Saks et al., 2010; Raian et al., 2016), the Dutch - funded OLFAR project Bentum et al., 2009; Rajan et al., 2011, 2016), the ESA - proposed projects SURO (Space - based Ultra - long wavelength Radio Observatory)(Blott et al., 2013) and DSL (Discovering the Sky at the Longest Wavelengths, together with a Chinese team) (Boonstra et al., 2016) and the ESA - funded CLE (CubeSat Low Frequency

Explorer)study (Bentum et al. , 2018). These studies made the foundation for the Dutch - Chinese NCLE (Netherlands - China Low Frequency Explorer) spacemission, which is a single -antenna satellite launched in 2018, located in the Moon - Earth L2 point, and which observes at low frequencies.

• In France, a NOIRE concept (Nanosatellites pour un Observatoire In - terferometrique Radio dans l' Espace/Nanosatellites for a Radio Interferometer Observatory in Space) was proposed by a team of several institutes led by the LESIA group (Cecconi et al. , 2018). The concept focused on the feasibility study of a swarm of nanosatellites equipped with ultra - low frequency radio astronomy instruments.

• A combined Swedish and UK team proposed the FIRST explorer (Bergman et al. , 2009), a constellation including six daughter spacecraft with radio astronomy antennas, and a mother spacecraft for science and metrology data processing and communications.

• A number of initiatives and design studies of various concepts of ultralow frequency missions have been presented over the past several decades in the USA. A comprehensive review of early studies is given by Kassim and Weiler (1990), Basart et al. (1997) and references therein. The most recent USA initiative, RELIC (Belov et al. , 2018) aims to deploy a multi - spacecraft constellation forming a decametric wavelength synthesized telescope on a selenocentric orbit. The science case of RELIC is presented extensively in about ten papers cited by (Belov et al. , 2018). The authors conclude that the proposed concept is viable for implementation on a rather short timescale of about five years. Another recent initiative in the USA, SunRISE (Kasper et al. , 2019) aims at studies of particle acceleration and transport in the inner heliosphere by means of radio interferometric imaging at frequencies from 100kHz to 25MHz. The concept assumes a formation of six satellites on a geosynchronous orbit with a typical baseline length between the satellites of about 10km. The key technological idea behind the SunRISE concept is in utilizing commercial access to space, including a launch of all six spacecraft to their target orbit as a piggyback payload in conjunction with a larger spacecraft intended for GEO positioning.

• Several design studies of a broad range of prospective Space VLBI missions have been addressed in China over the past decade, including those aiming to operate in the ultra - low frequency domain (An et al. , 2019). The NCLE experiment on the Chang'E - 4 mission (Prinsloo et al. , 2018) provided a new momentum for the ultra - low frequency studies which are envisioned in the framework of the DSL (Discovering the Sky at Longest wavelengths) initiative (Boonstra et al. , 2016; Chen et al. , 2019).

In Table 1, an overview of the initiatives is given, including the frequency range, maximum length of the baselines, the number of spacecrafts and the deployment location. All the initiatives, except NCLE, are project studies that have not proceeded to implementation stages. NCLE as well as OLFAR are discussed in more details in the following sections.

Table 1 History of space - based low frequency radio observatory studies

Name	Freq. range [MHz]	Baseline [km]	No. S/C	deploym. location	ref.
DARIS	1 - 30	⩽100	6 - 8	Dyn. Solar Orbit	(Saks et al., 2010)
FIRST	0.5 - 50	⩽30	7	Sun - Earth L2	(Bergman et al., 2009)
SURO	0.130	⩽30	8+	Sun - Earth L2	(Blott et al., 2013)
DSL	0.1 - 50	⩽100	8	Lunar Orbit	(Boonstra et al., 2016)
NOIRE	0.001 - 100	⩽100	various	Lunar Orbit	(Cecconi et al., 2018)
RELIC	0.1 - 30	⩽100	33	Lunar orbit	(Belov et al., 2018)
NCLE	0.08 - 80	0	1	Earth - Moon L2	(Prinsloo et al., 2018)
OLFAR	0.1 - 30	100	⩾50	TBD	(Rajan et al., 2016)

3 OLFAR - Orbiting Low Frequency Antennas for Radio Astronomy

Opening up the low frequency regime of 0.1 - 30MHz for radio astronomy requires a space - borne, in particular, Moon - based or Moon - orbiting, instrument as mentioned in the previous section. The objective of the OLFAR project is to design and build scalable autonomous satellite units to be used as an astronomical instrument at the ultra - low frequencies. In this frequency range, the sky brightness temperature can be as high at 10^7 K at around 1MHz. The OLFAR instrument should be able to measure weak astronomical signals, orders of magnitude less bright than the sky itself, and will therefore need a very high dynamic range. In subsection 3.2 the requirements of the OLFAR instrument are presented.

3.1 Science Case

A number of science case studies or ultra - low frequency astronomy have been published over the past decades, e.g. Gorgolewski (1966); GrieBmeier et al. (2007); Jester and Falcke (2009); Mimoun et al. (2012); Boonstra et al. (2016); Rajan et al. (2016); Saks et al. (2010); Cecconi et al. (2018); Zarka et al. (2012) and references therein. We briefly summarize major topics of these cases, not attempting a comprehensive review of the above mentioned publications.

3.1.1 Cosmology

The most popular high - profle science case for low frequency radio astronomy is the study of cosmology, especially mapping the so called Dark Ages, Jester and Falcke (2009) and references therein. It addresses the observation of the highly redshifted HI line, originated at the cosmological epoch before the formation of the first stars. Mapping and tomography of the HI distribution with adequate angular resolution, would require baselines of the order of tens of kilometers and a large number of antennas (Jester and Falcke, 2009).

3.1.2 Planetary radio emission

The magnetized planets in Solar System (Earth, Jupiter, Saturn, Uranus and Neptune) are emitting intense radio signals resulting from particle accelerations in various places of their magnetospheres. The auroral radio emission is emitted from above the magnetic poles of the planet through a cyclotron instability. The radiation belts (in the case of Jupiter and possibly the Earth) are emitting synchrotron radiation. The auroral radio emission, including AKR, carries imprints of the geophysical processes important for understanding the genesis and evolution of planets and their electromagnetic properties. Yet, comprehensive studies of AKR cannot be conducted from the Earth's surface due to the opaque ionosphere (Mutel et al., 2003) thus requiring a space - borne facility.

The planetary atmospheric discharges can also be the source of radio pulses, which are associated with lightning. With an ultra - low - frequency space - based array, observations of lightning from planets in Solar system will become possible, including the Earth - originated lightning emission too. In addition, signals from magnetized exoplanets could also be detected. Given the prominence of the exoplanetary studies in the modern astronomy agenda, a contribution from ultra - low - frequency astronomy should be seen as strong science driver (Grießmeier et al., 2007 ; Zarka et al., 2012).

3.1.3 Galactic and stellar astrophysics

The low frequency sky mapping and monitoring of the galactic and extragalactic sky address a multitude of science topics on radio galaxies, active galactic nuclei, large scale structures like clusters with radio halos. As demonstrated by, e.g., a number of studies of galactic and extragalactic targets at decameter wavelengths reviewed by Konovalenko et al. (2016) and the ongoing LOFAR Two - Metre Sky Survey (LoTTS, Shimwell et al. (2017)), the decameter to meter wavelength domain offers a very reach scientific harvest. There are no reasons to believe that pushing the frequency domain to even lower frequencies will be less productive. Of special interest for studies at ultra - low frequencies are extremely redshifted galaxies as beacons in the early cosmological epochs. While a number of ground - based radio observatories have produced sky maps at frequencies close but still above the ionosphere cut - off at about 10MHz, only very cursory sky mapping at frequencies below 10MHz is currently available. The latter observations were provided by the Interplanetary Monitoring Platform (IMP - 6) presented by Brown (1973), the Radio Astronomy Explorers RAE - 1 described by Alexander and Novaco (1974) and RAE - 2 Alexander et al. (1975); Novaco and Brown (1978). The need for deeper sky mapping at ultra - low frequencies, especially in polarised emission is discussed by Rajan et al. (2016) and references therein).

3.1.4 Space Weather

An important gap in the current space weather research capabilities can be filled by a space - based array. A terrestrial low - frequency array such as LOFAR can observe in the frequencies as low as 30MHz. However, at lower frequencies a space - based instrument is needed, which will enable a wide range of solar and space weather research topics. These include for example, tracing the initial launch of a CME (coronal mass ejections), detailed tracking of the solar wind and CMEs through interplanetary space and in - depth studies of micro - structure in the Earth's ionosphere (Oberoi and Kasper (2004)).

3.1.5 Ultra - high energy particles

Ultra - high energy particles are likely composed of atomic nuclei, protons and potentially photons or neutrinos, which are extremely hard to detect due to their small flux of less than 1 particle per square kilometer per century, requiring large detectors such as the LOFAR. However, if a large array of low - frequency antennas are placed on the Lunar surface, then an increased sensitivity to cosmic rays and neutrinos can be exploited, as compared to the LOFAR (Nelles et al. , 2015). In addition, the radio quietness at the Lunar surface would greatly benefit such measurements (Scholten et al. , 2006).

3.2 OLFAR high level system requirements

Table 2 lists the high level system requirements of OLFAR derived from the science cases (see (Rajan et al. , 2016)). The minimum distances between the satellites must be more than 10km and due to interstellar scattering the maximum baseline is limited to 100 km, giving a resolution of 1 arcminute at 10MHz. OLFAR will comprise of more than 50 satellites, each containing three linearly polarized antennas, observing the sky from 0.1 — 30MHz. Extrapolation to higher frequencies will increase data bandwidth and processing overload. On the other hand, the number of satellites could vary from 50 to larger than 10000, and meet a diverse set of aforementioned science cases, as discussed in (Jester and Falcke, 2009). The satellites will employ passive formation flying and yet maintain sufficient position stability for a given integration time. In the presence of a stable orbit and thus stable baseline, position estimates can be more precisely known and thus the integration time can be extended up to 1000 seconds and thereby reducing the down - link data rate (Rajan et al. , 2016).

Table 2 OLFAR system requirements

Number of satellites	≥50 (scalable)
Number of antennas	2 or 3
Observation frequency range	0.1—30MHz
Instantaneous bandwidth	≥1MHz
Spectral resolution	1kHz
Snapshot integration time	1 to 1000s
Maximum baseline between satellites	100km
Deployment location	Moon orbit, Earth trailing, L2

4　Roadmap

A roadmap towards implementing the OLFAR mission is illustrated in Figure 1 which provides a wide range of technology candidates and a development pathway for the oncoming decade. The long term goal is to enable the deployment of OLFAR satellites in a Lunar orbit by 2030. Furthermore, the roadmap outlines the high-level requirements of each phase, and the necessary technological developments, which will be incorporated into the mission along the road to realizing OLFAR.

The roadmap is divided into 4 distinct phases, as illustrated in Figure 1 to gradually build up on operation experience, validate newly developed technologies, and address diverse science cases. Furthermore, the proposed roadmap is in alignment with the long term planning of other major space agencies, which allows for gradual build up of the swarm in the final location of OLFAR.

In this section, we will outline all the four phases and provide a description, high level requirements, and the technologies that need to be developed to successfully realize the phase. We also mention the strategic importance of each phase in the context of the overall OLFAR mission. To this end, each phase description is divided into key subsections, as follows.

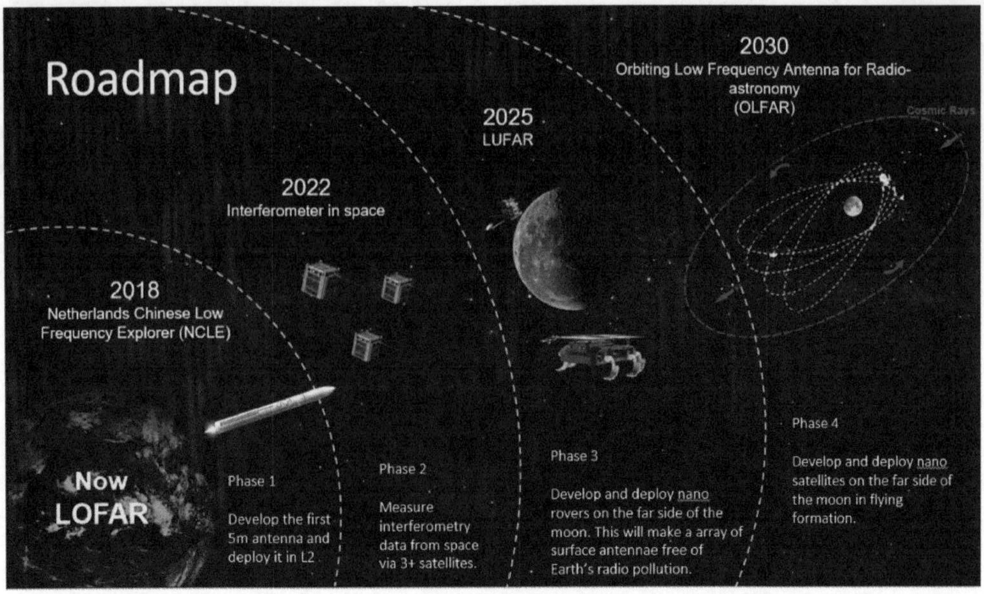

Figure 1　OLFAR Roadmap

1. Mission description
2. Technology challenges
3. High level requirements

4. Time to launch

4.1　Phase 1: Netherlands Chinese Low Frequency Erplorer (NCLE)

4.1.1　Mission description:

The NCLE is a low - frequency payload which is currently on - board the Chang'e 4 satellite (see Figure 2), which was launched on May 21th, 2018, and is located in a Lissajous orbit around the Earth - Moon L2 point (Boonstra et al. (2017)). The NCLE design concept involves 3 co - located, orthogonal, monopole antenna elements, each of ~5 meters in length. These 3 active antennas are mounted perpendicular to the upper side of the satellite body. The voltages between the monopoles and the spacecraft body are measured, digitized and send to Earth for further processing. The frequency range of NCLE is 80kHz to 80MHz, and overlaps with the Earth - based LOFAR LBA spectrum (van Haarlem et al., 2013), which could be useful for validation.

The NCLE is a pathfinder for a future space - based low - frequency radio interferometer, and thus one of the key objectives of NCLE is to characterize the lunar radio environment and to understand the RFI at ultra - low frequencies. In addition, NCLE will also enable the study of solar activity, space weather, and possibly a detection of bright pulsars and other radio transient phenomena at ultra - long wavelengths, although the latter is not warranted due to the sensitivity limitations.

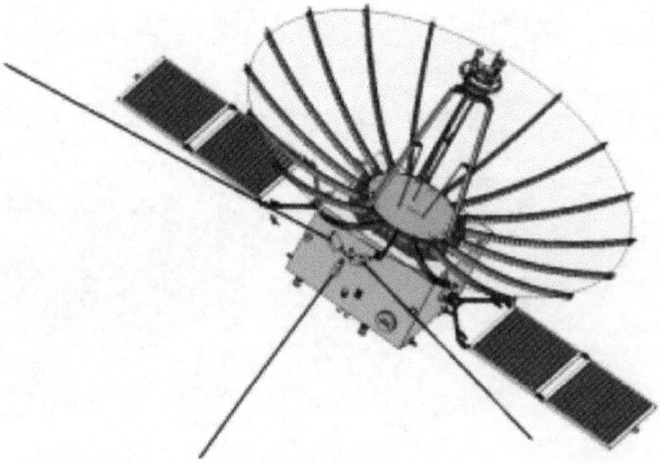

Figure 2　An illustration of the Chang'e 4 satellite with three monopole antennas mounted on the body of the satellite (Boonstra et al. (2017))

4.1.2　Technological Challenges

One of the fundamental challenge in the NCLE payload is the design and development of the wideband science antenna and the following low - noise amplifier (LNA) (Prinsloo et al. (2018)). The EMI modelling, system integration and calibration of the science front - end is a critical challenge in the NCLE mission. The analogue signals are digitized in a DSP system on which dedicated science modes are implemented in a flexible software - defined radio system, which is part of the NCLE payload. These modes for instance per - form fast Fourier transforms to create average radio spectra, allow triggering on transient radio events, or allow to retrieve direction of arrival information using beam - forming or goniopolarimetry techniques (Boonstra et al. (2017)). Finally, the data is relayed back Earth - based ground stations, by the Chang'e 4 satellite. The high level requirements of the NCLE Science instrument are listed in Table 4. 1. 2 (Prinsloo et al. , 2018).

Table 3　High Level Requirements of the NCLE payload

Req. ID	Requirement
NCLE1	The NCLE antennas operates in the frequency range of 80kHz to 80MHz.
NCLE2	Each of the three antennas is 5m in length.
NCLE3	NCLE instrument is adhere to electromagnetic interference specifications of MIL - STD - 461.
NCLE4	The location of the NCLE antennas on the spacecraft is not influencing the radiation performance of the antennas.

4. 2　Phase 2: LEO CubeSat Low Frequency Erplorer (LCLE)

4. 2. 1　Mission Description

The LCLE will be the first standalone mission to allow key interferometry technologies to be tested for OLFAR. The mission will consist of 3 cubesatellites in a low Earth orbit (LEO), each equipped with 3 dipoles. The low frequency measurements will take place every time the satellites are at the apogee point of orbit, which is followed by correlation and integration of the measured data. The processed data is sent to the ground station for further analysis, when the satellites are nearest to Earth.

4. 2. 2　Technological Challenges

There are several technical advancements that are needed, to make this mission feasible. The LCLE is by no means the first mission to have cubesats flying as a group of three but it will be the first to deploy a radio sensitive payload in LEO and fly in a very tight formation. The high accuracy needed to maintain flight within small position error means that the mission will heavily rely on an active attitude determination and control subsystem (ADCS) system. In the ADCS system, having a propulsion system which can provide enough delta V constantly maintaining the attitude and last for more than 3 - 4 years is one the biggest challenges to overcome for this mission. This also means a very accurate and reliable pointing algorithm is needed to make sure each satellite can independently locate its position and in relation to each other by sharing their own location details in the orbit. For inter - satellite communication, not only position data will be shared but also the raw measurements will be sent to each one of them to do distributed correlations. As each measurements will produce several Gigabits of data, a very high data inter-satellite communication is necessary which can be laser or RF based. The science front-end comprising of the antenna, antenna deployment mechanism, LNA and the digital signal processing system, will be miniaturized to enable power and mass efficiency. The high level requirements for the LCLE payload is listed in Table 4.

Table 4 High Level Requirements of LCLE

Req. ID	Requirement
LCLE1	The science instrument shall operate in the radio frequency range of 1MHz to 30MHz
LCLE2	Inter satellite distance shall be 100km at maximum
LCLE3	Observations shall take place at least for a year
LCLE4	Minimum inter-satellite communication data rate shall be 6 Mbps

4.2.3 Estimated time to launch

The mission is expected to be ready for launch in 2022/2023.

4.3 Phase 3: LUnar low Frequency Antennas for Radio Astronomy (LU-FAR)

4.3.1 Mission Description

LUFAR stands apart from the previously planned single-rover missions (see (Mimoun et al., 2012)), as it makes use of several individual nodes, like LOFAR (van Haarlem et al. (2013)), to make a large array of antennas. Where all other concepts deploy large static stretched out antennas from a central node, a group of nanorovers (swarm) will carry one antenna each for LUFAR. State-of-the art rovers will enable a flexible spatial sampling of the UVW plane for radio astronomy observations. As the array is made up of mobile platforms, formation is dynamic in nature and can adjust in shape and size to test a number of configurations.

The rovers will be deployed on the far side of the Moon onboard a lander which will land on the edge of the far side. This way, the lander can be in the field of view of the swarm on the far side and ground stations on Earth. This will allow the lander to act as a communication hub and relay data between the ground station, if a in-orbit relay spacecraft is not available. The rovers will take measurements and transfer raw data within the swarm for correlation via fibre optic cables. The processed data can be sent by the rovers to the relay system to be communicated to Earth. In the initial study, the number of rovers needed to make LUFAR successful for a wide range of science cases, is similar to OLFAR (50+ antenna systems). Therefore, the swarm will build itself gradually by adding 5 to 10 rovers each time a mission is flown to the Moon, if the ideal case of a dedicated mission is not feasible.

Given results from the LCLE, the initial formation algorithm and on-board correlation, LUFAR will be implementing the same technology but at a much larger scale which will be a much closer representation of the OLFAR mission. One of the biggest advantages of testing high accuracy navigation subsystems and techniques on LUFAR is that the change in attitude is very low and the nodes are in much more stable and predictable relative position. This allows for much less risk factor and focus every aspect of the mission to optimize for the large number of nodes and mass production. Besides testing systems for lunar environment, an autonomous self - organization ('swarm - intelligence') algorithm will be implemented for the first time on a large scale and provide vital experience for the next phase of OLFAR. LUFAR will also validate the science cases possible that can be carried out on the far side of the Moon as rovers can maintain continuous observations for a long period of time, which is vital to know if OLFAR satellites are optimized for the science cases described in Section 3. 1.

LUFAR can be extended to allow the rovers to be used as a navigation reference for OLFAR satellites on lunar ground and provide auxiliary data for validation of observations made in orbit. In case of emergency, LUFAR may also provide a communications relay for flying nanosatellites and a data backup location.

4. 3. 2　Technological Challenges

LUFAR rovers will be deployed in harsh lunar environment, and carry out very sensitive measurements over a long period of time, in full autonomy. This mission therefore brings a relatively new list of technical challenges compared to the last two phases but is also critical to the next and final phase.

For housekeeping and nominal operations of the rover, an autonomous swarming algorithm shall be developed which not only ensures the survival of the rover and swarm in very cold conditions but also navigates and maintains relative positioning. Antenna and its deployment system will also need to be developed to survive the abrasive and electrostatic nature of the dust particles. An 'intelligent' inter - rover communications will also be a demanding task to develop in order to prevent data congestion because a large amount of raw data will flow through this system. During night time, the rover should house a miniaturized very efficient power and active thermal system which will ensure critical system survive －180℃ and have minimal electromagnetic interference with the measurements. One major bottleneck that is unique to LUFAR mission is the communication relay system. Although the lander and in - orbit satellite provide a means to do so, the data communications systems on board the rover needs to handle a very high data rate while also being miniaturized enough for a shoe - box sized rover, which is currently not available off - the - shelf. In addition, the LUFAR will establish inter - rover communication links for exchanging raw data, and enable distributed processing for correlation in the swarm and navigation. Furthermore, the processed data must be downlinked to Earth via an rover on the near - side or using a relay satellite. The high level requirements for the LUFAR mission are listed in Table 5.

Table 5　High Level Requirements of LUFAR

Req. ID	Requirement
LUFAR1	The science instrument shall be able to operate in the radio frequency range of 0. 1MHz to 30MHz
LUFAR2	The distance between rovers shall be 10km at maximum
LUFAR3	Each rover shall have an internal operating temperature range of −40℃ to 50℃
LUFAR4	Each rover shall have the ability to be commanded from ground station individually
LUFAR5	The science instrument's antenna shall be 10m long
LUFAR6	Each rover shall not have a mass of more than 8kg
LUFAR7	The swarm shall be able to navigate and operate autonomously

4. 3. 3　Estimated time to launch

Delft university of technology and various Dutch partners are currently working on a concept of a Lunar Rover which will be a stepping stone for LUFAR. Given that NASA, ESA and other major agencies have similar plans and are up - scaling lunar missions in the coming decade, LUFAR has a good chance to be ready for flight by 2025, with full capability of swarm reached by 2030.

4. 4　Phase 4: Orbiting low Frequency Antennas for Radio Astronomy (OLFAR)

4. 4. 1　Mission Description:

Our dot on the horizon, OLFAR, aims to develop a space - based antenna array for radio astronomy comprising of 100's to 1000's of nano - satellites for radio astronomy observations at 0. 1 - 100MHz. The satellites will be spatially co - located within a maximum of 100km, which will result in skymaps of sub arc - minute resolutions at these wavelengths. This means that OLFAR will be 1 - 3 orders of magnitude more capable than its ground - based or space - borne predecessors in terms of sensitivity, dynamic range, and both angular and spectral resolution. The exact location of OLFAR is still under investigation. It could either be an orbit around the Moon or a location far far away from Earth to overcome the RFI from Earth.

4. 4. 2　Technological Challenges

To be able to launch OLFAR, a few technology breakthroughs will be needed. Many of the collaborators in aforementioned projects, including the NCLE, are part of our current consortium, offering a blend of expertise from radio astronomy, telecommunications, aerospace systems, embedded systems and navigation. A few of the fundamental science and technology challenges, which are listed in 5 distinct domains:

• *Array calibration and 3 - D imaging* Antenna calibration is one of the biggest challenges for any low - frequency phased arrays, and for terrestrial arrays, calibrating ionospheric phase fuctuations tops the list (van Haarlem et al. , 2013). The instrumental beam shape, and the frequency dependent gains of the antennas must be calibrated on the fly with minimal Earth - assistance. For low frequency observations, the requirements on positional stability and clock accuracies are in the order of sub - meters and sub - nanoseconds respectively, which drive the navigation requirements of OLFAR. Secondly, unlike terrestrial low - frequency arrays which are quasi - planar and have (at best) a hemispheric view, OLFAR will enable an instantaneous 3D view of the sky with a field - of - view of the full celestial sphere. 3D Imaging on a sphere with an interferometric array has been addressed by Carozzi (Carozzi, 2015) as well as by Vugt et al. (van Vugt et al., 2017).

• *Navigation* The knowledge of position, time and orientation of individual satellites is vital for radio - science interferometry and for communication. However, due to the distant deployment and the shear number of satellites, current dependence on GPS - satellites and Earth - based ground segment are either not scalable or too expensive. In particular, when OLFAR is orbiting the far side of the Moon, the satellite swarm is completely disconnected from Earth for few hours (Rajan et al., 2011). This drawback calls for self - navigation or relative navigation of OLFAR, which suffices the needs of radio science, intra - satellite communication and collision avoidance. The desired synchronization accuracy is related to the accuracy of the onboard clocks and the observation frequencies of the OLFAR network (Rajan et al. 2013a). Despite satellite mobilitv, clock synchronization can be achieved by emploving generalized two - way ranging between mobile satellites (Rajan and van der Veen, 2015). Recent studies show that joint relative localization and synchronization is feasible for such GPS - denied or reference - free networks (Rajan et al. 2015, 2019), however numerous questions still remain unanswered. For example, how can the relative time - varying position - time - orientation be estimated sufficiently accurate by a satellite swarm by only communicating with the neighbouring satellites? Can data fusion be applied to combine the satellite orbital dynamics and kinematics? Answering these questions will not only enable spacemissions such as OLFAR, it would also contribute to breakthroughs in terrestrial navigation.

• *Communications and Antenna Design* In the swarm of satellites there is a need for high - data - rate inter - satellite communication, both for localization and navigation purposes as well as for the exchange of data for the interferometric data processing. Multi satellite systems, operating as interferometers will generate so much data, that it needs to be processed to reduce the data volume to the essential information before it will be possible to relay the information onto the end user via a ground station. Apart from the need for high speed inter satellite communication, the need for high speed satellite - to - ground station communication is also rapidly growing. The total antenna system for the intended applications can be distinguished in three parts: the antenna system for the payload (the reception of astronomical signals), the antenna system for communication in the distributed space system and the antenna system for the data downlink to Earth.

• *Space systems engineering and miniaturization* To establish a swarm of satellites of perhaps hundreds of elements, miniaturization of the elements is required. This holds technology challenges on the level of individual elements, but also requires systems engineering breakthroughs in order to guarantee their efficient development, and a robust and efficient operations of the entire swarm and other mission elements.

Miniaturization may result in the design of small antennas. However, with smaller antennas, communication on lower frequencies is not possible. Higher frequencies are power hungry which might be a problem given that small harvesters provide limited energy. In the coming years, we want to address such problems both in terms of hardware and communication.

Attitude control is necessary for antenna pointing and orientation and energy harvesters towards the energy source. To this end, we want to explore the ways to miniaturize attitude control systems such as momentum wheels, magneto - torquers, and our existing micro - propulsion system.

• *Distributed processing* The OLFAR swarm will employ distributed architectures for navigation, communication and processing science data, to avoid single point of failure, optimize efficiency and to abide by power constraints imposed by nano - satellites (Rajan et al. , 2013b). While distributed computing in these individual domains are common in terrestrial applications, OLFAR will juggle between these tasks while adhering to the strict science requirements levied on the mission. OLFAR will employ on - board distributed correlations to process streaming payload data, in order to reduce the data - downlink back to Earth.

5 Conclusion

Space - based ultra - low - frequency radio astronomy will present a new milestone in our understanding of the Universe. It will open the frequency range below 30MHz, which will enable unique science. For that purpose, missions such as OLFAR are essential to get access to this low - frequency part of the spectrum. The road to OLFAR, however, includes both scientific opportunities, and challenges concerning technological development and programmatics. In this article, we briefly summarized the previous missions, and presented a roadmap to tackle the technological challenges for the dot on the horizon, a fully operational OLFAR.

Acknowledgements

We are grateful to the anonymous referees for very useful critique and constructive suggestions on the present paper.

References

[1] Alexander, J. , Kaiser, M. , Novaco, J. , Grena. , F. , and Weber, R. (1975).

Scientific instrumentation of the Radio - Astronomy - Explorer - 2 satellite. Astronomy and Astrophysics Journal, 40 (4): 365 - 371.

[2] Alexander, J. K., Brown, L. W., Clark, T. A., Stone, R. G., and Weber, R. R. (1969). The spectrum of the cosmic radio background between 0.4 and 6.5MHz. The Astrophysical Journal, 157: L163.

[3] Alexander, J. K. and Novaco, J. C. (1974). Survey of the galactic background radiation at 3.93 and 6.55MHz. Astron. J., 79: 777 - 785.

[4] An, T., Hong, X., Zheng, W., Ye, S., Qian, Z., Fu, L., Guo, Q., Jaiswal, S., Kong, D., Lao, B., Liu, L., Liu, Q., L, W., Mohan, P., Shen, Z., Wang, G., Wu, F., Wu, X., Zhang, J., Zhang, Z., Zheng, Z., and Zhong, W. (2019). Space Very Long Baseline Interferometry in China. Advances in Space Research, in press.

[5] Basart, J. P., Burns, J. O., Dennison, B. K., Weiler, K. W., Kassim, N. E., Castillo, S. P., and McCune, B. M. (1997). Directions for space - based low frequency radio astronomy: 1. system considerations. Radio Science, 32 (1): 251 -263.

[6] Belov, K., Branch, A., Broschart, S., Castillo - Rogez, J., Chien, S., Clare, L., Dengler, R., Gao, J., Garza, D., Hegedus, A., Hernandez, S., Herzig, S., Imken, T., Kim, H., Mandutianu, S., Romero - Wolf, A., Schaffer, S., Troesch, M., Wyatt, E. J., and Lazio, J. (2018). A space - based decametric wavelength radio telescope concept. Erperimental Astronomy, 46 (2): 241 - 284.

[7] Bentum, M., Alpay Koc, N., Fuster, S., Ravi Kumar, R., Prinsloo, D., Klein Wolt, M., Brinkerink, C., Boonstra, A., Verhoeven, C., and Verma, M. (2018). The cubesat low frequency explorer (cle) in lunar orbit. Presentation at the 42nd COSPAR Scientific Assembly, Pasadena, California, USA, July 2018.

[8] Bentum, M., Verhoeven, C., Boonstra, A. J. znd Gill, E., and van der Veen, A. (2009). A novel astronomical application for formation flying small satellites. In Proceedings of the 60th International Astronautical Congress, Daejeon, South Korea, October 2009, Daejeon, South Korea, IAC - 09. A3. 4. 3.

[9] Bentum, M. J. (2017). The search for exoplanets using ultra - long wavelength radio astronomy. In Proceedings of the IEEE Aerospace Conference, Big Sky, Montana, USA, March 2017. IEEE, inspec: 16932672.

[10] Bentum, M. J. (2018). Algorithms for direct radio detections of exoplanets in the neighbourhood of radiating host stars. In Proceedings of the IEEE Aerospace Conference, Big Sky, Montana, USA, March 2018. IEEE, inspec: 17877993.

[11] Bentum, M. J. and Boonstra, A. (2016). The RFI situation for a spacebased low - frequency radio astronomy instrument. In Proceedings of the 2016 RFI conference, Socorro, New Merixo, USA, October 2016. IEEE, inspec: 16620879.

[12] Bentum, M.J., van Vugt, P., Horlings, W., and Boonstra, A. (2019). The radio environment for a space - based low - frequency radio astronomy instrument. In Proceedings of the IEEE Aerospace Conference, Big Sky, Montana, USA, March 2019. IEEE, doi: 10.1109/AER0.2019.8741975.

[13] Bergman, J.E.S., Blott, R.J., Forbes, A.B., Humphreys, D.A., Robinson, D.W., and Stavrinidis, C. (2009). First explorer - an innovative low - cost passive formation - flying system. In Proceedings of CEAS 2009 European Air 8 Space Conference, Manchester, UK, October 2009, pages 1 - 14. Royal Aeronautical Society.

[14] Blott, R., Baan, W., Boonstra, A., Bergman, J., Robinson, D., Liddle, D., Navarathinam, N., Eves, S., Bridges, C., Gao, S., Bentum, M., Forbes, A.ad Humphreys, D., and Harroch, C. (2013). Space - based ultra - long wavelength radio observatory (low cost) suro - lc. Presentation at the European Planetary Science Congress, London, UK, September 2013.

[15] Boonstra, A., Wise, M.van der Marel, J., Ruiter, M.Arts, M., Prinsloo, D., Bast, J., Kruithof, G., Falcke, H., Klein - Wolt, M., et al. (2017). The Netherlands - China Low Frequency Explorer. Presentation at the 32nd International Union of Radio Science (URSI) General Assembly Scientific Symposium (GASS), Montreal, Canada, August 2017.

[16] Boonstra, A. -J., Garrett, M., Kruithof, G., Wise, M., van Ardenne, A., Yan, J., Wu, J., Zheng, J., Gill, E.K.A., Guo, J., Bentum, M., Girard, J.N., Hong, X., An, T., Falcke, H., Klein - Wolt, M., Wu, S., Chen, W. Koopmans, L., Rothkaehl, H., Chen, X., Huang, M., Chen, L., Gurvits, L., Zarka, P., Cecconi, B., and de Haan, H. (2016). Discovering the Sky at the Longest Wavelengths (DSL). In Proceedings of the IEEE Aerospace Conference, Big Sky, Montana, US, March 2016. IEEE, ispec: 16121612.

[17] Brown, L.W. (1973). The Galactic Radio Spectrum Between 130 and 2600kHz. ApJ, 180: 359 - 370.

[18] Carozzi, T.D. (2015). Imaging on a sphere with interferometers: the spherical wave harmonic transform. Monthly Notices of the Royal Astronomical Society: Letters, 451 (1): L6 - L10.

[19] Cecconi, B., Dekkali, M., Briand, C., Segret, B., Girard, J.N., Laurens, A., Lamy, A., Valat, D., Delpech, M., Bruno, M., Gelard, P., Bucher, M., Nenon, Q., Griesmeier, J. - M., Boonstra, A. - J., and Bentum, M. (2018). NOIRE study report: Towards a low frequency radio interferometer in space. In Proceedings of the IEEE Aerospace Conference, Big Sky, Montana, US, March 2018. IEEE, inspec: 17878057.

[20] Chen, X., Wu, J., Burns, J., Silk, J., Koopmans, L., Rothkael, H., Falanga,

M. , Bale, S. , Boonstra, A. - J. , Cecconi, B. , Chang, C. , Chen, L. , Deng, L. , Falcke, H. , Fan, Q. , Fang, G. , Fialkov, A. , Gurvits, L. , Ji, Y. , Kasper, J. , Li, K. , Mao, Y. , Mckinley, B. , Monsalve, R. , Pen, U. - L. , Peterson, J. , Ping, J. , Subrahmanyan, R. , Vedantham, H. , Wang, C. , Klein Wolt, M. , Wu, F. , Yan, Y. X. J. , and Yue, B. (2019). Discover the Sky by Longest Wavelengths with small satellite constellation, volume 14 of Taikong. International Space Science Institute - Beijing.

[21] French, F. , Huguenin, G. , Rodman, A. , and Rodman, A. (1967). A synthetic aperture approach to space - based radio telescopes. Journal of Spacecraft and Rockets, 4 (12): 1649 - 1656.

[22] Gorgolewski, S. (1966). Lunar radio astronomy observatory. In Malina, F. , editor, Proceedings of the First Lunar International Laboratory (LIL) Symposium Research in Geosciences and Astronomy, Athens, Greece, September 1965, pages 78 - 84. Springer: Berlin, Heidelberg.

[23] Grießmeier, J. - M. , Zarka, P. , and Spreeuw, H. (2007). Predicting lowfrequency radio fluxes of known extrasolar planets. Astronomy 8 Astrophysics, 475 (1): 359 - 368.

[24] Gurvits, L. I. (2019). Space VLBI: from first ideas to operational missions. Advances in Space Research, in press.

[25] Jester, S. and Falcke, H. (2009). Science with a lunar lowfrequency array: From the dark ages of the universe to nearby exoplanets. New Astronomy Reviews, 53 (1 - 2): 1 - 26.

[26] Jones, D. , Allen, R. , Basart, J. , Bastian, T. , Blume, W. , Bougeret, J. - L. Dennison, B. , Desch, M. , Dwarakanath, K. , Erickson, W. , Farrell, W. Finley, D. , Gopalswamy, N. , Howard, R. , Kaiser, M. , Kassim, N. , Kuiper, T. , MacDowall, R. , Mahoney, M. , Perley, R. , Preston, R. , Reiner, M. , Rodriguez, P. , Stone, R. , Unwin, S. , Weiler, K. , Woan, G. , and Woo, R. (2000). The ALFA medium explorer mission. Advances in Space Research, 26 (4): 743 - 746.

[27] Kasper, J. , Lazio, J. , Romero - Wolf, A. , Lux, J. , and Neilsen, T. (2019). The sun radio interferometer space experiment (sunrise) mission concept. In Proceedings of the IEEE Aerospace Conference, Big Sky, Montana, March 2019. IEEE, inspec: 18759562.

[28] Kassim, N. E. and Weiler, K. W. (1990). Low Frequency Astrophysics from Space, volume 362 of Lecture Notes in Physics. Springer - Verlag Berlin Heidelberg.

[29] Konovalenko, A. , Sodin, L. , Zakharenko, V. , Zarka, P. , Ulyanov, O. , Sidorchuk, M. , Stepkin, S. , Tokarsky, P. , Melnik, V. , Kalinichenko, N. Stanislavsky, A. , Koliadin, V. , Shepelev, V. , Dorovskyy, V. , Ryabov, V. , Koval, A. , Bubnov, I, Yerin, S. , Gridin, A. , Kulishenko, V. ,

Reznichenko, A. , Bortsov, V. , Lisachenko, V. , Reznik, A. , Kvasov, G. , Mukha, D. , Litvinenko, G. , Khristenko, A. , Shevchenko, V. V. , Shevchenko, V. A. , Belov, A. , Rudavin, E. , Vasylieva, I. , Miroshnichenko, A. , Vasilenko, N. , Olyak M. , Mylostna, K. , Skoryk, A. , Shevtsova, A. , Plakhov, M. , Kravtsov, I. , Volvach, Y. , Lytvinenko, O. , Shevchuk, N. , Zhouk, I. , Bovkun, V. , Antonov, A. , Vavriv, D. , Vinogradov, V. , Kozhin, R. , Kravtsov, A. , Bulakh, E. , Kuzin, A. , Vasilyev, A. , Brazhenko, A. , Vashchishin, R. , Pylaev, O. , Koshovyy, V. , Lozinsky, A. , Ivantyshin, O. , Rucker, H. O. , Panchenko, M. , Fischer, G. , Lecacheux, A. , Denis, L. , Coffre, A. , Grießmeier, J. M. , Tagger, M. , Girard, J. , Charrier, D. , Briand, C. , and Mann, G. (2016). The modern radio astronomy network in Ukraine: UTR - 2, URAN and GURT. Experimental Astronomy, 42 (1): 11 - 48.

[30] Mimoun, D. , Wieczorek, M. A. Alkalai, L. , Banerdt, W. B. , Baratoux, D. , Bougeret, J. - L. , Bouley, S. , Cecconi, B. , Falcke, H. , Flohrer, J. , Garcia, R. F. , Grimm, R. Grott, M. , Gurvits, L. , Jaumann, R. , Johnson, C. L. , Knapmeyer, M. , Kobayashi, N. , Konovalenko, A. , Lawrence, D. , Feuvre, M. L. , Lognonn, P. , Neal, C. , Oberst, J. , Olsen, N. , Röttgering, H. , Spohn, T. , Vennerstrom, S. , Woan, G. , and Zarka, P. (2012). Farside explorer: unique science from a mission to the farside of the moon. Erperimental Astronomy, 33 (2 - 3): 529 - 585.

[31] Mutel, R. , Gurnett, D. A. , Christopher, I. W. , and J. S. Pickett, a. M. S. (2003). Locations of auroral kilometric radiation bursts inferred from multispacecraft wideband cluster vlbi observations: 1. description of technique and initial results. Journal of Geophysical Research: Space Physics, 108 (A11), 1398.

[32] Nelles, A. , Hrandel, J. R. , Karskens, T. , Krause, M. , Buitink, S. , Corstanje, A. , Enriquez, J. E. , Erdmann, M. , Falcke, H. , Haungs, A. , Hiller, R. , Huege, T. , Krause, R. , Link, K. , Norden, M. J. , Rachen, J. P. , Rossetto, L. , Schellart, P. , Scholten, O. , Schrder, F. G. , ter Veen, S. , Thoudam, S. , Trinh, T. N. G. , Weidenhaupt, K. , Wijnholds, S. J. , Anderson, J. , Bhren, L. , Bell, M. E. , Bentum, M. J. , Best, P. , Bonafede, A. , Bregman, J. , Brouw, W. N. , Brggen, M. , Butcher, H. R. , Carbone, D. , Ciardi, B. , de Gasperin, F. , Duscha, S. , Eislffel, J. , Fallows, R. A. , Frieswijk, W. , Garrett, M. A. , van Haarlem, M. P, Heald, G. , Hoeft, M. , Horneffer, A. , Iacobelli, M. , Juette, E. , Karastergiou, A. , Kohler, J. , Kondratiev, V. I. , Kuniyoshi, M. , Kuper, G. , van Leeuwen, J. , Maat, P. , McFadden, R. , McKay - Bukowski, D. , Orru, E. , Paas, H. , Pandey - Pommier, M. , Pandey, V. N. , Pizzo, R. , Polatidis, A. G. , Reich, W. , Rttgering, H. , Schwarz, D. , Serylak, M. , Sluman, J. , Smirnov, O. , Tasse, C. , Toribio,

M. C. , Vermeulen, R. , van Weeren, R. J. , Wijers, R. A. M. J. , Wucknitz, O. , and Zarka, P. (2015). Calibrating the absolute amplitude scale for air showers measured at LOFAR. Journal of Instrumentation, 10 (11): P11005 - P11005.

[33] Novaco, J. C. and Brown, L. W. (1978). Nonthermal galactic emission below 10 megahertz. Astrophys. J. , 221: 114 - 123.

[34] Oberoi, D. and Kasper, J. (2004). Lofar: The potential for solar and space weather studies. Planetary and Space Science, 52 (15): 1415 - 1421.

[35] Prinsloo, D. , Ruiter, M, Arts, M. , Marel, J. , Boonstra, A. , Kruithof, G. Wise, M. , Falcke, H. , KleinWolt, M. , Rothkaehl, H. , Cecconi, B. , Dekkali, M. , and Ping, J. (2018). EMI modelling of an 80kHz to 80MHz wideband antenna and low - noise amplifier for radio astronomy in space. In Proceedings of the 12th European Conference on Antennas and Propagation (EuCAP 2018), London UK, April 2018. Institution of Engineering and Technology, doi: 10. 1049/cp. 2018. 0820.

[36] Rajan, R. T. , Bentum, M. , and Boonstra, A. - J. (2013a). Synchronization for space based ultra low frequency interferometry. In Proceedings of the IEEE Aerospace Conference, Big Sky, Montana, USA, March 2013. IEEE, inspec: 13502805.

[37] Rajan, R. T. , Bentum, M. , Gunst, A. , and Boonstra, A. (2013b). Distributed correlators for interferometry in space. In Proceedings of the IEEE Aerospace Conference, Big Sky, Montana, USA, March 2013. IEEE, inspec: 13502806.

[38] Rajan, R. T. , Boonstra, A. -J. , Bentum, M. , Klein - Wolt, M. , Belien, F. , Arts, M. , Saks, N. , and van der Veen, A. - J. (2016). Space - based aperture array for ultra - long wavelength radio astronomy. Experimental Astronomy, 41 (1 -2): 271 - 306.

[39] Rajan, R. T. , Engelen, S. , Bentum, M. , and Verhoeven, C. (2011). Orbit - ing low frequency array for radio astronomy. In Proceedings of the IEEE Aerospace Conference, Big Sky, Montana, USA, March 2011. IEEE, inspec: 11943731.

[40] Rajan, R. T. , Leus, G. , and van der Veen, A. - J. (2015). Joint relative position and velocity estimation for an anchorless network of mobile nodes. Signal Processing, 115: 66 - 78 .

[41] Rajan, R. T. , Leus, G. , and van der Veen, A. - J. (2019). Relative kinematics of an anchorless network. Signal Processing, 157: 266 - 279.

[42] Rajan, R. T. and van der Veen, A. - J. (2015). Joint ranging and synchronization for an anchorless network of mobile nodes. IEEE Transactions on Signal Processing, 63 (8): 1925 - 1940.

[43] Saks, N. , Boonstra, A. , Rajan, R. T. , Bentum, M. , Belien, F. , and van't Klooster, K. (2010). Daris, a fleet of passive formation flying small satellites for low frequency radio astronomy. In Proceedings of the Small Satellite Systems and Services - The 4S Symposium, Madeira, Portugal, June 2010, pages 1 - 15,

Funchal, Madeira, Portugal.

[44] Scholten, O., Bacelar, J., Braun, R., De Bruyn, A., Falcke, H., Stappers, B., and Strom, R. (2006). Optimal radio window for the detection of ultrahigh energy cosmic rays and neutrinos off the moon. Astroparticle Physics, 26 (3): 219 -229.

[45] Shimwell, T. W., Röttgering, H. J. A., Best, P. N., Williams, W. L., Dijkema, T. J., de Gasperin, F., Hardcastle, M. J., Heald, G. H., Hoang, D. N., Horneffer, A., Intema, H., Mahony, E. K., Mandal, S., Mechev, A. P., Morabito, L, Oonk, J. B. R., Rafferty, D., Retana - Montenegro, E., Sabater, J., Tasse, C., van Weeren, R. J., Brüggen, M., Brunetti, G., Chyży, K. T., Conway, J. E., Haverkorn, M. , Jackson, N. , Jarvis, M . J., McKean, J. P., Miley, G. K., Morganti, R., White, G. J., Wise, M. W., van Bemmel, I. M., Beck, R., Brienza, M., Bonafede, A., Calistro Rivera, G., Cassano, R., Clarke, A. O., Cseh, D., Deller, A., Drabent, A., van Driel, W., Engels, D., Falcke, H., Ferrari, C., Fröhlich, S., Garrett, M. A., Harwood, J. J., Heesen, V., Hoeft, M., Horellou, C., Israel, F. P., Kapinska , A. D., Kunert - Bajraszewska, M., McKay, D. J., Mohan, N. R., Orrú, E., Pizzo, R. F., Prandoni, I., Schwarz, D. J., Shulevski, A., Sipior, M., Smith, D. J. B., Sridhar, S. S., Steinmetz, M., Stroe, A., Varenius, E., van der Werf, P. P., Zensus, J. A., and Zwart, J. T. L. (2017). The LOFAR Two - metre Sky Survey. I. Survey description and preliminary data release. Astronomy 8 Astrophysics, 598: A104.

[46] van Haarlem, M. P., Wise, M. W., Gunst, A. W., Heald, G., McKean, J. P., Hessels, J. W. T., de Bruyn, A. G., Nijboer, R., Swinbank, J., Fallows, R., Brentjens, M., Nelles, A., Beck, R., Falcke, H., Fender, R., Hörandel, J., Koopmans, L. V. E., Mann, G., Miley, G., Röttgering, H., Stappers, B. W. , Wijers, R. A. M. J., Zaroubi, S., van den Akker, M., Alexov, A., Anderson , J., Anderson, K., van Ardenne, A., Arts, M., Asgekar, A., Avruch, I. M., Batejat, F., Bahren, L., Bell, M. E., Bell, M. R., van Bemmel, I., Bennema, P., Bentum, M. J., Bernardi, G., Best, P., Birzan, L., Bonafede, A. Boonstra, A., Braun, R., Bregman, J., Breitling, F., van de Brink, R. H., Broderick, J., Broekema, P. C., Brouw, W. N., Brüggen, M., Butcher, H. R., van Cappellen, W., Ciardi, B., Coenen, T., Conway, J., Coolen, A., Corstanje, A., Damstra, S., Davies, O., Deller, A. T., Dettmar, R., van Diepen, G., Dijkstra, K., Donker, P., Doorduin, A., Dromer, J., Drost , M., van Duin, A., Eislöffel, J., van Enst, J., Ferrari, C., Frieswijk, W., Gankema, H., Garrett, M. A., de Gasperin, F., Gerbers, M., de Geus, E., GrieBmeier, J., Grit, T., Gruppen, P., Hamaker, J. P., Hassall, T., Hoeft, M., Holties, H.,

Horneffer, A., van der Horst, A., van Houwelingen, A., Huijgen, A., Iacobelli, M., Intema, H., Jackson, N., Jelic, V., de Jong, A., Juette, E., Kant, D., Karastergiou, A., Koers, A., Kollen, H., Kondratiev, V. I., Kooistra, E., Koopman, Y., Koster, A., Kuniyoshi, M., Kramer, M., Kuper, G., Lambropoulos, P., Law, C., van Leeuwen, J., Lemaitre, J., Loose, M., Maat, P., Macario, G., Markoff, S., Masters, J., McKayBukowski, D., Meijering, H., Meulman, H., Mevius, M., Middelberg, E., Millenaar, R., Miller - Jones, J. C. A., Mohan, R. N., Mol, J. D., Morawietz, J., Morganti, R., Mulcahy, D. D., Mulder, E., Munk, H., Nieuwenhuis, L., van Nieuwpoort, R., Noordam, J. E., Norden, M., Noutsos, A., Offringa, A. R., Olofsson, H., Omar, A., Orrù, E., Overeem, R., Paas, H., PandeyPommier, M., Pandey, V. N., Pizzo, R., Polatidis, A., Rafferty, D., Rawlings, S., Reich, W., de Reijer, J., Reitsma, J., Renting, A., Riemers, P., Rol, E., Romein, J. W., Roosjen, J., Ruiter, M., Scaife, A., van der Schaaf, K., Scheers, B., Schellart, P., Schoenmakers, A., Schoonderbeek, G., Serylak, M., Shulevski, A., Sluman, J., Smirnov, O., Sobey, C., Spreeuw, H., Steinmetz, M., Sterks, C. G. M., Stiepel, H., Stuurwold, K., Tagger, M., Tang, Y., Tasse, C., Thomas, I., Thoudam, S., Toribio, M. C., van der Tol, B., Usov, O., van Veelen, M., van der Veen, A., ter Veen, S., Verbiest, J. P. W., Vermeulen, R., Vermaas, N., Vocks, C., Vogt, C., de Vos, M., van der Wal, E., van Weeren, R., Weggemans, H., Weltevrede, P., White, S., Wijnholds, S. J., Wilhelmsson, T., Wucknitz, O., Yatawatta, S., Zarka, P., Zensus, A., and van Zwieten, J. (2013). LOFAR: The LOw - frequency ARray. Astronomy & Astrophysics, 556: A2.

[47] van Vugt, P., Wijnholds, S., Meijerink, A., and Bentum, M. J. (2017). Frequency smearing in full 3d interferometry. In Proceedings of the IEEE Aerospace Conference, Big Sky, Montana, USA, March 2017. IEEE, inspec: 16946319.

[48] Weiler, K., Johnston, K.. Simon, R., Dennison, B., Erickson, W., Kaiser, M., Cane, H., and Desch, M. (1988). A low frequency radio array for space. Astronomy & Astrophysics, 372: 372 - 379.

[49] Zarka, P. (1998). Auroral radio emissions at the outer planets: Observations and theories. Journal of Geophysical Research: Planets, 103 (E9): 20159 - 20194.

[50] Zarka, P. (2007). Plasma interactions of exoplanets with their parent star and associated radio emissions. Planetary and Space Science, 55 (5): 598 - 617.

[51] Zarka, P., Bougeret, J. - L., Briand, C., Cecconi, B., Falcke, H., Girard, J., Grießmeier, J. - M., Hess, S., Klein - Wolt, M., Konovalenko, A., Lamy, L., Mimoun, D., and Aminaei, A. (2012). Planetary and exoplanetary low frequency radio observations from the Moon. Planetary and Space Science, 74: 156 - 166.

Photometric Normalization of Chang'e-4 Visible and Near-Infrared Imaging Spectrometer Datasets: A Combined Study of In-Situ and Laboratory Spectral Measurements

Xiaobin Qi[1], Zongcheng Ling[1,2], Jiang Zhang[1], Jian Chen[1], Haijun Cao[1], Changqing Liu[1], Le Qiao[1], Xiaohui Fu[1,2], Zhiping He[3], Rui Xu[3], Jianzhong Liu[4] and Yongliao Zou[5]

1. Shandong Key Laboratory of Optical Astronomy and Solar-Terrestrial Environment, School of Space Science and Physics, Institute of Space Sciences, Shandong University, Weihai 264209, China; xiaobinqi@mail.sdu.edu.cn (X. Q.); zhang_jiang@sdu.edu.cn (J. Z.); merchenj@mail.sdu.edu.cn (J. C.);
Caohj@mail.sdu.edu.cn (H. C.); liucq@mail.sdu.edu.cn (C. L.); leqiao.geo@gmail.com (L. Q.);
fuxh@sdu.edu.cn (X. F.)
2. CAS Center for Excellence in Comparative Planetology, Chinese Academy of Sciences, Hefei 230026, China
3. Key Laboratory of Space Active Opto-Electronics Technology, Shanghai Institute of Technical Physics, Chinese Academy of Sciences, Shanghai 200083, China; hzping@mail.sitp.ac.cn (Z. H.); xurui@mail.sitp.ac.cn (R. X.)
4. Institute of Geochemistry, Chinese Academy of Sciences, Guiyang 550002, China; liujianzhong@mail.gyig.ac.cn
5. State Key Laboratory of Space Weather, National Space Science Center, Chinese Academy of Sciences, Beijing 100190, China; zouyongliao@nssc.ac.cn

Abstract: Until 29 May 2020, the Visible and Near-Infrared Imaging Spectrometer (VNIS) onboard the Yutu-2 Rover of the Chang'e-4 (CE-4) has acquired 96 high-resolution surface in-situ imaging spectra. These spectra were acquired under different illumination conditions, thus photometric normalization should be conducted to correct the introduced albedo differences before deriving the quantitative mineralogy for accurate geologic interpretations. In this study, a Lommel-Seeliger (LS) model and Hapke radiative transfer (Hapke) model were used and empirical phase functions of the LS model were derived. The values of these derived phase functions exhibit declining trends with the increase in phase angles and the opposition effect and phase reddening effect were observed. Then, we discovered from in - situ and laboratory measurements that the shadows caused by surface roughness have significant impacts on reflectance spectra and proper corrections were introduced. The validations of different phase functions showed that the maximum discrepancy at 1500nm of spectra corrected by the LS model was less (~3.7%) than that by the Hapke model (~7.4%). This is the first time that empirical phase functions have been derived for a wavelength from 450 to 2395nm using in - situ visible and near - infrared spectral datasets. Generally, photometrically normalized spectra exhibit smaller spectral slopes, lower FeO contents and larger optical maturity parameter (OMAT) than spectra without correction. In addition, the band centers of the 1 and 2μm absorption features of spectra after photometric normalization exhibit a more concentrated distribution, indicating the compositional homogeneity of soils at the CE - 4 landing site.

Keywords: Chang'e - 4; visible and near - infrared imaging spectrometer; lunar soils; photometric correction; phase function; surface roughness

1 Introduction

On 3 January 2019, Chang' e - 4 (CE - 4) landed in the South Pole - Aitken (SPA) basin on the lunar farside and released the Yutu - 2 rover. The rover is equipped with four payloads including Panoramic Cameras (PCAMs), a Visible and Near - Infrared Imaging Spectrometer (VNIS), Lunar Penetrating Radars (LPRs), and international - joint Advanced Small Analyzed for Neutrals (ASANs), to investigate the morphology, mineral composition, subsurface structure, and energetic neutral atoms (ENAs) of the CE - 4 landing site and travel path[1]. The Yutu - 2 rover has driven a total of 447.68 m on the floor of the Von Kármán crater (Figure 1) in a north - westerly direction, heading for basalt or craters with high reflectance. Ninety - six hyperspectral measurements were performed (for 18 lunar days until 29 May 2020) over relatively uniform terrain and the rover is still active as of 23 September. This valuable dataset provides critical observations for uncovering the mineralogy and rock lithologies on the Von Kármán crater floor in South Pole - Aitken (SPA) basin of the lunar farside[2-6]. Note that these in - situ spectral datasets were acquired with different illumination conditions (e. g., incident angle (i), emission angle (e), and phase angle (g)). It is widely known that in addition to mineralogy, reflected light entering the spectrometer is also influenced by illumination conditions, causing albedo differences due to observation geometries[7]. Thus, photometric normalization should be conducted to normalize these spectra to a standard illumination condition ($i=30°$, $e=0$, $g=30°$) before deriving the quantitative mineralogy for accurate geologic interpretations.

Numerous investigations have been reported to eliminate the influence of illumination/observation geometry[8-12]. Existing lunar photometric models can be

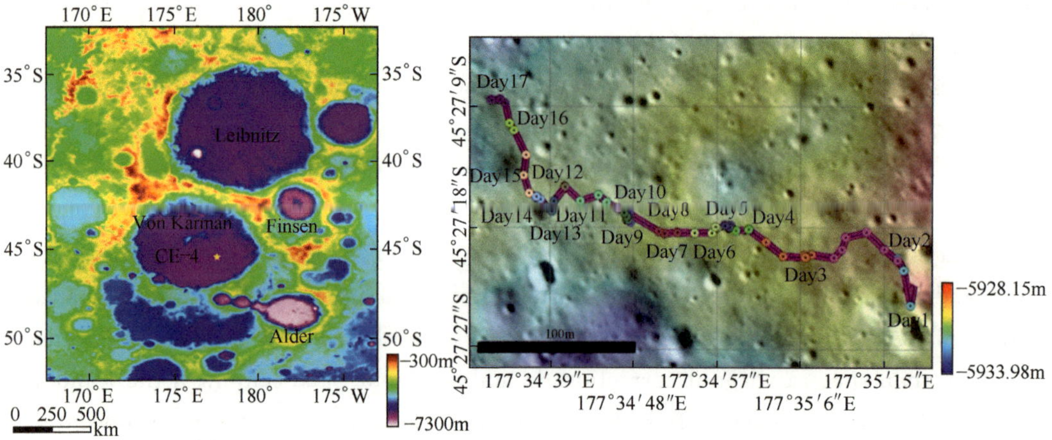

(a)　　(b)

Fig. 1　Orbital view of the Chang'e - 4 (CE - 4) landing site

(a) Nomenclature of major features surrounding the CE - 4 landing site on Lunar Reconnaissance Orbiter (LRO) Lunar Orbiter Laser Altimeter Digital Elevation Model; (b) the traverse map of Yutu - 2 in the first 17 lunar days (overlaid on Lunar Reconnaissance Orbiter Camera Narrow Angle Camera Digital Terrain Model). Yutu - 2 did not move in the eighteenth lunar day; each exploration site in different lunar days is marked by circles with different colors

classified as two distinct groups including empirical models[13,14] and models referenced to physical properties (e. g., radiative transfer models)[9,11,15]. Among the empirical models, the Lommel - Seeliger (LS) model has the advantage of simple analysis, and in many cases, it is an excellent first approximation to diuse reflection[16]. The LS model assumes that light attenuates exponentially as it penetrates the surface[16] and it has been widely applied to spectral datasets obtained from lunar spacecrafts (e. g., Lunar Reconnaissance Orbiter Camera[17], Chandrayaan - 1 Moon Mineralogy Mapper[10], Chang'e - 1 Imaging Interferometer[18-20], CE - 4 VNIS[5,21]). The Hapke radiative transfer (Hapke) model is based on the radiative transport equation of granular media and comprehensively considers the effects of mineralogy, surface roughness, grain sizes, compaction, and multiple scattering factors[9,11,22,23]. This model has been widely applied to the photometric normalization of reflectance spectra detected from various solar system objects (e. g., asteroids, Mercury, comets, Mars, Moon)[12,24-31].

For the above photometric models, phase functions describe the relationship between reflectance and phase angles[32]. Accurate phase functions of the above models are essential in photometric normalization, which is an important data processing step in the scientific analysis and application of Chang'e - 3 (CE - 3)[33,34] and CE - 4 VNIS data. However, most previous phase functions of the LS and Hapke model were obtained from orbital or Earth - based observations[7,14,20,35,36] and it was difficult to achieve the ideal observing conditions for the continuous detection of the same target. Besides, in - situ measurements of the lunar phase functions had never been carried out until the successful landing of the CE - 3 mission[37]. Images acquired by the CE - 3 PCAM were used to extract phase functions, but the limited three bands (470, 540, 640nm) may have constrained the application of these derived phase functions[37]. After CE - 4 landed, the in - situ phase functions of the LS model[21] were derived using the spectral datasets detected on the fourth lunar day. However, due to the limited observation data on the fourth lunar day, the derived phase functions only cover the spectral ranges of 470 - 945nm[21]. Using the Hapke model, Yang et al.[31] and Lin et al.[38] evaluated the effect of view geometry on CE - 4 spectral data and derived average Hapke parameters for the potential uses of a Legendre pronominal and Henyey - Greenstein single - scattering phase functions in the Hapke model.

In this study, we focus on the determination of the best models for photometric normalization of CE - 4 VNIS datasets using data acquired over 18 lunar days. We derived the fitness parameters of the phase functions (LS model) of VNIS datasets covering 450—2395nm using photometric experiments in the CE - 4 landing site on the tenth lunar day. Compared with photometrically corrected reflectance between the LS and Hapke models, this study aims to obtain the best - fit phase functions with the intent to get robust VNIS spectral datasets and better constrain the photometric properties of lunar soils at the CE - 4 landing site.

2 Data and Methods

2.1 VNIS Instrument

The VNIS spectrometer, mounted at the front of Yutu - 2 rover with ~0.69m altitude, comprises a visible and near - infrared (VNIR, 450—945nm) imaging detector and a non - imaging shortwave infrared detector (SWIR, 900 — 2395nm), with a scientific goal to detect the rock lithologies and mineral composition of the lunar soils and rocks at the CE - 4 landing site [39—41]. The spectral resolution of VNIS is 2~10nm for the VNIR range and 3~12nm for the SWIR, respectively, yet the spectral sampling interval is 5nm for both detectors[1,39,40]. The single point measurement of the SWIR spectrometer has a circular field of view (FOV) with a diameter of 107.6 pixels for the corresponding VNIR detector. The central point of the SWIR is located at the coordinate (98, 127.5) within the FOV of the VNIR image[2,39]. The radiance of the forward scattered sunlight from particles on the lunar surface are collected by the VNIS on a trapezoid with a nominal emission angle of ~45°[39,40]. The single - band spectral or image information pass through the Acousto - optic Tunable Filter (AOTF) to form quasi - monochromatic light with a specific wavelength and then converges to the Complementary Metal - Oxide Semiconductor (CMOS) detector[39,42]. The spectral or imaging data of all bands are obtained by changing the driving frequency of exerting on the AOTF in order to achieve wavelength selection[39,40].

2.2 Photometric Models

2.2.1 Lommel - Seeliger Model

As an empirical model, the LS model has the advantage of analytical simplicity and has been widely applied to various spectral datasets obtained from lunar spacecraft[7,12,21]. For low albedo solar system bodies such as the Moon, single scattering dominates the regolith albedo, and the variations of lunar surface reflectance with the illumination - viewing geometry can be approximated by the Lommel - Seeliger Law[5,10,14,19-21]:

$$r(i,\ e,\ g)=\mu_0/(\mu_0+\mu_1)\times f(g) \tag{1}$$

where r is bidirectional reflectance. μ_0 and μ_1 are the cosine values of the incident angle and emission angle, respectively. $f(g)$ describes reflectance variation with the phase angle.

A third - order polynomial was used in this work to model $f(g)$:

$$f(g)=p_1g^3+p_2g^2+p_3g+p_4 \tag{2}$$

The parameter values in Equation (2) were derived for each band using the non - linear least squares methods. The model parameters with the minimum Root Mean Square Error (RMSE)[43] are achieved and then VNIS data are normalized into the standard geometry ($i=30$, $e=0°$, $g=30°$) by Equation (3):

$$r_{\text{measure}}(30°,\ 0°,\ 30°)=r_{\text{model}}(30°,\ 0°,\ 30°)/r_{\text{model}}(i,\ e,\ g)r_{\text{measure}}(i,\ e,\ g) \tag{3}$$

2.2.2 Hapke Radiative Transfer Model

As a model tied to the physical properties of the lunar surface, the Hapke radiative transfer model is another way to characterize the photometric properties and normalize the photometric effects of lunar soils. The Hapke radiative transfer model often assumes that that all airless planetary materials in the media are an intimate mixture and the particle sizes are larger than wavelengths[11]. The Hapke model is derived from the radiative transport equation of the particle medium, and its parameters reflect the chemical and physical properties of the medium. The bidirectional reflectance as a function of i, e, g, and single scattering albedo (SSA) are approximated as follows[9,11,22,23]:

$$r(i,\ e,\ g)=K\frac{\omega}{4\pi}\frac{\mu_0}{\mu_0+\mu_0}\{[1+B(g)]P(g)+H(\mu_0/K)H(\mu_1/K)-1\} \tag{4}$$

where r is bidirectional reflectance (BREF). K is the porosity factor and can usually be calculated with the filling factor ϕ ($\phi=0.41$ for lunar soil)[11,22]: $K=-\ln(1-1.209\phi^{2/3})/(1.209\phi^{2/3})$. ω (SSA) describes the ratio of the energy loss due to the scattering of incident light through a particle medium to the total energy loss due to scattering and absorption. $B(g)$ is a backscattering function describing the shadow - hiding opposition effect and can be expressed as:

$$B(g)=\left(1+\frac{1}{h_s}\tan\frac{g}{2}\right)^{-1} \tag{5}$$

h_s[11] describes the angle width of backscattering and can be expressed as:

$$h_s \approx \frac{3\sqrt{3}}{8}\frac{K\phi}{\ln(1000)} \tag{6}$$

$H(x)$ describes the multiple scattering process and can be approximated as $H(x)=(1+2x)/(1+2\gamma x)$, where $\gamma=\sqrt{(1-\omega)}$[11].

The specific expression of the $P(g)$ is related to the particle size, orientation, electromagnetic properties, etc., and has analytical solutions only under special conditions. In practical applications, phase functions are usually expressed by a Legendre polynomial or empirical formulas such as the Henyey - Greenstein function[44]. In this work, a Legendre polynomial was used:

$$P(g)=1+b\times\cos(g)+c\times(1.5\times\cos^2(g)-0.5) \tag{7}$$

The coefficients b and c determine the general shape of phase functions (Equation (7)), where b describes the degree of forward and backward scattering and c the degree of side scatter[9]. Based on this model, the reflectance data at different phase angles were converted to SSA and then the bidirectional reflectance at a standard geometry was calculated from SSA with Equation (4). Then, bidirectional reflectance can be converted to reflectance factor (REFF) using Equation (8):

$$\text{REFF}=\text{BREF}\times\pi/\mu_0 \tag{8}$$

2.3 CE - 4 VNIS Datasets

The VNIS level 2B data used in this study have been underwent dark - current and scattering - background subtraction, flat - field and instrument temperature correction, and radiometric and geometric calibration[2,39,45] by the Ground Research and Application System (GRAS) of the Chinese Lunar Exploration Program. The level 2B data were then transformed into bidirectional reflectance of solar irradiance divided before normalization to a Sun - Moon distance of 1 AU[6]. For the view geometry of VNIS datasets, the illumination conditions (e. g., incident angle (i), emission angle (e), and phase angle (g)) of each pixel are calculated by the linear interpolation of the values of the four corner points and averaged over a 1° angle bin for the VNIR images. In addition, the central point of the FOV is used for the SWIR spectra to analyze. The detailed observation geometry is shown in Tab. 1. This calculation of illumination geometry assumes a horizontal surface and the incidence angle could be changed by $<2°$[38] due to the topographic slopes, while the phase angle remains unchanged.

Tab. 1　The observation geometry (incident angle (i), emission angle (e), and phase angle (g)) of the tenth lunar day

Observation	i/(°)	e/(°)	g/(°)	Observation	i/(°)	e/(°)	g/(°)
0068	76.543	48.273	79.376	0079	57.272	44.414	91.082
0069	75.946	48.273	78.433	0080	57.873	44.414	92.113
0070	71.319	48.273	71.432	0081	58.289	44.414	92.814
0071	70.102	48.273	69.544	0082	58.741	44.414	93.568
0072	69.576	48.273	68.722	0083	59.239	44.414	94.386
0073	68.867	48.273	67.607	0084	63.498	44.414	100.990
0074	68.299	48.273	66.708	0085	64.127	44.414	101.916
0075	63.709	48.273	59.185	0086	64.640	44.414	102.664
0076	62.823	48.273	57.672	0087	65.227	44.414	103.512
0077	62.015	48.273	56.269	0088	65.812	44.414	104.349
0078	61.149	48.273	54.737	0089	66.370	44.414	105.139
				0090	77.113	44.414	111.614

During the first 18 lunar days of the Yutu - 2 operation, 96 VNIS spectra datasets were obtained (Fig. 1) and three photometric experiments were conducted in the surrounding regions on the fourth lunar day and two sites on the tenth lunar day (Fig. 2). These observations provide a critical opportunity for the photometric studies of local lunar soils. On the fourth lunar day, the Yutu - 2 rover had remained stationary and conducted multiple spectral measurements with phase angles of 39°—97° for the surrounding lunar surface area of homogeneous composition[21]. Along the planned travel path, the VNIS observed another two targeted sites with phase angles of 55°—79° and 91° — 112° (Tab. 1), respectively, on the tenth lunar day. The spectrophotometric experiments on the tenth lunar day provide valuable in - situ data to investigate the photometric properties of CE - 4 lunar soils, thus we focus on this spectral dataset in this study (Tab. 1).

2.4　Laboratorial Spectral Measurements by a VNIS Replica

In order to better understand the photometric properties of the CE - 4 landing site and examine the effect of surface roughness on SWIR bands, we performed the spectral measurements in a laboratory using a VNIS replica. The mixed silicate powders were prepared as an analog of lunar regolith. The photometric measurements were conducted at the Shanghai Institute of Technical Physics, Chinese Academy of Sciences (SITP). The lunar regolith analog (Tab. 2) was prepared following the remote sensing results based on a Kaguya Multiband Imager[46,47], which suggest that lunar regolith at the landing site is mainly made up of ejecta from the Finsen crater[2-4,48]. Orthopyroxene (Opx), Clinopyroxene (Cpx), plagioclase (Pl) and olivine (Ol) were selected as pure mineral endmembers, and they were grounded and sieved into particle size of 20 — 100μm. The mineral proportions of the lunar regolith analog are similar to the Finsen crater ejecta, as shown in Tab. 2. The measurements were designed with instrument configuration as shown in Supplementary Materials Fig. S1. The vertical distance between the light source

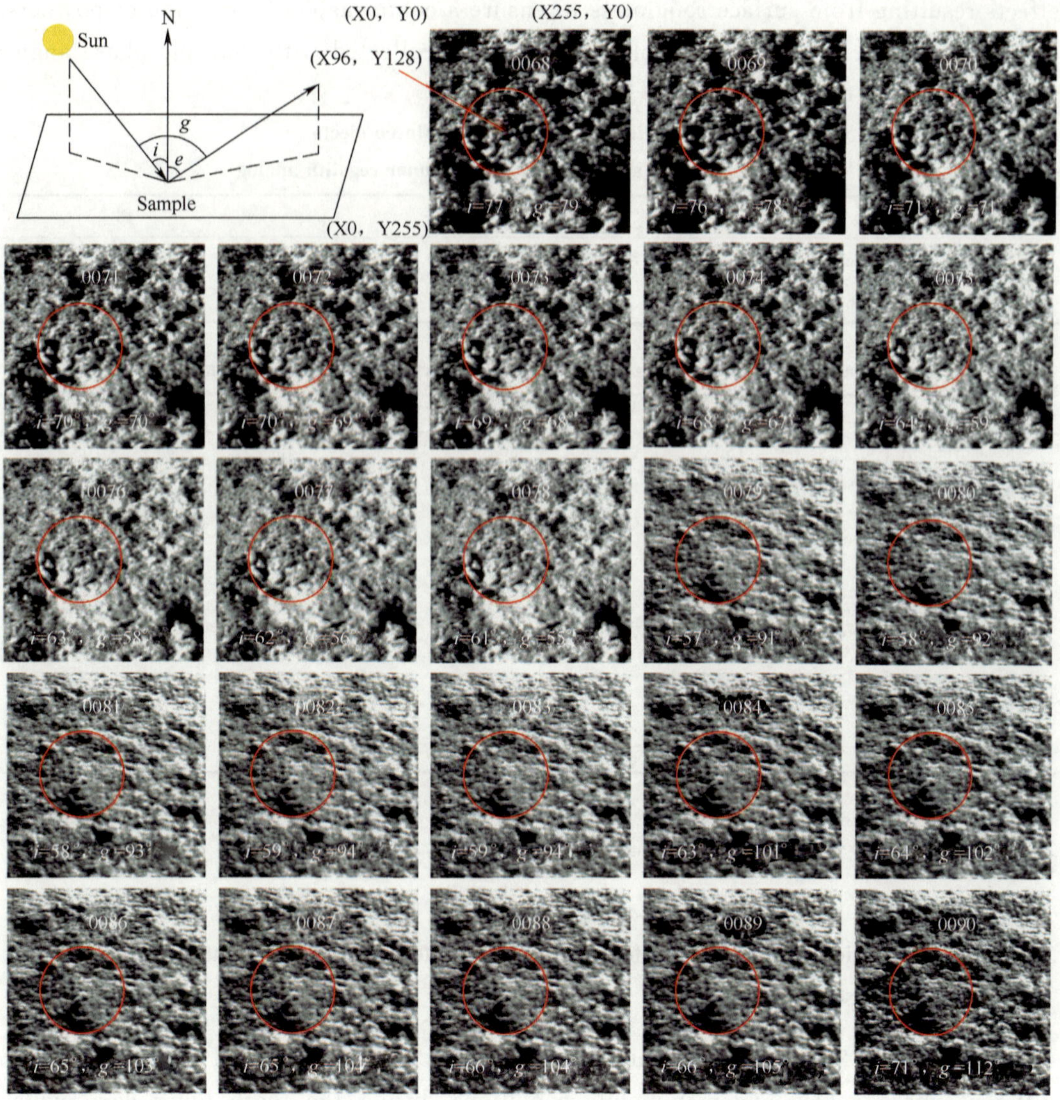

Fig. 2 Reflectance images at 750nm of the two detection targets measured in the tenth lunar day. Addition labels of pixels were added to the first Visible and Near-Infrared (VNIR) image; the red circles and number labels indicate Field of View (FOV) of Shortwave Infrared (SWIR) spectrometer and data filename released at the Data Release and Information Service System of China's Lunar Exploration Program; a schematic diagram of the relationship between incident, emission and phase angles is placed in the upper left corner of Fig. 2

and the experimental platform is 0.69m. The spectrometer was fixed (thus the emission angle was fixed) in order to reduce the error of instrument (Tab. 3). The phase angle was adjusted by changing the azimuth angle of the light source and the included angle between the light source and the normal of the experimental platform was maintained at 60° (thus the incident angle was maintained at 60°) (Tab. 3). The lunar regolith analog was flattened out before VNIS spectral measurements, aiming to suppress the shadowing effects resulting from surface roughness to ensure a conformable distribution of powders. The spectral data of the lunar regolith analog were collected with multiple phase angles varying from 28° to 103° (Tab. 3).

Tab. 2 Mineral proportions of Finsen ejecta (estimated from Kaguya MI data) and lunar regolith analog

Mineral (wt.%)	Ol	Opx	Cpx	Pl
Finsen ejecta	7±3	25±4	13±4	55±4
Lunar regolith analog	7	25	13	55

Tab. 3 The observation geometry of laboratory spectral measurements

Observation	Azimuth Angle/(°)	i/(°)	e/(°)	g/(°)
Measurement 1	150	60	45	103
Measurement 2	140	60	45	97
Measurement 3	130	60	45	76
Measurement 4	120	60	45	63
Measurement 5	110	60	45	56
Measurement 6	100	60	45	49
Measurement 7	90	60	45	42
Measurement 8	80	60	45	35
Measurement 9	70	60	45	28

3 Results

3.1 Phase Functions of the LS Model and Initial Photometric Normalization of CE-4 Datasets

The Yutu-2 rover performed the photometric measurements of two targets with different phase angles under immobile conditions during the tenth lunar day, which provides unprecedented opportunities for determining the accurate phase functions for lunar farside soils. As the first step, the phase curves of VNIR imaging bands and SWIR non-imaging bands are independently fitted using a third-order polynomial LS model as given above (Equation (2), Fig. 3). The derived best-fit values for phase functions of LS model are given in Table S1. The scattering intensities related to phase curves exhibit declining trends with the increase in phase angles shown in Fig. 3a, b. As expected for lunar regolith, when the phase angle is small, the values of phase curves rapidly increase with decreasing phase angles due to the opposition effect (Fig. 3)[24]. It is suggested that the phase curves of short-wavelength bands display faster declining trends with the decrease in phase angles than that of long-wavelength bands, resulting in a phase reddening effect[49]. In addition, the minimum phase angle used to fit phase curves in our work is 55° limited by the actual observations of VNIS (Tab. 1). We conducted an extrapolation correction to 30° for VNIS datasets, possibly importing additional errors and/or uncertainties.

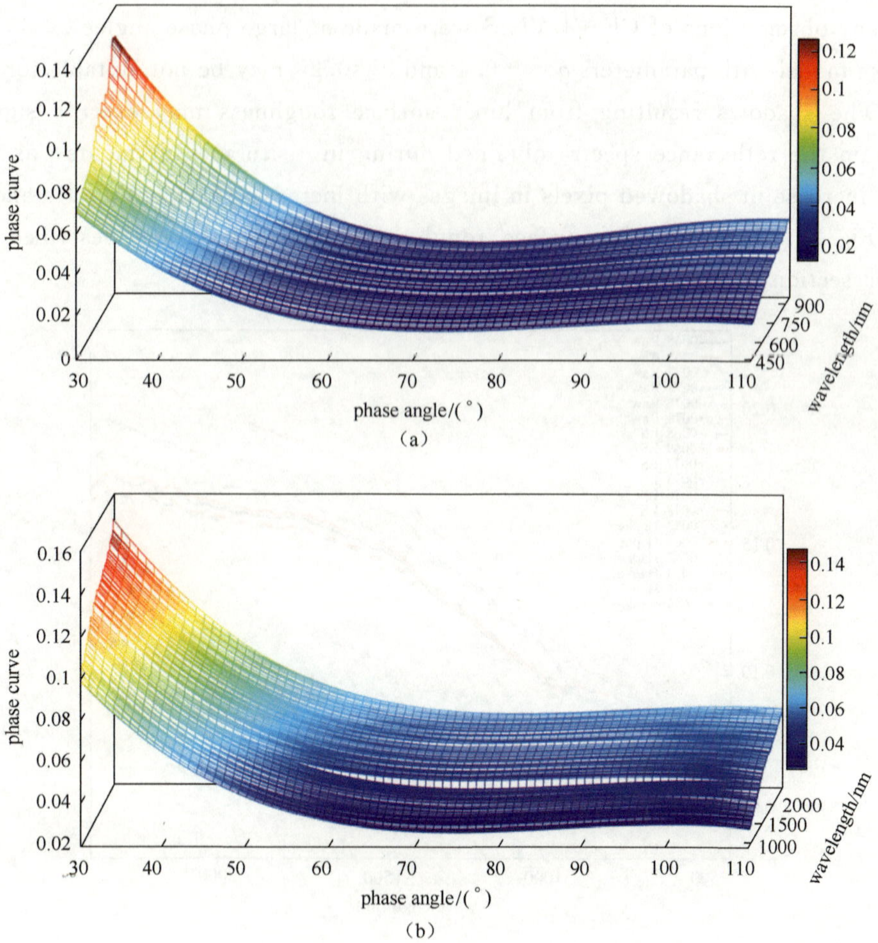

Fig. 3　Three-dimensional phase functions for the Lommel-Seeliger (LS) model derived from observations 0068—0090. (a) VNIR spectral regions; (b) SWIR spectral regions

For the Hapke model, the parameters of phase functions have been intensively studied. In this study, we preliminarily used $b=-0.4$, $c=0.25$ (named by Mustard parameters in this work) for the Legendre polynomials[50,51], as they are widely used in spectral unmixing analysis[4,43]. We then conducted a photometric correction using the Hapke model and LS model with the methods mentioned in Section 2.4. The initial photometric correction results of the LS model for VNIS datasets are compared with those of the Hapke model (Fig. 4), indicating that the corrected reflectance - using Hapke model decreases in contrary to the LS model. Besides, the spectra of the same target did not coincide after photometric corrections employing both the LS and Hapke model. Compared with raw VNIS datasets, spectral curves become more dispersed after the correction of the Hapke model (Fig. 4, the maximum discrepancy at 1500nm increase from 15.4% to 28.3%). As shown in Fig. S2, the values of the single-scattering phase function of the Hapke model are unrealistic when increasing with phase angles greater than ~60°. Many observations of CE-4 VNIS were made at large phase angles (>60°), thus the Hapke model with parameters $b=-0.4$ and $c=0.25$ may be not suitable for CE-4 spectra. The shadows resulting from lunar surface roughness may exert a significant influence on the reflectance spectra obtained during in-situ measurements, as evident from the increase in shadowed pixels in images with increasing incident angles and phase angles (Fig. 2). The effect of surface roughness was assessed and described in the discussion section.

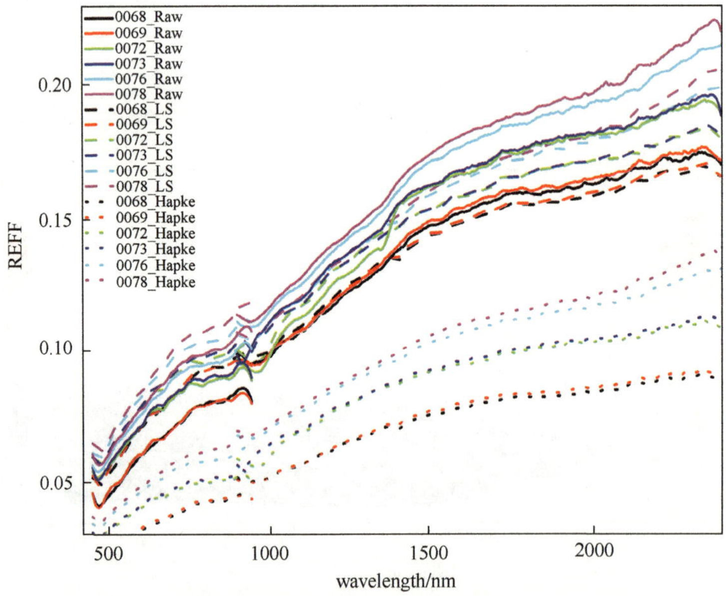

Fig. 4　Comparisons of initial photometric correction results of Hapke radiative transfer (Hapke) model and Lommel-Seeliger (LS) model

3.2　Laboratory Spectral Measurements of Lunar Regolith Analog

As shown in Fig. 5a, the nine spectra of the lunar regolith analog exhibit flat continuum and distinct 1 and 2μm absorption features due to the Fe^{2+} electronic transitions of mafic minerals. In general, the REFF values of nine spectra increase in accordance with phase angles in both the VNIR and SWIR spectral regions. Photometric corrections were performed (Fig. 5b, c) on these laboratory data in order to verify the photometric functions of the LS and Hapke models and help to constrain the photometric properties of the CE-4 landing site.

As indicated in Fig. 5b, the normalized spectra of the SWIR range using the LS model show a striking coincidence (maximum discrepancy was 8.2% at 1500nm). The spectra with very large phase angle ($g = 103°$) exhibit abnormal high REFF values before and after normalization, possibly due to the stray light contamination during VNIS observations. Compared with the normalized spectra using the LS model of CE - 4 lunar soils and regolith analog (Fig. 4 and 5b), CE - 4 spectra exhibit a larger spectral slope and smaller absorption depth caused by the formation and accumulation of submicroscopic metallic iron (SMFe), named the space weathering effect[52,53]. Furthermore, the normalized spectra of CE - 4 show a dispersed distribution (Fig. 4), while normalized laboratory spectra match with each other (Fig. 5b). We tentatively ascribe this to the fact that the lunar regolith analog underwent adequate grounding, sieving and mixing and the apparent surface for the measured sample was smooth at a large scale compared with the analog particle size (20—100μm). However, lunar soils at the CE - 4 landing site were composed of lumps of varying sizes and exhibited an uneven microtopography with shadows caused by surface roughness (Fig. 2). From our laboratory measurements, the surface roughness has a significant influence on the reflectance spectra acquired during the in - situ VNIS observations and its effects should not be ignored in the VNIS data processing. We will discuss the shadow effect in detail in the following section.

As shown in Fig. 5c, similar to CE - 4 VNIS spectra corrected by the Hapke model (mentioned in Section 3.1), the normalized laboratory spectra employing the Hapke model exhibit a similar dispersion. This may further enhance our recognition that the parameters ($b = -0.4$ and $c = 0.25$) in the phase functions of the Hapke model are not reliable when applying it to the CE - 4 VNIS spectra[31]. Instead, we would use the phase function parameters ($b = -0.17$ and $c = 0.70$, named by YZ parameters in this work) of the Hapke model obtained by CE - 4 in - situ measurements[31] to normalize the CE - 4 REFF spectra at different phase angles. Details will be given in following discussions.

4 Discussion

4.1 Evaluation of the Shadow Effects

As discussed above, different performances of the photometric normalization of lunar soils and lunar regolith analog were observed, leading to the necessity to evaluate shadow effects in CE - 4 in - situ VNIS measurements. As indicted by the VNIS images (Fig. 2) of the first target on the tenth lunar day, the numbers of shadowed pixels increase with the increase in incident and phase angles, revealing the influence of surface roughness.

The pixels of the VNIR image with REFF < 0.02 at 465nm were regarded as shadowed pixels. To evaluate the contribution of the shadowed pixels in VNIR and SWIR detectors, we first extracted the number of shadowed pixels in the VNIR images within regions of the FOV of the SWIR detector using a density slice function of ENVI (Fig. 6a, b). The proportions of shadowed pixels (indicated as red areas in Fig. 6b) related to the

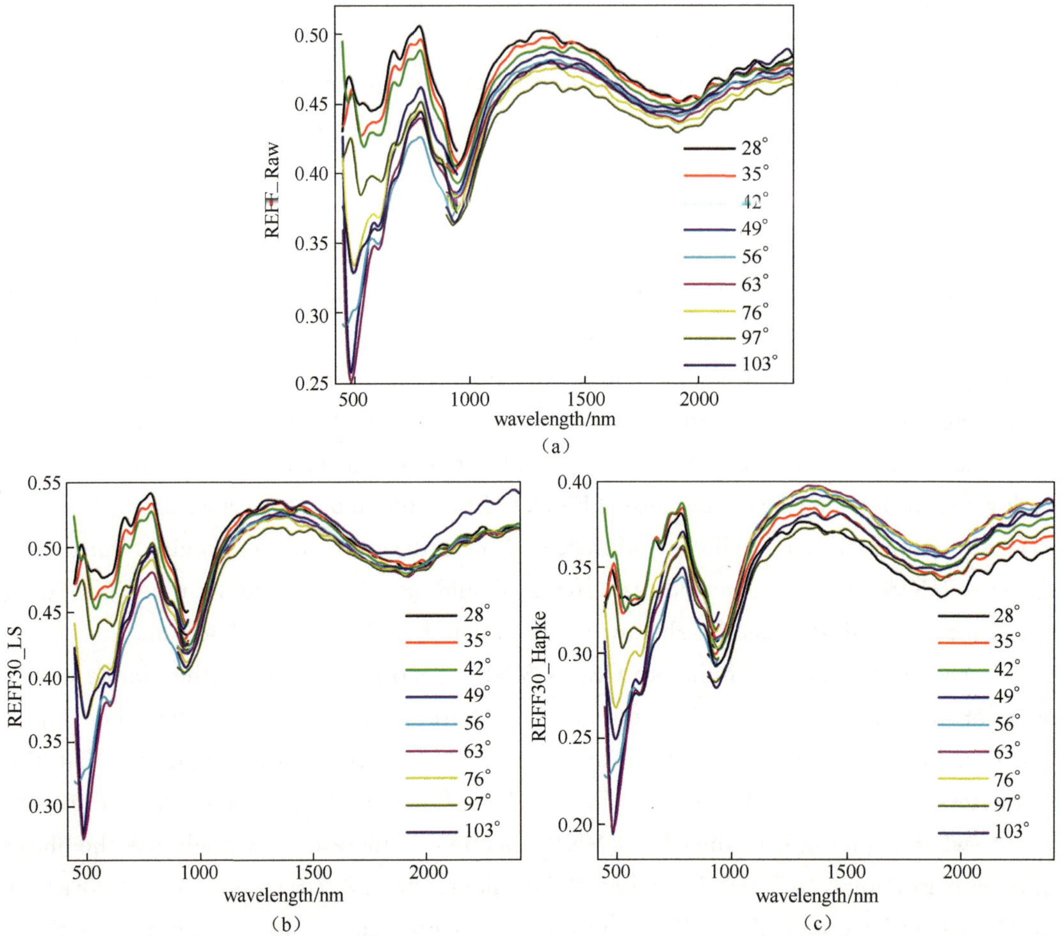

Fig. 5　Comparisons of the photometric correction results of (a) raw reflectance factor (REFF) spectra of lunar regolith analog obtained by VNIS replica at different phase angles; (b) spectra of lunar regolith analog corrected with LS model; (c) spectra of lunar regolith analog corrected with Hapke model

total pixels in the red circle for 750nm were obtained. We find a linear relationship between the proportions of shadowed pixels and REFF values for the eleven spectra obtained during the photometric experiments on the tenth lunar day. As given in Figure 6c, d, the squares of the determination coefficients are 0. 97 and 0. 98 for REFF at 750 and 1500nm, respectively. Therefore, the shadowed pixels will affect the total reflectance values varying with illumination conditions, possibly caused by the multiple scattering contribution to the shadowed pixels. This reveals that in - situ visible and near - infrared detection is susceptible to surface roughness, which should be corrected to get the authentic REFF. Conversely, for orbital images, these small shadows would not be resolvable.

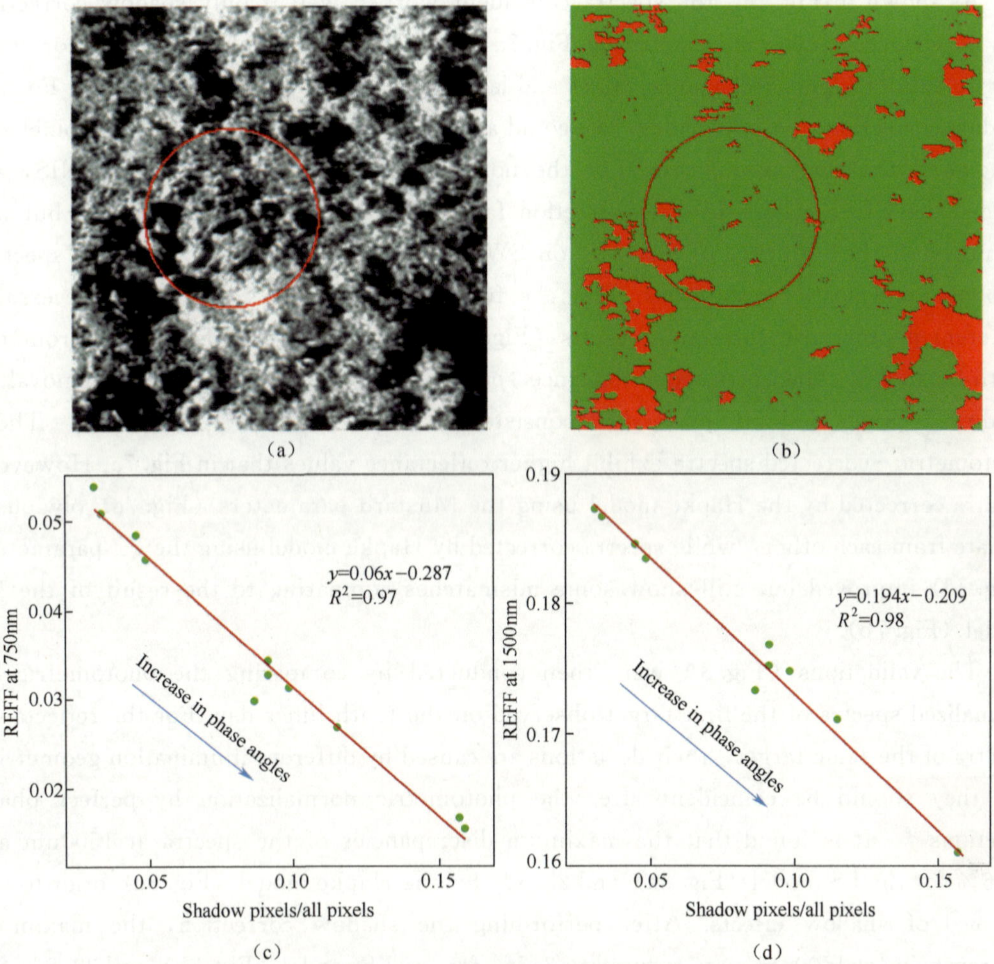

Fig. 6　(a) REFF images at 750nm of CE-4-VNIS-0068; (b) color mapping of CE-4-VNIS-0068 at 750nm; (c) the linear relationship between shadowed pixels in VNIR image and reflectance at 750nm (observations 0068—0079); (d) the linear relationship between shadowed pixels in VNIR image and reflectance at 1500nm (observations 0068—0079). The red circle indicates FOV of SWIR spectrometer; green: pixels with REFF (465nm) >0.02; red: pixels with REFF (465nm) >0.02, were regarded as shadowed pixels

4.2　Improvement and Validation of Photometric Models

To eliminate the shadow effects on the CE-4 VNIR data, all shadowed pixels in VNIR images were discarded. Then, the spectra of all VNIR non-shadowed pixels within the region of the FOV of the SWIR detector are averaged and used as the new REFF spectra from 450 to 950nm. To reduce the influence of shadows on SWIR spectra, the REFF values of all bands were divided by the correction factors that are defined by the proportions of the non-shadowed to all the pixels within the red circles (e.g., Fig. 6b) extracted from the corresponding VNIR images.

As shown in Fig. 7a, the spectra coincident worked well if only shadow correction was performed. However, comparing Fig. 7a with Fig. 7b, the photometric - corrected spectra exhibit higher reflectance values and larger spectral slopes in SWIR bands. For the shadow correction of VNIR bands, we could assess the reflectance value of each pixel and remove all the shadowed pixels. For the non - imaging SWIR detector of VNIS, we divided the REFF values by the correction factors. This could help to weaken but not eliminate all the influence of shadows on SWIR detector due to their different spectral responses and spatial resolutions. Thus, a further photometric correction is necessary. Different photometric correction results (Fig. 7) were compared for the data from the tenth lunar day. The corrected reflectance spectra using the LS model after removal of shadowed pixels exhibit preferable consistence with each other (Fig. 7b). These photometric - corrected spectra exhibit higher reflectance values that in Fig. 7a. However, spectra corrected by the Hapke model using the Mustard parameters (Fig. 7c) obviously deviate from each other, while spectra corrected by Hapke model using the YZ parameters (Fig. 7d) improved but still show some mismatches comparing to the result of the LS model (Fig. 7b).

The validations (Fig. 8) were then conducted by comparing the photometrically normalized spectra of the first target observed on the tenth lunar day. For the reflectance spectra of the same target, their deviations are caused by different illumination geometries and they should be coincident after the photometric normalization by perfect phase functions[36]. It is found that the maximum discrepancies of the spectra at 1500nm are 10.6% for the LS model (Fig. 8a) and 28.3% for the Hapke model (Fig. 8c) prior to the removal of shadow effects. After performing the shadow correction, the maximum discrepancies at 1500nm were reduced to 3.7% for the LS model (Fig. 8b) and 20.8% for the Hapke model using Mustard parameters (Figu. 7c and 8d). The relatively poor spectra coincidence of the Hapke model for Mustard parameters reveals that they are not suitable for spectra with large phase angles, as is the case for CE - 4 observations. In order to improve the performance of the Hapke model, the YZ parameters ($b = -0.17$ and $c = 0.70$), obtained by CE - 4 in - situ photometric experiment datasets[31], were employed to normalize the CE - 4 REFF spectra at different phase angles. Results show that the maximum discrepancy at 1500 nm was reduced to 7.4% (Fig. 7d and 8e). The spectra of the second target measured on the tenth lunar day were also shown to evaluate the performance of the phase functions of the LS and Hapke model (Fig. S3). However, correction results of both the LS and Hapke model show obvious trichotomy in terms of phase angles. This inconsistency could be caused by stray light contamination at very large phase angles, which was also observed during laboratory measurements using the CE - 4 VNIS replica. It is worth noting that the spectra measured at different phase angles are approximately parallel to each other after being corrected by the LS model (Fig. S3b, f), while they deviated from each other after being corrected by the Hapke model (Fig. S3c,

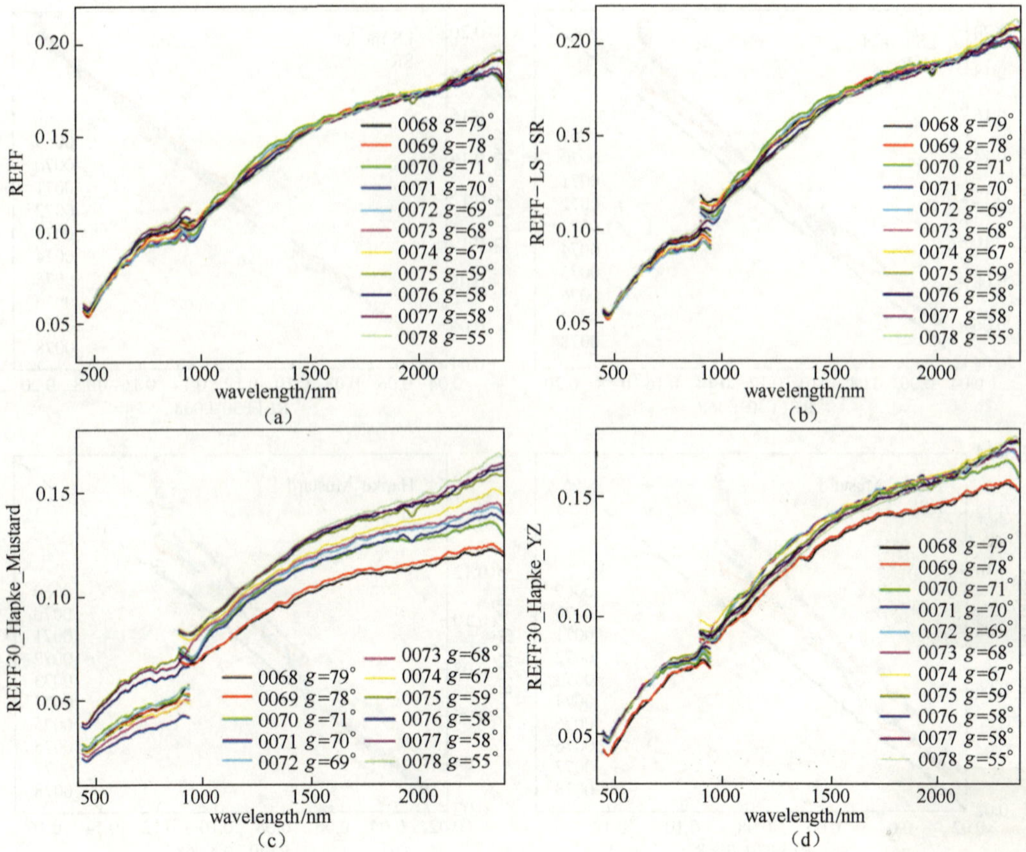

Fig. 7　Comparisons of photometric correction results of the CE - 4 VNIS spectra measured on the tenth lunar day (target 1). (a) Spectra that are only shadow - corrected; (b) spectra after shadow correction corrected by LS model; (c) spectra of the first target corrected by Hapke model ($b=-0.4$ and $c=0.25$); (d) spectra of the first target corrected by Hapke model ($b=-0.17$ and $c=0.70$)

d, g, h).

In summary, we find that in - situ measured spectra are greatly influenced by the shadows caused by surface roughness. For the in - situ photometric correction, proper shadow correction should be conducted first. We also find that the phase functions derived for the LS model are more suitable for the photometric normalization of CE - 4 VNIS datasets, which could help to gain an accurate interpretation of the relative quantitative mineralogy along the traverse route of the Yutu - 2 rover in the Von Kármán crater.

4.3　Comparison of the Spectral and Mineralogical Properties of the CE - 4 and CE - 3 Lunar Soils

The VNIS observations implemented by CE - 4 and CE - 3 were the only in - situ reflectance detections of the Moon so far, providing unique opportunities to comprehend the properties of lunar soils. In this work, the in - situ VNIS spectra of CE - 4 and CE - 3

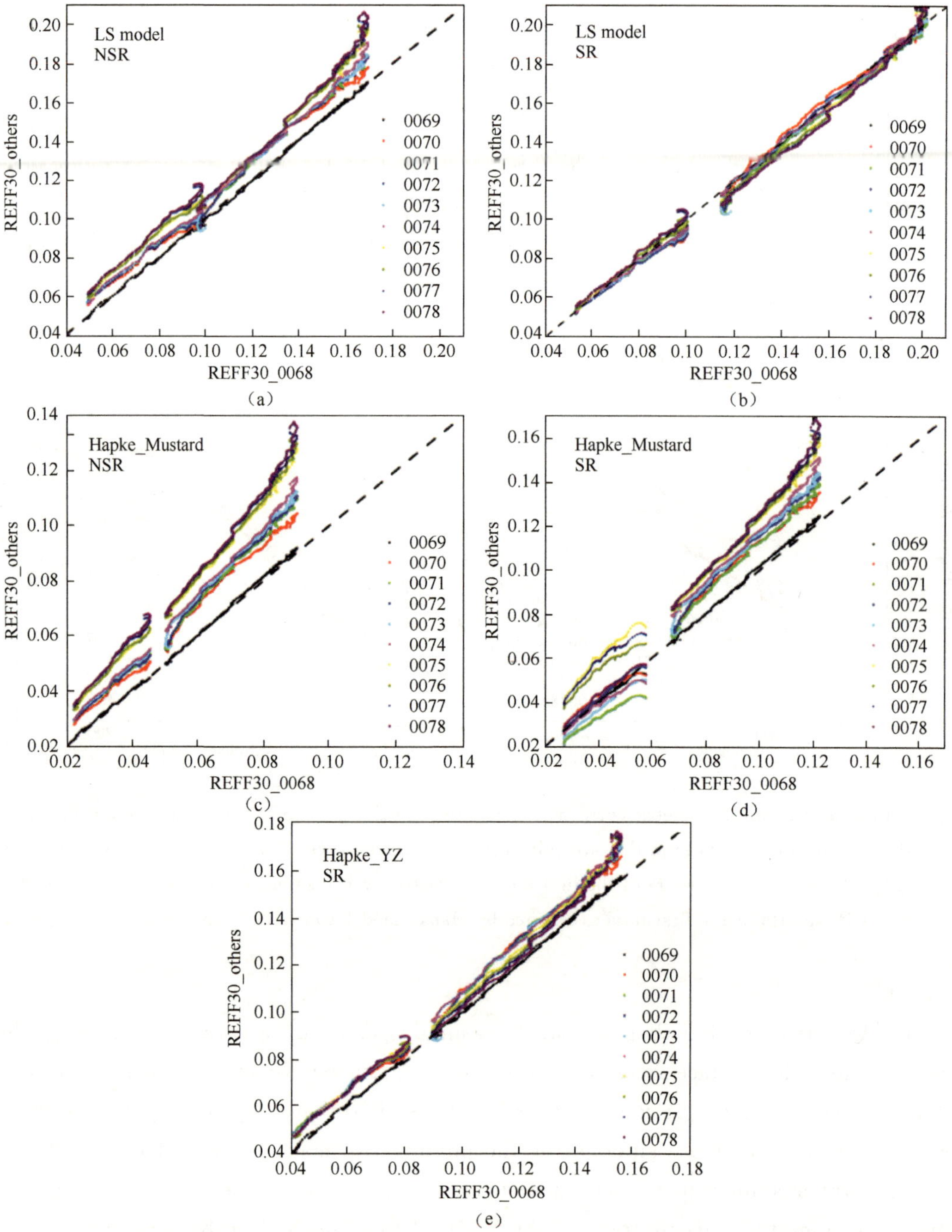

Fig. 8　Scatter plot between photometrically normalized VNIS spectra of the same target and at different illumination geometries. (a) Before shadow effect corrections (LS model); (b) after shadow effect corrections (LS model); (c) before shadow effect corrections (Hapke model, Mustard parameters); (d) after shadow effect corrections (Hapke model, Mustard parameters); (e) after shadow effect corrections (Hapke model, YZ parameters). The dashed line denotes the two spectra are coincident; NSR—no shadow effects removed; SR—shadow effects removed

landing site were normalized to the standard geometry using the LS model (Fig. 9a, b), ensuring that the corrected spectra are comparable with each other, devoid of the influence of solar illumination conditions. The VNIR spectra were multiplied by a factor to splice them with the SWIR spectra; then, a Savitzky - Golay filter [54] of order 3 that uses a frame size of 31 points was applied to smooth these spectra (Fig. 9a). Except some rock - related spectra (e. g., 0014, 0015, 0104 and 0107), the regolith spectra of the CE - 4 landing site show a weak absorption depth and consistent 1μm absorption band center. The spectra of the CE - 3 landing site exhibit lower reflectance values and a stronger absorption depth than that of CE - 4 spectra. Then, spectral continuums were removed and spectral parameters (e. g., band centers, integrated band depth (IBD) of 1μm absorption) were extracted using the data processing method of [6].

As shown in Fig. 9c, the normalized VNIS spectra of the CE - 4 landing site exhibited redder spectral slopes (R1600/R750) (Fig. 9c, average value of 1. 70) and a weaker IBD (Fig. 9d, average value of 16. 5) of the 1μm absorption than that of the CE - 3 landing site (average values of 1. 58 and 28. 1) due to different optical maturities. Then, the contents of FeO and the optical maturity parameter (OMAT)[53] were estimated (Tab. S2) from the photometric - normalized spectra using parameters provided by[21]. Results show that the estimated OMAT of CE - 4 soils are 0. 18±0. 03—smaller than the results of [21] (~0. 25). Besides, the soils at the CE - 4 landing site had a lower abundance of FeO (10. 8 ± 1. 06 wt.%) than that of the CE - 3 landing site (~22. 8 wt.%)[33], which is consistent with the lunar soils of the CE - 4 landing site as they exhibit higher maturity and are exposed to a longer duration of space weathering[55,56]. Moreover, previous studies suggest that the exposure ages of the CE - 4 and CE - 3 landing sites are about 3. 5—3. 6 Ga (Finsen crater ejecta)[56,57] and 27—80 My (Zi Wei crater)[58], and the maturity indexes I_s/FeO are ~82[59] and ~53[55], respectively.

In terms of mineral chemistries, the CE - 4 landing site is located near ballistic ejecta from the Imbrian - aged Finsen [56] and/or the Imbrian - aged Alder crater [4,47,60], which are rich in Low - Calcium Pyroxene (LCP) (HCP/LCP=~ 0. 7) and characterized by a more feldspathic composition[48]. However, the CE - 3 landing site is located on the young impact ejecta blankets of the Copernican ZiWei crater on the lunar nearside[58], which is rich in high - calcium pyroxene (HCP) (HCP/LCP=~ 2. 0), olivine (>20 wt.%) and characterized by medium - titanium lunar basalt exposure[33,61]. As shown in Fig. 9e, the band centers of the 1μm absorption are plotted against 2μm absorption to identify the composition of pyroxene. The band centers (1μm absorption) of CE - 3 soils fall in the range of 1004—1027nm, while CE - 4 regolith was concentrated in the range of 951—1006nm. In addition, the band centers (1μm absorption) of CE - 4 soils located at the position of non - mare clinopyroxene (CPX), while the pyroxene chemistries of the CE - 3 soils were more Fe, Ca - rich than that of CE - 4. Despite the large deviation of 2000nm band centers (with standard deviation of 154nm), the CE - 4 VNIS spectra are

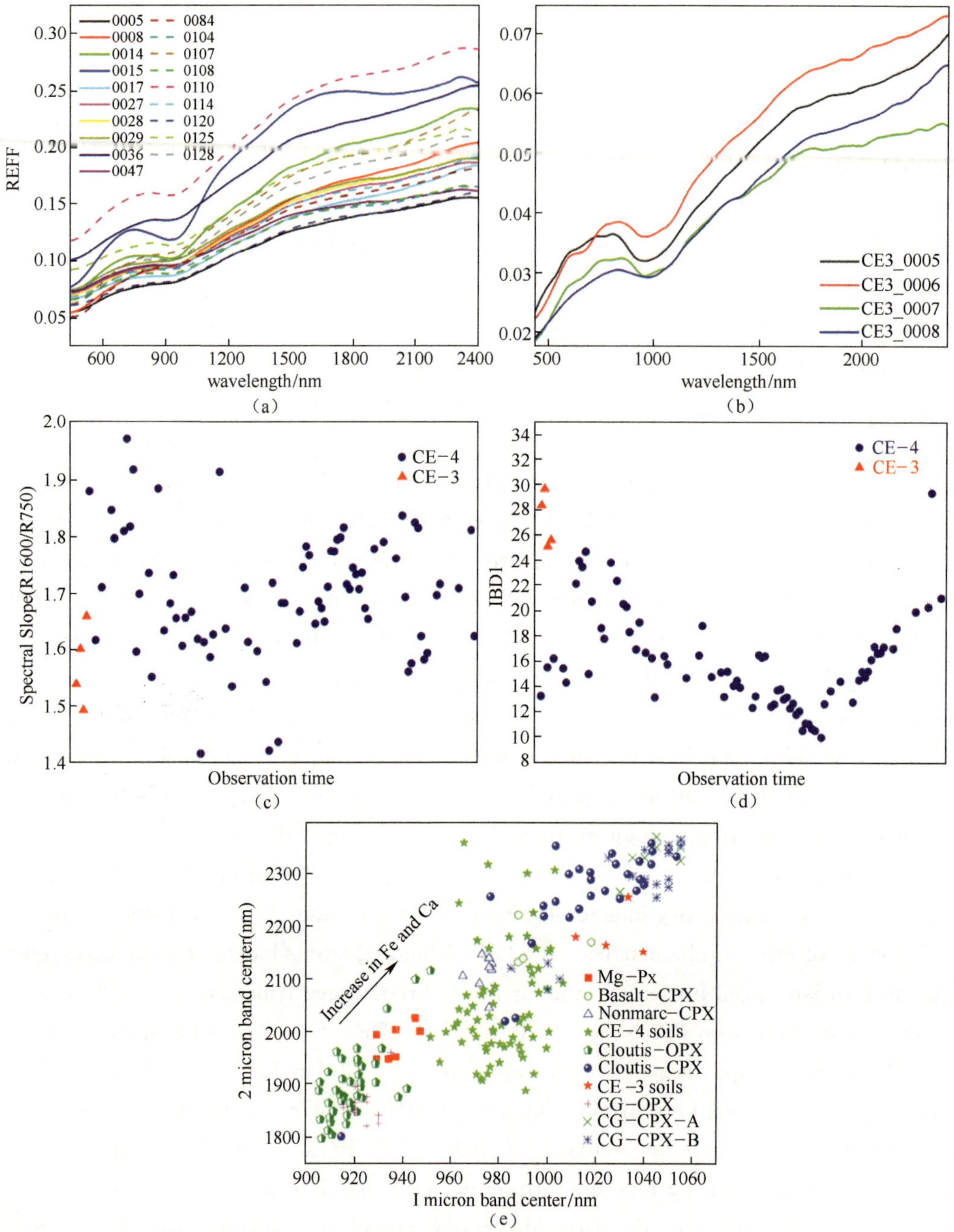

Fig. 9　(a) Part of the CE-4 VNIS spectra obtained during 1—18 lunar days; (b) photometric normalization results of CE-3 VNIR spectra using LS model; (c) spectral slopes of CE-3 and CE-4 measurements; (d) integrated band depth (IBD) of 1μm absorption (CE-3 and CE-4 measurements); (e) band center positions of the CE-3 and CE-4 VNIS detections overlain on band centers of pyroxene[6,62,63]. Mg-PX: Mg-rich pyroxene located between Mg - rich pyroxene and calcic CPX, which is similar to the results of [6]. The composition of lunar soils on the CE - 4 landing site are relatively homogeneous, indicating that the soils are well gardened and mixed.

A comparison of spectral parameters was conducted (Tab. S2) for the CE - 4 VNIS spectra before and after correction to better understand the influence of illumination conditions. Generally, normalized spectra exhibit smaller spectral slopes, lower FeO contents and thus a larger OMAT (Tab. S2). The standard deviations of these spectral parameters were also smaller than that of spectra without a photometric correction. Besides, the band centers of the 1 and 2μm absorption of spectra without a photometric correction exhibit a wider dispersion and the 1μm band centers range from 940 to 1041nm (Tab. S2). The difference in spectra shapes before and after photometric correction reveal that the variation of illumination conditions would influence the spectral characteristics and thus the geologic interpretation of the analyses of lunar materials. Thus, it is essential to acquire accurate phase functions and eliminate the effects of illumination conditions and shadows.

5 Conclusions

This work establishes empirical photometric functions to normalize the radiance variation of the lunar surface with incidence, emission, and phase angles, in order to eliminate the influence of illumination conditions. Major remarks are as follows:

(1) From in - situ and laboratory measurements, the shadows caused by surface roughness have significant impacts on the reflectance spectra acquired during lunar in - situ VNIS observations and proper corrections for shadows were introduced.

(2) We produced empirical wavelength - dependent phase functions of Lommel - Seeliger model from CE - 4 VNIS datasets. Values of phase functions exhibit declining trends with the increase in phase angles and an opposition effect and phase reddening effect were observed. This is the first time that empirical phase functions have been derived for wavelengths from 450 to 2395nm using in - situ visible and near - infrared spectral datasets.

(3) Photometric corrections of CE - 4 and laboratory VNIS spectra were conducted using the Lommel - Seeliger and Hapke radiative transfer models to determine the most suitable phase functions. Then, the validations of different phase functions conducted by comparing photometric - normalized spectra of the same target suggest that their discrepancies at 1500 nm are 3.7% using the Lommel - Seeliger model and 7.4% using the Hapke radiative transfer model (with parameters $b=-0.17$, $c=0.70$).

(4) Generally, photometrically normalized spectra exhibit smaller spectral slopes, lower FeO contents and thus a larger OMAT than that of spectra without correction. In addition, the band centers of the 1 and 2μm absorption of spectra after photometric

normalization exhibit a more concentrated distribution, indicating the compositional homogeneity of soils at the CE - 4 landing site after Yutu - 2 traveling ~448m.

(5) We prefer the phase functions derived from the LS model to apply to the photometric normalization of VNIS datasets and these empirical phase functions will contribute to the detection of, and accurate interpretation of, relative quantitative mineralogy along the traverse route of the Yutu - 2 rover in the Von Karman crater.

Acknowledgments

The scientific data are provided by China National Space Administration. The authors gratefully thank the whole Chang' e - 4 team for mission operations and the Ground Research and Application System (GRAS) of the Chinese Lunar Exploration Program for providing the data (http: //moon. bao. ac. cn). All original LRO data can be found in NASA's Planetary Data System (http: //pds. nasa. gov/). The authors are grateful to the editors and reviewers for their constructive reviews.

References

[1] Jia, Y.; Zou, Y.; Ping, J.; Xue, C.; Yan, J.; Ning, Y. The scientific objectives and payloads of Chang'E - 4 mission. Planet. Space Sci. 2018, 162, 207 -215.

[2] Li, C.; Liu, D.; Liu, B.; Ren, X.; Liu, J.; He, Z.; Zuo, W.; Zeng, X.; Xu, R.; Tan, X. Chang'E - 4 initial spectroscopic identification of lunar far - side mantle - derived materials. Nature 2019, 569, 378 - 382.

[3] Gou, S.; Di, K.; Yue, Z.; Liu, Z.; He, Z.; Xu, R.; Lin, H.; Liu, B.; Peng, M.; Wan, W. Lunar deep materials observed by Chang'e - 4 rover. Earth Planet. Sci. Lett. 2019, 528, 115829.

[4] Hu, X.; Ma, P.; Yang, Y.; Zhu, M. H.; Jiang, T.; Lucey, P. G.; Sun, L.; Zhang, H.; Li, C.; Xu, R. Mineral Abundances Inferred From In Situ Reflectance Measurements of Chang'E - 4 Landing Site in South Pole - Aitken Basin. Geophys. Res. Lett. 2019, 46, 9439 - 9447.

[5] Lin, H.; He, Z.; Yang, W.; Lin, Y.; Xu, R.; Zhang, C.; Zhu, M. - H.; Chang, R.; Zhang, J.; Li, C. Olivine - norite rock detected by the lunar rover Yutu - 2 likely crystallized from the SPA impact melt pool. Natl. Sci. Rev. 2019, 7, 913 - 920.

[6] Chen, J.; Ling, Z.; Qiao, L.; He, Z.; Xu, R.; Sun, L.; Zhang, J.; Li, B.; Fu, X.; Liu, C. Mineralogy of Chang'e - 4 landing site: Preliminary results of visible and near - infrared imaging spectrometer. Inf. Sci. 2020, 63, 1 - 140903.

[7] Sato, H.; Robinson, M.; Hapke, B.; Denevi, B.; Boyd, A. Resolved Hapke

parameter maps of the Moon. J. Geophys. Res. Planets 2014，119，1775 - 1805.

[8] Shkuratov，Y. G.；Grynko，Y. S. Light scattering by media composed of semitransparent particles of different shapes in ray optics approximation：Consequences for spectroscopy, photometry，and polarimetry of planetary regoliths. Icarus 2005，173，16 - 28.

[9] Hapke，B. Bidirectional reflectance spectroscopy：1. Theory. J. Geophys. Res. Solid Earth 1981，86，3039 - 3054.

[10] Besse，S.；Sunshine，J.；Staid，M.；Boardman，J.；Pieters，C.；Guasqui，P.；Malaret，E.；McLaughlin，S.；Yokota，Y.；Li，J. - Y. A visible and near - infrared photometric correction for Moon Mineralogy Mapper（M3）. Icarus 2013，222，229 - 242.

[11] Hapke，B. Theory of Reflectance and Emittance Spectroscopy；Cambridge University Press：Cambridge，UK，2012.

[12] Hillier，J. K.；Buratti，B. J.；Hill，K. Multispectral photometry of the Moon and absolute calibration of the Clementine UV/Vis camera. Icarus 1999，141，205 - 225.

[13] Minnaert，M. The reciprocity principle in lunar photometry. Astrophys. J. 1941，93，403 - 410.

[14] Hicks，M.；Buratti，B.；Nettles，J.；Staid，M.；Sunshine，J.；Pieters，C.；Besse，S.；Boardman，J. A photometric function for analysis of lunar images in the visual and infrared based on Moon Mineralogy Mapper observations. J. Geophys. Res. Planets 2011，116，E6.

[15] Shkuratov，Y.；Starukhina，L.；Homann，H.；Arnold，G. A model of spectral albedo of particulate surfaces：Implications for optical properties of the Moon. Icarus 1999，137，235 - 246.

[16] Fairbarin，M. B. Planetary Photometry：The Lommel - Seeliger Law. J. R. Astron. Soc. Can. 2005，99，92.

[17] Sato，H.；Boyd，A.；Robinson，M.；Denevi，B.；Hapke，B.；McEwen，A.；Speyerer，E. Photometric normalization of LROCWAC global color mosaic. epsc 2011，2011，636.

[18] Chen，C.；Qin，Q.；Zhang，N.；Wang，J.；Li，J.；Jiang，H. Research on photometric calibration and reflectance calculation of CE - 1 IIS data. Spectrosc. Spectr. Anal. 2011，31，1985 - 1990.

[19] Wu，Y.；Besse，S.；Li，J. - Y.；Combe，J. - P.；Wang，Z.；Zhou，X.；Wang，C. Photometric correction and in - flight calibration of Chang'E - 1 Interference Imaging Spectrometer（IIM）data. Icarus 2013，222，283 - 295.

[20] Zhang，J.；Ling，Z.；Zhang，W.；Ren，X.；Li，C.；Liu，J. Photometric modeling of the Moon using Lommel - Seeliger function and Chang'E - 1 IIM data. Chin. Sci. Bull. 2013，58，4588 - 4592.

[21] Lin，H.；Xu，R.；Yang，W.；Lin，Y.；Wei，Y.；Hu，S.；He，Z.；Qiao，

L.; Wan, W. In situ photometric experiment of lunar regolith with visible and near - infrared imaging spectrometer onboard the Yutu - 2 lunar rover. J. Geophys. Res. Planets 2020, 125, e2019JE006076.

[22] Hapke, B. Bidirectional reflectance spectroscopy: 4. The extinction coefficient and the opposition effect. Icarus 1986, 67, 264 - 280.

[23] Hapke, B. Space weathering from Mercury to the asteroid belt. J. Geophys. Res. Planets 2001, 106, 10039 - 10073.

[24] Helfenstein, P.; Veverka, J.; Hillier, J. The lunar opposition effect: A test of alternative models. Icarus 1997, 128, 2 - 14.

[25] Helfenstein, P.; Veverka, J.; Thomas, P. C. Uranus satellites: Hapke parameters from Voyager disk - integrated photometry. Icarus 1988, 74, 231 - 239.

[26] Domingue, D.; Hapke, B. Fitting theoretical photometric functions to asteroid phase curves. Icarus 1989, 78, 330 - 336.

[27] Guinness, E. A.; Arvidson, R. E.; Clark, I. H.; Shepard, M. K. Optical scattering properties of terrestrial varnished basalts compared with rocks and soils at the Viking Lander sites. J. Geophys. Res. Planets 1997, 102, 28687 - 28703.

[28] Warell, J. Properties of the Hermean regolith: IV. Photometric parameters of Mercury and the Moon contrasted with Hapke modelling. Icarus 2004, 167, 271 - 286.

[29] Li, J. - Y.; A'Hearn, M. F.; Belton, M. J.; Crockett, C. J.; Farnham, T. L.; Lisse, C. M.; McFadden, L. A.; Meech, K. J.; Sunshine, J. M.; Thomas, P. C. Deep Impact photometry of comet 9P/Tempel 1. Icarus 2007, 187, 41 - 55.

[30] Fernando, J.; Schmidt, F.; Ceamanos, X.; Pinet, P.; Douté, S.; Daydou, Y. Surface reflectance of Mars observed by CRISM/MRO: 2. Estimation of surface photometric properties in Gusev Crater and Meridiani Planum. J. Geophys. Res. Planets 2013, 118, 534 - 559.

[31] Yang, Y.; Lin, H.; Liu, Y.; Lin, Y.; Wei, Y.; Hu, S.; Yang, W.; Xu, R.; He, Z.; Zou, Y. The Effects of Viewing Geometry on the Spectral Analysis of Lunar Regolith as Inferred by in situ Spectrophotometric Measurements of Chang'E - 4. Geophys. Res. Lett. 2020, 47, e2020GL087080.

[32] Kreslavsky, M.; Shkuratov, Y. G. Photometric anomalies of the lunar surface: Results from Clementine data. J. Geophys. Res. Planets 2003, 108.

[33] Ling, Z.; Jolliff, B. L.; Wang, A.; Li, C.; Liu, J.; Zhang, J.; Li, B.; Sun, L.; Chen, J.; Xiao, L. Correlated compositional and mineralogical investigations at the Chang'e - 3 landing site. Nat. Commun. 2015, 6, 1 - 9.

[34] Zhang, H.; Yang, Y.; Yuan, Y.; Jin, W.; Lucey, P. G.; Zhu, M. H.; Kaydash, V. G.; Shkuratov, Y. G.; Di, K.; Wan, W. In situ optical measurements of Chang'E - 3 landing site in Mare Imbrium: 1. Mineral abundances inferred from spectral reflectance. Geophys. Res. Lett. 2015, 42, 6945 - 6950.

[35] Buratti, B.; Hicks, M.; Nettles, J.; Staid, M.; Pieters, C.; Sunshine, J.; Boardman, J.; Stone, T. A wavelengthdependent visible and infrared spectrophotometric function for the Moon based on ROLO data. J. Geophys. Res. Planets 2011, 116, E6.

[36] Yokota, Y.; Matsunaga, T.; Ohtake, M.; Haruyama, J.; Nakamura, R.; Yamamoto, S.; Ogawa, Y.; Morota, T.; Honda, C.; Saiki, K. Lunar photometric properties at wavelengths 0.5 — 1.6μm acquired by SELENE Spectral Profiler and their dependency on local albedo and latitudinal zones. Icarus 2011, 215, 639 - 660.

[37] Jin, W.; Zhang, H.; Yuan, Y.; Yang, Y.; Shkuratov, Y. G.; Lucey, P. G.; Kaydash, V. G.; Zhu, M. H.; Xue, B.; Di, K. In situ optical measurements of Chang'E - 3 landing site in Mare Imbrium: 2. Photometric properties of the regolith. Geophys. Res. Lett. 2015, 42, 8312 - 8319.

[38] Lin, H.; Yang, Y.; Lin, Y.; Liu, Y.; Wei, Y.; Li, S.; Hu, S.; Yang, W.; Wan, W.; Xu, R. Photometric properties of lunar regolith revealed by the Yutu - 2 rover. Astron. Astrophys. 2020, 638, A35.

[39] He, Z. - P.; Wang, B. - Y.; Lü, G.; Li, C. - L.; Yuan, L. - Y.; Xu, R.; Liu, B.; Chen, K.; Wang, J. - Y. Operating principles and detection characteristics of the Visible and Near - Infrared Imaging Spectrometer in the Chang'e - 3. Res. Astron. Astrophys. 2014, 14, 1567.

[40] Li, C.; Wang, Z.; Xu, R.; Lv, G.; Yuan, L.; He, Z.; Wang, J. The scientific information model of Chang' e - 4 visible and near - IR imaging spectrometer (VNIS) and in - flight verification. Sensors 2019, 19, 2806.

[41] Wu, W.; Wang, Q.; Tang, Y.; Yu, G.; Liu, J.; Zhang, W. Design of Chang'E - 4 lunar farside soft - landing mission. J. Deep Space Explor. 2017, 4, 111 - 117.

[42] He, Z.; Li, C.; Xu, R.; Lv, G.; Yuan, L.; Wang, J. Spectrometers based on acousto - optic tunable filters for in - situ lunar surface measurement. J. Appl. Remote Sens. 2019, 13, 027502.

[43] Li, S.; Li, L. Radiative transfer modeling for quantifying lunar surface minerals, particle size, and submicroscopic metallic Fe. J. Geophys. Res. Planets 2011, 116, E9.

[44] Henyey, L. G.; Greenstein, J. L. Diuse radiation in the galaxy. Astrophys. J. 1941, 93, 70 - 83.

[45] Liu, B.; Li, C. - L.; Zhang, G. - L.; Xu, R.; Liu, J. - J.; Ren, X.; Tan, X.; Zhang, X. - X.; Zuo, W.; Wen, W. - B. Data processing and preliminary results of the Chang'e - 3 VIS/NIR Imaging Spectrometer in - situ analysis. Res. Astron. Astrophys. 2014, 14, 1578.

[46] Lemelin, M.; Lucey, P. G.; Miljkovi'c, K.; Gaddis, L. R.; Hare, T.; Ohtake, M. The compositions of the lunar crust and upper mantle: Spectral analysis of the

inner rings of lunar impact basins. Planet. Space Sci. 2019，165，230 - 243.

[47] Qiao，L.；Ling，Z.；Fu，X.；Li，B. Geological characterization of the Chang'e - 4 landing area on the lunar farside. Icarus 2019，333，37 - 51.

[48] Ling，Z.；Qiao，L.；Liu，C.；Cao，H.；Bi，X.；Lu，X.；Zhang，J.；Fu，X.；Li，B.；Liu，J. Composition，mineralogy and chronology of mare basalts and non - mare materials in Von Kármán crater：Landing site of the Chang'E - 4 mission. Planet. Space Sci. 2019，179，104741.

[49] Schröder，S. E.；Grynko，Y.；Pommerol，A.；Keller，H.；Thomas，N.；Roush，T. Laboratory observations and simulations of phase reddening. Icarus 2014，239，201 -216.

[50] Mustard，J. F.；Pieters，C. M. Photometric phase functions of common geologic minerals and applications to quantitative analysis of mineral mixture reflectance spectra. J. Geophys. Res. Solid Earth 1989，94，13619 - 13634.

[51] Lucey，P. G. Model near - infrared optical constants of olivine and pyroxene as a function of iron content. J. Geophys. Res. Planets 1998，103，1703 - 1713.

[52] Pieters，C. M.；Taylor，L. A.；Noble，S. K.；Keller，L. P.；Hapke，B.；Morris，R. V.；Allen，C. C.；McKAY，D. S.；Wentworth，S. Space weathering on airless bodies：Resolving amystery with lunar samples. Meteorit. Planet. Sci. 2000，35，1101 - 1107.

[53] Lucey，P. G.；Blewett，D. T.；Taylor，G. J.；Hawke，B. R. Imaging of lunar surface maturity. J. Geophys. Res. Planets 2000，105，20377 - 20386.

[54] Savitzky，A.；Golay，M. J. Smoothing and differentiation of data by simplified least squares procedures. Anal. Chem. 1964，36，1627 - 1639.

[55] Wang，Z.；Wu，Y.；Blewett，D. T.；Cloutis，E. A.；Zheng，Y.；Chen，J. Submicroscopic metallic iron in lunar soils estimated from the in situ spectra of the Chang'E - 3 mission. Geophys. Res. Lett. 2017，44，3485 - 3492.

[56] Gou，S.；Yue，Z.；Kaichang，D.；Cai，Z.；Liu，Z.；Niu，S. Absolute model age of lunar Finsen crater and geologic implications. Icarus 2020，354，114046.

[57] Ivanov，M.；Hiesinger，H.；van der Bogert，C.；Orgel，C.；Pasckert，J.；Head，J. Geologic History of the Northern Portion of the South Pole - Aitken Basin on the Moon. J. Geophys. Res. Planets 2018，123，2585 - 2612.

[58] Xiao，L.；Zhu，P.；Fang，G.；Xiao，Z.；Zou，Y.；Zhao，J.；Zhao，N.；Yuan，Y.；Qiao，L.；Zhang，X. A young multilayered terrane of the northern Mare Imbrium revealed by Chang'E - 3 mission. Science 2015，347，1226 - 1229.

[59] Gou，S.；Yue，Z.；Di，K.；Wan，W.；Liu，Z.；Liu，B.；Peng，M.；Wang，Y.；He，Z.；Xu，R. In situ spectral measurements of space weathering by Chang'e - 4 rover. Earth Planet. Sci. Lett. 2020，535，116117.

[60] Fu，X. - H.；Qiao，L.；Zhang，J.；Ling，Z. - C.；Li，B. The subsurface structure and stratigraphy of the Chang'E - 4 landing site：Orbital evidence from

small craters on the Von Kármán crater floor. Res. Astron. Astrophys. 2020, 20, 008.

[61] Zhao, J.; Huang, J.; Qiao, L.; Xiao, Z.; Huang, Q.; Wang, J.; He, Q.; Xiao, L. Geologic characteristics of the Chang'E-3 exploration region. Sci. China Phys. Mech. Astron. 2014, 57, 569-576.

[62] Cloutis, E. A.; Gaey, M. J. Pyroxene spectroscopy revisited: Spectral-compositional correlations and relationship to geothermometry. J. Geophys. Res. Planets 1991, 96, 22809-22826.

[63] Moriarty Iii, D.; Pieters, C. The character of south pole-Aitken basin: Patterns of surface and subsurface composition. J. Geophys. Res. Planets 2018, 123, 729-747.

Effect of Lunar Complex Illumination on In Situ Measurements Obtained Using Visible and Near – Infrared Imaging Spectrometer of Chang'E – 4

Jiafei Xu[1,2,3+], Meizhu Wang[1,+], Honglei Lin[4], Rong Wang[1], Qi Feng[2] and Xuesen Xu[5]

1. Key Laboratory of Space Active Opto – Electronics Technology, Shanghai Institute of Technical Physics, Chinese Academy of Sciences, Shanghai 200083, China; xujf2@shanghaitech.edu.cn (J. X.); mzhwang@mail.sitp.ac.cn (M. W.); wangrong@mail.sitp.ac.cn (R. W.)
2. Key Laboratory of Infrared System Detection and Imaging Technology, Shanghai Institute of Technical Physics, Chinese Academy of Sciences, Shanghai 200083, China
3. School of Information Science and Technology, ShanghaiTech University, Shanghai 201210, China
4. Key Laboratory of Earth and Planetary Physics, Institute of Geology and Geophysics, Chinese Academy of Sciences, Beijing 100029, China; linhonglei@mail.iggcas.ac.cn
5. School of Physics and Optoelectronic Engineering, Hangzhou Institute for Advanced Study, University of Chinese Academy of Sciences, Hangzhou 310024, China; xuxuesen@ucas.ac.cn

+These authors contributed equally to this work.

Abstract: In – situ measurements of the spectral information on the lunar surface are of significance to study the geological evolution of the Moon. China's Chang'E – 4 (CE – 4) Yutu – 2 rover has conducted several in – situ spectral explorations on the Moon. The visible and near – infrared imaging spectrometer (VNIS) onboard the rover has acquired a series of in – situ spectra of the regolith at the landing site. In general, the mineralogical research of the lunar surface relies on the accuracy of the in – situ data. However, the spectral measurements of the Yutu – 2 rover may be affected by shadows and stray illumination. In this study, we analyzed 106 CE – 4 VNIS spectra acquired in the first 24 lunar days of the mission and noted that six of these spectra were affected by the shadows of the rover. Therefore, a method was established to correct the effects of the rover shadow on the spectral measurements. After shadow correction, the FeO content in the affected area is corrected to 14.46 wt.%, which was similar to the result calculated in the normal regolith. Furthermore, according to the visible images, certain areas of the explored sites were noted to be unusually bright. Considering the reflectance, geometric information, and shining patterns of the multi - layer insulation (MLI), we examined the influence of the specular reflection of the MLI on the bright spot regionsd, and found that the five sets of data were likely not affected by the specular reflection of the MLI. The results indicated that the complex illumination considerably influences the in situ spectral data. This study can provide a basis to analyze the VNIS scientific data and help enhance the accuracy of interpretation of the composition at CE - 4 landing sites.

Keywords: moon; Chang'E - 4; VNIS; multi - layer insulation; shadow effects

1 Introduction

The reflectance spectra of lunar rocks and soils are used to identify the composition of minerals and their relative abundance. The Chang'E - 3 (CE - 3) mission was successfully launched on 2 December 2013, and the lander landed to the east of a 430m crater in northwestern Mare Imbrium (19.51°W, 44.12°N) on 14 December 2013 [1]. In the China's Chang' E - 4 (CE - 4) mission, the lander successfully landed in the Von Kármán crater of the South Pole - Aitken basin on the far side of the Moon on 3 January 2019 [2, 3]. Considering the rock composition identified based on the data acquired by the CE - 3 visible and near - infrared imaging spectrometer (VNIS), the researchers inferred that when ferropyroxene - ilmenite cumulates sank and mixed with the relatively deep ferrous olivine and orthopyroxene in a mixed mantle source, the basalt source rock was formed during the late magma - ocean differentiation [4]. Moreover, it was considered that materials at the CE - 4 landing site may represent deep - seated materials, delivered to this site by ejecta from the Finsen crater [2, 5 - 8].

Notably, the multi - layer insulation (MLI) wrapped around the outside of the rover and lander and the rover's shadow likely affect the accuracy of the quantitative analysis based on the VNIS data. The results of ground simulation experiments performed under a specific solar altitude angle and different azimuth angles highlighted the effects of the shadow, specular reflection, and diffuse reflection on the imaging data [9]. Currently, researchers have focused on CE - 3 VNIS data that in - situ detection can be constrained by the observation angles and surface roughness and unevenness, resulting in significant small - area shadow effects [10]. However, no studies have been conducted to examine the effects of large - area shadow occlusion caused by the illumination angle as well as the orientation of the rover. In the CE - 4 photometric analysis, Lin found differences between the data characteristics and the minerals, and speculated that they might be influenced by stray illumination introduced by the MLI [11]. In general, when the CE - 4 Yutu - 2 rover advances, the VNIS simultaneously selects suitable observation points for spectral detection [3]. During operation on the lunar surface, considering the routes and areas of interest, it is hard to avoid the influence of shadows and stray illumination on the observation results. At certain observation angles (solar altitude angle and relative azimuth angle), the MLI reflected solar radiation, as shown in Figure 1a, and its specular reflection produce abnormally bright areas. In this study, we mainly focused on the specular reflection paths, from MLI to the lunar surface which was detected by VNIS, providing the basic analyzation for the MLI effects, as shown in Figure 1b. Furthermore, the shadows of the rover body may project onto the lunar surface at certain geometric angles and may also enter the VNIS detection area.

(a)

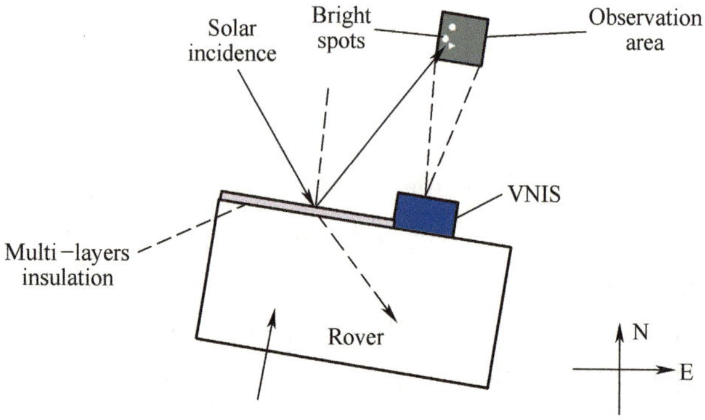

(b)

Figure 1　Image obtained using the obstacle avoidance camera on the fourth lunar day of Chang' E - 4 (CE - 4) [3]. (a) The area marked by the red circle shows the stray light introduced by the specular rflection of the multi - layer insulation (MLI). The area marked by the yellow circle indicates the projection of the shadow of the rover onto the lunar surface. (b) Light paths of bright spots introduced by MLI specular reflection

Considering these aspects, in this study, the data obtained by the VNIS and method to calculate the reflectance factor (RADF) [12] were considered. Because the rover shadows and MLI specular reflection spots can be easily distinguished in visible near - infrared (VIS/NIR) images, we focused on the spectral image data in the 450 - 945nm band. The spectra of the specular reflection and diffuse reflection of the MLI F46 films were obtained through laboratory testing. Subsequently, based on the statistics of the first 24 lunar days of CE - 4, we examined the effect of the large shadows generated by the rover on the VNIS in - situ data. In general, in the presence of considerable extensive shadows, the average reflectance of the image is low, which might affect the matter composition analysis based on the reflectance spectra. Finally, we proposed a method to determine whether the bright spot areas were affected by stray illumination introduced by the specular reflection of the MLI by considering the reflectance, geometric information, and shining patterns of the MLI.

2 Data and Methods

2.1 Chang' E-4 Spectral Data and Preprocessing

Under variations in the solar illumination and differences in the observation geometry, solar radiation may be reflected by the MLI wrapped around the rover, and shadows of the rover may be generated. These phenomena influence the data obtained from the detection area of the spectrometer. To clarify the effects of the complex illumination aspects, the angle information and spectral data must be considered. CE-3 and CE-4 adopt the same VNIS for target detection. On the first two lunar days, the CE-3 Yutu rover carried the VNIS, which obtained measurements in four areas. In the detection mode, four sets of data were obtained [13]. On 6 April 2021 (UTC), the CE-4 lander and Yutu-2 rover entered the 29th lunar day work period. The Yutu-2 rover has traveled approximately 682.8m. In the first 24 lunar days, VNIS obtained 106 sets of spectral data under different light geometries. The amount of data detected by the CE-3 VNIS is smaller, and no effect of the rover body or MLI can be observed in the VIS/NIR images. Therefore, we examined the data obtained by the CE-4 VNIS.

The CE-4 VNIS is installed at the front of the Yutu-2 rover at a height of approximately 0.7m, looking down at the lunar surface at 45°, as shown in Figure 2. The size of the observation area is ~15 × 21cm [14]. The device contains two channels: a VIS/NIR channel and a short-wave infrared (SWIR) channel. Single-band spectral or image information is passed through an acousto-optic tunable filter (AOTF) to form quasi-monochromatic light of a specific wavelength, which is then converged to the detector. The wavelength is selected by varying the drive frequency applied to the AOTF to obtain the spectral or imaging data in all wavelength bands [15, 16]. The VIS/NIR channel adopts a complementary metal-oxide semiconductor (CMOS) array detector to obtain data in the wavelength range of 450nm~945nm, with a sampling interval of 5nm; 100 bands of data are obtained at a spectral resolution of 2~10nm. The SWIR channel involves a InGaAs unit detector to obtain data in the wavelength range of 900nm~2395 nm wavelength range, with a sampling interval of 5nm; 300 bands of data are obtained at a spectral resolution of 3~12nm [15, 16]. The field of view of the SWIR detector is a circle with a diameter of 107.6 CMOS pixels, and the center located at sample 98, line 127.5 of the CMOS imager field of view [15]. The uncertainties in the CMOS imager and SWIR detector spectral calibrations are ±0.39nm and ±0.62nm, respectively [15, 17]. The uncertainties in the corresponding radiometric calibrations were 4.97% and 6.63% [15].

The VNIS 2B data used in this study were subjected to dark current and scattering background deductions, flat field and instrumental temperature corrections, and radio-

Figure 2 Image of Yutu - 2 rover at exploration site A. Visible and near - infrared imaging spectrometer (VNIS) mounted at the front of the rover, marked by the red rectangle in the picture. The VNIS detection area was marked by the white rectangle, which is approximately 1m below the spectrometer, and less than 5m from the center of the rover. This figure was obtained from the CE - 4 Terrain Camera (TCAM) [1]

metric and geometric corrections by using the Ground Research and Application System (GRAS) of the Chinese Lunar Exploration Program [2, 18]. The RADF data were calculated using a solar irradiance - based calibration method involving the following steps [19 - 21]. First, the scientific data were read frame by frame, and the average value was determined as the single - band radiance data. The irradiance data of the other bands were calculated and finally stitched into a complete VNIS irradiance spectral curve L, in units of $W \cdot m^{-2} sr^{-1} nm^{-1}$. Next, the reflectance was obtained by its ratio to the solar spectral irradiance data F by using Equation (1) [18, 22]. The distance d between the Sun and Moon was considered to be approximately 1 AU [23].

$$RADF(i) = \frac{\pi \times L(i) \times d^2}{F(i)}, \quad i = 1, 2, 3 \cdots 100 \tag{1}$$

The main angle information was obtained by reading the tag data corresponding to the auxiliary file, including the solar altitude angle (90° - solar incidence angle), solar incidence azimuth angle, emergence azimuth angle, and relative azimuth angle obtained of the two parameters.

Figure 2, shows that at certain geometric angles, the rover body produces shadows on the lunar surface, and the reflectance measurements exhibit significantly low values. In addition, the MLI exhibits specular reflection, and the reflected solar radiation enters the VNIS detection area. The reflectance measurement is thus biased, and notable brightness anomalies may be found in the VIS/NIR image.

In general, the MLI, applied to the exterior of the Yutu - 2 rover to reduce the heat exchange between the rover and solar radiation, lunar surface reflection, and cold space, consists of F46 auxiliary surface mirrors. The reflectivity of the MLI F46 film wrapped around the outside of Yutu - 2 was measured in the laboratory to provide a reference for subsequent studies. The test procedure is described in the following sections.

2.2 Laboratory Testing of F46 Film Spectrum

The experimental test system consisted of a halogen lamp, an optical fiber, an integrating sphere, a spectrometer, and the F46 film. A broad - spectrum halogen lamp was used as the light source, and a 600μm optical fiber transmitted the optical signal. The detection target, that is, the F46 film has a strong specular reflection and weak diffuse reflection, and thus, the reflected light was collected using a diffuse integrating sphere. The grating type spectrometer was used to evaluate the light emitted from the integrating sphere. The principle is shown in Figure 3: The incident light from the light source passed through the optical fiber from window - 1 into the integrating sphere. An exit window - 2 was present in the integrating sphere, symmetric with the incident window - 1. It was necessary to evaluate the diffuse reflection spectrum of the F46 film, as well as the superimposed spectrum of the diffuse reflection and specular reflection. When measuring the superposition of the specular reflectance and diffuse reflection, window - 2 was closed. In this configuration, the specularly reflected light together with the diffuse reflected light were collected by the integrating sphere by the outgoing fiber (from window - 3) into the spectrometer. To measure only the diffuse reflection of the F46 film, window - 2 was opened. In this configuration, the specular reflectance light was emitted directly from the integrating sphere, and only the diffuse spectrum of the F46 film was measured. Finally, we obtained the diffuse reflection spectra of the F46 film and the spectra of the specular reflection superimposed on the diffuse reflection in the band of 450—2300nm.

Furthermore, in the experiments, a diffuse reflectance integrating sphere was used to ensure traceability to a standard reflectance plate. The digital number (DN) value of the diffuse reflectance of the standard reflectance plate and that of the F46 film when the reflection occurred were measured separately. In addition, a set of dark noise data was tested without light incidence. The specular reflection and specular reflection superimposed on the diffuse reflection were measured by opening and closing window - 2. The final conversion from DN value to reflectance is achieved by the comparison method.

The experimental layout is shown in Figure 3.

To ensure the accuracy of the data, we conducted several experiments at different integration times to eliminate the errors caused by the nonlinearity of the detector.

Through the above mentioned experiment, we obtained the reflectance curves of the specular reflection and diffuse reflection of the F46 film. As shown in Figure 4, the

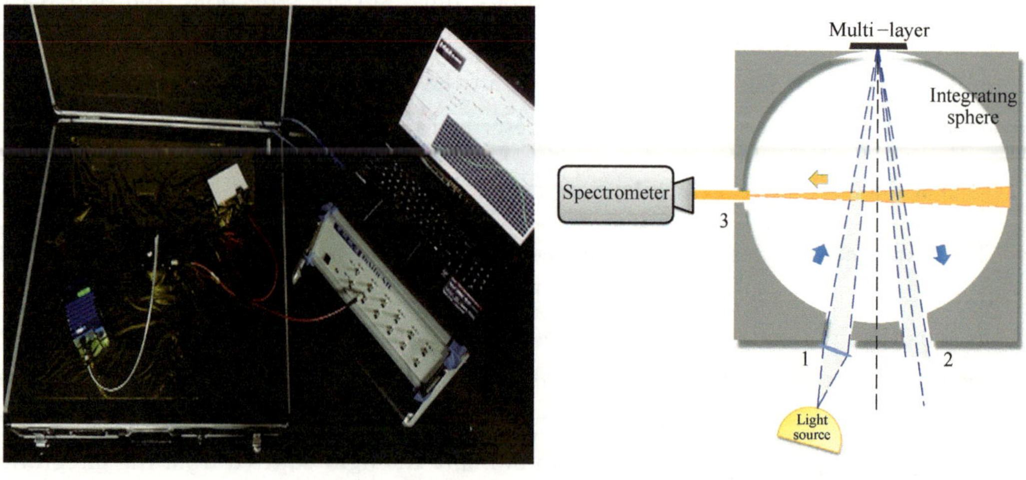

(a)　　(b)

Figure 3　(a) Laboratory test setup. (b) Schematic. Numbers 1, 2 and 3 correspond to the three windows. Light source from window 1 entering the integrating sphere

reflectance characteristics of the F46 film are relatively distinct. The reflectance curve tends to be flat in the visible band 0.4－0.7μm and infrared band. A distinct broad absorption appears at 0.75－0.85μm, and this feature is used to analyze the influence of the specular reflections from the F46 film on the spectral data collected during the first 24 lunar days by the VNIS.

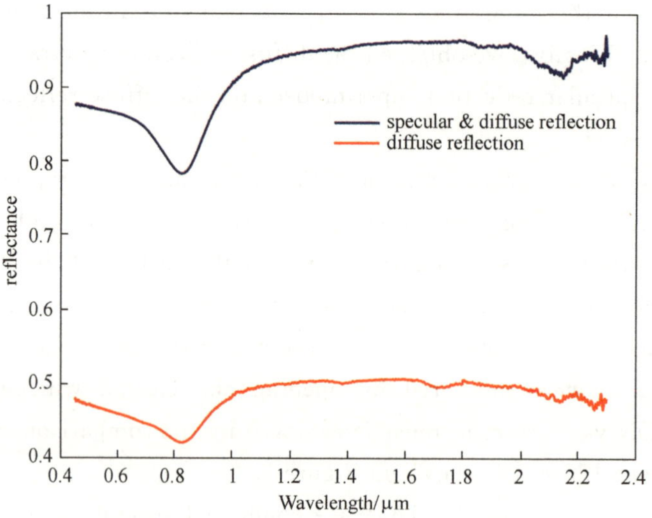

Figure 4　Diffuse reflection and specular reflection reflectance curve of the F46 film

3 Results and Discussion

Based on the laboratory test data and in－situ geometric angle and spectral data, we investigated the effect of the MLI specularly reflected solar radiation and rover shadow on the VNIS data. It was noted that the VNIS will be affected by the MLI and shadow at a specific angle, owing to which, the reflectance of the affected area is significantly different from that of the normal area. Moreover, these aspects considerably influence the quantitative analysis of the data. Therefore, we performed a statistical analysis of the geometric angle information obtained in the first 24 lunar days and examined the effect of shadows and stray illumination on the in－situ spectral data by analyzing the spectral images and data.

3.1 Statistics of Illumination and Observation Geometry

To investigate the angle relationship between the illumination and rover orientation, we conducted a statistical analysis based on the auxiliary information for the first 24 lunar days. There were 106 sets of data on the whole, and through statistical analysis, the solar altitude angle ranged from 12°～39° and the relative azimuth angle ranged from 56°～328°. The relative azimuth angle was calculated by the solar azimuth angle minus the emission azimuth angle, and the negative values added 360° to limit the range of the relative azimuth to 0～360°. The statistical histograms are shown in Figure 5.

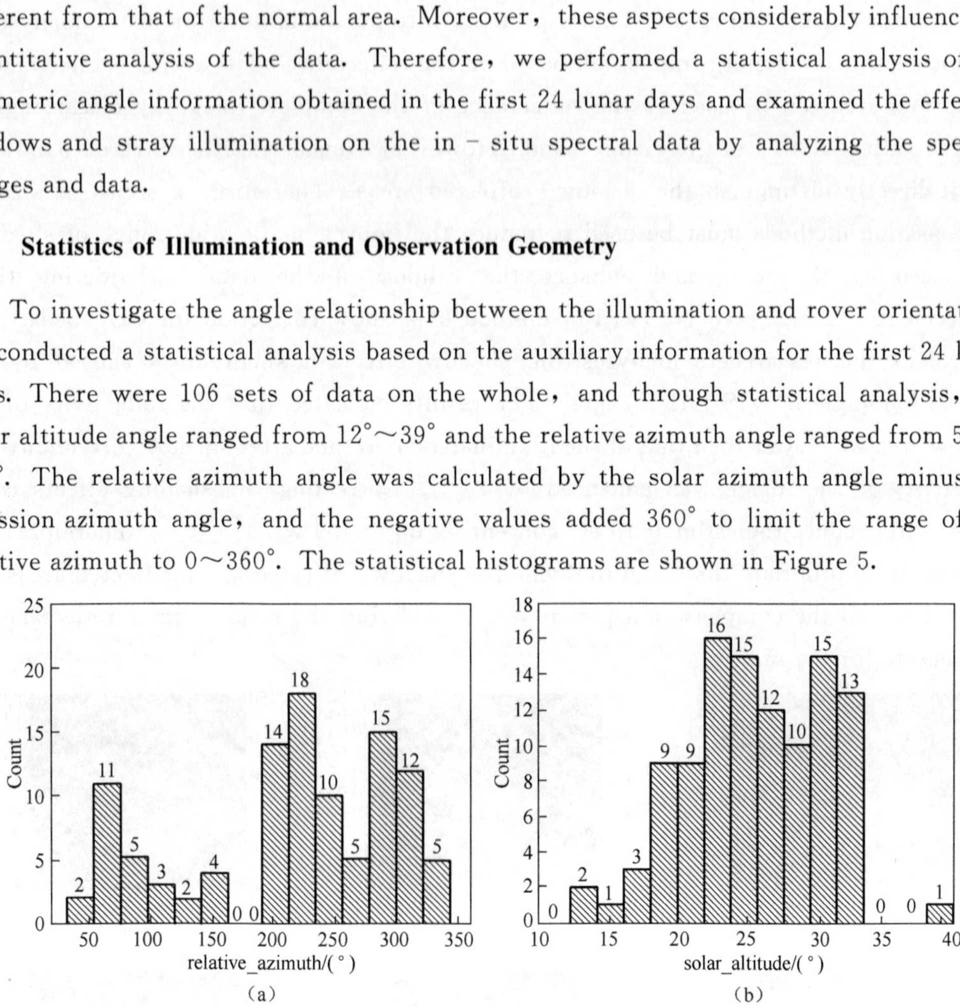

Figure 5　Histogram of the distribution of illuminance and observation angle in the first 24 lunar days. (a) Relative azimuth angle statistics, with an interval of 20°. The relative azimuth is the difference in the solar incident light azimuth and emission light azimuth. (b) Solar altitude angle statistics, with an interval of 2°

3.2 Shadow Effect

The observation area is affected by the shadow of the rover owing to the geometric relationship among the orientation of the rover, solar altitude angle, and relative azimuth angle of the incident light and emission light. Six sets of images exhibited notable large-area shadowing effects (Table 1). In this study, we primarily analyzed the large-area shadows caused by the rover occlusion, as shown in Figure 6.

In general, to calculate the reflectance of the VIS/NIR band, we considered the average value of the complete image as the reflectance of this band. Owing to the effect of the shadows in large areas, the calculated reflectance is extremely low and inconsistent with that for the normal areas. This aspect can introduce a large error in the analysis of the lunar surface mineral composition based on the reflectance. The average reflectance of the normal area can be used as the reflectance of the complete image by discarding the shadow-affected area in the visible band. However, considering the infrared band, we cannot directly distinguish the shadow-affected areas. Therefore, a series of shadow compensation methods must be used to reduce the impact of the wide range of shadows introduced by the rover and enhance the validity of the data. Considering these requirements, we adopted Hu's [10] method of shadow correction for CE-3 data. In particular, the researchers analyzes the shadow effect in small areas caused by the observation angle in VIS/NIR images. The results indicated that the reflectivity of the shadow area was lower than that of the illuminated area, and after shadow correction, the reflectivity of the image was enhanced by 24%. Correcting the shading effects could significantly reduce the estimated FeO content by up to 4.9 wt.% [10]. Therefore, we selected the appropriate threshold to divide the shadow-affected and unaffected areas and later calculated the compensation parameters considering the ratio of their reflectance to compensate for the impact.

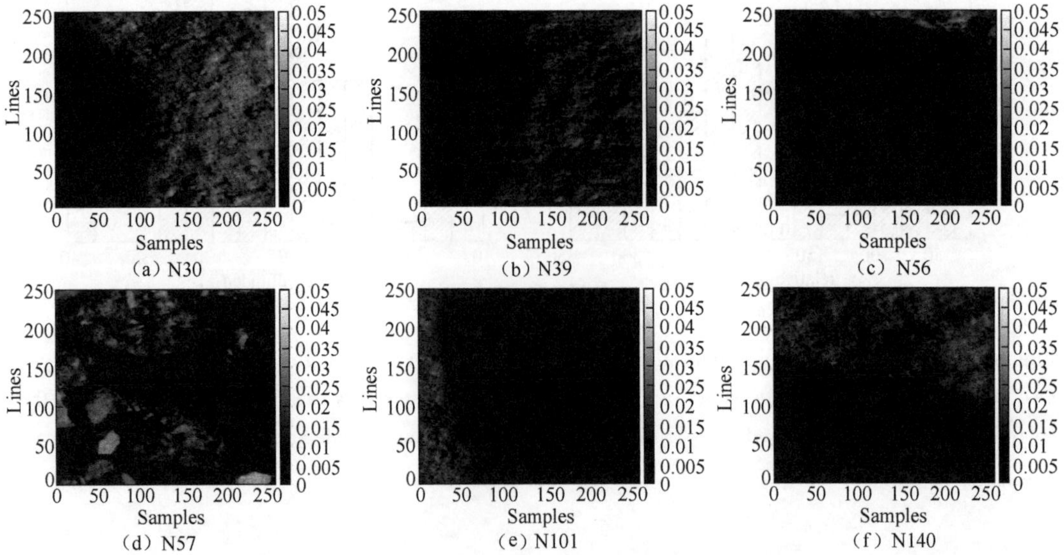

Figure 6 Radiance images of the lunar surface observed by the Yutu-2 rover at 750nm, affected by shadows. All the images are stretched from 0 to 0.05 to illustrate the albedo variation. (a-f) correspond to image IDs N30, N39, N56, N57, N101 and N140, respectively

Table 1　Observation angles of data affected by shadows

Data ID	Solar Altitude/(°)	Solar Azimuth Angle/(°)	Emission Azimuth Angle/(°)	Relative Azimuth Angle/(°)
N30	28.63	59.26	106.42	312.84
N39	12.25	78.84	116.83	320.01
N56	22.55	63.14	114.90	308.23
N57	24.55	60.36	111.47	308.90
N101	24.18	62.53	6.17	56.35
N140	21.00	64.94	118.14	306.80

We selected four sets of data, N30, N39, N56, and N140, as typical examples for the analysis. To distinguish the effect of the shadow areas, R_{750} (reflectance at 750nm) was considered to divide the complete image into shadow areas and normal areas. To determine the threshold value for the shadow region, N30 was considered as an example. The following Figure 7 shows the histogram of the reflectance distribution of the visible image at 750nm after photometric correction [21]. The most frequent occurrence pertains to the value 0, which represents a large area of notable shadow occlusion in the VIS/NIR image. The first trough appears at a reflectivity of 0.02, and the peak position occurs at a reflectivity of 0.058. Thus, the intermediate value (approximately 0.036) for the trough and peak was considered as the threshold to distinguish the shadow and normal areas. The reflectance range 0.02~0.036 exists as an excessive band between the shadow and normal areas, and this part was also affected by the shadow. Furthermore, the threshold value was chosen higher not only to remove the large shadowing area but also to remove the "micro" shadows induced by the regolith topography. As shown in Figure 6c, the N56 shadow area occupies almost the complete image, and the results of shadow compensation are not notable. Therefore, this set of data was not considered. The histogram results for the other two sets of data (N39 and N140) were compared using the same method. Reflectance values of 0.023 and 0.025 were selected as the thresholds to distinguish the shadow and normal areas. We obtained the coefficient of reflectivity enhancement by using Equation (2) [10].

$$k = \frac{1}{N}\sum_{i_\lambda=1}^{100}\frac{R_{illuminated}(\lambda)}{R(\lambda)} \tag{2}$$

where k represents the shadow correction factor, i_λ represents the ith band of the VIS/NIR band, and λ is the wavelength corresponding to this band, with an interval of 5nm. $R_{illuminated}$ (λ) denotes the average RADF of the remaining pixels at λ, excluding the shadow region; R (λ) is the average RADF of the complete image at λ; and N=100 denotes the total number of VIS/NIR bands. The correction factor for the whole detection area is wavelength-independent to eliminate the slight variation of different bands. The correction factor was calculated by averaging the factors of VIS/NIR 100 bands. The shadow correction coefficients (k) were calculated to be 1.63, 2.00, and 2.50 for the N30, N39, and N140 data, respectively. After shadow correction, the reflectance spectrum of the VNIS in the VIS/NIR band was significantly enhanced compared to the uncorrected one, as shown in Figure 7.

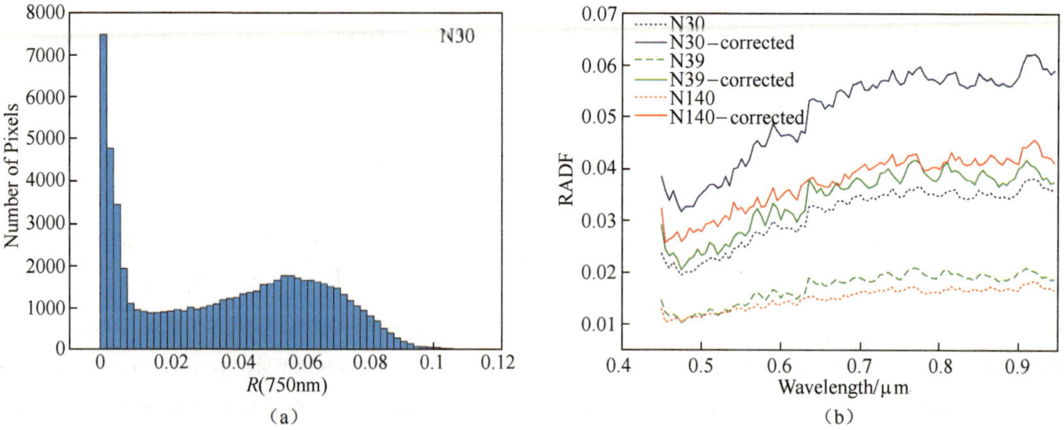

Figure 7 (a) Reflectance histogram of N30 detection area at 750nm. (b) Shadow-corrected and uncorrected spectra of N30, N39, N140 in the VIS/NIR band after photometric correction.

To verify the results of the shadow compensation correction, we compared the corrected average reflectance with the lunar surface soil reflectance in areas not affected by the shadows. We selected the normal area in N30, as shown in Figure 8a, marked through the red box. However, this area may contain fewer "micro" shadows induced by the regolith topography. So we picked 100 pixels for the red box outside the shadowed region randomly five times and calculated their mean reflectance and standard deviation. The reflectance after shadow correction was compared with that of the complete image, as shown in Figure 8b. The reflectance was considerably enhanced after the shadow correction, although its value was lower than that of the normal area. These findings indicate that researchers can exclude the shaded parts when performing the quantification of the matter composition based on VIS/NIR bands. The value for only the area not affected by the shadows must be considered as the average reflectance of the complete image. Nevertheless, shadow correction must be implemented in the infrared band. The set of data for images that are more than 80% shaded, such as N56 and N101, must be eliminated.

Furthermore, the FeO content can be calculated considering the VIS/NIR data. We calculated the FeO composition content in N30 based on Lucey's method [24] to analyze the effect of large shadows on the analysis of the mineral composition. θ_{Fe} for the Fe parameter can be calculated using Equation (3) [24, 25], and a linear relationship exists between FeO and θ_{Fe}, as shown in Equation (4) [21]. It is known from previous studies that the photometric correction of the reflectance are important for the retrieval of

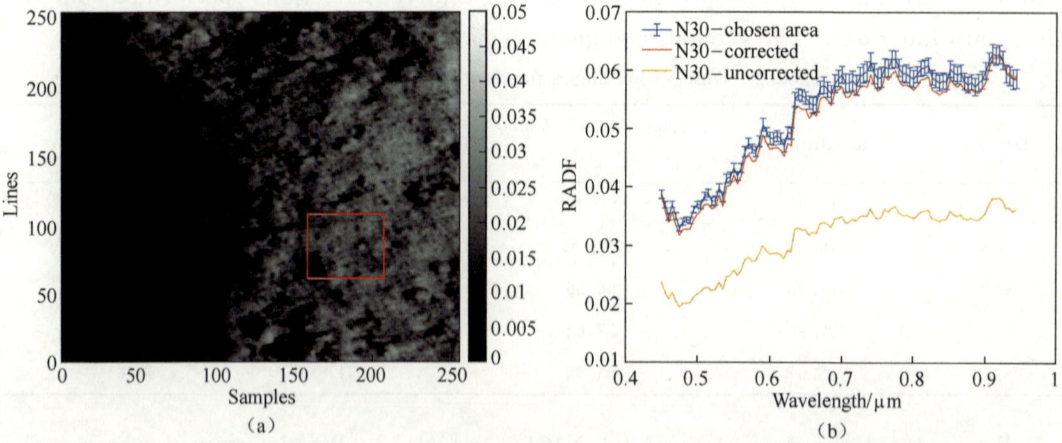

Figure 8　Comparison of reflectance after shadow correction and normal area reflectance. (a) Radiance image at 750nm; the red rectangle marks the area in which the normal lunar soil is not obscured by shadows. (b) We randomly chose 100 pixels of the normal areas marked in (a) five times. The blue curve shows the mean reflectance factor (RADF) curve of them and their standard deviation; the reflectance is significantly enhanced after shadow correction

the contents of FeO [24]. So the reflectance here were corrected photometrically to the standard viewing geometric angle (30°, 0°, 30°) based on the Lommel - Seeliger empirical model. We used the photometric function parameters obtained by Lin [21].

$$\theta_{Fe} = \arctan\left[\frac{R_{950}/R_{750} - 1.23}{R_{750} - 0.04}\right] \tag{3}$$

$$FeO = 14.42 \times \theta_{Fe} - 6.884 \tag{4}$$

where R_{950} and R_{750} represent the reflectance spectra of the VNIS at 950nm and 750nm, respectively. The calculated average reflectance was low in N30 when large shadows are present. Moreover, the presence of a negative denominator ($R_{750} - 0.04$) in the calculation of the Fe parameter θ_{Fe} precluded the analysis of its composition. After the shadow correction, the FeO content was 14.46 wt.%, calculated based on the average reflectance of the complete image. For the normal lunar soil area marked by the red rectangle in Figure 8, the FeO content was calculated as 14.60 wt.%. The two values were similar. The results showed that shadow correction is of significance, and the presence of large shadow areas may affect the analysis of the FeO contents.

3.3　MLI Effect

We analyzed the VIS/NIR images of 106 sets of data for the first 24 lunar days, among which five sets of data (N65, N66, N130, N131, and N132) were noted to have obvious bright spots or bright areas. We speculated that these spots could be attributed to the solar radiation reflected from the MLI entering the VNIS detection fleld. Note that N65 and N66 represent two consecutive detections performed on the ninth lunar day.

N130, N131, and N132 represent three consecutive detections performed on the nineteenth lunar day. The geometric angle information is summarized in Table 2.

Table 2 Observation angles for data involving bright spots

Data ID	Solar Altitude/(°)	Solar Azimuth Angle/(°)	Emission Azimuth Angle/(°)	Relative Azimuth Angle/(°)
N65	19.93	293.72	57.74	235.97
N66	19.00	292.55	57.73	234.82
N130	21.68	66.36	356.00	70.35
N131	27.90	57.64	356.28	61.36
N132	28.97	55.95	356.31	59.64

We analyzed the data sets N130, N131, and N132. We determined whether the detection area was affected by stray light from the MLI considering three aspects: (1) relationship between the illumination angle and vehicle body orientation; (2) comparison of the VIS/NIR images of nearby detections; (3) comparison of the average reflectance spectra of the abnormal region with the reflectance spectra of the MLI, measured in the laboratory. First, we compared the azimuth angles of the incident and emission light for the three sets of data and schematically illustrated the corresponding angle relationships. As shown in Figure 9, the Sun rays were incident from the rear left of the rover body. Consequently, the stray illumination introduced by the specular reflection of the MLI did not enter the detection area. Therefore, the bright spots were likely not caused by the stray illumination. To verify this aspect, we selected the bright spot areas and compared the average reflectance spectrum with the reflectance curve of the F46 film measured in the laboratory. If the stray light effects were indeed present, the reflectance spectrum of the bright spot region was expected to be superimposed with the reflectance characteristics of the F46 film. We selected these abnormal brightness regions and obtained their average RADF curves by using Equation (1). Next, we smoothed the reflectance data by using the cubic splines method with a smoothing parameter of 0.9999 [26]. Furthermore, we compared the reflectance through normalization ($R_{i-normalization} = R_i / \sum_{i=1}^{i=n} R_i$ [11], where i is the number of bands). According to Figure 10, the F46 film exhibits a wide absorption at 750 - 850nm, whereas the average reflectance spectrum of the bright spot region does not exhibit such an absorption feature. These two aspects demonstrate that the apparent bright spots in this region were not caused by the stray illumination introduced by the specular reflection of the MLI.

Next, we analyzed the geometric angles for N65 and N66 and noted that the solar illumination and VNIS detection directions were similar. The relevant values were calculated using the coordinate information of the vehicle and detection area pertaining to the 2BL, with the vehicle moving 7cm backward and forward for the two detections. The area marked by the red box in Figure 11 represented the area in which the bright spot appeared for two consecutive detections; it was observed that the bright spot appeared in the same lunar soil area. According to the analysis, the bright spot area was expected to shift if stray light was introduced by the MLI after the rover moved. We speculated that the bright spot area was not likely caused by the stray light introduced by the specular reflection of the MLI. In a previous study, researchers analyzed data from the ninth lunar day detection and reported the presence of glass in the pit fragments [11]. Consequently, we considered that this abnormally bright region in the VIS/NIR images may be the micro specular reflection of the glass. Notably, because of the solar incidence from the front of the rover, the impact of scattered light from the MLI could not be excluded in this area. The effect of MLI may also be superimposed on the glass spectra with high albedo features, resulting in a broad absorption feature near 800nm, as described in Lin [11].

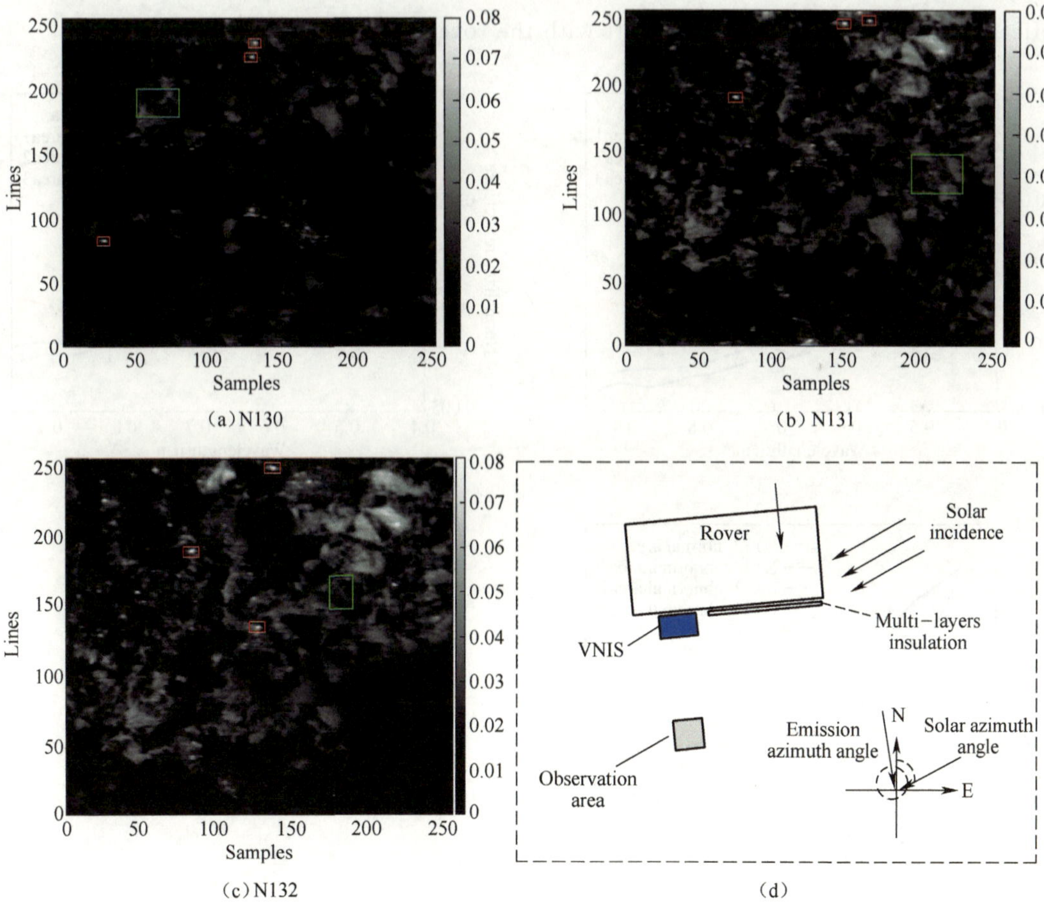

Figure 9　(a－c) Radiance images of lunar surface observed by the Yutu－2 rover. The images are stretched from 0 to 0.08 to illustrate the albedo variation. The red rectangles indicate the obvious bright spot areas. The normal areas were marked by the green rectangle. The image IDs are N130, N131, and N132. (d) Schematic of the angle relationships between the illumination angle and vehicle body orientation

In general, if notably bright areas are observed in a VIS/NIR image of the VNIS, it must be analyzed considering three aspects to established the influence of the specular reflection of the MLI. First, the relationship between the illumination angle and vehicle body orientation must be considered. Only when the solar light is incident from the front of the rover is there a possibility of the specular reflection of the MLI entering the detection area. Second, we must consider the absorption properties of the spectra. We measured the reflection spectra of the MLI in the laboratory. If specular reflection occurs, the absorption characteristics of the MLI should be superimposed on the area affected by the stray light. For example, as mentioned previously, the F46 film exhibits a wide absorption at 750—850nm. Third, in the presence of continuous detections such as N65 and N66 with similar illumination angles, the VIS/NIR images must be considered for the analysis. Under the same illumination conditions, if stray light is introduced by specular reflection, the bright spot should shift with the rover movement.

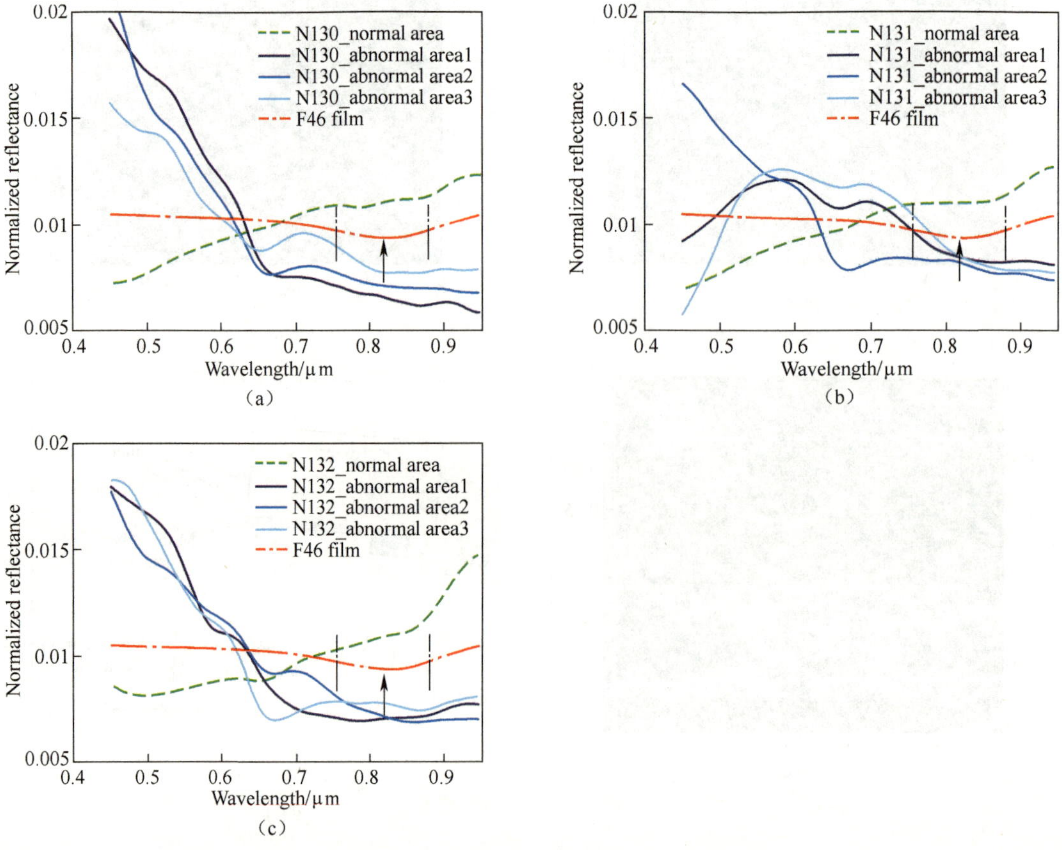

Figure 10　(a - c) Comparison of normalized reflectance in the VIS/NIR band of N130, N131, N132, and F46 film. The arrow indicates the wide absorption (750—850nm) of the F46 film measured in the laboratory. The bright region marked by the red rectangle in Figure 9 is considered to calculate the average reflectance of the abnormal region

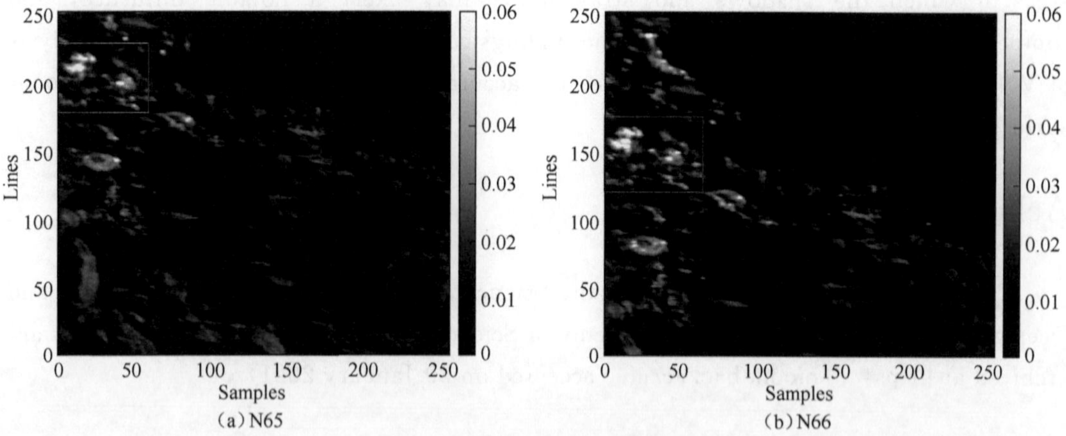

Figure 11　Radiance images of the lunar surface observed by the Yutu - 2 rover at 750nm. The images are stretched from 0 to 0. 06 to illustrate the albedo variation. The image IDs are N65 and N66. The area marked by the red rectangle is the same lunar soil area considered for image matching

4　Conclusions

The Yutu - 2 rover of the CE - 4 mission has performed notable explorations on the lunar surface and obtained considerable scientific data. We performed laboratory tests and in - flight data validation to demonstrate that the complex illumination in the lunar environment, including the shadows of the rover, affect the spectra data of the VNIS. This influence may introduce errors in the quantitative analysis of the VNIS spectra data. The following key conclusions were derived:

(1) Six sets of data pertaining to the VIS/NIR images obtained in the first 24 lunar days involved large shadows caused by the rover occlusion. The shadow correction significantly improved the reflectance of the affected images and influenced the interpretation of the FeO composition. We analyzed the data from statistical and geometric perspectives and concluded that the observation area is more likely to be influenced by the rover occlusion when the relative azimuth angle ranges from 300°～320°;

(2) Five sets of data exhibited bright spots that were significantly different from the normal areas. We proposed an analytical method that involved the comprehensive consideration of three aspects to determine the potential influence of the stray light from the specular reflections of the MLI. The results indicated that the five sets of data were likely not affected by the specular reflection of the MLI. However, diffuse reflection of MLI, specular reflection of MLI that entered the detector directly, and stray lights from rover wheels might also affect the spectral images, the influence of these portions needed further analysis in the future.

Overall, we calculated the spectral data for the first 24 lunar days, identified the areas in which the shadows and stray lights may exert a notable influence, and compensated for the shadow effects. The findings can provide a reference for the analysis of VNIS scientific data and help enhance the accuracy of the interpretation of the lunar surface mineral composition.

Acknowledgments

The scientific data are provided by the Science and Application Center for Moon and Deep Space Exploration, Chinese Academy of Science. The data reported in this work are archived at https://moon.bao.ac.cn (accessed on 20 January 2021).

References

[1] Li, C.; Liu, J.; Ren, X.; Zuo, W.; Tan, X.; Wen, W.; Li, H.; Mu, L.; Su, Y.; Zhang, H.; et al. The Chang'e 3 mission overview. Space Sci. Rev. 2015, 190, 85-101.

[2] Li, C.; Liu, D.; Liu, B.; Ren, X.; Liu, J.; He, Z.; Zuo, W.; Zeng, X.; Xu, R.; Tan, X.; et al. Chang' E-4 initial spectroscopic identification of lunar far-side mantle-derived materials. Nature 2019, 569, 378-382.

[3] Li, C.; Zuo, W.; Wen, W.; Zeng, X.; Gao, X.; Liu, Y.; Fu, Q.; Zhang, Z.; Su, Y.; Ren, X.; et al. Overview of the Chang'e-4 Mission: Opening the Frontier of Scientific Exploration of the Lunar Far Side. Space Sci. Rev. 2021, 217, 1-32.

[4] Ling, Z.; Jolliff, B.L.; Wang, A.; Li, C.; Liu, J.; Zhang, J.; Li, B.; Sun, L.; Chen, J.; Xiao, L.; et al. Correlated compositional and mineralogical investigations at the Chang'e-3 landing site. Nat. Commun. 2015, 6, 1-9.

[5] Hu, X.; Ma, P.; Yang, Y.; Zhu, M.H.; Jiang, T.; Lucey, P.G.; Sun, L.; Zhang, H.; Li, C.; Xu, R.; et al. Mineral abundances inferred from in situ reflectance measurements of Chang'E-4 landing site in South Pole-Aitken basin. Geophys. Res. Lett. 2019, 46, 9439-9447.

[6] Gou, S.; Di, K.; Yue, Z.; Liu, Z.; He, Z.; Xu, R.; Lin, H.; Liu, B.; Peng, M.; Wan, W.; et al. Lunar deep materials observed by Chang'e-4 rover. Earth Planet Sci. Lett. 2019, 528, 115829.

[7] Lin, H.; He, Z.; Yang, W.; Lin, Y.; Xu, R.; Zhang, C.; Zhu, M.H.; Chang, R.; Zhang, J.; Li, C.; et al. Olivine-norite rock detected by the lunar rover Yutu-2 likely crystallized from the SPA-impact melt pool. Natl. Sci. Rev. 2020, 7, 913-920.

[8] Huang, J.; Xiao, Z.; Xiao, L.; Horgan, B.; Hu, X.; Lucey, P.; Xiao, X.; Zhao, S.; Qian, Y.; Zhang, H.; et al. Diverse rock types detected in the lunar

South Pole - Aitken Basin by the Chang' E - 4 lunar mission. Geology 2020, 48, 723 - 727.

[9] Tao, D.; Jia, G.; Zhao, H. End - to - end simulation model of rover - based hyperspectral remote - sensing systems: Application to VNIS. Int. J. Remote Sens. 2014, 35, 7279 - 7302.

[10] Hu, S.; Lin, Y. T.; Liu, B.; Yang, W.; He, Z. P.; Xing, W. F. Reflectance calibration and shadow effect of VNIS spectra acquired by the Yutu rover. Res. Astron. Astrophys. 2015, 15, 1587.

[11] Lin, H.; Lin, Y.; Yang, W.; He, Z.; Hu, S.; Wei, Y.; Xu, R.; Zhang, J.; Liu, X.; Yang, J.; et al. New insight into lunar regolith - forming processes by the lunar rover Yutu - 2. Geophys. Res. Lett. 2020, 47, e2020GL087949.

[12] Hapke, B. Theory of Reflectance and Emittance Spectroscopy; Cambridge University Press: Cambridge, UK, 2012.

[13] He, Z.; Xu, R.; Li, C.; Lv, G.; Yuan, L.; Wang, B.; Shu, R.; Wang, J. Visible and near - infrared imaging spectrometer (VNIS) for in - situ lunar surface measurements. In Sensors, Systems, and Next - Generation Satellites XIX; International Society for Optics and Photonics: Toulouse, France, 2015; Volume 9639, p. 96391S.

[14] Yang, Y.; Lin, H.; Liu, Y.; Lin, Y.; Wei, Y.; Hu, S.; Yang, W.; Xu, R.; He, Z.; Zou, Y. The effects of viewing geometry on the spectral analysis of lunar regolith as inferred by in situ spectrophotometric measurements of Chang' E - 4. Geophys. Res. Lett. 2020, 47, e2020GL087080.

[15] Li, C.; Xu, R.; Lv, G.; Yuan, L.; He, Z.; Wang, J. Detection and calibration characteristics of the visible and near - infrared imaging spectrometer in the Chang'e - 4. Rev. Sci. Instrum. 2019, 90, 103106.

[16] He, Z.; Li, C.; Xu, R.; Lv, G.; Yuan, L.; Wang, J. Spectrometers based on acousto - optic tunable filters for in - situ lunar surface measurement. J. Appl. Remote Sens. 2019, 13, 027502.

[17] He, Z. P.; Wang, B. Y.; Lv, G.; Li, C. L.; Yuan, L. Y.; Xu, R.; Liu, B.; Chen, K.; Wang, J. Y. Operating principles and detection characteristics of the Visible and Near - Infrared Imaging Spectrometer in the Chang'e - 3. Res. Astron. Astrophys. 2014, 14, 1567.

[18] Liu, B.; Li, C. L.; Zhang, G. L.; Xu, R.; Liu, J. J.; Ren, X.; Tan, X.; Zhang, X. X.; Zuo, W.; Wen, W. B. Data processing and preliminary results of the Chang'e - 3 VIS/NIR Imaging Spectrometer in - situ analysis. Res. Astron. Astrophys. 2014, 14, 1578.

[19] Gueymard, C. A. The sun's total and spectral irradiance for solar energy applications and solar radiation models. Sol. Energy 2004, 76, 423 - 453.

[20] Wu, Y.; Wang, Z.; Cai, W.; Lu, Y. The absolute reflectance and new calibration site of the Moon. Astron. J. 2018, 155, 213.

[21] Lin, H.; Xu, R.; Yang, W.; Lin, Y.; Wei, Y.; Hu, S.; He, Z.; Qiao, L.; Wan, W. In situ photometric experiment of lunar regolith with visible and near-infrared imaging spectrometer on board the Yutu-2 lunar rover. J. Geophys. Res. Planets 2020, 125, e2019JE006076.

[22] Liu, B.; Liu, J. Z.; Zhang, G. L.; Ling, Z. C.; Zhang, J.; He, Z. P.; Yang, B. Y.; Zou, Y. L. Reflectance conversion methods for the VIS/NIR imaging spectrometer aboard the Chang' E-3 lunar rover: Based on ground validation experiment data. Res. Astron. Astrophys. 2013, 13, 862.

[23] Qi, X.; Ling, Z.; Zhang, J.; Chen, J.; Cao, H.; Liu, C.; Qiao, L.; Fu, X.; He, Z.; Xu, R.; et al. Photometric Normalization of Chang'e-4 Visible and Near-Infrared Imaging Spectrometer Datasets: A Combined Study of In-Situ and Laboratory Spectral Measurements. Remote Sens. 2020, 12, 3211.

[24] Lucey, P. G.; Blewett, D. T.; Jolliff, B. L. Lunar iron and titanium abundance algorithms based on final processing of Clementine ultraviolet-visible images. J. Geophys. Res. Planets 2000, 105, 20297-20305.

[25] Lin, H.; Lin, Y.; Wei, Y.; Xu, R.; Liu, Y.; Yang, Y.; Hu, S.; Yang, W.; He, Z. Estimation of Noise in the In Situ Hyperspectral Data Acquired by Chang' E-4 and Its Effects on Spectral Analysis of Regolith. Remote Sens. 2020, 12, 1603.

[26] De Boor, C.; De Boor, C. A Practical Guide to Splines; Springer: New York, NY, USA, 1978; Volume 27.

Analysis of plume - lunar surface interaction and soil erosion during the Chang'E - 4 landing process

Jilin You[1,2], Xiaoping Zhang[1,2], Haiyan Zhang[3,8], Cunhui Li[3,7], Yi Xu[1,2], Qi Yan[1,2], Hsinchen Yu[1,2], Jiahui Liu[1,2], Yong Li[1,2], Yi Wang[3], Chengxuan Zhao[3], He Zhang[4], Yingqiao Xu[4], Liping Chen[4], Hongyu Lin[5], Qingfei Fu[6], Yupeng Gao[6], Yuming Wang[7], Weidong Wang[8], Qijun Zhi[9]

1. State Key Laboratory of Lunar and Planetary Sciences, Macau University of Science and Technology, Macau 999078, China
2. CNSA Macau Center for Space Exploration and Science, Macau 999078, China
3. Science and Technology on Vacum Technology and Physics Laboratory, Lanzhou Institute of Physics, Lanzhou 73000, China
4. Beijing Institute of Spacecraft System Engineering China Academy of Space Technology, Beijing 100094, China
5. Beijing Institute of Space Mechanics and Electricity, China Academy of Space Technology, Beijing 100094, China
6. School of Astronautics, Beihang University, Beijing 100191, China
7. School of Earth and Space Science, University of Science and Technology of China, Hefei 230026, China
8. School of Mechano - Electronic Engineering Xidian University, Xi'an 710071, China
9. School of Physics and Electronic Science, Guizhou Normal University, Guiyang 550001, China

Abstract: When the lander approaches the lunar surface, the rocket exhaust will erode the surface and eject massive highspeed dust, which could damage the payloads on the lander or interfere with subsequent surface operations. This process also provides a unique opportunity to study the plume - dust interaction and the wind erosion on the surface of airless body such as the Moon. Previously, limited knowledge was obtained from Apollo landing images and a simplified erosion model is built to study the erosion process. With high - quality data acquired in the landing process of Chang'E - 4 (CE - 4) mission, for the first time, we aim to quantify the dust ejection angle and erosion depth for a low - thrust lander, test the effectiveness of the current erosion model, and provide reference information for future landing missions. We use image measurement methods to extract key parameters such as dust ejection angle and erosion depth from the landing image of CE - 4. For comparison with the observation, we also employ an erosion model to estimate the erosion thickness. The dust appears at about 7s before the lander touch the surface, with a height of 13m above the surface. When the nozzle altitude was about 0. 6m, the dust ejection angle is measured to be above 7°, while the dust ejection angle calculated by the particle trajectory method is less than 4. 3° for particle sizes smaller than 1mm. For the first time, an average erosion depth of 0. 7cm is measured between 1. 12m and 2m from the nozzle centerline for a low - thrust lander. The measurement corresponds to an erosion time duration from 1. 1s before landing to 0. 4s after landing. The measured erosion range using CE - 4 data is much larger than the estimated erosion radius given by the previous erosion model. It indicates that the dust's angular distribution, local topography of the landing area, and other factors are important in improving the precision of the erosion model.

Keywords: Wind erosion; Plume - dust interaction; Erosion depth; Dust ejection angle; Erosion model

1 Introduction

Wind erosion is a fundamental surface process on terrestrial planets with airflows. Dust on the planetary surface is usually transported by winds, which causes landform variation, environmental or economic damages. As an airless celestial body, it is generally impossible to study wind erosion on the Moon. However, during the landing process of space vehicles, the engine exhaust of a lunar lander impacts the lunar surface, resulting in fast surface dust transportation and surface erosion. Therefore, the landing process provides a unique opportunity to study the plume - dust interaction and the wind erosion process on the Moon's surface. It also provides useful data for extracting the properties of dust itself, such as its size distribution or mechanical properties.

Ejected high - speed dust could damage the landing instruments and interfere with subsequent surface operations[1,2]. The dust is tiny, usually tens of microns in diameter[3], so it is easy to adhere to the instrument's surface or penetrate its interior, e. g. joints of extravehicular activity (EVA) suits during Apollo mission[4,5]. Mpagazehe et al. show that even below the medium speed (about $105ms^{-1}$), the moving JSC - 1AF lunar dust simulant could reduce the transmittance and reflectance of the metallic and acrylic specimen's surface by up to ~70%[6] and ductile plastics are more vulnerable than brittle materials[7]. The dust's impact range is not limited to the lander itself but also distant instruments. For example, Immer et al. found that pits with a density of 103 pits cm^{-2} and diameter of about 80μm occurred on the equipment of Surveyor 3 were caused by particles ejected during the Apollo 12 landing[8]. The device is about 155m away from the Apollo 12 landing site. Furthermore, dust is toxic to astronauts' lung and cardiac health[9-12]. Due to the harmfulness of dust, "Dust is the number one priority for returning to the moon", Apollo astronaut John Young declared. Therefore, studying the characteristics of dust during landing will help plan and implement future lunar exploration missions.

The erosion process was studied with laboratory experiments, numerical simulations, and images of Apollo missions. Laboratory experiments using jets of gas were implemented, and it showed that soil erosion rate in near vacuum conditions was faster than that in ambient atmosphere conditions[13]. This experiment also showed that the dependence of erosion rate on various parameters is more complicated than expected. It is still challenging to observe the plumesoil interaction process in a vacuum chamber to model the environment of the Moon. A direct simulation Monte Carlo (DSMC) method[14] was used to analyze the interaction process between plume and lunar soil under the vacuum condition and the rarefied gas flowing out of the nozzle[15-17]. Another method[18] used the conservation laws in the Eulerian - Lagrangian framework to investigate near - field plumesurface interaction. Using landing images of Apollo missions, Immer et al. estimated that the ejected dust density by the exhaust was about $10^8 - 10^{15}$ particles m^{-3} when Apollo 11 and 16 approached the lunar surface[19]. Lane et al. found that 2600 kg of lunar soil was ejected during Apollo 12 landing process[20].

At present, the in - situ observations on lunar surface erosion during the landing are mainly based on Apollo's landing images. At 2: 26 (UTC) on January 3, 2019, CE - 4 successfully landed in the Von Kármán crater in the South - Pole Aitken basin on the farside of the Moon. While CE - 4 lander was approaching the lunar surface, a large amount of dust was ejected by the rocket plume, which was recorded by high resolution landing images. Here we analyze the plume erosion on the lunar soil during the CE - 4 landing. CE - 4 performed a vertical landing while the Apollo mission made an oblique landing. Besides, the engine thrusts of CE - 4 and Apollo missions are different. Therefore, CE - 4 collected a unique set of observation data and show clear dust transportation. Some quantitative information was determined for erosion model development: the height of CE - 4 lander when the ejected dust just appeared, the dust ejection angle, and the eroded lunar soil's thickness by the plume, etc. We find that results obtained from CE - 4 landing process are different from the estimations given by the erosion theoretical model[20], providing new insights into erosion model development and future exploration missions.

2 CE - 4 landing images

During the landing, the landing camera captured images of the lunar surface at a ten fps rate. Fig. 1 plots the average pixel value and standard deviation of the histogram of the images of lunar surface during the eleven seconds before the landing. In Fig. 1, the negative sign on the lower axis indicates the time before CE - 4 landed, and 0 is the time when CE - 4 just touched the lunar surface. In Fig. 1 and elsewhere in this paper, all CE - 4 landing images are gray scale images with 8 bits per sampled pixel. The contrast ranges from black at the weakest intensity (0) to white at the strongest (255). The exposure time of landing images is 2.2ms before −2s and to 1.6ms within 2s before landing. The image at −4.6s is not plotted due to image distortion. From −11s to about −7s, the average brightness of the images remains almost unchanged. After −7s, the average brightness shows some fluctuations because dust rose into the field of view of the landing camera or exposure time changed.

According to detailed analysis, landing images within 8.8s before landing are divided into six stages, as shown in Table 1. In stage Ⅰ, the lunar surface's brightness remains nearly unchanged. In stage Ⅱ, this brightness shows a continuous rise when the lander got closer to the lunar surface. However, it is difficult to see the dust by eye in this stage. The dust can be observed starting at −7.1s. At this time, the corresponding nozzle altitude is about 13m. The DSMC simulation[21] also showed that the plume of CE−4 has almost no effect on the lunar surface when the nozzle is above 20m from the surface. From the third stage to the sixth stage, more dust appear and show a radially spreading pattern.

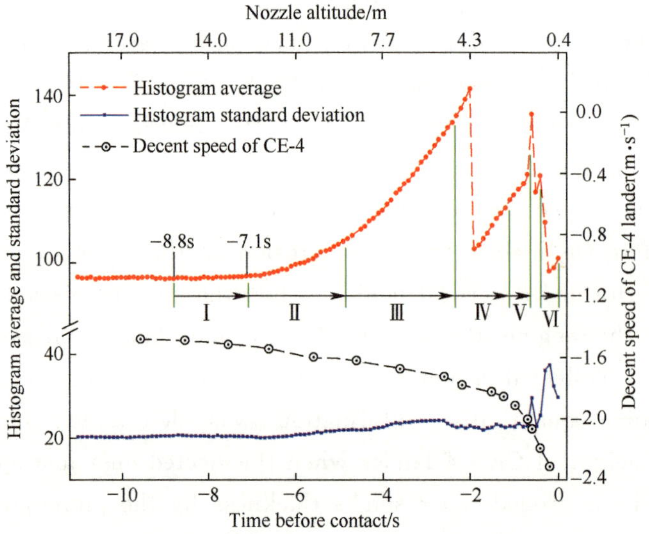

Fig. 1　Histogram average, the standard deviation for the photoed lunar surface in each landing image, and decent speed versus the nozzle altitude and the time before contact. The nozzle altitude denotes the distance of the nozzle exit plane from the lunar surface. See Appendix A for the detailed descriptions of altitude measurement

3　Measurement results

3.1　Ejection angle of dust

In Image F3372 (see Fig. 2 (b)), one can observe that dust covers the footpad of CE−4.

The footpad brightness for the adjacent image is shown in Fig. 2 (c). As the dust in F3372 covers the footpad, the footpad brightness suddenly increases.

A schematic diagram of the lander in F3372 is shown in Fig. 2 (d). If the dust starts to rise from the nozzle centerline, the calculated ejection angle of dust is about 7° and treated as a minimum when the altitude of footpad is 0. 24m and its distance to the nozzle centerline is 1. 95m. CE - 4 landed on an inclined surface (about 5° measured by Guidance Navigation and Control[21]; about 4° measured by Wu et al. [22]). In the case of an inclined surface of 5° (Fig. 2 (e)), the dust ejection angle is determined to be 6. 89° , which is quite close to the result deriving from a flat ground . Considering the small difference

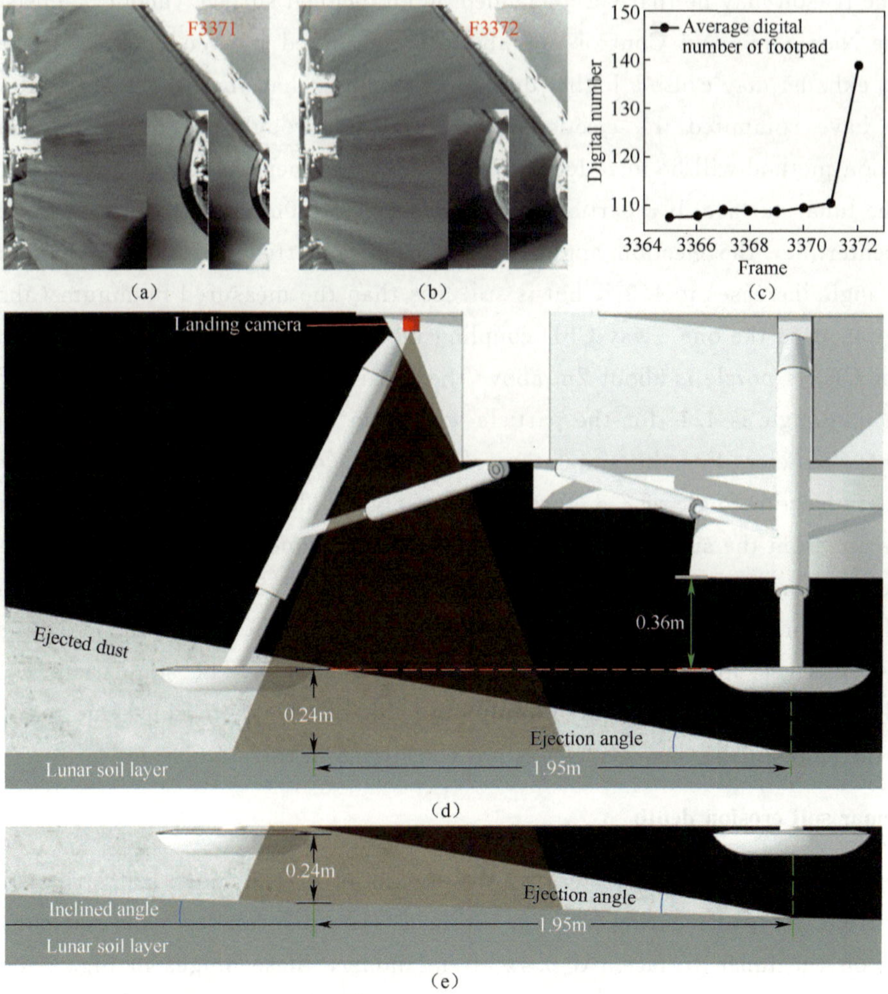

Fig. 2 (a) Dust does not cover the surface of the footpad in F3371. (b) Dust covers the footpad in F3372. (c) Average brightness change on the footpad in images (F3365 - F3372). (d) Schematic diagram of dust covering the footpad in F3372. (e) Similar to the panel (d), but the lunar surface is inclined. The altitude of 0. 24m is measured according to Appendix A. Other values are obtained from the design drawing of CE - 4 between an ejection angle of 7° obtained in the case of a flat surface and 6.89°, the slope's influence on measuring the minimum ejection angle of dust can be ignored. When calculating the ejection angle of dust, we assumed that all dust was emitted from the nozzle centerline. In fact, the dust is emitted at a certain distance from the nozzle centerline, so the actual ejection angle of dust will be larger. Immer et al. measured the ejection angle of dust when Apollo (excluding Apollo 12) landed[19]. The measured average ejection angle is 3.3° for these Apollo landings, the maximum ejection angle is 8.1° for Apollo 15. Immer et al. analyzed that the difference was due to the landing on an inclined surface of about 11°. By comparison, the ejection angle for CE-4 is relatively high. One reason may be that CE-4 landed on an inclined surface (about 5° measured by Guidance Navigation and Control[21]; about 4° measured by Wu et al. [22]), and the engine's exhaust may cause a higher dust ejection angle than that of the horizontal plane.

We have computed the ejection angle of dust through the numerical simulation (simulation method will be introduced in Appendix F) when the nozzle of CE-4 is 0.6m above the lunar surface. If a particle with a diameter of 70μm starts 0.4m away from the nozzle centerline, the ejection angle is 2.7°. For a particle with 1mm diameter, the ejection angle increases to 4.3°, but is still less than the measured minimum (about 7°). Zhang et al. used the one-way CFD coupling method to simulate the dust ejection angle when the CE-4 nozzle is about 2m above the lunar surface[23]. They computed that the dust ejection angle is 4.1° for the particle with 1μm diameter, and 2.3° for the particle with 1mm diameter. For the simulation of Apollo's dust ejection angle, Morris et al. obtained an ejection angle of 2.8° through a loosely coupled continuum-DSMC when the nozzle is 5m from the surface[24]. Although Apollo's plume parameters and nozzle height are different from that for CE-4, the simulated angle of 2.8° is close to CE-4 simulated value of 2.7° for the particle size of 70μm. At the same height of 5m, the maximum angle does not exceed 4° for particle sizes smaller than 50μm[18]. The simulation results are generally smaller than the measured minimum value of CE-4 mission. One reason could be that the slope of the lunar surface is not considered in the simulation.

3.2 Lunar soil erosion depth

Here, the erosion depth refers to the thickness of the lunar soil stripped by the exhaust from the lander during landing. In Fig. 3 (a), one can see some soil bulges with shadows on the lunar surface. At 0.4s after landing, these bulges in Fig. 3 (c) were almost entirely blown away by the exhaust. The thickness of these bulges approximates the lunar soil depth eroded by the exhaust. The estimated erosion depth from Appendix C is shown in Fig. 4 (a). Additionally, we plot the elevation map in Fig. 4 (b) to show the correlation between erosion depth and terrain elevation. Note that the elevation of zone D′ in Fig. 4 (b) may have a large error due to the influence of moving dust on the measurement process (see Appendix D). However, the elevation of other areas is not significantly different from the landing image.

From Fig. 4 (a), there is relatively severe erosion in the four marked areas (A - D). The erosion depth in these four areas is around 1cm. Among them, areas B, C, and D have higher elevations in Fig. 4 (b). The erosion depth and elevation are relatively high around those erosion points 1, 2, and 3. From Fig. 3 (d), the lunar soil surface near these erosion points becomes relatively smooth, which also indicates that severe erosion has occurred around these erosion points, which may be due to relatively high terrain. In the cyan area of Fig. 4 (a), the erosion depth is reduced to around 0.4cm. The corresponding elevation is relatively low.

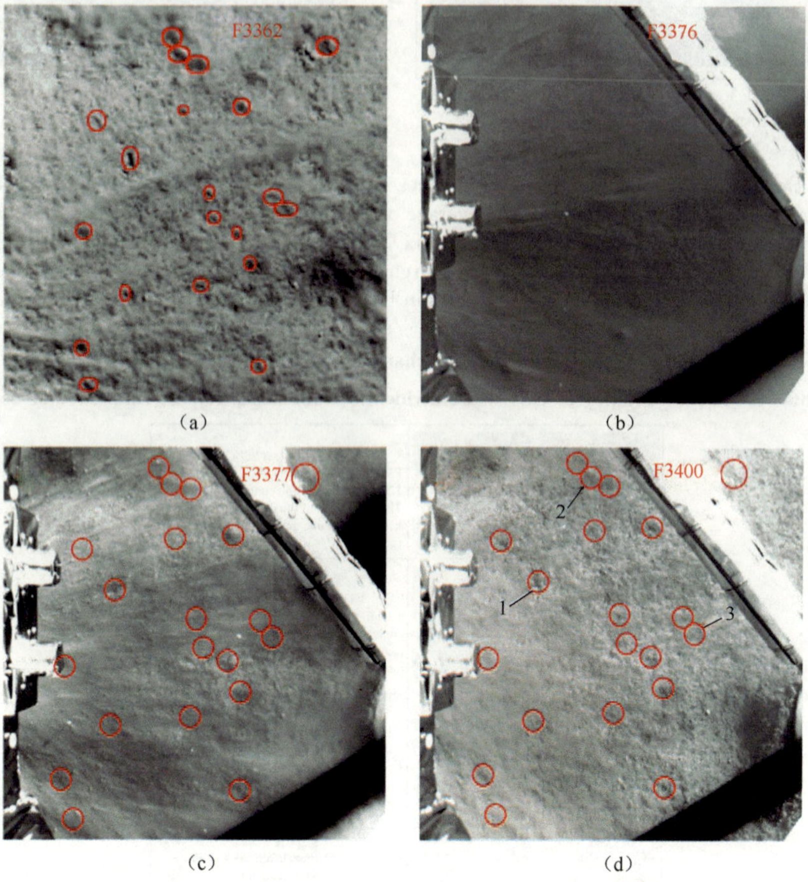

Fig. 3　Comparison of erosion in the same area before and after landing. The image at the touchdown moment is F3373. The red circles' positions in panels (a), (c), and (d) correspond one to one. Contrast enhanced for F3362, F3377, and F3400

The erosion depth is not only related to topography but also affected by other factors. For example, it can be seen from Figs. 3 (a) and 3 (b) that massive dust appears in some radial directions. The dust caused secondary erosion on the lunar surface. By comparing Figs. 4 (a) and 4 (b), it can be found that the terrain of belt A is relatively low, but the

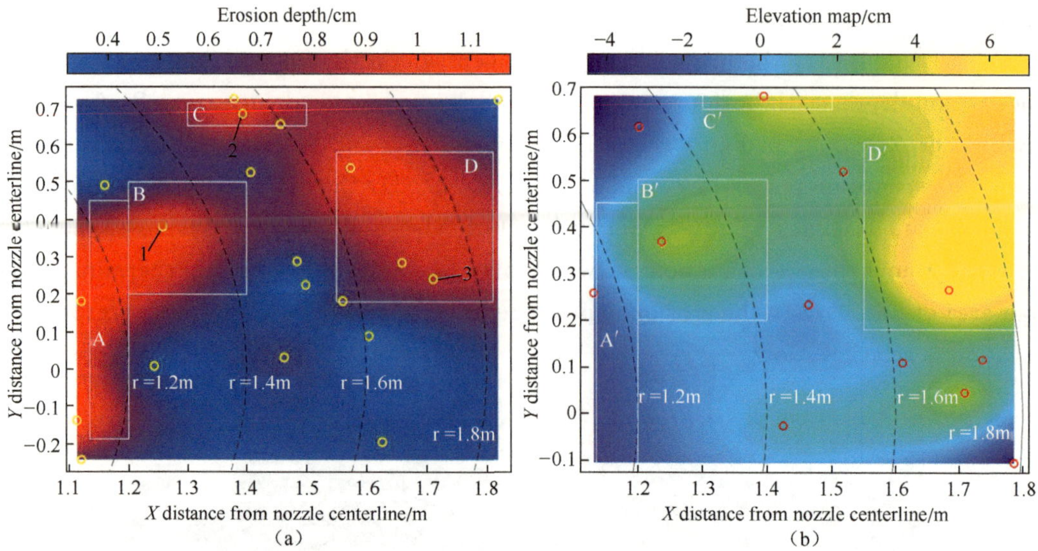

Fig. 4 (a) Erosion depth obtained from image measurement (open yellow circles), and the depth for other areas are obtained by interpolation. The average of these measurements is 0.71 ± 0.20cm (details of the uncertainty estimation could be found in Appendix C), 0.67cm for the interpolation area. (b) Elevation map obtained from the interpolation on image measurement (open red circles). The base plane of the map is the local slope. This map corresponds to the terrain in F3377 (see Fig. 3 (c)). The estimation method for the elevation is from Appendix D

Tab. 1 According to the dust characteristics, the landing images (F3286 - F3373, F: Frame) of CE - 4 are divided into six stages for the final 8.8s of descent

	Stage Ⅰ F3286–F3303.(nozzle altitudes: 15.23m–12.67m; time before contact:−8.8s—7.1s). The brightness of the images in this stage keeps almost unchanged.
	Stage Ⅱ F3304–F3325.(nozzle altitudes: 12.51m–9.22m; time before contact: −7s—4.9s).The brightness of the images in this stage gradually increases. However, dust is not visible in most images.
	Stage Ⅲ F3326–F3349.(nozzle altitudes: 9.04m–5.06m; time before contact: −4.8s—2.4s). In this stage, the dust can be clearly seen. Thedust starts from a certain position and spreads symmetrically on the lunar surface.
	Stage Ⅳ F3350–F3361.(nozzle altitudes: 4.90m–2.88m; time before contact: −2.3s—1.2s). In this stage, the distribution of dust begins to show asymmetry, and the inner erosion radius(r_0) decreases as the nozzle altitude decreases.
	Stage Ⅴ F3362–F3366. (nozzle altitudes: 2.67m–1.92m; time before contact: −1.1s—0.7s). In this stage, the amount of dust continues to increase, and the distribution of dust is radial.
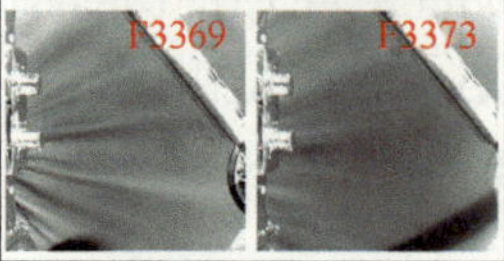	Stage Ⅵ F3369–F3373. (nozzle altitudes: 1.31m–0.38m; time before contact:−0.4s–0s).A large amount of dust is blown up radially, almost completely covering the lunar surface.

erosion depth is still high. Because belt A is close to the nozzle, the erosion effect is related with the distance to the nozzle. The smooth surface on the left side of Fig. 3 (d) also hints that the erosion in this area is relatively serious. Besides, the cohesion of the lunar soil is also a factor affecting the erosion depth.

Above comparisons between Figs. 4 (a) and 4 (b) indicate that the lander landing in a flat place can effectively reduce the impact of dust on the equipment. With more data in the future, it is expected that some quantitative relationships will be established among erosion depth, elevation map, and so on, which helps select an optimal landing site with the least amount of ejected dust during the landing process.

4 Theoretical calculation of erosion depth

The theoretical values obtained from the erosion model proposed by Lane and Metzger[20] are compared with the measured erosion depth in this paper. In the model, the histogram matching method (HMM) and the particle trajectory method (PTM) are carried to estimate the extinction factor and the speed of dust, respectively. This model has not been verified by in - situ observations since it was put forward. Here, we make some improvements to the HMM based on CE - 4 data.

The erosion model[20] is used to estimate both erosion rate and total mass of soil thrown from the lunar surface during the landing. The formula to calculate mass erosion rate is written as follows[20]:

$$\dot{m}(t)=\frac{\rho_L \pi \phi}{6s(f)}M_2(t)\frac{\int_0^\infty S(D)D^3 dD}{\int_0^\infty S(D)\left(\frac{D^2}{v_\perp(D)}\right)dD} \tag{1}$$

$$M_2(t)=\frac{4\alpha}{\pi Q_e}$$

where $\dot{m}(t)$ is mass erosion rate (in kg s^{-1} m^{-2}) at time t, ρ_L is the density of lunar soil particles, ϕ is the geometry factor which describes the transport pattern for the dust or rainfall, $s(f)$ is the particle shape factor, α is the optical extinction factor, Q_e is the scattering efficiency factor, D is the diameter of particle (in m), $S(D)$ is the normalized soil size distribution, $v_\perp(D)$ is the vertical component of the average speed of the dust.

In this paper, these quantities used in Eq. (1) are as follows [20]: $\rho_L = 3100\text{kgm}^{-3}$, $\phi = 2$ for radially spreading dust, $s(f) = 1$ for spherical particles, $Q_e = 2$ for visible light scattering on the lunar particles ($D > 0.5\mu\text{m}$). The diameter of large particles is a few millimeters from CE - 4 landing images. For small particles, the diameter is mainly greater than 1μm[3]. Therefore, the selected integral range in Eq. (1) is from 10^{-6} m to 10^{-2} m for most lunar dust.

In Eq. (1), $v_\perp(D)$ and α are essential quantities for calculating mass erosion rate.

The computation about $v_{\perp}(D)$ will be introduced in Appendix F. α can be determined by the following formula from the histogram matching method[20]:

$$p'_{ij} = p_{ij}\,\mathrm{e}^{-2\alpha x} + p_0(1-\mathrm{e}^{-\alpha x}), \tag{2}$$

where p'_{ij} and p_{ij} denote output and input image pixel, respectively, p_0 is a fitting parameter, x is the path length of effective optical - dust. This length is determined by the formula:

$$x = \begin{cases} x_D & r \leqslant a_0 \\ x_D\left(\dfrac{a_0}{r}\right)^2 & r > a_0 \end{cases} \tag{3}$$

where x_D denotes the length of the reflected light through the dust, and it is indicated by the dashed green line in Fig. 5 (a), r is the distance from a point on the lunar surface to the engine nozzle centerline, a_0 is the erosion radius in Fig. 5 (a).

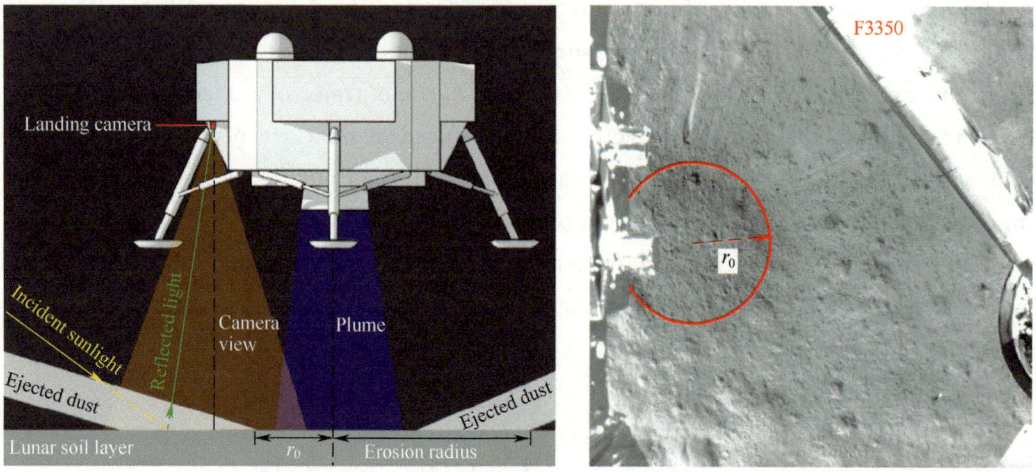

(a)　　(b)

Fig. 5　(a) Geometry of the erosion model. (b) Most dust rises from the area outside the radius r_0. (For interpretation of the references to color in this figure legend, the reader is referred to the web version of this article.)

The process of determining the extinction factor can be divided into three steps: selecting a reference image as p_{ij}; bring p_{ij} into Eq. (2) and outputting an image p'_{ij} under given parameters α and p_0; comparing the thistogram of p'_{ij} with that of the image covered by dust. The optimal parameter α is determined when the average and standard deviation of the histogram of p'_{ij} are equal to that of the image covered by dust, respectively. In the first step, we select the image (F3286, see also in stage I in Tab. 1) as the reference image for determining α. The fluctuation of the average and standard deviation for F3286 is minimal in adjacent images. Therefore, F3286 can be a reference image according to the histogram matching method. More details about this method are available in [20].

In this paper, we made three improvements to Eqs. (2) and (3) . The specific reasons and process of improvement are listed in Appendix E. The improved formula to determine the extinction factor α is as follows:

$$R'_{ij} = R_{ij} e^{-\alpha(x+d)} + R_0(1 - e^{-\alpha x})$$
$$x = \begin{cases} x_D & r_0 \leqslant r \leqslant a_0 \\ x_D \left(\dfrac{a_0 - r_0}{r - r_0}\right)^2 & r > a_0 \end{cases} \tag{4}$$

where R_{ij} is the equivalent radiance corresponding to a pixel on the reference image, d denotes the path length of effective optical - dust for the incident light, R_0 is still a fitting parameter. r_0 is marked in Fig. 5. Within $r_0 \leqslant r \leqslant a_0$, d refers to the dashed yellow line in Fig. 5 (a) . Outside the range of a_0, d is also multiplied by a factor $((a_0 - r_0)/(r - r_0))^2$. Calculating length x for the reflected light can be found in [20]. The relevant camera parameters are listed in Supplementary Table S2. As for the length d, the solar altitude and azimuth angle are used, which have been estimated with the method introduced in Appendix B. When calculating the lengths x and d, the dust ejection is set to be 7°.

The radius r_0 is determined by Eq. (4) for these images of F3300 - F3316, (corresponding nozzle altitudes: 13.13m - 10.66m). For subsequent images of F3317 - F3360 (corresponding nozle altitudes: 10.50m - 3.04m), r_0 is directly measured. In the last batch of images of F3361 - F3373 (corresponding nozzle altitudes: 2.88m - 0.38m), as the radius r_0 is not visble and dust distribution in asymmertic, Eq. (4) is not suitable for determining r_0. We will estimate r_0 by simulation introduced in Appendix F.

5 Discussions

5.1 Extinction factor and erosion radius

The extinction factor and erosion radius for CE - 4 and Apollo 12 landing processes are shown in Fig. 6. For both CE - 4 and Apollo 12, the extinction factor increases with decreasing nozzle altitude, and an inverse association is found between extinction factor and erosion radius. The erosion radius for Apollo 12 is about twice as large as that for CE - 4 at about 13m altitude, and the ratio rises to 5 at about 2m altitude. The main reason for the difference in this erosion radius is that Apollo 12 has a larger nozzle radius and pressure on the nozzle outlet plane.

Because of the thrust difference, it is expected that the extinction factor for CE - 4 should be relatively low, however, Fig. 6 shows that extinction factors of CE - 4 are generally higher than Apollo 12. One reason is that the angles (ψ) of the landing cameras' optical axes relative to vertical are different. The optical axis of CE - 4 landing camera is perpendicular to the lunar surface. As for Apollo 12, ψ is 33° . As ψ decreases, the observed ground position is closer to the nozzle center, resulting in a higher amount of observed dust.

Fig. 6 shows the variation of radius r_0 with the nozzle altitude. Above 10.66m altitude, the radius r_0 is obtained by Eq. (4) in the improved HMM. For altitudes in the range of 3 to 10.66m, r_0 is determined by measuring images. r_0 within these two ranges show the same linear change. Within the final 3m altitude, r_0 is estimated by the simulation method presented in Appendix F, which is between 0.2m and 0.3m.

When the nozzle of CE－4 was about 2m above the lunar surface, the engine shut down, but the nozzle was still exhausting gas. From Fig. 6, above 2m altitude, the extinction factor increases by about 0.79m^{-1} for every one－meter reduction in nozzle height. However, below 2m altitude, the extinction factor's growth rate slows down to zero. This change implies that turning off the engine before landing helps reduce the impact of dust on equipment.

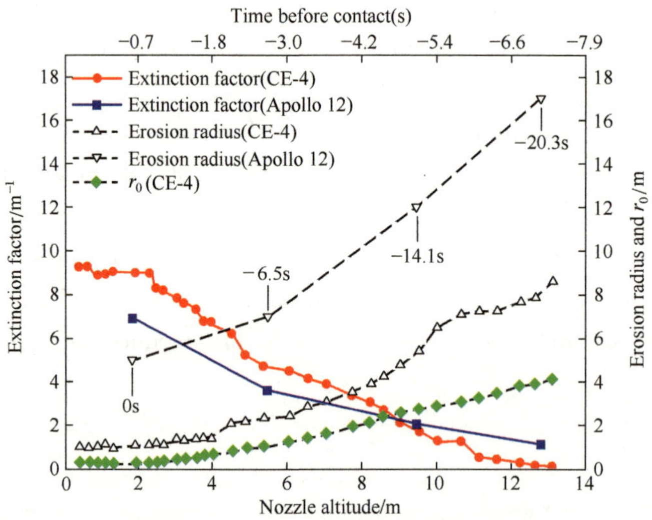

Fig. 6　Extinction factor and erosion radius for CE－4 and Apollo 12[20] versus nozzle altitude and time before contact. The upper axis (time before contact) is only applicable to CE－4, the time for Apollo 12 is marked in the figure

5.2　Particle velocity

Fig. 7 plots the average velocity of particles passing through the observation area, which is determined by the simulation in Appendix F. In Fig. 7 (a), before the engine shut down (nozzle altitude is higher than 2m), particle velocity basically increases with the decreasing nozzle altitude or particle diameter. Below 2m altitude, the velocity decreases due to the lower thrust. At about 2m altitude, as the engine starts to shut down, the dust speed decreases rapidly. When the nozzle altitude decreases further, the dust velocity increases again, resulting in a minimum velocity at 1.3m altitude. Fig. 7 (c) shows the variation of particle average velocity with particle diameter for several nozzle altitudes. From this figure, the speed of 1μm particles can reach about $200ms^{-1}$, the speed of 70μm particles is about 30 – $60ms^{-1}$, and the speed of millimeter particles is basically lower than $10ms^{-1}$.

As shown in Figs. 7 (b) and 7 (d), we consider different cases to determine the speed weighting factor η of Eq. (F. 3) in Appendix F. The factor η is different before and after the engine is shut down. Before the shutdown point (nozzle altitude $H > 1.92$m), four η values in Fig. 7 (b) are very close to one. After the shutdown point, another four η values in Fig. 7 (d) are around 0. 6. Extensive calculations show that the factor η is also around 0. 6 for the nozzle altitude after shutdown ($H = 1.31$m, 1. 07m, 0. 60m, and 0. 38m). Therefore, we use the average of the four η values in Fig. 7 (d) to calculate the soil erosion after the engine shutdown. Before the shutdown point ($H > 1.92$m), the average of the four η values in Fig. 7 (b) is used. Here, the weighting factor η is used for the following reason. In Eq. (1), $v_{\perp}(D)$ refers to the vertical speed of all particles with diameter D in the observation area. However, the particles may be ejected from different initial locations r, resulting in different speeds in the observation area. Therefore, their speeds need to be weighted based on their initial locations, and the weighting factor obtained is substituted into Eq. (F. 4) to obtain $v_{\perp}(D)$.

5.3 Mass erosion rate

The following parameters and formulae are used to calculate the erosion rate: the average factor $\eta = 1.06$ from Fig. 7 (b) for nozzle altitude $H > 1.92$m, the average factor $\eta = 0.63$ from Fig. 7 (d) for nozzle altitude $H \leqslant 1.92$m, and the minimum ejection angle of dust $\theta = 7°$, the declarative parameters used in Eq. (1), Eqs. (1) and (F. 4) in Appendix F, and Figs. 7 (a) and 8 (a). Through the numerical calculation of these data, we obtained the mass erosion rate plotted in Fig. 8 (b). By contrast, these two kinds of Apollo samples cause little difference in erosion rates. Above 6m, the erosion rate increases by 0. 28kg s^{-1} m^{-2} for every one – meter reduction in altitude. This factor rises to 0. 5kg s^{-1} m^{-2} for altitudes in the range of 6 to 1. 9m. At the stage below 1. 9m altitude, the factor increases to 1. 7kg s^{-1} m^{-2} even though the simulation's thrust is reduced to 322 N. So the erosion rate's growth is relatively large at low altitude.

The total erosion mass m_t is obtained by the summation formula:

$$
\begin{aligned}
m_t &= \sum_{n=2}^{32} \Delta m_n \\
&= \sum_{n=2}^{32} \frac{1}{2}(\dot{m}_n \pi(a_{0n}^2 - r_{0n}^2) + \dot{m}_{n-1}\pi(a_{0n-1}^2 - r_{0n-1}^2))(t_n - t_{n-1}),
\end{aligned} \tag{5}
$$

where t_1 represents the first moment when dust appears, and t_{32} represents the last moment before landing. Substituting the 32 sets of data in Figs. 6 and 8 (b) into Eq. (5), the soil erosion mass during the whole landing process can be calculated. The erosion mass of different types of samples are computed as follows: 212kg for 60010 samples, 214kg for

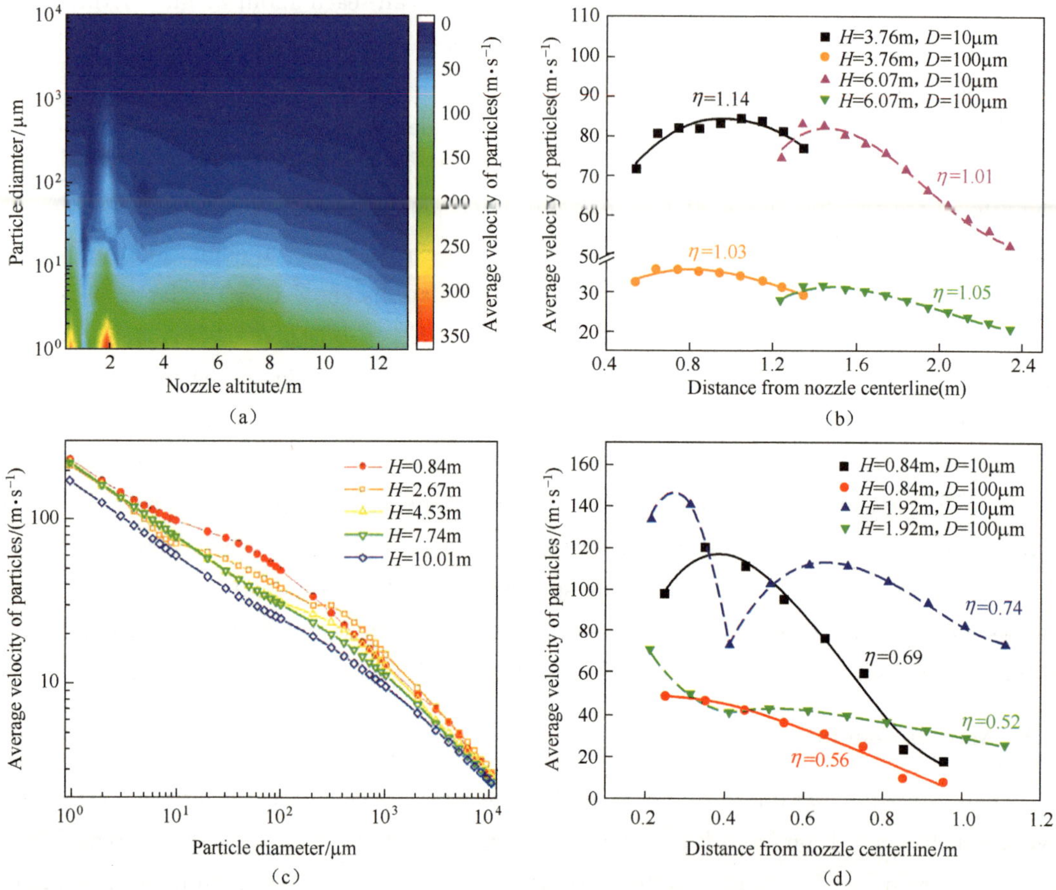

Fig. 7　(a) Average velocity $\overline{V}_t$ (r_0) of particles passing through the observation area. The initial altitude of the particles from the ground in the simulation is set to 0.005m, the initial horizontal distance of the particles from the nozzle centerline is equal to the corresponding r_0 for a nozzle altitude. (b) Average velocity $\overline{V}_t$ (r) as a function of the initial distance from nozzle centerline, these fitting curves are power polynomial functions. H and D represent the nozzle altitude and particle diameter, respectively. (c) Average velocity $\overline{V}_t$ (D) as a function of particle diameter D. The initial position of the particles is the same as in panel (a). (d) The same as panel (b), but for those two cases ($H=1.92$m), piecewise fitting is taken, and the dividing line is at $r=0.4$m

10084 and 70051 samples. Although two types of samples have a very different particle size, in the range of tens of microns (see Fig. 8 (a)), the erosion mass difference is only 1%.

Lane and Metzger estimated an erosion mass of 2600kg for Apollo 12[20]. This result is more than ten times higher than that of CE-4 mission (212kg and 214kg). It is caused by three quantities: erosion range on the lunar surface, erosion duration, and engine thrust. The erosion range for Apollo 12 is at least twice that for CE-4, and the erosion duration increases by a factor of seven. Besides, the pressure on the nozzle outlet plane of

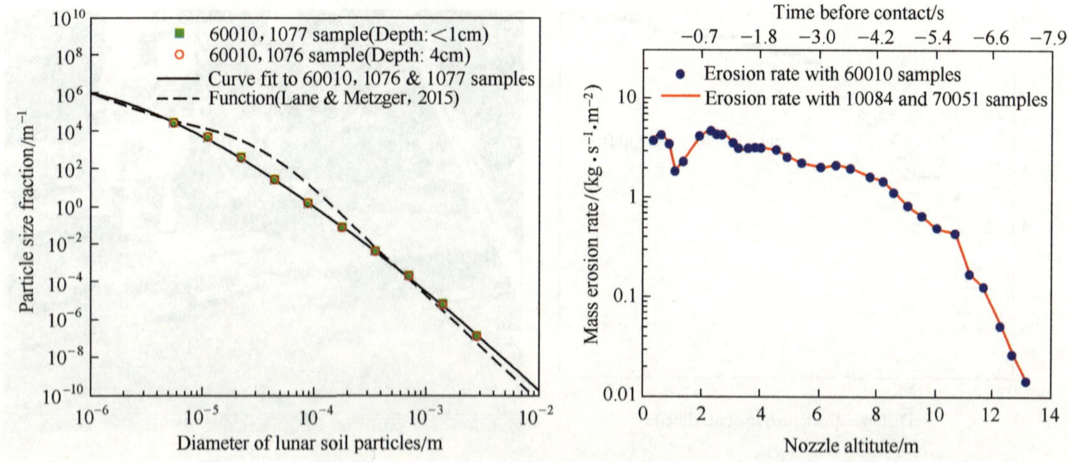

(a)　　(b)

Fig. 8　(a) Normalized particle size fraction $S(D)$ of lunar soil samples for 60010 (Apollo 16) at two different depths (< 1cm and 4cm). The dotted gray line denotes the curve fit to samples 10084 (Apollo 11) and 70051 (Apollo 17), this line from Eq. (9) in the article[20]. (b) Mass erosion rate under the various lunar soil samples

Apollo 12 is about 1.5 times that of CE - 4. Also, we use the average velocity of the particles in the observation area to calculate the erosion mass, not the maximum velocity of the particles[20]. If the maximum speed is used, the speed in Fig. 7 (a) will increase nearly twice, which will also lead to a significant increase in erosion rate and erosion mass. Therefore, the erosion mass of CE - 4 calculated in this paper is much smaller than that of Apollo 12.

5.4 Erosion depth for lunar soil

The erosion mass in each time step can be calculated by Eq. (5), and then divided by the bulk density of lunar soil ($1330 kg\ m^{-3}$[25]) to obtain the corresponding volume. This part of the volume is divided by the corresponding erosion area to obtain the erosion depth in this time step. In our calculation process, the erosion rate used in Eq. (5) is the average of the two sets of data in Fig. 8 (b), and the erosion radius a_0 and r_0 of CE - 4 can be obtained from Fig. 6. The estimated erosion depth during the entire landing is shown in Fig. 9 (a). In this figure, a 0.6cm depth crater appears at about 1m distance from the nozzle centerline. Within the radius of 0.2m, the erosion depth is reduced to zero. This erosion area under the nozzle has a radius of 0.4m.

Lane and Metzger used the erosion model to estimate the erosion depth produced by the Apollo 12 engine[20]. The estimated depth reaches 1.5cm below the LM. However, the radius r_0 was not considered, the erosion depth of 1.5cm was predicted in the area under the nozzle. Through the DSMC simulation, Morris et al. estimated a crater depth of about 2.2cm between 1m and 2m from the nozzle centerline of Apollo[24]. This crater depth is

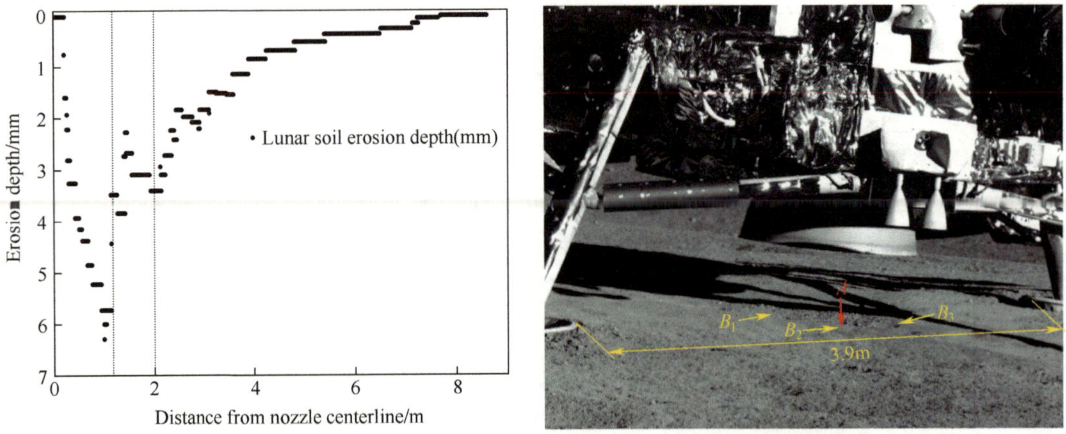

(a)　　(b)

Fig. 9 (a) Erosion depth obtained from the erosion model. (b) Erodent traces on the lunar soil surface under the nozzle of CE-4. The distance (3.9m) between the two diagonal footpads is obtained by the design drawing of CE-4

about 3.7 times larger than that in Fig. 9 (a). Both Fig. 9 (a) and the result of Morris et al. show that the largest soil removal is not under the nozzle, but 1m to 2m away from the centerline of the nozzle. The Apollo landings also show that the largest soil removal is located 1m or more away from the plume's impingement point[26]. Especially in the landing of Apollo 11, there is an annular area around the nozzle, where the soil is eroded by the plume[27].

After Apollo 11 landed, the soil surface under the lander show some layered shapes, not a smooth surface. These soil layers gradually descend moving away from the nozzle as far outward as the footpads[27]. However, both Fig. 9 (a) and the result of Morris et al. [24] do not reproduce the layered structure of the soil surface. This is because these two models do not consider the abrupt change of lunar soil properties with lunar soil depth. In fact, the lunar soil properties (such as soil compaction) of the upper layer are different from those of the lower layer. The terraces in the Apollo landings are caused by natural geological strata[27]. Note that the stepped erosion depth in Fig. 9 (a) is only caused by the assumption of a constant erosion rate within the corresponding erosion range and time step. In future model improvements, the geological structure of the lunar soil is a factor worth considering to estimate a more reasonable erosion depth.

In Fig. 9 (b), we can find some new craters created by the plume under the nozzle. The images taken by the panoramic camera of Yutu-2 rover are used to measure the depth of these craters (here, the depth refers to the distance from a point to the base plane where points B_1, B_2, and B_3 in Fig. 9 (b) are). At point A in Fig. 9 (b), which is 0.33m from the nozzle centerline, the measured depth reaches a maximum that is equal to 1.84cm. However, the erosion model calculation in Fig. 9 (a) shows that the erosion depth is about 0. 3cm at 0. 33m from the nozzle centerline. The possible reason for this difference is that the following factors are not considered in the numerical simulation: the topography of the lunar surface under the CE - 4 nozzle, the cohesion of the lunar soil, the porosity of the lunar soil, and so on. Another possible reason is that those craters under the nozzle are not erosion craters, but craters created by the diffused gas eruption (DGE[28]). The gas from the nozzle is injected into the pores of the soil under the nozzle and forms the gas cones with high pressure. When the engine shuts down, the gas cones rush out and blow away the lunar soil, finally leaving some craters on the lunar surface. Within one second before the CE - 4 landed, the engine discharged non - combusted gas with higher density, which was more likely to cause the DGE process to occur.

The descriptions of Apollo 11 astronauts and photographic evidence show that a crater of 2. 54cm deep eroded by the exhaust appeared beneath the engine after Apollo 11 landed[29]. This crater depth for Apollo 11 is 38% larger than the depth of 1. 84cm for CE - 4. The speed of Apollo 11 approaching the lunar surface is about 0. 5m s^{-1}[29], which is 1/4 times that of CE - 4 (around 2m s^{-1} for CE - 4 measured by Appendix A). In other words, Apollo 11 takes more time to complete the landing than CE - 4. Therefore, the exhaust from Apollo 11 can erode deeper crater on the lunar surface in a longer period of time. We note that no dust movement within r_0 is an ideal assumption in the improved erosion model in this paper. In the real situation, the erosion area under the nozzle needs to be further established and studied.

In Fig. 9 (a), the average depth is 0. 3cm for distances in the range of 1. 12m to 2m. The measured value is about 0. 7cm in the same range shown in Fig. 4 (a). The calculated value is 2. 3 times smaller than the measured value, and the corresponding erosion time is not at the same stage. The measured depth (0. 7cm) was produced from 1. 1s before landing to 0. 4s after landing. During this time, the erosion radius estimated by HMM[20] is less than 1. 13m (see Fig. 6). Hence, almost no erosion is predicted to occur in the range of 1. 12m to 2m during this period. The predicted erosion depth within the same region plotted in Fig. 9 (a) occurred at the earlier stage. Therefore, the HMM predictions are inconsistent with the observations in the same zone and at the same time.

The predicted erosion radius based on the HMM method is already smaller than the observed one. In the process of deriving this prediction, the ejection angle of dust used is the measured value of 7°obtained in Section 3. 1. However, this ejection angle is a lower limit, and the actual ejection angle may be higher. To show whether the calculated erosion radius for higher angle changes the prediction result, we substituted some higher inclination angles into the HMM method to obtain the corresponding erosion radius. When the nozzle altitude of CE - 4 is 0. 84m (time before contact: —0. 2s), increasing the dust inclination angle from 7° to 22° will reduce the erosion radius from 1. 02m to 0. 51m. When the nozzle altitude is 1. 92m (time before contact: —0. 7s), the same angle change reduces the erosion radius from 1. 12m to 0. 54m. From these two results, when the dust inclination angle increases, the predicted erosion radius is even smaller than the observed one in the range of 1.12m to 2m.

The main reason for this deviation is that the assumptions used in the simulation model are that all the dust is ejected within the erosion radius and with the same ejection angle. Outside the erosion radius, the model assumes that no erosion occurs. However, our calculation shows that there is severe erosion outside this radius, indicating the estimated erosion radius is too small and the erosion region depends on local topography and other factors. Additionally, the ejection angles of particles should not be a single value and satisfy certain distribution as CE-4 landing images show that some dust even has zero ejection angle. Therefore, the erosion model could be improved by considering the ejection angle of dust particles and local topography of the landing site and other essential factors.

6 Conclusions

The landing images of CE-4 were used to analyze plume erosion on the lunar surface during the landing process. Quantitative information of plume effects on the lunar dust was measured. When the nozzle was at least about 13m altitude, the plume began to erode the surface and affect the visibility of landing camera. When the nozzle was at an altitude of 0.6m, the estimated ejection angle of dust is at least 7°, which is higher than the measured average of 3.3° for Apollo landings (excluding Apollo 12). One possible reason is that CE-4 landed on an inclined surface of about 4°-5°, which may result in a larger dust ejection angle than Apollo's average value. Zhang et al. obtained a maximum ejection angle of 4.3° through the one-way CFD coupling calculation when the CE-4 nozzle is about 2m above the horizontal lunar surface. However, the calculated dust ejection angle is much smaller than we measured.

The dust is ejected from the edge at a certain radius r_0, for the first time, which is measured for a low-thrust lander. This radius r_0 reduces linearly with the decrease of the nozzle altitude above 3m altitude (constant thrust phase of CE-4). With the panorama from Yutu-2, we measure that the CE-4 engine scoured a crater of 1.84cm deep at 0.33m from the nozzle centerline, which is 0.7cm smaller than the measured value for Apollo 11. This is because the speed of Apollo 11 approaching the lunar surface is 1/4 times that of CE-4, so Apollo 11 has a relatively long erosion time and erodes deeper crater.

In addition, we measure from the CE-4 landing images that the lunar soil's average erosion depth is about 0.7cm from 1.1s before landing (corresponding nozzle altitudes: 2.67m) to 0.4 s after landing. The corresponding erosion area is between 1.12m and 2m from the nozzle centerline. However, the erosion simulation model shows almost no erosion in the same zone and at the same time. The inconsistency suggests that the erosion model could be improved by considering the distribution of the ejection angle of dust particles and local topography of the landing site and other essential factors. As CE-4 data suggest that the erosion depth depends on the terrain, distance to the nozzle, cohesions of lunar regolith, etc., the landing on a flat place can effectively reduce the impact of dust on the equipment.

The physical quantities related to the lunar dust measured in this paper are important references for future extraterrestrial landing missions. For example, the precise landing position selection process for a CE - 4 - like lander should be performed at an altitude above 13m to avoid the dust obscuration effect on the images of landing camera. The measured radius r_0 and dust ejection angle could be used to estimate the trajectory of ejected dust, which is helpful for dust protection. Dust - proof equipment should be installed in this impact range, and key equipment on the lander (such as optical equipment that is susceptible to dust) should be placed outside the impact range. In addition, these measurements (including the depth of lunar soil erosion caused by the plume) can stimulate the development of theoretical models.

Most previous simulations used a horizontal lunar surface to simulate the plume field, ignoring the local topography of the lunar surface. However, the analysis of lunar soil erosion shows that the depth of lunar soil erosion caused by the plume is positively correlated with the elevation of the lunar surface. In a words, the erosion of the lunar soil by the plume is significantly related to the topography of the lunar surface. Therefore, future simulation works are worth considering the local topography of the lunar surface.

In future work, we will combine observations to make more comprehensive improvements to this erosion model. It is expected that more high - quality data will provide precise physical details of exhaust impingement on the lunar soil and further verify the theoretical models, which is essential to identify key factors to minimize the effect of dust emission on the landing process.

Acknowledgments

Scientific data of Chang'e missions are provided by the China National Space Administration (CNSA). We are grateful for the support from the team members of the Ground Research and Application System (GRAS), who contributed to data receiving and preprocessing. The landing images used in this work are archived at http://moon.bao.ac.cn. This work was supported by the Science and Technology Development Fund (FDCT) of Macau (Grants 008/2017/AFJ and 0042/2018/A2) and National Natural Science Foundation of China (Grants 11761161001, 41941001, 42004157 and 12075108), the Key Laboratory of Science and Technology on Vacuum Technology and Physics, China (No. 6142207200202).

Appendix A　Measurement of camera altitude from the lunar surface

With the available camera parameters listed in Supplementary Table S2, we determine the camera altitude above the lunar surface by measuring the variation of the pixel distance of two fixed feature points. The measurement procedure can be divided into three steps: determining the camera altitude above the lunar surface after the landing, selecting the reference feature points which do not move, and calculating the camera altitude in each image.

The altitude of the camera can be determined by the image F3376 in Fig. A. 1 (a). In F3376, the edge of the footpad is almost level with the soil surface. With the Z - coordinate of the camera in Supplementary Tab. S2, we can determine that the camera altitude from the surface is approximately 1. 34m in this image.

Then we select three groups of feature point pairs in the images to determine the altitude. As shown in Fig. A. 1, we select feature points 1 and 2 as group I in the image F3364 - F3376. For image F3349 - F3363: feature points 2 and 3 are selected as group II. For images taken before F3349, feature points 4 and 5 are selected as group III. This division method considers the limitations caused by the increase in altitude: the distance between two feature points shortens, and the feature points themselves become more blurred. These two limitations will result in bigger errors in the measured altitude. As the landing area is an inclined surface (about 5° measured by Guidance Navigation and Control[21]; about 4° measured by Wu et al.[22]), the two points' midpoint should be as close as possible to the image center. This rule can minimize errors caused by terrain elevation.

The camera altitude corresponding to F3376 was determined, which is about 1. 34m. With this altitude and other parameters in Supplementary Tab. S2, one can estimate the actual distance between points 1 and 2 in F3376 by the formula

$$d_{12} = d_0 \sqrt{(X_1 - X_2)^2 + (Y_1 - Y_2)^2} \qquad \text{(A. 1)}$$

Scripts 1 and 2 represent the corresponding feature points, X and Y denote the horizontal pixel index and the vertical pixel index, respectively. In Eq. (A. 1), $(X_1, Y_1) = (812, 536)$, $(X_2, Y_2) = (771, 686)$. d_0 is the actual size of a pixel in an image, it is calculated by

$$d_0 = \frac{h \times w}{F} \qquad \text{(A. 2)}$$

h denotes the camera altitude above the lunar surface, the pixel size w and the focal length F are listed in Supplementary Table S2. With the camera altitude of 1. 34m, d_{12} is equal to 0. 163m calculated from Eqs. (A. 1) and (A. 2). To calculate the camera altitude in F3364 or other images, we integrate Eqs. (A. 1) and (A. 2) into the following formula:

$$h = \frac{d_{12} F}{w \sqrt{(X_1 - X_2)^2 + (Y_1 - Y_2)^2}}. \qquad \text{(A. 3)}$$

From Eq. (A. 3), one only needs to re - determine the feature points' pixel coordinates to get the camera and nozzle altitude in each image. The obtained nozzle height and descent speed of CE - 4 lander are plotted in Fig. A. 2.

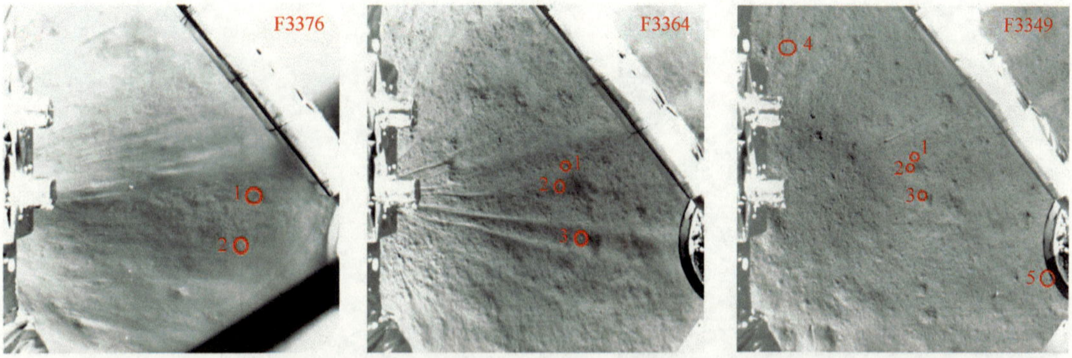

(a)　　(b)　　(c)

Fig. A. 1　Feature points to determine the camera altitude above the lunar surface. Contrast enhanced for F3376 and F3364

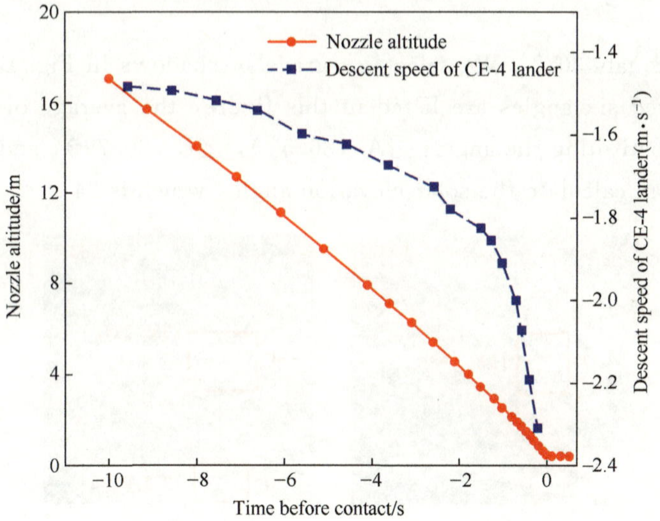

Fig. A. 2　The nozzle altitude and descent speed of CE－4 lander from image measurement

Appendix B　Solar elevation angle relative to the lunar surface

When CE－4 landed, the solar elevation angle can be calculated using three angles in Fig. B. 1 (b): $\angle A$, $\angle B$, and $\angle C$. $\angle A$ is the gravel shadow angle relative to x－axis. Here, $\angle A$ can represent the azimuth angle of sunlight. $\angle B$ is the foot shadow angle relative to x－axis. $\angle C$ is the foot angle relative to x－axis. The elevation angle is calculated by the geometric relation:

$$\text{Solar elevation angle}=\arctan\left(\tan(\angle C)\ \frac{\sin(\angle A-\angle B)}{\sin(\angle B)}\right). \tag{B. 1}$$

According to the design map of CE－4 lander, $\angle C$ is equal to 66°. $\angle B$ is determined from

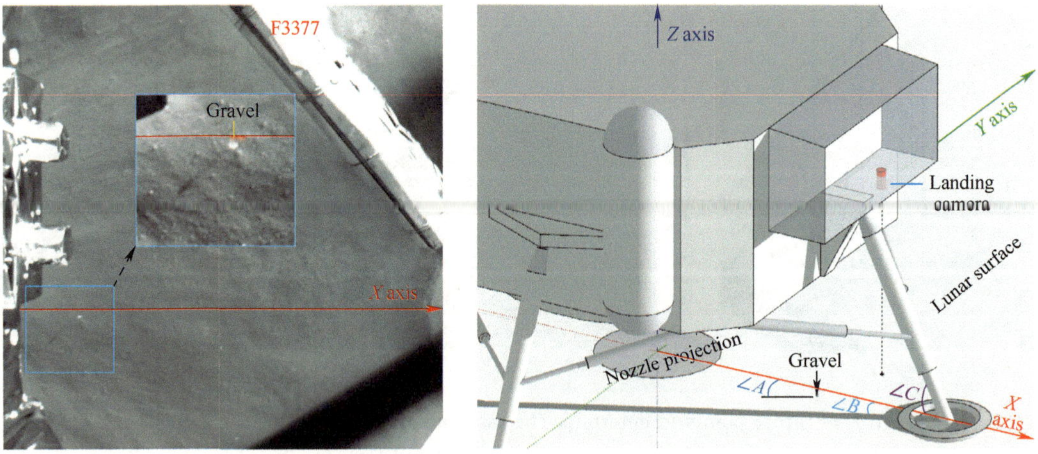

(a)　　(b)

Fig. B. 1　Schematic diagram of the angles for the gravel shadow, foot, and foot shadow relative to the x - axis

image F3377, it equals 30. 7°. We select six gravels' shadows in Fig. B. 2 to determine $\angle A$. The measured six angles are listed in this figure, the average of these six angles equals 36. 5° . Substituting the angles $\angle A$ (36. 5°), $\angle B$ (30. 7°), and $\angle C$ (66°) into Eq. (B. 1), one can calculate the solar elevation angle, which is 24°.

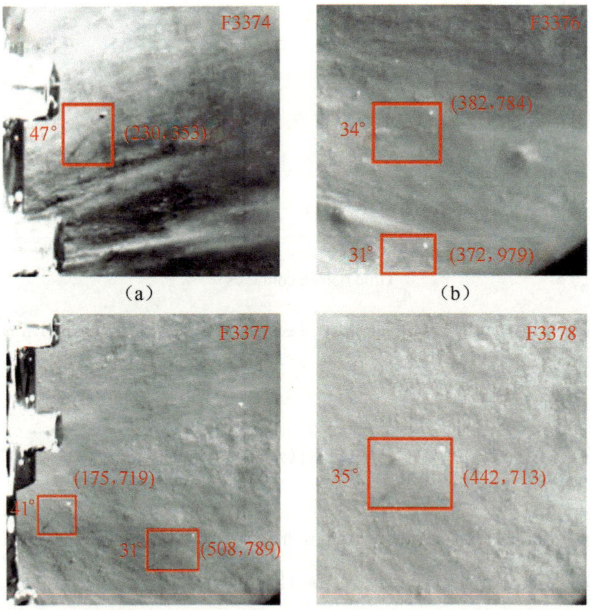

(a)　(b)

(c)　(d)

Fig. B. 2　The angles between the gravel shadow and the horizontal direction (x - axis). The values in parentheses are the pixel coordinates of each gravel. Contrast enhanced for all imanges

From the WebGeocalc (website: https: //wgc. jpl. nasa. gov: 8443/we bgeocalc/

#IlluminationAngles), the solar elevation angle relative to horizontal is about 32° at the time of landing. This angle is 8° larger than the angle measured from the image analysis. There are two main reasons for this difference: (1) the inclined angle of the landing area is about 4° - 5° (about 5° measured by Guidance Navigation and Control[21]; about 4° measured by Wu et al. [22]), the topography on the right of the image is relatively high; (2) the local terrain of the landing site is uneven, which is seen from those images in Fig. B. 2.

Appendix C　Lunar soil erosion depth

The thickness D_l of the bulges can be estimated using the given solar elevation angle and shadow length of the bulge. The estimated thicknesses of the bulges approximate the lunar soil depth eroded by the exhaust. The geometry map of calculating the bugle thickness is plotted in Fig. C. 1. The relationship between thickness D_l and other quantities in Fig. C. 1 can be written as follows:

$$\left(\frac{L}{D_l}\right)^2 = \tan^2(\varphi) + \frac{1}{\tan^2(\angle D)} - 2\frac{\tan(\varphi)\cos(\angle E)}{\tan(\angle D)}$$

$$L = d_0\sqrt{(X_A - X_B)^2 + (Y_A - Y_B)^2}$$

$$\tan(\phi) = \frac{w\sqrt{(X_A - 512)^2 + (Y_A - 512)^2}}{F} \qquad (C.1)$$

$$\cos(\angle E) = \frac{e_a e_b}{|e_a||e_b|}$$

$$e_a = \left(X_A - \frac{N}{2}, \frac{N}{2} - Y_A\right)$$

$$e_b = (-1, -\tan(\angle A))$$

Here, d_0, w, F, X, and Y are the actual size of a pixel in an image, the pixel size on the matrix CCD, the focal length, the horizontal pixel index, and the vertical pixel index, respectively. The solar azimuth angle $\angle D$ is equal to 36.5° (see Appendix B). N is the total number of horizontal or vertical pixels, which is equal to 1024. Before using Eq. (C. 1), one needs to measure the pixel coordinates of points A and B in Fig. C. 1. Point A is a point on the bulging edge in the landing image, and point B is the corresponding shadow of point A.

It is necessary to estimate the uncertainty of the measured erosion depth D_l by Eq. (C. 1). The uncertainty formula of the indirect measurement is as follows:

$$\Delta y = \sqrt{\sum_{k=1}^{n}\left(\frac{\partial f}{\partial x_k}\Delta_k\right)^2\Bigg|_{x_k=\bar{x}_k}} \qquad (C.2)$$

Here, x_k represents the input measurements in Eq. (C. 1) (pixel coordinates, solar altitude and azimuth angle, and the size of a pixel in the image), and Δ_k is the

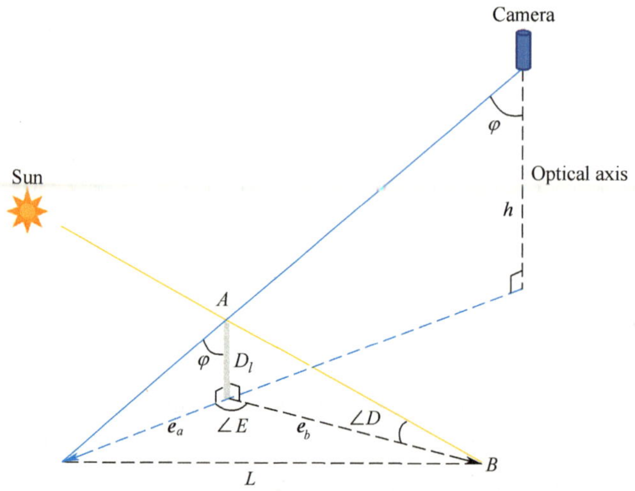

Fig. C. 1 Gemerty map for calcuating the soil erosion depth. The solid gray line denotes the soil bulge, and D_l is the thickness of the bulge. $\angle D$ is the solar elevation angle relative to the lunar surface (about 24° from Appendix B)

corresponding uncertainty. The uncertainties of these input measurements are estimated as follows:

1. Pixel coordinates (X_A, X_B, Y_A, and Y_A)

The measurement accuracy of pixel coordinates depends on the resolution of the image. We take the size of a pixel in the landing image as the uncertainty of the pixel coordinates.

2. Solar altitude and azimuth angles ($\angle D$ and $\angle A$)

From Eq. (B. 1), the uncertainty of the solar altitude depends on the accuracy of three angles: $\angle A$, $\angle B$ and $\angle C$. $\angle A$ is shown in Figs. B. 1 (b) and B. 2, and $\angle B$ and $\angle C$ are shown in Fig. B. 1 (b). We use the six angles in Fig. B. 2 to calculate the uncertainty of $\angle A$. The calculation formula is as follows:

$$\Delta \angle A = \frac{t}{\sqrt{n}} \sqrt{\frac{\sum (\angle A_i - \angle \overline{A})^2}{n-1}}, \tag{C. 3}$$

where t is equal to 2.57 ($n=6$), the average $\angle \overline{A}$ is 36.5°. Through calculation, the uncertainty of $\angle A$ is about 7°. $\angle B$ can be better measured according to the image F3377 (Fig. B. 1 (a)), and the uncertainty is relatively low. $\angle C$ is shown in Fig. B. 1 (b) that is provided by the CE-4 drawing, and the uncertainty of this angle is also relatively low. Therefore, we only consider the influence of the solar azimuth angle $\angle A$ on the solar altitude. Substituting the uncertainty of $\angle A$ and Eq. (B. 1) into Eq. (C. 2), one can obtain the uncertainty of the solar altitude, and the result is 24°.

3. The size of a pixel in the image (d_0)

(3a) According to Eq. (A. 2), the uncertainty of d_0 mainly depends on that of the

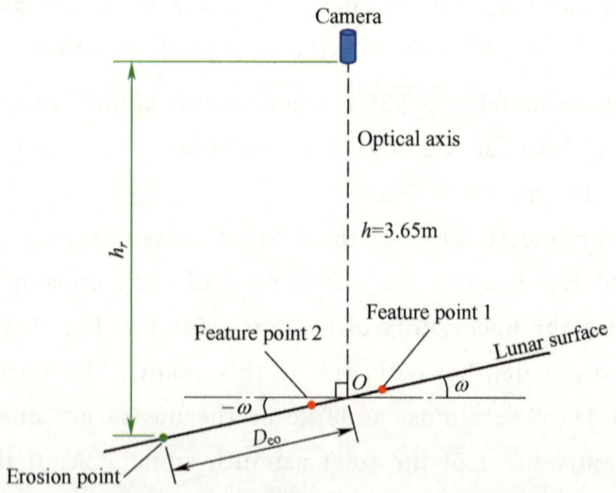

Fig. C. 2　Camera height calculation including the swing of the optical axis and the slope of the landing site. Feature points 1 and 2 are the same as those in Fig. A. 1. Point O is located at the midpoint of the line connecting feature points 1and 2. The erosion point here refers to an erosion point selected in Fig. 3. D_{eo} is the distance from point O to the erosion point. The camera height $h = 3.65\text{m}$ is calculated by Eq. (A. 3), this camera height corresponds to the landing image F3362

camera height h. h is obtained by Eqs. (A. 3) and (A. 1), these two equations include the measurement of pixel coordinates. We still assume that the uncertainty of the pixel coordinates is the size represented by one pixel. Substituting Eqs. (A. 3), (A. 1) and the uncertainty of pixel coordinates into Eq. (C. 2), one can calculate the uncertainty of the camera height h. Taking the image F3362 as an example (because this image is selected to measure the erosion depth, see Fig. 3 (a)), the uncertainty of the corresponding camera height is 0. 03m.

(3b) Strictly speaking, the camera altitude formula (Eq. (A. 3)) does not take into account the swing angle of the CE－4 lander during the landing and the slope of the landing site. These two factors also have an effect on the measurement of camera height. Also taking the image F3362 as an example, we roughly estimate the effect of these two factors on the corresponding camera height. The camera height h_r from a erosion point is shown in Fig. C. 2, h_r is written as

$$h_r = h - D_{eo}\sin(-\omega) \tag{C. 4}$$

For a certain erosion point, the value range of $|\omega|$ does not exceed 6° (5°is the slope, and 1° is the maximum swing angle of the camera's optical axis). Using Eq. (C. 2), the uncertainty of h_r is expressed as follows:

$$\Delta h_r = \sqrt{\Delta h^2 + (D_{eo}\cos(\omega)\Delta\omega)^2 + (\Delta D_{eo}\sin(\omega))^2} \tag{C. 5}$$

The first term Δh has been calculated, it is equal to 0. 03m. The maximum sum of the other two terms can be estimated. Using the image F3400 after landing, it can be calculated that the maximum value of D_{eo} is equal to 0. 73m, and the corresponding ΔD_{eo}

is taken as the size of one pixel (0.0011m). The value of $\Delta\omega$ is taken as 6°. When ω is within the value range [−6°, 6°], the maximum sum of the other two terms is 0.006m^2. Therefore, Δh_r is approximately $\sqrt{0.03^2+0.006}$ m = 0.08m. Finally, substituting Δh_r = 0.08m and Eq. (A. 2) into Eq. (C. 2), one can obtain the uncertainty of the pixel size d_0, the result is 6×10^{-5} m.

By substituting the uncertainties of these input measurements and the erosion depth D_l in Eq. (C. 1) into Eq. (C. 2), the uncertainty of each erosion measurement can be obtained. For example, the uncertainty of erosion point 1 in Fig. 4 (a) is 1.27cm, which is larger than the erosion depth (1.10cm) at this point. The uncertainty of the other measurements in Fig. 4 (a) is almost as large as the measurement itself. This is mainly caused by the uncertainty (7°) of the solar azimuth angle $\angle A$. If the uncertainty of the solar azimuth angle can be reduced to 3°, and then the uncertainty of each measured erosion depth is almost reduced by half. This reduction also means that if the solar azimuth angle on each erosion point is accurately determined, then the erosion depth can be better calculated.

Here, we also calculate the uncertainty of the average erosion depth (that is, the average erosion depth measured in Fig. 4 (a), 0.71cm). The uncertainty formula of the average erosion depth is derived by using Eq. (C. 2) as follows:

$$\Delta\overline{D_l} = \frac{1}{n}\sqrt{\sum_{i=1}^{n}\Delta D_{li}^2} \tag{C. 6}$$

Here, n represents the number of measurements of erosion depth in Fig. 4 (a), there are 20 in total. Substituting the uncertainty of each measured erosion depth into Eq. (C. 6), one can obtain $\Delta\overline{D_l}$, and the result is equal to 0.20cm. Therefore, the average erosion depth with uncertainty is 0.71 ± 0.20cm.

Appendix D　Elevation map after landing

Within one second after CE-4 touched the ground, the fuselage moved a short distance laterally on the lunar surface. In Fig. D. 1 (a), the optical axis pointed to point A. After 0.3s, the optical axis pointed to point A'. The distance between points A and A' can be calculated by Eq. (A. 1) in Appendix A. From this distance, one can estimate the feature point's elevation in Fig. D. 1 (a). The calculation principle is similar to using a pair of eyes to an object's three-dimensional coordinates.

The elevation of the feature point can be calculated by Fig. D. 1 and the following formula:

$$\Delta h = \frac{\Delta L}{\tan(\varphi)} - h$$

$$\Delta L = \Delta S\,\frac{\sin(\beta)}{\sin(\pi-\beta-\gamma)}$$

$$\tan(\varphi)=\frac{w\sqrt{(X-512)^2+(Y-512)^2}}{F},\tag{D. 1}$$

where Δh denotes the feature point's altitude relative to point A in F3377. h, w, F, X, and Y are the camera altitude above the lunar surface, the pixel size on the matrix CCD, the focal length, the horizontal pixel index of feature point, the vertical pixel index of feature point, respectively. The angle γ and β are measured by Figs. D. 1 (a) and D. 1 (b), respectively. Based on Eq. (D. 1) and Fig. D. 1, we have obtained the corresponding elevation map (see Fig. 4 (b)).

There are some limitations to utilize Eq. (D. 1) to estimate the elevation of Fig. D. 1 (a). It can be seen from Fig. D. 1 (a) that the soil was being eroded, which makes it difficult to determine the locations of some feature points accurately. For example, point B in Fig. D. 1 (a) is partially covered by dust, and it is difficult to determine its location. This occlusion may result in higher calculation results. However, the elevations of other feature points could be determined more precisely than point B .

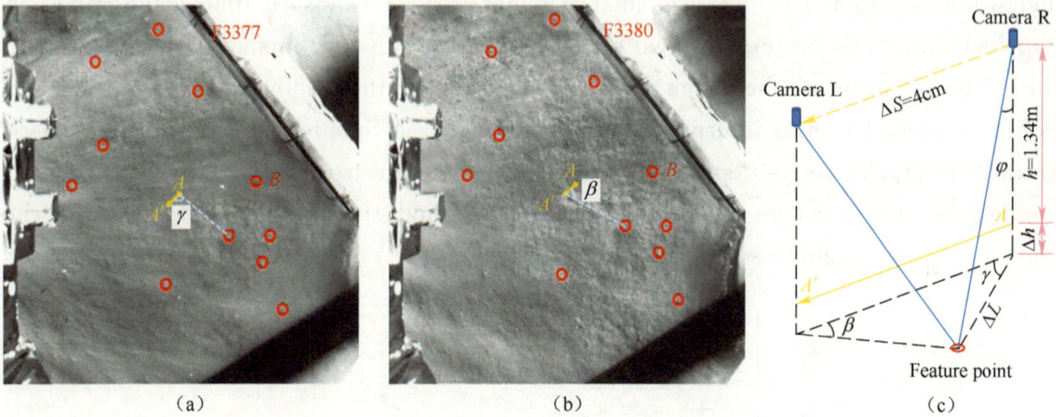

Fig. D. 1　(a) and (b) Selected feature points to calculate the elevation map. Points A and A' are the center points of images F3377 and F3380, respectively. Contrast enhanced for F3377 and F3380. (c) Geometry diagram for calculating the elevation of the feature point. The solid yellow line is on the lunar surface. Points A' and A on the lunar surface are shown in panels (a) and (b). The angles γ and β are marked in panels (a) and (b), respectively. (For interpretation of the references to color in this figure legend, the reader is refrred to the web version of this article)

Appendix E　Improvements on the histogram matching method

HMM is improved in three aspects in this paper: (1) utilizing the equivalent radiance at the entrance pupil of the landing camera instead of the digital number in Eq. (2); (2) subtracting a part of the area within the erosion radius a_0 in Eq. (3); (3) considering the light' s complete path length through the dust. The reasons for the three improvements are given below.

First of all, the first improvement is to address the issue that Eq. (2) fails for the time - varying exposure time during the landing. Eq. (2) uses the reference image p_{ij} and outputs an image p'_{ij}, and then match the histogram of p'_{ij} to that of the image covered by dust. However, this match can be implemented under the condition that the exposure time of the reference image p_{ij} is equal to that of the image covered by dust. If the exposure time for the image covered by dust is relatively long, which causes the dust brightness of the image becomes higher. Consequently, it will result in a higher dust amount. In turn, the relatively short exposure time will give a lower dust amount. To avoid this problem, we use the equivalent radiance at the landing camera's entrance pupil, not the digital number. This equivalent radiance represents the actual brightness emitted by the object, which improves the precision of the estimated amount of dust in an image. In this paper, we use the equivalent radiance to determine the dust's extinction factor α due to the time - varying exposure time during the landing of CE - 4.

To obtain the equivalent radiance corresponding to the landing image, we convert the digital number to the equivalent radiance under various exposure time using values in Supplementary Tab. S3. For the two kinds of exposure time, 2.2ms and 1.6 ms, used in the reference image and the image covered by dust during the landing, the converting rules can be obtained by linear interpolation of values listed in Supplementary Table S3. Based on these interpolation results plotted in Fig. E. 1, two linear relations between the digital number and the equivalent radiance were found. Through this way, one can determine the object's actual brightness appeared in the image.

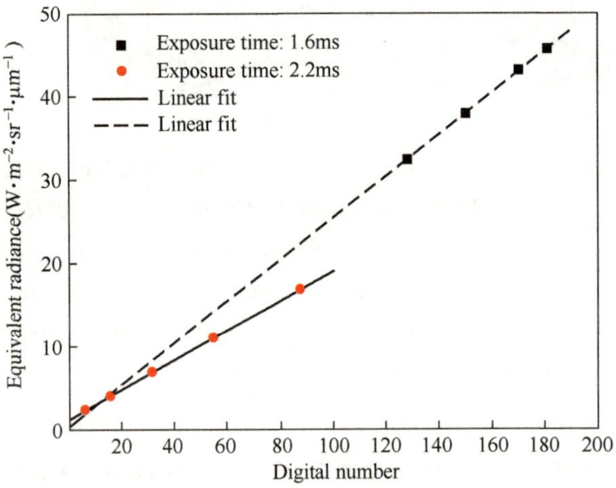

Fig. E. 1　Equivalent radiance from the linear interpolation on the data in Supplementary Tab. S3

The second improvement is for Eq. (3). According to the histogram matching method, lunar dust can be ejected from the lunar surface within the erosion radius a_0. However, one can observe that the dust hardly exists within the radius r_0 from the landing images (see stages III and IV in Tab. 1). Also, theoretically, the lunar soil under the nozzle is hard to rise because the shear stress of the plume on the lunar soil is less than the critical shear stress. Hence, we assume that all dust is emitted from outside the radius r_0. When calculating the effective path length x in Eq. (2), the radius r_0 needs to be subtracted from the erosion radius a_0.

Finally, the length of light path through the dust is considered. Lane et al. used $2x$ of e^{-2ax} in Eq. (2) to represent the length of light path [20]. Here, the length of the incident light through the dust is assumed to be equal to the length x of the reflected light, which is a good approximation for a thin layer of elevated dust and a large solar elevation angle. To better determine the dust's extinction factor, we directly calculated the incident light's length in this paper.

Appendix F　Velocity simulation for dust particles

The dust particles' velocity is computed from the particle trajectory model[30]. This model uses the gas properties (density, temperature, etc.) output from Computation Fluid Dynamics (CFD) or Direct Simulation Monte Carlo (DSMC) simulation. In this paper, we use ANSYS FLUENT (Version 19.0) to simulate the gas field. The plume simulation for CE-4 is divided into these two stages: before and after engine shutdown. The simulation parameters are listed in Supplementary Table S1. In the CFD simulation, we selected 28 nozzle altitudes from the 32 nozzle altitudes corresponding to the extinction factor data points in Fig. 6 as the nozzle altitudes in the simulation. The minimum height interval is 0.18m, and the maximum height interval is 0.96m. For the remaining 4 nozzle altitudes, the corresponding particle velocities are obtained by linear interpolation. The diameter of the particles used is from 1μm to 10000μm, and 37 types of particles are selected according to the following rules: 1μm, 2μm ... 10μm, 20μm ... 100μm, 200μm ...

The particles' movement in the fluid field is caused by the following forces: drag and lift generated by the gas, and lunar gravity. Drag and lift are obtained as follows[30]

$$\boldsymbol{a}_{\mathrm{D}}=\frac{3C_{\mathrm{D}}\rho_{\mathrm{g}}}{4D\rho_{\mathrm{p}}}\left|\boldsymbol{u}_{\mathrm{g}}-\boldsymbol{u}_{\mathrm{p}}\right|(\boldsymbol{u}_{\mathrm{g}}-\boldsymbol{u}_{\mathrm{p}}) \tag{F.1}$$

$$\boldsymbol{a}_{\mathrm{L}}=\frac{3C_{\mathrm{L}}\rho_{\mathrm{g}}}{2\rho_{\mathrm{p}}}\left|\boldsymbol{u}_{\mathrm{g}}-\boldsymbol{u}_{\mathrm{p}}\right|_{\mathrm{horiz}}\frac{\partial}{\partial x}\left|\boldsymbol{u}_{\mathrm{g}}-\boldsymbol{u}_{\mathrm{p}}\right|_{\mathrm{horiz}}, \tag{F.2}$$

where C_{D} and C_{L} are the coefficients of drag and lift, respectively, ρ is the density, $\boldsymbol{u}$ is the velocity, subscripts g and p represent the gas and particle, respectively. D is the diameter of a particle. 20 is adopted for the lift coefficient, and C_{D} is taken from[31] that considered the effects of compressibility and rarefaction on drag. We add the missing term $2\sqrt{\pi}/(3s)$ in (25b) in Loth' s article, as pointed out by Parmar et al. [32]. To obtain the velocity $v_{\perp}(D)$ in Eq. (1), the steps are as follows: (1) computing average velocity of particles passing through the observation area after being thrown from the radius r_0, here, the average velocity is the arithmetic average; (2) determining the speed weighting factor η under certain nozzle altitudes,

$$\eta = \frac{2\pi \int_{r_0}^{a_0} \overline{V}_t(r) r \mathrm{d}r}{2\pi \overline{V}_t(r_0) \int_{r_0}^{a_0} r \mathrm{d}r}. \tag{F. 3}$$

In Eq. (F. 3), $\overline{V}_t(r_0)$ represents the average speed mentioned in the first step. We select this average speed for the following reasons. The selected speed of a particle can at least represent the speed of this particle in the observation area (area in the landing image). Since the particle appearing in the image is captured at a certain moment, the speed of the particle in the image is the instantaneous speed. Finally, all instantaneous velocities of a particle passing through the observation area are averaged, and the average is taken as the velocity of this particle in the observation area. As η is assumed to be a constant for various particle sizes under some nozzle altitudes, then η obtained from Eq. (F. 3) can be utilized in the landing. Based on these two steps and the ejection angle of dust θ, one can determine $v_{\perp}(D)$ by the formula for a given nozzle altitude and particle diameter:

$$v_{\perp}(D) = \overline{V}_t(r_0) \eta \sin(\theta). \tag{F. 4}$$

In Eq. (F. 4), $v_{\perp}$ and $\overline{V}_t$ are the vertical velocity and total velocity, respectively. The initial position of the particles is critical for determining $\overline{V}_t(r_0)$, such as the height from the ground and the distance r_0 from nozzle centerline. The height of all particles from the lunar surface is set to 0.005m. The reason for this setting is as follows. According to Eqs. (F. 1) and (F. 2), when the particle is resting on the surface, it will not be blown up. Because the gas velocity in CFD simulation is assumed to be zero on the surface, this means the forces affecting the particle are zero. Therefore, the initial height of the particles should be greater than zero. On the other hand, the landing images of CE-4 show that dust near the nozzle centerline is close to the lunar surface. Therefore, the initial height of the particles is a small value. As the landing images' pixel size can reach the millimeter level, the particles' initial height should be below 0.01m (0.005m selected in the present work).

r_0 can be obtained from the method presented at the end of Section 4. However, r_0 is not visible in the last batch of images (F3361-F3373). For these images, we determine r_0 by the following method. In the images where r_0 is visible, we first use CFD simulation to determine the plume parameters (shear stress and pressure) at r_0, and then use the Mohr-Coulomb failure criterion to determine the mechanical parameters when the lunar soil is broken. Using these mechanical parameters and CFD simulation again, we can find the position of r_0 for the images where r_0 is invisible. The specific implementation process is described as follows.

In the images where r_0 can be seen, the shear stress and pressure of the plume at r_0 can be computed by the CFD simulation, and satisfy the Mohr-Coulomb failure criterion:

$$\tau^* = c + P\tan\varphi_c, \tag{F. 5}$$

where τ^* is threshold shear stress, c is the cohesive stress, P is the pressure, and φ_c is the friction angle. In the CFD simulation, the inclination of the lunar surface is set to be 0, and the simulation parameters used are in Supplementary Table S1. Fig. F. 1 shows the threshold shear stress and pressure at r_0 for some images where r_0 can be seen.

For the images (F3361 - F3373) where r_0 is invisible, the shear stress and pressure on the lunar surface can also be obtained under the corresponding nozzle altitude through the same CFD simulation. We can find the specific radial location where the shear stress and pressure satisfy the linear relationship shown in Fig. F. 1, and then this radial location can be regarded as r_0. Through this method, the obtained r_0 for the images of F3361 - F3373 (corresponding nozzle altitudes: 2. 88m - 0. 38m) is plotted in Fig. 6.

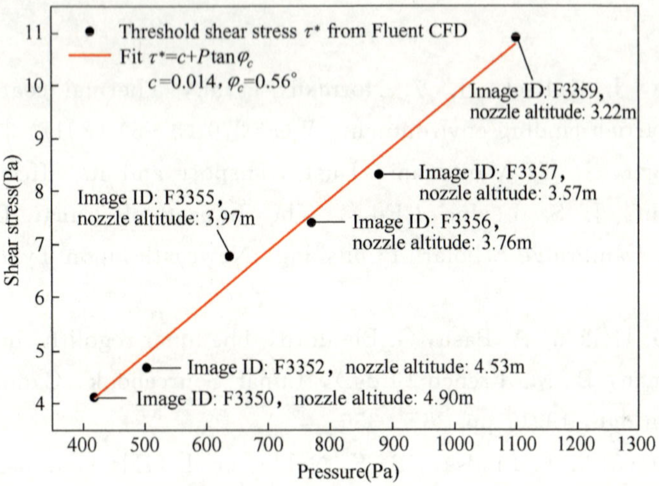

Fig. F. 1　The shear stress and pressure of the plume at r_0 from the CFD simucation for the images where r_0 can be seen.

After obtaining $v_\perp(D)$ of the 37 particles under the 28 nozzle altitudes, we use interpolation to obtain $v_\perp(D)$ of other nozzle altitudes and particles. The other four nozzle altitudes are 3. 04m, 3. 97m, 8. 23m, and 11. 13m respectively. For these four altitudes, $v_\perp(D)$ of the 37 particles is obtained by linearly interpolating $v_\perp(D)$ of adjacent nozzle altitudes. Then, $v_\perp(D)$ of particles of any size can be obtained by interpolating $v_\perp(D)$ of 32 particles for a certain nozzle altitude. Here, we use the piecewise cubic Hermite interpolation polynomial (PCHIP) to interpolate $v_\perp$ (D) of 10, 000 particles (10,000 particles are uniformly taken from the particle diameter range of 1μm to 10,000μm). For a certain nozzle altitude, $v_\perp$ (D) of these 10, 000 particles is substituted into the following summation formula to obtain the integral term $\int_0^{0.01} S(D)\left(\frac{D^2}{v_\perp(D)}\right)dD$ in Eq. (1):

$$\int_0^{0.01} S(D)\left(\frac{D^2}{v_\perp(D)}\right)dD$$

$$\approx \sum_{n=1}^{9999} \frac{1}{2}\left(S(D_n)\left(\frac{D_n^2}{v_\perp (D_n)}\right)+S(D_{n+1})\left(\frac{D_{n+1}^2}{v_\perp (D_{n+1})}\right)\right)\Delta D \tag{F. 6}$$

where the unit of particle diameter D is meter, the value of $S(D)$ is obtained from the fitting curve given in Fig. 8 (a), $\Delta D = 10^{-6}$ m.

Appendix G Supplementary data

Supplementary material related to this article can be found online at https://doi.org/10.1016/j.actaastro.2021.05.009.

References

[1] A. Sengupta, J. Kulleck, J. V. Norman, et al., Thermal coating erosion in a simulated Martian landing environment, Wear 270 (5 - 6) (2011) 335 - 343.

[2] P. T. Metzger, J. T. Mantovani, Dust transport and its effect due to landing spacecraft, in: J. S. Levine (Ed.), The Impact of Lunar Dust on Human Exploration, Cambridge Scholars Publishing, Newcastle upon Tyne, 2021, pp. 88 - 111.

[3] D. McKay, G. Heiken, A. Basu, G. Blanford, The lunar regolith, in: G. H. Heiken, D. T. Vaniman, B. M. French (Eds.), Lunar Sourcebook, Cambridge University Press, Cambridge, 1991, pp. 285 - 356.

[4] R. Christoffersen, J. F. Lindsay, S. K. Noble, et al. (Eds.), Lunar Dust Effects on Spacesuit Systems: insights from the Apollo Spacesuits, NASA/TP - 2008 - 000000, 2008.

[5] S. Wagner (Ed.), The Apollo Experience Lessons Learned for Constellation Lunar Dust Management, NASA/TP - 2006 - 213726, 2006.

[6] J. N. Mpagazehe, K. W. Street, I. R. Delgado, et al., An experimental study of lunar dust erosive wear potential using the JSC - 1AF lunar dust simulant, Wear 316 (1 - 2) (2014) 79 - 91.

[7] C. A. Bühler, Experimental investigation of lunar dust impact wear, Wear 342 (2015) 244 - 251.

[8] C. Immer, P. Metzger, P. E. Hintze, et al., Apollo 12 lunar module exhaust plume impingement on Lunar Surveyor III, Icarus 211 (2) (2011) 1089 - 1102.

[9] C. Darquenne, G. K. Prisk, Particulate deposition in the human lung under lunar habitat conditions, Aviat. Space Environ. Med. 84 (6) (2013) 190 - 195.

[10] G. K. Prisk, Effects of partial gravity on the function and particle handling of the human lung, Curr. Pathobiol. Rep. 6 (3) (2018) 159 - 166.

[11] Y. Sun, J. G. Liu, X. P. Zhang, et al., Mechanisms involved in inflammatory pulmonary fibrosis induced by lunar dust simulant in rats, Environ. Toxicol. 34 (2)

(2019) 131 - 140.

[12] Y. Sun, L. Zhang, J. G. Liu, et al., Effects of lunar dust simulant on cardiac function and fibrosis in rats, Toxicol. Res. 8 (4) (2019) 499 - 508.

[13] P. T. Metzger, Rocket exhaust blowing soil in near vacuum conditions is faster than predicted by continuum scaling laws, in: Earth and Space 2016: Engineering for Extreme Environments, American Society of Civil Engineers, Reston, VA, 2016, pp. 58 - 66.

[14] M. A. Gallis, J. R. Torczynski, D. J. Rader, An approach for simulating the transport of spherical particles in a rarefied gas flow via the direct simulation Monte Carlo method, Phys. Fluids 13 (11) (2001) 3482 - 3492.

[15] Z. Y. Tang, B. J. He, G. B. Cai, Investigation on a coupled Navier - Stokes - direct simulation Monte Carlo method for the simulation of plume flowfield of a conical nozzle, Internat. J. Numer. Methods Fluids 76 (2) (2014) 95 - 108.

[16] K. R. Al Khasawneh, M. Al Tarawneh, C. P. Cai, Comprehensive analytical study of rarefied jet flow of annular exit impingement at vertical flat plate, Int. J. Fluid Mech. Res. 42 (5) (2015) 404 - 417.

[17] K. R. Al Khasawneh, D. Kharouf, Analytical study of noncontinuum flow impingement onto a flat plate with an arbitrary angle, Int. J. Fluid Mech. Res. 45 (2) (2018) 153 - 162.

[18] A. Rahimi, O. Ejtehadi, K. H. Lee, et al., Near - field plume - surface interaction and regolith erosion and dispersal during the lunar landing, Acta Astronaut. 175 (2020) 308 - 326.

[19] C. Immer, J. Lane, P. Metzger, et al., Apollo video photogrammetry estimation of plume impingement effects, Icarus 214 (1) (2011) 46 - 52.

[20] J. E. Lane, P. T. Metzger, Estimation of Apollo lunar dust transport using optical extinction measurements, Acta Geophys. 63 (2) (2015) 568 - 599.

[21] Z. Z. Sun, H. Zhang, X. Y. Wu, et al., Flight results of Chang'E - 4 lander summary and evaluation, Sci. Sin. Technol. 49 (12) (2019) 1397 - 1407.

[22] B. Wu, Y. Li, W. C. Liu, et al., Centimeter - resolution topographic modeling and fine - scale analysis of craters and rocks at the Chang'E - 4 landing site, Earth Planet. Sci. Lett. 553 (2021) 116666.

[23] H. Y. Zhang, J. L. You, Y. Wang, et al., The study on lunar dust kinetic characteristics caused by lander landing in Chang'E - 4 mission using CFD method, Astron. Comput. (2021) submitted for publication.

[24] A. B. Morris, D. B. Goldstein, P. L. Varghese, et al., Plume impingement on a dusty lunar surface, AIP Conf. Proc. 1333 (1) (2011) 1187 - 1192.

[25] J. L. Lai, Y. Xu, X. P. Zhang, et al., Comparison of dielectric properties and structure of lunar regolith at Chang'E - 3 and Chang'E - 4 landing sites revealed by ground - penetrating radar, Geophys. Res. Lett. 46 (22) (2019) 12783 - 12793.

[26] NASA Manned Space Center, Apollo 14 technical air - to - ground voice transcription,

1971, pp. 381, 449 - 50. 464 - 6, 582.

[27] P. T. Metzger, J. Smith, J. E. Lane, Phenomenology of soil erosion due to rocket exhaust on the Moon and the Mauna Kea lunar test site, J. Geophys. Res. 116 (2011) E06005.

[28] R. F. Scott, H. - Y. Ko, Transient rocket - engine gas flow in soil, AIAA J. 6 (2) (1968) 258 - 264.

[29] C. C. Mason, Comparison of actual versus predicted lunar surface erosion caused by Apollo 11 descent engine, Geol. Soc. Am. Bull. 81 (6) (1970) 1807 - 1812.

[30] J. E. Lane, P. T. Metzger, C. D. Immer, et al., Lagrangian trajectory modeling of lunar dust particles, in: Earth and Space 2008: Engineering, Science, Construction, and Operations in Challenging Environments, American Society of Civil Engineers, Reston, VA, 2008, pp. 1 - 9.

[31] E. Loth, Compressibility and rarefaction effects on drag of a spherical particle, AIAA J. 46 (9) (2008) 2219 - 2228.

[32] M. Parmar, A. Haselbacher, S. Balachandar, Improved drag correlation for spheres and application to shock - tube experiments, AIAA J. 48 (6) (2010) 1273 - 1276.

Ground experiments and performance evaluation of the Low – Frequency Radio Spectrometer onboard the lander of Chang'e – 4 mission

Xin – Ying Zhu[1], Yan Su[1,3], Yi – Cai Ji[2,3], Hong – Bo Zhang[1,3], Bo Zhao[2], Jun – duo Li[1], Shun Dai[1], Xi – Ping Xue[1], Chun – Lai Li[1,3]

1. Key Laboratory of Lunar and Deep Space Exploration, National Astronomical Observatories, Chinese Academy of Sciences, Beijing 100101, China; suyan@bao.ac.cn
2. Institute of Electronics, Chinese Academy of Sciences, Beijing 100190, China; ycji@mail.ie.ac.cn
3. University of Chinese Academy of Sciences, Beijing 100049, China

Abstract: The Low – Frequency Radio Spectrometer (LFRS) is a scientific payload onboard the Chang'e – 4 lunar lander launched in December 2018. The LFRS provides in – situ measurements of the low – frequency radio phenomena on the far – side of the Moon for the first time in human history. To evaluate the performance of the LFRS, a series of ground experiments are conducted using an engineering model of the LFRS. It is not easy to perform the experiments because the Electro Magnetic Interference (EMI) from the Chang'e – 4 lunar lander itself and the environment is very intense. The results after EMI mitigation show that the sensitivity of the LFRS may be 10^{-18} W m^{-2} Hz^{-1}.

Keywords: Moon; far – side – space vehicles; instruments – techniques; spectroscopic

1 INTRODUCTION

The atmosphere is not perfectly transparent at any radio frequency. Radio waves are scattered and absorbed by Earth's atmosphere. Low – frequency astronomical observation from the ground is limited by severe ionospheric distortions below 50MHz and complete reflection of radio waves below 10 – 30MHz (Jester & Falcke 2009). Even from an orbiter around the Earth, human – made interference from the Earth and natural radio emission from the Sun turn out to be too overwhelming for any observations in this frequency range. For these reasons, the lowfrequency end is indeed one of the last portions of the electromagnetic spectrum to remain terra incognita in astrophysics (Boonstra et al. 2010; Bentum et al. 2011; Jester & Falcke 2009; Takahashi 2003; Wolt et al. 2012).

Chang'e – 4 probe is the first mission landed on the far – side of the Moon in human history. The probe was launched at 18: 23 (UTC) on 2018 December 7, and landed successfully in the Von Kármán crater within the South Pole - Aitken (SPA) basin at 2: 26 (UTC) on 2019 January 3. It will attempt to collect new evidence from the most massive crater in the solar system to determine the age and composition of an unexplored region of the Moon, as well as develop technologies required for the later stages of the program (Li et al. 2019; Wu et al. 2017, 2019). The Low - Frequency Radio Spectrometer (LFRS) is a scientific payload onboard the Chang'e - 4 lunar lander. The primary motivation for the LFRS is to learn about the universe through low - frequency spectral window. The Moon can be utilized as a shield against unwanted radiations from the Earth. By taking advantage of the unique environment, many astrophysical topics of interest such as cosmology with HI line emission, solar and planetary radio bursts, local plasma environment above the Moon's surface, ultra - high energy particle detection, meteoritic impacts could be studied through low - frequency observations (Jester & Falcke 2009; Lazio et al. 2011; Wolt et al. 2012). According to the characteristics of the Chang'e - 4 mission, the main scientific objectives of the LFRS are to probe solar radio bursts and local plasma environment above the Moon's surface in the frequency range 0.1 -40MHz.

The Chang'e - 4 probe was initially built as a backup for Chang'e - 3 and became operational after Chang'e - 3 successfully landing in 2013. According to Electro Magnetic Compatibility (EMC) test results of Chang'e - 4, the Electro Magnetic Interference (EMI) from the Chang'e - 4 lunar lander itself is very intense, so all target radio emissions are hidden in the EMI noise. A method is proposed in order to suppress the significant interference from the lander. Experiments on the ground (On Earth) are an essential tool for evaluating the payload performance, so a series of ground experiments were performed for this reason. In addition, the EMI mitigation method was also verified during the ground experiments.

2 INSTRUMENT DESCRIPTION

The LFRS which is mounted on the top of the Chang'e - 4 Lander, was designed by Institute of Electronics, Chinese Academy of Sciences (IECAS) together with National Astronomical Observatories, Chinese Academy of Sciences (NAOC), and manufactured by IECAS. The actual working environment of the LFRS on the Moon is shown in Figure 1. This picture was taken by the Chang'e - 4 cruiser.

The LFRS consists of sensors, pre - amplifiers, electronics unit and cable assemblies. The sensors and pre - amplifiers are installed outside the lander cabin, while the electronics unit is installed inside. The sensors are three 5 - meter - long antennas which are mounted on the top of the lander orthogonally (Arts et al. 2010). Another 20cm long antenna is mounted near the root of the long antenna as an auxiliary reference antenna.

Fig. 1　The actual working environment of the LFRS on the Moon.
This photo was taken by the Chang'e - 4 cruiser

The signals of the above four antennas are pre - amplified and fed to the electronics unit. The electronics unit includes a control unit, a four - channel radio receiver covering the full band of 0. 1—40MHz, and a calibration unit, etc (Ji et al. 2017).

The three 5 - meter - long antennas receive both the scientific radio signals and the EMI of the lander, while the 20cm long antennas only receive the EMI of the lander. The data of the 20cm long antennas will be used to suppress the EMI in the off - line data processing stage. The specifications of the LFRS are summarized in Table 1.

Table 1　Specifications of the LFRS Instrument

No.	Parameter	Characteristic
1	Frequency	0. 1—40MHz
2	Receiver sensitivity	≤10nV/ $\sqrt{Hz}$
3	Dynamic Range	≥75dB
4	Frequency Resolution	≤10kHz (0. 1—2. 0MHz) ≤200kHz (1. 0—40MHz)
5	Max allowed bit rate	5Mbps
6	Power	≤24W

3　GROUND EXPERIMENT

The main purposes of the ground experiments were to evaluate the performance of the prototype model of the LFRS and verify the off - line data processing method. According to the EMC test results of the Chang'e - 4 lander, the detection capability of the LFRS is related to its technical performance and depends on the EMI mitigation method for noise suppression of the lander. Therefore, the EMI mitigation method is also tested in the ground experiments.

3.1 Experimental Setup

The engineering model which is identical, in form, fit and function to the flight model, was tested in the ground experiments. The experimental settings are shown in Figure 2. The EMI from the Chang'e - 4 lunar lander was simulated by an arbitrary waveform generator which reproduced the lander noise recorded during the EMC test. Both the sinusoidal and wide - band signals generated by a vector signal generator were used to emulate the sky signals. The format of the emulated sky signals is listed in Table 2.

Fig. 2　Photograph of the LFRS antennas and the pre - amplifier

Table 2　Format of the Emulated Sky Signals

No.	Signal Mode	Frequency	Amplitude	Remark
1	Sinusoidal signals	1.8MHz	24dBm	Sinusoidal signals are generated by arbitrary waveform generator
2		5MHz	18dBm	
3		8MHz	18dBm	
4		10MHz	4dBm	
5		22.5MHz	−8dBm	
6		38MHz	−8dBm	
7	Wide - band signals	1.75MHz	20dBm	Bandwidth: 100kHz generated by vector signal generator
8		5MHz	20dBm	Bandwidth: 1MHz generated by vector signal generator
9		20MHz	20dBm	Bandwidth: 1MHz generated by vector signal generator

3.2 Experimental Method

As can be seen in Table 2, nine experiments were performed. Different rows in Table 2 mean different configurations of the emulated sky signal. Each configuration should be verified by an experiment. Each experiment was done in two steps. First, the signals were recorded by the LFRS while the simulated noise was off. Next, the simulated noise was turned on, both the simulated noise and the emulated sky signal were recorded by the LFRS.

A method named adaptive interference cancellation is used to suppress the EMI from the lander (Fridman & Baan 2001). A separate, dedicated reference channel is designed in order to obtain an independent estimate of the EMI from lander. The 20cm long antenna for the reference channel is installed very close to the lander, near the root of the 5 - meter - long antenna, as shown in Figure 2. The received signal from the short antenna is almost the EMI from the lander, because of the limited sensitivity for external signals. The signals received by the 5 meter long antenna, are be corrected using the signals received by the short reference antenna to suppress the EMI from the lander, the emulated sky signals with EMI and without EMI are compared to evaluate EMI mitigation. For the data of every experiment, the first 50 groups of data are averaged as correction coefficient. For every subsequent signal received by the antenna with a length of 5 meter is corrected using the corresponding signal received by the short reference antenna and the estimated correction coefficient. 75 groups of data (about 5 minutes) are averaged as the result of the experiments. The above method we used to suppress the EMI from the lander is shown in Equation (1).

$$\begin{cases} C_a(f) = \dfrac{\sum_{i=1}^{50} U_i^a(f)}{\sum_{i=1}^{50} U_i^d(f)}, \\ UC_{50+j}^a(f) = U_{50+j}^a(f) - C_a(f) \times U_{50+j}^d(f), \\ UO^a(f) = \dfrac{1}{75}\sum_{j=1}^{75} UC_{50+j}^a(f), \end{cases} \tag{1}$$

where $U_i^a(f)$ is the ith group data output by LFRS which is an amplitude spectrum array of 5 - meter - long antenna a, $U_i^d(f)$ is the ith group data of reference antenna d, $C_a(f)$ is the array of correction coefficient of antenna a, $UC_{50+j}^a(f)$ is the $50 + j$th group of amplitude spectrum array which is corrected by reference antenna data and the array of correction coefficient, $UO^a(f)$ represents the final amplitude spectrum array of 5 -meter - long antenna a.

The advantage of adaptive interference cancellation lies in keeping the structure of the signal - of - interest intact, the subtraction of an EMI estimate should not affect the wanted radio signal. This kind of EMI cancellation is especially useful for the LFRS observations where the EMI and the signal - of - interest occupy the same frequency domain, but the effectiveness of the method also depends on the temporal stability of correction coefficient. Equation (1) is workable under the assumption of constant correction coefficient during the later averaging. Furthermore, the effectiveness of this adaptive interference cancellation also depends on the sensitivity of the reference channel. There will be a loss of signal - of - interest after processing, if the auxiliary reference is sensitive enough to receive the external signals.

3.3 Experimental Procedure

In order to prevent the low - frequency RFI from environment and simulate an open area test site, a semianechoic chamber located in China Electronic standardization Institute (CESI) was selected as the site of the ground experiments as shown in Figure 3 (a) - (b). The semi - anechoic chamber consists of a 23m (Length) × 14m (Width) × 9m (Height) shielded enclosure. The chamber is lined with hybrid absorbers (model number IP - 130BLB) which consist of carbon loaded polystyrene foam absorber bonded to the ferrite absorber backing. The guaranteed level of shielding effectiveness in the chamber would provide over the frequency range of 100kHz to 18GHz is 100dB. Due to good electromagnetic shielding effectiveness of the semi - anechoic chamber, desired results of the ground experiments were achieved. The results and performance analysis are presented in the following section.

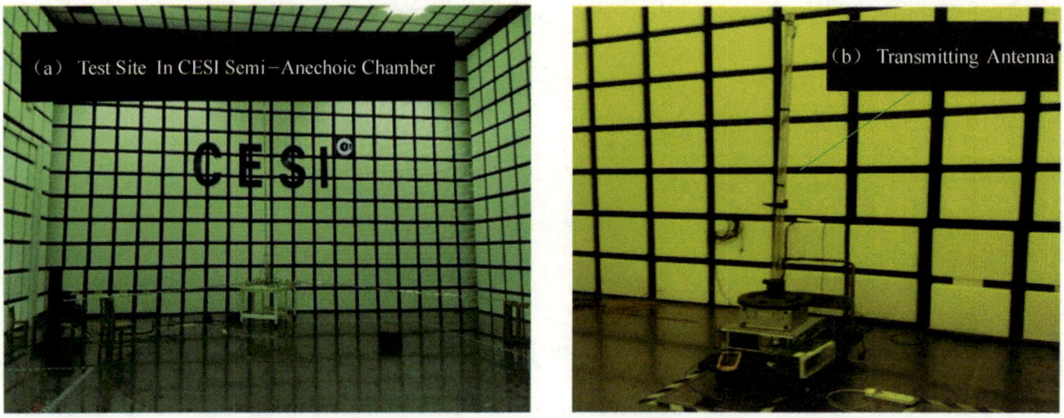

Fig. 3　Photographs of the test site CESI semi anechoic chamber

4 RESULTS AND PERFORMANCE ANALYSIS

4.1 Results

At the beginning of the experiment, the background noise of the site was tested.

Figure 4 shows the background signal spectra received by the LFRS's 5 - meter - long antenna in the semi - anechoic chamber. The Y - axis represents the amplitude of the signal spectrum in dBm, while the X - axis shows the frequency of the signal spectrum in MHz. Three tests were performed to check the characteristics of the background signals (black, red and green lines in Fig. 4). As you can see by Figure 4, although there are still RFI from the computers, test instruments, power supply, etc. , the power and frequency of the RFI do not appreciably change over observable time. For the above the reasons, the results from the experiments in the semi - anechoic chamber are used for analysis and performance evaluation of the LFRS.

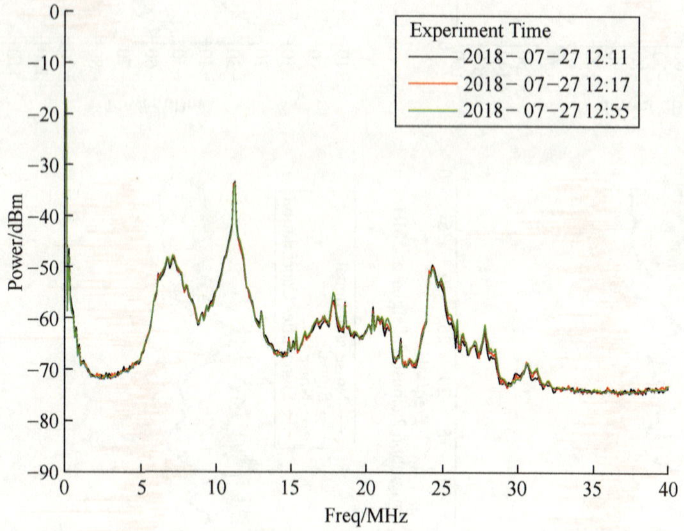

Fig. 4　Background signal spectra on site CESI semi - anechoic chamber

Figure 5 shows the typical results of sinusoidal signals tests. According to the configuration of Table 2, six tests of sinusoidal signals mode were carried out. The method that is briefly described in Section 3. 2 was used for EMI mitigation. In Figure 5 the blue lines indicate the spectrum of the generated signal described in Table 2. The green lines show the spectra of lander noise recorded during the EMC test. The black lines represent mixed spectra of signal and lander noise. The red lines reflect the spectra after EMI mitigation. The position of the red line relative to the position of the black line could be used to see the effectiveness of EMI mitigation. As shown in Figure 5, about 20 to 40dB noise reduction could be received 0. 1 — 40MHz after EMI mitigation, while the sinusoidal signals are not severely distorted.

Figure 6 show the typical results of broadband signal tests. As shown in Figure 6, about 20 to 40dB noise reduction could be received in 0. 1—40MHz after EMI mitigation, while the broadband signals are not severely distorted.

4.2　Performance Analysis

Average and peak spectra of the various planetary magnetosphere and atmospheric

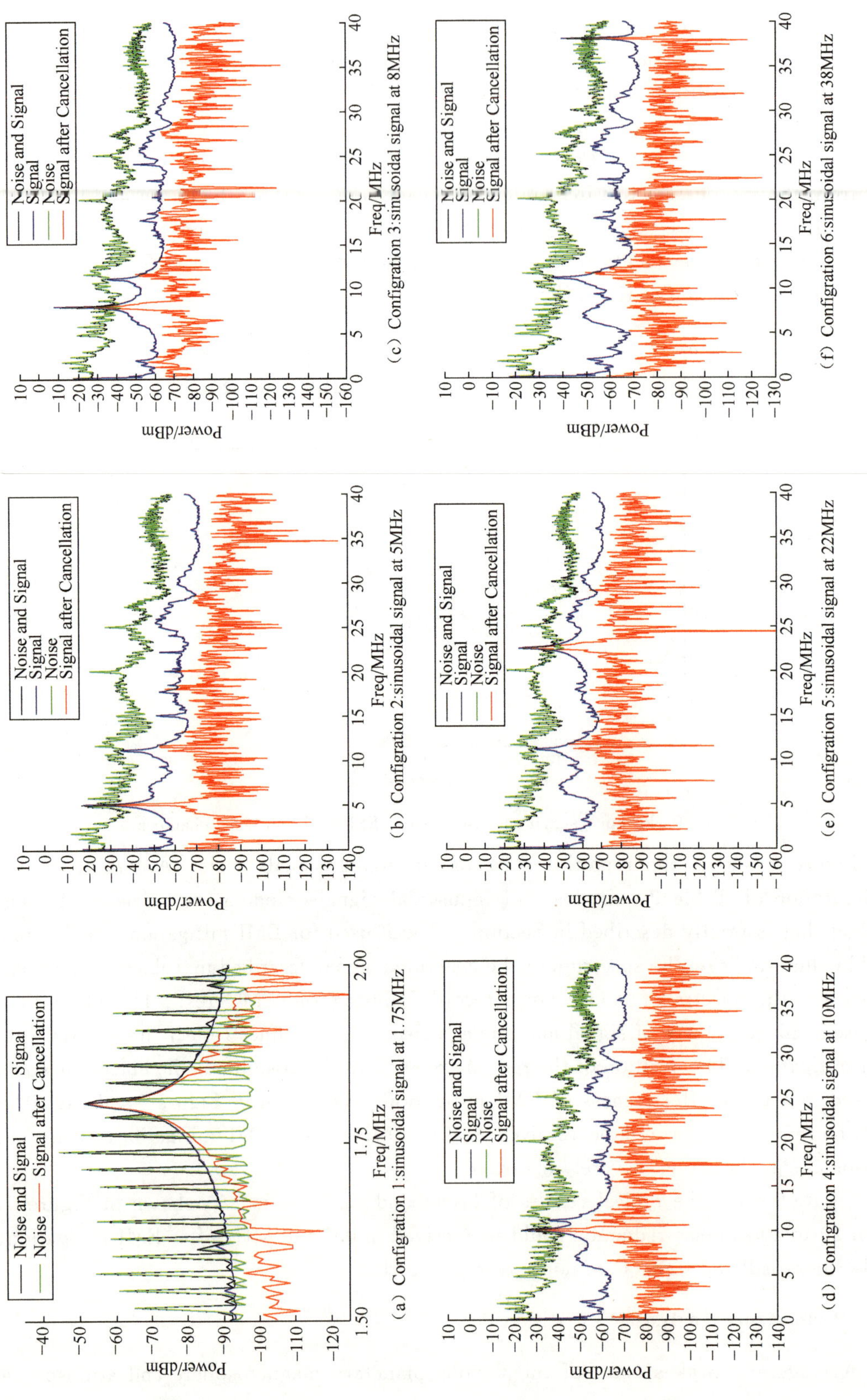

（a）Configration 1:sinusoidal signal at 1.75MHz

（b）Configration 2:sinusoidal signal at 5MHz

（c）Configration 3:sinusoidal signal at 8MHz

（d）Configration 4:sinusoidal signal at 10MHz

（e）Configration 5:sinusoidal signal at 22MHz

（f）Configration 6:sinusoidal signal at 38MHz

Fig. 5　Typical results of sinusoidal signals tests

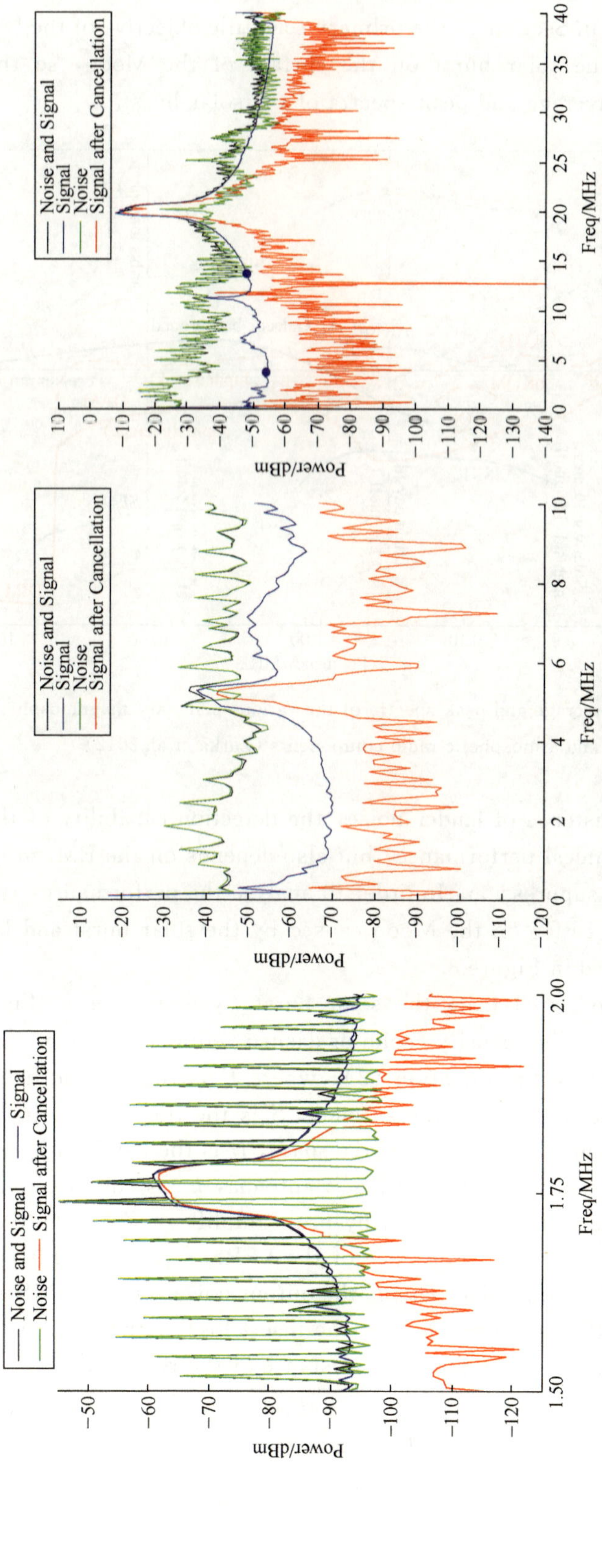

（a）Configration 7:BroadBand Signal at 1.75MHz

（b）Configration 8:BroadBand Signal at 5MHz

（c）Configration 9:BroadBand Signal at 20MHz

Fig. 6　Typical results of broadband signals tests

radio components as they would be measured from the Moon surface or orbit are displayed in Figure 7. As stated in Section 1, the primary scientific objective of the LFRS is to probe radio emission from the solar burst on the surface of the Moon, so the performance analysis focus on the average and peak spectra of the solar burst.

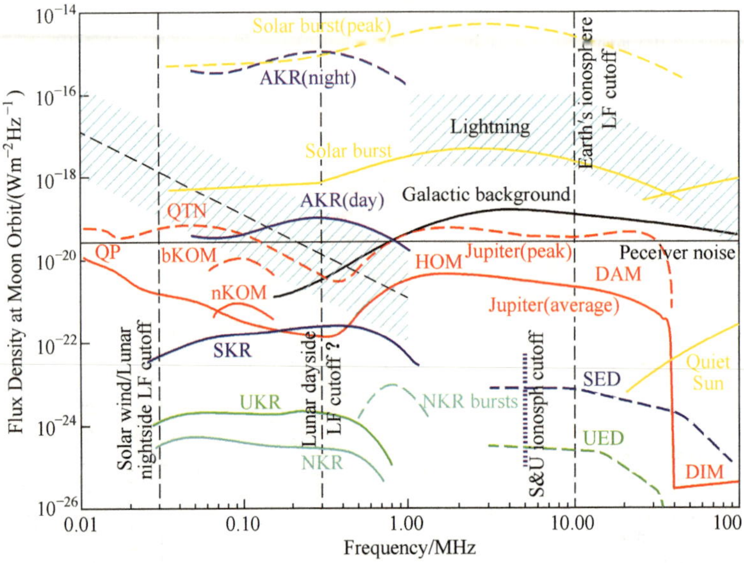

Fig. 7 Average and peak spectra of the various planetary magnetosphere and atmospheric radio components (Zarka et al. 2012)

Because of the existence of lander noise, the detection capability of the LFRS is not only related to its technical performance, but also depends on the EMI mitigation method for the lander's noise suppression. In order to analyse the performance, the power of the signal received by the LFRS on the Moon caused by the solar burst and lander noise are estimated and compared in Figure 8.

The power of the signal from the solar burst (yellow lines in Fig. 8) could be estimated by solar burst flux density (solid or dashed yellow line in Fig. 7) and the LFRS technical parameters using the Equation (2), where E is the electric field strength that produces voltage at the terminal of the antenna, S is the flux density in $\mathrm{Wm}^{-2}\ \mathrm{Hz}^{-1}$ of the wave, η is the intrinsic impedance of free space, B is the bandwidth of the LFRS, U_a is the voltage at the terminal of the LFRS's antenna, h_e is the effective length of the antenna, U_{preamp} is the voltage outputted by pre-amplifier, Z_a is the impedance of the LFRS's antenna, Z_{preamp} is the impedance of the LFRS's pre-amplifier, G_{preamp} is the gain of the LFRS's pre-amplifier, P_{out} is the output power of the LFRS's receiver, G_{rec} is the gain of the LFRS's receiver, L_{line} is the insert loss of the cable between the pre-amplifier and the receiver, Z_L is the load resistance of the receiver.

$$E = \sqrt{2\eta BS}$$

$$U_a = E \times h_e$$

$$U_{\text{preamp}} = \frac{U_a Z_{\text{preamp}} G_{\text{preamp}}}{Z_a + Z_{\text{preamp}}}$$

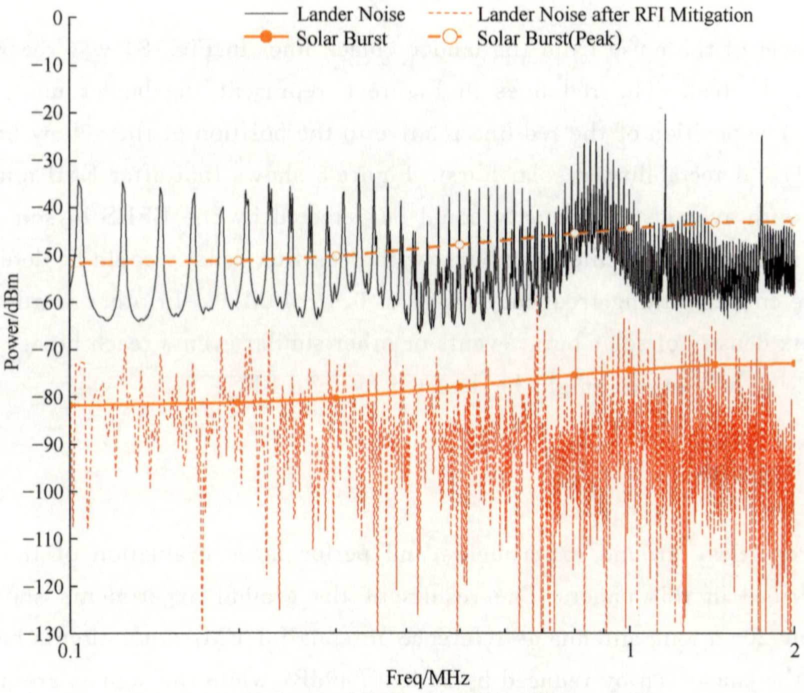

（a）Frequency range：0.1～2MHz RBW：1.22kHz

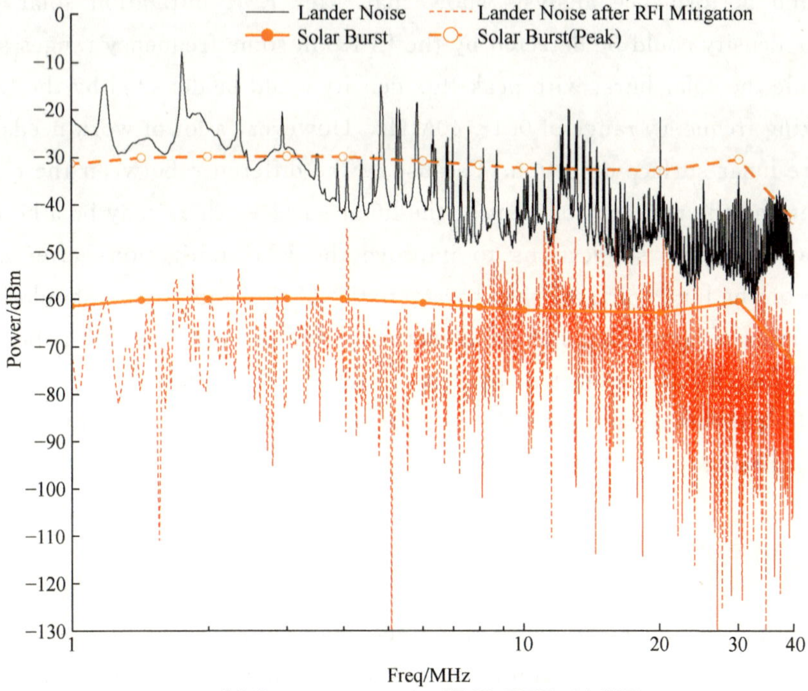

（b）Frequency range：1～40MHz RBW：24.42kHz

Fig. 8　Power spectrum received by the LFRS on the Moon caused by the solar burst and lander noise

$$P_{out} = \frac{(U_{preamp} G_{rec} L_{line})^2}{Z_L} \tag{2}$$

The power of the noise from the lander (black lines in Fig. 8) was recorded during the lander EMC test. The red lines in Figure 8 represent the lander noise after EMI mitigation. The position of the red line relative to the position of the yellow line could be used to analyse detectability of solar burst. Figure 8 shows that after EMI mitigation the solar burst with average flux density could be detected by the LFRS in some frequency range such as 1—6MHz, while solar burst with peak flux density could be detected by the LFRS in the entire working frequency range of 0.1—40MHz. In other words, when the radiation flux density of solar burst events or other similar events reach higher than 10^{-18} W m^{-2} Hz^{-1}, these events might be detected by the LFRS.

5 Conclusions

The properties, ground experiments and performance evaluation of the LFRS are briefly described in this paper. The results of the ground experiments show that the method using 20cm long antenna as reference antenna for EMI mitigation is helpful. The noise from the lander can by reduced by 20dB to 40dB, while the signals are not severely distorted in the frequency range of 0.1—40MHz with the above method. The detection capability and performance analysis show that after EMI mitigation solar burst with average flux density could be detected by the LFRS in some frequency ranges such as 1—6MHz, while the solar burst with peak flux density could be detected by the LFRS in the entire working frequency range of 0.1—40MHz. However, a lot of work needs to be done in the future lunar surface detection. Because of the difference between the experimental environment and the real environment of the lunar surface, there may be a better way for EMI mitigation. Some suggestions to improve the EMI mitigation: after a period of observation on the lunar surface, the characteristics of the lander noise should be carefully studied. A more accurate model of the lander noise will be very helpful for the process and analysis of the spectra collected by the LFRS. Furthermore, more advanced and complicated algorithms can be tried to improve the detection capability of the LFRS (Fridman & Baan 2001).

Acknowledgements

This work was funded by the Youth Innovation Promotion Association of Chinese Academy of Sciences (CAS) (Grant No. 2015334) and the National Natural Science Foundation of China (Grant No. 11941002). The authors thank the Ground Application System of Lunar Exploration, National Astronomical Observatories and Institute of Electronics, CAS, for their valuable and efficient assistance with providing the data and data calibration.

References

[1] Arts, M., Wal, E. V. D., & Boonstra, A. J. 2010, in ESA Antenna Workshop on Antennas for Space Applications, ed. H. Lacoste, Vol. 32 (Noordwijk: ESA), 5

[2] Bentum, M. J., Boonstra, A. J., & Baan, W. 2011, in General Assembly and Scientific Symposium (Piscataway: IEEE), 1

[3] Boonstra, A. J., Saks, N., Bentum, M. J., Klooster, V. T. K., & Falcke, H. 2010, in Cospar Scientific Assembly (Paris: COSPAR), 11

[4] Fridman, P. A., & Baan, W. A. 2001, Journal of Astronomy and Astrophysics, 378, 327

[5] Jester, S., & Falcke, H. 2009, New Astronomy Reviews, 53, 1

[6] Ji, Y., Zhao, B., Fang, G., et al. 2017, Journal of Deep Space Exploration, 4, 150

[7] Lazio, T. J. W., Macdowall, R. J., Burns, J. O., et al. 2011, Advances in Space Research, 48, 1942

[8] Li, C., Wang, C., Wei, Y., & Lin, Y. 2019, Science, 365, 238

[9] Takahashi, Y. D. 2003, Advances in Space Research, 31, 2473

[10] Wolt, M. K., Aminaei, A., Zarka, P., et al. 2012, Planetary and Space Science, 74, 167

[11] Wu, W., Wang, Q., Tang, Y., et al. 2017, Journal of Deep Space Exploration, 4, 111

[12] Wu, W., Li, C., Zuo, W., et al. 2019, Nature Geoscience, 12, 222

[13] Zarka, P., Bougeret, J. L., Briand, C., et al. 2012, Planetary and Space Science, 74, 156

嫦娥四号着陆点区域高分辨率数字高程模型构建

陈　昊，刘世杰，童小华，胡　明
同济大学测绘与地理信息学院，上海 200092

摘要： 嫦娥四号登月探测器成功着陆于冯·卡门撞击坑内，实现了人类历史上首次月背软着陆。嫦娥四号登月探测器着陆区的高分辨率数字高程模型（DEM）对后续任务顺利开展至关重要，着陆点区域的三维地形可为月球探测提供关键空间信息支撑，但在嫦娥四号着陆点区域，月球轨道激光高度计生成的 DEM/LOLA DEM 分辨率仅有 30m，未见公布高分辨率 DEM。基于高分辨率的月球勘测轨道飞行器窄角相机（LROC NAC）影像，利用摄影测量法和阴影恢复形状方法（SFS）针对嫦娥四号着陆点区域，生成了着陆点区域的高分辨率 DEM。结果表明，SFS 法生成的 DEM 分辨率更高，重建的地形更加精细。

关键词： 嫦娥四号登月探测器；冯·卡门撞击坑；数字高程模型；月球勘测轨道飞行器窄角相机（LROC NAC）；阴影恢复形状

High - Resolution Digital Elevation Model Generation for Chang'e - 4 Landing Site Area

CHEN Hao，LIU Shijie，TONG Xiaohua，HU Ming

College of Surveying and Geo - Informatics，Tongji University，
Shanghai 200092，China

Abstract： The Chang'e - 4 lunar probe lands on the Von Karman crater successfully，achieving the first soft landing on the back of the moon throughout the human history. It is very important for the successful implementation of follow - up missions with the high - resolution digital elevation model（DEM）for the Chang'e - 4 landing site area. Three - dimensional topography of the landing site area can provide critical spatial information for lunar exploration. In the Chang'e - 4 landing area，the limited LOLA DEM resolution is only around 30m per pixel，and the high - resolution DEM is not reported in any document. In this paper, high - resolution DEMs of the landing site area are generated by the photogrammetry method and the shape from shading (SFS) method based on the highresolution lunar reconnaissance orbiter camera narrow angle camera (LROC NAC) images. The results show that the DEM generated by the SFS method has higher resolution, with more elaborate reconstructed terrain.

Keywords: Chang ' e - 4 lunar probe; Von Karman crater; digital elevation model (DEM); lunar reconnaissance orbiter camera narrow angle camera (LROC NAC); shape from shading

0　引言

2019 年 1 月 3 日，我国的嫦娥四号登月探测器成功着陆于冯·卡门撞击坑内，实现了人类历史上首次月背软着陆。冯·卡门撞击坑中心经纬度为（44.8°S, 175.9°E），直径约为 180km，位于艾肯盆地内，具有重要的科学研究价值[1]。

月表三维地形的恢复是月球探测的基础研究之一，目前主要利用摄影测量手段对月球影像进行立体匹配生成三维地形，并利用激光测高计数据恢复月表形状[2]。立体匹配需基于高度重叠的影像，但因为月表材质较为均匀，纹理信息缺乏，通过匹配方法往往获取不到足够的匹配点[3]。激光数据生成的 DEM（digital elevation model）具有较高的高程精度，但水平方向分辨率较低，因此无法满足对分辨率要求较高的需求，如：着陆区选取、巡视器路径规划等[4]。阴影恢复形状（SFS）理论是基于假设的光照模型和物体表面反射模型建立图像的反射图方程，仅靠单幅影像或多幅影像恢复物体表面像素级地形信息，月表相对简单的地表反照率有利于该技术的实现[5]。

利用单幅影像进行阴影恢复形状（SI - SFS）是 SFS 进行表面重建的最基本形式，通过反射模型建立三维形状与图像强度之间的关系，然而这种关系仅恢复了表面斜率，而非真实的三维形状。HORN[6]利用单幅影像 SFS 恢复的斜率生成表面三维形状，如 DEM 等。这些 SFS 算法在行星测绘和科学领域也有一定的应用，例如，KRIK 和 BEYER 分别利用 SFS 法生成 DEM，用于美国 MER 和 MSL 任务的着陆区分析[7-8]。然而，这些方法仅在小尺度场景下适用。GRUPE 等[4,9]将低分辨率的 DEM 结合到 SFS 算法中，成功提高了大尺度场景下地形重建的精度。WOODHAM[10]首次提出利用不同光照条件下获取的多幅影像进行阴影恢复形状（MI - SFS）。HEIPKE[11]将 MI - SFS 运用到地形测绘中，WOHLER 等[12-13]进一步研究了其在行星表面三维重建方面的应用。MI - SFS 在无纹理区域处给出了合理的结果，在高分辨率表面建模中具有显著的价值[14]，Lzu 等[15]也研究了 MI - SFS 与影像间光照差异的关系。

随着月球探测任务越来越多，人类获取的月球遥感影像也越来越丰富。美国月球勘测轨道器（LRO）任务上搭载的窄角相机（NAC）可获得分辨率为 0.5～1.5m/pixel 月表高分辨率影像，是目前月球分辨率最高的轨道器影像[16]，其拍摄的影像基本覆盖全月球，对月球南北极完全覆盖。LRO 上搭载的 LOLA 数据最高可获取 1 024pixel/(°) 全月 DEM[2]。中国的嫦娥二号登月探测器可获取 7m 分辨率影像，在嫦娥三号着陆区月球虹湾区域可获取 1.5m/pixel 高分辨率影像[17]，利用 7m 分辨率影像生成 20m/pixel 全月

DEM[18]。日本月亮女神月球探测器的地形相机可获取 10 m/pixel 的影像[19]。

嫦娥四号着陆器精确着陆位置为（177.588°E，45.457°S）[20]，目前还未有公开发表的着陆点附近高分辨率 DEM。本文基于月球分辨率最高的 LROC NAC 影像，分别利用摄影测量法及 SFS 法生成着陆点附近高分辨率 DEM，并对结果进行比对，其中以摄影测量法生成的 DEM 作为 SFS 法的初始 DEM。本文生成的高分辨率 DEM 可用于嫦娥四号登月探测器着陆点区域地形地貌分析、着陆区适宜性分析等方面。

1　区域和数据

100m/pixel WAC GLOBAL MOSAIC 影像能对全月球进行覆盖[21]，嫦娥四号登月探测器着陆于月球背面冯·卡门撞击坑内，其在 WAC GLOBAL MOSAIC 上的显示如图 1 所示。由图可见，冯·卡门撞击坑坑壁比较陡峭，而坑底除中央峰外，其余区域均比较平坦。因此，冯·卡门撞击坑内适合嫦娥四号着陆器进行软着陆及玉兔二号巡视探测。美国 LROC NAC 影像是目前在嫦娥四号登月探测器着陆区冯·卡门撞击坑区域可获取的分辨率最高的遥感影像，因此本文利用 LROC NAC 影像对着陆点附近进行高分辨率 DEM 生成。

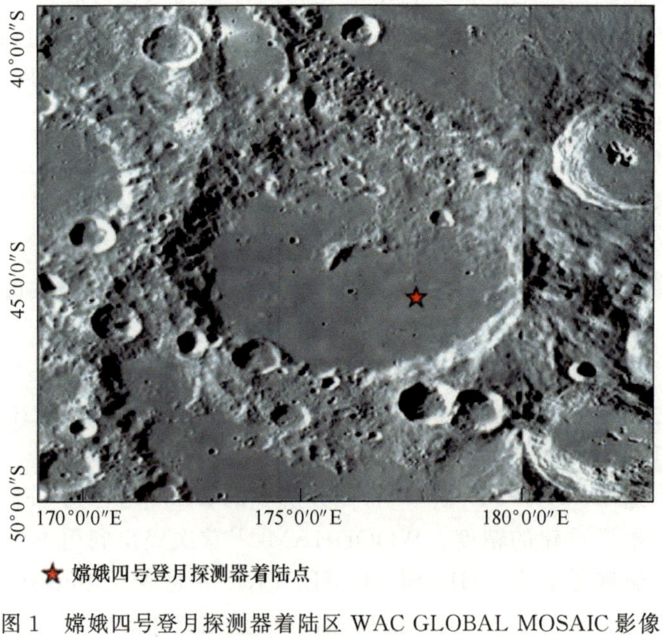

图 1　嫦娥四号登月探测器着陆区 WAC GLOBAL MOSAIC 影像

Fig. 1　WAC GLOBAL MOSAIC image of Chang'e－4 luner probe landing area

2　DEM 生成方法

2.1　摄影测量方法

摄影测量方法发展比较成熟，通常分为 3 个步骤：1）构建严密几何模型或有理函数模型，对立体影像进行平差，获取更优的影像姿态[22]。2）利用密集匹配进行视差图生成，再通过三角测量生成点云。对于密集匹配，SGM 算法的精度与效率均比较好，在行星摄影测量中应用较为广泛[23]；3）将点云生成 DEM。摄影测量法的核心是密集匹配，对于较少纹理、无光照的阴影区域，密集匹配结果较差。

2.2　阴影恢复形状方法

摄影测量法生成的 DEM 分辨率为影像的 2～3 倍，SFS 法生成的 DEM 分辨率可达到像素级。SFS 最早由 HORN 于 20 世纪 70 年代提出的 1 种单幅影像恢复三维信息的方法，自此出现了许多方法[24]。SFS 方法可分为 4 类[25]，本文采用最小值法。图像强度和表明形状之间的关系为

$$I_n(\phi)(x, y)=E_nA(x, y)R_n(\phi)(x, y) \tag{1}$$

式中：(x, y) 为像素坐标；$\phi(x, y)$ 为 (x, y) 对应的月表实际高程；$I_n(\phi)(x, y)$ 是 $\phi(x, y)$ 对应的第 n 幅图像像素值；E_n 是第 n 幅图像曝光值；$A(x, y)$ 是图像中 (x, y) 的反照率；$R_n(\phi)(x, y)$ 是第 n 幅图像的反射率。$R_n(x, y)$ 由月球反射模型构成，主要有 Lambertian，Lunar - Lambertian，Hapk 模型等[26-28]，本文采用 Lunar - Lambertian 反射模型。

考虑到已知影像强度与由反射模型所确定的物体表面强度之间可能存在误差，将强度方程转化为误差函数的形式，即

$$\iint\sum_n[I_n(\phi)(x, y)-E_nA(x, y)R_n(\phi)(x, y)]^2\mathrm{d}x\,\mathrm{d}y \tag{2}$$

平滑约束主要是避免生成的 DEM 不连续，通过最小化曲面的二阶导数实现平滑约束，即

$$\iint\|\nabla^2\phi(x, y)\|^2\mathrm{d}x\,\mathrm{d}y \tag{3}$$

式中：$\|\nabla^2\phi(x, y)\|^2$ 是 ϕ 所有二阶偏导数的平方和。

月球的初始 DEM 一般由摄影测量进行立体匹配生成或 LOLA 生成，高程方向具有较高的精度。高程约束是保证生成的 DEM 与初始 DEM 保持一致性，即

$$\iint[\phi(x, y)-\phi_0(x, y)]^2\mathrm{d}x\,\mathrm{d}y \tag{4}$$

因此，本文进行 SFS 生成高分辨率 DEM 除了 1 幅或多幅影像作为输入外，还应有 1 个初始 DEM 作为输入。

SFS 的最小化函数为

$$\begin{aligned}\iint\sum_n[I_n(\phi)(x, y)-T_nA(x, y)R_n(\phi)(x, y)]^2+\\ \mu\ \|\nabla^2\phi(x, y)\|^2+\\ \lambda\ [\phi(x, y)-\phi_0(x, y)]^2\mathrm{d}x\,\mathrm{d}y\end{aligned} \tag{5}$$

式中：μ 是平滑项系数；λ 是确定输出的 DEM 保持与输入地形的接近程度。

3　结果与讨论

利用嫦娥四号登月探测器着陆点附近的两幅影像，通过摄影测量法生成该区域高分辨

率 DEM，结果如图 2 所示。生成的 DEM 分辨率为 2.28m/pixel，其中密集匹配方法采用半全局匹配法。从图 2 中得出着陆点附近地势较为平坦，但无法精细地展示出地形细节。

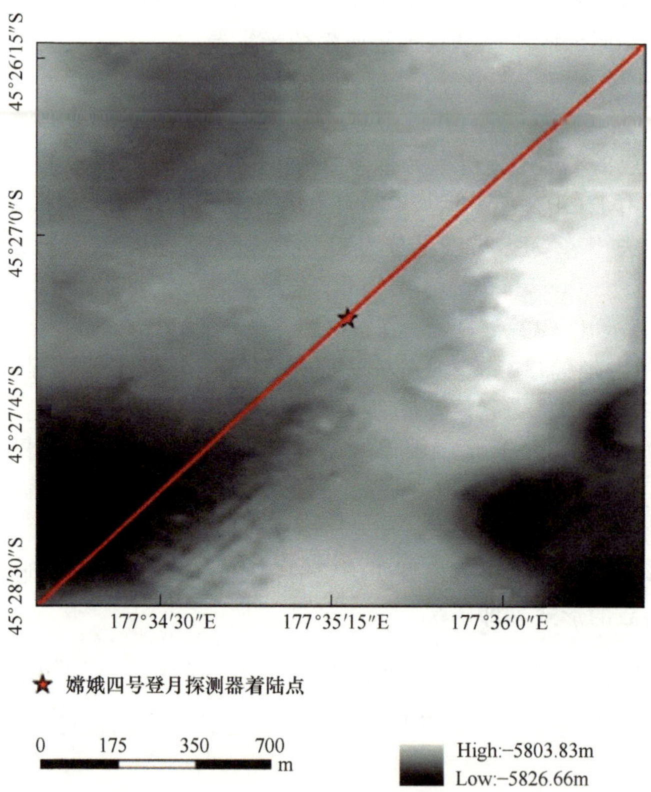

图 2　摄影测量法生成 2.28m/pixel DEM

Fig. 2　2.28m/pixel DEM generated by the photogrammetry method

摄影测量法生成的 DEM 分辨率为 2.28m/pixel，而嫦娥四号登月探测器着陆点附近另外 1 幅 LROC NAC 影像的分辨率为 1.4m/pixel，本文通过 SFS 法生成 1.5m/pixel 高分辨率 DEM。对摄影测量法生成的 2.28m/pixel DEM 升采样为 1.5m/pixel，作为 SFS 法初始 DEM。图 3 为利用单幅 LROC NAC 影像与初始 DEM 进行 SFS 生成的 1.5m/pixel 高分辨率 DEM。由于着陆点附近的坑均比较浅，摄影测量法生成的 2.28m/pixel DEM 无法显示细致的地形信息，因此无法通过视觉从图 2 中看到嫦娥四号着陆点附近的浅坑。由图 3 可以看出，SFS 生成的 1.5m/pixel DEM 能够更好显示地形细节，着陆点附近的浅坑可以明显显示出来。

摄影测量法精度较高，本文以摄影测量法生成的 DEM 作为参考，对 SFS 法生成的 DEM 与摄影测量法生成的 DEM 进行剖面分析，剖面线见图 2 中的红线，剖面分析结果见图 4。由图可见，SFS 法生成的 DEM 与摄影测量法生成的 DEM 具有较好的拟合，因此 SFS 法生成的 DEM 不仅分辨率更高，而且精度也较高，可将其应用于分辨率要求较高的场景下。

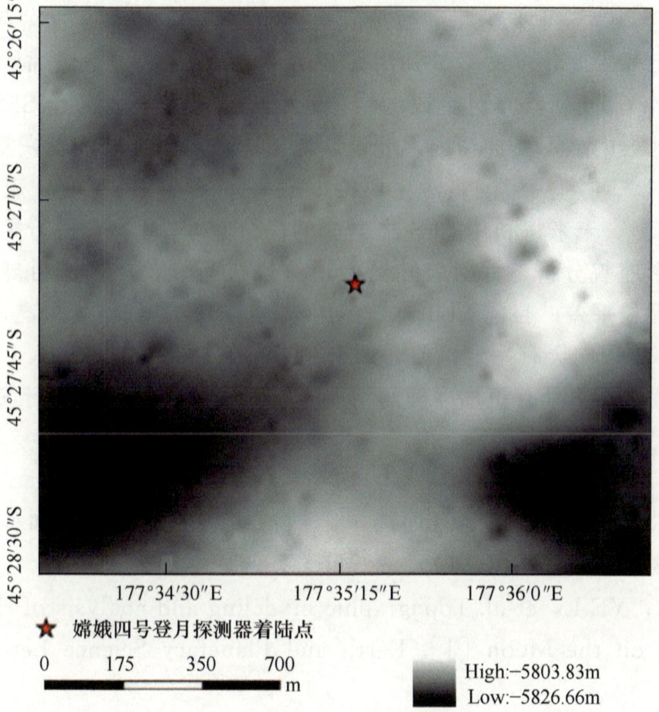

图 3　SFS 生成 1.5m/pixel DEM

Fig. 3　1.5m/pixel DEM generated by the SFS

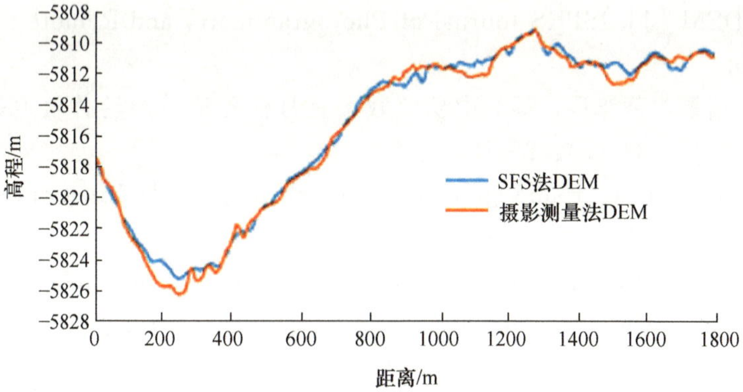

图 4　SFS 法 DEM 与摄影测量法 DEM 剖面分析

Fig. 4　Profile analysis of SFS DEM and photogrammetry DEM

4　结束语

月表三维地形重建方法有摄影测量法、激光测高法、SFS 法三种。其中，摄影测量法和激光测高法比较成熟，但都有相应的缺点。SFS 法虽然提出的时间比较晚，发展不成熟，运行效率较低，尚不可实现大范围月表地形重建，但 SFS 法的优点能很好弥补前 2 种方法存在的缺点。因此，本文尝试利用 SFS 法与摄影测量法生成嫦娥四号登月探测器着陆点附近高分辨率 DEM，并对生成的结果进行对比。结果表明，SFS 法生成的 DEM 分辨率更高，对地形的显示更加精细，能很好地将着陆点附近的浅坑恢复出来。同时，利用 SFS 法生成的 DEM 与摄影测量法生成的 DEM 具有较好的拟合，表明 SFS 法生成的 DEM 精度也比较高。未来将利用 SFS 法生成的高分辨率 DEM，进行嫦娥四号登月探测器着陆点区域地形地貌分析，并对着陆区的适宜性进行评价。由于月球缺乏高精度控制点，本文采用了无控平差。目前 LOLA 数据生成的 DEM 高程精度最高，生成的 DEM 与 LOLA DEM 在高程方向会有一定偏差，未来在进行平差时可考虑引入 LOLA 点作为控制点，以保证生成的 DEM 具有更加准确的精度，且对利用多幅影像进行 SFS 有一定帮助。

参考文献

[1] 孟治国，李翠，平劲松，等．月面冯·卡门撞击坑的着陆选址和科学探测目标浅析[J]. 深空探测学报，2018，5 (1)：3 - 11.

[2] WU B，LI F，YE L，et al. Topographic modeling and analysis of the landing site of Chang'E - 3 on the Moon [J]. Earth and Planetary Science Letters，2014，405：257 -273.

[3] 徐辛超，刘少创，徐爱功．梯度约束 SFS 的月面地形重构 [J]. 中国图象图形学报，2014，19 (5)：739 - 745.

[4] WU B，LIU W C，GRUMPE A，et al. Construction of pixel - level resolution DEMs from monocular images by shape and albedo from shading constrained with low - resolution DEM [J]. ISPRS Journal of Photogrammetry and Remote Sensing，2018，140：3 - 19.

[5] 贺波勇，顾绍景，黄海兵，等．中秋节载人登月任务窗口与转移轨道设计研究 [J]. 上海航天，2017，34 (5)：9 - 15.

[6] HORN B K P. Height and gradient from shading [J]. International Journal of Computer Vision，1990，5 (1)：37 - 75.

[7] KIRK，RANDOLPH L. High - resolution topomapping of candidate MER landing sites with Mars orbiter camera narrow - angle images [J]. Journal of Geophysical Research Atmospheres，2003，108 (12)：8088.

[8] BEYERRA，MCEWEN A，KIRK R L. Meter - scaleslopes of candidate MSL landing sites from point photoclinometry [J]. Space Science Reviews，2003，108：775 -791.

[9] GRUMPE A，BELKHIR F，WÖHLER C. Construction of lunar DEMs based on reflectance modelling [J]. Advances in Space Research，2014，53 (12)：1735 - 1767.

[10] WOODHAM R J. Photometric method for determining surface orientation from multiple images [J]. Optical Engineering，1992，19 (1)：191139.

[11] HEIPKE C. Integration of digital image matching and multi image shape from

shading [J]. Mustererkennung，1992：186 - 198.

[12] LOHSE V，HEIPKE C. Multi - image shape - fromshading：derivation of planetary digital terrain models using clementine images [J]. International Archives of the Photogrammetry，Remote Sensing，2004，35：828 - 833.

[13] WÖHLER C，GRUMPE A，BEREZHNOY A，et al. Integrated topographic，photometric and spectral analysis of the lunar surface：application to impact melt flows and ponds [J]. Icarus，2014，235：86 - 122.

[14] LIU W C，WU B，WÖHLER C. Effects of illumination differences on photometric stereo shape - and - albedo - from - shading for precision lunar surface reconstruction [J]. ISPRS Journal of Photogrammetry and Remote Sensing，2018，136：58 - 72.

[15] LIU W C，WU B. Photometric stereo shape - and - albedo - from - shading for pixel -level resolution lunar surface reconstruction [J]. International Archives of the Photogrammetry，Remote Sensing and Spatial Information Sciences，2017，42 (3/W1) .

[16] ROBINSONA M S，SPEYERERA E J，BOYDA A，et al. Exploring the moon with the lunar reconnaissance orbiter camera [J]. International Archives of the Photogrammetry，Remote Sensing and Spatial Information Sciences，2012：501 -504.

[17] 陈卓宁 . 嫦娥二号高分辨率月球影像图集的设计与制作 [J]. 科技资讯，2013 (31)：3 - 4.

[18] 李春来，刘建军，任鑫，等 . 基于嫦娥二号立体影像的全月高精度地形重建 [J]. 武汉大学学报（信息科学版），2018，43 (4)：485 - 495.

[19] HARUYAMA J，OHTAKE M，MATSUNAGA T，et al. Selene (Kaguya) terrain camera observation results of nominal mission period [C] //Lunar and Planetary Science Conference，2009，1123 - 1127.

[20] 邸凯昌，刘召芹，刘斌，等 . 多源数据的嫦娥四号着陆点定位 [J]. 遥感学报，2019，23 (1)：177 - 184.

[21] WAGNER R V，SPEYERER E J，ROBINSON M S，et al. New mosaicked data products from the LROC team [C] //Lunar and Planetary Science Conference. 2015，46：1473.

[22] 徐青，耿迅，蓝朝桢，等 . 火星地形测绘研究综述 [J]. 深空探测学报，2014，1 (1)：28 - 35.

[23] HIRSCHMULLER H. Stereo processing by semiglobal matching and mutual information [J]. IEEE Transactions on Pattern Analysis and Machine Intelligence，2008，30 (2)：328 - 341.

[24] RATIKA P，GHOSE M K，JEYARAM A. Extraction of depth elevation model (DEM) from high resolution satellite imagery using shape from shading approach [J]. International Journal of Computer Applications，2010，7 (12)：40 - 46.

[25] 廖熠，赵荣椿 . 从明暗恢复形状（SFS）的几类典型算法分析与评价 [J]. 中国图象

图形学报，2001，6（10）：953－961.

[26] MCEWEN A S. Photometric functions for photoclinometry and other applications [J]. Icarus，1991，92（2）：298－311.

[27] 彭祺擘，李桢，李海阳．载人登月飞行方案研究 [J]. 上海航天，2012，29（5）：14－19，72.

[28] FERNANDO J，SCHMIDT F，CEAMANOS X，et al. Surface reflectance of Mars observed by CRISM/MRO：estimation of surface photometric properties in gusev crater and meridiani planum [J]. Journal of Geophysical Research Atmospheres，2013，118（3）：534－559.

基于空间矢量天线的太阳低频射电爆发探测研究

陈林杰[1]，颜毅华[1,2]，谭宝林[1,2]

1. 中国科学院 国家天文台 太阳活动重点实验室，北京 100101
2. 中国科学院大学 天文学与空间科学学院，北京 100047

摘要： 目前已有空间探测器对太阳的甚低频射电爆发进行的探测主要是频谱观测，针对太阳低频射电爆发的成像观测仍然是空白，利用空间矢量天线可以对太阳低频射电爆发（包括Ⅱ型和Ⅲ型暴）进行空间定位和一定的成像观测。研究针对空间（甚）低频射电探测器对太阳射电爆发的探测，提出了利用三极子矢量天线对太阳爆发进行定位的算法；基于部分空间甚低频设备的观测数据，在不同条件下对算法进行了仿真，同时针对“嫦娥四号”上的甚低频射电探测器（同样采用了三极子矢量天线），分析了其探测太阳低频射电爆发的能力。结果表明：当信号的信噪比达到 10dB 以上时，矢量天线可以实现度量级的定位精度，并进一步分析了空间低频探测器探测太阳低频射电爆发的能力。

关键词： 甚低频；空间探测器；太阳射电爆发；矢量天线

Study of Very Low Frequency Solar Radio Emission Detection with Space Vector Antenna

CHEN Linjie[1]，YAN Yihua[1,2]，TAN Baolin[1,2]

1. CAS Key Laboratory of Solar Activity，National Astronomical Observatories of Chinese Academy of Sciences，Beijing 100101，China
2. School of Astronomy and Space Science，University of Chinese Academy of Sciences，Beijing 100047，China

Abstract: Solar winds，Coronal Mass Ejections (CME)，interplanetary shock waves and high - energy particle events，induced by the solar radio bursts，are called as solar electromagnetic storms. The propagation，acceleration，and evolution of the CMEs，solar winds，interplanetary shock waves and high - energy particle phenomena mainly happen at the frequency range from tens of kHz to tens of MHz，which is generally considered as the frequency band of Very Low Frequency (VLF). Though, there are already few space instruments that have investigated the VLF solar radio emissions, most of the observations are spectral measurements, and only occupied part of the whole VLF band, while the imaging observations of the VLF solar radio bursts are still blank. Based on the characteristics of the vector antenna, it can estimate the Direction of Arrival (DOA) of the incoming waves, which can be used to locate and image the solar radio bursts (Type Ⅱ & Ⅲ). So all the latest space VLF radio instruments adopt the tripole antenna can do the observations. In this paper, in order to investigate the detection of the solar radio bursts with the space VLF radio explorer, different algorithms are proposed to detect the solar radio bursts with a tripole vector antenna, and simulations have been done with some space VLF radio observations for these algorithms under different conditions. The simulation results show that the vector antenna can locate the radio bursts with a degree-level accuracy. The capability of detecting the solar radio bursts is also analyzed in further for the space VLF radio explorer.

Keywords: very low frequency; space explorer; solar radio burst; vector antenna

0　引言

由太阳爆发活动引起的太阳风、日冕物质抛射、行星际激波和高能粒子事件被称为太阳电磁风暴，超级太阳爆发将会引起一系列的空间灾害性影响。因此，探索和掌握这些事件的发生、发展和活动规律，不仅在空间科学领域，在人类社会发展方面也有着重要的意义。

从日球层密度模型可得出，太阳射电辐射的频率与日冕高度之间具有直接的联系：辐射频率越高，其辐射源区越接近太阳表面，频率越低其源区则距离太阳表面越远。来自太阳的日冕物质抛射（Coronal Mass Ejection，CME）、高能粒子流和太阳风暴等现象，当它们离开太阳表面附近空间后，随着等离子体密度的迅速降低，其辐射频率也将迅速降低，从而使光学、红外、紫外、X射线、微波及米波射电探测都失去了作用。我们无法知道这些高速扰动源在离开太阳表面附近以后在日地之间广阔空间里是如何传播和演化的。目前在轨运行的各种空间太阳探测器中，其探测波段主要为可见光、紫外线、X射线等波段，这些波段的太阳辐射源区基本上均位于距离太阳表面1～2个太阳半径以内的空间范围。这类探测对于离开太阳表面2个太阳半径以外到地球附近的广阔空间里所发生的物理过程几乎无能为力，而甚低频射电探测则有可能在这个领域里取得明显突破。

射电天文学中的甚低频（≤30MHz），作为最后几个未被实质观测的频谱窗口之一，一直受到天文学家的重视。然而，在低于30MHz的频段，大量人为的射电干扰严重限制了人们对宇宙低频射电辐射的观测，而且地球电离层的截止频率白天很少会低于10MHz，即使夜晚也不会低于几MHz，因此基于地基射电望远镜对宇宙甚低频射电辐射进行观测很难或者几乎不可能。为了实现对这一频段的射电观测，空间射电望远镜成为唯一的选择。

近年来，随着空间科学探测的进一步发展，包括我国在内的一些国家已经将空间大型甚低频射电探测项目列入未来的空间科学计划之中，并为此开展了一系列的研究。为了尽可能地减小地球干扰的影响，空间甚低频射电探测可以选择不同的探测地点。月球由于巨大的体积能够遮挡地球的干扰，因此其背面提供了近乎理想的甚低频射电观测条件[1-2]。另一方面，随着距离的增加，射电干扰也会相应地衰减。研究表明，当远离地球约 100 万 km 以上时，地球射电干扰将会衰减到宇宙背景噪声以下，达到研究可以接受的水平。因此，在不同的探测地点，其受到地球射电干扰的影响也是不同的。地球轨道的探测器由于受到的干扰比较强，很难开展实质的甚低频射电天文观测。月球轨道的甚低频探测器由于可以部分时间被月球遮挡，因此也可以开展一些灵敏度要求高的射电天文研究，且探测器的实施相对于月球背面来说比较容易[3]。而其他轨道如日-地朗格朗日点（如 L2 、L4 或 L5 点），由于其特殊的位置，它们能够提前一定时间对太阳电磁风暴进行观测，对于日地空间天气研究来说具有重要意义。

在空间甚低频射电探测方面，最新的研究是基于我国探月计划“嫦娥四号”开展的甚低频探测器项目，这些探测器将在月球背面或是月球轨道对空间甚低频射电辐射进行探测。在所有已开展的研究中，大部分探测器采用了三极子 Tripole 矢量天线作为基本的探测单元。Tripole 天线由 3 个相互正交的阵子天线组成，具有相同的长度，中心馈电并向 3 个方向伸展。它能够测量三维电场分量，并且能够抑制干扰，不同于空间分布的天线阵列，利用单个天线就能够实现对入射电磁波的达波方向估计。这些都使得 Tripole 天线成为空间甚低频观测天线的理想选择。

电磁波的达波方向估计有着广泛的应用。目前的算法大部分使用阵列天线技术[4-5]，也有一些是基于矢量天线来实现。这方面的研究主要有 2 个方向，提高解析度和降低运算量。一些被称为超解析度的技术，如以往空间谱估计采用的 MUSIC，Root-MUSIC 以及 ESPRIT 算法，它们基于特征分解技术，通过空间平滑来使信号源的协方差矩阵满足满秩的要求，以此实现。这些算法虽然能够对窄频信号实现很好的解析度，但是算法的计算量很大，不易实现。Sarkar 等[4]提出了运用矩阵束（MP）的方法来实现达波方向的估计，这一算法不需要进行空间平滑，因此算法的计算量有所降低。Haardt 在文献［5］中将一个单位变换引入到 ESPRIT 方法中，通过将一个复数矩阵变为一个实数矩阵来减少频率估计中的计算量。Nuri Yilmazer 又将类似的变换运用到矩阵束方法中，同样降低了频率估计的复数计算量[6]。为了分析三维电磁场的极化特性，Carozzi 等人引入了归一化的斯托克斯参数以及三维电磁场的替代描述[7]。通过三维极化特性参数计算入射波的三维极化椭圆方向，多个来自不同方向的信号源可以被同时进行测量，这就为利用矢量天线（例如 Tripole 天线）对多个信号源的达波方向估计提供了理论基础。

本文将重点讨论利用三极子矢量天线探测太阳甚低频射电Ⅱ和Ⅲ型爆发的能力。其中，第 1 节主要介绍三极子天线的达波方向计算；第 2 节重点介绍达波方向估计算法；第 3 节对太阳射电爆发的达波方向估计进行分析；第 4 节为“嫦娥四号”低频射电频谱仪数据的初步处理；第 5 节为结论及讨论。

1 三极子天线的达波方向计算

Tripole 天线包括 3 个相互垂直的共中心阵子天线，它可以被看做是一个由 3 个天线组成的小型天线阵，如图 1 所示。为了便于分析，建立如图所示的坐标系，每一个阵子天线与 xy 平面的夹角为 35.3°。对于这一种天线配置，如果 Tripole 天线被建在一个与 xy 平面平行的介质表面，则可以看出 3 个阵子的条件完全相同，必然有着相同的性能，这将使得 Tripole 天线的后续校准非常简单。这一点，对于空间基的天线来说尤为重要。

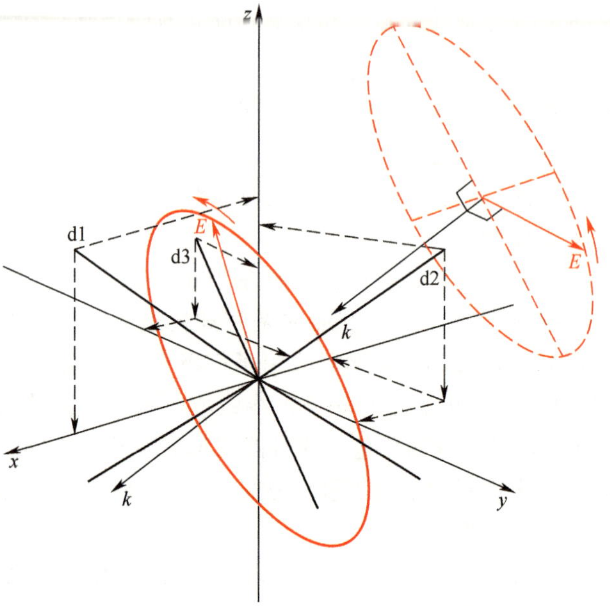

图 1　Tripole 天线对电场所扫偏振椭圆进行 3D 采样示意图

Fig. 1　A radio wave sensed by a tripole antenna propagates with a wave vector k sweeping out the polarization ellipse

为了研究 Tipole 天线上入射波的极化特性，电场矢量可以由 3 个与阵子平行的电场分量表示

$$\boldsymbol{E}(t)=(A_1^{\mathrm{m}}\mathrm{e}^{i\theta_1^{\mathrm{m}}},\ A_2^{\mathrm{m}}\mathrm{e}^{i\theta_2'^{\mathrm{m}}},\ A_3^{\mathrm{m}}\mathrm{e}^{i\theta_3^{\mathrm{m}}})\mathrm{e}^{i\omega t} \tag{1}$$

这里 A_1、A_2、A_3 和 θ_1、θ_2、θ_3 分别代表 3 个电场分量的幅度和相位。为了对图 1 所示的天线配置进行计算，将这 3 个电场分量在 xyz 坐标系下进行投影，得到新坐标系下的分量

$$\begin{bmatrix} E_x \\ E_y \\ E_z \end{bmatrix}=\boldsymbol{T}\times\begin{bmatrix} E_1 \\ E_2 \\ E_3 \end{bmatrix},$$

$$\boldsymbol{T}=\begin{bmatrix} \frac{\sqrt{6}}{3} & \frac{-\sqrt{6}}{6} & -\frac{\sqrt{6}}{6} \\ 0 & \frac{\sqrt{2}}{2} & -\frac{\sqrt{2}}{2} \\ \frac{\sqrt{3}}{3} & \frac{\sqrt{3}}{3} & \frac{\sqrt{3}}{3} \end{bmatrix} \tag{2}$$

这里 $\boldsymbol{T}$ 为转换矩阵。引入一个三维相干张量[7]如下

$$EE^{\mathrm{H}}=\begin{pmatrix} E_xE_x^* & E_xE_y^* & E_xE_z^* \\ E_yE_x^* & E_yE_y^* & E_yE_z^* \\ E_zE_x^* & E_zE_y^* & E_zE_z^* \end{pmatrix} \tag{3}$$

式（3）称之为频谱张量。通常情况下，相干张量用来对电场的时域极化特性进行分析，对应的频域张量被称为频谱张量，为了使用相干张量，要求式中的信号为准单色波，但是对于频谱张量，则没有这一限制。

斯托克斯参数中 I 描述的是场的总强度，这里等于上述张量的迹，

$$\boldsymbol{I}=|E_x|^2+|E_y|^2+|E_z|^2 \tag{4}$$

参照斯托克斯参数 V，引入一个三维广义矢量 $\mathbf{V}$，其中 Im 表示求虚部

$$\mathbf{V}=-2\mathrm{Im}(E_yE_z^*\hat{x}+E_zE_x^*\hat{y}+E_xE_y^*\hat{z}) \tag{5}$$

分析这一矢量可知，它平行于入射波的波矢量 $\boldsymbol{k}$，如图 1 所示，因此，通过计算矢量 $\mathbf{V}$ 的方向就可以得到入射波的波达方向。在球面极坐标系下，归一化的矢量 $\mathbf{V}$ 可以表示为

$$\boldsymbol{V}/\boldsymbol{I}=v\cdot(\sin\theta\cos\varphi\hat{x}+\sin\theta\sin\varphi\hat{y}+\cos\theta\hat{z}) \tag{6}$$

其中：归一化的幅度 $v=|\boldsymbol{V}|/\boldsymbol{I}$，表示入射波的圆极化度，零为线极化，而一则为圆极化；θ 和 φ 分别指俯仰角和方位角。于是，达波方向参数可以由式（6）在极坐标系下进行计算[8]

$$\theta=\arccos(V_z/\sqrt{V_x^2+V_y^2+V_z^2})$$

$$\varphi=\begin{cases}\arctan(V_y/V_x) & \text{如果} \quad V_x>0 \\ \arctan(V_y/V_x)+\pi & \text{如果} \quad V_x<0\end{cases} \tag{7}$$

2　射电信号达波方向估计算法

在月基甚低频射电天文研究中，一个重要的科学目标是对太阳低频射电爆发的探测。而我们知道，太阳的射电爆发信号不管是Ⅱ型暴、Ⅲ型暴或是Ⅳ型暴，其都是一个宽带信号。除此之外，有高能宇宙射电或者中微子撞击月球表面也会产生宽带的射电脉冲信号。为了确定宽带射电信号的来源，基于 Tripole 天线，对其的达波方向估计进行了研究。由于信号的宽频带特性，文献［8］和［9］中的频率估计算法将不再有效。但对于太阳爆发信号，大部分情况下，观测到的信号都明显大于背景噪声，达到 10～20dB 以上，这就意味着对于不同频率的信号仅有一个来波方向需要确定。

对于单个或是多个准单色波，信号的频率分量可以由基于矩阵束和最小二乘法的算法进行估计[9]。由于具有滤除噪声的功能，因此相比与快速傅里叶变换（Fast Fourier Transform，FFT），上述算法能够实现更高的频率估计精度和效率，从而保证了达波方向估计的精度。而对于宽带信号，上述估计算法则是无效的，这里考虑通过 FFT 来获得信号的频谱，然后用公式（7）来计算每一个频率分量所对应的达波方向。由于不同频率的信号仅有一个方向，因此考虑对相邻的不同频率分量的达波方向进行平均，以此来提高方向估计的精度[1]，具体的算法图 2 所示。

在具体的算法中，对采集的太阳射电爆发信号首先进行频谱变换或是信道化，得到不

同频率的子带信号，然后按上节所述的方法建立针对不同频率的广义三维矢量 **V**，以此求得不同频率信号的达波方向，并对其在一定频率带宽内进行平均，以此得到不同频率更为准确的达波方向。

理论上，该算法仅适用于理想情况下/自由空间的小天线，然而，大多情况下，天线并不是工作在自由空间，而是安装在一个卫星平台之上或是在一个大的介质表面（如月面）上。卫星平台或是介质会影响天线的辐射方向图，如图 3 所示：天线的长度为 2.5m，天线距离地面 1m，其中的月面电参数为 $\varepsilon_r = 3 + j0.01$，$\sigma = 10^{-14}$S/m；沙漠的电参数为 $\varepsilon_r = 3$ & $\sigma = 5\times10^{-5}$S/m；纯净水为 $\varepsilon_r = 80 + j0.001$ & $\sigma = 0.001$S/m；田地电参数为 $\varepsilon_r = 15$ & $\sigma = 0.02$S/m；以及理想地面。ε_r 和 σ 分别表示相对介电常数以及电导率。图中不同地面介质的电参数取值都为其典型值，所示的方向图（$\varphi = 0°$）为 Tripole 天线中的单个阵子的方向图。可以看出，月球表面的结果和沙漠上的几乎重合在一起，并且接近于自由空间里的方向图，这是因为月面和沙漠对电磁波的损耗比较小。然而，其他的方向图，由于介质损耗比较大，则与自由空间的差别非常大。将上述算法运用到不同的地面介质，可发现不同方向图与自由空间的差别决定了达波方向的估计误差，其中方向图的差别越大，DOA 估计的误差也就越大。为此，需要对天线的方向图相对于自由空间的方向图进行校正，从而得到准确的信号达波方向。

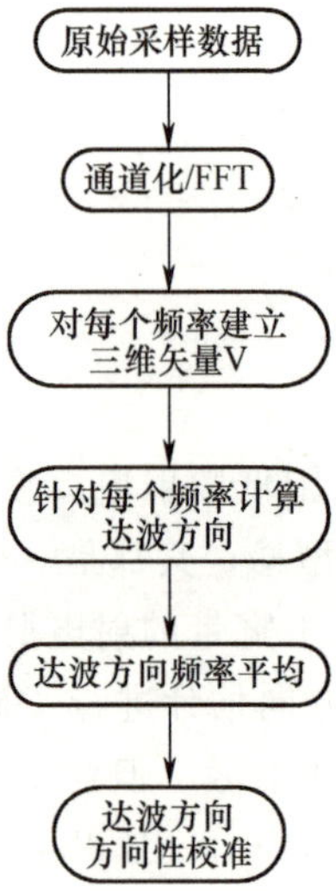

图 2 宽带脉冲信号的达波方向估计算法

Fig. 2 DOA Estimation for broad - band pulse signal

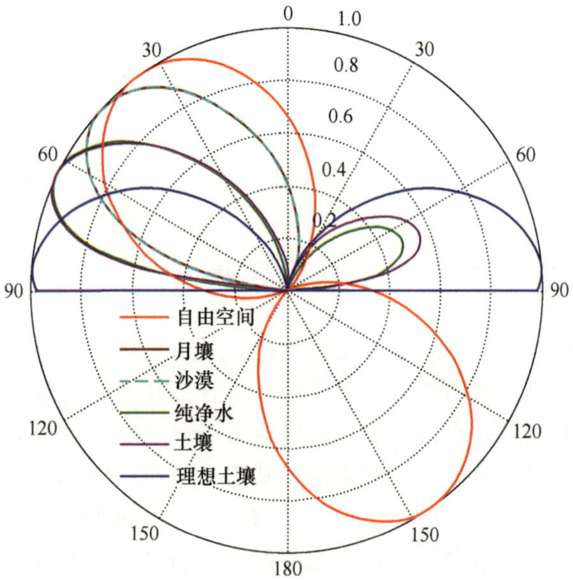

图 3 Tripole 天线 5MHz 的归一化方向图

Fig. 3 Normalized far - field patterns at 5MHz Tripole

对于天线方向图的校准，在自由空间和有平台或是介质情况下建立统一的坐标系，通过天线方向图的仿真数据或是测试数据来对信号 DOA 估计的误差进行确定，在不同方向建立一一对应的误差校正关系。当然，如果利用仿真的数据进行校正，仿真结果的准确性将直接影响校正的效果。另外，如果利用测试的数据进行校正，测试的精度也将直接影响最终 DOA 估计的精度。

3　太阳射电爆发的达波方向估计

太阳活动产生的高能粒子流在甚低频段主要表现为具有快速频率漂移的射电Ⅲ型爆发，其主要特点是辐射频率快速从高频向低频漂移。通过对甚低频射电Ⅲ爆的定位和监测，将对高能粒子是否存在二次加速、加速机制以及加速源区等问题给出新的解释，并为高能粒子的到达时间的预报给出可靠的依据。而 ICME 和行星际激波在射电甚低频段主要表现为射电Ⅱ型爆发。与射电Ⅲ型爆发不同，Ⅱ型爆发具有缓慢的频漂率。通过监测射电Ⅱ型爆发可以追踪 ICME 和行星际激波的位置、空间结构、传播规律和演变过程，从而为灾害性空间天气扰动事件的预报提供可靠依据。如图 4 所示，太阳的Ⅲ型暴表现出快速的频漂特性。而我们知道太阳射电辐射的不同频率对应的辐射源区也是不同的，也就是说我们探测到的太阳不同频率的辐射可能来自不同的方向，这就要求在探测太阳射电爆发的时候需要针对不同的频率进行不同的方向估计。通过对不同频率太阳射电辐射的定位和跟踪，将能够实现对其在行星际空间的动力学过程以及传播特性进行研究，并为预报高能粒子的到达时间提供更为可靠的依据。

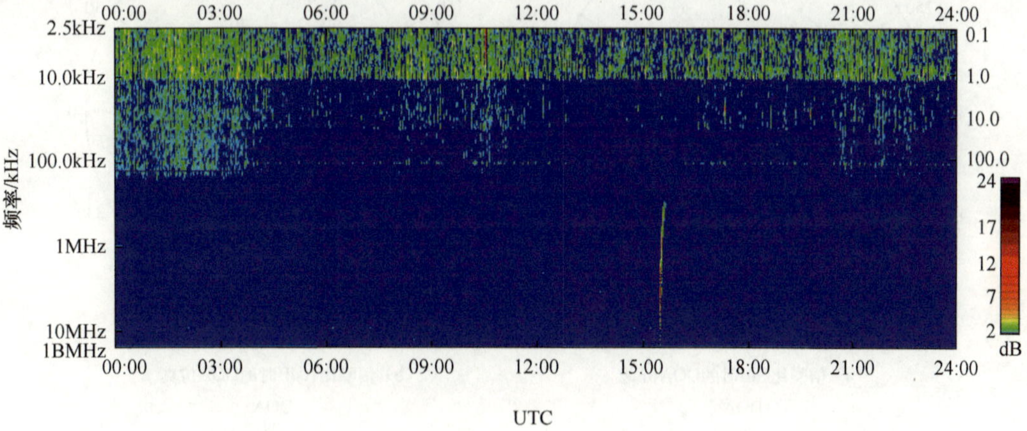

图 4　STEREO/WAVES 观测到的太阳Ⅲ型射电爆发事件（20190208）

Fig. 4　Solar type Ⅲ radio burst observed by STEREO/WAVES on Feb. 8，2019

为了评估三极子 Tripole 天线对太阳射电爆发的定位和方向估计能力，对其进行了仿真。根据太阳射电爆发信号的特性（如图 5），模拟太阳射电爆发的射电信号，在时间域其可以看做是一个脉冲型的宽带信号，而在不同频率其到达时间有一定的时间差，且不同频率的来波方向也是不一样的。理论上，太阳射电爆发的甚低频射电辐射位于约 2 个太阳半径（30MHz）到十几个太阳半径（1MHz），也就是说距离太阳约 1°～5°的范围。对于仿真来说此范围太小，不能够清楚地反映出信号达波方向估计的精度和适用性，为此在仿真时将不同频率信号的来波方向进行随机地设置，在不同信噪比下对其的达波方向估计进行了仿真，如图 6 所示，图中（a）、（b）、（c）和（d）分别为信噪比 5、10、20 及 30dB 下的 DOA 仿真。图中黑点为每次仿真的 DOA 估计，蓝色为方向图校准后的 DOA 结果，红色为信号的理想来波方向。不同信噪比下的 DOA 估计误差见表 1。

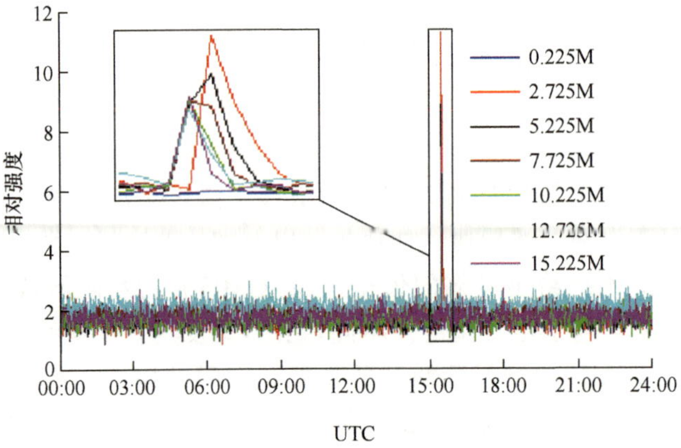

图 5　STEREO/WAVES 观测到的太阳Ⅲ型射电爆发频谱

Fig. 5　Dynamic spectrum of a solar type Ⅲ radio burst observed by STEREO/WAVES

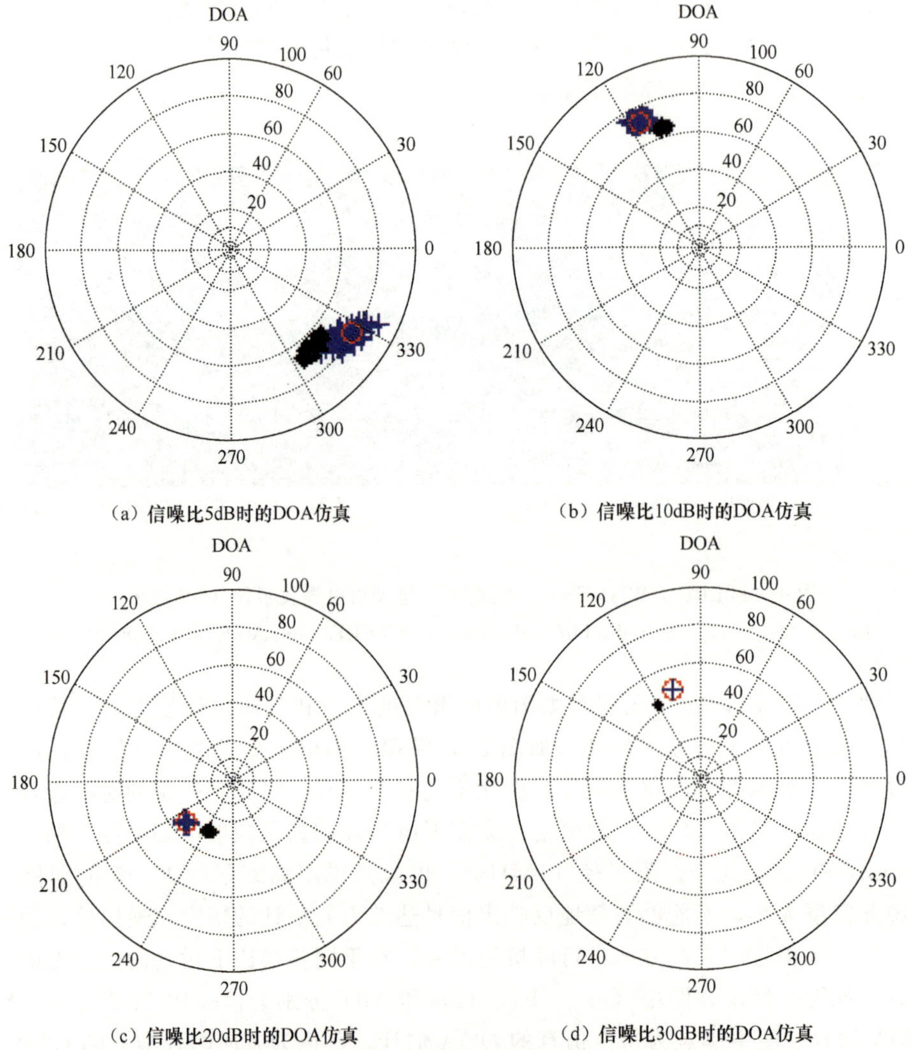

图 6　月面不同信噪比下的太阳射电爆发在 20MHz 时达波方向估计

Fig. 6　DOA estimation results of the solar radio burst at 20MHz on the Moon surface

表 1　不同信噪比下的 DOA 估计误差

Table 1　DOA estimation errors for different SNR

DOA 估计误差/dB		DOA	
		方位角/(°)	俯仰角/(°)
信噪比	5	4.189 9	2.835 6
	10	1.812 8	1.421 3
	20	0.651 9	0.632 1
	30	0.347 8	0.262 7

由以上仿真结果可以看出，在信号的信噪比达到 20dB 以上时，经过对方向图的校正(针对卫星平台或是月面的影响)，利用空间甚低频探测器可以实现小于 1°的达波方向估计精度，当然这要求方向图的校准要实现度量级的精度。这一结果也为研究太阳射电爆发在日地空间的传播特性提供了可能。

此外，由于不同频率天线的方向图是不一样的，而且随着频率的增加，方向图可能会产生裂瓣。为此，利用空间三极子天线在不同频率下对太阳的射电爆发进行了达波方向估计仿真。

图 7 及表 2 显示了 4 个频率的达波方向估计结果，图 7 中黑点为每次仿真的 DOA 估计，蓝色为方向图校准后的 DOA 结果，红色为理想来波方向。可以看出，相比于信噪比，不同频率对达波方向估计的精度影响不大，且比较稳定，精度在 1.5°左右。

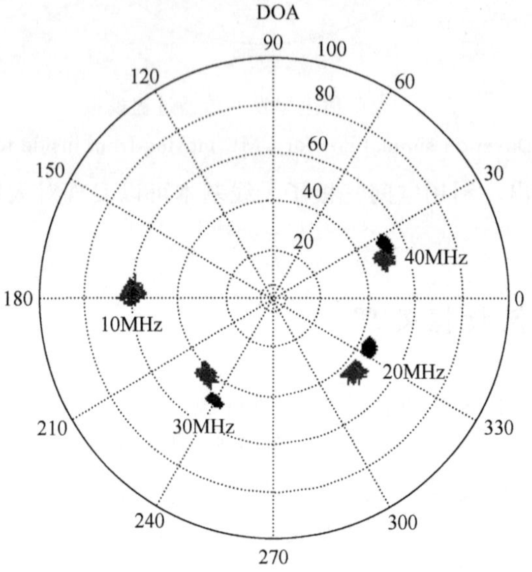

图 7　不同频率太阳射电爆发的达波方向估计

Fig. 7　DOA estimation results of solar radio burst for different frequencies

为了进一步直观地分析利用空间甚低频天线对太阳射电Ⅱ型或Ⅲ型爆发的定位能力，模拟了太阳射电爆发的传播过程，如图 8 所示，图中的每一个点为每次仿真的结果，圆圈为理想的来波方向，不同的颜色代表不同频率及不同位置的达波方向估计。其中信号的信噪比为 20dB；不同频率的爆发来自相互靠近的不同位置，间隔 3°左右。

表 2　不同频率的 DOA 估计误差

Table 2　DOA estimation errors for different frequencies

DOA 估计误差/MHz		DOA	
		方位角 /(°)	俯仰角 /(°)
频率	10	1.488 2	0.972 6
	20	1.212 2	0.716 7
	30	1.053 2	0.712 0
	40	1.417 8	0.783 2

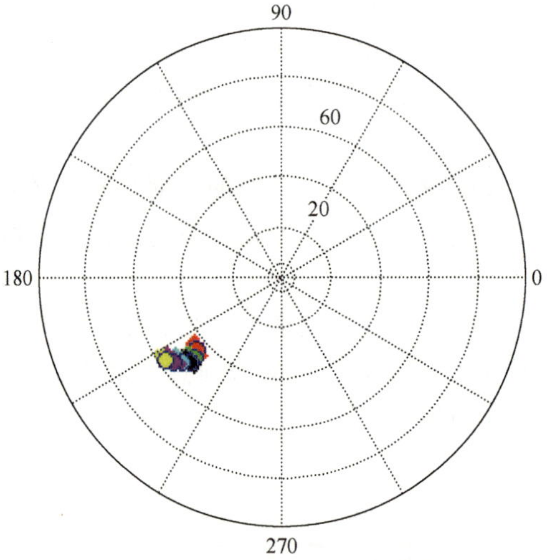

图 8　太阳射电爆发的传播过程模拟

Fig. 8　Direction simulations for CME moving from inside to outside

由以上结果可以看出，利用空间三极子天线基本可以实现对太阳射电爆发在行星际空间的定位和跟踪。

4　低频射电频谱仪数据处理

“嫦娥四号”任务的中继星和着陆器分别搭载有中-荷低频探测仪（Netherlands-China Low frequency Explorer，NCLE）和低频射电频谱仪（Low Frequency Radio Spectrometer，LFRS）两台设备，两者都可以对宇宙的甚低频射电辐射进行探测。其中 NCLE 刚完成了天线的在轨展开，还未开展正式的科学观测。而 LFRS 已经开展了十几个月昼的科学观测，并释放了相关的观测数据。

低频射电频谱仪 LFRS 工作在 0.1～40MHz，其天线是由 3 个 5m 长的单极子构成的 Tripole 天线。理论上 LFRS 完全可以利用上述 DOA 估计的方法来对太阳射电爆发进行定位。但由于“嫦娥四号”着陆器采用的是“嫦娥三号”的平台，并没有针对低频射电观测进行专门的电磁兼容设计，因此平台本身存在着大量的低频射电干扰，对射电观测带来了严重的影响。为了利用观测数据开展科学研究，必须要对数据进行干扰去除的处理。

在实际的数据处理中，采用了将 2 个单极子天线组合成偶极子的方法，来消除两个单极子天线上的共模分量。由于来自平台的干扰基本是近场辐射信号，在不同的天线上会产生一定的共模信号，因此这种方法可以在一定程度上减小来自平台的干扰，但这需要天线接收到的原始信号。

对 LFRS 在 2019 年 2 月 8 日观测的数据进行了处理，如图 9 所示：图中蓝色为 3 个单极子天线 A、B、C 的频谱，红色为 3 个组合的偶极子的频谱，其中左边显示的是低频段的结果，右边为高频段的结果。可以看到经过偶极子组合后的频谱中干扰信号在整个频段均得到了一定程度的抑制，在某些频段甚至达到 20～30dB。但同时也看到，由于 LFRS 单次采集仅得到一小段傅里叶变换点数的原始数据，时间太短，导致数据的灵敏度不够，因此频谱中的噪声起伏很大。为此，对一天的数据进行了积分，获得了相当于在低频段 2s 和高频段 100ms 积分时间的频谱，如图 10 所示：图中蓝色为 3 个单极子天线 A、B、C 的频谱，红色为 3 个组合的偶极子的频谱，其中左边显示的是低频段的结果，右边为高频段的结果。

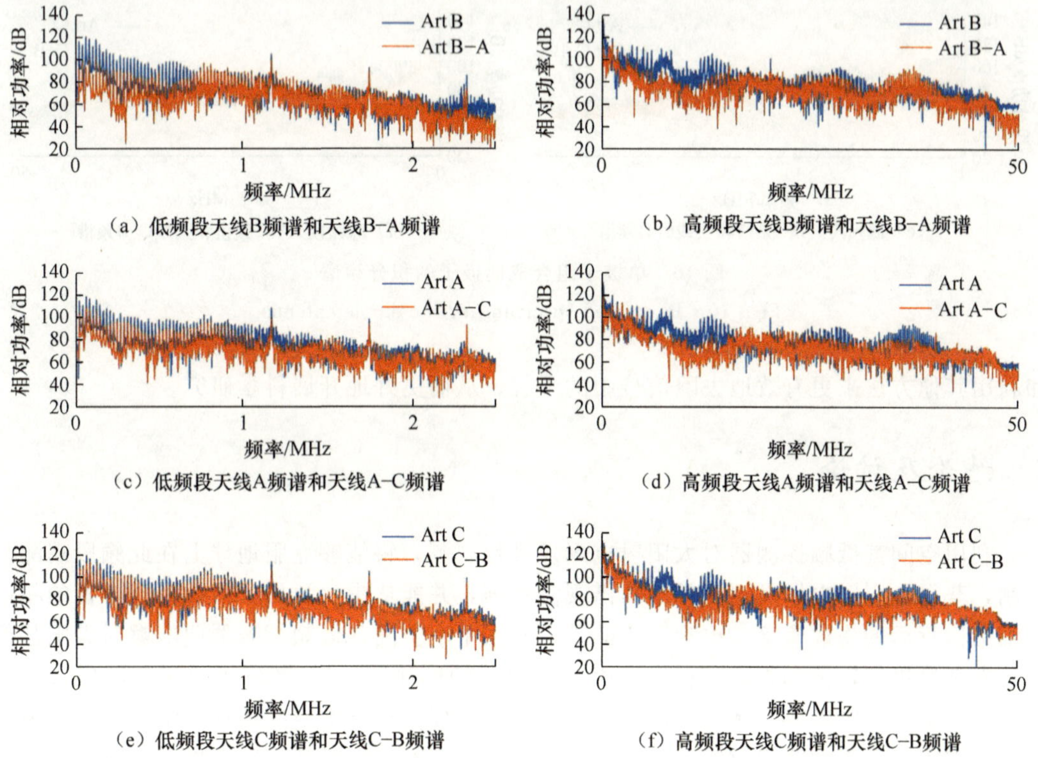

（a）低频段天线B频谱和天线B−A频谱　（b）高频段天线B频谱和天线B−A频谱

（c）低频段天线A频谱和天线A−C频谱　（d）高频段天线A频谱和天线A−C频谱

（e）低频段天线C频谱和天线C−B频谱　（f）高频段天线C频谱和天线C−B频谱

图 9　单极子组合成偶极子的共模抑制频谱

Fig. 9　Common - mode RFI rejections by combining two monopoles to get dipole antenna

可以看出，积分后频谱的灵敏度得到了明显的提高，频谱曲线变得平滑，尤其是高频频段。虽然通过长时间的积分，可以明显改善灵敏度，但也损失了时间分辨率。对于太阳爆发这种快变信号来说，很难利用这种方法来处理。但对于研究宇宙背景辐射等这类稳定的谱信号，这种处理非常有效。另外看到在整个频谱中仍然有大量的干扰，这主要是平台的低频干扰产生的二次和三次谐波。后续，将对观测数据做进一步的处理，同时也将改善

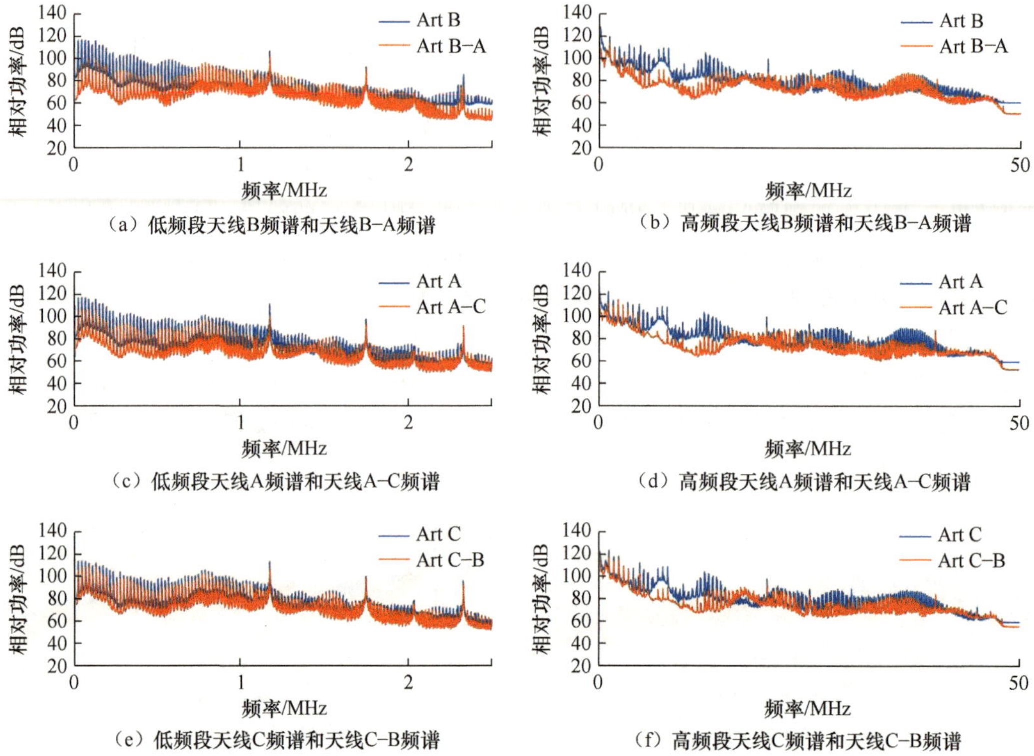

图 10　单极子组合成偶极子的积分频谱

Fig. 10　Integrated spectrum for the dipole antenna

和利用其他方法来更有效地去除信号中的干扰，从而更好地开展科学研究。

5　结论及讨论

利用空间甚低频探测器对太阳射电爆发进行观测，将能够克服地球上在此频段观测的限制，获得太阳射电爆发超宽带的频谱观测数据，并能从中反演出比以往更为丰富的频谱精细结构背后的物理信息。更为重要的是，利用单个空间甚低频探测器便能够对太阳甚低频射电爆发进行定位和跟踪。在信噪比大于 20dB 时，能够实现度量级的定位精度，基本满足进一步研究的需要，从而为通过研究太阳射电爆发在日地空间的动力学传播特性来进行空间灾害性天气的预报奠定一定的基础。目前，针对“嫦娥四号”上搭载的甚低频探测设备，其观测数据的处理和分析工作正在进行中。

另一方面，也要看到，虽然空间单个甚低频探测器能够在一定程度上实现对太阳射电爆发的空间定位，但是其只能实现爆发源区重心的定位，不能反映爆发内部的结构信息，而这些爆发的源区一定存在内部结构。对这种源区结构特征的探测和研究，必然需要空间甚低频射电成像设备。为此，需要将多个空间探测器组合构成甚低频阵列才能够实现这一目标。目前基于“嫦娥四号”上单个甚低频探测设备进行研究，大型的月基甚低频射电阵列也已经被提出，并进入实质性研究阶段，在其完成并发射升空后，太阳甚低频射电爆发的研究必将迎来新的阶段！

参考文献

[1] 陈林杰．月基甚低频射电干涉仪技术研究［D］．北京：中国科学院大学，2011.
CHEN L J. Research on Moon - based very low frequency radio interferometer ［D］. Beijing：School of Astronomy and Space Science，2011.

[2] 纪奕才，赵博，方广有，等．在月球背面进行低频射电天文观测的关键技术研究［J］．深空探测学报（中英文），2017，4（2）：150 - 157.
JI Y C，ZHAO B，FANG G Y，et al. Key technologies of very low frequency radio observations on the lunar far - side ［J］. Journal of Deep Space Exploration，2017，4（2）：150 - 157.

[3] 张锦绣，陈学雷，曹喜滨，等．月球轨道编队超长波天文观测微卫星任务［J］．深空探测学报（中英文），2017，4（2）：158 - 165.
CAO X B，ZHANG J X，CHEN X L，et al. Formation flying around lunar for ultra - long wave radio interferometer mission ［J］. Journal of Deep Space Exploration，2017，4（2）：158 - 165.

[4] SARKAR T K，PEREIRA O. Using matrix pencil method to estimatethe parameters of a sum of complex exponentials ［J］. IEEE Antennas Propag. Mag.，1994，37（1）：48 - 54.

[5] HAARDT M，NOSSEK J A. Unitary ESPRIT：how to obtain increased estimation accuracy with a reduced computational burden ［J］. IEEE Trans. Signal Processing，1995，43（5）：1232 - 1242.

[6] YILMAZER N，KOH J. Utilization of a unitary transform for efficient computation in the matrix pencil method to find the direction of arrival ［J］. IEEE Transactions on Antennas and Propagation，2006，，54（1）：171 - 181.

[7] CAROZZI T，KARLSSON R，BERGMAN J. Parameters characterizing electro - magnetic wave polarization ［J］. Phys Rev E Stat Phys Plasmas Fluids Relat Interdiscip Topics，2000，61（2）：2024 - 2028.

[8] DALDORFF L K S. TURAGA D S，VERSCHEURE O，et al. Direction of arrival estimation using single tripole radio antenna ［C］//Proceedings of the 2009 IEEE ICASSP. ［S. l.］：IEEE，2009.

[9] CHEN L，AMINAEI A，FALCKE H，et al. Optimized estimation of the direction of arrival with single tripole antenna ［C］//2010 Loughborough Antennas & Propagation Conference. ［S. l.］：IEEE，2010.

月球表面原位光谱探测技术研究与应用

何志平，李春来，吕刚，袁立银，徐睿，王建宇

中国科学院空间主动光电技术重点实验室　中国科学院上海技术物理研究所，上海 200083

摘要：探测月球与行星表面物质化学成分是了解其起源及演化历史的关键，而光谱探测技术则是物质成分识别与定量反演研究的重要手段。原位（In－Situ）光谱探测有别于空间探测中的环绕器遥感及采样返回探测，是指在目标现场进行的近距离光谱探测。我国“嫦娥三号”任务科学研究与资源勘查，需要开展月球表面原位光谱探测技术研究，突破凝视型时序扫描的新型声光光谱探测关键技术，研发适应表面恶劣环境的高性能、轻小型、高可靠仪器，在国际上率先实现月球表面光谱原位探测及分析。论文结合以“嫦娥三号、四号”为典型应用的红外成像光谱仪，介绍据此发展起来的月球表面原位光谱探测技术，包括探测机制、工作模式及仪器的功能、性能与应用；最后，也简要介绍了即将应用的“嫦娥五号”月球矿物光谱分析仪。

关键词：月球探测；光谱技术；原位探测

Research and applications of In－Situ lunar surface spectral detection technology

He Zhiping，Li Chunlai，Lv Gang，Yuan Liyin，Xu Rui，Wang Jianyu

Key Laboratory of Space Active Opto－electronics Technology，
Shanghai Institute of Technical Physics，Chinese Academy of Sciences，Shanghai 200083，China

Abstract: Exploring the chemical composition of materials on the surface of the moon and planets is the key to research its origin and evolution history. Spectral detection technology is the major methods of material composition recognition and quantitative inversion In－Situ spectral detection is different from remote sensing and sampling return detection，it refers to the short distance spectral detection performed at the target site. The lunar science research and resource exploration of “Chang’e－3” mission need to implement the first In -Situ spectral detection technology on the lunar surface, by breaking through the key technologies of the new acousto - optic spectroscopic detection of staring time series scanning, we have developed high - performance, light and small, highly reliable instruments, adapted to the harsh surface environment, and become the first in the world to realize In - Situ spectral detection and analysis on the lunar surface of the moon The paper takes "Chang' e - 3 and 4" as typical applications, and introduces the In - Situ spectral detection technology of the lunar surface, including detection mechanism, working mode and the function, performance and application of the instrument; Finally, the "Chang'e - 5" lunar mineral spectrometer was introduced which will be applied soon.

Keywords: lunar surface detection; spectral technology; In - Situ spectral detection

0　引言

月球与行星科学主要研究天体的运动和轨道、表面形貌、物质组成、内部结构、空间环境等[1-2]。月球表面主要是由大小不同且形状各异的矿物组成，如辉石、斜长石、橄榄石和钛铁矿等[3]。这些矿石在可见近红外谱段表现出独特的光谱特征，获取其光谱信息一直是月球表面矿物成分识别与定量反演的重要手段之一，科学成果也很丰富[4-5]。

原位（In - Situ）光谱探测，是指在目标现场进行的近距离光谱探测[6.7]。相对环绕探测及采样分析而言，原位光谱探测因现场、实时探测且同时具备高分辨、高性能且不损伤目标等优势，成为当前各航天大国竞相攻关，重点突破的核心技术。

面向我国探月工程中“嫦娥三号”任务的科学研究与资源勘查的战略需求，我国开展了月球表面原位光谱探测技术研究，克服了月球表面原位场景中低光照、宽温度等极端恶劣空间环境影响，研制了“嫦娥三号”“嫦娥四号”红外成像光谱仪，在国际上首次成功实现月球表面原位光谱探测。

论文结合红外成像光谱仪，重点介绍了原位光谱技术的探测机制、工作模式及典型应用，并对后续应用的“嫦娥五号”月球矿物光谱分析仪的功能、性能及探测特性进行了说明，最后介绍了其在月球表面的应用状况。

1　月球表面原位光谱探测仪器

1.1　红外成像光谱仪

红外成像光谱仪（Visible and Near - Infrared Imaging Spectrometer, VNIS）是“嫦娥三号、四号”巡视器（月球车）载有效载荷[4,8]，红外成像光谱仪由光谱仪探头及电控箱组成，通过电缆组件连接，其中探头安装于巡视器舱外，而电控箱部分安装在巡视器箱内。红外成像光谱仪的主要任务是在巡视器停止时，对月表目标进行可见近红外光谱成像及短波红外光谱探测，获取月表目标的光谱和几何图像数据，完成巡视区探测点矿物组成、含量（丰度）和与化学成分的原位分析。

红外成像光谱仪如图 1 所示[4]，由可见近红外光谱成像通道（0.45～0.95μm）和短波红外通道（0.9～2.4μm）组成，通过接收月表目标漫反射的太阳辐射来完成对目标光谱的探测，主要技术指标如表 1 所示[4]。红外成像光谱仪采用声光可调滤光器（AOTF）作为核心分光元件，其系统框图如图 2 所示，具体包括声光调制分光光学系统、超声驱动定标防尘隔热一体化组件、轻型复合结构、数据获取及处理模块，主控系统等[5,9]。月球表面矿物反射的太阳辐射，进入成像光谱仪，由仪器首先准直成平行光束，再由声光调制滤光器（AOTF）调制，形成某一波长的准单色光，会聚到探测器上，由此得到月表观测目标的单波段光谱及图像。红外成像光谱仪采用轻型超声电机实现定标、防尘、隔热功能，满足仪器在月面的昼、夜生存、防尘等环境适应性以及轻小型化要求。

图 1　红外成像光谱仪

Fig. 1　VNIS

表 1　红外成像光谱仪主要技术指标

Tab. 1　Key technical parameters of VNIS

No	Parameters	Spectra	
		VNIR	SWIR
1	Spectral range/nm	450—950	900—2400
2	Resolution	2—7	3—12
3	FOV/(°)	8.5×8.5	Φ3.6
4	SNR/dB	≥31@9%　Solar reflectance	
5	Power consumption/W	19.8	
6	Weight/kg	4.675（probe）～0.7（electronics）	

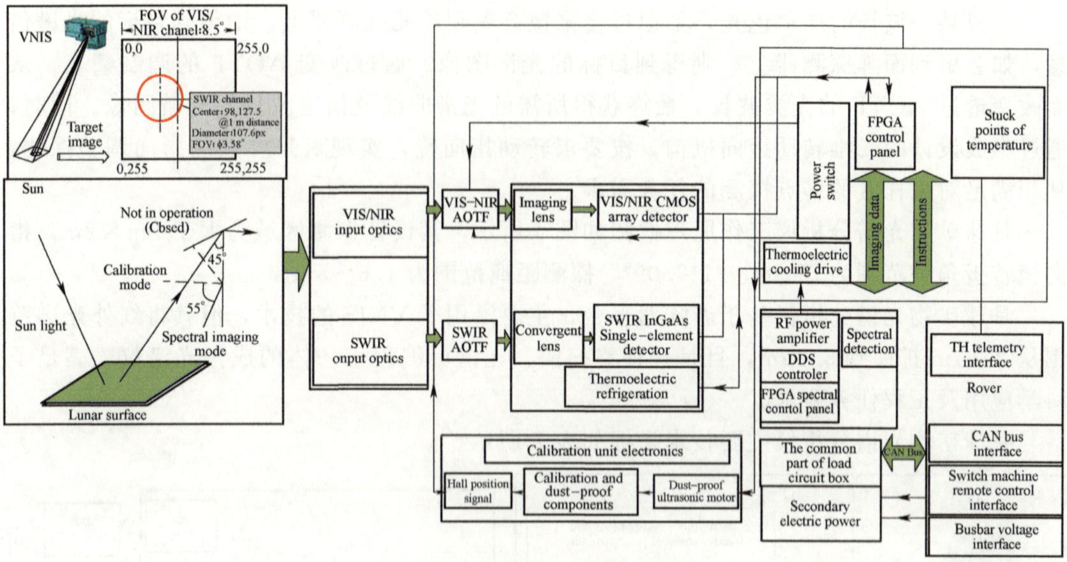

图 2　红外成像光谱仪系统框图

Fig. 2　Block diagram of VNIS system

1.2　月球矿物光谱分析仪

月球矿物光谱分析仪（Lunar Mineralogical Spectrometer，LMS）是“嫦娥五号”任务探测器上的主要有效载荷，用以完成对月面目标的光谱探测和分析任务，为采样区月表物质成分和资源勘察，以及样品实验室测量结果比对研究提供科学数据。

月球矿物光谱分析仪照片如图 3 所示[8]，其探测原理为：来自月球外的太阳辐射经

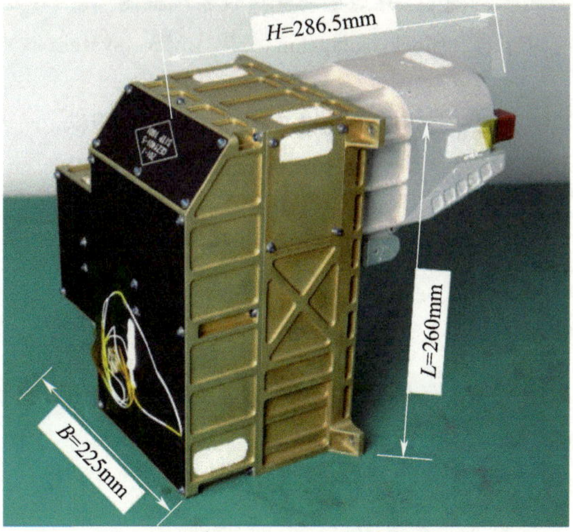

图 3　月球矿物光谱分析仪照片

Fig. 3　Photo of LMS

月球表面的矿物反射后，入射到光谱分析仪的视场内，经声光可调滤光器分光（AOTF）后，形成某一波长的准单色光，如通过会聚镜会聚到单元探测器上，则得到目标的光谱信息，如会聚到面阵探测器上，则得到目标的光谱图像。通过改变 AOTF 的调制频率，从而改变透过 AOTF 的光波波长，最终获得所需的光谱曲线及指定波长的光谱图像。同时，通过集成设计的二维转动指向机构，按要求转动指向镜，实现对采样点的指向调整探测，从而满足对采样区的多点探测的任务需求。

月球矿物光谱分析仪工作原理框图如图 4 所示，其设计探测区域为大于 2m×2m，指向镜转折角度范围为 57.09°～117.09°，探测距离范围为 1.6～5m。

月球矿物光谱分析仪基于着陆器平台，主要沿用了 VNIS 的技术，但其将红外光谱范围从 2.4μm 扩展到 3.2μm，且是一种集扫描、光谱可编程于一体的成像光谱仪，满足了科学应用及工程任务要求。

月球矿物光谱分析仪主要技术指标如表 2 所示[8]。

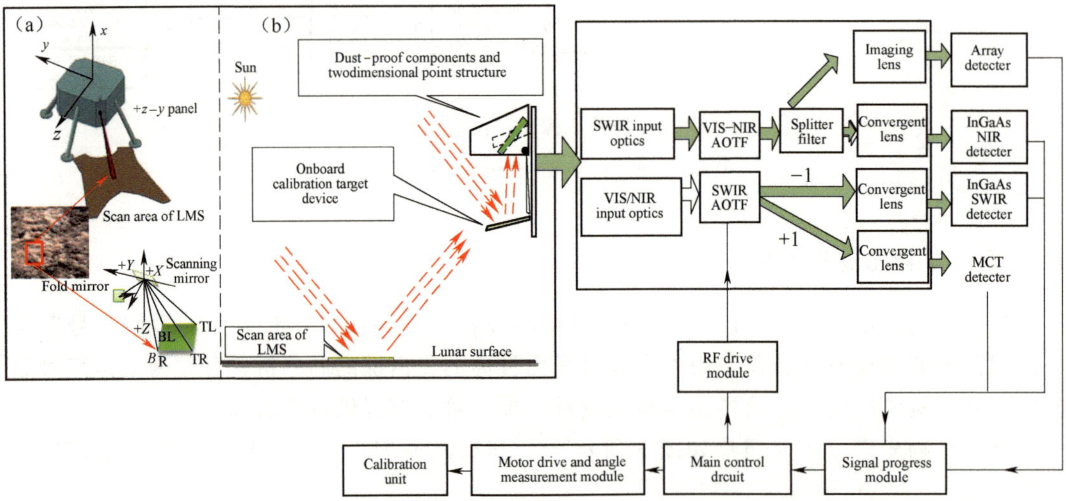

图 4　月球矿物光谱仪系统框图

Fig. 4　Block diagram of LMS system

表 2　月球矿物光谱分析仪主要技术指标

Tab. 2　Key technical parameters of LMS

Description	Specification requirements	
	VIS/NIR	SWIR
Spectral coverage/nm	480—950	900—3200
Spectral resolution/nm	5.0—25.0	5.0—25.0
FOV/(°)	≥ 3.0×3.0	≥ 3.0×3.0
Effective pixels	256×256	1
Quantization/bits	≥10	≥10
SNR (dB, albedo is 9% and solar incident angle is 45°)	≥30	≥30
Spectral sampling interval/nm	5	
Power consumption/W	16.0	
Weight/kg	5.6	

2　月球表面探测应用验证

2.1　嫦娥三号应用

红外成像光谱仪安装在“玉兔号”月球车的前面，它由可见近红外成像光谱仪(450～950nm）和短波红外光谱仪光谱仪（900～2400nm）组成，实现我国基于声光可调滤光器分光的凝视型成像及光谱探测技术，以及超声电机驱动的月面定标防尘一体化集成技术的首次空间应用。“嫦娥三号”红外成像光谱仪于北京时间 2013 年 12 月 23 日上午 10 点 10 分首次开机进行探测及定标，随后获取了多组不同探测点的月壤光谱图像及光谱数据，用于巡视区矿物和化学成分综合研究。红外成像光谱仪月面获取的探测与定标图像，以及光谱数据为原始的光响应信号，经过数据处理后供科学应用。图 5 为红外成像光谱仪获取的月壤初步探测结果[5]，图（a）为单带灰度图像，图（b）为白圆为 ROI 的 DN 曲线平均值，图（c）为 SWIR 通道的 DN 曲线，图（d）为入瞳的理论光谱辐射，初步分析其光谱具有辉石与橄榄石矿物月壤的典型吸收峰特征。

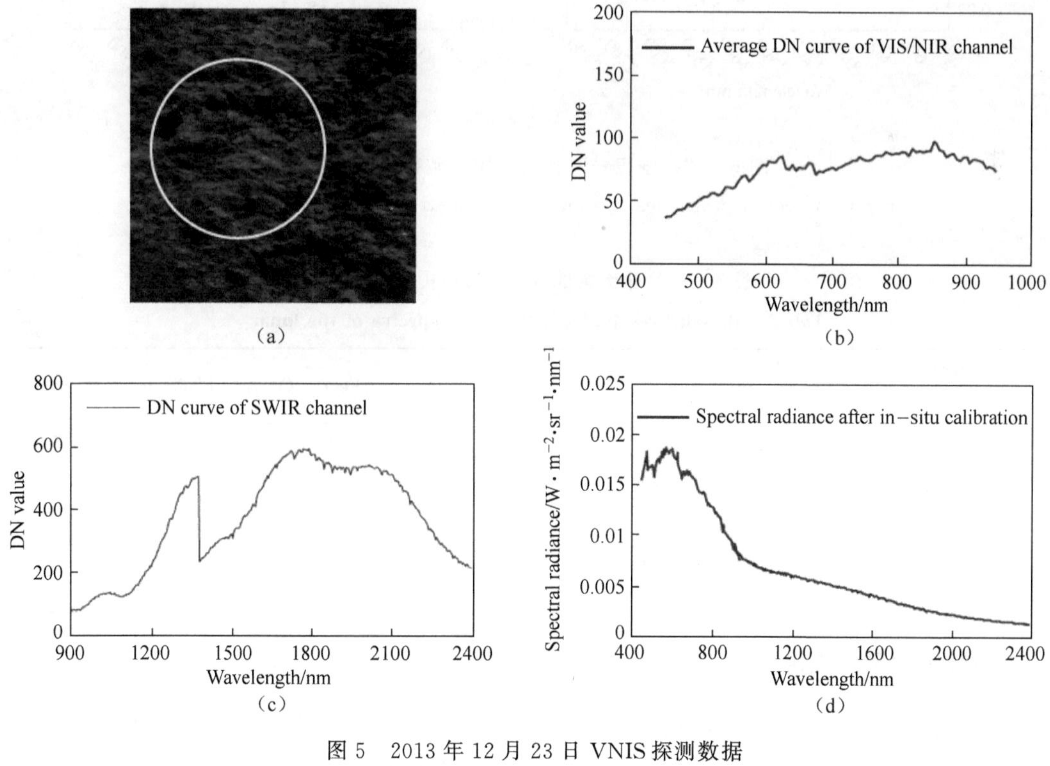

图 5　2013 年 12 月 23 日 VNIS 探测数据

Fig. 5　Detection data of VNIS on Dec. 23，2013

红外成像光谱仪分别于月面的 4 个位置进行了探测，获取了宝贵的科学数据，部分科学应用成果已形成论文在自然子刊（Nature. communication）[10]、美国科学院院刊（PNAS）杂志[11]、RAA[5]等发表。

如图 6 所示，凌宗成团队通过对“嫦娥三号”着陆地点及在着陆地点附近的新陨石坑中岩石和土壤的原位分析，揭示了该地区年轻月海玄武岩的化学成分特征（含有较为丰富的橄榄石和钛铁矿矿物）。此项研究成果不仅证实了之前遥感研究的推测，而且更为确切地表明橄榄石矿物成分趋向于富铁的矿物端元[10]。

林扬挺团队利用被修正的高斯模型（MGM）法从 4 个 VNIS 光谱数据中提取出了月球土壤的矿物模态组成，如表 3 所示，对其矿物和化学成分进行研究后发现大约 25 亿年前月球还泛滥着火山喷发的熔岩[12]，而科学家此前普遍认为月球在 31 亿年前就逐渐沉寂而少有地质活动了。

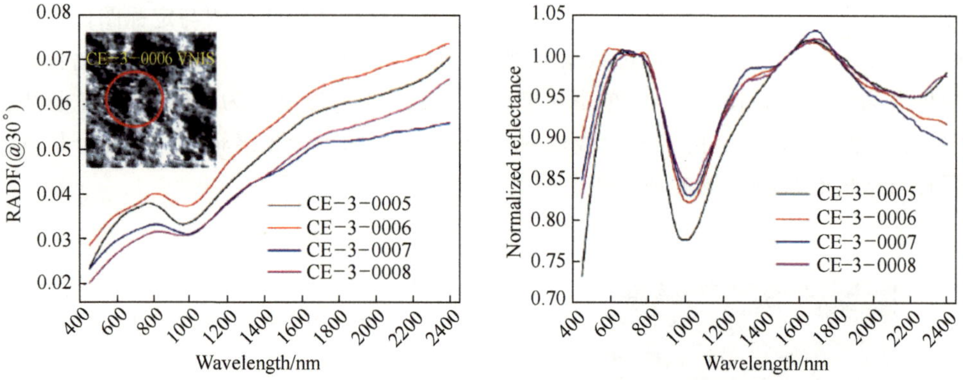

（a）　　　　（b）

图 6　“嫦娥三号”光谱性质和矿物化学特性

Fig. 6　Spectral properties and mineral chemistry of “Chang’e - 3” soils

表 3　从月球土壤的 VNIS 谱中提取的结果

Tab. 3　Decoded results from the VNIS spectra of the lunar

	Distance /m	R				ΘTi	TiO_2	ΘFe	FeO	OMAT	Plagioclase	Pyroxene
		450nm	415nm	750nm	950nm							
CD005	19.79	0.043	0.043	0.070	0.0611	1.175	5.8%	1.487%	18.8	0.312	15%	20.6%
CD006	32.06	0.033	0.033	0.030	0.057	1.154	5.3%	1.482%	18.7	0.157	16.6%	20.3%
CD007	38.72	0.027	0.027	0.025	0.047	1.195	6.4%	1.534%	19.5	0.158	17.5%	17.8%
CD008	40.89	40.89	0.032	0.029	0.051	1.274	9.0%	1.489%	18.8	0.098	16.3%	13%
Avg							6.6%		18.9	0.181	16.4%	17.9%

2.2　“嫦娥四号”应用

“嫦娥四号”红外成像光谱仪在“嫦娥三号”（CE-3）的基础上[13]，进一步开展月球背面巡视勘测任务。2019 年 1 月 4 日，“嫦娥四号”红外成像光谱仪于月球背面展开工作，获得了月球背面近距离高分辨率的光谱数据，实现了人类历史上首次对月球背面的软着陆原位探测。“嫦娥四号”（CE-4）红外成像是光谱仪前 3 个月昼共在 12 个探测点进行了探测，其中 8 个探测点进行了同步在轨定标（目前已工作 17 个月昼，获取了约 100 个探测点的数据)，如图 7 所示。基于“嫦娥四号”的探测数据研究结果，成功揭示了月球背面的物质组成，进一步丰富了人类对月球形成与演化的认识。

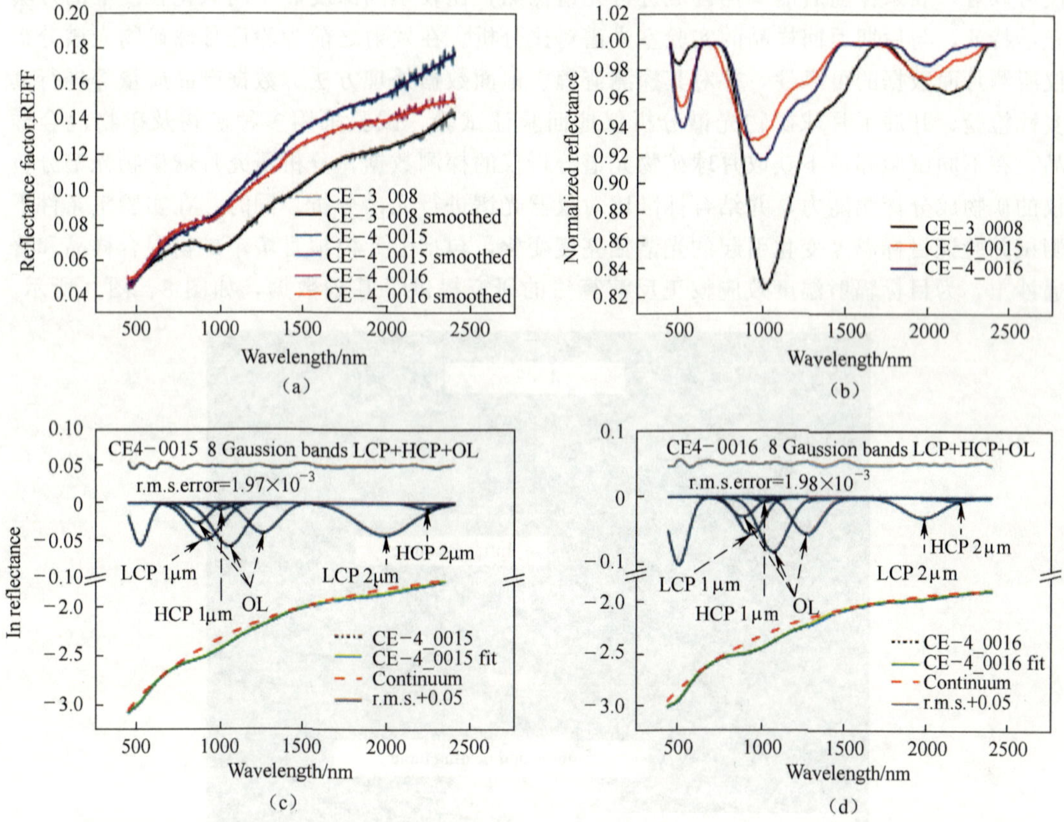

图 7　第一个月昼，“嫦娥四号”VNIS 获得的反射率光谱

Fig. 7　“Chang’e - 4” reflectance spectra acquired by the VNIS during the first lunar day

“嫦娥四号”着陆于南极-艾特肯（SPA）盆地，为探测月球内部的组成提供了难得的机会。“嫦娥四号”红外成像光谱仪科学数据研究成果之一（Nature 期刊）：李春来团队利用光谱初始观测结果推断出，月球背面存在以低钙辉石和橄榄石矿物为主的月球深部物质[13]，不仅为解答长期困扰国内外学者的有关月幔物质组成问题提供了直接证据，也为完善月球形成与演化模型提供了支撑。

林杨挺团队利用玉兔二号在第 4 个月昼测量的光谱，首次推导出覆盖可见光－近红外波段的月表原位光度函数，获得了更准确的月壤 FeO 含量和成熟度。研究发现月球风化层岩石的光谱在 1μm 和 2μm 处显示最大吸收，可以被用来区分低钙辉石和高钙辉石检测得出月壤中橄榄石存在的可能性[14]。

3　“嫦娥五号”地面验证

“嫦娥五号”月球矿物光谱分析仪突破了红外宽谱段的分光器性及探测器性能提升，将光谱范围从“嫦娥三号、四号”红外成像光谱仪的 450～2400nm 扩展到了 480～

3200nm，除了能探测月球表面主要矿物辉石、橄榄石等，还具备探测 3000nm 附近的羟基吸收峰特征的能力。此外，结合“嫦娥五号”月面取样的工作任务，月球矿物光谱分析仪可以对月表采样前后的采样区域进行光谱探测，比较不同深度、不同风化程度下的月壤光谱特征，与后期返回样品的实验室光谱对比分析。在发射之前为验证月球矿物光谱分析仪所测月面数据的可靠性，并对其探测流程、地面数据处理方法、数据产品质量等进行真实性检验，开展了月球矿物光谱分析仪地面验证试验。试验采用多种矿物及矿物混合样品，在不同试验环境下获取月球矿物光谱分析仪的探测数据，分析研究月球矿物光谱分析仪的矿物成分探测能力，并结合标准比对仪器光谱进行光谱分析。同时，在实验室条件下测试并验证目标温度变化引起的光谱辐亮度变化，包括七元模拟月壤单矿物混合样品和普通沙土，为目标辐射温度效应校正反演算法的研究提供了基础数据，如图 8、图 9 所示。

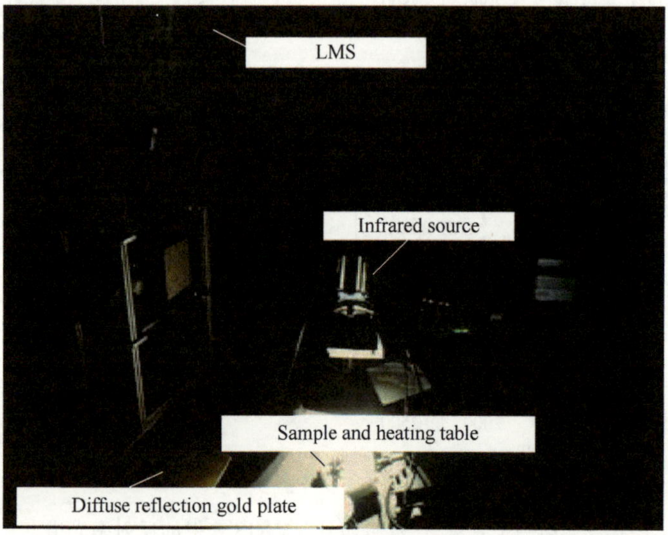

图 8　地面验证试验实物图

Fig 8　Photo of ground verification test

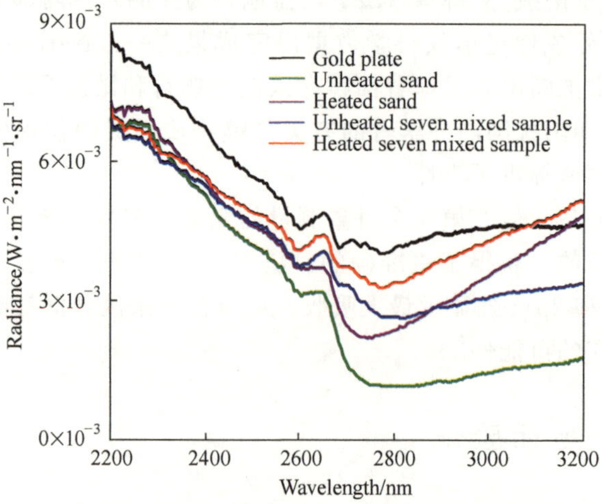

图 9　不同温度下金板/样品的光谱响应结果

Fig. 9　Spectral response result of the gold board/sample at different temperatures

实验表明月球矿物光谱分析仪具有较好的光谱数据质量，也验证了其光谱数据分析识别月球表面物质的矿物组成的能力，包括含水矿物，为月球的形成过程、月球地质演变及岩石一水交互作用的研究提供了数据支撑。

4　结论

月球表面原位光谱探测技术研究是围绕我国探月工程重大战略需求而开展的。该研究突破了凝视型时序扫描的新型声光光谱探测关键技术，解决了月球表面光照、温度变化的复杂环境中原位光谱高性能探测的技术问题及难点。据此研发了可适应表面恶劣环境的高性能、轻小型、高可靠的红外成像光谱仪，应用于“嫦娥三号”、“嫦娥四号”任务，在国际上率先成功实现了月球正面、背面原位光谱探测。目前，该技术进一步突破了红外宽谱段谱段的分光器性及探测器性能提升，完成的月球矿物光谱分析仪研制，将光谱范围从“嫦娥三号”、“嫦娥四号”红外成像光谱仪的 450～2400nm 扩展到了 480～3200nm，将为我国“嫦娥五号”任务提供新型仪器，继续不断应用于我国后续月球及深空探测任务。该技术也正推广应用于其他军、民领域，预计将不断产生经济及社会效益。

参考文献

[1] Ouyang Ziyuan, Zou Yongliao, Li Chunlai, et al. Bulletin of mineralogy, petrology and geochemistry [J] . Lunar Exploration and Containable Development for Human Society, 2003 (4): 328 - 333. (in Chinese)

[2] Ye Peijian, Peng Jing. Deep space exploration and its prospect in China [J]. Engineering Science, 2006 (10): 13 - 18 (in Chinese)

[3] 欧阳自远，邹永廖. 月球的地质特征和矿产资源及我国月球探测的科学目标 [J] 国土资源情报，2004 (1): 36 - 39.

[4] He Z P, Wang B Y, Lv G, et al. Visible and near - infrared imaging spectrometer and its preliminary results from the Chang'E - 3 project [J] The Review of Scientific Instruments, 2014, 85 (8): 1.

[5] Zhiping He, Binyong Wang, Gang Lv, et al. Operating principles and detection characteristics of the visible and near - Infrared imaging spectrometer in the Chang'e - 3 [J] . Research in Astronomy and Astrophysics, 2014, 14 (12): 1567 - 1577.

[6] Zhenchao Wang, Yunzhao Wu, Xiaomeng Zhang, et al. Mineralogy of northern nearside mare basalts [J] . Research in Astronomy and Astrophysics, 2019, 19 (4): 35 - 42.

[7] Wu Y, Hapke B Spectroscopic observations of the Moon at the lunar surface [J]. Earth & Planetary Science Letters, 2018, 484: 145 - 153.

[8] He Z, Li C, Xu R, et al The spectrometers based on AOTF for in - situ lunar surface measurement [J] . Journal of Applied Remote Sensing, 2018, 12: 2324437.

[9] He Zhiping, Shu, Rong, Wang JianYu Imaging spectrometer based on AOTF and its prospects in deep - space exploration application [J] . Proceedings of Spie the

International Society for Optical Engineering, 2011, 8196 (3): 819625.

[10] Ling Z, Jolliff B L, Wang A, et al. Correlated compositional and mineralogical investigations at the Chang'e－3 landing site [J] . Nature Communications, 2015, 6: 8880.

[11] Zhang J, Yang W, Hu S, et al. Volcanic history of the imbrium basin: A close－up view from the lunar rover Yutu [J] . Proceedings of the National Academy of Sciences, 2015, 112 (17): 5342－5347.

[12] He Zhiping, Li Chunlai, Xu Rui, et al. Spectrometers based on acousto－optic tunable filters for in－situ lunar surface measurement [J] . International Society for Optics and Photonics, 2019, 13 (2): 27502.

[13] Li C, Liu D, Liu B, et al Chang'E－4 initial spectroscopic identification of lunar far－side mantle－derived materials [J] . Nature, 2019, 569 (7756): 378－382 .

[14] Lin Honglei, He Zhiping, Yang Wei, et al. Olivine－norite rock detected by the lunar rover Yutu－2 likely crystallized from the SPA impact melt pool [J] . National Science Review, 2019 (14): 183.

嫦娥四号低频射电频谱仪 2C 数据对宇宙黑暗时代的探测能力分析

焦康[1]，王明远[2]，张同杰[1]，黄茂海[2]，平劲松[2]

1. 北京师范大学天文系，北京 100875
2. 中国科学院国家天文台，北京 100101

摘要： 嫦娥四号着陆器搭载的低频射电频谱仪（Low Frequency Radio Spectrometer，LFRS）已成功开展观测，每月把观测数据传回地面，并且把经过一定预处理后的 2C 级别的频谱数据释放给科学团队。旨在评估该级别数据是否满足探测宇宙黑暗时代 21cm 整体谱的要求，对生成的 2C 级别数据进行了长时间积分，并模拟仿真银河系前景射电辐射。通过对两套数据进行对比分析发现，2C 数据减弱 40～50dB，几乎不包含银河系前景的远场辐射信号，这种大幅度的减弱进而影响了该载荷对宇宙黑暗时代的探测能力。

关键词： 低频射电频谱仪；嫦娥四号；黑暗时代

Analysis of the Capability of Chang'E－4 Low Frequency Radio Spectrometer 2C Data in Detecting the Dark Ages

Jiao Kang[1]，Wang Mingyuan[2]，Zhang Tongjie[1]，Huang Maohai[2]，Ping Jinsong[2]

1. Department of Astronomy，Beijing Normal University，Beijing 100875，China，Email：tjzhang@bnu. edu. cn
2. National Astronomical Observatories，Chinese Academy of Sciences，Beijing 100101，China，Email：jsping@bao. ac. cn

Abstract： The low－frequency radio spectrometer carried by Chang'E－4 lander has successfully carried out observation. The observational data are continuously sent back to the ground data center monthly. After certain pretreatment，the 2C level spectrum data is released to scientific teams. The main purpose of this article is to evaluate whether this level of data meets the requirements for detecting the sky - averaged 21cm global spectrum of the cosmic dark ages. We make a long - term integration of the 2C level data, and simulate the reception of the Galactic foreground radio radiation by the lander. Through the comparison and analysis of the two sets of data, we find that the 2C data has been reduced by 40 - 50dB, which almost does not contain the far - field radiation signal of the Galactic foreground. Such a large reduction will affect the detection ability of the payload for the dark ages.

Keywords: low frequency radio spectrometer; Chang'E - 4; dark ages

现代宇宙学中，基于对宇宙微波背景（Cosmic Microwave Background，CMB）辐射温度图中的小扰动的观测，我们建立了宇宙大尺度结构形成理论。但是对于从宇宙微波背景辐射最后散射面（$z\sim1100$）到宇宙完全电离（$z\sim6$）这段漫长的宇宙演化历史，很多理论尚未经过观测验证。一般认为，从最后散射面到宇宙中最早的发光天体第 1 代恒星形成（$z\sim 30$）之前，宇宙处于完全透明的黑暗时代（Dark Ages，DA）。第 1 代恒星产生之后，宇宙黎明（Cosmic Dawn，CD）开始，期间第 1 代恒星产生的辐射显著改变了宇宙的辐射背景、气体的电离度和温度等。之后，随着第 1 代星系的批量形成（$z\sim15$），再电离时期（Epoch of Reionization，EoR）开始。在星系和类星体等发出的电离光子的驱动下，星系际介质（Intergalactic Medium，IGM）快速电离和加热，一直到 $z\sim6$ 左右，几乎所有的星系际介质再次回到电离状态，宇宙中残余的中性氢（Neutral Hydrogen，HI）主要在星系内部，再电离时期结束。

对于黑暗时代、宇宙黎明和再电离时期的观测是低频射电天文研究的重要内容。在过去 20 年里，地基低频射电天文学的研究主要在 30MHz 以上，在该频率以下，无线电波受到电离层影响而严重畸变，在 10MHz 以下则几乎完全被反射。因而，对于理论上位于 30MHz 以下的黑暗时代吸收谷的观测，只能通过空间射电望远镜进行。月球的背面为射电天文观测提供了一个独特的射电静谧区，它不受地面射频干扰，且电离层非常薄。作为空间低频射电天文的探路者，搭载于嫦娥四号着陆器的低频射电频谱仪已于 2019 年 1 月 3 日成功着陆，并观测获得了首批数据。本文评估 2C 级数据对黑暗时代信号的探测能力。

1 21cm 整体谱

中性氢原子存在电子自旋和核自旋平行（0 能级）与反平行（1 能级）的超精细能级结构，能级跃迁会吸收或辐射 21cm 波长谱线。我们定义 $T_* \equiv h_c/k_B\lambda_{21cm}=0.068\text{K}$，两个能级氢原子的数密度比可以由中性氢的自旋温度 T_S描述，

$$n_1 / n_0 = (g_1 / g_0)\exp(-T_* / T_S), \tag{1}$$

统计简并度比 $g_1/g_0=3$。根据辐射转移理论可以得到 21cm 线微分亮温度

$$\begin{aligned}\delta T_b &= \frac{T_S - T_R}{1+z}(1-e^{-\tau_\nu}) \\ &\approx \frac{T_S - T_R}{1+z}\tau\end{aligned}$$

$$\approx 21x_{\mathrm{HI}}(1+\delta_{\mathrm{b}})\left(\frac{\Omega_{\mathrm{b}}h^2}{0.023}\right)\left(\frac{0.15}{\Omega_{\mathrm{m}}h^2}\frac{1+z}{10}\right)^{1/2}\left(\frac{T_{\mathrm{S}}-T_{\mathrm{R}}}{T_{\mathrm{S}}}\right)\left[\frac{\partial_{\mathrm{r}}v_{\mathrm{r}}}{(1+z)H(z)}\right]\mathrm{mK}, \quad (2)$$

其中，T_{R}为背景辐射温度；τ_ν为谱线光深；z 为宇宙学红移；x_{HI}为中性氢电离度；δ_{b}为重子密度涨落比分数；Ω_{b}和 Ω_{m}分别为重子物质与暗物质的密度参量；$H(z)$ 为哈勃参量，约化哈勃参量 $h=\frac{H(z)}{100}$ ；$\partial_{\mathrm{r}}v_{\mathrm{r}}$ 为视线方向的速度梯度[1]。只有当 T_{S} 偏离 T_{R}，才存在可探测的 21cm 信号，即当 $T_{\mathrm{S}}>T_{\mathrm{R}}$ 时表现为辐射，当 $T_{\mathrm{S}}<T_{\mathrm{R}}$时表现为吸收。

自旋温度主要由 3 个物理过程共同决定：(1) 从（向）背景辐射吸收（辐射）21cm 光子；(2) 与其他中性原子及电子的碰撞；(3) 莱曼 α 光子的共振散射导致通过中间激发态的自旋翻转。这些过程远快于退激发时间，所以通过平衡态方程可以近似给出自旋温度

$$T_{\mathrm{S}}^{-1}=\frac{T_\gamma^{-1}+x_\alpha T_\alpha^{-1}+x_{\mathrm{c}}T_{\mathrm{K}}^{-1}}{1+x_\alpha+x_{\mathrm{c}}}, \quad (3)$$

其中，T_γ为光子的环境温度，近似地 $T_\gamma=T_{\mathrm{CMB}}$；$T_\alpha$为莱曼 α 辐射场色温度；T_{K}为气体动力学温度；x_α与 x_{c}为相应过程的耦合系数。当总耦合系数 $x_{\mathrm{tot}}\equiv x_\alpha+x_{\mathrm{c}}\gtrsim 1$ 时，自旋温度强耦合于气体温度，当 $x_{\mathrm{tot}}\ll 1$ 时趋于背景辐射温度。

从观测角度，背景辐射源主要有两类：宇宙微波背景辐射或射电噪点源，后者对应于“21cm 森林”的观测，本文主要讨论前者。当 $T_{\mathrm{R}}=T_{\mathrm{CMB}}$时，21cm 信号表现为宇宙微波背景辐射黑体谱在射电波段的微小扰动。对 21cm 信号的观测分为两类：(1) 考虑空间扰动的功率谱测量或成像测量；(2) 对全天所有方向平均的整体谱测量，即测量

$$\overline{\delta T_{\mathrm{b}}}(\nu)=\int \mathrm{d}\Omega\delta T_{\mathrm{b}}(\hat{r},\ \nu). \quad (4)$$

功率谱测量需要望远镜有足够高的空间分辨率，通常由射电阵列观测。通过嫦娥四号中继通信卫星上搭载的中荷低频射电探测仪（The Netherland - China Low Frequency Explorer，NCLE）与低频射电频谱仪进行空间低频射电干涉测量[2]，有望探测到 21cm 信号的功率谱。整体谱的观测可通过单天线实现，本文评估低频射电频谱仪的 2C 级数据对于宇宙黑暗时代 21cm 信号整体谱的探测能力。

21cm 信号蕴含丰富的信息，它由宇宙学参数与天体物理学参数共同决定。考虑到星系际介质密度、电离状态和热状态的演化以及天体的形成，不同阶段的 21cm 整体谱信号表现出不同的特征，不同模型（参数）的信号差异也较大，其理论值可由纯流体动力学模拟[3]、半解析模拟[4]或基于神经网络的仿真器[5]给出。图 1 给出不同模型下的 21cm 整体谱示例（详见文献［6］），横坐标分别为中性氢谱线的观测频率及对应的宇宙学红移，纵坐标为 21cm 线微分亮温度。图中的黑色虚线对应标准宇宙学模型；灰色区域是重子物质与暗物质相互作用的参数化模型；阴影区域分别是与“边缘”实验（Experiment to Detect the Global Epoch of reionization Signature，EDGES）观测结果[7]一致的 1σ 和 2σ 置信区间；蓝色曲线假设早于标准模型的绝热冷却；红色曲线冷却更早且冷却率更低；紫色曲线假设冷却速率非单调下降，而是存在一个过冷却时期。现有模型普遍认为 $z>30$ 存在黑暗时代吸收谷，$30>z>15$ 存在宇宙黎明吸收谷，$15>z>6$ 存在再电离时期发射线，信号大小为 $10\sim10^2$ mK 量级。

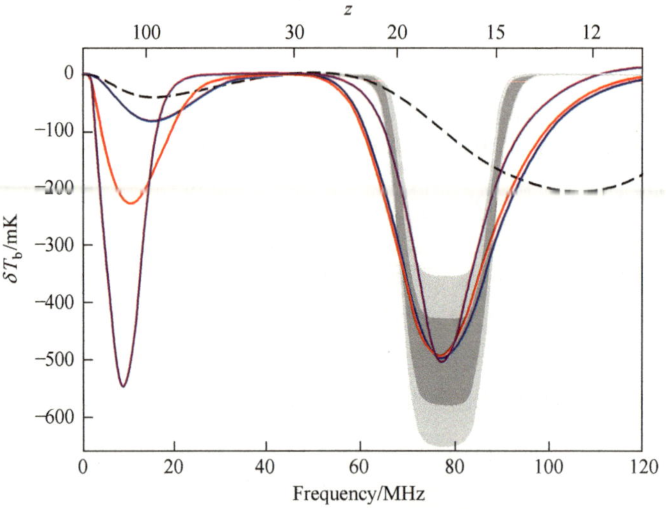

图 1　不同模型的 21cm 整体谱示例[6]

Fig. 1　Examples of 21cm global spectrum of different models[6]

2　前景辐射

21cm 信号观测充满挑战，主要在于将微弱信号从高数个量级的前景辐射中分离。在低频射电波段，银河系的同步辐射占主导，一个合理的亮温度模型近似为

$$T_{\text{syn}}(\nu)=A_{\text{syn}}\left(\frac{\nu}{\nu_0}\right)^{-\alpha_{\text{syn}}-\Delta\alpha_{\text{syn}}\ln(\nu/\nu_0)},\tag{5}$$

其中，ν 为观测频率；谱指数 $\alpha_{\text{syn}}\approx2.8$；谱指数跑动 $\Delta\alpha_{\text{syn}}\approx0.1$；$\nu_0$ 为（任意的）中心频率，与特定频率的总体振幅简并[8]。当 $\nu_0=150\text{MHz}$ 时，$A_{\text{syn}}=335.4\text{K}$。对于银河系的自由一自由辐射（在 150MHz 处约占总体前景辐射的 1%），我们同样采用跑动幂率谱近似，

$$T_{\text{ff}}(\nu)=A_{\text{ff}}\left(\frac{\nu}{\nu_0}\right)^{-\alpha_{\text{ff}}-\Delta\alpha_{\text{ff}}\ln(\nu/\nu_0)},\tag{6}$$

其中，$\alpha_{\text{ff}}\approx2.15$；$\Delta\alpha_{\text{ff}}\approx0.01$；$A_{\text{ff}}=33.5\text{K}$。此外，明亮射电源、未知点源以及尘埃辐射等组成的前景污染同样不容忽视。

然而，由于星系际介质在视线上的密度、电离状态和温度的不均匀性，21cm 信号在光谱上是高度变化的，而非热前景辐射光谱则是平滑的，这为分离前景辐射与信号提供了便利，但前提是需要消除任何仪器效应造成前景扰动带来的污染[9]。

3　2C 数据频谱

2019 年 1 月 3 日，嫦娥四号着陆器成功着陆月球背面南极一艾特肯盆地冯·卡门撞击坑。嫦娥四号着陆器搭载的低频射电频谱仪安装在着陆器顶板，包括 4 根接收天线、前置放大器和电子学单元。4 根接收天线中，A，B 和 C 天线互相正交且长度均为 5m，D 天线

为长度 20cm 的短天线。低频射电频谱仪的工作频率为 100kHz～2MHz 和 1～40MHz 两个频段，频率分辨率为 5kHz 和 100kHz。在低频射电频谱仪的数据处理过程中，首先对 4 根天线的原始电压观测量进行快速傅里叶变换，根据一定时间间隔内每根长天线与天线 D 频谱幅值间的比值确定对消系数，再以长天线频谱幅值减去天线 D 的频谱幅值乘以对消系数，得到 2C 观测数据[10]。

在低频射电波段，辐射亮温度定义为

$$T_b = \frac{I_\nu}{2k}\frac{c^2}{\nu^2}, \tag{7}$$

其中，k 为玻尔兹曼常数；c 为光速；ν 为谱线频率；辐射强度 I_ν 由单位立体角的流量密度决定，即

$$I_\nu = \frac{S_\nu}{d\Omega}. \tag{8}$$

根据有效面积的定义[11]，我们可由谱功率 P_ν 得到总流量密度为

$$S_\nu = \frac{2P_\nu}{\langle A_{eff}\rangle}, \tag{9}$$

其中，P_ν 由每个频率通道的接收功率除以谱频分辨率得到。对于所有的无损失天线，平均有效接收面积正比于波长 λ^2[11]，即

$$\langle A_{eff}\rangle = \frac{\lambda^2}{4\pi}. \tag{10}$$

通常低频单机子天线在高频（High Frequency，HF）频带的天线效率 η 取值的动态范围 10%～80%，我们将 2C 数据基线功率频谱转化为亮温度频谱，即

$$\overline{T}_b = \frac{4\pi P_\nu}{\eta k}. \tag{11}$$

基于上述过程，我们对 2019 年 1～4 月的 2C 级别数据做了长时间积分，并分别给出 A，B 和 C 天线测量的亮温度频谱上下限。结果如图 2 中阴影区域，横纵坐标分别表示观测频率和亮温度。假设天线效率 10%～80%，图 2 给出 A，B 和 C 天线测量的亮温度上下

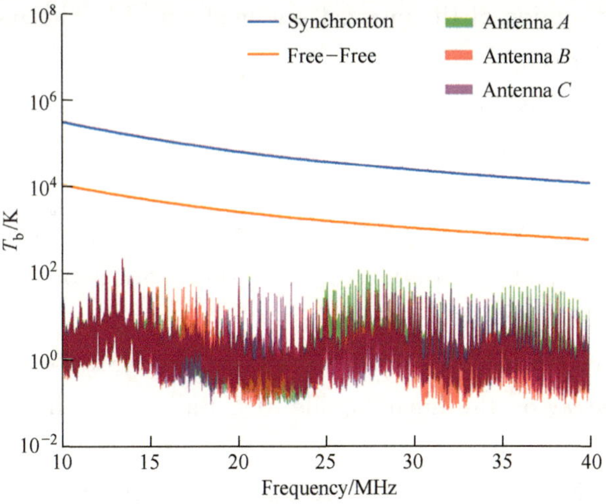

图 2　低频射电频谱仪 2C 数据（2019 年 1～4 月）基线频谱

Fig. 2　The baseline spectrum of the Low Frequency Radio Spectrometer 2C data (Jan. - Apr. 2019)

限（阴影区域），蓝色和橙色曲线为银河系同步辐射和自由-自由辐射的理论值（分别由（5）式和（6）式计算得出）。图 2 同时给出银河系前景辐射的理论曲线用于比较，结果显示，2C 数据基线频谱低于前景辐射 4～5 个数量级。

4 结论与讨论

上述结果表明，2C 数据基线频谱低于银河系前景辐射 4～5 个数量级，因而几乎不包含前景辐射信号，也无法使用前景辐射的仿真数据对 2C 数据进行定标校准。因此，我们需要更低级别的数据，从中寻找造成该结果的原因，并对更低级别的数据进行多种数据处理方法的对比研究，以便进一步评估嫦娥四号低频射电频谱仪对宇宙黑暗时代的探测能力。

致谢

我们感谢张志嵩在数据分析部分的深度讨论和建议，感谢陈林杰和张墨关于文稿的修改建议，也感谢所有参与本项目并做出贡献的同事们。

参考文献

[1] PRITCHARD J R, LOEB A. 21cm cosmology in the 21st century [J]. Reports on Progress in Physics, 2012, 75 (8): 086901.

[2] 武宇翔，温卫斌，平劲松，等. 利用 Halo 轨道仿真开展嫦娥四号中继星与月面设备干涉基线的研究 [J]. 天文研究与技术，2021，18（1）：69－76.
WU Y X, WEN W B, PING J S, et al. Research of interferometry baselines between CE－4 rover and lunar facility using Halo orbit simulation [J]. Astronomical Research & Technology, 2021, 18 (1): 69－76.

[3] TRAC H, CEN R, LOEB A. Imprint of inhomogeneous hydrogen reionization on the temperature distribution of the intergalactic medium [J]. The Astrophysical Journal, 2008, 689 (2): L81.

[4] MESINGER A, FURLANETTO S, CEN R. 21CMFAST: a fast, seminumerical simulation of the high－redshift 21－cm signal [J]. Monthly Notices of the Royal Astronomical Society, 2011, 411 (2): 955－972.

[5] COHEN A, FIALKOV A, BARKANA R, et al. Emulating the global 21－cm signal from Cosmic Dawn and reionization [J]. Monthly Notices of the Royal Astronomical Society, 2020, 495 (4): 4845－4859.

[6] BURNS J O, HALLINAN G, LUX J, et al. FARSIDE: a low radio frequency interferometric array on the lunar farside [J]. Bulletin of the AAS, 2019, 51 (7): 1－13.

[7] BOWMAN J D, ROGERS A E E, MONSALVE R A, et al. An absorption profile centred at 78 megahertz in the sky－averaged spectrum [J]. Nature, 2018, 555 (7694): 67－70.

[8] WANG X M, TEGMARK M, SANTOS M G, et al. 21cm tomography with foregrounds [J]. The Astrophysical Journal, 2006, 650: 529-537.
[9] KERN N S, DILLON J S, PARSONS A R, et al. Absolute calibration strategies for the Hydrogen Epoch of Reionization Array and their impact on the 21cm power spectrum [J]. The Astrophysical Journal, 2020, 890 (2): 122.
[10] 张韬，苏彦. 嫦娥四号低频射电频谱仪降低背景噪声方法的研究 [J]. 天文研究与技术，2019，16 (3)：312-320.
ZHANG T, SU Y. Research of the method for reducing background of very low frequency radio spectrumon Chang'E-4 [J]. Astronomical Research & Technology, 2019, 16 (3): 312-320.
[11] CONDON J J, RANSOM S M. Essential radioastronomy [M]. Princeton: Princeton University Press, 2016: 75-76.

嫦娥四号冯·卡门撞击坑着陆区地形及月面遮挡分析

李飞[1]，吴波[2]，杨眉[1]，赵洋[1]，张熇[1]，吴学英[1]，许映乔[1]，叶青[1]

1. 北京空间飞行器总体设计部，北京 100094
2. 香港理工大学土地测量与地理资讯学系，香港 999077

摘要： 嫦娥四号任务实现了人类首次月球背面软着陆。在任务设计阶段，为制定着陆器动力下降控制策略和设计月面工作程序，根据嫦娥二号、LRO 等月球卫星探测得到的图像、激光测高等数据，对冯·卡门撞击坑着陆区的高程起伏、坡度、撞击坑和石块分布密度，以及着陆区内地形对光照和中继链路遮挡进行了详细分析。本文对嫦娥四号冯·卡门撞击坑着陆区的地形地貌情况进行了总结，对地形地貌及遮挡分析所使用的方法进行了详细说明，同时将嫦娥四号着陆区同嫦娥三号虹湾着陆区的地形地貌进行了对比分析，最后对实际在轨着陆点进行了精确定位并分析了着陆点周围地形的遮挡情况。所形成的系统性分析方法在轨得到了全面验证，并为后续月球及深空探测软着陆任务奠定了坚实的基础。

关键词： 嫦娥四号；冯·卡门撞击坑；着陆区；地形地貌；遮挡

Topographic and occlusion analysis of the Chang'E－4 landing site in the Von Kármán crater

LI Fei[1], WU Bo[2], YANG Mei[1], ZHAO Yang[1], ZHANG He[1], WU XueYing[1], XU YingQiao[1] & YE Qing[1]

1. Beijing Institute of Spacecraft System Engineering, China Academy of Space Technology, Beijing 100094, China
2. Department of Land Surveying and Geo－Informatics, The Hong Kong Polytechnic University, Hong Kong 999077, China

Abstract: The Chang'E－4 lunar probe successfully made a soft landing on the far side of the Moon, the first such landing in the history. To determine the powered－descent control strategy and operational programs for the lunar surface, a detailed analysis was carried out on the key factors of the surface topography in the Von Kármán cater. It included the elevations, slopes, distribution densities of craters and rocks, as well as the terrain occlusions to sun illumination and communication signals between the probe and the relay satellite. The analysis was based on images collected by the Chang'E－2 satellite and the images and laser range measurements acquired by the Lunar Reconnaissance Orbiter. This paper presents the topographic and occlusion analysis results of the Chang'E－4 landing site in the Von Kármán crater. The methods for topographic and occlusion analysis are described in detail. This paper also presents a comparative study of the topographic features in the Chang'E－4 landing area and the Chang'E－3 landing area in Sinus Iridum, on the near side of the Moon. Finally, the precise landing location of Chang'E－4 is presented, and the above－mentioned terrain occlusions are analyzed. The analytical methods described in this paper have demonstrated their effectiveness in real operations and can benefit future missions involving soft landings on celestial bodies.

Keywords: Chang'E－4; Von Kármán crater; landing site; topography; occlusion

1　引言

2019 年 1 月 3 日 10 时 26 分，嫦娥四号探测器成功着陆于月球背面的南极-艾特肯盆地（South－Pole Aitken Basin，SPA）的冯·卡门（Von Kármán）撞击坑内，实现了人类首次月球背面软着陆，并相继开展了就位与巡视探测，获取了大量的数据并通过中继星将探测数据传输回地球[1]。

SPA 是太阳系中已知的规模最大、最古老的撞击盆地，具有极高的科学研究价值[2]，其中心纬度位于 40°～60°S，中心经度位于 180°附近，直径达 2000～2600km，平均深度约 12km。SPA 的撞击坑较大且分布密集，地势比较崎岖，但在大型的撞击坑底部以及部分月海地区也存在相对较平坦的区域。

冯·卡门撞击坑位于 SPA 的中西部，属于 SPA 内的典型地貌类型，其内部的月球物质成分和地质年代具有明显的代表性。冯·卡门撞击坑底部地形相对平坦，嫦娥四号着陆区选择在其坑底东南部区域。

本文系统地总结了冯·卡门撞击坑内着陆区的地形地貌特点，对坡度、撞击坑和石块分布密度情况进行了分析；并结合嫦娥四号探测器和“鹊桥”号中继星的轨道设计，对着陆区内光照和测控遮挡情况进行了详细分析；在嫦娥四号成功着陆后，对实际着陆点进行了精确定位并分析了着陆点周围地形的遮挡情况。本文还对嫦娥四号月背着陆区与嫦娥三号着陆的月球正面的虹湾地区进行了对比分析。

2　冯·卡门撞击坑地形地貌概况

冯·卡门撞击坑位于 SPA 盆地的中部，其地质年代属于前酒海纪，中心坐标为(44.8°S，175.9°E)，平均直径约 186km，东西向约 200km，南北向约 160km，平均海拔约－6km。坑底中部存在一个中央峰，长约 45.5km，宽约 13.7km，高约 1.6km；坑底南部地形起伏平缓，撞击坑数量相对较少；北部地形较为复杂，存在多个较大撞击坑[3]，如图 1 所示。

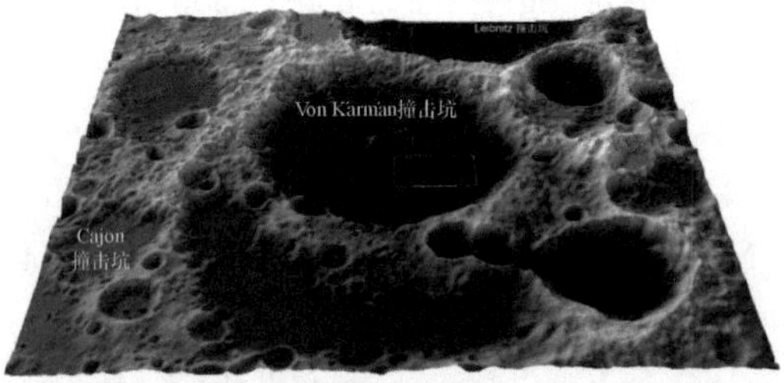

图 1　（网络版彩图）冯·卡门撞击坑示意图（高程放大 2 倍效果图）。方框所示为嫦娥四号预选着陆区

Fig. 1　(Color online) The elevation map of Von Kármán impact crater (elevation double magnification). The square shows the primary landing area of Chang'E－4

3　基于数字地形模型及影像的地形分析方法

3.1　数据来源简介

3.1.1　月球勘测轨道器数据简介

美国 NASA 于 2009 年 6 月成功发射了月球勘测轨道器（lunar reconnaissance orbiter，LRO）。LRO 上载有一个激光测高仪（lunar orbiter laser altimeter，LOLA）与广角相机（wide－angle camera，WAC）和窄角相机（narrowangle camera，NAC）两个数字相机，用来收集月球表面地形数据与影像数据。

LRO LOLA 是一个五点 X－模式脉冲激光测高仪，可以同时量测从轨道器到月球表面 5 个点的距离。五点模式使得每一轨 LOLA 数据包括 5 个相邻的测距轨迹，两两之间横向相隔 10～12m，总计覆盖 50～60m 的宽度范围（图 2）。这五点测距模式专为精准确定月表局部范围（50～60m 的着陆区域）的地形而设计，通过直接五点距离量测获取局部范围的坡度与表面粗糙度信息。

由于 LRO 轨道的定轨精度高，LOLA 数据经轨间交叉点平差处理后的几何精度高，在高程方向约为 0.5m，平面约为 25m[4,5]。LOLA 数据是目前世界各国月球探测任务所获取的地形数据中精度最高、覆盖最密集的激光测高数据[5,6]。

LRO 的 WAC 和 NAC 分别获取空间分辨率为 100 和 0.5m 的月球影像。LRO WAC 影像覆盖全月，LRO NAC 影像在月球的南北两极提供完整的立体覆盖，但在其他区域仅

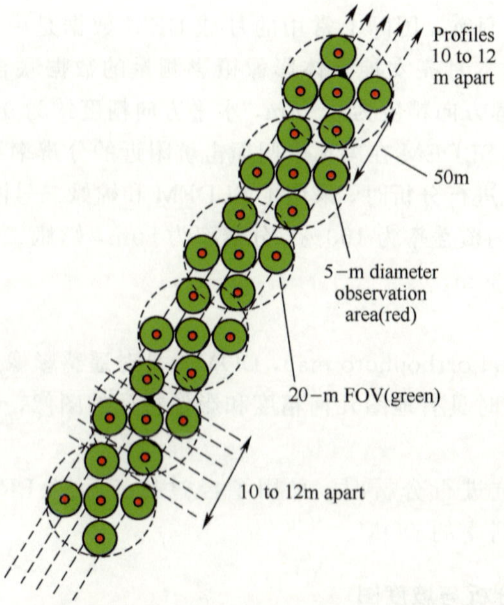

图 2　（网络版彩图）LRO LOLA 五点 X-模式激光测距[4]

Fig. 2　(Color online) LRO LOLA 5-point X-mode laser ranging[4]

覆盖较小的范围。LRO NAC 影像的单轨影像大小为 5000 像素×50000 像素，覆盖 2.5km×25km 的范围。LRO NAC 对阿波罗时代美国和苏联安置在月表的多个激光反射器与实验装置进行了摄像，通过比较其几何位置发现，LRO NAC 影像的平面定位精度（geo-reference）达到 20m[7]。

3.1.2　嫦娥二号数据简介

嫦娥二号的 CCD（charge coupled device）相机是一个包含两条线阵 CCD 的推扫式相机，两条线阵 CCD 分别组装在一个统一的焦平面上，沿飞行方向同时扫描获取前视和后视的影像构成立体像对，相机的焦距为 144.3mm. 前视影像的视角为 8°，后视影像的视角为 17.2°，两者构成交会角为 25.2°的立体像对[8]。

嫦娥二号有两种轨道高度，其一为 100km×100km 的圆形轨道上，在此轨道高度获取的影像的分辨率为 7m；其二为 100km×15km 的椭圆轨道，在此轨道高度获取的影像的分辨率优于 1.5m（未覆盖月球背面区域）。嫦娥二号单轨影像大小为 6144 像素×60640 像素，对应 7m 分辨率的影像覆盖范围约为 43km× 424km。

利用嫦娥二号影像的内方位元素（焦距、像主点位置等）与外方位元素（相机投影中心在月心坐标系的位置与姿态），基于摄影测量的方法，可以由前视与后视影像组成的立体像对实现三维量测，并进而获得数字地形信息，如数字高程模型等。

3.2　数字高程模型与数字正射影像

3.2.1　数字高程模型

数字高程模型（digital elevation model，DEM）在平面上即为一张规则的格网，格网中单元格的大小即为 DEM 的分辨率。DEM 的数据结构中，各单元格角点的平面位置由格网的起始点与单元格在格网中的相对位置来确定，其高程则由前述摄影测量或激光测高得到的三维信息内插得到。目前，国际上常用的月球 DEM 数据是由 LRO 上搭载的 LOLA 获取的激光测高数据与日本月亮女神立体影像摄影测量的数据联合生成的数字高程模型（SLDEM2015）[9]，其高程方向精度为 3～4m，水平方向精度约为 90m，有效分辨率为 512 PPD（pixel per degree），SLDEM 在冯·卡门撞击坑附近的分辨率约为 50m/像素。

在对冯·卡门撞击坑进行分析时，采用了 SLDEM 和嫦娥二号图像生成的 DEM 数据，其中，SLDEM 在着陆区内覆盖率为 100%，分辨率为 50m，嫦娥二号 DEM 在着陆区内覆盖率为 100%，分辨率为 30m。

3.2.2　数字正射影像

数字正射影像（digital orthophoto map，DOM）是对遥感影像进行数字微分纠正和镶嵌后的数字影像，它是同时具有地图几何精度和影像特征的图像，可直接利用其进行几何量测与特征提取。

在对冯·卡门撞击坑进行分析时，采用了经过校正后的 LRO NAC 1m 分辨率的 DOM 和嫦娥二号 7m 分辨率的 DOM。

3.3　基于 DEM 的坡度分析与坡度图

基于 DEM，可以有效分析 DEM 范围内的坡度情况，生成与 DEM 范围对应的坡度图。

坡度的计算方法[10]：以图 3（a）中 DEM 格网上任一格网点处的坡度计算为例，以该格网点（5 号点）为中心，取一定大小的坡度计算窗口，如图 3（a）所示为 2×2 的坡度计算窗口，若该 DEM 的分辨率为 20m，则该坡度计算窗口的大小为 40m×40m。坡度计算窗口也称为坡度计算的基线（Baseline 或者 Scale），对应于如图 3（a）所示的坡度计算基线长为 40m。图 3（a）中坡度计算窗口包含 9 个格网点的高程值，分别用 Z_1，…，Z_9 代表这 9 个格网点的高程值，则其 X 方向的坡度与 Y 方向的坡度可以通过式（1）和（2）计算得到：

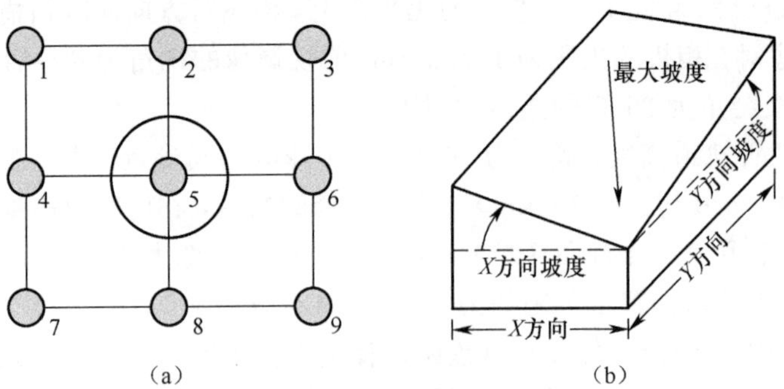

（a）　　（b）

图 3　基于 DEM 的坡度分析示例

（a）DEM 规则格网点；（b）坡度计算三维示例

Fig. 3　Examples of slope analysis based on DEM

(a) DEM regular grid points; (b) 3D examples of slope calculation

$$X_{坡度} = (Z_3 + 2Z_6 + Z_9 - Z_1 - 2Z_4 - Z_7)/8D \tag{1}$$

$$Y_{坡度} = (Z_1 + 2Z_2 + Z_3 - Z_7 - 2Z_8 - Z_9)/8D \tag{2}$$

式中，D 为单元格的大小（即 DEM 的分辨率，如 20m）。利用 X 方向与 Y 方向的坡度值，可以由式（3）计算得到该坡度计算窗口范围内的最大坡度值 S：

$$\tan S=\sqrt{X_{坡度}^2+Y_{坡度}^2} \tag{3}$$

对 DEM 中的每一个格网点都进行上述计算，可得到以每一个格网点为中心，坡度计算窗口范围内的最大坡度，将各格网点处的坡度值取代 DEM 中各格网点处的高程值，即可以生成与 DEM 对应的坡度图。

3.4 基于 DEM 及 DOM 的撞击坑、石块识别

基于 DEM 和 DOM，可以有效识别、分析撞击坑和石块。

撞击坑的识别步骤如下[6]：

（1）用机器学习的方法基于 DOM 进行撞击坑自动提取[11]；

（2）在格网（格网单元格的大小为撞击坑的最小识别尺寸）的辅助下进行人工检查，查漏补缺；

（3）撞击坑三维信息从对应 DEM 上自动获取。

石块的识别基于 LRO NAC 1m 分辨率的 DOM 来进行，其步骤如下[12]：

（1）根据 DOM 图像中光照方向的灰度梯度变化，检测提取属于石块的像素；

（2）提取石块像素的边界，并圆化拟合输出；

（3）对比 DOM，获得石块位置信息及其外包络直径信息；

（4）人工检查，删除可能错误的提取结果，如图像噪点等因素，添加可能遗漏的石块。

3.5 基于 DEM 的地形遮挡分析

由于冯·卡门撞击坑内地形高低起伏，撞击坑边缘或着陆点周围复杂的地形可能导致在月面工作的着陆器和巡视器存在光照和通信被遮挡的可能，如发生遮挡，对任务影响较大，甚至带来灾难性的风险。因此需要开展遮挡分析，以评估月面工作的风险，并制定相应的应对策略。

遮挡分析主要是指地形的高低起伏对着陆点所造成的地形遮挡情况的分析。为准确地描述着陆点的遮挡情况，本文引入了“遮挡角”作为度量指标，定义为以着陆点（由于着陆前着陆点未知，因此应覆盖着陆区内所有点）为起点，在某一方位的高程剖面上的每一点为终点，计算起点与终点的连线相对于水平面的仰角，仰角最大值称为该方位上的遮挡角，如图 4 所示。

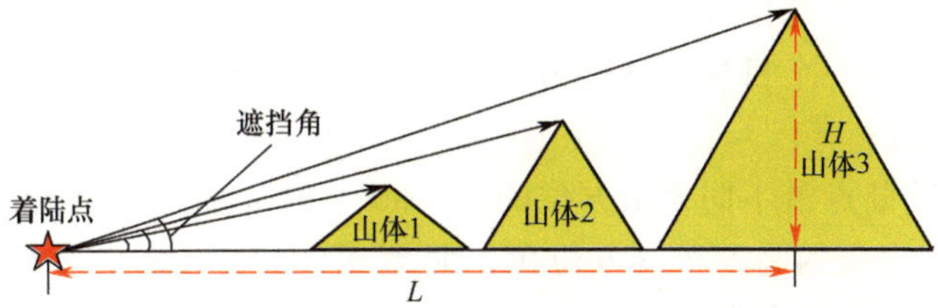

图 4　（网络版彩图）着陆点某一方位“遮挡角”定义示意图

Fig. 4　(Color online) The schematic of the definition of “occlusion angle” in an orientation of the landing site

假设冯·卡门撞击坑内某一点为着陆点，按 1°的方位角间隔，着陆点周围 360°遮挡角分布三维显示如图 5（a）所示，遮挡角统计结果如图 5（b）所示。

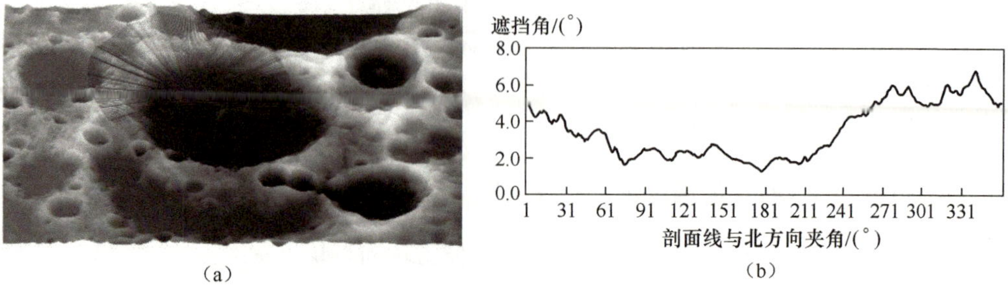

图 5　（网络版彩图）着陆区遮挡分析计算示意图

(a) 360°方位遮挡角三维显示图；(b) 360°方位遮挡角计算结果

Fig. 5 (Color online) The schematic of occlusion analysis and calculation in landing area

(a) 3D display of 360° azimuth occlusion angle; (b) calculation result of 360° azimuth occlusion angle

经分析，在考虑月面曲率影响的情况下，当遮挡山体距离着陆点大于 150km 时，只要该山体高度不超过 6.5km，该山体即位于着陆点的水平面以下，对着陆点不构成遮挡（如图 6 所示）。冯·卡门撞击坑周边最大高差约为 6km，因此遮挡角的计算，只需要在目标着陆点的周围以不超过 150km 的半径搜索遮挡山体。

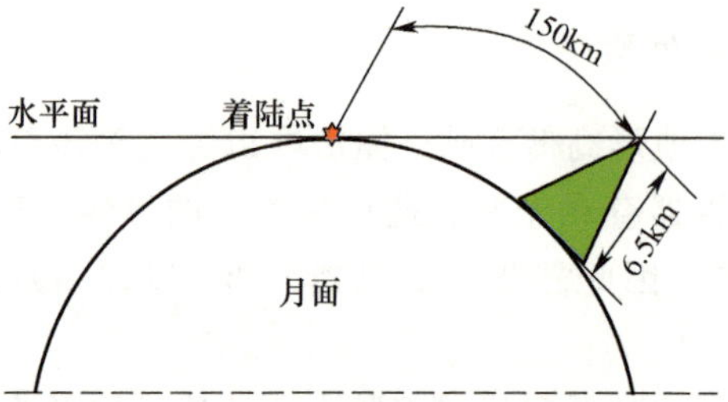

图 6　月面曲率影响示意图

Fig. 6　The schematic of the influence of lunar curvature

以遮挡角分析为基础，结合着陆点太阳和中继星的方位角、高度角变化情况，可以分别对着陆区内光照和测控遮挡进行分析，步骤如下：

（1）对着陆区内的一个目标点设置为着陆点，按前述方法计算得到沿不同方位（0°～360°）的最大遮挡角；

（2）计算得到着陆点沿不同太阳方位的（－90°～90°）太阳高度角以及不同中继星方位（－180°～180°）对应的高度角（方位角正北为 0°，逆时针减少，从 90°到－90°为由东往西，代表月球表面一天的从早到晚）；

（3）给定一个遮挡程度参数 $O=0$，比较不同方位角下对应的遮挡角与太阳高度角/中继星高度角，如出现一次遮挡角大于太阳高度角/中继星高度角的情况，则 $O=O+1$，直至比较完所有方位角；

(4) 将计算得到的遮挡程度参数 O 值赋给该点，同时记录下遮挡角大于太阳高度角/中继星高度角的各个方位角；

(5) 对着陆区内的所有点进行上述 4 个步骤的处理，进而获得着陆区范围内的遮挡程度图。

4　嫦娥四号着陆区地形地貌及遮挡分析

4.1　嫦娥四号着陆区

根据任务设计，嫦娥四号着陆区范围为 177.6°±1.2°E，45.5°±0.5°S[13]，位于冯·卡门撞击坑东南部。根据第 3 节提出的方法，对着陆区的高程、坡度、撞击坑、石块的分布特点以及光照和测控遮挡情况进行了分析。

4.2　嫦娥四号着陆区高程地势特点

基于 50m 分辨率的 SLDEM 得到着陆区范围内的高程统计值为：最小值为 −6089.5m，最大值为 −5768m，平均值为 −5926.14m，标准差为 20.44m. 着陆区高程分布情况如图 7 所示。

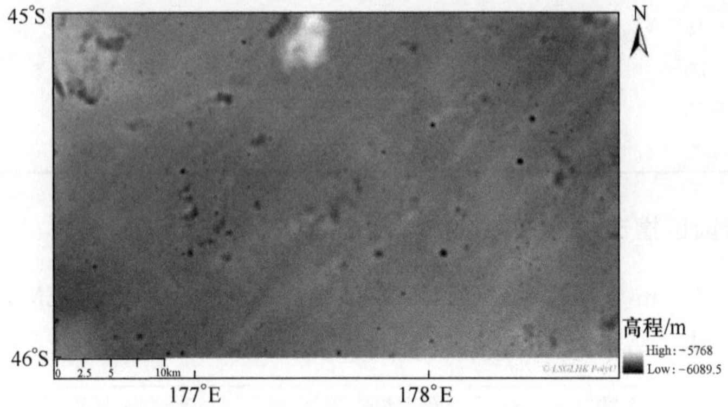

图 7　着陆区 DEM 的二维显示图

Fig. 7　2D display map of DEM in landing area

4.3　嫦娥四号着陆区坡度特点

根据 50m 分辨率的 SLDEM 数据统计着陆区内的坡度分布：坡度＜8°占整个着陆区的 96.57%，最小坡度为 0°，最大值为 48.97°，平均值为 2.51°，标准差为 2.26°。

坡度主要集中在 8°以内，说明着陆区整体趋势平坦，坡度较大的地区集中在撞击坑处，坑的外沿坡度在 10°左右，部分坑壁坡度超过 20°，着陆区坡度统计结果如图 8 和表 1 所示。

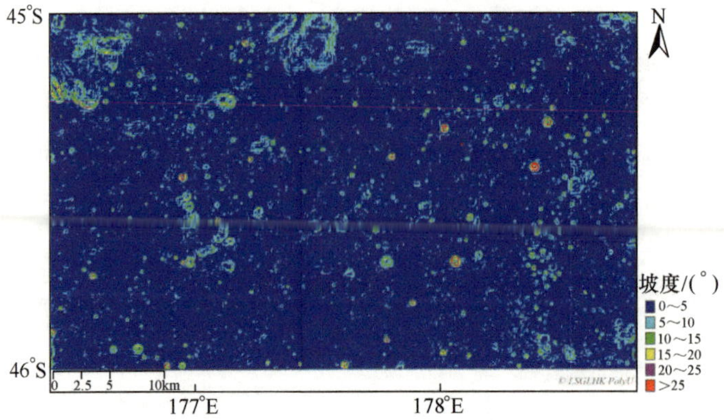

图 8　着陆区坡度统计结果（坡度计算基线：100m）

Fig. 8　Statistical results of landing slope (baseline: 100m)

表 1　着陆区坡度统计表

Tab. 1　Statistical table of landing slope

坡度范围	所占面积百分比（%）
0°~2°	50.90
2°~4°	33.44
4°~6°	9.21
6°~8°	3.02
8°~10°	1.33
10°~15°	1.24
15°~20°	0.34
20°~25°	0.12
>25°	0.41

4.4　嫦娥四号着陆区撞击坑分布特点

基于 LRO NAC 1m 分辨率的 DOM 和嫦娥二号 30m 分辨率的 DEM，进行撞击坑的识别与三维信息的提取，结果如图 9 所示。

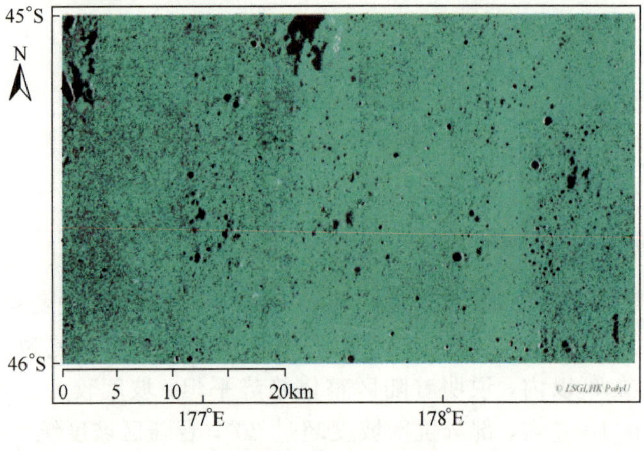

图 9　着陆区撞击坑统计结果

Fig. 9　Statistical results of impact craters in landing area

在着陆区内识别的撞击坑个数约为 410272 个，其中撞击坑最大直径为 1581m，最小直径为 14m。根据每 $10km^2$ 区域撞击坑数量分布情况，得到数量密度图，如图 10 所示。密度最大约为 5056 个/$10km^2$。

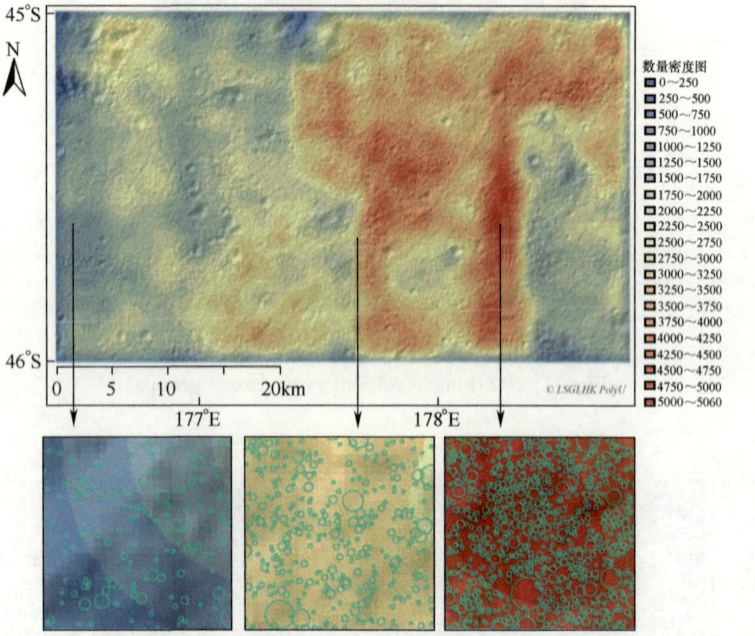

图 10　着陆区撞击坑数量密度统计结果

Fig. 10　Statistical results of number density of impact craters in landing area

着陆区撞击坑数量最密集的区域集中在 177.5°～178.5°E 区域内，该区域内有大量直径较小的坑密集分布。撞击坑数量密度的低值主要分布在 176°～177°E 区域内。

基于提取的撞击坑的平面信息（中心位置与直径），利用 DEM 来获取这些撞击坑的深度、边缘高度以及坑壁坡度信息。由于嫦娥二号的 DEM 分辨率为 30m/像素，所以取直径大于或等于 300m 的撞击坑进行分析。分析结果表明，绝大多数撞击坑属于比较浅平的撞击坑，97.64%的撞击坑深度与直径比小于 0.1，所有撞击坑的深度与直径比都不大于 0.2，绝大多数撞击坑的坑壁平均坡度小于 10°，所有撞击坑坑壁的平均坡度都不大于 20°。

4.5　嫦娥四号着陆区石块分布特点

基于 LRO NAC 1m 分辨率的 DOM 在着陆区内提取出 20031 个石块，其中石块外包络最大直径为 32.52m，最小直径为 1.23m。典型石块群标记结果如图 11 所示。

对着陆区每平方千米内的石块个数进行统计，计算出石块数量密度，如图 12 所示，其中着陆区每平方公里内平均有 5 个外包络直径≥1m 的石块，覆盖面积约占整个区域的 0.0046%。

根据统计结果，着陆区内 3～4m 范围内的石块数量最多，占总数的约 40.22%，直径≤5m 的石块占总数的约 86%。随着石块外包络直径的增大，石块的数量也逐渐减少。

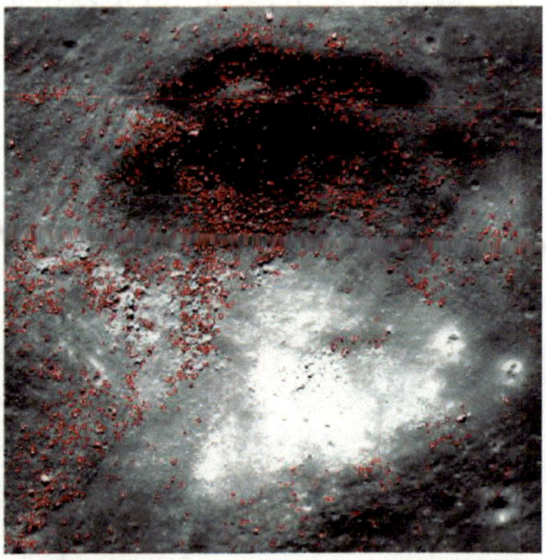

图 11　典型石块群标记结果

Fig. 11　Marking results of typical rock groups

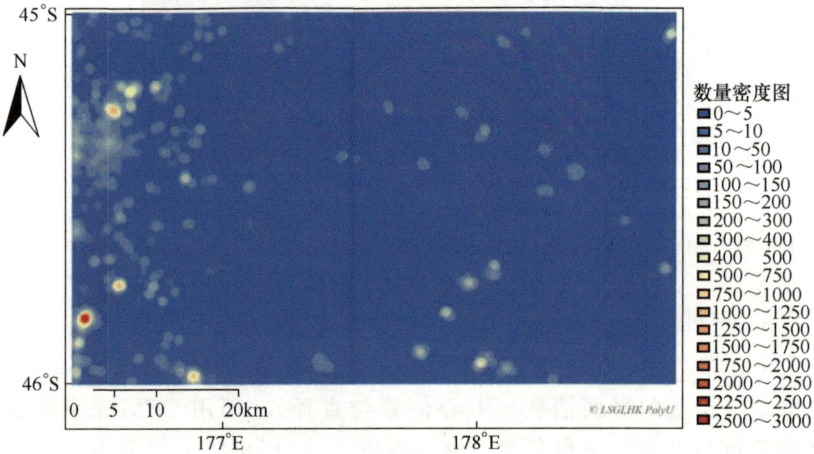

图 12　着陆区石块数量密度分布图

Fig. 12　Number density distribution map of rocks in landing area

4.6　嫦娥四号着陆区光照及测控遮挡

4.6.1　着陆区地形遮挡角计算

对着陆区基于 SLDEM 来进行遮挡分析。遮挡分析的参数取 1°的方位角间隔，150km 的搜索半径，50m 的搜索步长，生成的区域整体最大遮挡角分布如图 13 所示，其中每个点的不同颜色代表该点在 360°范围内的最大遮挡角。

着陆区地形最大遮挡角的统计结果：最小值为 2.07°，最大值为 56.22°，平均值为 4.15°，标准差为 1.98°。着陆区西北大部分区域地形遮挡角在 2°～4°，东南部分及东北角由于靠近外围大型撞击坑的边缘，遮挡角度逐渐变大到 4°～6°，6°～8°，直至 8°～10°。地形遮挡角最大的区域（图中红色或褐色区域）主要在小型撞击坑的内部，这是由于撞击坑

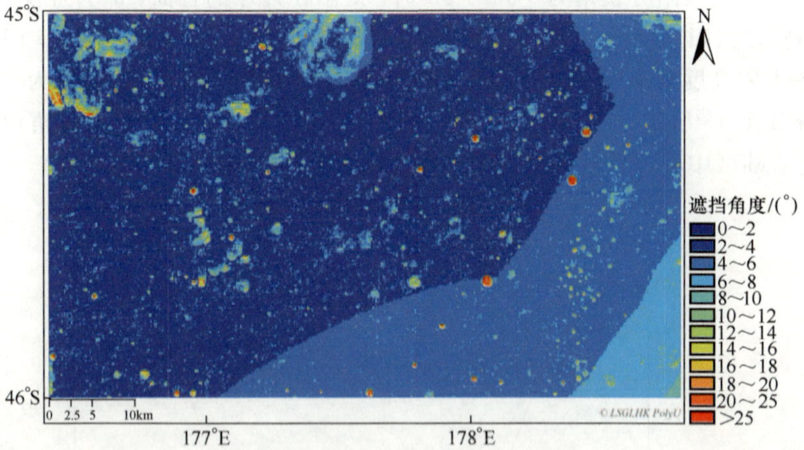

图 13 着陆区最大地形遮挡角分布图

Fig. 13 Distribution map of maximum terrain occlusion angle in landing area

自身地形造成的遮挡。在着陆区中部往北及西北角的隆起地带，由地形隆起造成的遮挡角超过 10°，局部区域的遮挡角甚至更大。

图 14 所示为与上述最大地形遮挡角分布图对应的方位角（即各点最大遮挡角对应方位角）分布图。可以看出，着陆区中南部大部分区域的遮挡角所在方位为东南方向（135°～180°，图中浅黄色区域），即其遮挡主要由东南方向的冯·卡门撞击坑的边缘所造成。东南部与西南角的遮挡角的方位主要为东南偏东方向（90°～135°，图中黄色区域），着陆区东北角的遮挡角的方位主要为西北向（270°～315°，315°～360°，图中青色及蓝色区域），着陆区东北部的遮挡角的方位主要为东北向（0°～45°，45°～90°，图中褐色及红色区域）。

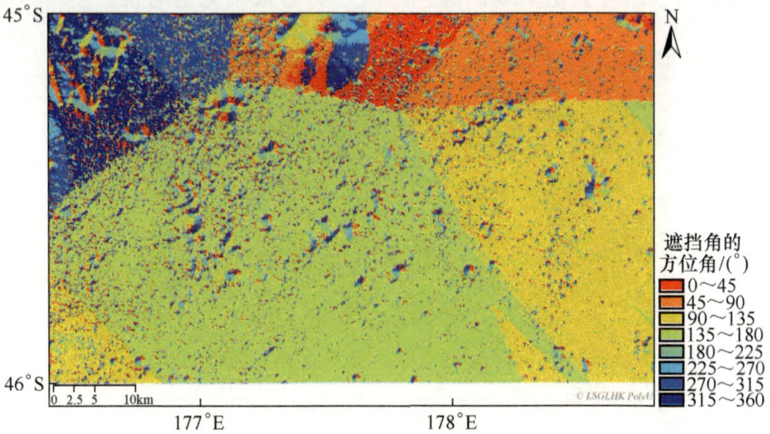

图 14 着陆区遮挡角对应的方位角分布图

Fig. 14 Azimuth distribution map related to occlusion angle in landing area

4.6.2 着陆区光照遮挡情况

本文比较了着陆区内各点在不同太阳方位的太阳高度角与对应的方位角下的地形对光

照遮挡的情况。当太阳方位角为－90°～90°时，太阳在月球的背面，即为月夜，因此不参与比较。着陆器在太阳高度角小于10°时不工作（探测器处于月夜休眠状态），因此在分析计算时不考虑太阳高度角小于10°的情况。以着陆区中心点177.6°E，45.5°S为例进行计算光照遮挡情况如图15所示，其中在太阳高度角大于10°的情况下，太阳高度角均大于地形遮挡角，因此着陆区中心点在整个月昼无光照遮挡。

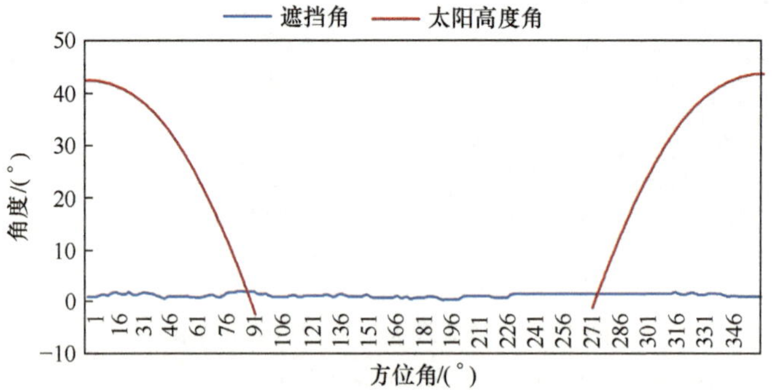

图15 着陆区中心点太阳高度角与遮挡角的比较结果

Fig. 15 Comparison of sun elevation angle and occlusion angle at central point of landing area

图16所示为着陆后6个月着陆区内太阳高度角与遮挡角的比较结果分布情况。其中每个点的数值表示各点在着陆后6个月内（太阳高度角大于10°）得到太阳照射的时间百分比，如果比值小于1，则表示光照存在遮挡。由于探测器在太阳高度角小于10°时不工作，因此在分析计算时没有统计太阳高度角小于10°的情况。颜色越绿的区域代表太阳的通视情况越好，颜色越红的区域代表太阳的通视情况越差。在嫦娥四号着陆区内，太阳照射时间百分比最小值为86%，即对应的点6个月内当中有约14%的时间不能得到太阳的照射。统计得到着陆区内有99.803%的面积（所有数值为0的绿色区域）为对太阳完全通视。

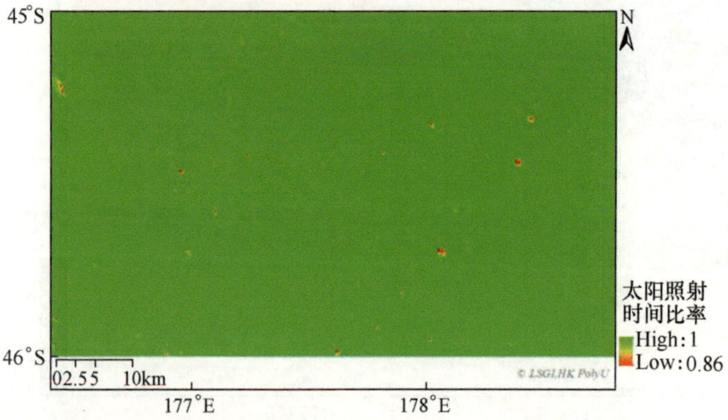

图16 着陆区太阳高度角与遮挡角的比较结果分布

Fig. 16 Comparison map of sun elevation angle and occlusion angle in landing area

4.6.3 着陆区测控遮挡情况

结合中继星轨道预报，比较了着陆区内各点在探测器6个月寿命内不同方位的中继星相对于当地地理坐标系的高度角与对应的方位角下的地形对测控通信链路遮挡情况。着陆区在着陆后 6 个月对中继卫星的测控遮挡面积占总面积的百分比为 0.0165%，主要集中在撞击坑的内壁及左上角的山脚处，即着陆区 99.9835%的面积内不存在测控遮挡。着陆区测控遮挡结果如图 17 所示。

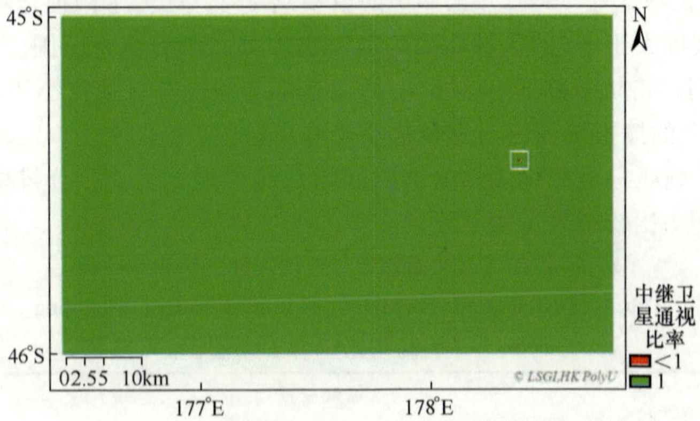

图 17　着陆区测控遮挡区域（红色像素点）

Fig. 17　Communication occlusion in landing area (red pixels)

测控遮挡最严重点的位置为（−45.4309°，178.3903°），即图 17 中黄色框内所标示的点。其总计遮挡时间为 79h，占总工作时间（太阳高度角 10°以上）的 4.2%。该点在着陆后 6 个月内的地形-太阳-中继卫星通视关系如图 18 所示。

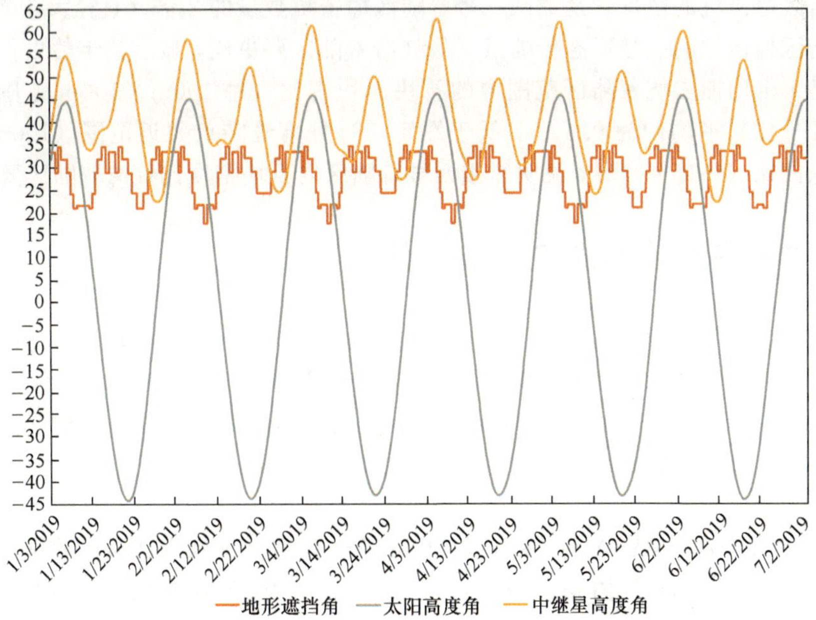

图 18　着陆后 6 个月内测控遮挡最严重点（−45.4309°，178.3903°）的地形-太阳-中继卫星通视关系

Fig. 18　Terrain – Sun – Relay Satellite visual relation at the most occluded point (−45.4309°, 178.3903°) within 6 months after landing

4.7 冯·卡门撞击坑着陆区与虹湾着陆区地形对比

根据分析结果，相较于嫦娥三号虹湾着陆区[14]：

(1) 冯·卡门撞击坑着陆区小于10°的坡分布概率大于97%，大于10°的坡不到3%；虹湾地区小于10°的坡大于98%，大于10°的坡不到2%。冯·卡门撞击坑和虹湾坡度分布情况较类似，且坡度大于20°的区域主要为撞击坑内部的部分坑壁范围。

(2) 冯·卡门撞击坑着陆区97%以上的撞击坑的深度与直径比小于0.1，嫦娥三号虹湾着陆区84%以上的撞击坑深度与直径比小于0.1。

为便于比较，嫦娥三号和嫦娥四号着陆区均采用嫦娥二号7m分辨率的DOM提取的撞击坑直径和数量，比较结果见表2。

表2 嫦娥三号和嫦娥四号着陆区撞击坑分布比较

Tab. 2 Distribution comparison of impact craters in landing areas of Chang'E-3 and Chang'E-4

撞击坑直径 D	嫦娥三号 虹湾着陆区 (个/100km^2)	嫦娥四号 冯·卡门撞击坑着陆区 (个/100km^2)
$140m < D \leq 300m$	56	152
$300m < D \leq 500m$	14	47
$D > 500m$	2	9

通过上述分析，嫦娥四号着陆区的撞击坑分布较嫦娥三号着陆区要更多、更密集，嫦娥四号着陆区撞击坑的数量约是嫦娥三号着陆区撞击坑数量的3倍（直径大于140m）；嫦娥四号着陆区撞击坑的三维形态较嫦娥三号虹湾着陆区要更浅，显示其年龄更古老。

(3) 冯·卡门撞击坑着陆区范围内的石块面积占比（每100m×100m范围内石块覆盖面积的百分比）平均约为0.5%，低于嫦娥三号虹湾着陆点邻近范围（100m×100m）的石块面积占比（1.15%）[10]，说明嫦娥四号着陆区内石块密度比虹湾着陆区要低。

5 嫦娥四号实际着陆点定位与遮挡分析

2019年1月3日10时26分02秒，嫦娥四号探测器成功着陆在冯·卡门撞击坑着陆区内。立即利用着陆器的降落相机所拍摄的序列影像与LRO NAC影像（影像ID：M134022629L）进行配准，如图19（a）所示，实现了嫦娥四号着陆器的精确定位。定位结果显示，嫦娥四号着陆点的经纬度为177.5892°E，45.4561°S，距平均海拔平面高度为−5926m。

2019年2月1日，NASA的LRO以82km轨道高通过了嫦娥四号的着陆区，并利用其NAC对着陆区进行了拍摄，图19（b）所示为其拍摄得到的影像（影像ID：M1303619844LR）。该LRO NAC影像的分辨率为0.85m/像素，在影像上嫦娥四号着陆器（竖直方向箭头所指处带斜长阴影的白色点）与玉兔二号月球车（水平方向箭头所指处的黑点）清晰可见。比较图19（a）与（b），可以发现我们利用下降影像的定位结果与LRO实际拍摄到的嫦娥四号着陆点位置一致。

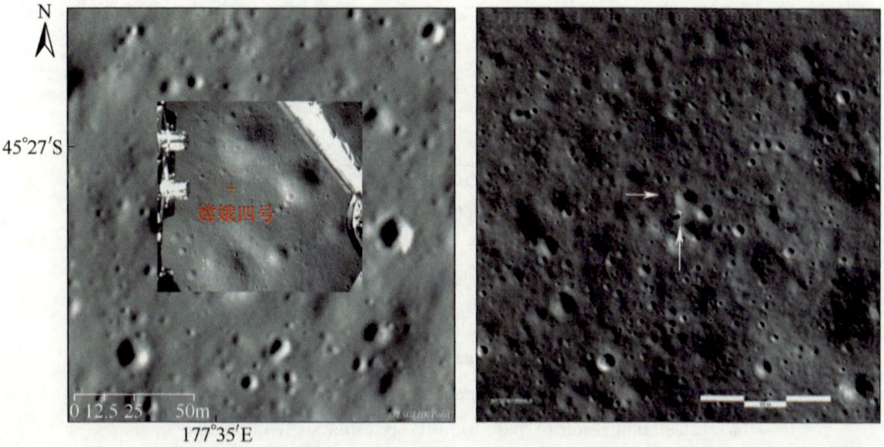

(a) (b)

图 19 嫦娥四号实际着陆点位置（177.5892°E，45.4561°S）

(a) 嫦娥四号着陆器上的降落相机拍摄的实际着陆点；(b) LRO NAC 影像拍摄的嫦娥四号（白色箭头所指处）

Fig. 19 The actual landing site of Chang'E-4 (177.5892°E, 45.4561°S)

(a) The actual landing site taken by landing camera on Chang'E-4 lander;

(b) Chang'E-4 (indicated by white arrow) taken by LRO NAC

基于上述着陆点定位结果，对着陆点周围的地形遮挡情况进行了分析，着陆点地形遮挡角分析结果如图 20 所示。图 20（a）中红色点所示为着陆点的位置，蓝色线代表着陆点周围 360°范围内的各个方位角，线的长度指向造成遮挡的位置（如冯·卡门撞击坑的边缘及中央峰的高处），线末端的数字显示了遮挡角的大小。着陆点周围 360°范围内各遮挡角的大小见图 20（b）所示，最大的地形遮挡角为 3°。

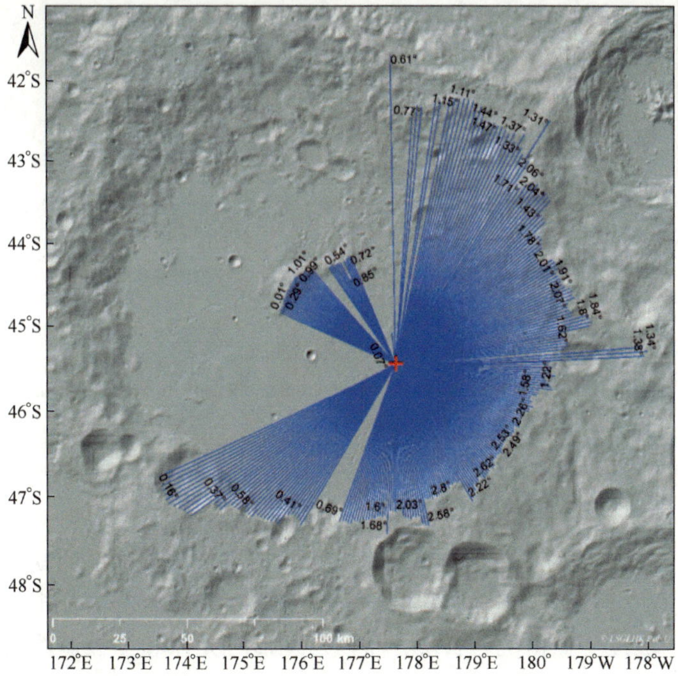

(a)

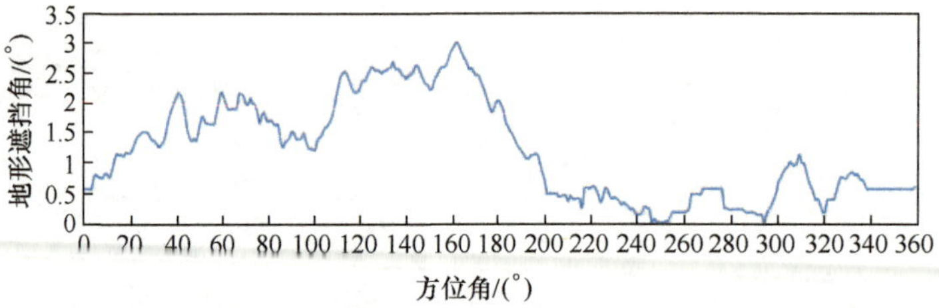

(b)

图 20　着陆点遮挡角示意图（红色十字为着陆点）

(a) 着陆点 360°方位遮挡角三维显示图；(b) 着陆点 360°方位遮挡角计算结果

Fig. 20　Map of occlusion angle of landing site (red cross)

(a) 3D display map of 360° azimuth occlusion angle of landing site

(b) calculation result of 360° azimuth occlusion angle of landing site

着陆后 6 个月，着陆点光照遮挡和测控遮挡情况如图 21 所示，在月昼期间，着陆点对太阳和中继星全部通视，无遮挡情况。该结果同任务分析时得到的在该位置无光照和测控遮挡的结果一致。

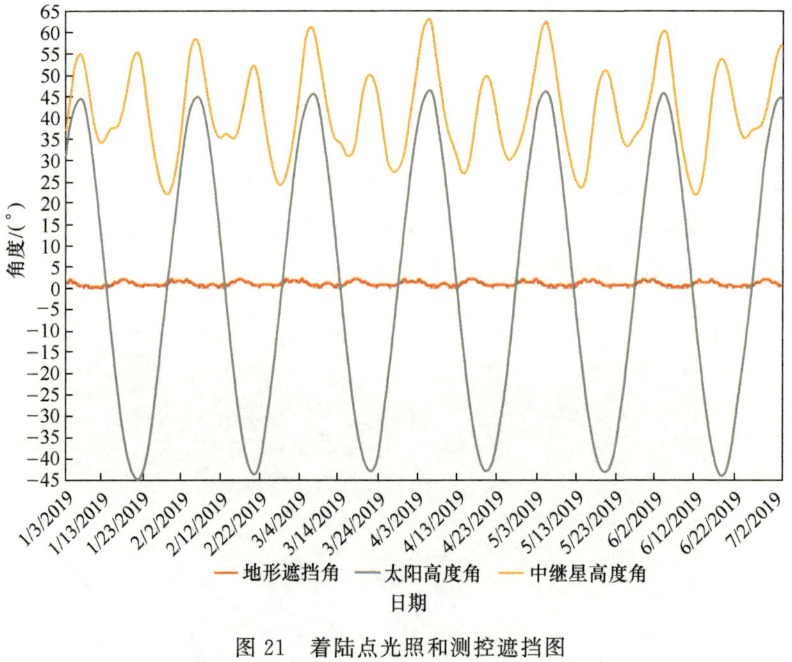

图 21　着陆点光照和测控遮挡图

Fig. 21　Illumination and communication occlusion map of landing site

6　结论

嫦娥四号任务成功实现了在月球背面冯·卡门撞击坑内着陆区的软着陆任务，实现了精准着陆并完成了月面工作。在任务设计阶段，对冯·卡门撞击坑着陆区的地形地貌进行分析，包括高程、坡度、撞击坑、石块分布以及地形对光照和测控遮挡的影响等。由于嫦娥四号着陆区地势崎岖，经分析在极小范围内存在光照遮挡和测控遮挡；将嫦娥四号着陆区与嫦娥三号着陆区地形特征进行比较，坡度基本一致，嫦娥四号着陆区撞击坑数量更多，石块较少。在嫦娥四号成功着陆后，对着陆点的位置进行了精确定位，并进而分析了着陆点周围地形对光照和测控遮挡的情况，结果表明着陆点在 6 个月寿命期内无光照和测控遮挡。

嫦娥四号在轨成功的实践结果表明，在着陆区地形及月面遮挡分析过程中采用的方法有效，可以将成功经验充分应用到后续月球探测及地外天体软着陆任务中。

参考文献

[1] Ye P J, Sun Z Z, Zhang H, et al. Mission design of Chang'E - 4 probe system (in Chinese). Sci Sin Tech, 2019, doi: 10. 1360/SST - 2019 - 0106

[2] Kring D A, Durda D D. A global lunar landing site study to provide the scientific context for exploration of the Moon. Technical Report. LPI - JSC Center for Lunar Science and Exploration, 2012

[3] Snape J F, Fagan A L, Ennis M E, et al. Science - rich mission sites within South Pole - Aitken basin, part 2: Von Karman crater. In: Lunar and Planetary Science Conference. The Woodlands, 2010

[4] Smith D E, Zuber M T, Jackson G B, et al. The lunar orbiter laser altimeter investigation on the lunar reconnaissance orbiter mission. Space Sci Rev, 2010, 150: 209 - 241

[5] Mazarico E, Rowlands D D, Neumann G A, et al. Orbit determination of the lunar reconnaissance orbiter. J Geod, 2012, 86: 193 - 207

[6] Wu B, Hu H, Guo J. Integration of Chang'E - 2 imagery and LRO laser altimeter data with a combined block adjustment for precision lunar topographic modeling. Earth Planet Sci Lett, 2014, 391: 1 - 15

[7] Speyerer E J, Wagner R V, Robinson M S, et al. Pre - flight and on - orbit geometric calibration of the lunar reconnaissance orbiter camera. Space Sci Rev, 2016, 200: 357 -392

[8] Zhao Bao C, Yang J F, Wen D S, et al. Chang'E - 2 lunar orbiter CCD stereo camera design and validation (in Chinese). Spacecr Eng, 2010, 20: 2 - 14

[9] Barker M K, Mazarico E, Neumann G A, et al. A new lunar digital elevation model from the lunar orbiter laser altimeter and SELENE terrain camera. Icarus, 2016, 273: 346 - 355

[10] Tang J, Pilesjö P. Estimating slope from raster data: A test of eight different algorithms in flat, undulating and steep terrain. In: Transactions on Ecology and the Environment, Vol 146. Southampton: WIT Press, 2011. 143 - 154

[11] Wang Y, Wu B. Active machine learning approach for crater detection from planetary imagery and digital elevation models. IEEE Trans Geosci Remote Sens, 2019, 1 - 13

[12] Li Y, Wu B. Analysis of rock abundance on lunar surface from orbital and descent

images using automatic rock detection. J Geophys Res Planets, 2018, 123: 1061 - 1088

[13] Wu W R, Wang Q, Tang Y H, et al. Design of Chang'E - 4 lunar farside soft - landing mission (in Chinese). J Deep Space Expl, 2017, 4: 111 - 117

[14] Wu B, Li F, Ye L, et al. Topographic modeling and analysis of the landing site of Chang'E - 3 on the Moon. Earth Planet Sci Lett, 2014, 405: 257 - 273

用 CLEAN 算法对嫦娥四号低频射电频谱仪信号进行干扰抑制

刘晨迪[1,2,3]，周建锋[4]，苏彦[1,2,3]

1. 中国科学院国家天文台，北京 100101
2. 中国科学院大学，北京 100049
3. 中国科学院月球与深空探测重点实验室，北京 100101
4. 清华大学工程物理系，北京 100084

摘要： 嫦娥四号低频射电频谱仪放置在月球背面，其天然观测条件得天独厚。然而，嫦娥四号平台存在约 10^{-15} W/(m^2·Hz) 级别的强干扰，并且干扰在每道时域数据中都存在明显的差异，这大大削弱了低频射电频谱仪的天文观测灵敏度。为此，本论文从两组信号的相关性出发，提出基于 CLEAN 算法，借助互相关功率谱、傅立叶级数展开等工具，把低频射电频谱仪 A、B、C 天线的时域观测数据切分为强相关的 CLEAN 模型信号和部分相关的残余信号两个部分。其中 CLEAN 模型信号主要由平台干扰信号和可能的低频强射电爆发组成；残余信号由接收机噪声、未扣除干净的平台干扰信号和常规的低频射电信号组成。将该方法应用到实际数据中，结果表明嫦娥四号低频射电频谱仪的未积分灵敏度可以提高约 8 个数量级，达到 10^{-23} W/(m^2·Hz) 水平。在此基础上，基于对平台干扰信号中确定性成分和宽带随机成分的分别处理，再借助低频射电爆发信号和平台干扰信号在功率谱上不同的表现，以及常规低频射电天文信号受月球自转调制等信息，将来科学分析工作的重点是进一步处理 CLEAN 模型信号和残余信号，以期发现低频强射电天文爆发信号，乃至对全天区进行粗略的成像。

关键词： 嫦娥四号低频射电频谱仪；平台干扰分离；CLEAN 算法

Using CLEAN Algorithm to Suppress the Interference of Chang'e-4 Low-frequency Radio Spectrum Analyzer

Liu Chendi[1,2,3], Zhou Jianfeng[4], Su Yan[1,2,3]

1. National Astronomical Observatories, China Academy of Sciences, Beijing 100101, China

2. University of China Academy of Sciences, Beijing 100049, China
3. Key Laboratory of Lunar and Deep - Space Exploration, National Astronomical Observatories, Chinese Academy of Sciences, Beijing 100101, China
4. Department of Engineering Physics, Tsinghua University, Beijing 100084, China

Abstract: The Chang'e - 4 low - frequency radio spectrometer is placed on the back of the moon, and its natural observation conditions are unique. However, the Chang'e - 4 platform has strong interference of about 10^{-15} W/(m^2 · Hz) level, and there are obvious differences in the interference in each channel of time domain data, which greatly weakens the astronomical observation sensitivity of the low - frequency radio spectrum analyzer. To this end, starting from the correlation of the two sets of signals, this paper proposes to divide the time - domain observation data of the low - frequency radio spectrum analyzer A, B, and C antennas into strongly correlated CLEAN model signals and partial correlated residual signal based on the CLEAN algorithm, with the help of cross - correlation power spectrum, Fourier series expansion and other tools. The CLEAN model signal is mainly composed of platform interference signals and possible low - frequency strong radio bursts; the residual signal is composed of receiver noise, undeducted platform interference signals and conventional low - frequency radio signals. Applying this method to actual data, the results show that the unintegrated sensitivity of the Chang'e - 4 low - frequency radio spectrum analyzer can be increased by about 8 orders of magnitude, reaching a level of 10^{-23} W/(m^2 · Hz). On this basis, based on the separate processing of the deterministic components and broadband random components in the platform interference signal, and then relying on the different performance of the low frequency radio burst signal and the platform interference signal in the power spectrum, and the conventional low frequency radio astronomy signal modulated by the moon rotation And other information, the focus of future scientific analysis work is to further process the CLEAN model signal and residual signal, with a view to discovering low - frequency strong radio astronomy burst signals, and even roughly imaging the entire sky.

Keywords: Chang'e - 4 low - frequency radio spectrum analyzer; Platform interference removal; CLEAN algorithm

1　引言

嫦娥四号低频射电频谱仪（Low frequency radio spectrum analyzer，简称 LFRS）放置在月球背面。由于月球屏蔽了大量来自地球的无线电干扰，因此 LFRS 具有得天独厚的观测条件，可以对太阳等宇宙天体的射电信号进行观测。[1]但是，由于着陆器平台干扰信号比接收到的低频射电天文信号大很多，如果不分离平台干扰信号，将无法得到有效的低频射电天文信号数据。因此，需要提出一种有效分离平台干扰信号的方法，从而实现对干扰的抑制。

在射电天文领域，传统抑制干扰的方法有很多，包括对空白脉冲干扰的时间限制[2]，自适应频率选择滤波[3]，参数信号相减[4]，自适应波束成形[5]，时间自适应滤波[6]，带有参考天线的频域后相关处理[7]，子空间投影的空间滤波[8]等等。到目前为止，运用比较广泛的算法是单通道总功率变化检测器，它使得相关器的输出得到减小。Friedman (1996) 基于对功率变化的检测，在 RATAN600 上实现了改进的功率检测器[9]。Weber 等人（1997 年）提出在所有滞后的情形下使用量化相关性来测试干扰的存在[10]。Maslakovic 等人 1996 年提出了一种基于小波分解的检测器[11]。这些都是单通道检测器，它们不利用干扰的空间特性分析。Kasper，Chute 和 Routledge 在 1982 年提出了将多个望远镜结合使用以改善探测质量的方法，此方法用于低频干涉测量，在这之中还提出了一种基于交叉频谱时间行为的鲁棒数据检查方法[12]。Barnbaum&Bradley (1998) 考虑了自适应过滤技术，他们提议使用参考天线和最小均方（LMS）类型算法消除 Green - Bank 射电望远镜的干扰[13]。W. A. BAAN 提出了一种针对韦斯特博克综合射电望远镜的射电干扰消除子系统[14]。这些都是在射电干扰消除方面典型的例子。

上面提到的射电干扰信号消除算法主要分为三类：第一类试图去通过线性的方式来区分射电干扰信号的特点，比如奇异值分解（SVD）[15]，主成分分析（PCA）[16]等等。如果射电干扰在时间和频率上表现出重复的模式，这些方法效果很好，但无法处理更多随机信号，例如由卫星引起的不规则信号。第二类是基于阈值的算法，例如 CUMSUM[8] 和 SUMTHRESHOLD[4]，尽管它们很简单，但是这些方法相当可靠并且非常有效。第三类使用传统的监督机器学习技术（例如 K 最近邻和高斯混合模型）对射电干扰信号进行聚类[17]。

嫦娥四号 LFRS 所收到的平台干扰信号很强，和太阳爆发峰值的量级 $10^{-15}\,W/(m^2 \cdot Hz)$ 相当，比普通的太阳爆发信号 $10^{-18}\,W/(m^2 \cdot Hz)$ 大三个量级，比太阳宁静时的信号 $10^{-22}\,W/(m^2 \cdot Hz)$（1S. F. U）大七个量级。之前仪器设计时推荐使用的谱减法[18]，虽然可以将信号的信噪比提升约 30dB，但是仍然不足以探测有效的射电天文信号。为此，本文提出了基于 CLEAN 算法的平台干扰信号分离方法，利用 LFRS 三根天线同时观测信号存在相关性的特点，并结合天线之间数据的互相关功率谱、傅立叶级数展开等工具，有效地分离平台干扰信号，从而实现对干扰的抑制。

本文的第 2 小节主要介绍 LFRS 的工作原理及接收信号构成成分，第 3 小节介绍 LFRS 平台干扰信号分离的方法及算法流程，第 4 小节展示新算法对 LFRS 实际数据的处理结果及效果分析，第 5 小节是讨论，最后的第 6 小节是结论。

2　LFRS 的工作原理及接收信号的构成

2.1　LFRS 简要介绍

太阳和宇宙中的其它来波信号是一个矢量信号。根据矢量场的理论，来波信号可以分解为任意三个互相垂直方向的分量，只要将这三个分量接收下来，进行合成，就可以得到矢量的大小和方向。通过后期处理，可以获得电场的频谱和时变信息。

LFRS 的天线分布如图 1 所示：三分量有源天线接收宇宙的来波电场。三根天线分别标注为“天线 A”、“天线 B”和“天线 C”，两两正交，还有一根短的“天线 D”，主要用于接收平台干扰信号。为天线数据处理做一个参考。

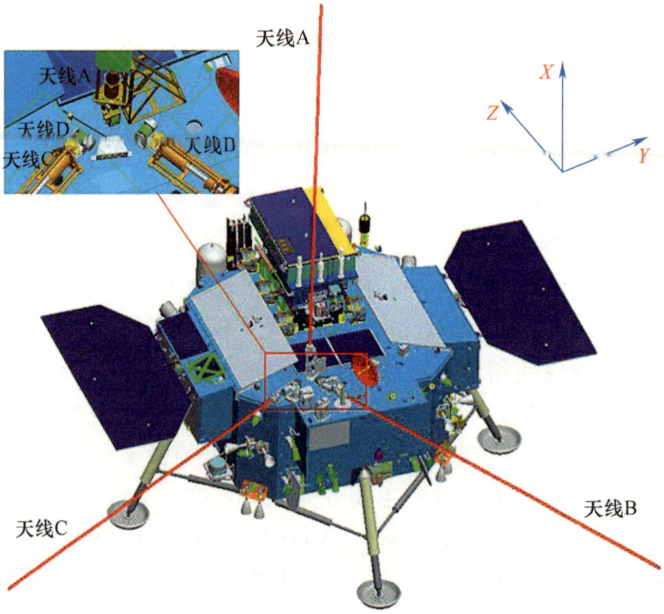

图 1 着陆器低频射电频谱仪的天线位置分布

Fig. 1 Antenna location distribution of lander low - frequency radio spectrum analyzer

LFRS 具有高频和低频两个频率接收波段，高频对应的频率范围是 1 - 40MHz，低频对应的频率范围是 0.1～2MHz。接收机灵敏度为 $6nV/\sqrt{Hz}$ ，高频的频率分辨率为 100kHz，低频的频率分辨率为 5kHz。

LFRS 主要由四根接收天线、前置放大器、电子学单元、电缆组件等部分组成。有源天线安装于卫星本体外，其余组件安装于卫星本体内。电子学单元由控制器、配电器、基准时钟模块、四通道接收机、内定标模块、存储单元、通信接口、数传接口等组成。LFRS 内部各单元之间及与数管分系统、供配电分系统之间的接口关系如图 2 所示。

LFRS 有六种工作模式，分别为 1）待机工作模式 2）内定标工作模式 3）时频对比工作模式 4）频谱巡查工作模式 5）三天线工作模式 6）时域工作模式。其中运用最广泛的是三天线工作模式，在此模式下，LFRS 分别采集四根天线四个通道的科学数据，本文进行读取处理的数据均对应三天线工作模式。

LFRS 的科学数据分为 0A 级、0B 级、01 级、2A 级、2B 级、2C 级共六级。其中预处理数据经过分路解帧和分包，生成 0A 级数据；0A 级数据经过排序、去重复、两站优化拼接、去源包包头、根据采样周期形成 LFRS 数据块，生成 0B 级数据产品；多个 0B 级数据文件拼合成一个探测周期，生成 01 级数据产品；对 01 级数据进行傅立叶变化、定标等操作，生成 2A 级数据产品；对 2A 级数据加入太阳位置等信息，生成 2B 级数据产品；对 2B 级数据进行对消噪声处理，生成 2C 级数据产品。其中 1 级数据为时域数据，2 级数据为频域数据。本论文处理所用的原始数据均为 1 级数据。

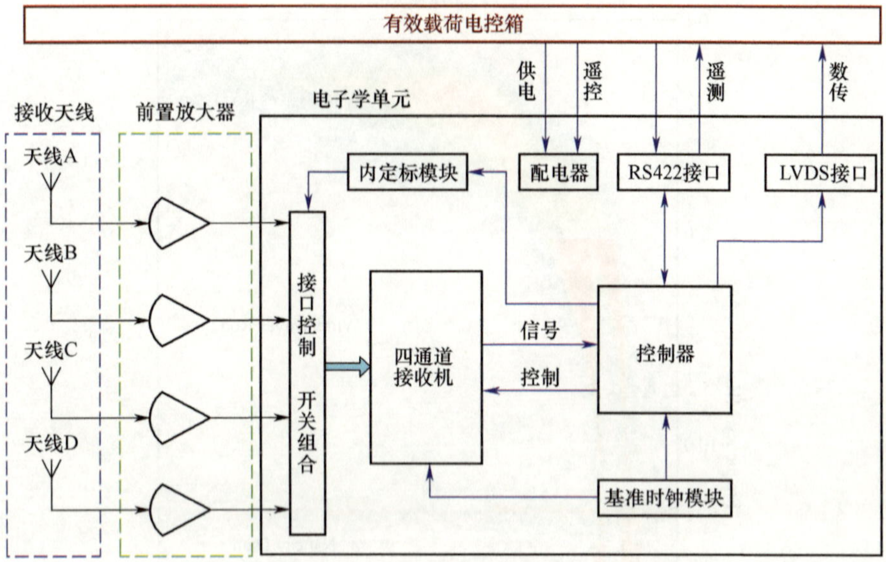

图 2　低频射电频谱仪系统组成及接口关系示意图

Fig. 2　Schematic diagram of low - frequency radio spectrum analyzer, with system composition and interface relationship

2.2　接收信号的构成成分

LFRS 收到的信号主要由三部分构成，一部分是平台干扰信号，一部分是接收机噪声，最后是低频射电天文信号。

平台干扰信号的特点是幅度相对较强，两两天线之间的相关性高。如果干扰源位置确定，各天线接收信号之间的强度比例不随时间变化。且来自电子设备的干扰信号，一般可分为固定频率的发射信号，如通信发射机，时钟信号等，另一部分是宽带的随机噪声信号，这是由于 EMC 兼容问题带来的。最终的平台干扰信号如图 6 所示，既有确定性变化的成分，也有宽带随机性噪声成分。

低频射电天文信号也具有一定的相关性，除此之外，还有一些天体源所特有的性质。比如，低频太阳射电信号其具有瞬时性和频谱漂移等特点，为此其具有很强的频谱可识别特征[19]，如图 3 所示。再比如，快速射电爆（Fast Radio Burst，FRB）FRB200428 的持续时间约 0.61ms，能量通量（Fluence）达到 1.5MJy ms[20]，其辐射强度足以被 LFRS 探测到。另外，由于仪器指向随月球的自转而改变，低频射电天文信号在 LFRS 各天线之间的强度比例随时间呈现有规律的变化。所有这些特性，都有助在强平台干扰的背景下，识别有用的低频射电天文信号。

接收机噪声的特点是天线两两之间无相关性。可以利用该特性，通过计算天线之间数据的相关性来评估平台干扰信号的分离程度。

图 4 展示了 LFRSB 天线观测到的一个频谱实例。由图可以得知，频谱在 1MHz 频率附近的流量密度大约为 10^{-15} W/(m^2 · Hz)，与太阳爆发峰值流量密度 10^{-15} W/(m^2 · Hz) 相当，比太阳爆发流量密度 10^{-18} W/(m^2 · Hz) 大了 2 - 3 个量级，说明着陆器平台

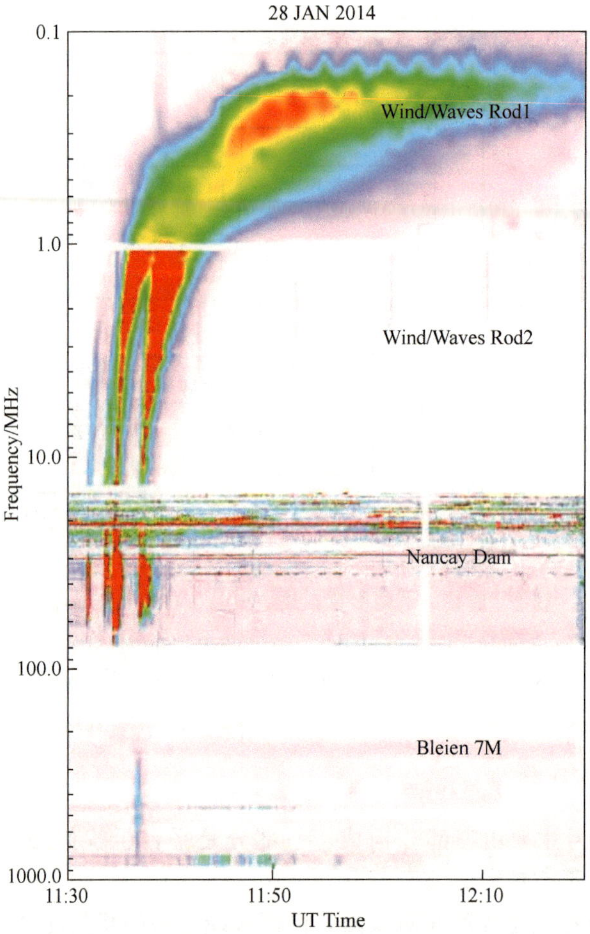

图 3　低频太阳射电信号频谱

Fig. 3　Low frequency spectrum of solar radio signal

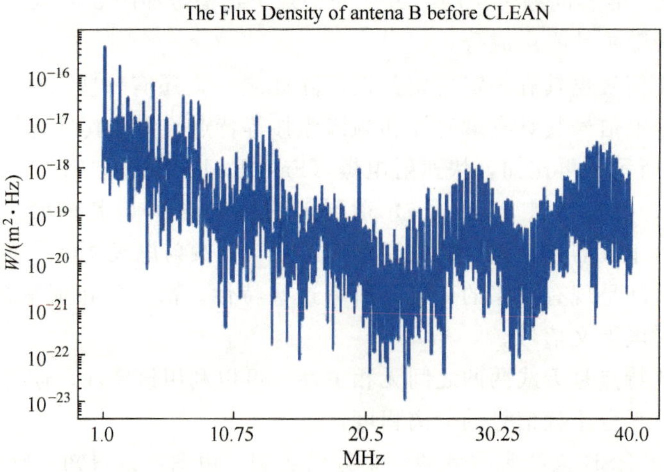

图 4　B 天线采集到的高频波段的一个频谱实例，对应第 23 月昼第二道数据

Fig. 4　An example of the spectrum of antenna B at the high - frequency band, corresponding to the second channel data on the 23rd moon day

干扰比我们希望探测的太阳爆发信号要大许多。如果不通过一定的方法分离平台干扰信号，将无法有效地提取出想要的低频射电天文信号。

3　LFRS 平台干扰信号的分离方法

低频射电源与平台干扰源会在 LFRS 的 3 个天线上激发出互相关联的接收信号。本论文的目的是从两组信号的相关性出发，借助 CLEAN 方法，以及互相关功率谱、傅立叶级数展开等工具，把 LFRS 观测信号中的强相关成分与部分相关成分进行分离，以便进行后续的科学分析。分离之后，原始的 LFRS 时域信号被切分为两个成分：即强相关的 CLEAN 模型信号和部分相关的残余信号。在这个过程中，原始信号的信息没有丢失或损伤。

CLEAN 模型信号主要由平台干扰信号组成。当低频强射电爆发信号到来时，也会进入 CLEAN 模型信号。由于数传的限制，LFRS 大约以 1 秒钟为间隔采集一道时域数据，采集间隔并不严格相等。每道数据由 4096 个数据组成，采样频率为 100MHz，采样持续时间只有大约 40 微秒。因此，不同道时域数据有明显的差别（参见图 6 与图 7）。

为了寻找有效的低频强射电爆发信号，需要提取每一道时域数据的 CLEAN 模型信号，并对其中的确定性成分和宽带随机成分分别进行整理。例如，对于 CLEAN 模型信号中的确定性时域变化结构或其频谱成分，可以把它们当作某种系统误差，同样借助 CLEAN 算法，将其扣除。剩下的 CLEAN 模型信号，由宽带随机信号组成，可以通过平均多道数据的方式，降低其涨落水平，提高探测低频强射电天文爆发的灵敏度。

残余信号主要由不相关的接收机噪声、未分离出去的平台干扰信号，以及常规低频射电天文信号组成。残余信号的特征类似于随机白噪声，里面没有确定性的信号成分，可以借助于数理统计工具来进行分析。例如，可以借鉴微波辐射计的思想，通过积分平均的方式进一步提高残余信号的探测灵敏度。

我们的最终目标是用嫦娥四号 LFRS 的观测数据来搜索低频射电天文信号，并进行深入的科学研究。此时，无论是 CLEAN 模型信号还是残余信号都是缺一不可的。

时标短（分钟到天量级）、幅度强（MJy 量级）的低频强射电爆发往往隐藏在 CLEAN 模型信号中。对这类射电爆发信号的提取离不开宁静状态的 CLEAN 模型信号的分类整理。特别强的低频射电天文爆发，在单道数据中就有明显异常的表现。此时，可以借助交叉相关，直接扣除该道数据对应的宁静状态的 CLEAN 模型信号，获得射电爆发信号的时域、频谱特征。如果射电爆发强度与平台干扰中的宽带随机成分接近，那么先扣除 CLEAN 模型信号中的确定性成分，再通过多道数据积分平均的方式，提高探测灵敏度，最终获得这类低频射电天文爆发信号的时域、频谱特征。

一些时标长（超过月量级）、幅度弱的低频射电天文信号则隐藏在残余信号中。为了探测这类弱信号，首先要借助星载定标装置或其他外部稳定的定标源，确认接收机工作稳定。之后，可以直接从残余信号出发，同样借助微波辐射计的思路，通过积分平均，提高残余信号的探测灵敏度。理论上讲，LFRS 三个天线残余信号的积分值会随着月球自转而发生调制变化。这样的调制信息甚至可以用来对全天进行粗略的成像。

3.1 用 CLEAN 算法分离平台干扰信号的基本思路

设 LFRS 中一个天线收到的信号为 $S(t)$，它包含了平台干扰信号 $I(t)$，宇宙射电信号 $C(t)$，以及接收机噪声 $N(t)$。这些信号之间的关系如下：

$$
\begin{aligned}
S_A(t) &= \alpha_A(t)I(t) + \beta_A(t)C(t) + N_A(t) \\
S_B(t) &= \alpha_B(t)I(t) + \beta_B(t)C(t) + N_B(t) \\
S_C(t) &= \alpha_C(t)I(t) + \beta_C(t)C(t) + N_C(t)
\end{aligned} \tag{1}
$$

其中 $S_A(t)$、$S_B(t)$、$S_C(t)$ 分别为天线 A、B、C 接收到的信号；$N_A(t)$、$N_B(t)$、$N_C(t)$ 分别为天线 A、B、C 对应接收机的本地噪声，它们之间相互独立；$\alpha_A(t)$、$\alpha_B(t)$、$\alpha_C(t)$ 为平台干扰信号在天线 A、B、C 上的投射系数，$\beta_A(t)$、$\beta_B(t)$、$\beta_C(t)$ 为宇宙射电信号在天线 A、B、C 上的投射系数。

在 2.2 小节中提到，平台干扰信号比低频射电天文信号大约 3～10 个量级。因此，分离平台干扰是进行有效天文观测的一个基本前提。

经典的 CLEAN 算法在脏图或残余图搜索最大值，此位置处包含部分真实目标，用减点源扩散函数（PSF）的方式将这部分真实目标提取出来，如此不断迭代，直到残余图中只剩噪声结构为止[21]。CLEAN 算法同样可以应用到时域信号上，比如利用 CLEAN 方法对时间序列进行频谱分析[22]，利用时域 CLEAN 识别声源[23]，以及利用 CLEAN 方法提取一些特定结构的信号，比如周期信号、准周期信号等等[24]。

我们利用 CLEAN 算法提取 LFRS 的平台干扰信号。算法的输入数据为一组（2 个以上）天线的观测数据。首先，计算每个天线的频谱，以及每一对天线的互相关功率谱，并将所有的互相关功率谱叠加起来。由于每个天线的接收机噪声 $N(t)$ 是相互独立的，因此互相关功率谱中的噪声水平会降低，而叠加后的互相关功率谱的信噪比会进一步增强，这有利于提取平台干扰信号。

叠加的互相关功率谱包含了平台干扰信号对应的频谱成分，同时囊括了振幅和相位信息。根据这些信息，我们借助 CLEAN 算法和傅立叶级数展开公式，可以逐一将某个频率成分对应的三角函数信号从时域数据中扣除。考虑到每个天线及其接收机的增益并不一致，在扣除时每个天线的 CLEAN 因子是不一样的。

离散傅立叶变化所对应的时域及频域信号都是周期信号。如果信号首尾的值并不一致，就会在傅立叶变换时出现频谱泄漏，导致频率估计出现偏差。通常，实际采集到的信号并不满足首尾连续的条件，正如图 6 中 LFRS 的 A、B、C 天线信号所示的那样。一般可以采用加窗函数的方式，来避免频谱泄漏。但是加窗函数后会导致信号发生扭曲，幅度值发生变化，假如我们直接采用傅立叶反变换的方式来进行信号分离，则会导致较大的偏差。为此，我们引入 CLEAN 方法，通过加窗函数的方式，来获得准确的频谱估计；然后在信号分离时，逐步扣除某一个谐波分量，可以避免扭曲信号，从而把首尾不连续的干扰信号精确的分离出来。

当大部分平台干扰信号被 CLEAN 算法提取出来后，残余信号主要由接收机噪声所主导，此时可以停止 CLEAN。来自宇宙的射电信号，其幅度往往小于噪声水平，加上小部分平台干扰信号，仍然隐藏在接收机噪声中。之后，可以通过对时域信号积分平均的手段，提高残余信号的信噪比，并利用信号随时间的变化特性来进一步辨别、分离平台干扰信号和低频射电天文信号，从而实现对干扰的抑制。

3.2 CLEAN 算法的流程

按照 3.1 子小节的思路，我们给出用 CLEAN 算法分离平台干扰信号的流程图。为了更简明清晰的展示算法，我们先考虑两根天线（分别记为 A 和 B）的情形，并且只考虑连续时域信号。定义在某一道观测数据中，天线 A 的时域数据记为 $f^A(t)$，天线 B 的时域数据记为 $f^B(t)$。

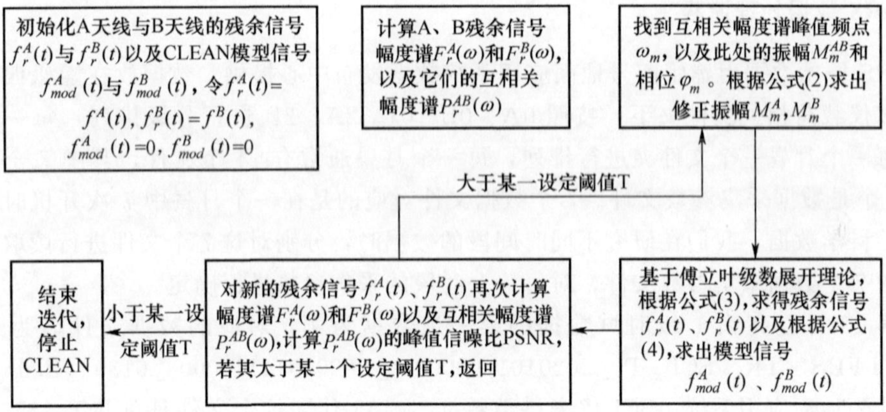

图 5　CLEAN 算法流程图

Fig. 5　CLEAN algorithm flow chart

相关的公式如公式（2）、（3）、（4）所示

$$M_m^A = \frac{\overline{M^A}}{\overline{M^{AB}}} M_m^{AB}, \quad M_m^B = \frac{\overline{M^A}}{\overline{M^{AB}}} M_m^{AB} \tag{2}$$

其中 $\overline{M^A}$ 和$\overline{M^B}$ 分别为 $F_r^A(\omega)$ 与 $F_r^B(\omega)$ 的平均振幅，$\overline{M^{AB}}$ 为 $P_r^{AB}(\omega)$ 的平均振幅。

$$\begin{aligned} f_r^A(t) &= f_r^A(t) - \delta M_m^A \cos(\omega_m t + \varphi_m) \\ f_r^B(t) &= f_r^B(t) - \delta M_m^B \cos(\omega_m t + \varphi_m) \end{aligned} \tag{3}$$

$$\begin{aligned} f_{mod}^A(t) &= f_{mod}^A(t) + \delta M_m^A \cos(\omega_m t + \varphi_m) \\ f_{mod}^B(t) &= f_{mod}^B(t) + \delta M_m^B \cos(\omega_m t + \varphi_m) \end{aligned} \tag{4}$$

本算法使用了修正振幅 M_m^A 和 M_m^B，没有直接采用频谱 $F_r^A(\omega)$ 和 $F_r^B(\omega)$ 在频点 ω_m 处的振幅，其原因是当信噪比较低时，频谱的振幅由噪声谱振幅主导，而互相关功率谱 $P_r^{AB}(\omega)$ 在频点 ω_m 处的振幅 M_m^{AB} 受噪声的影响较小。

在这里，我们假定天线 A 和 B 的时域信号 $f^A(t)$ 和 $f^B(t)$ 是同步的。若由于线路补偿没有做好，两个时域信号之间存在时间延迟，那么需要通过交叉相关等方法计算该时间延迟，并对时域信号做延迟修正，然后再用 CLEAN 算法做平台干扰信号的提取。

通过 CLEAN 算法分离平台干扰信号之后，得到的残余信号主要成分是接收机噪声，以及隐藏其中的宇宙射电信号，还有少部分没有扣除干净的平台干扰信号。此时两个天线残余信号之间的相关系数比较低，例如绝对值小于 0.3。这样，通过计算信噪比发现，残余信号中的平台干扰信号和接收机噪声基本在同一个量级水平上。再往后，可以通过积分平均的方式来提高残余信号的信噪比，这与传统辐射计的工作思路一致。至于平台干扰信号和宇宙射电信号的区分，则要借助三个天线信号随频谱、时间和来波方向变化的规律。

4 LFRS 实例数据处理

4.1 LFRS 数据处理流程

LFRS 原始数据由嫦娥四号地面应用系统数据发布中心提供。数据放在嫦娥四号低频射电频谱仪载荷相关文件夹下，按照 0A、0B、01、2A、2B 进行分级排列，每一级数据均按照每一个月昼一个文件夹进行排列，每一个月昼通常有 14 个文件，其中 7 个是数据文件，7 个是数据采集参数文件。7 个数据文件对应的是在一个月昼中 7 次开机时间段所采集到的科学数据，我们在研究不同时间段的数据时，分别对这 7 个文件进行读取，并结合相应的 7 个数据采集参数文件，对每一个时段的采集参数进行确定。

以第 23 月昼高频 1 级时域数据的第一道数据为例。对应的数据文件名为 CE4_GRAS_LFRS-TR_SCI_P_20201014020000_20201014140000_0135_B.01，其中 GRAS 代表地面应用系统，SCI 代表科学数据，0135 代表这个文件是总共第 135 个数据文件，数据结尾的 01 扩展名代表数据是 1 级数据，为时域数据。从文件名可以看出采集的数据时间段是从 2020 年 10 月 14 日 2 点 0 分 0 秒到 2020 年 10 月 14 日 14 点 0 分 0 秒，这个时间是世界时，转换为北京时间还需要加 8 小时进行换算。

在数据文件中，数据按照道的顺序依次排列，每一道数据由数据头（79 字节）和实际科学数据（8192 字节）组成。其中数据头包含了仪器的相关参数，科学数据共有 4096 个科学数据点，每个数据 2 个字节，由 16 位无符号整数组成，高字节在前，低字节在后。读取后还需要将整数电压转化为浮点数电压值，具体数据形式可参考文档《探月工程嫦娥四号任务着陆器低频射电频谱仪数据预处理方法设计》[25]。第一道数据一共采集了 4096 个点，高频数据的采样间隔为 0.01μs，低频数据的采样间隔为 0.2μs。

我们是通过 python 编程的方式，将数据按照道的顺序依次读取出来，将每一道 4096 个点排成一行，生成一个科学数据矩阵。处理的时候，提取每一行（每一道）数据，分别用第 3.2 子小节所述的 CLEAN 方法进行处理。

4.2 LFRS 部分处理结果展示

在本子小节，我们将分别展示 CLEAN 前后天线时域数据（图 6、图 11），频域数据（图 8、图 13），流量密度随频率分布的曲线（图 9），CLEAN 后 model 信号（图 12），互相关功率谱（图 10、图 14），相关系数随迭代次数变化曲线（图 15），并对相应的结果进行说明。

从 CLEAN 前的时域数据（图 6）中可以看到，A 天线和 B 天线的信号值范围基本接近，均为−0.5V～0.5V 左右。C 天线数据比 A、B 天线略大。三个天线的信号随时间变化的结构具有较高的相似性，A、B 之间的相关系数高达 0.965，B、C 之间的相关系数为

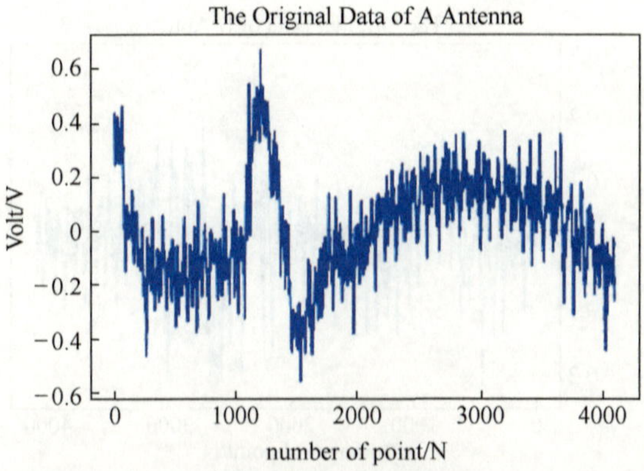

（a）A天线时域数据

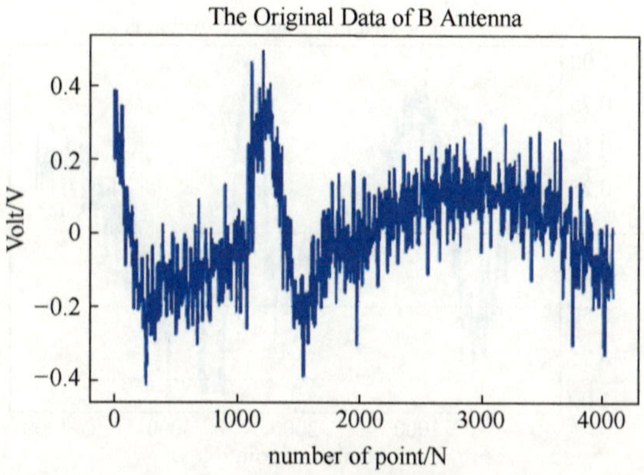

（b）B天线时域数据

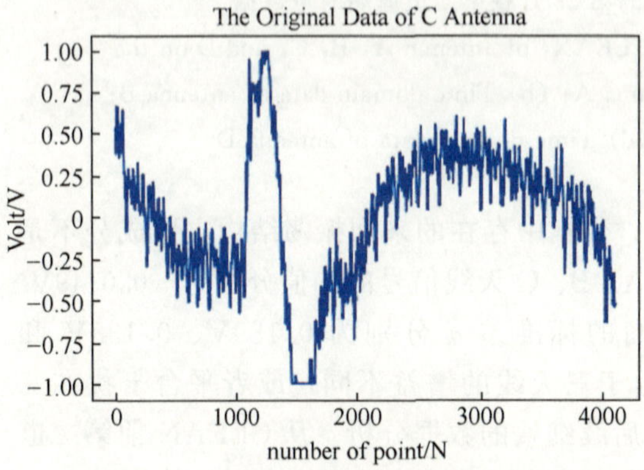

（c）C天线时域数据

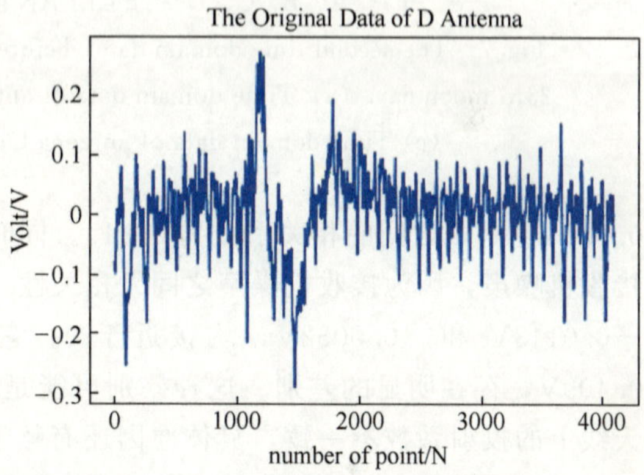

（d）D天线时域数据

图 6　A、B、C、D 天线 CLEAN 前的第 23 月昼第一道原始时域数据

Fig. 6　The first time - domain data, before CLEAN, of antenna A, B, C, and D on the 23rd moon day.
（a）Time domain data of antenna A;（b）Time domain data of antenna B;
（c）Time domain data of antenna C;（d）Time domain data of antenna D

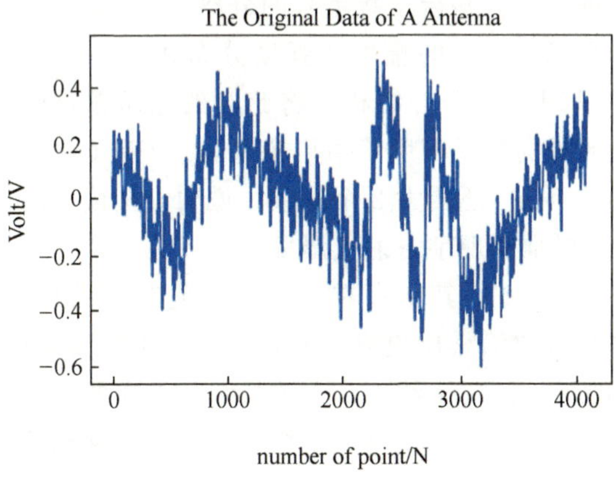

（a）A天线时域数据

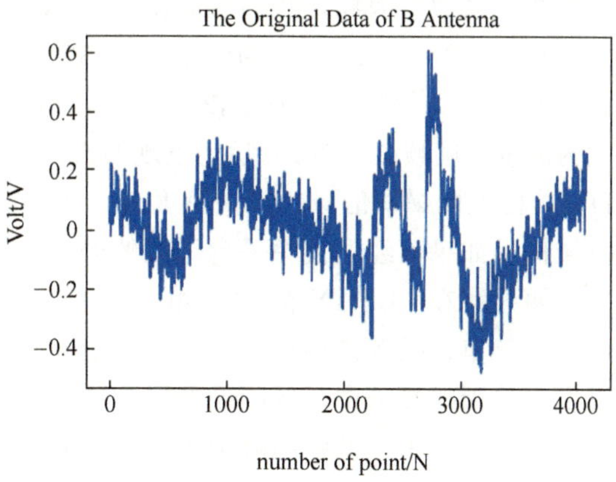

（b）B天线时域数据

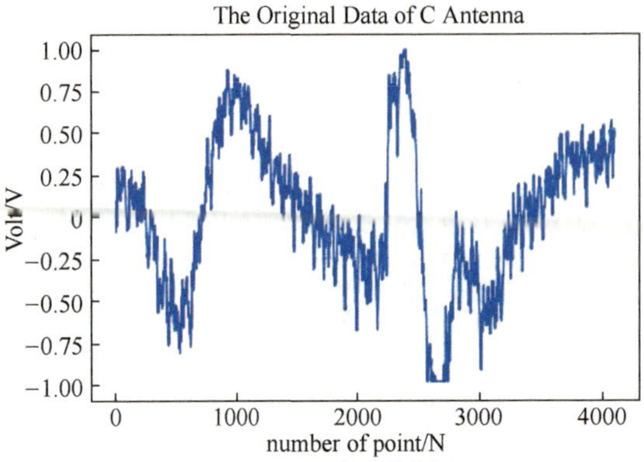

（c）C天线时域数据

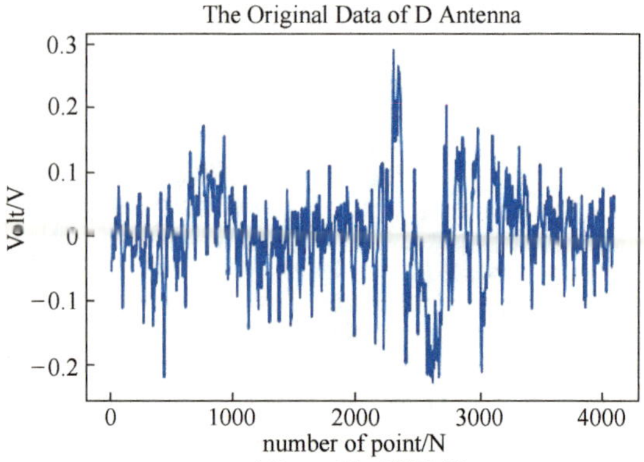

（d）D天线时域数据

图 7　A、B、C、D 天线 CLEAN 前的第 23 月昼第二道原始时域数据

Fig. 7　The second time domain data, before CLEAN, of antenna A, B, C, and D on the 23rd moon day.（a）Time domain data of antenna A;（b）Time domain data of antenna B;（c）Time domain data of antenna C;（d）Time domain data of antenna D

0.823，A、C 之间的相关系数为 0.915。同时，数据中存在的来回振荡结构主要成分不是接收机噪声，因为接收机噪声之间无相关性。A、B、C 天线信号的均值分别为−0.0149V、−0.0143V 和−0.0032V，均接近于 0。它们的标准差 σ 分别为 0.189V、0.138V 和 0.409V，存在明显的差别。这种差别可能是由于三天线的增益不同，或者平台干扰在三天线上的投射系数不一致，具体原因还有赖于后续细致的数据分析。从 CLEAN 前第二道时域数据（图 7）中可以看到，第一道数据与第二道数据时域信号有较为明显的差别。

从四根天线的数据来看，发现 D 天线的响应值比其余三根天线小很多，这是由于 D 天线长度只有 20cm，比 A、B、C 长度（5m）小很多造成的。同时，D 天线信号与其余三根天线信号在形状上也具有明显的差别，说明 D 天线的频谱响应明显异于其余三根天线，并且这种差异是非线性的。从天线之间的相关系数也可以看出，A、B、C 三根天线与 D 天线的数据相关系数分别为 0.440，0.397 和 0.298，显著低于 A、B、C 天线信号之间的相关系数。因此，我们放弃使用 D 天线数据来进行平台干扰信号的分离。

从两天线 CLEAN 前的频谱（图 8）可以看到，频谱的幅度谱强度分布在 10^{-1}V～10^{-5}V 之间，低频成分强度较大，并且频谱中呈现出复杂的结构。将此时的电压频谱转化为流量密度频谱（图 9）后，发现两天线较低频段的流量密度峰值可超过 10^{-16}W/(m^2·Hz)，比太阳爆发信号的流量密度 10^{-18}W/(m^2·Hz) 还要大 2 个量级。说明直接读取的时域信号如果不经过去除平台干扰的处理，我们很难发现宇宙射电爆发信号。

两天线在 CLEAN 前的互相关功率谱（图 10）显示其中存在一些较为复杂的结构。此外，与单天线频谱相比较，两天线互相关功率谱的信噪比更高。对比图 10 和图 8，可以发现单天线频谱中，低频的峰值大约比高频的平均值高了约 1 个多数量级，而在两天线互相关功率谱中，低频的峰值大约比高频的平均值高了约 2 个多数量级。因此，从互相关功率谱出发，更有利于把平台干扰信号给提取出来。

我们进行 CLEAN 时选取的 δ 增益因子为 0.2，迭代次数设置为 10 万次。从经过 CLEAN 之后，天线 A 和天线 B 的残余信号显示在图 8 中。可以看出，CLEAN 前天线时

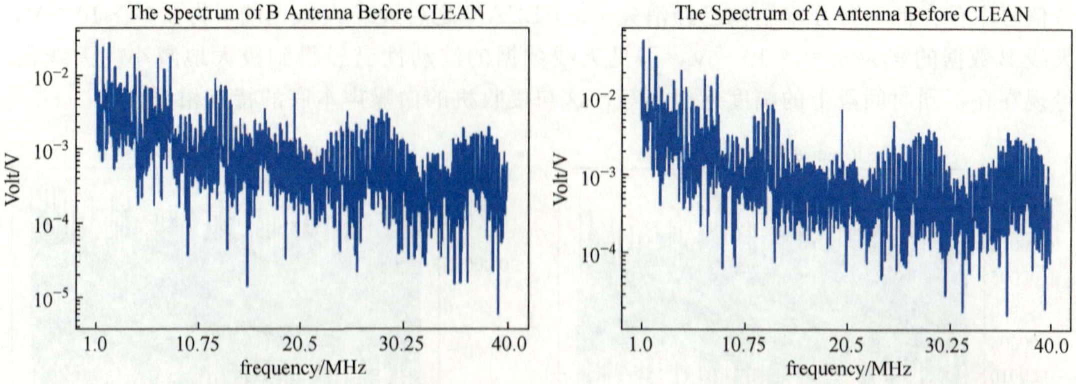

（a）A天线时域数据CLEAN前的频谱　　（b）B天线时域数据CLEAN前的频谱

图 8　A、B 天线 CLEAN 前第 23 月昼第一道数据的频谱

Fig. 8　The spectrum of the first data, before CLEAN, of antenna A and B on the 23rd moon day
(a) The spectrum, before CLEAN, of antenna A; (b) The spectrum, before CLEAN, of antenna B

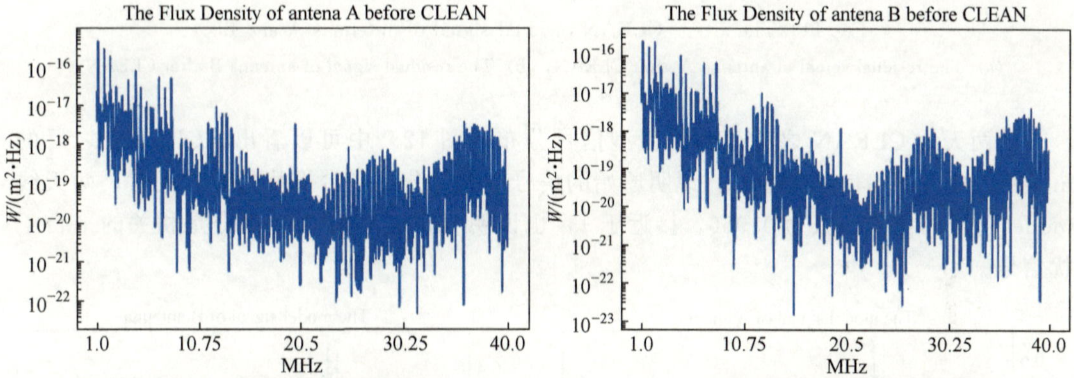

（a）A天线CLEAN前流量密度　　（b）B天线CLEAN前流量密度

图 9　A、B 天线第 23 月昼第一道数据 CLEAN 前的流量密度

Fig. 9　The first flux density data, before CLEAN, of antennas A and B on the 23rd moon day
(a) Flux density of antenna A before CLEAN; (b) Flux density of antenna B before CLEAN

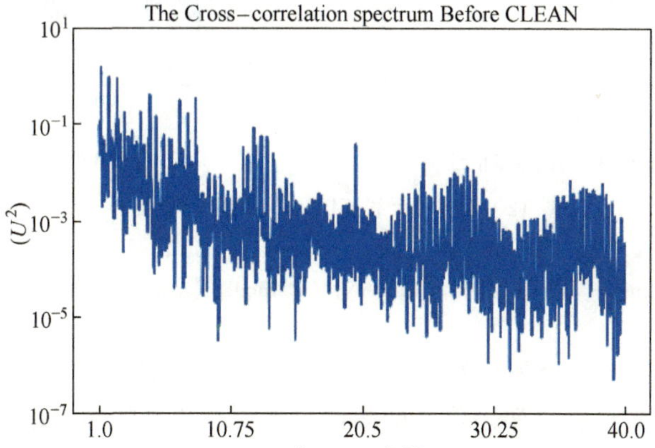

图 10　A、B 天线 CLEAN 前的互相关功率谱

Fig. 10　Cross power spectrum of antennas A and B before CLEAN

域信号中存在的一些强峰结构已经消失了。CLEAN 之后天线 A 数据的 σ 为 6.58×10^{-5} V，天线 B 数据的 σ 为 5.36×10^{-5} V，可见天线数据的波动性已经得到极大地减小。天线信号现在在不同时间段上的幅度近似相等，这与接收机的白噪声本底的特点相符合。

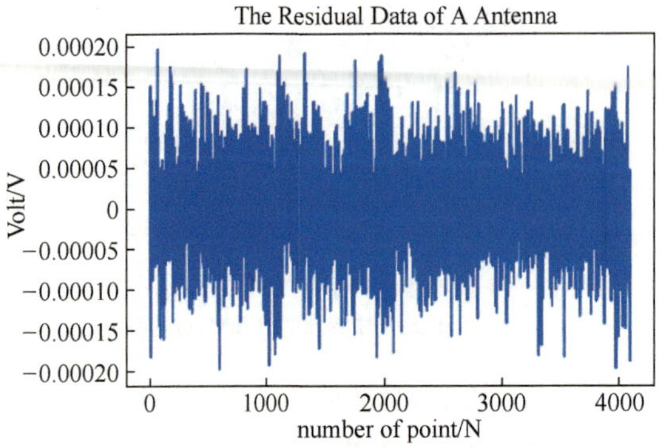

(a) A天线CLEAN后的残余信号

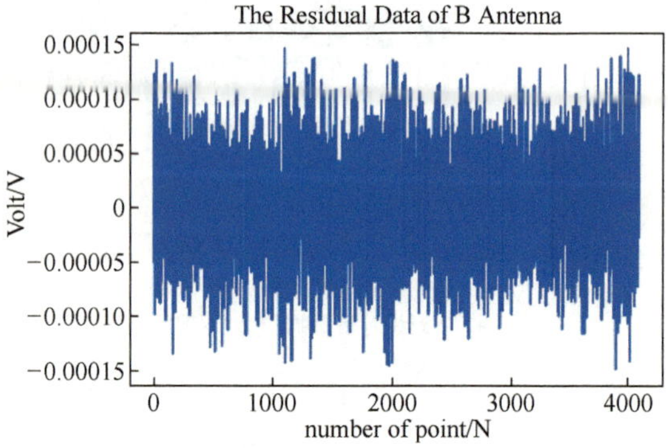

(b) B天线CLEAN后的残余信号

图 11　A、B 天线 CLEAN 后的残余信号

Fig. 11　The after - CLEAN residual signal of antennas A and B

(a) The residual signal of antenna A after CLEAN; (b) The residual signal of antenna B after CLEAN

从两天线 CLEAN 之后得到的模型信号分布（图 12）中可以看出，CLEAN 之后的 model 信号与初始信号很接近，说明原始的信号中大部分是平台干扰信号。同时，两天线 model 信号的相关系数为 0.966，接近于 1，也说明 model 信号中大部分是相关的平台干扰信号。

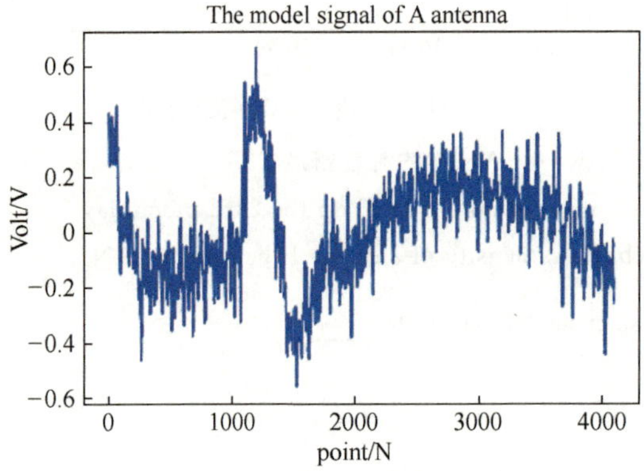

(a) A天线CLEAN后的模型信号

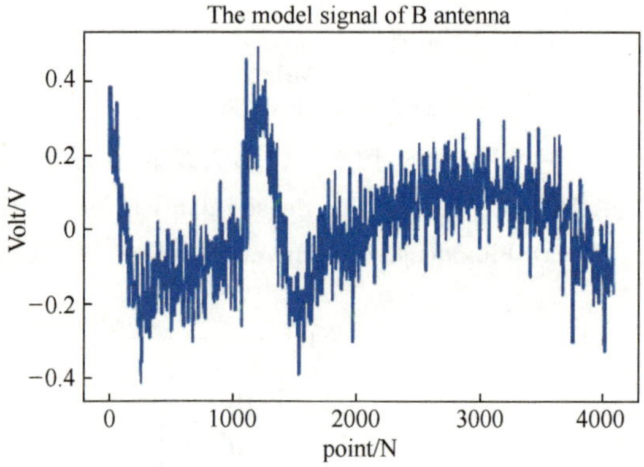

(b) B天线CLEAN后的模型信号

图 12　A、B 天线 CLEAN 后的模型信号

Fig. 12　The model signal of antenna A and B after CLEAN

(a) Model signal of antenna A after CLEAN; (b) Model signal of antenna B after CLEAN

CLEAN 之后两天线残余信号的频谱（图 13）幅度峰值由原来的约 10^{-1} V 降低到约 10^{-5} V，降低约 4 个数量级。残余信号的频谱也不再有明显的频谱结构，在各个频率分量的强度大致相等，可以看作白噪声谱。这说明，天线信号中的强平台干扰成分得到了非常有效的去除。

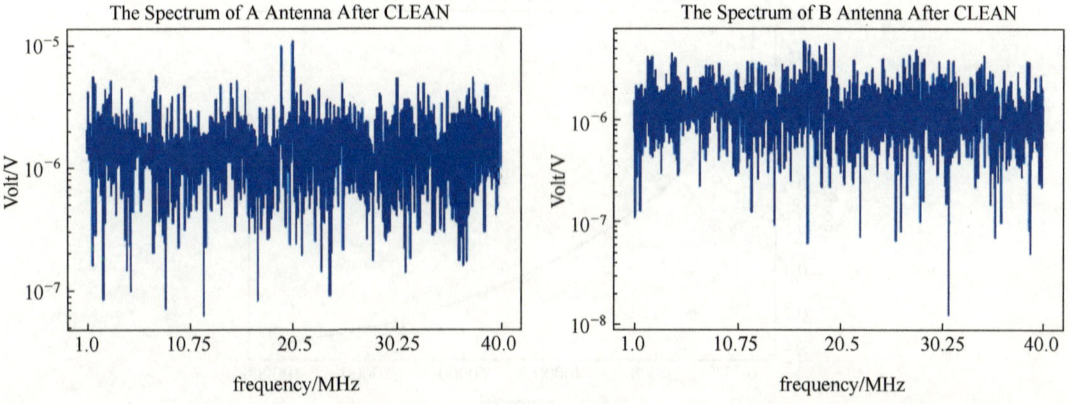

（a）A天线CLEAN后频谱　　（b）B天线CLEAN后频谱

图 13　A、B 天线直接 CLEAN 后频谱

Fig. 13　Spectrum of antennas A and B after CLEAN

（a）Spectrum of antenna A after CLEAN；（b）Spectrum of antenna B after CLEAN

CLEAN 后的互相关功率谱（图 14）同样印证了算法的有效性。可以看出，两天线的互相关功率谱中已经没有明显的峰值结构，各个频段的互相关功率谱峰值都接近于 10^{-9} V^2，对比图 7 中的结果，整整降低了约 9 个数量级。

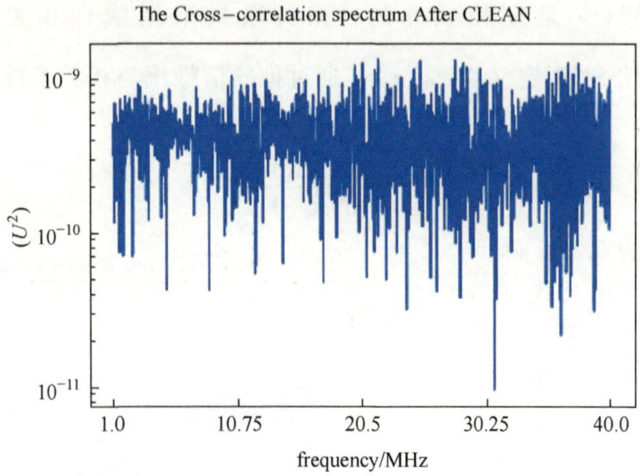

图 14　A、B 天线 CLEAN 后的末态互相关功率谱

Fig. 14　The cross - power spectrum of antennas A and B after CLEAN

我们给出了两天线信号之间的相关系数随 CLEAN 迭代次数变化的曲线（图 15）。CLEAN 之前初始的相关系数为 0.965，由强平台干扰信号所主导；CLEAN 十万次之后的相关系数接近于 0.2，根据此相关系数计算出的信噪比，可以反映出此时残余信号中的平台干扰信号和接收机噪声基本在同一个水平上，此时信号由接收机噪声所主导。当然，CLEAN 结束之后，两天线信号的相关系数并没有显著接近于 0。这说明，残余信号中可能包含有来自宇宙的低频射电信号，以及没有彻底清除的平台干扰信号。

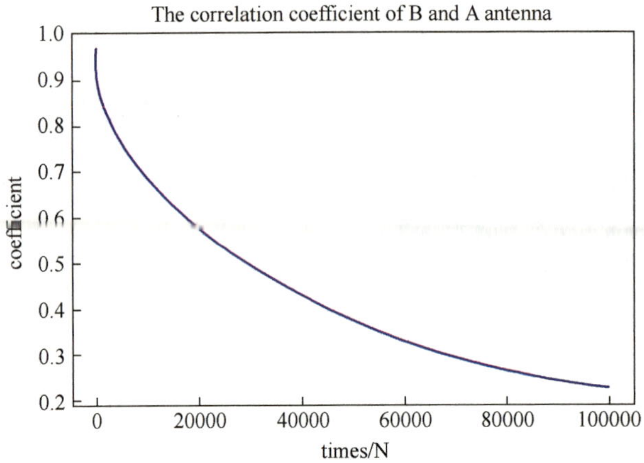

图 15　A、B 天线 CLEAN 时相关系数随迭代次数分布曲线

Fig. 15　Distribution curve of correlation coefficient between antennas A and B vs. CLEAN iteration.

4.3　CLEAN 方法效果分析

我们假设 A、B 两个天线的时域数据分别表示为 $f^A(t)=S(t)+N^A(t)$ 和 $f^B(t)=S(t)+N^B(t)$，其中 $S(t)$ 为完全相同或相关的信号成分，设其均值为 0，σ 记为 σ_S，$N^A(t)$ 和 $N^B(t)$ 分别表示 A 和 B 天线的接收机噪声信号，其均值也都为 0，σ 都记为 σ_N。但是，$N^A(t)$ 和 $N^B(t)$ 是互相独立的。相关系数采用线性相关系数的定义 $r=\frac{\mathrm{cov}(A,\ B)}{\sigma_A \cdot \sigma_B}$，信噪比 SNR 定义为 σ_S/σ_N。通过推导信噪比与相关系数之间的关系，可以得到如下公式：

$$r=\frac{SNR^2}{SNR^2+1} \tag{5}$$

当相关系数从最初的 0.965 降低到最后接近 0.2 时，由公式 (5) 计算得到的信噪比 SNR 从 5.3 降低到了 0.5。说明在末态残余信号中，相干的信号成分和不相干的噪声成分几乎在同一个水平上。

通过将末态信号转化为流量密度，得到 B 天线末态信号的流量密度随频率的分布（图 16）。从中可见，末态流量密度基本上降低到 10^{-23} W/(m^2 · Hz) 以下，与太阳宁静时的信号水平相当。后续可以在此基础上对强射电爆发进行研究。

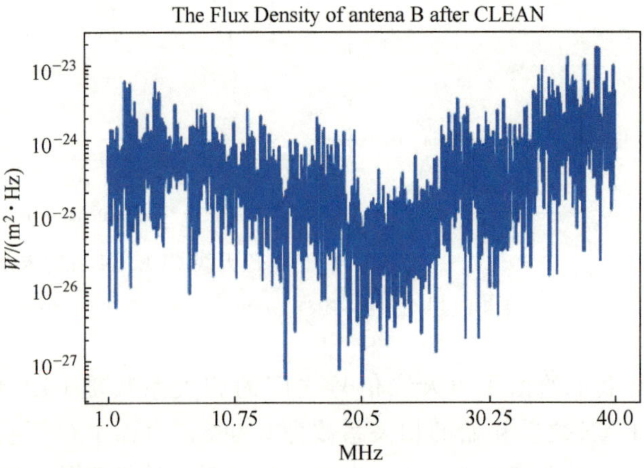

图 16　B 天线 CLEAN 后残余信号的流量密度

Fig. 16　The residual flux density of antenna B after CLEAN

5　讨论

经过 CLEAN 算法处理之后，嫦娥四号低频射电频谱仪各天线的探测灵敏度（未积分）从 10^{-15} W/(m^2 · Hz) 提高到 10^{-23} W/(m^2 · Hz)，整整 8 个数量级。在此灵敏度下，可以对太阳爆发进行监测。太阳爆发的流量密度一般在 10^{-18} W/(m^2 · Hz)，极端爆发可以达到 10^{-15} W/(m^2 · Hz)。除此之外，这样的灵敏度也足以探测部分强的快速射电爆（Fast Radio Burst，FRB），其辐射流量密度可以达到几十万央斯基（Jy），1Jy=10^{-26} W/(m^2 · Hz)。如果再对时域数据进行积分平均，那么可以进一步提高探测灵敏度。例如，如果对一道 4096 个数据进行积分平均，则灵敏度可以额外提高 64 倍。这将大大拓展低频射电频谱仪的科学探测目标。

一系列用 CLEAN 算法分离出来的平台干扰信号也很有用。首先是被用来分析平台干扰信号的频谱、空间和时间变化特征，并确定其来源，以便在今后开发类似设备的时候，改进电磁兼容设计，大幅度降低平台干扰水平。其次，在全面了解平台干扰信号特征的基础上，充分利用这些知识，发现那些特征异常的信号，这些异常信号往往来自宇宙。

分离并抑制干扰信号的最终目的是进行有效的天文观测。那么，如何在强烈的平台干扰背景下寻找并确认宇宙射电信号呢？主要的思路是寻找异常信号。例如，频谱异常。平台干扰往往有固定的一组频谱特征，一旦有例外的频谱信号出现，可能意味着记录到了来自宇宙的射电爆发信号。如果这些频谱还具备色散等特征，那么几乎可以肯定它来自宇宙。再例如，A、B、C 天线信号的强度比例系数。若平台干扰信号产生的空间分布是固定的，那么不管其在时域如何变化，信号之间的强度比例系数也是固定的。倘若这些比例系数有明显的改变，则往往意味着天线接收到了来自宇宙的强射电爆发信号。若射电爆发信号的时标长达几天（此处 1 天=24 小时）的量级，那么这些比例系数还会随着月球自转，发生规律性变化，有助于更加可靠的确定爆发信号的来源。当然，具体如何搜寻异常信号，还得依赖于后续细致的数据分析工作。

本论文提出的方法也可以应用到其他射电观测设备，尤其是地基的综合射电干涉阵。目前地基干涉阵受到越来越多的来自地面设备以及空间卫星的无线电干扰。借助 CLEAN 算法，我们利用两两天线之间的互相关功率谱，在信号层面把干扰信号提取出来，再对残余数据做相关获得可见度数据，对可见度数据作成图等处理工作。同样，也可以对提取出的干扰信号进行类似的操作，获得其详细的频谱、空间分布、时域变化等信息。

6　结论

本论文提出一种方法，它从两组信号的相关性出发，基于 CLEAN 算法，并借助互相关功率谱、傅立叶级数展开等工具，把低频射电频谱仪 A、B、C 天线的时域观测数据分为强相关的 CLEAN 模型信号和部分相关的残余信号两个部分。其中 CLEAN 模型信号主要由平台干扰信号和可能的低频强射电爆发组成；残余信号由接收机噪声、未扣除干净的平台干扰信号和常规的低频射电信号组成。将该方法应用到实际数据中，结果表明嫦娥四号低频射电频谱仪的未积分灵敏度可以提高约 8 个数量级，达到 10^{-23} W/(m^2 · Hz)

水平。

在此基础上，基于对平台干扰信号中确定性成分和宽带随机成分的分别处理，再借助低频射电爆发信号和平台干扰信号在功率谱上不同的表现，以及常规低频射电信号受月球自转调制等信息，将来工作的重点是进一步分析 CLEAN 模型信号和残余信号，以期搜索低频强射电爆发信号，乃至对全天区进行粗略的成像。

参考文献

[1] 梅丽，苏彦，周建锋．极低频射电天文观测现状与未来发展 [J]. 天文研究与技术，2018，15 (2)：127 - 149.

[2] Leshem A , Veen A V D . Detection and blanking of GSM interference in radio - astronomical observations [C] // IEEE Workshop on Signal Processing Advances in Wireless Communications. IEEE，2002.

[3] Fridman P A , Baan W A . RFI mitigation methods in radio astronomy [J]. Journal of Astronomy & Astrophysics，2001，378 (1)：327 - 344.

[4] Ellingson S W , Bunton J D , Bell J F . Removal of the GLONASS C/A Signal from OH Spectral Line Observations Using a Parametric Modeling Technique [J]. The Astrophysical Journal Supplement Series，2000，135 (1).

[5] Ellingson S W . Beamforming and interference canceling with very large wideband arrays [J]. IEEE Transactions on Antennas and Propagation，2003，51 (6)：1338 - 1346.

[6] Cecilia B , Bradley R F. A New Approach to Interference Excision in Radio Astronomy：Real - Time Adaptive Cancellation [J]. Astronomical Journal，1998，116 (5)：2598.

[7] Briggs F H , Bell J F , Kesteven M J . Removing radio interference from contaminated astronomical spectra using an independent reference signal and closure relations [J]. Astronomical Journal，2000，120 (6)：3351 - 3361.

[8] Leshem A , van der Veen，Alle - Jan，Boonstra，Albert - Jan. Multichannel interference mitigation methods in radio astronomy [J]. Astrophysical Journal Supplement，2000，131 (1)：355.

[9] Friedman，P. [IEEE Comput. Soc. Press 8th Workshop on Statistical Signal and Array Processing - Corfu，Greece (24 - 26 June 1996)] Proceedings of 8th Workshop on Statistical Signal and Array Processing - A change point detection method for elimination of industrial interference in radio astronomy receivers [J] 1996：264 - 266.

[10] Weber R , Faye C , Biraud F , et al. Spectral detector for interference time blanking using quantized correlator [J]. Astronomy & Astrophysics Supplement，1997，126 (1)：161 - 167.

[11] Maslakovic S , Linscott I R , Oslick M , et al. [IEEE Third International Symposium on Time - Frequency and Time - Scale Analysis (TFTS - 96) - Paris，France (18 - 21

June 1996)] Proceedings of Third International Symposium on Time - Frequency and Time - Scale Analysis (TFTS - 96) - Excising radio frequency interference using the discrete wavelet transform [J]. 1996: 349 - 352.

[12] Kasper B L , Chute F S , Routledge D . Excising terrestrial radio interference in low frequency radio astronomy [J]. Monthly Notices of the Royal Astronomical Society (2): 345 - 354.

[13] Cecilia B , Bradley R F . A New Approach to Interference Excision in Radio Astronomy: Real - Time Adaptive Cancellation [J]. Astronomical Journal, 1998, 116 (5): 2598.

[14] Baan W A , Fridman P A , Millenaar R P . Radio Frequency Interference Mitigation at the Westerbork Synthesis Radio Telescope: Algorithms, Test Observations, and System Implementation [J]. Astronomical Journal, 2004, 128 (2): 933 - 949.

[15] Offringa A R , De B A G , Biehl M , et al. Post - correlation radio frequency interference classification methods [J]. Monthly Notices of the Royal Astronomical Society, 2010, 405 (1): 155 - 167.

[16] Zhao, J, Zou, et al. WindSat Radio - Frequency Interference Signature and Its Identification Over Greenland and Antarctic [J]. IEEE Transactions on Geoscience & Remote Sensing, 2013, 51 (9): 4830 - 4839.

[17] C. J. Wolfaardt, Machine learning approach to radio frequency interference (rfi) classification in radio astronomy [D]. Stellenbosch University (2016).

[18] 张韬，苏彦 . 嫦娥四号低频射电频谱仪降低背景噪声方法的研究 [J]. 天文研究与技术，2019，16 (3)：312 - 320.

[19] Bastian, T. S , Benz, et al. Radio emission from solar flares. [J]. Annual Review of Astronomy & Astrophysics, 1998, 36: 131 - 188.

[20] Bochenek C D, Ravi V, Belov K V, et al. A fast radio burst associated with a Galactic magnetar [J]. Nature, 2020, 587 (7832): 59 - 62.

[21] Hogbom J A. Aperture Synthesis with a Non - Regular Distribution of Interferometer Baselines [J]. Astronony Astrophys Suppl, 1974, 15: 417.

[22] Heslop D, Dekkers M J. Spectral analysis of unevenly spaced climatic time series using CLEAN: signal recovery and derivation of significance levels using a Monte Carlo simulation [J]. Physics of the earth and Planetary Interiors, 2002, 130 (1 - 2): 103 - 116.

[23] Cousson R, Leclere Q, Pallas M A, et al. A time domain CLEAN approach for the identification of acoustic moving sources [J]. Journal of Sound and Vibration, 2019, 443: 47 - 62.

[24] Kulpa K . The CLEAN type algorithms for radar signal processing [M]. IEEE, 2008.

[25] 朱新颖，探月工程嫦娥四号任务着陆器低频射电频谱仪数据预处理方法设计 [R]. 北京 . 2018.

基于嫦娥四号月球车图像的地形遮挡高精度预报试验研究

马友青[1]，彭松[2]，温博[2]，贾阳[2]，申振荣[2]，亓晨[3]，张烁[1]，杨欢[3]，司马兵[1]，王楷文[1]，刘少创[1]

1. 中国科学院遥感与数字地球研究所，北京 100101
2. 北京空间飞行器总体设计部，北京 100094
3. 武汉大学测绘学院，武汉 430072

摘要：为进行天际线遮挡角的准确预报，提出了月球车图像的地形遮挡角高精度计算方法。首先阐述了月球车图像的地形遮挡角预报流程，然后重点分析了月球车相机在轨几何标定、基于闭合区域网平差的月球车相机在轨位姿参数估计、月球车位姿参数高精度计算和天际线自动提取等关键技术。最后，通过我国嫦娥四号月球车的专项试验，验证了月球车图像的地形遮挡角高精度计算方法具有较高的地形遮挡角预测精度和应用稳定性。

关键词：嫦娥四号月球车；地形遮挡角；在轨几何标定；天际线自动提取；闭合区域网平差

Experimental research of precise terrain occlusion prediction algorithm based on image sequences of the Chang'E－4 lunar rover

MA YouQing[1], PENG Song[2], WEN Bo[2], JIA Yang[2], SHEN ZhenRong[2], QI Chen[3], ZHANG Shuo[1], YANG Huan[3], SIMA Bing[1], WANG KaiWen[1] & LIU ShaoChuang[1]

1. Institute of Remote Sensing and Digital Earth, Chinese Academy of Sciences, Beijing 100101, China
2. Beijing Institute of Spacecraft System Engineering, Beijing 100094, China
3. School of Geodesy and Geomatics, Wuhan University, Wuhan 430072, China

Abstract: An accurate terrain occlusion prediction algorithm based on image sequences obtained by the Chang'E (CE - 4) lunar rover stereo cameras is proposed. First, the proposed algorithm is briefly introduced. Next, the key procedures, including on - orbit camera calibration, high - precision camera pose estimation based on a loop—closed bundle block adjustment model, high - precision rover pose estimation, and automatic extraction of the skyline, are explained. Lastly, the conclusions are drawn based on the experimental results of China's Chang'E - 4 lunar rover. The proposed algorithm can achieve high accuracy and application stability in the occlusion prediction of the skyline.

Keywords: Chang'E - 4 lunar rover; terrain occlusion; on - orbit camera calibration; automatic skyline extraction; loop - closed bundle block adjustment

1 引言

在我国探月工程中，嫦娥四号由"鹊桥"中继星、着陆器和月球车组成。其中，中继星于 2018 年 5 月 21 日发射升空，着陆器携带月球车计划于 2018 年 12 月登陆月球背对地球一面的南极附近的艾特肯（South Pole Aitken，SPA）盆地[1]。SPA 盆地作为嫦娥四号探测器最合适的着陆区，区域内最大高程差 6518m，坡度大于 8°的区域占总区域面积比为 18.5%，且分布着若干个形成于酒海纪或前酒海纪的较为古老的撞击坑和许多后期叠加的撞击坑或盆地。嫦娥四号探测器在 SPA 盆地工作时，周围由山脉包围，极易处在地势较低的地方。通过星历可以准确计算太阳和中继星在某个时刻北东地系下的方位，当中继星的方位（方位角和高度角）处于月球车天线指向角度范围外，或者太阳边缘处于月球车（可视为质点）与山脉连线的下方，即月球车处于光照阴影区域时，将影响月球车的能源获取和通信[2]为制定地形遮挡应对策略，需研究地形遮挡预报方法。利用月球车遥测数据准确预报地形遮挡情况，有利于月球车太阳翼展开收拢、月球车天线转向、月球车加电断电和移动转向等策略的生成，有利于节省月球车能源、提高工作效率和降低工程任务风险。

嫦娥四号月球车与嫦娥三号月球车结构类似，其立体相机系统均由一对导航相机、一对全景相机和一对避障相机构成，搭载在相机云台的两端，云台安装在桅杆上节点处，云台与桅杆构成一个整体，安装在月球车上[3,4]。立体相机系统感知周围地形信息，获取全景序列图像，如图 1 所示。因此，在嫦娥四号月球车遥测数据中，只有相机序列图像包含周围地形信息，可以为地形遮挡角计算提供解决思路。

由于应用领域的局限性，国内外开展基于月球车图像数据进行地形遮挡角的高精度计算的研究极少。李飞等人[2]利用月球数字地形模型（digital elevation model，DEM）和月球车位置，以及相机对周围地形拍摄照片的分析，初步预测出可能存在光照遮挡的方位和遮挡的时长。Cozman 等人[5]利用行星巡视器相机图像、姿态角信息和轨道器影像，提出基于概率框架模型的匹配方法，获得巡视器高精度的位置与姿态角信息。Oiri 等人[6]利用行星巡视器全景相机图像和 DEM 的匹配，获得巡视器的全局位置。Ahmad 等人[7]、Bouyssounouse 等人[8]、Nefian 等人[9]提出了利用行星月球车图像边缘信息和局部特征点提取天际线（地形与天空的交界线）的方法，然后与 DEM 图像中的虚拟天际线进行配准，实现无 GPS 或路标点信息的月球车定位定向。Ahmad 等人[7]提出利用边缘图像和局部特征匹配点（SIFT 匹配像点）等信息，然后利用机器学习方法对全景图像进行水平天际线的检测。Bouyssounouse 等人[8]通过构建行星巡视器全景图像和 DEM 图像中天际线的匹配最优化函数，实现巡视器的姿态角的全局计算。Nefian 等人[9]利用视觉里程计、DEM 和天际线匹配的综合结果，实现行星巡视全局位置与姿态角的预估。Grelsson 等人[10]通过对鱼眼相机图像进行天际线的检测实现飞机的姿态角的精确计算。Fefilatyev 等人[11]针对雾天和多云情境下的图像，利用机器学习的方法检测天际线。针对图像边缘信息二值化和易出现空洞的情况，Ahmad 等人[12,13]提出在分类后的图像上利用动态规划法进行天际线提取的方法。Ahmad 等人[14]认为天际线提取是一种分类问题，探讨了局部纹理特征点提取算子的适用性。Grishin 和 Maslov[15]针对用于目标物导航的水平线不稳定性，提出一种水平线检测和测量的方法。在其他如无人机、小型飞行器和海事无人艇等领域[16—19]，通过图像边缘检测、图像分割（聚类）和图像分类等多种方法的组合应用，实现天际线或者地平线的自动检测。然而，这些方法在计算地形遮挡角时，地形遮挡角精度严重受制于 DEM 图像和天际线匹配的精度，较低精度的 DEM 无法保证地形遮挡角的计算精度和准确性。

为提高月球车在轨工作的效率，本文提出了基于图像和月球车部分结构控制信息的地形遮挡角高精度计算方法。与基于 DEM 图像和月球车图像的天际线匹配算法相比较，新方法具有以下优点：(1) 对相机内方位元素[20]重新进行了计算，避免受到外太空环境、月球车降落过程中不规则震颤等因素的影响发生变化时，仍采用地面几何标定值导致地形遮挡角计算精度降低的情况；(2) 消除了月球车桅杆机构转角的测量误差（0.3°，3σ），提高了相机遮挡角的计算精度；(3) 进一步抑制了月球车 IMU 设备获取遥测姿态角的测量误差（1°，3σ）；(4) 提高了天际线检测精度。在给出嫦娥四号月球车图像地形遮挡角预报流程的基础上，本文从相机在轨几何标定、月球车相机位姿参数和月球车位姿参数高精度计算以及天际线的自动提取等关键技术进行原理的阐述。通过嫦娥四号月球车（验证器）图像遮挡角试验性的研究，表明了新方法具有可行性和高精度的特性。

2 月球车图像地形遮挡预报流程

以月球车左导航相机图像为例，在图 1 中，首先通过天际线自动提取算法提取图像里的天际线，获得若干个天际线特征点 A 对应的像点 a，其像素坐标 x，y 通过相机标定，获得内方位元素（主距 f，像主点坐标 x_0，y_0）、畸变系数（k_1，k_2，p_1，p_2），然后对像点坐标进行畸变纠正，补偿像点坐标系统误差[3]（Δx，Δy），获得纠正后的像点坐标 $\underline{x}=x-\Delta x$，$\underline{y}=y-\Delta y$。构建在相机系下的遮挡角方向矢量 γ_S：

$$\gamma_S=[\cos\eta_S\sin\tau_S \quad \sin\eta_S \quad -\cos\eta_S\cos\tau_S]^{\mathrm{T}} \tag{1}$$

其中，天际线特征点的高度角

$$\eta_S=\mathrm{atan}((\underline{y}-y_0)/\sqrt{f^2+(\underline{x}-x_0)^2})$$

方位角 $\tau_S=\mathrm{atan}((\underline{x}-x_0)/f)$，符号 sin，cos 和 atan 分别表示正弦、余弦和反正切函数。

在月球车一个工作任务周期内，将在周期起始时刻建立工作坐标系（即北东地系），并进行若干次移动，月球车共行驶距离一般为 20～30m。相较于相机与天际线特征点的

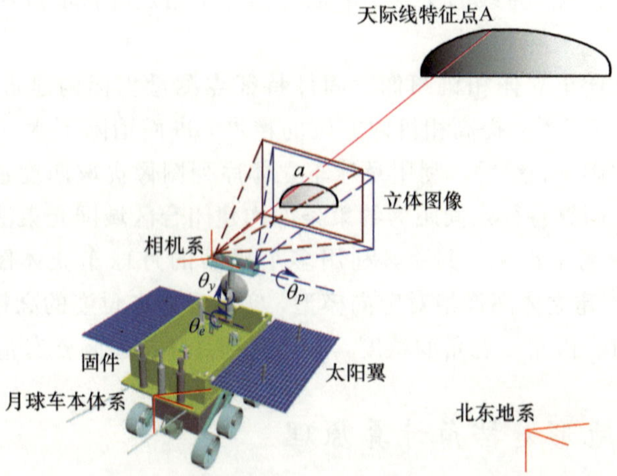

图 1　（网络版彩图）月球车导航相机工作示意图

Fig. 1　(Color online) Work principle sketch map of the lunar rover's stereo navigation cameras system

距离（10km 级），相机与月球车车体的距离（1.2m）和月球车质心与工作坐标系原点的距离可忽略不计。当月球车姿态和桅杆姿态均处于零位状态，相机系原点与特征点 A 构成的方向矢量可近似为月球车本体系原点与特征点 A 的方向矢量。也可近似为工作坐标系原点与特征点 A 的方向矢量。通过上述分析，在北东地系下的遮挡角方向矢量 γ 通过月球车坐标系统转换框架[3]、月球车姿态角（偏航角 κ_R、俯仰角 ω_R、滚动角 φ_R）建立如下关系：

$$\gamma = \mathbf{R}_{\mathrm{NED-R}} \mathbf{R}_{\mathrm{S-R}} \gamma_S \tag{2}$$

其中，$\mathbf{R}_{\mathrm{NED-R}}$ 表示月球车本体系与北东地系的旋转矩阵，$\mathbf{R}_{\mathrm{S-R}}$ 表示相机系与月球车本体系的旋转矩阵，$\mathbf{R}_{\mathrm{S-R}} = \mathbf{R}_{\mathrm{expansion}} \mathbf{R}_{\mathrm{yaw}} \mathbf{R}_{\mathrm{pitch}}$。$\mathbf{R}_{\mathrm{expansion}}$ 表示机械臂绕桅杆展开系 Y 轴旋转形成的转换矩阵，展开角为 θ_e，$\mathbf{R}_{\mathrm{yaw}}$ 表示桅杆绕偏航坐标系 Z 轴旋转形成的旋转矩阵，偏航角为 θ_y，$\mathbf{R}_{\mathrm{pitch}}$ 为桅杆绕桅杆俯仰坐标系 Y 轴旋转形成的旋转矩阵，俯仰角为 θ_p，具体表达式请参看文献 [3]。由方向矢量 $\boldsymbol{\gamma} = (\boldsymbol{\gamma}_X \quad \boldsymbol{\gamma}_Y \quad \boldsymbol{\gamma}_Z)$ 计算天际线特征点在北东地系下的高度角 η 和方位角 τ：

$$\begin{cases} \eta = -\mathrm{asin}(\gamma_Z / \sqrt{\gamma_X^2 + \gamma_Y^2}) \\ \tau = \mathrm{atan}(\gamma_Y / \gamma_X) \end{cases} \tag{3}$$

其中，符号 asin 表示反正弦函数，方位角以北偏东为正，北偏西为负，高度角以水平向上为正，向下为负。式（3）表明，高度角 η 和方位角 τ 的计算精度受到天际线像点坐标、相机内方位元素、桅杆机构转角以及月球车姿态角的影响。桅杆机构转角（θ_e，θ_y，θ_p）和 IMU 设备获取月球车姿态角的理论测量精度分别为 0.3°（3σ）和 1.0°（3σ）。相较于相机内方位元素和天际线像点坐标的精度，对高度角和方位角的影响程度更高。在图 1 中，月球车车体后方上部分结构（固件）处于立体相机视场范围内，且在发射前已精测出其在月球车本体系下的坐标，这些固件与月球车车体关系保持不变。在月球车能源不充足的情况下，太阳翼将展开至一定角度，尽可能接受最大范围的太阳光照射。太阳翼由排列整齐的长方形电池片构成，电池片直线的交点、电池片的 4 个角点等特征点在发射前已精测出其长度。

本文首先利用月球车立体相机图像、固件特征点测量坐标构建自检校光束法平差模型，获得相机在轨几何参数，提高相机内方位的精度，进而消除不规则震颤等因素对相机几何参数和遮挡角的影响；然后，利用月球车立体序列图像提取四度重叠像点，并以固件特征点测量坐标和太阳翼特征点间距为约束条件构建闭合区域网光束法平差模型，消除月球车桅杆机构转角的测量误差。其次，利用多个摄站的月球车立体图像，采用 SIFT 算子[21]提取连接点，构建立体图像绝对定向模型，实现月球车位姿的高精度计算，提高月球车当前摄站相对于前一摄站姿态角的精度，有效抑制月球车遥测姿态角的误差累积效应。

3　月球车图像地形遮挡角计算原理

3.1　月球车相机在轨几何标定

月球车等深空巡视探测器视觉系统在轨几何标定的主要方法包括：Di 和 Li[22]使用了一种基于 4 个矢量 **CAHV** 的相机模型，每个字母分别代表模型的一个矢量。其中，矢量 **C** 为物方坐标系到相机光心的矢量，**A** 为垂直于像平面的单位矢量，**H** 和 **V** 分别为描述水平信息和垂直信息的矢量。**CAHV** 模型应用在美国的勇气号和机遇号的导航相机上，解决了物方坐标系向像方坐标系的变换问题。在 **CAHV** 模型的基础上，Gennery[23]增加了一个向量和三元组 **OR** 来描述畸变参数，并将 **CAHVOR** 模型运用在勇气号和机遇号的全景相机模型中。其中，**O** 为光轴单位矢量，描述透镜畸变，**R** 为物方坐标系下径向畸变系数组成的三元组向量。为了对避障相机进行建模，Gennery 设计了 **CAHVORE** 模型，重新描述光线束在像平面的投影位置。Heikkila[24]提出了非对称性投影的校正方法，具有一定的适用性。张烁等人[25]提出了基于月球车全景相机的正交变换投影模型，一定程度上克服了透视投影模型相机标定中的病态性。刘少创等人[26]、张烁等人[27]提出了基于标定控制场的月球车立体相机检校方法，可以实现相机检校参数的高精度标定，但需要控制点信息的支持，因此限制了该方法在地外天体环境下的应用。针对此问题，马友青等人[3]利用同名匹配像点采用加权总体最小二乘平差计算月球车立体相机相对方位元素，不依赖于控制点。

为充分利用月球车结构上的控制信息，本文提出月球车相机在轨几何标定模型，以式(4) 共线条件方程为基础，以像点坐标、固件（控制点）测量坐标和相机外方位元素的初始值为输入。

$$\begin{cases} x - x_0 - \Delta x = -f\dfrac{a_1(X-X_S)+b_1(Y-Y_S)+c_1(Z-Z_S)}{a_3(X-X_S)+b_3(Y-Y_S)+c_3(Z-Z_S)} \\ y - y_0 - \Delta y = -f\dfrac{a_2(X-X_S)+b_2(Y-Y_S)+c_2(Z-Z_S)}{a_3(X-X_S)+b_3(Y-Y_S)+c_3(Z-Z_S)} \end{cases} \tag{4}$$

式中，X_S，Y_S，Z_S 为相机图像在月球车本体系下的外方位线元素，a_1，a_2，a_3，b_1，b_2，b_3，c_1，c_2，c_3 构成旋转矩阵 $\mathbf{R}_{S-R}$，其独立元素为 φ，ω 和 κ，X，Y，Z 为控制点三维坐标，在计算过程中视为真值。对共线方程式进行线性化，如下所示：

$$
\begin{cases}
v_x = \dfrac{\partial x}{\partial X_S}\Delta X_S + \dfrac{\partial x}{\partial Y_S}\Delta Y_S + \dfrac{\partial x}{\partial Z_S}\Delta Z_S + \dfrac{\partial x}{\partial \varphi}\Delta \varphi \\
\quad + \dfrac{\partial x}{\partial \omega}\Delta \omega + \dfrac{\partial x}{\partial \kappa}\Delta \kappa + \dfrac{\partial x}{\partial f}\Delta f \\
\quad + \dfrac{\partial x}{\partial x_0}\Delta x_0 + \dfrac{\partial x}{\partial y_0}\Delta y_0 + \dfrac{\partial x}{\partial k_1}\Delta k_1 \\
\quad + \dfrac{\partial x}{\partial k_2}\Delta k_2 + \dfrac{\partial x}{\partial p_1}\Delta p_1 + \dfrac{\partial x}{\partial p_2}\Delta p_2 - l_x \\
v_y = \dfrac{\partial y}{\partial X_S}\Delta X_S + \dfrac{\partial y}{\partial Y_S}\Delta Y_S + \dfrac{\partial y}{\partial Z_S}\Delta Z_S + \dfrac{\partial y}{\partial \varphi}\Delta \varphi \\
\quad + \dfrac{\partial y}{\partial \omega}\Delta \omega + \dfrac{\partial y}{\partial \kappa}\Delta \kappa \\
\quad + \dfrac{\partial y}{\partial x_0}\Delta x_0 + \dfrac{\partial y}{\partial y_0}\Delta y_0 + \dfrac{\partial y}{\partial k_1}\Delta k_1 \\
\quad + \dfrac{\partial y}{\partial k_2}\Delta k_2 + \dfrac{\partial y}{\partial p_1}\Delta p_1 + \dfrac{\partial y}{\partial p_2}\Delta p_2 - l_y
\end{cases}
\tag{5}
$$

式中，v_x 和 v_y 表示像点坐标的改正量；ΔX_S，ΔY_S，ΔZ_S，$\Delta\varphi$，$\Delta\omega$，$\Delta\kappa$ 分别表示外方位线元素和角元素的增量；Δf，Δx_0，Δy_0 分别表示主距和像主点的增量；Δk_1，Δk_2，Δp_1，Δp_2 表示畸变系数的增量；l_x，l_y 表示残差。将式（5）改写为

$$\mathbf{V}=\mathbf{B}\hat{\mathbf{x}}-\mathbf{l}-\mathbf{P} \tag{6}$$

式中，$\mathbf{V}=[v_x, v_y]^{\mathrm{T}}$ 表示像点坐标的改正量向量，$\mathbf{l}=[l_x, l_y]^{\mathrm{T}}$ 表示残差向量，$\hat{\mathbf{x}}=[\mathbf{x_E}\ \mathbf{x_I}]^{\mathrm{T}}$ 为相机待标定参数的增量向量，$\mathbf{x_E}=[\Delta X_S\quad \Delta Y_S\quad \Delta Z_S\quad \Delta\varphi\quad \Delta\omega\quad \Delta\kappa]$ 为相机外方位元素的增量向量，$\mathbf{x_I}=[\Delta f\quad \Delta x_0\quad \Delta y_0\quad \Delta k_1\quad \Delta k_2\quad \Delta p_1\quad \Delta p_2]$ 为相机内方位元素的增量。

$$\mathbf{B}=\left[\frac{\partial x}{\partial[\mathbf{x_E}\quad \mathbf{x_I}]}\quad \frac{\partial y}{\partial[\mathbf{x_E}\quad \mathbf{x_I}]}\right]^{\mathrm{T}}$$

为系数矩阵，具体表达式见文献［20］。$\mathbf{P}$ 表示权矩阵。定义位置 S_i 的左相机系在月球车本体系下的外参数（位置向量 $\mathbf{T}_i^{\mathrm{l}}$ 和姿态角矩阵 $\mathbf{R}_i^{\mathrm{l}}$），右相机系在月球车本体系下的外参数（位置向量 $\mathbf{T}_i^{\mathrm{r}}$ 和姿态角矩阵 $\mathbf{R}_i^{\mathrm{r}}$），则相对位姿关系有如下约束条件：

$$\mathbf{E}=\begin{bmatrix}((\mathbf{R}_i^{\mathrm{l}})^{\mathrm{T}}\mathbf{R}_i^{\mathrm{r}})_{21}\\ ((\mathbf{R}_i^{\mathrm{l}})^{\mathrm{T}}\mathbf{R}_i^{\mathrm{r}})_{31}\\ ((\mathbf{R}_i^{\mathrm{l}})^{\mathrm{T}}\mathbf{R}_i^{\mathrm{r}})_{32}\\ (\mathbf{R}_i^{\mathrm{l}})^{\mathrm{T}}(\mathbf{T}_i^{\mathrm{r}}-\mathbf{T}_i^{\mathrm{l}})\end{bmatrix}-\mathbf{C}_1=\mathbf{0} \tag{7}$$

其中，$\mathbf{C}_1$ 为常数向量，$((\mathbf{R}_i^{\mathrm{l}})^{\mathrm{T}}\mathbf{R}_i^{\mathrm{r}})_{jk}$ 为矩阵 $(\mathbf{R}_i^{\mathrm{l}})^{\mathrm{T}}\mathbf{R}_i^{\mathrm{r}}$ 的第 j 行第 k 列数值。为与式（6）表示一致，函数 $\mathbf{E}$ 对外方位参数线性化：

$$\mathbf{J}\hat{\mathbf{x}}+\mathbf{W}=\mathbf{0} \tag{8}$$

其中，$\mathbf{W}=-\mathbf{E}^0-\mathbf{C}_1$，$\mathbf{J}=[\mathbf{J}_i \quad \underset{7\times7}{\mathbf{0}}]^{\mathrm{T}}$，$\mathbf{E}^0$ 表示三维坐标参数为当前迭代值下的函数值，$\mathbf{J}_i=\delta\mathbf{E}/\delta\boldsymbol{x}_{\mathbf{E}_i}$，$\delta$ 为导数符号. 通过附有限制条件的间接平差[20]得到

$$\hat{\mathbf{x}}=(\mathbf{N}_{\mathbf{BB}}^{-1}-\mathbf{N}_{\mathbf{BB}}^{-1}\mathbf{J}^{\mathrm{T}}\mathbf{N}_{\mathbf{JJ}}^{-1}\mathbf{J}\mathbf{N}_{\mathbf{BB}}^{-1})\mathbf{W}_x-\mathbf{N}_{\mathbf{BB}}^{-1}\mathbf{J}^{\mathrm{T}}\mathbf{N}_{\mathbf{JJ}}^{-1}\mathbf{W} \tag{9}$$

其中，$\mathbf{W}_x=\mathbf{B}^{\mathrm{T}}\mathbf{Pl}$，$\mathbf{N}_{\mathbf{BB}}=\mathbf{B}^{\mathrm{T}}\mathbf{PB}$，$\mathbf{N}_{\mathbf{JJ}}=\mathbf{J}\mathbf{N}_{\mathbf{BB}}^{-1}\mathbf{J}^{\mathrm{T}}$，单位权方差为 $\sigma_0=\sqrt{\mathbf{V}^{\mathrm{T}}\mathbf{PV}/(n-u+s)}$，$n$ 表示观测值总数，u 表示待估计参数的个数，s 表示约束条件的数目。参数中误差为 $\sigma_{\hat{x}\hat{x}}=\sigma_0\sqrt{\mathbf{Q}_{\hat{x}\hat{x}}}$，$\mathbf{Q}_{\hat{x}\hat{x}}=\mathbf{N}_{\mathbf{BB}}^{-1}-\mathbf{N}_{\mathbf{BB}}^{-1}\mathbf{J}^{\mathrm{T}}\mathbf{N}_{JJ}^{-1}\mathbf{J}\mathbf{N}_{\mathbf{BB}}^{-1}$，表示待标定参数的协因数矩阵。通过参数中误差的计算，可以获得所有待标定参数的理论精度数值。

月球车在轨工作过程中，应将相机云台下俯一定角度，使固件尽可能多地出现在相机视场范围内。

3.2　月球车相机在轨位姿参数高精度计算

为提高月球车相机图像遮挡角计算的精度，相机系在月球车本体系下的位姿参数精度是极其重要的。由于月球车在轨工作时，导航相机工作频率高于全景相机，且视场角为46°，相机图像能同时包含天空和月球车车体固件和太阳翼部分信息。而全景相机序列图像视场难以同时包括天空和固件，需要调整相机成像的俯仰角。为抑制月球车桅杆机构转角误差对天际线特征点遮挡角的影响，本文提出利用月球车序列立体图像、固件特征点坐标和太阳翼电池片长度控制信息构建闭合区域网光束法平差模型，实现相机在轨位姿参数的高精度计算，如图 2 所示。

如图 2 所示，月球车立体相机将以一定角度间隔（在 10°～20°区间）从月球车正后方为起始点逆时针旋转拍照，获取序列立体图像. 以相机位置 S_1 的相机系为起始基准坐标系。在相机位置 S_1 和位置 S_2，利用 SIFT 算子首先进行左图像的匹配，获得部分匹配像点集合 p_1^1 和 p_2^1。然后利用核线平行对齐的限制条件，通过相关系数匹配算法[20]获得右图像的同名像点 p_1^{r} 和 p_2^{r}。最后再次利用相关系数匹配算法对 p_1^{r} 和 p_2^{r} 进行匹配和检核，剔除误匹配点，得到四度重叠像点。通过若干个（6 个以上）分布均匀的四度重叠像点，将位置 S_1 和位置 S_2 的 4 幅图像关联起来，并建立光束法平差模型，从而获得位置 S_2 的相机系与位置 S_1 的相机系的位姿关系（旋转矩阵 $\mathbf{R}_1^2$ 和平移向量 $\mathbf{T}_1^2$）。以此类推，最后一个位置 S_n 相机系可通过位置 S_{n-1} 相机系关联。由于存在像点提取误差、四度重叠区域较小的情况，导致区域网光束法平差模型的参数解算精度较低，后续相机位姿参数将偏离真实值，且所有的相机系均在起始基准坐标系下。

在图 2 中 S_n，S_1，S_2 等相机成像位置，月球车固件可提供在月球车本体下的三维坐标，如图 2 十字丝所示的固件控制点。在月球车左右方向的相机位置 S_i 和 S_j，太阳翼电池片能提供长度控制信息，即电池片角点间的距离，且角点位于图像四度重叠像点中. 将这些控制信息融入区域网光束法平差模型中，不仅可以将相机系和起始基准坐标系纳入月球车本体系下，而且将未闭合的区域网平差模型强制转换为闭合区域网光束法平差模型。

以 3.1 节月球车在轨相机标定参数为已知值，利用图像四度重叠点作为桥梁，将式（5）可以改写为

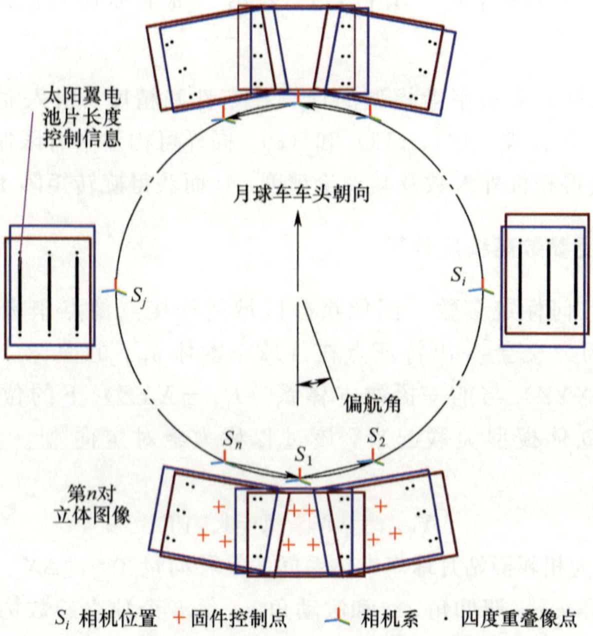

图 2　（网络版彩图）相机图像闭合区域网平差

Fig. 2　(Color online) Visual loop－closed bundle block adjustment

$$
\begin{cases}
v_x = \dfrac{\partial x}{\partial X_S}\Delta X_S + \dfrac{\partial x}{\partial Y_S}\Delta Y_S + \dfrac{\partial x}{\partial Z_S}\Delta Z_S \\
\quad + \dfrac{\partial x}{\partial \varphi}\Delta\varphi + \dfrac{\partial x}{\partial \omega}\Delta\omega + \dfrac{\partial x}{\partial \kappa}\Delta\kappa \\
\quad + \dfrac{\partial x}{\partial X}\Delta X + \dfrac{\partial x}{\partial Y}\Delta Y + \dfrac{\partial x}{\partial Z}\Delta Z - l_x \\
v_y = \dfrac{\partial y}{\partial X_S}\Delta X_S + \dfrac{\partial y}{\partial Y_S}\Delta Y_S + \dfrac{\partial y}{\partial Z_S}\Delta Z_S \\
\quad + \dfrac{\partial y}{\partial \varphi}\Delta\varphi + \dfrac{\partial y}{\partial \omega}\Delta\omega + \dfrac{\partial y}{\partial \kappa}\Delta\kappa \\
\quad + \dfrac{\partial y}{\partial X}\Delta X + \dfrac{\partial y}{\partial Y}\Delta Y + \dfrac{\partial y}{\partial Z}\Delta Z - l_y
\end{cases}
\tag{10}
$$

其中，ΔX，ΔY，ΔZ 表示特征点三维坐标的增量。当月球车车体固件上的特征点三维坐标作为真值时，ΔX，ΔY，ΔZ 设为 0，对应的像点坐标依旧按照式（10）构建误差方程式。当太阳翼电池片提取 n 个特征点 $M_i=$（$i=1$，…，n）时，具有如下关系式：

$$(X_i - X_j)^2 + (Y_i - Y_j)^2 + (Z_i - Z_j)^2 = C_2 \tag{11}$$

其中，$(X,Y,Z)_i$ 和 $(X,Y,Z)_j$ 分别表示特征点 M_i 和 M_j 的三维坐标，C_2 为常数，代表太阳翼电池片的长度信息，由地面测量获得。令式（11）表示为

$$F = (X_i - X_j)^2 + (Y_i - Y_j)^2 + (Z_i - Z_j)^2 - C_2 = 0 \tag{12}$$

函数 F 对三维坐标参数 $\overline{\mathbf{X}}=[\Delta X, \Delta Y, \Delta Z]^{\mathrm{T}}$ 线性化：

$$\overline{\mathbf{H}}\hat{\mathbf{x}} + \overline{\mathbf{W}} = \mathbf{0} \tag{13}$$

其中，$\overline{\mathbf{W}}=-\mathbf{F}^0-\mathbf{C}_2$，$\overline{\mathbf{H}}=[\underset{6\times6}{\mathbf{0}}\quad \mathbf{H}_i]^{\mathrm{T}}$，$\mathbf{F}^0$ 表示三维坐标在当前迭代值下的函数值，$\mathbf{H}_i=\delta F/\delta\overline{\mathbf{X}}_i$。

为提高闭合区域网光束法平差模型的计算稳定性和精度，引入立体相机严格几何关系，如式（7）所示。联合式（10），(13）和（8），同样可构建附有限制条件的间接平差模型。通过最小二乘获得相机外参数及其理论精度，从而获得旋转矩阵 $\mathbf{R}_{\mathrm{S-R}}$。

3.3 月球车在轨位姿参数高精度计算

通过立体相机几何标定参数、图像重叠区域的连接点像点坐标和相机在月球车本体系下的外参数，前方交会[3]出特征点在月球车本体系下的坐标。为获得当前摄站月球车本体系（R_2-XYZ）与前一摄站本体系（R_1-XYZ）下的位姿关系，需借助连接点将相邻摄站的立体模型关联起来，该过程称为绝对定向[3]。立体图像绝对定向模型的表达式如下：

$$\mathbf{Y}_{n\times3}=\rho\mathbf{A}_{n\times3}\boldsymbol{\xi}+\mathbf{1}_{n\times1}\mathbf{T} \tag{14}$$

其中，绝对定向参数为相邻摄站月球车本体系间的平移向量 $\mathbf{T}=[\Delta X\quad \Delta Y\quad \Delta Z]^{\mathrm{T}}$、$3\times3$ 的旋转矩阵 $\boldsymbol{\xi}$（偏航角 κ_R、俯仰角 ω_R 和滚动角 φ_R 等 3 个独立参数构成）和比例系数 ρ。在式（14）中，下标 n 为连接点个数，$\mathbf{A}_{n\times3}$ 和 $\mathbf{Y}_{n\times3}$ 分别为 n 个连接点在坐标系 R_2-XYZ，R_1-XYZ 下的三维坐标构成的矩阵，$\mathbf{1}_{n\times1}$ 为元素为 1 的向量. 将连接点三维坐标视为伪观测值时，令其含有随机误差 $\mathbf{E_A}$ 和 $\mathbf{E_Y}$。采用正交 Procrustes 和总体最小二乘平差[28,29]获得绝对定向参数：

$$\left.\begin{array}{l}\hat{\boldsymbol{\xi}}=\mathbf{U}\mathbf{V}^{\mathrm{T}}\\ \mathrm{tr}(\mathbf{Y}^{\mathrm{T}}\mathbf{A}')\hat{\rho}^2+\mathrm{tr}(\mathbf{A}^{\mathrm{T}}\mathbf{A}-\mathbf{Y}^{\mathrm{T}}\mathbf{Y})\hat{\rho}-\mathrm{tr}(\mathbf{Y}^{\mathrm{T}}\mathbf{A}')=0\\ \hat{\mathbf{T}}=(\mathbf{1}_{n\times1}^{\mathrm{T}}\mathbf{W}_n^2\mathbf{1}_{n\times1})^{-1}\mathbf{1}_{n\times1}^{\mathrm{T}}\mathbf{W}_n^2(\mathbf{Y}-\hat{\rho}\mathbf{A}')\end{array}\right\} \tag{15}$$

其中，矩阵 $\mathbf{A}'=\mathbf{A}\boldsymbol{\xi}$，$\mathbf{W}_n$ 反映的是［$\mathbf{A}\quad\mathbf{Y}$］行向量的权矩阵，tr（）为矩阵的迹. 令矩阵 $\boldsymbol{\chi}=\mathbf{1}_{n\times1}-(\mathbf{1}_{n\times1}^{\mathrm{T}}\mathbf{W}_n^2\mathbf{1}_{n\times1})^{-1}(\mathbf{1}_{n\times1}\mathbf{1}_{n\times1}^{\mathrm{T}}\mathbf{W}_n^2)$，$\overline{\mathbf{A}}=\mathbf{W}_n\boldsymbol{\chi}\mathbf{A}$，$\overline{\mathbf{Y}}=\mathbf{W}_n\boldsymbol{\chi}\mathbf{Y}$，对矩阵 $\overline{\mathbf{A}}^{\mathrm{T}}\overline{\mathbf{Y}}$ 进行奇异值分解(singular value decomposition， SVD)， SVD（$\overline{\mathbf{A}}^{\mathrm{T}}\overline{\mathbf{Y}}$）$=\mathbf{U}\boldsymbol{\Sigma}\mathbf{V}^{\mathrm{T}}$，$\mathbf{U}$ 和 $\mathbf{V}$ 分别为对应的左奇异正交矩阵、右奇异正交矩阵，$\boldsymbol{\Sigma}$ 为奇异值构成的对角线矩阵，具体表达式请参看文献［3，4］。观测值误差表达式如下：

$$\begin{cases}\mathbf{E_Y}=\mathbf{W}_n^{-1}(\mathbf{Y}-\hat{\rho}\mathbf{A}\hat{\boldsymbol{\xi}})/(1+\hat{\rho}^2)\\ \mathbf{E_A}=-\hat{\rho}\mathbf{E_Y}\boldsymbol{\xi}^{\mathrm{T}}\end{cases} \tag{16}$$

文献［28］指出总体最小二乘平差是有偏的。单位权方差为 $\sigma_0=\sqrt{\mathbf{e}^{\mathrm{T}}\mathbf{W}_n\mathbf{e}/(n-u)}$，其中，$\mathbf{e}=[\mathrm{vec}(\mathbf{E_Y})\ \mathrm{vec}(\mathbf{E_A})]$，vec 为矩阵拉直运算，$n$ 表示观测值总数，u 表示待估计参数的个数。通过理论证明[30]，式（15）所示的绝对定向表达式等价于基于高斯牛顿法的总体最小二乘解。因此，旋转矩阵 $\hat{\boldsymbol{\xi}}$ 中误差可通过近似函数法[31]表示为

$$\sigma_{\hat{\xi}\hat{\xi}}=\sigma_0\sqrt{((\mathbf{A}-\mathbf{E_A})^{\mathrm{T}}(\mathbf{B}\mathbf{Q_{ll}}\mathbf{B}^{\mathrm{T}})^{-1}(\mathbf{A}-\mathbf{E_A}))^{-1}} \tag{17}$$

其中，$\mathbf{Q_{ll}}=\begin{bmatrix}\mathbf{Q_{AA}} & \mathbf{Q_{AY}}\\ \mathbf{Q_{YA}} & \mathbf{Q_{YY}}\end{bmatrix}=\mathbf{W}_n^{-1}$ 为［$\mathbf{A}\ \mathbf{Y}$］的协因数矩阵，矩阵 $\mathbf{B}=[\hat{\boldsymbol{\xi}}^{\mathrm{T}}\otimes\mathbf{I_n},\ -\mathbf{I_n}]$。

式（17）表示月球车姿态角参数的理论精度。

将月球车移动过程视为一个整体，可获得当前摄站月球车本体系在第一个摄站月球车本体系下的位姿关系，即在工作坐标系下。其中，工作坐标系在全局坐标系的位姿信息设为真值。式（14）中的旋转矩阵 ξ 即式（2）中的 $\mathbf{R}_{\mathrm{NED-R}}$。

3.4　天际线提取

通过检测图像中所有灰度值变化较大的点，并将其连接起来构成了若干线条，即可完成图像的边缘检测。由于月面环境无空气，导致太阳光为直射照射，不发生漫反射和散射等地球、火星表面常见的现象。因此，月球车相机图像中的天际线表现为天空较为黑暗，地面区域较亮，且分界较为明显。本文首先对相机图像进行 3×3 模板的去噪，然后进行二值化，阈值设为 10～30；其次，采用 Canny 算子[32]进行边缘检测，轮廓阈值设为 3，选取边缘点的 Sobel 算子，内核设为 3×3 矩阵；最后，在边缘检测后的图像中逐列采用自上而下判断像素是否发生变化，若发生变化，则认为该点是天际线特征点。由于天际线为连续的曲线，需加入天际线相邻特征点的像素高度差应小于 2 个像素的限制条件。嫦娥三号玉兔号月球车导航相机未拍照包含天空的图像，图 3 和 4 为其全景相机图像利用本节方法提取的天际线。

（a）　（b）　（c）

图 3　（网络版彩图）玉兔号月球车全景相机图像天际线提取示例一

（a）原始图像；（b）边缘检测后的图像；（c）天际线提取效果图，曲线为天际线

Fig. 3　(Color online) The first visual skyline extraction example in the "YUTU" lunar rover's panoramic camera system

(a) Original panoramic image; (b) panoramic image after edge detection; (c) skyline extraction map, the curve denotes the skyline

（a）　（b）　（c）

图 4　（网络版彩图）玉兔号月球车全景相机图像天际线提取示例二

（a）原始图像；（b）边缘检测后的图像；（c）天际线提取效果图，曲线为天际线

Fig. 4　(Color online) The second visual skyline extraction example in the "YUTU" lunar rover's panoramic camera system

(a) Original panoramic image; (b) panoramic image after edge detection; (c) skyline extraction map, the curve denotes the skyline

4　试验验证

4.1　相机几何标定测试

在测试相机几何标定时，利用月球车在探测点 A2 的遥测数据。导航相机在车尾 3 个位置的图像提取 15 个控制点，且部分控制点成像在两幅或三幅图像上，构成二度和三度重叠像点，重叠度分别为 60%和 20%，如图 5 中十字丝标记所示。桅杆机构转角如表 1 所示。为便于标记和测量，控制点均位于月球车机构拐角、直线段交点位置，因此，本文采用基于 Harris 算法的角点检测方法[33]提取控制点对应的图像特征点，精度优于 1 pixel。

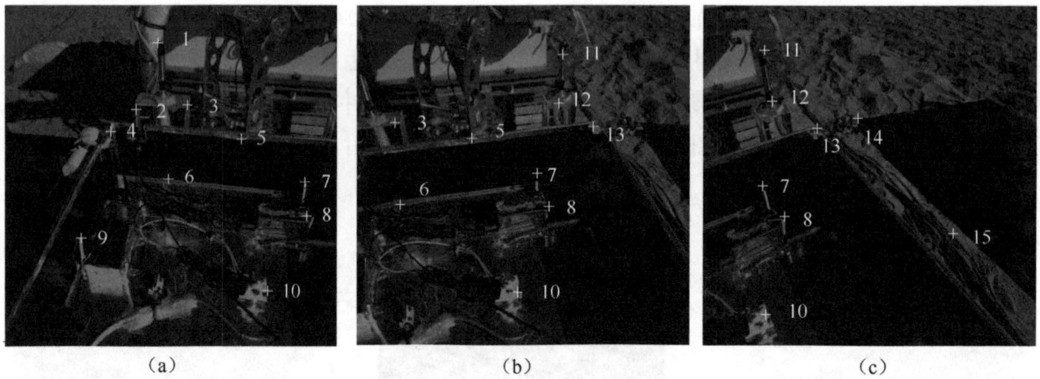

图 5　在探测点 A2 月球车左导航相机图像控制点的像点提取分布

Fig. 5　Distribution of selected image points in the left navigation camera at location A2

表 1　探测点 A2 月球车机构转角数据

Tab. 1　Rotation angles of the mast at location A2

转角	图 5（a）	图 5（b）	图 5（c）
桅杆俯仰角	−39.905°	−39.905°	−39.905°
桅杆偏航角	−169.903°	169.881°	149.885°
桅杆展开角	0.066°	0.066°	0.066°

本节对光束法平差模型进行最小二乘平差（简称为常规解，仅包含式（6））和本文提出的新方法求解（简称为新方法，包含式（6）和（8）），统计算法收敛（收敛次数为 5）时的单位权方差 σ_0，常规解 $\sigma_0=0.408$ pixel，新方法 $\sigma_0=0.380$ pixel。

图 6 表示的是探测 A2 点位置的左右导航相机图像外参数（线元素 X_S，Y_S，Z_S 和角元素 φ，ω，κ）和左导航相机内参数的精度。通过分析常规解和新方法的单位权方差与图 6 中相机内外参数精度，可说明加入立体相机固定几何关系后的光束法平差模型解算的相机内外参数精度更高。在新方法中，左导航相机畸变系数 k_1，k_2，p_1，p_2 的精度分别是 5.23×10^{-9}，1.17×10^{-14}，5.26×10^{-7}，4.78×10^{-7}。

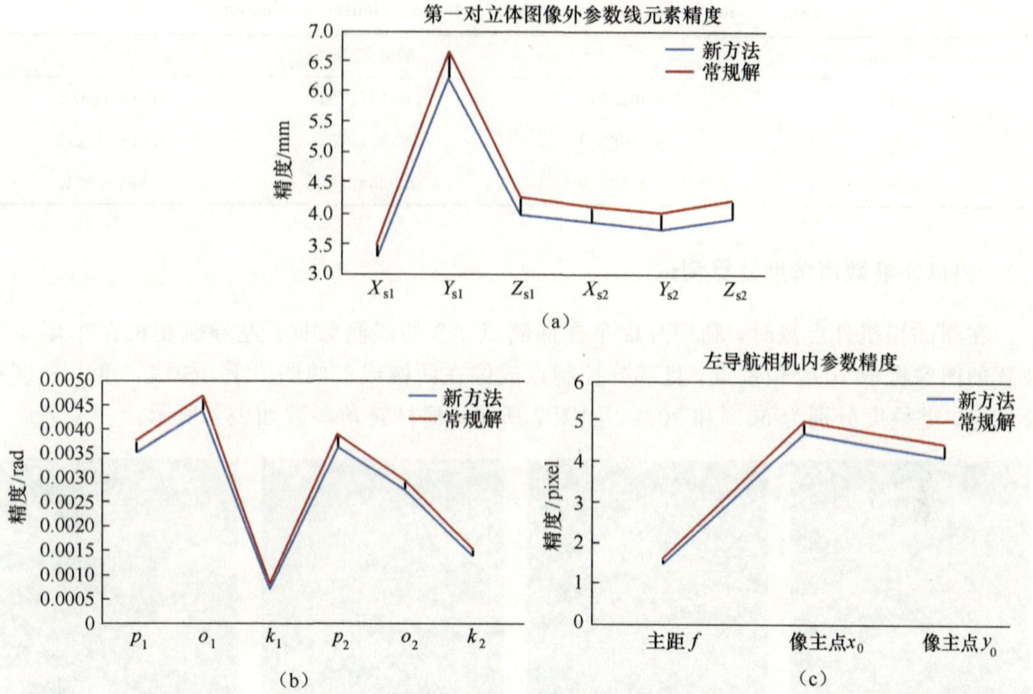

图 6　（网络版彩图）探测 A2 点位置的左右导航相机图像外参数和左导航相机内参数的精度

(a) 导航相机第一对图像外方位线元素精度，横坐标表示的是左相机图像外方位线元素（X_{S1}，Y_{S1}，Z_{S1}）和右相机图像外方位线元素（X_{S2}，Y_{S2}，Z_{S2}）；

(b) 角元素精度，横坐标（p_1，o_1，k_1，p_2，o_2，k_2）依次为左右相机图像角元素（φ，ω，κ）；

(c) 左导航相机内参数精度，横坐标依次为主距 f 和像主点坐标 x_0，y_0

Fig. 6　(Color online) Precision of the stereo navigation camera calibration and pose estimation at location A2. (a) Corresponds to the first stereo camera position (X_{S1}, Y_{S1}, Z_{S1}, X_{S2}, Y_{S2}, Z_{S2}), respectively; (b) corresponds to the first stereo camera altitude angle p_1, o_1, k_1, p_2, o_2, k_2, respectively; (c) represents the calibration parameters f, x_0, y_0 of the left navigation camera

由新方法计算出嫦娥四号月球车验证器导航相机内方位元素如表 2 所示，立体几何关系如表 3 所示。上述几何标定参数将作为后续月球车相机外参数高精度计算和月球车位姿高精度计算的基础数据。

表 2　月球车导航相机内方位元素

Tab. 2　Stereo navigation cameras' interior orientation elements

	主距/pixel	像主点坐标/pixel	径向畸变系数		切向畸变系数	
			k_1	k_2	p_1	p_2
左相机	1181.4	3.07，11.48	-2.15×10^{-8}	3.11×10^{-14}	1.46×10^{-7}	-3.13×10^{-7}
右相机	1196.6	−1.0，−18.4	-2.22×10^{-8}	3.09×10^{-14}	-5.32×10^{-9}	-2.85×10^{-7}

表 3　月球车导航相机立体几何关系

Tab. 3　Stereo navigation cameras' exterior orientation elements

位置向量/mm	旋转矩阵		
262.447	0.999991964	0.001372791	0.00376672
0.496	−0.001396945	0.999978434	0.006417225
0.859	−0.003757829	−0.006422435	0.999972315

4.2　相机外参数高精度计算测试

在测试相机外参数时，利用月球车在探测点 A2 的遥测数据。左导航相机在车尾 4 个位置的图像提取 10 个控制点，且部分控制点成像在两幅或三幅图像上，构成二度、三度重叠像点，重叠度分别为 60%和 20%，如图 7 所示，桅杆转角参数如表 4 所示。

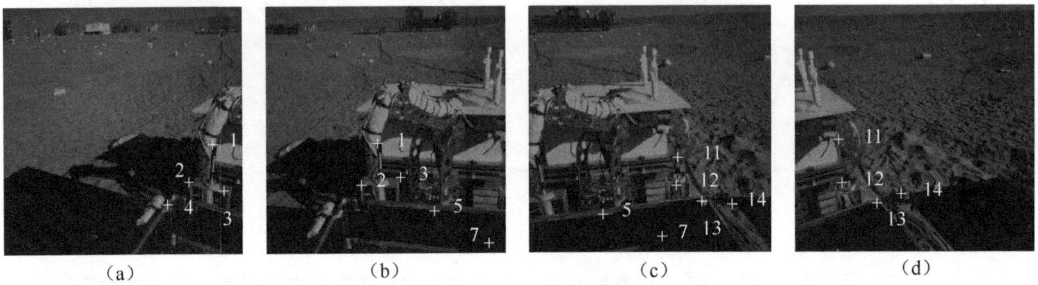

图 7　在探测点 A2 月球车左导航相机图像控制点的像点提取分布

Fig. 7　Distribution of selected image points in the left navigation camera at location A2

表 4　探测点 A2 月球车机构转角数据

Tab. 4　Rotation angles of the mast at location A2

转角	图 7 (a)	图 7 (b)	图 7 (c)	图 7 (d)
桅杆俯仰角	−19.93°	−19.93°	−19.93°	−19.93°
桅杆偏航角	−150.126°	−169.903°	169.881°	149.885°
桅杆展开角	0.066°	0.066°	0.066°	0.066°

经相机视场角分析，在月球车桅杆俯仰角为−24°～−10°的范围内，月球车导航相机图像能同时拍摄到天空和月球车固件控制点。在月球车具体工作中，环拍时桅杆俯仰角为 0°，当立体相机向后成像时，桅杆俯仰角可调至−20°，但需要保证不同俯仰角的序列图像间包含接近 20%的重叠区域。在探测点 A2，立体导航相机从后方起按逆时针成像 18 次，相邻相机位置的立体图像间自动匹配出 6 对同名像点，共计 118 个特征点，其中 10 个特征点如图 7 所示，108 个像点为四度重叠像点。在月球车左右方向，选择太阳翼电池片的长度作为特征点的约束条件，共 20 个。设置收敛次数为 5 次。控制点坐标在月球车本体系下，相机外参数初始值依据表 5 的桅杆姿态角通过本体系与相机系的坐标转换获得。使用闭合区域网平差模型，单位权方差为 1.20 pixel，36 个位置相机外参数精度如图 8 所示。

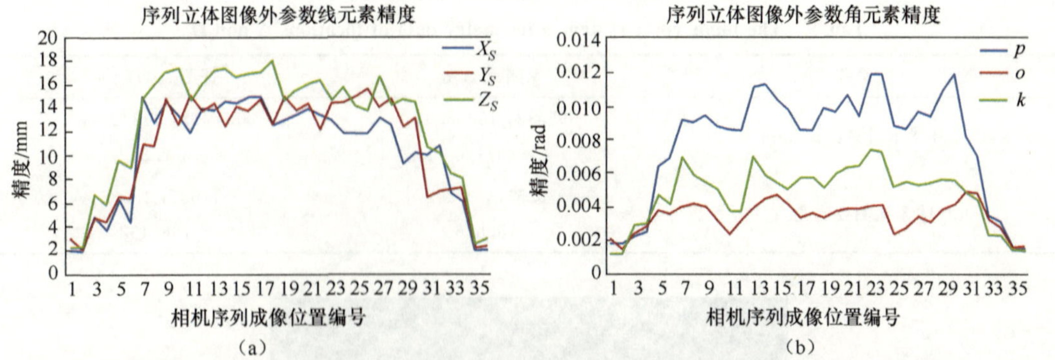

图 8　(网络版彩图) 在探测点 A2 导航相机序列立体图像外参数线元素 (X_S, Y_S, Z_S) 精度 (a) 和角元素 (p, o, k 对应图像外参数角元素 φ, ω, κ) 精度 (b)。横坐标为相机序列成像位置的编号

Fig. 8　(Color online) Precision of the stereo navigation camera pose estimation at location A2: (a) corresponds to the stereo camera position X_S, Y_S, Z_S, respectively; (b) corresponds to the stereo camera altitude angle p, o, k, respectively The horizontal axis represents image number, which is from 1 to 36

由于桅杆机构转角的遥测精度为 0.3°，即 0.0052rad，从图 8 (b) 中可说明在图像编号 1～6 (桅杆偏航角为＋170°～＋130°) 和编号 32～36 (桅杆偏航角为－170°～－130°) 中的相机位置下，利用相机图像闭合区域网平差模型可提高相机外参数 (在月球车本体系下) 的精度，其角元素精度均高于 0.0052rad。在具体的天际线遮挡角计算过程中，则采用本文方法的计算结果，替代月球车桅杆机构转角。在图像编号 7～31 (桅杆偏航角为－110°～＋110°) 位置，由于受到四度重叠点分布较为集中的因素导致区域网平差模型稳定性较差，相机外参数角元素精度低于 0.0052rad。在具体计算相机外参数时，则应该采用桅杆展开角、偏航角以及相机云台俯仰角的遥测数据。从图 8 (a) 中分析可知，在图像编号 7～31 的相机外参数线元素精度低于 10mm。因此，减小相机序列成像间隔，提高相邻立体图像间的重叠区域比例，以及尽可能将四度重叠点均匀分布在图像重叠区域和提供四度重叠像点的数量，将提高区域网平差的稳定性和外参数解算精度。

在月球车图像天际线遮挡角计算过程中，需要对每个位置的相机外参数角元素精度进行判断。当某个位置图像外参数角元素精度均高于 0.3°时，说明本文提出的相机图像闭合区域网平差模型能有效地提升相机系与月球车本体系的姿态角精度，消除月球车桅杆机构转角测量误差对计算天际线遮挡角的影响。

4.3　天际线遮挡角高精度计算测试

本节采用嫦娥四号月球车验证器在地面的试验数据对地形遮挡预测算法进行验证。验证器工作在近似月面环境中的室内试验场内，用两台全站仪联测出所有控制点在北东地坐标系下的坐标，仪器的测角精度是 2.0″，点位测量精度为 0.12mm。月球车导航控制系统 GNC 同步记录导航相机成像时刻的月球车遥测位姿参数，如表 5 所示。左导航相机在探测点 A 点拍摄的第 9 幅图像如图 9 所示。

表 5 在探测点 A 处和 D 处的月球车遥测位姿数据

Tab. 5 The lunar rover system's telemetry data at locations A and D

参数	探测点 A 处	探测点 D 处
月球车遥测位置/mm	956.2，15898.0，−371.0	38254.8，44343.4，1802.24
月球车遥测姿态角/(°)	−0.537，0.802，115.576	−1.687，1.086，−168.462

图 9 （网络版彩图）探测点 A 处月球车左导航相机第 9 幅图像及特征点显示

Fig. 9 (Color online) Distribution of the extracted image points in the ninth image taken by the left navigation camera at location A

在实际工作环境中，天际线特征点离月球车较远，导致无法量测其在北东地系下的真实坐标。在室内试验场环境中，利用月球车验证器相机视场内的特征点模拟天际线特征点，并获得其在北东地系下的实测坐标。式 (2) 成立的前提是特征点距离较远，因此式 (2) 在室内试验场环境下无法成立。为此，本文利用月球车在北东地系下的姿态角旋转矩阵 $\mathbf{R}_{\mathrm{NED-R}}$、室内 IGPS 测量的月球车位置参数和旋转矩阵 $\mathbf{R}_{\mathrm{S-R}}$，通过月球车坐标转换框架[3]将特征点坐标从北东地系转到相机坐标系下，并计算在相机系下的高度角 $\eta_{\mathrm{S}} = \mathrm{atan}(Y/\sqrt{Z^2+X^2})$ 和方位角 $\tau_{\mathrm{S}} = \mathrm{atan}(X/Z)$。其中，$X$，$Y$，$Z$ 为特征点在相机系下的坐标。在上述过程中，月球车位置参数可视为真值，高度角和方位角的计算过程是式 (2) 的逆过程，其精度只受到旋转矩阵 $\mathbf{R}_{\mathrm{NED-R}}$ 和 $\mathbf{R}_{\mathrm{S-R}}$ 误差的影响。利用式 (1) 计算出特征点的高度角与方位角信息，统计方位角和高度角与上述过程遮挡角的差值。旋转矩阵 $\mathbf{R}_{\mathrm{NED-R}}$ 和 $\mathbf{R}_{\mathrm{S-R}}$ 元素的误差越大，方位角和高度角的对应差值越大，反之，则越小。

获得旋转矩阵 $\mathbf{R}_{\mathrm{NED-R}}$ 和 $\mathbf{R}_{\mathrm{S-R}}$ 的方式有以下两种：一种是由月球车遥测姿态角参数和桅杆结构转角直接获取；另一种是本文 3.2 和 3.3 节。因此，遮挡角差值是考察 3.2 和 3.3 节能否替代月球车遥测姿态角参数和桅杆结构转角和提高遮挡角精度的有效方式。两种方式均采用相同的相机内参数。

在图 9 中，利用 Forstner 点特征提取算子[34]提取 11 个特征点，用十字丝标记，像点提取精度为 0.5pixel，通过式（1）计算出在相机系下的高度角。利用特征点在北东地系下的测量坐标和月球车遥测姿态角和桅杆结构转角参数计算出在相机系下的高度角和方位角。统计遮挡角的差值，称之为“遥测”，如表 6 第 2 列和第 4 列所示。月球车立体导航相机在探测点 A 点获得 18 对立体图像，间隔 20°。通过闭合区域网光束法平差和立体图像绝对定向模型，计算旋转矩阵 $\mathbf{R}_{NED-R}$ 和 $\mathbf{R}_{S-R}$，然后利用特征点在北东地系下的测量坐标计算出在相机系下的高度角和方位角，并统计遮挡角的差值，称之为“新方法”，如表 6 第 3 列和第 5 列所示。

闭合区域网光束法平差模型中，在车体固件上量测 9 个控制点，相邻立体图像的连接点选择 12 对，特征点共计 225 个。在太阳翼电池片上选择 15 个特征点间长度信息的约束条件，迭代次数为 5 次。单位权方差为 0.95pixel，在当前摄站第 9 对图像位置时的相机外参数角元素 φ，ω，κ 为 0.0061，0.0032 和 0.0035rad。36 个相机位置的相机外参数角元素精度如图 10 所示。在立体图像绝对定向模型中，需设定起始摄站月球车位姿参数为真值，相邻摄站图像重叠区域选择 25 个连接点，即四度重叠像点。单位权方差为 0.223mm，当前摄站月球车姿态角精度分别为 0.0164，0.0087 和 0.0078rad，均高于月球车的遥测姿态角精度（1°），即 0.0175rad。

表 6 的最后一行为高度角差值和方位角差值的平均值。高度角差值的平均值在“遥测”和“新方法”两种方法下分别是−0.23°和−0.18°，其绝对值的最大值分别是 0.42°和 0.20°。方位角差值的平均值在“遥测”和“新方法”两种方法下分别是−0.18°和−0.16°，其绝对值的最大值分别是 0.32°和 0.26°。说明“新方法”计算的遮挡角差值绝对值小于“遥测”方法计算的遮挡角差值，表明本文提出的遮挡角计算方法的精度更高。在月球车连续移动过程中，其 IMU 记录的遥测位姿信息将会出现误差累计甚至出现野值的情况，本文提出的方法能更好地计算出地形遮挡角，且能保证精度。

表 6　地面特征点遮挡角差值计算

Tab. 6　Difference in the altitude angles and azimuthal angles of the control points in the camera frame

编号	高度角差值		方位角差值	
	遥测	新方法	遥测	新方法
1	−0.27°	−0.17°	−0.09°	−0.08°
2	−0.02°	−0.18°	−0.20°	−0.16°
3	−0.17°	−0.18°	−0.04°	−0.26°
4	−0.10°	−0.18°	−0.16°	−0.16°
5	−0.32°	−0.19°	−0.16°	−0.16°
6	−0.21°	−0.18°	−0.23°	−0.22°
7	−0.26°	−0.18°	−0.11°	−0.10°
8	−0.23°	−0.19°	−0.06°	−0.05°
9	−0.41°	−0.20°	−0.28°	−0.20°
10	−0.27°	−0.19°	−0.34°	−0.19°

续表 6

编号	高度角差值		方位角差值	
	遥测	新方法	遥测	新方法
18	−0.22°	−0.19°	−0.33°	−0.18°
平均值	−0.23°	−0.18°	−0.18°	−0.16°

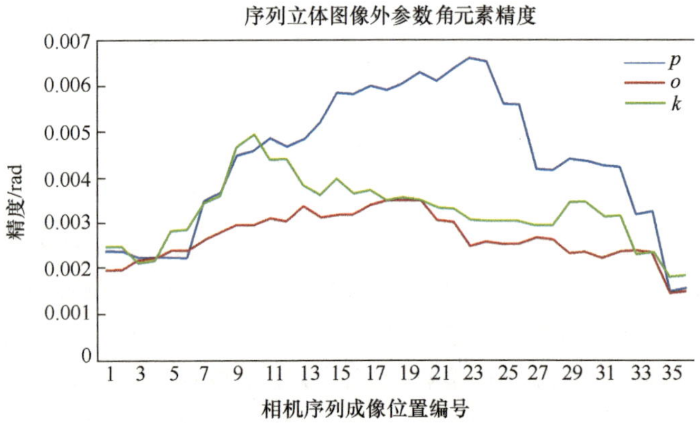

图 10 （网络版彩图）导航相机序列立体图像外参数角元素

（p，o，k 对应图像外参数角元素 φ，ω，κ）精度。横坐标为相机序列成像位置的编号

Fig. 10 (Color online) Precision of the stereo navigation camera pose estimation: the stereo camera altitude angle p, o, k, respectively

The horizontal axis represents image number, which is from 1 to 36

为进一步考察在野外环境下本文方法的稳定性和精度，利用图 11 所示探测点 D 的序列立体图像。月球车遥测位置数据如表 5 所示。

图 11 （网络版彩图）基于探测点 D 导航相机序列图像的特征点分布

Fig. 11 (Color online) Distribution of the extracted image points in the panoramic image taken by the left navigation camera at location D

在探测点 D，月球车导航相机从后方起逆时针以间隔 20°成像 18 次，云台俯仰角为 −19.93°。相邻相机位置的立体图像间自动匹配出 20～30 对同名像点，共计 378 个特征点，其中 10 个特征点如图 7 所示，110 个像点为六度重叠像点（相邻 3 个位置立体相机图像均有成像），其余为四度重叠像点。在月球车左右方向的序列图像中，选择太阳翼电池片的长度作为特征点的约束条件，共 10 个。设置收敛次数为 5 次。在立体图像绝对定向模型中，需设定起始摄站月球车位姿参数为真值，相邻摄站图像重叠区域选择 32 个连接点，即四度重叠像点。单位权方差为 0.209mm，当前摄站月球车姿态角精度分别为 0.0099，0.0156 和 0.0109rad，均高于月球车的遥测姿态角精度（1°），即 0.0175rad。

在实际布设控制点时，利用全站仪测量出天际线部分特征点在北东地坐标系下的天顶角和方位角，仪器的测角精度是 2.0″，点位测量精度为 0.12mm。如图 11 所示，采用基于 Harris 算法的角点检测方法和手动复核的方式，提取山峰与天空分界线处包含明显特征（如山尖、山峰与平地交界处等）的特征点，像点提取精度优于 1pixel。将天顶角转换为高度角后，和方位角的测量结果一并视为真值。在本次实验的“遥测”和“新方法”两种方法中，均采用表 2 和 3 所示的相机标定参数。在“遥测”方法中，月球车相机外参数计算和月球车姿态角计算两个步骤均采用遥测数据，包括月球车遥测姿态角和桅杆结构转角参数。而在“新方法”中，月球车相机外参数计算和月球车姿态角计算两个步骤均采用本文提出的方法，图 12 表示的是在探测点 D 导航相机序列立体图像外参数线元素和角元素。两种方法均采用式（3）计算出天际线特征点在北东地系下的高度角和方位角，然后与真值进行求差，其计算结果如表 7 所示。

在图 11 中，天际线特征点“1”，“2”，“3”和“4，5”分别位于编号“1，2”，“35，36”，“29，30”，“21，22”的立体图像中。如图 12（b）所示，天际线特征点“1”，“2”和“3”对应的相机外参数角元素精度均高于 0.0052rad。而天际线特征点“4，5”对应的相机外参数角元素精度均低于 0.0052rad，说明包含控制点的相机外参数角元素的精度高于不含控制点的相机外参数角元素。在图 12（a）中，编号“1，2”，“35，36”的相机外参数线元素精度要高于编号“29，30”和“21，22”。

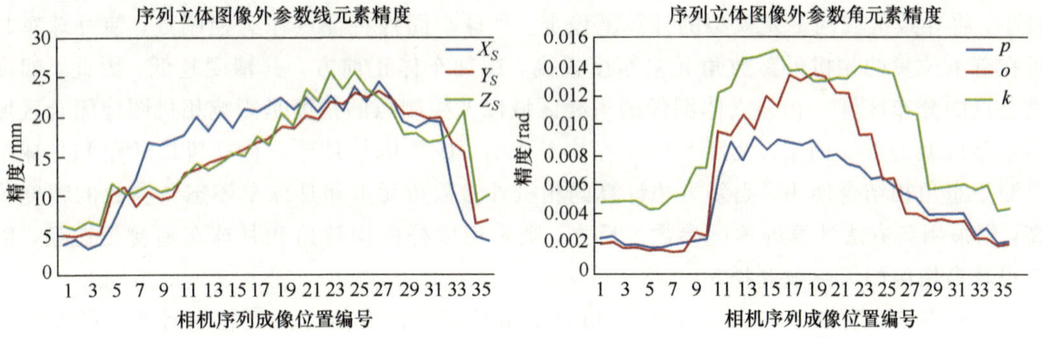

（a）　　（b）

图 12　（网络版彩图）在探测点 D 导航相机序列立体图像外参数线元素（X_S，Y_S，Z_S）精度（a）和角元素（p，o，k 对应图像外参数角元素 φ，ω，κ）精度（b）横坐标为相机序列成像位置的编号

Fig. 12　(Color online) Precision of the stereo navigation camera pose estimation at location D (a) Corresponds to the stereo camera position X_S, Y_S, Z_S, respectively; (b) corresponds to the stereo camera altitude angle p, o, k, respectively. The horizontal axis represents image number, which is from 1 to 36

在表 7 中，相较于“遥测”方法，天际线特征点“1”，“2”和“3”在“新方法”中的方位角和高度角更接近于真值。然而，天际线特征点“4”和“5”在“遥测”方法中的方位角和高度角更接近于真值。因此，通过精度对比，是判断相机外参数角元素和月球车姿态角参数用于计算天际线遮挡角的重要依据。

5　结果与讨论

本文以我国嫦娥四号月球车在轨任务中的天际线遮挡角预测问题为研究对象，提出了基于月球车图像的天际线遮挡角预测算法，并利用模拟试验数据进行验证。得出以下结论。

(1) 基于月球车图像的天际线遮挡角预测算法能够自动识别出天际线并能较高精度地计算地形遮挡角（方位角和高度角），解决了在模拟月面环境下的天际线遮挡预测问题。新方法采用基于固件控制点的相机标定方法，加入立体相机严格几何关系作为约束条件，获得较高精度的相机内参数，精度为 0.38pixel，可实现月球车相机在轨高精度几何标定。新方法提出了基于控制点的相机图像闭合区域网平差模型，并结合立体相机严格几何关系和太阳翼电池片长度信息，能提升当前摄站相机系相对于月球车本体系的姿态参数精度，有效消除月球车桅杆机构转角的测量误差（$0.3°$，3σ）。新方法采用立体图像绝对定位模型，月球车本体系相对于北东地系下的姿态角精度高于 $1°$，能进一步抑制月球车遥测姿态角的测量误差，提高天际线遮挡角计算精度。

(2) 由于涉及立体图像匹配，月球车天际线遮挡角预测算法与四度重叠像点的分布、月球车固件控制点精度等密切相关。在相机图像闭合区域网平差模型中，图像内分布范围较小的四度重叠点将严重影响相机外参数角元素的精度。另外，相邻立体图像的重叠区域较小，将导致区域网平差模型的计算不稳定。月球车固件特征点作为控制点，使月球车相机在首末位置的相机外参数角元素精度极高，越往车体正前方，其精度越低。因此，四度重叠点的分布区域、相邻立体图像的重叠区域以及控制点的精度是影响相机图像闭合区域网平差模型的重要因素。实际月球车在轨工作时，应考虑与月球车桅杆机构转角和月球车遥测姿态角相结合使用，当新方法计算的相机外参数角元素和月球车姿态角参数的精度较高，则采用新方法计算的角度参数。反之，则采用桅杆机构转角和月球车遥测姿态角，以获得较高精度的天际线遮挡角。

(3) 在本文算法 3.3 节中，实现相邻摄站月球车位姿信息的高精度计算，可以消除 IMU 设备测量月球车姿态角的误差累积效应。然而，本文方法需要假定月球车器初始位姿信息为真值，无法修正月球车在全局坐标系下的位姿信息。因此，研究月球车在全局坐标系下的高精度位姿信息是后续的研究和工程实践的一个重要方向。

(4) 本文研究结果具有较高的精度和应用稳定性，已应用于嫦娥四号月球车地面模拟试验环境下的工程实践中，对月球车在轨任务具有一定的指导意义。

参考文献

[1] Jia Y Z, Zou Y L, Xue C B, et al. Scientific objectives and payloads of Chang'E-4 mission (in Chinese). Chin J Space Sci, 2018, 38: 118-130.

[2] Li F, Zhang H, Wu X Y, et al. Influence analysis of terrain of the farside of the moon on softlanding (in Chinese). J Deep Space Explor, 2017, 4: 143-149.

[3] Ma Y Q, Liu S C, Jia Y, et al. Experimental research of navigation and localization algorithm based on stereo images for the lunar rover (in Chinese). Sci Sin Tech,

2014, 44: 1097 - 1104.

[4] Qing M Y, Chuang L S, Bo W, et al. Weighted total least squares for the visual localization of a planetary rover. Photogramm Eng Remote Sens, 2018, 84: 605 - 618.

[5] Cozman F, Krotkov E, Guestrin C. Outdoor visual position estimation for planetary rovers. Auton Robot, 2000, 9: 135 - 150.

[6] Oiri A, Nagatani K, Yoshida K. Global positioning for planetary rovers based on panoramic skyline image. In: Proceedings of the 2010 JSME Conference on Robotics and Mechatronics. 2010.

[7] Ahmad T, Bebis G, Regentova E E, et al. A machine learning approach to horizon line detection using local features. In: International Symposium on Visual Computing. Berlin, Heidelberg: Springer, 2013. 181 - 193.

[8] Bouyssounouse X, Nefian A V, Thomas A, et al. Horizon based orientation estimation for planetary surface navigation. In: IEEE International Conference on Image Processing. Phoenix: IEEE, 2016. 4368 - 4372.

[9] Nefian A, Bouyssounousse X, Edwards L, et al. Infrastructure free rover localization. In: International Symposium on Artificial Intelligence, Robotics and Automation in Space (i—SAIRAS). 2014.

[10] Grelsson B, Felsberg M, Isaksson F. Highly accurate attitude estimation via horizon detection. J Field Robot, 2016, 33: 967—993.

[11] Fefilatyev S, Smarodzinava V, Hall L O, et al. Horizon detection using machine learning techniques. In: International Conference on Machine Learning and Applications. Orlando: IEEE Computer Society, 2006. 17 - 21.

[12] Ahmad T, Bebis G, Nicolescu M, et al. An edge - less approach to horizon line detection. In: International Conference on Machine Learning and Applications. Miami: IEEE, 2015. 1095 - 1102.

[13] Ahmad T, Bebis G, Regentova E, et al. Coupling dynamic programming with machine learning for horizon line detection. Int J Artif Intell Tools, 2015, 24: 1540018.

[14] Ahmad T, Bebis G, Regentova E, et al. An experimental evaluation of different features and nodal costs for horizon line detection. In: International Symposium on Visual Computing. Cham: Springer, 2014.

[15] Grishin V A, Maslov I A. Horizon line stability observations over the sea. J Navigat, 2018, 71: 437 - 453.

[16] McGee T G, Sengupta R, Hedrick K. Obstacle detection for small autonomous aircraft using sky segmentation. In: Proceedings of the 2005 IEEE International Conference on Robotics and Automation. Barcelona: IEEE, 2005. 4679 - 4684.

[17] Thurrowgood S, Soccol D, Moore R J D, et al. A vision based system for attitude estimation of UAVs. In: 2009 IEEE/RSJ International Conference on Intelligent Robots and Systems. St. Louis: IEEE, 2009. 5725 - 5730.

[18] De Croon G, De Wagter C, Remes B D W, et al. Sky segmentation approach to obstacle avoidance. In: 2011 Aerospace Conference. Big Sky: IEEE, 2011. 1 - 16.

[19] Dumble S J, Gibbens P W. Efficient terrain - aided visual horizon based attitude estimation and localization. J Intell Robot Syst, 2015, 78: 205 - 221.

[20] Zhang J Q, Pan L, Wang S G. Photogrammetry (in Chinese). 2nd ed. Wuhan: Wuhan University Press, 2009. 54 - 72.

[21] Lowe D G. Distinctive image features from scale - invariant keypoints. Int J Comput Vision, 2004, 60: 91 - 110.

[22] Di K, Li R. CAHVOR camera model and its photogrammetric conversion for planetary applications. J Geophys Res, 2004, 109: E04004.

[23] Gennery D B. Generalized camera calibration including fish - eye lenses. Int J Comput Vision, 2006, 68: 239 - 266.

[24] Heikkila J. Geometric camera calibration using circular control points. IEEE Trans Pattern Anal Mach Intell, 2000, 22: 1066 - 1077.

[25] Zhang S, Li Q Z, Wen B, et al. Orthogonal projection model of jade Rabbit lunar rover Pancam. Geomat Inform Sci Wuhan Univ, 2014, 39: 804 - 808.

[26] Liu S C, Jia Y, Ma Y Q, et al. High precision localization of the Chang'E - 3 lunar rover (in Chinese). Chin Sci Bull, 2015, 60: 372 - 378.

[27] Zhang S, Xu Y M, Liu S C, et al. Calibration of Chang'E - 3 lunar rover stereo - camera system based on control field (in Chinese). Geomat Inform Sci Wuhan Univ, 2015, 40: 1509 - 1513.

[28] Fang X. Weighted total least squares: Necessary and sufficient conditions, fixed and random parameters. J Geod, 2013, 87: 733 - 749.

[29] Mahboub V. On weighted total least - squares for geodetic transformations. J Geod, 2012, 86: 359 - 367.

[30] Ma Y Q. Research on navigation and localization of CE - 3 lunar rover (in Chinese). Dissertation for Doctoral Degree. Wuhan: Wuhan University, 2014. 32 - 40.

[31] Zhao Y W. Research on precision estimation method for total least squares (in Chinese). Dissertation for Doctoral Degree. Nanchang: East China University of Technology, 2017. 8 - 10.

[32] Canny J. A computational approach to edge detection. IEEE Trans Pattern Anal Mach Intell, 1986, PAMI - 8: 679 - 698.

[33] Harris C G, Stephens M. A combined corner and edge detector. In: Alvey Vision Conference. 1988. 147 - 151.

[34] Förstner W, Gülch E. A fast operator for detection and precise location of distinct points, corners and centres of circular features. In: Proceeding of the ISPRS Intercommission Workshop on Fast Processing of Photogrammetric Data. Interlaken, 1987. 281 - 305.

嫦娥四号月面巡视器全景相机制图方法

司马兵[1,2]，彭松[3]，马友青[1]，温博[3]，刘少创[1]

1. 中国科学院空天信息创新研究院，北京 100094
2. 中国科学院大学，北京 100049
3. 中国空间飞行器总体设计部，北京 100094

摘要：作为嫦娥四号月面巡视器的科学载荷，全景相机不仅承担着探测月面地质的任务，同时可以用于制图和地形重建，直接服务于巡视器的导航规划工作。在试验场选用一种模型方法用于前期标定实验，但为了后期制图工作的方便往往需要选用另一种模型，基于此，本文提出了一种基于畸变模型参数转换的去畸变方法，技巧性地解决了标定模型和制图模型不一致的问题，并融合支持像点（support points）稠密匹配、空间前方交会、最小二乘光束法平差、离散点云生成、德劳内三角化等多项技术，将全景相机影像用于制图工作。基于月面巡视器在 LE00210 站点拍摄的全景相机影像，在实验室构建控制场对本文提出的畸变模型参数转换方法进行了可行性和精度测试，结果表明畸变模型转换后定位精度优于 1.6cm，相对距离精度优于 4%，并成功应用于嫦娥四号巡视器全景相机制图。

关键词：嫦娥四号；月面巡视器；地形重建；去畸变；遥操作

Research on a mapping method of Chang'E－4 lunar rover's panoramic camera

SIMA Bing[1,2], PENG Song[3], MA YouQing[1], WEN Bo[3] & LIU ShaoChuang[1]

1. Aerospace Information Research Institute, Chinese Academy of Sciences, Beijing 100094, China
2. University of Chinese Academy of Sciences, Beijing 100049, China
3. Beijing Institute of Spacecraft System Engineering, Beijing 100094, China

Abstract: As a scientific payload of Chang'E－4 lunar rover, the panoramic camera not only detects lunar geology but can be directly used for mapping and constructing the lunar terrain in the lunar rover's navigation and planning. We choose one model for calibrating the camera early; however, we often need to choose another model for the convenience of mapping work afterward. This paper proposes a distortion removal method based on the model parameter transformation algorithm, which overcomes the mismatch problem between the mapping model and the calibration model. The support points dense matching, spatial forward intersection, least－square bundle adjustment, discrete point cloud generation, and Delaunay triangulation technique are used together to achieve the topographic mapping work of panoramic camera images. A control field is constructed in the laboratory and used to test the feasibility and accuracy of the proposed distortion model parameter transformation method using the images taken by Chang'E－4 lunar rover's panoramic camera located in Site LE00210. The result shows that after distortion model transformation, the position accuracy was higher than 1. 6 cm, while the relative distance accuracy was higher than 4%. The proposed algorithm has been successfully applied to the mapping work of Chang'E－4 lunar rover's panoramic camera.

Keywords: Chang'E－4, lunar rover, terrain reconstruction, distortion removal, teleoperation

1 引言

嫦娥四号于 2018 年 12 月 8 日在西昌卫星发射中心发射，并于 2019 年 1 月 3 日在月球背面冯·卡门撞击坑（45.5°S，177.56°E）成功软着陆，是人类探测器首次着陆月球背面，实现了在地月拉格朗日 L2 点的中继通讯。全景相机影像制图，尤其是重建出远处的地形地貌，能够有效地为巡视器路径规划和导航定位提供参考。在月面特殊的成像环境下，如低对比度、缺乏纹理、阴影遮挡等，加之全景相机有畸变大、视场小等特性，导致月面巡视器全景相机制图相较于导航相机制图有更大难度。目前地形重建算法主要有基于静态阴影的单幅图像重建技术[1]，基于原始地形和等高线数据的多分辨率拼接技术[2]，基于动态规划算法[3]、图割算法[4,5]和人工智能方法[6]进行匹配的视觉重建方法等，而基于多幅影像的重建方法对月面全景相机制图至关重要。近年来，国内外一些学者针对月面制图对算法进行了不同程度的改进。李明磊等人[7]采用基于动态规划的立体匹配算法，在嫦娥三号巡视器制图中得到了较好的效果。Jie 和 Huang[8]提出了基于运动的长基线双目立体视觉模型，实现了适用月球表面的地形三维重建算法。

全景相机成像时存在几何畸变，需要通过去畸变恢复实际的成像信息。由于后期基于开放性编程库制图时采用的畸变模型与在正样试验场标定时使用的畸变模型不同，标定后的畸变模型参数不能直接应用于制图模型，给制图工作带来不便。标定模型结果经由畸变模型参数的数学转换，能够得到一套可靠的制图模型结果，从而应用于后续制图工作。基于以上分析，本文提出了畸变模型的数学转换方法，采用基于支持像点（support points）的立体匹配算法，介绍了全景相机制图的操作流程，并成功应用于嫦娥四号月面巡视器全景相机影像，实现了全景相机的高精度制图。

2 全景相机制图流程

嫦娥四号月面巡视器上安装有 1 对全景相机，装于相机桅杆上，是一对窄视场相机，月面距离为 1.5m，基线长为 270mm，像元物理尺寸为 7.4μm，如图 1 所示。全景相机是科学载荷，主要用于对地质、地形和地貌的科学研究。

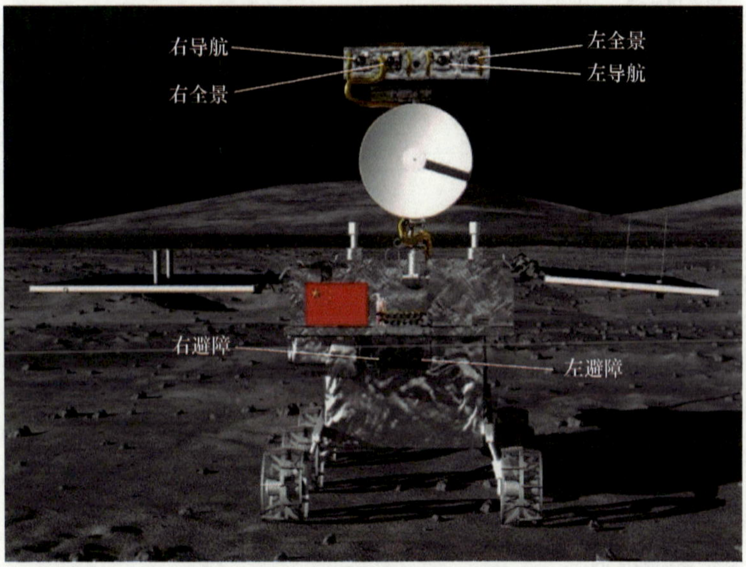

图 1　月面巡视器上的相机

Fig. 1　Cameras equipped on the lunar rover

OpenCV 是开放性计算机视觉库（open source computer vision library）的简称，丰富的 API 接口函数和强大的图像处理能力使它在制图等领域有着广泛的应用，因此采用基于 OpenCV 库的算法进行制图工作。月面巡视器全景相机制图的目的是通过影像预处理，畸变模型转换和畸变改正，密集匹配生成视差图，离散点云生成，模型重建等一系列流程，高精度地生成月面数字高程模型（digital elevation model，DEM）和具有图像纹理信息的正射影像图（digital orthophoto map，DOM）。算法要求有一定的重叠度，因此月面巡视器在某一位置通过桅杆的左右旋转和上下俯仰，使全景相机在不同姿态角成像时，要保证两个相机的影像间应有一定的重叠度，相邻影像间也要有一定的重叠度。月面巡视器全景相机制图流程如图 2 所示。

2.1　影像预处理

首先对全景相机采集的影像进行质量检查，然后进行高斯滤波、伽马变换以及直方图均衡化等预处理，去除影像噪点，减弱光照条件的影响，增强图像对比度。预处理工作能够提高影像质量，减小数据源本身的误差。

2.2　相机标定和去畸变

相机标定通常采用两种方式：通过相机多角度对特定标定板成像的方式和采用布设高精度控制场的方式。常用的去畸变模型有摄影测量畸变模型[9]、Photometrix 畸变模型[10]和 Tsai 畸变模型[11]等。在嫦娥四号月面巡视器正样标定实验时，通过布设高精度的控制

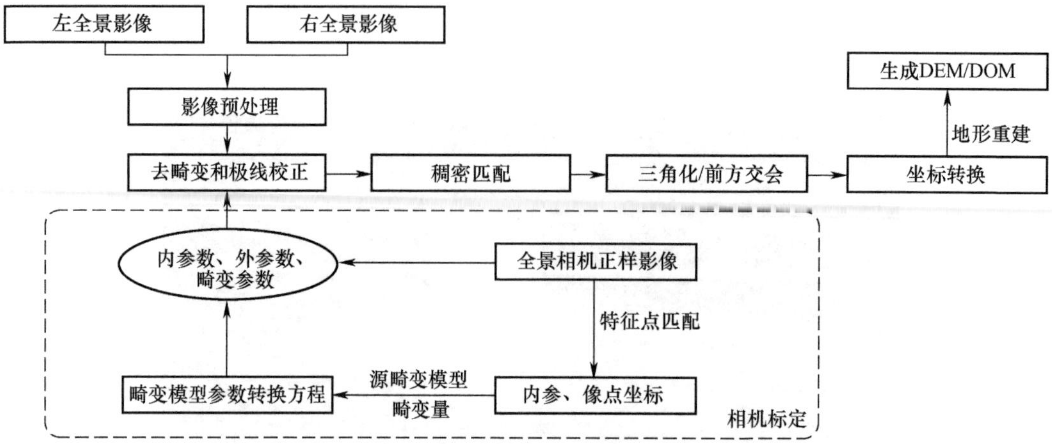

图 2　全景相机制图流程图

Fig. 2　Pancam mapping flow chart

场，基于 Photometrix 畸变模型对全景相机内参数、外参数以及畸变系数进行了高精度标定。本节对 3 种畸变模型进行简要介绍，并给出其数学表达式。

(1) 摄影测量畸变模型

摄影测量是利用摄影技术进行测量，主要目的是要保证测量的精度和准确性，在数字化成像时代，随着 CCD/CMOS 的数字化航空成像仪的普及，摄影测量开始使用桶形畸变和枕形畸变的概念以解决畸变问题，不断发展出基于径向畸变和切向畸变的畸变模型。摄影测量畸变模型的数学形式为

$$\begin{aligned}\Delta x = &(x - x_0)\cdot(k_1 r^2 + k_2 r^4 + k_3 r^6 \\ &+ k_4 r^8 + k_5 r^{10}) + p_1\cdot[r^2 + 2(x - x_0)^2] \\ &+ 2p_2\cdot(x - x_0)\cdot(y - y_0),\end{aligned}\tag{1}$$

$$\begin{aligned}\Delta y = &(y - y_0)\cdot(k_1 r^2 + k_2 r^4 + k_3 r^6 \\ &+ k_4 r^8 + k_5 r^{10} + p_2\cdot[r^2 + 2(y - y_0)^2] \\ &+ 2p_1\cdot(x - x_0)\cdot(y - y_0),\end{aligned}\tag{2}$$

$$r^2 = (x - x_0)^2 + (y - y_0)^2.\tag{3}$$

式 (1) ～ (3) 表示摄影测量模型在 X 方向和 Y 方向上坐标的畸变量。其中，(x_0，y_0) 为主点坐标，k_1，k_2，k_3，k_4，k_5，p_1，p_2 为模型畸变系数，由标定结果给出。(x，y) 为畸变前像点坐标，(Δx，Δy) 为像点相应的畸变差。坐标采用标准化的像素单位。

(2) Tsai 畸变模型

畸变量可分为在 X 方向和 Y 方向上，这种畸变量只考虑径向畸变，根据径向畸变的线性关系，即畸变后坐标与畸变前坐标具有比例约束关系，Tsai 于 1987 年发展出了先线性求解后非线性优化的 Tsai 畸变模型。Tsai 畸变模型的数学形式为

$$x = \frac{u(\text{undistorted}) - c_x}{f_x},\tag{4}$$

$$y = \frac{v(\text{undistorted}) - c_y}{f_y},\tag{5}$$

$$r^2 = x^2 + y^2, \tag{6}$$

$$x' = x\,\frac{1 + k_1 r^2 + k_2 r^4 + k_3 r^6}{1 + k_4 r^2 + k_5 r^4 + k_6 r^6} + 2p_1 xy + p_2(r^2 + 2x^2) + s_1 r^2 + s_2 r^4, \tag{7}$$

$$y' = y\,\frac{1 + k_1 r^2 + k_2 r^4 + k_3 r^6}{1 + k_4 r^2 + k_5 r^4 + k_6 r^6} + 2p_2 xy + p_1(r^2 + 2y^2) + s_1 r^2 + s_2 r^4, \tag{8}$$

$$u(\text{distorted}) = f_x^* x' + c_x, \tag{9}$$

$$v(\text{distorted}) = f_y^* y' + c_y. \tag{10}$$

式（4）～（10）表示 Tsai 模型在 X 方向和 Y 方向上坐标的畸变量。其中（c_x，c_y）为主点坐标，k_1，k_2，k_3，k_4，k_5，k_6，p_1，p_2，s_1，s_2 为模型畸变系数，f_x 与 f_y 分别为 X 方向和 Y 方向上的主距，由标定结果给出。（u(undistorted)，v(undistorted)）为畸变前像点坐标，（u（distorted)，v（distorted)）为去畸变后像点坐标，即畸变差等于畸变后坐标与畸变前坐标之差。坐标和主距采用标准化的像素单位。

(3) Photometrix 畸变模型

澳大利亚 Photometrix 公司的主要业务是通过摄影图像高精度地完成 3D 地形重建，同时发展出了一套基于 Photometrix 畸变模型的用于非工业近景摄影测量领域的标定系统，该模型与 Fryer 和 Brown[12] 用于校正镜头畸变的 10 参数矫正法类似，在径向畸变和切向畸变的基础上还考虑了仿射和非正交化的畸变。Photometrix 畸变模型的数学形式为

$$x = x(\text{distorted}) - x_p, \tag{11}$$

$$y = y(\text{distorted}) - y_p, \tag{12}$$

$$r^2 = x^2 + y^2, \tag{13}$$

$$\mathrm{d}r = K_1 r^3 + K_2 r^5 + K_3 r^7, \tag{14}$$

$$x(\text{undistorted}) = x + x\,\frac{\mathrm{d}r}{r} + P_1(r^2 + 2x^2) + 2P_2 xy + B_1 x + B_2 y, \tag{15}$$

$$y(\text{undistorted}) = y + y\,\frac{\mathrm{d}r}{r} + P_2(r^2 + 2y^2) + 2P_1 xy. \tag{16}$$

式（11）～（16）表示 Photometrix 模型在 X 方向和 Y 方向上坐标的畸变量。其中，（x_p，y_p）为主点坐标，K_1，K_2，K_3，P_1，P_2，B_1，B_2 为畸变系数，由标定结果给出。（x(distorted)，y(distorted)）为畸变前像点坐标，（x(undistorted)，y(undistorted)）为畸变量。由于该模型使用非标准化的像素单位，因此畸变前坐标以 mm 为单位表示，畸变量采用标准化的像素单位。

2.3　畸变模型参数转换方法

在使用 OpenCV 库进行大多数的制图工作时，畸变模型通常采用 Tsai 模型的简单形式：只需要 k_1，k_2，p_1，p_2 4 个畸变系数即可。令 k_3，k_4，k_5，k_6，s_1，s_2 为 0，式（7）和（8）右式变为线性方程，即

$$x' = x(1 + k_1 r^2 + k_2 r^4) + 2p_1 xy + p_2(r^2 + 2x^2), \tag{17}$$

$$y' = y(1 + k_1 r^2 + k_2 r^4) + 2p_2 xy + p_1(r^2 + 2y^2). \tag{18}$$

线性形式给了我们一个高精度求解 Tsai 畸变模型参数方法的启发。第一种方法是基于特定的标定板或控制场，提取像点坐标和物方点坐标，排列成特定的坐标数据结构，依据基于 Tsai 畸变模型的张正友平板标定法[13]进行解算，亦可直接基于共线方程建立数学表达式进行解算，但该方法解算相对复杂。第二种方法是一种技巧性方法，输入像点坐标，依据一种畸变模型计算出畸变量，然后通过式（19）～（22）所示的畸变模型参数数学变换方法得到 Tsai 畸变模型参数，此后便可在 OpenCV 库中应用这些参数以期完成制图工作。需要注意的是，在此过程中假定内参数保持不变。给定像点坐标和畸变量，将等式（7）和（8）合并改写为矩阵形式有：

$$\begin{pmatrix} x'-x \\ y'-y \end{pmatrix}_i = \begin{pmatrix} xr^2 & xr^4 & 2xy & r^2+2x^2 \\ yr^2 & yr^4 & r^2+2y^2 & 2xy \end{pmatrix}_i \begin{pmatrix} k_1 \\ k_2 \\ p_1 \\ p_2 \end{pmatrix}, \tag{19}$$

$$(x,\ y)_i = \left(\frac{u(\text{before})-c_x}{f_x},\ \frac{v(\text{before})-c_y}{f_y}\right)_i, \tag{20}$$

$$(x',\ y')_i = \left(\frac{u(\text{after})-c_x}{f_x},\ \frac{v(\text{after})-c_y}{f_y}\right)_i, \tag{21}$$

$$(r^2)_i = (x^2+y^2)_i, \tag{22}$$

其中，下标 i 表示第 i 个像点坐标和相应的畸变量。建立如式（19）所示的线性方程组 $\boldsymbol{AX}=\boldsymbol{B}$，方程共有 4 个未知数（4 个畸变参数 k_1，k_2，p_1，p_2），一共可列出 $2i$ 个方程（通常取 $i>10$），因此可以通过最小二乘或拟最小残差方法求解畸变参数。

同样地，由于各个畸变模型的线性性质，Photometrix 模型能够转换至摄影测量模型，摄影测量模型也能够转换至 Tsai 模型，在转换过程中，线性方程组 $\boldsymbol{AX}=\boldsymbol{B}$ 的组成矩阵 $\boldsymbol{AB}$ 将根据目标模型的数学形式发生改变，但本质上与式（19）类似。为了验证模型转换精度，根据 Photometrix 模型求解出 Tsai 模型和摄影测量模型参数，并计算了绝对坐标差值和相对距离差值，结果将在实验结果分析部分展示。

2.4　立体匹配

在左右图像间自动匹配一些可以稳健稠密匹配的支持像点[14]，然后以这些点为顶点，进行德劳内三角化，再通过三角化得到的平面，对三角形内部的像素点视差进行插值，从而获得稠密匹配点和视差图。算法流程示意图如图 3 所示。

支持像点选择：对原始立体图像进行 9×9 窗口的竖直和水平方向的 Sobel 滤波，获得待选像点集的向量，并对该向量进行 1 范数计算，获得足够数量的稳健的支持像点。采用最优匹配和次优匹配的匹配阈值，剔除错误的匹配支持像点。

构建基于支持像点和待匹配图像的概率模型 $p(d_n,\ o_n^{(l)},\ o_n^{(r)},\ S)$：基于支持像点的像素坐标和视差以及左图像像素坐标，构建先验概率函数 $p(d_n \mid S,\ o_n^{(l)})$，获得左图像视差；然后构建图像稠密匹配的相似函数 $p(o_n^{(r)} \mid o_n^{(l)},\ d_n)$，基于左图像视差和

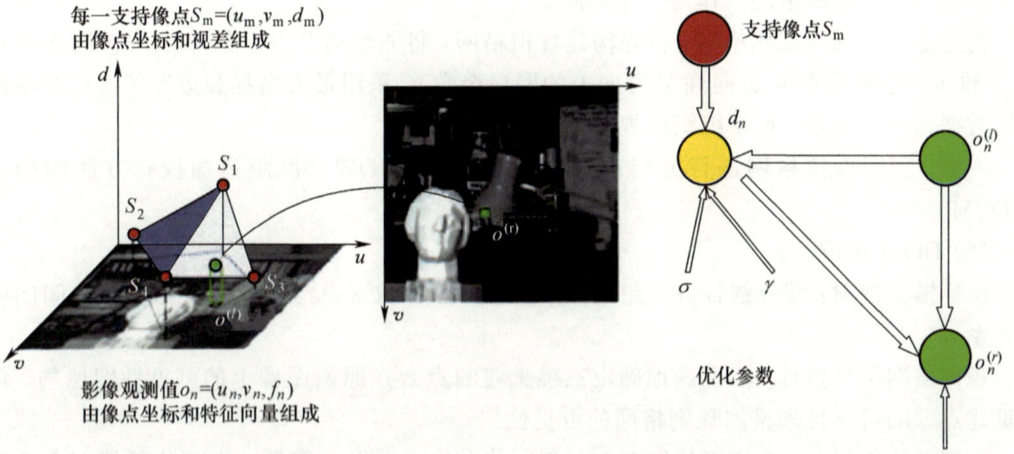

图 3　(网络版彩图) 立体匹配算法示意图

Fig. 3　(Color online) Illustration of stereo matching algorithm

左图像的像素坐标，获得右图像的匹配像素坐标。对左图像每个像素重复上述步骤 100 次以上，最后对多次计算得到的右图像像素坐标取平均，以得到精确匹配的像素坐标。

视差估计：在上一步骤的基础上，构建基于极线校正的立体图像像素坐标和像素值、支持像点的最大后验估计：$d_n^* = \operatorname{argmax} p(d_n \mid o_n^{(l)}, o_n^{(r)}, \cdots, o_N^{(r)}, S)$。

匹配验证：立体图像分别通过上述步骤获得相应视差图，然后对其左右视差图像进行连续性检查。若在一定阈值范围内，可视为视差估计最优。

2.5　离散点云生成

为了能够整体平差计算，需要进行影像匹配，建立多幅影像之间的联系。首先通过划分影像内的区域格网，限制提取的特征点数量，保留更为明显的目标，然后基于几何相对关系和附加约束条件的 Affine - SIFT 特征匹配方法[15]获取连接点匹配点对。在完成匹配计算后，依据处理数据的特殊性，采用唯一性约束、极线几何约束、仿射变换约束和人工检查 4 种约束。基于中心投影的共线方程，将标定参数和连接点坐标作为输入，并引入惯性测量单元 (inertial measurement unit，IMU) 数据，建立加权总体最小二乘的立体图像条带网光束法平差模型，计算出月面巡视器高精度的位姿坐标[16]。

利用获取的视差图，通过前方交会 (三角测量) 获得在相机坐标系下的三维点云，然后通过多摄站月面巡视器的高精度位置坐标和坐标转换框架，将三维点云纳入全局坐标系下，完成多摄站点云的生成和拼接。

2.6　DEM 和 DOM 生成

DEM 表示地形三维向量的有限序列，DOM 存储图像的纹理特征。利用多摄站的点云数据，通过构建德劳内三角网和栅格化，可以得到 DEM 和 DOM。

(1) DEM 生成

点云预处理：对离散点云的高程进行横切分层，统计每个分割区域内的点云数量，将分割区内小于一定阈值的点作为粗差剔除。

构建虚拟格网与规则格网：首先构建虚拟格网，将原始离散点云分为 $N_x \times N_y$ 个网格（N_x 和 N_y 分别代表 X 方向和 Y 方向上的网格个数），采用最大高程和最近邻方法对格网内的数据做内插处理，重新组织成规则格网。

内插：利用虚拟格网进行分层次滤波，然后依据 IDW（反距离加权）方法内插生成 DEM。

（2）DOM 生成

坐标模型映射：建立线性方程组，解算正射影像各像素点与离散点云点坐标之间的映射关系。

规则格网可见性计算：准确地确定三维重建的点云在原始图像上的可见性图像集，借助重建点云的可见性来推测规则格网的可见性。

规则面片选择：一个规则格网对应正射影像的一个面片，但是一个面片可能在多个原始图像上可见，因此最终的正射影像的面片对应于哪个图像需要一种选择机制。建立基于马尔科夫随机场的能量函数，考虑影像成像条件和规则格网边界，通过能量函数的最小化，得到各格网最优的标记号，保证相邻面片的纹理一致性。

匀光：基于图像梯度场，在图像空间以图像梯度差值实现全局匀光，保持目标区域原有的梯度变化，同时最小化目标图像边界的灰度差异，可以最大限度地消除格网边界的拼接缝。

3　结果

在嫦娥四号正样标定场进行标定实验，将两台全景相机固定在约 270mm 的基线上，对高精度控制场进行成像，然后基于 Photometrix 畸变模型的标定算法，对全景相机左右相机内外参数以及畸变系数进行了精确的标定，标定图像如图 4 所示（图像大小为 1176 像素×865 像素），内参数和畸变系数标定结果如表 1 所示（主点和主距以标准化像素为单位）。

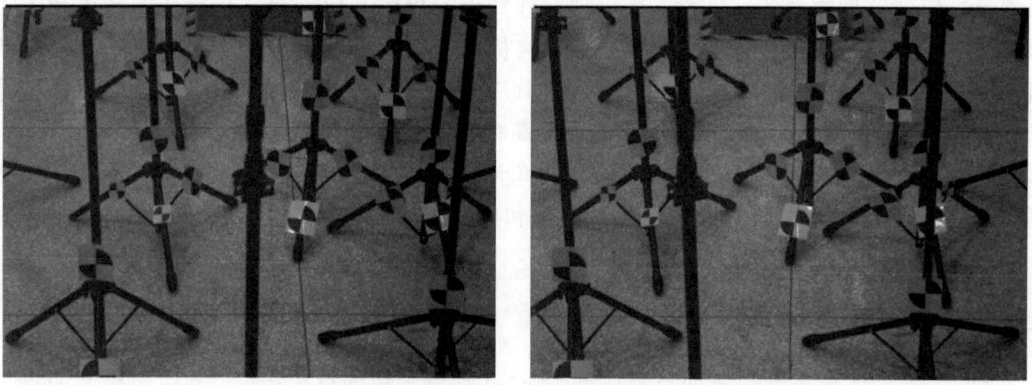

图 4　（网络版彩图）标定影像

Fig. 4　(Color online) A pair of image for calibration

需要注意的是，该模型定义的像平面坐标系是以影像中心为原点，X 轴指向右侧，Y 轴指向下侧。通过自主开发的点位坐标提取软件，可以从标定影像中提取出高精度的同名像点坐标，软件定义的像平面坐标系是一般意义下的像平面坐标系，即以影像左上角为原点，X 轴指向右侧，Y 轴指向下侧。因此在实际畸变差计算中，需将提取出的像点坐标平移至以影像中心，即（588，432.5）为原点，然后由畸变模型数学表达式和标定参数值计算出畸变差，如表 2 所示。

表 1　基于 Photometrix 畸变模型的标定结果

Table 1 Calibration results based on Photometrix distortion model

系数	B 相机（左）	A 相机（右）
xp	1.6562838	−1.965878
yp	0.6054595	−24.473649
f_x	3397.972973	3396.891892
f_y	3397.972973	3396.891892
K_1	5.6613×10^{-7}	-8.7695×10^{-7}
K_2	3.5207×10^{-8}	6.3169×10^{-8}
K_3	-1.3835×10^{-10}	-2.640×10^{-10}
P_1	1.9053×10^{-6}	1.6107×10^{-6}
P_2	1.4390×10^{-6}	-5.0584×10^{-6}
B_1	1.9856×10^{-5}	4.2724×10^{-5}
B_2	1.8739×10^{-5}	-3.9761×10^{-5}

表 2　像点坐标及其畸变差

Table 2　Image coordination and distortion

	Image point coordinate				Photometrix distortion			
	L_x	L_y	R_x	R_y	ΔL_x	ΔL_y	ΔR_x	ΔR_y
1	221.8462	601.4231	92.9231	604.2692	−2.73396	1.24108	−3.63457	1.63318
2	1060.8462	659.6538	931.8462	665.1923	3.47466	1.67206	2.57356	2.08405
3	271.6154	430.4231	183.4615	434.5000	−2.36569	−0.02432	−2.96457	0.37696
4	377.9231	484.7307	281.0000	488.6538	−1.57903	0.37755	−2.24278	0.77771
5	451.7692	408.5000	262.3846	317.2692	−1.03260	−0.18656	−2.38052	−0.49052
6	357.5385	312.9615	577.9231	370.2692	−1.72989	−0.89353	−0.04547	−0.09830
7	659.6923	366.1923	742.2308	369.9615	0.50599	−0.49964	1.17046	−0.10059
8	825.3077	365.8846	833.6154	455.9615	1.73154	−0.50191	1.84671	0.53581
9	925.5385	451.5769	935.3846	337.6538	2.47327	0.13222	2.59990	−0.33969
10	1042.3846	487.6538	341.5385	186.1154	3.33780	0.39920	−1.79474	−1.46107
11	1032.9231	333.1923	409.4615	136.5000	3.26792	−0.74384	−1.29208	−1.82824
12	405.3846	181.5000	865.9231	236.4231	−1.37586	−1.86634	2.08590	−1.08880
13	467.0769	132.2692	881.4615	17.9615	−0.91935	−2.23065	2.20100	−2.70552
14	932.0769	232.5000	696.0769	38.9615	2.52163	−1.48897	0.82903	−2.55008

在标定完成并计算出畸变差后，由于畸变模型的不一致性，需要由 Photometrix 模型参数转换为基于 Tsai 畸变模型的 OpenCV 参数，以便在后期制图时使用。首先将原始像点坐标和相应的畸变量代入畸变模型参数转换方程，依据式（19）～（22）组成矩阵 $\boldsymbol{A}$ 和矩阵 $\boldsymbol{B}$，采用最小二乘或拟最小残差方法分别求解出摄影测量畸变模型参数和基于 Tsai 模型的 OpenCV 畸变参数。“玉兔二号”巡视器在月面移动时，全景相机在 LE00210 摄站点拍摄了 28 对立体影像（此时，月面巡视器桅杆的俯仰角为 $-18.898°$），影像缩略图如图 5 所示。从图上可以看出，左右影像的重叠度很高。受全景相机视场和畸变的影响，在制图前应检查左右影像的编码是否与实际相符，然后将参数转换结果用于在轨的真实立体影像。依据图 2 的操作流程，即去畸变和极线校正后，经由立体稠密匹配、定位和影像拼接重建出离散点云，最后生成 DEM 和 DOM 等地形产品，如图 6 所示。

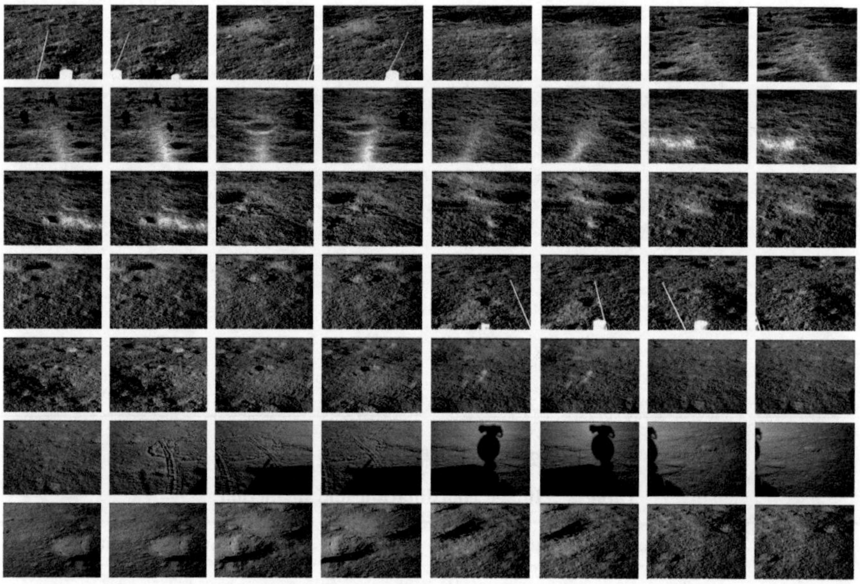

图 5　全景相机在 LE00210 点的在轨影像（28 对，左图为左相机影像，右图为右相机影像）

Fig. 5　PanCam images located in Site LE00210 (28 pairs; Left: Left images, Right: Right images)

在巡视器本体坐标系下生成的 DEM 和 DOM 上方出现了两块空白区域，结合全景相机在该站点的第 22～24 对原始影像，认为是巡视器太阳翼电池板和桅杆的阴影遮挡了地形。“玉兔二号”巡视器行走到该站点时，接收到的遥测坐标为（18.481，-64.557，0.387），这是巡视器在着陆北东地坐标系下的坐标。从 DOM 上选取空白区域的一个点，如 A 点，其在着陆北东地坐标系下的坐标为（17.091，-58.605，0.477）. XY 坐标之间的相对关系符合坐标系指向，即以巡视器中心为原点，X 轴指向东，Y 轴指向北，Z 轴指向地面。

在求得畸变模型参数后，结果发现各个模型畸变差之间的绝对差值有 95％都在 0.5 像素以内。考虑到模型间的差异性，选取一对正样场影像，提取出左右影像上的同名像点坐标，通过空间前方交会算法获得在控制场坐标系下的三维点坐标，并与在控制场坐标系下由高精度全站仪测量出的实际三维坐标进行比较。同时，从全景相机在轨影像中无法实际测量三维点坐标，因此使用两对同名像点间的距离作为评判标准，分析各个模型计算结果

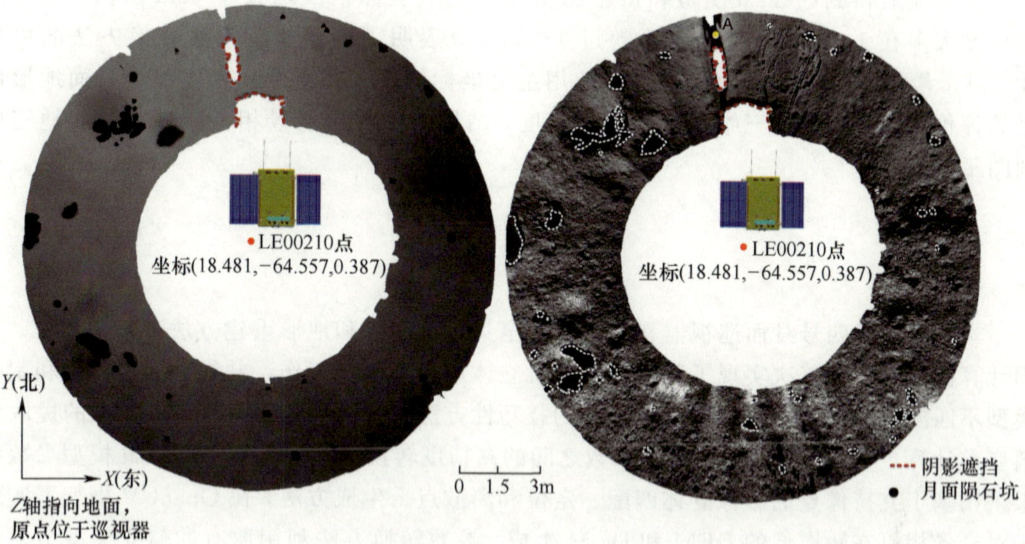

图 6　（网络版彩图）生成的 DEM 和 DOM

Fig. 6　(Color online) Generated DEM and DOM

的相对距离精度，以此作为去畸变模型转换和生成地形产品的精度，结果如表 3 所示。

表 3　坐标结果和相对距离结果比较[a)]

Table 3　Comparison of coordination results and relative distance[a)]

	实际坐标（mm）			模型差异 Δ（mm）			距离	相对距离精度（%）		
	X	Y	Z	Photometrix	Tsai	Photogrammetry	(mm)	Photometrix	Tsai	Photogrammetry
1	−557.825	687.619	3646.955	4.423	5.746	4.709	904.209	0.09	0.35	0.14
2	343.939	746.791	3616.734	6.210	13.413	10.292	1177.619	0.20	1.32	0.08
3	−556.502	880.442	4363.815	1.412	14.783	4.629	221.992	1.64	2.37	2.72
4	−414.181	862.738	4194.370	4.750	7.434	3.256	216.517	0.04	0.23	0.11
5	−439.398	688.835	4320.866	1.718	7.464	1.327	498.075	0.82	3.03	2.30
6	−51.654	892.351	4558.174	8.858	15.685	15.081	223.342	1.07	1.42	1.74
7	165.924	868.847	4513.577	3.627	5.315	12.902	251.745	0.20	1.68	0.15
8	275.143	870.423	4286.764	2.822	10.709	13.269	225.402	0.10	0.19	3.78
9	393.091	682.073	4249.103	3.068	11.090	3.171	1151.959	0.72	1.12	0.55
10	−402.524	856.494	5063.709	8.930	8.669	11.279	211.177	3.41	2.35	3.38
11	−315.112	864.964	5255.759	0.531	2.944	2.860	736.760	0.16	0.61	0.39
12	357.473	888.036	4955.905	1.375	11.188	3.451	313.095	0.36	0.05	0.71
13	380.353	613.794	5105.224	8.022	2.671	14.978	690.201	0.95	0.32	1.90
14	125.568	887.394	5685.400	1.294	5.208	1.102				

a) $\Delta=\sqrt{\Delta X^2+\Delta Y^2+\Delta Z^2}$，$\Delta X$，$\Delta Y$，$\Delta Z$ 分别表示模型得到的物方点坐标与实际坐标在 XYZ 方向上的差值。

从表 3 可以看出，使用初始的 Photometrix 畸变模型得到的绝对坐标精度在 1cm 以内，经转换后得到的 Tsai 模型和摄影测量模型绝对坐标精度均在 1.6cm 以内；相对距离精度大多在 1.5%以内，最差不超过 4%，有效表明了畸变模型参数转换方法的可靠性。对于最终生成的 DEM 和 DOM，使用绝对坐标精度和相对距离精度作为月面地形恢复的评价因子，反映了三维地形的重建精度，结果表明使用全景相机能够高精度地完成制图工作。

4　结论

本文对嫦娥四号月面巡视器科学载荷的全景相机制图和地形重建方法进行了研究，采用计算机视觉方法首次实现了基于全景相机立体影像的制图工作，并针对标定模型和制图模型不匹配的问题提出了模型参数转换的技巧性方法。通过对立体影像同名像点的提取和畸变差计算，可实现多种畸变模型参数之间的高精度转换，利用转换后的 Tsai 模型参数结果采用基于支持像点的影像立体匹配、定位和离散点云生成方法，在 OpenCV 框架下实现了对全景相机在轨影像的 DEM 和 DOM 生成。参数转换方法利用原有的畸变模型结果可实现向另一畸变模型的数学变换，提供了解决“标定时使用的畸变模型与后期制图时使用的畸变模型不同，从而导致无法直接使用标定后畸变参数”这一问题的一个解决方法，该方法的可行性和高精度性也为相机标定提供了新的思路。通过模拟试验表明，模型转换后重建出的三维点绝对坐标精度在 1.6cm 以内，两点间的相对距离精度最差不超过 4%，并且该方法已成功应用于嫦娥四号月面巡视器的全景相机制图工作，顺利保障了路径规划和导航定位的进行。

致谢

特别感谢中国航天科技集团第五研究院总体部的各位专家教授，为本文提供行文思路和数据支持，同时感谢课题组同学的大力帮助，使得本文顺利完成。

参考文献

[1] Domrongwatthanayothin B, Roeksabutr A, Chongchaewchamnanand M, et al. A simple static - shadow based algorithm for 3D complex - terrain surface reconstruction. In: IEEE International Geoscience and Remote Sensing Symposium. Seoul, 2005. 1738 - 1741.

[2] David M. Reconstruction of surface from planar contours. Dissertation for Doctoral Degree. Seattle: University of Washington, 1994.

[3] Zhao M, Li X B, Lang R L. Improved adaptive stereo matching using double dynamic programming with correlation of row and column. In: 2010 2nd International Conference on Signal Processing Systems. Dalian, 2010. V3 - 95 - V3 - 100.

[4] Wang H, Ray N, Zhang H. Graph - cut optimization of the ratio of functions and its application to image segmentation. In: 2008 15th IEEE International Conference on Image Processing. Las Vegas, 2008. 749 - 752.

[5] Faliu Y, Moon I. Image segmentation: A survey of graph - cut methods. In: 2012

International Conference on Systems and Informatics. Shanghai，2012. 1936 - 1941.

[6] Žbontar J，Lecun Y. Stereo matching by training a convolutional neural network to compare image patches. J Mach Learn Res，2015，17：1 - 32.

[7] Li M L，Liu S C，Peng S，et al. Improved dynamic programming in the lunar terrain reconstruction (in Chinese). Opto - Electron Eng，2013，40：6 - 11 [李明磊，刘少创，彭松，等. 基于改进动态规划法的月面地形三维重建. 光电工程，2013，40：6 - 11].

[8] Jie M，Huang X L. A lunar terrain reconstruction method using long base - line stereo vision. In：Chinese Control Conference. Zhangjiajie，2007. 488 - 492.

[9] Zhang Z X，Zhang J Q. Digital Photogrammetry (in Chinese). Wuhan：Wuhan University Press，2002 [张祖勋，张建清. 数字摄影测量学. 武汉：武汉大学出版社，2002].

[10] Photometrix. The Australis Image Coordinate Correction Model. Melbourne，Australia：Photometrix Pty Ltd，2004.

[11] Tsai R Y. An efficient and accurate camera calibration technique for 3D machine vision. In：Proceedings of IEEE Conference on Computer Vision and Pattern Recognition. Miami Beach，1986. 364 - 374.

[12] Fryer J G，Brown D C. Lens distortion for close - range photogrammetry. Photogramm Eng Remote Sens，1986，52：51 - 58.

[13] Zhang Z Y. A flexible new technique for camera calibration. IEEE Trans Pattern Anal Machine Intell，2000，22：1330 - 1334.

[14] Geiger A，Roser M，Urtasun R. Efficient large - scale stereo matching. In：Asian Conference on Computer Vision. Berlin，Heidelberg：Springer，2010.

[15] Yu G，Morel J M. A fully affine invariant image comparison method. In：Proceeding of the IEEE International Conference on Acoustics，Speech，and Signal Processing (ICASSP). Taipei，2009.

[16] Qing M Y，Chuang L S，Bo W，et al. Weighted total least squares for the visual localization of a planetary rover. Photogramm Eng Remote Sens，2018，84：605 - 618.

基于视觉的嫦娥四号探测器着陆点定位方法

王镓[1]，万文辉[3]，赵焕洲[1]，王保丰[1,2]，彭嫚[3]，荣志飞[1]

1. 北京航天飞行控制中心，北京 100094
2. 航天飞行动力学技术重点实验室，北京 100094
3. 中国科学院遥感与数字地球研究所，北京 100101

摘要：针对探测器月球背面着陆点快速高精度定位问题，利用嫦娥四号探测器动力下降过程中的降落相机序列图像以及着陆区高精度地形图，提出了一种基于高精度图像匹配和几何变换的探测器着陆点高精度快速定位方法。考虑到嫦娥四号、三号任务在动力下降阶段的相似性，利用嫦娥三号降落相机相关数据进行了着陆点定位实验与精度验证。其中降落相机序列影像间的匹配精度达到子像素；降落相机变分辨率图像的拼接优于 1 个像素。该方法将降落相机序列图像进行拼接，有效扩大着陆区域范围，提高匹配成功率与计算效率，实时性较好，已应用于嫦娥四号探测器月面着陆定位。

关键词：嫦娥四号；月球探测；着陆定位；图像匹配

Vision - Based Positioning Method for Chang'e - 4 Lander

WANG Jia[1], WAN Wenhui[3], ZHAO Huanzhou[1], WANG Baofeng[1,2], PENG Man[3], RONG Zhifei[1]

1. Beijing Aerospace Control Center, Beijing 100094, China
2. State Key Laboratory of Remote Sensing Science, Beijing 100094, China
3. Institute of Remote Sensing and Digital Earth, Chinese Academy of Sciences, Beijing 100101, China

Abstract: To realize the fast and high - precision positioning of the landing location of the probe on the far side of the lunar surface, a positioning method was proposed on the basis of high - precision image matching and geometric transformation using the descent sequence images during the powered descent phase and the high - precision Digital

Orthophoto Map of the landing zone. Considering the similarity of Chang'e－3 and Chang'e－4 in the powered descent phase, the landing location experiment and the accuracy verification were carried out using the relevant Chang'e－3 descent data. The matching accuracy between sequent descent images was sub－pixel, and the stitching precision of the multi－resolution descent images was higher than 1 pixel. The experiments showed that the method proposed in this paper could effectively extend the range of the landing area, improve the mapping accuracy and the computation efficiency. It has been applied to Chang'e－4 mission.

Keywords: Chang'e－4; lunar exploration; landing positioning; image matching

1 引言

月球作为深空探测的前哨站和转运站，是人类探索宇宙、各国开展深空探测的首选目标[1]。我国月球探测工程始于 20 世纪 90 年代，2013 年 12 月 14 日嫦娥三号月球探测器在月球虹湾地区成功着陆，实现了我国航天器首次地外天体软着陆。2018 年 12 月 8 日发射的嫦娥四号月球探测器，实施世界上第一次月球背面软着陆和巡视探测，着陆区选定在月球背面南极-艾特肯盆地内的卡门撞击坑内[2]。探测器着陆点的快速高精度定位是实现这一过程的重要技术环节，也是地外天体探测器开展各项工作的重要前提。

目前国际上月球探测器着陆点的定位方法主要有基于地基测量和基于视觉图像两种[3]。基于地基测量的着陆定位技术是指利用地面跟踪站对着陆器进行测距测速和干涉测量（Very Long Baseline Interferometry，VLBI），结合月球运动信息，将测量数据进行综合平差处理，从而确定探测器在天体中的位置[4]。基于视觉图像的定位是指利用探测器着陆过程序列图像与轨道器卫星影像的配准关系和轨道器所携带相机的精密成像模型，结合已有的高精度数字高程模型来实现着陆器落点的定位[3]。嫦娥四号任务着陆区为月球背面，由于遮挡，传统测距、测速及 VLBI 测角技术在探测器动力下降段无法发挥作用，因此依靠视觉系统完成探测器着陆点的快速、高精度定位是一种主要的技术手段。

基于视觉图像的着陆定位技术在月球、行星、小行星和彗星的着陆与采样返回任务中均得到了应用[5]。深空 1 号（DS－1）小行星探测任务首次对该方法进行了试验验证，完全利用视觉图像的方法实现了自主轨道确定[6-7]。在火星探路者任务中，Golombek 等利用 Viking 卫星影像与着陆器影像上人工选取的 5 对特征点通过方位角计算进行着陆点定位，定位精度达到 100m[8]。我国在嫦娥三号任务中也采用了视觉图像处理的方法实现探测器着陆点的定位。万文辉[9]、贾阳[10]等利用降落图像、导航相机图像和嫦娥二号 DOM 影像对着陆点位置进行解算，其定位结果均是在探测器落月后数小时完成的。Wagner[11]、刘斌[12]等利用月球勘测轨道器（Lunar Reconnaissance Orbiter，LRO）再次飞越着陆区上方获取的图像，以事后处理的方式对定位结果的精度进行了评估。目前，应用于工程中的着陆点视觉定位方法实时性不高，不能完全满足工程任务的实际需求。

本文在充分分析降落相机序列图像成像特点的基础上，针对相邻图像间的尺度与旋转变化特点引入 SIFT 匹配方法，基于相邻图像间获得的匹配点集建立同名像点的映射模型。结合着陆区高精度正射影像图（DOM），通过两类影像间的撞击坑匹配建立几何关系，将降落序列影像上的着陆点投影至着陆区 DOM 上并获取相应的地理位置，以提升定位效率。

2 着陆过程分析

相对于嫦娥三号任务，嫦娥四号任务的着陆区地形起伏程度明显增加且波动明显，因此在动力下降策略中进行了如下优化[2]：在主减速段结束后探测器调整姿态，由倾斜运动轨迹改为垂直向下运动轨迹，以实现测距波速指向与着陆点位置基本一致；同时，引入测距敏感器对导航滤波算法进行修正以保证高度方向导航结果的正确性，实现安全着陆。嫦娥四号任务探测器动力下降过程示意如图 1 所示。

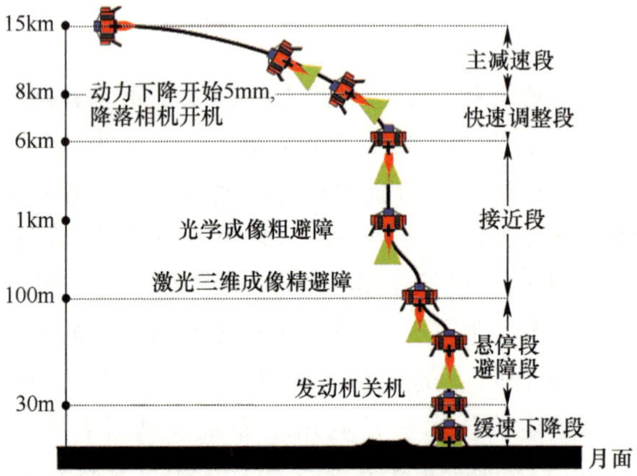

图 1 嫦娥四号动力下降轨迹示意图[2]

Fig. 1 Powered descent of Chang'e-4[2]

同嫦娥三号任务一样，嫦娥四号探测器底部安装有降落相机，是 1024×1024 像素的面阵相机。在动力下降开始 5min 后降落相机将开机随后开始抽帧下传降落相机的图像数据，直至探测器着陆月面后，降落相机关机。降落相机相关参数与嫦娥三号任务一致，在 1×1（km^2）范围内，影像分辨率高于 1m；在以着陆点为中心的 50×50（m^2）范围内分辨率可达到 0.1m[13]。由此可见，通过获取的降落相机图像，不仅能够连续反映探测器落月过程月球表面的变化范围，还能实时监视探测器的降落过程，并且影像的重叠度较高，是基于视觉的探测器着陆定位的重要数据源。

嫦娥四号任务主要采用国防科技工业局发布的由嫦娥二号月球探测器获得的分辨率为 7m 的全月球影像图作为着陆点定位的基础数据。

3 着陆点定位方法

探测器着陆点的定位流程如图 2 所示，主要分为以下四步：首先利用降落相机下传的抽帧序列图像进行相邻图像重叠区域的特征提取与匹配，根据得到的匹配特征点集计算降落相机初帧图像和末帧图像的坐标变换关系，从而求解探测器落点在初帧图像中的像坐标；接着根据相邻图像间的坐标变换关系，建立成像面对应地面水平的约束条件，对图像进行单应变换，实现降落序列图像的着陆区图像拼接；然后，根据惯性导航提供的相关位置数据对预选着陆区 DOM 影像进行裁剪；最后，通过 MeanShift 算法依次完成撞击坑的提取、撞击坑的匹配，将拼接影像与裁剪的 DOM 进行配准，从而实现着陆点的高精度定位。

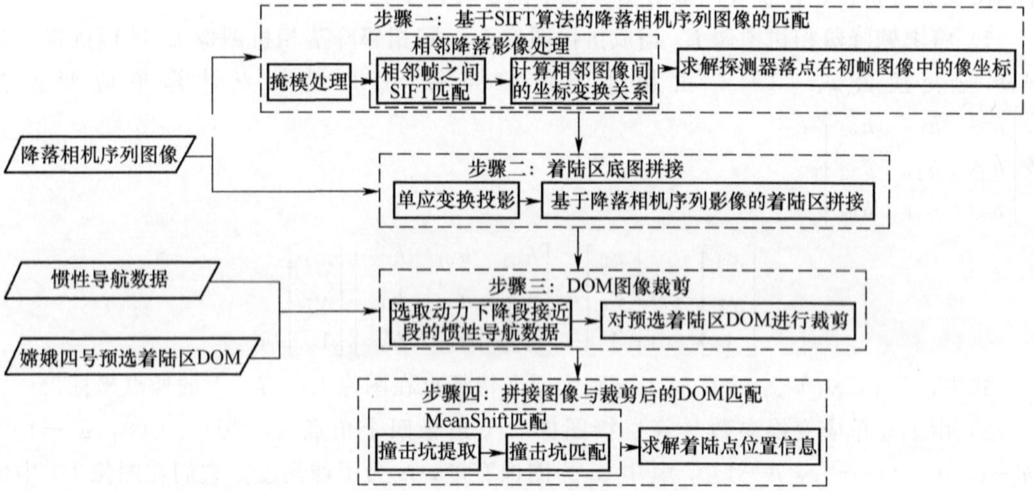

图 2　着陆点定位实现流程图

Fig. 2　Flowchart of landing positioning

3.1　基于 SIFT 的降落相机相邻影像间的匹配

由于相邻两幅降落相机序列图像存在缩放、部分旋转等仿射变形，为获得稳健的匹配结果，选用 SIFT 特征提取作为图像仿射不变特征提取方法，并基于提取的特征实现图像间的匹配。具体步骤如下：

1）生成 SIFT 特征向量：获取降落相机在探测器着陆过程中拍摄的序列图像，分别生成相邻序列图像的 SIFT 特征向量。为了增强匹配的稳健性，对每个关键点使用 4×4 共 16 个种子点来描述。这样 SIFT 特征向量已经去除了尺度变化、旋转等几何因素的影响，后续再将特征向量的长度归一化，便可以进一步去除光照变化的影响。

2）特征匹配：采用关键点特征向量的欧式距离来作为两幅图像中关键点的相似性判定度量，如果最近的距离除以次近的距离小于某个比例阈值，则接受这一对匹配点。降低这个比例阈值，SIFT 匹配点数目会减少，但会更加稳定。应用中为了获得足够多的特征点，一般取 0.75～0.85。

3）双向匹配策略剔除粗差：采用双向匹配策略剔除匹配的粗差，具体如下：令 (x_{ref}, y_{ref}) 为参考图像上的特征点，(x_{mat}, y_{mat}) 为该点在其匹配图像上的同名特征点。再将同名点 (x_{mat}, y_{mat}) 作为目标点，在参考图像上匹配获得另一同名点 (x'_{ref}, y'_{ref})，这一过程称为逆向匹配。若 (x_{ref}, y_{ref}) 与 (x'_{ref}, y'_{ref}) 间的距离小于 1 个像素，则确定 (x_{mat}, y_{mat}) 为 (x_{ref}, y_{ref}) 的同名点；否则，将其作为误匹配点剔除。

3.2 基于降落相机图像的着陆区底图拼接

降落相机序列图像覆盖探测器从动力下降段主减速段至着陆月面过程中平扫过的月球表面纹理。为了便于实现与预选着陆区底图 DOM 进行配准，设计中将变分辨率降落相机序列图像进行拼接，以扩大图像数据区域范围。具体步骤如下：

1）将末帧降落相机图像 L_1 当成拼接图像 L_{1M} 同相邻降落相机图像 L_2 进行拼接。根据单应变换公式（1），由相邻图像的同名 SIFT 特征点计算单应变换参数 $\begin{bmatrix} h_{00} & h_{01} & h_{02} \\ h_{10} & h_{11} & h_{12} \\ h_{20} & h_{21} & h_{22} \end{bmatrix}$。

$$\begin{bmatrix} x_2 \\ y_2 \\ 1 \end{bmatrix} = \boldsymbol{H} \begin{bmatrix} x_1 \\ y_1 \\ 1 \end{bmatrix} = \begin{bmatrix} h_{00} & h_{01} & h_{02} \\ h_{10} & h_{11} & h_{12} \\ h_{20} & h_{21} & h_{22} \end{bmatrix} \begin{bmatrix} x_1 \\ y_2 \\ 1 \end{bmatrix} \tag{1}$$

式中，(x_1, y_1)、(x_2, y_2) 分别为同名特征点在图像 L_1、L_2 上的像点坐标。

2）由上述单应变换参数计算拼接图像 L_{1M} 边缘四个角点（0，0）、（0，$w-1$）、（$h-1$，0）、（$w-1$，$h-1$），其中 w 为图像宽度，h 为图像高度。它们在图像 L_2 中的对应坐标范围记为 $\boldsymbol{R}_{1-2}$。根据图像 L_{1M} 边缘四个角点坐标计算拼接图像 L_{1M} 的坐标范围，记为 $\boldsymbol{R}_{2M}$，同时计算它的起始坐标，记为 $O(w_{2M}, h_{2M})$。

3）依据拼接图像 L_{1M} 的范围 $\boldsymbol{R}_{2M}$ 构建拼接图像，记为 L_{2M}。图像 L_{2M} 中 p_m 的像素值由 L_{1M} 与 L_2 之间的单应变换参数 $\begin{bmatrix} h_{00} & h_{01} & h_{02} \\ h_{10} & h_{11} & h_{12} \\ h_{20} & h_{21} & h_{22} \end{bmatrix}$ 计算对应 L_{1M} 中位置 p_i 的像素值确定。如果 p_i 的位置超出 L_{1M} 的有效范围，则依据 $\boldsymbol{R}_{2M}$ 计算对应在图像 L_2 中位置 p_j，并将 L_2 中 p_j 的像素值赋予 p_m。

4）依次传入后续降落相机图像 L_3 至 L_N，同 L_{2M} 通过上述第 1～3 步构建 L_{3M} 至 L_{NM}，最终可获得经 N 张降落相机序列图像拼接的图像 L_{NM}。

3.3 基于 MeanShift 算法的拼接图像与着陆区 DOM 影像的匹配

拼接图像与着陆区 DOM 影像在成像时间、太阳高度角、太阳方位角均不相同，而这些因素又会导致两幅影像基于同一目标区域的纹理出现较大差异，常见的基于点特征的匹配算法均无法适用。实际上，月球表面又会布满大大小小的撞击坑，由于地形凹陷，这些撞击坑在影像中多会呈现出两侧坑沿明暗对比明显的特点。

MeanShift 算法通过检测局部图像的特征矢量能够处理图像的缩放、旋转等变形带来的匹配困难的问题。依据现有的经验，目前该算法可适用于观测图（如拼接图像）与基准图（如着陆区 DOM 影像）6 倍以内缩放的自动匹配。因此，在实施过程中可首先通过 MeanShift 提取两幅影像中撞击坑的中心位置[14]而得到相应撞击坑的分布情况，然后计算两幅影像中撞击坑分布的几何关系，从而实现影像间的特征匹配，具体的实现方法参见文献［9］。

4　实验与分析

考虑嫦娥四号与嫦娥三号任务的动力下降过程具有一定相似性，选用嫦娥三号任务接近段至缓速下降段的 37 幅降落相机序列图像数据、相关惯性导航数据及虹湾地区 DOM 影像进行着陆点定位实验，中间的处理结果见本章各节。经解算，嫦娥三号探测器着陆点位置信息为（44.1210°N，19.5112°W），与目前国内外公布的定位结果精度一致，见表 1。

表 1　嫦娥三号探测器着陆点定位结果统计

Tab. 1　Statistics of landing location positioning of Chang'e - 3 probe

实验数据	经度/(°)	纬度/(°)	出　处
降落相机序列图像、CE - 2 DOM	−19.5125	44.1189	万文辉，航天器工程，2014[9]
降落相机序列图像、导航相机图像及 CE - 2 DOM	−19.5126	44.1188	贾阳，科学通报，2014[10]
着陆后 1 幅 LRO NAC 影像	−19.5116	44.1214	LRO 团队，定位结果由 M1142582775R 计算[15]
着陆前后 5 幅 LRO NAC 影像	−19.5115	44.1213	Wagner，The 45th Lunar and Planetary Science Conference，2014[11]
着陆前后 14 幅 LRO NAC 影像	−19.5113	44.1219	刘斌，科学通报，2015[12]
降落相机序列图像、基于 NAC 影像重建的 DOM	−19.5112	44.1210	本文方法，DOM 的分辨率约为 1.6m/pixel

4.1　降落相机序列图像间的匹配

受降落相机安装位置影响，每幅降落相机图像中均有一部分固定区域被探测器支架遮挡。如果直接使用原始图像进行匹配，遮挡区域会产生一定数量的无效特征点。因此，在进行相邻序列降落图像的匹配运算之前，先对每幅图像进行预处理，图 3（a）为一幅降落相机的原始图像，图 3（b）为用于标记无效区域的掩模模板。

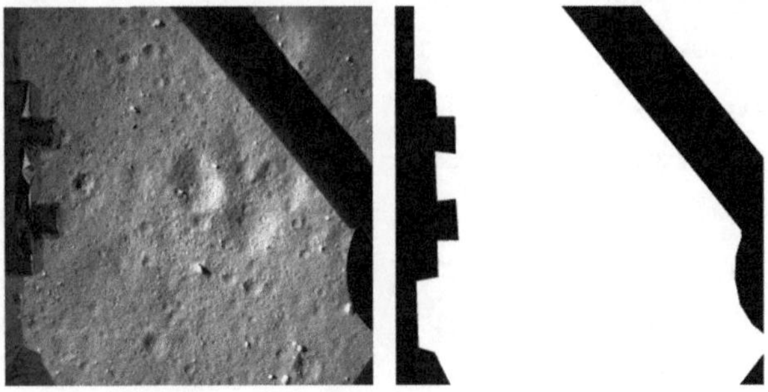

(a)　(b)

图 3　降落相机原始图像与掩模影像

（a）原始降落影像；（b）掩模板

Fig. 3　Original descent imageand its mask template

（a）Original descent image；（b）Mask template

利用 SIFT 算法进行相邻序列降落相机图像的匹配，其中降落相机图像共 37 幅，图像间隔约为 5s，具体的成像情况见表 2 所示。其中成像高度信息由惯性导航系统经由遥测信息下传提供给地面控制中心，是一个概略值，与实际情况存在一定的偏差。由表 2 中可以看出，图像 23 至图像 32 探测器在进行精避障，之后探测器发动机关机，图像 33 至图像 37 探测器缓速下降至月面，距月面高度接近于 0，与设计的下降轨道基本一致。

记降落相机视场角为 θ，探测器垂直月面的高度为 h，月面视场距离为 L，像素个数为 s，图像分辨率为 r。根据嫦娥三号任务降落相机相关参数，相机视场角 $\theta = 46.4°$、像素个数 $s = 1024$，可将降落相机图像分辨率 r 表示如式（2）：

$$r = L/s = 2\frac{h}{s}\tan\left(\frac{\theta}{2}\right) \tag{2}$$

表 2　降落相机成像情况

Tab. 2　Imaging of the landing camera

图像序号	成像时间（北京时间）	成像高度/m	分辨率/(m/pixel)	着陆点像坐标	
				x	y
0001	2013/12/14 21：08：12.875	2591	2.1	510.866 276	406.885 933
0002	2013/12/14 21：08：20.000	2307	1.8	527.915 607	760.319 688
0003	2013/12/14 21：08：23.000	2182	1.7	541.847 233	1 127.465 892
0004	2013/12/14 21：08：29.000	1930	1.5	516.474 361	1 296.504 672
0005	2013/12/14 21：08：38.000	1671	1.3	507.964 664	1 319.501263
0006	2013/12/14 21：08：45.000	1448	1.1	505.697 817	1 326.907 137
0007	2013/12/14 21：08：49.000	1324	1.0	501.920 699	1 320.780 323
0008	2013/12/14 21：08：54.000	1225	1.0	503.783 839	1 318.329 69
0009	2013/12/14 21：08：58.875	1069	0.8	513.200 984	1 313.918 092
0010	2013/12/14 21：09：03.875	941	0.7	517.197 608	1 306.921 963
0011	2013/12/14 21：09：10.000	817	0.6	514.211 303	1 301.598 874
0012	2013/12/14 21：09：14.875	718	0.5	510.970 57	1 296.616 597
0013	2013/12/14 21：09：19.875	623	0.5	513.849 015	1 286.735 254
0014	2013/12/14 21：09：23.875	556	0.4	511.013 175	1 268.240 32
0015	2013/12/14 21：09：29.875	463	0.3	507.924 105	1 250.425 092
0016	2013/12/14 21：09：33.875	400	0.3	508.230 924	1 188.898 039
0017	2013/12/14 21：09：39.875	333	0.2	515.762 653	1 153.623 205
0018	2013/12/14 21：09：44.875	269	0.2	501.451 95	1137.871 641
0019	2013/12/14 21：09：50.000	238	0.1	484.219 639	1104.131 241
0020	2013/12/14 21：09：56.000	145	0.1	487.820 423	1029.677 975
0021	2013/12/14 21：09：59.000	145	0.1	485.207 644	971.924 109
0022	2013/12/14 21：10：05.875	113	0.0	471.606 057	867.105 503
0023	2013/12/14 21：10：09.000	82	0.0	477.337 046	734.661 373
0024	2013/12/14 21：10：15.000	82	0.0	462.769 635	578.458 091

续表 2

图像序号	成像时间（北京时间）	成像高度/m	分辨率/（m/pixel）	着陆点像坐标	
				x	y
0025	2013/12/14 21：10：21.000	79	0.0	420.212 165	436.762 815
0026	2013/12/14 21：10：25.000	79	0.0	409.940 362	336.512 892
0027	2013/12/14 21：10：30.875	79	0.0	425.848 012	526.645 417
0028	2013/12/14 21：10：34.875	108	0.0	433.356 942	454.794 298
0029	2013/12/14 21：10：41.000	83	0.0	438.123 814	459.821 033
0030	2013/12/14 21：10：44.875	51	0.0	445.252 1	461.493 215
0031	2013/12/14 21：10：50.000	51	0.0	479.913 28	473.078 207
0032	2013/12/14 21：10：54.875	51	0.0	498.109 816	487.444 233
0033	2013/12/14 21：10：59.875	−5	0.0	490.172 738	499.288 338
0034	2013/12/14 21：11：05.875	−5	0.0	489.683 81	498.560 08
0035	2013/12/14 21：11：09.875	−5	0.0	500.948 022	497.506 076
0036	2013/12/14 21：11：15.875	23	0.0	504.926 99	500.659 308
0037	2013/12/14 21：11：20.000	23	0.0	512	512

根据每幅降落图像距离月面的高度计算其分辨率，结果见表 2。由于图像 23 至图像 37 的距离月面高度不是很准确（表 2 中标红处），因此这几幅图像的分辨率也不是很准确。经事后分析，图像 37 的成像高度约为 2m，分辨率约为 0.008m/pixel。根据第 3.1 节给出的相邻序列图像的匹配算法，37 幅降落相机相邻图像的（部分）结果如图 4 所示，计算出的探测器着陆点在每幅降落相机图像中的像点坐标见表 2 右侧两列。

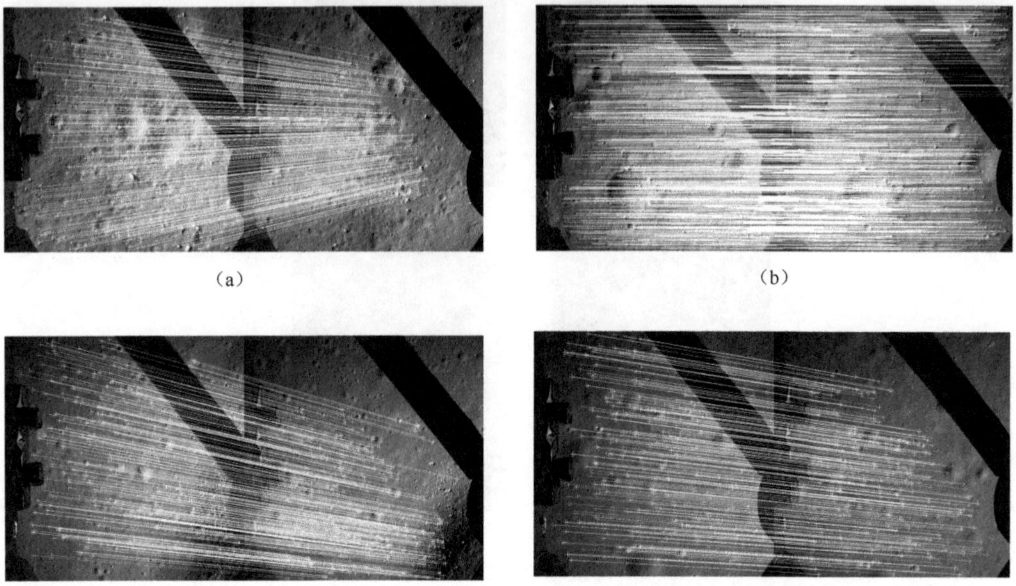

（a）　　（b）　　（c）　　（d）

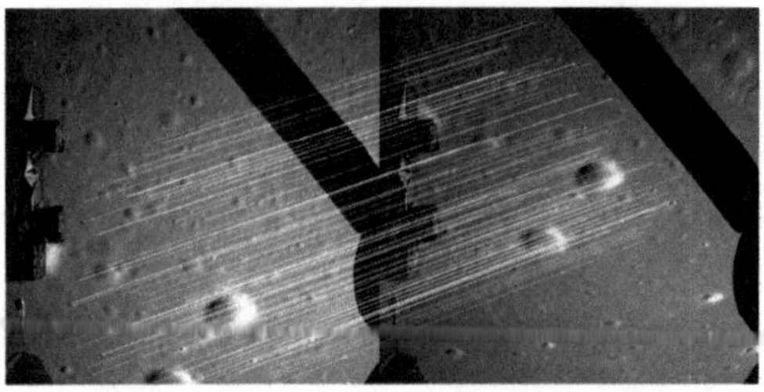

(e)

图 4　相邻图像的 SIFT 匹配结果

(a) 降落图像 0037 与 0036 间的匹配，590 个同名点；(b) 降落图像 0030 与 0029 间的匹配，1735 个同名点；(c) 降落图像 0020 与 0019 间的匹配，594 个同名点；(d) 降落图像 0010 与 0009 间的匹配，452 个同名点 (e) 降落图像 0002 与 0001 间的匹配，154 个同名点

Fig. 4　Matching results of adjacent descent imagesbased on SIFT

(a) Matching between 0037—0036，590homnyms；(b) Matching between 0030—0029，1735homnyms；(c) Matching between 0020—0019，594homnyms；(d) Matching between 0010—0009，452homnyms；(e) Matching between 0002 - 0001，154 homonyms

4.2　着陆区底图拼接

单应投影变换虽有 9 个待求解元素，实际上只具有 8 个自由度，理论上只需要有 4 对以上的同名特征点，就可以根据最小二乘原理求解出单应矩阵对应的各元素。由 4.1 节匹配结果看出，相邻图像的匹配结果经 RANSAC 剔除粗差点之后，剩余特征点对个数均大于 100。由于拼接计算的位置可能是非整数的像素，在实际拼图过程中采用双线性内插方法进行像素点重采样。

图 5 给出了降落相机图像的拼接结果，其中白色的十字叉代表了着陆点在该幅拼接图

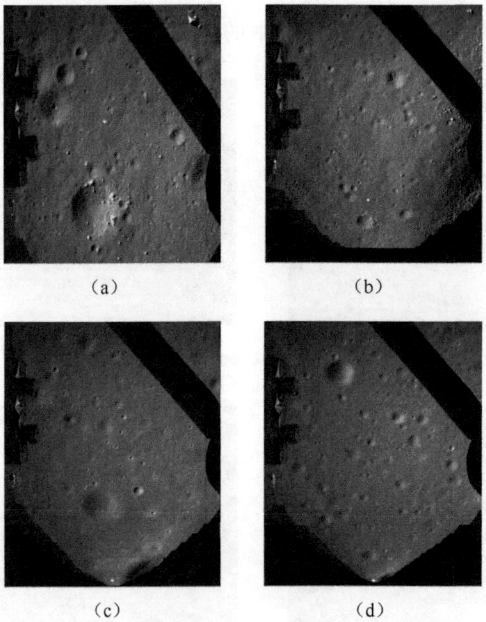

图 5　降落相机图像的拼接结果

(a) 7 幅影像的拼接结果；(b) 17 幅影像的拼接结果；(c) 27 幅影像的拼接结果；(d) 34 幅影像的拼接结果

Fig. 5　Resultant Mosaic of the Landing Area

(a) Mosaic image using 7 images；(b) Mosaic image using 17 images；(c) Mosaic image using 27 images；(d) Mosaic image using 34 images

像中的位置情况。可以看出，经拼接处理后能明显扩大图像的数据区域范围，有利于与预选着陆区底图 DOM 进行图像匹配。

4.3　拼接图像与着陆区 DOM 图的匹配

将 4.2 节获得的着陆区拼接图像与 CE-3 虹湾地区 DOM 图进行图像匹配。其中，CE-3 着陆区 DOM 图由 Lunar Orbital Data Explorer 下载获得[16]，选用的图像名称为：NAC_DTM_CHANGE3_M1144922100_160CM. tif，分辨率为 1.6m/pixel。

首先根据惯性导航提供的探测器的概略位置对该 DOM 进行裁剪，生成的图像记为 I_{DOM}。然后采用 MeanShift 算法，将拼接图像（记为 I_{mosaic}）与裁剪后的 DOM 图像（即 I_{DOM}）进行匹配，两幅影像的匹配结果如图 6 所示。

实验采用主频为 3.0GHz、操作系统为 64 位麒麟系统的图形工作站进行，其中 37 幅降落相机序列图像的匹配及拼接在 2min 内完成全自动处理，精度优于 1pixel，相比文献 [9]、[12] 的方法在处理效率上有提升。由于成像时刻不同，降落相机图像与着陆区 DOM 影像纹理差异较大，匹配时加入适量的人工交互，匹配精度优于 1pixel。

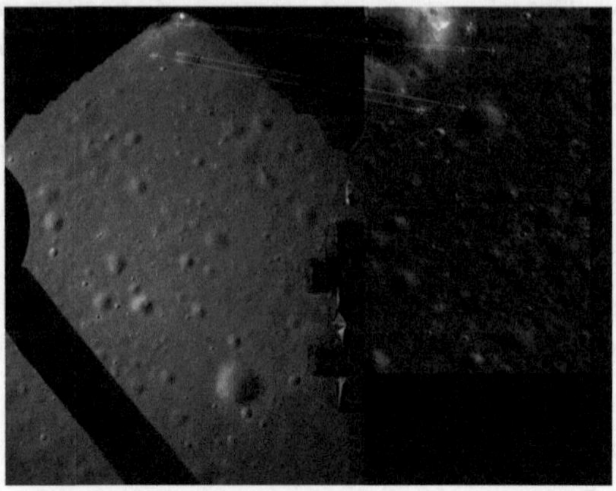

图 6　拼接图像与 DOM 影像的配准结果

Fig. 6　Matching result between the mosaic imageand the tailored DOM

5　结论

本文采用计算机视觉的相关方法在高精度图像匹配和几何变换的基础上实现了探测器月面着陆定位。使用嫦娥三号任务相关影像开展了探测器着陆点定位实验与精度验证，解算结果与目前国内外公布的各类定位结果精度一致。该方法将降落相机序列图像进行拼接，能有效扩大着陆区域范围，提高匹配成功率，实时性也较好。但其定位结果是一个相对值，定位精度与着陆区底图 DOM 的制图精度密切相关。下一步将开展“月球多源遥感数据的高精度着陆区制图”研究，为基于视觉的着陆点定位方法提供覆盖范围更广、精度更高的制图产品，同时也能为探测器落月后的路径规划和安全导航提供基础数据支持。

参考文献

[1] 欧阳自远. 月球探测的进展与我国月球探测的科学目标 [J]. 贵州工业大学学报，2003，32 (6)：1—17.
OUYANG Ziyuan. International lunar exploration progress and scientific objectives of Chinese lunar exploration program [J]. Journal of Guizhou University of Technology，2003，32 (6)：1—17. (in Chinese)

[2] 李飞，张熇，吴学英，等. 月球背面地形对软着陆探测的影像分析 [J]. 深空探测学报，2017，4 (2)：143—149.
Li F，Zhang H，Wu X，et al. Influence analysis of terrain of the farside of the moon on softlanding [J]. Journal of Deep Space Exploration，2017，4 (2)：143—149. (in Chinese)

[3] 王镓，王保丰，曹建峰，等. 月球探测着陆点定位综述 [C] //中国测绘地理信息学会大地测量与导航专业委员会2017年学术会议论文集，2017，432—440.
Wang J，Wang B，Cao J，et al. A review of positioning methods for lunar exploration [C]//CSGPC Geodetic Survey and Navigation Professional Committee 2017 Academic Annual Meeting，432—440. (in Chinese)

[4] 黄勇，昌胜骐，李培佳，等. 嫦娥三号月球探测器的轨道确定和月面定位 [J]. 科学通报，2014，59：2268—2277.
Huang Y，Chang S Q，Li P J，et al. Orbit determination of Chang'e-3 and position of the lander and the rover [J]. Chin Sci Bull，2014，59：3858—3867. (in Chinese)

[5] 李俊峰，崔文，宝音贺西. 深空探测自主导航技术综述 [J]. 力学与实践，2012，34 (2)：1—9.
LI Junfeng，CUI Wen，BAOYIN Hexi. A survey of autonomous navigation for deep space exploration [J]. Mechanics in Engineering，2012，34 (2)：1—9. (in Chinese)

[6] Bhaskaran S，Nandi S，Broschart S B，et al. Small body landings using autonomous onboard optical navigation [J]. Journal of the Astronautical Sciences，2011，58 (3)：409—427.

[7] 邵巍. 基于图像信息的小天体参数估计及探测器自主导航研究 [D]. 哈尔滨：哈尔滨工业大学，2009.
SHAO Wei. Image-Based Small Body Parameters Estimation and Spacecraft Autonomous Navigation [D]. Harbin：Harbin Institute of Technology，2009. (in Chinese)

[8] 李建国. 火星探测器自主光学着陆导航方法研究 [D]. 哈尔滨：哈尔滨工业大学，2014.
LI Jianguo. Research on Autonomous Vision Navigation Methods for Spacecraft Landing on Mars [D]. Harbin：Harbin Institute of Technology，2014. (in Chinese)

[9] 万文辉，刘召芹，刘一良，等. 基于降落图像匹配的嫦娥三号着陆点位置评估 [J]. 航天器工程，2014，23 (4)：5—12.

WAN Wenhui, LIU Zhaoqin, LIU Yiliang, et al. Descent image matching based position evaluation for Chang'e-3 landing point [J]. Spacecraft Engineering, 2014, 23 (4): 5-12. (in Chinese)

[10] 贾阳，刘少创，李明磊，等. 利用降落影像序列实现嫦娥三号系统着陆点高精度定位 [J]. 科学通报，2014，59：1838—1843.

Jia Y, Liu S C, Li M L, et al. Chang'e-3 system pinpoint landing localization based on descent image sequence [J]. Chin SCI Bull (Chin Ver), 2014, 59: 1838—1843. (in Chinese)

[11] Wagner R V, Robinson M S, Speyerer E J, et al. Locations of anthropogenic sites on the Moon [C]//45th Lunar & Planetary Science Conference. Lunar and Planetary Science Conference, 2014, 45: 225.

[12] 刘斌，邸凯昌，王保丰，等. 基于 LRO NAC 影像的嫦娥三号着陆点高精度定位与精度验证 [J]. 科学通报，2015，60：2750—2757.

LIU Bin, DI Kaichang, WANG Baofeng, et al. Positioning and precision validation of Chang'e-3 Lander based on multiple LRO NAC images [J]. Chin SCI Bull (Chin Ver), 2015, 60: 2750—2757. (in Chinese)

[13] 刘斌，徐斌，刘召芹，等. 基于降落相机图像的嫦娥三号着陆轨迹恢复 [J]. 遥感学报，2014，18 (5)：981—987.

LIU Bin, XU Bin, LIU Yiliang, et al. Descending and landing trajectory recovery of Chang'e-3 lander using descent images [J]. Journal of Remote Sensing, 18 (5): 981—987.

[14] 丁萌，曹云峰，吴庆宪. 一种从月面图像检测陨石坑的方法 [J]. 宇航学报，2009，30 (3)：1243—1248.

DING Meng, CAO Yunfeng, WU Qingxian. A method of craters detection from the surface imagery of Moon [J]. Journal of Astronautics, 30 (3): 1243—1248. (in Chinese)

[15] NASA Images of Chang'e-3 Landing Site [EB/OL]. NASA/GSFC/Arizona State University. [2014-01-23]. http: //www. nasa. gov/content /nasa-images-of-change-3-landing-site/.

[16] PDS Geosciences Node. Lunar Orbital Data Explorer [EB/OL]. Washington University in St. Louis. (2018) [2019]. http: //ode. rsl. wustl. edu/moon/.

基于视觉的嫦娥四号探测器着陆点定位

王镓[1,2]，吴伟仁[3]，李剑[1]，邸凯昌[4]，万文辉[4]，谢剑锋[1]，彭嫚[4]，王保丰[1,2]，刘斌[4]，贾萌娜[4]，席露华[1]，赵瑞[1]

1. 北京航天飞行控制中心，北京 100094
2. 航天飞行动力学技术重点实验室，北京 100094
3. 探月与航天工程中心，北京 100037
4. 中国科学院遥感与数字地球研究所，北京 100101

摘要： 我国于 2018 年 12 月 8 日发射嫦娥四号月球探测器，实施世界上首次月球背面软着陆和巡视探测。探测器着陆点的快速高精度定位是其中的一个重要技术环节，也是着陆器和巡视器开展月面探测工作的重要前提。本文基于高精度图像匹配和几何变换方法，结合实际工程任务需求，首先利用探测器动力下降期间的近实时高压缩比降落序列影像完成了着陆点的快速初始定位；其次利用回放的低压缩比降落影像对该位置进行了精化，最终解算的嫦娥四号探测器着陆点位置为（177.5884°E，45.4565°S）。该定位方法及结果被成功应用于实际工程任务中，为嫦娥四号任务着陆区的地形分析及后续遥操作的任务规划提供了重要支持。

关键词： 嫦娥四号；月球探测；着陆点定位；降落影像

Vision based Chang'E－4 landing point localization

WANG Jia[1,2], WU WeiRen[3], LI Jian[1], DI KaiChang[4], WAN WenHui[4], XIE JianFeng[1], PENG Man[4], WANG BaoFeng[1,2], LIU Bin[4], JIA MengNa[4] & XI LuHua[1], ZHAO Rui[1]

1. Beijing Aerospace Control Center, Beijing 100094, China
2. National Key Laboratory of Science and Technology on Aerospace Flight Dynamics, Beijing 100094, China
3. Lunar Exploration and Space Engineering Center, Beijing 100037, China
4. State Key Laboratory of Remote Sensing Science, Institute of Remote Sensing and Digital Earth, Chinese Academy of Sciences, Beijing 100101, China

Abstract: On December 8, 2018, China launched the Chang'E - 4 lunar probe and implemented soft landing and patrol exploration on the lunar farside for the first time. Fast and high - precision positioning of the landing point is a critical step of the mission, and also an important prerequisite for surface operations of the lander and rover. Based on the high - precision image matching and geometric transformation methods, and considering the engineering requirements, the landing point location of the Chang'E - 4 lander was initially determined using the near real - time high compression ratio descent sequence images. Then, the position was refined using the replayed low compression ratio descent images, and the landing point position was calculated to be (177.5884°E, 45.4565°S). The landing point localization method and result have been successfully applied to actual engineering tasks for the first time in China, thus providing important support to the topographic analysis of the landing area and mission planning of the follow - up teleoperations.

Keywords: Chang'E - 4; lunar exploration; landing position; descent images

1 引言

北京时间2019年1月3日10时26分，嫦娥四号月球探测器顺利在预选着陆区月球背面南极-艾特肯（South Pole - Aitken，SPA）盆地内的冯·卡门（Von Karman）撞击坑成功着陆[1]。1月11日16时，嫦娥四号巡视器与着陆器在“鹊桥”中继星的支持下，成功实施了两器互拍，标志着我国嫦娥四号任务圆满成功。此次任务实现了人类历史上首次航天器在月球背面软着陆和巡视勘察，首次月球背面与地球的中继通信，开启了人类探索宇宙奥秘的新篇章。着陆点的精确位置信息是建立巡视器工作坐标系的重要依据，为着陆区的地形分析和后续任务规划及科学探测提供位置信息和基础数据，是着陆器和巡视器开展各项月面探测工作的重要前提[2-4]。

基于视觉的着陆定位技术在月球、行星、小行星及彗星的着陆与采样返回任务中均得到了应用[2-5]。我国在嫦娥三号任务中也采用了视觉图像处理的方法实现探测器着陆点的定位[6-7]，万文辉等人[2]、贾阳等人[3]利用降落图像、导航相机图像和嫦娥二号数字正射影像（digital orthophoto map，DOM）对着陆点位置进行解算，其定位结果均是在探测器落月后数小时完成的。Wagner 等人[8]、LRO 团队[9]、刘斌等人[4]利用月球勘测轨道器（lunar reconnaissance orbiter，LRO）再次飞越着陆区上方获取的图像，通过对比着陆前后影像以离线处理的方式对定位结果的精度进行了评估。目前看来，提高现有的应用于工程中的着陆点视觉定位方法实时性，才能更好地满足工程任务的实际需求。此外，嫦娥三号着陆区为月球正面，借助嫦娥一号X频段UXB（unified X - band）与甚长基线干涉测量VLBI（very long baseline interferometry）等观测数据可为着陆点定位提供服务[10]。而嫦娥四号着陆区为月球背面，由于月球遮挡，传统测距、测速及VLBI技术在探测器动力下降阶段无法发挥作用，因而整个降落过程中无法获取无线电定位结果。因此，在嫦娥四号任务中，基于视觉影像的定位技术成为着陆点快速、高精度定位的主要技术手段[7]。另外，受“鹊桥”中继传输链路的带宽限制，嫦娥四号探测器动力下降阶段采取了高压缩比（1∶64）的方式对降落影像进行压缩传输（嫦娥三号采用的是 1∶8 压缩比）[11]，使得原始影像斑块效应明显，为基于降落影像的图像匹配增加了难度。

着陆区大范围、高分辨率的 DOM 是基于视觉的着陆点定位的参考基准。目前覆盖全月、分辨率最高的立体影像数据为嫦娥二号轨道器影像，国家天文台已面向全球发布了全月 7m/pixel 的 DOM 及 20m/pixel 的数字高程模型（digital elevation model，DEM）产品[12]。北京航天飞行控制中心联合中国科学院遥感与数字地球研究所遥感科学国家重点实验室，利用 LRO 窄角相机（narrow angle camera，NAC）[13]的 100 幅影像，制作了着陆区分辨率约为 1m/pixel 的高精度无缝 DOM，与嫦娥二号 DOM 一起用作着陆点定位的底图。

本文基于嫦娥四号序列降落影像的成像特点及下传约束，利用近实时下传的高压缩比（1∶64）降落影像实现了着陆点的快速初始定位；利用回放的低压缩比（1∶8）的降落影像实现了着陆点的高精度定位。为了解决高压缩比导致原始降落影像的斑块效应，在匹配定位处理前先对原始影像采用“去块效应”进行预处理增强，为后续图像匹配奠定基础。针对序列影像间的尺度与旋转变化特点引入 SIFT 匹配方法及构建几何映射模型，并同步完成着陆点位置信息的传递，有效提升了定位效率。利用回放的低压缩比降落影像，选择序列高分辨率图像进行二次精细定位解算，有效保证了最终着陆点定位结果的精度及可靠性。

2　探测器动力下降过程分析

嫦娥四号探测器的动力下降过程经历着陆准备段、主减速段、快速调整段、接近段、悬停段、（精）避障段以及缓速下降段共 7 个阶段[1,11,14]，由距离月面 15km 高度至接触月球表面，整个过程约为 687s。利用下传的遥测信息绘制了嫦娥四号探测器动力下降阶段空间位置变换曲线，如图 1 所示。

嫦娥四号着陆区为月球背面，航迹的高度起伏相对于月球正面明显增大且波动明显，因此需要优化着陆过程中的分段控制目标和导航信息的引入时间，同时对导航算法进行优化[11]。探测器在距离月面高度为 6km 之后就迅速调整着陆姿态，运动轨迹改为垂直向下[11,14]；同时，为了增加着陆安全性，嫦娥四号探测器还采取了接力避障的方式：即在距离月面高度约为 2km 时，探测器首先进行一次光学粗避障；在距离月面高度为 100m 左右时，探测器利用激光敏感器实现精细避障。

同嫦娥三号任务一样，嫦娥四号探测器底部以垂直向下的视角安装了降落相机，像幅大小为 1024 像素×1024 像素，视场角约为 45°[1]。动力下降段开始后约 328s 降落相机开机，随后开始抽帧下传降落影像数据，影像分辨率由低至高，直至探测器着陆月面。在 1km×1km 范围内，降落影像的分辨率优于 1m/pixel；在以着陆点为中心的 50m×50m 范围内，降落影像的分辨率可达到 0.1m/pixel[15]。在探测器动力下降阶段采取近实时、抽帧、压缩方式共下传 59 幅 1∶64 高压缩比的降落影像，这些数据是用于着陆点初始定位的重要数据。当探测器落月顺利完成两器分离后，通过回放的方式回传给地面约 5441 幅降落原始影像，其中 1∶8 低压缩比的影像约为 5300 幅，这些数据是用于着陆点精确定位的主要数据。图 2（a）为探测器下传的首幅降落影像，是 1∶64 高压缩比的近实时影像，距

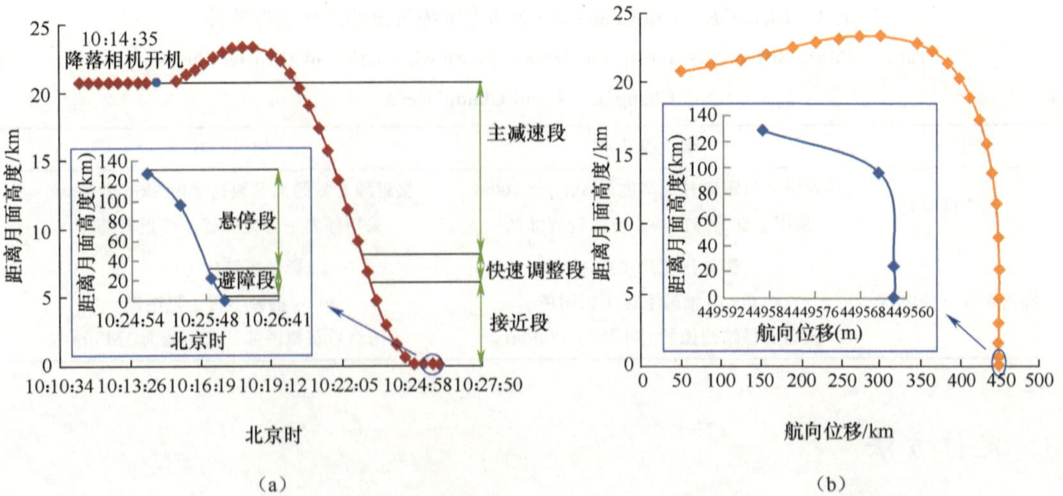

图 1　（网络版彩图）Chang'E－4 探测器动力下降阶段空间位置变化

（a）月面高度-时间关系图；（b）月面高度-航向位移关系图

Fig. 1　(Color online) Position of Chang'E－4 during the descent phase with action of thrusts

(a) Lunar height－time; (b) lunar height－heading displacement

离月面高度为 3km 以上；图 2（b）为探测器落月后的首幅影像，图像压缩比为 1∶8。

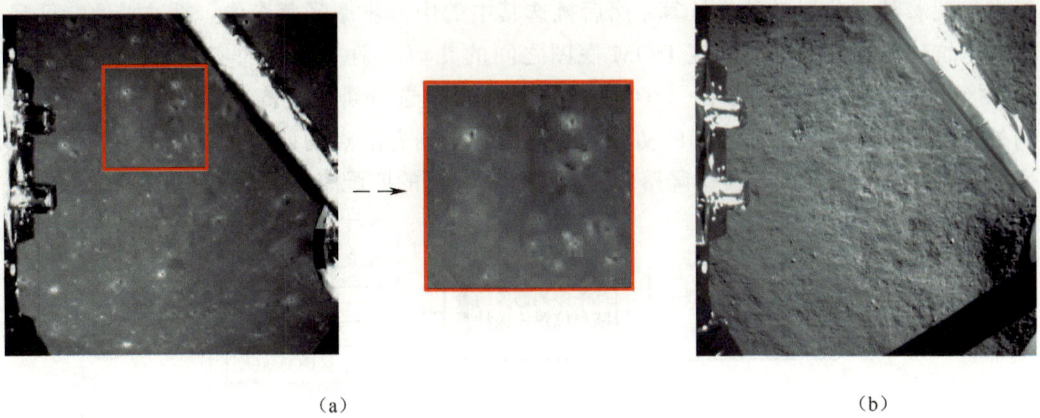

图 2　（网络版彩图）地面接收的首幅降落影像和落月后影像

（a）地面接收到的首幅降落影像（北京时间 2019－01－03T10：23：03，帧号 1624）；

（b）地面接收到的首幅落月后影像（北京时间 2019－01－03T11：44：33，帧号 5416）

Fig. 2　(Color online) The first descent and landing image received by the ground

(a) The first descent image (2019－01－03T10：23：03, Beijing time, Frame No. 1624);

(b) the first landing image (2019－01－03T11：44：33, Beijing time, Frame No. 5416)

由以上分析可知，相对于嫦娥三号任务，嫦娥四号任务在动力下降阶段主要有两方面的变化：一是下降轨迹；二是受传输带宽的限制，降落影像的下传受到诸多限制，具体见表 1 所示。

表 1　Chang'E－4 和 Chang'E－3 动力下降阶段的主要状态差异

Tab. 1　Main differences during the descent phase with action of thrusters between Chang'E－4 and Chang'E－3

变化项	嫦娥四号	嫦娥三号
动力下降轨迹	接近段开始距离月面高度由 6km 至 100m 采用垂直下降方式逐步接近着陆区	接近段开始距离月面高度由 3km 至 100m 采用倾斜下降方式逐步接近着陆区
降落图像下传方式	影像压缩比 1∶64 约 6～7s 抽帧下传 1 幅图像 压缩后每幅图像的大小约为 128kbits	影像压缩比 1∶8 约 5s 抽帧下传 1 幅图像 压缩后每幅图像的大小约为 1Mbits

3　定位方法

为了兼顾工程任务对着陆点定位的时效性与高精度要求，本文提出由粗至精的着陆点定位方法，首先是利用动力下降阶段近实时抽帧下传的高压缩比（1∶64）降落影像实现着陆点的初始（粗）定位；其次是利用回放的低压缩比（1∶8）降落影像，通过高分辨率降落影像与着陆区 DOM 底图匹配来实现着陆点的精确（精）定位，具体实现流程如图 3 所示。

如图 3 所示，着陆点定位的算法流程主要可以分为以下 5 个步骤：首先采用“去块效应”预处理算法对原始图像进行增强处理，提升图像纹理质量；再使用序列图像获取的特征点实现着陆点位置的传递计算；然后挑选其中的中分辨率降落影像，通过影像特征显著的撞击坑建立该影像与着陆区 DOM 底图之间的几何映射关系从而实现着陆点的初始定位；接着利用回放的低压缩比（1∶8）降落影像，挑选分辨率较高的影像，采用之前类似的处理方法完成着陆点像坐标的初始传递；最后根据着陆点初始定位结果对着陆区 DOM 底图进行裁剪，通过高分辨率降落影像与 DOM 底图的匹配来实现着陆点的精确定位。

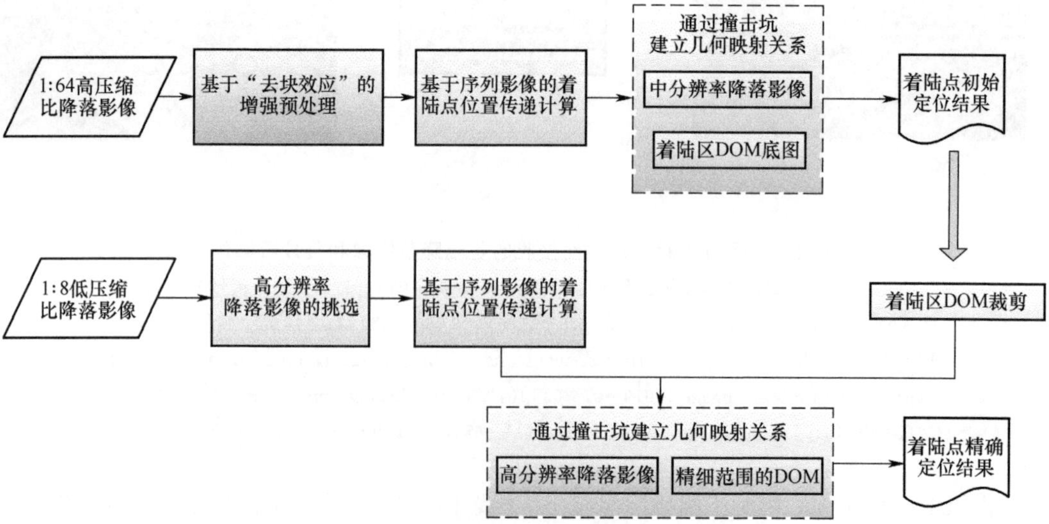

图 3　嫦娥四号着陆点定位流程图

Fig. 3　Flowchart of landing location for Chang'E－4

3.1　高压缩比降落影像的增强预处理

高压缩比降落影像在近实时下传过程中，采用了基于分块的离散余弦变换（BDCT）的压缩技术，由于没有考虑像素在相邻块中的相关性，粗糙量化使得相邻块的 DCT 系数取样落在不同的量化区间，从而导致在块边界上产生“块效应”[16]。为了提升高压缩比降落影像质量，首先需对其进行图像增强预处理，总体算法流程如图 4 所示。

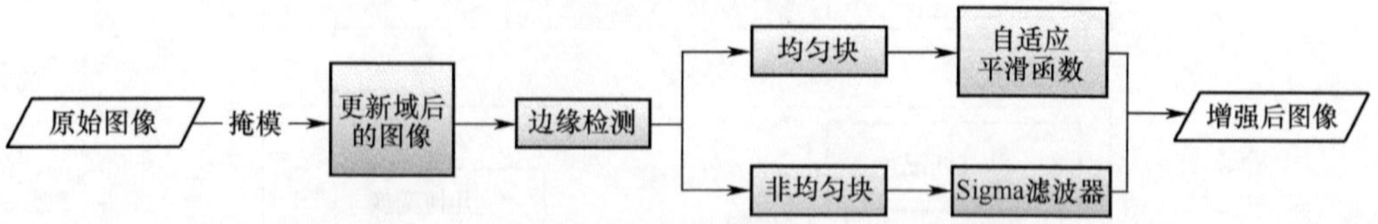

图 4　去除块效应的图像增强处理流程

Fig. 4　Flowchart of image enhancement to remove block effect

受降落相机安装位置影响，每幅降落影像中均有一部分固定区域被探测器支架遮挡，而这部分遮挡区域未包含有效的着陆区纹理信息。基于这个考虑，首先设计了一个掩模模板对其中的无效区域进行标记；为了提高图像增强效果，结合 DCT 特性及月球表面纹理特征，将图像分为均匀块和非均匀块；然后对均匀块使用自适应平滑函数增强，对非均匀块用 Lee 提出的 Sigma 滤波器，使用 5×5 大小的窗口在需要平滑的区域处理；最后获得增强后的图像。

3.2　基于降落序列影像的着陆点位置传递

基于降落序列影像的着陆点位置传递流程如图 5 所示，使用序列图像进行基于 SIFT 的特征提取与匹配，根据得到的特征点集求解降落初帧图像和末帧图像的坐标变换关系，从而解算着陆点在初帧图像中的像坐标。

序列降落影像（包括高、低压缩比）存在缩放、部分旋转等几何变形，为获得稳健的匹配结果，选用 SIFT 特征提取作为图像仿射不变特征提取方法，并基于提取的特征实现图像间的高精度匹配. 采用关键点特征向量的欧式距离作为两幅影像关键点的相似性判定度量，应用中为了获得足够多的特征点，相似性测度阈值取 0.75～0.85。考虑到探测器着陆点小范围一般地势起伏较为平坦，本文采用如式（1）的单应变换模型计算相邻影像间的几何关系[17]。

$$\begin{bmatrix} x_{n+1}^k \\ y_{n+1}^k \\ 1 \end{bmatrix} = H_n^{n+1} \begin{bmatrix} x_n^k \\ y_n^k \\ 1 \end{bmatrix},\ (n=0,\ 1,\ \cdots,\ m,\ 3 \leqslant k < j) \tag{1}$$

其中，H_n^{n+1} 为单应变换系数，通常用 $\begin{bmatrix} h_{00} & h_{01} & h_{02} \\ h_{10} & h_{11} & h_{12} \\ h_{20} & h_{21} & h_{22} \end{bmatrix}$ 9 个参数来表示，$(x_n^k,\ y_n^k)$ 及 $(x_{n+1}^k,\ y_{n+1}^k)$ 分别为相邻降落影像同名特征点的像坐标。

针对相邻影像间 SIFT 的误匹配，采用选权迭代法检测匹配的中粗差，即首先通过式

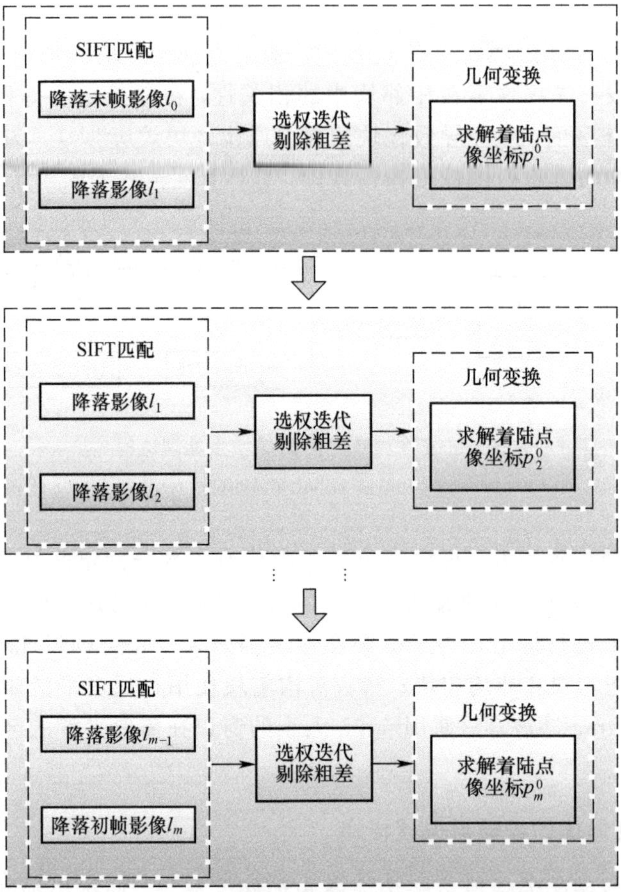

图 5　基于降落序列影像的着陆点位置传递流程

Fig. 5　Flowchart of landing location transfer based on descent sequence images

(1) 模型计算每一对匹配点的模型残差，剔除其中不满足残差阈值的点对；然后，重复上述步骤，直至参与计算的所有点对的残差均满足阈值要求；最后输出粗差剔除后的正确匹配结果进行几何关系计算。

3.3　降落影像与 DOM 底图的几何关系建立

降落影像与 DOM 底图影像的成像时间、太阳高度角、太阳方位角均不相同，这些因素会导致两类影像基于同一目标区域的纹理出现较大差异，常见的基于特征的匹配算法及基于区域的匹配算法均无法适用。实际上，月球表面会布满大大小小的撞击坑，由于地形凹陷，这些撞击坑在影像中多会呈现出两侧坑沿明暗对比明显的特点。

基于以上分析，首先通过降落影像中的撞击坑找到 DOM 底图上的大致区域，然后在人工干预下确认 4 组以上的同名特征点，最后采用仿射模型求得它们之间的几何关系。为确保人工干预的准确性，在实施过程中采取多次实验取平均值的策略。

4　定位结果及分析

嫦娥四号任务执行过程中，在接收到全部高压缩比（1∶64）影像后约 20min，通过图像增强处理、序列影像间的匹配及中分辨率降落影像与 DOM 底图等一系列处理运算，实现了着陆点的初始定位。在探测器着陆后第二天，接收到低压缩比降落影像后约 10min，北京航天飞行控制中心利用高分辨率的降落影像与着陆区 DOM 进行特征匹配，进一步优化了着陆点的定位结果。本文中的中分辨率影像指的是距离月面高度为 1.0～1.5km 时的降落相机影像，分辨率为 0.8 ～1.2m/pixel，该分辨率同着陆区底图分辨率较接近（约 1.0m/pixel）；高分辨率影像指的是距离月面高度为 0.2km 以下时的降落相机影像，分辨率均高于 0.2m/pixel。

4.1　着陆点初始定位

嫦娥四号任务在动力下降阶段（即北京时间 2019 年 1 月 3 日 10 时 23 分 03 秒至 10 时 29 分 21 秒），共实时接收到 1∶64 压缩比的降落影像 59 幅。其中前 27 幅为“接近段”至“缓速下降段”的影像，后 32 幅均为探测器落月后的影像。因此，在任务实施过程中选用前 27 幅影像进行着陆点的初始定位计算。

4.1.1　高压缩比影像预处理效果

图 6 为处理前后的对比效果图。经过增强处理后，不仅提高了原始图像的纹理质量，而且还有效保留了边缘和图像细节，为后续图像匹配奠定了基础。采用批处理方式全自动运行，处理 27 幅图像的时间约为 15s。

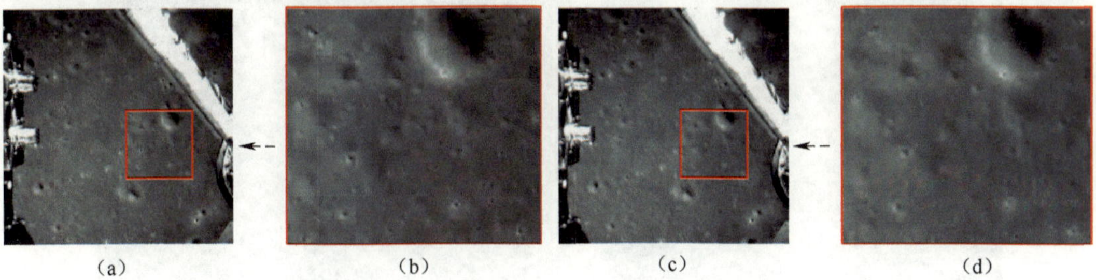

图 6　（网络版彩图）原始图像及增强图像对比图（帧号 1754）

（a）原始图像；（b）原始图像局部细节；（c）增强后的图像；（d）增强后图像局部细节

Fig. 6　(Color online) Comparison of the original image and the enhanced image (Frame No. 1754)

(a) Original image; (b) local details of the original image; (c) enhanced image; (d) local details of the enhanced image

4.1.2　序列影像匹配效果

图 7 为相邻序列影像的部分匹配情况，其中图 7（a）两幅影像的距离月面高度约为 200m，图 7（b）两幅影像的距离月面高度约为 1.6km。经过图像增强处理后，软件能够实现全自动匹配；基于获得的匹配特征点集建立映射模型并完成着陆点的像坐标传递，无需人工干预，处理 27 幅图像的时间约为 2min。

4.1.3　中分辨率降落影像与 DOM 的匹配结果

首先根据惯导提供的探测器着陆点概略位置（177.5°E，45.5°S）对着陆区 DOM 底图

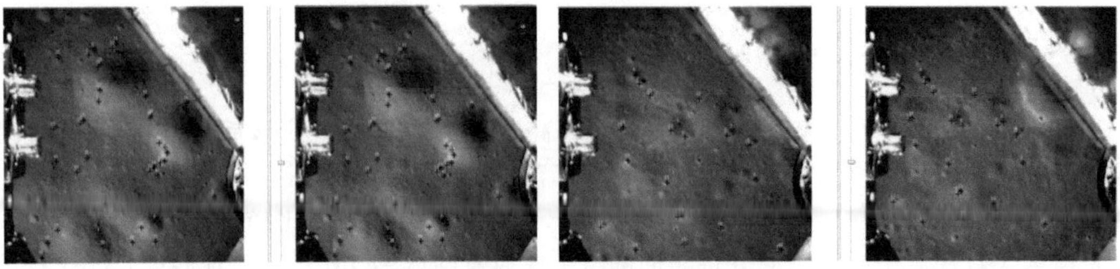

(a) (b)

图 7 （网络版彩图）相邻序列影像的 SIFT 匹配结果

（a）降落影像（帧号 3054－2989）匹配结果，共 44 个同名点；

（b）降落影像（帧号 2339－2274）匹配结果，共 24 个同名点

Fig. 7 (Color online) SIFT matching results of adjacent sequence images

(a) Matching result of the descent images (Frame Nos. 3054 and 2989), 44 matched points included

(b) matching result of the descent images (Frame Nos. 2339 and 2274), 24 matched points included

进行裁剪，然后选用中分辨率降落影像与裁剪后的 DOM 图像进行匹配。为确保匹配正确性，需人工干预，处理时间约为 15min。两幅影像的匹配结果如图 8 所示，中分辨率降落影像距离月面高度约为 1km，图像分辨率约为 0.8m/pixel。着陆点的初始定位结果为(177.5876°E，45.4564°S)。

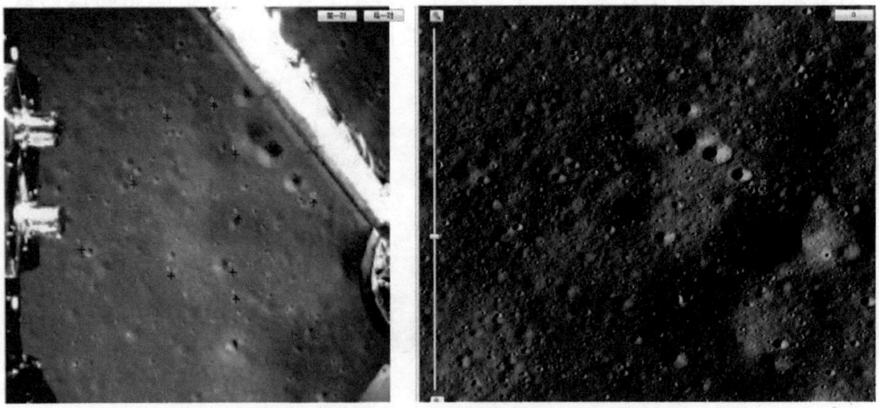

(a) (b)

图 8 （网络版彩图）中分辨率降落影像与裁剪后的 DOM 图像的匹配结果

（a）降落影像（帧号 2469）；（b）裁剪后的 DOM 图像（I_{DOM}）

Fig. 8 (Color online) Matching results of the medium－resolution descent image and the clipped DOM image

(a) Descent image (Frame No. 2469); (b) clipped DOM image (I_{DOM})

4.2 着陆点精确定位

在探测器落月后约 15h，即北京时间 2019 年 1 月 04 日 07 时 22 分至 10 时 37 分，嫦娥四号探测器开始回放低压缩比（1∶8）降落影像，其中约有 2600 幅图像为探测器动力下降过程的图像，距离月面高度约为 2～3km，选用其中的 10 幅高分辨降落影像进行着陆点

的精确定位。

4.2.1　高分辨率降落影像的最优选择结果

选取探测器缓速下降段的降落影像（帧号 3396）作为末帧图像，距离月面高度约为 30m；接近段的降落影像（帧号 2946）作为初帧图像，距离月面高度约为 200m。实验中，依照 50 帧的均匀间隔选取接近段至缓速下降段共 10 幅图像，相邻图像的成像间隔为 5s，具体的成像情况见表 2 所示。其中成像帧序号、成像时间包含在降落影像的描述文件中，成像时探测器的星下点经度、星下点纬度、高度信息可根据成像时间由相应的遥测信息对查到。受传输时延的影响，这 3 类信息是个概略值，与实际情况存在一定偏差。同时，由降落相机相关参数及影像距离月面高度，还可计算求得每幅图像的分辨率。

表 2　降落影像成像情况

Tab. 2　Details of descent images

图像帧号	成像时刻（北京时间）	星下点经度/(°)	星下点纬度/(°)	距离月面高度/m	图像分辨率/(m/pixel)	着陆点像坐标	
						x	y
3396	2019/01/03 10：25：49	—	—	—	—	512	512
3346	2019/01/03 10：25：44	—	—	—	—	492.95	463.55
3296	2019/01/03 10：25：39	177.5744	−45.4599	146	0.12	487.16	459.04
3246	2019/01/03 10：25：34	—	—	—	—	477.07	473.87
3196	2019/01/03 10：25：29	—	—	—	—	466.01	496.09
3146	2019/01/03 10：25：24	—	—	—	—	456.40	494.75
3096	2019/01/03 10：25：19	—	—	—	—	425.60	526.62
3046	2019/01/03 10：25：14	177.5764	−45.4598	200	0.17	398.73	549.98
2996	2019/01/03 10：25：09	—	—	—	—	393.82	563.82
2946	2019/01/03 10：25：04	—	—	—	—	393.37	554.10

a）“—”项由于无法根据成像时刻从遥测信息中查到相应的信息。

4.2.2　序列影像匹配效果

相邻图像的部分匹配结果如图 9 所示，计算的着陆点在每幅影像中的像坐标见表 2 右侧两列。

4.2.3　高分辨率降落影像与 DOM 的匹配结果

根据第 4.1 节计算出的着陆点初始定位结果对着陆区 DOM 图像进行更精确范围的裁剪，然后选用高分辨率降落影像（帧号 2946）与裁剪后的 DOM 图像（记为 I'_{DOM}）进行匹配。本过程需适当人工干预，为确保同名特征点的准确性，共进行 3 组实验，总处理时间约为 5min。3 组实验的影像匹配结果如图 10～图 12 所示，其中图 10（b）～图 12（b）的圆圈代表着陆点在影像中的位置，具体的同名点信息见表 3。依据 3 组实验结果，取它们的均值作为精确定位结果。

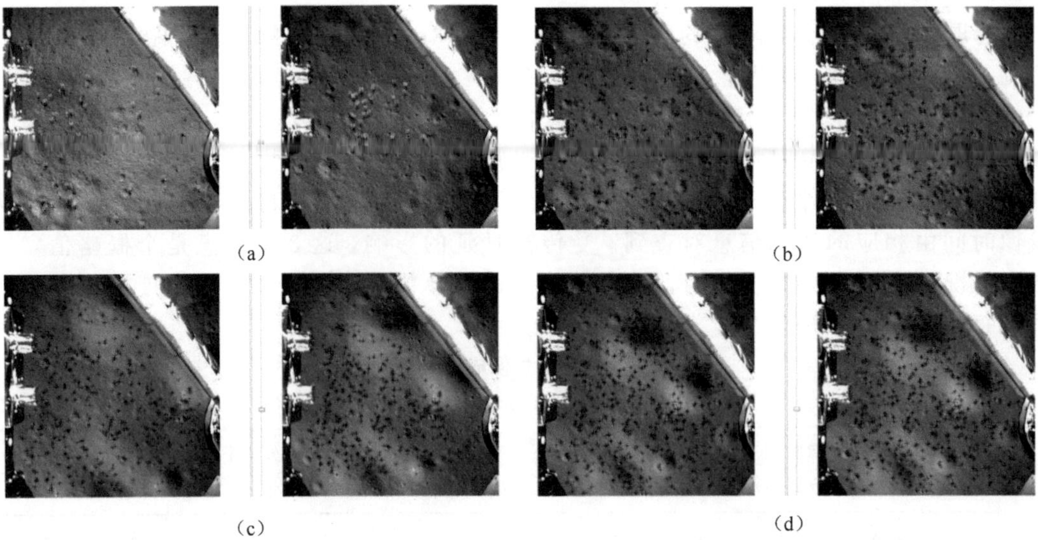

图 9 （网络版彩图）部分相邻图像的 SIFT 匹配结果

(a) 降落影像 3396－3346 匹配结果，共 59 个同名点；(b) 降落影像 3296－3246 匹配结果，共 495 个同名点；(c) 降落影像 3146－3096 匹配结果，共 555 个同名点；(d) 降落影像 2996－2946 匹配结果，共 871 个同名点

Fig. 9 (Color online) SIFT matching results of partial adjacent images

(a) Matching result of descent images (Frame Nos. 3396 and 3346), 59 matched points included

(b) matching result of descent images (Frame Nos. 3296 and 3246), 459 matched points included

(c) matching result of descent image (Frame Nos. 3146 and 3096), 555 matched points included

(d) matching result of descent image (Frame Nos. 2996 and 2946), 871 matched points included

图 10 （网络版彩图）第 1 组结果—降落影像（帧号 2946）与 I'_{DOM}图像

(a) 原始影像；(b) 局部放大

Fig. 10 (Color online) Group 1－descent image (Frame No. 2946) and I'_{DOM}

(a) Original images; (b) local details

图 11　（网络版彩图）第 2 组结果—降落影像（帧号 2946）与 I'_{DOM} 图像

(a) 原始影像；(b) 局部放大

Fig. 11　(Color online) Group 2—descent image (Frame No. 2946) and I'_{DOM}

(a) Original images; (b) local details

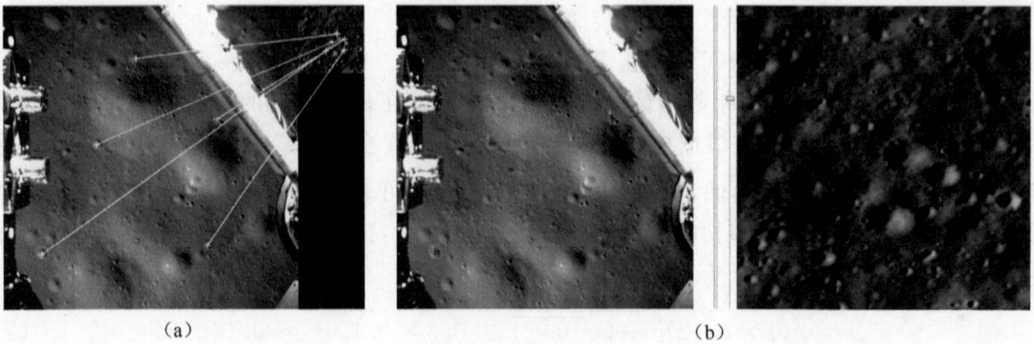

图 12　（网络版彩图）第 3 组结果—降落影像（帧号 2946）与 I'_{DOM} 图像

(a) 原始影像；(b) 局部放大

Fig. 12　(Color online) Group 3—descent image (Frame No. 2946) and I'_{DOM}

(a) Original images; (b) local details

表 3　3 组实验的匹配结果

Tab. 3　Matching results of three groups

点位	影像	第 1 组		第 2 组		第 3 组	
		x	y	x	y	x	y
同名点 1	降落影像	467.24	467.24	483.50	105.34	458.16	176.01
	DOM 图像	122.43	122.43	137.76	88.63	133.54	95.49
同名点 2	降落影像	224.41	224.41	802.59	596.51	130.84	838.45
	DOM 图像	104.89	104.89	173.59	131.88	94.80	152.11
同名点 3	降落影像	649.48	649.48	678.57	839.25	319.66	467.88
	DOM 图像	135.93	135.93	154.79	151.57	117.78	122.44
同名点 4	降落影像	655.88	655.88	323.36	467.24	706.11	808.45
	DOM 图像	136.61	136.61	117.16	122.43	158.22	149.49

续表 3

点位	影像	第 1 组		第 2 组		第 3 组	
		x	y	x	y	x	y
同名点 5	降落影像	839.72	839.72	312.17	655.88	741.40	383.18
	DOM 图像	152.13	152.13	115.69	136.61	159.68	113.42
着陆点像坐标		(125.03，129.69)		(127.58，128.42)		(125.37，128.14)	
着陆点地理坐标		(177.58835，−45.45652)		(177.58844，−45.45648)		(177.58836，−45.45647)	

4.3　精度评估

4.3.1　算法理论精度评估

依据第 3 节所述的着陆点定位实施算法流程及策略可知，其绝对定位精度主要取决于两个方面：一是图像的匹配精度；二是着陆区底图 DOM 的制图精度。其中，图像匹配的误差主要来源于以下两点。

（1）着陆点在序列降落影像间的传递误差

着陆点在序列降落影像间的传递误差主要来源于以下两个方面：一是降落影像间的匹配误差，基于 SIFT 算法的相邻影像的匹配精度可控制在传递影像的 1 个像素[6,18]；二是由影像间的传递模型引入，根据控制点的拟合残差，精度在亚像素[18]，本文中取 0.3 个像素，假设有 n 幅序列降落影像，那么模型间的传递误差可表示为$\sqrt{n-1}\times 0.3$ 个像素。取传递影像的最大分辨率（即初帧影像）计算，为 0.12m/pixel，n 取 10，那么，着陆点在序列影像间的传递误差 α 表示为

$$a=\sqrt{(1\times 0.12)^2+(\sqrt{10-1}\times 0.3\times 0.12)^2}\approx 0.16\text{m} \tag{2}$$

（2）降落影像与着陆区 DOM 底图的匹配

记匹配误差为 β，主要是由人工在降落影像与着陆区 DOM 底图之间确定同名特征点造成，难以给出具体的定性指标，采取多次实验取平均值的方法可将 β 控制在 2pixel。记底图 DOM 的分辨率为 res，本文取 res 为 1m/pixel，那么降落影像与 DOM 的匹配误差 $\beta=2\times 1=2\text{m}$。

因此，图像匹配的精度 pre1 可表示为 $\text{prel}=\sqrt{0.16^2+2^2}\approx 2\text{m}$。

另一方面，就是底图 DOM 的制图精度，记为 pre2。其误差主要取决于传感器本身的定位误差，包括传感器的定轨、定姿以及内方位的测量误差，据文献[18]介绍，LRO 定轨精度约为 20m。

那么本文的绝对定位误差精度 pre 可表示为

$$\text{pre}=\sqrt{\text{pre1}^2+\text{pre2}^2}=\sqrt{2^2+20^2}\approx 20.1\text{m} \tag{3}$$

需要说明的是，由于 LRO 的定轨精度是大量轨道数据的统计精度，定轨误差会导致每幅 LRO NAC 影像的定位精度不尽相同；同时 LRO NAC DOM 底图的精度还会受到影像传感器姿态精度的影响，因此式（3）中得到的绝对定位精度仅是一种理想状态下的大致估计。

4.3.2　定位结果比对

中国科学院遥感与数字地球研究所的研究团队基于嫦娥二号正射影像、LROC NAC

DOM、嫦娥四号降落相机及监视相机影像等多源数据，利用影像特征匹配定位和单像视觉测量定位技术，确定了嫦娥四号探测器着陆点的位置[19]；NASA/GSFC/ASU 于探测器着陆后，利用着陆前的 1 幅 LRO NAC 影像，也公布了嫦娥四号探测器着陆点的位置信息[20]。这些定位结果与本文定位结果的比对见表 4，可以看出本文的精确定位结果与目前国内外公布的结果的一致性较好，而初始定位的偏差主要来源于高压缩比下图像纹理信息的缺失以及抽帧下传方式导致降落序列图像的不连续。

表 4　嫦娥四号着陆点定位结果比对

Tab. 4　Comparison of Chang'E - 4 landing position results

数据	经度/(°)	纬度/(°)	说　明
27 幅（高压缩比）降落影像，LRO NAC DOM，CE - 2 DOM	177.5876	−45.4564	本文方法，着陆点初始定位结果
10 幅（低压缩比）降落影像，LRO NAC DOM，CE - 2 DOM	177.5884	−45.4565	本文方法，着陆点精确定位结果
LRO NAC 影像 M1298916428LR	177.589	−45.457	NASA/GSFC/ASU[20]
CE - 2 DOM，LRO NAC DOM，降落影像，监视相机影像	177.588	−45.457	邸凯昌等，遥感学报，2019[19]

基于探测器着陆点的快速、高精度定位结果，北京航天飞行控制中心成功实施了嫦娥四号两器互拍的周期规划，见图 13 所示，其中 A 点、C′点为两器互拍点，S1 点为第 1 月昼巡视器的休眠点，底图是用多幅降落影像制作的高分辨率 DOM。

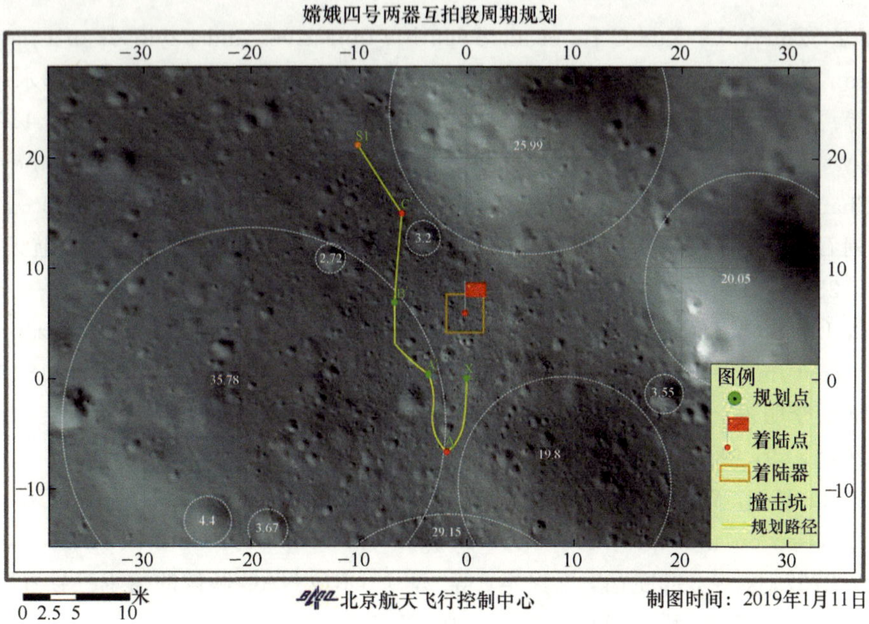

图 13　（网络版彩图）嫦娥四号两器互拍段周期规划示意图

Fig. 13　(Color online) Schematic diagram of periodic planning for rover and lander mutual - photograph of CE - 4

5　结论

本文采用计算机视觉方法在高精度图像匹配和几何变换的基础上实现了嫦娥四号探测器着陆点绝对定位。利用动力下降过程中高压缩比的降落影像，在探测器着陆月面后的20min内实现了着陆点的初始定位；之后通过回放的低压缩比降落影像，进一步精化了着陆点的位置信息。该结果与目前国内外公布的各类定位结果精度一致，探测器着陆点的高精度定位结果有效地支持了嫦娥四号后续遥操作任务。

本文的定位方法是相对着陆区底图的结果，定位精度与底图DOM的制图精度密切相关，下一步将开展月球多源遥感数据的高精度着陆区制图研究，为基于视觉的着陆点定位方法提供覆盖范围更广、精度更高的制图产品，同时也能为探测器落月后的路径规划和安全导航提供基础数据支持。另外，由于目前覆盖嫦娥四号着陆区的高精度LRO NAC影像数量还比较少，随着后期影像数据的累积，之后还可以开展基于多重覆盖遥感影像的高精度定位研究。

致谢

感谢北京航天飞行控制中心刘磊工程师、卜彦龙工程师为本文写作提供的帮助和支持。

参考文献

[1] Jia Y Z, Zou Y L, Xue C B, et al. Scientific objectives and payloads of Chang'E－4 mission (in Chinese). Chin J Space Sci, 2018, 38: 118－130 [贾瑛卓，邹永廖，薛长斌，等. 嫦娥四号任务科学目标和有效载荷配置. 空间科学学报，2018，38：118－130].

[2] Wan W H, Liu Z Q, Liu Y L, et al. Descent image matching based position evaluation for Chang'E－3 landing point (in Chinese). Space Craft Eng, 2014, 23: 5－12 [万文辉，刘召芹，刘一良，等. 基于降落图像匹配的嫦娥三号着陆点位置评估. 航天器工程，2014，23：5－12].

[3] Jia Y, Liu S C, Li M L, et al. Chang'E－3 system pinpoint landing localization based on descent image sequence (in Chinese). Chin Sci Bull, 2014, 59: 1838－1843 [贾阳，刘少创，李明磊，等. 利用降落影像序列实现嫦娥三号系统着陆点高精度定位. 科学通报，2014，59：1838－1843].

[4] Liu B, Di K C, Wang B F, et al. Positioning and precision validation of Chang'E－3 lander based on multiple LRO NAC images (in Chinese). Chin Sci Bull, 2015, 60: 2750－2757 [刘斌，邸凯昌，王保丰，等. 基于LRO NAC影像的嫦娥三号着陆点高精度定位与精度验证. 科学通报，2015，60：2750－2757.

[5] Li J F, Cui W, Baoyin H X. A survey of autonomous navigation for deep space exploration (in Chinese). Mech Eng, 2012, 34: 1－9 [李俊峰，崔文，宝音贺西. 深空探测自主导航技术综述. 力学与实践，2012，34：1－9].

[6] Wu W R, Zhou J L, Wang B F, et al. Key technologies in the teleoperation of Chang'E－3

"Jade Rabbit" rover (in Chinese). Sci Sin Inf，2014，44：425 - 440 [吴伟仁，周建亮，王保丰，等. 嫦娥三号"玉兔号"巡视器遥操作中的关键技术. 中国科学：信息科学，2014，44：425 - 440.

[7] Wang J，Wan W H，Zhao H Z，et al. Vision - based positioning method for Chang'E - 4 lander (in Chinese). Manned Spaceflight，2019，25：12 - 18 [王镓，万文辉，赵焕洲，等. 基于视觉的嫦娥四号探测器着陆点定位方法. 载人航天，2019，25：12 - 18].

[8] Wagner R V，Robinson M S，Speyerer E J，et al. Locations of anthropogenic sites on the Moon. In：45th Lunar & Planetary Science Conference. The Woodlands，2014. 2259.

[9] NASA/GSFC/Arizona State University. NASA Images of Chang'E - 3 Landing Site. [2014 - 01 - 23]. http：//www. nasa. gov/content/nasa - images - of - change - 3 - landing - site/.

[10] Cao J F，Zhang Y，Hu S J，et al. An analysis of precise positioning and accuracy of the CE - 3 lunarlander soft landing (in Chinese). Geom Inform Sci Wuhan Univ，2016，41：274 - 278 [曹建峰，张宇，胡松杰，等. 嫦娥三号着陆器精确定位与精度分析. 武汉大学学报：信息科学版，2016，41：274 - 278].

[11] Ye P J，Sun Z Z，Zhang H，et al. Mission design of Chang'E - 4 probe system (in Chinese). Sci Sin Tech，2019，49：1 - 14 [叶培建，孙泽洲，张熇，等. 嫦娥四号探测器系统任务设计. 中国科学：技术科学，2019，49：1 - 14].

[12] Li C L，Liu J J，Ren X，et al. Lunar global high - precision terrain reconstruction based on Chang'E - 2 stereo images (in Chinese). Geom Inform Sci Wuhan Univ，2018，43：485 - 495 [李春来，刘建军，任鑫，等. 基于嫦娥二号立体影像的全月高精度地形重建. 武汉大学学报：信息科学版，2018，43：485 - 495].

[13] Robinson M S，Brylow S M，Tschimmel M，et al. Lunar Reconnaissance Orbiter Camera (LROC) instrument overview. Space Sci Rev，2010，150：81 - 124.

[14] Li F，Zhang H，Wu X Y，et al. Influence analysis of terrain of the farside of the moon on soft landing (in Chinese). J Deep Space Expl，2017，4：143 - 149 [李飞，张熇，吴学英，等. 月球背面地形对软着陆探测的影像分析. 深空探测学报，2017，4：143 - 149].

[15] Liu B，Xu B，Liu Z Q，et al. Descending and landing trajectory recovery of Chang'E - 3 lander using descent images (in Chinese). J Remote Sens，2014，18：981 - 987 [刘斌，徐斌，刘召芹，等. 基于降落相机图像的嫦娥三号着陆轨迹恢复. 遥感学报，2014，18：981 - 987].

[16] Xie S L，Zhou Z H. A new adapt reduction algorithm of blocking artifacts (in Chinese). Acta Electron Sin，2005，33：1897 - 1900 [谢胜利，周智恒. 一种自适应消除块效应的新算法. 电子学报，2005，33：1897 - 1900].

[17] Wang J，Peng M，Wan W，et al. Positioning method of Chang'E - 4 lander based on multi - source images. In：69th International Astronautical Congress (IAC)，2018. 4321 - 4332.

[18] Mazarico E, Rowlands D D, Neumann G A, et al. Orbit determination of the Lunar Reconnaissance Orbiter. J Geod, 2012, 86: 193 - 207.

[19] Di K C, Liu Z Q, Liu B, et al. Chang'E - 4 lander localization based on multi - source data (in Chinese). J Remote Sens, 2019, 23: 177 - 184. [邸凯昌, 刘召芹, 刘斌, 等. 多源数据的嫦娥四号着陆点定位. 遥感学报, 2019, 23: 177 - 184].

[20] NASA/GSFC/Arizona State University. Chang'E 4 Lander Coordinates. [2019 01 11]. http: //lroc. sese. asu. edu/posts/1087/.

利用Halo轨道仿真开展嫦娥四号中继星与月面设备干涉基线的研究

武宇翔[1,2,3]，温卫斌[1,2]，平劲松[1,2,3]，朱新颖[1,2]，张洪波[1,2,3]，孔德庆[1,2,3]，戴舜[1,2]，薛喜平[1,2]，李臣[1,2]

1. 中国科学院国家天文台，北京 100101
2. 中国科学院月球与深空探测重点实验室，北京 100101
3. 中国科学院大学，北京 100049

摘要： 初步建立了嫦娥四号中继星的晕轨道（Halo轨道）数值模型，将其应用于空间低频射电天文观测的干涉测量仿真，以地月拉格朗日平动点 L_2 区域的Halo轨道为基础，以发布的嫦娥四号中继星理论轨道为参考，校准和调整模型参数，通过时间和空间参考系统的变换，把轨道数据转换到 L_2 点旋转坐标系。然后对比Halo轨道模型与理论轨道数据之间的差别，对比模拟和理论数据分别与月面设备形成基线的长度，并对结果进行分析。两者在所分析的运行阶段，与月面设备联成基线的长度差在60km以内，在高频频带基本满足干涉测量条纹搜索对基线初值精度的需求。

关键词： Halo轨道；平动点；参考系；仿真

Research of Interferometry Baselines Between CE－4 Rover and Lunar Facility Using Halo Orbit Simulation

Wu Yuxiang[1,2,3], Wen Weibin[1,2], Ping Jinsong[1,2,3], Zhu Xinying[1,2], Zhang Hongbo[1,2,3], Kong Deqing[1,2,3], Dai Shun[1,2], Xue Xiping[1,2], Li Chen[1,2]

1. National Astronomical Observatories, Chinese Academy of Sciences, Beijing 100101, China, Email: jsping@bao. ac. cn
2. Key Laboratory of Lunar and Deep Space Exploration, Chinese Academy of Sciences, Beijing 100101, China
3. University of Chinese Academy of Sciences, Beijing 100049, China

Abstract: An orbit model for Chang'E - 4 relay satellite is established which is supposed to be used in a simulation of space low - frequency radio interferometry. Based on classic Lagrange libration point L_2 Halo orbit, some parameters are calibrated and modified by the theoretical orbit of Chang'E - 4 relay satellite, converting theoretical orbit to a rotation orbit in the vicinity of L_2 point through time and coordinate reference transferring. Comparisons between the Halo orbit model and the theoretical orbit model are carried out, according to which, their baselines to lunar - surface facility are calculated respectively and analyzed. In the analysis, differences between the baselines are below 60km, which satisfy the needs of interferometry fringe searching in the HF band.
Keywords: Halo orbit; Libration point; Reference system; Simulation

1　Halo 轨道和中继卫星

我国的嫦娥工程已经进行了多次任务，嫦娥四号探测器首次实现了月球背面着陆探测。由于月球背面的着陆器无法与地面直接进行通信，为了建立着陆器、巡视器与地球之间的联系，我国发射了中继通信卫星“鹊桥号”，通过它把获得的科学数据传回地球。这颗中继星运行在地月拉格朗日点 L_2 附近的使命轨道上[1-4]。中继通信卫星搭载的低频射电探测仪和月球着陆器上搭载的射电频谱仪为空间低频射电干涉测量创造了条件[5]。

月球背面屏蔽了来自地球的各种信号，给低频射电观测提供了宁静的环境，可以在 0.1～40MHz 的频段开展低频射电天文观测[6]。由于低频射电波的波长较长，为达到较高的空间分辨率，可参照甚长基线干涉测量（Very Long Baseline Interferometry，VLBI）技术，借助月面、中继星和地面的低频射电观测设备联合进行协同观测，甚至是干涉测量。为了考察这种空间干涉测量的可行性，我们建立理论和仿真模型进行模拟实验，其中的一个重点是模拟中继星到月面设备的基线并计算基线长度。为此，本文讨论了 Halo 轨道模型在空间低频射电干涉测量模拟中的适用性。

嫦娥四号任务选择南极-艾特肯盆地（Southpole - Atiken）作为着陆区，着陆器及巡视器在这一地区开展科学探测，并获得了有关月幔物质的探测数据[7]。通过开展形貌分析、物质组成、月壤和月表浅层结构的综合探测，有望获得诸多有关月球形成的新证据[8]。2018 年 5 月 21 日，中继通信卫星“鹊桥号”成功发射，进入地月 L_2 点附近的使命轨道，其中，L_2 点是指限制性三体问题（Restricted Three Body Problem，RTBP）的 5 个特解，被称作拉格朗日平动点中位于较小大天体外侧的共线点，文献［9］进行过系统性的总结。利用地月 L_2 点附近轨道部署中继卫星开展月球背面的探索这一构想由文献［10］提出，文献［11 - 12］研究丰富了平动点轨道力学的理论体系，文献［13 - 15］给出了 Halo 轨道的三阶近似解析解，广泛应用于三体问题，在文献［16］提出在平动点建立空间站的设想之后，文献［17］提出了中继通信链路和月球采样高速公路。第一次平动点轨道探测是日地 L_1 的 ISEE - 3，美国的 ARTEMIS 任务首先进行了地月平动点的验证飞行；嫦娥二号探访了日地 L_2 点的 Lissajous 轨道；嫦娥 5T1 月球再入返回试验任务进入了地月 L_2 点 Lissajous 轨道；嫦娥四号中继星则使用地月 L_2 点附近的 Halo 轨道[18-19]，国内也有多项与中继星使命轨道相关的建模仿真以及实际测量研究[20-21]。

为了开展空间低频射电干涉测量系统的相关研究，本文构建了地月 L_2 点附近的 Halo 轨道，并比较模型与中继星实际使用的理论轨道之间的关系，对模型与理论轨道的偏差进行了比较，并分析了模型对星上设备与月面设备形成基线长度的影响，评价模型的仿真效果。

2　平动点轨道的相关理论

限制性三体问题描述了一个质量小的物体在两个大天体附近空间受万有引力作用的情况，在特殊情况下，小物体受到来自大天体的引力能满足自身向心加速度所需，从而稳定运行。这种稳定平动的 5 个特解被称为“拉格朗日平动点”，其中，L_2点是位于两个大天体连线上、较小的天体外侧的平动点，在这一点上可以找到一些轨道使得小物体能以较低的能耗、较长的时间维持在稳定的位置。然而 L_2 点并不是一个稳定点，所以这类轨道都不是完美的周期轨道，依然需要时常进行轨道维持机动，常见的 L_2 点轨道有 Lyapunov 轨道、Lissajous 轨道和 Halo 轨道。

在地月系空间定义一个右手直角坐标系，并称之为旋转坐标系（Rotation System），它的原点始终是地月连线的瞬时平动点 L_2，x 轴方向与地球指向月球的方向相同，z 轴与地月之间的旋转角速度方向相同。在这个坐标系内，Lissajous 轨道和 Halo 轨道的运动都可以描述为“$A_z = 0$”的平面内与平面外的两种振动的叠加。平动点轨道的一般表达式为

$$\begin{cases} x = -A_x\cos(\lambda t + \phi) \\ y = A_y\sin(\lambda t + \phi) \\ z = A_z\sin(\nu t + \psi) \end{cases}, \tag{1}$$

其中，A_y 和 A_z 分别为旋转坐标系下面内和面外振幅；λ 为面内振动频率；ν 为垂直于面的振动频率；ϕ 和 ψ 分别为面内外振动的初始相位。

一种 Lissajous 轨道的示例如图 1（a），这种轨道在地月旋转面内周期与面外周期并不相同，所以轨道呈现 Lissajous 曲线的形式，卫星在经过中央位置时，有可能被月球遮挡，不能建立对地球的联系。而 Halo 轨道（如图 1（b））的尺寸较大，且面内周期与面外周期相同，它不会被遮挡，能够保证在全程保持同时对月背和地球的直视，保证地面—中继星—月球背面的通讯连续性。Halo 轨道的表达式为

$$\begin{cases} x = -A_x\cos(\lambda t + \phi) \\ y = A_y\sin(\lambda t + \phi) \\ z = A_z\sin(\lambda t + \phi) \end{cases}, \tag{2}$$

此时，面内振动和面外振动的频率相等且为 λ。描述 Halo 轨道的尺寸一般用 A_x 和 A_z 两个参数，分别是面内振幅和面外振幅。Halo 轨道实际上是一个集合，被称为轨道族（Family），因此，对 Halo 轨道的模拟还需要进一步简化，并且 Halo 轨道的稳定性有限，随着时间的变化，航天器会逐渐偏离标称轨道（Standard Orbit）。文献［12］给出了一个用准周期（Quasi-periodic）轨道描述 Halo 轨道的数值构造方法，这种方法计算量较小，且能够较好地描述 Halo 轨道的特征。它的基本形式是

$$\begin{cases} x = m^{1/2}x_1 + mx_2 + m^{3/2}x_3 + m^2x_4 \\ y = m^{1/2}y_1 + my_2 + m^{3/2}y_3 + m^2y_4 \\ z = m^{1/2}z_1 + mz_2 + m^{3/2}z_3 + m^2z_4 \end{cases} \tag{3}$$

各参数意义见文献［12］。在描述振幅时，也可以用 y 方向的面内振幅 A_y 与面外振幅 A_z 两个参数。为了保证面内与面外的频率相等，A_y 和 A_z 应该有以下关系[12]：

$$A_y^2 = 1.176726A_z^2 + 3.361330, \tag{4}$$

利用这个关系式可以得到 A_y 的最小值是 46 793km，由此可以得到 Halo 轨道的一般表达式与 A_z 之间的关系。

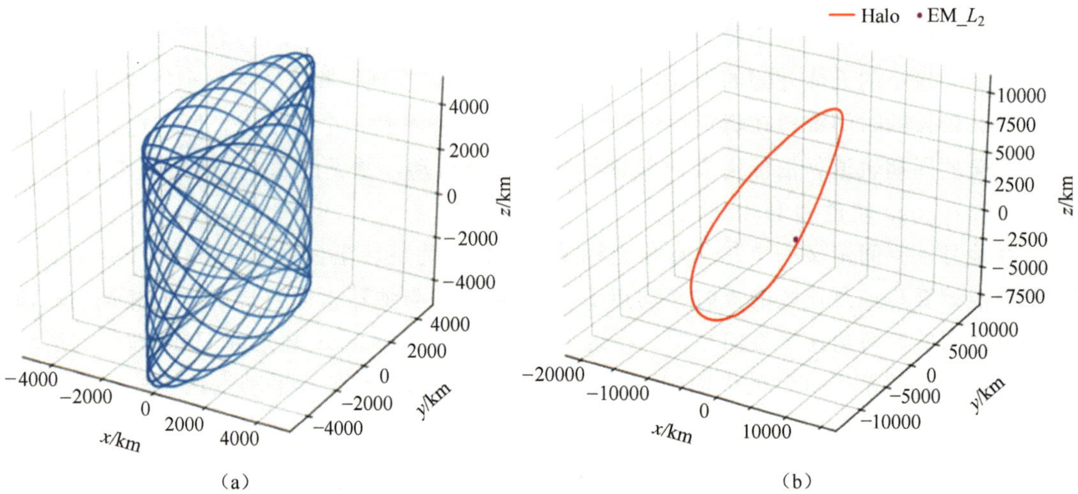

图 1　(a) Lissajous 轨道和 (b) Halo 轨道示例

Fig. 1　Examples of (a) Lissajous orbit and (b) Halo orbit

3　利用中继星轨道的验证实验

为检验 Halo 轨道模型，选取嫦娥四号中继星"鹊桥号"于 2019 年 12 月 22 日 0 时 0 分 0 秒到 12 月 23 日 23 时 59 分 59 秒实际使用的理论轨道[19]（以下简称实际轨道），每个数据点的间隔为 1s。实际轨道数据包含北京时间、地心坐标系统下中继星的位置坐标（km）以及速度矢量（km/s）。根据嫦娥四号中继星所在使命轨道的相关参数，调整模型内的面外振幅 A_z和初始相位 θ，使得模型与实际轨道匹配[20,21]。对 Halo 轨道的模拟实际上是对使命轨道的标称轨道进行模拟，卫星进行多次轨道机动（Orbit Maneuver）以确保卫星不远离标称轨道，但综合考虑轨道稳定性，并节省推进剂以延长任务寿命，实际轨道与标称轨道也有所差别。

为了清晰地表现轨道的特征，选择旋转坐标系对实际轨道数据和模型数据进行对比。首先将实际轨道的时间系统和空间坐标系进行转换，将时间统一到地球时（Terrestrial Time，TT），能够兼顾地球附近时间和行星历表的精度要求[22]。在进行空间坐标系转换时选用喷气推进实验室（Jet Propulsion Laboratory，JPL）的 DE430 行星历表。

首先，将时间统一到地球时。由于中继星轨道处于地月附近的空间，为获取相应时刻的行星历表数据，应当采用地球时代替历书时读取历表，在保证精度的前提下统一模型和实际轨道数据的时间基准。首先将北京时间转换到格林尼治时间的协调世界时，然后使用国际地球自转服务（International Earth Rotation Service，IERS）发布的地球指向参数（Earth Orientation Parameters，EOP）将协调世界时转换到 UT1 时间，计算跳秒（Leap Seconds）后将时间转换到地球时即可。图 2 是这一过程的示意图。

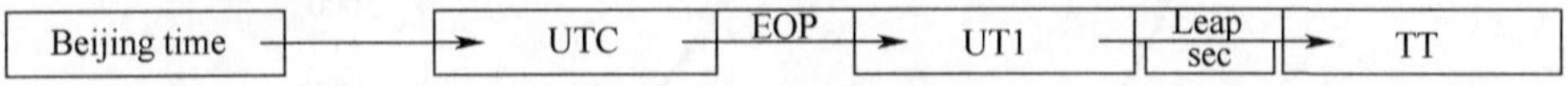

图 2　时间系统转换

Fig. 2　Transforming of time system

通过地面深空站测距、测速和 VLBI 测量数据估算和预报中继星实际轨道的位置和速度，这种信息采用的参考系应为地心地固坐标系。若要使用实际的测量数据，应该将这种坐标转化到合适的空间坐标系。地心地固坐标系包含地球自转、章动岁差等因素，且这些因素与月球附近卫星运动以及月球本身的天平动无关。在地心地固坐标系中描述中继卫星平台和月面着陆器平台的运动规律复杂且不能体现特征，需要将其转换到合适的坐标系。为了呈现 Halo 轨道的主要特征（近似于一个竖立的圆环），应将轨道数据转换到旋转坐标系中，直观地对轨道数据和模型进行比较。

坐标转换的流程如图 3，地固坐标系是原点位于地球质心，且固结于地球，随地球一同运动的坐标系。首先将地心坐标系转换到国际地球参考系（International Terrestrial Reference System，ITRS）；再利用国际地球自转服务发布的地球指向参数数据，调整极移、地球自转角和章动-岁差，将坐标转换到地心天球参考系（Geocentric Celestial Reference System，GCRS）以及国际天球参考系（International Celestial Reference System，ICRS）。

图 3　空间坐标系统转换

Fig. 3　Transforming of reference system

下一步将坐标转换到旋转坐标系。由于旋转坐标系与地月有关，首先得到某时刻地球和月球的质心，由两者连线得到旋转坐标系的 x 轴，月球围绕地球公转的角速度方向即为 z 轴，根据右手螺旋法则可以得到 y 轴的方向，由此可以将国际天球参考系下中继星的位置转换到旋转坐标系，转化结果如图 4 所示。图 4 中，蓝色粗线是实际轨道数据，从 12 月 22 日 0 时至 12 月 23 日 24 时中继星在旋转坐标系中的轨迹，称为观测数据（图中“O”）；红色细线是与这段时间的实际轨道对应的 Halo 轨道模型在一个整周期内的轨迹，称为计算数据（图中“C”）。由图 4 可以看出在这段时间内，实际轨道与 Halo 轨道模型符合较好。实际数据与 Halo 轨道模型重合的部分做差，并对差值取模，可以得到图 5。

从图 5 可以看到，这段数据总体符合较好，总的误差值不超过 700km，说明模型参数与实际情况接近。但观测数据和模型数据之差在中段的变化趋势发生了变化，说明实际轨道有更加复杂的运动形式没有被模型完全描述。

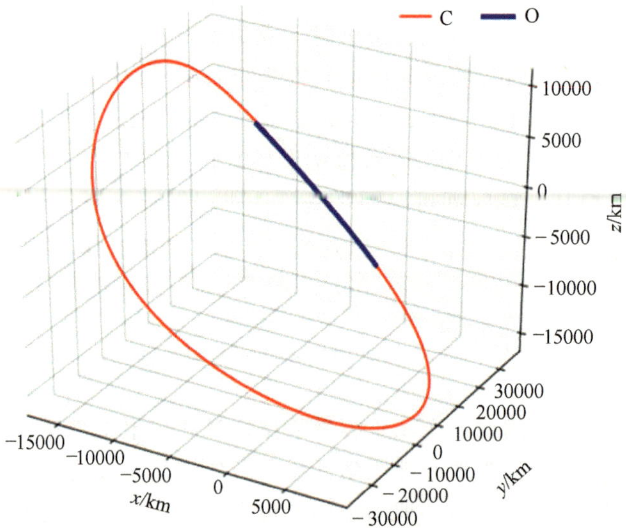

图 4　Halo 轨道模型与部分中继星轨迹

Fig. 4　Halo model and part of relay satellite orbit

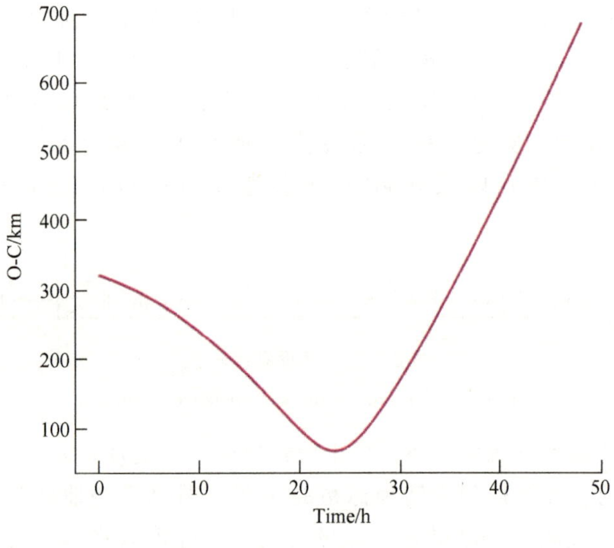

图 5　观测轨道与模型轨道的差

Fig. 5　Differences between observation and model orbits

这里的 Halo 轨道模型将应用于空间射电干涉测量的研究，所以需要考察模型与理论轨道在干涉测量中的差别。首先考察中继星与月面设备干涉时形成的基线。基线是指干涉测量中两个天线单元之间的距离，随着基线的变化，两个天线单元对同一目标的观测信号能够形成干涉，为了顺利开展干涉测量，需要基线的长度以及基线与目标源的方向。

如果在月球背面位于地月连线上的一点有一个低频探测设备，它与中继卫星上的低频探测设备进行干涉测量，那么它们之间的基线是从月面设备指向中继卫星的向量，长度就是这个向量的模。利用观测数据和模型数据可以分别求出这段时间内的基线及长度，基线长度随着时间的变化如图 6。图 6 是从 2019 年 12 月 22 日开始，基于观测数据和模型数据分别计算基线长度随时间变化的情况。其中横坐标是从 12 月 22 日 0 时起经过的时间（单位：小时），纵坐标是此时此刻基线的长度；蓝色线条代表观测数据得到的基线长度，红色线条代表模型数据得到的基线长度。总体来说它们的变化趋势一致，为了比较两者的差异，对两条基线长度做差得到观测数据和模型数据的基线长度之差随时间的变化关系如图 7，从图 7 可以看到，两种情况下基线长度的差距不大，普遍在几十千米范围内。甚长基线干涉测量的相关处理需要时延的误差在一定范围内，时延误差的重要组成部分是几何时延误差，即对基线长度的计算应与真实值尽可能一致。相关处理对时延误差的要求可以描述为[22]

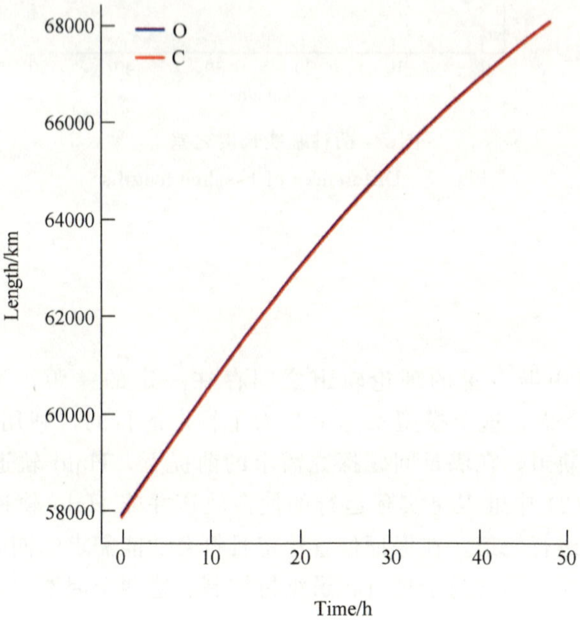

图 6　观测基线与模型基线长度

Fig. 6　Baseline lengths of observation and calculation

$$\Delta\tau \leqslant \frac{N}{2B}, \tag{5}$$

其中，N 为谱通道个数；B 为带宽。如果以嫦娥四号着陆器低频射电频谱仪的相关性能参数为参照，进行甚长基线干涉测量相关处理需要的时延误差应低于 0.819ms，对应的基线长度误差应低于 245km。几何时延虽然是总时延的主要部分，其他的因素（如传播介质造成的时延）也会对最终时延的估计产生影响，所以，基线长度估计误差造成的时延估计误差应该小于上述计算值。如图 7，随着卫星继续运行，它的轨道逐渐回到标称轨道附近，后半段的差值最大只有 50.94km，平均值 30.00km，可以认为在这一阶段，模型对实际轨道的符合效果较好。

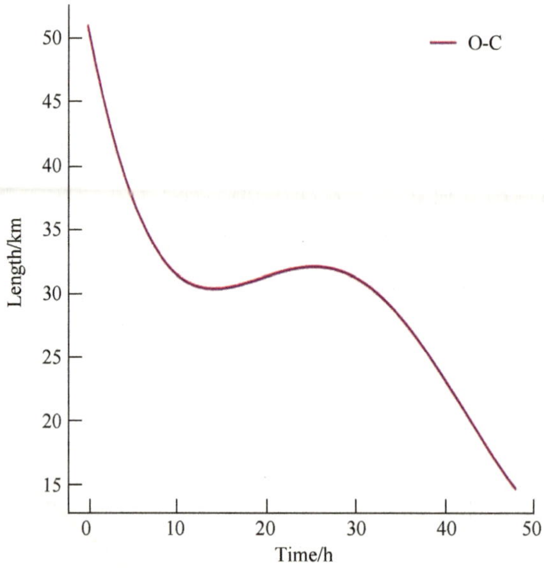

图 7　两种基线长度之差

Fig. 7　Differences of baseline lengths

4　讨论和展望

数据显示模型与中继卫星的理论轨道之间存在一定的差距，可能的原因有两个：(1) 出于计算简便的考虑，这个模型实际上是为了特定的目的，利用一个周期化的 Halo 轨道代替实际的使命轨道，在满足问题探究需求的前提下，Halo 轨道模型与理论轨道之间存在较小的差异；(2) 中继卫星实际运行的使命轨道并非 Halo 轨道族，而是以理论轨道为标称轨道设计的运行轨道，在保证位置稳定且符合功能需求的同时，尽可能降低推进剂消耗，延长任务寿命，同时易于进行轨道维持控制。这两个因素会显著影响模型与实际轨道的符合程度。

本文采用一个准周期的 Halo 轨道模型，与嫦娥四号中继卫星“鹊桥号”的理论轨道数据进行了比对，两者与月面着陆器形成的基线长度的差异较小，两种基线长度差平均值为 30.00km。

在高频频带，干涉测量拟采用条纹搜索方法，对基线精度需求是波长的数千倍，即百千米左右，可以认为这种模型基本满足空间低频射电干涉模拟实验的要求。在空间低频射电干涉测量得以实施之前，可以通过建立模型、数值模拟的方法对干涉测量系统的组成、方法和性能进行计算和预测，计算结果将为实施空间低频射电干涉测量提供参考。

参考文献

[1] 吴伟仁，刘继忠，唐玉华，等. 中国探月工程 [J]. 深空探测学报，2019，6 (5)：405 - 416.

WU W R，LIU J Z，TANG Y H，et al. China lunar exploration program [J].

Journal of Deep Space Exploration，2019，6（5）：405 - 416.

[2] 张立华，熊亮，王鹏，等."嫦娥 4 号"中继星任务分析与系统设计 [J]. 深空探测学报，2018，5（6）：515 - 523.

ZHANG L H，XIONG L，WANG P，et al. The mission analysis and system design of Chang'e - 4 lunar relay communication satellite [J]. Journal of Deep Space Exploration，2018，5（6）：515 - 523.

[3] 段建锋，李勰，李翠兰，等."嫦娥 4 号"中继星使命轨道段定轨计算与分析 [J]. 深空探测学报，2019，6（3）：247 - 253.

DUAN J F，LI X，LI C L，et al. Orbit determination and analysis of Chang'E - 4 relay satellite on mission orbit [J]. Journal of Deep Space Exploration，2019，6（3）：247 - 253.

[4] 叶培建，孙泽洲，张熇，等. 嫦娥四号探测器系统任务设计 [J]. 中国科学：技术科学，2019，49（2）：124 - 137.

YE P J，SUN Z Z，ZHANG H，et al. Mission design of Chang'e - 4 probe system [J]. Scientia Sinica：Technologica，2019，49（2）：124 - 137.

[5] 贾瑛卓，邹永廖，薛长斌，等. 嫦娥四号任务科学目标和有效载荷配置 [J]. 空间科学学报，2018，38（1）：118 - 130.

JIA Y Z，ZOU Y L，XUE C B，et al. Scientific objectives and payloads of Chang'E - 4 mission [J]. Chinese Journal of Space Science，2018，38（1）：118 - 130.

[6] 梅丽，苏彦，周建锋. 极低频射电天文观测现状与未来发展 [J]. 天文研究与技术，2018，15（2）：127 - 139.

MEI L，SU Y，ZHOU J F. The history and development of the low - frequency radio observation [J]. Astronomical Research & Technology，2018，15（2）：127 - 139.

[7] LI C L，LIU D W，LIU B，et al. Chang'E - 4 initial spectroscopic identification of lunar far - side mantle - derived materials [J]. Nature，2019，569：378 - 382.

[8] 吴伟仁，王琼，唐玉华，等."嫦娥 4 号"月球背面软着陆任务设计 [J]. 深空探测学报，2017，4（2）：111 - 117.

WU W R，WANG Q，TANG Y H，et al. Design of Chang'e - 4 Lunar farside soft - landing mission [J]. Journal of Deep Space Exploration，2017，4（2）：111 - 117.

[9] SZEBEHELY V. Theory of orbit [M]. New York：Academic Press，1967.

[10] HOWELL K C，BREAKWELL J V. Almost rectilinear halo orbits [J]. Celestial Mechanics，1984，32（1）：29 - 52.

[11] BREAKWELL J V，BROWN J V. The 'Halo' family of 3 - dimensional periodic orbits in the Earth - Moon restricted 3 - body problem [J]. Celestial Mechanics，1979，20（4）：389 - 404.

[12] FARQUHAR R W，KAMEL A A. Quasi - periodic orbits about the translunar libration point [J]. Celestial Mechanics & Dynamical Astronomy，1973，7（4）：458 - 473.

[13] RICHARDSON D L. Halo orbit formulation for the ISEE - 3 mission [J]. Journal of

Guidance Control & Dynamics, 1979, 3 (6) : 543 - 548.

[14] RICHARDSON D L. Analytic construction of periodic orbits about the collinear points [J]. Celestial Mechanics, 1980, 22 (3) : 241 - 253.

[15] 卢松涛，赵育善. Halo 轨道 Richardson 三阶近似解析解的改进 [J]. 宇航学报，2009，30 (3) : 863 - 869.

LU S T, ZHAO Y S. The improvement of Richardson's three order approximate analytical solution of Halo orbit [J]. Journal of Astronautics, 2009, 30 (3): 863 - 869.

[16] FARQUHAR R W, MUHONEN D P, RICHARDSON D L. Mission design for a halo orbiter of the earth [J]. Journal of Spacecraft & Rockets, 1976, 14 (3): 170 - 177.

[17] LO M W. The InterPlanetary Superhighway and the origins program [C]//Proceedings of the Aerospace Conference. 2002.

[18] 曹鹏飞，李维国，王俊彦，等. 高精度模型下 Halo 轨道设计研究 [J]. 深空探测学报，2019，6 (3) : 277 - 283.

CAO P F, LI W G, WANG J Y, et al. Study on Halo orbit design under high precision model [J]. Journal of Deep Space Exploration, 2019, 6 (3) : 277 - 283.

[19] 刘磊，陈明，张哲，等. 地月平动点轨道应用与研究进展 [J]. 宇航学报，2019，40 (8): 849 - 860.

LIU L, CHEN M, ZHANG Z, et al. Progress on application and research of Earth - Moon libration orbits [J]. Journal of Astronautics, 2019, 40 (8) : 849 - 860.

[20] 高珊，周文艳，张磊，等. 嫦娥四号中继星任务轨道设计与实践 [J]. 中国科学：技术科学，2019，49 (2) : 156 - 165.

GAO S, ZHOU W Y, ZHANG L, et al. Trajectory design and flight results for Chang'e 4 - relay satellite [J]. Scientia Sinica: Technologica, 2019, 49 (2) : 156 - 165.

[21] 高清鹏，李春晓，李荣旺，等. 鹊桥卫星激光测距时间窗口及测距成功概率分析 [J]. 天文研究与技术，2019，16 (4): 422 - 430.

GAO Q P, LI C X, LI R W, et al. Magpie bridge satellite laser ranging time window and distance probability analysis [J]. Astronomical Research & Technology, 2019, 16 (4) : 422 - 430.

[22] 舒逢春，张秀忠. 相关处理软件系统中模型计算的精度分析 [J]. 中国科学院上海天文台年刊，2001 (22) : 100 - 106.

SHU F C, ZHANG X Z. Accuracy analysis on model calculation in correlator software package [J]. Annals of Shanghai Observatory Academia Sinica, 2001 (22) : 100 - 106.

嫦娥四号低频射电频谱仪降低背景噪声方法的研究

张韬[1,2,3]，苏彦[1,2,3]

1. 中国科学院大学，北京 100049
2. 中国科学院国家天文台，北京 100101
3. 中国科学院月球与深空探测重点实验室，北京 100101

摘要： 嫦娥四号着陆器将搭载低频射电频谱仪在月球背面进行低频射电天文观测，低频射电频谱仪的观测波段为 0.1～40MHz。根据着陆器在中国空间技术研究院的微波暗室进行的电磁兼容性试验结果，着陆器平台在该频段内自身存在非常强的噪声，其强度甚至淹没大部分来自太阳爆发的信号，难以探测有效信号，实现预期的科学目标。通过模拟仿真分析谱减法、维纳滤波及自适应滤波 3 种方法对着陆器噪声消除的效果，从而选择更为有效的噪声消除方法，为低频射电频谱仪在轨探测任务的数据处理提供依据。

关键词： 低频射电频谱仪；嫦娥四号；谱减法；维纳滤波；自适应滤波

Research of the Method for Reducing Background of Very Low Frequency Radio Spectrumon Chang'E－4

Zhang Tao[1,2,3]，Su Yan[1,2,3]

1. University of Chinese Academy of Sciences，Beijing 100049，China，Email：zhangtao@nao. cas. cn
2. National Astronomical Observatories，Chinese Academy of Sciences，Beijing 100101，China
3. Key Laboratory of Lunar and Deep Space Exploration，Chinese Academy of Sciences，Beijing 100101，China

Abstract： Chang'E－4 will carry a Very Low Frequency Radio Spectrometer（VLFRS），which will land on the far－side of the moon forRadio observation. The VLFRS has an observation frequency of 0.1 — 40MHz. According to Electromagnetic Compatibility (EMC) test results in the Microwave darkroom of China Academy of Space Technology, the lander platform is quite noisy in this band, and the intensity of the noise is even stronger than that of most signals from the sun burst. In this pape, we simulate and optimize background elimination through improved spectral subtraction, adaptive filtering and wiener filtering, and choose a more effective method foreliminating background.

Keywords: Very low frequency radio spectrometer; Chang'E - 4; Improved spectral subtraction; Wiener filtering; Adaptive filtering

美国著名工程师卡尔，央斯基在 1930 年用他自己建造的“旋转木马”首次观测到来自银河系的射电辐射[1]，自此展开了对射电天文的广泛研究。天文学家的研究发现，不同天体因为辐射机制的不同产生不同波段的电磁辐射，因此，对宇宙进行全波段观测是一件非常有意义的事，但是，地球大气的存在使很多射电波段的信号特别是频率低于 10MHz 的甚低频波段难以到达地面，然而，低频波段的探测对于研究天体的演化与起源有至关重要的作用，为了弥补甚低频波段的观测空白，许多国家或组织发射卫星到太空进行观测，比如太阳长期探测卫星（WIND）和尤利西斯号卫星（Ulysses）。

由于地球以及众多在轨卫星的电磁干扰，空间低频射电观测也面临挑战。因为月球自身可以遮挡来自地球方向的射电干扰，所以月球背面的射电宁静区是十分理想的观测地点。欧洲航天局于 2015 年 1 月提出在 2025 年向月球背面发射一颗载有低频射电频谱仪着陆器的月球背面探测任务（Farside Explorer）计划[2]，我国提出嫦娥四号探月计划，嫦娥四号是世界首颗在月球背面软着陆和巡视探测的航天器。2018 年 5 月 21 日，嫦娥四号中继星“鹊桥”成功发射，着陆器和巡视器将于年底发射，着陆器上搭载了一台科学载荷低频射电频谱仪，是人类在月球背面首次实现对宇宙和太阳的低频段（0.1～40MHz）观测。

根据着陆器整器在微波暗室的电磁兼容性试验结果，虽然来自外界的信号干扰可以很好地被屏蔽，但是着陆器本身搭载的电子设备以及其它的载荷对低频观测存在电磁干扰，由于无法对平台进行工程上的调整，所以只能通过数据处理的方式消除平台的干扰。目前降低信号的噪声处理方法主要有 3 种：谱减法、维纳滤波及自适应滤波方法。本文结合低频射电频谱仪的科学目标，对比、分析信号增强的效果，得到最优化的方法，为低频射电频谱仪在轨探测任务的数据处理提供依据。

1　低频射电频谱仪的工作原理以及噪声来源

低频射电频谱仪系统由 4 根接收天线（3 根 5m 的长天线 A，B，C 和 1 根 20cm 的短天线 D）、前置放大器以及电子学单元组成。3 根长天线 A，B，C 相互垂直[3-4]（如图 1），接收空间电磁信号在三个方向上的分量，根据电磁波的传播理论计算接收信号的强度以及方向。短天线 D 与长天线 B，C 处于同一个平面，根据电磁波的传播理论以及不同长度天线的增益不同，A，B，C 长天线可以同时接收来自宇宙的远场信号和来自着陆器本身的近场信号，而天线 D 因天线短且离着陆器近，主要接收来自近场的着陆器噪声信号[3]。

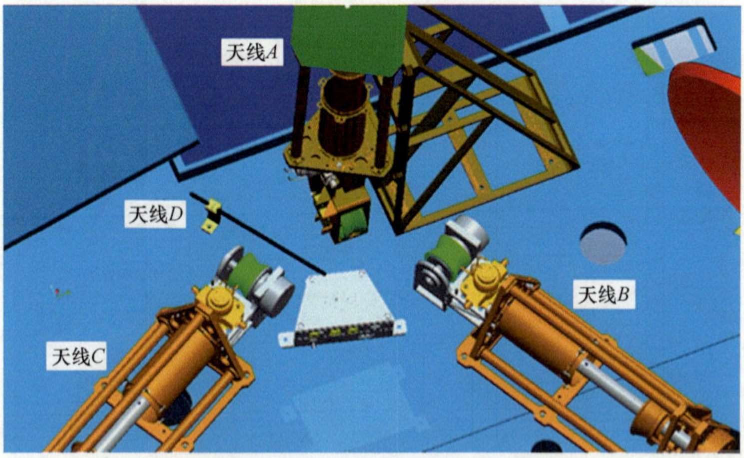

图 1　低频射电频谱仪的 4 根天线[3]

Fig. 1　The four antennas of the very low frequency radio spectrometer[3]

嫦娥四号搭载的低频射电频谱仪的工作频率包括低频（100kHz～2MHz）和高频（1～40MHz）两个频段，频谱仪的主要技术参数如表 1。

在低频射电频谱仪工作的同时，着陆器上其他部分载荷以及一些电子模块也处于运行状态，比如电源控制器、数传调制器、发射段 RE102、数传控制电源测量单元（Source Measure Unit，SMU）等，图 2 给出了月球表面的电磁波通量密度，天线附近区域的电场强度和接收机输出的电压值根据（1）式计算：

$$U_{out} = \frac{G_{preamp}\ G_{Receiver}\ IL_{coax}}{1 + Z_{\alpha}\ /\ Z_{preamp}}\ \overrightarrow{h_e}\overrightarrow{E} \tag{1}$$

表 1　低频射电频谱仪主要技术参数[3]

Table 1　Main technical parameters of Very low frequency radio spectrometer[3]

项　目	设计参数	实测结果
工作频率	100kHz～40MHz	100kHz～40MHz
接收机灵敏度/(nV / Hz)	6	8.1
频率分辨率	低频：5kHz	4.88kHz
	高频：100kHz	97.66kHz
最大数据率/bps	4.5M	4.34M
功耗/W	24	24

其中，U_{out} 为接收机的输出电压值；G_{preamp} 和 $G_{Receiver}$ 分别为前置放大器增益和接收机增益；IL_{coax} 为高频电缆插损；Z_{α} 和 Z_{preamp} 分别为天线阻抗和前置放大器阻抗；$\overrightarrow{h_e}$ 为天线的有效长度；$\overrightarrow{E}$ 为天线处的电场强度。

（1）式中各项参数根据低频射电频谱仪生产方提供的定标报告获取，结合各项参数以及（1）式将接收机测得的电压数据反演出空间电场以及通量密度。图 2 给出了各天体到

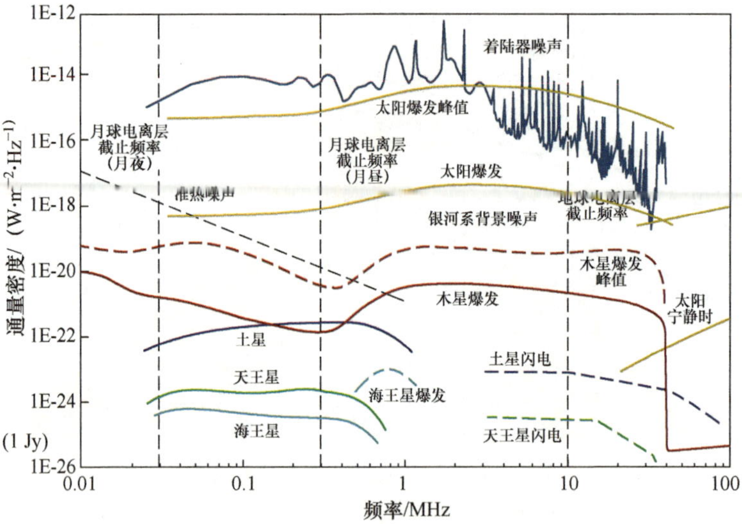

图 2　月球表面的电磁波通量密度[2]

Fig. 2　Radio emission levels on the surface of moon[2]

达月球表面的通量密度。图中的着陆器噪声是根据中国空间技术研究院微波暗室中对嫦娥四号着陆器整器进行的电磁兼容性试验测得工况三（着陆器电源控制器主份、数管管理单元主份、数传调制器主份、载荷电控箱主份、低频射电频谱仪等依次加电连续工作）天线 A 接收的高频段数据反演得到的。从图 2 可以发现，着陆器的噪声通量密度大于普通的太阳爆发产生的通量密度，与太阳爆发峰值通量密度相当，所以不对接收到的信号进行降噪处理，难以反演出太阳爆发的真实强度。

2　低频射电频谱仪有效信号提取方法

由于低频射电频谱仪的 3 根主要天线 A，B，C 收到的目标信号都包含来自着陆器自身的噪声信号，所以在对信号进行提取之前需要通过降噪算法抑制噪声并最大限度地分离出有用的信号。用 $x(n)$ 表示观测到的信号，$s(n)$ 表示有用信号，$v(n)$ 表示噪声信号。它们之间的关系表示为

$$x(n)=s(n)+v(n) \tag{2}$$

在本文的模拟仿真中，将一组强度与太阳爆发强度接近的模拟信号作为有用信号 $s(n)$，用电磁兼容性试验测得的工况三的着陆器噪声作为噪声信号 $v(n)$，通过谱减法、维纳滤波及自适应滤波的方法进行仿真处理。

2.1　谱减法

谱减法是信号增强的一种常见方法，在频域上用信号加噪声的幅值减去噪声的幅值得到有用信号的幅值。文献 [5] 于 1979 年改进了该方法，称为改进的谱减法。改进的谱减法在谱减的过程中使用噪声的平均功率，有效地减小剩余的“音乐噪声”。改进的谱减法

用公式表示为

$$|S(n)|^{\gamma}=\begin{cases}|X(n)|^{\gamma}-aD(n); & |X(n)|^{\gamma}\geqslant aD(n)\\ bD(n); & |X(n)|^{\gamma}<aD(n)\end{cases} \tag{3}$$

其中，$D(n)$ 为噪声的平均幅值，

$$D(n)=\frac{1}{N}\sum_{i=1}^{N}|D_i(n)|^{\gamma} \tag{4}$$

(3) 式、(4) 式中，γ 为 1 时，幅值相减；γ 为 2 时，功率相减。a ($a\geqslant 1$) 为过减因子，b($b>0$) 为补偿因子，通过调节两个参数减少周期性的误差。结合低频射电频谱仪的结构特点以及对科学数据的要求，用长天线 A，B，C 收到信号的平均幅值减去短天线 D 收到信号的平均幅值获取信号的幅值。

由于天线 A，B，C 与天线 D 的长度及材质不一样导致其增益不同，所以在进行谱减时需要补充不同增益带来的信号强度差异，具体来说，先取一小段时间内的信号计算它们的对消系数，然后在后续的处理过程中将天线 D 收到信号的平均幅值乘以对消系数获取噪声幅值。结合低频射电频谱仪的结构特点，运用改进的谱减法进行降噪处理，以天线 A 和天线 D 为例，用公式表示为

$$|S(f)|=\begin{cases}|V_A(f)|-|P(f)V_D(f)|; & |V_A(f)|\geqslant|P(f)V_D(f)|\\ b|P(f)V_D(f)|; & |V_A(f)|<|P(f)V_D(f)|\end{cases} \tag{5}$$

其中，$V_A(f)$ 为长天线 A 收到的信号进行傅里叶变换后在相应频点的平均值；$V_D(f)$ 为短天线 D 收到的信号在相应频点的平均值；γ 取 1；$P(f)$ 取太阳宁静时一小段时间内 4 个接收机接收到的信号分别算出：

$$P_A(f)=\frac{V_A(f)}{V_D(f)} \tag{6}$$

处理流程如图 3 所示。

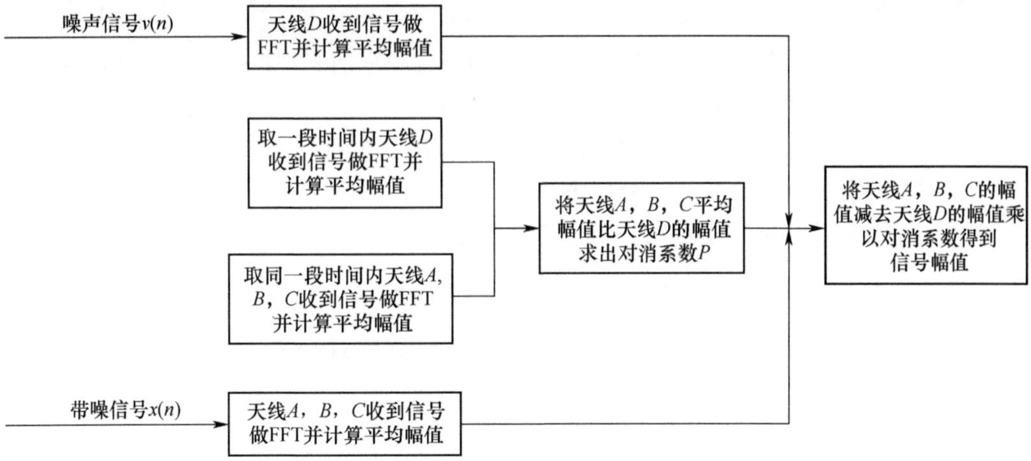

图 3　改进谱减法运算示意图

Fig. 3　Improved spectral subtraction diagram

下面3种数据处理方法以天线 A 与天线 D 接收到的数据进行仿真为例，谱减法仿真结果如图4所示。

图4展示了快速傅里叶变换后的频域信号，横轴表示频率（单位为MHz），纵轴为幅度（单位为dBuV）。图4中黑色曲线是带有噪声的信号，蓝色曲线是经过改进谱减法处理后的信号，红色曲线是有用信号。从图4可以看出，该方法在整个波段对噪声进行了有效抑制，在10MHz以下频段内噪声幅度下降了20dB左右，在其他频段噪声降低10dB左右，且处理后的信号波形基本能还原原本的有用信号，在背景噪声较低的频段（20～40MHz）处理效果更好。

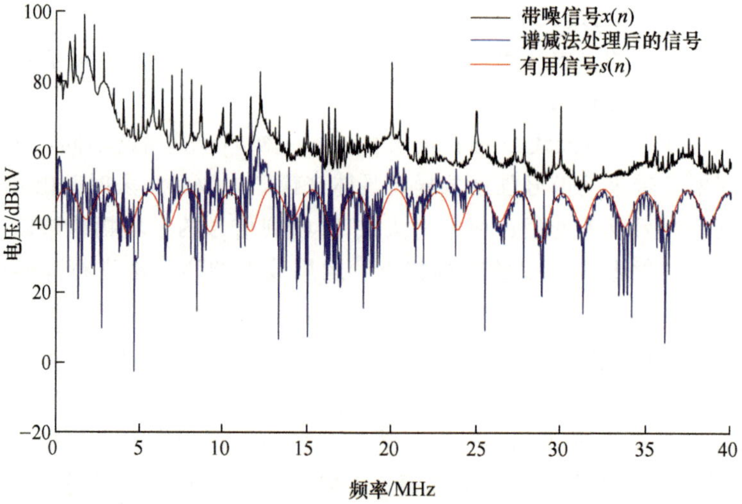

图4　改进谱减法仿真结果

Fig. 4　Simulation results by improved spectral subtraction

2.2　维纳滤波

从（2）式可知，进行信号处理的目标是得到不含噪声的有用信号 $s(n)$。但是在实际信号处理过程中，求得的有用信号并不完全等于 $s(n)$，只是 $s(n)$ 的逼近值或估计值，因此对信号的处理可以看作对 $s(n)$ 的估计，处理信号就是找到一个最佳的估计器。维纳滤波是一个寻找最佳估计器来估计信号的方法[6-7]。

图5为维纳滤波的原理示意图，用公式表示为

$$y(n)=x(n)*h(n) \tag{7}$$

其中，$h(n)$ 为维纳滤波器的滤波参数；$y(n)$ 为对有用信号 $s(n)$ 的估计值。

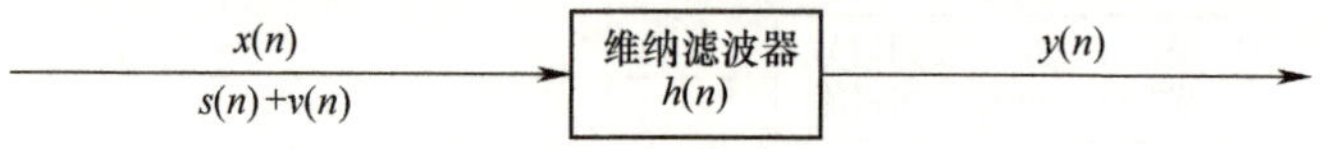

图5　维纳滤波器

Fig. 5　Wiener filter

根据最小误差准则使 $\varepsilon = E[\{s(n) - y(n)\}^2]$ 达到最小可以算出最佳滤波参数，

$$h(n) = \frac{E[s(n) * x(n)]}{E[x^2(n)]} \tag{8}$$

将上式进行傅里叶变换可以导出，

$$H(k) = \frac{P_{sx}(k)}{P_x(k)} \tag{9}$$

其中，$P_{sx}(k)$ 为 $s(n)$ 与 $x(n)$ 的互功率谱密度；$P_x(k)$ 为 $x(n)$ 的功率谱密度，由于有用信号 $s(n)$ 与噪声信号 $v(n)$ 互不相关，即有 $P_{sv}(k)=0$，则可以得到：

$$P_{sx}(k) = P_s(k) \tag{10}$$

$$P_x(k) = P_s(k) + P_d(k) \tag{11}$$

将（10）式和（11）式代入（9）式可以得到：

$$H(k) = 1 - \frac{1}{\gamma(k)} \tag{12}$$

（12）式中，$\gamma(k)$ 为傅里叶变换后对应频点处带噪信号的功率谱与噪声功率谱的比值[8]。根据天线 D 和天线 A 接收到的数据计算出 $\gamma(k)$，从而计算出最佳的滤波参数。图 6 为仿真结果。

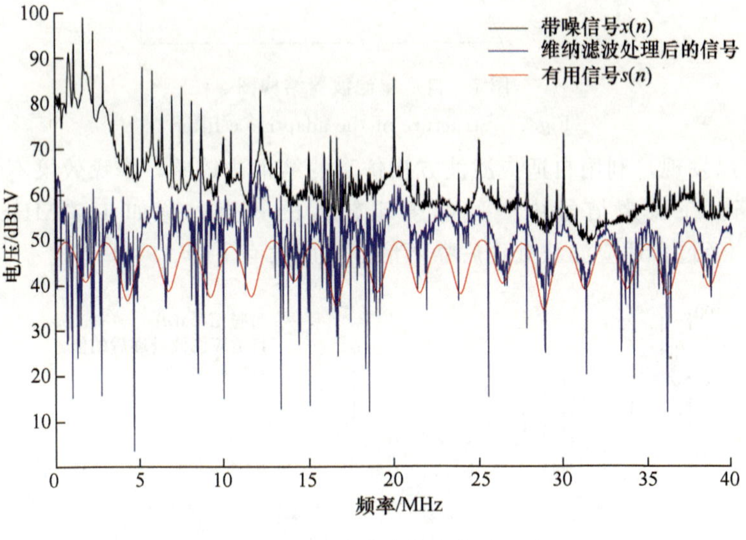

图 6　维纳滤波仿真结果

Fig. 6　Simulation results by Wiener filter

从图 6 可以看出，维纳滤波能够降低噪声，但是相对于谱减法的降噪效果要差一些，基本能还原有用信号的波形，但是处理后的信号在背景噪声比较大的波段（0～20MHz）仍难以分辨有用信号。

2.3　自适应滤波

不同于维纳滤波，自适应滤波器根据收到的信号不断地自动调节自身的参数，最大限度地降低噪声信号[9]。自适应滤波已经广泛应用在系统识别、信道均衡、信号增强以及信号预测方面。本文主要运用信号增强的功能，结构示意图如图 7。自适应滤波算法至少需要两套接收设备，其中一套用来接收带有噪声的信号，一套用来接收噪声，然后通过自适应算法求得有用信号，而低频射电频谱仪正好有这样两套接收设备。

图 7 为一个自适应降噪滤波器的结构示意图。其中，$x(n)$ 指接收机收到的信号（包括有用信号 $s(n)$ 以及噪声信号 $v(n)$）；$v_1(n)$ 为另一个接收机收到的噪声信号；$e(n)$ 为误差信号，它们之间的关系表示为

$$e(n)=x(n)-\sum_{i=1}^{N}[w(i)\,v_1(n-i)] \tag{13}$$

其中，$w(n)$ 为滤波器的加权系数，该系数通过最小均方误差准则或者最小二乘法评估 $E[e^2(n)]$ 来调整，使滤波器工作在最佳状态[10-12]。下面利用最小二乘法进行仿真。

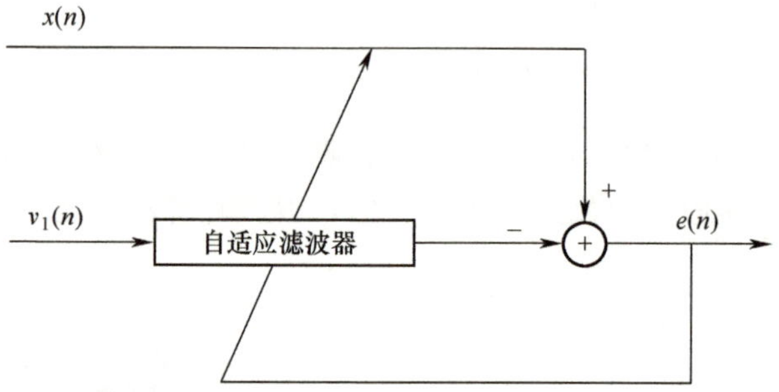

图 7 自适应滤波器结构图

Fig. 7 Structure of the adaptive e filter

从图 8 可以发现，利用自适应滤波方法仿真的结果比较差，降噪效果不明显，对信号的还原度不高。在噪声信号中一些幅度比较大的频点处（如 1.73MHz，2.32MHz，2.9MHz 等），经过自适应滤波处理后噪声信号幅度没有下降。

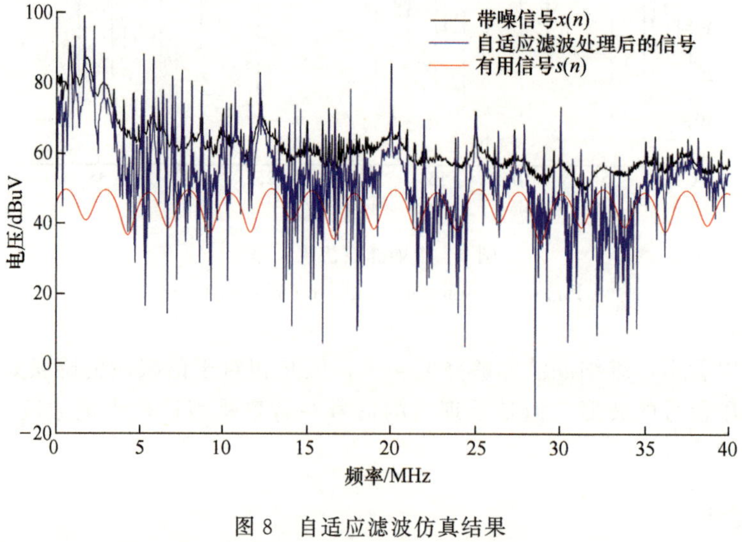

图 8 自适应滤波仿真结果

Fig. 8 Simulation results byadaptive filter

3　低频射电频谱仪背景噪声消除结果分析

前文通过 3 种不同的数据处理方法对着陆器噪声进行了仿真处理，其中自适应滤波方法是在时域上的处理方法，改进的谱减法和维纳滤波则是在频域上的处理方法。表 2 给出这 3 种信号增强方法对信号信噪比提升效果的对比。

从表 2 可以发现，改进谱减法和维纳滤波的处理结果明显好于自适应滤波，且其中改进谱减法的信噪比提升略大于维纳滤波，自适应滤波降噪能力受输入端的噪声信号 $v_1(n)$ 与带噪信号中的噪声信号相关度影响很大。在低频射电频谱仪设计中用于接收信号的长天线 A，B，C 和用于接收着陆器噪声信号的短天线 D 的参数不同，仿真信号中有用信号的信噪比非常低等因素导致自适应滤波降噪的效果变差。长天线 A，B，C 与短天线 D 收到噪声信号的不同可以通过评估短天线与长天线在不同频点处的增益差异来抵消，所以在频域上处理的两种算法在降噪方面取得比较好的效果。

表 2　3 种方法处理数据前后的信噪比

Table 2　SNR before and after the data processing by three different algorithms

数据处理方法	改进谱减法	维纳滤波	自适应滤波
处理前信噪比/dB	−27.64	−27.64	−27.64
处理后信噪比/dB	1.99	−4.64	−25.35
信噪比提升/dB	29.63	23.00	2.29

为了验证低频射电频谱仪的性能以及数据处理方法，在中国科学院国家天文台密云观测站食堂楼顶安装低频射电频谱仪及接收天线进行外场试验，并且在距离接收天线 100m 左右的宿舍楼里用任意波形发生器连接天线发射模拟信号。着陆器噪声信号采用电磁兼容性工况三的试验结果，由一个放在接收天线下方 1m 长的小天线连接波形发生器发出，图 9 为试验天线 A 接收到的一组数据利用改进的谱减法处理后的结果，其中发射的有用信号为 1.8MHz 点频信号。

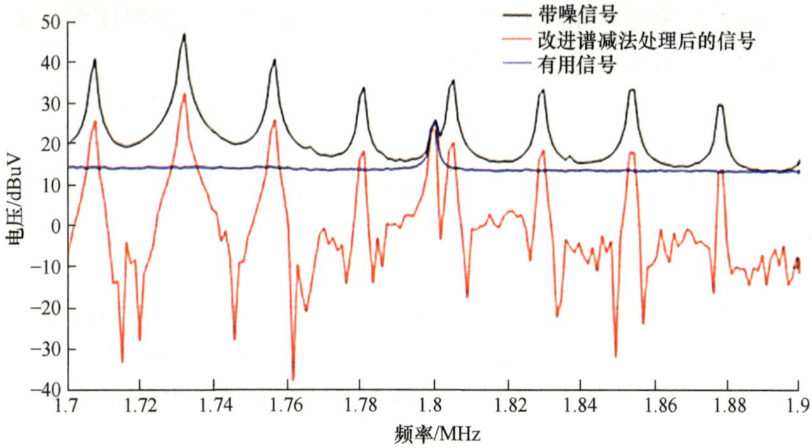

图 9　改进谱减法处理结果

Fig. 9　Noise reductionby improved spectral subtraction

图 9 截取了 1.7～1.9MHz 的信号，从图中可以看出，经过改进谱减法处理后，着陆器的噪声被有效地抑制，幅度下降 10dB，而有用信号的幅度并没有降低，信噪比从 −25.23dB 提升到 −2.60dB。通过改进谱减法处理后的信号信噪比提升了 −22.63dB，说明改进的谱减法在外场实验中对着陆器背景噪声信号的降低也是有效的。

4　总结

在月球背面进行极低频（0.1～40MHz）探测非常有科学意义[13]，同时由于该频段信号微弱，信噪比低，所以提取有用信号十分困难，本文利用电磁兼容性测试的着陆器噪声以及嫦娥四号搭载的低频射电频谱仪对月球背面接收到的目标信号强度进行了仿真实验。利用改进的谱减法、维纳滤波以及自适应滤波这 3 种常见的信号增强算法对仿真信号进行提取，其中改进的谱减法对信号的提取效果显著，经处理后信噪比有明显提升，可以直观地从处理后的波形中看到目标信号波形。改进的谱减法对在密云进行外场试验的数据处理也有很好的效果。因此，将改进的谱减法作为低频射电频谱仪主要的预处理方法，实现对太阳爆发较为准确的观测。低频射电频谱仪在月球背面进行探测的过程中，温度等环境的变化可以引起对消系数的改变，所以在实际探测过程中需要结合环境情况，每隔一段时间计算一次对消系数。

参考文献

[1] JANSKY K G. Directional studies of atmospherics at high frequencies [C]// Proceedings of the Institute of Radio Engineers. 1932，20：1920 - 1932.

[2] MIMOUN D，WIECZOREK M A，ALKALAI L，et al. Farside explorer：unique science from a mission to the farside of the moon [J]. Experimental Astronomy，2012，33 (2/3)：529 - 585.

[3] 纪奕才，赵博，方广有，等．在月球背面进行低频射电天文观测的关键技术研究 [J]. 深空探测学报，2017，4 (2)：150 - 157.

[4] 施硕彪，董亮，高冠男，等．米波太阳射电频谱仪的科学目标和技术方案 [J]. 天文研究与技术——国家天文台台刊，2011，8 (3)：229 - 235.

[5] BOLL S. Suppression of acoustic noise in speech using spectral subtraction [J]. IEEE Transactions on Acoustics Speech & Signal Processing，1979，27 (2)：113 - 120.

[6] BROWN R G，HWANG P Y C. Introduction to random signals and applied Kalman filtering：with MATLAB exercises and solutions [M]. Hoboken：John Wiley & Sons，1997.

[7] WIENER N. Extrapolation，interpolation，and smoothing of stationary time series：with engineering applications [M]. Hoboken：John Wiley & Sons，1949.

[8] 宋知用．MATLAB 语音信号分析与合成 [M]. 北京：北京航空航天大学出版社，2017.

[9] 耿妍，张端金．自适应滤波算法综述 [J]. 信息与电子工程，2008，6 (4)：315 - 320.

[10] DINIZ P S R. Adaptive filtering: algorithms and practical implementation [M]. 4th ed. [S. l.]: Kluwer Academic Publishers, 2008.
[11] 沈福民．自适应信号处理 [M]. 西安：西安电子科技大学出版社，2001.
[12] 王树棠，王蕴红．自适应滤波器阶数与稳态均方误差 [J]. 电路与系统学报，1997 (1)：65－67.
[13] 梅丽，苏彦，周建锋．极低频射电天文观测现状与未来发展 [J]. 天文研究与技术，2018，15 (2)：127－149.

第 3 章　基于光谱学的着陆区物质组成研究

月表物质成分的种类和空间分布是由月球早期的全球岩浆洋结晶和后期火山喷发、小天体撞击等内外动力地质过程共同作用的产物，是月球形成和地质演化历史最重要的研究载体之一。嫦娥四号探测器着陆在月球背面的南极-艾特肯盆地，由于该盆地具有大且深的特点（深达 12km），研究着陆区物质组成不仅可以帮助理解月表演化过程，更是提供了探测月球深部成分信息的机会，这对于限定月球的形成与演化历史至关重要。

嫦娥四号任务中承担物质成分探测科学目标的载荷主要为搭载于“玉兔二号”月球车上的红外成像光谱仪，通过获取月表的反射光谱反演着陆区的物质类型。此外，月表的光度特性、太空风化作用和热辐射特征也会在光谱上有所响应，并与矿物的光谱特征互相耦合。通过解耦分析，也可获得着陆区月壤的光学物理特性、成熟度和温度等信息。

本章将对嫦娥四号着陆巡视区的光谱数据科学研究成果进行综述，着重介绍“玉兔二号”月球车在月表巡视探测过程中获取的重要成果，主要包含着陆区月壤光度特性、太空风化作用、热辐射特征、物质组成和撞击残留物等方面的研究成果。

3.1　月面物质光度特性

月球基本无大气，因此月壤层是月球与空间环境发生直接作用的界面，研究月壤的光度特性有助于了解月壤的演化过程，是提高对月球认识的重要途径。利用可见-近红外反射光谱可以对月表的主要矿物类型进行有效识别。但是除了矿物成分外，光照与观测几何角度也会对所测光谱产生较大影响。光度学研究的正是物质在不同角度的光散射特性，光度特征主要受月壤颗粒尺寸、孔隙度、透明度、表面粗糙度等物理特性影响。通过开展光度学研究，既能构建光度校正模型，将不同角度下获取的反射光谱数据校正到标准的角度，以提高光谱解译准确度，又能反推出观测目标的物理特性。月壤的光度特性可通过 Hapke 辐射传输模型中的相函数参数来描述，较常用的相函数形式包括 Henyey-Greenstein（HG）相函数和 Legendre 多项式形式的相函数。

嫦娥三号月球车所携带的红外成像光谱仪只获得了 4 条不同地点的光谱，不足以推导月壤光度参数，因此只从其全景相机所拍摄图像中提取了红、绿、蓝三个波段覆盖 2°～141°相位的相位曲线，该方法同样应用于嫦娥四号月球车的全景图像观测。嫦娥四号原位的红外成像光谱观测获取了波段范围更宽、光谱分辨率更高、目标更为准确的多角度数据，为深入理解月壤的光度特性提供了更好的机会。

在嫦娥四号探测器着陆后的第十个月昼，“玉兔二号”月球车利用红外成像光谱仪对月表的同一个区域进行了一次原位光度测量试验，利用太阳高度与方位的变化，月球车在不同时间段对准同一月表区域进行了连续的光谱观测，成功获取了相同目标在不同观测几何角度条件下的光谱，为推导月壤的光度特性提供了前所未有的数据集。对比不同太阳高度角条件下测量的相同目标的光谱特征，发现随着太阳高度角的降低，光谱数据质量逐渐下降，而且会影响物质组成的反演准确性（图 3－1）。

利用“玉兔二号”开展的原位光度实验数据，结合相机重建出的地形数据对测量区域的光谱进行了地形校正，将观测几何校正到真正的局地角度，然后提取出月壤的相位曲线，最终推导出 HG 相函数中的两个参数。结果表明，这些参数随波长而变化，表明着陆区原始月壤表现出更强的前向散射特性。利用相同的数据研究 Legendre 相函数参数，同样表明嫦娥四号着陆区月壤具有前向散射特性。这些研究获取的月壤光度特性参数为光谱的光度校正和更加准确的解混提供了重要信息和参数输入。

图 3-1　“玉兔二号”月球车在月面上进行光度测量

科学团队利用“玉兔二号”月球车在第四和第十个月昼定点绕圈观测获取的光谱数据，获得了光谱光度校正经验函数和 Hapke 辐射传输模型参数，对反射率数据进行光度校正，有效消除了光照观测几何的影响。利用推导出的光度函数将光谱校正到标准的几何角度（入射角 30°，出射角 0°，相位 30°），并利用校正后的光谱计算了月壤的 FeO 含量和光学成熟度（OMAT），OMAT 表征月壤因暴露于太空而发生物理变化程度的相对量度，较小的值表示较大的光学成熟度。未经光度函数校正的反射率光谱，在反演小区域内（直径<5m）月壤 FeO 含量和 OMAT 时有显著的离散，FeO 含量由 9.0 wt.%变化到 10.89 wt.%，OMAT 值由 0.11 到 0.18。基于光度函数校正后的光谱，获得的 FeO 含量和 OMAT 值都相当一致，标准偏差分别从 13.5 wt.%（FeO）和 35.7%（OMAT）降低到 0.7%和 7.1%。结果也显示出反射光谱光度校正的重要性，经过光度校正后的光谱，可以获得更加可靠的解译结果。

3.2　月面物质太空风化特性

太空风化是天体表面暴露在严苛的空间环境中所经历的一系列变化过程的总称，太空风化作用已经在月球、水星以及一些小行星上被广泛地观测到。月球几乎没有大气，陨石、微陨石、太阳风、宇宙射线等长期作用于裸露于空间环境的月球表面，太空风化作用使得月表物质的物理化学性质以及光学性质发生改变，会改变月壤的反射率光谱，影响月壤矿物含量的遥感反演准确度。

研究证实，月壤颗粒表面存在很薄的一层由太空风化作用产生的亚微米尺度无晶体结构的玻璃质物质，还分布有几纳米大小的亚微观金属铁颗粒，正是这种亚微观金属铁导致月壤光谱变暗、变红、变平。随着暴露在空间环境中的时间增加，月壤会受到更多的太空风化作用，新鲜的月壤逐渐变为成熟的月壤。“玉兔二号”月球车在着陆区开展的一系列原位光谱测量，为开展真实月表环境下的太空风化作用研究提供了珍贵的数据。

根据地质背景分析可以知道，嫦娥四号着陆区表面物质主要源于芬森撞击坑的溅射物，芬森撞击坑中央峰的新鲜物质应该与其溅射物成分一致且未受太空风化影响或者影响很小。利用印度月船一号（Chandrayaan-1）轨道器搭载的月球矿物制图仪（Moon Mineralogy Mapper，M^3）数据，在1500～2200nm的光谱波段范围内，将芬森撞击坑中央峰新鲜物质的M^3光谱重采样到“玉兔二号”光谱仪的波长范围和波段中心位置，作为着陆区尚未遭受太空风化的月壤的基准光谱（假定亚微观铁含量为0）；通过Hapke辐射传输模型，生成月壤表面沉积不同含量亚微观铁的模拟光谱；然后根据模拟光谱与“玉兔二号”原位测量光谱的最大相似度，估算出嫦娥四号着陆区月壤的亚微观铁含量约为0.32±0.06wt.%。月壤成熟度是描述太空风化效应的常用指标，可结合亚微观铁含量和FeO含量计算，嫦娥四号着陆区FeO含量约为12.6±0.6wt.%，计算显示着陆区月壤成熟度为82±15，根据月壤成熟度的划分方案，表明嫦娥四号着陆区主要源于芬森撞击坑溅射物的月壤大概率属于成熟月壤（图3-2）。

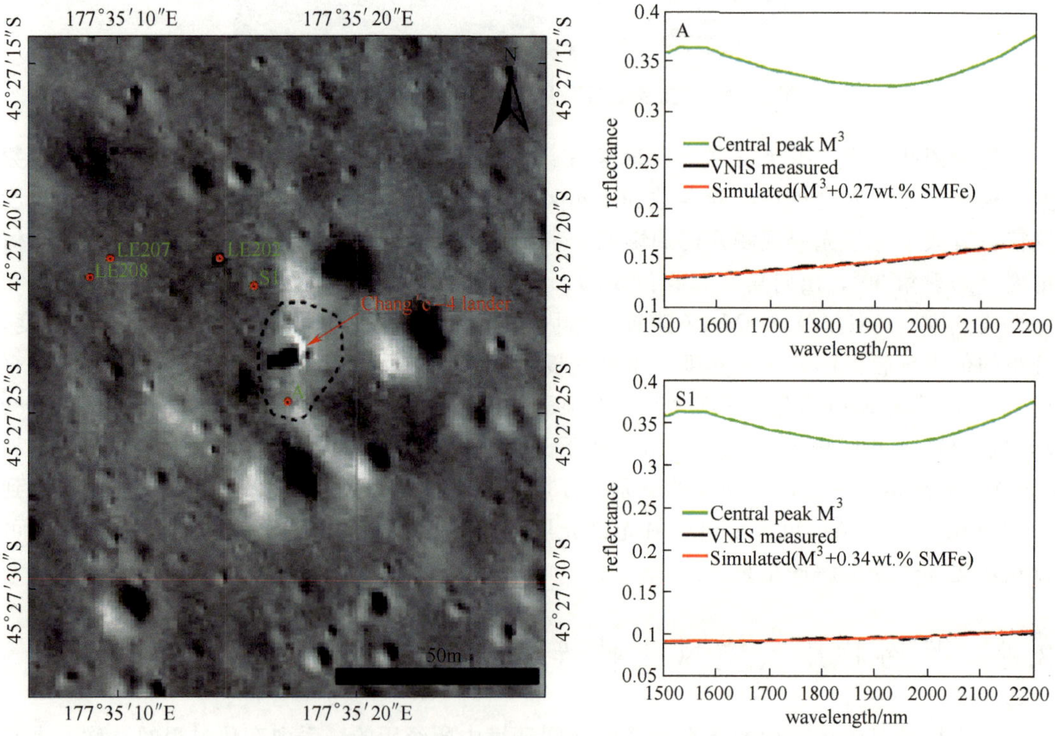

图3-2　嫦娥四号着陆区太空风化效应研究。左图为“玉兔二号”月球车巡视区背景图，右图为太空风化光谱拟合示意图

利用测月雷达数据对 24m 深度内的月壤成熟度进行量化分析表明，0～2m 的表层、2～12m 的细粒月壤、12～24m 的粗粒喷出物月壤的平均成熟度分别为 68.98%、82.61% 和 72.33%。月球表面低成熟度的区域可能是由于局部撞击形成的致密物质或远处撞击产生的岩石碎片造成的；在 12～18m 深处新发现的低成熟度的夹层结构是由古火山口的异质风化作用形成的（图 3-3）。

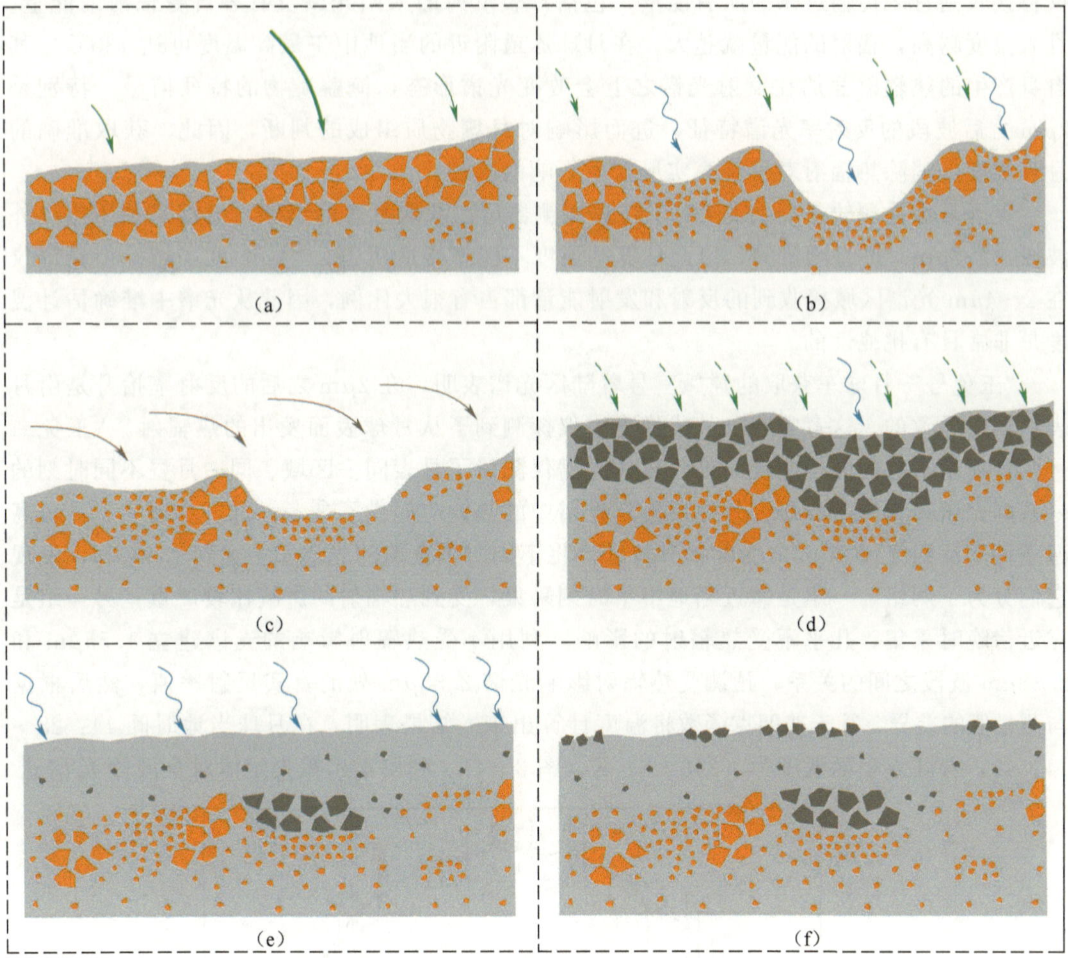

图 3-3　嫦娥四号着陆点风化层精细演化

(a) 不同程度的撞击事件重塑了粗粒风化层的结构；(b) 侵蚀和小规模撞击不断地改变着月球表面；(c) 芬森喷出物的覆盖；(d) 细粒风化层的形成；(e) 远端撞击事件造成的碎片输送；(f) 现在的风化层结构，岩石的颗粒代表风化层的成熟度，橙色异常表示附近多个喷出物的叠加，灰色异常表示芬森喷出物。绿色箭头和虚线箭头分别表示单个和多个撞击事件；波浪形箭头和灰色箭头分别表示多源物质和芬森喷出物的堆积

嫦娥三号的原位测量光谱表明，其着陆区的月壤属于亚成熟月壤，这与嫦娥三号着陆区的年轻玄武岩地层一致；月面最表层未受着陆器降落时羽流影响的月壤与受影响区域月壤相比表现出更强烈的太空风化特征。而嫦娥四号着陆区的月壤属于成熟月壤，这与着陆区已经历 30 多亿年太空风化作用的地质背景相一致。但是，月壤的亚微观铁含量随月球车与着陆器距离远近并未产生明显变化，这说明嫦娥四号着陆区月壤的快速形成过程充分混合了最表层已高度成熟的月壤。这项研究对于开展月球正面和背面太空风化作用比较分

析，以及深刻认识太空风化作用的全球性均具有重要意义。

3.3 月面物质热辐射特征

任何高于绝对零度（0K）的物体自身都会向外辐射能量，“玉兔二号”月球车的光谱仪将接收到月表传输过来的所有能量，包括月表反射的太阳光以及月表自身辐射的能量。月表温度越高，辐射的能量就越大。在月球赤道附近的当地中午最高温度可达 130℃，其自身产生的热辐射叠加在反射光谱之上会改变光谱形态，掩盖矿物的特征信号，特别是 2μm 之后波段的反射率光谱特征，进而影响对月壤物质组成的判断，因此，获取准确的月表温度是消除热辐射对月壤真实反射率影响的关键。

美国月球勘测轨道器上的 DIVINER 探测器可以获取月表的温度，但都是利用热红外波段（>7μm）进行测量的，以此估算的温度不能直接适用于 2～4μm 的区域。但光谱仪在 2～4μm 光谱区域接收到的反射和发射能量都占有很大比例，因此从光谱中准确估计温度是非常具有挑战性的。

“玉兔号”月球车获取的嫦娥三号着陆区光谱表明，在 2μm 之后的反射率抬升是由月表热辐射导致的，这意味着红外成像光谱仪探测到了从月球表面发出的热辐射。“玉兔二号”月球车搭载的同样配置的红外成像光谱仪测量了月表同一区域、同一月昼不同时刻的一系列光谱，为进一步研究月球反射率光谱中的热效应提供了参考，进一步证实了月表高温条件下反射光谱 2μm 之后的抬升特征。将不同时刻获取的光谱归一化到 1.55μm，发现它们分为了两组，一组是靠近当地中午时刻采集，受到热辐射的影响比较严重；另一组是靠近傍晚时采集，几乎不受热辐射的影响。利用不受热辐射影响的光谱建立 1.55μm 和 2.38μm 波段之间的关系，预测受热辐射影响光谱 2.38μm 处的真实反射率值，然后根据与测量值的差异，基于普朗克函数将温度计算出来。结果表明，在月球当地时间 14：28～14：41，着陆点表面温度为（346±8）K（图 3-4）。利用基尔霍夫定律对全波段光谱进

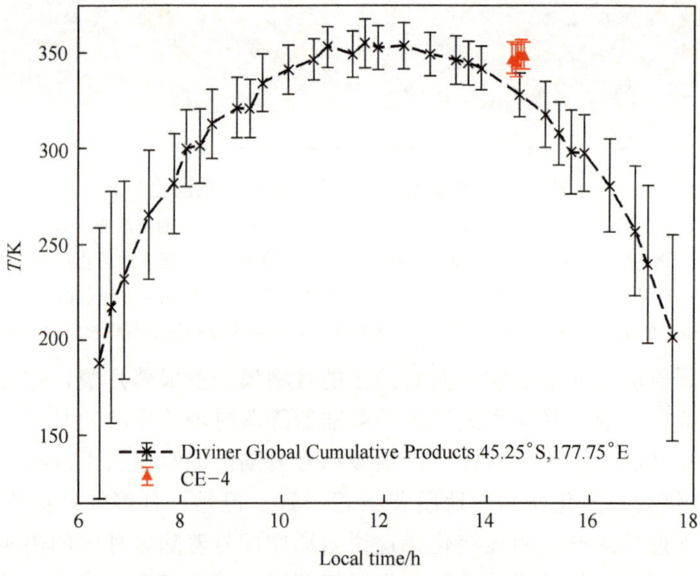

图 3-4 嫦娥四号着陆区月表温度

行热校正，校正前后对比发现，反射光谱中的热贡献会显著增大辉石和/或玻璃丰度的估算误差。此外，利用蒙特卡洛反演方法直接求解了月壤温度与发射率，进而对原始光谱进行了热校正，发现除了物质成分外，热辐射特性受微尺度粗糙度显著影响，与平面模型温差可达7～10K。

过去对月表温度特性的研究主要依靠宏观遥感数据以及热红外波段，嫦娥四号在微观尺度和短波方面为月壤温度研究提供了数据补充。

3.4　深部物质组成特性

南极艾特肯盆地是月球上最大、最深、最古老的撞击盆地，是月球下月壳甚至上月幔的天然窗口，具有月球表面天然的深度剖面，是认识月球深部物质组成的佳地。通过月球深部物质成分和结构的研究，可以获取月球月壳不同深度乃至撞击挖掘出的深层月幔的矿物岩石信息，这对于细致研究月球岩浆洋结晶分异过程和月幔反转机制等均具有重要价值，是解决月球岩浆洋假说和月球壳-幔分异机制重大科学争议问题的关键。嫦娥四号着陆区物质成分的探测对于理解南极艾特肯盆地基底乃至月球的深部物质组成非常关键。嫦娥四号在此区域进行光谱探测，可以获取到巡视区域的橄榄石、辉石、斜长石等矿物的相对含量以及矿物中钙、镁等元素的含量，为理解着陆巡视区物质来源与地质演化历史提供重要线索。

冯·卡门撞击坑位于南极艾特肯盆地的镁辉石环单元内，覆盖在着陆区最上面的一层是来自着陆点东北方向芬森撞击坑的溅射物，FeO 等化学成分分布图表明着陆区表面可能存在未混合均匀的月海（冯·卡门撞击坑底部玄武岩）与非月海（芬森等撞击坑溅射物）物质。

冯·卡门撞击坑底部的玄武岩主要为低钛玄武岩（1.5～3 wt.% TiO_2），富含单斜辉石，橄榄石含量较低，冯·卡门撞击坑底部的玄武岩单元受到芬森等撞击坑溅射物影响，具有明显的成分不均一性，且 FeO 含量（13～18 wt.%）相比月海玄武岩样品较低。在着陆点西部存在一个新鲜的撞击坑，被命名为织女撞击坑，织女撞击坑挖掘出了未受芬森撞击坑溅射物影响的冯·卡门本地玄武岩，这些玄武岩物质的成分特征随深度有所变化，表明冯·卡门撞击坑底部至少经历过三次熔岩流活动，形成了三期玄武岩，这三期玄武岩虽然在不同类型矿物的比例上具有一定的变化，但其中橄榄石的镁值［镁/（铁＋镁）］比较均一，表明三个期次火山活动的源区可能具有相似的成分，或源自一个熔体成分随时间演化的岩浆房。来自芬森等撞击坑的非月海物质具有不同的物质成分，芬森撞击坑中央峰具有富集高镁、低钙辉石的成分特征，与南极艾特肯盆地镁辉石环单元相似，而芬森撞击坑壁与其溅射物则表现为相对更高钙、富铁的辉石成分，与南极艾特肯成分异常区单元矿物学特征一致。

自 2019 年 1 月 3 日嫦娥四号成功着陆以来，“玉兔二号”月球车对巡视区域的月壤、石块、车辙等目标开展了高光谱探测，获得了大量的有效探测数据（图 3 - 5）。基于这些光谱数据，人们在巡视区月球深部物质成分特征与岩石类型等方面采用了多种模型和方法，进行了深入的研究。

1）通过光谱参数分析、光谱解混、光谱匹配等手段对嫦娥四号红外成像光谱仪数据

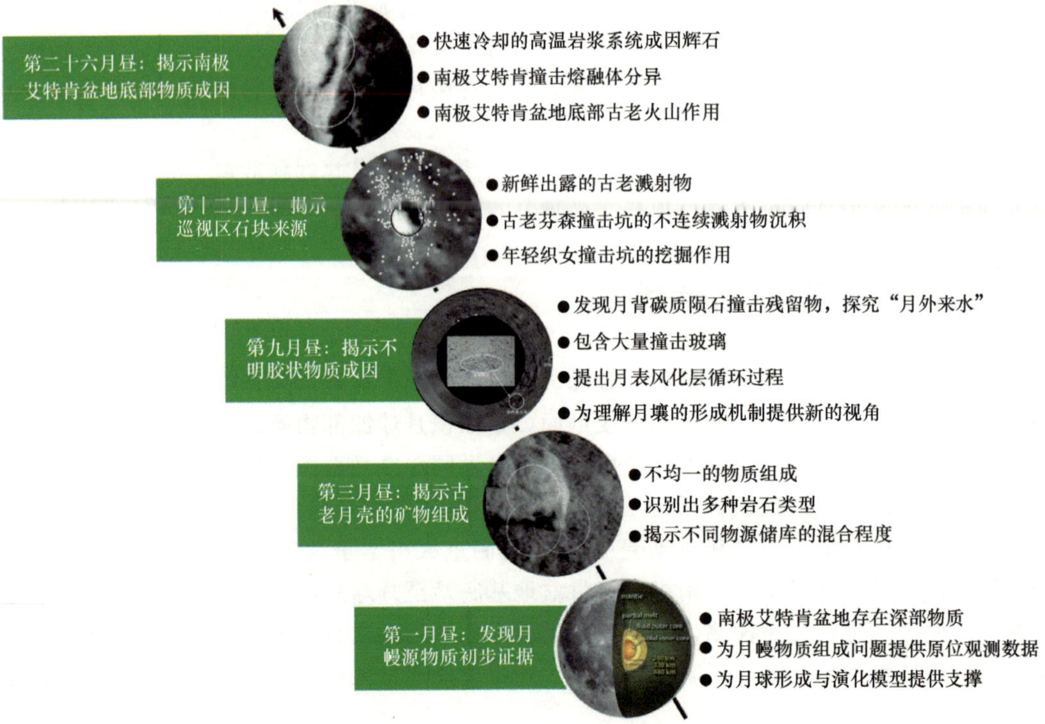

图 3-5　“玉兔二号”月球车在月面的物质组成探测

的分析结果表明，“玉兔二号”巡视路线上的不同探测目标可能具有不均一的物质组成，目前已识别出斜长岩、苏长岩/辉长岩等多种岩石类型，揭示了不同物源储库（冯·卡门玄武岩、南极艾特肯撞击熔融、月幔物质、壳源溅射物）的混合程度。

2）采用修正高斯模型对“玉兔二号”在第一月昼的前两个探测点光谱的镁铁质矿物解混，结果表明探测目标中橄榄石含量最高，低钙辉石次之，高钙辉石最少，两个探测点的橄榄石、低钙辉石和高钙辉石矿物的相对含量比例分别为 48∶42∶10 和 55∶38∶7。基于 Hapke 辐射传输模型建立的光谱查找表匹配结果发现，前两个月昼的探测目标中斜长石含量很高，大约占 60%，镁铁质矿物含量较低。基于阿波罗样品构建的经验模型表明前三个月昼的月壤目标普遍包含接近 60%的黏结集块岩，橄榄石等镁铁质矿物含量较少，这与着陆区经历了较长时间的太空风化作用的预期一致。根据吸收位置、波段面积比等光谱参数的组合分析发现，前三个月昼月壤的目标光谱的镁铁质矿物以橄榄石和单斜辉石特征为主，辉石成分介于高钙辉石与低钙辉石之间，这种辉石成分与岩浆的快速冷却结晶一致。

3）“玉兔二号”月球车在第三月昼首次对石块目标开展了光谱探测，石块光谱的修正高斯模型拟合结果表明其中的富镁橄榄石和富镁斜方辉石矿物含量基本相当，而基于 Hapke 辐射传输模型的分析结果发现石块约含有 48%斜长石、38%低钙辉石和 14%的橄榄石，对应含橄榄石的苏长质岩石。该石块具有中等细粒的结构，不大可能是深成岩，而与南极艾特肯盆地撞击形成的岩浆湖的快速结晶条件一致。石块光谱与 NASA RELAB 数据库中月岩、月球陨石光谱的匹配结果则表明该石块富集斜长石矿物（60%～80%），可能是来自织女撞击坑的溅射物。

4）采用 Hapke 模型与稀疏解混算法相结合，获得了前 21 个月昼探测目标的矿物含量与化学成分信息。矿物组成方面，以斜长石为主，辉石次之，橄榄石较少，对应苏长质/辉长质的矿物学特征，大部分（50/86）探测目标中低钙辉石含量超过高钙辉石；化学成分方面，前 21 个月昼的探测目标 FeO 含量变化较大（7.42～18.82 wt.%），TiO_2含量普遍较低（1.48～2.1 wt.%），而镁值中等偏低（约 55）。

5）通过红外成像光谱仪备份样机，对模拟月壤的测试结果进行分析，建立了估算月表实测目标中单斜辉石、斜方辉石、橄榄石等镁铁质矿物相对含量的方法，并应用于月面探测数据。结果表明，前 26 个月昼的探测目标中镁铁质矿物以单斜辉石为主，橄榄石含量有限。矿物成分方面，辉石具有中等的钙、铁含量，橄榄石具有中等镁值。4 个石块目标的光谱具有较强的 1.2μm 吸收强度。根据辉石光谱学的研究经验，1.2μm 吸收带的强度可以用于指示辉石形成时的岩浆结晶温度和冷却速率。分析结果表明这些石块中的辉石可能形成于快速冷却的高温（780～1300℃）岩浆系统，如南极艾特肯撞击熔融体的分异或盆地底部的火山作用。

综合目前对嫦娥四号着陆区原位光谱的分析结果，大部分研究认为该地区主要矿物为斜长石，辉石次之（低钙辉石多于高钙辉石），而橄榄石较少。基于原位光谱数据反演得到的嫦娥四号着陆巡视区的物质组成以及岩石结构特征，南极艾特肯盆地基底很可能不是简单挖掘暴露出的原始月球深部物质或撞击熔融湖中分异结晶等单一成因的物质，而是多个物源的混合产物，说明了南极艾特肯盆地成分和地质形成过程的复杂性。嫦娥四号获得的这些认识为理解月球的形成和演化提供了重要线索，也为后续的南极艾特肯盆地探测任务设计和实施提供了重要参考。

3.5　月壤中的撞击玻璃

月壤物质中含有大量的撞击玻璃，其中玻璃微珠是月球撞击玻璃的重要组成分布，形成于气化物凝结和熔融物冷凝。和月球火山玻璃微珠一样，月球撞击玻璃珠的直径大多远小于 1mm。撞击玻璃微珠是反演月壤物质成分、记录月表撞击历史的关键载体。使用玉兔二号导航相机获取的数据，在月球上发现了数粒厘米级直径的半透明的玻璃球，这是国际首次发现的手标本大小的、半透明的月球玻璃球。使用全景相机数据分析这些透明玻璃球的颜色、形态和产出特征，发现这些玻璃球不是月球火山活动或外来撞击体直接降落的产物，它们是成分特殊的撞击玻璃（图 3-6），对照主要月球岩石的熔体在冷却时发生结晶的临界降温速率，组成月球高地月壳的斜长岩具有很低的铁含量，其熔体的临界降温速率小于 5K/s，是形成厘米级直径的透明玻璃珠的最佳原岩，进而通过限定这些玻璃球沉积时的弹道速度，并使用月球表面的矿物和元素丰度数据，在相应的范围内寻找可能的母坑。结果发现，着陆点附近很多撞击事件可能形成了此类玻璃球，它们可能是月球高地月壤上常见的物质。这项发现对撞击机理研究具有重要的指示意义，它们是记录月球高地早期撞击历史的关键信息载体，例如，这些玻璃球的形成过程可能和地球上广泛分布的远撞击玻璃（俗称玻璃陨石）相似，将证明远撞击玻璃是类地天体上撞击事件的固有产物；月球高地上形成的古老撞击盆地可能也形成了大量的此类玻璃球，其形成年龄是反演内太阳系早期撞击历史的重要信息。

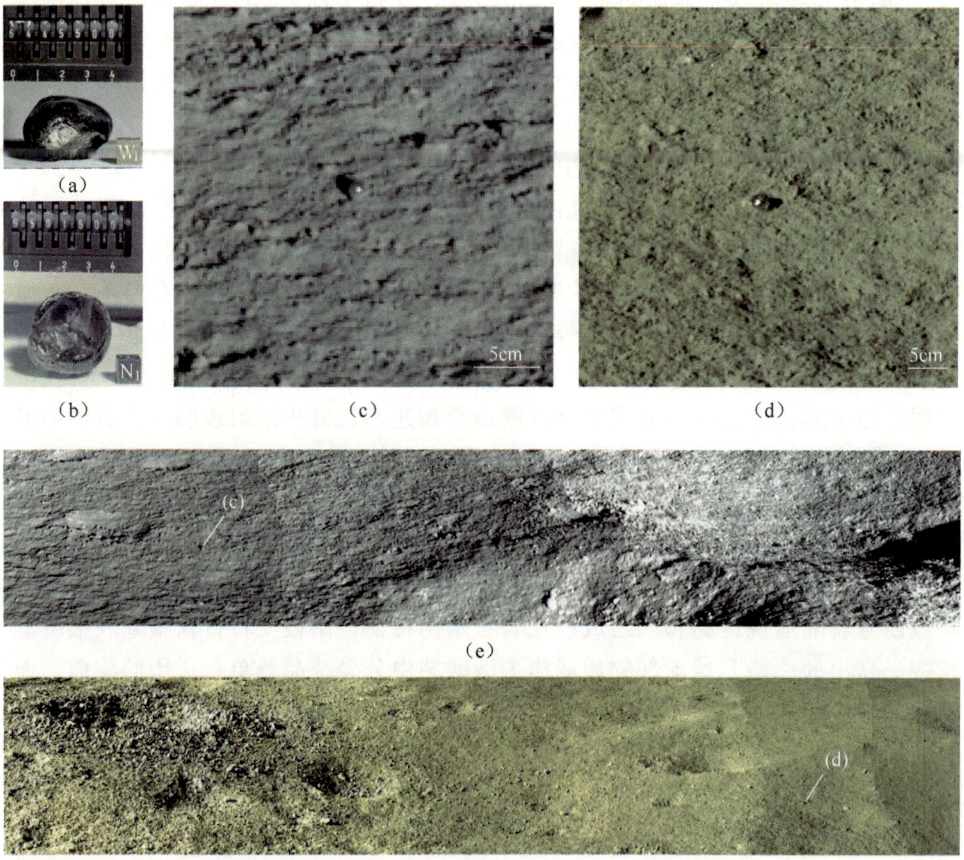

(a) (b) (c) (d) (e) (f)

图 3-6 “玉兔二号”沿途发现的 2 个厘米级粒径的透明玻璃珠

3.6 发现月面陨石碎片

撞击作用贯穿了月球乃至太阳系其他类地天体的形成与演化历史。在过去的四十几亿年里，小天体撞击产生的粉碎、熔融、黏合和混合等物理过程在很大程度上改变了月球表面。月球由于其特殊的空间环境使得部分撞击体碎片有可能保留在月表，可作为地-月系统撞击体成分演变的直接证据。寻找月壤中与撞击相关的物质，对理解月壤的形成和演化过程至关重要。然而由于持续不断的撞击翻耕作用，撞击产生的物质与月壤高度混合，因此很难在月表对它们的物质特性直接进行研究。

“玉兔二号”装备的全景相机和红外成像光谱仪可以获取超高分辨率（< 1mm/pixel）的月表影像与高光谱数据。月球车在巡视过程中发现月球表面有许多充满表土角砾的小撞击坑，与被细粒月壤覆盖的普通撞击坑明显不同，仅在前 12 个月昼里，在行驶路径上就观测到了超过 45 个随机分布的小撞击坑，它们直径从几十厘米至几米不等。在第 8～9 个月昼的探测过程中，月球车抵近了其中一个充满表土角砾的小坑，进行了详细

探测，发现了具有特殊光泽的神秘“胶状”物质（图 3-7）。对该“胶状”物质光谱进行了初步分析，结果表明其与编号为 70019 和 15546 的阿波罗撞击熔融角砾岩类似。详细的光谱特征分析表明该物质与坑内外的典型月壤、岩石碎块的光谱呈现明显不同的特征，包含大量的撞击玻璃，并据此提出了月壤形成中的循环过程，认为月壤是在反复撞击砸细又重新胶结成岩的过程中不断演化成熟，这为理解月壤的形成机制提供了新的视角。

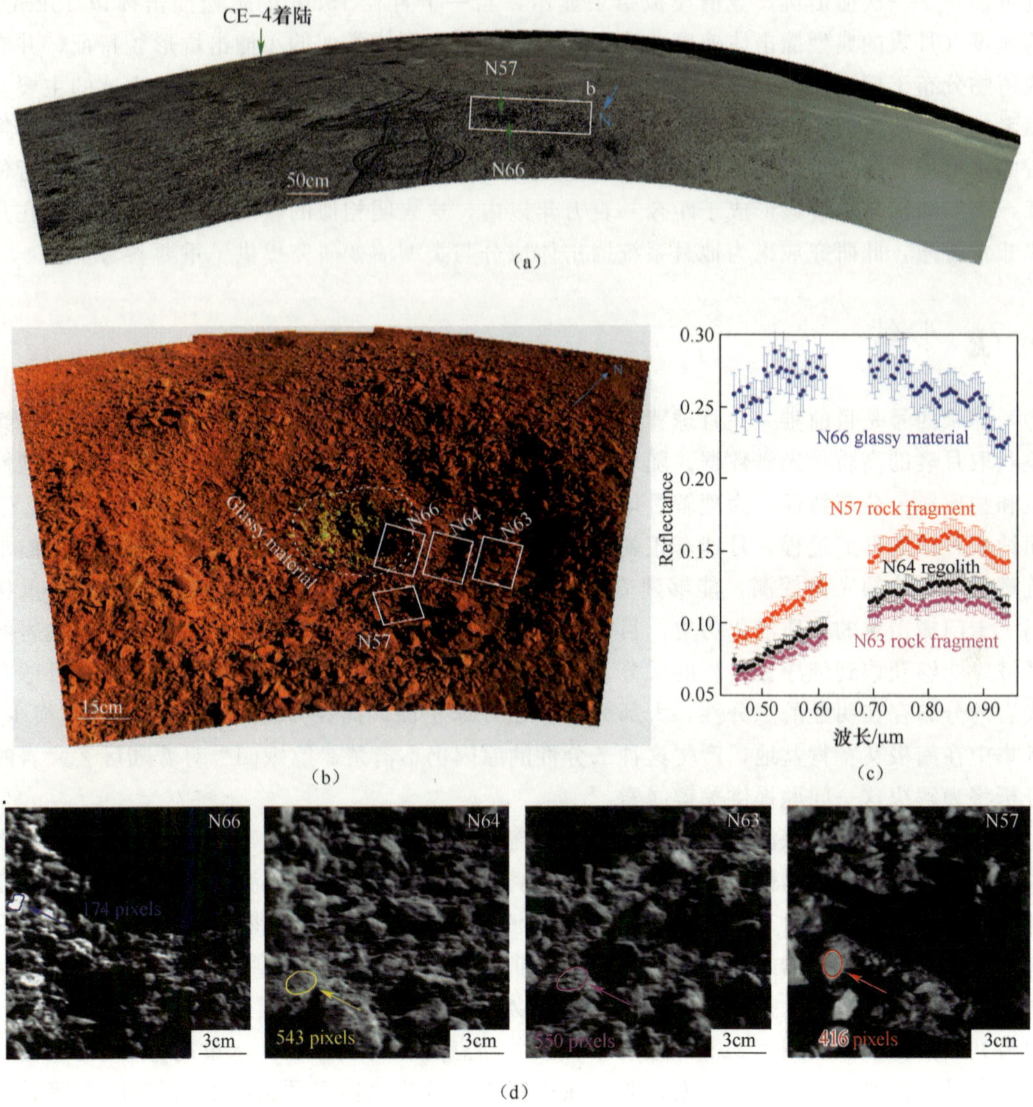

图 3-7　“玉兔二号”对巡视路径上“偶遇”的一个小型新鲜撞击坑进行详细的光谱探测

(a) 由全景相机影像拼接而成的全景图，可以看到撞击坑与着陆器的相对位置；(b) 全景相机近距离拍摄的小撞击坑全貌，该图为假彩色图。撞击坑中心呈现的黄绿色部分为反射率较高的神秘“胶状”物质。白框指示的是进行成像光谱观测的区域；(c) 从成像光谱观测区域中提取的高反射率物质、月壤、岩石碎块的反射光谱；(d) 成像光谱仪获取的观测区域影像

在“胶状”物质光谱的形态表面中还可能存在外来碳质小行星撞击体物质的混入。通

过与大量碳质球粒陨石光谱的对比发现，“胶状”物质与碳质陨石光谱特征具有很高的相似度。采用辐射传输模型的光谱定量反演结果显示，该“胶状”物质中碳质陨石组分占比达到 40 wt. %以上。为了进一步确认上述分析结果，基于全景相机获取的立体像对，利用摄影测量方法构建的高精度数字高程模型对该小撞击坑进行了详尽的形貌分析，表明该小撞击坑的深度与直径比值是着陆区二次撞击坑的深度与直径比的 1.7～3.4 倍，该撞击坑可能属于一次撞击坑。数值模拟结果显示，当一个直径 15cm 的疏松撞击体以 15km/s 的速度（月表的典型撞击体速度）撞击月表时，可以形成类似的小撞击坑形貌特征，并有残留物分布于撞击坑中心。撞击输运过程被认为是月球表面水及永久阴影区水冰的主要贡献者之一，而碳质小行星是小天体中相对比较富含水及挥发分的一类，在撞击过程中其携带的水可能有部分得以保留在月表。基于撞击坑退化模型对该小撞击坑的形成年龄进行估算，表明该撞击坑应该形成于距今一百万年以内，这表明相似的碳质陨石残留物可能在月表非常普遍。此研究成果为地月系统撞击体成分与类型演变研究提供了重要参考。

3.7　小结

嫦娥四号是目前唯一在月球背面运行的人类探测器，“玉兔二号”月球车仍然在持续地获取月表的高质量光谱数据。随着巡视区域的扩大，它能够提供更大范围的月壤物理性质和物质成分分布特征，为理解月球南极艾特肯盆地的形成和演化提供更丰富的信息，进而约束月球的形成过程。月球车正朝着陆点的西北方向行进，通过对经过的不同深度撞击坑溅射物石块的光谱探测，能够建立物质成分剖面，结合测月雷达等载荷的结果，重构冯·卡门撞击坑的地质演化历史。目前规划的月球车行驶路径将会经过一片裸露玄武岩的区域，能够获取到纯净玄武岩的成分，将完成首次对月球背面玄武岩的成分探测。月球玄武岩的分布存在明显的二分性，大部分分布在月球正面，而在月球背面玄武岩分布很少，且集中在南极艾特肯盆地，产生这种二分性的原因仍不清楚，嫦娥四号对着陆区玄武岩的分析将为解决这一问题提供关键线索。

未来的月球光谱探测正向宽光谱范围、高空间分辨率方向发展，研究更小尺度上的物质成分特征及与形貌的耦合关系，理解区域演化历史。嫦娥五号探测器成功地从风暴洋北部采集回 1.731kg 的月球样品。嫦娥六号探测器在不久的未来也将从月球极区或背面采集样品并返回地球。嫦娥五号和嫦娥六号着陆器上都搭载了月球矿物光谱仪，在采样返回之前对月表物质组成进行原位探测，月球矿物光谱仪主要继承了嫦娥四号光谱仪的技术，但将其光谱范围扩展到 3.2μm，这对于检测着陆点的矿物成分和含水量更为有效。但是，如果在月表高温条件下采集数据，那么在进行光谱分析之前必须去除 2μm 波段之后的热影响，嫦娥四号在热辐射研究方面的成果为嫦娥五/六号的光谱分析提供了重要参考。嫦娥七号轨道器上将搭载一个宽谱段成像光谱仪，光谱覆盖 0.4～10.5μm，热红外波段的加入不仅能够更准确地反演月表物质组成，而且提供了研究月表热特征的机会，基于嫦娥四号建立起的数据方法可为嫦娥七号提供参考，“玉兔二号”月球车也可与嫦娥七号进行联动，为轨道遥感的光谱探测提供标定。

在嫦娥四号着陆区发现的撞击体残留物距今只有一百万年左右，而从月表一个年龄约 20 亿年的玄武岩单元里采集返回的嫦娥五号样品中，将有很大概率发现类似的撞击体残留物，结合主微量元素与同位素年代学的月壤分析，将可以对撞击体成分与通量演变进行更好的限定。将来利用更高空间分辨率的遥测光谱数据，将有可能在月表更多地方发现类似的撞击体残留物分布，从而进一步加深对月球水来源与分布的认识。

采集嫦娥四号着陆区中新鲜小撞击坑内部的疑似熔融物以及月壤集块的样品，将是验证基于雷达数据反演的月壤孔隙闭合效应的最佳方法；钻取嫦娥四号浅表层的岩芯柱样品，将为厘定雷达数据揭示的浅表层结构及其物源提供更确切的约束；返回月球表面手标本大小的透明玻璃珠，将是验证天体表面远撞击溅射玻璃（tektites）的成因机理和月球撞击历史的绝佳载体；月球背面月海玄武岩的矿物成分，以及和正面月海玄武岩的异同，还有月球背面可能暴露在表面的月幔物质的矿物组合等也将是探测和研究的主要内容。因此，结合国内、外深空探测领域的技术和科学发展趋势，综合有人和无人的遥感观测和样品返回任务，将更高效地推动月球科学研究的进步。

主要成果附录

[1] Chen J, Ling Z, Jolliff B, et al. Radiative Transfer Modeling of Chang'e-4 Spectroscopic Observations and Interpretation of the South Pole-Aitken Compositional Anomaly [J]. The Astrophysical Journal Letters, 2022, 931, L24.

[2] CHEN J, LING Z C, QIAO L, et al. Mineralogy of Chang'e-4 landing site: preliminary results of visible and near-infrared imaging spectrometer [J]. Science China (Information Sciences), 2020, 63 (4): 195-206.

[3] Gou S, Di K, Yue Z, et al. Forsteritic olivine and magnesium-rich orthopyroxene materials measured by Chang'e-4 rover [J]. Icarus, 2020b, 345: 113776.

[4] Gou S, Di K, Yue Z, et al. Lunar deep materials observed by Chang'e-4 rover [J]. Earth and Planetary Science Letters, 2019, 528: 115829.

[5] Gou S, Yue Z, Di K, et al. Impact melt breccia and surrounding regolith measured by Chang'e-4 rover [J]. Earth and Planetary Science Letters, 2020c, 544: 116378.

[6] Gou S, Yue Z Y, Di K C, et al. Mare basalt flooding events surrounding Chang'e-4 landing site as revealed by Zhinyu crater ejecta [J]. Icarus, 2021a, 360.

[7] Gou S, Yue Z, Di K, et al. Absolute model age of lunar Finsen crater and geologic implications [J]. Icarus, 2021b, 354: 114046.

[8] Gou S, Yue Z, Di K, et al. In situ spectral measurements of space weathering by Chang'e-4 rover [J]. Earth and Planetary Science Letters, 2020a, 535: 116117.

[9] Hu X, Ma P, Yang Y, et al. Mineral Abundances Inferred From In Situ Reflectance Measurements of Chang'E-4 Landing Site In South Pole-Aitken Basin [J]. Geophysical Research Letters, 2019, 46, 9439-9447.

[10] Huang J, Xiao Z, Xiao L, et al. Diverse rock types detected in the lunar South Pole-Aitken Basin by the Chang'E-4 lunar mission [J]. Geology, 2020, 48: 723-727.

[11] Jiang T, Hu X, Zhang H, et al. In situ lunar phase curves measured by Chang'E-4 in the Von Kármán Crater, South Pole-Aitken basin [J]. Astronomy & Astrophysics, 2021, 646, A2.

[12] Li C L, Liu D W, Liu B, et al. Chang'E - 4 initial spectroscopic identification of lunar far - side mantle - derived materials [J]. Nature, 2019, 569 (7756) .

[13] Lin H, He Z, Yang W, et al. Olivine - norite rock detected by the lunar rover Yutu - 2 likely crystallized from the SPA - impact melt pool [J]. National Science Review, 2020c, 7: 913 - 920.

[14] Lin H, Li S, Lin Y, et al. Thermal Modeling of the Lunar Regolith at the Chang'E - 4 Landing Site [J]. Geophysical Research Letters, 2021, 48 (6), e2020GL091687.

[15] Lin H, Lin Y, Yang W, et al. New Insight Into Lunar Regolith - Forming Processes by the lunar Rover Yutu - 2 [J]. Geophysical Research Letters, 2020d, 47, e2020GL087949. https://doi.org/10.1029/2020GL087949.

[16] Lin H, Xu R, Yang W, et al. In Situ Photometric Experiment of Lunar Regolith With Visible and Near - Infrared Imaging Spectrometer On Board the Yutu - 2 Lunar Rover [J]. Journal of Geophysical Research: Planets, 2020b, 125 (2), e2019JE006076.

[17] Lin H, Yang Y, Lin Y, et al. Photometric properties of lunar regolith revealed by the Yutu - 2 rover [J]. Astronomy & Astrophysics, 2020a, 638, A35.

[18] Lin H L, Lin Y T, Wei Y, et al. Estimation of Noise in the In Situ Hyperspectral Data Acquired by Chang'E - 4 and Its Effects on Spectral Analysis of Regolith [J]. Remote Sensing, 2020, 12 (10) .

[19] Ling Z C, Qiao L, Liu C Q, et al. Composition, mineralogy and chronology of mare basalts and non - mare materials in Von Kármán crater: Landing site of the Chang'E - 4 mission [J]. Planetary and Space Science, 2019, 179.

[20] Ma P, Sun Y, Zhu M H, et al. A plagioclase - rich rock measured by Yutu - 2 Rover in Von Kármán crater on the far side of the Moon [J]. Icarus, 2020, 350: 113901.

[21] Wu Y, Kührt E, Grott M, et al. Chang'E - 4 Rover Spectra Revealing Micro - scale Surface Thermophysical Properities of the Moon [J]. Geophysical Research Letters, 2021, 48 (4), e2020GL089226.

[22] Xiao Z, Yan P, Wu B, et al. Translucent glass globules on the Moon [J]. Science Bulletin, 2021. doi: https://doi.org/10.1016/j.scib.2021.11.004.

[23] Yang Y, Li S, Zhu M H, et al. Impact remnants rich in carbonaceous chondrites detected on the Moon by the Chang'e - 4 rover [J]. Nat Astron 6, 207—213 (2022). https://doi.org/10.1038/s41550 - 021 - 01530 - w.

[24] Yang Y, Lin H, Liu Y, et al. The Effects of Viewing Geometry on the Spectral Analysis of Lunar Regolith as Inferred by *in situ* Spectrophotometric Measurements of Chang'E - 4 [J]. Geophysical Research Letters, 2020, 47 (8): e2020GL087080.

[25] Zeng Q, Chen S, Zhang Y, et al. Mineralogical and chemical properties inversed from 21 - lunar - day VNIS observations taken during the Chang'E - 4 mission [J].

Sci Rep 11, 15435 (2021).

[26] Zhou H, Feng X, Ding C, et al. Heterogeneous Weathering Process of Lunar Regolith Revealed by Polarimetric Attributes Analysis of Chang'E - 4 Lunar Penetrating Radar Data Acquired During the Yutu - 2 Turnings [J]. Geophysical Research Letters, 2022, 49, e2022GL099207. https://doi.org/10.1029/2022GL099207.

[27] 王振超，柳稼航，盛庆红，等．嫦娥四号着陆点原位高光谱的太空风化研究 [J]. 红外与激光工程，2020，49 (5)：108 - 112.

[28] 阳梅萍，岳宗玉，邸凯昌，等．基于全景相机数据的嫦娥四号着陆区次级坑统计分析 [J]. 矿物岩石地球化学通报，2021，40 (3)：720 - 729.

Radiative Transfer Modeling of Chang'e – 4 Spectroscopic Observations and Interpretation of the South Pole – Aitken Compositional Anomaly

Jian Chen[1], Zongcheng Ling[1], Bradley L. Jolliff[2], Lingzhi Sun[3], Le Qiao[1], Jianzhong Liu[4], Xiaohui Fu[1], Jiang Zhang[1], Bo Li[1], Changqing Liu[1], Xiaobin Qi[1], Xuejin Lu[1], Zhiping He[5], and Rui Xu[5]

1. Shandong Key Laboratory of Optical Astronomy and Solar – Terrestrial Environment, Institute of Space Sciences, Shandong University, Weihai, 264209, People's Republic of China; zcling@sdu. edu. cn
2. Department of Earth & Planetary Sciences and the McDonnell Center for the Space Sciences, Washington University, St . Louis, MO 63130, USA
3. Hawai 'i Institute of Geophysics and Planetology, Department of Earth Sciences, University of Hawai 'i at Mānoa, Honolulu, HI 196826, USA
4. Institute of Geochemistry, Chinese Academy of Sciences, Guiyang, 550002, People's Republic of China
5. Key Laboratory of Space Active Opto – Electronics Technology, Shanghai Institute of Technical Physics, Chinese Academy of Sciences, Shanghai, 200083, People's Republic of China

Abstract: Chang'e – 4 landed in the Von Kármán basaltic zone affected by Finsen ejecta from the South Pole – Aitken (SPA) "compositional anomaly" (SPACA), providing an opportunity to untangle the origin of the SPACA. We developed a mineral abundance unmixing model and built a spectral library composed of mineral mixtures of plagioclase, low – Ca pyroxene, high – Ca pyroxene, and olivine. Based on the mineral abundances returned by our model, the average plagioclase abundance (60. 4 vol. %) of SPACA is beyond the high – plagioclase end of the mineralogical range of SPA impact melt differentiation products and would require an additional contribution from crustal materials. We suggest that either the crust was not entirely removed in the SPA – forming impact or the ejecta from post – SPA basins fed feldspathic materials back onto the SPA basin floor. Our unmixing analyses indicate that the surface regolith of the SPACA is a mixture of 70. 2% crustal materials and 29. 8% Fe – and incompatible – element enriched materials.

1 Introduction

The South Pole - Aitken (SPA) basin, located on the Moon's farside southern hemisphere, is the largest confirmed impact structure and one of the most diversified geochemical provinces on the Moon (Spudis et al. 1994). The SPA basin holds key to unlock the secrets of lunar evolution with its unique characteristics: a spatial location around the antipode of the nearside Procellarum KREEP terrane, as a potential cause of lunar global asymmetry, its vast size as a boundary condition for modeling and simulation of lunar impact processes, its antiquity as an ideal anchor for the early history of the Moon, and unusual materials as possible outcrops of lunar deep strata (Jolliff et al. 2006). These characteristics make the SPA basin an excellent target for investigating the evolution of the Moon. Among its various characteristics, the nonmare mafic anomaly associated with the interior of the SPA basin has attracted wide attention from the lunar scientific community since it was fully mapped by the Clementine mission (Spudis et al. 1994).

The formation of the SPA basin is thought to have penetrated into the lunar interior and exposed deep (lower crust and/or upper mantle) materials (Potter et al. 2012). This interpretation is supported by elevated concentrations of mafic components and incompatible elements such as thorium inside the basin compared with adjacent feldspathic highlands (Figure A1). Various orbital spectroscopic data sets indicate that the mafic mineralogy of the SPA basin floor is dominated by pyroxene (e.g., Pieters et al. 1997), and few olivine - rich exposures were observed within the SPA basin (e.g., Yamamoto et al. 2010). Heterogeneous pyroxene compositions were subsequently observed within the SPA basin using hyperspectral remote - sensing data sets (Moriarty 2016). Based on the pyroxene compositions, the interior of the SPA basin was further divided into several subunits, one of which exhibits unusual pyroxene compositions and was defined as the SPA compositional anomaly (SPACA). Different scenarios were proposed to explain the characteristics of the SPACA region; however, the origin of the unusual mineralogy remains uncertain with limited knowledge of the actual pyroxene composition.

In addition to the possible direct exposure of materials from the deep interior, the unusual materials within the basin have also been considered as products of the SPA impact melt sheet. The energetic SPA impact event is expected to have produced a large volume of melt (Melosh et al. 2017). The pervasive noritic lithology in the SPA basin floor has been suggested to be either an undifferentiated impact melt sheet, reflecting the bulk composition of the lower crust (e.g., Pieters et al. 1997) or the top layer of a differentiated melt beginning from an ultramafic composition representing the upper mantle (Nakamura et al. 2009). Numerous contributions have documente observations and simulations to constrain models of SPA impact and melting processes; however, the composition of the melt and source materials is still under debate with regard to the SPACA.

The Chang'e - 4 (CE - 4) landing region was initially selected using technical constraints, i.e., a safe location (Jia et al. 2018) on the lunar farside. Hence, the smooth mare plains in the Von Kármán crater were chosen with the scientific objective to explore the farside volcanism that occurred in the SPA basin, whereas subsequent geological and dspectroscopic characterizations of the CE - 4 mission target tregion from orbit trevealed that the basaltic deposits in the Von Kármán crater floor have been mostly obscured by nonmare crater ejecta (Qiao et al. 2019). Given that the unusual pyroxene compositions of the SPACA were also observed on the wall of the Finsen crater (Moriarty 2016), the spectral measurements of materials ejected from the Finsen crater wall (FW) by the CE - 4 visible and nearinfrared imaging spectrometer (VNIS) with an unprecedented spatial resolution are expected to help explore in detail the mineralogy in the SPACA.

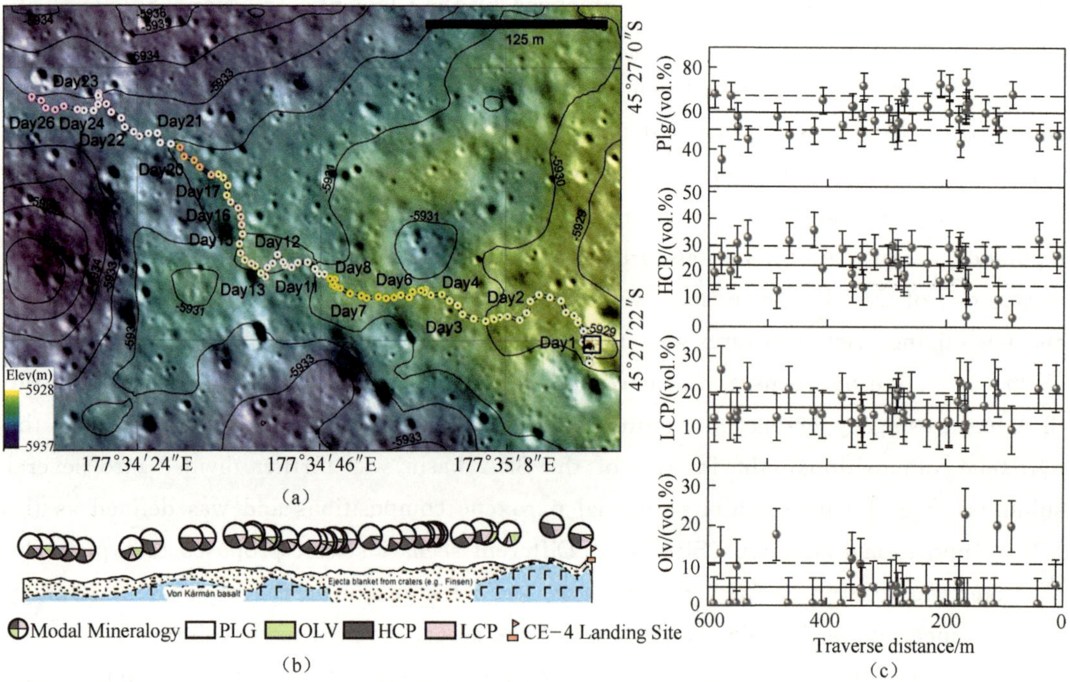

Figure 1　Yutu - 2 traverse map and mineralogical variations along the path. (a) Traverse path of the Yutu - 2 rover and locations of the exploration sites overlaid on the image taken by LROC NAC (frame M1303619844) and the color - coded digital terrain mode (DTM) derived from stereo pairs of NAC images (https://wms.lroc.asu.edu/lroc/view _ rdr/NAC _ DTM _ CHANGE4). Topographic contours are shown as black solid lines with the elevation values labeled. (b) Shallow subsurface stratigraphy along the traverse path. The thicknesses of different deposit layers reflect the mixing ratios. The vertical scale is zoomed 10×. (c) Estimated mineralogy along the traverse path. The solid and ashed lines indicate the mean values and standard deviations of mineral proportions. (The data used to create this figure are available.)

Since the successful landing of CE - 4 and the deployment of the Yutu - 2 rover in 2019 January (Di et al. 2019), a number of morphological and mineralogical results have been reported (Gou et al . 2019; Hu et al. 2019; Li et al. 2019; Huang et al. 2020; Chen et al. 2020; Lin et al. 2020; Ma et al. 2020) . There is consensus that the surface regolith at the CE - 4 exploration region is affected by Finsen crater ejecta; however, the mineralogy and petrogenetic origin of these materials remains controversial.

To constrain the regolith mineralogy at the CE - 4 landing region and understand the processes that generated the unusual materials in the SPACA, we carried out comprehensive mineralogical analyses utilizing data collected by the VNIS in the first 26 lunar days off the CE - 4 mission (Figure 1) . The modal mineralogy and compositions of major silicate minerals (i. e. , plagioclase (Plg), low - Ca pyroxene (LCP), high - Ca pyroxene (HCP), and olivine (Olv)) in the regolith were determined using a spectral look - up table (LUT) constructed using the Hapke radiative transfer model (RTM) . Unmixing analysis was further conducted based on the mineralogical and compositional results, aiming to evaluate the contributions of different potential reservoirs to the observed mineralogy in the regolith at the CE - 4 landing region as well as in the SPACA.

2 Data and Methods

The Yutu - 2 rover traversed ~628m in the first 26 lunar days (Figure 1) . The VNIS aboard the Yutu - 2 rover collected 120 hyperspectral visible and near - infrared (VNIR) images (450~950nm) and single - pixel short - wave infrared (SWIR) spectra (900~2400nm) for 54 soil - dominated targets and four boulders. The complete figure set (120 images of Yutu - 2 VNIS targets) is available. The VNIS raw data were processed to level 2B radiance through a series of processing pipelines including dark - current, scattering - background subtractions, flat - field, instrument temperature corrections, and radiometric and geometric calibrations (Liu et al. 2014) . The data were then distributed by the Ground Research and Application System (GRAS) of the Chinese Lunar Exploration Program. For the convenience of readers, VNIS images and spectra are labeled by the sequence numbers as shown in the filenames released by GRAS (e. g. , 0001 for CE4 _ GRAS _ VNIS* N _ yyyymmddhhmiss _ YYYYMMDDMISS _ 0001* . 2B data) rather than the engineering labels used by other publications. The level 2B VNIS radiance data were further processed (e. g. , solar distance correction) to derive a credible reflectance factor (REFF) for quantitative mineralogical analyses. A detailed description of processing procedures can be found in the appendices and our previous publication (Chen et al. 2020) .

2.1 Correction of Instrumental Artifacts

An artifact absorption around 820nm (Figure A2) is noticed in VNIS spectra and is probably caused by the F46 films in the multilayer insulation (Xu et al. 2021) on the body and/or aluminum metal in the wheels of the Yutu - 2 rover. This artifact is more obvious in VNIS spectra of darker pixels where the sunlight is shielded by local topography while the reflected and/or scattered sunlight with 820nm absorption from the rover serves as a dominant light source. Besides introducing an additional absorption, the artifact is expected to affect the shortwavelength shoulder position of 1μm absorption in VNIS spectra (Figure A2). Furthermore, spectra with such an artifact used for matching with the spectral library would tend to generate a modal mineralogy with more olivine because olivine has a broader absorption than pyroxene (Figure A2). Thus, in order to identify and remove this artifact, the integrated band depth (IBD) of the 820nm feature (IBD820) was calculated by fifitting a linear continuum over the feature and then integrating the band depths relative to the continuum at each wavelength within the regions (Table A1). The pixels with IBD820 greater than 0.01 were referred to as affected by this artifact.

In addition, signifificant darkening effects on the reflectance owing to shadows in the VNIR image caused by surface roughness were observed and relevant correction is also necessary. The pixels with REFF at 465nm less than 0.02 were regarded as shadows. Both bright pixels with 820nm artifacts and dark/shadow pixels are removed when calculating the average VNIR reflectance in the FOV (field of view) of SWIR detectors.

2.2 Photometric Correction

As the CE - 4 VNIS measurements were performed at large phase angles (~40°-117°), photometric correction is important for comparisons of reflectance spectra taken under different illumination conditions and further quantitative analysis of mineral abundances for local regolith. The standard deviations of mineral abundances of the same target derived from raw spectra (without photometric correction) and photometrically corrected spectra would be as large as 12.3 vol.%, 9.2 vol.%, 4.8 vol.%, and 7.6 vol.% for Plg, HCP, LCP, and Olv, respectively. Photometric models for local regolith at the CE - 4 landing region have been obtained from VNIS data collected on the 10th lunar day when spectral measurements of the same target at various illumination geometry were done by the Yutu - 2 rover. The VNIS spectra of local soils at the CE - 4 landing region were corrected to REFF at standard illumination geometry ($i = 30°$, $e = 0°$, and $g = 30°$) using the Lommel - Seeliger (LS) model (Qi et al. 2020). Although the LS model is a highly simplified solution of the Hapke RTM with the assumption of isotropic single scattering without multiple scattering, this empirical model somehow works better for photometric correction of CE - 4 VNIS data sets than Hapke's RTM. After the photometric correction, the spectral discrepancy of the same target collected at different viewing geometries was suppressed to ~3.7% with the LS model and ~7.4% with the Hapke model (Qi et al. 2020). The residual errors of mineralogy (standard deviations of the same target) derived from LS - model photometrically corrected spectra are 1. 1 vol. %, 0. 6 vol. %, 0. 5 vol. %, and 2. 0 vol. % for Plg, HCP, LCP, and Olv, respectively.

2. 3 Modal Mineralogy and Mafic Mineral Compositions

The LUT of modeled REFF spectra using Hapke's RTM was constructed to compare with the observed REFF spectra by CE - 4 VNIS. Abundances of mineral end - members composing the CE - 4 regolith were derived via minimizing the rms errors (RMSE) and spectral angle (SA) between observed VNIS spectra and modeled spectral LUT. The detailed description of Hapke's RTM used in this work can be found in AppendixA. 1. The modeled spectral LUT contains REFF spectra of 323, 449, 000 mixtures of Olv (Fo_{75} ~ Fo_{64}), LCP ($Fs_{27}Wo_{1}En_{72}$ ~ $Fs_{31}Wo_{20}En_{49}$), HCP ($Fs_{16}Wo_{40}En_{44}$ ~ $Fs_{31}Wo_{20}En_{49}$), and Plg ($An_{95}Ab_{5}$) . The uncertainties of this model in estimating mineral abundance are better than 4 vol. % for LCP and HCP, better than 5 vol. % for Olv, and better than 6 vol. % for Plg. In this work, the model is validated using 19 Apollo soil samples (Figure 2) .

The agglutinitic glass is the dominant (30 ~ 70 vol. %; Taylor et al. 2001) component in lunar soils, while the absorption characteristics in VNIS spectra of soils at the CE - 4 landing region resemble crystalline phases. Mafic glass and mafic minerals exhibit similar 1μm absorption bands, the position and strength of which depend on the composition and cooling rate. If impact melts cooled slowly enough for recrystallization to occur, their spectra would resemble those of mafic minerals with similar compositions (Tompkins & Pieters 2010) . Hence, it is expected that a proportion of the mafic mineral abundances reported in CE - 4 regolith is probably recrystallized from impact melts. Given the overlapping 1μm absorption features of agglutinitic glass and olivine, the incorporation or exclusion of glass may have an impact on the LUT model - retrieved results of olivine abundance. Compared to the spectral LUT without glass, the spectral LUT with glass improved the abundance retrieval accuracy of olivine (4. 2 vol. % → 1. 9 vol. %) for LSCC (Lunar Soil Characterization Consortium) samples while reducing the accuracies of LCP (3. 8 vol. % → 4. 7 vol. %), HCP (3. 4 vol. % → 4. 1 vol. %), and Plg (5. 4 vol. %→8. 0 vol. %) abundances as well as the total performance of the model (RMSE 6. 5 vol. % → 8. 4 vol. %) . Thus our spectral LUT only includes abundances of Plg, LCP, HCP, and Olv. The space weathering effects were modeled with npFe and SMFe abundances.

3 Observations

3. 1 Spectral Properties

The CE - 4 VNIS spectra of soils are similar to the spectra of the CE - 4 landing region, remotely observed by orbital spectrometers (Figure 3). Both orbital and ground spectra of the CE-4 exploration region are composed of reddish continuum and weak 1μm and 2μm absorption bands, suggesting that the soils are heavily space-weathered products from mafific-mineral-bearing lithologies. Among the detection targets, the spectra of boulders exhibit relatively higher reflectance and stronger 1μm and 2μm absorption band strengths, which are consistent with their immature nature.

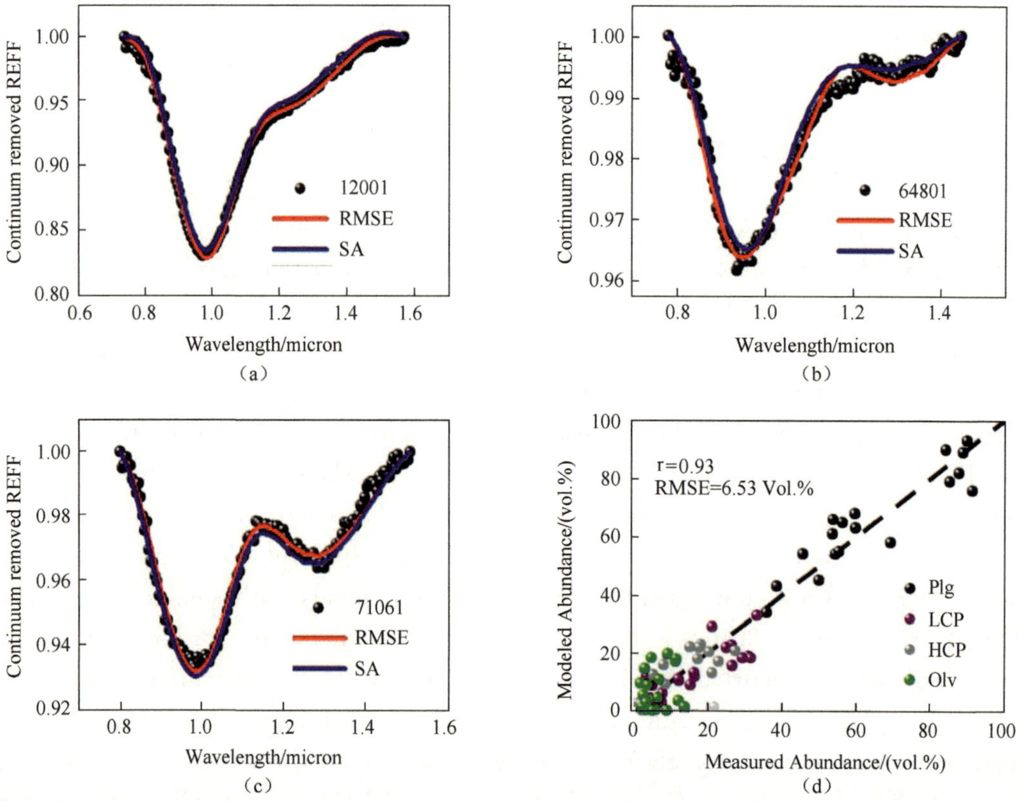

Figure 2　Comparison between the modeled and measured data. (a) — (c) LSCC spectra of samples 12001, 64804, and 71061 and their best-matched modeled spectra. (d) Mineral abundances modeled with our LUT vs. measured modal mineralogy of LSCC samples. (The data used to create this figure are available.)

Multiple measurements on different parts of boulders (Figure A3) that demonstrated spectral variations with maturity were conducted. The more mature parts of the three boulders exhibit lower reflectance and weaker mafic absorption bands than their immature parts (Figure 3). These trends are consistent with the common observations of lunar space weathering: As the boulder is weathered to form regolith, it becomes darker and the mafic absorption strength decreases (Pieters et al. 2000). However, a difference is noted wherein the spectral slope of boulders decreases with increasing maturity. In other words, the more mature regolith tends to be bluer than less mature boulders, contrasting with the reddening effects of nanophase metallic Fe (npFe, tens of nanometers) residing in rims of mineral grains in lunar - style space weathering (e. g. , Pieters et al. 2000) . An alternative explanation to the observation of bluer spectra for more mature soils is the effects of larger - sized (>50nm) submicroscopic metallic Fe (SMFe) in the interior of agglutinates that mainly darken but do not redden the host materials (Lucey & Riner 2011) . Combined with reddish spectra of boulders, a two - stage space - weathering model could be inferred: (1) During space weathering of freshly exposed rocks, the decrease in the filling factor (increase of the porosity) decreased the reflectance (Hapke 2008) and the npFe formation further darkened the rock and reddened the spectral slope. (2) As the metallic Fe aggregates, larger - sized SMFe particles began to dominate the space weathering effects that generated dark and flat spectra for highly mature soils. The average abundances of npFe and SMFe in CE - 4 soils were estimated to be, respectively, 0. 22 wt. % and 0. 23 wt. %, which are comparable to some Apollo and Luna soils (e. g. , 15601; Morris 1980) . The less mature parts of the boulders exhibit higher npFe abundances (0. 14 wt. %) and lower SMFe abundances than their more mature parts (0. 11 wt. % npFe and 0. 03 wt. % SMFe), which is consistent with the redder and brighter spectra.

3. 2　Modal Mineralogy

The CE - 4 VNIS spectra and LUT - derived modal mineralogy of soils exhibit extensive mixing between the continuous ejecta of Finsen (hypothesized SPACA component excavated from FW) and local Von Kármán basalts (VKB, Figures 1, A4, A5, and A6), which was interpreted by Huang et al. (2020) as lithologic diversity. The average modal mineralogy of soils in the CE - 4 landing region is 57. 0 vol. % Plg, 15. 8 vol. % LCP, 22. 7 vol. % HCP, and 4. 4 vol. % Olv, which are consistent with some previous researchers working on CE - 4 VNIS data (56%～72% Plg, 9%～28% LCP, 4%～19% HCP, and 2%～12% Olv, Hu et al. 2019) within the model uncertainties.

3. 3　Mafic Phases

The mafic mineralogy in the CE - 4 regolith returned from LUT spectral matching further supports our preliminary conclusion based on spectral and mineralogical indices (Chen et al. 2020) . The mafic assemblages in soils are dominated by pyroxenes (38. 5 vol. %, 8～10 times the olivine proportion (4. 4 vol. %)) . The relative proportions of LCP and HCP in soils are comparable to each other. It is also supported by the intermediate band centers (970. 0 ± 5. 2 nm) between LCP and HCP. Pyroxenes in soil targets exhibit extensive compositional distribution, and the most abundant pyroxene compositions are $Fs_{18.2}$ $Wo_{48.0}$ $En_{33.8}$ and $Fs_{34.7}$ $Wo_{1.0}$ $En_{64.4}$ for LCP and HCP, respectively.

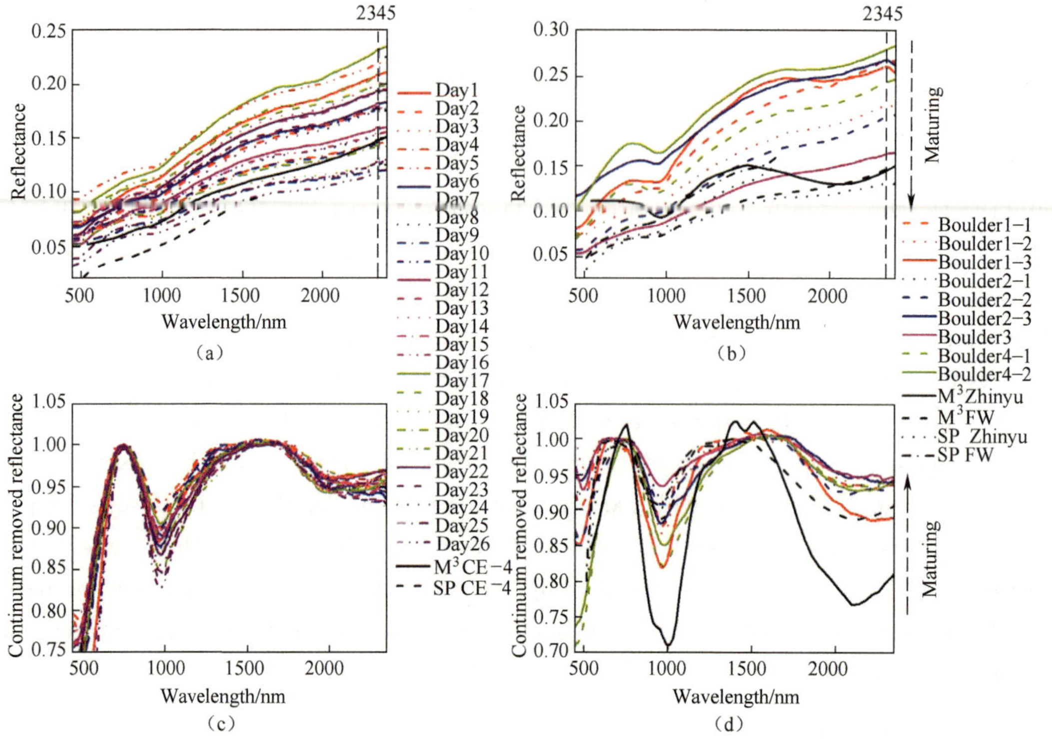

Figure 3　Comparison of orbital and ground spectra. (a) — (b) Average CE - 4 VNIS spectra of soils and boulders compared with M^3 and SP spectra of the CE - 4 landing region, Zhinyu crater, and FW. The reflectance is the REFF calibrated to the RELAB standard illumination geometry ($i=30°$, $e=0°$, and $g=30°$. (c) - (d) Continuum - removed spectra. The arrows indicate the trend of maturation by space weathering. The average spectrum of the multiple measurements in each lunar day is compared here. The complete spectral figure set (54 images) is available online. (The data used to create this figure are available.) (The complete figure set (54 images) is available.)

The olivine - rich mineralogy as reported in previous studies on soil spectra (e. g. , Li et al. 2019) is likely to be an artifact caused by the inappropriate continuum removal method and the 820nm artifact. The convex - hull method was used by Li et al. (2019) to estimate the continuum of VNIS spectra where the last bands (2345～2395 nm) are on the hull, while these bands should not be on the continuum because the 2μm absorption was not completely measured by VNIS (Figure A2) . Such a continuum, connecting the short - wavelength edge of 2μm absorption and the endpoint of VNIS spectra, significantly attenuates the true absorption strength at 2μm and generates the illusion that the 2μm absorption of VNIS spectra is weak and the soil of the CE - 4 landing site is rich in olivine. In other words, the convex - hull method is not suitable for continuum removal with respect to the 2μm absorption of VNIS spectra. Although the convex - hull method works well to remove the continuum of the 1μm absorption, the resulting long - wavelength edges of 1μm absorption in CE - 4 VNIS spectra are located at ～1400 -

1500nm, which is different from that of olivine - dominated spectra (typically at ~1700nm).

4 Discussion

The compositional and mineralogical remote sensing of SPACA indicates that plagioclase (58.4~60.4 vol.%) is more abundant than mafic minerals (39.6~41.6 vol.%, Lemelin et al. 2016) and imparts a crustal signature to the SPACA regolith. The plagioclase - rich mineralogy was interpreted to be from the plagioclase - rich crust formed by the differentiated SPA impact melt sheet (Hu et al. 2019). However, the SPACA is located within the assumed transient cavity formed by the SPA impact event (Figure A1) where all of the ~45km thick and feldspathic crust was thought to have been completely removed (Potter et al. 2012) and deposited far (>200km distance) beyond the SPACA. Beginning from a mafic - rich bulk composition (23.3~49.1 wt.% FeO+MgO, 0.9~13.2 wt.% Al_2O_3), the differentiation of the SPA impact melt sheet is not expected to fractionate plagioclase from the residual melt and produce a secondary plagioclase - rich (Plg > 60 vol.%) crust (Hurwitz & Kring 2014). Instead, the final mineral assemblage after fractional crystallization is likely plagioclase + ferroan (Mg# = 5.1 - 68.7) mafific minerals that form noritic (HCP/LCP = 0.16 ~ 0.24) lithologies with relatively low Al compositions (10.9~14.5 wt.%). The average plagioclase proportion for SPACA is~60.4 vol.%, which is beyond the high - Plg end of the mineralogical range for SPA impact melt differentiation products (Figure 4, 3.0~55.2 vol.% Plg). The observed composition and modal mineralogy of the SPACA would require either a more Al - enriched starting composition of SPA impact melt differentiation or additional contribution from post - SPA processes (e.g., feldspathic ejecta from post - SPA basins).

More Al - enriched initial compositions for SPA impact melt would indicate that more crustal materials went into the melt. During the transient crater collapse stage of large impact events, the crust would migrate back in toward the basin center (Freed et al. 2014). Hence, it is suggested that the crust was not entirely removed in the SPA - forming impact (Trowbridge et al. 2020). As shown by the elliptical structure of topographical and compositional distribution of the SPA surface expression (Garrick - Bethell & Zuber 2009), the SPA formation process was likely an oblique impact although the vertical case is preferred in most SPA impact simulations (e.g., Potter et al. 2012). Recent three - dimensional models have attempted to take the elliptical nature of the SPA impact into account (e.g., Melosh et al. 2017); an oblique SPA impact would have incorporated a greater proportion of feldspathic crustal materials into the melt (Vaughan & Head 2014). Contamination (~50%) of crustal materials into the SPA impact melt has been examined, while the modeled differentiation results of a contaminated SPA

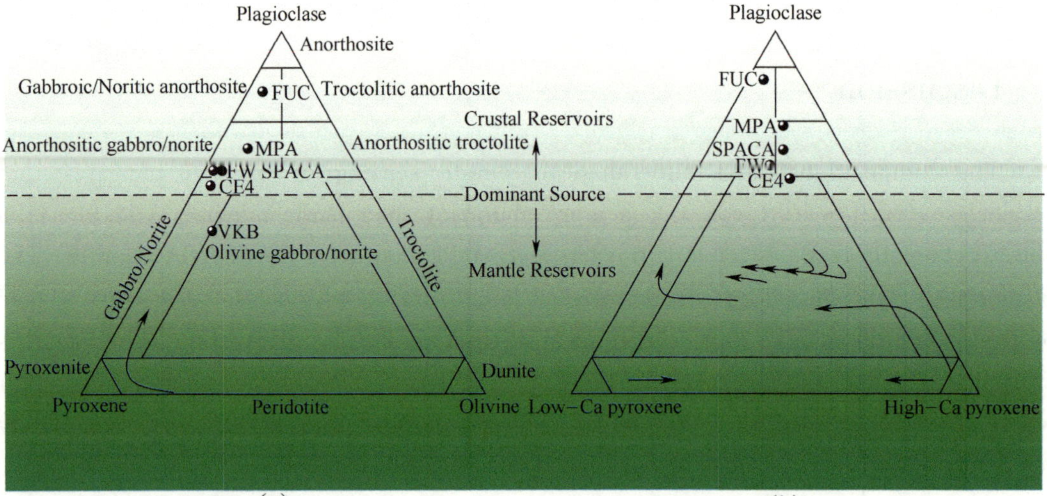

Figure 4　Mineralogy of CE－4 VNIS exploration sites.（a）Modal mineralogy plotted on a plagioclase－pyroxene－olivine ternary.（b）Modal mineralogy plotted on a Plg－LCP－HCP ternary（Olv ＜ 10 vol. %）. The modeled SPA impact melt differentiation and mineralogic evolution are indicated by arrows（Hurwitz & Kring 2014）.（The data used to create this figure are available.）

impact melt reveal that the additional Al may not affect the top（～5km thick）noritic layer. Instead, these added crustal materials resulted in earlier saturation of plagioclase and a（～3km）thicker layer of deep－seated plagioclase－bearing lithologies（Hurwitz & Kring 2014）. Unless the SPA impact melt did not differentiate and the crustal materials directly contributed to the compositions of solidified surface, the observed Al－rich composition of the SPACA surface is more likely from post－SPA processes.

Considering that the SPA region experienced numerous impact events in the long history（＞4.3Ga）since its formation（Wilhelms et al. 1987）, materials ejected from more feldspathic regions outside the basin floor may account for a considerable proportion of the SPACA. The nearest potential source is the so－called Mg－pyroxene annulus（MPA）（Moriarty 2016）, where the pyroxene composition is more Mg rich（64.8 Mg#）compared to the SPACA（58.2 Mg#）. The mineral map products from the Kaguya MI data sets（Lemelin et al. 2016）reveal that MPA is more feldspathic（66.1 vol. % Plg）than SPACA（60.4 vol. % Plg）and meets the requirement as a candidate. The irregular boundary of SPACA is obviously shaped by those adjacent basins and large craters in MPA（Figure A1）. The ejecta of these large impacts is believed to have contaminated the SPACA with MPA materials. Besides MPA, another potential source of feldspathic materials is the highlands outside the SPA basin. Although the feldspathic upper crust may have been removed by the SPA impact, distal ejecta from post－SPA basins could feed feldspathic materials back into the SPA basin floor（e.g., Xiao et al. 2021）. This issue (contribution from post - SPA bombardments) was also discussed in the work of Petro & Pieters (2004, 2006). They found that some portion (20%~50%) of material at the surface of the SPA basin floor is introduced from the exterior of SPA, while it would not be enough to mask the SPA impact melt products. Ultimately, only the large (e. g., Imbrium) or adjacent basins (e. g., Apollo, Ingenii) could move the appreciable amount of material into the interior of SPA. Our unmixing results also suggest that the proportion of materials from the feldspathic upper crust outside the SPA is lower than 4. 5 vol. % (versus 41. 4 - 65. 7 vol. % materials from MPA).

The thickness distribution of post - SPA impact ejecta provides further evidence of SPACA being contaminated by more feldspathic materials. Based on the transient cavity size and impact ejecta thickness estimation models (Pike 1974), it yields a post - SPA ejecta thickness of up to~1. 5km in the current range of SPACA (Figure A1).

Assuming that the initial SPACA surface is formed in the SPA impact melt sheet, the surface regolith observed today could be regarded as a mixture of initial SPA impact melt products and post - SPA ejecta. Given the predicted compositions of the initial top noritic layer of the differentiated SPA impact melt sheet (Hurwitz & Kring 2014), results from applying linear unmixing procedures to the observed mineralogy show that post - SPA ejecta accounts for 70. 2 vol. % of the materials of SPACA. With a significant contribution from feldspathic (MPA and highlands) end - members, unmixing analysis of average soil mineralogy indicates that another endmember for mixing is Fe, ITE (incompatible trace elements) - enriched materials (16. 9~18. 9 wt. % FeO, 3. 1~6. 7 ppm Th). Accordingly, the compositions of the CE - 4 landing site (C) and the SPACA (S) are not represented by any known lithologies in the Apollo and Luna collections (Figure A7).

5　Conclusion and Perspective

Based on the VNIS data obtained from the CE - 4 mission, we investigated the mineralogical properties of regolith encountered by the Yutu - 2 rover and their implications for SPACA. The major outcomes of this study are as follows. The average plagioclase abundance of SPACA is beyond the highplagioclase end of the mineralogical range of SPA impact melt differentiation products and would require an additional contribution from crustal materials. We suggest that either the crust was not entirely removed in the SPA - forming impact or the ejecta from post - SPA basins fed feldspathic materials back onto the SPA basin floor. These analyses highlight the significance of plagioclase and crustal materials in mineralogical investigations and future explorations of the SPA basin (and the Moon).

We thank the Chang'e - 4 mission of China's Lunar Exploration Program and the team members of the GRAS, who contributed to receiving and preprocessing raw VNIS data. This work was supported by the National Natural Science Foundation of China (grant Nos. 11941001, 41972322), China Postdoctoral Science Foundation (grant No. 2020M682164), the State Scholarship Fund (grant No. 201706220310), National Key Research and Development Program of China (grant Nos. 2020YFE0202100, 2019YFE0123300), Strategic Priority Research Program, Chinese Academy of Sciences (grant No. XDB41000000), Pre-research project on Civil Aerospace Technologies, CNSA (grant Nos. D020102, D020204). This is the SDU-CPS publication #96.

Appendix

A.1 Hapke Radiative Transfer Model

The single scattering albedo single scattering albedo (SSA; ω) of mafic minerals in the VNIS spectral range (450～2395nm) are modeled based on the refractive index ($n+ik$) and particle diameter (D) as follows (Hapke 2012):

$$\omega = S_e + (1-S_e)(1-S_i)\frac{\Theta}{1-S_i\Theta} \tag{1}$$

$$S_e = \frac{(n-1)^2}{(n+1)^2} + 0.05 \tag{2}$$

$$S_i = 1 - \frac{4}{n(n+1)^2} \tag{3}$$

$$\Theta = e^{-\alpha\langle D\rangle} \tag{4}$$

$$\alpha = 4\pi k/\lambda \tag{5}$$

$$\langle D\rangle = \frac{2}{3}\left[n^2 - \frac{1}{n}(n^2-1)^{3/2}\right]D \tag{6}$$

The grain size (D) is held to a constant value of 17μm as VNIR spectral properties of lunar soils are dominated by the 10～20μm size fraction (Pieters et al. 1993).

The real part of the refractive index (n) for mafic minerals dependent on the mineral compositions (represented by Mg# (100 x molar Mg/(Mg+Fe))) is calculated using Equations (7) ～ (9):

$$n_{\mathrm{LCP}} = 1.768 - 0.118 \times \mathrm{Mg\#}/100 \tag{7}$$

$$n_{\mathrm{HCP}} = 1.726 - 0.082 \times \mathrm{Mg\#}100 \tag{8}$$

$$n_{\mathrm{Olv}} = 1.827 - 0.192 \times \mathrm{Mg\#}/100 \tag{9}$$

The imaginary part of the refractive index (k) for mafic minerals as a function of mineral composition is calculated with the use of parameters derived from the least-squares regression of modified gaussian model (MGM) fitting results from Trang et al. (2013).

For Plg end-members, the optical constant n dependent on An# (molar Ca/(Ca+Na)) could be calculated based on Equation (12) from Warell & Davidsson (2010) as follows:

$$n_{Plg} = 1.523 + 0.0227 \times An\# + 0.0264 \times (An\#)^2 \tag{10}$$

Given the narrow compositional range of lunar Plg, the n of Plg is set to a fixed value (1.57) in this study. The k of Plg was calculated from the RELAB (Reflectance Experiment LABoratory) reflectance spectrum with the n and D of the lunar plagioclase sample (LR - CMP - 171 from the Apollo sample 15555, 965 ($D < 125\mu m$)) and the USGS (United States Geological Survey) spectral library of a terrestrial anorthite sample (HS201.3B, Sun & Lucey 2021).

The space - weathering effects of npFe and SMFe were quantified with the theories described in Lucey & Riner (2011):

$$\alpha_{sw} = \alpha_h + \alpha_c + \alpha_g = \frac{4\pi k_h}{\lambda} + \frac{36\pi z M_c \rho_h}{\lambda \rho_{Fe}} + \frac{3 q_a M_g \rho_h}{D_{Fe} \rho_{Fe}} \tag{11}$$

$$z = \frac{n_{Fe} k_{Fe} n_h^3}{(n_{Fe}^2 - k_{Fe}^2 + 2n_h^2)^2 + (2 n_{Fe} k_{Fe})^2} \tag{12}$$

where α_{sw} is the absorption coefficient of space - weathered minerals as a sum of absorption coefficients of host materials (α_h) and metallic Fe in the coatings (α_c) and grains (α_g). n_h and k_h are optical constants of host materials. λ is wavelength in units of micron. The optical constants of metallic Fe recently reported by Cahill et al. (2019) are linearly interpolated to the CE - 4 VNIS bands. M_c and M_g are the mass fractions of metallic Fe in the grain and coating. ρ_h and ρ_{Fe} are the densities of host materials and metallic Fe. The density of metallic Fe and Plg is set to 7.6 and 2.75, respectively. The solid solution density of mafic minerals is calculated as the weighted average of the end - member density based on Webmineral data as follows:

$$\rho_{Olv} = 3.22 \times Mg\#/100 + 4.39 \times (100 - Mg\#)/100, \tag{13}$$

$$\rho_{LCP} = 3.21 \times Mg\#/100 + 3.96 \times (100 - Mg\#)/100, \tag{14}$$

$$\rho_{HCP} = 3.22 \times Mg\#/100 + 3.56 \times (100 - Mg\#)/100. \tag{15}$$

q_a is the absorption efficiency of metallic Fe computed from Mie theory (Bohren & Huffman 2008). The particle diameter of metallic Fe (D_{Fe}) in grains is adjusted to $0.8\mu m$.

After the treatment of space - weathering effects, the total SSA of the mineral mixture was calculated as the weighted average of the end - member SSA of space - weathered minerals as follows (Hapke 2012):

$$\omega = \left(\sum_j \frac{M_j \omega_j}{\rho_j D_j}\right) \Big/ \left(\sum_j \frac{M_j}{\rho_j D_j}\right) \tag{16}$$

where M_j, ω_j, ρ_j, and D_j are mass fraction, SSA, density, and grain size of the jth end - member.

The reflectance spectra of mineral mixtures could be predicted from the average SSA of mineral mixtures by forward modeling of the Hapke RTM.

The RTM used in this work is approximated as follows (Hapke 2012; Sun & Lucey

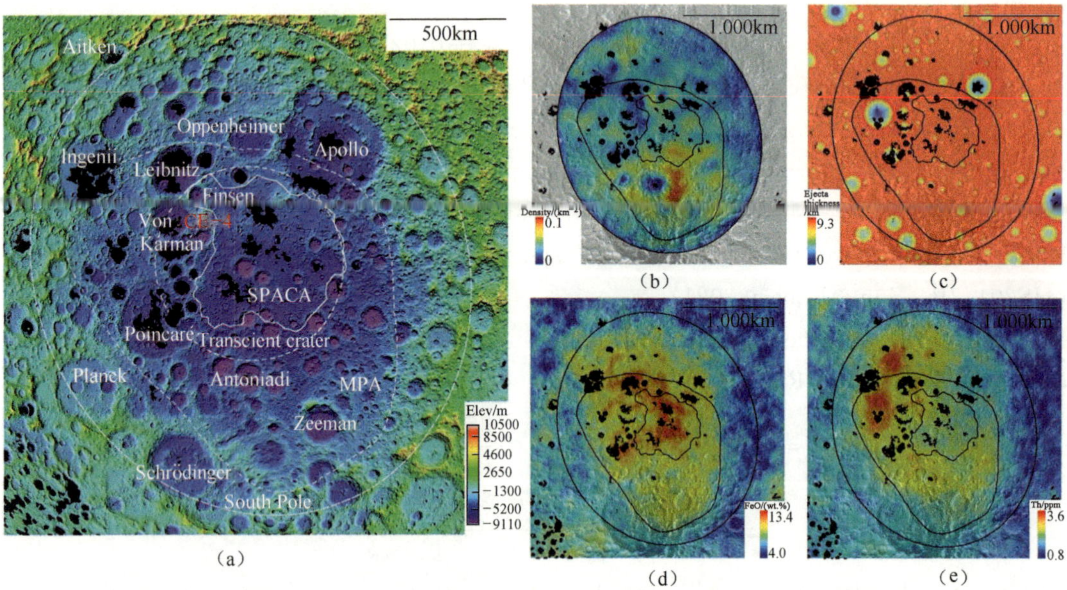

Figure A1　SPA region as seen in different data sets. (a) LOLA color shaded relief map (https://astrogeology.usgs.gov/search/map/Moon/LMMP/LOLA-derived/Lunar_LRO_LOLA_ClrShade_Global_128ppd_v04) centered on Mons Marguerite. The dashed circle is determined by the diameter (820km) of the transient cavity formed by the SPA impact event from Hurwitz & Kring (2014), that is perfectly outlining the SPACA region identified by Moriarty (2016). The irregular dashed polygon is the extent of the so-called MPA. (b) Impact crater density calculated based on the lunar crater database from Robbins (2019). Note the high-density annulus of the secondary craters surrounding the Imbrian crater Antoniadi. (c) Ejecta thickness distribution of post-SPA impacts (>15km). (d)-(e) LPGRS Half Degree Abundance Data of FeO and Th (https://pds-geosciences.wustl.edu/missions/lunarp/reduced_special.html). The boundary of SPACA identified by Moriarty (2016) is shown as an irregular polygon. The elliptical lines are the SPA basin ring. The black mask represents mare basalt regions modified from Nelson et al. (2014). (The data used to create this figure are available.)

2021):

$$\mathrm{REFF}(i,\ e,\ g)=K\frac{\omega}{4}\frac{1}{\mu_0+\mu}\{p(g)[1+B_s(g)]+H(\mu_0/K)H(\mu/K)-1\},\tag{17}$$

$$K=\frac{-\ln(1-1.209\phi^{2/3})}{1.209\phi^{2/3}},\tag{18}$$

where $\phi=0.41$ for lunar regolith (Sun & Lucey 2021),

$$\mu_0=\cos(i)\tag{19}$$

$$\mu=\cos(e)\tag{20}$$

$$p(g)=1+b\cos(g)+c(1.5\cos^2(g)-0.5).\tag{21}$$

The values of $b=-0.4$ and $c=0.25$ were approximated for photometrically corrected spectra ($g=30°$) (Sun & Lucey 2021),

$$B_s(g)=\left(1+\frac{1}{h_s}\tan\frac{g}{2}\right)^{-1}, \tag{22}$$

$$h_S=\frac{3\sqrt{3}}{8}\frac{K\phi}{\ln(1000)}, \tag{23}$$

$$H(x)=\frac{1+2x}{1+2\gamma x}, \tag{24}$$

$$\gamma=\sqrt{1-\omega}. \tag{25}$$

To verify the robustness of our spectral LUT, the RELAB spectra of 19 Apollo soil samples were matched with the spectral LUT to estimate the mineralogy and compare with their modal mineralogy characterized by LSCC (Taylor et al. 2001). Two different criteria (RMSE and SA) were used to find the best match between the spectral LUT and LSCC spectra. The RMSE and SA are calculated as follows:

$$\text{RMSE}=\sqrt{\frac{1}{m}\sum_{i=1}^{m}(y_{i,\text{ measured}}-y_{i,\text{ modeled}})^2}, \tag{26}$$

$$\text{SA}=\cos^{-1}\left(\frac{y^T_{\text{measured}}y_{\text{modeled}}}{\sqrt{y^T_{\text{measured}}y_{\text{measured}}}\sqrt{y^T_{\text{modeled}}y_{\text{modeled}}}}\right) \tag{27}$$

The superscript "T" represents transposition. The spectral matching returned excellent results (Figure 2) that the RMSE and SA between the modeled and observed spectra are lower than 0.005. The mineral abundances of 18 soil samples (except 71061) modeled with our LUT exhibit a correlation of 0.96 and an RMSE of 6.53 vol.% versus their modal mineralogy quantified by LSCC. The RMSE/SA between the modeled and CE-4 VNIS spectra are lower than 0.03 for soil targets.

A.2 Estimation of Impact Ejecta Thickness

Ejecta thickness of post-SPA impacts were calculated using the model proposed by Pike (1974):

$$T_E=0.033R_t(r/R_t)^{-3.0} \tag{28}$$

where T_E is the ejecta thickness at a distance r in units of meters, and R_t is the rim radius of the transient cavity in meters and could be inverted as follows (Kring 1995):

$$R_t=^{1.07}\sqrt{R_C/0.86}, \tag{29}$$

where R_C is the current radius of the crater/basin. r is the distance from the crater/basin center in meters and is defined as the spherical distance as follows:

$$D(x1,\ y1,\ x2,\ y2)=R_M^*\cos^{-1}(\sin(x1)*\sin(x2)+\cos(x1)*\cos(x2)*\cos(y1-y2)), \tag{30}$$

where $(x1,\ y1)$ and $(x2,\ y2)$ are the longitudes and latitudes of two points, and R_M is the radius of the Moon (1737.4km).

A.3 Effects of the SG Smoothing Algorithm on BC and IBDR

The CE-4 VNIS reflectance spectra present noisy (fluctuant) variations with wavelength (e. g. , Figure A2) . Therefore, the Savitzky – Golay (SG) smoothing algorithm (Savitzky & Golay 1964) was used to reduce the effects of local noises on the automatic detection of tie points for continuum removal with the tangency line method and on the calculation of the band center (BC) and IBD values (Table A1) . The order of the SG smoothing algorithm is set to 5 because the VNIS spectrum has nominally four minimum/maximum points (i. e. , the shortwavelength maximum of 1μm absorption at ~700 – 750nm, minimum of 1μm absorption, maximum between 1μm and 2μm absorptions at ~1400 – 1700nm, minimum of 2μm absorption) . We tested the resulting BC and IBD values with all possible numbers of points in the smoothing window (i. e. , 7 – 389 points, Figure A8) . As shown in Figure A8, when the number of points is between 8 and 40, the SG smoothing algorithm has negligible effects on the BC1 values (966. 0± 0. 6) . Furthermore, the BC2 (2204. 6±2. 3) and IBDR (0. 97±0. 04) values exhibit relative uncertainties below 4%. Compared to the BC and IBDR values derived from the unsmoothed spectrum, the difference in BC1 values (~1. 0nm) is negligible. However, the differences in the BC2 and IBDR values between unsmoothed and smoothed spectra are large (~32. 4 nm for BC2 and 0. 36 for IBDR, respectively) . In summary, we suggest that an SG smoothing algorithm with number of points between 8 and 40 and an order of 5 would improve the spectral quality of 1μm absorption in VNIS data but bring in a relative difference below 4% for 2μm absorption and a larger relative difference of 37% for IBDR values in comparison with those from unsmoothed spectra.

A. 4 Mixing Effects in CE – 4 Soils

Because CE – 4 landed in the VKB zone affected by FE, ballistic ejecta from the Finsen crater should mix to some degree with local Von Kármán materials during emplacement (Oberbeck 1975) . Unmixing analyses were used to identify the FE component in CE – 4 VNIS data (Figure A4) . Given that the basaltic deposits in the Von Kármán crater floor have been mostly obscured by nonmare crater ejecta, the VKB end-member for mixing was defined by the basaltic ejecta of the Zhinyu crater, which presents the highest FeO contents (>15 wt. %, pure mare area, Gillis et al. 2003) in the floor of the Von Kármán crater. The modal mineralogy and FeO contents derived from orbital observations using data from the Kaguya multiband imager (MI) (Lemelin et al. 2016) reveal continuous compositional and mineralogical variations along the section line Zhinyu – CE4 – Finsen (Figure A4) . The Plg proportion increases from Zhinyu to the continuous ejecta that surrounds the Finsen crater rim (1 ~ 2 radius) while the mafic mineral abundances and FeO contents both decrease from Zhinyu to Finsen. The central peak of Finsen exhibits a different mineralogy (LCP – richer, Figure A4) from FW and FE, hence SPACA data sets (not associated with Finsen) were also included for comparison (Figure A6) . The composition and mineralogy of SPACA (not associated with Finsen) resemble those exposed on the Finsen crater wall and continuous ejecta surrounding the

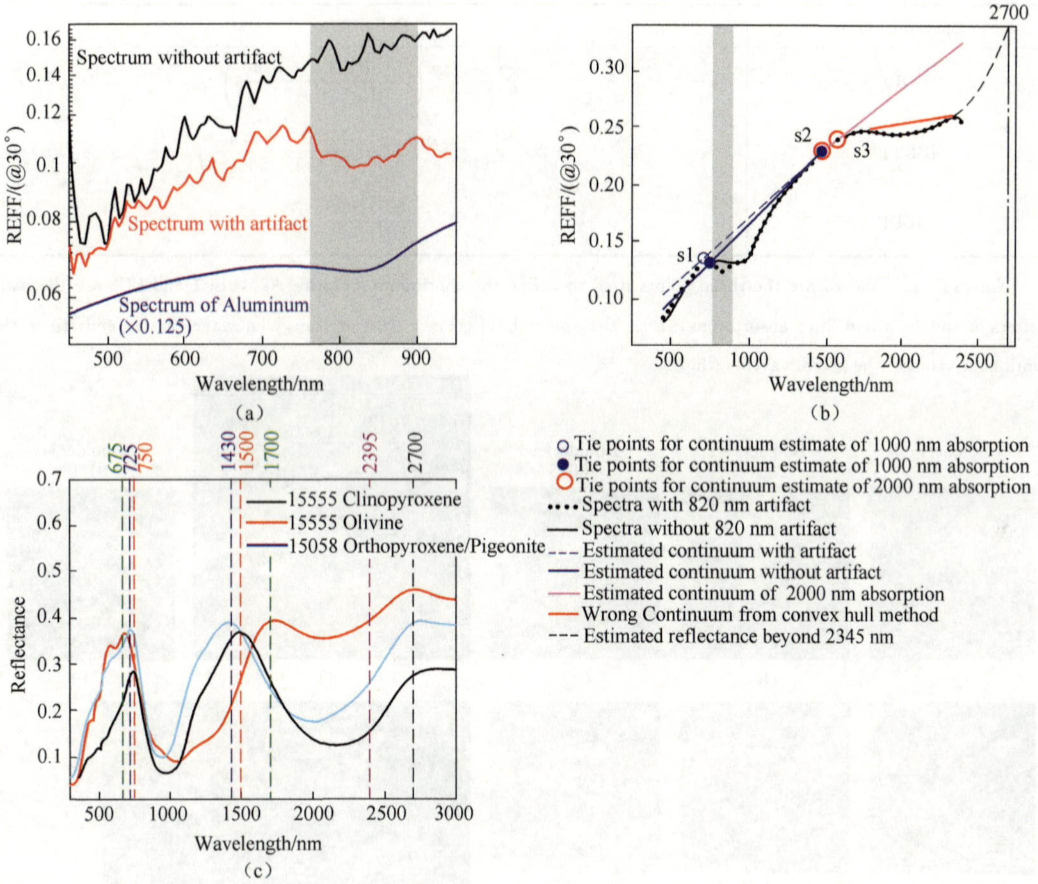

Figure A2　The 820nm artifact in VNIS spectra and its effects on continuum removal. (a) The artifact (gray region) between 760nm and 900nm. (b) The artifact affects the positions of the tie point used for continuum removal of CE－4 VNIS data. (c) VNIR spectra of lunar clinopyroxene, olivine, and orthopyroxene/pigeonite collected and characterized by the Apollo missions (Data are from RELAB database: clinopyroxene (LR－CMP－168 from Apollo 15 sample 15555), olivine (LR－CMP－169 from Apollo 15 sample 15555), and orthopyroxene (LR－CMP－173 from Apollo 15 sample 15058). The endpoint wavelength (2395nm) of VNIS spectra is marked by a magenta dashed line to illustrate that the 2μm absorption band was not completely measured by VNIS. (The data used to create this figure are available.)

crater rim.

Table A1　Definitions of Spectral Indices Used in This Work

Spectral Index	Definition
BD940	$\frac{R(740\text{nm})}{R(940\text{nm})}$
IBD820	$\sum_{n=0}^{28}\left(1-\frac{R(760+5\times n)}{R_C(760+5\times n)}\right)$

Continued Table A1

Spectral Index	Definition
IBD1000	$\sum_{n=0}^{(s2-s1)/5}\left(1-\frac{R(s1+5\times n)}{R_C(s1+5\times n)}\right)$
IBD2000	$2\times\sum_{n=0}^{(BC2-s3)/5}\left(1-\frac{R(s3+5\times n)}{R_C(s3+5\times n)}\right)$
IBDR	$\frac{IBD2000}{IBD1000}$

Note. $s1$, $s2$, and $s3$ are the three points used to define the continuum (Figure A2), BC1 and BC2 are the band centers of the 1μm and 2μm absorptions, R is the spectral reflectance data at the given wavelength, and R_C is the continuum value at the given wavelength.

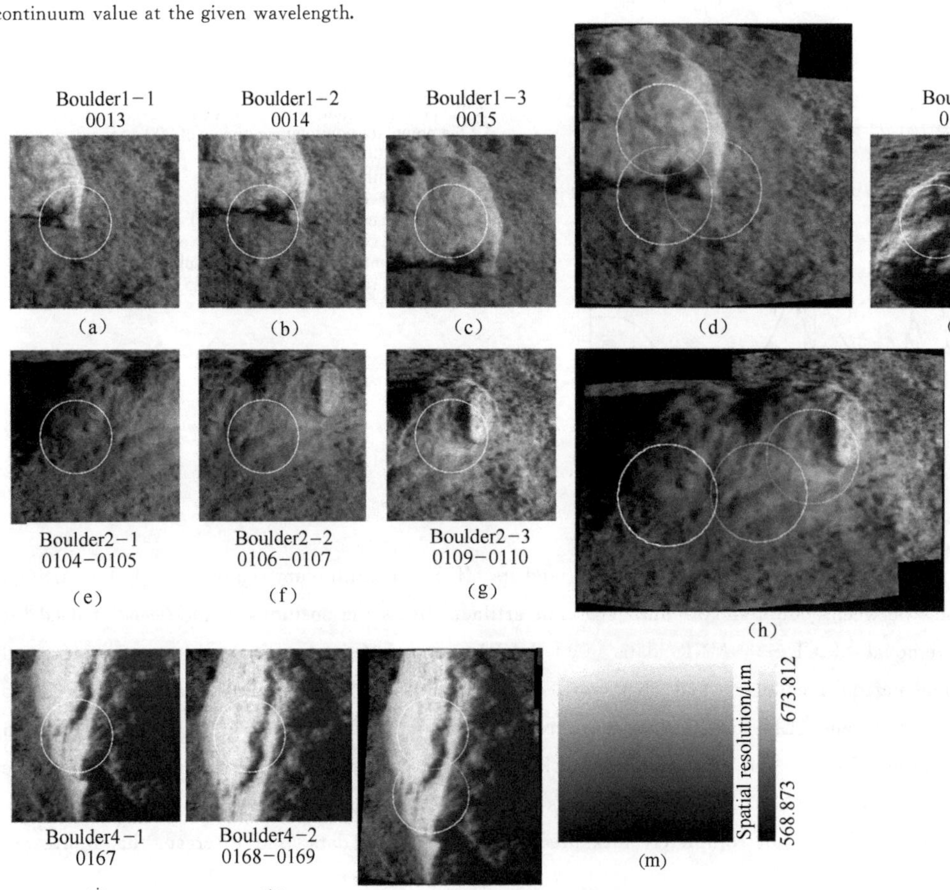

Figure A3　900nm band images of VNIS detections for boulders. (a) — (c) images of the boulder measured on the third lunar day. (d) Mosaic of a—c. (e - g) images of the boulder measured on the 13th lunar day. (h) Mosaic of e - g. (i) image of the boulder measured on the 23rd lunar day. (j) — (k) Images of the boulder measured on the 26th lunar day. (l) mosaic of j - k. (m) spatial resolutions of each pixel in the images. The white circles show the FOV (with a diameter of ~65mm projected onto the lunar surface) of the SWIR detectors. The VNIS spectra of different parts of boulders are shown in Figures 3b and 3d. (The data used to create this figure are available.) (The complete figure set (120 images) is available.)

Although the soils of the CE - 4 landing region seen by both orbital and ground spectrometers are mixtures, an advantage of the CE - 4 mission is that the VNIS aboard the Yutu - 2 rover can make measurements on centimeter - sized boulders, which are not resolvable for orbital spectrometers. Individual boulders encountered by the Yutu - 2 rover would commonly be Von Kármán basalts excavated by younger local craters (e. g., Zhinyu), but the first boulder with LCP - rich mafic mineralogy might be surviving blocks ejected from Finsen and represent an end - member for the mixing in local soils.

The modal mineralogy of boulders could be semiquantitatively determined based on the pixel classification assuming that the area fraction of pixels containing diagnostic 1μm mafic absorption is equal to the volume percentage of mafic phases within the boulders. The pixels without obvious 1μm absorptions could be regarded as a pure Plg pixel. This method is only valid for coarse - grained rocks with grain size larger than the pixel resolution (~600μm; Figure A3) and may not be suitable for estimating Plg and mafic mineral abundances in fine - grained rocks. For fine - grained rocks, most pixels are Plg intergrown with mafic minerals so that they exhibit mafic absorptions and the proportion of Plg will be underestimated based on the pixel classification. The fraction of pixels without mafic absorptions in VNIR images can be used as a lower estimate for Plg proportion in boulders.

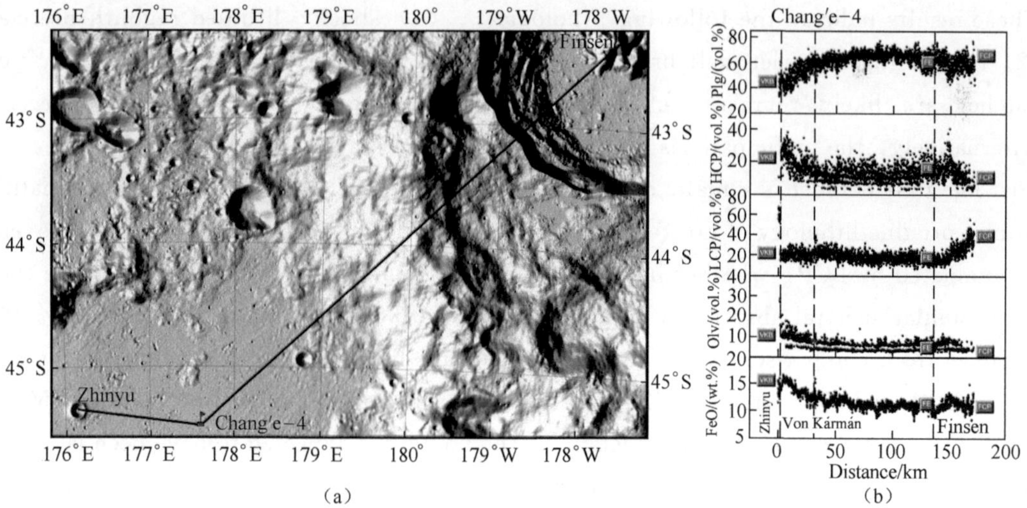

Figure A4　Compositional and mineralogical profile along the section line Zhinyu - CE4 - Finsen. (a) Section line overlain on the shaded relief derived from LRO + Kaguya merged topography products (https://astrogeology.usgs.gov/search/map/Moon/LRO/LOLA/Lunar _ LRO _ LOLAKaguya _ Shade _ 60N60S _ 512ppd). (b) Compositional and mineralogical variations along the traverse path. FeO and mineralogy data are from Kaguya MI global map products. Three dashed lines indicate the locations of the Zhinyu crater rim, CE - 4 landing site, and Finsen crater rim on the section line. The red stars mark the mineralogy derived from CE - 4 VNIS data sets. The dark marbles represent the three end - members for mixing, i. e., VKB, FE, and FCP (Finsen central peak). (The data used to create this figure are available.)

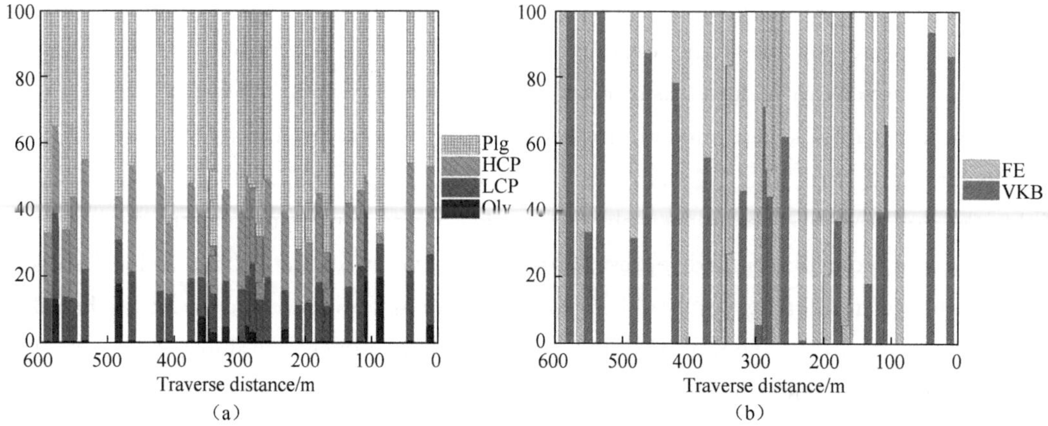

Figure A5　Modal analyses of CE－4 soils. (a) Mineralogy.
(b) Mixing ratio. (The data used to create this figure are available.)

The VNIR images could be also used to analyze the texture of boulders. As shown in Figure A9, the pixels with a strong 1μm mafic absorption feature are probably mafic phenocrysts or clasts. The largest mafic grain/clast in the first boulder is about 65 pixels (~3.9cm) in the longest dimension. The texture (i.e., large mafic clast/phenocrysts in a fine－grained matrix) of the boulder would exclude the possibility of a plutonic origin. These results indicate the following lithologies: (1) shock－lithified regolith breccia; (2) recrystallized impact melt breccia; and (3) mafic－phyric basalt. The mafic－rich patches are likely remnants of pristine mafic lithologies in regolith/melt breccia. Alternatively, the mafic patches also could be phenocrysts in a basalt.

The first boulder is a mafic clast－bearing melt/regolith breccia probably originating from a pristine lithology of norite/gabbro or mafic－phyric basalt. Although the pyroxene is dominated by LCP, whether it is orthopyroxene or pigeonite is unknown because they share similar spectral absorption features. A feldspathic mineralogy (60－80 vol.% Plg obtained from comparison with the RELAB spectral library of returned lunar rocks and meteorites) and source region (Zhinyu crater) were reported for the first boulder (Ma et al. 2020). As a distal ejecta from Zhinyu, the boulder is expected to have been launched from the target surface of the Zhinyu impact, which is either mare basalt or Finsen ejecta according to the stratigraphic studies (e.g., Qiao et al. 2019). The Plg－rich mineralogy of the boulder reported by Ma et al. (2020) is obviously not basaltic, but the other scenario of relaunching Finsen ejecta from the Zhinyu crater to the CE－4 landing region is possibly responsible for adding mafic clasts into the fine－grained matrix.

A.5　Mantle Materials Identified by CE－4?

Although basin－forming events could breach the lunar crust and may have exposed mantle materials (e.g., Yamamoto et al. 2010), few mantle samples have been

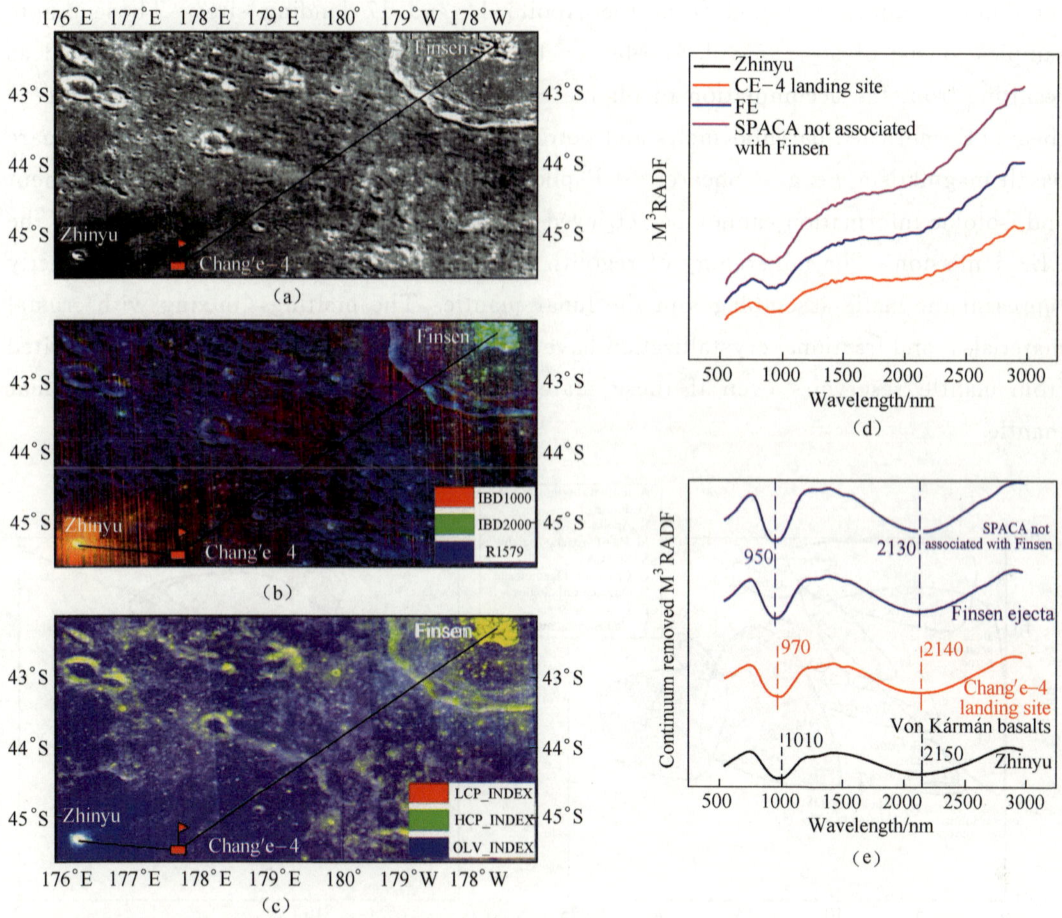

Figure A6　Spectral and mineralogical profile along the section line Zhinyu - CE4 - Finsen. (a) Mosaic of the Zhinyu - CE4 - Finsen region at 540nm from the M^3 data sets (https://pdsimage2.wr.usgs.gov/archive/chandrayaan_1/m3/). (b) M^3 Standard color composite. (c) Composite of mineral indices. IBD parameters and mineral indices are defined in Table A1 and Chen et al. (2020). (d) M^3 RADF reflectance spectra of the Von Kármán basalts (VKB exposed by Zhinyu), CE - 4 landing site, Finsen ejecta (FE), and SPACA region (not associated with Finsen). (e) Continuum - removed spectra. (The data used to create this figure are available.)

identified in the Apollo and Luna samples and in lunar meteorites up to now. Apollo 17 sample 74275 may contain mantle clasts (Shearer et al. 2015) but their mantle origin did not arrive at a consensus. The excavated mantle materials have been contaminated by crustal signatures and diluted in postbasin processes.

The soils of the CE - 4 landing site were interpreted by some early researchers as originating from the lunar mantle (Li et al. 2019); however, their conclusion is based only on the presence of low - Ca pyroxene and olivine and lacks the key constraint from plagioclase abundance. Moreover, even if the soils measured by CE - 4 VNIS were ultramafic, they cannot be interpreted as mantle materials. Typical counterexamples are the dunite samples collected from the Apollo 14 and 17 landing sites. Those dunite samples consist of 90 – 95 vol. % Mg – rich Olv (Fo_{86-90}) yet they are interpreted as resulting from the accumulation of olivine crystallized in crustal environments owing to their ITE – enriched parental melts and petrogenetic relationships to Mg – suite and mare basalt magmatism (e. g. , Shearer & Papike2005) . Unfortunately, the trace element and isotopic information cannot be retrieved by the VNIR spectroscopy employed by the CE – 4 mission. The mineralogy of regolith at the CE – 4 landing region cannot directly represent the mafic assemblages in the lunar mantle. The melting, mixing with crustal materials, and fractional crystallization have altered the petrogenetic information inherited from mantle reservoirs even if these materials were initially sourced from the lunar mantle.

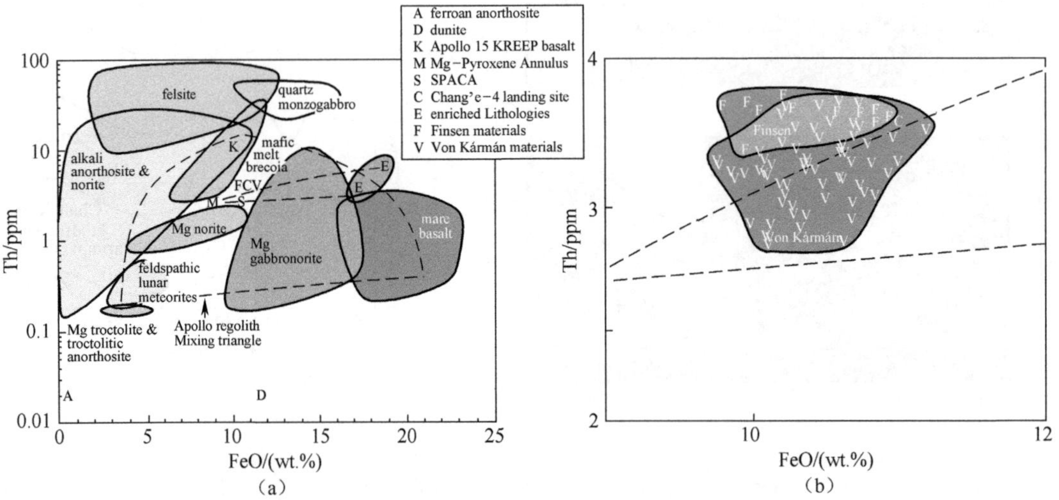

Figure A7　FeO – Th diagrams of lunar materials. (a) FeO – Th compositions of lunar rock samples modified from Jolliff et al. (2006) . The fields with different grayscales represent compositional ranges of different lithologies that occurred in Apollo and Luna rock samples. The "A" point represents the highly feldspathic ferroan anorthosite sample 15418. The "C" point represents the regolith composition of the CE – 4 landing site and the traverse area as seen by the Lunar Prospector gamma – ray spectrometer. The "D" point represents the Apollo 17 dunite sample 72415 – 7. The "E" points represent the enriched end – member lithologies for the mixing in SPACA. The "K" point represents the average compositions of Apollo 15 KREEP basalt samples. The "S" point represents the average compositions of SPACA. The dotted curves indicate mixing effects in lunar regolith samples among three end – members (i. e. , feldspathic highlands materials, mare basalt, and KREEP – bearing materials) . (b) Data used to unmix the Finsen materials ("F" points) and Von Kármán materials ("V" points) . (The data used to create this figure are available.)

The end – member lithologies with Fe and ITE enrichments (mafic clasts in the first boulder) are consistent with a possible origin from differentiated SPA impact melt. The subsequently erupted basalts also could be mixed with the enriched materials by lava flow

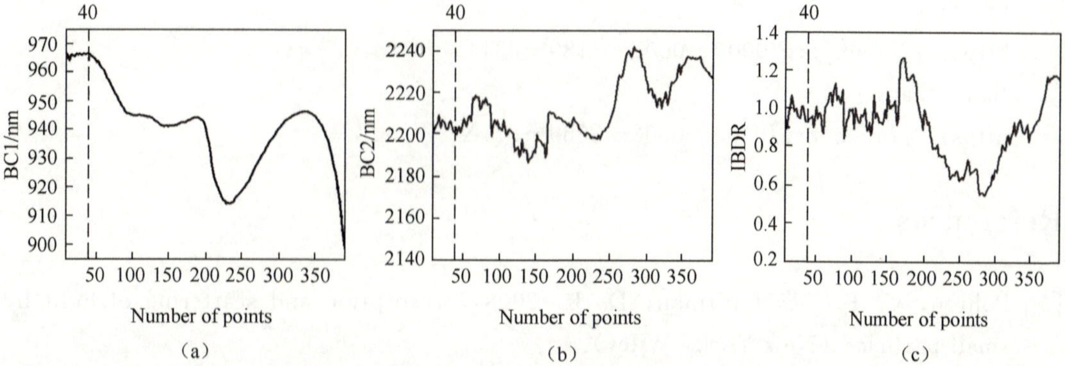

Figure A8　Variations of BC1, BC2, and IBDR values with different SG smoothing windows. (The data used to create this figure are available.)

assimilation and continuous impact gardening. These enriched materials were diluted by more feldspathic ejecta, and a less mafic upper regolith layer was developed in SPACA and observed by orbital spectrometers. Some of these mixing products were delivered by the Finsen impact event and observed by the CE - 4 VNIS. The enriched end - member lithologies are recorded as mafic clasts in a feldspathic matrix (as shown in the first boulder) . To further reveal the nature of these mafic lithologies, future orbital spacecraft with highenough spatial resolution to resolve these ancient fragments or landed missions with more advanced in - situ analytical techniques are required. Undoubtedly, returning samples from the SPACA would significantly contribute to a better under - standing of the impact melt evolution and mare volcanism in this enormous basin and of the underlying lunar mantle.

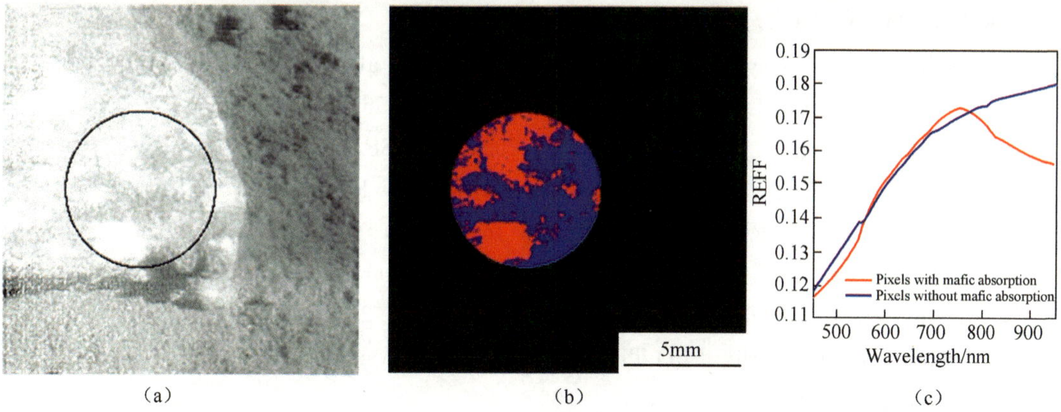

Figure A9　Spectral classification of VNIS images of the first boulder (0015) . The circles indicate the FOV of the SWIR detector. (a) BD940 image. (b) Binary classification results with a pixel resolution of ~600μm. (c) Mean spectra (normalized at 750nm) of blue pixels (with mafic absorptions) and red pixels (without mafic absorptions) . (The data used to create this figure are available.)

ORCID iDs
Le Qiao
https：//orcid. org/0000—0002—6180—2344
Jiang Zhang
https：//orcid. org/0000—0002—8369—871X

References

[1] Bohren，C. F.，& Huffman，D. R. 2008，Absorption and scattering of light by small particles（New York：Wiley）.
[2] Cahill，J. T.，Blewett，D. T.，Nguyen，N. V.，et al. 2019，Icar，317，229.
[3] Chen，J.，Ling，Z.，Qiao，L.，et al. 2020，Sci. China-Inf. Sci，63，140903.
[4] Di，K.，Liu，Z.，Liu，B.，et al. 2019，JRemS，23，177.
[5] Freed，A. M.，Johnson，B. C.，Blair，D. M.，et al. 2014，JGRE，119，2378.
[6] Garrick-Bethell，I.，& Zuber，M. T. 2009，Icar，204，399.
[7] Gillis，J. J.，Jolliff，B. L.，& Elphic，R. C. 2003，JGRE，108，5009.
[8] Gou，S.，Di，K.，Yue，Z.，et al. 2019，E&PSL，528，115829.
[9] Hapke，B. 2008，Icar，195，918.
[10] Hapke，B. 2012，Theory of Reflectance and Emittance Spectroscopy（Cambridge：Cambridge Univ. Press）.
[11] Hu，X.，Ma，P.，Yang，Y.，et al. 2019，GeoRL，46，9439.
[12] Huang，J.，Xiao，Z.，Xiao，L.，et al. 2020，Geo，48，723.
[13] Hurwitz，D. M.，& Kring，D. A. 2014，JGRE，119，1110.
[14] Jia，Y.，Zou，Y.，Ping，J.，et al. 2018，P&SS，162，207.
[15] Jolliff，B. L.，Wieczorek，M. A.，Shearer，C. K.，et al. 2006，RvMG，60，1.
[16] Kring，D. A. 1995，JGRE，100，16979.
[17] Lemelin，M.，Lucey，P. G.，Gaddis，L. R.，et al. 2016，LPSC，47，2994.
[18] Li，C.，Liu，D.，Liu，B.，et al. 2019，Natur，569，378.
[19] Lin，H.，He，Z.，Yang，W.，et al. 2020，NatSR，7，913.
[20] Liu，B.，Li，C. L.，Zhang，G. L.，et al. 2014，RAA，14，1578.
[21] Lucey，P. G.，& Riner，M. A. 2011，Icar，212，451.
[22] Ma，P.，Sun，Y.，Zhu，M. H.，et al. 2020，Icar，350，113901.
[23] Melosh，H. J.，Kendall，J.，Horgan，B.，et al. 2017，Geo，45，1063.
[24] Moriarty，D. P. 2016，PhD Thesis，Brown University.
[25] Morris，R. V. 1980，LPSC，11，1697.
[26] Nakamura，R.，Matsunaga，T.，Ogawa，Y.，et al. 2009，GeoRL，36，L22202.
[27] Nelson，D. M.，Koeber，S. D.，Daud，K.，et al. 2014，LPSC，45，2861.
[28] Oberbeck，V. R. 1975，RvGeo，13，337.

[29] Petro, N. E., & Pieters, C. M. 2004, JGRE, 109, E06004.
[30] Petro, N. E., & Pieters, C. M. 2006, JGRE, 111, E09005.
[31] Pieters, C. M., Fischer, E. M., Rode, O., & Basu, A. 1993, JGR, 98, 20817.
[32] Pieters, C. M., Taylor, L. A., Noble, S. K., et al. 2000, M&PS, 35, 1101.
[33] Pieters, C. M., Tompkins, S., Head, J. W., et al. 1997, GeoRL, 24, 1903.
[34] Pike, R. J. 1974, E&PSL, 23, 265.
[35] Potter, R. W., Collins, G. S., Kiefer, W. S., et al. 2012, Icar, 220, 730.
[36] Qi, X., Ling, Z., Zhang, J., et al. 2020, RemS, 12, 3211.
[37] Qiao, L., Ling, Z., Fu, X., et al. 2019, Icar, 333, 37.
[38] Robbins, S. J. 2019, JGRE, 124, 871.
[39] Savitzky, A., & Golay, M. J. 1964, AnaCh, 36, 1627.
[40] Shearer, C. K., Burger, P. V., Bell, A. S., et al. 2015, M&PS, 50, 1449.
[41] Shearer, C. K., & Papike, J. J. 2005, GeCoA, 69, 3445.
[42] Spudis, P. D., Gillis, J. J., & Reisse, R. A. 1994, Sci, 266, 1848.
[43] Sun, L., & Lucey, P. G. 2021, JGRE, 126, e06691.
[44] Taylor, L. A., Pieters, C. M., Keller, L. P., et al. 2001, JGRE, 106, 27, 985.
[45] Tompkins, S., & Pieters, C. M. 2010, M&PS, 45, 1152.
[46] Trang, D., Lucey, P. G., Gillis-Davis, J. J., et al. 2013, JGRE, 118, 708.
[47] Trowbridge, A. J., Johnson, B. C., Freed, A. M., et al. 2020, Icar, 352, 113995
[48] Vaughan, W. M., & Head, J. W. 2014, P&SS, 91, 101.
[49] Warell, J., & Davidsson, B. J. R. 2010, Icar, 209, 164.
[50] Wilhelms, D. E., John, F., & Trask, N. J. 1987, The Geologic History of the Moon, US Geological Survey Professional Paper, 1348
[51] Xiao, Z., Ding, C., Xie, M., et al. 2021, GeoRL, 48, e90935.
[52] Xu, J., Wang, M., Wang, R., et al. 2021, RemS, 13, 2359.
[53] Yamamoto, S., Nakamura, R., Matsunaga, T., et al. 2010, NatGe, 3, 533.

Mineralogy of Chang'e-4 landing site: preliminary results of visible and near-infrared imaging spectrometer

Jian CHEN[1], Zongcheng LING[1*], Le QIAO[1], Zhiping HE[2], Rui XU[2], Lingzhi SUN[3], Jiang ZHANG[1], Bo LI[1], Xiaohui FU[1], Changqing LIU[1] & Xiaobin QI[1]

1. Shandong Key Laboratory of Optical Astronomy and Solar-Terrestrial Environment, School of Space Science and Physics, Institute of Space Sciences, Shandong University, Weihai 264209, China;

2. Key Laboratory of Space Active Opto-Electronics Technology, Shanghai Institute of Technical Physics, Chinese Academy of Sciences, Shanghai 200083, China;

3. Hawai'i Institute of Geophysics and Planetology, Department of Earth Sciences, University of Hawai'i at Mānoa, Honolulu HI 96826, USA

Abstract: The exploration of mafic anomaly in South Pole-Aitken (SPA, the largest confirmed) basin on the Moon provides important insights into lunar interior. The landing of Chang'e-4 (CE-4) and deployment of Yutu-2 rover on the discontinuous ejecta from Finsen crater enabled *in-situ* measurements of the unusual mineralogy in the central portion of SPA basin with visible and near-infrared imaging spectrometer (VNIS). Here we present detailed processing procedures based on the level 2B data of CE-4 VNIS and preliminary mineralogical results at the exploration area of Yutu-2 rover. A systematic processing pipeline is developed to derive credible reflectance spectra, based on which several spectral and mineral indices are calculated to constrain the mafic mineralogy. The mafic components in the soils and boulder around CE-4 landing site are concluded as clinopyroxene-bearing with intermediate composition and probably dominated by pigeonite although the possibility of mixing orthopyroxen (OPX) and calcic clinopyroxene (CPX) also exists. These mineralogical results are more consistent with a petrogenesis that the CE-4 regolith and rock fragment are derived from rapid-cooling magmatic systems and we interpret that the materials at the CE-4 landing site ejected from Finsen crater are probably recrystallized from impact melt settings.

Keywords: Chang'e-4; visible and near-infrared spectroscopy; mineralogy; impact melt; the Moon

Citation Chen J, Ling Z C, Qiao L, et al. Mineralogy of Chang'e-4 landing site: preliminary results of visible and near-infrared imaging spectrometer. Sci China Inf Sci, 2020, 63 (4): 140903, https://doi.org/10.1007/s11432-019-2768-1

1 Introduction

The first-ever landing on the farside of the Moon (Fig. 1) was realized by Chang'e-4 (CE-4) mission with its successful touchdown in the South Pole-Aitken (SPA) basin on 3 January 2019[1-3]. As the largest confirmed impact basin on the Moon[4], SPA basin is thought to have penetrated into lunar interior and excavated deep materials from lower crust and upper mantle during its formation[5-7]. Since global remote sensing enabled observations of the lunar farside, numerous researches have been performed on SPA basin with various orbital datasets. Observations of the SPA basin floor indicate that it is characterized by mafic anomaly [5-11] and elevated concentrations of incompatible elements (e. g., Th,[12-14]) and is dominated by low-Ca pyroxene (LCP, orthopyroxene (OPX) or pigeonite)-rich noritic materials with disperse high-Ca pyroxene (HCP)-rich gabbroic or basaltic lithologies (e. g.,[5,7-9]) compared to the feldsparthic highlands outside of the basin. Among diverse aspects of SPA region, the nature of mafic anomaly in the floor of SPA basin, which exhibits distinct geochemical and spectral characteristics from returned lunar samples and currently known lunar meteorites[5-14], is regarded as an ideal indicator of the composition and mineralogy of lunar interior (lower crust or upper mantle).

The CE-4 landing site is located on the floor of ancient Pre-Nectarian Von Kármán crater, where it is flooded by low-TiO_2 mare basaltic lava flows and then mostly overprinted by discontinuous ejecta from the nearby Finsen crater in the northeast[15-22]. The ejecta, characterized by non-mare mafic components in orbital remote sensing datasets, exhibits pervasive mineralogical similarities to the mafic anomaly observed in the central portion of SPA basin[8]. Different from all previous studies based on orbital datasets in the past, the *in-situ* spectroscopic observation datasets of Finsen ejecta around the CE-4 landing site acquired by the visible and near-infrared imaging spectrometer (VNIS) onboard the Yutu-2 rover provide a valuable opportunity to improve our understanding of the unusual mineralogy in the floor of SPA basin. This study aims to develop a systematic VNIS data processing strategy and report preliminary analysis results in terms of mineralogical information of the exploration region of Yutu-2 rover in the first three lunar days.

2 Data and methods

2.1 VNIS data

The VNIS collected hyperspectral visible and near-infrared (VNIR) images (450—

950nm) and single - pixel short - wave infrared (SWIR) spectra (900 — 2400nm) for fourteen detection targets during the first three lunar days of CE - 4 mission. The locations of the VNIS measurements are shown in Fig. 1. The VNIS raw data are reduced into

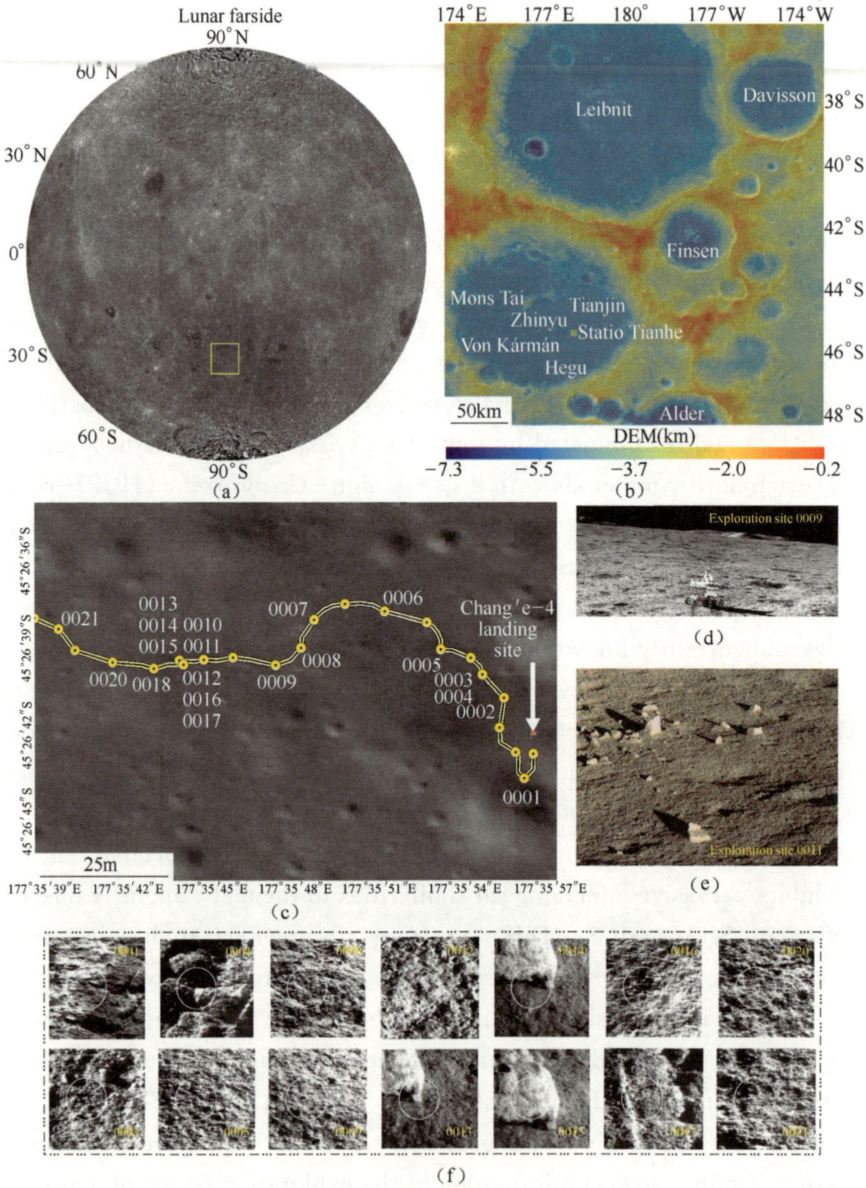

Fig. 1　(Color online) Global to ground view of the CE - 4 landing site, exploration region of Yutu - 2 rover, and the VNIS detections. (a) CE - 1 digital orthophoto map (DOM) of lunar farside. The yellow indicator shows the extent of frame (b). (b) Nomenclature surrounding CE - 4 landing site on CE - 2 digital elevation model (DEM) and DOM. (c) Traverse path of Yutu - 2 rover and location of exploration sites during the first three lunar days on image taken by the CE - 4 landing camera (LCAM), each exploration site is marked by yellow circles and its number which corresponds that in panel (f). (d) Image of Yutu - 2 rover at exploration site 0009 acquired by the CE - 4 terrane camera (TCAM). (e) Boulders viewed by the CE - 4 panoramic camera (PCAM) at the exploration site 0011. (f) 900 - nm - band images of 14 VNIS detections. The white circles show the field of view of short - wave infrared (SWIR) detectors

level 2B radiance through a series of processing pipelines including dark - current, scattering - background subtractions, flat - field, instrument temperature corrections, and radiometric and geometric calibrations[23, 24], and then distributed by the Ground Research and Application System (GRAS) of Chinese Lunar Exploration Program. The level 2B VNIS radiance data are further processed to extract diagnostic absorption characteristics and identify the surface mineralogy of exploration sites in this study. The processing procedures based on level 2B VNIS radiance data are shown in Fig. 2.

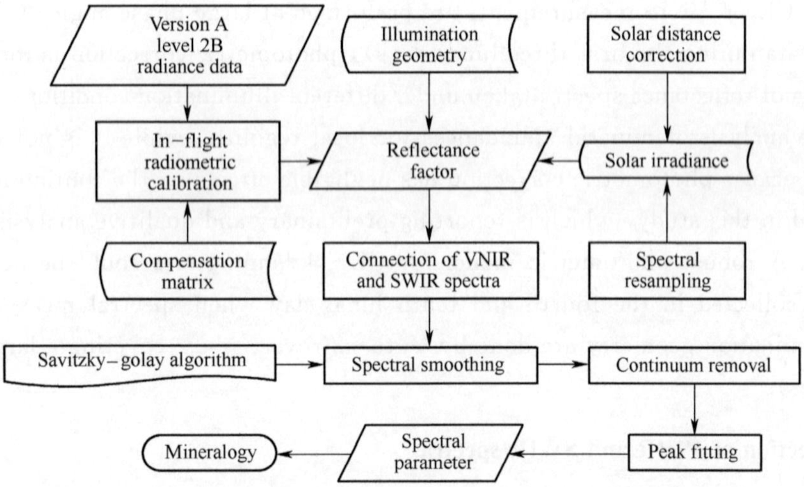

Fig. 2　Processing procedures based on CE - 4 VNIS level 2B radiance data

2.2　Reflectance calculation

Because the VNIS version A level 2B radiance data distributed by GRAS are generated with the radiometric calibration matrix obtained from ground/pre - flight experiments, a compensation matrix from in - flight radiometric calibration experiments is applied to VNIS level 2B radiance data to correct various effects owing to the change of working environment of the instrument. Afterwards, the corrected radiance data are converted to reflectance factor by (1) to eliminate the Fraunhofer lines inherited from the incident sunlight and emphasize the absorption features related to minerals on the lunar surface.

$$\mathrm{REFF}(i, e, g, \lambda) = \frac{\pi L(\lambda)}{\cos(i)\,\frac{1}{D^2}\int F(\lambda)T(\lambda)\mathrm{d}\lambda} \tag{1}$$

where REFF is the reflectance factor. i, e, and g are the incidence, emission, and phase angles. $L(\lambda)$ is radiometrically corrected radiance data. D is the Sun - Moon distance in a unit of astronomical unit (AU) and is calculated for the VNIS data according to the observation time of each band, with the NASA NAIF SPICE Toolkit and SPICE kernels naif0012. tls, pck00010. tpc, and DE438. bsp[1]. $\frac{1}{D^2}\int F(\lambda)T(\lambda)d\lambda$ is spectral resampling and distance correction for solar irradiance, where $F(\lambda)$ is the extraterrestrial solar irradiance data without terrestrial atmospheric absorptions [25] and $T(\lambda)$ is the spectral response function of detectors. The solar irradiance data are reported at a spectral resolution distinct from that of VNIS and a standard distance of 1 AU[25], hence, the solar irradiance data must be resampled with the spectral response function of VNIS and corrected to the Sun - Moon distance during VNIS detections by a factor of D^2.

As the CE - 4 VNIS measurements are performed at large phase angles (~65°—105° for SWIR data during the first three lunar days), photometric correction is important for comparisons of reflectance spectra taken under different illumination conditions and further quantitative analysis of mineral abundances for local regolith, while it is not included in this study because photometric correction has negligible effects on the continuum removed spectra used in this study, which is reporting preliminary and qualitive analysis results of mineralogy. A robust photometric model for CE - 4 landing site could be derived from VNIS data collected in the fourth and tenth lunar day when spectral measurements at various illumination geometry are done by Yutu - 2 rover (i. e., covering a large range of phase angles).

2.3 Connection of VNIR and SWIR spectra

The field of view (FOV) of the single - pixel SWIR detector is corresponding to a circular region with a diameter of 107. 6 pixels (a radius of 53. 8 pixels) and centered at the coordinate (98, 127. 5) within the FOV of the VNIR image[26]. The distance of each pixel in VNIR image to the circle center is calculated as follows:

$$\text{Distance} = \sqrt{(x-98)^2+(y-127.5)^2} \tag{2}$$

where x and y are the center coordinates of each pixel. If the distance is less than the radius (53. 8), the pixel is considered as in the FOV of SWIR detector. In this study, 9104 pixels in the VNIR hyperspectral image are concluded to be covered by the FOV of SWIR detector. For each of the fourteen detections, VNIR spectral data of these 9104 pixels are averaged and smoothed using the Savitzky - Golay (SG) algorithm[27], which essentially performs a polynomial regression and convolution to the spectral data points in a moving window. The smoothed VNIR spectrum is then connected with the SWIR spectrum measured from the circle FOV. To fix the step artifacts resulted from the response differences of the detectors, factors obtained at 900nm and 1375—1380nm of the SWIR spectrum are used to correct the VNIR spectra in 450—900nm region and the SWIR spectra in 900—1375nm region, respectively. The description and correction of artifacts at 1375—1380nm and 900—945nm are presented in Appendix A. The combined spectra are

---

http://naif.jpl.nasa.gov.

then treated with SG smoothing procedure, because CE - 4 VNIS reflectance spectra present noisy (fluctuant) variations with wavelength (e. g., Figure S1). SG smoothing is used to suppress the effects of local noises on the automatic detection of tie points for continuum removal with tangency line method and on the calculation of spectral parameters afterwards. The uncertainties introduced by SG smoothing are discussed in Appendix B.

2.4 Continuum removal

The absorption bands of mafic minerals in lunar reflectance spectra are usually superimposed on a continuum, which is induced by a combination of complex effects from various mineralogical and nonmineralogical factors (e. g., SMFe, submicron Fe produced from space weathering[28, 29]). Therefore, an important step in mineralogical analysis based on reflectance spectra is to estimate and remove the continuum from the raw reflectance spectra, leaving only the absorption bands, of which positions and widths are more dependent on mineralogy (e. g., [30-32]). Because the exact form of continuum as a function of wavelength or energy (wavenumber) is unknown, it is difficult to model the continuum[33]. This study employs tangent line method (Appendix C) to remove the continuum of CE - 4 reflectance spectra. The tangent line method is chosen because it automatically detects tie points of continuum whose wavelength ranges vary with mineralogy (e. g., ~725nm and ~1430nm for LCP, ~750nm and ~1500nm for HCP, and ~675nm and ~1700nm for olivine in Appendix D). Therefore, tangency line method is more sensitive to the potential mineralogical differences among different sites explored by Yutu - 2 rover compared to those methods which set tie points at fixed wavelengths. Convex hull method is popular for continuum removal of other lunar VNIR hyperspectral remote sensing datasets (e. g., Moon Mineralogy Mapper (M^3) [34-36]) while it is not suitable for continuum removal of 2000 nm absorption in VNIS data, which is not completely measured (Appendixes C and D).

2.5 Spectral fitting

To identify mafic mineralogy at the CE - 4 landing site, several spectral indices (e. g., band centers, integrated band depth (IBD) of 1000 and 2000nm absorptions, IBD ratio (IBDR), and mineral indices of LCP, HCP, olivine (OL)) sensitive to mineralogy, composition, and mineral mixtures in lunar surface spectra were applied on VNIS data[37, 38]. Following the removal of continuum, the band centers of 1000nm and 2000nm (BC1 and BC2) are determined by Gaussian fitting, and other spectral indices are calculated according to the definitions given in Tab. 1[39]. The primary indices used to detect mafic minerals in VNIS spectra are LCP INDEX, HCP INDEX, and OL INDEX, which are based on searching for lower reflectance in selected wavelengths inside the absorption bands of their target minerals compared to references outside of the absorption bands (Tab. 1). The LCP INDEX is defined to detect the 1000nm and 2000nm absorptions of LCP (e. g., OPX or pigeonite) via reduced reflectance at 1080nm and 1820nm relative to the reflectance at the spectral shoulder near 1330nm, and the HCP INDEX is similarly defined to detect HCP (e. g., augite) via reduced reflectance at 1080 and 2070nm relative to the reflectance at 1470m[40,41]. The OL INDEX is defined to detect olivine via reduced reflectance at 1050, 1210, 1330, and 1470nm relative to the continuum value at those wavelengths[41]. Although these mineral indices are designed for compact reconnaissance imaging spectrometer for Mars (CRISM) data, they also could be used on remote sensing spectroscopic data on the Moon because these mafic minerals (LCP, HCP, and OL) share the same VNIR absorption properties on different bodies. While the wavelengths of these mineral indices could be optimized for lunar data, unoptimized indices already prove the capability to distinguish terrestrial LCP, HCP, and OL. As seen in Fig. 3[8, 34, 42−44], OPX exhibits high LCP _ INDEX and low HCP INDEX, and cpxclinopyroxene (CPX) presents high HCP _ INDEX and low LCP _ INDEX while olivine shows high OL _ INDEX, low HCP _ INDEX, and low LCP _ INDEX.

Tab. 1 Definitions of spectral indices used in [39][a)]

Spectral index	Definition
IBD1000	$\sum_{n=0}^{s2-s1} 1-\frac{R(s_1+n)}{R_c(s_1+n)}$
IBD2000	$2\times\sum_{n=0}^{bc2-s3} 1-\frac{R(s_3+n)}{R_c(s_3+n)}$
IBDR	$\frac{\text{IBD2000}}{\text{IBD1000}}$
LCP_INDEX	$100\times\left[\frac{R(1330)-R(1080)}{R(1330)+R(1080)}\right]\times\left[\frac{R(1330)-R(1815)}{R(1330)+R(1815)}\right]$
HCP_INDEX	$100\times\left[\frac{R(1470)-R(1080)}{R(1470)+R(1080)}\right]\times\left[\frac{R(1470)-R(2067)}{R(1470)+R(2067)}\right]$
OL_INDEX	$1-\left[0.1\times\frac{R(1054)}{R_C(1054)}+0.1\times\frac{R(1211)}{R_C(1211)}\right]+0.4\times\frac{R(1329)}{R_C(1329)}+0.4\times\frac{R(1474)}{R_C(1474)}$

a) $s1$, $s2$, and $s3$ are the three points used to define continuum (Figure S3), $bc2$ is the band center of 2000nm, R is the spectral reflectance data at the given wavelength, and R_C is the continuum value at the given wavelength.

3 Results and discussion

The VNIS spectra of soils at CE - 4 landing site are composed of reddish continuum and weak 1000nm and 2000nm absorption bands (Fig. 3 (a)), suggesting that the soils are heavily space weathered products from primary mafic - bearing lithologies[28,29]. Among the fourteen detection targets, the boulder (0013—0015 in Fig. 1 (f), 3 (a) and 3 (b)) exhibits the highest reflectance and the strongest 1000nm and 2000nm absorption band strength, which is consistent with its immature nature. Regardless of the different space weathering degrees between soils and the boulder, their 1000nm absorption band centers are fallen in a relatively narrow range between 962. 1—981. 3nm (Fig. 3 (c)), indicating that they are likely cogenetic and mineralogically similar and that the local soils are developed from ever existed boulders by space weathering.

The diagnostic 1000nm and 2000nm absorption features within the VNIS spectra can be interpreted as the presence of pyroxene within soils and boulder around the CE - 4 landing site. The olivine - rich mineralogy as reported in previous studies on CE - 4 VNIS spectra[16,18] is likely to be an illusion caused by inappropriate continuum removal method. The convex hull method was used by Li et al. [18] and Gou et al.[16] to estimate the continuum of VNIS spectra where the last band (2395nm) is on the hull, while the last band should not be on the continuum because the 2000nm absorption is not completely measured by VNIS (see more basic knowledges of lunar VNIR reflectance spectra in Appendix D). Such a continuum connecting the short - wavelength edge of 2000nm absorption and the endpoint of VNIS spectra significantly attenuates the true absorption strength at 2000nm (Fig. S3) and generates the illusion that the 2000nm absorption of VNIS spectra is weak and the mineralogy of CE - 4 landing site is olivine - rich. In a word, convex hull method is not suitable for continuum removal of 2000nm absorption of VNIS spectra.

Although the convex hull method works well to remove continuum of 1000nm absorption, the resulting long - wavelength edges of 1000nm absorption in CE - 4 VNIS spectra are located at ~1400—1500nm (Fig. 2 (b) in [16,18]), which is very different from that of true olivine (typically at ~1700nm, Fig. S4). This difference between CE - 4 VNIS spectra and true olivine - rich spectra observed by CE - 3 VNIS is also obvious in Fig. 2 (b) of[18] while those authors ignored that and still claimed olivine - rich lithologies for CE - 4 landing site. In this study, the long - wavelength edges of 1000nm absorption in continuum removed CE - 4 VNIS spectra detected by tangency line method are located at~ 1430 — 1500nm, which is comparable to those derived from other continuum removal methods (e. g., convex hull or tie points at fixed wavelengths) and is consistent with the spectral characteristics of pyroxene (Appendix D). Hence, we interpret that the CE - 4 VNIS spectral properties are dominated by pyroxene and that olivine is little at CE - 4 landing site.

The 1000nm and 2000nm absorption band centers of the continuum removed spectra are utilized to further identify the composition of pyroxene. Compared with the absorption band centers of a suite of pyroxenes with composition varying from Mg - rich pyroxene (OPX or pigeonite) to Ca - and Fe - rich CPX (Fig. 3 (c)), the absorption bands of CE - 4 VNIS spectra are located at intermediate wavelengths between those of OPX and calcic CPX from experiments[42], and also between those of Mg - rich pyroxene and calcic CPX - bearing mare basalt remotely observed by M^3, but resemble those of non - mare CPX in the central SPA mafic anomaly[8] and the spectra of pigeonites with relative enrichments of Fe^{2+} ions in M2 sites[45]. Although the 2000nm absorption band centers of CE - 4 VNIS spectra exhibit large variation from 2110—2264nm, and slightly deviate from the pyroxene cluster, it is plausible because the long - wavelength edge (typically beyond 2500nm, Appendix D) of 2000nm absorption band of pyroxene is not covered by VNIS spectral range (450 — 2395nm) so that the continuum removal and spectral fitting of 2000nm absorption band are subject to large uncertainties (Appendixes B, C, and E).

The band centers of the 1000nm absorption are plotted against the ratio of IBD2000 to IBD1000 (IBDR, Fig. 3 (d)) to determine the mafic mineral assemblages that possibly exist at the CE - 4 landing site. Data point of the boulder is fallen into the field defined by CPX - bearing lithologies, however the spectra of soils present apparently larger variation of IBD2000 to IBD1000 ratio relative to that of boulder (Fig. 3 (d)), implying larger uncertainties (Appendixes B and C) of the mitigated absorption bands by strong space weathering or shadow - hiding effects in soil spectra.

Pyroxene and olivine spectral indices (Tab. 1) are used to evaluate the effects of subtle mineralogical variations in VNIS data. The pyroxene indices of VNIS data reveal that the CE - 4 soils and boulder are mineralogically distinct from common OPX and CPX (naturally occurred pyroxenes except pigeonite), and the VNIS data cluster is in a fashion close to the trend line of olivine (Fig. 3(e)). However, the OL _ INDEX values of VNIS spectra are much lower than those of olivine spectra (Fig. 3 (f)). A special case of CPX points occurs within the field of VNIS data (Fig. 3 (e)), in support of the fact that these spectral parameters are generally compromised by minerals with atypical compositions. Hence, based on the spectral inspections above, the dominant spectral properties of the soils at the CE - 4 landing site are more consistent with atypical pyroxene composition (e. g., pigeonite).

Besides of the prominent peak at 962 — 981nm, an absorption at around 1200nm contributing to the asymmetric 1000nm band is observed in CE - 4 VNIS spectra. This absorption is also observed in relatively pure anorthosite where small amounts of Fe^{2+} are incorporated into the structure of crystalline plagioclase (e. g., [46-49]). However, this absorption is sensitive to impact events and would disappear under high shock pressures [50, 51]. Hence, the ~1200nm absorption in CE - 4 VNIS spectra is not likely contributed by the presence of plagioclase given that the CE - 4 landing site is dominated by Finsen's ejecta, which is product of impact events, although we cannot exclude the possibility that recrystallized plagioclase exists in the local soils. Furthermore, the ~1200nm feature from plagioclase would be effectively masked by stronger absorptions from mafic minerals when they are present even in small amounts (>15%) (e. g., [52]). On the other hand, the 1200nm absorption band could be resulted from spinallowed crystal - field transition of Fe^{2+} in the six - coordinated M1 octahedral crystallographic site of Fe - bearing LCP[49,53,54]. It seems incredible that LCP - dominated lithologies present ferroan compositions considering that LCP - rich non - mare lunar samples returned by Apollo and Luna missions as well as meteorites are virtually magnesian and belong to so called

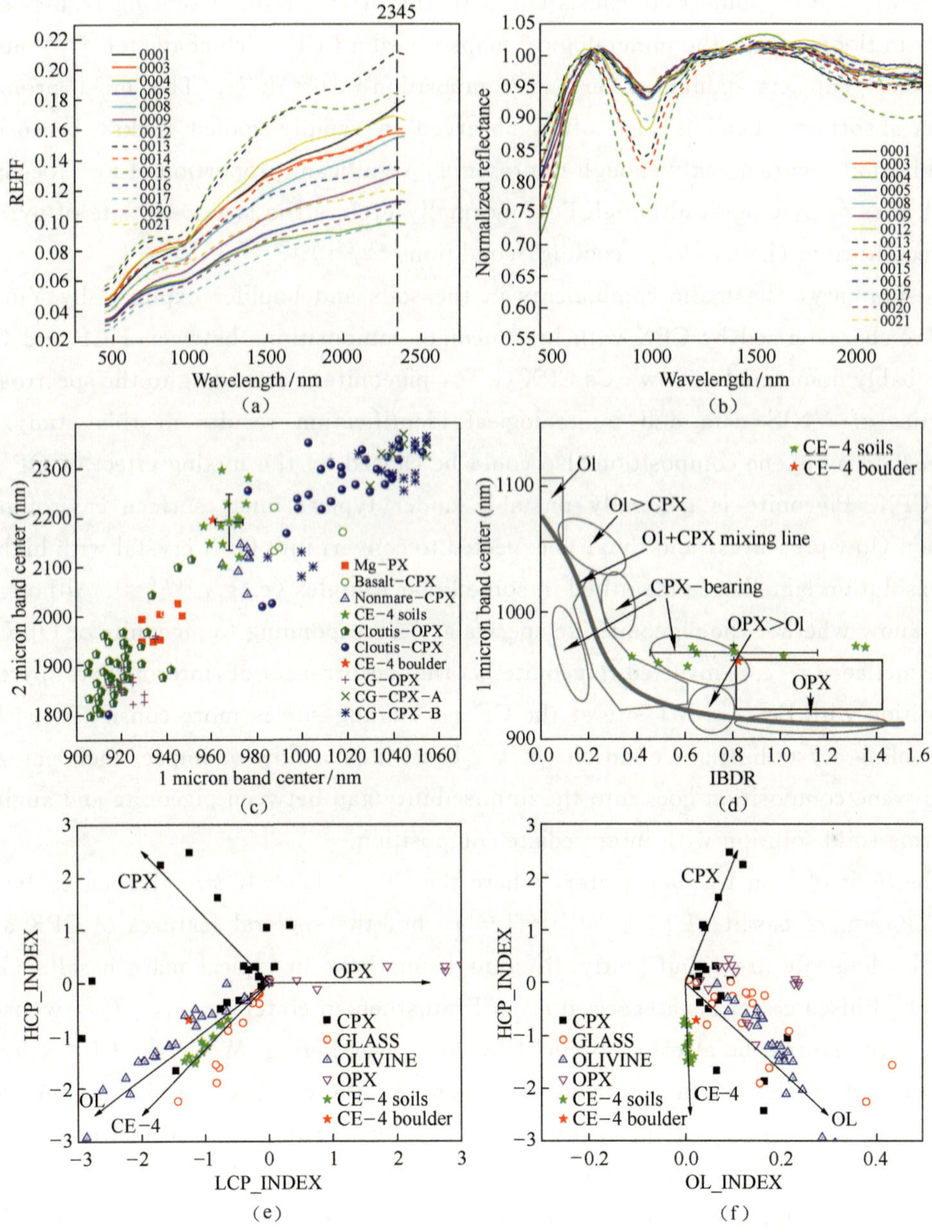

Fig. 3　(Color online) VNIS spectra and mineral compositions of Yutu-2 detection sites. (a) Combined VNIS spectra (450—2395nm) from 14 sites. (b) VNIS spectra after continuum removal. (c) Band center positions of the CE-4 VNIS detection targets overlain on band centers of pyroxene from [8, 42]. (d) BC1-IBDR plot for the fourteen CE-4 VNIS detections. The polygon and elliptic fields represent mineral combinations[34, 43, 44]. The data points with error bars in (c) and (d) represent the average and standard deviations of VNIS spectral parameters. (e) HCP INDEX vs. LCP INDEX plot for laboratory spectra[39] of clinopyroxene (black squares), orthopyroxene (purple triangles), olivine (blue triangles), and glass (red circles) and VNIS spectra. (f) HCP INDEX vs. OL INDEX plot for the same datasets in (e). Mg-PX: Mg-rich pyroxene; Ol: olivine.

"Mg - suite"[55-57], while it is consistent with the orbital remote sensing results of the SPA basin floor, where the mineralogical maps reveal a LCP - rich character[58, 59] and the geochemical datasets exhibit a ferroan composition[5, 6, 8, 10, 11]. The most prominent 1200nm absorption band is also often observed in rapidly cooled rocks, because the crystallization occurs quickly enough to preserve a significant proportion of Fe^{2+} occupying the M1 sites of pyroxene, although Fe^{2+} normally prefers the larger M2 site of pyroxene under equilibrium (i. e., slow - cooling) conditions[49, 53, 54].

In summary, the mafic components in the soils and boulder explored by Yutu - 2 rover are characterized by CPX with intermediate compositions between HCP and OPX, and probably dominated by low - Ca CPX (i. e., pigeonite), according to the spectroscopic properties of VNIS data and mineralogical identification results in this study. The intermediate pyroxene composition also could be caused by the mixing effect of OPX and calcic CPX. Pigeonite is generally unstable under typical lunar surface environmental condition (low pressures) and thus, is expected to convert into OPX crystal with high - Ca CPX exsolution lamellae as identified in some lunar samples (e. g., [60-63]). Although we do not know whether the pigeonitelike spectra are corresponding to pigeonite or OPX with CPX lamellae (i. e., inverted pigeonite), the occurrence of intermediate pyroxene composition with Fe^{2+} in M1 site at the CE - 4 landing site is more consistent with the fast - cooling crystallization condition (e. g., basaltic lava flow or impact melt) at which the pyroxene composition goes into the immiscibility gap between pigeonite and augite[64] and forms solid solution with intermediate composition.

The floor of Von Kármán crater, where the CE - 4 landed, was flooded by Imbrian low - TiO_2 mare basalt (LT, [9,15,17,19,21,22]), but the spectral features of CPX at the CE - 4 landing site are significantly different from those in typical mare basalts (Fig. 3 (c)). The Finsen crater is categorized as an Eratosthenian crater (e. g., [8]), whose ray system is thought to be already obscured by space weathering. While the CE - 4 landing site is located on the bright ray from Finsen crater, the observed prominent reflectance pattern of the Finsen rays is probably mainly related to compositional contrast with the background mare on the floor of Von Kármán crater, rather than difference of optical maturity[65]. The mineralogy at the CE - 4 landing site also shows a strong similarity to those pervasively exposed outcrops on the wall of the Finsen crater (e. g., [5, 8, 21]). Recent spectral and geological investigations using M^3 observation data produced mineralogical maps at an exceptional level of detail and revealed obviously different units (LT and Finsen's ejecta (FE)) on the floor of Von Kármán crater and suggested that the CE - 4 landing site has been affected by additional mafic non - mare materials ejected from the wall of Finsen crater (e. g., [15 - 18]). The dominant mineralogy of FE has been described as LCP - rich (e. g., [20 - 22]), which is consistent with what the VNIS observes during the *in - situ* explorations.

Considering that SPA region experienced numerous impact events in the long geological history since its formation (e. g., [66]), materials derived from geologic settings of impact melt maybe account for a large proportion of the SPA basin floor. Several previous studies have proposed that the SPA basin forming impact event has extensively melted the mafic cumulates and generated a local magma sea in the floor of the SPA[5, 8, 67-69], in which Mg - rich ultramafic cumulate sank to the base of SPA basin and the modestly evolved residual melt is subsequently deposited as relatively ferroan lithologies at the central portion of SPA. In addition to the ancient SPA basin - forming event, the parental magma of CE - 4 soils and boulder also could be sourced from other impact events occurred in SPA region such as Leibnitz crater (Fig. 1 (b)). Previous studies[16, 18] suggested an origin of lunar mantle materials for CE - 4 regolith and rock fragment, while the conclusion is mostly based on the illusion of olivine - rich lithologies as discussed above. According to the perspectives of mineralogy and petrology, the regolith and rock fragment observed by Yutu - 2 rover are more likely to be products of rapid - cooling magmatic episodes, rather than fragments of plutonic cumulates from either mantle or the base of a differentiated SPA impact melt sheet. The melting, potential mixing with crustal materials, and recrystallization have blurred the genes inherited from mantle reservoir even if these materials are initially sourced from lunar mantle.

4 Conclusion and future work

We developed a systematic processing pipeline of CE - 4 VNIS spectra based on the version A level 2B data to derive credible reflectance spectra. Based on several spectral and mineral indices calculated from the VNIS reflectance spectra, the mafic components in the soils and boulder at CE - 4 landing site are concluded as clinopyroxene - bearing and probably dominated by pigeonite, although mixing effect of OPX and calcic CPX cannot be excluded. Such a mafic mineralogy with intermediate pyroxene composition (pigeonite) is more consistent with a petrogenesis that the CE - 4 regolith is derived from rapid - cooling magmatic systems (basaltic lava flow or impact melt). The LCP - rich mafic mineralogy and residence on the Finsen's discontinuous ejecta suggest that the soils and boulder observed by Yutu - 2 rover are not likely from basaltic lava flows that infill Von Kármán crater, and are probably recrystallized from impact melt settings and delivered to CE - 4 landing site by Finsen impact event.

Future work will focus on building a spectral library to derive quantitative modal abundances and mineral compositions (e. g., Mg # (Molar Mg/(Mg + Fe)) of mafic minerals) as well as space weathering degree (abundance of SMFe) with spectral matching. By the extraction of complete mineralogical information from VNIS data, the (crustal or mantle in question) origin of the materials encountered by Yutu - 2 rover will be better constrained.

Acknowledgements

This work was supported by National Natural Science Foundation of China (Grant Nos. 11941001, 41972322, U1931211), Natural Science Foundation of Shandong Province (Grant No. ZR2019MD008), Qilu Young Scholar (TANG SCHOLAR) Program of Shandong University, Weihai (Grant No. 2015WHWLJH14), Key Research Program of Frontier Sciences, Chinese Academy of Sciences (Grant No. QYZDY - SSW - DQC028), and Pre - research Project on Civil Aerospace Technologies Funded by China National Space Administration (CNSA) (Grant No. D020102).

References

[1] Di K, Liu Z, Liu B, et al. Chang'e - 4 lander localization based on multi - source data. J Remote Sens, 2019, 23: 177 - 184.

[2] Wu W, Li C, Zuo W, et al. Lunar farside to be explored by Chang'e - 4. Nat Geosci, 2019, 12: 222 - 223.

[3] Liu J, Ren X, Yan W, et al. Descent trajectory reconstruction and landing site positioning of Chang'e - 4 on the lunar farside. Nat Commun, 2019, 10: 4229.

[4] Garrick - Bethell I, Zuber M T. Elliptical structure of the lunar South Pole - Aitken basin. Icarus, 2009, 204: 399 - 408.

[5] Hurwitz D M, Kring D A. Differentiation of the South Pole - Aitken basin impact melt sheet: implications for lunar exploration. J Geophys Res Planets, 2014, 119: 1110 - 1133.

[6] Lucey P G, Taylor G J, Hawke B R, et al. FeO and TiO_2 concentrations in the South Pole - Aitken basin: implications for mantle composition and basin formation. J Geophys Res, 1998, 103: 3701 - 3708.

[7] Pieters C M, Tompkins S, Head J W, et al. Mineralogy of the mafic anomaly in the South Pole - Aitken basin: implications for excavation of the lunar mantle. Geophys Res Lett, 1997, 24: 1903 - 1906.

[8] Moriarty Iii D P, Pieters C M. The character of South Pole - Aitken basin: patterns of surface and subsurface composition. J Geophys Res Planets, 2018, 123: 729 - 747.

[9] Pieters C M, Head Iii J W, Gaddis L, et al. Rock types of South Pole - Aitken basin and extent of basaltic volcanism. J Geophys Res, 2001, 106: 28001 - 28022.

[10] Lawrence D J, Feldman W C, Elphic R C, et al. Iron abundances on the lunar surface as measured by the lunar prospector Gamma - ray and neutron spectrometers. J Geophys Res Planets, 2002, 107: 5130.

[11] Prettyman T H, Hagerty J J, Elphic R C, et al. Elemental composition of the lunar surface: analysis of gamma ray spectroscopy data from Lunar Prospector. J Geophys

Res, 2006, 111: E12007.

[12] Lawrence D J, Feldman W C, Barraclough B L, et al. Thorium abundances on the lunar surface. J Geophys Res, 2000, 105: 20307 - 20331.

[13] Kobayashi S, Karouji Y, Morota T, et al. Lunar farside Th distribution measured by Kaguya gamma - ray spectrometer. Earth Planet Sci Lett, 2012, 337 - 338: 10 - 16.

[14] Jolliff B L, Wieczorek M A, Shearer C K, et al. New Views of the Moon. Gottingan: Mineralogical Society of America, 2018. 60.

[15] Fu X H, Qiao L, Zhang J, et al. The subsurface structure and stratigraphy of the Chang'e - 4 landing site: orbital evidence from small craters on the Von Kármán crater floor. Res Astron Astrophys, 2020, 20: 008.

[16] Gou S, Di K, Yue Z, et al. Lunar deep materials observed by Chang'e - 4 rover. Earth Planet Sci Lett, 2019, 528: 115829.

[17] Huang J, Xiao Z, Flahaut J, et al. Geological characteristics of Von Kármán crater, northwestern South Pole - Aitken basin: Chang'e - 4 landing site region. J Geophys Res Planets, 2018, 123: 1684 - 1700.

[18] Li C, Liu D, Liu B, et al. Chang'e - 4 initial spectroscopic identification of lunar far - side mantle - derived materials. Nature, 2019, 569: 378 - 382.

[19] Ling Z, Jolliff B L, Liu C, et al. Composition, mineralogy, and chronology of mare basalts in Von Kármán crater: a candidate landing site of Chang'e - 4. In: Proceedings of the 49th Lunar and Planetary Science Conference, Houston, 2018. 1939.

[20] Ling Z, Jolliff B L, Liu C, et al. A close view of Chang'e - 4 landing site and science questions to be answered by Yutu - 2. In: Proceedings of the 50th Lunar and Planetary Science Conference, Houston, 2019. 2330.

[21] Ling Z, Qiao L, Liu C, et al. Composition, mineralogy and chronology of mare basalts and non - mare materials in Von Kármán crater: landing site of the Chang'e - 4 mission. Planet Space Sci, 2019, 179: 104741.

[22] Qiao L, Ling Z, Fu X, et al. Geological characterization of the Chang'e - 4 landing area on the lunar farside. Icarus, 2019, 333: 37 - 51.

[23] He Z, Xu R, Li C, et al. Visible and near - infrared imaging spectrometer (VNIS) for *in - situ* lunar surface measurements. In: Proceedings of Sensors, Systems, and Next - Generation Satellites XIX, 2015. 9639.

[24] Liu B, Li C L, Zhang G L, et al. Data processing and preliminary results of the Chang'e - 3 VIS/NIR imaging spectrometer *in - situ* analysis. Res Astron Astrophys, 2014, 14: 1578 - 1594.

[25] Gueymard C A. The Sun's total and spectral irradiance for solar energy applications and solar radiation models. Sol Energy, 2004, 76: 423 - 453.

[26] Li C, Wang Z, Xu R, et al. The scientific information model of Chang'e - 4 visible and near - IR imaging spectrometer (VNIS) and in - flight verification. Sensors,

2019, 19: 2806.

[27] Savitzky A, Golay M J E. Smoothing and differentiation of data by simplified least squares procedures. Anal Chem, 1964, 36: 1627 - 1639.

[28] Hapke B. Space weathering from Mercury to the asteroid belt. J Geophys Res, 2001, 106: 10039 - 10073.

[29] Pieters C M, Taylor L A, Noble S K, et al. Space weathering on airless bodies: resolving a mystery with lunar samples. Meteoritics Planet Sci, 2000, 35: 1101 -1107.

[30] Clark R N, Roush T L. Reflectance spectroscopy: quantitative analysis techniques for remotesensing applications. J Geophys Res, 1984, 89: 6329 - 6340.

[31] Clark R N, Swayze G A, Livo K E, et al. Imaging spectroscopy: earth and planetary remote sensing with the USGS Tetracorder and expert systems. J Geophys Res, 2003, 108: 5131.

[32] McCord T B, Clark R N, Hawke B R, et al. Moon: near - infrared spectral reflectance, a first good look. J Geophys Res, 1981, 86: 10883 - 10892.

[33] Sunshine J M, Pieters C M, Pratt S F. Deconvolution of mineral absorption bands: an improved approach. J Geophys Res, 1990, 95: 6955 - 6966.

[34] Kaur P, Bhattacharya S, Chauhan P, et al. Mineralogy of Mare Serenitatis on the near side of the Moon based on Chandrayaan - 1 Moon Mineralogy Mapper (M^3) observations. Icarus, 2013, 222: 137 - 148.

[35] Wu Y, Li L, Luo X, et al. Geology, tectonism and composition of the northwest Imbrium region. Icarus, 2018, 303: 67 - 90.

[36] Zhang X Y, Ouyang Z Y, Zhang X M, et al. Study of the continuum removal method for the Moon Mineralogy Mapper (M^3) and its application to Mare Humorum and Mare Nubium. Res Astron Astrophys, 2016, 16: 115.

[37] Besse S, Sunshine J M, Staid M I, et al. Compositional variability of the Marius Hills volcanic complex from the Moon Mineralogy Mapper (M^3). J Geophys Res, 2011, 116: 13.

[38] Cheek L C, Donaldson Hanna K L, Pieters C M, et al. The distribution and purity of anorthosite across the Orientale basin: new perspectives from Moon Mineralogy Mapper data. J Geophys Res Planets, 2013, 118: 1805 - 1820.

[39] Horgan B H N, Cloutis E A, Mann P, et al. Near - infrared spectra of ferrous mineral mixtures and methods for their identification in planetary surface spectra. Icarus, 2014, 234: 132 - 154

[40] Pelkey S M, Mustard J F, Murchie S, et al. CRISM multispectral summary products: parameterizing mineral diversity on Mars from reflectance. J Geophys Res, 2007, 112: 14.

[41] Salvatore M R, Mustard J F, Wyatt M B, et al. Definitive evidence of Hesperian basalt in Acidalia and Chryse planitiae. J Geophys Res, 2010, 115: E07005.

[42] Cloutis E A, Gaffey M J. Pyroxene spectroscopy revisited - spectral - compositional correlations and relationship to geothermometry. J Geophys Res, 1991, 96: 22809 -22826.

[43] Gaffey M J, Bell J F, Brown R H, et al. Mineralogical variations within the S - type asteroid class. Icarus, 1993, 106: 573 - 602.

[44] Varatharajan I, Srivastava N, Murty S V S. Mineralogy of young lunar mare basalts: assessment of temporal and spatial heterogeneity using M^3 data from Chandrayaan - 1. Icarus, 2014, 236: 56 - 71

[45] Pieters C M, Englert P A J. Remote Geochemical Analysis, Elemental and Mineralogical Composition. Cambridge: Cambridge University Press, 1993.

[46] Adams J B, Goullaud L H. Plagioclase feldspars - visible and near infrared diffuse reflectance spectra as applied to remote sensing. In: Proceedings of the 9th Lunar and Planetary Science Conference, 1978. 2901 - 2909.

[47] Donaldson Hanna K L, Cheek L C, Pieters C M, et al. Global assessment of pure crystalline plagioclase across the Moon and implications for the evolution of the primary crust. J Geophys Res Planets, 2014, 119: 1516 - 1545.

[48] Ohtake M, Matsunaga T, Haruyama J, et al. The global distribution of pure anorthosite on the Moon. Nature, 2009, 461: 236 - 240.

[49] Tompkins S, Pieters C M. Spectral characteristics of lunar impact melts and inferred mineralogy. Meteoritics Planet Sci, 2010, 45: 1152 - 1169.

[50] Adams J B, Horz F, Gibbons R V. Effects of shock - loading on the reflectance spectra of plagioclase, pyroxene, and glass. In: Proceedings of the 10th Lunar and Planetary Science Conference, 1979. 1 - 3.

[51] Johnson J R, Hörz F. Visible/near - infrared spectra of experimentally shocked plagioclase feldspars. J Geophys Res, 2003, 108: 5120.

[52] Crown D A, Pieters C M. Spectral properties of plagioclase and pyroxene mixtures and the interpretation of lunar soil spectra. Icarus, 1987, 72: 492 - 506.

[53] Klima R L, Pieters C M, Dyar M D. Spectroscopy of synthetic Mg - Fe pyroxenes I: spin - allowed and spin - forbidden crystal field bands in the visible and near - infrared. Meteoritics Planet Sci, 2007, 42: 235 - 253.

[54] Klima R L, Pieters C M, Dyar M D. Characterization of the 1. 2μm M1 pyroxene band: extracting cooling history from near - IR spectra of pyroxenes and pyroxene - dominated rocks. Meteoritics Planet Sci, 2008, 43: 1591 - 1604.

[55] Gross J, Treiman A H, Mercer C N. Lunar feldspathic meteorites: constraints on the geology of the lunar highlands, and the origin of the lunar crust. Earth Planet Sci Lett, 2014, 388: 318 - 328.

[56] Shearer C K, Elardo S M, Petro N E, et al. Origin of the lunar highlands Mg - suite: an integrated petrology, geochemistry, chronology, and remote sensing perspective. Am Miner, 2015, 100: 294 - 325.

[57] Shearer C K, Hess P C, Wieczorek M A, et al. Thermal and magmatic evolution of the Moon. Rev Mineral GeoChem, 2006, 60: 365 - 518.

[58] Lemelin M, Lucey P G, Gaddis L R, et al. Global map products from the Kaguya multiband imager at 512 ppd: Minerals, FeO, and OMAT. In: Proceedings of the 47th Lunar and Planetary Science Conference, 2016. 2994.

[59] Lucey P G. Mineral maps of the Moon. Geophys Res Lett, 2004, 31: L08701.

[60] Bickel C E, Warner J L, Phinney W C. Petrology of 79215 - Brecciation of a lunar cumulate. In: Proceedings of Lunar and Planetary Science Conference, 1976. 7: 1793 -1819.

[61] Dymek R F, Albee A L, Chodos A A. Comparative petrology of lunar cumulate rocks of possible primary origin - Dunite 72415, troctolite 76535, norite 78235, and anorthosite 62237. In: Proceedings of Lunar and Planetary Science Conference, 1975. 6: 301 - 341.

[62] Takeda H, Miyamoto M. Inverted pigeonites from lunar breccia 76255 and pyroxene -crystallization trends in lunar and achondritic crusts. In: Proceedings of Lunar and Planetary Science Conference, 1977. 8: 2617 - 2626.

[63] Warner J L, Simonds C H, Phinney W C. Apollo 17, Station 6 boulder sample 76255 -Absolute petrology of breccia matrix and igneous clasts. In: Proceedings of Lunar and Planetary Science Conference, 1976. 7: 2233 - 2250.

[64] Nakamura Y, Kushiro I. Equilibrium relations of hypersthene, pigeonite and augite in crystallizing magmas: microprobe study of a pigeonite andesite from Weiselberg, Germany. American Mineralogist, 1970, 55: 1999 - 2015.

[65] Hawke B R, Blewett D T, Lucey P G, et al. The origin of lunar crater rays. Icarus, 2004, 170: 1 - 16.

[66] Wilhelms D E, John F, Trask N J. The Geologic History of the Moon. United States Geological Survey, 1987.

[67] Morrison D A. Did a thick South Pole - Aitken basin melt sheet differentiate to form cumulates? In: Proceedings of the 29th Lunar and Planetary Science Conference, 1998. 1657.

[68] Vaughan W M, Head J W, Wilson L, et al. Geology and petrology of enormous volumes of impact melt on the Moon: a case study of the Orientale basin impact melt sea. Icarus, 2013, 223: 749 - 765.

[69] Vaughan W M, Head J W. Impact melt differentiation in the South Pole - Aitken basin: Some observations and speculations. Planet Space Sci, 2014, 91: 101 - 106.

Forsteritic olivine and magnesium – rich orthopyroxene materials measured by Chang'e – 4 rover

Sheng Gou[1,2], Kaichang Di[1,2,3], Zongyu Yue[1,3], Zhaoqin Liu[1], Zhiping He[4], Rui Xu[4], Bin Liu[1,2], Man Peng[1], Wenhui Wan[1], Yexin Wang[1], Jianzhong Liu[5]

1. State Key Laboratory of Remote Sensing Science, Aerospace Information Research Institute, Chinese Academy of Sciences, Beijing 100101, China
2. State Key Laboratory of Lunar and Planetary Sciences, Macau University of Science and Technology, Macau, China
3. CAS Center for Excellence in Comparative Planetology, Hefei 230026, China
4. Key Laboratory of Space Active Opto – Electronics Technology, Shanghai Institute of Technical Physics, Chinese Academy of Sciences, Shanghai 200083, China
5. Institute of Geochemistry, Chinese Academy of Sciences, Guiyang 550002, China

Abstract: China's Chang'e – 4 probe achieved the first soft landing within the South Pole – Aitken (SPA) basin, which is the oldest, largest, and deepest basin on the lunar farside. The deployed Chang'e – 4 rover made in situ spectral measurements along the rover traverse during a nominal three – month mission period. Spectral analyses imply that materials at the Chang'e – 4 landing site have a forsteritic olivine (OL) and magnesium (Mg) – rich orthopyroxene (OPX) assemblage in almost equal fractions. The Chang'e – 4 rover measured materials, which were essentially mixture of multiple sources, were primarily the weathering products of Finsen crater ejecta. Plagioclase (PLG), which has significant implication for the provenance, is often spectrally transparent or featureless in the wavelength range (450 – 2395nm) of the in situ measured spectrum. Depending on the possible absence or presence of abundant PLG, the materials likely originated from a differentiated SPA impact melt pool or from an Mg – suite pluton in the lunar lower crust, and were less likely to originate from the lunar upper mantle.

Keywords: Chang'e – 4 rover; In situ spectral measurements; Finsen crater ejecta; Differentiated impact melt pool; Mg – suite pluton

1 Introduction

The South Pole - Aitken (SPA) basin is recognized as the largest, deepest, and oldest impact basin on the Moon (Stuart - Alexander, 1978; Wilhelms et al., 1979; Spudis et al., 1994). Although studies have suggested the SPA - forming impact was perhaps very oblique, and thus maybe have excavated only shallowly into the lunar crust (Garrick - Bethell and Zuber, 2009; Schultz and Crawford, 2011), recent simulation works revealed that the basin - forming impact should have penetrated the lunar crust and exposed the lunar mantle materials (Miljkovic et al., 2015; Melosh et al., 2017; Zhu et al., 2018). Because it provides both a unique window to study the structure and constitution of the lunar interior (Ennis et al., 2012) and a portal to the early history of the solar system, especially relating to postaccretional heavy bombardment (Jolliff et al., 2010), exploring the SPA basin ranks as a high priority for lunar and solar system science by the National Research Council (NRC, 2011).

China's Chang'e - 4 probe successfully landed in the Von Kármán crater inside the SPA basin on January 3, 2019, making the first soft landing of a probe on the lunar farside (Wu et al., 2019). A lunar rover equipped with several scientific payloads was deployed on the same day. The Chang'e - 4 rover conducted a series of in situ measurements along its traverse. Interpretations of the measured datasets provide first - hand information on the geomorphology, mineral composition, and subsurface structure within the Von Kármán crater and could shed light on the understanding of the lunar evolution. This study analyzes the in situ measured spectra on lunar surface materials during the nominal three - month mission period. Mineral compositions and relative fractions of the measured materials are inferred from spectral deconvolution outputs.

2 Instrument and measurement

2.1 Specifications of the instrument

Three payloads are installed on the Chang'e - 4 rover: a panoramic camera, a lunar - penetrating radar, and a Visible and Near - Infrared Imaging Spectrometer (VNIS) (Wu et al., 2019). This study focuses on spectral datasets measured by the VNIS. The VNIS is mainly composed of a visible/near - infrared (VIS/NIR) imaging spectrometer, a shortwave infrared (SWIR) spectrometer, and a white panel for calibration and dust - proofing (He et al., 2019; Li et al., 2019). The spectral range is 450 - 2395nm with a default sampling interval of 5 nm, and the total number of spectral channels is 400 (with 10 overlapping channels between 900 and 945nm) (He et al., 2019; Li et al., 2019). The effective image sizes of the VIS/NIR imaging spectrometer are 256×256 pixels, and the spatial resolution is approximately 1 mm/pixel from a typical 1m detection distance. The field of view (FOV) of the SWIR spectrometer corresponds to a circular region in the VIS/NIR image with the center located at (96, 128) and a radius of 54 pixels (He et al., 2019; Li et al., 2019).

2.2 In situ spectral measurements

Since the successful deployment on the lunar surface, the Chang'e-4 rover traversed about 163m during its nominal three-month mission period. The VNIS made 13 measurements on lunar regolith and 1 measurement on an about 20-cm rock fragment along the rover traverse (Fig. 1 and Supplementary Table 1).

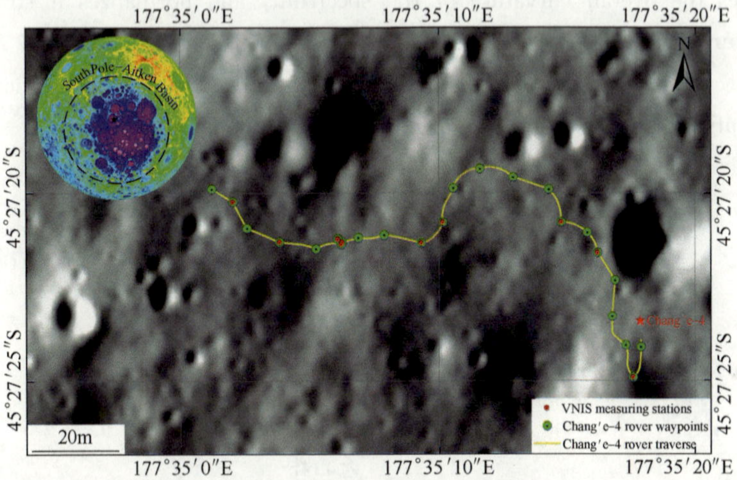

Fig. 1　Traverse map of the Chang'e-4 rover during the first 3 lunar days. The base map is a high resolution (0.9 m/pixel) digital orthophoto map (DOM) (Liu et al., 2019) generated from Lunar Reconnaissance Orbiter Camera (LROC) Narrow Angle Camera (NAC) imagery (Robinson et al., 2010). The Lunar Orbiter Laser Altimeter (LOLA) elevation (Smith et al., 2010) inset shows Chang'e-4 landing site within the SPA basin (black star). Colors represent different heights in the inset, with red being the highest and purple being the lowest

3　Methods

3.1 Full-spectrum reflectance derivation

The spectral analysis is based on the Level 2B (L2B) radiance datasets, which are produced and released online by the Ground Research and Application System (GRAS). The L2B radiance datasets are first corrected by onboard radiometric calibration coefficients for both the VIS/NIR and SWIR spectrometers. These calibration coefficients are obtained from the onboard synchronous calibration experiments that were performed during the three lunar days and are provided by the VNIS manufacturer. The reflectance (reflectance factor, REFF) is then derived in this study by the solar irradiance calibration method (Wu et al., 2018).

A gap exists at the overlapping bands of the VIS/NIR and SWIR spectra (Fig. 2a). Because the SWIR spectrum has a higher signal - to - noise ratio and stability than the average VIS/NIR spectrum (He et al., 2019; Li et al., 2019), the VIS/NIR spectrum in the FOV of the SWIR spectrometer is adjusted to the SWIR spectrum by moving the VIS/NIR spectrum around with an offset (Fig. 2b). The offset is the difference between the average reflectance of the overlapping bands from 900 to 945nm. The full - range spectrum from 450 to 2395nm is thus constructed and it is smoothed by a simple 11 - points moving average method (Fig. 2b). Continuum is removed during spectral deconvolution by the following descried Modified Gaussian model (MGM). The process removes the overall curvature of the spectrum, and normalizes it so that individual absorption feature of interest is compared from a common baseline. Therefore, the influence of moving the spectrum around to derive full - range spectrum has minor influence on making relative, qualitative assessments of general mineralogical trends in this study.

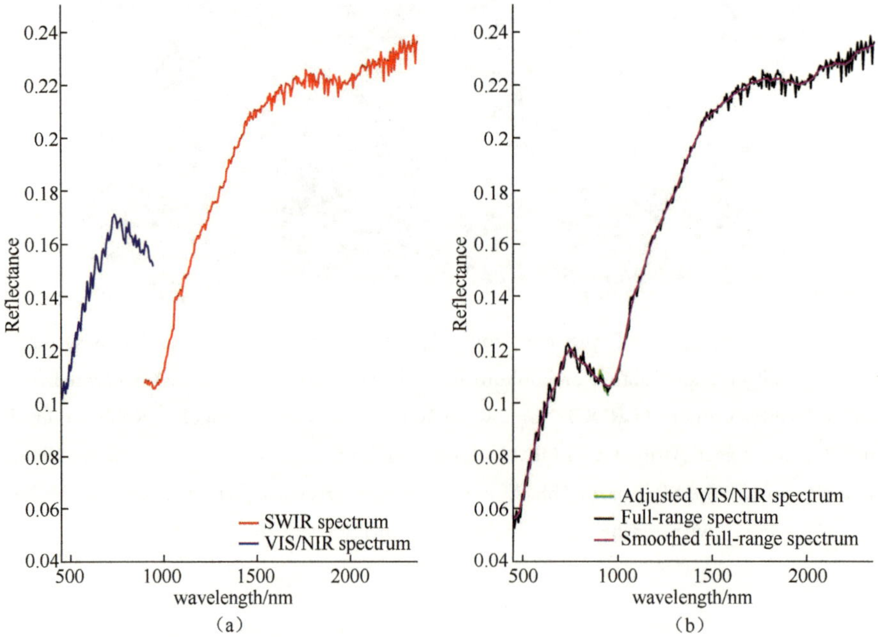

Fig. 2　Spectrum measured at rover station LE00303 for a rock fragment

(a) Illustration of spectrum discontinuity; (b) Illustration of eliminating discontinuity and smoothing spectrum

3.2　Mineral composition identification

The primary rock - forming minerals on the lunar surface are pyroxene (PYX), olivine (OL), and plagioclase (PLG). PYX is roughly classified as orthopyroxene (OPX) and clinopyroxene (CPX) based on chemistry and crystallography (Burns, 1993). PLG dominates the lunar highlands, while OPX, CPX, and OL dominate the mare regions (Pieters, 1986; Spudis and Davis, 1986). These minerals can be uniquely distinguished by their specific spectral characteristics in the wavelength range of the Chang'e-4 rover measured spectrum. Both OPX and CPX spectra exhibit Gaussian shape absorption features. Their absorption positions shift with the content of Fe/Ca/Mg. OPX has absorptions at ~900nm and ~1900nm, respectively. CPX has absorptions at ~1000nm and ~2200nm, respectively (Adams, 1974; Denevi et al., 2007). Typical pure OL spectrum has three overlapping absorption features at approximately 850, 1050, and 1250nm, forming what appears to be a single, broad, and composite absorption feature near 1050nm (Burns, 1970; Sunshine and Pieters, 1998). The overall absorption center shifts towards longer wavelengths with increasing Fe (Burns, 1970), and the absorption band of fayalite (Fe-rich OL) is broader and more flat-bottomed than that of forsterite (Mg-rich OL) (Sunshine and Pieters, 1998). PLG is often considered as a spectrally transparent or featureless mineral. On one hand, due to incorporation of minor amounts of FeO (<1.0 wt%), iron-bearing crystalline PLG spectrum has only one weak and broad absorption near 1250nm (Adams and Goullaud, 1978; Burns, 1993). However, this feature is easily masked by a few to several vol% of mafic minerals in a mixture. Cheek and Pieters (2014) showed that as little as 2 vol% of OL or PYX completely masks the PLG signal. On the other hand, due to shock-induced loss of crystal structure, many lunar PLGs are spectrally featureless (Adams et al., 1979; Bruckenthal and Pieters, 1984). Therefore, PLG cannot be assessed with Chang'e-4 rover measured spectral datasets alone. Nevertheless, the possible presence or absence of abundant PLG in the Chang'e-4 rover measured materials, which has important implications, is discussed in this study.

In addition, other components in the regolith, such as opaque minerals (e. g., ilmenite), and maturation products (e. g., agglutinitic glass), also contribute to the overall reflectance and have significant implications. Ilmenite is highly absorbing and strongly affects the reflectance spectrum when its abundance is high (e. g., ilmenite-rich (>17%) Apollo 17 bulk rock samples), especially for the fine particle size samples (Moriarty and Pieters, 2016). Compared with the ilmenite-rich Apollo 17 samples, TiO_2 abundance at Chang'e-4 landing site estimated from Selenological and Engineering Explorer (SELENE) multiband imager (MI) data is about 1.42±0.24 wt% (Otake et al., 2012) (Supplementary Fig. 1), thus the ilmenite does not strongly affect the overall continuum compared with PYX and OL. Iron-bearing glass (GLS) spectrum exhibits two broad and symmetric absorptions typically centered between 1070 and 1200nm, and between 1090 and 2050nm (Horgan et al., 2014). However, the 1000nm broad iron band absorption of GLS is easily suppressed by iron-bearing mafic minerals and nanophase iron that are ubiquitous in the lunar regolith (Cannon and Mustard, 2015). Therefore, possible presence of these components is not discussed here either.

3.3 MGM deconvolution and composition prediction

Modified Gaussian model (MGM), originally proposed by Sunshine et al. (1990) and revised by Clénet et al. (2011), is an analytical tool to deconvolve overlapping absorptions of mafic mineral spectrum into their fundamental absorption components (Sunshine et al., 1990). One Gaussian function (characterized by band center, width and strength) or combination of multiple Gaussian functions permits the identification of a mineralogical absorption band, indicative of the presence and composition of the predominant mafic minerals (Sunshine and Pieters, 1998). MGM is used in this study with 7 different configurations (Supplementary Table 2) to deconvolve overlapping absorptions of each VNIS measured spectrum into their fundamental absorptions. For example, The 'OL - OPX' configuration (Supplementary Table 2) has 6 Gaussians, including 1 for strong absorption centered around 450nm in ultraviolet (UV) region (UV450), 3 for OL absorptions (OL850, OL1050, and OL1250), and 2 for OPX absorptions (OPX900 and OPX1900) (Clénet et al., 2011). The spectrum continuum is fitted by a second order polynomial with anchoring points adjusted on the local maxima along the spectrum and initial parameters of each Gaussian are made on the basis of laboratory results (Clénet et al., 2011). As one spectrum may be fitted with multiple candidate configurations, the fitting with the least root mean square error (RMSE) is selected as the optimum among candidate configurations that have physical meaning (i. e., satisfy the spectroscopic criteria).

MGM outputs are informative of the chemical composition of OL and PYX (Sunshine and Pieters, 1998; Kanner et al., 2007). To predict the Mg number (Fo#) of a given spectrum that is modeled with OL - related configurations, the MGM modeled three OL - related band centers are fitted to a laboratory derived compositional trend line by finding the least error (Sunshine and Pieters, 1998; Isaacson and Pieters, 2010). It should be noted here that the uncertainty of predicted Fo# arises from slight variations in initial spectral parameters and other errors (e. g., spectrum with lower signal - to - noise ratio) is likely to be on the order of 5 - 10% Fo# range (Isaacson and Pieters, 2010; Pinet et al., 2018; Pinet et al., 2019). Kanner et al. (2007) assessed relationship between normalized band strength ratio (NBSR) that calculated from MGM derived band depths (also called absorption strengths) and the known mineral fractions in an OPX and CPX mixture. The estimated OPX and CPX fraction errors are constrained to ±10%, which is still robust in the presence of up to 10 - 15% OL (Kanner et al., 2007). The concept of NBSR including OL has been subsequently introduced by Clénet et al. (2011). MGM is able to model simple and complex mafic mineralogies, including binary and ternary OL - PYX mixtures, for a large range of grain sizes (Clénet et al., 2011). Five NBSR values (if applicable) are calculated in this study to estimate relative mineral fractions (Tab. 1).

Table 1　Formulae of normalized band strength ratio (NBSR)

NBSR		Definition
1000nm absorption feature	NBSR (OL)	$\frac{BD1000(OL)^{*}}{BD1000(OL)+BD1000(OPX)+BD1000(CPX)}$
	NBSR (OPX)	$\frac{BD1000(OPX)}{BD1000(OL)+BD1000(OPX)+BD1000(CPX)}$
	NBSR (CPX)	$\frac{BD1000(CPX)}{BD1000(OL)+BD1000(OPX)+BD1000(CPX)}$
2000nm absorption feature	NBSR (OPX)	$\frac{BD2000(OPX)}{BD2000(OPX)+BD2000(CPX)}$
	NBSR (CPX)	$\frac{BD2000(CPX)}{BD2000(OPX)+BD2000(CPX)}$

* BD1000 and BD2000 are MGM derived band depths (also called absorption strengths) for OL, OPX, and CPX.

4　Results

4.1　Spectroscopic features of lunar surface materials

According to space weathering studies (Pieters and Noble, 2016), the reflectance spectra measured by the Chang'e-4 rover should have suffered from space weathering. The in situ measured reflectance spectra exhibit typical features of an increasing spectral slope towards longer wavelengths (reddening), a darkening of the overall reflectance and a weakening of the absorption features (Fig. 3a). Two absorption features at approximately

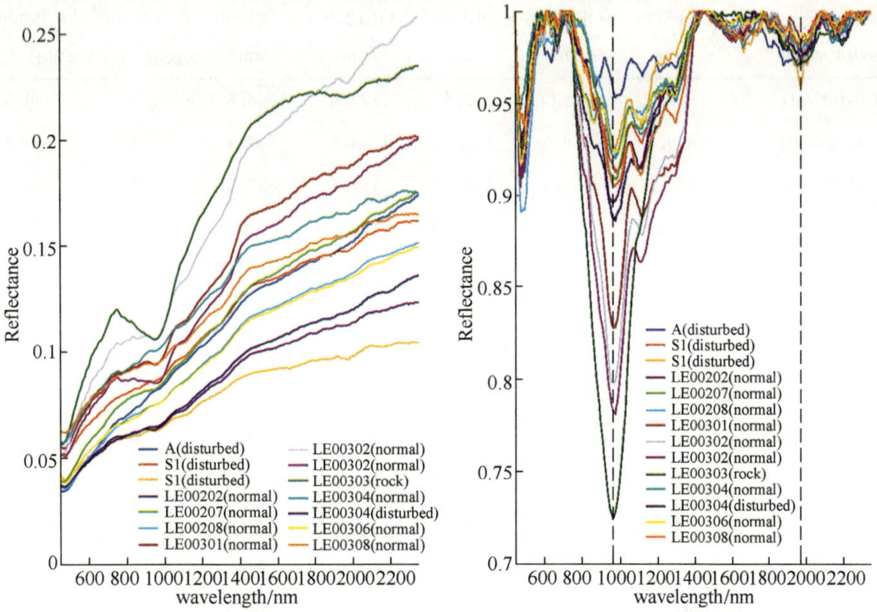

(a)　　　　　　　　　　　　　　　　(b)

Fig. 3　Spectra measured by the Chang'e-4 rover at 11 different stations during the first three lunar days: (a) the original reflectance and (b) the continuum-removed spectrum. The dashed lines are located at 965 nm and 1975nm. The discontinuity in some spectra in this figure near 1400nm is related to signals that arise from artifacts of the instrument design

1000nm and 2000nm are discernible from the continuum - removed spectra (Fig. 3b). The band depths of the 1000nm absorption for PYX - and OL - bearing minerals are relatively stronger than that of the 2000nm absorption for PYX.

4.2 Mineral compositions of lunar regolith

The MGM fitting results show that 10 out of the 13 lunar regolith spectra can be optimally fitted with the 'OL - OPX' configuration (Supplementary Fig. 2 to 11), indicating that the Chang'e - 4 rover measured regolith has an OL and OPX assemblage. The modeled OPX - related 1000nm absorption centers vary from 900.4nm to 969.8nm (939.8±26.5nm) (Table 2), and the 2000nm absorption centers vary from 1816.2nm to 1915.2nm (1862.9±37.2nm). The relative short band centers indicate that PYX in the regolith at the Chang'e - 4 landing site is chiefly Mg - rich orthopyroxene (OPX). NBSR values (OL: OPX is 50%: 50%) (Tab. 2) indicate that OL and OPX are present in almost equal fractions in the regolith. The predicted magnesium number (Fo# 87±10) (Tab. 2), with an uncertainty on the order of 10, indicates that the OL is forsteritic (Mg - rich) OL. It should be noted here that the complex OL absorption feature not only varies with the iron content but also with grain size and abundance. As demonstrated by Clénet et al. (2011), a forsterite with very large grains and a fayalite with smaller grains may have similar spectral features. Thus high Fo# value is very tentative here and should be treated with caution for geological interpretation.

Table 2　MGM derived band centers, NBSR, and Fo# values

Rover station (regolith state)	OL850 /nm	OPX900 /nm	OL1050 /nm	OL1250 /nm	OPX1900 /nm	1000nm absorption		Fo#
						NBSR (OL)	NBSR (OPX)	
S1 (disturbed)	840.3	940	1066.7	1182.6	1866.6	0.5	0.5	96
S1 (disturbed)	849.8	918.9	1107.5	1213.3	1837.4	0.6	0.4	65
LE00202 (normal)	845.9	954.5	1110.1	1194.1	1833.3	0.5	0.5	76
LE00301 (normal)	834.5	969.8	1070.9	1188	1848.1	0.5	0.5	94
LE00302 (normal)	850.1	967	1074.5	1182.7	1915.2	0.5	0.5	89
LE00302 (normal)	839.5	962.6	1071.7	1187.8	1910.4	0.6	0.4	92
LE00304 (normal)	844.4	906.5	1088	1179.7	1824.6	0.5	0.5	90
LE00304 (disturbed)	850	918.9	1078.5	1197.8	1908.2	0.5	0.5	80
LE00306 (normal)	850	900.4	1053	1183.7	1868.6	0.5	0.5	94
LE00308 (normal)	832.8	959.8	1087.3	1180.9	1816.2	0.5	0.5	95
Mean	843.7	939.8	1080.8	1189.1	1862.9	0.5	0.5	87
Standard deviation	6.6	26.5	17.8	10.3	37.2	0.0	0.0	10

Note that NBSR values represent relative mineral fractions, and the uncertainties are constrained to±10%.

4.3 Mineral composition of a rock fragment

MGM with the 'OL - OPX - CPX' configuration provides the best modeling results in terms of mineral detection for the rock spectrum (Fig. 4), and the modeled OL band centers predict that the Mg - number (Fo #) is 79, pointing at a Mg - rich composition. The short OPX absorption bands (920. 6nm and 1927. 8nm) indicate it is Mg - rich. NBSR values (Table 3) for the 1000 - nm absorption feature predict that the relative fractions of OPX and CPX are equal (OL: OPX: CPX is 20%: 40%: 40%). NBSR values for the 2000 - nm absorption features also predict that OPX and CPX are present in equal fractions (OPX: CPX is 50%: 50%), indicating self - consistency in the mineral fractions. The OL and OPX compositions of the rock fragment predicted from MGM fitting results are in agreement with that of the regolith, indicating that the rock fragment has a high probability of having the same source as the surrounding lunar regolith. Because of minerals' overlapping spectral absorption features and above stated grain size effects, it's also should be noted here further MGM modeling experiments should be carried out in well - constrained conditions to better understand the influence of the presence of possible other components (e. g., PLG) and mineral grain sizes on assessing mineral compositions and estimating relative fractions (Pinet, 2019).

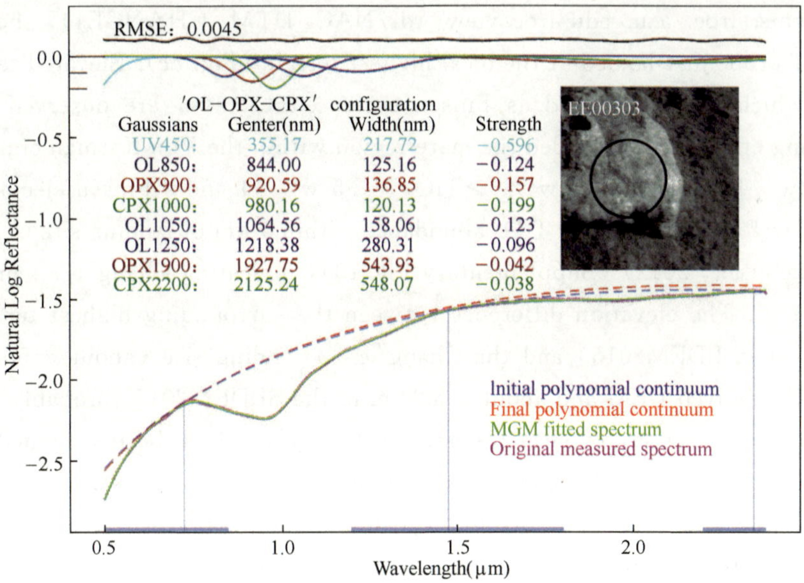

Fig. 4　Rock spectrum modeled with 'OL - OPX - CPX' configuration.
The VIS/NIR image inset is taken at station LE00303
(the circle corresponds to the FOV of SWIR spectrometer)

Table 3　NBSR values of the rock spectrum

NBSR	1000nm absorption			2000nm absorption	
	NBSR (OL)	NBSR (OPX)	NBSR (CPX)	NBSR (OPX)	NBSR (CPX)
Value	0. 2	0. 4	0. 4	0. 5	0. 5

Note: The NBSR values represent relative mineral fractions.

5 Discussion

5.1 Primary source of the in situ spectrally measured materials

The materials measured by Chang'e-4 rover were essentially mixtures of many origins, especially in a complex, ancient terrane such as the SPA basin. Obvious NE-SW trending linear features, which radiate from the ~73-km diameter Finsen crater (Fig. 5b), are observed from the elevation map of Chang'e-4 landing area (Huang et al., 2018). In addition, the MI reflectance mosaic (Supplementary Fig. 12) and false color composite (Supplementary Fig. 13) show that Chang'e-4 landed on a low-albedo region that was covered by old mixed/gardened ejecta materials. Compared with surrounding highest terrain in the Chang' e-4 landing area, topographic profiles, which are extracted from both an improved lunar digital elevation model (DEM) SLDEM2015 (512 pixels/degree) (Barker et al., 2016) and a Lunar Reconnaissance Orbiter Camera (LROC) Narrow Angle Camera (NAC) -derived Digital Terrain Model (DTM, about 5 m/pixel) (http: //wms. lroc. asu. edu/lroc/view _rdr/NAC _DTM _CHANGE4), show that the Chang'e-4 probe just landed at the local highest terrain (Fig. 5c). Elevated terrains with intervals, which are interpreted as Finsen crater ejecta rays, are observed from both profiles. Ling et al. (2018) divided the mare region within the Von Kármán crater into two units: A low-Ti unit (LT, 3 wt% $>TiO_2>$1.5 wt %) and a Finsen ejecta unit (FE, $TiO_2<$1.5 wt%). MI-derived TiO_2 abundance within 5 km of landing site is 1.42 ± 0.24 wt% (Otake et al., 2012) (Supplementary Fig. 13), indicating Chang'e-4 probe landed on the FE unit. The elevation difference between the surrounding highest terrain (about −5915m in the SLDEM2015) and the Chang'e-4 landing site (about −5925m in the LROC NAC-derived DEM and about −5930m in the SLDEM2015) probably reflects the thickness of the Finsen crater ejecta is uneven. Chang'e-4 probe just landed in a place where the ejecta thickness of the ejecta is relatively thin. On the basis of these analyses, it is proposed that the local mare basalt deposits at the Chang'e-4 landing site were extensively covered by ejecta from Finsen crater. And the Chang'e-4 rover measured forsteritic OL and Mg-rich OPX materials are primarily lunar deep-seated materials excavated from Finsen crater.

5.2 Constrain for the provenance of the materials

Topographic and geologic context analysis reveals that the Chang'e-4 rover measured materials was primarily ejecta from Finsen crater. MGM deconvolution of the in situ measured spectra implies that the materials at the landing site have a forsteritic OL and Mg-rich OPX assemblage in almost equal fractions. It cannot be ruled out the possi-bility that PLG is present , which has significant implication for the provenance of the

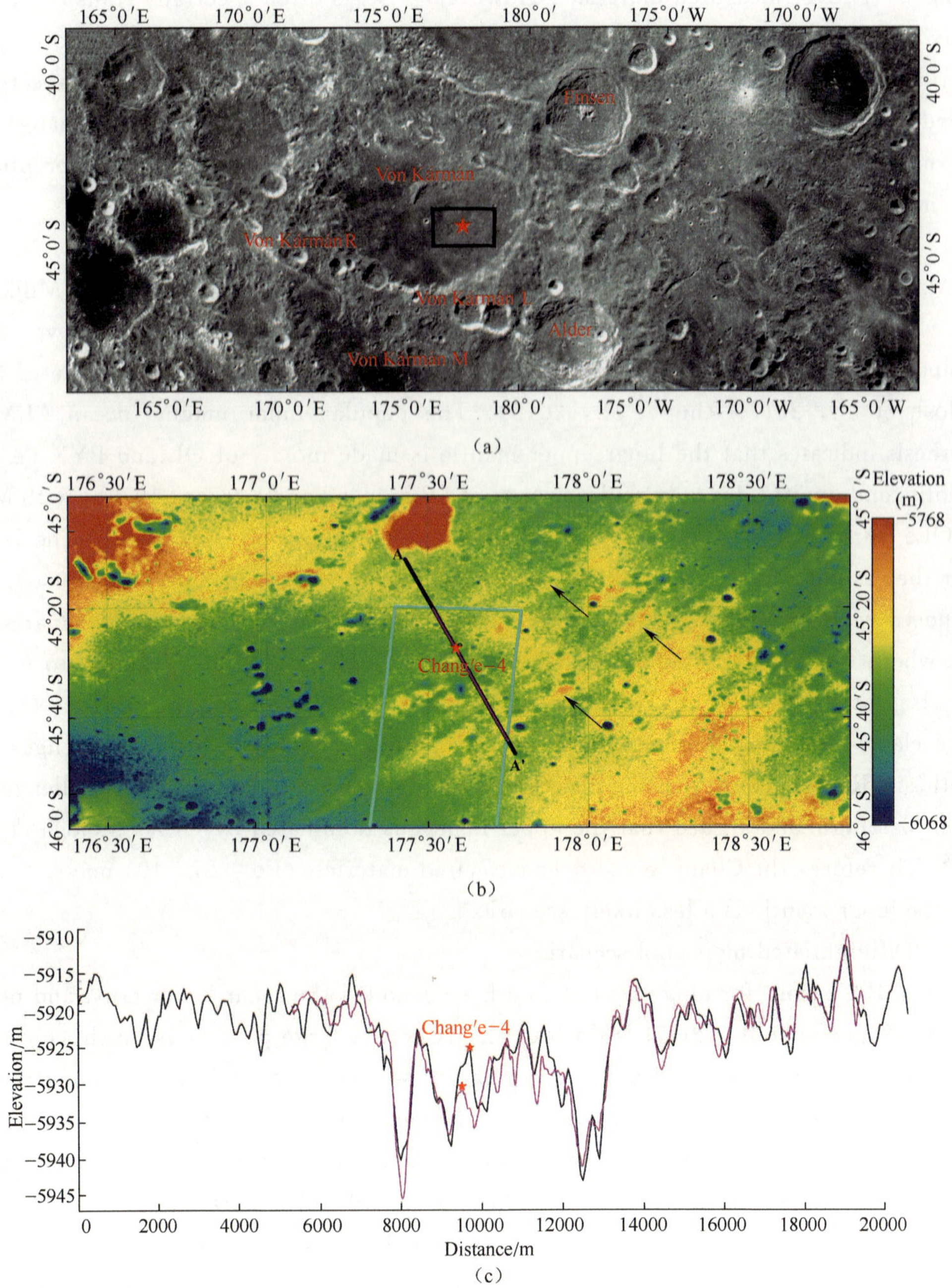

Fig. 5　(a) Chang' e - 4 probe landing site (red star) inside the Von Kármán crater and surrounding craters within the SPA basin. The black rectangle marks the extent of the elevation map shown in b. (b) SLDEM2015 elevation map in the Chang' e - 4 probe landing area. Black arrows indicate the NE - SW trend features. The cyan polygon is the footprint of LROC NAC - derived DTM. (c) Topographic profiles of A - A' that cross the Chang' e - 4 probe landing site. Elevation data of black profile are extracted from SLDEM2015 and the magenta are from LROC NAC - derived DTM. The offset between the two stars that mark the Chang' e - 4 probe landing site along the x - axis is caused by the two elevation datasets are not strictly co - registrated. The offset along the y - axis is caused by the elevation accuracy of the two datasets

Chang'e - 4 rover measured materials. However, PLG is often spectrally transparent due to mix with other mafic minerals or featureless due to shock - induced loss of crystal structure in the wavelength range (450 — 2395nm) of the in situ measured spectrum. Depending on the possible absence or presence of a large amount of PLG, the Chang'e - 4 rover measured materials may originate from: (1) lunar upper mantle; (2) differentiated SPA impact melt pool; (3) Mg - suite pluton in the lunar lower crust.

5.2.1　Lunar upper mantle scenario

Mantle materials are expected to be exposed by the basin - forming events (Miljkovic et al., 2015). The SPA basin - forming impact should have sampled the lunar lower crust and upper mantle even if it's a very oblique impact (e. g., impact at an angle of 20°) (Melosh et al., 2017; Zhu et al., 2018). The popular lunar magma ocean (LMO) hypothesis indicates that the lunar upper mantle is made mostly of OL and PYX (e. g., peridotite and dunite) due to cumulate overturn (Ringwood and Kesson, 1976), with Mg - rich OLs being believed to be the most abundant constituent (Fo is expected to be >90) under the assumption that all planets usually are like the Earth (Gaul et al., 2000). OL has indeed been remotely detected on several large - impact basin rims on the lunar sur - face, where the crust was substantially thinned through excavation (Yamamoto et al., 2010; Isaacson et al., 2011). However, the almost complete absence of evidence (e. g., dunite clasts) in the returned lunar samples (Shearer et al., 2015) challenges the hypothesis. Recent remote sensing observations and numeric simulations also challenge the current paradigm and argued that the upper mantle is dominated by OPX (Melosh et al., 2017). Therefore, the Chang'e - 4 rover measured materials (Fo# 87±10) may originate from the lunar mantle is a less likely scenario.

5.2.2　Differentiated melt pool scenario

The SPA basin - forming event should have sampled the lunar lower crust and upper mantle (Miljkovic et al., 2015; Melosh et al., 2017). A large portion as much as 60%～80% of the impact - generated melts was retained within the SPA basin, forming a melt pool. The melt pool likely differentiated and formed a noritic layer with a thickness of ～6km and possibly as thick as ～13km (Hurwitz and Kring, 2014; Vaughan and Head, 2014). The estimated transient cavity radius of the SPA basin ranges from 840km to 1400km (Wieczorek and Phillips, 1999; Petro and Pieters, 2002; Potter et al., 2012). Finsen crater is just about 370km from the center of the SPA basin. Therefore, the materials measured by Chang'e - 4 rover were probably primarily the weathering products of the noritic rocks that mainly excavated from Finsen crater 3.61 billion years ago (Ivanov et al., 2018).

5.2.3　Mg - suite pluton scenario

OL that is present in mare basalt is usually Fe - rich (Papike et al., 1976), while a Mg - rich composition suggests a relatively primitive plutonic source (Isaacson et al., 2011). Despite the relative high Fo value, Apollo samples rich in OL have been confirmed to not necessarily contain mantle components. For example, OLs in Type C basalt 74,275 have forsterite contents ranging from 68.2—83.2, and 72,415 contains OL with forsterite contents >88 (Shearer et al., 2015). In addition, the definitive identification of SPA derived impact melt on the basin floor, either by remote sensing or via photogeology, is extremely difficult due to the number of subsequent impacts and volcanic activities (Petro, 2011). And the Finsen crater is on the margin of the proposed impact melt pool. Therefore, the possibility that the Chang'e-4 rover measured forsteritic OL and Mg-rich OPX materials may originate from an Mg-suite pluton in the lunar lower crust cannot be excluded in this study.

Based on these analyses, the Chang'e-4 rover measured forsteritic OL and Mg-rich OPX materials likely originated from the differentiated SPA impact melt pool or from an Mg-suite pluton in the lunar lower crust, and are less likely to originate from the lunar mantle. As SPA basin has not been sampled either by previous lunar sample return missions (Apollo, Luna) or by lunar meteorites, a definitive answer to the sources of VNIS measured materials can only be addressed in the future through detailed analysis on returned samples. Specifically, trace element geochemistry or the amount and composition of minor minerals (including plagioclase) are needed to fully constrain the petrologic history of these materials.

6 Conclusion

Chang'e-4 rover made in situ spectral measurements during a nominal three-month mission period within the SPA basin on the lunar farside. MGM deconvolution on the in-situ measured spectra reveals materials at Chang'e-4 landing site have a Fo-rich OL and high-Mg OPX assemblage in almost equal fractions. The materials measured by the rover were actually a highly mixture of multiple sources, which were primarily the weathering products of Finsen crater ejecta. PLG, which has significant implication for the provenance of the measured materials, is often spectrally transparent or featureless in the wavelength range (450—2395nm) of the in situ measured spectrum. Depending on the possible absence or presence of abundant PLG, the Chang'e-4 rover measured forsteritic OL and Mg-rich OPX materials likely originated from the differentiated SPA impact melt pool or from an Mg-suite pluton in the lunar lower crust, and were less likely to originate from the lunar upper mantle.

Acknowledgments

The authors gratefully thank the whole Chang'e-4 team for making the mission a great success and for providing the data (http://moon.bao.ac.cn). This research is supported by the B-type Strategic Priority Program of the Chinese Academy of Sciences (grant No. XDB41000000), the National Natural Science Foundation of China (grant Nos. 41702354, 41941003, and 41972321), Macau Young Scholars Program (grant No. AM201902), and the Science and Technology Development Fund of Macau (grant No. 131/2017/A3). The authors greatly appreciated two anonymous reviewers for constructive comments that improved the manuscript quality substantially.

Appendix A. Supplementary data

Supplementary data to this article can be found online at https://doi.org/10.1016/j.icarus.2020.113776.

References

[1] Adams, J. B., 1974. Visible and near - infrared diffuse reflectance spectra of pyroxenes as applied to remote sensing of solid objects in the solar system. J. Geophys. Res. 79 (32), 4829 - 4836.

[2] Adams, J. B., Goullaud, L. H., 1978. Plagioclase feldspars - visible and near infrared diffuse reflectance spectra as applied to remote sensing. In: 9th Lunar and Planetary Science Conference, pp. 2901 - 2909.

[3] Adams, J. B., Horz, F., Gibbons, R. V., 1979. Effects of shock - loading on the reflectance spectra of plagioclase, pyroxene, and glass. In: 10th Lunar and Planetary Science Conference. Pergamon Press, Houston, TX, pp. 1 - 3.

[4] Barker, M. K., Mazarico, E., Neumann, G. A., Zuber, M. T., Haruyama, J., Smith, D. E., 2016. A new lunar digital elevation model from the lunar orbiter laser altimeter and SELENE terrain camera. Icarus 273, 346 - 355.

[5] Bruckenthal, E. A., Pieters, C. M., 1984. Spectral Effects of Natural Shock on Plagioclase Feldspar. 15th Lunar and Planetary Science Conference, Houston, TX, pp. 96 - 97.

[6] Burns, R. G., 1970. Crystal field spectra and evidence of cation ordering in olivine minerals. Am. Mineral. 55, 1608 - 1632.

[7] Burns, R. G., 1993. Mineralogical Applications of Crystal Field Theory, 2nd edition. Cambridge University Press, New York.

[8] Cannon, K. M., Mustard, J. F., 2015. Preserved glass - rich impactites on Mars. Geology 43 (7), 635 - 638.

[9] Cheek, L. C., Pieters, C. M., 2014. Reflectance spectroscopy of plagioclase - dominated mineral mixtures: implications for characterizing lunar anorthosites remotely. Am. Mineral. 99 (10), 1871 - 1892.

[10] Clénet, H., Pinet, P., Daydou, Y., Heuripeau, F., Rosemberg, C., Baratoux, D., Chevrel, S., 2011. A new systematic approach using the modified Gaussian model:

insight for the characterization of chemical composition of olivines, pyroxenes and olivine - pyroxene mixtures. Icarus 213 (1), 404 - 422.

[11] Denevi, B. W., Lucey, P. G., Hochberg, E. J., Steutel, D., 2007. Near - infrared optical constants of pyroxene as a function of iron and calcium content. J. Geophys. Res. Planets 112 (E5), E05009.

[12] Ennis, M., Fagan, A., Pogue, J., Porter, S., Snape, J., 2012. Feasibility assessment of all science concepts within south pole - Aitken Basin. In: Kring, D. A., Durda, D. D. (Eds.), A Global Lunar Landing Site Study to Provide the Scientific Context for Exploration of the Moon. Lunar and Planetary Institute, Houston, pp. 477 - 562.

[13] Garrick - Bethell, I., Zuber, M. T., 2009. Elliptical structure of the lunar south pole - Aitken basin. Icarus 204 (2), 399 - 408.

[14] Gaul, O. F., Griffin, W. L., O'Reilly, S. Y., Pearson, N. J., 2000. Mapping olivine composition in the lithospheric mantle. Earth Planet. Sci. Lett. 182 (3), 223 - 235.

[15] He, Z., Li, C., Xu, R., Lv, G., Yuan, L., Wang, J., 2019. Spectrometers based on acousto optic tunable filters for in - situ lunar surface measurement. J. Appl. Remote. Sens. 13 (2), 27502.

[16] Horgan, B. H. N., Cloutis, E. A., Mann, P., Bell, J. F., 2014. Near - infrared spectra of ferrous mineral mixtures and methods for their identification in planetary surface spectra. Icarus 234, 132 - 154.

[17] Huang, J., Xiao, Z., Flahaut, J., Martinot, M., Head, J., Xiao, X., Xie, M., Xiao, L., 2018. Geological characteristics of Von Kármán crater, northwestern south pole - Aitken Basin: Chang'E - 4 landing site region. J. Geophys. Res. Planets 123 (7), 1684 -1700.

[18] Hurwitz, D. M., Kring, D. A., 2014. Differentiation of the south pole - Aitken basin impact melt sheet: implications for lunar exploration. J. Geophys. Res. Planets 119 (6), 1110 - 1133.

[19] Isaacson, P. J., Pieters, C. M., 2010. Deconvolution of lunar olivine reflectance spectra: implications for remote compositional assessment. Icarus 210 (1), 8 - 13.

[20] Isaacson, P. J., Pieters, C. M., Besse, S., Clark, R. N., Head, J. W., Klima, R. L., Mustard, J. F., Petro, N. E., Staid, M. I., Sunshine, J. M., 2011. Remote compositional analysis of lunar olivine - rich lithologies with moon mineralogy mapper (M3) spectra. J. Geophys. Res. Planets 116 (E6), E00G11.

[21] Ivanov, M. A., Hiesinger, H., van der Bogert, C. H., Orgel, C., Pasckert, J. H., Head, J. W., 2018. Geologic history of the northern portion of the south pole - Aitken Basin on the moon. J. Geophys. Res. Planets 123 (10), 2585 - 2612.

[22] Jolliff, B., Alkalai, L., Pieters, C., Head, J., Papanastassiou, D., Bierhaus, E., 2010. Sampling the south pole - Aitken Basin: Objectives and site selection criteria.

In: 41st Lunar and Planetary Science Conference, p. 2450.

[23] Kanner, L. C., Mustard, J. F., Gendrin, A., 2007. Assessing the limits of the modified Gaussian model for remote spectroscopic studies of pyroxenes on Mars. Icarus 187 (2), 442 - 456.

[24] Li, C. L., Xu, R., Lv, G., Yuan, L. Y., He, Z. P., Wang, J. Y., 2019. Detection and calibration characteristics of the visible and near - infrared imaging spectrometer in the Chang'e - 4. Rev. Sci. Instrum. 90 (10), 103106.

[25] Ling, Z., Jolliff, B. L., Liu, C. Q., Bi, X. Y., Liu, L., Qiao, L., Zhang, J., Fu, X., Liu, J., 2018. Composition, mineralogy, and chronology of Mare basalts in Von Kármán crater: A candidate landing site of Chang'e - 4. In: 49th Lunar and Planetary Science Conference, the WoodLands, Texas, p. 1939.

[26] Liu, B., Niu, S., Xin, X., Jia, M., Di, K., Liu, Z., Peng, M., Yue, Z., 2019. High precision DTM and DOM denerating using multi - source orbital data on Chang'e - 4 landing site. In: Int. Arch. Photogramm. Remote Sens. Spatial Inf. Sci, XLII - 2/W13. Copernicus Publications, Enschede, The Netherlands, pp. 1413 - 1417.

[27] Melosh, H. J., Kendall, J., Horgan, B., Johnson, B. C., Bowling, T., Lucey, P. G., Taylor, G. J., 2017. South pole - Aitken basin ejecta reveal the Moon' s upper mantle. Geology 45 (12), 1063 - 1066.

[28] Miljkovic, K., Wieczorek, M. A., Collins, G., C. Solomon, S., E. Smith, D., T. Zuber, M., 2015. Excavation of the mantle in basin - forming events on the moon. Earth Planet. Sci. Lett. 409, 243 - 251.

[29] Moriarty, D. P., Pieters, C. M., 2016. Complexities in pyroxene compositions derived from absorption band centers: examples from Apollo samples, HED meteorites, synthetic pure pyroxenes, and remote sensing data. Meteorit. Planet. Sci. 51 (2), 207 -234.

[30] NRC, 2011. Vision and Voyages for Planetary Science in the Decade 2013 - 2022. National Research Council, Washington, D. C.

[31] Otake, H., Ohtake, M., Hirata, N., 2012. Lunar Iron and titanium abundance algorithms based on SELENE (Kaguya) multiband imager data. In: 43rd Lunar and Planetary Science Conference, the Woodlands, Texas, p. 1905.

[32] Papike, J. J., Hodges, F. N., Bence, A. E., Cameron, M., Rhodes, J. M., 1976. Mare basalts: crystal chemistry, mineralogy, and petrology. Rev. Geophys. 14 (4), 475 -540.

[33] Petro, N. E., 2011. Volume of impact melt generated by the formation of the south pole - Aitken Basin. In: Annual Meeting of the Lunar Exploration Analysis Group, Houston, Texas, p. 2051.

[34] Petro, N. E., Pieters, C. M., 2002. The size and location of the transient crater of the south pole - Aitken Basin. In: 33rd Lunar and Planetary Science Conference,

Houston，Texas，p. 1848.

[35] Pieters，C. M.，1986. Composition of the lunar highland crust from near - infrared spectroscopy. Rev. Geophys. 24 (3)，557 - 578.

[36] Pieters，C. M.，Noble，S. K.，2016. Space weathering on airless bodies. J. Geophys. Res. Planets 121 (10)，1865 - 1884.

[37] Pinet，P.，2019. The Moon's mantle unveiled. Nature. Nature. 569，338 - 339.

[38] Pinet，P. C.，Chevrel，S. D.，Daydou，Y. H.，2018. Characterization of the olivine/ plagioclase mineralogy at Copernicus crater from MGM Deconvolution of M3 observations. In：49th Lunar and Planetary Science Conference，the Woodlands，Texas.

[39] Pinet，P. C.，Chevrel，S. D.，Daydou，Y. H.，2019. Reassessing the relationship between olivine composition and reflectance spectroscopy from advanced MGM deconvolution. In：50th Lunar and Planetary Science Conference，the Woodlands，Texas.

[40] Potter，R. W. K.，Collins，G. S.，Kiefer，W. S.，McGovern，P. J.，Kring，D. A.，2012. Constraining the size of the south pole - Aitken basin impact. Icarus 220 (2)，730 -743.

[41] Ringwood，A. E.，Kesson，S. E.，1976. A dynamic model for mare basalt petrogenesis. In：7th Lunar Science Conference. Pergamon Press，Houston，Texas，pp. 1697 - 1722.

[42] Robinson，M. S.，Brylow，S. M.，Tschimmel，M.，Humm，D.，Lawrence，S. J.，Thomas，P. C.，Denevi，B. W.，Bowman - Cisneros，E.，Zerr，J.，Ravine，M. A.，Caplinger，M. A.，Ghaemi，F. T.，Schaffner，J. A.，Malin，M. C.，Mahanti，P.，Bartels，A.，Anderson，J.，Tran，T. N.，Eliason，E. M.，McEwen，A. S.，Turtle，E.，Jolliff，B. L.，Hiesinger，H.，2010. Lunar reconnaissance orbiter camera (LROC) instrument overview. Space Sci. Rev. 150 (1 - 4)，81 - 124.

[43] Schultz，P. H.，Crawford，D. A.，2011. Origin of nearside structural and geochemical anomalies on the moon. Geol. Soc. Am. Spec. Pap. 477，141 - 159.

[44] Shearer，C. K.，Burger，P. V.，Bell，A. S.，Guan，Y.，Neal，C. R.，2015. Exploring the Moon's surface for remnants of the lunar mantle 1. Dunite xenoliths in mare basalts. A crustal or mantle origin? Meteorit. Planet. Sci. 50 (8)，1449 - 1467.

[45] Smith，D. E.，Zuber，M. T.，Neumann，G. A.，Lemoine，F. G.，Mazarico，E.，Torrence，M. H.，McGarry，J. F.，Rowlands，D. D.，Head III，J. W.，Duxbury，T. H.，Aharonson，O.，Lucey，P. G.，Robinson，M. S.，Barnouin，O. S.，Cavanaugh，J. F.，Sun，X.，Liiva，P.，Mao，D. - d.，Smith，J. C.，Bartels，A. E.，2010. Initial observations from the lunar orbiter laser altimeter (LOLA). Geophys. Res. Lett. 37 (18)，L18204.

[46] Spudis，P. D.，Davis，P. A.，1986. A chemical and petrological model of the lunar

crust and implications for lunar crustal origin. J. Geophys. Res. Solid Earth 91 (B13), E84 - E90.

[47] Spudis, P. D., Gillis, J. J., Reisse, R. A., 1994. Ancient multiring basins on the moon revealed by Clementine laser altimetry. Science 266 (5192), 1848 - 1851.

[48] Stuart - Alexander, D. E., 1978. Geologic map of the central far side of the moon IMAP 1047. http://pubs. er. usgs. gov/publication/i1047.

[49] Sunshine, J. M., Pieters, C. M., 1998. Determining the composition of olivine from reflectance spectroscopy. J. Geophys. Res. 103 (E6), 13675 - 13688.

[50] Sunshine, J. M., Pieters, C. M., Pratt, S. F., 1990. Deconvolution of mineral absorption bands - an improved approach. J. Geophys. Res. 95 (B5), 6955 - 6966.

[51] Vaughan, W. M., Head, J. W., 2014. Impact melt differentiation in the south pole - Aitken basin: some observations and speculations. Planet. Space Sci. 91, 101 - 106.

[52] Wieczorek, M. A., Phillips, R. J., 1999. Lunar multiring basins and the cratering process. Icarus 139 (2), 246 - 259.

[53] Wilhelms, D. E., Howard, K. A., Wilshire, H. G., 1979. Geologic map of the south side of the moon. In: U. S. G. S. Misc. Inves. Ser. Map I - 1162, 1 : 5M Scale.

[54] Wu, Y., Wang, Z., Cai, W., Lu, Y., 2018. The absolute reflectance and new calibration site of the moon. Astron. J. 155 (5), 213.

[55] Wu, W., Li, C., Zuo, W., Zhang, H., Liu, J., Wen, W., Su, Y., Ren, X., Yan, J., Yu, D., Dong, G., Wang, C., Sun, Z., Liu, E., Yang, J., Ouyang, Z., 2019. Lunar farside to be explored by Chang'e - 4. Nat. Geosci. 12 (4), 222 - 223.

[56] Yamamoto, S., Nakamura, R., Matsunaga, T., Ogawa, Y., Ishihara, Y., Morota, T., Hirata, N., Ohtake, M., Hiroi, T., Yokota, Y., Haruyama, J., 2010. Possible mantle origin of olivine around lunar impact basins detected by SELENE. Nat. Geosci. 3, 533 - 536.

[57] Zhu, M. - H., Wünnemann, K., Morbidelli, A., Artemieva, N. A., 2018. South pole - Aitken basin may contribute insignificantly to the late accretion of the moon. In: 49th Lunar and Planetary Science Conference, the WoodLands, Texas, p. 1955.

Lunar deep materials observed by Chang'e – 4 rover

Sheng Gou[1,2], Kaichang Di[1,2,3], Zongyu Yue[1,3], Zhaoqin Liu[1], Zhiping He[4], Rui Xu[4], Honglei Lin[5], Bin Liu[1,2], Man Peng[1], Wenhui Wan[1], Yexin Wang[1], Jianzhong Liu[6]

1. State Key Laboratory of Remote Sensing Science, Institute of Remote Sensing and Digital Earth, Chinese Academy of Sciences, Beijing 100101, China

2. State Key Laboratory of Lunar and Planetary Sciences, Macau University of Science and Technology, Macau, China

3. CAS Center for Excellence in Comparative Planetology, Hefei 230026, China

4. Key Laboratory of Space Active Opto – Electronics Technology, Shanghai Institute of Technical Physics, Chinese Academy of Science, Shanghai 200083, China

5. Key Laboratory of Earth and Planetary Physics, Institute of Geology and Geophysics, Chinese Academy of Sciences, Beijing 100029, China

6. Institute of Geochemistry, Chinese Academy of Sciences, Guiyang 550002, China

Abstract: China's Chang'e – 4 spacecraft achieved the first ever soft – landing within the South Pole – Aitken (SPA) basin on the farside of the Moon. The Chang'e – 4 rover, named Yutu – 2, made in – situ spectral observations on lunar regolith and a rock fragment at 11 locations during a nominal three – month mission period. The lunar regolith has a relative high olivine/pyroxene ratio, with the pyroxene being chiefly Mg – rich Low – Ca pyroxene (LCP). The rock fragment has a similar Mg – rich composition to that of the regolith. According to the surrounding topographic and geologic context, though originating from the lower base of a differentiated melt pool cannot be excluded here, the rover observed regolith and rock fragment are very likely to be lunar mantle materials excavated from nearby Finsen crater.

Keywords: Chang'e – 4 rover; in – situ observations; regolith and rock fragment; lunar mantle materials; differentiated melt pool

1 Introduction

Chang'e – 4 successfully landed in Von Kármán crater inside the South Pole – Aitken

(SPA) basin on January 3, 2019 (Li et al., 2019), marking the first soft landing of human spacecraft on the farside of the Moon. The SPA basin is the largest (~2500km in diam - eter), oldest (Pre - Nectarian, formed between ~3. 8 and ~4. 3 Ga ago) and deepest (~13km deep) recognized impact basin on the Moon (Stuart - Alexander, 1978; Spudis et al., 1994). Several studies have suggested the SPA - forming impact was perhaps very oblique. The identified evidences include elliptical ring structures (Garrick - Bethell and Zuber, 2009), an observed Fe anomaly and Thenhancement within the basin interior (Lawrence et al., 1998), and a far offset between the SPA antipode and the nearside center of the lunar mare concentrations (Schultz and Crawford, 2011). These evidences suggest the SPA impact would not have excavated as deep into the lunar interior, maybe only excavating shallowly into the lunar crust, which is even favored by observations of the feldspathic ejecta of SPA. However, according to the numerical impact simulations, lunar mantle material should be excavated by some of the largest lunar impact events (Miljkovic et al., 2015). Melosh et al. (2017) showed that, even in the most conservative simu - lation case of a 200 - km - diameter impactor striking the Moon at 45°, the SPA basin should have sampled the lunar lower crust and upper mantle to a maximum excavation depth of 105km. Zhu et al. (2018) showed in the case of a 630 - km - diameter impactor with an impact velocity of 18km/s hitting the Moon at an angle of 20°produces a transient cavity with a diameter of ~1400km, excavates up to a depth of ~140km and exposes the mantle within the basin.

The ~186km diameter pre - Nectarian Von Kármán crater lies in the northwestern portion of the Mg - Pyroxene Annulus of the SPA basin (Stuart - Alexander, 1978; Moriarty and Pieters, 2018), and pre - impact crust beneath Von Kármán is likely to have already been thinned by the SPA impact. The estimated approximate impact excavation and melting depth *Hexc* of Von Kármán crater is 9. 3—12. 05km, which is calculated from the equation $Hexc \approx 0.05-0.065D$ (D is the final complex crater diameter) (Melosh, 1989). The newly acquired high - resolution gravity data from the Gravity Recovery and Interior Laboratory (GRAIL) mission showed Von Kármán crater has a relatively thin crust that is less than 5km (Wieczorek et al., 2013). Von Kármán is thus one candidate crater, with an impact melting and central peak source depth clearly exceeding the modeled lunar crustal thickness, to investigate possible excavated mantle materials. The interior of Von Kármán was heavily bombarded by numerous impacts and subsequently flooded with one or several periods of basalt flows and the ejecta from the nearby craters during the Imbrian period (Huang et al., 2018), which have greatly modified the topography and surface compositions. In order to have the maximum likelihood of con - ducting in - situ observations on the occurrence of deep seated, possibly mantle - derived materials within Von Karman crater, Chang'e - 4 landed on the relatively smooth mare basalt plain emplaced at the south floor of the crater (Wu et al., 2019). Therefore, interpretations of the in - situ measured spectra, which is collected by the Visible and Near -

Infrared Imaging Spectrometer (VNIS) onboard Chang'e - 4 rover, should shed light on the mineralogical compositions of the possible lunar deep interior's materials that has long puzzled lunar geologists. For example, olivine is expected to be abundant in the lunar mantle, however, orbital spectral measurements have only found mere traces of olivine in limited places within SPA basin (Yamamoto et al., 2012). These results would thereby provide favorable or unfavorable evidences for previous suggestions on the compositions of the possible lunar deep materials excavated from the mantle.

This study analyzes characteristics of spectra collected by in - situ observations from VNIS during a nominal three - month mission period, and infers mineral compositions of the regolith and rock fragment at the Chang'e - 4 landing site. Low band area ratio (BAR) values (0.27 ± 0.10) suggest a relatively high olivine/pyroxene ratio within local lunar regolith and rock fragment, and short 1000nm and 2000nm absorption centers (979.6 ± 15.5nm and 1990.7 ± 35.0nm, respectively) indicate pyroxenes are chiefly Mg - rich orthopyroxenes (low - Ca pyroxene, LCP). According to analysis on surrounding topography and geologic background, the Chang'e - 4 landing site is on the ejecta of the ~72.98km diameter Finsen crater, and it would make sense that the local lunar regolith and rock fragment be essentially composed of lunar upper mantle materials excavated from Finsen.

2 Spectrometers and spectral products

2.1 Description of VNIS

The Chang'e - 4 mission includes a communication relay satellite, a lander, and a rover (Jia et al., 2018). Six scientific payloads have been installed on Chang'e - 4 spacecraft to carry out planned objectives and tasks, with three payloads on the lander: the descent camera, the terrain camera, and the low frequency spectrometer, and three payloads on the rover (named Yutu - 2): the panoramic camera, the lunar penetrating radar, and the Visible and Near - Infrared Imaging Spectrometer (Jia et al., 2018).

VNIS was designed and manufactured by Shanghai Institute of Technical Physics, Chinese Academy of Sciences. It's composed of a visible/near - infrared (VIS/NIR) imaging spectrometer, a shortwave infrared (SWIR) single - pixel non - imaging spectrometer, and a white panel for calibration and dustproofing (He et al., 2014). The spectral range is 450 - 2395nm with a default sampling interval of 5nm, and the total spectral channels is 400 (100 channels from 450 to 945nm for the VIS/NIR imaging spectrometer and 300 channels from 900 to 2395nm for the SWIR spectrometer, with 10 overlap channels between 900—945nm) (He et al., 2014). The VNIS is mounted at the front of the Yutu - 2 rover and is inclined 45°from a height of 0.69m above the lunar surface (He et al., 2014). VNIS is capable of synchronously acquiring the full spectrum of lunar surface objects and in - situ calibration. The effective pixels of the VIS/NIR imaging spectrometer are 256×256, and the spatial resolution is about 1mm/

pixel. The field of view of the SWIR spectrometer corresponds to a circular region in the VIS/NIR image with the center located at (96, 128) and a radius of 54 pixels (Supplementary Fig. 1) (He et al., 2014).

2.2 VNIS in-situ observations

Since the successful deploying on the lunar surface, the Yutu-2 rover has traversed 163m (odometer reading) during a nominal three-month mission period and made in-situ observations using its suite of science instruments (Supplementary Tab. 1). These observations provide first-ever datasets to study the geomorphology, mineral composition, and subsurface structure within Von Kármán crater. VNIS made 14 in-situ observations of different targets at 11 sites along the rover traverse, including 13 observations for lunar regolith and 1 observation for a rock fragment (Supplementary Fig. 1).

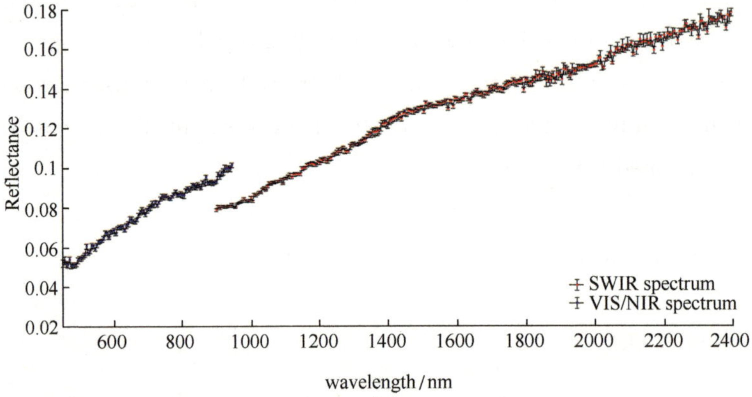

(a)

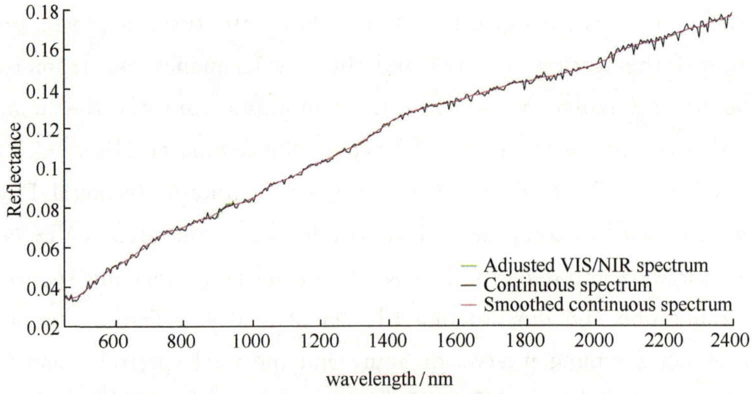

(b)

Fig. 1　The first in-situ observed spectrum on the lunar farside at site A

(a) Illustration of spectrum discontinuity; (b) Eliminating discontinuity and smoothing spectrum.

The error bars in (a) represent standard deviation of each spectral channel

(For interpretation of the colors in the figure (s), the reader is referred to the web version of this article.)

3 Methods

3.1 Reflectance derivation

Details of processing and calibration for producing VNIS L2B spectral radiance data from original L0 data can be found in He et al. (2014) and Liu et al. (2013). The reflectance properties, which are crucial for mineral identification and quantification, are controlled by the physical and chemical state of the surface regolith, the surface roughness and the geometric circumstances (e. g. solar incidence angle). The L2B radiance data is firstly corrected by on-board radiometric calibration coefficients for both VIS/NIR and SWIR spectrometers in this study, which is provided by the VNIS manufacturer after on-board synchronous calibration experiments during the three lunar days. Reflectance (Bidirectional Reflectance Factor, BRF) is then derived by the solar irradiance calibration method (Eq. (1)) (Wu et al., 2018).

$$BRF(\lambda, \theta_i, \varphi_i, \theta_r, \varphi_r) = \frac{\pi \cdot I(\lambda, \varphi_i, \varphi_i, \theta_r, \varphi_r) \cdot D^2}{E_0(\lambda) \cdot \cos(\theta_i)} \quad (1)$$

where λ, θ_i, φ_i, θ_r, and φ_r rare the wavelength, solar zenith angle, solar azimuth angle, viewing zenith angle, and viewing azimuth angle, respectively. I is the VNIS L2B spectral radiance after radiometric correction. E_0 (λ) is the solar irradiance at 1 AU, which is adopted from Coddington et al. (2016) and resampled to VNIS spectral channels. D is the Sun-Moon distance in kilometers at the observation time divided by the standard Sun-Earth distance (149, 597, 870.7km).

3.2 Spectrum discontinuity elimination

The VIS/NIR spectrum is derived firstly by averaging 9145 pixels in a circle area that corresponds to the field of view of the SWIR spectrometer (the center of the circle is located at (96, 128) of the VIS/NIR image and the radius is 54 pixels). The discontinuity at the boundary of derived average VIS/NIR spectrum and SWIR spectrum is clear, both for the calibration white panel and the target on lunar surface (e. g., regolith) (Fig. 1a). Typical average signal-to-noise ratios (SNR) of the derived VIS/NIR spectrum and SWIR spectrum that calculated from in-flight calibration are 37.4dB and 44.0dB, respectively. Average standard deviations of each measurement for VIS/NIR and SWIR channels are 2.3% and 2.7%, respectively, demonstrating absorption features with depth greater than 3% can be uniquely distinguished from the noise. Considering the SWIR spectrum has higher signal-to-noise ratio and stability, the derived VIS/NIR spectrum should be adjusted to the SWIR spectrum. There is no obvious slope difference on the overlapping 10 bands from 900 to 945nm between the derived VIS/NIR spectrum and SWIR spectrum, thus there is only an offset between the two spectra. A linear regression fitting is applied on the overlapping bands between VIS/NIR and SWIR spectrum for the purpose of keeping the shape of the VIS/NIR spectrum. The equation has the form Y = X + b, where Y is SWIR spectrum, X is VIS/NIR spectrum, and b is the offset. The calculated offset is then added to the VIS/NIR spectrum and the mean spectrum value of the overlapping bands is also calculated to construct full - range continuous spectrum (Fig. 1b). The full - range continuous spectrum is further smoothed with a 3 - point unweighted sliding - average algorithm to reduce noise before calculating spectral parameters (Fig. 1b).

3.3 Spectral parameter calculation

Absorption features are a function of mineralogy and their spectral shapes are controlled by chemical compositions, modal proportions, and space weathering. Pyroxene and olivine are major mafic minerals on planetary bodies, which are distinguished by their specific spectral characteristics. Pyroxene exhibits simple gaussian shaped absorption features at ~1000nm and ~2000nm, whereas olivine has three overlapping absorption features at approximately 850, 1050, and 1250nm, forming what appears to be a single, broad, and composite absorption feature at ~1050nm (Sunshine and Pieters, 1998). Plagioclase is also an important mineral component and has significant implications for the origin of the VNIS observed lunar regolith and rock fragment. Due to the incorporation of minor amounts of FeO (<1. 0 wt.%), the iron - bearing plagioclase spectrum has a weak and broad absorption near 1250 nm. However, this absorption feature is easily masked by a few to several vol.% of mafic minerals in a mixture. Cheek and Pieters (2014) showed that as little as 2 vol.% of olivine or pyroxene completely masks this absorption. Nevertheless, implications of possible presence of large amounts of plagioclase in the interpreted spectral signature are discussed in this study.

Fundamental spectral parameters (band center (BC), band depth (BD) and band area (BA)) and derived parameters (integrated band depth (IBD), integrated band depth ratio (IBDR) and band area ratio (BAR)) are calculated in this study. All of them depict spectral variabilities among the VNIS measured in - situ spectra and therefore reveal mineral compositions of lunar regolith, which are of great significance to the interpretation of geological evolution. Band center (BC) is the wavelength position at the deepest absorption in a given wavelength range. BC positions of 1000nm (BC1000) and 2000nm (BC2000) absorption features are estimated in this study to differentiate between low - Ca pyroxenes (~900nm and 1900nm) and high - Ca pyroxenes (~1000nm and 2200nm). Band Depth (BD) represents the absorption strength at a particular wavelength. Integrated band depth (IBD) is the cumulative sum of the absorption strengths over a wavelength range. IBD1000 and IBD2000 that refer to integrated depths of the 1000nm and 2000nm absorption feature were introduced by Moon Mineralogy Mapper (M^3) science team, to characterize spectral variations modulated by Fe - bearing mineral distribution and space weathering (Mustard et al., 2011). Integrated band depth ratio (IBDR) is defined as the ratio of IBD2000 to IBD1000. Band area (BA) is the area covered by an absorption feature in a given wavelength range. BA1000 and BA2000 refer to the area covered by the 1000nm and 2000nm absorptions respectively. They are used to derive band area ratio (BAR), which is the ratio of BA2000 to BA1000. As IBD is proportional to the band area of an absorption feature, BAR is calculated and discussed in this study. In accordance with the scheme adopted by the M^3 team (Mustard et al., 2011) and considering the spectral resolution of VNIS data, the wavelength ranges used for calculating BA1000 and BA2000 are the same as that of IBD1000 (between 790 and 1310nm) and IBD2000 (between 1660 and 2395nm), respectively. Though they are wider than the actual absorption features owing to pyroxene and olivine, they provide a common baseline for comparing individual absorption features.

Before calculating spectral parameters, the spectrum has been spectrally subset to the region containing the absorption features of interest, and its continuum is removed, which helps minimizing influences from different observation backgrounds (e. g., various light conditions). The continuum is a convex hull fit over the top of a spectrum using straight-line segments that connect local spectra maxima. The first and last bands are on the hull, and their spectral values in the continuum-removed spectrum are equal to 1. 0 (Clark and Roush, 1984). In addition, spectral slope for the visible wavelength region (VISslope), near-infrared wavelength region (NIRslope), shortwave infrared wavelength region (SWIRslope), and full-range spectral slope (FSslope) are calculated. The definitions and formulae of these parameters are stated in Supplementary Table 2.

3.4 MGM and NBSR analysis

The principle of the Modified Gaussian Model (MGM), originally proposed by Sunshine et al. (1990), is to deconvolve overlapping absorptions of mafic mineral spectra into their fundamental absorption components (Sunshine et al., 1990; Sunshine and Pieters, 1993). One gaussian function (characterized by band center, width and strength) or combination of multiple gaussian functions permits the identification of a mineralogical absorption band, indicative of the presence and composition of the predominant mafic minerals (Sunshine and Pieters, 1998). Based on Clénet et al. (2011) and Pinet et al. (2016), MGM is used here with 7 configurations to deconvolve overlapping absorptions of the Chang'e-4 VNIS spectrum. The numbers of gaussians in these configurations are listed in Supplementary Table 3. For example, The 'OL-OPX-CPX' configuration (last column in Supplementary Table 3) has 8 gaussians, including 1 for a strong absorption centered around 450nm in the ultraviolet (UV) region (UV450), 3 for olivine (OL) absorptions (OL850, OL1050, OL1250), 2 for orthopyroxene (OPX) absorptions (OPX900, OPX1900), and 2 for clinopyroxene (CPX) absorptions (CPX1000, CPX2200) (Clénet et al., 2011, 2013). The continuum is fit by a second order polynomial with anchoring points adjusted on the local maxima along the spectrum and initial parameters of each gaussian are made on the basis of laboratory results (Clénet et al., 2011).

Kanner et al. (2007) found there is a strong positive relationship between normalized band strength ratio (NBSR) calculated from MGM derived band depths (also called absorption strengths) and the known mineral fractions in a two - pyroxene mixture. The concept of NBSR including olivine has been first introduced by Clénet et al. (2013). Five NBSR values are calculated in this study to predict the relative fractions of the mafic mineral mixtures, including NBSR $(OL)_{1000nm}$, NBSR $(OPX)_{1000nm}$, and NBSR $(CPX)_{1000nm}$ for the 1000nm absorption region, NBSR $(OPX)_{2000nm}$ and NBSR $(CPX)_{2000nm}$ for the 2000nm absorption region. The formulae for them are stated in Supplementary Table 4.

4 Results and discussions

4.1 Lunar regolith reflectance properties

The 13 lunar regolith reflectance spectra observed by VNIS at the Chang'e - 4 landing site has typical characteristics caused by space weathering. These in - situ measured reflectance spectra exhibit common features of increasing reflectance with increasing wavelength (reddening) (Fig. 2a). Two absorption features near 1000nm and 2000nm are discernible from continuum removed spectra (Fig. 2b). The averages and standard deviations of band depths for the 1000nm and 2000nm absorption features of the 13 regolith spectra are 0.12 ±0.05 and 0.04 ±0.01, respectively, indicating the existence of pyroxene. The lack of significant 2000nm absorptions in these regolith spectra indicates they have a very high degree of space weathering. A broad olivine absorption feature centered near 1050nm also can be distinguished in continuum removed spectra. These in - situ spectral measurements are consistent with remote sensing observations of the SPA basin (Huang et al., 2018; Moriarty and Pieters, 2018).

All spectral slope values, regardless of type, have a general decreasing trend with Yutu - 2 rover traversing away from the lander. This phenomenon is attributed to lander rocket plume, which blows away the uppermost mature regolith and exposes the less - mature regolith. Surface modifications by plumes, which changes reflectance within blast zone, have been documented by Surveyor, Luna, Apollo and Chang'e - 3 on the Moon, and by Viking, Curiosity and especially Phoenix on Mars (Clegg et al., 2014). The full - range spectral slope from 450nm to 2350nm has a value of 0.07 ±0.02, the visible slope from 450 to 750nm is 0.13 ±0.05, the near - infrared slope from 750 to 1500nm is 0.09 ±0.04, and the shortwave infrared slope from 1500 to 2350nm is 0.04 ±0.01. Both the mean and the standard variation of the slope value decrease gradually from visible to shortwave infrared wavelength region, which indicate that lander rocket exhaust effects has the largest influence on the visible wavelength region.

4.2 Spectral characteristics of regolith

Band area ratio (BAR) is a sensitive indicator for interpreting the spectral behavior of olivine - orthopyroxene mixtures, which is nearly independent of particle size and mineral composition (Cloutis et al., 1986). BAR contains information about the olivine abundance in olivine - pyroxene mixtures, which is inversely proportional to the abundance of olivine but increases linearly with the increase in pyroxene abundance (Cloutis et al., 1986). Gaffey et al. (1993) found the relationship between BC1000 and BAR can reduce the error caused by the presence of clinopyroxenes in the mixture. The low BAR values (0.27 ± 0.10) seen from BAR - BC1000 scatter plot (Fig. 3a) are in favor of a mixture of olivine - orthopyroxene and suggest a relatively high olivine/pyroxene ratio within minerals at the Chang'e - 4 landing site (Clénet et al., 2011), though olivine absorptions are not obvious in the original spectra (Fig. 2a).

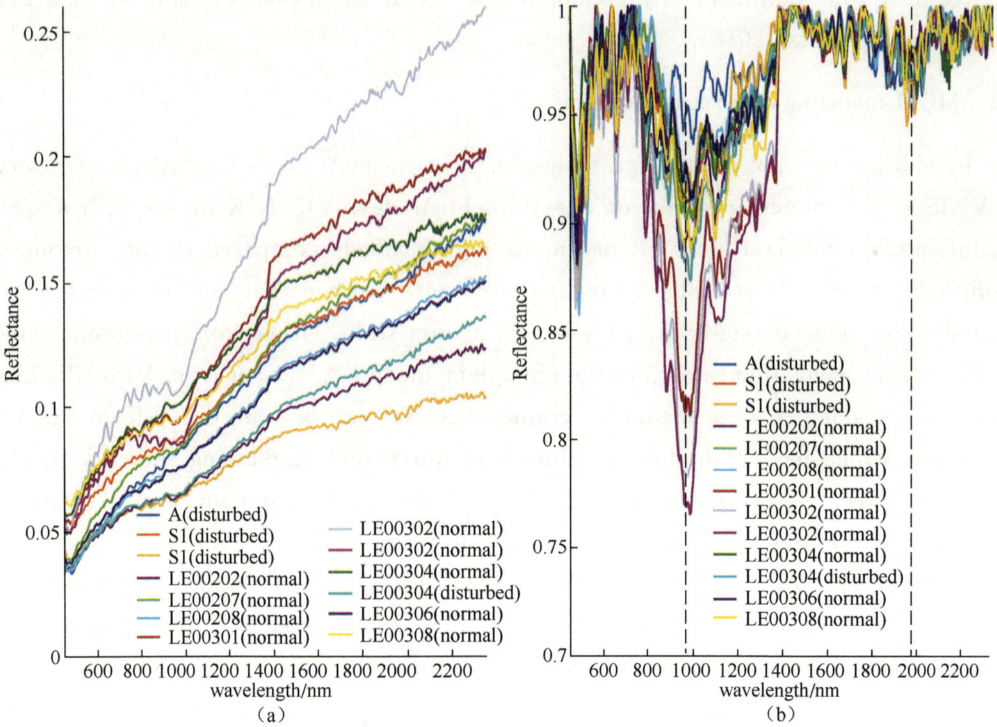

Fig. 2　Lunar regolith spectra observed by VNIS during the first three lunar days: (a) original; (b) continuum removed. The dashed lines in (b) are located at 965nm and 1975nm, respectively. The regolith state (normal or disturbed) of each observation is marked in the legend. The discontinuity in some spectra in this figure near 1400nm is related to signals that arise from artifacts within the VNIS design

Extensive laboratory studies of terrestrial and synthetic pyroxenes provide the basis to correlate the 1000nm and 2000nm absorption positions with their chemical compositions. Orthopyroxene absorption bands (BC1000 and BC2000) shift to longer wavelengths with increasing amounts of iron and calcium and shift to shorter wavelengths with increasing magnesium, whereas clinopyroxene absorption bands shift to longer wavelengths with increasing calcium content and shift to shorter wavelengths with increasing iron (Adams, 1974; Cloutis et al., 1986; Klima et al., 2007; Clénet et al., 2011). Thus, the relationship between band centers and iron and calcium content has been widely used to infer pyroxene composition from reflectance spectrum. The BC2000 -BC1000 plot shows BC1000 varies from 949.8nm to 998.1nm (979.6 ±15.5nm), and BC2000 varies from 1935.3nm to 2028.0nm (1990.7 ± 35.0 m) (Fig. 3b). Short 1000nm and 2000nm absorption centers indicate pyroxenes at Chang'e - 4 landing site are chiefly Mg - rich orthopyroxenes (low - Ca pyroxene, LCP). These in - situ spectral observations are consistent with remote sensing observations from Moon Mineralogy Mapper and SELENE/Kaguya that suggests that Von Kármán crater (where Chang'e - 4 landed) is situated in a zone dominated by abundant Mg - rich pyroxenes (Huang et al., 2018; Moriarty and Pieters, 2018).

4.3 MGM modeling of a rock spectrum

In addition to regolith, the spectrum of a 20 - centimeter rock fragment was observed by VNIS at 1.2 meters distance on the third lunar day. VIS/NIR image with a spatial resolution of 1mm/pixel shows it has much higher albedo compared to the surrounding regolith (Fig. 4a). It is really quite hard to figure out whether this rock fragment is crystallized or non - crystallized at a sub - cm or mm scale, either from panoramic camera image or from navigation camera image (Supplementary Fig. 2), or from VIS/NIR image (Fig. 4a). A small to even medium - grained textures can be seen dimly from the VIS/NIR image which has the highest resolution (1mm/pixel), indicating the rock is likely plutonic. It is conservatively estimated that the rock fragment lacks discernible coarsegrained textures (>5mm crystals visible to naked eyes).

The rock spectrum has a low BAR value of 0.17, and the 1000nm and 2000nm mafic mineral absorption features are obvious (Fig. 4b). Because fitting with other configurations either have extreme weak or even positive absorption strengths (e. g., fitting with 'OL - CPX' configuration produces a positive absorption strength for olivine - related absorption at 1050nm, Supplementary Fig. 3), or have unreasonable wavelength centers/widths (e. g., fitting with 'OL' configuration produces an unreasonable wavelength center for olivine - related absorption at 1000nm, Supplementary Fig. 4), or have relative large fitting errors (e. g., fitting with 'OPX - CPX' con - figuration produces a relative large error, Supplementary Fig. 5), MGM fitting with 'OL - OPX - CPX' configuration (initial and final fitting parameters are listed in Supplementary Table 5) provides the optimal modeling in terms of mineral detection for this rock spectrum

(Fig. 5). Then, in order to predict Mg number (Fo#) (Sunshine and Pieters, 1998), the MGM modeled three olivine - related band centers estimates are fit to a laboratory derived compositional trend line by finding the least error (Sunshine and Pieters, 1998). However, the precise determination of the overall uncertainty on Mg number is not an easy task (Isaacson and Pieters, 2010; Pinet et al., 2018, 2019). Relying on these studies, a first order estimate of the uncertainty is likely to be on the order of 10 - 15% Fo# range. Accordingly, depending on the minimization used, the predicted olivine Fo# appears to be in the range 80—84, with an uncertainty on the order of 10, pointing at a Mg - rich composition (forsterite). It should be noted here that the complex olivine absorption feature not only varies with the iron content but also with grain size and abundance. As demonstrated by Clénet et al. (2011), a forsterite with very large grains and a fayalite with smaller grains may have similar spectral features. Everything considered, this high Fo# value is very tentative here and should thus be treated with caution.

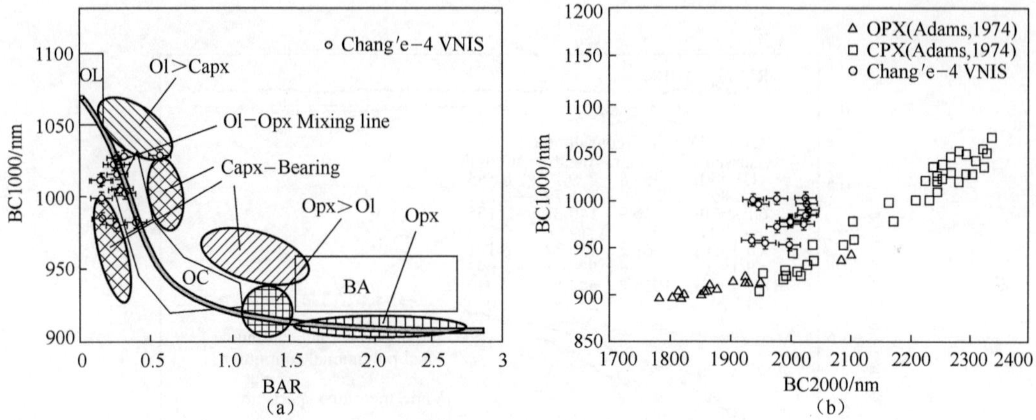

Fig. 3 Scatter plots of spectral parameters

(a) BAR - BC1000; (b) BC2000 - BC1000. Reference compositional groups are from Cloutis et al. (1986) and Adams (1974), respectively. The error bars in each plot represent a first - order approximation for the uncertainty determined by the law of propagation of uncertainty. Here OC - ordinary chondrites and BA - pyroxene dominated basaltic achondrites

NBSR values (Table 1) for the 1000nm absorption feature predict olivine accounts for the largest fraction, followed by orthopyroxene and then clinopyroxene (olivine: orthopyroxene: clinopyroxene is 38%: 33%: 29%). NBSR values for the 2000nm absorption features also predict orthopyroxene has a bit higher fraction than clinopyroxene (orthopyroxene: clinopyroxene is 51%: 49%), indicating self - consistent in mineral fractions. The mineral composition and fraction of the rock fragment predicted from MGM fitting and NBSR values are in agreement with that inferred from the spectral parameter scatter plots for the regolith, indicating the rock fragment may have a high probability that it originates from the same material as the surrounding lunar regolith.

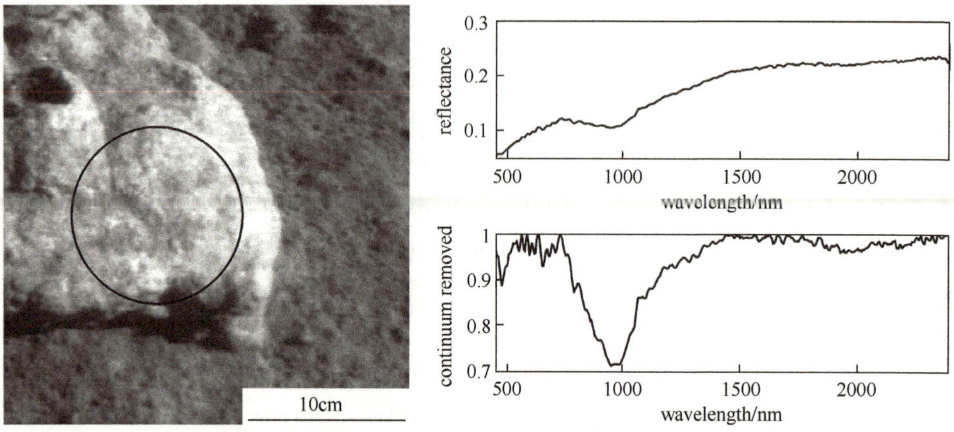

(a)　　(b)

Fig. 4　A rock fragment detected along the Yutu - 2 rover traverse

(a) VIS/NIR image (1 mm/pixel) @ 750nm, the black circle corresponds to the field of view of the SWIR spectrometer.
(b) Rock spectrum (top: original, bottom: continuum removed)

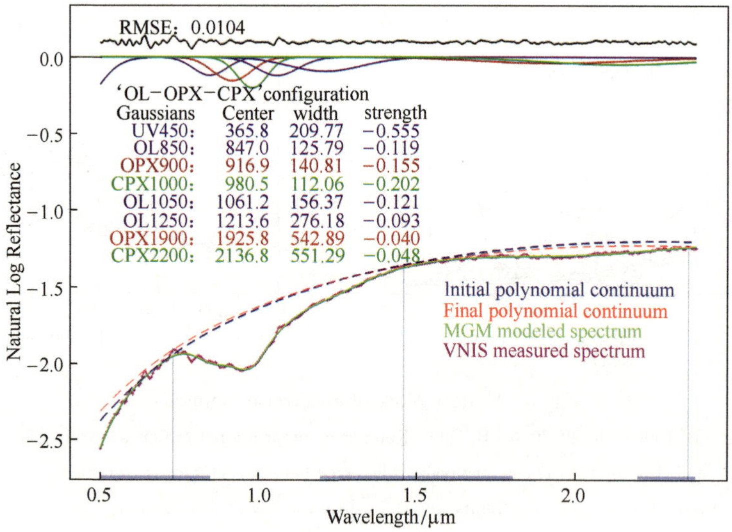

Fig. 5　Rock spectrum modeled with 'OL - OPX - CPX' configuration by MGM

Table 1　NBSR values of the rock spectrum

NBSR*	1000nm absorption			2000nm absorption	
	NBSR (OL)	NBSR (OPX)	NBSR (CPX)	NBSR (OPX)	NBSR (CPX)
Value**	0.38	0.33	0.29	0.51	0.49

* The definitions of the NBSR are stated in supplementary material.

* * These values represent relative fractions.

Lunar Reconnaissance Orbiter Diviner observations show that thermal infrared (TIR) emissivity spectra at a spatial resolution of ~250m/pixel indicate mixtures of plagioclase and olivine (Arnold et al., 2016) where olivine - rich sites (e. g., locations around the SPA basin) have been identified with visible to near infrared (VNIR) observations at a resolution of ~500m/pixel (Yamamoto et al., 2010). The discrepancy between this study and previous work was probably caused by: (1) different observation methods (orbital vs. in - situ); (2) different observation bands (TIR vs. VNIR); (3) great resolution differences (e. g., non - imaging single - pixel vs. hundreds of meters per pixel). In addition, because of minerals' overlapping spectral absorption features (e. g., mafic minerals and plagioclase), large amounts of plagioclase in the rock (if exists) can be masked in the spectral signature of the olivine - and pyroxene - dominated spectrum that are being interpreted here. Considering this scenario and above stated grain size effects, it should also be noted that further MGM modeling experiments should be carried out to better understand the influence of the possible presence of large amounts of plagioclase and mineral grain sizes on estimating relative mineral fractions and assessing olivine compositions (Pinet, 2019), which has huge implications for the origin of the VNIS observed materials (e. g., large amounts of plagioclase in a rock would completely change the type of rock being observed, which indicate the rock would not be expected to originate from the mantle, but rather likely from the crust).

4.4 Origin of the VNIS observed regolith and rock fragment

Obvious NE - SW trending linear features cut across this region, which is radiated from the relatively young ~72.98km diameter Finsen crater (Fig. 6). Topographic profile extracted from an improved lunar digital elevation model (DEM) SLDEM2015 (Barker et al., 2016), which has a resolution of 512 pixels per degree (~42m/pixel at the Chang'e - 4 landing site), shows Chang'e - 4 landed on these NE - SW trending elevated streaks which have an interval of about 3km (Fig. 6). SELENE/Kaguya Multiband Imager (MI) reflectance mosaic and false color composite images show Chang'e - 4 landed on the low albedo regions (Supplementary Fig. 6), which are likely ejecta materials (Supplementary Fig. 7). Therefore, the Chang'e - 4 landing site is in the area affected by ejecta materials coming from Finsen crater, which is 370km away from the center of the transient cavity of SPA basin (Potter et al., 2012). The local mare basalt deposits at the Chang'e - 4 landing site should be extensively mixed with or covered by ejecta materials from Finsen crater. Ling et al. (2018) divided the mare region within Von Kármán crater into two units: low - Ti unit (LT, 3 wt.% $>TiO_2>$1.5wt.%) and Finsen ejecta unit (FE, $TiO_2<$1.5wt.%). Chang'e - 4 just landed at the FE unit (1.42 ±0.24 wt.% TiO_2). Combined with above information and the rock fragment may have a plutonic origin, the VNIS observed Mg - rich olivine (Fo# 80 - 84 +/-10) and LCP - bearing regolith and rock fragment are essentially lunar deep - seated materials excavated from the nearby Finsen crater at a depth H_{exc} of 3.65—4.74km.

The SPA basin forming impact event had substantially thinned the lunar crust

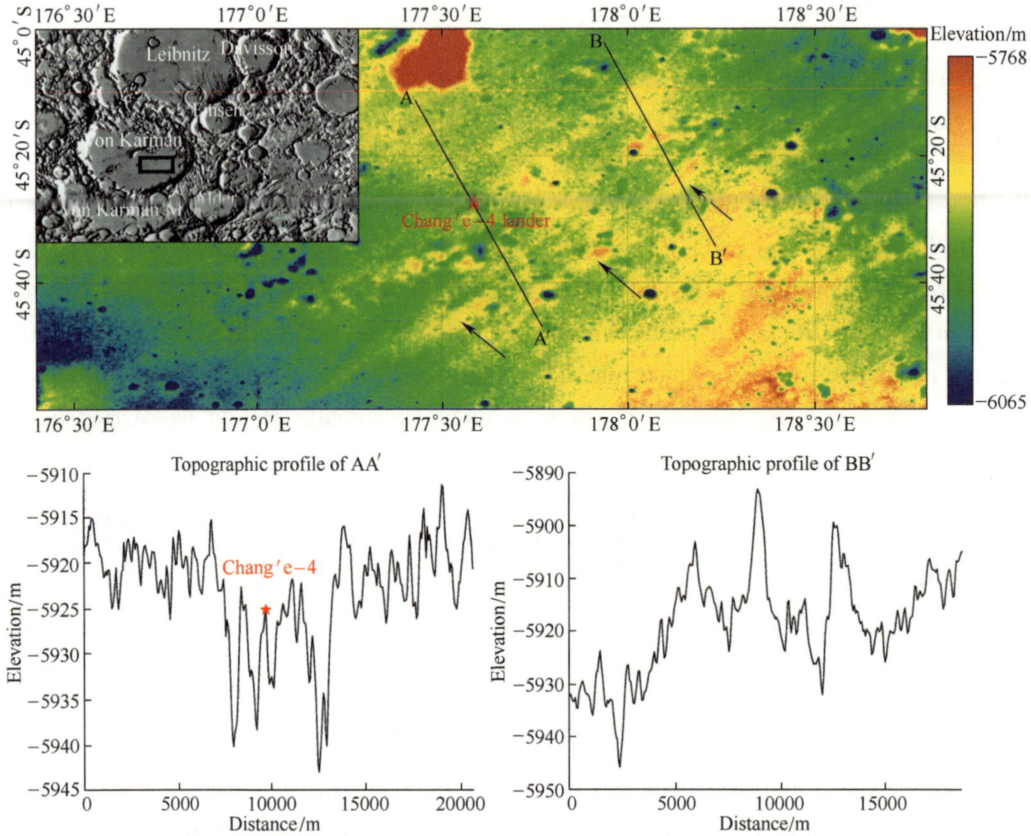

Fig. 6　Top panel: Elevation map at Chang'e-4 landing site. Black arrows indicate the NE-SW trend features and black rectangle in the shaded relief inset in the upper left indicates the location of the elevation map within the Von Kármán crater. Bottom panel: Topographic profile of A-A′ that crosses the Chang'e-4 landing site, and that of B-B′shows topographic variations with notable intervals

through excavation and the event can generate an impact melt pond within the basin. Hence, the sources of these VNIS observed deep-seated materials may originate from: (1) the lunar lower crust; (2) the lunar upper mantle; (3) a differentiated melt pool. Previous works have suggested that perhaps only the lower crust was exposed during the SPA impact event if it was an oblique impact (Pieters et al., 2001; Garrick-Bethell and Zuber, 2009), however, recent numerical modeling works suggest that mantle material should be excavated by the SPA forming impact even if it's an very oblique impact (e. g., impact at an angle of 20°) (Melosh et al., 2017; Zhu et al., 2018). Melosh et al. (2017) proposed the SPA basin should have exposed the Moon's upper mantle materials, which is dominated by LCP. Remote sensing observations suggested the presence of a thick LCP-bearing layer in the SPA subsurface (Pieters et al., 1997). LCP-bearing materials indeed have been detected using orbital spectral data at many central peaks within SPA basin, including Finsen crater (Moriarty et al., 2013). Therefore, the VNIS observed Mg-rich

LCP - bearing lunar regolith and rock fragment (with high Fo# for the olivine mineral) are very likely to be lunar mantle materials excavated from nearby Finsen crater. Alternatively, most materials within the SPA basin are melted due to impact - produced high pressure (Hurwitz and Kring, 2014; Vaughan and Head, 2014). However, the definitive identification of SPA derived impact melt on the basin floor, either by remote sensing or via photogeology, is extremely difficult due to the number of subsequent impacts and volcanic activities (Petro, 2011). It cannot be excluded here that the VNIS observed local materials might originate from the lower base of a differentiated melt pool. As the SPA basin has not been sampled either by previous lunar sample return missions (Apollo, Luna) or by lunar meteorites, a definitive answer to the sources of VNIS observed materials can only be addressed in the future through detailed analysis on returned samples.

5 Conclusion

The first in - situ spectral observations on the lunar farside reveal the regolith at the Chang'e - 4 landing site has a relative high olivine/pyroxene ratio; with the pyroxene being mainly Mg - rich LCP. The observed rock fragment has a composition similar to that of the surrounding regolith, with a rather Mg - rich olivine composition (Fo# 80 - 84 +/-10). Topographic and geologic analysis show Chang'e - 4 just landed on the ejecta of nearby ~72.98km diameter Finsen crater. Though originating from the lower base of a differentiated melt pool cannot be excluded here, the VNIS observed lunar regolith and rock fragment are very likely to be lunar upper mantle materials excavated from nearby Finsen crater. The in - situ spectral observations confirm previous orbital observations and numerical impact simulations in the SPA basin, and complement the initial results recently reported by Li et al. (2019) that Chang'e - 4 may have discovered material from the Moon's mantle.

Acknowledgements

The authors gratefully thank the whole Chang'e - 4 team for making the mission a great success and providing the data. This research is supported by Key Research Program of the Chinese Academy of Sciences (Grant No. XDPB11), National Natural Science Foundation of China (grant No. 41702354 and 41941003), Macau Young Scholars Program (grant No. AM201902), and Science and Technology Development Fund of Macau (grant No. 131/2017/A3). Constructive reviews by Patrick Pinet and an anonymous reviewer improved the quality of this manuscript substantially, and are greatly appreciated.

Appendix A　Supplementary material

Supplementary material related to this article can be found online at https://doi.org

/10.1016 /j. epsl . 2019. 115829.

References

[1] Adams, J. B., 1974. Visible and near - infrared diffuse reflectance spectra of pyroxenes as applied to remote sensing of solid objects in the solar system. J. Geophys. Res. 79, 4829 - 4836.

[2] Arnold, J. A., Glotch, T. D., Lucey, P. G., Song, E., Thomas, I. R., Bowles, N. E., Greenhagen, B. T., 2016. Constraints on olivine - rich rock types on the Moon as observed by Diviner and M3: implications for the formation of the lunar crust. J. Geophys. Res., Planets 121, 1342 - 1361.

[3] Barker, M. K., Mazarico, E., Neumann, G. A., Zuber, M. T., Haruyama, J., Smith, D. E., 2016. A new lunar digital elevation model from the Lunar Orbiter Laser Altimeter and SELENE Terrain Camera. Icarus 273, 346 - 355.

[4] Cheek, L. C., Pieters, C. M., 2014. Reflectance spectroscopy of plagioclase - dominated mineral mixtures: implications for characterizing lunar anorthosites remotely. Am. Mineral. 99, 1871 - 1892.

[5] Clark, R. N., Roush, T. L., 1984. Reflectance spectroscopy: quantitative analysis techniques for remote sensing applications. J. Geophys. Res. 89, 6329 - 6340.

[6] Clegg, R. N., Jolliff, B. L., Robinson, M. S., Hapke, B. W., Plescia, J. B., 2014. Effects of rocket exhaust on lunar soil reflectance properties. Icarus 227, 176 -194.

[7] Clénet, H., Pinet, P., Ceuleneer, G., Daydou, Y., Heuripeau, F., Rosemberg, C., Bibring, J. P., Bellucci, G., Altieri, F., Gondet, B., 2013. A systematic mapping procedure based on the Modified Gaussian Model to characterize magmatic units from olivine/pyroxenes mixtures: application to the Syrtis Major volcanic shield on Mars. J. Geophys. Res. 118, 1632 - 1655.

[8] Clénet, H., Pinet, P., Daydou, Y., Heuripeau, F., Rosemberg, C., Baratoux, D., Chevrel, S., 2011. A new systematic approach using the modified Gaussian model: insight for the characterization of chemical composition of olivines, pyroxenes and olivine -pyroxene mixtures. Icarus 213, 404 - 422.

[9] Cloutis, E. A., Gaffey, M. J., Jackowski, T. L., Reed, K. L., 1986. Calibrations of phase abundance, composition, and particle size distribution for olivine - orthopyroxene mixtures from reflectance spectra. J. Geophys. Res., Solid Earth91, 11641 - 11653.

[10] Coddington, O., Lean, J., Pilewskie, P., Snow, M., Lindholm, D., 2016. A solar irradiance climate data record. Bull. Am. Meteorol. Soc. 97, 1265 - 1282.

[11] Gaffey, M. J., Bell, J. F., Brown, R. H., Burbine, T. H., Piatek, J. L., Reed, K. L., Chaky, D. A., 1993. Mineralogical variations within the S - type asteroid

class. Icarus 106, 573 - 602.

[12] Garrick - Bethell, I., Zuber, M. T., 2009. Elliptical structure of the lunar South Pole - Aitken basin. Icarus 204, 399 - 408.

[13] He, Z., Wang, B. Y., Lv, G., Li, C. L., Yuan, L. Y., Xu, R., Chen, K., Wang, J., 2014. Visible and near - infrared imaging spectrometer and its preliminary results from the Chang'e 3 project. Rev. Sci. Instrum. 85, 083104.

[14] Huang, J., Xiao, Z., Flahaut, J., Martinot, M., Head, J., Xiao, X., Xie, M., Xiao, L., 2018. Geological characteristics of von Kármán Crater, Northwestern South Pole - Aitken Basin: Chang'e - 4 landing site region. J. Geophys. Res., Planets 123, 1684 -1700.

[15] Hurwitz, D. M., Kring, D. A., 2014. Differentiation of the South Pole - Aitken basin impact melt sheet: Implications for lunar exploration. J. Geophys. Res., Planets 119, 1110 -1133.

[16] Isaacson, P. J., Pieters, C. M., 2010. Deconvolution of lunar olivine reflectance spectra: implications for remote compositional assessment. Icarus 210, 8 - 13.

[17] Jia, Y., Zou, Y., Ping, J., Xue, C., Yan, J., Ning, Y., 2018. The scientific objectives and payloads of Chang'e - 4 mission. Planet. Space Sci. 162, 207 - 215.

[18] Kanner, L. C., Mustard, J. F., Gendrin, A., 2007. Assessing the limits of the modified Gaussian model for remote spectroscopic studies of pyroxenes on Mars. Icarus 187, 442 - 456.

[19] Klima, R. L., Pieters, C. M., Dyar, M. D., 2007. Spectroscopy of synthetic Mg - Fe pyroxenes I: spin - allowed and spin - forbidden crystal field bands in the visible and near - infrared. Meteorit. Planet. Sci. 42, 235 - 253.

[20] Lawrence, D. J., Feldman, W. C., Barraclough, B. L., Binder, A. B., Elphic, R. C., Maurice, S., Thomsen, D. R., 1998. Global elemental maps of the Moon: the lunar prospector gamma - ray spectrometer. Science 281, 1484 - 1489.

[21] Li, C., Liu, D., Liu, B., Ren, X., Liu, J., He, Z., Zuo, W., Zeng, X., Xu, R., Tan, X., Zhang, X., Chen, W., Shu, R., Wen, W., Su, Y., Zhang, H., Ouyang, Z., 2019. Chang'e - 4 initial spectroscopic identification of lunar far - side mantle - derived materials. Nature 569, 378 - 382.

[22] Ling, Z., Jolliff, B. L., Liu, C. Q., Bi, X. Y., Liu, L., Qiao, L., Zhang, J., Fu, X., Liu, J., 2018. Composition, mineralogy, and chronology of mare basalts in von Kármán Crater: a candidate landing site of Chang'e - 4. In: 49th Lunar and Planetary Science Conference. The Woodlands, Texas, p. 1939.

[23] Liu, B., Liu, J. -Z., Zhang, G. -L., Ling, Z. C., Zhang, J., he, Z., Yang, B. - Y., Zou, Y. - L., 2013. Reflectance conversion methods for the VIS/NIR imaging spectrometer aboard the Chang'e - 3 lunar rover: based on ground validation experiment data. Res. Astron. Astrophys. 13, 862 - 874.

[24] Melosh, H. J., 1989. Impact Cratering: A Geologic Process. Oxford University

Press.

[25] Melosh, H. J., Kendall, J., Horgan, B., Johnson, B. C., Bowling, T., Lucey, P. G., Taylor, G. J., 2017. South Pole - Aitken basin ejecta reveal the Moon' s upper mantle. Geology 45, 1063 - 1066.

[26] Miljkovic, K., Wieczorek, M. A., Collins, G. S., Solomon, S. C., Smith, D. E., Zuber, M. T., 2015. Excavation of the lunar mantle by basin - forming impact events on the Moon. Earth Planet. Sci. Lett. 409, 243 - 251.

[27] Moriarty, D. P., Pieters, C. M., 2018. The character of South Pole - Aitken Basin: patterns of surface and subsurface composition. J. Geophys. Res., Planets 123, 729 -747.

[28] Moriarty, D. P., Pieters, C. M., Isaacson, P. J., 2013. Compositional heterogeneity of central peaks within the South Pole - Aitken Basin. J. Geophys. Res., Planets 118, 2310 -2322.

[29] Mustard, J. F., Pieters, C. M., Isaacson, P. J., Head, J. W., Besse, S., Clark, R. N., Klima, R. L., Petro, N. E., Staid, M. I., Sunshine, J. M., 2011. Compositional diversity and geologic insights of the Aristarchus crater from Moon Mineralogy Mapper data. J. Geophys. Res., Planets 116.

[30] Petro, N. E., 2011. Volume of impact melt generated by the formation of the South Pole -Aitken Basin. In: Annual Meeting of the Lunar Exploration Analysis Group. Houston, Texas.

[31] Pieters, C. M., Head III, J. W., Gaddis, L., Jolliff, B., Duke, M., 2001. Rock types of South Pole - Aitken basin and extent of basaltic volcanism. J. Geophys. Res., Planets 106, 28001 - 28022.

[32] Pieters, C. M., Tompkins, S., Head, J. W., Hess, P. C., 1997. Mineralogy of the Mafic Anomaly in the South Pole - Aitken Basin: implications for excavation of the lunar mantle. Geophys. Res. Lett. 24, 1903 - 1906.

[33] Pinet, P., 2019. The Moon' s mantle unveiled. Nature 569, 338 - 339.

[34] Pinet, P. C., Chevrel, S. D., Daydou, Y. H., 2018. Characterization of the olivine/plagioclase mineralogy at Copernicus Crater from MGM deconvolution of M3 observations. In: 49th Lunar and Planetary Science Conference. The Woodlands, Texas, LPI Contrib. No. 2083.

[35] Pinet, P. C., Chevrel, S. D., Daydou, Y. H., 2019. Reassessing the relationship between olivine composition and reflectance spectroscopy from advanced MGM deconvolution. In: 50th Lunar and Planetary Science Conference. The Woodlands, Texas.

[36] Pinet, P. C., Glenadel - Justaut, D., Daydou, Y., Ceuleneer, G., Gou, S., Launeau, P., Chevrel, S., Carli, C., 2016. MGM deconvolution of complex mafic mineralogy rock slab spectra from visible - near infrared imaging spectroscopy: implications for the characterization of the terrestrial ocean crust and of the lunar

crust. In: The 8th Workshop on Hyperspectral Image and Signal Processing: Evolution in Remote Sensing. WHISPERS, Los Angeles, USA.

[37] Potter, R. W. K., Collins, G. S., Kiefer, W. S., McGovern, P. J., Kring, D. A., 2012. Constraining the size of the South Pole - Aitken basin impact. Icarus 220, 730 -743.

[38] Schultz, P. H., Crawford, D. A., 2011. Origin of nearside structural and geochemical anomalies on the Moon. Geol. Soc. Am. Spec. Pap. 477, 141 - 159.

[39] Spudis, P. D., Gillis, J. J., Reisse, R. A., 1994. Ancient multiring basins on the Moon revealed by Clementine Laser Altimetry. Science 266, 1848 - 1851.

[40] Stuart - Alexander, D. E., 1978. Geologic map of the central far side of the Moon. IMAP 1047.

[41] Sunshine, J. M., Pieters, C. M., 1993. Estimating modal abundances from the spectra of natural and laboratory pyroxene mixtures using the modified Gaussian model. J. Geophys. Res. 98, 9075 - 9087.

[42] Sunshine, J. M., Pieters, C. M., 1998. Determining the composition of olivine from reflectance spectroscopy. J. Geophys. Res. 103, 13675 - 13688.

[43] Sunshine, J. M., Pieters, C. M., Pratt, S. F., 1990. Deconvolution of mineral absorption bands - an improved approach. J. Geophys. Res. 95, 6955 - 6966.

[44] Vaughan, W. M., Head, J. W., 2014. Impact melt differentiation in the South Pole - Aitken basin: some observations and speculations. Planet. Space Sci. 91, 101 -106.

[45] Wieczorek, M. A., Neumann, G. A., Nimmo, F., Kiefer, W. S., Taylor, G. J., Melosh, H. J., Phillips, R. J., Solomon, S. C., Andrews - Hanna, J. C., Asmar, S. W., Konopliv, A. S., Lemoine, F. G., Smith, D. E., Watkins, M. M., Williams, J. G., Zuber, M. T., 2013. The crust of the Moon as seen by GRAIL. Science 339, 671 -675.

[46] Wu, W., Li, C., Zuo, W., Zhang, H., Liu, J., Wen, W., Su, Y., Ren, X., Yan, J., Yu, D., Dong, G., Wang, C., Sun, Z., Liu, E., Yang, J., Ouyang, Z., 2019. Lunar farside to be explored by Chang'e - 4. Nat. Geosci. 12, 222 - 223.

[47] Wu, Y., Wang, Z., Cai, W., Lu, Y., 2018. The absolute reflectance and new calibration site of the Moon. Astron. J. 155, 213.

[48] Yamamoto, S., Nakamura, R., Matsunaga, T., Ogawa, Y., Ishihara, Y., Morota, T., Hirata, N., Ohtake, M., Hiroi, T., Yokota, Y., Haruyama, J., 2010. Possible mantle origin of olivine around lunar impact basins detected by SELENE. Nat. Geosci. 3, 533.

[49] Yamamoto, S., Nakamura, R., Matsunaga, T., Ogawa, Y., Ishihara, Y., Morota, T., Hirata, N., Ohtake, M., Hiroi, T., Yokota, Y., Haruyama, J., 2012. Olivine - rich exposures in the South Pole - Aitken Basin. Icarus

218, 331 - 344.

[50] Zhu, M. - H., Wünnemann, K., Morbidelli, A., Artemieva, N. A., 2018. South Pole -Aitken basin may contribute insignificantly to the late accretion of the Moon. In: 49th Lunar and Planetary Science Conference, p. 1955.

Impact melt breccia and surrounding regolith measured by Chang'e - 4 rover

Sheng Gou[1,2], Zongyu Yue[1,3], Kaichang Di[1,2,3], Jia Wang[4], Wenhui Wan[1], Zhaoqin Liu[1], Bin Liu[1,2], Man Peng[1], Yexin Wang[1], Zhiping He[5], Rui Xu[5]

1. State Key Laboratory of Remote Sensing Science, Aerospace Information Research Institute, Chinese Academy of Sciences, Beijing 100101, China

2. State Key Laboratory of Lunar and Planetary Sciences, Macau University of Science and Technology, Macau, China

3. CAS Center for Excellence in Comparative Planetology, Hefei 230026, China

4. Science and Technology on Aerospace Flight Dynamics Laboratory, Beijing Aerospace Control Center, Beijing 100094, China

5. Key Laboratory of Space Active Opto - Electronics Technology, Shanghai Institute of Technical Physics, Chinese Academy of Sciences, Shanghai 200083, China

Abstract: Chang'e - 4 rover discovered a dark greenish and glistening impact melt breccia in a crater during its traverse on the floor of Von Kármán crater within the South Pole Aitken (SPA) basin on the lunar farside. The discovered breccia, being 52 × 16cm, resembles the lunar impact melt breccia samples 15466 and 70019 that returned by the Apollo missions. It was formed by impact - generated welding, cementing and agglutinating of lunar regolith and breccia. Clods surrounds the breccia - hosting crater were crushed into regolith powders by the rover's wheels, indicating the regolith may be compacted slightly and becomes blocky and friable. Relative mineral fractions are estimated from the in situ measured spectra by a Hapke model - based unmixing algorithm. Unmixing reveals that plagioclase (PLG, 45 ± 6%) is dominant in the regolith, followed by almost equal fractions of pyroxene (PYX, 7 ± 1%) and olivine (OL, 6 ± 2%), indicating the regolith is likely related to noritic rocks. The regolith measured by Chang'e - 4 rover was actually a highly mixture of multiple sources, with ejecta from Finsen crater being primary and possible contributions from Alder crater. Finsen and Alder craters are on the margin of the proposed impact melt pool produced by the SPA basin - forming event. Therefore, the ultimate source of the regolith might originate from a differentiated melt pool or from a suite of igneous rocks.

Keywords: impact melt breccia; Hapke model; relative mineral fractions; noritic rocks;

differentiated melt pool; igneous rocks

1 Introduction

The lunar regolith mainly consists of unconsolidated rock and mineral fragments, breccias, glasses, and agglutinates (e. g., Simon et al., 1981; Houck, 1982; Simon et al., 1982). It was formed by meteorite impact and gardening on the lunar surface over bil - lions of years. It covers nearly the entire lunar surface (McKay et al., 1991), except the underlying bedrocks occasionally exposed on lava channels, the walls of steep - sided craters (e. g., Galilei crater), or the central peaks of complex craters (e. g., Copernicus crater). The average thickness of the lunar regolith, determined by remote sensing techniques or in situ measurements, varies over mare and highland areas from 4 - 5m to proximately 10 - 15m, respectively (McKay et al., 1991). For example, the regolith thicknesses at the Apollo 11, 12, 14, 15, 16 and 17 sites estimated from the in situ seismic experiments are about 4. 4, 3. 7, 8. 5, 4. 4, 12. 2 and 4m, respectively (Cooper et al., 1974; Nakamura et al., 1975). Recent in situ penetrating radar measurements revealed that the regolith thicknesses at the Chang'e - 3 and Chang'e - 4 probe landing sites are about 3. 2m and 11. 1m, respectively (Lai et al., 2016, 2019). The composition of lunar regolith, in addition to provide potential resources for future human lunar explorations, records both the lunar geological evolution and its interaction with the dynamic space environment (Lucey et al., 2006).

According to the popular crystallization models of the Lunar Magma Ocean (LMO) hypothesis, olivine and pyroxene are rich in the lunar upper mantle due to the cumulate mantle overturn (Ringwood and Kesson, 1976). In order to constrain the lunar evolution, it is critically important to investigate the composition and structure of the lunar deep - seated materials. The SPA basin, stretching about 2500km across and 13km deep, is the largest, oldest and deepest recognized impact basin on the Moon (Stuart - Alexander, 1978). The SPA basin - forming impact should have sampled the lunar lower crust and upper mantle (Miljkovic et al., 2015; Melosh et al., 2017). The study on the possible mantle - derived materials may provide clues to the lunar evolution. China's Chang'e - 4 probe successfully landed within Von Kármán crater (186km in diameter, 176. 2°E, 44. 5°S) inside the South Pole - Aitken (SPA) basin on January 3, 2019 (Fig. 1) (Di et al., 2019). A lunar rover was deployed on the same day and conducted a series of in situ measurements during its travels on the lunar farside with onboard science payloads, e. g., a visible and near - infrared imaging spectrometer (VNIS) and a panoramic camera (Pancam) (Jia et al., 2018). Various targets on the lunar surface were measured by the VNIS, e. g., plume - disturbed regolith, wheel - trenched regolith, fresh crater ejecta and a rock fragment. Initial spectral analysis revealed that Chang'e - 4 rover may have discovered material from the lunar mantle, or from an impact - produced differentiated melt sheet (Gou et al., 2019; Hu et al., 2019; Li et al., 2019b). These interpretations perhaps would shed light on the long-standing mysteries about the Moon's formation and evolution, e. g., if the rover measured regolith was confirmed to originate from the mantle, the cumulate overturn model is supported.

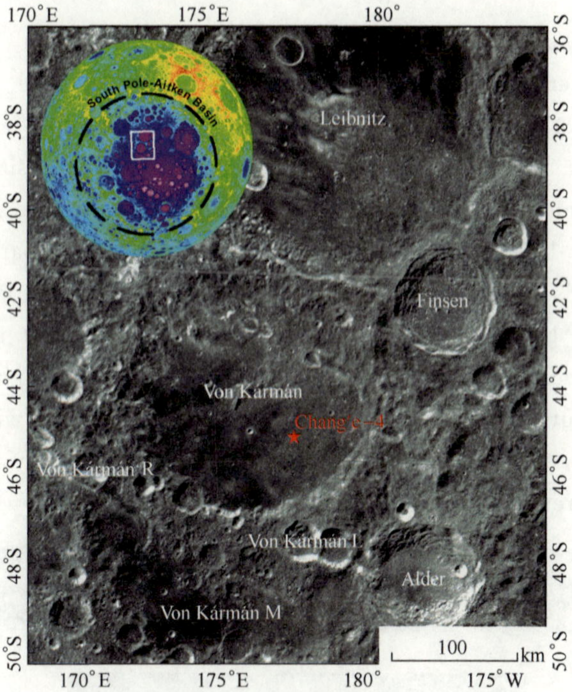

Fig. 1　The regional geologic setting of the Chang'e-4 probe landing site (red star). The base map shown in orthographic projection is Chang'e-2 high resolution (7m/pixel) digital orthophoto map. The Lunar Orbiter Laser Altimeter (LOLA) topography inset shows extent of the figure within the SPA basin (white box). Colors represent different heights in the inset, with red being the highest and purple being the lowest (For interpretation of the colors in the figure (s), the reader is referred to the web version of this article)

Chang'e-4 rover found an unusual substance in the bottom of a small crater during the eighth lunar day (July 26 to August 7, 2019). As small pieces of the unusual substance scattered in the crater, it seems that the substance is actually a distribution of multiple pieces of the same material. Its outermost fragments appear fairly similar to that of the surrounding materials. However, the unusual substance differs from the surrounding regolith blocks in shape, color and texture. It is dark greenish and glistening. The apparent length and width of its main part, which has an elongated mass-like configuration, are about 52cm and 16cm, respectively. Detailed in situ spectral measurements were conducted on the crater ejecta and the unusual substance during the ninth lunar day (August 24 to September 6). This study presents a detailed analysis on the unusual substance and its surrounding regolith at Chang'e-4 landing site, and infers the fundamental source of the regolith.

2 Datasets

2.1 Pancam images

The Pancam system is composed of two cameras with identical functions, performances and interfaces. Each camera has the ability to take color (red: 640nm, green: 540nm, blue: 470nm) and panchromatic (420—700nm) images with a field of view (FOV) of 19.6°×14.5° (Jia et al., 2018). The effective image size is 2,352 ×1,728 pixels in color mode and 1,176 ×864 pixels in panchromatic mode. As the camera has a resolving power of 0.15 mrad (about 120 pixels/degree), the spatial resolution of the Pancam image depends strongly on distance from the camera. The Pancam can acquire a 360° panoramic view by taking 56 pairs of images (28 different azimuth angles at 2 elevation angles). Pancam has a stereo base of 270mm and the stereo images can be used to generate high - resolution topographic maps of the lunar surface with photogrammetric techniques (Jia et al., 2018). Pancam imagery is used in this study for visual perception and geometric measurement of the unusual substance (Fig. 2).

Fig. 2 Top panel: The 360° Pancam panchromatic mosaic produced from the left - eye images taken at rover station LE00803. Bottom panel: Pancam color image of the unusual substance. The arrow in the Pancam mosaic marks the location of the crater that hosts the unusual substance

2.2 VNIS spectra

The VNIS onboard the Chang'e - 4 rover mainly consists of a visible/near - infrared (VIS/NIR) imager with an effective image size of 256 ×256 pixels (~1mm/pixel), a shortwave infrared (SWIR) single - pixel detector, and a white panel for calibration and dust - proofing (He et al., 2018; Li et al., 2019a). The VIS/NIR imager works from 450 to 945nm with a spectral resolution of 2 - 7nm, and the SWIR detector that works from 900 to 2,395nm has a spectral resolution of 3 - 12nm. The FOV of the SWIR (ϕ3.58°) corresponds to a circular region within the FOV of the VIS/NIR (8.5°×8.5°). It which has a diameter of 108 pixels on the VIS/NIR image from a detection distance of 1m (He et al., 2018; Li et al., 2019a), with the center being located at the coordinate (96, 128). In situ measured VNIS data are preprocessed and released to the science community as L2B radiance data by the ground application and research system (GRAS, http://moon.bao.ac.cn/). The L2B radiance datasets are further processed in this study to derive fullrange (450—2,395nm) reflectance (reflectance factor, REFF) spectra by the solar irradiance calibration method (Gou et al., 2019). The spectra measured on the lunar regolith and the unusual substance at different rover stations are shown in Fig. 3.

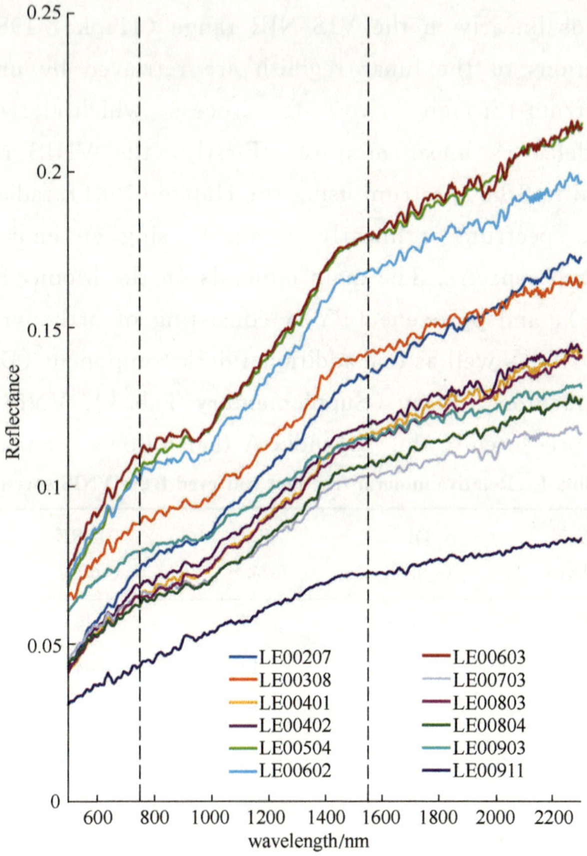

Fig. 3　In situ measured VNIS spectra. The targets of spectra measured at stations LE00903 and LE00911 are the regolith in crater interior and the unusual substance, respectively. The targets of the rest spectra are the surrounding lunar regolith. The dashed lines located at 750nm and 1550nm mark the wavelength range used for spectral unmixing to exclude thermal effects

2.3　Hazcam images

In addition, images acquired by the hazard avoidance camera (Hazcam) are also used in this study to infer the property of the lunar regolith at the Chang'e-4 landing site. The Hazcam is a stereo camera system, and each camera has a fisheye lens that offers a 170-degree FOV. The Hazcam acquires 1024 ×1024 pixels stereo images, which are used to obtain 3 dimension terrain information in front of the rover to detect hazards (e. g., large rocks and craters) along the rover's pathways.

3 Relative mineral fraction determination

The regolith reflectance spectrum measured by VNIS is a highly non-linear combination of the spectra of all components in the VIS/NIR wavelength range. However, the single-scattering albedo (SSA) spectrum of each component, which can be derived from reflectance, adds linearly in the VIS/NIR range (Hapke, 1981). Therefore, the relative mineral fractions of the lunar regolith are retrieved by unmixing the in situ measured VNIS spectrum through a two-step process, which allows nonlinearly mixed spectrum to be modeled as linear mixtures. Firstly, the VNIS reflectance (REFF) spectrum is converted to SSA spectrum using the Hapke (1981) radiative transfer model. Secondly, each SSA spectrum is linearly unmixed using an endmember library that consists of multiple SSA spectra. The main minerals on the Moon, including plagioclase (PLG), olivine (OL), and pyroxene (PYX, consisting of orthopyroxene (OPX), and clinopyroxene (CPX)), as well as one additional dark component (DARK), are selected to construct the endmember library (Supplementary Tab. 1). VNIS data reduction and unmixing details are presented in the Appendix A that follows the main text.

Table 1 Relative mineral fractions retrieved from VNIS spectra

Rover stations	PYX (vol.%)	OL (vol.%)	PLG (vol.%)	DARK (vol.%)	RMSE (%)
LE00207	10	5	45	40	0.2
LE00308	7	7	42	44	0.3
LE00401	8	6	40	46	0.3
LE00402	7	6	40	46	0.2
LE00504	8	7	57	28	0.3
LE00602	9	7	51	33	0.3
LE00603	8	8	54	30	0.3
LE00703	6	6	35	53	0.3
LE00803	7	5	44	43	0.2
LE00804	5	7	40	48	0.2
LE00903	7	2	47	43	0.2
LE00911	0	0	38	62	0.2

4 Results and discussions

4.1 Mineral fractions of the lunar regolith

The relative mineral fractions and root - mean - square error (RMSE) retrieved from the spectral unmixing algorithm are summarized in Tab. 1, and the best fitting results for these spectra are shown in Supplementary Fig. 1. Results show the regolith is dominated by PLG (45 ±6%), followed by PYX (7 ±1%) and OL (6 ±2%). These in situ spectral measurements generally are consistent with orbital observations that Chang'e - 4 landed in the PYX - abundant zone (Moriarty et al., 2013; Moriarty Ⅲ and Pieters, 2018). Though the inclusion of a dark endmember precludes this study from determining absolute mineral abundances, the relative abundant PLG and almost equation fractions of PYX (consisting of 6 ±1% OPX and 2 ±1% CPX) and OL indicate the regolith is likely related to noritic rocks.

Compared with the regolith, various particle sizes (e. g., 10μm, 20μm, 30μm), which have influences on effective optical path, are tested in the unmixing model for the unusual substance spectrum. However, only PLG (38%) and DARK (62%) components are detected. The unmixing result is unlikely in line with the actual situation. Probable physical reasons for the retrieved unrealistic mineral component fractions are: (1) The rover measured the unusual substance under a bad illumination condition that shadow accounts for about 42% of the FOV of VNIS (Supplementary Fig. 1), resulting in the VNIS measured spectrum is unreliable. (2) The composition of the unusual substance is the same as that of the regolith; however, the formation process changed its spectral properties in a way that it becomes insensitive in the VNIS spectral range, for example, loss of crystal structure through glass formation; (3) The composition of the unusual substance is different from that of the regolith but cannot be detected by VNIS, for example, incorporation of opaque materials results in spectrally neutral in the VNIS wavelength range.

4.2 Possible formation mechanisms

The Hazcam images, acquired when the rover approached the crater rim (Fig. 4a), made in situ spectral measurements (Fig. 4b), and withdrew from the crater rim (Fig. 4c), revealed that various sizes of blocks around the crater had been crushed into powder particles by the rover's wheels. The particles can even be clearly observed on the wheel (inset in Fig. 4b), indicating these blocks are clods that composed of fine regolith particles, rather than solid rock fragments/breccias. The generation of clods indicates the local regolith at Chang'e - 4 landing site may be compacted slightly and becomes blocky and friable; therefore, the regolith may be broken up and fractured easily by impacts on

the lunar surface.

(a)　　(b)　　(c)

Fig. 4　Hazcam images show the rover's locations

(a) Approached the crater rim; (2) Made in situ spectral measurements on the unusual substance; (3) Withdrew from the crater rim. The inset in the upper right of (b) shows powdered regolith particles stick on the wheel

Crater morphology and its surrounding boulder's size/shape distribution (if any) reflect the degradation and provide insights into the relative age dating (Basilevskii, 1973). Due to gravity and weathering processes, continued degradation causes the crater rim to lower, infilling produces a broad and flat crater floor, and physical weathering disintegrates surrounding boulders. Compared with the very gently sloping craters around, the preservation of the clods indicates that the crater didn't suffer from severe weathering. The crater's current depth and diameter are 0.3m and 2m, respectively. The relative high depth to diameter ratio (0.15) indicates it's a rather fresh crater (Pike, 1974).

The dark greenish and shiny features observed from Pancam color image (Fig. 2) are signs of possible presence of glasses. Glasses in the lunar regolith are usually sourced from impact melts or from volcanic eruptions. For example, the orange lunar soil sample 74220 returned by Apollo 17 mission in 1972 is rich in orange glass spheres and fragments, and they are thought to be recrystallized pyroclastic glasses that sprayed from an explosive volcanic eruption 3.71 $\pm$ 0.06 billion years ago (Husain and Schaeffer, 1973). The volcanic eruption within Von Kármán crater ceased about 3.35 billion years ago (Haruyama et al., 2009). It is unlikely that the unusual substance is a volcanic glass. Impact melts, including glass coatings, glass spherules, glassy breccias, and crystalline melt rocks, can be formed by hypervelocity impacts (Melosh, 1989). The Pancam color image (Fig. 2) gives clues to the genesis of the Chang'e - 4 rover discovered unusual substance. In appearance, it especially resembles lunar sample 15466 (Fig. 5a), which was sampled by the Apollo 15 mission from the rim of Spur crater (80m in diameter). Sample 15466 is a collection of breccia fragments cemented together with a vesicular black glass (Meyer, 2011). Due to limited imagery acquired by the Chang'e - 4 rover, it cannot be judged from other morphological evidences that the unusual substance is an entirely impact melt breccia. It may only be a coating on unmelted rocks that may only account for a small component of the observed total volume. It also resembles lunar sample 70019 (Fig. 5b), which was collected by the Apollo 17 mission from the bottom of a 3m fresh crater. Sample 70019 is a glass - coated soil breccia, and can be considered as a large "agglutinate" made up of dark regolith microbreccia cemented together with black, shiny glasses (Meyer, 2011). Thus, the best petrologic interpretation of the unusual substance discovered by the Chang'e - 4 rover in the bottom of the fresh crater is an impact melt breccia or a glass - coated agglutinitic regolith breccia.

(a)　　(b)

Fig. 5　Photos of lunar impact melt breccia samples

(a) S1 face of sample 15466; (b) T1 face of sample 70019. The cubes in (a) and (b) are 1 inch and 1cm, respectively.

(Photo source: NASA/JSC; IDs: S71 - 44184 and S73 - 15333)

There may be two sources of the unusual substance: (1) entirely impact melt emplaced in a different event; (2) glass coating on unmelted rocks emplaced in the crater. In the first scenario, assuming a spherical impactor with density of $3000kg/m^3$ and velocity of 17km/s vertically impacts the porous lunar regolith that has density of $1500kg/m^3$, according to the scaling laws (Melosh, 1989), the estimated diameter of the impactor that creates the 2m fresh crater is about 2cm, and the estimated diameter of the initial impact melt is about 6cm (see Supplementary Text 1 for the estimation). As the estimation is made on normal conditions, it can be considered as a minimum estimate only to a limited extent. The length (~52cm) and width (~16cm) of the Chang'e - 4 discovered melt breccia are several times larger than the predicted melt diameter even if the melt didn't disperse in subsequent crater collapse. Hence, the impact melt breccia was not likely formed in situ in the crater where it was found, but was very likely emplaced in a different event and was ejected to the 2m fresh crater. However, it is difficult to determine whether the breccia impacted the lunar surface at relative low velocity and formed the fresh crater, or if the fresh crater formed earlier and the breccia happened to be ejected and deposited onto the floor of the crater later. In the second scenario, a meteorite (about 2cm in diameter) impacted the lunar surface and formed relatively dark, glass - rich impact melt (6cm in diameter) in the fresh crater. The glassy melts coated the unmelted rocks, agglutinated the surrounding clods, and formed the current observed glass - coated agglutinitic breccia. Nevertheless, on the basis of the above comprehensive analysis, one most likely scenario for the formation of the Chang'e - 4 discovered impact melt breccia was that an impact partially melted local regolith, small drops of the dark greenish melt cooled rapidly, became shiny and cemented. The breccia formation process may darken the reflectance spectrum and rendered the regolith insensitive in the VNIS wavelength range.

4.3 Fundamental source of the regolith

Regolith is typically mixtures of materials with many origins, especially in a complex, ancient terrane such as the SPA basin (53°S, 169°W). Di et al. (2019) revealed that the regolith measured by the Chang'e - 4 rover was ejecta primarily from Finsen crater (42.3°S, 177.732°W) with possible contributions from Alder crater (48.6°S, 177.9°W), rather than the underlying mare basalt. Many hypotheses have been put forward regarding the fundamental source of the regolith measured by the Chang'e - 4 rover, including from the lunar upper mantle, or from a differentiated impact melt pool (Hu et al., 2019; Li et al., 2019b).

Mantle materials are expected to be exposed by the basin - forming events (Miljkovic et al., 2015), particularly the SPA basin inevitably spewed them onto the farside highlands (Melosh et al., 2017). The LMO hypothesis indicates that the lunar upper mantle is rich in OL and PYX due to cumulate overturn (Ringwood and Kesson, 1976). However, recent remote sensing observations and numeric simulations challenge the current paradigm and argue that the upper mantle is dominated by OPX (Melosh et al., 2017). For example, strong OPX signatures in the outer basin and rim of the SPA basin were spectrally observed (Moriarty Ⅲ and Pieters, 2018), while only mere traces of OL in limited places within the SPA basin were detected (Yamamoto et al., 2012). Based on the relative mineral fractions (abundant PLG) derived in this study, the hypothesis of possible mantle - derived regolith is not supported.

Numerical simulations suggested that the SPA basin should have sampled the lunar lower crust and upper mantle (Miljkovic et al., 2015; Melosh et al., 2017). The basin - forming event might generate an impact melt pond which differentiated and formed noritic layers with a thickness of ~6km and possibly as thick as ~13km (Hurwitz and Kring, 2014; Vaughan and Head, 2014). The estimated transient cavity radius of the SPA basin ranges from 840km to 1400km (Moriarty et al., 2013). Finsen and Alder craters are just about 370km and 210 away from the center of the SPA basin, respectively. Therefore, the lunar regolith at Chang'e - 4 landing site was probably primarily the physical weathering products of the noritic rocks that mainly excavated from Finsen crater billion years ago.

However, as the Finsen and Alder craters are on the margin of the proposed impact melt pool, there is another possibility that the regolith measured by Chang'e - 4 rover was produced from a suite of igneous rocks. In addition to the relative mineral fractions of the regolith derived in this study, the rock fragments encountered by the rover during the third lunar day (Fig. 6) at rover station LE303 provide supplemental evidence. Given that one of the measured fragment exhibits strong absorption bands (Fig. 6), it is likely that these rock fragments are crystallized. However, it is hard to figure out whether they are aphanitic or phaneritic. They lack discernible coarse - grained textures, e. g., no grains ≥3mm can be unambiguously recognized on the rock fragment surface, indicating the melt that formed the igneous rocks may cool quickly that crystals didn't have enough time to grow to sizes easily seen by the unaided eye. Spectral unmixing reveal the rock fragment contains 40% PYX (29% OPX and 11% CPX), 11% OL, and 43% PLG, indicating it can be classified as norite.

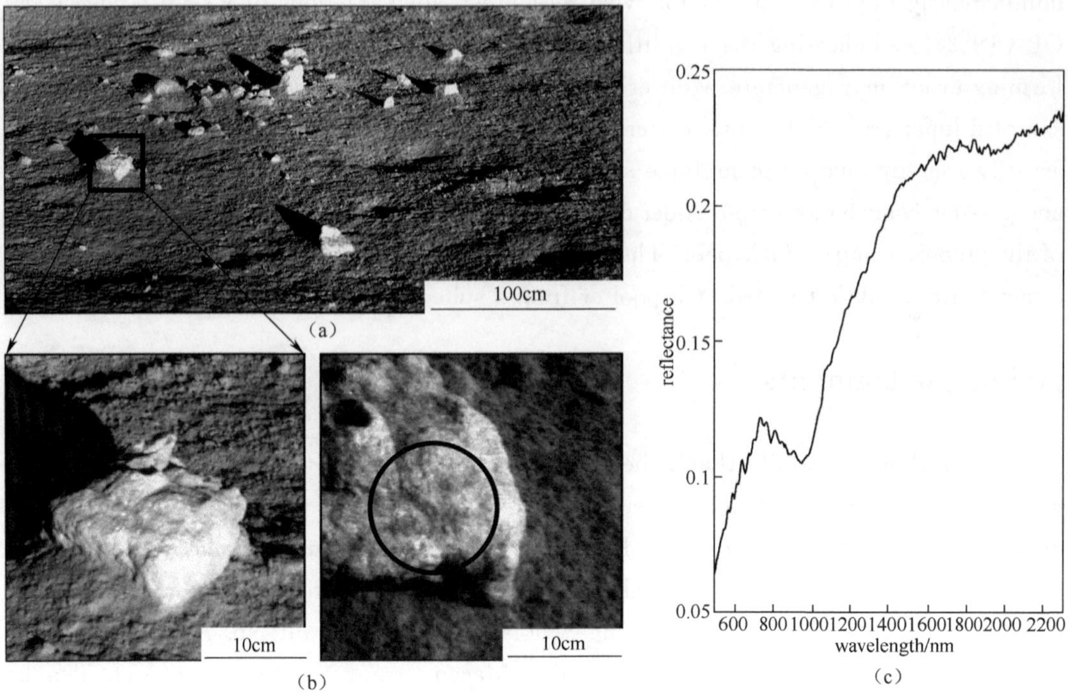

Fig. 6　(a) Pancam image of rock fragments encountered by Chang'e - 4 rover; (b) VIS/NIR image (~1mm/pixel) @ 750nm, the black circle corresponds to the field of view of the SWIR spectrometer; (c) VNIS spectrum of the rock fragment

The regolith measured by Chang'e - 4 rover was actually a highly mixture of multiple sources. It was primarily the weathering products of Finsen crater ejecta. This study supports the hypotheses that the fundamental source of the regolith might originate from a differentiated impact melt pool or from a suite of igneous rocks.

5 Conclusion

Chang'e - 4 rover discovered an impact melt breccia at the bottom of a small fresh crater along its traverse within Von Kármán crater. The length and width of the dark greenish and glistening breccia are about 52cm and 16cm, respectively. The discovered breccia resembles the lunar breccia samples 15466 and 70019 that returned by the Apollo missions. The breccia may be an entirely impact melt emplaced in a different event, or it may only be a coating on unmelted rocks emplaced in the crater. In either case, it was formed by impact - generated welding, cementing and agglutinating of lunar regolith and breccia.

A Hapke model - based spectral unmixing algorithm was applied on the in situ measured spectra. Unmixing results imply the regolith at the Chang'e - 4 landing site is dominated by PLG (45±6%), followed by almost equal fractions of PYX (7±1%) and OL (6±2%), indicating the regolith is likely related to noritic rocks. The SPA basin - forming event may generate a differentiated impact melt pool via thorough mixing the sampled lunar crust and mantle materials. The regolith measured by Chang'e - 4 rover was actually a highly mixture of multiple sources, with ejecta from Finsen crater being primary and possible contributions from Alder crater. Both Finsen and Alder craters are on the margin of the proposed impact melt pool. Therefore, the fundamental source of the regolith might originate from a differentiated melt pool or from a suite of igneous rocks.

Acknowledgements

The authors gratefully thank the whole Chang'e - 4 team for making the mission a great success and providing the data (http://moon.bao.ac.cn). This research was supported by the Strategic Priority Research Program of Chinese Academy of Sciences (grant No. XDB41000000), National Natural Science Foundation of China (grant Nos. 41702354 and 41941003), Macao Young Scholars Program (grant No. AM201902), and Science and Technology Development Fund of Macau (grant No. 131/2017/A3). Thanks for Kevin Cannon and Honglei Lin for discussions about Hapke model. Constructive reviews by Daniel P. Moriarty Ⅲ and Axel Wittmann improved the quality of this manuscript substantially, and are greatly appreciated.

Appendix A Methods

A. 1 VNIS data reduction

Major steps for the VNIS data reduction includes: (a) extraction of a VIS/NIR average spectrum, (b) elimination of spectrum discontinuity, (c) construction of a full-range spectrum, and (d) spectrum smoothing. More details on the data processing can be found in Gou et al. (2019). To avoid any thermal emission effects, the wavelength range is limited from 750 to 1550nm in this study. Therefore, only the absorption feature around 1000nm is essentially interpreted in this study. This feature is highly nonunique and can be attributed to numerous lunar materials, including volcanic glass, olivine, crystalline plagioclase, and pyroxene mixtures. Usually, these minerals can be spectroscopically distinguished by their diagnostic absorption centers. Orthopyroxene (OPX) spectrum exhibits a relatively narrow and symmetrical absorption near 900nm, and a broad absorption near 1900nm; Clinopyroxene (CPX) spectrum exhibits a narrow, symmetrical absorption near 1000nm, and a broad absorption near 2200nm; Olivine (OL) spectrum exhibits three overlapping absorption features at approximately 850, 1050, and 1150nm, forming what appears to be a single, broad, and asymmetric absorption near 1050nm; Iron-bearing crystalline plagioclase (PLG) spectrum has a weak and broad absorption near 1250nm; Iron-bearing Glass (GLS) spectrum exhibits two wide absorptions centered near 1060nm and 1900nm (Horgan et al., 2014).

A. 2 Spectral unmixing

A. 2. 1 Spectral endmember selection

The primary minerals on the lunar surface are pyroxene (PYX), olivine (OL) and plagioclase (PLG) (Spudis and Davis, 1986). It is understood from Apollo samples that natural lunar materials exhibit a multitude of diverse PYX compositions even within individual rocks and soils (Isaacson et al., 2011; Moriarty Ⅲ and Pieters, 2016). Though spectral measurements cannot sufficiently reflect their diversity, PYX is roughly classified as orthopyroxene (OPX) and clinopyroxene (CPX) based on chemistry and crystallography (Burns, 1993). And the relative abundances of OPX and CPX in the lunar regolith can be distinguished appropriately by spectral deconvolution (Noble et al., 2006). OL and PLG also have diverse compositions.

The aim of this study is to make relative, qualitative assessments of general mineralogical trends, therefore, SSA spectra of four lunar components, including OL, OPX, CPX and crystalline PLG, as well as one additional dark component (DARK) that is physically represented by opaque materials (e. g., ilmenite), and actual shade in the FOV of VNIS, are selected as spectral endmembers in this study. All the mineral endmember spectra are from reflectance experiment laboratory (RELAB, http://www.planetary.brown.edu/relab) and their details are listed in Supplementary Tab. 1. Note that ilmenite is not a generic opaque material. It is highly absorbing and strongly affects the reflectance spectrum when its abundance is high, especially for the fine particle size samples (Moriarty Ⅲ and Pieters, 2016). TiO_2 abundance at Chang'e-4 landing site estimated from Kaguya multiband imager data is about 1.42 ±0.24 wt.%. Thus the influence of ilmenite is not so strong on the overall continuum and pyroxene absorption band properties as that of the ilmenite - rich (>17%) Apollo 17 bulk rock sample.

The shiny features of the breccia (Fig. 2) show the possible presence of glass (GLS). However, the broad spectral absorption of GLS at approximately 1.06μm is easily suppressed by mafic minerals and nanophase iron that is ubiquitous in the lunar regolith (Cannon and Mustard, 2015). To test the sensitivity of the inclusion of a GLS component to the unmixing results, the VNIS measured breccia spectrum is modeled with and without a GLS endmember, multiple GLS endmembers (RELAB IDs: KC - JFM - 038, KC - JFM - 039, KC - JFM - 040, KC - JFM - 041, KC - JFM - 042, KC - JFM - 043) are tested in the unmixing model to check whether one composition of GLS provides a better fit. However, none of them can provide reasonable result and F - test result indicates the inclusion of a GLS component does not have significant influence on the modeled spectrum at the 95% confidence level (See Supplementary Text 1 for the introduction of F - test). Hence, a GLS component is not included in the final endmember library.

A. 2. 2 Single - scattering albedo determination

The core of the unmixing algorithm is the Hapke model (Hapke, 1981), which offers an approximate solution to the radiative transfer theory. Due to the definite physical meaning of the model parameters, the simplicity of model expression, and the fast com - putation, Hapke model is widely applied in the planetary science community to characterize the properties of planetary surface materials, e. g., grain size, morphology, internal structure and surface roughness (Schmidt and Fernando, 2015). When the particle sizes of all components in a mixture are known or fixed (similar as - sumption is made in this study), the uncertainties of relative mineral fractions estimated from reflectance spectrum by the simplified Hapke model are commonly <5% (Mustard and Pieters, 1987).

With the assumptions that all components within the lunar regolith are mixed intimately and the regolith particles are much larger than the spectral wavelength, the relationship between VNIS reflectance (REFF) and single - scattering albedo (SSA) can be expressed in equation (1) (Hapke, 1981).

$$\mathrm{REFF}(i, e, g) = \frac{\omega_{\mathrm{avg}}}{4(\mu_0 + \mu)}[(1 + B(g))P(g) + H(\mu_0, \omega_{\mathrm{avg}})H(\mu, \omega_{\mathrm{avg}}) - 1] \tag{1}$$

where μ_0 and μ are the cosine of the incidence angle i and emission (viewing) angle e, respectively; g is phase angle. ω_{avg} is the average single scattering albedo (SSA) of all the components. $B(g)$ is the backscatter function that describes opposition effect. $P(g)$ is the single - particle phase function. $H(x, \omega_{\mathrm{avg}})$ is the isotropic scattering function approximation (equation (2)) (Hapke, 2002).

$$H(x, \omega_{\mathrm{avg}}) = \frac{1 + 2x}{1 + 2x\sqrt{1 - \omega_{\mathrm{avg}}}} \tag{2}$$

Because the phase angle of each VNIS spectrum analyzed in this study is larger than 15° (see supplementary material for VNIS measure parameters at each station), the opposite effect is negligible and $B(g)=0$ (Mustard and Pieters, 1987). Assuming the lunar surface scatters light equally in all directions (isotropic), the single - particle phase function $P(g)=1.0$. The SSA spectrum can thus be solved from REFF spectrum by the simplified equation (3).

$$\mathrm{REFF}(i, e, g)=\frac{\omega_{\mathrm{avg}}}{4(\mu_0+\mu)}H(\mu_0, \omega_{\mathrm{avg}})H(\mu, \omega_{\mathrm{avg}}) \tag{3}$$

A. 2. 3　Space weathering simulation

To consider the space weathering effect, spectra of endmembers with different weathering levels are simulated by adding different mass fraction of submicroscopic iron (SMFe) to the host materials of the lunar regolith (Hapke, 1981, 1984, 1986, 2001) (equations (4) - (12)). The major steps are stated below.

a) Resample the endmember spectra to the VNIS bands. The spectral resolution of the endmember and VNIS spectra are different, thus the endmember spectra are resampled to the VNIS bands to ensure both have the same spectral resolution (band number). The spectra are limited to the wavelength range from 750 to 1,550nm to avoid any thermal emission effects.

b) Derive SSA from reflectance. SSA ω is derived from VNIS reflectance (REFF) spectrum (equations (4) - (5)).

$$\sqrt{1-\omega}=\frac{[(\mu_0+\mu)^2\Gamma^2+(1+4\mu_0\mu\Gamma)(1-\Gamma)]^{1/2}-(\mu_0+\mu)\Gamma}{1+4\mu_0\mu\Gamma} \tag{4}$$

$$\Gamma=4\frac{\mu_0+\mu}{\mu_0}\frac{1}{(1+2\mu_0)(1+2\mu)}\mathrm{REFF} \tag{5}$$

where μ_0 and μ are the cosine of the incidence and emission angles, respectively.

c) Derive absorption coefficient from SSA. Absorption coefficient α is derived from optical constants of each endmember (equation (6)).

$$\alpha=\frac{1}{\langle D\rangle}\ln\left[S_{\mathrm{i}}+\frac{(1-S_{\mathrm{e}})(1-S_{\mathrm{i}})}{\omega-S_{\mathrm{e}}}\right] \tag{6}$$

where $\langle D\rangle$ is the average distance traveled by all rays during a single transit of the particle. It has various approximation expressions (Hapke, 2001). Area weighted mean size (equation (7)) is adopted in this study and the derived $\langle D\rangle$ is listed in Supplementary Table 1. S_{e} and S_{i} are the average Fresnel reflection coefficients for externally and internally incident light, respectively. For $k^2\ll1$ and $1.2\leqslant n\leqslant2.2$, where n and k are the real and imagery parts (refractive index and extinction coefficient) of the endmembers' complex refractive index (also known as optical constants), respectively, S_{e} and S_{i} are empirically approximated by equations (8) and (9) (Hapke, 2012).

$$\langle D\rangle=D_{\mathrm{L}}\ln\frac{D_U}{D_L} \tag{7}$$

where D_U and D_L are the effective upper and lower limit of the particle size

$$S_e \approx \frac{(n-1)^2}{(n+1)^2} + 0.05 \tag{8}$$

$$S_i \approx 1 - \frac{4}{n(n+1)^2} \tag{9}$$

d) Simulate different space weathering level. Assuming the space weathering levels are proportional to the mass fraction of SMFe, the effect of SMFe that produced by space weathering on spectrum is simulated by modification on the absorption coefficient (equation (10)).

$$\alpha_w = \alpha_h + \alpha_{Fe} \tag{10}$$

where α_w, α_h, α_{Fe} are the absorption coefficients of the weathered host materials, unweathered host materials (equation (6)) and SMFe (equation (11)), respectively.

$$\alpha_{Fe} = \frac{36\pi f \rho_h}{\lambda \rho_{Fe}} \cdot \frac{n_h^3 n_{Fe} k_{Fe}}{(n_{Fe}^2 - k_{Fe}^2 + 2n_h^2)^2 + 4n_{Fe}^2 k_{Fe}^2} \tag{11}$$

where f is mass fraction of SMFe; λ is wavelength; n_*, k_*, ρ_* ($*$ denotes h or Fe) are the real and imagery parts of the complex refractive index (also known as optical constants), and densities of host materials or SMFe, respectively. The optical constants of Fe are cited from Johnson and Christy (1974), and that of OL, OPX and CPX are determined by Dorschner et al. (1995), and that of PLG are from Anthony et al. (2003).

e) Derive SSA from modified absorption coefficient. The relationship between absorption coefficient and SSA is expressed in equation (12) (See supplementary Fig. 2 for the simulated endmember spectra with the inclusion of different SMFe mass fraction).

$$\omega = \left[S_e + \frac{(1-S_e)(1-S_i)}{1-S_i\Theta}\Theta \right] \tag{12}$$

where $\Theta = \exp(-\alpha \langle D \rangle)$. For the Apollo lunar regolith sample, the effective upper and lower limit of the mineral particle sizes are 1000μm and 5μm, hence the approximate effective optical path is 26μm. This $\langle D \rangle$ value is adopted for the regolith at Chang'e-4 landing site for deriving SSA from modified absorption coefficient α_w in this study. Because the unusual substance is present in a relatively young crater and appears physically different than the surrounding regolith, its particle size should be different from the Apollo regolith. Therefore, various particle sizes (e. g., 10μm, 20μm, 30μm) are tested for the unusual substance in the unmixing model.

A. 2. 4 Unmixing of the VNIS SSA spectra

Endmember components in a mixture combine linearly in SSA space (Hapke, 1981). After converting into the SSA space, VNIS SSA spectra are deconvolved by a linear spectral unmixing model to retrieve each endmember's fraction (equation (13)). The fractions are estimated using a least squares solution with non-negativity and sum-to-one constraints (equation (14)).

$$\omega_{\text{mix}} = \omega F + E = \sum_{i=1}^{N} \omega_i F_i + E \tag{13}$$

$$\sum_{i=1}^{M} F_i = 1 (0 \leqslant F_i \leqslant 1) \tag{14}$$

where ω_{mix} is the $N \times 1$ VNIS SSA spectrum; ω is an $N \times M$ matrix of M endmember spectra with N spectral channels, and each column ω_i ($i = 1, 2, \ldots, M$) represents an endmember spectrum); F is a $M \times 1$ fraction vector, and each term F_i ($i = 1, 2, \ldots, M$) represents the spectral contribution of the ith endmember to the mixture; E is a $N \times 1$ residual error vector, and each term E_i ($i = 1, 2, \ldots, N$) represents the discrepancy between the measured and modeled SSA at the ith spectral channel.

According to Hapke (1981), the resolved F_i is the relative geometric cross - section of each endmember i, which is related to bulk - density M_i, solid density ρ_i and particle diameter D_i (equation (15)). It's assumed here that all endmember components have approximately the same particle size (26μm) after considering the contribution of SMFe inclusions. Hence, F_i can be considered as relative volume fraction of the ith endmember component in this study when ignoring pore space (equation (16)).

$$F_i = \frac{M/\rho_i D_i}{\sum_i^M M_i/\rho_i D_i} \tag{15}$$

$$F_i = \frac{M/\rho_i}{\sum_i^M M_i/\rho_i} \tag{16}$$

Appendix B　Supplementary material

Supplementary material related to this article can be found online at https: //doi. org /10. 1016/j. epsl. 2020. 116378.

References

[1] Anthony, J. W., Bideaux, R. A., Bladh, K. W., Nichols, M. C., 2003. Handbook of Mineralogy. Mineralogical Society of America, Chantilly, VA, USA. 20151 - 1110. http: //www. handbookofmineralogy. org/.

[2] Basilevskii, A. T., 1973. On the evolution rate of small lunar craters. In: 7th Lunar Science Conference. Pergamon Press, Houston, TX, pp. 1005 - 1020.

[3] Burns, R. G., 1993. Mineralogical Applications of Crystal Field Theory, 2nd edition. Cambridge University Press, New York.

[4] Cannon, K. M., Mustard, J. F., 2015. Preserved glass - rich impactites on Mars. Geol - ogy43, 635 - 638.

[5] Cooper, M. R., Kovach, R. L., Watkins, J. S., 1974. Lunar near - surface

structure. Rev. Geophys. 12, 291 - 308.

[6] Di, K., Zhu, M. - H., Yue, Z., Lin, Y., Wan, W., Liu, Zhaoqin, Gou, S., Liu, B., Peng, Man, Wang, Y., Niu, S., Zhang, J., Li, J., Xie, J., Xi, L., Yang, J., Xue, B., 2019. Topographic evolution of von Kármán crater revealed by the lunar rover Yutu -2. Geophys. Res. Lett. 46, 12764 - 12770.

[7] Dorschner, J., Begemann, B., Henning, T., Jager, C., Mutschke, H., 1995. Steps toward interstellar silicate mineralogy. II. Study of Mg - Fe - silicate glasses of variable composition. Astron. Astrophys. 300, 503 - 520.

[8] Gou, S., Di, K., Yue, Z., Liu, Z., He, Z., Xu, R., Lin, H., Liu, B., Peng, M., Wan, W., Wang, Y., Liu, J., 2019. Lunar deep materials observed by Chang'e - 4 rover. Earth Planet. Sci. Lett. 528, 115829.

[9] Hapke, B., 1981. Bidirectional reflectance spectroscopy: 1. Theory. J. Geophys. Res., Solid Earth 86, 3039 - 3054.

[10] Hapke, B., 1984. Bidirectional reflectance spectroscopy: 3. Correction for macroscopic roughness. Icarus 59, 41 - 59.

[11] Hapke, B., 1986. Bidirectional reflectance spectroscopy: 4. The extinction coefficient and the opposition effect. Icarus 67, 264 - 280.

[12] Hapke, B., 2001. Space weathering from Mercury to the asteroid belt. J. Geophys. Res. 106, 10039 - 10073.

[13] Hapke, B., 2002. Bidirectional reflectance spectroscopy: 5. The coherent backscatter opposition effect and anisotropic scattering. Icarus157, 523 - 534.

[14] Hapke, B., 2012. Theory of Reflectance and Emittance Spectroscopy, second edition. Cambridge University Press, Cambridge.

[15] Haruyama, J., Ohtake, M., Matsunaga, T., Morota, T., Honda, C., Yokota, Y., Abe, M., Ogawa, Y., Miyamoto, H., Iwasaki, A., Pieters, C. M., Asada, N., Demura, H., Hirata, N., Terazono, J., Sasaki, S., Saiki, K., Yamaji, A., Torii, M., Josset, J. - L., 2009. Long - lived volcanism on the lunar farside revealed by SELENE terrain camera. Science 323, 905 - 908.

[16] He, Z., Li, C., Xu, R., Lv, G., Yuan, L., Wang, J., 2018. Spectrometers based on acoustooptic tunable filters for in - situ lunar surface measurement. J. Appl. Remote Sens. 13, 027502.

[17] Horgan, B. H. N., Cloutis, E. A., Mann, P., Bell, J. F., 2014. Near - infrared spectra of ferrous mineral mixtures and methods for their identification in planetary surface spectra. Icarus 234, 132 - 154.

[18] Houck, K. J., 1982. Petrologic variations in Apollo 16 surface soils. J. Geophys. Res., Solid Earth 87, A197 - A209.

[19] Hu, X., Ma, P., Yang, Y., Zhu, M. - H., Jiang, T., Lucey, P. G., Sun, L., Zhang, H., Li, C., Xu, R., He, Z., Lin, H., Huang, C., Sun, Y., 2019. Mineral abundances inferred from in - situ reflectance measurements of Chang' E - 4 landing

site in South Pole - Aitken basin. Geophys. Res. Lett. 46, 9439 - 9447.

[20] Hurwitz, D. M., Kring, D. A., 2014. Differentiation of the South Pole - Aitken basin impact melt sheet: implications for lunar exploration. J. Geophys. Res., Planets 119, 1110 - 1133.

[21] Husain, L., Schaeffer, O. A., 1973. Lunar volcanism: age of the glass in the Apollo 17 orange soil. Science 180, 1358 - 1360.

[22] Isaacson, P. J., Sarbadhikari, A. B., Pieters, C. M., Klima, R. L., Hiroi, T., Liu, Y., Taylor, L. A., 2011. The lunar rock and mineral characterization consortium: deconstruction and integrated mineralogical, petrologic, and spectroscopic analyses of mare basalts. Meteorit. Planet. Sci. 46, 228 - 251.

[23] Jia, Y., Zou, Y., Ping, J., Xue, C., Yan, J., Ning, Y., 2018. The scientific objectives and payloads of Chang'E -4 mission. Planet. Space Sci. 162, 207 - 215.

[24] Johnson, P. B., Christy, R. W., 1974. Optical constants of transition metals: Ti, V, Cr, Mn, Fe, Co, Ni, and Pd. Phys. Rev. B9, 5056 - 5070.

[25] Lai, J., Xu, Y., Zhang, X., Tang, Z., 2016. Structural analysis of lunar subsurface with Chang'E - 3 lunar penetrating radar. Planet. Space Sci. 120, 96 - 102.

[26] Lai, J., Xu, Y., Zhang, X., Xiao, L., Yan, Q., Meng, X., Zhou, B., Dong, Z., Zhao, D., 2019. Comparison of dielectric properties and structure of lunar regolith at Chang'e - 3 and Chang'e - 4 landing sites revealed by ground penetrating radar. Geophys. Res. Lett. 46, 12783 - 12793. https://doi.org/10.1029/2019GL084458.

[27] Li, C., Wang, Z., Xu, R., Lv, G., Yuan, L., He, Z., Wang, J., 2019a. The scientific information model of Chang'e - 4 visible and near - IR imaging spectrometer (VNIS) and in - flight verification. Sensors 19, 2806.

[28] Li, C., Liu, D., Liu, B., Ren, X., Liu, J., He, Z., Zuo, W., Zeng, X., Xu, R., Tan, X., Zhang, X., Chen, W., Shu, R., Wen, W., Su, Y., Zhang, H., Ouyang, Z., 2019b. Chang'E - 4 initial spectroscopic identification of lunar far - side mantle - derived materials. Nature 569, 378 - 382.

[29] Lucey, P., Korotev, R. L., Gillis, J. J., Taylor, L. A., Lawrence, D., Campbell, B. A., Elphic, R., Feldman, B., Hood, L. L., Hunten, D., Mendillo, M., Noble, S., Papike, J. J., Reedy, R. C., Lawson, S., Prettyman, T., Gasnault, O., Maurice, S., 2006. Understanding the lunar surface and space - Moon interactions. Rev. Mineral. Geochem. 60, 83 - 219.

[30] McKay, D. S., Heiken, G., Basu, A., Blanford, G., Simon, S., Reedy, R., French, B. M., Papike, J., 1991. The lunar regolith. In: Heiken, G., Vaniman, D., French, B. M. (Eds.), Lunar Sourcebook. Cambridge University Press, pp. 285 - 356.

[31] Melosh, H. J., 1989. Impact Cratering: A Geologic Process. Oxford University Press.

[32] Melosh, H. J., Kendall, J., Horgan, B., Johnson, B. C., Bowling, T., Lucey, P.

G., Taylor, G. J., 2017. South Pole - Aitken basin ejecta reveal the Moon's upper mantle. Geology 45, 1063 - 1066.

[33] Meyer, C., 2011. Lunar sample compendium. https://curator.jsc.nasa.gov/lunar/lsc/index.cfm.

[34] Miljkovic, K., Wieczorek, M. A., Collins, G., Solomon, S. C., Smith, D. E., Zuber, M. T., 2015. Excavation of the mantle in basin - forming events on the Moon. Earth Planet. Sci. Lett. 409, 243 - 251.

[35] Moriarty, D. P., Pieters, C. M., Isaacson, P. J., 2013. Compositional heterogeneity of central peaks within the South Pole - Aitken basin. J. Geophys. Res., Planets 118, 2310 - 2322.

[36] Moriarty Ⅲ, D. P., Pieters, C. M., 2016. Complexities in pyroxene compositions derived from absorption band centers: examples from Apollo samples, HED meteorites, synthetic pure pyroxenes, and remote sensing data. Meteorit. Planet. Sci. 51, 207 -234.

[37] Moriarty Ⅲ, D. P., Pieters, C. M., 2018. The character of South Pole - Aitken basin: patterns of surface and subsurface composition. J. Geophys. Res., Planets 123, 729 -747.

[38] Mustard, J. F., Pieters, C. M., 1987. Quantitative abundance estimates from bidirec - tional reflectance measurements. J. Geophys. Res. 92, E617 - E626.

[39] Nakamura, Y., Dorman, J., Duennebier, F., Lammlein, D., Latham, G., 1975. Shallow lunar structure determined from the passive seismic experiment. Moon 13, 57 - 66.

[40] Noble, S. K., Pieters, C. M., Hiroi, T., Taylor, L. A., 2006. Using the modified Gaussian model to extract quantitative data from lunar soils. J. Geophys. Res., Planets 111.

[41] Pike, R. J., 1974. Depth/diameter relations of fresh lunar craters: revision from space - craft data. Geophys. Res. Lett. 1,291 - 294.

[42] Ringwood, A. E., Kesson, S. E., 1976. A dynamic model for mare basalt petro - genesis. In: 7th Lunar Science Conference. Pergamon Press, Houston, Texas, pp. 1697 - 1722.

[43] Schmidt, F., Fernando, J., 2015. Realistic uncertainties on Hapke model parameters from photometric measurement. Icarus 260,73 - 93.

[44] Simon, S. B., Papike, J. J., Laul, J. C., 1981. The lunar regolith: comparative studies of the Apollo and Luna sites. Petrology of soils from Apollo 17, Luna 16, 20, and 24. In: 12th Lunar and Planetary Science Conference. Pergamon Press, Houston, TX, pp. 371 - 388.

[45] Simon, S. B., Papike, J. J., Laul, J. C., 1982. The Apollo 14 regolith: petrology of cores 14210/14211 and 14220 and soils 14141, 14148, and 14149. J. Geophys. Res., Solid Earth 87, A232 - A246.

[46] Spudis, P. D., Davis, P. A., 1986. A chemical and petrological model of the lunar crust and implications for lunar crustal origin. J. Geophys. Res., Solid Earth 91, E84 -E90.

[47] Stuart - Alexander, D. E., 1978. Geologic map of the central far side of the Moon IMAP 1047. http: //pubs . er. usgs . gov /publication /i1047.

[48] Vaughan, W. M., Head, J. W., 2014. Impact melt differentiation in the South Pole - Aitken basin: some observations and speculations. Planet. Space Sci. 91, 101 -106.

[49] Yamamoto, S., Nakamura, R., Matsunaga, T., Ogawa, Y., Ishihara, Y., Morota, T., Hirata, N., Ohtake, M., Hiroi, T., Yokota, Y., Haruyama, J., 2012. Olivine - rich exposures in the South Pole - Aitken basin. Icarus218, 331 - 344.

Mare basalt flooding events surrounding Chang'e-4 landing site as revealed by Zhinyu crater ejecta

Sheng Gou[1,2,3], Zongyu Yue[2,4], Kaichang Di[2,4], Roberto Bugiolacchi[1,3,6], Meng-Hua Zhu[1,3], Patrick C. Pinet[5], Zhanchuan Cai[1,3]

1. State Key Laboratory of Lunar and Planetary Sciences, Macau University of Science and Technology, Macau, China
2. State Key Laboratory of Remote Sensing Science, Aerospace Information Research Institute, Chinese Academy of Sciences, Beijing 100101, China
3. CNSA Macau Center for Space Exploration and Science, Macau, China
4. Center for Excellence in Comparative Planetology, Chinese Academy of Sciences, Hefei 230026, China
5. Institut de Recherche en Astrophysique et Planétologie, University of Toulouse, CNRS, UPS, CNES, UT3, Toulouse, France
6. Earth Sciences, University College London, London, UK

Abstract: Von Kármán crater's floor was flooded with mare basalts during the Imbrian period. This site is the target of China's ongoing Chang'e-4 mission that includes the Yutu-2 rover and its instrumental payload. The Zhinyu crater, one of the largest craters within a few tens of kilometers from the landing site, is the product of a fresh impact that excavated subsurface, basaltic materials onto the surface. The compositional characteristics of the continuous ejecta around the crater vary radially with distance, suggesting possible mineral heterogeneity at depth. At least three main mare flooding phases could have occurred within Von Kármán, producing a basalt layer of at least 320 m in thickness. However, a broadly consistent olivine composition (Fo#: 55) of the three proposed basalt layers suggests that the sources of the mare basalts would have been similar in composition, perhaps even consisting of a single magma chamber where magma had time to evolve. The Yutu-2 rover that is edging its way westwards, could in principle test this hypothesis, thus further constraining the thermal history of the Von Kármán crater.

Keywords: Zhinyu crater; Von Kármán crater; Mare flooding; Homogenous composition; Single magma chamber

1 Introduction

The Von Kármán crater (~186km in diameter), lying in the north - west of South Pole - Aitken (SPA) basin on the lunar farside, was formed 3.97 Ga ago in the pre - Nectarian period (Stuart - Alexander, 1978; Yingst et al., 2017). During the Imbrian period, the floor of Von Kármán crater was flooded with several basaltic lava flows (or mare basalts) (Huang et al., 2018; Pasckert et al., 2018; Ling et al., 2019). China's Chang'e - 4 probe touched down successfully in the Von Kármán crater at 177.588°E, 45.457°S on January 3, 2019, and released the Yutu - 2 rover to conduct in situ measurements on this previously unexplored region (Di et al., 2019a; Pinet, 2019). The surface of the Chang'e - 4 landing site is generally thought to be mantled by ejecta from the Finsen crater (e. g., Gou et al., 2019; Hu et al., 2019; Gou et al., 2020; Huang et al., 2020; Lin et al., 2020), though the estimation of ejecta thickness from Finsen crater at the landing site is debated (Huang et al., 2018; Di et al., 2019b; Fu et al., 2020).

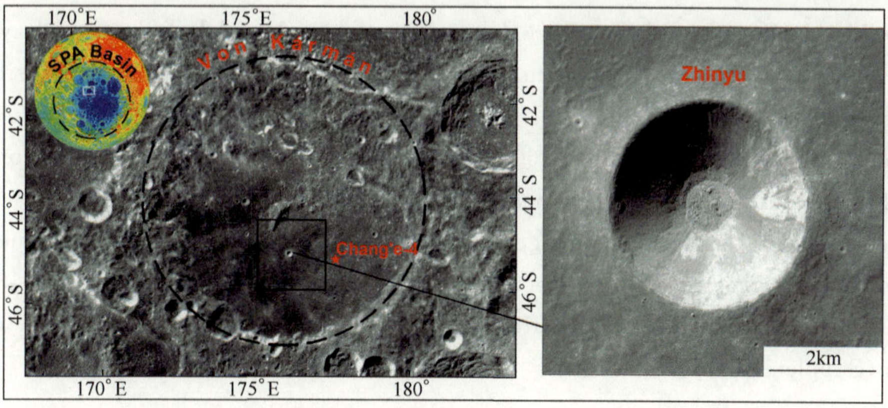

Fig. 1 Location of the Zhinyu crater inside the Von Kármán crater. The base map is a digital orthophoto map mosaic (7m/pixel) generated from Chang'e - 2 images (Ren et al., 2014). Different colors in the inset represent different heights, with red being the highest and blue being the lowest. The white box shows the boundaries of Fig. 1. The black box frames the area of Figs. 2, 4 and 5. (For interpretation of the references to colour in this figure legend, the reader is referred to the web version of this article.)

Zhinyu (176.15°E, 45.34°S), a fresh, simple crater with a diameter of ~3.8km and a depth of ~800m (Fig. 1) is one of the largest impacts near the Chang'e - 4 landing site, lying about 30km away to the west. The Zhinyu impact penetrated the mare basalts within the Von Kármán crater, thus it opens an exploratory window into the composition and stratigraphy of the subsurface. Hypervelocity impacts exhume materials from the subsurface and deposit them both within and outside the newly formed crater, allowing for investigation. For instance, Fu et al. (2020) targeted 40 dark - haloed craters and 77 bright ray craters around the Chang'e - 4 landing area and constructed a possible stratigraphy (including a low - Ti mare basalt layer) based on the excavation depths of these craters and their distances from the Finsen crater. Further, observed differences in the ejecta layers would represent geological variations with depth at the target point. For example, Ding et al. (2020) proposed that several fresh craters surrounded by blocky fragments encountered by the Yutu - 2 rover might be secondaries of the Zhinyu crater. Since the primary source for the regolith at the Chang'e - 4 landing site is reckoned to represent Finsen crater ejecta (compositionally different from mare basalt), caution must be applied in interpreting the in situ measured spectra around these secondary craters. Therefore, the Zhinyu crater ejecta provides a rare opportunity to explore the sub - regolith mare basalts, and should also contribute to the correct interpretation of the in situ measured data by the Yutu - 2 rover.

2 Data

2.1 Moon mineralogy mapper spectra

The Moon Mineralogy Mapper (M^3), which flew onboard India's Chandrayaan - 1 lunar orbiter, was a push - broom imaging spectrometer operating from the visible into the near - infrared (0.42 — 3.0μm). M^3 observed the lunar surface both in a high - resolution target mode (80m/pixel; 260 spectral channels) and in a low - resolution global mode (140m/pixel; 85 spectral channels) from a 100km orbit (Pieters et al., 2009). The M^3 Level 2 spectral reflectance images (REFIMG), which have been photometrically calibrated and thermal emission corrected (Clark et al., 2011; Besse et al., 2013), are used in this study to extract spectra from regions of interest (ROIs) for further detailed analysis (Supplementary Fig. S1).

2.2 Multiband imager products and images

The multiband imager (MI) was a high - resolution imaging camera onboard the Japanese Kaguya lunar orbiter that included separate visible and near infrared sensors (Ohtake et al., 2008). It acquired push - broom images in five bands for the visible and four bands for the near - infrared spectral ranges (Ohtake et al., 2008). The spatial resolution of the MI images is 20m/pixel for the visible bands and 62m/pixel for the near - infrared bands taken at a nominal orbit altitude of 100km (Ohtake et al., 2008). The mineral abundance products derived from topographically - corrected MI reflectance data (Lemelin et al., 2016) are used in this study to evaluate compositional variabilities of the mare basalts. As concentrations of TiO_2 and FeO are useful in classifying major lunar rock types (Lucey et al., 1998), derived abundances around the Zhinyu crater were produced using the algorithm of Otake et al. (2012) to investigate possible differences.

3 Methods

3.1 Basalt layer identification

There is evidence that the lava infill of the Von Kármán crater occurred during the Imbrian (Im) period (Huang et al., 2018). Inferred TiO_2 content is commonly used as a classifier for basalts: very low Ti (<1.5 wt% TiO_2), low-Ti (1.5–6 wt% TiO_2), and high-Ti (>6 wt% TiO_2) (Hiesinger and Jaumann, 2014). The overall TiO_2 content of the continuous ejecta of Zhinyu crater is relatively low (see Section 4.1), making it not suitable for subdivision. Consequently, to evaluate the mineral composition of the ejecta around Zhinyu crater, mineral maps (Fig. 2) of clinopyroxene (CPX), orthopyroxene (OPX), olivine (OL) and plagioclase (PLG) were extracted from the MI derived global map products (Lemelin et al., 2016). It emerges that three rings with varied mineral abundance, in particular for the OL (Fig. 2a), are distributed concentrically around the Zhinyu crater. The boundaries of these three rings are broadly outlined by the OL variation (Lai et al., 2020) and defined in this study as Im1, Im2 and Im3, in the order from the rim outward (Fig. 2). Previous studies have proposed buried mare basalts fields inside the Von Kármán crater (Pasckert et al., 2018; Ling et al., 2019). Here the spectral/mineralogical variations suggest there might have been at least three distinct mare flooding events with a degree of compositional differences.

3.2 Olivine composition analysis

The olivine fraction content of igneous rocks is highly diagnostic for interpreting the magma source and the degree of petrologic evolution (Isaacson et al., 2011). Based on the established relation between laboratory spectral absorption features and OL composition (Sunshine and Pieters, 1998; Isaacson and Pieters, 2010), the relative Mg/(Mg+Fe) ratio, as known as the forsterite number (Fo#), can be estimated from the remotely-sensed spectra. Therefore, the average spectrum of each layer was extracted from M^3 images (Supplementary Fig. S1) and deconvolved by the Modified Gaussian Model (MGM) (Supplementary Figs. S2–S4) (Sunshine and Pieters, 1993) with different mineral combination configurations (Supplementary Tab. S1) to estimate mineral modes (Clénet et al., 2011; Clénet et al., 2013). The OL composition is determined by fitting the MGM-derived OL-related absorption centers to a laboratory-derived compositional trend line (Sunshine and Pieters, 1998; Isaacson and Pieters, 2010). The uncertainty of the predicted Fo# values that arises from model initial conditions and other error sources, e.g., the low signal-to-noise ratio and low resolution of M^3 spectrum, is estimated to be about 10% Fo# and possibly more in the case of ternary mineral mixtures (Isaacson and Pieters, 2010; Pinet et al., 2018; Pinet et al., 2019).

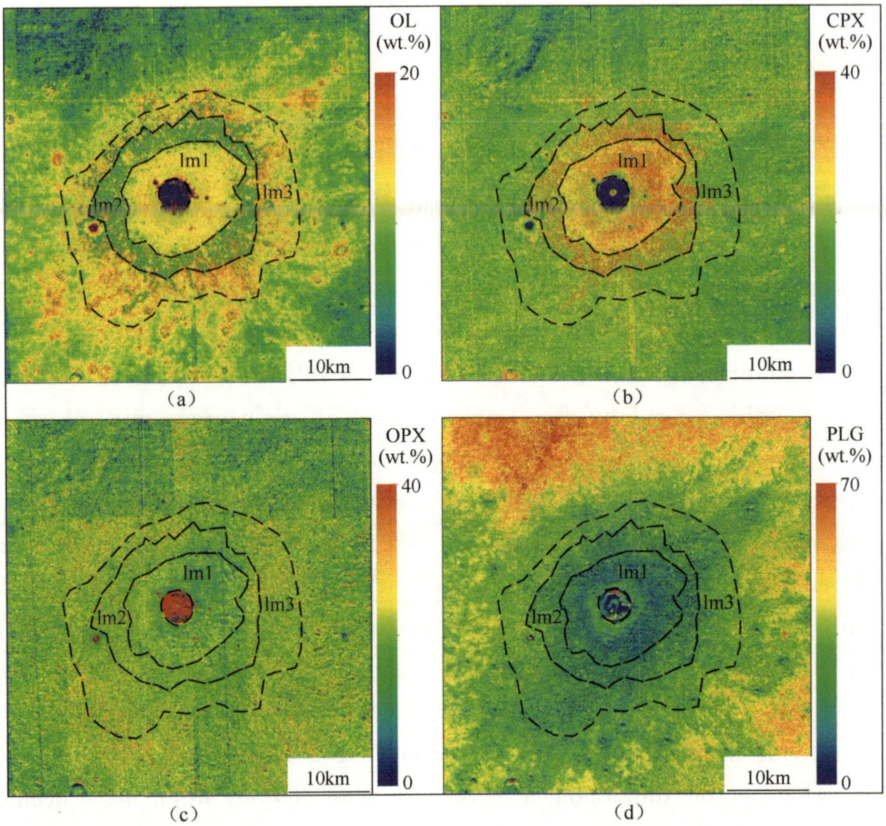

Fig. 2 Three mineral abundance variation rings distributed concentrically around the Zhinyu crater. (a) OL; (b) CPX; (c) OPX; (d) PLG. These MI derived products are from Lemelin et al. (2016). Upper range boundaries differ to highlight variations

3.3 Basalt layer thickness estimation

For a lunar simple crater with a diameter smaller than 15km, the maximum excavation depth (H_{exc}) is approximately 1/10 of the transient crater diameter (D_t), which is about 0.84 times the rim-to-rim diameter (D), i. e., $D_t = 0.84D$ (Melosh, 1989). The ejecta apron of an impact crater is composed of target materials excavated during the excavation stage (Melosh, 1989). They are deposited around the crater following an inverted preimpact stratigraphy of the underlying bedrock: the materials from the shallower depths tend to be thrown farther away from the crater rim than materials excavated from deeper layers (Thomson et al., 2009). The ejecta becomes thinner and more mixed with local mature regolith with increasing radial distance (Melosh, 1989). The average radius of continuous ejecta blanket (R_{ce}) can be estimated from crater radius (R) using the relation $R_{ce} = (2.348^{+0.564}_{-0.454}) R^{1.006}$ (Moore et al., 1974).

Although many laboratory experiments and numerical simulations have studied impacts into layered targets (e. g., Stöffler et al., 1975; Senft and Stewart, 2007;

Prieur et al., 2018), little work has been done to specifically address the relationship between the radial position of ejecta on the upper surface of the continuous ejecta blanket and depth of excavation in layered targets (e. g., Baratoux et al., 2007; Thomson et al., 2009). As a first approximation in this study, it's assumed that continuous ejecta at the farthest distance (D_{max}) comes from the surface, and the radial position of the continuous ejecta blanket is inversely related to the excavation depth, i. e., $(H_{exc}-S_{ce})/H_{exc}=D/D_{max}$. Therefore, the source depth of continuous ejecta blanket (S_{ce}) at a given radial position (D) is estimated by equation $S_{ce}=H_{exc}-(D^{*}H_{exc}/D_{max})$.

4　Results and discussions

4.1　Composition variations

Results of the MI-derived mineral abundances (Lemelin et al., 2016) for the three layers (in the order Im1, Im2, and Im3) show that the PLG abundance increases from 38.2% to 43.1%, the CPX decreases from 26.3% to 18.8%, the OPX increases from 22.6% to 25.2%, and the OL decreases from 12.9% to 10.5% and then increases to 13.2%. The mineral abundances almost fall within each other's standard deviation (Fig. 3), suggesting variations to be marginal. For example, the average CPX abundances for Im1, Im2 and Im3 are ~26.3 ±5.3%, ~24.8 ±6.9% and ~18.8 ±6.2%, respectively. Also, the average OL abundances are ~12.9 ±2.6%, ~10.5 ±4.7% and ~13.2 ±3.4%, respectively. MGM fitting results reveal that Fo# values derived from OL-related absorptions are ~55, suggesting that the OL to be relatively homogenous within the sampled units.

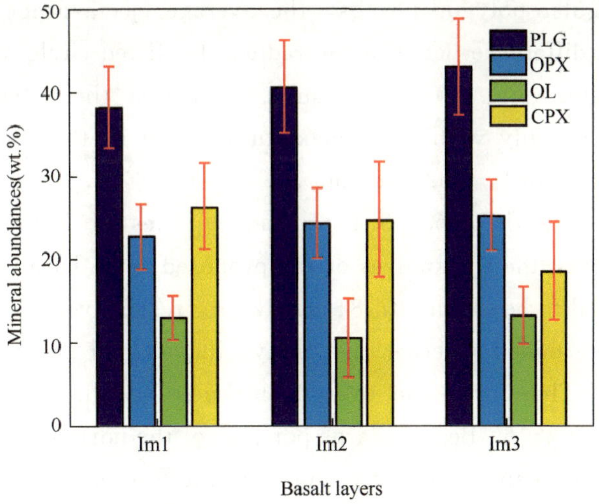

Fig. 3　Mineral abundances of the identified three basalt layers.
The error bars show the standard deviation of each mineral abundance

The average FeO contents for Im1, Im2 and Im3 are ~17.1±0.4%, ~16.6±0.4% and ~15.9±0.5%, respectively. Also, the average TiO_2 contents are ~1.5 ±0.4%, ~1.7 ±0.3% and ~1.6 ±0.3%, respectively (Fig. 4). Compared with lunar basalts sampled by Apollo and Luna missions or in situ measured by a rover (e. g., the FeO and TiO_2 contents of basalt measured by the Chang'e-3 rover are ~22.8 wt% and ~5.0 wt%, respectively (Ling et al., 2015), the identified three layers generally have relatively low FeO and TiO_2 contents.

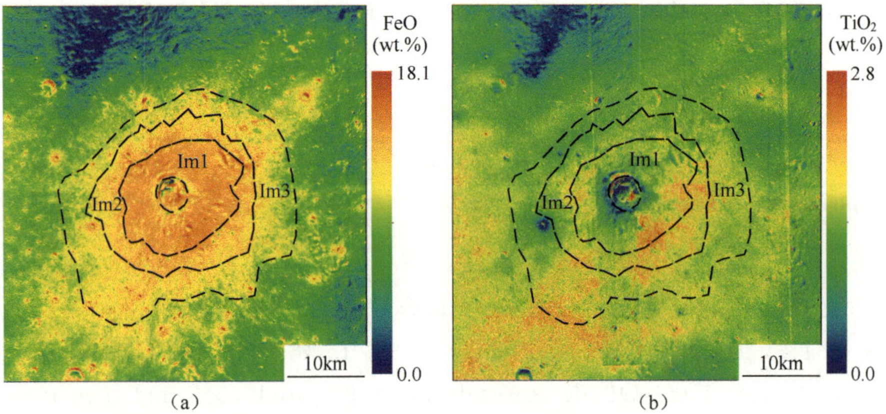

Fig. 4　FeO and TiO_2 contents of identified three mare basalt layers

4.2　The thickness of the basalt layers

Based on the excavation depth of impact crater (Section 3.3), the maximum excavation depth (H_{exc}) and radius of continuous ejecta blanket (R_{ce}) of Zhinyu cater are about 320m and 11.4km, respectively. Because the extents of the identified three layers are outlined by irregular polygonal rings, the average ejecta range from each layer is approximated by the difference between the radius of a fitted circle of each layer and the radius of Zhinyu crater (Fig. 5). The results show that the extents of all the three identified layers are roughly within the maximum extent of the continuous ejecta. The average ejecta ranges of Im1, Im2 and Im3 are about 4.6, 7.6 and 11.4km, indicating the extent of each layer is about 3.4, 5 and 7 times of Zhinyu crater's radius, respectively. Therefore, the thicknesses of the proposed three basalt layers (from top to bottom) are estimated to be obout 107, 84 and 129m, respectively. It is important to note that it cannot be determined whether the Zhinyu impact had penetrated the entire Im1 layer in this study. Therefore, the overall thickness of three mare basalt layers is estimated to be at least 320m. Because large portions of the northeastern region of the Von Kármán crater floor are mantled by the ejecta from the Finsen crater, and the area of the mare unit is about 6145km^2 (Pasckert et al., 2018), a minimum mare volume is thus estimated to be about 1966km^3.

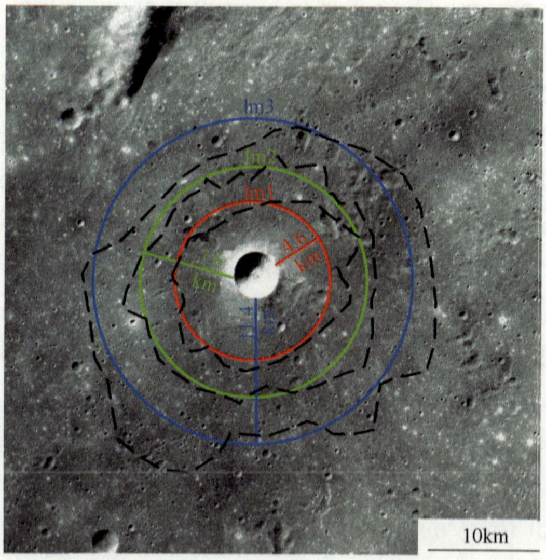

Fig. 5　Approximation of ejecta range around Zhihyu crater

Ling et al. (2019) divided the mare region inside the Von Kármán crater into Low - Ti unit (1.5 wt% < TiO_2 < 3 wt%) and Finsen ejecta unit (TiO_2 < 1.5 wt%). Crater size - frequency distribution (CSFD) measurements of the Low - Ti mare unit show three absolute model ages (AMAs), i. e., 3.38Ga, 3.57Ga, and 3.72Ga (Ling et al., 2019). These AMAs indicate there might be older buried lava flows in the Von Kármán crater (Pasckert et al., 2018; Ling et al., 2019), and the composition variations (Fig. 3) highlighted in this study indicate the recent erupted basalt (Im3) might have a lower CPX/OPX ratio than the older basalts (Im1). The kink (deflection diameter) between the youngest (3.38Ga) and oldest (3.72) isochron curves is ~1km, suggesting the thickness of buried lava flows to be >100m (Ling et al., 2019). The estimated total basalt thickness (≥320m) in this study agrees with this estimate. As a result, a refined stratigraphic sequence for the low - Ti mare basalts (Ling et al., 2019; Qiao et al., 2019; Fu et al., 2020) is proposed in this study, from top to bottom: Im3 (~107m), Im2 (~84m), and Im1 (≥129m).

A Lunar Penetrating Radar (LPR), with two work frequency channels (channel 1, 60MHz; channel 2, 500MHz), is part of the instrumentation on the Yutu - 2 rover to study the subsurface structure of its traversing area (Fang et al., 2014; Jia et al., 2018). Initial interpretations of the data indicate that there are several subsurface layers/units, pointing to a complex thermal history. For example, the high - frequency LPR data reveal that the shallow subsurface at the landing area is roughly composed of three layers (Lai et al., 2019; Li et al., 2020; Zhang et al., 2020a; Zhang et al., 2020b): a topmost fine - grained regolith layer (depth, 0 to 12m), an intermediate layer consisting of coarse materials with embedded rocks (12 to 24m), and a bottom layer composed of alternating coarse and fine materials (24 to 40m). As of November 22, 2020, the Yutu - 2 rover had already operated for 24 lunar days and traversed about 590m on the Finsen ejecta unit within the Von Kármán crater. The rover is currently about 30km away from the Zhinyu crater. The collected low - frequency LPR channel has the maximum detection depth of ~330m (Lai et al., 2020), possibly up to~500 m (Zhang et al., 2020a). However, due to the low thickness resolution (~10m) of the LPR low - frequency data and different processing procedures, the deep multi - layered stratigraphic structure revealed by the low - frequency LPR data is not entirely consistent between most of the studies, including Lai et al. (2020), Zhang et al. (2020a), and Zhang et al. (2020b). Nevertheless, the basaltic stratigraphic sequence at the Zhinyu crater proposed in this study roughly corresponds to the D and E layers in the work of Lai et al. (2020). The rover will continue to traverse westward and it is planned to reach the mare basalt area in the future. The proposed stratigraphic sequence and basalt thickness are expected to be validated by the LPR data, which will provide ground truth and should contribute to a fuller understanding of the basalt thermal history of the Von Kármán crater.

4.3 Implications for magma evolution

The ejecta under investigation is composed of materials exhumed by the Zhinyu crater during formation, a process that inverts the target's stratigraphical sequence. Therefore, the variation of ejecta composition along the radial distances around Zhinyu crater reflects the composi - tional variation of the basalt source at different stratigraphic depths. A fairly homogenous OL composition (Fo#: 55) of the three layers indicates the sources of these mare basalts (≥320m in thickness) are almost homogenous in composition, or might even be from a single magma chamber. Generally, for primary magmas, forsteritic (high Fo#) OL is indicative of a primitive source, and more fayalitic (low Fo#) OL is indicative of an evolved source (Basaltic Volcanism Study Project, 1981). The predicted Fo# value points to a moderate magma evolution. The small mineral abundance differences (Fig. 3) among layers may indicate there were at least three mare flooding events occurred inside Von Kármán crater during the Imbrian period, with two dome - like structures being the possible latest volcanic vents of these basalts (Qiao et al., 2019).

5 Conclusion

China's Chang'e - 4 probe landed inside the Von Kármán crater, whose floor was flooded with mare basalts during the Imbrian period. Zhinyu, a fresh impact crater, is the largest close to the landing site. The materials exhumed by the blast would have distributed the ejecta around the crater, exposing the stratigraphy sequence of the underlying basaltic rock layer (s). The mineral abundances derived from multispectral data, vary radially and discontinuously from the rim craters outwards, forming compositionally distinguished annuli. In particular, variations in the mineral olivine and clinopyroxene abundances support the idea that there could be at least three distinct basalt layers (named Im1, Im2, and Im3 from bottom to top) within the Von Kármán crater. The thickness of each layer is estimated by an empirical relation according to the processes of impact excavation and ejecta deposition, i. e., the shallow material deposits at the further distance whereas the deep material deposits around the crater. As it cannot be determined in this study whether the Zhinyu impact penetrated through the entire Im1 layer, the total thickness of the three basalt layers is estimated to be at least 320m. However, the minor differences noted in the mineral abundances and the fairly consistent olivine compositions (Fo#: 55) of these three layers indicate the sources of these mare basalts to be comparable in composition, perhaps even from a single magma chamber where the magma underwent only moderate evolution. The westward traversing Yutu - 2 rover may provide ground truth for the findings in this study, which would contribute to the understanding of the thermal history of the Von kármán crater.

Acknowledgments

The M^3 data are from the PDS Geoscience Node (https://pds - geosciences. wustl. edu/). The MI data are from the Selene Data Archive (http://l2db. selene. darts. isas. jaxa. jp/). The MI - derived mineral maps are from the Astrogeology Science Center (https://astrogeology. usgs. go v/maps/lunar - kaguya - multiband - imager - mosaics). This work was supported by the Strategic Priority Research Program of Chinese Academy of Sciences (grant No. XDB41000000), National Natural Science Foundation of China (grant Nos. 41702354 and 41941003), Macao Young Scholars Program (grant No. AM201902), and Science and Technology Development Fund of Macau (grant No. 131/2017/A3).

Appendix A　Supplementary data

Supplementary data to this article can be found online at https://doi. org/10. 1016/j. icarus. 2021. 114370.

References

[1] Baratoux, D., Pinet, P., Gendrin, A., Kanner, L., Mustard, J., Daydou, Y., Vaucher, J., Bibring, J. - P., 2007. Mineralogical structure of the subsurface of Syrtis Major from OMEGA observations of lobate ejecta blankets. J. Geophys. Res.: Planets 112 (E8), E08S05.

[2] Basaltic Volcanism Study Project, 1981. Basaltic Volcanism on the Terrestrial Planets. Pergamon Press, New York.

[3] Besse, S., Yokota, Y., Boardman, J., Green, R., Haruyama, J., Isaacson, P., Mall, U., Matsunaga, T., Ohtake, M., Pieters, C., Staid, M., Sunshine,

J., Yamamoto, S., 2013. One Moon, many measurements 2: photometric corrections. Icarus 226 (1), 127 - 139.

[4] Clark, R. N., Pieters, C. M., Green, R. O., Boardman, J. W., Petro, N. E., 2011. Thermal removal from near - infrared imaging spectroscopy data of the Moon. J. Geophys. Res.: Planets 116 (E6), E00G16.

[5] Clénet, H., Pinet, P. C., Daydou, Y., Heuripeau, F., Rosemberg, C., Baratoux, D., Chevrel, S., 2011. A new systematic approach using the modified Gaussian model: insight for the characterization of chemical composition of olivines, pyroxenes and olivine - pyroxene mixtures. Icarus 213 (1), 404 - 422.

[6] Clénet, H., Pinet, P., Ceuleneer, G., Daydou, Y., Heuripeau, F., Rosemberg, C., Bibring, J. P., Bellucci, G., Altieri, F., Gondet, B., 2013. A systematic mapping procedure based on the modified Gaussian model to characterize magmatic units from olivine/ pyroxenes mixtures: application to the Syrtis Major volcanic shield on Mars. J. Geophys. Res.: Planets 118 (8), 1632 - 1655.

[7] Di, K., Liu, Z., Liu, B., Wan, W., Peng, M., Wang, Y., Gou, S., Yue, Z., Xn, X., Jia, M., Niu, S., 2019a. Chang'e - 4 lander localization based on multi - source data. J. Remote Sens. 23 (1), 177 - 184.

[8] Di, K., Zhu, M. - H., Yue, Z., Lin, Y., Wan, W., Zhaoqin, Liu, Gou, S., Liu, B., Man, Peng, Wang, Y., Niu, S., Zhang, J., Li, J., Xie, J., Xi, L., Yang, J., Xue, B., 2019b.

[9] Topographic evolution of Von Kármán crater revealed by the lunar rover Yutu - 2. Geophys. Res. Lett. 46 (22), 12764 - 12770.

[10] Ding, C., Xiao, Z., Wu, B., Li, Y., Prieur, N. C., Cai, Y., Su, Y., Cui, J., 2020. Fragments delivered by secondary craters at the Chang'E - 4 landing site. Geophys. Res. Lett. 47 (7), e2020GL087361.

[11] Fang, G. - Y., Zhou, B., Ji, Y. - C., Zhang, Q. - Y., Shen, S. - X., Li, Y. - X., Guan, H. - F., Tang, C. - J., Gao, Y. - Z., Lu, W., Ye, S. - B., Han, H. - D., Zheng, J., Wang, S. - Z., 2014. Lunar penetrating radar onboard the Chang'e - 3 mission. Res. Astron. Astrophys. 14 (12), 1607 - 1622.

[12] Fu, X. - H., Qiao, L., Zhang, J., Ling, Z. - C., Li, B., 2020. The subsurface structure and stratigraphy of the Chang'E - 4 landing site: orbital evidence from small craters on the Von Kármán crater floor. Res. Astron. Astrophys. 20 (1), 8 (pp. 12).

[13] Gou, S., Di, K., Yue, Z., Liu, Z., He, Z., Xu, R., Lin, H., Liu, B., Peng, M., Wan, W., Wang, Y., Liu, J., 2019. Lunar deep materials observed by Chang'e - 4 rover. Earth Planet. Sci. Lett. 528, 115829.

[14] Gou, S., Di, K., Yue, Z., Liu, Z., He, Z., Xu, R., Liu, B., Peng, M., Wan, W., Wang, Y., Liu, J., 2020. Forsteritic olivine and magnesium - rich orthopyroxene materials measured by Chang'e - 4 rover. Icarus 345, 113776.

[15] Hiesinger, H., Jaumann, R., 2014. Chapter 23 - the moon. In: Spohn, T., Breuer, D., Johnson, T. V. (Eds.), Encyclopedia of the Solar System, Third edition. Elsevier, Boston, pp. 493 - 538.

[16] Hu, X., Ma, P., Yang, Y., Zhu, M. - H., Jiang, T., Lucey, P. G., Sun, L., Zhang, H., Li, C., Xu, R., He, Z., Lin, H., Huang, C., Sun, Y., 2019. Mineral abundances inferred from in - situ reflectance measurements of Chang'E - 4 landing site in south pole - Aitken basin. Geophys. Res. Lett. 46, 9439 -9447.

[17] Huang, J., Xiao, Z., Flahaut, J., Martinot, M., Head, J., Xiao, X., Xie, M., Xiao, L., 2018. Geological characteristics of Von Kármán crater, northwestern south pole - Aitken Basin: Chang'E - 4 landing site region. J. Geophys. Res.: Planets 123 (7), 1684 - 1700.

[18] Huang, J., Xiao, Z., Xiao, L., Horgan, B., Hu, X., Lucey, P., Xiao, X., Zhao, S., Qian, Y., Zhang, H., Li, C., Xu, R., He, Z., Yang, J., Xue, B., He, Q., Zhong, J., Lin, H., Huang, C., Xie, J., 2020. Diverse rock types detected in the lunar south pole - Aitken Basin by the Chang'E - 4 lunar mission. Geology 48, 723 - 727.

[19] Isaacson, P. J., Pieters, C. M., 2010. Deconvolution of lunar olivine reflectance spectra: implications for remote compositional assessment. Icarus 210 (1), 8 - 13.

[20] Isaacson, P. J., Pieters, C. M., Besse, S., Clark, R. N., Head, J. W., Klima, R. L., Mustard, J. F., Petro, N. E., Staid, M. I., Sunshine, J. M., 2011. Remote compositional analysis of lunar olivine - rich lithologies with Moon Mineralogy Mapper (M^3) spectra. J. Geophys. Res.: Planets 116 (E6), E00G11.

[21] Jia, Y., Zou, Y., Ping, J., Xue, C., Yan, J., Ning, Y., 2018. The scientific objectives and payloads of Chang'E - 4 mission. Planet. Space Sci. 162, 207 - 215.

[22] Lai, J., Xu, Y., Zhang, X., Xiao, L., Yan, Q., Meng, X., Zhou, B., Dong, Z., Zhao, D., 2019. Comparison of dielectric properties and structure of lunar regolith at Chang'e - 3 and Chang'e - 4 landing sites revealed by ground penetrating radar. Geophys. Res. Lett. 46, 12783 - 12793.

[23] Lai, J., Xu, Y., Bugiolacchi, R., Meng, X., Xiao, L., Xie, M., Liu, B., Di, K., Zhang, X., Zhou, B., Shen, S., Xu, L., 2020. First look by the Yutu - 2 rover at the deep subsurface structure at the lunar farside. Nat. Commun. 11 (1), 3426.

[24] Lemelin, M., Lucey, P. G., Gaddis, L. R., Hare, T., Ohtake, M., 2016. Global Map Products From the Kaguya Multiband Imager at 512 ppd: Minerals, FeO and OMAT. 47th Lunar and Planetary Science Conference, The Woodlands, Texas (pp. 2994).

[25] Li, C., Su, Y., Pettinelli, E., Xing, S., Ding, C., Liu, J., Ren, X., Lauro, S. E., Soldovieri, F., Zeng, X., Gao, X., Chen, W., Dai, S., Liu,

D., Zhang, G., Zuo, W., Wen, W., Zhang, Z., Zhang, X., Zhang, H., 2020. The Moon's farside shallow subsurface structure unveiled by Chang'E - 4 lunar penetrating radar. Sci. Adv. 6 (9), eaay6898.

[26] Lin, H., He, Z., Yang, W., Lin, Y., Xu, R., Zhang, C., Zhu, M. - H., Chang, R., Zhang, J., Li, C., Lin, H., Liu, Y., Gou, S., Wei, Y., Hu, S., Xue, C., Yang, J., Zhong, J., Fu, X., Wan, W., Zou, Y., 2020. Olivine - norite rock detected by the lunar rover Yutu - 2 likely crystallized from the SPA impact melt pool. Natl. Sci. Rev. 7, 913 - 920.

[27] Ling, Z., Jolliff, B. L., Wang, A., Li, C., Liu, J., Zhang, J., Li, B., Sun, L., Chen, J., Xiao, L., Liu, J., Ren, X., Peng, W., Wang, H., Cui, X., He, Z., Wang, J., 2015. Correlated compositional and mineralogical investigations at the Chang'e - 3 landing site. Nat. Commun. 6, 8880.

[28] Ling, Z., Qiao, L., Liu, C., Cao, H., Bi, X., Lu, X., Zhang, J., Fu, X., Li, B., Liu, J., 2019. Composition, mineralogy and chronology of mare basalts and non - mare materials in Von Kármán crater: landing site of the Chang'E - 4 mission. Planet. Space Sci. 179, 104741.

[29] Lucey, P. G., Taylor, G. J., Hawke, B. R., Spudis, P. D., 1998. FeO and TiO_2 concentrations in the south pole - Aitken basin: implications for mantle composition and basin formation. J. Geophys. Res.: Planets 103 (E2), 3701 - 3708.

[30] Melosh, H. J., 1989. Impact Cratering: A Geologic Process. Oxford University Press, New York.

[31] Moore, H. J., Hodges, C. A., Scott, D. H., 1974. Multiringed Basins - Illustrated by Orientale and Associated Features. 5th Lunar Science Conference, Houston, Texas, pp. 71 - 100.

[32] Ohtake, M., Haruyama, J., Matsunaga, T., Kodama, S., Morota, T., Yokota, Y., 2008. Scientific objectives and specification of the SELENE multiband imager. Adv. Space Res. 42 (2), 301 - 304.

[33] Otake, H., Ohtake, M., Hirata, N., 2012. Lunar Iron and Titanium Abundance Algorithms Based on SELENE (Kaguya) Multiband Imager Data. 43rd Lunar and Planetary Science Conference, The Woodlands, Texas (pp. 1905).

[34] Pasckert, J. H., Hiesinger, H., van der Bogert, C. H., 2018. Lunar farside volcanism in and around the south pole - Aitken basin. Icarus 299, 538 - 562.

[35] Pieters, C. M., Boardman, J., Buratti, B., Chatterjee, A., Clark, R., Glavich, T., Green, R., Head, J., Isaacson, P., Malaret, E., 2009. The moon mineralogy mapper (M3) on Chandrayaan - 1. Curr. Sci. 96 (4), 500 - 505.

[36] Pinet, P. C., 2019. The Moon's mantle unveiled. Nature 569 (7756), 338 - 339.

[37] Pinet, P. C., Chevrel, S. D., Daydou, Y. H., 2018. Characterization of the Olivine/ Plagioclase Mineralogy at Copernicus Crater from MGM Deconvolution of M3 Observations. 49th Lunar and Planetary Science Conference, The Woodlands,

Texas (pp. 1899).

[38] Pinet, P. C., Chevrel, S. D., Daydou, Y. H., 2019. Reassessing the Relationship Between Olivine Composition and Reflectance Spectroscopy From Advanced MGM Deconvolution. 50th Lunar and Planetary Science Conference, The Woodlands, Texas (pp. 1806).

[39] Prieur, N. C., Rolf, T., Wünnemann, K., Werner, S. C., 2018. Formation of simple impact craters in layered targets: implications for lunar crater morphology and regolith thickness. J. Geophys. Res.: Planets 123 (6), 1555 - 1578.

[40] Qiao, L., Ling, Z., Fu, X., Li, B., 2019. Geological characterization of the Chang'e - 4 landing area on the lunar farside. Icarus 333, 37 - 51.

[41] Ren, X., Liu, J. J., Wang, F. F., Wang, W. R., Mu, L. L., Li, H. H., 2014. A New Lunar Global Topographic Map Products From Chang'E - 2 Stereo Camera Image Data. European Planetary Science Congress, Cascais, Portugal (pp. EPSC2014 - 344).

[42] Senft, L. E., Stewart, S. T., 2007. Modeling impact cratering in layered surfaces. J. Geophys. Res.: Planets 112 (E11), E11002.

[43] Stöffler, D., Gault, D. E., Wedekind, J., Polkowski, G., 1975. Experimental hypervelocity impact into quartz sand: distribution and shock metamorphism of ejecta. J. Geophys. Res. 80 (29), 4062 - 4077.

[44] Stuart - Alexander, D. E., 1978. Geologic Map of the Central Far Side of the Moon.

[45] Sunshine, J. M., Pieters, C. M., 1993. Estimating modal abundances from the spectra of natural and laboratory pyroxene mixtures using the modified Gaussian model. J. Geophys. Res. 98 (E5), 9075 - 9087.

[46] Sunshine, J. M., Pieters, C. M., 1998. Determining the composition of olivine from reflectance spectroscopy. J. Geophys. Res. 103 (E6), 13675 - 13688.

[47] Thomson, B. J., Grosfils, E. B., Bussey, D. B. J., Spudis, P. D., 2009. A new technique for estimating the thickness of mare basalts in Imbrium Basin. Geophys. Res. Lett. 36 (12), L12201.

[48] Yingst, R. A., Chuang, F. C., Berman, D. C., Mest, S. C., 2017. Geologic Mapping of the Planck Quadrangle of the Moon (LQ - 29). 48th Lunar and Planetary Science Conference, The Woodlands, Texas (pp. 1680).

[49] Zhang, J., Zhou, B., Lin, Y., Zhu, M. - H., Song, H., Dong, Z., Gao, Y., Di, K., Yang, W., Lin, H., Yang, J., Liu, E., Wang, L., Lin, Y., Li, C., Yue, Z., Yao, Z., Ouyang, Z., 2020a. Lunar regolith and substructure at Chang'E - 4 landing site in south pole - Aitken basin. Nat. Astron. https://doi. org/10. 1038/s41550 - 020 - 1197 - x.

[50] Zhang, L., Li, J., Zeng, Z., Xu, Y., Liu, C., Chen, S., 2020b. Stratigraphy of the Von Kármán crater based on Chang'E - 4 lunar penetrating radar data. Geophys. Res. Lett. 47 (15), e2020GL088680.

Absolute model age of lunar Finsen crater and geologic implications

Sheng Gou[1,2], Zongyu Yue[2,3], Kaichang Di[2,3], Zhanchuan Cai[1], Zhaoqin Liu[2], Shengli Niu[1]

1. State Key Laboratory of Lunar and Planetary Sciences, Macau University of Science and Technology, Macau, China

2. State Key Laboratory of Remote Sensing Science, Aerospace Information Research Institute, Chinese Academy of Sciences, Beijing 100101, China

3. Center for Excellence in Comparative Planetology, Chinese Academy of Sciences, Hefei 230026, China

Abstract: As the primary source of materials measured inside of Von Kármán crater by the Chang'e-4 rover, the absolute model age (AMA) of Finsen crater is derived by the crater size-frequency distribution (CSFD) measurements. Both cumulative and differential fits reveal an AMA of ~3.5Ga, indicating Finsen crater is Imbrium-aged. Based on thickness estimation of the Finsen crater ejecta-sourced fine-grained regolith at Chang'e-4 landing site (~12m) (Lai et al., 2019; Li et al., 2020), the average regolith growth rate at the landing site is estimated to be about 3.4m/Gyr. The current depths of 25 largest craters on the floor of Finsen crater are calculated by the profile-average-depth method, and their original depths are estimated under the assumption that the initial depth-to-diameter (d/D) ratio is 0.2. Depth differences between current and estimated original depths, together with the AMA of Finsen crater, derive the average crater degradation/infilling rate within Finsen crater is about 21 ±3m/Gyr.

Keywords: Finsen crater; Absolute model age; Chang'e-4 landing site; Regolith growth rate; Crater degradation/infilling rate

1 Introduction

China's Chang'e-4 probe, including a lander and a rover, successfully touched down within the 185-km-wide Von Kármán crater inside the South Pole-Aitken (SPA) basin on January 3, 2019, making the first-ever soft landing on the lunar farside (Di et al.,

2019a) (Fig. 1). The rover Yutu - 2 was released from the lander and started surface exploration on the same day. As of July 27, 2020, the rover had already operated for 20 lunar days and traversed about 490. 9m, making it the longest - working rover on the lunar surface in human history that continues to set the record. The Chang'e - 4 rover has conducted a series of in situ measurements along traverse with the equipped scientific payloads: the Panoramic Camera (Pancam), the Visible and Near - Infrared Imaging Spectrometer (VNIS), and the Lunar Penetrating Radar (LPR) (Jia et al., 2018). The measured datasets enable scientists to study the unexplored area in more detail. LPR radargrams indicate that the stratigraphy of shallow subsurface structure at the landing area can be roughly divided into three units (Lai et al., 2019; Li et al., 2020): a shallower unit consisting of fine - grained regolith (depth interval, 0 to 12m), a underlying unit consisting of coarser materials with embedded rocks (12 to 24m), and a bottom unit containing alternating layers of coarse and fine materials (24 to 40m). Pancam images demonstrate that surface regolith measured by the rover are lunar deep materials excavated mainly by Finsen crater (northeast of Von Kármán) with possible contributions from Alder crater (southeast of Von Kármán), and numerical simulations predict ~32 and ~35m ejecta deposited at the landing site from Finsen and Alder, respectively (Di et al., 2019b). VNIS spectra reveal the regolith at the landing site is mature (Gou et al., 2020a), and has a forsteritic olivine and Mg - rich orthopyroxene assemblage in almost equal fractions (Gou et al., 2019; Gou et al., 2020b; Gou et al., 2020c).

As primary source of Chang'e - 4 rover measured materials (e. g., Gou et al., 2019; Di et al., 2019b; Gou et al., 2020b), the formation time of Finsen crater is still unclear. A geologic map of the lunar central farside originally created in the 1970s interpreted Finsen crater as an Eratoshenian - aged crater (Stuart - Alexander, 1978; Fortezzo et al., 2020). Recently compiled geologic map of the northern portion of the SPA basin registered Finsen crater as an Imbrium - aged crater (Ivanov et al., 2018). Determination of the absolute model age (AMA) of Finsen crater as accurate as possible is of great importance for understanding the evolution of regolith at landing site. For example, the estimated fine - grained regolith thickness at the Chang'e - 4 landing site is about 12m (Lai et al., 2019; Li et al., 2020), if Finsen crater formed in the Eratosthenian period (range from about 3. 2 to 1. 1 billion years), the regolith formation rate should be between 3. 8m/Gyr and 10. 9m/Gyr. In this scenario, this rather high value would challenge our current knowledge on the regolith formation rate, or otherwise would indicate the lunar farside is different. The AMA of the Finsen crater is derived by the crater size - frequency distribution (CSFD) measurements in this study to conduct a comprehensive analysis, including regolith growth rate at Chang'e - 4 landing site and crater degradation/infilling rate within Finsen crater.

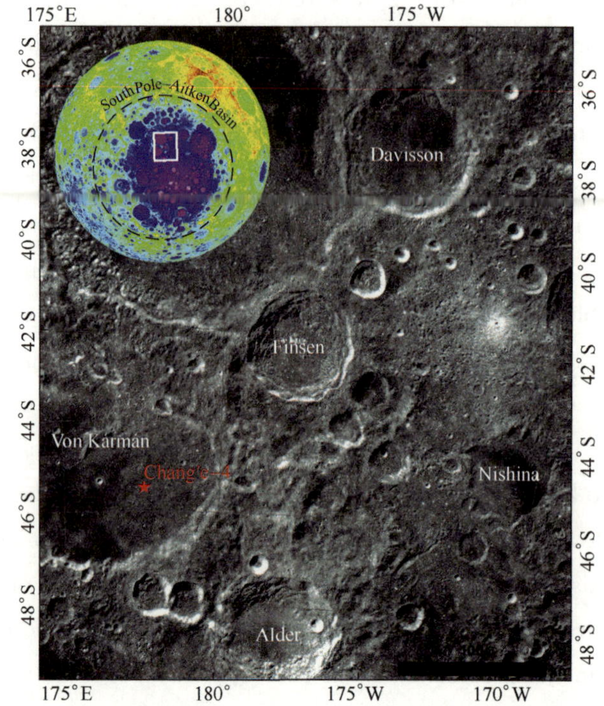

Fig. 1 Context map of Chang'e-4 landing site. The basemap is Chang'e-2 image mosaic (Ren et al., 2014). The Lunar Orbiter Laser Altimeter (LOLA) elevation (Smith et al., 2010) inset shows topography within the SPA basin. Colors represent different heights in the inset, with red being the highest and purple being the lowest. The white box frames the extent of Fig. 1. (For interpretation of the references to colour in this figure legend, the reader is referred to the web version of this article)

2 Data and context

A digital orthophoto map (DOM) mosaic and a digital elevation model (DEM) mosaic generated from images collected by Chang'e-2 mission (Ren et al., 2014) are used as base maps in this study. Chang'e-2 was China's second probe to orbit the Moon, whose main objective was to image the entire lunar surface with a resolution of 7m/pixel, to provide high resolution imagery (1.5m /pixel) of possible landing site (Sinus Iridum) for future Chang'e-3 mission, and to test some crucial techniques for the follow-up lunar and deep space explorations (Zhao et al., 2011). The returned images were processed with radiometric, photometric, and geometric calibrations, to produce a global DOM product with a resolution of 7m/pixel and a global DEM product with a grid spacing of 20m using photogrammetric techniques (Ren et al., 2014). Each product is divided into 844 map subdivisions, which are stored in the Data Release and Information Service System of China's Lunar Exploration Program (http://moon.bao.ac.cn) (Ren et al., 2014).

Both DOM and DEM map subdivisions that have a full coverage over the Chang'e – 4 landing site and its surroundings were downloaded and mosaicked in this study for crater identification and crater depth measurement.

The location of the Chang'e – 4 lander was determined to be 177. 588°E, 45. 457°S using visual localization techniques (Di et al., 2019a). Finsen crater (42. 29°S, 177. 72°W, D =73km), located about 135km northeast of Chang'e – 4 landing site (Fig. 1), is a well – preserved complex crater that has a prominent central peak. Representative spectra from both the central peak and crater floor exhibit similar and clear mafic absorptions, indicating the presence of low – Ca pyroxene (Moriarty et al., 2011). Further spectral analysis revealed the pyroxene composition differs significantly from mare basalt mixtures (Moriarty III and Pieters, 2018). Based on the absence of apparent volcanic sources and the morphology of the deposits (e. g., hummocky terrain and viscous flow), the floor of the 73 – km – wide Finsen crater plausibly represent impactmelt deposits. Finsen crater and the melt deposits were formed almost simultaneously in a single impact event, therefore, a flat and homogenous area, which is outlined on the floor of Finsen crater in Fig. 2, is selected as crater count area for CSFD measurements.

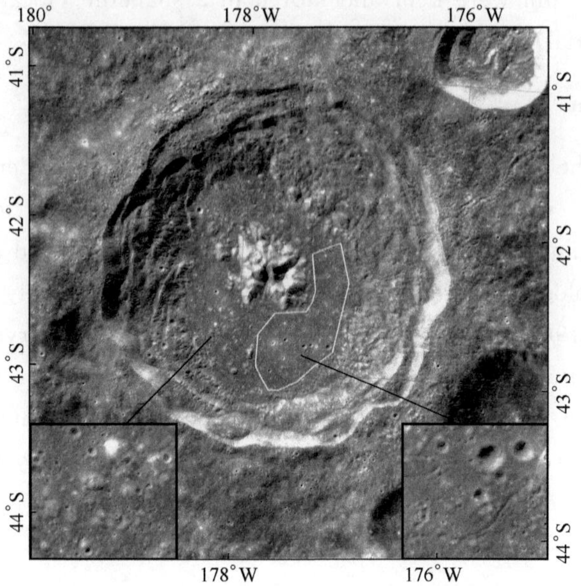

Fig. 2　Context map of Finsen crater (Chang'e – 2 mosaic as base map), with crater count area being outlined in white polygon. The lower – left and lower – right insets show hummocky terrain and viscous flow, respectively

3　Methods

3.1　Crater identification

Craters are identified by their unique characteristics (e. g., presence of rim, circularity of the rim, etc.), and they are mapped by an almost evenly distributed three-point rim fitting method to eliminate any possible errors (Kneissl et al., 2011; Yue et al., 2019). The crater mapping work is done with an ArcGIS add-in "CraterTools", which automatically adjusts for map-projection distortions to prevent mismeasurement of crater diameters and count area (Kneissl et al., 2011). During the crater mapping, obvious secondary craters, which exhibit distinctive morphologies such as shallow, irregular shapes and occurrence in chains and clusters, sometimes with distinctive herring-bone patterns (e. g., McEwen and Bierhaus, 2006), are excluded to reduce contamination. In order to investigate the spatial distribution and variability of the mapped craters within the measurement area, two individual randomness analyses, standard deviation of adjacent area (SDAA) and mean 2nd-closest neighbor distance (M2CND), are performed in the widely used Craterstats software (Michael et al., 2012). Due to the inherent uncertainty in diameter measurements, which gets larger as craters approach the resolution limit, and the possibility that craters near the resolution limit will be missed, only craters with diameters larger than about 10 pixels of the basemap (here is 70m) (Robbins et al., 2014; Wang et al., 2020) are kept for further AMA derivation. As a result, 216 craters with diameter larger than 70m were kept and stored in a shapefile (locations are provided in supplementary material).

3.2 AMA derivation

The AMA of the outlined unit within the Finsen crater is derived using standard crater chronologies, which combines information from crater formation rates and radiometric ages of lunar rock samples. This method is widely used and well accepted to obtain AMAs of geological units on planetary bodies from remotely sensed data (e. g., Shoemaker et al., 1962; Baldwin, 1964; Hartmann, 1965; Arvidson et al., 1979; Neukum, 1983). After mapping craters on the outlined count area, the obtained crater size-frequency distribution (CSFD) is plotted using pseudo-log binning with Craterstats (Michael and Neukum, 2010). The AMA of the area is derived by both cumulative and differential fit, with chronology function (CF) and production function (PF) from Neukum (1983).

3.3 Crater depth measurement

The depths of identified craters are calculated from the DEM by the profile-average-depth method (Vincent et al., 2014; Gou et al., 2018; Robbins et al., 2018). The major steps are described as follows. First, 8 profiles crossing the initial circle center of each crater are drawn starting from north at a fixed interval angle 22.5°. Second, each profile is updated by finding local topographic inflection point (local maxima). Third, the depth of the crater is calculated as the average elevation difference between the rim and the floor of each profile. More detail about crater depth measurement can be found in Gou et al. (2018).

4 Result and geologic implications

4.1 AMA of Finsen crater

Spatial randomness analysis of the mapped craters falls within $\pm 3\sigma$ of the expectation value derived from Monte Carlo simulation for all the diameters (Fig. 3), indicating craters in the size range examined show a random spatial distribution. AMAs of Finsen crater derived from cumulative and differential fits, which are based on CSFD measurements on an representative area of impact - melt deposits on the floor of Finsen crater, are $3.47^{+0.067}_{-0.12}$ Ga and $3.51^{+0.10}_{-0.39}$ Ga, respectively (Fig. 3). These two AMAs are within each other' s error bars, suggesting they are in excellent agreement that reveal an AMA of ~3.5 Ga. AMAs were also computed with the newer PF and CF values (Neukum et al., 2001) and found to produce the same model age, indicating the determined AMA (~3.5Ga) is robust. This model age is consistent with the AMA ($3.61^{+0.02}_{-0.02}$ Ga) derived by Ivanov et al. (2018), indicating Finsen crater is Imbrium - aged. The slight model age difference between this study and Ivanov et al. (2018) is probably due to factors such as the difference in count area selection and crater identification (Robbins et al., 2014). The

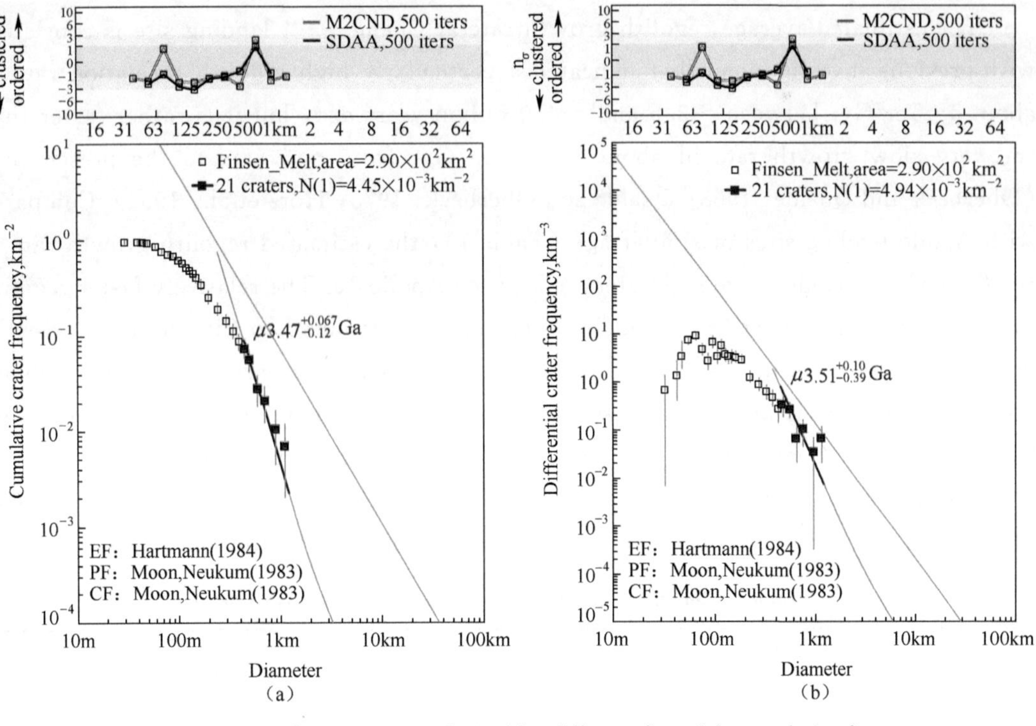

Fig. 3　AMA of Finsen crater derived by different fits: (a) cumulative fit, (b) differential fit. Craters for fitting are marked as red filled squares. (For interpretation of the references to colour in this figure legend, the reader is referred to the web version of this article)

derived AMA (~3.5Ga) of Finsen crater is consistent with the geology context that ejecta from Finsen covers the southeastern and northeastern part of the interior floor of Leibnitz (~3.88Ga) and Von Kármán (~3.97Ga) crater, respectively (Stuart—Alexander, 1978; Yingst et al., 2017).

4.2 Regolith growth rate at Chang'e-4 landing site

Previous works suggested the Chang'e-4 landing site is chiefly covered by ejecta excavated by Finsen crater (e. g., Gou et al., 2019; Di et al., 2019b; Lin et al., 2020). The surface of the landing site has then been bombarded by meteorites for billions of years. Therefore, a finegrained layer termed "regolith", which is mainly composed of unconsolidated materials, accumulated gradually by the continuous impacting and gardening of the lunar surface materials. The variations of permittivity with depth and the LPR radargrams indicate that the thickness of the Finsen crater ejecta-sourced fine-grained regolith at Chang'e-4 landing site is about 12m (Lai et al., 2019; Li et al., 2020). Simplifying the actual dynamic regolith growth process, we assume the lunar regolith at Chang'e-4 landing site accumulated immediately and steadily after the formation of Finsen crater about 3.5 billion years ago, the average regolith growth rate at Chang'e-4 landing site is thus estimated to be about 3.4m/Gyr (12m/3.5Ga).

The estimated average regolith growth rate at Chang'e-4 landing site is consistent with previous investigations that indicated a particularly high regolith formation rate of about 3—5m/Gyr between ~3.5 and ~4.0 billion years ago, but it is rather faster than the very slow growth rate of about 1—1.5m/Gyr from ~3.5Ga to the present day (Oberbeck and Quaide, 1968; Quaide and Oberbeck, 1975; Hörz et al., 1991). Compared with Apollo landing sites of similar age (Table 1), the estimated regolith growth rate at the Chang'e-4 landing site is greater, except for Apollo 16. The relatively faster regolith growth rate at the Chang'e-4 landing site may be attributed to Finsen ejecta possibly having a low weathering resistivity that is easily weathered by harsh space environment. However, this rate is incomparable with rather rapid growth rate of new regolith (about 5—10m/Gyr) at Chang'e-3 landing site (Table 1) that results from very fast space weathering for young fresh surfaces (~100Ma) (Fa et al., 2015; Lai et al., 2016), indicating a great regional variation of regolith formation rate on the lunar surface.

Table 1 Summary of estimated average regolith growth rate

Missions	Regolith thickness* /m	Age** / Ga	Growth rate/ (m/Gyr)	References
Apollo 11	4.4	3.58	1.23	(Nakamura et al., 1975; Stöffler and Ryder, 2001)
Apollo 12	3.7	3.15	1.17	(Nakamura et al., 1975; Stöffler and Ryder, 2001)

Continued

Missions	Regolith thickness* /m	Age** / Ga	Growth rate/ (m/Gyr)	References
Apollo 14	8.5	3.94	2.16	(Cooper et al., 1974; Snape et al., 2016)
Apollo 15	4.4	3.30	1.33	(Nakamura et al., 1975; Stöffler and Ryder, 2001)
Apollo 16	12.2	3.74	3.26	(Cooper et al., 1974; Stöffler and Ryder, 2001)
Apollo 17	4	3.75	1.07	(Cooper et al., 1974; Stöffler and Ryder, 2001)
Chang'e-3	0.95	0.1	5—10***	(Fa et al., 2015; Lai et al., 2016)
Chang'e-4	12	3.5	3.4	(Lai et al., 2019; Li et al., 2020)

* The regolith thicknesses are in situ measured values.

** The ages for Apollo missions are absolute ages determined by radiometric dating, the ages for Chang'e-3/4 are AMAs determined from CSFD measurements.

*** This estimation combines an uncertainty of 30 Ma in age and 5% in depth.

4.3 Crater degradation/infilling rate within Finsen crater

Small craters are easily destroyed with time by erosion of crater rim/ ejecta and infilling of crater interior by geologic processes (e. g., wall collapse and mass wasting) (e. g., Barlow, 2015). After 3 Gyr of evolution, a 300m crater is almost imperceptible, while a 1 km crater still has significant relief (Fassett and Thomson, 2014). Therefore, variations of depth for craters within Finsen crater can provide constraints on the crater degradation/infilling rate on the lunar surface. The initial morphometry of fresh simple craters has been well-studied, with depth-to-diameter (d/D) ratio as a key parameter for describing/quantifying crater morphology. On the Moon, fresh, strength-regime dominated craters (diameters less than 400m) exhibit d/D ratios of ~0.11, ~0.12, ~0.13, ~0.15, and ~0.17 (Stopar et al., 2017) due to varying crater shapes (flat floor, normal bowl, and central mound) and diameter bin examined (200—400m, 100—200m, and 40—100m); whereas larger-scale craters (diameters 400m—10km) have d/D ratios averaging around 0.21 (Stopar et al., 2017).

The twenty-five largest craters within the count area, whose diameters range from 407m to 1175m (details are provided in supplementary material), are selected for average degradation/infilling rate estimation in this study. To simplify the analysis here, according to the scaling law (Melosh, 1989), the depth of a simple fresh crater is about 1/ 5 of its final diameter when it forms after the modification stage is complete. The original depths of the selected twenty-five largest craters within Finsen crater are estimated to range from 81m to 235m. Their current crater depths that are calculated by the above described profile - average - depth method vary from 4m to 157m (details are provided in supplementary material). Therefore, the differences between the estimated original depths and the calculated current depths, as well as the AMA of Finsen crater are used to estimate crater degradation/infilling rates. Results show that the average crater degradation/infilling rate within Finsen crater is about 21 ±3m/Gyr, which has the same order of magnitude to that on the lunar maria (about 32m/Gyr) (Fassett and Thomson, 2014). The slight discrepancy of average degradation/infilling rates may be related to many potential factors, such as target material that affects resistance to degradation (impact melt inside Finsen crater vs. mare basalts for lunar maria) (Melosh, 1989), degradation time implied by the AMAs (~3.5Ga for craters on the floor of Finsen crater vs. ~3Ga for craters on lunar maria), and location that affects degradation intensity (inside Finsen crater vs. inside impact basin for lunar maria), etc. However, this estimated degradation/infilling rate is much slower than that on other airless rocky bodies. For example, average degradation/infilling rate on Vesta is 350m/Gyr (Vincent et al., 2014), and on Gaspra is 100—1000m/Gyr (Carr et al., 1994). One of the most likely reasons is that craters on asteroids are easily degraded or even erased by mass movements (e.g., landslides and crater erasures) caused by impact - induced seismic shaking (e.g., Carr et al., 1994; Richardson et al., 2004; Jutzi et al., 2013; Matsue et al., 2020).

5 Conclusion

As the primary source of materials measured inside of Von Kármán crater by the Chang'e - 4 rover, the absolute model age (AMA) of Finsen crater is derived from the crater size - frequency distribution (CSFD) measurements in this study. Both cumulative and differential fits using the Neukum chronology (Neukum, 1983) reveal an AMA of ~3.5Ga, indicating Finsen crater is Imbrium - aged.

Based on thickness estimation of the Finsen crater ejecta - sourced fine - grained regolith at Chang'e - 4 landing site (~12m) (Lai et al., 2019; Li et al., 2020), the average regolith growth rate at the landing site is estimated to be about 3.4m/Gyr, which is consistent with previous investigations that indicated a particularly high regolith formation rate of about 3—5m/Gyr between ~3.5 and ~4.0 billion years ago.

The current depths of 25 largest craters on the floor of Finsen crater are calculated by the profile - average - depth method, and their original depths are estimated under the assumption that the initial depth - to - diameter (d/D) ratio is 0.2. Differences between the estimated original depths and the calculated current average depths, as well as the AMA of Finsen crater indicate the average crater degradation/infilling rate within Finsen crater is about 21 ±3m/Gyr, which has the same order of magnitude to that on the lunar maria (about 32m/Gyr).

Acknowledgements

This work was supported by the Strategic Priority Research Program of Chinese Academy of Sciences (grant No. XDB41000000), National Natural Science Foundation of China (grant Nos. 41702354 and 41941003), Macao Young Scholars Program (grant No. AM201902), and Science and Technology Development Fund of Macau (grant No. 131/2017/A3). The authors thank Jun Du for discussions about crater degradation process, and greatly appreciate Mikhail Ivanov and Michelle Kirchoff for constructive reviews that improved the manuscript quality substantially.

Appendix A　Supplementary data

Supplementary data to this article can be found online at https: //doi. org/10. 1016/j. icarus. 2020. 114046.

References

[1] Arvidson, R. E., Boyce, J., Chapman, C., Cintala, M., Fulchignoni, M., Moore, H., Neukum, G., Schultz, P., Soderblom, L., Strom, R., Woronow, A., Young, R., 1979. Standard techniques for presentation and analysis of crater size - frequency data. Icarus. 37 (2), 467 - 474.

[2] Baldwin, R. B., 1964. Lunar crater counts. Astron. J. 69, 377.

[3] Barlow, N. G., 2015. Constraining geologic properties and processes through the use of impact craters. Geomorphology. 240, 18 - 33.

[4] Carr, M. H., Kirk, R. L., Mcewen, A. S., Veverka, J., Thomas, P., Head, J. W., Murchie, S. L., 1994. The geology of Gaspra. Icarus. 107 (1), 61 - 71.

[5] Cooper, M. R., Kovach, R. L., Watkins, J. S., 1974. Lunar near - surface structure. Rev. Geophys. 12 (3), 291 - 308.

[6] Di, K., Liu, Z., Liu, B., Wan, W., Peng, M., Wang, Y., Gou, S., Yue, Z., Xn, X., Jia, M., Niu, S., 2019a. Chang'e - 4 lander localization based on multi - source data. J. Remote Sens. 23 (1), 177 - 184.

[7] Di, K., Zhu, M. - H., Yue, Z., Lin, Y., Wan, W., Liu, Zhaoqin, Gou, S., Liu, B., Peng, Man, Wang, Y., Niu, S., Zhang, J., Li, J., Xie, J., Xi, L., Yang, J., Xue, B., 2019b.

[8] Topographic evolution of Von Kármán crater revealed by the lunar rover Yutu - 2. Geophys. Res. Lett. 46 (22), 12764 - 12770.

[9] Fa, W., Zhu, M. - H., Liu, T., Plescia, J. B., 2015. Regolith stratigraphy at the Chang'E - 3 landing site as seen by lunar penetrating radar. Geophys. Res. Lett. 42

(23), 10179 - 10187.

[10] Fassett, C. I., Thomson, B. J., 2014. Crater degradation on the lunar maria: topographic diffusion and the rate of erosion on the moon. J. Geophys. Res: Planet. 119 (10), 2255 - 2271.

[11] Fortezzo, C. M., Spudis, P. D., Harrel, S. L., 2020. Release of the Digital Unified Global Geologic Map of the Moon At 1 : 5,000,000 - Scale. In: 51st Lunar and Planetary Science Conference, p. 2760 the Woodlands, Texas.

[12] Gou, S., Yue, Z. Y., Di, K. C., Liu, Z. Q., 2018. A global catalogue of Ceres impact craters ≥1km and preliminary analysis. Icarus. 302, 296 - 307.

[13] Gou, S., Di, K., Yue, Z., Liu, Z., He, Z., Xu, R., Lin, H., Liu, B., Peng, M., Wan, W., Wang, Y., Liu, J., 2019. Lunar deep materials observed by Chang'e - 4 rover. Earth Planet. Sci. Lett. 528, 115829.

[14] Gou, S., Yue, Z., Di, K., Wan, W., Liu, Z., Liu, B., Peng, M., Wang, Y., He, Z., Xu, R., 2020a. In situ spectral measurements of space weathering by Chang'e - 4 rover. Earth Planet. Sci. Lett. 535, 116117.

[15] Gou, S., Di, K., Yue, Z., Liu, Z., He, Z., Xu, R., Liu, B., Peng, M., Wan, W., Wang, Y., Liu, J., 2020b. Forsteritic olivine and magnesium - rich orthopyroxene materials measured by Chang'e - 4 rover. Icarus. 345, 113776.

[16] Gou, S., Yue, Z., Di, K., Wang, J., Wan, W., Liu, Z., Liu, B., Peng, M., Wang, Y., He, Z., Xu, R., 2020c. Impact melt breccia and surrounding regolith measured by Chang' e - 4 rover. Earth Planet. Sci. Lett. 544, 116378.

[17] Hartmann, W. K., 1965. Terrestrial and lunar flux of large meteorites in the last two billion years. Icarus. 4 (2), 157 - 165.

[18] Hörz, F., Grieve, R. A. F., Heiken, G., Spudis, P. D., Binder, A. B., 1991. Lunar surface processes. In: Heiken, G., Vaniman, D., French, B. M. (Eds.), Lunar Sourcebook. Cambridge University Press, New York, pp. 61 - 120.

[19] Ivanov, M. A., Hiesinger, H., van der Bogert, C. H., Orgel, C., Pasckert, J. H., Head, J. W., 2018. Geologic history of the northern portion of the south pole - Aitken Basin on the moon. J. Geophys. Res: Planet. 123 (10), 2585 - 2612.

[20] Jia, Y., Zou, Y., Ping, J., Xue, C., Yan, J., Ning, Y., 2018. The scientific objectives and payloads of Chang'E - 4 mission. Planet. Space Sci. 162, 207 - 215.

[21] Jutzi, M., Thomas, N., Benz, W., El Maarry, M. R., Jorda, L., Kührt, E., Preusker, F., 2013. The influence of recent major crater impacts on the surrounding surfaces of (21) Lutetia. Icarus. 226 (1), 89 - 100.

[22] Kneissl, T., Van Gasselt, S., Neukum, G., 2011. Map - projection - independent crater size - frequency determination in GIS environments—new software tool for ArcGIS. Planet. Space Sci. 59 (11), 1243 - 1254.

[23] Lai, J., Xu, Y., Zhang, X., Tang, Z., 2016. Structural analysis of lunar subsurface with Chang'E - 3 lunar penetrating radar. Planet. Space Sci. 120, 96 - 102.

[24] Lai, J., Xu, Y., Zhang, X., Xiao, L., Yan, Q., Meng, X., Zhou, B., Dong, Z., Zhao, D., 2019. Comparison of dielectric properties and structure of lunar regolith at chang'e - 3 and chang'e - 4 landing sites revealed by ground penetrating radar. Geophys. Res. Lett. 46, 12783 - 12793.

[25] Li, C., Su, Y., Pettinelli, E., Xing, S., Ding, C., Liu, J., Ren, X., Lauro, S. E., Soldovieri, F., Zeng, X., Gao, X., Chen, W., Dai, S., Liu, D., Zhang, G., Zuo, W., Wen, W., Zhang, Z., Zhang, X., Zhang, H., 2020. The Moon's farside shallow subsurface structure unveiled by Chang'E - 4 lunar penetrating radar. Sci. Adv. 6 (9), eaay6898.

[26] Lin, H., He, Z., Yang, W., Lin, Y., Xu, R., Zhang, C., Zhu, M. - H., Chang, R., Zhang, J., Li, C., Lin, H., Liu, Y., Gou, S., Wei, Y., Hu, S., Xue, C., Yang, J., Zhong, J., Fu, X., Wan, W., Zou, Y., 2020. Olivine - norite rock detected by the lunar rover Yutu - 2 likely crystallized from the SPA impact melt pool. Natl. Sci. Rev. 7, 913 - 920.

[27] Matsue, K., Yasui, M., Arakawa, M., Hasegawa, S., 2020. Measurements of seismic waves induced by high - velocity impacts: implications for seismic shaking surrounding impact craters on asteroids. Icarus. 338, 113520.

[28] McEwen, A. S., Bierhaus, E. B., 2006. The importance of secondary cratering to age constraints on planetary surfaces. Annu. Rev. Earth Planet. Sci. 34 (1), 535 - 567.

[29] Melosh, H. J., 1989. Impact Cratering: A Geologic Process. Oxford University Press, New York.

[30] Michael, G. G., Neukum, G., 2010. Planetary surface dating from crater size - frequency distribution measurements: partial resurfacing events and statistical age uncertainty. Earth Planet. Sci. Lett. 294 (3), 223 - 229.

[31] Michael, G. G., Platz, T., Kneissl, T., Schmedemann, N., 2012. Planetary surface dating from crater size - frequency distribution measurements: spatial randomness and clustering. Icarus. 218 (1), 169 - 177.

[32] Moriarty III, D. P., Pieters, C. M., 2018. The character of south pole - Aitken Basin: patterns of surface and subsurface composition. J. Geophys. Res: Planet. 123 (3), 729 -747.

[33] Moriarty, D. P., Pieters, C. M., Nettles, J., Isaacson, P. J., Cheek, L., Head, J. W., Tompkins, S., Petro, N., 2011. Finsen and Alder: A Compositional Study of Lunar Central Peak Craters in the South Pole - Aitken Basin. In: 42nd Lunar and Planetary Science Conference, p. 2564. The Woodlands, Texas.

[34] Nakamura, Y., Dorman, J., Duennebier, F., Lammlein, D., Latham, G., 1975. Shallow lunar structure determined from the passive seismic experiment. Moon. 13 (1), 57 - 66.

[35] Neukum, G., 1983. Meteoriten bombardement und Datierung planetarer

Oberflächen (German original). Habilitation Dissertation for Faculty Membership. University of Munich.

[36] Neukum, G., Ivanov, B. A., Hartmann, W. K., 2001. Cratering records in the inner solar system in relation to the lunar reference system. Space Sci. Rev. 96, 55 - 86.

[37] Oberbeck, V. R., Quaide, W. L., 1968. Genetic implications of lunar regolith thickness variations. Icarus. 9 (1), 446 - 465.

[38] Quaide, W., Oberbeck, V., 1975. Development of the mare regolith: some model considerations. Moon 13 (1), 27 - 55.

[39] Ren, X., Liu, J. J., Wang, F. F., Wang, W. R., Mu, L. L., Li, H. H., 2014. A new lunar global topographic map products from Chang'E - 2 stereo camera image data. In: European Planetary Science Congress. Centro de Congressos do Estoril, Cascais, Portugal. pp. EPSC2014 - 344.

[40] Richardson, J. E., Melosh, H. J., Greenberg, R., 2004. Impact - induced seismic activity on asteroid 433 Eros: a surface modification process. Science. 306 (5701), 1526 -1529.

[41] Robbins, S. J., Antonenko, I., Kirchoff, M. R., Chapman, C. R., Fassett, C. I., Herrick, R. R., Singer, K., Zanetti, M., Lehan, C., Huang, D., Gay, P. L., 2014. The variability of crater identification among expert and community crater analysts. Icarus. 234, 109 - 131.

[42] Robbins, S. J., Watters, W. A., Chappelow, J. E., Bray, V. J., Daubar, I. J., Craddock, R. A., Beyer, R. A., Landis, M., Ostrach, L. R., Tornabene, L., Riggs, J. D., Weaver, B. P., 2018. Measuring impact crater depth throughout the solar system. Meteorit. Planet. Sci. 53 (4), 583 - 637.

[43] Shoemaker, E. M., Hackman, R. J., Eggleton, R. E., 1962. Interplanetary correlation of geologic time. Adv. Astronaut. Sci. 8, 70 - 79.

[44] Smith, D. E., Zuber, M. T., Neumann, G. A., Lemoine, F. G., Mazarico, E., Torrence, M. H., McGarry, J. F., Rowlands, D. D., Head III, J. W., Duxbury, T. H., Aharonson, O., Lucey, P. G., Robinson, M. S., Barnouin, O. S., Cavanaugh, J. F., Sun, X., Liiva, P., Mao, D. - d., Smith, J. C., Bartels, A. E., 2010. Initial observations from the lunar orbiter laser altimeter (LOLA). Geophys. Res. Lett. 37 (18), L18204.

[45] Snape, J. F., Nemchin, A. A., Grange, M. L., Bellucci, J. J., Thiessen, F., Whitehouse, M. J., 2016. Phosphate ages in Apollo 14 breccias: resolving multiple impact events with high precision U - Pb SIMS analyses. Geochim. Cosmochim. Acta 174, 13 - 29.

[46] Stöffler, D., Ryder, G., 2001. Stratigraphy and isotope ages of lunar geologic units: chronological standard for the inner solar system. Space Sci. Rev. 96 (1), 9 - 54.

[47] Stopar, J. D., Robinson, M. S., Barnouin, O. S., McEwen, A. S., Speyerer, E.

J., Henriksen, M. R., Sutton, S. S., 2017. Relative depths of simple craters and the nature of the lunar regolith. Icarus. 298, 34 - 48.

[48] Stuart - Alexander, D. E., 1978. Geologic map of the central far side of the Moon. http://pubs. er. usgs. gov/publication/i1047.

[49] Vincent, J. - B., Schenk, P., Nathues, A., 2014. Crater depth - to - diameter distribution and surface properties of (4) vesta. Planet. Space Sci. 103, 57 - 65.

[50] Wang, Y., Xie, M., Xiao, Z., Cui, J., 2020. The minimum confidence limit for diameters in crater counts. Icarus. 341, 113645.

[51] Yingst, R. A., Chuang, F. C., Berman, D. C., Mest, S. C., 2017. Geologic Mapping of the Planck Quadrangle of the Moon (LQ - 29). In: 48th Lunar and Planetary Science Conference, p. 1680 the Woodlands, Texas.

[52] Yue, Z., Di, K., Liu, Z., Michael, G., Jia, M., Xin, X., Liu, B., Peng, M., Liu, J., 2019. Lunar regolith thickness deduced from concentric craters in the CE - 5 landing area. Icarus. 329, 46 - 54.

[53] Zhao, B., Yang, J., Wen, D., Gao, W., Chang, L., Song, Z., Xue, B., Zhao, W., 2011. Overall scheme and on - orbit images of Chang'E - 2 lunar satellite CCD stereo camera. Sci. China Technol. Sci. 54 (9), 2237.

In situ spectral measurements of space weathering by Chang'e – 4 rover

Sheng Gou[1,2], Zongyu Yue[1,3], Kaichang Di[1,2,3], Wenhui Wan[1], Zhaoqin Liu[1], Bin Liu[1,2], Man Peng[1], Yexin Wang[1], Zhiping He[4], Rui Xu[4]

1. State Key Laboratory of Remote Sensing Science, Aerospace Information Research Institute, Chinese Academy of Sciences, Beijing 100101, China
2. State Key Laboratory of Lunar and Planetary Sciences, Macau University of Science and Technology, Macau, China
3. CAS Center for Excellence in Comparative Planetology, Hefei 230026, China
4. Key Laboratory of Space Active Opto – Electronics Technology, Shanghai Institute of Technical Physics, Chinese Academy of Sciences, Shanghai 200083, China

Abstract: Space weathering introduces confounding effects on visible and near infrared reflectance spectra of airless bodies, which considerably darkens the reflectance, reddens the continuum slope and suppresses absorption features. It's mainly attributed to the gradual formation and accumulation of submicroscopic metallic iron (SMFe) on regolith grains. In situ spectral measurements from Chang'e – 4 rover provide a unique opportunity to investigate the space weathering effects on the intact lunar farside regolith. SMFe abundance at the landing site, which is 0.32 ± 0.06 wt.%, is retrieved from in situ measured reflectance spectra by using Hapke model. The derived Is/FeO maturity index (82 ± 15) indicates the Finsen crater ejecta – sourced regolith is mature, which is consistent with the geologic background that it had experienced about 3.7 Ga space weathering.

Keywords: in situ measurements; space weathering; Hapke model; SMFe abundance; Chang'e – 4 rover

1 Introduction

Space weathering is the gradual alteration of materials when they are exposed to a variety of natural processes that occur in the space environment, which is ubiquitous on the Moon and other airless bodies (Pieters and Noble, 2016). The space weathering process results in the visible and near infrared reflectance spectrum of the regolith, which covers many planetary bodies, becomes considerably dark, exhibiting a reddened continuum and even almost completely suppressed absorption features (Pieters and Noble, 2016). The optical effects of space weathering, which has already been successfully simulated in the laboratory (Sasaki et al., 2001; Noble et al., 2007), are mainly caused by the gradual formation and accumulation of submicroscopic metallic iron (SMFe) that imbedded within or deposited on the regolith particles (Pieters et al., 2000; Hapke, 2001; Noble et al., 2001; Lucey and Noble, 2008; Lucey and Riner, 2011). Smaller SMFe particles (<10nm in diameter) greatly redden spectrum in the visible (VIS) wavelengths while leaving the near - infrared (NIR) region largely unaffected; Larger particles (>40nm in diameter) darken spectrum across the VIS/NIR range with little change in the overall shape of the continuum; And intermediate particles impact the spectrum in a distinct pattern that varies with total iron abundance (Noble et al., 2007; Lucey and Noble, 2008). Consequently, the SMFe abundance in the regolith is an important indicator for evaluating space weathering and regolith evolution.

The Moon is a unique body that already has been remotely observed by orbiters, in situ measured by rovers and sampled by manned/unmanned spacecraft. The chemical and spectral data collected during these missions, especially those from the well - studied Apollo and Luna samples, provide valuable and critical ground - truth information to perform quantitative analyses on space weathering from various optical remote sensing techniques, which are the only means to study the vast area of the lunar surface. However, the sampling process may disturb the primitive state of the very top of lunar regolith, the optical properties of returned lunar samples may not be representative of the undisturbed lunar surface (Wang et al., 2017). Sampling the uppermost surface while preserving its structure (e. g. porosity, grain - to - grain contacts) has not been achieved on any sample return mission to date (Noble et al., 2011). Therefore, in situ spectral measurements, along with experiences and expertise from weathering simulation theory (Hapke, 2001; Lucey and Riner, 2011) and practice (Sasaki et al., 2001; Noble et al., 2007), provide a feasible means to study the space weathering effects caused by SMFe on the lunar surface.

China's Chang'e - 4 spacecraft successfully landed inside Von Kármán crater within the South Pole - Aitken (SPA) basin on the lunar farside on January 3, 2019 (Di et al., 2019). The Visible and Near - Infrared Imaging Spectrometer (VNIS) (Li et al., 2019), one of the scientific payloads on the deployed Chang'e - 4 rover, had made in situ spectral measurements on the lunar regolith along the rover traverse (Fig. 1). These reflectance spectra provide a unique opportunity to investigate the space weathering effects on the intact lunar farside regolith, which is accomplished by evaluating the SMFe abundance via Hapke model (Hapke, 1981; 2001; Lucey and Riner, 2011) in this study.

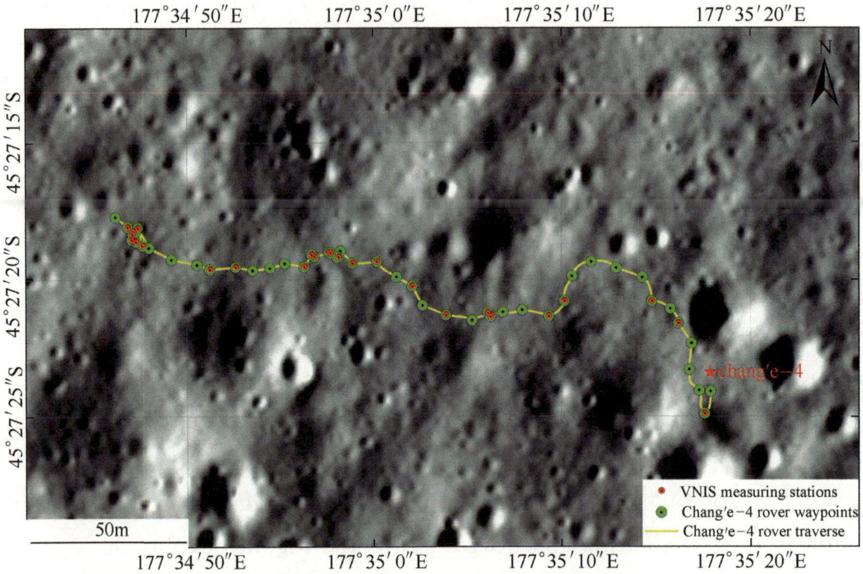

Fig. 1 Traverse map of the Chang'e-4 rover during the first 9 lunar days. The base map is a high resolution (0.9 m/pixel) digital orthophoto map (DOM) (Liu et al., 2019) generated from Lunar Reconnaissance Orbiter Camera (LROC) Narrow Angle Camera (NAC) imagery (Robinson et al., 2010). A detailed regional geologic setting map of the Chang'e-4 landing site is shown in Fig. 3

2 Data

2.1 VNIS spectra

The VNIS instrument consists of a VIS/NIR imager with an effective 256 ×256pixels (~1mm/pixel) and a shortwave infrared (SWIR) single-pixel detector (Li et al., 2019). The VIS/NIR imager works from 450 to 945nm with a spectral resolution of 2—7nm, and the SWIR detector works from 900 to 2395nm with a spectral resolution of 3—12nm (Li et al., 2019). The VNIS made in situ measurements on the lunar regolith along the traverse at different stations between the first and the ninth lunar days (January 3 — September 6, 2019). These measured raw data are preprocessed and released to the scientific team members as L2B radiance spectra by the team of ground application and research system (GRAS). The L2B radiance spectra are further processed in this study by the solar irradiance calibration method (Gou et al., 2019) to derive full-range reflectance spectra to study the space weathering effects (Fig. 2).

Indoor radiance calibration at typical temperature (20℃) shows the signal-to-noise ratio (SNR) values of VNIS are greater than 50dB in the wavelength range of 1000 nm to 2200nm (He et al., 2019). Inflight radiance calibration shows that the SNR values are greater than 40dB in the wavelength range of 1200nm to 2400nm. On one hand, averaging

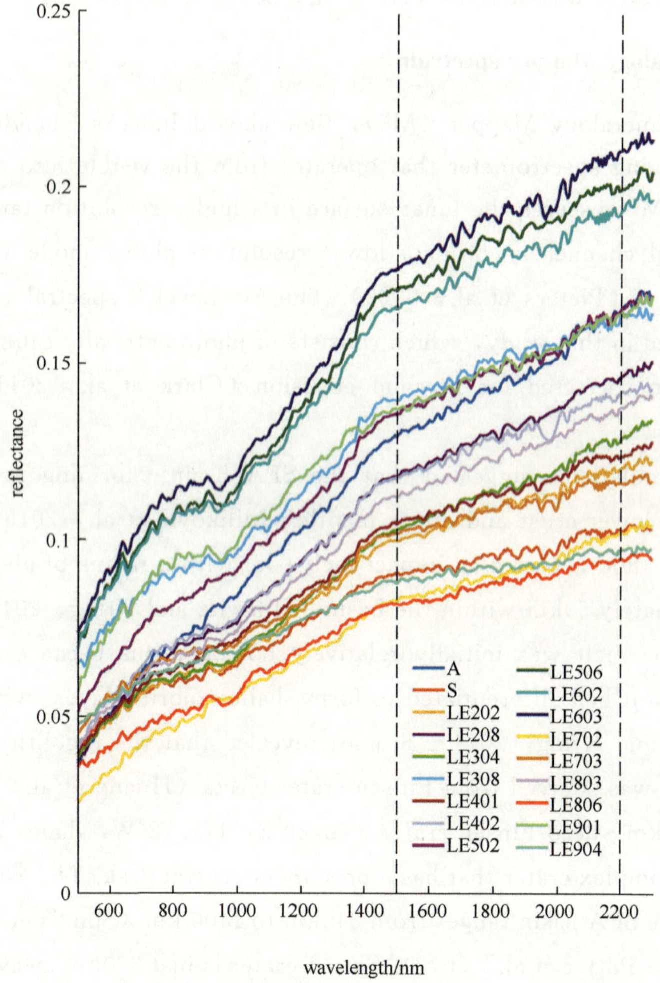

Fig. 2　Examples of in situ measured VNIS spectra on the lunar regolith at different rover stations. Only two of the multiple spectra measured in each lunar day are shown here. For example, spectra A and S are measured during the first luanr day. All the spectra analyzed in sthis study are shown in the supplementary material. The dashed lines located at 1500nm and 2200nm mark the wavelength range used for simulation in this study to avoid thermal emission effects at longer wavelengths (Li and Milliken, 2016) and low signal - to - noise ratio (SNR) at shorter wavelengths (He et al., 2019; Li et al., 2019). (For interpretation of the colors in the figure (s), the reader is referred to the web version of this article.)

is an effective way to improve SNR, and the SNR values of the VNIS at VIS/NIR wavelengths shown in both He et al. (2019) and Li et al. (2019) were calculated by averaging the pixels within the field of view of SWIR, however the SNR values of. VIS/NIR are still lower than that of SWIR. On the other hand, due to the design principle of the VNIS instrument (He et al., 2019), in situ measured spectrum may have a spectral kink around 1400nm. Therefore, VNIS spectra from 1500nm to 2200nm are used in this study to avoid the thermal emission effects at longer wavelengths (Li and Milliken, 2016) and low SNR at shorter wavelengths (He et al., 2019; Li et al., 2019).

2.2 Moon Mineralogy Mapper spectrum

The Moon Mineralogy Mapper (M^3), flew aboard India's Chandrayaan - 1, is a push - broom imaging spectrometer that operates from the visible into the near infrared (0.42—3.0μm). M^3 observes the lunar surface in a high - resolution targetmode (80m/pixel; 260 spectral channels), or in a low - resolution global mode (140m/pixel; 85 spectral channels) (Pieters et al., 2009). The M^3 Level 2 spectral reflectance image (REFIMG) is used in this study, which consists of photometrically calibrated reflectance data that has been corrected for thermal emission (Clark et al., 2011; Besse et al., 2013).

Numerical simulations suggested that the SPA basin - forming event should have sampled the lunar lower crust and upper mantle (Miljkovic et al., 2015; Melosh et al., 2017). The event could generate an impact melt pool with a radius of about 200km and a depth of approximately 50km within the basin (Hurwitz and Kring, 2014; Vaughan and Head, 2014). The melt was initially relatively homogeneous because of the thorough mixing, however, it had differentiated to form shallow noritic layers with a thickness of ~6km (Hurwitz and Kring, 2014). Studies revealed that the regolith observed by the Chang'e - 4 rover was sourced from Finsen crater ejecta (Huang et al., 2018; Di et al., 2019). The 3.9 - km - deep Finsen crater (42.29°S, 177.72°W, diameter =73km) is a well - preserved complex crater that has a prominent central peak. The estimated transient cavity radius of the SPA basin ranges from 840km to 1400 km (Spudis et al., 1994; Petro and Pieters, 2002; Potter et al., 2012). Finsen crater is just 370km away from the center of the SPA basin. The estimated approximate maximum excavation depth H_{exc} of Finsen crater, which can be calculated from $H_{exc} \approx 0.05—0.065D$ (D is the final complex crater diameter) (Melosh, 1989), is about 3.65—4.74km. Therefore, the Finsen crater ejecta were very likely completely sourced from the shallow noritic layers. In this case, the shallower material that was ejected from the Finsen crater and the relative deeper material in the central peak would have the same composition. The central peak bedrock of Finsen crater has steep walls that prevent the accumulation of a space - weathered regolith layer, thus preserves optical immaturity (Tompkins and Pieters, 1999). A large portion of the ejecta was from the excavated material during the crater formation, hence, M^3 spectrum of the exposed central peak of Finsen crater, which is resampled to the VNIS bands to ensure both have the same spectral resolution (band number), is used as the base spectrum in this study. The spectrum clearly exhibits strong mafic absorptions at short wavelengths, indicating the relatively pristine state (optical immaturity) was kept since the central peak formed and the presence of Mg - rich low - Ca pyroxene (LCP) (Fig. 3).

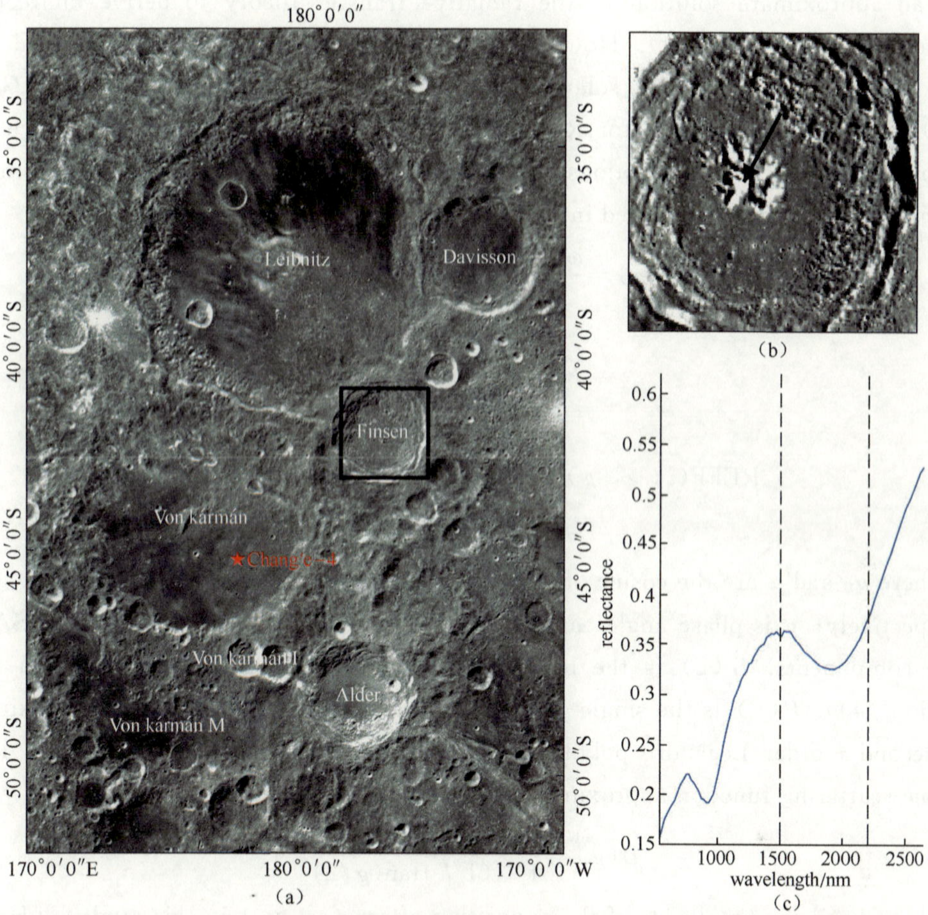

Fig. 3　(a) Chang'e－2 image (Ren et al., 2014) of Chang'e－4 landing site inside the Von Kármán crater and surrounding craters. (b) M^3 reflectance image @750nm of Finsen crater, with location indicated by the black box in (a) (image ID: M3G20090815T202602); (c) M^3 spectrum corresponding to the central peak of Finsen crater, with location indicated by the black arrow in (b). The spectrum is derived from a 3 × 3pixel average

3　Quantifying SMFe abundance in lunar regolith

Hapke model (Hapke, 1981; 2001; Lucey and Riner, 2011) is used in this study to quatify the SMFe mass fraction in the lunar regolith at Chang'e－4 landing site from in situ measured VNIS spectra. Details of assessing the SMFe abundance are described be－low.

3.1　Single scattering albedo determination

With the assumptions that all materials within the media are mixed intimately and the particle size is much larger than the spectral wavelength, Hapke model (Hapke, 1981)

offers an approximate solution to the radiative transfer theory to derive bidirectional reflectance r (equation (1)). However, reflectance factor (REFF) is derived from VNIS calibrated radiance by the solar irradiance calibration method in this study (Gou et al., 2019). The reflectance factor REFF is related to the bidirectional reflectance r by equation (2). Hence, the relationship between reflectance factor REFF and single-scattering albedo can be expressed in equation (3).

$$r(i, e, g)=\frac{\omega_{avg}}{4\pi}\frac{\mu_0}{(\mu_0+\mu)}[(1+B(g))P(g)+H(\mu_0, \omega_{avg})H(\mu, \omega_{avg})-1] \quad (1)$$

$$\mathrm{REFF}=\frac{\pi * r}{\mu_0} \quad (2)$$

$$\mathrm{REFF}(i, e, g)=\frac{\omega_{avg}}{4(\mu_0+\mu)}[(1+B(g))P(g)+H(\mu_0, \omega_{avg})H(\mu, \omega_{avg})-1] \quad (3)$$

where μ_0 and μ are the cosine of the incidence angle i and emission (viewing) angle e, respectively; g is phase angle. ω_{avg} is the average single scattering albedo (SSA) of all the components. $B(g)$ is the backscatter function that describes opposition effect (equation (4)). $P(g)$ is the single-particle phase function, which can be approximated by a second-order Legendre polynomial series (equation (6)). $H(x, \omega_{avg})$ is the isotropic scattering function approximation (equation (7)) (Hapke, 2002).

$$B(g)=\frac{B_0}{1+(1/h)\tan(g/2)} \quad (4)$$

where B_0 is the amplitude of the opposition effect (set to 1 in this study) (Hapke, 1981), and h is the angular width parameter of the opposition effect, which is approximated by equation (5).

$$h=-\frac{3}{8}\ln(1-\varphi) \quad (5)$$

where φ is the filling factor (decreasing porosity), which is set to 0.41 for lunar regolith (Bowell et al., 1989).

$$P(g)=1+b\cos(g)+c(1.5\cos^2(g)-0.5) \quad (6)$$

b and c are set to -0.4 and 0.25, respectively (Mustard and Pieters, 1989).

$$H(x, \omega_{avg})=\left[1-\omega x\left[r_0+1\frac{-2r_0x}{2}\ln\left(\frac{1+x}{x}\right)\right]\right]^{-1} \quad (7)$$

$$\text{where } r_0=\frac{1-\sqrt{1-\omega_{avg}}}{1+\sqrt{1-\omega_{avg}}}$$

When equations (4) – (7) are inserted into equation (3), a nonlinear relationship between REFF and ω_{avg} can be established. Therefore, the average single scattering albedo ω_{avg} of the lunar regolith at Chang'e-4 landing site can be solved from the in situ measured VNIS reflectance spectrum.

3.2 Absorption coefficient determination

Theoretically, ω_{avg} can be computed by equation (8) (Hapke, 2012).

$$\omega_{avg} = S_e + (1 - S_e)\frac{(1 - S_i)\Theta}{1 - S_i\Theta} \tag{8}$$

where S_e and S_i are the average Fresnel reflection coefficients for externally and internally incident light, respectively. Θ is the internal transmission coefficient of the particle without internal scatters. S_e and S_i are empirically approximated by equations (9) and (10) (Hapke, 2012). Θ is calculated from equation (12).

$$S_e = 0.0587 + 0.8453R(0) + 0.08707R(0)^2 \tag{9}$$

$$S_i = 1 - \frac{1}{n^2}[0.9413 - 0.8453R(0) + 0.08707R(0)^2] \tag{10}$$

$$R(0) = [(n-1)^2 + k^2]/[(n+1)^2 + k^2] \tag{11}$$

where n and k are the real part (refractive index) and imagery part (extinction coefficient) of the complex refractive index (also known as optical constant), respectively. For the lunar regolith, the refractive index n is 1.78 (Hapke, 1994).

$$\Theta = e^{-\alpha<D>} \tag{12}$$

where α is the absorption coefficient, $<D>$ is the average distance traveled by all rays during a single transit of the particle, which has various approximation expressions (Hapke, 2001). Area weighted mean size (equation (13)) is adopted in this study

$$\langle D\rangle = D_L \ln\frac{D_U}{D_L} \tag{13}$$

where D_U and D_L are the effective upper and lower limit of the particle size. As for the Apollo lunar regolith sample, the effective upper and lower limit of the mineral particle sizes are 1000μm and 5μm, hence the approximate effective optical path is about 26μm and this value is adopted for deriving absorption coefficient from SSA in this study.

When equation (12) is inserted into equation (8), the relationship between absorption coefficient of the lunar regolith α and average single scattering albedo ω_{avg} can be expressed in equation (14).

$$\alpha = \frac{1}{\langle D\rangle}\ln\left[S_i + \frac{(1 - S_e)(1 - S_i)}{\omega_{avg} - S_e}\right] \tag{14}$$

3.3 Space weathering simulation

Assuming the space weathering being proportional to the mass fraction of SMFe, Hapke (2001) successfully reproduced the reddening and darkening effects caused by fine - grained SMFe particles, Lucey and Riner (2011) proposed a new formulation to model the effects of particles at any size by using Mie theory (equation (15)).

$$\alpha_w = \alpha_h + \alpha_g + \alpha_c \tag{15}$$

where α_w and α_h are the absorption coefficients of the weathered host material, unweathered host material (equation (14)), respectively; α_g and α_c represent the additional absorption due to SMFe within grains (equation (16)) and in coatings (equation (17)), respectively.

$$\alpha_g = \frac{36\pi M_g \rho_h}{\lambda \rho_{Fe}} \cdot \frac{n_h^3 n_{Fe} k_{Fe}}{(n_{Fe}^2 - k_{Fe}^2 + 2n_h^2)^2 + 4n_{Fe}^2 k_{Fe}^2} \tag{16}$$

$$\alpha_c = \frac{36\pi M_c \rho_h}{\lambda \rho_{Fe}} \cdot \frac{n_h^3 n_{Fe} k_{Fe}}{(n_{Fe}^2 - k_{Fe}^2 + 2n_h^2)^2 + 4n_{Fe}^2 k_{Fe}^2} \tag{17}$$

where M_g and M_c are the mass fraction of SMFe in the grain and coating relative to the host material. λ is wavelength; n_*, k_*, ρ_* ($*$ denotes h or Fe) are the real and imagery parts of the complex refractive index, and density of host material or SMFe, respectively. The optical constants of Fe used in this study are from Querry (1985).

The REFF spectrum with different mass fraction of SMFe incorporated on the host material can thus be simulated by inserting equation (15) into (8), and then into (3).

3.4 SMFe abundance estimation

The mass fraction of SMFe in the lunar regolith is estimated by comparing the spectral shape similarity between in situ measured VNIS spectrum and simulated M^3 spectrum by considering the contribution of SMFe. The simulated M^3 spectrum is created by adding various amounts of SMFe (0.0—2.0wt.%) with an increment of 0.001wt.%. The similarity between these two spectra is measured through the spectral angle parameter θ (in radians) by spectral angle mapper (SAM) technique (equation (18)) (Kruse et al., 1993). Smaller angles indicate higher similarity between the in situ measured VNIS spectrum and the simulated M^3 spectrum. The reason for choosing SAM technique is that it concentrates on spectral shape and is insensitive to absolute reflectance related to illumination or albedo effects. As stated in the section 2.1, the comparison is performed over the wavelength range 1500—2200nm in this study to avoid the thermal emission effects at longer wavelengths (Li and Milliken, 2016) and low SNR at shorter wavelengths (He et al., 2019; Li et al., 2019). The optimal estimation for the mass fraction of SMFe in the lunar regolith at Chang'e-4 landing site is determined when the spectral angle between the in situ measured VNIS spectrum and the simulated M^3 spectrum becomes the lowest.

$$\theta = \arccos\left(\frac{\sum_{\lambda=1}^{N} \mathrm{REFF}_\lambda \mathrm{REFF}'_\lambda}{\sqrt{\sum_{\lambda=1}^{N} \mathrm{REFF}_\lambda^2}\sqrt{\sum_{\lambda=1}^{N} \mathrm{REFF}'^2_\lambda}}\right) \tag{18}$$

where REFF_λ and REFF'_λ ($\lambda = 1, 2, \ldots, N$) are the in situ measured VNIS spectrum and simulated M^3 spectrum, respectively.

4　Result and discussion

4.1　SMFe abundance

The method used in this study successfully simulates the spectral shape of the in situ measured VNIS spectrum from the base M^3 spectrum. For example, the best match for a regolith spectrum measured by VNIS at station LE902 is found to be the M^3 spectrum plus 0.42 wt.% SMFe (Supplementary Fig. 1), which is typical for mature lunar regolith (Warell and Blewett, 2004). The average SMFe abundance in the lunar regolith at Chang'e-4 landing site, which is derived from 34 in situ measured spectra, is 0.32 ±0.06 wt.%. Trang and Lucey (2019) produced a SMFe abundance map through radiative transfer modeling of Kaguya multiband imager (MI) data. The SMFe abundance at Chang'e-4 landing site reported in their work is about 1.31 wt.%, which is the sum of the nanophase (particles <100nm in diameter) and microphase iron (particles >100nm in diameter) abundances. The possible main reason for the discrepancy is that the SMFe particles in the regolith at Chang'e-4 landing site may be chiefly in nanophase form. In this scenario, the slight difference between the estimated SMFe abundance (~0.32 wt.%) in this study and the nanophase iron abundance (~0.43 wt.%) reported by Trang and Lucey (2019) may attributed to: (1) different observation modes, i.e., in situ observation for the VNIS and orbital observation for the MI. (2) huge resolution differences, i.e., ~1mm/pixel from a detection distance of 1m for the VNIS and ~30.3km/pixel for the reported map. Wang et al. (2017) estimated the SMFe abundance (0.368 wt.%) and inferred the regolith at Chang'e-3 landing site to be submature after analyzing a spectrum measured by VNIS on the minimally disturbed regolith at Chang'e-3 landing site (Mare Imbrium). Although the regoliths at Chang'e-3 and Chang'e-4 landing sites have similar SMFe abundance, the SMFe saturation limit generally increases with FeO content (Trang and Lucey, 2019), which is important for estimating the maturity degree. The FeO contents at Chang'e-3 and Chang'e-4 landing sites are ~22.8 wt.% (Ling et al., 2015) and ~12.6wt.% (Supplementary Fig. 2), respectively. The absolute model ages of the two landing sites are 27—80 Ma (Xiao et al., 2015) and 3.7Ga (Supplementary Text 1), respectively. Considering the large differences in both FeO content and absolute model age between the two landing sites, the lunar regolith maturity degree at Chang'e-4 landing site is mature (discussed in the next section) and different from that at Chang'e-3 landing site.

By comparing images acquired both before and after the Chang'e-4 probe landing by the LROC NAC, it is found that the lander rocket plume had modified the surface. This effect had also been observed in previous missions (Clegg et al., 2014). The severely affected region, which was disturbed by Chang'e-4 probe descent engine during soft landing, is approximately 450m² (Fig. 4). The rover station A, which has an abundance of 0.27wt.% and is about 12.9m from the lander, is within the most significantly affected region. Because rocket plume can blow away the very uppermost, finest, and highly mature particles, previous study reported that the SMFe abundance in the regolith may vary significantly with distance from the lander (Wang et al., 2017). However, the phenomenon is not observed at Chang'e-4 landing site. For example, the VNIS made in situ measurements sequentially at rover stations S1, LE202, LE207 and LE208 just immediately after station A (Fig. 4). These stations are not within the extent of the possible severely affected region, and they are 19.2m, 28.6m, 50.6m and 53.5m away from the lander. The SMFe abundances estimated from VNIS spectra at these stations are 0.34wt.%, 0.30wt.%, 0.27wt.% and 0.28wt.%, respectively (Supplementary Fig. 1). The possible reason for the insignificant SMFe abundance difference is that the relative high regolith growth rate (3.1m/Gyr) at Chang'e-4 landing area (Lai et al., 2019). The fast gardening and impacting processes that formed the regolith had sufficiently mixed the uppermost, finest, and highly mature particles.

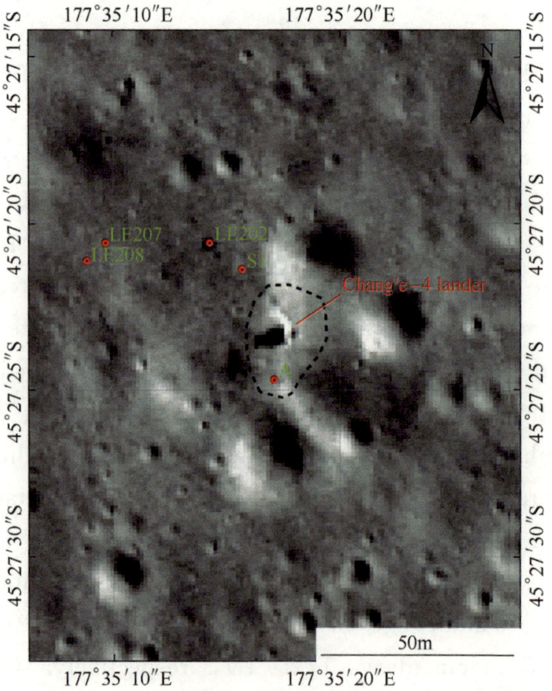

Fig. 4　The region severely affected by rocket engine plume during soft landing.
It is outlined by the black dotted line and is about 450m² (LROC NAC image ID: M1303619844LE)

4.2　Regolith maturity

Morris (1980) deduced the relationship between SMFe abundance, FeO content, and maturity index Is/FeO from Apollo and Lunar regolith samples, and classified the regolith into three categories: immature (0.0 <Is/FeO <29.0), submature (30.0 <Is/FeO <59.0), mature (Is/FeO >60.0). According to the products derived from Kaguya lunar multiband imager (Lemelin et al., 2016), the FeO abundance at Chang'e-4 landing site (Supplementary Fig. 2) is 12.6 ±0.6wt.%. Hence, the calculated Is/FeO value is 82 ±15, indicating the regolith at Chang'e-4 landing site is mature. As noted by Wang et al. (2017), the ferromagnetic resonance technique used by Morris (1980) to determine the Is value is only sensitive to iron particles in the diameter range about 4—33nm. However, the influence of SMFe particles >33nm, which are not sensed by magnetic resonance, is taken into consideration (equation (15)) by the spectral simulation process in this study. As a result, the use of estimated SMFe abundance in this study tends to slightly overestimate the Is/FeO value. However, it should be noted that the SMFe content of the base M^3 spectrum of the central peak bedrock is assumed to be zero, which can cause an underestimation because of the existence of possible little patina on the surface of the rock. Consequently, the overestimation of Is/FeO value should be balanced to some extent by the underestimation of SMFe content in the base M^3 spectrum.

The maturity of lunar regolith is a function of weathering environment, surface properties, exposure time, and gardening rates (Pieters and Noble, 2016). To simplify the actual dynamic space weathering process, assume the lunar regolith at Chang'e-4 landing site accumulated SMFe immediately and steadily after its formation, thus the maturity index Is/FeO can be treated as an indicator of exposure age to the harsh space environment. For example, radiometric dating revealed that many of the Apollo 17 samples are 3.6—4.2 billion years old, and Is/FeO values indicate they are mature (e.g., 93 for sample 78221, 82 for sample 72150) (Morris, 1978). The Von Karman crater was formed about 3.97 billion years ago (Yingst et al., 2017), and Finsen crater ejecta that had covered/sculptured the crater floor mare plain at the landing site happened around 3.7 billion years ago (Supplementary Text 1). Hence, the high Is/FeO of the regolith at the Chang'e-4 landing site is generally consistent with the geologic background that the regolith had experienced 3.7 billion years of space weathering.

5 Conclusion

Chang'e-4 rover made in situ spectral measurements on the lunar farside regolith, which provides a unique chance to study space weathering effects on an airless planetary body. In order to simulate the space weathering effects that arise from SMFe in the visible and the near infrared region, i.e., reflectance darkening, continuum slope reddening and absorption features flattening, Hapke model is used in this study to investigate the SMFe abundance in the regolith at the Chang'e-4 landing site. Simulation results reveal that the average SMFe abundance in the regolith is 0.32 ±0.06 wt.%. The derived Is/FeO maturity index, which is 82 ±15, indicates the regolith at Chang'e-4 landing site is mature. The mature state is consistent with the geologic background that the Finsen crater ejecta - sourced regolith, which was measured by Chang'e - 4 rover, had experienced about 3.7 billion years space weathering that caused by solar wind, galactic cosmic rays, micrometeorites and larger meteors.

Acknowledgements

The authors gratefully thank the whole Chang'e - 4 team for making the mission a great success and providing the data (http://moon.bao.ac.cn/). This research is supported by the B - type Strategic Priority Program of the Chinese Academy of Sciences (grant No. XDB41000000), National Natural Science Foundation of China (grant Nos. 41702354 and 41941003), Macau Young Scholars Program (grant No. AM201902), and Science and Technology Development Fund of Macau (grant No. 131/2017/A3). Constructive reviews by anonymous reviewers improved the quality of this manuscript substantially, and are greatly appreciated.

Appendix A Supplementary material

Supplementary material related to this article can be found online at https://doi.org/10.1016/j.epsl.2020.116117.

References

[1] Besse, S., Yokota, Y., Boardman, J., Green, R., Haruyama, J., Isaacson, P., Mall, U., Matsunaga, T., Ohtake, M., Pieters, C., Staid, M., Sunshine, J., Yamamoto, S., 2013. One Moon, many measurements 2: photometric corrections. Icarus 226, 127 - 139.

[2] Bowell, E., Hapke, B., Domingue, D., Lumme, K., Peltoniemi, J., Harris, A., 1989. Application of photometric models to asteroids. In: Binzel, R. P., Gehrels, T., Shapley Matthews, M. (Eds.), Asteroids II. The University of Arizona Press, Tucson, AZ, pp. 524 - 556.

[3] Clark, R. N., Pieters, C. M., Green, R. O., Boardman, J. W., Petro, N. E., 2011. Thermal removal from near - infrared imaging spectroscopy data of the Moon. J. Geophys. Res., Planets 116, E00G16.

[4] Clegg, R. N., Jolliff, B. L., Robinson, M. S., Hapke, B. W., Plescia, J. B., 2014. Effects of rocket exhaust on lunar soil reflectance properties. Icarus 227, 176 - 194.

[5] Di, K., Zhu, M. - H., Yue, Z., Lin, Y., Wan, W., Zhaoqin, Liu, Gou, S., Liu, B., Man, Peng, Wang, Y., Niu, S., Zhang, J., Li, J., Xie, J., Xi,

L., Yang, J., Xue, B., 2019. Topographic evolution of Von Kármán crater revealed by the lunar rover Yutu - 2. Geophys. Res. Lett. 46, 12764 - 12770.

[6] Gou, S., Di, K., Yue, Z., Liu, Z., He, Z., Xu, R., Lin, H., Liu, B., Peng, M., Wan, W., Wang, Y., Liu, J., 2019. Lunar deep materials observed by Chang'e - 4 rover. Earth Planet. Sci. Lett. 528, 115829.

[7] Hapke, B., 1981. Bidirectional reflectance spectroscopy: 1. Theory. J. Geophys. Res., Solid Earth 86, 3039 - 3054.

[8] Hapke, B., 1994. The refractive index of the regolith of mercury. In: 25th Lunar and Planetary Science Conference. Houston, TX, pp. 507 - 508.

[9] Hapke, B., 2001. Space weathering from Mercury to the asteroid belt. J. Geophys. Res. 106, 10039 - 10073.

[10] Hapke, B., 2002. Bidirectional reflectance spectroscopy: 5. The coherent backscatter opposition effect and anisotropic scattering. Icarus157, 523 - 534.

[11] Hapke, B., 2012. Theory of Reflectance and Emittance Spectroscopy, second edition. Cambridge University Press, Cambridge.

[12] He, Z., Li, C., Xu, R., Lv, G., Yuan, L., Wang, J., 2019. Spectrometers based on acoustooptic tunable filters for in - situ lunar surface measurement. J. Appl. Remote Sens. 13, 027502.

[13] Huang, J., Xiao, Z., Flahaut, J., Martinot, M., Head, J., Xiao, X., Xie, M., Xiao, L., 2018. Geological characteristics of Von Kármán Crater, Northwestern South Pole - Aitken Basin: Chang'E - 4 landing site region. J. Geophys. Res., Planets 123, 1684 - 1700.

[14] Hurwitz, D. M., Kring, D. A., 2014. Differentiation of the South Pole - Aitken basin impact melt sheet: implications for lunar exploration. J. Geophys. Res., Planets 119, 1110 - 1133.

[15] Kruse, F., Lefkoff, A., Boardman, J., Heidebrecht, K., Shapiro, A., Barloon, P., Goetz, A., 1993. The spectral image processing system (SIPS) - interactive visualization and analysis of imaging spectrometer data. Remote Sens. Environ. 44, 145 -163.

[16] Lai, J., Xu, Y., Zhang, X., Xiao, L., Yan, Q., Meng, X., Zhou, B., Dong, Z., Zhao, D., 2019. Comparison of dielectric properties and structure of lunar regolith at Chang'e - 3 and Chang'e - 4 landing sites revealed by ground penetrating radar. Geophys. Res. Lett. 46, 12783 - 12793. https://doi.org/10.1029/2019GL084458.

[17] Lemelin, M., Lucey, P. G., Gaddis, L. R., Hare, T., Ohtake, M., 2016. Global map products from the Kaguya Multiband Imager at 512 ppd: minerals, FeO and OMAT. In: 47th Lunar and Planetary Science Conference. Lunar and Planetary Institute. The Woodlands, Texas, p. 2994.

[18] Li, S., Milliken, R. E., 2016. An empirical thermal correction model for Moon

Miner – alogy Mapper data constrained by laboratory spectra and Diviner temperatures. J. Geophys. Res., Planets121, 2081 – 2107.

[19] Li, C. L., Xu, R., Lv, G., Yuan, L. Y., He, Z. P., Wang, J. Y., 2019. Detection and calibration characteristics of the visible and near – infrared imaging spectrometer in the Chang'e – 4. Rev. Sci. Instrum. 90, 103106.

[20] Ling, Z., Jolliff, B. L., Wang, A., Li, C., Liu, J., Zhang, J., Li, B., Sun, L., Chen, J., Xiao, L., Liu, J., Ren, X., Peng, W., Wang, H., Cui, X., He, Z., Wang, J., 2015. Correlated compositional and mineralogical investigations at the Chang'e – 3 landing site. Nat. Commun. 6, 8880.

[21] Liu, B., Niu, S., Xin, X., Jia, M., Di, K., Liu, Z., Peng, M., Yue, Z., 2019. High precision DTM and DOM generating using multi – source orbital data on Chang'e – 4 landing site. In: Int. Arch. Photogramm. Remote Sens. Spatial Inf. Sci. Copernicus Publications, Enschede, the Netherlands, pp. 1413 – 1417.

[22] Lucey, P. G., Noble, S. K., 2008. Experimental test of a radiative transfer model of the optical effects of space weathering. Icarus197, 348 – 353.

[23] Lucey, P. G., Riner, M. A., 2011. The optical effects of small iron particles that darken but do not redden: evidence of intense space weathering on Mercury. Icarus 212, 451 – 462.

[24] Melosh, H. J., 1989. Impact Cratering: A Geologic Process. Oxford University Press.

[25] Melosh, H. J., Kendall, J., Horgan, B., Johnson, B. C., Bowling, T., Lucey, P. G., Taylor, G. J., 2017. South Pole – Aitken basin ejecta reveal the Moon's upper mantle. Geology 45, 1063 – 1066.

[26] Miljkovic, K., Wieczorek, M. A., Collins, G. S., Solomon, S. C., Smith, D. E., Zuber, M. T., 2015. Excavation of the mantle in basin – forming events on the Moon. Earth Planet. Sci. Lett. 409, 243 – 251.

[27] Morris, R. V., 1978. The surface exposure (maturity) of lunar soils: some concepts and IS/FeO compilation. In: 9th Lunar and Planetary Science Conference. Pergamon Press, Houston, TX, pp. 2287 – 2297.

[28] Morris, R. V., 1980. Origins and size distribution of metallic iron particles in the lu – nar regolith. In: 11th Lunar and Planetary Science Conference. Pergamon Press, Houston, TX, pp. 1697 – 1712.

[29] Mustard, J. F., Pieters, C. M., 1989. Photometric phase functions of common geologic minerals and applications to quantitative analysis of mineral mixture reflectance spectra. J. Geophys. Res., Solid Earth 94, 13619 – 13634.

[30] Noble, S. K., Keller, L. P., Christoffersen, R., 2011. Sampling the uppermost surface of airless bodies. In: The Importance of Solar System Sample Return Missions to the Future of Planetary Science. The Woodlands, Texas, p. 5008.

[31] Noble, S. K., Pieters, C. M., Keller, L. P., 2007. An experimental approach

to understanding the optical effects of space weathering. Icarus 192，629 - 642.

[32] Noble，S.，Pieters，C.，Taylor，L.，Morris，R.，Allen，C.，McKay，D.，Keller，L.，2001. The optical properties of the finest fraction of lunar soil：implications for space weathering. Meteorit. Planet. Sci. 36，31 - 42.

[33] Petro，N. E.，Pieters，C. M.，2002. The size and location of the transient crater of the South Pole - Aitken basin. In：33rd Lunar and Planetary Science Conference. Hous - ton，Texas，p. 1848.

[34] Pieters，C. M.，Boardman，J.，Buratti，B.，Chatterjee，A.，Clark，R.，Glavich，T.，Green，R.，Head，J.，Isaacson，P.，Malaret，E.，2009. The Moon mineralogy mapper（M3）on Chandrayaan - 1. Curr. Nanosci. 96，500 - 505.

[35] Pieters，C. M.，Noble，S. K.，2016. Space weathering on airless bodies. J. Geophys. Res.，Planets 121，1865 - 1884.

[36] Pieters，C. M.，Taylor，L. A.，Noble，S. K.，Keller，L. P.，Hapke，B.，Morris，R. V.，Allen，C. C.，McKay，D. S.，Wentworth，S.，2000. Space weathering on airless bodies：resolving a mystery with lunar samples. Meteorit. Planet. Sci. 35，1101 - 1107.

[37] Potter，R. W. K.，Collins，G. S.，Kiefer，W. S.，McGovern，P. J.，Kring，D. A.，2012. Constraining the size of the South Pole - Aitken basin impact. Icarus 220，730 - 743.

[38] Querry，M. R.，1985. Optical Constants. Contractor Report CRDC - CR - 85034.

[39] Ren，X.，Liu，J. J.，Wang，F. F.，Wang，W. R.，Mu，L. L.，Li，H. H.，2014. A new lunar global topographic map products from Chang'E - 2 stereo camera image data. In：European Planetary Science Congress. Centro de Congressos do Estoril，Cascais，Portugal，pp. EPSC2014 - EPSC2344.

[40] Robinson，M. S.，Brylow，S. M.，Tschimmel，M.，Humm，D.，Lawrence，S. J.，Thomas，P. C.，Denevi，B. W.，Bowman - Cisneros，E.，Zerr，J.，Ravine，M. A.，Caplinger，M. A.，Ghaemi，F. T.，Schaffner，J. A.，Malin，M. C.，Mahanti，P.，Bartels，A.，Anderson，J.，Tran，T. N.，Eliason，E. M.，McEwen，A. S.，Turtle，E.，Jolliff，B. L.，Hiesinger，H.，2010. Lunar Reconnaissance Orbiter Camera（LROC）instrument overview. Space Sci. Rev. 150，81 - 124.

[41] Sasaki，S.，Nakamura，K.，Hamabe，Y.，Kurahashi，E.，Hiroi，T.，2001. Production of iron nanoparticles by laser irradiation in a simulation of lunar - like space weathering. Nature 410，555 - 557.

[42] Spudis，P. D.，Gillis，J. J.，Reisse，R. A.，1994. Ancient multiring basins on the Moon revealed by Clementine laser altimetry. Science 266，1848 - 1851.

[43] Tompkins，S.，Pieters，C. M.，1999. Mineralogy of the lunar crust：results from Clementine. Meteorit. Planet. Sci. 34，25 - 41.

[44] Trang，D.，Lucey，P. G.，2019. Improved space weathering maps of the lunar

surface through radiative transfer modeling of Kaguya multiband imager data. Icarus 321, 307 - 323.

[45] Vaughan, W. M., Head, J. W., 2014. Impact melt differentiation in the South Pole - Aitken basin: some observations and speculations. Planet. Space Sci. 91, 101 - 106.

[46] Wang, Z., Wu, Y., Blewett, D. T., Cloutis, E. A., Zheng, Y., Chen, J., 2017. Submicroscopic metallic iron in lunar soils estimated from the in situ spectra of the Chang'E - 3 mission. Geophys. Res. Lett. 44, 3485 - 3492.

[47] Warell, J., Blewett, D. T., 2004. Properties of the Hermean regolith: V. New optical reflectance spectra, comparison with lunar anorthosites, and mineralogical modelling. Icarus 168, 257 - 276.

[48] Xiao, L., Zhu, P., Fang, G., Xiao, Z., Zou, Y., Zhao, J., Zhao, N., Yuan, Y., Qiao, L., Zhang, X., Zhang, H., Wang, J., Huang, J., Huang, Q., He, Q., Zhou, B., Ji, Y., Zhang, Q., Shen, S., Li, Y., Gao, Y., 2015. A young multilayered terrane of the northern Mare Imbrium revealed by Chang'E - 3 mission. Science347, 1226 - 1229.

[49] Yingst, R. A., Chuang, F. C., Berman, D. C., Mest, S. C., 2017. Geologic mapping of the Planck quadrangle of the Moon (LQ - 29). In: 48th Lunar and Planetary Science Conference. The Woodlands, Texas, p. 1680.

Mineral Abundances Inferred From In Situ Reflectance Measurements of Chang'E - 4 Landing Site in South Pole - Aitken Basin

Xiaoyi Hu[1], Pei Ma[1], Yazhou Yang[1], MengHua Zhu[2,3], The Jiang[1], Paul G. Lucey[4], Lingzhi Sun[4], Hao Zhang[1,3], Chunlai Li[5], Rui Xu[5], Zhiping He[5], Hongyu Lin[6], Changning Huang[6], and Yuxue Sun[1]

1. Planetary Science Institute, School of Earth Sciences, China University of Geosciences, Wuhan, China
2. State Key Laboratory of Lunar and Planetary Sciences, Macau University of Science and Technology, Taipa, Macau
3. CAS Center for Excellence in Comparative Planetology, Hefei, China
4. Hawaii Institute of Geophysics and Planetology, University of Hawai'i atMānoa, Honolulu, HI, USA
5. Key Laboratory of Space Active Opto - Electronics Technology, Shanghai Institute of Technical Physics, Chinese Academy of Sciences, Shanghai, China
6. Beijing Institute of Space Mechanics and Electricity, China Academy of Space Technology, Beijing, China

Abstract: During its first two lunar day measurements, the visible and near - infrared spectrometer on board the Yutu - 2 Rover of the Chang'E - 4 mission acquired six in situ reflectance spectra from the floor of Von Kármán crater within the South Pole - Aitken basin. A spectral lookup table search has shown that the regolith at the landing site contains 56% — 72% plagioclase, 9% — 28% orthopyroxene, 4% — 19% clinopyroxene, and 2%—12% olivine, in relative abundance, in agreement with the results from Kaguya Multiband Imager mineral mapping. A mineral composition analysis using measurements made by the Moon Mineralogy Mapper and the Multiband Imager has shown that the surface materials of the landing site may have been ejected from the nearby Finsen crater. This result is also supported by principal component analysis which shows a strong correlation between the compositions of the landing area and Finsen crater.

Keypoints: We report the in situ visiblenear - infrared reflectance spectra measured by Chang'E - 4 during the first two lunar days

We obtained the relative mineral abundances at the landing site by using a spectral lookup table

The surface materials of the Chang'E - 4 landing site are plagioclase rich and may have come from the nearby Finsen crater

Plain Language Summary: With its successful soft landing in South Pole - Aitken basin on the lunar far side, the Chang'E - 4 mission measured six reflectance spectra during the first two lunar days. With these measurements, we obtained the mineral compositions of the landing site and identified the possible source location of regolith materials at the landing site, which may help us to uncover the mysteries of the formation of the South Pole - Aitken basin.

1 Introduction

The South Pole - Aitken (SPA) basin is the largest and oldest impact basin on the Moon (Garrick - Bethell & Zuber, 2009; Zuber et al., 1994), and the basin - forming impact may have exposed materials from the lower crust and even the upper mantle (e. g. Lucey et al., 1998, Melosh et al., 2017, Pieters et al., 1997, Potter et al., 2012, Wieczorek et al., 2012, Wieczorek & Phillips, 1998). Since the basin materials contain information regarding the early evolution of the Moon (Jolliff et al., 2006; Moriarty & Pieters, 2018; Ohtake et al., 2014; Pieters et al., 2001), the accurate mapping and validation of its mineral types, distribution, and abundance can provide important clues for understanding the compositional distribution of Moon's lower crust or upper mantle and the geological history of early Moon.

China's Chang'E - 4 (CE - 4) spacecraft landed at 177.588°E and 45.4578°S (Di et al., 2019) in Von Kármán crater inside the SPA basin on 3 January 2019 (Fig. 1) and thus has provided a unique opportunity to perform in situ investigations on the materials within the SPA basin. As the backup spacecraft of the Chang'E - 3 (CE - 3) mission, the CE - 4 released the Yutu - 2 Rover that carries a nearly identical payload of instruments as the CE - 3 including the panorama cameras, a visible and near - infrared spectrometer (VNIS), and a groundpenetrating radar (Wu et al., 2019). In this study, we report the VNIS measurements during the first two lunar days and the relevant mineral results.

2 Instrument, Measurement, and Data Reductions

Similar to the Yutu - 1 Rover of CE - 3, the VNIS on board the Yutu - 2 Rover consists of a complementary metal - oxide - semiconductor (CMOS) imager with 256 by 256 pixels and a short - wavelength infrared (SWIR) detector with a single pixel. The VNIS is installed on the front surface of the rover and looks down at the lunar surface at a nominal 45° viewing angle at a height of 0.69m. The CMOS imager works from 450 to 945nm with a spectral resolution of 2—7nm, and the SWIR detector works from 900 to 2,395nm with a spectral resolution of 3—12nm (He et al., 2019; Li, Wang, et al., 2019; Liu et al., 2014). The CMOS camera has a field of view (FOV) of 8.5° by 8.5° and

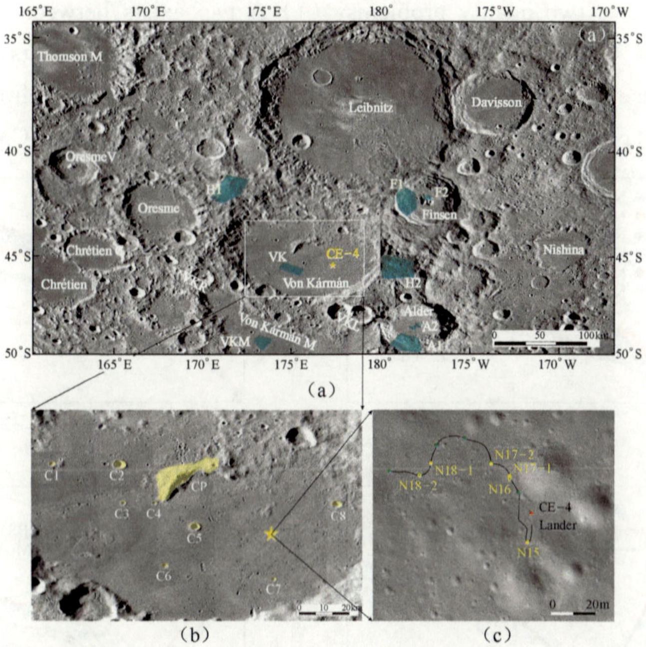

Fig. 1　The CE－4 landing area and Yutu－2 Rover's path during the first two lunar days:
(a) Lunar Reconnaissance Orbiter Camera Wide Angle Camera mosaic of the CE－4 landing area with the landing site indicated by the yellow star. (b) Enlarged view of the Von Kármán crater area indicated by the white box shown in (a);
(c) The path of the Yutu－2 Rover during the first two lunar days together with the visible and near－infrared spectrometer measurement sites (yellow dots), the locations for panorama camera (green dots), and the landing site (red triangle). The image of the landing area shown in (c) was taken by the landing camera on board the CE－4 spacecraft. Green and yellow zones in (a) and (b) are selected regions of interest (ROI) for detailed mineral abundance analysis in (a) and (b), respectively. The Moon Mineralogy Mapper spectra of each ROI shown in this work were averages of the typical spectra for that region: for ROIs highlighted in green in (a), we averaged the spectra of all pixels within the highlighted area. For the eight relatively fresh craters highlighted in yellowin (b), we only averaged the spectra of those brighter pixels to avoid the shadows and boundaries. CE－4 = Chang'E－4 landing site; VK = mare background of Von Kármán; F1 = crater wall of Finsen; F2 = central peak of Finsen; VKM = mare background of Von Kármán M; A1 = crater wall of Alder; A2 = central peak of Alder; H1 and H2 = typical high－elevation units; C1 to C8 = eight relatively fresh craters distributed uniformly across the Von Kármán crater; CP = central peak of the Von Kármán crater. The shadowed regions in each ROI were excluded when averaging the data.

roughly images an area of 15 by 21cm. The FOV of the single－pixel SWIR detector is 3.58°, and the measured area is a circle ~7cm in diameter within the FOV of the CMOS camera. More details of the VNIS can be found in He et al. (2014) and Zhang et al. (2015). Figure 1c shows the CE－4 landing area and Yutu－2 Rover's path with six VNIS measurement sites during the first two lunar days, during which the rover moved a distance of ~54m away from the lander. Starting from the raw data, we obtained calibrated radiance by following the standard data reduction procedures including dark current subtraction, temperature correction, and radiometric calibration (Liu et al., 2014). To obtain reflectance (reflectance factor; Hapke, 2012), we used two approaches: the solar irradiance calibration method and the onboard calibration target method (supporting information [SI] Text S1). Due to some unidentified glitches, the obtained reflectance spectra have two quality problems: (1) A gap exists between the CMOS (450－945nm) and SWIR (900－2,395nm) spectra in several measurements; and (2) spectra obtained using these two calibration methods do not coincide with each other for four out of the six spectra (Fig. S2). For the first problem, we multiplied the CMOS spectra by a

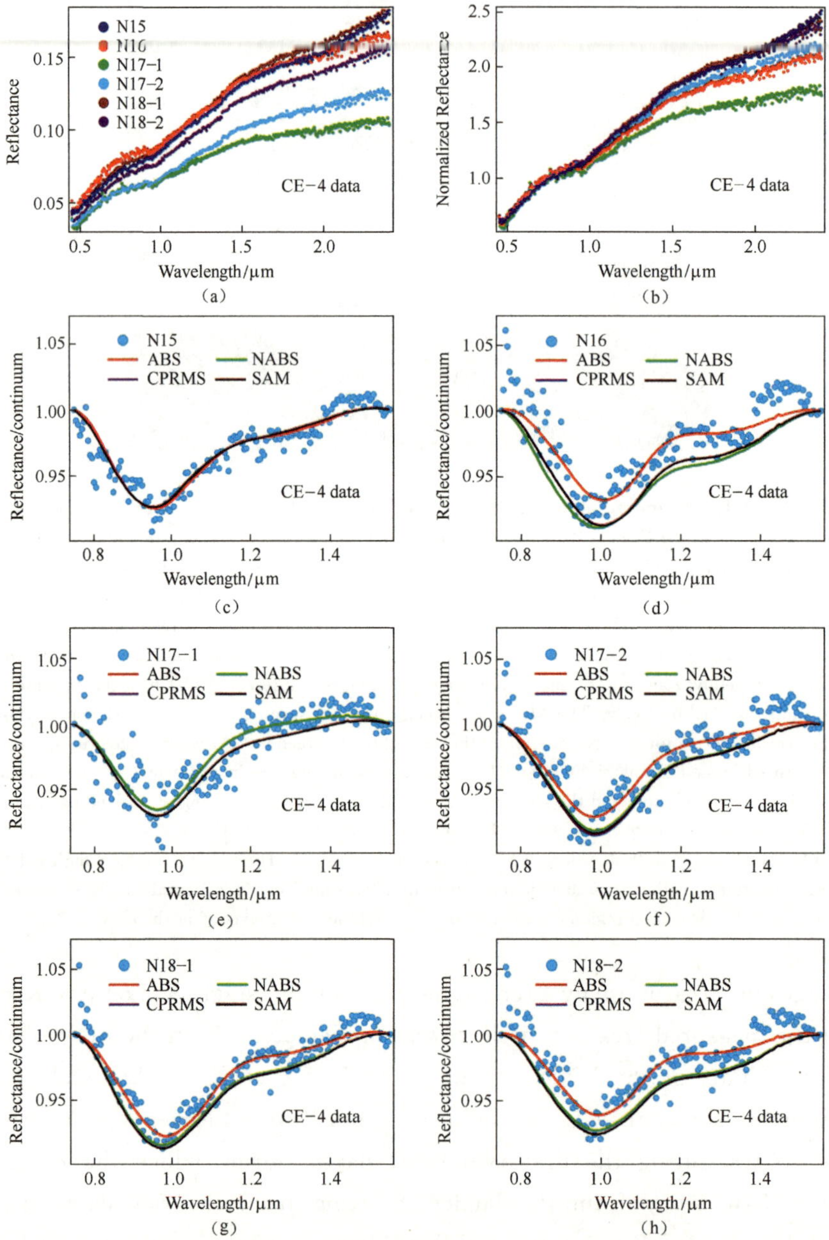

Fig. 2　(a) Original in situ visible and near - infrared spectrometer reflectance spectra expressed in reflectance factor measured by CE - 4 during the first two lunar days; (b) the six spectra normalized at 750nm. Comparisons of continuum - removed measurement spectra of six different sites with the corresponding best matched lookup table spectra obtained using four different fitting metrics are shown in (c) — (h). (c) For N15, (d) for N16, (e) for N17 - 1, (f) for N17 - 2, (g) for N18 - 1, and (h) for N18 - 2. CE - 4 = Chang'E - 4; ABS = minimum absolute difference; NABS = normalized absolute difference; CPRMS = centered pattern root mean square; SAM = spectral angle mapper factor to connect to the SWIR spectra in the overlapping region (900—945nm), following the payload team suggestions (Fig. S1). For the second one, as all analysis presented in this work uses continuum - removed spectra, the inferred mineral abundances by two calibration methods are close (SI Text S1). Therefore, we present the reflectance spectra calibrated by the solar irradiance method. The final spectra (450—2,395nm) of the six measurements are displayed in Fig. 2a.

Among the six spectra, N17 - 1 has the lowest average reflectance, while N15, N16, and N18 - 1 have the highest reflectance values. It is noted that the N17 - 1 site was on the rover track and more than half of the detection area was in shadow (Fig. S5 and S6 in SI), thus causing its lower apparent reflectance. The spectra normalized at 750nm shown in Fig. 2b have similar shapes below 1μm and start to diverge above 1μm.

Tab. 1 Mineral Abundance (wt.%) Obtained by Comparing the CE - 4 Data and the LUT Spectra, and the Errors Were Obtained by Using Equation (S8)

Data	First lunar day		Second lunar day				Average
	N15	N16	N17 - 1	N17 - 2	N18 - 1	N18 - 2	
OLV	9.0 ± 1.0	11.1 ± 3.9	2.1 ± 2.1	7.9 ± 2.3	11.8 ± 2.2	10.9 ± 2.5	8.8 ± 3.3
OPX	27.5 ± 3.4	9.2 ± 4.5	24.0 ± 0.6	17.8 ± 1.1	20.3 ± 2.5	10.9 ± 2.5	18.3 ± 6.6
CPX	4.1 ± 0.3	8.5 ± 0.3	14.0 ± 1.7	18.8 ± 0.7	10.6 ± 3.1	6.8 ± 1.2	10.4 ± 4.8
PLG	59.5 ± 2.6	71.3 ± 8.1	60.0 ± 1.0	55.5 ± 2.7	57.3 ± 1.6	71.5 ± 3.8	62.5 ± 6.4
i (°)	50.88	72.06	73.39	64.51	64.56	66.23	
e (°)	46.93	42.89	47.97	46.53	46.30	45.48	
α (°)	93.97	89.55	77.81	91.87	90.82	101.93	

Note. *i*, *e*, and *α* are solar zenith angle, sensor viewing zenith angle, and the solar phase angle, respectively. LUT = lookup table; OLV = olivine; OPX = orthopyroxene; CPX = clinopyroxene; PLG = plagioclase.

3　Mineral Abundance Analysis

To infer the major mineral abundances from measurements, we employed a synthesized lunar spectral mineral lookup table (LUT) based on radiative transfer models (SI Text S2; Lucey, 2004; Denevi et al., 2008; Lucey et al., 2014). Based on a study by Pieters et al. (1993), Lucey (2006) concluded that the spectra of bulk lunar regolith are extremely similar to that of soils with grain sizes between 10 and 20μm, and thus the effective grain diameter was fixed at 17μm. The Mg number in mole percent was fixed at 65 as this has been found to be a representative value of lunar regolith (Lucey et al., 2014). The effects of space weathering were simulated by adding different concentrations of submicroscopic iron to the grains. The LUT did not include glass, as the presence of glass (cagglutinitic and/or volcanic) based on its broad 1 - μm absorption (e. g., Cannon et al., 2017) is not apparent in the CE - 4 spectra. Glass absorptions due to agglutinates may be suppressed by large amounts of nanophase iron that is ubiquitous in agglutinate glass. The absence of glass in the models still enables mineral abundances errors less than 10wt.%.

Only spectra of olivine (OLV), clinopyroxene (CPX), orthopyroxene (OPX), and plagioclase (PLG) were considered, and the results presented in this work are relative mineral abundances. Lunar pyroxene (PYX) chemistries, abundances, and their spectral consequences are extremely diverse (Moriarty & Pieters, 2016), and spectral measurements do not contain sufficient information to accurately represent this diversity. However, using the simplification of representing PYX assemblages as being composed of only two PYXs, with a single shared Mg number of 65, still enables derivation of mineral abundances with errors less than 10wt.% (Lucey et al., 2014).

To avoid any thermal emission effects, we limited our analysis to the wavelength range from 750 to 1, 550nm (Lucey, 2004; Lucey et al., 2014). We first removed the continuum of both the measurement and the LUT spectra by dividing the spectra by a straight line that connects the first (750-nm) and the last (1,550-nm) data points. Then we searched the best match between measurement and the LUT spectra by using four different fitting metrics (Zhang et al., 2015): minimum absolute difference, normalized absolute difference, centered pattern root mean square (Taylor, 2001), and spectral angle mapper (SI Text S3). The mineral compositions of the best matched LUT spectra are used as the retrieved relative mineral abundances. The results of these four fitting algorithms were averaged to obtain the mean abundance, and the standard deviation was used to represent the fitting error (SI Text S3).

3.1 Results of CE-4 Spectra

The best fitting results for the six measurements are displayed in Fig. 2c—2h, and the mineral abundances with uncertainties, obtained by averaging the four results, are summarized in Tab. 1. These results indicate that the landing site is dominated by PLG (56%—72%), followed by OPX (9%—28%), CPX (4%—19%), and OLV (2%—12%). The ratio of PYX and OLV is consistent with results derived from the remote sensing observations (e. g., Moriarty et al., 2013; Moriarty & Pieters, 2018).

3.2 Results From Moon Mineralogy Mapper Spectra

Stratigraphy analysis suggests that the landing site has three major layers. The topmost one is mainly deposited ejecta from the nearby Finsen crater. The second layer is the ejecta from Alder and Leibnitz craters, and the bottom one consists of mare basalts (Huang et al., 2018). To compare the mineral concentrations with the surrounding areas, in particular with the Finsen and Alder craters, we used the Moon Mineralogy Mapper (M^3) spectral data to get the average spectrum within a square of 1, 120m by 1, 120m with 16 pixels of the landing site (SI Text S4). We also collected the spectra of Von Kármán crater's central peak and eight relatively fresh craters that are uniformly distributed within the Von Kármán crater (Fig. 1b) to verify the differences. Moreover, we collected the spectra of typical high - elevation regions (H1 and H2), Finsen (F1 and F2), Alder (A1 and A2) and Von Kármán M (VKM) for analyzing the possible mineral sources of the landing areas (Fig. 1a). All averaged M^3 spectra (SI Text S4) of selected areas (Fig. 1) are displayed in Fig. 3.

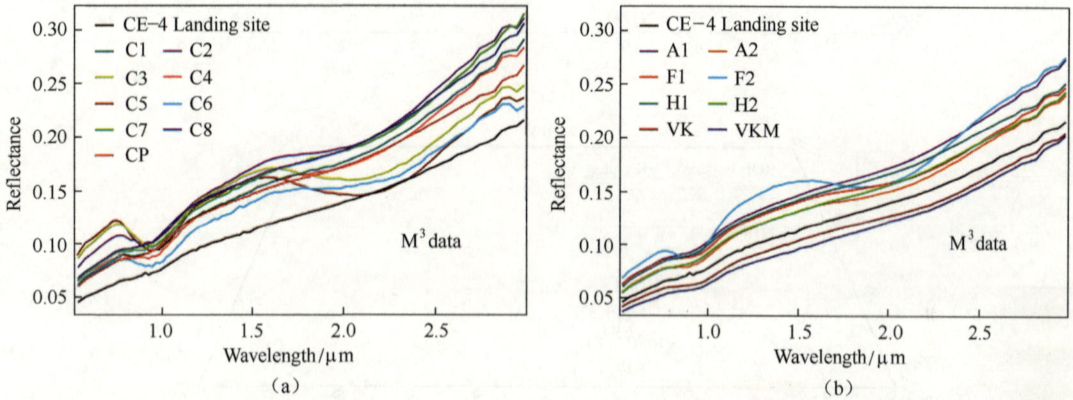

Fig. 3　Average M^3 spectra of selected typical units. The locations of selected areas in (a) and (b) are marked in Fig. 1b and 1a, respectively. CE - 4 = Chang'E - 4

Compared to those fresh craters, the CE - 4 landing site has the weakest absorption features and lowest reflectance (Fig. 3a), whereas the spectra of typical high - elevation regions (H1 and H2) and the crater wall of Finsen (F1) are more similar to that of the landing site. The mineral abundances of these regions extracted from the LUT are plotted in the ternary OLV - PYX - PLG abundance diagram as shown in Fig. 4 (see Tab. S2 for exact abundance values and Fig. S8 and S9 for the best matched LUT spectra). From Fig. 4a and 4b, we can find that the landing site has similar mineral abundances (M^3 results) to that of the central peak (CP) of Von Kármán, the crater wall of Finsen (F1), typical high - elevation units (H1 and H2), and crater 8 (C8), whereas results from CE - 4 in situ spectra show higher OLV and lower PLG abundances than that of M^3 spectra. It is possible that any OLV - rich patches in small scales (approximately centimeters) that have been detected by in situ measurements cannot be resolved by remote sensing measurements ($\sim$280m/pixel for M^3). Such difference in spatial resolutions may have caused the measured mineral abundances difference.

3.3　Mineral Mapping Using Kaguya Lunar Multiband Imager Data

To interpret the CE - 4 spectra at a larger geological setting scale, we plot the mineral abundance derived from the Kaguya Lunar Multiband Imager (MI) data (SI Text S5; Lemelin et al., 2015, 2016) in a region with the same size of Fig. 1a in Fig. 5. The average mineral abundances of the regions of interest shown in Fig. 1a and 1b are also extracted

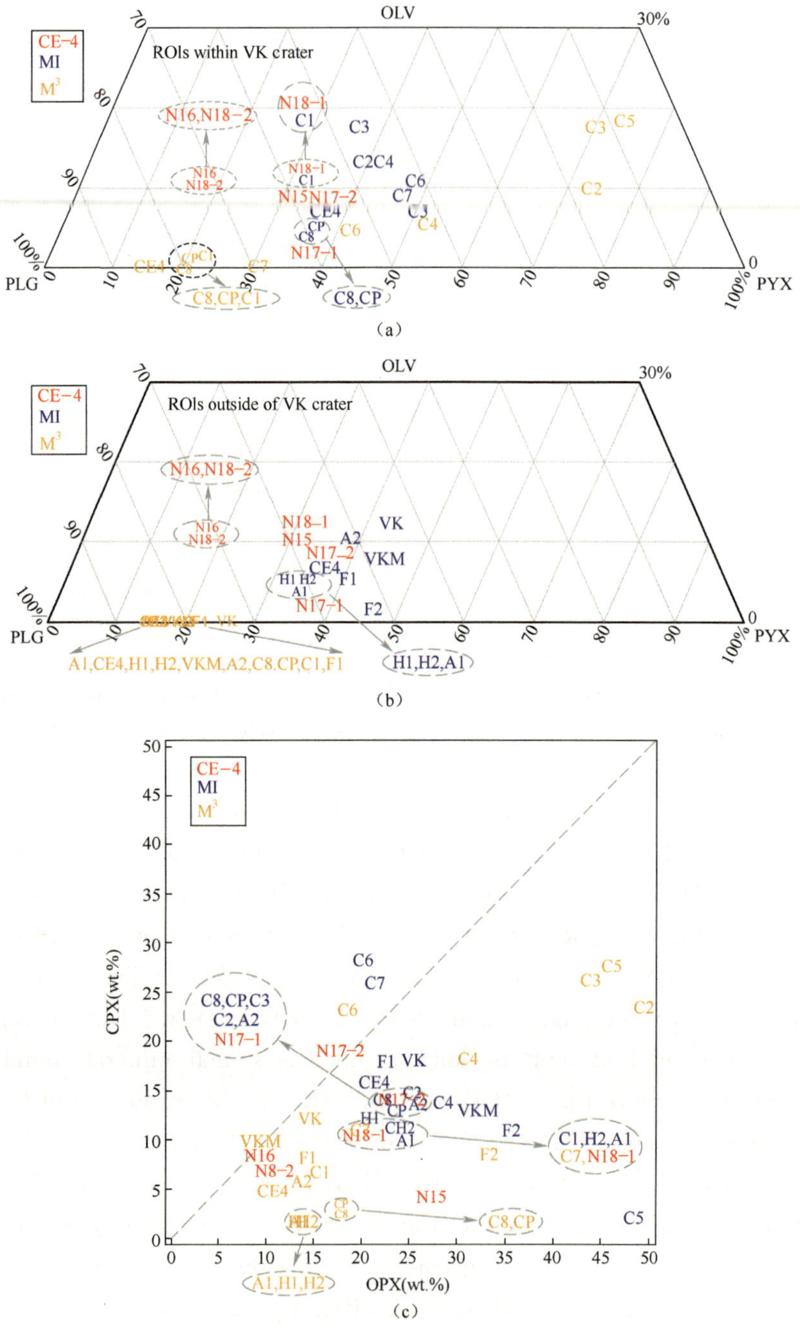

Fig. 4　The ternary diagram of PLG - PYX - OLV abundances of (a) the CE4 landing site and selected ROIs within the Von Kármán (VK) crater; (b) the CE4 landing site and selected ROIs outside the VK crater extracted from the CE - 4, M^3, and Kaguya Lunar Multiband Imager (MI) data; (c) the plot of CPX abundance versus OPX abundance of the CE4 landing site and selected ROIs extracted from the CE - 4, M^3, and MI data. Note the abundances are relative. The different results of M^3 and Kaguya/CE - 4 are probably caused by their different spatial resolutions. PLG = plagioclase; PYX = pyroxene; OLV = olivine; CE - 4 = Chang'E - 4; ROI = region of interest; CPX = clinopyroxene; OPX = orthopyroxene

from this mapping. By averaging a square of 300m by 300m with 25 pixels, the mineral abundances of the landing site are found to be 50% − 60% PLG and 5% − 10% OLV (Fig. 4 and Tab. S3), in agreement with the CE - 4 results. It should be noted that the MI data used here did not consider contributions from glass either. However, this should not affect the quantitative comparison with the CE - 4 results as both data sets have used similar LUT spectra.

It is readily seen from Fig. 5a and 5b that the crater rays from Finsen crater extend to the CE - 4 landing site, consistent with the fact that the landing site and F1 have similar mineral compositions (Fig. 4 and Tab. S3). Although M^3 and MI give different mineral abundance results, they are consistent in showing that the ejecta of Finsen crater is the major contributor to the surficial materials at the CE - 4 landing site.

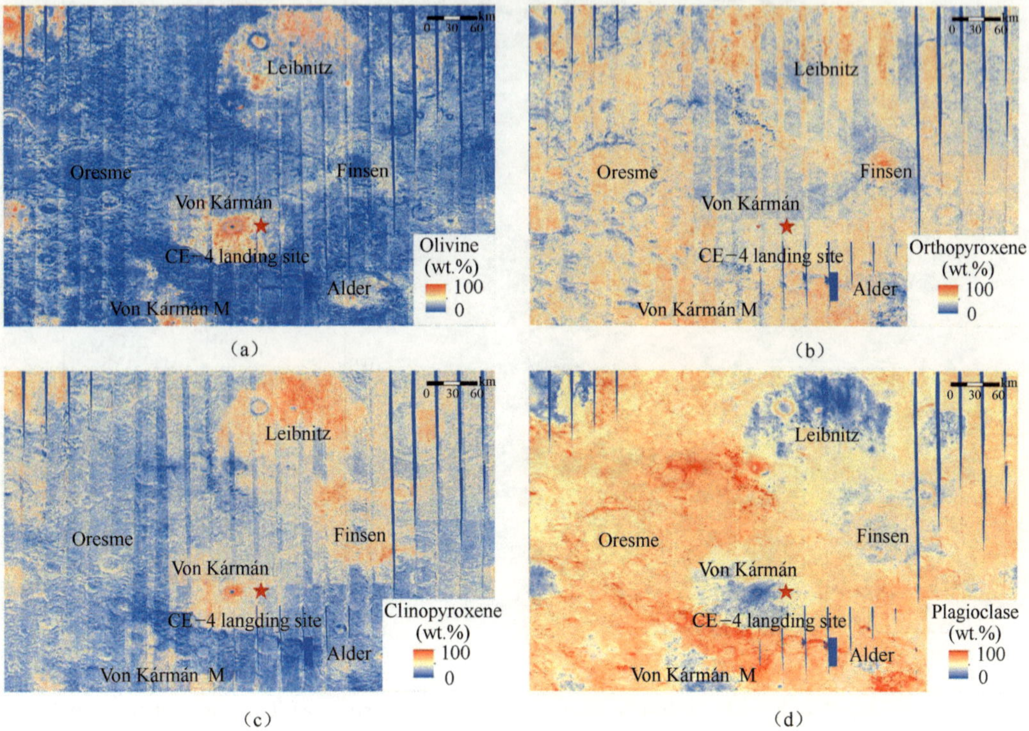

Fig. 5　Kaguya Lunar Multiband Imager mineral mapping results.

(a) Olivine mineral abundance, (b) orthopyroxene mineral abundance,

(c) clinopyroxene mineral abundance, and (d) plagioclase mineral abundance

3.4　Principal Component Analysis Using M^3 Spectra

To further identify the locations of source materials of the CE - 4 landing site, we applied principal component analysis (PCA) to the M^3 spectra, and the major results are shown in Fig. 6. The PCA can convert a hyperspectral data set into a set of orthogonal variables known as principal components (PCs) and hence reduce the data set dimension

(Jensen, 2015). As a result, the 85 - band M^3 spectra were reduced to 85 PCs with the first four PCs accounting for most of (> 99.7%) the entire spectral data variances (Fig. S10). The first component (PC1) accounts for 97.0% of the entire spectral variances and is essentially correlated to the albedo (Kramer et al., 2011). Although PC2, PC3, and PC4 only account for 2.27%, 0.35%, and 0.08% of the total variances, respectively, and there appear to be no specific physical meaning for them, they can be very useful in identifying different units. For example, the PC4 image of the CE - 4 landing area (Fig. 6a) clearly shows the crater rays extending from the Finsen crater to Von Kármán, indicating a strong correlation between the materials at the landing site and the Finsen crater (the other PCs images are shown in Fig. S11).

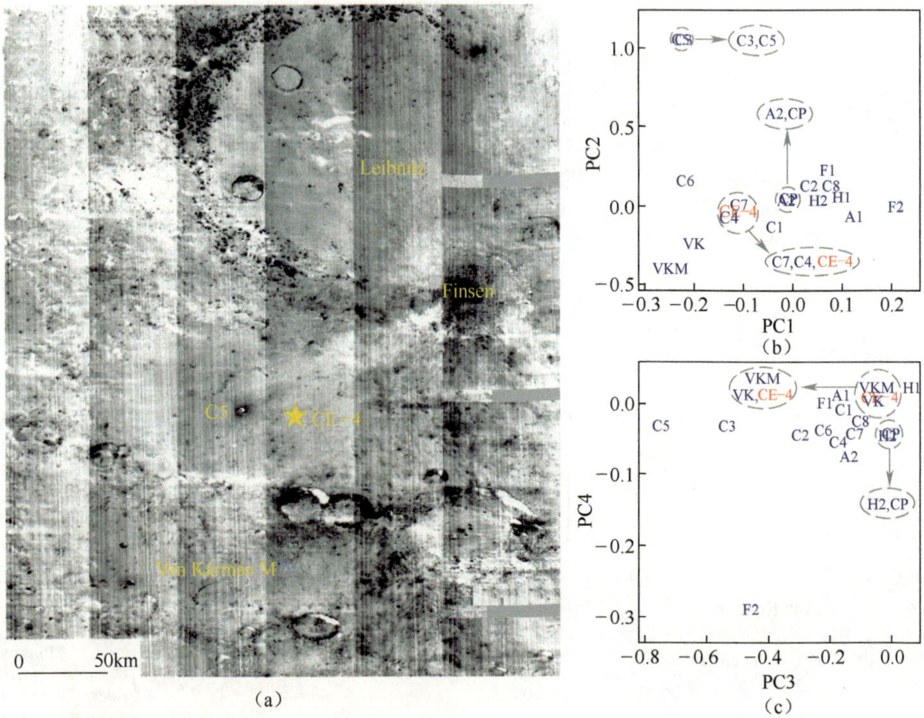

Fig. 6　Principal component analysis results of CE - 4 landing site. (a) The image of the fourth principal component (PC4) of the M^3 data. (b) PC2 vs PC1 and (c) PC4 vs PC3 are distributions of the typical PC values of selected regions of interest calculated from the M^3 spectra in the PC space. CE - 4 = Chang'E - 4

Fig. 6b and 6c show the distributions of the first four PC values of those selected regions of interest in the PC space. The C1—C8 craters show large variations in PC1 and PC2, while C3 and C5 have very similar PC values which are consistent with their mineral compositions as shown in Fig. 4. In contrast, the central peak of Finsen (F2) shows obviously different mineral compositions from its crater wall (F1) (Moriarty & Pieters, 2018).

4 Discussion and Conclusions

The PLG abundance at the CE - 4 landing site (56%—72%) is slightly lower than the average PLG abundance of the farside highlands of the Moon (~75%, e. g., Lucey, 2004) but is similar to the values of SPA high - elevation regions (Fig. 4). The abundances of OPX and CPX are much lower than those for the mare basalts throughout the SPA basin. The OLV abundance is similar to the average OLV abundance for the mare basalts throughout the SPA basin (<10 wt.%, Lucey, 2004) but far lower than the value of fresh mare basalts on the nearside of the Moon as measured by CE - 3 (Zhang et al., 2015). The high PLG abundance at the landing site may indicate the associated OLV is inherent of anorthosites, rather than the Mg - suite troctolites erupted from the interior of the Moon (Lucey, 2004). So far, the materials at the landing site are unlikely OLV - and PYX - rich materials from the lunar mantle as found by Li, Liu, et al. (2019). This discrepancy was probably caused by the fact that the modified Gaussian method used by Li, Liu, et al. (2019) to estimate the mineral abundance is insensitive to the presence of PLG.

The mineral composition analysis in the landing area based on M^3 and the Kaguya MI data have shown that the regolith materials of CE - 4 landing region may have come from the nearby Finsen crater. This is corroborated by the PCA that shows crater rays extending from the Finsen crater to the landing site.

The CE - 4 landing site has similar concentrations of the dominant minerals with the central peak (CP) of Von Kármán, the crater wall of Finsen (F1), the typical high - elevation units (H1 and H2), and the crater wall of Alder crater (A1; Figures 3—5). This suggests an insignificant vertical or horizontal compositional variation for the area surrounding the CE - 4 landing site. According to the numerical simulations on the formation of the SPA basin, this area is located at the boundary of the transient crater (Melosh et al., 2017; Wieczorek et al., 2012). In light of the impact cratering process, the materials deposited on this area were probably excavated from a depth of 90 — 140km (Melosh et al., 2017; Potter et al., 2012 ; Zhu et al., 2018), dominated by the PLG - poor but OLV - and PYX - rich upper mantle materials (Khan et al., 2007). However, the expected mantle materials are not observed at the CE - 4 landing site, indicating that the materials of the present - day crust (at least the upper crust) at the landing area is not the breccia deposits derived from the ejecta of the SPA basin (Ohtake et al., 2014; Pieters et al., 2001). The CE - 4 landing area is located within the impact - induced melt pool (Potter et al., 2012; Zhu et al., 2018). The thick impact sheet (>50km; Potter et al., 2012; Zhu et al., 2018) produced during the impact cratering process may differentiate and form the PLG - rich but OLVand PYX - poor crust (Hurwitz & Kring, 2014; Uemoto et al., 2017; Vaughan & Head, 2014), as measured at the landing site by CE - 4 on the ejecta deposits excavated from the Finsen crater (Huang et al., 2018).

Acknowledgments

We thank Kevin Cannon and Daniel Moriarty for helpful reviews that improved the quality of this paper. We also thank the Chang'E－4 payload teams for mission operations and China National Space Administration for providing the Chang'E－4 data that made this study possible. This work was supported by the National Natural Science Foundation of China (11773023, 11941001, U1631124, and 11773087), the Lunar and Planetary Science Laboratory, Macau University of Science and Technology Partner Laboratory of Key Laboratory of Lunar and Deep Space Exploration, Chinese Academy of Sciences (039/2013/A2), the Fundamental Research Funds for the Central Universities, China University of Geosciences (Wuhan), and by the Science and Technology Development Fund, Macau (0079/2018/A2). The Chang'E－4 data used in this work can be downloaded online (https: //zenodo. org/record/3361004#. XUjxKi2B1TY).

References

[1] Cannon, K. M., Mustard, J. F., Parman, S. W., Sklute, E. C., Dyar, M. D., & Cooper, R. F. (2017). Spectral properties of Martian and other planetary glasses and their detection in remotely sensed data. Journal of Geophysical Research: Planets, 122, 249－268. https://doi. org/10. 1002/2016JE005219.

[2] Denevi, B. W., Lucey, P. G., & Sherman, S. B. (2008). Radiative transfer modeling of near－infrared spectra of lunar mare soils: Theory and measurement. Journal of Geophysical Research, 113, E02003. https://doi. org/10. 1029/2007JE002929.

[3] Di, K., Liu, Z., Liu, B., Wan, W., Peng, M., Wang, Y., et al. (2019). Chang'E－4 lander localization based on multi－source data. Journal of Remote Sensing, 23 (1). https://doi. org/10. 11834/jrs. 20199015.

[4] Garrick－Bethell, I., & Zuber, M. T. (2009). Elliptical structure of the lunar South Pole－Aitken basin. Icarus, 204 (2), 399－408. https://doi. org/10. 1016/j. icarus. 2009. 05. 032.

[5] Hapke, B. (2012). Theory of reflectance and emittance spectroscopy. Cambridge, UK: Cambridge University Press.

[6] He, Z., Li, C., Xu, R., Lv, G., Yuan, L., & Wang, J. (2019). Spectrometers based on acousto－optic tunable filters for in－situ lunar surface measurement. Journal of Applied Remote Sensing, 13 (02). https://doi. org/10. 1117/1. JRS. 13. 027502.

[7] He, Z. －P., Wang, B. Y., Lü, G., Li, C. L., Yuan, L. Y., Xu, R., et al. (2014). Operating principles and detection characteristics of the Visible and Near－Infrared Imaging Spectrometer in the Chang－3. Research in Astronomy and Astrophysics, 14

(12), 1567 - 1577. https://doi.org/10.1088/1674-4527/14/12/006.

[8] Huang, J., Xiao, Z., Flahaut, J., Martinot, M., Head, J., Xiao, X., et al. (2018). Geological characteristics of Von Kármán Crater, northwestern South Pole - Aitken basin: Chang'E - 4 landing site region. Journal of Geophysical Research: Planets, 123, 1684 - 1700. https://doi.org/10.1029/2018JE005577.

[9] Hurwitz, D. M., & Kring, D. A. (2014). Differentiation of the South Pole - Aitken basin impact melt sheet: Implications for lunar exploration. Journal of Geophysical Research: Planets, 119, 1110 - 1133. https://doi.org/10.1002/2013JE004530.

[10] Jensen, J. R. (2015). Introductory digital image processing: A remote sensing perspective (4th ed.). Upper Saddle River, NJ, USA: Prentice Hall Press.

[11] Jolliff, B. L., Wieczorek, M. A., & Shearer, C. K. (2006). New views of the Moon. Reviews in Mineralogy and Geochemistry, 60 (1), v - xv. https://doi.org/10.2138/rmg.2006.60.0.

[12] Khan, A., Connolly, J. A. D., Maclennan, J., & Mosegaard, K. (2007). Joint inversion of seismic and gravity data for lunar composition and thermal state. Geophysical Journal International, 168 (1), 243 - 258. https://doi.org/10.1111/j.1365246X.2006.03200.

[13] Kramer, G. Y., Besse, S., Dhingra, D., Nettles, J., Klima, R., Garrick - Bethell, I., et al. (2011). M - 3 spectral analysis of lunar swirls and the link between optical maturation and surface hydroxyl formation at magnetic anomalies. Journal of Geophysical Research, 116, E00G18. https://doi.org/10.1029/2010je003729.

[14] Lemelin, M., Lucey, P., Gaddis, L., Hare, T., & Ohtake, M. (2016). Global map products from the Kaguya multiband imager at 512 ppd: Minerals, FeO, and OMAT (Vol. 47, p. 2994). Presented at the Lunar and Planetary Science Conference.

[15] Lemelin, M., Lucey, P. G., Song, E., & Taylor, G. J. (2015). Lunar central peak mineralogy and iron content using the Kaguya Multiband Imager: Reassessment of the compositional structure of the lunar crust. Journal of Geophysical Research: Planets, 120, 869 - 887. https://doi.org/10.1002/2014JE004778.

[16] Li, C., Liu, D., Liu, B., Ren, X., Liu, J., He, Z., et al. (2019). Chang'E - 4 initial spectroscopic identification of lunar far - side mantle - derived materials. Nature, 569 (7756), 378 - 382. https://doi.org/10.1038/s41586-019-1189-0.

[17] Li, C., Wang, Z., Xu, R., Lv, G., Yuan, L., He, Z., & Wang, J. (2019). The scientific information model of Chang'E - 4 Visible and Near - IR Imaging Spectrometer (VNIS) and in - orbit verification. Sensors, 19 (12), 2806. https://doi.org/10.3390/s19122806.

[18] Liu, B., Li, C. L., Zhang, G. L., Xu, R., Liu, J. J., Ren, X., et al. (2014). Data processing and preliminary results of the Chang'E - 3 VIS/NIR Imaging Spectrometer in - situ analysis. Research in Astronomy and Astrophysics, 14 (12), 1578 - 1594. https://doi.org/10.1088/1674-4527/14/12/007.

[19] Lucey, P. G. (2004). Mineral maps of the Moon. Geophysical Research Letters, 31, L08701. https://doi.org/10.1029/2003GL019406.

[20] Lucey, P. G. (2006). Radiative transfer modeling of the effect of mineralogy on some empirical methods for estimating iron concentration from multispectral imaging of the Moon. Journal of Geophysical Research, 111, E08003. https://doi.org/10.1029/2005JE002661.

[21] Lucey, P. G., Norman, J. A., Crites, S. T., Taylor, G. J., Hawke, B. R., Lemelin, M., & Melosh, H. J. (2014). A large spectral survey of small lunar craters: Implications for the composition of the lunar mantle. American Mineralogist, 99 (11 - 12), 2251 - 2257. https://doi.org/10.2138/am-2014-4854.

[22] Lucey, P. G., Taylor, G. J., Hawke, B. R., & Spudis, P. D. (1998). FeO and TiO_2 concentrations in the South Pole - Aitken basin: Implications for mantle composition and basin formation. Journal of Geophysical Research, 103 (E2), 3701 - 3708. https://doi.org/10.1029/97JE03146.

[23] Melosh, H. J., Kendall, J., Horgan, B., Johnson, B. C., Bowling, T., Lucey, P. G., & Taylor, G. J. (2017). South Pole - Aitken basin ejecta reveal the Moon's upper mantle. Geology, 45 (12), 1063 - 1066. https://doi.org/10.1130/G39375.1.

[24] Moriarty, D. P. III, & Pieters, C. M. (2016). Complexities in pyroxene compositions derived from absorption band centers: Examples from Apollo samples, HED meteorites, synthetic pure pyroxenes, and remote sensing data. Meteoritics & Planetary Science, 51 (2), 207 - 234. https://doi.org/10.1111/maps.12588.

[25] Moriarty, D. P. III, & Pieters, C. M. (2018). The character of South Pole - Aitken basin: Patterns of surface and subsurface composition. Journal of Geophysical Research: Planets, 123, 729 - 747. https://doi.org/10.1002/2017JE005364

[26] Moriarty, D. P., Pieters, C. M., & Isaacson, P. J. (2013). Compositional heterogeneity of central peaks within the South Pole - Aitken basin. Journal of Geophysical Research: Planets, 118, 2310 - 2322. https://doi.org/10.1002/2013JE004376.

[27] Ohtake, M., Uemoto, K., Yokota, Y., Morota, T., Yamamoto, S., Nakamura, R., et al. (2014). Geologic structure generated by large - impact basin formation observed at the South Pole - Aitken basin on the Moon. Geophysical Research Letters, 41, 2738 - 2745. https://doi.org/10.1002/2014GL059478.

[28] Pieters, C. M., Fischer, E. M., Rode, O., & Basu, A. (1993). Optical effects of space weathering: The role of the finest fraction. Journal of Geophysical Research, 98 (E11), 20, 817 - 20, 824. https://doi.org/10.1029/93JE02467.

[29] Pieters, C. M., Head, J. W., Gaddis, L., Jolliff, B., & Duke, M. (2001). Rock types of South Pole - Aitken basin and extent of basaltic volcanism. Journal of

Geophysical Research, 106 (E11), 28, 001 - 28, 022. https://doi.org/10.1029/2000JE001414.

[30] Pieters, C. M., Tompkins, S., Head, J. W., & Hess, P. C. (1997). Mineralogy of the mafic anomaly in the South Pole - Aitken basin: Implications for excavation of the lunar mantle. Geophysical Research Letters, 24 (15), 1903 - 1906. https://doi.org/10.1029/97GL01718.

[31] Potter, R. W. K., Collins, G. S., Kiefer, W. S., McGovern, P. J., & Kring, D. A. (2012). Constraining the size of the South Pole - Aitken basin impact. Icarus, 220 (2), 730 - 743. https://doi.org/10.1016/j.icarus.2012.05.032.

[32] Taylor, K. E. (2001). Summarizing multiple aspects of model performance in a single diagram. Journal of Geophysical Research, 106 (D7), 7183 - 7192. https://doi.org/10.1029/2000JD900719.

[33] Uemoto, K., Ohtake, M., Haruyama, J., Matsunaga, T., Yamamoto, S., Nakamura, R., et al. (2017). Evidence of impact melt sheet differentiation of the lunar South Pole - Aitken basin. Journal of Geophysical Research: Planets, 122, 1672 - 1686. https://doi.org/10.1002/2016JE005209.

[34] Vaughan, W. M., & Head, J. W. (2014). Impact melt differentiation in the South Pole - Aitken basin: Some observations and speculations. Planetary and Space Science, 91, 101 - 106. https://doi.org/10.1016/j.pss.2013.11.010.

[35] Wieczorek, M. A., & Phillips, R. J. (1998). Potential anomalies on a sphere: Applications to the thickness of the lunar crust. Journal of Geophysical Research, 103 (E1), 1715 - 1724. https://doi.org/10.1029/97JE03136.

[36] Wieczorek, M. A., Weiss, B. P., & Stewart, S. T. (2012). An impactor origin for lunar magnetic anomalies. Science, 335 (6073), 1212 - 1215. https://doi.org/10.1126/science.1214773.

[37] Wu, W., Li, C., Zuo, W., Zhang, H., Liu, J., Wen, W., et al. (2019). Lunar farside to be explored by Chang'E - 4. Nature Geoscience, 12 (4), 222 - 223. https://doi.org/10.1038/s41561-019-0341-7.

[38] Zhang, H., Yang, Y., Yuan, Y., Jin, W., Lucey, P. G., Zhu, M. - H., et al. (2015). In situ optical measurements of Chang'E - 3 landing site in Mare Imbrium: 1. Mineral abundances inferred from spectral reflectance. Geophysical Research Letters, 42, 6945 - 6950. https://doi.org/10.1002/2015GL065273.

[39] Zhu, M. - H., Wünnemann, K., Morbidelli, A., & Artemieva, N. (2018). South Pole - Aitken basin may contribute insignificantly to the late accretion of the Moon (Vol. 49, p. 2083). Presented at the Lunar and Planetary Science Conference.

[40] Zuber, M. T., Smith, D. E., Lemoine, F. G., & Neumann, G. A. (1994). The shape and internal structure of the Moon from the Clementine mission. Science, 266 (5192), 1839 -1843. https://doi.org/10.1126/science.266.5192.1839.

Diverse rock types detected in the lunar South Pole – Aitken Basin by the Chang'E – 4 lunar mission

Jun Huang (黄俊)[1,2], Zhiyong Xiao (肖智勇)[2,3,4], Long Xiao (肖龙)[1,2,4], Briony Horgan[5], Xiaoyi Hu (胡晓依)[1], Paul Lucey[6], Xiao Xiao (肖潇)[1], Siyuan Zhao (赵思源)[1], Yuqi Qian (钱煜奇)[1], Hao Zhang (张昊)[1], Chunlai Li (李春来)[7], Rui Xu (徐睿)[7], Zhiping He (何志平)[7], Jianfeng Yang (杨建峰)[8], Bin Xue (薛彬)[8], Qi He (何琦)[1], Jie Zhong (钟杰)[9], Hongyu Lin (林宏宇)[10], Changning Huang (黄长宁)[10], Jianfeng Xie (谢剑锋)[11]

1. State Key Laboratory of Geological Processes and Mineral Resources, Planetary Science Institute, School of Earth Sciences, China University of Geosciences, Wuhan, Hubei 430074, China

2. Chinese Academy of Sciences (CAS) Center for Excellence in Comparative Planetology, Hefei, Anhui 230026, China

3. Planetary Environmental and Astrobiological Laboratory, School of Atmospheric Sciences, Sun Yat – sen University, Zhuhai, Guangdong 519000, China

4. State Key Laboratory of Lunar and Planetary Sciences, Space Science Institute, Macau University of Science and Technology, Macau 999078, China

5. Department of Earth, Atmospheric and Planetary Sciences, Purdue University, West Lafayette, Indiana 47907, USA

6. Hawai 'i Institute of Geophysics and Planetology, University of Hawai 'i at Manoa, Honolulu, Hawai 'i 96822, USA

7. Key Laboratory of Space Active Opto – Electronics Technology, Shanghai Institute of Technical Physics, Chinese Academy of Sciences, Shanghai 200083, China

8. Xi' an Institute of Optics and Precision Mechanics, Chinese Academy of Sciences, Xi' an, 710119, China

9. Institute of Optics and Electronics, Chinese Academy of Sciences, Chengdu 610209, China

10. Institute of Space Mechanics and Electricity, China Academy of Space Technology, Beijing 100086, China

11. Beijing Aerospace Flight Control Center, Beijing 100094, China

Abstract: The South Pole – Aitken (SPA) basin, located between the South Pole and Aitken crater on the far side of the Moon, is the largest confirmed lunar impact structure.

The pre-Nectarian SPA basin is a 2400 × 2050 km elliptical structure centered at 53°S, 191°E, which should have exposed lower crust and upper mantle due to the enormous excavation depth. Olivine, the dominant mineral in Earth's mantle, has only been identified in small and localized exposures in the margins of the SPA basin, and the dominant mafic component is, instead, pyroxene. These mineralogical characteristics could be explained by the recent hypothesis that the lunar upper mantle is dominated by low-calcium pyroxene, not olivine. Here, we present observations from imaging and spectral data from China's Chang'E-4 (CE-4) lunar mission in the first 4 synodic days, especially the first *in situ* visible/near-infrared spectrometer observations of an exposed boulder. We identified a variety of rock types, but not the recently reported olivine-rich materials in the landing region. The results are consistent with orbital observations. The obtained mineralogical information provides a better understanding of the nature and origin of SPA materials.

Introduction

The surface materials of the South Pole-Aitken (SPA) basin show distinct chemical (Lucey et al., 1998; Jolliff et al., 2000) and mineralogical (Pieters et al., 2001; Ohtake et al., 2014; Moriarty and Pieters, 2018) characteristics compared to other compositional regions on the Moon. The vast scale of the SPA basin-forming event is modeled as having produced a thick ejecta blanket, which excavated materials from throughout the crust and down into the upper mantle (Melosh et al., 2017). However, orbital studies of SPA basin have revealed very few exposures of Earth-like olivine-rich mantle material (Yamamoto et al., 2012), and they instead show abundant low-calcium pyroxene (LCP; Melosh et al., 2017; Moriarty and Pieters, 2018) and some high-calcium pyroxene (HCP; Ohtake et al., 2014; Moriarty et al., 2019). SPA also should have generated thick impact melt, which has been hypothesized to have undergone igneous differentiation (Morrison, 1998; Nakamura et al., 2009; Hurwitz and Kring, 2014; Vaughan and Head, 2014). Modeling has indicated that the approximate diameter of the differentiated melt sheet is ~400km (Hurwitz and Kring, 2014), which is roughly spatially consistent with the South Pole-Aitken compositional anomaly (SPACA; Moriarty and Pieters, 2018), a central HCP-dominated spectral unit (Fig. 1a) with distinctive topographic and slope properties compared to the rest of the SPA basin (Ohtake et al., 2014). Assuming an olivine-rich upper mantle, a possible top-to-bottom compositional stratigraphy of the differentiated melt is norite, pyroxenite, and dunite (Hurwitz and Kring, 2014; Vaughan and Head, 2014).

The Chang'E-4 (CE-4) lunar spacecraft mission (Chinese Lunar Exploration Program) was planned to provide *in situ* observations to address the fundamental question about the possible results of SPA melt differentiation and the composition of the lunar

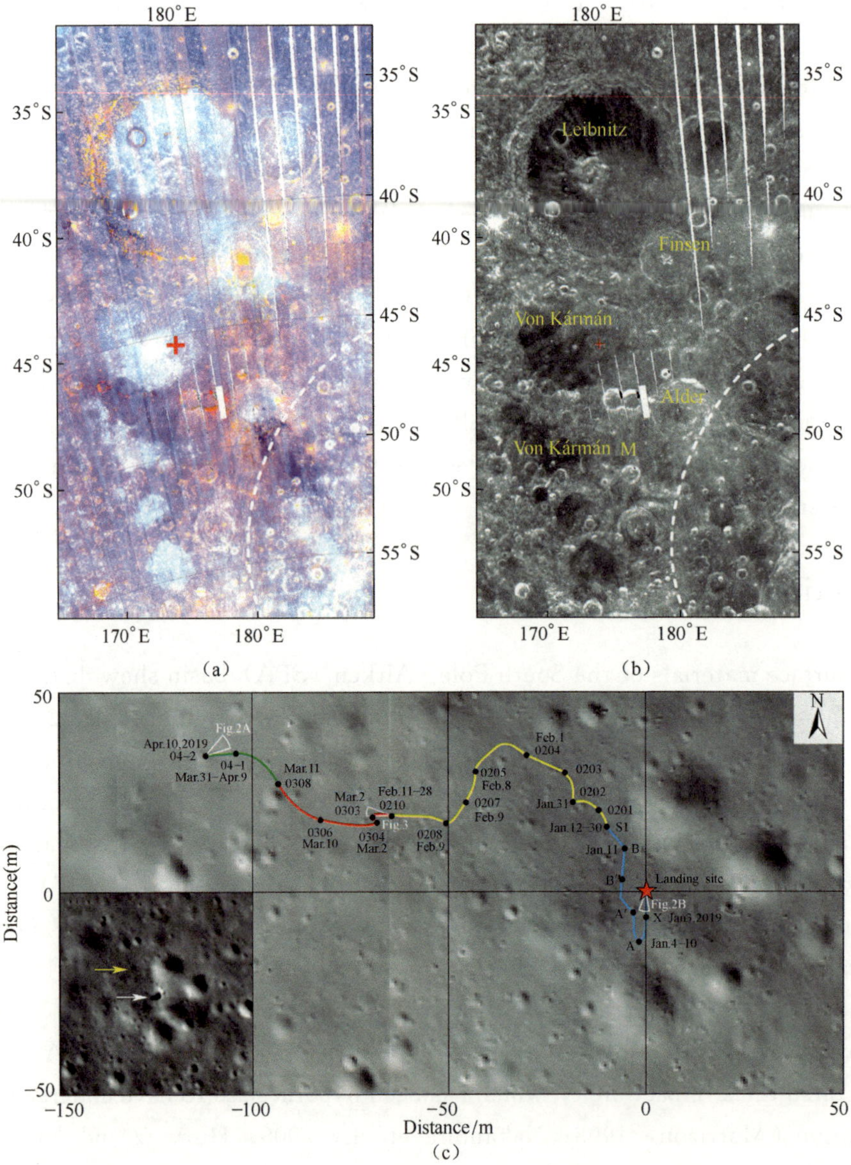

Fig. 1　Geologic context of the Chang'E - 4 (CE - 4) lunar spacecraft mission (Chinese Lunar Exploration Program) landing region and *in situ* exploration traverse route of rover Yutu - 2. (a) Multispectral Imager (MI) color composite mosaic of the Von Kármán region; continuum - removed reflectance spectra of bands at 900nm, 1050nm, and 1250nm are assigned to red, green, and blue channels, respectively. CE - 4 landing point is marked with red cross. (b) MI mosaic of 750nm reflectance. Von Kármán M, Von Kármán, Alder, Finsen, and Leibnitz craters are labeled in yellow. White dashed line is part of outer margins of hypothesized South Pole - Aitken (SPA) impact melt pond (Ohtake et al., 2014). (c) Traverse of Yutu - 2 in first 4 synodic days on landing camera (LCAM) mosaic. Blue, yellow, red, and green lines indicate traverse of synodic day 1, 2, 3, and 4, respectively. Letters and numbers besides black dots are names of navigation points. Red star marks landing point. (d) NASA Lunar Reconnaissance Orbiter Narrow Angle Camera (NAC) image for lander (yellow arrow) and Yutu - 2 (white arrow)

upper mantle (Huang et al., 2018). With the help of a relay satellite (Que Qiao) at the Earth - Sun Lagrangian 2 point (1.5×10^6 km directly 'behind' Earth as viewed from the Sun), the CE - 4 lander and the rover (Yutu - 2) have been conducting successful exploration of the lunar far side for the first time (Wu et al., 2019). The landing site is located in a relatively high - albedo region within Von Kármán crater (Fig. 1b), which can be tracked as LCP - bearing ejecta from Finsen crater, based on orbital reflectance spectra data (Huang et al., 2018). Yutu - 2 traveled 178.9m in the first 4 synodic days, and 24 observations were made by the visible/near - infrared spectrometer (VNIS) along the route (Fig. 1c), of which 18 were analyzed here. Here, we report diverse rock types using the VNIS data and the panorama camera (PCAM) on Yutu - 2 and the landing camera (LCAM) and the terrain camera (TCAM) on the lander.

Imagine Results

The stratigraphic framework of the landing area features a surface layer dominated by the ejecta deposits from Finsen crater and other various - aged nearby craters and the underlying mare basalts (Huang et al., 2018). Images obtained by the LCAM and PCAM show that the landing region is relatively free of large boulders (>1m), indicating a relatively long time of degradation, which is consistent with the estimated thicknesses of regolith and impact crater ejecta (Huang et al., 2018). Craters with diameters of centimeters to tens of meters (Figs. 1c and 2a) were observed. We also identified clusters of centimeter - size rock fragments in the rover exploration region (Fig. 2a), which are most likely shattered clusters of ejecta launched from other impact craters (Schultz and Gault, 1985). Similar occurrences of pebble - decorated small and shallow craters were also observed near the NASA Surveyor 7 landing site (Shoemaker et al., 1969). A few rocks around the lander exhibit lighter or darker tones compared to the regolith and most other surface rocks (Fig. 2b), where the relatively light - toned rocks resemble feldspar - rich rocks (e. g., anorthosite), and the relatively dark rocks resemble mafic - mineral - dominated lunar rocks (e. g., basalt) or meteorites. Unfortunately, the Yutu - 2 did not perform VNIS observations of these interesting targets.

The PCAM instrument observed a group of relatively light - toned rocks near a local depression (Fig. 3a). One of the rocks, which we informally name as "Qi Yuan" ("unexpected encounter" in Chinese), is ~15—20cm in size, with an irregular shape. Qi Yuan looks like a coherent igneous rock, and the bright spots on its relatively fresh surface are possibly either plagioclase phenocrysts or pits created by micrometeorite impacts (Fig. 3b). However, we cannot rule out a breccia origin due to the limited resolution of the image. The VNIS instrument made four attempts to take a spectrum of Qi Yuan, of which only the fourth was successful (Fig. 3c). It is the first *in situ* visible/near - infrared spectrum of a lunar rock (Fig. 4a).

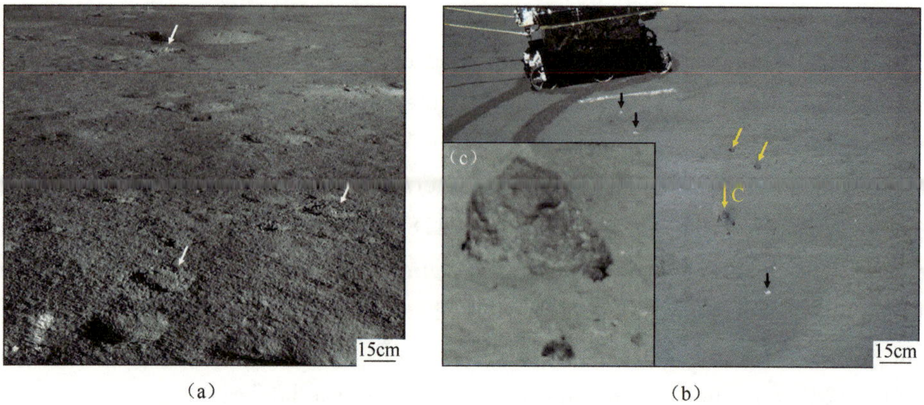

Fig. 2　Panorama camera (PCAM) and terrain camera (TCAM) observations in the Chang'E - 4 (CE - 4) landing region. (a) Craters with different sizes. Clusters of rock fragments (white arrows) were frequently identified. (b) Relatively dark - toned (yellow arrows) and light - toned (black arrows) rocks. (c) Close - up view of dark - toned rock. Light - toned spots are plagioclase phenocrysts or pits created by micrometeorite impacts

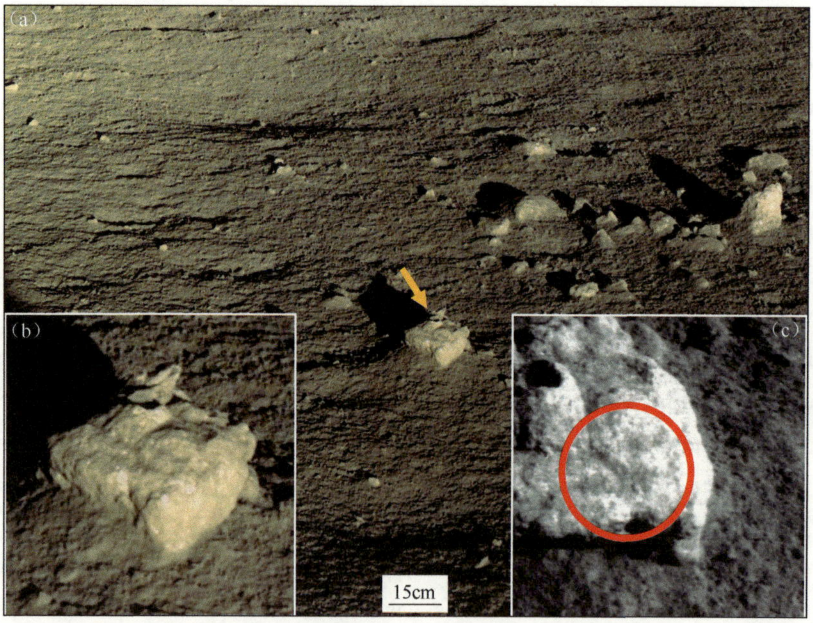

Fig. 3　Panorama camera (PCAM) and visible/near - infrared spectrometer (VNIS) observations of Qi Yuan - the first lunar rock for which a VNIS spectrum was obtained. (a) Group of relatively light - toned rocks. Half - buried rocks were likely exposed due to long - term mass wasting. Qi Yuan is indicated by yellow arrow. (b) Close - up view of Qi Yuan, which looks like a coherent igneous rock; bright spots on its relatively fresh surface are plagioclase phenocrysts or pits created by micrometeorite impacts. (c) VNIS visible image of Qi Yuan. Red circle indicates footprint of VNIS short - wave infrared detector

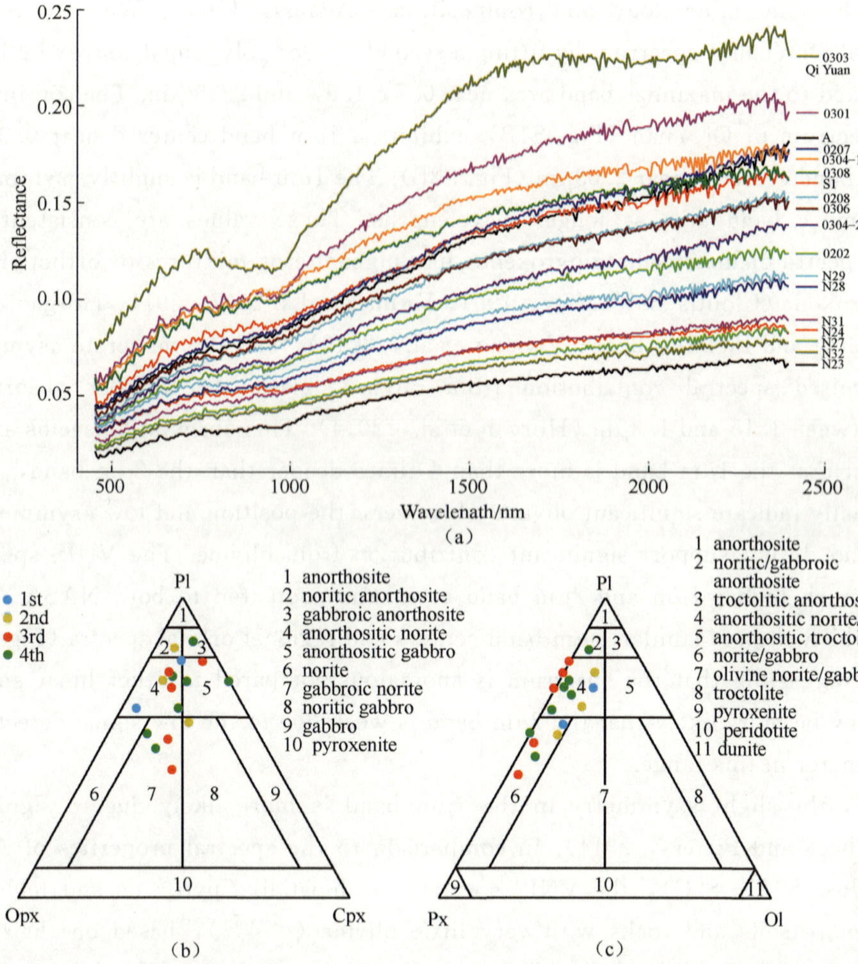

Fig. 4 Visible/near - infrared spectrometer (VNIS) - obtained spectra and look - up table (LUT) mineral abundance of lunar regolith. (a) Spectra of all VNIS observations in first 4 synodic days. (b) LUT mineral abundances on orthopyroxene - clinopyroxene - plagioclase (Opx - Cpx - Pl) ternary diagram. (c) LUT mineral abundances on pyroxene - olivine - plagioclase (Px - Ol - Pl) ternary diagram

In situ Spectroscopy Results

The spectrum of Qi Yuan shows more prominent 1μm and 2μm spectral features (Fig. S1A in the Supplemental Material[1]) than those of the regolith (Fig. 4a), but all of the VNIS - obtained spectra exhibit red slopes due to space weathering (Pieters et al., 2000; Hapke, 2001). The spectral features around 1μm and 2μm are due to an Fe^{2+} crystal field absorption (Burns, 1993), and the center position and shape of the 1 and 2μm bands are

[1] Supplemental Material. Supplemental information and figures. Please visit https: //doi . org/10. 1130/GEOL. 26213S. 12114414 to access the supplemental material, and contact editing@geosociety. org with any questions.

diagnostic of mafic mineralogy and composition (Adams, 1968). We removed the continuum of Qi Yuan's spectrum by fitting a second-order polynomial convex hull to tie points adjusted to the maximize band area near 0.7, 1.5, and 2.35 μm. The continuum-removed spectrum of Qi Yuan (Fig. S1B) exhibits a 1μm band centered near 0.96μm, and a 2 μm band centered near 2.03μm (Fig. S2B). The 1μm band is slightly asymmetric, with 18% more band area at longer wavelengths. These values are consistent with laboratory spectra of intermediate pyroxenes like pigeonite or mixtures of orthopyroxene and clinopyroxene (Cloutis and Gaffey, 1991; Klima et al., 2007, 2011; Horgan et al., 2014; Moriarty and Pieters, 2016; Viviano et al., 2019), and the moderate asymmetry suggests limited spectral contribution from olivine, glass, or feldspar absorptions centered between 1.15 and 1.3μm (Horgan et al., 2014). This observation seems at odds with the fact that the 1μm band is more than 4 times deeper than the 2μm band, which would normally indicate significant olivine. However, the position and low asymmetry of the 1μm band do not support significant contribution from olivine. The VNIS spectrum exhibits a much higher 1μm and 2μm band area ratio compared to both NASA Apollo samples, which showed similar 1 μm band centers and regional orbital spectra (Figs. S2C and S2F), suggesting that the spectrum is anomalous compared to other lunar spectra. Thus, it may be more likely that the 2μm band is weak due to the low signal detected by the spectrometer in this range.

Instead, the slight asymmetry in the 1μm band is more likely due to significant feldspar (Cheek and Pieters, 2014). In comparison to the spectral properties of Apollo samples (Figs. S2A - S2C), the VNIS spectrum is most like pyroxene and feldspar-dominated mare soils and rocks with very little olivine (<4%), based on their band centers and asymmetries (Fig. S1B). Thus, the spectral properties of Qi Yuan appear to be consistent with a pyroxene and feldspar-dominated assemblage with at most minor olivine, where the pyroxene is composed of either pigeonite or a mixture of clinopyroxene and orthopyroxene.

For all the 17 observation of the lunar regolith in the landing region, we removed the continuum of the spectra by subtracting the connecting line of reflectance values between 750 and 1550 nm. Then, we generated a lunar spectral mineral look-up table (LUT) with 17μm grain size (Pieters et al., 1993) and Mg number of 65 (Lucey et al., 2014) using a radiative transfer model (Lucey, 2004; Denevi et al., 2008; Lucey et al., 2014) at the wavelength range of 750—1550nm. Finally, we fit the spectra following the procedures applied to the Chang'E-3 VNIS data (Zhang et al., 2015) to estimate the abundances of plagioclase (Pl), orthopyroxene (Opx), clinopyroxene (Cpx), and olivine (Ol). The method is detailed in the supplementary material. Mineral abundances are the mean values of a comprehensive set of fitting results (Tab. S1). We did not derive the mineral abundances for Qi Yuan using the LUT method, because this method was constructed using only lunar regolith samples, and it might induce large errors and uncertainties in estimation of mineral abundance due to grain size and/or phase angle effects.

The modeled olivine concentrations of all observations were less than 12%, and we plotted the estimated mineral abundances on two ternary diagrams. In the Px - Ol - Pl (Px - pyroxene) ternary diagram (Fig. 4b), 10 out of 17 VNIS observations fall into the field of anorthositic norite/gabbro for intrusive igneous rocks, which have elevated plagioclase concentrations. Six out of 17 VNIS observations fall into norite/gabbro. Only one observation falls into noritic/gabbroic anorthosite with highest plagioclase concentration. In the Opx - Cpx - Pl ternary diagram (Fig. 4c), we can see most observations have higher Opx concentrations than Cpx. These results are consistent with our interpretation for Qi Yuan based on band parameters above, perhaps suggesting a similar origin for Qi Yuan and at least some of the regolith materials. Interestingly, the 7 VNIS spectra on the fourth synodic day were obtained while Yutu - 2 rotated heading without changing its location, so these spectra are all from a very small area. The large compositional variation in an area less than a few square meters is consistent with the idea that the landing region has received multiple sources of various - aged impact ejecta (Huang et al., 2018).

Interpretations and Implications

Various types of rock identified using spectral and imaging data in the landing site are expected from orbital observation (Huang et al., 2018). The dark - toned rocks (Fig. 2b) could originate from the local mare infill excavated by craters penetrating through the overlying ejecta, while the light - toned rocks (Fig. 2b) could have been delivered by craters that impacted the heterogeneous annulus (Moriarty and Pieters, 2018) unit (Fig. 1a). Orbital spectra of the rim and peak of Finsen crater are generally spectrally similar to both VNIS spectra of Qi Yuan and orbital spectra of the landing site (Figs. S1C - S1D, S2D - S2F), which is consistent with an origin for Qi Yuan (and spectrally similar regolith targets nearby) as an ejected block of plagioclase - and pyroxene - rich materials from Finsen crater that has been exposed by mass wasting (Xiao et al., 2013) or a subsequent impact event (Fig. 1d). Alternatively, Qi Yuan could be a shattered ejecta block (Schultz and Gault, 1985) that was delivered by another impact crater in the heterogeneous annulus unit.

The target material of Finsen crater may include both SPA ejecta sourced from the crust and upper mantle (Melosh et al., 2017) as well as SPA impact melts. In addition, the target material has experienced modification by subsequent impact cratering (e. g., the ejecta from Von Kármán and Leibnitz craters). These processes could have led to significant heterogeneity of the target materials for Finsen crater, which is also supported by the compositional variations of the Leibnitz crater rim (Fig. 1a). However, the location of Finsen crater is ~200km away from the outer margins of the hypothesized SPA impact melt pond (Fig. 1b; Pieters et al., 2000; Hurwitz and Kring, 2014; Ohtake et al., 2014). Therefore, the SPA impact melt should have been negligible in the target lithology where the Finsen impact occurred. In the Multispectral Imager color composite, the surface materials (Fig. 1a) at the CE - 4 landing region are more akin to heterogeneous annulus unit materials. Thus, our interpretation of the pyroxene - and plagioclase - rich CE - 4 spectral properties for Qi Yuan and nearby regolith suggests that they may contain a significant amount of highlands crust (e. g., anorthosite) and an SPA ejecta component.

In summary, CE - 4 observations provide the first *in situ* characterization of various types of SPA materials. No olivine - rich mantle materials (Li et al., 2019) have been detected so far, and, instead, the only ultramafic component is Opx mixed with highlands - like materials. Rock samples from the local mare infill and mantle - originated materials should be high - priority targets as Yutu - 2 continues roving to the west.

Acknowledgments

This work was supported by the Strategic Priority Research Program of the Chinese Academy of Sciences (grant XDA17010403), the Natural Science Foundation of China (grants 41773061, 41772050, 41830214, 41773063), the China National Space Administration (CNSA) Pre - research Project on Civil Aerospace Technologies (grant D020101), and the Science and Technology Development Fund of Macau (grants 121/2017/A3, 0042/2018/A2). We thank editor Mark Quigley for editorial handling, and Daniel Moriarty, Hap McSween, and an anonymous reviewer for their constructive comments. CNSA Lunar Exploration and Space Program Center and the Chinese Academy of Sciences Lunar and Deep Space Exploration General Department coordinated data access.

References Cited

[1] Adams, J. B., 1968, Lunar and Martian surfaces: Petrologic significance of absorption bands in the near - infrared: Science, v. 159, p. 1453 - 1455.

[2] Burns, R. G., 1993, Mineralogical Applications of Crystal Field Theory: Cambridge, UK, Cambridge University Press, 224 p., https://doi.org/10.1017/CBO9780511524899.

[3] Cheek, L. C., and Pieters, C. M., 2014, Reflectance spectroscopy of plagioclase - dominated mineral mixtures: Implications for characterizing lunar anorthosites remotely: The American Mineralogist, v. 99, p. 1871 - 1892, https://doi.org/10.2138/am-2014-4785.

[4] Cloutis, E. A., and Gaffey, M. J., 1991, Spectral - compositional variations in the constituent minerals of mafic and ultramafic assemblages and remote sensing

implications: Earth, Moon, and Planets, v. 53, p. 11 - 53, https://doi . org/10.1007/BF00116217.

[5] Denevi, B., Lucey, P., and Sherman, S., 2008, Radiative transfer modeling of near - infrared spectra of lunar mare soils: Theory and measurement: Journal of Geophysical Research: Planets, v. 113, E02003, https://doi . org/10.1029/2007JE002929.

[6] Hapke, B., 2001, Space weathering from Mercury to the asteroid belt: Journal of Geophysical Research: Planets, v. 106, E5, p. 10039 - 10073, https://doi . org/10.1029/2000JE001338.

[7] Horgan, B. H. N., Cloutis, E. A., Mann, P., and Bell, J. F., 2014, Near - infrared spectra of ferrous mineral mixtures and methods for their identification in planetary surface spectra: Icarus, v. 234, p. 132 - 154, https://doi . org/10.1016/j. icarus. 2014.02.031.

[8] Huang, J., Xiao, Z., Flahaut, J., Martinot, M., Head, J., Xiao, X., Xie, M., and Xiao, L., 2018, Geological characteristics of Von Kármán Crater, northwestern South Pole - Aitken basin: Chang'E - 4 landing site region: Journal of Geophysical Research: Planets, v. 123, p. 1684 - 1700, https: //doi . org/10.1029/2018JE005577.

[9] Hurwitz, D. M., and Kring, D. A., 2014, Differentiation of the South Pole - Aitken basin impact melt sheet: Implications for lunar exploration: Journal of Geophysical Research: Planets, v. 119, p. 1110 - 1133, https://doi . org/10.1002/2013JE004530.

[10] Jolliff, B. L., Gillis, J. J., Haskin, L. A., Korotev, R. L., and Wieczorek, M. A., 2000, Major lunar crustal terranes: Surface expressions and crust - mantle origins: Journal of Geophysical Research: Planets, v. 105, E2, p. 4197 - 4216, https://doi . org/10.1029/1999JE001103.

[11] Klima, R. L., Pieters, C. M., and Dyar, M. D., 2007, Spectroscopy of synthetic Mg - Fe pyroxenes I: Spin - allowed and spin - forbidden crystal field bands in the visible and near - infrared: Meteoritics & Planetary Science, v. 42, p. 235 - 253, https://doi . org/10.1111/ j. 1945 - 5100.2007.tb00230.x.

[12] Klima, R. L., Dyar, M. D., and Pieters, C. M., 2011, Near - infrared spectra of clinopyroxenes: Effects of calcium content and crystal structure: Meteoritics & Planetary Science, v. 46, p. 379 - 395, https://doi . org/10.1111/ j. 1945 - 5100.2010.01158.x.

[13] Li, C., Liu, D., Liu, B., Ren, X., Liu, J., He, Z., Zuo, W., Zeng, X., Xu, R., and Tan, X., 2019, Chang'E - 4 initial spectroscopic identification of lunar far - side mantle - derived materials: Nature, v. 569, p. 378 - 382, https://doi . org/10.1038/s41586 - 019 - 1189 - 0.

[14] Lucey, P. G., 2004, Mineral maps of the Moon: Geophysical Research Letters, v. 31, L08701, https://doi . org/10.1029/2003GL019406.

[15] Lucey, P. G., Taylor, G. J., Hawke, B. R., and Spudis, P. D., 1998, FeO and TiO_2 concentrations in the South Pole - Aitken basin: Implications for mantle composition and basin formation: Journal of Geophysical Research: Planets, v. 103, E2, p. 3701 - 3708, https://doi . org/10. 1029/97JE03146.

[16] Lucey, P. G., Norman, J. A., Crites, S. T., Taylor, G. J., Hawke, B. R., Lemelin, M., and Melosh, H. J., 2014, A large spectral survey of small lunar craters: Implications for the composition of the lunar mantle: The American Mineralogist, v. 99, p. 2251 - 2257, https://doi . org/10. 2138/am - 2014 - 4854.

[17] Melosh, H. J., Kendall, J., Horgan, B., Johnson, B. C., Bowling, T., Lucey, P. G., and Taylor, G. J., 2017, South Pole - Aitken basin ejecta reveal the Moon's upper mantle: Geology, v. 45, p. 1063 - 1066, https://doi . org/10. 1130/G39375. 1.

[18] Moriarty, D. P., and Pieters, C. M., 2016, Complexities in pyroxene compositions derived from absorption band centers: Examples from *Apollo* samples, HED meteorites, synthetic pure pyroxenes, and remote sensing data: Meteoritics & Planetary Science, v. 51, p. 207 - 234, https://doi . org/10. 1111/maps. 12588.

[19] Moriarty, D. P., and Pieters, C. M., 2018, The character of South Pole - Aitken basin: Patterns of surface and subsurface composition: Journal of Geophysical Research: Planets, v. 123, p. 729 - 747, https://doi . org/10. 1002/2017JE005364.

[20] Moriarty, D. P., Watkins, R. N., Valencia, S. N., Kendall, J. D., and Petro, N. E., 2019, Mineralogy of thorium - enhanced materials within the South Pole - Aitken basin: Possible traces of the lunar upper mantle, in 50th Lunar and Planetary Science Conference: Houston, Texas, Lunar and Planetary Institute #2874.

[21] Morrison, D. A., 1998, Did a thick South Pole - Aitken basin melt sheet differentiate to form cumulates?, in 29th Lunar and Planetary Science Conference: Houston, Texas, Lunar and Planetary Institute #1657.

[22] Nakamura, R., et al., 2009, Ultramafic impact melt sheet beneath the South Pole - Aitken basin on the Moon: Geophysical Research Letters, v. 36, L22202, https://doi . org/10. 1029/2009GL040765.

[23] Ohtake, M., Uemoto, K., Yokota, Y., Morota, T., Yamamoto, S., Nakamura, R., Haruyama, J., Iwata, T., Matsunaga, T., and Ishihara, Y., 2014, Geologic structure generated by large - impact basin formation observed at the South Pole - Aitken basin on the Moon: Geophysical Research Letters, v. 41, p. 2738 - 2745, https://doi . org/10. 1002/2014GL059478.

[24] Pieters, C. M., Fischer, E. M., Rode, O., and Basu, A., 1993, Optical effects of space weathering: The role of the finest fraction: Journal of Geophysical Research: Planets, v. 98, E11, p. 20817 - 20824, https://doi . org/10. 1029/93JE02467.

[25] Pieters, C. M., Taylor, L. A., Noble, S. K., Keller, L. P., Hapke, B., Morris, R. V., Allen, C. C., McKay, D. S., and Wentworth, S., 2000, Space weathering on airless bodies: Resolving a mystery with lunar samples: Meteoritics & Planetary Science, v. 35, p. 1101 - 1107, https://doi . org/10. 1111/ j. 1945 - 5100. 2000. tb01496. x.

[26] Pieters, C. M., Head, J., Gaddis, L., Jolliff, B., and Duke, M., 2001, Rock types of South Pole - Aitken basin and extent of basaltic volcanism: Journal of Geophysical Research: Planets, v. 106, E11, p. 28001 - 28022, https://doi . org/10. 1029/2000JE001414.

[27] Schultz, P. H., and Gault, D. E., 1985, Clustered impacts: Experiments and implications: Journal of Geophysical Research: Solid Earth, v. 90, B5, p. 3701 - 3732, https://doi . org/10. 1029/JB090iB05p03701.

[28] Shoemaker, E. M., Batson, R. M., Holt, H. E., Morris, E. C., Rennilson, J., and Whitaker, E. A., 1969, Observations of the lunar regolith and the Earth from the television camera on Surveyor 7: Journal of Geophysical Research, v. 74, p. 6081 - 6119, https://doi . org/10. 1029/JB074i025p06081.

[29] Vaughan, W. M., and Head, J. W., 2014, Impact melt differentiation in the South Pole - Aitken basin: Some observations and speculations: Planetary and Space Science, v. 91, p. 101 - 106, https://doi . org/10. 1016/ j. pss. 2013. 11. 010.

[30] Viviano, C., Murchie, S. L., Daubar, I. J., Morgan, M. F., Seelos, F. P., and Plescia, J. B., 2019, Composition of Amazonian volcanic materials in Tharsis and Elysium, Mars, from MRO/CRISM reflectance spectra: Icarus, v. 328, p. 274 - 286, https://doi . org/10. 1016/ j. icarus. 2019. 03. 001.

[31] Wu, W., et al., 2019, Lunar farside to be explored by Chang'E - 4: Nature Geoscience, v. 12, p. 222 - 223, https://doi . org/10. 1038/s41561 - 019 - 0341 - 7.

[32] Xiao, Z., Zeng, Z., Ding, N., and Molaro, J., 2013, Mass wasting features on the Moon - How active is the lunar surface?: Earth and Planetary Science Letters, v. 376, p. 1 - 11, https://doi . org/10. 1016/ j. epsl. 2013. 06. 015.

[33] Yamamoto, S., Nakamura, R., Matsunaga, T., Ogawa, Y., Ishihara, Y., Morota, T., Hirata, N., Ohtake, M., Hiroi, T., and Yokota, Y., 2012, Olivine - rich exposures in the South Pole - Aitken basin: Icarus, v. 218, p. 331 - 344, https://doi. org/10. 1016/ j. icarus. 2011. 12. 012.

[34] Zhang, H., Yang, Y., Yuan, Y., Jin, W., Lucey, P. G., Zhu, M. H., Kaydash, V. G., Shkuratov, Y. G., Di, K., and Wan, W., 2015, In situ optical measurements of Chang'E - 3 landing site in Mare Imbrium: 1. Mineral abundances inferred from spectral reflectance: Geophysical Research Letters, v. 42, p. 6945 - 6950, https://doi . org/10. 1002/2015GL065273.

In situ lunar phase curves measured by Chang'E 4 in the Von Kármán Crater, South Pole – Aitken basin

Te Jiang[1], Xiaoyi Hu[1], Hao Zhang[1,2], Pei Ma[1], Chunlai Li[3], Xin Ren[3], Bin Liu[3], Dawei Liu[3], Jianfeng Yang[4], Bin Xue[4], Weidong Jin[5], Meng – Hua Zhu[6], Changning Huang[7], and Hongyu Lin[7]

1. Planetary Science Institute, School of Earth Sciences, China University of Geosciences, Wuhan, PR China

e—mail: zhanghao@cug. edu. cn

2. CAS Center for Excellence in Comparative Planetology, Hefei, PR China

3. Key Laboratory of Lunar and Deep Space Exploration, National Astronomical Observatories, Chinese Academy of Sciences, Beijing, PR China

e—mail: liudw@nao. cas. cn

4. Xi'an Institute of Optics and Precision Mechanics, Chinese Academy of Sciences, Xi'an, PR China

5. Department of Physics & Astronomy, The University of Alabama, Tuscaloosa, AL, USA

6. State Key Laboratory of Lunar and Planetary Sciences, Macau University of Science and Technology, Taipa, Macau

7. Beijing Institute of Space Mechanics and Electronics, China Academy of Space Technology, Beijing, PR China

Abstract: Context. The Yutu – 2 rover of the Chang'E – 4 (CE – 4) mission measured the lunar phase curves in the Von Kármán crater, South Pole – Aitken basin.

Aims: We aim to study the photometric properties of the regolith at CE – 4's landing site and compare them with those of Chang'E – 3 (CE – 3) in order to understand the regolith physical properties of the two landing sites.

Methods: We extracted the in situ lunar phase curves measured by CE – 4 with a very wide phase angle coverage (1°—144°) and performed photometric model inversions using both the Hapke model and the Lumme – Bowell model.

Results: Compared with the CE – 3 measurement taken in Mare Imbrium, the CE – 4 phase curves show the colorimetric opposition effect and have a steeper and narrower opposition spike. The surface regolith at the CE – 4 site is much darker, more porous, more forward scattering, and has a larger slope angle (Hapke model) than that of CE – 3.

Conclusions: The CE – 4 site may have experienced more space weathering alterations than the CE – 3 site, which is consistent with their different surface model ages (~3.6 Ga for CE – 4 and ~3 Ga for CE – 3).

Keywords: Moon – planets and satellites; surfaces – radiative transfer – techniques; photometric

1　Introduction

Compared to reflectance spectroscopy, which is sensitive to mineralogy and chemical compositions, the phase curve of an airless body is more sensitive to regolith physical properties such as particle shape and transparency, size distribution, packing density, and macroscopic and microscopic surface roughness. The nonlinear increase of surface brightness in the anti – solar direction (the opposition effect (OE)), caused by shadow hiding and/or coherent back – scattering, is useful when inferring the regolith opaqueness, porosity, and particle size distribution (e. g., Helfenstein et al. 1997; Abe et al. 2006; Mishchenko 2008; Shkuratov et al. 2011; Hapke 2012). Various photometric models have been developed to describe light scattering properties of the surface regolith (e. g., Lumme & Bowell 1981; Muinonen 2004; Hapke et al. 2012; Mishchenko et al. 2015), and it has been recognized that a wide coverage of the solar phase angle is a crucial to reliably retrieving model parameters from the measurement.

Since the successful landing of the Chang'E – 4 (CE – 4) spacecraft in the Von Kármán crater (177.588°E, 45.4578°S) on the floor of the South Pole – Aitken basin (Fig. A. 1) on Jan. 3, 2019 (Di et al. 2019a), the panorama cameras (PCAMs) onboard the Yutu – 2 rover have extensively imaged the landing area. This has provided a good opportunity to study the regolith photometric properties in addition to its spectroscopic properties (e. g., Li et al. 2019; Hu et al. 2019). From this work, we extracted the in – situ lunar phase curves using the PCAM images and compared them with the Chang'E – 3 (CE – 3) measurement made in Mare Imbrium (Jin et al. 2015), which was the first of its kind since the Apollo era. By performing photometric model inversions using both the Hapke model (Hapke 2012) and the Lumme – Bowell model (Lumme & Bowell 1981, LB model hereafter), we infer the regolith physical properties and compare them with those of CE – 3.

2　Instrument and data descriptions

As its backup payload, the Yutu – 2 rover has a PCAM system identical to that of CE – 3 (Jin et al. 2015). As shown in Fig. A. 2, the left PCAM (PCAML) and right PCAM (PCAMR) are located on top of the mast and can rotate horizontally (0°—360° yaw) and vertically (—30°—0°pitch), with their horizontal pointing directions turned toward each other by 1°. The PCAMs can work in color and panchromatic modes with a field of view of 19.6° by 14.5° (Jia et al. 2018).

A more detailed description of the PCAM data product can be found in Ren et al. (2014). In brief, each level 2B (L2B) digital number (DN) image was obtained by applying a series of operations on level 0 raw data including format processing, dark current subtraction, and flat fielding correction. The gain, exposure time, and geometric information are also included in each 2B image as the header file. The geometric information includes the solar zenith and azimuth angles, the exterior orientation elements (T1), and the inner orientation elements (T2). Each L2B frame in color mode is 2352 by 1728 pixels and consists of four color bands with a Bayer filter mosaic arrangement: red (R, 640nm), green (G1, 540nm), green (G2, 540nm), and blue (B, 470nm). As the PCAM data with higher levels above L2B have undergone color interpretations and are released as color photos, they are not suitable for quantitative photometric analysis. Unfortunately, most PCAM data measured so far have been taken in panchromatic mode and thus we have selected the L2B data collected on the third lunar day for phase curve extractions.

Fig. 1 shows a panoramic view of the study area, which is about 75 m west of the lander. This picture is a mosaic of 43 overlapping images taken by the PCAMR during the

Fig. 1　Panoramic view of the study area imaged by the PCAMR with image IDs listed in Tab. C. 1. The opposition effect is seen centred around the shadow of the PCAMR. The average solar zenith angle for these images is 80°

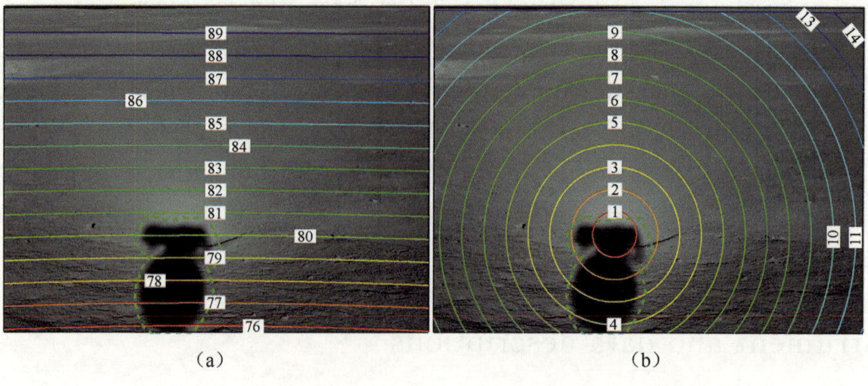

Fig. 2　Example contour plots of the calculated viewing zenith angles (a) and phase angles (b) in phase curve extractions. Regions surrounded by dashed green lines are treated as shadows and are not used in phase curve extractions

third lunar day, and the image IDs are listed in Tab. C. 1. When selecting these images, we avoided pictures with the presence of obvious solar glares (Fig. A. 3). A topographic analysis of the landing site (Di et al. 2019b) shows that the terrain is generally higher in the northeast than in the southwest, with an elevation drop of ~8m over 350m distance, corresponding to a slope angle of about 1°. Although there are decimeter-sized craters in the scene, the study area around the rover is otherwise flat, and therefore we can treat the surface as macroscopically flat with an identical solar zenith angle of 80° for all images in the mosaic. The actual solar zenith angle in various images ranges from 79.6635° to 80.2584°.

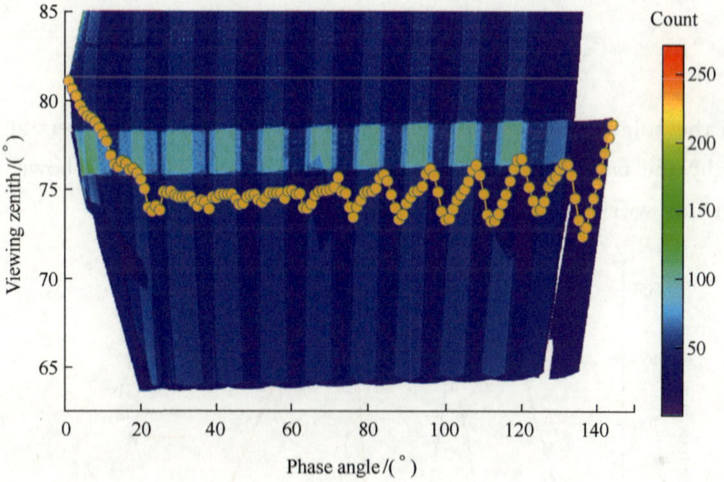

Fig. 3　Extracted viewing zenith angles for all useful pixels (color dots represent pixel densities with a bin width of 0.15 in the phase angle axis and 0.021 in the viewing zenith axis) and the averaged values within a 1° phase-angle bin (yellow circles)

3　Phase-curve extraction

We followed the same procedure to extract the in situ lunar phase curves as we did for the CE-3 data (Jin et al. 2015) as summarized below:

First, we used T1 and T2 to calculate the vector (V1) from a pixel to the PCAMs through the collinearity equations and obtained the viewing zenith angle (e) for each pixel (Fig. 2a). Second, we used the solar zenith angle and solar azimuth angle in each image's header file to calculate the vector (V2) from the pixel to the Sun. Then, the angle between V1 and V2, or the solar phase angle (α), can be obtained using vector algebra (Fig. 2b). Third, we removed pixels from the shadow of the rover (in the rover image) and pixels with e larger than 85° in all frames due to their lower collection efficiencies. The retrieved values are plotted as gray dots in Fig. 3, and those averaged within a 1° phase-angle bin are shown as the yellow dots. Fourth, we converted the L2B DN value of each pixel in the frame to radiance (I_s) with the brightness normalization parameter A_{Cali} and the absolute radiation calibration coefficients C_{Rad} provided by the payload team:

$$I_s = \frac{DN}{A_{Cali} \cdot C_{Rad}} \tag{1}$$

where A_{Cali} is recorded in each header file, and C_{Rad} (in Sr nm m^2/W) is 7.667×10^5, 9.893×10^5, 9.893×10^5, and 9.705×10^5 for the R, $G1$, $G2$, and B bands, respectively. Fifth, we averaged the radiance values of pixels within a 1° phase angle to reduce albedo variations caused by surface heterogeneities. Finally, we used the solar irradiance $J(\lambda)$ (Gueymard 2004) to convert radiance to reflectance factor (*REFF*, r_s, Hapke 2012) as follows:

$$r_s = \frac{\pi I_s}{\cos i \int J(\lambda) S(\lambda) d\lambda} \tag{2}$$

where i is the solar zenith angle, and $S(\lambda)$ is the normalized spectral response of the sensor provided by the payload team (the original response curves are shown in Fig. 4). All phase curves in this work are presented in terms of the *REFF*.

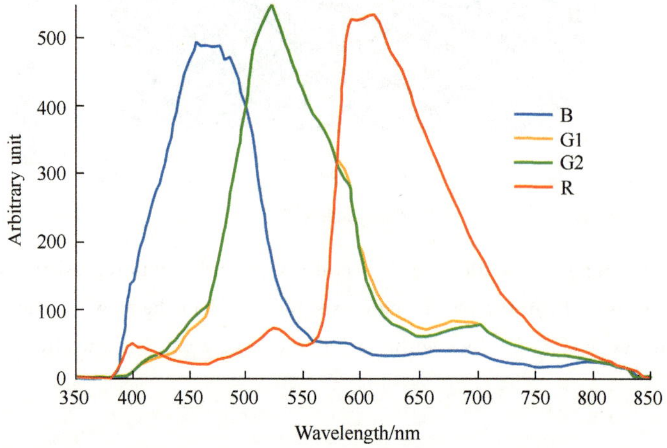

Fig. 4　Spectral response functions of PCAM sensors provided by the payload team. B, G1, G2, and R represent the blue, green 1, green 2, and red pixels, respectively. The area of each of these functions was normalized to 1 when deriving the reflectance factor using Eq. (2)

Fig. 5 shows the comparisons of the CE－3 (Jin et al. 2015) and CE－4 phase curves in red, the CE－4 phase curves in R, G, and B, and the color ratios of CE－4. The gray dots in Fig. 5a are the original reflectance values, and the red curve with error bars is the averaged one within a 1° phase－angle bin for CE－4. Compared with that of the landing site of CE－3 in Mare Imbrium, the CE－4 phase curve has a lower absolute value (Fig. 5a) and a narrower opposition surge (Fig. 5b), implying the CE－4 site is more porous and/or the average CE－4 regolith is more transparent (e. g., Abe et al. 2006).

To remove the effects of different incident and view zenith angles between CE－3 and CE－4, in Figs. 5c－d we plot the reduced reflectance r_r (Hapke 2012) as

$$r_r(i, e, \alpha) = \frac{I/F}{\mu_0/(\mu_0+\mu)} = REFF \cdot (\mu_0+\mu) \tag{3}$$

where I/F is the radiance factor, and μ_0 and μ are the cosine of i and e, respectively. Fig. 5c shows that the reduced reflectance of CE - 3 is about four - times higher than that of CE - 4. After normalizing at 2° phase angle, the two curves still have marked differences (Fig. 5d), implying the different regolith physical properties of the two sites.

Many airless bodies have a steeper phase curve at shorter wavelength or have a redder reflectance spectrum at larger phase angles. This phase reddening effect has been observed for the Moon by Earth - based telescopes (e. g., Gehrels et al. 1964; McCord 1969), and orbiter (e. g., Hapke et al. 2012) and in situ measurements (Jin et al. 2015), and it has also been observed in laboratory experiments (e. g., Adams & Filice 1967; Johnson et al. 2013; Pommerol et al. 2013; Schröder et al. 2014). The color ratio spectra of CE - 4 shown in Fig. 5f clearly show the phase reddening effect: the curves first drop from 1° to 3° and then increase sharply from 3° to about 40° and gradually approach a peak around 100° before decreasing again. The increase at phase angles larger than ~130° is similar to what happens in the CE - 3 case (Jin et al. 2015) but is more pronounced. The phase curves in the G and B bands have error levels commensurate with the red one, which for clarity are not shown in Fig. 5e. The uncertainties of the color ratio curves (R/G and R/B), estimated from error propagation equations, are better than ~40%, which is the upper limit in case the covariances between the R, G, and B bands are zero.

4 Photometric model inversions

In order to compare the CE - 3 and CE - 4 landing - site photometric properties, we used the Hapke model (Hapke 2012) and the LB model (Lumme & Bowell 1981) to invert the reflectance phase curve. Although the physical significance of the model parameters has been a constant subject of debate (Mishchenko 1994; Hapke 1996, 2013; Hapke et al. 2009; Zhang & Voss 2011; Shkuratov et al. 2012), by comparing the fitting results of CE - 3 and CE - 4, useful conclusions in comparative planetology may be made. A comparison of similar parameters from different models may also help to verify the modeling results.

4.1 The Hapke model

The Hapke model in reflectance factor ($REFF$) is given as (Hapke 2012; Hapke et al. 2012)

$$r_s(i, e, \alpha) = K\frac{\tilde{\omega}_0}{4\mu_0}\frac{\mu_{0e}}{\mu_{0e}+\mu_e}\{[1+B_S(\alpha)]P(\alpha)$$
$$+H(\mu_{0e}/K)H(\mu_e/K)-1\}[1+B_C(\alpha)]S(i, e, \psi, \bar{\theta}) \tag{4}$$

where K is the porosity coefficient, $\tilde{\omega}_0$ is the single - scattering albedo (SSALB),

u_0 is the cosine of the incident angle (i), and μ_{0e} and μ_e are cosines of the effective incident angle and the effective viewing zenith angle, respectively. To account for both backward - and forward - scattering lobes, a two - term Henyey - Greenstein phase function is used to approximate single - scattering phase function $P(\alpha)$ as

$$P(\alpha)=\frac{1+c}{2}\frac{1-b^2}{(1-2b\cos\alpha+b^2)^{3/2}}+\frac{1-c}{2}\frac{1-b^2}{(1+2b\cos\alpha+b^2)^{3/2}} \tag{5}$$

with a corresponding asymmetry parameter $\zeta=-bc$. The Ambartsumian - Chandrasekhar H function is approximated by

$$H(x)\approx\frac{1}{1-\tilde{\omega}_0 x\left(r_0+\frac{1-2r_0x}{2}\ln\left(\frac{1+x}{x}\right)\right)} \tag{6}$$

where r_0 is the diffusive reflectance given by

$$r_0=\frac{1-\sqrt{1-\tilde{\omega}_0}}{1+\sqrt{1-\tilde{\omega}_0}}. \tag{7}$$

The functions $B_S(\alpha)$ and $B_C(\alpha)$ describe the shadow - hiding opposition effect (SHOE) and the coherent - back - scattering opposition effect (CBOE) at small phase angle region as

$$B_S(\alpha)=\frac{B_{S0}}{1+\frac{1}{h_S}\tan\left(\frac{\alpha}{2}\right)} \tag{8}$$

$$B_C(\alpha)=B_{C0}\frac{1+[1-\exp(-\gamma)]/\gamma}{2(1+\gamma)^2} \tag{9}$$

$$\gamma=\frac{\tan(\alpha/2)}{h_C} \tag{10}$$

and $S(i, e, \psi, \bar{\theta})$ is the surface roughness factor with a mean slope angle $\bar{\theta}$ (Hapke 2012). For equant particles much larger than the radiation wavelength, K can be expressed as

$$K=-\frac{\ln(1-1.209\phi^{2/3})}{1.209\phi^{2/3}} \tag{11}$$

where the ϕ (= 1 − porosity) is the filling factor. For equant particles larger than the radiation wavelength and with a narrow particle size distribution, Eq. (9.26) from Hapke (2012) can be used to estimate the filling factor from h_S:

$$h_S=\frac{3}{8}K\phi \tag{12}$$

With the assumption that the regolith obeys a power - law size distribution (McKay et al. 1974), h_S can be used to estimate the particle size distribution as

$$h_S=-0.537\frac{\ln(1-1.209\phi^{2/3})}{\ln(d_l/d_s)}\phi^{1/3} \tag{13}$$

where d_l and d_s are the diameters of the largest and smallest grains, respectively.

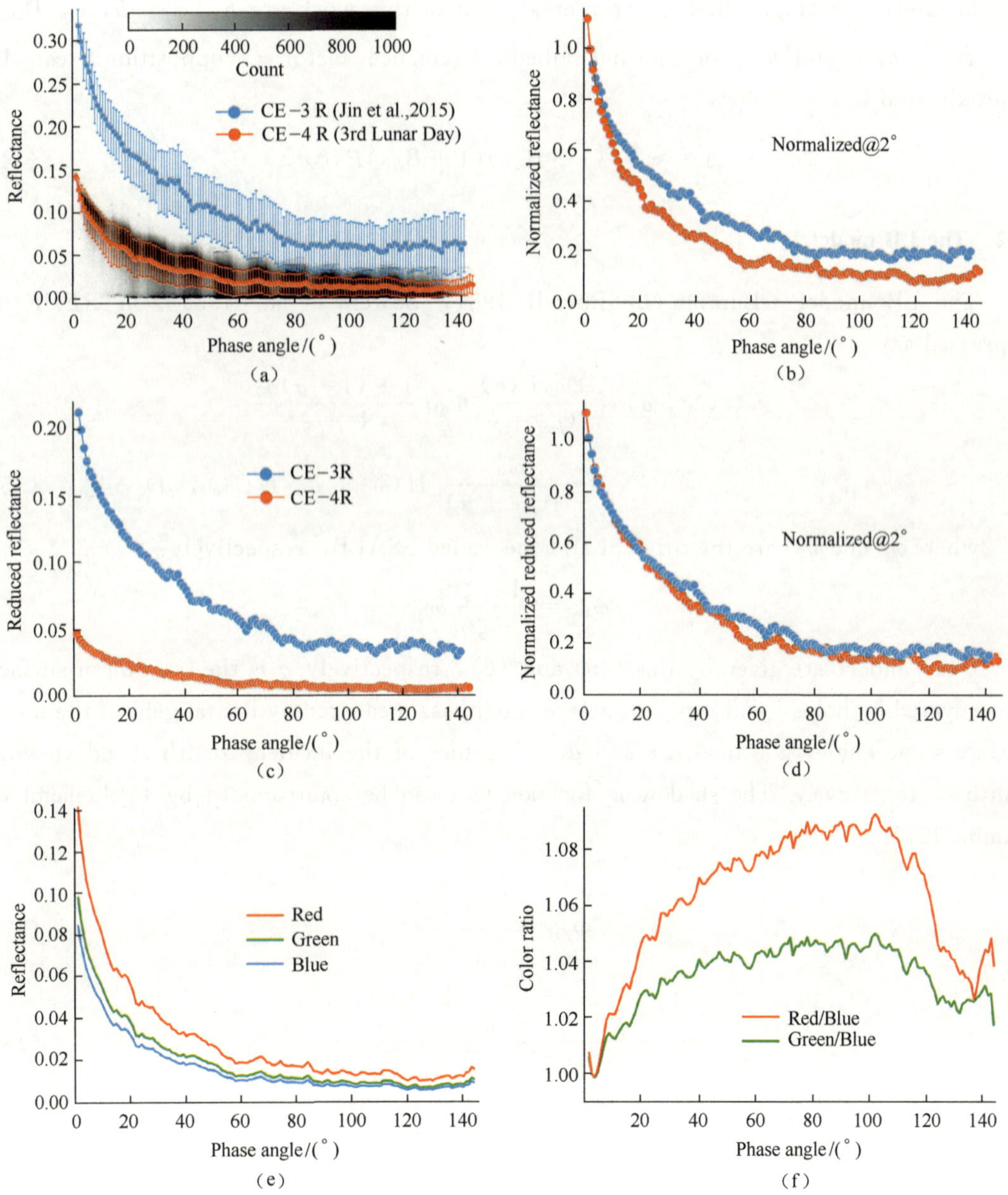

Fig. 5　Extracted phase curves and color ratio curves

(a) Comparisons of the CE－4 (orange, this work) and CE－3 (blue, Jin et al. 2015) phase curves in the *R* band (640 nm). For CE－4 data, the gray color dots represent pixel densities (with a bin width of 0.375 on the phase－angle axis and 0.000375 on the reflectance axis) of the original reflectance values and the red curve with error bars is the averaged value within a 1° phase angle; (b) reflectance in the *R* band (640 nm) normalized at a 2°phase angle. We note that the minimum phase angle of the CE－4 phase curve is 1°, while that of CE－3 is 2°; (c) Reduced reflectance values obtained from Eq. (3); (d) reduced reflectance normalized at a 2°phase angle; (e) extracted phase curves in three colors for CE－4: red (R, 640 nm), green (G, the average of two 540 nm channels G1 and G2), and blue (B, 470 nm); (f) color ratios of red over blue and green over blue for CE－4

In brief, the Hapke fitting parameters used in this work are K, $\tilde{\omega}_0$, b, c, B_{S0}, h_S, B_{C0}, h_C, and $\bar{\theta}$. The normal albedo (radiance factor at opposition) can be approximated by

$$A_N \approx \frac{\tilde{\omega}_0}{8}(1+B_{C0})(1+B_{S0})P(0) \tag{14}$$

4.2 The LB model

The LB model (Lumme & Bowell 1981; Bowell et al. 1989) in *REFF* is expressed as:

$$r_s(i, e, \alpha)=\frac{2\bar{\omega}_0 P(\alpha)}{4(\mu_0+\mu)}\Phi_{SH}\frac{1+(1-\sigma)\rho\xi}{1+\rho\xi}+\frac{\tilde{\omega}_0^*}{4(\mu_0+\mu)}[H(\tilde{\omega}_0^*, \mu_0)H(\tilde{\omega}_0^*, \mu)-1] \tag{15}$$

where $\tilde{\omega}_0$ and $\tilde{\omega}_0^*$ are the original and the scaled SSALB, respectively:

$$\tilde{\omega}_0^* = \frac{1-\zeta}{1-\zeta\tilde{\omega}_0}\tilde{\omega}_0 \tag{16}$$

$P(\alpha)$ and H are given by Eqs. (5) and (6), respectively. σ is the fraction of surface area covered by holes, and ρ is the surface roughness measured by the tangent of the mean surface slope angle, $\rho=\tan\bar{\theta}$. μ_0 and μ are cosines of the incident zenith i and viewing zenith e, respectively. The shadowing function Φ_{SH} can be approximated by (Peltoniemi & Lumme 1992):

$$\Phi_{SH} = \frac{y+\frac{3}{4}}{y+\frac{3}{2}} \tag{17}$$

$$y=\frac{\phi}{2.38}\frac{\mu_0+\mu}{\sqrt{\mu_0^2+\mu^2-2\mu_0\mu\cos\alpha}} \tag{18}$$

$$\xi=\frac{\sqrt{\mu_0^2+\mu^2-2\mu_0 \cdot \mu\cos\alpha}}{\mu_0\mu}$$

where ϕ is the filling factor. The LB model in this form has six free parameters: $\tilde{\omega}_0$, b, c, ϕ, ρ, and σ.

4.3 Model fitting and results

Since curve fitting to multivariable models can be highly dependent on initial guess values, we used the following procedures to find the most likely fitting parameters.

For the Hapke model (Appendix B), we first fit the data in the phase-angle range of 20°−144°, where the opposition effects are negligible with regard to obtaining the initial guess values of K, $\tilde{\omega}_0$, b, c, and $\bar{\theta}$. We find that the fit K value was consistently below 1, which is outside its valid domain (Hapke 2012), and therefore we set the constraint $K \geqslant 1$ in all following steps. Next, we extended the fitting to the full range [1°—144°] to include the opposition surge [1° — 20°] to estimate all nine parameters including B_{S0}, h_S, B_{C0}, and h_C. The resultant negative B_{C0} value (−0.05) implies that the CBOE may have no contribution to the OE. Also, as is shown in Fig. B. 5, multiple scattering contributes less than 14% to total reflectance at small phase angles. Thus we removed CBOE and the model parameter number was reduced to seven. Then a grid search took place with the initial guess values derived from the above procedures as ([lower, upper; step]; Appendix B) $K = [1.0, 1.3; 0.1]$, $\tilde{\omega}_0 = [0.05, 0.20; 0.05]$, $b = [0.3, 0.5; 0.1]$, $c = [-0.4, 0.3; 0.1]$, $\bar{\theta} = [26°, 38°; 2°]$, and $B_{S0} = [0.4, 0.9; 0.1]$, $h_S = [0.01, 0.15; 0.02]$, with constraints $K \geqslant 1$, $0 < b < 1$, and $-1 < c < 1$. The histograms of the seven resultant parameters are shown in Figs. 6a—g. The narrow histograms are likely caused by the constraint we set for K. Finally, the peak position values of the histograms were used as the initial guess values and the converged fitting results are as follows: $K = 1.0 \pm 2.9$, $\tilde{\omega}_0 = 0.11 \pm 0.28$, $b = 0.35 \pm 0.04$, $c = -0.21 \pm 0.06$, $\bar{\theta} = 30.0° \pm 4.5°$, $B_{S0} = 0.28 \pm 0.16$, $h_S = 0.023 \pm 0.013$, and $A_N = 0.026 \pm 0.068$, and the fit curve is shown in Fig. 6h. In comparison, the nine-parameter fitting results for the CE-3 data are as follows: $K = 1.47^{+0.16}_{-0.07}$, $\tilde{\omega}_0 = 0.19^{+0.1}_{-0.3}$, $b = 0.30 \pm 0.1$, $c = 0.08^{+0.12}_{-0.08}$, $\bar{\theta} = 17°^{+1°}_{-6°}$, $B_{S0} = 0.52^{+0.3}_{-0.2}$, $B_{C0} = 0.47^{+0.15}_{-0.17}$, $h_S = 0.059 \pm 0.01$, $h_C = 0.12 \pm 0.01$, and $A_N = 0.15^{+0}_{-0.05}$ (Jin et al. 2015).

For the LB model, we first applied a grid search with initial values of $\tilde{\omega}_0 = [0.15, 0.35; 0.02]$, $b = [0.1, 0.5; 0.1]$, $c = [-0.5, 0.5; 0.1]$, $\phi = [0.1, 1; 0.1]$, slope $= [3°, 35°; 3°]$, $\sigma = [0.1, 1; 0.1]$ for CE-3 and $\tilde{\omega}_0 = [0.01, 0.15; 0.02]$, $b = [0.1, 0.5; 0.1]$, $c = [-0.5, 0.5; 0.1]$, $\phi = [0.1, 1; 0.1]$, slope $= [3°, 35°; 3°]$, and $\sigma = [0.1, 1; 0.1]$ for CE-4. We also set the following constraints: $0 < \tilde{\omega}_0 < 1$, $0 < D < 1$, $0 < \sigma < 1$, according to their physical meanings. The histogram distributions of each resulting parameter are shown in Figs. 7a—f.

The most probable values obtained are $\tilde{\omega}_0 = 0.25$; $b = 0.31$; $c = 0.38$; $\phi = 0.12$; slope$=17.8°$; $\sigma = 0.99$ for CE-3, and $\tilde{\omega}_0 = 0.040$, $b = 0.32$, $c = 0.40$, $\phi = 0.0010$, slope$=14.8°$, $\sigma = 0.99$ for CE-4. These values were then used as initial guess values to fit the data, and the final fitting results are as follows: $\tilde{\omega}_0 = 0.22 \pm 0.01$, $b = 0.32 \pm 0.01$, $c = 0.58 \pm 0.02$, $\phi = 0.17 \pm 0.11$, $\rho = 2.0 \pm 1.9$ (with a corresponding slope angle of $63.4°^{+12.2°}_{-57.7°}$), $\sigma = 0.34 \pm 0.06$ for CE-3 and $\tilde{\omega}_0 = 0.080 \pm 0.017$; $b = 0.32 \pm 0.04$; $c = 0.42 \pm 0.27$; $\phi = 0.0095 \pm 0.0112$; $\rho = 0.32 \pm 0.25$ (with a corresponding slope angle of $17.7°^{+12.0°}_{-13.7°}$), $\sigma = 0.95 \pm 0.36$ for CE-4 as shown in Figs. 7h and 7i. The factor that caused the drop at small phase angles for CE-4 (Fig. 7i) is unknown. We noticed that this dip would disappear if the incident zenith angle were changed slightly (e. g., from 80° to 80.5°), but then it would appear at other slightly different values (e. g., 80.4°). Therefore, we did not attempt to make the fit look better by fine-tuning the incident angle, but we fixed the incident angle at 80°.

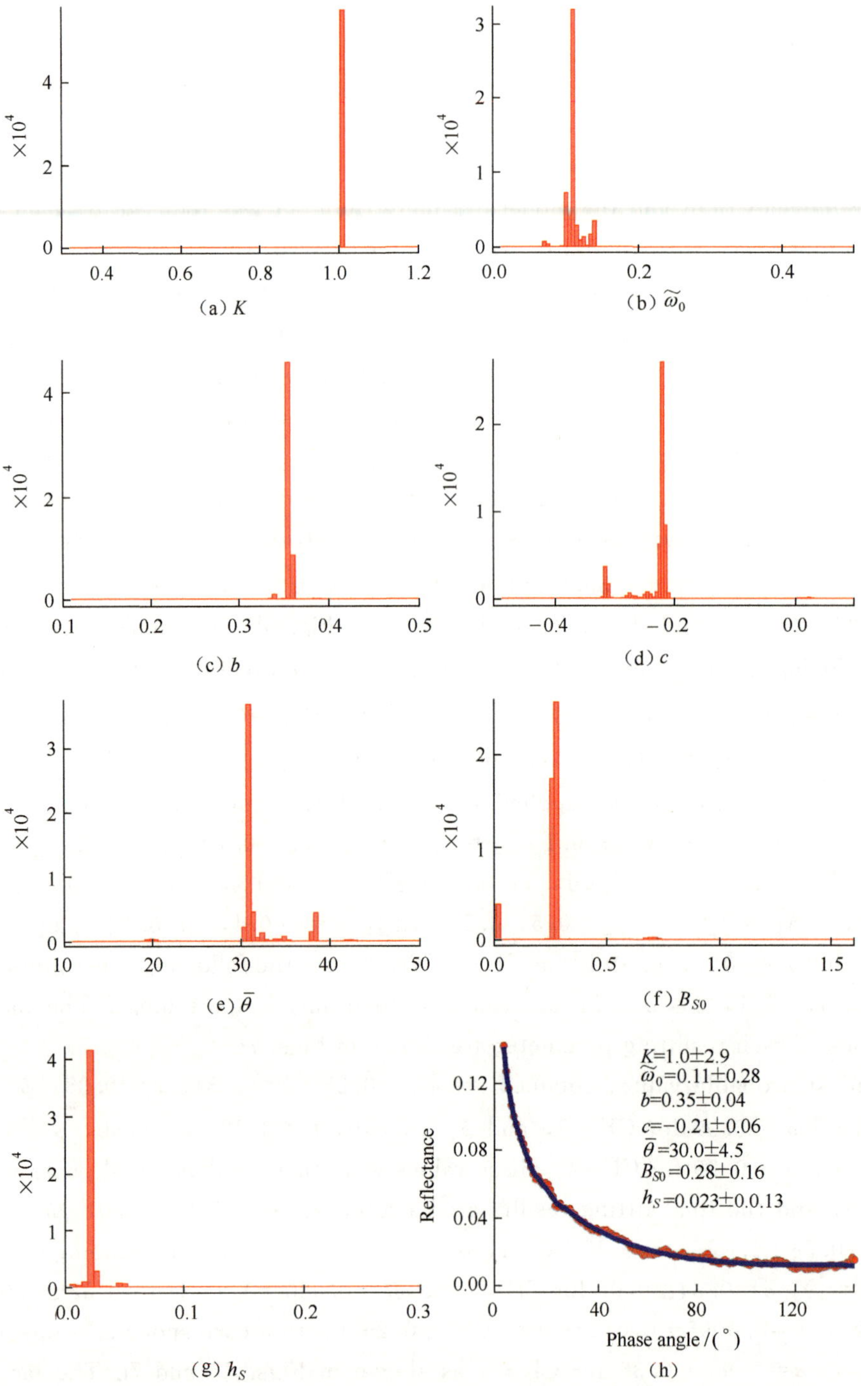

Fig. 6　Histograms of the fitted parameters of the seven-parameter Hapke model over the full phase angle range (1°—144°) with no CBOE contribution ($B_C(\alpha)=0$) and with the constraint $K \geqslant 1$: (a) porosity factor K; (b) single scattering albedo $\widetilde{w}_0$; (c) and (d) are b and c in Eq. (5); (e) slope angle $\bar{\theta}$; (f) and (g) are B_{S0} and h_S in Eq. (8), respectively. The peak values of the (a) — (g) are then used as the initial guess values and the fitting results are shown in (h)

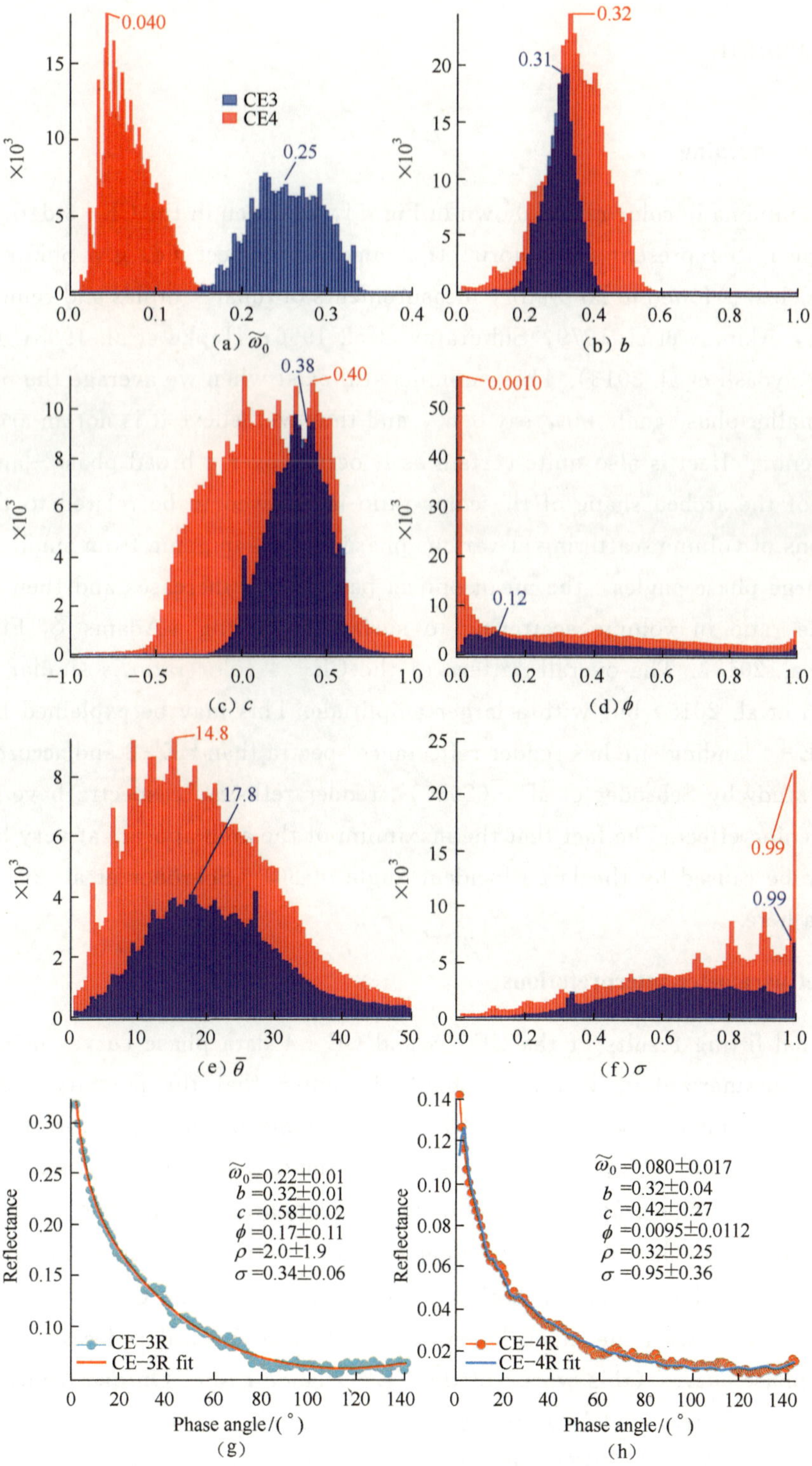

Fig. 7　Fitting results of Lumme - Bowell model on CE - 3's and CE - 4's R - band data. Histograms of the grid searching results are shown in (a) to (f) as (a) $\tilde{\omega}_0$, (b) b, (c) c, (d) ϕ, (e) slope angle (°), and (f) σ. The fitting curves of (h) CE - 3 and (i) CE - 4 are also shown. ρ is the tangent of the slope angle, $\rho = \tan\overline{\theta}$

5 Discussion

5.1 Phase reddening

The 3° minima in color ratios shown in Fig. 5f, not seen in the CE - 3 data (Jin et al. 2015), appear to represent the colorimetric opposition effect (e. g., Shkuratov et al. 2011), which was found in laboratory measurements of lunar samples and remote sensing data (e. g., Akimov et al. 1979; Shkuratov et al. 1996; Hapke et al. 1998; Johnson et al. 2013; Kaydash et al. 2013). These minima still exist when we average the pixel values within a smaller phase angle bin, say 0.5°, and thus we believe it is not an artifact. The phase reddening effect is also quite certain as it occurs over a broad phase - angle range. The cause of the arched shape of the color ratio is believed to be related to the various contributions of volume scattering at various phase angles: in going from small to medium and then large phase angles, the mean optical path length increases and then decreases, so does the ratio of volume scattering to surface scattering (Adams & Filice 1967; Johnson et al. 2013). The overall pattern of the CE - 4 color ratio is similar to that of CE - 3 (Jin et al. 2015) but with a larger amplitude. This may be explained by the fact that the CE - 4 landing site has redder reflectance spectra than CE - 3 and according to the laboratory study by Schröder et al. (2014), redder reflectance spectra have a stronger phase reddening effect. The fact that the maximum of the arch appears at very large phase angles may be caused by the large incident angle of 80° (Schröder et al. 2014) for the CE - 4 data here.

5.2 Model parameter interpretations

The final fitting results of the CE - 3 and CE - 4 data phase curves using the two models are summarized in Tab. 1. It should be noted that the porosity coefficient K retrieved by the Hapke model for CE - 4 is 1.0, leading to a filling factor of 0 according to Eq. (11). Therefore, Eq. (12) was used to estimate ϕ for CE - 4. Tab. 1 shows that three out of the four parameters that both models contain, $\tilde{\omega}_0$, ϕ, and ζ, have similar differential behaviors despite their different absolute values: although the LB $\tilde{\omega}_0$ differ from the Hapke $\tilde{\omega}_0$ by 20—30%, both are lower for CE - 4 than CE - 3 by about 50%, which is consistent with the much lower values of both spectral reflectance (Hu et al. 2019) and phase curves (this work) of the CE - 4 landing site. On one hand, we notice that many photometric modelings using the Hapke model retrieved similar low SSALB values (e. g., Beck et al. 2012; Sato et al. 2014). On the other hand, we recognize that these SSALB values may be too low to be considered realistic, as Mie calculations can hardly produce SSALB lower than 0.6 or so, even for highly absorbing grains. Therefore, the retrieved SSALB values should be interpreted more in a comparative sense. The filling factors ϕ retrieved by both models are consistently much lower for the CE－4 site (~6% by Hapke and ~1% by LB) than for the CE－3 (~32% by Hapke and ~17% by LB), indicating that the regolith of the former is more porous than that of the latter. Due to the long－term space weathering processes and the Moon's low gravity, the porosity of the uppermost regolith layer is believed to be very high. Hapke & Sato (2016) confirmed the high porosity values (74－87%) of the uppermost layer of lunar regolith derived by Ohtake et al. (2010) and concluded this may be a lower limit. Although our retrieval results (porosity >90%) are basically commensurate with these values, as the rover did not photograph any (nearly) levitating particles on the surface, our retrieved porosity values should be more of comparative significance. The asymmetry parameters ζ retrieved by the two models have opposite signs for both the CE－3 and CE－4 landing sites; however, both are more positive for CE－4 than CE－3, indicating the CE－4 regolith is more forward scattering. With a fit $h_S \sim 0.023$, the ratio of the largest and smallest particle diameters d_l/d_s can be estimated by Eq. (13) as ~7, in contrast to ~46 for CE－3, suggesting a much narrower size distribution at the CE－4 site than the CE－3 (The d_l/d_s estimated from Eq. (11) for CE－3 is ~180). Even the absolute accuracy of these model parameters cannot be evaluated or justified, the semi－quantitative and qualitative agreement between two model parameters can give us confidence in comparing the model parameters between the CE－3 and CE－4 landing sites. The lower BS_0 value by the Hapke model, 0.28 for CE－4 and 0.52 for CE－3, indicates the CE－4 regolith is more transparent and this larger transparency may have caused the regolith layer to be more forward scattering.

The conclusion that coherent back－scattering (CBS) is absent at CE－4's landing site comes mainly from the fact that the Hapke model fitting did not return physically possible CBS parameters. Considering the fact that CE－4's landing site is very dark and the solar zenith angles when taking the PCAM images were very large (~80°), it is possible that multiple scattering is weak, and thus the CBS contribution is small. However, recent numerical simulations (e. g., Muinonen et al. 2012, 2018; Markkanen et al. 2018) and laboratory measurements (Hapke et al. 1993; Nelson et al. 1998, 2000) show that CBS could be present on dark surfaces. Therefore, the possible presence of the CBS in CE－4 phase curves cannot be completely ruled out, and its exclusion may have led to the very high porosity value derived from the Hapke model.

As the landing site surface ages of CE－3 and CE－4 have been estimated to be ~3 Ga and ~3.6 Ga, respectively (e. g., Hiesinger et al. 2000; Fa et al. 2015; Pasckert et al. 2018), the surface material of the latter may have experienced a higher level of space weathering including meteorite bombardment and comminution, and thus have become darker and finer with a narrower size distribution. The much lower normal albedo value at the CE - 4 site (~0.03) compared to that of CE - 3 (0.15), together with the fact that the CE - 4 site shows no signs of albedo increase caused by descent engine blow①, also support this argument. The higher transparency of the CE - 4 regolith could be caused by the higher concentrations of glassy materials generated by space weathering, but this speculation would need further investigation, as other factors such as different mineralogies (Zhang et al. 2015; Hu et al. 2019) may also play a role.

Tab. 1 Comparisons of fitting parameters of the Hapke model and the Lumme - Bowell model for CE - 3 and CE - 4 data

Parameter	CE - 3		CE - 4	
	Hapke	Lumme - Bowell	Hapke	Lumme - Bowell
SSALB	$0.19^{+0.1}_{-0.3}$	0.22±0.01	0.11±0.28	0.080±0.017
Filling factor (ϕ)	$0.32^{+0.11}_{-0.05}$	0.17±0.11	0.061±0.180	0.0095±0.0112
B_{S0}	$0.52^{+0.3}_{-0.2}$	N/A	0.28±0.16	N/A
h_S	0.059±0.01	N/A	0.023±0.013	N/A
Asymmetry parameter (ζ)	−0.024±0.031	−0.19±0.01	0.074±0.023	−0.13±0.09
Slope angle (θ)	$17.0°^{+1.0°}_{-6.0°}$	$62.9°^{+12.7°}_{-61.1°}$	30.0°±4.5°	$17.7°^{+12.0°}_{-13.7°}$

Notes. The Hapke model results for CE - 3 are from Jin et al. (2015), and the rest are from this work. The filling factor ϕ in the Hapke model for CE - 4 was calculated using Eq. (12), and its uncertainty was estimated by using the error propagation equation.

Tab. 2 Comparisons of the three - and five - parameter Hapke model parameters derived from this study and that by Sato et al. (2014)

Model parameters	CE - 4 at 640 nm				Data from Sato et al. (2014)		
	Three - parameter fitting		Five - parameter fitting		Highlands	Mares	All area
	[1°—144°]	[1°—97°]	[1°—144°]	[1°—97°]			
ϖ_0	0.086	0.032	0.082	0.094	0.42	0.24	0.38
b	0.26	0.016	0.34	0.39	0.23	0.26	0.24
c	—	—	−0.085	−0.29	0.37	0.08	0.32
B_{S0}	—	—	0.52	0.47	1.7	2.0	1.7
h_S	0.0037	0.26	0.037	0.020	0.080	0.05	0.070

Notes. The fittings performed in this study were done in both the full phase - angle range [1°—144°] and [1°—97°], which is the same as that of Sato et al. (2014).

For slope angle $\bar{\theta}$, the two models did not return consistent values between the CE - 3 and CE - 4 sites. Although neither model specified the valid spatial scales of its model roughness, the scales of the "macroscopic roughness" (Hapke 2012) and "depth - to -

① https://www.nasa.gov/feature/goddard/2019/chang-e-landers-a-closer-look

radius ratio of a hole" (Lumme & Bowell 1981) are both likely below ~8cm (Helfenstein & Shepard 1999). In comparison, Rozitis & Green (2011) derived the root mean square (RMS) slope of the Moon's entire surface to be ~32° at scales from thermal skin depth (~1cm) to observational spatial resolution (~18km). A topographical study using Kaguya's Terrain Camera dataset with a 30 m baseline found the CE-4 landing area is flat, with 90% of the surface having slopes of less than 5° (Qiao et al. 2019). By studying the Apollo Lunar Surface Closeup Camera pictures, Helfenstein & Shepard (1999) found the RMS slope at a 1mm scale is ~16° for the lunar mare and ~25° for the highland. The Hapke $\bar{\theta}$ values (~17° for CE-3 and ~30° for CE-4) are close to that of Helfenstein & Shepard (1999) and these are consistent with the fact that the lunar highlands are photometrically rougher than the mares. Since the spatial scale that $\bar{\theta}$ characterizes may also depend on surface albedo (Helfenstein & Shepard 1999), and the CE-3 regolith is much brighter with stronger multiple scattering, it is possible that CE-3's $\bar{\theta}$ has a larger spatial scale than that of CE-4. The LB $\bar{\theta}$ (~63°) for CE-3 is probably too large, as in Fig. 7e most $\bar{\theta}$ values are concentrated below 20°. In short, the Hapke $\bar{\theta}$ values for CE-3 and CE-4 may be realistic on millimeter and centimeter scales and are not inconsistent with the result from Qiao et al. (2019) as they were made at different spatial scales.

5.3 Comparison with remote sensing measurement

To compare with remote sensing measurements of lunar highland at 643nm made by Sato et al. (2014), we also fit the Hapke model to the CE-4 data with fewer parameters. Since the maximum phase angle of Sato et al. (2014) is 97°, we carried out fittings over the full phase angle range (1°—144°) and over (1°—97°) as well (Appendix B. 2, Figs. B. 7—B. 10, Tab. 2). The principal results are as follows: (1) our $\tilde{\omega}_0$, 0.082, is smaller than their value, 0.42, which is consistent with the darker nature of the CE-4 landing site; (2) our b (0.34) and c (−0.085) are close to the typical highland crystalline in the "hockey-stick" graph (Fig. 8); (3) our h_S, 0.037, is smaller than their 0.080, indicating we are seeing a narrower size distribution (this does not contradict the above CE-3/CE-4 comparison, however); (4) our B_{S0}, 0.52, is smaller than their value of 1.7, suggesting that we observe more transparent grains.

Sato et al. (2014) found that, in general, the lunar surface with higher level of maturity is more forward scattering, though mineralogy must be considered when making comparisons (e.g., the anorthosite-rich but ilmenite-poor highland material and the ilmenite-rich mare material may have different scattering properties). On the other hand, in our laboratory photometric study of analog materials including lunar regolith simulant, silicates, and ilmenite before and after simulated space weathering, we found that the back-scattering lobes of five out of the six samples became stronger with more negative asymmetry parameters of the single-scattering phase functions after laser irradiations (Jiang et al. 2019). Reflectance spectroscopy

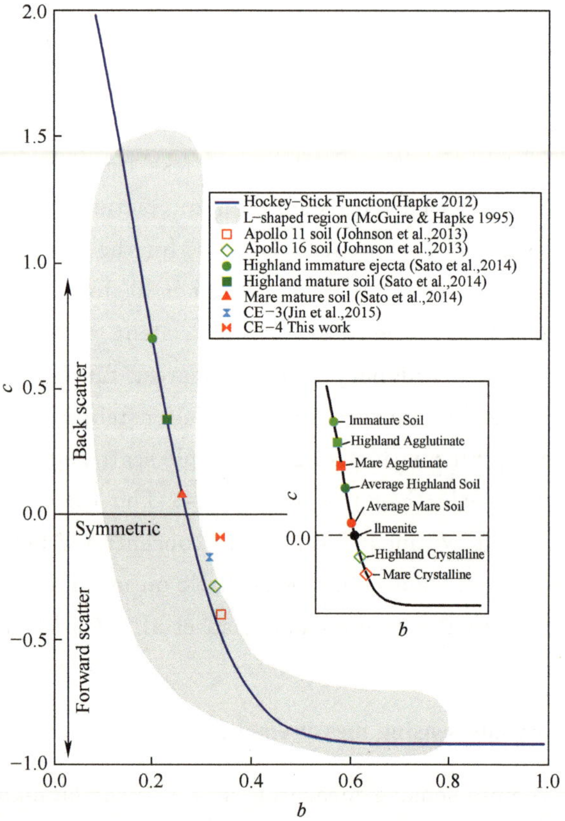

Fig. 8 Typical values of b and c (Henyey - Greenstein phase - function parameters) for the mature mare (low and high Ti content), the mature highland, and the immature highland ejecta in 643 nm band parameter maps by Sato et al. (2014). The red horizontal double triangle indicates the b and c values derived in B. 2. The blue vertical double triangle represents the b and c values of CE - 3 in Jin et al. (2015)

shows that the CE - 4 regolith is redder with more subdued absorption features than the CE - 3 regolith (Zhang et al. 2015; Hu et al. 2019), indicating the former is more mature and therefore, should be more back scattering. In this study, both photometric models have returned more positive asymmetry parameters for the CE - 4 site than CE - 3 (Tab. 1). Thus, according to the result of Sato et al. (2014), the CE - 4 site should be more mature than that of CE - 3, consistently with the spectroscopic results. The contradiction with our laboratory measurement is expected to be solved by performing more laboratory and modeling work in the future.

6 Summary and conclusion

We extracted the phase curves of CE - 4's landing site using the panorama camera images measured by the Yutu - 2 rover and carried out photometric inversions using the

Hapke model and the Lumme - Bowell model. We obtained the following results: compared with the in situ lunar phase curves measured by CE - 3, the CE - 4 phase curves have lower reflectance values and are relatively steeper, have a stronger phase - reddening effect, and show the colorimetric opposition effect. The Hapke model and the Lumme - Bowell model returned qualitatively consistent results in a single - scattering albedo, porosity, and an asymmetry parameter, which shows that the CE - 4 regolith is much darker, more porous, and more forward scattering than that of CE - 3. The Hapke model shows the CE - 4 site is rougher than the CE - 3 site. These features indicate that the surface materials at the CE - 4 landing site are more mature than those of CE - 3.

Even concerning parameters for which the two models yielded consistent results, their absolute values are not guaranteed to be physically correct. Therefore, more focus should be placed on their interpretations in comparative analysis.

Acknowledgements

We thank Karri Muinonen for critical comments that improved the quality of the manuscript. We thank the Chang'E payload team for mission operations and China National Space Administration for providing the Chang'E data that made this study possible. This work was supported by the National Natural Science Foundation of China (11773023, 11941001, U1631124, 11941002) and the Civil Aerospace Pre - research Project (D020302). M. - H. Z. was supported by the Science and Technology Development Fund, Macau (0079/2018/A2) and the pre - research project on Civil Aerospace Technologies No. D020202 funded by China National Space Administration (CNSA). The Chang'E data used in this work were processed and produced by the Ground Research and Application System of China's Lunar Exploration Program and can be downloaded at http://moon.bao.ac.cn/searchOrder_dataSearchData.search.

References

[1] Abe, M., Takagi, Y., Kitazato, K., et al. 2006, Science, 312, 1334.

[2] Adams, J. B., & Filice, A. L. 1967, J. Geophys. Res., 72, 5705.

[3] Akimov, L. A., Antipova - Karataeva, I. I., & Shkuratov, T. G. 1979, in Lunar and Planetary Science Conference, 9.

[4] Beck, P., Pommerol, A., Thomas, N., et al. 2012, Icarus, 218, 364.

[5] Bowell, E., Hapke, B., Domingue, D., et al. 1989, in Asteroids II, eds. R. P.

[6] Binzel, T. Gehrels, & M. S. Matthews (Tucson: University of Arizona Press), 524.

[7] Di, K., Liu, Z., Liu, B., et al. 2019a, J. Rem. Sensing, 23, 177.

[8] Di, K., Zhu, M. - H., Yue, Z., et al. 2019b, Geophys. Res. Lett., 46,

12, 764.

[9] Fa, W., Zhu, M. H., & Liu, T. 2015, in European Planetary Science Congress, EPSC2015-190.

[10] Gehrels, T., Coffeen, T., & Owings, D. 1964, AJ, 69, 826.

[11] Gueymard, C. A. 2004, Sol. Energy, 76, 423.

[12] Hapke, B. 1996, J. Quant. Spectr. Rad. Transf., 55, 837.

[13] Hapke, B. 2012, Theory of reflectance and emittance spectroscopy (Cambridge: Cambridge university press).

[14] Hapke, B. 2013, J. Quant. Spectr. Rad. Transf., 116, 184.

[15] Hapke, B., & Sato, H. 2016, Icarus, 273, 75.

[16] Hapke, B. W., Nelson, R. M., & Smythe, W. D. 1993, Science, 260, 509.

[17] Hapke, B., Nelson, R., & Smythe, W. 1998, Icarus, 133, 89.

[18] Hapke, B. W., Shepard, M. K., Nelson, R. M., Smythe, W. D., & Piatek, J. L. 2009, Icarus, 199, 210.

[19] Hapke, B., Denevi, B., Sato, H., Braden, S., & Robinson, M. 2012, J. Geophys. Res. Planets, 117, E00H15.

[20] Helfenstein, P., & Shepard, M. K. 1999, Icarus, 141, 107.

[21] Helfenstein, P., Veverka, J., & Hillier, J. 1997, Icarus, 128, 2.

[22] Hiesinger, H., Jaumann, R., Neukam, G., & Head, J. W. 2000, J. Geophys. Res., 105, 29239.

[23] Hu, X., Ma, P., Yang, Y., et al. 2019, Geophys. Res. Lett., 46, 9439.

[24] Jia, Y., Zou, Y., Ping, J., et al. 2018, Planet. Space Sci., 162, 207.

[25] Jiang, T., Zhang, H., Yang, Y., et al. 2019, Icarus, 331, 127.

[26] Jin, W., Zhang, H., Yuan, Y., et al. 2015, Geophys. Res. Lett., 42, 8312.

[27] Johnson, J. R., Shepard, M. K., Grundy, W. M., Paige, D. A., & Foote, E. J. 2013, Icarus, 223, 383.

[28] Kaydash, V., Pieters, C., Shkuratov, Y., & Korokhin, V. 2013, J. Geophys. Res. Planets, 118, 1221.

[29] Li, C., Liu, D., Liu, B., et al. 2019, Nature, 569, 378.

[30] Lumme, K., & Bowell, E. 1981, AJ, 86, 1694.

[31] Markkanen,J., Väisänen, T., Penttilä, A., & Muinonen, K. 2018, Opt. Lett., 43, 2925.

[32] McCord, T. B. 1969, J. Geophys. Res., 74, 3131.

[33] McGuire, A. F., & Hapke, B. W. 1995, Icarus, 113, 134.

[34] McKay, D., Fruland, R., & Heiken, G. 1974, Lunar Planet. Sci. Conf. Proc., 5, 887.

[35] Mishchenko, M. I. 1994, J. Quant. Spectr. Rad. Transf., 52, 95.

[36] Mishchenko, M. I. 2008, Rev. Geophys., 46, RG2003.

[37] Mishchenko, M. I., Dlugach, J. M., Chowdhary, J., & Zakharova, N. T. 2015,

J. Quant. Spectr. Rad. Transf., 156, 97.
[38] Muinonen, K. 2004, Waves Random Media, 14, 365.
[39] Muinonen, K., Mishchenko, M. I., Dlugach, J. M., et al. 2012, ApJ, 760, 118.
[40] Muinonen, K., Markkanen, J., Väisänen, T., Peltoniemi, J., & Penttilä, A. 2018, Opt. Lett., 43, 683.
[41] Nelson, R. M., Hapke, B. W., Smythe, W. D., & Horn, L. J. 1998, Icarus, 131, 223.
[42] Nelson, R. M., Hapke, B. W., Smythe, W. D., & Spilker, L. J. 2000, Icarus, 147, 545.
[43] Ohtake, M., Matsunaga, T., Yokota, Y., et al. 2010, Space Sci. Rev., 154, 57.
[44] Pasckert, J. H., Hiesinger, H., & van der Bogert, C. H. 2018, Icarus, 299, 538.
[45] Peltoniemi, J. I., & Lumme, K. 1992, J. Opt. Soc. Am. A, 9, 1320.
[46] Pommerol, A., Thomas, N., Jost, B., et al. 2013, J. Geophys. Res. Planets, 118, 2045.
[47] Qiao, L., Ling, Z., Fu, X., & Li, B. 2019, Icarus, 333, 37.
[48] Ren, X., Li, C. - L., Liu, J. - J., et al. 2014, Res. Astron. Astrophys., 14, 1557.
[49] Rozitis, B., & Green, S. F. 2011, MNRAS, 415, 2042.
[50] Sato, H., Robinson, M. S., Hapke, B., Denevi, B. W., & Boyd, A. K. 2014, J. Geophys. Res. Planets, 119, 1775.
[51] Schröder, S. E., Grynko, Y., Pommerol, A., et al. 2014, Icarus, 239, 201.
[52] Shkuratov, Y. G., Melkumova, L. Y., Opanasenko, N. V., & Stankevich, D. G. 1996, Sol. Syst. Res., 30, 71.
[53] Shkuratov, Y., Kaydash, V., Korokhin, V., et al. 2011, Planet. Space Sci., 59, 1326.
[54] Shkuratov, Y., Kaydash, V., Korokhin, V., et al. 2012, J. Quant. Spectr. Rad. Transf., 113, 2431.
[55] Zhang, H., & Voss, K. J. 2011, Icarus, 215, 27.
[56] Zhang, H., Yang, Y., Yuan, Y., et al. 2015, Geophys. Res. Lett., 42, 6945.

Appendix A　More information on the landing site and images

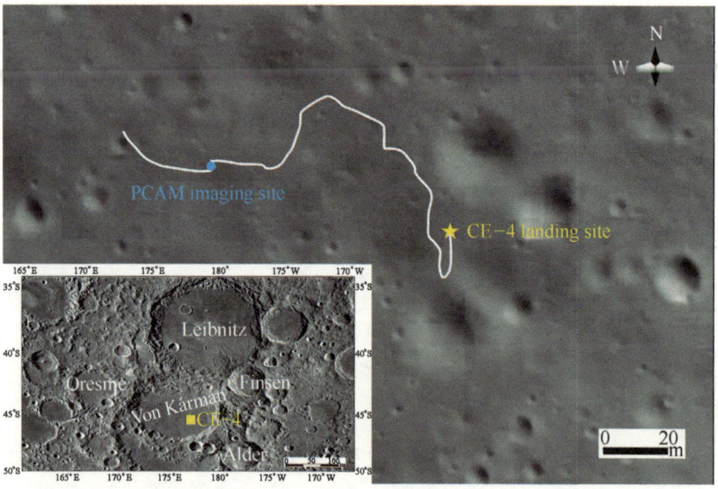

Fig. A. 1　Location of CE-4 landing site and PCAM measurement site. The white line indicates the Yutu-2 Rover's path during the first three lunar days. The base image was photographed by the landing camera onboard the CE-4 spacecraft. The inset shows the Lunar Reconnaissance Orbiter Camera Wide Angle Camera mosaic of the surrounding region of the CE-4 landing site in the South Pole-Aitken basin

Fig. A. 2　Arrangement of PCAMs on CE-4's rover Yutu-2. PCAML and PCAMR are indicated by yellow arrows

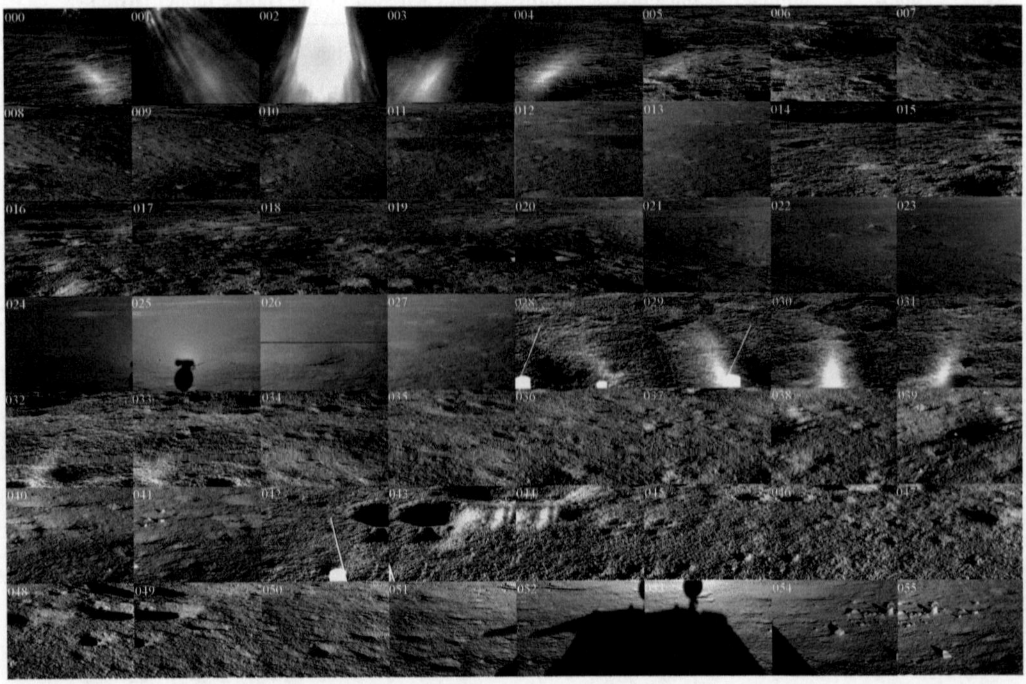

Fig. A. 3　All images taken by PCAMR in color mode during the third lunar day. Obviously, images with numbers in yellow are identified as glare-contaminated, and thus are not used in mosaicing

Appendix B　Detailed procedures for fitting the Hapke model

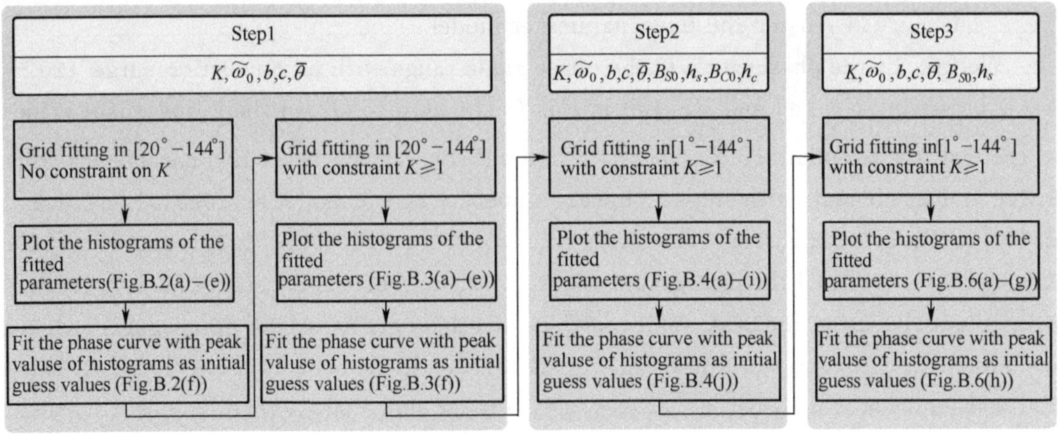

Fig. B. 1　Flowchart of Hapke model fitting process

Tab. B. 1　Summary of parameters obtained from each step of the Hapke model fitting

Phase - angle range		Step 1 [20°—144°]		Step 2 [1°—144°]	Step 3 [1°—144°]
Constraint on K		no	$K\geqslant1$	$K\geqslant1$	$K\geqslant1$
Initial guess values [lower, upper; step]	K	[0.3, 1.8; 0.2]	[0.3, 1.8; 0.2]	[1, 1.3; 0.1]	[1, 1.3; 0.1]
	$\tilde{\omega}_0$	[0.05, 0.5; 0.1]	[0.05, 0.5; 0.1]	[0.05, 0.2; 0.05]	[0.05, 0.2; 0.05]
	b	[0, 0.5; 0.1]	[0, 0.5; 0.1]	[0.1, 0.5; 0.1]	[0.3, 0.5; 0.1]
	c	[−0.5, 0.5; 0.3]	[−0.5, 0.5; 0.3]	[−0.5, 0.1; 0.2]	[−0.4, 0.3; 0.1]
	$\bar{\theta}$	[10°, 35°; 3°]	[10°, 35°; 3°]	[25°, 40°; 5°]	[26°, 38°; 2°]
	B_{S0}	N/A	N/A	[0.4, 1; 0.2]	[0.4, 0.9; 0.1]
	h_S	N/A	N/A	[0.05, 0.15; 0.05]	[0.01, 0.15; 0.02]
	B_{C0}	N/A	N/A	[0.4, 1; 0.2]	N/A
	h_C	N/A	N/A	[0.05, 0.2; 0.05]	N/A
Corresponding figure		Fig. B. 2	Fig. B. 3	Fig. B. 4	Fig. B. 6

B. 1　A piecewise fitting approach

To avoid any possible initial guess value - dependent fitting uncertainties, we performed a grid search in parameter space to obtain a more reliable fitting result. The fitting steps are summarized in Fig. B. 1 and are described as follows.

B. 1. 1　Step 1: fitting of the phase curve in a phase - angle range with no opposition surge (20°—144°) using the five - parameter model

We first fit the phase curve in the phase angle range with no opposition surge (20°—144°) by setting B_S (α) and B_C (α) in Eq. (4) to zero to obtain the initial guess values of the rest five parameters: K, $\tilde{\omega}_0$, b, c, and $\bar{\theta}$. By varying these five parameters from lower to upper bounds with steps ([lower, upper; step]) as: $K=$ [0.3, 1.8; 0.2], $\tilde{\omega}_0=$ [0.05, 0.5; 0.1], $b=$ [0, 0.5; 0.1], $c=$ [−0.5, 0.5; 0.3], and $\bar{\theta}=$ [10°, 35°; 3°], we obtained the histograms (Figs. B. 2a—e) of the distributions of these five parameters. By using the peak values in these five histograms as the initial guess values, we obtain the fitting result shown in Fig. B. 2f. It can be seen that the fitted K value 0.074 is unphysical as by definition (Eq. (11)) it should be larger than or equal to 1. Fig. B. 2a also shows that the porosity factor K has large amount of values concentrated in regions of smaller than 1. To make the fitting results more physically plausible, we repeated the above step with K constrained to be larger or equal to 1, and the results are shown in Fig. B. 2. The peak values of the histograms in Figs. B. 3a—e, $K=1.0$, $\tilde{\omega}_0=0.11$, $b=0.35$, $c=-0.23$, $\bar{\theta}=32.5°$, were then used as the initial fitting parameters, and the resultant fitting is shown in Fig. B. 3f. Although now the fit K has a larger spread, the central value 1.0 does not contradict its definition.

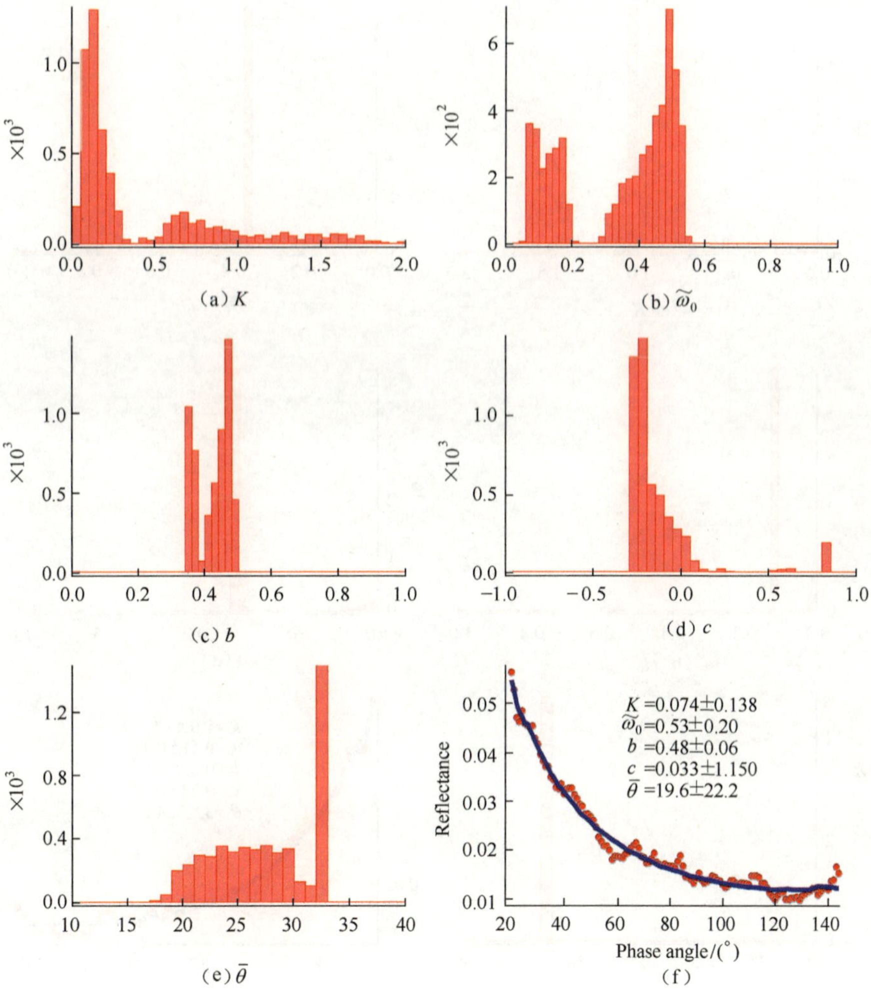

Fig. B. 2　Histograms of fit parameters ((a) porosity factor K; (b) singlescattering albedo $\tilde{\omega}_0$; (c) and (d) are b and c in Eq. (5); and (e) slope angle $\bar{\theta}$) of the Hapke model with no SHOE and CBS terms over the phase - angle range (20°—144°). The final fitting result (f) was obtained using the peak values retrieved from (a) to (e) as initial guess values

B. 1. 2　Step 2: fitting the phase curve using the nine - parameter Hapke model over the full phase - angle range (1°—144°)

In this step, we performed the fitting using the full model including the shadow - hiding (B_{S0}, h_S) and coherent - back - scattering (B_{C0}, h_C) contributions over the full phase - angle range (1°—144°). Based on the results shown in Figs. B. 3a—f, we chose the grid values as follows: [lower, upper; step] as: K = [1, 1.3; 0.1], $\tilde{\omega}_0$ = [0.05, 0.2; 0.05], b = [0.1, 0.5; 0.1], c = [−0.5, 0.1; 0.2], $\bar{\theta}$ = [25°, 40°;

5°], B_{S0}= [0.4, 1; 0.2], h_S = [0.05, 0.15; 0.05], B_{C0}= [0.4, 1; 0.2], and h_C= [0.05, 0.2; 0.05]. With these 245, 760.

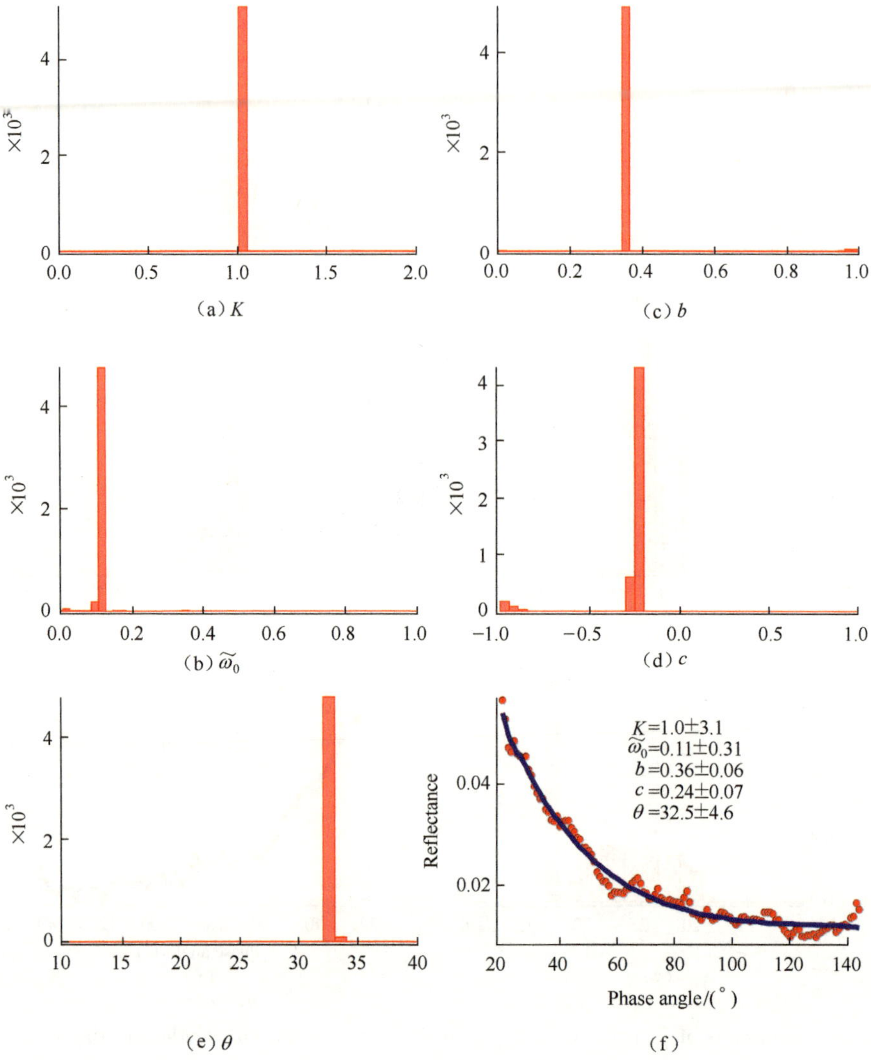

Fig. B. 3　Histograms of fit model parameters((a) porosity factor K; (b) single - scattering albedo $\tilde{\omega}_0$; (c) and (d) are b and c in Eq. (5); (e) slope angle $\bar{\theta}$) of the five - parameter Hapke model fitting with no opposition terms (by setting B_S (α) =0 and B_C (α) =0 in Eq. (4) over the phase angle range (20°—144°) with K constrained to be larger than or equal to 1. The peak values of the parameters in (a) — (e) were then used as the initial guess values in the fitting and the final results are shown in (f)

combinations as initial guess values, nonlinear least square fittings of Eq. (4) were performed, and the resultant parameter histograms are shown in Figs. B. 4a - i. With these peak values as the initial fitting guess values, the nine - parameter fitting was performed again, and the fitting result is shown in Fig. B. 4j.

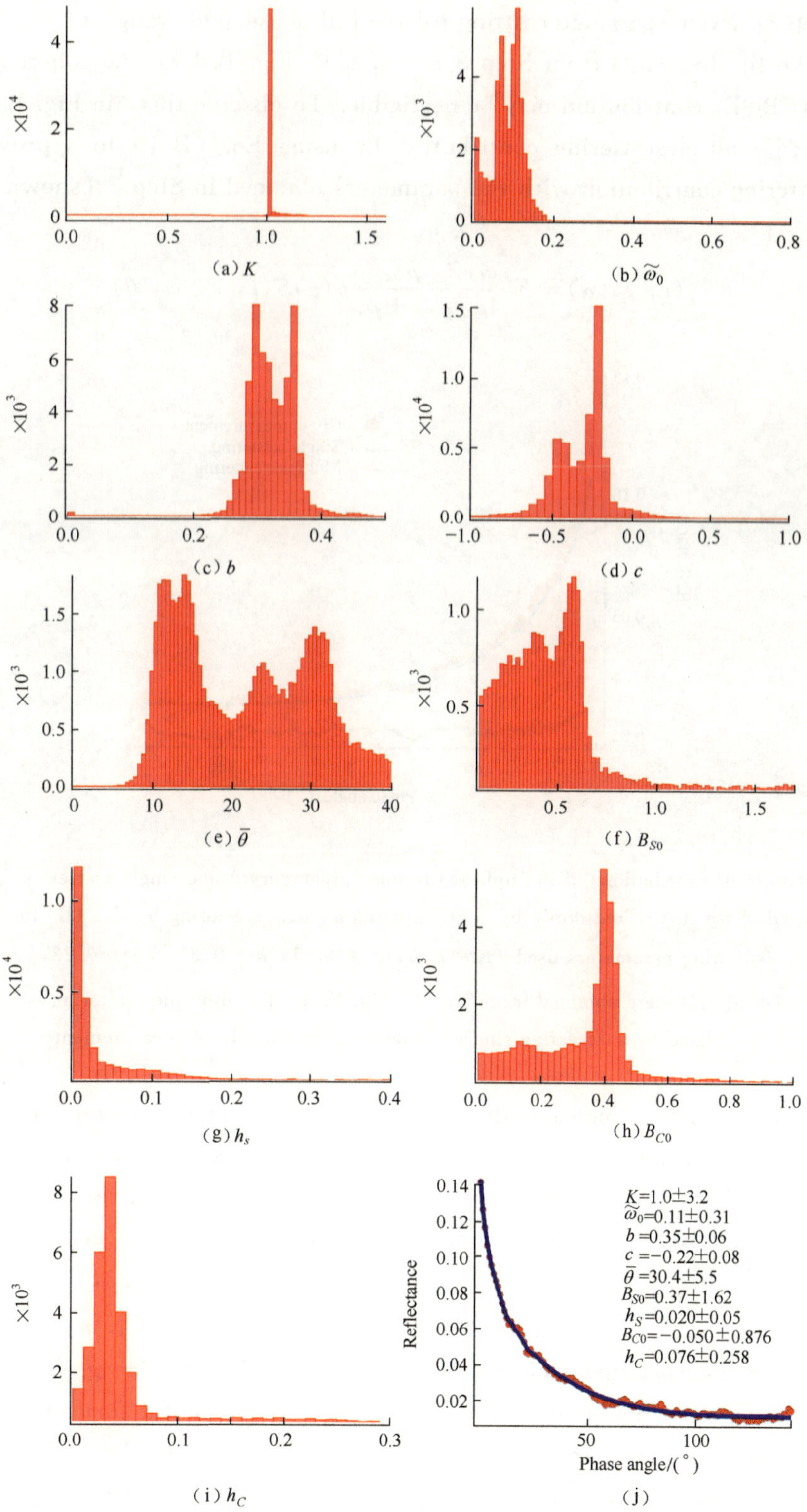

Fig. B. 4　Histograms of fit parameters((a) porosity factor K; (b) singlescattering albedo $\tilde{\omega}_0$; (c) and (d) are b and c in Eq. (5); (e) slope angle $\bar{\theta}$; (f) and (g) are B_{S0} and h_S in Eq. (8); and (h) and (i) are B_{C0} and h_C in Eq. (9)) of the nine - parameter Hapke model over the full phas - angle range (1°—144°) with K constrained to be larger than or equal to 1. The peak values of (a) — (i) were then used as the initial guess values, and the fitting results are shown in (j)

B. 1. 3 Step 3: seven - parameter fitting for the full phase angle range (1°- 144°)

Since the fit B_{C0} value from Step 2 is -0.050 (Fig. B. 4j), the coherent - back - scattering (CBOE) contribution may be negligible. To observe this, in Fig. B. 5 we plot the estimated multiplescattering contribution by using Eq. (B. 1) to approximate the single - scattering contribution with the parameters obtained in Step 2 (shown in Fig. B. 4j):

$$r_s(i, e, \alpha) = K \frac{\tilde{\omega}_0}{4\mu_0} \frac{\mu_{0e}}{\mu_{0e} + \mu_e} p(\alpha) S(i, e, \psi, \bar{\theta}) \tag{B. 1}$$

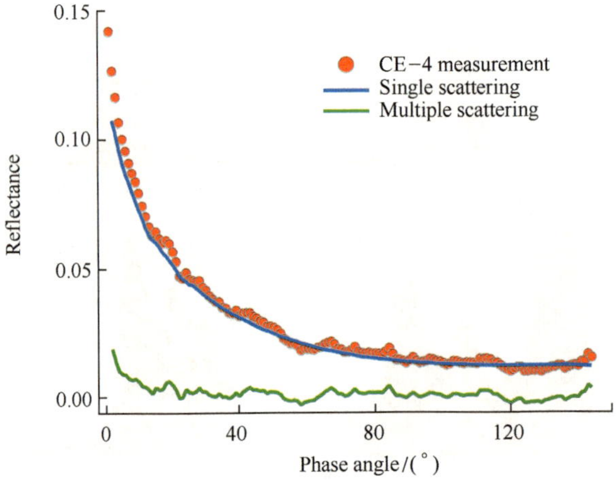

Fig. B. 5 Estimated contributions of multiple scattering (green curve) and single scattering (blue) to the CE - 4 measured phase curve (red dots) by approximating a single scattering by Eq. (B. 1). The single - scattering parameters used (K=1. 0, $\tilde{\omega}_0$=0. 11, b= 0. 35, c=−0. 22, $\bar{\theta}$= 30. 4°) were obtained from Step 2, Fig. B. 4j. The multiple scattering is calculated by subtracting single scattering from the CE - 4 measurement

Obviously, Fig. B. 5 indicates that multiple scattering plays a minor role in the CE - 4 phase curve, and as a result we can safely ignore the CBS contribution. By removing the CBOE terms in Eq. (4), the nine - parameter model has now been reduced to a seven - parameter version. By supplying the initial guess values ([lower, upper; step]) as follows: K= [1, 1. 3; 0. 1], $\tilde{\omega}_0$= [0. 05, 0. 2; 0. 05], b= [0. 3, 0. 5; 0. 1], c= [−0. 4, 0. 3; 0. 1], $\bar{\theta}$= [26°, 38°; 2°], B_{S0}= [0. 4, 0. 9; 0. 1], and h_S= [0. 01, 0. 15; 0. 02], the fit parameter histograms are shown in Figs. B. 6a—g. The peak values in Figs. B. 6a - g were then used as the initial guess values, and the final fitting results of the seven - parameter model are obtained as K=1. 0±2. 9, $\tilde{\omega}_0$=0. 11±0. 28, b= 0. 35± 0. 04, c=−0. 21±0. 06, $\bar{\theta}$= 30. 0°±4. 5°, B_{S0}=0. 28±0. 16, and h_S=0. 023±0. 013. During this final fitting step, the parameters $\tilde{\omega}_0$, b, c, $\bar{\theta}$ are quite stable and appear plausible. With K constrained to be larger than or equal to 1, the fit K always stays near

1. Since the fit B_{S0} value is below 1, the shadow-hiding term alone should be sufficient to describe the opposition effect. The very narrow histograms were probably a consequence of constraining K in the seven-parameter model fitting. The fitting results obtained in each of the above procedures are summarized in Tab. B. 1.

B. 2 Comparisons with Sato's Hapke fitting to lunar global data

Sato et al. (2014) mapped the Hapke parameters over the Moon using a simplified version with three parameters ($\tilde{\omega}_0$, b, B_{S0}, and h_S, but with $\tilde{\omega}_0$ and B_{S0} correlated through Eq. (B. 3) below). They correlated b with c through Eq. (B. 2) and B_{S0} with $\tilde{\omega}_0$ through Eq. (B. 3) and fixed the slope angle to $\bar{\theta} = 23.4°$. These simplifications allowed quicker computations of large amounts of data:

$$c = 3.29e^{-17.4b^2} - 0.908, \tag{B. 2}$$

$$B_{S0} \approx \frac{S(0)}{\tilde{\omega}_0 p(0)}, \tag{B. 3}$$

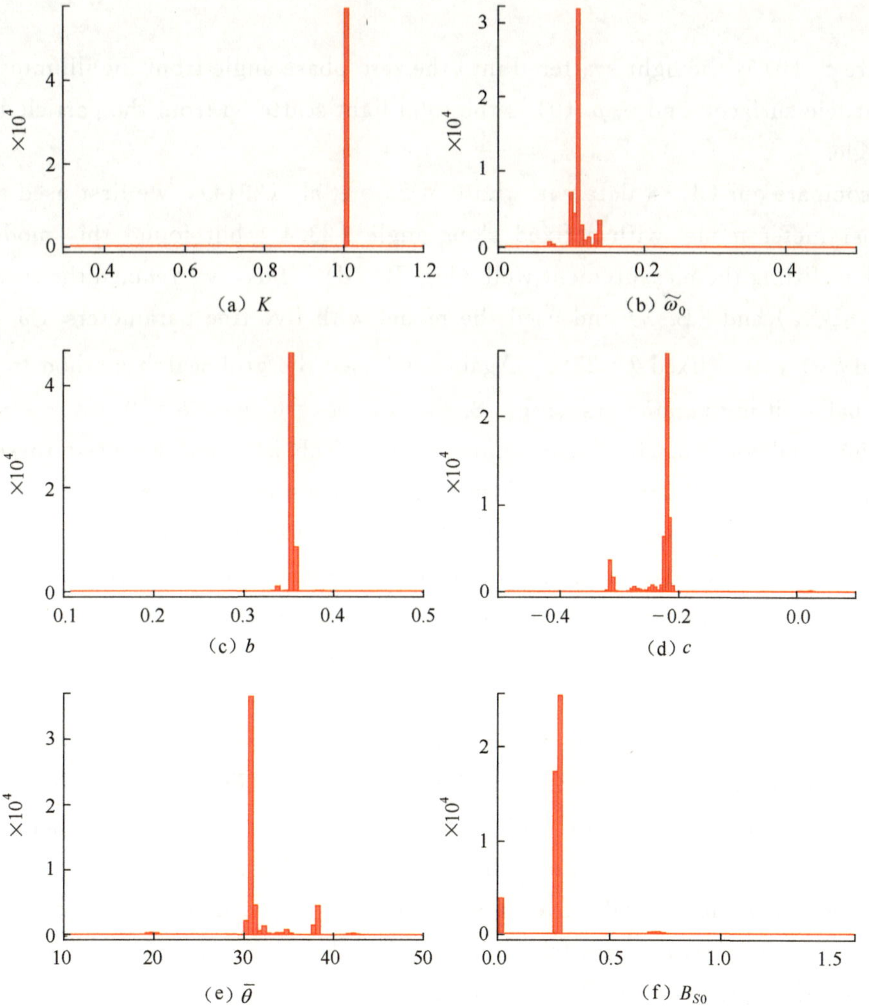
(a) K　(b) $\tilde{\omega}_0$　(c) b　(d) c　(e) $\bar{\theta}$　(f) B_{S0}

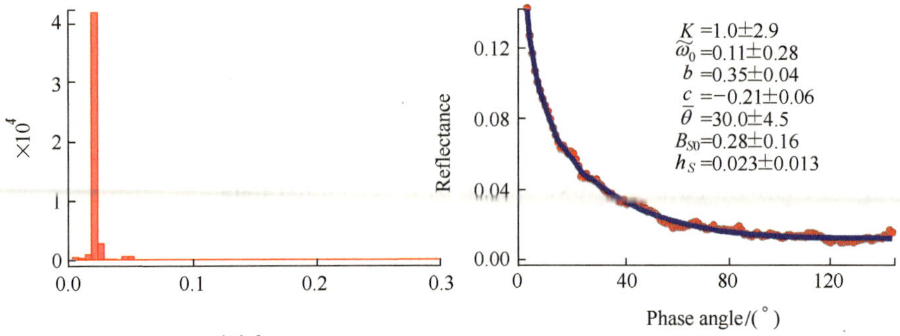

(g) h_S　　　　(h)

Fig. B. 6　Histograms of fit parameters((a) porosity factor K; (b) singlescattering albedo $\tilde{\omega}_0$; (c) and (d) are b and c in Eq. (5); (e) slope angle $\bar{\theta}$; and (f) and (g) are B_{S0} and h_S in Eq. (8)) of the seven - parameter Hapke model over the full phase - angle range (1°—144°) with no CBOE contribution ($B_C(\alpha)=0$) and with K constrained to be larger than or equal to 1. The peak values of (a) — (g) are then used as the initial guess values and the fitting results are shown in (*h*)

where S (0) is the light scattered into the zero phase angle from the illuminated part of the particle surface, and $\tilde{\omega}_0 p$ (0) is the total light scattered from the particle into zero phase angle.

To compare our CE - 4 data with those of Sato et al. (2014), we first used the same three - parameter model with a fixed slope angle (23.4°) but found this model to be incapable of fitting the measurement well (Fig. B. 7d). Thus, we relaxed the correlations of Eqs. (B. 2) and (B. 3) and used the model with five free parameters ($\tilde{\omega}_0$, b, c, B_{S0}, and h_S) with a fixed $\bar{\theta}=23.4°$. Again, we used the grid search method to find the most probable fitting parameters (Figs. B. 8a—e): $\tilde{\omega}_0=0.090$, $b=0.35$, $c=-0.090$, $B_{S0}=0.50$, and $h_S=0.040$ (also summarized in Tab. 2), and we used these as the initial guess values to obtain the final fitting result shown in Fig. B. 8f.

It should be noted that the maximum phase angle in the data of Sato et al. (2014) is 97°, while ours is 144°. To understand if such a difference in phase angle coverage would affect the fitting results, we also fit the above three - parameter and five - parameter models within a 97° phase angle, and the results are shown in Figs. B. 9 and B. 10 and summarized in Tab. 2.

Obviously, the three - parameter model fitting over 1°—97° is still less satisfactory (Fig. B. 9d), and the five - parameter fitting over 1°—97° (Fig. B. 10f) produced very a similar result to that over 1°—144° (Fig. B. 8f), with the lorgest difference being in the c parameter. To be consistent with our CE - 4 data, we chose the b and c values derived from the five - parameter model over 1 - 144 to compare with the results of Sato et al. (2014). The b and c parameters for both CE - 3 and CE - 4 are added to the "hockey - stick" plot (McGuire & Hapke 1995) in Fig. 8.

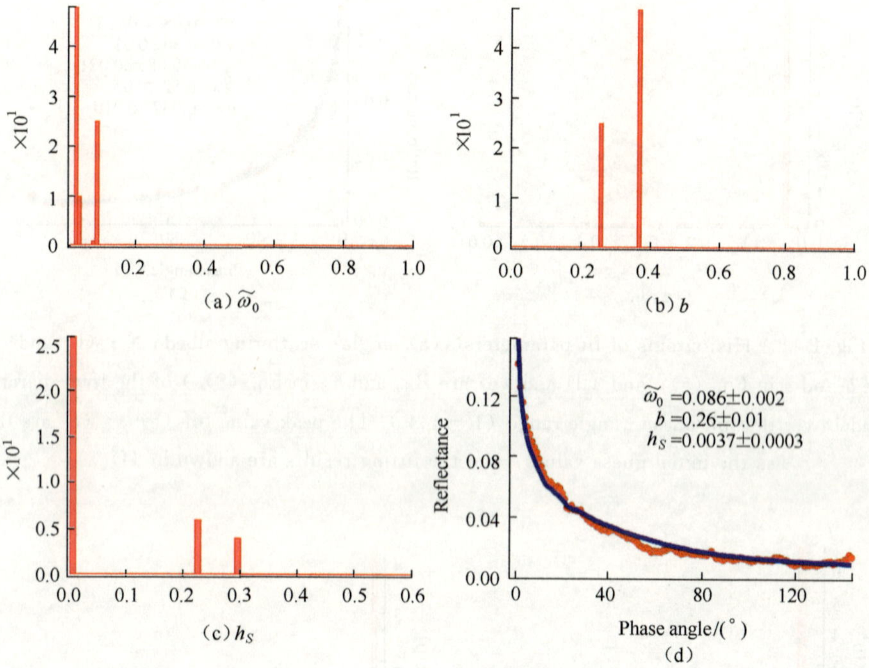

Fig. B. 7　Histograms of fit parameters((a) single-scattering albedo $\widetilde{\omega}_0$; (b) b; and (c) h_S) of the simplified Hapke model over the full phaseangle range (1°—144°). The peak values of (a) — (c) were then used as the initial guess values, and the fitting results are shown in (d)

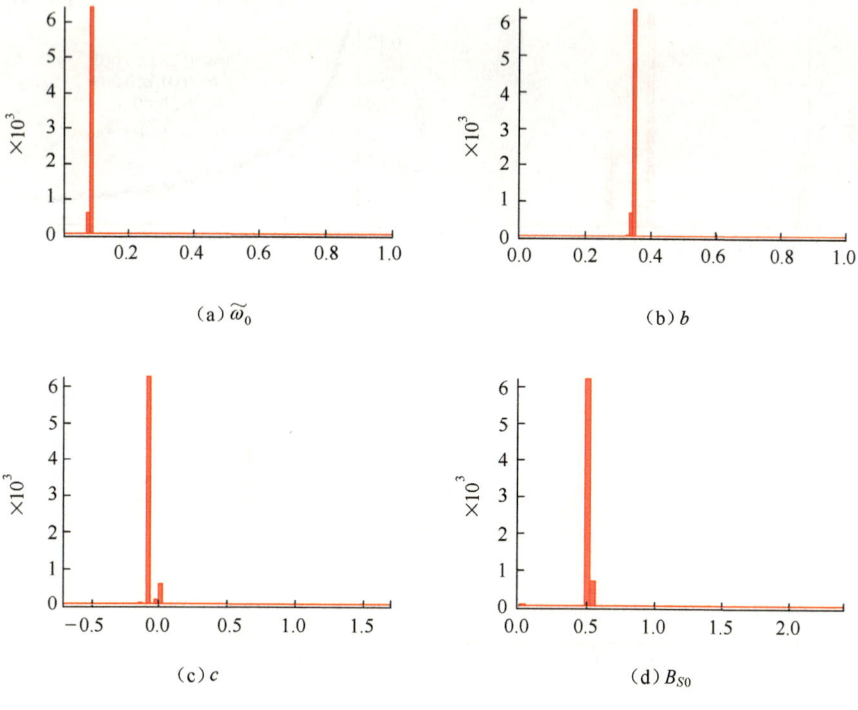

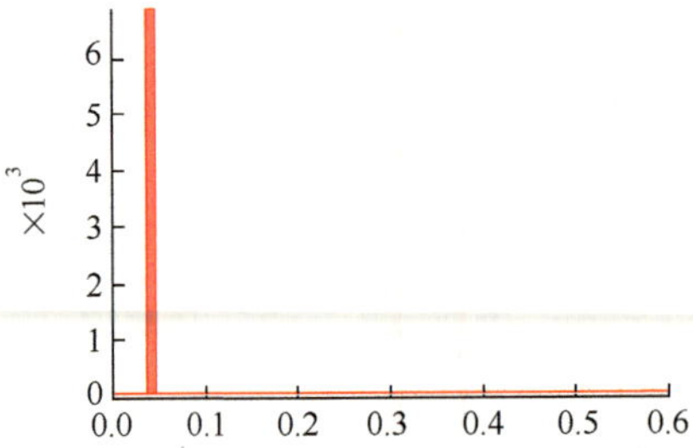

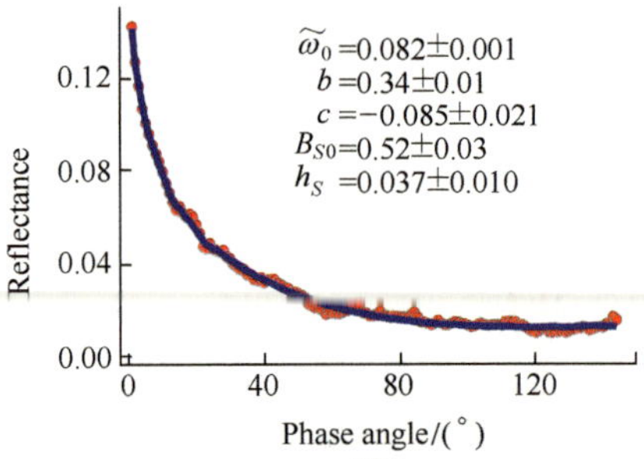

(e) h_S　　(f)

Fig. B. 8　Histograms of fit parameters((a) single-scattering albedo $\widetilde{\omega}_0$; (b) and (c) are b and c in Eq. (5); and (d) and (e) are B_{S0} and h_S in Eq. (8)) of the five-parameter Hapke model over the full phase-angle range (1°—144°). The peak values of (a) — (e) are then used as the initial guess values, and the fitting results are shown in (f)

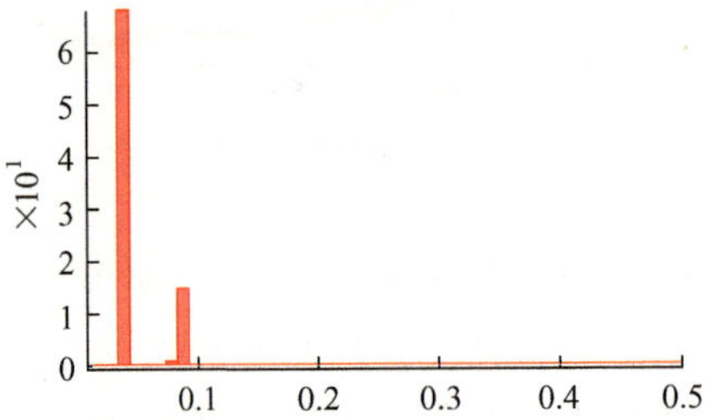

(a) $\widetilde{\omega}_0$

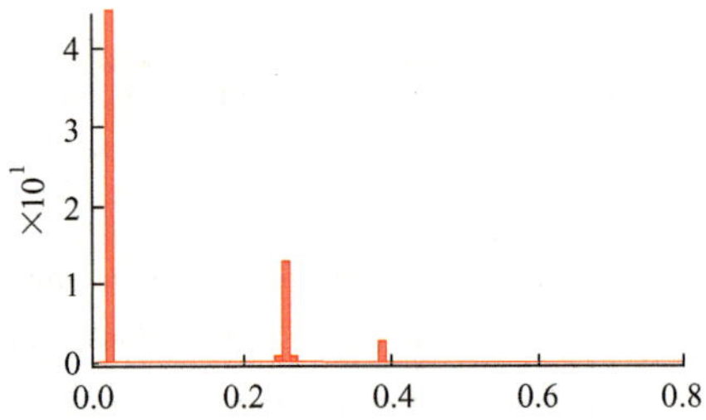

(b) b

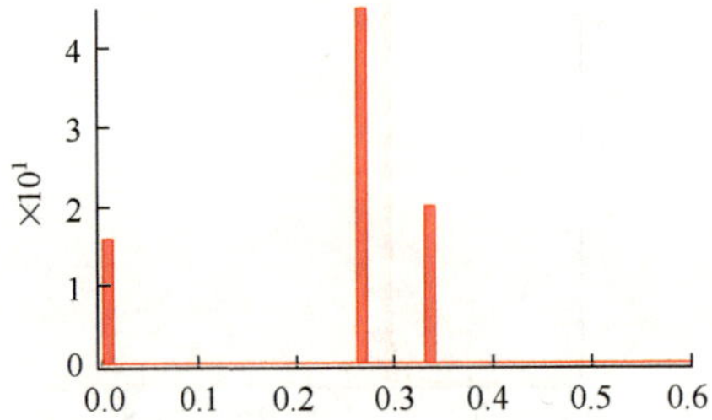

(c) h_S

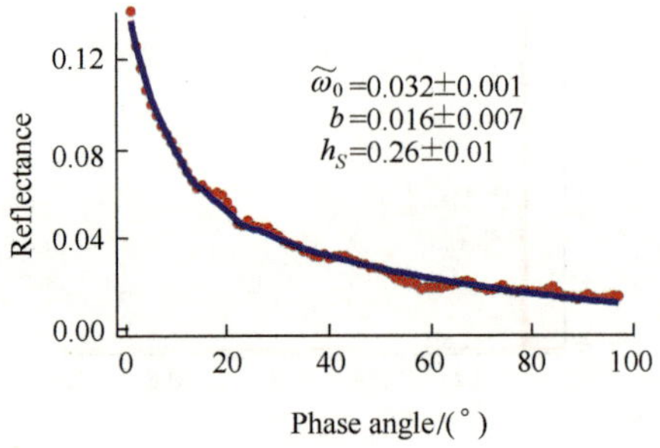

(d)

Fig. B. 9　Same as in Fig. B. 7 but over the phase—angle range (1°—97°)

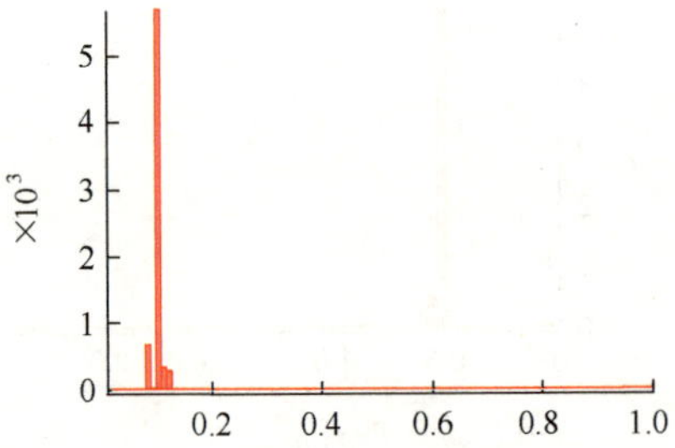

(a) $\widetilde{\omega}_0$

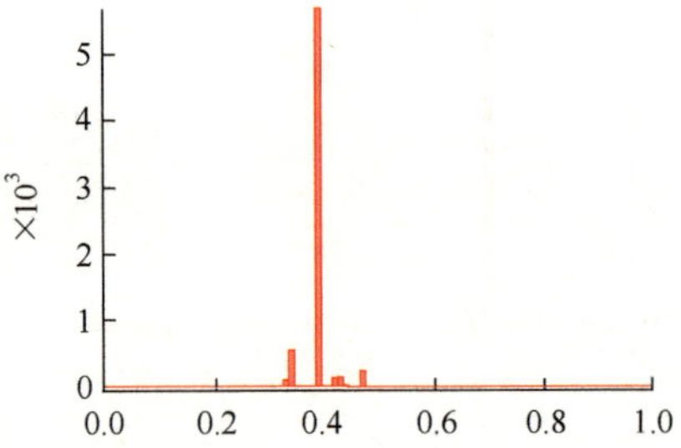

(b) b

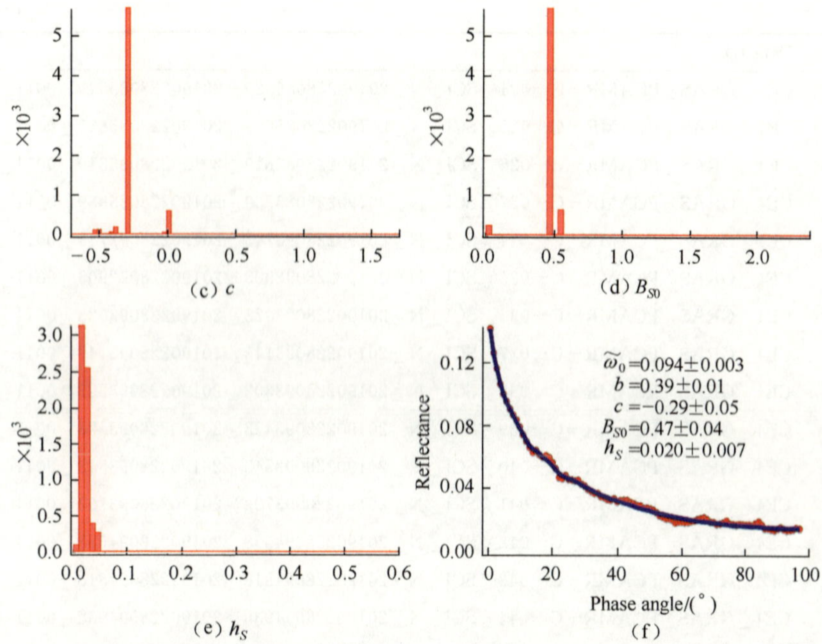

Fig. B. 10　Same as in Fig. B. 8 but over the phase-angle range (1°—97°)

Appendix C　List of all data file names used in this work

Tab. C. 1　Data IDs used in this work

Figure number	Data ID
Fig. 1	CE4_GRAS_PCAMR-C-006_SCI_N_20190228082303_20190228082303_0011_B.2B
	CE4_GRAS_PCAMR-C-007_SCI_N_20190228082423_20190228082423_0011_B.2B
	CE4_GRAS_PCAMR-C-008_SCI_N_20190228082543_20190228082543_0011_B.2B
	CE4_GRAS_PCAMR-C-009_SCI_N_20190228082703_20190228082703_0011_B.2B
	CE4_GRAS_PCAMR-C-010_SCI_N_20190228082823_20190228082823_0011_B.2B
	CE4_GRAS_PCAMR-C-011_SCI_N_20190228082943_20190228082943_0011_B.2B
	CE4_GRAS_PCAMR-C-012_SCI_N_20190228083103_20190228083103_0011_B.2B
	CE4_GRAS_PCAMR-C-013_SCI_N_20190228083223_20190228083223_0011_B.2B
	CE4_GRAS_PCAMR-C-015_SCI_N_20190228084339_20190228084339_0011_B.2B
	CE4_GRAS_PCAMR-C-016_SCI_N_20190228084459_20190228084459_0011_B.2B
	CE4_GRAS_PCAMR-C-017_SCI_N_20190228084619_20190228084619_0011_B.2B
	CE4_GRAS_PCAMR-C-018_SCI_N_20190228084739_20190228084739_0011_B.2B
	CE4_GRAS_PCAMR-C-019_SCI_N_20190228084859_20190228084859_0011_B.2B
	CE4_GRAS_PCAMR-C-020_SCI_N_20190228085019_20190228085019_0011_B.2B
	CE4_GRAS_PCAMR-C-021_SCI_N_20190228085139_20190228085139_0011_B.2B
	CE4_GRAS_PCAMR-C-022_SCI_N_20190228085259_20190228085259_0011_B.2B
	CE4_GRAS_PCAMR-C-023_SCI_N_20190228085419_20190228085419_0011_B.2B

Continued

Figure number	Data ID
	CE4 _ GRAS _ PCAMR－C－024 _ SCI _ N _ 20190228085539 _ 20190228085539 _ 0011 _ B. 2B
	CE4 _ GRAS _ PCAMR－C－025 _ SCI _ N _ 20190228085659 _ 20190228085659 _ 0011 _ B. 2B
	CE4 _ GRAS _ PCAMR－C－026 _ SCI _ N _ 20190228085819 _ 20190228085819 _ 0011 _ B. 2B
	CE4 _ GRAS _ PCAMR－C－027 _ SCI _ N _ 20190228085939 _ 20190228085939 _ 0011 _ B. 2B
Fig. 1	CE4 _ GRAS _ PCAMR－C－034 _ SCI _ N _ 20190228092743 _ 20190228092743 _ 0011 _ B. 2B
	CE4 _ GRAS _ PCAMR－C－035 _ SCI _ N _ 20190228092903 _ 20190228092903 _ 0011 _ B. 2B
	CE4 _ GRAS _ PCAMR－C－036 _ SCI _ N _ 20190228093023 _ 20190228093023 _ 0011 _ B. 2B
	CE4 _ GRAS _ PCAMR－C－037 _ SCI _ N _ 20190228093143 _ 20190228093143 _ 0011 _ B. 2B
	CE4 _ GRAS _ PCAMR－C－038 _ SCI _ N _ 20190228093303 _ 20190228093303 _ 0011 _ B. 2B
	CE4 _ GRAS _ PCAMR－C－039 _ SCI _ N _ 20190228093423 _ 20190228093423 _ 0011 _ B. 2B
	CE4 _ GRAS _ PCAMR－C－040 _ SCI _ N _ 20190228093543 _ 20190228093543 _ 0011 _ B. 2B
	CE4 _ GRAS _ PCAMR－C－041 _ SCI _ N _ 20190228093703 _ 20190228093703 _ 0011 _ B. 2B
	CE4 _ GRAS _ PCAMR－C－042 _ SCI _ N _ 20190228094658 _ 20190228094658 _ 0011 _ B. 2B
	CE4 _ GRAS _ PCAMR－C－043 _ SCI _ N _ 20190228094818 _ 20190228094818 _ 0011 _ B. 2B
	CE4 _ GRAS _ PCAMR－C－044 _ SCI _ N _ 20190228094938 _ 20190228094938 _ 0011 _ B. 2B
	CE4 _ GRAS _ PCAMR－C－045 _ SCI _ N _ 20190228095058 _ 20190228095058 _ 0011 _ B. 2B
	CE4 _ GRAS _ PCAMR－C－046 _ SCI _ N _ 20190228095218 _ 20190228095218 _ 0011 _ B. 2B
	CE4 _ GRAS _ PCAMR－C－047 _ SCI _ N _ 20190228095338 _ 20190228095338 _ 0011 _ B. 2B
	CE4 _ GRAS _ PCAMR－C－048 _ SCI _ N _ 20190228095458 _ 20190228095458 _ 0011 _ B. 2B
	CE4 _ GRAS _ PCAMR－C－049 _ SCI _ N _ 20190228095618 _ 20190228095618 _ 0011 _ B. 2B
	CE4 _ GRAS _ PCAMR－C－050 _ SCI _ N _ 20190228095738 _ 20190228095738 _ 0011 _ B. 2B
	CE4 _ GRAS _ PCAMR－C－051 _ SCI _ N _ 20190228095858 _ 20190228095858 _ 0011 _ B. 2B
	CE4 _ GRAS _ PCAMR－C－052 _ SCI _ N _ 20190228100018 _ 20190228100018 _ 0011 _ B. 2B
	CE4 _ GRAS _ PCAMR－C－053 _ SCI _ N _ 20190228100138 _ 20190228100138 _ 0011 _ B. 2B
	CE4 _ GRAS _ PCAMR－C－054 _ SCI _ N _ 20190228100258 _ 20190228100258 _ 0011 _ B. 2B
	CE4 _ GRAS _ PCAMR－C－055 _ SCI _ N _ 20190228100418 _ 20190228100418 _ 0011 _ B. 2B
Fig. 2	CE4 _ GRAS _ PCAMR－C－025 _ SCI _ N _ 20190228085659 _ 20190228085659 _ 0011 _ B. 2B
Fig. A. 2	CE4－La _ GRAS _ TCAM－I _ SCI _ P _ 20190110061721 _ 20190111011837 _ 0016 _ A. 0B
Fig. A. 3	CE4 _ GRAS _ PCAMR－C－000 _ SCI _ N _ 20190228081503 _ 20190228081503 _ 0011 _ B. 2B
	CE4 _ GRAS _ PCAMR－C－001 _ SCI _ N _ 20190228081623 _ 20190228081623 _ 0011 _ B. 2B
	CE4 _ GRAS _ PCAMR－C－002 _ SCI _ N _ 20190228081743 _ 20190228081743 _ 0011 _ B. 2B
	CE4 _ GRAS _ PCAMR－C－003 _ SCI _ N _ 20190228081903 _ 20190228081903 _ 0011 _ B. 2B
	CE4 _ GRAS _ PCAMR－C－004 _ SCI _ N _ 20190228082023 _ 20190228082023 _ 0011 _ B. 2B
	CE4 _ GRAS _ PCAMR－C－005 _ SCI _ N _ 20190228082143 _ 20190228082143 _ 0011 _ B. 2B
	CE4 _ GRAS _ PCAMR－C－006 _ SCI _ N _ 20190228082303 _ 20190228082303 _ 0011 _ B. 2B
	CE4 _ GRAS _ PCAMR－C－007 _ SCI _ N _ 20190228082423 _ 20190228082423 _ 0011 _ B. 2B
	CE4 _ GRAS _ PCAMR－C－008 _ SCI _ N _ 20190228082543 _ 20190228082543 _ 0011 _ B. 2B
	CE4 _ GRAS _ PCAMR－C－009 _ SCI _ N _ 20190228082703 _ 20190228082703 _ 0011 _ B. 2B
	CE4 _ GRAS _ PCAMR－C－010 _ SCI _ N _ 20190228082823 _ 20190228082823 _ 0011 _ B. 2B
	CE4 _ GRAS _ PCAMR－C－011 _ SCI _ N _ 20190228082943 _ 20190228082943 _ 0011 _ B. 2B

Continued

Figure number	Data ID
	CE4 _ GRAS _ PCAMR－C－012 _ SCI _ N _ 20190228083103 _ 20190228083103 _ 0011 _ B. 2B
	CE4 _ GRAS _ PCAMR－C－013 _ SCI _ N _ 20190228083223 _ 20190228083223 _ 0011 _ B. 2B
	CE4 _ GRAS _ PCAMR－C－014 _ SCI _ N _ 20190228084219 _ 20190228084219 _ 0011 _ B. 2B
	CE4 _ GRAS _ PCAMR－C－015 _ SCI _ N _ 20190228084339 _ 20190228084339 _ 0011 _ B. 2B
	CE4 _ GRAS _ PCAMR－C－016 _ SCI _ N _ 20190228084459 _ 20190228084459 _ 0011 _ B. 2B
	CE4 _ GRAS _ PCAMR－C－017 _ SCI _ N _ 20190228084619 _ 20190228084619 _ 0011 _ B. 2B
	CE4 _ GRAS _ PCAMR－C－018 _ SCI _ N _ 20190228084739 _ 20190228084739 _ 0011 _ B. 2B
	CE4 _ GRAS _ PCAMR－C－019 _ SCI _ N _ 20190228084859 _ 20190228084859 _ 0011 _ B. 2B
	CE4 _ GRAS _ PCAMR－C－020 _ SCI _ N _ 20190228085019 _ 20190228085019 _ 0011 _ B. 2B
	CE4 _ GRAS _ PCAMR－C－021 _ SCI _ N _ 20190228085139 _ 20190228085139 _ 0011 _ B. 2B
	CE4 _ GRAS _ PCAMR－C－022 _ SCI _ N _ 20190228085259 _ 20190228085259 _ 0011 _ B. 2B
	CE4 _ GRAS _ PCAMR－C－023 _ SCI _ N _ 20190228085419 _ 20190228085419 _ 0011 _ B. 2B
	CE4 _ GRAS _ PCAMR－C－024 _ SCI _ N _ 20190228085539 _ 20190228085539 _ 0011 _ B. 2B
	CE4 _ GRAS _ PCAMR－C－025 _ SCI _ N _ 20190228085659 _ 20190228085659 _ 0011 _ B. 2B
	CE4 _ GRAS _ PCAMR－C－026 _ SCI _ N _ 20190228085819 _ 20190228085819 _ 0011 _ B. 2B
	CE4 _ GRAS _ PCAMR－C－027 _ SCI _ N _ 20190228085939 _ 20190228085939 _ 0011 _ B. 2B
	CE4 _ GRAS _ PCAMR－C－028 _ SCI _ N _ 20190228091633 _ 20190228091633 _ 0011 _ B. 2B
	CE4 _ GRAS _ PCAMR－C－029 _ SCI _ N _ 20190228092103 _ 20190228092103 _ 0011 _ B. 2B
	CE4 _ GRAS _ PCAMR－C－030 _ SCI _ N _ 20190228092223 _ 20190228092223 _ 0011 _ B. 2B
	CE4 _ GRAS _ PCAMR－C－031 _ SCI _ N _ 20190228092343 _ 20190228092343 _ 0011 _ B. 2B
	CE4 _ GRAS _ PCAMR－C－032 _ SCI _ N _ 20190228092503 _ 20190228092503 _ 0011 _ B. 2B
	CE4 _ GRAS _ PCAMR－C－033 _ SCI _ N _ 20190228092623 _ 20190228092623 _ 0011 _ B. 2B
	CE4 _ GRAS _ PCAMR－C－034 _ SCI _ N _ 20190228092743 _ 20190228092743 _ 0011 _ B. 2B
	CE4 _ GRAS _ PCAMR－C－035 _ SCI _ N _ 20190228092903 _ 20190228092903 _ 0011 _ B. 2B
	CE4 _ GRAS _ PCAMR－C－036 _ SCI _ N _ 20190228093023 _ 20190228093023 _ 0011 _ B. 2B
	CE4 _ GRAS _ PCAMR－C－037 _ SCI _ N _ 20190228093143 _ 20190228093143 _ 0011 _ B. 2B
	CE4 _ GRAS _ PCAMR－C－038 _ SCI _ N _ 20190228093303 _ 20190228093303 _ 0011 _ B. 2B
	CE4 _ GRAS _ PCAMR－C－039 _ SCI _ N _ 20190228093423 _ 20190228093423 _ 0011 _ B. 2B
	CE4 _ GRAS _ PCAMR－C－040 _ SCI _ N _ 20190228093543 _ 20190228093543 _ 0011 _ B. 2B
	CE4 _ GRAS _ PCAMR－C－041 _ SCI _ N _ 20190228093703 _ 20190228093703 _ 0011 _ B. 2B
	CE4 _ GRAS _ PCAMR－C－042 _ SCI _ N _ 20190228094658 _ 20190228094658 _ 0011 _ B. 2B
	CE4 _ GRAS _ PCAMR－C－043 _ SCI _ N _ 20190228094818 _ 20190228094818 _ 0011 _ B. 2B
	CE4 _ GRAS _ PCAMR－C－044 _ SCI _ N _ 20190228094938 _ 20190228094938 _ 0011 _ B. 2B
	CE4 _ GRAS _ PCAMR－C－045 _ SCI _ N _ 20190228095058 _ 20190228095058 _ 0011 _ B. 2B

Continued

Figure number	Data ID
	CE4 _ GRAS _ PCAMR−C−046 _ SCI _ N _ 20190228095218 _ 20190228095218 _ 0011 _ B. 2B
	CE4 _ GRAS _ PCAMR−C−047 _ SCI _ N _ 20190228095338 _ 20190228095338 _ 0011 _ B. 2B
	CE4 _ GRAS _ PCAMR−C−048 _ SCI _ N _ 20190228095458 _ 20190228095458 _ 0011 _ B. 2B
	CE4 _ GRAS _ PCAMR−C−049 _ SCI _ N _ 20190228095618 _ 20190228095618 _ 0011 _ B. 2B
	CE4 _ GRAS _ PCAMR−C−050 _ SCI _ N _ 20190228095738 _ 20190228095738 _ 0011 _ B. 2B
	CE4 _ GRAS _ PCAMR−C−051 _ SCI _ N _ 20190228095858 _ 20190228095858 _ 0011 _ B. 2B
	CE4 _ GRAS _ PCAMR−C−052 _ SCI _ N _ 20190228100018 _ 20190228100018 _ 0011 _ B. 2B
	CE4 _ GRAS _ PCAMR−C−053 _ SCI _ N _ 20190228100138 _ 20190228100138 _ 0011 _ B. 2B
	CE4 _ GRAS _ PCAMR−C−054 _ SCI _ N _ 20190228100258 _ 20190228100258 _ 0011 _ B. 2B
	CE4 _ GRAS _ PCAMR−C−055 _ SCI _ N _ 20190228100418 _ 20190228100418 _ 0011 _ B. 2B

Chang'E – 4 initial spectroscopic identification of lunar far – side mantle – derived materials

Chunlai Li[1], Dawei Liu[1], Bin Liu[1], Xin Ren[1], Jianjun Liu[1], Zhiping He[2], Wei Zuo[1], Xingguo Zeng[1], Rui Xu[2], Xu Tan[1], Xiaoxia Zhang[1], Wangli Chen[1], Rong Shu[2], Weibin Wen[1], Yan Su[1], Hongbo Zhang[1] & Ziyuan Ouyang[1,3]

1. Key Laboratory of Lunar and Deep Space Exploration, National Astronomical Observatories, Chinese Academy of Sciences, Beijing, China

2. Key Laboratory of Space Active Opto – Electronics Technology, Shanghai institute of Technical Physics, The Chinese Acadmy of Sciences, Shanghai, China

3. Institute of Geochemistry, Chiness Academy of Sciences, Guiyang, China

Over 60 years of spacecraft exploration has revealed that the Earth's Moon is characterized by a lunar crust[1] dominated by the mineral plagioclase, overlying a more mafic (richer in iron and magnesium) mantle of uncertain composition. Both crust and mantle formed during the earliest stages of lunar evolution when late – stage accretional energy caused a molten rock (magma) ocean, flotation of the light plagioclase, sinking of the denser iron – rich minerals, such as olivine and pyroxene, and eventually solidification[2]. Very large impact craters can potentially penetrate through the crust and sample the lunar mantle. The largest of these craters is the approximately 2, 500 – kilometre – diameter South Pole – Aitken (SPA) basin[3] on the lunar far side. Evidence obtained from orbiting spacecraft shows that the floor of the SPA basin is rich in mafic minerals[4], but their mantle origin is controversial and their in situ geologic settings are poorly known. China's Chang'E – 4 lunar far – side lander recently touched down in the Von Kármán crater[5,6] to explore the floor of the huge SPA basin and deployed its rover, Yutu – 2. Here we report on the initial spectral observations of the Visible and Near Infrared Spectrometer (VNIS)[7] onboard Yutu – 2, which we interpret to represent the presence of low – calcium (ortho) pyroxene and olivine, materials that may originate from the lunar mantle. Geological context[6] suggests that these materials were excavated from below the SPA floor by the nearby 72 – km – diameter Finsen impact crater event, and transported to the landing site. Continued exploration by Yutu – 2 will target these materials on the floor of the Von Kármán crater to understand their geologic context, o –

rigin and abundance, and to assess the possibility of sample-return scenarios.

The detailed nature of the lunar mantle has eluded investigators since the Moon was first found to be differentiated and the crust to have formed by late-stage accretional energy and plagioclase flotation—a process different from the one that formed Earth's crust[2]. However, the initial composition and structure of the lunar mantle and how these constrain models of the formation and evolution of the Moon remain unknown. No direct evidence (such as unequivocal samples of the lunar mantle) was found in material returned by the Apollo and Luna missions[8,9]. Indirect information has come from Apollo seismic experiments and the nature of mare basalts sourced from the lunar mantle, but fundamental questions remain[8]. Predictions ranged from a mantle dominated by olivine, an orthopyroxenite with trace amounts of olivine, or some other combination of these component minerals[8]. After recognizing that impact craters form 'drill holes' into the lunar subsurface, investigators turned to orbital elemental and mineralogical measurements to probe the nature of the lunar crust and mantle, finding evidence of stratification in the lunar crust[10]. Evidence of the presence of olivine—a known major component of planetary mantles—was surprisingly sparse on the lunar surface; the mineral was found mostly around impact basins of moderate to large size, but even there it was not abundant or widespread[11]. Attention turned to the largest and oldest impact basin, the about 2,500-km-diameter SPA basin, which is located on the lunar far side[12,13]. A basin this large would be the best candidate to have penetrated below the crust and deep into the lunar interior, excavating lunar mantle and distributing it on the surface in the form of ejecta and impact melt deposits. Recent refinement (downward) of estimates of lunar crustal thickness from the GRAIL mission[1] further supported the likelihood that the SPA impact should have excavated olivine-rich mantle material. Orbital remote-sensing measurements revealed that, despite not being flooded by mare basalts, the SPA interior was clearly more iron-rich than the lunar highland[13]. In addition, although there is abundant evidence of the presence of pyroxene[12] in the SPA interior, olivine was found in only two places, inside the Schrödinger basin and the Zeeman crater - both later events superposed near the edges of the SPA interior[14]. The lack of evidence of widespread and abundant olivine in the largest impact basin on the Moon led to an alternative interpretation: that the SPA basin event was probably an oblique impact[15], excavating pyroxene-rich lower-crust material (norites)[3,12] instead of olivine-bearing mantle material. The picture was further complicated by the growing awareness that large impact basins would produce huge quantities of impact melt, which is predicted to pond and possibly undergo differentiation in the basin interior[16,17]. What would these products look like and how can we distinguish these from samples of the lunar mantle? Furthermore, how might emplacement of post-SPA-basin maria and cryptomaria[18] influence remote-sensing data?

Recent developments in impact-basin hydrocode modelling enabled a more detailed prediction of the nature of an SPA - scale basin - forming event[19,20]. These models show that the SPA event should have penetrated to depths of several hundred kilometres, well below the crustal thickness estimated by the GRAIL data[1], and would have clearly exca - vated mantle material and redistributed it on the surface within and in the vicinity of the SPA basin. However, the absence of abundant olivine in the SPA interior remains a conundrum. Could the predictions of an olivine - rich lunar mantle[8,9] be incorrect? A partial solution to this problem has been recently proposed[20]: perhaps the mantle was composed predominantly of orthopyroxine, not olivine. Could orthopyroxene[4] actually be the dominant material in the upper lunar mantle and olivine a minor component?

These questions were of such fundamental importance that they were prominently featured in documents reporting national space exploration strategies[21,22]. However, exploration infrastructure technology had confined the several dozen in situ lunar missions to land on the lunar near side. Further insight required in situ analyses and sample return from the floor of the SPA basin, on the lunar far side. In an effort to address these fundamental questions, the Chinese Lunar Exploration Program (CLEP) developed and implemented the QueQiao relay satellite to facilitate lunar far - side in situ surface exploration and samplereturn missions.

Such a mission[5] is Chang'E - 4, which was targeted at the floor of the SPA basin. Paramount in the selection of the landing site were scientific goals and objectives, as well as landing safety. To have the highest likelihood of sampling lunar - mantle material[6], the landing site was chosen on the relatively smooth mare floor of the post - SPA - basin Von Kármán crater (diameter of about 186km). The geologic setting of the landing site[6] is a region of impact craters superposed on the floor of the SPA basin (the 225 - km - diameter Von Kármán M crater and the 245 - km - diameter Leibnitz crater), and other nearby craters that are likely to deliver SPA - basin - subfloor material to the landingsite. The youngest of these is the 72 - km - diameter crater Finsen, which is located to the northeast of the landing site and clearly delivers ejecta to the landingsite region. The fact that the central peaks of Finsen have provided the most prominent and purest detection of orthopyroxene in the SPA basin[23] strongly suggests that mantle material has been delivered to the site by the Finsen cratering event. Thus the regionally smooth Von Karman floor mare deposits enhance landing safety, and the extensive ejecta delivered by Finsen and other nearby deeply excavating post - SPA craters provide multiple possible samples of lunar mantle[6].

Chang'E - 4 successfully landed in the Von Kármán crater of the SPA basin at 10: 26 (utc＋8) on 3 January 2019. The Chang'E - 4 landing site (called Statio Tianhe) is 177.5991°E, 45.4446°S, with a - 5, 935 - m surface elevation (relative to a 1,737.4 - km - radius lunar spheroid). The VNIS[7] onboard the Yutu - 2 rover enabled the first in situ reflectance measurements of the far side of the Moon. The VNIS acquires hyper - spectral reflectance data at wavelengths of 450—2, 400nm with spectral resolutions of 2.4

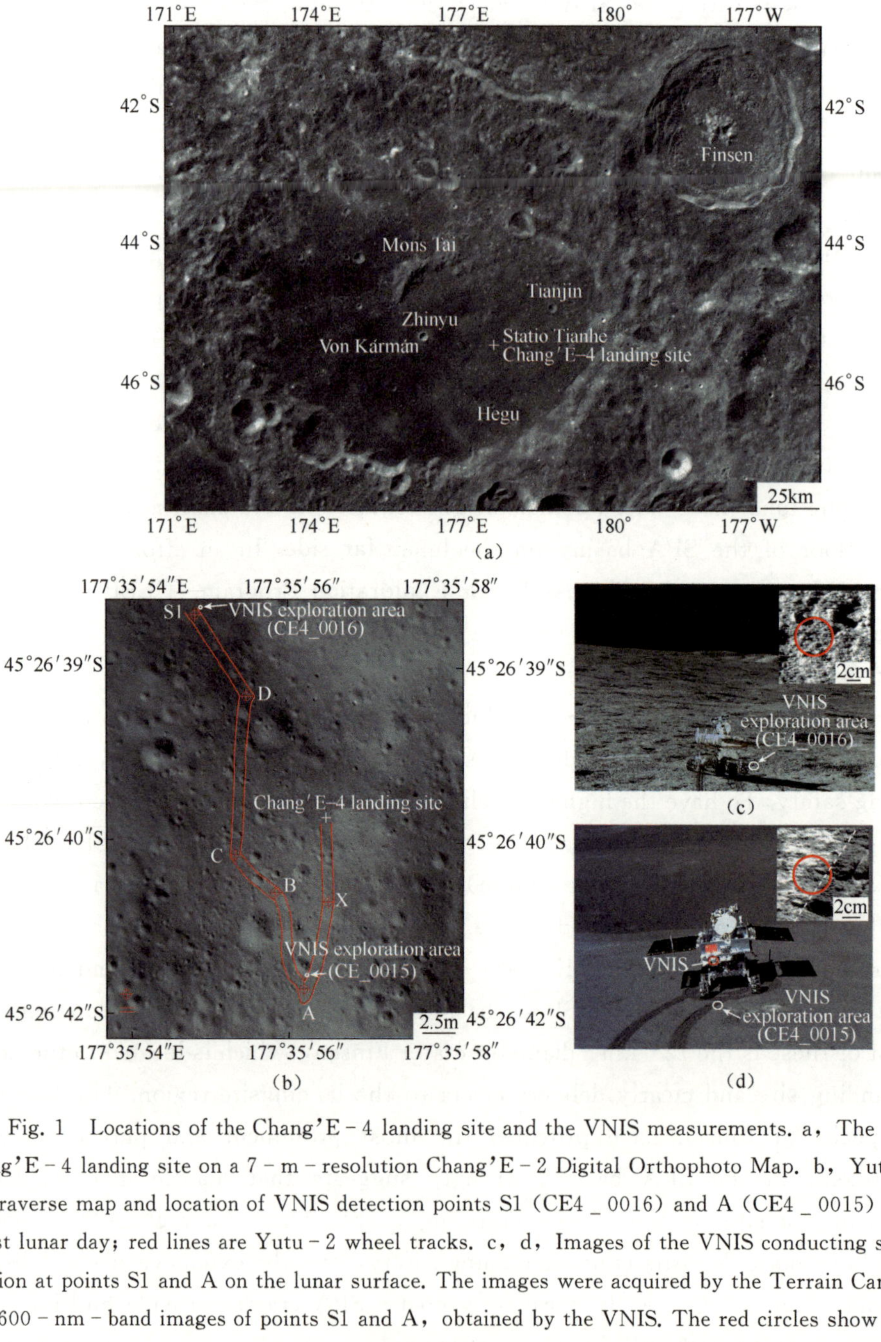

Fig. 1　Locations of the Chang'E - 4 landing site and the VNIS measurements. a, The Chang'E - 4 landing site on a 7 - m - resolution Chang'E - 2 Digital Orthophoto Map. b, Yutu - 2 rover traverse map and location of VNIS detection points S1 (CE4 _ 0016) and A (CE4 _ 0015) during the first lunar day; red lines are Yutu - 2 wheel tracks. c, d, Images of the VNIS conducting spectral detection at points S1 and A on the lunar surface. The images were acquired by the Terrain Camera[5]. Insets, 600 - nm - band images of points S1 and A, obtained by the VNIS. The red circles show the field view of the SWIR detector, and the white dashed line is the boundary between the rover wheel track and the lunar surface

—6.5nm for visible/near - infrared (VIS/NIR) and 3.6 — 9.6 nm for the short wave infrared spectrometer (SWIR; see Methods).

The radiance spectra of two sites adjacent to the lander (Fig. 1) were collected during the first lunar day. These radiance spectra were first divided by the solar irradiance and smoothed to obtain the reflectance factor (REFF) (Fig. 2a). Continuum - removal and parabola fits were then applied to acquire an accurate absorption - band centre for the two reflectance spectra (Fig. 2b; see Methods).

The two spectra obtained by Chang'E - 4 are obviously different from Chang'E - 3 spectra of mare soils in Mare Imbrium[24] (Fig. 2) and those of mare samples measured by the Lunar Soils Characterization Consortium (LSCC; Extended Data Fig. 1a). They are characterized by an attenuated absorption - band strength and a red - sloped continuum, indicating a stronger space weathering effect[25,26] than that of the Chang'E - 3 samples. In addition, the centres of the 1 - μm and 2 - μm absorption bands in the Chang'E - 4 spectra move to shorter wavelengths, showing apparent spectral features of low - Ca pyroxene (LCP) and olivine. The overall spectral shape and absorption - band positions are more similar to those of lunar highland materials (Extended Data Fig. 1b). These results suggest

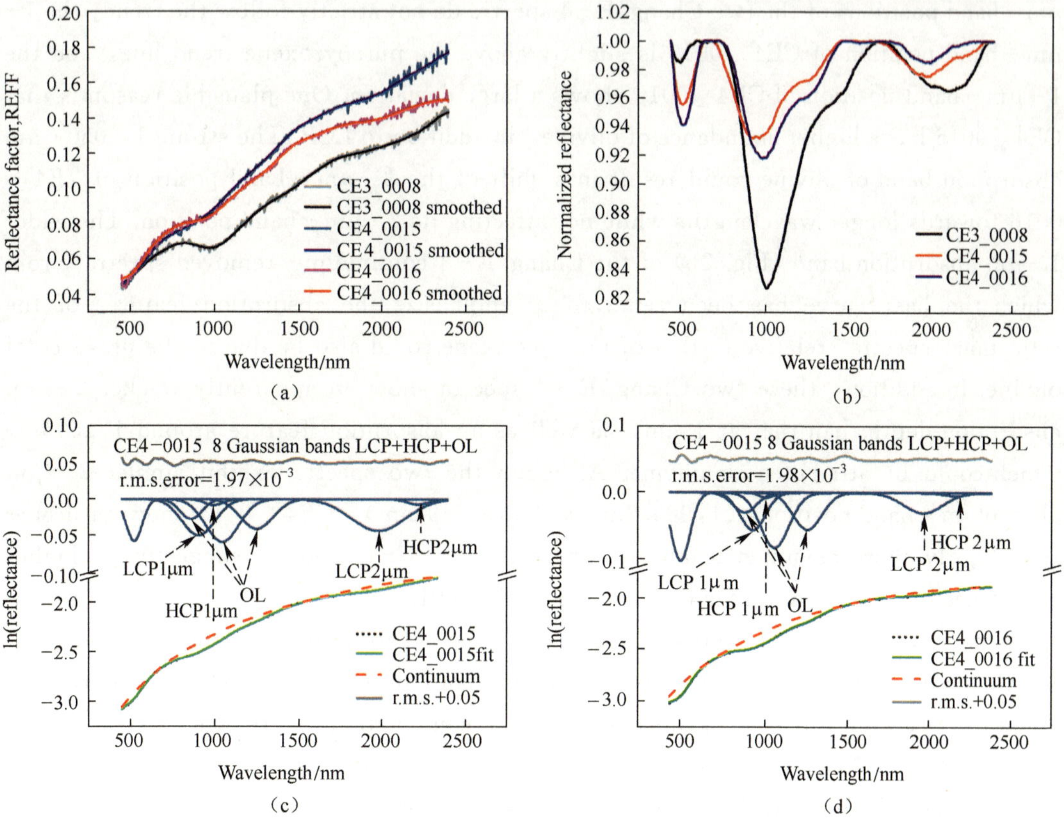

Fig. 2　Chang'E - 4 reflectance spectra acquired by the VNIS during the first lunar day. a, Reflectance (REFF) spectra obtained by Chang'E - 4 VNIS (CE4 _ 0015 and CE _ 0016) and the Chang'E - 3 VNIS detection point (CE3 _ 0008). b, Continuum - removed spectra of a. c, d, MGM - fitting results for CE4 _ 0015 and CE4 _ 0016 using endmember LCP, HCP and olivine

that the materials at the Chang'E - 4 landing site are enriched in mafic components distinct from mare basalt.

The characteristic 1 - μm and 2 - μm absorption features of the two VNIS observations are attributed to the presence of pyroxene - bearing lithologies in the soils of the Chang'E - 4 landing site. We plotted the positions of the 1 - μm and 2 - μm bands of the two spectra (Extended Data Fig. 2) on the trend line (Fig. 3) defined by pyroxene of varying compositions (from Mg - rich LCP to Ca - and Fe - rich pyroxene)[4] to further classify the type of pyroxene and determine whether other possible minerals might exist within the Chang'E - 4 landing site. Although CE4 _ 0015 and CE4 _ 0016 have a weak 2 - μm absorption, it can still be seen that the centres of their 2 - μm absorption bands are below 2,000nm, suggesting that the materials at the landing site exhibit an LCP signature[27,28] (CE4 _ 0015 and CE4 _ 0016 represent the 15th and 16th navigation points of the Yutu - 2 rover, respectively, where the two spectra were acquired). However, the 1 - μm and 2 - μm - band positions of the two Chang'E - 4 spectra do not strictly follow the trend; the 1 - μm - band position of CE4 _ 0015 is slightly above the purepyroxene trend line, and the 1 - μm - band position of CE4 _ 0016 shows a large deviation. One plausible reason is that CE4 _ 0016 has a higher abundance of olivine, in addition to LCP. The strong 1, 050 - nm absorption band of olivine could result in a shift of the 1 - μm - band position of CE4 _ 0016 towards longer wavelengths while not affecting its 2 - μm - band position. The wider 1 - μm absorption band (Fig. 2b) of the Chang'E - 4 continuum - removed spectra (from which the base curve has been removed to emphasize the absorption features of the reflectance spectra) relative to that of pure pyroxene could also be due to the presence of olivine. In addition, these two Chang'E - 4 spectra show an apparently weaker 2 - μm absorption depth than that at 1 μm, as well as an absorption feature around 1,250nm, which could be attributed to olivine. Although the two spectra exhibit similar 2 - μm absorption - band positions (1,985.9nm and 1,984.9 nm), CE4 _ 0016 shows a deeper and slightly more asymmetric absorption relative to CE4 _ 0015, suggesting a higher abundance of olivine within CE4 _ 0016 than CE4 _ 0015.

The modified Gaussian model (MGM) was applied to deconvolve the two spectra of the Chang'E - 4 landing site to constrain the minerals present and estimate their relative amounts[29,30]. Four different groups of mineral combinations were used in the MGM deconvolution: (1) LCP, HCP and olivine (OL); 2) LCP, HCP and plagioclase (Plag); 3) LCP and Plag; 4) LCP and OL. The results show that the continuum - removed reflectance spectra of CE4 _ 0015 can be accurately fitted by the mineral combination of LCP+ HCP + OL, with the lowest root - mean - square (r. m. s.) deviation relative to the other three groups of mineral combination (Fig. 2c, Extended Data Fig. 3). The significance test results also show that HCP+ LCP + OL can achieve the best MGM fitting results (see Methods). Further analysis on the deconvolved 1 - μm - band depth of LCP, HCP and OL suggests that for CE4 _ 0015 the abundances for LCP: HCP: OL are

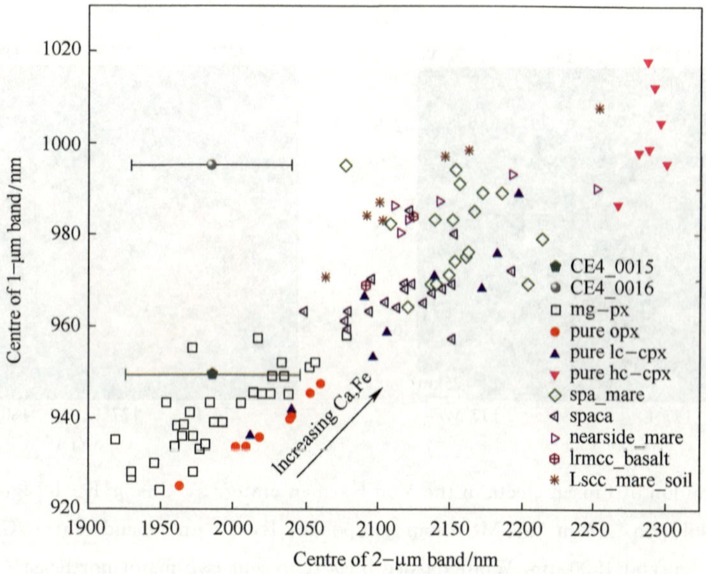

Fig. 3　Absorption - band positions for CE _ 0015 and CE _ 0016. Positions of the 1 - μm and 2 - μm bands of the two Chang'E - 4 spectra compared with the trend line defined by pyroxene of varying compositions (data are from figure 8b of ref.[4]; annotation as in ref. [4]). The solid green polygon and black circle represent CE4 _ 0015 and CE4 _ 0016, respectively. The green and black error bars show the wavelength range used for the 2 - μm absorption parabola fitting

42%: 10%: 48% (see Methods), with the highest abundance for OL, followed by LCP and the lowest abundance for HCP. MGM deconvolution of CE4 _ 0016 (Fig. 2d) indicates that LCP: HCP: OL 38%: 7%: 55%. CE4 _ 0016 is dominated by LCP and OL and has a greater abundance of OL than CE4 _ 0015 and a very small amount of HCP. In addition, Moon Mineralogy Mapper (M^3) spectra of materials at the Chang'E - 4 landing site also exhibit spectral features of LCP[6] (Extended Data Fig. 6), in agreement with the analysis results of the Chang'E - 4 data.

The Chang'E - 4 landing site is located in the Von Kármán crater, which flooded after its formation by lava flows of basaltic composition[6]. However, the Chang'E - 4 spectra have considerable differences from spectral features typical of mare basalts. Analysis of the Chang'E - 1 image data reveals that the Chang'E - 4 landing site is located in a bright northeast - southwest linear feature, which, together with another linear feature to the north, converges towards the location of the Finsen crater[6] (Fig. 4a). Therefore, we consider the Chang'E - 4 landing site to have been modified by the emplacement of ejecta from Finsen[6]. The ejecta of the Finsen crater could represent deep - seated materials emplaced at the time that the SPA basin formed. Spectral analysis of M^3 data also indicates an obvious compositional difference between the bright ejecta and mare materials within the Von Kármán crater[4,6] (Fig. 4b). Moreover, extensions of these two

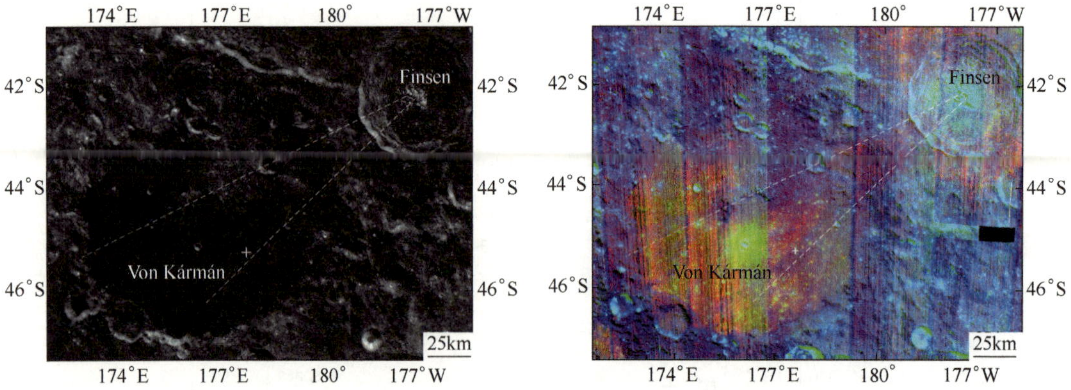

(a)　　(b)

Fig. 4　Distribution of Finsen ejecta in the Von Kármán crater. a, Chang'E－1 Digital Orthophoto Map; spatial resolution, 120m. b, M^3 colour composite; R, 2－μm－band centre; G, 2－μm－band depth; B, reflectance at 1580 nm. White dashed lines represent two major northeast－southwest ejecta rays of the Finsen crater converging towards its central peak. The white cross is the Chang'E－4 landing site, which is located on the ejecta material of the Finsen crater. The 2－μm－band centre can be used to distinguish between LCP and HCP. As the content of Fe and Ca in the pyroxene increases, the 2－μm－band centre shifts towards longer wavelengths (redder). The 2－μm－band depth indicates the relative amounts of mafic minerals. The 1580－nm reflectance represents the brightness of the lunar soils (which can be affected by plagioclase content). LCP－bearing materials appear light blue, and HCP－bearing materials appear green

bright linear features within the Von Kármán crater converge to the central peak of the Finsen crater, the dominant composition of which has been shown to be LCP[4,23,31]. This is consistent with what we observe in the CE4 _ 0015 and CE _ 0016 data. On the basis of these analyses, we propose that materials in the immediate Chang'E－4 landing site are not dominated solely by mare basalt, but that they also contain additional mafic material ejected from the Finsen crater[6]. The materials of the Chang'E－4 landing site show a strong resemblance to those exposed in the mafic central peak of the Finsen crater and therefore could readily originate from deep－seated materials excavated at the time of the formation of the SPA basin and subsequently re－excavated by the Finsen crater event and delivered to the Chang'E－4 landing site on the floor of the Von Kármán crater.

Several studies indicate that the impact event that formed the SPA basin might not have excavated through the lunar crust and exposed the mantle material on the surface of the Moon[12,18]. Our results, however, imply that the Chang'E－4 landing site is characterized by mafic components that are dominated by LCP and olivine, with a very small amount of HCP, suggesting the presence of deep－seated material[32] from the upper mantle[20], rather than the surface mare basalts on the floor of the Von Kármán crater. Alternatively, these mafic components might originate from the base of a differentiated melt sheet[17].

In summary, we found that materials at the Chang'E-4 landing site are obviously different from most sampled mare basalts and possess mafic components that are interpreted to be dominated by LCP and olivine, with very small amounts of HCP. These may represent deep-seated materials potentially from the lunar mantle, delivered to the Chang'E-4 landing site by ejecta from the Finsen crater. Moreover, our results suggest that the predictions of an olivine-rich lunar mantle cannot be ruled out. The lunar upper mantle could be composed predominantly of both LCP and olivine, rather than simply dominated by LCP with minor olivine content. Future operations of the Chang'E-4 Yutu-2 rover will focus on validating these initial results, understanding the origin of the olivine and analysing the distribution, geologic settings, characteristics and origin of these mafic candidate mantle materials on the floor of the Von Kármán crater.

Online content

Any methods, additional references, Nature Research reporting summaries, source data, statements of data availability and associated accession codes are available at https://doi.org/10.1038/s41586-019-1189-0.

References

[1] Wieczorek, M. A. et al. The crust of the Moon as seen by GRAIL. Science 339, 671-675 (2013).

[2] Wood, J. A., Dickey, J. S., Marvin, U. B. & Powell, B. N. Lunar anorthosites and a geophysical model of the Moon. Geochim. Cosmochim. Acta 1, 965-988 (1970).

[3] Stuart-Alexander, D. E. Geologic Map of the Central Far Side of the Moon I-1047 (US Geological Survey, Denver, 1978).

[4] Moriarty, D. P. III & Pieters, C. M. The character of South Pole Aitken basin: patterns of surface and subsurface composition. J. Geophys. Res. Planets 123, 729-747 (2018).

[5] Li, C. L. et al. The Chang'e 3 mission overview. Space Sci. Rev. 190, 85-101 (2015).

[6] Huang, J. et al. Geological characteristics of Von Kármán crater, northwestern South Pole Aitken basin: Chang'E 4 landing site region. J. Geophys. Res. Planets 123, 1684-1700 (2018).

[7] He, Z. P. et al. Operating principles and detection characteristics of the visible and near-infrared imaging spectrometer in the Chang'E-3. Res. Astron. Astrophys. 14, 1567-1577 (2014).

[8] Wieczorek, M. A. et al. The constitution and structure of the lunar interior. Rev. Mineral. Geochem. 60, 221 - 364 (2006).

[9] Shearer, C. et al. Thermal and magmatic evolution of the Moon. Rev. Mineral. Geochem. 60, 365 - 518 (2006).

[10] Tompkins, S. & Pieters, C. M. Mineralogy of the lunar crust: results from Clementine. Meteorit. Planet. Sci. 34, 25 - 41 (1999).

[11] Yamamoto, S. et al. Possible mantle origin of olivine around lunar impact basins detected by SELENE. Nat. Geosci. 3, 533 - 536 (2010).

[12] Pieters, C. M., Tompkins, S., Head, J. W. & Hess, P. C. Mineralogy of the mafic anomaly in the South Pole - Aitken basin: implications for excavation of the lunar mantle. Geophys. Res. Lett. 24, 1903 - 1906 (1997).

[13] Prettyman, T. H. et al. Elemental composition of the lunar surface: analysis of gamma ray spectroscopy data from Lunar Prospector. J. Geophys. Res. 111, E12007 (2006).

[14] Yamamoto, S. et al. Olivine - rich exposures in the South Pole - Aitken basin. Icarus 218, 331 - 344 (2012).

[15] Garrick - Bethell, I. & Zuber, M. T. Elliptical structure of the lunar South Pole - Aitken basin. Icarus 204, 399 - 408 (2009).

[16] Vaughan, W. M., Head, J. W., Wilson, L. & Hess, P. C. Geology and petrology of enormous volumes of impact melt on the Moon: a case study of the Orientale basin impact melt sea. Icarus 223, 749 - 765 (2013).

[17] Vaughan, W. M. & Head, J. W. Impact melt differentiation in the South Pole - Aitken basin: some observations and speculations. Planet. Space Sci. 91, 101 - 106 (2014).

[18] Pieters, C. M., Head, J. W., Gaddis, L., Jolliff, B. & Duke, M. Rock types of South Pole Aitken basin and extent of basaltic volcanism. J. Geophys. Res. Planets 106, 28001 - 28022 (2001).

[19] Potter, R. W. K., Collins, G. S., Kiefer, W. S., McGovern, P. J. & Kring, D. A. Constraining the size of the South Pole - Aitken basin impact. Icarus 220, 730 - 743 (2012).

[20] Melosh, H. J. et al. South Pole - Aitken basin ejecta reveal the Moon's upper mantle. Geology 45, 1063 - 1066 (2017).

[21] Guo, H. & Wu, J. in Space Science and Technology in China: A Roadmap to 2050 (ed. Guo, H.) 99 (Chinese Academy of Sciences Science Press, Beijing, 2009).

[22] US National Research Council. The Scientifc Context for Exploration of the Moon (National Academies Press, Washington DC, 2007).

[23] Moriarty, D. P. et al. Finsen and Alder: a compositional study of lunar central peak craters in the South Pole - Aitken basin. Lunar Planet. Sci. Conf. 42, 2564 (2011).

[24] Ling, Z. et al. Correlated compositional and mineralogical investigations at the

Chang'E-3 landing site. Nat. Commun. 6, 8880 (2015).

[25] Hapke, B. Space weathering from Mercury to the asteroid belt. J. Geophys. Res. Planets 106, 10039-10073 (2001).

[26] Pieters, C. M. et al. Space weathering on airless bodies: resolving a mystery with lunar samples. Meteorit. Planet. Sci. 35, 1101-1107 (2000).

[27] Klima, R. L., Dyar, M. D. & Pieters, C. M. Near infrared spectra of clinopyroxenes: effects of calcium content and crystal structure. Meteorit. Planet. Sci. 46, 379-395 (2011).

[28] Klima, R. L., Pieters, C. M. & Dyar, M. D. Spectroscopy of synthetic Mg-Fe pyroxenes I: spin-allowed and spin-forbidden crystal feld bands in the visible and near-infrared. Meteorit. Planet. Sci. 42, 235-253 (2007).

[29] Sunshine, J. M., Pieters, C. M. & Pratt, S. F. Deconvolution of mineral absorption bands: an improved approach. J. Geophys. Res. Solid Earth 95, 6955-6966 (1990).

[30] Sunshine, J. M. & Pieters, C. M. Estimating modal abundances from the spectra of natural and laboratory pyroxene mixtures using the modifed Gaussian model. J. Geophys. Res. Planets 98, 9075-9087 (1993).

[31] Nakamura, R. et al. Ultramafc impact melt sheet beneath the South Pole-Aitken basin on the Moon. Geophys. Res. Lett. 36, L22202 (2009).

[32] Neumann, G. & Head, J. Geophysical characteristics of Von Kármán crater: Chang'E 4 landing site region. Lunar Planet. Sci. Conf. 50, 2132 (2019).

Acknowledgements

This research was funded by the Chang'E-4 mission of CLEP. We thank the team members of the Ground Application and Research System (GRAS), who contributed to data receiving and preprocessing.

METHODS

The VNIS. As one of the four scientific payloads on the Chang'E-4 Yutu-2 rover, the main task of the VNIS is in situ mineral detection on the lunar surface. The technical characteristics of the Chang'E-4 VNIS (Extended Data Table 1) are consistent with those of Chang'E-3[7,33]. The VNIS spectral range is 450-2,400nm and consists of one hyperspectral imaging detector(VIS/NIR;450-950nm)and one single-point spectrometer (SWIR; 900—2,400nm). The VNIS is mounted on the front of the rover at a height of 0.69m above the ground and detects the lunar surface at a nominal angle of 45°, leading to an isosceles trapezoid detection area on the surface (height, 20.6cm and two parallel sides of 13.5cm and 15.7cm)[34,35]. The single-point SWIR spectrometer has a circular field of view, the centre of which corresponds to (X98, Y128) of the VIS/NIR image. The radius of the circle covers 53.8 pixels of the VIS/NIR image (pixel size, 25μm). In addition, the VNIS carries a calibration panel composed of polytetrafluoroethylene.

VNIS data processing and analysis. The VNIS spectral data processing began with level 2B data, which underwent dark-current subtraction, scattering-background correction, flat-field and instrument temperature corrections, and radiometric and geometric calibration[7,33]. First, the radiance data of the VIS/NIR and SWIR bands of CE4_0015 and CE4_0016 were divided by the solar irradiance, multiplied by π, and further divided by the cosine of the incident angle to obtain the reflectance factor, REFF, of the two detection points. Second, for each detection point, the reflectance spectra (450—950nm) of 9,082 pixels from the circle shown in Fig. 1c, d within the imaged area were averaged to obtain one spectrum (450—950nm), which was then connected with the point spectrum (900—2,400nm) measured in the same circular area. However, this process requires the correction of a step artefact caused by the different spectral responses of the two detectors. Considering the small spectral jitter and the stability of the point spectral data, we took the SWIR data as the standard and multiplied the VIS/NIR data by an adjustment factor, which was derived from the ratio of SWIR 900-nm to the VIS-NIR 900-nm reflectance data, to obtain one continuous spectrum. Finally, the continuous spectra were smoothed using the Savitzky-Golay method[24,36] (Fig. 2a) and the spectral continuum was removed using the ENVI software (convex hull fit) (Fig. 2b). After removal of the continuum, the positions of the 1-μm and 2-μm absorption bands of the two spectra were identified through parabola fitting (Extended Data Fig. 2). The results show that the positions of the 1-μm band for CE4_0015 and CE4_0016 are 949.2nm and 995.1nm, and those of the 2-μm band are 1,985.9nm and 1,984.9nm, respectively. A photometric correction has not yet been performed in this study owing to the lack of a suitable correction model for large phase angles (>90°). However, this does not affect the comparison among CE4_0015, CE4_0016 and CE3_0008 data because of the similar phase angles (~90°) of the three spectra.

Mineral assemblage analysis using MGM deconvolution. MGM was applied to deconvolve the CE4_0015 and CE4_0016 spectra to constrain their mineral composition[30,37]. The software developed by the Reflectance Experiment Laboratory (Relab; http://www.planetary.brown.edu/relab/) was used. The identified absorption-band positions of CE4_0015 and CE4_0016 imply the presence of pyroxene. Both spectra have wide and strong 1-μm absorption but weak 2-μm absorption, resulting in a higher area ratio between the 1-μm and 2-μm bands[33,38,39]. This suggests the presence of a high amount of olivine within the Chang'E-4 landing site in addition to the pyroxene components. However, we cannot rule out the existence of plagioclase. Therefore, we used four different mineral assemblages (LCP+HCP+OL, LCP+HCP+Plag, LCP+Plag, LCP+OL) and different numbers of Gaussian bands (8, 6, 4, 6) to deconvolve the CE4_0015 spectra[24]. The initial parameter settings for MGM are shown in Extended Data Table 2, and Extended Data Fig. 3 shows the MGM modelling results for different mineral assemblages. Generally, the derived 1 - μm - band depth ratios of LCP/HCP should be similar to their 2 - μm - band depth ratios[30]. Moreover, when olivine is considered in the MGM deconvolution process, the Fo value (the relative content of forsterite in olivine, Fo= { [Mg]/([Mg] + [Fe]) } ×100, where square brackets denote concentrations in mol%) derived from the three absorption bands of olivine around 1μm should be consistent[37]. Following these criteria, the best mineral combination is LCP+ HCP + OL and its MGM deconvolution result reveals the lowest r. m. s. error (1.97×10^{-3}). We also analysed the statistical significance of the difference between the MGM fit and the observed spectral data for all of these mineral assemblages using two - factor analysis of variance without replication to further evaluate the MGM fitting results. The significance test (see Extended Data Table 3) shows that HCP+ LCP + OL has the lowest F value (the value on the F distribution) and the largest P value compared to the other three groups, which means that this mineral assemblage can achieve the best MGM fitting results. The same mineral assemblage (LCP + HCP + OL) was also used to perform MGM deconvolution on CE4 _ 0016, resulting in an r. m. s. error of 1.98×10^{-3}. The initial parameter settings for CE4 _ 0016 are shown in Extended Data Table 4, and the deconvolution result is shown in Fig. 2d.

The content ratio of HCP/LCP can be calculated using their band - depth ratios at 1μm or 2μm, which are derived from MGM deconvolution[27,28,30,40,41]. The ratio of the 1 - μm - band depth for HCP to the 1. 25 - μm - band depth for olivine can be used to estimate their relative content[40]. The MGM deconvolution results of the two spectra are shown in Fig. 2c,d, Extended Data Figs. 4,5. The results show that for CE4 _ 0015 the content ratio LCP : HCP is 81% : 19% (1μm position) and HCP : OL is 17% : 83%. Therefore, LCP : HCP : OL is 42% : 10% : 48%. For CE4 _ 0016, the ratio LCP : HCP is 84% : 16% (1μm position) and HCP/OL is 12% : 88%. Thus, LCP : HCP : OL is 38% : 7% : 55%. According to the relationship between the Fo value and the absorption - band positions of olivine[37] (Extended Data Fig. 5), the derived Fo values for CE4 _ 0015 and CE4 _ 0016 are 57% and 28%, respectively. CE4 _ 0016 has more fayalite than CE4 _ 0015. Currently, we cannot explain the large deviation between the olivine Fo values of these two detection points. In a previous study[40] it was shown that large - grain - size forsterites have absorption characteristics that may mimic the absorption features of fayalite. For olivine of the same composition, the variation in Fo value derived from MGM deconvolution can be up to about 30% when the grain size changes from 45—75μm to 250—500μm. Interpretation of the MGM deconvolution results should consider the effect of changing grain size. In this study, we are unable to accurately constrain the grain size distribution of the landing site. Therefore, the Fo value of olivine derived here could be affected by grain size. The difference between the CE4 _ 0015 and CE4 _ 0016 Fo values may not result from the chemical variation of olivine—it may be also due to the changing grain size of the landing site.

[33] Liu, B. et al. Data processing and preliminary results of the Chang'E-3 VIS/NIR imaging spectrometer in-situ analysis. Res. Astron. Astrophys. 14, 1578-1594 (2014).

[34] Wu, Y. Z. & Hapke, B. Spectroscopic observations of the Moon at the lunar surface. Earth Planet. Sci. Lett. 484, 145-153 (2018).

[35] Wu, Y. Z. et al. The absolute refectance and new calibration site of the Moon. Astron. J. 155, 213 (2018).

[36] Savitzky, A. & Golay, M. J. Smoothing and differentiation of data by simplifed least squares procedures. Anal. Chem. 36, 1627-1639 (1964).

[37] Sunshine, J. M. & Pieters, C. M. Determining the composition of olivine from refectance spectroscopy. J. Geophys. Res. Planets 103, 13675-13688 (1998).

[38] Staid, M. I. et al. The mineralogy of late stage lunar volcanism as observed by the Moon Mineralogy Mapper on Chandrayaan-1. J. Geophys. Res. Planets 116, E00G10 (2011).

[39] Besse, S. et al. Compositional variability of the Marius Hills volcanic complex from the Moon Mineralogy Mapper (M^3). J. Geophys. Res. Planets 116, E00G13 (2011).

[40] Clénet, H. et al. A new systematic approach using the Modifed Gaussian Modal: insight for the characterization of chemical composition of olivines, pyroxenes and olivine-pyroxene mixtures. Icarus 213, 404-422 (2011).

[41] Kanner, L. C., Mustard, J. F. & Gendrin, A. Assessing the limits of the Modifed Gaussian Model for remote spectroscopic studies of pyroxenes on Mars. Icarus 187, 442-456 (2007).

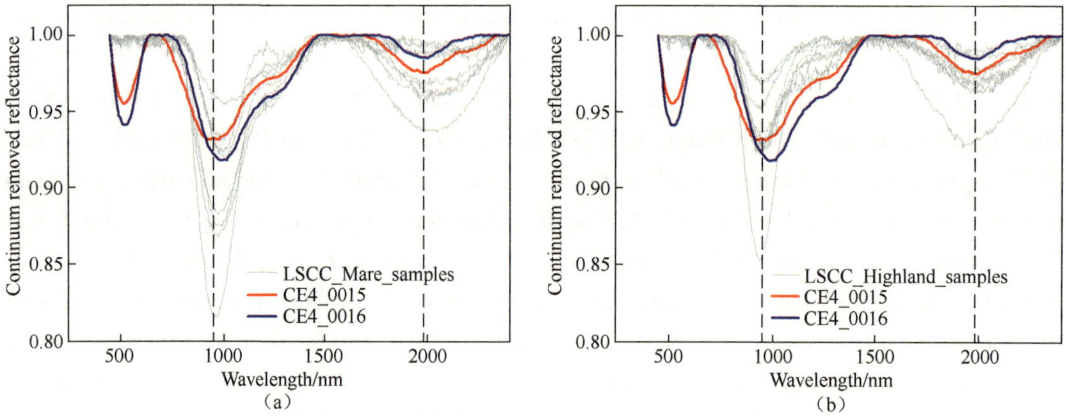

Extended Data Fig. 1　Spectral comparison between CE4_0015, CE4_0016 and LSCC samples (continuum removed). The red and blue lines represent CE4_0015 and CE4_0016. (a) Comparison of CE4_0015 and CE4_0016 with mare soil (particle size<45μm) samples: 10084, 12001, 12030, 15041, 15071, 70181, 71501, 79221. The 1-μm and 2-μm absorption bands of CE4_0015 and the 2-μm absorption band of CE4_0016 (dashed lines) obviously shifts to shorter wavelengths and differ considerably from the spectral features of the mare samples. (b) Comparison of CE4_0015 and CE4_0016 with highland samples (particle size<45μm): 14141, 14163, 14259, 14260, 61141, 61221, 62231, 64801, 67461, 67481. The 1-μm and 2-μm absorption bands of CE4_0015 and the 2-μm absorption band (dashed lines) of CE4_0016 resemble those of the highland samples. However, the 1-μm absorption band of CE4_0016 is more similar to that of basaltic soil, which can be attributed to elevated abundance of olivine (see main text)

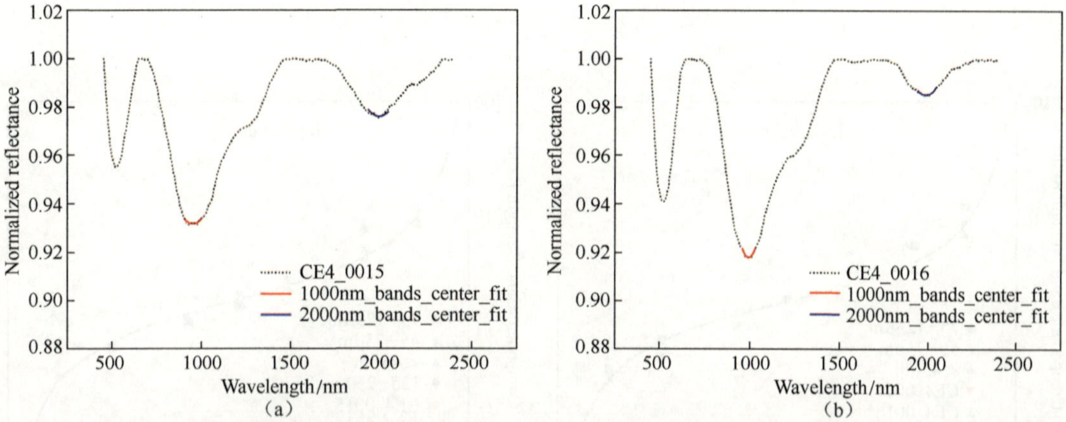

Extended Data Fig. 2　Parabola - fitting results for CE4 _ 0015 and CE4 _ 0016 1 - μm and 2 - μm absorption - band positions. (a) Reflectance of CE4 _ 0015; the positions of the 1 - μm and 2 - μm bands are 949. 2nm and 1, 985. 9nm, respectively. (b) Reflectance of CE4 _ 0016; the 1 - μm - and 2 - μm - band positions are 995. 1nm and 1, 984. 9nm, respectively. The 2 - μm - band centre assignment is tentative because of its weak absorption. The red and blue lines represent the wavelength ranges used for the parabola fitting of the 1 - μm - and 2 - μm - band centres, respectively

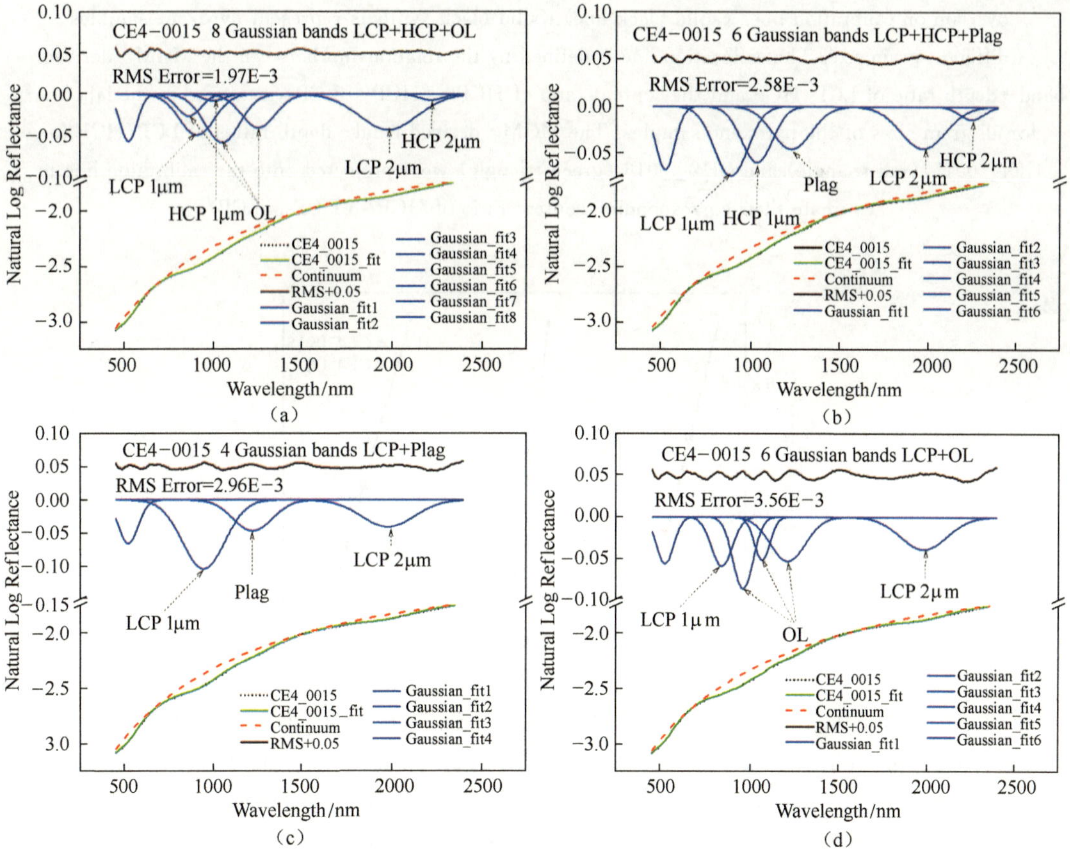

Extended Data Fig. 3　MGM deconvolution results of CE4 _ 0015 spectra using four different mineral assemblages. (a) LCP + HCP + OL. (b) LCP + HCP + Plag. (c) LCP + Plag. (d) LCP + OL

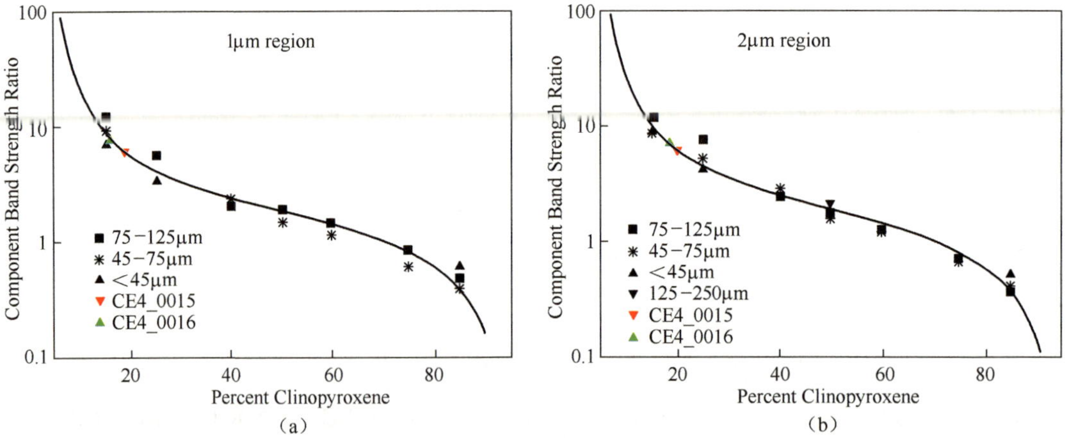

Extended Data Fig. 4　Content ratio HCP/(HCP + LCP) for CE4 _ 0015 and CE4 _ 0016, calculated by MGM deconvolution. (a) 1 - μm result: 19% (CE4 _ 0015) and 16% (CE4 _ 0016). (b) 2 - μm results: 20% (CE4 _ 0015) and 18% (CE4 _ 0016). Data are from figure 7 of ref.[30], overlain on calibration lines (solid black line). Solid black symbols represent pyroxene samples of different grain size. The calibration line is defined by the relationship between the MGM - derived band - depth ratio of LCP/HCP and the content ratio of HCP/ (HCP+ LCP) measured in the laboratory for all grain sizes of the pyroxene samples. The MGM - derived band - depth ratios of LCP/HCP for CE4 _ 0015 (red triangle) and CE4 _ 0016 (green triangle) were projected onto the calibration line to estimate their corresponding content ratio of HCP/ (HCP+ LCP)

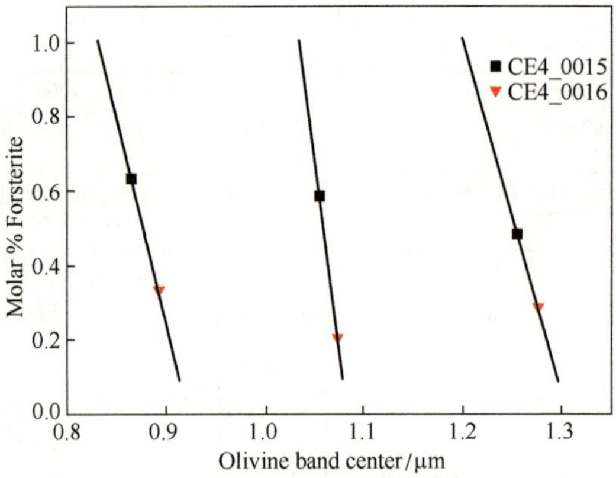

Extended Data Fig. 5　Molar fraction of forsterite in olivine in CE4 _ 0015 and CE4 _ 0016. The results were derived from VNIS spectra, overlain on calibration lines defined by the relationship between the Fo value and the absorption - band positions of olivine of various compositions (data used to plot the calibration lines are from figure 4 of ref.[37])

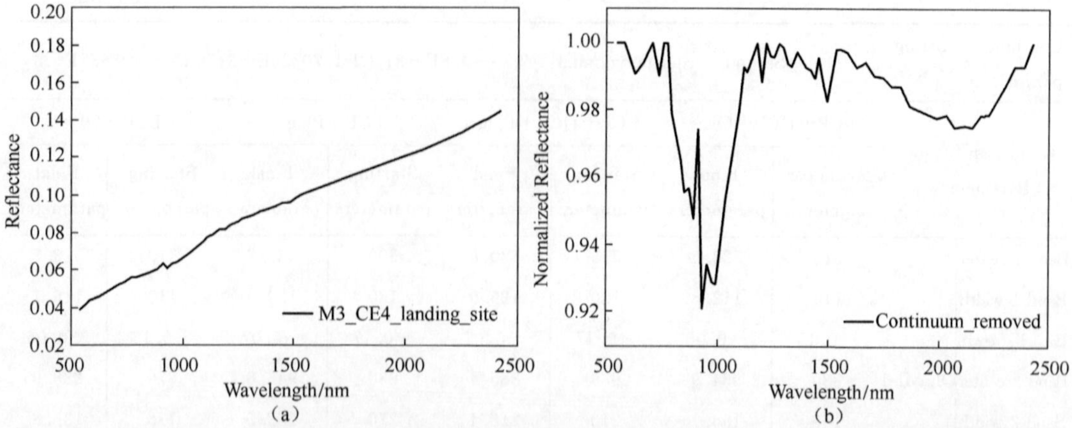

Extended Data Fig. 6　M^3 reflectance spectrum of Chang'E－4 landing site. (a) Average M^3 reflectance spectra obtained using five adjacent pixels with central pixel position (64, 4, 927). (b) Continuum－removed spectrum; 1－μm－band position, 930.1 nm; 2－μm－band position, 2,058.66nm

Extended Data Tab. 1　Performance characteristics of Chang'E－4 VNIS

Description	Specification	
	VIS/NIR	SWIR
Spectral range/nm	450～950	900～2400
Spectral resolution/nm	2.4～6.5	3.6～9.6
Field of view/(°)	8.5° * 8.5°	Φ3.58°
Effective pixel counts	256 * 256	1
Bit resolution	10	16
Signal to noise ratio	43 (maximum SNR) 33 (albedo＝0.09, and solar elevation angle＝45°)	46 (maximum SNR) 31 (albedo＝0.09, and solar elevation angle＝45°)
Detection distance/m	0.7～1.3	
Power consumption/W	16.95	
weight/kg	4.65 (probe part) and 0.48 (electronics part)	
Work modes	Detection mode、calibration mode and dust proofing mode	

SNR, Signal－to－noise ratio; Φ, circular field of view of SWIR.

Extended Data Tab. 2　MGM modelling parameters and errors of CE4 _ 0015 spectra using different mineral components

Continuum fitting parameters	Second－order polynomial: C0＝－3.8E－3; C1 1.20223E－3; C2＝－1.8657E－8							
Gauussian fitting Parameters	LCP＋HCP＋OL		LCP＋HCP＋Plag		LCP＋Plag		LCP＋OL	
	Startingparameters	Final parameters	Starting parameters	Final parameters	Starting parameters	Final parameters	Starting parameters	Final parameters
No. of Bands	8	8	6	6	4	4	6	6

Continued

Continuum fitting parameters	Second - order polynomial：C0＝－3. 8E－3；C1 1. 20223E－3；C2＝－1. 8657E－8								
Gauussian fitting Parameters		LCP＋HCP＋OL		LCP＋HCP＋Plag		LCP＋Plag		LCP＋OL	
		Startingparameters	Final parameters	Starting parameters	Final parameters	Starting parameters	Final parameters	Starting parameters	Final parameters
Band 1 center		515	523. 5	515	520. 8	500	521. 2	510	523. 3
Band 1 width		110	112. 5	120	126. 9	130	129. 3	110	115. 7
Band 1 strength		−0. 15	−0. 06	−0. 17	−0. 07	−0. 17	−0. 07	−0. 15	−0. 06
Band 2 center	OL M1-1	880	864. 9	930	885. 4	930	946. 6	870	838. 9
Band 2 width		155	156. 1	210	218. 1	270	282. 5	115	136. 8
Band 2 strength		−0. 1	−0. 03	−0. 25	−0. 07	−0. 25	−0. 1	−0. 08	−0. 06
Band 3 center	LCP	915	909. 3	980	1029. 6	1300	1224. 7	915	956. 0
Band 3 width		230	231. 0	200	205. 3	260	259. 7	120	132. 4
Band 3 strength		−0. 25	−0. 05	−0. 11	−0. 06	−0. 02	−0. 05	−0. 25	−0. 09
Band 4 center	HCP	1000	1001. 5	1290	1225. 2	1980	1977. 6	1050	1063. 0
Band 4 width		135	134. 9	260	258. 5	340	343. 3	80	105. 6
Band 4 strength		−0. 15	−0. 008	−0. 02	−0. 05	−0s. 1	−0. 04	−0. 08	−0. 05
Band 5 center	OL M2	1050	1052. 1	1980	1968. 2			1250	1208. 5
Band 5 width		235	234. 6	290	299. 2			225	235. 6
Band 5 strength		−0. 1	−0. 06	−0. 1	−0. 04			−0. 1	−0. 05
Band 6 center	OL M1−2	1250	1257. 2	2250	2244. 5			1950	1976. 7
Band 6 width		225	225. 0	160	161. 3			300	310. 4
Band 6 strength		−0. 15	−0. 04	−0. 1	−0. 01			−0. 1	−0. 04
Band 7 center	LCP	1985	1978. 9						
Band 7 width		380	380. 4						
Band 7 strength		−0. 1	−0. 04						
Band 8 center	HCP	2215	2215. 8						
Band 8 width		150	149. 9						
Band 8 strength		−0. 07	−0. 007						
RMS Error		1. 97E−3		2. 58E−3		2. 96E−3		3. 56E−3	

RMS，root mean square.

Extended Data Tab. 3 Significance test for MGM deconvolution of CE4 _ 0015

Source of variation	F	P - value	F crit
fit _ 0015 _ HCP+LCP+OL	4. 94774E−07	0. 999439	3. 865475
fit _ 0015 _ LCP+OL	0. 000416932	0. 98372	3. 865475
fit _ 0015 _ LCP+plag	0. 000215657	0. 988291	3. 865475
fit _ 0015 _ HCP+LCP+Plag	0. 001506455	0. 969059	3. 865475

'F crit' is the critical *F* value, which can be obtained from the *F* distrbution table.

Extended Data Tab. 4 MGM modelling parameters and errors for CE4 _ 0016

Continuum fitting parameters	Second−order polynomial: C0=−3. 8E−3; C1=1. 20223E−3; C2=−1. 8657E−8									
Gauussian fitting Parameters		LCP+HCP+OL		LCP+HCP+Plag		LCP+Plag		LCP+OL		
		Startingparameters	Final parameters	Starting parameters	Final parameters	Starting parameters	Final parameters	Starting parameters	Final parameters	
No. of Bands		8	8	6	6	4	4	6	6	
Band 1 center		515	523. 5	515	520. 8	500	521. 2	510	523. 3	
Band 1 width		110	112. 5	120	126. 9	130	129. 3	110	115. 7	
Band 1 strength		−0. 15	−0. 06	−0. 17	−0. 07	−0. 17	−0. 07	−0. 15	−0. 06	
Band 2 center	OL M1-1	880	864. 9	930	885. 4	930	946. 6	870	838. 9	
Band 2 width		155	156. 1	210	218. 1	270	282. 5	115	136. 8	
Band 2 strength		−0. 1	−0. 03	−0. 25	−0. 07	−0. 25	−0. 1	−0. 08	−0. 06	
Band 3 center	LCP	915	909. 3	980	1029. 6	1300	1224. 7	915	956. 0	
Band 3 width		230	231. 0	200	205. 3	260	259. 7	120	132. 4	
Band 3 strength		−0. 25	−0. 05	−0. 11	−0. 06	−0. 02	−0. 05	−0. 25	−0. 09	
Band 4 center	HCP	1000	1001. 5	1290	1225. 2	1980	1977. 6	1050	1063. 0	
Band 4 width		135	134. 9	260	258. 5	340	343. 3	80	105. 6	
Band 4 strength		−0. 15	−0. 008	−0. 02	−0. 05	−0s. 1	−0. 04	−0. 08	−0. 05	
Band 5 center	OL M2	1050	1052. 1	1980	1968. 2			1250	1208. 5	
Band 5 width		235	234. 6	290	299. 2			225	235. 6	
Band 5 strength		−0. 1	−0. 06	−0. 1	−0. 04			−0. 1	−0. 05	
Band 6 center	OL M1−2	1250	1257. 2	2250	2244. 5			1950	1976. 7	
Band 6 width		225	225. 0	160	161. 3			300	310. 4	
Band 6 strength		−0. 15	−0. 04	−0. 1	−0. 01			−0. 1	−0. 04	
Band 7 center	LCP	1985	1978. 9							
Band 7 width		380	380. 4							
Band 7 strength		−0. 1	−0. 04							
Band 8 center	HCP	2215	2215. 8							
Band 8 width		150	149. 9							
Band 8 strength		−0. 07	−0. 007							
RMS Error		1. 97E−3		2. 58E−3		2. 96E−3		3. 56E−3		

Olivine – norite rock detected by the lunar rover Yutu – 2 likely crystallized from the SPA – impact melt pool

Honglei Lin[1], Zhiping He[2], Wei Yang[1], Yangting Lin[1], Rui Xu[2], Chi Zhang[1], MengHua Zhu[3], Rui Chang[1], Jinhai Zhang[1], Chunlai Li[2], Hongyu Lin[4], Yang Liu[5], Sheng Gou[6], Yong Wei[1], Sen Hu[1], Changbin Xue[7], Jianfeng Yang[8], Jie Zhong[9], Xiaohui Fu[10], Weixing Wan[1] and Yongliao Zou[5]

1. Key Laboratory of Earth and Planetary Physics, Institute of Geology and Geophysics, Chinese Academy of Sciences, Beijing 100029, China
2. Key Laboratory of Space Active Opto – Electronics Technology, Shanghai Institute of Technical Physics, Chinese Academy of Sciences, Shanghai 200083, China
3. State Key Laboratory of Lunar and Planetary Sciences, Macau University of Science and Technology, Macau, China
4. Beijing Institute of Space Mechanics and Electricity, Beijing 100076, China
5. State Key Laboratory of Space Weather, National Space Science Center, Chinese Academy of Sciences, Beijing 100190, China
6. State Key Laboratory of Remote Sensing Science, Institute of Remote Sensing and Digital Earth, Chinese Academy of Sciences, Beijing 100101, China
7. Key Laboratory of Electronics and Information Technology for Space System, National Space Science Center, Chinese Academy of Sciences, Beijing 100190, China
8. Xi' an Institute of Optics and Precision Mechanics, Chinese Academy of Sciences, Xi'an 710119, China
9. Institute of Optics and Electronics, Chinese Academy of Sciences, Chengdu 610209, China
10. Shandong Provincial Key Laboratory of Optical Astronomy and Solar – Terrestrial Environment, Institute of Space Sciences, Shandong University, Weihai 264209, China

Abstract: Chang'E – 4 landed in the South Pole – Aitken (SPA) basin, providing a unique chance to probe the composition of the lunar interior. Its landing site is located on ejecta strips in Von Kármán crater that possibly originate from the neighboring Finsen crater. A surface rock and the lunar regolith at 10 sites along the rover Yutu – 2 track were measured by the onboard Visible and Near – Infrared Imaging Spectrometer in the first three lunar days of mission operations. *In situ* spectra of the regolith have peak band positions at 1 and 2μm, similar to the spectral data of Finsen ejecta from the Moon Mineralogy Mapper,

which confirms that the regolith's composition of the landing area is mostly similar to that of Finsen ejecta. The rock spectrum shows similar band peak positions, but stronger absorptions, suggesting relatively fresh exposure. The rock may consist of 38.1±5.4% low - Ca pyroxene, 13.9±5.1% olivine and 48.0±3.1% plagioclase, referred to as olivine - norite. The plagioclase - abundant and olivine - poor modal composition of the rock is inconsistent with the origin of the mantle, but representative of the lunar lower crust. Alternatively, the rock crystallized from the impact - derived melt pool formed by the SPA - impact event via mixing the lunar crust and mantle materials. This scenario is consistent with fast - cooling thermal conditions of a shallow melt pool, indicated by the fine to medium - sized texture (<3mm) of the rock and the SPA - impact melting model.
Keywords: impact basin; lunar interior composition; visible and near - infrared spectra; lunar rover Yutu - 2; Chang'E - 4 mission

Introduction

According to the Lunar Magma Ocean (LMO) hypothesis[1-3], the Moon was, early in its history, covered by global silicate magma up to 800km thick[4], which crystallized into the modern crust and mantle. Several model compositions of the lunar mantle and crust have been proposed, with the lowermost cumulates dominated by Mg - rich olivine, followed by orthopyroxene (OPX) and finally by Ca pyroxene and plagioclase[5,6]. Furthermore, the lunar - mantle composition may have been modified by the overturn process, with the heavy top mantle materials sinking down due to gravitational instability[7]. In order to constrain the Moon's formation history, it is critically important to determine the composition of the lunar deep interior. The South Pole - Aitken (SPA) basin is the largest impact basin on the Moon, 2200km in mean diameter and 13km in depth[8,9], theoretically opening a window into the lunar lower crust and likely into the upper mantle[10-12]. However, compositional information of the SPA basin has mainly been obtained from orbital remote sensing[13-15] and there have been no *in situ* measurements of the region before the first landing on the far side of the Moon by Chang'E - 4.

On 3 January 2019, the Chinese lunar mission Chang'E - 4 (CE - 4), the backup of Chang'E - 3 (CE - 3), landed at 45.457°S, 177.588°E in Von Kármán crater (Fig. 1a), which is within the Mg - Pyroxene Annulus [13] of the SPA basin. Von Kármán crater has been filled with mare basalt and partially covered by ejecta from nearby impact craters. The landing site is located on ejecta strips radiating from Finsen crater, which lies ~ 135km to the northeast (Fig. 1b). The lunar surface at the landing site consists of very homogenous regolith overlain by a few scattered rocks (Fig. 1c). During the first three lunar days of mission operations, the Visible and Near - Infrared Imaging Spectrometer (VNIS) on board the rover Yutu - 2 measured nine hyperspectral images of the lunar regolith and one hyperspectral image of a rock boulder along the 163 - m rover track (Fig.

1d). These *in situ* measurements of the lunar regolith and overlying rock reveal the fine-scale composition of the SPA basin at unprecedented spatial resolution, shedding light on the composition of the lunar interior and the LMO crystallization.

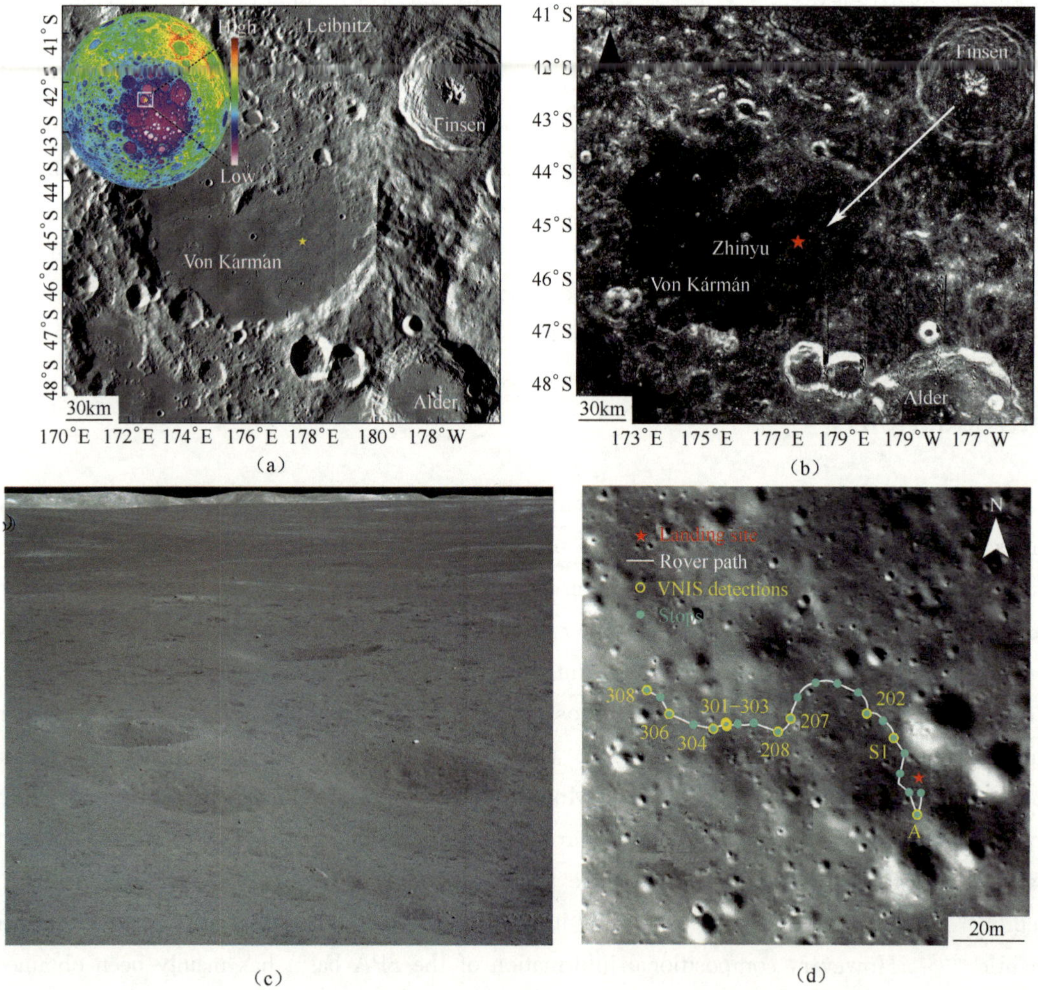

Fig. 1　CE-4 landing site and the rover track. (a) The geologic setting of the CE-4 landing site (yellow star) seen in a Lunar Reconnaissance Orbiter Camera (LROC) Wide Angle Camera (WAC) image. The inset is SLDEM (SELENE and LRO DEM 2015) [16], stretched from -8000 to 10000m. (b) The ejecta strips radiating from Finsen crater. Zhinyu crater, the largest crater close to the landing site, is also labeled. (c) The typical landscape of the CE-4 landing site, showing very few rocks overlying the lunar regolith. (d) The Yutu-2 rover traverse during the first 3 lunar days of mission operations. The context image was taken by the descent camera on CE-4. The red star is the location of CE-4 landing site and the scale bar is 20m

Results

The radiance spectra measured by VNIS on board Yutu-2 were converted to reflectance via division of the solar irradiance spectrum and then were photometrically corrected using an empirical photometric function (see 'Materials and methods'). The spectra were smoothed twice with a boxcar average of 17 pixels followed by another boxcar average of 7 pixels. All 10 VNIS spectra, including 1 of the rock boulder and 9 of the lunar regolith, are shown in Fig. 2 alongside a close-up image of the rock and an image of the lunar regolith. The other images of the regolith are given in the Supplementary Materials (Supplementary Fig. 5). The surface around the CE-4 landing site is distinctly brighter than the basaltic soils at the Chang'E-3 landing site, which is exhibited by their different albedos (Fig. 2a).

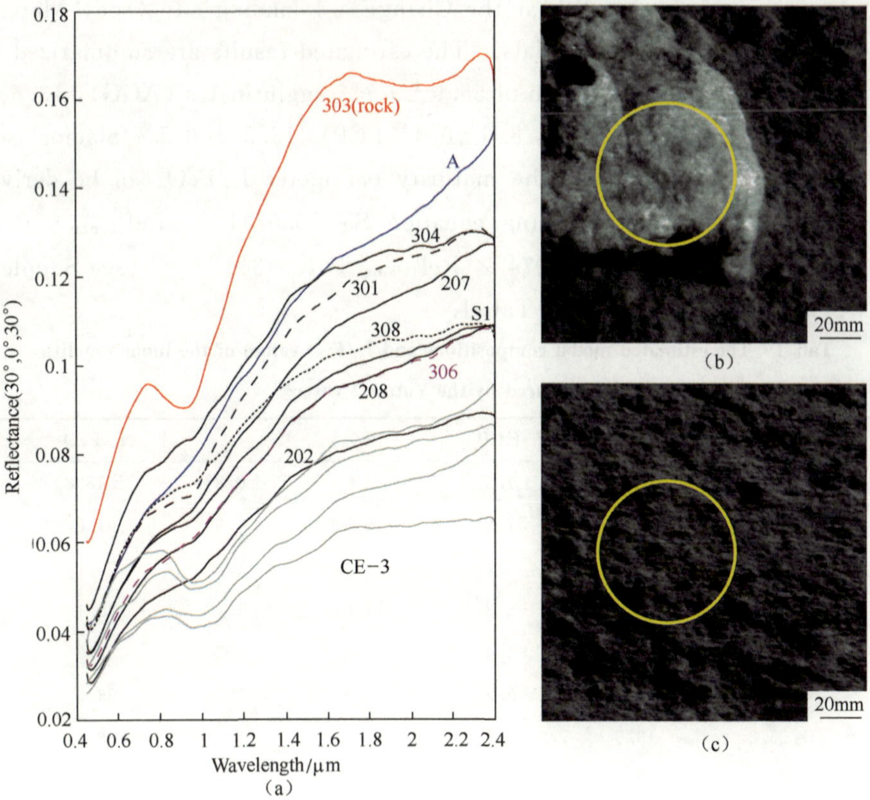

Fig. 2　(a) VNIS spectra of the rock and lunar regolith measured by Chang'E-4. The analysis labels are the same as in Fig. 1d. Four other spectra (in gray color) of the basaltic lunar regolith measured by Chang'E-3[17] are shown for comparison. (b) and (c) are the CMOS images of the rock (labeled as 303) and the lunar soil (labeled as 207) at 0.75μm observed by VNIS imaging spectrometer, respectively. The yellow circle is the SWIR field and the scale bar is 20mm

The spectra of the lunar regolith and the rock show maximum absorptions at ~1 and ~2μm (Fig. 2a and Supplementary Fig. 10). The peak positions of the ~1- and ~2-μm bands can be used to distinguish low-calcium pyroxene (LCP, centered at ~0.95 and ~1.9μm) from high-calcium pyroxene (HCP, centered at ~1.0 and ~2.0μm)[18] and to detect the possible presence of olivine (centered at ~1.1μm). The precise peak positions were determined by searching for the minimum reflectance[19] of the bands in continuum - removed [20] spectra (see 'Materials and methods'). As shown in Fig. 3, the pyroxene composition of the lunar - surface materials measured by Yutu - 2 is close to LCP, which is obviously different with the basalts excavated from Zhinyu crater.

The abundances of the constituents (i. e. agglutinates, total pyroxene, augite, olivine and plagioclase) of lunar regolith show strong correlation with the spectral parameters (empirical combination of albedos in the principal spectral bands) [25,26]. As such, we used a Lunar Soil Characterization Consortium (LSCC) dataset[27] to estimate the modal compositions of the regolith at the ChangE - 4 landing site (see 'Materials and methods' and Supplementary Materials). The estimated results are summarized in Table 1, with an averagemodal composition of 55.9±0.7% agglutinates (AGG), 13.5± 0.9% pyroxene (PYX) (4.9±0.5% HCP, 8.6±0.4%LCP), 13.6 ± 0.7% plagioclase (PLG) and 1.4 ± 0.3% olivine (OL). The maturity parameter I_s/FeO can be derived from VNIS reflectance (Ref) spectra using equation S5 ($0.0341 \times Ref_{500nm} - 0.2421 \times Ref_{750nm} - 0.0201 \times Ref_{900nm} + 0.2074 \times Ref_{1000nm} + 1.9430$) [25,28] (see Supplementary Materials). The results are listed in Tab. 1.

Tab. 1　The estimated modal compositions and I_s/FeO values of the lunar regolith measured by the Yutu - 2 rover[a]

Site	AGG	PYX	HCP	PLG	OL	I_s/FeO	*D*/m
Uncertainty	±7.7%	±3.1%	±1.3%	±4.4%	±0.8%	±18.8	—
A	57.1	11.5	3.8	15.0	1.6	97.0	11.5
S1	55.6	14.2	5.2	13.0	1.5	78.7	20.1
202	55.6	14.4	5.3	13.6	1.7	85.1	29.4
207	56.7	12.9	4.4	14.1	1.7	89.2	51.3
208	56.3	13.9	5.2	12.9	1.7	86.0	54.5
301	55.2	13.6	4.8	14.1	1.3	76.6	71.3
304	55.4	13.1	4.9	13.6	0.9	76.3	68.9
306	56.1	13.6	4.9	13.4	1.4	86.8	85.8
308	55.2	14.2	5.5	12.9	1.1	74.0	98.1

[a] AGG, agglutinates; PYX, pyroxene; HCP, augite; PLG, plagioclase; OL, olivine. *D* is the straight - line distance from the analysis position to the lander.

The rock analysed by Yutu - 2 is >20cm in size, sitting on the regolith surface. No grains can be unambiguously recognized on the surface, suggesting a fine - or medium - grain - size texture (<3mm) based on the 0.6 - mm/pixel spatial resolution of the image (Fig. 2b). In addition, the rock shows no lithic clasts of lunar - impact breccia (Supplementary Fig. 15), which usually vary in brightness and are embedded in a dark fine - grained matrix. The rock shows deep absorptions at ~1 - and ~2 - μm bands due to the low degree of space weathering. Hapke radiative transfer modeling [29], which has been validated and applied to lunar samples and meteorites[30,31] (see Supplementary Materials), was used to estimate the modal composition of the rock and our modeling results suggest 38.1 ± 5.4% LCP, 13.9 ± 5.1% olivine and 48.0 ± 3.1% plagioclase in this rock sample. As such, it is referred to as olivine-norite.

Discussion and Conclusion

The spectral features at site A are distinct from those at other sites (Fig. 2a), resulting in different estimated modal mineral compositions and maturity (Tab. 1). Site A shows the highest agglutinates and I_s/FeO value, indicating greater maturity than the other sites. Comparison to images acquired by the Lunar Reconnaissance Orbiter (LRO) Narrow Angle Camera (NAC) shows that the surface disturbed by CE-4 descent engines is about 1400m^2 (Supplementary Fig. 3), which is in agreement with the relationship between lander mass and blast-zone area constrained by previousmissions[32]. Site A, about 11.5m from the lander, should have been most significantly affected by the rocket exhaust. The higher maturity value at site A is therefore probably caused by deposition of the finest dust particles that were transported from the topmost surface below the lander blown by the rocket exhaust[33].

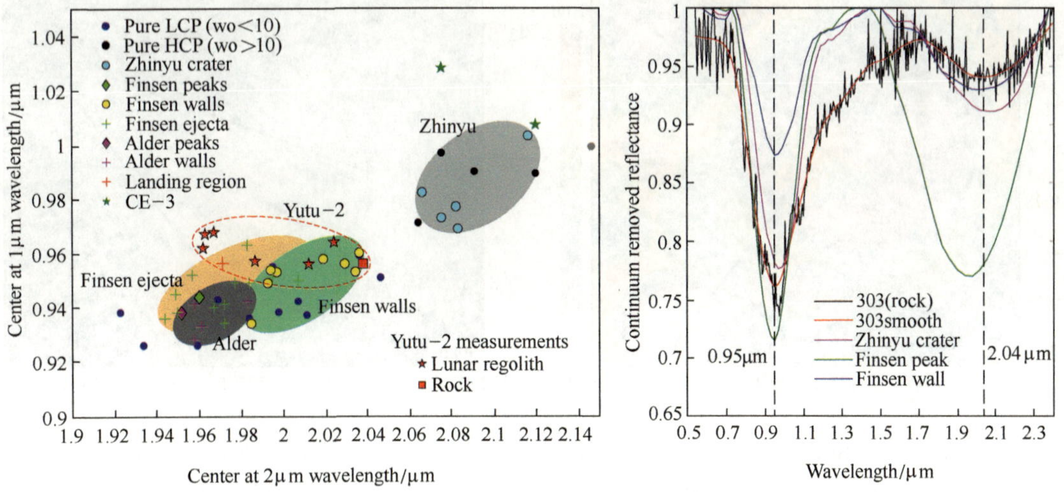

(a)　　　　　　　　　　　　(b)

Fig. 3　Spectral features of the lunar-surface materials observed by Yutu-2. (a) The peak center positions of the 1- and 2-μm bands, which are compared with those from spectra measured by CE-3 and the Moon Mineralogy Mapper (M^3) of nearby craters. The pure LCP and HCP data are from[21]. The locations of all the M^3 spectra are labeled in Supplementary Fig. 14. (b) Comparison of the rock spectrum with spectra from Zhinyu crater, Alder crater peak, Finsen crater peak and Finsen crater wall. The spectral calibration error of VNIS onboard CE-3 and CE-4 is <2nm[22,23] and that of M^3 is <1nm[24]

The regolith at the CE-4 landing site is dominated by plagioclase and pyroxenes with more LCP than HCP (1.8 : 1) (Tab. 1), consistently with the findings of a recent study[34]. The estimated modal composition of the lunar regolith is inconsistent with mare basalts (which would contain more pyroxene than plagioclase, with higher HCP than LCP)[35], but suggests its main source was norite[36]. Figure 3 also shows that band positions from *in situ* spectra of the lunar regolith at the landing site fall within the ranges of Finsen crater materials (the ejecta and crater walls), which are distinct from the underlying mare basalt excavated by Zhinyu crater (Supplementary Fig. 13) and the basaltic regolith on the nearside of the Moon that was analysed by Yutu-1[17].

Our observations are also supported by topographic features. The landing site is located on NE-SW ejecta strips radiating from Finsen crater (Fig. 4), which superpose the SE-NW dome-like ridge directed towards Alder crater (Fig. 4a). Furthermore, the elevation difference between the landing site and the edge of the SE-NW dome-like ridge was estimated to be ~70m, suggesting a very thick deposit of ejecta from Finsen crater and probably also from Alder crater. Previous studies also reported that mare basalts were deeply buried by ejecta from nearby craters, with aminimum burial depth of 33m[37,38]. Hence, the surface materials at the landing site are predominantly ejecta from neighboring craters, with little contribution from the underlying mare basalt.

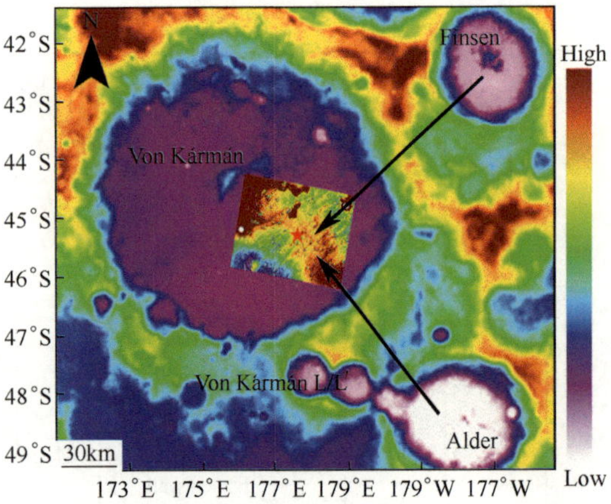

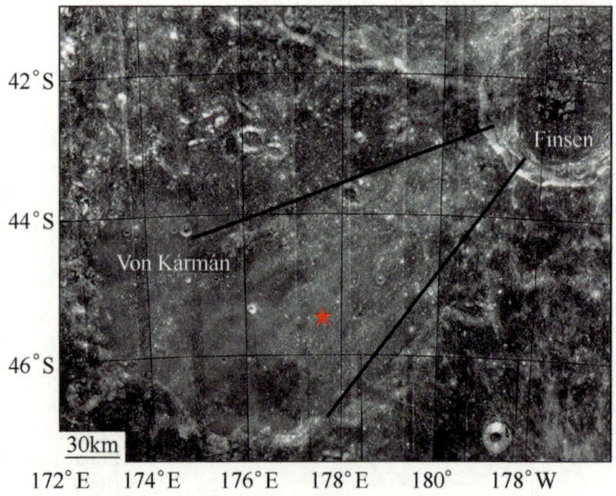

Fig. 4 (a) SLDEM[16]. The image is stretched from -7581 to -158m. The inset image is stretched from -6000 to -5900m. (b) MI band ratio of 415 and 750nm[39] (stretched from 0.53 to 0.55). The landing site is marked as a red star

The rock analysed by Yutu-2 is olivine-norite, not basalt. This is consistent with its setting on the surface of the regolith. Its modal composition is close to that of the regolith, but has higher modal abundances of olivine and LCP. It is possible that the rock was also delivered to its current location from Finsen crater. This is consistent with the similar peak positions of 1- and 2-μm bands (Fig. 3a). Alternatively, this rock may have been excavated from Alder crater. In this case, it would have been covered by ejecta from

Finsen crater and then subsequently excavated from depth. However, this uncertainty makes no difference for constraining the composition of SPA basin, because both Finsen and Alder craters are located within the Mg - Pyroxene Annulus[13] at a similar distance (~350km) from the center of SPA basin. Hence, this rock is likely representative of the original bedrock in the Mg - Pyroxene Annulus of SPA basin.

The SPA - impact event excavated the lunar deep interior and exposed lunar lower crust and/or mantle materials[10,12]. The bedrock of SPA basin could be the original plutonic rocks crystallized directly from LMO or may have formed from crystallization of the SPA - impact melt pool. The plagioclase - abundant and olivine - poor modal compositions of the materials measured by Yutu - 2 rover are consistent with the origin of the lower crust rather than the mantle. The deep interior origin would also expect coarse - grained textures, typical of plutonic rocks (3mm or larger)[40,41]. However, the observed fine - to medium - grain - sized texture of the rock suggests a fast - cooling thermal condition, which is consistent with crystallization from the SPA - impact melt pool[10,15,42]. Furthermore, the numerical simulations of impacting suggest that the SPA - scale event could generate a transient cavity with a diameter of 840km and a melt pond with ~50 - km depth (Fig. 5a)[10,42,43]. The SPA - impact melt would have been a mixture of the lunar crust and upper - mantle materials. Both Finsen and Alder craters are located close to the margin of the melt pond (Fig. 5) [9,10]. Hence, the sources of the regolith and the rock boulder analysed by Yutu - 2 are unlikely to be representative of the original lunar lower crust or the mantle as claimed by the previous study[44], but rather are the differentiated rocks from the impact melt pond. It is likely that the Finsen - impact event excavated shallow - layer materials crystallized from the SPA melt pond and delivered them to Von Kármán crater, where they were measured by the rover Yutu - 2 (Fig. 5).

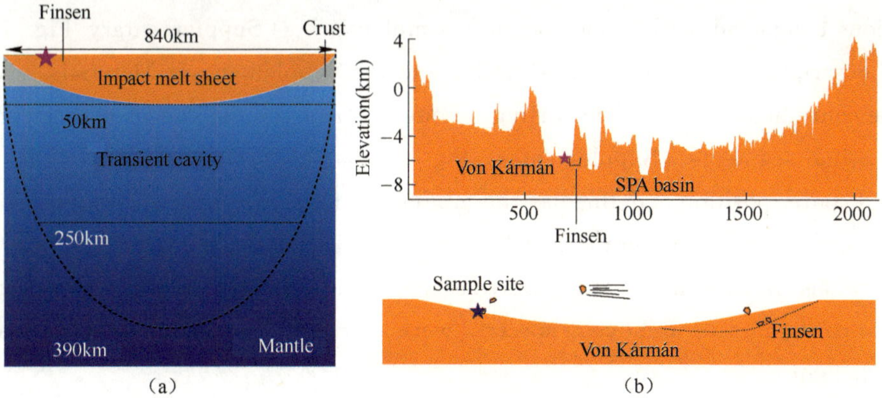

Fig. 5　Schematic illustration for the origin of the rock and regolith analysed by Yutu - 2. (a) The impact event of SPA formed a transient crater that was ~840km in diameter and ~390km in depth[10,43]. Hydrocode models suggest that the upper ~250km of the lunar interior was melted[43] and the melts gathered at the bottom of the SPA basin to form a~50 - km - deep melt pond after impact[10,43]. (b) The topographic profile from A to A' shown in Supplementary Fig. 2. (c) The schematic illustration of the sputtering process of the rock

In summary, the *in situ* spectral measurements of the rock and regolith at the Chang'E-4 landing site result in similar estimates for mineral compositions and the rock is likely olivine-norite. These surface materials were delivered mainly from the neighboring Finsen crater, with possible additional contributions from Alder crater, but not from the underlying mare basalts. The fine-to medium-grain-size texture of the rock suggests fast crystallization, probably from the impact melt pond produced viamelting the lunar lower crust and mantle materials by the SPA basin-forming event. This scenario is also consistent with the SPA-impact melting models[10]. These observations shed light on the composition of the SPA basin floor and the formation of the SPA basin.

Materials and Methods

Data preprocessing

The VNIS on board the rover Yutu-2 consists of a Complementary Metal-Oxide Semiconductor (CMOS) imager (450-950nm) with 256 × 256 pixels and a short-wavelength near-infrared (SWIR) detector (950—2395nm) with single-pixel [22,45]. VNIS is installed on the front of the rover and measures the lunar surface from a height of ~1m above the lunar surface at a 45°emission angle. The SWIR field is centered at pixel 97.5, 127.5 of the CMOS field with a diameter of 107.6 pixels (Supplementary Fig. 1). The spectral-sampling interval of the CMOS and SWIR detectors is 5nm, and the wavelength and FWHM (full width at half maximum) are shown in Supplementary Table 6.

The data were calibrated in flight and converted to reflectance using solar irradiance[46] and were then photometrically corrected [47,48] to common viewing geometry (incidence angle=30°, emission angle = 0° and phase angle = 30°) based on illumination simulations conducted with lunar regolith simulants[49] (Supplementary Fig. 7). The detailed description of this process is available in the Supplementary Materials.

The continuum is derived by fitting a convex hull over the top of a spectrum using straight-line segments (Supplementary Fig. 9) that connect local spectra maxima[20]. The band center position is defined as the minimum of a third-order polynomial fit to the channels of the band minimum[19]. All the Moon Mineralogy Mapper (M^3) spectra and laboratory spectra used in this study were resampled to VNIS bands with a Gaussian model and the continuum-removed spectra were derived using the spectral range of 0.545—2.395μm.

Estimation of the mineral abundances from the regolith spectra

The statistical formulations linking spectral properties (empirical combination of spectral bands) with each mineral abundance were optimized with LSCC soils and have been successfully applied to Clementine UVVIS data[25,26]. We resampled the LSCC spectra to the spectral resolutions of Yutu - 2 VNIS bands using a Gaussian model and parameterized the formulations (Supplementary Tab. 4) to estimate the mineralogy and maturity parameter of the regolith at the CE - 4 landing site. The detailed descriptions are available in the Supplementary Materials.

Estimation of the mineral abundance from the rock spectrum

The rock is much less affected by space weathering than the soil and has typical spectral characteristics. Thus, amore rigorous non - linear model, i. e. a Hapke radiative transfermodel[29], was applied to retrieve the mineral abundances of the rock. The endmembers used in this study are high - calcium pyroxene (HCP), low - calcium pyroxene (LCP), olivine (OL) and plagioclase (PLG) (Supplementary Fig. 11 and Supplementary Tab. 5). More details about the methods are available in Supplementary Materials.

Acknowledgements

The Chang'E - 4 mission was carried out by the Chinese Lunar Exploration Program and the data were provided by the Science and Application Center for Moon and Deep Space Exploration, Chinese Academy of Sciences. The authors are grateful to the editor and three anonymous reviewers for their constructive reviews.

References

[1] Warren PH. The Magma Ocean concept and lunar evolution. Annu Rev Earth Planet Sci 1985; 13: 201 - 40.

[2] Wood JA, Dickey JS Jr and Marvin UB et al. Lunar anorthosites and a geophysical model of the moon. Geochim Cosmochim Acta 1970; 1: 965 - 88.

[3] Smith J, Anderson A and Newton R et al. Petrologic history of the moon inferred from petrography, mineralogy and petrogenesis of Apollo 11 rocks. Geochim Cosmochim Acta 1970; 1: 897 - 925.

[4] Elkins - Tanton LT, Burgess S and Yin Q - Z. The lunar magma ocean: reconciling the solidification process with lunar petrology and geochronology. Earth Planet Sci Lett 2011; 304: 326 - 36.

[5] Rapp J and Draper D. Fractional crystallization of the lunar magma ocean: updating the dominant paradigm. Meteorit Planet Sci 2018; 53: 1432 - 55.

[6] Lin YH, Tronche EJ and Steenstra ES et al. Evidence for an early wet moon from experimental crystallization of the lunar magma ocean. Nat Geosci 2017; 10: 14 - 8.

[7] Hess PC and Parmentier E. A model for the thermal and chemical evolution of the Moon's interior: implications for the onset of mare volcanism. Earth Planet Sci Lett

1995; 134: 501 - 14.

[8] Smith DE, Zuber MT and Neumann GA et al. Initial observations from the lunar orbiter laser altimeter (LOLA). Geophys Res Lett 2010; 37: L18204.

[9] Garrick - Bethell I and Zuber MT. Elliptical structure of the lunar South Pole - Aitken basin. Icarus 2009; 204: 399 - 408.

[10] Potter RWK, Collins GS and Kiefer WS et al. Constraining the size of the South Pole - Aitken basin impact. Icarus 2012; 220: 730 - 43.

[11] James PB, Smith DE and Byrne PK et al. Deep structure of the lunar South Pole - Aitken basin. Geophys Res Lett 2019; 49: 5100 - 6.

[12] Melosh H, Kendall J and Horgan B et al. South Pole - Aitken basin ejecta reveal the Moon's upper mantle. Geology 2017; 45: 1063 - 6.

[13] Moriarty DP and Pieters CM. The character of South Pole - Aitken basin: patterns of surface and subsurface composition. J Geophys Res Planets 2018; 123: 729 - 47.

[14] Ohtake M, Uemoto K and Yokota Y et al. Geologic structure generated by largeimpact basin formation observed at the South Pole - Aitken basin on the moon. Geophys Res Lett 2014; 41: 2738 - 45.

[15] Uemoto K, Ohtake M and Haruyama J et al. Evidence of impact melt sheet differentiation of the lunar South Pole - Aitken basin. J Geophys Res Planets 2017; 122: 1672 - 86.

[16] Barker MK, Mazarico E and Neumann GA et al. A new lunar digital elevation model from the lunar orbiter laser altimeter and SELENE terrain camera. Icarus 2016; 273: 346 - 55.

[17] Zhang JH, Yang W and Hu S et al. Volcanic history of the Imbrium basin: a close - up view from the lunar rover Yutu. Proc Natl Acad Sci USA 2015; 112: 5342 - 7.

[18] Adams JB. Visible and near - infrared diffuse reflectance spectra of pyroxenes as applied to remote sensing of solid objects in the solar system. J Geophys Res 1974; 79: 4829 - 36.

[19] Horgan BHN, Cloutis EA and Mann P et al. Near - infrared spectra of ferrous mineral mixtures and methods for their identification in planetary surface spectra. Icarus 2014; 234: 132 - 54.

[20] Clark RN. Water frost and ice: The near - infrared spectral reflectance 0. 65—2. 5μm. J Geophys Res Solid Earth 1981; 86: 3087 - 96.

[21] Klima RL, Pieters CM and Dyar MD. Spectroscopy of synthetic mg - Fe pyroxenes I: spin - allowed and spin - forbidden crystal field bands in the visible and nearinfrared. Meteorit Planet Sci 2007; 42: 235 - 53.

[22] He ZP, Wang BY and Lu G et al. Operating principles and detection characteristics of the visible and near - infrared imaging spectrometer in the Chang'e - 3. Res Astron Astrophys 2014; 14: 1567 - 77.

[23] Li CL, Wang ZD and Xu R et al. The scientific information model of Chang'e - 4

visible and near - IR imaging spectrometer (VNIS) and in - flight verification. Sensors 2019; 19: 2806.

[24] Green RO, Pieters C and Mouroulis P et al. The Moon Mineralogy Mapper (M^3) imaging spectrometer for lunar science: Instrument description, calibration, onorbit measurements, science data calibration and on - orbit validation. J Geophys Res Planet 2011; 116: E00G19.

[25] Pieters C, Shkuratov Y and Kaydash V et al. Lunar soil characterization consortium analyses: pyroxene and maturity estimates derived from Clementine image data. Icarus 2006; 184: 83 - 101.

[26] Shkuratov YG, Kaydash VG and Pieters CM. Lunar clinopyroxene and plagioclase: surface distribution and composition. Sol Syst Res 2005; 39: 255 - 66.

[27] Taylor LA, Pieters CM and Keller LP et al. Lunar Mare soils: space weathering and the major effects of surface - correlated nanophase Fe. J Geophys Res Planet 2001; 106: 27985 - 99.

[28] Shkuratov YG, Kaydash VG and Starukhina LV et al. Lunar surface agglutinates: mapping composition anomalies. Sol Syst Res 2007; 41: 177 - 85.

[29] Hapke B. Theory of reflectance and emittance spectroscopy. Cambridge: Cambridge University Press, 2012.

[30] Li S and Li L. Radiative transfer modeling for quantifying lunar surface minerals, particle size, and submicroscopic metallic Fe. J Geophys Res 2011; 116: E09001.

[31] Li S and Milliken RE. Estimating the modal mineralogy of eucrite and diogenite meteorites using visible - near infrared reflectance spectroscopy. Meteorit Planet Sci 2015; 50: 1821 - 50.

[32] Clegg - Watkins R, Jolliff B and Boyd A et al. Photometric characterization of the Chang'e - 3 landing site using LROC NAC images. Icarus 2016; 273: 84 - 95.

[33] Clegg RN, Jolliff BL and Robinson MS et al. Effects of rocket exhaust on lunar soil reflectance properties. Icarus 2014; 227: 176 - 94.

[34] Hu X, Ma P and Yang Y et al. Mineral abundances inferred from in situ reflectance measurements of Chang'E - 4 landing site in South Pole - Aitken basin. Geophys Res Lett 2019; 46: 9439 - 47.

[35] Jolliff BL, Wieczorek MA and Shearer CK et al. New views of the Moon. Berlin: Walter de Gruyter GmbH & Co KG; 2018.

[36] NASA. Exploring the Moon—A Teacher's Guide with Activities. https://www.nasa.gov/pdf/180557main_ETM.Rock.Fact.Sheet.pdf. 1997.

[37] Huang J, Xiao ZY and Flahaut J et al. Geological characteristics of Von Karman Crater, Northwestern South Pole - Aitken basin: Chang'E - 4 Landing Site Region. J Geophys Res Planets 2018; 123: 1684 - 700.

[38] Qiao L, Ling Z and Fu X et al. Geological characterization of the Chang'e - 4 landing area on the lunar farside. Icarus 2019; 333: 37 - 51.

[39] Ohtake M, Haruyama J and Matsunaga T et al. Performance and scientific objectives of the SELENE (KAGUYA) multiband imager. Earth Planet Space 2008; 60: 257-64.

[40] Bickel CE and Warner J. Survey of lunar plutonic and granulitic lithic fragments. In: Lunar and Planetary Science Conference Proceedings, Houston, 1978.

[41] Warner J and Bickel C. Lunar plutonic rocks; a suite of materials depleted in trace siderophile elements. Am Mineral 1978; 63: 1010-5.

[42] Vaughan WM and Head JW. Impact melt differentiation in the South Pole-Aitken basin: some observations and speculations. Planet Space Sci 2014; 91: 101-6.

[43] Hurwitz DM and Kring DA. Differentiation of the South Pole-Aitken basin impact melt sheet: implications for lunar exploration. J Geophys Res Planets 2014; 119: 1110-33.

[44] Li CL, Liu DW and Liu B et al. Chang'E-4 initial spectroscopic identification of lunar far-side mantle-derived materials. Nature 2019; 569: 378-82.

[45] Liu B, Li CL and Zhang GL et al. Data processing and preliminary results of the Chang'e-3 VIS/NIR imaging spectrometer in-situ analysis. Res Astron Astrophys 2014; 14: 1578-94.

[46] Gueymard CA. The sun's total and spectral irradiance for solar energy applications and solar radiation models. Sol Energy 2004; 76: 423-53.

[47] Besse S, Sunshine J and Staid M et al. A visible and near-infrared photometric correction for moon mineralogy Mapper (M^3). Icarus 2013; 222: 229-42.

[48] Wu YZ, Besse S and Li JY et al. Photometric correction and in-flight calibration of Chang'E-1 Interference Imaging Spectrometer (IIM) data. Icarus 2013; 222: 283-95.

[49] Zheng YC, Wang SJ and Ouyang ZY et al. CAS-1 lunar soil simulant. Adv Space Res 2009; 43: 448-54.

Thermal Modeling of the Lunar Regolith at the Chang'E - 4 Landing Site

Honglei Lin[1], Shuai Li[2], Yangting Lin[1], Yang Liu[3,4], Yong Wei[1], Wei Yang[1], Yazhou Yang[3], Sen Hu[1], Xing Wu[3], Rui Xu[5], Chunlai Li[5], and Zhiping He[5]

1. Key Laboratory of Earth and Planetary Physics, Institute of Geology and Geophysics, Chinese Academy of Sciences, Beijing, China
2. Department of Earth Sciences, University of Hawai' i at Manoa, Honolulu, HI, USA
3. State Key Laboratory of Space Weather, National Space Science Center, Chinese Academy of Sciences, Beijing, China
4. Center for Excellence in Comparative Planetology, Chinese Academy of Sciences, Hefei, China
5. Key Laboratory of Space Active Opto - Electronics Technology, Shanghai Institute of Technical Physics, Chinese Academy of Sciences, Shanghai, China

Abstract: Accurate assessments of surface temperatures on the Moon are important for understanding the physical properties of the lunar surface regolith and thermal effects on the reflectance spectra beyond 2μm. The local time resolved spectral measurements of the same region were used to estimate the surface temperature at the Chang'E - 4 landing site. The results show that the surface temperatures at the landing site at lunar local time 14: 28—14: 41 are 346 ± 8K. The spectral absorptions at ~2μm of the lunar surface are overestimated without sufficient thermal removal, and consequently may have resulted in overestimation of pyroxene and/or glasses at the landing site. The results play a foundation for correcting thermal effects of reflectance data that will be acquired by Chang'E - 5 and - 6 and possible other future Chang'E missions.

Plain Language Summary: Surface temperature is an important physical parameter of the lunar surface and it reflects the thermal state of the lunar regolith. Though some orbital observations have been conducted to measure the surface temperature, the in situ detections are very limited. We derived the surface temperature at the Chang'E - 4 landing site using the in situ spectral measurements acquired at different local time. In addition, we found that the thermal contributions on reflectance would induce significant errors on estimating pyroxene and/or glass abundances. The thermal correction model is critical for removing thermal contributions from reflectance data that will be acquired by Chang'E - 5 and - 6.

Key Points:

• We modeled the thermal effect of the near infrared reflectance spectra acquired by the Yutu - 2 rover near the Chang'E - 4 landing site

• Our model results suggest that thermal emissions at the Chang'E - 4 landing site could substantially alter the reflectance spectra beyond 2 μm

• This work plays a foundation for correcting thermal effects of infrared reflectance spectra that will be acquired by Chang'E - 5/6 missions

1 Introduction

Remote estimation of surface temperatures on the Moon is challenged by the surface roughness and depth - dependent thermal properties of lunar regolith in the sensing column (Bandfield et al., 2011; Cremers et al., 1971; Keihm et al., 1973; Lucas et al., 1967; Vasavada et al., 1999, 2012). Temperature of the topmost layer of the lunar surface regolith can reach over 400K at local noon near the equator (Cremers et al., 1971; Keihm et al., 1973; Vasavada et al., 1999, 2012). Thermal emissions at such high temperatures can substantially alter the reflectance features at wavelengths beyond around 2μm (e. g., Bandfield et al., 2018; Clark, 1979; Li & Milliken, 2016; Wohler et al., 2017). Accurate surface temperatures are required to remove the contamination of thermal emissions at long wavelengths, which is critical to derive the true reflectance. However, it is very challenging to accurately estimate temperatures from the spectral regions where both reflected and emitted energies can take significant portions of remotely acquired radiance (e. g., ~2-4μm) (Bandfield et al., 2018; Clark, 2011; Li & Milliken, 2016; Wohler et al., 2017). Temperatures estimated from spectral bands dominated by thermal emissions at long wavelengths (e. g., >6μm) are not immediately applicable to those at around 2-4μm regions due to different sensing depth and anisothermality of the lunar surface (e. g., Bandfield et al., 2018; Keihm et al., 1973; Paige et al., 2010).

Spectral bands between ~2 and ~3 μm of the Moon Mineralogy Mapper (M^3) data, particularly those acquired at low latitudes and near local noon, received severe thermal contamination (Clark et al., 2011). Essential efforts were made to remove thermal effects of those M^3 bands (Bandfield et al., 2018; Clark et al., 2011; Li & Milliken, 2016; Wohler et al., 2017). Two different types of approaches were proposed to estimate the temperatures to remove the thermal contamination in M^3 data. One type of the approaches assumes thermal equilibrium and predicts temperature of each facet of a rough surface (i. e., a M^3 pixel) (Bandfield et al., 2018; Wohler et al., 2017). Slope angles of those facets are commonly described with a Gaussian roughness model (Bandfield et al., 2015; Helfenstein & Shepard, 1999). The total thermal emission can then be calculated as a weighted average of Planck radiance of each facet in that surface. Thermal emissions can be then removed from the acquired radiance of the lunar surface (e. g., M^3 data). It is suggested that a ray - tracing based cast shadowing is necessary to improve the accuracy of estimated temperature (Bandfield et al., 2018). High - resolution Digital Terrain Model (DTM) data and high computation power are needed in this type of models to achieve high accuracy of estimated temperatures (e. g., ~1 K uncertainties) (Bandfield et al., 2018). The other type of models estimate surface temperatures based on the difference between orbitally measured radiance and the reflected solar radiance that is predicted from empirical models (Clark et al., 2011; Li & Milliken, 2016). Any excess radiance of the orbital measurements is attributed to thermal emissions and temperatures can then be estimated from the Planck function (Clark et al., 2011; Li & Milliken, 2016). The prediction of reflected solar radiance is based on empirical relationships of reflectance between spectral bands that cannot be affected by thermal emissions (e. g., 1. 5 μm) and spectral bands that may contain thermal emissions (e. g., >2μm) on the Moon (Clark et al., 2011; Li & Milliken, 2016). The derived temperatures are used to remove thermal emissions from other spectral bands that may have also been contaminated by thermal emissions. Models in Bandfield et al. (2018) and Li and Milliken (2016) are validated with temperatures measured by the Diviner Lunar Radiometer Experiment (Diviner) on board the Lunar Reconnaissance Orbiter (LRO). Although both models claim high accuracies in estimated temperatures for correcting M^3 data, they arise controversial interpretations about the absorption near 3μm due to the presence of water (OH/H_2O) on the lunar surface. Bandfield et al. (2018) suggests that almost identical absorption strengths of 3μm bands are observed on the global lunar surface and thus similar water content across the lunar surface is suggested. In contrast, Li and Milliken (2017) shows that latitudinal and diurnal variations of the 3μm absorption, which is a proxy of water content, are observed in their global water map. Further investigations are needed to resolve the thermal effects of spectral bands (e. g., ~2－4μm) that are possibly dominated by both reflected and emitted energies on the Moon. One approach can be in situ observations of the diurnal variations of thermal bands on the lunar surface. The temperatures of the lunar surface near dawn or dusk are lower than 300K (lower than room temperatures) (Cremers et al., 1971; Keihm et al., 1973; Paige et al., 2010; Vasavada et al., 1999, 2012; Williams et al., 2017). Reflectance spectra acquired under such conditions can be used as "thermal - free" references for those acquired close to local noon or any local time when surface temperatures are high enough to contribute notable thermal emissions to modify reflectance features of spectral bands between ~2 and 4μm. One or more thermal bands at this wavelength range (i. e., ~2－4μm) can be used to estimate temperatures. Selections of thermal bands should be out of major absorptions on the Moon, such as spectral bands near 2. 5－2. 6 and 4μm (Li & Milliken, 2016), which is crucial to minimize effects from absorptions of Fe^{2+} in the M2 site near 2μm and water

(OH/H_2O) absorptions near 3μm, when assessing thermal emissions. Excess radiance at the selected thermal band (s) between thermally contaminated measurements and "thermal - free" references can be attributed to thermal emission. Temperatures can then be derived from the Plank function in conjunction with the excess radiance at the selected thermal band (s). Thermal effects of other spectral bands can be removed with the estimated temperatures.

On December 14, 2013, Chang'E - 3 landed on the nearside of the Moon and measured the in situ reflectance spectra from 0.4 to 2.4μm (Ling et al., 2015; Wu & Hapke, 2018; Zhang et al., 2015). Unfortunately, Chang'E - 3 rover stopped working because of the mechanical breakdown after collecting four spectra, which could not provide enough information for the purpose of our study (i. e., studying thermal effects of 2—2.4μm). On January 3, 2019, Chang'E - 4 spacecraft successfully landed in the South Pole - Aitken (SPA) basin on the lunar farside and its rover Yutu - 2 has been carrying out detailed surveys since then. Chang'E - 4 landed at 45.457°S, 177.588°E in the Von Kármán crater. The landing site is located within the ejecta blanket from nearby Finsen crater (e. g., Lin, He, et al., 2020). The rover is equipped with a Visible and Near - Infrared Imaging Spectrometer (VNIS) to collect reflectance spectra from 0.4 to 2.4μm on the lunar surface (He et al., 2019; Wu et al., 2019). In this study, a series of local - time resolved spectral measurements at the same spot (an area of ~15cm by ~21cm) are acquired in one lunation and are used to study the thermal effects of spectral bands between ~2 and 2.4μm. The VNIR spectrometers on future Chang'E missions will include more IR bands. For instance, the spectrometers on Chang'E - 5 and - 6 landers will measure reflectance spectra up to 3.2μm on the lunar surface (He et al., 2019). This work will play a foundation for correcting thermal effects of reflectance data that will be acquired by Chang'E - 5 and - 6 and possible other future Chang'E missions.

2 Data and Methods

The VNIS on board the Yutu - 2 rover consists of a Complementary Metal - Oxide Semiconductor (CMOS) imager (450—950nm with a spectral resolution of 2—7nm) and a short - wavelength near - infrared (SWIR) detector (950 — 2,395nm with a spectral resolution of 3—12nm) (He et al., 2019). In this study, we only used SWIR bands because the CMOS bands are almost unaffected by thermal emissions. On the tenth lunar day of the Chang'E - 4 mission, the Yutu - 2 rover acquired 12 spectra (N79 - N90, Figure S1) at exactly the same spot on the lunar surface (Lin, Yang, et al., 2020; Yang et al., 2020). The spectrum labeled N90 was acquired at a very low solar altitude and received substantial effects from shadows and poor instrument performance due to weak solar illumination (Yang et al., 2020), and is excluded in this study. We derived the reflectance (radiance factor I/F) using the solar spectrum in Gueymard (2004) and

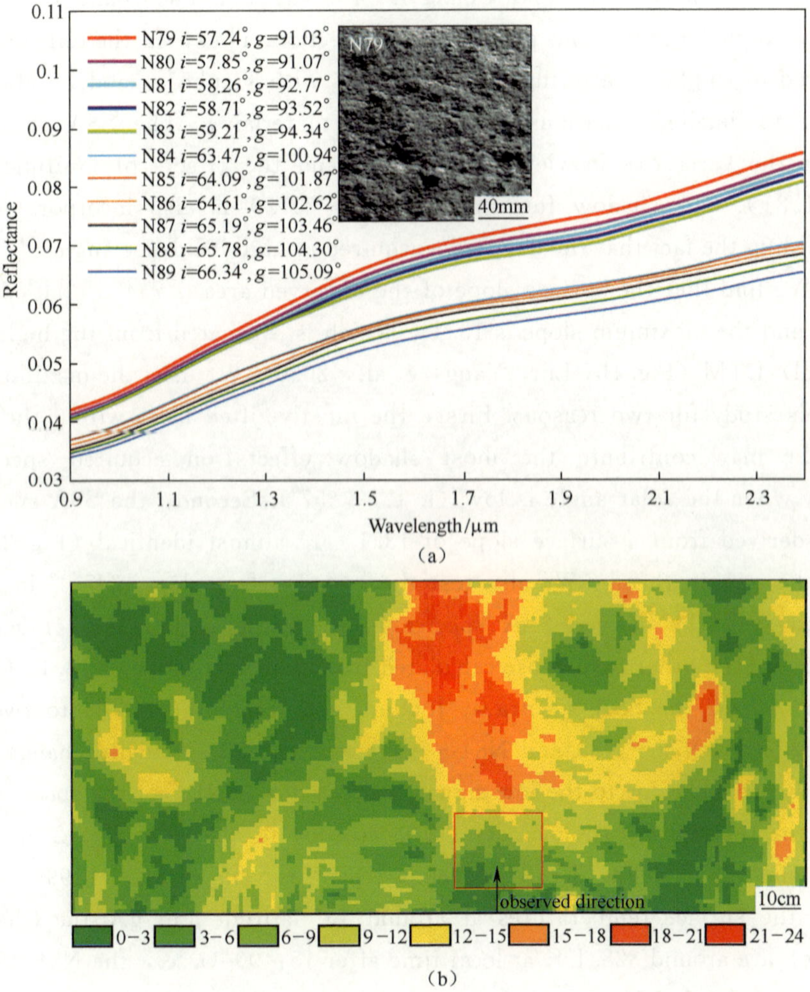

Fig. 1　(a) The infrared spectra at the same spot acquired by the Visible and Near-Infrared Imaging Spectrometer (VNIS) on board the Yutu-2 rover of Chang'E-4; the respective solar incidence angle (i) and phase angle (g) of each spectrum are listed in the legend; (b) The surface slope (°) of the observed area (red rectangle) at the Chang' E-4 landing site

showed them in Figure 1a. All the spectra were smoothed using the cubic spline method with a smoothing parameter of 0.999 (De Boor, 1978). The original spectra can be found in Figure S1.

A similar empirical model as that in Li and Milliken (2016) is used to estimate the surface temperatures. Li and Milliken (2016) established the relationships between the reflectance at 1.55 and 2.54μm of "thermal-free" samples (e.g., measured under room temperatures) and then predicted the "true" reflectance of a "thermal" band (i.e., 2.54μm). The excess radiance of M^3 data in comparison to the predicted "thermal-free" radiance at 2.54μm is used to estimate the surface temperatures. However, the Chang'E-4 data only cover spectral range up to 2.4μm. We use the "thermal" band at 2.38μm to assess the possible thermal effects of Chang'E - 4 VNIS data. The Chang'E - 4 reflectance data have yet to undergo rigorous photometric correction. Thus, all the reflectance spectra are converted to single - scattering albedo (SSA) with Hapke's model (Hapke et al., 2012; refer to detailed equations in supporting materials). The SSA spectra clearly account for the variations in viewing geometry and the effects of multiple scattering (Hapke, 1981). The shadow function (Hapke, 2012) is also incorporated into our model regarding the fact that the data were acquired at high incidence angles (~57°—66°) (Fig. 1a). We find that the surface slope of the observed area is 7.4 ± 3.0° (1 standard deviation) and the maximum slope is 13.1°, which is estimated from the high resolution (1mm/pixel) DTM (Fig. 1b; Lin, Yang, et al., 2020). We used the maximum slope of 13.1° in this study for two reasons. First, the mostly tilted facet within the FOV of a spectrometer may contribute the most shadow effects on acquired spectral data, particularly when the solar angle is low (i. e., >57°). Second, the SSA spectra of the same spot derived from a surface slope of 13.1° are almost identical (Fig. 2a), which should be the case in theory. We also test the mean surface slope of 7.4° in our Hapke model. However, the derived SSA spectra at the same region show much stronger deviation from each other than those calculated from a surface slope of 13.1° (Fig. S2).

We normalized the SSA spectra at 1.55μm and found they fall into two different groups (Fig. 2b), N79 - N83 and N84 - N89, but do not gradually change over local time. This is probably due to the low signal - noise ratio (SNR) of the spectrometer. We performed the thermal diffusion model (0.08 albedo, 0.99 emissivity, 20cm first layer thickness, and 1m modeling depth) developed by Vasavada et al. (1999, 2012) and found that the surface temperatures at around 45° latitude (e. g., the Chang'E - 4 landing site) are around 328.1 K at local time after 15:00 (i. e., the N84 - 89 group), while it is around 341.5 K at local time around 14:30 (i. e., the N79 - 83 group). The SNR of the Chang'E - 4 VNIS is ~40 at reflectance of 0.09 (Li, Wang, Xu, et al., 2019). The reflectance spectra can be affected by the thermal emission when the surface temperature T less than 330 K at the wavelength of 2.38μm. But we simulated the thermal effect on the Chang'E - 4 spectra (Fig. S3) and found that the VNIS may not be able to capture the variation of received energies when the thermal emission is small, such as under a temperature of ~330K (Fig. S3). The difference between the spectra measured at<330 K and >340K seems distinguishable under the SNR level of Chang'E - 4 VNIS (Fig. S3), which may explain why our spectra fall into two groups other than showing gradual changes. Thus, we assumed the N84 - 89 group as "thermal - free" reference to correct the thermal effects of the N79 - 83 group. It is noteworthy that spectra in the N84 - 89 group may also have been affected by thermal emission. However, the low SNR (~40) of Chang'E - 4 data make it impossible to resolve thermal emissions at surface temperature less than around 330K (Fig. S3). The error bar of the derived temperatures (Figure 3) includes the deviations caused by this assumption (Fig. 2c). The relationship

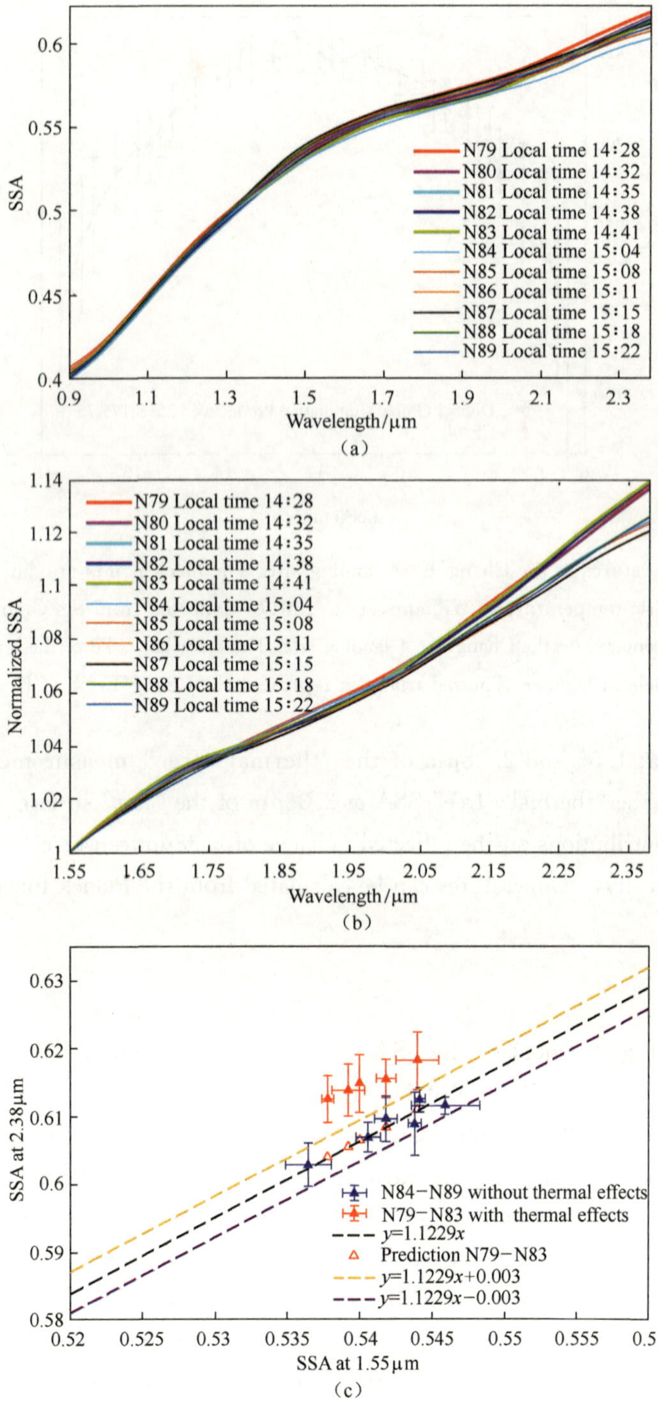

Fig. 2　(a) Single-scattering albedo (SSA) spectra derived with Hapke's model and respective viewing geometries of all spectra in Fig. 1. (b) SSA spectra normalized at 1.55μm. (c) Correlation between SSA at 1.55 and 2.38μm. The error bar is the measurement noise

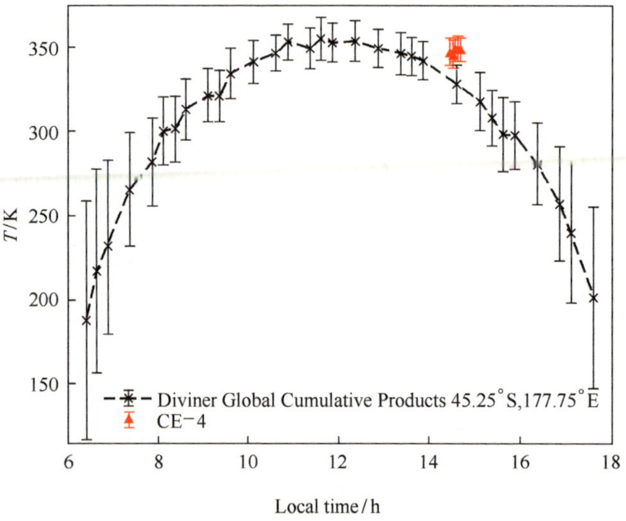

Fig. 3　The temperatures at the Chang'E - 4 landing site. The black points are diurnal variations of Diviner bolometric temperatures (Williams et al., 2017) collected from the Global Cumulative Product for an area centered on the Chang'E - 4 landing site. The error bar is 2δ of the temperatures derived from different channels of Diviner. The red triangles represent the Chang'E - 4 (CE - 4) measurements

between the SSA at 1.55 and 2.38μm of the "thermal - free" measurements (N84 - N89) was used to predict the "thermal - free" SSA at 2.38μm of the "hot" surface (N79 - N83) and then the thermal contributions on the reflected radiance of 2.38μm can be estimated (supporting information S1). Finally, temperatures can be calculated from the Planck function.

3　Results

As shown in Fig. 2, most of the SSA spectra are very similar but the measurements close to the noon (N79 - N83) uplift beyond 2μm (Fig. 2b). This is consistent with the findings of Wu et al. (2021). We assumed N84 - N89 as the reference spectra without thermal effects to model surface temperatures at the measurement time of N79 - N83. The plot of SSA between 2.38 and 1.55 μm are shown in Fig. 2c. The relationship between SSA at 1.55 and 2.38μm can be fitted by the linear equation $SSA_{2.38} = 1.1229SSA_{1.55}$. The "true" SSA without thermal contribution at 2.38μm of N84 - N89 can be predicted using this empirical relationship. Making a 0.003 (average error bar of N84 - N89 SSA values at 2.38μm, Fig. 2c) offset to the fitted line, the error bar of temperature estimation can be obtained (Fig. 3 and S4). The temperatures derived from the spectra of N79, N80, N81, N82, N83, N84 are 345^{+6}_{-10} K, 344^{+7}_{-10} K, 347^{+6}_{-8} K, 347^{+6}_{-9} K, 347^{+6}_{-8} K, respectively. The superscript and subscript are the temperature ranges determined by the measurement errors ($SSA_{2.38} = 1.1229SSA_{1.55} - 0.003$ and $SSA_{2.38} = 1.1229SSA_{1.55} + 0.003$). The results show that the surface temperatures at the Chang'E - 4 landing site in lunar local time 14：28—14：41 are 346 ± 8K (mean + 1 standard deviation of the five measurements) (Fig. 3). The reflectance spectra were then thermally corrected using our modeled surface temperature. After thermal correction, spectra beyond 2μm bands are consistent with those unaffected by thermal emission (Fig. 4a).

4 Discussions

4.1 Uncertainties of the Thermal Model

By assuming a perfectly black surface, the temperatures at local time 14：28—14：41 calculated using Stefan - Boltzmann law are 333—338K. The direct Diviner observations at the same location and same local time are not available at the Chang'E - 4 landing site. Thus, we also compared the results with Diviner bolometric temperatures (BOT) collected from the Global Cumulative Product (Williams et al., 2017) at the Chang'E - 4 landing site (Fig. 3). Our modeled surface temperatures (346 ± 8 K) are higher than those from Stefan - Boltzmann law (333—338K) and the mean Diviner BOT (~331K). However, it matches well with more realistic thermal diffusion model (341.5 K) that was developed based on thermal properties of Apollo samples (Vasavada et al., 1999) and was refined by Diviner data (Vasavada et al., 2012). Additionally, the surface roughness is one of the factors that may lead to differences in the derived temperatures. The surface bolometric temperature was derived from Diviner channels 3—9 (7.55—8.05 μm, 100—200μm), which is less sensitive to the surface components than the shorter wavelength because of the surface roughness. The discrepancy between different wavelengths caused by surface roughness can be much larger than 5K (Bandfield et al., 2015). There are at least other three factors to account for the inconsistency between our result with Stefan - Boltzmann law and Diviner BOT: (1) The Diviner BOT used in this study is derived with 0.5° spatial resolution (Williams et al., 2017) while the in situ measurements have the spatial resolution of ~ 15 × 21cm. The short wavelength observations have brightness temperatures that approach the highest temperature within the field - of - view (Bandfield et al., 2015). (2) The penetration depth of the shorter wavelength is smaller than that of longer wavelength. The subsurface temperature could be much lower than the topmost layer due to the strong thermal gradient of the lunar regolith (Conel, 1969; Hanna et al., 2012; Vasavada et al., 1999). (3) The uncertainties of photometric corrections (~10%, Hapke et al., 2012) and the low SNR levels of the Chang'E - 4 spectra result in a wide temperature range. The lower bound of our results match with Diviner mean values and those modeled from the Stefan - Boltzmann model. However, it is impossible to know why we systematically over - estimate the surface temperature for a few degrees without high quality rover data.

It is ideal to use the longer wavelength (e.g., 2.54μm) to estimate the surface temperature (Li & Milliken, 2016). This is because the spectral band at ~2.38μm can be affected by absorption near 2μm introduced by Fe^{2+} in pyroxene and glasses. However, the Chang'E - 4 data only cover spectral range up to 2.4μm. All our spectra near the Chang'E - 4 landing site exhibit very weak absorption near 2μm (Fig. 1a), which indicates that our estimation of temperatures from the thermal band at 2.38μm may receive little/negligible effects from Fe^{2+} absorptions. The parameters in the empirical model derived in this study may not be universal for all the lunar spectra, especially those regions showing strong absorption near 2μm. But these Chang'E - 4 spectra that were acquired at exactly the same spot on the lunar surface provide an ideal validation condition for the empirical thermal correction model. A thermal band at 2.54μm will be available in the data acquired by the Lunar Mineralogical Spectrometer (LMS) on board the Chang'E - 5/6 landers (He et al., 2019), which expands the Chang'E - 4 VNIS to 3.2μm. Chang'E - 5 was launched in November 23, 2020 and successfully returned spectral data and soil samples from the lunar surface. It provides the opportunity to further verify the empirical thermal models.

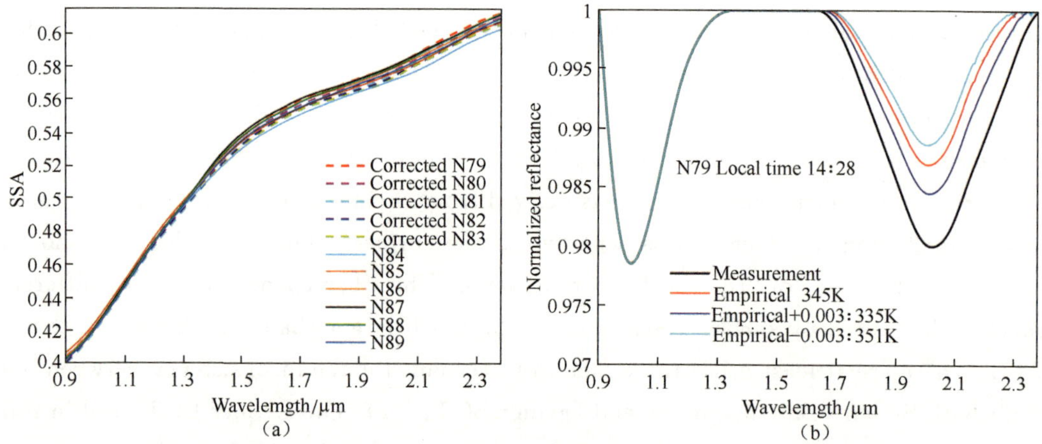

Figure 4　(a) The spectra after thermal correction. (b) The continuum - removed spectra of N79. The continuumremoved spectra of N80 - N89 are provided in Figure S5.

4.2 Effects of Thermal Correction on Spectral Analysis

Radiance acquired by the VNIS at wavelengths beyond 2μm includes both reflectance and thermal emission from the lunar surface. The thermal contributions changed the spectral shapes, as shown in Fig. 4. Although the Yutu - 2 rover only works in the local morning and afternoon to avoid high temperature environment, many spectral observations close to the local noon are strongly affected by the thermal emission. If the thermal contributions are not removed, the spectral features at ~2μm will be altered. As shown in Fig. 4b, the absorption at ~2μm after thermal correction becomes shallower. The insufficient thermal correction will result in the misestimates of mineral abundances because the absorption depths are proportional to mineral abundance. It is expected that the absorption depth of ~2μm would be more overestimated from the spectra acquired near local noon because the temperatures are higher. The spectral absorptions of the lunar surface at ~2μm are mainly related to pyroxenes and/or some glasses. Although the exact relationship between the mineral abundance and absorption depth is difficult to estimate due to the effects of complex factors (e. g., particle size, space weathering) on reflectance spectra, it is undoubtedly that the pyroxene and/or glass abundance will be significantly overestimated using the uncorrected spectrum. Besides the thermally emitted radiation, absorption bands that cause dips in the reflectance could also cause peaks in the absorption spectra according the Kirchhof's law.

4.3 Implications for Chang'E-5/6 Spectral Exploration

Chang'E-5 spacecraft was launched and successfully returned 1.731kg lunar samples from northern Oceanus Procellarum (Zeng et al., 2017) in late 2020. Chang'E-6 spacecraft will return samples from the lunar south polar area (Li, Wang, Wei, et al., 2019). Both Chang'E-5 and Chang'E-6 landers equip with a Lunar Mineralogical Spectrometer (LMS) (He et al., 2019) to investigate the lunar surface mineralogy before sampling. The LMS primarily inherits the technology of the VNIS of Chang'E-4 but expands its spectral range to 3.2μm (He et al., 2019), which is more useful to detect the mineral compositions and water contents at the landing site. However, the thermal contributions on reflectance at the longer wavelengths must be removed before spectral analysis if the data will be acquired at high temperatures. This study provides a foundation for analyzing spectral data from Chang'E-5/6 and possibly future missions. The joint analysis using the returned samples and in situ detections will help to better understand the thermal properties of the lunar regolith at spectral region near 3μm.

Similar experiments in this study can be conducted by the Chang'E-5/6 landers to accurately remove the thermal contributions from the reflectance spectra and then obtain the mineral abundance and water content at the landing sites. Lunar samples returned by Chang'E-5/6 spacecrafts will be analyzed in the laboratory and thus the accurate mineral abundances can be quantified, providing the validation for the in situ detections. Moreover, the results from both lander measurements and returned samples will help to validate orbital observations, which will improve the globally mineral and water mapping results on the lunar surface.

5 Conclusions

The Yutu-2 rover of the Chang'E-4 mission measured a series of infrared spectra of the same lunar surface in an extremely high spatial resolution (centimeter scale), providing an unprecedented opportunity to understand the thermal effects of reflectance data on the lunar surface. The surface temperatures at the Chang'E - 4 landing site were derived from the time resolved spectral measurements of the same spot. Our results show that the surface temperatures at lunar local time 14: 28 — 14: 41 are 346 ± 8K. The spectral absorption depths at ~ 2μm of the lunar surface will be overestimated if the thermal correction is insufficient for our data, which will lead to the overestimation of pyroxene and/or glass abundance. The estimation error for absorption depth of 2μm will be larger when the spectra are acquired at higher temperatures. Our results provide a fundamental reference to the Chang'E - 5 and - 6 spectral analysis, and the returned lunar samples by Chang'E - 5 and - 6 will help validate the empirical thermal model and provide ground truth for validating orbital observations.

Acknowledgments

The Chang'E - 4 mission was carried out by the Chinese Lunar Exploration Program, and the scientific data are provided by China National Space Administration. The authors are grateful to the Editor and Prof. Bruce Hapke and an anonymous reviewer for their constructive reviews. This work was supported by NSFC (11941001, 41902318, 41430105, 41490631, and 41525016), the Beijing Municipal Science and Technology Commission (Z181100002918003, Z191100004319001) and Key Research Program of Frontier Sciences, CAS (QYZDJ - SSW - DQC001). H. Lin also acknowledges the support of the Key Research Program of the Institute of Geology and Geophysics, CAS (IGGCAS - 201905).

References

[1] Bandfield, J. L., Ghent, R. R., Vasavada, A. R., Paige, D. A., Lawrence, S. J., & Robinson, M. S. (2011). Lunar surface rock abundance and regolith fines temperatures derived from LRO Diviner Radiometer data. Journal of Geophysical Research, 116. https://doi.org/10.1029/2011JE003866.

[2] Bandfield, J. L., Hayne, P. O., Williams, J., Greenhagen, B. T., & Paigeet, D. A. (2015). Lunar surface roughness derived from LRO Diviner Radiometer observations. Icarus, 248, 357 - 372. https://doi.org/10.1016/j.icarus.2014.11.009

[3] Bandfield, J. L., Michael, J. P., Rachel, L. K., & Christopher, S. E. (2018). Widespread distribution of OH/H_2O on the lunar surface inferred from spectral data. Nature Geoscience, 11 (3), 173 - 177. https://doi.org/10.1038/s41561-018-0065-0

[4] Clark, R. N. (1979). Planetary reflectance measurements in the region of planetary thermal emission. Icarus, 40, 94 - 103.

[5] Clark, R. N., Pieters, C. M., Green, R. O., Boardman, J. W., & Petro, N. E. (2011). Thermal removal from near - infrared imaging spectroscopy data of the Moon. Journal of Geophysical Research, 116, E00G16. https://doi.org/10.1029/2010JE003751.

[6] Conel, J. E. (1969). Infrared Emissivities of Silicates - Experimental Results and a Cloudy Atmosphere Model of Spectral Emission from Condensed Particulate Mediums. Journal of Geophysical Research, 74, 1614 - 1634. https://doi.org/10.1029/JB074i006p01614

[7] Cremers, C. J., Birkebak, R. C., & White, J. E. (1971). Lunar surface temperatures from Apollo 12. The Moon, 3, 346 - 351.

[8] De Boor, C. (1978). A practical guide to splines. Springer - Verlag. https://doi.org/10.1007/978-1-4612-6333-3.

[9] Gueymard, C. A. (2004). The sun's total and spectral irradiance for solar energy applications and solar radiation models. Solar Energy, 76, 423 - 453. https://doi.org/10.1016/j.solener.2003.08.039.

[10] Hanna, K. L. D., Wyatt, M. B., Thomas, I. R., Bowles, N. E., Greenhagen, B. T., Maturilli, A., et al. (2012). Thermal infrared emissivity measurements under a simulated lunar environment: Application to the diviner lunar radiometer experiment. Journal of Geophysical Research, 117, E00h0510. https://doi.org/10.1029/2011JE003862.

[11] Hapke, B. (1981). Bidirectional reflectance spectroscopy: 1. Theory. Journal of Geophysical Research, 86, 3039 - 3054.

[12] Hapke, B. (2012). Theory of reflectance and emittance spectroscopy. Cambridge University Press.

[13] Hapke, B., Denevi, B., Sato, H., Braden, S., & Robinson, M. (2012). The wavelength dependence of the lunar phase curve as seen by the Lunar Reconnaissance Orbiter wide - angle camera. Journal of Geophysical Research, 117, E00H15. https://doi.org/10.1029/2011JE003916.

[14] He, Z., Li, C., Xu, R., Lv, G., Yuan, L., & Wang, J. (2019). Spectrometers based on acousto - optic tunable filters for in - situ lunar surface measurement. Journal of Applied Remote Sensing, 13, 027502. https://doi.org/10.1117/1.JRS.13.027502.

[15] Helfenstein, P., & Shepard, M. K. (1999). Submillimeter - scale topography of the lunar regolith. Icarus, 141, 107 - 131. https://dx.doi.org/10.1006/icar.1999.6160.

[16] Keihm, S. J., Peters, K., Langseth, M. G., & Chute, J. L., Jr (1973). Apollo 15 measurement of lunar surface brightness temperatures thermal conductivity of the upper 1 1/2 meters of regolith. Earth and Planetary Science Letters, 19, 337 - 351.

[17] Li, C., Wang, C., Wei, Y., & Lin, Y. (2019). China's present and future

lunar exploration program. Science, 365, 238 - 239.

[18] Li, C., Wang, Z., Xu, R., Lv, G., Yuan, L., He, Z., & Wang, J. (2019). The scientific information model of Chang'e - 4 visible and near - IR imaging spectrometer (VNIS) and in - flight verification. Sensors, 19, 2806. https://doi.org/10.3390/s19122806.

[19] Li, S., & Milliken, R. E. (2016). An empirical thermal correction model for Moon Mineralogy Mapper data constrained by laboratory spectra and Diviner temperatures. Journal of Geophysical Research: Planet, 121, 2081 - 2107. https://doi.org/10.1002/2016je005035.

[20] Li, S., & Milliken, R. E. (2017). Water on the surface of the Moon as seen by the Moon Mineralogy Mapper: Distribution, abundance, and origins. Science Advances, 3, e1701471. https://doi.org/10.1126/sciadv.1701471.

[21] Lin, H., He, Z., Yang, W., Lin, Y., Xu, R., Zhang, C., et al. (2020a). Olivine - norite rock detected by the lunar rover Yutu - 2 likely crystallized from the SPA - impact melt pool. National Science Review, 7, 913 - 920. https://doi.org/10.1093/nsr/nwz183.

[22] Lin, H., Yang, Y., Lin, Y., Liu, Y., Wei, Y., Li, S., et al. (2020b). Photometric properties of lunar regolith revealed by the Yutu - 2 rover. Astronomy and Astrophysics, 638. https://doi.org/10.1051/0004 - 6361/202037859.

[23] Ling, Z. C., Jolliff, B. L., Wang, A., Li, C., Liu, J., Zhang, J., et al. (2015). Correlated compositional and mineralogical investigations at the Chang'e - 3 landing site. Nature Communications, 6, 888010. https://doi.org/10.1038/ncomms9880.

[24] Lucas, J. W., Conel, J. E., Hagemeyer, W. A., Garipay, R. R., & Saari, J. M. (1967). Lunar surface thermal characteristics from Surveyor 1. Journal of Geophysical Research, 72, 779 - 789.

[25] Paige, D. A., Foote, M. C., Greenhagen, B. T., Schofield, J. T., Calcutt, S., Vasavada, A. R., & Jakosky, B. M. (2010). The lunar reconnaissance orbiter diviner lunar radiometer experiment. Space Science Reviews, 150, 125 - 160. https://doi.org/10.1007/s11214 - 009 - 9529 - 2.

[26] Vasavada, A. R., Bandfield, J. L., Greenhagen, B. T., Hayne, P. O., Siegler, M. A., Williams, J. P., & Paige, D. A. (2012). Lunar equatorial surface temperatures and regolith properties from the Diviner Lunar Radiometer experiment. Journal of Geophysical Research, 117. https://doi.org/10.1029/2011JE003987.

[27] Vasavada, A. R., Paige, D. A., & Wood, S. E. (1999). Near - surface temperatures on Mercury and the Moon and the stability of polar ice deposits. Icarus, 141, 179 - 193.

[28] Williams, J. P., Paige, D. A., Greenhagen, B. T., & Sefton - Nash, E. (2017). The global surface temperatures of the moon as measured by the diviner lunar

radiometer experiment. Icarus, 283, 300 - 325. https://doi.org/10.1016/j.icarus.2016.08.012.

[29] Wohler, C., Grumpe, A., Berezhnoy, A. A., Feoktistova, E. A., Evdokimova, N. A., Kapoor, K., & Shevchenko, V. V. (2017). Temperature regime and water/hydroxyl behavior in the crater Boguslawsky on the Moon. Icarus, 285, 118 - 136. https://doi.org/10.1016/j.icarus.2016.12.026.

[30] Wu, W., Li, C., Zuo, W., Zhang, H., Liu, J., Wen, W., et al. (2019). Lunar farside to be explored by Chang'e - 4. Nature Geoscience, 12, 222 - 223. https://doi.org/10.1038/s41561-019-0341-7.

[31] Wu, Y., & Hapke, B. (2018). Spectroscopic observations of the Moon at the lunar surface. Earth and Planetary Science Letters, 484, 145 - 153. https://doi.org/10.1016/j.epsl.2017.12.003.

[32] Wu, Y., Kührt, E., Grott, M., Jin, Q., Xu, T., Helbert, J., et al. (2021). Chang'E - 4 rover spectra revealing micro - scale surface thermophysical properties of the Moon. Geophysical Research Letters, 48, e2020GL089226. https://doi.org/10.1029/2020GL089226.

[33] Yang, Y. Z., Lin, H. L., Liu, Y., Lin, Y. T., Wei, Y., Hu, S., et al. (2020). The effects of viewing geometry on the spectral analysis of Lunar Regolith as inferred by in situ spectrophotometric measurements of Chang'E - 4. Geophysical Research Letters, 47, e2020GL08708010. https://doi.org/10.1029/2020GL087080.

[34] Zeng, X., Zuo, W., Zhang, Z., Liu, Y., & Li, C. (2017). Topographic and geologic analysis of the pre - selection landing sites for Chang'E 5 (CE - 5) lunar sample returning mission of China. Paper Presented at 19th EGU General Assembly. EGU.

[35] Zhang, H., Yang, Y., Yuan, Y., Jin, W., Lucey, P. G., Zhu, M. H., et al. (2015). In situ optical measurements of Chang'E - 3 landing site in Mare Imbrium: 1. Mineral abundances inferred from spectral reflectance. Geophysical Research Letters, 42, 6945 - 6950. https://doi.org/10.1002/2015gl065273.

New Insight Into Lunar Regolith – Forming Processes by the Lunar Rover Yutu – 2

Honglei Lin[1], Yangting Lin[1], Wei Yang[1], Zhiping He[2], Sen Hu[1], Yong Wei[1], Rui Xu[2], Jinhai Zhang[1], Xiaohui Liu[3], Jianfeng Yang[4], Yan Xing[5], Chengwu Yu[5], and Yongliao Zou[6]

1. Key Laboratory of Earth and Planetary Physics, Institute of Geology and Geophysics, Chinese Academy of Sciences, Beijing, China
2. Key Laboratory of Space Active Opto – Electronics Technology, Shanghai Institute of Technical Physics, Chinese Academy of Sciences, Shanghai, China
3. Beijing Aerospace Control Center, Beijing, China
4. Xi'an Institute of Optics and Precision Mechanics, Chinese Academy of Sciences, Xi'an, China
5. Beijing institute of control engineering, Beijing, China
6. State Key Laboratory of Space Weather, National Space Science Center, Chinese Academy of Sciences, Beijing, China

Key Points: • Unique fragment – filled pits distinct from the meteorite impact craters were observed at the Chang'E – 4 landing site

• Impact glass in the fragment – filled pits was identified in situ for the first time

• The discoveries of the Yutu – 2 rover provide important evidence for the recycling of the lunar regolith

Abstract: The Yutu – 2 rover of the Chang'E – 4 spacecraft observed many meter – sized shallow pits fully covered with small fragments, distinct from the typical rock – free impact craters with comparable sizes in the landing area. The unique morphology of the pits and the visible and near – infrared spectra of the fragments suggest that the fragments are broken pieces of impact melt – conglutinated regolith breccia projectiles, which were excavated from preexisting craters. The rareness of rock on the landing area surface suggests that the preexisting craters were probably small in size (e. g., <60m in diameter), not large enough to penetrate the thick regolith (~ 12m) and to excavate the beneath rock breccia and/or bedrock. The presence of the impact melts was confirmed by the glass – like spectra of some fragments with unusually high albedo and blue – green tint in the centers of the pits. These observations reveal the gardening and consolidating processes on the Moon.

Plain Language Summary: There are very rare outcrops of bedrock on the lunar surface because the Moon is covered by the lunar regolith. The detailed formation processes of the lunar regolith are important for understanding the orbiting and landing exploration data, and for planning future sample return missions. The lunar rover Yutu - 2 discovered fragment - filled pits and identified in situ impact glass in the fragments. The glass - bearing and regolith - like fragments were probably broken pieces of lunar regolith breccia projectiles, which were excavated from the consolidated bottom and walls of preexisting small craters within the lunar regolith (<12 — 15m in thickness). These observations contribute to our understanding of the formation processes of the lunar regolith.

1 Introduction

Physical processes such as small body shock - induced impact, agglutination, and mixing have largely modified the lunar surface over the last four billion years (e. g., Joy et al., 2016; Mckay et al., 1974, 1978), forming a widespread layer of fine - grained regolith about 2 — 15m thick, with a variable abundance of rock clasts (Shkuratov & Bondarenko, 2001). The solar wind irradiation, cosmic ray exposure, and asteroid impact history have been recorded in the depth profile of the lunar regolith (Crawford et al., 2010; Shuster et al., 2010; Wieler, 1998). The surface of the Moon was observed remotely by lunar orbiters and rovers, and samples were collected by Apollo missions and the Luna missions. However, there are few in situ observations that can be directly related to the formation processes of the lunar regolith. These processes are important for the interpretation of data obtained by orbiting and landing spacecraft and by laboratory measurements of the lunar samples, and for understanding the evolution of the regolith on the Moon and other planetary bodies (National Research Council, 2007).

The Chinese robotic spacecraft Chang'E - 4 (CE - 4) landed in the Von Kármán crater (45.457°S, 177.577°E, with a diameter of ~186km) on the farside of the Moon (Fig. S1) on 3 January 2019 (Li et al., 2019). After landing, the Yutu - 2 rover has been exploring the lunar surface with an onboard panorama camera (PCAM), lunar penetrating radar (LPR), navigation camera (NAVCAM), hazard avoidance camera (HAZCAM), and visible and near - infrared imaging spectrometer (VNIS). The exploration by Yutu - 2 provides a unique perspective of the lunar surface features that preserve the records of its evolutional history. Previous studies of the first three lunar day exploration data indicate that CE - 4 landed on the ejecta delivered from the Finsen crater (D = 72km, ~135km to the northeast) and likely other craters (e. g., Alder crater, D=77km, ~139km to the southeast), covering on the beneath basalt (Gou et al., 2019; Hu et al., 2019; Li et al., 2019; Lin et al., 2019). Based on the crater dating of the landing area, the surface age was estimated as ~3.6Ga (Huang et al., 2018; Ling et al., 2019), which would expect a thick layer of regolith because of the long gardening period (Quaide & Oberbeck, 1975). The LPR results revealed a Finsen ejecta layer of about 36 -m - thick, including a top layer of 12 - m - thick fine - grained regolith (Ding et al., 2020; Lai et al., 2019; Li et al., 2020). The regolith surface observed by CE - 4 is relatively homogeneous with many small craters (< 10m in diameter) and very few scattered rocks (Lin et al., 2019). Most of the craters along the Yutu - 2 rover traverse have a smooth surface without rocks (Fig. S2), as noted in previous reports (Lin et al., 2019; Wu et al., 2019). The images taken by the PCAM, NAVCAM, and HAZCAM during the first 12 lunar days of the mission showed more than 45 pits with diameters up to several meters, covered fully with abundant fragments (Fig. S3). These pits are distributed randomly along the traverse and are distinct from the typical rock - free craters with the comparable sizes. The in situ observations of these fragment - rich pits with high spatial resolution conducted by the Yutu - 2 rover provide unique and important information of the lunar surface processes.

2　Exploration Data and Methods

The PCAM, NAVCAM, and HAZCAM onboard the Yutu - 2 rover acquired lots of images with a high spatial resolution (millimeters per pixel). The NAVCAM and HAZCAM are engineering payloads onboard the rover, providing the gray images for the rover's progress (Liu et al., 2015). The PCAM consists of gray and color modes (Jin et al., 2015); these two modes can be switched according to the purposes. The high spatial - resolution images were used to identify the pits and craters around the CE - 4 landing site. A complementary metal - oxide semiconductor (CMOS) imager of the VNIS instrument onboard the rover acquired hyperspectral images of fragments and soil in the pit with spectral resolutions of 2.4—6.5nm and spatial resolution of ~0.6mm at wavelengths of 450—950nm with 256 × 256 pixels. The CMOS observed the surface area at a height of ~0.75m with the fields - of - view of 8.5°. The size of the analyzed area was ~15cm × 21cm (Yang et al., 2020). The initial spectral data reduction was at level 2B data, which was radiometric calibrated radiance data (He et al., 2019; Li et al., 2019). The Yutu - 2 data used in this study are listed in Tab. S1 and S2. The reflectance spectra of the Apollo samples (Tab. S3) were collected from the reflectance experiment laboratory (RELAB) spectral database for comparison with Yutu - 2 spectral data. The RELAB spectra used in this work covers 450—945nm with a spectral resolution of 5nm.

In order to observe the morphologic features and determine the composition of the fragments in the pits, the Yutu - 2 rover approached one of the pits (Pit - 33) and carried out in situ measurements in the ninth lunar day of the mission. A high - resolution digital elevation model (DEM) was created from the stereo HAZCAM images through image matching, 3 - D coordinate calculation of the matched points, and DEM interpolation (Di et al., 2019; Liu et al., 2015). The reflectance (reflectance factor) spectra were obtained using the solar irradiance calibration method (Gueymard, 2004), and then smoothed using the cubic spline method with the smoothing parameter of 0.999 (De Boor, 1978). All the reflectance spectra were normalized ($r_i - normalization = r_i / \sum_{i=1}^{i=n} r_i$, i is band number) for comparison.

3 Results

3.1 Morphology of Fragment - Filled Pits and Small Craters

The surface at the landing site is composed of fine - grained regolith with very few rocks (Fig. S2). This is consistent with the very thick regolith (~12m) revealed by the LPR onboard the rover (Ding et al., 2020; Lai et al., 2019; Li et al., 2020). Observations of the Yutu - 2 rover show that most of the craters around the landing site are almost rock -free, whereas a large amount of fragments were observed in many pits along the rover's path (Fig. 1 and S3). Similarly, no large amount of rocks were observed around the pits (Figure S2. The DEM created using HAZCAM images indicates that the Pit - 33 is

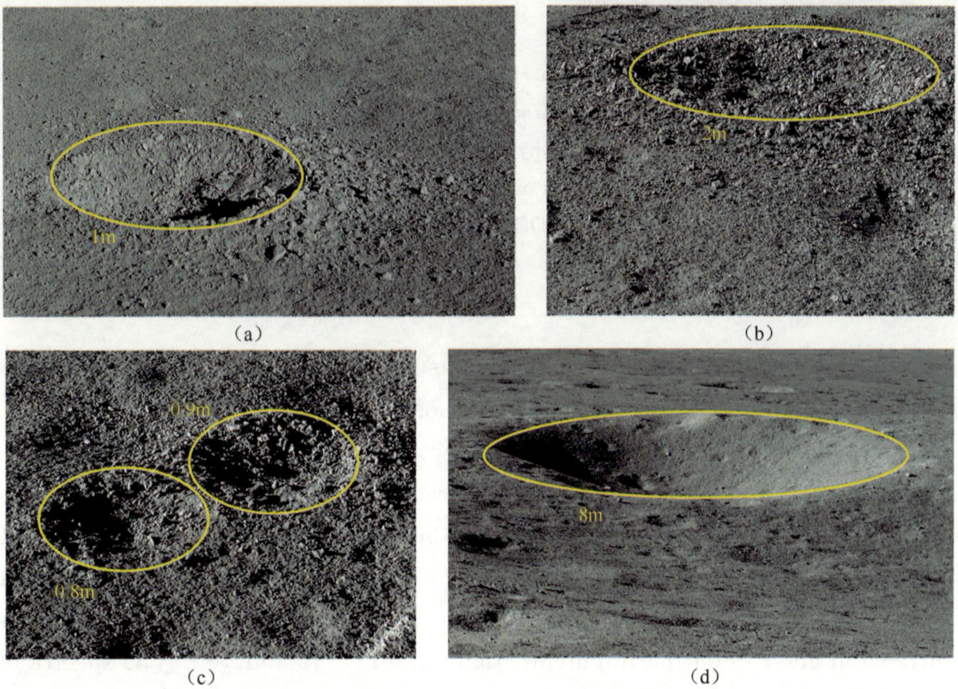

Fig. 1　The panoramic images of typical small craters and pits along the traverse of Yutu - 2 rover. (a - c) The small pits filled with abundant fragments. More images of the fragment - filled pits are shown in Fig. S3. (d) A typical small crater in the Chang'E - 4 landing area, noting the rock -free surface of this crater and other smaller ones. The image catalog numbers of panels (a) — (d) are PCAM - 13 (Pit - 20), PCAM - 26 (Pit - 33), PCAM - 2 (pits 6 and 7), and PCAM - 36 (Tab. S1), respectively

~2m in diameter and ~30cm deep (Fig. 2). The pit exhibits a central depression at the bottom (Fig. 2c). Most of the fragments in the pit are smaller than 5cm.

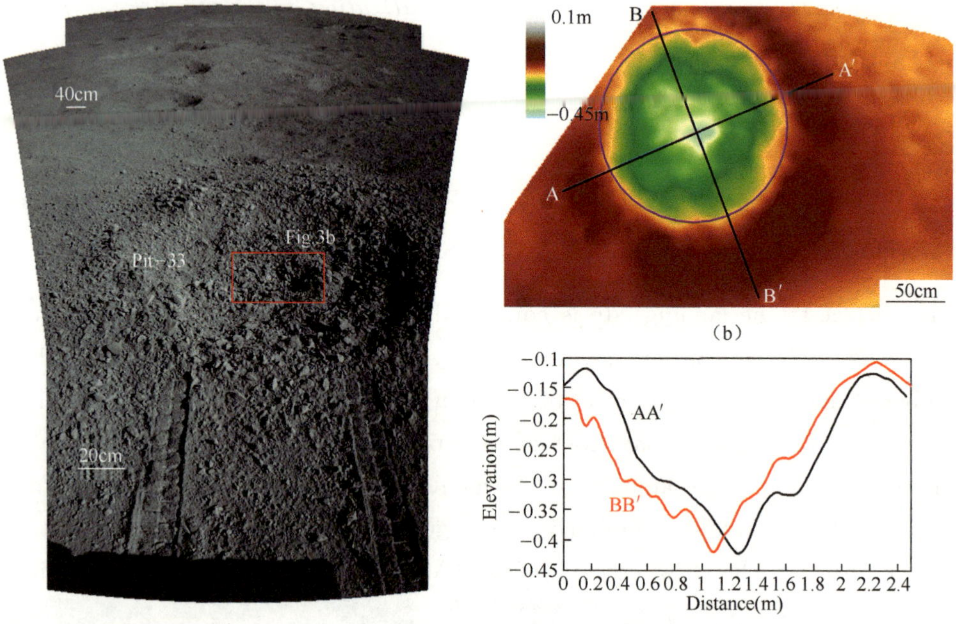

Fig. 2　The morphology of a small fragment-filled pit (Pit-33). (a) The NAVCAM image of the small pit. The distance between the rover tracks is 0.85m (center to center). The width of each wheel is 0.15m. This image was generated by mosaicking the NAVCAM images listed in Tab. S1. (b) The DEM of the small fragment-filled pit, generated using HAZCAM images. The pink circle indicates the edge of the pit in Fig. 2a. (c) DEM profiles of the lines are indicated in Fig. 2b

3.2　Spectral Features of the Fragments

The spectra of the pit fragments were measured by the onboard VNIS and showed characteristics of the regolith rather than those of fresh boulder rocks (Fig. 3). Furthermore, the spectra derived from hyperspectral images of the soil and fragments in the same pit have similar spectral shapes, indicating similar compositions. The spectral slope (945nm/750nm) of the soil (1.06) in the pit is slightly greater than that of the fragments (0.94). In the center of the Pit-33, the fragments have very high albedo with blue-green tint (Fig. 3), likely the glassy fragments. These glassy fragments are concentrated in clusters on the fragment-rich pits (Fig. 3a and 3b). The spectra of the glassy fragments are different from those of the soil in the same pit and the normal regolith surface, those of the fragments with normal albedo and those of the rock boulders (Fig. 3c). The spectra of the glassy fragments exhibit a reflection peak at ~600nm, which is significantly different from those of the rock and regolith at the landing site (Fig. 3c).

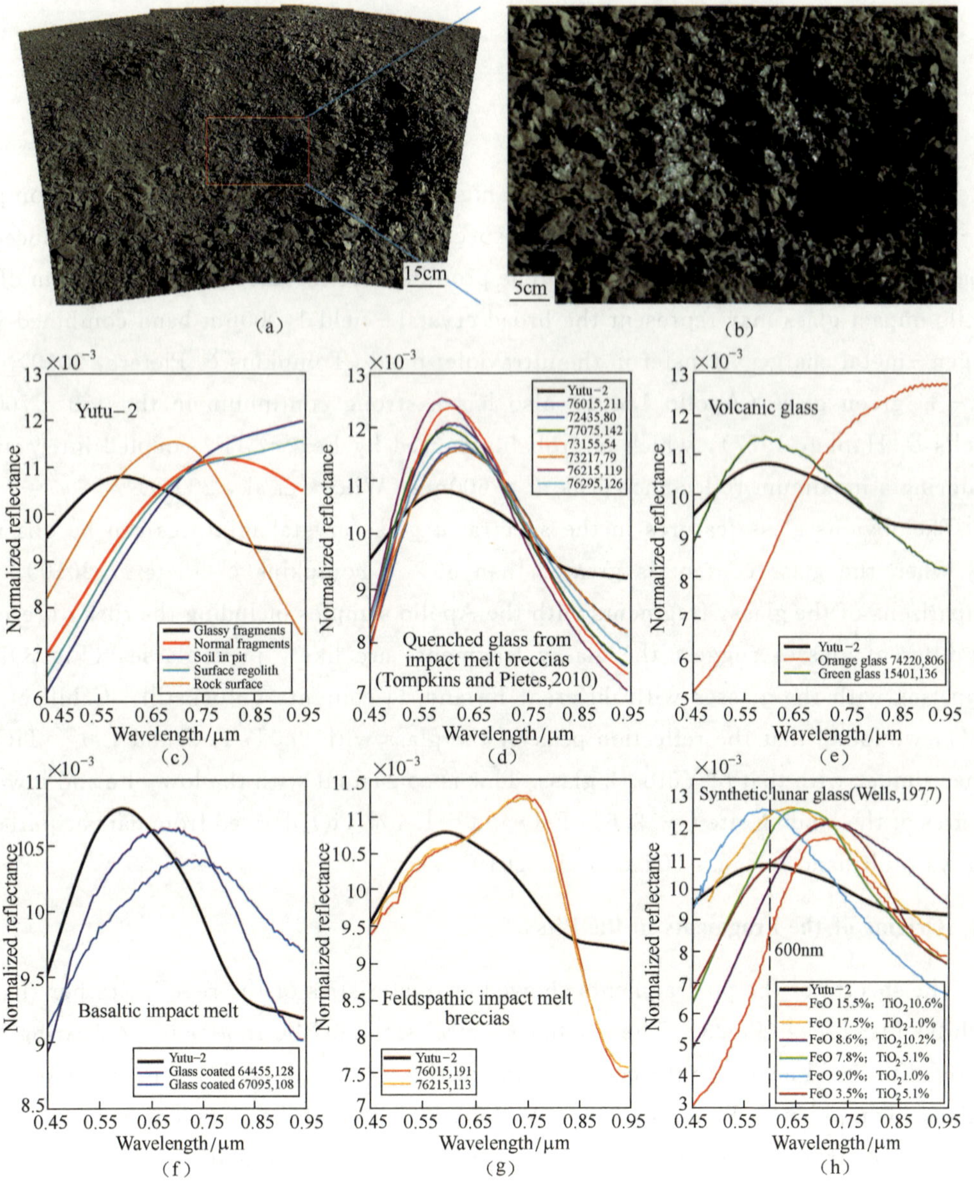

Fig. 3　Visible and near - infrared spectra of fragments from Pit - 33. (a) Close - up view of the small pit in Fig. 2a. The colored image is the mosaic of PCAM 39 - 44 in Tab. S1 in the supporting information. (b) Enlarged image of the rectangular area in Fig. 3a. The image was stretched to highlight the glassy fragments. (c) Spectra of the glassy fragments, compared with those of soil in the same pit, the rock and surface regolith observed by Yutu - 2 (Lin et al., 2019). The location of these spectra in the CMOS image are shown in Fig. S4. The absorption around 825nm of the glassy fragments may be artifacts caused by light reflected from the protective layer attached to the rover (Fig. S5). (d - h) The comparison of the glassy fragments with glass quenched from (d) Apollo impact melt breccias (Tompkins & Pieters, 2010), (e) Apollo volcanic glass, (f) Apollo basaltic impact melt, (g) Apollo feldspathic impact melt breccias, and (h) the synthetic lunar glass collected from Wells and Hapke (1977), respectively. The reflectance was normalized for comparison

4 Discussions

4.1 Presence of Glass in the Pit Fragments

The glassy fragments have the unusual high albedos and show obvious reflection peak at ~600nm, which are consistent with the spectral features of Apollo glasses produced by impact melting or explosive volcanism. The prominent reflectance peak at ~600nm of the Apollo impact glass may represent the broad crystal - field 1,000nm band combined with oxygen - metal charge - transfer in the ultraviolet band (Tompkins & Pieters, 2010). The low - Ti green glass (Apollo 15401) also has a strong continuum in the 450 — 700nm (Wells & Hapke, 1977), which is probably caused by Fe^{2+} — Ti^{4+} coupled interaction, producing a maximum reflectance peak at ~600nm (Wilcox et al., 2006).

The obvious glass features in the spectra of glass/crystal mixtures can be observed only when the glass content is greater than 60% (Tompkins & Pieters, 2010). The comparisons of the glassy fragments with the Apollo samples including the rock, breccias, and different glasses suggest the glassy fragments are likely pure glasses (Fig. 3). By comparing with the glasses with different Fe and Ti contents (Stockstill - Cahill et al., 2014), we found that the reflection peak of the glass with 9.0% FeO and 1.0% TiO_2 is rather similar with that of Yutu - 2 glass. This is consistent with the low - Fe and low - Ti features of the landing site (~12.6% FeO and ~1.4% TiO_2 derived from data acquired by Kaguya Multiband Imager) (Qiao et al., 2019).

4.2 Origins of the Fragments in the Pits

The spectra of the pit fragments showed characteristics of the regolith rather than of fresh boulder rocks (Fig. 3). The smaller spectral slope of the fragments comparing with the regolith could be due to the consolidation of the regolith (Fig. S6) and/or the presence of glass (Tompkins & Pieters, 2010). A scenario is that these fragments were probably broken pieces from single large rocks that were excavated from bedrock by an impact and then hit on the surface with a low speed (< 2.38km/s). However, this origin is inconsistent with the distribution of abundant small fragments (<5cm) in these pits, which indicates that the projectiles were fragile or loose assemblages instead of hard igneous rocks. In addition, the placement of the fragments on the surface is different from the presence of the rock boulders around the landing site (Fig. S7). Although the similarity between the spectra of the fragments and the regolith can also be explained by the covering of lunar soil on rock clasts, this possibility is inconsistent with the observation of high - albedo glass in the fragment - filled pits. In addition, heterogeneous distribution of lunar soil on the fragments should have been observed, with presence of some soil - free surfaces. In fact, the single scattered rock boulders on the landing area were not completely covered by the lunar soil (see the rock boulder analyzed by Yutu - 2, Lin et al., 2019).

The topography observed by Yutu - 2 rover shows the rock - free features of most surface of the landing site, including the surroundings of the fragment - rich pits. The lunar surface at the landing site is covered by the ~12 - m - thick regolith. Impacting simulations on the lunar regolith target (Holsapple & Keith, 2003) (https://www.lpi.usra.edu/lunar/tools/lunarcratercalc/) show that a crater >60m in diameter is needed to penetrate such thick regolith. This has two important limitations: (1) The projectiles were mainly delivered from small impact craters (<60m in diameter) within the regolith, but not from those larger ones that penetrated the overlying regolith and excavated the beneath rock - rich ejecta layers. (2) The regolith - like projectiles were mainly excavated from the impact melt - conglutinated and consolidated walls and bottoms of preexisting small craters within the regolith, and the preexisting craters probably did not penetrate the overlying regolith and exposure the beneath rock - rich ejecta layers either. The small bodies that hit the lunar regolith with high - speed would fuse the regolith into breccias and melt part of the regolith at the bottom and walls of the craters because of the high shock pressures (Kieffer, 1975). The scenario implies that the fragment - filled pits were produced by small meteorite - impact processes on the Moon.

The observed spectra of the glass - bearing fragments are similar to those of some Apollo volcanic glass, especially the green glass (Fig. 3e). However, the volcanic origin of the glass - bearing fragments can be excluded because the mare floor around the CE - 4 landing site was covered by a ~70 - m - thick layer of ejecta (Di et al., 2019) from the adjacent craters (e. g., Finsen crater and Alder crater) (Fig. S1). In addition, the glasses at the CE - 4 landing site are concentrated in clusters on the small fragment - rich pits rather than widely distributed on the surface. The presence of impact glass is consistent with the origin of the regolith breccias, with the impact glass playing a role of conglutination.

4. 3　Different Origins of the Fragment - Filled Pits and the Typical Rock - Free Craters

Most craters observed by the Yutu - 2 rover are almost rock - free (Fig. S2), whereas the pits are typically fragment rich (Fig. 3). The rock - free craters were likely formed by the hypervelocity impacts of small bodies, with the impactors completely destroyed. They are small in size, not penetrating the thick regolith layer and excavating the beneath rocks. These small craters are common and the typical feature of the Moon. The fragment - filled pits were likely formed by low - velocity impacts, because the projectiles were merely broken into small pieces. Hence, these fragment - filled pits are likely the secondary craters, and the impact velocity was <2. 38km/s. As discussed above, the projectiles were probably the impact melt conglutinated and consolidated regolith breccia, instead of stone boulders excavated from bedrocks by large asteroid impact events.

4.4 A Model for Recycling of the Lunar Regolith

The above observations reveal repeated impacting and mixing processes in the lunar regolith. The lunar regolith was impacted by asteroids with high velocity. The compression and temperature distribution data obtained from impact simulation studies (e. g., Pierazzo & Melosh, 2000) indicate that impact events shattered not only rock boulders on the surface and in the regolith but also consolidated and melted part of the regolith (Cintala, 1992; Kieffer, 1975; Schaal & Horz, 1980), mainly in the bottom and walls of the craters (Fig. 4a). Subsequent impact then excavated the conglutinated regolith breccias from the existing crater walls and bottoms, and ejected them to a distance (Fig. 4b). The observed pits and their abundant fragments were likely produced by hits of the impact melt - conglutinated regolith breccia projectiles (Fig. 4c). The ejected regolith breccia could be loosely packed due to short diagenesis time, hence they disintegrated into small fragments, as observed by Yutu - 2. The presence of the impact melts was confirmed by the observations of glass fragments in the pits. The impact melts were formed when the primary craters were created. These observations also explain the formation of the lunar regolith breccia (Fig. S8) commonly found in Apollo lunar samples (e. g., Wolfe, 1981), as well as lunar meteorites (e. g., Gnos et al., 2004; Valencia et al., 2014). The PCAM onboard the Yutu rover also observed several fragment - filled pits emplaced with the possible glass at the Chang'E - 3 landing site (Fig. S9), providing the broader evidence for our model. However, the rocky surface, shallow regolith layer, and the limited detections (Yutu rover only traveled 114m) at the Chang'E - 3 landing site prevented us from finding as many fragment - filled pits as the Chang'E - 4 landing site.

Alternatively, the fragments and glass were derived from superheated blobs of impact melt that has splashed onto the area from Finsen crater or other large craters ($D >$ 60m) in Von Kármán crater (Fig. S1). If so, the rocks should be widely found around the landing site. But this is not the case. In addition, the origin of impact melt from Finsen crater can be excluded, because the event took place about 3.6Ga ago (Huang et al., 2018) and the ejecta from Finsen crater should have been highly processed. Therefore, we prefer the hypothesis that the abundant fragments in the pits are broken pieces of impact melt - conglutinated regolith breccias projectiles excavated from the preexisting craters with diameters less than 60m as discussed above.

Space weathering, including asteroid and micrometeorite impacting and solar wind irradiation, has modified the morphology, subsurface structure, and composition of the surface of nonatmospheric celestial bodies, which has great effects on orbital and rover explorations and the returned samples. The Moon, which is almost completely covered by about 2—15m of lunar regolith (Shkuratov & Bondarenko, 2001), is an ideal target to study the surface processes of similar celestial bodies. Previous studies attributed the formation of the lunar regolith to meteoroid impact and irradiation by particles from the

Sun and cosmic rays (Jolliff et al., 2006). However, the mechanism involved in the formation process of the lunar regolith layer is still unclear. The discovery of the fragment - filled pits by Yutu - 2 rover provides new evidence for the recycling processes of the lunar regolith.

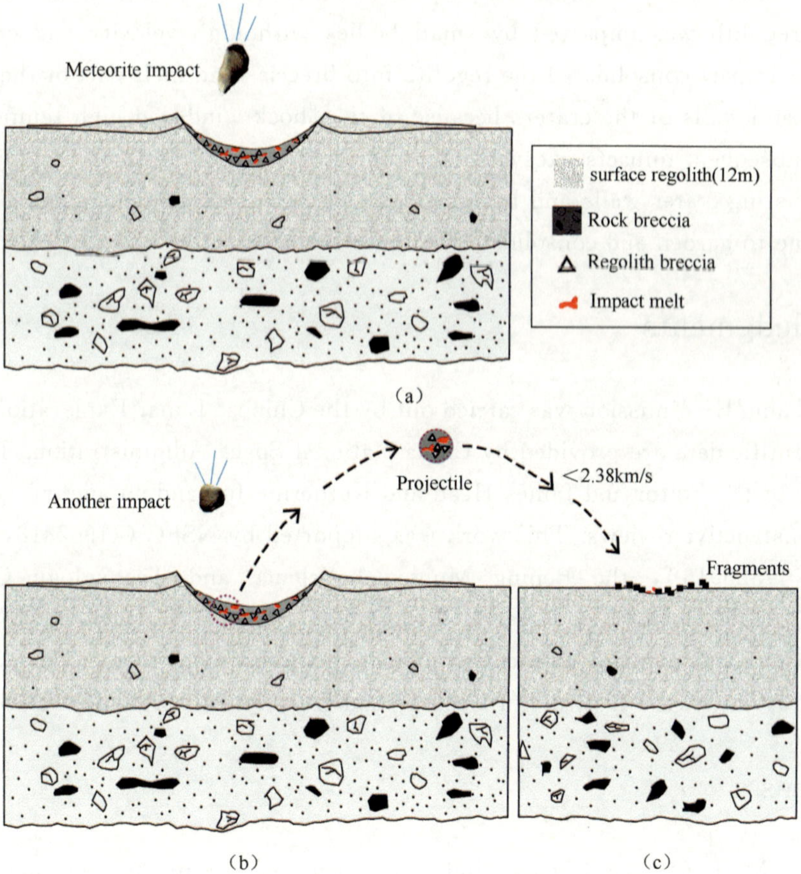

Fig. 4　Illustration of lunar regolith reforming processes

(a) The formation of the primary crater, producing consolidated and conglutinated bottom and wall. Part of the regolith was melted. (b) Another impact hit the preexisting crater, excavated the consolidated and conglutinated materials, and ejected them to a distance. (c) The formation of the small fragment - filled pits. The conglutinated regolith breccia projectile was broken into pieces when it hit the surface

5　Conclusions

The Yutu - 2 rover observed many meter - sized shallow pits fully covered with small fragments in a thick layer of regolith at the CE - 4 landing site. The visible and near - infrared spectra of these fragments are regolith - like rather than the rock boulders. These observations suggest that the fragments are regolith breccias rather than the igneous rocks or rock breccias excavated from the ejecta or bedrock layers beneath the regolith. The regolith breccia projectiles were mainly delivered from small impact craters (<60m in diameter) within the regolith. In addition, the impact glass was in situ found in the pits for the first time. These observations support that the impact glass played a role of conglutination in the regolith breccias and provide new evidence for the recycling processes of the lunar regolith. Accordingly, we proposed a model for recycling of the lunar regolith. The lunar regolith was impacted by small bodies with high velocity and craters were formed. The impact consolidated the regolith into breccias and melt part of the regolith at the bottom and walls of the craters because of the shock - induced high temperature and pressure. Subsequent impacts excavated the impact melt - conglutinated regolith breccias from the existing crater walls and bottoms, and ejected them to a distance. This process may continue to garden and consolidate the lunar surface.

Acknowledgments

The Chang'E - 4 mission was carried out by the Chinese Lunar Exploration Program, and the scientific data are provided by China National Space Administration. The authors are grateful to the Editor and James Head and Katherine Joy and an anonymous reviewer for their constructive reviews. This work was supported by NSFC (41902318, 41430105, 41490631, 41525016), the Beijing Municipal Science and Technology Commission (Z181100002918003) and Key Research Program of Frontier Sciences, CAS (QYZDJ - SSW - DQC001). Honglei Lin and Sen Hu also acknowledge the support of the Key Research Program of the Institute of Geology and Geophysics, CAS (IGGCAS - 201905).

References

[1] Cintala, M. J. (1992). Impact - induced thermal effects in the lunar and Mercurian regoliths. Journal of Geophysical Research, 97 (E1), 947 - 973. https://doi.org/10.1029/91JE02207

[2] Crawford, I. A., Fagents, S. A., Joy, K. H., & Rumpf, M. E. (2010). Lunar palaeoregolith deposits as recorders of the galactic environment of the solar system and implications for astrobiology. Earth, Moon, and Planets, 107 (1), 75 - 85. https://doi.org/10.1007/s11038 - 010 - 9358 - z de Boor, C. (1978). A practical guide to splines (Vol. 27). New York, 1978: Springer - Verlag. https://doi.org/10.1007/978 - 1 - 4612 - 6333 - 3

[3] Di, K., Zhu, M. H., Yue, Z., Lin, Y., Wan, W., Liu, Z., et al. (2019). Topographic evolution of Von Kármán crater revealed by the lunar rover Yutu - 2. Geophysical Research Letters, 46, 12, 764 - 12, 770. https://doi.org/10.1029/2019GL085252

[4] Ding, C., Xiao, Z., Wu, B., Li, Y., Prieur, N. C., Cai, Y., et al. (2020).

Fragments delivered by secondary craters at the Chang'E－4 landing site. Geophysical Research Letters，47，e2020GL087361. https：//doi. org/10. 1029/2020GL087361

[5] Gnos，E.，Hofmann，B. A.，al－Kathiri，A.，Lorenzetti，S.，Eugster，O.，Whitehouse，M. J.，et al. (2004). Pinpointing the source of a lunar meteorite：Implications for the evolution of the moon. Science，305 (5684)，657－659. https：//doi. org/10. 1126/science. 1099397

[6] Gou，S.，di，K.，Yue，Z.，Liu，Z.，He，Z.，Xu，R.，et al. (2019). Lunar deep materials observed by Chang'e－4 rover. Earth and Planetary Science Letters，528，115829. https：//doi. org/10. 1016/j. epsl. 2019. 115829

[7] Gueymard，C. (2004). A. The sun's total and spectral irradiance for solar energy applications and solar radiation models. Solar Energy，76 (4)，423－453. https：//doi. org/10. 1016/j. solener. 2003. 08. 039

[8] He，Z.，Li，C.，Xu，R.，Lv，G.，Yuan，L.，& Wang，J. (2019). Spectrometers based on acousto－optic tunable filters for in－situ lunar surface measurement. Journal of Applied Remote Sensing，13 (02)，1. https：//doi. org/10. 1117/1. JRS. 13. 027502

[9] Holsapple，C.，& Keith，A. (2003). Theory and equations for "Craters from Impacts and Explosions". St. Louis，MO：Washington University.

[10] Hu，X.，Ma，P.，Yang，Y.，Zhu，M. H.，Jiang，T.，Lucey，P. G.，et al. (2019). Mineral abundances inferred from in situ reflectance measurements of Chang'E－4 landing site in south pole－Aitken Basin. Geophysical Research Letters，46，9439－9447. https：//doi. org/10. 1029/2019GL084531

[11] Huang，J.，Xiao，Z.，Flahaut，J.，Martinot，M.，Head，J.，Xiao，X.，et al. (2018). Geological characteristics of Von Karman crater，northwestern south pole－Aitken Basin：Chang'E－4 landing site region. Journal of Geophysical Research：Planet，123，1684－1700. https：//doi. org/10. 1029/2018JE005577

[12] Jin，W.，Zhang，H.，Yuan，Y.，Yang，Y.，Shkuratov，Y. G.，Lucey，P. G.，et al. (2015). In situ optical measurements of Chang'E－3 landing site in Mare Imbrium：2. Photometric properties of the regolith. Geophysical Research Letters，42，8312－8319. https：//doi. org/10. 1002/2015GL065789

[13] Jolliff，B. L.，Wieczorek，M. A.，Shearer，C. K.，& Neal，C. R. (2006). New views of the Moon. Reviews in Mineralogy & Grochemistry，60.

[14] Joy，K. H.，Crawford，I. A.，Curran，N. M.，Zolensky，M.，Fagan，A. F.，& Kring，D. A. (2016). The Moon：An archive of small body migration in the solar system. Earth，Moon，and Planets，118 (2－3)，133－158. https：//doi. org/10. 1007/s11038－016－9495－0

[15] Kieffer，S. W. (1975). From regolith to rock by shock. The Moon，13 (1－3)，301－320. https：//doi. org/10. 1007/BF00567522

[16] Lai，J.，Xu，Y.，Zhang，X.，Xiao，L.，Yan，Q.，Meng，X.，et al. (2019).

Comparison of dielectric properties and structure of lunar regolith at Chang'e - 3 and Chang' e - 4 landing sites revealed by ground penetrating radar. Geophysical Research Letters, 46, 12783 - 12793. https://doi.org/10.1029/2019GL084458

[17] Li, C., Liu, D., Liu, B., Ren, X., Liu, J., He, Z., et al. (2019). Chang'E - 4 initial spectroscopic identification of lunar far - side mantle - derived materials. Nature, 569 (7756), 378 - 382. https://doi.org/10.1038/s41586-019-1189-0

[18] Li, C., Su, Y., Pettinelli, E., Xing, S., Ding, C., Liu, J., et al. (2020). The Moon's farside shallow subsurface structure unveiled by Chang'E - 4 lunar penetrating radar. Science Advances, 6 (9).

[19] Li, C., Wang, C., Wei, Y., & Lin, Y. (2019). China's present and future lunar exploration program. Science, 365 (6450), 238 - 239. https://doi.org/10.1126/science.aax9908

[20] Li, C., Wang, Z., Xu, R., Lv, G., Yuan, L., He, Z., & Wang, J. (2019). The scientific information model of Chang'E - 4 visible and near - IR imaging spectrometer (VNIS) and in - orbit verification. Sensors, 19 (12), 2806. https://doi.org/10.3390/s19122806

[21] Lin, H., He, Z., Yang, W., Lin, Y., Xu, R., Zhang, C., et al. (2019). Olivine - norite rock detected by the lunar rover Yutu - 2 likely crystallized from the SPA impact melt pool. National Science Review, 7 (5), 913 - 920. https://doi.org/10.1093/nsr/nwz183

[22] Ling, Z., Qiao, L., Liu, C., Cao, H., Bi, X., Lu, X., et al. (2019). Composition, mineralogy and chronology of mare basalts and non - mare materials in Von Kármán crater: Landing site of the Chang'E - 4 mission. Planetary and Space Science, 179, 104741. https://doi.org/10.1016/j.pss.2019.104741

[23] Liu, Z., di, K. C., Peng, M., Wan, W. H., Liu, B., Li, L. C., et al. (2015). High precision landing site mapping and rover localization for Chang'e - 3 mission. Science China Physics, Mechanics & Astronomy, 58 (1), 1 - 11. https://doi.org/10.1007/s11433-014-5612-0

[24] McKay, D., Fruland, R. & Heiken, G. (1974), In lunar and planetary science conference proceedings (Vol. 5, pp. 887 - 906).

[25] McKay, D., Heiken, G. & Waits, G. (1978), In lunar and planetary science conference proceedings (Vol. 9, pp. 1913 - 1932).

[26] National Research Council (2007). The scientific context for exploration of the Moon. Washington, DC: The National Academies Press.

[27] Pierazzo, E., & Melosh, H. (2000). Melt production in oblique impacts. Icarus, 145 (1), 252 - 261. https://doi.org/10.1006/icar.1999.6332

[28] Qiao, L., Ling, Z., Fu, X., & Li, B. (2019). Geological characterization of the Chang'e - 4 landing area on the lunar farside. Icarus, 333, 37 - 51. https://doi.org/10.1016/j.icarus.2019.05.029

[29] Quaide, W., & Oberbeck, V. (1975). Development of the mare regolith: Some model considerations. The Moon, 13 (1 - 3), 27 - 55. https://doi.org/10.1007/BF00567506

[30] Schaal, R. B., & Horz, F. (1980). Experimental shock metamorphism of lunar soil. In Lunar and Planetary Science Conference Proceedings, 11, 1679 - 1695.

[31] Shkuratov, Y. G., & Bondarenko, N. V. (2001). Regolith layer thickness mapping of the Moon by radar and optical data. Icarus, 149 (2), 329 - 338. https://doi.org/10.1006/icar.2000.6545

[32] Shuster, D. L., William, G. B., Cassata, S., Fernandes, V. A., Garrick - Bethell, I., & Weiss, B. P. (2010). A record of impacts preserved in the lunar regolith. Earth and Planetary Science Letters, 290 (1 - 2), 155 - 165. https://doi.org/10.1016/j.epsl.2009.12.016

[33] Stockstill - Cahill, K. R., Blewett, D. T., Cahill, J. T., Denevi, B. W., Lawrence, S. J., & Coman, E. I. (2014). Spectra of the Wells lunar glass simulants: New old data for reflectance modeling. Journal of Geophysical Research: Planets, 119, 925 - 940. https://doi.org/10.1002/2013JE004600

[34] Tompkins, S., & Pieters, C. M. (2010). Spectral characteristics of lunar impact melts and inferred mineralogy. Meteoritics and Planetary Science, 45 (7), 1152 - 1169. https://doi.org/10.1111/j.1945 - 5100.2010.01074.x

[35] Valencia, S., Carpenter, P., Jolliff, B., & Korotev, R. (2014). Using electron - probe microanalysis and quantitative compositional mapping to study lithic clasts in lunar meteorites NWA 2727 and NWA 3170. Microscopy and Microanalysis, 54 (9), 2083 - 2115. https://doi.org/10.1111/maps.13370

[36] Wells, E., & Hapke, B. (1977). Lunar soil: Iron and titanium bands in the glass fraction. Science, 195 (4282), 977 - 979. https://doi.org/10.1126/science.195.4282.977

[37] Wieler, R. (1998). The solar noble gas record in lunar samples and meteorites. Space Science Reviews, 85 (1/2), 303 - 314. https://doi.org/10.1023/A: 1005166904225

[38] Wilcox, B., Lucey, P., & Hapke, B. (2006). Radiative transfer modeling of compositions of lunar pyroclastic deposits. Journal of Geophysical Research, 111, E09001. https://doi.org/10.1029/2006JE002686

[39] Wolfe, E. W. (1981), The geologic investigation of the Taurus - Littrow valley, Apollo 17 landing site. Washington: US Govt. Print. Off., 1981.

[40] Wu, W., Li, C., Zuo, W., Zhang, H., Liu, J., Wen, W., et al. (2019). Lunar farside to be explored by Chang'e - 4. Nature Geoscience, 12 (4), 222 - 223. https://doi.org/10.1038/s41561 - 019 - 0341 - 7

[41] Yang, Y., Lin, H., Liu, Y., Lin, Y., Wei, Y., Hu, S., et al. (2020). The effects of viewing geometry on the spectral analysis of lunar regolith as inferred by in situ spectrophotometric measurements of Chang'E - 4. Geophysical Research Letters, 47, e2020GL087080. https://doi.org/10.1029/2020GL087080

In Situ Photometric Experiment of Lunar Regolith With Visible and Near – Infrared Imaging Spectrometer On Board the Yutu – 2 Lunar Rover

Honglei Lin[1], Rui Xu[2], Wei Yang[1], Yangting Lin[1], Yong Wei[1], Sen Hu[1], Zhiping He[2], Le Qiao[3], and Weixing Wan[1]

1. Key Laboratory of Earth and Planetary Physics, Institute of Geology and Geophysics, Chinese Academy of Sciences, Beijing, China
2. Key Laboratory of Space Active Opto – Electronics Technology, Shanghai Institute of Technical Physics, Chinese Academy of Sciences, Shanghai, China
3. Shandong Provincial Key Laboratory of Optical Astronomy and Solar – Terrestrial Environment, Institute of Space Sciences, Shandong University, Weihai, China

Abstract: China's Chang'E – 4 (CE – 4) mission successfully landed in Von Kármán crater within South Pole – Aitken basin of the Moon. The Visible and Near – Infrared Imaging Spectrometer (VNIS) on board Yutu – 2 rover investigated the photometric properties of lunar regolith. Seven VNIS measurements were conducted on a small lunar surface with a diameter <5m by the rover rotating at the center, with the phase angles from 39.6 to 97.1° obtained in the similar observational geometry of solar altitude and observation angle. The phase function, which varies in different wavelength, is derived using a third – order polynomial fitting, in combination with the calibration and comparison of orbital/in situ VNIS data at the Chang'E – 4 landing site and the same regions. After the photometrical correction of the spectra with the phase function, the derived FeO contents and optical maturity parameters of the regolith reduce much of their deviations, which is consistent with the homogeneity of the regolith and hence demonstrates the significance of the photometric correction on the VNIS spectra.

Plain Language Summary: For mineral identification and mapping with optical data, the photometric correction is important and challenging. The rover Yutu – 2 conducted experiments on the lunar surface to measure how the spectra change with phase angles. We derived a photometric function using seven measurements obtained by the Visible and Near –Infrared Imaging Spectrometer on board Yutu – 2. The photometric correction of reflectance spectra is critical for estimations of the FeO content and the spectral maturity, which relate to geologic interpretations. The photometric function is also important for the calibration, validation, and comparison of different observations acquired by the same/different instruments.

Key Points:

• In situ photometric experiment was conducted by Yutu－2, and a visible and near－infrared photometric function of lunar regolith was derived

• The FeO content and optical maturity parameter derived with corrected spectra are more reliable

1 Introduction

The reflectance spectra were routinely used to identify minerals and their relative abundances on the lunar surface, especially remotely from the orbit. The in situ measurements of the lunar regolith by Visible and Near－Infrared Imaging Spectrometer (VNIS) were so far conducted only by China's Chang'E－3 (CE－3) and Chang'E－4 (CE－4) mission. CE－3 landed in the northern Mare Imbrium on 14 December 2013 and ceased to transmit data in March 2015. During the mission, the VNIS on board the rover Yutu observed four spectra of the lunar surface (Ling et al., 2015; Wu & Hapke, 2018; Zhang et al., 2015; Zhang et al., 2015). CE－4, the backup of CE－3, landed at 45.457°S, 177.588°E in the Von Kármán crater within the South Pole－Aitken basin on 3 January 2019, and the rover Yutu－2 has tracked 190.66m during the first five lunar days of mission operations. The Von Kármán crater was filled with mare basalt, but the CE－4 landing site was covered by thick ejecta on the mare basalt, which can be noted as light gray strips irradiating from the neighbor Finsen crater and could represent the material of the South Pole－Aitken floor (Hu et al., 2019; Huang et al., 2018; Li et al., 2019; Lin et al., 2019; Qiao et al., 2019). The thickness of the ejecta was estimated at least 33m including ～7.5m regolith (Huang et al., 2018) or ～30m of Finsen ejecta plus additional ～35m of the beneath ejecta layer from the Alder crater (Di et al., 2019). In the fourth lunar day after the CE－4 landing, the in situ photometric experiments of lunar regolith were conducted with the VNIS on board Yutu－2 (Fig. 1), providing a new perspective to understand the photometric properties of the lunar surface.

An accurate photometric function is very important for correcting the effects of observation geometry (incidence angle, emission angle and phase angle) and wavelength－dependent phase function on spectra (Besse et al., 2013; Hicks et al., 2011; Wu et al., 2013; Yokota et al., 2011). Many studies have been reported for correcting the orbital data, such as the Spectral profiler on board the SELENE mission, Moon Mineralogy Mapper (M^3) on board Chandrayaan－1 mission, and Interference Imaging Spectrometer (IIM) on board Chang'E－1 mission (Besse et al., 2013; Besse, Sunshine, et al.,

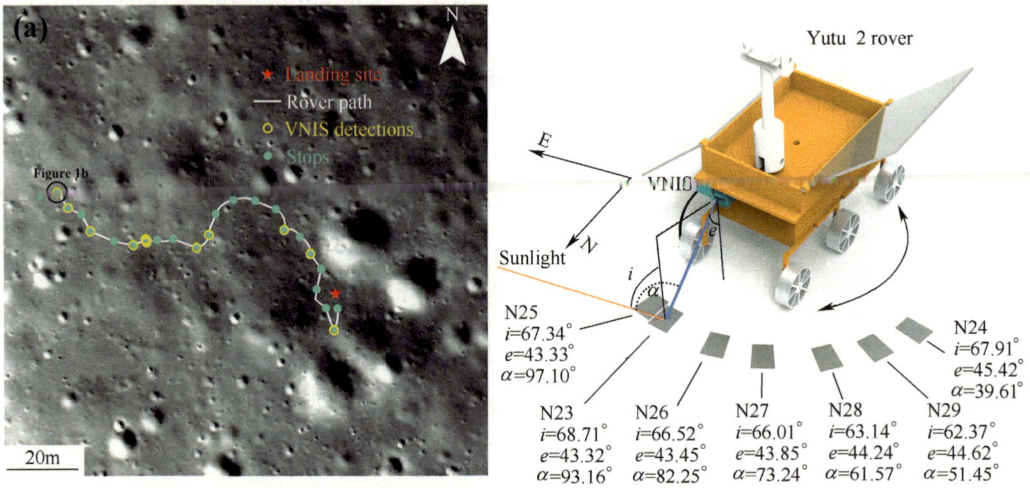

(a)　　(b)

Fig. 1 The photometric experiment in the fourth lunar day of mission operations. (a) The path of Yutu - 2 rover in the first four lunar days. The background is the image acquired by landing camera (LCAM). The black circle is the location of photometric experiment; (b) The schematic of photometric experiment conducted by VNIS. The VNIS measures the regolith from a height of ~1m above the lunar surface. The rover rotated within a circle with radius of 2 - 3m, and the gray rectangle is the measured region. The distance from the rover center to each VNIS measurement is less than 5m. The incidence angle i, emission angle e, and phase angle α are labeled. The labels of VNIS measurements are corresponding to Fig. 3

2013; Wu et al., 2013; Yokota et al., 2011). Ideally, the accurate photometric function should be derived from a set of measurements on the surface with the same/similar composition. However, this requirement is difficult to meet from the orbit even though the regions can be separated based on some thresholds (e. g., FeO content). The phase curves of the regolith at Chang' E - 3 landing site have been extracted using the observations of panorama cameras, and the photometric properties were also determined (Jin et al., 2015). Even though covering the wide phase angles (from 2° to 140°), the small band numbers and the large area of likely heterogeneous surface as well as the low signal - to - noise ratio in the observations with large phase angles limit its applications. In order to investigate the effect of the phase angle, a set of VNIS analyses of the lunar regolith were carried out by the rover Yutu - 2. A smooth and boulder - free area was selected for this experiment, which was expected homogenous (Fig. 1a and 2). The rover rotated in the center, with phase angle changing from 39.6° to 97.1° (Fig. 1b). The analyzed zone of the regolith is expected homogeneous within such a small area (< 5m in diameter), because the regolith at the landing site was well mixed for a very long time (up to 3.6 Ga) (e. g., Huang et al., 2018; Ling et al., 2019). This is also confirmed by the smooth and rocky clast - free surface as shown in the panorama images (Fig. 2).

The homogeneous surface at the landing site provides a unique perspective to conduct the lunar photometric experiments. In this study, we derived the in situ phase curve and photometric function covering 470 to 945 nm with a spectral interval of 5nm.

Fig. 2　The mosaic panorama image around the regions measured by VNIS instrument in the fourth lunar day of mission operations. The brightness variation in the images is probably caused by the surface relief

2　Data and Methods

The VNIS consists of a Complementary Metal - Oxide Semiconductor (CMOS) imager (100 bands covering 450 − 945nm) with 256 by 256 pixels and a short - wavelength near - infrared (SWIR) detector (300 bands covering 900−2,395nm) with one pixel (He et al., 2019). The spectral resolutions are 2.4−6.5nm in the VIS/NIR bands and 3.6 − 9.6nm in the SWIR bands. The uncertainty of spectral calibration of CMOS imager is ±0.39nm and that of SWIR detector is ±0.62nm (He et al., 2014; Li et al., 2019). In laboratory radiometric calibration, the relations between the spectral radiance and digital number (DN) were modeled using the integrating sphere, which provides the standard radiation source. The uncertainty of the radiometric calibration results is 4.97% in the VIS/NIR channel and 6.36% in the SWIR channel (Li, Xu, et al., 2019). Eight measurements (Fig. 3) were conducted for photometric study, covering the phase angle of 39.61° (CE - 4 N24), 40.42° (CE - 4 N30), 51.45° (CE - 4 N29), 61.57° (CE - 4 N28), 73.24° (CE - 4 N27), 82.25° (CE - 4 N26), 93.16° (CE - 4 N23), and 97.10° (CE - 4 N25). The incidence angle and emission angle of these measurements are similar (Tab. 1). We only focus on the CMOS data to derive the photometric functions because of the lack of reliable SWIR spectra. Even though a Lambertian reference plate is on board the rover, it has not been measured in the fourth lunar day for engineering reasons. However, the radiance used in this study was in flight calibrated using the valid measurements of the Lambertian reference plate (Xu et al., 2014). The reflectance (radiance factor, RADF) was retrieved from radiance by divided by the solar irradiance (Gueymard, 2004; Wu et al., 2018). The Sun - Moon distance is approximately 1 AU. It should be noted that the pixels with the reflectance value less than 0.01 were referred as shadow and removed for calculating average reflectance (Tab. S1 in the supporting information). The last measurement with phase angle of 40.42° before Yutu - 2 "noon nap" was shadowed by the rover body and it has the very similar phase angle with the first measurement (39.61°) in the fourth lunar day. Thus, this measurement was not considered in this study. The spectra before 470 nm, which were discarded in this study, show unreasonable uplift (Fig. 3). For the same region of the surface, the reflectance varies in local observation geometry as well as a wavelength - dependent phase function $f(\alpha)$. The photometric function based on Lommel - Seeliger model was employed in this study (Besse, Sunshine, et al., 2013; Hicks et al., 2011; Wu et al., 2013).

Tab. 1 The Observation Geometry of the Data Used in This Study

Observation	Incidence angle	Emission angle	Phase angle
CE - 4 N23	68.71	43.32	93.16
CE - 4 N24	67.91	45.42	39.61
CE - 4 N25	67.34	43.33	97.10
CE - 4 N26	66.55	43.45	82.25
CE - 4 N27	66.01	43.85	73.24
CE - 4 N28	63.14	44.24	61.57
CE - 4 N29	62.37	44.62	51.45
CE - 4 N30	61.44	45.23	40.42
M^3 - 1	53.00	5.47	47.60
M^3 - 2	52.05	12.97	57.87
S1	72.06	42.89	89.55
LE00202	64.51	46.53	91.88
LE00207	64.56	46.30	90.82
LE00301	70.86	43.36	66.89
LE00304	61.39	44.37	73.58

$$RADF = \frac{\cos(i)}{\cos(i) + \cos(e)} f(\alpha)$$

The phase function $f(\alpha)$ can be expressed an empirical equations as described in Hicks et al. (2011) and Wu et al. (2013). A fourth order polynomial (Besse, Sunshine, et al., 2013) of phase function has been applied to M^3 data (Pieters et al., 2009), and a combination of exponential function and fourth - order polynomial (Wu et al., 2013) has been used to fit Interference Imaging Spectrometer (IIM) data. The phase curve (Fig. S1) derived with fourth - order polynomial cannot represent the trend of phase function well because only seven samples can be used and the phase angles are relatively

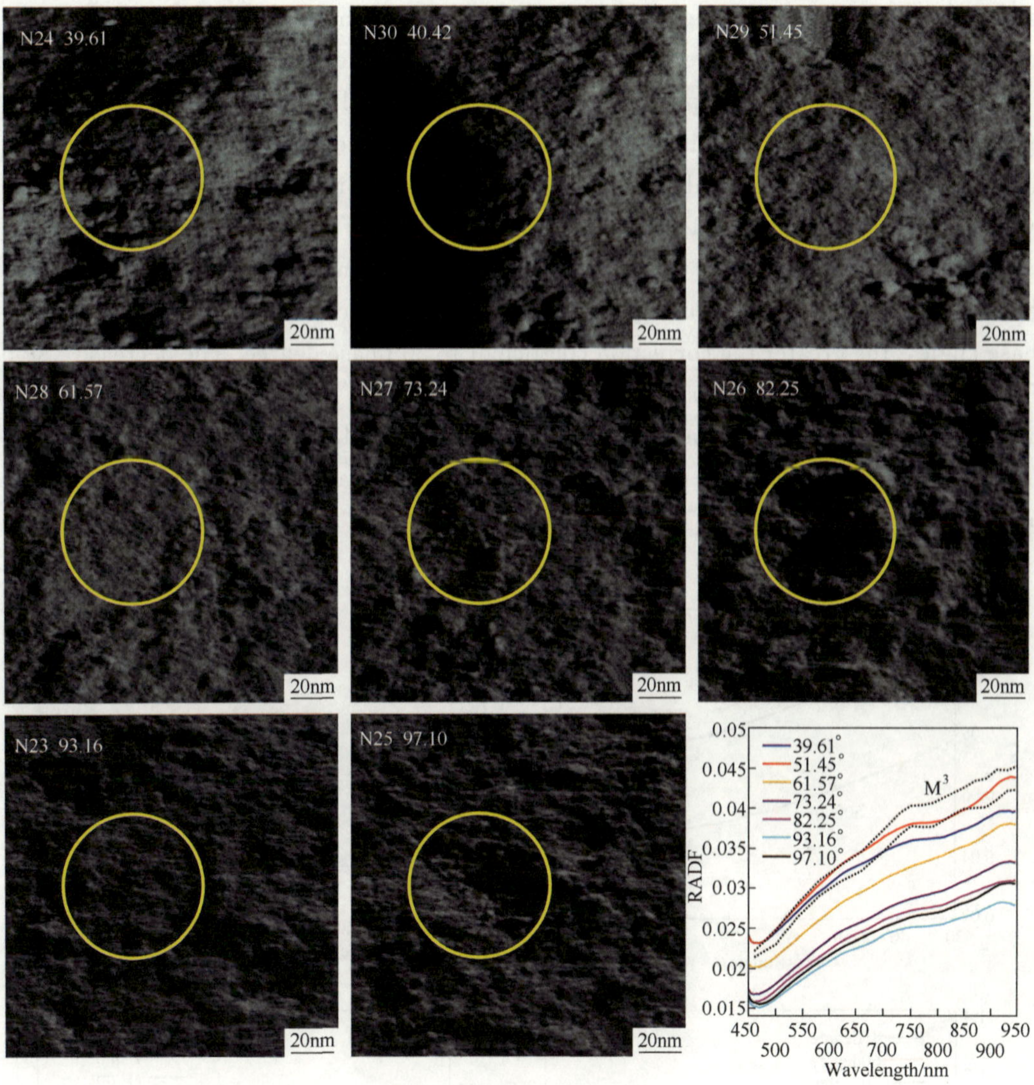

Fig. 3　Radiance images at 750nm and mean reflectance spectra of measurements obtained at different phase angle (°). The yellow circle is SWIR field. The M^3 spectra (dotted lines) around CE－4 landing site are extracted from M3G20090719T200620 (7×7 pixels average centered at X: 64, Y: 4935) and M3G20090720T003411 (7 × 7 pixels average centered at X: 254, Y: 27699). The M^3 radiance data were downloaded from Planetary Data System and then converted to radiance factor (RADF) using the same method with CE－4 data. The location of CE－4 landing site in M^3 images was determined by registering the same features in LROC Narrow Angle Camera image

larger. Thus, we applied a third－order polynomial to fit Yutu－2 data and the photometric function parameter set will be obtained.

3　Results

The phase functions of all wavelengths are fitted independently. The third－order polynomial fitted phase function of 500 and 750nm are shown in Fig. 4a as examples, indicating a good correlation between the phase function and phase angle. The opposition effect of lunar regolith is apparent. The threedimensional demonstration of the phase functions at all wavelengths is shown in Fig. 4b, and the parameter set fitted to the photometric function is available in Tab. S2 of the supporting information. The phase function varies with wavelength in magnitude. The plot of the wavelength versus phase function (Fig. 4c) shows that the correction coefficient from a specific phase angle to 30 phase degrees decreases with increasing wavelength. For each wavelength, the larger the phase angle, the greater the correction coefficient to 30 phase degree. The decrease (slope of phase function) will be smaller when the phase angle is large enough. We normalized the phase function to 750nm at phase angle of 30° (Fig. 4d), indicating that, for two wavelengths with a difference of 80nm, the maximum relative error between the phase functions is ~28%.

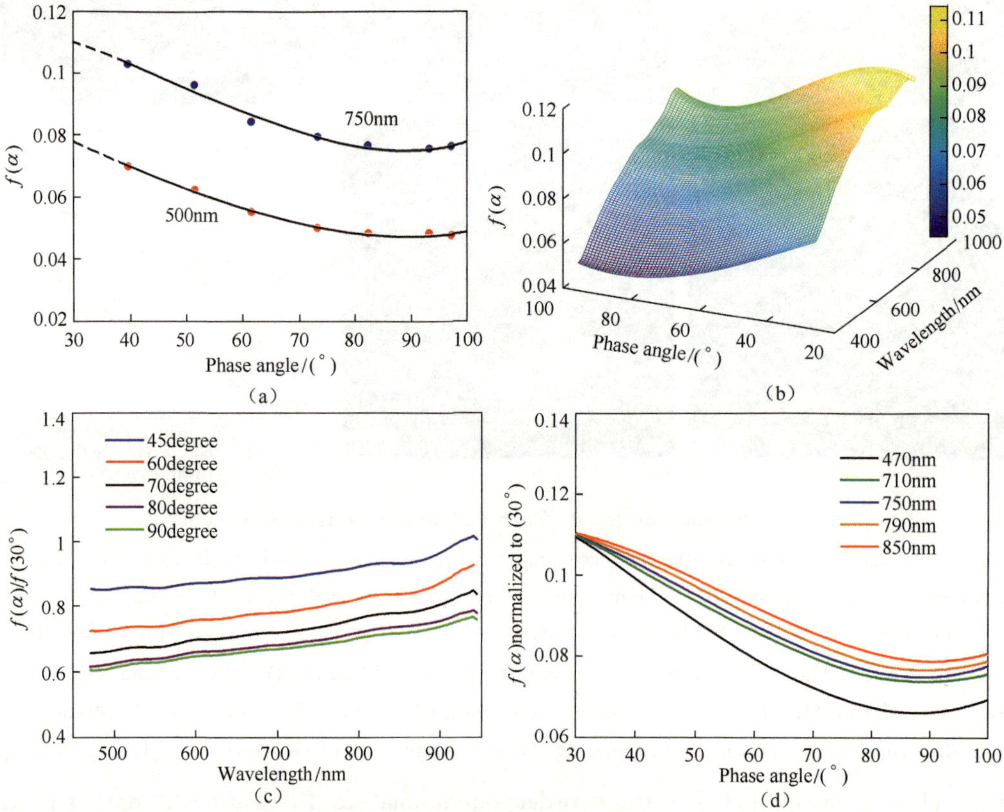

Fig. 4 Phase function derived from CMOS data obtained by Yutu-2 rover. (a) The fits of phase function at 500 and 750nm; (b) Three-dimensional phase functions at all wavelengths; (c) The comparison of phase functions between different wavelengths for five exampled phase angles; (d) The phase functions were normalized to 750nm at phase angle of 30°

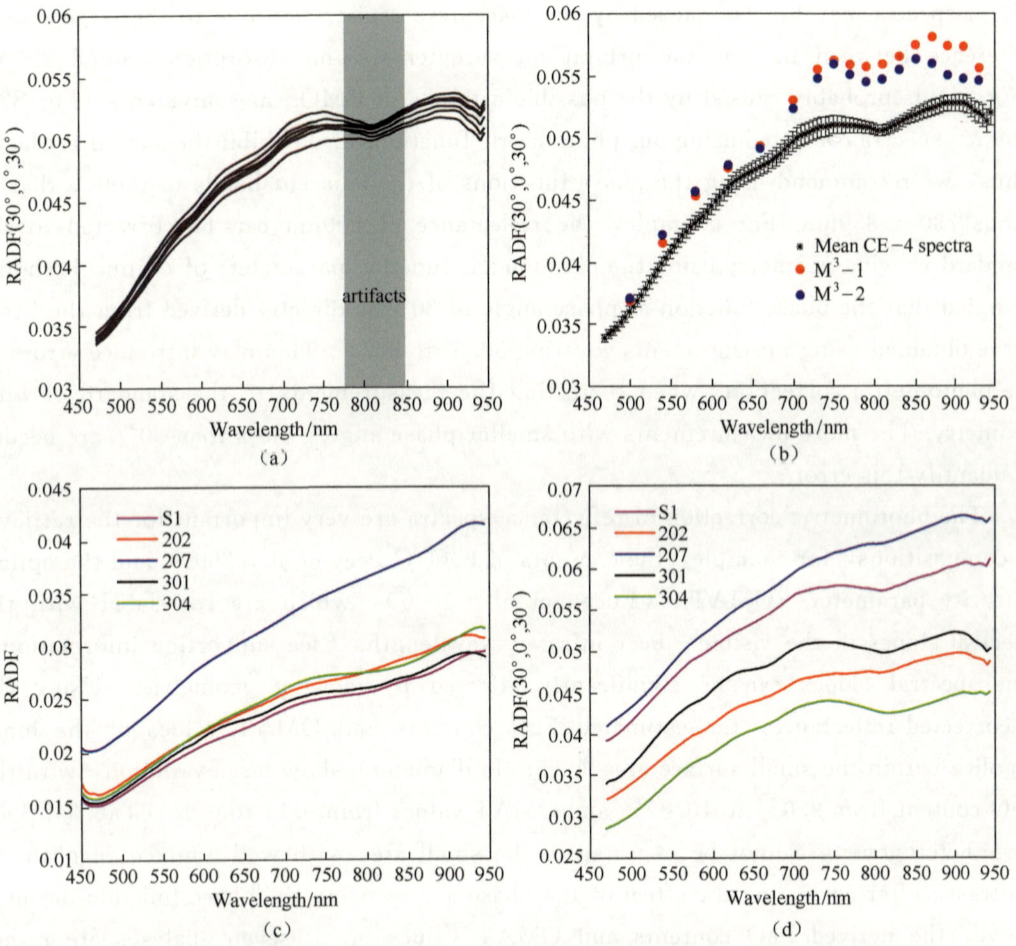

Fig. 5　The photometrically corrected spectra. (a) The corrected CE－4 spectra. The absorption between 780 and 850nm is probably caused by artifacts of instrument; (b) The mean spectrum of CE－4 and the corrected M^3 spectra of the CE－4 exploration area; (c) The RADF of measurements observed in first three lunar days. The phase angles of these measurements are within 39.6° to 97.1°; (d) The corrected CE－4 RADF spectra of the measurements of first three lunar days

4　Discussion

The seven measurements of Yutu－2 were photometrically corrected to the standard viewing geometry, that is, $i=30°$, $e=0°$, $\alpha=30°$, as shown in Fig. 5a. The corrected reflectance is matched very well, and the error bar with standard deviation is provided in Figure 5b. Two spectra extracted from different scenes obtained by Moon Mineralogy Mapper (M^3) at the CE－4 landing site region were also corrected to the standard geometry. The spectra of two different M^3 scenes are matched after correction with our model. Overall speaking, the CMOS spectra acquired by Yutu－2 rover are comparable with M^3 spectra. The CMOS spectra are slightly darker than M^3 spectra, possibly because of the contributions of darker pixels in CMOS images. The difference between M^3 and

CE - 4 spectra may be also caused by the inaccurate registration due to the large - scale difference between in situ and orbital measurements. The absorption around 825nm (Fig. 5a) is probably caused by the possible artifacts of CMOS measurements (Fig. S2). The M^3 spectra corrected using our photometric functions also exhibit the similar artifacts. Thus, we recommend using the phase functions of the adjacent bands to replace that of bands 780 — 850nm. For example, the reflectance at 800nm can be corrected to the standard viewing geometry using the photometric function parameters of 775nm. It should be noted that the phase function at phase angle of 30°was directly derived from the fitted curve obtained using measurements covering 39.6° to 97.1°. This may introduce errors to the photometric corrections when correcting the measurements to the standard viewing geometry. The more measurements with smaller phase angles (less than 30°) are needed to quantify this error.

The photometric correction of reflectance spectra are very important for the retrieval of compositions, for example, the contents of FeO (Lucey et al., 2000) and the optical maturity parameter (OMAT) (Lucey et al., 2000), which are correlated with the spectral slopes in the visible - bear infrared wavelengths (see supporting information). The spectral slopes can be significantly affected by viewing geometries. Using the uncorrected reflectance, the estimated FeO contents and OMAT values of the lunar regolith within the small surface area (<5m in diameter) show large variation, with the FeO content from 9.0% to 10.89% and OMAT values from 0.11 to 0.18 (Tab. 2). Such large heterogeneity cannot be expected in the small area with well - mixed regolith. In contrast, after correcting the effect of the phase angles using the phase function depicted above, the derived FeO contents and OMAT values of all seven analyses are rather homogeneous (Tab. 2). The standard deviation of the measurements significantly reduces to (2σ) 0.7% and 7.1% from 13.5% (FeO) and 35.7% (OMAT) with the correction, respectively, convincingly demonstrating the significance of the photometrical correction on the reflectance spectra. The FeO contents at different locations along the rover track (~ 178 m) also exhibit small deviations, but the OMAT values are somewhat heterogeneous. The variation of OMAT was probably caused by the subsurface regolith excavated by the small impact craters, whereas the FeO contents of the regolith were not affected by space weathering. Another effect of the photometrical correction is enhancement of the FeO contents, from 9.87 ± 1.33 to 12.21 ± 0.08 wt % (Tab. 2). The higher FeO contents of the regolith at the landing site of Chang'E - 4 suggest a deeper origin (e. g., Warren, 1990), probably from the lower crust or even the upper mantle. This is consistent with the recent studies (Gou et al., 2019; Hu et al., 2019; Li, Liu, et al., 2019; Lin et al., 2019) that the surface materials at the CE - 4 landing site could originate from the lunar deep. The mineral fractions in these recent studies (Gou et al., 2019; Hu et al., 2019; Li, Liu, et al., 2019; Lin et al., 2019) are inconsistent, exhibiting the systematic biases of the different approaches. However, the retrieved mineralogy at multiple measurement sites are similar in each published work when considering their own model errors, consistent with our hypothesis for the local compositional homogeneity of the CE - 4 exploration area. Lin et al. (2019) have corrected the VNIS spectra using the photometric function derived from the laboratory measurements of the lunar simulant with the backup CE - 4 VNIS instrument and concluded that the materials at the CE - 4 landing site are olivine - bearing norite.

Tab. 2 The FeO Contents and OMAT Derived From the Original and Photometric Corrected CMOS Spectra

Site	Phase angle/(°)	Original FeO (wt%)	Corrected FeO (wt %)	Original OMAT	Corrected OMAT
N24	39.61	9.68	12.22	0.178	0.246
N29	51.45	9.00	12.17	0.128	0.250
N28	61.57	9.41	12.27	0.115	0.250
N27	73.24	9.68	12.18	0.116	0.239
N26	82.25	10.62	12.18	0.136	0.246
N23	93.16	10.89	12.26	0.152	0.263
N25	97.10	9.80	12.18	0.113	0.236
	average	9.87	12.21	0.13	0.25
	2SD	1.33	0.08	0.05	0.02
	2SD%	13.5	0.7	35.7	7.1
S1	89.55	7.84	11.77	0.094	0.220
LE00202	91.87	10.80	12.32	0.104	0.228
LE00207	90.82	6.83	11.42	0.074	0.202
LE00301	66.89	10.10	12.27	0.123	0.268
LE00304	105.39	11.00	12.50	0.103	0.239
	average	9.31	12.06	0.10	0.23
	2SD	3.74	0.89	0.04	0.05
	2SD%	40.2	7.4	35.7	21.2

Note. The two standard deviations (SD) are calculated.

5 Conclusions

The in situ photometric experiment was conducted by rover Yutu - 2, which is important to achieve correctly the compositions of the farside of the Moon from the VNIS data. Seven measurements with different phase angles were used to derive the photometric functions and the parameter sets for fitting phase functions. The magnitude of the phase function varies with the wavelength, and the correction coefficient from a specific phase angle to 30° decreases with the increase of wavelength (for example, 1.64 at 500nm to 1.36 at 900nm when correcting phase angle from 90° to 30°). At each wavelength, the greater correction coefficient is needed to correct the larger phase angle to 30°. Phase functions can be substituted for each other within 80nm centered at a given wavelength with maximum relative error of ~28%. The derived FeO contents and optical maturity parameters of the regolith using the photometrically corrected spectra reduced much of their deviations, which is consistent with the well - mixed regolith at the small surface area (<5m in diameter). The in situ photometric function supply with the reference for the calibration, validation, and comparison of VNIS cameras and spectrometers and the retrieval of compositions.

Another better - controlled experiment of SWIR sensor is needed because the mineral absorptions are mostly located in SWIR range, which will supply with the mineral identification and mapping. The comprehensive comparison with orbital data, for example, M^3, Spectral Profiler, Interference Imaging Spectrometer, and Clementine, will be also analyzed in the next work.

Acknowledgments

The Chang'E - 4 mission was carried out by the Chinese Lunar Exploration Program, and the data were supplied by the Science and Application Center for Moon and Deep Space Exploration, Chinese Academy of Sciences. We thank the Editor David Baratoux and the reviewers Bruce Hapke and Jun Huang for their efforts to the peer review of this work. This work was supported by the NSFC (41902318, 41490631, 41525016, 61605231, and 11941002) and the Key Research Program of Frontier Sciences, CAS (QYZDJ - SSW - DQC001). Le Qiao also acknowledges support from a Preresearch Project on Civil Aerospace Technologies funded by CNSA (D020205). The data reported in this work were archived at http: //moon. bao. ac. cn/searchOrder _ dataSearchData. search. The detailed data ID of each observation is listed in Table S3. The authors declare no competing interests.

References

[1] Besse, S., Sunshine, J., Staid, M., Boardman, J., Pieters, C., Guasqui, P., et al. (2013). A visible and near - infrared photometric correction for Moon Mineralogy Mapper (M - 3). Icarus, 222 (1), 229 - 242. https: //doi. org/10. 1016/j. icarus. 2012. 10. 036

[2] Besse, S., Yokota, Y., Boardman, J., Green, R., Haruyama, J., Isaacson, P., et al. (2013). One Moon, many measurements 2: Photometric corrections. Icarus, 226 (1), 127 - 139. https: //doi. org/10. 1016/j. icarus. 2013. 05. 009

[3] Di, K., Zhu, M. H., Yue, Z., Lin, Y., Wan, W., Liu, Z., et al. (2019). Topographic evolution of Von Kármán crater revealed by the lunar rover Yutu - 2. Geophysical Research Letters., 46, 12, 764 - 12, 770. https: //doi. org/10. 1029/2019

[4] Gou, S., Di, K., Yue, Z., Liu, Z., He, Z., Xu, R., et al. (2019). Lunar deep materials observed by Chang'E - 4 rover. Earth and Planetary Science Letters, 528, 115, 829. https://doi.org/10.1016/j.epsl.2019.115829

[5] Gueymard, C. A. (2004). The Sun's total and spectral irradiance for solar energy applications and solar radiation models. Sol Energy, 76 (4), 423 - 453. https://doi.org/10.1016/j.solener.2003.08.039

[6] He, Z., Li, C., Xu, R., Lv, G., Yuan, L., & Wang, J. (2019). Spectrometers based on acousto - optic tunable filters for in - situ lunar surface measurement. Journal of Applied Remote Sensing, 13 (2), 027502. https://doi.org/10.1117/1.JRS.13.027502

[7] He, Z. P., Wang, B. Y., Lu, G., Li, C. L., Yuan, L. Y., Xu, R., et al. (2014). Operating principles and detection characteristics of the visible and near - infrared imaging spectrometer in the Chang'E - 3. Research in Astronomy and Astrophysics, 14 (12), 1567 - 1577. https://doi.org/10.1088/1674 - 4527/14/12/006

[8] Hicks, M. D., Buratti, B. J., Nettles, J., Staid, M., Sunshine, J., Pieters, C. M., et al. (2011). A photometric function for analysis of lunar images in the visual and infrared based on Moon Mineralogy Mapper observations. Journal of Geophysical Research, 116, EO0G15. https://doi.org/10.1029/2010je003733

[9] Hu, X., Ma, P., Yang, Y., Zhu, M. H., Jiang, T., Lucey, P. G., et al. (2019). Mineral abundances inferred from in - situ reflectance measurements of Chang'E - 4 landing site in South Pole - Aitken basin. Geophysical Research Letters, 46, 9439 - 9447. https://doi.org/10.1029/2019GL084531

[10] Huang, J., Xiao, Z. Y., Flahaut, J., Martinot, M., Head, J., Xiao, X., et al. (2018). Geological characteristics of Von Karman crater, northwestern South Pole - Aitken basin: Chang'E - 4 landing site region. Journal of Geophysical Research: Planet, 123, 1684 - 1700. https://doi.org/10.1029/2018je005577

[11] Jin, W., Zhang, H., Yuan, Y., Yang, Y., Shkuratov, Y. G., Lucey, P. G., et al. (2015). In situ optical measurements of Chang'E - 3 landing site in Mare Imbrium: 2. Photometric properties of the regolith. Geophysical Research Letters, 42, 8312 - 8319. https://doi.org/10.1002/2015GL065789

[12] Li, C., Liu, D., Liu, B., Ren, X., Liu, J., He, Z., et al. (2019). Chang'E - 4 initial spectroscopic identification of lunar far - side mantle - derived materials. Nature, 569 (7756), 378 - 382. https://doi.org/10.1038/s41586 - 019 -1189 -0

[13] Li, C., Xu, R., Lv, G., Yuan, L., He, Z., & Wang, J. (2019). Detection and calibration characteristics of the visible and near - infrared imaging spectrometer in the Chang'E - 4. Review of Scientific Instruments, 90 (10). https://doi.org/10.1063/1.5089737

[14] Lin, H., He, Z., Yang, W., Lin, Y., Xu, R., Zhang, C., et al. (2019).

Olivine - norite rock detected by the lunar rover Yutu - 2 likely crystallized from the SPA impact melt pool. National Science Review. https://doi.org/10.1093/nsr/nwz183

[15] Ling, Z., Jolliff, B. L., Wang, A., Li, C., Liu, J., Zhang, J., et al. (2015). Correlated compositional and mineralogical investigations at the Chang'E - 3 landing site. Nature Communications, 6 (1), 8880. https://doi.org/10.1038/ncomms9880

[16] Ling, Z., Qiao, L., Liu, C., Cao, H., Bi, X., Lu, X., et al. (2019). Composition, mineralogy and chronology of mare basalts and non - mare materials in Von Kármán crater: Landing site of the Chang'E - 4 mission. Planetary and Space Science, 179. https://doi.org/10.1016/j.pss.2019.104741

[17] Lucey, P. G., Blewett, D. T., & Jolliff, B. L. (2000). Lunar iron and titanium abundance algorithms based on final processing of Clementine ultraviolet - visible images. Journal of Geophysical Research, 105 (E8), 20, 297 - 20, 305. https://doi.org/10.1029/1999je001117

[18] Lucey, P. G., Blewett, D. T., Taylor, G. J., & Hawke, B. R. (2000). Imaging of lunar surface maturity. Journal of Geophysical Research, 105 (E8), 20, 377 - 20, 386. https://doi.org/10.1029/1999je001110

[19] Pieters, C. M., Boardman, J., Buratti, B., Chatterjee, A., Clark, R., Glavich, T., et al. (2009). The Moon Mineralogy Mapper (M^3) on Chandrayaan - 1. Current Science, 500 - 505. https://www.jstor.org/stable/24105459

[20] Qiao, L., Ling, Z., Fu, X., & Li, B. (2019). Geological characterization of the Chang'E - 4 landing area on the lunar farside. Icarus, 333, 37 - 51. https://doi.org/10.1016/j.icarus.2019.05.029

[21] Warren, P. H. (1990). Lunar anorthosites and the magma - ocean plagioclase - flotation hypothesis; importance of FeO enrichment in the parent magma. American Mineralogist, 75 (1 - 2), 46 - 58.

[22] Wu, Y., Wang, Z., Cai, W., & Lu, Y. (2018). The absolute reflectance and new calibration site of the Moon. The Astronomical Journal, 155 (5), 213.

[23] Wu, Y. Z., Besse, S., Li, J. Y., Combe, J. P., Wang, Z. C., Zhou, X. F., & Wang, C. (2013). Photometric correction and in - flight calibration of Chang' E - 1 Interference Imaging Spectrometer (IIM) data. Icarus, 222 (1), 283 - 295. https://doi.org/10.1016/j.icarus.2012.11.010

[24] Wu, Y. Z., & Hapke, B. (2018). Spectroscopic observations of the Moon at the lunar surface. Earth and Planetary Science Letters, 484, 145 - 153. https://doi.org/10.1016/j.epsl.2017.12.003

[25] Xu, R., G. Lv, Y. - h. Ma, and J. Wang (2014), Calibration of Visible and Near - infrared Imaging Spectrometer (VNIS) on lunar surface, paper presented at Multispectral, Hyperspectral, and Ultraspectral Remote Sensing Technology, Techniques and Applications V, International Society for Optics and Photonics.

[26] Yokota, Y., Matsunaga, T., Ohtake, M., Haruyama, J., Nakamura, R., Yamamoto, S., et al. (2011). Lunar photometric properties at wavelengths 0.5—1.6 mu m acquired by SELENE Spectral Profiler and their dependency on local albedo and latitudinal zones. Icarus, 215 (2), 639 - 660. https://doi.org/10.1016/j.icarus.2011.07.028

[27] Zhang, H., Yang, Y., Yuan, Y., Jin, W., Lucey, P. G., Zhu, M. H., et al. (2015). In situ optical measurements of Chang'E - 3 landing site in Mare Imbrium: 1. Mineral abundances inferred from spectral reflectance. Geophysical Research Letters, 42, 6945 - 6950. https://doi.org/10.1002/2015gl065273

[28] Zhang, J., Yang, W., Hu, S., Lin, Y., Fang, G., Li, C., et al. (2015). Volcanic history of the Imbrium basin: A close - up view from the lunar rover Yutu. Proceedings of the National Academy of Sciences of the United States of America, 112 (17), 5342 - 5347. https://doi.org/10.1073/pnas.1503082112

Photometric properties of lunar regolith revealed by the Yutu – 2 rover

Honglei Lin[1], Yazhou Yang[2], Yangting Lin[1], Yang Liu[2,3], Yong Wei[1], Shuai Li[4], Sen Hu[1], Wei Yang[1], Wenhui Wan[5], Rui Xu[6], Zhiping He[6], Xiaohui Liu[7], Yan Xing[8], Chengwu Yu[8], and Yongliao Zou[2]

1. Key Laboratory of Earth and Planetary Physics, Institute of Geology and Geophysics, Chinese Academy of Sciences, Beijing 100029, PR China

e—mail: linyt@mail. iggcas. ac. cn

2. State Key Laboratory of Space Weather, National Space Science Center, Chinese Academy of Sciences, Beijing 100190, PR China

e—mail: yangliu@nssc. ac. cn

3. Center for Excellence in Comparative Planetology, Chinese Academy of Sciences, Hefei 230026, PR China

4. Department of Earth Sciences, University of Hawai 'i at Mānoa, Honolulu, HI, USA

5. State key Laboratory of Remote Sensing Science, Aerospace Information Research Institute, Chinese Academy of Sciences, Beijing 100101, PR China

6. Key Laboratory of Space Active Opto – Electronics Technology, Shanghai Institute of Technical Physics, Chinese Academy of Sciences, Shanghai 200083, PR China

7. Beijing Aerospace Control Center, Beijing 100094, PR China

8. Beijing Institute of Control Engineering, Beijing 100094, PR China

Abstract: *Context*. The surface composition of the Moon has mainly determined based on the visible and near – infrared spectra achieved from orbits and/or landing sites, and the spectroscopic analysis is based on photometric properties of the topmost lunar regolith. However, the lack of a ground truth for the photometric parameters of the undisturbed lunar surface has limited accurate applications of spectral observations.

Aims: Here we report the photometric properties of the small – scale (i. e., centimeter level) undisturbed lunar regolith around the Chang'E – 4 landing site, determined from a series of photometric experiments conducted by the rover Yutu – 2.

Methods: The simplified Hapke model was used to derive the photometric properties. The micro – topographic effect on the spectral measurements was corrected for the first time in the in situ photometric investigations on the Moon, which improves the accuracy of the derived photometric parameters.

Results: The single - scattering albedo w and two parameters (b, c) of the Henyey - Greenstein phase function were derived, and they show a wavelength dependence. The regolith at the Chang'E - 4 landing site exhibits strong forward scattering according to the retrieved c values, and the higher asymmetry parameter indicates that the regolith here is more strongly forward scattering than the Apollo lunar soil samples. The derived photometric parameters can serve as ground truth and can be used in the radiative transfer modeling analysis of the orbital remote - sensing data.
Keywords: instrumentation: photometers - planets and satellites: surfaces - infrared: planetary systems - radiative transfer

1 Introduction

The global and in situ mineral compositions of the lunar surface as well as the space weathering of the lunar soil can be derived from the spectral measurements, with the known photometric properties of the lunar surface, including the porosity K, the single - scattering albedo w, and the two parameters (b, c) of the Henyey - Greenstein phase function. The photometric properties of the lunar surface play key roles in interpreting spectroscopic measurements (e. g., Shkuratov et al. 2011; Hapke 2012a; Jin et al. 2015). Accurate photometric parameters of lunar regolith are essential to characterize its spectroscopic properties and to precisely estimate the mineral abundances. In the past decades, the photometric properties of lunar regolith have been investigated by laboratory measurements of the Apollo lunar samples and simulated samples, Earth - based observations, and orbital remote sensing. For example, a few photometric experiments have been conducted with both synthetic and natural terrestrial granular samples (Mcguire & Hapke 1995; Souchon et al. 2011) and the Apollo samples (Johnson et al. 2013). Although these measurements provided important clues on understanding the photometric properties of lunar regolith, it is difficult to simulate the pristine lunar surface with laboratory measurements. The reflectance data acquired by the Lunar Reconnaissance Orbiter (LRO) have been used to calculate the photometric parameters (Hapke et al. 2012; Sato et al. 2014; Liu et al. 2018) of lunar regolith. Orbital constraints generally do not allow observing a given target with many different viewing geometries (Lin et al. 2020a). Chang'E - 3 (CE - 3) only acquired four spectra at different sites, which is insufficient to derive the photometric parameters. The three - band panoramic camera (PCAM) images acquired by CE - 3 have been used to extract the phase curve and determine the photometric parameters of the lunar surface (Jin et al. 2015), but the lack of measurements at longer wavelengths may have constrained the application of these derived phase functions to other spectroscopic observations. In addition, microtopographic irregularities, which might introduce large uncertainties in the spectral properties of the surface regolith, were ignored in previous studies.

In January 2019, the Chang'E - 4 (CE - 4) mission successfully landed in the Von Kármán crater within the South - Pole Aitken (SPA) basin, and the rover Yutu - 2 has been exploring the lunar surface since then. In order to solve the problems discussed above, the in situ photometric experiments (Fig. 1a) with the Visible and Near - Infrared Imaging Spectrometer (VNIS) on board the Yutu - 2 rover were conducted on the tenth lunar day of the mission operations. The spectra of the same area were acquired at different viewing geometries (i. e., incidence angle i, emission angle e, and phase angle g), providing an unprecedented dataset (Fig. 2) and a unique perspective to understand the photometric properties of the lunar surface. The CE - 4 landing site is suggested to be covered by the ejecta from the neighboring Finsen crater (Gou et al. 2019; Hu et al.

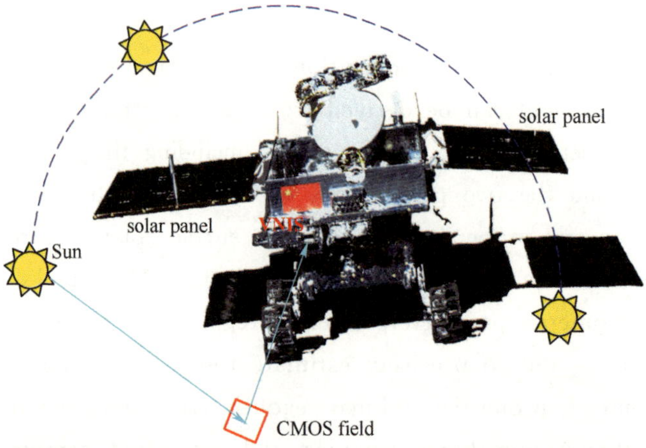

(a)

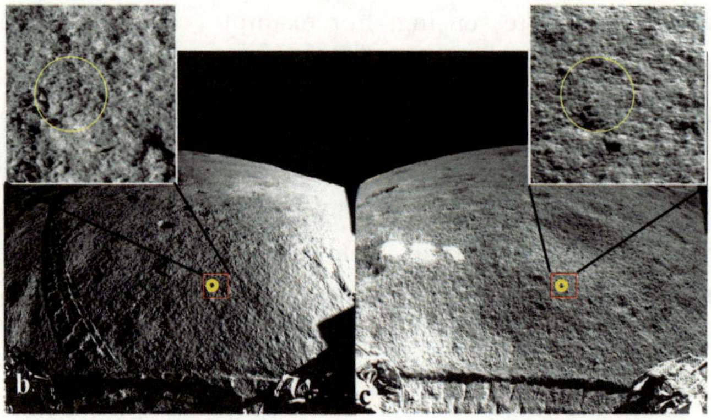

(b)　　(c)

Fig. 1　Spectrophotometric measurements of lunar regolith by the Yutu - 2 rover. Panel a: illustration of photometric experiments conducted by the Yutu - 2 rover. The VNIS collected spectral data at the same location, but different solar angles. Two adjacent sites were measured in the morning and afternoon, respectively. Panels b and c: images acquired by the hazard - avoidance camera (HAZCAM) at sites 1 and 2, respectively. The red rectangle is the viewing field of the CMOS imager. The inset images within panels b and c show a close - up view of lunar regolith imaged by the CMOS, and the yellow circle is the field of view of SWIR

2019; Li et al. 2019a; Lin et al. 2020b; Qiao et al. 2019). The digital elevation model (DEM) with a resolution of 5 cm generated using PCAM images indicates that the landing site is about 70m higher than the mare basalts within the Von Kármán crater (Di et al. 2011). The measurements conducted by the lunar penetrating radar (LPR) on board the Yutu - 2 rover suggest an about 11m thick regolith layer on the top of the crater ejecta (Lai et al. 2019). The modeled age of the landing region is up to 3. 6Ga (e. g., Huang et al. 2018; Ling et al. 2019). The regolith of the uppermost layer may have been churned often enough to be homogenized in composition both vertically and laterally at the landing site. Therefore, the surface at the landing region is likely rather homogeneous in composition, providing an ideal site for conducting a series of photometric experiments. The spectra were acquired at site 1 (Fig. 1b) and site 2 (Fig. 1c). The phase angles of these spectra range from 54. 79° to 111. 57°. The stereo pair with high spatial resolutions were also acquired during the spectral measurements, which allow us to study the topographic effects on the spectral observations.

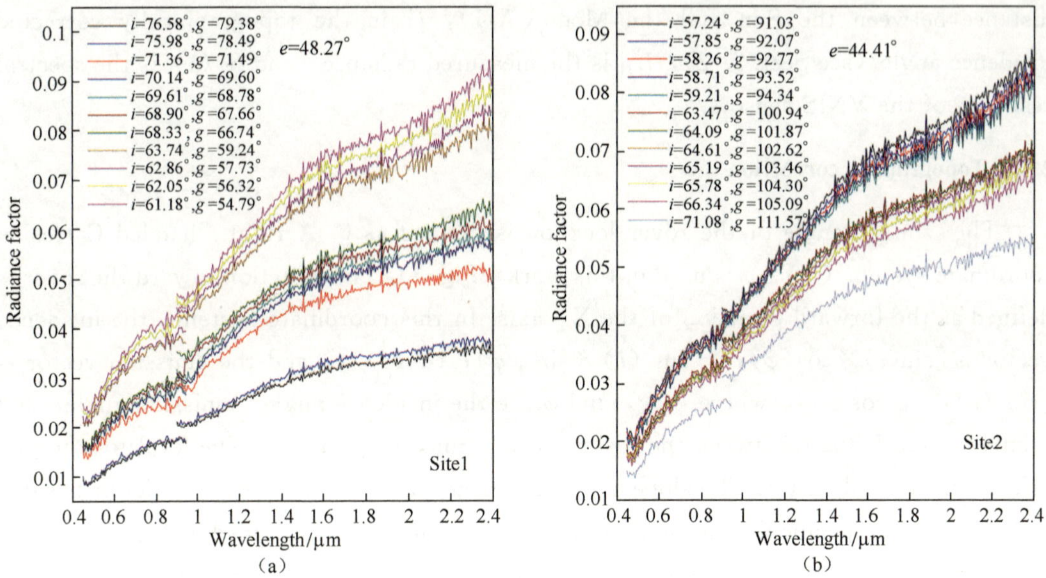

Fig. 2　VNIS spectra acquired by Yutu - 2. Panel a: site 1 and panel b: site 2

2　Data and methods

2.1　VNIS instrument and data preprocessing

The VNIS instrument on board the Yutu - 2 rover measures the lunar surface from a height of about 1m. It has two detectors that include a complementary metal - oxide semiconductor (CMOS) imager and a short - wavelength near - infrared (SWIR) detector (He et al. 2019). The spectral resolution of the CMOS imager is 2.4－6.5nm from 450 to 945nm and that of the SWIR detector is 3.6－9.6nm from 900 to 2395nm (He et al. 2014; Li et al. 2019b). The spatial calibration between the CMOS and SWIR field was made in the laboratory. The CMOS image is 15cm×21cm, and the SWIR field covers a circular region with a radius of ~3.5cm within the CMOS image (Fig. 1).

The DN values acquired by the VNIS instrument are converted into radiance data by dark reduction, background removal, thermal correction, and radiometric calibration (He et al. 2014; Liu et al. 2014; Li et al. 2019b). The radiance used in this study was in-flight calibrated using the valid measurements of the Lambertian reference plate on board the rover (Xu et al. 2014). The reflectance (reflectance factor, REFF) was retrieved from radiance by dividing the solar irradiance (Gueymard 2004; Wu et al. 2018),

$$\mathrm{REFF} = \frac{\pi L_{\mathrm{H}} D^2}{\cos(i') \int J(\lambda) S(\lambda) \mathrm{d}(\lambda)} \tag{1}$$

where $J(\lambda)$ is the solar irradiance (Gueymard 2004) at wavelength λ, D is the distance between the Sun and the Moon (AU), i' is the topographically corrected incidence angle (see Sect. 2.2), L_{H} is the measured radiance, and $S(\lambda)$ is the spectral response of the VNIS sensors.

2.2 Topographic correction

The azimuth angle of the rover location is defined as 0. A right-handed Cartesian coordinate system (XYZ) is used in this work (Fig. 3). The direction toward the rover is defined as the forward direction of the X-axis. In this coordinate system, the incidence vector is $[\sin(i)\times\cos(\varphi), -\sin(i)\times\sin(\varphi); \cos(i)]$, and the emission vector is $[\sin(e), 0, \cos(e)]$, where i, e, and φ are the incidence angle, emission angle, and relative azimuth angle between the rover and the Sun, respectively. The local topography affects the normal vector. The slope along the rover observation direction is defined as θ_1, and the slope perpendicular to the rover observation direction is defined as θ_2. The length of the projection of the normal vector onto the YOZ plane is defined as $p_1 = \sqrt{\cos(\theta_1)^2/(1-\sin(\theta_1)^2\sin(\theta_2)^2)}$, onto the XOZ plane defined as $p_2 = \sqrt{1-p_1^2\sin(\theta_2)^2}$, and onto the Z-axis, it is defined as $p_3=\sqrt{(1-p_2^2\sin(\theta_1)^2)-p_1^2\sin(\theta_2)^2}$. In this case, the new normal vector that accounts for the local topography can be written as $[p_2\times\sin(\theta_1), p_1\times\sin(\theta_2), p_3]$. Thus, the corrected incidence angle i' and the corrected emission angle e' can be calculated using the vector product principle with the derived new normal vector. The angle between the incidence vector and emission vector is the phase angle g, which does not change with topography. The slope angles can be derived from the DEM (Di et al. 2011), which was generated using the images acquired by the HAZCAM.

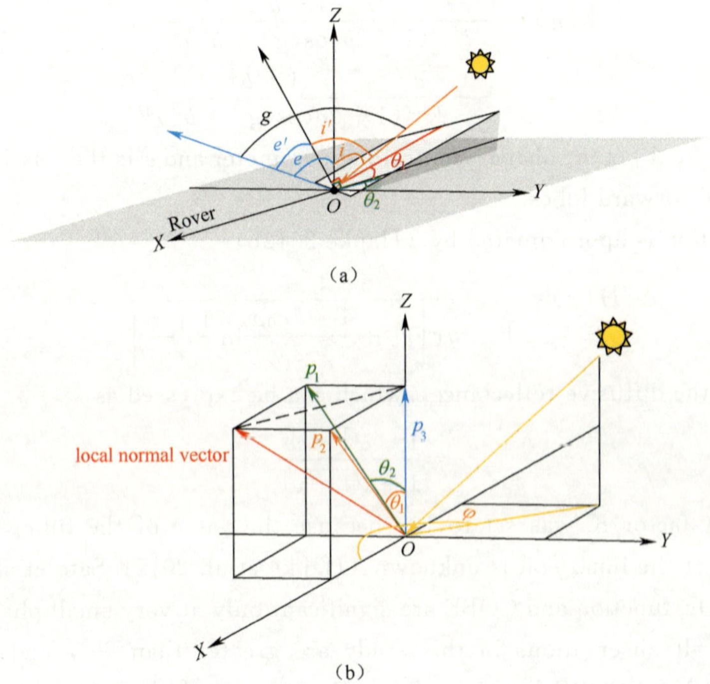

(a)

(b)

Fig. 3　Illustration of viewing geometries and effect of microtopography
Panel a: relations between the viewing angles. i, e, and g are the incidence angle, emission angle, and phase angle, respectively, and i' and e' are the incidence and emission angles after topographical correction. θ_1 and θ_2 are the slope angles along and perpendicular to the rover viewing direction, respectively. Panel b: corrected angles. φ is the relative azimuth angle between the rover and the Sun, p_1 is the length of the projection of the normal vector onto the YOZ plane, p_2 is the length of the projection of the normal vector onto the XOZ plane, and p_3 is the length of the projection of the normal vector onto the Z-axis

2.3　Photometric model for parameter determination

We employed the Hapke model to derive the photometric properties of lunar regolith. The reflectance r (reflectance factor, REFF) can be written as (Hapke 2012b)

$$r=\frac{Kw}{4(\mu_0+\mu_e)}\left\{p(g)[1+B_{S0}B_S(g)]+H\left(\frac{\mu_0}{K}\right)H\left(\frac{\mu_e}{K}\right)-1\right\}\times[1-B_{C0}B_C(g)] \tag{2}$$

where K is the porosity factor, w is the single-scattering albedo, μ_0 and μ_e are the cosines of the incidence angle (i) and emission angle (e), respectively, $p(g)$ is the phase function, B_{S0} is the amplitude of the shadow-hiding opposition effect (SHOE) function and $B_S(g)$ is the SHOE function, B_{C0} is the amplitude of the coherent backscattering opposition effect (CBOE), and $B_C(g)$ is the CBOE function, and finally, $H(x)$ is the Ambartsumian-Chandrasekhar H function.

The two-parameter Henyey-Greenstein phase function $p(g)$ was employed,

$$p(g)=\frac{1+c}{2}\frac{1-b^2}{(1-2b\cos(g)+b^2)^{3/2}}+\frac{1-c}{2}\frac{1-b^2}{(1+2b\cos(g)+b^2)^{3/2}} \tag{3}$$

where $b\in[0, 1]$ is the shape - controlling parameter and c is the relative strength of the backward and forward lobes.

The H function is approximated by (Hapke 2012b)

$$H(x)=\frac{1}{1-wx\left[r_0+\frac{1-2r_0x}{2}\ln\frac{1+x}{x}\right]} \tag{4}$$

where r_0 is the diffusive reflectance, which can be expressed as

$$r_0=\frac{1-\sqrt{1-w}}{1+\sqrt{1-w}} \tag{5}$$

The porosity factor K was set to 1.0 because the value of the filling factor of the uppermost layer of the lunar soil is unknown (Hapke et al. 2012; Sato et al. 2014). The value of the SHOE function and COBE are significant only at very small phase angle. The phase angles of all observations in this study are greater than 50°, and we therefore ignored the SHOE and COBE functions. Then Eq. (2) would have three free parameters (w, b, and c). We note that the parameters c and b in the Henyey - Greenstein phase function may be correlated with each other, and they can be expressed as an empirical hockey - stick relation (Hapke 2012a),

$$c=3.29\exp(-17.4b^2)-0.908 \tag{6}$$

This allows us to compare the relationship between the c and b values derived in this study to the hockey - stick relation.

The nonlinear least - squares method was employed to solve Eq. (2). The determination of initial values of the free parameters has significant effects on the final results. Therefore we combined a heuristic piecewise approach and a grid - search method (Hapke et al. 2012; Jin et al. 2015) to fit the phase curves. To avoiding random initial values, we conducted a grid search in parameter space of the free parameters for each model to determine the parameter values of the first band. The lower bounds, steps, and upper bounds were set as $w=[0.1, 0.1, 1]$, $b=[0.1, 0.1, 1]$, and $c=[-1, 0.1, 2]$. The model fit with the minimum RMSE was accepted. The final parameter values of the band were set as the initial values of the next band. To solve the nonlinear least - squares algorithm, we constrained the values of w and b to between 0 and 1.

3　Results and discussions

3.1　Topographic features of the explored area

Of the two sites that were explored on the tenth day of the rover operations, site 1 is located on a sloped surface, whereas site 2 is located at the edge of a small crater, as is shown in the highresolution images (Fig. 1). The spatial resolution of the derived DEM is better than 2cm. The slope angles of site 1 are 7.4° and 1.9° in the rover viewing direction and perpendicular to the rover viewing direction, respectively (Fig. 4a). The slope angles of site 2 are −1.5° and 4.7° tilted in the rover viewing direction and perpendicular to the rover viewing direction, respectively (Fig. 4b). The topographic slopes would bias the true values of the viewing geometry angles, which might affect the spectral properties and thus the geological interpretations. We therefore derived the local viewing geometry angles for all the spectral measurements using the method described in Sect. 2.2 to account for the topographic effects (Fig. 4c). The corrected angles are listed in Tab. 1.

Tab. 1　Incidence angle, emission angle, and phase angle of all measurements for the original and the corrected angles that take the topography into account

Data ID	i	Corrected i	e	Corrected e	g	φ
N68	76.58	74.51	48.27	40.91	79.38	272.36
N69	75.98	73.82	48.27	40.91	78.49	273.04
N70	71.36	68.55	48.27	40.91	71.49	278.52
N71	70.14	67.15	48.27	40.91	69.60	280.04
N72	69.61	66.54	48.27	40.91	68.78	280.72
N73	68.90	65.73	48.27	40.91	67.66	281.64
N74	68.33	65.07	48.27	40.91	66.76	282.39
N75	63.74	59.75	48.27	40.91	59.24	288.91
N76	62.86	58.72	48.27	40.91	57.73	290.29
N77	62.05	57.77	48.27	40.91	56.32	291.58
N78	61.18	56.74	48.27	40.91	54.79	293.02
N79	57.24	59.72	44.41	46.10	91.03	133.42
N80	57.85	60.24	44.41	46.10	92.07	134.62
N81	58.26	60.59	44.41	46.10	92.77	135.43
N82	58.71	60.97	44.41	46.10	93.52	136.30
N83	59.21	61.40	44.41	46.10	94.34	137.22
N84	63.47	65.08	44.41	46.10	100.94	144.37
N85	64.09	65.62	44.41	46.10	101.87	145.33
N86	64.61	66.08	44.41	46.10	102.62	146.09
N87	65.19	66.59	44.41	46.10	103.46	146.96
N88	65.78	67.10	44.41	46.10	104.30	147.80
N89	66.34	67.60	44.41	46.10	105.09	148.59
N90	71.08	71.79	44.41	46.10	111.57	154.84

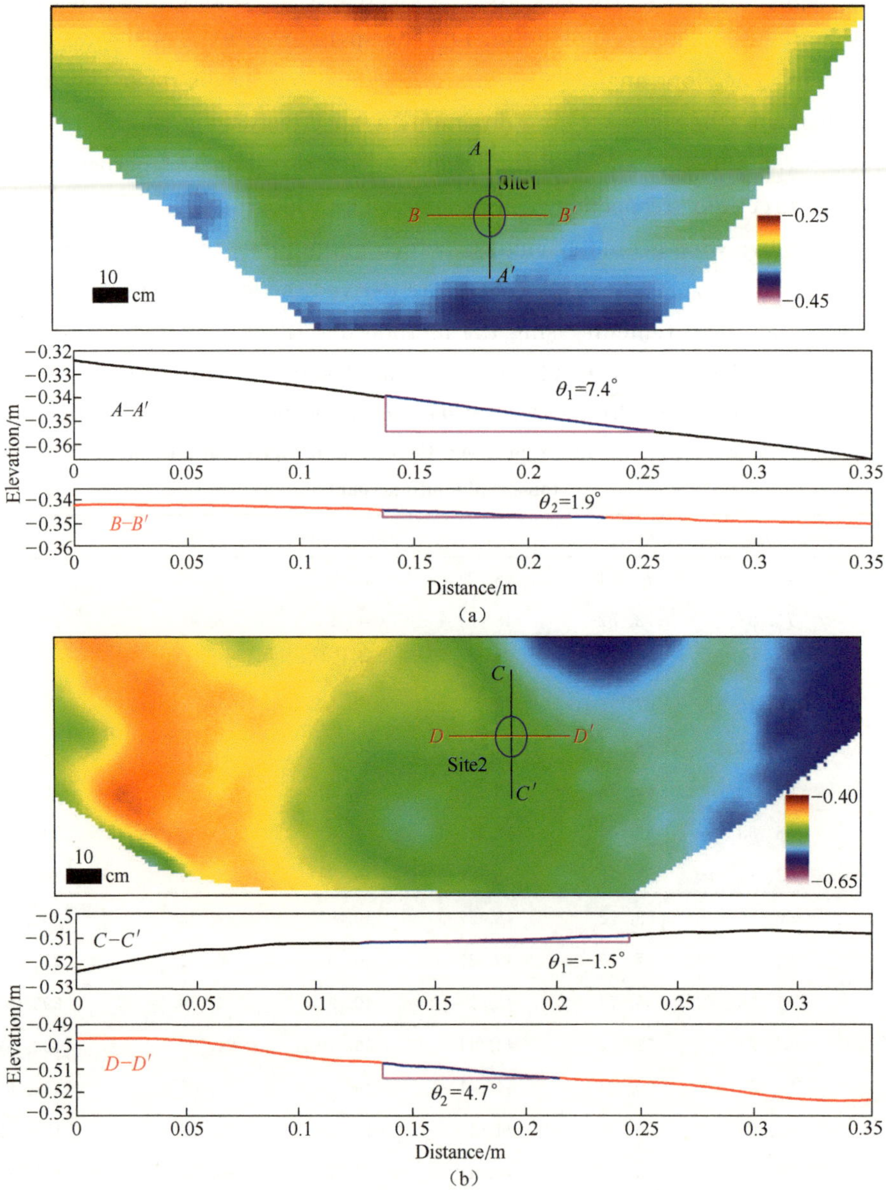

Fig. 4　Microtopographic features of explored area and the corrected measurement angles. Panel a: DEM generated using the images acquired by HAZCAM at site 1. The context of the region is shown in Fig. 1b. Topographic profiles from A to A′ and B to B′ are indicated. Panel b: DEM generated using the images acquired by HAZCAM at site 2. The context of the region is shown in Fig. 1c. Topographic profiles from C to C′ and D to D′ are also indicated. The blue line is the field of view of SWIR

3.2　Phase curves and photometric parameters of lunar regolith

The reflectance was obtained from the radiance using the solar irradiance calibration method (Eq. (1)). The albedo at the same wavelength changed after topographic correction, indicating that the topography affects the shapes of the phase curves. The phase curves of the representative wavelengths, which were extracted from the spectra, are shown in Fig. 5. Overall, the reflectance increases with increasing wavelength, which is consistent with the reddish spectra of the Moon. This may be caused by space weathering, which produces abundant agglutinates and submicroscopic metallic iron (SMFe; Pieters et al. 2000, 2016). The reflectance values decrease with the increase in phase angles, and then it increases at phase angles greater than 80° (Fig. 5b).

The reflectance variation of the same location at different phase angles in Fig. 5 is dominantly controlled by the phase function of lunar regolith. The phase curves corrected for the topographic effect were fit using the three - parameter (i.e., singlescattering albedo w, phase function parameters b and c) version of Hapke's model (Eq. (2)) to obtain the photometric parameters. The phase - curve fitting at the representative wavelengths is shown in Fig. 6a. The model gives a high fitting quality with a small root - mean - square error (RMSE). Most RMSE values are lower than 10^{-2}. However, the curve fitting seems worse at longer wavelengths, likely because the reflectance increases, the noise level is higher, and because of thermal emission at longer wavelengths. The strongest deviation between the observed and modeled values occurs at a phase angle of 80° (Fig. 6a), which might be caused by the effects of the shadows caused by the low solar altitude.

The discontinuities in CE - 4 spectra and in these Hapke parameters are caused by the differences between the spectral responses of the CMOS and SWIR detectors. The differences between the CMOS and the SWIR measurements over 900—945nm may vary with the change of measurement geometries. We therefore carried out photometric modeling of the CMOS and SWIR data separately. As shown in Fig. 6, the results suggest that these parameters are wavelength dependent. The single - sattering albedo of the regolith derived from the spectral measurement before and after topographic correction is similar in the CMOS bands, whereas the variations of single - scattering albedo at longer wavelength are evident (Fig. 6b). The phase function parameters b and c show large variations before and after topographic correction. Interestingly, the relations between b and c for the uncorrected and corrected case are not consistent with those derived in laboratory measurements (Fig. 7), although the corrected case is closer to the L - shaped region (i. e, to the hockey - stick relation). The c values are comparable with the Aplollo samples, but the b values are much higher, suggesting that the scattered lobes of the pristine lunar surface are higher and narrower. The scattering properties of the samples vary with the grain sizes, shapes, the state of sample compaction, etc. (Shepard & Helfenstein 2007; Souchon et al. 2011). It is difficult to simulate the stacking structure of the lunar surface, such as the fairy - castle structure (Jaffe 1965), with laboratory samples. The phase curves show that light is more scattered at large phase angles after topographic correction (Fig. 5), indicating strong forward - scattering properties of the regolith. This is consistent with the negative c values in the Henyey - Greenstein phase

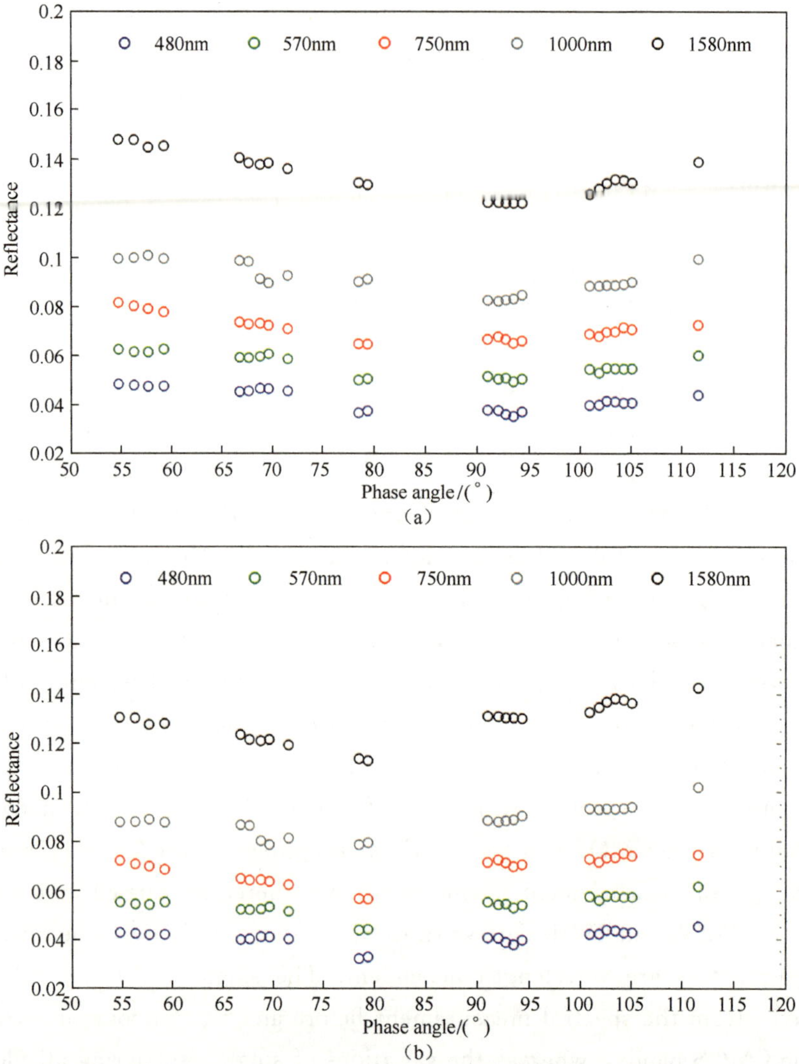

Fig. 5 Phase curves at five different wavelengths derived from in situ spectral observations. Panel a: variation of the reflectance with phase angles at representative wavelengths derived from spectral measurements at a wide range of phase angles from 54.79° to 111.57°. Panel b: same dataset as in panel a, but after correction for the microtopography effect

function (e. g., Hapke 2012a). The regolith at the CE－4 landing site exhibits forward scattering that is similar to that of the Apollo samples (Johnson et al. 2013) (Fig. 7). When the fitted b and c values are used, the asymmetry parameter $\langle \cos\theta \rangle = -b \times c$ (Hapke 2012b) can be obtained. The asymmetry parameter of CE－4 (0.39 ± 0.06) is larger than those of the Apollo samples (0.12 ± 0.02), suggesting a stronger forwardscattering lobe (Fig. 7). We note that the asymmetry parameters between CE－3 and CE－4 are also diffrenent, although they were both derived from the measurements of undisturbed regolith. Several factors may have caused this difference. First, the composition of the CE - 3 and CE - 4 landing sites is different. The regolith at the CE - 3 landing site is high - titanium basalt, while that at the CE - 4 landing site may consist of norite (Lin et al. 2020b). Second, the difference in space - weathering degree may have caused the difference. The CE - 4 landing area (~3.6Ga) is older than that of CE - 3 (~2.96 Ga), indicating that the regolith at the CE - 4 landing area is more mature (Huang et al. 2018). Alternatively, the difference might be due to the photometric measurement method. The photometric data of CE - 3 were obtained using the panorama camera (PCAM), which covers a large circular area (~20m diameter) (Jin et al. 2015), where the measurements were not repeatedly conducted over the same area. In contrast, the VNIS instrument on board the Yutu - 2 rover of CE - 4 measured exactly the same area under varied viewing geometries, which essentially provided more reliable photometeric measurements. The photometric parameters derived from these in situ measurements provide important insight on the scattering properties of the lunar regolith. Our analysis also indicates that the topography has strong effects on the photometric values derived from the in situ spectral measurements of the lunar regolith, and these effects need to be corrected for. The photometric parameters to which the topographic correction is applied are critical for obtaining more accurate information on the mineral abundance at the landing site. The wavelength - dependent b and c values in the two - parameter Henyey - Greenstein phase function, derived from CE - 4 in situ meaurements, can serve as ground truth and can be used in the radiative transfer modeling analysis of the orbital remote - sensing data.

4　Conclusions

A series of in situ photometric measurements were conducted by the Yutu - 2 rover. They provide an excellent opportunity to study the photometric properties of the lunar regolith. Microtopographic correction was performed on these measurements, and the phase curves were derived. We then estimated the single - scattering albedo w and two parameters b and c in the two - parameter Henyey - Greenstein phase function using the simplified Hapke model. Our results show that these parameters show a strong wavelength dependence. The regolith at the CE - 4 landing site exhibits forward scattering based on the negative c values. Compared to the Apollo samples, the regolith at the CE - 4 landing site has stronger forward scattering, as indicated by the higher asymmetry parameter. The derived photometric parameters are critical for a quantitative analysis of the mineral abundances at the landing site from the VNIR spectra collected in situ, and they also provide critical ground truth for the photometric correction of orbital data.

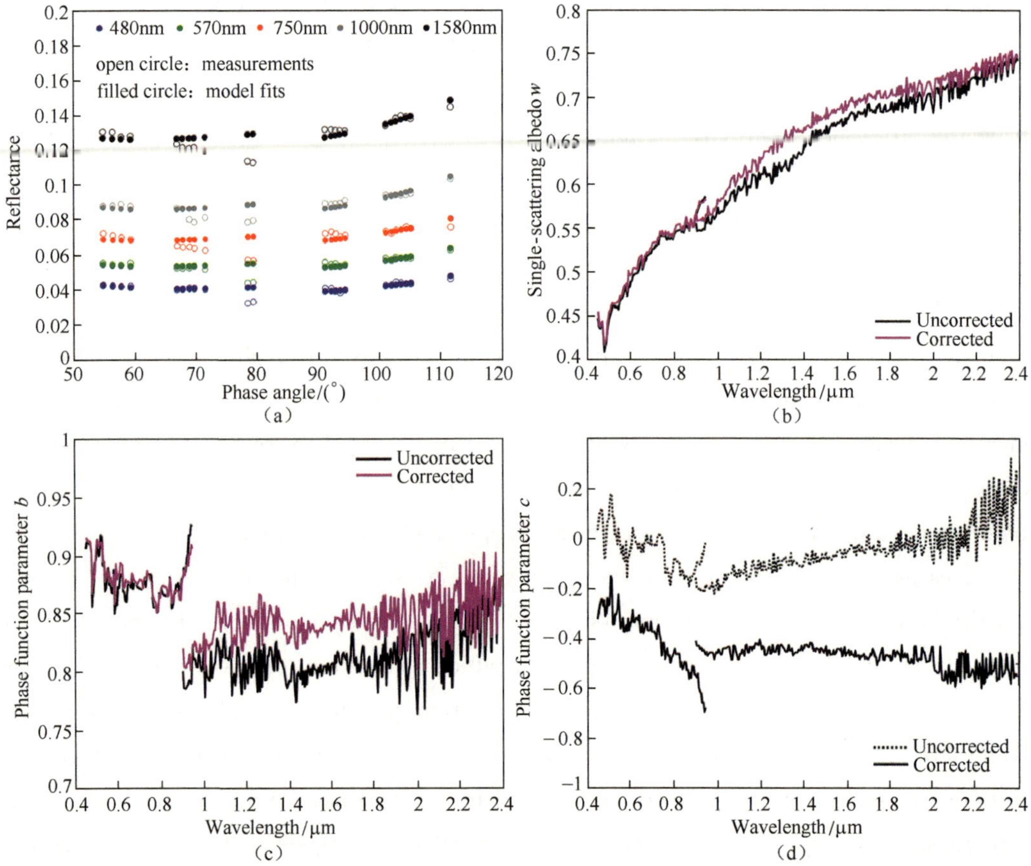

Fig. 6 Derived Hapke model parameters of lunar regolith. Panel a: model fits of the phase curves at 480, 570, 750, 1000, and 1580 nm. The open circles show measurements, and the solid circles show the modeled values. Panel b: variation of the single-scattering albedo w with wavelength. Panel c: variation of the phase function parameter b with wavelength. Panel d: variation of the phase function parameter c with wavelength.

Acknowledgements

The Chang'E-4 mission was carried out by the Chinese Lunar Exploration Program, and the data was provided by the Science and Application Center for Moon and Deep Space Exploration, Chinese Academy of Sciences. We thank Stéphane Le Mouélic for the constructive comments that significantly improved the content and clarity of the manuscript. This work was supported by the Strategic Priority Research Program of Chinese Academy of Sciences (Grant No. XDB 41000000), the National Natural Science Foundation of China (41902318, 11941001, 41490631, and 41525016), the Key Research Program of Frontier Sciences, CAS (QYZDJ-SSW-DQC001), the preresearch project on Civil Aerospace Technologies (No. D020101 and D020102) funded by China National Space Administration (CNSA) and the Beijing Municipal Science and Technology Commission (Z181100002918003 and Z191100004319001). H. Lin also thanks the support of the Key Research Program of the Institute of Geology and Geophysics, CAS (IGGCAS-201905) and Y. Yang acknowledges the support from the China Postdoctoral Science Foundation (No. 2019TQ0323). The data reported in this work are archived at http://moon.bao.ac.cn. The authors declare no competing interests.

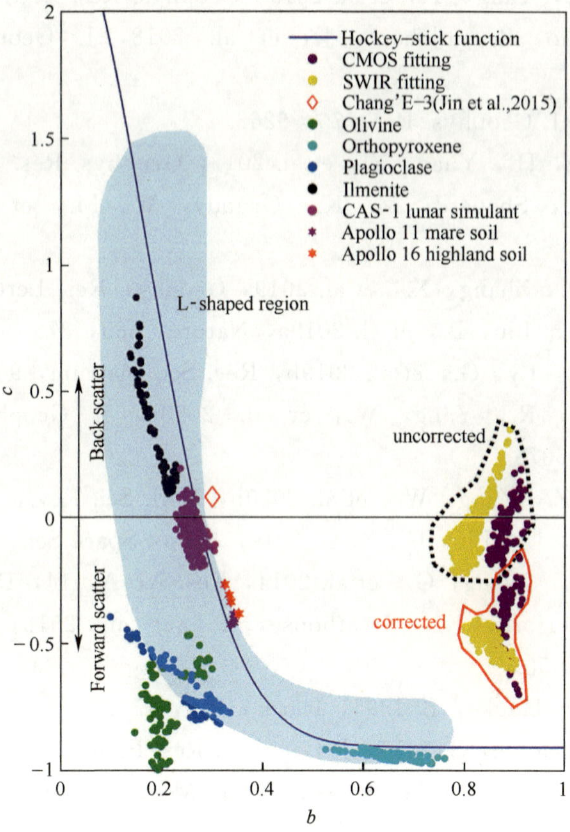

Fig. 7　Relation between phase function parameters b and c (modified from Sato et al. 2014). The phase function parameters of pure minerals were calculated using the data in Yang et al. (2018). We show the lunar simulated regolith (Zheng et al. 2009; Lin et al. 2020b), and regolith at the Chang'E-3 landing site as well as Apollo soil (Johnson et al. 2013) for comparison. The areas marked with the dashed black and red lines are the data points derived using the CE-4 reflectance without and with topographic correction, respectively

References

[1] Di, K., Zhu, M. H., Yue, Z., et al. 2019, Geophys. Res. Lett., 46, 12764.

[2] Gou, S., Di, K., Yue, Z., et al. 2019, Earth Planet. Sci. Lett., 528, 115829.
[3] Gueymard, C. A. 2004, Sol. Energy, 76, 423.
[4] Hapke, B. 2012a, Icarus, 221, 1079.
[5] Hapke, B. 2012b, Theory of Reflectance and Emittance Spectroscopy (Cambridge: Cambridge University Press).
[6] Hapke, B., Denevi, B., Sato, H., et al. 2012, J. Geophys. Res. Planet, 117.
[7] He, Z., Wang, B., Lu, G., et al. 2014, Res. A&A, 14, 1567.
[8] He, Z., Li, C., Xu, R., et al. 2019, J. Appl. Remote Sens., 13, 027502.
[9] Hu, X., Ma, P., Yang, Y., et al. 2019, Geophys. Res. Lett., 46, 9439.
[10] Huang, J., Xiao, Z., Flahaut, J., et al. 2018, J. Geophys. Res. Planet, 123, 1684.
[11] Jaffe, L. 1965, J. Geophys. Res, 70, 6268.
[12] Jin, W., Zhang, H., Yuan, Y., et al. 2015, Geophys. Res. Lett., 42, 8312.
[13] Johnson, J. R., Shepard, M. K., Grundy, W. M., et al. 2013, Icarus, 223, 383.
[14] Lai, J., Xu, Y., Zhang, X., et al. 2019, Geophys. Res. Lett., 46, 12783.
[15] Li, C., Liu, D., Liu, B., et al. 2019a, Nature, 569, 378.
[16] Li, C., Xu, R., Lv, G., et al. 2019b, Rev. Sci. Instrum, 90, 103106.
[17] Lin, H., Xu, R., Yang, W., et al. 2020a, J. Geophys. Res. Planets, 125, e2019JE006076.
[18] Lin, H., He, Z., Yang, W., et al. 2020b, Natl. Sci. Rev., 7, 913.
[19] Ling, Z., Qiao, L., Liu, C., et al. 2019, Planet Space Sci., 179, 104741.
[20] Liu, B., Li, C., Zhang, G., et al. 2014, Res. A&A, 14, 1578.
[21] Liu, Y., Retherford, K., Greathouse, T., et al. 2018, J. Geophys. Res. Planets, 123, 2550.
[22] Mcguire, A., & Hapke, B. 1995, Icarus, 113, 134.
[23] Pieters, C., & Noble, S. 2016, J. Geophys. Res. Planets, 121, 1865.
[24] Pieters, C., Taylor, L., Noble, S., et al. 2000, Meteorit. Planet. Sci., 35, 1101.
[25] Qiao, L., Ling, Z., Fu, X., et al. 2019, Icarus, 333, 37.
[26] Sato, H., Robinson, M., Hapke, B., et al. 2014, J. Geophys. Res. Planets, 119, 1775.
[27] Shepard, M., & Helfenstein, P. 2007, J. Geophys. Res. Planets, 112, E3.
[28] Shkuratov, Y., Kaydash, V., Korokhin, V., et al. 2011, Planet Space Sci., 59, 1326.
[29] Souchon, A., Pinet, P., Chevrel, S., et al. 2011, Icarus, 215, 313.
[30] Wu, Y., Wang, Z., Cai, W., et al. 2018, AJ, 155, 213.
[31] Xu, R., Lv, G., Ma, Y., et al. 2014, SPIE Conf. Ser., 926315.
[32] Yang, Y., Li, S., Milliken, R., et al. 2018, J. Geophys. Res. Planets, 124, 31.
[33] Zheng, Y., Wang, S., Ouyang, Z., et al. 2009, Adv. Space Res., 43, 448.

Estimation of Noise in the In Situ Hyperspectral Data Acquired by Chang'E – 4 and Its Effects on Spectral Analysis of Regolith

Honglei Lin[1], Yangting Lin[1], Yong Weil[1], Rui Xu[2], Yang Liu[3], Yazhou Yang[3], Sen Hu[1], Wei Yang[1] and Zhiping He[2]

1. Key Laboratory of Earth and Planetary Physics, Institute of Geology and Geophysics, Chinese Academy of Sciences, Beijing 100029, China; linhongleiOmail. iggcas. ac. cn (H. L); weiy@ mail. iggcas. ac. cn (Y. W.); husen@ mailiggcas. ac. cn (S. H.); yangw@ mailiggcas. ac. cn (W. Y.)
2. Key Laboratory of Space Active Opto – Electronics Technology, Shanghai Institute of Technical Physics, Chinese Academy of Sciences, Shanghai 200083, China; xurui@ mail. sitp. ac. cn (R. X.); hzping@ mail. sitp. ac. cn (Z. H.)
3. State Key Laboratory of Space Weather, National Space Science Center, Chinese Academy of Sciences, Beijing 100190, China; yangliu@ nssc. ac. cn (YL.); yangyazhou@ nssc. ac. cn (Y. Y.)

Abstract: The Chang'E – 4 (CE – 4) spacecraft landed successfully on the far side of the Moon on 3 January 2019, and the rover Yutu – 2 has explored the lunar surface since then. The visible and near – infrared imaging spectrometer (VNIS) onboard the rover has acquired numerous spectra, providing unprecedented insight into the composition of the lunar surface. However, the noise in these spectral data and its effects on spectral interpretation are not yet assessed. Here we analyzed repeated measurements over the same area at the lunar surface to estimate the signal – noise ratio (SNR) of the VNIS spectra. Using the results, we assessed the effects of noise on the estimation of band centers, band depths, FeO content, optical maturity (OMAT), mineral abundances, and submicroscopic metallic iron (SMFe). The data observed at solar altitudes $<20°$ exhibit low SNR (25dB), whereas the data acquired at 20°– 35°exhibit higher SNR (35 – 37dB). We found differences in band centers due to noise to be ~6. 2 and up to 28. 6nm for 1 and 2μm absorption, respectively. We also found that mineral abundances derived using the Hapke model are affected by noise, with maximum standard deviations of 6. 3%, 2. 4%, and 7. 0% for plagioclase, pyroxene, and olivine, respectively. Our results suggest that noise has significant impacts on the CE – 4 spectra, which should be considered in the spectral analysis and geologic interpretation of lunar exploration data.

Keywords: the Moon; Chang'E – 4; hyperspectral data; noise estimation; spectral interpretation

1 Introduction

Visible and near-infrared (VNIR) reflectance spectra are the primary data sources used in analysis of the composition of the materials and processes on the lunar surface. Hyperspectral remote sensing sensors, including the Moon Mineralogy Mapper (M^3), the Spectral Profilers, and the Interference Imaging Spectrometer, have allowed the accumulation of extensive spectral datasets describing the entire surface of the Moon [1-3]. The VNIR spectral signature of the lunar surface is a complex function of parameters including surface composition, particle size, scattering property, space weathering, and viewing geometry [4-7]. Previously, the spectral features of lunar regolith were studied in a laboratory setting using the Apollo samples to help decode remote sensing data [6, 8]. However, laboratory measurements are typically obtained under controlled conditions (i. e., with fixed incidence and viewing angles), and differences between the acquired spectra are generally associated with differences in the chemical and physical properties of the target. Moreover, it is difficult to reproduce the undisturbed lunar surface and environment in a laboratory; any such reproduction may have a significant effect on the optical features obtained. These experimental issues can be addressed by conducting in situ measurements of the lunar surface.

On 3 January 2019, the Chang'E-4 (CE-4) spacecraft landed successfully in the South Pole-Aitken (SPA) basin on the far side of the Moon. The visible and near-infrared imaging spectrometer (VNIS) onboard the Yutu-2 rover obtained numerous VNIR hyperspectral images, providing unprecedented insight into the spectrophotometric properties and compositions of lunar regolith and rocks [9-13]. However, the ability of these hyperspectral data to represent the real lunar environment is degraded by noise, which is related to factors such as the instrument system used and the measurement environment [14]. In particular, instrumental noise can introduce uncertainties in spectral bands to varying degrees; moreover, the received radiance can be degraded by imaging conditions such as illumination [14]. Thus, the effect of noise on observed data must be estimated in the spectral analysis of lunar surface characteristics.

Lunar regolith and rocks are composed of varying quantities of dominant minerals, including plagioclase, pyroxene, olivine, and ilmenite. The accurate identification of minerals on the lunar surface is critical to our understanding of the surface processes and the geological history of the Moon [15-18]. Absorption related to electronic transitions, vibrational modes, and charge transfer processes of specific minerals can be used to characterize mineral types [4]. For example, pyroxenes have diagnostic band centers at ~1 and ~2μm, with wavelength varying as a function of the Ca/Fe/Mg ratio [19, 20]. Similarly, absorption occurs at ~1.05μm for olivine [21]. Additionally, band depth (absorption strength) is related to mineral abundance and space weathering. Noise is known to affect the estimation of spectral features such as band center and band depth. In this context, the estimation of spectral features may be affected by the signal - noise ratio (SNR) of hyperspectral data. Accurate estimation of the abundances of each of these minerals in lunar regolith and rocks would help further constrain the Moon's evolution [22]. Typically, mineral abundances can be derived from spectra of lunar regolith and rocks using the radiative transfer models. For example, the Hapke model is used widely to estimate mineralogical information from VNIR reflectance spectra of lunar materials [9, 11, 22 - 24]. However, the deconvolution of such spectra is known to be sensitive to noise. Therefore, quantitative assessment of how noise affects spectral data, and the mineralogical information derived from such data, is important for robust geologic interpretation.

Space weathering is the primary surface process on the Moon which can produce the submicroscopic metallic iron (SMFe) [25 - 27]. The estimation of the SMFe abundance in the lunar regolith is undoubtedly important in evaluating space weathering and hence the regolith evolution [27, 28]. The existence of SMFe reduces the albedo and band depth of the spectra of the lunar regolith, and physical modeling was developed to derive the SMFe abundance from the acquired spectra [25, 29]. However, how noise of CE - 4 spectral data affects the estimations of SMFe abundance needs to be quantitatively assessed.

Noise in in situ hyperspectral data and its effects on the spectral features of lunar

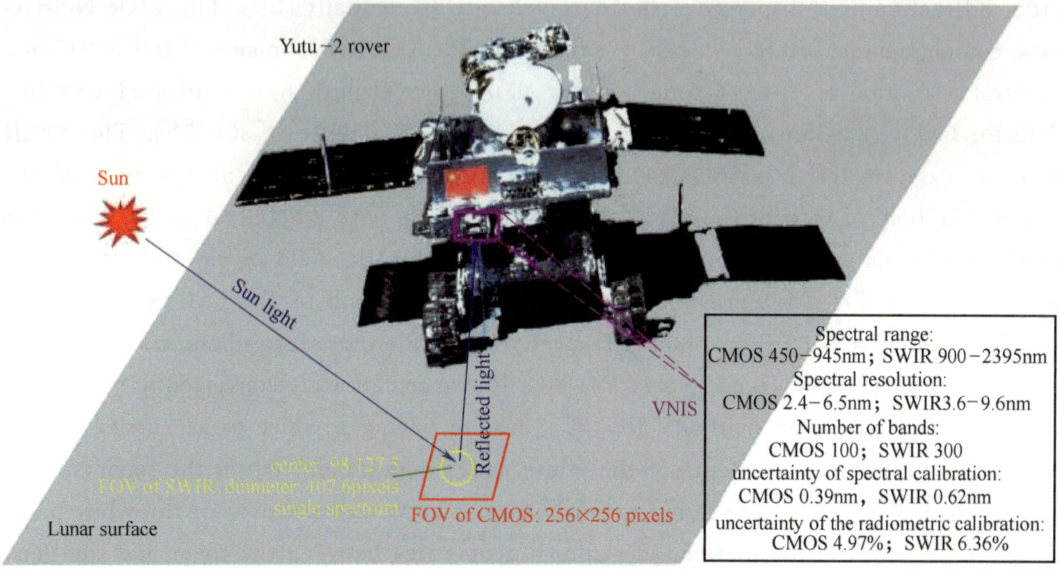

Figure 1　Schematic of working mode of the Yutu - 2 rover at lunar surface and the major specifications of visible and near - infrared (VNIR) imaging spectrometer [30]. The field of view (FOV) of the complementary metal - oxide - semiconductor (CMOS) imager is ~15cm × 21cm because of the different resolution in the horizontal and vertical directions. The FOV of the short - wavelength near - infrared (SWIR) detector is a circle with a diameter of 107. 6 CMOS pixels and is centered at sample 98, line 127. 5 within the FOV of the CMOS imager. Only a spectrum can be obtained by SWIR detector

regolith can be determined by observing the same areas repeatedly. On the tenth and thirteenth lunar days of the CE - 4 mission, the VNIS onboard the Yutu - 2 rover conducted experiments to measure the same target repeatedly at various sites (Figure 1). In this study, we first estimated the noise level of the VNIS measurements (i. e, the SNR of the hyperspectral images obtained) and analyzed their variations with solar altitude and incident energy. We also analyzed the effect of noise on the estimation of band center and depth. Finally, we assessed variations in mineral abundance and SMFe caused by the noise. These results are critical in the interpretation of the in situ spectral data acquired by CE - 4.

2 Data and Methods

2.1 Chang'E - 4 Spectral Data and Preprocessing

The VNIS onboard the Yutu - 2 rover measures the lunar surface at a height of ~0.75m (Figure 1). The VNIS uses an acousto - optic tunable filter (AOTF), which is a spectroscopic device based on the principle of acousto - optic diffraction, used to discriminate the light wavelength [30, 31]. The VNIS detectors acquire the spectral information by rapidly driving frequency scanning on the AOTF, to change the first - order diffraction light dispersed with the wavelength sequentially [30]. The VNIS consists of a complementary metal - oxide - semiconductor (CMOS) imager (450 - 945nm, spectral resolution of 2.4 - 6.5nm) and a short - wavelength near - infrared (SWIR) detector (900 - 2395nm, spectral resolution of 3.6 - 9.6nm) [30, 31]. The SWIR detector (the material is InGaAs) does not image but acquires a spectrum of the target. The field of view of the SWIR detector is a circle with a diameter of 107.6 CMOS pixels and is centered at sample 98, line 127.5, within the field of view of the CMOS imager (Figure 1). The uncertainties for spectral calibration of the CMOS imager and SWIR detector are 0.39 and 0.62nm, respectively, while those for radiometric calibration are 4.97% and 6.36%, respectively [10, 30]. Radiance data were converted to reflectance using a solar calibration method [10, 32] despite having a white panel carried by the rover. Only a few measurements of the white panel were conducted on the lunar surface because of the rigorous engineering requirements. A good measurement of the white panel relies on the appropriate relations of azimuth geometries between the rover and the Sun, otherwise the white panel would be shadowed. In addition, the rover's protective layer scatters the light on the white panel in some observation geometries, which could cause uncertainties in the calibration. Fortunately, the released radiance data were calibrated in fight using the valid measurements of white panel. The CMOS and SWIR channels have an overlap region from 900 to 945nm; differences within this overlapping region can be attributed to response differences between the CMOS imager and SWIR detector [30, 31,

33, 34]. Accordingly, we synchronized the CMOS and SWIR channels using the reflectance at 900nm [9, 10]. In addition, the spectra acquired by the SWIR detector have a gap between 1375 and 1380nm due to an unexpected spectral response. We filled this gap by extending the spectra before 1375nm to 1380nm.

On the tenth and thirteenth lunar days of mission operations, the Yutu - 2 rover conducted two different experiments using the VNIS (Figure 1): (1) measuring the spectra of the same regolith area at different solar altitudes and (2) measuring the spectra of the same rock - bearing area continuously at the same viewing geometry. The measurement angles for each site are listed in Table 1 and the corresponding CMOS images are shown in Figure 2. Regolith characteristics were measured at different solar altitudes for the same area to investigate errors caused by varying illumination conditions (Figure 2a). Similarly, rock outcrops were measured repeatedly at the same viewing geometry (ie., with little variation in solar altitude) which could help assess the measurement uncertainty. Measurements were obtained for three rock - bearing areas (Figure 2b).

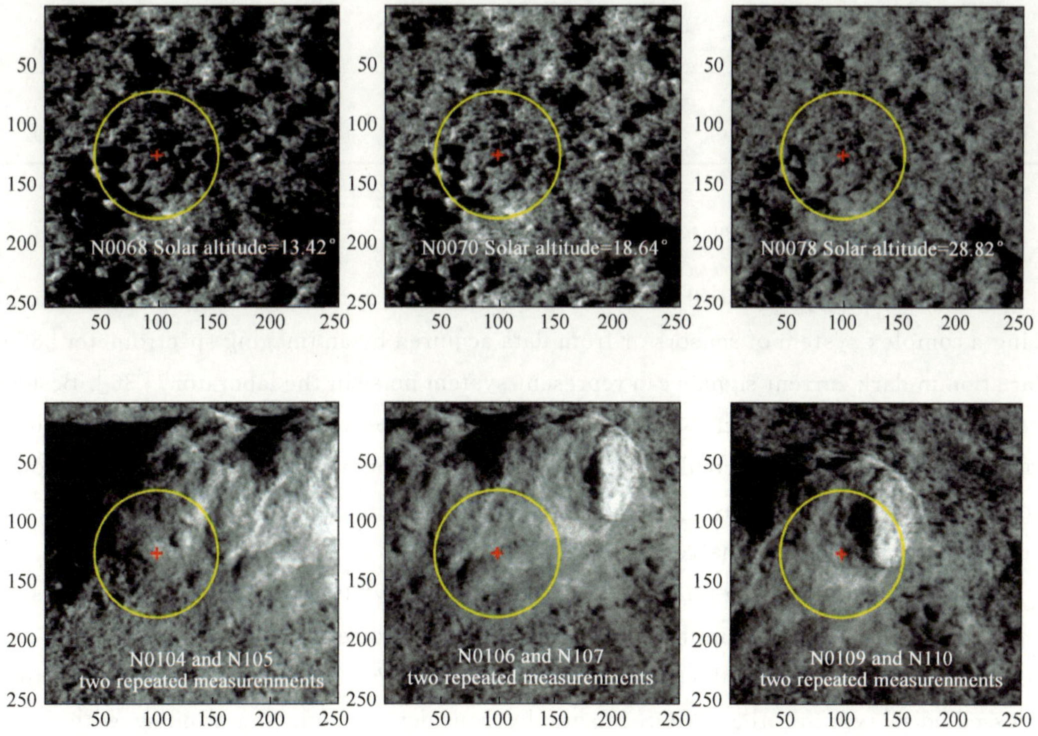

Figure 2　The CMOS images of lunar surface observed by Yutu - 2 rover.

(a) The images of the same regolith area measured at different solar altitudes.

(b) The repeated measurements of rock - bearing areas. The size of CMOS image is～15×21cm.

The yellow circle is the field of view of the SWIR detector

Table 1　Observation angles of the repeated measurements

	Data ID	Incidence Angle (°)	Solar Altitude (°)	Emission Angle (°)	Phase Angle (°)	Incidence Angle Variation	Phase Angle Variation
Figure 2a (regolith)	N0068	76.58	13.42	[illegible]	79.38		
	N0069	75.98	14.02	48.27	78.49	0.60	0.89
	N0070	71.36	18.64	48.27	71.49	4.62	7.0
	N0071	70.14	19.86	48.27	69.60	1.22	1.89
	N0072	69.61	20.39	48.27	68.78	0.53	0.82
	N0073	68.90	21.10	48.27	67.66	0.71	1.12
	N0074	68.33	21.67	48.27	66.76	0.57	0.9
	N0075	63.74	26.26	48.27	59.24	4.59	7.52
	N0076	62.86	27.14	48.27	57.73	0.88	1.51
	N0077	62.05	27.95	48.27	56.32	0.81	1.41
	N0078	61.18	28.82	48.27	54.79	0.87	1.53
Figure 2b (rock - bearing area)	N0104	58.83	31.17	43.23	49.70		
	N0105	59.53	30.47	43.23	50.94	0.70	1.24
Figure 2c (rock - bearing area)	N0106	60.24	29.76	43.08	54.06		
	N0107	60.35	29.65	43.08	54.26	0.11	0.20
Figure 2d (rock - bearing area)	N0109	67.40	22.60	42.93	66.76		
	N0110	67.52	22.48	42.93	66.95	0.12	0.19

2.2　Estimation of Signal - Noise Ratio

The SNR of hyperspectral remote sensing systems can be estimated in two ways: using a complex system of sensors or from data acquired by an imaging spectrometer [35]. Variation in dark current signals can represent system noise in the laboratory [36]. Besides the dark current of the detector, the thermal and readout effects also contribute to the data noise [31]. Furthermore, conditions in the lunar environment are complex which could induce additional noise. Thus, it is important to have an effective way to estimate the noise level in the spectral analysis of the CE - 4 data.

Noise in remote sensing images can be divided into two types based on the relationship between noise and signal, multiplicative and additive noises, which exhibit strong and weak correlation with signal strength, respectively [37, 38]. In this study, we first assessed noise type for the CMOS images by considering a moving window with 5×5 pixels as the local area. The local mean radiance value and local standard deviation were assumed to represent the signal and noise, respectively [38]. The correlation coefficients of the images were found to be <0.3 for all bands, suggesting that the noise in the CMOS images is additive (Figure 3).

Thus, we used the residual - scaled local standard deviations (RLSD) method,

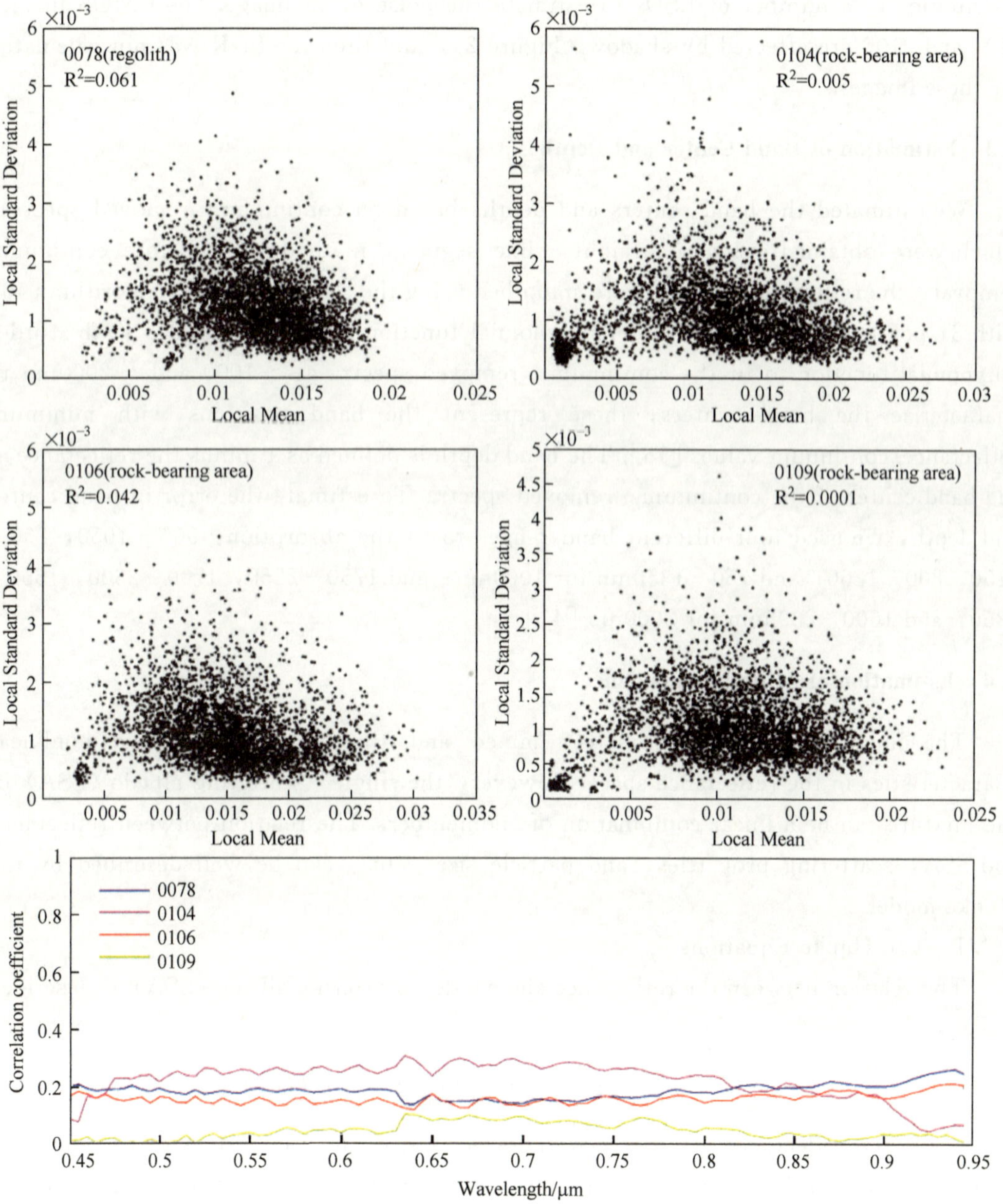

Figure 3　The scatterplot of the local mean radiance values and the local standard deviations at 0.945μm and their correlation coefficients at all bands

which assumes that the noise in hyperspectral images is additive, to estimate the SNR [35]. The RLSD method has been applied successfully to hyperspectral images, including those from the interference imaging spectrometer of the Chang'E-1 spacecraft [39]. This method is based on the concept of strong band correlation of hyperspectral images and the local standard deviation (LSD) of small imaging patches. In particular, the RLSD method uses multiple linear regression to obtain noise-like residuals and calculate the distribution of a number of LSDs to estimate the noise of an image. The CMOS images 0104 and 0105 are affected by shadow (Figure 2), and thus the SNR was not calculated for these images.

2.3 Estimation of Band Center and Depth

We estimated the band centers and depths based on continuum - removed spectra, which were obtained using a straight - line segment method [4]. Before continuum removal, the reflectance spectra were smoothed using the Savitzky - Golay algorithm [40] with 31 points and a second - order polynomial function. Then, we used a sixth - order polynomial function to fit the continuum - removed spectra at ~1000 and ~2000nm to characterize the band centers; these represent the band positions with minimum reflectance/continuum values [15]. The band depth is defined as 1 minus the reflectance at the band center of the continuum - removed spectra. To estimate the error in band center and depth, we used four different band ranges to fit the absorption: 900 - 1050, 850 - 1150, 800 - 1250, and 750 - 1350nm for 1000nm; and 1750 - 2250, 1700 - 2300, 1650 - 2350, and 1600 - 2395nm for 2000nm [41].

2.4 Estimation of Mineral Abundance

The lunar regolith is intimately mixed and the endmembers have nonlinear characteristics in the reflectance space. However, the single - scattering albedo (SSA) of the mixture can be a linear combination of endmembers. The relation between reflectance and SSA, scattering properties, and particle size, etc., can be well described by the Hapke model.

2.4.1 The Hapke Equations

The relation between the reflectance and single - scattering albedo (SSA) is described as [5]:

$$r = \frac{\omega}{4(\mu_0 + \mu_e)}\{P(g)[1 + B(g)] + H(\mu_0)H(\mu_e) - 1\} \tag{1}$$

where r is the reflectance of lunar materials; μ_0 and μ_e are the cosines of the angles of incidence and emission, respectively; and g is phase angle. $B(g)$ is the backscattering function; $P(g)$ is the phase function, which can be expressed with Legendre polynomials; $H(x)$ is a multiple - scattering function; and ω is the average SSA, which is the approximately linear combination of endmembers, SSAs. The equations and parameters of $B(g)$, $P(g)$, $H(\mu_0)$, and $H(\mu_e)$ are the same as those in [22]:

$$P(g) = 1 - 0.4\cos(g) + 0.25[1.5\cos^2(g) - 0.5] \tag{2}$$

$$B(g) = 1/[1 + \tan(g/2)/h] \tag{3}$$

$$h = -\frac{3}{8}\ln(1 - \phi) \tag{4}$$

$$H(\mu_0) = 1/\{1 - (1 - \sqrt{1 - \omega})\mu_0[r_0 + (1 - 0.5r_0 - r_0\mu_0)\ln(1 + \mu_0)/\mu_0]\} \tag{5}$$

$$r_0 = (1 - \sqrt{1-\omega})/(1 + \sqrt{1-\omega}) \tag{6}$$

where h is the angular width parameter of opposition effect. The equation for $H(\mu_e)$ is same as for $H(\mu_0)$, but μ_0 is replaced with μ_e.

2.4.2 Single-Scattering Albedo of Mineral Endmember

The endmembers used in this study are pyroxene (PYX), olivine (OL), plagioclase (PLG), and ilmenite (ILM), which are the dominant minerals of the Moon. Considering the spectral variations, at least two spectra of each endmember were used in this study (Figure 4). These endmember spectra were collected from the Lunar Rock and Mineral Characterization (LRMCC) database [42]. The endmember SSA can be calculated using the Hapke model with given optical constants and grain size:

$$\omega = S_e + (1 - S_e)\frac{(1 - S_i)\theta}{1 - S_i\theta} \tag{7}$$

$$S_e = \frac{(n-1)^2 + k^2}{(n+1)^2 + k^2} + 0.05 \tag{8}$$

$$S_i = 1 - \frac{4}{n(n+1)^2} \tag{9}$$

$$\theta = e^{-\alpha\langle D\rangle} \tag{10}$$

where n and k are optical constants of the minerals. S_i and S_e are the Fresnel reflectivity for internally and externally incident light, respectively. θ is the internal transmission coefficient of the particle without internal scatter. The variable $\langle D\rangle$ is the average distance traveled by transmitted rays during one traverse of a particle, which is related to partidle size D:

$$\langle D\rangle = \frac{2}{3}\left[n^2 - \frac{1}{n}(n^2-1)^{3/2}\right]D \tag{11}$$

where $\alpha = 4\pi k/\lambda$ is the absorption coefficient of the mineral. For considering the effects of space weathering on the lunar surface, the spectral contribution of submicroscopic metallic iron (SMFe) is taken into account by modeling the absorption coefficient:

$$\alpha = \frac{4\pi k}{\lambda} + \frac{36\pi z M_{Fe}\rho}{\lambda\rho_{Fe}} \tag{12}$$

$$z = \frac{n^3 n_{Fe} k_{Fe}}{(n_{Fe}^2 - k_{Fe}^2 + 2n^2)^2 + 4n_{Fe}^2 k_{Fe}^2} \tag{13}$$

where ρ is density of host mineral; n_{Fe}, k_{Fe}, and ρ_{Fe} are optical constants and the density of SMFe. M_{Fe} is the mass fraction of SMFe.

The endmember's optical constant k was calculated using Equations (1-11) with known reflectance and particle size. The endmembers' SSAs were calculated by setting the particle size as (low boundary, step, upper boundary) = (10, 5, 60) and SMFe abundance as (0, 0.001, 0.005). Thus, a SSA spectral library was constructed. Theoretically, there are only a small number of materials in a mixture, which is sparse compared with the hundreds of spectra in our spectral library. Therefore, a sparse unmixing algorithm [43] that was applied to Martian hyperspectral data [44 – 46] was used to obtain the optimal solution. The sparse regression problem is written as [43]:

$$\min_{x} \|x\|_0 \text{ Subject to } \|Y - Ax\|_2 \leqslant \delta, \quad x \geqslant 0 \tag{14}$$

where $\delta \geqslant 0$ is the modeling error tolerance, x is the mineral abundance, Y is the SSA of mixture, and A is the spectral library. Variable $\|x\|_0$, the L_0 norm, denotes the number of nonzero components of x. Due to the non – deterministic polynomial (NP) – hard problem of the L_0 norm, it is typically replaced with the L_1 norm [43]. Therefore, the objective function of the optimal problem can be formulated as:

$$\min_{x} \frac{1}{2} \|Ax - Y\|_2^2 + \gamma \|X\|_1 \tag{15}$$

where γ is regularization parameter. A sparse unmixing via variable splitting augmented Lagrangian (SunSAL) algorithm proposed by Iordanche et al. [43] was applied to retrieve the mineral abundances.

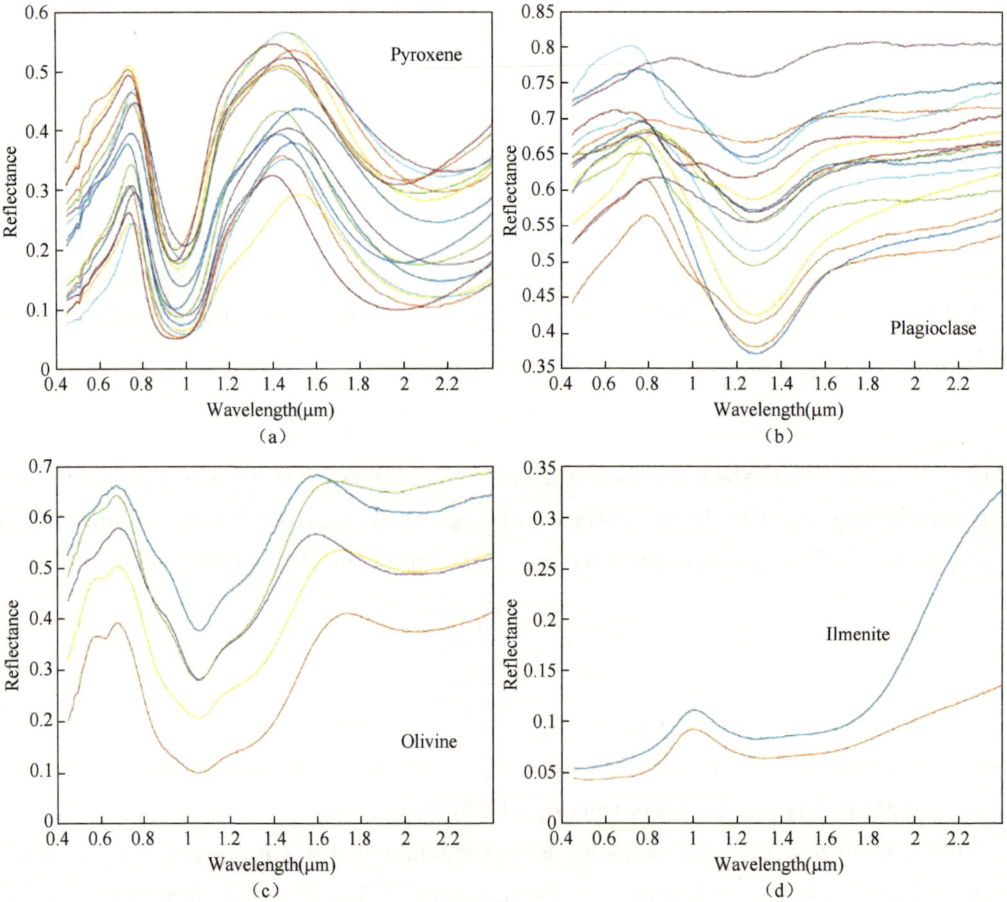

Figure 4　Endmember spectra used in this study.

(a) Pyroxene. (b) Plagioclase. (c) Olivine. (d) Ilmenite. The minerals were separated from the Apollo samples, which were collected in Lunar Rock and Mineral Characterization Consortium (LMRCC) database [42]

2.5 Estimation of SMFe Abundance

The SMFe abundance of the lunar regolith can be modeled using the Hapke model [22] by accounting for the space weathering. As shown in Equations (12) and (13), the absorption coefficient of the lunar regolith can be modeled by adding a different mass fraction of SMFe incorporated on the host material, which did not experience space weathering [25]. A rock with fresh exposure (Figure 5) was measured by the Yutu - 2 rover at the third lunar day of the mission, which was suggested to be a norite [9]. The lunar regolith was believed to have the same origin with this rock, based on the similar spectral features [9]. Even though part of the rock surface was masked by the lunar regolith, the fresh rock surface can be clearly identified from the CMOS image [9]. Therefore, the mean spectrum of the pixels collected from the fresh rock surface was considered as the host material in Equation (12). The optical constant n was set as 1.78 [28] and the mean grain size of the rock was assumed to be 80μm [9]. In this case, the optical constant k of the host material can be calculated using Equations (1 - 11). The mass fraction of SMFe and particle size of the lunar regolith were then estimated by fitting the regolith spectra using Equations (1 - 13) with least squares algorithm [25]. The initial values of the SMFe abundance and particle size of lunar regolith in the algorithm were set as 0.1 wt.% and 17μm, respectively [24].

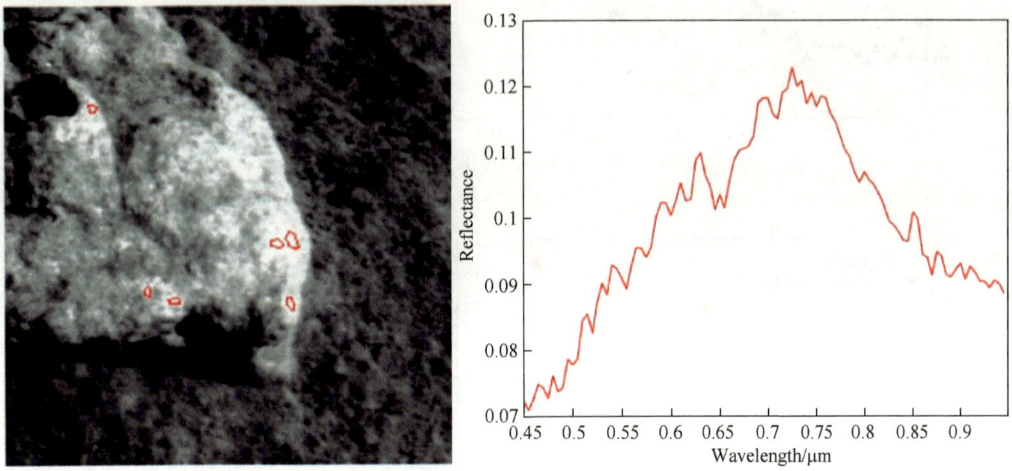

(a)　　(b)

Figure 5　The rock measured by Yutu - 2 rover.

(a) The CMOS image of the rock. The red regions are the fresh rock surface.

(b) The average reflectance spectrum of the fresh rock surface

3 Results and Discussion

3.1 SNR of CMOS Image and Spectral Difference

We found the SNR to increase with increasing solar altitude (Figure 6), which indicates that the quality of the hyperspectral remote sensing data acquired by CE-4 was affected significantly by the illumination conditions at the lunar surface. The irradiance of solar energy also increases with solar altitude, in conjunction with increasing SNR in our dataset (Figure 6c). In general, the SNR increases with wavelength (Figure 6a). The SNR is lower than 25dB for solar altitudes<20° (Figure 6b); this could affect spectral interpretation, particularly for short wavelengths (10-15dB). The low SNR of the image acquired at low solar altitude was primarily caused by the low solar incident energy and the shadows produced by the uneven lighting. Therefore, we suggest that future measurements should be obtained at solar altitudes >20° to improve the quality of

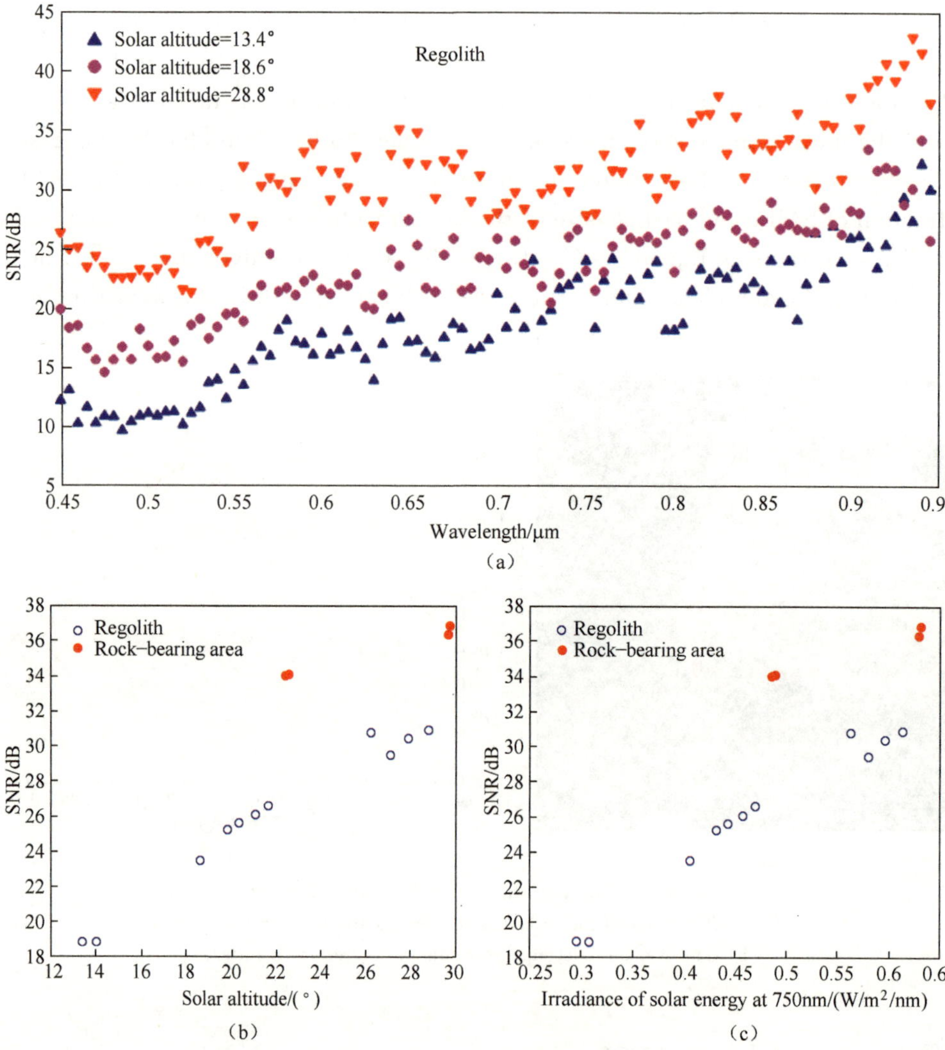

Figure 6　Signal-noise ratio (SNR) of the CMOS image.

(a) The SNR versus wavelength at different solar altitude.

(b) The SNR versus solar altitude. (c) The SNR versus irradiance of solar energy at 0.75μm obtained lunar spectra. Compared with the regolith, the rock - bearing areas measured at a similar solar altitude were found to exhibit higher SNR (Figure 6b), suggesting that fresher materials exhibit a higher SNR due to their stronger reflectance.

We found differences in the paired measurements between 0106 and 0107, as well as between 0109 and 0110, despite that these observations were collected for the same areas and with similar viewing geometries. Note that the differences in incidence angles were smaller than 0.12°. We attribute these differences to measurement error or noise. We found considerable variability in the spectra at wavelengths longer than 1.8μm (Figure 7), suggesting that variability could affect the 2μm spectral features. This variability is particularly pronounced for the two observations collected at the lowest solar altitudes (Figure 7b, d).

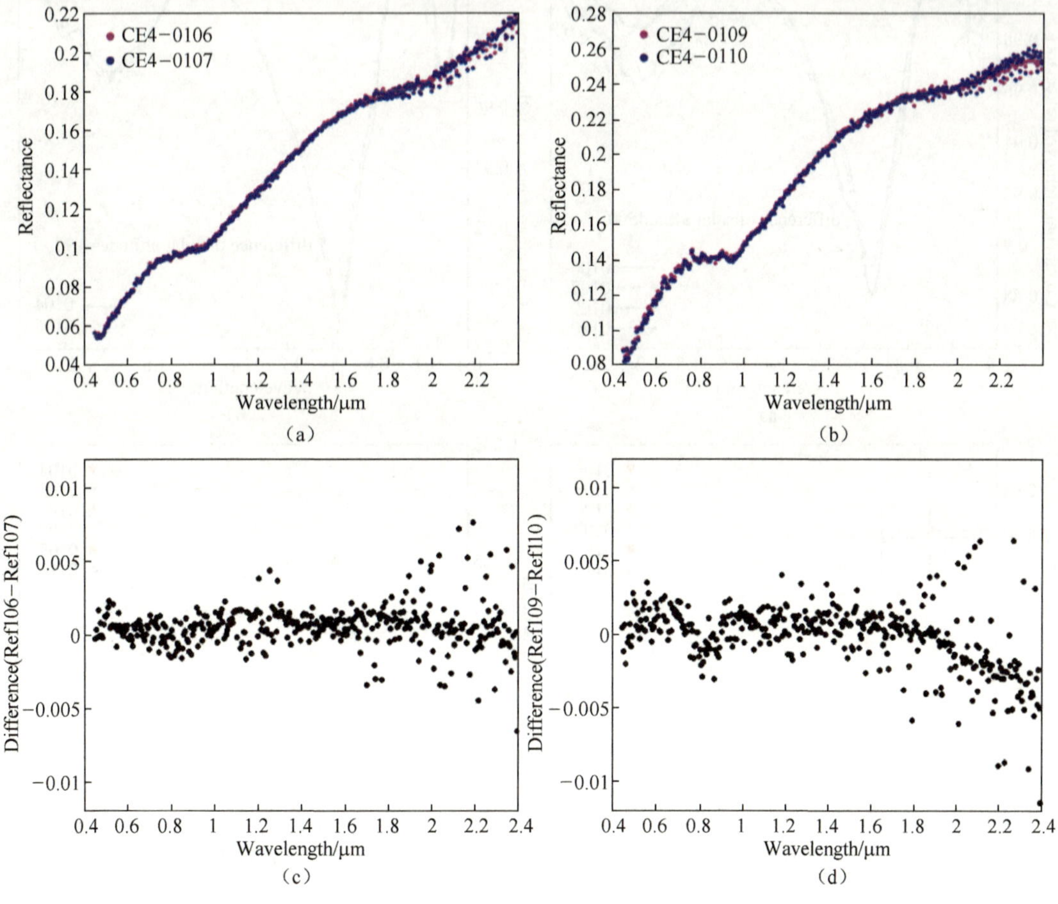

Figure 7　The spectra and their differences acquired from the same rock - bearing area.
(a) The spectra of 0106 and 0107. (b) The spectra of 0109 and 0110. (c) The difference of the spectra between 0106 and 0107. (d) The difference of the spectra between 0109 and 0110

3.2 Effects of Noise on Band Center and Depth

The effects of noise on band center and band depth were also investigated. Spectral variations in the continuum-removed reflectance spectra between measurements acquired under the same conditions are evident (Figure 8), particularly for the 2μm absorption. This is consistent with the large differences observed at ~2μm in the reflectance spectra. We observed the following differences in band centers. For 0106 and 0107, differences were 2.6 and 10.2nm for the 1 and 2μm bands, respectively. For 0109 and 0110, differences were 2.7 and 1.7nm for the 1 and 2μm bands, respectively. Finally, for 0104 and 0105, differences were 1.9 and 16.7nm for the 1 and 2μm bands, respectively. These findings suggest that the effects of noise on band center estimation are

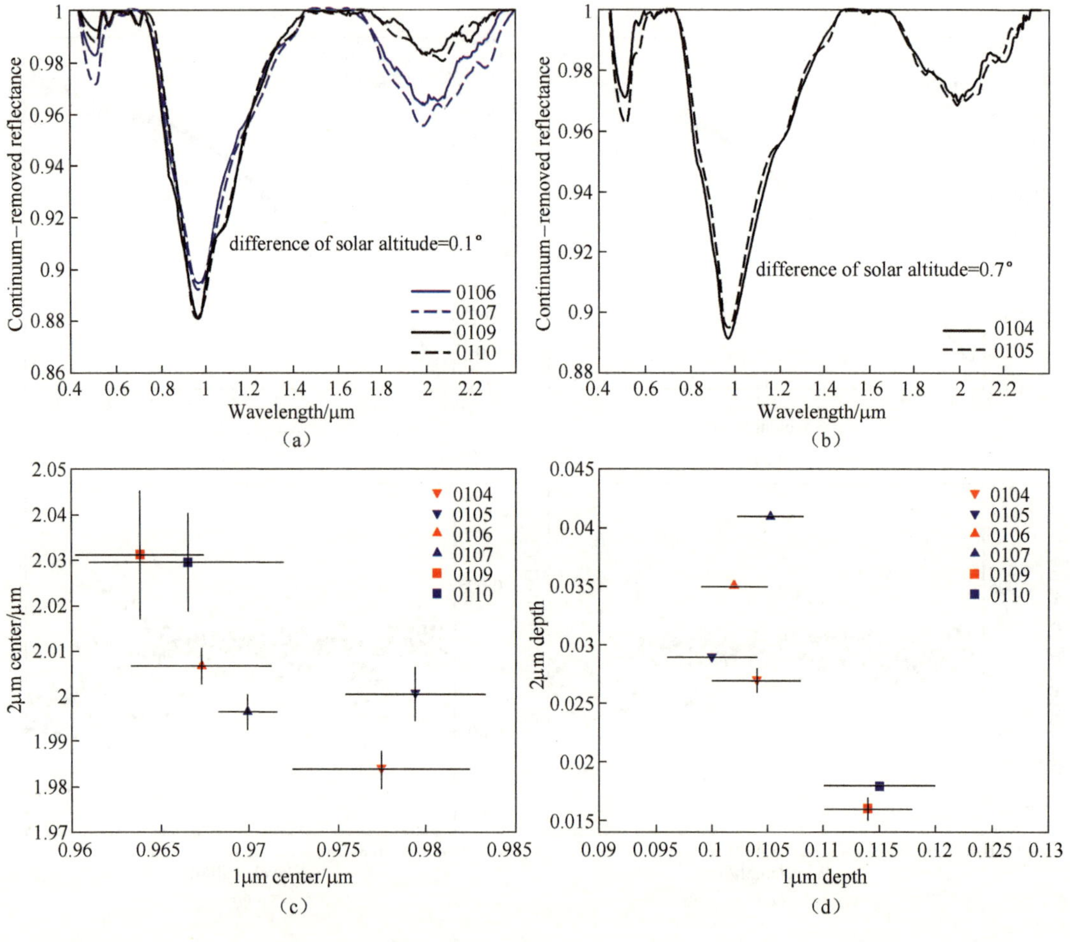

Figure 8　The continuum-removed spectra and band centers and depths of rock-bearing areas.
(a) The continuum-removed spectra of the paired measurements with small difference of solar altitudes.
(b) The continuum-removed spectra of the paired measurements with large difference of solar altitudes.
(c) The positions of the band centers. (d) The band depths. The lines in (c) and (d) are error bars

random, with variation of 1.9 - 2.7 and 1.7 - 16.7nm for the 1 and 2μm absorptions, respectively. Moreover, the noise caused by solar altitude was found to have significant effects on the band centers, with uncertainties of 9.6 and 46.0nm for the 1 and 2μm bands, respectively. Removal of measurements acquired at solar altitudes $<20°$ reduced these uncertainties to 6.2 and 28.6nm for the 1 and 2μm absorptions, respectively. Band depth, which is related to mineral abundance, was also affected by noise (Figure 8d). Similarly, we found noise to induce variations in estimation of band depth, and these variations are 0.001 - 0.004 and 0.002 - 0.006 for the 1 and 2μm absorptions, respectively. The tie points of the spectral continuum were changed by the noise and thus introduced significant variations when estimating the band center and depth.

The CE - 4 landing site is located in Von Kármán crater of the SPA basin. Von Kármán crater was filled with mare basalts [47]. However, the ejecta blanket from Finsen crater, which is to the northeast of the CE - 4 landing site, can be seen from the orbital images [9, 11 - 13, 47]. In this case, the regolith at the landing site could represent the original materials of the SPA basin floor excavated by an asteroid impact [9]. It is critical to determine the source of the materials around the landing site because their formation conditions would be varied. We estimated the band centers of numerous regolith spectra acquired by the Yutu - 2 rover and compared our results with the Apollo samples from the Lunar Soil Characterization Consortium (LSCC) database [8, 48]. As shown in Figure 9, the maximum shifts of the band centers of each regolith spectrum affect the determination of mineral types, thus such shifts caused by the noise need to be considered in the geologic interpretation.

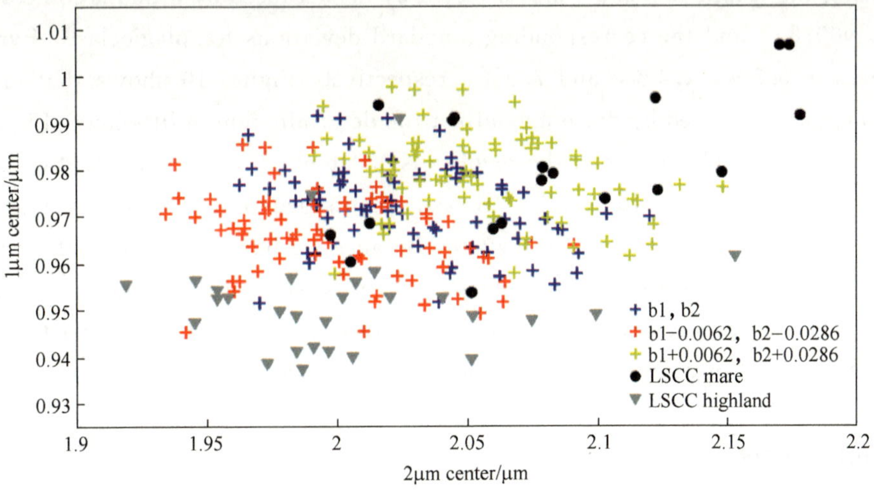

Figure 9　The band centers of the regolith spectra measured by Yutu - 2 rover and their shifts caused by the data noise. The bl and b2 indicate the 1μm band center and 2μm band center, respectively. The black dots and gray triangles represent spectral features of the Apollo samples included in the LSCC database. For obtaining the accurate band centers of Apollo samples, only the spectra with strong absorptions were plotted (48 of 76 samples)

3.3 Effects of Noise on Estimations of Compositions

The noise changed the spectral shapes (band centers, band depths, and spectral slopes) and thus could affect the estimation of compositions, as the composition derived from the spectra relies on the fitting of the spectral shape and absorption features. Therefore, we analyzed the influence of noise on the interpretation of FeO content, optical maturity (OMAT), and mineral abundances, as well as the SMFe abundance. The spectra obtained were photometrically corrected with photometric functions derived using the CE-4 spectra, which were acquired on the fourth lunar day of the mission [10]. Then, FeO content and OMAT were calculated using spectral parameters [10, 49, 50]. The variation in OMAT (i. e., absolute difference/OMAT value) between the two measurements acquired under the same conditions was found to be 0.9%—2.7%, whereas the uncertainty induced by noise related to illumination was found to be 6.8%. The variation in FeO content (i. e., absolute difference/FeO content) between two measurements acquired under the same conditions was found to be 3.9%-8.1%. In contrast, variation in FeO content caused by illumination-related noise was found to be 11.2%.

The mineral abundances derived from the VNIR spectra using the Hapke model are shown in Figure 10, showing that the abundance values vary between two sets of measurements acquired under the same conditions. In particular, we found the abundances could vary as much as 0%—6%, 1%—5%, and 0%—4%for pyroxene, plagioclase, and olivine, respectively. Furthermore, we found the mineral abundances derived from CE-4 spectra using the Hapke model are also affected by noise caused by illumination conditions (N0068-N0078), and the corresponding standard deviations for plagioclase, Pyroxene, and olivine are 6.3%, 2.4%, and 7.0%, respectively. Figure 10 shows that errors of mineral abundances caused by noise may affect the determination of lithology (Figure 10).

The mean SMFe abundance of the regolith is 0.096 wt.%, with standard variation of 0.025 wt.% (Figure 11a). Removal of measurements acquired at solar altitudes $<20°$ redued this standard variation to 0.011 wt.%. Because we have photometrically corrected the spectra to the standard viewing geometry, this standard variation is attributed primarily to the noise. Similarly, our modeling results show that the mean particle size of the regolith is 45.73μm, with a standard deviation of 4.75μm (Figure 11b).

4 Conclusions

The Yutu-2 rover of the CE-4 mission has explored the far side of the Moon since 3 January 2019, providing a unique perspective on the composition of the lunar surface. The rover conducted a series of in situ spectral measurements in two specific areas. We estimated the noise inherent in the obtained hyperspectral data, analyzed its effects on the

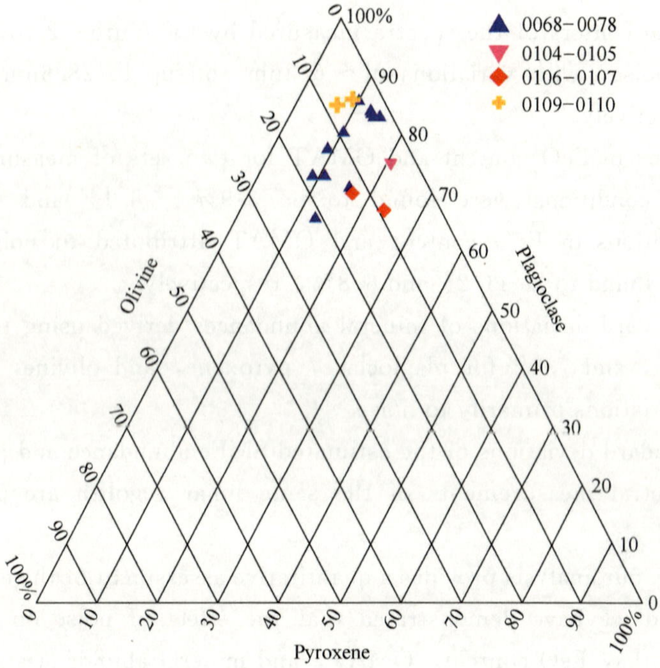

Figure 10　The variation of mineral abundances caused by the noise in the hyperspectral data acquired by CE−4

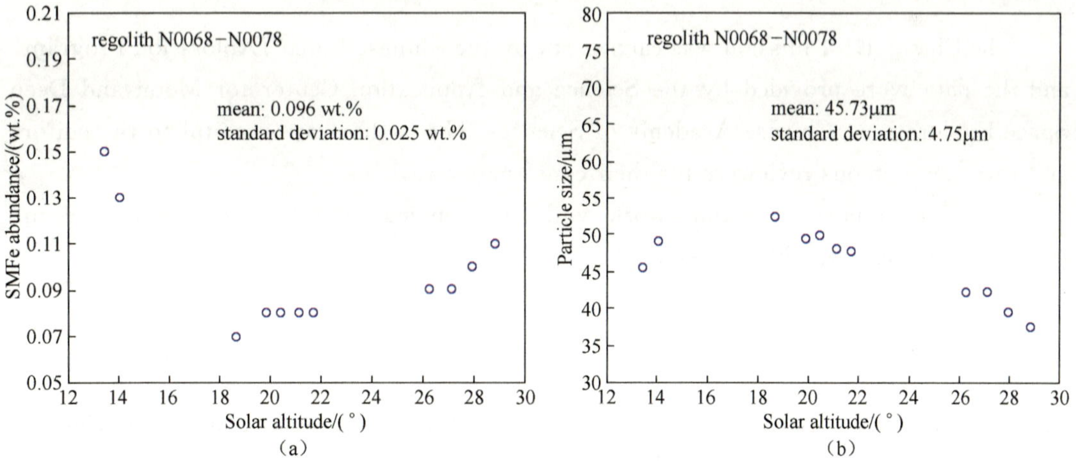

Figure 11　The variations of (a) SMFe abundance and (b) particle size of lunar regolith caused by the noise in the hyperspectral data acquired by CE−4

band centers and band depths, and discussed the effects of the noise on the derived FeO content, OMAT, and mineral abundances of lunar regolith and rocks. The main findings can be summarized as follows:

(1) Data observed at solar altitudes $<20°$ exhibit low signal−noise ratio (SNR)

(25dB). In contrast, those acquired at 20°—35°exhibit higher SNR (35—37dB).

(2) The band centers of the spectra measured by the Yutu - 2 rover were affected significantly by noise, with variations of ~ 6.2nm and up to 28.6nm for 1 and 2μm absorption, respectively.

(3) Variations in FeO content and OMAT for two sets of measurements acquired under the same conditions were found to be 3.9% — 8.1% and 0.9% — 2.7%, respectively. Variations in FeO content and OMAT attributed to noise due to weak illumination were found to be 11.2%and 6.8%, respectively.

(4) The standard deviations of mineral abundances derived using the Hapke model are 6.3%, 2.4%, and 7.0% for plagioclase, pyroxene, and olivine, respectively. We attribute these variations primarily to noise.

(5) The standard deviations of the estimated SMFe abundance and particle size from the different spectral measurements of the same lunar regolith are 0.025 wt. % and 4.75μm, respectively.

In summary, our analysis provides a quantitative assessment of the effects of noise on spectral data, and we have demonstrated that the effect of noise on estimating band centers, band depths, FeO content, OMAT, and mineral abundances is significant and could mislead the geologic interpretation of the evolution history of the CE - 4 landing site.

Acknowledgments

The Chang'E - 4 mission was carried out by the Chinese Lunar Exploration Program, and the data were provided by the Science and Application Center for Moon and Deep Space Exploration, Chinese Academy of Sciences. The authors are grateful to the editors and three anonymous reviewers for their constructive reviews.

The data reported in this work will be archived at http://moon.bao.ac.cn/searchOrder_dataSearchData.search.

References

[1] Pieters, C. M.; Boardman, J.; Buratti, B.; Chatterjee, A.; Clark, R.; Glavich, T.; Green, R.; Head, J., Ⅲ; Isaacson, P.; Malaret, E. The Moon Mineralogy Mapper (M^3) on Chandrayaan - 1. Curr. Sci. 2009, 96, 500 - 505.

[2] Wu, Y. Z.; Zheng, Y. C.; Zou, Y. L.; Chen, J.; Xu, X. S.; Tang, Z. S.; Xu, A. A.; Yan, B. K.; Gan, F. P.; Zhang, X. A preliminary experience in the use of Chang'E - 1 IIM data. Planet. Space Sci. 2010, 58, 1922 - 1931.

[3] Yamamoto, S.; Matsunaga, T.; Ogawa, Y.; Nakamura, R.; Yokota, Y.; Ohtake, M.; Haruyama, J.; Morota, T.; Honda, C.; Hiroi, T.; et al. Calibration of NIR 2 of Spectral Profiler Onboard Kaguya/SELENE. IEEE Trans. Geosci. Remote

Sens. 2014, 52, 6882 - 6898.

[4] Clark, R. N.; Roush, T. L. Reflectance Spectroscopy - Quantitative - Analysis Techniques for Remote - Sensing Applications. J. Geophys. Res. 1984, 89, 6329 - 6340.

[5] Hapke, B. Bidirectional reflectance spectroscopy: 1. Theory. J. Geophys. Res. Solid Earth 1981, 86, 3039 - 3054.

[6] Pieters, C. M.; Taylor, LA.; Noble, S. K.; Keller, L. P.; Hapke, B.; Morris, R. V.; Allen, C. C.; McKay, D. S.; Wentworth, S. Space weathering on airless bodies: Resolving a mystery with lunar samples. Meteorit. Planet. Sci. 2000, 35, 1101 - 1107.

[7] Johnson, J. R.; Shepard, M. K.; Grundy, W. M.; Paige, D. A.; Foote, E. J. Spectrogoniometry and modeling of martian and lunar analog samples and Apollo soils. Icarus 2013, 223, 383 - 406.

[8] Taylor, L. A.; Pieters, C. M.; Keller, L. P.; Morris, R. V.; McKay, D. S. Lunar Mare Soils: Space weathering and the major effects of surface - correlated nanophase Fe. J. Geophys. Res. Planets 2001, 106, 27985 - 27999.

[9] Lin, H.; He, Z.; Yang, W.; Lin, Y.; Xu, R.; Zhang, C.; Zhu, M. - H.; Chang, R.; Zhang, J.; Li, C. Olivine - norite rock detected by the lunar rover Yutu - 2 likely crystallized from the SPA impact melt pool. Natl. Sci. Rev. 2019.

[10] Lin, H.; Xu, R.; Yang, W.; Lin, Y.; Wei, Y.; Hu, S.; He, Z.; Qiao, L.; Wan, W. In situ photometric experiment of lunar regolith with visible and near - infrared imaging spectrometer onboard the Yutu - 2 lunar rover. J. Geophys. Res. Planets 2020, 125, e2019JE006076.

[11] Hu, X.; Ma, P.; Yang, Y.; Zhu, M. H.; Jiang, T.; Lucey, PG.; Sun, L.; Zhang, H.; Li, C.; Xu, R. Mineral Abundances Inferred From In Situ Reflectance Measurements of Chang'E - 4 Landing Site in South Pole - Aitken Basin. Geophys. Res. Lett. 2019, 46, 9439 - 9447.

[12] Li, C. L.; Liu, D. W.; Liu, B.; Ren, X; Liu, J. J.; He, Z. P.; Zuo, W.; Zeng, X. G.; Xu, R.; Tan, X.; et al. Chang'E - 4 initial spectroscopic identification of lunar far - side mantle - derived materials. Nature 2019, 569, 378 - 382.

[13] Gou, S.; Di, K.; Yue, Z.; Liu, Z.; He, Z.; Xu, R.; Lin, H.; Liu, B.; Peng, M.; Wan, W. Lunar deep materials observed by Chang'e - 4 rover. Earth Planet. Sci. Lett. 2019, 528, 115829.

[14] Rasti, B.; Scheunders, P.; Ghamisi, P.; Licciardi, G.; Chanussot, J. Noise Reduction in Hyperspectral Imagery: Overview and Application. Remote Sens. 2018, 10, 482.

[15] Horgan, B. H. N.; Cloutis, E. A.; Mann, P.; Bell, J. F. Near - infrared spectra of ferrous mineral mixtures and methods for their identification in planetary surface spectra. Icarus 2014, 234, 132 - 154.

[16] Wu, Y.; Hapke, B. Spectroscopic observations of the Moon at the lunar surface. Earth Planet. Sci. Lett. 2018, 484, 145 - 153.

[17] Moriarty, D.; Pieters, C.; Isaacson, P. Compositional heterogeneity of central peaks within the South Pole - Aitken Basin. J. Geophys. Res. Planets 2013, 118, 2310 - 2322.

[18] Moriarty, D. P.; Pieters, C. M. The Character of South Pole - Aitken Basin: Patterns of Surface and Subsurface Composition. J. Geophys. Res. Planets 2018, 123, 729 - 747.

[19] Adams, J. B. Visible and near - infrared diffuse reflectance spectra of pyroxenes as applied to remote sensing of solid objects in the solar system. J. Geoplrys. Res. 1974, 79, 4829 - 4836.

[20] Klima, R. L.; Dyar, M. D.; Pieters, C. M. Near - infrared spectra of clinopyroxenes: Effects of calcium content and crystal structure. Meteorit. Planet. Sci. 2011, 46, 379 - 395.

[21] Sunshine, J. M.; Pieters, C. M. Determining the composition of olivine from reflectancespectroscopy. J. Geoplhys. Res. Planets 1998, 103, 13675 - 13688.

[22] Li, S. A.; Li, L. Radiative transfer modeling for quantifying lunar surface minerals, particle size, and submicroscopic metallic Fe. J. Geophys. Res. Planets 2011, 116.

[23] Denevi, B. W.; Lucey, P. G.; Sherman, S. B. Radiative transfer modeling of near - infrared spectra of lunar mare soils: Theory and measurement. J. Geophys. Res. Planets 2008, 113.

[24] Zhang, H.; Yang, Y. Z.; Yuan, Y.; Jin, W. D.; Lucey, P. G.; Zhu, M. H.; Kaydash, V. G.; Shkuratov, Y. G.; Di, K. C.; Wan, W. H.; et al. In situ optical measurements of Chang'E - 3 landing site in Mare Imbrium: 1. Mineral abundances inferred from spectral reflectance. Geophys. Res. Lett. 2015, 42, 6945 - 6950.

[25] Hapke, B. Space weathering from Mercury to the asteroid belt. J. Geoplhys. Res. Planets 2001, 106, 10039 - 10073.

[26] Pieters, C. M.; Noble, S. K. Space weathering on airless bodies. J. Geophys. Res. Planets 2016, 121, 1865 - 1884.

[27] Wang, Z.; Wu, Y.; Blewett, D. T.; Cloutis, E. A.; Zheng, Y.; Chen, J. Submicroscopic metallic iron in lunar soils estimated from the in situ spectra of the Chang'E - 3 mission. Geophys. Res. Lett. 2017, 44, 3485 - 3492.

[28] Gou, S; Yue, Z. Y.; Di, K. C.; Wan, W. H.; Liu, Z. Q.; Liu, B.; Peng, M.; Wang, Y. X.; He, Z. P.; Xu, R. In situ spectral measurements of space weathering by Chang'e - 4 rover. Earth Planet. Sci. Lett. 2020, 536.

[29] Lucey, P. G.; Noble, S. K. Experimental test of a radiative transfer model of the optical effects of space weathering. Icarus 2008, 197, 348 - 353.

[30] Li, C. ; Xu, R. ; Lv, G. ; Yuan, L. ; He, Z. ; Wang, J. Detection and calibration characteristics of the visible and near - infrared imaging spectrometer in the Chang' e - 4. Rev. Sci. Instrum. 2019, 90, 103106.

[31] He, Z.; Li, C.; Xu, R. ; Lv, G.; Yuan, L.; Wang, J. Spectrometers based on acousto - optic tunable flters for in - situ lunar surface measurement. J. Appl. Remote Sens. 2019, 13, 027502.

[32] Gueymard, C. A. The sun's total and spectral irradiance for solar energy applications and solar radiation models. Sol. Energy 2004, 76, 423 - 453.

[33] He, Z. P.; Wang, B. Y.; Lu, G. ; Li, C. L. ; Yuan, L. Y; Xu, R. ; Liu, B. ; Chen, K. ; Wang, J. Y. Operating principles and detection characteristics of the Visible and Near - Infrared Imaging Spectrometer in the Chang'e - 3. Res. Astron. Astrophys. 2014, 14, 1567 - 1577.

[34] Li, C. L.; Wang Z. D.; Xu, R.; Lv, G.; Yuan, L. Y.; He, Z. P.; Wang, J. Y. The Scientific Information Model of Chang'e - 4 Visible and Near - IR Imaging Spectrometer (VNIS) and In - Flight Verification. Sensors 2019, 19, 482.

[35] Gao, L.; Wen, J. ; Ran, Q. Residual - Scaled Local Standard Deviations Method for Estimating Noise in Hyperspectral Images. In Proceedings of the MIPPR 2007: Multispectral Image Processing, Wuhan, China, 15 - 17 November 2007; p. 678713.

[36] Van der Meer, F. D. ; De Jong, S. M. Imaging Spectrometry: Basic Principles and Prospective Applications; Springer: Berlin/Heidelberg, Germany, 2011.

[37] Corner, B. R.; Narayanan, R. M.; Reichenbach, S. E. Noise estimation in remote sensing imagery using data masking. Int. J. Remote Sens. 2003, 24, 689 - 702.

[38] Gao, B. C. An Operational Method for Estimating Signal to Noise Ratios from Data Acquired with Imaging Spectrometers. Remote Sens. Environ. 1993, 43, 23 - 33.

[39] Xia, Z.; Tong, S. ; Dong, Z. Assessment of the information quality of hyperspectral data from interference imaging spectrometer. J. Infrared Millim. Waves 2012, 31, 143 - 147.

[40] Savitzky, A. ; Golay, M. J. E. Smoothing + Differentiation of Data by Simplified Least Squares Procedures. Anal. Chem. 1964, 36, 1627 - 1639.

[41] Yang, Y.; Lin, H.; Liu, Y.; Lin, Y.; Wei, Y.; Hu, S.; Yang, W.; Xu, R.; He, Z; Zou, Y. The effects of viewing geometry on the spectral analysis of lunar regolith as inferred by in situ spectrophotometric measurements of Chang'E - 4. Geophys. Res. Lett. 2020, 47, e2020GL087080.

[42] Isaacson, P. J. ; Sarbadhikari, A. B. ; Pieters, C. M. ; Klima, R. L. ; Hiroi, T.; Liu, Y.; Taylor, L. A. The lunar rock and mineral characterization consortium: Deconstruction and integrated mineralogical, petrologic, and spectroscopic analyses of mare basalts. Meteorit. Planet. Sci. 2011, 46, 228 - 251.

[43] lordache, M. D. ; Bioucas - Dias, J. M.; Plaza, A. Sparse Unmixing of Hyperspectral Data. IEEE Trans. Geosci. Remote Sens. 2011, 49, 2014 - 2039.

[44] Lin, H. L.; Mustard, J. F.; Zhang, X. A methodology for quantitative analysis of hydrated minerals on Mars with large endmember library using CRISM near - infrared data. Planet. Space Sei. 2019, 165, 124 - 136.

[45] Lin, H. L; Zhang, X. Retrieving the hydrous minerals on Mars by sparse unmixing and the Hapke model using MRO/CRISM data. Icarus 2017, 288, 160 - 171.

[46] Lin, H. L.; Zhang, X.; Shuai, T.; Zhang, L. F.; Sun, Y. L. Abundance retrieval of hydrous minerals around the Mars Science Laboratory landing site in Gale crater, Mars. Planet. Space Sci. 2016, 121, 76 - 82.

[47] Huang, J.; Xiao, Z. Y.; Flahaut, J.; Martinot, M.; Head, J.; Xiao, X.; Xie, M. G.; Xiao, L. Geological Characteristics of Von Karman Crater, Northwestern South pole - Aitken Basin: Chang' E - 4 Landing Site Region. J. Geophys. Res. Planets 2018, 123, 1684 - 1700.

[48] Pieters, C.; Shkuratov, Y.; Kaydash, V.; Stankevich, D.; Taylor, L. Lunar soil characterization consortium analyses: Pyroxene and maturity estimates derived from Clementine image data. Icarus 2006, 184, 83 - 101.

[49] Lucey, P. G.; Blewett, D. T.; Joliff, B. L. Lunar iron and titanium abundance algorithms based on final processing of Clementine ultraviolet - visible images. J. Geophys. Res. Planets 2000, 105, 20297 - 20305.

[50] Lucey, P. G.; Blewett, D. T.; Taylor, G. J.; Hawke, B. R Imaging of lunar surface maturity. J. Geophys. Res. Planets 2000, 105, 20377 - 20386.

Composition, mineralogy and chronology of mare basalts and non-mare materials in Von Kármán crater: Landing site of the Chang'E-4 mission

Zongcheng Ling[1,2], Le Qiao[1], Changqing Liu[a], Haijun Cao[1], Xiangyu Bi[1], Xuejin Lu[1], Jiang Zhang[1], Xiaohui Fu[1], Bo Li[1,3], Jianzhong Liu[4]

1. Shandong Provincial Key Laboratory of Optical Astronomy and Solar-Terrestrial Environment, Institute of Space Sciences, Shandong University, Weihai, 264209, China
2. Center for Excellence in Comparative Planetology, Chinese Academy of Sciences, Hefei, 230026, China
3. College of Geoexploration Science and Technology, Jilin University, Changchun, 130026, China
4. Institute of Geochemistry, Chinese Academy of Sciences, Guiyang, 550002, China

Abstract: As the first lunar farside landed mission, China's Chang'E-4 (CE-4) has safely landed on the volcanically resurfaced floor of Von Kármán crater in South Pole-Aitken (SPA) basin. The CE-4 mission will conduct rover exploration by Yutu-2 similar to what Chang'E-3 did in the northern Imbrium basin. Here we present a comprehensive study of the composition, mineralogy, and chronology of the basaltic and nonmare units in this crater, with the intent to provide context for the compositional properties of the landing site. Compositional maps (FeO, TiO_2, Mg #, and Th) demonstrate the heterogeneous distribution of mare and nonmare materials. Using a spectral survey of 1899 small impact craters (<1km in diameter), we present a mineral map of high-Ca to low-Ca pyroxene ratio that is useful to untangle the effects of spectral mixing of these materials. Mare basalts with low-Ti composition and model ages of ~3.7 Ga and ~3.6Ga have distinct mineral compositions that differ from the Chang'E-3 landing site, including no olivine enrichment. The non-mare materials have Mg-rich pyroxene, and Finsen crater materials (and Finsen ejecta) have higher Ca/Fe ratios than other parts of the rim of Von Kármán crater, indicating compositional anomaly of the SPA basin. The Von Kármán mare basalts show obvious compositional heterogeneity and complicated mixing effects with non-mare ejecta from subsequent impacts (e. g., the Eratoshenian-aged Finsen impact). We also present a geological map of the landing site, which reveals the geologic history of the landing region and will help to guide the surface exploration

route of the CE - 4's Yutu - 2 rover. We also put forward some science questions to be answered by Yutu - 2, to aid mission planning and scientific research.
Keywords: Moon; Chang' E - 4; Von Kármán crater; Mare volcanism; Nonmare materials; Lunar exploration

1 Introduction

The enormous SPA basin is generally regarded as the largest (Wilhelms, 1987; Head et al., 1993; Garrick - Bethell and Zuber, 2009) and oldest impact basin (older than 4 Ga; e. g., Fassett et al., 2012) recognized on the Moon. Von Kármán crater (centered at 44.45°S, 176.3°E; diameter: 186km) is a degraded Nectarian or pre - Nectarian impact crater, flooded with mare basalt units on the floor (Stuart - Alexander, 1978; Yingst et al., 2017; Pasckert et al., 2018). January 3, 2019, China's Chang'E - 4 (CE - 4) spacecraft, consisting of a lander and rover (Yutu - 2) safely landed on the lunar farside, i. e., the scheduled landing site of the Von Kármán crater in the South Pole - Aitken (SPA) basin (Wu et al., 2017; Huang et al., 2018). After landing, CE - 4 released the Yutu - 2 rover to traverse around and perform detailed geologic investigations. Surface *in - situ* exploration missions (e. g., CE - 4) into the central portion of SPA basin have great potential to advance scientific understanding in key areas including knowledge of very large impact basin formation processes, late heavy bombardment history of the Earth - Moon system, composition of the lower crust and possibly upper mantle, and lateral and vertical compositional diversity of the lunar interior (Duke et al., 2000; National Research Council, 2007; Jolliff et al., 2017).

Here we report the geologic context, spectroscopic, compositional and chronological investigations of the mare basalt and nonmare units inVon Kármán crater, with an intent to recognize their scientific potential from a remote sensing view, which should be better understood during the CE - 4 mission's operations. In this study, we provide geologic context for the landing site within Von Kármán crater, focusing on the mineralogical and compositional properties of the mare basalts and nonmare materials that occur in the region of the CE - 4 lander and rover. The geologic map and compositional studies provide critical constraints and guidance for the CE - 4 Visible - Near Infrared Spectrometer (VNIS) and Lunar Ground Penetration Radar (LPR) studies in early 2019. We also provide context for science questions to be answered by Yutu - 2, to aid mission planning and scientific research.

2 Data and methods

We employ the Chang' E - 1 CCD images (120m/pixel), Lunar Reconnaissance Orbiter Camera (LROC) Wide Angle Camera (WAC) mosaic (100m/pixel) for regional geologic context and geomorphological analyses (Li et al., 2010; Robinson et al., 2010) to give the geologic context, geomorphology of SPA basin, Von Kármán crater and the CE-4 landing site. Topographic information of the studied region is derived from the merged digital elevation model (DEM) that is derived from the LRO Lunar Orbiter Laser Altimeter (LOLA; Smith et al., 2010) and KaguyaTerrain Camera (TC) data (SLDEM: 59m/pixel; Barker et al., 2016). We produced a slope map for the CE-4 landing region using SLDEM topography at a baseline of ~180m. Detailed results are given in section 3 and Figs. 1-2.

To evaluate the compositional properties of the mare basalts in Von Kármán crater, we use Kaguya Multi-band Imager (MI) data (20m/pixel; Ohtake et al., 2010) to map the TiO_2 and FeO distributions in Von Kármán crater with algorithms from Otake et al. (2012) and Lemelin et al. (2015), respectively. Those algorithms are derived based on the "ground truth" from returned Apollo and Luna samples, in combination with the multispectral imaging data sets of corresponding sampling sites. They are as follows:

$$TiO_2(\text{wt. }\%) = 0.72 \times \theta_{Ti}^{14.964} \quad (1)$$

$$\theta_{Ti} = \arctan(((R_{415}/R_{750}) - 0.208)/(R_{750} + 0.108)) \quad (2)$$

$$FeO(\text{wt. }\%) = (1.0708 \times \theta_{Fe2}) - 0.3986 \quad (3)$$

$$\theta_{Fe2} = 0.0656e^{(3.6681 \times \theta_{Fe1})} \quad (4)$$

$$\theta_{Fe1} = -\arctan(((R_{950}/R_{750}) - 1.39)/(R_{750} - 0.04)) \quad (5)$$

The Mg# (mole percent MgO/(MgO+FeO)) map of the region is derived from Chang'E-1 Imaging Interferometer (IIM; 200m/pixel) data. Similar to the methods of FeO and TiO_2 mapping using Kaguya MI data, we employ Mg# for Apollo and Luna sample analyses as ground truth to map Mg#. The IIM data was binned into 4×4 pixels to suppress the random noise and improve the signal-to-noise ratio. Thorium (Th) content distribution map of the CE-4 landing site from Lunar Prospector has a spatial resolution of 0.5°/pixel from Lunar Prospector data (Prettyman et al., 2006). Detailed results are given in section 4 and Fig. 3.

We use orbital hyperspectral imaging data from Chandrayaan-1 Moon Mineralogy Mapper (M^3) data (Green et al., 2011) to evaluate the spectral and mineralogical variations of the mare units and the nonmare materials. Moon Mineralogy Mapper (M^3), a guest payload on Chandrayaan-1, has been employed to detect lunar global and regional mineralogy with high spatial and spectral resolution (Green et al., 2011; Boardman et al., 2011). We select ten scenes M^3 data that are of optical period OP2C according to the range of Von Karman in the NASA PDS Geosciences Node (http://ode.rsl.wustl.edu/). OP2C data is collected with a spatial resolution of ~280m/pixel at a spacecraft altitude of ~200km (Cheek et al., 2011). Ground truth value correction and geometry correction are employed on the data. We then mosaic all the M^3 data (Tab. 1) to give a full view of the Von Kármán crater. We acquired the spectra of small craters with average values of 2×2 pixels. All the spectra extracted are only shown from a single optical period

(OP2C) for direct spectral comparisons. The spectra are smoothed by Savitzky - Golay (SG) convolution method. The continuum of each individual spectrum is removed using two straight lines with three points defined by the local maximum reflectance between 0.66 ~ 0.91μm and 1.35 ~ 1.82μm, and the right endpoint position at 2.62μm. Continuum removal of the spectra lead to more prominent absorption features to help identify the mineral chemistries.

Tab. 1　Information of Moon Mineralogy Mapper (M^3) datasets used in this study

Data File Name	Date	Optical Period	Orbit Altitude/km	Resolution/ (m/pixel)
M3G20090622T082242_V01_RFL	2009/6/22	OP2C	200	280
M3G20090622T124051_V01_RFL	2009/6/22	OP2C	200	280
M3G20090623T012541_V01_RFL	2009/6/23	OP2C	200	280
M3G20090719T160300_V01_RFL	2009/7/19	OP2C	200	280
M3G20090719T200620_V01_RFL	2009/7/19	OP2C	200	280
M3G20090720T003411_V01_RFL	2009/7/20	OP2C	200	280
M3G20090720T043741_V01_RFL	2009/7/20	OP2C	200	280
M3G20090720T090521_V01_RFL	2009/7/20	OP2C	200	280
M3G20090816T005433_V01_RFL	2009/8/16	OP2C	200	280
M3G20090816T045833_V01_RFL	2009/8/16	OP2C	200	280

Integrated Band Depth (IBD) is a useful parameter to evaluate the properties of the whole absorption band of mafic minerals. Following the method by Thiessen et al. (2014), we created color composite images using IBD1000 (red channel, IBD from 789 to 1308nm), IBD2000 (green channel, IBD from 1658 to 2458nm) and R750 (blue channel, reflectance of 750nm). Note that here we use the reflectance of 750nm instead of the reflectance of 1589nm of the M^3 science team (e. g., Besse et al., 2011) for the blue channel due to its limited absorptions of mafic minerals at 750nm (Thiessen et al., 2014). On the contrary, the 1250nm absorption feature of plagioclase and 2000nm absorption feature oflow - Ca pyroxene may affect the 1589nm reflectance from their endmember spectra (Fig. 1 from Pieters et al., 2014). In order to evaluate the spatial variations of mineralogy in Von Kármán crater, we conduct a systematic spectral survey of small, fresh craters superposed on the basaltic and nonmare units. We collect spectra of 1899 small craters less than 1km in diameter, which present fresh materials with sharp spectral absorption peaks. We estimate the mineral modes of these spectra using the Modified Gaussian Model (MGM) deconvolution method (Sunshine and Pieters, 1993, 1998). Detailed results are given in section 5 and Fig. 4 - 5.

We estimate the absolute model ages of mare basaltic units and Finsen crater using crater size - frequency distribution (CSFD) measurements. We count all superposed impact craters ≥200m in diameter using the CraterTools extension in ESRI's ArcGIS environment (Kneissl et al., 2011) on Kaguya TC images (~10m/pixel). During the

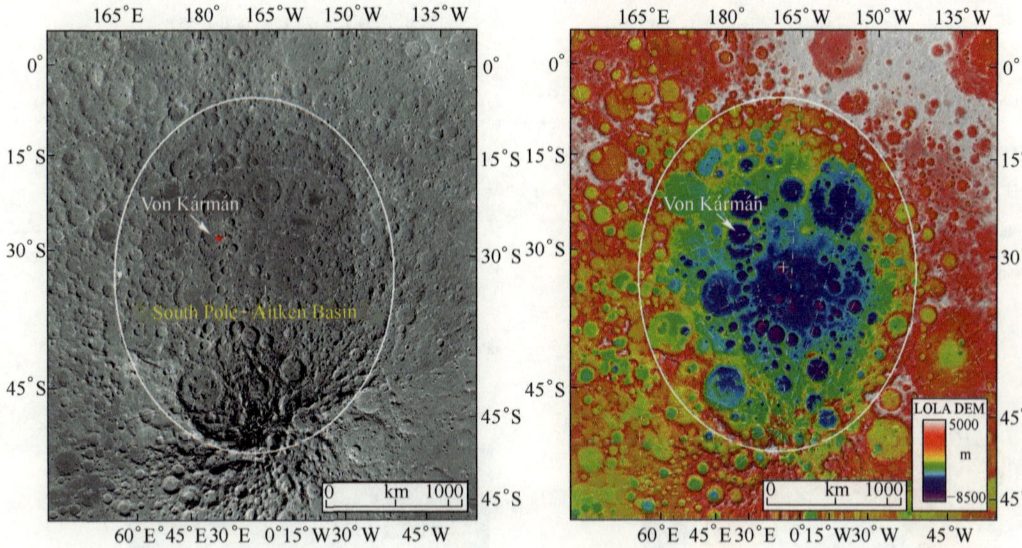

(a)　　　　(b)

Fig. 1　The context of Von Kármán crater within the South Pole - Aitken (SPA) basin. (a) LROC WAC image mosaic and (b) LOLA DEM topography (128 pixel/degree) overlaid on WAC images. The white outline in panels (a) and (b) represents the outer ellipse of SPA basin. The location of Von Kármán crater is pointed out by the white arrow. The Chang'E - 4 landing site is shown by the red star. The white cross marks in (b) the ellipse center of SPA basin (Garrick - Bethell and Zuber, 2009). The projections are stereographic centered at 191°E, 53°W, and north is up. (For interpretation of the references to color in this figure legend, the reader is referred to the Web version of this article.)

crater counting investigations, special care has been taken to eliminate contamination by secondary impact craters and endogenous pits according to their morphologic characteristics (e. g., Shoemaker, 1962; Oberbeck and Morrison, 1974). We calculate absolute model ages by fitting the crater counting results on the basis of the lunar chronology function (CF) and production function (PF) of Neukum et al. (2001) using CraterStats software (Michael and Neukum, 2010; Michael et al., 2016). Detailed results are given in section 6 and Fig. 6.

3　Geologic context of Von Kármán crater

The Von Kármán crater is centered at 44.45°S, 176.25E (based on LOLA 2011 control network), located in the central region of SPA basin (Fig. 1). The exact landing site of CE - 4 has been confirmed by multisource data (e. g., digital orthophoto map of Chang'E - 2, LROC NAC images, CE - 4 descent camera images and monitoring camera images) and is given as the red star in Fig. 1 at 45.457°S, 177.588°E (Di et al., 2019). The Von Kármán crater shows degraded topographic and morphologic patterns, and geological mapping studies have designated it as a Nectarian - aged or pre - Nectarian

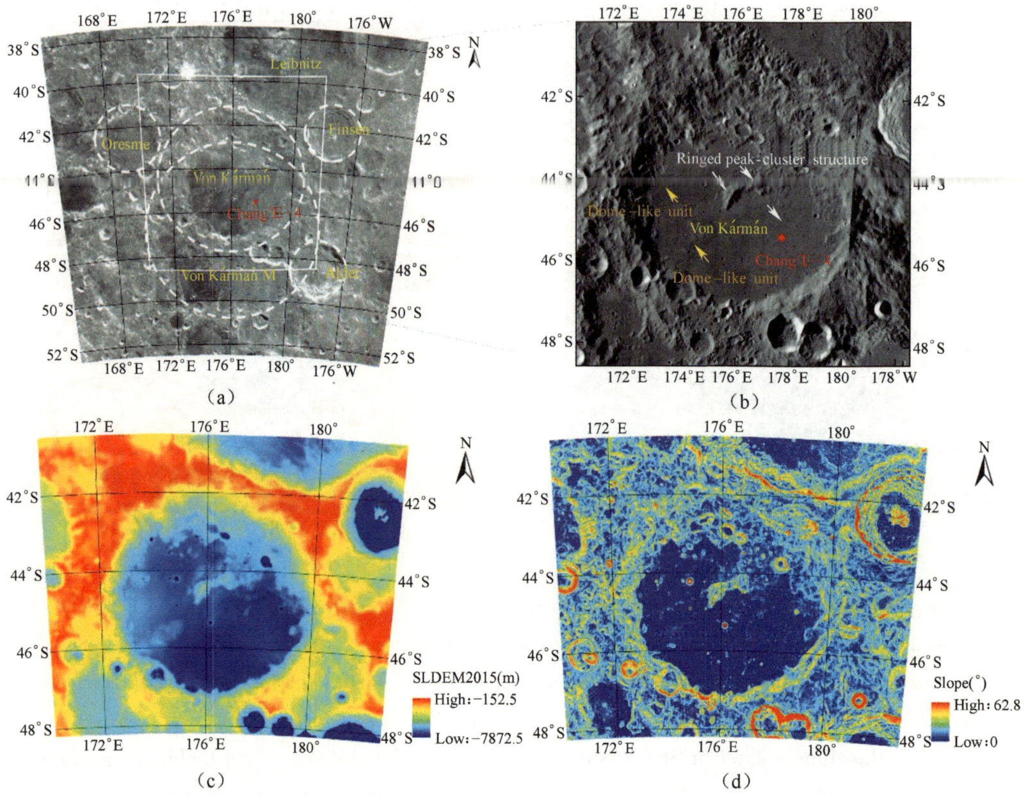

Fig. 2　The regional context of Von Kármán crater. (a) Chang'E-1 CCD image; (b) LROC WAC image, the three white arrows point to the ringed peak-cluster and the two white arrows point to the dome-like structures, respectively; (c) SLDEM2015 topography (~60m horizontal resolution) overlaid on LOLA shaded relief image; (d) SLDEM2015-derived slope map (baseline ~180m) overlaid on LOLA shaded relief image. The dotted white circles in panel (a) marks the rim locations of Von Kármán, Von Kármán M, Finsen, Alder, and Oresme craters

crater (Stuart-Alexander, 1978; Yingst et al., 2017).

Regionally, the Von Kármán crater is encircled by several other post-SPA impact basins and craters; these basins/craters have complex stratigraphic relationships (Fig. 2a and b; Qiao et al., 2019). The Von Kármán crater clearly superposes (and is thus younger than) the pre-Nectarian Von Kármán M cater. Geomorphologic investigations show the northern portion of the rim structures of Von Kármán M crater has been largely obliterated and buried during the Von Kármán impact event. The Von Kármán M crater has an estimated diameter of ~245km (Head et al., 2010). A much larger impact crater, Leibnitz, with a diameter of ~240km, lies very close to the north rim of Von Kármán crater. It appears that this larger crater postdates Von Kármán crater, and the rugged and knobby terrains of the northern interior of Von Kármán are probably ejecta deposits from the Leibnitz impact crater.

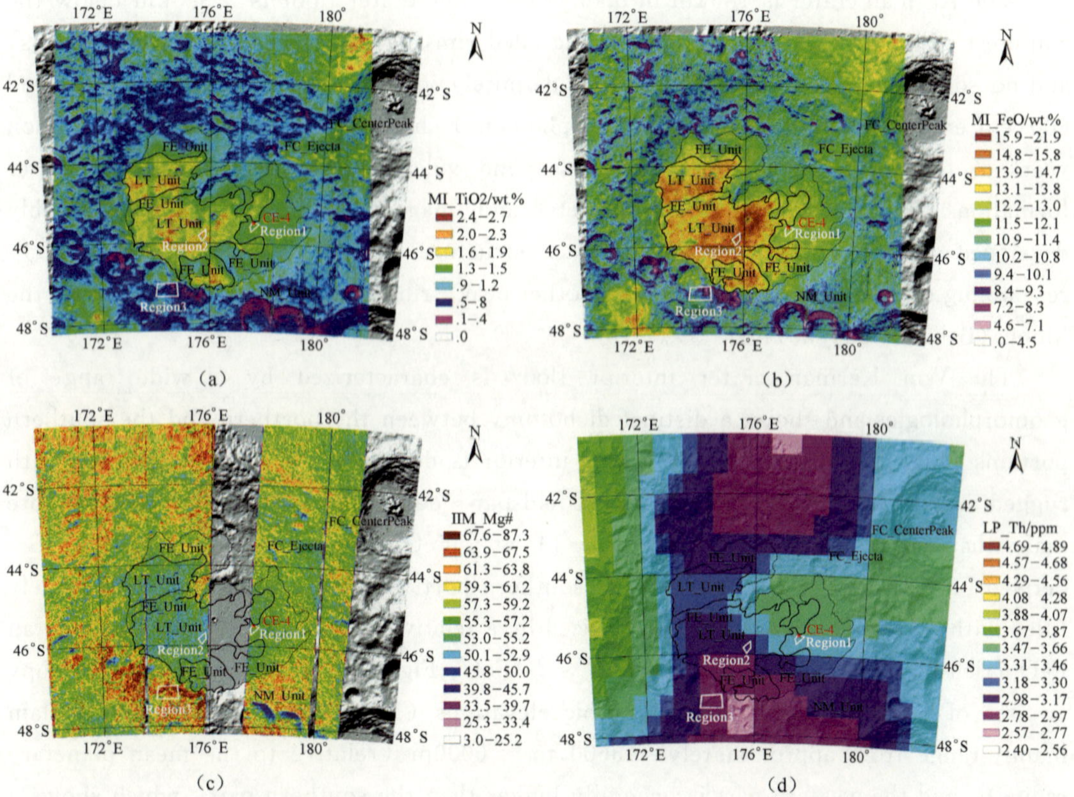

Fig. 3　Chemical compositions of Von Kármán crater. (a) TiO_2 map and (b) FeO map using Kaguya Multiband Imager (MI) data; (c) Mg# ($100\times MgO/(MgO+FeO)$) map using Chang'E-1 IIM data, and (d) Th maps from Lunar Prospector data. The basemap is LOLA shaded relief image. Black polygons are the unit division results: LT: low-titanium mare unit, FE: Finsen ejecta unit

The most prominent large impact crater in the regional context is the 73-km-diameter Finsen crater to the northeast of Von Kármán crater. Earlier geologic mapping has interpreted it as an Eratoshenian-aged crater (Stuart-Alexander, 1978). On optical images and reflectance maps, extensive ray patterns and secondary craters radial to Finsen crater center are clearly observed. Imaging spectrometer mapping analyses show that, however, these bright ray patterns are mainly due to the compositional difference between highlands-rich Finsen ejecta (brighter) and local mare basalts (darker), not due to the presence of fresh, high-albedo materials (Hawke et al., 2004). Thus, these ray patterns associated with Finsen crater are probably not reliable indications for the Copernian crater ages (Hawke et al., 2004). In addition, Finsen is also not registered as a Copernian crater by several prior investigations (Werner and Medvedev, 2010; Ravi et al., 2016; Mazrouei et al., 2019). The mare deposits within Von Kármán are also extensively overprinted by ejecta materials from the Finsen crater (Pasckert et al., 2018) (see more in sections 4-5).

Von Kármán crater is 186km in diameter and the crater floor is ~3.5km below the rim crest (Fig. 2c). This crater shows degraded rim, wall terraces and ejecta terrains, and no solidified impact melts have been definitely identified in the crater interior and adjacent exterior; all these degraded morphological characteristics reflect several billion years of topographic degradation processes and volcanic resurfacing after the crater formation. High - resolution LRO altimetry and image data show this basin probably developed a peak - ring or peak - cluster structure (poorly preserved) (Fig. 2b), resembling central structures observed in other peak - ring and peak - cluster basins on the Moon (Baker et al., 2011).

The Von Kármán crater interior floor is characterized by a wide range of geomorphologies and shows a distinct dichotomy between the northern and the southern portions (Fig. 2b). The northern crater interior is dominated by ejecta deposits, with rugged surface textures (Fig. 2b) and a feldspar - dominated composition that is more mafic than typical lunar highlands (10—14 wt% FeO, using Clementine UV - VIS - derived FeO (Lucey et al., 2000). The southern portion of the crater interior is largely filled with mare lava deposits, characterized by a relatively flat surface (Fig. 2d) and an FeO - rich basaltic composition (13—18 wt% FeO; Fig. 3b). These mare plains occupy an area of $1.063 \times 10^4 km^2$. Topographic elevations (Fig. 2c) of the mare lava plain mainly range from approximately −5600 to −6000m (relative to the mean planetary radius), and the western part is generally higher than the southern part, which shows a topographic decline of ~350m across a distance of ~80km. Two bulgeor dome - like structures, ~15—40km in diameter and up to ~350m high, occur in the western margin of the interior mare units (Pasckert et al., 2018), which may represent the source vent areas for the later volcanic flows in the western Von Kármán floor.

4 Chemical compositional properties of mare basalts and surrounding non - mare materials

As shown in Fig. 3a and b, many fresh craters excavated relatively FeO - and TiO_2 - richer underlying basaltic materials than those of surface materials, indicating heavy reworking and impact gardening of the surface basaltic materials. The contamination and mixing of Finsen ejecta with local mare materials is evident by light - colored linear patterns (Fig. 2a) with relatively low FeO and TiO_2 values pointing toward Finsen crater (Fig. 3 a, b). Using TiO_2 content as the discriminator for lava units, we divide the mare region in Von Kármán crater into two units, i. e., Low - Ti unit (LT, 3 wt % > TiO_2 > 1.5wt%) and Finsen ejecta unit (FE, TiO_2 < 1.5wt%), as shown by the polygons in Fig. 3.

Except mare basalts in the Von Kármán crater, the nonmare materials are ubiquitous from nearby terrains of SPA interior floors, which are derived from SPA impact event rather than the later mare volcanisms that formed the mare basalts. For those nomare materials (NM units) surrounding the mare basalts in Von Kármán crater, we find they generally have lower FeO and TiO_2 contents, except for the ejecta of Finsen crater, which seems to be slightly richer in FeO and TiO_2 than other nonmare regions (Fig. 3a and b). The Mg# (mole percent MgO/ (MgO/FeO)) map (Fig. 3c) derived from Chang'E - 1 Imaging Interferometer (IIM) data provide the detailed heterogeneous Mg# distribution of this region. We find that the nonmare materials from Finsen impact ejecta and Von Kármán crater rim (represented by Regions 1 and 3 in Fig. 3, which include the Chang'E - 4 landing site) are slightly more magnesian (Mg# = ~56±2 for Region 1 and Mg# = ~64±4 for Region 3) than the mare materials (represented by Region 2 in Fig. 3, Mg# = ~52±2). Lunar Prospector gamma - ray data for Th (Fig. 3d) (Prettyman et al., 2006) indicates that the ejecta of Finsen crater has a higher Th content (~3. 6ppm) than that of mare basalts (~2. 8ppm).

The mare surface in Von Kármán crater is relatively low in FeO (~13—18 wt %) and TiO_2 (~1—3 wt %) in comparison with basalt at the Chang'E - 3 landing site (i. e., ~22. 8 wt%FeO and ~5. 0 wt%TiO_2) (Ling et al., 2015). It is clear that the typical mare basalt (Region 2) belong to a low - Fe (15. 3±0. 6 wt%) and low - Ti (2. 0±0. 3 wt%) endmember, while the Mg composition suggest they are relatively high in Mg# values (~52±2). Compared to the various mare basalt types of Apollo and Luna samples (Fig. 4), CE - 4 mare basalt stands out as a special type of basalt with similar Fe, Ti contents and Mg# values with the very high K basalt and high aluminum basalt from Apollo 14 (e. g., Apollo 14053, 14168, 14305, 14321 samples) and Luna 16 missions. The characteristic chemical compositions of CE - 4 basalt may imply a unique view of farside volcanism.

5 Spectral properties and mineralogical composition

As shown in Fig. 5a, mare basaltic materials show hues between yellow and red, indicating they are rich in mafic minerals. Ejecta from Finsen crater make the mixed mare materials appear bluish in color, suggesting the contamination of feldspathic materials. We select seven small, fresh craters or steep slopes (coordinates of locations are given in Tab. 2) with sharp spectral features among different units as shown in Fig. 5b, and extract their spectra. As shown in Fig. 5c and d, we find that the LT unit generally has longer 1 and 2μm absorption centers than those of the FE unit, indicating the existence of more Ca - and Fe - rich pyroxenes. The central peak of Finsen crater shows its characteristic spectral absorption features approaching ~900nm and ~1900nm, indicating more magnesian compositions than its ejecta (i. e., FC _ ejecta, blue line), consistent with the findings of Moriarty and Pieters (2018). Similarly, nonmare material also show spectral features, indicating a more Mg - rich composition of the pyroxene.

The result of spectral survey of small craters suggests that the LT unit has more high -

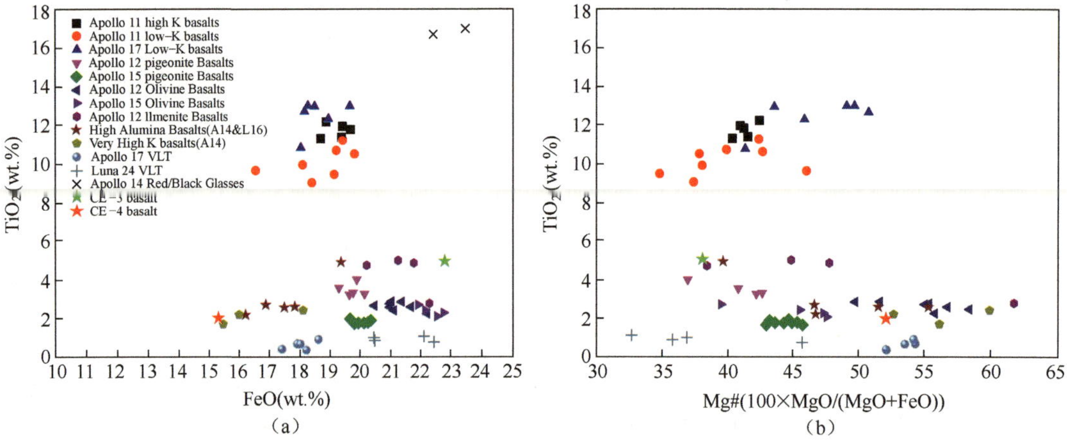

Fig. 4　Comparison of chemical variations between Chang'E-4 mare basalt and other mare basalts of Apollo and Luna samples (Papike et al., 1998) as well as Chang'E-3 mare basalt (Ling et al., 2015). (a) FeO versus TiO_2; (b) Mg# (100×MgO/(MgO+FeO)) versus TiO_2

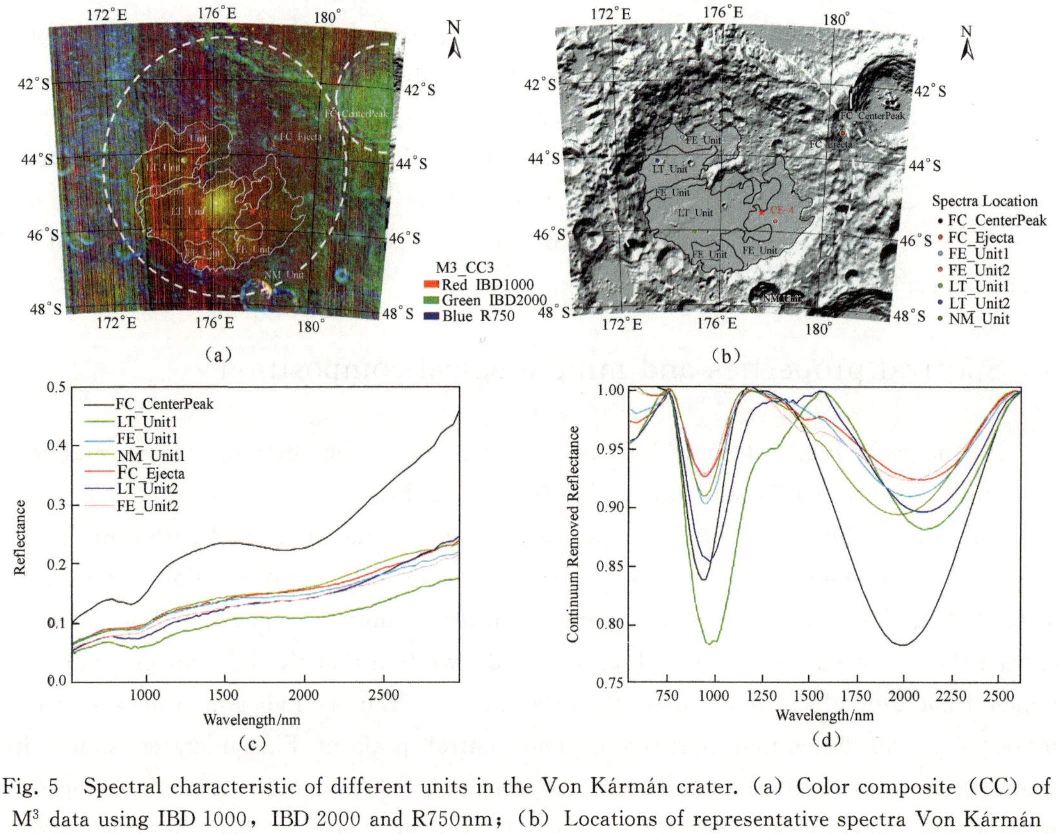

Fig. 5　Spectral characteristic of different units in the Von Kármán crater. (a) Color composite (CC) of M^3 data using IBD 1000, IBD 2000 and R750nm; (b) Locations of representative spectra Von Kármán crater; (c) Representative spectra of Von Kármán crater; (d) Continuum-removed representative spectra of Von Kármán crater. Points in panel (b) indicate area where the representative spectra are taken out of LT, FE and nonmare (NM) units. The basemap is LOLA shaded relief image. (For interpretation of the references to color in this figure legend, the reader is referred to the Web version of this article.)

Ca pyroxene (HCP) relative to low - Ca pyroxene (LCP) than the FE unit. We calculate the mineral absorption centers using MGM, which present quantitative mineralogy information than conventional spectral profile analyses. The spatial distribution of these results is given in Fig. 6a. The volume percentage ratio of HCP and LCP (HCP/LCP) is estimated using their band - strength ratios of 1 and 2μm bands from MGM deconvolution of M^3 spectra. Using a Kriging interpolation technique, we obtain the mineral (HCP/LCP ratio) map shown in Fig. 6b, which provides better discrimination of the compositional boundaries.

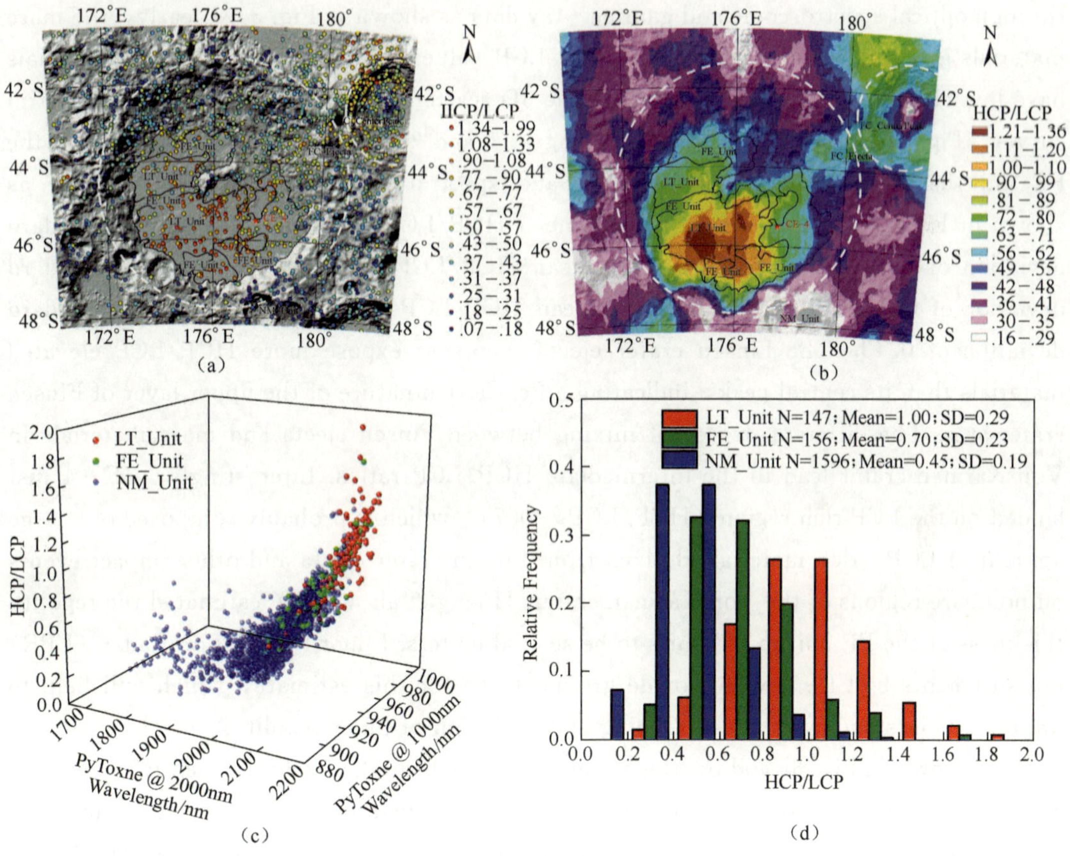

Fig. 6　Spectral survey of small craters of Von Kármán crater. (a) Mineral ratios of High - Ca pyroxene (HCP) and Low - Ca pyroxene (LCP) derived from MGM method; (b) Mineral map based on the interpolation of (a) using Kriging interpolation technique; (c) Pyroxene VNIS peak positions and HCP/LCP ratios of Von Kármán crater materials; (d) Histogram of HCP/LCP ratios derived from the spectral survey of small craters. The basemap in (a) and (b) is LOLA shaded relief image

Tab. 2　Coordinates of locations of representative spectra in Von Kármán crater area

Spectrum ID	Latitude/(°)	Longitude/(°)	Unit Name
Spec _ 01	−42. 360178	182. 133365	FC _ CenterPeak
Spec _ 02	−43. 283169	180. 551836	FC _ Ejecta
Spec _ 03	−46. 046955	174. 967225	LT _ Unit1

Continued Tab. 2

Spectrum ID	Latitude/(°)	Longitude/(°)	Unit Name
Spec _ 04	−44.149121	173.561998	LT _ Unit2
Spec _ 05	−43.532065	174.516101	FE _ Unit1
Spec _ 06	−45.720279	178.145839	FE _ Unit2
Spec _ 07	−48.131462	177.040111	NM _ Unit

This mineral map has advantages of probing the subsurface mineral variations and minimizing the effects of impact mixing in the upper meter of soils (such as those obtained through optical spectroscopic and gamma - ray data as shown in Fig. 3). Clearly, the mare materials generally have the greatest HCP/LCP values, whereas the nonmare materials have low HCP/LCP values. As given by the 3D scatter plot of the results of the spectral survey (Fig. 6c), the LT unit has the longest 1 and 2μm absorption feature, while the FE unit has intermediate compositional and spectral characteristics. Statistically, as shown in Fig. 6d, the LT units have a mean HCP/LCP value of 1.0, with a standard deviation of 0.29; the FE unit has a mean HCP/LCP value of 0.7, with a standard deviation of 0.23; the NM unit has a mean HCP/LCP value of 0.45, with a standard deviation of 0.19. The Finsen crater ejecta seems to expose more HCP/LCP elevated materials than its central peak, indicating a Fe/Ca rich nature of the upper layer of Finsen crater area. The different degree of mixing between Finsen ejecta and mare materials in Von Kármán crater lead to the intermediate HCP/LCP ratios. Interestingly, CE - 4 just landed on the LCP rich region (HCP/LCP ~0.7), which is probably composed of a large amount of LCP - rich materials derived from Finsen crater ejecta and other impact events on nonmare regions of the Von Kármán crater. Huang et al. (2018) estimated the regolith thickness of the FE unit in LT units to be several meters. Lunar Penetration Radar (LPR) measurements by CE - 4 will provide ground truth for this estimate, which will help to improve understanding of the gardening and evolution of lunar regolith.

As shown in Fig. 6a and b, the nonmare materials are generally Mgrich for pyroxene. Specifically, pyroxene on the southern rim of Von Kármán crater is even more Mg - rich than those pyroxene in the ejecta of Finsen crater. This is consistent with the Mg# map (Fig. 3c) which showing the more Mg - rich southern rim of the crater than the Finsen ejecta and the mare basalts. Interestingly, the central peak of Finsen crater exhibits low HCP/LCP values, indicating a more Mg - rich pyroxene composition. The northern and southern rim of Von Kármán crater appear to expose more Mg - rich pyroxenes. This observation is consistent with Moriarty and Pieters (2018), who suggested the interior of SPA could be divided into three composition zones as SPA compositional anomaly (SPACA), Mg - Pyroxene Annulus, and Heterogeneous Annulus. According to Moriarty and Pieters (2018), Finsen crater and its ejecta belong to the SPACA unit, whereas other nonmare regions in Von Kármán crater belong to the Mg - Pyroxene Annulus. In addition, olivine was not observed in the spectral survey, in agreement with previous spectral studies of materials in SPA basin which is expected to have excavated the lunar upper mantle materials onto the surface (Yamamoto et al., 2012; Melosh et al., 2017), supporting the hypothesis that the lunar upper mantle was probably enriched in low-calcium pyroxene (LCP) but not olivine. As have been noted by Moriarty and Pieters (2018), the origin of Mg-rich pyroxene in SPA and the nature of unusual SPACA surface materials are unanswered questions which are keys for unraveling the complex geologic history of SPA basin. In this sense, Von Kármán crater is a good place to check the compositional and geologic differentiation of the SPA basin, if these relatively high Ca/Fe nonmare pyroxenes might be encountered by the CE-4 rover and detected with its VNIS payload.

In order to present detailed geological and compositional context and provide remote sensing observation constraints for Yutu-2 operations, we show a more close view of the Chang'E-4 landing site (Fig. 7). As shown in Fig. 7a and b, Von Kármán mare basalt is relatively low in FeO (~13—18 wt %) and TiO_2 (~1—3wt %). Relatively FeO and TiO_2-rich underlying basaltic materials are exposed to surface, indicating the degree of regolith mixing and gardening are high. CE-4 landing area is a topographically very flat region on the mare plain (Fig. 7c), which is on an average elevation of −5928±6m. As

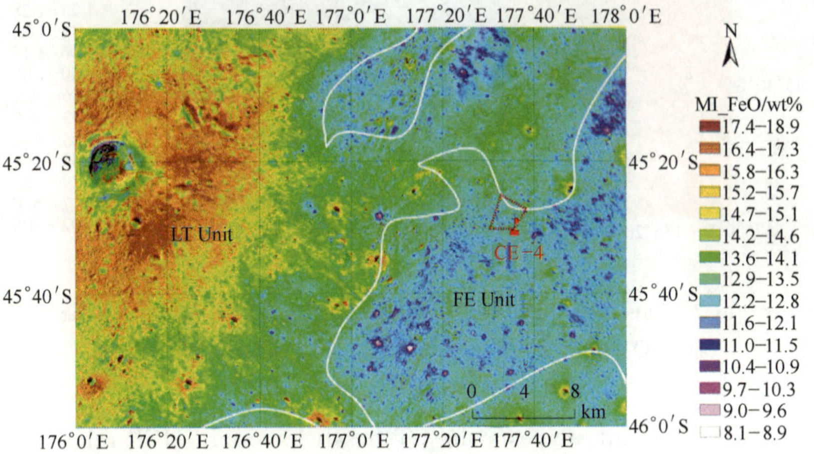

(a)

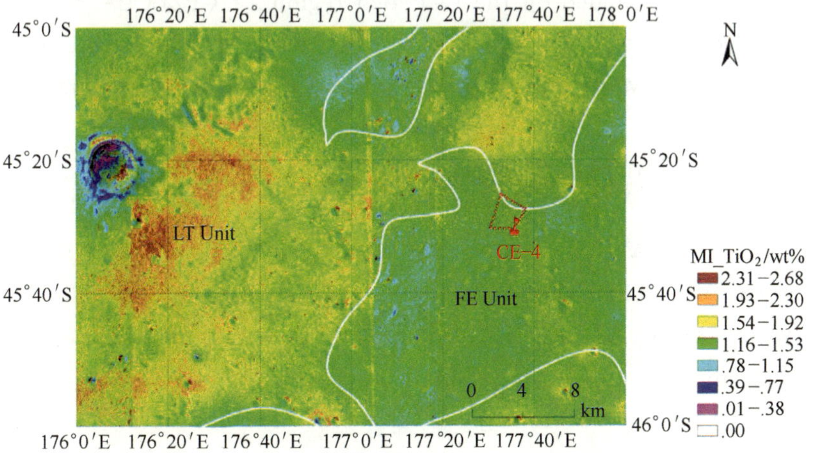

(b)

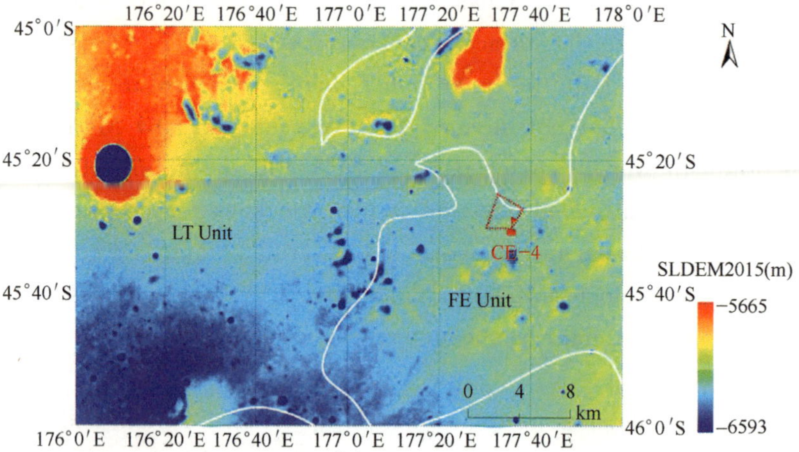

(c)

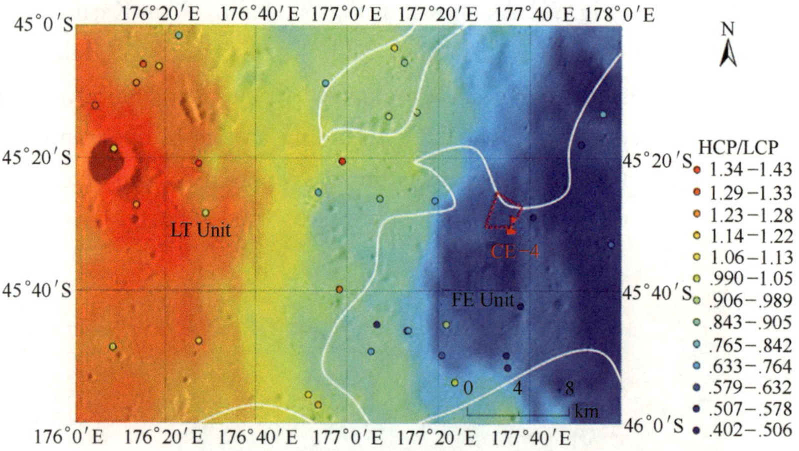

(d)

Fig. 7　Close view of Chang'E-4 landing site in Von Kármán crater
(a) FeO map; (b) TiO_2 map; (c) SLDEM2015 topography;
(d) HCP/LCP mineral ratios.

shown in Fig. 2d, the result suggests the LT basalt has more HCP relative to LCP than the FE unit. CE-4 just landed on the LCP rich region (HCP/LCP=~0.7), which is probably composed of a large amount of LCP-rich materials derived from Finsen crater ejecta and other impact events on nonmare regions of the Von Kármán crater.

6　Chronology of mare basalts

The chronology of Von Kármán crater basalts is conducted by CSFD using areas shown in Fig. 8a. We separate the LT region into two areas (Fig. 8a, LT_1 and LT_2) for age analysis according to their different mineral distributions (i.e., HCP/LCP) as indicated in Fig. 5b.

There are two previous chronological studies of mare basalts in Von Karman crater floor using CSFD measurements. Haruyama et al. (2009) obtained a model age of 3.35 Ga for these mare deposits, whereas a recent study by Pasckert et al. (2018) derived an age of 3.15 Ga and inferred a buried lava flow to be 3.75 Ga old. Here we contribute to the chronological investigation using our updated mare basalt unit division (section 4). As shown in Fig. 8c, The CSFD curve of the LT _ 1 units is featured by an characteristic deflection (knee) at ~1km diameter, which can be interpreted to the presence of two mare flow units separated in time: an older mare flow unit and a younger mare flow resurfacing unit (preferentially covering smaller craters (with crater depths smaller than the resurfacing flow thickness) and resulting in deficit of smaller crater) (Hiesinger et al., 2002). The age of LT _ 1 is determined to be ~3.38±0.03 Ga by fitting craters from ~300m to ~1km in diameter for the younger resurfacing lava flows, similar to that of Haruyama et al. (2009), and the larger craters (>1km) yield an age of ~3.72 Ga for the buried lava flows. The depth of craters (approximately 1/10 of the crater diameter; Melosh, 1989) of the deflection diameter (~1km) can be used to constrain the depth of the older ~3.72Ga buried lava flow, i. e., >100m. The more recent volcanic event may have a higher HCP/LCP ratio than the earlier volcanic event (~3.72Ga), indicating different composition of their mantle sources. The age of LT _ 2 (Fig. 8d) is dete rmined

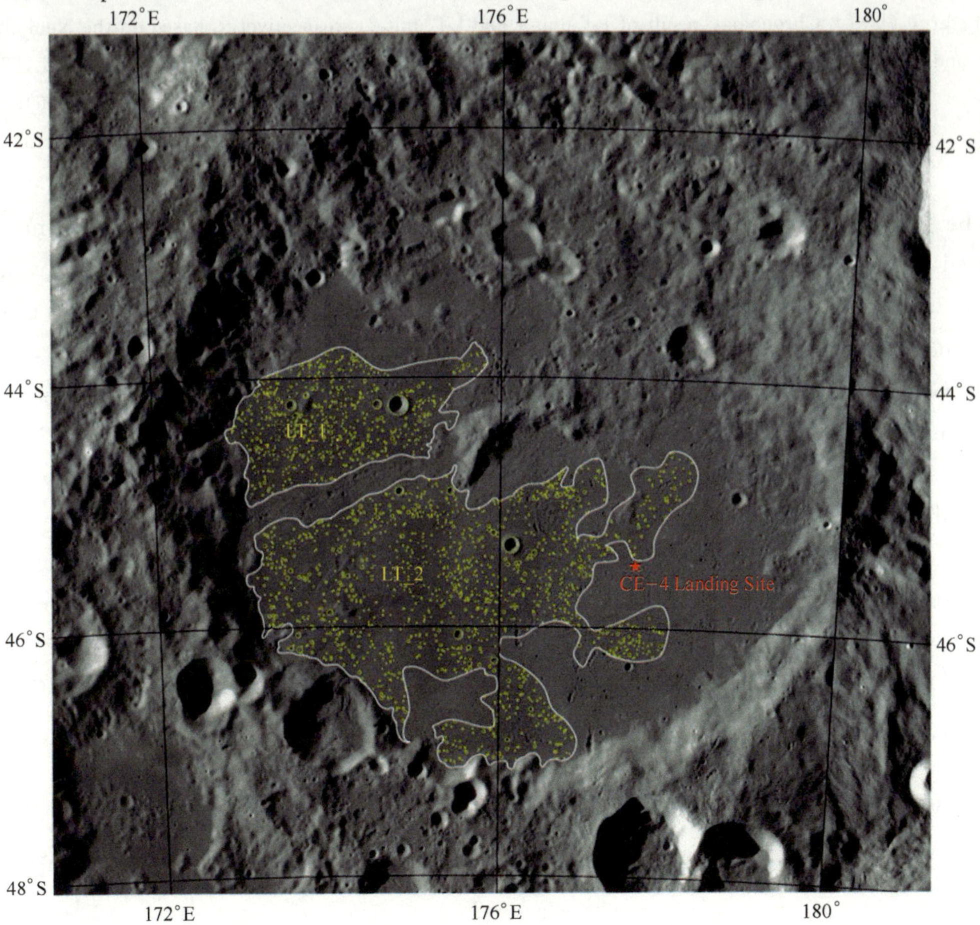

(a)

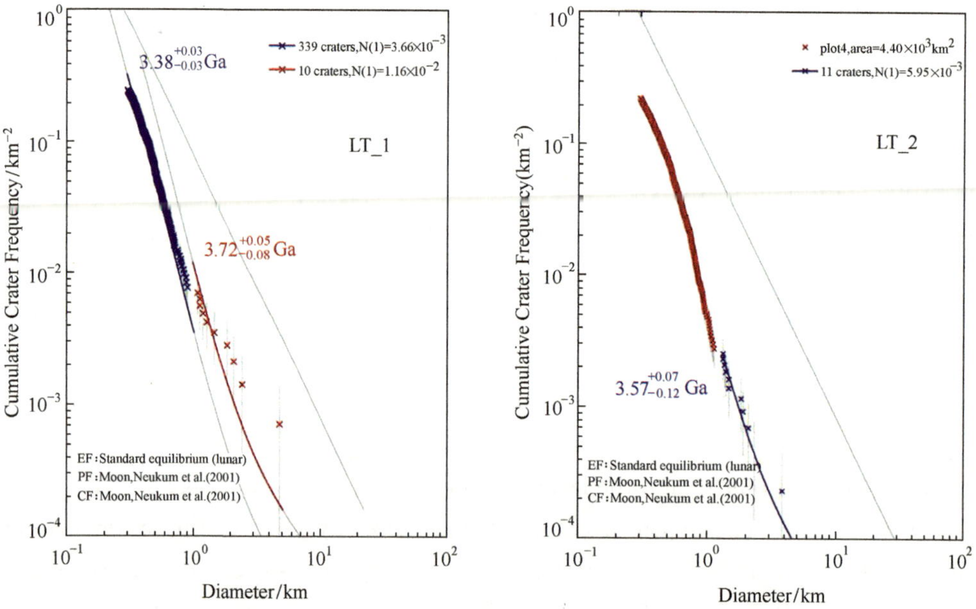

(b)　　　　　　　　　　　　(c)

Fig. 8　Chronology of Von Karman crater basalts and Finsen crater. (a) Locations of crater - counting area for LT unit 1 (LT _ 1), LT unit 2 (LT _ 2), yellow points are the counted impact craters >200m in diameters; (b - c) Chronology result of LT unit 1 and LT unit 2, respectively, based on the Neukum lunar PF and CF, using the CraterStats software package (Michael and Neukum, 2010; Michael et al., 2016). The gray line on the right is the lunar equilibrium function (EF) curve from Trask (1966). (For interpretation of the references to color in this figure legend, the reader is referred to the Web version of this article.)

to be ~3. 57 (+0. 07/−0. 12) Ga by fitting craters >1. 3km. The Finsen crater has been dated to be ~3. 6Ga through crater population study of Finsen crater floors (Ivanov et al., 2018). However, this reported age will place Finsen even older than parts of the Von Kármán floor maria (3. 35Ga, Haruyama et al., 2009; 3. 15Ga, Pasckert et al., 2018; 3. 38Ga, this study), which is in consistent with the pervasively observed superposition of Finsen ejecta materials on the Von Kármán floor maria (Figs. 2a, 3a and 3b). We suggest the calculated CSFD age of Finsen crater is probably biased by many factors including target properties (different from mare surface), hummocky topography and its effects on mass wasting and crater identification. We refer to the geologic mapping results by Stuart - Alexander (1978) and interpret Finsen it as an Eratoshenian - aged crater.

7　Geologic history of CE - 4 landing site and science questions to be answered by Yutu - 2

We create a new geologic map of Von Kármán crater region (Fig. 9), which illustrate the regional stratigraphic characteristics and geological evolution history as follows:

1. The South Pole－Aitken basin was formed more than four billion years ago（e. g., Hiesinger et al., 2012; Garrick－Bethell and Miljković, 2018).

2. Multiple large impact craters formed during the post－SPA period, creating a complicated stratigraphic sequence. The impact event that formed the 186－km－diameter Von Kármán crater occurred during the Nectarian period（Ncf: crater floor units, Ncw: crater wall units and Ncrh: hummocky rim units), which has largely destroyed the northern part of the previously formed 245－km－diameter Von Kármán Mcrater. Then, the 240－km－diameter Leibnitz crater formed north of Von Kármán crater and destroyed part of the northern rim structure of Von Kármán crater and ejected material into the northern part of the Von Kármán crater floor（Ncf units). These ancient, pre－mare crater related materials are generally characterized by very low FeO and TiO_2 contents, while a slightly more magnesian composition（Fig. 3a, b, c).

3. During the Imbrian period, regional lava－infilling events took place within the SPA basin interior, which generated small mare plains（compared with extensive mare plain within nearside basins), including those within the Von Kármán interior and southern part of the Von Kármán M crater floor（Im1 and Im2: Imbrian mare and Imd: Imbrian dome－like units). The Von Kármán floor surface mare deposits are dated to be ～3. 4—3. 6Ga, with buried mare flows to be ～3. 7 Ga lava flow at a depth of ＞100m. Mare basalts emplaced at the Von Kármán crater floor are characterized by a Ca－and Fe－rich pyroxene composition, with relatively low iron and titanium contents, but relatively higher magnesium.

4. During the post－mare period, the major surface modification processes have been mainly degradation due to ongoing impact cratering, which formed many smaller craters（Ic, Icw, Icr: Imbrian crater materials and Ec: Eratoshenian crater materials). One of most prominent features formed during this period is the Eratoshenian－aged Finsen crater（Ecr: Eratoshenian crater rim materials and Ccw: Eratoshenian crater wall materials), whose ejecta materials extensively overprinted the pre－emplaced mare deposits within Von Kármán crater floor, generating many bright secondary crater chains. These Finsen ejecta materials are characterized by elevated FeO, TiO_2, and LCP content than typical lunar highlands.

Within this geologic context, we propose the following scientific questions related to the traverse and in－situ exploration by the Yutu－2 rover:

(1) Composition and origin of nonmare materials. From the landing site, Yutu－2 will be able to measure in－situ the visible and near－infrared spectra of materials mainly composed of Finsen ejecta（perhaps with just a small percentage of basaltic material). To－be－explored questions include: What are the mineral chemistry and assemblage of Finsen ejecta? How much might the surface represent the SPA Composition Anomaly（SPACA) and indicate its origin（Moriarty and Pieters, 2018)? Could Yutu－2 help to find some clues relating to lunar mantle components（i. e., does it contain olivine or Mg－

rich LCP)?

(2) Composition and origin of lunar farside mare basalts. If Yutu - 2 traverses to the west, it will encounter HCP - rich Von Kármán mare basalts. To - be - explored questions include: What are the mineral chemistry and assemblage of these basalts? How do they differ from nearside mare basalts and what are the implications for lunar volcanism?

(3) Mixing effects of lunar soils and their evolution. As discussed above, the CE - 4 landing site has been greatly affected by post - volcanism impacts. To - be - explored questions include: What's the thickness of nonmare ejecta on basaltic units? What's the variability of soil compositions? Whether the apparent compositional variations are due mainly to contamination of ejecta materials or variations in the basalts?

We believe that Yutu - 2 rover will combine information including morphologic (via panorama cameras), mineralogical (via VNIS), and subsurface structures (via LPR), all helping to answer these questions and provide new ground truth for the lunar farside for the first time.

8 Conclusions

We presented a detailed study of the composition, mineralogy, and chronology of basaltic and nonmare materials in Von Kármán crater. Mare basalts in this location have relatively low Fe (13 — 18 wt% FeO) and low Ti (1 — 3 wt% TiO_2), and the Mg composition suggests they are relatively high in MgO (~10 wt %) and Mg# (~55). By a spectral survey of 1899 small craters, we derived a mineral (HCP/LCP ratio) map of the region of the landing site, which helps to discriminate the basaltic and nonmare units by suppressing the upper layer soils mixing and disturbance by impacts and gardening. This region includes nonmare Mg - rich pyroxene materials, nonmare HCP materials, and mare basalts. The nonmare materials are characterized by more magnesian pyroxene compositions compared to the basalts, with Finsen crater material (and its ejecta) more Ca/Fe rich than other parts of the rim of Von Kármán crater. The mare basalts show different reworking and mixing effects. The mixing and evolution of lunar regolith and the layers of mare basalts in this region will benefit from detailed in - situ exploration by the CE - 4 rover. In addition, we dated the mare basalts at two model ages of ~3.6Ga and ~3.4Ga. We finally draw a new geologic map of this landing region based on above studies, which demonstrate the geologic evolution of landing region and would help to guide the surface roving exploration route of the CE - 4 mission.

Our study provides some mineralogical and geological information about the basaltic units in Von Kármán crater. The in - situ analysis of Yutu - 2 will help to evaluate the mineral chemistries and subsurface structures in - situ by VNIS and LPR, respectively. We anticipate that the Chang'E - 4 mission will bring back new ground truth from the lunar farside surface.

Acknowledgments

We thank the Editor and three reviewers for their comments and suggestions, which helped us to improve the quality of the manuscript. We also thank Bradley L. Jolliff for for helpful discussions on the geology of CE - 4 landing site. This work is supported by the National Natural Science Foundation of China (11941001, 41972322, U1931211 and 41703063), the Natural Science Foundation of Shandong Province (ZR2019MD008), Qilu (Tang) Young Scholars Program of Shandong University, Weihai (2015WHWLJH14), the China Postdoctoral Science Foundation (No. 2017M610421), the Program for JLU Science and Technology Innovative Research Team (JLUSTIRT, 2017TD - 26), the Focus on research and development plan in Shandong province (2018GGX101028), the Open Fund of State Key Laboratory of Remote Sensing Science (OFSLRSS201802). All original WAC, SLDEM2015 and M^3 data can be found in the NASA's Planetary Data System (http://pds.nasa.gov/), the original MI data are archived at SELENE Data Archive (https://darts.isas.jaxa.jp/planet/pdap/selene/) and the original IIM data are available from the Data Publishing and Information Service System of China's Lunar Exploration Program (http://moon.bao.ac.cn).

References

[1] Baker, D. M., Head, J. W., Fassett, C. I., Kadish, S. J., Smith, D. E., Zuber, M. T., Neumann, G. A., 2011. The transition from complex crater to peak - ring basin on the Moon: new observations from the Lunar Orbiter Laser Altimeter (LOLA) instrument. Icarus 214, 377 - 393.

[2] Barker, M., Mazarico, E., Neumann, G., Zuber, M., Haruyama, J., Smith, D., 2016. A new lunar digital elevation model from the lunar orbiter laser altimeter and SELENE terrain camera. Icarus 273, 346 - 355.

[3] Besse, S., Sunshine, J. M., Staid, M. I., Petro, N. E., Boardman, J. W., Green, R. O., Head, J. W., Isaacson, P. J., Mustard, J. F., Pieters, C. M., 2011. Compositional variability of the Marius Hill volcanic complex from the Moon mineralogy mapper (M^3). J. Geophys. Res. 116, E6.

[4] Boardman, J. W., Pieters, C. M., Green, R. O., Lundeen, S. R., Varanasi, P., Nettles, J., Petro, N., Isaacson, P., Besse, S., Taylor, L. A., 2011. Measuring moonlight: an overview of the spatial properties, lunar coverage, selenolocation, and related Level 1B products of the Moon Mineralogy Mapper. J. Geophys. Res. 116 https://doi.org/10.1029/2010je003730.

[5] Cheek, L. C., et al., 2011. Goldschmidt crater and the Moon's north polar region:

results from the Moon mineralogy mapper (M^3). J. Geophys. Res. 116 https: //doi. org/10. 1029/2010je003702.

[6] Di, K. C., Liu, Z. Q., Liu, B., Wan, W. H., Peng, M., Wang, Y. X., Gou, S., Yue, Z. Y., Xin, X., Jia, M. N., Niu, S. L., 2019. Chang'e - 4 lander localization based on multi - source data. J. Remot. Sens. 23 (1), 177 - 184,

[7] Duke, M., Agee, C., Bogard, D., Carrier, W., Coombs, C., Gaddis, L., Head, J. I., Jolliff, B., Lofgren, G., Papanastassiou, D., Papike, J., Pieters, C., Ryder, G., 2000. South Pole - Aitken basin sample return mission. In: Proceedings of the Fourth International Conference on Exploration and Utilization of the Moon. European Space Agency, ESTEC, Noordwijk.

[8] Fassett, C. I., Head, J. W., Kadish, S. J., Mazarico, E., Neumann, G. A., Smith, D. E., Zuber, M. T., 2012. Lunar impact basins: stratigraphy, sequence and ages from superposed impact crater populations measured from Lunar Orbiter Laser Altimeter (LOLA) data. J. Geophys. Res. Planets 117, E00H06.

[9] Garrick - Bethell, I., Miljković, K., 2018. Age of the Lunar South Pole - Aitken Basin. 49th Lunar and Planetary Science Conference. Lunar and Planetary Institute, The Woodlands, Texas. Abstract #2633.

[10] Garrick - Bethell, I., Zuber, M. T., 2009. Elliptical structure of the lunar South Pole - Aitken basin. Icarus 204, 399 - 408.

[11] Green, R. O., Pieters, C., Mouroulis, P., Eastwood, M., Boardman, J., Glavich, T., et al., 2011. The Moon Mineralogy Mapper (M^3) imaging spectrometer for lunar science: instrument description, calibration, on - orbit measurements, science data calibration and on - orbit validation. J. Geophys. Res. Planets 116, E00G19.

[12] Haruyama, J., Ohtake, M., Matsunaga, T., Morota, T., Honda, C., Yokota, Y., et al., 2009. Long - lived volcanism on the lunar farside revealed by SELENE terrain camera. Science 323, 905 - 908.

[13] Hawke, B. R., Blewett, D. T., Lucey, P. G., Smith, G. A., Bell, J. F., Campbell, B. A., Robinson, M. S., 2004. The origin of lunar crater rays. Icarus 170, 1 - 16.

[14] Head, J. W., Murchie, S. L., Mustard, J. F., Pieters, C., Neukum, G., McEwen, A., Greeley, R., Nagel, E., Belton, M. J. S., 1993. New data from the western limb and far side (Orientale and South Pole - Aitken basins) from the first Galileo flyby. J. Geophys. Res. 98, 17149 - 171181.

[15] Head, J. W., Fassett, C. I., Kadish, S. J., Smith, D. E., Zuber, M. T., Neumann, G. A., Mazarico, E., 2010. Global distribution of large lunar craters: implications for resurfacing and impactor populations. Science 329, 1504 - 1507.

[16] Hiesinger, H., Head III, J., Wolf, U., Jaumann, R., Neukum, G., 2002. Lunar mare basalt flow units: Thicknesses determined from crater size - frequency

distributions. Geophys. Res. Lett. 29 (8), 89.

[17] Hiesinger, H., van der Bogert, C. H., Pasckert, J. H., Schmedemann, N., Robinson, M. S., Jolliff, B., Petro, N., 2012. New crater size - frequency distribution measurements of the south Pole - Aitken basin. Lunar and Planetary Sci. Conf. 43. Abstract #2863.

[18] Huang, J., Xiao, Z., Flahaut, J., Martinot, M., Head, J., Xiao, X., et al., 2018. Geological characteristics of Von Kármán crater, northwestern South Pole - Aitken Basin: Chang'E - 4 landing site region. J. Geophys. Res. Planets 123, 1684-1700.

[19] Ivanov, M. A., Hiesinger, H., van der Bogert, C. H., Orgel, C., Pasckert, J. H., Head, J. W., 2018. Geologic history of the northern portion of the south Pole - Aitken basin on the Moon. J. Geophys. Res.: Planets 123, 2585 - 2612.

[20] Jolliff, B. L., Shearer, C. K., Papanastassiou, D. A., Liu, Y., the MoonRise Science Team, 2017. Why do we need samples from the Moon's South Pole - Aitken basin and what would we do with them? Lunar Planet. Sci. 48. Abstract #1300.

[21] Kneissl, T., van Gasselt, S., Neukum, G., 2011. Map - projection - independent crater size - frequency determination in GIS environments - new software tool for ArcGIS. Planet. Space Sci 59, 1243 - 1254.

[22] Lemelin, M., Lucey, P. G., Song, E., Taylor, G. J., 2015. Lunar central peak mineralogy and iron content using the Kaguya Multiband Imager: reassessment of the compositional structure of the lunar crust. J. Geophys. Res. Planets 120, 869 - 887.

[23] Li, C., Liu, J., Ren, X., Mou, L., Zou, Y., Zhang, H., et al., 2010. The global image of the Moon obtained by the Chang'E - 1: data processing and lunar cartography. Sci. China Earth Sci. 53 (8), 1091 - 1102.

[24] Ling, Z., Jolliff, B. L., Wang, A., Li, C., Liu, J., Zhang, J., et al., 2015. Correlated compositional and mineralogical investigations at the Chang'e - 3 landing site. Nat. Commun. 6 (1), 8880.

[25] Lucey, P. G., Blewett, D. T., Jolliff, B. L., 2000. Lunar iron and titanium abundance algorithms based on final processing of Clementine ultraviolet - visible images. J. Geophys. Res. 105 (E8), 20297 - 20305.

[26] Mazrouei, S., Ghent, R. R., Bottke, W. F., Parker, A. H., Gernon, T. M., 2019. Earth and Moon impact flux increased at the end of the Paleozoic. Science 363, 253.

[27] Melosh, H. J., 1989. Impact Cratering: A Geologic Process. Research Supported by NASA. Oxford University Press, New York. Oxford Monographs on Geology and Geophysics, No. 11.

[28] Melosh, H. J., Kendall, J., Horgan, B., Johnson, B. C., Bowling, T.,

Lucey, P. G., Taylor, G. J., 2017. South Pole - Aitken basin ejecta reveal the Moon's upper mantle. Geology 45 (12), 1063 - 1066.

[29] Michael, G. G., Neukum, G., 2010. Planetary surface dating from crater size - frequency distribution measurements: partial resurfacing events and statistical age uncertainty. Earth Planet. Sci. Lett. 294, 223 - 229.

[30] Michael, G. G., Kneissl, T., Neesemann, A., 2016. Planetary surface dating from crater size - frequency distribution measurements: Poisson timing analysis. Icarus 277, 279 - 285.

[31] Moriarty III, D. P., Pieters, C. M., 2018. The character of South Pole - Aitken Basin: patterns of surface and subsurface composition. J. Geophys. Res. Planets 123, 729 - 747.

[32] National Research Council, 2007. The Scientific Context for the Exploration of the Moon. National Academies Press, Washington, D. C., p. 120

[33] Neukum, G., Ivanov, B. A., Hartmann, W. K., 2001. Cratering records in the inner solar system in relation to the lunar reference system. Space Sci. Rev. 96, 55 - 86.

[34] Oberbeck, V., Morrison, R., 1974. Laboratory simulation of the herringbone pattern associated with lunar secondary crater chains. The Moon 9, 415 - 455.

[35] Ohtake, M., Matsunaga, T., Yokota, Y., Yamamoto, S., Ogawa, Y., Morota, T., et al., 2010. Deriving the absolute reflectance of lunar surface using SELENE (Kaguya) multiband imager data. Space Sci. Rev. 154 (1 - 4), 57 - 77.

[36] Otake, H., Ohtake, M., Hirata, N., 2012. Lunar iron and titanium abundance algorithms based on SELENE (Kaguya) Multiband Imager data. Lunar and Planetary Sci. Conf. 43rd. Abstract 1905.

[37] Papike, J. J., Ryder, G., Shearer, C. K., 1998. Lunar samples (Ch. 5). In: Papike, J. J. (Ed.), Planetary Materials. Mineralogical Society of America, Washington, D. C., USA, pp. 1 - 234.

[38] Pasckert, J. H., Hiesinger, H., van der Bogert, C. H., 2018. Lunar farside volcanism in and around the South Pole - Aitken basin. Icarus 299, 538 - 562.

[39] Pieters, C. M., Hanna, K. D., Cheek, L., Dhingra, D., Prissel, T., Jackson, C., et al., 2014. The distribution of Mg - spinel across the Moon and constraints on crustal origin. Am. Mineral. 99 (10), 1893 - 1910.

[40] Prettyman, T. H., Hagerty, J. J., Elphic, R. C., Feldman, W. C., Lawrence, D. J., Mckinney, G. W., Vaniman, D. T., 2006. Elemental composition of the lunar surface: analysis of gamma ray spectroscopy data from Lunar Prospector. J. Geophys. Res. Planets 111, E12007.

[41] Qiao, L., Ling, Z., Fu, X., Li, B., 2019. Geological characterization of the Chang'e - 4 landing area on the lunar farside. Icarus 333, 37 - 51.

[42] Ravi, S., Mahanti, P., Meyer, H., Robinson, M. S., 2016. On the Usefulness

of Optical Maturity for Relative Age Classification of Fresh Craters. American Geophysical Union Fall Meeting, San Francisco, California abstract ＃P53A－2166.

[43] Robinson, M. S., Brylow, S. M., Tschimmel, M., Humm, D., Lawrence, S. J., Thomas, P. C., et al., 2010. Lunar reconnaissance orbiter camera (LROC) instrument overview. Space Sci. Rev. 150 (1－4), 81－124.

[44] Shoemaker, E. M., 1962. Interpretation of lunar craters. In: Kopal, Z. (Ed.), Physics and Astronomy of the Moon. Academic Press, New York, pp. 283－360.

[45] Smith, D. E., Zuber, M. T., Neumann, G. A., Lemoine, F. G., Mazarico, E., Torrence, M. H., et al., 2010. Initialobservations from the lunar orbiter laser altimeter (LOLA). Geophys. Res. Lett. 37, L18204.

[46] Stuart－Alexander, D. E., 1978. Geologic Map of the Central Far Side of the Moon. U. S. Geol. Surv., Map I－1047, Reston, VA.

[47] Sunshine, J. M., Pieters, C. M., 1993. Estimating modal abundances from the spectra of natural and laboratory pyroxene mixtures using the modified Gaussian model. J. Geophys. Res. 98, 9075－9087.

[48] Sunshine, J. M., Pieters, C. M., 1998. Determining the composition of olivine from reflectance spectroscopy. J. Geophys. Res. 103, 13675－13688.

[49] Thiessen, F., Besse, S., Staid, M. I., Hiesinger, H., 2014. Mapping lunar mare basalt units in mare Imbrium as observed with the Moon Mineralogy Mapper (M^3). Planet. Space Sci. 104, 244－252.

[50] Trask, N. J., 1966. Size and spatial distribution of craters estimated from the Ranger photographs. Jet Propul. Lab. Tech. Rep., 32－700 Pasadena, California, pp. 252－263.

[51] Werner, S. C., Medvedev, S., 2010. The Lunar rayed－crater population－characteristics of the spatial distribution and ray retention. Earth Planet. Sci. Lett. 295, 147－158.

[52] Wilhelms, D. E., 1987. The Geologic History of the Moon. U. S. Geol. Surv. Prof. Pap. 1348.

[53] Wu, W. R., Wang, Q., Tang, Y. H., Yu, G. B., Liu, J. Z., Zhang, W., et al., 2017. Design of Chang'E－4 lunar farside soft－landing mission (in Chinese). Journal of Deep Space Explor. 4 (2), 111－117.

[54] Yamamoto, S., Nakamura, R., Matsunaga, T., Ogawa, Y., Ishihara, Y., Morota, T., et al., 2012. Olivine－rich exposures in the south Pole－Aitken basin. Icarus 218 (1), 331－344. Yingst, R., Chuang, F., Berman, D., Mest, S., 2017. Geologic mapping of the planck quadrangle of the Moon (LQ－29). Lunar Planet. Sci. 48. Abstract ＃1680.

A plagioclase – rich rock measured by Yutu – 2 Rover in Von Kármán crater on the far side of the Moon

Pei Ma[1], Yuxue Sun[1], MengHua Zhu[2,3], Yazhou Yang[4], Xiaoyi Hu[1], Te Jiang[1], Hao Zhang[1,3], Paul G. Lucey[5], Rui Xu[6], Chunlai Li[6], Zhiping He[6], Bin Xue[7], Jianfeng Yang[7], Changning Huang[8], Hongyu Lin[8]

1. Planetary Science Institute, School of Earth Sciences, China University of Geosciences, Wuhan, China
2. State Key Laboratory of Lunar and Planetary Sciences, Macau University of Science and Technology, Taipa, Macau
3. CAS Center for Excellence in Comparative Planetology, Hefei, China
4. State Key Laboratory of Space Weather, National Space Science Center, Chinese Academy of Sciences, Beijing, China
5. Hawaii Institute of Geophysics and Planetology, University of Hawai' i, Honolulu, HI, USA
6. Key Laboratory of Space Active Opto – Electronics Technology, Shanghai Institute of Technical Physics, Chinese Academy of Sciences, Shanghai, China
7. Xi'an Institute of Optics and Precision Mechanics, Chinese Academy of Sciences, Xi'an, China
8. Beijing Institute of Space Mechanics and Electricity, China Academy of Space Technology, Beijing, China

Abstract: The Chang'E – 4 spacecraft successfully landed in Von Kármán crater inside the South Pole – Aitken basin on the lunar far side on January 3, 2019 and the Yutu – 2 Rover has performed explorations on the lunar surface for nine lunar days as of September 2019. Our earlier analysis of the visible and near – infrared spectrometer measurements made by the Yutu – 2 rover during the first two lunar days shows that the regolith of the landing site may have come from the nearby Finsen crater and is dominated by plagioclase with lesser amount of mafic minerals. During its third lunar day explorations, the Yutu – 2 photographed a small piece of lunar rock and measured its reflectance spectra. Compared with the spectra of its surrounding regolith, this rock's spectrum has deeper absorption features, indicating its fresher nature. To obtain the mineralogy of the rock, we compared the rock's spectrum with the spectral library data of NASA's reflectance experiment laboratory of returned lunar rocks and lunar meteorites. We found that this rock is also plagioclase – rich with a possible plagioclase abundance of 60 – 80 vol%. A source region analysis using the Moon Mineralogy Mapper's remote sensing observations indicates this rock was ejected from the Zhinyu crater, about 30 km west of the landing site, rather than directly from the Finsen crater. Numerical simulations of the Zhinyu crater on the impact cratering process and ejecta thickness distribution confirmed our findings and imply that the surficial materials at the CE - 4 landing site experienced a complicated evolution rather than simply retaining the pristine or primordial ejecta directly from the Finsen crater.

Keywords: Spectroscopy; Moon; Surface; Mineralogy

1 Introduction

Since its successful soft landing on the far side of the Moon on January 3, 2019, the Chang'E - 4 mission's Yutu - 2 rover has roamed the lunar surface for consecutive nine lunar days as of early September 2019, thus providing a unique opportunity to study the compositions of the lunar surface in - situ. Located within the Von Kármán crater [Di et al., 2019a], the landing site is covered by a thick layer of mare basalts [Huang et al., 2018] with some parallel linear - structures on top that were considered to be possibly originated from the 80 - km - diameter Finsen crater which is ~140km east of the landing site [Qiao et al., 2019; Di et al., 2019b]. Early spectral investigations using the modified Gaussian method show that the surficial materials at the landing site are olivine - and pyroxene - rich and therefore may represent the compositions of lunar mantle materials excavated by Finsen crater [Li, Liu et al., 2019; Gou et al., 2019]. However, by analyzing the reflectance measured in the first two lunar days using a lunar mineral reflectance lookup table (LUT) [Lucey, 2004; Denevi et al., 2008; Lucey et al., 2014], Hu et al. [2019] found that the regolith at the landing site is dominated by plagioclase (56—72 wt%) with low olivine (2—12 wt%) concentrations and thus these materials are unlikely of lunar mantle origins. Alternatively, as the CE - 4 landing site is located within the impact - induced melt pool of the South Pole - Aitken basin [Potter et al., 2012; Zhu et al., 2018], it is highly possible that the surficial materials at the landing site are the secondary product of re - crystallization of impact melt pool produced by the South Pole - Aitken forming impact event [Hu et al., 2019; Lin et al., 2019]. The thick impact melt sheet (>50km, Potter et al., 2012; Zhu et al., 2018) may differentiate and form the plagioclase - rich, but olivine - and pyroxene - poor crust [Hurwitz and Kring, 2014; Vaughan and Head, 2014], as measured at the landing site on the ejecta deposits excavated from Finsen crater. As the rover has roamed further away from the landing site, the diversity of the surface materials may have begun to emerge and may provide more clues on the origins of surficial material at the landing site. For example, during its third lunar day exploration, the Yutu - 2 rover measured the reflectance spectra of a piece of rock located about 70m away from the landing site. Since the degree of space weathering of the rock is much lower than that of mature regolith, its reflectance spectra may provide more direct information on the mineralogy of the surface materials.

In this work, we report the in-situ reflectance measurement of this rock ("the Rock" hereafter) and the neighboring regolith during the first five lunar days. By comparing the CE-4 measurement with NASA reflectance experiment laboratory (RELAB)'s spectral data of returned lunar rocks and lunar meteorites, we attempt to determine the mineral abundance of the Rock. An analysis using the Moon Mineralogy Mapper (M^3) remote sensing observations has identified the possible source region of the Rock. Numerical simulations of impact cratering and ejecta thickness distribution have been conducted to validate our findings.

2 Instruments, observations, and data reductions

2.1 Instruments

The Yutu-2 rover carries a suite of instruments including a visible and near-infrared imaging spectrometer (VNIS), a panoramic camera (PCAM) and a ground-penetrating radar for analyzing the topography, regolith, rock and mineral compositions of the landing and roving sites [Wu et al., 2019]. The VNIS is installed on the front surface of the rover to view the lunar surface at a fixed 45° zenith angle at a height of 0.69m. The VNIS consists of a complementary metal-oxide-semiconductor (CMOS) imager (450—945nm) with 256 by 256 pixels and a shortwavelength infrared (SWIR) detector (900—2395nm) with a single pixel. The spectral resolution of the CMOS imager and the SWIR detector varies between 2.4—6.5nm and 3.6—9.6nm, respectively [Li et al., 2019b, 2019a]. The field of views (FOVs) of the CMOS imager and SWIR detector are 8.5° and 3.58°, respectively, and the SWIR FOV is their overlapping FOV (see yellow circles in Fig. 2 below). The CE-4 VNIS measurement uncertainties have been estimated to be 5% for the CMOS and 7% for the SWIR, respectively [Li et al., 2019]. More details about the VNIS instrumentation and calibrations can be found in C. Li et al., 2019; C. L. Li et al., 2019 and He et al. [2019].

2.2 Observations

The roving path of the Yutu-2 and the measurement sites during the first five lunar days are shown in Fig. 1. During this period, the Yutu-2 moved a total path length of ~190m and collected 28 reflectance spectra. The CMOS images of all 28 measurements are shown in Fig. 2 and the scattering configurations of these measurements including the solar and sensor zenith angles, azimuth angles, and solar phase angles are summarized in Table A1 in Appendix A. During its third lunar day measurement, the Yutu-2 came across the Rock located ~70m away from the spacecraft touchdown point. The VNIS collected three reflectance spectra of the Rock and its surrounding regolith (N_0013, N_0014, and N_0015)

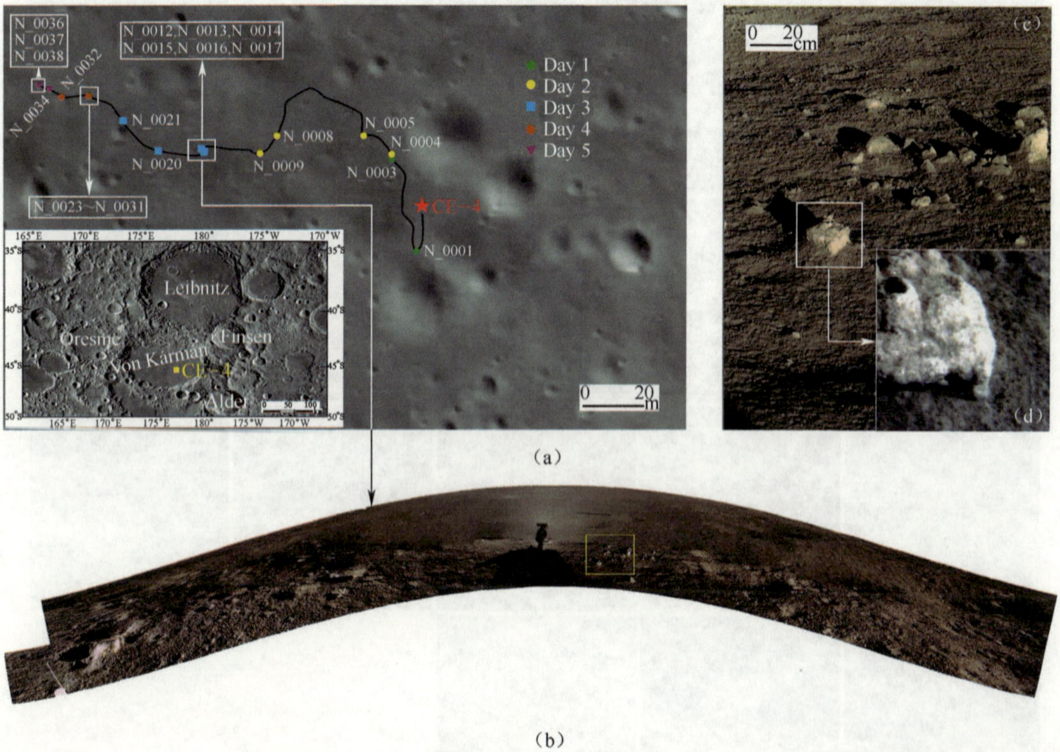

Fig. 1　Images of the CE－4 landing site and the location of the Rock

(a) Yutu－2 rover's path during the first five lunar days (Day 1 to Day 5). The base image was photographed by the landing camera onboard the CE－4 spacecraft. The sub－window of (a) is the Lunar Reconnaissance Orbiter Camera Wide Angle Camera mosaic of the CE－4 landing area and the landing site is marked by the yellow square. (b) A mosaic of 43 PCAM images near measurement site N_0015 taken during the third lunar day. The yellow box indicates the location of the Rock. (c) The original PCAM image of the Rock. (d) A pseudo color picture of the Rock synthesized by using CMOS images at 450, 550, and 635nm (For interpretation of the references to color in this figure legend, the reader is referred to the web version of this article)

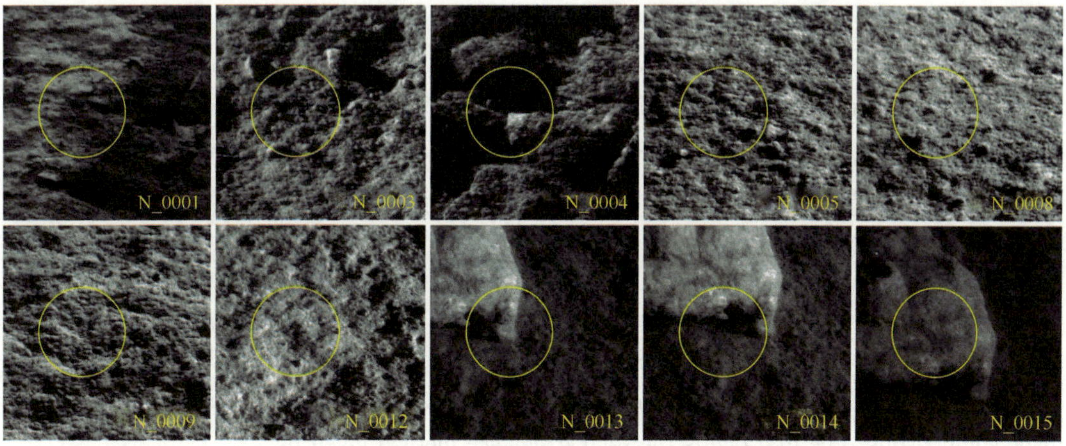

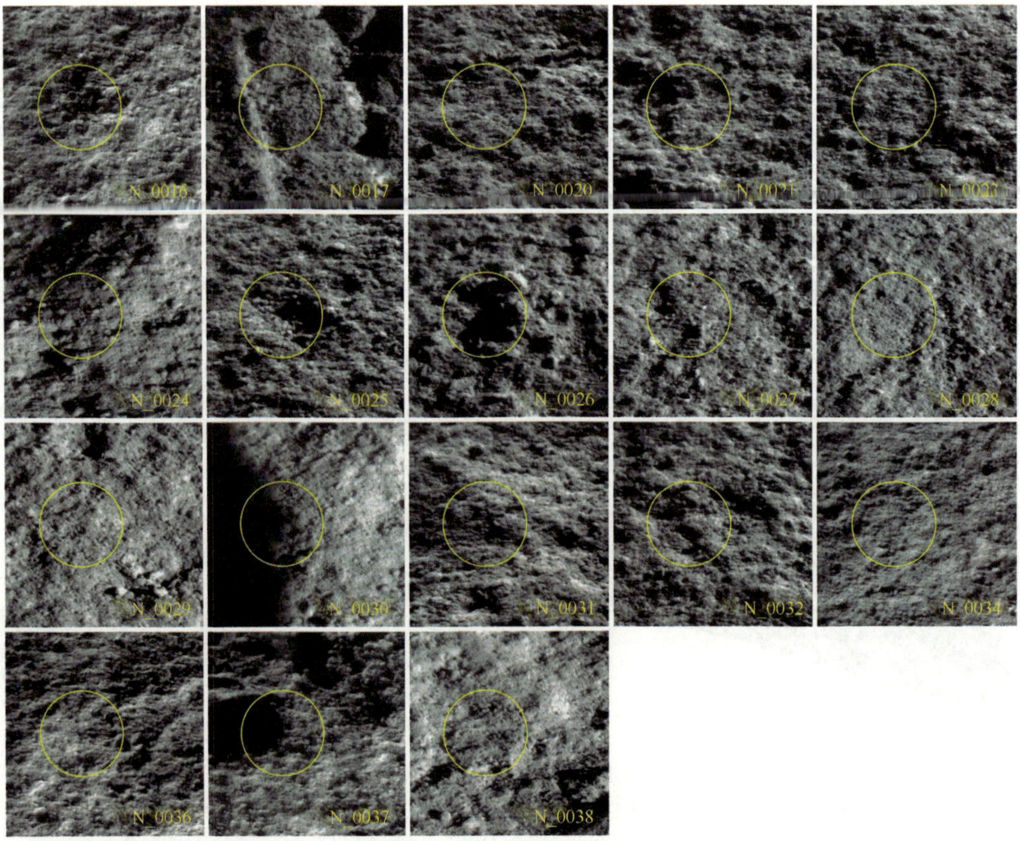

Fig. 2　CMOS images at 600nm of the lunar surface at 28 measurement sites. The overlapping FOVs of the CMOS and SWIR detectors are marked as yellow circles. The FOVs of N _0004, N _0026, N _0030, and N _0037 are partly filled with shadows. The FOVs of measurements N _0013 and N _0014 contain both the Rock and regolith, and only that of N _0015 is totally filled up with the Rock. Measurements N _0004 and N _0017 were made on Yutu-2 rover's track. (For interpretation of the references to color in this figure legend, the reader is referred to the web version of this article.)

but only the FOV of N _0015 was totally filled up by the Rock (Fig. 2). Since the FOV of the CMOS imager is 8.5°and roughly images an area of 15 by 21 cm at a viewing angle of 45° [Wu and Hapke, 2018; Li et al., 2019b, 2019a], the Rock is estimated to have a dimension of ~20 by ~20cm. Compared with the Chang'E-3 landing site in Mare Imbrium that is cluttered with boulders and pebbles [Zhang et al., 2015; Jin et al., 2015; Di et al., 2016], the CE-4 landing site has lower abundance of rocks as evidenced by Fig. 1 (b) and (c).

2.3 Data reductions

In this work, we present all reflectance measurements in reflectance factor (*REFF*, e. g., Hapke, 2012). Since the VNIS carries a polytetrafluoroethylene - type calibration target, we converted the VNIS measured lunar surface radiance to *REFF* using two alternative methods and made comparisons. The first one is the solar irradiance method given by

$$REFF = \frac{\pi}{\cos(i)} \frac{I(\lambda)}{\int J(\lambda) S(\lambda) d\lambda} \frac{D^2}{D_{AU}^2} C_m(\lambda) \quad (1)$$

where $I(\lambda)$, $J(\lambda)$ and $S(\lambda)$ are radiance reflected from the lunar surface, the solar irradiance at 1 AU [Gueymard, 2004], and the detector's spectral response function approximated by Gaussian function [Liu et al., 2014; He et al., 2019], respectively. The parameters of i, D, D_{AU}, $C_m(\lambda)$, represent the solar incident zenith angle, the actual Sun - Moon distance in AU, 1 AU, and the correction array provided by the payload team, respectively.

Alternatively, the *REFF* can be obtained by using

$$REFF = \frac{I}{I_{plaque}} REFF_{plaque} \quad (2)$$

where I and I_{plaque} are the radiance reflected from the lunar surface and the onboard calibration target, respectively; $REFF_{plaque}$ is the pre - flight laboratory measured *REFF* values of the calibration target. Among the total 28 measurements, only 11 had measurements on the calibration target. After a detailed comparison of the data reduced by these two methods (Appendix B), we chose method one (Eq. (1)) to present all data. Since the FOVs of measurements N _0004, N _0026, N _0030, and N _0037 contain large amount of shadows (Fig. 2), and the SWIR spectrum of N _0024 has a gap at 0.98μm, we only analyzed the remaining 23 spectra out of the total 28 measurements.

3 Results and discussion

Fig. 3 (a) — (d) shows the selected 23 reflectance spectra in *REFF* and Fig. 3 (e) and (f) is the normalized (at 750nm) and continuum - removed spectra, respectively. The continuum - removed spectra were obtained by dividing the *REFF* spectra by straight - line segments [Clark and Roush, 1984] as shown in the sub - window of Fig. 3 (f). As different continuumremoval methods may result in different absorption features, we have also tried the straight line method and a three - point polyline method [Appendix C]. Although the resultant continuum - removed spectra have slightly different band width and band depth, such difference did not affect the mineralogical analysis as long as the same method is applied to both the CE - 4 measurements and the library spectra.

Among all the 23 continuum - removed spectra (Fig. 3 (f)), the Rock's spectrum N _0015 has the widest and deepest 1μm absorption, indicating its fresher nature as compared to the more mature neighboring regolith. As the N _0013 and N _0014 spectra measured some portion of the rock, their 1μm absorptions are also deeper than that of the surrounding regolith.

3.1 Comparisons with reflectance spectra of returned lunar rocks and lunar meteorites

In our analysis of CE - 4's VNIS measurements made during the first two lunar days,

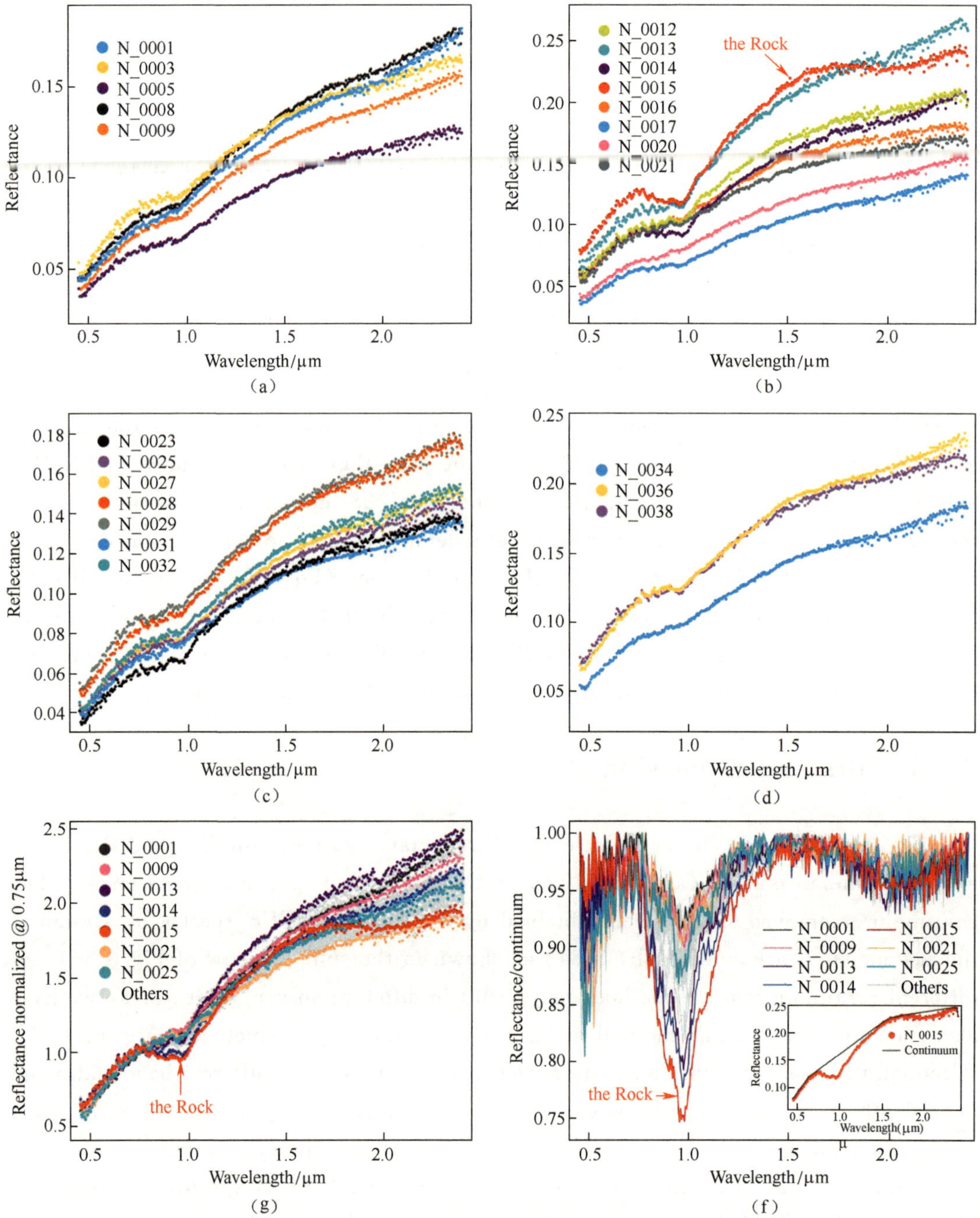

Fig. 3　The selected 23 reflectance spectra measured by Yutu-2 during the first five lunar days used in this work. (a-d) The reflectance spectra measured during different lunar days

(a) Day 1 and Day 2; (b) Day 3; (c) Day 4; (d) Day 5. (e) The spectra normalized at 750nm and (f) the continuum-removed spectra. The REFF spectrum of the Rock (N_0015) and the continuum are shown in the sub-window of (f). To make the figure clear, in (e) and (f) only seven representative spectra are shown in colors and all others are shown in gray. (For interpretation of the references to color in this figure legend, the reader is referred to the web version of this article.)

we employed a mineral reflectance LUT [Lucey, 2004; Denevi et al., 2008; Lucey et al., 2014] to retrieve the relative mineral abundances and found that the landing site regolith is enriched in plagioclase, followed by pyroxene and olivine [Hu et al., 2019]. Although lunar rocks should contain more direct information on mineral abundance than the more space weathered regolith, there appears to be no well-developed approach similar to the LUT method to retrieve their mineral abundance. Therefore, to infer the mineral abundances of the Rock, we collected all available RELAB reflectance spectra of returned lunar rocks and lunar meteorites and searched the best matches between CE-4 spectra and the library spectra. Altogether a total of 755 RELAB spectra of returned lunar rocks and lunar meteorites that covered CE-4's spectral range (450-2395nm) have been downloaded and searched. These 755 RELAB spectra were measured in ambient environment and thus may differ from real lunar environment measurement results. However, various laboratory studies on olivine, pyroxene, and Apollo samples have indicated that changes in spectral features and reflectance values at various temperatures are relatively minor compared to compositional variations (e. g., Singer and Roush, 1985; Roush and Singer, 1986, 1987; Hinrichs and Lucey, 2002). The vacuum condition has virtually no effects on the visible and near-infrared reflectance spectra within 0.5—2.5μm. Therefore, the comparison study in this work should be valid.

We first resampled the RELAB spectra to CE-4 wavelength range and spectral steps using cubic spline interpolation and then removed the continua of both spectra using the straight-line segments method (Appendix C contains comparisons of all three continuum-removal methods). To find the best matches between the two datasets, we employed four searching criteria to evaluate their similarities: the absolute difference (ABS), the normalized absolute difference (NABS), the Spectral Angle Mapper (SAM, [Kruse et al., 1993]) and the Centered Pattern Root Mean Square (CPRMS, [Taylor, 2001]), as these four algorithms have been found to be able to describe spectral patterns well [Zhang et al., 2015; Hu et al., 2019]. The ABS, NABS, SAM, and CPRMS are defined as:

$$\mathrm{ABS}=\sum_{i=1}^{n}\left|t_i-r_i\right| \tag{3}$$

$$\mathrm{NABS}=\sum_{i=1}^{n}\left|\frac{t_i}{\bar{t}}-\frac{r_i}{\bar{r}}\right| \tag{4}$$

$$\mathrm{SAM}=\frac{\sum_{i=1}^{n}t_i r_i}{\sqrt{\sum_{i=1}^{n}t_i^2}\sqrt{\sum_{i=1}^{n}r_i^2}} \tag{5}$$

$$\mathrm{CPRMS}=\sqrt{\frac{1}{n}\sum_{i=1}^{n}[(t_i-\bar{t})-(r_i-\bar{r})]^2} \tag{6}$$

where the t_i and r_i, are the ith data point for test spectrum and reference spectrum, and $\bar{t}$ and $\bar{r}$ are the averages of test and reference spectra, respectively. Here the test and reference spectra are CE-4 N_0015 and RELAB spectra, respectively. As all four methods produced very similar searching results (Appendix C), we use the CPRMS result in Fig. 4 to represent the best matches between N _0015 and the RELAB spectra.

The RELAB spectrum of the returned lunar rock that best matched the Rock's N _0015 spectrum is C4LS23 (spectrum ID) with sample ID LSCMP - 023 which is a thin section numbered with 15565 # 57. According to the RELAB sample and spectrum descriptions, sample 15565 # 57 consists of anorthosite and norite - rich pristine and the regolith breccia. The measurement area of spectrum C4LS23 was located in the regolith breccia region of the thin section. Sample 15565 was returned by the Apollo 15 mission and it was probably sampled near station 9 [Meyer, 2009]. The thin section 15565 # 57 used in reflectance measurement was a small piece chipped off from a special white clast from 15565 # 11 [Meyer, 2009]. Although its accurate mineral compositions and abundances are not available, this white clast is described by Warren et al. [1982] as an anorthositic norite which means plagioclase is the dominant mineral phase. Therefore, the regolith breccia from this white clast is also likely dominated by plagioclase.

The RELAB spectrum of the lunar meteorite that best matched the Rock's spectrum is a large feldspathic lunar meteorite Yamato - 86032 with spectrum ID BKR1LM013F and sample ID LM - LAM - 013 - F possibly from the lunar far side [Takeda et al., 1989; Takeda et al., 1990; Takeda et al., 2002; Yamaguchi et al., 2004; Nyquist et al., 2006]. The RELAB measurement was made on a piece numbered with Yamato - 86032 # 117 that was ground and sieved into a particle size distribution of smaller than 45μm. The Yamato - 86032 # 117 is an anorthositic clast [Nyquist et al., 2002, 2006] which implies the plagioclase abundance is likely >50 vol%. The 1μm absorption feature indicates that this sample should also contain pyroxenes. As will be shown below, the high plagioclase abundances of the two RELAB samples can be justified by a semi - quantitative comparison method.

One can see from Fig. 4 (a) that except for the band depth of 1μm absorption for Yamato - 86032 # 117, all three spectra have similar absorption features near 1μm and 2μm. Compared with Yamato - 86032 # 117, the N _0015 has a deeper 1μm absorption band and this may indicate that the Rock has higher abundance of pyroxene, or Yamato - 86032 # 117's fine grains caused its subdued absorption feature. It is noticed that the CE - 4 N _0015 has redder reflectance spectrum as compared with that of Yamato - 86032 # 117 (Fig. 4 (c)), possibly caused by the presence of splashes of impact melt on the surface of the Rock.

To search for any possible correlations between the CE4 - RELAB spectral similarities and the sample plagioclase abundance, we picked out the RELAB spectra of samples with documented plagioclase (together with olivine and pyroxene) abundances and plotted the CPRMS values (Eq. (6)) between the RELAB spectra and CE - 4 measurement versus plagioclase abundance in Fig. 5 (a). From this plot we can see that overall the similarity increases with increasing plagioclase abundance (Note: a lower CPRMS value corresponds to a higher similarity). By fitting a three - term polynomial to the data, the plagioclase

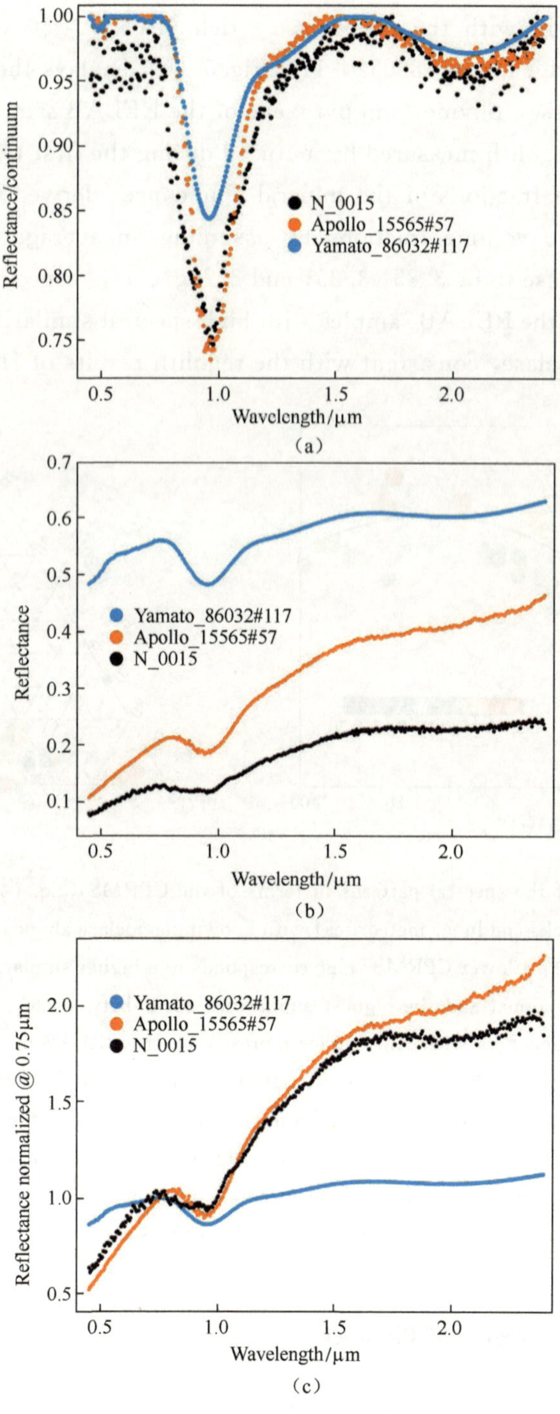

Fig. 4　Comparisons of CE－4 measurement N_0015 and the best－matched RELAB spectra of lunar meteorite Yamato－86032＃117 and returned lunar rock Apollo－15565＃57 using the CPRMS algorithm

(a) Continuum－removed spectra using the straight－line segments. (b) Reflectance in REFF and (c) spectra normalized at 750nm abundance of the Rock can be estimated to be somewhere between 60 vol% and 80 vol% (Fig. 5 (a)), consistent with the plagioclase - rich (i. e., >50 vol%) nature of the Apollo - 15565#57 and Yamato - 86032#117. Fig. 5 (b) displays the normalized mineral abundances of plagioclase, olivine, and pyroxene of the RELAB samples mentioned above and that of the lunar regolith measured by Yutu - 2 during the first two lunar days [Hu et al., 2019]. The weight fractions of the mineral abundances derived in Hu et al. [2019] have been converted to volume fractions by assuming an average density for olivine, pyroxene, and plagioclase to be 3. 85, 3. 25, and 2. 76g/cm^3, respectively. As can be seen in Fig. 5 (b), most of the RELAB samples with high spectral similarities to the Rock have high contents of plagioclase, consistent with the regolith results of Hu et al. [2019].

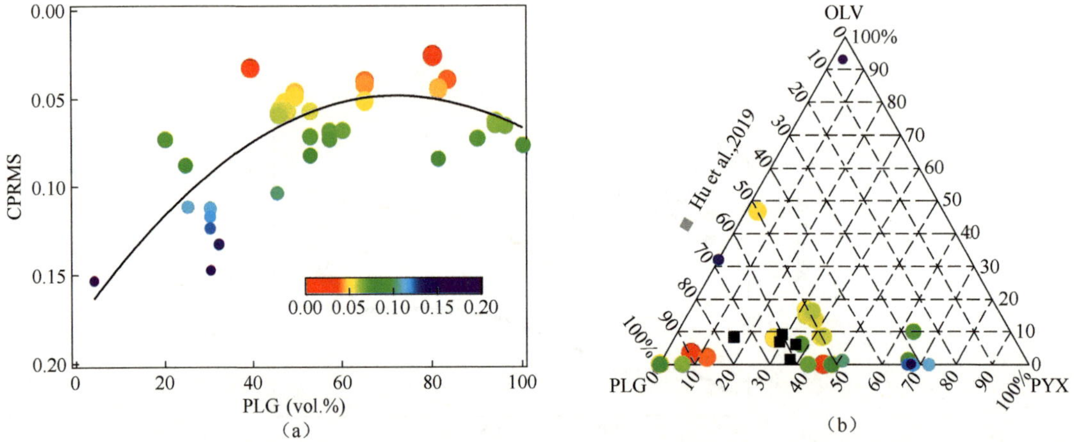

Fig. 5　(a) Similarities of the spectral patterns in terms of the CPRMS (Eq. (4)) between the RELAB samples (returned lunar rocks and lunar meteorites) with known plagioclase abundances and the Rock versus plagioclase abundance. Note, a lower CPRMS value corresponds to a higher similarity. The data points were fitted with a 3 - term polynomial and the highest similarities occur between 60 vol% and 80 vol%. (b) Ternary plot of mineral abundances with color circles representing the RELAB modal mineralogy data used in this work. Black squares are CE - 4 regolith data from Hu et al. [2019] but converted from weight fractions to volume fractions. The colors of the circles in (a) and (b) represent the CPRMS values as shown in the color bar in (a). OLV, PLG, and PYX represent olivine, pyroxene, and plagioclase, respectively. The RELAB sample information is summarized in Appendix E (For interpretation of the references to color in this figure legend, the reader is referred to the web version of this article)

3. 2　The possible source region of the Rock

3. 2. 1　Similarity between CE - 4 and M^3 spectra

To identify the possible source region of the Rock, we performed a comparison study between the CE - 4 and the M^3 spectra over the Von Kármán crater area (see Fig. 6). Although the CE - 4 measurements were made on samples with centimeter scales while remote sensing measurements were made at hundred - meter scales, if the M^3 spectrum of a region shows strong similarities to the spectrum of the Rock, it is very likely that this region has very similar mineralogy as the Rock.

We selected the CE - 4 and M^3's overlapping spectral region between 540nm and 2375nm and resampled the CE - 4 data to M^3's spectral steps. After removing the continua of the CE - 4 and M^3 data using straight - line segments, a search was performed to evaluate the similarity between the two datasets by using the ABS (Eq. (3)), NABS (Eq. (4)), SAM (Eq. (5)), and CPRMS (Eq. (6)) algorithms. It turned out that the searches using these four criteria have produced similar results (Fig. D1 in Appendix D) and the CPRMS result is shown in Fig. 6 (pixels in colors). Note, since a lower CPRMS value corresponds to a higher similarity, those pixels with CPRMS values higher than 0.035 were made transparent to enhance the comparison effects.

From Fig. 6 we can find that an annular region of Zhinyu crater has the highest similarity to the Rock's spectrum (N _0015) and the similarity decreases outwards gradually from the crater rims. No other areas within or surrounding the Von Kármán crater shows the high similarity to the Rock. Thus the Rock is likely to have originated from the Zhinyu crater which has a diameter of ~4km and is about 30km from the CE - 4 landing site. Although the landing site is far beyond the boundary of the ejecta blanket (e.g., 1 - 2 radii beyond the crater rim) of the Zhinyu crater, the fragments from the shallow layer near the impact site may have high ejecta velocities and may deposit in regions far away from the impact site [Shuvalov, 2011; Shuvalov and Dypvik, 2013] (see more detailed discussions in Section 3.2.2). This is consistent with the high similarity of the Rock with the ejecta around crater rims that are mainly from the shallow layer as well, but ejected a little further from the impact site, possibly owing to the different launch velocities and angles.

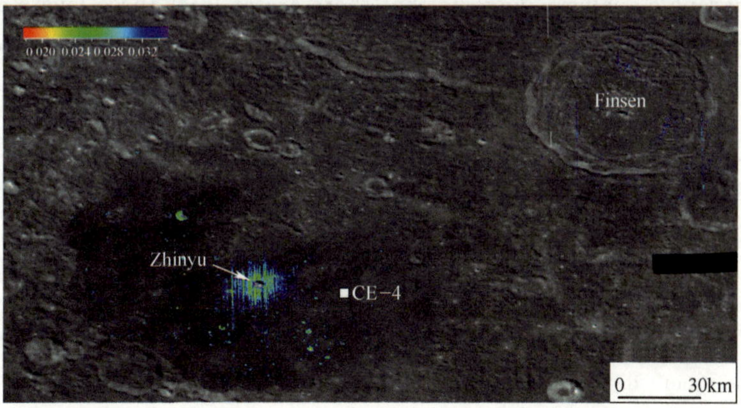

Fig. 6　Image of the CPRMS values (Eq. (6)) between the Rock's spectrum (N _0015) and the spectra of all M^3 pixels to show the similarities between the CE - 4 and M^3 spectra. The base map is an M^3 image of the Von Kármán crater area. The CE - 4 landing site (CE - 4), Zhinyu, and Finsen craters are indicated. The distance between the center of Zhinyu crater and CE - 4 landing site is ~30km

3.2.2 Ejecta thickness from Zhinyu crater at CE－4 landing site predicted by numerical simulations

In order to validate the above finding, we simulated the impact cratering process and ejecta distribution of Zhinyu crater by using the iSALE shock－physics code [Collins et al., 2004; Wünnemann et al., 2006]. The iSALE code has been used to simulate the impact cratering process and the ejecta thickness distribution for both small－scale labo－ratory experiments [Wünnemann et al., 2016] and large－scale lunar basins [Zhu et al., 2015, 2017, 2019; Luther et al., 2018]. In our simulations, we assumed a vertical impact of an asteroid made of dunite [Pierazzo et al., 1998] at a typical lunar impact velocity of 15km/s [Le Feuvre and Wieczorek, 2011] on a basalt target. Our simulations indicate that an asteroid with diameter of 200m can produce a crater with diameter of ~4km, the analog to Zhinyu crater (Fig. 7 (a)). The impact excavates the materials from an excavation zone with a maximum depth of ~350m and a width of~3km (Fig. 7 (b)). The excavated materials deposit near the crater rim is ~0.1km－thick and decreases with the increase of the radial distance from the crater center as shown in Fig. 7 (c). The variation of ejecta thickness with distance beyond the crater follows the power law scaling function with an exponent of －3.2 which is consistent with laboratory experiments [Stöffler et al., 1975; Wünnemann et al., 2016] and ejecta thickness observations for lunar crater [McGetchin et al., 1973; Xie and Zhu, 2016]. At CE－4's landing site, about 30km from Zhinyu crater, the *average ejecta* would be ~4cm thick according to the impact cratering simulation. However, as the CE－4 landing site is beyond the ejecta blanket zone, the ejecta deposited at the landing site are very rare, mostly are solely fragments or rocks likely with the sizes of the observed rocks (~20cm), consistent with therock poor nature of the landing site from in－situ observations (Fig. 1).

The stratigraphic analysis suggests that there are at least three layers at the landing site [Huang et al., 2018]. The topmost one is mainly the deposited ejecta from the nearby Finsen crater with a diameter of ~80km. The second layer is the ejecta from Alder (~80km in diameter) and Leibnitz (~220km) crater, and the bottom one consists of mare basalts [Di et al., 2019b]. The Zhinyu crater is ~160km and ~155km from the crater center of Finsen and Alder, respectively. According to the impact cratering simulations for Finsen and Alder crater in Di et al. [2019b], the ejecta thickness at the Zhinyu crater are estimated to be ~25m and ~20m from Finsen and Alder crater, respectively. Our impact simulation for Zhinyu crater formation indicates that the initial positions of the ejecta deposited at the landing site, as shown in Fig. 7 (b), are from the excavation zone with a depth smaller than ~200m. The possible source of the Rock, or the initial positions of the ejecta deposited on the topmost surface at landing site, are mostly from a depth smaller than ~100m (see blue squares in Fig. 7 (b)). There are two possible initial depths of the Rock. One is that the Rock is the ejecta from Finsen or Alder but was excavated again by the formation of the Zhinyu crater and deposited at the landing site.

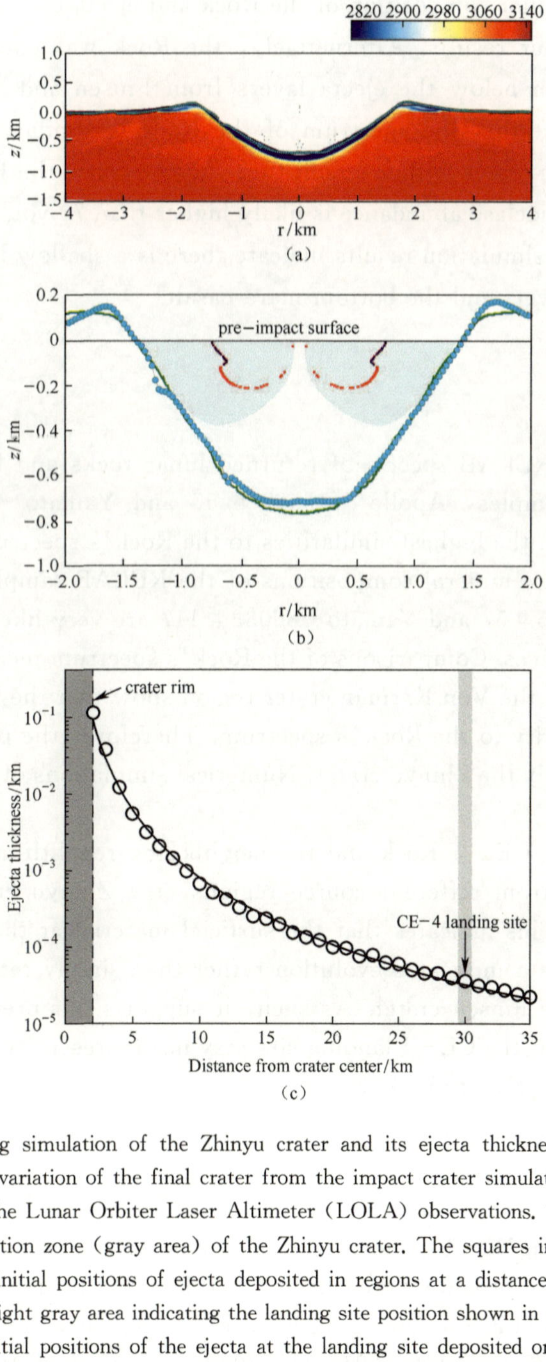

Fig. 7　Impact cratering simulation of the Zhinyu crater and its ejecta thickness distribution. (a) The snapshot of the density variation of the final crater from the impact crater simulation and the crater profile (points) derived from the Lunar Orbiter Laser Altimeter (LOLA) observations. The color bar has unit of kg/m^3. (b) The excavation zone (gray area) of the Zhinyu crater. The squares in red and blue within the gray area represent the initial positions of ejecta deposited in regions at a distance of 30 ± 0.5km from the impact site, i. e., the right gray area indicating the landing site position shown in (c). Specifically, the blue squares represent the initial positions of the ejecta at the landing site deposited on top of the surface layer while the red squares represent the initial positions of ejecta at the landing site deposited below the blue squares. The cyan points and green line represent the crater profile derived from LOLA observations and impact cratering simulation, respectively. (c) Ejecta thicknesses of the simulated crater. The gray area on the left indicates the dimensions of the crater (which equals the size of the Zhinyu crater). The gray area on the right marks the distance of the CE‑4 landing site to the center of the Zhinyu crater. The vertical axis is in logarithm scale. The points are simulation results and the line is a fitted curve with power law function with an exponent of −3.2 (For interpretation of the references to color in this figure legend, the reader is referred to the web version of this article)

This would imply a similar composition of the Rock and ejecta of Finsen or Alder crater, which is contrary to our results. Alternatively, the Rock was excavated from a depth between 60m and 100m below the ejecta layers from Finsen and Alder crater. This is consistent with the fact that the spectrum of the Rock is similar to that of the ejecta surrounding the Zhinyu crater within 2 radii (see Fig. 6). Since the Rock is dominated by plagioclase and its plagioclase abundance is likely higher than 70 vol%, it is unlikely mare basalt. Therefore, our simulation results indicate there is a shallow layer between the top two deposited ejecta layers and the bottom mare basalt.

4 Conclusions

By searching the RELAB spectra of returned lunar rocks and lunar meteorites, we found two RELAB samples, Apollo - 15565 # 57 and Yamato - 86032 # 117, have reflectance spectra with the highest similarities to the Rock's spectrum measured by CE-4. Although the accurate mineral compositions of the RELAB samples are not available, both the Apollo-15565#57 and Yamato-86032#117 are very likely to be plagioclase-rich and contain pyroxenes. Comparisons of the Rock's spectrum measured by CE-4 with M^3 measurements over the Von Kármán crater region show that the Zhinyu crater has the highest spectral similarity to the Rock's spectrum. Therefore, the possible source region of the Rock is very likely the Zhinyu crater. Numerical simulations of impact cratering also support our findings.

Although both the CE-4 Rock and the neighboring regolith are plagioclase-rich, they may have come from different source regions, the Zhinyu crater and the Finsen craters, respectively. This indicates that the surficial material at the CE-4 landing site may have experienced a complicated evolution rather than simply retaining the primordial ejecta directly from the Finsen crater. As such, it supports our previous conclusion that the surficial materials at the CE-4 landing site may not represent the compositions of the lunar mantle materials [Hu et al., 2019].

Acknowledgment

We thank the Chang'E-4 payload teams for mission operations and China National Space Administration for providing the Chang'E-4 data that made this study possible. M. Z. acknowledges the iSALE developers. This work was supported by the National Natural Science Foundation of China (11941001, 11773023, 11773087, U1631124), the Science and Technology Development Fund, Macau (0079/2018/A2), the Preresearch Project on Civil Aerospace Technologies funded by Chinese National Space Administration (D020302), and the Beijing Municipal Science and Technology Commission (Z181100002918003). The Chang'E-4 data used in this work can be requested from China's Lunar and Deep Space Exploration (CLEP) web site at: http://moon.bao.ac.cn/searchOrder_dataSearchData.search.

Appendix A. Summary of scattering configurations of all 28 in-situ measurements during the first five lunar days

Tab. A1　Scattering configurations of all 28 measurements made during the first five lunar days. Site names in *italic* indicate measurements that included calibration target measurements

Day number	Site	Sun		Rover		Phase angle/(°)
		Zenith/(°)	Azimuth/(°)	Zenith/(°)	Azimuth/(°)	
Day 1	N_0001	50.88	33.19	46.93	185.12	93.97
	N_0003	72.06	289.48	42.89	179.83	89.55
Day 2	N_0004	73.39	73.02	47.96	344.61	77.81
	N_0005	64.51	61.85	16.53	181.97	91.87
	N_0008	64.56	297.76	46.30	56.22	90.82
	N_0009	66.23	295.47	45.48	74.06	101.93
Day 3	N_0012	70.86	71.19	43.35	147.45	66.89
	N_0013	68.58	68.42	43.55	144.91	65.51
	N_0014	65.20	64.04	43.20	143.42	65.14
	N_0015	63.35	61.49	42.91	143.32	65.48
	N_0016	62.11	59.69	47.54	266.64	105.39
	N_0017	61.39	58.60	44.37	154.17	73.58
	N_0020	64.54	296.66	46.26	83.03	104.24
	N_0021	71.41	287.95	45.28	58.93	102.55
Day 4	N_0023	68.69	69.18	43.15	189.35	93.16
	N_0024	67.91	68.21	45.43	108.11	39.62
	N_0025	67.33	67.48	43.34	197.10	97.10
	N_0026	66.52	66.44	43.47	170.58	82.24
	N_0027	65.96	65.70	43.86	156.19	73.24
	N_0028	63.13	61.84	44.24	137.69	61.58
	N_0029	62.39	60.78	44.63	122.65	51.44
	N_0030	61.40	59.31	45.25	106.42	40.42
	N_0031	60.73	301.71	45.76	88.54	100.49
	N_0032	67.89	291.82	45.16	57.09	96.54
Day 5	N_0034	57.74	53.28	45.05	265.97	97.26
	N_0036	65.58	295.08	38.92	146.48	99.59
	N_0037	66.74	293.56	40.57	145.73	101.88
	N_0038	72.11	287.01	51.86	344.37	53.60

Appendix B　Data reduction descriptions

A comparison of the *REFF* of the CMOS (450—945nm) and SWIR (900—2395nm) spectra obtained by using Eqs. (1) and (2) has shown two problems.

The first problem is that two gaps exist in the reduced *REFF* spectra. Gap 1 exists in the overlapping wavelength range (900—945nm) of the CMOS and SWIR spectra, and Gap 2 exists between 1375nm and 1380nm as shown in Fig. B1. For Gap 1, as the VNIS payload team suggested that the SWIR measurement should be more reliable than the CMOS measurement, we multiplied the CMOS spectra (450—900nm) by a factor to connect to the SWIR spectra (900—2395nm). We noticed that Gap 2 was not apparent in the first two lunar day's data and has become more apparent from the third lunar day. It was likely caused by a change of the SWIR spectral resolution between spectral ranges (900—1380nm) and (1380—2395nm) [Li et al., 2019b, 2019a]. To deal with this problem, we multiplied the spectra over 450—1375nm by a factor to connect to the 1380—2395nm part to obtain the final full range spectra (450—2395nm). It should be noted that such scaling would not change the absorption features and since all spectral analysis work presented here uses continuum - removed spectra, the absolute *REFF* values are not critically important.

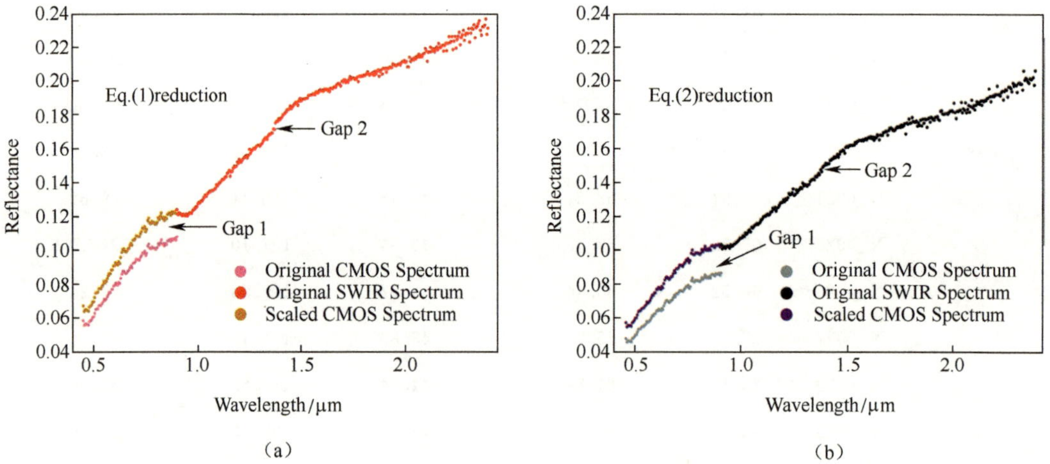

Fig. B1. The original CMOS (450—900nm) and SWIR (900—2395nm) *REFF* spectra obtained using the two calibration methods and the scaled CMOS (450—900nm) spectra for N_0036 measurement. (a) *REFF* spectra obtained by using Eq. (1). (b) *REFF* spectra obtained by using Eq. (2). The two gaps are indicated by arrows. Note that gap 2 in (b) is not apparent as in (a) for this dataset.

The second one is that significant differences exist in most of the data obtained by using Eqs. (1) and (2). Fig. B2 shows the ratio of the *REFF* spectra obtained by using Eqs. (1) and (2). If the two calibration methods produced identical spectra, this ratio should be 1. However, from Fig. B2 one can see that except for N _0001, N _0008, and N _0020 that are quite close to 1, ratios of other spectra deviate significantly from 1. Nonetheless, most of the ratio spectra do not deviate from a horizontal straight line too much which means that the two calibration methods have yielded similar spectral patterns. As an example, the case for N _0036 is shown in Fig. B3. By inspecting the corresponding plaque calibration images (Fig. B4), it is found that the ratio spectra that significantly deviate from 1 appear to be those having non - uniform illuminations in the calibration target. For this reason, we used the solar irradiance calibration method (Eq. (1)) to reduce all spectra used in this work.

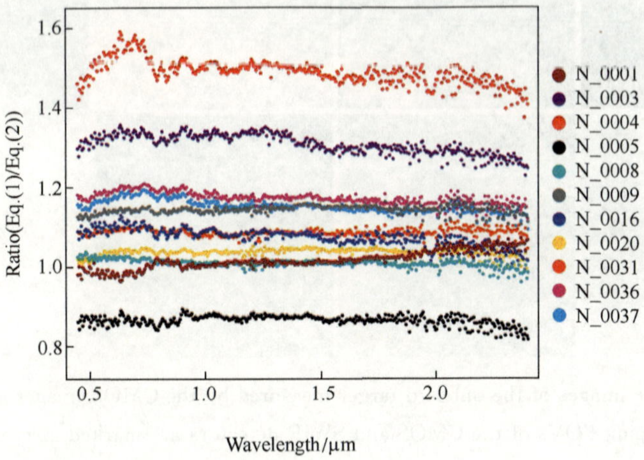

Fig. B2　The ratio of spectra obtained by using Eqs. (1) and (2)

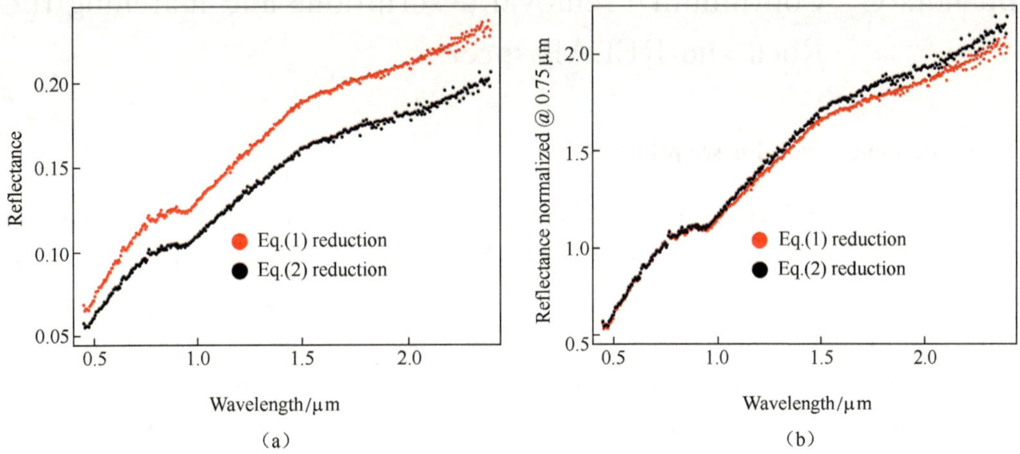

Fig. B3　Comparisons of the *REFF* spectra obtained by using Eqs. (1) and (2) for N _0036 measurement. (a) Spectra obtained by using Eq. (1) and (b) spectra obtained by using Eq. (2). Gap 1 and Gap 2 existing in the original spectra as shown in Fig. B1 have been eliminated by scaling

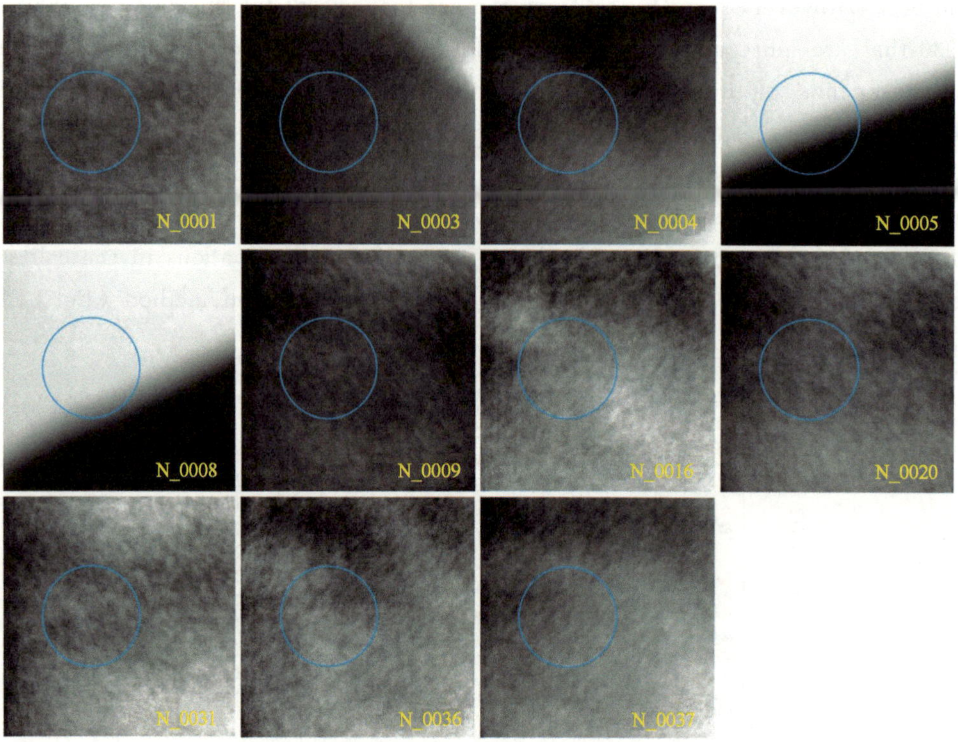

Fig. B4　The images of the onboard target measured by the CMOS imager at 600nm. The overlapping FOVs of the CMOS and SWIR detectors are marked as cyan circles.

Appendix C　Continuum - removal descriptions and matching the Rock and RELAB spectra

C. 1　Continuum removal descriptions

To compare the effects of using different continuum - removal methods, we chose three typical continuum functions (CF) as shown in Fig. C1 for the case of CE - 4 N _ 0015 measurement. The first one (CF1 in Fig. C1 (a)) is a straight line that connects the first (450nm) and the last (2395nm) data points in the spectra. In case the last data point (2395nm) drops down significantly and thus must have lower data quality, we chose a nearby data point that appears more reliable. The second one (CF2 in Fig. C1 (a)) uses a 2 - segment broken line that passes the maximum points around 750, 1550, and 2395nm. The third one (CF3 in Fig. C1 (a)) is straight - line segments generated in the following steps with the condition that the continuum should be larger than reflectance everywhere by at least - 1E - 6. The first line segment for trial was generated by connecting the 1st and 2nd data points (450nm and 455nm) and then extending to the last data point (2395nm). If the condition was violated then the next data point (3rd data point) was connected to the 1st data point and then extended to 2395nm, and so forth. In the case of N _ 0015, the first line segment that met the condition was found by connecting the 1st and 13th data points as shown in Fig. C1(b). The second line segment was generated in a similar way by starting with the 13th and 14th data points as the initial trial and the segment that met the condition was found to be the straight line that passes the 13th and 37th data points (Fig. C1(b)). The final continuum is shown as the gray line in Fig. C1 (a).

The continuum - removed spectra were obtained by dividing the CE - 4 spectra by the continuum function and the results for N _0015 are shown in Fig. C1(c). It is evident that the continuum - removed spectra using different continuum functions have different absorption features (Fig. C1(d)). However, in the following we will show that as long as the CF is used consistently for both in - situ measurement and library spectra, the matching results for different continuum - removal methods agree with each other well.

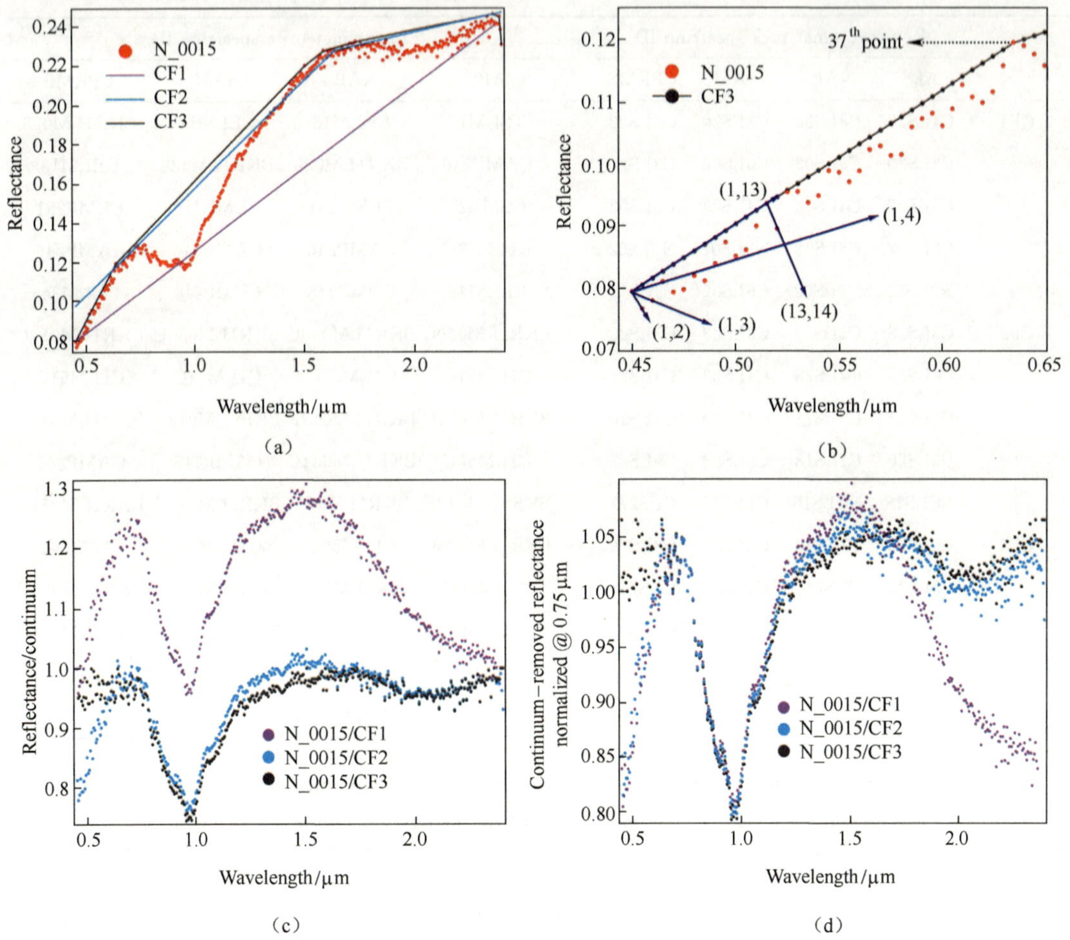

Fig. C1　Different continuum functions (CF) and the corresponding continuum - removed spectra for CE - 4 N _0015measurement.

(a) *REFF* reflectance of N _0015 and the three continuum functions described in text, CF1, CF2, and CF3.

(b) The first two line segments of the third continuum function (CF3).

(c) Continuum - removed spectra of N _ 0015, and (d) the normalized (at 750nm) continuum - removed spectra

C.2　Search of best matched RELAB spectra

The Rock's spectrum (N _0015) and 755 RELAB spectra of lunar meteorites and returned lunar rocks (resampled to the sampling interval of CE - 4 measurements) were continuum - removed using three CFs and then compared using four algorithms (Eqs. (3) — (6)). Tables C1 and C2 summarize the top five best matches and Fig. C2 shows the spectra and the normalized spectra. The best matched spectra with their continua removed using different continuum - removal methods are shown in Fig. C3. Since the three continuum - removal methods produced similar results, the spectra produced by using CF3 were chosen to present the results.

Tab. C1　The top five best matched RELAB spectral identifier (spectrum ID) using three continuum - removal methods and four searching algorithms (Eqs. (3) — (6))

	Returned lunar rock spectrum ID				Lunar meteorite spectrum ID			
	ABS	NABS	SAM	CPRMS	ABS	NABS	SAM	CPRMS
CF1	C4LS23	C3LS20	C3LS20	C5LS02	TSLM13	CELM13	CELM13	BKR1LM028
	C3LS20	C5LS02	C5LS02	C3LS20	CAMP246	BKR1LM028	BKR1LM028	CELM13
	C5LS02	C4LS23	C4LS23	C6LS02	CCMB29	CEMP251	CEMP251	CEMP251
	C6LS02	C6LS02	N2LS02	N2LS02	C1LM51C	CAMP246	CAMP246	CAMP246
	N2LS02	N2LS02	C6LS02	S3LS02	CELM34	CDMP248	CDMP248	CELM34
CF2	C3LS23	C3LS23	C3LS23	C3LS23	BKR1LM050C	BKR1LM050C	BKR1LM050C	BKR1LM050C
	C4LS23	C4LS23	C4LS23	C4LS23	C1LM51C	C1LM51C	C1LM51C	C1LM51C
	C5LS23	C5LS23	C3LS20	C3LS20	BKR1LM051C	BKR1LM013F	BKR1LM013F	BKR1LM013F
	C3LS20	C1LS18	C5LS23	C5LS23	C1LM50C	BKR1LM051C	CAMP245	CAMP245
	NCLS18	NCLS18	CLS277	CLS277	BKR1LM013F	BKR1LM028	BKR1LM051C	BKR1LM051C
CF3	C5LS23	C3LS23	C3LS23	C3LS23	BKR1LM050C	C1LM50C	C1LM50C	C1LM50C
	C3LS23	CLS750	C4LS23	CLS750	C1LM50C	BKR1LM050C	BKR1LM050C	BKR1LM050C
	C4LS23	C4LS23	CLS750	C4LS23	BKR1LM028	BKR1LM013F	BKR1LM013F	BKR1LM013F
	C3LU06	C5LS23	C5LS23	C5LS23	C1LM50	C1LM51C	C1LM51C	C1LM51C
	CLS741	C3LU06	C3LU06	C3LU06	C1LM51C	BKR1LM051C	BKR1LM028	BKR1LM051C

Tab. C2　The rock types of returned lunar rocks (Moon - Ret) and meteorites (Moon - Met) corresponding to the RELAB spectral identifier (Spectrum ID) listed in Tab. C1

Spectrum ID	Source	Rock type
C5LS02, C6LS02, N2LS02, S3LS02	Moon - Ret	Basalt
C1LS18, NCLS18	Moon - Ret	Impact Melt Breccia
C3LS20	Moon - Ret	Matrix Breccia/Gabbroic Anorthosite
C3LS23, C4LS23, C5LS23	Moon - Ret	Regolith Breccia
CLS741	Moon - Ret	Mare Basalt
CLS277	Moon - Ret	Regolith Breccia

Continued Tab. C2

Spectrum ID	Source	Rock type
C3LU06	Moon - Ret	Mare Breccia
CLS750	Moon - Ret	Glass
CELM34	Moon - Met	Lunar Feldspathic Breccia
C1LM50	Moon - Met	Anorthositic Impact Melt Breccia
BKR1LM050C，C1LM50C	Moon - Met	Anorthositic Impact Melt Breccia
BKR1LM051C，C1LM51C	Moon - Met	Impact Melt Breccia
CELM13	Moon - Met	Anorthositic Clast
BKR1LM013F	Moon - Met	Anorthositic Clast
BKR1LM028	Moon - Met	Anorthosite - bearing Basaltic Regolith Breccia
CCMB29	Moon - Met	Gabbroic
CAMP245	Moon - Met	Basaltic Regolith
CAMP246	Moon - Met	Anorthositic Fragmental
CDMP248	Moon - Met	Anorthositic Fragmental
CEMP251	Moon - Met	Basaltic Regolith

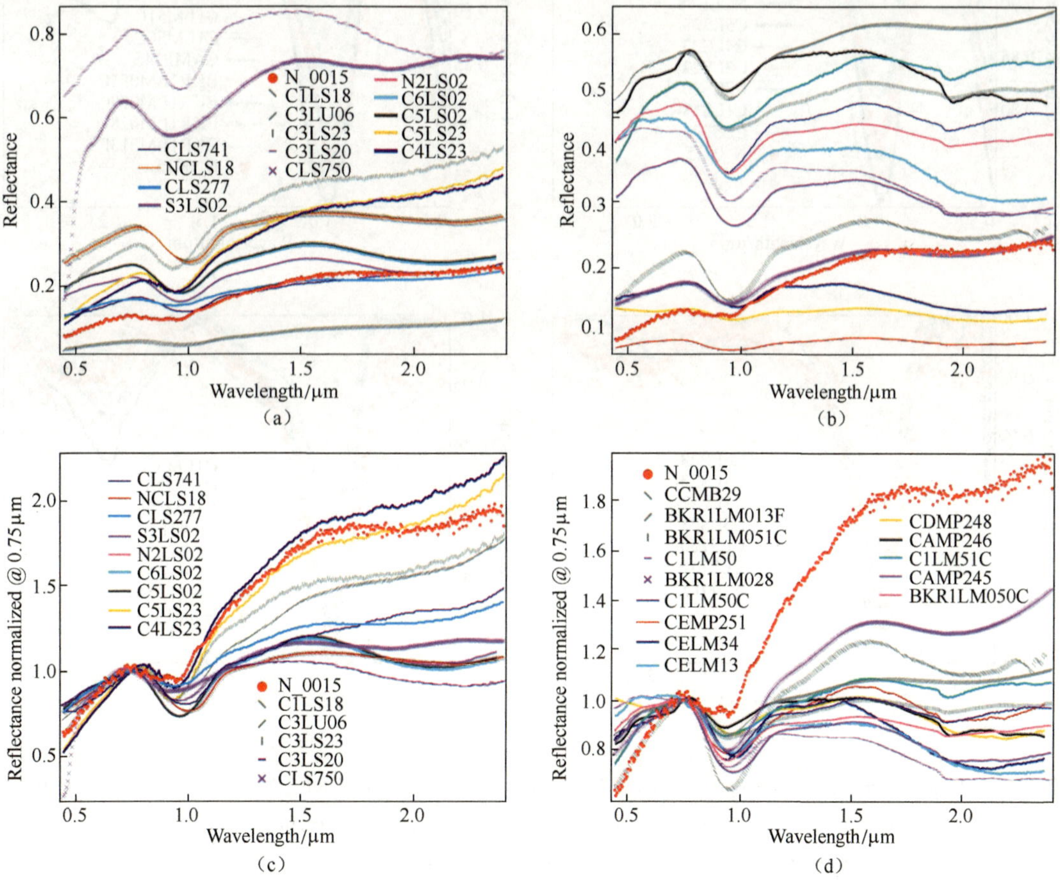

Fig. C2. The best matched RELAB spectra of returned lunar rocks ((a) and (c)) and lunar meteorites ((b) and (d)) and the Rock's spectrum N _0015. (a) — (b) Reflectance in *REFF*. (c) — (d) *REFF* normalized to 750nm. (b) and (d) share the same legend

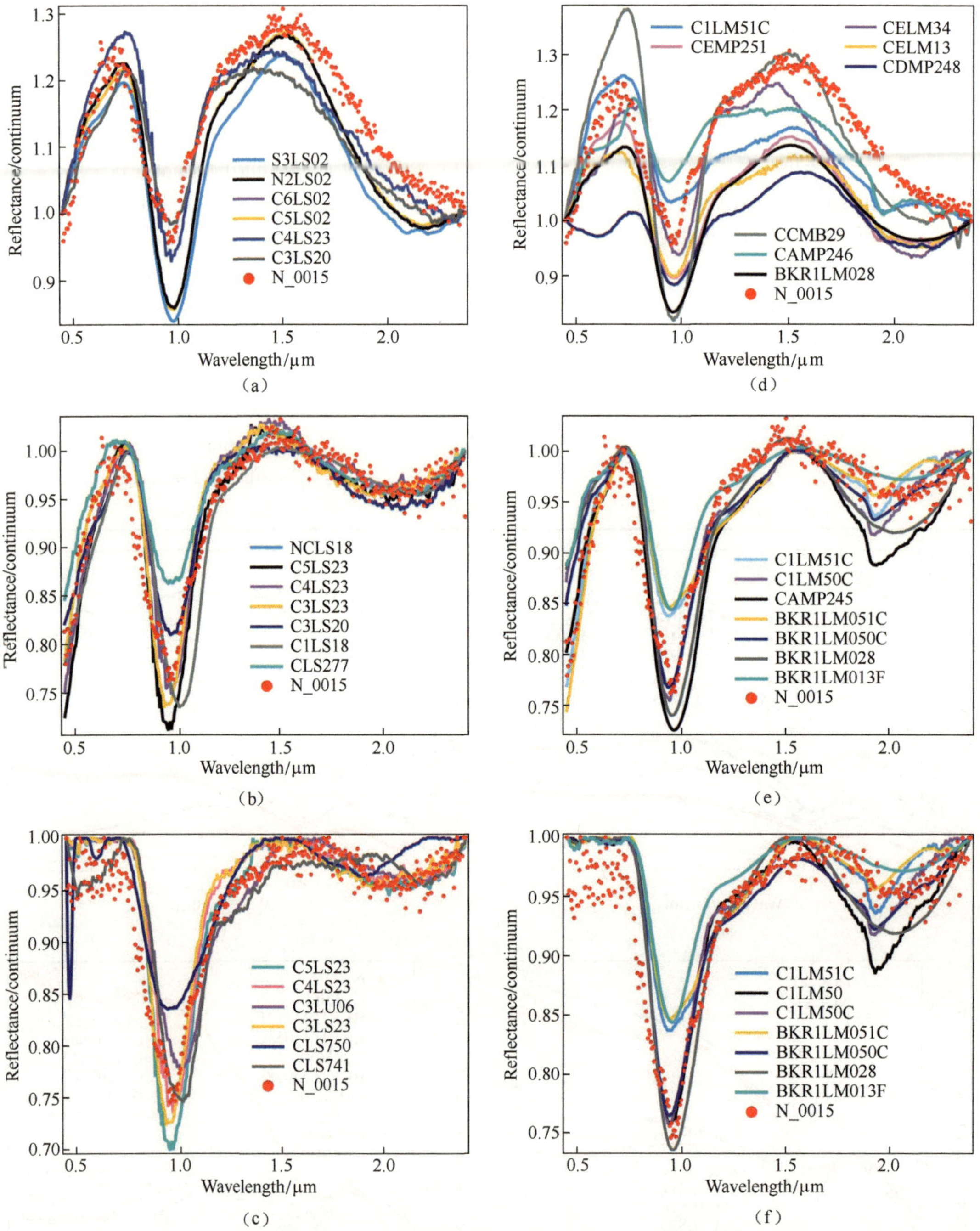

Fig. C3　Continuum - removed spectra of the Rock and the best matched RELAB spectra of returned lunar rocks and lunar meteorites. (a) — (c) Continuum - removed spectra of returned lunar rocks and the CE - 4 N_0015 using continuum functions CF1, CF2, and CF3, respectively. (d) — (f) Continuum - removed spectra of lunar meteorites and the Rock's spectrum (CE - 4 N_0015) using continuum functions CF1, CF2, and CF3, respectively

Appendix D　Similarities between the Rock's spectrum and the M^3 measurements

We used four algorithms, the ABS, NABS, SAM and CPRMS (Eqs. (3)－(6)), to calculate the similarities between the Rock's spectrum and the M^3 measurements as shown in Fig. D1. The base image of Fig. D1 is a mosaic of five M^3 images of the study area and the colored pixels are calculated similarities. Similar to the matching results of the Rock's spectrum with RELAB data, the similarities using the four algorithms are nearly identical. Since higher SAM values and lower ABS, NABS, CPRMS values correspond to a higher similarity, those pixels with ABS, NABS values higher than 2 (Fig. D1 (a), (b)), and those pixels with SAM values lower than 0.9993 (Fig. D1 (c)), and those with CPRMS values higher than 0.035 (Fig. D1 (d)) are made transparent to enhance the comparison effects.

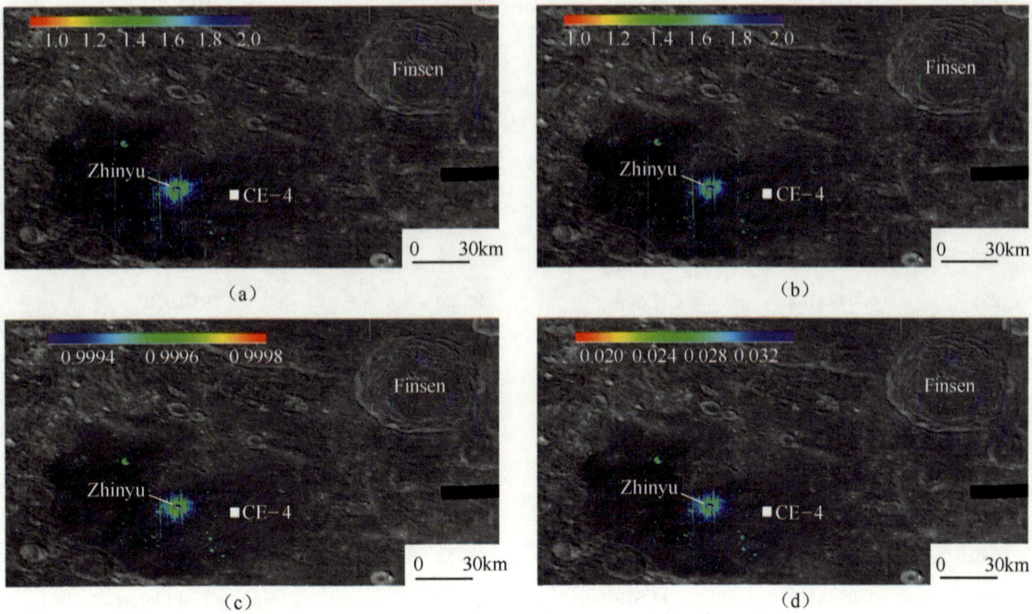

Fig. D1　Similarities between CE-4 measured the Rock's spectrum and M^3 spectra in the Von Kármán crater area. The base image is a mosaic of five M^3 images of the Von Kármán crater area. The CE-4 landing site (CE-4), Zhinyu and Finsen craters are indicated. The distance between the center of Zhinyu crater and CE-4 landing site is about 30km. From (a) to (d): the ABS, NABS, SAM, and CPRMS values between the N_0015 spectrum and all M^3 pixels, respectively

Appendix E. List of the CPRMS values between the RELAB spectra and CE-4 spectrum of the Rock, the plagioclase (PLG) abundances of the corresponding RELAB samples, and the spectral identifiers (Spectrum ID) of the RELAB spectra. Since a smaller CPRMS value corresponds to a higher similarity in spectral patterns, the sorting is from higher to lower similarities

Similarity order	CPRMS	Spectrum ID	PLG (vol%)	Reference
4	0.0257597	C1LM50C	80.00	Daubar et al., 2002

Continued

Similarity order	CPRMS	Spectrum ID	PLG (vol%)	Reference
5	0.0262912	BKR1LM050C	80.00	Daubar et al. , 2002
25	0.0329855	NCLS18	39.30	Tompkins and Pieters, 2010
26	0.0329855	C1LS18	39.30	Tompkins and Pieters, 2010
27	0.0330035	CMLS18	39.30	Tompkins and Pieters, 2010
29	0.0330486	C2LS18	39.30	Tompkins and Pieters, 2010
70	0.0395156	CDLS08	83.30	RELAB
73	0.0395652	SDLS08	83.30	RELAB
82	0.0401335	C2LS15	65.00	RELAB
84	0.0402705	SG1L15	65.00	RELAB
98	0.0418935	S3LS15	65.00	RELAB
104	0.0424179	C3LS15	65.00	RELAB
109	0.0425195	C1GL15	65.00	RELAB
132	0.0442049	CLS646	81.30	RELAB
154	0.0453983	CLS647	81.30	RELAB
179	0.0466227	CA2LS8	49.20	RELAB
180	0.0466297	SA2LS8	49.20	RELAB
228	0.0492364	C2LS08	49.20	RELAB
231	0.0493168	C1LS08	49.20	RELAB
271	0.0515916	S1LS15	65.00	RELAB
286	0.0521651	C1LS15	65.00	RELAB
289	0.0522906	NCLS77	46.97	Tompkins and Pieters, 2010
290	0.0522906	C1LS77	46.97	Tompkins and Pieters, 2010
319	0.0537172	CMLS77	46.97	Tompkins and Pieters, 2010
342	0.0545563	NCLS79	46.52	Tompkins and Pieters, 2010
343	0.0545563	C1LS79	46.52	Tompkins and Pieters, 2010
346	0.0546850	CMLS79	46.52	Tompkins and Pieters, 2010
349	0.0547421	C2LS79	46.52	Tompkins and Pieters, 2010
374	0.0560190	NCLS39	46.01	Tompkins and Pieters, 2010
375	0.0560190	C1LS39	46.01	Tompkins and Pieters, 2010
382	0.0565961	C2LS29	46.17	Tompkins and Pieters, 2010
387	0.0567739	CMLS29	46.17	Tompkins and Pieters, 2010
390	0.0567975	CMLS39	46.01	Tompkins and Pieters, 2010
396	0.0569610	C1LS29	46.17	Tompkins and Pieters, 2010
397	0.0569610	NCLS28	47.57	Tompkins and Pieters, 2010
409	0.0573019	CG5L15	52.60	RELAB

Continued

Similarity order	CPRMS	Spectrum ID	PLG (vol%)	Reference
413	0.0574264	CZLS28	47.57	Tompkins and Pieters, 2010
428	0.0580586	NCLS35	45.59	Tompkins and Pieters, 2010
429	0.0580586	C1LS35	45.59	Tompkins and Pieters, 2010
442	0.0586659	CMLS35	45.59	Tompkins and Pieters, 2010
445	0.0588032	NCLS29	46.17	Tompkins and Pieters, 2010
452	0.0594429	C2LS35	45.59	Tompkins and Pieters, 2010
500	0.0627427	CBLS08	93.90	RELAB
527	0.0650772	C5LS08	93.90	RELAB
528	0.0651102	S5LS08	93.90	RELAB
531	0.0655497	CLS060	96.00	RELAB
558	0.0681171	CEMP248	60.00	Hiroi et al., 2016
562	0.0683954	CG2L15	57.00	RELAB
573	0.0691920	SG2L15	57.00	RELAB
589	0.0714723	CG3L15	52.60	RELAB
590	0.0715770	SG3L15	52.60	RELAB
601	0.0726136	C1LR235	90.00	Dixon and Papike, 1975
603	0.0729721	C2GL15	57.00	RELAB
604	0.0732169	CLS687	20.00	Brown et al., 1975; Dymek et al., 1975; Longhi et al., 1974; McGee et al., 1977
616	0.0766419	CAMP248	100.00	Hiroi et al., 2016
640	0.0822571	C3GL15	52.60	RELAB
642	0.0842465	CLS648	81.30	RELAB
653	0.0878354	C1LR158	24.50	https://curator.jsc.nasa.gov/lunar/lsc/70035.pdf
654	0.0878354	BKR1LR158	24.50	https://curator.jsc.nasa.gov/lunar/lsc/70035.pdf
692	0.1036250	C1LM47	45.30	Righter et al., 2005
693	0.1036250	BKR1LM047	45.30	Righter et al., 2005
701	0.1113650	C1DD36	25.00	Boyd and Smith, 1971; Brown et al., 1971; Klein Jr et al., 1971; McGee et al., 1977; Neal et al., 1994; Papike et al., 1976
702	0.1119540	CLS067	31.20	https://curator.jsc.nasa.gov/lunar/lsc/15065r.pdf
704	0.1123120	BKR1LR204	30.10	RELAB
705	0.1123120	C1LR204	30.10	RELAB
706	0.1129370	C2LR204	30.10	RELAB
710	0.1169020	CLS465	31.20	https://curator.jsc.nasa.gov/lunar/lsc/15065r.pdf

Continued

Similarity order	CPRMS	Spectrum ID	PLG (vol%)	Reference
719	0.1235040	C2MB29	30.00	Oba and Kobayashi, 2001
731	0.1325590	C1LR154	32.00	https: //curator. jsc. nasa. gov/lunar/lsc/15555. pdf
732	0.1325590	BKR1LR154	32.00	https: //curator. jsc. nasa. gov/lunar/lsc/15555. pdf
742	0.1472160	C1LR153	30.10	RELAB
743	0.1472160	BKR2LR153	30.10	RELAB
747	0.1529810	S2LS03	4.00	Dymek et al., 1975
748	0.1530930	C2LS03	4.00	Dymek et al., 1975
749	0.1532220	S1LS03	4.00	Dymek et al., 1975
750	0.1534480	C1LS03	4.00	Dymek et al., 197

References

[1] Boyd, F., Smith, D., 1971. Compositional zoning in pyroxenes from lunar rock 12021, Oceanus Procellarum. J. Petrol. 12, 439 - 464.

[2] Brown, G., Emeleus, C., Holland, J., Peckett, A., Phillips, R., 1971. Picrite basalts, ferrobasalts, feldspathic norites, and rhyolites in a strongly fractionated lunar crust. In: Lunar and Planetary Science Conference Proceedings.

[3] Brown, G., Peckett, A., Emeleus, C., Phillips, R., Pinsent, R., 1975. Petrology and mineralogy of Apollo 17 mare basalts. In: Lunar and Planetary Science Conference Proceedings.

[4] Clark, R. N., Roush, T. L., 1984. Reflectance spectroscopy: quantitative analysis techniques for remote sensing applications. J. Geophys. Res. 89, 6329 - 6340. https: // doi. org/10. 1029/JB089iB07p06329.

[5] Collins, G. S., Melosh, H. J., Ivanov, B. A., 2004. Modeling damage and deformation in impact simulations. Meteorit. Planet. Sci. 39, 217 - 231. https: //doi. org/10. 1111/j. 1945 - 5100. 2004. tb00337. x.

[6] Daubar, I. J., Kring, D. A., Swindle, T. D., Jull, A. T., 2002. Northwest Africa 482: a crystalline impact - melt breccia from the lunar highlands. Meteorit. Planet. Sci. 37, 1797 - 1813.

[7] Denevi, B. W., Lucey, P. G., Sherman, S. B., 2008. Radiative transfer modeling of near - infrared spectra of lunar mare soils: theory and measurement. J. Geophys. Res. 113, E02003. https: //doi. org/10. 1029/2007JE002929.

[8] Di, K., Xu, B., Peng, M., Yue, Z., Liu, Z., Wan, W., Li, L., Zhou, J., 2016. Rock size - frequency distribution analysis at the Chang' E - 3 landing site. Planetary and Space Science 120, 103 - 112. https: //doi. org/10. 1016/j. pss. 2015. 11. 012.

[9] Di, K., Liu, Z., Liu, B., Wan, W., Peng, M., Wang, Y., Gou, S., Yue, Z., Xin, X., Jia, M., Niu, S., 2019a. Chang'e - 4 lander localization based on multi - source data. Journal of Remote Sensing 23, 177 - 184. https://doi.org/10.11834/jrs.20199015.

[10] Di, K., Zhu, M. - H., Yue, Z., Lin, Y., Wan, W., Liu, Z., Gou, S., Liu, B., Peng, M., Wang, Y., Niu, S., Zhang, J., Li, J., Xie, J., Xi, L., Yang, J., Xue, B., 2019b. Topographic evolution of Von Kármán crater revealed by the lunar rover Yutu - 2. Geophys. Res. Lett. 46, 12764 - 12770. https://doi.org/10.1029/2019GL085252.

[11] Dixon, J. R., Papike, J., 1975. Petrology of anorthosites from the Descartes region of the moon - Apollo 16. In: Lunar and Planetary Science Conference Proceedings.

[12] Dymek, R., Albee, A., Chodos, A., 1975. Comparative petrology of lunar cumulate rocks of possible primary origin - Dunite 72415, troctolite 76535, norite 78235, and anorthosite 62237. In: Lunar and Planetary Science Conference Proceedings.

[13] Gou, S., Di, K., Yue, Z., Liu, Z., 2019. Lunar deep materials observed by Chang'e - 4 rover. Earth Planet. Sci. Lett. 528, 115829 https://doi.org/10.1016/J.EPSL.2019.115829.

[14] Gueymard, C. A., 2004. The sun's total and spectral irradiance for solar energy applications and solar radiation models. Sol. Energy 76, 423 - 453. https://doi.org/10.1016/j.solener.2003.08.039.

[15] Hapke, Bruce, 2012. Theory of Reflectance and Emittance Spectroscopy. Cambridge University Press.

[16] He, Z., Li, C., Xu, R., Lv, G., Yuan, L., Wang, J., 2019. Spectrometers based on acousto - optic tunable filters for in - situ lunar surface measurement. JARS 13, 027502. https://doi.org/10.1117/1.JRS.13.027502.

[17] Hinrichs, J. L., Lucey, P. G., 2002. Temperature - dependent near - infrared spectral properties of minerals, meteorites, and lunar soil. Icarus 155, 169 - 180. https://doi.org/10.1006/ICAR.2001.6754.

[18] Hiroi, T., Kaiden, H., Yamaguchi, A., Kojima, H., Uemoto, K., Ohtake, M., Arai, T., Sasaki, S., 2016. Visible and near - infrared spectral survey of lunar meteorites recovered by the National Institute of Polar Research. Polar Science. https://doi.org/10.1016/j.polar.2016.06.004.

[19] Hu, X., Ma, P., Yang, Y., Zhu, M., Jiang, T., Lucey, P. G., Sun, L., Zhang, H., Li, C., Xu, R., He, Z., Lin, H., Huang, C., Sun, Y., 2019. Mineral abundances inferred from in situ reflectance measurements of Chang'E - 4 landing site in South Pole - Aitken Basin. Geophys. Res. Lett. 46 https://doi.org/10.1029/2019GL084531.

[20] Huang, J., Xiao, Z., Flahaut, J., Martinot, M., Head, J., Xiao, X., Xie, M.,

Xiao, L., 2018. Geological characteristics of Von Kármán crater, northwestern south pole - Aitken Basin: Chang'E - 4 landing site region. Journal of Geophysical Research: Planets 123, 1684 - 1700. https://doi.org/10.1029/2018JE005577.

[21] Hurwitz, D. M., Kring, D. A., 2014. Differentiation of the South Pole - Aitken basin impact melt sheet: implications for lunar exploration. J. Geophys. Res. 110, 1110 - 1133. https://doi.org/10.1002/2013JE004530.

[22] Jin, W., Zhang, H., Yuan, Y., Yang, Y., Shkuratov, Y. G., Lucey, P. G., Kaydash, V. G., Zhu, M. - H., Xue, B., Di, K., Xu, B., Wan, W., Xiao, L., Wang, Z., 2015. In situ optical measurements of Chang'E - 3 landing site in Mare Imbrium: 2. Photometric properties of the regolith. Geophys. Res. Lett. 42, 8312 - 8319. https://doi.org/10.1002/ 2015GL065789.

[23] Klein Jr., C., Drake, J., Frondel, C., 1971. Mineralogical, petrological, and chemical features of four Apollo 12 lunar microgabbros. In: Lunar and Planetary Science Conference Proceedings.

[24] Kruse, F., Lefkoff, A., Boardman, J., Heidebrecht, K., Shapiro, A., Barloon, P., Goetz, A., 1993. The spectral image - processing system (sips) - interactive visualization and analysis of imaging spectrometer data. Remote Sens. Environ. 44, 145 -163. https:// doi.org/10.1016/0034 - 4257 (93) 90013 - N.

[25] Le Feuvre, M., Wieczorek, M. A., 2011. Nonuniform cratering of the moon and a revised crater chronology of the inner solar system. Icarus 214, 1 - 20. https://doi.org/10.1016/j.icarus.2011.03.010.

[26] Li, C., Xu, R., Lv, G., Yuan, L. Y., He, Z. P., Wang, J. Y., 2019a. Detection and calibration characteristics of the visible and near - infrared imaging spectrometer in the Chang'e - 4. Rev. Sci. Instrum. 90, 103106 https://doi.org/10.1063/1.5089737.

[27] Li, C. L., Liu, D., Liu, B., Ren, X., Liu, J., He, Z., Zuo, W., Zeng, X., Xu, R., Tan, X., Zhang, X., Chen, W., Shu, R., Wen, W., Su, Y., Zhang, H., Ouyang, Z., 2019b. Chang'E - 4 initial spectroscopic identification of lunar far - side mantle - derived materials. Nature 569, 378 - 382. https://doi.org/10.1038/s41586 - 019 -1189 - 0.

[28] Li, C., Wang, Z., Xu, R., Lv, G., Yuan, L., He, Z., Wang, J., 2019b. The scientific information model of Chang' e - 4 visible and near - IR imaging spectrometer (VNIS) and in - flight verification. Sensors 19, 2806. https://doi.org/10.3390/s19122806.

[29] Lin, Honglei, He, Z., Yang, W., Lin, Y., Xu, R., Zhang, C., Zhu, M. - H., Chang, R., Zhang, J., Li, C., Hongyu, Lin, Liu, Y., Gou, S., Wei, Y., Hu, S., Xue, C., Yang, J., Zhong, J., Fu, X., Wan, W., Zou, Y., 2019. Olivine - norite rock detected by the lunar rover Yutu - 2 likely crystallized from the SPA impact melt pool. National Science Review nwz 183. https://doi.org/10.1093/nsr/nwz183.

[30] Liu, B., Li, C., Zhang, G., Xu, R., Liu, J., Ren, X., Tan, X., Zhang, X., Zuo,

W., Wen, W., 2014. Data processing and preliminary results of the Chang'e - 3 VIS/NIR imaging spectrometer in - situ analysis. Res. Astron. Astrophys. 14, 1578 -1594. https://doi.org/10.1088/1674-4527/14/12/007.

[31] Longhi, J., Walker, D., Grove, T., Stolper, E., Hays, J., 1974. The petrology of the Apollo 17 mare basalts. In: Lunar and Planetary Science Conference Proceedings.

[32] Lucey, P. G., 2004. Mineral maps of the moon. Geophys. Res. Lett. 31, L08701 https://doi.org/10.1029/2003GL019406.

[33] Lucey, P. G., Norman, J. A., Crites, S. T., Taylor, G. J., Hawke, B. R., Lemelin, M., Melosh, H. J., 2014. A large spectral survey of small lunar craters: implications for the composition of the lunar mantle. Am. Mineral. 99, 2251 - 2257. https://doi.org/10.2138/am-2014-4854.

[34] Luther, R., Zhu, M. - H., Collins, G., Wünnemann, K., 2018. Effect of target properties and impact velocity on ejection dynamics and ejecta deposition. Meteorit. Planet. Sci. 53, 1705 - 1732. https://doi.org/10.1111/MAPS.13143.

[35] McGee, P. E., Warner, J. L., Simonds, C., 1977. Introduction to the Apollo Collections: Part 1. Lunar Igneous Rocks.

[36] McGetchin, T. R., Settle, M., Head, J. W., 1973. Radial thickness variation in impact crater ejecta: implications for lunar basin deposits. Earth Planet. Sci. Lett. 20, 226 - 236. https://doi.org/10.1016/0012-821X(73)90162-3.

[37] Meyer, C., 2009. https://www.lpi.usra.edu/lunar/samples/atlas/compendium/15565.pdf.

[38] Neal, C. R., Hacker, M. D., Snyder, G. A., Taylor, L. A., Liu, Y., Schmitt, R. A., 1994. Basalt generation at the Apollo 12 site, part 1: new data, classification, and re - evaluation. Meteoritics 29, 334 - 348.

[39] Nyquist, L. E., Bogard, D. D., Shih, C. Y., Wiesmann, H., 2002. Negative εNd in anorthositic clasts in Yamato 86032 and MAC88105: evidence for the LMO?. In: Presented at the Lunar and Planetary Science Conference, p. 1289.

[40] Nyquist, L., Bogard, D., Yamaguchi, A., Shih, C. - Y., Karouji, Y., Ebihara, M., Reese, Y., Garrison, D., McKay, G., Takeda, H., 2006. Feldspathic clasts in Yamato -86032: remnants of the lunar crust with implications for its formation and impact history. Geochim. Cosmochim. Acta 70, 5990 - 6015. https://doi.org/10.1016/j.gca.2006.07.042.

[41] Oba, T., Kobayashi, Y., 2001. The mineral assemblage of symplectites in lunar meteorite Asuka - 881757. Antarctic Meteorite Research 14, 21 - 27.

[42] Papike, J. J., Hodges, F. N., Bence, A., Cameron, M., Rhodes, J., 1976. Mare basalts: crystal chemistry, mineralogy, and petrology. Rev. Geophys. 14, 475 - 540.

[43] Pierazzo, E., Kring, D. A., Melosh, H. J., 1998. Hydrocode simulation of the Chicxulub impact event and the production of climatically active gases. Journal of Geophysical Research: Planets 103 (E12), 28, 607 - 28, 625. https://doi.org/

10.1029/98je02496.

[44] Potter, R. W. K., Collins, G. S., Kiefer, W. S., McGovern, P. J., Kring, D. A., 2012. Constraining the size of the South Pole - Aitken basin impact. Icarus 220, 730 - 743. https://doi.org/10.1016/J.ICARUS.2012.05.032.

[45] Qiao, L., Ling, Z., Fu, X., Li, B., 2019. Geological characterization of the Chang'e - 4 landing area on the lunar farside. Icarus 333, 37 - 51. https://doi.org/10.1016/j.icarus.2019.05.029.

[46] Righter, K., Collins, S., Brandon, A., 2005. Mineralogy and petrology of the LaPaz Icefield lunar mare basaltic meteorites. Meteorit. Planet. Sci. 40, 1703 - 1722.

[47] Roush, T. L., Singer, R. B., 1986. Gaussian analysis of temperature effects on the reflectance spectra of mafic minerals in the 1 - μm region. J. Geophys. Res. 91, 10301 - 10308. https://doi.org/10.1029/JB091IB10P10301.

[48] Roush, T. L., Singer, R. B., 1987. Possible temperature variation effects on the interpretation of spatially resolved reflectance observations of asteroid surfaces. Icarus 69, 571 - 574. https://doi.org/10.1016/0019-1035(87)90026-1.

[49] Shuvalov, V., 2011. Ejecta deposition after oblique impacts: an influence of impact scale: ejecta deposition after oblique impacts. Meteorit. Planet. Sci. 46, 1713 - 1718. https://doi.org/10.1111/j.1945-5100.2011.01259.x.

[50] Shuvalov, V., Dypvik, H., 2013. Distribution of ejecta from small impact craters. Meteorit. Planet. Sci. 48, 1034 - 1042. https://doi.org/10.1111/maps.12127.

[51] Singer, R. B., Roush, T. L., 1985. Effects of temperature on remotely sensed mineral absorption features. J. Geophys. Res. 90, 12434 - 12444. https://doi.org/10.1029/JB090IB14P12434.

[52] Stöffler, D., Gault, D. E., Wedekind, J., Polkowski, G., 1975. Experimental hypervelocity impact into quartz sand: distribution and shock metamorphism of ejecta. J. Geophys. Res. 80, 4062 - 4077. https://doi.org/10.1029/JB080i029p04062.

[53] Takeda, H., Kojima, H., Nishi, F., Yanai, K., Lindstrom, M. M., 1989. Preliminary report on the Yamato - 86032 lunar meteorite: I. Recovery sample descriptions, mineralogy and petrography. Antarctic Meteorite Research 2, 3 - 14.

[54] Takeda, H., Miyamoto, M., Mori, H., Wentworth, S. J., McKay, D. S., 1990. Mineralogical comparison of the Y86032 - type lunar meteorites to feldspathic fragmental breccia 67016. In: Presented at the Lunar and Planetary Science Conference Proceedings, pp. 91 - 100.

[55] Takeda, H., Nyquist, L. E., Kojima, H., 2002. Mineralogical study of a gray anorthositic clast in the Yamato 86032 lunar meteorite: windows to the far - side highland. In: Presented at the Lunar and Planetary Science Conference.

[56] Taylor, K. E., 2001. Summarizing multiple aspects of model performance in a single diagram. J. Geophys. Res. 106, 7183 - 7192. https://doi.org/10.1029/2000JD900719.

[57] Tompkins, S., Pieters, C. M., 2010. Spectral characteristics of lunar impact melts

and inferred mineralogy. Meteorit. Planet. Sci. https://doi.org/10.1111/j.1945-5100.2010.01074.x.

[58] Vaughan, W. M., Head, J. W., 2014. Impact melt differentiation in the South Pole-Aitken basin: some observations and speculations. Planetary and Space Science 91, 101-106. https://doi.org/10.1016/J.PSS.2013.11.010.

[59] Warren, P. H., Taylor, G. J., Keil, K., Marshall, C., Wasson, J. T., 1982. Foraging westward for pristine nonmare rocks - complications for petrogenetic models. In: Lunar and Planetary Science Conference Proceedings, pp. 21-40.

[60] Wu, Y., Hapke, B., 2018. Spectroscopic observations of the Moon at the lunar surface. Earth Planet. Sci. Lett. 484, 145-153. https://doi.org/10.1016/j.epsl.2017.12.003.

[61] Wu, W., Li, C., Zuo, W., Zhang, H., Liu, J., Wen, W., Su, Y., Ren, X., Yan, J., Yu, D., Dong, G., Wang, C., Sun, Z., Liu, E., Yang, J., Ouyang, Z., 2019. Lunar farside to be explored by Chang'e-4. Nat. Geosci. 12, 222-223. https://doi.org/10.1038/s41561-019-0341-7.

[62] Wünnemann, K., Collins, G. S., Melosh, H. J., 2006. A strain-based porosity model for use in hydrocode simulations of impacts and implications for transient crater growth in porous targets. Icarus 180, 514-527. https://doi.org/10.1016/j.icarus.2005.10.013.

[63] Wünnemann, K., Zhu, M.-H., Stöffler, D., 2016. Impacts into quartz sand: crater formation, shock metamorphism, and ejecta distribution in laboratory experiments and numerical models. Meteorit. Planet. Sci. 51, 1762-1794. https://doi.org/10.1111/maps.12710.

[64] Xie, M., Zhu, M.-H., 2016. Estimates of primary ejecta and local material for the Orientale basin: implications for the formation and ballistic sedimentation of multi-ring basins. Earth Planet. Sci. Lett. 440, 71-80. https://doi.org/10.1016/J.EPSL.2016.02.012.

[65] Yamaguchi, A., Takeda, H., Nyquist, L. E., Bogard, D. D., Ebihara, M., Karouji, Y., 2004. The orign and impact history of lunar meteorite Yamato 86032. In: Presented at the Lunar and Planetary Science.

[66] Zhang, H., Yang, Y., Yuan, Y., Jin, W., Lucey, P. G., Zhu, M.-H., Kaydash, V. G., Shkuratov, Y. G., Di, K., Wan, W., Xu, B., Xiao, L., Wang, Z., Xue, B., 2015. In situ optical measurements of Chang'E-3 landing site in Mare Imbrium: 1. Mineral abundances inferred from spectral reflectance. Geophys. Res. Lett. 42, 6945-6950. https://doi.org/10.1002/2015GL065273.

[67] Zhu, M.-H., Wünnemann, K., Potter, R. W. K., 2015. Numerical modeling of the ejecta distribution and formation of the Orientale basin on the Moon. J. Geophys. Res. 120, 2118-2134. https://doi.org/10.1002/2015JE004827.

[68] Zhu, M.-H., Wünnemann, K., Artemieva, N., 2017. Effects of Moon's thermal

state on the impact basin ejecta distribution. Geophys. Res. Lett. 44, 11292 - 11300. https://doi.org/10.1002/2017GL075405.

[69] Zhu, M. H., Wünnemann, K., Morbidelli, A., Artemieva, N., 2018. South Pole - Aitken Basin may contribute insignificantly to the late accretion of the Moon. In: Lunar and Planetary Science Conference, p. 1955.

[70] Zhu, M. - H., Wünnemann, K., Potter, R. W. K., Kleine, T., Morbidelli, A., 2019. Are the Moon's nearside - farside asymmetries the result of a giant impact? J. Geophys. Res. 124, 2117 - 2140. https://doi.org/10.1029/2018JE005826.

Chang'E - 4 Rover Spectra Revealing Micro - scale Surface Thermophysical Properties of the Moon

Yunzhao Wu[1,2,3], Ekkehard Kührt[4,5], Matthias Grott[6], Qi Jin[2], Tianyi Xu[2], Jörn Helbert[6], Maximilian Hamm[6,7], Nannan Qin[1,8], Jinsong Ma[9], Jean - Baptiste Vincent[6], and Paul Hayne[10]

1. Key Laboratory of Planetary Sciences, Purple Mountain Observatory, Chinese Academy of Sciences, Nanjing, China
2. Space Science Institute, Macau University of Science and Technology, Macau, China
3. Center for Excellence in Comparative Planetology, Chinese Academy of Science, Hefei, China
4. Institute of Optical Sensor Systems, German Aerospace Center, Berlin, Germany
5. Qian Xuesen Laboratory of Space Technology, China Academy of Space Technology, Beijing, China
6. Institute of Planetary Research, German Aerospace Center, Berlin, Germany
7. Institute for Mathematics, University of Potsdam, Potsdam - Golm, Germany
8. Jiangsu Center for Collaborative Innovation in Geographical Information Resource Development and Application, Nanjing, China
9. School of Geographic and Oceanographic Sciences, Nanjing University, Nanjing, China
10. Laboratory for Atmospheric & Space Physics, and Astrophysical & Planetary Sciences Department, University of Colorado Boulder, Boulder, CO, USA

Abstract: Lunar surface temperature variations provide key information about the thermophysical properties of the regolith. To date, temperatures have been measured using telescopes and orbiter instruments, providing information from mid - infrared to long infrared (IR) wavelengths. Here, we report on temperature measurements in the short - wavelength IR at centimeter scales observed in situ by the Chang'E - 4 rover. These local observations are an important complement to the existing large - scale data. We show that even at 2.39μm, where reflected radiation dominates the spectrum, thermal information can be retrieved from the data. The observed thermal radiances and derived temperatures depend on the observation geometry, specifically the relative azimuth angle between Sun and detector. This indicates that surface roughness on subresolution (millimeter) scales causes a non - Lambertian radiation pattern of the emitted flux that is per definition independent of the observation angle. This behavior must be considered when deriving temperatures and thermal properties of lunar regolith.

Plain Language Summary: The Chang'E－4 spacecraft landed in the Von Kármán Crater in the Moon's South Pole－Aitken Basin. A Visible and Near Infrared Spectrometer (VNIS) was carried on the "Yutu－2" rover. The VNIS measured thermal emission on the surface of the Moon at short wavelengths and temperature variations at centimeter scales were retrieved during the first, second, and tenth day of rover operations on the lunar surface. These in situ observations are an important complement to the existing large－scale measurements taken from orbit and reveal that the millimeter scale roughness plays an important role when deriving temperatures and physical properties of the lunar surface regolith.

Key Points:

• The Chang' E－4 VNIS observed the lunar surface in situ in the 0.45－2.4μm wavelength range

• Emitted flux at centimeter scales is non－Lambertian and depends on surface roughness and observation geometry

• Surface roughness as described by the bidirectional RMS slope was found to be 22°－24°

1 Introduction

The Moon's surface temperature holds important information about the thermophysical properties of the lunar regolith. It has been measured by Earth－based telescopes (Hayne et al., 2015; Maghrabi, 2016; Pettit & Nicholson, 1930; Shorthill, 1962; Sinton, 1962), orbiting spacecraft (Lawson et al., 2000; Paige et al., 2010; Vasavada et al., 2012; J. P. Williams et al., 2017), as well as in situ during the Apollo missions (Grott et al., 2010; Keihm & Langseth, 1973). The surface brightness temperature of the Moon and other atmosphereless bodies strongly depends on surface roughness, and roughness effects have been studied extensively (e. g., Bandfield et al., 2015; Buhl et al., 1968; Davidsson et al., 2015; Kührt et al., 1992; Lagerros, 1997; Rosenburg et al., 2011, 2015; Rozitis & Green, 2011; Rubanenko et al., 2020; Smith, 1967; Spencer, 1990; Vasavada et al., 1999).

For many atmosphereless objects, it was found that surface brightness temperatures derived from remote sensing data were significantly higher than predicted by theoretical models. This discrepancy has been explained in terms of so－called "beaming" (Lebofsky et al., 1986), a combination of beaming and self－heating (Spencer, 1990), a "rough surface" (Palmer & Sykes, 2014), and microscale roughness (Bandfield et al., 2015). When interpreting infrared flux in terms of thermophysical properties, care must be taken to account for surface features below the resolution of the digital terrain model, as this roughness at small scales is not known and must be modeled (see, e. g., Davidsson et al., 2015, for a detailed review).

Furthermore, as pointed out by Davidsson et al. (2015), studies at short wavelength are rare but important to learn how the degree of roughness and dependency on the type of topography may differ compared to those in the thermal infrared. While temperature measurements were made with the Clementine longwave infrared camera as well as the Lunar Reconnaissance Orbiter (LRO) Diviner Radiometer at 8.75μm (Lawson et al., 2000) and beyond 13μm (Paige et al., 2010), we here report on temperature measurements in the relatively unexplored wavelength region around 2.39μm by the Visible and Near Infrared Imaging Spectrometer (VNIS) onboard the Chang'E-4 (CE-4) Yutu-2 rover. In this paper, we analyze the thermal information contained in the measured lunar spectra at 2.39μm and prove by a statistical test that considering surface roughness significantly improves the agreement of the surface temperatures derived from the data and the thermophysical model.

2 Data and Methods

2.1 Data

At 2:25:30 UTC on January 3, 2019, the CE-4 spacecraft landed in Von Kármán Crater in the South Pole-Aitken Basin. The VNIS, one of its rover's instruments, consists of a visible and near infrared imaging spectrometer (0.45—0.95 μm) and a shortwave infrared (SWIR) spectrometer (0.9—2.4μm). It is mounted on the front of the rover and observes lunar surface from a height of 0.69 m above the surface at 45° emission angle. The nominal spatial resolution of the imaging spectrometer is 0.53—0.80mm/pixel and the field of view (FOV) is an isosceles trapezoid with a height of 20.6cm and two parallel sides of 13.5cm and 15.7cm. The FOV of the SWIR spectrometer is oval with diameters of 6.14cm and 8.68cm. Details of the instrument are provided in Wu et al. (2018).

Fig. 1 (a) and 1 (b) show the raw spectra measured by the CE-4 VNIS during the first two lunar days of operation. The y axes, I/F, is the normalized radiance which is calculated as the radiance from the Moon divided by the radiance from a perfect Lambert surface illuminated by the Sun. The spectral upturns at wavelengths beyond 2μm are consistent with the angles of solar incidence. The two spectra showing a marked upturn of radiance (#1 and #8) correspond to local times having low incidence angles, and the two spectra which do not show a clear upturn (#3 and #4) correspond to local times having high incidence angles. The observations from the CE-4 VNIS confirms Wu and Hapke (2018) which firstly revealed that the Chang'E-3 (CE-3) VNIS captured thermal emission.

To investigate the microscale temperature variations, the CE-4 team performed

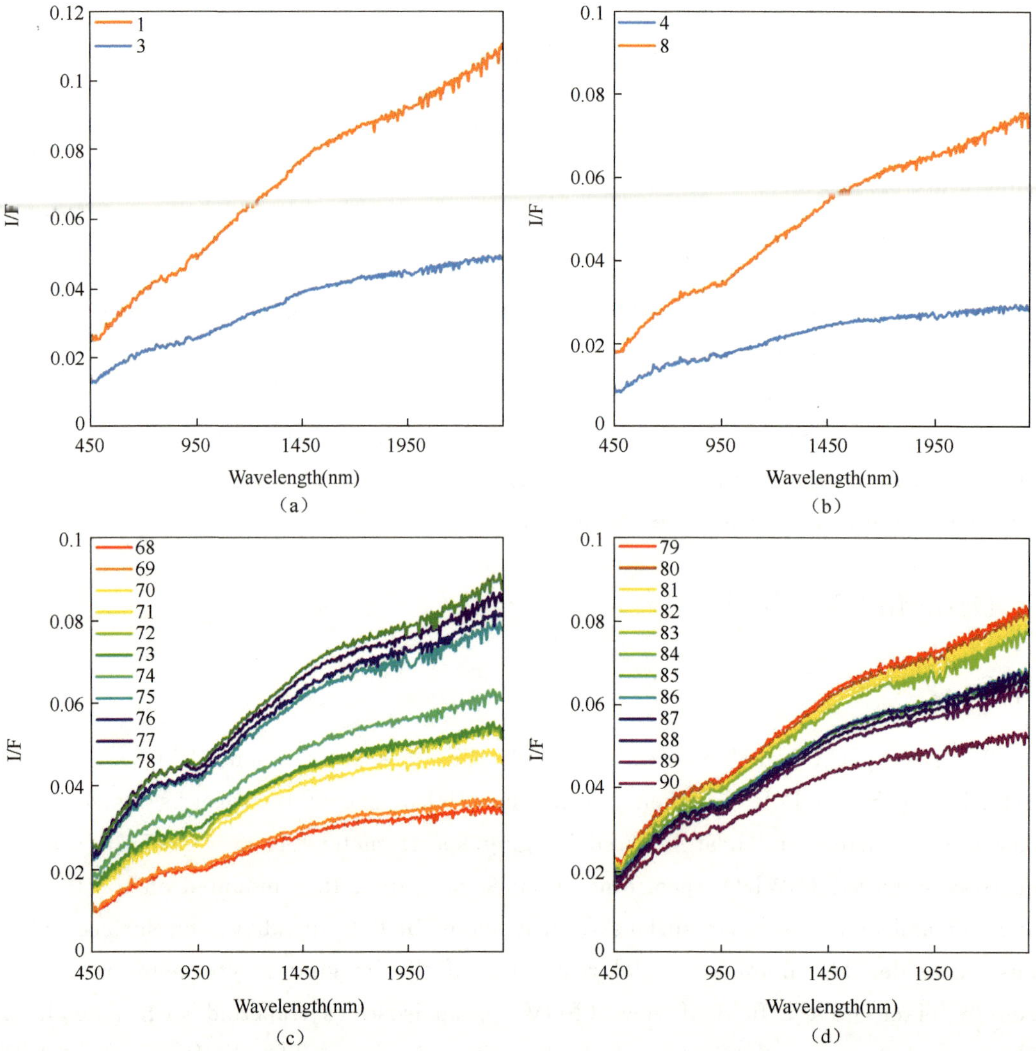

Fig. 1　Normalized radiances of the lunar regolith measured by the VNIS for the first (a), second (b), and tenth (c) and (d) lunar day of operation. For the morning measurements (c), the spectral upturn beyond 2μm gradually increases, while in the afternoon (d) the spectral upturn beyond 2μm gradually decreases. Numbers in the legend relate to the number # of spectral measurement. See Table S1 for detailed information on observation time and geometry

directional spectral measurements during a single lunar day (the tenth lunar day of surface operation). The regolith was observed while the rover was stationary under varying solar incidence angles. Due to the necessity to shut down the rover in a particular orientation during the lunar noon, two adjacent sites rather than one site were observed in the lunar morning (#68－78) and afternoon (#79－90). The sequential spectra taken on the tenth lunar day are shown in Fig. 1 (c) and 1 (d) and high - resolution images of two representative sites for the morning and afternoon observations are shown in Fig. 2. Shadows at submillimeter scales are clearly visible in the images, indicating a rough surface at small scales. Details of the observing conditions can be found in Tab. S1.

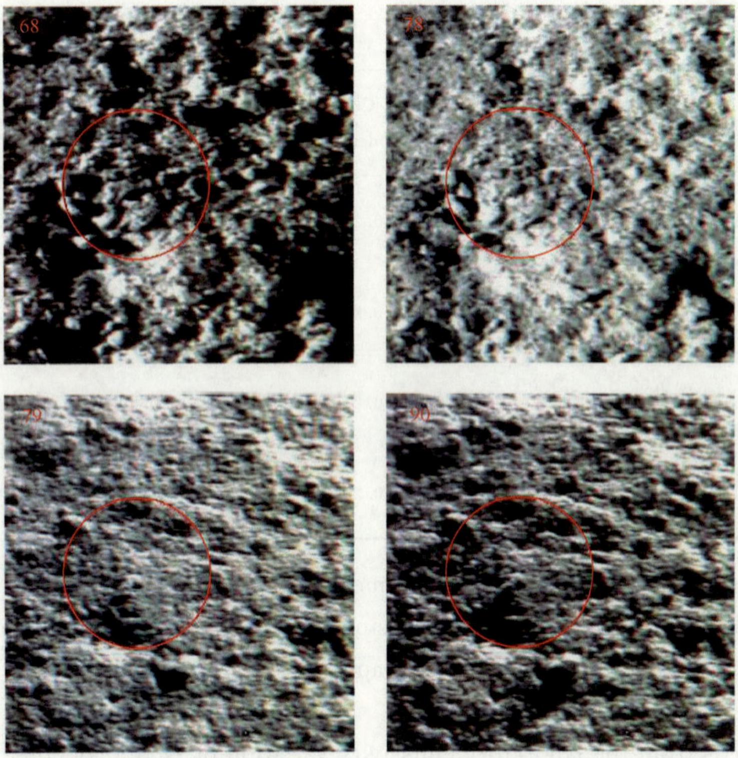

Fig. 2　VNIS images of two sites acquired during the tenth lunar day of operation. Images were taken in the lunar morning (68: 7: 23 local time, 78: 9: 02 local time) and afternoon (79: 14: 28 local time, 90: 16: 02 local time), respectively. The red circle in the images indicates the SWIR FOV

Tab. 1　Model Parameters Retrieved From VNIS Measurements Without and With Applying a Roughness Correction

Lunar operation day	Measurement #	VNIS results			Roughness model						Thermal model
		Emissivity[a]	Temperature /K[b]	Shadow fraction[c]	Crater density	Crater angle /(°)	Shadow fraction	Correction factor	Corrected temperature/K	RMS slope/(°)	Temperature/K
1st/morning	1	0.85±0.02	344.5±1.5	0.17	0.20	190	0.17	0.90	346.6	22	346.3
2nd/afternoon	8	0.80±0.02	319.2±1.7	0.17	0.20	190	0.17	1.22	315.9	22	314.8
10th/morning	75	0.80±0.02	320.3±1.7	0.20	0.20	270	0.19	1.25	316.5	24	313.1
	78	0.80±0.02	322.9±2.1	0.18	0.20	270	0.18	1.28	318.7	24	320.0

Continued

Lunar operation day	Measurement #	VNIS results			Roughness model						Thermal model
		Emissivity[a]	Temperature /K[b]	Shadow fraction[c]	Crater density	Crater angle /(°)	Shadow fraction	Correction factor	Corrected temperature/K	RMS slope/ (°)	Temperature/K
10th/ afternoon	79	0.80±0.02	330.9±1.9	0.19	0.20	190	0.19	0.94	332.0	22	330.8
	82	0.81±0.02	324.3±2.3	0.19	0.20	190	0.19	0.91	326.0	22	327.4
	83	0.80±0.02	321.0±2.0	0.19	0.20	190	0.19	0.90	322.8	22	326.3
	84	0.80±0.02	313.1±1.5	0.19	0.20	190	0.20	0.87	315.4	22	315.5

Temperatures predicted by the thermal model are given for reference. Out of a total of 27 measurements within the three lunar days of operation, eight measurements for which spectral upturns beyond 2 μm are very apparent were chosen for the temperature inversions.

[a] Error bar is from the laboratory emissivity variation of Apollo 14 soils (Salisbury et al., 1997). [b] Error bar indicate 1σ confidence limit. [c] Method of deriving the shadow fraction from images is provided in Text S2 in the supporting information.

2.2 Retrieval of Surface Temperature

We interpret surface temperature as a modified brightness temperature derived from the thermal flux that takes the deduced emissivity into account (usually brightness temperature is defined for a black body). Temperature retrieval from the spectral measurements is complicated by the fact that the set of equations is underdetermined for the given set of observations, as data are available at N channels but there are N + 1 unknowns (N emissivities plus one temperatures). For the VNIS (and other infrared data sets up to about 5μm), another difficulty is that reflected and emitted radiation overlap and must be separated. The spectral sampling interval of VNIS is 5nm, which provides an opportunity for finding the emissivity and reflected fraction of solar radiation simultaneously using multiple channels by assuming the same temperature for the longest adjacent channels. We use the multiband inversion approach from Wu and Hapke (2018) to find the emissivity, reflected sunlight fraction, and temperature simultaneously. Reflected sunlight fraction R'_i, emissivity ε_λ, and temperature T are then related by

$$\frac{R_i - R'_i}{R_i} \times L_i = \varepsilon_\lambda \times \frac{c_1 \lambda^{-5}}{e^{c_2/\lambda T} - 1} \tag{1}$$

where C_1 and C_2 are Planck's constants $C_1 = 1.191 \times 10^8\,\mathrm{W \cdot \mu m^4/(sr \cdot m^2)}$ and $C_2 = 1.439 \times 10^4\,\mathrm{\mu m \cdot K}$. Parameters R_i and R_i' are the measured reflectance and the reflected sunlight fraction of band i, respectively. ε_λ is the wavelength dependent emissivity, and L_i is the measured radiance of band i. The solution of the optimization was judged using the relative error (RE), defined as

$$RE = | r_L - r_R | / r_L \tag{2}$$

in which r_L and r_R are values of the left and right hand side of Equation 1, respectively. An iterative numerical approach was used to calculate R', ε, and T by setting initial parameter values with a step of $\varepsilon = 0.01$ and $T = 0.01$ and using the longest three channels (2.385 – 2.395 μm). For given small RE, there are a large number of possible solutions. To deal with the inherent multiple solutions of the inversion problem, Wu et al. (2009) developed a Monte Carlo inversion approach which is also adopted here. With this strategy, unique meaningful solutions can be determined by choosing the mode of the data set, that is, those parameter values that occur most often in the family of solutions. The temperature inversion used the smoothed spectra by B spline fitting.

2.3 Thermophysical Model

To compare retrieved temperatures with predictions, a one-dimensional thermal model was used. The model calculates surface and subsurface temperatures solving the heat conduction equation subject to the boundary conditions of insolation at the surface and zero heat flow at depth. It is given by

$$\rho(z)C(T)\frac{\partial T(z,t)}{\partial t} = \frac{\partial}{\partial z}\left[k(z,T)\frac{\partial T(z,t)}{\partial z}\right] \tag{3}$$

$$\varepsilon\sigma T_0^4(t) = F(t) + k(z,T)\frac{\partial T(z,t)}{\partial z}\bigg|_{z=0} \tag{4}$$

$$F(t) = \frac{S}{d^2}\cos(i)(1-A) \tag{5}$$

where z is depth, T is temperature, T_0 is the surface temperature, t is time, S is the solar constant, d is the Sun-Moon distance, A is the bolometric albedo, i is the solar incidence angle, ε is the wavelength-integrated hemispherical emissivity, and σ is the Stefan-Boltzmann constant ($5.6704 \times 10^{-8}\ \mathrm{W/m^2/K^4}$). $F(t)$ represents the solar insolation and the last term of Equation 4 represents the conduction of heat into the subsurface. For the thermal conductivity k (z, T), heat capacity C (T), and density ρ (z) of lunar soils, we adopt the model of Hayne et al. (2017). They are given by

$$K(z,T) = [3.4 \times 10^{-3} - (3.4 \times 10^{-3} - 7.4 \times 10^{-4})e^{-z/H}]\left[1 + 2.7\left(\frac{T}{350}\right)^3\right] \tag{6}$$

$$\rho(z) = 1{,}800 - 700e^{-z/H} \tag{7}$$

$$C(T) = (-3.6125 + 2.7431T + 2.3616 \times 10^{-3}T^2 - 1.234 \times 10^{-5}T^3 + 8.9093 \times 10^{-9}T^4)(1+\beta) \quad (8)$$

where H is the parameter governing the increase of density and conductivity with depth and β is the parameter controlling the fluctuation range of heat capacity after comparing the data from Hemingway (1973). Temperatures are modeled by varying $A = 0.11 \pm 0.02$, $\varepsilon = 0.96 \pm 0.02$, $H = 0.06 \pm 0.04$, and $\beta = \pm 0.17$ to cover a range typically used for sites in similar geological settings of the CE－4 landing site (Bandfield et al., 2015; Bauch et al., 2014; Hayne et al., 2017; Hemingway, 1973; Racca, 1995; Vasavada et al., 2012; Wöhler et al., 2017).

2.4 Surface Roughness Description

To account for the effects of surface roughness, several methods have been described in the literature. Surface features can be modeled as spherical segments, sinusoidal trenches, random Gaussian features, or fractals (for an overview with references, see Davidsson et al., 2015). In this study, we model a rough surface by spherical crater segments in a smooth environment. We use the model developed by Giese and Kührt (1990) and by Kührt et al. (1992) that is an extension of the approach by Buhl et al. (1968) by allowing emissivities <1 and by considering multireflected radiation within the crater that contributes to self－heating. Davidsson et al. (2015) demonstrated (see Fig. 14 in their paper) that spherical segments and a Gaussian surface roughness model approach both produce very similar results with a parameter set similar to our study (small IR wavelengths, solar incidence angles of 50°—70°, root mean square (RMS) slope ~20°, see below) and both models describe the measured data well. Heat conduction along the crater walls is neglected since the dominating heat transport mechanism in craters with a size >1 mm is self－heating by radiation that is included in the model (a justification is given in Text S1). The roughness model is defined by two free parameters, the crater density ρ_c as a measure of the cratered to total (cratered + smooth) surface and the crater opening angle γ that is directly related by the depth/diameter ratio of the spherical crater ($\gamma = 180°$ for a half sphere with depth/diameter$=0.5$). It should be noted that the absolute diameter does not play a role for the roughness effect in spherical craters (Buhl et al., 1968).

To constrain the two free parameters, we determine shadow fractions both from the images (see Fig. 2 and Text S2) and from the model by a common raytracing method, and ρ_c and γ are chosen such that the modeled shadows fit the measured ones best at the given solar insolation and observation conditions. For this purpose, ρ_c is varied in 0.05 steps from 0 to 1 and γ in steps of 5° from 0° to 300° and the best fit in this 2D parameter space is selected. A further condition is that measurements at same site like, for example, #75 and #78 as well as #79－#84, must be described by the same roughness parameters. Tab. 1 summarizes the results together with the derived RMS slopes derived from the parameterized model by applying Equation 1 in Davidsson et al. (2015).

With the fixed crater opening angle γ, the inner crater surface is divided into ~100 facets and the temperature of each facet is calculated by considering direct solar insolation, reradiation, and self - heating by multiple reflections and emissions in the spherical segments. Finally, the Planck radiation from all facets with their specific temperature that are seen by the detector is summed up together with the radiation from the smooth part of the surface defined by ρ_c.

The ratio of total radiance from the rough surface to that of a smooth Lambertian surface provides a correction factor that gives the deviation of the measured radiance from that of a Lambertian radiator. This factor can be smaller (reduced radiance) or larger (beaming effect) than unity. Eventually, the temperature mixture on a rough surface is responsible for the non - Lambertian behavior. However, when such a rough surface is isothermal (which may or may not be the case at nighttime, depending on the time constant for cooling down and on possible subsurface heat sources), the radiation pattern is Lambertian and the correction factor is 1.

The derived correction factor depends on the roughness model, the heliocentric distance, albedo and emissivity, the solar zenith angle, the angle between surface normal and observation direction, as well as the angle between projections of the Sun vector and the observation direction onto the surface (relative azimuth). By correcting the measured thermal radiance with the calculated correction factor one gets the radiance of a Lambertian radiator at the same insolation conditions that can be described by the calculated temperature (see Section 2. 3) and the Planck law.

3 Results and Discussion

The two sequential observations at the same locations on the tenth lunar day of operations show that as the Sun rises in the morning the spectral upturn gradually increases, whereas in the afternoon, as the Sun slowly sets, the spectral upturn gradually decreases. The consistency of the in situ spectra in Fig. 1 for all the three lunar days of operation indicates that though the wavelength range of the VNIS is much shorter than for a typical thermal radiometer, it does capture thermal emission from the surface of the Moon. This provides an opportunity to understand the temperature response of lunar regolith to insolation at centimeter scales in combination with high - resolution images.

Temperature variations measured by the CE - 4 VNIS with and without taking surface roughness into account are shown as a function of local time in Fig. 3 along with the thermal models. Dashed lines in Fig. 3 indicate model temperature uncertainty (see Section 2. 3). While temperatures retrieved ignoring surface roughness differ from modeled values by up to 7 K, discrepancies can be significantly reduced if surface roughness is taken into account. Differences between predicted and inverted temperatures are largest for

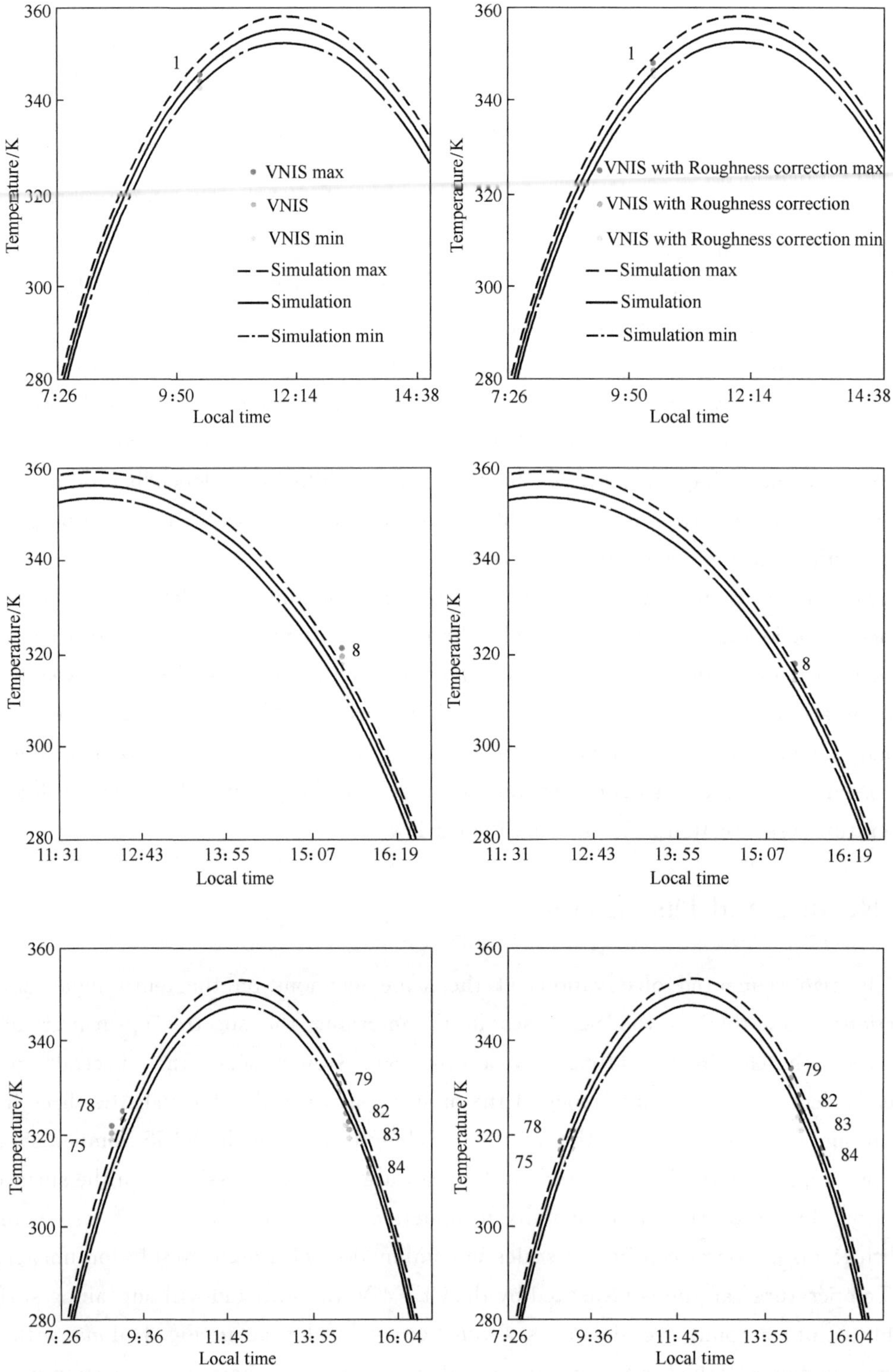

Fig. 3　Retrieved and modeled surface temperatures. The left and right columns are the retrieved temperature variations without and with taking the surface roughness into account. Model uncertainties (dashed lines) were calculated from the parameter variations of the thermal model. The error bars of the retrieved temperatures (blue and green dots) indicate 1σ confidence limit

#75, which was observed at a local time of 8: 44, while being smallest for #79, which was observed at a local time of 14: 28. This clearly indicates that roughness effects become most important for large solar zenith angles, as would be expected.

Tab. 1 shows the numerical results. Columns 1 and 2 give the day and number of the measurements, columns 3 – 5 contain emissivity, temperature, and shadow fraction derived from the VNIS measurements, and columns 6 and 7 give the input parameters for the roughness model which were derived such that the calculated shadow fraction (column 8) for the given observation geometry fits best to the measured one (column 5). Resulting flux correction factor and corrected temperature are shown in columns 9 and 10, where the corrected temperature is the temperature an isothermal Lambertian radiator would have under the same insolation conditions. This temperature can be directly compared with temperatures derived from the thermal model, which are given in column 12. Furthermore, the RMS slopes resulting from the two given roughness parameters are shown in column 11 for reference.

Surface emissivity was evaluated at a wavelength of 2.39μm and is 0.8 ± 0.02 for all measurements except for #1, where a value of 0.85 ± 0.02 was determined. This indicates that soil properties are similar among these sites. Site 1 is an outlier and the reason is unknown but perhaps because it was disturbed by the rover track (Fig. S1). The emissivity of the CE - 4 regolith is a little smaller than that of CE - 3 regolith which is 0.85—0.90 (Wu & Hapke, 2018) and a little larger than that of Apollo 14 samples measured in the laboratory which is 0.79 ± 0.02 (Salisbury et al., 1997). This is consistent with their albedo characteristics. The CE - 3 landing site is among the darkest regions of the Moon, corresponding to a highest emissivity. The CE - 4 landing site, similar to that of Apollo 14, has an albedo between those found for Maria and Highlands. The reflectance of lunar samples measured in the laboratory is larger than that measured on the surface of the Moon because of the differences in roughness, compactness, maturity, and particle size.

Fig. S2 shows the thermal emission fraction, original spectrum, and thermal removed spectrum. Both the reflectance and the emitted fraction increase with wavelength. The emitted fraction is in general positively correlated with temperature. At a local time of 10: 19 and latitude of 45.45°, the emitted fraction can be ~1.5% at 2.38μm.

The correction factors vary between 0.87 and 1.28 depending on the observing conditions and are kept fixed for the same measuring conditions. The bidirectional RMS slopes derived from the roughness parameters (calibrated by shadow measurements) range from 22° to 24°. Bandfield et al. (2015) found bidirectional RMS slopes between 20° and 35° from LRO Diviner data and multiple emission angle observations what is in good agreement. Helfenstein and Shepard (1999) found from Apollo data undirectional RMS slope angles of $\sim 16^{\circ}{}^{+4^{\circ}}_{-3^{\circ}}$ to $25^{\circ} \pm 1^{\circ}$. When correcting this result by a factor of $\sqrt{2}$ to meet our bidirectional slope definition one gets a range of 18°—37° in the bidirectional case what is also in agreement with our values.

The effects of surface roughness are also clearly visible in Fig. 4, where the difference between modeled and observed temperature is shown as a function of relative azimuth angle (azimuth difference between Sun and detector). The observed temperature of a rough surface is higher than that of a Lambertian surface when viewing from an azimuth toward the solar azimuth because the detector receives strong radiation from the sunlit, hot areas in the FOV (beaming effect). The opposite effect occurs at large relative azimuth angles when the detector looks mostly on shadowed areas in the roughness features.

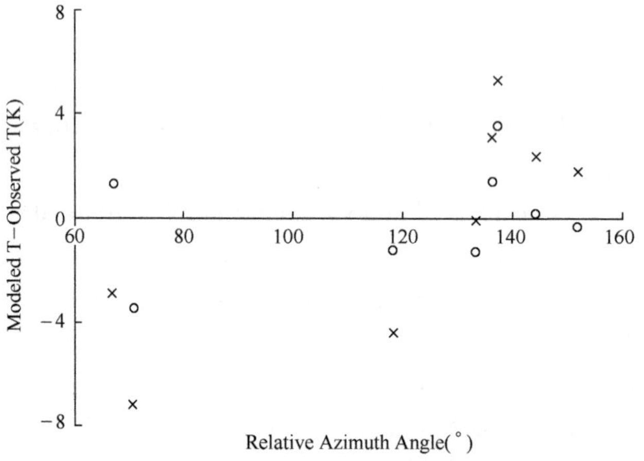

Fig. 4　Difference between modeled and observed temperature as a function of relative azimuth angle without (×) and with (○) roughness correction

Fig. 4 shows that the roughness corrected temperatures fit much better to the model. An F-test was applied to quantify this visual impression. The variances of the roughness corrected and uncorrected observed temperature sets from the expected (modeled) ones given in Table 1 are 4.37 and 18.05, respectively. The resulting F-value as the quotient of the variances is 4.13 and larger than the critical value of 3.79 in the F-distribution table for seven degrees of freedom and 5% significance level. This means that the null hypothesis (no significant statistical difference between roughness corrected and uncorrected data sets) can be rejected with a probability >95%. Consequently, applying the roughness corrections gives a better agreement between the measured and expected temperatures also in a statistical sense.

4　Summary and Conclusions

The VNIS, though initially designed for mineral detection, captured thermal emission from the surface of the Moon in the short-wavelength IR. VNIS data offer an opportunity for investigating temperature variations at centimeter scales. We applied a multiband inversion approach to determine emissivity, reflected sunlight fraction, and temperature from the measured flux, which we corrected for surface roughness using a spherical section crater model. Taking surface roughness into account significantly reduces the misfit between inverted and modeled temperatures, indicating that roughness is important even at subcentimeter scales.

An *F* test proved the statistical significance of considering the surface roughness for minimizing the difference between modeled temperatures and temperatures derived from measurements. Directional thermal emission at short IR wavelengths was measured by CE－4's in situ. These measurements illustrate the dependence of radiance on roughness and observation conditions, where large relative azimuth angles result in significant deviations from the emission of a Lambertian surface. Our study shows that roughness acts also on small scales and creates specific non－Lambertian radiation characteristics. It must be considered even on mm and cm scales and at short IR wavelength ranges.

The wavelength range investigated here carries contributions from both reflected sunlight and thermal emission, and reflectance spectra are often investigated to learn about the mineralogical composition of the Moon. As surface roughness can have a significant effect on thermal emission, care must be taken when interpreting the wavelength range beyond 2μm. In particular, the depth of mineral absorptions around 2 μm could be affected by the non－Lambertian emission characteristics.

Acknowledgments

The authors thank the whole Chang'E－4 team for making the mission a success. This research was supported by the Strategic Priority Research Program of Chinese Academy of Sciences (XDB 41000000), Preresearch project on Civil Aerospace Technologies by CNSA (D020201), the National Natural Science Foundation of China (42050202), the Macau Science and Technology Development Fund (0042/2018/A2 and 0090/2020/A), and Minor Planet Foundation of Purple Mountain Observatory. M. Hamm was financially supported by Geo. X the research network for geosciences in Berlin and Potsdam (SO _ 087 _ GeoX). The authors acknowledge the useful comments of two anonymous reviewers that helped to improve the paper a lot.

References

[1] Bandfield, J. L., Hayne, P. O., Williams, J. P., Greenhagen, B. T., & Paige, D. A. (2015). Lunar surface roughness derived from LRO Diviner Radiometer observations. Icarus, 248, 357－372. https://doi.org/10.1016/j.icarus.2014.11.009

[2] Bauch, E. K., Hiesinger, H., Helbert, J., Robinson, S. M., & Scholten, F. (2014). Estimation of lunar surface temperatures and thermophysical properties: Test

of a thermal model in preparation of the MERTIS experiment onboard BepiColombo. Planetary and Space Science, 101, 27 - 36. https://doi.org/10.1016/j.pss.2014.06.004

[3] Buhl, D., Welch, W. J., & Rea, D. G. (1968). Reradiation and thermal emission from illuminated craters on the lunar surface. Journal of Geophysical Research, 73 (15), 5281 - 5295. https://doi.org/10.1029/JB073i016p05281

[4] Davidsson, B. J. R., Rickman, H., Bandfield, J. L., Groussin, O., Gutierrez, P. J., & Wilska, M. (2015). Interpretation of thermal emission. I. The effect of roughness for spatially resolved atmosphereless bodies. Icarus, 252, 1 - 21. https://doi.org/10.1016/j.icarus.2014.12.029 Giese, B., & Kührt, E. (1990). Theoretical interpretation of infrared measurements at Deimos in the framework of crater radiation. Icarus, 88, 372 - 379. https://doi.org/10.1016/0019-1035(90)90088-Q

[5] Grott, M., Knollenberg, J., & Krause, C. (2010). Apollo lunar heat flow experiment revisited: A critical reassessment of the in situ thermal conductivity determination. Journal of Geophysical Research, 115, E11005. https://doi.org/10.1029/2010JE003612

[6] Hayne, P. O., Bandfield, J. L., Siegler, M. A., Vasavada, A. R., Ghent, R. R., Williams, J., et al. (2017). Global regolith thermophysical properties of the Moon from the Diviner Lunar Radiometer Experiment. Journal of Geophysical Research: Planets, 122, 2371 - 2400. https://doi.org/10.1002/2017JE005387

[7] Hayne, P. O., Lucey, P. G., Swindle, T. R., Bandfield, J. L., Siegler, M. A., & Paige, D. A. (2015). Thermal infrared observations of the moon during lunar eclipse using the air force Maui space surveillance system. (Vol. 46, p. 1997). Paper presented at 46th Lunar and Planetary Science Conference, Houston, TX.

[8] Helfenstein, P., & Shepard, M. K. (1999). Submillimeter - scale topography of the lunar regolith. Icarus, 141, 107 - 131. https://doi.org/10.1006/icar.1999.6160

[9] Hemingway, B. S., Robie, R. A., & Wilson, W. H. (1973). Specific heats of lunar soils, basalt, and breccias from the Apollo 14, 15, and 16 landing sites, between 90 and 350°K. (Vol. 4, p. 2481). Paper presented at 4th Proceedings of the Lunar Science Conference. Houston, TX.

[10] Keihm, S. J., & Langseth, M. G. J. (1973). Surface brightness temperature at the Apollo 17 heat flow site: Thermal conductivity of the upper 15 cm of regolith. (Vol. 3, pp. 2503 - 2513). Paper presented at 3rd Proceedings of the Lunar Science Conference, Houston, TX.

[11] Kührt, E., Giese, B., Keller, H. U., & Ksanfomality, L. V. (1992). Interpretation of the KRFM - infrared measurements on Phobos. Icarus, 96, 213 - 218. https://doi.org/10.1016/0019-1035(92)90075-I.

[12] Lagerros, J. S. (1997). Thermal physics of asteroids. III. Irregular shapes and albedo variegations. Astronomy and Astrophysics, 325, 1226 - 1236.

[13] Lawson, S. L., Jakosky, B. M., Park, H., & Mellon, M. T. (2000). Brightness temperatures of the lunar surface: Calibration and global analysis of the Clementine long - wave infrared camera data. Journal of Geophysical Research, 105, 4273 - 4290. https://doi.org/10.1029/1999JE001047

[14] Lebofsky, L. A., Sykes, M. V., Tedesco, E. F., Veeder, G. J., Matson, D. L., Brown, R. H., et al. (1986). A refined 'standard' thermal model for asteroids based on observations of 1 Ceres and 2 Pallas. Icarus, 68 (2), 239 - 251. https://doi.org/10.1016/0019 - 1035 (86) 90021 - 7

[15] Maghrabi, A. H. (2016). Infrared radiometric measurements of lunar disk temperatures during lunar eclipse on 15th June 2011. Advances in Space Research, 58 (6), 1044 - 1049. https://doi.org/10.1016/j.asr.2016.05.040

[16] Paige, D. A., Foote, M. C., Greenhagen, B. T., Schofield, J. T., Calcutt, S., Vasavada, A. R., et al. (2010). The lunar reconnaissance orbiter diviner lunar radiometer experiment. Space Science Reviews, 150, 125 - 160. https://doi.org/10.1007/s11214 - 009 - 9529 - 2

[17] Palmer, E. E., & Sykes, M. V. (2014). The observational bias of thermal spectra due to subpixel variations. Paper presented at 45th Lunar and Planetary Science Conference. Houston, TX.

[18] Pettit, E., & Nicholson, S. B. (1930). Lunar radiation and temperatures. The Astrophysical Journal, 71 (71), 102 - 135. https://doi.org/10.1086/143236

[19] Racca, G. D. (1995). Moon surface thermal characteristics for moon orbiting spacecraft thermal analysis. Planetary and Space Science, 43, 835 - 842. https://doi.org/10.1016/0032 - 0633 (95) 00067 - F

[20] Rosenburg, M. A., Aharonson, O., Head, J. W., Kreslavsky, M. A., Mazarico, E., Neumann, G. A., et al. (2011). Global surface slopes and roughness of the Moon from the Lunar Orbiter Laser Altimeter. Journal of Geophysical Research, 116, E02001. https://doi.org/10.1029/2010JE003716

[21] Rosenburg, M. A., Aharonson, O., & Sari, R. E. (2015). Topographic power spectra of cratered terrains: Theory and application to the Moon. Journal of Geophysical Research: Planets, 120, 177 - 194. https://doi.org/10.1002/2014JE004746

[22] Rozitis, B., & Green, S. F. (2011). Directional characteristics of thermal - infrared beaming from atmosphereless planetary surfaces - a new thermophysical model. Monthly Notices of the Royal Astronomical Society, 415 (3), 2042 - 2062. https://doi.org/10.1111/j.1365 - 2966.2011.18718.x

[23] Rubanenko, L., Schorghofer, N., Greenhagen, B. T., & Paige, D. A. (2020). Equilibrium temperatures and directional emissivity of sunlit airless surfaces with applications to the Moon. Journal of Geophysical Research: Planets, 125,

e2020JE006377. https://doi.org/10.1029/2020JE006377
[24] Salisbury, J. M., Basu, A., & Fischer, E. M. (1997). Thermal infrared spectra of lunar soils. Icarus, 130 (1), 125 - 139. https://doi.org/10.1006/icar.1997.5809
[25] Shorthill, R. W. (1962). Measurements of lunar temperature variations during an eclipse and throughout a lunation, Seattle, WA: Boeing Scientific Research Laboratories.
[26] Sinton, W. M. (1962). Eclipse temperatures of the lunar crater Tycho. International Astronomical Union, 469 - 471.
[27] Smith, B. G. (1967). Lunar surface roughness: Shadowing and thermal emission. Journal of Geophysical Research, 72, 4059 - 4067. https://doi.org/10.1029/JZ072i016p04059
[28] Spencer, J. R. (1990). A rough - surface thermophysical model for airless planets. Icarus, 83 (1), 27 - 38. https://doi.org/10.1016/0019 - 1035 (90) 90004 - S
[29] Vasavada, A. R., Bandfield, J. L., Greenhagen, B. T., Hayne, P. O., Siegler, M. A., Williams, J. P., & Paige, D. A. (2012). Lunar equatorial surface temperatures and regolith properties from the Diviner Lunar Radiometer Experiment. Journal of Geophysical Research, 117, E00H18. https://doi.org/10.1029/2011JE003987
[30] Vasavada, A. R., Palage, D. A., & Wood, S. E. (1999). Near - surface temperatures on Mercury and the Moon and the stability of polar ice deposits. Icarus, 141 (2), 179 - 193. https://doi.org/10.1006/icar.1999.6175
[31] Williams, J. P., Paige, D. A., Greenhagen, B. T., & Sefton - Nash, E. (2017). The global surface temperatures of the moon as measured by the diviner lunar radiometer experiment. Icarus, 283, 300 - 325. https://doi.org/10.1016/j.icarus.2016.08.012
[32] Wöhler, C., Grumpe, A., Berezhnoy, A. A., Feoktistova, E. A., Evdokimova, N. A., Kapoor, K., & Shevchenko, V. V. (2017). Temperature regime and water/hydroxyl behavior in the crater Boguslawsky on the Moon. Icarus, 285, 118 - 136. https://doi.org/10.1016/j.icarus.2016.12.026
[33] Wu, Y. Z., Gong, P., Liu, Q., & Chappell, A. (2009). Retrieving photometric properties of desert surfaces in China using the Hapke model and MISR data. Remote Sensing of Environment, 113, 213 - 233. https://doi.org/10.1016/j.rse.2008.09.006
[34] Wu, Y. Z., & Hapke, B. (2018). Spectroscopic observations of the Moon at the lunar surface. Earth and Planetary Science Letters, 484, 145 - 153. https://doi.org/10.1016/j.epsl.2017.12.003
[35] Wu, Y. Z., Wang, Z. C., Cai, W., & Lu, Y. (2018). The absolute reflectance and new calibration site of the Moon. The Astronomical Journal, 155, 213.

https：//doi. org/10. 3847/1538 - 3881/aabaf5

References From the Supporting Information

Wood, S. E. (2020). A mechanistic model for the thermal conductivity of planetary regolith: 1. The effects of particle shape, composition, cohesion, and compression at depth. Icarus, 352, 113964. https：//doi. org/10. 1016/j. icarus. 2020. 113964

Yu, S., & Fa, W. (2016). Thermal conductivity of surficial lunar regolith estimated from Lunar Reconnaissance Orbiter Diviner Radiometer data. Planetary and Space Science, 124, 48 - 61. https：//doi. org/10. 1016/j. pss. 2016. 02. 001

Translucent glass globules on the Moon

Zhiyong Xiao[1,2], Pan Yan[1], Bo Wu[3], Chunyu Ding[4], Yuan Li[3,5], Yiren Chang[6], Rui Xu[1], Yunhua Wu[1], Yichen Wang[1], Yizhen Ma[7], Jun Cui[1,2]

1. Planetary Environmental and Astrobiological Research Laboratory, School of Atmospheric Sciences, Sun Yat - sen University, Zhuhai 519000, China
2. Center for Excellence in Comparative Planetology, Chinese Academy of Sciences, Hefei 230026, China
3. Department of Land Surveying and Geo - Informatics, Hong Kong Polytechnic University, Hong Kong, China
4. Department of Earth and Space Sciences, Southern University of Science and Technology, Shenzhen 518055, China
5. School of Geospatial Engineering and Science, Sun Yat - sen University, Zhuhai 519082, China
6. Shanghai Key Laboratory for Astrophysics, Shanghai Normal University, Shanghai 200233, China
7. School of Earth Sciences, China University of Geosciences (Wuhan), Wuhan 430074, China

Glass is ubiquitous in lunar regolith, and volcanism and hypervelocity impacts are the major mechanisms of forming lunar glasses. Volcanic glasses on the Moon occur as quenched skin of basaltic rocks or as glass spherules in pyroclastic deposits. Lunar volcanic glass spherules are less than 1mm in diameters, and most are less than 300μm [1]. Impact glasses on the Moon are formed by cooling of impact melt and/or condensation of impact vapor, and impact glass spherules have rotational shapes, smooth surfaces, and largely clast - free interiors[2]. Recognized impact glass spherules on the Moon are dominantly sub - millimeter in sizes, and few are up to ~8mm in diameter[3]. Glass spherules record important information about the mantle composition and the history of lunar volcanism and impact cratering[4]. Transparent and translucent glasses on the Moon are less than 1mm in diameters, and larger ones are dark and opaque[5]. Hitherto discovered macro - sized glass globules on the Moon (up to ~4cm in diameter) are opaque impact glass (Fig. 1a, b) that usually contains hollows and lithic clasts[6].

Anorthosites are abundant in the lunar highland, and petrophysical modelling revealed that lunar anorthosites are excellent raw materials to form macroscopic transparent glasses[7]. Impact cratering in the lunar highland can form widespread anorthositic melt, some of which was shocked to temperatures way above the liquidus, so instant cooling could form distal glass ejecta. For comparison, tektites and microtektites on Earth are distal impact glasses sourced from highly shocked shallow targets[8]. Although their detailed formation mechanism (s) in terms of shock physics has not been completely settled (e.g., quench from impact melt, and/or condense from impact vapor), the required pressure and temperature to account for their geochemical characteristics are not special in planetary impacts[8]. The large crater density and the small weathering rate on the lunar highland indicate that transparent anorthositic glass globules formed due to impact should be common on the Moon. However, while abundant highresolution field data have been obtained by multiple long-range surface explorations (such as the Apollo and Luna missions), tektite-like holohyaline objects have not been discovered on the Moon[3], possessing a potential question about the possible ubiquitous of tektite-equivalents across terrestrial bodies.

The Chinese Chang'E-4 mission is exploring the South Pole-Atkien basin at the lunar farside. Since being deployed from the lander, the Yutu-2 rover has traveled over 700m on the floor of the Von Kármán crater. In images obtained by the panorama camera, two translucent glass globules are recognized along the route (Fig. 1). No compositional data were obtained for these globules, but their special morphology and local context suggest they are most likely impact glass quenched from anorthositic impact melt.

The traverse area of the rover is located on the debris deposits that cover the mare unit in Von Kármán crater. Reflectance spectra obtained by both lunar orbiters and the Yutu-2 rover reveal that the debris cover is dominantly anorthositic and gabbroic in compositions, which are mainly composed by plagioclase and lowcalcium pyroxene[9]. Stratigraphic study revealed that the ~40m thick debris cover is mainly composed of distal impact ejecta[10]. The lunar penetrating radar onboard the rover revealed an ~11m thick regolith in the upper section of the debris cover[10], where numerous small craters but few rock fragments are visible (Fig. 2b).

The discovered glass globules are located adjacent to the route of Yutu-2 (Fig. 2b) where image resolution is better. Perching on the surface regolith, the gray-brownish glasses are ~1.5—2.5cm in diameters (Fig. 1c, d), and they have intact spherical shapes as seen at millimeter-scale digital terrain models that are constructed from panorama camera stereo pairs (Text S1 and Fig. S1 online). The globules are dramatically different from both the surrounding blocky fragments (Fig. 1e, f) and similar-sized glass spheres sampled by the Apollo missions (Fig. 1a, b), as the spherical globules are translucent or semi-transparent and they exhibit a vitreous luster. Four other similar globules are also identified based on their vitreous luster and spherical shapes, but the available image resolution is not adequate to resolve whether or not they are translucent (Fig. S2 online).

The globules were formed or exposed recently as evidenced by their intact shapes and surface exposure. Surface materials on the Moon are subjected to continual abrasion and collisional disruption due to impacts. Updated growth rate of lunar regolith suggested that the top 2—3cm thick regolith is thoroughly overturned in <100000 years[11], indicating

Fig. 1　(Color online) Centimeter - sized glass globules collected by the Apollo 16 missions (a, b) and those observed by the Chang'E - 4 mission (c, d). Note that the Apollo 16 mission was landed in the lunar highland, but the landing region was dominated by distal ejecta from the nearside. (e, f) The two globules are accompanied by similar - sized fragments excavated by fresh impact craters nearby

the globules might be extremely young. However, due to the stochastic and repeated impact excavation and burial, the exact formation age of a single fragment on lunar surfaces cannot be constrained based only on their surface occurrences. Close to each of the globules, irregularly - shaped fragments excavated by meter - sized fresh cratersare visible (Fig. 1e, f). The fragments are collocated with the globules and they have similar sizes but different colors and shapes. Therefore, it is alternatively possible that the globules may be existed in the shallow regolith and later were exposed by the small craters.

Hypervelocity impacts are propellant for interplanetary exchange of materials, as up to ~0.5 vol% of the top ~10m thick lunar regolith may be ejected from the Earth[12]. However, no translucent meteorites were noticed on the Earth. *In* vacuo impact experiments in regolith - like target showed that cm - sized spherical glass projectiles are readily shattered at velocities down to ~1km/s[13], far less than the minimum impact velocity on the Moon. The heterolithic glass globules have intact spheric shapes (Fig. 1c, d), suggesting they are not meteorites delivered from the other planetary bodies.

Glass formation on the Moon requires either endogenic or exogenic melting. The critical cooling rate of crystallization plays a central role in the formation of glasses, as lower critical cooling rates mean better glass - forming ability. Geochemical data (e. g., content of meteoritic components) are usually required to differentiate impact and volcanic micro - glasses on the Moon[2]. However, no composition data were obtained for the globules. The spheric shape of the globules indicates a cooling and solidification history in flight. Lunar basalts are poor glass - former because of their high iron and titanium contents and also the large critical cooling rate (~24 — 47K/s), thus a pronounced tendency of crystallization[5]. In radiative cooling on the Moon, basaltic droplets with masses larger than ~0. 1 — 0. 3g (equals to ~4 — 6mm) would show indications of crystallization, and those larger than about 3g (equals to ~12mm diameter) should be completely crystallized[14]. Therefore, the globules are not likely mafic volcanic or impact glasses. While silicic volcanic deposits on the Moon have relatively high reflectances, they are mainly located within and near the Oceanus Procellarum[15], not occurring around the landing region.

By excluding a potential volcanic origin, it can be ascertained that the globules are impact glasses, and the precursor materials are not mafic materials. Among known lithologies on the Moon, anorthositic melt droplets are semi - transparent, and their critical cooling rate is much less than 5K/s[14]. Upon free - flight cooling in lunar vacuum, anorthositic melt >5cm in diameter could form transparent holohyaline globules[7]. Returned lunar anorthosites contain coarse - grained plagioclases that are as large as 3 cm, e. g., sample 15415. Pure anorthosites that contain >98 wt% of plagioclase have also been identified at dozens of locations on the Moon (Fig. 2a), including several areas within the less than 1km/s ballistic range around the landing site (Fig. 2c — e). Collectively, the peculiar morphology, geometry, and local context of the glass globules are consistent with being anorthositic impact glasses.

Macroscopic glass globules collected by Apollo (Fig. 1a, b) were traced back to nearby small craters, and their heterogeneous compositions are mainly caused by unmelted clasts and schlieren[6]. On the contrary, distal impact glasses are generally more homogeneous, because their precursor materials had a higher shock pressure and shorter passage paths through excavation flows. Theoretical calculations predict that during typical impact cratering on the Earth, the volume of impact melt that could form tektites is about that of the impactor[8]. Therefore, the translucent globules might be tektite - like glasses on the Moon, but this speculation could only be testified with compositional data.

As the first discovery of macroscopic and translucent glass globules on the Moon, this study predicts that such globules should be abundant across the lunar highland,

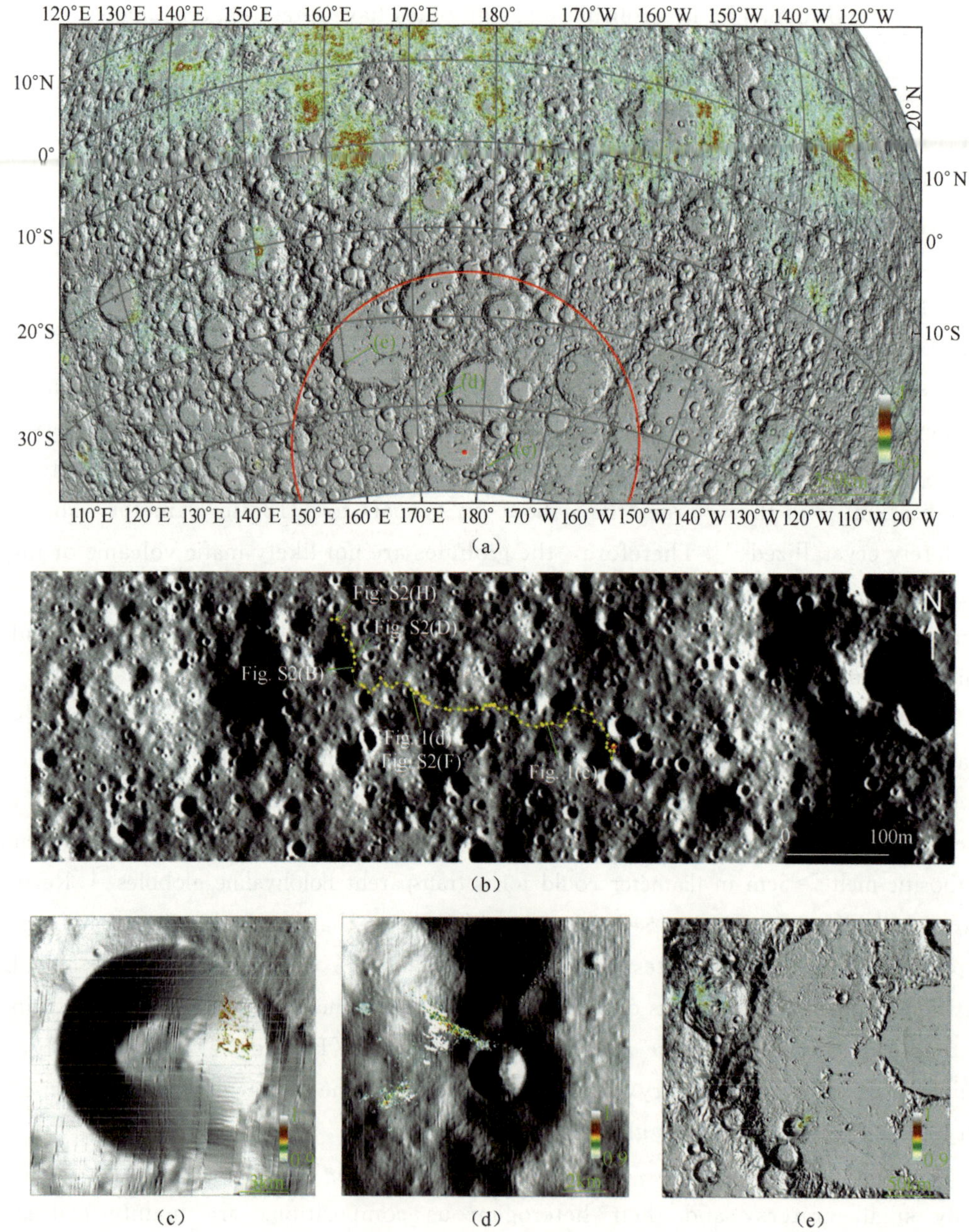

Fig. 2　Locations of glass globules along the Yutu - 2 rover path and distribution of anorthosites at the farside highland. (a) Content of plagioclase at latitudes above 50°s of the lunar farside. Red dot marks the location of the Chang'E - 4 mission, and red circle denotes ballistic ranges with a velocity of 1km/s from the landing site. (b) Route of the Yutu - 2 rover before July 2020, and the landing site is marked in red. Tips of the green arrows point to the locations of the confirmed and possible glass globules. (c - e) Within the red circle shown in (a), many craters have excavated anorthosites

providing promising sampling targets to reveal the early impact history of the Moon.

Acknowledgments

This work was supported by the Strategic Priority Program of the Chinese Academy of Sciences（XDB41000000），the Pre－research Project on Civil Aerospace Technologies（D020101，D020202）of China National Space Administration，the National Natural Science Foundation of China（41773063），the Key Research Program of Frontier Sciences，Chinese Academy of Sciences（QYZDY－SSWDQC028），and the Fundamental Research Funds for the Central Universities. Comments and suggestions provided by the four anonymous reviewers are appreciated. The Chang'E－4 mission was carried out by the Chinese Lunar Exploration Program，and the scientific data were provided by China National Space Administration.

Appendix A　Supplementary materials

Supplementary materials to this article can be found online at https：//doi.org/10.1016/j.scib.2021.11.004.

References

［1］Ryder G，Delano J W，Warren P H，et al. A glass spherule of questionable impact origin from the Apollo 15 landing site：unique target mare basalt. Geochim Cosmochim Acta 1996；60：693－710.

［2］Delano J W. Pristine lunar glasses：criteria，data，and implications. J Geophys Res 1986；91：201－13.

［3］Glass B P. Lunar sample 14425：not a lunar tektite. Geochim Cosmochim Acta 1986；50：111－3.

［4］Zellner N E B. Lunar impact glasses：probing the Moon's surface and constraining its impact history. J Geophys Res 2019；124：2686－702.

［5］Arndt J，Flad K，Feth M. Radiative cooling experiments on lunar glass analog. In：Lunar and Planetary Science Conference. Houston：Lunar and Planetary Institute；1979. p. 355－73.

［6］See T H，Hörz F，Morris RV. Apollo 16 impact－melt splashes：petrography and major－element composition. J Geophys Res 1986；91：E3－E20.

［7］Isard J O. The formation of spherical glass particles on the lunar surface. In：Proceedings of the Lunar Science Conference. 1971；3：2003－8.

［8］Johnson B J，Melosh HJ. Formation of melt droplets，melt fragments，and accretionary impact lapilli during a hypervelocity impact. Icarus 2017；228：347－63.

[9] Lin H, He Z, Yang W, et al. Olivine - norite rock detected by the lunar rover Yutu - 2 likely crystallized from the SPA - impact melt pool. Natl Sci Rev 2020; 7: 913 - 20.

[10] Zhang J, Zhou B, Lin Y, et al. Lunar regolith and substructure at Chang' E - 4 landing site in South Pole - Aitken basin. Nat Astron 2021; 5: 25 - 30.

[11] Costello E S, Ghent R R, Lucey PG. The mixing of lunar regolith: vital updates to a canonical model. Icarus 2018; 314: 327 - 44.

[12] Crawford I A, Baldwin E C, Taylor EA, et al. On the survivability and detectability of terrestrial meteorites on the Moon. Astrobiology 2008; 8: 242 - 52.

[13] Cintala M J, Berthoud L, Hörz F. Ejection - velocity distributions from impacts into coarse - grained sand. Meteorol Planet Sci 1999; 34: 605 - 23.

[14] Arndt J, von Engelhardt W. Formation of Apollo 17 orange and black glass beads. J Geophys Res 1987; 92: E372 - 6.

[15] Wilson J T, Eke V R, Massey R J, et al. Evidence for explosive silicic volcanism on the Moon from the extended distribution of thorium near the Compton - Belkovich Volcanic Complex. J Geophys Res 2015; 120: 92 - 108.

Impact remnants rich in carbonaceous chondrites detected on the Moon by the Chang'e – 4 rover

Yazhou Yang[1], Shuai Li[2], MengHua Zhu[3], Yang Liu[1,4], Bo Wu[5], Jun Du[1,6], Wenzhe Fa[3,4,6], Rui Xu[7], Zhiping He[7], Chi Wang[1], Bin Xue[8], Jianfeng Yang[8] and Yongliao Zou[1]

1. State Key Laboratory of Space Weather, National Space Science Center, Chinese Academy of Sciences, Beijing, China
2. Hawi'i Institute of Geophysics and Planetology, University of Hawaii, Honolulu, HI, USA
3. State Key Laboratory of Lunar and Planetary Sciences, Macau University of Science and Technology, Taipa, China
4. Center for Excellence in Comparative Planetology, Chinese Academy of Sciences, Hefei, China.
5. Department of Land Surveying and Geo – Informatics, The Hong Kong Polytechnic University, Hong Kong, China
6. Institute of Remote Sensing and Geographical Information System, School of Earth and Space Sciences, Peking University, Beijing, China
7. Key Laboratory of Space Active Opto – Electronics Technology, Shanghai Institute of Technical Physics, Chinese Academy of Sciences, Shanghai, China
8. Xi' an Institute of Optics and Precision Mechanics, Chinese Academy of Sciences, Xi'an, China

The Moon has experienced an intense bombardment history since its formation[1]. Fragments of the impactor can remain on the lunar surface[2-4] and can provide evidence of the evolution of the impactor composition and impact population in the Earth – Moon system[3-5]. However, the retained impactor fragments previously identified in the Apollo samples have been well mixed into bulk lunar regolith due to the subsequent impact gardening, and their properties cannot be easily isolated[3,6,7]. Here we report observations of a two – metre – sized crater that formed less than one million years ago obtained by the Yutu – 2 rover of Chang'e – 4. Hyperspectral images in the visible and near – infrared range (0.45 – 0.945μm) with a spatial resolution less than 1mm per pixel highlight the presence of glassy material with high concentration (47%) of carbonaceous chondrites. We identify this material as remnants of the original impactor that was not entirely vaporized by the impact. Although carbonaceous chondrite fragments have been found in Apollo samples[8,9], no carbonaceous chondrite remnant had been directly observed on the lunar surface by remote sensing exploration. We suggest that carbonaceous chondrite – like bodies may still provide one of the sources of water to the present Moon.

The Chang'e - 4 mission has been exploring the lunar surface since its successful landing at Von Kármán crater (Extended Data Fig. 1) in January 2019. The Yutu - 2 rover has returned valuable spectra and images of the lunar regolith and rocks that improve our

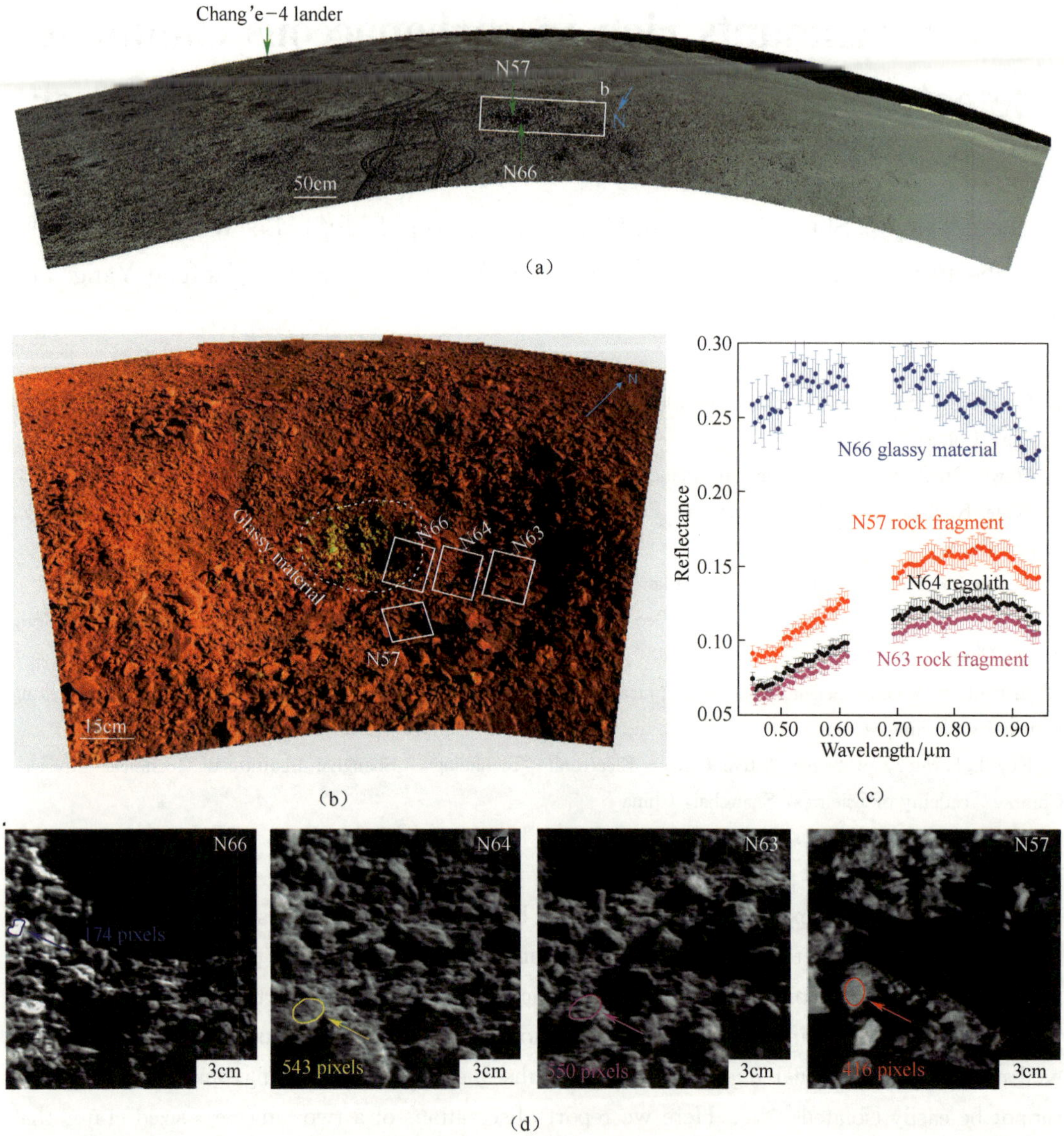

Fig. 1　Images and reflectance spectra of the targeted small crater. (a) False - colour mosaic of 85 images obtained by the Chang'e - 4 PCAM showing the crater and surrounding lunar surface. The blue arrow indicates north direction, and two typical spectral measurement sites (N57 and N66) are indicated by green arrows. The width of the rover track is 15cm. (b) Enlarged view of the crater investigated (outlined by the white box in a). The white boxes here indicate the areas imaged by the CMOS camera. The colour contrast of the image was enhanced with Photoshop (hue, - 38; saturation, 100; lightness, - 11). The area of glassy materials is marked with a dashed white circle. (c) CMOS spectra extracted from ROIs shown in d. The uncertainties were estimated based on the lower limit of the signal - to - noise ratio of 25dB in the CMOS channels[37]. (d) Calibrated CMOS images at wavelength of 560nm obtained at sites N66, N64, N63 and N57 (left to right). All of the images are stretched from 0 to 0.2 (that is, scaling the reflectance values in all the images to be between 0.0 and 0.2) to show albedo variations between images. The coloured circles show the representative ROIs used for spectral analysis

understanding of the evolution history of the landing region[10-13]. On the ninth lunar day of the mission's operation, Yutu - 2 encountered a small fresh crater (177.590575°E, 45.443274°S; ~2m in diameter) after travelling ~290m (Fig. 1). The high - resolution image, obtained by the rover's panoramic camera (PCAM), shows that this crater is fresh because areas within and surrounding the centre contain abundant fragments, whereas the landing region is relative flat and free of rocks (Fig. 1a). The close - up view of the crater, in an enhanced false - colour mode, shows distinct 'glassy' materials (yellowish colour) around the central pit of the crater with a size of ~21×16cm^2 (Fig. 1b and Extended Data Fig. 2). These glassy materials clearly exhibit stronger specular reflection than the surrounding regolith and rocks (Extended Data Fig. 2).

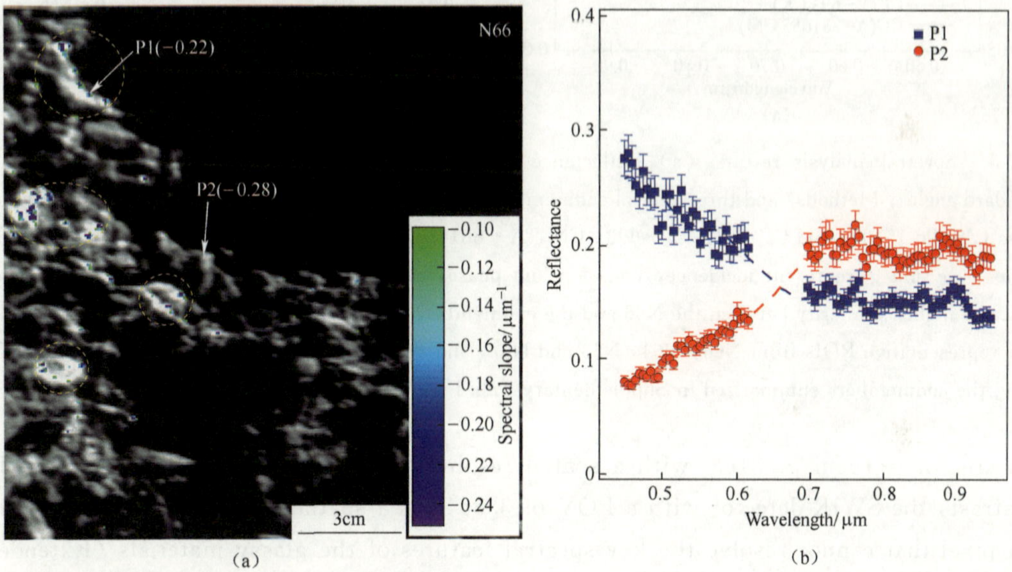

Fig. 2　Spectral slope analysis results. (a) Spectral slope (Methods) value for each pixel superimposed on the CMOS reflectance image of N66. Pixels with slope values larger than −0.1 are not shown. Areas rich in blueing (that is, reflectance decreases with increasing wavelength) pixels are marked out with yellow circles. White arrows show the positions of two pixels with different spectral slopes, implying different compositions (P1 and P2). (b) Comparison of the spectra with their slope values for P1 and P2. The error bars were generated based on the lower limit of the signal - to - noise ratio of 25dB of the CMOS channel[37]. The dashed lines represent the smoothed spectra

Yutu - 2 conducted a series of spectral measurements over the wall and the centre of the crater using the visible and near - infrared imaging spectrometer (VNIS), to investigate the composition of the radially distributed materials associated with this crater, including the glassy materials at the central pit (Extended Data Fig. 3). The VNIS consists of a complementary metal oxide semiconductor (CMOS) imager (wavelength: 0.45−0.945μm) and a single - pixel short - wavelength infrared (SWIR) detector (wavelength: 0.9−2.395μm) (see Methods and Extended Data Fig. 4 for the details of these two detectors). The CMOS imager has a field of view (FOV) of 8.5° and can capture

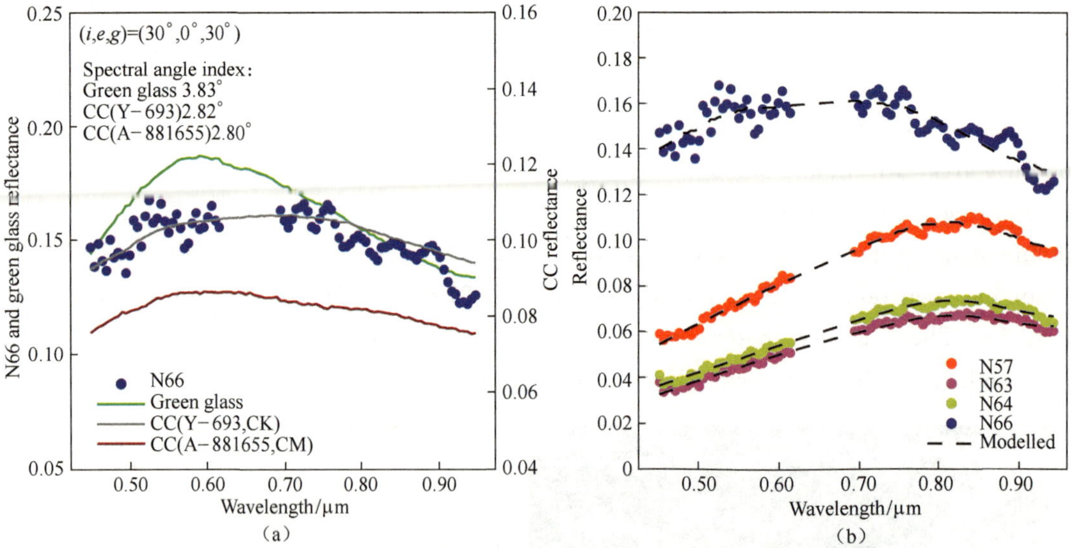

Fig. 3　Spectral analysis results. (a) Reflectance spectra of the glassy material in N66 (corrected to standard angles; Methods) and three typical endmembers (Methods and Supplementary Tab. 1): lunar green glass (Apollo 15401) and CC (CK, Y-693; CM, A-881655, RELAB database file ID: c1mp55). The angles i, e and g represent incidence, emission and phase, respectively. A smaller spectral angle index indicates higher similarity between the N66 and the endmember spectra (Methods). (b) Averaged spectra of four representative ROIs from N66, N64, N63 and N57, and the corresponding best-fit modelling results using the endmembers summarized in Supplementary Tab. 1

an area of ~15cm × 21cm with a spatial resolution of better than 1mm per pixel. In contrast, the SWIR detector with a FOV of 3.58° has a spatial resolution of only ~7cm per pixel that cannot resolve the key spectral features of the glassy materials (Extended Data Fig. 5)[14]. Four representative CMOS images (N66, N64, N63 and N57) are shown in Fig. 1d. As shown in Fig. 1b and d, CMOS image N66 contains more glassy materials than the other three images (N57, N63 and N64), mostly containing typical background regolith and rock fragments. Image N66 clearly contains materials with much higher reflectance than the typical regolith and rock fragments. In addition, the reflectance spectra between 0.45 and 0.945μm of these bright glassy materials show distinct blue spectral slopes (that is, reflectance decreases with increasing wavelength) and shapes, which strongly suggests that these materials are compositionally different from the surrounding regolith (N64) and the rock fragments (N63 and N57) with red spectral slopes (Fig. 1c and Extended Data Fig. 6). The spectral slope of each pixel of all 12 CMOS images (Extended Data Fig. 3b) around this crater were calculated (Methods and Extended Data Fig. 7) to map the pixels with blue spectral slopes. Interestingly, the blue spectral slopes are only observed at the exposures of glassy materials around the central pit of the crater (Fig. 2a and Extended Data Fig. 7). Repeated measurements on exactly the same spot with a wide range of viewing geometries (for example, phase angles varying

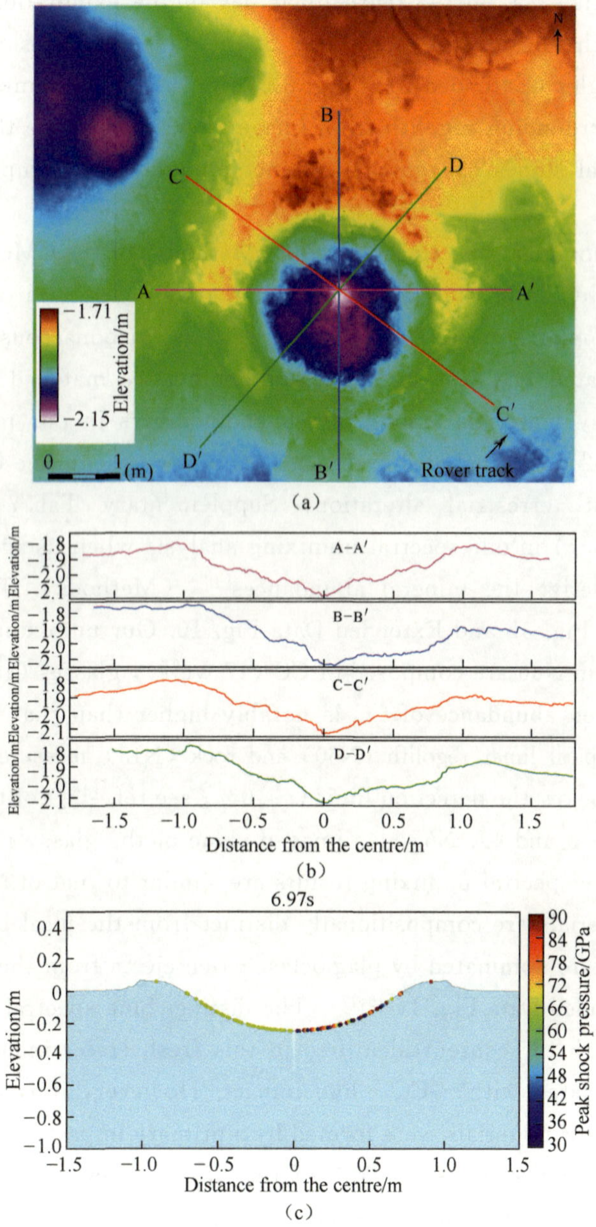

Fig. 4　The digital elevation model of the 2m crater and the iSALE simulation of crater formation. (a) Digital elevation model derived from PCAM stereo image pairs with a spatial resolution of 5mm per pixel. The profiles shown in b are indicated with coloured lines. (b) Topographic profiles for the 2m crater that show the structure of the crater. (c) Distribution of impactor relics based on numerical modelling for a dunite impactor on a basalt layer. The distribution of impact relics is shown on the left half of the simulated crater profile, whereas the peak shock pressures of these impact relics experienced during the impact cratering process are symmetrically shown on the right half of the profile (colour scale). The 6.97s label represents the modelling time for the formation of the crater

from 55° to 112°; ref. [15]) on the tenth lunar day of the exploration show similar red spectral slopes, which rules out the possibility of this blue spectral slope being caused by the measurement angles (Extended Data Fig. 8). Fig. 2b shows example spectra of glassy materials and the surrounding regolith from image N66. We can see that the two spectra have opposite spectral slopes, suggesting distinct mineralogical compositions (Extended Data Fig. 9).

We performed nonlinear spectral unmixing (Methods) of the CMOS spectra of glassy and non-glassy materials in the crater. We found that the spectra of lunar green glass (Apollo 15401; Supplementary Tab. 1), as well as carbonaceous chondrites (CC; including both hydrated and less hydrated endmembers), matched the best with the spectra of N66 glassy materials among all 13 endmembers in our model (Fig. 3a and Supplementary Tab. 2). We included green glass and a representative CC (Y-693, a CK type with the least terrestrial alteration; Supplementary Tab. 7) as endmembers (Supplementary Tab. 1) in our spectral unmixing analysis when using Hapke's radiative transfer model to derive the mineral abundances[16] (Methods). The spectral fitting results are shown in Fig. 3b and Extended Data Fig. 10. Our unmixing results show that the glassy materials in N66 are composed of CC (47 wt%), glasses (45 wt%) and olivine (8 wt%). The derived abundance of CC is notably higher than that derived from N57, N63, N64 and the typical lunar regolith (N60) and rock (N15) measured along the rover's path, which are all below the detection limit (~10%; see ref. [16]) of the unmixing model (Supplementary Tab. 3 and 4). N65 also imaged some of the glassy materials (Extended Data Fig. 3b) and the spectral unmixing results are similar to that of N66.

The glassy materials are compositionally distinct from the local background regolith and from rocks that are dominated by plagioclase-rich ejecta from the nearby Finsen and Alder craters (Extended Data Fig. 1)[10,17]. The distinct blue spectral slope of the glassy materials and the high CC content identified in this fresh crater indicate that the crater formation was associated with a CC-like impact. However, it is difficult to identify whether these impactor materials were formed by a primary impact, or transported from a nearby impact. The depth: diameter ratio of this fresh crater is around 1 : 5 with the central pit and 1 : 10 without the central pit (Fig. 4 and Supplementary Tab. 5), higher than ratios of 1 : 17 of secondary craters at the landing site[18], suggesting a primary impact. Our impact cratering simulation, using the state-of-the-art iSALE shock physics code[19], shows that a 15-cm-diameter porous impactor hitting the lunar surface with a typical velocity of 15km s^{-1} can produce the observed crater structure and impact remnants at the crater centre, with most remnants experiencing peak shock pressures larger than 30GPa (Fig. 4c). However, laboratory hypervelocity experiments show that a depth: diameter ratio between 1 : 5 and 1 : 10 can also be produced by low-velocity impact[20]. According to the previous observation of impact melts in secondary craters on the Moon[21], a secondary impact (for example, from nearby Zhinyu crater) forming this fresh crater cannot be ruled out. Analysis from the crater degradation model[22,23] indicates that the age of this fresh crater might be less than 1 Ma (Supplementary Fig. 1).

Extended Data Fig. 1　Geological context of the Chang'e - 4 landing area. The mosaic of images obtained by the Lunar Reconnaissance Orbiter Camera Wide Angle Camera of the Chang'e - 4 landing area. The landing site is indicated by the yellow star. The data was obtained from https://quickmap. lroc. asu. edu/

The impact delivery is considered to be one of the major contributors to lunar surface water and water ice in the permanently shaded regions[24-28]. A notable amount of water in the CC impactors may have been retained at the lunar surface[8]. Hypervelocity impact experiments in the laboratory suggest that up to 30wt% water in projectiles can be retained in impact melt and remnants[29]. Previous studies suggest that even the less hydrated CK - type CC endmember (that is, Y - 693) contains around 0.2wt% (2,000ppm) water, and some CK - type CC could contain up to 3wt% water[30]. Thus, this glassy material containing 47wt% CC remnants could have retained at least 280ppm of water on the Moon, notably higher than that observed (<100ppm) in the mid - and low latitudes of the Moon[31]. These CC impactors may be important sources of water on the illuminated surface[32] and of the water ice found in the permanently shaded lunar regions[31].

Impactors striking the Moon since the crust formed have provided an indelible record of the impactor population. Geochemical analysis of elements abundant in meteorites in returned lunar soil samples shows that ~2wt% CC materials have been added to the lunar surface after billions of years' exposure to space[33,34]. Recent analyses of the compositions of Apollo impact melt coats and anorthositic regolith breccia meteorites suggest that the impactors striking the Moon changed from ordinary chondrite - like bodies to CC - like impacts with time[6,7]. Our direct detection of CC impactor remnants in the fresh crater by the Chang'e - 4 rover suggests that CC - like impactors may not be uncommon in the modern population of impactors on the Moon. Similar CC remnants might be numerous on the lunar surface and their fragments are expected to be observed in the samples collected by the Chang'e - 5 mission from a relatively young mare unit.

Methods

Instruments and data description. *VNIS*. VNIS consists of a CMOS imager and a SWIR detector. It uses an acousto - optic tunable filter as the dispersive component. It is installed on the front of the rover and points to the lunar surface at a nominal angle of 45° at a height of ~0.7m. The CMOS imager has 256 × 256 pixels and can image the lunar surface at 100 different bands from 450 to 945nm with a spectral resolution of 2.4 – 6.5nm. The FOV of the CMOS imager is 8.5°, hence it can image an area of ~15cm × 21cm. The SWIR detector works from 900 to 2,395nm with a spectral resolution of 3.6 – 9.6nm (ref. [14]). It has only one pixel with a FOV of 3.58° (ref. [14]), which can measure a circular area with a diameter of ~7cm within the FOV of the CMOS image (Extended Data Fig. 4). The SWIR field is centred at pixel 98,128 of the CMOS field with a radius of 53.8 pixels. Both of the detectors have a sampling frequency of 5nm. Detailed information on VNIS can be found in ref. [35]. The uncertainty of the wavelength is ±0.39nm (ref. [36]), and the signal - to - noise ratio of the CMOS channel is better than 25 dB based on a repeated measurement on exactly the same area[37].

In this work, we used the high - spatial and - spectral - resolution CMOS data to analyse millimetre - scale differences in the composition of the targeted small crater. The SWIR detector has only one pixel and the spatial resolution (~7cm per pixel) is much lower than that of the CMOS imager. As shown in Extended Data Fig. 5, we compared the average spectra within the SWIR FOV and other smaller regions, the spectral features of some glassy materials are obscured due to the lower spatial resolution.

PCAMs. Yutu - 2 is equipped with two PCAMs on the top of its mast (Extended Data Fig. 4), and the horizontal pointing directions of the two cameras are always tilted towards angled away from each other by 1° so that stereo images can be collected[38]. Both PCAMs have a FOV of 19.7° × 14.5° and two working modes: colour and greyscale. The frame pixels are 2,352 × 1,728 for the colour mode and 1,176 × 864 for the greyscale mode. With these high - resolution PCAM stereo images, we constructed a digital elevation model (DEM) for the small crater using photogrammetric methods[39]. The DEM was then refined using the shape - from - shading method[40] and the final derived DEM has an effective resolution of 5mm per pixel.

Hazard - avoidance cameras (HAZCAMs). There are two HAZCAMs installed aside the VNIS, and they have a FOV of 58°, which is much larger than the CMOS imager. The positions of both the VNIS and HAZCAMs are fixed, hence the relative positions of the imaged areas of the two cameras are also fixed (Extended Data Fig. 4). The area of CMOS imager is located approximately at the centre of the HAZCAM FOV. With these large - view HAZCAM images, we could locate the position of each VNIS measurement on the lunar surface accurately (Supplementary Fig. 2).

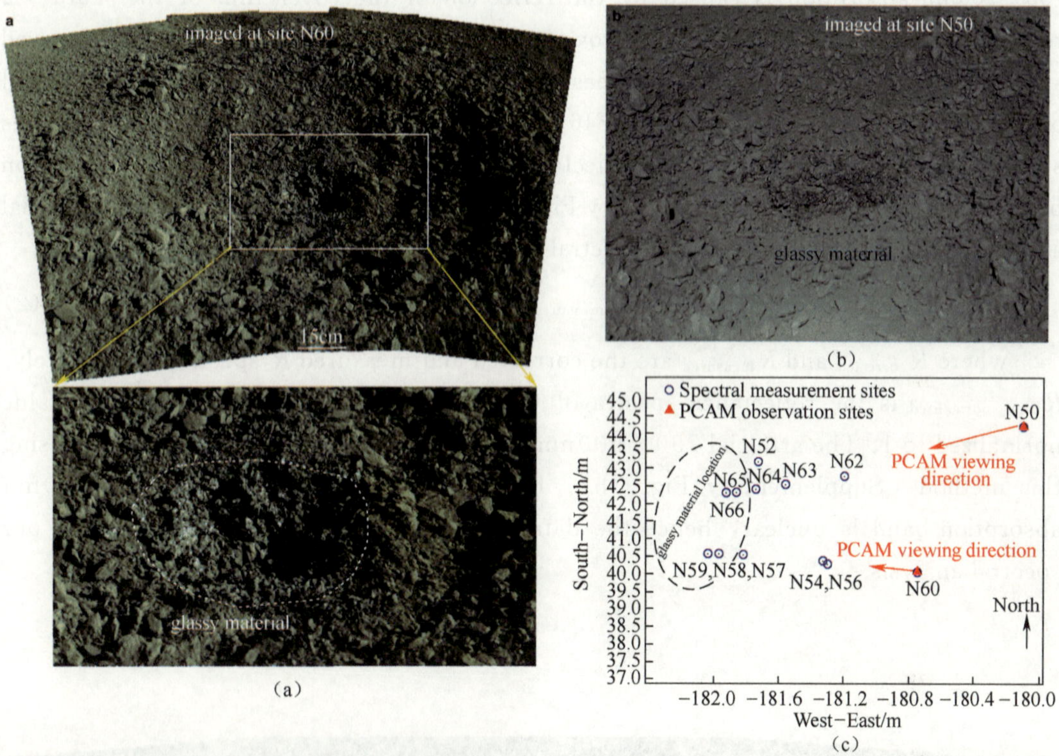

Extended Data Fig. 2　The "glassy" materials imaged by the rover's panorama camera. (a) Color mosaic of PCAM images of the targeted small crater observed on the 9th lunar day at site N60. The enlarged image at bottom shows the observed glassy material. (b) The PCAM image of the crater centre observed from another direction on the 8th lunar day at site N50. (c) The relative locations of the observed "glassy" material under the rover - lander coordinate system and the two PCAM observation sites of (a) and (b)

Absolute reflectance calibration. We started the data reduction from the level 2B data, which are the radiometric calibrated radiance data. The absolute reflectance was obtained by following[10,12,41]:

$$R(\lambda)=\frac{\pi}{\cos(i)}\frac{I(\lambda)}{\int J(\lambda)F(\lambda)\mathrm{d}\lambda}D_{\mathrm{au}}^{2} \tag{1}$$

where $R(\lambda)$ is reflectance factor[42] at wavelength λ, i is angle of incidence, $I(\lambda)$ is the radiance obtained by the VNIS spectrometer, $J(\lambda)$ is the solar irradiance at 1 au (ref. [43]), $F(\lambda)$ is the spectral response function of the VNIS detector, approximated by Gaussian function with a given band centre and full - width at half - maximum for each band[14,15], and D_{AU} is the real - time Sun - Moon distance in astronomical units obtained from https://ssd.jpl.nasa.gov/horizons.cgi#top.

Stray light correction. During the analysis of the CMOS spectra, we found that there were two absorption features around 620－690nm and 750－900nm. After careful validation, we determined that these two absorption bands are artificial features. The 750－900nm absorption is caused by the reflection of the cover film of the Yutu－2 rover[44]. The light reflected from the rover cover may have been reflected by the observed surface and entered the detector. We measured the reflectance spectra of the cover material of the rover, which is a type of F46 film, using an Analytical Spectral Devices spectrometer in the laboratory. The reflectance spectra of the film show clearly absorption between 750 and 900nm (Supplementary Fig. 3). To correct for the effect of this artificial absorption feature on the measured spectral data, we employed the following method:

$$R_{\text{corrected}}(\lambda)=R_{\text{measured}}(\lambda)/R_{\text{film\_normalized}}(\lambda) \tag{2}$$

where $R_{\text{corrected}}$ and R_{measured} are the corrected and measured R spectra, respectively. $R_{\text{film\_normalized}}$ is the reflectance spectra of the film material with the maximum value normalized to 1. The artificial 750－900nm absorption can be corrected effectively using this method (Supplementary Fig. 3b). However, the cause of the 620－690nm absorption band is unclear, hence the data in this band regions were masked in our spectral analysis.

(a)

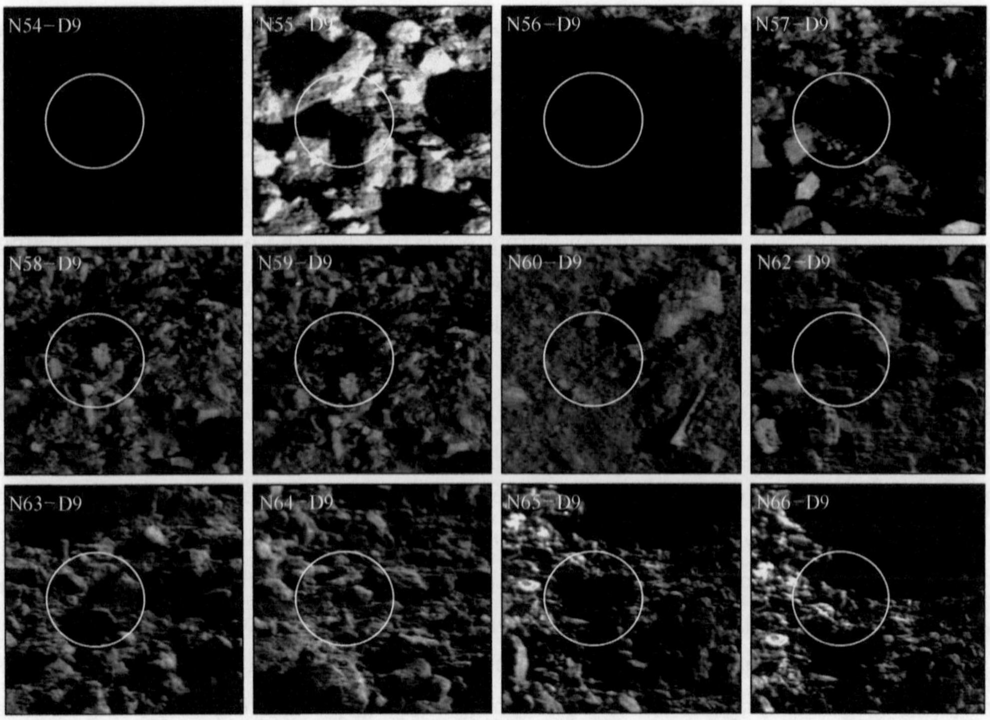

(b)

Extended Data Fig. 3 The VNIS measurement sites on Day 9 and the corresponding CMOS images. (a) A PCAM image mosaic showing the rover tracks and the crater rim, and all the VNIS measurement sites are marked out with green dots and the corresponding site IDs are given in yellow numbers. The blue arrow indicates the north direction. (b) The 12 reflectance images at wavelength of 560nm obtained by the CMOS imager at the sites shown on (a), all the images are stretched from 0 to 0.2. The white circles show the field of view of the SWIR detector

Evaluation of the spectral slope. For each CMOS spectrum, the spectral slope was calculated using the following procedure: first, the bands smaller than 500nm and larger than 900nm were excluded, and the bands between 620 and 690nm were also masked due to potential artificial absorption here. Then, a linear function ($y = ax + b$) was employed to fit to the remaining part of the spectrum, and the best-fit slope value (a) was adopted as the spectral slope.

Spectral angle algorithm for similarity quantification. The spectral angle can be used to quantify the similarity between two spectra following the equation below[45]:

$$\alpha = \arccos\left(\frac{\sum_{i=1}^{n} x_i y_i}{\sqrt{\sum_{i=1}^{n} x_i^2}\sqrt{\sum_{i=1}^{n} y_i^2}}\right) \tag{3}$$

where x and y represent two spectra, respectively, α is the spectral angle between the two spectra that are treated as two m - dimensional vectors, m is the total band number in the wavelength domain. α varies between 0° and 90°; the closer the angle to 0°, the higher similarity of the two spectra. The results from the spectral angle analysis between the selected endmembers and the N66 region of interest (ROI) spectrum are listed in Supplementary Tab. 2.

Spectral unmixing. Reflectance model. We employed the Hapke radiative transfer model for spectral unmixing[46]. For an intimate mixture like the lunar regolith, mixing in the reflectance dimension is highly nonlinear. However, the average single scattering albedo (SSA) of a mixture can be taken as the linear combinations of each endmember's SSA[46]. The mixed SSA spectra can then be modelled through several iterations with a least squares approach to produce the best mathematical and visual fit by optimizing for the lowest root mean squared error between the observed and the modelled SSA, providing mineral abundances. Therefore, the key to our spectral unmixing is to convert the reflectance of both the observed mixture and endmembers into SSA. SSA is a function of the complex refractive index (n: real part; k: imaginary part) and the optical path length $<D>$ that is related to particle size D. The optical constants (n and k) of all endmembers can be derived using equation (4) from their reflectance spectra following the method used in ref. [16]. The relationship between reflectance and these parameters can be expressed as:

$$R(i, e, g, n, k, D, \varphi, \lambda)$$
$$=\frac{\omega_{\text{ave}}(n, k, D, \lambda)}{4\pi}\frac{\mu_0}{\mu_0+\mu}\{[1+B(g, \varphi)](P(g)+H(\mu_0, \omega_{\text{ave}})H(\mu, \omega_{\text{ave}})-1\} \tag{4}$$

where i, e and g are the incidence, emission and phase angles, respectively; ω_{ave} is the average SSA of the endmembers used in the unmixing weighted by solid density and particle size; Φ is the porosity; μ_0 and μ are the cosines of the incidence and emission angles, respectively; $B(g, \varphi)$ represents the opposition effect; $P(g)$ is the phase function and the Ambartsumian - Chandrasekhar H function accounts for the multiple scattering. We set $\omega_{\text{ave}}=\sum_i M_i\omega_i$, where M_i represents the fraction of cross - sectional area for the i th endmember. The detailed parameterizations can be found in refs. [16,47].

Endmember selection. To build an endmember spectral library, we collected spectra of typical lunar mineral separates, including olivine, orthopyroxene, clinopyroxene, plagioclase and ilmenite, as the mineral spectral endmembers from the Lunar Rock and Mineral Characterization Database (http: //www. planetary. brown. edu/LRMCC)[48]. These are common minerals on the lunar surface that are essential for spectral modelling of lunar surface materials. We also included the spectra of four typical lunar glass samples (agglutinate, green glass, orange glass and black spherules)[49]. It has been suggested that there is around 30—70 wt% agglutinate in Apollo samples[50]. Therefore, agglutinate was added in our endmember pool. Recent studies suggested that agglutinates at different Apollo sampling sites exhibit similar spectral features[51], which strengthens the case for the use of agglutinate as an endmember for spectral unmixing. So far, only three distinct volcanic glasses (green and orange glasses and black beads) have been identified in Apollo samples. These are also included in our endmember library.

Traditionally, the spectra of lunar rock or regolith samples have a reddened slope in the visible to near-infrared wavelength region due to space weathering. In contrast, as shown in Fig. 1c, the spectrum of the observed glassy material shows distinctly different spectral slopes to those of the nearby lunar regolith and rock fragments. We found that the spectra of this region do not match with the spectra of any known minerals and glasses in Apollo samples. Therefore, there must be some other exotic materials that have blue spectral slopes mixed with the local surface materials forming the observed glassy materials. The most common S-type meteorites show similar spectral features to lunar surface silicate minerals, which have already been included in our endmember pool. Another type of common impactor shared by the Earth and Moon is CC meteorites, which have similar blue spectral slopes[30,52]. We performed a thorough survey and collected 183 spectra from a variety of CC meteorites. To select the closest spectral match with the endmembers in our spectral unmixing analysis, we performed spectral angle analysis to characterize the similarity between CC meteorite and the glassy material. We found that the spectra of some CK- and CM-type CC match well with Chang'e-4 spectra (see Supplementary Tab. 6, where a smaller spectral angle means higher similarity). The comparisons of these spectra and the corresponding sample information are provided in Supplementary Tab. 7. Some samples could have been contaminated under terrestrial conditions (represented by a weathering index); we found that only Y-693 had a weathering index of 0 (Supplementary Tab. 6) among these nine CK samples. Therefore, we used Y-693 as the representative CC endmember for our spectral unmixing analysis. The reflectance spectra of all endmembers in the COMS band region are shown in Extended Data Fig. 9, and the detailed spectral and sample information are listed in Supplementary Tab. 1.

Incorporating the space weathering effect. The spectral effect of space weathering was also considered in our analysis by incorporating the absorption effect of $npFe^0$ (nanophase iron or submicrometre iron, the primary product of space weathering that accounts for the optical effect) into that of host materials. As proposed by Hapke[42], the absorption coefficient of host materials containing $npFe^0$ can be expressed as:

$$\alpha = \frac{4\pi n_{\mathrm{h}} k_{\mathrm{h}}}{\lambda} + \frac{36\pi z M_{\mathrm{Fe}} \rho_{\mathrm{h}}}{\lambda \rho_{\mathrm{Fe}}}$$

$$z = \frac{n_{\mathrm{h}}^3 n_{\mathrm{Fe}} k_{\mathrm{Fe}}}{(n_{\mathrm{Fe}}^2 - k_{\mathrm{Fe}}^2 + 2n_{\mathrm{h}}^2)^2 + 4n_{\mathrm{Fe}}^2 k_{\mathrm{Fe}}^2} \tag{5}$$

where n and k are the optical constants, ρ is the solid density and M is the mass fraction. The subscripts h and Fe indicate the host material and $npFe^0$, respectively. The detailed parameterization of Hapke's model for accommodating $npFe^0$ effects is the same as that of ref. [16].

The performance of Hapke's model in deriving the mineral abundances of lunar materials has been thoroughly studied in refs. [16,47,53,54]. We employed the same method and code as used in refs. [16,47]. According to their results, the uncertainty of this technique for estimating mineral abundances of Apollo samples in the LSCC dataset is within ±10wt% (refs. [16,54]). The HED meteorites show similar mineral endmembers as lunar basalts, dominantly clinopyroxene, orthopyroxene, plagioclase and olivine[47]. Similar uncertainty (±10 wt%) in estimating minerals in HED meteorites is seen in their Hapke modelling work[47]. We have confidence in the modelling results if the estimated endmember abundance is notably higher than 10wt%. We also performed sensitivity tests on whether the CC (rather than agglutinat) was required to reproduce the spectral features observed by the Chang'e-4 rover using spectral unmixing model: (1) with all endmembers used, (2) with only the CC endmember removed and (3) with only the agglutinate endmember removed. The results show that the model failed to reproduce the observed spectral features without the CC endmember. In contrast, the model could still fit the Chang'e-4 spectra well even without including the agglutinate endmember when the CC endmember was retained (Supplementary Fig. 4).

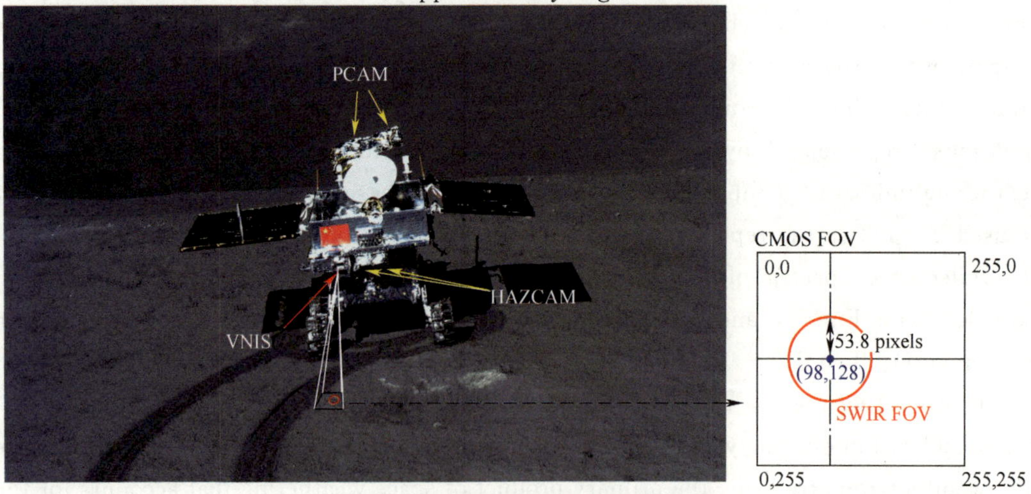

Extended Data Fig. 4　A front view of the Yutu-2 Rover taken by the terrain camera on board the Chang'e-4 lander. The panorama cameras (PCAMs) are on top of the rover mast and the hazard avoidance cameras (HAZCAMs) are installed on the front side of the rover, close to the visible and near-infrared spectrometer (VNIS). The CMOS imager has a field of view (FOV) of 8.5° and looks down to the lunar surface from a height of ~0.7m with a nominal fixed viewing angle of 45°, hence it can image an area of ~15cm×21cm. The SWIR detector only has one pixel with a FOV of 3.58° and the relative position of the SWIR field within the CMOS image is shown in the right schematic diagram. The black rectangle on the right figure represents the FOV of CMOS and the red circle within represents the FOV of SWIR detector

In an effort to quantify the effects of spectral uncertainty on the modelled abundances, we performed spectral unmixing on the average spectrum of N66, the spectrum with uncertainty added, and the spectrum with uncertainty subtracted, respectively. The modelled CC abundances were 47wt%, 15wt% and 25wt%, respectively, with an average of 29 ± 16wt%. We also randomly selected a series of subgroups of pixels from the N66 ROI and performed spectral unmixing over the average spectrum of each subgroup. We generated 300 averaged spectra by randomly selecting 140 pixels out of the original N66 ROI (174 pixels; Fig. 1d) and the modelling results show that the derived average abundance of CC was 43 ± 25wt%. These sensitivity tests confirm the presence of CC when uncertainties are included in the spectral unmixing modelling.

Conversion of Chang'e - 4 reflectance at standard angles. To compare with the endmember spectra, the Chang'e - 4 spectra were converted into the spectra measured at the standard measurement angles, that is $i = 30°$, $e = 0°$, $g = 30°$. With given measurement angles (Supplementary Tab. 8) and reflectance values, the corresponding SSA for each spectrum could be derived by solving equation (4). By inserting these standard angles and SSA into equation (4), we obtained the corrected reflectance spectra. Since the phase angles of the Chang'e - 4 measurements were all larger than 40°, the $B(g)$ term in equation (4) that accounts for the opposition effect was ignored in our modelling.

Impact cratering simulation. We used the multimaterial, multirheology 2D iSALE (Dellen version) shock physics code[19,55] to model the impact cratering process of the observed fresh crater. iSALE has been used to simulate both small - scale laboratory experiments[56] and large - scale lunar basin formation[57-62].

In our simulation, we used the analytic equation of state (ANEOS) for dunite[63,64] and basalt[65] to represent the compositions of impactor and target, respectively. It would have been more accurate to model the CC impactor as forsterite or serpentine in our simulation, but the shock equations of state for forsterite and serpentine are not included in the current version. However, the ANEOS and Hugoniot curves for forsterite and dunite are almost similar[64], so our use of dunite for the representative CC impactor did not affect our conclusion. We assigned a porosity of ~55% to the dunite impactor[55] so that the average density of the impactor was ~1,500kg · m^{-3}, approaching the average density for C - type asteroids[66]. Material strength and damage were accounted for using the model of refs. [19,67]. The computational domain covered an area of 1.5m in the lateral and vertical directions, with a cell size of 0.05m × 0.05m in the high - resolution zone. The resolution of 20 cells per projectile radius is suitable for impact crater processing and tracing of the peak shock pressure of the impactor relics[61,62]. The simulation assumed an average impact velocity of 15km · s^{-1}. We assigned a lunar surface gravitational acceleration of 1.62m · s^{-2} in our model. Owing to the axisymmetric nature of the 2D code, all simulations considered vertical impacts. Although vertical impacts are highly unlikely, they provide a reasonable proxy for the most common angle of impact (45°) on

the lunar surface.

We tracked the impactor physical states by using Lagrangian tracers in iSALE, which were initially placed in each computational cell and represented the matter originally in that cell throughout the simulation. We recorded the position and peak shock pressure the impactor experienced during the impact cratering process in the same manner as refs. [61,68]. Fig. 4c shows the structure of the final crater for the simulation with an impactor diameter of 15cm. This simulation produces a simple crater with diameter of ~2m, similar to the size of observed crater. Our simulation shows that ~85% of the impactor accreted on the lunar surface, of which ~60% of relics are located within the crater. These impactor relics within the crater experienced the peak shock pressure ranging from ~30GPa to ~90GPa (Fig. 4c).

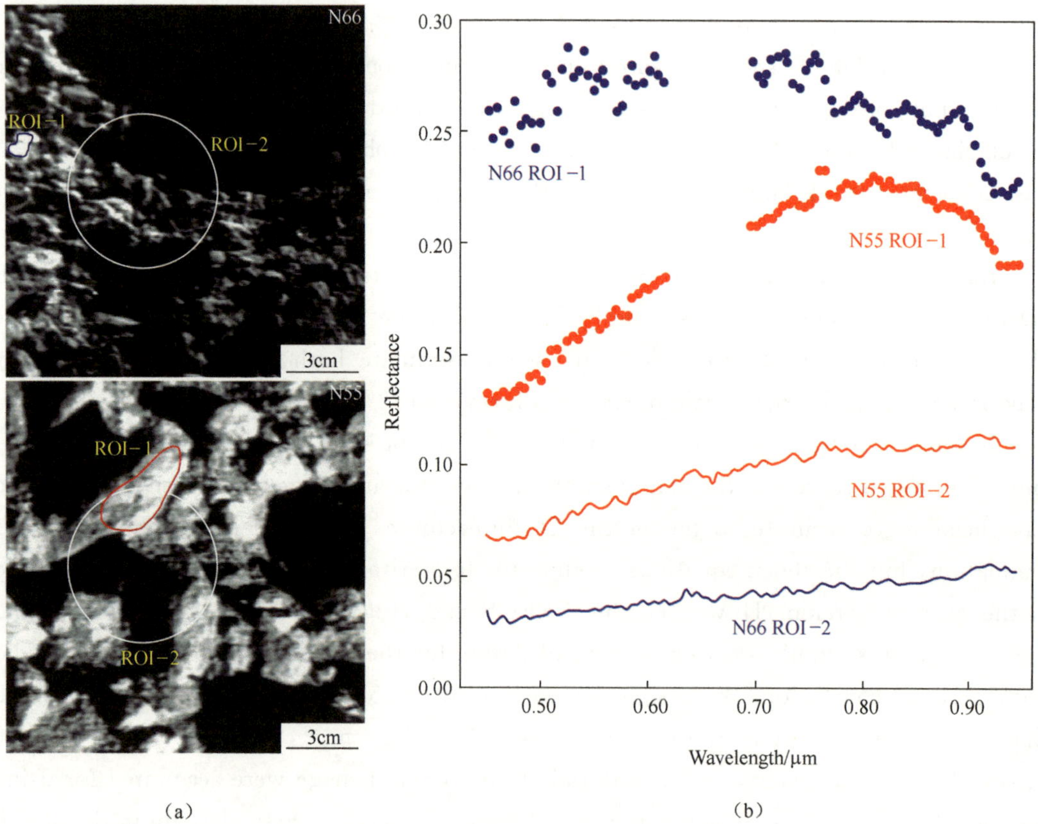

Extended Data Fig. 5　Comparison of the average CMOS reflectance spectra of the different ROIs. The average CMOS reflectance spectra within the FOV of SWIR detector. (a) The image of N66 and N55 at a wavelength of 560nm, the white circles indicate the FOV of SWIR detector. (b) The average spectra of the corresponding ROIs shown in (a). In this work we only used the CMOS data, as the spatial resolution of the SWIR data is too low to resolve those "glassy" materials and shadow regions

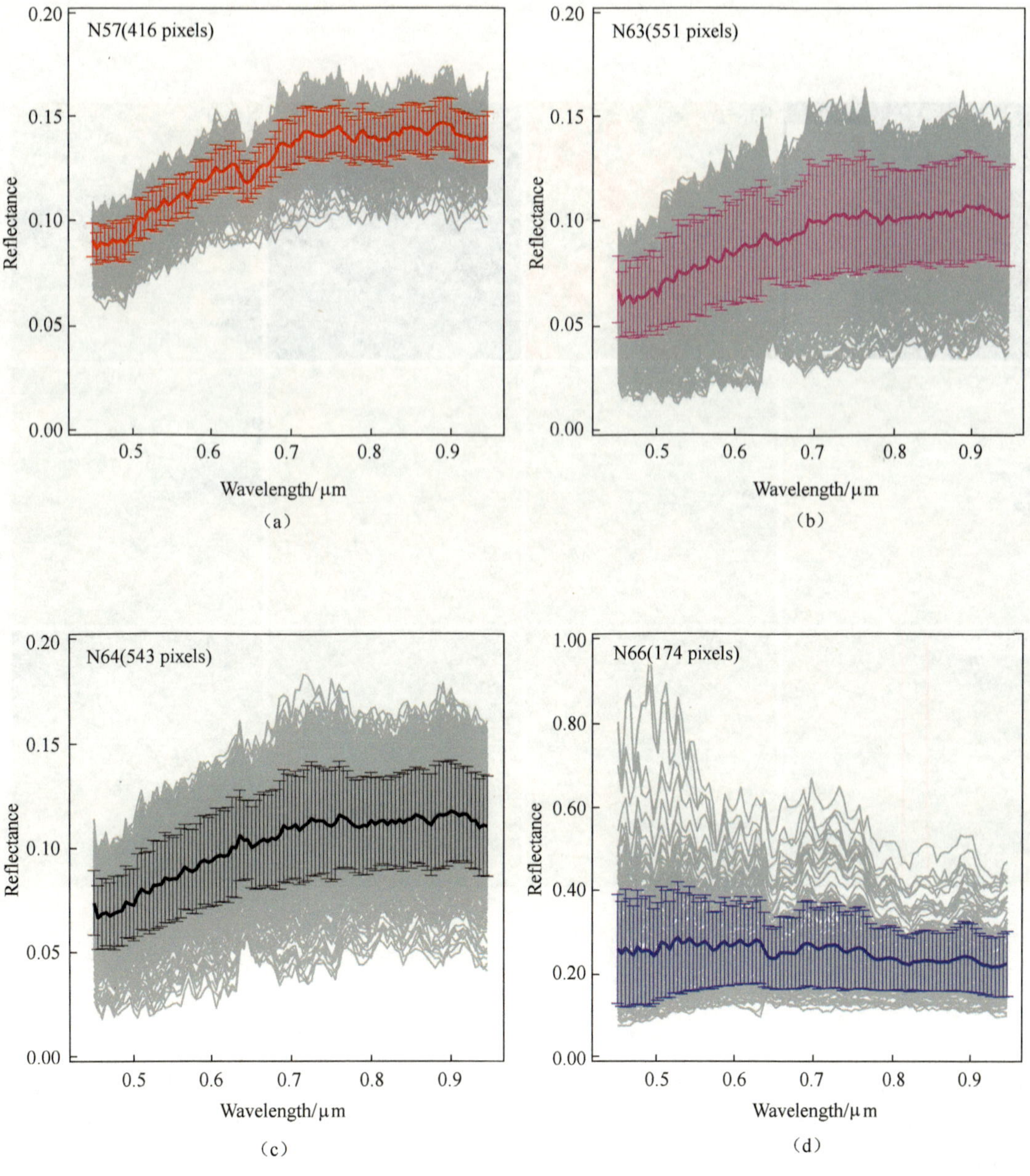

Extended Data Fig. 6　The spectra of all pixels within each ROI shown in Fig. 1d. The spectra of single pixel are shown in gray, and the average spectra of each ROI are shown in color lines with $\pm 1\sigma$ uncertainties determined from the statistic of the data within each ROI, respectively

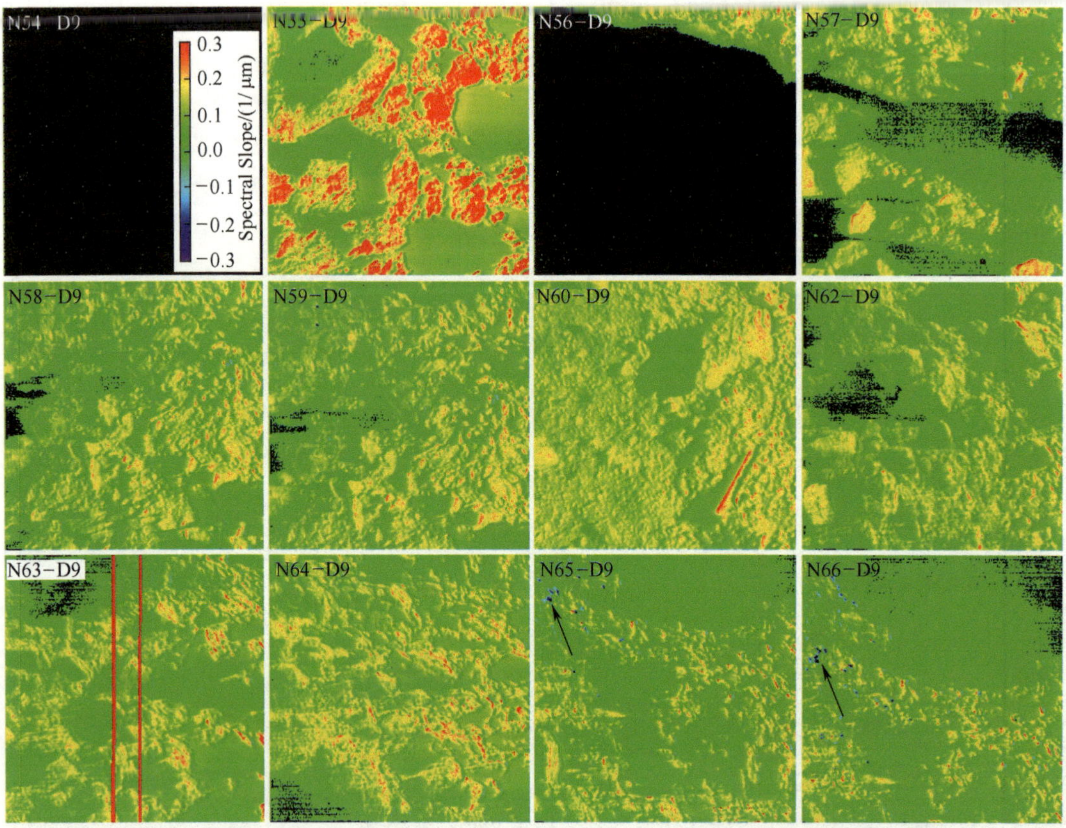

Extended Data Fig. 7　Estimated spectral slopes of all the 12 CMOS measurements on Day 9. Estimated spectral slopes of all the 12 CMOS measurements on Day 9. On sites N65 and N66, the CMOS camera captured the "glassy" materials occurred at the crater centre. Some pixels in N65 and N66 show clearly blueing spectral trend

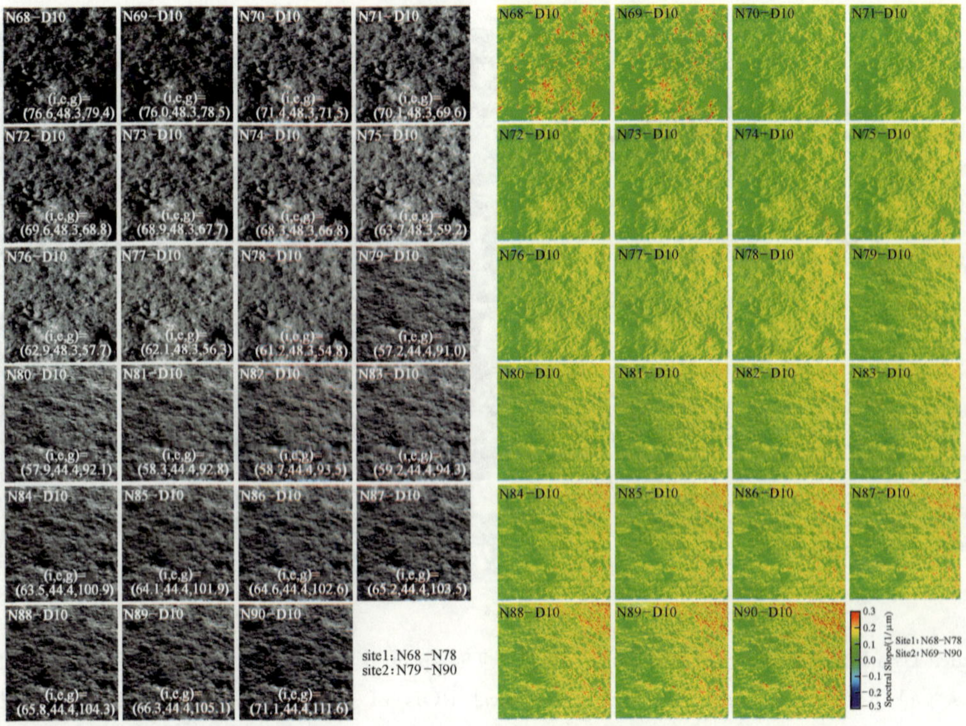

Extended Data Fig. 8　The CMOS images obtained on Day 10 and corresponding spectral slope maps. (a) The CMOS images at wavelength 560nm obtained on Day 10. N68 - N78 and N79 - N90 represent images of two sites taken at varied measurement angles, respectively (ref. [15]). (b) Estimated spectral slopes of the CMOS images

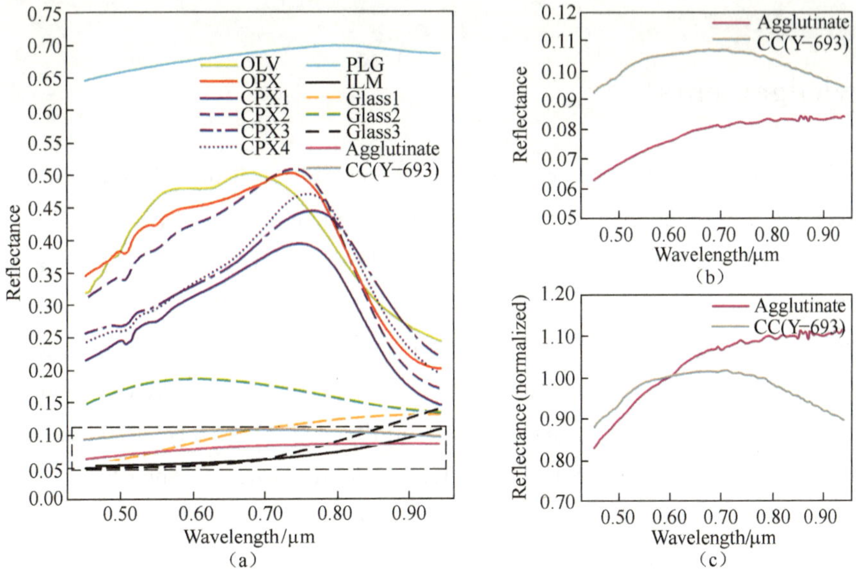

Extended Data Fig. 9　Reflectance spectra of endmembers used for spectral unmixing. (a) Reflectance spectra of endmembers used for spectral unmixing obtained from the RELAB spectral database. The detailed information of these endmembers is listed in Supplementary Tab. 1. (b) Enlarged view of the dashed box area shown in panel (a). (c) Reflectance spectra of agglutinate and CC normalized to 1 at 0.6μm

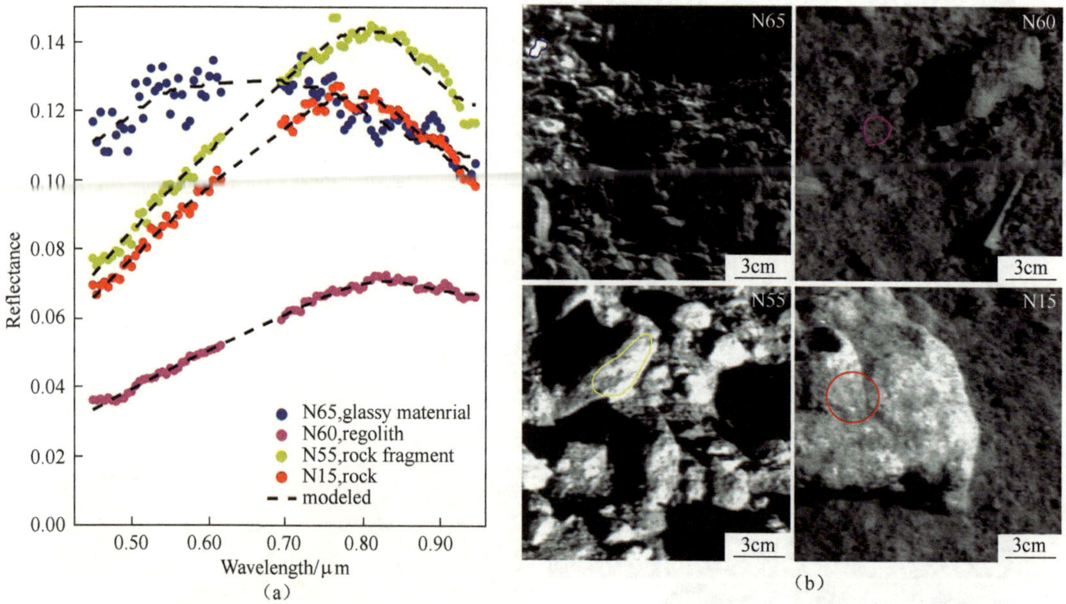

Extended Data Fig. 10　The spectral fitting results. The spectral fitting results and corresponding CMOS images. (a) Reflectance spectra of the four typical ROIs of N65，N60，N55，and N15，and the corresponding best－fit modeling results. (b) CMOS images at wavelength of 560 nm with ROIs of N60，N55，N16，and N15. All the images are stretched from 0 to 0.2. The measurement locations of N65，N60，and N55 are shown in Extended Data Fig. 3a. The ROI of N65 is the same material as that of N66 (Fig. 1d). The ROI of N60 represents the background regolith outside the crater. The ROI of N15 represents a fresh rock found on the 3rd lunar day of the mission's operation

Acknowledgements

The Chang'e－4 mission was carried out by the Chinese Lunar Exploration Program，and the data were provided by the China National Space Administration. We thank the Chang'e－4 engineering team for their tremendous efforts to make such a successful mission possible. This work was supported by the National Natural Science Foundation of China (grant numbers 11941001，42002306 and 42072337)，the Strategic Priority Research Program of Chinese Academy of Sciences (grant number XDB41000000) and the Civil Aerospace Pre－research Project (grant numbers D020201，D020203 and D020204). Y. Y. also acknowledges support from the China Postdoctoral Science Foundation (grant number 2019TQ0323) and the Pandeng Program of the National Space Science Center, Chinese Academy of Sciences. M.－H. Z. was supported by the Science and Technology Development Fund，Macau (grant numbers 0079/2018/A2 and 0020/2021/A1) and the Civil Aerospace Presearch Project (grant number D020202). B. W. was supported by a grant from the Research Grants Council of Hong Kong (RIF project number R5043－19).

We acknowledge the developers of iSALE and the pySALEPlot visualization package.

References

[1] Neukum, G., Ivanov, B. & Hartmann, W. Cratering records in the inner Solar System in relation to the lunar reference system. Space Sci. Rev. 1, 55 - 86 (2001).

[2] Yue, Z. et al. Projectile remnants in central peaks of lunar impact craters. Nat. Geosci. 6, 435 - 437 (2013).

[3] Joy, K. H. et al. Direct detection of projectile relics from the end of the lunar basin - forming epoch. Science 336, 1426 - 1429 (2012).

[4] Joy, K. H. et al. The Moon: an archive of small body migration in the Solar System. Earth Moon Planets 118, 133 - 158 (2016).

[5] Strom, R. G., Malhotra, R., Ito, T., Yoshida, F. & Kring, D. A. The origin of planetary impactors in the inner Solar System. Science 309, 1847 - 1850 (2005).

[6] McIntosh, E. C., Day, J. M., Liu, Y. & Jiskoot, C. Examining the compositions of impactors striking the Moon using Apollo impact melt coats and anorthositic regolith breccia meteorites. Geochim. Cosmochim. Acta 274, 192 - 210 (2020).

[7] Puchtel, I. S., Walker, R. J., James, O. B. & Kring, D. A. Osmium isotope and highly siderophile element systematics of lunar impact melt breccias: implications for the late accretion history of the Moon and Earth. Geochim. Cosmochim. Acta 72, 3022 - 3042 (2008).

[8] Joy, K. H. et al. The isotopic composition of volatiles in the unique bench crater carbonaceous chondrite impactor found in the Apollo 12 regolith. Earth Planet. Sci. Lett. 540, 116265 (2020).

[9] Zolensky, M. E., Weisberg, M. K., Buchanan, P. C. & Mittlefehldt, D. W. Mineralogy of carbonaceous chondrite clasts in HED achondrites and the Moon. Meteorit. Planet. Sci. 31, 518 - 537 (1996).

[10] Hu, X. et al. Mineral abundances inferred from in situ reflectance measurements of Chang'e - 4 landing site in South Pole - Aitken Basin. Geophys. Res. Lett. 46, 9439 - 9447 (2019).

[11] Gou, S. et al. Lunar deep materials observed by Chang'e - 4 rover. Earth Planet. Sci. Lett. 528, 115829 (2019).

[12] Lin, H. et al. Olivine - norite rock detected by the lunar rover Yutu - 2 likely crystallized from the SPA - impact melt pool. Natl Sci. Rev. 7, 913 - 920 (2020).

[13] Huang, J. et al. Diverse rock types detected in the lunar South Pole - Aitken Basin by the Chang'e - 4 lunar mission. Geology https://doi.org/10.1130/G47280.1 (2020).

[14] Li, C. et al. Detection and calibration characteristics of the visible and near - infrared imaging spectrometer in the Chang'e - 4. Rev. Sci. Instrum. 90, 103106 (2019).

[15] Yang, Y. et al. The effects of viewing geometry on the spectral analysis of lunar regolith as inferred by in situ spectrophoto－metric measurements of Chang'e－4. Geophys. Res. Lett. 47, e2020GL087080 (2020).
[16] Li, S. & Li, L. Radiative transfer modeling for quantifying lunar surface minerals, particle size, and submicroscopic metallic Fe. J. Geophys. Res. Planets 116, E00001 (2011).
[17] Huang, J. et al. Geological characteristics of Von Kármán crater, northwestern South Pole－Aitken Basin: Chang'e－4 landing site region. J. Geophys. Res. Planets 123, 1684－1700 (2018).
[18] Wu, B. et al. Centimeter－resolution topographic modeling and fine－scale analysis of craters and rocks at the Chang'e－4 landing site. Earth Planet. Sci. Lett. 553, 116666 (2020).
[19] Collins, G. S., Melosh, H. J. & Ivanov, B. A. Modeling damage and deformation in impact simulations. Meteorit. Planet. Sci. 39, 217－231 (2004).
[20] Schultz, P. H. & Gault, D. E. Clustered impacts: experiments and implications. J. Geophys. Res. 90, 3701－3732 (1985).
[21] Schultz, P. H. Evidence and mechanisms for the non－local contribution to ejecta deposits. In Workshop on Apollo16 (eds James, O. & Hörz, F.) Technical Report 81－01 (Lunar and Planetary Institute, 1981).
[22] Basilevskii, A. On the evolution rate of small lunar craters. Lunar Planet. Sci. Proc. 7, 1005－1020 (1976).
[23] Du, J. et al. Thickness of lunar mare basalts: new results based on modeling the degradation of partially buried craters. J. Geophys. Res. Planets 124, 2430－2459 (2019).
[24] Watson, K., Murray, B. C. & Brown, H. The behavior of volatiles on the lunar surface. J. Geophys. Res. 66, 3033－3045 (1961).
[25] Pieters, C. M. et al. Character and spatial distribution of OH/H_2O on the surface of the Moon seen by M^3 on Chandrayaan－1. Science 326, 568－572 (2009).
[26] Sunshine, J. M. et al. Temporal and spatial variability of lunar hydration as observed by the Deep Impact spacecraft. Science 326, 565－568 (2009).
[27] Clark, R. N. Detection of adsorbed water and hydroxyl on the Moon. Science 326, 562－564 (2009).
[28] Li, S. et al. Direct evidence of surface exposed water ice in the lunar polar regions. Proc. Natl Acad. Sci. USA 115, 8907－8912 (2018).
[29] Daly, R. T. & Schultz, P. H. The delivery of water by impacts from planetary accretion to present. Sci. Adv. 4, eaar2632 (2018).
[30] Cloutis, E., Hudon, P., Hiroi, T. & Gaffey, M. Spectral reflectance properties of carbonaceous chondrites: 7. CK chondrites. Icarus 221, 911－924 (2012).
[31] Li, S. & Milliken, R. E. Water on the surface of the Moon as seen by the Moon

Mineralogy Mapper: distribution, abundance, and origins. Sci. Adv. 3, eaba1940 (2017).

[32] Hurley, D. et al. Anatomy of the lunar water exosphere. In 50th Lunar and Planetary Science Conference 2132 (Lunar and Planetary Institute, 2019).

[33] Anders, E., Ganapathy, R., Krähenbühl, U. & Morgan, J. W. Meteoritic material on the moon. Moon 8, 3 - 24 (1973).

[34] Wasson, J. T., Boynton, W. V., Chou, C. - L. & Baedecker, P. A. Compositional evidence regarding the influx of interplanetary materials onto the lunar surface. Moon 13, 121 - 141 (1975).

[35] He, Z. et al. Spectrometers based on acousto - optic tunable filters for in - situ lunar surface measurement. J. Appl. Remote Sens. 13, 027502 (2019).

[36] Lin, H. et al. In situ photometric experiment of lunar regolith with visible and near - infrared imaging spectrometer on board the Yutu - 2 lunar rover. J. Geophys. Res. Planets 125, e2019JE006076 (2020).

[37] Lin, H. et al. Estimation of noise in the in situ hyperspectral data acquired by Chang'e - 4 and its effects on spectral analysis of regolith. Remote Sens. 12, 1603 (2020).

[38] Jin, W. et al. In situ optical measurements of Chang'e - 3 landing site in Mare Imbrium: 2. Photometric properties of the regolith. Geophys. Res. Lett. 42, 8312 - 8319 (2015).

[39] Liu, W. C. & Wu, B. An integrated photogrammetric and photoclinometric approach for illumination - invariant pixel - resolution 3D mapping of the lunar surface. ISPRS J. Photogramm. Remote Sens. 159, 153 - 168 (2020).

[40] Wu, B., Liu, W. C., Grumpe, A. & Wöhler, C. Construction of pixel - level resolution dems from monocular images by shape and albedo from shading constrained with low - resolution dem. ISPRS J. Photogramm. Remote Sens. 140, 3 - 19 (2018).

[41] Ma, P. et al. A plagioclase - rich rock measured by Yutu - 2 rover in Von Kármán crater on the far side of the Moon. Icarus 350, 113901 (2020).

[42] Hapke, B. Theory of Reflectance and Emittance Spectroscopy (Cambridge Univ. Press, 2012).

[43] Gueymard, C. A. The Sun's total and spectral irradiance for solar energy applications and solar radiation models. Sol. Energy 76, 423 - 453 (2004).

[44] Lin, H. et al. New insight into lunar regolith - forming processes by the lunar rover Yutu - 2. Geophys. Res. Lett. 47, e2020GL087949 (2020).

[45] Taylor, K. E. Summarizing multiple aspects of model performance in a single diagram. J. Geophys. Res. Atmos. 106, 7183 - 7192 (2001).

[46] Hapke, B. Bidirectional reflectance spectroscopy: 1. Theory. J. Geophys. Res. Solid Earth 86, 3039 - 3054 (1981).

[47] Li, S. & Milliken, R. E. Estimating the modal mineralogy of eucrite and diogenite meteorites using visible - near infrared reflectance spectroscopy. Meteorit. Planet. Sci. 50, 1821 - 1850 (2015).

[48] Isaacson, P. J. et al. The lunar rock and mineral characterization consortium: deconstruction and integrated mineralogical, petrologic, and spectroscopic analyses of mare basalts. Meteorit. Planet. Sci. 46, 228 - 251 (2011).

[49] Tompkins, S. & Pieters, C. M. Spectral characteristics of lunar impact melts and inferred mineralogy. Meteorit. Planet. Sci. 45, 1152 - 1169 (2010).

[50] Heiken, G. H., Vaniman, D. T. & French, B. M. Lunar Sourcebook: A User's Guide to the Moon (Cambridge Univ. Press, 1991).

[51] Denevi, B., Yasanayake, C., Jolliff, B., Lawrence, S. & Hiroi, T. The spectral properties of lunar agglutinates. Lunar Planet. Sci. Conf. 51, 2026 (2020).

[52] Cloutis, E. et al. Spectral reflectance properties of carbonaceous chondrites: 6. CV chondrites. Icarus 221, 328 - 358 (2012).

[53] Denevi, B., Lucey, P. & Sherman, S. Radiative transfer modeling of near - infrared spectra of lunar mare soils: theory and measurement. J. Geophys. Res. Planets 113, E02003 (2008).

[54] Lucey, P. G. et al. A large spectral survey of small lunar craters: implications for the composition of the lunar mantle. Am. Mineral. 99, 2251 - 2257 (2014).

[55] Wünnemann, K., Collins, G. & Melosh, H. A strain - based porosity model for use in hydrocode simulations of impacts and implications for transient crater growth in porous targets. Icarus 180, 514 - 527 (2006).

[56] Wünnemann, K., Zhu, M. - H. & Stöffler, D. Impacts into quartz sand: crater formation, shock metamorphism, and ejecta distribution in laboratory experiments and numerical models. Meteorit. Planet. Sci. 51, 1762 - 1794 (2016).

[57] Melosh, H. et al. The origin of lunar mascon basins. Science 340, 1552 - 1555 (2013).

[58] Potter, R. W., Collins, G. S., Kiefer, W. S., McGovern, P. J. & Kring, D. A. Constraining the size of the South Pole - Aitken Basin impact. Icarus 220, 730 - 743 (2012).

[59] Miljkovic, K. et al. Asymmetric distribution of lunar impact basins caused by variations in target properties. Science 342, 724 - 726 (2013).

[60] Zhu, M. - H., Wünnemann, K. & Potter, R. W. Numerical modeling of the ejecta distribution and formation of the Orientale basin on the Moon. J. Geophys. Res. Planets 120, 2118 - 2134 (2015).

[61] Zhu, M. - H., Wünnemann, K. & Artemieva, N. Effects of Moon's thermal state on the impact basin ejecta distribution. Geophys. Res. Lett. 44, 11292 - 11300 (2017).

[62] Zhu, M. - H., Wünnemann, K., Potter, R. W., Kleine, T. & Morbidelli, A. Are

the Moon' s nearside - farside asymmetries the result of a giant impact? J. Geophys. Res. Planets 124, 2117 - 2140 (2019).

[63] Ahrens, T. J. & Okeefe, J. D. in Impact and Explosion Cratering: Planetary and Terrestrial Implications (eds Roddy D. J. et al.) 639 - 656 (1977).

[64] Benz, W., Cameron, A. & Melosh, H. The origin of the Moon and the single - impact hypothesis III. Icarus 81, 113 - 131 (1989).

[65] Pierazzo, E. & Melosh, H. Hydrocode modeling of oblique impacts: the fate of the projectile. Meteorit. Planet. Sci. 35, 117 - 130 (2000).

[66] Britt, D. & SJ, G. C. Modeling the structure of high porosity asteroids. Icarus 152, 134 - 139 (2001).

[67] Ivanov, B., Melosh, H. & Pierazzo, E. Basin - forming impacts: reconnaissance modeling. Spec. Pap. Geol. Soc. Am. 465, 29 - 49 (2010).

[68] Potter, R. W., Kring, D. A., Collins, G. S., Kiefer, W. S. & McGovern, P. J. Numerical modeling of the formation and structure of the Orientale impact basin. J. Geophys. Res. Planets 118, 963 - 979 (2013).

The Effects of Viewing Geometry on the Spectral Analysis of Lunar Regolith as Inferred by *in situ* Spectrophotometric Measurements of Chang'E – 4

Yazhou Yang[1], Honglei Lin[2], Yang Liu[1,3], Yangting Lin[2], Yong Wei[2], Sen Hu[2], Wei Yang[2], Rui Xu[4], Zhiping He[4], and Yongliao Zou[1]

1. State Key Laboratory of Space Weather, National Space Science Center, Chinese Academy of Sciences, Beijing, China
2. Key Laboratory of Earth and Planetary Physics, Institute of Geology and Geophysics, Chinese Academy of Sciences, Beijing, China
3. Center for Excellence in Comparative Planetology, Chinese Academy of Sciences, Hefei, China
4. Key Laboratory of Space Active Opto – Electronics Technology Shanghai Institute of Technical Physics, Chinese Academy of Sciences, Shanghai, China

Abstract: The visible and near – infrared imaging spectrometer on board the Yutu – 2 rover of Chang'E – 4 mission has conducted 2 sets of spectrophotometric measurements at two sites on its 10th lunar day. At each site, the reflectance spectra were measured on the same lunar regolith at different geometric configurations. The experiments reveal an increasing trend of the spectral reddening effect with the increasing phase angles, which may cause significant uncertainties on interpreting the maturity of the lunar regolith. Furthermore, the phase angles have significant effects on the 2 – μm band center and the 1 – μm band depth of the spectra, consequently on the estimation of mineral composition of the regolith. Finally, the derived average Hapke parameters using the *in situ* spectrophotometric measurements provide important ground – truth for remote sensing observations.

Plain Language Summary: Olivine, pyroxenes and plagioclase, the major constituent minerals of the surface of the Moon, have diagnostic absorption features in their visible and near – infrared reflectance spectra. These features, including the wavelength position of the absorption band center and the depth of the absorption band, can be used to identify the mineralogy of the observed surface. The slope of the reflectance spectral curve is another feature that can be used to describe the degree of how the surface materials interact with space environment . The Yutu – 2 rover on board the China ' s Chang ' E – 4 mission has been equipped with a visible to near - infrared spectrometer, to carry out *in situ* spectral measurements at varied illumination and azimuth angles. These measurements provided us an excellent chance to study the effects of measurement angles on these spectral features, which can help us better understand the mineralogical information of lunar surface.

Key Points:

• In situ spectrophotometric measurements by Chang'E - 4 Yutu - 2 rover show viewing geometry may affect band centers and depth of the spectra

• Phase reddening coupled with space weathering effects on the spectra may cause significant uncertainties on interpreting the maturity of lunar regolith

• The derived Hapke parameters from the in situ spectrophotometric measurements provide ground truth for both in situ and orbital observations

1 Introduction

Visible and near - infrared (VNIR) reflectance spectroscopy has been widely used in characterizing the surface mineralogical compositions of airless bodies (e.g., Chapman, 1996; Gaffey, 2010). Numerous previous studies have shown that viewing geometry may change the spectral features, including band center, band depth, and spectral slope, and thus complicate the interpretation of the reflectance spectra (Sanchez et al., 2012; Ruesch et al., 2015; Izawa et al., 2016). For example, with increasing phase angles, the reflectance spectra of some planetary bodies show a reddening trend (Schröder et al., 2014), that being said, the reflectance spectra at larger phase angles exhibit a steeper spectral slope. This effect is so called phase reddening, which has been widely observed on airless bodies (Abe et al., 2006; Sanchez et al., 2012; Hapke, 2012; Liu et al., 2018), must be considered in the evaluation of lunar regolith maturity.

The average single particle scattering albedo (SSA), defined as the ratio of scattering efficiency to the sum of scattering and absorption efficiencies, is an important parameter in quantitative spectral analysis. As demonstrated by Hapke (2012), the average SSA of a mixture can be taken as the linear combination of each endmember's SSA, which allows the retrieval of mineral abundances using spectral unmixing technique (e.g., Li & Li, 2011; Liu et al., 2018; Yang et al., 2019). In theory, SSA should be independent of viewing geometries, as it is only dependent on optical constants and grain size (Hapke, 2012). However, in remote sensing analysis the SSA of a mixture cannot be directly calculated, because the optical constants and grain size of the mixture cannot easily be obtained. Instead, SSA of the mixture for lunar regolith can be derived from Hapke model by assuming known phase function parameters (Lucey, 1998), and in this case SSA may become dependent on viewing geometries. Alternatively, SSA and phase function parameters can be solved through spectrophotometric measurements (e.g.,

Mustard & Pieters, 1989; Hapke, 2012). The description of these two methods are detailed in section 2.5. The *in situ* measurements with varying viewing geometries can help evaluate the two methods and thus can provide more accurate SSA values for spectral unmixing analysis.

The Chang'E-4 (CE-4) has successfully landed on the far side of the Moon on January 3, 2019. The Visible and Near-infrared Imaging Spectrometer (VNIS) on board the Yutu-2 rover provides us an excellent opportunity to study the spectral features of the lunar surface through *in situ* measurements. On the mission's 10th lunar day, the VNIS carried out a series of spectrophotometric measurements at two sites along the traverse for the first time on lunar surface. During the measurements at each site, the position of the rover was kept unchanged and the spectra were collected at different time of the day with fixed emission angle at ~45°. Since the azimuth and zenith of the Sun changed with time, a series of spectra over the same area with different viewing geometries were obtained. These *in situ* spectrophotometric measurements are critical to study both the photometric properties and the spectral features of the lunar regolith.

In this contribution, we study how the spectral features, including the absorption band centers and depths, the spectral slope, and the derived SSA vary as a function of viewing geometries, and discuss how these variations influence our interpretation on the composition and maturity of lunar regolith.

2 Instruments, Measurements, and Data Reductions

2.1 VNIS Instruments

The VNIS consists of a complementary metal-oxide semiconductor (CMOS) imager (256×256 pixels, 450~945nm, spectral resolution: 2~7nm) and a single pixel short-wavelength infrared (SWIR) detector (900~2395nm, spectral resolution: 3~12nm) (He et al., 2019). It is fixed on the front of the rover and looks down at the lunar surface at a nominal 45° at a height of ~0.7m. The fields-of-view of the CMOS imager and SWIR detector are 8.5° and 3.58°, respectively (Li, Wang, et al., 2019; Li, Xu, et al., 2019). The CMOS imager captures an area of ~15cm × 21cm, and the SWIR detector measures a circular area with a radius of ~3.5cm within the area imaged by the CMOS imager. More details about the VNIS instrumentation and calibrations can be found in Li, Wang, et al. (2019), Li, Xu, et al. (2019) and He et al. (2019).

2.2 VNIS Measurements

On the mission's 10th lunar day, the Yutu-2 rover firstly collected 11 spectra (spectral ID N68-N78) over the exactly same area at site 1. Then, it rotated for ~102° and collected another set of 12 spectra (N79-N90) over site 2 which is adjacent to site 1.

The CMOS images of the lunar regolith for these measurements are shown in Fig. 1. The distance between these two sites is 2.133 m and the precise locations of the rover and the lander are listed in Table S4. The incidence, emission, azimuth and phase angles of these measurements are listed in Table S1. It takes ~11 to 13 seconds for one spectral measurement depending on the altitude of the sun (Fig. S1).

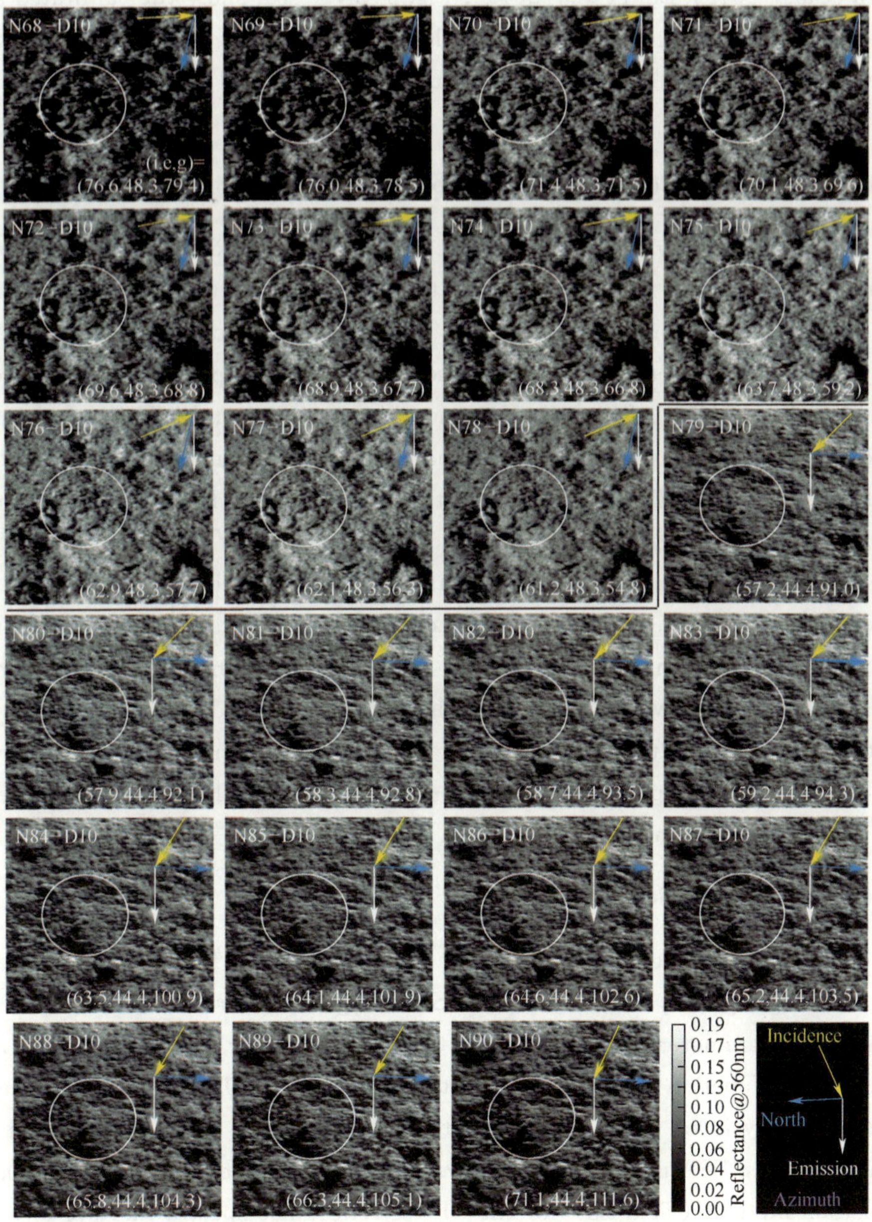

Fig. 1　The calibrated CMOS images at wavelength of 560nm. N68 - N78 and N79 - N90 represent images of site 1 and site 2 taken at varied measurement angles, respectively. The yellow, blue, and white arrows indicate the relative azimuths of incident light, north direction, and emission light, respectively. All the images are stretched from 0 to 0.19 to show the albedo variations. The white circle in each CMOS image shows the viewed area of the SWIR detector. The corresponding incidence, emission, and phase angles are given in the bottom of each image

2.3 Absolute Reflectance

We started the data reduction from the radiometric calibrated radiance data. To obtain the absolute reflectance (defined as reflectance factor in Hapke, 2012), we used the solar irradiance (Gueymard, 2004) calibration method (see Text S1 for detail) (Zhang et al., 2015; Hu et al., 2019; Lin et al., 2019). Fig. 2 (a)—(b) show the calibrated absolute reflectance spectra of site 1 and site 2, and the spectra normalized at 0.7μm are shown in Fig. 2 (c)—(d). For both sites, their absolute reflectance and spectral slopes vary with the measurement angles.

2.4 Estimation of Band Center and Band Depth

We first used the Savitzky - Golay algorithm (Savitzky & Golay, 1964) with 31 points and 2nd polynomial order to smooth those CE - 4 spectra. Then we employed a straight - line segments method (Clark & Roush, 1984) to obtain the continuum - removed spectra for characterizing the band center and band depth. Fig. 2 (e) shows the reflectance spectra of N78 before and after smoothing as an example. The continuum of the smoothed spectrum is also included. To estimate the band center and depth, we employed a 6th order polynomial fitting to the absorption regions around band1 (~1 - μm) and band2 (~2 - μm) of the continuum - removed spectra (Wu et al., 2019), respectively. We have set four different band ranges for the polynomial fitting for band1 and band2, respectively (Fig. S2). The average center and depth values obtained from the four fitting results are taken as the final results and the corresponding standard deviations are taken for the error bar. As shown in Fig. 2(f), the band center is defined as the wavelength having the minimum reflectance/continuum value calculated using the best - fit polynomial function, and the band depth is defined as $1 - R_{minimum}$, where $R_{minimum}$ is the reflectance/continuum value at the band center.

2.5 Derivation of Single Scattering Albedo

Here we employed the simplified Hapke model (Hapke, 1981) that is widely used for spectral unmixing analysis to derive the SSA (e. g., Mustard & Pieters, 1989; Li & Li, 2011). The parameter values of $b = -0.4$, $c = 0.25$ for the two - term Legendre polynomial phase function (Eq. S3) suggested by Lucey (1998) have been widely used for unmixing analysis of lunar spectral data (e. g., Denevi et al., 2008; Li & Li, 2011). In this case, SSA is the only unknown and can be solved from Eq. (S2) using the false - position method (Text S3) for the CE - 4 reflectance spectra that are measured at given viewing geometries.

Alternatively, SSA can be derived by fitting to the CE - 4 spectrophotometric data with Eq. (S2). In this case, the phase function parameters b and c are free parameters and can be solved simultaneously . We used the same iterative least square method

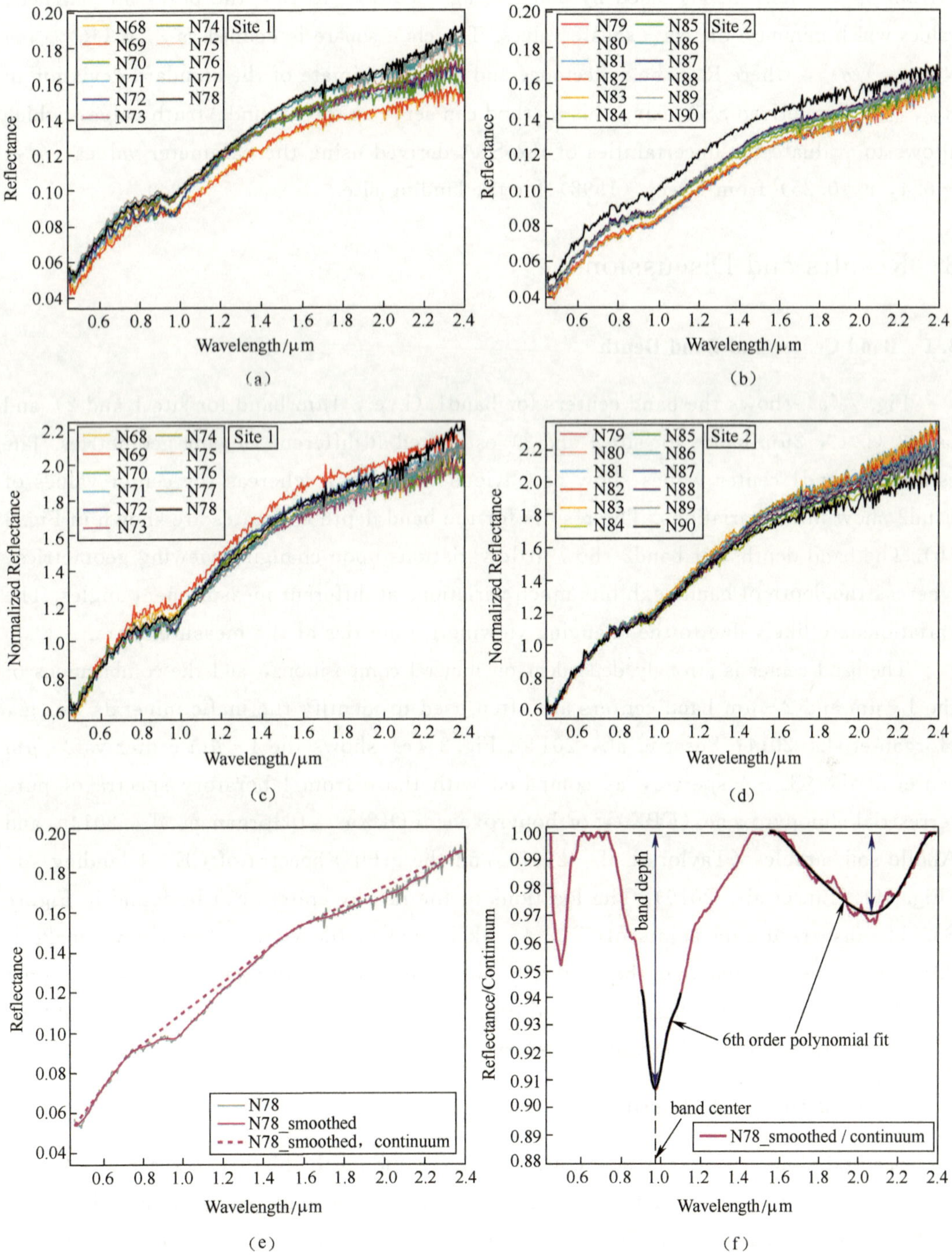

Fig. 2　Reflectance spectra obtained by Yutu-2 rover on its 10th lunar day

(a) and (b) are reflectance spectra of site 1 and site 2, respectively. (c) and (d) are reflectance spectra normalized at 0.7μm of site 1 and site 2, respectively. (e) reflectance spectra of N78 before and after smoothing with the continuum shown as dashed line. (f) continuum-removed spectrum of N78.

(Atkinson & Han, 2005) used by Yang et al. (2019) to find the best - fit parameter values which minimize the chi - square values. The chi - square is defined as: $\Sigma((R_{modeled} - R_{measured})/\sigma)^2$, where R is the reflectance and σ is an estimate of the standard deviation of $R_{measured}$. The derived SSA using this method can serve as the ground - truth value, which allows to evaluate the uncertainties of the SSA derived using the parameter values ($b = -0.4$, $c = 0.25$) from Lucey (1998) for the landing site.

3 Results and Discussions

3.1 Band Center and Band Depth

Fig. 3 (a) shows the band centers for band1 (i. e., 1μm band for site 1 and 2) and band2 (i. e., 2μm bands for site 1 and 2) estimated at different viewing geometries. The estimated band1 center values show no obvious variations, whereas the center values of band2 show much variations. The results for the band depth estimates are shown in Fig. 3 (b). The band depths for band2 show little variations upon changing viewing geometries, whereas the depth of band1 exhibits much variations at different measurement angles. The variations are likely due to the changing viewing geometries at the measurements.

The band center is strongly dependent on mineral compositions, and the combinations of the 1 - μm and 2 - μm band centers are often used to identify the mafic minerals (e. g., Horgan et al., 2014; Besse et al., 2014). Fig. 3 (c) shows the 1 - μm center vs 2 - μm center of the CE - 4 spectra, as compared with those from laboratory spectra of pure terrestrial clinopyroxene (CPX), orthopyroxene (OPX) (Horgan et al., 2014) and Apollo soil samples (Taylor et al., 2001) and the orbital spectra of CE - 4 landing site (Fig. S4) (Lin et al., 2019). The locations of the nearby craters can be found in Figure S5. The dispersed data points of CE - 4, likely due to the viewing geometry, make it difficult to conclude whether the materials observed were the orginal SPA floor material from Finse ejecta or mare basalt ecavated from Zhinyu crater (Lin et al., 2019), and thus affect our interpretation of evolution history of the area.

3.2 Spectral Slope and Maturity

Spectral slope provides critical clues for understanding the history of space weathering (e. g., Chapman, 1996; Pieters & Noble, 2016). In general, spectral slope increases with increasing space weathering degree, i. e., more matured lunar soil shows redder slope (Taylor et al., 2001). Here the spectral slope is defined as the ratio of the reflectance at 1600 nm to that at 700 nm (R1600/R700) which can well characterize the overall slope of the VNIR spectra. We also analyzed the changes of reflectance ratio R950 - nm/R750 - nm which is often used for deriving the optical maturity parameter of the lunar soil (Lucey et al., 2000). To account for the effects of random noise on the spectral

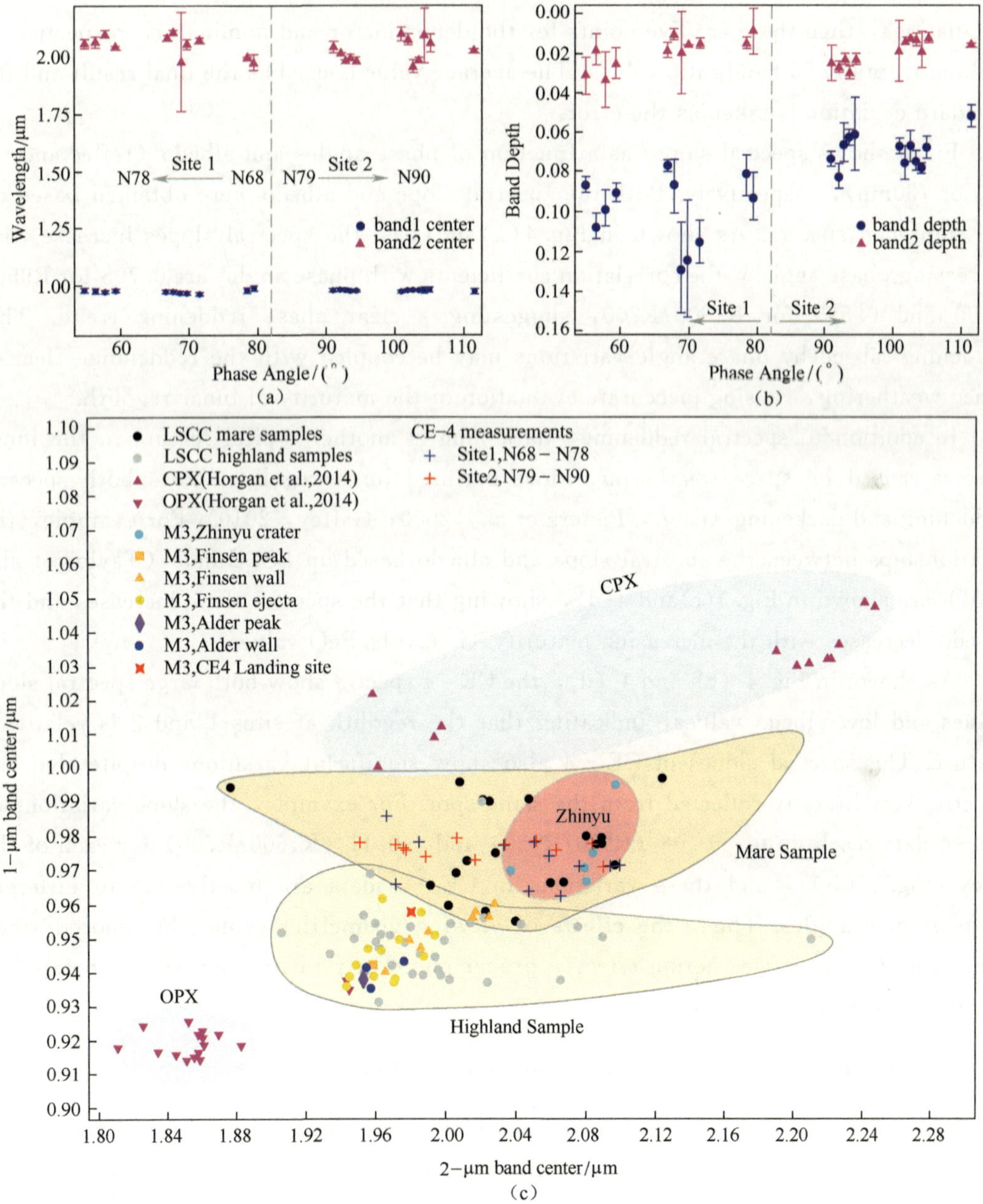

Fig. 3　Estimated band centers and band depths

(a) the band centers and (b) band depths of the 1－μm and 2－μm absorption bands for spectra measured at site 1 and site 2; (c) the 1－μm band centers vs. 2－μm band centers for CE－4 spectra, spectra of pure minerals and lunar soil samples measured in laboratory, and spectra collected by the Moon Mineralogy Mapper (M^3) of craters surrounding the CE－4 landing site. The spectra data of CPX and OPX are from Horgan et al. [2014]; the locations of all the M^3 spectra are labeled in Figure S4 in Lin et al. [2019]

slopes, we estimated the uncertainty of each slope following the method used by (Sanchez et al., 2012). We set a range of ± 10 nm around the bands that used for slope calculations, then there are five points for the denominator and nominator, respectively, making it totally 25 band ratio values. The average value is used as the final result and the standard deviation is taken as the error.

Fig. 4 shows spectral slopes as a function of phase angles and albedo (reflectance at 700 or 750nm), respectively. Both the spectral slope and albedo were obtained based on the smoothed spectra. As shown in Fig. 4(a) and (b), the spectral slopes increase with increasing phase angles, the correlation coefficients with phase angles are 0.793 for R950/R750 and 0.597 for R1600/R700, suggesting a clear phase reddening trend. This reddening caused by phase angle variations may be coupled with the reddening effect of space weathering, causing inaccurate evaluation of the maturity of lunar regolith.

In addition to spectral reddening, darkening is another typical feature in the lunar spectra caused by space weathering. More matured lunar regolith shows both spectral reddening and darkening (e. g., Pieters et al., 2000; Gaffey, 2010). For example, the relationships between the spectral slope and albedo based on LSCC data (Taylor et al., 2001) are shown in Fig. 4(c) and 4(d), showing that the spectral slope increases and the albedo decreases with the increasing maturity (i. e., Is/FeO values).

As shown in Fig. 4 (c) and 4 (d), the CE-4 spectra show both large spectral slope values and low albedo values, indicating that the regolith at sites 1 and 2 is relatively mature. The spectral slopes of CE-4 also show significant variations despite that the spectra were literally collected from the same spot. For example, the slope variations in CE-4 data reach up to ~0.06 (R950/R750) and ~0.11 (R1600/R700) for each of the sites (Fig. 4 (b)), and these variations in CE-4 data are mostly due to different measurement angles. Thus, the effects of viewing geometries cannot be ignored when evaluating the space weathering effects, proper photometric correction should be applied whenever possible.

3.3 Photometric Properties of Lunar Regolith at Landing Site

The *in situ* spectrophotometric measurements with varying geometries provide an opportunity to study the photometric properties of lunar regolith (Lin et al., 2020). Photometric modeling was performed using the datasets from site 1 and site 2 as well as the combined datasets from the two sites (i. e., site 1 + site 2), respectively. Fig. 5 (a) shows phase curves extracted at five typical wavelengths varied from 450 to 2000nm. The three sets of derived SSAs are shown in Fig. 5 (b). The derived phase function parameter c as a function of b is displayed in Fig. 5 (c). Note that the derived SSA spectra for site 1 and site 2 both show strong spikes (Fig. 5 (b)), and the b, c values for both sites are very dispersive (Fig. 5 (c)). This may due to the narrow phase angle ranges used for site 1 (i. e., ~24.5°) and site 2 (i. e., ~20.5°), which have caused large uncertainties in

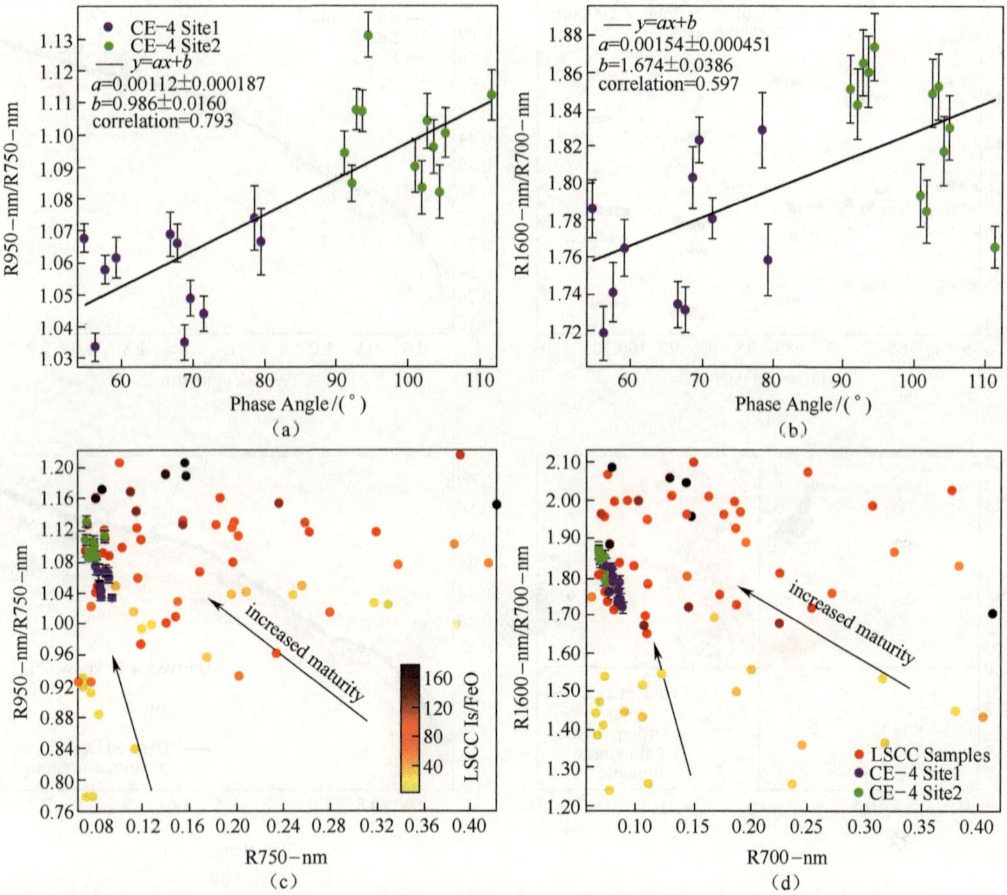

Fig. 4　The variation trends of spectral slope towards phase angle and albedo. Spectral slopes of (a) R950-nm/R750-nm and (b) R1600-nm/R700-nm vs phase angles for the CE-4 spectra; (c) R950-nm/R750-nm vs R750-nm and (d) R1600-nm/R700-nm vs R700-nm for the CE-4 spectra and spectra of returned lunar soil samples from the LSCC database. The Is/FeO values are from [Pieters et al., 2006]. Please note that the LSCC spectra are all measured under a standard geometry ($i=30°$, $e=0°$, $g=30°$)

photometric modeling. In contrast, the derived SSA spectrum using the combined datasets from both site 1 and site 2 (covering a range of the phase angle of ~57°) looks much smoother (Fig. 5 (b)), and the derived b and c values converged at −0.17 and 0.70, respectively (Fig. 5 (c)). Thus, using the combined datasets from the two sites can provide more reasonable values for photometric parameters.

The derived b and c values in this work are different from that in Lucey (1998). The Lucey values are widely used in the derivation of SSAs and the subsequent spectral unmixing analysis to retrieve mineral abundances from remote sensing data. To test if the b and c values from Lucey (1998) are applicable for CE-4 landing site, with these values we derived the SSAs from the 11 and 12 spectra collected over the same areas for

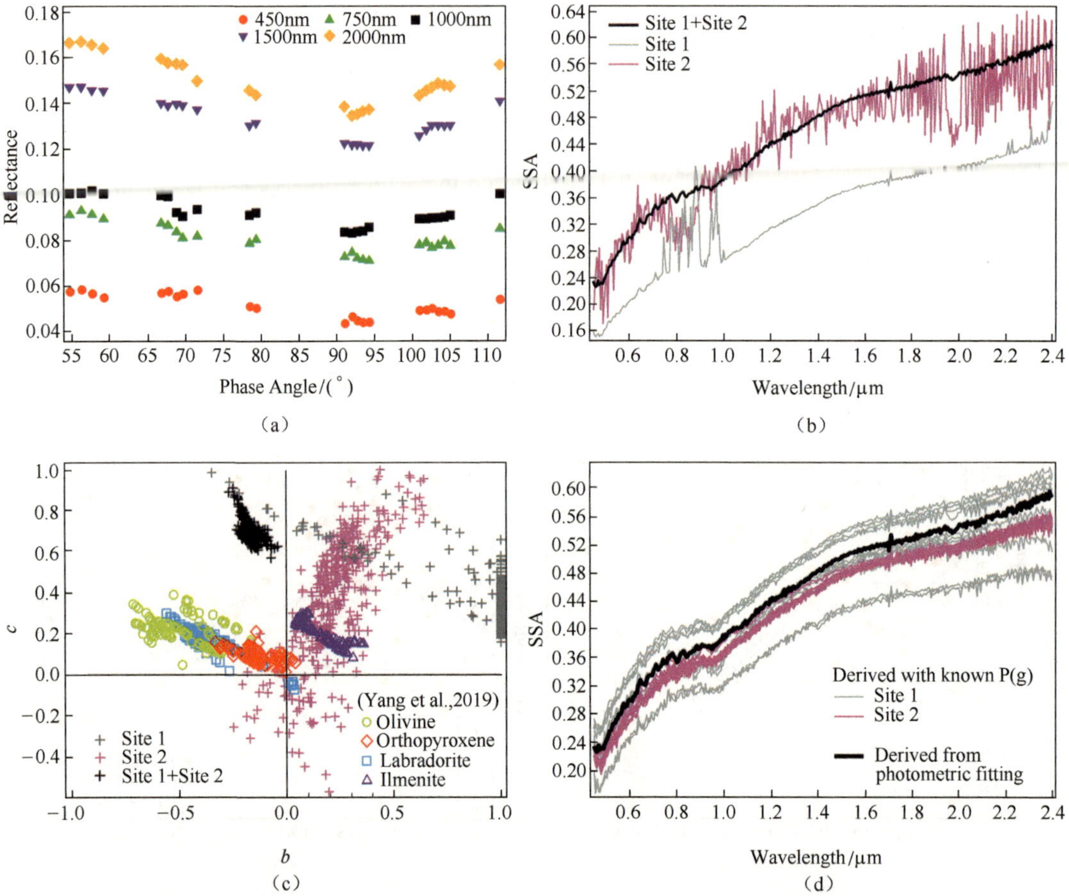

Fig. 5　SSA and the Legendre polynomial phase function parameters b, c for different wavelengths derived by fitting the photometric spectral data. (a) phase curves at five typical wavelengths; (b) SSA derived using the data measured at site 1, site 2, and the combination of site 1 + site 2, respectively; (c) phase function parameters b and c derived for site 1, site 2, and site 1 + site 2 at all VNIR wavelengths. The b, c values of olivine, orthopyroxene, labradorite, and ilmenite obtained by Yang et al. [2019] are shown for comparison; (d) the SSA derived from measured reflectance of site 1 and site 2 using the Hapke model (Eq. S2) with fixed b, c values ($b=-0.4$, $c=0.25$), and the ground - truth SSA derived from photometric fitting to the data of site 1 + site 2

site 1 and site 2, respectively. Our results indicate that the derived SSAs show great variations with viewing geometries (especially for site 1), which is contrary to the fact that SSA should be independent of viewing geometry. Thus, the b, c values from Lucey (1998) may not applicable for CE - 4 landing site, and the average values of $b=-0.17$ and $c=0.70$, derived from photometric modeling using the data from the *in situ* spectrophotometric measurements by Yutu - 2 rover, may be used as the typical values in Legendre pronominal phase function for the CE - 4 landing site. Therefore, these values can be used to retrieve SSAs over other sites along Yutu - 2 rover's traverse, and spectral unmixing of the SSAs derived from the ground - truth - based b and c values could provide more reliable mineral abundance information.

4 Conclusions

The Yutu - 2 rover on board the CE - 4 mission has carried out a series of *in situ* spectrophotometric measurements over the same area at different viewing geometries on its 10th lunar day. Based on these measurements, we analyzed the effects of measurement geometries on band centers, band depths, spectral slopes, and photometric properties of lunar regolith, and we discussed how these affect our interpretations on the composition, marurity, and minerals abundances of lunar soil from the spectra. Here summarizes our main findings:

1. Band center (i. e., ~2 μm region) and band depth (i. e., ~1 μm region) in the spectra measured by Yutu - 2 rover show significant variations, likely due to the changing viewing geometries, which could mislead our interpretation of composition of lunar regolith and thus the evolution history of the region.

2. Spectral slopes and absolute reflectance of the lunar regolith are strongly affected by phase angle variations. Reddening caused by increasing phase angles may be coupled with space weathering effects to result in overestimation of the maturity of lunar reglith.

3. The values of b and c should be derived from fitting phase curves whenever possible for the photometric correction and analysis of the given data sets. The derived average values of $b = -0.17$, $c = 0.70$ in the two - term Legendre phase function, using the *in situ* spectrophotometric measurements, can serve as the ground - truth values in the retrieval of SSAs in CE - 4 landing site and the subsequent spectral unmixing analysis.

Acknowledgments

The Chang'E - 4 mission was carried out by the Chinese Lunar Exploration Program, and the data were provided by the Science and Application Center for Moon and Deep Space Exploration, Chinese Academy of Sciences. This work was supported by the National Natural Science Foundation of China (11941001, 41902318), the Strategic Priority Research Program of Chinese Academy of Sciences (Grant No. XDB 41000000), the Beijing Municipal Science and Technology Commission (Z181100002918003) and Key Research Program of Frontier Sciences, CAS (QYZDJ - SSW - DQC001). Yang Y. also acknowledges the support from the China Postdoctoral Science Foundation (No. 2019TQ0323). The authors declare no competing interests. The data reported in this work is archived at http: //moon. bao. ac. cn/searchOrder _ dataSearchData. search.

References

[1] Abe, M., Takagi, Y., Kitazato, K., Abe, S., Hiroi, T., Vilas, F., et al. (2006). Near - infrared spectral results of asteroid Itokawa from the Hayabusa spacecraft. Science, 312 (5778), 1334 - 1338. https://doi.org/10.1126/science.1125718.

[2] Atkinson, K., & Han, W. (2005). Theoretical Numerical Analysis. Berlin: Springer.

[3] Besse, S., Sunshine, J. M., & Gaddis, L. R. (2014). Volcanic glass signatures in spectroscopic survey of newly proposed lunar pyroclastic deposits. Journal of Geophysical Research: Planets, 119 (2), 355 - 372.

[4] Chapman, C. R. (1996). S - type asteroids, ordinary chondrites, and space weathering: The evidence from Galileo's fly - bys of Gaspra and Ida. Meteoritics & Planetary Science, 31 (6), 699 - 725.

[5] Clark, R. N., & Roush, T. L. (1984). Reflectance spectroscopy: Quantitative analysis techniques for remote sensing applications. Journal of Geophysical Research, 89 (B7), 6329. https://doi.org/10.1029/JB089iB07p06329.

[6] Denevi, B., Lucey, P., & Sherman, S. (2008). Radiative transfer modeling of near - infrared spectra of lunar mare soils: Theory and measurement. Journal of Geophysical Research: Planets, 113 (E2), 1991 - 2012.

[7] Gaffey, M. J. (2010). Space weathering and the interpretation of asteroid reflectance spectra. Icarus, 209 (2), 564 - 574. https://doi.org/10.1016/j.icarus.2010.05.006.

[8] Gueymard, C. A. (2004). The sun's total and spectral irradiance for solar energy applications and solar radiation models. Solar Energy, 76 (4), 423 - 453. https://doi.org/10.1016/j.solener.2003.08.039.

[9] Hapke, B. (1981). Bidirectional reflectance spectroscopy: 1. Theory. Journal of Geophysical Research, 86 (B4), 3039. https://doi.org/10.1029/JB086iB04p03039.

[10] Hapke, B. (2012). Theory of Reflectance and Emittance Spectroscopy. Cambridge: Cambridge University Press.

[11] He, Z., Li, C., Xu, R., Lv, G., Yuan, L., & Wang, J. (2019). Spectrometers based on acousto - optic tunable fifilters for in - situ lunar surface measurement. Journal of Applied Remote Sensing, 13 (02). https://doi.org/10.1117/1.JRS.13.027502.

[12] Horgan, B. H., Cloutis, E. A., Mann, P., & Bell, J. F. III (2014). Near - infrared spectra of ferrous mineral mixtures and methods for their identification in planetary surface spectra. Icarus, 234, 132 - 154. https://doi.org/10.1016/j.icarus.2014.02.031

[13] Hu, X., Ma, P., Yang, Y., Zhu, M. H., Jiang, T., Lucey, P. G., et al. (2019). Mineral abundances inferred from in situ reflectance measurements of Chang'E - 4 landing site in south pole - Aitken Basin. Geophysical Research Letters. https://doi.org/10.1029/2019gl084531

[14] Izawa, M. R. M., Schäfer, T., Pietrasz, V. B., Cloutis, E. A., Mann, P., Nathues, A., et al. (2016). Effects of viewing geometry, aggregation state, and particle size on reflectance spectra of the Murchison CM2 chondrite deconvolved to Dawn FC band passes. Icarus, 266, 235 - 248. https://doi.org/10.1016/j.icarus.2015.10.029

[15] Li, C., Wang, Z., Xu, R., Lv, G., Yuan, L., He, Z., & Wang, J. (2019). The scientific information model of Chang'E - 4 visible and near - IR imaging spectrometer (VNIS) and in - orbit verifification. Sensors, 19 (12), 2806. https://doi.org/10.3390/s19122806

[16] Li, C. L., Xu, R., Lv, G., Yuan, L. Y., He, Z. P., & Wang, J. Y. (2019). Detection and calibration characteristics of the visible and near - infrared imaging spectrometer in the Chang'e - 4. Review of Scientific Instruments, 90 (10), 103, 106.

[17] Li, S., & Li, L. (2011). Radiative transfer modeling for quantifying lunar surface minerals, particle size, and submicroscopic metallic Fe. Journal of Geophysical Research, 116 (E9). https://doi.org/10.1029/2011je003837.

[18] Lin, H., He, Z., Yang, W., Lin, Y., Xu, R., Zhang, C., et al. (2019). Olivine - norite rock detected by the lunar rover Yutu - 2 likely crystallized from the SPA impact melt pool. National Science Review. https://doi.org/10.1093/nsr/nwz183.

[19] Lin, H., Xu, R., Yang, W., Lin, Y., Wei, Y., Hu, S., et al. (2020). In situ photometric experiment of lunar regolith with visible and near - infrared imaging spectrometer on board the Yutu - 2 lunar rover. Journal of Geophysical Research: Planets, 125, e2019JE006076. https://doi.org/10.1029/2019JE006076.

[20] Liu, Y., Retherford, K. D., Greathouse, T. K., Raut, U., Mandt, K. E., Hendrix, A. R., et al. (2018). The far ultraviolet wavelength dependence of the lunar phase curve as seen by LRO LAMP. Journal of Geophysical Research: Planets, 123 (10), 2550 - 2563. https://doi.org/10.1029/2018JE005580

[21] Lucey, P. G. (1998). Model near - infrared optical constants of olivine and pyroxene as a function of iron content. Journal of Geophysical Research: Planets, 103 (E1), 1703 - 1713.

[22] Lucey, P., Blewett, D., Taylor, G., & Hapke, B. (2000). Imaging of lunar surface maturity. Journal of Geophysical Research, 105 (E8), 20, 377 - 20, 386.

[23] Mustard, J. F., & Pieters, C. M. (1989). Photometric phase functions of common geologic minerals and applications to quantitative analysis of mineral mixture

reflectance spectra. Journal of Geophysical Research: Solid Earth, 94 (B10), 13, 619 - 13, 634. https: //doi. org/10. 1029/JB094iB10p13619

[24] Pieters, C., Shkuratov, Y., Kaydash, V., Stankevich, D., & Taylor, L. (2006). Lunar soil characterization consortium analyses: Pyroxene and maturity estimates derived from Clementine image data. Icarus, 184 (1), 83 - 101.

[25] Pieters, C. M., & Noble, S. K. (2016). Space Weathering on Airless Bodies. Journal of Geophysical Research: Planets. https: //doi. org/10. 1002/2016je005128

[26] Pieters, C. M., Taylor, L. A., Noble, S. K., Keller, L. P., Watson, B., Morris, R. V., et al. (2000). Space weathering on airless bodies: Resolving a mystery with lunar samples. Meteoritics & Planetary Science, 35, 1101 - 1107.

[27] Ruesch, O., Hiesinger, H., Cloutis, E., Le Corre, L., Kallisch, J., Mann, P., et al. (2015). Near infrared spectroscopy of HED meteorites: Effects of viewing geometry and compositional variations. Icarus, 258, 384 - 401. https: // doi. org/10. 1016/j. icarus. 2015. 06. 034

[28] Sanchez, J. A., Reddy, V., Nathues, A., Cloutis, E. A., Mann, P., & Hiesinger, H. (2012). Phase reddening on near - earth asteroids: Implications for mineralogical analysis, space weathering and taxonomic classification. Icarus, 220 (1), 36 - 50. https: //doi. org/10. 1016/j. icarus. 2012. 04. 008

[29] Savitzky, A., & Golay, M. J. E. (1964). Smoothing and differentiation of data by simplified least squares procedures. Analytical Chemistry, 36 (8), 1627 - 1639.

[30] Schröder, S. E., Grynko, Y., Pommerol, A., Keller, H. U., Thomas, N., & Roush, T. L. (2014). Laboratory observations and simulations of phase reddening. Icarus, 239, 201 - 216. https: //doi. org/10. 1016/j. icarus. 2014. 06. 010

[31] Taylor, L. A., Pieters, C. M., Keller, L. P., Morris, R. V., & McKay, D. S. (2001). Lunar Mare soils: Space weathering and the major effects of surface - correlated nanophase Fe. Journal of Geophysical Research, 106 (E11), 27, 985. https: //doi. org/10. 1029/2000je001402

[32] Wu, Y. Z., Wang, Z. C., & Lu, Y. (2019). Space weathering of the moon from in situ detection. Research in Astronomy and Astrophysics, 19 (4), 51.

[33] Yang, Y., Li, S., Milliken, R. E., Zhang, H., Robertson, K., & Hiroi, T. (2019). Phase Functions of Typical Lunar Surface Minerals Derived for the Hapke Model and Implications for Visible to Near - Infrared (VNIR) Spectral Unmixing. Journal of Geophysical Research: Planets. https: //doi. org/10. 1029/2018je005713

[34] Zhang, H., Yang, Y., Yuan, Y., Jin, W., Lucey, P. G., Zhu, M. - H., et al. (2015). In situ optical measurements of Chang'E - 3 landing site in Mare Imbrium: 1. Mineral abundances inferred from spectral reflectance. Geophysical Research Letters, 42 (17), 6945 - 6950. https: //doi. org/10. 1002/2015gl065273

Mineralogical and chemical properties inversed from 21 – lunar – day VNIS observations taken during the Chang'E – 4 mission

Qinghong Zeng[1], Shengbo Chen[1,2], Yuanzhi Zhang[3], Yongling Mu[1], Rui Dai[1], CongyuYang[1], Anzhen Li[1] & Peng Lu[1]

1. College of Geo – Exploration Science and Technology, Jilin University, Changchun, China
2. CAS Center for Excellence in Comparative Planetology, Hefei, China
3. Key Laboratory of Lunar and Deep – Space Exploration, National Astronomical Observatories, Chinese Academy of Sciences, Beijing, China

We report on the mineralogical and chemical properties of materials investigated by the lunar roverYutu – 2, which landed on the Von Kármán crater in the pre – Nectarian South Pole – Aitken (SPA) basin. Yutu – 2 carried several scientific payloads, including the Visible and Near – infrared Imaging Spectrometer (VNIS), which is used for mineral identification, offering insights into lunar evolution. We used 86 valid VNIS data for 21 lunar days, with mineral abundance obtained using the Hapke radiative transfer model and sparse unmixing algorithm and chemical compositions empirically estimated. The mineralogical properties of the materials at the Chang'E – 4 (CE – 4) site referred to as norite/gabbro, based on findings of mineral abundance, indicate that they may be SPA impact melt components excavated by a surrounding impact crater. We find that CE – 4 materials are dominated by plagioclase and pyroxene and feature little olivine, with 50 of 86 observations showing higher LCP than HCP in pyroxene. In view of the effects of space weathering, olivine content may be underestimated, with FeO and TiO_2 content estimated using the maturity – corrected method. Estimates of chemical content are 7.42—18.82 wt% FeO and 1.48—2.1 wt% TiO_2, with a low – medium Mg number (Mg # ~ 55). Olivine-rich materials are not present at the CE – 4 landing site, based on the low-medium Mg #. Multi-origin materials at the CE – 4 landing site were analyzed with regard to concentrations of FeO and TiO_2 content, supporting our conclusion that the materials at CE – 4 do not have a single source but rather are likely a mixture of SPA impact melt components excavated by surrounding impact crater and volcanic product ejecta.

Through October 25, 2020, the Yutu – 2 rover of the Chang'E – 4 (CE – 4) mission had toured 519.29m up to 600 days (for 21 lunar days), with the Visible and Near – Infrared Image Spectrometer (VNIS) on board the rover acquiring 106 in situ reflectance measurements (Fig. 1). This in situ scientific dataset provides a unique perspective from which to understand the mineralogical and chemical properties of the Von Kármán crater inside the South Pole – Aitken (SPA) basin on the far side of the moon[1,2]. A range of remote - sensing and in situ data, some based on spectral observations, analysis of geomorphology, and the subsurface structure within Von Kármán crater, have helped develop understandings of the moon's history and evolution.

After the CE - 4's safe landing, researchers have reported initial results by interpreting the first lunar - day VNIS data showing that the materials along the rover route are dominated by olivine and low - calcium pyroxene (LCP)- suggesting that the Yutu - 2 rover may have detected the moon's mantle material[3,4]. These results are consistent with orbital data, confirming that the floor of the SPA is rich in mafic materials[5].

However, additional data from the Yutu - 2 rover have not indicated the presence of expected mantle materials at the CE - 4 landing site, where materials are dominated by plagioclase (Plag) and pyroxene but include little olivine (OL), for the first three lunar days[6]. As exploration continues to produce more scientific data, we can arrive at new and different understandings. Several inversion methods could cause the controversial components found at the CE - 4 landing site. Findings of magnesium - rich materials obtaining using the modified Gaussian method (MGM) may overstate olivine content, because plagioclase and olivine have very similar spectral features around 1250nm, making it difficult to confirm the abundance of plagioclase by MGM deconvolution[7,8]. Further studies have considered additional minerals (such as plagioclase and agglutinates) for inversion through the Hapke radiation transmission model, but different unmixing methods (lookup tables[9] and other empirical approaches[6]) may produce unexpected results. Accordingly, more lunar - days' worth of data are needed to research mineralogical properties along the rover's route and support use of an objective unmixing method based on radiative transfer models.

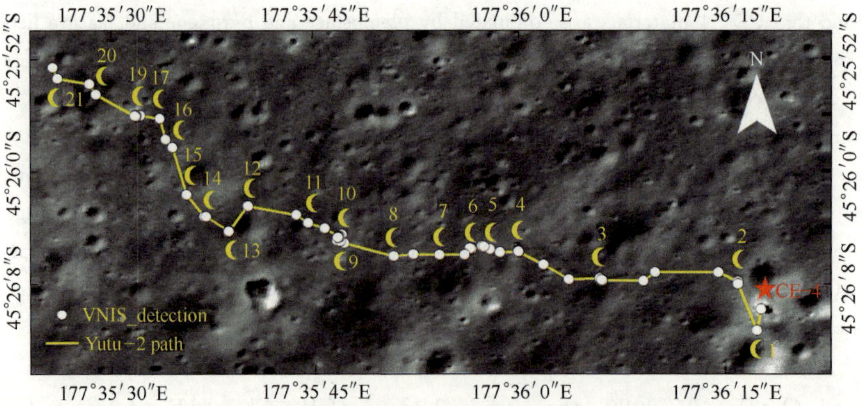

Fig. 1 The routing path and the VNIS detections of the Yutu - 2 rover during the 21 lunar days (except for the eighteenth lunar day when the Yutu - 2 did not move). The number upper the point is the first VNIS detection of a new lunar day. The base map is from the LROC NAC image (image number M1311886645RC and M1311886645LC). This map was created in ESRI ArcMap 10. 2 (https://desktop. arcgis. com/zh - cn/arcmap/)

By contrast, the lunar surface material is affected mainly by space weathering, a natural process that affects spectral band depths[10], hindering precise modeling of lunar mixtures[10,11]. Notably, various mature materials such as rock and regolith were found along the Yutu - 2 rover's route[6,7,12]. The regolith is more mature than the surrounding rocks[13]. Olivine and pyroxene, the major mafic minerals of lunar material, have distinctive spectral characteristics in the visible and near - infrared (VIS/NIS) and short - wave infrared (SWIR) regions[8,14]: for example, pyroxene exhibits absorption at ~1000nm and ~2000nm, whereas olivine has three overlaps in absorption near 1000nm but no absorption at 2000nm. Olivine and pyroxene differ in their sensitivity to space weathering[15], making it diffcult to ascertain the extent to which space weathering and mineral composition influence the spectrum. Thus chemical characteristics (such as iron and titanium content) of the lunar surface are needed to be concerned except for lunar minerals.

Iron and titanium are two elements of particular interest on the moon and are useful for understanding its origin and geological evolution. The major rock types can be distinguished by concentrations of FeO and TiO_2 content[16] supplemented and compared with the results of mineral inversion. Although optical maturity strongly influences spectral reflectance[10,17-19], there are ways of reducing the effect of space weathering when inversion of FeO and TiO_2 content. Generally speaking, two types of models are used, based on spectral data. One is based on the statistical relation of reflectance spectra and compositions, such as principal component analysis[20,21], the support vector machine model[21], and partial least squares regression[22], mathematical methods that show potential for estimating element content by spectra. However, purely mathematical statistics are limited by mature differences in the training data, producing artifacts when applied to other regions. The other algorithms were maturity - corrected by parameterizing spectral properties of sensitivity to iron or titanium and soil maturity (e. g., Lucey et al. 1995[18], 1998[16], 2000[19]; Le Mouélic et al. 2000[23]), an approach that can normalize the effects of space weathering, optimized to remove mature effects. Thus FeO and TiO_2 content and the optical maturity parameter (OMAT) of Yutu - 2 spectral data can be estimated using Lucey et al.'s method[19] to help explain the moon's geological background.

In this study, we analyze in situ spectra observed by VNIS spectral data during the first 21 synodic days, with a view to reporting mineralogical and chemical information for the region of the Yutu - 2 rover and offering insights into the nature and origin of the SPA basin. Based on analysis of the spectral data, we attempt to determine the mineralogical and chemical properties of the materials at the landing site.

Results

Mineralogy properties observed by Yutu - 2 rover. The VNIS spectra measured by the

Yutu - 2 rover show absorption regions around 1000nm and 2000nm (Fig. 2b). These absorption characteristics can be used to distinguish the primary minerals on the lunar surface, including pyroxene (PYX), olivine (OL), and plagio - clase (Plag), because their absorption positions vary with Fe/Ca/Mg content[24-26]. As Fig. 2 shows, similarities in absorption peak position notwithstanding, the band depth of the rock (LE00303) is larger than for other nearby regolith, indicating similarities of components but differing degrees of space weathering.

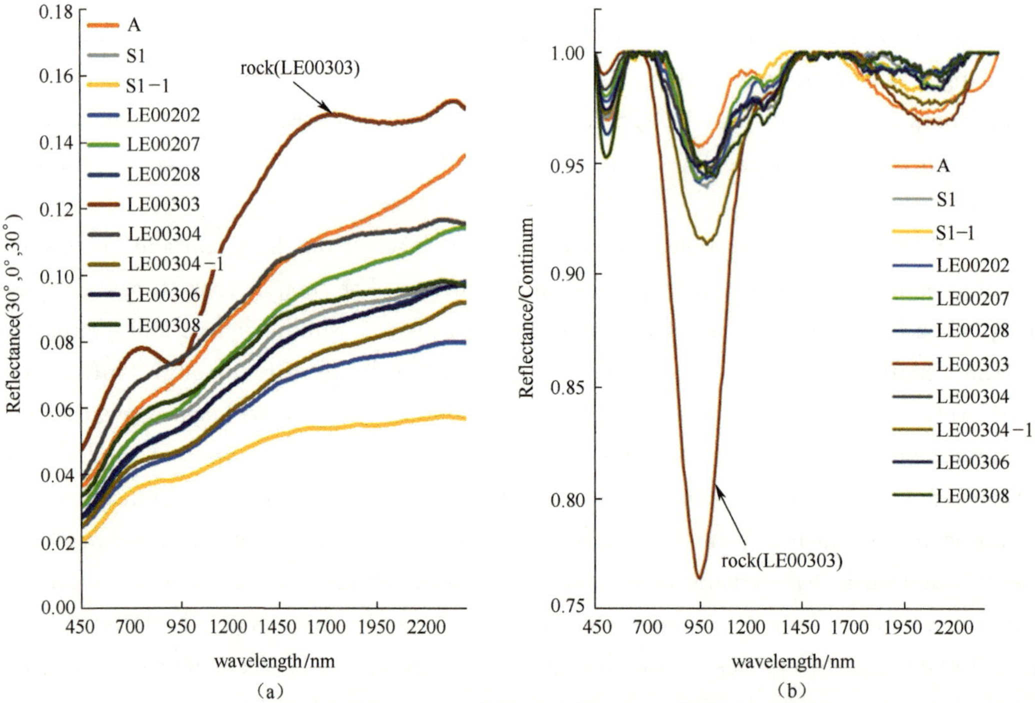

Fig. 2 Examples of VNIS spectra measured by Yutu - 2 rover

(a) The reflectance data with photometric correction and smoothing. (b) The spectra after continuum removed

To quantitatively infer the abundance of major minerals from the VNIS spectra measured by the Yutu - 2 rover, we use the Hapke radiative model, in combination with a sparse unmixing algorithm, to estimate their relative content. In view of the space weathering effect, we add the agglutinates as the products of space weathering to estimate the relative contents of the primary minerals[27]. Our estimates show an average modeled composition of 50.5% agglutinates (AGG), 16.3% low - calcium pyroxene (LCP), 14.8% high - calcium pyroxene (HCP), 6.1% olivine (OL), and 12.3% plagioclase (PLG) (Fig. 3a). The modeled mineral abundance of the CE - 4 observations suggests that their main sources are norite/gabbro and olivine norite/gabbro. Furthermore, we see higher LCP than HCP at most observation sites, consistent with the findings of the recent study[7].

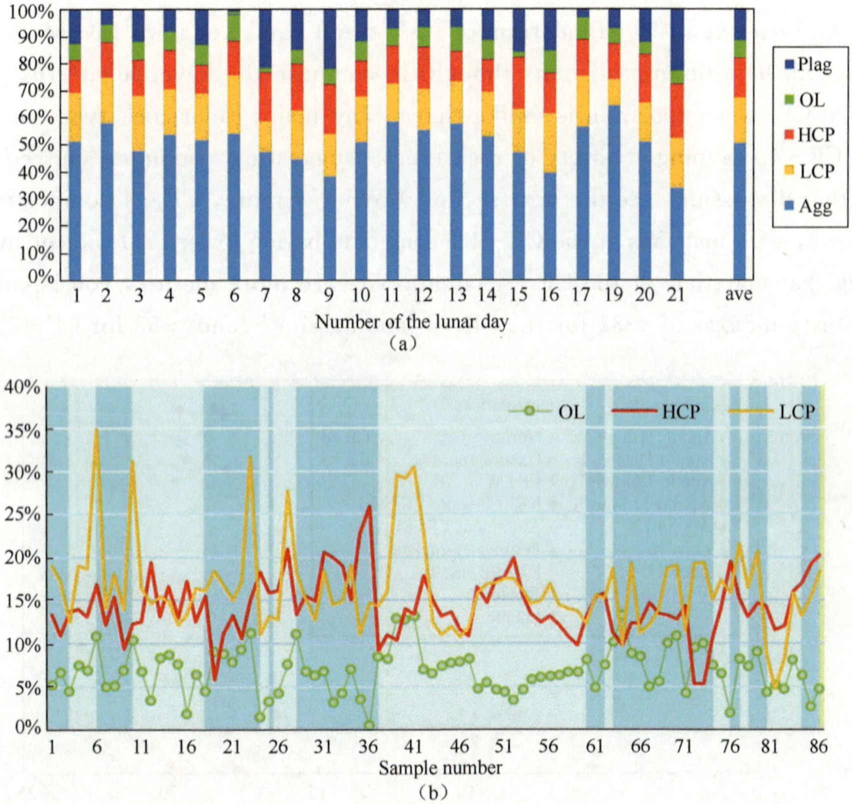

Fig. 3 (a) The abundance of lunar materials during the first 21 lunar days. The estimated results show an average modal composition of 50.5% agglutinates (AGG), 16.3% LCP, 14.8% HCP, 6.1% olivine (OL) and 12.3% plagioclase (PLG). The ave means the average value during the first 21 lunar days. (b) The detailed abundance information for every VNIS detection during the first 21 lunar days. We use the background of the color bars to distinguish the observations during a lunar day

Fig. 3b shows detailed abundance information for every VNIS detection. Most observations (50 of 86) show higher LCP concentrations than HCP, indicating that composition variation within a small area of the Yutu - 2 route is consistent with that suggested by Huang et al.[7] for the CE - 4 landing region, based on multiple sources.

Chemistry properties observed by Yutu - 2 rover. From VNIS spectral data, adopting the method used by Luccy et al.[19], we derived FeO and TiO_2 content for Yutu - 2 in situ measurements, with estimated content of 7.42 — 18.82 wt% FeO and 1.48 — 2.1 wt% TiO_2 Fig. 4. FeO + TiO_2 is consistent with estimates (11 — 19 wt%) based on remote sensing data[19,28,29]. The results show that surface materials have higher FeO content but lower TiO_2 content.

Elaborating on the chemical properties of the CE - 4 landing site, Fig. 4 plots CE - 4 FeO versus TiO_2 for major lunar rock type fields and CE - 3 landing site materials. CE - 3 data are from Ling et al.[30], and lunar rock (mare basalt and nonmare rock) data are reported by Lucey et al.[16], Lindstrom et al.[31], and Papike et al.[32] LKFM (low-K Fra Mauro), with basaltic impact melts thought to compose the lower lunar crust[33]. Fig. 4 shows that CE-4 composition is unlike that of any major lunar rock type, but with the range of CE-4 spanning a variety of rock types - suggesting a complex source for CE-4. (For further discussion, see the next section.) What's more, CE-4 composition differs from CE-3, with materials at the CE-4 landing site having lower FeO content than CE-3, indicating that materials at the CE-4 landing site are more mature, consistent with Is/FeO maturity indexes of ~82 for the CE-4 landing site[13] and ~53 for CE-3[34].

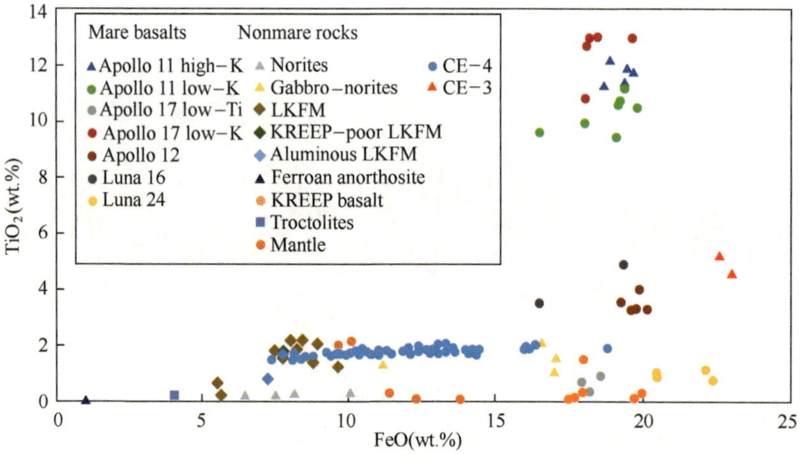

Fig. 4 Chemical properties of CE-4 measured materials. The LKFM (11.4—20 wt% FeO, 0.1—0.34 wt% TiO_2) is a rock type thought to compose the lower lunar crust. The lunar mantle (11.4—17.5 wt% FeO and 0.1—0.34 wt% TiO_2) compositions are inferred by the experiments at high temperatures and pressure of mare basalt source regions in the lunar mantle which may lead to the elevation of TiO_2. Mare basalts shows the range of 16.5—22.4 wt% FeO and 0.36—12.99 wt% TiO_2

Mg number (Mg # = mole percent MgO/(MgO + FeO)) is one of the chemical indexes used to discuss the origin of lunar samples. The Mg number (Mg # = mole percent MgO/(MgO + FeO)) at the CE-4 site was empirically estimated by the 1000-nm and 2000-nm band centers, as Fig. 5 shows. We seek to identify mafic compositional trends and assess whether they are related to band centers, with the magnesium ratio classified as high (Mg # > 75), medium (Mg # 50—75), or low (Mg # < 50). The materials at the CE-4 site have a relatively low-medium Mg # (Fig. 5), consistent with the Mg # of ~55 derived from the Chang'E-1 Imaging Interferometer (IIM) data[29].

Discussions

Space weathering introduces changes to visible and near-infrared reflectance

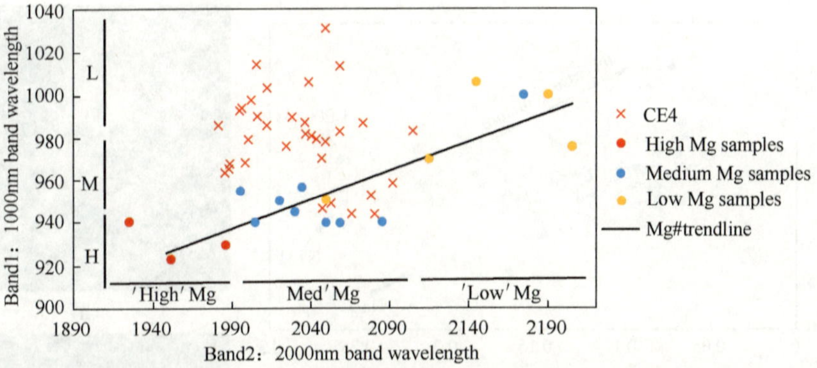

Fig. 5 Comparison of the 1000nm and 2000nm band centers for spectra observed by Yutu - 2 with the lunar highland and mare samples. The Mg # trend line was estimated using Apollo mare and highland samples. The Mg # of lunar samples were calculated using chemical information of MgO and FeO

spectra[35]; because the moon is an airless body, its surface is altered through irradiation by solar wind and galactic cosmic rays[36], a process that reduces the strength of mineralogical absorption bands, including the mafic minerals pyroxene and olivine. Various types of pits are on the path of the Yutu - 2, and the fragments that fill them show varying degrees of space weathering (Fig. S1), providing a unique perspective on space weathering's influence on mineral abundance, based on in situ spectra on the far side of the moon.

Optical maturity (OMAT) is a spectral parameter used to estimate the degree of space weathering for lunar soils, based on reflectance properties. We calculated the optical maturity parameter[19] and produced a scatterplot (Fig. 6) of the relationship between OMAT and lunar mafic materials (HCP, LCP, and OL), showing that the three minerals have different weathering resistances. OL has decreased tendency with increasing maturity, whereas OL has weak resistance to weathering. LCP has better weathering resistance, whereas HCP shows relative increases with maturity, because they are in the relative content system, indicating that HCP has better weathering resistance. Because of the effects of space weathering, olivine may be underestimated, so further discussion of the sources of materials at CE - 4 should be combined with the chemical properties.

The CE - 4 landing site is in the Von Kármán crater, which formed in the pre - Nectarian era and subsequently underwent a series of complex geological evolutions[37-40]. The materials at the CE - 4 landing site measured by Yutu - 2 are dominated by pyroxenes and plagioclase, with little olivine seen - suggesting that the rock type is norite/gabbro. The norite - like surface materials at the CE - 4 landing site may contain materials of the noritic layer during SPA melt pool formation[41,42], excavated by the surrounding impact craters (e. g., Finsen crater and Alder crater). Furthermore, no olivine - rich materials were found at the CE - 4 landing site with low - medium Mg#, suggesting that the source

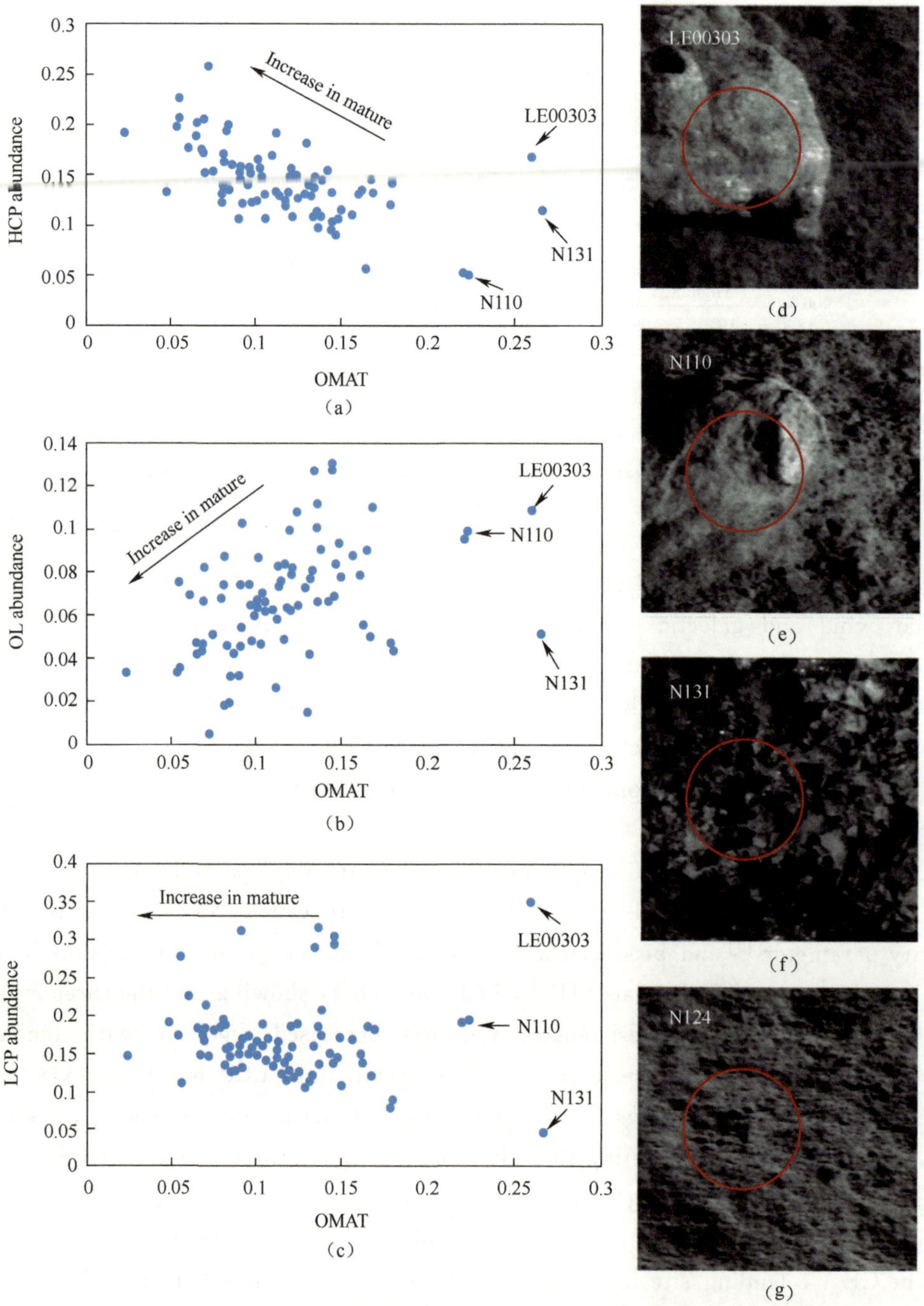

Fig. 6　The relation between the OMAT and the main lunar minerals (HCP, LCP and OL). (a—c) are the scatterplots showing the relationship between the OMAT and the main lunar materials. (d—g) are the different type of lunar materials during the CE - 4 routine path, (d, e) are the rock types, (f) is the fragment filled in the pit, (g) is the normal type of the lunar regolith nearby the Yutu - 2, the red circle is the field of view (FOV) for SWIR. The VNIS images (d—g) were created by our own script in MATLAB 2013a (https://www. mathworks. com/produ cts/matlab. html)

might not be the lunar mantle, although the possibility of the mixture's coming from the lower crust cannot be ruled out.

By contrast, elevated CE - 4 FeO content (Fig. 4) may result from exposure from lunar mantle[16], but TiO_2 content is significantly higher than the expected mantle composition. The correlation of FeO with TiO_2 of CE - 4 shown in Fig. 4 could result from the differentiation of the large melt sheet, with the differentiation materials perhaps containing parts of LKFM (lower crustal composition).

The CE - 4 materials have a distinctive chemical characteristic compared with other various mare basalt (Fig. 4), and the chemical composition at CE - 4 may offer insights into the mixing effects of impact melt ejecta and volcanic products.

Wc note multiple lava - inffling events[45] within the Von Kármán crater, ~70km west of the CE - 4 landing site. Volcanic activity may have peaked in the Late Imbrium and could have continued into the Eratosthenian epoch[45], with multiple lava flooding products excavated by the fresh Zhinyu crater (~ 4km diameter), ~30km west of the CE - 4 landing site[12]. The orbital data indicate the presence of a slight impact ray to the CE - 4 landing site from the Zhinyu crater, based on color ratio imaging from Clementine UVVIS data (Fig. 7).

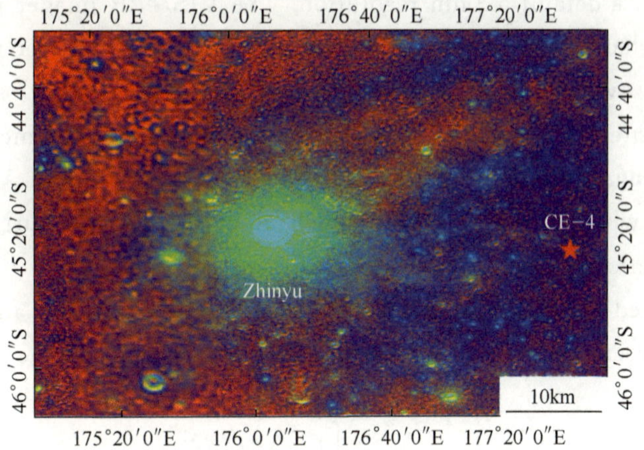

Fig. 7　The color ratio image[43,44] from Clementine UVVIS data at the CE - 4 landing site, showing a slight impact ray to the CE - 4 landing site (red star) from the Zhinyu crater. The map was created in ESRI ArcMap 10.2 (https://desktop.arcgis.com/zh-cn/arcmap/)

As already discussed, the CE - 4 landing site has been affected by the mixing effects of ejecta and volcanic products. Accordingly, multiple lavas were probably infilled in the bottom of Zhinyu crater and the CE - 4 landing site before the Zhinyu impact event, containing the ejecta and volcanic products from the Zhinyu crater brought to the surface of the CE - 4 landing site.

This paper investigates the mineralogical and chemical properties by in situ spectra as measured by Yutu - 2, finding that the materials detected by Yutu - 2 were norite/gabbro.

No olivine - rich mantle materials[3] have been detected thus far, but plagioclase and pyroxene have been. Agglutinates are abundant, based on the results of mineral abundance, suggesting that the materials have undergone space weathering effects and reached a mature state. The chemical compositions of materials at the CE - 4 landing site were empirically estimated using in situ spectra. Chemical properties show that the surface materials have higher FeO (7.42 — 18.82 wt%) and lower TiO_2 (1.48 — 2.1 wt%) content and low - medium Mg number (Mg # ~ 55). Based on analysis of their mineralogical and chemical properties, the materials measured by Yutu - 2 likely originated from SPA impact melt components excavated by the surrounding impact crater and from the mixing effects of SPA ejecta and volcanic products.

Methods

VNIS spectra. The Visible and Near - infrared Imaging Spectrometer (VNIS) is composed of a 256 - by 256 - pixel VIS/NIR imager and a short - wave infrared (SWIR) single - pixel non - imaging spectrometer[3,14]. The VNIS is fixed on the front of the rover, offering a 45° viewing angle at a height of 0.69m. Its spectral range is 450 - 2395nm, with 400 spectral channels at a default 5 - nm resolution. The VIS/NIR imager works from 450 to 945nm, with 300 channels, and the 100 - channel SWIR spectrometer works from 900 to 2395nm, with 10 overlap channels[1,14]. The SWIR spectrometer's field of view is circular, with a center of (96, 128) in the VIS/NIR image and a diameter of 107 pixels.

VNIS data processing. This section describes VNIS data processing. VNIS spectral data processing began with level 2B data, which underwent dark current, flat - field, temperature corrections and radiometric and geometric calibration[3,46-48]. L2B radiance data were converted to reflectance data by solar irradiance, for an airless planetary surface, with bidirectional reflectance commonly given as radiance factor (RADF) I/F,

$$I/F(\lambda_i)=\frac{\pi I_s(\lambda_i)}{\int J(\lambda)S(\lambda)d\lambda} \tag{1}$$

where $J(\lambda)$ is the solar irradiance[49], $S(\lambda)$ the spectral response of the VNIS sensor, and $I_s(\lambda_i)$ the radiance measured by Yutu - 2 at wavelength λ_i.

To reduce the influence of the geometric circumstances, the photometric correction described in Lin et al.[6] was used to correct the data to fit common viewing geometry (incidence angle = 30°, emission angle = 0°, phase angle = 30°).

Owing to differences in signal - to - noise ratios (SNR) between the VIS/NIR spectrum (37.4dB) and SWIR spectrum (44.0dB)[8], a gap is seen at the boundary of the two spectra. With no obvious slope difference at the overlapping region (900 - 950nm), and taking into account SWIR's higher SNR, we multiplied the VIS/NIR spectrum by a factor (reflectance ratio of SWIR 900 - nm to the VNIS 900 - nm) to connect to the SWIR spectrum and obtain a continuous spectrum. Finally, Savitzky - Golay method was performed during the spectral data smoothing, shown in Fig. 2.

Estimation of absorption band center. In lunar mineralogical studies, the absorption band center is an important spectral parameter[50], especially for mafic minerals (olivine and pyroxene), which have two obvious absorption bands, band 1 (~1000nm) and band2 (~2000nm). The properties of absorption band centers are related to mineral structure and chemical composition[51].

To calculate the band center, we first used the straight - line continuum method[52] to obtain the continuum - removed spectra. We then used a sixth - order polynomial to fit the bottom of the continuum - removed absorption feature around band 1 (~1000nm) and band2 (~2000nm)[48]. The band center is defined as the wavelength at the minimum point on the polynomial fitted curve. We set three different spectral ranges for band 1 and band 2 when polynomial fitting, and take the average center values as the final results.

Estimation of the mineral abundances from the spectra. In this study, we used the radiative transfer model[53,54] (see Supplementary Text S1) and the sparse unmixing algorithm[55,56] to estimate mineral abundance from the VNIS spectra. To avoid any thermal emission effects, we limited our analysis to the wavelength range 950—1500nm. First we calculated the imaginary part (k) of the endmembers (shown in Tab. S1) using the Hapke model. Next we computed the single scattering (SSA) for each endmember with different particle sizes (ranging from 5 to 200nm, with an interval of 5nm) using the Hapke slab model, constructed as the endmember library. Finally, we used the sparse unmixing algorithm to unmix the SSA of the Yutu - 2 spectra derived by the Hapke model.

Derivation of iron and titanium abundance. To evaluate the chemical properties of the materials detected by Yutu - 2, we calculated FeO and TiO_2 content using the algorithms described by Luccy et al.[19]. These approaches were based on returned *Apollo* and *Luna* samples. FeO and TiO_2 content was calculated from the best fit curve,

$$\text{FeO (wt\%)} = 31.875\theta_{\text{Fe}} - 17.895 \tag{2}$$

$$\theta_{\text{Fe}} = -\arctan[(R_{950}/R_{750} - y_{0\text{Fe}})/(R_{750} - x_{0\text{Fe}})] \tag{3}$$

$$\text{TiO}_2\text{(wt\%)} = 1.836 \times \theta_{\text{Ti}}^{9.779} \tag{4}$$

$$\theta_{\text{Ti}} = \arctan[(R_{415}/R_{750} - y_{0\text{Ti}})/(R_{750} - x_{0\text{Ti}})] \tag{5}$$

where the origin (x_{0Fe}, y_{0Fe}) and (x_{0Ti}, y_{0Ti}) were set to (−0.1, 1.39) and (−1.08, 0.208), optimized to maximize the correlation coefficient between content and θ_{Fe} using the *Apollo* lunar samples[10,57]. R_{415} was determined by $R_{415} = 0.95R_{415} - 0.0013$ ($R^2 = 0.99$).

Acknowledgements

The authors thank the whole Chang'E - 4 team for the successful mission and providing the scientific data. This study is supported by the Strategic Priority Research

Program of Chinese Academy of Sciences (XDB 41000000) and National Natural Science Foundation of China (41772346).

References

[1] Li, C. L. et al. Detection and calibration characteristics of the visible and near - infrared imaging spectrometer in the Chang'E - 4. Rev. Sci. Instrum. 90 (10), 103 - 106 (2019).

[2] Chen, J. et al. Mineralogy of Chang' E - 4 landing site: Preliminary results of visible and near - infrared imaging spectrometer. Sci. China Inform. Sci. 63 (4), 1869 - 1919 (2020).

[3] Li, C. L. et al. Chang'E - 4 initial spectroscopic identification of lunar far - side mantle - derived materials. Nature 569 (7756), 378 - 382 (2019).

[4] Gou, S. et al. Forsteritic olivine and magnesium - rich orthopyroxene materials measured by Chang'E - 4 rover. Icarus 345, 113776 (2020).

[5] Moriarty, D. P. & Pieters, C. M. The character of South Pole - Aitken Basin: Patterns of surface and subsurface composition. J. Geophys. Res. Planet 123 (3), 729 - 747 (2018).

[6] Lin, H. L. et al. Olivine - norite rock detected by the lunar rover Yutu - 2 likely crystallized from the SPA - impact melt pool. Natl. Sci. Rev. 7 (5), 913 - 920 (2020).

[7] Huang, J. et al. Diverse rock types detected in the lunar South Pole - Aitken Basin by the Chang'E - 4 lunar mission. Geology 48 (7), 723 - 727 (2020).

[8] Gou, S. et al. Lunar deep materials observed by Chang'E - 4 rover. Earth Planet. Sci. Lett. 528, 115829 (2019).

[9] Hu, X. Y. et al. Mineral abundances inferred from in situ reflectance measurements of Chang'E - 4 landing site in South Pole - Aitken Basin. Geophys. Res. Lett. 46 (16), 9439 - 9447 (2019).

[10] Pieters, C. M. et al. Space weathering on airless bodies: Resolving a mystery with lunar samples. Meteor. Planet. Sci. 35 (5), 1101 - 1107 (2000).

[11] Moriarty, D. P. & Pieters, C. M. The character of South Pole - Aitken Basin: Patterns of surface and subsurface composition. J. Geophys. Res. Planets 123 (3), 729 - 747 (2018).

[12] Ma, P. et al. A plagioclase - rich rock measured by Yutu - 2 Rover in Von Karman crater on the far side of the Moon. Icarus 350, 569 - 580 (2020).

[13] Gou, S. et al. In situ spectral measurements of space weathering by Chang'E - 4 rover. Earth Planet. Sci. Lett. 535, 116117 (2020).

[14] He, Z. P. et al. Spectrometers based on acousto - optic tunable filters for in situ lunar surface measurement. J. Appl. Remote Sens. 13 (2), 027502 (2019).

[15] Weber, I. et al. Space weathering by simulated micrometeorite bombardment on natural olivine and pyroxene: A coordinated IR and TEM study. Earth Planet. Sci. Lett. 530, 230 - 242 (2020).

[16] Lucey, P. G., Taylor, G. J., Hawke, B. R. & Spudis, P. D. FeO and TiO_2 concentrations in the South Pole - Aitken basin: Implications for mantle composition and basin formation. J. Geophys. Res. Atmos. 103 (E2), 3701 - 3708 (1998).

[17] Fischer, E. M. & Pieters, C. M. Remote determination of exposure degree and iron concentration of lunar soils using VIS - NIR spectroscopic methods. Icarus 111 (2), 475 - 488 (1994).

[18] Lucey, P. G., Taylor, G. J. & Malaret, E. Abundance and distribution of iron on the Moon. Science 268 (5214), 1150 - 1153 (1995).

[19] Lucey, P. G., Blewett, D. T. & Jolliff, B. L. Lunar iron and titanium abundance algorithms based on final processing of Clementine ultraviolet - visible images. J. Geophys. Res. 105 (E8), 20297 - 20305 (2000).

[20] Pieters, C. M., Stankevich, D. G., Shkuratov, Y. G. & Taylor, L. A. Statistical analysis of the links among lunar mare soil mineralogy, chemistry, and reflectance spectra. Icarus 155 (2), 285 - 298 (2002).

[21] Zhang, X. Y., Li, C. L. & Lü, C. Quantification of the chemical composition of lunar soil in terms of its reflectance spectra by PCA and SVM. Chin. J. Geochem. 28, 204 - 211 (2009).

[22] Li, L. Partial least squares modeling to quantify lunar soil composition with hyperspectral reflectance measurements. J. Geophys. Res. 111 (E4), pp E04002 (2006).

[23] Mouélic, S. L. et al. Discrimination between maturity and composition of lunar soils from integrated Clementine UV - visible/near - infrared data: Application to the Aristarchus Plateau. J. Geophys. Res. 105 (E4), 9445 - 9455 (2000).

[24] Adams, J. B. & Goullaud, L. H. Plagioclase feldspars: Visible and near infrared diffuse reflectance spectra as applied to remote sensing. Geochim. Cosmochim. Acta 10 (1978).

[25] Bowell, E. et al. Application of photometric models to asteroids. Asteroids II (C), 524 - 556 (1988).

[26] Sunshine, J. M. & Pieters, C. M. Determining the composition of olivine from reflectance spectroscopy. J. Geophys. Res. 103 (E6), 13675 - 13688 (1998).

[27] Trang, D. & Lucey, P. G. Improved space weathering maps of the lunar surface through radiative transfer modeling of Kaguya multiband imager data. Icarus 321, 307 - 323 (2019).

[28] Otake, H., Ohtake, M. & Hirata, N. Lunar iron and titanium abundance algorithms based on SELENE (Kaguya) multiband imager data. In 43rd Lunar and Planetary Science Conference, Vol. 43, 1905 (2012).

[29] Ling, Z. et al. Composition, mineralogy and chronology of mare basalts and non-mare materials in Von Kármán crater: Landing site of the Chang'E-4 mission. Planet. Space Sci. 179, 104741 (2019).

[30] Ling, Z. et al. Correlated compositional and mineralogical investigations at the Chang'E-3 landing site. Nat. Commun. 6, 1-9 (2015).

[31] Lindstrom, M. M., Marvin, U. B., Holmberg, B. B. & Mittlefehldt, D. W. Apollo 15 KREEP-poor impact melts. In Lunar and Plan-etary Science Conference 20th (1989).

[32] Papike, J. J., Ryder, G. & Shearer, C. K. Lunar samples. Rev. Mineral. Geochem. 36 (1), 51-5234 (1998).

[33] Spudis, P. D. Apollo 16 site geology and impact melts: Implications for the geologic history of the lunar highlands. J. Geophys. Res. Solid Earth 89 (1), 95-107 (1984).

[34] Xiao, L. et al. A young multilayered terrane of the northern Mare Imbrium revealed by Chang'E-3 mission. Science 347 (6227), 1226-1229 (2015).

[35] Pieters, C. M. & Noble, S. K. Space weathering on airless bodies. J. Geophys. Res. Planets 121 (10), 1865-1884 (2016).

[36] Brett, W. D. et al. Characterization of space weathering from lunar reconnaissance orbiter camera ultraviolet observations of the Moon. J. Geophys. Res. Planets 119 (5), 976-997 (2014).

[37] Fu, X. H., Qiao, L., Zhang, J., Ling, Z. & Li, B. The subsurface structure and stratigraphy of the Chang'E-4 landing site: Orbital evidence from small craters on the Von Kármán crater floor. Res. Astron. Astrophys. 20 (1), 59-70 (2020).

[38] Huang, J. et al. Geological characteristics of Von Kármán Crater, Northwestern South Pole-Aitken Basin: Chang'E-4 landing site region. J. Geophys. Res. Planets 123 (7), 1684-1700 (2018).

[39] Paskert, J. H., Hiesinger, H. & van der Bogert, C. H. Lunar far side volcanism in and around the south pole-Aitken basin. Icarus 299, 538-562 (2018).

[40] Wilhelms, D. E., Howard, K. A. & Wilshire, H. G. Geologic map of the south side of the Moon. In IMAP (1979).

[41] Hurwitz, D. M. & Kring, D. A. Differentiation of the South Pole-Aitken basin impact melt sheet: Implications for lunar exploration. J. Geophys. Res. Planets 119 (6), 1110-1133 (2014).

[42] Vaughan, W. M. & Head, J. W. Impact melt differentiation in the South Pole-Aitken basin: Some observations and speculations. Planet. Space Sci. 91, 101-106 (2014).

[43] McEwen, A. S. & Robinson, M. S. Mapping of the Moon by Clementine. Adv. Space Res. 19 (10), 1523-1533 (1997).

[44] Pieters, C. M. & Tompkins, S. Tsiolkovsky crater: A window into crustal processes

on the lunar farside. J. Geophys. Res. 14 (E9), 21935 - 21949 (1999).

[45] Lai, J. L. et al. First look by the Yutu - 2 rover at the deep subsurface structure at the lunar farside. Nat. Commun. 11, 1 - 9 (2020).

[46] He, Z. P. et al. Visible and near - infrared imaging spectrometer and its preliminary results from the Chang'E - 3 project. Rev. Sci. Instrum. 85 (8), 083 - 104 (2014).

[47] Liu, B. et al. Reflectance conversion methods for the VIS/NIR imaging spectrometer aboard the Chang'E - 3 lunar rover: based on ground validation experiment data. Res. Astron. Astrophys. 13 (7), 862 - 874 (2013).

[48] Wu, Y. & Hapke, B. Spectroscopic observations of the Moon at the lunar surface. Earth Planet. Sci. Lett. 484, 145 - 153 (2018).

[49] Gueymard, C. Direct solar transmittance and irradiance predictions with broadband models. Part I: detailed theoretical performance assessment. Sol. Energy 76 (4), 513 (2004).

[50] Zhang, X. et al. Study of the continuum removal method for the Moon Mineralogy Mapper (M3) and its application to Mare Humorum and Mare Nubium. Res. Astron. Astrophys. 16, 7 - 15 (2016).

[51] Klima, R. L. et al. New insights into lunar petrology: Distribution and composition of prominent low - Ca pyroxene exposures as observed by the Moon Mineralogy Mapper (M3). J. Geophys. Res. Planets 116 (E6), 2156 - 2202 (2011).

[52] Clark, R. N. & Roush, T. L. Reflectance spectroscopy: Quantitative analysis techniques for remote sensing applications. J. Geophys. Res. Solid Earth 89 (B7), 6329 - 6340 (1984).

[53] Hapke, B. Bidirectional reflectance spectroscopy: 1. J. Geophys. Res. 86, 3039 - 3054 (1981).

[54] Hapke, B. Theory of Reflectance and Emittance Spectroscopy 528 (Cambridge University Press, 2012).

[55] Iordache, M. - D., Bioucas - Dias, J. M. & Plaza, A. Sparse unmixing of hyperspectral data. IEEE Trans. Geosci. Remote Sens. 49 (6), 2014 - 2039 (2011).

[56] Lin, H. et al. Retrieval of the mineral abundance and particle size distribution at the landing site of Yutu rover with hyperspectral remote sensing data (Article). J. Remote Sens. 23 (5), 831 - 840 (2019).

[57] Taylor, L. A., Pieters, C. M., Keller, L. P., Morris, R. V. & McKay, D. S. Lunar mare soils: Space weathering and the major effects of surface - correlated nanophase Fe. J. Geophys. Res. 106 (E11), 27985 - 27999 (2001).

Heterogeneous Weathering Process of Lunar Regolith Revealed by Polarimetric Attributes Analysis of Chang'E-4 Lunar Penetrating Radar Data Acquired During the Yutu-2 Turnings

Haoqiu Zhou[1], Xuan Feng[1], Chunyu Ding[2],
Zejun Dong[1], Cai Liu[1], and Wenjing Liang[1]

1. College of Geo-exploration Science and Technology, Jilin University, Changchun, China;
2. Institute of Moon-base Exploration and Observation, Institute of Advance Study, Shenzhen University, Shenzhen, China

Abstract: This paper provides analyses to quantify the maturity of regolith within 24m depth using a novel polarimetric attributes analysis of lunar penetrating radar (LPR) data. The results demonstrate that LPR signals were mainly from subsurface rock fragments of different sizes and were rarely from strata interfaces. The averages and ranges of maturities for the near-surface (0~2m), the fine-grained regolith (2~12m), and the coarse-grained ejecta (12~24m) are 68.98% (0%~99.82%), 82.61% (41.52% ~ 96.56%), and 72.33% (5.91% ~ 96.66%), respectively. Local regions with anomalously low maturities on the lunar surface may be due to the dense materials formed by local impacts or the rock fragments from distal impacts; a newly discovered sandwich structure with low maturity at a depth between 12 and 18m is formed by the heterogeneous weathering in a paleo-crater. Based on these new insights, we infer a heterogeneous weathering process of the regolith.

Plain Language Summary: The entire lunar surface is considered to be covered by a layer of regolith. The study of regolith weathering degree almost focused on the near-surface because the traditional techniques cannot detect the regolith at large depths. The Chang' E-4 lunar penetrating radar (LPR) can detect the regolith structure within dozens of meters depth. However, previous publication of LPR results only use the macro layering model to interpret the regolith structure. This study innovatively extracts new properties from the LPR data acquired while the rover was turning and estimated the quantitative maturity of regolith within ～24m depth at lunar farside. We found that the LPR can mainly detect the subsurface rock fragments that survived weathering and rarely the interfaces of strata because the materials of different strata are mixed at interfaces and make it gradational. Our results also reveal the spatial difference of weathering in the regolith. We investigate the formations and material compositions of several interesting regions with unusual weathering degrees. Based on these new insights, we establish a model to illustrate the weathering process of the regolith at the CE－4 landing site.

Key Points:

• We quantify the maturity of the lunar regolith within 24 m depth by analyzing the lunar penetrating radar (LPR) data acquired during the rover turnings

• LPR signals were mainly from subsurface rock fragments of different sizes and were rarely from strata interfaces

• Inferring the heterogeneous weathering process by comprehensively investigating the regions with anomalous maturities in the regolith

1 Introduction

The entire lunar surface is considered to be covered by a layer of regolith which is formed by meteorite impacts and gardening and consists of unconsolidated rock and mineral fragments, breccias, glasses, and agglutinates (Gou et al., 2019; McKay et al., 1991). Most of the knowledge of regolith such as optical maturity and surface roughness was acquired by remote sensing and mainly characterize the regolith properties at the near－surface. The regolith over the full depth records the lunar geological evolution. It is generally believed that regolith at larger depths contains more and larger rock fragments than shallower regolith (Bart & Melosh, 2010; McKay et al., 1991). However, the regolith at larger depths is rarely studied as it is beyond the detection of remote sensing techniques.

The Chinese Chang' E－4 probe landed on the floor of the Von Kármán crater in the South Pole－Aitken basin (Liu et al., 2019) and the lunar penetrating radar (LPR) (Fang et al., 2014; Li et al., 2020; Su et al., 2014) carried by the Yutu－2 rover are now mapping the fine structure of subsurface at lunar farside. The Von Kármán crater was covered by several periods of basalt flows and the ejecta from the adjacent craters during the Imbrian and Eratoshenian periods (Head III et al., 2012; Huang et al., 2018; Stuart－Alexander, 1978), which have resulted in complex subsurface compositions and structures.

The previous LPR sounding results have revealed the macro layering structure of lunar regolith. The top unit 1 is a fine－grained regolith with a depth of ～0－12m and is interpreted to be the ejecta from the nearby Finsen crater (Gou et al., 2021; Hu et al., 2019; Li et al., 2019); the middle unit 2 with a depth of 12－24m corresponds to strong reflections and it is considered to be the superposition of multiple ejecta sources from nearby craters; the bottom unit 3 has a depth of 24 – 40m and contains alternating layers of fine – grained materials and rocks (Lai et al. , 2019; Li et al. , 2020; Zhang, Zhou et al. , 2021) . However, these impressive works only provide the macro layering structure of regolith and its maturity remains unknown.

In other studies, the LPR data are applied to the analysis of subsurface properties and rock quantity in the regolith (Dong et al. , 2020; Lai et al. , 2021a, . 2021b; Zhang, Xu et al. , 2021; Zhou, Feng, Ding, et al. , 2021) . However, the rock localization using LPR data mainly depends on the reflection signals from the subsurface and the results in previous works cannot determine whether these signals are from horizontally distributed rocks or continuous geological interfaces (Zhou, Feng, Ding, et al. , 2021) because the regolith maturity over the full depth has not been obtained.

The LPR system onboard the Yutu – 2 rover adopted the time – trigger mode, which means the LPR transmitted and received signals by a certain time interval. However, the rover did not move at a constant speed (Fang et al. , 2014; Feng et al. , 2017; Su et al. , 2014) . The obtained LPR data are therefore related to three different situations, namely: the data acquired when the rover was moving, stopped, and was turning. In previous works, only the first type of data was selected to analyze the regolith (Li et al. , 2020; Zhou, Feng, Dong, et al. , 2021) . However, the LPR data acquired during the rover turnings is also a type of data that worth analysis. The rings distributed along the Yutu – 2 path (See Figures 1c – 1e) indicate that the LPR acquired data with different orientations at the same positions and these data can be used to obtain polarimetric attributes and better characterize the subsurface properties.

In this paper, we extract and analyze the polarization information from LPR data acquired when the rover was turning during the 15 lunar days. The results provide unprecedented quantitative maturity information and newly discovered fine structures of regolith within ~24m depth at lunar farside. The results reveal the heterogeneity of space weathering on the Moon and produce a fine evolution process of the regolith at Chang'E – 4 landing site.

2 Methods

2.1 Analysis of the LPR Data Acquired During the Rover Turnings

When the rover was turning, the collected LPR data is a function of the rotation angles. The differences in the LPR data of different measuring angles are related to the Sinclair matrix of the subsurface (Cloude & Pottier, 1996) . We select the rover heading direction when the rover started to turn as the x – axis, and the direction perpendicular to the x – axis as the y – axis as Figure 1e shows. Supposing that the turning angle of the rover/antennas is θ, the acquired turning data is given by (Dong et al., 2021, 2022; Sinclair, 1950; van Zyl et al., 1987):

$$M(t,\ s,\ \theta)=E_r^{\mathrm{T}}SE_t=\begin{bmatrix}\sin\theta\\ \cos\theta\end{bmatrix}^{\mathrm{T}}\begin{bmatrix}S_{hh}(t,\ s) & S_{hv}(t,\ s)\\ S_{vh}(t,\ s) & S_{vv}(t,\ s)\end{bmatrix}\begin{bmatrix}\sin\theta\\ \cos\theta\end{bmatrix} \tag{1}$$

where E_t and E_r are the polarization modes of the transmitting and receiving antennas and are the same for the LPR system. S represents the Sinclair matrix of the subsurface. t and s denote the travel time of an LPR signal and the turning position along the Yutu - 2 route, respectively. (1) can be further simplified as follows:

$$M(t,\ s,\ \theta)=\frac{1}{2}[\gamma(t,\ s)\sin 2\theta-\beta(t,\ s)\cos 2\theta+\alpha(t,\ s)] \tag{2}$$

where $\alpha=S_{hh}+S_{vv}$, $\beta=S_{hh}-S_{vv}$, and $\gamma=2S_{hv}$ are the Pauli components to be obtained (Cloude & Pottier, 1996). The acquired rover turning LPR data is composed of sine and cosine functions and a constant threshold. Subsequently, the least - squares fitting is performed as follows:

$$\arg\min L(\alpha,\ \beta,\ \gamma)=\arg\min \| D-M \|^2 \tag{3}$$

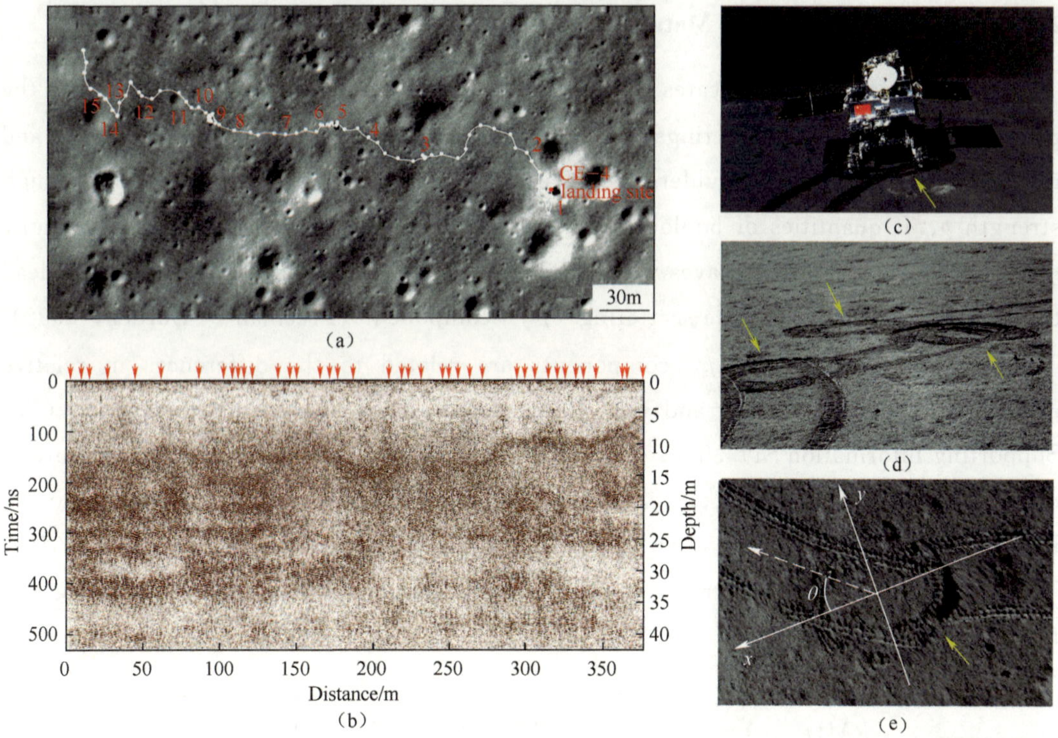

Figure 1　Yutu - 2 route and Lunar Penetrating Radar data. (a) The Yutu - 2 route within the first 15 lunar days. (b) The acquired CH - 2B LPR data. The red arrows indicate the positions to extract rover turning data. (c — e) Show the details of the rover route, the yellow arrows indicate the turning positions. A coordinate is established in (e) to analyze the acquired data when the rover was turning. The rover heading direction when the rover started to turn is selected as the x - axis

where D represents the acquired rover turning LPR data. By computing the minimum value of L, the polarization information S_{hh}, S_{vv}, and S_{hv} can be extracted.

2.2 Freeman Decomposition

Freeman decomposition (FD) is a type of classic polarimetric analysis technique, which is a physical model-based, three-component decomposition method (Freeman & Durden, 1998). The key idea of FD method is that for each scattering point, its scattering mechanism is contributed by three basic scattering components, which refer to a surface-like scattering (p_s), a double-bounce scattering (p_d), and a volume scattering (p_v) (Feng et al., 2017; Freeman & Durden, 1998; Zhou, Feng, Dong, Liu, Liang, & An, 2022, Zhou, Feng, Dong, Liu, & Liang, 2022). By decomposing the polarimetric scattering covariance matrix, the power of the three components can be obtained (See Supporting Information S1). In this study, the p_s and p_v components are adopted to derive the regolith maturity.

2.3 Quantitation of Regolith Maturity

The Freeman theory indicates that the surface-like scattering (p_s) characterizes the power of surface-like scatterings, such as the signals from the surfaces of strata and relatively coarse-grained boulders. Relatively fine-grained rocks can not generate high strength p_s. If quantities of boulders or rock fragments concentrate in a certain region, the electromagnetic (EM) waves will scatter many times in this region, which will lead to the high strength of volume scattering (p_v) component (Freeman & Durden, 1998). The amplitudes of p_s and p_v components are related to the differences in relative permittivity, the grain size, and the clutter degree of rocks within the regolith (See Supporting Information S1). The grain size and the clutter degree of rocks is a measure of regolith maturity (McKay et al., 1991), whereas relative permittivity is related to the FeO and TiO_2 content (Dong et al., 2017, 2020; Feng et al., 2017; Lai et al., 2019). Therefore, we can adopt the relevance of the two components to characterize the regolith maturity as follows:

$$RM(t, x) = \left[1 - \sqrt{\frac{p_s(t, x)p_v(t, x)}{\max[p_s(t, x)p_v(t, x)]}}\right] \times 100\% \tag{4}$$

where RM can be used to represent the regolith maturity. The verification can be found in Supporting Information S1. The assumption of Equation 4 is that the ejecta layer with a depth of 12 — 24m at the CE-4 landing site contains totally unweathered materials. Otherwise, the RM represents the relative maturity of the investigated zone.

3 Results and Discussions

3.1 Quantitative Regolith Maturity

The LPR is equipped with two types of channels (CH - 1 and CH - 2), with center frequencies of 60 and 500MHz and bandwidths of about 40 and 450MHz, respectively. CH - 2 channel contains one transmitting and two receiving bow - tie antennas, known as CH - 2A and CH - 2B, with different offsets of 0.16 and 0.32m (Fang et al., 2014). The CH - 1 antenna is located at the tail of the rover, whereas the CH - 2 antenna is at the bottom of the rover (Fang et al., 2014; Su et al., 2014). To obtain accurate polarization information, we should select the LPR data which are acquired at the same position for each turning, which means the center of transmitting and receiving antennas should be close to the center of the rover. Therefore, CH - 1 data is excluded and the CH - 2B data within 24m depth with a higher signal - to - noise ratio is adopted in this study.

Figure 2a is the instantaneous amplitude profile of migrated CH - 2B LPR image within 24m depth. Figures 2b and 2c present the surface - like and volume scattering components (p_s and p_v) which correspond to the models in Figures 2d - 2g, respectively. p_s mainly characterizes the strata interfaces and the surface of boulders whereas p_v mainly demonstrates the rock fragments of different grain sizes. The results demonstrate that the regolith is mainly composed of rock fragments that survived weathering whereas continuous strata interfaces are rare in the regolith. A reasonable interpretation is that the materials of different strata are mixed at interfaces which makes the boundary gradational and invisible (Giannakis et al., 2021). The received LPR signals are mainly from the dense fragments located at the bottoms of strata and survived the weathering (Zhang, Xu et al., 2021).

Subsequently, the p_s and p_v components can be applied to estimating the regolith maturity along the Yutu - 2 route in Figure. 3a. The subsurface can be divided into three units by depths. The near - surface region with a depth of 0 - 2m has variable maturities ranging from 0% to 99.82% with an average of 68.98%; the maturities of the fine - grained regolith (2 - 12m) vary from 41.52% to 96.56% with an average of 82.61%; coarse - grained ejecta layer (12 - 24m) has a variable maturity of 5.91%—96.66% with an average of 72.33%.

Although the 0 - 10m is convinced to be the fine - grained Finsen ejecta, the 0—2m region presents distinct lower maturities than the 2—10m region. A possible reason is that the surface of the CE - 4 landing area may consist of lithic material from secondary cratering whereas the 2 — 10m region below consists mainly of native materials which experienced longer time weathering (Ding et al., 2021). Besides, both maturities of near - surface and coarse ejecta regions have significant spatial differences whereas that of the fine - grained regolith layer is generally of higher degrees and lower variations.

3.2 Heterogeneous Lunar Surface Along the Yutu - 2 Route

The surface - like scattering component dominates at the top of Figure 2b but it is not uniform along the Yutu - 2 route. Figure 3b presents average curves of the two scattering components and average regolith maturity within 0 — 2m. The red arrows indicate the regions where surface - like scatterings are greater than the volume scatterings. Regions 2—4 show the high strengths of both two scatterings, demonstrating their high relative permittivities and roughness of the lunar surface. Therefore, the regolith in these regions

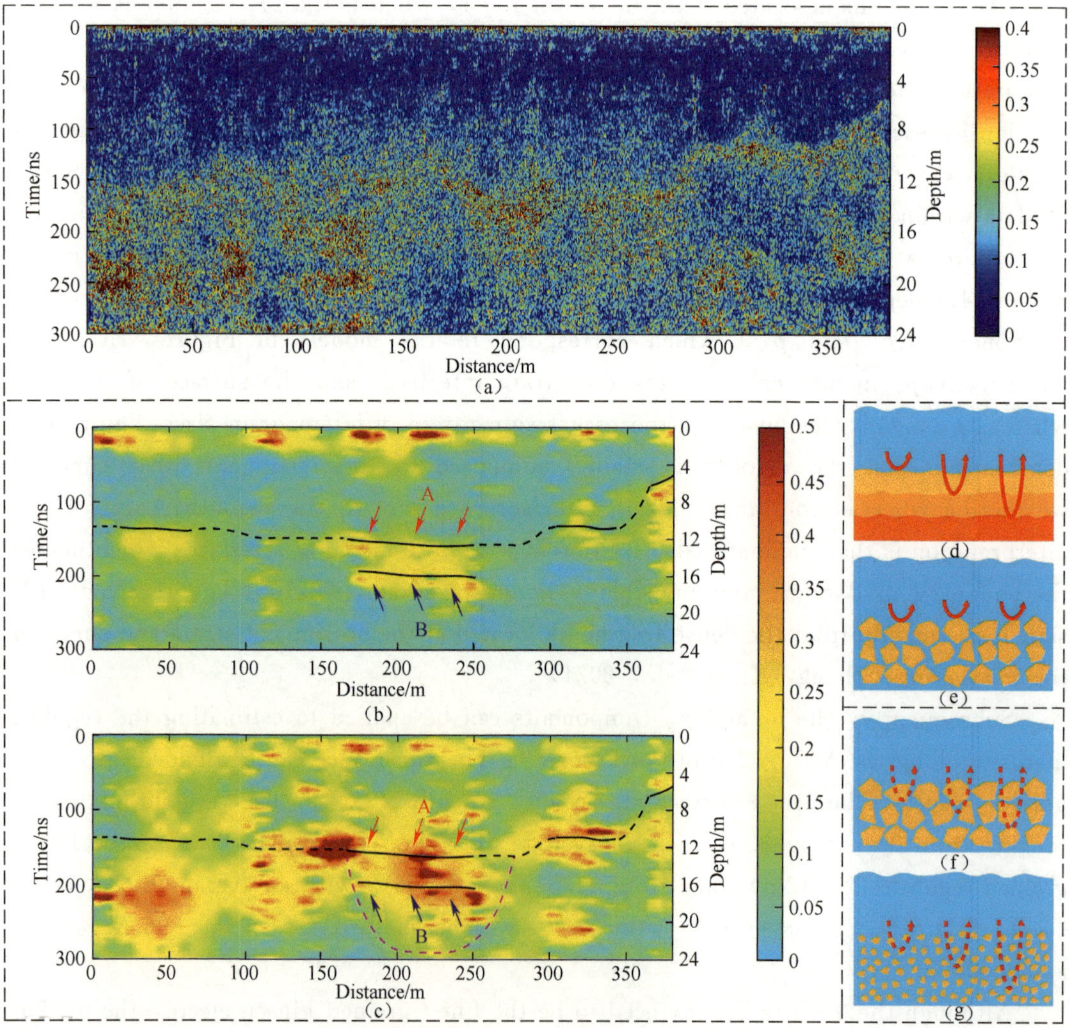

Figure 2　The Lunar Penetrating Radar polarimetric attributes processing results. (a) The instantaneous amplitude profile of migrated LPR image. (b) The surface - like scattering component (p_s) corresponding to the signals from continuous rough interfaces in (d) or the surfaces of boulders in (e). (c) The volume scattering component (p_v) corresponding to the signals from different sizes of boulders of strata interiors in (f and g). The curved arrows in (d — g) represent the schematic propagating paths of electromagnetic waves transmitted by LPR. A and B indicate the boundaries of the newly discovered sandwich structure in the paleo - crater

shows lower maturities. These regions are likely to contain dense material on the rough surface (Ding et al., 2021) and can be described by the hybrid model composed of Figures 2d and 2e. Regions 1, 5, and 6 exhibit moderate strengths of the surface - like scattering and low strengths of volume scattering, which indicate the moderate relative permittivities and smoother surface and correspond to moderate regolith maturities. These regions are likely to contain a smooth surface with small fragments or without fragments and can be described by the hybrid model composed of Figures 2d and 2g or the single model in Figure 2d.

Besides, the blue arrows indicate the regions where volume scatterings dominate and are with low strengths of both the two scattering components. Therefore, regions 7 - 11 demonstrate a lunar surface with low relative permittivities, moderate roughness, and high maturities, also corresponding to the hybrid model composed of Figures 2d and 2g.

The results reveal the heterogeneous lunar surface along the Yutu - 2 route and this heterogeneity is mainly due to the uneven space weathering. The regions with high maturities are likely to experience more impacting events whereas the regions with low maturities or high dielectric constants may compose of fresh ejecta from local impact or dense materials like melt - bearing impact breccias. Besides, the distal impact events can also deliver rock fragments to the region (Melosh, H. J., 1989), some of which formed small and fresh secondary craters and increased the roughness of the lunar surface (Ding et al., 2020, 2021; Gou et al., 2019; Lin et al., 2020). These events can also produce melt welded impact breccia in the near - surface of fine - grained regolith and resulting in the horizontal heterogeneity of the lunar surface (Ding et al., 2020, 2021).

3.3 Newly Discovered Sandwich Structure in the Ejecta Layer

Figure 3c presents average curves of the two scattering components and regolith maturity within 12—24m. In this depth range, volume scattering dominates along the Yutu - 2 route and three anomalies (12—14) are observed. Regions 12 and 13 are two separated areas of concentrated rocks; different strengths of surface - like scatterings indicate smaller rock sizes and higher regolith maturity in region 13. Besides, region 14 has been interpreted as a buried crater with large quantities of rocks (Lai, Cui, et al., 2021; Zhang, Xu et al., 2021; Zhou, Feng, Ding, et al., 2021), presenting the minimum of regolith maturity along the Yutu - 2 route. The regolith below 12m depth also shows horizontal heterogeneity. Therefore, we use the discontinuous black curves in Figures 2b and 2c to illustrate the bottom of the fine - grained regolith. The dashed lines denote the regions with weaker surface - like scatterings and higher regolith maturities.

It is interesting that we newly observe another continuous interface B within the buried crater (the blue arrows in Figures 2b and 2c). The newly discovered interface is of higher strengths compared to boundary A between the buried crater and the fine - grained regolith, which indicates that the materials on the two sides of interface B have a higher

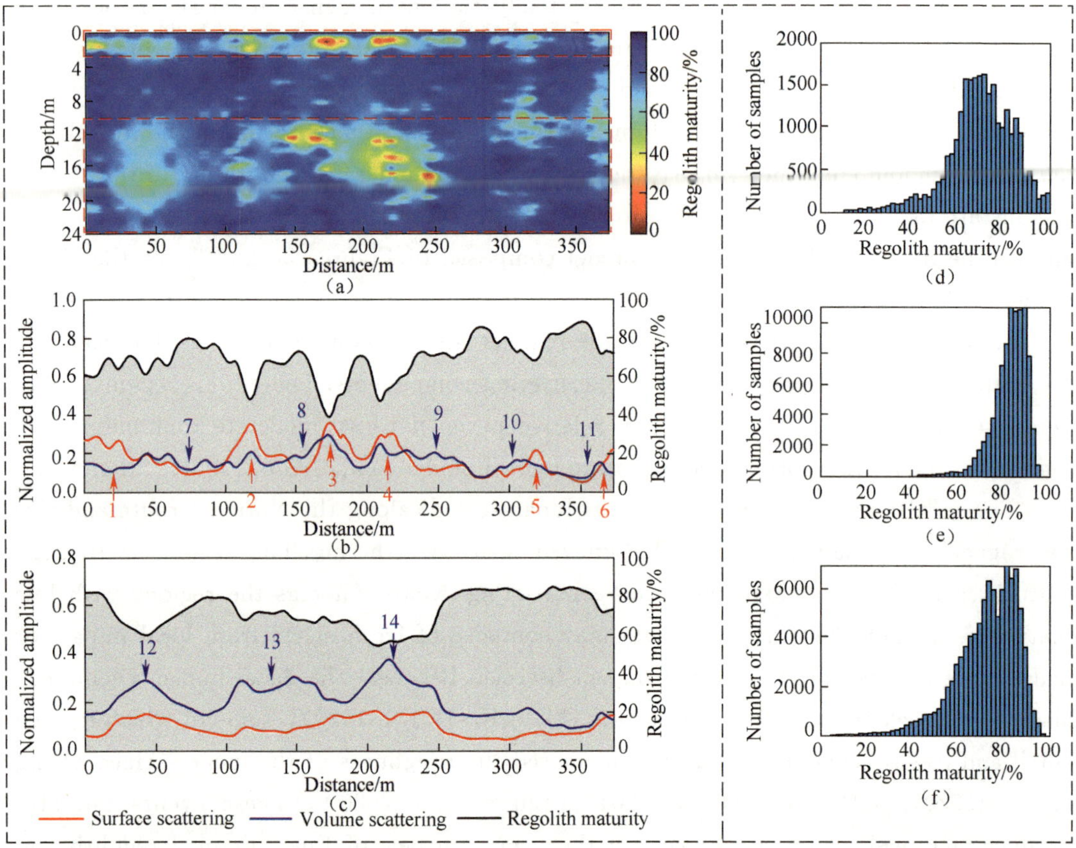

Figure 3 The regolith maturity and analyses. (a) The regolith maturity within 0—24m depth. The black curve in (b and c) represent the average maturity of 0—2m and 12—24m depths along the Yutu-2 route, respectively. The red and blue curves in (b and c) denote the average p_s and p_v within the same depth ranges. The red numbers 1—6 indicate the regions with unusual strong surface-like scattering and the blue numbers 7—14 mark the regions with anomalously strong volume scattering. (d—f) Are the distribution of regolith maturity of three-layered regions of different depths. (d) The near-surface region. (0—2m) (e) the fine-grained regolith layer (2—12m). (f) The coarse-grained ejecta layer (12—24m)

difference in dielectric constants than those of interface A. In other words, the compositions of the materials between A and B are more close to those of the fine-grained regolith. The fine-grained regolith is convinced to be the ejecta from the nearby Finsen crater, so the materials between A and B are also the Finsen ejecta but with lower maturity. These materials are likely to have survived the gardening on the lunar surface, resulting in their unique properties compared to both the materials of the underlying and overlying regolith. Besides, B is the bottom of the buried crater when it was covered by the Finsen ejecta. Compared to the excavated depth of the buried crater observed on radargram, it is likely to have experienced degradation before the coverage of Finsen ejecta.

3.4 Indications for Regolith Formation

Based on the regolith maturity and the analysis before, a hypothesis of the weathering process for the regolith within ～24m depth at the CE－4 landing site is possibly established as follows (Figure 4):

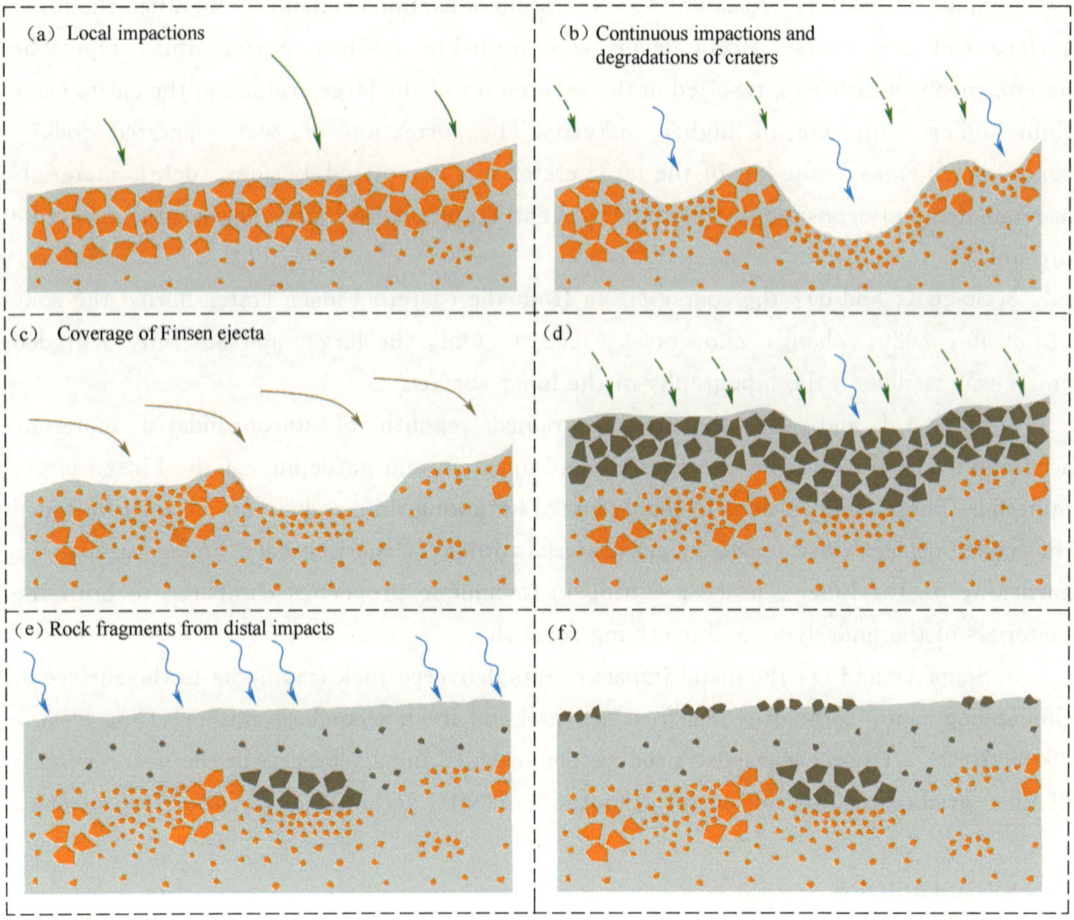

Figure 4　Fine Evolution of the Regolith at the CE－4 landing site. (a) Impact events of different extents reshaped the structure of coarse－grained regolith. (b) Erosion and small－scale impactions constantly modified the lunar surface. (c) Coverage of Finsen ejecta. (d) Formation of fine－grained regolith. (e) Delivery of fragments by distal impact events. (f) Current regolith structure. The grains of the rocks represent the maturities of regolith. The orange－colored anomalies represent the superposition of multiple ejecta from nearby caters and the gray－colored anomalies denote the Finsen ejecta. The green arrows and the dashed arrows represent the single and multiple impact events, respectively; whereas the undulated arrows and the gray arrows indicate the accumulations of multi－source materials and Finsen ejecta, respectively

1. Stage (a and b) postdated the superposition of multiple coarse ejecta from nearby caters. During this period, the lunar surface was likely to experience multiple and important impacts events of different extents that reshaped the structure of the

regolith. These impacting events resulted in the undulation and discontinuity of the lunar surface. Today, we can only recognize the largest crater from the LPR radargram located at ～170—290m on the Yutu - 2 route (Zhang, Xu et al., 2021; Zhou, Feng, Ding, et. al., 2021). However, the heterogeneous rock sizes and maturities of the rest regions (0—170m and 290—370m) probably indicate other craters that have degraded.

2. Stage (b and c): erosion and small - scale impactions constantly modified the lunar surface and the coarse ejecta layer was probably grained during this stage. The heterogeneous weathering resulted in the asymmetry of the large crater and the ejecta in its right sliding wing are of higher maturity. The formation of two separated rock - concentrated areas at the left of the large crater was completed. Besides, debris materials accumulated and deposited in the craters; small craters were highly degraded and became invisible.

3. Stage (c and d): the coarse ejecta from the eastern Finsen crater buried the zone (Li et al., 2020; Zhang, Zhou et al., 2021). Only the larger and not fully degraded crater was visible on the topography of the lunar surface.

4. Stage (d and e): the fine - grained regolith of unconsolidated materials accumulated gradually from the continuous impaction and gardening of the Finsen ejecta materials. These modifications probably lasted for a long time and simultaneously flattened the lunar surface. Most coarse ejecta materials filled in the degraded crater survive the gardening on the lunar surface resulting in its unique properties compared to both the materials of the underlying and overlying regolith.

5. Stage (e and f): the distal impact events delivered rock fragments to the surface of the landing zone, some of which formed small and fresh secondary craters (Ding et al., 2020, 2021). These events also produce melt welded impact breccia in the near - surface of fine - grained regolith and resulting in the horizontal heterogeneity of the lunar surface.

4 Conclusions

This study provides quantitative maturity and fine structure of regolith at lunar farside using polarimetric attributes analysis of CE - 4 LPR data. The polarization information is extracted from LPR data acquired during the rover turnings. Local regions with anomalously low maturities are observed on the lunar surface which may be due to the dense materials formed by local impact or rock fragments delivered from distal impacts. A sandwich structure with low maturity is newly discovered within the ejecta layer which is formed by the heterogeneous weathering in a paleo - crater. We also found that LPR signals were mainly from subsurface rock fragments of different sizes and were rarely from strata interfaces. Based on these new insights, we infer a heterogeneous weathering process of the regolith.

The polarimetric information is important and useful to characterize the subsurface and is applied to planetary exploration for the first time in this paper. Both the Chinese Tianwen - 1 Mars mission and the ESA's ExoMars 2022 mission are equipped with polarimetric rover radar. Our results are pivotal and instructive for the interpretation of polarimetric data from future Mars rover radar.

Acknowledgments

We thank the China National Space Administration for providing the scientific data used in this study. This work was supported by the National Key Research and Development Program of China under Grant 2021YFC1523401 and the National Natural Science Foundation of China under Grant 41974129.

References

[1] Bart, G. D., & Melosh, H. J. (2010). Distributions of boulders ejected from lunar craters. Icarus, 209 (2), 337 - 357.

[2] Cloude, S. R., & Pottier, E. (1996). A review of target decomposition theorems in radar polarimetry. IEEE Transactions on Geoscience and Remote Sensing, 34 (2), 498 - 518.

[3] Ding, C., Xiao, Z., Wu, B., Li, Y., Prieur, N. C., Cai, Y., et. al. (2020). Fragments delivered by secondary craters at the Chang'e 4 landing site. Geophysical Research Letters, 47, e2020GL087361. https://doi. org/10. 1029/2020gl087361.

[4] Ding, C., Xiao, Z., Wu, B., Li, Z., Su, Y., Zhou, B., et. al. (2021). Rock fragments in shallow lunar regolith: Constraints by the lunar penetrating radar onboard the Chang' E - 4 mission. Journal of Geophysical Research: Planets, 126, e2021JE006917. https://doi. org/10. 1029/2021je006917.

[5] Dong, Z., Fang, G., Ji, Y., Gao, Y., Wu, C., & Zhang, X. (2017). Parameters and structure of lunar regolith in Chang' E - 3 landing area from lunar penetrating radar (LPR) data. Icarus, 282, 40 - 46. https://doi. org/10. 1016/j. icarus. 2016. 09. 010.

[6] Dong, Z., Feng, X., Zhou, H., Liu, C., Lu, Q., & Liang, W. (2022). Assessing the effects of induced field rotation on water ice detection of Tianwen - 1 full - polarimetric Mars rover penetrating radar. IEEE Transactions on Geoscience and Remote Sensing, 60, 1 - 13. https://doi. org/10. 1109/tgrs. 2021. 3138463.

[7] Dong, Z., Feng, X., Zhou, H., Liu, C., & Sato, M. (2021). Effects of induced field rotation from rough surface on H - alpha decomposition of full - polarimetric GPR. IEEE Transactions on Geoscience and Remote Sensing, 59 (11), 9192 - 9208.

[8] Dong, Z., Feng, X., Zhou, H., Liu, C., Zeng, Z., Li, J., & Liang, W.

(2020) . Properties analysis of lunar regolith at Chang' E－4 landing site based on 3D velocity spectrum of lunar penetrating radar. Remote Sensing, 12, 629. https: //doi. org/10. 3390/rs12040629

[9] Fang, G. , Zhou, B. , Ji, Y. , Zhang, Q. , Shen, S. , Li, Y. , et al. (2014). Lunar penetrating radar onboard the Chang'E－3 mission. Research in Astronomy and Astrophysics, 14 (12), 1607－1622.

[10] Feng, J. , Su, Y. , Ding, C. , Xing, S. , Dai, S. , & Zou, Y. (2017). Dielectric properties estimation of the lunar regolith at CE－3 landing site using lunar penetrating radar data. Icarus, 284, 424－430. https: //doi. org/10. 1016/j. icarus. 2016. 12. 005

[11] Freeman, A. , & Durden, S. L. (1998) . A three－component scattering model for polarimetric SAR Data. IEEE Transactions on Geoscience and Remote Sensing, 36 (3), 963－973.

[12] Giannakis, I. , Zhou, F. , Warren, C. , & Giannopoulos, A. (2021) . Inferring the shallow layered structure at the Chang' E－4 landing site: A novel interpretation approach using lunar penetrating radar. Geophysical Research Letters, 48 (16), e2021GL092866. https: //doi. org/10. 1029/2021gl092866.

[13] Gou, S. , Di, K. , Yue, Z. , Liu, Z. , He, Z. , Xu, R. , et al. (2019). Lunar deep materials observed by Chang'e－4 rover. Earth and Planetary Science Letters, 528 (15), 115829. https: //doi. org/10. 1016/j. epsl. 2019. 115829.

[14] Gou, S. , Yue, Z. , Di, K. , Cai, Z. , Liu, Z. , & Niu, S. (2021) . Absolute model age of lunar Finsen crater and geologic implications. Icarus, 354 (15), 115829. https: //doi. org/10. 1016/j. icarus. 2020. 114046

[15] Head, J. W. , III, Fassett, C. I. , Kadish, S. J. , Smith, D. E. , Zuber, M. T. , Neumann, G. A. , & Mazarico, E. (2012) . Global distribution of large lunar craters: Implications for resurfacing and impactor populations. Science, 329 (5998), 1504－1507.

[16] Hu, X. , Ma, P. , Yang, Y. , Zhu, M. , Jiang, T. , Lucey, P. G. , et al. (2019) . Mineral abundances inferred from in situ reflectance measurements of Chang' E－4 landing site in South Pole－Aitken basin. Geophysical Research Letters, 46, 9439－9447. https: //doi. org/10. 1029/2019gl084531

[17] Huang, J. , Xiao, Z. , Flahaut, J. , Martinot, M. , Head, J. , Xiao, X. , et al. (2018) . Geological characteristics of Von Kármán crater, northwestern South PoleAitken basin: Chang' E－4 landing site region. Journal of Geophysical Research: Planets, 123, 1684－1700. https: //doi. org/10. 1029/2018je005577

[18] Lai, J. , Cui, F. , Xu, Y. , Liu, C. , & Zhang, L. (2021) . Dielectric properties of lunar materials at the Chang'e－4 landing site. Remote Sensing, 13, 4056. https: //doi. org/10. 3390/rs13204056

[19] Lai, J. , Xu, Y. , Bugiolacchi, R. , Wong, H. K. , Xu, L. , Zhang, X. , et al.

(2021). A complex paleo - surface revealed by the Yutu - 2 rover at the lunar farside. Geophysical Research Letters, 48, e2021GL095133. https://doi.org/10.1029/2021gl095133

[20] Lai, J., Xu, Y., Zhang, X., Xiao, L., Yan, Q., Meng, X., et al. (2019). Comparison of dielectric properties and structure of lunar regolith at Chang'e - 4 landing site in South Pole - Aitken basin. Geophysical Research Letters, 46, 12783 - 12793. https://doi.org/10.1029/2019gl084458

[21] Li, C., Liu, D., Liu, B., Ren, X., Liu, J., He, Z., et al. (2019). Chang' E - 4 initial spectroscopic identification of lunar far side mantle - derived materials. Nature, 569 (7756), 378 - 382.

[22] Li, C., Su, Y., Pettinelli, E., Xing, S., Ding, C., Liu, J., et. al. (2020). The Moon's farside shallow subsurface structure unveiled by Chang'E - 4 lunar penetrating radar. Science Advances, 6 (9), eaay6898. https://doi.org/10.1126/sciadv.aay6898

[23] Lin, H., Lin, Y., Yang, W., He, Z., Hu, S., Wei, Y., et al. (2020). New insight into lunar regolith forming processes by the lunar rover Yutu - 2. Geophysical Research Letters, 47, e2020GL087949. https://doi.org/10.1029/2020gl087949

[24] Liu, J., Ren, X., Yan, W., Li, C., Zhang, H., Jia, Y., et al. (2019). Descent trajectory reconstruction and landing site positioning of Chang' E - 4 on the lunar farside. Nature Communications, 10, 4229. https://doi.org/10.1038/s41467 - 019 - 12278 - 3.

[25] McKay, D. S., Heiken, G., Basu, A., Blanford, G., Simon, S., Reedy, R., et al. (1991). The lunar regolith. In G. H. Heiken, D. T. Vaniman, & B. M. French (Eds.), Lunar sourcebook: A user's guide to the Moon (pp. 285 - 356). Cambridge University Press.

[26] Melosh, H. J. (1989). Impact cratering: A geologic process. Oxford Univ. Press.

[27] Sinclair, G. (1950). The transmission and reception of elliptically polarized waves. Proceedings of the IRE, 38, 148 - 151. https://doi.org/10.1109/jrproc.1950.230106.

[28] Stuart - Alexander, D. E. (1978). Geologic map of the central far side of the Moon I -1047. US Geological Survey.

[29] Su, Y., Fang, G., Feng, J., Xing, S., Ji, Y., Zhou, B., et al. (2014). Data processing and initial results of Chang'E - 3 lunar penetrating radar. Research in Astronomy and Astrophysics, 14 (12), 1623 - 1632.

[30] van Zyl, J. J., Zebker, H. A., & Elachi, C. (1987). Imaging radar polarization signatures: Theory and observation. Radio Science, 22 (4), 529 - 543.

[31] Zhang, J., Zhou, B., Lin, Y., Zhu, M., Song, H., Dong, Z., et al. (2021). Lunar regolith and substructure at Chang' E - 4 landing site in South Pole -

Aitken basin. Nature Astronomy, 5, 25 - 30. https://doi.org/10.1038/s41550-020-1197-x.

[32] Zhang, L., Xu, Y., Bugiolacchi, R., Hu, B., Liu, C., Lai, J., et al. (2021). Rock abundance and evolution of the shallow stratum on Chang'e-4 landing site unveiled by lunar penetrating radar data. Earth and Planetary Science Letters, 564, 116912. https://doi.org/10.1016/j.epsl.2021.116912.

[33] Zhou, H., Feng, X., Ding, C., Dong, Z., Liu, C., Zhang, Y., & Meng, Z. (2021). Yutu-2 radar sounding evidence of a buried crater at Chang'E-4 landing site. IEEE Transactions on Geoscience and Remote Sensing, 60, 1-19. https://doi.org/10.1109/tgrs.2021.3090528.

[34] Zhou, H., Feng, X., Dong, Z., Liu, C., & Liang, W. (2021). Application of denoising CNN for noise suppression and weak signal extraction of lunar penetrating radar data. Remote Sensing, 13 (4), 779. https://doi.org/10.3390/rs13040779.

[35] Zhou, H., Feng, X., Dong, Z., Liu, C., & Liang, W. (2022). Multi-parameter adaptive target classification using full-polarimetric GPR: A novel approach to landmine detection. IEEE Journal of Selected Topics in Applied Earth Observations and Remote Sensing, 15, 2592 - 2606. https://doi.org/10.1109/jstars.2022.3159305.

[36] Zhou, H., Feng, X., Dong, Z., Liu, C., Liang, W., & An, Y. (2022). Local Freeman decomposition for robust imaging and identification of subsurface anomalies using misaligned full-polarimetric GPR data. Remote Sensing, 14 (3), 804. https://doi.org/10.3390/rs14030804.

嫦娥四号着陆点原位高光谱的太空风化研究

王振超[1]，柳稼航[1]，盛庆红[1]，吴昀昭[2]

1. 南京航空航天大学航天学院，江苏南京 210016
2. 中国科学院紫金山天文台，江苏南京 210034

摘要： 月球表面没有大气层的保护，岩石矿物长期在太空风化的作用下，逐渐演化为月壤。太空风化过程是月壤产生、成熟的过程，由于物质形态结构的改变，导致月表光谱特征产生变化。因而，作为风化产物的月壤的光谱特性，包含了月表的太空风化信息。月壤成熟度，是描述太空风化程度的重要指标，利用高光谱遥感数据进行亚微观铁（SMFe）的反演进而获取月壤成熟度，是目前研究月表太空风化的主要手段。原位探测数据，由于没有受到其他因素的干扰，获得的反演结果相对更加准确、可靠。我国嫦娥四号（CE-4）玉兔二号巡视器搭载了能直接获取月表原位高光谱数据（450～2395nm）的科学载荷红外成像光谱仪，为研究月表的太空风化提供了很好的机会。选取了 CE-4 卫星着陆器登陆点附近的两处光谱数据，采用 Hapke 模型和光谱角匹配法对 CE-4 卫星登陆点附近月壤的 SMFe 进行了反演。根据 Morris 模型和 FeO 含量进一步反演 CE-4 卫星登陆点附近月壤的成熟度。结果表明，该处月壤成熟度为 11.5，较大概率为不成熟月壤。

关键词： 月球高光谱；Hapke 模型；嫦娥四号；亚微观铁含量；成熟度

Study on space weathering of Chang'e－4 landing site by in situ spectra

Wang Zhenchao[1]，Liu Jiahang[1]，Sheng Qinghong[1]，Wu Yunzhao[2]

1. College of Astronautics，Nanjing University of Aeronautics and Astronautics，Nanjing 210016，China
2. Purple Mountain Observatory，Chinese Academy of Sciences，Nanjing 210034，China

Abstract：Without atmosphere，the lunar surface is gradually evolved into lunar soil by the space weathering. The space weathering generates lunar soil and matures it. This progress changes the morphological structure and the spectral characteristics. Thus the spectral characteristics of lunar soil contains information about the space weathering of the moon's surface. Lunar soil maturity is an important indicator to describe the degree of the space weathering. Using hyperspectral data to produce the submicroscopic metallic iron (SMFe) and to obtain lunar maturity is an important method for studying the space weathering Since the in situ detection data is not disturbed，the results obtained are relatively more accurate and reliable Therefore，the Chang'e－4 (CE－4) Yutu－2 rover was equipped with a detection instrument that directly obtained the in situ hyperspectral data (450－2395nm) of the lunar surface，which provided a good opportunity for studying the lunar weathering Two spectral data near the landing site of the CE－4 lander were selected，using the Hapke model and the spectral angle matching method，SMFe content was produced. The maturity of the lunar soil near the landing site of the CE－4 was further inferred based on the Morris model and FeO content The results show that the lunar soil's maturity is 11 5，indicating that it may be immature lunar soil.

Keywords：lunar hyperspectra；Hapke model；Chang'e－4；SMFe；maturity

0 引言

早期认为表面几乎没有大气的月球是不存在演化的[1]，而事实上，在长期遭受宇宙射线和太阳风的照射、陨石撞击、昼夜热循环等作用，导致表面物质的物理化学性质发生改变，这种过程即太空风化。太空风化显著地改变了月球表面的光谱特征，使得月表反射率总体变低、光谱斜率增大（变红）、矿物诊断性吸收谱带普遍减弱[2-4]。产生这种变化，主要是由于微陨石的撞击和太阳风的溅射过程中汽化的铁粒子在表层的沉积作用。由于几十甚至几百纳米粒径的铁粒子对太空风化作用同样具有影响，所以早期将该铁粒子称为纳米相铁（$npFe^0$）并不合理，而应该称为亚微观铁（SMFe）。该风化过程，随着时间的增加而变得更加显著，即成熟度越来越高。太空风化对月球表面过程、光谱特性、物质成分反演等都具有重要意义。成熟度是描述太空风化的常用指标，利用 SMFe 含量与氧化亚铁含量可以反演得到成熟度[5]。采用高光谱遥感数据结合月壤的相关参数可以反演得到月壤中 SMFe 的含量[1,6-7]。

月球原位光谱数据不足且月球代表性样品不完备，付晓辉[8]、赵哲[9]等利用 Hapke 模型对实验模拟光谱进行了太空风化分析，是对太空风化研究的重要补充。研究表明，近距离获取原位高光谱数据对于月表太空风化反演的结果更加准确可靠[7]。绕月卫星距离月面较远，由于空间分辨率较低不可避免地存在混合像元现象，进而导致光谱混合像元，使得反演结果不准确；采样返回的月球样品在采用过程中，不可避免地会发生翻动、搅动混合和孔隙度变化等物理变化，从而使得实验室测得的光谱很难代表月球表面真实光谱信息。

我国 2018 年底发射的嫦娥四号（CE－4）卫星巡视器玉兔二号搭载的红外成像光谱仪所获取的原位光谱数据，结合相关参数以及 FeO 的含量，对 CE－4 着陆点附近月壤进行太空风化研究，是继嫦娥三号（CE－3）太空风化研究后的又一绝佳机会。相比 CE－3，CE－4 是人类首次获取月球背面原位光谱数据，对月球背面太空风化研究对人类对月球的总体完整认识具有重要意义。

1　嫦娥四号高光谱数据

2019 年 1 月 3 日嫦娥四号着陆器与巡视器玉兔二号成功地于月球背面（远离地球的一侧）软着陆[10]。玉兔二号目前已经正常工作至第八个月昼，搭载的红外成像光谱仪已经获取了大量的月球原位高光谱数据。红外成像光谱仪的相关参数见表 1。红外成像光谱仪包括两个通道，短波红外 SWIR 通道和可见近红外 VNIS 通道，其中 SWIR 所探测的区域为 VNIS 所探测区域的中间偏左侧内部的圆形区域。由于光谱由两个光谱通道观测范围不一致，所以需要对二者进行统一。该研究对 VNIS 范围内与 SWIR 观测范围一致部分进行平均，对于原始数据去噪、平滑以及二者的整合等预处理参照参考文献［11］和［12］。由于在小于 530nm 的短波范围内光谱的信噪比较低，而在大于 2200nm 的长波范围内受热辐射影响较大，故该研究只采用 530～2200nm 波长范围内的光谱数据。

表 1　嫦娥四号，玉兔二号红外成像光谱仪技术参数

Tab. 1　Parameters of the CE－4，Yutu－2 visible and near－IR imaging spectrometer（VNIS）and short－wave infrared（SWIR）spectrometer

Parameters	VNIS	SWIR
Wavelength/nm	450—950	900—2400
Band resolution/nm	2—10	3—12
View field/(°)	8.5×8.5	ϕ3.58
Numbers of valid. pixel	≥256×256	1
Detective distance/m	0.7—1.3	
Detective time/min	～2	

在嫦娥四号的观测点中，第三个月昼共获得了十个点位的光谱数据（点 12～点 21）。为了研究嫦娥四号登陆点附近的月壤太空风化情况，选取了两个观测点的光谱数据。通过可见光近红外成像光谱仪的影像以及短波红外的光谱数据研究发现，嫦娥四号登录后的第三月昼的所探测的点 15 的光谱为岩石光谱（见图 1）。尚未风化的岩石表面的 SMFe 含量为 0[7]，如果得到其他点的月壤 SMFe 含量与该点的差值，便可以得到相应光谱对应月壤的 SMFe 含量。为了得到可以代表 SMFe 原位赋存状态的光谱，该研究筛选掉了玉兔二号车轮碾压过的月壤光谱以及表面碎石较多或者太阳斜度较大造成表面阴影较为严重的光谱，最终选取点 20 的光谱（图 2）。由于记录的角度信息均为假设被拍摄面为水平面，而点 15 为岩石的侧上方，故该研究重新估算了点 15 的几何角度，点 15 与点 20 光谱的几何信息见表 2。

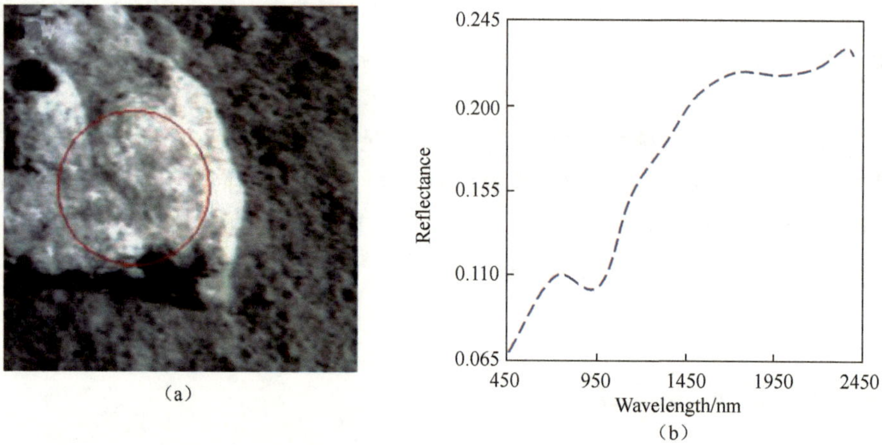

图 1　嫦娥四号 15 号观测点影像及其光谱。(a) 成像数据 (450～945nm)，红色圆圈为短波红外数据探测范围；(b) 红圈范围内光谱

Fig. 1　Image and spectra of site 15 of Chang'e-4. (a) Imaging data (450～945nm), red circle is the scope of the SWIR data; (b) spectra inside the red circle scope

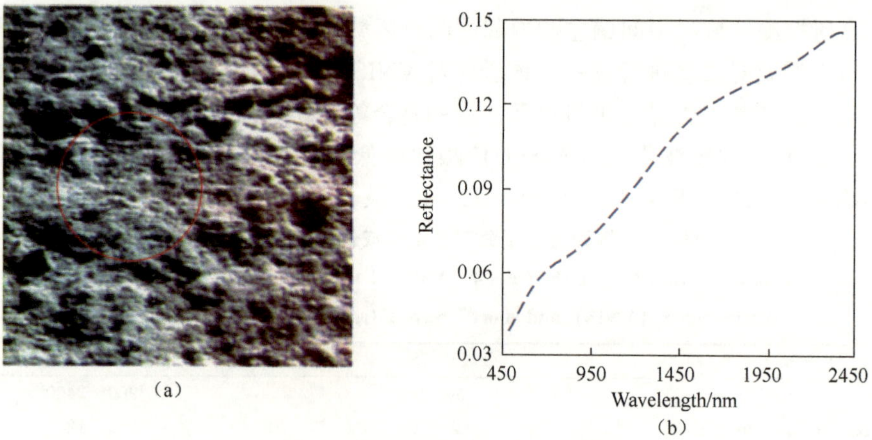

图 2　嫦娥四号 20 号观测点影像及其光谱。(a) 成像数据 (450～945nm)，红色圆圈为短波红外数据探测范围；(b) 红圈范围内光谱

Fig. 2　Image and spectra of site 20 of Chang'e-4. (a) Imaging data (450～945nm), red circle is the scope of the SWIR data; (b) spectra inside the red circle scope

表 2　玉兔二号获取的点 15 与点 20 光谱的几何信息

Tab. 2　Data acquisition conditions of Site 15 and Site 20 by Yutu - 2

Parameters	Point 15	Point 20
Local time	08：32：00	15：35：00
Distance to the Sun/(AU)	0.98914637	0.99153951
View angle/(°)	17	46.26
Incident angle/(°)	70.5	64.54
View azimuth angle/(°)	143.32	83.03
Incident azimuth angle/(°)	61.48	296.66
Phase angle/(°)	65.48	104.24

2　反演模型

基于辐射传输理论的二相性反射传输模式的 Hapke 模型，主要表达被研究物质成分、颗粒大小以及观测几何角度的光学特性等方面的内容[1]，其公式如下：

$$REFF=\frac{\omega_{\text{ave}}}{4(\mu_0+\mu)}\{[1+B(g)]P(g)+H(\mu_0,\ \omega_{\text{ave}})H(\mu,\ \omega_{\text{ave}})-1\} \tag{1}$$

式中：g 为观测相位角；μ_0 为入射角 i 的余弦；μ 为观测角 e 的余弦；$B(g)$ 为后向散射函数

$$B(g)=\frac{B_0}{1+(1/h)\tan(g/2)} \tag{2}$$

式中：B_0 为后向效应幅度，可以近似为 1[13]；h 为后向效应的宽度参数，与颗粒的粒径、介质的填充因子等相关，其表达式为：

$$h=-\frac{3}{8}\ln(1-\varphi) \tag{3}$$

式中：φ 为填充因子，月壤可以约等于 0.41[14]。$P(g)$ 为单个粒子的相函数，可以由以拉格朗日多项式来定义：

$$P(g)=1+b\cos(g)+c(1.5\cos^2(g)-0.5) \tag{4}$$

式中：$b=-0.4$，$c=0.25$[15]。ω_{ave} 为所有成分的平均单次散射反照率（average Single Scattering Albedo，SSA）。其表达式如下：

$$\omega_{\text{ave}}=S_e+(1-S_e)\frac{(1-S_i)\Theta}{1-S_i\Theta} \tag{5}$$

$$S_e=\frac{(n_h-1)^2+k_h^2}{(n_h+1)^2+k_h^2}+0.05\approx\frac{(n_h-1)^2}{(n_h+1)^2}+0.05 \tag{6}$$

$$S_i=1.014-\frac{4}{n_h(n_h+1)^2} \tag{7}$$

$$\Theta=e^{-\alpha_w<D>} \tag{8}$$

式中：n_h 和 k_h 分别为基质折射率指数（refractive. index）的实部（即折射率）和虚部（即消光系数），月壤的折射率 n_h 为 1.7[3]，而消光系数 k_h 则由模型反演得到；α_ω 为吸收系数（absorption coefficient），其大小为 $4\pi nk/\lambda$，λ 为波长；$\langle D\rangle$为光在颗粒内部的平均路径长度，在月壤中约为 30μm[16]。H（μ，ω_{ave}）为各向同性散射函数[5]：

$$H(\mu,\ \omega_{ave})=\left\{1-(1-\sqrt{1-\omega_{ave}})*\mu\left[r_0+\left(1-\frac{r_0}{2}-r_0\mu\right)\ln\frac{1+\mu}{\mu}\right]\right\}^{-1} \tag{9}$$

其中，

$$r_0=\frac{2}{1+\sqrt{1-\omega_{ave}}}-1 \tag{10}$$

空间风化的光谱效应是计算吸收系数 α 的因素[4]：

$$\alpha=\frac{4\pi n_h k_h}{\lambda}+\frac{36\pi z f\rho_h}{\lambda\rho_{Fe}} \tag{11}$$

其中，

$$z=\frac{n_h^3 n_{Fe} k_{Fe}}{(n_h^2-k_{Fe}^2+2n_h^2)^2+(2n_{Fe}k_{Fe})^2} \tag{12}$$

该研究中铁的光学常数为 Paquin[17]实验数据。ρ_h，ρ_{Fe}则分别代表基体和铁单质的密度，基体的密度本章假定为 1.6.g/cm^3[18]；铁的密度为 7.87g/cm^3[19]；f 为 SMFe 在月壤颗粒表面的质量比例。

在该研究中，首先利用 Hapke 模型反演得到尚未风化的岩石矿物光谱（即点 15 的光谱)对应的光谱消光系数 k，此时输入该光谱对应的各种观测角度，而吸收系数以 $4\pi nk/\lambda$ 计算。然后利用计算所得的波长范围内 k 值在其他光谱（即点 20 的光谱）测定时的测光角度信息等模拟计算，此时，吸收系数 α 增加空间风化因子。从 0 逐步增加 SMFe 的含量，计算模拟光谱与点 20 光谱的光谱角，当二者最为相等或者接近时的 f 便是点 20 月壤的 SMFe 含量。

3　结果与分析

经过模型反演，当未风化的岩石光谱（点 15 光谱）增加 SMFe 含量为 0.048wt.％时，与点 20 的光谱拟合结果达到最佳（见图 3)。由于点 15 的 SMFe 含量为 0，故点 20 的 SMFe 含量为 0.048wt.％。利用 Morris[5]的公式（4）定义的成熟度为：

$$(I_s/\mathrm{FeO})=f/(3.2*10^{-4}*\mathrm{FeO}) \tag{13}$$

结合该点的 FeO 含量 13wt.％[20]，可以得出该点的成熟度为 11.5，可能属于不成熟月壤。

需要说明的是，该研究中假定岩石矿物的表面 SMFe 含量为 0。而事实上，岩石在风化前表面也可能存在包浆并不为 0，所以反演的 SMFe 含量可能与实际相比偏低。利用 Hapke 模型反演得到的亚微观铁粒径范围为不大于 50nm，而 Morris 模型中采用的核磁共振敏感的亚微观铁范围仅为 4～33nm，这就使得在反演成熟度时，结果会偏高，这在一定程度上缓解了上述假设的偏差。

前人的研究目前主要是月球正面月海的研究，而对月球背面月海高地过渡区的相关参数是否需要调整仍需进一步研究与分析。

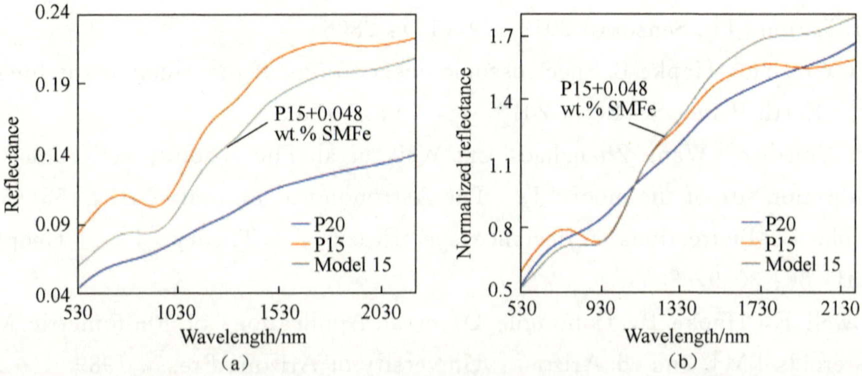

图 3　模拟光谱与所选取光谱的（a）原始数据与（b）在 1100nm 波长处归一化后的对照图

Fig. 3 (a) Original and (b) the normalized diagram at 1100nm wavelength of the modeled spectra and the selected spectra

参考文献

[1] Hapke B. Theory of Reflectance and Emittance Spectroscopy [M] . 2nd ed. Cambridge: Cambridge University Press, 2012.

[2] Fischer E M, Pieters C M. Remote determination of exposure degree and iron concentration of lunar soils using VIS - NIR spectroscopic methods [J] . Icarus, 1994, 111: 475 - 488.

[3] Lucey P G, Taylor G J, Malaret E. Abundance and distribution of iron on the Moon [J] . Science, 1995, 268 (5214): 1150 - 1153.

[4] Hapke B. Space weathering from Mercury to the asteroid belt [J] . J Geophys Res, 2001, 106: 10039 - 10073.

[5] Morris R V. Origins and size distribution of metallic ironparticles in the lunar regolith [C] //Lunar Planet Sci Conf, 1980: 1697 - 1712 .

[6] Liu Dawei, Li Lin, Sun Ying. An improved radiative transfer model for estimating mineral abundance of immature and mature lunar soils [J]. Icarus, 2015, 253: 40 - 50.

[7] Wang Zhenchao, Wu Yunzhao, Blewett D T, et al. Submicroscopic metallic iron inlunar soils estimated from the in situ spectra of the Chang'E - 3 mission [J]. Geophys Res Lett, 2017, 44 (3485): 3492.

[8] Fu Xiaohui, Zou Yongliao, Zheng Yongchun, et al. Space weathering processes and effects on the Moon [J] . Chin J Space Sci, 2011, 31 (6): 705 - 715.

[9] Zhao Zhe, Zhou Ping, Yan Bokun, et al. Spectra simulation of lunar regolith based on the Hapke radiative transfer model. Front [J]. Earth Sci, 2016, 23 (3): 266 - 278.

[10] Li Chunlai, Wang Zhendong, Xu Rui, et al. The scientific information model of

Chang'e - 4 visible and near - IR imaging spectrometer (VNIS) and in - flight verification [J]. Sensors, 2019, 19 (12): 2806.

[11] Wu Yunzhao, Hapke B. Spectroscopic observations of the Moon at the lunar surface [J]. Earth Planet Sci Lett, 2018, 484: 145 - 153.

[12] Wu Yunzhao, Wang Zhenchao, Cai Wei, et al. The absolute reflectance and new calibration site of the moon [J]. The Astronomical Journal, 2018, 155 (5): 213.

[13] Hapke B. Bidirectional reflectance spectroscopy: 1 Theory [J]. J Geophys Res, 1981, 86: 3039 - 3054.

[14] Bowell E, Hapke B, Domingue D, et al. Applications of Photometric Models to Asteroids [M]. 2nd ed. Arizona: University of Arizona Press, 1989.

[15] Mustard J, Pieters C. Photometric phase functions of common geologic minerals and applications to quantitative analysis of mineral mixture reflectance spectra [J]. Geophys Res, 1989, 94 (13): 619 - 634.

[16] Mckay D S, Cooper B L, Riofrio L M. New measurements of the particle size distribution of Apollo 11 lunar soil [C] //Lunar Planet Sci Conf, 2009.

[17] Paquin R A. Handbook of Optics [M]. New York: McGraw - Hill, Inc, 1995.

[18] Houston W N, Mitchell J K, Carrier III W D. Lunar soil density and porosity [C] // Proc Lunar Planet Sci Conf, 1974: 2361 - 2364.

[19] Askeland D R, Phule P P. The Science and Engineering of Materials [M]. 4th ed. USA: Brooks/Cole Publishing/Thompson Learning, 2003.

[20] Wu Yunzhao, Xue Bin, Zhao Baochang, et al. Global estimates of lunar iron and titanium contents from the Chang'E - 1 IIM data [J]. J Geophys Res, 2012, 117: E02001.

基于全景相机数据的嫦娥四号着陆区次级坑统计分析

阳梅萍[1,2]，岳宗玉[1,4]，邸凯昌[1,4]，万文辉[1]，刘建忠[3,4]，史珂[1,2]

1. 中国科学院空天信息创新研究院，遥感科学国家重点实验室，北京 100101
2. 中国科学院大学，北京 100049
3. 中国科学院地球化学研究所，贵阳 550081
4. 中国科学院比较行星学卓越创新中心，合肥 230026

摘要： 玉兔二号月球车在嫦娥四号着陆区发现了许多坑缘呈破碎石块状的小型撞击坑，其直径大多在亚米级，过去对于这种小尺度撞击坑的研究大多源于 Apollo、Surveyor、Ranger 系列任务，然而当时获取的影像质量较低，且并未覆盖月球背面。小尺度撞击坑在月球表面分布十分广泛，与月球最表层的演化密切相关，因此对其展开研究具有重要的意义。本文利用嫦娥四号任务获取的高分辨率全景影像，统计了 8、9、11～15 月昼影像中新鲜撞击坑的直径、频数、深度等信息，并根据这些撞击坑和周围石块的形态特征，推测这些坑应该形成于不同方向的二次撞击事件，这为揭示玉兔二号月球车行进途中观测物质的来源提供了重要信息。小型二次撞击坑从深径比中难以与一次撞击坑区分，且一般处于饱和状态，因此应避免使用小尺寸的撞击坑进行定年。

关键词： 嫦娥四号；全景影像；撞击坑统计

Statistical Analysis of Secondary Craters in the Chang'E - 4 Landing Area Based on Panoramic Camera Data

YANG Mei - ping[1,2]，YUE Zong - yu[1,4]，DI Kai - chang[1,4]，WAN Wen - hui[1]，LIU Jian - zhong[3,4]，SHI Ke[1,2]

1. State Key Laboratory of Remote Sensing Science，Aerospace Information Research Institute，Chinese Academy of Sciences，Beijing 100101，China

2. University of Chinese Academy of Sciences, Beijing 100049, China
3. Institute of Geochemistry, Chinese Academy of Sciences, Guiyang 550081, China
4. CAS Center for Excellence in Comparative Planetology, Hefei 230026, China

Abstract: Many fresh small sized (mostly in diameter of sub - meter) impact craters which have blocky - rim exposed in the regolith were encountered by the Yutu - 2 rover in the Chang'E - 4 landing area. Due to the limitation of low resolution of images, such kind of small sized impact craters were mostly found in images from the Ranger, Surveyor, Apollo missions in the past, nevertheless those images of low quality did not cover the area of lunar farside. Small sized impact craters are widely distributed in the lunar surface and they are closely related to the evolution of lunar regolith. Therefore, it is significant to study these craters in fine scales. In this paper, the very high resolution images acquired by the CE - 4 Panoramic camera (Pancam) in 8th, 9th, 11th - 15th lunar days are used to analyse information including the diameter, frequency, and depth of these fresh impact craters. Combining with the morphological characteristics of these impact craters and boulders, it is inferred that these craters could be mostly formed by secondary impact events in different directions. This provides important information for revealing the source of the material observed during the journey of the Yutu - 2 rover. Small sized secondary impact craters are difficult to be distinguished from primary impact craters in the ratio of depth to diameter, and they are generally in a saturated state. Therefore, small sized craters should be avoided to be used for the sizefrequency dating.
Keywords: Chang'E - 4; panoramic image; craters

0　引言

撞击坑是月球表面最主要的地貌形态，它在月球科学研究中具有重要意义。例如，根据撞击坑的物质成分特征可以推测月球的内部结构信息（Pieters，1982；Pinet et al.，1993；Yamamoto et al.，2010），依据撞击坑的大小-频率分布规律可以分析月表的地质年龄（Hartman et al.，1981；Neukum，1983；Neukum et al.，2001）等。此外，许多研究认为小型撞击坑的大小-频率分布规律与大型撞击坑也有所不同，不能用于撞击坑统计定年（Hartmann，1984；Xiao and Strom，2012；Xiao and Werner，2015）。深径比是描述撞击坑形态的一个重要参数，已经积累了大量的研究，典型的一次坑深径比约 0.2，二次坑约 0.1（Baldwin，1963，1965；Bouška，1968；Pike，1974；Chappelow，2013；Daubar et al.，2014）。小型撞击坑主要形成于月球浅表层，其溅射物是月壤的主要来源（Melosh，1989），极大地影响了月壤的全球分布，因此小规模撞击坑的研究正受到越来越多的重视。

亚米级的撞击坑及其溅射物通常在轨道影像上难以识别，因此与其相关的研究主要来源于 Ranger、Surveyor、Apollo 等系列任务（Trask，1966；Shoemaker et al.，1967；Aldrin et al.，1969；Shoemaker et al.，1969；Swann et al.，1971）。已有的研究表明，百米级以下的撞击坑累积频数－直径分布规律近似于 $N(D)-D^{-2}$［其中 D 表示撞击坑直径，N（D）表示直径≥D 的撞击坑累积个数]，当直径超过数百米时，分布规律发生了变化，这两种分布曲线的交点 C_s 则代表了小型撞击坑稳定状态的直径上限。然而，这些任务获取的影像大多质量较低，且任务区域不覆盖月球背面，因此使用最新的高分辨率对月球背面的小型撞击坑进行统计具有重大意义。

2019 年 1 月 3 日，嫦娥四号成功着陆月球南极艾特肯盆地（SPA）的冯·卡门撞击坑，实现了历史上首次月球背面软着陆。艾特肯盆地是月球上规模最大最古老的盆地，其直径约为 2500km，深约 13km（Shevchenko et al.，2007），根据直径与最大挖掘深度的关系（Melosh，1989），艾特肯盆地理论上已经穿透了上月壳，并且很有可能挖掘出了上月幔的物质（Melosh et al.，2017）。冯·卡门撞击坑直径约 186km，深约 2.4km（Huang et al.，2018），形成于古老的前酒海纪（约 3.97Ga），坑底被多期雨海纪玄武岩覆盖。在月海形成之后，坑底又被附近大型撞击坑溅射物覆盖，玉兔二号探测器光谱数据显示，嫦娥四号着陆区的风化层主要来源于 Finsen 撞击坑溅射物（Gou et al.，2020a，2020b）。整体来说，嫦娥四号着陆区十分平坦，很适合玉兔二号月球车行驶。然而，在探测器行进途中，发现了许多的高反射率的小型撞击坑，其直径在几厘米到几米不等，坑缘上堆积了许多石块，与周围暗色的风化层和高度退化的撞击坑形成了鲜明的对比，暗示这些石块是在近期的地质事件中形成的。由于月表持续外来侵蚀作用以及影像分辨率的限制，小型撞击坑及其厘米级溅射物很难被记录下来。因此这些被玉兔二号搭载的全景相机记录下来的撞击坑和石块，为小型撞击坑的研究提供了宝贵的契机，对于揭示嫦娥四号着陆区表面物质来源有着重要的意义。

为了减小阴影和其他遮挡物对撞击坑识别的影响，本文使用 8、9、11～15 月昼的高分辨率全景相机数据，圈定和识别了玉兔二号月球车路途中遇到的亚米级-米级撞击坑 371 个，统计和分析了它们的尺寸、空间分布特征，并根据这些撞击坑和周围石块的形态特征，推测这些坑应该形成于不同的二次撞击事件，同时还分析了可能的撞击方向，这为揭示玉兔二号月球车行进途中观测物质的来源提供了重要信息。根据本文的统计结果，小型二次撞击坑从深径比中难以与一次撞击坑区分，且一般处于饱和状态，因此应避免使用小尺寸的撞击坑进行定年。

1　数据和方法

1.1　数据

玉兔二号搭载的全景相机可在彩色成像和全色成像两种模式之间切换，彩色模式的影像分辨率是 2352×1728 像素，全色成像模式的影像分辨率是 1176×864 像素，相机视场角为 19.6°×14.5°（贾瑛卓等，2018）。全景相机安装在玉兔二号巡视器的桅杆上，依靠桅杆的上下左右转动，全景相机拍摄站点周围 360°的图像，每转一圈可以拍下 56 幅影像（在两个高度角上不同方位的 28 对影像），从而实现着陆区和巡视区的地形地貌的三维立体成像（代树武等，2014）。玉兔二号全景相机可实现从巡视器脚下到视野范围无穷远处清晰成像，在近距离观测的空间分辨率可以达到毫米量级。截至 2020 年 9 月 24 日，嫦娥四号已完成第 22 月昼工作，为了避免阴影面积过大而遮挡撞击坑，本文使用的数据为 8、

9、11～15 月昼的全景影像。另外文中还使用了空间分辨率为 0.9m 的 LRO NAC 影像作为研究区域的底图（影像编号为 M1331814485LE）。

1.2　方法

本文的总体技术路线如图 1 所示。本文利用目视解译的方法识别全景影像中具有破碎石块状坑缘的新鲜撞击坑，由于不同月昼得到的全景影像存在重叠的区域，为了消除统计中重复的撞击坑，使用全景影像数据生成了研究区域数字正射影像图（DOM），将测得的所有撞击坑投影到 DOM 上，对重叠的撞击坑进行判别，删除重复识别的撞击坑，并贴合

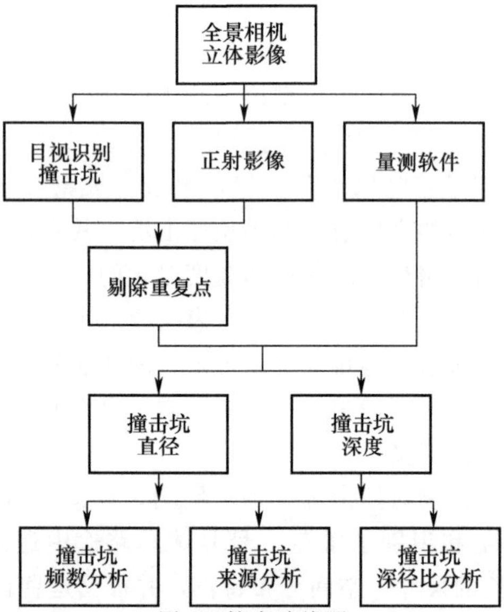

图 1　技术路线图

Fig. 1　The chart of technical procedure

撞击坑的分布范围选取统计区域，面积为 0.004km^2。之后使用月球摄影测量制图软件（Di et al.，2016）对撞击坑进行测量，首先在左像对撞击坑左右两端坑缘上选取点，右像对会自动匹配出同名点，同时根据影像内外方位参数和计算出对应点的三维地面真实坐标，这两点的距离即为撞击坑直径 D（Barnouin et al.，2012）。最小石块的尺寸至少占三个像素，约 0.4cm，而最小能被肉眼识别的撞击坑直径在 2cm 左右。而深度测量仅对影像中露出坑底的撞击坑进行，由于陡坡上的撞击坑深度测量误差较大，不将这类撞击坑纳入深度统计范围以内。为了减小由 DEM 插值带来的二次误差，高度值同样来源于月球摄影测量制图软件中所得到的三维地面坐标，通过对撞击坑中心多个点进行高度的测定，选取最深的点，计算该点与坑缘高度平均值之差即为深度 d。

2　结果

2.1　撞击坑直径－频数关系

本文共测量了 371 个撞击坑的直径，将其叠加在 LRONAC 影像底图上得到它们的空

间分布图（图 2）（由于并未使用前几个月昼的数据，因此撞击坑只集中在八月昼之后）。撞击坑的频数-直径图和累积频数-直径图如图 3 所示。统计结果表明只有 12 个撞击坑直径大于 1m，50%以上的撞击坑直径为 0.1～0.3m。在全景影像中，存在许多微小的边缘

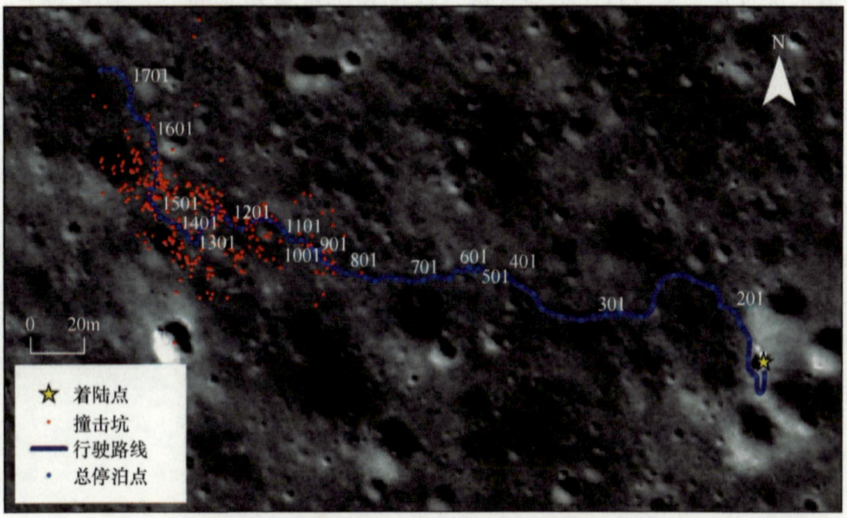

底图为 LRO NAC 正射影像，蓝线为前 17 个月昼玉兔二号的行驶路线，沿途停泊点旁的数字标记以月昼序号开头，代表每个月昼的首个站点

图 2　着陆区测量撞击坑分布图

Fig. 2　The distribution of impact craters in the Chang'E－4 landing area

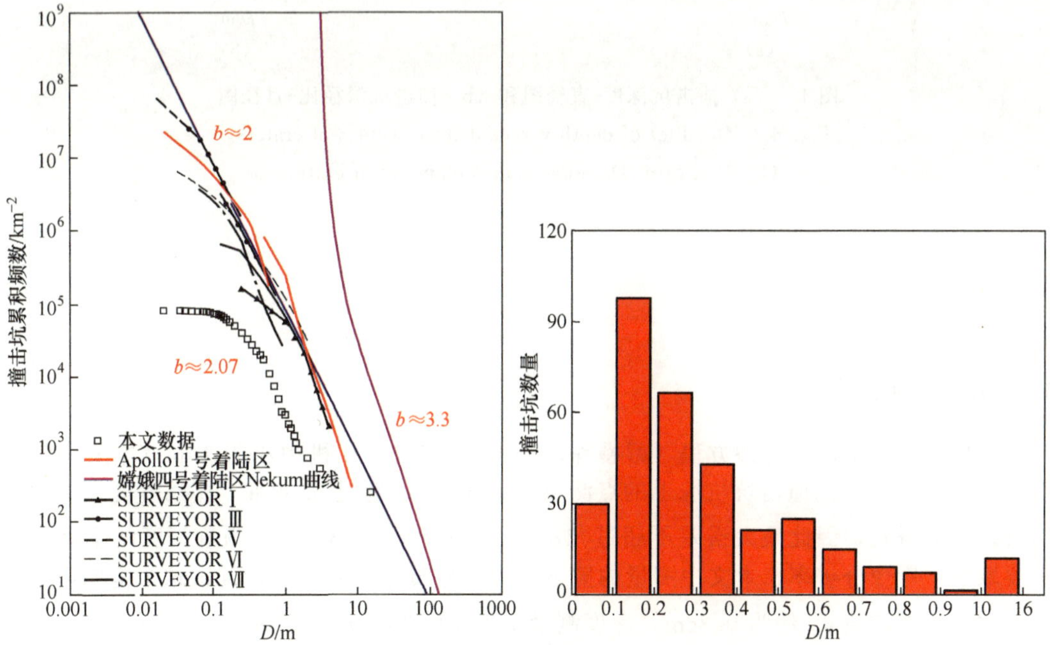

图 3　(a) 撞击坑累计频数-直径图和 (b) 撞击坑频数-直径图

Fig. 3　(a) The diagram of cumulative frequencies versus diameters of impact craters and (b) histogram of frequencies versus diameters of craters

带有细粒石块的撞击坑，但是由于分辨率的限制，无法清晰辨认和测量直径，导致直径小于 0.1m 的撞击坑数量大幅小于直径为 0.1～0.2m 的撞击坑数量。整体来说，撞击坑数量随着直径的增加而减少。撞击坑累积频数和直径的关系近似满足 $N(D)=AD^{-b}$ [其中 D 表示撞击坑直径，$N(D)$ 表示直径≥D 的撞击坑累积个数，A 是系数，b 是指数，对于本文中嫦娥四号着陆区撞击坑，b 约等于 2.07]。

2.2　撞击坑深度-直径关系

本文共测量了 63 个撞击坑的深度，深度与直径的关系如图 4a 所示。从图中可以看出深度和直径呈明显的线性关系。深径比为 0.03～0.18，均值为 0.12，方差约 0.03。超过 78%的撞击坑深径比为 0.10 ～0.16。

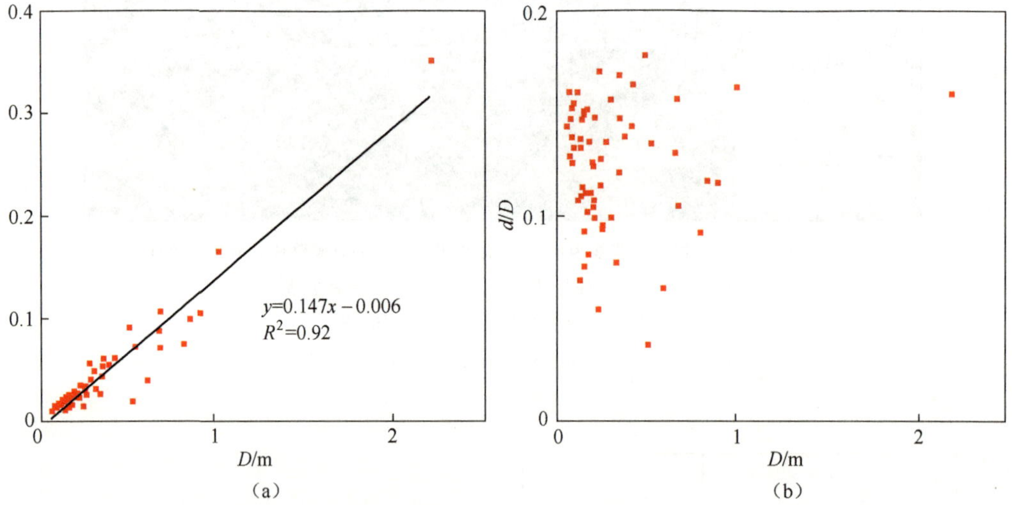

图 4　(a) 撞击坑深度-直径图和 (b) 撞击坑深径比-直径图

Fig. 4　(a) Plot of depth versus diamter values of craters, (b) Plot of d/D values versus diamters of craters

3　讨论

3.1　测量不确定度

根据摄影测量误差传播方程（彭嫚等，2014），在全景相机中小于 15m（本文中最大撞击坑约为 15m）的撞击坑坑缘上的点测量误差 σ_{pmax} 不超过 0.09m。直径由坑缘上两点的直线距离得到，因此最大误差不超过 $\sqrt{2}\sigma_{pmax}$，约等于 0.13m。而测量深度的不确定度主要取决于影像分辨率，本文中测量的撞击坑大多分布在巡视器的近处，这个范围内全景影像的分辨率均值 σ_h 约为 0.3cm，深度测量不确定度为 $\sqrt{2}\sigma_d$，约等于 0.4cm。

3.2　小型撞击坑的二次撞击特征

着陆区整体地形平坦，颜色黯淡，区域内的小型撞击坑侵蚀程度较低、反射率较高，说明这些撞击坑是最近才暴露在月表的，根据其形成速率（Neukum and Ivanov，1994；Neukum et al.，2001），短时间内很难形成如此高空间密度的一次撞击坑，所以这些撞击坑极有可能是二次撞击形成的（Dinget al.，2020），它们的平均深径比（～0.12）也小于新鲜的简单一次坑的深径比（～0.2）（Pike，1974），说明成坑效率较低（Schmidt and Holsapple，1980）。除此之外，这些撞击坑在形态上也表现出与二次坑相同的特征：①撞击坑附近的石块分布不均匀，这种不均匀体现在石块数量不均匀（图 5 和图 8）和大小不均匀（图 6）上，通常来说，下靶区的石块溅射物速度更大，分布范围更远（Schultz and Gault，1985）；②撞击坑沿着撞击方向呈拉长的椭圆形（McEwenand Bierhaus，2006），图 7a 撞击坑西北-东南向长轴约 17.5m，西南-东北向短轴约 13.5m，扁率为 0.23；图 8c 中撞击坑西北-东南向长轴约 0.44m，西南-东北向短轴约 0.36m，扁率为 0.18；③撞击坑坑缘高低起伏很大，存在明显的某侧坑缘抬升或者凹陷（图 7）。

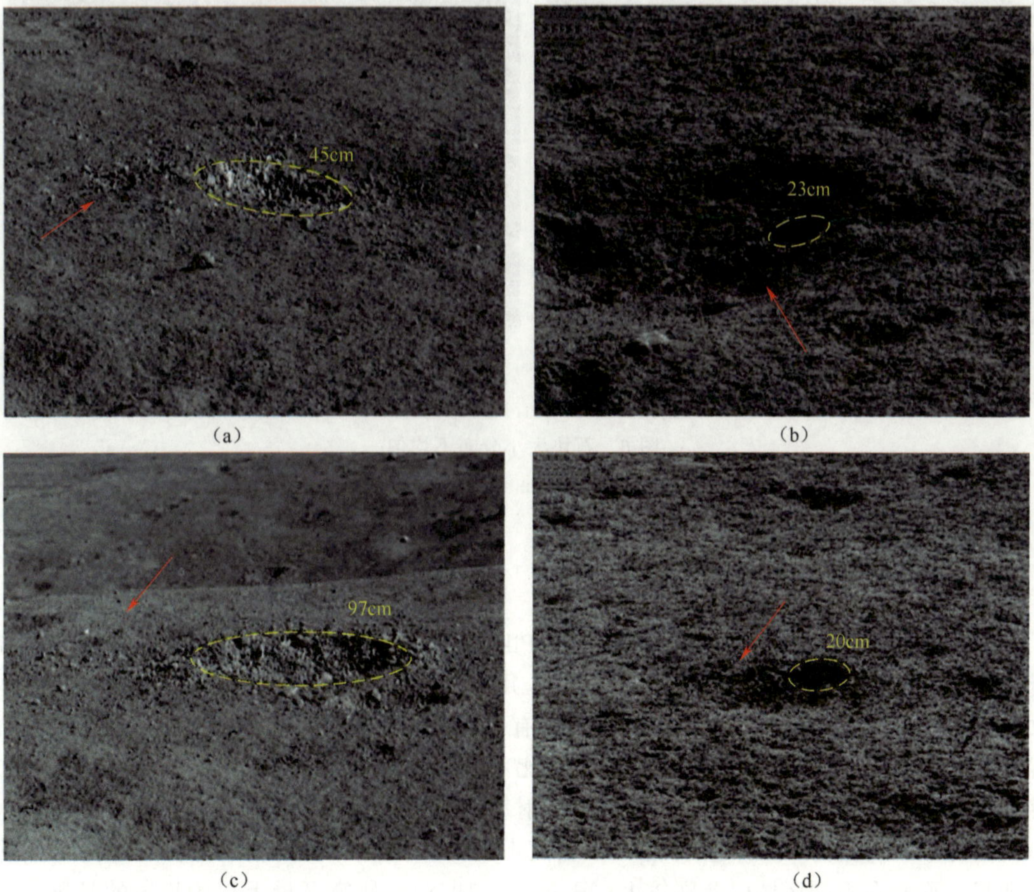

红色箭头所指处石块数量多于其他方向，并且（a）（c）红色箭头所指的石块明显溅射得更远。

图 5　石块数量分布不均匀

Fig. 5　The nonuniformly distributed bonlders around different sized craters

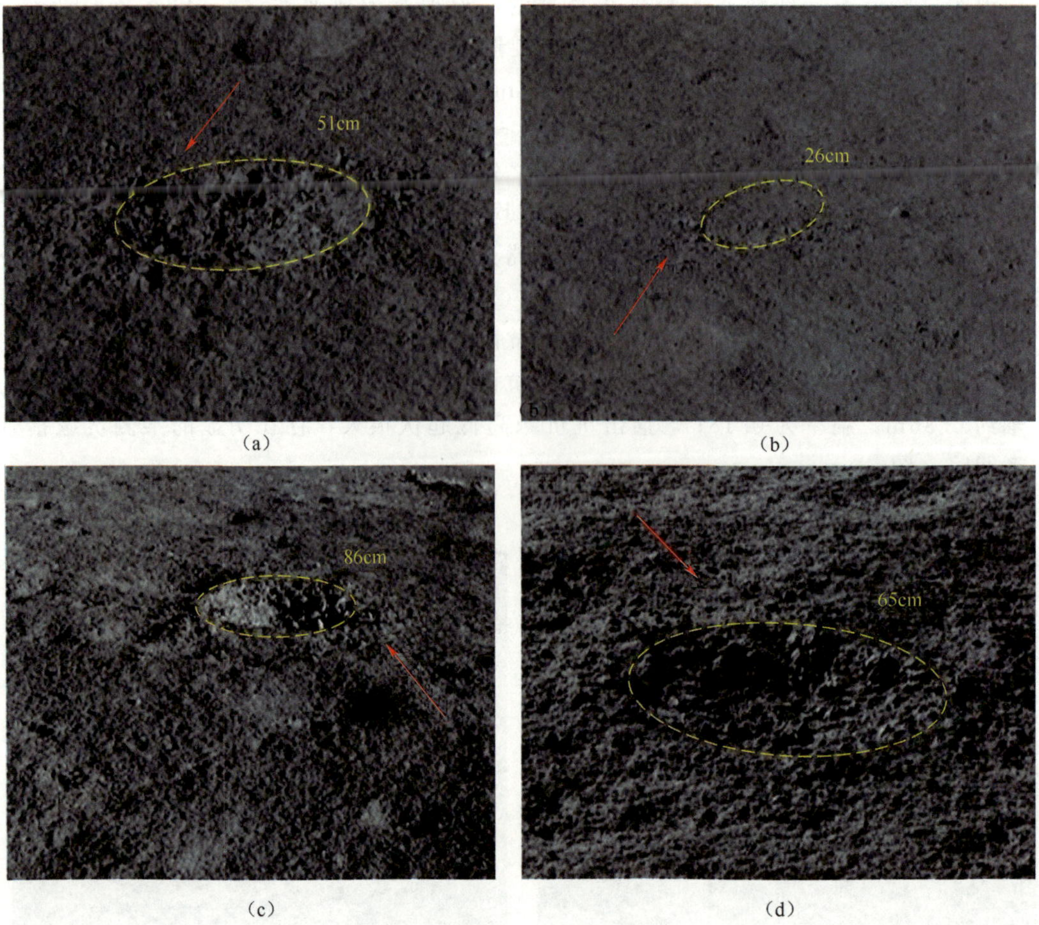

箭头所指处石块直径大于其他方向

图 6　石块大小分布不均匀

Fig. 6　The nonuniformly distributed various sized bonlders around different sized craters

3.3　研究区撞击坑统计特征

月球上典型二次坑的深径比约为 0.1（Pike and Wilhelms，1978），略低于本文的统计结果（0.12），这是由于撞击坑会在风化作用下逐渐变浅（Basi - levsky，1976），而本文选取的撞击坑大多退化程度不高，其中具有锐利的抬升坑缘且堆积了大量石块的撞击坑深径比大多在 0.15 左右，与一次坑的深径比（0.2）相近（Pike，1974）；Jia 等（2021）使用 NAC 数据对嫦娥四号着陆区具有高反射率溅射物且坑缘明显的撞击坑进行了测量，得到的深径比均值为 0.14。实际上，小型撞击坑的深径比在一次坑和二次坑之间差别并不明显，而是在不同区域内差别较大（Bouška，1968）。研究区最上层为松散的月壤，高孔隙度使得撞击坑的直径减小（Wünnemann et al.，2005），从而增大了深径比；新鲜小型一次坑的退化速度很快（Fassett and Combellick，2014），这些原因导致观察到的一次撞击坑和二次撞击坑深径比十分接近，难以辨认，所以还需要结合撞击坑及其溅射物的平面形态特征来识别二次坑。

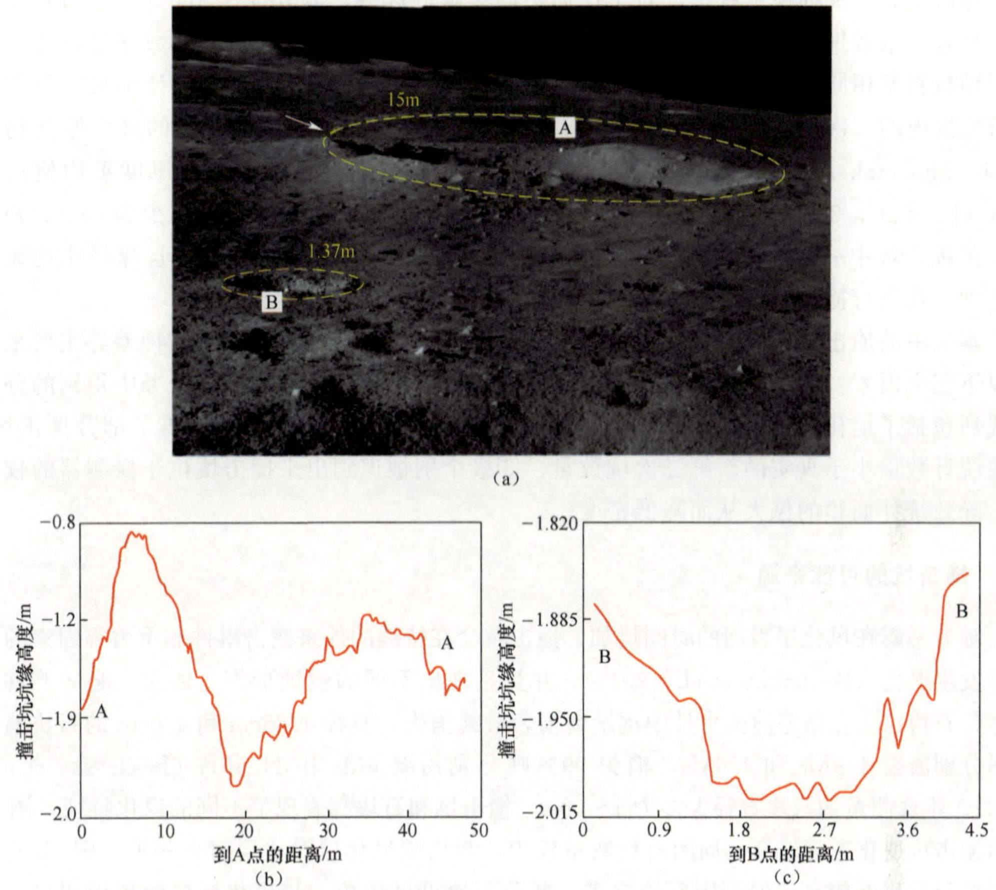

(a) 图白色箭头所指撞击坑西北-东南向长轴约 17.5m，西南-东北向短轴约 13.5m

图 7　(a) 第十三月昼的撞击坑全景相机影像图；(b) 图 (a) 中 A 为起始点的坑缘高程图；(c) 以 (a) 中 B 点为起始点的坑缘高程图

Fig. 7 (a) Pancam image of crater in13th lunar day;

(b) Height variation of the crater rim with startingpoint A in the image (a);

(c) Height variation of the crater rim with starting point B in the image (a)

分析撞击坑的大小-频数分布（SFD）是获取行星表面年龄的重要手段（Neukum and Ivanov，1994），原理是年龄越大的表面，其遭受撞击的次数越多，表面累积的撞击坑密度也越高。直径-累积频数（CSFD）曲线通常是一个幂律函数，服从 $N(D) = AD^{-b}$ 的分布关系。Neukum 等（2001）统计得到全月的 CSFD 分布曲线在直径 100m 处的斜率约为 3.3，随着直径的减小，曲线也变得更陡（图 3a）；通常来说，一次坑的斜率要小于二次坑的斜率（Shoemaker，1965；Wilhelms et al.，1978）。本文得到的研究区域的 b 值约为 2.07，远小于 Neukum 等（2001）得到的撞击坑 CSFD 曲线斜率。而 Shoemaker 等（1969）对 Surveyor 7 任务中 0.13～4m 的撞击坑进行统计得到 $b \approx 2$，十分接近本文的结果；而从 Lunar Orbiter 5 获取的 Surveyor 7 区域 10m 以上的撞击坑中解译得到的 $b \approx 2.93$，Surveyor 其他任务以及 Apollo 11 区域的撞击坑同样也有两种分布规律（Shoemaker et al.，1970）。这两种分布曲线的交点被定义为 C_s，当撞击坑的直径 $\leqslant C_s$

时，撞击坑处于一种稳定状态，对于任何不同年龄的区域，其 $b \approx 2$；Jia 等（2021）使用第一月昼全景数据对嫦娥四号着陆区直径小于 25m 的一次坑进行直径-频数增量统计（与累积频数斜率相同），也得到了相同的斜率（～2）。因此，结合本文的统计结果，我们认为研究区内的二次撞击已经达到了饱和，满足 $N \sim D^{-2}$ 的分布。风化层的最大厚度约为 $C_s/4$（Shoemaker et al.，1970），雷达数据获得的嫦娥四号区域风化层厚度平均值约为 12m（Li et al.，2020），因此嫦娥四号着陆区处于稳定状态坑群的上限至少为 48m，$D \leqslant$ 48m 的撞击坑中一次坑与二次坑的直径-频数曲线分布与表面年龄无关，且深径比均无显著差别，在进行撞击坑定年时应当避免使用小尺寸撞击坑。

本文中的撞击坑密度略小于勘探者号任务的撞击坑密度，我们推测这种差异主要来源于以下三个因素：①本文中仅统计了带有石块的新鲜撞击坑，而勘探者任务中得到的分布曲线还包括了退化的撞击坑；②全景影像中的阴影以及坡度起伏可能掩盖了部分撞击坑，导致统计数量小于真实的新鲜二次坑数量。③极个别被识别出来撞击坑位于探测器的较远处，导致统计面积的偏大从而降低密度。

3.4 撞击坑的可能来源

随着暴露在风化层表面的时间增加，撞击坑会在持续的外来撞击事件和重力等因素的作用下发生退化（Basilevsky et al.，2013），并且呈现出不同的侵蚀形态（Basilevsky，1976）。同样，石块也会在微型撞击事件中逐渐磨损，最终消失。直径 1.85cm 和 4.0cm 的石块消失 50%分别需要 1.9Ma 和 4.6Ma，消失 99%则分别需要 6Ma 和 14.5Ma（Basilevsky et al.，2013）。本文测量的石块直径大多为 1～5cm，撞击坑和石块都表现了不同的退化特征，图 9a 所示撞击坑退化程度较高，周围石块数量较少，与周围风化层对比反射率相近。图 9b 所示撞击坑退化程度较高，但周围石块略多，部分石块可见棱角，与风化层接触界面明显。图 9c、9d 所示撞击坑十分新鲜，且深径比在 0.17 左右，周围分布大量石块，形状不规则，棱角明显，反射率高，和周围灰暗的风化层对比十分明显。根据撞击坑和石块不同的退化特征，我们推测它们可能是不同的撞击事件产生的二次撞击坑。

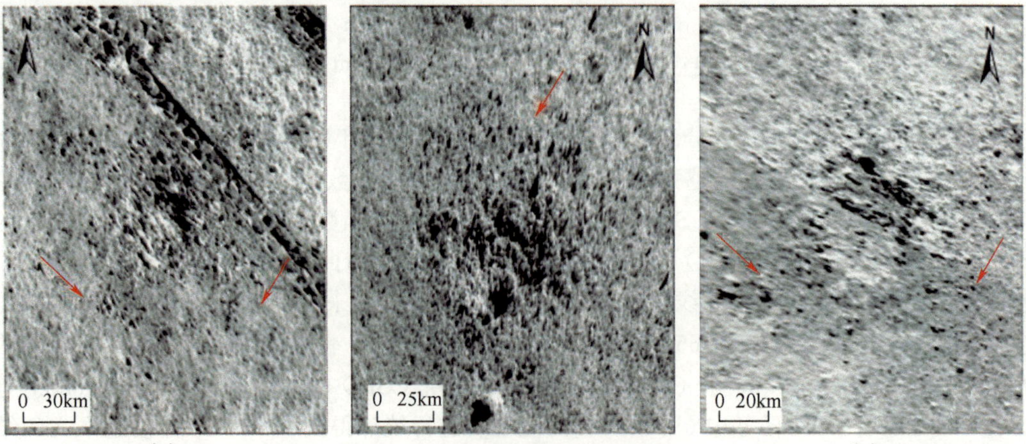

红色箭头所指处溅射范围更广

图 8 二次撞击坑正射影像图

Fig. 8 Digital orthophotos of sencondary impact craters

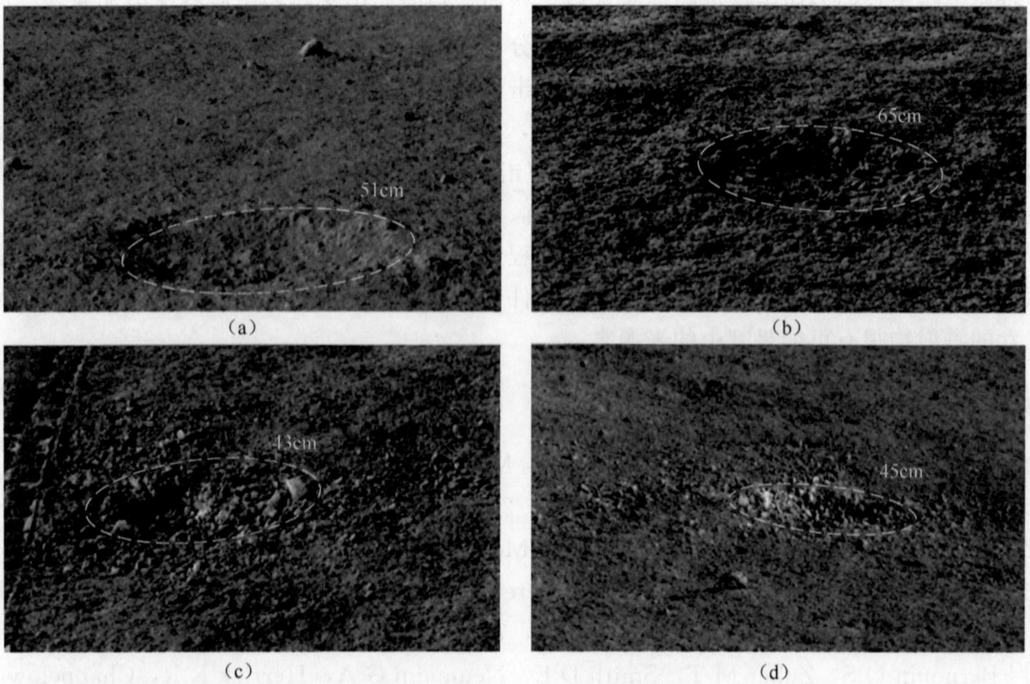

(a) 撞击坑退化程度较高，周围石块数量较少，与周围风化层对比反射率相近；(b) 较高退化程度撞击坑，但周围石块略多，磨圆程度中等，部分石块可见棱角，与风化层接触界面明显；(c)(d) 十分新鲜的撞击坑，周围分布大量石块，形状不规则，棱角明显，反射率高，和周围灰暗的风化层对比十分明显

图 9　不同退化程度的撞击坑

Fig. 9　Images showing impact craters with different degradation degrees

研究区的地质年代十分古老，表面被多次撞击事件的溅射物覆盖（张健等，2011），这使得找到这些二次坑的来源难度较大。图 8a 撞击坑长轴大致为南北向，溅射物向东南方向延伸，应该形成于西北-东南向的二次撞击。图 8b 所示的撞击坑西南侧坑缘只有少量石块，而东北方向有大量石块堆积，形成了一个溅射扇，溅射距离较远，同时可以发现石块的直径随着到撞击坑的距离增大而减小，根据 Krishna 和 Kumar（2016）对 Censorinus 撞击坑周围石块的分布特征统计结果，上靶区的石块经历了更加复杂的变形，撞击方向附近会形成一个溅射物的“封闭区”，在其周围石块的分布范围较小且石块较集中，下靶区周围石块的分布范围较大离散程度高，这是由于下靶区溅射物速度更大而导致的（Mc Ewen and Bierhaus，2006），由此推断，该撞击坑应该形成于原始西南-东北向的二次撞击。Ding 等（2020）发现织女坑的一条溅射线穿过了嫦娥四号着陆区，所以织女溅射物到达嫦娥四号着陆区时，很可能形成了一些二次坑。

4　结论

(1) 大多数的撞击坑直径为 0.1～0.3m，整体上来说，撞击坑的数量随着撞击坑的直径增大而减少。深度和直径呈明显的线性关系，深径比大多数为 0.10～0.16。

(2) 本文统计的撞击坑周围富集石块，反射率高，与周围成熟的暗色月壤形成了鲜明的对比，暗示它们形成于最近的暴露事件。然而，短时间内无法产生研究区内如此高密度的撞击坑，结合撞击坑呈椭圆形、周围石块分布不均匀、某侧坑缘明显抬升或凹陷的形态特征，我们推测它们极有可能形成于二次撞击事件。

（3）数十米的撞击坑一般处于饱和状态，在不同年龄的区域表面其直径-频数曲线基本相同，且从深径比中很难区分在小尺寸撞击坑中的一次坑和二次坑，因此在撞击坑直径-频数定年统计时应尽量避免使用小撞击坑。

（4）撞击坑和石块均表现出不同的退化程度，二次坑特征也指示了不同方向的撞击来源，因此它们应该形成于不同的二次撞击事件。织女坑的一条溅射线穿过了研究区，因此织女的溅射物很有可能是其中的来源之一。

参考文献

[1] Aldrin E E Jr，Armstrong N A，Collins M. 1969. Cre wobservations. In：Apollo 11 Preliminary Science Report. NASA SP－214，35－40.

[2] Baldwin R B. 1963. The Measure of the Moon. University of Chicago Press，Chicago.

[3] Baldwin R B. 1965. The crater diameter － depth relationship from ranger VII photographs. The Astronomical Journal，70（8）：545－547.

[4] Barnouin O S，Zuber M T，Smith D E，Neumann G A，Herrick R R，Chappelow J E，Murchie S L，Prockter L M. 2012. The morphology of craters on Mercury：Results from MESSENGER flybys. Icarus，219（1）：414－427.

[5] Basilevsky A T. 1976. On the evolution rate of small lunar craters. In：Proceedings of the 7th Lunar Science Conference. Houston T X，1005－1020.

[6] Basilevsky A T，Head J W，Horz F. 2013. Survival times of meter－sized boulders on the surface of the Moon. Planetary and Space Science，89：118－126.

[7] Bouška J. 1968. Crater diameter－depth relationship from ranger lunar charts. Bulletin of the Astronomical Institute of Czechoslovakia，19：229.

[8] Chappelow J E. 2013. Simple impact crater shape determination from shadows. Meteoritics and Planetary Science，48（10）：1863－1872.

[9] Daubar I J，Atwood－Stone C，Byrne S，McEwen A S，Russell P S. 2014. The morphology of small fresh craters on Mars and the Moon. Journal of Geophysical Research：Planets，119（12）：2620－2639.

[10] Di K C，Xu B，Peng M，Yue Z Y，Liu Z Q，Wan W H，Li L C，Zhou J L. 2016. Rock size － frequency distribution analysis at the Chang'E － 3 landing site. Planetary and Space Science，120：103－112.

[11] Ding C Y，Xiao Z Y，Wu B，Li Y，Prieur N C，Cai Y Z，Su Y，Cui J. 2020. Fragments delivered by secondary craters at the Chang'E － 4 landing site. Geophysical Research Letters，47（7）：e2020GL087361.

[12] Fassett C I，Combellick J R. 2014. The rate of crater degradation and topographic evolution on the Moon：Results from the Maria and Initial comparisons with the highlands. LPI，（1777）：1429.

[13] Gou S, Di K C, Yue Z Y, Liu Z Q, He Z P, Xu R, Liu B, Peng M, Wan W H, Wang Y X, Liu J Z. 2020a. Forsteritic olivine and magnesium - rich orthopyroxene materials measured by Chang'E - 4 rover. Icarus, 345: 113776.

[14] Gou S, Yue Z Y, Di K C, Wan W H, Liu Z Q, Liu B, Peng M, Wang Y X, He Z P, Xu R. 2020b. In situ spectral measurements of space weathering by Chang'E - 4 rover. Earth and Planetary Science Letters, 535: 116117.

[15] Hartmann W K, Strom R G, Weidenschilling S, Blasius K. R, Voronow A, Dence M R, Grieve R A F, Diaz J, Chapman C R, Shoemaker E M, Jones K L. 1981. Chronology of planetary volcanism by comparative studies of planetary craters. In: Basaltic Volcanism on the Terrestrial Planets. Elmsford N Y: Pergamon Press, 1050 - 1127.

[16] Hartmann W K. 1984. Does crater "saturation equilibrium" occur in the solar system? Icarus, 60 (1): 56 - 74.

[17] Huang J, Xiao Z Y, Flahaut J, Martinot M, Head J, Xiao X, Xie M G, Xiao L. 2018. Geological characteristics of Von Kármán crater, northwestern south pole - Aitken Basin: Chang'E - 4 landing site region. Journal of Geophysical Research: Planets, 123 (7): 1684 - 1700.

[18] Jia M N, Di K C, Yue Z Y, Liu B, Wan W H, Niu S L, Liu J Z, Cheng W M, Lin Y T. 2021. Multi - scale morphologic investigation of craters in the Chang'E - 4 landing area. Icarus, 355: 114164.

[19] Krishna N, Kumar P S. 2016. Impact spallation processes on the Moon: A case study from the size and shape analysis of ejecta boulders and secondary craters of Censorinus crater. Icarus, 264: 274 - 299.

[20] Li C L, Su Y, Pettinelli E, Xing S G, Ding C Y, Liu J J, Ren X, Lauro S E, Soldovieri F, Zeng X G, Gao X Y, Chen W L, Dai S, Liu D W, Zhang G L, Zuo W, Wen W B, Zhang Z B, Zhang X X, Zhang H B. 2020. The Moon's farside shallow subsurface structure unveiled by Chang ' E - 4 Lunar Penetrating Radar. Science Advances, 6 (9): eaay6898.

[21] McEwen A S, Bierhaus E B. 2006. The importance of secondary cratering to age constraints on planetary surfaces. Annual Review of Earth and Planetary Sciences, 34 (1): 535 - 567.

[22] Melosh H J. 1989. Impact cratering: A geologic process. New York: Oxford University Press

[23] Melosh H J, Kendall J, Horgan B, Johnson B C, Bowling T, Lucey P G, Taylor G J. 2017. South Pole - Aitken basin ejecta reveal the Moon ' s upper mantle. Geology, 45 (12): 1063 - 1066.

[24] Neukum G. 1983. Meteoritenbombardement Und Datierung Planetarer Oberflachen. Habilitation Thesis for Faculty Membership. Germany: University of Munich.

[25] Neukum G, Ivanov B A. 1994. Crater size distributions and impact probabilities on

Earth from lunar, terrestrial - planet, and asteroid cratering data. In: Hazards due to Comets and Asteroids. Tucson A Z: University of Arizona Press, 359 - 416.

[26] Neukum G, Ivanov B A, Hartmann W K. 2001. Cratering records in the inner solar system in relation to the lunar reference system. Space Science Reviews, 96 (1): 55 - 86.

[27] Pieters C M. 1982. Copernicus crater central peak: Lunar mountain of unique composition. Science, 215 (4528): 59 - 61.

[28] Pike R J. 1974. Depth /diameter relations of fresh lunar craters: Revision from spacecraft data. Geophysical Research Letters, 1 (7): 291 - 294.

[29] Pike R J, Wilhelms D E. 1978. Secondary - impact craters on the Moon: Topographic form and geologic process. Lunar and Planetary Science IX, 9: 907 -909.

[30] Pinet P C, Chevrel S D, Martin P. 1993. Copernicus: A regional probe of the lunar interior. Science, 260 (5109): 797 - 801.

[31] Schmidt R M, Holsapple K A. 1980. Theory and experiments on cenrifuge cratering. Journal of Geophysical Research: Solid Earth, 85 (B1): 235 - 252.

[32] Schultz P H, Gault D E. 1985. Clustered impacts: Experiments and implications. Journal of Geophysical Research: Solid Earth, 90 (B5): 3701 - 3732.

[33] Shevchenko V V, Chikmachev V I, Pugacheva S G. 2007. Structure of the South Pole - Aitken lunar basin. Solar System Research, 41 (6): 447 - 462.

[34] Shoemaker E M. 1965. Preliminary analysis of the fine structure of the lunar surface in Mare Cognitum. In: The Nature of the Lunar Surface. Baltimore, MD USA: Johns Hopkins Press, 23.

[35] Shoemaker E M, Batson R M, Holt H E, Morris E C, Rennilson J J, Whitaker E A. 1967. Surveyor V: Television pictures. Science, 158 (3801): 642 - 652.

[36] Shoemaker E M, Batson R M, Holt H E, Morris E C, Rennilson J J, Whitaker E A. 1969. Observations of the lunar regolith and the Earth from the television camera on Surveyor 7. Journal of Geophysical Research, 74 (25): 6081 - 6119.

[37] Shoemaker E M, Hait M H, Swann G A, Schleicher D L, Dahlem D H, Schaber G G, Sutton R L. 1970. Lunar regolith at Tranquillity Base. Science, 167 (3918): 452 - 455.

[38] Swann G A, Bailey N G, Batson R M, Eggleton R E, Hait M H, Holt H E, Larson K B, McEwen M C, Mitchell E D, Schaber G G, SchaferJ B, Shepard A B, Sutton R L, Trask N J, Ulrich G E, Wilshire H G, Wolfe E W. 1971. Preliminary geologic investigations of the Apollo 14 landing site. In: Apollo 14. Washington, DC: NASA, 39 - 86.

[39] Trask N J. 1966. Size and spatial distribution of craters estimated from the Ranger photographs. In: Ranger VIII and IX, part II, Experimenters' analyses and interpretations. Pasadena, California: Jet Propulsion Laboratory, 252 - 264.

[40] Wilhelms D E, Oberbeck V R, Aggarwal H R. 1978. Size - frequency distributions

of primary and secondary lunar impact craters. In：Proceedings of the 9th Lunar and Planetary Science Conference. New York：Pergamon Press，3735－3762.

[41] Wünnemann K，Collins G S，Melosh H J. 2006. A strain－based porosity model for use in hydrocode simulations of impacts and implications for transient crater growth in porous targets. Icarus，180（2）：514－527.

[42] Xiao Z Y，Strom R G. 2012. Problems determining relative and absolute ages using the small crater population. Icarus，220（1）：254－267.

[43] Xiao Z Y，Werner S C. 2015. Size－frequency distribution of crater populations in equilibrium on the Moon. Journal of Geophysical Research：Planets，120（12）：2277－2292.

[44] Yamamoto S，Nakamura R，Matsunaga T，Ogawa Y，Ishihara Y，Morota T，Hirata N，Ohtake M，Hiroi T，Yokota Y，Haruyama J. 2010. Possible mantle origin of olivine around lunar impact basins detected by SELENE. Nature Geoscience，3（8）：533－536.

[45] 代树武，吴季，孙辉先，张宝明，杨建峰，方广有，王建宇，王焕玉，安军社. 2014. 嫦娥三号巡视器有效载荷. 空间科学学报，34（3）：332－340.

[46] 贾瑛卓，邹永廖，薛长斌，平劲松，严俊，宁远明. 2018. 嫦娥四号任务科学目标和有效载荷配置. 空间科学学报，38（1）：118－130.

[47] 彭嫚，万文辉，吴凯，刘召芹，李力，邸凯昌，李立春，苗毅，詹磊. 2014. 嫦娥三号导航相机测图能力分析及地形重建. 遥感学报，18（5）：995－1002.

[48] 张健，缪秉魁，廖庆园，荆元，王辉. 2011. 月球南极艾特肯盆地的地质特征：探索月球深部的窗口. 矿物岩石地球化学通报，30（2）：234－240，244.

中国探月工程科学成果（第二辑）
航天科技图书出版基金资助出版

嫦娥四号科学研究创新成果
（中册）

刘继忠　严　俊　邹永廖　唐玉华　等　编著

中国宇航出版社
·北京·

目　　录

---

① 为了保持与原刊载媒体一致，本书沿用了每篇论文发表时对该词的表达方式，全书未做统一，far－side 等词同理。

第 4 章　形貌构造与地质演化研究

嫦娥四号探测器着陆在月球背面南极艾特肯盆地的冯·卡门撞击坑底部。冯·卡门撞击坑形成于前酒海纪，坑底受后期溅射物和玄武岩填充；着陆区表面地貌整体平坦，主要受月海形成后的撞击事件改造；各类大小、形态和保存状态各异的撞击坑记录了丰富的撞击历史；着陆区的表面石块丰度和撞击坑的退化状态指示了成熟的月壤覆盖，局部密集分布石块与新鲜的小撞击坑紧密相关；玉兔二号携带的测月雷达数据揭示了清晰的次表层结构，显示了多层厚度不均一的层状结构，这些结构与区域的撞击和火山活动历史紧密相关。

4.1　地形地貌特征

4.1.1　整体地貌特征

地形粗糙度是描述一定水平尺度（称为基线）上地形变化的地貌参数。月球和行星表面的地形粗糙度受不同尺度的地质过程的控制，是认识表面地质过程和演化历史的重要信息。地形粗糙度也是与光学反射率、热发射率和雷达反射率等物理性质有关的主要参数，在遥感探测分析中有重要意义。基于高分辨率、高精度月表 DEM，在百米基线上，冯·卡门撞击坑的坑底非常平坦，但在一些较年轻的撞击坑内部和周围，局部坡度可达约 10°；着陆区附近约 10km 范围内的双向坡度的中值为 2°，着陆点所在像元为 0.5°。在十米基线上，着陆点附近双向坡度有较大的变化，一些撞击坑内部和周围的最大坡度可大于 20°，区域中值坡度为 2.6°，着陆点所在像元的坡度为 3.7°。在米级和几十米级基线上，着陆区的双向坡度和月海相近（2.0°～3.4°），但显著低于该基线上的高地（7.5°～8.2°）。上述表明嫦娥四号着陆区的地形特征可能主要受早期填充的月海控制，局部地形受较年轻的撞击作用影响。

使用全景相机数据重建获得的毫米分辨率的 DEM，详细地研究了玉兔二号月球车的巡视区域的地貌与地形粗糙度特征，得到了月表微米到千米尺度的双向坡度。结果表明在 7mm 基线上，巡视路线的双向坡度的中位数约为 10°，高于米级到十米级尺度的 2°～3°。双向坡度从微米尺度的 45°～60°降低到千米尺度的 1°～4°。该结果填补了前人数据的空白，结果也揭示在小尺度和大尺度上，着陆区内不同地质单元的地形粗糙度都存在差异。嫦娥四号探测区域 125mm 窗口的均方根高度在 2mm 到大于 15mm 之间变化，中值约为 4mm。由 Fraunhofer 判据分析表明，月表对 LRO Mini - RF 雷达电磁波（波长 12.6cm）

是光滑的，因此回波信号主要以体散射为主。月表粗糙度与月表地质背景和地质单元的演化有密切关系。月表毫米尺度的粗糙度主要受小型撞击坑和撞击碎屑、石块，以及月壤性质控制。该结果首次比较了月表微米至千米量级的粗糙度，研究了大尺度范围粗糙度的变化特征和影响粗糙度的关键参数（图 4－1）。

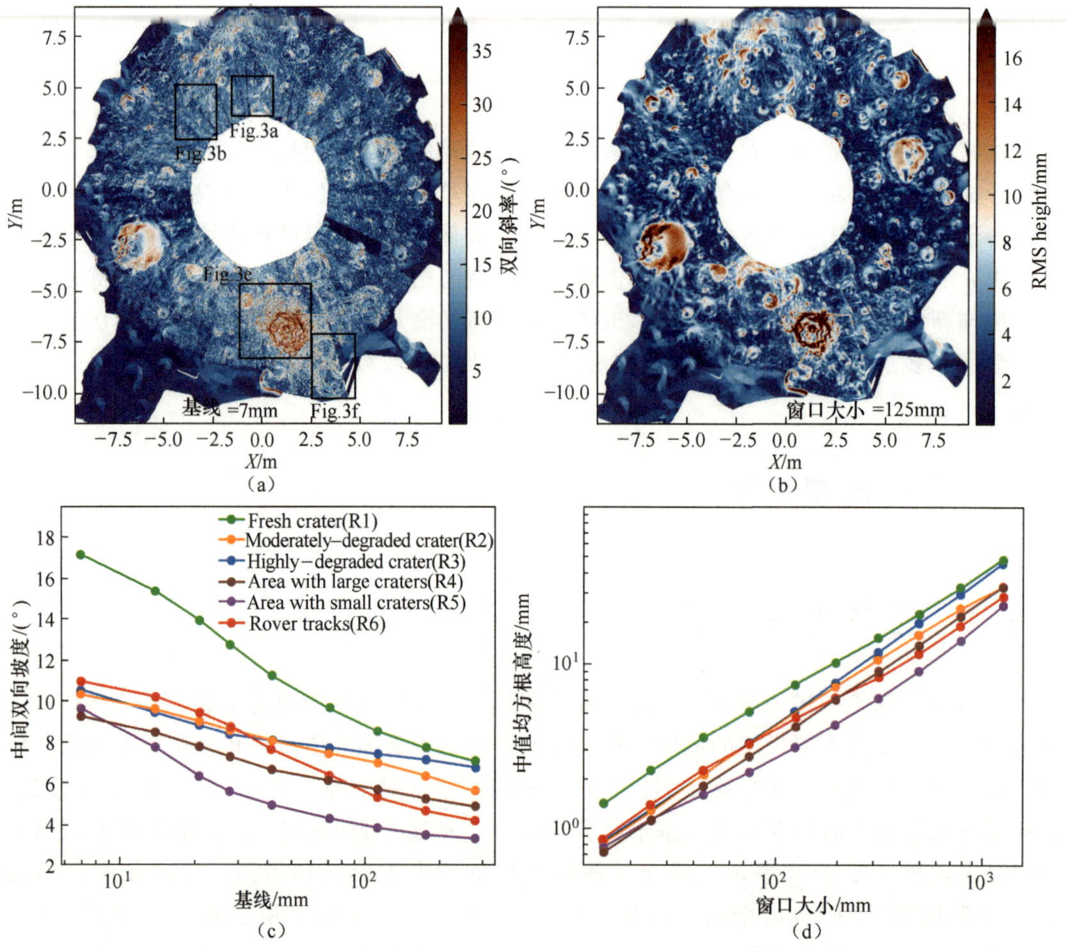

图 4－1　“玉兔二号”巡视路线上的微尺度地形粗糙度特征

4.1.2　撞击溅射物的地貌特征

冯·卡门坑内被玄武岩所充填，整体上地形平坦，然而，全景立体影像上不仅发现这里有大量的小撞击坑，地形趋势也崎岖不平。利用玉兔二号在多个停泊点拍摄的全景立体影像制作了 5cm 分辨率的 DEM（覆盖范围 380m×390m），发现着陆区的整体地形东北高、西南低，在东南至西北方向有波浪状的起伏，进一步解译判断，这些地形起伏不是由小的撞击坑引起的，而可能是堆积在平坦玄武岩基底之上的至少两层来自周围撞击坑的溅射物。

结合已有的中低分辨率 DEM，进一步识别和量测着陆区的线性溅射物地形特征

(图 4-2)，发现着陆区所覆盖的溅射物厚达 60～70m（溅射物表面高程相对于冯·卡门撞击坑底部的玄武岩基底高程的差值），最上层的东北-西南向分布的线性溅射物来自芬森坑，叠加在东南-西北向的艾德勒坑的溅射物上。撞击过程数值模拟表明，来自芬森坑和艾德勒坑的溅射物在着陆点的厚度分别为约 30m 和 35m。溅射物厚度累计值与着陆点和冯·卡门坑底玄武岩的高差一致，验证了在玄武岩基底之上叠加了这两层溅射物解译的合理性。芬森坑和艾德勒坑是在 SPA 形成之后撞击挖掘形成的，它们内部并没有后期玄武岩的填充，因此这两个坑的内部物质及其溅射物实际上来自南极艾特肯盆地底部，是月球深部物质。

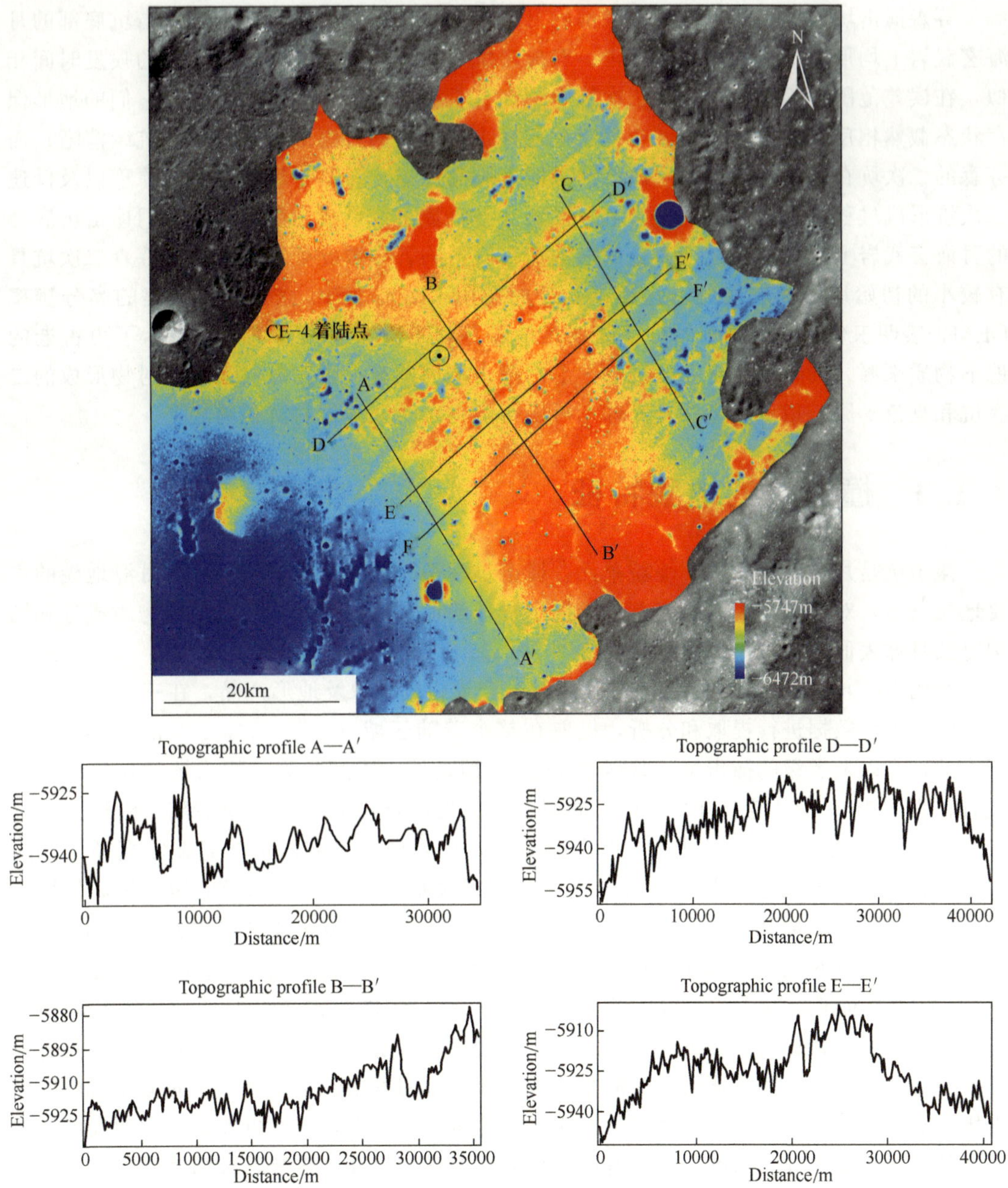

图 4-2　嫦娥四号着陆区的撞击溅射物的地貌特征

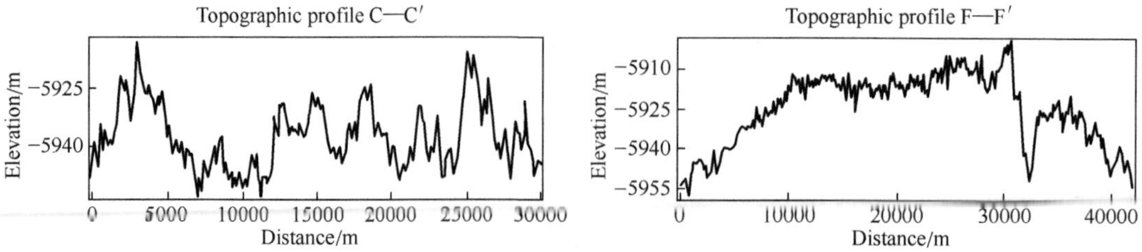

图 4－2　嫦娥四号着陆区的撞击溅射物的地貌特征（续）

芬森撞击坑在冯·卡门撞击坑底部的月海玄武岩和其北侧的莱布尼茨撞击坑底部的月海玄武岩上均形成了大量的二次坑串。对撞击坑统计发现这两处月海玄武岩的喷发时间相似，在误差范围内相当，这两处二次坑串距离芬森撞击坑的距离相当，但是它们的地形保存状态截然相反。芬森的二次坑在莱布尼茨撞击坑底部的月海上保存完好，坑缘清晰；而芬森的二次坑在冯·卡门撞击坑底部的月海上则高度退化。基于地形退化的原理以及低速二次坑形成过程中受靶体物性的影响，发现在芬森撞击坑形成之前，冯·卡门撞击坑底部的月海玄武岩上可能已填充了大量的松散撞击溅射物，导致此处后续叠加的芬森二次坑具有较小的初始地形和较快的退化速率。进而使用 LROC NAC 立体相对重建的高分辨率 DEM，表明玉兔二号月球车的行径路线是沿着芬森二次坑串内的坑缘，指示了更古老的地下物质来源。同时，有研究发现冯·卡门撞击坑周围有大量东海盆地的溅射物形成的二次坑和亮色平原，指示玉兔二号月球车可能行走在东海盆地溅射沉积物之上。

4.1.3　撞击坑的地貌特征

撞击坑是月球表面的主要地貌单元，撞击过程是改造嫦娥四号着陆区的月海地貌的主要地质营力，对撞击地貌的研究有助于揭示月球的地质历史、月球浅表层的物质成分和结构以及月球表面改造作用等地质特性（图 4－3）。

利用 LROC NAC 正射影像图、玉兔二号全景相机影像及地形产品，在三个尺度上对着陆区撞击坑数据进行提取和分析，提取在整个着陆区域（2.4°×1°，51km×30km）范围内的百米以上量级的撞击坑，共得到 4333 个撞击坑，求解得到该撞击坑数据集的完整性直径为 127m，其中直径 127m 以上的撞击坑的数量为 2298 个。对于着陆点附近 3.2km×3.0km 的区域，共提取得到 11787 个撞击坑，计算得到该数据集的完整性直径为 9m，大于该直径的撞击坑数量为 5108 个。对于更小尺度如米级甚至亚米级的撞击坑，使用玉兔二号全景相机影像及其地形产品来进行研究，在距相机 60m 范围以内共提取得到 738 个撞击坑。通过对撞击坑尺寸-频率分布图的分析，可以推断出 2.4°×1°区域内直径小于 250m 的撞击坑和 3.2km×3.0km 范围内直径小于 90m 的撞击坑已经达到了饱和均衡状态。对撞击坑深径比的分析表明，以上三个直径段的撞击坑的深度和直径间的关系具有较好的一致性。但是，其深度直径比的分布很分散，说明其中包含了处于不同退化阶段的撞击坑。总体来看，撞击坑的尺寸越大，其深径比也相对更大，这很可能是由月面表层和次表层的物质特性差异导致的。与已有的研究对比发现，研究区内缺乏大深径比（≥0.18）的撞击坑，这可以部分归因于该地区古老的地质年龄，以及在漫长岁月中不断遭受的后期撞

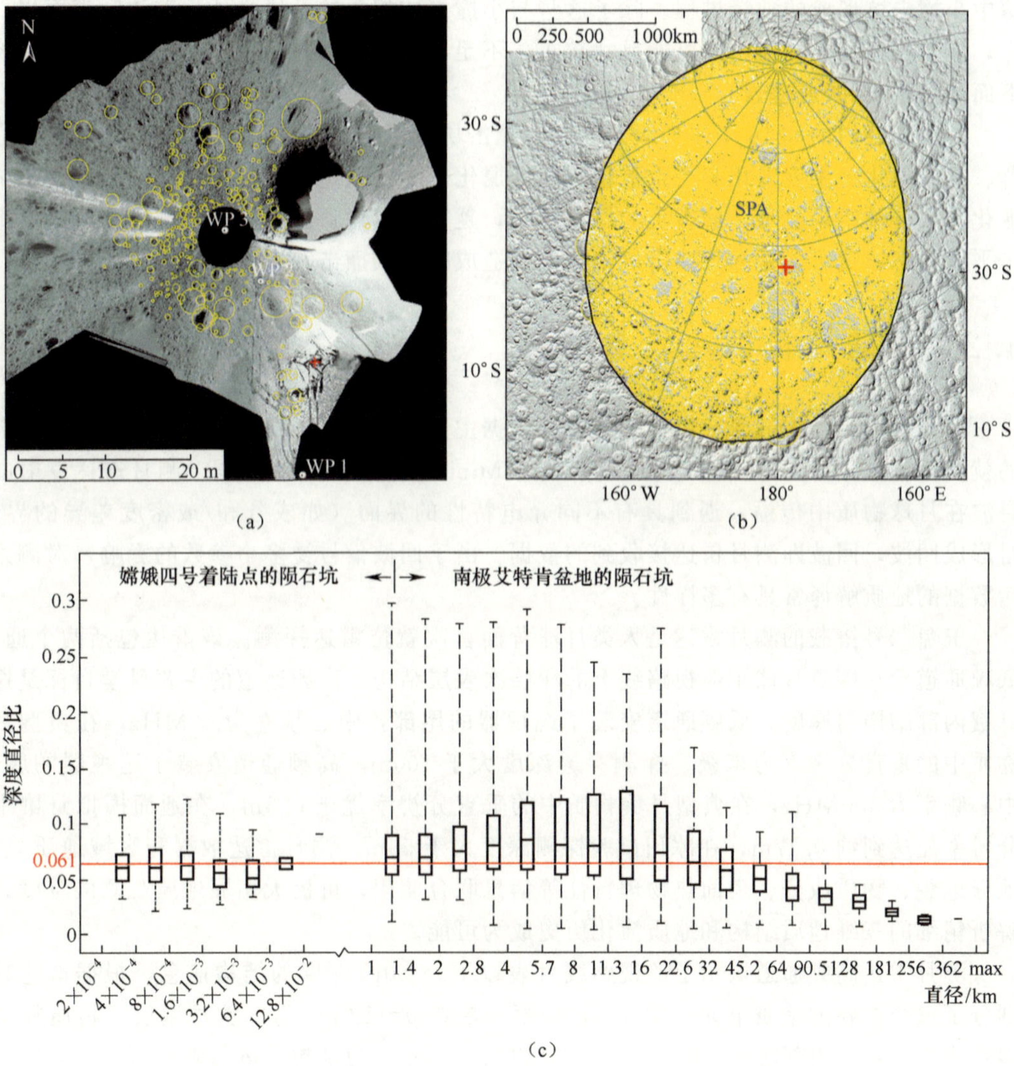

图 4-3　嫦娥四号着陆点周围的撞击坑及其与 SPA 盆地撞击坑的地貌对比

击导致的撞击坑退化加速。

利用全景相机立体影像制作了着陆点附近 1cm 分辨率的 DEM 和正射影像图（覆盖范围约 50m×50m），并识别出 1023 个直径大于 10cm 的撞击坑。经统计分析发现，这些小撞击坑的大小频率分布与 100m 以上的撞击坑一样呈现类似的幂律模式。另外统计了 310 个小撞击坑的深径比，其范围为 0.02～0.12，均值为 0.061，明显小于千米级撞击坑的深径比。

沿着玉兔二号的行径路线还散布着一些新鲜的撞击坑，直径从<1m 到 3m 不等。这些撞击坑的坑缘和坑底散布着大量棱角分明的土块，它们的长轴大多小于 5cm，混杂堆积。在这些撞击坑底部还可见一些反照率较高的胶状物质，使用玉兔二号获取的光谱数据推测这些物质可能是撞击熔融物。通过使用全景相机立体影像制作的高分辨率 DEM 和正射影像图像，统计发现这些石块的大小-频率分布与其他着陆点略有不同，指示了成熟月

壤中小撞击坑形成的独特过程。除了这些与小撞击坑相关的石块之外，着陆区表面岩石稀少，表明先前存在的撞击坑可能尺寸较小，不足以穿透较厚的风化层（约 12m），并挖掘下面的角砾岩和/或基岩。

对着陆区内的 6316 个厘米级到米级的撞击坑进行研究，它们的形态在影像上分为新鲜、轻微退化、中度退化和严重退化等四个退化等级，从定性和定量的角度验证撞击坑的退化等级分类结果显示，撞击坑的尺寸越小，越容易退化；撞击坑尺寸越小，越难达到撞击平衡状态，这可能是邻近撞击坑的喷出物造成的二次撞击和表面重铺的结果。

4.2 着陆巡视区次表层结构

对于次表层结构的研究，可用的观测数据主要是玉兔二号搭载的双通道测月雷达获取的就位探测数据，以及 LRO 轨道器搭载的 Mini－RF 雷达探测数据。测月雷达发射的电磁波在月球物质中传播，遇到具有不同介电特性的界面（如成分和/或密度差异的界面）时形成回波，回波即测月雷达接收到的数据。由于回波信号受多个参数的影响，对测月雷达数据的地质解译常具有多样性。

玉兔二号搭载的测月雷达是人类月球背面首次就位雷达探测。该雷达包括两个通道，低频通道旨在探测月球车巡视路线上的深层次表层结构，高频通道的主要科学目标是探测月壤内部结构与厚度。低频通道安装于巡视器的尾部，中心频率为 60MHz，在典型月壤物质中的垂直分辨率为米级，探测穿透深度大于 100m。高频通道安装于巡视器的底部，中心频率为 500MHz，在典型月壤物质中的垂直分辨率优于 0.3m，在地面模拟月壤中的分辨率能达到约 0.17m，在冰川试验探测深度大于 50m。测月雷达数据和区域地质概况、地形地貌、物质成分、表面热物理特性等信息联合使用，可极大拓展地质观测的维度，使解析精细的三维地质结构和地质演化历史成为可能。

测月雷达高频通道揭示了巡视路线月表以下＞ 40m 深度的结构信息，根据雷达数据划分了三个主要的地质单元（图 4－4）。第一层约 0～12m，为细粒月壤层，物质粒径大多数小于 1cm，内部的石块较少；第二层约 12～20m，为细粒月壤与破碎石块混合的溅射物层，其中夹杂着大量的石块；第三层约 20～45m，由不连续互层的、回波信号差异较大的物质构成，表现为粗粒与细粒物质的交替叠置。高频测月雷达数据揭示的三层地下结构得到多个团队研究的交叉检验，但是，每层物质的上、下界线及其物源解译存在较大的差异，可能的原因是不同团队对着陆区的地质背景的细节理解的角度不同，因此在解译上存在差异。例如，高频测月雷达数据是否已有效探测到玄武岩基岩的位置还不确定；着陆区覆盖的溅射物厚度的理论推测值和基于高频测月雷达数据解译的溅射物厚度还存在不吻合；雷达数据揭示的第二层和第三层物质内一些弧形界面被解释为埋藏的撞击坑或非水平的分层界线，且每层溅射物的来源和沉积方式上还存在争议等。在随后的研究中，高频雷达剖面显示，在导航地点 4－11 和 16－31 下方有两个不同直径的埋藏撞击坑。总体而言，高频测月雷达数据揭示的精细地下结构，充分反映了较年轻的月球地质单元上的复杂地质历史，其地质多解性说明了未来开展月球钻取采样的必要性。

测月雷达低频通道穿透至月表下＞400m 的深度，雷达回波信号揭示了至少三层较深（＞40m）的次表层结构（图 4－5）。在具体层位划分和物源解译方面，根据不同的数据分

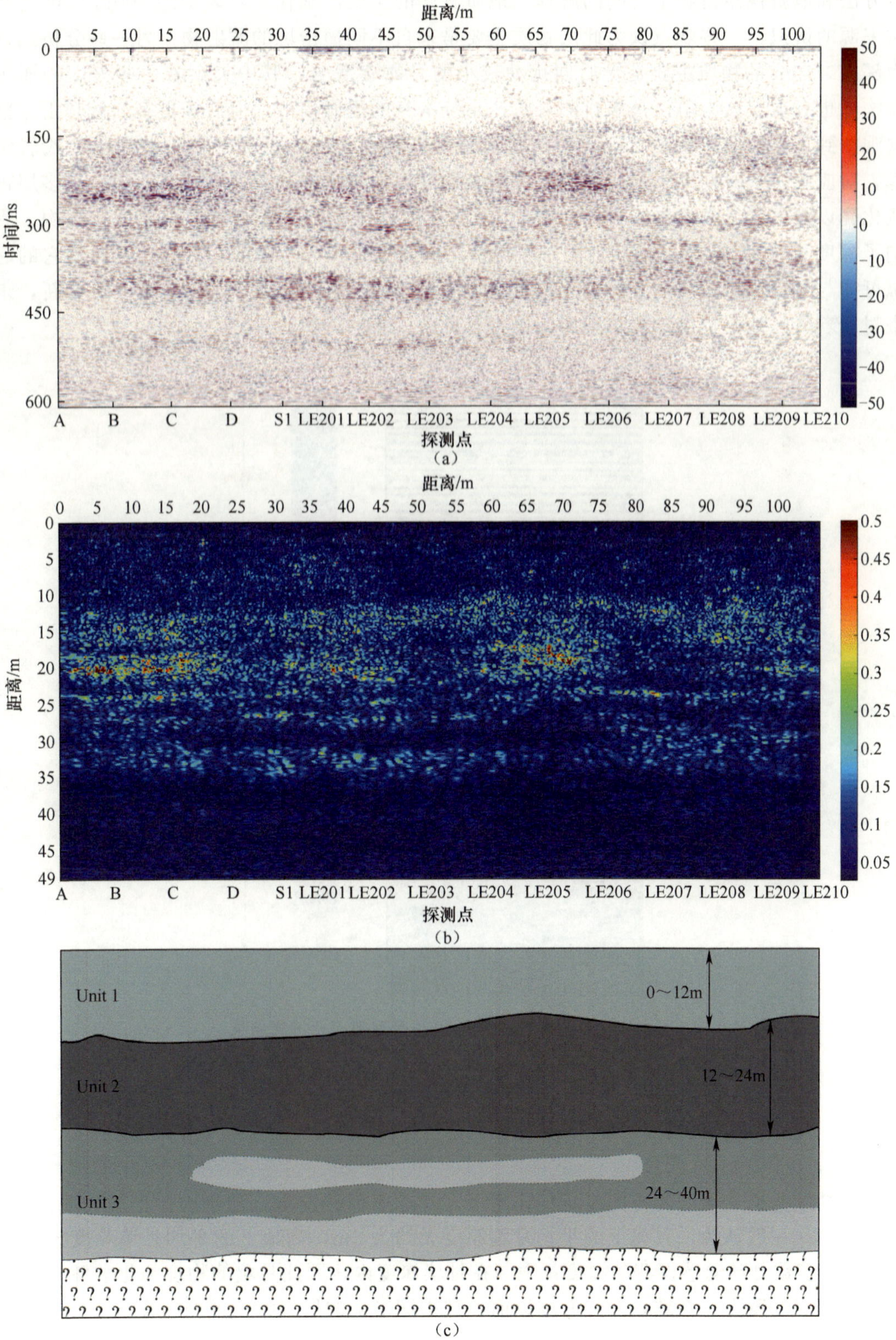

图 4-4　玉兔二号高频测月雷达获取的浅表层结构信息

析方法和地质视角得到了不同的解释，地质解译的主要区别在于对表层约 100m 以内的物质来源的认识，一些观点认为此层物质主要是来自不同撞击坑的溅射物；另一些分析则认为深度＞40m 的物质主要是受后期扰动较小的月海玄武岩，其中夹杂着一些火山碎屑沉积物；也有观点认为争议是来自测月雷达系统本身的虚假信号，但后续更多的低频雷达数据观测到了界面的起伏变化，证明这些信号是真实存在的。更深的物质由多层玄武岩组成，对应雨海纪时期冯·卡门撞击坑内的多期月海岩浆活动。玄武岩层之间夹杂了多层古风化壳以及更古老的撞击溅射物。在 LPR 低频数据约 300m 处出现了一个分叉结构，影响了深度 90～310m 之间的反射面，根据该结构的形态特征、厚度、深度和方向，它的起源可能与薛定谔撞击坑的溅射物沉积有关。随着 LPR 雷达数据处理技术的不断更新，未来对于着陆区地下分层结构与地质历史将有更进一步的认识。

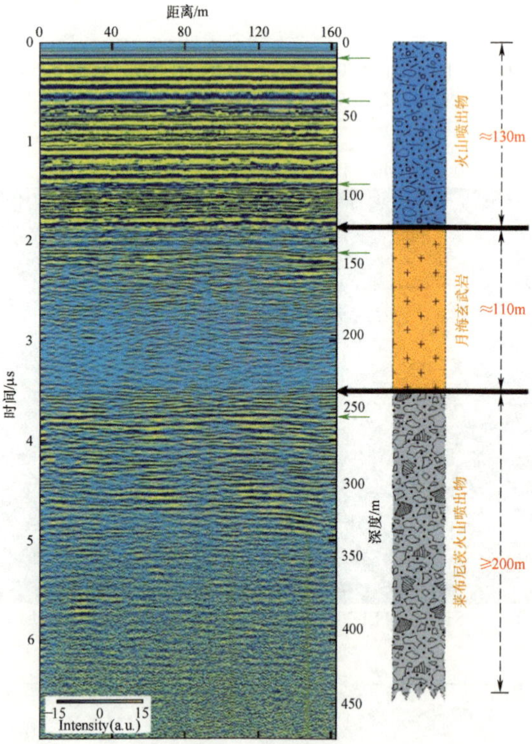

图 4－5　测月雷达低频通道的探测剖面及解译结果

4.3　月壤特性

在持续的撞击碾磨和空间风化的作用下，月球表面普遍被一层细粒的松散物质覆盖，称为月壤。一般认为，月海上的平均月壤厚度大约为 5m，高地上的平均月壤厚度大约为 10m。大多数月球遥感探测的深度都限制在月壤深度内，因此月壤的物理和化学性质是月球科学研究的重点，在嫦娥四号着陆以前，利用撞击坑的形态特征推测的着陆区月壤厚度大约为 2.5～7.5m。

玉兔二号搭载的测月雷达高频通道获取的数据中存在大量的双曲线形的回波信号，它们可能是由较细粒的月壤中相对致密的石块的散射所致，这些双曲线形的回波信号是推测上覆物质的平均介电常数的重要方法。使用三维速度追踪法，得到了玉兔二号月球车行径路线以下约 50m 深度内物质的介电常数，发现其与典型的月壤物质类似。估算高频测月雷达约在 400ns 之内的损耗角正切，约为 0.005，表明表层＞ 40m 的物质的介电特性和典型的月壤物质类似。和嫦娥三号着陆点相比，嫦娥四号着陆点的浅表层物质具有更小的介电常数，因此电磁波随深度衰减得更慢，测月雷达在嫦娥四号着陆区的探测深度更大。

玉兔二号的高频测月雷达探测发现，嫦娥四号着陆区月壤中的石块含量远远低于嫦娥三号。基于玉兔二号在前 12 个月昼的高频雷达数据，估算了巡视路线上月表约 4cm 之内的介电常数的均值为 3.11±0.085，其中，约有 93 处介电常数 ＞ 4，显示高密度非月壤特性（图 4－6）。利用 2019 年 1 月到 2020 年 9 月的雷达数据，生成了 540m 雷达图，分析着陆点的月壤具有 2.64～3.85 的介电常数和 0.0032～0.0044 的损耗正切，表明当地的月壤与 CE－3 着陆区相比是由细粒的、低损耗的材料组成的。通过高精度三维表面模型并模拟计算高频雷达回波信号的结果表明，这些成熟月壤中的高介电常数区域并不由表面地形所导致，而是由表面暴露或者埋藏的石块所引起的高反射回波幅值，小型撞击坑形成过程中也可以通过冲击压实和熔融表层月壤导致其孔隙度降低，从而导致密度增大，月表岩石碎块的分布为解译月壤的形成和演化提供了新的线索。

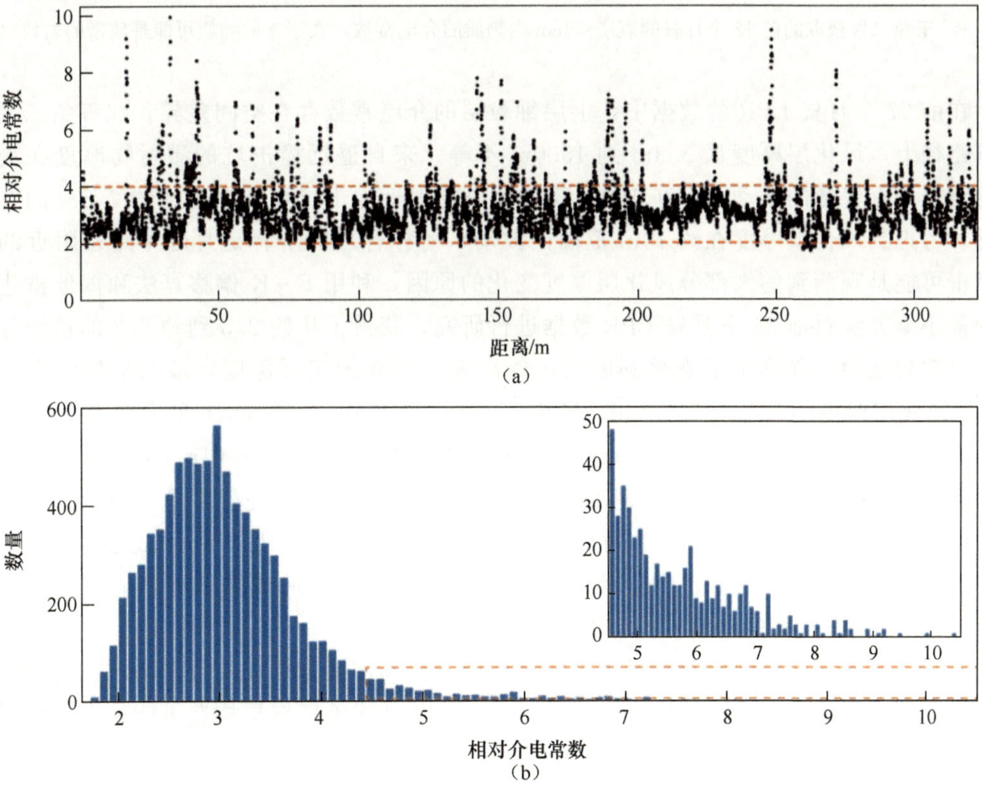

图 4－6　玉兔二号获取的前 12 个月昼的表层＜4cm 内物质的介电常数，大于 4 的物质可能是致密的岩块

(c)

图 4-6 玉兔二号获取的前 12 个月昼的表层<4cm 内物质的介电常数，大于 4 的物质可能是致密的岩块（续）

在前 27 个月昼 LPR 的数据中，上层细粒层的介电常数存在空间差异。在玉兔二号探测车路径上，风化层厚度从 5.3m 到 15.0m 不等。来自遥远撞击坑的溅射物厚度在漫游者的路径上有轻微的波动，而在 100m 缓冲区内的附近撞击坑可能会造成约 2.5m 的厚度变化。考虑到风化层厚度在约 10m 范围内波动，一种合理的解释是月球车路径附近的古代撞击可能是观测到的大部分风化层厚度变化的原因。利用 F-K 偏移算法和探地雷达信号的最小熵方法对前 31 个月昼 LPR 数据进行研究，获得了从约 2.3 到约 6.3 的精细分辨率介电常数剖面，并确定了观察到的突出的次表层界面的实际深度，最大平均深度约为 38m。风化层厚度约为 5.7～15.6m，平均 11.8m。近地层（<30cm）的介电常数为 2.78±0.01，密度为（1.57±0.01）g/cm^3，接近阿波罗 15 号着陆区 1.61g/cm^3的结果。

利用 LPR 数据的极化属性，对 24m 深度内风化层的成熟度进行了量化分析。结果表明，LPR 信号主要来自不同大小的地下岩石碎片，很少来自地层界面。近地表（0 ～ 2m）、细粒风化层（2 ～ 12m）和粗粒溅射物（12 ～ 24m）的成熟度平均值和范围分别为 68.98%（0% ～ 99.82%）、82.61%（41.52% ～ 96.56%）和 72.33%（5.91% ～ 96.66%）。月球表面局部成熟度异常低的区域可能由局部撞击形成的致密物质或远端撞击溅射形成的岩石碎片组成；新发现的 12 ～ 18m 深的低成熟度夹层构造可能是由古撞击坑非均匀风化形成的。

在垂向剖面上，月壤物质的平均粒径随深度增加，因此不同探测方法约束的月壤厚度和整体物理性质存在差异。在测月雷达高频通道获取的数据中，表层 0～12m 物质的回波信号强度与更深的物质截然不同，表明巡视路线上的月壤厚度约为 12m，表明嫦娥四号着陆点的月壤厚度所对应的月壤生成速率可能远大于月球表面正常的月壤生成速率。使用全景相机获取的立体影像制作高分辨率 DEM，分析玉兔二号的车辙在月壤中留下的地形痕迹，并采用岩土力学方法估算嫦娥四号着陆点的月壤物质的强度，发现嫦娥四号着陆点的月壤成熟度很高，月壤的强度比嫦娥三号和 Luna 17 号等较年轻的探测区域高，揭示了局部地形和空间风化速率对月壤强度的相对影响。通过玉兔二号在前两年运行过程中的工程数据，发现月球车经历了不同程度的轻度滑移和打滑，说明该地区地形大尺度相对平坦，但零星分布着局部缓坡。粘在轮子上的土壤表明月壤的凝聚力比在其他月球着陆点遇到的更强。进一步的研究结果表明，着陆区风化层在承载性能上类似于地球上的干砂和沙壤土。

利用数值模拟的方法，对 CE－4 着陆区小型撞击事件影响月球风化层的演变进行了定量分析。结果表明，月球风化层的形成受控于较大的撞击事件，而月球风化层的表面翻转和混合主要是由较小的撞击事件引起的。此外，由于当时的撞击通量较大，月球风化层大多形成于较早时期。CE－4 着陆区的绝对模式年龄约为 3.5Ga，而月球风化层大多在 3.0Ga 之前形成，之后变化缓慢。模型超出了玉兔二号测月雷达的观测能力，结果有助于更好地理解区域月球风化层的形成和演化。

4.4　碎石特性

月壤中岩石碎块的尺寸-深度的频率分布和月球表面撞击历史直接相关，这对了解月壤物质的来源具有指示意义。嫦娥四号着陆区的整体岩石丰度很低，是成熟月壤的典型特征，对比阿波罗和 Luna 任务着陆点的表面岩石丰度，统计结果显示在年龄为约 3.6Ga 形成的月表岩石丰度通常小于 1%，如阿波罗 16 号为 0.11%、阿波罗 11 号为 0.28%，嫦娥四号地区为 0.21%；更年轻的表面分布着更多的石块，如 Luna 17 约 1.5%、嫦娥三号约 1.2%。测月雷达高频通道对一个小型新鲜撞击坑的溅射物进行了两个剖面探测，和周围正常月壤的雷达数据对比表明，该新鲜撞击坑的溅射物及其浅表层物质和周围月壤的物理性质基本类似，未见大于 1cm 的致密石块的信号。雷达数据仿真研究表明这些土块状的溅射物不是撞击挖掘的致密石块，而是由疏松的月壤在撞击作用下发生瞬间孔隙闭合而成。

4.5　月球背面冯•卡门撞击坑地区区域地质演化

前人利用阿波罗任务的返回样品和撞击坑统计划分了月球的地质年代，从老到新分别为：前酒海纪（39.2 亿年之前），主要事件为“形成月壳和大块月壤角砾、火山和侵入活动、月盆和月坑垮塌”；酒海纪（距今 39.2 亿～38.5 亿年），主要事件为“形成酒海、其他盆地与一些月坑垮塌，形成浅色的平原”；早雨海纪（距今 38.5 亿～37.5 亿年），主要事件为“形成雨海、东海盆地、凯利平原，一些月坑垮塌”；晚雨海纪（距今 37.5 亿～32 亿年），主要事件为“大部分月海物质形成”；艾拉托辛纪（距今 32 亿～12 亿年），主要事件为“轻微垮塌的月坑，大量的月海物质”；哥白尼纪（距今 12 亿年至今），主要事件为“新鲜的有辐射线的月坑，少量月海物质”。

前人根据光谱特性的研究将南极-艾特肯盆地细分为四个不同的成分区：中心约

700km 宽的南极艾特肯盆地组分异常区，它不同于典型的月海玄武岩，展示了强烈的钙辉石特征、以富镁辉石为特征的镁辉石环区；显示局部富辉石岩单元和长石物质的混合的异质环区；无镁铁和以长石物质为主的南极艾特肯盆地的外部区。

根据空间内撞击坑的地层关系和南极艾特肯盆地组分异常中央峰暴露情况，镁辉石单元位于南极艾特肯盆地组分异钙辉石单元之下，镁辉石环的物质代表着形成南极艾特肯盆地事件挖掘和熔融的主要物质，目前尚未在其内部发现橄榄石的广泛分布，表明上月幔的主要成分可能为辉石。冯·卡门撞击坑位于南极艾特肯盆地的西北部的镁辉石环单元内，主要撞击坑包括冯·卡门 M 撞击坑、冯卡门撞击坑、莱布尼茨撞击坑、芬森撞击坑、阿尔德撞击坑、冯·卡门 L 和 L′撞击坑(图 4－7)。冯·卡门撞击坑被一次或数次的月海玄武岩淹没，之后周边地区众多撞击坑的溅射物被抛射到其内部。

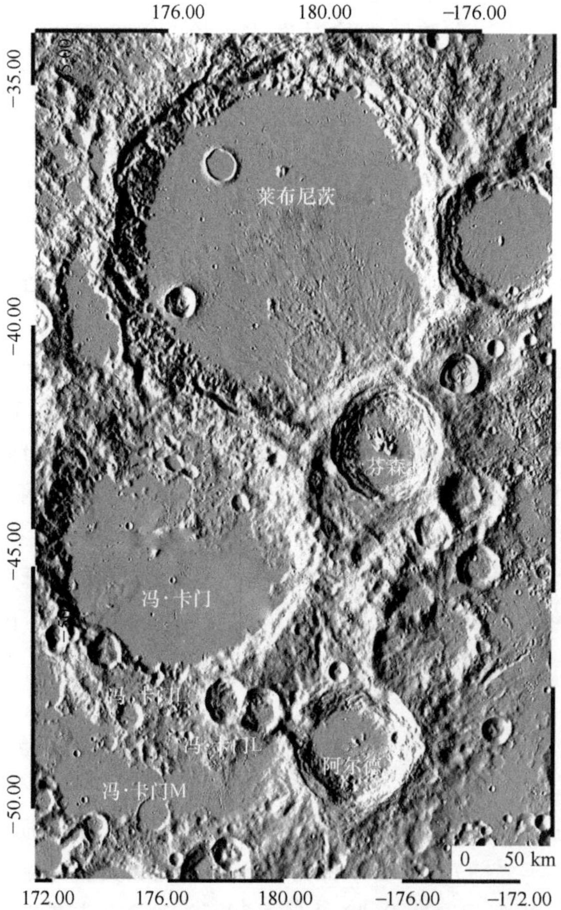

图 4－7 嫦娥四号着陆点（绿色星点）及其周围的主要撞击坑

目前，玉兔二号尚未走出芬森撞击坑的溅射物范围，而芬森撞击坑溅射物来自南极艾特肯组分异常区单元，芬森撞击坑靶体性质的复杂性直接导致了其溅射物成分的多样性和复杂性。玉兔二号携带的全景相机、可见光-近红外光谱仪和探月雷达获得了大量科学数据，研究表明着陆点和巡视区的表面物质主要来自芬森撞击坑，其下为多期次的撞击坑溅射物和冯·

卡门撞击坑内的多期次玄武岩，揭示了着陆和巡视区的地形地貌特征、表面物质组成以及浅部地下结构，为认识冯·卡门撞击坑地区区域地质演化特征提供了重要支撑。

可见光-近红外光谱仪观察到了月球表面的多种矿物组合，包括斜长岩、苏长岩/辉长岩、撞击熔融角砾岩等多种岩石类型，这些成分特征可能代表了深部月壳，甚至部分的上部月幔，南极艾特肯撞击事件的冲击熔融席和冯·卡门坑内填充的月海玄武岩不同程度混合；表层的月壤受到了长期的空间风化作用，着陆巡视区的月壤属于成熟的月壤，并且包含了部分月外的物质。新的研究认为，由于溅射物沉积和灰尘静电悬浮，光谱仪观测到的岩石表面有不同程度的灰尘覆盖，灰尘覆盖较少的岩石表面的光谱表明，这些岩石主要由镁铁质物质组成，这表明它们可能不是来自月球上地壳，为着陆区的地质演化提供了新的线索。

基于玉兔二号低频雷达特征，将着陆区的浅层结构划分为三大基本单元，由上往下依次为强反射单元、弱反射单元和中等反射单元。结合区域地质和大型撞击坑的空间分布等约束，对着陆区地质解译如下：表层强反射单元总厚度约 130m，为临近多个撞击坑的溅射物堆积（包括芬森、阿尔德、冯·卡门 L 和 L′等撞击坑）和底部的玄武岩角砾层；弱反射层总厚度约 110m，为多次喷发的玄武岩层；中等反射单元厚度不小于 200m，为着陆区北部莱布尼茨撞击坑的溅射物。高频雷达信号给出了强反射单元上部的精细结构，其特征为顶部约为 12m 的月壤层，基本不含大石块，其下为约 22m 的条带状溅射物，它们均是来自芬森撞击坑的抛射物，总厚度达 34m，这与前期利用遥感数据解译得到的结果基本一致。

依据嫦娥二号光学影像、SLDEM 高程数据以及含量分布数据等，利用撞击坑统计定年法得到嫦娥四号着陆区各个玄武岩地质单元的年龄在 3.42Ga～2.28Ga 之间。基于撞击坑形成过程中挖掘深度、颗粒大小分布、溅射席厚度分布等经验公式，利用蒙特卡罗数值模拟方法建立了月球表面月壤演化模型，进而依据嫦娥四号着陆区的撞击坑产率函数定量分析了嫦娥四号着陆区石块大小、月壤厚度等重要信息随时间的演化关系，结果表明月壤的形成受较大的撞击事件约束，而月壤的表层翻转和混合主要由较小的撞击事件引起，月壤主要是在早期形成的，CE-4 着陆区的绝对模型年龄约为 3.5Ga，而月壤大多在 3.0Ga 之前形成，之后变化缓慢。

基于遥感解译的结果和嫦娥四号就位探测的结果，着陆巡视区的地质演化由老到新的序列如下：冯·卡门撞击坑形成于前酒海纪，之后莱布尼茨撞击坑的溅射物覆盖了冯·卡门撞击坑坑底的部分区域；大约在距今 32 亿～37 亿年之间，月海玄武岩填充了冯·卡门撞击坑内部的部分区域；在这些玄武岩填充期间，很可能混入了阿尔德撞击坑的溅射物；玄武岩填充完成后，受到了可能包括东海盆地在内的远端溅射物的覆盖；在此之后，芬森撞击坑的溅射物覆盖到了着陆巡视区的最表层；这些表层物质经历了长时间的后期改造，部分区域混入了月球外部的物质。

4.6　小结

嫦娥四号着陆器和玉兔二号月球车依然在月球背面稳定地工作，持续传回科学数据。玉兔二号目前正在向西行进，目标是达到溅射物覆盖较薄的月海区域。着陆巡视区的地质特征非常复杂，使用玉兔二号就位探测获取的地形地貌和地下结构信息，和已有的阶段性科学认识对比和互补，尽管对于部分现象的解释仍然具有一定的争议，但仍有望进一步揭示着陆区的表面地形地貌演化和三维地质结构。特别地，溅射物覆盖较薄的玄武岩月海区域上的撞击坑的整体形态和内部结构特征，可用于对比研究靶体物性对撞击过程的影响；表面和地下石块的三维分布信息和地下结构等信息将对进一步了解着陆区的地质演化、修正和约束已有的地质解译提供对比的依据。

嫦娥四号任务在更长的表面巡视路线上证明月球表面的地质过程高度丰富和复杂，轨道探测和就位探测在地形地貌的分辨尺度上存在数量级的差别，因此不同地质视角的观测带来了不同的地质解译，也进一步丰富了基于遥感探测的月球地貌和地质演化的认识。基于嫦娥四号任务的科学数据得到的地质认识和讨论，深刻地证明了遥感探测和样品分析相结合的必要性和探测发展趋势。

主要成果附录

[1] Cao H，Xu Y，Xu L，et al. From Schrödinger to Von Kármán：An Intriguing New Geological Structure Revealed by the Chang'e - 4 Lunar Penetrating Radar [J]. Geophysical Research Letters，2023，50 (2)：e2022GL101413.

[2] Chang R Y，Xiao Z Y，Wang Y C，et al. An updated constraint on the local stratigraphy at the Chang'E - 4 landing site [J] . Earth and Planetary Physics，2021，5 (1)：19 - 31.

[3] Chisenga Chikondi，Yan Jianguo，Zhao Jiannan，et al. Density Structure of the Von Kármán Crater in the Northwestern South Pole - Aitken Basin：Initial Subsurface Interpretation of the Chang'E - 4 Landing Site Region [J] . Sensors (Basel，Switzerland)，2019，19 (20) .

[4] Di K，Zhu M H，Yue Z，et al. Topographic Evolution of Von Kármán Crater Revealed by the Lunar Rover Yutu - 2 [J]. Geophysical Research Letters，2019，46. https：//doi. org/10. 1029/ 2019GL085252.

[5] Di K C，Liu Z Q，Wan W H，et al. Geospatial technologies for Chang'e - 3 and Chang'e - 4 lunar rover missions [J]. Geospatial Information Science，2020，23 (1)：87 - 97，DOI：10. 1080/10095020. 2020. 1718002.

[6] Ding C，Xiao Z，Su Y，et al. Hyperbolic reflectors determined from peak echoes of ground penetrating radar [J] . Icarus，2020，358：114280.

[7] Ding C，Xiao Z，Wu B，et al. Fragments Delivered by Secondary Craters at the Chang' E - 4 Landing Site [J] . Geophysical Research Letters，2020，47，e2020GL087361. https：//doi. org/ 10. 1029/2020GL087361.

[8] Ding C，Xiao Z，Wu B，et al. Rock Fragments in Shallow Lunar Regolith：Constraints by the Lunar Penetrating Radar Onboard the Chang'E - 4 Mission [J]. Journal of Geophysical Research：Planets，2021，126，e2021JE006917. https：// doi. org/10. 1029/2021JE006917.

[9] Ding L，Zhou R，Yuan Y，et al. A 2 - year locomotive exploration and scientific investigation of the lunar farside by the Yutu - 2 rover [J]. Science robotics，2022，7

(62), eabj6660.

[10] Dong Z, Fang G, Zhao D, et al. Dielectric Properties of Lunar Subsurface Materials. Geophysical Research Letters, 2020a47, e2020GL089264. https://doi.org/ 10.1029/2020GL089264.

[11] Dong Z, Fang G, Zhou B, et al. Properties of Lunar Regolith on the Moon's Farside Unveiled by Chang'E-4 Lunar Penetrating Radar [J]. Journal of Geophysical Research: Planets, 2021, 126, e2020JE006564. https://doi.org/10.1029/2020JE006564.

[12] Dong Z, Feng X, Zhou H, et al. Properties Analysis of Lunar Regolith at Chang'E-4 Landing Site Based on 3D Velocity Spectrum of Lunar Penetrating Radar [J]. Remote Sens, 2020b, 12, 629; doi: 10.3390/rs12040629.

[13] Feng J, Siegler M A, White M N. Dielectric properties and stratigraphy of regolith in the lunar South Pole-Aitken basin: Observations from the Lunar Penetrating Radar [J]. Astron and Astrophys, 2022, 661, A47.

[14] Feng Y, Chen S, Tong X, et al. Exploring the Lunar Regolith's Thickness and Dielectric Properties Using Band-Limited Impedance at Chang'E-4 Landing Site [J]. Journal of Geophysical Research: Planets, 2023, e2022JE007540.

[15] Fu X H, Qiao L, Zhang J, et al. The subsurface structure and stratigraphy of the Chang'E-4 landing site: Orbital evidence from small craters on the Von Kármán crater floor [J]. Research in Astronomy and Astrophysics, 2020, 20 (1): 59-70.

[16] Giannakis I, Zhou F, Warren C, et al. Inferring the Shallow Layered Structure at the Chang'E-4 Landing Site: A Novel Interpretation Approach Using Lunar Penetrating Radar [J]. Geophysical Research Letters, 48, e2021GL092866. https://doi.org/10.1029/2021GL092866.

[17] Gou S, Yue Z, Di K, et al. Geologically Old but Freshly Exposed Rock Fragments Encountered by Yutu-2 Rover [J]. Journal of Geophysical Research: Planets, 2021, 126, e2020JE006565. https://doi.org/10.1029/2020JE006565.

[18] Guo D, Fa W, Zeng X, et al. Geochemistry of the Von Kármán crater floor and thickness of the non-mare ejecta over the Chang'e-4 landing area [J]. Icarus, 2021a, 359, 114327.

[19] Guo D, Fa W, Wu B, et al. Millimeter- to Decimeter-Scale Surface Roughness of the Moon at the Chang'E-4 Exploration Region [J]. Geophysical Research Letters, 2021, 48, e2021GL094931. https://doi.org/10.1029/2021GL094931.

[20] Hu T, Yang Z, Kang Z Z, et al. Population of Degrading Small Impact Craters in the Chang'E-4 Landing Area Using Descent and Ground Images [J]. Remote Sensing, 2022, 14, 3608.

[21] Jia M, Di K, Yue Z, et al. Multi-scale morphologic investigation of craters in the Chang'e-4 landing area [J]. Icarus, 2021, 355, 114164.

［22］ Lai J，Cui F，Xu Y，et al. Dielectric Properties of Lunar Materials at the Chang'e - 4 Landing Site ［J］. Remote Sens，2021，13，4056.

［23］ Lai J，Xu Y，Bugiolacchi R，et al. A Complex Paleo - Surface Revealed by the Yutu - 2 Rover at the Lunar Farside ［J］. Geophysical Research Letters，2021a，48，e2021GL095133.

［24］ Lai J，Xu Y，Bugiolacchi R，et al. First look by the Yutu - 2 rover at the deep subsurface structure at the lunar farside ［J］. Nat Commun 2020，11，3426. https：//doi. org/10. 1038/s41467 - 020 - 17262 - w.

［25］ Lai J，Xu Y，Zhang X，et al. Comparison of Dielectric Properties and Structure of Lunar Regolith at Chang'e - 3 and Chang'e - 4 Landing Sites Revealed by Ground - Penetrating Radar ［J］. Geophysical Research Letters，2019，46，12，783 - 12，793. https：//doi. org/10. 1029/ 2019GL084458.

［26］ Li B，Yue Z，Zhang J，et al. High - Resolution Terrain Analysis for Lander Safety Landing and Rover Path Planning Based on Lunar Reconnaissance Orbiter Narrow Angle Camera Images：A Case Study of China's Chang'e - 4 Probe ［J］. Earth and Space Science，2019，6，398 - 410. https：//doi. org/10. 1029/2018EA000507.

［27］ Li C，Su Y，Pettinelli E，et al. The Moon's farside shallow subsurface structure unveiled by Chang'E - 4 Lunar Penetrating Radar ［J］. Sci Adv，6，eaay6898 (2020).

［28］ Li C，Zhang J. Preserving signal during random noise attenuation through migration enhancement and local orthogonalization ［J］. Geophysics，2022，87，V451 - V466. https：//doi. org/ 10. 1190/GEO2021 - 0385. 1.

［29］ Li C，Zhang J. Velocity Analysis Using Separated Diffractions for Lunar Penetrating Radar Obtained by Yutu - 2 Rover ［J］. Remote Sens，2021，13，1387. https：// doi. org/10. 3390/rs13071387.

［30］ Liu W C，Wu B. An integrated photogrammetric and photoclinometric approach for illumination - invariant pixel - resolution 3D mapping of the lunar surface ［J］. ISPRS Journal of Photogrammetry and Remote Sensing，2019，159 (2020)：153 - 168.

［31］ Liu Z Q，Di K C，Li J，et al. Landing site topographic mapping and rover localization for Chang'e - 4 mission ［J］. Sci China Inf Sci，2020，63 (4)：140901，https：//doi. org/10. 1007/s11432 - 019 - 2796 - 1.

［32］ Lu Y，Wu Y Z，Gregory G Michael，et al. Chronological sequence of Chang'E - 4 landing zone within Von Kármán crater ［J］. Icarus，Volume 354，2021，114086. https：//doi. org/10. 1016/j. icarus. 2020. 114086.

［33］ Lv W M，Li C，Song H J，et al. Comparative analysis of reflection characteristics of lunar penetrating radar data using numerical simulations ［J］. Icarus，2020，350 (C).

［34］ Ma Y，Liu S，Sima B，et al. A PRECISE VISUAL LOCALISATION METHOD

FOR THE CHINESE CHANG'E-4 YUTU-2 ROVER [J]. Photogram Rec, 2020, 35: 10-39. https: //doi. org/10. 1111/phor. 12309.

[35] Ma Y Q, Peng S, Jia Y, et al. Prediction of terrain occlusion in Chang'e-4 mission [J]. Volume 152, 2020, 107368. https: //doi. org/10. 1016/j. measurement. 2019. 107368.

[36] Peng M, Di K, Wang Y, et al. A Photogrammetric-Photometric Stereo Method for High-Resolution Lunar Topographic Mapping Using Yutu-2 Rover Images [J]. Remote Sens, 2021, 13, 2975. https: //doi. org/10. 3390/rs13152975.

[37] Pettinelli E, Lauro S E, Mattei E, et al. Stratigraphy versus artefacts in the Chang'e-4 low-frequency radar [J]. Nat Astron, 2011, 5: 890-893.

[38] Qiao L, Ling Z, Fu X, et al. Geological characterization of the Chang'e-4 landing area on the lunar farside [J]. Icarus, 2019, 333, 37-51.

[39] Ruonan Chen, Yi Xu, Minggang Xie, et al. Sub-surface stratification and dielectric permittivity distribution at the Chang'E-4 landing site revealed by the lunar penetrating radar [J]. Astronomy & Astrophysics, 664: A35, Aug. 2022. DOI: https: //doi. org/10. 1051/0004-6361/202142677.

[40] Shi K, Yue Z, Di K, et al. The gardening process of lunar regolith by small impact craters: A case study in Chang'E-4 landing area [J]. Icarus, 2022, 377, 114908.

[41] Song H J, Li C, Zhang J H, et al. Rock Location and Property Analysis of Lunar Regolith at Chang'E-4 Landing Site Based on Local Correlation and Semblance Analysis [J]. Remote Sensing, 2020, 13 (1).

[42] Tang Z, Liu J, Wang X, et al. Physical and Mechanical Characteristics of Lunar Soil at the Chang'E-4 Landing Site [J]. Geophysical Research Letters, 2020, 47, e2020GL089499. https: //doi. org/ 10. 1029/2020GL089499.

[43] Wang R, Su Y, Ding C, et al. A Novel Approach for Permittivity Estimation of Lunar Regolith Using the Lunar Penetrating Radar Onboard Chang'E-4 Rover [J]. Remote Sens, 2021, 13, 3679. https: //doi. org/10. 3390/rs13183679.

[44] Wang X, Liu J, Liu D, et al. Dusty Mafic Rocks Along the Path of Chang'E-4 Rover: Initial Analysis of the Image Cubes of the Onboard Visible and Near-Infrared Imaging Spectrometer [J]. Geophysical Research Letters, 2022 (2): 49.

[45] Wang Y, Wan W, Gou S, et al. Vision-Based Decision Support for Rover Path Planning in the Chang'e-4 Mission [J]. Remote Sensing, 2020, 12, 624.

[46] Wu B, Li F, Hu H, et al. Topographic and Geomorphological Mapping and Analysis of the Chang'E-4 Landing Site on the Far Side of the Moon [J]. Photogrammetric Engineering and Remote Sensing, 2020, 86 (4).

[47] Wu B, Li Y, Liu W, et al. Centimeter-resolution topographic modeling and fine-scale analysis of craters and rocks at the Chang'E-4 landing site [J]. Earth and Planetary Science Letters, 2021, 553, 116666.

[48] Xiao Z, Ding C, Xie M, et al. Ejecta From the Orientale Basin at the Chang'E-4

Landing Site [J]. Geophysical Research Letters, 2021, 48, e2020GL090935. https: //doi. org/10. 1029/2020GL090935.

[49] Xu C, Zhang G, Zhang J, et al. Time - Frequency Attribute Analysis of Channel 1 Data of Lunar Penetrating Radar [J]. Appl Sci, 2020, 10, 535. https: //doi. org/10. 3390/app10020535.

[50] Xu L, Zhang X, Qiao L, et al. Evaluating the Thickness and Stratigraphy of Ejecta Materials at the Chang'e - 4 Landing Site [J]. The Astronomical Journal, 2021, 162: 29.

[51] Yuan Y, Zhu P, Xiao L, et al. Intermittent volcanic activity detected in the Von Kármán crater on the farside of the Moon [J]. Earth and Planetary Science Letters, 2021, 569, 117062.

[52] Yue Z, Shi K, Michael G, et al. Chronology of the Basalt Units Surrounding Chang'e - 4 Landing Area [J]. Remote Sensing, 2022, 14, 49.

[53] Zhang J, Zhou B, Lin Y, et al. Lunar regolith and substructure at Chang'E - 4 landing site in South Pole - Aitken basin [J]. Nat Astron, 2021, 5, 25 - 30. Http: doi. org/10. 1038/s41550 - 020 - 1197 - x.

[54] Zhang J, Zhou B, Lin Y. Reply to: Stratigraphy versus artefacts in the Chang'e - 4 low - frequency radar [J]. Nat Astron 5, 894 - 897 (2021) . https: //doi. org/10. 1038/s41550 - 021 - 01484 - z.

[55] Zhang L, Li J, Zeng Z, et al. Stratigraphy of the Von Kármán Crater Based on Chang'E - 4 Lunar Penetrating Radar Data [J]. Geophysical Research Letters, 2020, 47, e2020GL088680. https: //doi. org/ 10. 1029/2020GL088680.

[56] Zhang L, Xu Y, Bugiolacchi R, et al. Rock abundance and evolution of the shallow stratum on Chang'e - 4 landing site unveiled by lunar penetrating radar data [J]. Earth and Planetary Science Letters, 2021, 564, 116912.

[57] Zhang X, Lv W, Zhang L, et al. Self - Organization Characteristics of Lunar Regolith Inferred by Yutu - 2 Lunar Penetrating Radar [J]. Remote Sens, 2021, 13, 3017. https: //doi. org/10. 3390/rs13153017.

[58] Zhou H, Feng X, Ding C, et al. Heterogeneous Weathering Process of Lunar Regolith Revealed by Polarimetric Attributes Analysis of Chang'E - 4 Lunar Penetrating Radar Data Acquired During the Yutu - 2 Turnings [J]. Geophysical Research Letters, 2022, 49 (13): e2022GL099207.

[59] Zhou H, Feng X, Ding C, et al. Yutu - 2 Radar Sounding Evidence of a Buried Crater at Chang'E - 4 Landing Site [J]. IEEE TRANSACTIONS ON GEOSCIENCE AND REMOTE SENSING. 10. 1109/TGRS. 2021. 3090528.

[60] Zhou H, Feng X, Dong Z, et al. Application of Denoising CNN for Noise Suppression and Weak Signal Extraction of Lunar Penetrating Radar Data [J]. Remote Sens, 2021, 13, 779. https: //doi. org/10. 3390/rs13040779.

[61] 李圣卉，刘建忠，刘敬稳，等. 嫦娥 4 号着陆区冯 · 卡门撞击坑溅射物的识别及年

代分析 [J]．岩石学报，2022，38 (9)：2889 - 2899.
[62] 史珂，岳宗玉，邸凯昌，等．Apollo 11 和嫦娥四号着陆区撞击坑退化对比分析 [J]．矿物岩石地球化学通报，2021，40 (3)：711 - 719.

From Schrödinger to Von Kármán: An Intriguing New Geological Structure Revealed by the Chang'e－4 Lunar Pentrating Radar

Huaqing Cao[1], Yi Xu[1], Luyuan Xu[1], Ling Zhang[1,2],
Roberto Bugiolacchi[1,3], and Feng Zhang[4]

1. State Key Laboratory of Lunar and Planetary Sciences, Macau University of Scienceand Technology, Macau, China
2. School of Earth Science and Geological Engineering, Sun Yat－sen University, Guangzhou, China
3. Earth Sciences, University College London, London, UK, 4State Key Laboratory of Space Weather, National Space Science Center, Chinese Academy of Sciences, Beijing, China

Abstract: Since early 2019, the Lunar Penetrating Radar (LPR) onboard Chang'e－4 (CE－4)'s Yutu－2 rover has been gathering data relating to the subsurface structure of the Von Kármán crater within the South Pole－Aitken Basin (SPA) on the lunar farside. Low－frequency radar data have the potential of carrying geological information of about 300 m worth of strata below the traversed path. Forty－two days' data have revealed a bifurcated structure within the layered structure beneath the CE－4 surveying area for the first time, affecting the overlying reflectors between 90 and 310m. This study suggests that, based on the morphological characteristics, thickness, depth (timing sequence) and direction of the newly found structure, its origin might be linked to the deposition of ejecta from the Schrödinger impact. The local stratigraphy is interpreted as consisting of distinct geological layers, corresponding to the superposition of ejecta from different impact craters, paleo－regolith, and basaltic lava flows.

KeyPoints: • The CH－1 Lunar Penetrating Radar data of Chang'e－4 reveals ～300 m deep subsurface stratigraphy of the lunar farside

• A unique bifurcated structure that resembles an ejecta discontinuity affects the overlying and underlying strata

• The geologic evolution of the Von Kármán crater and the ejecta layer might be connected to the Schrödinger basin

Plain Language Summary: A radar (the Lunar Penetrating Radar, or LPR), built to look at the rocks below the surface, is presently traveling across an area of the lunar farside onboard a rover, called Yutu - 2 sent by China to shed some light on the striking difference between the Earth - facing side of the Moon and its farside. So far, the radar has revealed layered structures interpreted as distinct volcanic flows, thus helping to unravel their history. The rover has been surveying the floor of the Von Kármán crater inside one of the oldest and largest impact structures of the Moon, the South Pole - Aitken (SPA) Basin. By analyzing 42 lunar days of data, we found a new layer with a hunched upper surface and a relatively flat bottom inserted into a multilayered structure. Based on the morphology, thickness, depth, and direction of the new feature, we argue that its source could represent the ejecta deposition from one of the ancient lunar impact basins: Schrödinger. The subsurface structure revealed by the radar shows a complicated geologic history on the farside including multiple lava eruption events, the deposition of ejecta from different impact craters and weathering layers that develop during the quiescent stage.

1 Introduction

On 3 January 2019 the Chang'e - 4 (CE - 4) mission achieved the first soft landing on the farside of the Moon (Liu et al., 2019) targeting the Von Kármán crater (44.45°S, 176.25°E, 186km diameter) within the South Pole - Aitken (SPA) basin. The SPA is the largest basin on the Moon and one of the oldest impact structures (Ohtake et al., 2014). Impact scaling considerations based on the present crater outline (~2500km) suggest that the excavation blast may have reached and exhumed materials from the upper mantle, allowing a rare opportunity of studying its composition (Huang et al., 2018; Melosh et al., 2017).

The impact that excavated the Von Kármán crater occurred during the pre - Nectarian period (Losiak et al., 2009; Yingst et al., 2017). Its morphology was reshaped by several and often concurring geological processes, some constructive, for example, effusive processes, the emplacement, excavation, and mixing of subsequently deliv - ered ejecta (e.g., Finsen crater, Von Kármán L crater, Von Kármán L′ crater) (Ling et al., 2019; Moriarty & Pieters, 2018; Xu et al., 2021; Yuan et al., 2021), and some destructive, mostly due to the bombardment of subse quent impacts (affecting the cohesion of the surface materials and their mineralogical attributes). The instrumen - tation onboard the Yutu - 2 rover allows for the surface and subsurface investigation of these materials. The Lunar Penetrating Radar (LPR) on board the CE - 4 rover (Yutu - 2) sounds the subsurface below the exploration path, adding a third dimension to our geological quest.

Regolith and rock fragments′ spectra from this region have been acquired by the Visible and Near Infrared Spectrometer (VNIS) onboard the Yutu - 2 rover, allowing for a more detailed estimation of their mineralogy which is beyond the capabilities of remote sensing from orbital altitudes. Current geological models, in common with other terrestrial planetary bodies, argue for a lunar upper mantle dominated by Mg - rich olivine and low - calcium pyroxenes (LCP), which are progressively substituted by Ca -, and Al -richer silicates at lower pressures (Lin et al., 2017; Rapp & Draper, 2018). This is a very broad scenario with many caveats and expected localized "anomalies," but by and large, the detection of olivine - rich mineralogies on the lunar surface is generally inter - preted as exhumed from depth.

Li et al. (2019) and Gou et al. (2019) presented evidence that the materials at the CE - 4 landing site contain Mg - rich olivine and pyroxene (LCP mostly). However, Hu et al. (2019) and Lin et al. (2019) based on reflectance spectra also acquired by the CE - 4 estimated that the surface materials are rich in plagioclase and pyroxene but poor in olivine, arguing that the materials could only represent the inferred composition of the lunar lower crust, not the upper mantle. Likewise, Huang et al. (2020) did not observe olivine - rich materials; however, since they consider the lunar upper mantle to be dominated by low - calcium pyroxene, not olivine, they did not rule out a noncrustal origin.

The LPR on board the CE - 4 rover (Yutu - 2) allows the study of the subsurface strata and the corresponding sequences of geological events, potentially providing a glimpse into the evolution of the lunar farside (Pasckert et al., 2018). The LPR is a dual - frequency GPR system, operating at 60MHz (CH1) and 500MHz (CH2), with a frequency bandwidth of 40 - 80MHz and 250 - 750MHz, respectively (Su et al., 2014). Due to the longer wavelength, CH1 can reach deeper structures, while CH2 has a higher resolution although a shallower penetra tive potential. CH2 data detect the relative permittivity of the top mature regolith and fragments to analyze the origin of regolith and fragments (Ding et al., 2021). CH1 data have revealed several discontinuities beneath the Yutu - 2 traverse path: these can be attributed to at least three main subsurface layers to a depth of approximately 350 m, including regolith associated with ejecta from Finsen crater, multistage mare basalts, and ejecta from other craters (Lai et al., 2020; Zhang, Zhou, Lin, et al., 2021). In more detail, it appears that distinct types of lava flows occurred during the Imbrian and Eratosthenian periods suggesting heterogeneous source regions. The infill of the Von Kármán was interspersed by the production of effusive materials, the formation of paleo - regolith punctuated by the deposition of ejecta from other impacts. From the bottom up, The thicknesses of the basalt units gradually decrease, suggesting that the internal energy weakened with time (Lai et al., 2020; Zhang et al., 2020). The intermittent lava flows are accompanied by two (detectable) periods of volcanic activity within the Von Kármán crater (Yuan et al., 2021). However, there is some debate about the consistency of the data from the CH1, in particular, whether the systematic errors result in a low SNR (Li et al., 2018; Pettinelli et al. , 2021), which may swamp the actual subsurface signals. The LPR data were processed using different approaches, including a bandpass filter and background removal to mitigate the system noises (e. g. , Zhang, Zhou, & Lin, 2021) . However, the limited amount of data collected during just three lunar days was not enough to determine the reliability of the processed information beyond reasonable assumptions; indeed, since the horizontal reflectors in the radargram show few variations, they appear virtually indistinguishable from the system noise.

By the end of April 2022, the Yutu - 2 rover had traveled 42 lunar days and collected data over an accumulative distance of more than 1, 100 m. In this study, the characteristics of several recognized subsurface layers including those overlying and underlying a bifurcated structure have been analyzed and discussed. We offer new insights into the geologic evolution of the Von Kármán crater and investigate whether the bifurcated structure might be connected to the ejecta of the Schrödinger basin.

2 Data and Method

2.1 The CE - 4 LPR

The Yutu - 2 rover is equipped with the same LPR system as the original Yutu rover, part of the Chang'e - 3 mission to Mare Imbrium on the lunar nearside (Fang et al. , 2014; Su et al. , 2014) . LPR is a nanosecond imaging radar that is carrier free and operates

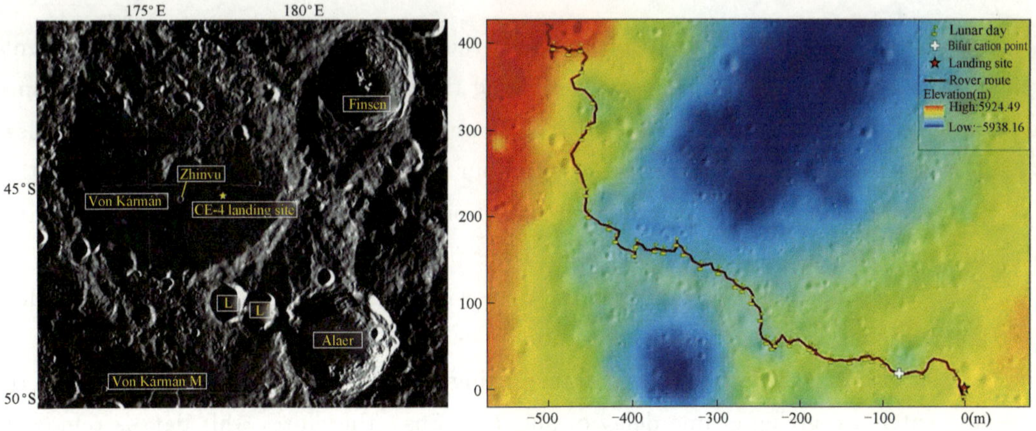

(a)　　　　　　　　(b)

Figure 1　(a) Lunar Reconnaissance Orbiter Camera (LROC) Wide - Angle Camera (WAC) image of the Von Kármán crater. The yellow star indicates the landing site of the CE - 4. (b) The route of the Yutu - 2 rover during the first 42 lunar days. The red line represents the route, the red five - point star represents the CE - 4 landing site, the yellow points represent the first points at each lunar day, and the white cross represents the starting point of the "Bifurcation" feature. The base images are the LROC Narrow Angle Camera (NAC) image (ID: M1324767262LE) and a DEM generated from the LROC NAC stereo pairs images (ID: M1303619844LR, M1303640934LR) in the time domain (Fang et al. , 2014) . It operates on two channels: one low frequency (CH1) with a spatial resolution of several meters, and high penetrative potential, and CH2, operating at a higher frequency with a fine spatial resolution of 30cm but with a much shallower penetrative potential.

As the Yutu - 2 rover travels, the LPR sends pulse signals into the lunar subsurface which are subsequently reflected accordingly to the dielectric properties of the encountered materials at various depths, with the ampli tude of the reflections being proportional to the differences in the relative permittivity (Slyuta, 2014) . Following signal processing steps, eventual subsurface boundaries are highlighted and interpreted in the regional geological context (Fang et al. , 2014) . The results from the Ground Penetrating Radar (GPR) reveal strong reflections at the boundaries between the fine - grained regolith, ejecta layer, and the brecciated zone of bedrock, while the reflections within the fine - grained regolith and homogeneous basalt layers are weak (Zhang, Zhou, Lin, et al. , 2021) .

Radar data are stored by tracks, which are amplitude profiles of the reflections versus time or depth. Next, we obtain the radargram by combining all the LPR tracks along the rover path. CH1 has 8, 192 echoes and CH2 has 2, 048 echoes per track, and the sampling interval is 2. 5ns (CH1) and 0. 3125ns (CH2), respectively. The time window is 20,480 and 640ns, respectively. From January 2019 to May 2022, the Yutu - 2 rover traveled 42 lunar days, as shown in Figure 1b. We used the whole 42 lunar days of LPR CH1 data, for a total of 7375 raw valid data tracks.

2. 2 Data Processing

Compared with the CE - 3 site, the CE - 4 landing area has lower ilmenite content, which results in a much smaller signal loss along the propagation path and a better Signal to Noise Ratio (SNR) (Lai et al. , 2019) . To improve the quality of the radar profiles, we followed the same processing steps used in our previous work (Zhang et al. , 2020):

1. *Data reading*. The radar data and position information are read from the initial files and converted to a binary format for further processing and analysis.

2. Stitching data files. Since the raw data were stored by tracks in different files, they are combined and sorted by time to obtain an integrated radar profile.

3. Zero timing correction. The transmitting antenna is turned on later than the receiving antenna, giving a time delay of about 57. 5ns, therefore, this data is removed.

4. Deduplication. Because the Yutu - 2 rover does not traverse at a constant speed, it stops collecting other scientistic information while still collecting radar data; in this step, we select the valid tracks and delete low SNR data.

5. Data range selection. Because of the loss of energy, only the first 5, 000 ns data is useful and deleted the rest.

6. Data gain. Because the electromagnetic wave energy decreases with the propagation and the reflected signal becomes weaker, the deeper information is not

prominent compared to the shallow signal and needs to be highlighted. We use automatic gain control (AGC) to compensate for deeper data.

7. Band - pass filtering. The high - frequency and low - frequency noise may interfere with the true signal; conse quently, we apply a bandpass filter to reduce the noise. The bandwidth of the bandpass filter is between 10 and 80MHz, which filters out high and low - frequency noise.

The actual depth of the lunar subsurface d can be converted from the two - way traveling time t of the radar signal and the relative permittivity ε of the substrate,

$$d = \frac{ct}{(2\sqrt{\varepsilon})} \tag{1}$$

We use $\varepsilon = 4.5$, when the depth is less than 52m, which is considered the boundary between the basalt and the transition zone, and $\varepsilon = 6.5$ when the depth is more than 52m (Lai et al., 2019). The error of depth value is $\pm$ 2m. After these processing procedures, a complete radar profile is obtained. Figure 2 displays the subsurface structure in the CE - 4 surveying area.

To construct the surface and bottom layers of the observed bifurcated structure, we extracted the depth values of corresponding reflectors at each waypoint and used a polynomial fitting method to interpolate and extrapolate the trend surface as shown in Figure S3 in Supporting Information S1.

3 Results

Figure 2 shows the amplitude of the reflections of the LPR CH1 data. Adistinctive bifurcated structure at Stratum C with a depth between 115 and 155m emerges. Within 0 - 300m from the lander, Stratum C appears as a very thin layer (less than 2m thick) with a strong reflection at the depth of about 130m. As the Yutu - 2 rover traveled farther from the lander the LPR data reveals that the main reflector at depth split into separate strong reflective boundaries, here interpreted as representing the upper and lower interfaces of Stratum C. About 200m along the exploration path, the upper boundary gradually raises from a depth of 130 to 125m, while the depth of the lower boundary descends rapidly from 130 to 155m. At a distance greater than 570m the reflection related to the upper surface gradually fades, probably due to increased roughness, which increases from 125 to 115m, resulting in a thicker Stratum C. The reflector interpreted as the lower boundary changes its strength when the traverse distance is between 300 and 500m. Its depth is around 150m and then raises from 155 to 150m at a distance of 600m. When the traverse distance is 300m from the end, the reflection within Stratum C appears significantly weaker than those at the two boundaries, and the thickness of this stratum remains uneven during the whole rover path. The average thickness of Stratum C is 30m, with the thickest part at the distance between 660 and 690m (Table S1 in Supporting

Information S1) .

Since Stratum B overlies Stratum C, the upper and lower interfaces are affected by the fluctuation of the upper surface of Stratum C, rising at the distance between 300 and 450 m. The bottom depth of Stratum B changes from 130 to 110 m. The reflections within Stratum B are much weaker than those in Stratum C, indicating more homogeneity in this layer. The average thickness of Stratum B is 16m.

The whole Stratum A has a significant strong reflection at the depth between 85 and 110 m. The upper and bottom boundaries have the same rising trend, the upper boundary raises from 95 to 85 m, and the bottom boundary goes up from 110 to 103 m. Although the horizontal variation is relatively small, Stratum A still follows the trend of Stratum B and C rising at a distance between 300 and 400 m, the same position as the starting point of the bifur cated structure. Because the reflections within this layer have similar characteristics, and the thickness of Stratum A is consistent throughout the path, we deduce that the materials were deposited by a single event. The reflections within Stratum A and C are relatively strong, possibly due to inhomogeneous components, so we interpreted these two layers as ejecta deposits combining local materials and remote impact craters. The average thickness of Stratum A is 15m. The reflectors above Stratum A have no obvious horizontal variations, which are not the focus of this work and have been analyzed and discussed in prior works (Ding et al. , 2021; Yuan et al. , 2021; Zhang et al. , 2020; Zhang, Xu, et al. , 2021) .

The deeper Stratum D has a relatively strong reflection at a depth between 210 and 230 m throughout the Yutu - 2 rover path, and the average thickness of Stratum D is 56 m. The boundary between stratum C and stratum D has a strong reflection, implying a density/porosity change at the interface. We observed a similarly sharp reflector relating to a buried crater wall, as shown in the CE - 4 CH2 radargram, further, the inversed relative permittivity is consistent with the breccia samples (Lai et al. , 2021) . It also indicates different dielectric properties between strata C and D, where Stratum D is interpreted as basalt materials. The reflections within stratum D are different from those of stratum B, probably relating to distinct basalt units. Because this stratum has a relatively stronger reflection than stratum B, the stratum probably contains more pyroclastic rocks (the rock clasts are larger than the resolution of the LPR) . The reflector at the bottom of Stratum D also rises at the distance between 300 and 600 m. This could represent the change of the average permittivity caused by the insertion of Stratum C, see Equation 1. The ε of the ejecta deposits is smaller than the basalt of Stratum *D*, so the real thickness of Stratum C is presumably closer to 35 m using $\varepsilon = 5.8$ (Lai et al. , 2021), larger than the value shown in Figure 2 which uses the uniform ε of 6. 5. If the variation of ε for Strata C and D is considered, the fluctuations of the bottom of Stratum D will be reduced. Table S1 in Supporting Information S1 shows the properties of each layer at the CE - 4 landing site.

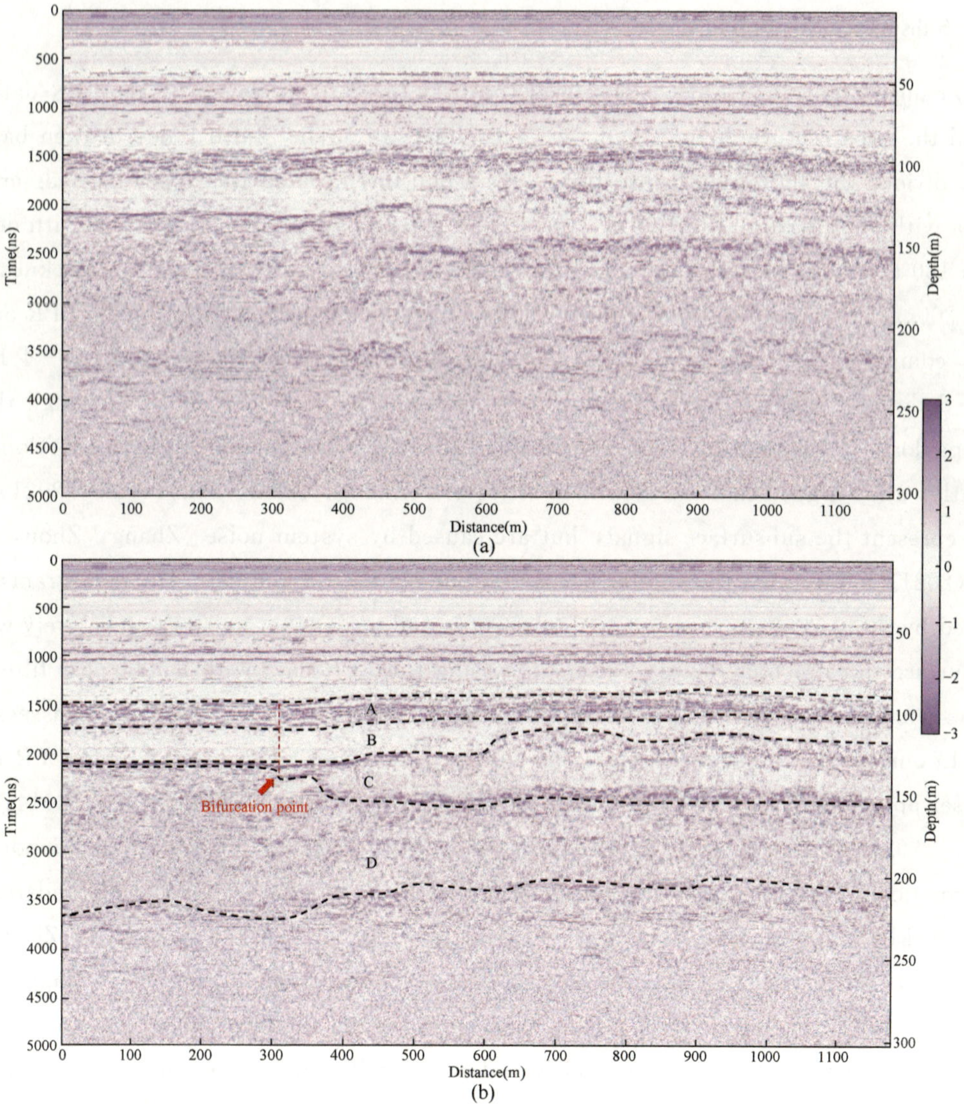

Figure 2　(a) Radar profile of the Lunar Penetrating Radar (LPR) CH1 data at the Chang'e - 4 landing site. The Yutu - 2 rover travel is over 1, 100 m long. The color bar represents the relative reflectance amplitude strength. (b) Interpreted radar profile of the LPR CH1 data at the Chang'e - 4 landing site. Black dash lines represent identified boundaries between strata A - D. The red vertical line represents the start position of the bifurcation structure, corresponding to the "Bifurcation point" in Figure 1b. The color barrepresents the relative amplitude strength of the LPR reflections.

4 Discussion

4.1 Subsurface Stratums

Zhang, Zhou, and Lin (2021) used the first three lunar days of CH1 LPR data to reveal the substructure at the Chang'e - 4 landing site in the South Pole - Aitken basin. They divided the 450m depth into three layers and inter preted the upper layer as crater ejecta with depths from 0 to 130 m, the second layer as a mare basalt stratum with depth from 130 to 240m, and the third layer as ejecta from Leibnitz crater with a thickness of over 200m. However, Pettinelli et al. (2021) questioned the reliability of the LPR data. They compared the cross sections and average spectrograms of CE - 3 and CE - 4 LPR CH1 data and found similar horizontal lines appearing in both LPR data, although these two payloads measured different sites located in distinctive geological settings. Conse - quently, they argued that the identified reflectors (Zhang, Zhou, Lin, et al., 2021) do not represent the subsurface signal, but are caused by system noise. Zhang, Zhou, and Lin (2021) agreed that the similar horizontal lines in CE - 3 and CE - 4's radargrams are caused by the strong electromagnetic interference and noises that mask the rela tively weak subsurface signals; nonetheless, after processing the LPR data with band - pass filtering in the range of 10 - 80MHz, the noise was removed and the real signals could be assessed.

In our study, we adopted the experience of processing CE - 3 and CE - 4 LPR data and set up a series of procedures to process the CE - 4 LPR data (Fang et al., 2014; Lai et al., 2020; Wu et al., 2019; Zhang et al., 2019, 2021). Figure S1 in Supporting Information S1 summarizes the subsurface reflectors of CE - 4 CH1 LPR radargram identified by our work and prior work (Lai et al., 2020; Yuan et al., 2021; Zhang et al., 2020; Zhang, Zhou, Lin, et al., 2021). As shown in Figure S1 in Supporting Information S1, the majority of previous studies agree on the locations of most reflectors with one or two exceptions. The small variations of reflectors are mainly due to different processing methods and parameters. The uncertainties of ε also lead to different depth/thickness values of the subsurface layer although the same time delay t is used. Most of our identified reflectors are consistent with the other studies as shown in Figure S1 in Supporting Information S1: indeed, they share the same reflectors at 1,400 - 1,600 ns, 2,000 - 2,500 ns, and 3,700 ns. Unlike straight interfaces obtained in prior studies, the reflectors after 1,000 ns (the focus of this study) show distinct horizontal variations (using 42 lunar days of LPR data) that gradually resemble the outline of a buried landform. This contrasts with inherent systematic noises which usually do not change along the traveling path and appear as periodic straight lines in the radargram. In particular, the upper surface of Stratum C shows a bulge-like structure, resembling the rough surface of potential ejecta. In addition, the unique bifurcated feature of Stratum C is not observed in CE-3 LPR CH1 radargram. Layers A, B, and D have distinct undulations, influenced by the insertion of Layer C. The consistent parallel trend of these four layers confirms the reliability of recognized reflectors in the radargram following the removal of system noises.

In our study, the CH-1 LPR data reveal the subsurface structure of the CE-4 landing site within a depth of more than 300 m, using the range from 1,000 to 4,000 ns. Starting from 1,400 ns, the materials in layer A may represent a lava flow containing numerous pyroclastic rocks (Yuan et al., 2021) or ejecta from the nearby crater, for instance, the Alder crater (Lai et al., 2020). The second layer B has weak reflections; therefore, this layer is probably relatively homogenous and enriched with mare basalt. The third layer C might be a mixture of paleo-regolith and ejecta. In this layer, the uppermost part probably represents paleo-regolith just less than two m in thickness; with time, under the effect of impact bombardment, a fine-grained layer would have developed and subsequently been buried by other materials (Head & Wilson, 2020). The mean regolith production rate at late Imbrian is ~7.3-20.5m/Ga (Zhu et al., 2021), hence, it took about 0.1 Ga to form this layer. The lower part is probably ejecta from a large crater/basin. Alternatively, it might represent a ghost crater formed by the interaction of molten lavas with preexisting impact craters (Figure S2 in Supporting Information S1). Around a given crater, the lava will rise around the crater's topographic-high rim zone, reducing the elevation difference between the crater rim and the surroundings, but the elevation of the crater rim would remain higher than the surroundings (inside and outside the crater) (Baldwin, 1968; Cruikshank et al., 1973). However, the morphology of Stratum C is uncharacteristic of a ghost crater so far that the rim part is lower than the crater's floor. It also lacks the surface expression of ghost crater-like rings above the buried crater rims. Another scenario might be one where the flow crust formed above a moving layer of magma in the low-lying terrain, which was fractured and entrained by surges in the eruption flow (Campbell et al., 2009). This process can produce rugged fragments several meters thick, probably caused by an increased lava viscosity due to differences in the process of lava transport over the surface (e.g., Belousov & Belousova, 2018); however, other lava layers in the local stratigraphy shown in the radargram do not share similar geomorphological characteristics. Hence, in these cases, the emplacement of ejecta is the most likely scenario.

The bottom layer D detected from the deepest signals has no obvious reflections, probably because the materials at this depth are relatively homogenous. Therefore, it could represent the basalt bedrock (Lai et al., 2020; Yuan et al., 2021; Zhang et al., 2020). The radar signals of layers B and D differ, and this could be due to the presence of mare units from different periods with distinctive geomorphological characteristics.

The ejecta of the nearby crater Zhinyu (3.8km in diameter) has shown distinctive spectral signatures interpreted as different olivine content to support the presence of at least three subsurface layers with different compositions (Lai et al., 2020), consistent with the observations of CE-4 LPR CH-1.

4.2 The Origin of the Bifurcated Structure

If we consider the hypothesis that the bifurcated structure of Stratum C represents an ejecta deposits layer created by a large crater/basin (see Section 4.1), the source crater/basin could be inferred by applying the following criteria:

1. The ejecta from the Alder crater, formed around the Early Imbrian Epoch, should be located above the layer of Stratum C, according to recent literature (Lai et al., 2020; Yuan et al., 2021), and the layer beneath Stratum C (i.e., Stratum *D*) has been interpreted as basalt bedrock belonging to the Imbrian Period. Therefore, the potential source crater/basin should be around the Early Imbrian Epoch (i.e., older than the Orientale basin and younger than the Imbrium basin), with a formation age slightly older than the Alder crater.

2. The ejecta deposit thickness of the potential source crater/basin ought to be comparable with the thickness of Stratum C (~30 m on average and ~41 m maximum), at least falling within the same order of magnitude. Figure S3 in Supporting Information S1 shows the extrapolated planes of the upper and lower boundaries of Stratum C using the LPR data points along the rover path. Table S1 in Supporting Information S1 gives the maximum measured thickness of Stratum C to be 41 m; however, the Yutu-2 rover might only have scanned the edge part of the bifurcated structure as hinted in Figure S3 in Supporting Information S1. If extrapolated data were adopted, the thickness of the interesting feature could reach over 60 m.

After vetting all lunar craters/basins ($D > 100$ km) formed within the Imbrian Period (Table 1 and Figure S4 in Supporting Information S1), we conclude that the Schrödinger basin of the Early Imbrian Epoch is the only source crater/basin that could fit the criteria set above. The Schrödinger basin (diameter 321.5 km, 74.8°S, 133.1°E; Robbins, 2019) is located southwest of the CE-4 landing site, and the traverse of Yutu-2 is broadly perpendicular to the radial direction of this basin. Xu and Xie (2020) measured the ejecta thickness of the Schrödinger basin and obtained a distribution model with a value of ~0.74km at the basin rim, decaying with a power law expo nent of $b = 3.0$. Taking the distance and deposition corrections into consideration, as in Xu et al.

(2021), the model - derived (Xu & Xie, 2020) estimation of the primary ejecta delivered from Schrödinger basin to the CE - 4 landing site is ~2. 6 m in thickness. Besides the emplacement of the primary ejecta itself, the excavation of local materials also contributes to the total thickness of the ejecta deposits. The ratio (μ') of the local material to the primary ejecta thickness produced by the Schrödinger impact can be estimated by the distance between the basin center to the CE - 4 landing site, by adopting the mixing model of Petro and Pieters (2004, 2006, 2008), which is a modification of Oberbeck et al. (1975) (see Xu et al. , 2021 for further details) . As a result, we estimate that the Schrödinger's ejecta also excavated a local materials layer of 16. 6m, with a ratio μ' of 6. 4, and thus created a combined ejecta deposits layer of ~19m thick. Incidentally, the actual thickness of distal ejecta is probably in the order of several times the model - derived value given that the ejecta model assumes an ideal case of a uniform distribution of primary ejecta. However, the ejecta outside the continuous ejecta blanket of a center crater/basin are discrete, thus, the actual thickness of Schrödinger's ejecta could create an ejecta deposits layer as thick as 30 - 60m in the vicinity of the CE - 4 landing site. Assuming that the original topography of the right part of Stratum D can be obtained by extrapolating from the left side data, the estimated ratio μ' from the radargram is 2. 3. The uncertainties of the original topography and the actual bottom of the ejecta deposits might contribute to the discrepancy in μ' between the model and radar data. According to the radar image (Figure 2), even the local material beneath the well-defined lower interface of Stratum C might be disturbed by the impact caused by the ejecta, leading to the underestimation of the thickness of the excavated local materials.

In addition, we find no craters/basins ($D>100$ km) of Late Imbrian Epoch could create an ejecta deposits layer as thick as 10m level, and either the Orientale basin or Imbrium basin may create over 100 m thick in the vicinity of CE - 4 landing site (Table 1). These considerations suggest that the Schrödinger basin is the most likely source for the bifurcated structure of Stratum C, of course, assuming that this layer was indeed created by impact ejecta. Based on the vertical placement of Stratum A and Stratum C, we could constrain the timing of the Schrödinger impact that occurred earlier than the Alder formation event dated to $\sim 3.5^{+0.01}_{-0.2}$ Ga (Lu et al. , 2021) or 3. 8 Ga (Chang et al. , 2021) .

Table 1 Estimations of the Thicknesses of Primary Ejecta, Local Materials, and Ejecta Deposits of Possible Crater/Basins ($D>100$ km) in the Vicinity of the CE-4 Landing Site

Periods (Fortezzo et al., 2020)		Source crater/ basin	Rim Diameter (km)[a]	Longitude (°)[a]	Latitude (°)[a]	Transient rim diameter (km)[b]	Spherical distance (km)	Primary ejecta (m)	Local materials (m)	Ejecta deposits (m)
		Humboldt	204.2	81.0	−27.2	143.0	2283.3	0.11	1.12	1.23
	↑	Tsiolkovskiy	183.7	129.0	−20.4	130.7	1421.4	0.24	1.82	2.07
Late Imbrian (Ic2)	Unknown order	Antoniadi	137.9	−173.0	−69.4	102.4	739.9	0.57	3.08	3.65
	↓	Aitken	133.0	173.1	−16.5	99.3	885.6	0.30	1.77	2.07
		Sklodowska	125.2	96.1	−18.0	94.4	2164.4	0.02	0.23	0.26
		Plaskett	121.7	177.6	81.6	92.1	3851.4	0.01	0.16	0.17
		Kovalevskaya	111.6	−129.4	30.8	85.5	2733.1	0.01	0.12	0.13
		Fizeau	106.8	−134.1	−58.2	82.4	958.8	0.11	0.70	0.81
		Plato	103.1	−9.4	51.6	80.0	5224.6	0.03	0.54	0.57
		Langemak	102.8	119.4	−9.9	79.8	1844.6	0.02	0.15	0.17
	The youngest →	Orientale	941.0	−95.2	−19.8	481.9	2245.8	13.88	139.33	153.22
	↑	**Schrödinger**	**321.5**	**133.1**	**−74.8**	**191.0**	**1061.4**	**2.58**	**16.55**	**19.13**
		Sinus Iridum	249.8	328.6	44.8	169.8	4840.1	0.23	4.00	4.24
Early Imbrian (Ic1)		Petavius	180.2	60.7	−25.4	128.6	2693.9	0.05	0.60	0.66
		Compton	165.6	104.4	55.8	119.7	3589.1	0.03	0.42	0.45
	Unknown order	Keeler	159.6	161.8	−9.8	116.0	1158.2	0.26	1.76	2.03
	↓	Letronne	114.9	−42.5	−10.7	87.7	3434.0	0.01	0.12	0.13
	The oldest →	Imbrium	1198.4	−17.1	34.3	484.3	4976.3	18.00	316.05	334.05

Note. The locations of those craters/basins are shown in Figure S4 in Supporting Information S1. The bold values are the information of Schrödinger basin that is the most possible.

[a]The crater diameters, longitudes, and latitudes are according to the crater database of Robbins (2019). [b]The transient rim diameters of craters ($D<300$ km) are converted from the final diameters based on Croft (1985), and the transient diameters of basins ($D>300$ km) Orientale, Schrödinger, and Imbrium are referred to the individual studies of Xie and Zhu (2016), Xu and Xie (2020), and Miljković et al. (2013). The ejecta thickness of the Orientale and Schrödinger basin is computed from individual measurements of Xie and Zhu (2016) and Xu and Xie (2020), and the others are computed from the ejecta distribution model of Pike (1974). Assuming the ejecta thickness distribution of each crater/basin to be symmetric.

4.3　Local Geologic History

Based on our interpretation of these layers, we propose the following geological history events of the deep subsur face structure of the CE－4 landing site (Figure 3):

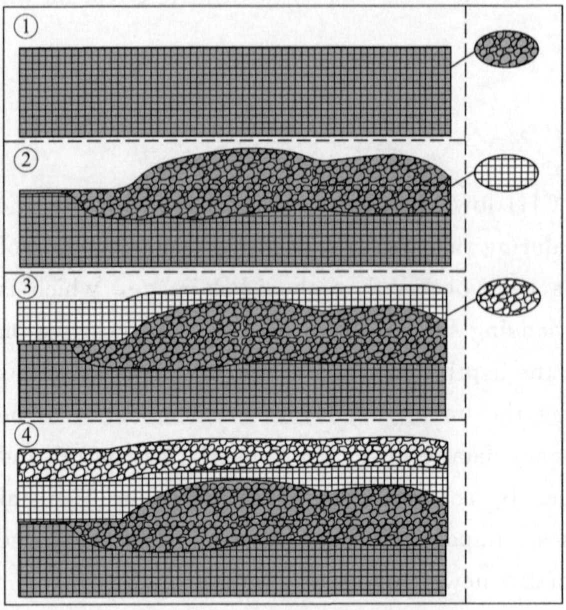

	Estimated age	Depth(±2m)	
		Upper boundary	Lower boundary
Basalt1	~3.60Ga	~139-161m	~211-220m
Ejecta 1	350-3.60Ga	~120-139m	~139-161m
Basalt2	~3.50Ga	~109-115m	~120-139m
Ejecta 2/ lave flow	~3.50Ga	~95-102m	~109-115m

Figure 3　Geological evolution model of the deep subsurface structure at the CE－4 landing site (The error of the depth is ±2m)

1. Early in its history, the interior of the crater was filled with a thick basaltic layer, with time, the exposed surface was subject to impact bombardment that broke down breccias and loose rocks forming a regolith layer of finer materials, the surface was also marginally modified by the arrival of ejecta from local impacts and those from outside Von Kármán (① in Figure 3).

2. Ejecta from the Schrödinger impact deposited in the CE－4 landing site area. The ejecta scoured the basalt unit and formed a mixed layer of both autochthonous and allochthonous materials. The exposed materials were thereafter at the mercy of impact bombardment. Therefore, the ejecta produced two substratigraphic structures: the lower materials represented by coarser ejecta with no obvious strong reflections, and the upper materials consisting of finer－grained particles causing strong reflection (from ① to ② in Figure 3). Due to the stochastic deposition mechanism, the thickness of this layer is uneven, with pronounced depressions in the lower part and bulges in the upper part. Affected by the undulating nature of the upper ejecta, the paleo－regolith atop also parallels the trend.

3. Following the formation of crater ejecta deposits and the development of the paleo－

regolith, the area was resurfaced by a mare flow. This basalt unit is different from the lower basalt unit (from ② to ③ in Figure 3), younger and more homogenous than the lower basalt unit.

4. The uppermost part also appears to represent ejecta from another impact and follows the underlying topogra phy (from ③ to ④ in Figure 3). Relating to this layer, the ejecta did not mix at depth but settled over the mare basalt, with little effect on the local topography.

5 Conclusions

In this study, we have processed the CH1 low - frequency LPR data collected by the Yutu - 2 rover on the farside of the Moon during the first 42 lunar days. By mitigation of system noises, a bifurcated structure has been observed in the LPR image, which is unique from the LPR results at the CE - 3 landing site. We identified four layers based on the location of the subsurface reflectors at the depth range of 90 - 310 m. Each layer has distinct undulations and was influenced by the underlying layers. It demonstrates the feasibility of using CE - 4 LPR low - frequency data to unveil the subsurface structure and hopefully resolve the trustworthiness issue. By comparing the modeling value of distal ejecta thickness of each possible large - scale impact event that occurred later than the Alder impact, the most probable source of the newly found structure is the Schrödinger basin. With Yutu - 2 rover moving toward the northwestern direction in the future, we expect the observed thickness of Stratum C will eventually reduce to ~2 m and its upper surface will be flat again. LPR data can obtain the full cross section of the "Bifurcation" feature for further analysis of its geomorphological parameters including length, symmetry, thickness variation, etc. to test the hypothesis of its origin.

The subsurface structure and stratigraphy within a depth of more than 300m at the CE - 4 landing area obtained with LPR low - frequency data can help with the interpretation of the geological activity events within the Von Kármán crater. Based on the results, the geological - historical events of the Von Kármán crater at the locality of the landing area include multiple phases of lava flows, punctuated by the arrival of ejecta materials from different impact events, while in the interim the exposed materials were subjected to impact bombardment. It also constrains the timing sequence of the major impact events that delivered ejecta to the landing area. The deviation between the observed and empirical modeling results of distal ejecta thickness calls for a more sophisticated prediction model, for example, considering the properties of target material, impact angle, etc.

References

[1] Baldwin, R. B. (1968). Rille pattern in the lunar crater Humboldt. Journal of Geophysical Research, 73 (10), 3227 - 3229. https://doi.org/10.1029/JB073i010p03227.

[2] Belousov, A., & Belousova, M. (2018). Dynamics and viscosity of 'a' a and pahoehoe lava flows of the 2012 - 2013 eruption of Tolbachik volcano, Kamchatka (Russia). Bulletin of Volcanology, 80 (1), 1 - 23. https://doi.org/10.1007/s00445-017-1180-2.

[3] Campbell, B. A., Hawke, B. R., Carter, L. M., Ghent, R. R., & Campbell, D. B. (2009). Rugged lava flows on the Moon revealed by Earth - based radar. Geophysical Research Letters, 36, L22201. https://doi.org/10.1029/2009GL041087.

[4] Chang, Y., Xiao, Z., Wang, Y., Ding, C., Cui, J., Cai, Y., et al. (2021)..

[5] An updated constraint on the local stratigraphy at the Chang'E-4 landing site. Earth and Planetary Physics, 5 (1), 19 - 31. https://doi.org/10.26464/epp2021007.

[6] Croft, S. K. (1985). The scaling of complex craters. Journal of Geophysical Research, 90 (S02), C828 - C842. https://doi.org/10.1029/JB090iS02p0C828.

[7] Cruikshank, D. P., Hartmann, W. K., & Wood, C. A. (1973). Moon: 'Ghost' craters formed during mare filling. The Moon, 7 (3 - 4), 440 - 452. https://doi.org/10.1007/BF00564645.

[8] Ding, C., Xiao, Z., Wu, B., Li, Z., Su, Y., Zhou, B., et al. (2021). Rock fragments in shallow lunar regolith: Constraints by the lunar penetrating radar onboard the Chang'E-4 mission. Journal of Geophysical Research: Planets, 126, e2021JE006917. https://doi.org/10.1029/2021JE006917.

[9] Fang, G., Zhou, B., Ji, Y., Zhang, Q., Shen, S., Li, Y. X., et al. (2014). Lunar penetrating radar onboard the Chang'e-3 mission. Research in Astronomy and Astrophysics, 14 (12), 1607 - 1622. https://doi.org/10.1088/1674-4527/14/12/009.

[10] Fortezzo, C. M., Spudis, P. D., & Harrel, S. L. (2020). Release of the digital unified global geologic map of the Moon at 1 : 5, 000, 000 - Scale. In 51st Lunar and Planetary Science Conference.

[11] Gou, S., Di, K., Yue, Z., Liu, Z., He, Z., Xu, R., et al. (2019). Lunar deep materials observed by Chang'e-4 rover. Earth and Planetary Science Letters, 528, 115829. https://doi.org/10.1016/j.epsl.2019.115829.

[12] Head, J. W., & Wilson, L. (2020). Rethinking lunar mare basalt regolith formation: New concepts of lava flow protolith and evolution of regolith thickness and internal structure. Geophysical Research Letters, 47, e2020GL088334. https://doi.org/10.1029/2020GL088334.

[13] Hu, X., Ma, P., Yang, Y., Zhu, M., Jiang, T., Lucey, P. G., et al. (2019). Mineral abundances inferred from in situ reflectance measurements of Chang'E - 4 landing site in south pole - Aitken Basin. Geophysical Research Letters, 46, 9439 - 9447. https://doi.org/10.1029/2019GL084531.

[14] Huang, J., Xiao, Z., Flahaut, J., Martinot, M., Head, J., Xiao, X., et al. (2018). Geological characteristics of Von Kármán crater, northwest ern south pole - Aitken Basin: Chang'E - 4 landing site region. Journal of Geophysical Research: Planets, 123, 1684 - 1700. https://doi.org/10.1029/2018JE005577.

[15] Huang, J., Xiao, Z., Xiao, L., Horgan, B., Hu, X., Lucey, P., et al. (2020). Diverse rock types detected in the lunar South Pole - Aitken Basin by the Chang' E - 4 lunar mission. Geology, 48 (7), 723 - 727. https://doi.org/10.1130/G47280.1.

[16] Lai, J., Xu, Y., Bugiolacchi, R., Meng, X., Xiao, L., Xie, M., et al. (2020). First look by the Yutu - 2 rover at the deep subsurface structure at the lunar farside. Nature Communications, 11 (1), 3426. https://doi.org/10.1038/s41467 - 020 - 17262 - w.

[17] Lai, J., Xu, Y., Bugiolacchi, R., Wong, H. K., Xu, L., Zhang, X., et al. (2021). A complex paleo - surface revealed by the Yutu - 2 rover at the lunar farside. Geophysical Research Letters, 48, e2021GL095133. https://doi.org/10.1029/2021GL095133.

[18] Lai, J., Xu, Y., Zhang, X., Xiao, L., Yan, Q., Meng, X., et al. (2019). Comparison of dielectric properties and structure of lunar regolith at Chang'e - 3 and Chang'e - 4 landing sites revealed by ground - penetrating radar. Geophysical Research Letters, 46, 12783 - 12793. https://doi.org/10.1029/2019GL084458.

[19] Li, C., Liu, D., Liu, B., Ren, X., Liu, J., He, Z., et al. (2019). Chang'E - 4 initial spectroscopic identification of lunar far - side mantle - derived materials. Nature, 569 (7756), 378 - 382. https://doi.org/10.1038/s41586 - 019 - 1189 - 0.

[20] Li, C., Xing, S., Lauro, S. E., Su, Y., Dai, S., Feng, J., et al. (2018). Pitfalls in GPR data interpretation: False reflectors detected in Lunar radar cross sections by Chang'e - 3. IEEE Transactions on Geoscience and Remote Sensing, 56 (3), 1325 - 1335. https://doi.org/10.1109/TGRS.2017.2761881.

[21] Lin, H., He, Z., Yang, W., Lin, Y., Xu, R., Zhang, C., et al. (2019). Olivine - norite rock detected by the lunar rover Yutu - 2 likely crystallized from the SPA impact melt pool. National Science Review, 7 (5), 15 - 920. https://doi.org/10.1093/nsr/nwz183.

[22] Lin, Y., Tronche, E. J., Steenstra, E. S., & van Westrenen, W. (2017). Evidence for an early wet Moon from experimental crystallization of the lunar magma ocean. Nature Geoscience, 10 (1), 14 - 18. https://doi.org/10.1038/

ngeo2845.

[23] Ling, Z., Qiao, L., Liu, C., Cao, H., Bi, X., Lu, X., et al. (2019). Composition, mineralogy and chronology of mare basalts and non-mare mate rials in Von Kármán crater: Landing site of the Chang'E-4 mission. Planetary and Space Science, 179, 104741. https://doi.org/10.1016/j.pss.2019.104741.

[24] Liu, J., Ren, X., Yan, W., Li, C., Zhang, H., Jia, Y., et al. (2019). Descent trajectory reconstruction and landing site positioning of Chang'E-4 on the lunar farside. Nature Communications, 10 (1), 4229. https://doi.org/10.1038/s41467-019-12278-3.

[25] Losiak, A., Wilhelms, D. E., Byrne, C. J., Thaisen, K. G., Weider, S. Z., Kohout, T., et al. (2009). A new lunar impact crater database. In 40th Annual Lunar and Planetary Science Conference.

[26] Lu, Y., Wu, Y., Michael, G. G., Ma, J., Cai, W., & Qin, N. (2021). Chronological sequence of Chang'E-4 landing zone within Von Kármán crater. Icarus, 354, 114086. https://doi.org/10.1016/j.icarus.2020.114086.

[27] Melosh, H. J., Kendall, J., Horgan, B., Johnson, B. C., Bowling, T., Lucey, P. G., & Taylor, G. J. (2017). South Pole-Aitken basin ejecta reveal the Moon's upper mantle. Geology, 45 (12), 1063-1066. https://doi.org/10.1130/G39375.1.

[28] Miljkovic, K., Wieczorek, M. A., Collins, G. S., Laneuville, M., Neumann, G. A., Melosh, H. J., et al. (2013). Asymmetric distribution of lunar impact basins caused by variations in target properties. Science, 342 (6159), 724-726. https://doi.org/10.1126/science.1243224.

[29] Moriarty, D. P., & Pieters, C. M. (2018). The character of South Pole-Aitken Basin: Patterns of surface and subsurface composition. Journal of Geophysical Research: Planets, 123, 729-747. https://doi.org/10.1002/2017JE005364.

[30] Oberbeck, V. R., Hörz, F., Morrison, R. H., Quaide, W. L., & Gault, D. E. (1975). On the origin of the lunar smooth-plains. The Moon, 12 (1), 19-54. https://doi.org/10.1007/BF02626332.

[31] Ohtake, M., Uemoto, K., Yokota, Y., Morota, T., Yamamoto, S., Nakamura, R., et al. (2014). Geologic structure generated by large-impact basin formation observed at the South Pole-Aitken basin on the Moon: Rock and Melt Pool within the SPA Basin. Geophysical Research Letters, 41, 2738-2745. https://doi.org/10.1002/2014GL059478.

[32] Pasckert, J. H., Hiesinger, H., & van der Bogert, C. H. (2018). Lunar farside volcanism in and around the South Pole-Aitken basin. Icarus, 299, 538-562. https://doi.org/10.1016/j.icarus.2017.07.023.

[33] Petro, N. E., & Pieters, C. M. (2004). Surviving the heavy bombardment:

Ancient material at the surface of south Pole - Aitken Basin. Journal of Geophysical Research, 109, E06004. https: //doi. org/10. 1029/2003JE002182.

[34] Petro, N. E., & Pieters, C. M. (2006). Modeling the provenance of the Apollo 16 regolith. Journal of Geophysical Research, 111, E09005. https:// doi. org/10. 1029/2005JE002559.

[35] Petro, N. E., & Pieters, C. M. (2008). The lunar - wide effects of basin ejecta distribution on the early megaregolith. Meteoritics & Planetary Science, 43 (9), 1517 - 1529. https: //doi. org/10. 1111/j. 1945 - 5100. 2008. tb01025. x.

[36] Pettinelli, E., Lauro, S. E., Mattei, E., Cosciotti, B., & Soldovieri, F. (2021). Stratigraphy versus artefacts in the Chang'e - 4 low - frequency radar. Nature Astronomy, 5 (9), 890 - 893. https://doi. org/10. 1038/s41550 - 021 - 01432 - x.

[37] Pike, R. J. (1974). Ejecta from large craters on the Moon: Comments on the geometric model of McGetchin et al. Earth and Planetary Science Letters, 23 (3), 265 - 271. https://doi. org/10. 1016/0012 - 821X (74) 90114 - 9.

[38] Rapp, J. F., & Draper, D. S. (2018). Fractional crystallization of the lunar magma ocean: Updating the dominant paradigm. Meteoritics & Planetary Science, 53 (7), 1432 - 1455. https: //doi. org/10. 1111/maps. 13086.

[39] Robbins, S. J. (2019). A new global database of lunar impact craters > 1 - 2km: 1. Crater locations and sizes, comparisons with published data - bases, and global analysis. Journal of Geophysical Research: Planets, 124, 871 - 892. https: // doi. org/10. 1029/2018JE005592.

[40] Slyuta, E. N. (2014). Physical and mechanical properties of the lunar soil (a review). Solar System Research, 48 (5), 330 - 353. https: //doi. org/10. 1134/ S0038094614050050.

[41] Su, Y., Fang, G., Feng, J., Xing, S., Ji, Y., Zhou, B., et al. (2014). Data processing and initial results of Chang'e - 3 lunar penetrating radar. Research in Astronomy and Astrophysics, 14 (12), 1623 - 1632. https: //doi. org/10. 1088/ 1674 - 4527/14/12/010.

[42] Wu, W., Li, C., Zuo, W., Zhang, H., Liu, J., Wen, W., et al. (2019). Lunar farside to be explored by Chang' e - 4. Nature Geoscience, 12 (4), 222 - 223. https: //doi. org/10. 1038/s41561 - 019 - 0341 - 7.

[43] Xie, M., & Zhu, M. (2016). Estimates of primary ejecta and local material for the Orientale basin: Implications for the formation and ballistic sedimentation of multi - ring basins. Earth and Planetary Science Letters, 440, 71 - 80. https: // doi. org/10. 1016/j. epsl. 2016. 02. 012.

[44] Xu, L., & Xie, M. (2020). Ejecta thickness distribution of the Schrödinger basin on the Moon. Journal of Geophysical Research: Planets, 125, e2020JE006506. https: //doi. org/10. 1029/2020JE006506.

[45] Xu, L., Zhang, X., Qiao, L., & Lai, J. (2021). Evaluating the thickness and stratigraphy of ejecta materials at the Chang'e-4 landing site. The Astronomical Journal, 162 (1), 29. https://doi.org/10.3847/1538-3881/abf8b0.

[46] Yingst, R. A., Chuang, F. C., Berman, D. C., & Mest, S. C. (2017). Geologic mapping of the Planck Quadrangle of the Moon (LQ-29). In 48th Lunar and Planetary Science Conference.

[47] Yuan, Y., Zhu, P., Xiao, L., Huang, J., Garnero, E. J., Deng, J., et al. (2021). Intermittent volcanic activity detected in the Von Kármán crater on the farside of the Moon. Earth and Planetary Science Letters, 569, 117062. https://doi.org/10.1016/j.epsl.2021.117062.

[48] Zhang, L., Li, J., Zeng, Z., Xu, Y., Liu, C., & Chen, S. (2020). Stratigraphy of the Von Kármán crater based on Chang'E-4 lunar penetrating radar data. Geophysical Research Letters, 47, e2020GL088680. https://doi.org/10.1029/2020GL088680.

[49] Zhang, L., Xu, Y., Bugiolacchi, R., Hu, B., Liu, C., Lai, J., et al. (2021). Rock abundance and evolution of the shallow stratum on Chang'e-4 landing site unveiled by lunar penetrating radar data. Earth and Planetary Science Letters, 564, 116912. https://doi.org/10.1016/j.epsl.2021.116912.

[50] Zhang, L., Zeng, Z., Li, J., Huang, L., Huo, Z., Zhang, J., & Huai, N. (2019). A story of regolith told by Lunar Penetrating Radar. Icarus, 321, .

[51] 148-160. https://doi.org/10.1016/j.icarus.2018.11.006.

[52] Zhang, J., Zhou, B., & Lin, Y. (2021). Reply to: Stratigraphy versus artefacts in the Chang'e-4 low-frequency radar. Nature Astronomy, 5 (9), 894-897. https://doi.org/10.1038/s41550-021-01484-z.

[53] Zhang, J., Zhou, B., Lin, Y., Zhu, M., Song, H., Dong, Z., et al. (2021). Lunar regolith and substructure at Chang'E-4 landing site in South Pole-Aitken basin. Nature Astronomy, 5 (1), 25-30. https://doi.org/10.1038/s41550-020-1197-x.

[54] Zhu, T., Zhang, J., & Lin, Y. (2021). Ultra-thick paleoregolith layer detected by lunar penetrating radar: Implication for fast regolith formation between 3.6 and 2.35 Ga. Geophysical Research Letters, 48, e2021GL095282. https://doi.org/10.1029/2021GL095282.

An updated constraint on the local stratigraphy at the Chang'E－4 landing site

YiRen Chang[1], ZhiYong Xiao[2,3], YiChen Wang[2], ChunYu Ding[2], Jun Cui[2,3,4], and YuZhen Cai[2]

1. State Key Laboratory of Lunar and Planetary Sciences, Macau University of Science and Technology, Macau 999078, China
2. Planetary Environmental and Astrobiological Research Laboratory, School of Atmospheric Sciences, Sun Yat－sen University, Zhuhai, Guangdong 519000, China
3. Center for Excellence in Comparative Planetology, Chinese Academy of Sciences, Hefei 230000, China
4. National Astronomical Observations of China, Chinese Academy of Sciences, Beijing 100000, China

Abstract: The Chang'E－4 mission has been exploring the lunar farside. Two scientific targets of the rover onboard are (1) resolving the possible mineralogy related to the South Pole－Aitken basin and (2) understanding the subsurface processes at the lunar farside. Publications to date that are based on the reflectance spectra and radar data obtained by the rover have shown a persistent inconsistency about the local stratigraphy. To explain both the abnormal surface topography at the landing site and the unexpected radargram observed by the rover, the Alder crater has been frequently reported to be older than the mare basalts at that landing site. However, this argument is not supported by earlier geological mapping nor recent crater statistics. Resolving this controversy is critical for a full understanding of the geological history of the landing area and for correct interpretations of the scientific data returned. Employing detailed crater statistics, rigorous statistical analyses, and an updated crater chronology function, this study is determined to resolve the relative ages of the Alder crater, Finsen crater, and the mare basalts on the floor of Von Kármán. Our results reveal that while background secondaries and local resurfacing have widely occurred in the study area, affecting age determinations, the statistics are significant enough to conclude that the Alder crater is the oldest among the three targets. This independent constraint is consistent with both the crosscutting relationships of different terrains in this area and global stratigraphic mapping. Our results exclude Alder as a possible contributor of the post－mare deposits at the landing site, appealing for a more systematic stratigraphy study to resolve the provenances of these deposits.

Keywords: Moon; Chang'E－4; impact craters; absolute model age; stratigraphy

Key Points:

• Detailed crater statistics were performed for key stratigraphic markers at the Chang'E－4 landing region.

• Alder is older than the mare basalts at the landing site, and Finsen is the youngest.

• Minor contributions of ejecta by Alder should exist at the shallow subsurface.

1 Introduction

The Chang'E－4 mission, consisting of a relay satellite, a lander, and a rover, has been successfully deployed to the floor of the Von Kármán crater (diameter $D = 185$km; 46.4°S, 176.5°E). The landing area is located within the South Pole－Aitken (SPA) basin (Wu WR et al., 2017), which is the largest impact structure on the Moon. The formation of the SPA basin has excavated ancient lunar crustal and even upper mantle materials (Potter et al., 2012). However, the possible composition of the lunar mantle has been elusive in previous studies that were based on orbital composition data and numerical simulations of the SPA impact (Melosh et al., 2017). Two important reasons for the debates are that (1) simulations of large impact basins depend heavily on unconstrained parameters, and (2) interpretations of planetary remote sensing normally have multiplicity. Therefore, sampling the SPA has been a high priority target for furture surface explorations on the Moon.

A key scientific target of the Chang'E－4 mission (CE－4) is to explore the composition of surface materials within the SPA, finding information that might be traced to possible mantle materials excavated by this ancient basin (Wu WR et al., 2017). However, this is not a straightforward task. Besides the ambiguous initial crustal configuration after the SPA basin was formed (e.g., whether or not the hypothesized thick melt sheet has been differentiated), postimpact modification has been extensive, as numerous impact basins and multiple episodes of volcanic eruptions have occurred at different places within the SPA (Moriarty Ⅲ and Pieters, 2018). Resolving the provenance of surface materials and detecting the primordial materials formed by the SPA is critical to finding the mantle materials that were excavated.

Based on reflectance spectra returned by the rover, most interpretations have suggested that the surface regolith is enriched in calcium－rich pyroxenes, and that olivine is a minor component (Lin HL et al., 2020; Ma P et al., 2020; Huang J et al., 2020), which are consistent with orbital observations (Huang J et al., 2018; Ling ZC et al., 2019). An obvious impact ray is visible at the landing site, which was delivered by the Finsen crater ($D = 73$km; 42.3°S, 177.7°W) to the northeast of Von Kármán. Therefore, the provenance of the surface regolith is usually ascribed to the distal ejecta delivered by Finsen (Figure 1). On the other hand, morphological studies have shown that the entire landing area has been occupied by secondaries with various sizes and degradation states, which were delivered by craters from different directions (Huang J et al., 2018). Mixing of surface materials should be pronounced (Figure 1c), which is evident by the different compositions exposed from various depths by relatively large impact craters (Huang J et al., 2018; Li CL et al., 2019). Indeed, both basaltic (Huang J et al., 2020) and feldspathic rocky fragments (Ma P et al., 2020) have been found by the rover. Therefore, understanding the major source (s) of regolith at the landing site is critically important to interpret the compositional data returned. Specifically, besides Finsen, which other crater (s) contributed additional ejecta to the post-mare deposits?

Following the first successful exterrestrial ground-penetrating radar, performed by the Chang'E-3 mission (Xiao L et al., 2015; Fa WZ et al., 2015; Zhang JH et al., 2015), the Yutu-2 rover of Chang'E-4 is equipped with a similar payload to investigate the propagation of electromagnetic waves in lunar farside materials. A scientific target is to understand the formation of regolith at the lunar farside. In the radargram obtained, materials deeper than ~40-45m are believed to be mare basalts that filled in the floor of Von Kármán; materials shallower than ~10-13m (Lai JL et al., 2019, 2020; Li CL et al., 2020; Ding CY et al., 2020; Zhang L et al., 2020) have dielectric properties similar to typical lunar regolith.

The possible provenance of the >~40m thick porous deposits is obscure, and widespread discrepancies exist in the interpretations. Referring to published stratigraphic ages of impact structures around the landing site (Figure 1a), earlier studies predicted that Finsen was the major contributor of the post-mare deposits (Li CL et al., 2019). Employing an analytical model for the thickness of impact ejecta as a function of radial distances to the parent crater, the total thickness of ejecta delivered to the landing site by all significant post-mare craters is less than 14 meters (Huang J et al., 2018; Lai JL et al., 2020). However, numerical simulations for the formation of Finsen- and Alder-sized craters predicted an ejecta thickness of >60meters at the landing sites (Di KC et al., 2019), which is much larger than the thickness of postmare deposits observed by the lunar penetrating radar (Lai JL et al., 2019; Lai JL et al., 2020; Li CL et al., 2020). On the other hand, the thickness of post-mare deposits constrained by the radargram is consistent with orbital observations, which were based on the reflectance spectra of small craters that have exposed subsurface mare basalts (Huang J et al., 2018; Ling ZC et al., 2019; Li CL et al., 2019). After scanning all the possible source craters that may have delivered ejecta to the landing site (Huang J et al., 2018; Lai JL et al., 2020), the key question now is whether Alder is older or younger than the mare basalts at the landing site (Figure 1a).

The above literature review summarizes the key question targeted in this paper, which is to resolve the relative ages of the Alder crater, the Finsen crater, and the mare basalts on the floor of Von Kármán. While we are aware that earlier studies (Ivanov et al., 2018; Pasckert et al., 2018; Huang J et al., 2018; Ling ZC et al., 2019; Gou S et al., 2020;

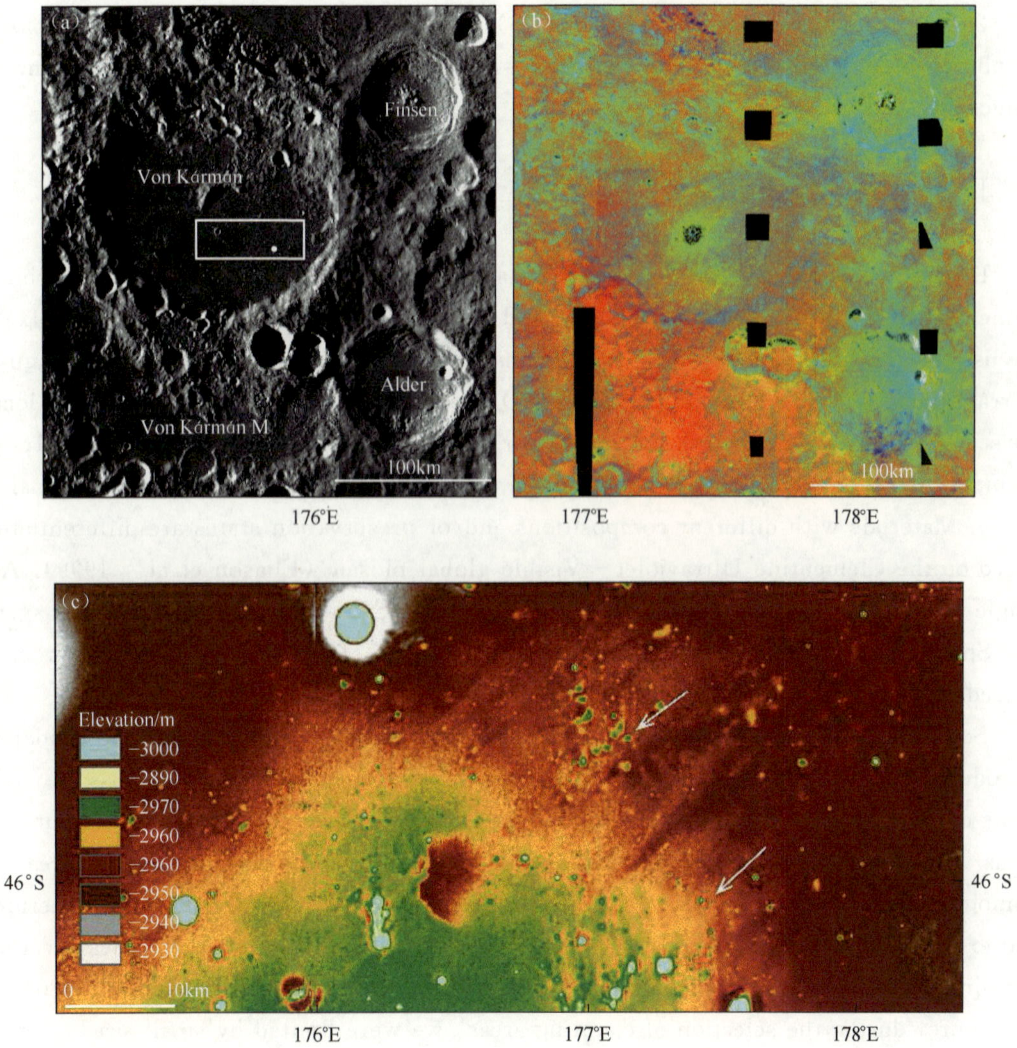

Figure 1　Geological context of the Chang'E－4 landing area and the landing site is marked as a white circle. (a) The landing area is located on a mare unit, and key craters in the surrounding area are denoted in white. The central coordinates of this area are 46°S, 178°E. (b) Composition and maturity difference of materials in the landing area and impact rays delivered by the Finsen crater are the blueish diffuse materials. (c) The regional topography of the landing site is dominated by the northeast－to－southwest－trending lineations that were formed by the Finsen crater. Note that numerous younger secondaries are visible in this map, e. g. , white arrows. The location of this area is denoted as the white box in panel (a)

Lu et al. , 2021) have tried to estimate the model ages for some of our targets via crater statistics, this paper is built on more focused studies of various crater populations for each of the targets. Strictly following technical suggestions to perform crater counts (Michael and Neukum, 2010) and analyze crater sizefrequency distributions (SFD; Robbins et al. , 2018), we avoid caveats known to the crater count community, such as obvious secondaries (Xiao ZY and Strom, 2012; Xiao ZY, 2018), image resolution (Wang YC et al., 2020), and slopes (Xiao ZY et al., 2013). Multiple counting areas are selected based on photogeology study for each of the targets, and crater populations at various diameter ranges are studied.

2 Data and Method

Different versions of global mosaics obtained by the Lunar Reconnaissance Orbiter Camera (LROC; Robinson et al., 2010) and Wide Angle Camera (WAC; 100 m/pixel) are used as the base map to study the regional context. Images obtained by the Kaguya Terrain Camera (TC; Haruyama et al., 2008) are used to analyze the morphology of local areas. The LOLA digital elevation model coregistered with SELENE data (SLDEM; 59 m/pixel; Barker et al., 2016) is employed to analyze the topography (Barker et al., 2016). Materials with different compositions and/or preservation states are differentiated based on the Clementine Ultraviolet - Visible global mosaic (Eliason et al., 1999). All single frame imagery data are calibrated using the USGS Integrated Software for Imagers and Spectrometers (http://isis.astrogeology.usgs.gov/) following the standard process procedure. Table S1 summarizes the data sources used in this study.

Selection of counting areas was based on the above data, following the guidance introduced by Michael and Neukum (2010). Continuous ejecta deposits and possible melt pools on both crater rims and crater floors were taken as the counted areas. The counting areas voided complicated topographies and shadows, and each counting area is homogeneous in terms of surface morphology (Xiao ZY et al., 2013). Both the interiors and exteriors of Alder and Finsen have been heavily contaminated by secondaries (section 3), similar to the situation on the mare surface in Von Kármán (Huang J et al., 2018). Therefore, during the selection of counting areas, we were limited by small sizes of ideal counting areas. To enhance statistical significance, we also selected larger areas that include shallow slope regions, but only craters that are much larger than half of the topographic wavelength are included, so that potential negative influences caused by mass movements are avoided (Xiao ZY et al., 2013). This approach allows us to perform cross comparisons of results derived for multiple counting areas of a single target. Table S2 in the appendix shows detailed information of the counting areas.

To avoid incomplete recognition, images with a solar incidence angle of $> 60°$ were used to collect craters in the counting areas (Ostrach et al., 2011). Table S3 shows information of the base images used for the counts. The crater counts were performed by exhaustive searching of craters larger than 3 pixels of the base image, but only craters larger than 10 pixels are considered completely included (Wang YC et al., 2020). The counting process has excluded obvious chains and clusters of secondaries (McEwen and Bierhaus, 2006; Xiao ZY and Strom, 2012).

The novel technique of analyzing crater size - frequency distribution (SFD) advocated by Robbins et al. (2018) is used here. A Gaussian kernel density estimator (i. e. KDE) is used to derive the crater SFD, and the mean value and bandwidth of the KDE are the measured crater diameter (D) and 0.1D, respectively. The hybrid - bootstrap method is used to estimate the 2σ confidence intervals of the SFD (Robbins et al., 2018). Compared to traditional methods of analyzing crater SFD, the new technique is more rigorous in terms of statistic reliability (Robbins et al., 2018). The dedicated crater counts performed here are characterized by small samples of usable craters (Section 3), because shadows normally occupy large areas of images obtained for such high latitude areas, and abundant secondaries are recognized in these areas, limiting both the size of usable counting areas and the number of usable craters (Section 3). The advantage of this new statistical technique is especially apparent for such small samples. The crater chronology function proposed by Robbins (2014) is updated from that of Neukum et al. (2001), and it is used here together with the production function proposed by Neukum et al. (2001) to derive the absolute model ages.

3 Results

3.1 Crater Size - Frequency Distribution at the Finsen Crater

Finsen is a typical complex crater that features faint impact rays (Figure 1b). It has been taken as a Copernican - aged crater (Ling et al., 2018), but the preservation of impact rays is not strictly correlated with the stratigraphic age of the parent crater (Hawke et al., 2004). In geologic maps published by both Wilhelms et al. (1979) and Fortezzo et al. (2020), Finsen was mapped as an Eratosthenian - aged crater. Previous crater counts for Finsen (Ivanov et al., 2018; Gou et al., 2020) yielded a model age of ~3.5 billion years (Ga), which is not so different compared to the ~3.6 Ga old mare basalts in the floor of Von Kármán (Huang J et al., 2018; Ling ZC et al., 2019; Pasckert et al., 2015). After the submission of this paper, a new publication on the model age of Finsen emerged, which was based on crater counts on the southern floor and ejecta, yielding numbers of ~3.0 and 3.1 Ga, respectively (Lu et al., 2021).

We take two strategies of crater statistics to investigate the model age of the Finsen crater. Four populations of relatively large craters at several locations of both the continuous ejecta deposits and crater floors are selected (Figure 2), and 7 populations of relatively small impact craters on several smaller areas are selected (Figure 3). The two sets of crater populations are complementary to each other, as they cover different diameter ranges. Therefore, possible contributions caused by background secondaries (Xiao ZY, 2018) and heterogeneous resurfacing processes can be investigated. A ~30km diameter is located on the northeastern crater rim of Finsen (Figure 1a), so its secondaries and the impact - induced resurfacing have heavily affected nearby regions. Therefore, the selected counting areas are away from the northern and eastern rims, and craters on the floor of Finsen are used with special cautions (Figures 2 and 3).

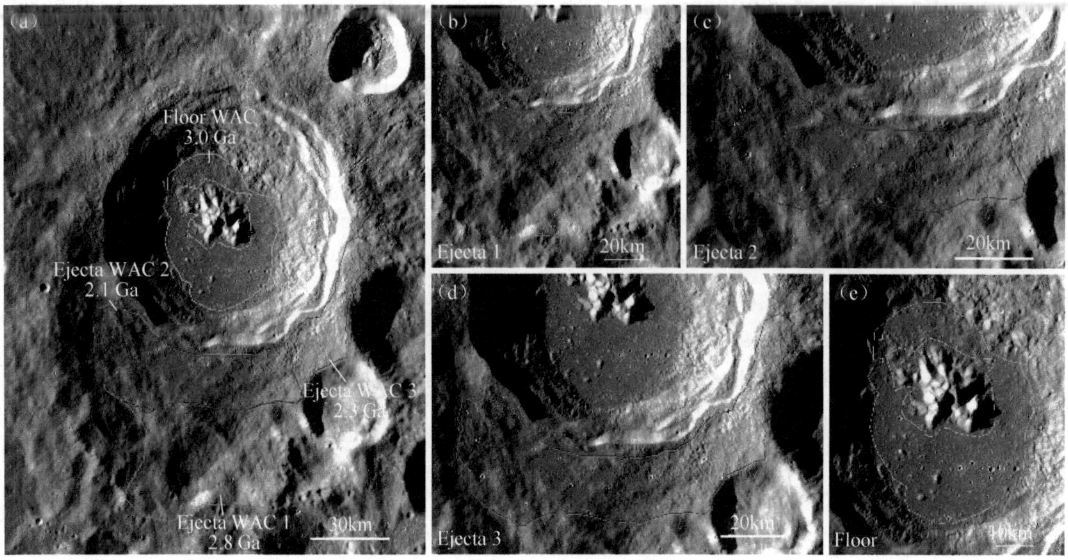

Figure 2　Collections of relatively large craters on the continuous ejecta deposits and floor of Finsen. Panel (a) shows the locations of the counting areas, and panels (b) - (e) show the detailed morphology of the counting areas and craters used to estimate absolute model ages. Craters as small as several pixels were included in counts, but only those larger than chains and clusters of obvious secondaries are used to derive the crater densities (marked as yellow circles). For crater populations collected in panels (b) - (e), the minimum diameter used for age determination is 1. 1, 0. 7, 0. 7 and 0. 8km, respectively. Names of the counting areas are denoted nearby, and WAC means that the count was based on LROC WAC mosaics

The absolute model ages derived from the four populations of relatively large craters and those derived from the seven populations of relatively small craters are shown in Figures 4 and 5, respectively. The results differ from each other by a maximum factor of >2, i. e. , the $D>0.4$km small crater population on the crater floor (Figure 5c) versus the $D>0.15-0.45$km crater populations on the continuous ejecta deposits (Figure 5a). Within error bars, most model ages of the 11 populations are around ~2. 0Ga (Figures 4c, 4d, 5b and 5d). The population of relatively large craters on the crater floor yields a comparable model age with that derived by Lu et al. (2021), Gou et al. (2020), and Ivanov et al. (2018), testifying a good repeatability of our crater collection. In conclusion, the majority of the absolute ages modelled for Finsen are less than 3. 0Ga.

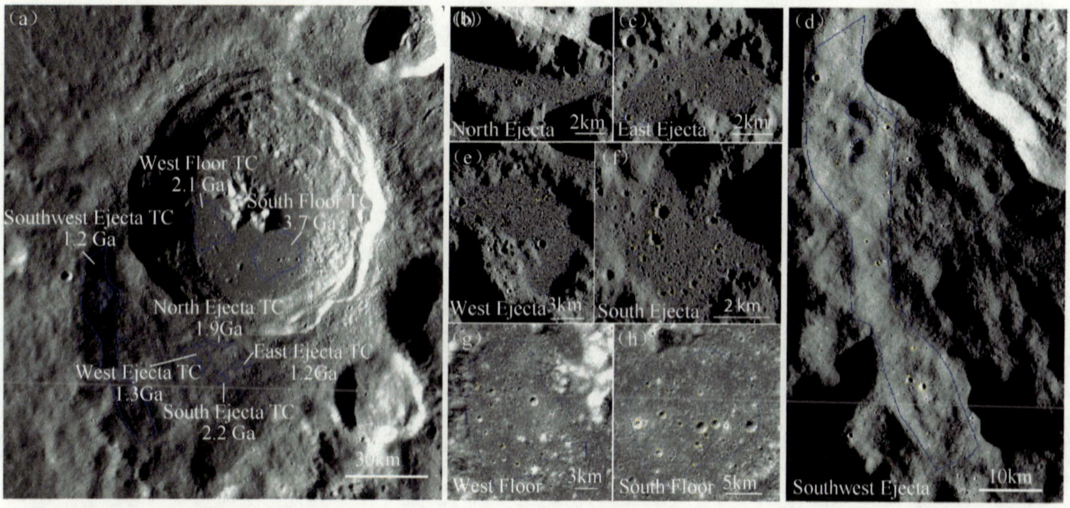

Figure 3　Collections of relatively small craters on the continuous ejecta deposits and floor of Finsen. Panel (a) shows the locations of the counting areas, and panels (b) - (h) show the detailed morphology of the counting areas and craters used to estimate absolute model ages. Craters as small as several pixels were included in counts, but only those larger than chains and clusters of obvious secondaries are used to derive the crater densities (marked as yellow circles). For crater populations collected in (b) - (h), the minimum diameter used for age determination is 0.14, 0.15, 0.45, 0.2, 0.15, 0.5, and 0.4km, respectively. Names of the counting areas are denoted nearby; TC means the count was based on Kaguya TC images

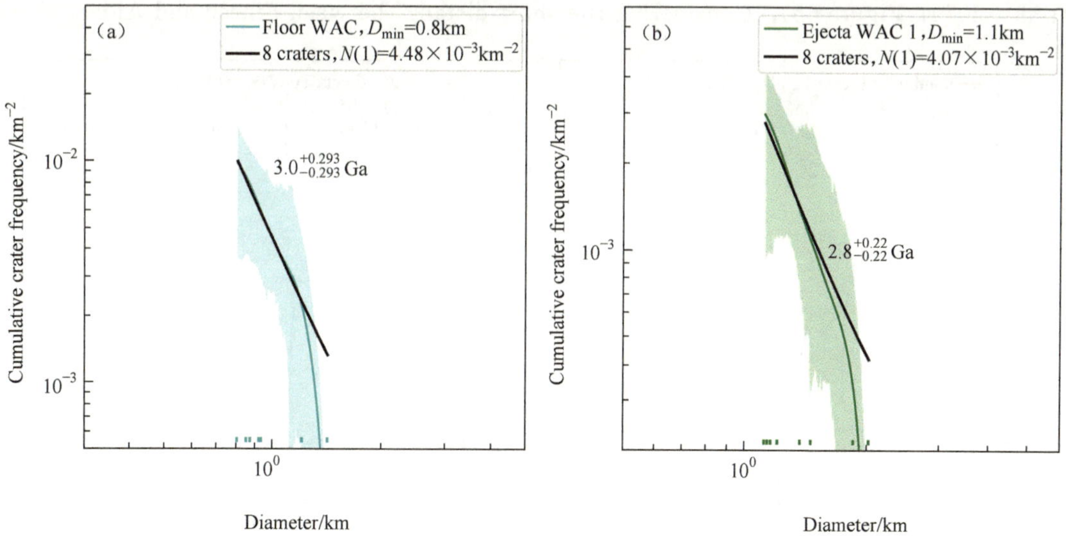

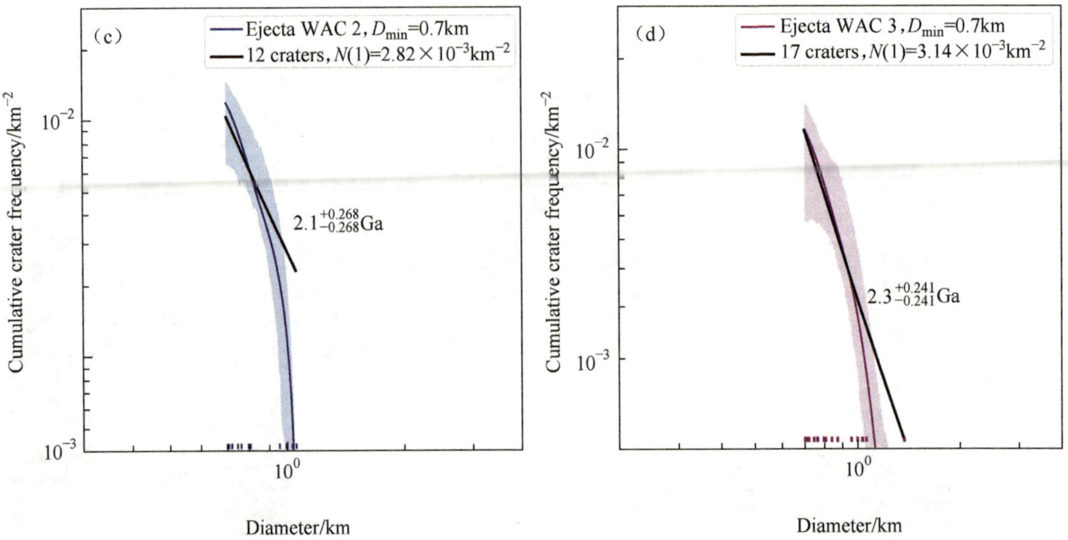

Figure 4　Cumulative crater size - frequency distribution and absolute model ages for the four populations of relatively large craters collected on the crater floor and ejecta deposits of Finsen. The colors and names of the counting areas correspond to those shown in Figure 2. Shaded colors are the errors for the crater size - frequency distributions

3.2 Crater Size - Frequency Distribution at the Alder Crater

The overlapping relationship between ejecta formed by Finsen and Alder suggests that Finsen is younger than Alder, which is consistent with the updated geologic map (Fortezzo et al., 2020; Figure 6b). Considering the short distance between Finsen and Alder, the

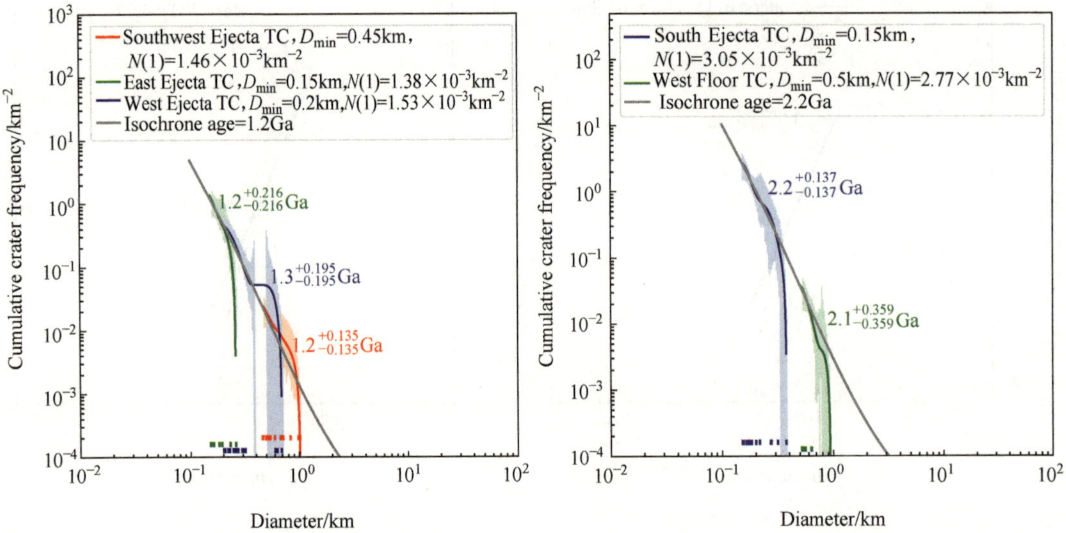

(a)　　(b)

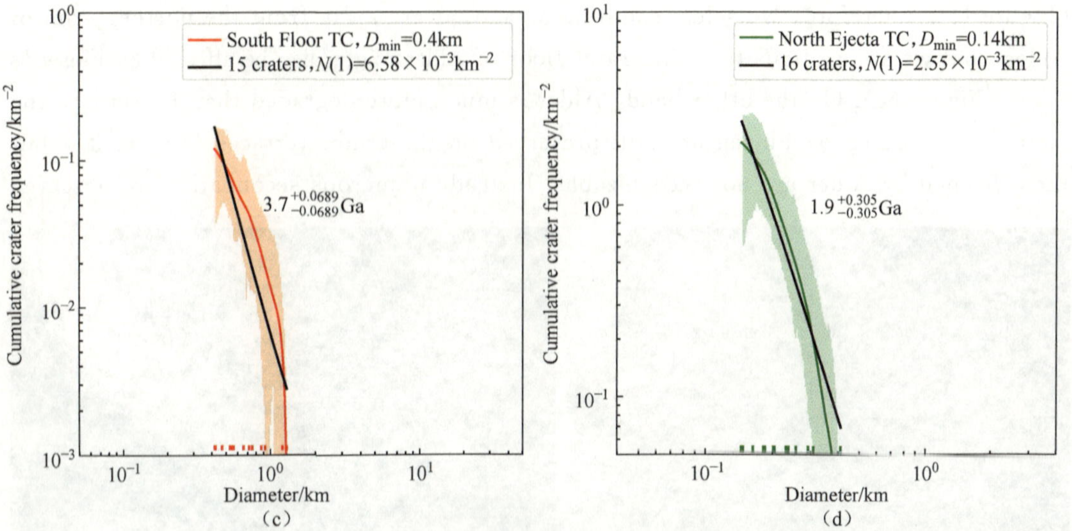

Figure 5　Cumulative crater size - frequency distribution and absolute model ages for the seven populations of relatively small craters collected on the crater floor and ejecta deposits of Finsen. The names of the counting areas correspond to those shown in Figure 3. Shaded colors are the errors for the crater size - frequency distributions

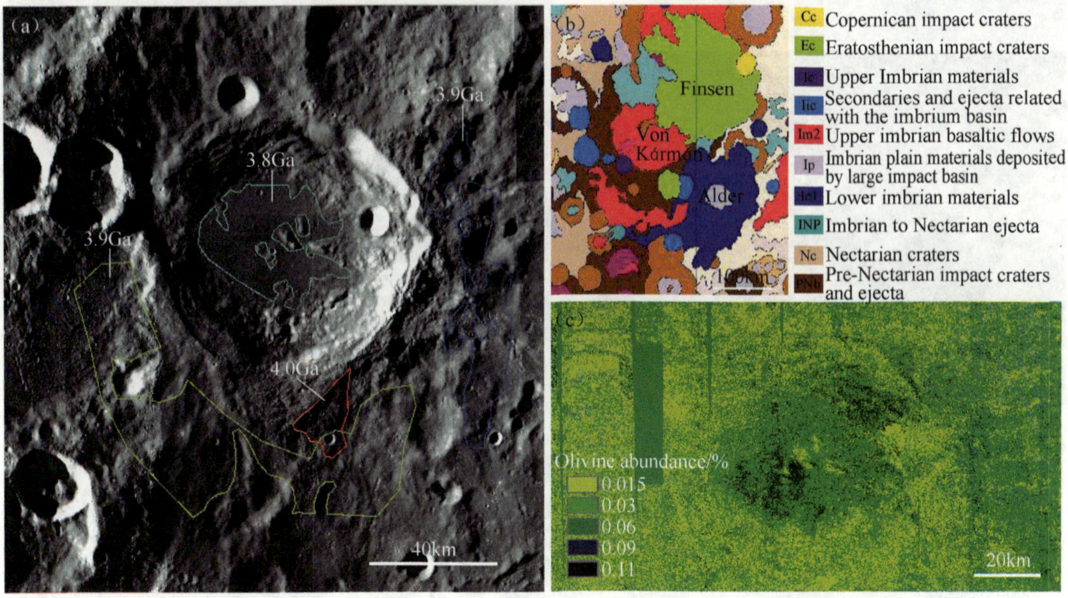

Figure 6　Morphological context of the Alder crater and the counting areas.

(a) The four counting areas on the floor and continuous ejecta deposits of Alder are marked in different colors. (b) Geologic map for the area of interest around the landing site (Fortezzo et al., 2020). Stratigraphic units with different colors are explained in the figure legend. (c) Contents of olivine within and around the Alder crater derived from the Kaguya Multiband Imager (Lemelin et al., 2016)

Finsen impact has substantially modified the northern rim and floor of Alder, evidenced by the strong modification of the mare basalts on the floor of Von Kármán caused by Finsen's ejecta

(Figure 1c). Therefore, we select counting areas that are away from the northern rim of Alder, and we are cautious that the crater floor of Alder might also be affected by Finsen's ejecta (Figure 6a). On the other hand, Alder is much more degraded than Finsen, as the melt sheets formed by Finsen are still preserved on the crater terraces (Figure 3), but those formed by Alder are not recognizable. Instead, numerous secondaries are observed

Figure 7　Detailed morphology of the 4 counting areas selected within and around Alder. The color and name of the counting areas correspond to those shown in Figure 6. Yellow circles in each panel mark the location of craters that are used to estimate the crater densities

overlapping Alder' s ejecta deposits (Figure 7). Therefore, the heavy topographic degradation and large abundance of superposed secondaries at Alder prohibit us from using populations of small craters to estimate the model ages. In total, three areas on the continuous ejecta deposits and the entire crater floor are selected to investigate the absolute model age of Alder (Figures 6 and 7).

The absolute ages modeled for the four populations of impact craters collected at Alder are shown in Figure 8. Considering error bars, ejecta on the southern ejecta (Figure 8d) is significantly older and that on the crater floor is significantly younger (Figure 8b). The ages are clustered around ~3. 9Ga with a much smaller range than the 11 ages derived for Finsen (Figure 5). As such, the modeled ages of Alder are uniformly larger than 3. 8Ga.

4　Discussion

4. 1　Age Relationships Between the Mare Basalts in Von Kármán and the Alder and Finsen Craters

4. 1. 1　Inconsistencies in ages modelled for different crater populations

For both Finsen and Alder, the absolute ages modelled for the different crater populations exhibit statistically significant differences (Figures 4, 5 and 8). The crater counts have strictly followed standard technical requirements (Michael and Neukum, 2010; Xiao ZY and Strom, 2012; Wang YC et al., 2020), so that each of the crater populations collected was designed to represent the age of the target crater.

Considering error bars, 5 of the 11 crater populations collected on both the floor and ejecta deposits of Finsen are clustered around 2. 0Ga (Figures 4c, 4d, 5d); 2 are close to ~3Ga (Figures 4a and 4b); 3 are around ~1. 2Ga (Figure 5a); and 1 is ~3. 7Ga (Figure 5c). The 3 crater populations collected on the floor yield model ages of 3. 0Ga (Figure 4a), 2. 1Ga (Figure 5b), and 3. 7 Ga (Figure 5c), indicating a significantly non - random spatial distribution. Figure 2e shows that on the crater floor of Finsen, craters larger than 0. 8km are obviously clustered at the southeast, where the 3. 7 Ga model age was derived (Figures 3h and 5c). The eastern crater floor is completely depleted of craters larger than 0. 8km (Figure 2e). However, when employing a prevalent algorithm to estimate the spatial distribution of impact craters (Michael et al., 2012), the statistical test does not show obvious clustering for the $D>0.8$km population (Figure S1 in the appendices). This echoes a previous test, which showed that the spatial distribution algorithms were not accurate (Xiao ZY and Werner, 2015). Our observations suggest that the $D>1.1$km crater population on the crater floor is heavily contaminated by background secondaries that cannot be differentiated from similar - sized primaries based on morphology (Xiao ZY, 2018). Therefore, the crater densities derived do not represent the true value for the crater floor.

Notably, the 3.68±0.03Ga, ~3.5Ga, ~3.0Ga ages of Finsen reported by Ivanov et al. (2018), Gou et al. (2020), and Lu et al. (2021), respectively, were all located on the crater floor. Of the four melt pools investigated on the southern rim of the Finsen crater, three exhibit clustered model ages around 1.2Ga (Figure 5a), while the remaining one is ~2.0Ga (Figure 5b). We do not have an explanation for the 1.2Ga age, which is substantially younger than the majority of ages (i.e., ~2.0Ga). Normally, crater populations that are ~2.0Ga and younger have not reached an equilibrium state at $D > 0.2$km (Xiao ZY and Werner, 2015), so the small ages are not due to crater obliteration, as demonstrated by their sparse spatial distribution. Contamination by secondaries is serious at smaller diameters, as shown by obvious secondaries chains at smaller diameters Figure 3). However, contamination by background secondaries would cause densities larger than the average value, which is inconsistent with the smaller densities observed. Two possible explanations exist, but we are not able to find evidence to confirm or exclude either: (1) the Finsen impact was accompanied by immediate but uneven deposition of self-secondaries (Xiao ZY et al., 2016), such that a quick equilibrium was achieved on the melt pools that would have been otherwise impossible on normal mare surfaces (Xiao ZY and Werner, 2015); (2) as evidenced by the numerous impact rays that pass the southern rim of Finsen (Figure 1b), post-impact resurfacing by distal ejecta might have occurred on the southern rim of Finsen, which would have preferentially removed some of the small crater populations there.

The inconsistencies in the four ages modelled for Alder are relatively small, but still statistically significant (Figure 8). Past and current geology maps both assigned a stratigraphic age of Early Imbrian (i.e., 3.85-3.75 Ga; Stöffler et al., 2006) for the Alder impact, and the floor materials were mapped as younger basin ejecta that have a stratigraphic age of Imbrian (Figure 6b). The content of olivine mapped by the Kaguya Multiband Imager (Lemelin et al., 2016) shows that the flat floor, northeastern crater wall, and eastern ejecta deposits of Alder all exhibit an elevated abundance compared to the surrounding materials (Figure 6c). Morphology study does not reveal potential volcanic deposits on the floor, suggesting that the olivine-rich materials might be distal ejecta that postdate Alder. Therefore, the 3.8 Ga age derived for the crater floor is consistent with a partial resurfacing event by distal ejecta (Robbins, 2014). The counting area on the southern ejecta deposits (Figure 7c) has a model age of ~4.0Ga, and the largest crater collected is ~4km, which is 1/4 of the width of the counting area. The effect of small sizes of counting areas on age determination might be a reason for the slightly-larger density derived (Van Der Bogert et al., 2015).

4.1.2 Alder is older than the mare basalts on the floor of Von Kármán

The mare basalts on the floor of Von Kármán have been rigorously constrained via crater counts on three small areas and also the entire floor, yielding a robust age of ~3.6Ga (Huang J et al., 2018). Within error bars, this value is consistent with both

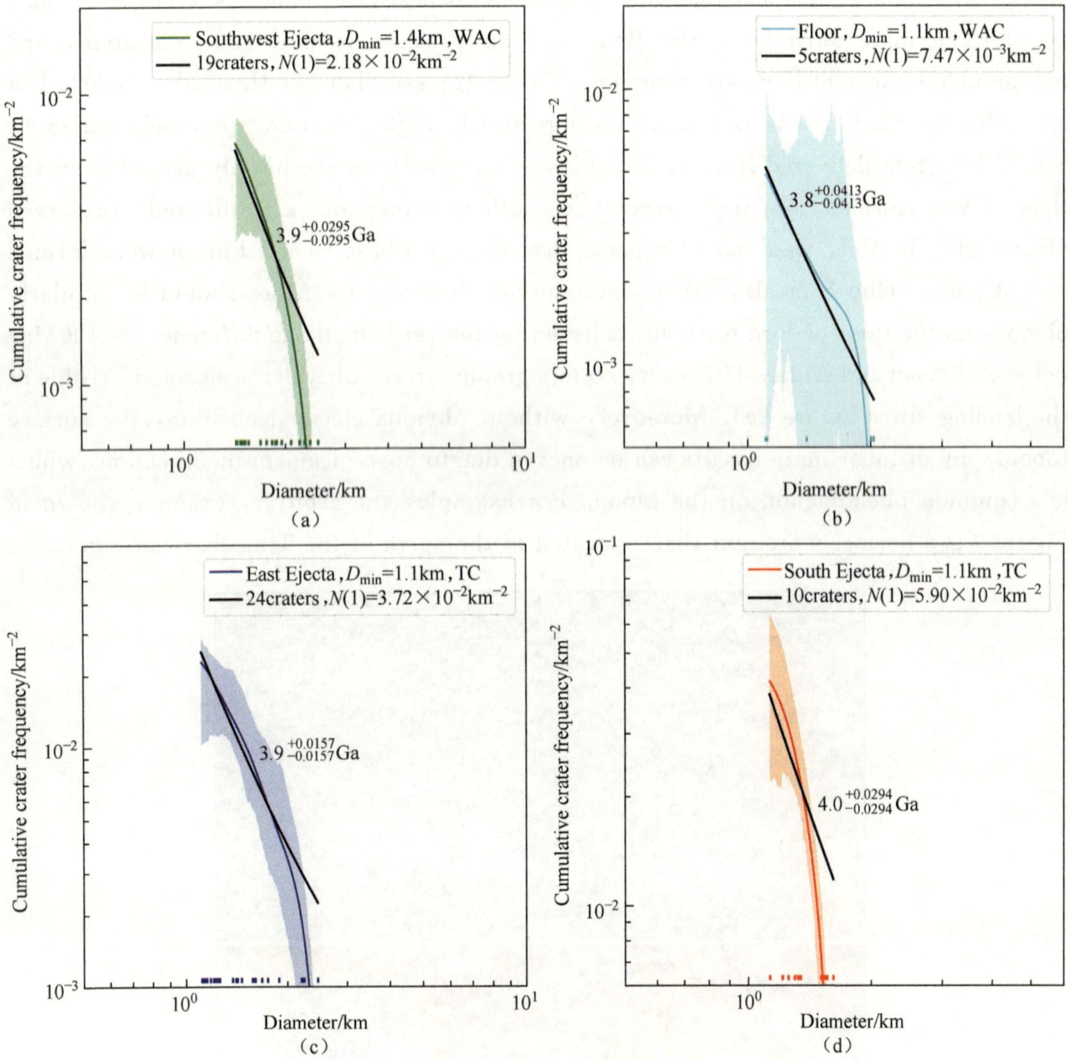

Figure 8　Cumulative crater size - frequency distribution and absolute model ages for the four populations of craters collected on the crater floor and ejecta deposits of Alder. The names of the counting areas correspond to those shown in Figure 7. Shaded colors are the errors for the crater size - frequency distributions

earlier (Pasckert et al., 2015) and later (Ling ZC et al., 2019) independent determinations. For comparison, here we have placed a solid constraint on the maximum and minimum model ages for Finsen and Alder, respectively, which are ～2Ga and ～3.9Ga according to the chronology function of Robbins (2014). Therefore, the debate about whether or not the Alder crater has contributed ejecta to the mare surface in Von Kármán can be settled.

4.2　Indications to Interpretations of Scientific Data Returned by the Yutu - 2 Rover

The surface topography at the landing site was once regarded as critical evidence to support that Alder was older than the mare basalts beneath the landing site (Di KC et al., 2019). The mare surface on the floor of Von Kármán is uneven in elevation, and topographic highs and lows are common (Figure 1c; see also Lai JL et al., 2020). For example, the landing site is located on a relatively high elevation area, believed to be caused by ejecta delivered from Alder. Finsen's secondaries are heavily degraded on the floor of Von Kármán, but the northeast to southwest lineations are still well-preserved (Figure 2b). If Alder were indeed younger than the mare basalts and Finsen were formed ~3.5Ga ago (Gou S et al., 2020), secondaries delivered by Alder should be similarly obvious on the floor of Von Kármán, considering the very small age difference (<100Ma) between Finsen and Alder. However, no topographic traces of Alder's ejecta are visible at the landing site (Figure 2c). Moreover, without obvious ejecta deposition, the surface topography of lunar mare basalts can be uneven due to postvolcanism modification, which is a common phenomenon on the Moon. For example, the arbitrary example shown in Figure 9 is a normal mare unit that is located to the north of the Timocharis crater.

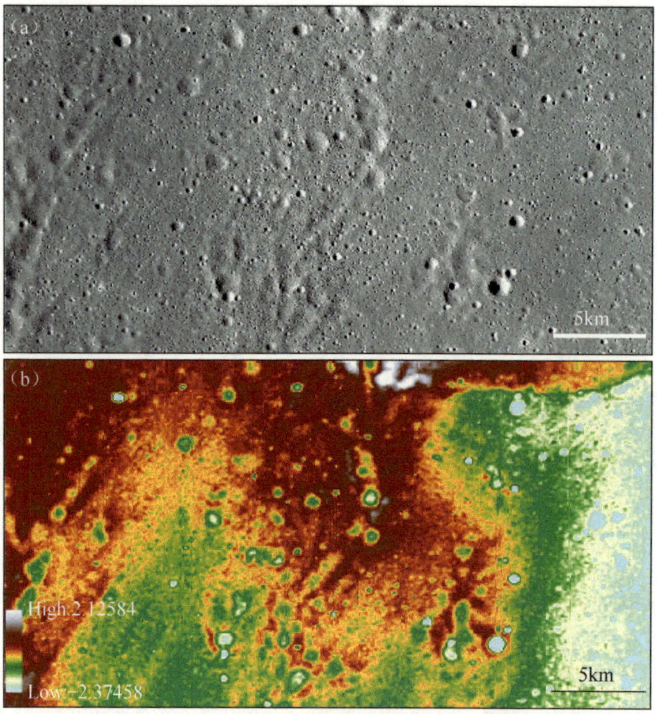

Figure 9　Mare surface on the Moon frequently exhibits uneven elevations, which are not obviously correlated with ejecta deposits. A nearside mare unit (central coordinates of the frame are 28.6°N, 12.0°W) close to the Timocharis crater is used here as an example, and such phenomena are not rare on the Moon. (a) A Kaguya Terrain Camera mosaic shows the general morphology of the mare units, where many degraded secondaries are visible. (b) SLDEM shows the regional topography of the mare unit, which features uneven elevations, but no obvious ejecta deposits are associated with the high topography. The unit of elevation here is kilometer

After reviewing recent advances in the local stratigraphy at the Chang'E－4 landing site, this work pins down the age relationship between Alder and the mare basalts at the landing site. The result confirms earlier orbital observations (Huang J et al., 2018; Ling ZC et al., 2019) and supports past and recent geologic maps with quantified model ages. Implications of the results to the interpretation of data obtained by Yutu－2 await further studies, but we mention two immediate indications here: (1) for the post－mare deposits, the lunar penetrating radar would not detect substantial ejecta delivered by Alder; and (2) for the surface regolith detected by the Yutu－2 rover, the reflectance spectra obtained contain little information about Alder's ejecta.

This result leaves a question: Besides Finsen, which craters are the other provenances of materials in the post－mare deposits? Theoretical predictions suggest that Finsen's ejecta should have dominated the post－mare ejecta, which is on the order of ～8m (Huang J et al., 2018; Lai JL et al., 2020). However, the lunar penetrating radar resolves post－mare deposits up to ～40m thick (Lai JL et al., 2020; Li CL et al., 2020). More systemic work is warranted on the possible contributors of distal ejecta to the landing site. On the other hand, the ballistic sedimentation model that was used to predict ejecta thickness was built on empirical and limited observations, and the predictions may be biased compared to real world impacts, such as complexities caused by various impact angles and pre－impact topography.

5 Conclusion

We perform rigorous crater statistics for different populations of impact craters that are superposed on various ejecta facies of the Finsen and Alder crater. The results are cross－compared to check replicability and reliability. Together with the revealed heterogeneous modification history of the different facies, discrepancies in crater densities are analyzed, concurring in the significant contribution of uneven depositions of background secondaries and resurfacing by distal ejecta. Compared to the model age of mare basalts in the floor of Von Kármán, our results unambiguously show that Alder is older than the basalts. Thus, their geological signature is not recorded in the surface topography, nor in subsurface structures revealed by the lunar penetrating radar, nor in the reflectance spectra obtained by the Yutu－2 rover. While this study has reshaped the possible provenances of the post－mare ejecta deposit in Von Kármán, we appeal for more focused studies on the post－mare impact history around the impact site.

Supplementary Materials

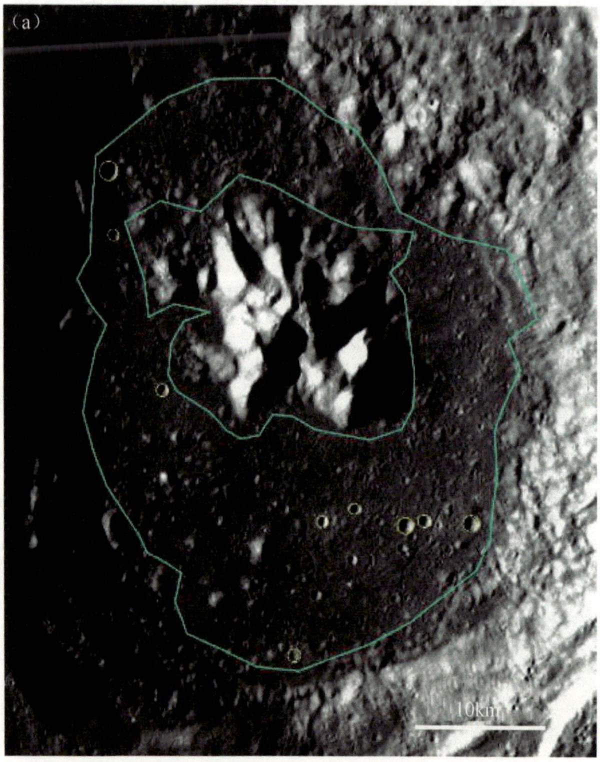

(a)

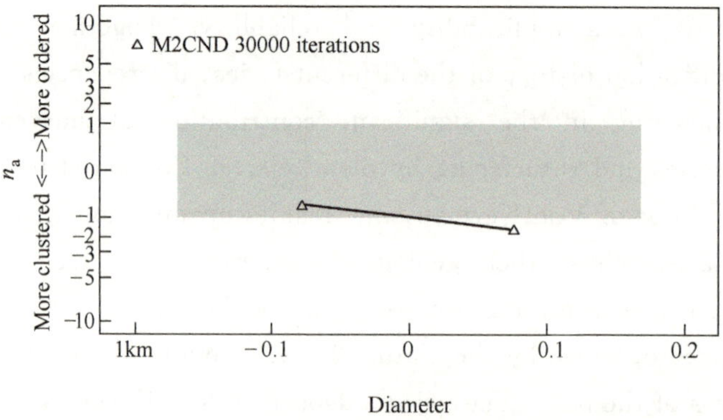

(b)

Figure S1　(a) Crater counts on the crater floor of Finsen. (b) The spatial distribution of craters larger than 1.1km collected on the floor of Finsen as represented by the standard deviation of adjacent area (SDAA). Using this algorithm (Michael et al., 2012), the spatial distribution of the collected craters is not significant different from a random distribution

Table S1　ID of imagery data used in this study

Figure ID	Image ID	Pixel scale
Figure 1a	LROC _ WAC _ Farside _ Lowsun	100m/pixel
Figure 1b	Clem _ mineral _ ration _ 200m	200m/pixel
Figure 1c	SLDEM2015 _ 512 _ 60S _ 30S _ 135 _ 180	59m/pixel
Figure 2a	M118681152M	111m/pixel
Figure 2b	M118681152M	111m/pixel
Figure 2c	M118681152M	111m/pixel
Figure 2d	M118681152M	111m/pixel
Figure 2e	M118681152M	111m/pixel
Figure 3a	M118681152M	111m/pixel
Figure 3b	TCO _ MAP _ 04 _ S42E177S45E180SC	7.4m/pixel
Figure 3c	TCO _ MAP _ 04 _ S42E177S45E180SC	7.4m/pixel
Figure 3d	TCO _ MAP _ 04 _ S42E177S45E180SC	7.4m/pixel
Figure 3e	TCO _ MAP _ 04 _ S42E177S45E180SC	7.4m/pixel
Figure 3f	TCO _ MAP _ 04 _ S42E177S45E180SC	7.4m/pixel
Figure 3g	TCO _ MAP _ 02 _ S42E177S45E180SC	7.4m/pixel
Figure 3h	TCO _ MAP _ 02 _ S42E177S45E180SC	7.4m/pixel
Figure 6a	M118640443M	75m/pixel
	M118653839M	75m/pixel
	M118660799M	75m/pixel
	M118667400M	75m/pixel
	M118681152M	78m/pixel
Figure 6c	Lunar _ Kaguya _ MIMap _ MineralDeconv _ OlivinePercent _ 50N50S	
Figure 7a	M118653839M	75m/pixel
Figure 7b	M118660799M	75m/pixel
	M118667400M	75m/pixel
Figure 7c	TCO _ MAP _ 04 _ S48E177S51E180SC	7.4m/pixel
Figure 7d	TCO _ MAP _ 04 _ S48E174S51E177SC	7.4m/pixel
Figure 9a	TCO _ MAP _ 04 _ N27E345N30E348SC	7.4 m/pixel
	TCO _ MAP _ 04 _ N27E348N30E351SC	
Figure 9b	SLDEM2015 _ 512 _ 00N _ 30N _ 315 _ 360	59m/pixel

Table S2　Information of the counting areas performed in this study

Counting areas	Central Latitude (°N)	Central Longitudes (°E)	Area (km^2)
Finsen – Southwest Ejecta	−43.6	−179.5	598.2
Finsen – North Ejecta	−43.6	−178.2	7.1
Finsen – West Ejecta	−43.7	−178.4	51.4
Finsen – South Ejecta	−43.9	−178.12	6.6
Finsen – East Ejecta	−43.8	−177.9	8.1
Finsen – West Floor	−42.5	−178.3	167.8
Finsen – South Floor	−42.3	−177.8	127
Finsen – Ejecta 1	−44.3	−177.5	2698.7
Finsen – Ejecta 2	−43.8	−178.1	1006.5
Finsen – Ejecta 3	−43.8	−178.1	1447.8
Finsen – Floor	−42.3	−177.8	822
Alder – Southwest Ejecta	−50.6	−179.2	2819.5
Alder – East Ejecta	−49	−174.8	1058.2
Alder – South Ejecta	−50.4	−177.4	240.8
Alder – Floor	−48.6	−178.3	1026

Table S3　Basic information of base images used for the crater counts

Counting areas	Base image	Pixel scale	Angle (°)
Finsen – North Ejecta	TCO_MAP_04_S39E177S42E180SC	7.4m/pixel	
Finsen – West Ejecta	TCO_MAP_04_S42E177S45E180SC		
Finsen – South Ejecta			
Finsen – East Ejecta			
Finsen – Southwest Ejecta			
Finsen – West Floor	TCO_MAP_02_S42E177S45E180SC	7.4m/pixel	
Finsen – South Floor			
Finsen – Ejecta 1	M118681152M	75m/pixel	76.2
Finsen – Ejecta 2			
Finsen – Ejecta 3			
Finsen – Floor			
Alder – Floor	M118640443M	75m/pixel	75.2
Alder – Southwest Ejecta	M118653839M	75m/pixel	77.9
	M118660799M	75m/pixel	75.2
	M118667400M	75m/pixel	77.8
	M118681152M	78m/pixel	76.2

Continued Table

Counting areas	Base image	Pixel scale	Angle (°)
Alder - East Ejecta	TCO _ MAP _ 04 _ S45E 174S48E177SC	7. 4m/pixel	
Alder - South Ejecta	TCO _ MAP _ 04 _ S48E 174S51E177SC		
	TCO _ MAP _ 04 _ S45E 177S48E180SC		
	TCO _ MAP _ 04 _ S48E 177S51E180SC		

Acknowledgments

This work is supported by the B - type Strategic Priority Program of the Chinese Academy of Sciences (Grant No. XDB41000000), the Science and Technology Development Fund of Macau (0042/2018/A2), the National Natural Science Foundation of China (No. 41773063), and the pre - research Project on Civil Aerospace Technologies (No. D020201 and D020202) that is funded by Chinese National Space Administration. The two reviewers are thanked for their suggestions and comments.

References

[1] Barker, M. K., Mazarico, E., Neumann, G. A., Zuber, M. T., Haruyama, J., and Smith, D. E. (2016). A new lunar digital elevation model from the Lunar Orbiter Laser Altimeter and SELENE Terrain Camera. Icarus, 273, 346 - 355. https://doi.org/10.1016/j.icarus.2015.07.039.

[2] Di, K. C., Zhu, M. H., Yue, Z. Y., Lin, Y. T., Wan, W. H., Liu, Z. Q., Gou, S., Liu, B., Peng, M., … Xue, B. (2019). Topographic Evolution of Von Kármán Crater Revealed by the Lunar Rover Yutu - 2. Geophys. Res. Lett., 46 (22), 12764 - 12770. https://doi.org/10.1029/2019GL085252.

[3] Ding, C. Y., Xiao, Z. Y., Wu, B., Li, Y., Prieur, N. C., Cai, Y. Z., Su, Y., and Cui, J. (2020). Fragments delivered by secondary craters at the Chang'e - 4 landing site. Geophys. Res. Lett., 47 (7), e2020GL087361. https://doi.org/10.1029/2020GL087361.

[4] Eliason, E. M., McEwen, A. S., Robinson, M. S., Lee, E. M., Becker, T., Gaddis, L., Weller, L. A., Isbell, C. E., Shinaman, J. R., … Malaret, E. (1999). Digital processing for a global multispectral map of the moon from the clementine UVVIS imaging instrument. In Proceedings of the 30th Annual Lunar and

Planetary Science Conference. Houston.

[5] Fa, W. Z., Zhu, M. H., Liu, T. T., and Plescia, J. B. (2015). Regolith stratigraphy at the Chang'E-3 landing site as seen by lunar penetrating radar. Geophys. Res. Lett., 42 (23), 10179-10187. https://doi.org/10.1002/2015GL066537.

[6] Fortezzo, C. M., Spudis, P. D., and Harrel, S. L. (2020). Release of the digital unified global geologic map of the moon at 1∶5, 000, 000-Scale. In Proceedings of the 51nd Lunar and Planetary Science Conference. Houston, TX: Lunar and Planetary Institute.

[7] Gou S., Yue, Z. Y., Di, K. C., Cai, Z. C., Liu, Z. Q., and Niu, S. L. (2020). Absolute model age of lunar Finsen crater and geologic implications. Icarus, 354, 114046. https://doi.org/10.1016/j.icarus.2020.114046.

[8] Haruyama, J., Matsunaga, T., Ohtake, M., Morota, T., Honda, C., Yokota, Y., Torii, M., Ogawa, Y., and LISM Working Group. (2008). Global lunar-surface mapping experiment using the Lunar Imager/Spectrometer on SELENE. Earth Planets Space, 60 (4), 243-255. https://doi.org/10.1186/BF03352788.

[9] Hawke, B. R., Blewett, D. T., Lucey, P. G., Smith, G. A., Bell III, J. F., Campbell, B. A., and Robinson, M. S. (2004). The origin of lunar crater rays. Icarus, 170, 1-16. https://doi.org/10.1016/j.icarus.2004.02.013.

[10] Huang, J., Xiao, Z. Y., Flahaut, J., Martinot, M., Head, J., Xiao, X., Xie, M. G., and Xiao, L. (2018). Geological characteristics of Von Kármán Crater, northwestern South Pole-Aitken basin: Chang'E-4 landing site region. J. Geophys. Res.: Planets, 123 (7), 1684-1700. https://doi.org/10.1029/2018JE005577.

[11] Huang, J., Xiao, Z. Y., Xiao, L., Briony, H., Hu, X. Y., Lucey, P., Xiao, X., Zhao, S. Y., Qian, Y. Q., … Xie, J. F. (2020). Diverse rock types detected in the lunar South Pole-Aitken Basin by the Chang'E-4 lunar mission. Geol. Soc. Am., 48 (7), 723-727. https://doi.org/10.1130/G47280.1.

[12] Ivanov, M. A., Hiesinger, H., Van Der Bogert, C. H., Orgel, C., Pasckert, J. H., and Head, J. W. (2018). Geologic history of the northern portion of the South Pole-Aitken Basin on the moon. J. Geophys. Res.: Planets, 123 (10), 2585-2612. https://doi.org/10.1029/2018JE005590.

[13] Lai, J. L., Xu, Y., Zhang, X. P., Xiao, L., Yan, Q., Meng, X., Zhou, B., Dong, Z. H., and Zhao, D. (2019). Comparison of dielectric properties and structure of lunar regolith at Chang'e-3 and Chang'e-4 landing sites revealed by groundpenetrating radar. Geol. Res. Lett., 46 (12), 12783-12793. https://doi.org/10.1029/2019GL084458.

[14] Lai, J. L., Xu, Y., Bugiolacchi, R., Meng, X., Xiao, L., Xie, M. G., Liu, B., Di, K. C., Zhang, X. P., … Xu, L. Y. (2020). First look by the Yutu-2 rover at the deep subsurface structure at the lunar farside. Nature Communications, 11

(1), 3426. https://doi.org/10.1038/s41467-020-17262-w.

[15] Lemelin, M., Lucey, P. G., Gaddis, L. P., Hare, T., and Ohtake, M. (2016). Global map products from the kaguya multiband imager at 512 ppd: minerals, FeO, and OMAT. In Proceedings of the 47th Lunar and Planetary Science Conference. The Woodlands, Texas.

[16] Li, C. L., Liu, D. L., Liu, B., Ren, X., Liu, J. J., He, Z. P., Zuo, W., Zeng, X. G., Xu, R., …Ouyang, Z. Y. (2019). Chang'E-4 initial spectroscopic identification of lunarfar-side mantle-derived materials. Nature, 569 (7756), 378-382. https://doi.org/10.1038/s41586-019-1189-0.

[17] Li, C. L., Su, Y., Pettinelli, E., Xing, S. G., Ding, C. Y., Liu, J. J., Ren, X., Lauro, S. E., Soldovieri, F., … Zhang, H. B. (2020). The Moon's farside shallow subsurface structure unveiled by Chang'E-4 lunar penetrating Radar. Sci. Adv., 6 (9), eaay6898. https://doi.org/10.1126/sciadv.aay6898.

[18] Lin, H. L., He, Z. P., Yang, W., Lin, Y. T., Xu, R., Zhang, C., Zhu, M. H., Chang, R., Zhang, J. H., … Zou, Y. L. (2020). Olivine-norite rock detected by the lunar rover Yutu-2 likely crystallized from the SPA-impact melt pool. Natl. Sci. Rev., 7 (5), 913-920. https://doi.org/10.1093/nsr/nwz183.

[19] Ling, Z. C., Jolliff, B. L., Liu, C. Q., Bi, X. Y., Liu, L., Qiao, L., Li, B., Zhang, J., Fu, X. H., and Liu, J. Z. (2018). Composition, mineralogy, and chronology of mare basalts in Von Kármán crater: a candidate landing site of Chang'E-4. In Proceedings of the 49th Lunar and Planetary Science Conference. The Woodlands, Texas.

[20] Ling, Z. C., Qiao, L., Liu, C. Q., Cao, H. J., Bi, X. Y., Lu, X. J., Zhang, J., Fu, X. H., Li, B., and Liu, J. Z. (2019). Composition, mineralogy and chronology of mare basalts and non-mare materials in Von Kármán crater: landing site of the Chang'E-4 mission. Planet. Space Sci., 179, 104741. https://doi.org/10.1016/j.pss.2019.104741.

[21] Lu, Y., Wu, Y. Z., Michael, G. G., Ma, J. S., Cai, W., and Qin, N. N. (2021). Chronological sequence of Chang'E-4 landing zone within Von Kármán crater. Icarus, 354, 114086. https://doi.org/10.1016/j.icarus.2020.114086.

[22] Ma, P., Sun, Y. X., Zhu, M. H., Yang, Y. Z., Hu, X. Y., Jiang, T., Zhang, H., Lucey, P.G., Xu, R., … Lin, H. Y. (2020). A plagioclase-rich rock measured by Yutu-2 Rover in Von Kármán crater on the far side of the Moon. Icarus, 350, 113901. https://doi.org/10.1016/j.icarus.2020.113901.

[23] McEwen, A. S., and Bierhaus E. B. (2006). The importance of secondary cratering to age constraints on planetary surfaces. Annu. Rev. Earth Planet., 34, 535-567. https://doi.org/10.1146/annurev.earth.34.031405.125018.

[24] Melosh, H. J., Kendall, J., Horgan, B., Johnson, B. C., Bowling, T., Lucey, P. G., and Taylor, G. J. (2017). South Pole-Aitken basin ejecta reveal the Moon's

upper mantle. Geology, 45 (12), 1063 - 1066. https://doi.org/10.1130/G39375.1.

[25] Michael, G. G., and Neukum, G. (2010). Planetary surface dating from crater sizefrequency distribution measurements: partial resurfacing events and statistical age uncertainty. Earth Planet. Sci. Lett., 294 (3 - 4), 223 - 229. https://doi.org/10.1016/j.epsl.2009.12.041.

[26] Michael, G. G., Platz, T., Kneissl, T., and Schmedemann N. (2012). Planetary surface dating from crater size - frequency distribution measurements: spatial randomness and clustering. Icarus, 218 (1), 169 - 177. https://doi.org/10.1016/j.icarus.2011.11.033.

[27] Moriarty Ⅲ, D. P., and Pieters, C. M. (2018). The character of South Pole - Aitken Basin: patterns of surface and sub - surface composition. J. Geophys. Res.: Planets, 123 (3), 729 - 747. https://doi.org/10.1002/2017JE005364.

[28] Neukum, G., Ivanov, B. A., and Hartmann, W. K. (2001). Cratering records in the inner solar system in relation to the lunar reference system. Space Sci. Rev., 96 (1 - 4), 55 - 86. https://doi.org/10.1023/A:1011989004263.

[29] Ostrach, L. R., Robinson, M. S., Denevi, B. W., and Thomas, P. C. (2011). Effects of incidence angle on crater counting observations. In Proceedings of the 42nd Lunar and Planetary Science Conference.

[30] Pasckert, J. H., Hiesinger, H., and Van Der Bogert, C. H. (2015). Small - scale lunar farside volcanism. Icarus, 257, 336 - 354. https://doi.org/10.1016/j.icarus.2015.04.040.

[31] Pasckert, J. H., Hiesinger, H., and Van Der Bogert, C. H. (2018). Lunar farside volcanism in and around the south pole - Aitken basin. Icarus, 299, 538 - 562. https://doi.org/10.1016/j.icarus.2017.07.023.

[32] Potter, R. W. K., Collins, G. S., Kiefer, W. S., McGovern, P. J., and Kring, D. A. (2012). Constraining the size of the South Pole - Aitken basin impact. Icarus, 220 (2), 730 - 743. https://doi.org/10.1016/j.icarus.2012.05.032

[33] Robbins, S. J. (2014). New crater calibrations for the lunar crater - age chronology. Earth Planet. Sci. Lett., 403, 188 - 198. https://doi.org/10.1016/j.epsl.2014.06.038.

[34] Robbins, S. J., Riggs, J. D., Weaver, B. P., Bierhaus, E. B., Chapman, C. R., Kirchoff, M. R., Singer, K. N., and Gaddis, L. R. (2018). Revised recommended methods for analyzing crater size - frequency distributions. Mateoritics Planet. Sci., 53 (4), 891 - 931. https://doi.org/10.1111/maps.12990.

[35] Robinson, M. S., Brylow, S. M., Tschimmel, M., Humm, D., Lawrence, S. J., Thomas, P. C., B. W. Denevi, E. Bowman - Cisneros, J. Zerr, … Hiesinger, H. (2010). Lunar Reconnaissance Orbiter Camera (LROC) instrument overview. Space Sci. Rev., 150 (1 - 4), 81 - 124. https://doi.org/10.1007/s11214 -

010 - 9634 - 2.

[36] Stöffler, D., Ryder, G., Ivanov, B. A., Artemieva, N. A., Cintala, M. J., and Grieve, R. A. F. (2006). Cratering history and lunar chronology. Rev. Mineral. Geochem., 60 (1), 519 - 596. https://doi.org/10.2138/rmg.2006.60.05.

[37] Van Der Bogert, C. H., Michael, G., Kneissl, T., Hiesinger, H., and Pasckert, J. H. (2015). Effects of count area size on absolute model ages derived from random crater size - frequency distributions. In Proceedings of the 46th Lunar and Planetary Science Conference. The Woodlands, Texas.

[38] Wang, Y. C., Xie, M. G., Xiao, Z. Y., and Cui, J. (2020). The minimum confidence limit for diameters in crater counts. Icarus, 341, 113645. https://doi.org/10.1016/j.icarus.2020.113645.

[39] Wilhelms, D. E., Howard, K. A., and Wilshire. H. G. (1979) Geologic map of the south side of the Moon. United States Geological Survey. https://doi.org/10.3133/i1162.

[40] Wu, W. R., Wang, Q., Tang, Y. H., Yu, G. B., Liu, J. Z., Zhang, W., Ning, Y. M., and Lu, L. L. (2017). Design of Chang'E - 4 lunar farside soft - landing mission. J. Deep Space Explor. (in Chinese), 4 (2), 111 - 117. https://doi.org/10.15982/j.issn.2095 - 7777.2017.02.002.

[41] Xiao, L., Zhu, P. M., Fang, G. Y., Xiao, Z. Y., Zou, Y. L., Zhao, J. N., Zhao, N., Yuan, Y. F., Qiao, L., … Gao, Y. Z. (2015). A young multilayered terrane of the northern Mare Imbrium revealed by Chang'E - 3 mission. Science, 347 (6227), 1226 - 1229. https://doi.org/10.1126/science.1259866.

[42] Xiao, Z. Y., and Strom R. G. (2012). Problems determining relative and absolute ages using the small crater population. Icarus, 220 (1), 254 - 267. https://doi.org/10.1016/j.icarus.2012.05.012.

[43] Xiao, Z. Y., Zeng, Z. X., Ding, N., and Molaro, J. (2013). Mass wasting features on the moon - how active is the lunar surface?. Earth Planet Sci. Lett., 376, 1 - 11. https://doi.org/10.1016/j.epsl.2013.06.015.

[44] Xiao, Z. Y., and Werner S. C. (2015). Size - frequency distribution of crater populations in equilibrium on the moon. J. Geophys. Res.: Planets, 120 (12), 2277 - 2292. https://doi.org/10.1002/2015JE004860.

[45] Xiao, Z. Y., Prieur, N. C., and Werner, S. C. (2016). The self - secondary crater population of the Hokusai crater on mercury. Geophys. Res. Lett., 43 (14), 7424 - 7432. https://doi.org/10.1002/2016GL069868.

[46] Xiao, Z. Y. (2018). On the importance of self - secondaries. Geosci. Lett., 5 (1), 17. https://doi.org/10.1186/s40562 - 018 - 0116 - 9.

[47] Zhang, J. H., Yang, W., Hu, S., Lin, Y. T., Fang, G. Y., Li, C. L., Peng, W. X., Zhu, S. Y., He, Z. P., … Ouyang, Z. Y. (2015). Volcanic history of the Imbrium basin: a close - up view from the lunar rover Yutu. Proc. Natl. Acad.

Sci. U S A, 112 (17), 5342 - 5347. https://doi.org/10.1073/pnas.1503082112.

[48] Zhang, L., Li, J., Zeng, Z.F., Xu, Y., Liu, C., and Chen, S.B. (2020). Stratigraphy of the Von Kármán crater based on Chang'E - 4 lunar penetrating radar data. Geophys. Res. Lett., 47(15), e2020GL088680. https://doi.org/10.1029/2020GL088680.

Density Structure of the Von Kármán Crater in the Northwestern South Pole - Aitken Basin: Initial Subsurface Interpretation of the Chang'E - 4 Landing Site Region

Chikondi Chisenga[1,2], Jianguo Yan[1], Jiannan Zhao[3,4], Qingyun Deng[1] and Jean - Pierre Barriot[1,5]

1. State Key Laboratory of Information Engineering in Surveying, Mapping and Remote Sensing (LIESMARS), Wuhan University, Box 129, Luoyu Road, Wuhan 430070, China; cchisenga@must.ac.mw (C. C.); dengqy@whu.edu.cn (Q. D.); jean-pierre.barriot@upf.pf (J.-P. B.)
2. Department of Earth Sciences, Ndata School of Climate and Earth Sciences, Malawi University of Science and Technology, Limbe P. O. Box 5196, Malawi
3. State Key Laboratory of Geological Process and Mineral Resources, Planetary Science Institute, China University of Geosciences, Wuhan 430074, China; jnzhao@cug.edu.cn
4. School of Environmental Studies, China University of Geosciences, Wuhan 430074, China
5. Geodesy Observatory of Tahiti, BP 6570, Faa'a 98702, Tahiti, French Polynesia

Abstract: The Von Kármán Crater, within the South Pole - Aitken (SPA) Basin, is the landing site of China's Chang'E - 4 mission. To complement the in situ exploration mission and provide initial subsurface interpretation, we applied a 3D density inversion using the Gravity Recovery and Interior Laboratory (GRAIL) gravity data. We constrain our inversion method using known geological and geophysical lunar parameters to reduce the non - uniqueness associated with gravity inversion. The 3D density models reveal vertical and lateral density variations, 2600 - 3200kg/m^3, assigned to the changing porosity beneath the Von Kármán Crater. We also identify two mass excess anomalies in the crust with a steep density contrast of 150kg/m^3, which were suggested to have been caused by multiple impact cratering. The anomalies from recovered near surface density models, together with the gravity derivative maps extending to the lower crust, are consistent with surface geological manifestation of excavated mantle materials from remote sensing studies. Therefore, we suggest that the density distribution of the Von Kármán Crater indicates multiple episodes of impact cratering that resulted in formation and destruction of ancient craters, with crustal reworking and excavation of mantle materials.

Keywords: Von Kármán Crater; density structure; gravity inversion; Chang'E - 4 mission

1 Introduction

The Von Kármán Crater is a pre - Nectarian crater with a diameter of ~186km [1] within the largest known impact structure on the farside of the Moon, the South Pole - Aitken (SPA) Basin (Figure 1) [2] . The SPA is the oldest impact basin with a well - preserved impact history, which suggests that the earliest lunar history and processes related to impact cratering can be studied from it [3, 4] . The SPA is subdivided into four distinct compositional zones [4], with the Von Kármán Crater situated within the Mg - Pyroxene Annulus zone. The Mg - Pyroxene Annulus contains the main materials that were excavated and melted by the SPA - forming event as indicated by its relatively uniform composition, great area, depth, and thickness. Subsequently, the Von Kármán Crater was filled with mare basalt around 3. 6 Ga in the Imbrian period [2, 5, 6] . The Von Kármán Crater is further characterized by a very thin crust with an average of 10km [7] . It also shows evidence of differentiated impact melts during its formation [3], low crater - floor elevation values with an average of 5926m [2], relative to a mean lunar radius of 1737. 4km [8] and high Bouguer gravity values in the southern part [9] . The thin crust inferred from a gravity - derived crustal thickness model [7] implies that the mantle is uplifted in this region [7, 9, 10], with a possibility of upper mantle materials exposed on the surface [4] . These characteristics of the Von Kármán Crater make it a good candidate for subsurface studies using gravity data, as density distribution beneath the Von Kármán Crater could complement the geological and spectral studies.

The China's Chang'E - 4 lunar exploration spacecraft successfully landed on the Moon inJanuary 2019 [11], and started the first in - situ exploration of the farside of the Moon. The landing region was located in the southern part of the Von Kármán Crater, within the mare basalt region [12] . A detailed geological and spectral analysis in the Von Kármán Crater indicates compositional and mineralogical differences for basaltic rocks compared to the nearside of the Moon. The surface materials in this region are characterized by a low titanium content, few mafic minerals and iron - depleted mare basalts [2, 13], which forms a basis for further interpretation of the in - situ exploration. Some previous studies, however, have indicated that mantle materials are possibly exposed in the SPA region [4], while some studies cast doubt on such occurrences [13 - 15] . However, the first published results from the Chang'E - 4 mission [16] show that the surface of the Von Kármán basin is characterized by mafic components that are dominated by low - Ca pyroxene (LCP) and olivine, with a very small amount of high - Ca pyroxene (HCP), which suggest a deep - seated, upper

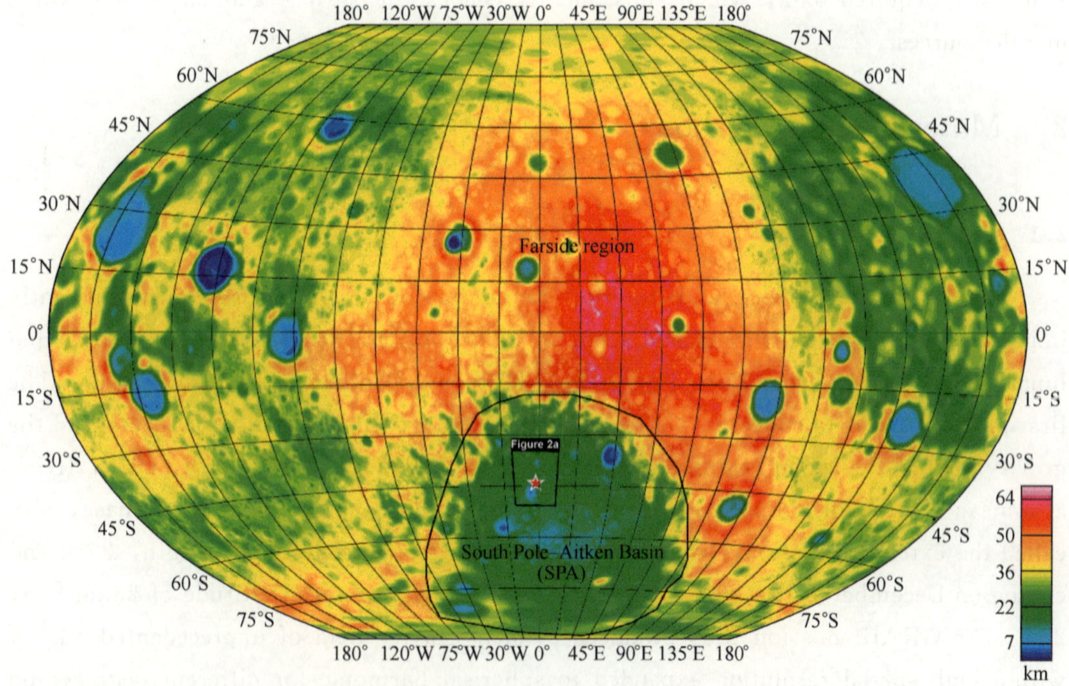

Figure 1 The crustal thickness map of the Moon, centered on the farside (Winkel tripel projection [7]). Also shown are the locations of the South Pole - Aitken Basin and the Von Kármán Crater. The black polygon shows the extent of the study area, which includes the Von Kármán Crater and surrounding craters. The red pentagram indicates the location of the Chang'E - 4 landing site (177.59°E, 45.46°S; [11, 12])

mantle origin [17]. Despite the presence of mare and other basaltic materials on the landing site of the Change'E - 4 in the Von Kármán Crater [4], it is suggested that these excavated mantle materials originated from the Finsen Crater in the northeast [13, 16], with a possibility that they originated from the base of the differentiated melt sheet [18]. These studies, however, do not consider the subsurface extent of the surficial manifestation of the mare basalt and possible mantle excavated materials. Our study therefore correlates the surface basaltic materials with the subsurface density distribution. We investigate the geophysical characteristics of the subsurface structures for the Von Kármán Crater using GRAIL gravity data. The presence of small - scale features revealed by the GRAIL gravity data [9, 19] was correlated to the near surface materials, its porosity and compositional differences within the Von Kármán Crater [19 - 22]. We then identify the existence and possible extent of differentiation of the SPA melt sheet [18, 23] and the possible subsurface extension of the surficial basaltic material [2, 13] based on density models using density inversion method. The approach gives a 3D subsurface characteristic of the Von Kármán Crater that furthers our understanding of the origin of the basaltic surface materials. Therefore, it complements the geological and spectral analysis of the Chang'E - 4 in - situ acquired data, which subsequently echoes the surface anomalies of possible mantle sources.

2 Materials and Methods

2.1 GRAIL Gravity Data

The Gravity Recovery and Interior Laboratory (GRAIL) mission significantly improved the gravity resolution and accuracy of the Moon, adding value to the existing lunar gravity models from Luna Prospector, Kaguya and other existing data [24]. The first phase of the data collection, also called the Primary Mission, of GRAIL mapped the gravity field of the Moon beginning from March 1st, 2012 up to May 29th, 2012, flying at a mean altitude of 55km [25, 26]. This was followed up by the second phase, also called the extended mission, that mapped the Moon starting on August 30th, 2012 and ending on December 14th, 2012, collecting data at a lower mean altitude of 23km [25, 26]. The GRAIL mission thus produced a lunar gravity field of unprecedented quality with a high spatial resolution expanded to spherical harmonic for different degrees and orders, e.g., 660, 900, and 1200 [9, 27, 28]. Such improvement has increased the spatial resolution by a factor of 3 to 4, with the RMS power of the GRAIL data error ~5 orders of magnitude smaller than the RMS power error in previous models [27]. High resolution data have the ability to resolve smaller anomalies on the Moon [19], which improved the results obtained from previous gravity models.

To achieve the objective of the study, we used higher spatial resolution (~4.5 km) GRAIL gravity data expanded to a spherical harmonics degree and order 1200, extracted from the GRGM1200A model [9]. The Bouguer anomaly data were calculated using SHTOOLS [29 - 31] at a height of 10km with a Moon reference radius of 1738 km. The bulk density of the South Pole - Aitken, in which the Von Kármán Crater is located, is ~2880 kg/m^3 [19]. The density value is higher than the average crustal density of 2550kg/m^3 for the feldspathic highlands crust [7]. It is also noticed that the typical SPA interior is more mafic than feldspathic material [32], which is consistent with the use of the higher average density value of ~2880 kg/m^3 [18]. Thus, we used this assumed density value for the South Pole - Aitken Basin in the gravity data calculation to obtain the Bouguer anomaly. The gravity data were then truncated between degrees 6 to 450 to avoid long - wavelength data from the deep mantle and noise that is contained in short - wavelength signals as a result of orbit parallel stripping at higher altitude (Figure 2). The resulting filtered data highlight mid - to short - wavelength crustal sources required for the subsurface analysis [25].

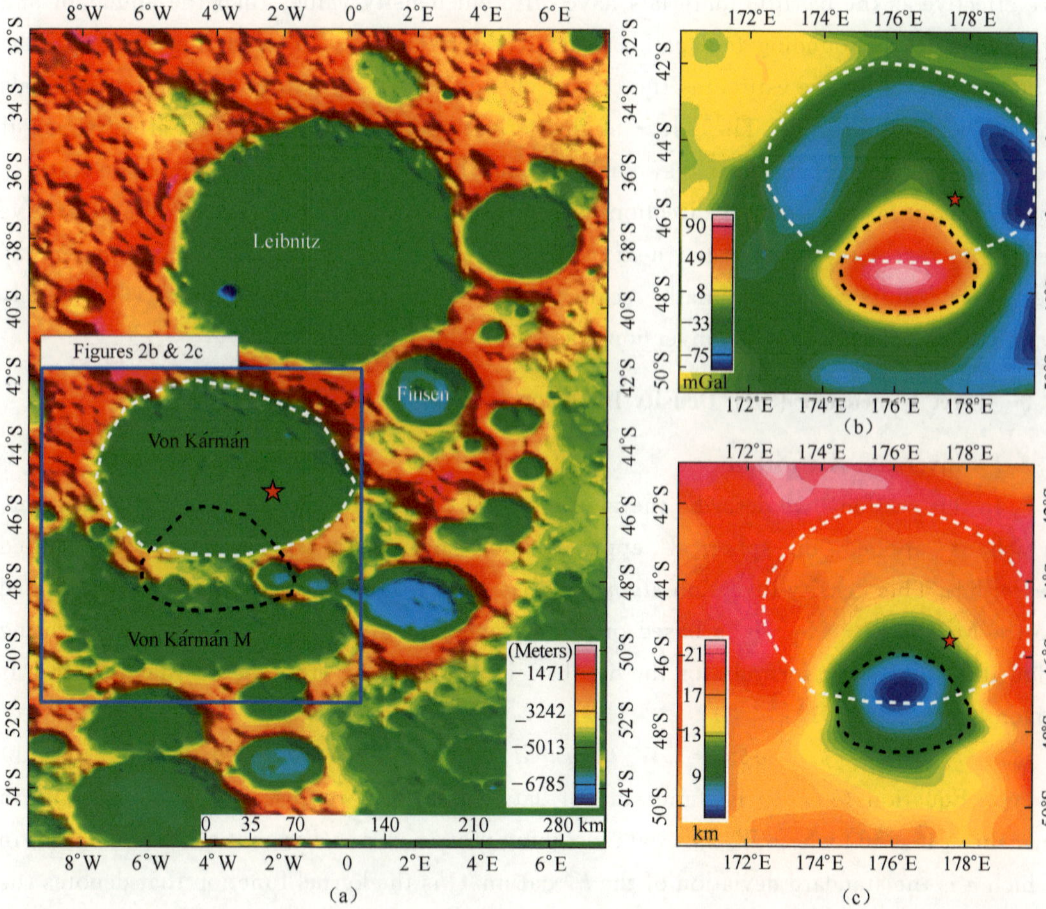

Figure 2　Study area. (a) Topographic map of the broader Von Kármán region and the surrounding craters, farside centred, after Barker et al. [8]. The Von Kármán Crater is enclosed by several other impact craters, namely, the Leibnitz crater in the north, the Finsen crater in the northeast and the Von Kármán M in the south. Also shown are the extent of our gravity inversion (Blue polygon), the boundary of the Von Kármán raised rim (White - dashed polygon) and the extent of high gravity region (Black - dashed polygon; Figure 2b). (b) Lunar Bouguer anomaly from the GRGM1200A gravity model available for the spherical harmonic degree and order 1200 [9], truncated between 6 - 450 degrees, nearside centred, after Goossens et al. [9]; (c) The crustal thickness map of the Von Kármán Crater, nearside centred, after Wieczorek et al. [7]. The red pentagram shows the location of the Chang'E - 4 landing site (177.59°E, 45.46°S; [11, 12]).

2.2　Gravity Derivative Calculation

In order to characterize the relationship between the surficial and near surface basaltic materials and the deep - seated structures, the Bouguer gravity data were subjected to first - order vertical derivative and tilt angle derivative calculations [33]. The calculations enhance the small - scale anomalies of the crustal structures [34, 35]. These techniques are effective as the basaltic materials have different density values than the abundant and wide - spread surrounding feldspathic crustal materials [36 - 40]. In this scenario, the first vertical derivative estimates the rate of change of density values of basaltic materials in the vertical direction. The tilt - angle derivative, however, is very effective as it normalizes all the density changes, such that subtle and less evident small - scale features in the horizontal and vertical directions are well enhanced. Thus, we used the derivative results to reveal the surface and near - surface basaltic materials. We also compared the derivative results with the small - scale features recovered from 3D density inversion, which could indicate that the inversion was successfully applied.

2.3 Three Dimension (3D) Density Inversion

2.3.1 Inversion Algorithm

In this section, we describe the mathematical description of the inversion algorithm, which is applied in spherical approximation by using spherical prisms, called tesseroids. This density inversion algorithm has been tested on synthetic data and applied to the Moon [41 - 44]. It is based on the Li and Oldenburg [45, 46] depth weighting algorithm. The basic formula for the depth weighting inverse problem for the algorithm is given by Equation (1):

$$\phi(m)=\phi_d+\mu\phi_m=\| W_d(Gm-d^{obs})\|_2^2+\mu\| W_m(m-m_{ref})\|_2^2 \quad (1)$$

In Equation (1), ϕ_d is a measure of data misfit. It is represented using a 2 - norm measure [45, 46] where W_d is a data weighing matrix with a diagonal elements of $1/\sigma_i$ in which σ is the standard deviation of the ith datum. G is the kernel function that denotes the relationship between the geological model (m) and the observed data (d^{obs}). The model objective function (ϕ_m), also called a model norm, measures the smoothness of the model. W_m is also a weighting matrix for the model objective function that defines the closeness and smallness between the recovered model (m) and the reference model (m_{ref}). Thus, the objective function ϕ (m) for the inversion combined the model objective function and the data misfit, and is controlled by a regularization parameter (μ). A regularization parameter [47] basically balances up the model objective function and data misfit. The best fit value of μ lies at the corner of the L _ curve for a plot of the model norm against data misfit [48 - 50]. The original objective function [45, 46] was designed to work in a Cartesian coordinate system. This poses a diffculty when applied to a large area with noticeable surface curvature on the Moon due to its small radius. To overcome this limitation, Liang et al. [44] extended the model objective function within the depth weighting algorithm for application in a spherical coordinate system [44]. They modified the depth weighting function in the algorithm using uniform prism cells to re - scalable prism cells along the radial direction from the surface going downwards designed for the spherical coordinate system (scs).

The solution of the inversion problem is obtained by minimizing the objective function (Equation (1)) following Equation (2). This equation is solved as the conjugate gradient (CG) matrix with an iterative algorithm,

$$[G^T W_d^T W_d G + \mu W_m^T W_m]m = G^T W_d^T W_d d^{obs} + \mu W_m^T W_m m_{ref} \tag{2}$$

In Equation (2) G^T, W_d^T and W_m^T are transpose matrices for the G, W_d and W_m that were described in Equation (1). We also include the Lagrangian multiplier [51, 52] method introduced by Zhang et al. [41] within the model objective function to fit the constraints information. This increases the reliability of the inversion results. For instance, the penalty factor compels the recovered model to be more reliable, whilst the geological constraints improve the inversion results [43]. The resulting extended formulae (Equation (3)) for this constrained inversion algorithm is described by Zhang et al. [43] and is given by:

$$\begin{aligned}&[G^T W_d^T W_d G + \mu W_m^T W_m + \frac{1}{2}M(F_1 + F_2 + F_3)]m \\ &= G^T W_d^T W_d d^{obs} + \mu W_m^T W_m m_{ref} - F_0 \lambda_0^T + F_1 \lambda_1^T + F_2 \lambda_2^T + \\ &\frac{1}{2}M(F_0 m_0 + F_1 m_1 + F_2 m_2) + \frac{1}{2}M(F_1 Z_1^2 - F_2 Z_2^2)\end{aligned} \tag{3}$$

In Equation (3), the diagonal of the matrix F_i denotes the index of the constrained information in each divided rectangle cell, M is the penalty factor, λ represents Lagrangian multipliers, and Z_i is the slack variable of the ith cell, where i is the index number for each cell that represents defined geological/geophysical bounds constraints.

2.3.2 Implementation

The dataset and the inversion space were prepared using the 3D mesh as illustrated in Table 1. The spatial resolution of the data was increased from ~4.5 km [9] to ~6 km (or 0.2°), given that 1° is equivalent to ~30.2 km on the Moon. The depth of the 3D mesh was constrained to 22 km, which represents the deepest point of the crust-mantle boundary in our study area [7]. The depth extent is consistent with the calculated Bouguer gravity data, which are mostly characterized by signals from the lunar crust. Then, the distance in the radial direction was discretized into cell sizes of 0.5km. The final 3D mesh had 98, 400 data cells (50 cells in the longitudinal direction × 48 cells in the latitudinal direction × 41 cells in the radial direction) that occupied a physical volume space of 26, 299, 735km³ (302km × 295.95km × 22km). The geophysical bound constraints, representing density values, within each cell in the 3D mesh were set to -580kg/m³ and 520kg/m³. These values represent the geophysical/geological upper and lower bound constraints, based on the crustal densities, relative to the average density of the South Pole-Aitken Basin. Normally, lunar crust densities vary between 2300kg/m³ and 2900kg/m³ with an average of 2550kg/m³ [7]. A 3D mesh with a flat base extending to 22km could include the mantle, since the greater part of the study area has crustal thicknesses less than 22km, with the thinnest crust at 5km. Furthermore, the region is known to have occurrences of basalts that have higher density values [2, 6, 16, 23]. Thus, we extended the density variation to 3300kg/m^3; hence, the values of the upper bound exceed the difference between the average density value of the Aitkin Basin, ~2880kg/m^3, and the average lunar crustal density, ~2550kg/m^3. As a result, the background density for each cell was set to zero (0) kg/m^3 that represents the 2880kg/m^3 in absolute density terms, to obtain the density contrast within the upper and lower bound constraints.

Table 1 The 3D mesh and dataset for inversion

Region	Inversion Range	Model		Data	
		Grid Size	Grid Number	Data Size	Data Number
Von Kármán	Longitude	0.2°	50	0.2°	50
	Latitude	0.2°	48	0.2°	48
	Depth		0~2km		0~50km
	Radial direction	0.5km	41	0.5km	41

During the inversion, some parameters, i.e., roughness factor, length scale, depth weighting parameter, were kept constant (Table 2), based on previously accepted studies and extensive test results [41, 45, 46]. However, a penalty factor was assigned to a small value of 1.0×10^{-6} (with an increment of 2) and iteratively increased until the penalty factor fit the constrained information [41]. The same was applied to the criteria for terminating the iteration by defining the convergence values for the inversion and conjugate gradient (CG) method (Table 2). Each of the iterative results was forward modeled in spherical approximation by numerically solving the Gauss - Legendre quadrature (GLQ) integration problem [53], thereby finding an optimal solution that minimizes the data misfit and model objective function. Despite this parameter control, the gravity inversion procedure is still an ill - posed problem that suers from non - uniqueness, which produces multiple solutions. We controlled this and identify a best fit solution by inverting the data with 14 different values of the regularization parameter (Table 2; Figure 3). The 14 values were obtained after experimenting with a number of ranges to produce 14 different models (Figures A1 and A2). The plotted L _ curve for the 14 regularizing parameters (Figure 3) identifies the value of 1.5 as a best fit regularization parameter, which enabled a construction of a desirable density model. This model was used for geophysical interpretation of the Von Kármán Crater as discussed in Section 3.

Table 2　Inversion parameters

Final Parameters for the Density Inversion		Determining the Optimal Value of the Regularization Parameter				
Parameter	Value	No	Tikhonov Parameter Exponent	Tikhonov Parameter	Data Misfit Value	Model Norm Value
Roughness factor	2，2，2	1	4.00	10，000.00	22714.7	9.55
Length scale	1×10^{-10}	2	3.00	1000.00	2146.65	14.49
Depth weighting parameter	2	3	2.50	316.20	879.65	16.63
Penalty factor	1×10^{-6}	4	2.00	100.00	437.87	19.09
Increase number for penalty factor	2	5	1.50	31.62	256.41	23.03
Convergence threshold of the inversion	1×10^{-4}	6	1.00	10.00	167.94	31.41
Maximum iteration number of the inversion	1000	7	0.50	3.16	114.40	49.92
Convergence threshold of the CG method	1×10^{-8}	8	0.00	1.00	75.13	91.54
Maximum iteration number of the CG method	500	9	0.05	1.12	78.55	85.74
		10	0.10	1.26	82.09	80.37
		11	0.15	1.41	85.75	75.41
		12	0.20	1.59	89.51	70.83
		13	0.30	1.59	89.51	70.83
		14	0.40	2.51	105.65	55.79

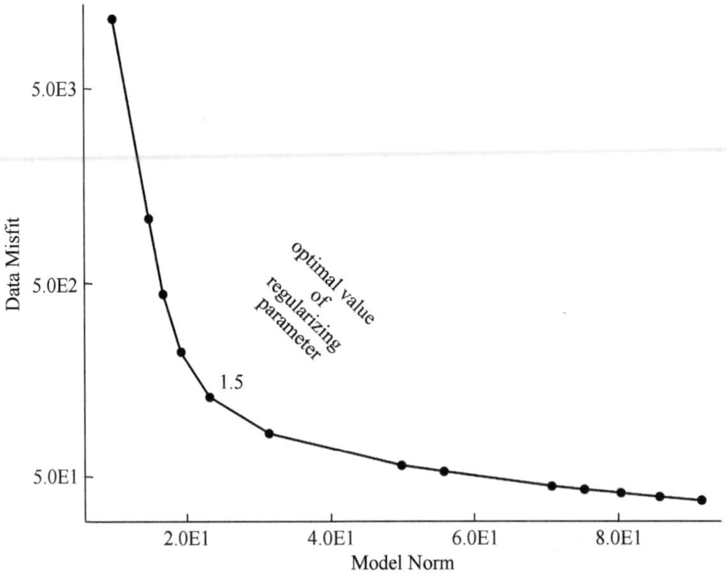

Figure 3 L _ curve solution for data misfit (ϕ_d) against model objective function (ϕ_m). The plot, called the L _ curve method, indicates the optimal value of regularization parameter as a value that lies at the corner of the L _ curve

3 Results

The recovered 3D density structures beneath the Von Kármán Crater indicate the presence of both lateral and horizontal density variations that range from 2600kg/m^3 to 3200kg/m^3 (Figures 4 and 5). We also noticed that the small - scale density anomalies are confined to the upper 8km of the crust. The anomalies observed on the density slice up to the depth of 5km can also be seen in the first vertical derivative and the tilt angle derivative (Figure 4a, b), which validates our inversion results that resolved the small - scale anomalies. The inversion results also show short - wavelength artifacts up to the depth of 7.5km (Figure 4c - f) that were introduced in the results primarily due to the use of a radial cell size of 0.5km. The crust - mantle boundary is generally thin in the study area, with high - density anomalies, $>$3000kg/m^3, located below this boundary. The density values increase with an increase in crustal depth beneath the high gravity regions. However, in other areas, especially the immediate area just below the crust - mantle boundary of the Von Kármán Crater, the density values decrease with an increase in crustal depth. The upper 2km indicates a relatively low - density value of ~2900kg/m^3 and conversely the region between 2km and 10km reveals a higher density value of greater than 3000kg/m^3, which could be due to presence of buried and thick near surface basaltic materials [2, 13] that overlay the feldspathic crust [54, 55]. The crustal model number

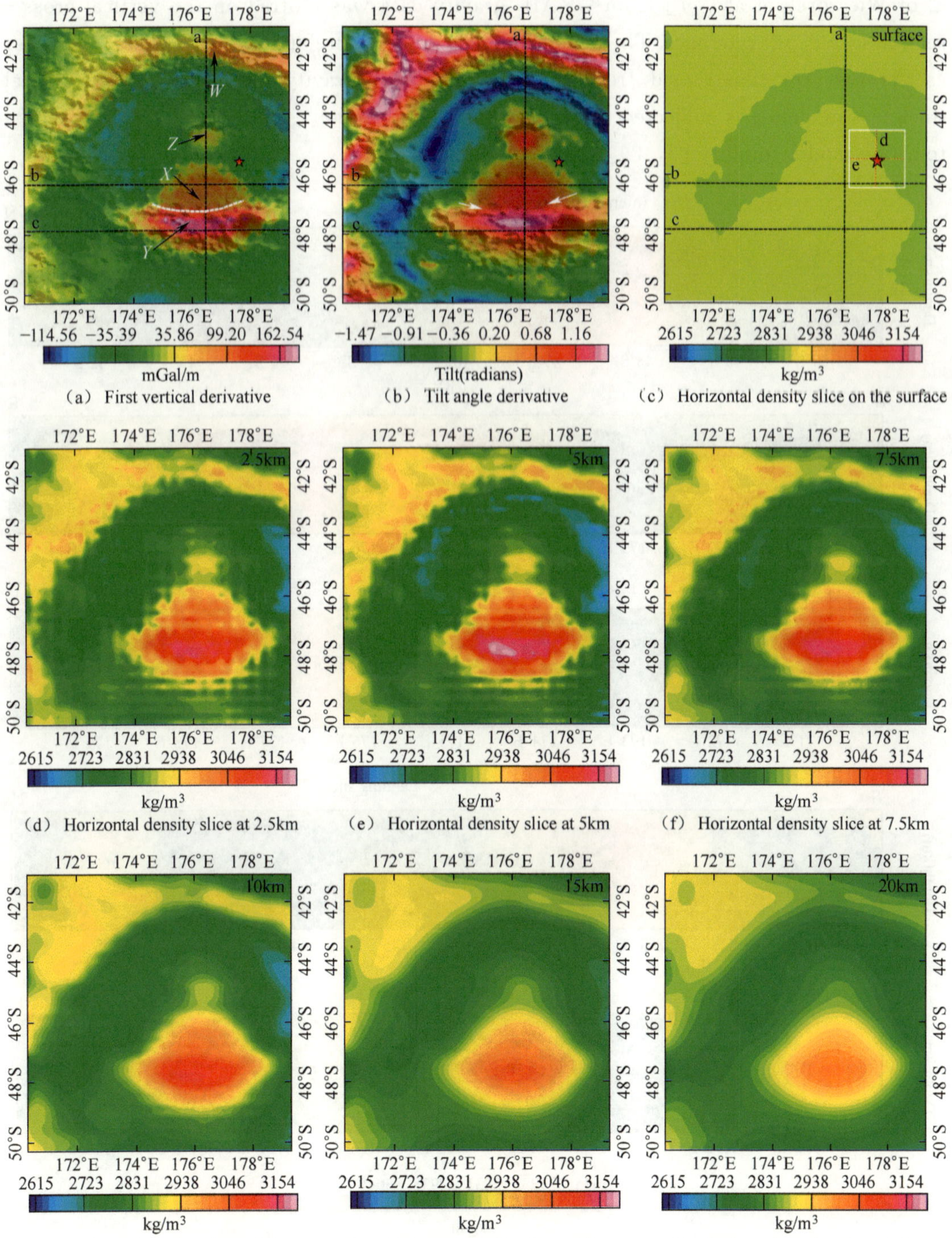

（a） First vertical derivative　（b） Tilt angle derivative　（c） Horizontal density slice on the surface

（d） Horizontal density slice at 2.5km　（e） Horizontal density slice at 5km　（f） Horizontal density slice at 7.5km

（g） Horizontal density slice at 10km　（h） Horizontal density slice at 15km　（i） Horizontal density slice at 20km

Figure 4　Derivative calculations results draped on a hill - shade topographic map，a - b；(a) First vertical derivative. The letters X，Y，Z and W represent anomalous areas in this region. The dashed - white line in Figure 4a represents a possible boundary that separates the high - density region from a relatively low - density region. and (b) Tilt angle derivative; and the horizontal density slices c - i，at the depth of (c) surface，(d) 2. 5km，(e) 5km，(f) 7. 5km，(g) 10km，(h) 15km and (i) 20km. The three black lines (a - c) in Figure 4a. c represent the location of the vertical cross - section shown in Figure 5，and the red dashed(d and e) in Figure 4c lines represents the location of the vertical cross - section across the landing site.

2 of Wieczorek et al. [7] from GRAIL gravity data was plotted on the vertical cross-section as a depth reference. The density models do not correlate with the crustal thickness as the two are constructed from different levels of gravity anomalies, since crustal thickness modeling is subjected to heavy filtering that removes most of the gravity signal to attain model stability [7, 29].

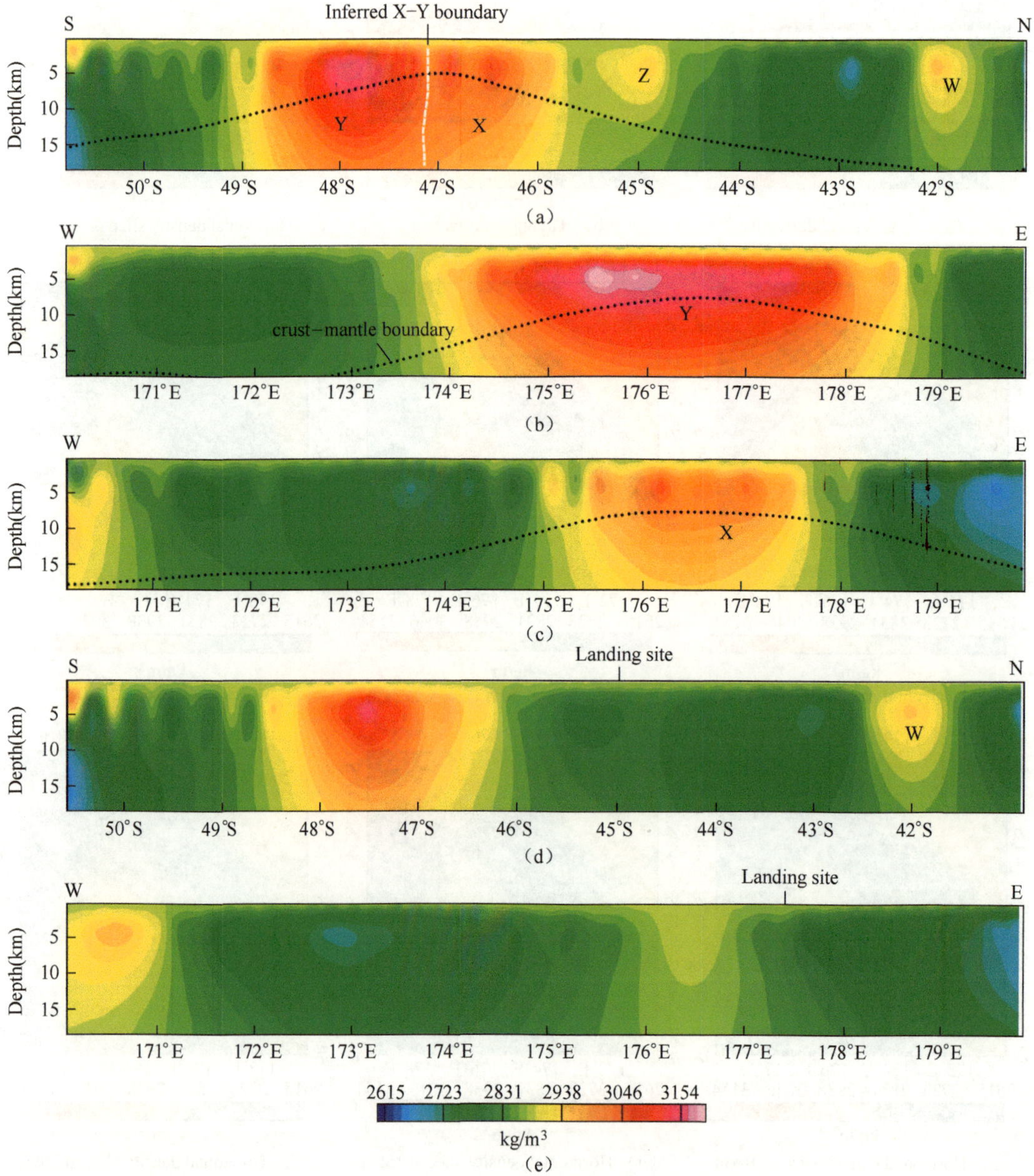

Figure 5　Vertical cross-section maps for the Von Kármán Crater. A vertical cross section (a-c), reveals the subsurface density distribution of locations shown in Figure 4 of the same letters, with a vertical exaggeration of 2. A vertical cross-section, (d, e), indicates the subsurface structure across the landing site (177.59°E, 45.46°S; [11, 12]). The black dotted line indicates the crust-mantle boundary from model 2 of Wieczorek et al. [7].

Figures 4 and 5 reveal both shallow and deep subsurface density structures of the Von Kármán Crater. The shallow subsurface structures are clearly visible even from the derivative maps（Figure 4a，b），which are also consistent with the recovered density model（Figure 4c – f）. The northern rim of the Von Kármán Crater，labeled W，reveals an anomalous region（Figure 4a，b），with a density value of ~2900kg/m^3 that extends to the depth of ~10km（Figure 5a）. This area is flanked by low gravity regions within the base of the elevated region，noticed in both the derivative maps and the density slices up to the depth of 7.5km. The southern rim of the Von Kármán Crater，however，reveals the highest density values，~2900kg/m^3 on the surface（Figure 5）that go up to 3150kg/m^3 on the crust – mantle boundary. Below the crust – mantle boundary，the density decreases to ~3000kg/m^3. The crust is noticeably thin in this region [7]. This anomaly forms an elliptic and elongated E – W trending feature along the elevated areas，apparently squeezed between the Von Kármán and the Von Kármán M Craters.

The high – density subsurface structures（labeled Y）show an obvious annulus of a low – density anomaly，which is more visible on the gravity anomaly than on density models. Unlike other regions with a similar high gravity center surrounded by an annulus of low gravity features [56 – 58]，the annulus in this region transcends two impact craters，the Von Kármán and the Von Kármán M. The low – density annulus are consistent with the thicker crust surrounding the high – density central part（Figures 2 and 3），which also transcend these two craters. Adjacent to this body，an extension of the high – density anomaly of ~2950kg/m^3 is located within a flat crater floor，also noticed on the topographic map（Figure 4a – c）. The derivative maps（Figure 4a，b）reveal a slightly low – gravity region between these two bodies（white dashed line on Figure 4a and pointed by two white arrows on Figure 4b）. The striking dissimilarity between these two anomalies，low – density northern part，labeled X，and relatively higher – density southern part，labeled Y，suggests that these two anomalies were possibly formed as a single body but underwent different post formation conditions. The possible boundary of these two anomalies is placed on the ~2900kg/m^3 region（dashed white line on Figure 4a）with the subsurface density structures，indicating two regions（Figure 4c – e，Figure 5b，c）.

The high – density anomaly ends on the edge of the Chang'E – 4 landing site（red star on Figure 4），which suggests that the landing site is directly above the feldspathic crust，overlain by considerable amount of mare basalt [13]. These characteristics are also revealed by the vertical cross – section on the landing site（Figure 5d，e）. The densities are slightly higher than the underlying crust，~ 2800kg/m^3. Another high – density anomaly，labeled Z，is located in the middle of the Von Kármán Crater. The anomaly indicates possible buried materials of the crater's central peak within the crust，with a density value of ~2850kg/m^3. The high density associated with this anomaly is entirely located above the crust – mantle boundary（Figure 5a）. The density values of this anomaly are slightly comparable with the values of the northern rim anomaly, labeled W, but lower than the southern anomalies, labeled X and Y.

4 Discussion

4.1 3D Density Model and Evolution for the Von Kármán Basin

The 3D gravity inversion has revealed a prominent mass excess on the southern rim of the Von Kármán Crater. The subsurface configuration of this feature, with a bowl-shaped anomaly, thin crust and a high-density region beneath the surface characterizes a buried impact basin [59-61]. The Bouguer gravity anomaly in this region (~85mGal) is relatively small to indicate uplifted mantle materials. Instead, it could be derived from the filled basalts in the impact crater [61]. However, the noticeable different subsurface structural configurations, with high-density asymmetry, could indicate an increase in density towards the southern part of the high-density region, which suggests a possible decrease in porosity or an increase in the mafic content [7, 19, 21]. We hypothesize that the structures were initially formed by a single impact cratering event. This is further evidenced by an annulus of low gravity values that surrounds the high-density anomaly. We also suggest that the high-density region predates the Von Kármán Crater, possibly formed during the formation of the Von Kármán M Crater. Stratigraphic sequences and dating techniques indicate that the Von Kármán Crater postdates and is superimposed on the Von Kármán M Crater, obscuring the crater morphology of the Von Kármán M Crater [13]. Thus, we discuss the formation and evolution of the mass excess on the southern rim of the Von Kármán Crater in relation to the Von Kármán M Crater.

We investigate the possible mode of formation for the mass excess related to our proposed multiple impact cratering. We propose that the mass excess beneath the Von Kármán and Von Kármán M Craters was formed by three impact events as a result of multiple impact cratering. The first major impact event, that occurred more than 4Ga years ago, in the multiple cratering scenario created the SPA that eventually excavated the lower crust and probably the upper mantle, resulting in a very thin crust region [7, 23]. Then, this event was followed by another impact event that created the Von Kármán M Crater, similar to the formation of an ancient mascon [56, 57, 62]. This event happened between 4 Ga and 3.97Ga years ago, as dating techniques indicate that the Von Kármán M Crater predates the Von Kármán Crater [13]. However, the mantle materials did not push the crust to attain isostatic stability in this region, as noticed from the lack of the excessive high-density anomaly beneath the crust-mantle boundary. This impact event probably excavated an already thinned crust forming a bowl-shaped geometry and possibly further exposing mantle materials in the Von Kármán M Crater. The Von Kármán M Crater was subsequently filled by mare basalts that created a buried impact basin [59,

61] with high - density and high gravity anomalies labeled X and Y. Finally, a third impact event reworked the Von Kármán M Crater and created the Von Kármán Crater. In the process, the existing high - density region beneath the Von Kármán M Crater was probably pushed southwards or thermally reworked during the impact. The other explanation is that the northern part of the mass excess (labeled X) was excavated by the impact event that created the Von Kármán Crater while the southern part (labeled Y) returned its basaltic thickness. This could explain the density asymmetry of the body. The presence of a central peak, labeled Z on Figure 4, suggests that the high - density anomalies, X and Y, predate the Von Kármán Crater and that the central peak is associated with the formation of Von Kármán Crater. Thus, the high - density region was created during later stages of multiple cratering and was not excavated during the formation of SPA. The Von Kármán Crater was also later filled by mare basalts, since multiple volcanic lava flows in the Von Kármán Crater region have been identified by remote sensing techniques to have occurred in different time periods, e. g. , ~3. 75 Ga [6], ~3. 35 Ga [63], 3. 15 Ga [6] .

4. 2 Implication of the Von Kármán Density Structure for the Chang'E - 4 Mission

The outstanding research question in the formation of SPA hinges on the possibility that the upper mantle was excavated and brought to the surface. The Chang'E - 4 mission explored this possibility with initial interpretation confirming the presence of mantle materials from the presence of olivine and LCP [16] . Li et al. [16] and Qiao et al. [13] suggested that the mantle materials, with the presence of olivine and LCP, originated from the Finsen Crater and were transported to the Von Kármán Crater. Our inversion results, however, reveal that the mass excess anomalies (Y and X on Figure 4) were created as the result of possible multiple impact cratering that reshaped and concentrated the anomaly in the southern end of the Von Kármán Crater. The mass excess anomaly labeled X has an apparent correlation with both surface mineralogy and mare volcanism [4, 13] . The floor of the Von Kármán Crater is characterized by mare basalts in the southern part and feldspathic materials in the northern part. This is evidenced in the recovered density models, with >2900kg/m^3 values localized in the southern part. Qiao et al. [13] also identified an additional possible underlying basaltic layer, beneath the mare basalt layer, at a depth up to ~ 30 - 90m, characterized by slightly higher iron and titanium contents.

In Section 4. 1, we indicated that the second impact event that created the Von Kármán M Crater excavated a thinned crust that probably contained upper mantle and lower crust materials. The mare basalts that were deposited in this region could have buried the mantle materials, which were eventually excavated by the third impact event that created the Von Kármán Crater. Since it is shown that the materials of the Chang'E - 4 landing site have a strong resemblance to the mantle originated mafic central peak of the

Finsen Crater, we suggest that the impact cratering that formed the SPA basin excavated mantle materials. Then, re – excavation by the impact forming the Von Kármán Crater exposed the mantle materials. This is evidenced by the absence of mantle materials in region Y, and high correlation of basaltic materials with the presence of olivine and LCP with region X [16] . Possibly, the mare basalts mixed with mantle materials, indicating a relationship between the mantle materials on the floor of the Von Kármán Crater and the Finsen Crater. We thus suggest that the impact forming the Von Kármán Crater contributed somewhat to the secondary excavation of mantle materials that were initially excavated by events that formed the SPA and were buried by the mare basalts that occurred after the impact event that created the Von Kármán M Crater.

5 Conclusions

We performed a 3 – D inversion of the GRAIL gravity data to produce a density model of the Von Kármán Crater. Our results reveal that the region is underlain by a mass excess anomaly beneath the Von Kármán M Crater with a density of ~3200kg/m^3, on the southern rim of the Von Kármán basin. A relatively high – density mass excess anomaly with a density value of up to 3100kg/m^3 connects to the first anomaly in the northern part. Basaltic rocks covering the Von Kármán basin show an extended body on the surface, with the northern part of the Von Kármán basin indicating a more feldspathic composition. We suggest that this was as the result of the buried mare basalts that created a buried impact basin in the Von Kármán M Crater. The inversion results are consistent with the spectral and geological results from remote sensing and gravity derivative data. The occurrence of high – density materials that extend to the lower crust correlates with excavated mantle materials observed on the floor of the Von Kármán basin, suggesting that the impact cratering could have brought upper mantle materials to the surface. We thus infer that the evolution of the Von Kármán basin was due to multiple episodes of impact cratering that resulted in crustal reworking and secondary excavation of mantle materials.

Acknowledgments

GRAIL gravity data, which we used, are available from NASA Planetary Data System (https://pds.nasa.gov/) .

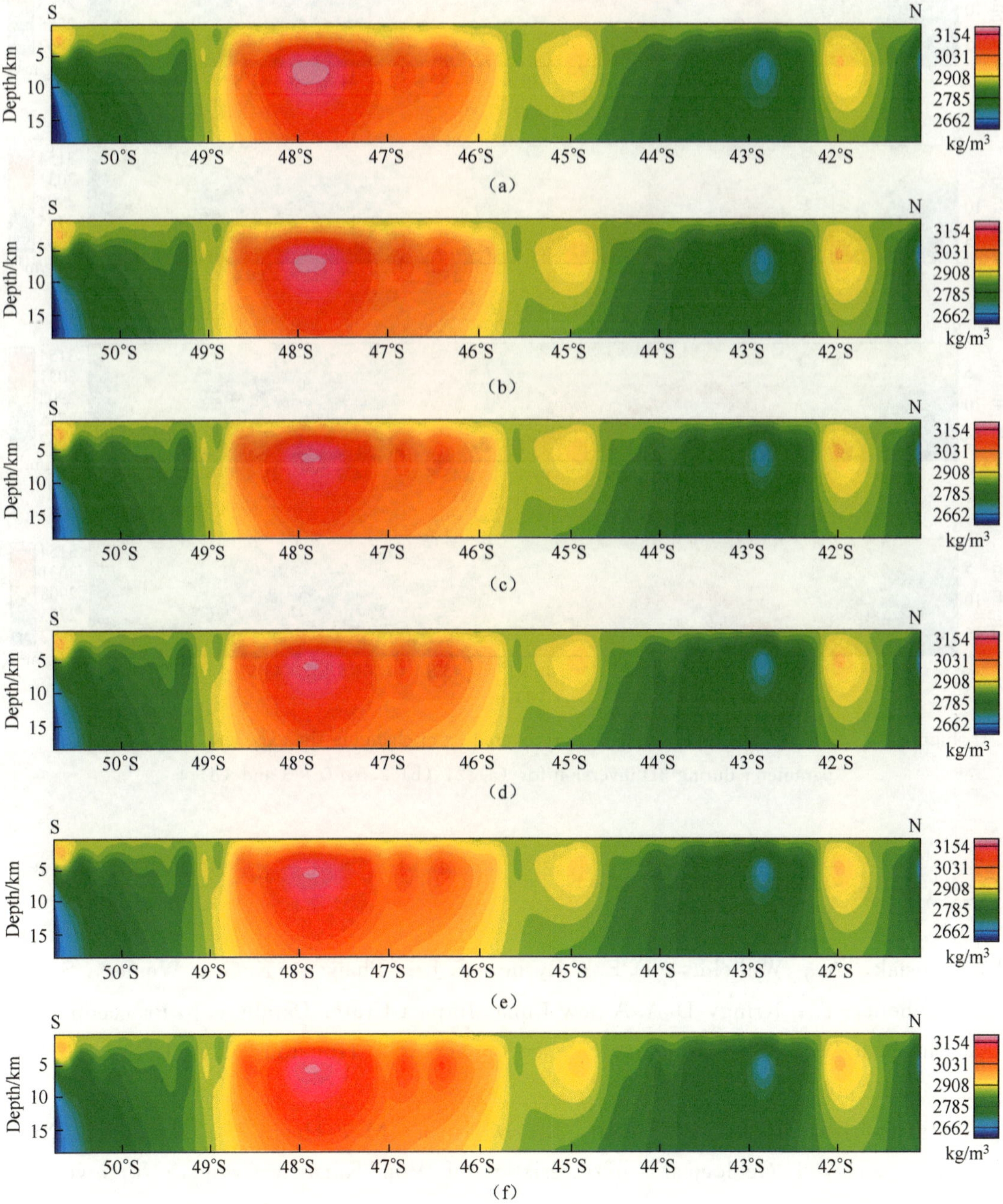

Figure A1　Some of the models that were recovered from changing the exponent of the Tikhonov parameter during 3D inversion for (a) 0.2, (b) 0.3, (c) 0.4, (d) 0.5, (e) 1 and (f) 1.5

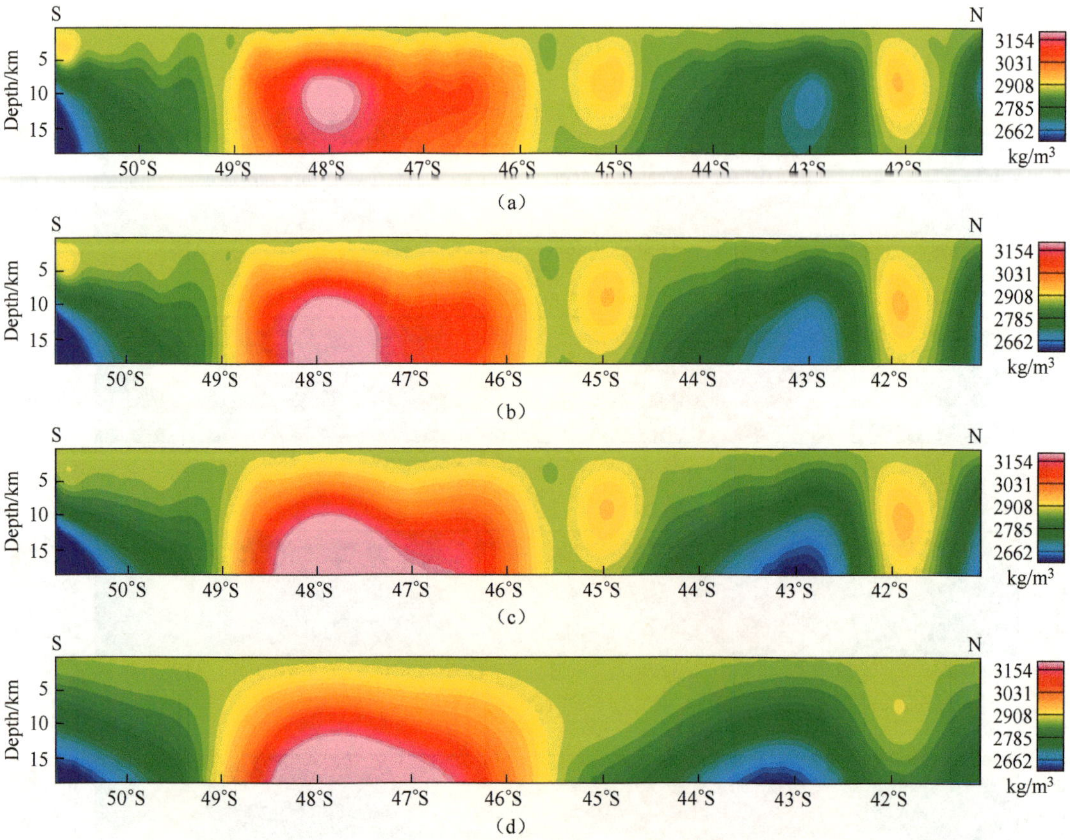

Figure A2　Some of the models that were recovered from changing the exponent of the Tikhonov parameter during 3D inversion for (a) 2, (b) 2.5, (c) 3 and (d) 4

References

[1] Losiak, A.; Wilhelms, D. E.; Byrne, C. J.; Thaisen, K. G.; Weider, S. Z.; Kohout, T.; Kring, D. A. A new Lunar Impact Crater Database. In Proceedings of the 40th Lunar and Planetary Science Conference, Houston, TX, USA, 23 - 27 March 2009.

[2] Huang, J.; Xiao, Z.; Flahaut, J.; Martinot, M.; Head, J.; Xiao, X.; Xie, M.; Xiao, L. Geological Characteristics of Von Kármán Crater, Northwestern South Pole - Aitken Basin: Chang'E - 4 Landing Site Region. J. Geophys. Res. Planets 2018, 123, 1684 - 1700.

[3] Potter, R. W. K.; Collins, G. S.; Kiefer, W. S.; Mcgovern, P. J.; Kring, D. A. Constraining the size of the South Pole - Aitken basin impact. Icarus 2012, 220, 730 - 743.

[4] Moriarty, D. P.; Pieters, C. M. The character of South PoleAitken basin: Patterns

of surface and subsurface composition. J. Geophys. Res. Planets 2018，123，729－747.

[5] Yingst，R. A.；Head，J. W.，III. Volumes of lunar lava ponds in South Pole－Aitken and Orientale Basins：Implications for eruption conditions，transport mechanisms，and magma source regions. J. Geophys. Res. 1997，102，10909－10931.

[6] Pasckert，J. G.；Hiesinger，H.；van der Bogert，C. H. Lunar farside volcanism in and around the South Pole－Aitken basin. Icarus 2018，599，538－562.

[7] Wieczorek，M. A.；Neumann，G. A.；Nimmo，F.；Kiefer，W. S.；Taylor，G. J.；Melosh，H. J.；Philips，R. J.；Solomon，S. C.；Andrews－Hanna，J. C.；Asmar，S. W.；et al. The Crust of the Moon as Seen by GRAIL. Science 2013，339，671－675.

[8] Barker，M. K.；Mazarico，E.；Neumann，G. A.；Zuber，M. T.；Haruyama，J.；Smith，D. E. A new lunar digital elevation model from the Lunar Orbiter Laser Altimeter and SELENE Terrain Camera. Icarus 2016，273，346－355.

[9] Goossens，S.；Lemoine，F. G.；Sabaka，T. J.；Nicholas，J. B.；Mazarico，E.；Rowlands，D. D.；Loomis，B. D.；Chinn，D. S.；Neumann，G. A.；Smith，D. E.；et al. A global degree and order 1200 model of the lunar gravity field using grail mission data. In 47th Lunar and Planetary Science Conference；Lunar and Planetary Institute：Houston，TX，USA，2016；pp. 13－14.

[10] Neumann，G. A.；Zuber，M. T.；Wieczorek，M. A.；Head，J. W.；Baker，D. M. H.；Solomon，S. C.；Smith，D. E.；Lemoine，F. G.；Mazarico，E.；Sabaka，T. J.；et al. Lunar impact basins revealed by gravity recovery and interior laboratory measurements. Sci. Adv. 2015，1，1－11.

[11] Di，K.；Liu，Z.；Liu，B.；Wan，W.；Peng，M.；Wang，Y.；Gou，S.；Yue，Z.；Xin，X.；Jia，M.；et al. Chang'E－4 lander localization based on multi－source data. J. Remote Sens. 2019，4619，177－184.

[12] Wu，W. R.；Wang，Q.；Tang，Y. H.；Yu，G. B.；Liu，J. Z.；Zhang，W. Design of Chang'E－4 lunar farside soft－landing mission. J. Deep Space Explor. 2017，4，111－117.（In Chinese）

[13] Qiao，L.；Ling，Z.；Fu，X.；Li，B. Geological characterization of the Chang'e－4 landing area on the lunar farside. Icarus 2019，333，37－51.

[14] Pieter，C. M.；Head，J. W.；Gaddis，L.；Jolliff.，B.；Duke，M. Rock types of South Pole－Aitken basin and extent of basaltic volcanism. J. Geophys. Res. 2001，106，28001－28022.

[15] Pieters，C. M.；Tompkins，S.；Head，J. W.；Hess，P. C. Mineralogy of the Mafic Anomaly in the South Pole－Aitken Basin：Implications for excavation of the lunar mantl. Geophys. Res. Lett. 1997，24，1903－1906.

[16] Li，C.；Liu，D.；Liu，B.；Ren，X.；Liu，J.；He，Z.；Wei，Z.；Zeng，X.；

Xu, R. ; Tan, X. ; et al. Chang'E - 4 initial spectroscopic identification of lunar far - side mantle - derived materials. Nature 2019, 569, 378 - 382.

[17] Melosh, H. J. ; Kendall, J. ; Horgan, B. ; Johnson, B. C. ; Bowling, T. ; Lucey, P. G. ; Taylor, G. J. South Pole - Aitken basin ejecta reveal the Moon' s upper mantle. Geology 2017, 45, 1063 - 1066.

[18] Vaughan, W. M. ; Head, J. M. Impact melt differentiation in the South Pole - Aitken basin: Some observations and speculations. Planet. SpaceSci. 2014, 91, 101 - 106.

[19] Jansen, J. C. ; Andrews - Hanna, J. C. ; Li, Y. ; Lucey, P. G. ; Taylor, G. J. ; Goossens, S. ; Lemoine, F. G. ; Mazarico, E. ; Head, J. W. , III; Milbury, C. ; et al. Small - scale density variations in the lunar crust revealed by GRAIL. Icarus 2017, 291, 107 - 123.

[20] Han, S. C. ; Schmerr, N. ; Neumann, G. ; Holmes, S. Global characteristics of porosity and density stratification within the lunar crust from GRAIL gravity and Lunar Orbiter Laser Altimeter topography data. Geophys. Res. Lett. 2014, 41, 882 - 1889.

[21] Huang, Q. ; Wieczorek, M. A. Density and porosity of the lunar crust from gravity and topography. J. Geophys. Res. 2012, 117.

[22] Besserer, J. ; Nimmo, F. ; Wieczorek, M. A. ; Weber, R. C. ; Kiefer, W. S. ; McGovern, P. J. ; Andrewa - Hanna, J. C. ; Smith, D. E. ; Zuber, M. T. GRAIL gravity constraints on the vertical and lateral density structure of the lunar crust. Geophys. Res. Lett. 2014, 47, 5771 - 5777.

[23] Nakamura, R. ; Matsunaga, T. ; Ogawa, Y. ; Yamamoto, S. ; Hiroi, T. ; Saiki, K. ; Hirata, A. ; Tomoko, A. ; Kohei, T. ; Hiroshi, T. ; et al. Ultramafic impact melt sheet beneath the South Pole - Aitken basin on the Moon. Geophys. Res. Lett. 2009, 36, L22202.

[24] Konopliv, A. S. ; Asmar, S. W. ; Carranza, E. ; Sjogren, W. L. ; Yuan, D. N. Recent gravity models as a result of the Lunar Prospector mission. Icarus 2001, 150, 1 - 18.

[25] Zuber, M. T. ; Smith, D. E. ; Lehman, D. H. ; Hoffman, T. L. ; Asmar, S. W. ; Watkins, M. M. Gravity Recovery and Interior Laboratory (GRAIL): Mapping the Lunar Interior from Crust to Core. Space Sci. Rev. 2013, 178, 2 - 27.

[26] Zuber, M. T. ; Smith, D. E. ; Watkins, M. M. ; Asmar, S. W. ; Konopliv, A. S. ; Lemoine, F. G. ; Melosh, H. J. ; Neumann, G. A. ; Philips, R. J. ; Solomon, S. C. ; et al. Gravity Field of the Moon from the Gravity Recovery and Interior Laboratory (GRAIL) Mission. Science 2013, 339, 668 - 671.

[27] Lemoine, F. G. ; Goossens, S. ; Sabaka, T. J. ; Nicholas, J. B. ; Mazarico, E. ; Rowlands, D. D. ; Loomis, B. D. ; Chinn, D. S. ; Neumann, G. A. ; Smith, D. E. ; et al. GRGM900C: A degree 900 lunar gravity model from GRAIL primary

and extended mission data. Geophys. Res. Lett. 2014，41，3382 – 3389.

[28] Konopliv, A. S.；Park，R. S.；Yuan，D. N.；Asmar，S. W.；Watkins，M. M.；Williams，J. G.；Fahnstock，E.；Kruizinga，G.；Paik，M.；Strekalov，D.；et al. High – resolution lunar gravity . fields from the GRAIL Primary and Extended Missions. Geophys. Res. Lett. 2014，41，452 – 1458.

[29] Wieczorek，M. A.；Phillips，R. J. Potential anomalies on a sphere' Applications to the thickness of the lunar crust. J. Geophys. Res. 1998，103，1715 – 1724.

[30] Wieczorek，M. A. SHTOOLS – Tools for working with spherical harmonics (v2. 9. 1) . Zenodo 2014.

[31] Wieczorek，M. A.；Meschede，M. SHTools – Tools for working with spherical harmonics. Geochem. Geophys. Geosyst. 2018，19，2574 – 2592.

[32] Jolliff，B. L.；Gillis，J. J.；Haskin，L. A.；Korotev，R. L.；Wieczorek，M. A. Major lunar crustal terranes：Surface expressions and crust – mantle origins. J. Geophys. 2000，105，4197 – 4216.

[33] Ma，Q.；Du，X. – J.；Li，L. – L.；Meng，L. – S. Interpretation of magnetic anomalies by horizontal and vertical derivatives of the analytic signal. Appl. Geophys. 2012，9，468 – 474.

[34] Miller，H. G.；Singh，V. Potentialfield tilt – A new concept for location of potentialfield sources. J. Appl. Geophys. 1994，32，213 – 217.

[35] Kolawole，F.；Atekwana，E. A.；Laó – Dávila，D. A.；Abdelsalam，M. G.；Chindandali，P. R.；Salima，J.；Kalindekafe，L. Active Deformation of Malawi Rift ' s North Basin Hinge Zone Modulated by Reactivation of Preexisting Precambrian Shear Zone Fabric. Tectonics 2018，37，683 – 704.

[36] Cheek, L. C.；Hanna，K. L. D.；Pieters，M.；Head，J. W.；Whitten，J. L. The distribution and purity of anorthosite across the Orientale basin：New perspectives from Moon Mineralogy Mapper data. J. Geophys. Res. Planets 2013，118，1805 – 1820.

[37] Grove，T. L.；Krawczynski，M. J. Lunar Mare Volcanism：Where Did the Magmas Come From? Elements 2009，5，29 – 34.

[38] James，W. Lunar Volcanism in Space and Time. Rev. Geophys. 1976，14，265 – 300.

[39] Neal，C.；Taylor，L. Petrogenesis of mare basalts：A record of lunar volcanism. Geochim. Cosmchimica Acta 1992，56，2177 – 2211.

[40] Wood, J. A. Petrology of the Lunar Soil and Geophysical Implications. J. Geophys. Res. 1970，75，6497 – 6513.

[41] Zhang，Y.；Yan，J.；Li，F.；Chen，C.；Mei，B.；Jin，S.；Dohm，J. H. A new bound constraints method for 3 – D potentialfield data inversion using Lagrangian multipliers. Geophys. J. Int. 2015，201，267 –275.

[42] Zhang，Y.；Wu，Y.；Li，H.；Zhang，X.；Yan，J.；Wu，G.；Wang，J. 3D

inversion of full gravity gradient tensor data in spherical coordinate system using local north – oriented frame. In CGS/SEG International Geophysical Conference; Society of Exploration Geophysicists: Tulsa, OK, USA, 2017; pp. 17 – 20.

[43] Zhang, Y.; Wu, Y.; Yan, J.; Wang, H.; Rodriguez, J. A. P.; Qiu, Y. 3D inversion of full gravity gradient tensor data in spherical coordinate system using local north – oriented frame. Earth Planets Space 2018, 70.

[44] Liang, Q.; Chen, C.; Li, Y. 3 – D inversion of gravitydata in spherical coordinates with application to the GRAIL data. J. Geophys. Res. Planets Res. 2014, 119, 1359 – 1373.

[45] Li, Y.; Oldenburg, D. Y. 3 – D inversion of magnetic data. Geophysics1996, 61, 394 – 408.

[46] Li, Y.; Oldenburg, D. Y. 3 – D inversion of gravity data. Geophysics1998, 63, 264 – 267.

[47] Tikhonov, A. N.; Arsenin, V. Y. Solutions of Ill. – Posed Problems; Winston and Sons: Washington, DC, USA, 1977.

[48] Lawson, C. L.; Hanson, R. J. Solving Least Squares Problem; Prentice – Hall: New Jersey, NJ, USA, 1974.

[49] Hansen, P. C. Analysis of discrete ill – posed problems by means of the L – curve. Soc. Ind. Appl. Math. Rev. 1992, 34, 561 – 580.

[50] Calvettia, D.; Morigib, S.; Reichelc, L.; Sgallarid, F. Tikhonov regularization and the L – curve for large discrete ill – posed problems. J. Comput. Appl. Math. 2000, 123, 423 – 446.

[51] Hestenes, M. R. Multiplier and Gradient Methods. J. Optim. TheoryAppl. 1969, 4, 303 – 320.

[52] Rockafellar, R. T. Convex Analysis; Princeton University Press: Princeton, NJ, USA, 1970.

[53] Asgharzadeh, M. F.; Von Frese, R. R. B.; Kim, H. R.; Leftwich, T. E.; Kim, J. W. Spherical prism gravity effects by Gauss – Legendre quadrature integration. Geophys. J. Int. 2007, 169, 1 – 11.

[54] Smith, J. V.; Anderson, A. T.; Newton, R. C.; Olsen, E. J.; Wyllie, P. J.; Crewe, A. V.; Isaacson, M. S.; Johnson, D. Petrologic history of the moon inferred from petrography, mineralogy, and petrogenesis of Apollo 11 rocks. In Apollo 11 Lunar Science Conference; Pergammon Press: New York, NY, USA, 1970; pp. 897 – 925.

[55] Wood, J. A.; Dickey, J. S.; Marvin, U. B.; Powell, B. N. Lunar anorthosites and a geophysical model of the moon. In Apollo 11 Science Conference; Pergammon Press: New York, NY, USA, 1970; pp. 965 – 988.

[56] Melosh, H. J.; Freed, A. M.; Johnson, B. C.; Blair, D. M.; Andrews – Hanna, J. C.; Neumann, G. A.; Philips, R. J.; Smith, D. E.; Solomon, S. C.;

Wieczorek, M. A. ; et al. The origin of lunar mascon basins. Science 2013, 340, 1552 - 1555.

[57] Dombard, A. J. ; Hauck, S. A. ; Balcerski, J. A. On the origin of mascon basins on the Moon (and beyond) . Geophys. Res. Lett. 2013, 40, 28 - 32.

[58] Wieczorek, M. A. ; Phillips, R. J. Lunar Multiring Basins and the Cratering Process. Icarus 1999, 259, 246 - 259.

[59] Deutsch, A. N. ; Neumann, G. A. ; Head, J. W. ; Wilson, L. GRAIL - identified gravity anomalies in Oceanus Procellarum: Insight into subsurface impact and magmatic structures on the Moon. Icarus 2019, 331, 192 - 208.

[60] Edgar, L. A. ; Frey, H. V. Buried impact basin distribution on Mars: Contributions from crustal thickness data. Geophys. Res. Lett. 2008, 35, 1 - 5.

[61] Evans, A. J. ; Soderblom, J. M. ; Andrews - Hanna, J. C. ; Solomon, S. C. ; Zuber, M. T. Identi. cation of buried lunar impact craters from GRAIL data and implications for the nearside maria. Geophys. Res. Lett. 2016, 43, 2445 - 2455.

[62] Freed, A. M. ; Johnson, B. C. ; Blair, D. M. ; Melosh, H. J. ; Neumann, G. A. ; Phillips, R. J. ; Solomon, S. C. ; Wieczorek, M. A. ; Zuber, M. T. The formation of lunar mascon basins from impact to contemporary form. J. Geophys. Res. E Planets 2014, 119, 2378 - 2397.

[63] Haruyama, J. ; Ohtake, M. ; Matsunaga, T. ; Morota, T. ; Honda, C. ; Yokota, Y. ; Abe, M. ; Ogawa, Y. ; Miyamoto, H. ; Iwasaki, A. ; et al. Long - Lived Volcanism on the Lunar Farside Revealed by SELENE Terrain Camera. Science 2009, 323, 905 - 908.

Topographic Evolution of Von Kármán Crater Revealed by the Lunar Rover Yutu - 2

Kaichang Di[1,2], Meng Hua Zhu[2], Zongyu Yue[1], Yangting Lin[3], Wenhui Wan[1], Zhaoqin Liu[1], Sheng Gou[1,2], Bin Liu[1], Man Peng[1], Yexin Wang[1], Shengli Niu[1], Jinhai Zhang[3], Jian Li[4], Jianfeng Xie[4], Luhua Xi[4], Jianfeng Yang[5], and Bin Xue[5]

1. State Key Laboratory of Remote Sensing Science, Institute of Remote Sensing and Digital Earth, Chinese Academy of Sciences, Beijing, China

2. State Key Laboratory of Lunar and Planetary Sciences, Macau University of Science and Technology, Macau, China

3. Key Laboratory of Earth and Planetary Physics, Institute of Geology and Geophysics, Chinese Academy of Sciences, Beijing, China

4. Beijing Aerospace Control Center, Beijing, China

5. Xi'an Institute of Optics and Precision Mechanics, Chinese Academy of Sciences, Xi'an, China

Abstract: Chang'e - 4 (CE - 4) achieved the frst farside landing in Von Kármán crater. In the landing site, linear features have been identified previously from SLDEM and considered to be ejecta from the neighboring Finsen crater. The 5cm grid spacing digital elevation model of the landing site, generated from the rover's panoramic images, provides more details of the rugged terrain. We further interpret the superimposition of NE - SW ejecta from Finsen crater on the underlying SE - NW dome - like surface relief from Alder crater. The landing site is ~70m higher than the mare basalts within Von Kármán crater. Numerical simulations predict ~30 and ~35 m ejecta deposited at the landing site from Finsen and Alder craters, respectively. The good agreement between the digital elevation model data and ejecta predicted thickness reveals the topographic evolution of Von Kármán crater, indicating that the rover - measured material is excavated from Finsen crater with possible contributions from Alder crater.

Plain Language Summary: Chang'e - 4 probe, including a lander and a rover, achieved the frstfarside landing in Von Kármán crater within the South Pole - Aitken basin (SPA). The SPA is the largestand oldest impact basin on the Moon and is thought to have exposed lunar mantle materials according to numerical simulations. The in situ explorations at the landing site by the Yutu - 2 rover can reveal information about lunar deep materials and shed light on the early impacting history of the Solar System. Here, we present a detailed topographic analysis of the landing site using a rover - image - derived fner grid spacing DEM, existing orbital - data - derived DEMs, and numerical simulations. The results reveal the topographic evolution of the SPA and demonstrate that the surface material explored by the Yutu - 2 rover is lunar deep interior material excavated from Finsen crater with possible contributions from Alder crater rather than the underlying mare basalt.

1 Introduction

As the oldest and largest impact basin on the Moon, the South Pole - Aitken (SPA) basin was thought to have exposed the lunar lower crust and probably upper mantle material according to the spectral analysis of remote sensing observations and numerical impact cratering simulations (e. g., Head et al., 1993; Ivanov et al., 2018; Lucey et al., 1998; Melosh et al., 2017; Ohtake et al., 2014; Pieters et al., 2001; Potter et al., 2012; Spudis et al., 1994). The basin was subsequently bombarded by numerous impacts and filled by multistage basaltic eruptions (Moriarty & Pieters, 2018; Wilhelms, 1987), which have greatly modified the topography and surficial compositions and buried the original structures of the SPA basin (Pieters et al., 2001; Prettyman et al., 2006; Spudis et al., 1994). The topographic evolution of the SPA basin is a key to tracking the origination of its surficial compositions (e. g., Moriarty & Pieters, 2016; Prettyman et al., 2006) and reconstructing the early impacting and volcanic history. The Chang'e - 4 (CE - 4), composed of a lander and a rover, landed at (177. 588°E, 45. 457° S) in Von Kármán crater within the SPA on 3 January 2019 (Di et al., 2019a), providing a unique opportunity for the in - situ exploration of the SPA basin (Jia et al., 2018; Wu et al., 2019). In this work, we used the rover's panoramic images to generate the finer grid spacing digital elevation model (DEM) of the landing area and conducted numerical impacting simulations of the neighboring craters. Based on morphology from the rover DEM and existing orbital DEMs, and crater modeling, we quantitatively deciphered the topographic structure at the CE - 4 landing site and the topographic evolution history of the Von Kármán crater. With the morphology at the CE - 4 landing site, we reevaluated the origination of surficial materials measured by the Yutu - 2 rover (Li et al., 2019; Lin et al., 2019).

2 Data and Methods

2.1 Data

The lunar rover Yutu - 2 (Jade Rabbit - 2) carries two pairs of stereo cameras, that is, a panoramic camera (Pancam) and a navigation camera (Navcam). These cameras were designed to obtain three - dimensional (3 - D) terrain information for the lunar surface, with Pancam mainly for topographic analysis and Navcam for rover navigation (Jia et al., 2018). The Pancam is a stereo camera mounted on the camera bar of the CE - 4 rover. It has a stereo base of 270mm and a field of view of 22. 9° × 16. 9°. The image size is 2,352 pixels × 1,728 pixels in color mode and 1, 176 pixels × 864 pixels in panchromatic mode (Jia et al., 2018). To acquire a 360° panorama, the rover takes 56 pairs of images with 28 different azimuth angles and two elevation angles. The panoramic mosaics (Figure 1) are automatically generated from the left - eye images taken at three locations. The Pancam stereo images are used to generate a high - resolution DEM of the landing site. The landing site is located in Von Kármán crater with a diameter of 186km, which is adjacent to Leibnitz cater (237km in diameter), Finsen crater (73km in diameter), and Alder crater (82km in diameter) (Figure S1 in the supporting information). To analyze the geologic context of the landing site, the SLDEM2015, which is the combined product of Lunar Orbiter Laser Altimeter (LOLA) data and DEMs generated from Kaguya's terrain camera images (Barker et al., 2016), was also employed. The SLDEM2015 has a resolution of 512 pixels per degree (an effective resolution of ~42m at the CE - 4 landing site).

In addition, we produced a 0. 9 - m - resolution seamless digital orthophoto map (DOM) mosaic of the CE - 4 landing site region covering 51km × 30km using 100 Lunar Reconnaissance Orbiter Camera (LROC) narrow - angle camera (NAC) images (Di et al., 2019b). This high - resolution DOM was used for detailed context analysis of the landing site. The Chang'e - 2 DOM with a resolution of 7m was also used as a base map for topographic analysis.

2.2 DEM Generation and Topographic Analysis

High - resolution DEMs at the three locations were automatically generated from the stereo images through image matching, 3 - D coordinate calculation of the matched points, and DEM interpolation (Liu et al., 2015). The three DEMs were merged into a larger, 380 × 390m DEM of the landing site based on accurate locations of the rover determined by the visual localization technique (see Figure 2). Note that the resolution of the rover data depends strongly on distance from the camera station, but even in the far range where it is worst, it is smaller than the 5cm grid spacing used, and reveals much smaller details than the previously available LROC and SLDEM2015 data sets.

The topographic analysis was performed at the CE - 4 landing site and its vicinity based on the Pancam DEM and the SLDEM2015. The Pancam DEM reveals the details of topography not associated with local craters, and the surface relief of the landing site may have preserved the uneven background terrain. This allowed us to analyze the topographic context using the SLDEM2015. In order to bridge the gap for the different resolution between the Pancam DEM and the SLDEM2015, a DEM with a grid spacing of 5m derived from LROC NAC stereo images (downloaded from http://wms. lroc. asu. edu/lroc/view _ rdr/NAC _ DTM _ CHANGE4) was incorporated (see Figure 2). The DEMs and elevation profiles were rendered using the ArcGIS software. Figure 3 shows the SLDEM2015 color - coded based on nonlinear stretching (see Figure 3) within a large region, in which the trend of topographic variation can be clearly seen.

2.3 Numerical Simulation of Impact Ejecta

In order to investigate whether the topography at the CE - 4 landing site is dominated by the ejecta of neighboring craters of Von Kármán crater, we used the multimaterial, multirheology iSALE - 2D (Dellen version) shock physics code (Collins et al., 2004; Wünnemann et al., 2006) to model the impact cratering process and ejecta distribution of Finsen and Alder craters. The iSALE is based on the SALE (simplified arbitrary Lagrangian Eulerian) hydrocode (Amsden et al., 1980) and has been used to simulate both small - scale laboratory experiments (Wünnemann et al., 2016) and large - scale lunar basin formation (Miljković et al., 2013; Potter et al., 2012; Potter et al., 2013; Yue et al., 2013; Zhu et al., 2015, 2017, 2019).

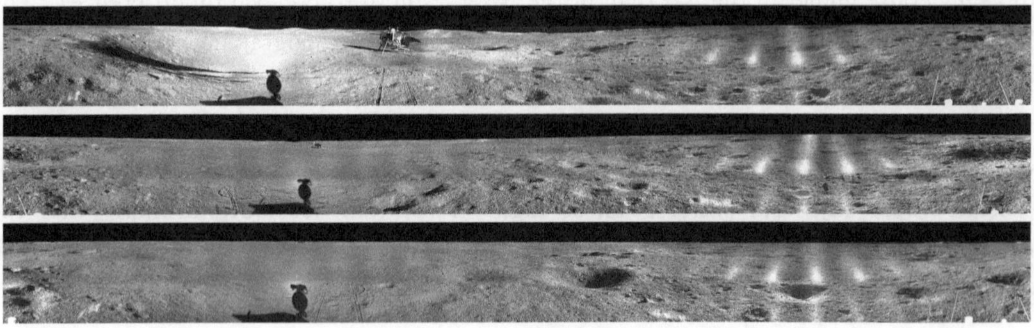

Figure 1　The three 360° image mosaics (left is north, clockwise) produced from the Pancam's left - eye images taken at the locations marked as black dots in Figure 2 labeled by S1, S2, and S3

In our simulations, we used the analytic equation of state for dunite (Benz et al., 1989; Ivanov et al., 2010) and the Tillotson equation of state for gabbroic anorthosite to represent the compositions of impactor and target, respectively. We accounted for the strength and damage models (Collins et al., 2004; Ivanov et al., 2010). The acoustic fluidization model (Melosh, 1979; Melosh & Ivanov, 1999) was considered as well to facilitate collapse in the impact cratering process. The parameters of materials used in the model are listed in Table S1 in the supporting information. We assumed a vertical impact with a typical impact velocity of 15km/s (Le Feuvre & Wieczorek, 2011) on a gabbroic target (Hurwitz & Kring, 2014; Potter et al., 2012). We tracked the ejecta trajectories during the impact cratering process and estimated the ejecta thickness at any given distance from the impact site (see Wünnemann et al., 2016; Zhu et al., 2015, 2017, 2019).

3 Results and Discussion

3.1 Local Variation of Topography From the Pancam DEM

The topography of the CE - 4 landing area was expected to be relatively flat, excluding numerous small craters, because it was filled with mare basalt, similar to that at the Chang' e - 3 landing site (Liu et al., 2015). However, the 360° panoramas achieved by the Pancam show a rugged landscape (Figure 1) that cannot be attributed to the presence of small craters but suggests a rolling topographic background. The DEM made from the rover data (Figure 2a) quantifies this assessment. For comparison, the LROC NAC DEM is also shown (see Figure 2b). The rover DEM is in landing site local coordinate system and the elevation is with respect to the landing point. The elevation of LROC NAC DEM is referenced to the average radius of the Moon, and simple cylindrical projection is used in Figure 2b. The elevation profiles from these two DEMs are shown in Figure 2c, from which we can find that these two DEMs are consistent but the Pancam DEM provides more details of the terrain. The DEMs show that the terrain of the landing site is generally higher in the NE than in the SW (~8m overall elevation difference). The elevation profiles in two directions clearly depict the local topographic variations. Local elevation changes in the NE and SW corners (~5m) and undulations in the NW - SE direction (~2 - 3m) can be noted. The comparison between the DEM and the LROC NAC DOM clearly shows that the local elevation changes and undulations are not associated with any local craters. These high - resolution observations imply that the surface relief has preserved the uneven background terrain, considerably more rugged than expected for a basalt flow with small craters superim posed, but consistent with the ejecta deposits of large, distant craters.

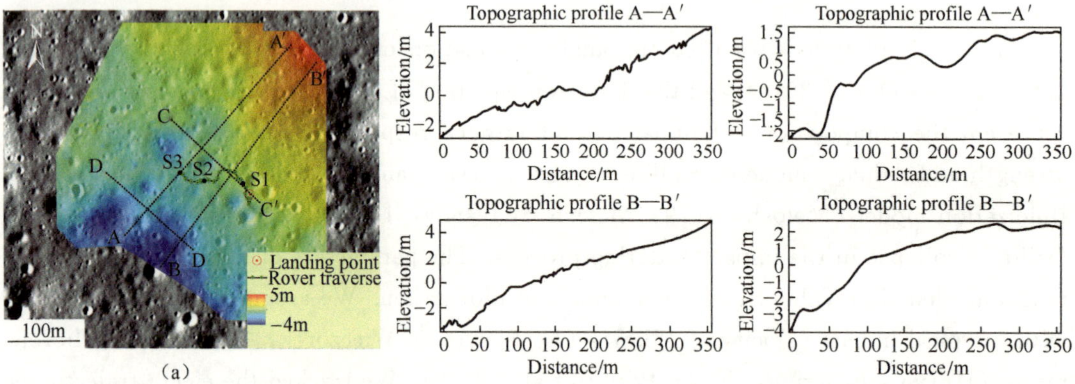

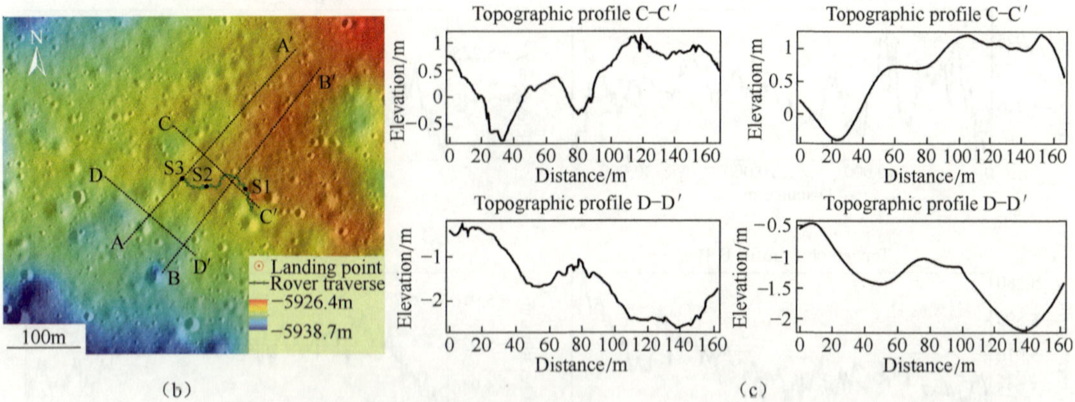

Figure 2　The DEMs and topographic profiles of the CE－4 landing area. The base map is the LROC NAC DOM in (a), with 5cm grid spacing DEM generated from Pancam stereo images, and the elevation data in (b) is from LRO NAC DEM. The elevation profiles of left and right panel in (c) correspond to lines in (a) and (b) with the same labels, respectively. The elevations in the right panel are referenced to the landing point so that consistent with the left panel

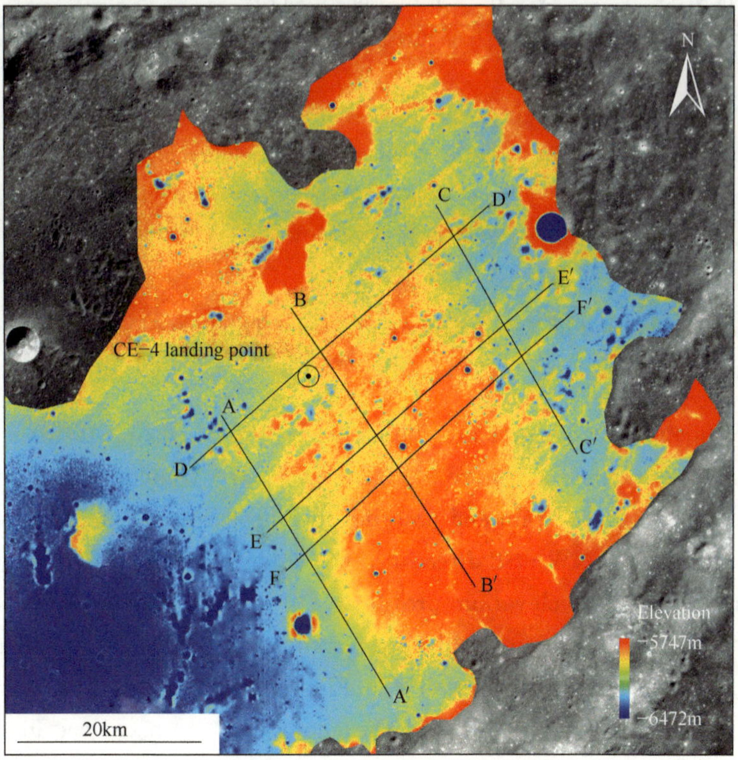

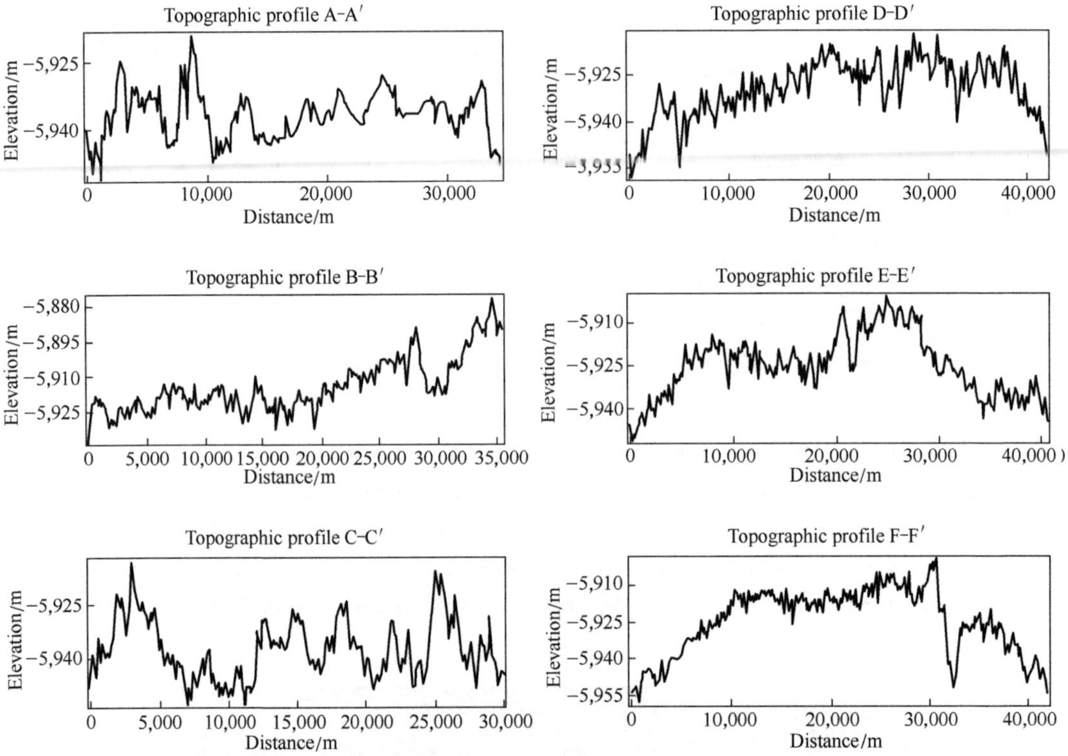

Figure 3　SLDEM2015 and topographic profiles of the context of the CE - 4 landing site. The base map is a DOM from the Chang'e - 2 image. The resolution of the DEM is approximately 42m in this region. Six profiles correspond to the lines on the DEM with the same labels. Simple cylindrical projection is used in the image

3.2　Large - Scale Variation of Topography and Ejecta Thickness From Finsen and Alder Crater

The above topographic features are confirmed by the SLDEM2015 for a large area of eastern Von Kármán crater (see Figure 3). The NE - SW - trending stripes, ~30 - 50 m higher than the background terrain and pointing to Finsen crater, strongly indicate ejecta from Finsen crater. In the NW - SE direction, the dome - like uplifts (e. g., profile BB′) point to Alder crater. The elevation profiles (D - D′, E - E′, and F — F′) show differences of ~40—50m, which could be referred to as the maximum thickness of the dome - like uplifts because the effect of the Finsen crater ejecta has been eliminated. According to the DEM measurements and the geolo gical context (see Figure S1), we interpret the rugged topography of the landing site as the overlapping of ejecta from Finsen crater on top of ejecta of Alder crater. The directions of ejecta from Fisen and Alder crater are approximately perpendicular.

At the CE - 4 landing site, the average elevation is ~70 m higher than the surface of

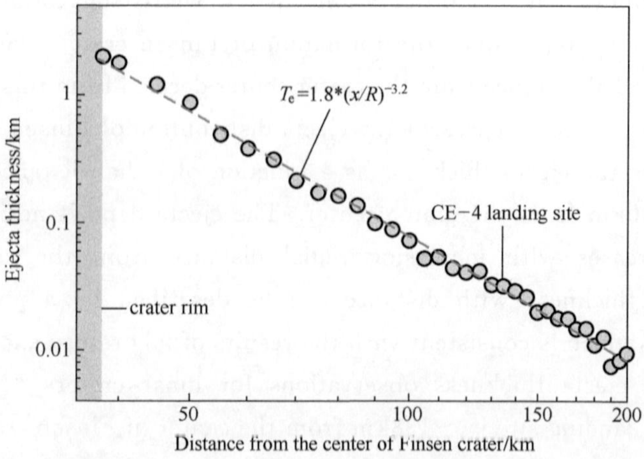

Figure 4　Ejecta thickness simulation for Finsen crater. The thickness of ejecta at the CE - 4 landing site from Finsen crater is ~32m. A similar simulation was also carried out for ejecta from Alder crater with a result of ~35m

background mare basalt measured at the center of the Von Kármán crater (see Figure 3). This suggests that the accumulated thickness of ejecta from Finsen and Alder craters at the landing site should be approaching to this value. Although we cannot completely exclude that the NW - SE dome - like uplifts was due to the eruption of mare basalt along a straight line fracture, there were no topographic features of basaltic lava flows. It is important to point out that the topographic variation at the landing area noted by previous works (e. g., Huang et al., 2018; Qiao et al., 2019) are the local elevation variations and undulations due to ridges of topmost layer, similar to our observations in the Pancam and LROC NAC DEMs (see Figure 2). The maximum value of the variations is at best a lower limit on the accumulated thickness of ejecta from the Finsen and Alder crater. In this study, the total ejecta thickness is measured as the height of the site above the floor of von Karman based on SEDEM2015 data rather than the amplitude of the ray - like ridges or of very local topography at the site.

3.3　Ejecta Thickness Predicted by Numerical Impact Cratering Simulations

From the impact cratering simulations, we find that asteroids with diameters of 6.2 and 6.5km can produce impact craters with diameters of ~73 and ~82km, respectively, the analogs to the diameters of Finsen and Alder craters (see the supporting information).

Figure S2 shows the structure of the final crater for the simulation with an impactor diameter of 6.2km and an impact velocity of 15km/s. This simulation produces a complex crater with a diameter of ~73km, similar to the size of Finsen crater. The central peak is ~2km below the preimpact surface, consistent with that of Finsen crater. We calculated the average profile of topography for Finsen crater from the LOLA observations in a crater - centered projection over the white region in Figure S3. The profiles of elevation derived from the impact simulation and observations match well, suggesting that our impact model could reasonably repro duce the formation of Finsen crater. The ejecta thickness varying along the radial distance from the crater center derived from this typical simulation (see Figure 4) can therefore represent the ejecta distribution of Finsen crater.

Figure 4 shows the ejecta thickness as a function of radial distance from the rim of Finsen crater to 200km from the crater center. The ejecta deposit near the crater rim is ~1.9km and decreases with increasing radial distance from the crater center. The decrease in ejecta thickness with distance can be described by a power law with an exponent of −3.2, which is consistent with the results of laboratory experiments (Stöffler et al., 1975) and ejecta thickness observations for lunar craters (McGetchin et al., 1973). The CE-4 landing site is ~138km from the center of Finsen crater. According to our simulation, ~32-m-thick ejecta from the Finsen crater was deposited at the CE-4 landing site, in agreement with the thickness derived from the topo— graphic variation. In addition, ~35-m-thick ejecta from Alder crater was estimated to deposit at the landing site based on the simulations. The total ejecta thickness of ~67m from both Finsen and Alder craters match well with the difference of topographic evaluation (~70m) between the background mare basalt (measured at the center of the Von Kármán crater) and the terrain at the landing site. These results directly support that the surficial materials at the CE-4 landing site are from ejecta of Finsen crater (Hu et al., 2019; Huang et al., 2018; Li et al., 2019; Lin et al., 2019), rather than the local mare basalts.

4 Conclusions

From the 5cm grid spacing DEM of the CE-4 landing site, LROC NAC DEM and the SLDEM2015, linear ejecta features were identified and measured. Thicknesses of ejecta deposited at the landing site from Finsen and Alder craters were estimated through numerical impact cratering simulations. The observations and numerical simulations reveal the topographic evolution of eastern Von Kármán crater within the SPA basin. Although Von Kármán crater was filled with mare basalt (Huang et al., 2018; Pieters et al., 2001), its surface was covered at least by two sets of ejecta with approximately perpendicular directions. The findings indicate that the surface materials explored by the lunar rover were delivered mostly from Finsen crater, but little from the deeply buried mare basalt. This study also demonstrated that topographic evolution is an important factor for probing the inherent compositions of the SPA basin in future missions to under-stand the early impacting history and the origination of surfical materials.

References

[1] Amsden, A. A., Ruppel, H. M., & Hirt, C. W. (1980). SALE: A simplified ALE computer program for fluid flow at all speedsRep. LA - 8095; TRN: 80 - 015132 United States 10.2172/5176006 TRN: 80 - 015132 NTIS, PC A06/MF A01. LANL English, Medium: ED pp, Los Alamos Scientific Lab., NM (USA).

[2] Barker, M. K., Mazarico, E., Neumann, G. A., Zuber, M. T., Haruyama, J., & Smith, D. E. (2016). A new lunar digital elevation model from the Lunar Orbiter Laser Altimeter and SELENE Terrain Camera. Icarus, 273, 346 - 355. https://doi.org/10.1016/j.icarus.2015.07.039.

[3] Benz, W., Cameron, A. G. W., & Melosh, H. J. (1989). The origin of the Moon and the single - impact hypothesis III. Icarus, 81 (1), 113 - 131. https://doi.org/10.1016/0019 - 1035 (89) 90129 - 2.

[4] Collins, G. S., Melosh, H. J., & Ivanov, B. A. (2004). Modeling damage and deformation in impact simulations. Meteoritics & Planetary Science, 39 (2), 217 - 231. https://doi.org/10.1111/j.1945 - 5100.2004.tb00337.x.

[5] Di, K., Liu, Z., Liu, B., Wan, W., Peng, M., Li, J., et al. (2019b). Topographic analysis of Chang'e - 4 landing site using orbital, descent and ground data, Int. Arch. Photogramm. Remote Sens. Spatial Inf. Sci., XLII - 2/W13, 1383 - 1387.

[6] Di, K., Liu, Z. Q., Liu, B., Wan, W. H., Peng, M., Wang, Y. X., et al. (2019a). Chang'e - 4 lander localization based on multi - source data. Journal of Remote Sensing, 23 (1), 177 - 184. .

[7] Head, J. W., Murchie, S., Mustard, J. F., Pieters, C. M., Neukum, G., McEwen, A., et al. (1993). Lunar impact basins: New data for the western limb and far side (Orientale and South Pole - Aitken Basins) from the first Galileo flyby. Journal of Geophysical Research, *98* (E9), 17, 149 - 17, 181. https://doi.org/10.1029/93JE01278.

[8] Hu, X., Ma, P., Yang, Y., Zhu, M. H., Jiang, T., Lucey, P. G., et al. (2019). Mineral abundances inferred from in situ reflectance measure - ments of Chang'E - 4 landing site in South Pole - Aitken basin. Geophysical Research Letters, *46*, 9439 - 9447. https://doi.org/10.1029/2019GL084531.

[9] Huang, J., Xiao, Z., Flahaut, J., Martinot, M., Head, J., Xiao, X., et al. (2018). Geological Characteristics of Von Kármán Crater, Northwestern South Pole - Aitken Basin: Chang'E - 4 Landing Site Region. Journal of Geophysical Research: Planets, *123*, 1684 - 1700. https://doi.org/10.1029/2018JE005577.

[10] Hurwitz, D. M., & Kring, D. A. (2014). Differentiation of the South Pole - Aitken basin impact melt sheet: Implications for lunar exploration. Journal of

Geophysical Research: Planets, 119, 1110 - 1133. https://doi. org/10. 1002/ 2013JE004530.

[11] Ivanov, B., Melosh, J., & Pierazzo, E. (2010). Basin - forming impacts: Reconnaissance modeling. In R. L. Gibson & W. U. Reimold (Eds.), Large Meteorite Impacts and Planetary Evolution IV (pp. 29 - 49). Boulder, CO: Geological Society of America.

[12] Ivanov, M. A., Hiesinger, H., van der Bogert, C. H., Orgel, C., Pasckert, J. H., & Head, J. W. (2018). Geologic History of the Northern Portion of the South Pole - Aitken Basin on the Moon. Journal of Geophysical Research: Planets, *123*, 2585 - 2612. https://doi.org/10.1029/2018JE005590.

[13] Jia, Y., Zou, Y., Ping, J., Xue, C., Yan, J., & Ning, Y. (2018). The scientific objectives and payloads of Chang'E - 4 mission. Planetary and Space Science, *162*, 207 - 215. https://doi. org/10. 1016/j. pss. 2018. 02. 011.

[14] Le Feuvre, M., & Wieczorek, M. A. (2011). Nonuniform cratering of the Moon and a revised crater chronology of the inner Solar System. Icarus, *214* (1), 1 - 20. https://doi. org/10. 1016/j. icarus. 2011. 03. 010.

[15] Li, C., Liu, D., Liu, B., Ren, X., Liu, J., He, Z., et al. (2019). Chang'E - 4 initial spectroscopic identification of lunar far - side mantle - derived materials. Nature, *569*(7756), 378 - 382.https://doi.org/10.1038/s41586 - 019 - 1189 - 0.

[16] Lin, H., He, Z., Yang, W., Lin, Y., Xu, R., Zhang, C., ... Zou, Y. (2019). Olivine - norite rock detected by the lunar rover Yutu - 2 likely crys - tallized from the SPA impact melt pool. National Science Review. https://doi. org/ 10. 1093/nsr/nwz183.

[17] Liu, Z., Di, K., Peng, M., Wan, W., Liu, B., Li, L., et al. (2015). High precision landing site mapping and rover localization for Chang'e - 3 mission. Science China - Physics Mechanics & Astronomy, *58* (1), 1 - 11. .

[18] Lucey, P. G., Taylor, G. J., Hawke, B. R., & Spudis, P. D. (1998). FeO and TiO_2 concentrations in the South Pole - Aitken basin: Implications for mantle composition and basin formation. Journal of Geophysical Research, 103 (E2), 3701 - 3708. https://doi. org/10. 1029/97JE03146.

[19] McGetchin, T. R., Settle, M., & Head, J. W. (1973). Radial thickness variation in impact crater ejecta: Implications for lunar basin deposits. Earth and Planetary Science Letters, *20* (2), 226 - 236. https://doi. org/10. 1016/0012 - 821X (73) 90162 - 3.

[20] Melosh, H. J. (1979). Acoustic fluidization: A new geologic process? Journal of Geophysical Research, 84 (B13), 7513 - 7520. https://doi. org/10.10291 JB084iB13p07513.

[21] Melosh,H. J., & Ivanov, B. A. (1999). Impact Crater Collapse. Annual Review of Earth and Planetary Sciences, 27 (1), 385 - 415. https://doi. org/10. 1146/

annurev. earth. 27. 1. 385.

[22] Melosh, H. J., Kendall, J., Horgan, B., Johnson, B. C., Bowling, T., Lucey, P. G., & Taylor, G. J. (2017). South Pole - Aitken basin ejecta reveal the Moon's upper mantle. Geology, 45 (12), 1063 - 1066. https://doi. org/10. 1130/G39375. 1.

[23] Miljković, K., Wieczorek, M. A., Collins, G. S., Laneuville, M., Neumann, G. A., Melosh, H. J., et al. (2013). Asymmetric Distribution of Lunar Impact Basins Caused by Variations in Target Properties. Science, 342 (6159), 724 - 726. https://doi. org/10. 1126/science. 1243224.

[24] Moriarty, D. P., & Pieters, C. M. (2018). The Character of South Pole - Aitken Basin: Patterns of Surface and Subsurface Composition. Journal of Geophysical Research: Planets, *123*, 729 - 747. https://doi. org/10. 1002/2017JE005364.

[25] Moriarty, D. P. III, & Pieters, C. M. (2016). Complexities in pyroxene compositions derived from absorption band centers: Examples from Apollo samples, HED meteorites, synthetic pure pyroxenes, and remote sensing data. Meteoritics & Planetary Science, 51 (2), 207 - 234. https://doi. org/10. 1111/maps. 12588.

[26] Ohtake, M., Uemoto, K., Yokota, Y., Morota, T., Yamamoto, S., Nakamura, R., et al. (2014). Geologic structure generated by large - impact basin formation observed at the South Pole - Aitken basin on the Moon. Geophysical Research Letters, *41*, 2738 - 2745. https://doi. org/10. 1002/2014GL059478.

[27] Pieters, C. M., Head, J. W. III, Gaddis, L., Jolliff, B., & Duke, M. (2001). Rock types of South Pole - Aitken basin and extent of basaltic volcanism. Journal of Geophysical Research, 106 (E11), 28, 001 - 28, 022. https://doi. org/10. 1029/2000JE001414.

[28] Potter, R. W. K., Collins, G. S., Kiefer, W. S., McGovern, P. J., & Kring, D. A. (2012). Constraining the size of the South Pole - Aitken basin impact. Icarus, 220 (2), 730 - 743. https://doi. org/10. 1016/j. icarus. 2012. 05. 032.

[29] Potter, R. W. K., Kring, D. A., Collins, G. S., Kiefer, W. S., & McGovern, P. J. (2013). Numerical modeling of the formation and structure of the Orientale impact basin. Journal of Geophysical Research: Planets, 118, 963 - 979. https://doi. org/10. 1002/jgre. 20080.

[30] Prettyman, T., Hagerty, J., Elphic, R., Feldman, W., Lawrence, D., McKinney, G., & Vaniman, D. (2006). Elemental composition of the lunar surface: Analysis of gamma ray spectroscopy data from Lunar Prospector. Journal of Geophysical Research, 111, E12007. https://doi. org/10. 1029/2005JE002656.

[31] Qiao, L., Ling, Z., Fu, X., & Li, B. (2019). Geological characterization of the Chang'e - 4 landing area on the lunar farside. Icarus, 333, 37 - 51. .

[32] Spudis, P. D., Gillis, J. J., & Reisse, R. A. (1994). Ancient Multiring Basins

on the Moon Revealed by Clementine Laser Altimetry. Science, *266* (5192), 1848 - 1851. https://doi. org/10. 1126/science. 266. 5192. 1848.

[33] Stöffler, D., Gault, D. E., Wedekind, J., & Polkowski, G. (1975). Experimental hypervelocity impact into quartz sand: Distribution and shockmetamorphism of ejecta. Journal of Geophysical Research, *80* (29), 4062 - 4077. https://doi. org/10. 1029/JB080i029p04062.

[34] Wilhelms, D. E. (1987). The geologic history of the Moon, U. S. Geologic Survey Prof. Paper 1348. .

[35] Wu, W., Li, C., Zuo, W., Zhang, H., Liu, J., Wen, W., et al. (2019). Lunar farside to be explored by Chang'e - 4. Nature Geoscience, *12* (4), 222 - 223. https://doi. org/10. 1038/s41561 - 019 - 0341 - 7.

[36] Wünnemann,K., Collins, G. S., & Melosh, H. J. (2006). A strain - based porosity model for use in hydrocode simulations of impacts and implications for transient crater growth in porous targets. Icarus, *180* (2), 514 - 527. https:// doi. org/10. 1016/j. icarus. 2005. 10. 013.

[37] Wünnemann, K., Zhu, M. - H., & Stöffler, D. (2016). Impacts into quartz sand: Crater formation, shock metamorphism, and ejecta distri - bution in laboratory experiments and numerical models. Meteoritics & Planetary Science, *51* (10), 1762 - 1794. https://doi. org/10. 1111/maps. 12710.

[38] Yue, Z., Johnson, B. C., Minton, D. A., Melosh, H. J., Di, K., Hu, W., & Liu, Y. (2013). Projectile remnants in central peaks of lunar impact craters. Nature Geoscience, *6* (6), 435 - 437. https://doi. org/10. 1038/ngeo1828.

[39] Zhu, M. - H., Wünnemann, K., & Artemieva, N. (2017). Effects of Moon's Thermal State on the Impact Basin Ejecta Distribution. Geophysical Research Letters, *44*, 11, 292 - 211, 300. https://doi. org/10. 1002/2017GL075405.

[40] Zhu, M. - H., Wünnemann, K., Potter, R. W., Kleine, T., & Morbidelli, A. (2019). Are the Moon's nearside - farside asymmetries the result of a giant impact? Journal of Geophysical Research: Planets, *124*, 2117 - 2140. https://doi. org/10. 1029/2018JE005826.

[41] Zhu, M. - H., Wünnemann, K., & Potter, R. W. K. (2015). Numerical modeling of the ejecta distribution and formation of the Orientale basinon the Moon. Journal of Geophysical Research: Planets, *120*, 2118 - 2134. https://doi. org/10. 1002/2015JE004827.

Geospatial technologies for Chang'e – 3 and Chang'e – 4 lunar rover missions

Kaichang Di, Zhaoqin Liu, Wenhui Wan, Man Peng,
Bin Liu, Yexin Wang, Sheng Gou & Zongyu Yue

State Key Laboratory of Remote Sensing Science, Institute of Remote Sensing and Digital Earth, Chinese Academy of Sciences, Beijing, China

Abstract: This paper presents a brief overview of the geospatial technologies developed and applied in Chang'e – 3 and Chang'e – 4 lunar rover missions. Photogrammetric mapping techniques were used to produce topographic products of the landing site with meter level resolution using orbital images before landing, and to produce centimeter – resolution topographic products in near real – time after landing. Visual positioning techniques were used to determine the locationsof the two landers using descent images and orbital basemaps immediately after landing. During surface operations, visual – positioning – based rover localization was performed routinely at each waypoint using Navcam images. The topographic analysis and rover localization results directly supported waypoint – to – waypoint path planning, science target selection and scientific investigations. A GIS – based digital cartography system was also developed to support rover teleoperation.

Keywords: Geospatial technology; lunar rover; landing site mapping; topographic analysis; rover localization and navigation

1 Introduction

Unmanned lunar and Mars rovers are in – situ robotic explorers that provide detailed information on the topography, geology and mineralogy of the landing sites with much higher resolutions than that from orbiters. So far, successfully landed rovers include Lunokhod – 1 and Lunokhod – 2 lunar rovers of the former Soviet Union, Sojourner, Spirit, Opportunity and Curiosity Mars rovers of the USA, Yutu and Yutu – 2 lunar rovers of China (Zakrajsek, Mckissock, and Woytach 2005; Golombek et al. 1999; Li et al. 2005; Grotzinger et al. 2012; Liu, Di, and Peng 2015; Wu, Li, and Zuo 2019; Di, Liu, and Liu 2019a).

Chang'e – 3 (CE – 3) probe, including a lander and a rover, successfully landed in northern Mare Imbrium of the moon on 14 December 2013 The Yutu (Jade Rabbit) rover was released to the lunar surface and began surface exploration on December 15. The rover traversed 114. 8m (cross - site visual localization result) on the lunar surface and accomplished a series of scientific investigations under teleoperation (Liu, Di, and Peng 2015; Xiao et al. 2015) .

Chang' e - 4 (CE - 4) mission was designed to be a backup of the CE - 3 mission. With the support of the Queqiao (Magpie Bridge) relay satellite orbiting Earth - Moon L2 point, the CE - 4 probe successfully landed on the far side of themoon inVon Kármán crater inside the South Pole - Aitken (SPA) basin on 3 January 2019. The Yutu - 2 rover was successfully released from the lander and touched on the lunar surface on the same day. The success of CE - 4 marks the first soft landing of human spacecraft on the far side of the moon and provides a unique opportunity for in - situ exploration of the SPA basin which is the oldest and largest impact basin on the moon (Jia et al. 2018; Huang, Xiao, and Flajault 2018; Wu, Li, and Zuo 2019; Di, Liu, and Liu 2019a; Di et al. 2019b) . As of 4 December 2019, the rover has traversed 345. 059m (odometer reading) (China Daily 2019) . Figure 1 shows the CE - 3 and CE - 4 rovers on the lunar surface photographed by the cameras on - board the landers.

Figure 1　Photos of the CE - 3 and CE - 4 rovers on the lunar surface taken by the landers

In CE - 3 and CE - 4 lunar rover missions, geospatial technologies were extensively used to support various scientific studies of the landing sites, and to support mission operations such as topographic mapping, hazard detection, path planning, rover localization and navigation, etc. (Wan et al. 2014; Peng, Wan, and Wu 2014; Liu, Di, and Peng 2015; Di et al. 2016; Liu, Di, Li 2019) . The Planetary Remote Sensing team of State Key Laboratory of Remote Sensing Science, Institute of Remote Sensing and Digital Earth (RADI), Chinese Academy of Sciences, in collaboration with the rover teleoperation team of Beijing Aerospace Control Center (BACC), developed and applied the technologies for teleoperation of the rovers at BACC during the missions. The paper presents a brief overview of the geospatial technologies and applications in these two lunar rover missions.

2 Data and technical framework

Geospatial technologies for landing site mapping and rover localization have been performed using multiple data sources, including orbital data acquired by previous or ongoing orbiter missions, descent images acquired by the lander during the entry, descent and landing (EDL) process, and rover images taken by the rover after landing.

In the early phase of a lunar rover mission, landing site selection is critical and is usually accomplished by careful evaluation of scientific requirements and constraints for engineering implementation (Zhao, Li, and Wu, 2019). Landing site topographic analysis is one of the key steps in the landing site selection process. This requires both digital elevation model (DEM) and digital orthophoto map (DOM) of the candidate landing sites. The products are primarily generated by photogrammetric mapping technology using highresolution orbital images.

The orbital data used in topographic mapping and characterization of CE - 3 and CE - 4 landing sites mainly include Chang'e - 2 (CE - 2) images and Lunar Reconnaissance Orbiter Camera (LROC) Narrow Angle Camera (NAC) images. CE - 2 orbiter acquired forward - and backward - looking stereo images (viewing angle 7.98 and - 17.2 degrees, respectively) in the same orbit through push - broom imaging; the stereo images cover the entire moon with a spatial resolution of 7m, and in the CE - 3 landing site 1.5 - m resolution images are available (Zuo, Li, and Zhang, 2014). Based on CE - 2 stereo images, DEM (20 - m grid spacing) and DOM (7 - m resolution) of the entire moon were produced and released by the Lunar and Deep Space Exploration Science Applications Center of National Astronomical Observatories, Chinese Academy of Sciences (NAOC) (Li, Liu, and Ren, 2018; Ren et al., 2019). LROC NAC images have spatial resolutions of 0.5 - 2 m and almost cover the entire moon surface (Robinson et al. 2010); NAC stereo images can be acquired from adjacent orbits using off - nadir slew, but the stereo coverage is very limited. In addition to CE - 2 and LROC images, the SLDEM2015 (Barker et al. 2016) and LROC Wide - Angle Camera (WAC) mosaic (Wagner et al., 2015) were also used for landing site topographic analysis and characterization.

Both CE - 3 and CE - 4 landers carry a downwardlooking descent camera with a Field Of View (FOV) of 45.4° and an image size of 1024×1024 pixels (Liu, Di, and Peng 2015). The descent camera took thousands of images (about 4600 in CE - 3 and 5300 in CE - 4 mission) with resolutions from meter level to millimeter level during the EDL process. The descent images were used in lander localization and landing site mapping.

Both Yutu and Yutu - 2 rovers carry three pairs of stereo cameras, i.e. panoramic camera (Pancam), navigation camera (Navcam), and hazard avoidance camera (Hazcam). Pancam and Navcam are mounted on the same camera bar atop the rover mast; Hazcam is fixed on the lower front of the rover body. Both Pancam and Navcam have a stereo base of 270mm. Pancam has an image size of 2352 pixels×1728 pixels, and a FOV of 22.9°×16.9° (Jia et al. 2018). Navcam has an image size of 1024×1024 pixels, a FOV of 46.6° (Liu, Di, and Peng 2015). Along the traverse in each mission, the rover obtains Navcam stereo images at every waypoint and acquires Pancam stereo images some waypoints. These stereo images have been routinely used in detailed topographic mapping and rover localization to support mission operations.

The geospatial technologies and derived products have directly contributed to the safe and efficient surface operations of the rovers. The mapping and localization products have also been frequently used to support various scientific investigations before and during the missions, e.g. to support data acquisition and analysis by other science instruments, such as the Visible and Near-Infrared Imaging Spectrometer (VNIS) and the Lunar Penetrating Radar (LPR) for investigations of mineralogical compositions and subsurface structures, respectively (Xiao et al. 2015; Di et al. 2016; Li et al. 2019; Gou et al. 2019; Di et al. 2019b).

Figure 2 shows the technical framework of the geospatial technologies developed and applied in CE-3 and CE-4 missions. Photogrammetric mapping using orbital images is performed before the launch of the landers and rovers for general topographic analysis of the landing sites. The mapping products include DEM and DOM, and some derived products such as slope map and contour map. The orbital DOM is used as a basemap to support lander localization, which is accomplished quickly (e.g. within 30min) after landing. During surface operations, detailed topographic mapping and rover localization are performed at the earth control center (i.e. BACC) in near real-time, i.e. a couple of minutes after the rover images are downlinked. More details of the key techniques and results are given in the following sections.

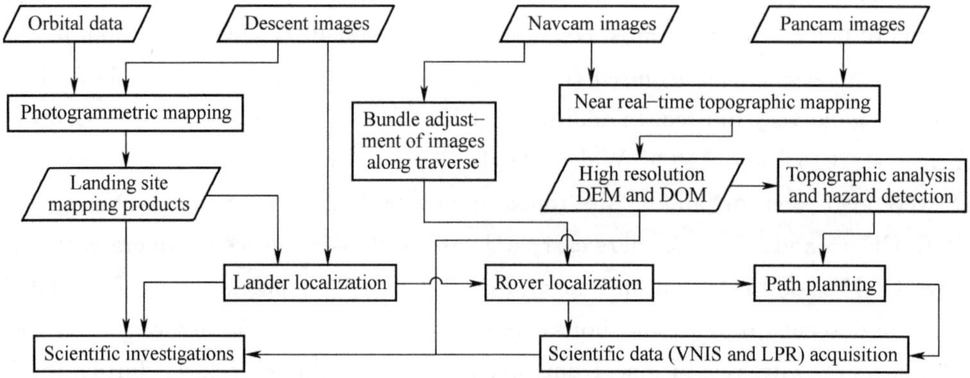

Figure 2　Technical framework of the geospatial technologies and applications in CE-3 and CE-4 missions

3　Photogrammetric mapping techniques and applications

3.1　Landing site mapping using orbital images

Photogrammetric mapping techniques, such as rigorous sensor modeling, bundle block adjustment, and image matching, have been developed in order to produce DEM and DOM from CE - 2 images (Di et al. 2014; Wu, Hu, and Guo 2014; Li, Liu, and Ren 2018; Liu et al. 2018; Ren et al. 2019). The DEM and DOM have been used in topographic characterization (e. g. slop and crater distributions) of CE - 3 and CE - 4 landing sites (Wu, Hu, and Guo 2014; Di et al. 2019b; Di, Liu, and Liu 2019c; Zhao, Li, andWu 2019).

CE - 2 orbiter acquired 1.5 - m resolution images at the pre - selected CE - 3 landing site (40°N ~50°N, 18°W ~35°W). Figure 3 shows a 4 - m - resolution DEM and a 1.5 - m - resolution DOM of CE - 3 landing site generated from the 1.5 - m - resolution CE - 2 stereo images of No. 0236 track by the NAOC team. The maps are rendered in the Mercator projection; the elevation value is with respect to the reference sphere with a radius of 1737.4km, and the red flag indicates the location of CE - 3 lander (Liu, Di, and Peng 2015).

Figure 4 shows the imagery and topography of CE - 4 landing site and surrounding area. The pre - selected landing site (white box in the lower - right image of Figure 4) is in the southeast of Von Kármán crater and covers the area of (176.4°E ~ 178.8°E, 45.0°S~ 46.0°S), approximately 52km×30km. The resolutions of the CE - 2 DOM and LROCWACmosaic are 7mand 100m, respectively. The resolution of the SLDEM2005 is about 42m in this region; the landing site region is generally flat in this scale.

Based on 100 high - resolution LROC NAC images of CE - 4 landing site, we produced a seamless DOM mosaic with 0.9 - m resolution using similar techniques for producing a seamless DOM mosaic of Chang'e - 5 landing region using LROC NAC images (Di et al. 2018, 2019d). Figure 5 shows the DOM mosaic in Lambert conformal conic projection. The DOM has an image size of 57,483 columns×34,084 rows. It reveals a lot of details of the topographic features, e. g. numerous small craters that are not discernable in the maps shown in Figure 1. In particular, there are many secondary crater clusters and ejecta materials which can be traced back to several craters in the vicinity (Di et al. 2019b; Di, Liu, and Liu 2019c). In addition to applications in landing site characterization, this high - resolution DOM, along with CE - 2 DOM, has been used as a basemap in lander localization and long - term rover traverse planning (Di, Liu, and Liu 2019a; Liu, Di, Li, 2019).

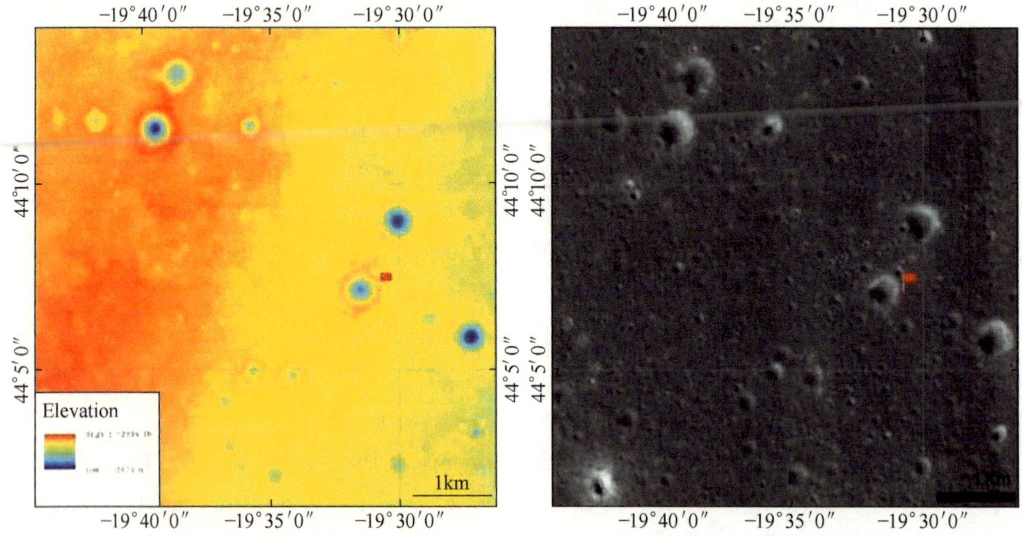

Figure 3　DEM (Left) and DOM (Right) of CE－3 landing site

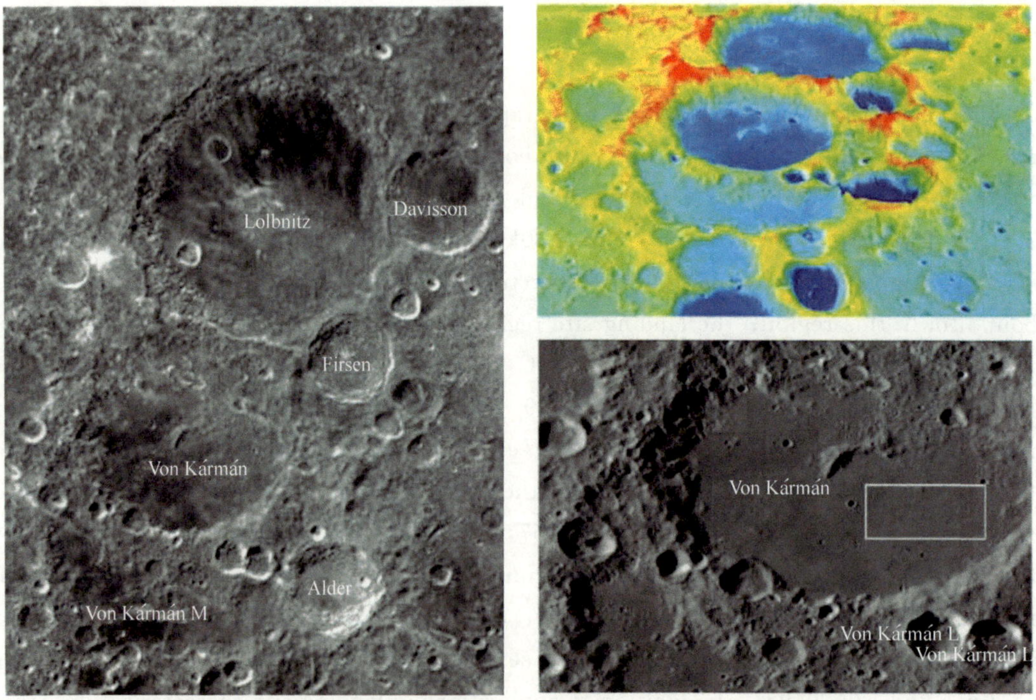

Figure 4　Imagery and topography of CE－4 landing site and surrounding area.

Left: CE－2 DOM, Upper－right: 3D perspective of colorcoded SLDEM2015, Lower－right: LROC WAC mosaic and pre－selected landing site (white box) of CE－4

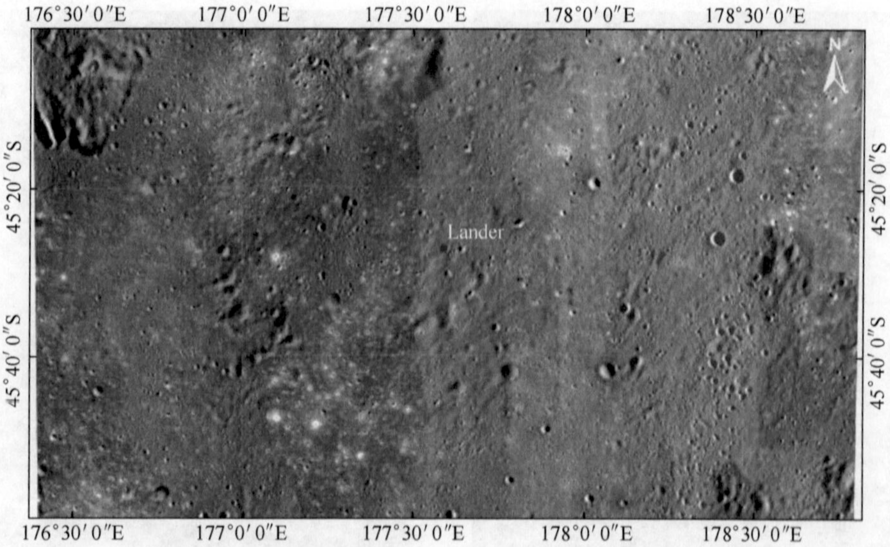

Figure 5　LROC NAC DOM mosaic of CE - 4 landing site with a resolution of 0. 9m（Liu，Di，Li 2019）

3. 2　Landing site mapping using descent images

The sequential descent images were utilized to produce topographic maps with higher resolutions than those generated from orbital images. In CE - 3 mission，180 descent images were selected at intervals from 2. 35km to 10 - m altitude，photogrammetric processing techniques，including feature matching，bundle adjustment，dense matching，space intersection，and DEM interpolation，were employed to produce DEM and DOM products of the landing site；26 Ground Control Points（GCPs）and 18 checkpoints were selected from the CE - 2 DEM and DOM for the bundle adjustment and accuracy assessment（Liu，Di，and Peng 2015）. DEM and DOM products with 0. 4 - m，0. 2 - m，and 0. 05 - m resolutions were generated，respectively. Figure 6 shows the DEM and DOM that cover an area of 97 × 115m with 0. 5 - m resolution. The 0. 4 - m resolution DEM and DOM cover a larger area of 780 × 1800m. Small features，e. g. rocks，can be identified from these maps. The maps were used as basemaps for rover localization and traverse planning.

In CE - 4 mission，80 descent images were selected and used for detailed landing site mapping，with GCPs selected from the LROC NAC DOM（Figure 5）. Figure 7 shows the DOM that covers an area of 211m × 187m with 0. 03 - m resolution. Small features，e. g. meter level craters，were identified and measured from this very high - resolution DOM. It was used for a detailed analysis of the landing site and utilized as one of the basemaps for overall rover traverse planning.

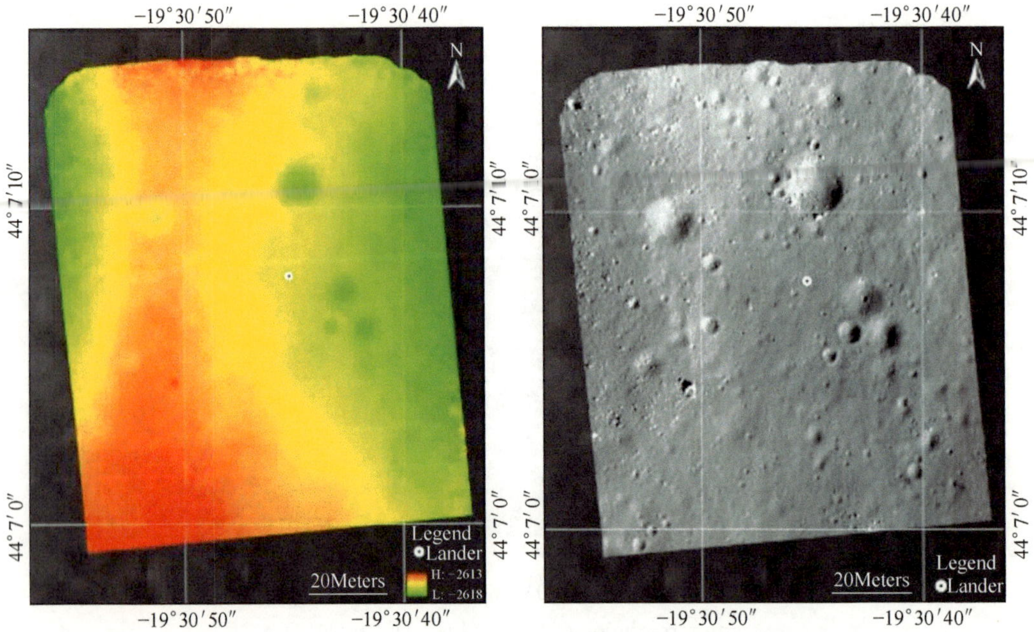

Figure 6 CE-3 landing site DEM and DOM (0.05-m resolution) generated from descent images (Liu, Di, and Peng 2015)

Figure 7 CE-4 landing site DOM (0.03-m resolution) generated from descent images (Di, Liu, and Liu 2019c)

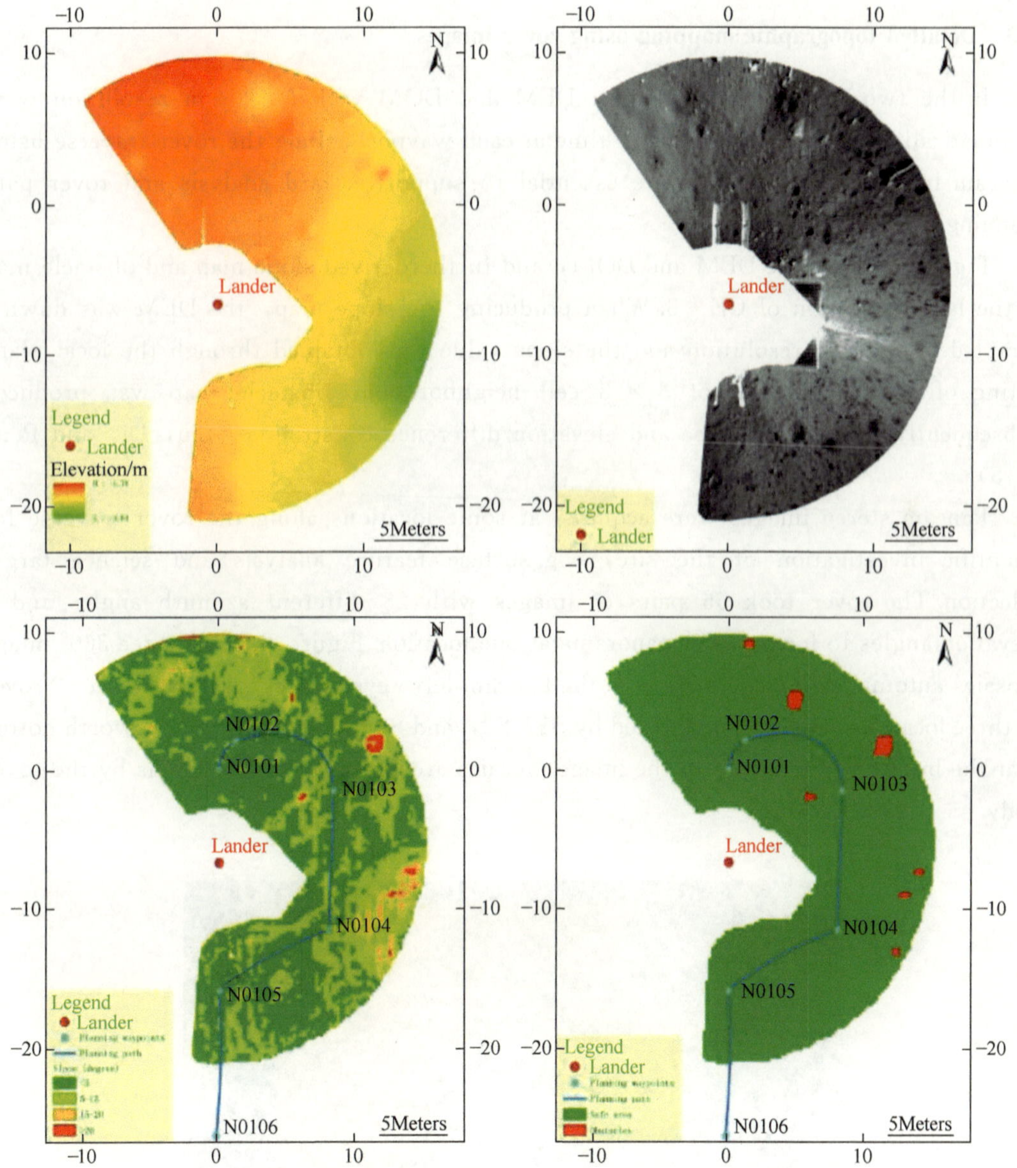

Figure 8　DEM (Upper - left), DOM (Upper - right), slope map (Lower - left), and obstacle map (Lower - right) derived from CE - 3 Navcam images at the lander location (Liu, Di, and Peng 2015)

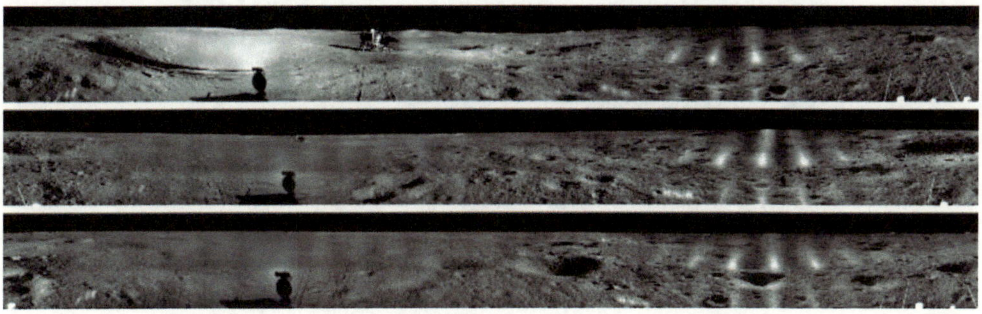

Figure 9　The 360° Pancam image mosaics generated from left - eye images taken by Yutu - 2 rover at three locations

3.3 Detailed topographic mapping using rover images

In the two rover missions, local DEM and DOM with 0.02 - m resolution were automatically produced in near real - time at each waypoint along the rover traverse using Navcam images. These maps were essential to support hazard analysis and rover path planning.

Figure 8 shows the DEM and DOM, and furtherderived slope map and obstacle map at the lander location of CE - 3. When producing the slope map, the DEM was down - sampled to 0.2 - m resolution and the slope value was obtained through the local plane fitting of the z - values of 3×3 cell neighborhood; obstacle map was produced subsequently considering slope and elevation difference constraints (Liu, Di, and Peng 2015).

Pancam stereo images were acquired at some locations along the rover traverse for scientific investigation of the site, e.g. surface feature analysis and science target selection. The rover took 56 pairs of images with 28 different azimuth angles and 2 elevation angles to form a 360° panorama at one location. Figure 7 shows three 360° image mosaics automatically generated from the Pancam left - eye images taken by Yutu - 2 rover at three locations (black dots labeled by S1, S2, and S3 in Figure 10). It is worth noting that the bright white stripes in the image mosaics are due to light reflections by the rover body.

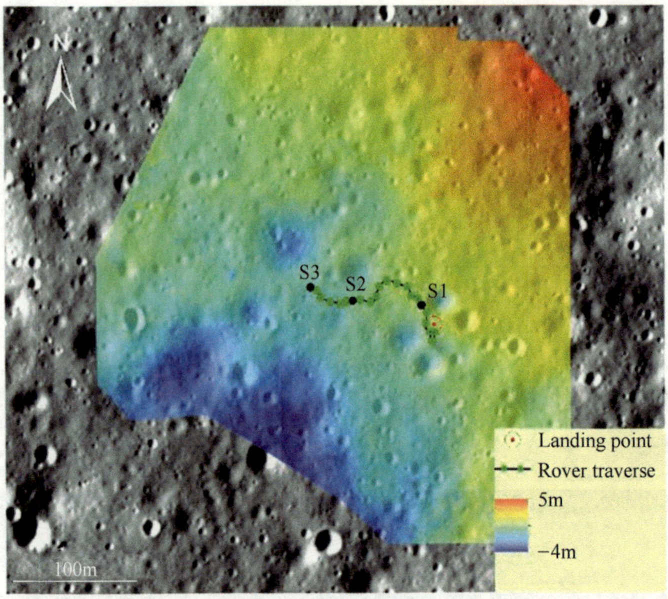

Figure 10 DEM of CE - 4 landing site derived from Pancam images taken at three locations (black dots)

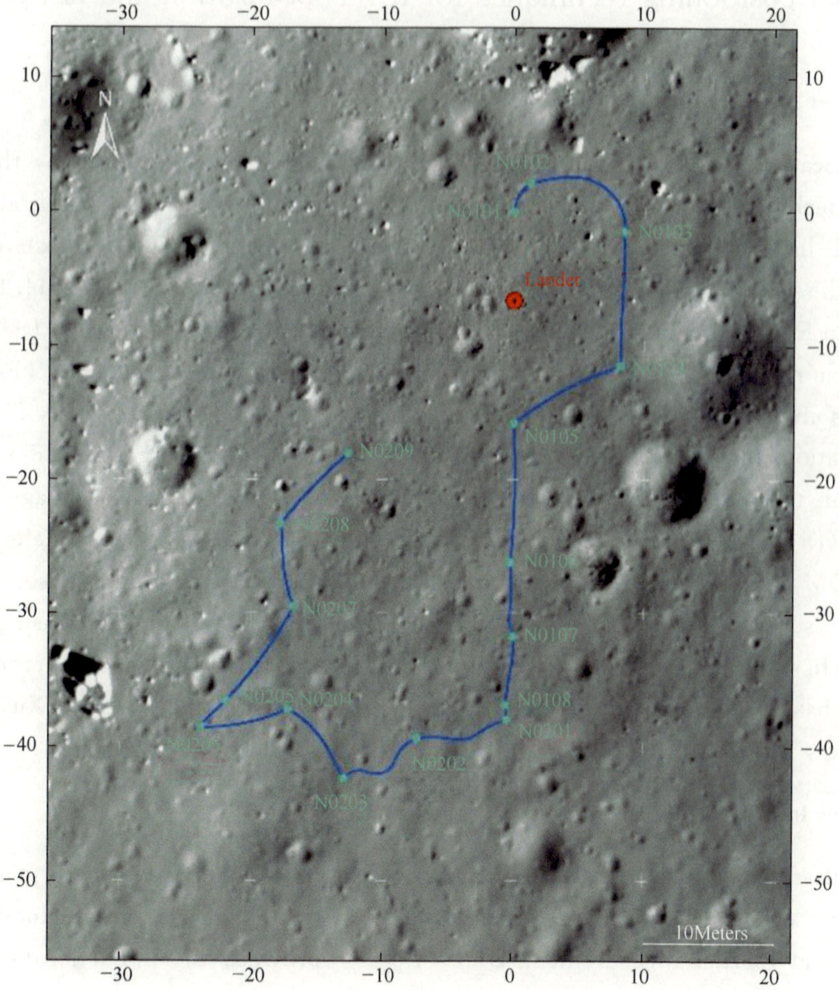

Figure 11　Yutu rover traverse map

Three DEMs (0.05 - grid spacing) at the three locations were automatically generated from the Pancam stereo images and then were merged into a larger DEM (380×390m) based on accurate locations of the rover determined by the visual localization technique (see section 4.2) . Figure 10 shows the merged DEM overlaid on the LROC NAC DOM. Figures 9 and 10 reveal that in finer resolution the terrain of CE - 4 landing site is much rugged and is generally higher in the NE than in the SW. This DEM contributed to the study of the topographic evolution of Von Kármán crater (Di et al. 2019b), which is significant to understanding the origin of the surface material measured by Yuyu - 2 rover (Li et al. 2019; Gou et al. 2019) .

4 Visual positioning techniques for rover localization and navigation

4.1 Lander localization

The locations of CE - 3 and CE - 4 landers were initially determined by the radio - tracking technique. To improve the positioning accuracy, lander localization was performed in each mission using visual positioning techniques immediately after landing. The CE - 3 lander location was determined to be (19.51256°W, 44.11884°N) (marked in Figures 3 and 6) through feature matching between the Navcam DOM (from images acquired when the rover was on the lander) and the descent DOM (Figure 4), which was already registered to CE - 2 DOM (Liu, Di, and Peng 2015). The CE - 4 lander location was determined to be (177.588°E, 45.457°S) (marked in Figures 5, 7 and 10) on the LROC NAC DOM through matching between descent images and the basemap (Di, Liu, and Liu 2019a). The CE - 4 lander location was also determined on CE - 2 DOM and SLDEM2015. High - precision lander localization in multi - source data is valuable to support the synergistic scientific investigation of the landing site using multiple data sets. In addition to lander localization, the descent and landing trajectory can be recovered based on bundle adjustment of sequential descent images (Liu, Xu, and Liu 2014; Wan, Liu, and Liu 2019; Liu, Ren, Li, 2019).

4.2 Rover localization

Localization of the rover with high accuracy is important for rover path planning and approaching science targets. The onboard dead - reckoning system, including wheel odometer and Inertial Measurement Unit (IMU), provides the initial locations of the rover along the traverse in real time. To reduce the localization error caused by wheel slippage and IMU drift, visual positioning techniques, namely cross - site visual localization and DOM matching, have been routinely applied to localize the rover with high accuracy using the Navcam images. The cross - site visual localization method is based on ASIFT feature matching and bundle adjustment of Navcam images acquired at adjacent sites (waypoints) (Wan et al. 2014); it was applied in near real - time (a couple of minutes after images downlinked) at each waypoint to support teleoperation of the rover. The DOM matching method is based on feature matching between the DOMs generated from Navcam stereo images and the basemaps generated from orbital or descent images (Liu, Di, and Peng 2015); it was applied at some waypoints where distinguishing features (e.g. small craters and large rocks) are available in the local areas.

In CE - 3 mission, the overall traverse distance from the lander is 114.8m from cross - site visual localization and 111.2m from DOM matching; both results are of higher accuracy than that from dead reckoning and the DOM matching result is of highest accuracy (Liu, Di, and Peng 2015). Figure 11 shows the Yutu rover traverse map with the traverse from the DOM matching method overlaid on the descent DOM.

In CE - 4 mission, Yutu - 2 rover traveled 190.66m from the lander according to wheel odometer by the end of the 5th lunar day (11 May 2019). The overall traverse distance was calculated to be 186.66m from visual localization, which should have corrected some wheel slippage (Liu, Di, and Li 2019). By the end of the 11th lunar day (21 November 2021), the rover traveled 318.62m (odometer reading) (CLEP, 2019). As of 4 December 2019, Yutu - 2 rover has traveled 345.059m (odometer reading) on the far side of the moon and accomplished the exploration tasks for the 12th lunar day (China Daily 2019). Visual localization of the rover has been continually performed to support the surface operations of CE - 4 mission. The latest results will be reported in the future.

4.3 Application in rover navigation

Rover navigation is critical for safe and efficient traversing, and science target approaching. Path planning is a key component of rover navigation. During the mission operations, path planning was implemented by the teams of BACC and China Academy of Space Technology. Topographic analysis and rover localization results directly contributed to the three levels of rover path planning, namely overall planning, exploration period planning, and navigation unit planning (Yu, Fei, and Li 2019). The high - accuracy rover location and topographic information were indispensable for the analysis of the important factors of path planning, such as terrain attainability, hazard, illumination condition, communication attainability, traverse cost, etc.

5 GIS technology and applications

The abovementioned photogrammetric mapping and visual positioning techniques were developed and implemented as the topographic mapping and rover localization sub - systems, which are important parts of the rover teleoperation system at BACC. Meanwhile, Geographic Information System (GIS) technology was also used in support of surface operations and scientific studies in the two missions. For example, the RADI team, in collaboration with the BACC team, developed a stand - alone digital cartography system for supporting rover teleoperation based on ArcGIS Engine - Developer Kit. Figure 12 shows the main interface of the system.

The major purpose of this GIS - based software system is to make and release standardized maps using multiple source data acquired or derived before and during the rover missions. It has three modules: (1) multi - source data integration, (2) data processing and analysis, and (3) cartographic product generation. The first module

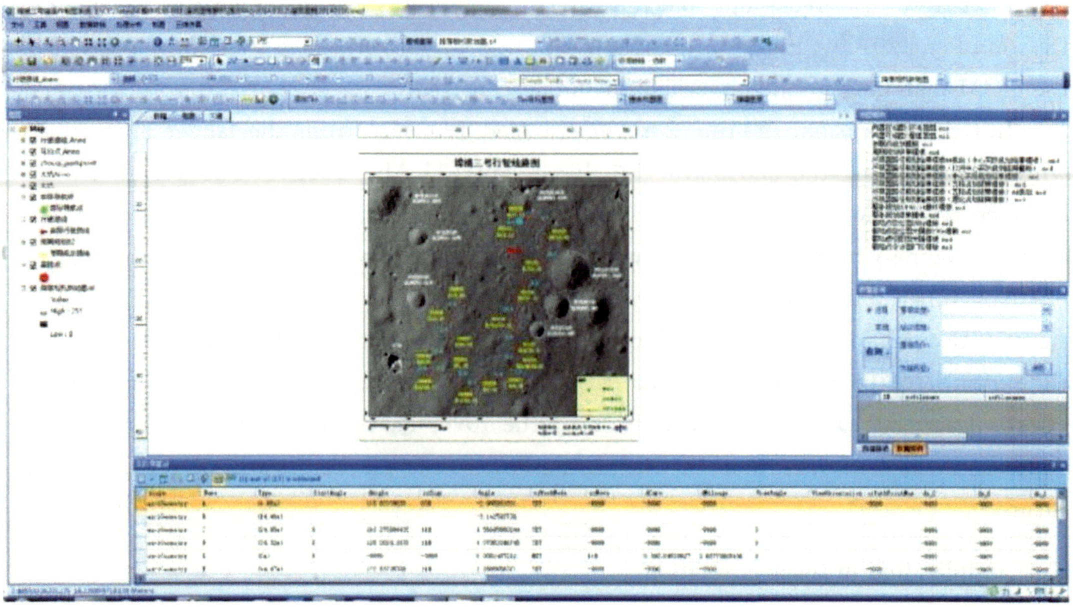

Figure 12　Main interface of the digital cartography system for supporting rover teleoperation

manages topographic maps (derived from orbital, descent and rover images), control point data, rover traverse data, path planning data, etc., and integrates them into the same spatial reference frame and coordinate system through data conversion, transformation, and editing. To meet the needs of rover teleoperation decision - making, the second module consists of a set of tools to derive specific maps or data, including slope, aspect, elevation profile, contour, 3D perspective view, line of sight for communication, skyline, illumination condition, projection of a camera's FOV on the surface, and so on. The third module provides specially designed map templates, map symbol database, and annotations for rover teleoperation, and produces standardized high - quality maps that can be used and released electronically or printed out on large format printers.

6　Conclusions

A series of geospatial technologies, i. e. photogrammetric mapping, visual positioning, and GIS, were developed and used in CE - 3 and CE - 4 lunar rover missions. The technologies greatly contributed to landing site mapping and characterization, lander and rover localization, hazard detection, and rover path planning during mission operations. In summary, the geospatial technologies are indispensable and critical in support of engineering operations and scientific investigations of the missions.

Future lunar and Mars rover missions may require landing in more challenging terrain, traversing longer distance within one navigation unit, acquiring more scientific data with a certain level of autonomy, etc. To meet such new requirements, the relevant geospatial technologies need to be expended and new technologies need to be developed not only to improve accuracy, robustness, and on - board real - time processing capability but also to increase the intelligence of the rover.

Acknowledgments

The authors gratefully acknowledge the Beijing Aerospace Control Center for the support and close collaboration during Chang'e - 3 and Chang - 4'e mission operations. We thank China National Space Administration for providing Chang'e - 3 and Chang'e - 4 data, and the Lunar and Deep Space Exploration Science Applications Center of the National Astronomical Observatories for providing the Chang'e - 2 DOM and DEM. We also thank those who worked on the Planetary Data System archive to make the LROC images and SLDEM2015 available, and Arizona State University for making the LROC WAC mosaic available.

References

[1] Barker, M. K., E. Mazarico, G. A. Neumann, M. T. Zuber, J. Haruyama, D. E. Smith. 2016. "A New Lunar Digital Elevation Model from the Lunar Orbiter Laser Altimeter and SELENE Terrain Camera". Icarus 273: 346 - 355. doi: 10. 1016/j. icarus. 2015. 07. 039.

[2] China Daily, 2019. "China's Lunar Rover Travels over 345 Meters on Moon's Far Side." 4 December 2019. http://www. cinadaily. com. cn/a/201912/04/WS5de77085a - 310cf3e3557bfc4. html

[3] CLEP (China's Lunar and Deep Space Exploration), 2019. "The Chang'e - 4 Lander and Rover Woke up and Entered the Twelfth Lunar Day." 22 November 2019. http://www. clep. org. cn/n5982341/c6808267/content. html

[4] Di, K., M. Jia, X. Xin, B. Liu, Z. Liu, M. Peng, Z. Yue, et al. 2018. "High Resolution Seamless DOM Generation over Chang'e - 5 Landing Area Using LROC NAC Images". International Archives of the Photogrammetry, Remote Sensing & Spatial Information Sciences XLII - 3: 271 - 276. doi: 10. 5194/isprs - archives - XLII - 3 - 271 - 2018.

[5] Di, K., M. Jia, X. Xin, J. Wang, B. Liu, J. Li, J. Xie et al. 2019d. "High - resolution Large - area Digital Orthophoto Map Generation Using LROC NAC Images." Photogrammetric Engineering and Remote Sensing 85 (7): 481 - 491. DOI: 10. 14358/PERS. 85. 7. 481.

[6] Di, K., Y. Liu, B. Liu, M. Peng, W. Hu. 2014. "A Selfcalibration Bundle Adjustment Method for Photogrammetric Processing of Chang'E - 2 Stereo Lunar

Imagery." IEEE Transactions on Geoscience and Remote Sensing 52 (9): 5432 - 5442. DOI: 10.1109/TGRS.2013.2288932.

[7] Di, K., Z. Liu, B. Liu. 2019a. "Chang'e - 4 Lander Localization Based on Multi - source Data." Journal of Remote Sensing 23 (1): 177 - 184.

[8] Di, K., Z. Liu, B. Liu. 2019c. "Topographic Analysis of Chang'e - 4 Landing Site Using Orbital, Descent and Ground Data." ISPRS Geospatial Week 2019 - GSW2019." International workshop on Planetary Remote Sensing and Mapping, June 10 - 14, 1383 - 1387. The Netherlands: Enschede.

[9] Di, K., B. Xu, M. Peng, Z. Yue, Z. Liu, W. Wan, L. Li, and J. Zhou. 2016. "Rock Size - frequency Distribution Analysis at the Cháng'e - 3 Landing Site." Planetary and Space Science 120: 103 - 112. doi: 10.1016/j.pss.2015.11.012.

[10] Di, K., M. - H. Zhu, Z. Yue, Y. Lin, W. Wan, Z. Liu, S. Gou, et al. 2019b. "Topographic Evolution of Von Kármán Crater Revealed by the Lunar Rover Yutu -2". Geophysical Research Letters 46: 12764 - 12770. doi: 10.1029/2019GL085252.

[11] Golombek, M., R. Anderson, J. Barnes, J. F. Bell, N. T. Bridges, D. T. Britt, J. Brückner et al. 1999. "Overview of the Mars Pathfinder Mission: Launch through Landing, Surface Operations, Data Sets, and Science Results." Journal of Geophysical Research: Planets 104 (E4): 8523 - 8553. DOI: 10.1029/98JE02554.

[12] Gou, S., K. Di, Z. Yue, Z. Liu, Z. He, R. Xu, H. Lin, et al. 2019. "Lunar Deep Materials Observed by Chang'e - 4 Rover". Earth and Planetary Science Letters 528: 115829. doi: 10.1016/j.epsl.2019.115829.

[13] Grotzinger, J. P., J. Crisp, A. R. Vasavada, R. C. Anderson, C. J. Baker, R. Barry, D. F. Blake, et al. 2012. "Mars Science Laboratory Mission and Science Investigation". Space Science Review 170: 5 - 56. doi: 10.1007/s11214 - 012 - 9892 - 2.

[14] Huang, J., Z. Xiao, J. Flajault. 2018. "Geological Characteristics of Von Kármán Crater, Northwestern South Pole - Aitken Basin: Chang'E - 4 Landing Site Region." Journal of Geophysical Research: Planets 123: 1684 - 1700.

[15] Jia, Y., Y. Zou, J. Ping, C. Xue, J. Yan, Y. Ning. 2018. "The Scientific Objectives and Payloads of Chang'E - 4 Mission". Planetary and Space Science 162: 207 - 215. doi: 10.1016/j.pss.2018.02.011.

[16] Li, C., D. Liu, B. Liu, X. Ren, J. Liu, Z. He, W. Zuo, et al. 2019. "Chang'e - 4 Initial Spectroscopic Identification of Lunar Far - side Mantle - derived Materials". Nature 569: 378 - 382. doi: 10.1038/s41586 - 019 - 1189 - 0.

[17] Li, C., J. Liu, X. Ren. 2018. "Lunar Global High - precision Terrain Reconstruction Based on Chang'e - 2 Stereo Images." Geomatics and Information Science of Wuhan University 43 (4): 485 - 489. (in Chinese with English abstract).

[18] Li, R., S. W. Squyres, R. E. Arvidson, B. A. Archinal, J. Bell, Y. Cheng, L. Crumpler et al. 2005. "Initial Results of Rover Localization and Topographic

Mapping for the 2003 Mars Exploration Rover Mission." Photogrammetric Engineering and Remote Sensing: Special Issue on Mapping Mars 71 (10): 1129-1142. DOI: 10.14358/PERS.71.10.1129.

[19] Liu, B., M. Jia, K. Di, J. Oberst, B. Xu, W. Wan. 2018. "Geopositioning Precision Analysis of Multiple Image Triangulation Using LROC NAC Lunar Images". Planetary and Space Science 162: 20-30. doi: 10.1016/j.pss.2017.07.016.

[20] Liu, B., B. Xu, Z. Liu. 2014. "Descending Ad Landing Trajectory Recovery of Chang'e-3 Lander Using Descent Images." Journal of Remote Sensing 18 (5): 988-994. (in Chinese with English abstract).

[21] Liu, J., X. Ren, W. Yan, C. Li, H. Zhang, Y. Jia, X. Zeng. 2019. "Descent Trajectory Reconstruction and Landing Site Positioning of Chang'E-4 on the Lunar Farside". Nature Communications 10: 4229. doi: 10.1038/s41467-019-12278-3.

[22] Liu, Z., K. Di, J. Li. 2019. "Landing Site Topographic Mapping and Rover Localization for Chang'e-4 Mission." Science China-Information Sciences 63. Accepted.

[23] Liu, Z., K. Di, M. Peng. 2015. "High Precision Landing Site Mapping and Rover Localization for Chang'e-3 Mission." Science China-physics Mechanics & Astronomy 58 (1): 1-11.

[24] Peng, M., W. Wan, K. Wu. 2014. "Topographic Mapping Capability Analysis of Chang'e-3 Navcam Stereo Images and Three-dimensional Terrain Reconstruction for Mission Operations." Journal of Remote Sensing 18 (5): 995-1002. (in Chinese with English abstract).

[25] Ren, X., J. Liu, C. Li, H. Li, W. Yan, F. Wang, W. Wang et al. 2019. "A Global Adjustment Method for Photogrammetric Processing of Chang'E-2 Stereo Images." IEEE Transactions on Geoscience and Remote Sensing 57 (9): 6832-6843. DOI: 10.1109/TGRS.36.

[26] Robinson, M.S., S.M. Brylow, M. Tschimmel, D. Humm, S.J. Lawrence, P.C. Thomas, B.W. Denevi, et al. 2010. "Lunar Reconnaissance Orbiter Camera (LROC) Instrument Overview". Space Science Reviews 150: 81-124. doi: 10.1007/s11214-010-9634-2.

[27] Wagner, R.V., E.J. Speyerer, M.S. Robinson, and L.R.O.C. Team. 2015. "New Mosaicked Data Products from the LROC Team." Proceedings of 46th Lunar and Planetary Science Conference, March 16-20 2015, 1473. The Woodlands, Texas.

[28] Wan, W., Z. Liu, K. Di, B. Wang, J. Zhou. 2014. "A Cross-Site Visual Localization Method for Yutu Rover". Int. Arch. Photogramm. Remote Sens. Spatial Inf. Sci. XL-4: 279-284. doi: 10.5194/isprsarchives-XL-4-279-2014.

[29] Wan, W., Z. Liu, B. Liu. 2019. "Descent Trajectory Recovery of Chang'e-4 Lander Based on Descent Images." ISPRS Geospatial Week 2019-GSW2019,

"International workshop on Planetary Remote Sensing and Mapping, June 10 - 14, 1457 - 1461. The Netherlands: Enschede.

[30] Wu, B., H. Hu, and J. Guo. 2014. "Integration of Chang'E - 2 Imagery and LRO Laser Altimeter Data with a Combined Block Adjustment for Precision Lunar Topographic Modelling." Earth and Planetary Science Letters 391: 1 - 15. doi: 10.1016/j.epsl.2014.01.023.

[31] Wu, W., C. Li, W. Zuo. 2019. "Lunar Farside to Be Explored by Chang'e - 4." Nature Geoscience 12 (4): 222 - 223. DOI: 10.1038/s41561 - 019 - 0341 - 7.

[32] Xiao, L., P. Zhu, G. Fang, Z. Xiao, Y. Zou, J. Zhao, N. Zhao et al. 2015. "A Young Multilayered Terrane of the Northern Mare Imbrium Revealed by Chang'E - 3 Mission." Science 347 (6227): 1226 - 1229. DOI: 10.1126/science.1259866.

[33] Yu, T., J. Fei, L. Li. 2019. "Study on Path Planning Method of Lunar Rover." Journal of Deep Space Exploration 6 (4): 384 - 390. (in Chinese with English abstract).

[34] Zakrajsek, J., D. Mckissock, J. Woytach. 2005. "Exploration Rover Concepts and Development Challenges." First Space Exploration Conference: Continuing the Voyage of Discovery, 1 - 23. Orlando: AIAA 2005.

[35] Zhao, Y., F. Li, B. Wu. 2019. "Precise Landing Site Selection and Evaluation System Design for Chang'e - 4 Probe." Spacecraft Engineering 28 (4): 22 - 30. (in Chinese with English abstract).

[36] Zuo, W., C. Li, and Z. Zhang. 2014. "Scientific Data and Their Release of Chang'E - 1 and Chang'E - 2." Chinese Journal of Geochemistry 33: 24 - 44. doi: 10.1007/s11631 - 014 - 0657 - 3.

Hyperbolic reflectors determined from peak echoes of ground penetrating radar

Chunyu Ding[1,2,3], Zhiyong Xiao[2,4], Yan Su[3,5], Jun Cui[2,3,4]

1. Department of Earth and Space Science, Southern University of Science and Technology, Shenzhen, 518055, China
2. Planetary Environmental and Astrobiological Research Laboratory, School of Atmospheric Sciences, Sun Yat-sen University, Zhuhai, 519000, China
3. Key Laboratory of Lunar and Deep Space Exploration, National Astronomical Observatories, Chinese Academy of Sciences, Beijing, 100020, China
4. Center for Excellence in Comparative Planetology, Chinese Academy of Sciences, Hefei, 230026, China
5. University of Chinese Academy of Sciences, Beijing, 100049, China

Abstract: Ground penetrating radar has recently been used to detect subsurface structures of extraterrestrial bodies, but in situ physical properties (permittivity, density, porosity, etc.) of planetary materials are not well constrained, so that geological interpretations for a given radar observation can be widely different. Hyperbolic reflectors in radargrams obtained by ground penetrating radars are caused by subsurface materials that have a different permittivity, and these reflectors are important agent to deduce the physical properties of materials at smaller depths. The geometry of hyperbolic reflectors is critical for such calculations. Manual extraction, image analyses, and peak echoes have been used to extract the geometry of hyperbolic reflectors, but uncertainties caused by these different methods have not been systematically investigated. Here we compare differences caused by the three methods based on both modelled and observed hyperbolic reflectors in radargrams. Our results suggest that minor differences in the extracted geometry could yield significant uncertainty in the estimated bulk permittivities, leading to dramatically different interpretations. For the interpretation of the top-most radargram that was obtained by the high-frequency lunar penetrating radar onboard the Chang'E-3 mission, uncertainties caused by the different extraction techniques are trivial. However, for radargrams obtained by both the Chang'E-4 and the first Chinese Mars missions, uncertainties of depth estimations can be tens of meters based on the different extraction methods. Such uncertainties should be considered as the minimum inherent uncertainty of radar detection, because hyperbolic reflectors in radargrams are formed by peak echoes of reflected electromagnetic waves, and in actual worlds, their geometry normally deviates from ideal forms of hyperbolas. We recommend extracting hyperbolas using the peak echo method.

Keywords: Moon; Regolith; Permittivity; Lunar penetrating radar; Radargram; Chang'E-3 mission; Chang'E-4 mission

1 Introduction

Ground penetrating radar (i. e., GPR) is a nondestructive detecting technique that utilizes the principle of scattering of electromagnetic waves in media, which has been widely used in Earth sciences to detect subsurface structures since about 100 years ago (Daniels, 2005). This technique has recently been applied to extraterrestrial bodies, i. e., the lunar penetrating radar onboard the Chinese Chang'E-3 and Chang'E-4 rovers (Fang et al., 2014; Jia et al., 2018). Radargrams obtained by signal processing of GPR data are affected by the combination of electrical conductivity of the media, the transmitted center frequency, and the radiated power of the radar signal (Daniels, 2005). Subsurface structures of various sizes, geometry, depths, and internal structures are frequently revealed as layers, scattered signals, and hyperbolic reflectors in radargrams.

Hyperbolic reflectors in radargrams are a key target to invert the shape and identity of buried structures that have different relative permittivitties, e.g., boulders in glacier (Brandt et al., 2007). Hyperbolic reflectors are formed due to different round-trip travel times of electromagnetic waves that were transmitted from the moving antenna, when the electromagnetic waves encounter a buried target (Chow and Rees, 1989; Davis and Annan, 1989). The geometry of hyperbolic reflectors in radargrams is critical to estimate the permittivity and other related physical properties of the subsurface structure, since the aperture of hyperbolas depends on the shape, size, and dielectric permittivity of the buried target (Al-Nuaimy et al., 2000; Al-Nuaimy et al., 2001). Uncertainties in extracting hyperbolic reflectors affect signal processing and interpretation. For example, a 1% uncertainty in the eccentricity of hyperbolas equals a difference of ~2 in the derived root mean square (i. e., RMS) bulk permittivity compared to an assumed bulk permittivity of 3 (section 3). Such uncertainties can cause substantial problems in geological interpretation.

Extracting the shape of hyperbolic reflectors is normally based on anchor points along the recognized reflector. Three methods have been used to extract the geometry of hyperbolic reflectors: (1) manual extraction based on subjective determination of the hyperbolic outlines (Windsor et al., 2005); (2) image analyses that extract hyperbolic outlines based on the albedo contrast using automatic or semi-automatic image processing methods such as pattern recognition (Dou et al., 2017; Maas and Schmalzl, 2013; Mertens et al., 2016; Muniappan et al., 2012; Noreen and Khan, 2017; Yuan et al.,

2018); (3) extract positions of peak echoes along hyperbolic reflectors (Ristic et al., 2009; Li and Zhou, 2007). While the three methods all have extensive applications, no systematical cross-comparison was done to evaluate their differences and possible uncertainties in the results.

The electrical permittivity of in situ planetary materials is barely known compared to terrestrial materials, so that hyperbolic reflectors in radargrams were frequently used to deduce their physical properties (Fa et al., 2015). Subsurface structures within lunar regolith and martian soil are both important scientific targets and engineering needs (e.g., Atkinson et al., 2019; Binnie et al., 2018), so that GPR is an invaluable application. So far, GPR experiments have been successfully performed on the Chang'E-3 and Chang'E-4 missions (Jia et al., 2018; Xiao et al., 2015). The coming Chang'E-5 and the first Chinese Mars mission will also carry a similar GPR payload (Xiao et al., 2019; Zhou et al., 2016). Previous studies applying the Chang'E-3 lunar penetrating radar (i.e., LPR) data have used different methods to extract hyperbolic reflectors, e.g., manual extraction (Feng et al., 2017) and image analyses (Lai et al., 2016), and different permittivity values and thus interpreted subsurface structures were derived (see Fig. S1 in the appendices). However, for the estimated relative permittivity of extraterrestrial materials, potential differences might exist if different extraction method of hyperbolic reflectors was employed, but such differences were not evaluated.

This paper first evaluates the applicability of the three different methods in extracting hyperbolic reflectors in radargrams (sections 2). Based on the high-frequency LPR data obtained by the Chang'E-3 mission and simulated radargrams, results derived using the three methods are slightly different from each other (section 3). We recommend the extraction method based on peak echoes, since this method follows the physical law of electromagnetic wave propagation in media, and the observed uncertainties should be regarded as intrinsic to radar detection. Interpreting subsurface structures using the GPR onboard the Chang'E-4 mission and the first Chinese Mars mission have to consider such uncertainties (section 4).

2 Method

The three methods used to extract hyperbolic reflectors all follow the same logical procedure in principle (Al-Nuaimy et al., 2001; Shihab and Al-Nuaimy, 2005): (1) subjectively determine possible hyperbolic reflectors, (2) set boundary for extraction, (3) locate anchor points along the hyperbolic reflector using different extraction methods, and (4) estimate the geometry of the hyperbola via best fit.

Manual extraction is done by subjective recognition of anchor points along the outline of a hyperbolic reflector (e.g., Windsor et al., 2005; Fig. 1b and 1f). Image analyses are a frequently-used technique for this purpose, and the basic theory is based on the albedo contrast between the hyperbolic reflector and the surrounding radargram. Here we utilize the skeletonization processing method of image analyses (Muniappan et al., 2012) to extract the shape of the hyperbolic reflectors (Fig. 1d and 1h).

As electromagnetic waves propagate into a medium, the transmitted pulse is reflected when encountering a buried structure that has different dielectric properties. The peak of the waveform corresponds to the strongest reflected signal (e. g., Ristic et al., 2009; Li and Zhou, 2007), which is approximately corresponded to the position of the buried object. Therefore, for a recognized hyperbolic reflector, the positions of peak echoes are selected as the anchor points, which are located along the hyperbolic reflector (Figs. 1c, g and 2).

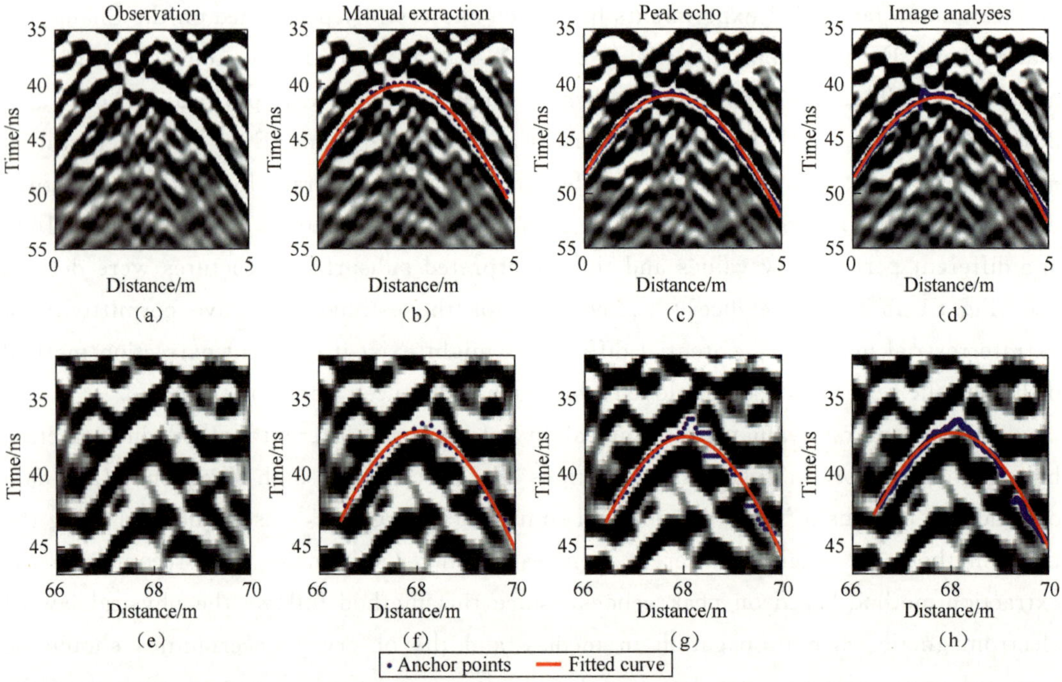

Fig. 1 Different extraction methods for the geometry of modelled and observed hyperbolic reflectors. The hyperbolic reflector in the first row is based on our modelled result, and that in the second row is observed by the Chang'E-3 LPR. Blue dots are the extracted anchor points that are used in the fitting, and the red lines are the fitted hyperbolas. (For interpretation of the references to colour in this figure legend, the reader is referred to the web version of this article.)

The mathematic equation of hyperbolas (Eq. 1) is used to fit the extracted anchor points via the least square method.

$$\frac{t^2}{\left(\frac{2h}{v}\right)^2}-\frac{(x-x_0)^2}{(h)^2}=1 \tag{1}$$

where x_0 is the horizontal position of the hyperbola, (x, t) is the coor dinate of anchored points, and v and h are RMS velocity and depth, respectively. The major $(2h/v)$ and minor semi-axis (h) of hyperbolic reflectors are derived using the best fit curve.

We use both simulated radargrams and those obtained by the Chang'E-3 LPR data to compare the three methods used to extract hyperbolic reflectors. Note that accomplishing our target does not depend on which GPR is simulated or which observational data are used, because the simulation purely employs the physics of electromagnetic wave propagation, and the high - frequency LPR data obtained by the Chang'E - 3 mission contains hyperbolic reflectors that have various complicated shapes (section 3. 2) .

The Chang ' E - 3 high - frequency LPR was operated at a central frequency of 500 MHz, and it has a range resolution better than 0. 3 min typical lunar regolith (Su et al. , 2014; Fang et al. , 2014) . The LPR data obtained on the lunar surface cover a transverse distance of ~107. 4m from N101 to N208 (Ding et al. , 2020), and different gain values were used along the route (Feng et al. , 2017) . To avoid uncertainties in the final radargram raised by the different gain values (Ding et al. , 2020), data obtained from N105 to N208 are used here. The data processing procedure for the raw radargram of the high - frequency channel contains removal of redundant data, time - zero correction, trace equalization, removal of direct current component, background subtraction, and band - pass filtering (Feng et al. , 2017; Ding et al. , 2020) . Fig. 2a shows the result.

A lunar regolith model (6. 5m × 5m) is constructed for the numerical simulation, which combines both knowledge gained from the Apollo core samples (Olhoeft and Strangway, 1975) and the layering structures observed within the radargarm (Fig. 3a). The regolith model has an assumed depth profile of relative permittivity (colour bar), which was built based on measurements of Apollo core samples (Olhoeft and Strangway, 1975) . Simulated radargrams are calculated using the Finite - Difference Time - Domain code (Teixeira et al. , 1998), and the operation mode of the high - frequency LPR onboard the Chang'E - 3 mission was used as the model input (Fang et al. , 2014) . To form a hyperbolic reflector, a 0. 3 m diameter basaltic boulder that has a relative permit - tivity of 7. 8 is implanted at a depth of ~2. 83 m in the regolith model. Layering structures in lunar regolith are simulated as curved lines that have a relative permittivity of 5, which is slightly larger than that of the background regolith (Ding et al. , 2020) . At depths less than the planted boulder, the average permittivity of the lunar regolith is ~3. 28 in the model.

3 Results

3. 1 Cross - comparison based on individual hyperbolic reflectors

For individual hyperbolic reflector in a radargram, the three methods of extracting hyperbolic reflectors yield similar but not identical geometries. We first compare the three methods using the case example shown in Fig. 3, which contains both simulated and observed radar grams of the Chang'E - 3 high - frequency LPR. It is evident that using the three methods, small differences exist in the geometries of the observed hyperbolas than those modelled (Fig. 1). For the simulated hyperbolic reflector (Fig. 3b), the derived geometry is not substantially different using the three methods, as the major semi-axis ($2h/v$) of the best fit curves are 41.04 ns, 41.08 ns, and 41.26 ns using the manual extraction, peak echo method, and image analyses, respectively. The RMS bulk permittivity (ε_{rms}) are calculated using the Eq. (1) and $\varepsilon_{rms} = (c/v)^2$, where c is speed of light in vacuum (Daniels, 2005), yielding ε_{rms} of 3.19 (Fig. 1b), 3.09 (Fig. 1c) and 3.17 (Fig. 1d), respectively. Compared to the average permittivity of the modelled materials above the boulder (~3.28), differences compared to the three estimated ε_{rms} are ~2.68% (manual), ~5.77% (peak echo) and ~3.45% (image analyses), respectively. On the other hand, radargrams for real geological terrains on the Moon are normally much more complicated than the dedicated regolith model (e.g., Fig. 3b and 3d). For the observed hyperbolic reflector (Fig. 3d), the major semi-axis of the fitted hyperbolas are 37.19 ns, 37.58 ns and 37.31 ns based on manual extraction, peak echo method, and image analyses, respectively. The three values derived for the observed hyperbolic reflector have slightly larger difference compared to the three values for the modelled reflector, which can be ascribed to the more complicated shape of the observed reflector (Fig. 3b and 3d). Similar comparisons are done for each of the 36 recognized hyperbolic reflector (Fig. S2 in the appendices).

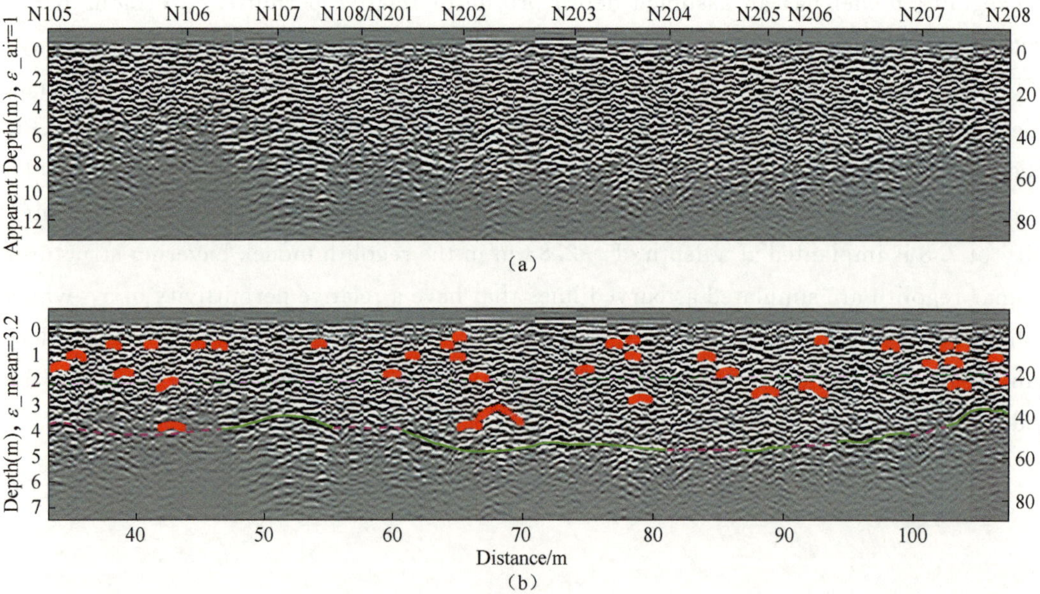

Fig. 2　Radargram and interpreted subsurface structures based on the high-frequency LPR onboard the Yutu rover of the Chang'E-3 mission. The data were obtained from navigation points N105 to N208, which correspond to 33.2 m to 107.4 m along the transverse route from the landing site (e.g., Ding et al., 2020). a. The radargram derived following standard signal processing (Feng et al., 2017). b. 36 hyperbolic reflectors are used in this study (red lines), and the cyan and green lines are the maximum depth of the studied radargram. (For interpretation of the references to colour in this figure legend, the reader is referred to the web version of this article.)

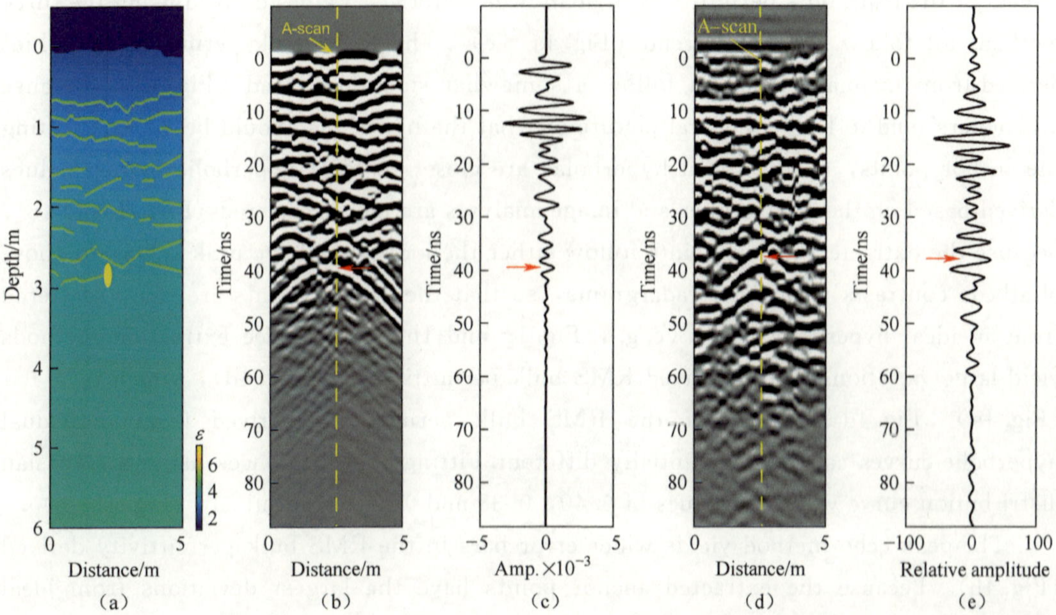

Fig. 3　Extracting hyperbolic reflectors based on peak echoes of reflected radar waves. a. Regolith model with layering structure obtained from LPR observations. b. Simulated radargram shows the hyperbolic reflector caused by the buried boulder. The yellow dotted line denotes the trace where the A - scan data are derived. c. A - scan waveform shows that the summit of the hyperbolic reflector corresponds to a local increase in the echo strength. d. A section of the radargram that was derived from the high frequency LPR data. The red arrow denotes a hyperbolic reflector recognized in the radargram. e. A - scan waveform along the yellow dashed line shown in panel (d). The outline of the hyperbolic reflector corresponds to the peak echo. Note that we compare the echo pattern along the A - scan to determine the position of echo peaks, and the absolute strength of the reflected signal is not compared between the simulated and modelled hyperbolic reflector. (For interpretation of the references to colour in this figure legend, the reader is referred to the web version of this article.)

3.2　Cross - comparison using the Chang'E - 3 LPR data

The effect of the different extraction methods on the geological interpretation of radargrams is more practical when dealing with a population of hyperbolic reflectors. Here we use the high - frequency LPR data returned by the Chang'E - 3 mission to estimate the RMS bulk permittivity (Fig. 2a). 36 hyperbolic reflectors are recognized within ～50 ns round - trip travel time of the radargram (Fig. 2b; Ding et al., 2020), and we use the above three methods to extract their geometries (Fig. 2b). Detailed information about the anchor points and the fitted hyperbolas based on the three methods are shown in the Fig. S2 of the appendices.

To better show the effect of the different extraction methods in geological interpretations, we compare the 3 sets of bulk RMS permit tivitties derived from the three extraction methods. For each set of the 36 hyperbolic reflectors, their geometries are used to deduce the RMS bulk permittivity of shallower materials. Values derived using the three methods all follow the same trend（Fig. 4a – c）. The RMS bulk permit tivity values derived from manual extraction follow a somewhat smoother trend（Fig. 4a）, because practioners tend to have a general picture of what the hyperbola should be when selecting the anchor points, so the fitted hyperbolas are closer to ideal hyperbolic forms. Values derived based on the peak echoes and image analyses are more scattered（Fig. 4b and 4c）, because the extracted anchor points follow either the locations of the peak echoes or those of albedo contrasts within the radargrams, so that the anchor points are more scattered from an ideal hyperbolic curve（e. g. , Fig. 1g and 1h）. The three extraction methods yield large variations in the derived RMS bulk permittivitties（Fig. 4d）, which is ～1. 1（Fig. 4e）. Fig. 4d shows that the RMS bulk permittivity derived from individual hyperbolic curves can be substantially different. Fitting the differences using a Gaussian distri bution curve yields 1δ values of 0. 40, 0. 33 and 0. 36, respectively（Fig. 4f – h）.

The peak echo method yields wider error bars in the RMS bulk permittivity derived（Fig. 4b）, because the extracted anchor points have the largest deviations from ideal hyperbolas（e. g. , Fig. 1g）. Nevertheless, we consider that the peak echo method is more suitable to extract the shape of hyperbolic reflectors compared to the other two methods, because the occurrences of the anchor points strictly follow the physical law of forming hyperbolic reflectors in radargrams（Chow and Rees, 1989; Li and Zhou, 2007）. Although the geometric parameters of hyperbolas derived using the peak echo method exhibit both wider dispersion and larger error bars（Fig. 4b）, they should be included in analyzing radargrams for actual geological materials, since the actual complexities of geological materials are much higher than the ideal simulations（e. g. , Fig. 1）.

Discussion and conclusion

For the top – most ～50 ns of the radargram obtained by the high – frequency LPR onboard the Chang'E – 3 mission（Fig. 2）, the three different methods used to extract hyperbolic reflectors yield small variations in deriving the RMS bulk permittivity（Fig. 4）. Previous studies usually assumed a relative permittivity of 3 for the lunar regolith when interpreting the high – frequency LPR data（Fa et al. , 2015; Xiao et al. , 2015; Zhang et al. , 2018）, so that the top ～50 ns of the radargram corresponds to a depth of ～4. 33 m orange line of Fig. 5a.

A better inversion of depth information in radargrams can be derived based on the depth profile of relative permittivity（Daniels, 2005）. Based on the hyperbolic reflectors extracted using the peak echo method（Fig. 4b）, the top – most ～50 ns of the radargram corresponds to a depth of ～3. 70 ±0. 19 m（black lines of Fig. 5a）after considering the uncertainties（1δ = 0. 36）. The difference of the interpreted depths（4. 33m versus 3. 70±0. 19m）is not substantial, since it is comparable with the wavelength of the high –

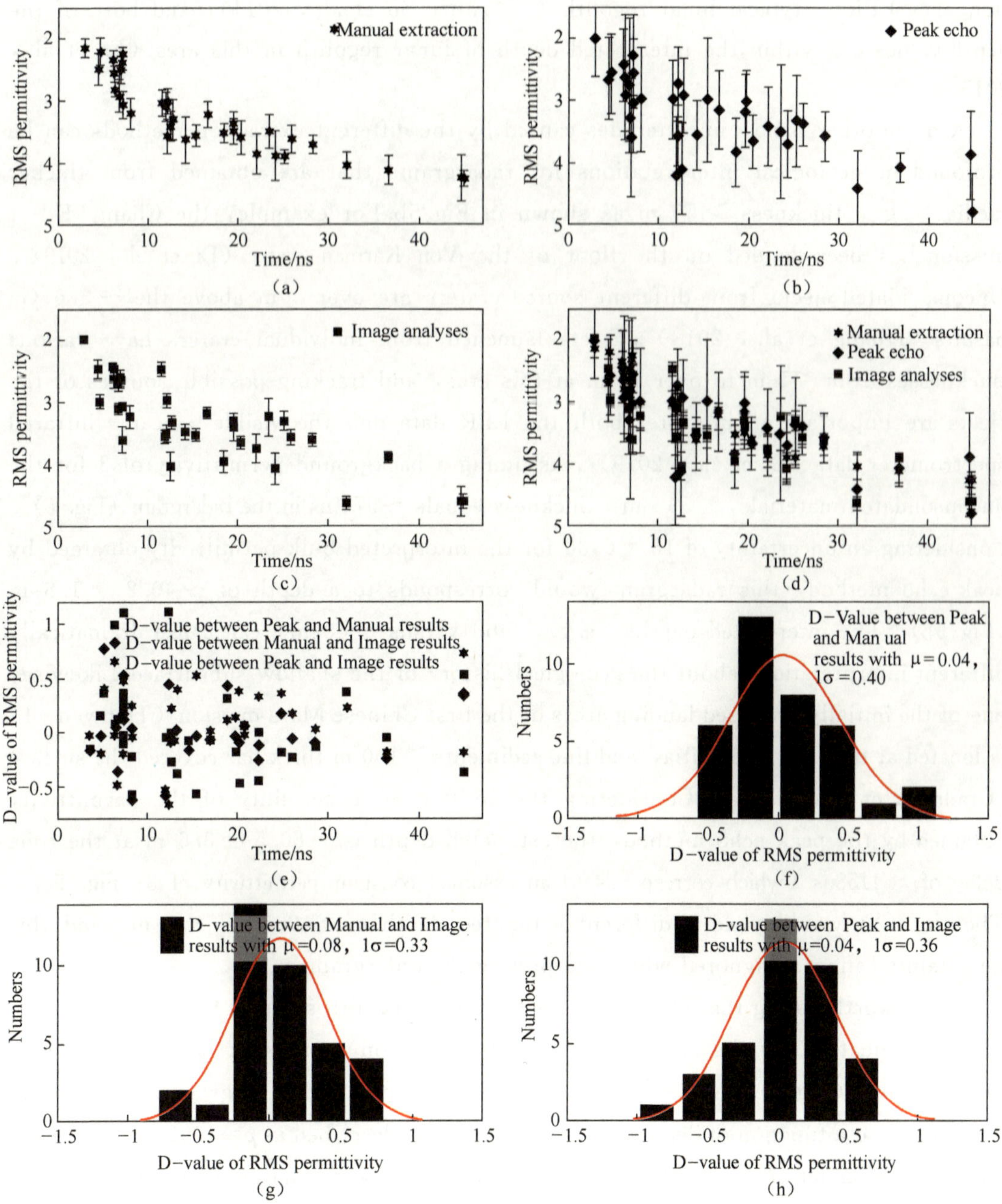

Fig. 4　Inversion of RMS bulk permittivity based on 36 hyperbolic reflectors observed in the radargam obtained by the Chang'E-3 mission. a-c are the results based on manual extraction, image analyses, and peak echos of the hyperbolic reflectors, respectively. d. Comparison of the results shown in a-c. e. Differences of RMS bulk permittivities estimated using the three methods. f-h. Histograms of the differences between the RMS bulk permittivities estimated based on the peak echo and manual extraction methods, manal extraction and image analyses, and peak echo and image analyses, respectively. The red lines are fitting curves following normal Gaussian distribution, and the average values and standard deviations are listed in the legend. (For interpretation of the references to colour in this figure legend, the reader is referred to the web version of this article.)

frequency LPR in typical lunar regolith (~0.3m; Su et al., 2014), and both of the depth values are within the interpreted depth of lunar regolith in this area (Fa et al., 2015).

On the other hand, uncertainties caused by the different extraction methods can be profound in geological interpretations for radargrams that are obtained from thicker media, e.g., thickness >50 m as shown in Fig. 5b. For example, the Chang'E-4 mission has been landed on the floor of the Von Kármán crater (Di et al., 2019). Unconsolidated ejecta from different source craters are over 50m above the ~3.6 Ga basalts (Huang et al., 2018). Ejecta launched from individual craters have various thicknesses from <1 m to over 10 m at this area, and tracking possible sources of the ejecta are important to interpret both the LPR data and the visible-near-infrared spectrometer data (Li et al., 2019). Assuming a background permittivity of 3 for the unconsolidated materials, a 50-m-thickness equals ~577 ns in the radargram (Fig. 5). Considering an uncertainty of $1\delta = 0.36$ for the interpreted bulk permittivity obtained by peak echo method, this radargram would corresponds to a depth of ~40.2 ±1.8 m (Fig. 5b). The interpreted depths (e.g., 50m versus ~40 m) can cause dramatically different interpretations about the geological history of the shallow subsurface. Likewise, one of the initially proposed landing areas of the first Chinese Mars mission (Tianwen-1) is located at the Chryse planitia, and fine sediments >100 m thick has covered the surface (Craddock et al., 1997). Considering the $1\delta = 0.36$ uncertainty of the permittivity obtained by the peak echo method, the estimated depth is ~80.5 ±3.6 m at the time delay of ~1158ns (which corresponds to an assumed constant permittivity of 3; Fig. 5c). Therefore, it would yield a different estimated depth of ~20.5 ±3.6 m, and this uncertainty cannot be ignored when interpreting the radargram.

It is worth noting that measurements of Apollo regolith samples have established an empirical equation for the relative permittivity (episilon) and bulk density (rtho) of lunar regolith (Fig. 5), as $\varepsilon = (1.93 \pm 0.17)^{\rho}$ (Olhoeft and Strangway, 1975). Bulk density is also a function of depths (d), which can be described as $\rho = 1.92\ (d+12.2)\ /\ (d+18)$ (Carrier et al., 1991). Therefore, the inverse estimation of the bulk density and porosity for lunar subsurface materials based on hyperbolic reflectors can also be seriously affected by the different extraction methods employed (Fig. 5b and c).

While we recommend the extraction method of hyperbolic reflectors based on peak echo of the radar signal, other errors raised by subjective determination of hyperbolic reflectors and boundary setting for extraction still exist. Therefore, we are cautious that the uncertainties in deducing the RMS bulk permittivity (thus the interpreted depths) using the different extraction methods should be considered as a lower limit for geological interpretation.

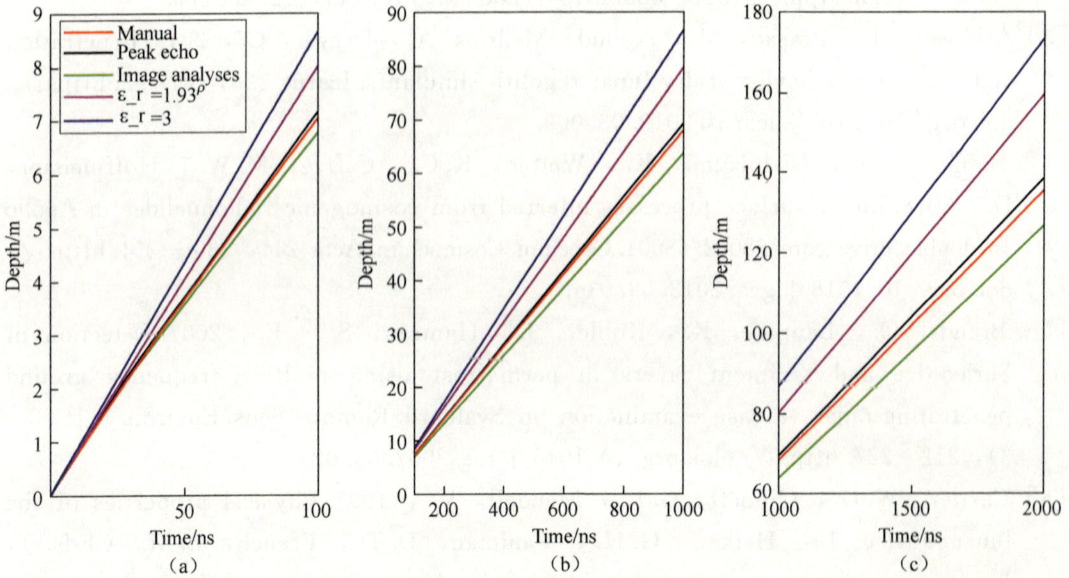

Fig. 5　The relationship between two - way travel time (ns) of radar waves and interpreted depths (m) assuming different depth profiles for the RMS bulk permittivity. The depth profiles of RMS bulk permittivity are derived using the three different methods (black, red and green lines). a - c. Extrapolation of the same relationship to two - way travel times of 0 - 100ns, 100 - 1000ns, and 1000 - 2000ns, respectively (For interpretation of the references to colour in this figure legend, the reader is referred to the web version of this article)

Acknowledgements

This work is supported by the B - type Strategic Priority Program of the Chinese Academy of Sciences (XDB41000000), National Natural Science Foundation of China (42004099, 41773063, 41525015, and 41830214), the Pre - research Project on Civil Aerospace Technologies (D020101, D020202) of CNSA, and the Opening Fund of the Key Laboratory of Lunar and Deep Space Exploration, Chinese Academy of Sciences (No. LDSE201908 and LDSE202005).

References

[1] Al - Nuaimy, W., Huang, Y., Nakhkash, M., Fang, M. T., Nguyen, V., Eriksen, A., 2000. Automatic detection of buried utilities and solid objects with GPR using neural networks and pattern recognition. J. Appl. Geophys. 43 (2 - 4), 157 - 165. https://doi.org/10.1016/s0926 - 9851 (99) 00055 - 5.

[2] Al - Nuaimy, W., Huang, Y., Eriksen, A., Nguyen, V. T., 2001. Automatic

detection of hyperbolic signatures in ground – penetrating radar data. Subsurf. Surf. Sens. Technol. Appl. Ⅲ 327 – 335. https://doi. org/10. 1117/12. 450177.

[3] Atkinson, J. , Prasad, M. , Abbud – Madrid, A. , Dreyer, C. , 2019. Penetration and relaxation behavior of dry lunar regolith simulants. Icarus 328, 82 – 92. https://doi. org/ 10. 1016/j. icarus. 2019. 03. 009.

[4] Binnie, S. A. , Nishiizumi, K. , Welten, K. C. , Caffee, M. W. , Hoffmeister, D. , 2018. Lunar surface processes inferred from cosmogenic radionuclides in Apollo 16 double drive core 68002/68001. Geochim. Cosmochim. Acta 244, 336 – 351. https://doi. org/ 10. 1016/j. gca. 2018. 09. 036.

[5] Brandt, O. , Langley, K. , Kohler, J. , Hamran, S. – E. , 2007. Detection of buried ice and sediment layers in permafrost using multi – frequency ground penetrating radar: a case examination on Svalbard. Remote Sens. Environ. 111 (2 – 3), 212 – 227. https:// doi. org/10. 1016/j. rse. 2007. 03. 025.

[6] Carrier, W. D. , Olhoeft, G. R. , Mendell, W. , 1991. Physical properties of the lunar surface. In: Heiken, G. H. , Vaniman, D. T. , French, B. M. (Eds.), Lunar Sourcebook: A user' s Guide to the Moon. Cambridge University Press, Cambridge, pp. 475 – 594.

[7] Chow, T. , Rees, H. W. , 1989. Identification of subsurface drain locations with ground – penetrating radar. Can. J. Soil Sci. 69 (2), 223 – 234. https://doi. org/ 10. 4141/cjss89— 023.

[8] Craddock, R. A. , Crumpler, L. S. , Aubele, J. C. , Zimbelman, J. R. , 1997. Geology of central Chryse Planitia and the Viking 1 landing site: implications for the Mars pathfinder mission. J. Geophys. Res. Planets 102 (E2), 4161 – 4183. https://doi. org/10. 1029/ 97je00058.

[9] Daniels, D. J. , 2005. Ground Penetrating Radar. Wiley Online Library.

[10] Davis, J. L. , Annan, A. P. , 1989. Ground – penetrating radar for high – resolution mapping of soil and rock stratigraphy 1. Geophys. Prospect. 37 (5), 531 – 551. https://doi. org/ 10. 1111/j. 1365 – 2478. 1989. tb02221. x.

[11] Di, K. , Liu, Z. , Liu, B. , Wan, W. , Peng, M. , Wang, Y. , Gou, S. , Yue, Z. , Xin, X. , Jia, M. , Niu, S. , 2019. Chang' E – 4 lander localization based on multi – source data. J. Remote Sens. 23 (1), 177 – 184. https://doi. org/ 10. 11834/jrs. 20199015.

[12] Ding, C. , Li, C. , Xiao, Z. , Su, Y. , Xing, S. , Wang, Y. , et al. , 2020. Layering structures in the porous material beneath the Chang'E – 3 landing site. Earth Space Sci. 7 https://doi. org/10. 1029/2019EA000862 e2019EA000862.

[13] Dou, Q. , Wei, L. , Magee, D. R. , Cohn, A. G. , 2017. Real – time hyperbola recognition and fitting in GPR data. IEEE Trans. Geosci. Remote Sens. 55 (1), 51 – 62. https://doi. org/ 10. 1109/tgrs. 2016. 2592679.

[14] Fa, W. , Zhu, M. , Liu, T. , Plescia, J. , 2015. Regolith stratigraphy at the

Chang'E – 3 landing site as seen by lunar penetrating radar. Geophys. Res. Lett. 42 (23), 10, 179 – 10, 187. https://doi.org/10.1002/2015gl066537.

[15] Fang, G., Zhou, B., Ji, Y., Zhang, Q., Shen, S., Li, Y., Guang, H., Tang, C., Gao, Y., Lu, W., Ye, S., Han, H., Zheng, J., Wang, S., 2014. Lunar penetrating radar onboard the Chang'E – 3 mission. Res. Astron. Astrophys. 14 (12), 1607 – 1622. https://doi.org/ 10.1088/1674 – 4527/14/12/009.

[16] Feng, J., Su, Y., Ding, C., Xing, S., Dai, S., Zou, Y., 2017. Dielectric properties estimation of the lunar regolith at CE – 3 landing site using lunar penetrating radar data. Icarus 284, 424 – 430. https://doi.org/10.1016/j.icarus.2016.12.005.

[17] Huang, J., Xiao, Z., Flahaut, J., Martinot, M., Head, J., Xiao, X., Xiao, L., 2018. Geological Characteristics of Von Kármán Crater, Northwestern South Pole – Aitken Basin: Chang'E – 4 Landing Site Region. J. Geophys. Res. Planets 123 (7), 1684 – 1700. https://doi.org/10.1029/2018je005577.

[18] Jia, Y., Zou, Y., Ping, J., Xue, C., Yan, J., Ning, Y., 2018. The scientific objectives and payloads of Chang'E – 4 mission. Planet. Space Sci. 162, 207 – 215. https://doi.org/ 10.1016/j.pss.2018.02.011.

[19] Lai, J., Xu, Y., Zhang, X., Tang, Z., 2016. Structural analysis of lunar subsurface with Chang'E – 3 lunar penetrating radar. Planet. Space Sci. 120, 96 – 102. https://doi.org/ 10.1016/j.pss.2015.10.014.

[20] Li, T., Zhou, Z., 2007. Fast extraction of hyperbolic signatures in GPR. In: 2007 International Conference on Microwave and Millimeter Wave Technology. Builin, pp. 1 – 3. https://doi.org/10.1109/ICMMT.2007.381489.

[21] Li, C., Liu, D., Liu, B., Ren, X., Liu, J., He, Z., Ouyang, Z., 2019. Chang'E – 4 initial spectroscopic identification of lunar far – side mantle – derived materials. Nature 569 (7756), 378 – 382. https://doi.org/10.1038/s41586 – 019 – 1189 – 0.

[22] Maas, C., Schmalzl, J., 2013. Using pattern recognition to automatically localize reflection hyperbolas in data from ground penetrating radar. Comput. Geosci. 58, 116 – 125. https://doi.org/10.1016/j.cageo.2013.04.012.

[23] Mertens, L., Persico, R., Matera, L., Lambot, S., 2016. Automated detection of reflection hyperbolas in complex GPR images with no a priori knowledge on the medium. IEEE Trans. Geosci. Remote Sens. 54 (1), 580 – 596. https://doi.org/ 10.1109/tgrs.2015.2462727.

[24] Muniappan, N., Hebsur, A. V., Rao, E. P., Venkatachalam, G., 2012. Radius estimation of buried cylindrical objects using GPR – A case study. In: 2012 14th International Conference on Ground Penetrating Radar (GPR), pp. 789 – 794. https://doi.org/ 10.1109/icgpr.2012.6254969.

[25] Noreen, T., Khan, U., 2017. Using pattern recognition with HOG to

automatically detect reflection hyperbolas in ground penetrating radar data. In: 2017 International Conference on Electrical and Computing Technologies and Applications, pp. 465 – 470. https://doi. org/10. 1109/ICECTA. 2017. 8252064.

[26] Olhoeft, G. R. , Strangway, D. , 1975. Dielectric properties of the first 100 meters of the moon. Earth Planet. Sci. Lett. 24, 394 – 404. https://doi. org/10. 1016/0012 – 821X (75) 90146 – 6.

[27] Ristic, A. V. , Petrovacki, D. , Govedarica, M. , 2009. A new method to simultaneously estimate the radius of a cylindrical object and the wave propagation velocity from GPR data. Comput. Geosci. 35, 1620 – 1630. https://doi. org/ 10. 1016/j. cageo. 2009. 01. 003.

[28] Shihab, S. , Al – Nuaimy, W. , 2005. Radius estimation for cylindrical objects detected by ground penetrating radar. Subsurf. Sens. Technol. Appl. 6, 151 – 166. https://doi. org/ 10. 1007/s11220 – 005 – 0004 – 1.

[29] Su, Y. , Fang, G. , Feng, J. , Xing, S. , Ji, Y. , Zhou, B. , Gao, Y. , Li, Han, Dai, S. , Xiao, Y. , Li, C. , 2014. Data processing and initial results of Chang' E – 3 lunar penetrating radar. Res. Astron. Astrophys. 14 (12), 1623 – 1632. https://doi. org/10. 1088/1674 – 4527/14/ 12/010.

[30] Teixeira, F. L. , Weng, C. C. , Straka, M. , Oristaglio, M. L. , Wang, T. , 1998. Finite – diference time – domain simulation of ground penetrating radar on dispersive, inhomogeneous, and conductive soils. IEEE Transac. Geosci. Remote Sens. 36, 1928 – 1937. https:// doi. org/10. 1109/36. 729364.

[31] Windsor, C. G. , Capineri, L. , Falorni, P. , 2005. The estimation of buried pipe diameters by generalized Hough transform of radar data. PIERS Online. 1 (3), 345 –349. https:// doi. org/10. 2529/PIERS041117130829.

[32] Xiao, L. , Zhu, P. , Fang, G. , Xiao, Z. , Zou, Y. , Zhao, J. , Zhao, N. , Yuan, Y. , Qiao, L. , Zhang, X. , Zhang, H. , Wang, J. , Huang, J. , Huang, Q. , He, Q. , Zhou, B. , Ji, Y. , Zhang, Q. , Shen, S. , Li, Y. , Gao, Y. , 2015. A young multilayered terrane of the northern Mare Imbrium revealed by Chang' E – 3 mission. Science 347 (6227), 1226 – 1229. https:// doi. org/10. 1126/science. 1259866.

[33] Xiao, Y. , Su, Y. , Dai, S. , Feng, J. , Xing, S. , Ding, C. , Li, C. , 2019. Ground experiments of Chang' E – 5 lunar Regolith penetrating radar. Adv. Space Res. 63, 3404 – 3419. https://doi. org/10. 1016/j. asr. 2019. 02. 001.

[34] Yuan, C. , Li, S. , Cai, H. , Kamat, V. R. , 2018. GPR signature detection and decomposition for mapping buried utilities with complex spatial configuration. J. Comput. Civ. Eng. 32 (4), 04018026 https://doi. org/10. 1061/ (asce) cp. 1943 – 5487. 0000764.

[35] Zhang, L. , Zeng, Z. , Li, J. , Lin, J. , Hu, Y. , Wang, X. , Sun, X. , 2018. Simulation of the lunar Regolith and lunar – penetrating radar data

processing. IEEE J. Select. Topics Appl. Earth Observ. Remote Sens. 11 (2), 655 - 663. https://doi. org/10. 1109/ jstars. 2017. 2786476.

[36] Zhou, B., Shen, S. X., Ji, Y. C., Lu, W., Zhang, F., Fang, G. Y., Su, Y., Dai, S., 2016. The subsurface penetrating radar on the rover of China's Mars 2020 mission. In: 2016 16th International Conference on Ground Penetrating Radar (GPR). https://doi. org/10. 1109/icgpr. 2016. 7572700.

Fragments Delivered by Secondary Craters at the Chang'E - 4 Landing Site

Chunyu Ding[1], Zhiyong Xiao[1,2,3], Bo Wu[4], Yuan Li[4], Nils C. Prieur[5], Yuzhen Cai[1], Yan Su[6], and Jun Cui[1,3,6]

1. Planetary Environmental and Astrobiological Laboratory, School of Atmospheric Sciences, Sun Yat - sen University, Zhuhai, China
2. State Key Laboratory of Lunar and Planetary Sciences, Space Science Institute, Macau University of Science and Technology, Macau, China
3. Center for Excellence in Comparative Planetology, Chinese Academy of Science, Hefei, China
4. Department of Land Surveying and Geo - Informatics, The Hong Kong Polytechnic University, Hong Kong
5. Centre for Earth Evolution and Dynamics, University of Oslo, Oslo, Norway
6. Key Laboratory of Lunar and Deep Space Exploration, National Astronomical Observatories, Beijing, China

Abstract: The Chang'E - 4 landing site is depleted with boulders seen from both orbit and surface. However, the Yutu - 2 rover came across thousands of concreted fragments in and around an abnormally fresh crater that has more elevated northwestern rims. The origin of the fragments is crucial to resolve the provenances of surface materials detected by the rover. The lunar penetrating radar performed two in - and - out scans for the blocky ejecta, revealing that the subsurface materials have indistinguishable radar permittivity with the surrounding regolith. Forward modeling of electromagnetic wave propagation shows that the fragments were not an original component in the subsurface. This crater is among the several fresh craters photoed by the rover, and they are located in an eastern extension ray of the Zhinyu crater. The small craters are likely secondaries of Zhinyu, and the fragments contain a mixture of shattered projectiles and most likely compacted regolith clumps formed during the secondary impacts.

Plain Language Summary: The Chang'E - 4 landing site lacks boulders seen from both orbit and surface, but several abnormally fresh craters were encountered by the Yutu - 2 rover. The blocky rims and interiors are in sharp contrast with the heavily degraded crater population in the landing area. These craters are less than 3 m in diameter, and the fragments are less than 10 cm long and appear earthy in color. The origin of the fragments needs an explanation because it is the basis for the interpretation of data returned by both the lunar penetrating radar and reflectance spectrometer onboard the rover. At the ninth lunar day, the rover was carefully driven into the blocky ejecta deposits of such a crater, and ground - penetrating radar detections were conducted. Radar measurements reveal that the physical properties beneath the blocky ejecta deposits are essentially the same with that of the surrounding regolith, and the observed fragments were not originated from the subsurface. The high spatial density, more pronounced northwestern rims, and colocation with an impact ray suggest that the small craters are most likely secondaries formed by the Zhinyu crater. The fragments are mainly formed by impact compaction of the preimpact target regolith, and the shattered secondaries - forming projectiles are a minor component.

Key Points:

- Yutu - 2 encountered several abnormally fresh craters surrounded by blocky fragments
- Radar permittivity of the blocky ejecta is not distinct from nearby normal regolith
- The small craters are likely secondaries, and impact compaction occurred in regolith

1 Introduction

The landing site of the Chang'E - 4 mission is located on an Imbrian - aged mare surface on the lunar farside (~3. 6 billion years old; Huang et al. , 2018), on the floor of the Von Kármán crater (Fig. 1a) . The location of the landing area within the largest impact basin on the Moon, the South Pole - Aitken basin, makes the study of surface composition particularly interesting. Secondaries delivered by various craters have covered the entire landing area, so that the Yutu - 2 rover might encounter not only the Imbrian - aged mare materials along its path but also highland materials ejected from impact craters

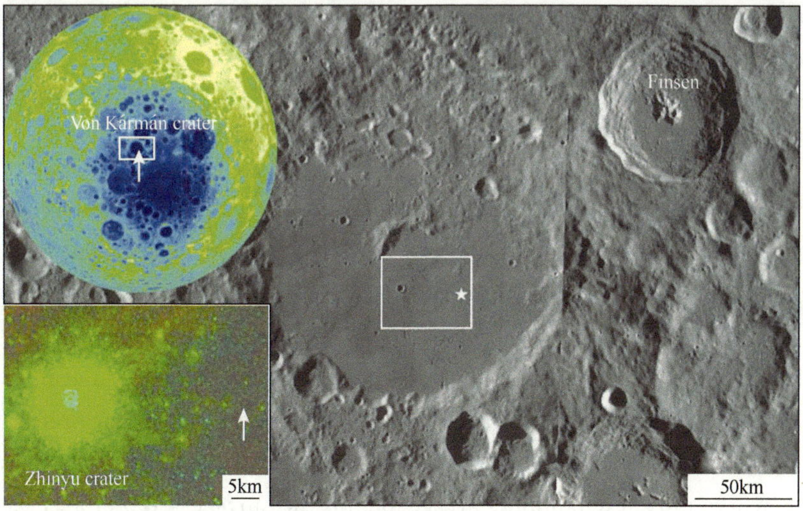

(a)

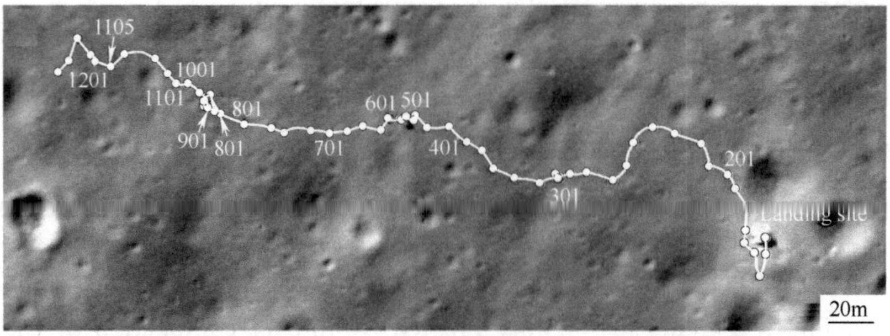

(b)

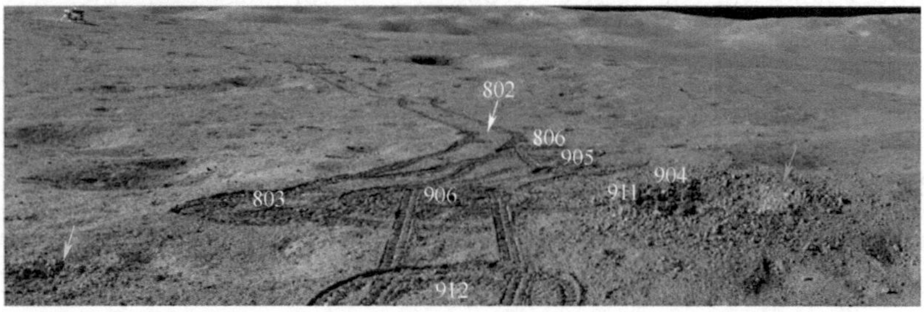

(c)

Fig. 1　Geomorphological context of the Chang'E－4 mission. (a) The Chang'E－4 landing site on the floor of the Von Kármán crater (white star; 177.588°E, 45.457°S). The landing site is positioned in the South Pole－Aitken basin on the farside of the Moon (white box in the upper－left inset; LROC WAC global DTM data set; Scholten et al., 2012). The lower－left inset (Clementine color mosaic; McEwen & Robinson, 1997) shows the Zhinyu crater to the west of the landing site, which has delivered a faint impact ray to the landing site (white arrow). North is up in all the mosaics. (b) The route of the Yutu－2 rover in the first 12 lunar days (white curve). Along the route, numbers along the dots are navigation points that are started with the number of lunar days counted since landing. The base image is from LROC NAC (image number M1311886645LE). North is up. (c) Images obtained at the ninth lunar day by the panoramic camera onboard the Yutu－2 rover. The imaging direction is southwest, and the lander is to the upper left of the panel. The targeted fresh crater is denoted by the green arrow, and another fresh crater is visible nearby (yellow arrow). Both of them are not visible from orbit. Width of wheel track is 15cm for scale. The denoted numbers are the navigation points around the target crater

outside of the Von Kármán crater (Huang et al., 2018). Finding surface materials that were sourced from lunar interior, analyzing their com position, and resolving their origins could be a key to reveal the possible structure and composition of the lunar interior (Wu et al., 2019).

Since being deployed from the lander to the lunar surface, the Yutu－2 rover has explored a distance of over 330 m across 12 lunar days (Fig. 1b). Images returned by both the lander and rover in the first 5 lunar days depict a landing area flat in topography (Di et al., 2019), where only few boulders are visible on the surface (Wu et al., 2019). This observation is consistent with orbital data, which depict a highly degraded topography, overall thick regolith cover (Huang et al., 2018), and low-thermal inertia (Fig. S1 in the supporting information). The boulder-scarce appearance is further verified by radar observations. The lunar penetrating radar (i. e., LPR; Fang et al., 2014) onboard the Yutu-2 rover has obtained high-quality data along the path. Using the LPR data obtained in the first few lunar days, Lai et al. (2019) and Li et al. (2020) noticed that electromagnetic waves decayed slowly in the top ~10—13m of the subsurface. Several hyperbolic reflectors were observed in the radargram, which are likely caused by subsurface boulders. The geometry of the hyperbolic reflectors was used to calculate the average relative permittivity and bulk density for this radar—transparent layer, and the results are comparable with those of typical lunar regolith (Lai et al., 2019; Li et al., 2020).

From the sixth lunar day, the rover suddenly encountered several fresh impact craters close to its path. These craters are less than 3 m in diameters and they have exposed thousands of small (less than 10 cm) fragments (e. g., Fig. 1c). Both the craters and fragments are in sharp contrast with the otherwise heavily degraded crater population, suggesting a recent emplacement on the surface. This observation is in contrast with both orbital and earlier *in situ* observations, provoking questions about the physical properties of subsurface materials at the landing site, such as potential heterogeneities within lunar regolith. The origin of the frag ments is also a basic question to be addressed for interpreting the radar and reflectance spectral data returned by the rover. For example, the rover has obtained reflectance spectra for both the surface regolith and rocky fragments, which were believed to be representative of Finsen's ejecta (Fig. 1a; Li et al., 2019; Lin et al., 2019). However, a layer of blocky materials might have been uniformly deposited to the landing area as possibly indicated by the fragments (Fig. 1c), so that the reflectance spectra collected are most likely for these distal materials.

The rover preformed a detailed investigation for such a fresh crater at the ninth day. The LPR obtained two profiles across the blocky ejecta along different azimuths, and the panoramic camera took multiple photos for the crater at mm-scale resolutions. Here we use data obtained by the LPR to investigate the radar permittivity of subsurface materials beneath and around the blocky ejecta. Images obtained by the panoramic cam-eras are used to construct millimeter-scale digital terrain models (DTM). Together with data obtained from orbit, the origin of the small fresh craters and the nature of the observed fragments are discussed. Indications to both the geological interpretation of the rover data and impact cratering mechanism are extended.

2 Data and Methods

The large-scale geological context of the landing area has been deciphered in previous studies (e. g. , Huang et al. , 2018; Ling et al. , 2019) . Here the general morphology of the landing area (Fig. 1a and 1b) is shown using images obtained by both the wide - angle camera (WAC) and narrow - angle camera (NAC) of the Lunar Reconnaissance Orbiter Camera (LROC; Robinson et al. , 2010) . The morphology and topography of the targeted small crater (Figure 1c) are studied using images obtained by the panoramic cameras on the Yutu - 2 rover. DTM constructed using photogrammetric methods requires multiple stereo images, and the spatial resolution is usually several times that of the original images, for example, the ~4m/pixel DTM derived from the ~1. 2m/pixel LROC NAC stereo pairs for the Chang'E - 4 landing site (Liu & Wu, 2020) . To derive the detailed topography for the target crater, a DTM was built using traditional photogrammetric methods based on stereo pairs obtained by the panoramic cameras (Fig. S2a in the supporting information) . Integrating this DTM as an input, the shape - from - shading (SfS) method (Horn, 1990) was employed to improve the spatial resolution. The effective resolution of the SfS - derived DTM is 5mm/pixel (Fig. 2b) . Methods of implementing SfS and the related uncertainties are introduced in detail by Wu et al. (2018) . Note that small blurry areas exist in the DTM because of shadows and difficult view geometries in the panoramic images (e. g. , east of the navigation point 906; Fig. 2c), where topography reconstruction is prohibited.

The LPR data obtained between the navigation points 904—905 and those between 906—911 are for the blocky ejecta deposit (Fig. 2c) . During this observation campaign, the rover was driven into the blocky ejecta until reaching the crater rim. Two endeavors of scanning from different directions were accomplished for the blocky ejecta by the LPR (Fig. 2c) . The LPR system onboard the Yutu - 2 rover has a similar design with that on the Yutu rover, and it has two subsystems that are operated at center frequencies of 60 and 500 MHz, respectively (Fang et al. , 2014) . We only use the high - frequency LPR data here, because the small craters are <3 m in diameter (Fig. 2) and considering their excavation depths are less than a meter, the top few meters of materials are interesting for the purpose of this work. The vertical resolution of the high - frequency LPR data is better than 0. 3m in lunar regolith (Su et al. , 2014) . The radar data are the level 2B data that were collected in the first 12 lunar days. The standard radar processing procedure (Feng et al. , 2017) is employed to process the data, that is, removal of redundant data, dewow filtering, amplitude compensation, background subtraction, and band - pass filtering.

To investigate whether or not the observed fragments originally existed in the subsurface and later been excavated by the small crater, we carry out numerical simulations for the propagation of radar waves in dedicated regolith models. The regolith models were built based on the detailed morphology and geometry of both the small craters and fragments. The 5 mm/pixel resolution DTM allows subtraction of the areal fraction and size frequency distribution of the fragments. Based on the SfS DTM and images obtained by the panoramic cameras, an orthoimage map is constructed for the

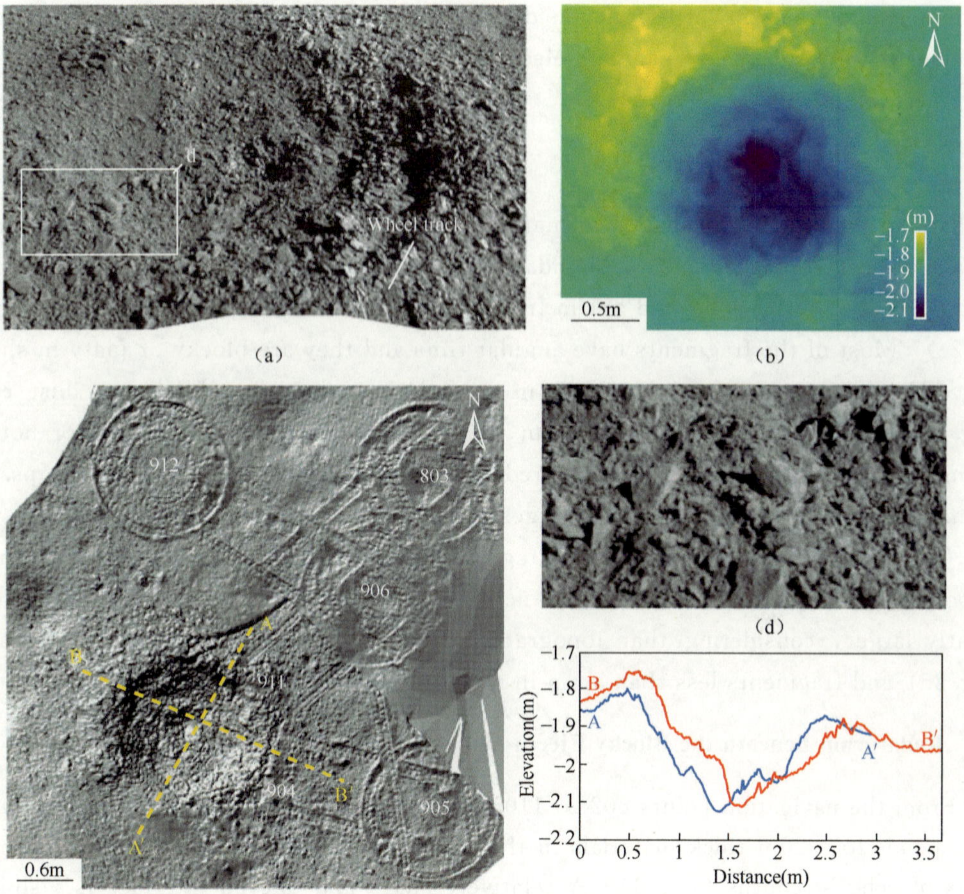

(a)　(b)　(c)　(d)　(e)

Fig. 2　Morphology and topography of the small blocky crater inspected by the rover. (a) Large amounts of fragments are visible within and around the small crater. The crater exhibits concentric walls, and high-reflectance materials are visible in the central pit. North is up. (b) High-resolution DTM constructed using the SfS method for the small crater shown in Fig. 2b. This crater is ~2.2m in diameter, and the rim-to-floor depth is ~35cm. (c) Shade relief map built for the small crater and the surrounding areas. The base DTM is built using the SfS method, and the crater pointed by the yellow arrow in Figure 1c is not covered. Denoted numbers are the navigation points. (d) Detailed morphology of the dusty fragments. This frame is an enlarged view for the polygonal areas marked in panel (a). (e) Topographic profiles for the small crater show the larger elevation at the northwestern rim. Locations of the two profiles are denoted in panel (c)

measurement (Fig. S2 in the supporting information). The grid size of the regolith models is set to 1cm, and the time window is set to 200ns. The transmitted electromagnetic waves in the models are identical with that transmitted by the high-frequency LPR (Fang et al., 2014). In the regolith models, the depth profile for the relative permittivity of the background regolith follows the method used by Lai et al. (2019). The finite-difference time-domain (Teixeira et al., 1998; Irving & Knight, 2006) method is employed for the numerical simulationwhich performs differential iteration calculations of the Maxwell's equations to simulate the propagation electromag netic waves in media.

3 Results

Unlike typical bowl - shaped primary craters, the target crater exhibits concentric walls and a central pit (Fig. 2) . Abundant fragments are visible both within and around this crater, but surfaces beyond ~1 m from the crater rim exhibit few fragments (Fig. 1c and 2c) . Most of the fragments have angular rims and they are blocky or platy in shapes (Fig. 2d) . The fragments are earthy in color due perhaps to the heavy dust cover (Fig. 2d) . It cannot be determined from the appearance alone that whether or not the fragments are crystalline rocks, sintered breccias, and/or just regolith clumps. The fragments are less than 10cm in the longest dimension, and most of them are less than 5cm. The areal fraction of recognizable fragments in the DTM is ~10% (Fig. S2b in the supporting information) . Note that the actual areal fraction of the fragments should be slightly larger, considering that topographic construction is imperfect in some areas (Fig. 2d) and fragments less than 5mm in dimension were not included.

3.1 Radargram Beneath the Blocky Ejecta and the Surrounding Regolith

From the navigation points 802 to 1105, the top ~250 ns of the radargram, which corresponds to ~20m thick materials in the subsurface can be divided into two layers in terms of echo strengths (Fig. 3) . A relatively sharp and uneven interface is visible at depths of ~10 — 13m in the radargram. Shallower materials have weak echoes and few scatters, and deeper materials exhibit overall stronger echoes (Fig. 3) . This pattern is generally consistent with that observed by Lai et al. (2019) and Li et al. (2020) for the LPR data obtained in the first five lunar days, supporting the interpretation that the top ~11m thick materials in the landing area are similar to fine - grained lunar regolith, and deeper materials are blockier.

Within the first layer of the radargram, materials beneath the blocky ejecta (i. e. , navigation points 904 — 905 and 906 — 911) are indistinguishable from the surrounding normal surfaces where fragments are few (Fig. 3) . Therefore, the relative permittivity of materials beneath the blocky ejecta is comparable with that of the surrounding normal regolith. The results indicate that either the subsurface materials along the entire rover path are composed of relative homogenous regolith - like materials (Lai et al. , 2019; Li et al. , 2020) or that small fragments might uniformly exist within the excavation depth of the small crater, but the small sizes of the fragments prohibited strong radar reflections.

3.2 Simulated Radargrams

Regolith models that have the same horizontal and vertical dimensions with the LPR segments for the blocky ejecta are constructed, that is, 6 × 15m. Referring to the subsurface structures revealed in the top ~200ns of the radargram (Fig. 3), the regolith models assumed different thicknesses of blocky materials in the top 12 - m - thick regolith, which is overlain on top of blocky materials at larger depths. The first model (Fig. 4a) assumed that the entire upper layer is fragment - free regolith; that is, the simulated materials have grain - sizes less than the grid size of the model (1cm). Considering that the excavation depth of the target crater is less than the apparent rim - to - floor depth (~35cm; Fig. 2e), the second to fourth regolith models assumed that the top ~30cm of the first layer is composed of fragment—rich materials (Fig. 4b—4d). In the three models, fragments in the shallow subsurface were generated with random shapes. Sizes of the modeled fragments follow a random distribution, and the longest dimensions are from 1—10cm, which are similar to the size - range of the observed fragments (Fig. S2 in the supporting information). The modeled fragments are assumed to have a total areal fraction of ~10% (Fig. 4b), 30% (Fig. 4c), and 60% (Fig. 4d), respectively. To account for the possibility that the fragmental layer is thicker than ~0.3 m, an additional regolith model is set up with a fragmental layer of 1m thick and the rest of the model design is the same with that shown in Fig. 4b. In all the models, the depth profile of relative permittivity is derived based on hyperbolic reflectors observed in the subsurface (Lai et al., 2019), and the fragments were assumed as basaltic in composition that have a typical permittivity of 7.5 (Chung et al., 1970). At depths larger than the fragment layer,

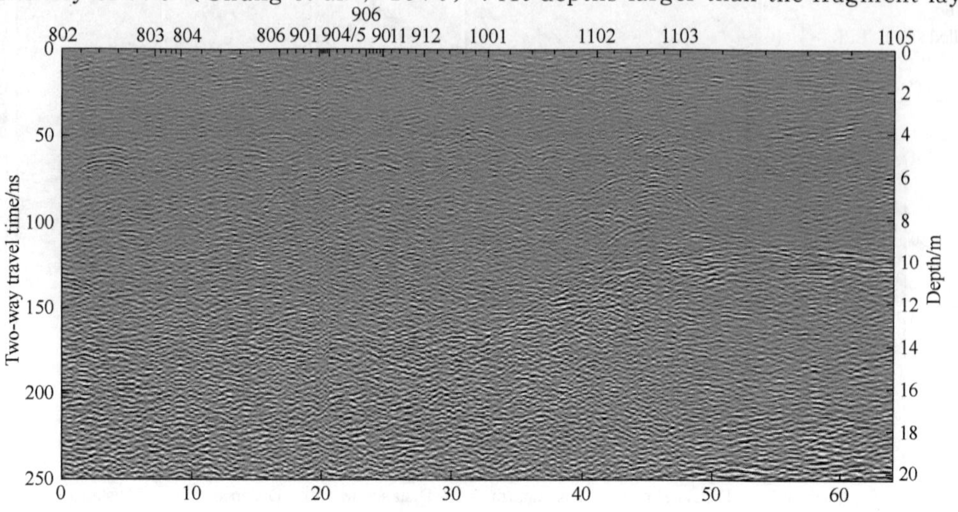

Figure 3　Radargram derived by the high - frequency LPR from the eighth to eleventh lunar day. The top ~10—13m thick materials have weaker echo strengths compared to deeper materials. Numbers at the upper x axis are the navigation points (red ticks). The sections at navigation points 904—905 and 906—911 are for the blocky ejecta (Fig. 2c), and the radar echoes are basically the same with those of the surrounding normal regolith. The lower x axis shows the distance sampled by the LPR, which is started from the navigation point 802. The left y axis shows the two - way travel time, and the right y axis denotes depths beneath ground

an irregularly shaped artificial block that is 10cm in the longest dimension is implanted as a tracer (Fig. 4) . This tracer is to produce a hyperbolic reflector in the radargram, and the echo strengths are compared to those formed by the fragment layers.

The observed radargram beneath the blocky ejecta deposits (Fig. 4f) is most similar with the first modeled radargram (Fig. 4g), which assumed that the first 12m was composed by fragment - free regolith. A fragment layer that has different assumed thicknesses and/or abundances of rock fragments would uniformly cause strong echoes at depths < ~50ns, which are comparable with those reflected from the buried artificial fragments (Fig. 4g 4k) . This comparison suggests that most materials beneath the blocky ejecta have grain sizes less than 1cm, and the observed fragments were not preexisting in the subsurface. The results are consistent with both the orbital observations at different wavelengths (Fig. 1b and S1) and the observations by the Yutu - 2 panoramic cameras (Fig. 1c) .

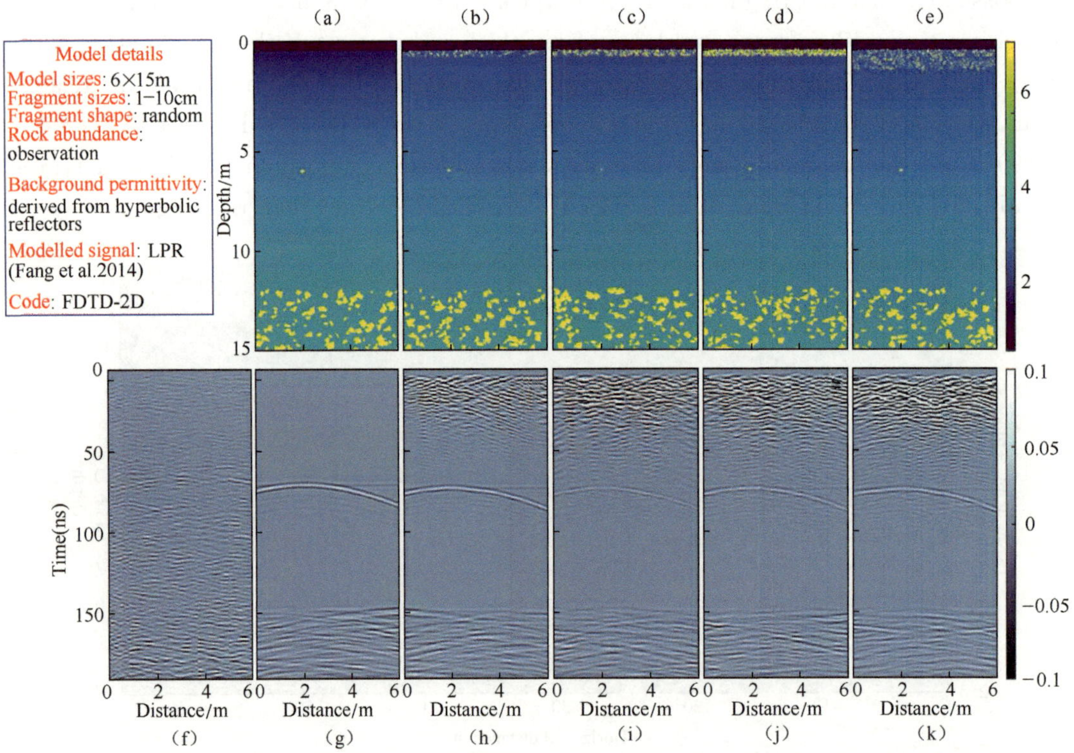

Fig. 4 Comparison of the observed radargram between the navigation points 904 and 911 and the simulated radargrams. The first row (a - e) shows the model setups, and the color bar denotes the relative permittivity used in the model. Below each model, the simulated radargrams (g - k) are listed, together with the observed radargram (f) . The scale bar shows the normalized echo strengths against the peak of the transmitted signal. The comparison suggests that subsurface materials beneath the blocky ejecta should be similar with fragment - free regolith

4 Discussion

4.1 Origin of the Small Fresh Craters at the Landing Region

The small fresh crater shown in Fig. 2 is not the only one that was encountered by the rover. From Days 6 to 12, at least four such craters are observed along the ~150m long path, and more are visible at larger distances from the path (Fig. S3 in the supporting information). The high spatial density of similar - sized fresh craters in such a small region is not expected for primary craters, considering that the fast overturn rate of surface regolith (Speyerer et al., 2016) and the quick breakdown of small fragments on the lunar surface (McKay et al., 1991) would form continuous transition in the degradation states of similar - sized primary craters. Fig. S4 shows the normal appearance of the heavily degraded crater population in this area.

The rover is located in a faint impact ray sourced from the ~3.8km diameter Zhinyu crater, which is located to the west of the landing site (Fig. 1a). Secondaries are frequently visible in impact rays on planetary bodies, and small impact craters on the Moon are efficient enough to deliver large amounts of ejecta to distances that are hundreds of radii of the parent crater (Speyerer et al., 2016). While the small craters shown in Fig. 2 and S3 cannot be resolved from orbit due to their small sizes (Fig. 1b), larger secondaries are observed in this faint ray of Zhinyu (Fig. S5). The high - resolution DTM shows that the northwestern rim of the small crater is higher in elevation than the rest of the crater rim (Fig. 2). The observations together suggest that the small blocky craters are most likely secondaries formed by the Zhinyu crater, so that enhanced ejection toward the downrange has caused the larger elevation at the uprange rim of secondaries (Schultz & Gault, 1985).

4.2 Origin of the Observed Fragments

The small crater shown in Fig. 2 is ~31 km from the center of the Zhinyu crater (Fig. 1). Assuming an ejection angle of ~45°, the impact velocity that formed the small crater is ~140 m/s. With such small impact velocities, cratering in porous target is mainly driven by the momentum instead of kinetic energy of the impactor (Schultz & Gault, 1985). The deep pit in the floor of the small crater (Fig. 2b) suggests that pene tration by projectiles was profound during the cratering process. Based on the different morphological aspects of lunar secondaries (e.g., asymmetric rims and central mounds), laboratory simulations of low - velocity impacts suggested that typical secondaries on the Moon are formed by clusters of fragments instead of single fragments, and most materials excavated by secondaries were local target materials, not the primary ejecta (Schultz & Gault, 1985). The ~0.15 depth - to - diameter ratio of the small crater (Fig. 2e) indicates that the projectile delivered from Zhinyu should be highly clustered or single fragment (s), see Figure 13 in Schultz and Gault (1985) for a comparison. Therefore, the shattered projectile (s) that formed the secondaries are a minor portion of the observed fragments (Schultz & Gault, 1985) .

During the formation of secondaries, compaction and then ejection of porous surface regolith could be an important source for the observed fragments. On the Moon, lithification of regolith occurs at a wide range of impact velocities due to dynamic loads (Ahrens & Cole, 1974; Kieffer, 1975) . At peak pressures larger than ~10 GPa, shock lithification occurs via melting, forming thermally sintered regolith breccia (Spray, 2016). At shock pressures less than ~5 GPa, lithification via melting in lunar regolith is weak, and the major effects of low shock are the collapse of pores and mechanical shortening of grains (Kieffer, 1975) . The impacts that formed the secondaries at the landing region were in the subsonic domain, so the peak pressure was restricted below the melting level (Kieffer, 1975), and shock lithification via melting should be very limited, if not at all. On the other hand, physical experiments show that porous materials exhibit inelastic behavior at impact velocities as low as ~20 m/s, and lithification via pore collapse is profound (Scott & Pearce, 1975) . Therefore, it is possible that most of the fragments might be formed by impact compaction during the landing of Zhinyu's ejecta, accounting for the earthy appearances (Fig. 2d) .

The interpretation is in line with previous interpretations that the top 10—13 m thick materials at the landing area are likely regolith (Lai et al. , 2019; Li et al. , 2020), negating the scenario that there might be a recent uniform deposition of fragments across the landing area. The visible - to - near - infrared spectrometer onboard the Yutu - 2 rover has acquired reflectance spectra around this crater. The proposed origin of the fragments could be tested and intergraded in the compositional analyses for the spectral data, for example, most fragments should have exhibit similar spectra with the surrounding regolith. On the other hand, the ubiquitous occurrences of secondaries on the Moon suggest that compaction caused by secondary impact might have been common in lunar history, hinting that secondary impacts might have been an efficient process to form fragments from lunar regolith, increasing the local porosity (Short, 1966) .

5 Conclusions

Several blocky craters are observed along the path of the Yutu - 2 rover. The rover performed a detailed observation for a blocky crater at the ninth lunar day, permitting detailed morphological, geometrical, and radar scattering studies for the fragments. Radargram obtained by the high - frequency lunar penetrating radar across the blocky ejecta are indistinguishable from the surrounding normal surfaces that are depleted with surface fragments. Numerical simulations for the propagation of radar waves in high -

resolution regolith models suggest that the fragments were not existing in the preimpact subsurface. The blocky craters are located in an impact ray from the Zhinyu crater, and the rims toward Zhinyu are more elevated. These observations are consistent with the high spatial density of this crater population, suggesting that the fresh craters are secondaries of Zhinyu. The fragments are likely dominated by impact - lithified regolith clumps, with additional contribution from the shattered primary ejecta that formed the secondaries. The origin of the fragments is indicative to the interpretation of both the regional stratigraphy and provenances of surface materials detected by the mission.

Acknowledgments

We thank the efforts of the engineering team of the Chang'E - 4 mission. The two anonymous reviewers provided helpful comments to improve the manuscript. The LRO data reported in this study are available from PDS. Data obtained by the LPR and panoramic camera in the first 2 lunar days have been released (http: //moon. bao. ac. cn/), and those used in this study will be released to the same archive after 12 months ofinternal investigation. The DTM constructed using the SfS method, the processed radar data, and the regolith models are available at http: //doi. org/10. 5281/zenodo. 3706469. The FDTD - 2D simulations follow the method of Irving and Knight (20062006). The authors are supported by the B - type Strategic Priority Program of the Chinese Academy of Sciences (XDB41000000), the National Natural Science Foundation of China (41773063, 41525015, and 41830214), the Science and Technology Development Fund of Macau (0042/2018/A2), and the Pre - research Project on Civil Aerospace Technologies (D020101) of CNSA.

References

[1] Ahrens, T., & Cole, D. M. (1974). Shock compression and adiabatic release of lunar fines from Apollo 17. Paper presented at the Lunar and Planetary Science Conference Proceedings.

[2] Chung, D. H., Westphal, W. B., & Simmons, G. (1970). Dielectric properties of Apollo 11 lunar samples and their comparison with Earth materials. Journal of Geophysical Research, 75 (32), 6524 - 6531. https://doi. org/10. 1029/JB075i032p06524.

[3] Di, K., Liu, Z., Liu, B., Wang, W., Peng, M., Wang, Y., et al. (2019). Chang'e - 4 lander localization based on multi - source data. Journal of Remote Sensing, 23 (1), 181 - 184. https://doi. org/10. 11834/jrs. 20199015.

[4] Fang, G., Zhou, B., Ji, Y., Zhang, Q., Shen, S., Li, Y., et al. (2014). Lunar penetrating radar onboard the Chang'e - 3 mission. Research in Astronomy and Astrophysics, 14 (12), 1607 - 1622. https://doi. org/10. 1088/1674 - 4527/14/

12/009.

[5] Feng, J., Su, Y., Ding, C., Xing, S., Dai, S., & Zou, Y. (2017). Dielectric properties estimation of the lunar regolith at CE-3 landing site using lunar penetrating radar data. Icarus, 284, 424-430. https://doi.org/10.1016/j.icarus.2016.12.005.

[6] Horn, B. K. P. (1990). Height and gradient from shading. International Journal of Computer Vision, 5 (1), 37-75. https://doi.org/10.1007/bf00056771.

[7] Huang, J., Xiao, Z., Flahaut, J., Martinot, M., Head, J., Xiao, X., et al. (2018). Geological characteristics of Von Kármán crater, northwestern south pole-Aitken Basin: Chang'E-4 landing site region. Journal of Geophysical Research: Planets, 123, 1684-1700. https://doi.org/10.1029/2018JE005577.

[8] Irving, J., & Knight, R. (2006). Numerical modeling of ground-penetrating radar in 2-D using MATLAB. Computers & Geosciences, 32 (9), 1247-1258. https://doi.org/10.1016/j.cageo.2005.11.006.

[9] Kieffer, S. W. (1975). From regolith to rock by shock. The Moon, 13 (1-3), 301-320. https://doi.org/10.1007/bf00567522.

[10] Lai, J., Xu, Y., Zhang, X., Xiao, L., Yan, Q., Meng, X., et al. (2019). Comparison of dielectric properties and structure of lunar regolith at Chang'e-3 and Chang'e-4 landing sites revealed by ground penetrating radar. Geophysical Research Letters, 46, 12783-12793. https://doi.org/10.1029/2019GL084458.

[11] Li, C., Liu, D., Liu, B., Ren, X., Liu, J., He, Z., et al. (2019). Chang'e-4 initial spectroscopic identification of lunar far-side mantle-derived materials. Nature, 569 (7756), 378-382. https://doi.org/10.1038/s41586-019-1189-0.

[12] Li, C., Su, Y., Pettinelli, E., Xing, S., Ding, C., Liu, J., et al. (2020). The Moon's farside shallow subsurface structure unveiled by Chang'e-4 lunar penetrating radar. Science Advances, 6 (9), eaay6898. https://doi.org/10.1126/sciadv.aay6898.

[13] Lin, H., He, Z., Yang, W., Lin, Y., Xu, R., Zhang, C., et al. (2019). Olivine-norite rock detected by the lunar rover Yutu-2 likely crystallized from the SPA impact melt pool. National Science Review. https://doi.org/10.1093/nsr/nwz183.

[14] Ling, Z., Qiao, L., Liu, C., Cao, H., Bi, X., Lu, X., et al. (2019). Composition, mineralogy and chronology of mare basalts and non-mare materials in Von Kármán crater: Landing site of the Chang'e-4 mission. Planetary and Space Science, 179, 104, 741. https://doi.org/10.1016/j.pss.2019.104741.

[15] Liu, W., & Wu, B. (2020). An integrated photogrammetric and photoclinometric approach for illumination-invariant pixel-resolution 3D mapping of the lunar surface. ISPRS Journal of Photogrammetry and Remote Sensing, 159, 153-168. https://doi.org/10.1016/j.isprsjprs.2019.11.017.

[16] McEwen, A., & Robinson, M. (1997). Mapping of the Moon by Clementine. Advances in Space Research, 19 (10), 1523 - 1533. https://doi.org/10.1016/S0273-1177 (97) 00365-7.

[17] McKay, D. S., Heiken, G., Basu, A., Blanford, G., Simon, S., Reedy, R., et al. (1991). The lunar regolith. In G. H. Heiken, D. T. Vaniman, & B. M. French (Eds.), Lunar sourcebook: A user's guide to the Moon, (pp. 285 - 356). Cambridge: Cambridge University Press.

[18] Robinson, M. S., Brylow, S. M., Tschimmel, M., Humm, D., Lawrence, S. J., Thomas, P. C., et al. (2010). Lunar Reconnaissance Orbiter Camera (LROC) instrument overview. Space Science Reviews, 150, 81 - 124.

[19] Scholten, F., Oberst, J., Matz, K., Roatsch, T., Wählisch, M., Speyerer, E., & Robinson, M. (2012). GLD100: The near-global lunar 100m raster DTM from LROC WAC stereo image data. Journal of Geophysical Research, 117, E00H17. https://doi.org/10.1029/2011JE003926.

[20] Schultz, P., & Gault, D. E. (1985). Clustered impacts: Experiments and implications. Journal of Geophysical Research, 90 (B5), 3701 - 3732. https://doi.org/10.1029/JB090iB05p03701.

[21] Scott, R., & Pearce, R. W. (1975). Soil compaction by impact. Géotechnique, 25 (1), 19 - 30. https://doi.org/10.1680/geot.1975.25.1.19.

[22] Short, N. M. (1966). Shock-lithification of unconsolidated rock materials. Science, 154 (3747), 382 - 384. https://doi.org/10.1126/science.154.3747.382.

[23] Speyerer, E., Povilaitis, R., Robinson, M., Thomas, P., & Wagner, R. (2016). Quantifying crater production and regolith overturn on the Moon with temporal imaging. Nature, 538 (7624), 215 - 218. https://doi.org/10.1038/nature19829.

[24] Spray, J. G. (2016). Lithification mechanisms for planetary regoliths: The glue that binds. Annual Review of Earth and Planetary Sciences, 44 (1), 139 - 174. https://doi.org/10.1146/annurev-060115-012203.

[25] Su, Y., Fang, G., Feng, J., Xing, S., Ji, Y., Zhou, B., et al. (2014). Data processing and initial results of Chang'e-3 lunar penetrating radar. Research in Astronomy and Astrophysics, 14 (12), 1623 - 1632. https://doi.org/10.1088/1674-4527/14/12/010.

[26] Teixeira, F., Weng, C., Straka, M., Oristaglio, M., & Wang, T. (1998). Finite-difference time-domain simulation of ground penetrating radaron dispersive, inhomogeneous, and conductive soils. IEEE Transactions on Geoscience and Remote Sensing, 36 (6), 1928 - 1937. https://doi.org/10.1109/36.729364.

[27] Wu, B., Liu, W. C., Grumpe, A., & Wöhler, C. (2018). Construction of pixel-level resolution DEMs from monocular images by shape and albedo from shading constrained with low-resolution DEM. ISPRS Journal of Photogrammetry

and Remote Sensing，140，3 - 19. https://doi. org/10. 1016/j. isprsjprs. 2017. 03. 007.

[28] Wu，W.，Li，C.，Zuo，W.，Zhang，H.，Liu，J.，Wen，W.，et al.（2019）. Lunar farside to be explored by Chang'e - 4. Nature Geoscience，12（4），222 - 223. https://doi. org/10. 1038/s41561 - 019 - 0341 - 7.

Rock Fragments in Shallow Lunar Regolith: Constraints by the Lunar Penetrating Radar Onboard the Chang'E－4 Mission

Chunyu Ding[1], Zhiyong Xiao[2,3], Bo Wu[4], Zhaojin Li[4], Yan Su[5,6], Bin Zhou[7,8], Kaijun Liu[1], and Jun Cui[2,3]

1. Department of Earth and Space Sciences, Southern University of Science and Technology, Shenzhen, China
2. Planetary Environmental and Astrobiological Research Laboratory, School of Atmospheric Sciences, Sun Yat－sen University, Zhuhai, China
3. Centre for Excellence in Comparative Planetology, Chinese Academy of Science, Hefei, China
4. Department of Land Surveying and Geo－Informatics, Planetary Remote Sensing Laboratory, The Hong Kong Polytechnic University, Hong Kong, China
5. Key Laboratory of Lunar and Deep Space Exploration, National Astronomical Observatories, Chinese Academy of Sciences, Beijing, China
6. University of Chinese Academy of Sciences, Beijing, China
7. Key Laboratory of Electromagnetic Radiation and Detection Technology, Chinese Academy of Sciences, Beijing, China
8. Aerospace Information Research Institute, Chinese Academy of Sciences, Beijing, China

Abstract: It is generally believed that the top centimeters of mature lunar regolith should be dominated by fine particles that contain few rocky fragments, but surface fragments are occasionally visible in mature lunar regolith. Based on amplitudes of surface echoes received by the Lunar Penetrating Radar onboard the Yutu－2 rover of the Chang'E－4 mission, we extracted regions with abnormally high relative permittivity that is larger than 4, quantifying the abundance of dense fragments within the top ~4 cm of the mature regolith along the rover path. We carried out high－resolution numerical simulations for the propagation of electromagnetic waves, showing that the abnormally high relative permittivity is not caused by local topography but due to the existence of dense materials. Using high－resolution images obtained by the panorama camera onboard the rover, we performed correlated observation for areas with abnormally high relative permittivity, confirming that some of these dense materials are exposed rocky fragments. However, most of the high relative permittivity areas are not correlated with surface fragments, and some are located beneath floors of highly degraded small impact craters. We found that regolith clumps formed due to impact compaction are not dense enough to cause the abnormally large relative permittivity. Alternatively, the detected dense materials may be melt - bearing impact breccias in the shallow regolith, which were formed by the host craters. The results revealed a highly heterogeneous structure at shallow depths of mature lunar regolith, revealing a new mechanism of sand - blasting impacts on the destruction of pores in mature lunar regolith.

Plain Language Summary: Interior structures of lunar regolith record important information about the surface dynamics on the Moon. It has been noticed that vertical and lateral heterogeneities widely exist in lunar regolith. However, the possible abundance of rocky fragments in mature regolith (e. g. , those in the top few centimeters) is not well constrained. Resolving this issue is important for the understanding of regolith formation. The Chang'E - 4 mission provides an excellent opportunity to study the interior structures of mature lunar regolith. The landing area is covered by ~12 - m - thick regolith, and the surface regolith has a high optical maturity index with few visible surface fragments. In this work, we detect the abundance of rock materials within a depth of ~4 cm in the mature regolith along the path of Yutu - 2, based on a combination of photogeology observation and numerical modeling. We found that the spatial distribution of rock materials in the shallow regolith is highly heterogenous, and >1/4 of the detected dense materials are located beneath small and degraded impact craters, where no fragments are visible on the surface. The results indicate that by producing melt - welded impact breccia, small impacts formed in mature regolith could substantially reduce the porosity of local regolith.

Key Points:

- Relative permittivity (ε) of the top ~4 - cm - thick mature regolith is derived along the Yutu - 2 path at the Chang'E - 4 landing area
- Some high permittivity areas along the path of Yutu - 2 are linked to exposed boulders while others probably feature buried rock fragments
- Sand - blasting impacts on the Moon can decrease the porosity of lunar regolith by compaction and melt welding

1 Introduction

The surface of the Moon is almost entirely covered by a layer of fine - grained debris (i. e. , regolith) that is formed by repeated impact cratering and space weathering (McKay et al. , 1991) . Regolith grains are mostly originated from local bedrocks, with a minor contribution from distal ejecta (McKay et al. , 1991) . Estimated regolith thickness at the lunar highland is ~10m and that at lunar mare is ~5 m on average (Shkuratov &

Bondarenko, 2001), both are much larger than detection depths of frequently used remote sensing techniques that are targeted on the lunar surface. Most of our knowledge about lunar geology was acquired by remote sensing, therefore regolith has been the direct observation target and thus the key information carrier.

Interior structures of lunar regolith record the history of surface dynamics on the Moon. Early observations reveal that older surfaces received more impacts, hosting thicker regolith and fewer interior rock fragments (Bandfield et al., 2011; Campbell et al., 2007; Oberbeck & Quaide, 1967), and regolith at larger depths con tains more and larger rock fragments than shallower regolith (Bart & Melosh, 2010; McKay et al., 1991). While traditional models for lunar subsurface structures usually include a surface regolith layer that overlies competent bedrocks, the interior structure of regolith is more complicated due to the long and chaotic reworking history during formation. Within a given geological unit, both the thickness and abundance of interior rock fragments exhibit large variations across short distances (Wilcox et al., 2005). A consensus has recently emerged that while the contact between regolith and deeper bedrocks is transitional rather than abrupt, the interior structure of regolith can be sketched as a gradually coarsening sequence of particles toward depths, which are interspersed with larger boulders (Wilcox et al., 2005).

The grain size - frequency distribution at different depths of lunar regolith is directly related with the impact bombardment history on the surface. This information is indicative to interpretations of provenances of regolith materials (Y. - H. Huang, Minton, et al., 2018; L. Li & Mustard, 2003). Thick regolith developed in old geological units has a high optical maturity caused by space weathering (Noble et al., 2001). It is generally believed that the top - most regolith materials are dominated by fine - grained particles that contain few rocky fragments (McKay et al., 1991). At Apollo and Luna landing regions that were emplaced >3.6 Ga, the areal percentage of surface rock fragments in mature regolith is usually less than 1%, such as that at the Apollo 11 (~0.28%; B. Li et al., 2017), Apollo 16 (0.11%; B. Li et al., 2017), and Apollo 17 (~1%; Owen, 1973) landing sites. Younger surfaces are naturally covered by a larger percent of rock fragments, such as the Apollo 15 (~1%; Swann et al., 1972) and Chang'E - 3 (~1.2%; Di et al., 2016; Y. Li & Wu, 2018) landing regions. However, regarding the entire thickness of a regolith layer, the particle size - frequency distribution has only been established for the few core samples returned by Apollo (McKay et al., 1991). While heterogeneities in regolith commonly exist among different locations, the particle size - frequency distribution at different depths of lunar regolith should be investigated across larger areas.

The Chang'E - 3 mission performed the first extraterrestrial experiment of ground penetrating radar (Fang et al., 2014), and radargrams obtained for the ejecta deposits of a recently formed small crater indeed have revealed abundant subsurface boulders (e. g.,

Fa et al. , 2015) and structural heterogeneities (e. g. , Ding, Xiao, Su, & Cui, 2021). The Chang'E-4's mission is performing the first surface exploration at the lunar far side (W. Wu et al. , 2019) . Featuring a model age of ~3. 63. 7 Ga (J. Huang, Xiao, et al. , 2018; Ling et al. , 2019), the landing area is covered by a thick and mature regolith that is ~12 m thick. The lunar penetrating radar (i. e. , LPR) onboard the Chang'E-4 rover (Yutu-2) has been scanning the interior structure of the subsurface (C. Li et al. , 2020; Xiao et al. , 2021) . The high-frequency channel of the LPR is capable of revealing subsurface materials with a resolution of ~0. 3 m in typical regolith (Su et al. , 2014). Based on hyperbolic echo patterns revealed in the radargram, dozens of competent rocks have been discovered in this regolith layer (Lai et al. , 2019; C. Li et al. , 2020; Xiao et al. , 2021), which are much fewer than those in the regolith-like ejecta deposits at the Chang'E-3 landing site (Ding, Cai, et al. , 2020; Fa et al. , 2015) . However, no subsurface boulders shallower than ~2 m have been detected in the regolith (Xiao et al. , 2021) . The top ~1 m of the regolith was revealed to be free of fragments that are larger than 1 cm, except for the limited occurrence of regolith clumps that were excavated by fresh and small impact craters (Ding, Xiao, et al. , 2020; Gou et al. , 2021; Lin et al. , 2020) . Nevertheless, rock fragments on the surface have been occasionally observed by the panorama camera onboard the Yutu-2 rover (B. Wu et al. , 2021), but the possible existence and distribution of fragments in the shallower regolith (≪1 m) is not constrained.

In this paper, we investigated the distribution of rocky fragments within the <~4-cm-thick regolith along the path of the Yutu-2 rover. The results bridged a gap in the grain size-frequency distribution at different depths of mature lunar regolith. More importantly, the revealed distribution of shallowly buried rock fragments provides new clues about the formation and modification mechanism of lunar regolith. Section 2 introduces the geological context of the Chang'E-4 landing region and data and method used in this study. The derived relative permittivity for the shallow regolith is introduced in Section 3. 1, and the possible effect of surface topography on the estimation of relative permittivity is investigated and then excluded in Section 3. 2. We investigated the surface morphology for regions that have abnormally large relative permittivity (Section 3), revealing that many such locations are correlated with exposed surface boulders, and most others are located on floors of small impact craters. After analyzing the errors that existed in the calculation method (Section 4. 1), we investigated the cause for the abnormally large relative permittivity using numerical simulations (Section 4. 2) . The origin of the buried dense materials is discussed based on their spatial distribution and numerical simulations, and the indications to lunar regolith evolution is discussed in Section 4. 3.

2　Data and Method

2.1　Geological Context of the Chang'E - 4 Landing Site

The Chang'E - 4 landing site is located on a layer of ejecta deposits that cover an Imbrian - aged mare unit (J. Huang, Xiao, et al., 2018; Fig. 1a). Based on the reflectance spectra obtained by both the Yutu - 2 rover and lunar orbiters, surface materials are believed to be dominated by plagioclase - rich materials that contain little mafic minerals (J. Huang et al., 2020). The content of $TiO_2 + FeO$ in this area is ~14 wt. % (Qiao et al., 2019), Suggesting that the Surface material has a relatively small loss tangent, which is verified by results (~0.005) derived from the decay rate of radar waves (Lai et al., 2019; C Li et al., 2020). The high - frequency LPR is capable of detecting subsurface materials up to a depth of ~45m, where 3 - 4 layers that have clearly differentiable echo patterns are recognized (Xiao et al., 2021). The top - most echo layer, including the surficial regolith, is ~12 - m - thick, and it features homogeneous echo patterns as seen in radargrams, suggesting a dominance of fine - grained materials

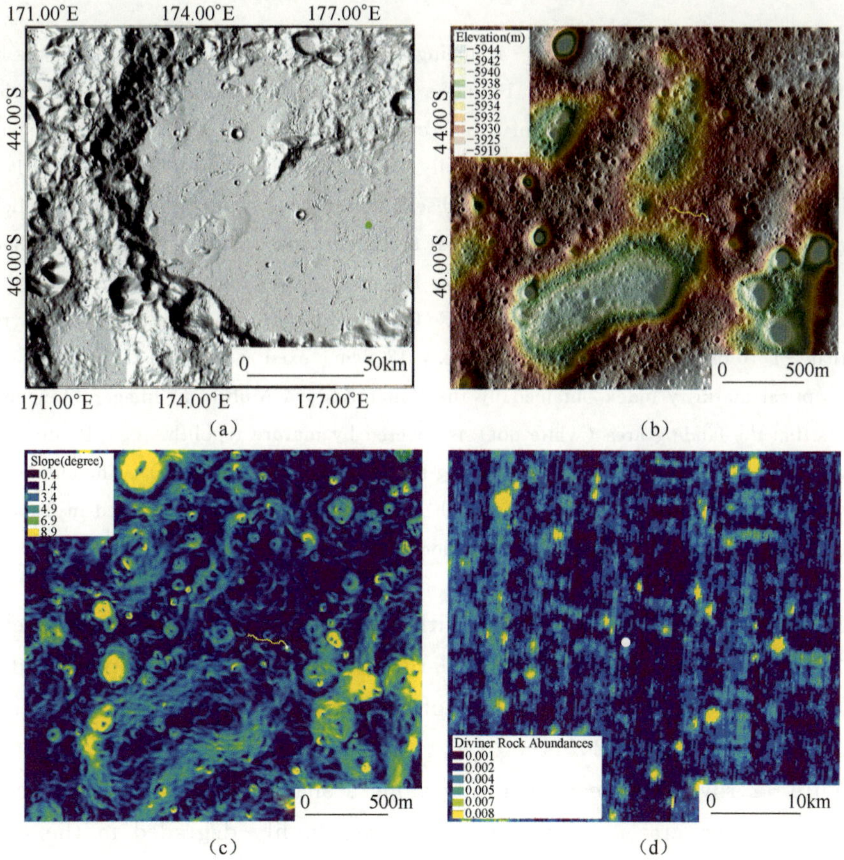

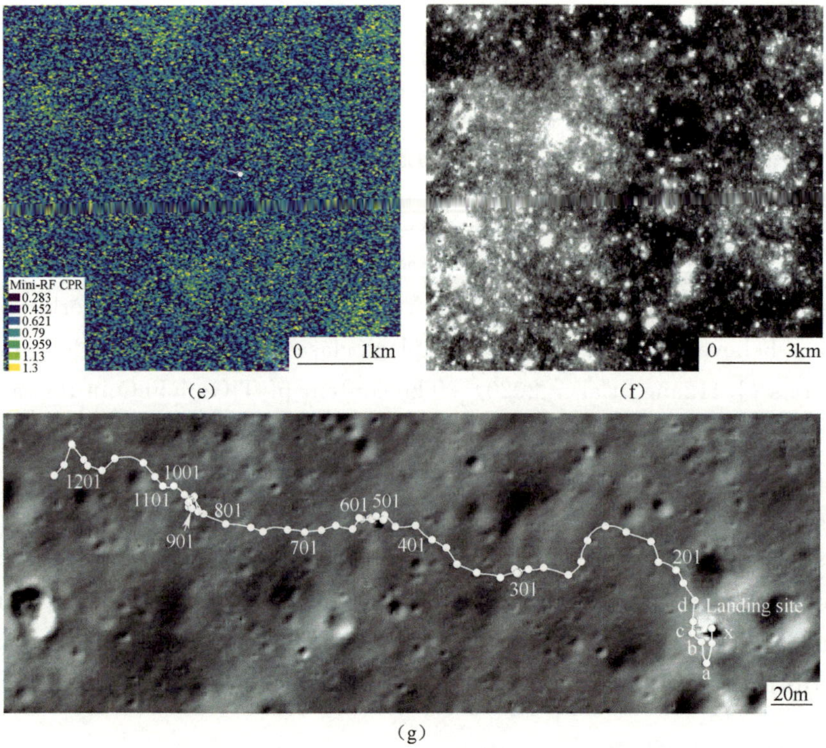

Fig. 1　Geological context of the Chang'E－4 landing area. (a) The lander (green dot) was landed on the floor of the Von Kármán crater. (b) Both the landing site (white star) and rover path (yellow curve) are located on top of a shared rim of two secondaries (Xiao et al., 2021). (c) Distribution of topographic slopes in the landing area shows generally flat surfaces. The white dot denotes the landing site and the yellow curve marks the rover path. (d) Observation by the Diviner radiometer (Bandfield et al., 2011) reveals a low abundance of surface rocks that are larger than 1 m in diameter at the landing site (white dot). (e) The traverse area by the rover has much lower circular polarization ratios (i. e., CPR) than the global average as seen by Mini－RF (Cahill et al., 2014), indicating that few rocks that are larger than ~1/10 of the radar wavelengths (i. e., 12.6cm) exist at both the surface and subsurface. (f) Map of optical maturity index obtained by the Lunar Kaguya Multiband Imager (Lemelin et al., 2019) shows that the landing area (white dot) is covered by mature regolith. (g) Route of Yutu－2 rover accomplished in first 12 lunar days after being released from the lander (x). The numbers denoted the navigation points of Yutu－2 rover. All orbital data used in this study are listed in Tab. S1 of the appendices, together with their IDs and available addresses.

(Ding, Xiao, et al., 2020). Together with the layering structures in the debris layer that were interpreted based on the LPR radargram, the local stratigraphic study showed that the debris cover was mainly composed of crustal materials that were delivered by the Orientale basin (Xiao et al., 2021).

The Yutu－2 rover has been exploring along a shared rim of two secondary craters formed by the Finsen crater (Fig. 1b), which are highly degraded in the topography (Fig. 1c). Seen from orbit, the traverse area of the Yutu－2 rover features a low rock abundance at both the surface (Fig. 1d) and subsurface (Fig. 1e), consistent with the large model age of this landing area (J. Huang, Xiao, et al., 2018) and the low rock abundances observed by the rover (B. Wu et al., 2021).

2.2 Method Used to Detect Dense Fragments in the Shallow Regolith

The Yutu-2 rover has been working for over 29 lunar days since been released from the lander to April 2021. The Lunar Penetrating Radar onboard the rover contains two radar systems, a low-frequency LPR that has a center frequency of 60MHz and a high-frequency LPR that has a center frequency of 500MHz (Fang et al., 2014; C. Li et al., 2020). Signals received by the low-frequency LPR might be affected by coupling waves from the rover body (C. Li et al., 2018), so they are not used here for our purpose. In this study, the level 2B data returned by the high-frequency LPR and images obtained by the panorama camera in the first 12 lunar days (from January 2019 to December 2019) are used, which are now in the public domain (Tab. S1 in the appendices list the IDs and available address for the data used in this study). Su et al. (2014) introduced the detailed process procedure for the LPR data, that is, from raw (level 0B) to the level 2B archive. Images obtained by the panorama camera are used to study the surface morphology for regions of interest (Section 3.2). High-resolution digital terrain models (i.e., DTM) were constructed for surfaces traversed by the rover, and they have pixel scales better than 1cm/pixel (B. Wu et al., 2021). The DTM was used to investigate geometries of encountered topography, such as small craters and surface boulders.

Laboratory measurements of dielectric properties of samples returned from the Moon revealed that the relative permittivity of lunar materials is mainly related with the bulk density (Olhoeft & Strangway, 1975). In this study, the relative permittivity is used to investigate the density distribution in the shallow regolith. The amplitudes of transmitted signals (A_i) and received echoes (A_r) are used to derive the relative permittivity of materials that are located within a maximum depth of 1/8 wavelength (λ) of the electromagnetic wave transmitted (Widess, 1973), that is, ~4 cm in typical lunar regolith by the high-frequency LPR. This method was initially applied for ground penetrating radars on Earth (Benedetto et al., 2017; Huisman et al., 2003), and it has also been employed for the high-frequency LPR onboard the Chang'E-3 mission and future WISDOM radar onboard the ExoMars mission (Dong et al., 2017; Herve et al., 2020).

The reflection coefficient (Γ) is defined as the amplitude ratio (Equation 1) between the reflected echo (A_r) and the incident signal (A_i).

$$\Gamma = \frac{A_r}{A_i} \tag{1}$$

Electromagnetic waves transmitted by the LPR propagate in vacuum before penetrating into the subsurface, so Γ can also be related with the intrinsic impedances of vacuum (η_v)

and lunar regolith (η_r), as shown in Equation 2 (Ulaby & Long, 2014).

$$\Gamma = \frac{\eta_r - \eta_v}{\eta_r + \eta_v} \tag{2}$$

Intrinsic impedances of media (η) is related with its permeability (μ), relative permittivity (ε), angular frequency of the electromagnetic wave (ω), and conductivity of loss materials in which the electromagnetic wave propagates (σ). Equation 3 shows the expression (Ulaby & Long, 2014).

$$\eta = \sqrt{\frac{\mu}{\varepsilon}}\left(1 + \mathrm{j}\frac{\sigma}{2\omega\varepsilon}\right) \tag{3}$$

Typical lunar regolith can be treated as nonmagnetic and low loss materials, and $\sigma/\omega\varepsilon < 10^{-2}$, so that the second term of Equation 3 can be ignored (Carrier et al., 1991; Ulaby & Long, 2014).

In combination Equations 1–3, the surface permittivity (ε_r) can be connected with the relative permittivity of vacuum ($\varepsilon_v = 1$), as rewritten in Equation 4. Note again that ε_r is the relative permittivity of materials within depths of less than $1/8 \times \lambda$.

$$\sqrt{\varepsilon_r} = \frac{\sqrt{\varepsilon_v} - \frac{A_r}{A_i}}{\sqrt{\varepsilon_v} + \frac{A_r}{A_i}} \tag{4}$$

The amplitude of reflected echo (A_r) was recorded in the level 2B data obtained by LPR. The amplitude of transmitted signals (A_i) by the high-frequency LPR was not reported in literatures (Dong et al., 2017; Fang et al., 2014).

Data obtained from ground tests of LPR that were conducted before the launch of the Chang'E-4 mission are used here to derive A_i. The prototype of the Yutu-2 rover preserves the observation geometry of the high-frequency LPR (Fig. 2). The transmitting antenna is 16cm from the receiving antenna (Fig. 2a). The high-frequency LPR is 30cm above the rover foot. The geometry of LPR suggests that the first peak signal received is direct wave that is transmitted by the antenna, since direct transmission is the shortest path compared to propagation path from the transmitting antenna to elsewhere and then to receiving antenna. An air observation was performed to determine the arrival time of the direct wave (~28 ns; black solid curve in Fig. 2c). The second observation was targeted on metallic plates that have a relative permittivity of $\varepsilon_r \approx +\infty$, so immediately following the direct wave, a total reflection was received. A trace of A-scan obtained in this experiment is shown in Fig. 2c (red solid curve). The time delay of the metallic reflection is ~31 ns, and the amplitude is 7,161 (a. u.), equal to A_i. During the operation of LPR on the lunar surface, echoes reflected from the surface are recognized as the second peak in the A-scan, immediately after the direct wave (see Section 4.1 for a discussion about uncertainties in this judgment). Fig. 2d shows the statistics of A_r received along the rover path in the first 12 lunar days, which are much less than

A_i. Fig. 2f - 2h show examples of identifying A_r in three traces of A - scan that were obtained during the operating LPR on the Moon. The above method of deriving the relative permittivity of surficial materials (ε_r) is affected by surface topography (Giannopoulos & Diamanti, 2008), that is, distances and angles of incidence and reflection between the LPR and lunar surface. To quantify the effect of surface topography on the results derived, we performed high - resolution two - dimensional numerical simulations for the propagation (i. e., reflection and transmission) of electromagnetic waves. The effect caused by three - dimensional surface roughness is discussed in Section 4. 1. The transmitted signals in the simulations have identical properties with those transmitted by the high - frequency LPR, that is, a quasi—Ricker wavelet that has a central frequency of 500MHz (Fang et al., 2014). The algorithm developed by Bitri and Grandjean (1998), which is in the frequency domain is employed here for the simulation. The radiation pattern of the modeled transmitted signal is approximated by an equivalent dip - filter, with optimum values of cut - off angles set to 60°—80° (Bitri & Grandjean, 1998). In the 2D and 3D models, the transmitter and receiver are at the same point, so the models did not record direct coupling waves. We build different surface models to test the effect of topography. The models consist of a vacuum layer that is 30— cm - thick from the original line to the simulated lunar surface, equivalent to the height of the high - frequency LPR from the rover foot. The lunar subsurface is modeled as homogenous regolith materials that have a dialectic constant of 3 (i. e., that of typical lunar regolith; Carrier et al., 1991) and a loss tangent of 0. 005 (equivalent to that of regolith at the Chang'E - 4 landing site; C. Li et al., 2020). The surface topography of the modeled regolith layer has different shapes, and they are constructed based on theoretical shapes of different - sized impact craters, stochastic Gaussian roughness, and extracted topography from high - resolution DTM that were constructed using stereo pairs obtained by the panorama camera (Section 3. 2). To investigate the possible cause of abnormally high relative permittivity derived, we further changed the geometry and dialectic properties of the modeled regolith to compare model results against observations (Section 4. 1).

(a)　　(b)

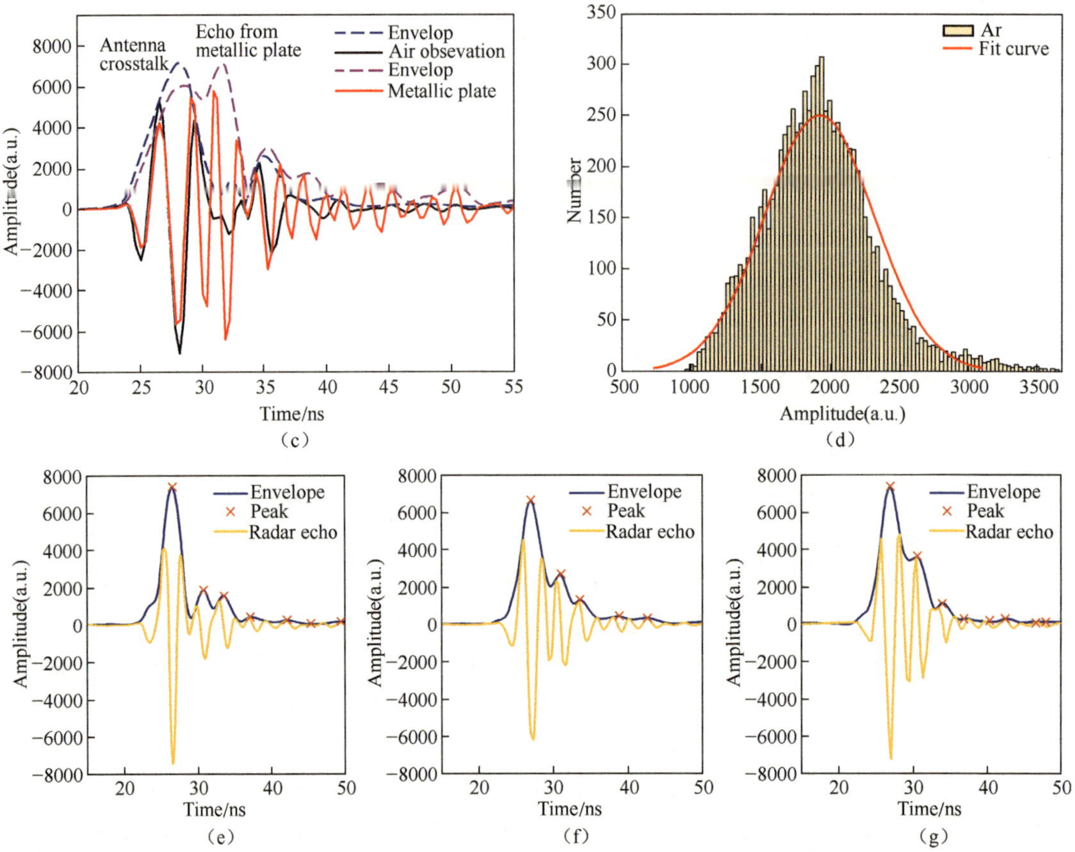

Fig. 2　Methods used to determine the strength of transmitted and received signals by the high - frequency lunar penetrating radar (LPR) during both ground test before flight and operations on lunar surface. (a) Air observation by the high - frequency LPR (green electronic components) was performed to determine the arrival time of direct wave. (b) Metallic plate observation was performed to determine the arrival time of total reflection and the strength of transmitted signals. (c) A - scans observed in ground tests during air and metallic plate observations. (d) Histogram of amplitudes of echo strengths reflected from the lunar surface as received by the high - frequency LPR during operation in the first 12 lunar days. (e - g) Three traces of A - scan are used here to illustrate the process of extracting the amplitude of echoes reflected from lunar surface. The yellow curves show signals in the time - domain, and the blue curves are envelopes of the surface echoes. Red crosses indicate peaks of signals. The first peak is for direct transmission by antenna crosstalk, and the second peak is the amplitude of echo strength reflected from the lunar surface, which are 1877.8, 2685.3, and 3602.5 (a. u.), respectively for the three examples. The amplitude of the incident wave is 7,161 (a. u.), so according to Equation 4, the estimated relative permittivity of surface permittivity is 2.93, 4.84, and 9.15, respectively, for the three A - scans

3　Results

3.1　Relative Permittivity of Shallow Regolith Along the Route of Yutu - 2

Eight thousand, five hundred seventy - three traces of A - scan were obtained by the high - frequency LPR during operation in the first 12 lunar days, over a rover path of

~326m long (Fig. 1g) . The relative permittivity derived for each trace of the A - scan (ε_r) is shown in Fig. 3a. The largest ε_r derived is 9. 47, and the smallest is 1. 74. 89. 49% of the derived ε_r is less than 4%, and 52. 49% is less than 3 (Fig. 3b) . This result is generally comparable with the relative permittivity of regolith samples returned from the Moon (Carrier et al. , 1991), also consistent with the top ~4 - cm - thick materials being dominated by typical regolith (Ding, Xiao, et al. , 2020), the low abundance of rocks

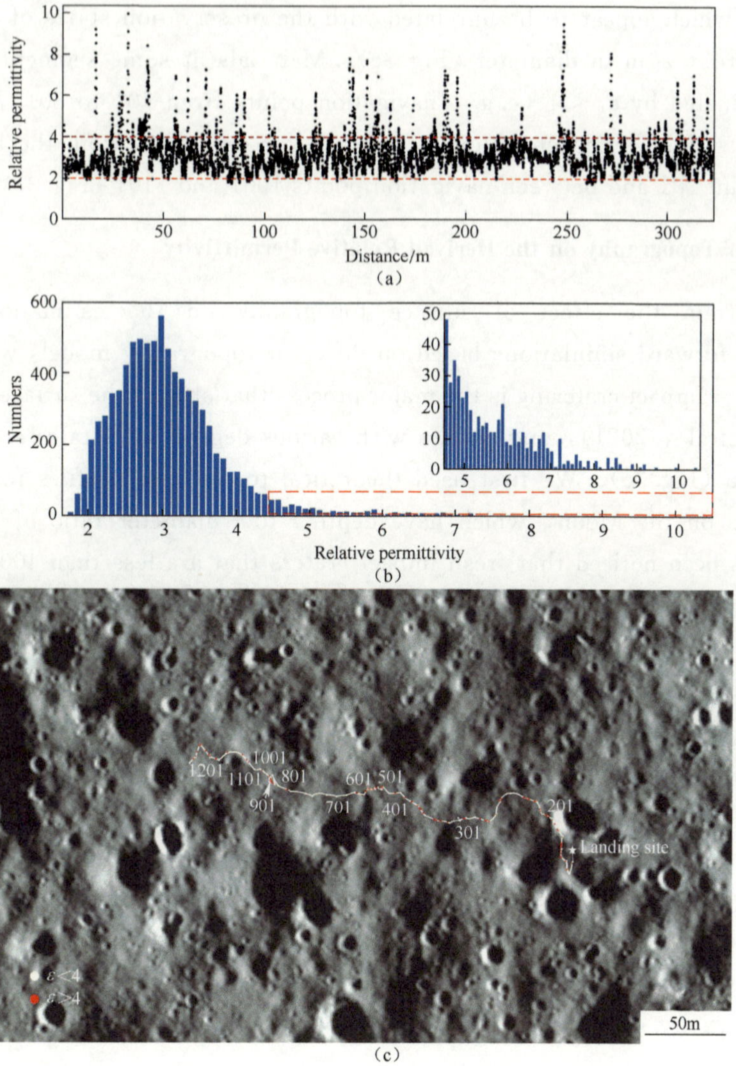

Fig. 3　Relative permittivity derived for the top - most (~4 cm thick) regolith along the path of Yutu - 2 in the first 12 lunar days. (a) Distribution of relative permittivity along the 326. 36m long path of the rover. The two dashed red lines are for relative permittivity of 2 and 4, respectively. (b) Histogram of the estimated surface permittivity. The inset figure shows the histogram of relative permittivity that is larger than 4. 5 (red box of the main figure). (c) Distribution of relative permittivity over high - resolution images along the traverse path of the rover. Major navigation points are annotated along the rover path. The white star denotes the location of the lander. White segments are for areas with relative permittivity less than 4, consistent with that of typical lunar regolith. Red segments are locations that have relative permittivity larger than 4, suggesting the existence of dense materials at depths <4cm

revealed by orbital observations (Fig. 1), and low areal percentage of surface rocks observed by the rover (B. Wu et al., 2021). On the other hand, 10.51% of the derived ε_r is larger than 4, and most of them are concentrated in ~93 prominent clusters (i.e., each cluster contains at least three continuous traces of A-scan that have derived ε_r larger than 4).

Projected over high-resolution images that were obtained with large incidence angles (i.e., enhanced shadows and topography), values of ε_r have a highly heterogenous distribution, which appear to be unrelated with the preservation states of nearby impact craters that are ~20m in diameter (Fig. 3c). Materials at some segments of the rover path are dominated by $\varepsilon_r < 4$ (e.g., navigation points from 701 to 801), and those at other segments show a concentration of $\varepsilon_r > 4$, for example, from the landing site to navigation point 201 and between navigation points 1001 and 1101 (Fig. 3c).

3.2 Effect of Topography on the Derived Relative Permittivity

To constrain the effect of surface topography on the estimation of relative permittivity, forward simulations based on different topography models were performed (see method). Impact cratering is the major process that shaped the surface of the landing area (B. Wu et al., 2021), and craters with various degradation states are visible along the rover path (Fig. 3c). We first used theoretical topography profiles for typical fresh impact craters on the Moon, which have depth-to-diameter ratio of 0.2 (Melosh, 1989). It has been noticed that fresh impact craters that are less than 400m in diameter exhibit large variations in the depth-to-diameter ratio (most are less than 0.2) due to the effect of target properties (Mahanti et al., 2017). A depth-to-diameter ratio of 0.2 is selected here to maximize the topography difference for the modeled regolith. The diameters of the modeled craters are set as 0.15 (Fig. 4a), 0.3 (Fig. 4b), 0.45 (Fig. 4c), and 0.6m (Fig. 4d), respectively, so different topographic amplitudes can be modeled. For a reference, the wheel rut of the Yutu-2 rover is 15cm wide, and the rover body is 75cm (ignoring the widths of solar panels). Therefore, the modeled topography should be regarded as extreme cases of surface topography that can be safely traversed by the rover. The modeled results are shown in Fig. 4e-4h. As shown in the models, when the signal-transmitter approaches and leaves craters on the surface, the model-derived relative permittivity exhibit ups and downs, but the derived values do not exceed 3.8 (Fig. 4e-4h).

Angles of incidence and reflection between the transmitted radar waves and local topography are more complicated than the models sketched in Fig. 4. Therefore, we built rougher topography models than those shown in Fig. 4a-4d to complicate the angles of incidence and reflection. The topography profiles shown in Fig. 4i-4l are constructed following stochastic Gaussian roughness, and the standard deviations of the modeled surface elevations are 0.25%, 0.5%, 1%, and 2%, respectively. The maximum topography difference in the model is ~5 cm. For these models, the derived ε_r is also uniformly less than 4 (Fig. 4m - 4p) . In each of the models, the derived ε_r does not strictly follow the surface topography, suggesting complicated reflection patterns caused by highly diverse angles of incidence and reflection.

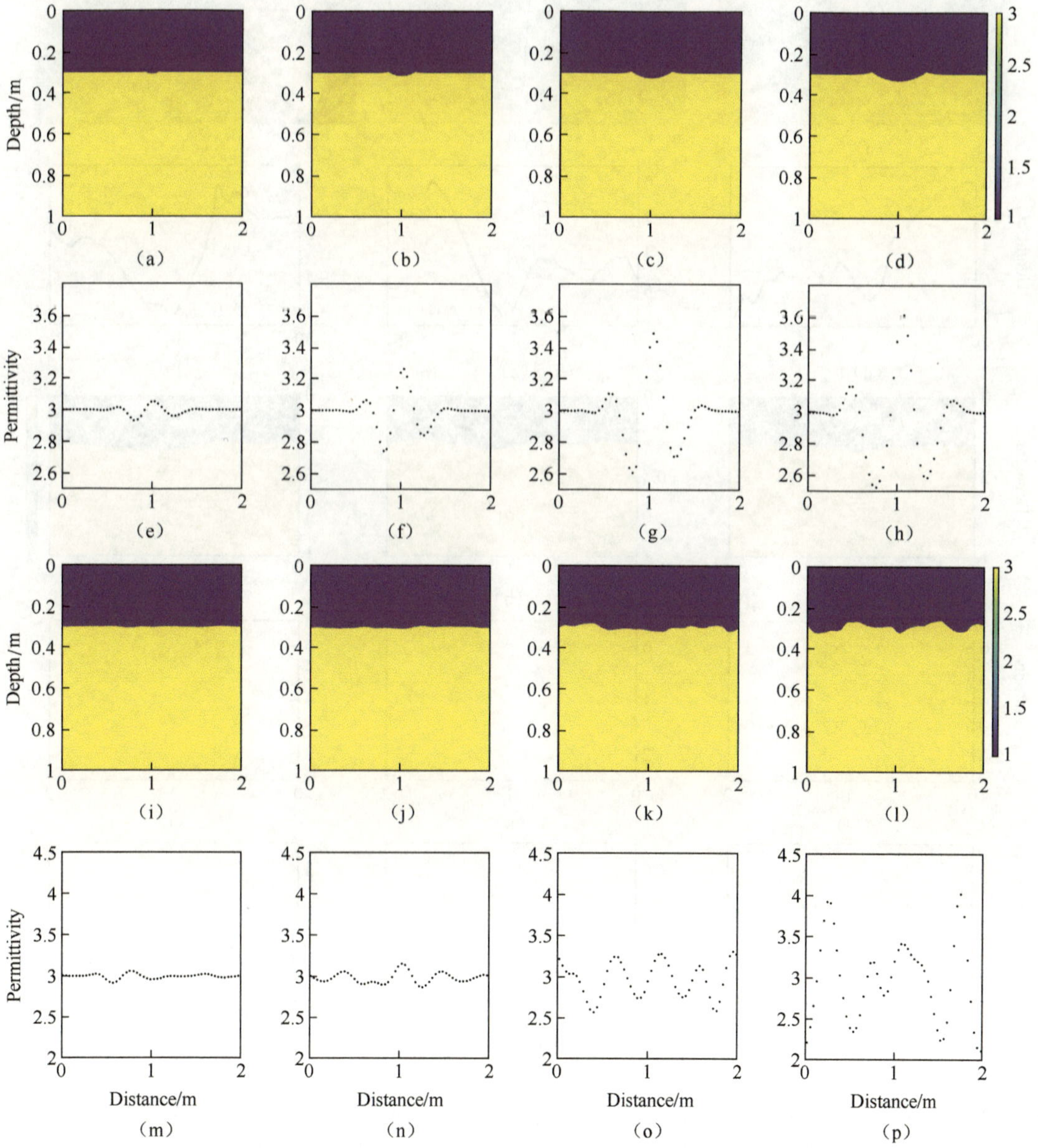

Fig. 4　Numerical simulations for the effect of surface topography on the estimation of relative permittivity. The sizes of the models are 1 × 2m, and the grid interval is set to 0.005 m. Color bars are for relative permittivity used in the models. (a - d) Regolith models have surface topography being consistent with impact craters that have rim - to - rim diameters 0.15, 0.3, 0.45, and 0.6 m, respectively. The depth - to - diameter ratios of the modeled craters are set to 0.2. (e and f) Relative permittivity of surface materials derived based on the topography models shown in panels (a - d), respectively. (i - l) Regolith models that have surface topography constructed based on stochastic Gaussian roughness. The modeled surface reliefs have standard deviations of 0.25%, 0.5%, 1%, and 2%, respectively. (m - p) Relative permittivity of surface materials derived based on the topography models shown in panels (i - l)

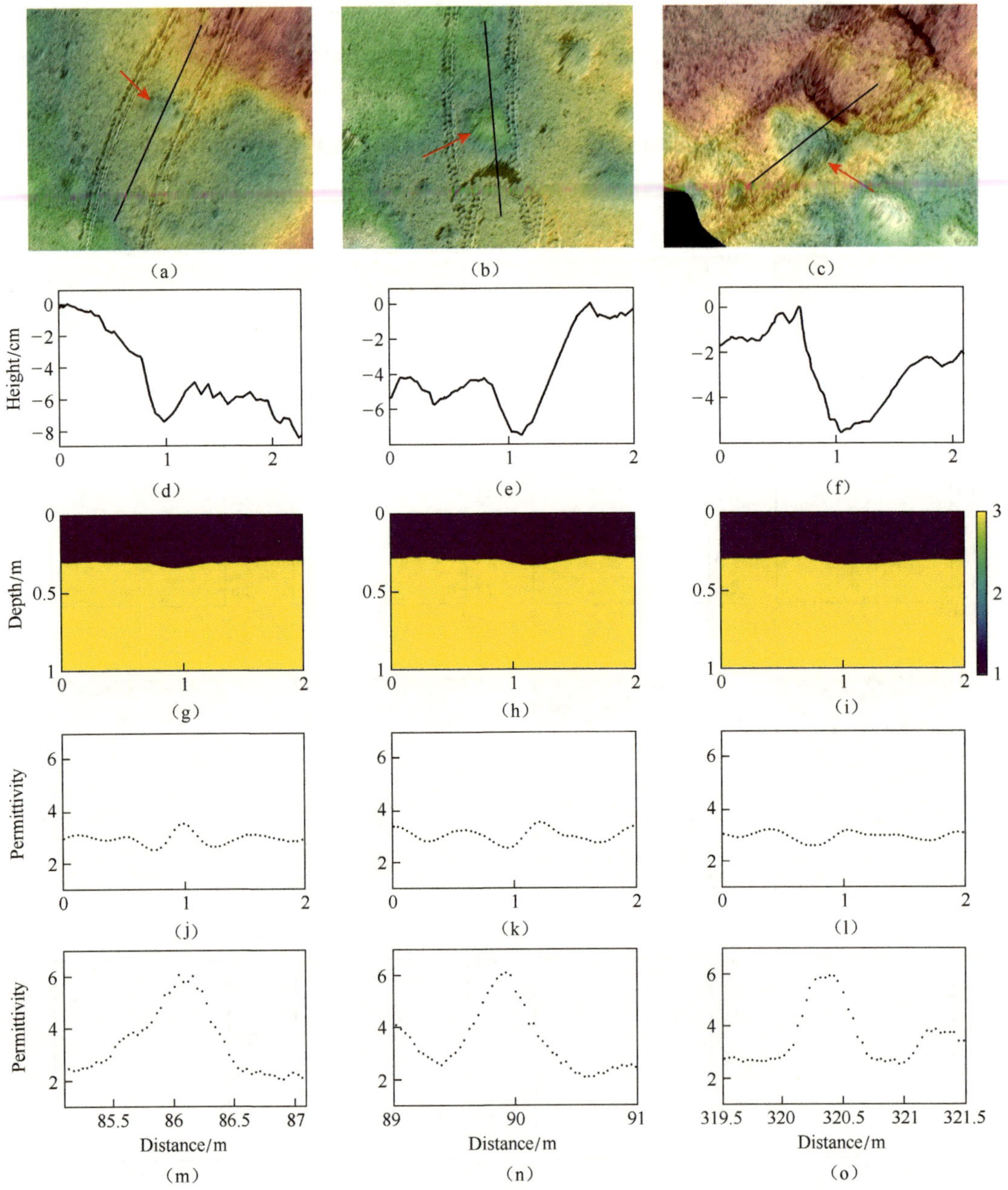

Fig. 5　Effect of actual lunar surface topography on the estimation of relative permittivity based on high-reflectance lunar penetrating radar (LPR) data. (a and b) Digital terrain models constructed for the rover path between the navigation points 207-208. Red arrows point to highly degraded small craters that were scanned by the LPR. (c) Digital terrain model constructed for rover path between the navigation points 1201-1204 (Fig. 1g). The red arrow points to a highly degraded small crater that was scanned by the LPR. The black lines are the locations of the extracted topography profiles, which are shown in panels (d-f), respectively. These topography profiles are used to build regolith models that are shown in the panels (g i). Note the different scales of elevations plotted in the second and third rows. The sizes of the models are 1×2m, and the grid interval is set to 0.005m. Color bars are for relative permittivity used in the model. (j-l) show the derived relative permittivity based on the models that assumed subsurface materials as typical lunar regolith ($\varepsilon=3$). (m-o) show the relative permittivity observed by the high-frequency LPR along the rover paths shown in the first row

We further modeled the effect of actual lunar surface topography on the derived relative permittivity, based on extracted high - resolution topographic profiles for paths traversed by the Yutu - 2 rover. Fig. 5a - 5c are three DTMs generated along the rover path using stereo pairs obtained by the panorama camera (B. Wu et al., 2021), and the extracted topography profiles are shown in Fig. 5d - 5f. Using typical lunar regolith as the model input, the model - derived ε_r is also uniformly less than 4. Interestingly, the rover paths shown in Fig. 5a - 5c do not contain surface boulders, but the derived ε_r at floors of small craters can be as large as 6 (Fig. 5m - 5o), similar to that of lunar rock samples returned by the Apollo missions (Carrier et al., 1991; Olhoeft & Strangway, 1975).

Simulation of the radar reflections from a random rough surface are determined by the coherent sum of echoes from individual small planar scatters on the surface (Nouvel et al., 2004). For spaceborne radars such as MARSIS and SHARAD, the resulting distribution of surface echo strengths is a Rayleigh probability density function, which is characterized by a long tail toward high values of strengths (Grima et al., 2012). However, the surface footprint of electromagnetic waves transmitted by high - frequency LPR is much less than 1m^2 (Dong et al., 2017). We performed three - dimensional simulation for the high - frequency LPR to in vestigate whether or not the observed $\varepsilon_r > 4$ was caused by random scattering caused by three - dimensional surface roughness. The algorithm designed for SHARAD radar by Ferro et al. (2013) was used here after being modified according to the observation geometry of the high - frequency LPR, that is, the radar transmitter was placed at a height of 30cm from the flat surface, and the Fresnel coefficient in this algorithm is changed to the reflection coefficient of the lunar surface. We used the 5mm/pixel DEM for the route between the navigation points N206 to N207 for the modeling (Fig. 6a and 6b). To investigate the effect of different amounts of topographic faces on the derivation of relative permittivity, the DEM was further down sampled to pixel scales of 1, 2, and 6cm for additional modeling. In the modeling, the surface footprint of the LPR radar waves was assumed to be 1 × 1m to maximize the amount of scatters caused by surface roughness (Fig. 6c). Figure 6d shows the relative permittivity of surficial materials along the rover path as derived from the three - dimensional simulation, and the majority of the derived values are less than 4, with a hand ful of A - scans yield $\varepsilon_r \approx 4$. For comparison, values of ε_r derived from the LPR data

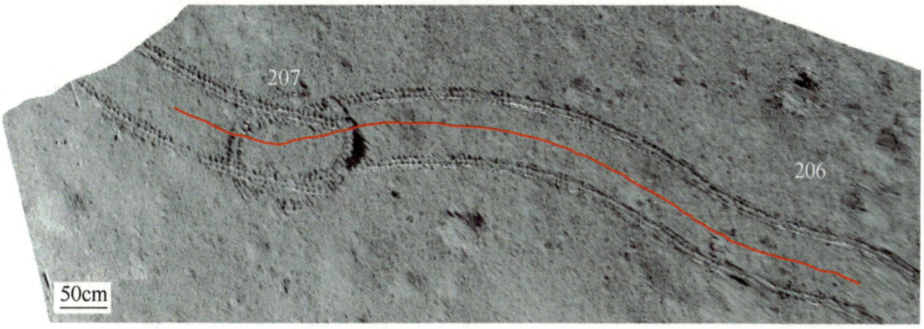

(a)

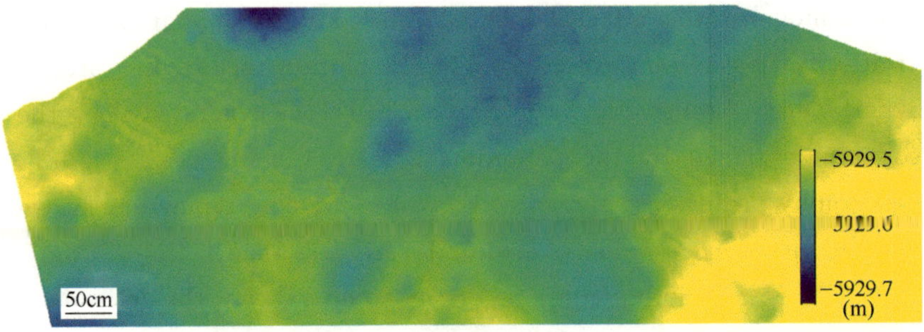

(b)

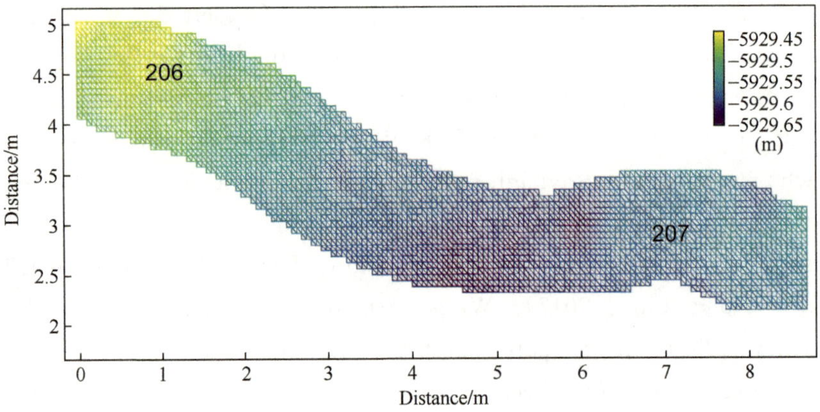

(c)

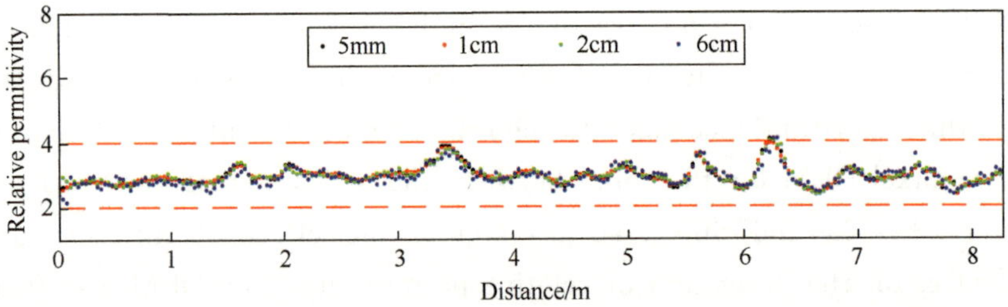

(d)

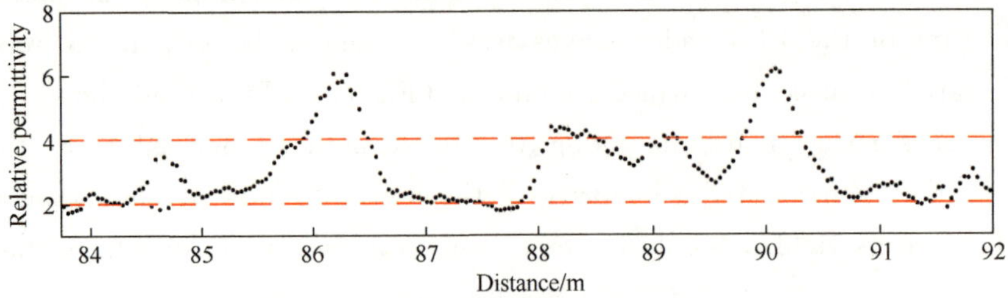

(e)

Fig. 6　Effect of millimeter and centimeter scale surface roughness on the estimation of relative permittivity. (a and b) are orthographic mosaic and DEM (5mm/pixel) for the rover path between the navigation point N206 and N207, respectively. (c) Triangulation of DEM for the surface footprint of signals transmitted by the high-frequency lunar penetrating radar (LPR) between the navigation point N206 and N207. Each footprint is assumed to be a 1 × 1m square so the number of scatters is maximized compared to the true geometry. (d) Relative permittivity of surface materials derived from our three-dimensional modeling. Different colors denote results derived using DEM that has different pixel scales (see the legend). (e) Relative permittivity of surface materials between the navigation point N206 and N207 as calculated based on the obtained LPR data along this route segment have several prominent peaks that are as large as 6 (Fig. 10e). Therefore, the observed values of $\varepsilon_r > 4$ are most likely caused by different properties of surface and subsurface materials, instead of being caused by random scattering caused by surface roughness.

In conclusion, surface topography of typical regolith materials is not the cause for the LPR - derived relative permittivity that is larger than 4. As evidenced by the rockless areas with $\varepsilon_r > 4$ (Fig. 5a - 5c and 6), materials denser than typical regolith should exist in the subsurface at depths $<\sim$4cm, causing abnormally large relative permittivity.

3.3 Surface Morphology of Areas With Abnormally High Relative Permittivity

Using images obtained by the panorama camera, we investigated the surface morphology of 93 areas that have abnormally high relative permittivity. Forty - nine of them have sufficient image coverage to resolve their detailed surface morphology. These areas are indexed in sequence according to their distances along the rover path that are sourced from the lander, that is, 1 - 49. Some of the examples are shown in Fig. 7 - 9, and the others are shown in the appendices (Fig. S1 - S8) .

After the rover was released to the lunar surface, it initially drove along a J - shaped route around the lander (navigation point NP - x) and arrived at the navigation point NP - d (Fig. 7a) . Five areas with $\varepsilon_r > 4$ exist along this ~32 - m - long route (Fig. 7b). Area 1 features $\varepsilon_r = \sim 4.54$, and it is located at ~0.87m from the lander (Fig. 7a). Small blocks that are ~3cm in dimensions are visible among a cluster of small craters that are less than ~20cm in diameter (Fig. 7c) . Area 2 features the largest relative permittivity in this route segment, where the rover traveled across a highly degraded crater that is ~70cm in diameter (Fig. 7d) . Resolvable fragments are not visible on the floor or wall of this crater, but the derived ε_r here is ~9 (Fig. 7b) . Similarly, areas 3 - 5 have an ε_r of ~5 - 6, but no discernable fragments are visible on the surface (Fig. 7e). Although the regolith surface is uneven along this rover route (Fig. 7e), topographic rough ness is not the cause for the abnormally high permittivity (Section 3.2), suggesting the existence of dense materials at depths of $<\sim$4 cm.

From the navigation points 603 to 701 (Fig. 3a), five prominent clusters of $\varepsilon_r > 4$ exist over the ~24 - m - long rover route and several smaller clusters are also visible (Fig. 8a) . Two of the five prominent areas (33 and 35) have surface boulders that are as large as ~8 cm in dimension (Fig. 8b and 8c) . The other three areas (34, 36, and 37) do not have discernable surface fragments, and both the areas 34 and 36 are correlated with highly degraded craters (Fig. 8b and 8c) . Interestingly, the derived ε_r at the area 33 is much smaller than that at area 35, but the surface fragment observed at area 33 is much larger (Fig. 8c) . Moreover, between the area 35 and the navigation point 605, two rocky fragments that are ~5cm in dimensions are visible (green arrows in Fig. 8c), but the derived ε_r is less than 4 (Fig. 8a) . The same is true for the small frag ment (~5cm in dimension) that is located near the location 37 (green arrow in Fig. 8d).

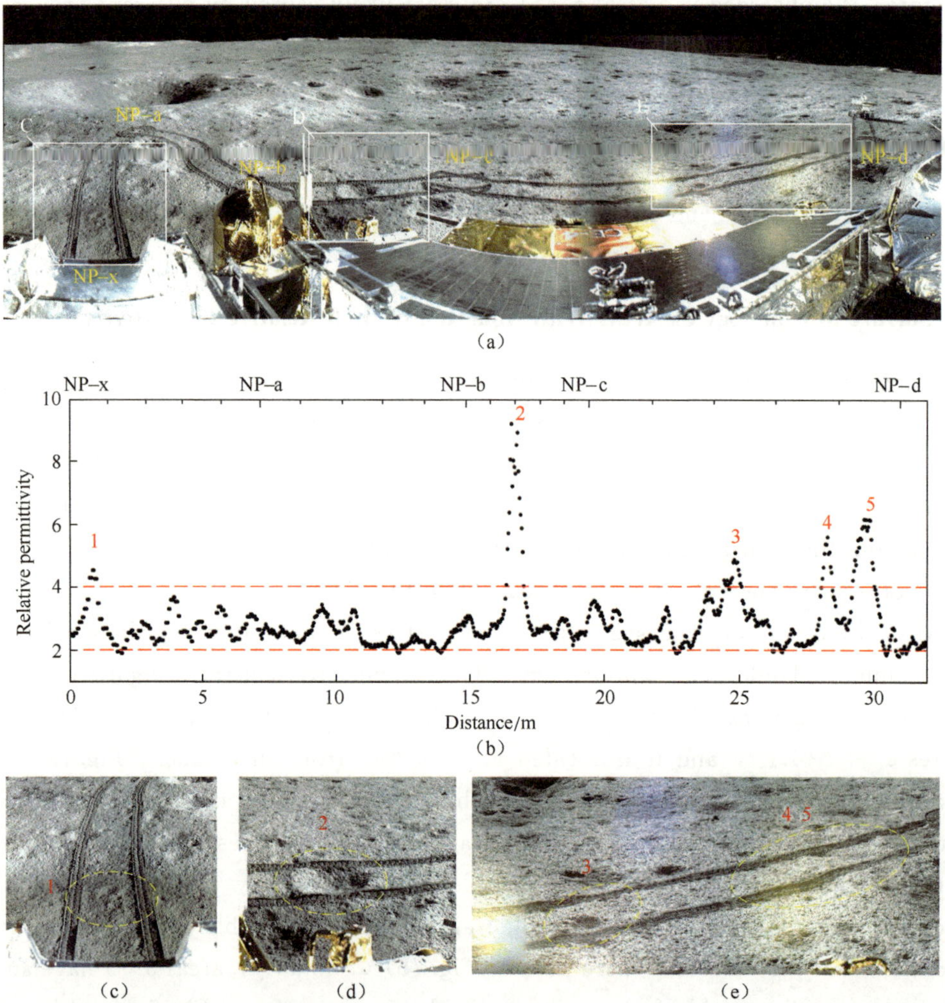

Fig. 7 Surface morphology of areas with abnormally high relative permittivity from the navigation points x to d along the rover path. (a) Overview of surface morphology adjacent to the lander. The navigation points are annotated in yellow as NP-x, NP-a, NP-b, NP-c, and NP-d along the rover path. (b) Distribution of derived relative permittivity along the rover path. The two dashed lines are relative permittivity of 2 and 4, respectively. (c-e) Surface morphology of the five areas (yellow ellipses) that have abnormally high relative permittivity. Areas of interest are outlined as yellow dashed circles and labeled in red

The traverse route of the rover in the first 12 lunar days is generally flat and few rocky fragments are visible along the path (B. Wu et al., 2021). However, from the navigation points 905 to 906 (Figure 3a), abundant rocky fragments are visible around an ~2.2m in diameter and ~35cm deep fresh crater (Fig 9a). The fragments are apparently impact ejecta excavated by this crater, and individual fragments can be up to 5 cm in sizes (Ding, Xiao, et al., 2020). The rover has approached the crater rim from two different

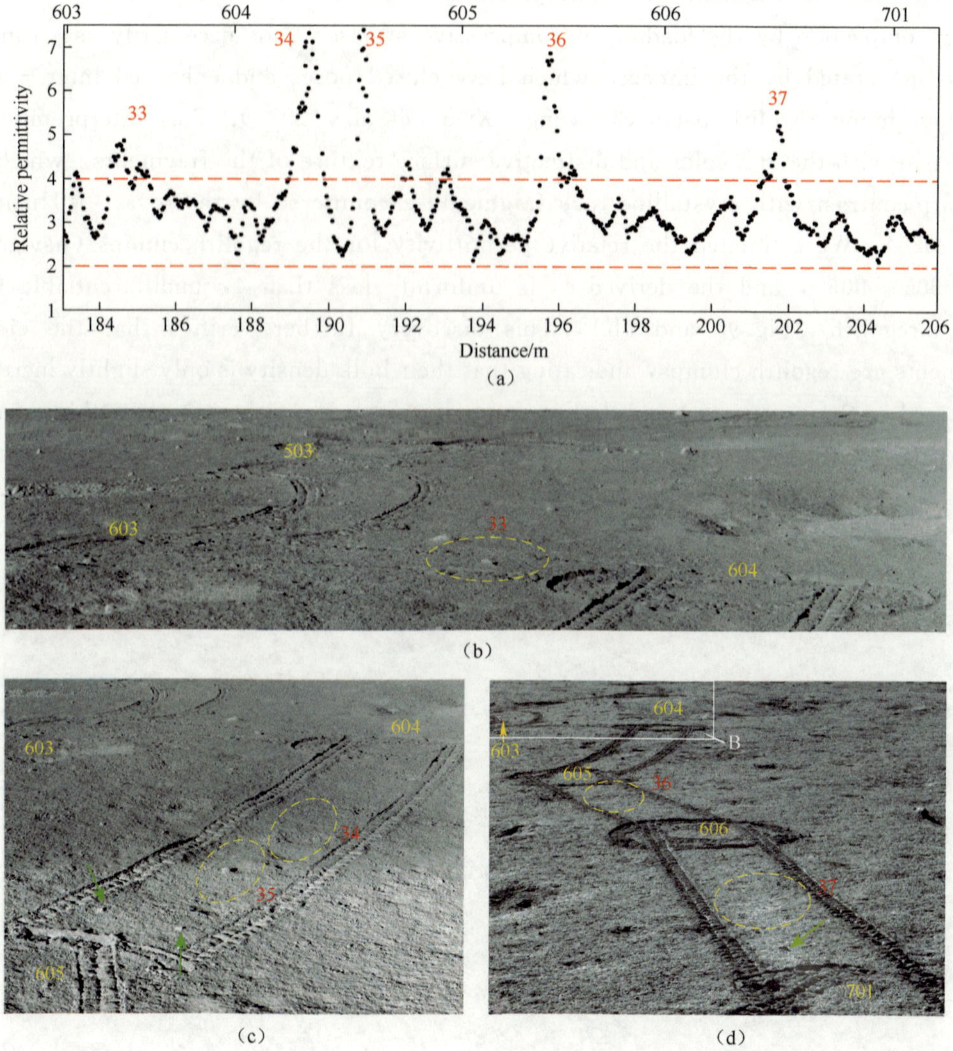

Fig. 8　Surface morphology of areas with abnormally high relative permittivity from the navigation points 603 and 701. (a) Distribution of relative permittivity along the rover path from navigation points 603 to 701. The two dashed lines are relative permittivity of 2 and 4, respectively. (b - d) Surface morphology of areas with abnormally high relative permittivity that are annotated in panel (a). The navigation points are annotated in yellow, and the areas of interest are outlined as yellow dashed circles and labeled in red. The green arrows point to discernable surface fragments along the rover path, but they did not induce abnormally large relative permittivity

paths (Fig. 9a), and the high - frequency LPR has thus performed continual profile investigations for the blocky ejecta (Ding, Xiao, et al., 2020). In the radargram, the blocky ejecta does not exhibit different radar reflections compared with the surrounding fragment - free regolith (Ding, Xiao, et al., 2020). Numerical simulations further verified that the blocky fragments were not preexisting in the subsurface when the impact occurred, and the fragments were interpreted as regolith clumps that were formed due to impact compaction by the loading of compressive stresses (not necessarily as strong as shock) generated by the impact, which have closed pores and enhanced inter - grain bonding among regolith particles (Ding, Xiao, et al., 2020). This interpretation is consistent with the dim color and distinctive surface texture of the fragments, which are in sharp contrast with crystalline rock fragments encountered by the rover (J. Huang et al., 2020). We estimated the relative permittivity for the regolith clumps (navigation point 905 - 906), and the derived ε_r is uniformly less than 4, undifferentiable from typical regolith (Fig. 9a and 9b). This discovery further verifies that the ejected fragments are regolith clumps, indicating that their bulk density is only slightly increased compared to the precursor regolith. For comparison, areas with $\varepsilon_r > 4$ in this area are related with cm - sized fragments that were interbedded in the regolith (e.g., locations 39, 40, and 41), these fragments are not excavated by the recent impact as evidenced by their spatial distribution and surface morphology (Fig. 9c and 9d).

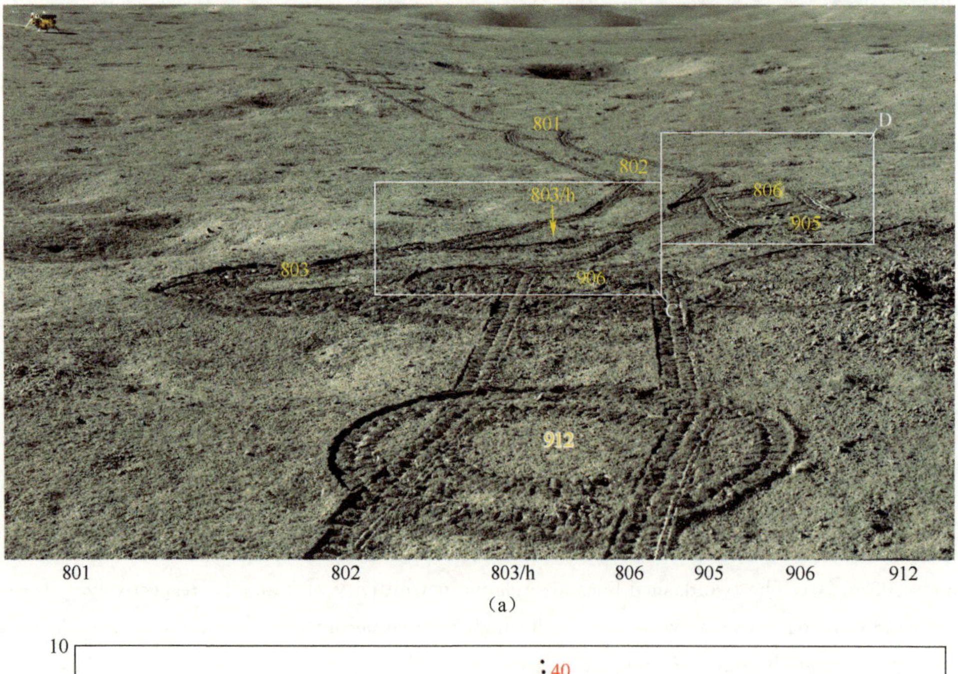

(a)

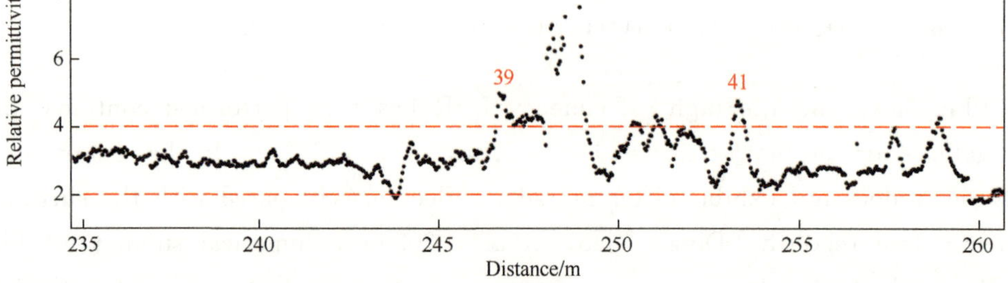

(b)

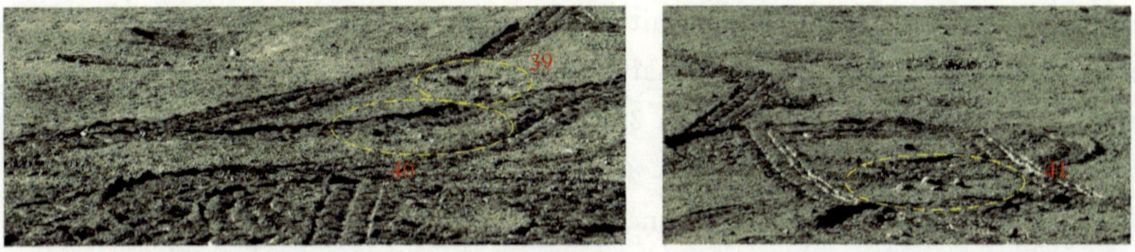

(c)　　　　　　　　　　　　　　(d)

Fig. 9　Surface morphology of areas with abnormally high relative permittivity from the navigation points 801 and 912. (a) Surface morphology of the rover route from navigation points 801 to 912. The lander is visible at the upper left corner. (b) Distribution of relative permittivity derived along the rover path. The two dashed lines are relative permittivity of 2 and 4, respectively. (c and d) Surface morphology of areas with abnormally high relative permittivity annotated in panel (a). The navigation points are annotated in yellow, and the areas of interest are outlined as yellow dashed circles and labeled in red

In conclusion, among the 49 areas that have both $\varepsilon_r > 4$ and resolvable surface morphology, 10 of them are obviously correlated with surface rocky fragments, 11 are located on floors of small craters where coarse fragments are not visible on the surface, and the rest 28 are not correlated with impact craters or visible surface fragments. Moreover, observed rocky fragments (e. g., Fig. 8) and small craters (Fig. S5 and S8 in the appendices) along the rover path do not uniformly cause $\varepsilon_r > 4$.

4　Discussion

4.1　Uncertainties in the Estimation of Relative Permittivity

Uncertainties from different sources existed in the method of estimating the relative permittivity, but they do not affect the overall result of this study. As shown in Fig. 2e－2g, the amplitude of the first envelope peak is different among traces of A－scan. Based on preflight ground test for the high－frequency LPR, we noticed that this issue was caused by a combination of inconstant transmitted signals (Fig. S9) and interaction with ground reflections (Fig. S10). This issue affected the derivation of relative permittivity based on echo strengths, and the uncertainties of results are constrained to be $\sim\pm0.31$ (Text S1). Also, there is a significant overlap between the direct wave traveling between the transmitter and receiver and the reflection from the ground surface (Fig. 2e－2g). The amplitude of the second envelope peak is affected by the direct wave, the second return and their relative phases and timing, and internal electronics. To evaluate this uncertainty, we referred to preflight ground test data of LPR on analog materials that have a relative permittivity (2.89 ± 0.02) similar with typical lunar regolith (Fig. 10a).

Note that the analog materials have a generally smooth surface while microtopography is abundant. We derived the relative permittivity for the surficial analog materials using the same method here, which are systematically less than the true value (Fig. 10b) . The average relative permittivity derived is 2. 49 (a standard deviation of ±0. 10; Fig. 10c), and it is ~0. 4 less than the true value.

We also evaluated the system uncertainty of LPR using data obtained during a repeated scan on the Moon. When the rover was investigating the fresh crater shown in Fig. 8a, the LPR repeatedly scanned the same path between the navigation points N803 and N805, and two sets of radar observations were obtained (Fig. 11) . We compared the two sets of ε_r derived, revealing good consistency (Fig. 11) . The differences between the two sets of ε_r have 1δ =0. 31, confirming that random scattering by surface roughness is not adequate to cause $\varepsilon_r > 4$ for typical lunar regolith.

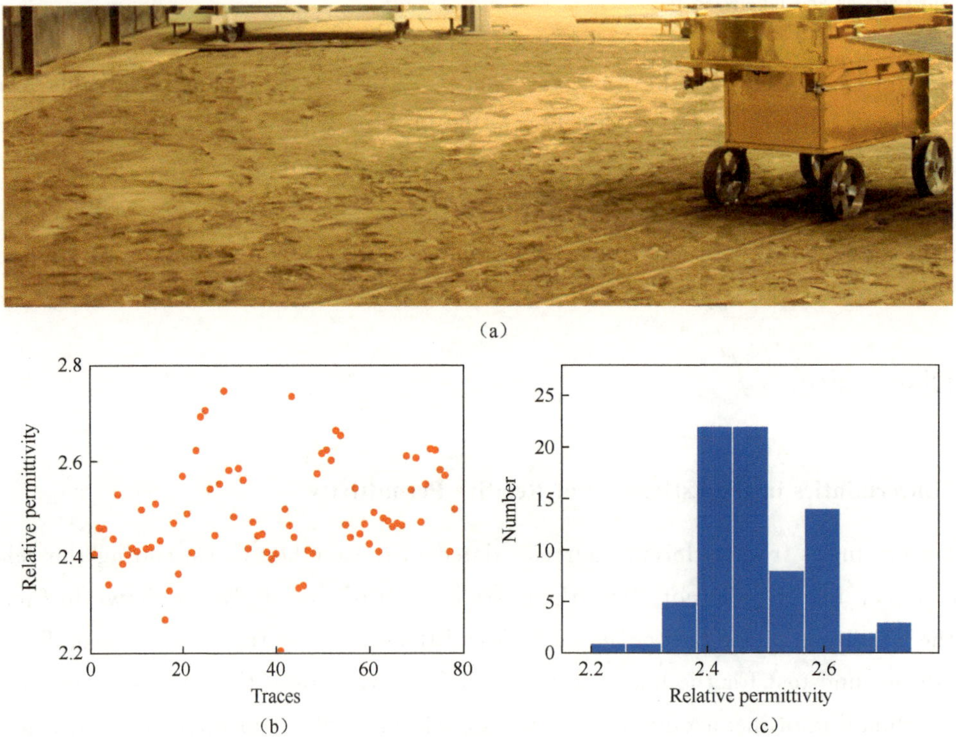

Fig. 10 Uncertainty of derived relative permittivity based on preflight test data of lunar penetrating radar. (a) The test was performed using a prototype of the Yutu rover over analog materials of lunar regolith that had a relative permittivity of 2. 89±0. 02. Seventy - nine traces of A - scan were obtained during this experiment. (b) Using the method of estimating relative permittivity for top - most target materials employed in this study, the relative permittivity of the analog materials were derived based on the experimental data. (c) Histogram of the relative permittivity derived from the experimental data

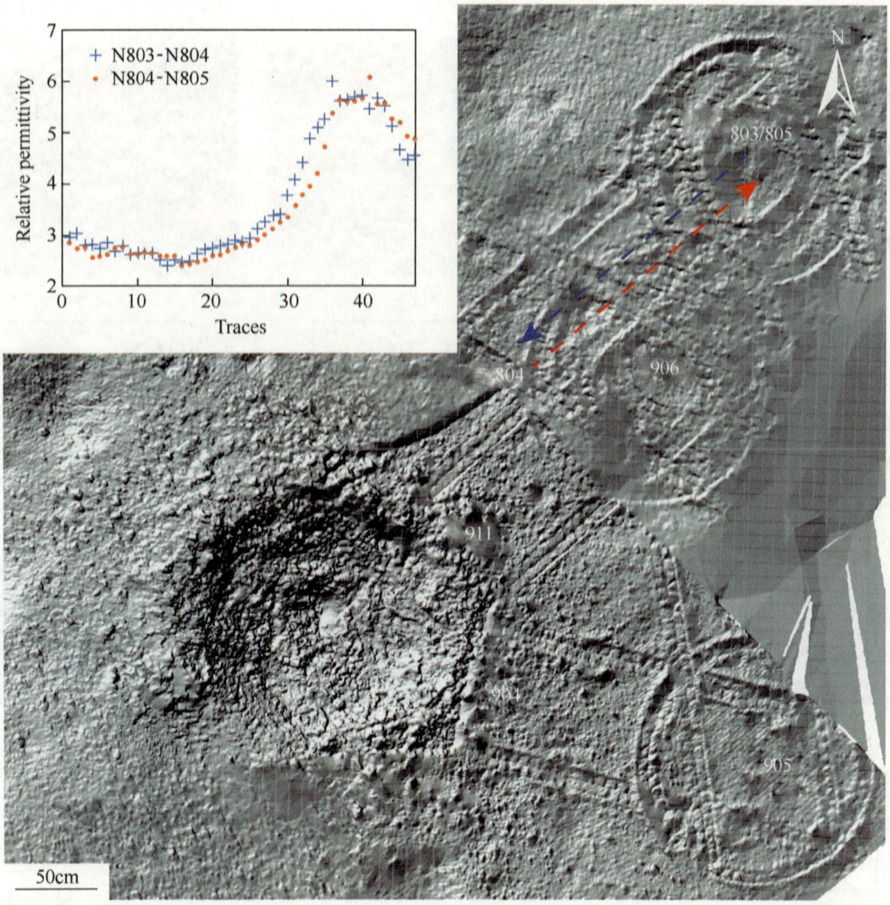

Fig. 11　Repeated scan of the same rover segment by lunar penetrating radar (LPR) revealing the effect of random scattering by surface topography on the estimation of relative permittivity. The Yutu－2 rover repeatedly scanned the segment between the navigation point N803 and N805. The upper－left inset compares the two sets of relative permittivity derived from the repeated scan of LPR. The base figure is referred from Ding，Xiao，et al. (2020)

4.2　Cause of the Abnormally Large Relative Permittivity

The 10 areas with $\varepsilon_r > 4$ that are associated with surface fragments can be explained by the larger relative permittivity of the fragments than typical lunar regolith. Fig. 12 shows the high－resolution DTM for three such examples. The fragments are perched on the surface regolith，their major and the minor semiaxis are 8.5 and 1.6cm，7.7 and 2.6cm，and 10.5cm and 2.3m，respectively (Fig. 12a－12c). We performed numerical simulations to investigate the cause of the abnormally large relative permittivity derived. The geometries of the surface fragments were used as model inputs (Fig. 12d－12f)，which were measured based on the high－resolution DTM. The relative permittivity of the surrounding regolith was set to 3 in the model and that of the fragments was set as a

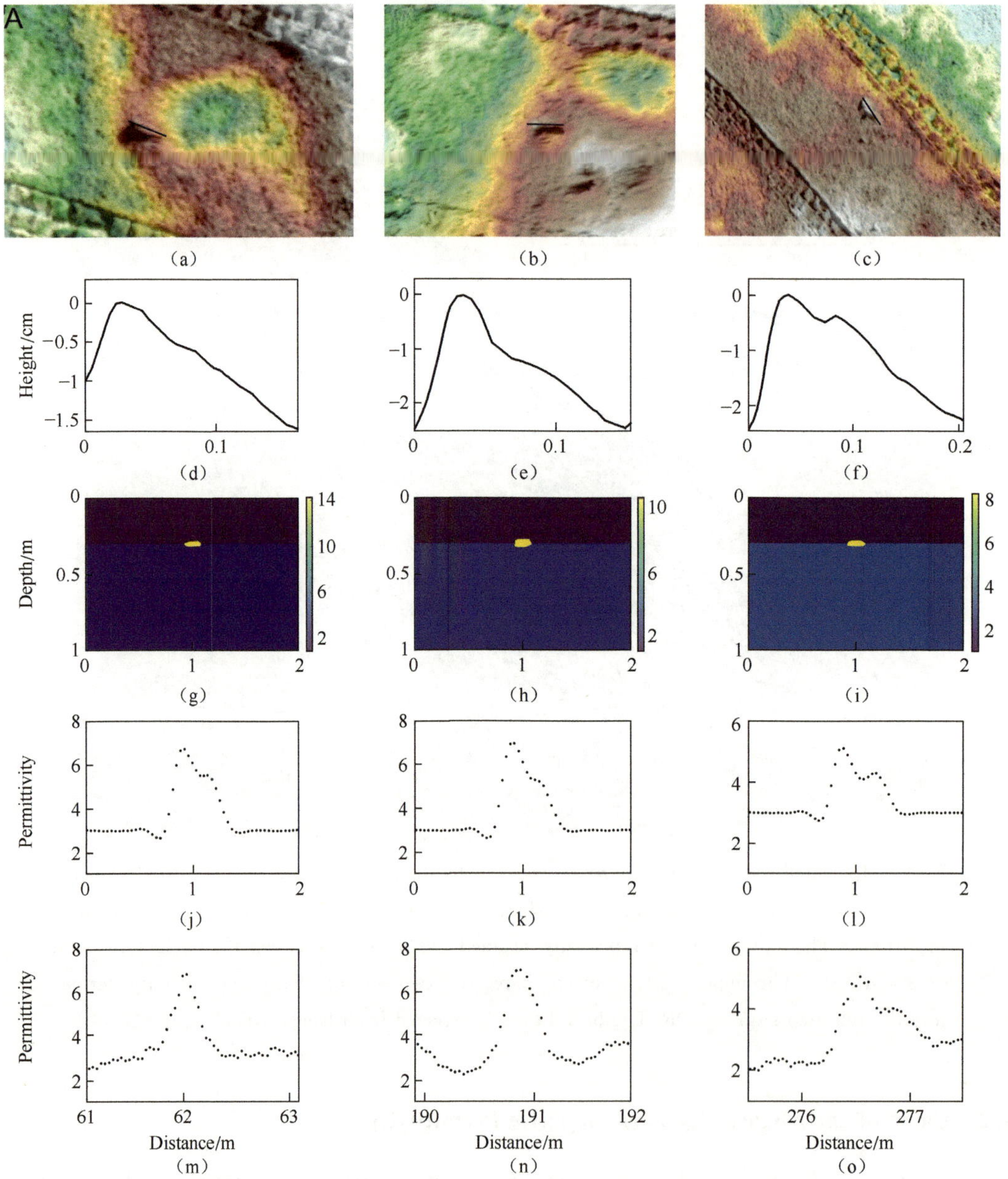

Fig. 12 Estimation of relative permittivity for the surface rocky fragments that have lunar penetrating radar (LPR) - derived relative permittivity larger than 4. (a - c) Digital terrain models for three fragments (indicated by black lines) at areas 16 (Fig. S1), 33 (Fig. 8), and 43 (Fig. S7), respectively. These fragments are correlated with LPR - derived relative permittivity of >4. (d - f) Topography profiles of the rock fragments along the black lines extracted from the DTMs. (g i) Two - dimensional models for the three fragments. The topography profiles are used as model inputs. The sizes of the models are 1 × 2m, and the grid interval is set to 0.005m. Color bars are for relative permittivity used in the models. (j - l) Profiles of modeled relative permittivity. These are the best - fit models compared to observations, and the assigned relative permittivity of the fragments in the models are ~14.2, 10.5, and 8.3, respectively (m - o) profiles of relative permittivity derived from traces of A - scan obtained by the high - frequency LPR along the rover paths as shownin the first row

free parameter (Fig. 12g - 12i) . By varying the relative permittivity of the fragments, different trial tests were performed to compare the modeled relative permittivity (Fig. 12j - 12l) and derived values (Fig. 12m - 12o) . The results show that when the relative permittivity of the fragments was set as ~14. 2 (Fig. 12j), ~10. 5 (Fig. 12k), and ~8. 3 (Fig. 12l), both the shape and magnitude of the profile of modeled ε_r fit the observations. For competent lunar rocks returned by Apollo missions, laboratory measurements revealed that their relative permittivity is 7 - 14 (Chung et al. , 1972) . Therefore, it can be confirmed that the rocky fragments observed are indeed competent rocks rather than regolith clumps. However, the possible rock types (e. g. , crystalline, igneous, and breccia) or their petrographic classifications cannot be inferred based on the derived ε_r, since different types of returned lunar rocks can have similar relative permittivity (Chung et al. , 1972) .

For areas with $\varepsilon_r > 4$ that are not associated with visible fragments on the surface, materials denser than typical regolith should exist in the subsurface. We are able to quantify the potential burial depth of the subsurface dense materials, answering the question of whether or not the dense materials could be icebergs of large boulders, that is, boulders that are much larger than 4 cm in dimensions, but their top edges have been buried by regolith. For this purpose, we performed simulations for the effect of buried fragments on the derived relative permittivity (Fig. 13) . In these simulations, the relative permittivity of the implanted rock fragments was set as 9 and that of the surrounding regolith was set as 3. The size and burial depth of the implanted boulders were varied in the models. Since fragments larger than 7cm can be confidently captured by the panorama camera (Fig. 7 - 9), a single fragment was set as a 7cm diameter sphere, and sizes of implanted rock fragments in the models were varied by adding or reducing the number of fragments (Fig. 13) . With respect to the wavelength of the transmitted electromagnetic waves by the high - frequency LPR ($\lambda = \sim 0.35$m in typical lunar regolith), the fragments were buried at various depths of the regolith models.

The model results show that while buried fragments could easily induce $\varepsilon_r > 3$ (first three columns of Fig. 13), $\varepsilon_r > 4$ is only yielded by fragments that were implanted at depths less than 1/16 of the radar wavelength (first and second column of Fig. 13) . For large boulders that were implanted at depths larger than 1/16 of the radar wavelength (e. g. , the 14 - and 21 - cm - long boulders modeled in Fig. 13c _ 4 and 13e _ 4), this method did not yield $\varepsilon_r > 4$ (Fig. 13d _ 4 and 13f _ 4) . The results are consistent with the theoretical constraint of this technique (Widess, 1973), as the first echo received is only sensitive to the relative permittivity of surficial materials that are located within ~1/8 wavelength of the transmitted radar wave. The results also show that with the same burial depth, larger subsurface boulders generally yield bigger ε_r. Furthermore, for a single rock fragment on the surface (e. g. , Fig 13a _ 1), the radar - derived relative permittivity can be less than 4 (e. g. , Fig. 13a _ 2) . This result explains the observation that some

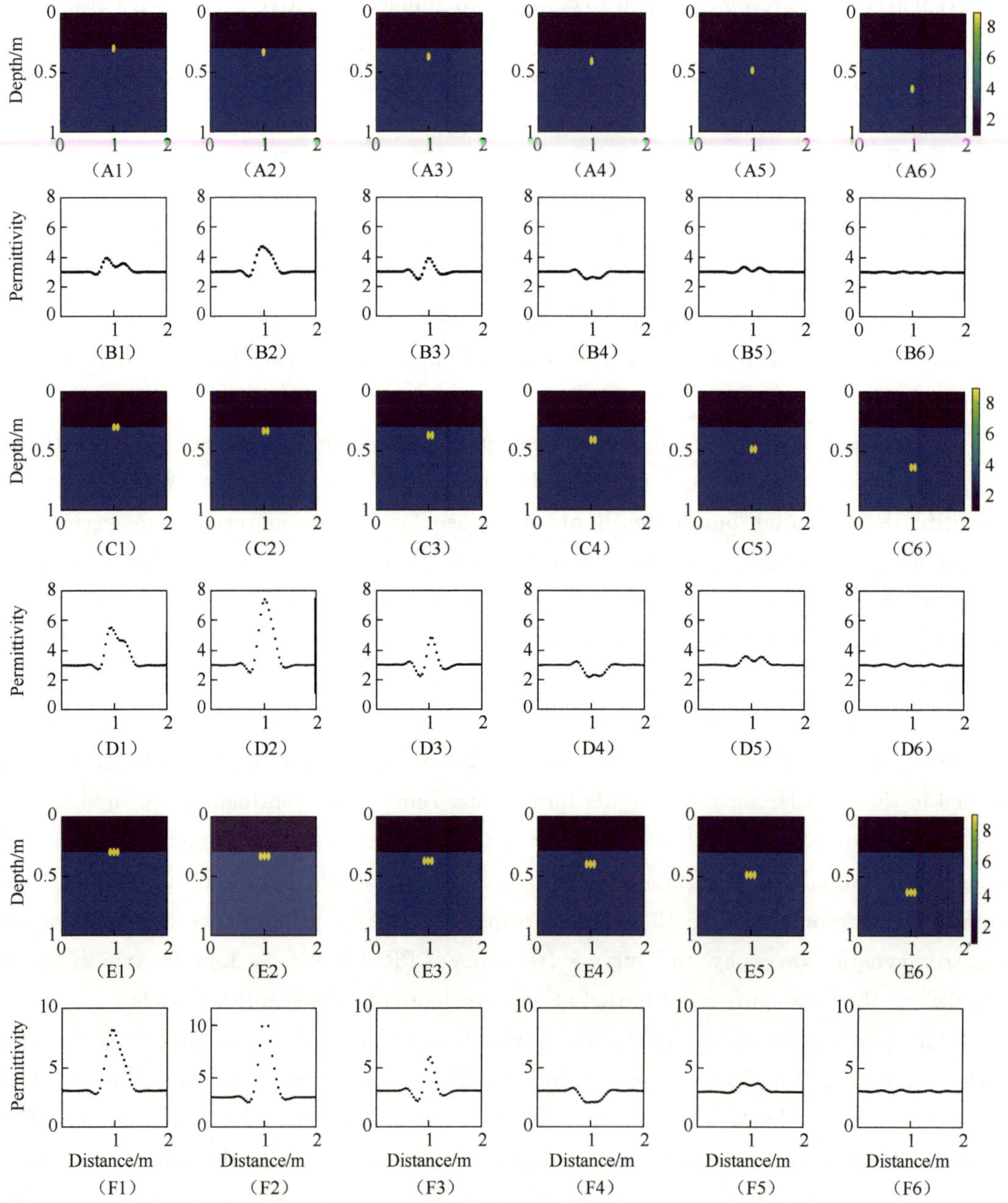

Fig. 13　Effects of the size and burial depth rock fragments in typical regolith on lunar penetrating radar-derived relative permittivity. All the models are 1 × 2m in dimension, the relative permittivity of regolith is set to 3 and that of the implanted rocks is 9. Each individual fragment has a cylindrical shape perpendicular to the profile that is 7cm in diameter, and the size of implanted fragments is enlarged by adding more fragments (the first, third, and fifth rows). The centers of the fragments are implanted at depths of 0 (first column), 1/16 (second column), 1/8 (third column), 1/4 (forth column), and 1/2 (fifth column) of the radar wavelength below the lunar surface. Below each model, the simulated profile of radar-derived relative permittivity is shown (the second, fourth, and sixth rows)

surface fragments are not associated with $\varepsilon_r > 4$ derived by LPR (green arrows in Fig. 8). It is notable that many factors of the implanted fragments (e. g., burial depth, shape, geometry, and relative permittivity) could affect the relative permittivity derived using this technique, and their influences are intertwined for a given value of observed ε_r. Fig. S11 shows that a single surface fragment (relative permittivity of 9) that is ~1-6 cm in dimension is not able to cause LPR-derived relative permittivity larger than 4, but a larger number of such small fragments could cause $\varepsilon_r > 4$ (Fig. S12). Therefore, we are not able to constrain the possible size and relative permittivity for the buried fragments.

4.3 Origin of Buried Rock Fragments and Indications for Lunar Surface Dynamics

Along the rover path accomplished in the first 12 lunar days, the amount of detected surface and subsurface rock fragments in this study is minor compared with the dominating porous regolith (Fig. 3). This result is generally consistent with the low abundance of rocks revealed by orbital observations (Fig. 1), statistics of surface rocks (B. Wu et al., 2021), and earlier interpretations for the entire regolith column based on the high-frequency LPR data (Ding, Xiao, et al., 2020). However, compared with earlier observations, this study provides unprecedented details about the heterogenous spatial distribution of rock fragments within the top ~4cm of the mature regolith (Fig. 1g), especially those located beneath small impact craters.

Thirty-nine of the 49 resolved areas with $\varepsilon_r > 4$ are not associated with surface fragments, and the above discussions support the existence of buried rock fragments in the shallow regolith. The overturn and accumulation history of lunar regolith is chaotic due to both random impacts by various-sized primary craters and sudden disturbances by distal impact ejecta (Y.—H. Huang, Minton, et al., 2018). In general, deeper materials are overturned less frequently compared to shallower materials, so surface materials have younger exposure ages (Hörz et al., 2020). However, the exact formation age of a single surface or subsurface fragment cannot be precisely constrained due to stochastic and repeated impact excavation and burial. Recent evidence gained from contents of radioactive isotopes in Apollo regolith samples (Costello et al., 2018), thermophysical properties of lunar regolith (Bandfield et al., 2014), estimated flux of lunar micrometeoroids (Hrz et al., 2020), and the observed formation rate of small impact craters and their ejecta (Speyerer et al., 2016) collectively suggest that the top 2—3cm of lunar regolith are thoroughly overturned in <80000 years. Therefore, the surface and subsurface fragments detected in this study can be simply interpreted as distal ejecta that were delivered recently to the landing region, possibly within the past ~80000 years. The possible origin of the buried fragments as distal ejecta is consistent with the heterogeneous distribution of areas with $\varepsilon_r > 4$ along the rover path (Fig. 1g).

The 11 buried fragments beneath floors of highly degraded small craters (e. g., Fig. 5

and 6) can alternatively be interpreted as impact breccias welded by melt. Along the rover path, small fresh craters featuring high - reflectance materials on crater floors have been noticed (Fig. S13) . There is no consensus yet about whether these small craters are primary (Lin et al. , 2020) or secondary craters (Ding, Xiao, et al. , 2020; Gou et al. , 2020), but reflectance spectra revealed that the high - reflectance materials were likely glass - rich impact breccias (Gou et al. , 2020; Lin et al. , 2020) . Melt - bearing impact breccias returned by the Apollo missions have comparable relative permittivity with crystalline and igneous rocks (Chung et al. , 1972), and topographic degradation of small craters on the Moon mainly occur as diffusive downslope movements on crater walls (Xie et al. , 2017) . Therefore, while regolith clumps formed by impact compaction are not dense enough to induce $\varepsilon_r > 4$, small impact craters formed in mature regolith could significantly reduce the porosity of local regolith by producing melt welded impact breccia in the crater floor, and the melt - bearing breccias may have a longer lifetime (due to burial) than the topography of the parent crater. The occurrence of many small melt - bearing breccias beneath crater floors could induce $\varepsilon_r > 4$ as revealed by numerical modeling (Fig. S12) .

The possible contribution of small impacts in reducing local regolith porosity brings new insights about surface dynamics on the Moon. Considering the high formation rate of small primary (Speyerer et al. , 2016) and secondary craters (Costello et al. , 2018) on the Moon, widespread patches of melt - bearing breccias should exist in the shallow subsurface of the Moon, although each patch covers only limited areas beneath the crater floor. This mechanism could explain the highly heterogeneous interior structures observed in lunar regolith. This mechanism also potentially occurs on the other airless bodies, such as Mercury.

5 Conclusion

Particle size - frequency distributions at different depths of lunar regolith record important information about surface dynamics on the Moon. Top centimeters of lunar regolith are generally believed to be composed of more - or - less homogeneous fine grains, considering their long overturn history. However, the abundance of interior fragments in the top - most regolith, which carries most information of lunar geology learned by remote sensing, is not adequately understood. The high - frequency Lunar Penetrating Radar onboard the Yutu - 2 rover of the Chang'E - 4 mission has been scanning the thick regolith at the lunar far side. Based on data obtained in the first 12 lunar days of surface operation, we used amplitudes of transmitted signals and those of the first received echoes by high - frequency Lunar Penetrating Radar to estimate the relative permittivity for the top - most ~4 - cm - thick regolith (ε_r) . The results show that ~89% of the surficial materials are composed of typical fine - grained regolith, consistent with both earlier orbital and in situ observations. Regolith clumps that were formed by impact compaction do not feature larger ε_r than typical regolith. Ninety - three clusters of abnormally high permittivity ($\varepsilon_r > 4$) were detected. We performed high - resolution numerical simulations for the propagation of electromagnetic waves in different regolith models, precluding surface topography as a cause for the abnormally large ε_r. Forty - nine areas with $\varepsilon_r > 4$ have been covered with sufficient resolution by the panorama camera onboard the rover. Among them, 10 of them are correlated with surface fragments, and our numerical simulations revealed that the surface fragments are rocks. The other 39 such areas do not have visible surface fragments, and our numerical simulations indicate that abnormally large relative permittivity is caused by dense materials within ~4cm of the regolith. Interestingly, 11 such buried dense materials are located on floors of small and highly degraded craters. While these dense materials could be rock fragments delivered by distal impacts, they are also possibly remanent of melt - bearing impact breccias formed by the parent crater. This mechanism could explain the widespread heterogeneous nature of the uppermost 4 cm of lunar regolith, which may also be valid on other airless bodies.

Acknowledgments

The Chang'E - 4 mission was carried out by the Chinese Lunar Exploration Program, and the scientific data are provided by the China National Space Administration. Comments and suggestions provided by Dr. Sarah Kruse and an anonymous reviewer helped clarifying and improving the manuscript. The authors are supported by the B - type Strategic Priority Program of the Chinese Academy of Sciences (grant XDB41000000), the National Natural Science Foundation of China (42004099 and 41773063), the SUSTech Presidential Postdoctoral Fellowship, the Opening Fund of the Key Laboratory of Lunar and Deep Space Exploration (No. LDSE202005) and the Key Research Program of Frontier Sciences of Chinese Academy of Sciences (No. QYZDY - SSW - DQC028), the pre - research Project on Civil Aerospace Technologies (D020201 and D020202) funded by the Chinese National Space Administration, a grant from the Research Grants Council of Hong Kong (Research Impact Fund—Project No: R5043 - 19), and Shenzhen Municipal Government Investment Project (No. 2106 - 440300 - 04 - 03 - 901272).

References

[1] Bandfield, J. L., Ghent, R. R., Vasavada, A. R., Paige, D. A., Lawrence, S. J., & Robinson, M. S. (2011). Lunar surface rock abundance and regolith fines temperatures derived from LRO Diviner Radiometer data. Journal of Geophysical Research, 116 (E12). https://doi.org/10.1029/2011JE003866.

[2] Bandfield, J. L., Song, E., Hayne, P. O., Brand, B. D., Ghent, R. R.,

Vasavada, A. R. , & Paige, D. A. (2014) . Lunar cold spots: Granular flow features and extensive insulating materials surrounding young craters. Icarus, 231, 221 - 231. https: //doi. org/10. 1016/j. icarus. 2013. 12. 017.

[3] Bart, G. D. , & Melosh, H. J. (2010) . Distributions of boulders ejected from lunar craters. Icarus, 209 (2), 337 - 357. https: //doi. org/10. 1016/j. icarus. 2010. 05. 023.

[4] Benedetto, A. , Tosti, F. , Ciampoli, L. B. , & D'amico, F. (2017) . An overview of ground - penetrating radar signal processing techniques for road inspections. Signal Processing, 132, 201 - 209. https: //doi. org/10. 1016/j. sigpro. 2016. 05. 016.

[5] Bitri, A. , & Grandjean, G. (1998) . Frequency - wavenumber modelling and migration of 2D GPR data in moderately heterogeneous dispersive media. Geophysical Prospecting, 46, 287 - 301. https: //doi. org/10. 1046/j. 1365 -2478. 1998. 00091. x.

[6] Cahill, J. T. (2014) . LRO Moon Mini - RF 5 global mosaic V1. 0 [Data set] . NASA Planetary Data System. https: //doi. org/10. 17189/1520656.

[7] Cahill, J. T. , Thomson, B. J. , Patterson, G. W. , Bussey, D. B. J. , Neish, C. D. , Lopez, N. R. , et al. (2014) . The miniature radio frequency instrument's (Mini - RF) global observations of Earth's Moon. Icarus, 243, 173 - 190. https: // doi. org/10. 1016/j. icarus. 2014. 07. 018.

[8] Campbell, B. A. , Campbell, D. B. , Margot, J. L. , Ghent, R. R. , Nolan, M. , Chandler, J. , et al. (2007) . Focused 70 - cm wavelength radar mapping of the Moon. IEEE Transactions on Geoscience and Remote Sensing, 45 (12), 4032 - 4042. https: //doi. org/10. 1109/TGRS. 2007. 906582 .

[9] Carrier, W. D. , Olhoeft, G. R. , & Mendell, W. (1991) . Physical properties of the lunar surface (pp. 475 - 594) . Lunar source book.

[10] Chung, D. H. , Westphal, W. B. , & Olhoeft, G. R. (1972) . Dielectric properties of Apollo 14 lunar samples. Paper presented at Lunar and Planetary Science Conference Proceedings (Vol. 3, p. 3161) .

[11] Costello, E. S. , Ghent, R. R. , & Lucey, P. G. (2018) . The mixing of lunar regolith: Vital updates to a canonical model. Icarus, 314, 327 - 344. https: // doi. org/10. 1016/j. icarus. 2018. 05. 023 .

[12] Di, K. C. , Xu, B. , Peng, M. , Yue, Z. Y. , Liu, Z. Q. , Wan, W. H. , et al. (2016) . Rock size - frequency distribution analysis at the Chang'e - 3 landing site. Planetary and Space Science, 120, 103 - 112. https: //doi. org/10. 1016/ j. pss. 2015. 11. 012 .

[13] Ding, C. Y. , Cai, Y. Z. , Xiao, Z. Y. , & Su, Y. (2020) . A rocky hill on the continuous ejecta of Ziwei crater revealed by the Chang'e - 3 mission. Earth and Planetary Physics, 4 (2), 105 - 110. https: //doi. org/10. 26464/epp2020016 .

[14] Ding, C. Y. , Xiao, Z. , Su, Y. , & Cui, J. (2021) . Hyperbolic reflectors determined from peak echoes of ground penetrating radar. Icarus, 358,

114280. https：//doi. org/10. 1016/j. icarus. 2020. 114280.

[15] Ding，C. Y.，Xiao，Z.，Wu，B.，Li，Y.，Prieur，N. C.，Cai，Y.，et al. (2020). Fragments delivered by secondary craters at the Chang'e 4 landing site. Geophysical Research Letters，47，e2020GL087361. https：//doi. org/10. 1029/2020GL087361.

[16] Ding，C. Y.，Xiao，Z.，Wu，B.，Li，Z.，Su，Y.，Zhou，B.，et al. (2021). Rock fragments in shallow regolith：Constraints by the lunar penetrating radar onboard the Chang'E-4 mission. Zenodo. https://doi. org/10. 5281/zenodo. 4964896.

[17] Dong，Z. H.，Fang，G. Y.，Ji，Y. C.，Gao，Y. Z.，Wu，C.，& Zhang，X. J. (2017). Parameters and structure of lunar regolith in Chang'e-3 landing area from lunar penetrating radar (LPR) data. Icarus，282，40 - 46. https：//doi. org/10. 1016/j. icarus. 2016. 09. 010.

[18] Fa，W. Z.，Zhu，M. H.，Liu，T. T.，& Plescia，J. B. (2015). Regolith stratigraphy at the Chang'e - 3 landing site as seen by lunar penetrating radar. Geophysical Research Letters，42，10179 - 10187. https：//doi. org/10. 1002/2015gl066537 .

[19] Fang，G.，Zhou，B.，Ji，Y.，Zhang，Q.，Shen，S.，Li，Y.，et al. (2014). Lunar Penetrating Radar onboard the Chang'e - 3 mission. Research in Astronomy and Astrophysics，14 (12)，1607 - 1622. https：//doi. org/10. 1088/1674 - 4527/14/12/009.

[20] Ferro，A.，Pascal，A.，& Bruzzone，L. (2013). A novel technique for the automatic detection of surface clutter returns in radar sounder data. IEEE Transactions on Geoscience and Remote Sensing，51 (5)，3037 - 3055. https：//doi. org/10. 1109/TGRS. 2012. 2219315.

[21] Giannopoulos，A.，& Diamanti，N. (2008). Numerical modelling of ground penetrating radar response from rough subsurface interfaces. Near Surface Geophysics，6 (6)，357 - 369. https：//doi. org/10. 3997/1873 - 0604. 2008024.

[22] Gou，S.，Yue，Z.，Di，K.，Bugiolacchi，R.，Wan，W.，Yang，M.，& Ye，L. (2021). Geologically old but freshly exposed rock fragments encountered by Yutu - 2 rover. Journal of Geophysical Research：Planets，126，e2020JE006565. https：//doi. org/10. 1029/2020JE006565.

[23] Gou，S.，Yue，Z.，Di，K.，Wang，J.，Wan，W.，Liu，Z.，et al. (2020). Impact melt breccia and surrounding regolith measured by Chang'e - 4 rover. Earth and Planetary Science Letters，544，116378. https：//doi. org/10. 1016/j. epsl. 2020. 116378.

[24] Grima，C.，Kofman，W.，Herique，A.，Orosei，R.，& Seu，R. (2012). Quantitative analysis of Mars surface radar reflectivity at 20MHz. Icarus，220 (1)，84 - 99. https：//doi. org/10. 1016/j. icarus. 2012. 04. 017.

[25] Herve，Y.，Ciarletti，V.，Le Gall，A.，Corbel，C.，Hassen - Khodja，R.，Benedix，W. S.，et al. (2020). The WISDOM radar on board the ExoMars 2022

Rover: Characterization and calibration of the flight model. Planetary and Space Science, 189, 104939. https://doi.org/10.1016/j.pss.2020.104939.

[26] Hörz, F., Basilevsky, A. T., Head, J. W., & Cintala, M. J. (2020). Erosion of lunar surface rocks by impact processes: A synthesis. Planetary and Space Science, 194, 105105. https://doi.org/10.1016/j.pss.2020.105105.

[27] Huang, J., Xiao, Z., Flahaut, J., Martinot, M., Head, J., Xiao, X., et al. (2018). Geological characteristics of Von Kármán crater, northwestern south pole Aitken Basin: Chang'E-4 landing site region. Journal of Geophysical Research: Planets, 123, 1684-1700. https://doi.org/10.1029/2018JE005577.

[28] Huang, J., Xiao, Z., Xiao, L., Horgan, B., Hu, X., Lucey, P., et al. (2020). Diverse rock types detected in the lunar South Pole-Aitken Basin by the Chang'E-4 lunar mission. Geology, 48 (7), 723-727. https://doi.org/10.1130/G47280.1.

[29] Huang, Y.-H., Minton, D. A., Zellner, N. E. B., Hirabayashi, M., Richardson, J. E., & Fassett, C. I. (2018). No change in the recent lunar impact flux required based on modeling of impact glass spherule age distributions. Geophysical Research Letters, 45, 6805-6813. https://doi.org/10.1029/2018GL077254.

[30] Huisman, J. A., Hubbard, S. S., Redman, J. D., & Annan, A. P. (2003). Measuring soil water content with ground penetrating radar: A review. Vadose Zone Journal, 2 (4), 476-491. https://doi.org/10.2136/vzj2003.4760.

[31] Lai, J., Xu, Y., Zhang, X., Xiao, L., Yan, Q., Meng, X., et al. (2019). Comparison of dielectric properties and structure of lunar regolith at Chang'e-3 and Chang'e-4 landing sites revealed by ground penetrating radar. Geophysical Research Letters, 46, 12783-12793. https://doi.org/10.1029/2019GL084458.

[32] Lemelin, M., Lucey, P. G., Miljković, K., Gaddis, L. R., Hare, T. M., & Ohtake, M. (2019). The compositions of the lunar crust and upper mantle: Spectral analysis of the inner rings of lunar impact basins. Planetary and Space Science, 165, 230-243. https://doi.org/10.1016/j.pss.2018.10.003.

[33] Li, B., Ling, Z., Zhang, J., & Chen, J. (2017). Rock size-frequency distributions analysis at lunar landing sites based on remote sensing and in-situ imagery. Planetary and Space Science, 146, 30-39. https://doi.org/10.1016/j.pss.2017.08.008.

[34] Li, C., Su, Y., Pettinelli, E., Xing, S., Ding, C., Liu, J., et al. (2020). The Moon's farside shallow subsurface structure unveiled by Chang'e-4 lunar penetrating radar. Science Advances, 6 (9), eaay6898. https://doi.org/10.1126/sciadv.aay6898.

[35] Li, C., Xing, S., Lauro, S. E., Su, Y., Dai, S., Feng, J., et al. (2018). Pitfalls in GPR data interpretation: False reflectors detected in lunar radar cross sections by Chang'e-3. IEEE Transactions on Geoscience and Remote Sensing, 56 (3),

1325 -1335. https：//doi. org/10. 1109/ tgrs. 2017. 2761881 .

[36] Li，L.，& Mustard，J. F. （2003）. Highland contamination in lunar mare soils：Improved mapping with multiple end - member spectral mixture analysis（MESMA）. Journal of Geophysical Research：Planets，108 （E6）. https：//doi. org/10. 1029/2002je001917.

[37] Li，Y.，& Wu，B. （2018）. Analysis of rock abundance on lunar surface from orbital and descent images using automatic rock detection. Journal of Geophysical Research：Planets，123，1061 - 1088. https：//doi. org/10. 1029/2017JE005496.

[38] Lin，H.，Lin，Y.，Yang，W.，He，Z.，Hu，S.，Wei，Y.，et al. （2020）. New insight into lunar regolith - forming processes by the lunar rover Yutu - 2. Geophysical Research Letters，47，e2020GL087949. https：//doi. org/10. 1029/2020GL087949.

[39] Ling，Z.，Qiao，L.，Liu，C.，Cao，H.，Bi，X.，Lu，X.，et al. （2019）. Composition，mineralogy and chronology of mare basalts and non - mare materials in Von Kármán crater：Landing site of the Chang'E - 4 mission. Planetary and Space Science，179，104741. https://doi. org/10. 1016/j. pss. 2019. 104741.

[40] Mahanti，P.，Robinson，M. S.，Thompson，T. J.，& Henriksen，M. R. （2017）. Small lunar craters at the Apollo 16 and 17 landing sites - Morphology and degradation. Icarus，299，475 - 501. https：//doi. org/10. 1016/j. icarus. 2017. 08. 018.

[41] McKay，D. S.，Heiken，G.，Basu，A.，Blanford，G.，Simon，S.，Reedy，R.，et al. （1991）. The lunar regolith. In G. H. Heiken，D. T. Vaniman，& B. M. French（Eds.），Lunar sourcebook：A user's guide to the Moon（pp. 285 - 356）. Cambridge University Press.

[42] Melosh，H. J. （1989）. Impact cratering：A geologic process. Oxford University Press.

[43] Noble，S. K.，Pieters，C. M.，Taylor，L. A.，Morris，R. V.，Allen，C. C.，McKay，D. S.，& Keller，L. P. （2001）. The optical properties of the finest fraction of lunar soil：Implications for space weathering. Meteoritics & Planetary Sciences，36，31 - 42. https：//doi. org/10. 1111/j. 1945 - 5100. 2001. tb01808. x .

[44] Nouvel，J. F.，Herique，A.，Kofman，W.，& Safaeinili，A. （2004）. Radar signal simulation：Surface modeling with the Facet Method. Radio Science，39（1），1 - 17. https：//doi. org/10. 1029/2003RS002903 .

[45] Oberbeck，V. R.，& Quaide，W. L. （1967）. Estimated thickness of a fragmental surface layer of Oceanus Procellarum. Journal of Geophysical Research，72（18），4697 - 4704. https://doi. org/10. 1029/JZ072i018p04697.

[46] Olhoeft，G. R.，& Strangway，D. （1975）. Dielectric properties of the first 100 meters of the moon. Earth and Planetary Science Letters，24（3），394 - 404. https：//doi. org/10. 1016/0012 - 821X（75）90146 - 6.

[47] Owen，G. M. （1973）. Lunar surface science（Apollo 17 mission report JSC -

079044, pp. 1 - 41) .

[48] Paige, D. A. (2019) . LRO DLRE 4 calibrated radiance V1. 0 [Data set] . NASA Planetary Data System. https: //doi. org/10. 17189/1520651.

[49] Qiao, L. , Ling, Z. , Fu, X. , & Li, B. (2019) . Geological characterization of the Chang'e - 4 landing area on the lunar farside. Icarus, 333, 37 - 51. https: //doi. org/10. 1016/j. icarus. 2019. 05. 029 .

[50] Shkuratov, Y. G. , & Bondarenko, N. V. (2001) . Regolith layer thickness mapping of the Moon by radar and optical data. Icarus, 149 (2), 329 - 338. https: //doi. org/10. 1006/icar. 2000. 6545.

[51] Speyerer, E. , Povilaitis, R. , Robinson, M. , Thomas, P. , & Wagner, R. (2016) . Quantifying crater production and regolith overturn on the Moon with temporal imaging. Nature, 538 (7624), 215 - 218. https: //doi. org/10. 1038/nature19829 .

[52] Su, Y. , Fang, G. Y. , Feng, J. Q. , Xing, S. G. , Ji, Y. C. , Zhou, B. , et al. (2014) . Data processing and initial results of Chang'e - 3 lunar penetrating radar. Research in Astronomy and Astrophysics, 14 (12), 1623 - 1632. https: //doi. org/10. 1088/1674 - 4527/14/12/010.

[53] Swann, G. A. , Bailey, N. G. , Batson, R. M. , Freeman, V. L. , Hait, M. H. , Head, J. W. , et al. (1972) . Preliminary geologic investigation of the Apollo 15 landing site (Apollo 15 Preliminary Science Report NASA SP - 289, pp. 5 - 1 - 5 - 112) .

[54] Ulaby, F. , & Long, D. (2014) . Microwave radar and radiometric remote sensing. The University of Michigan Press.

[55] Widess, M. B. (1973) . How thin is a thin bed? Geophysics, 38 (6), 1176 - 1180. https: //doi. org/10. 1190/1. 1440403 .

[56] Wilcox, B. B. , Robinson, M. S. , Thomas, P. C. , & Hawke, B. R. (2005). Constraints on the depth and variability of the lunar regolith. Meteoritics & Planetary Sciences, 40 (5), 695 - 710. https: //doi. org/10. 1111/j. 1945 - 5100. 2005. tb00974. x .

[57] Wu, B. , Li, Y. , Liu, W. C. , Wang, Y. , Li, F. , Zhao, Y. , & Zhang, H. (2021) . Centimeter - resolution topographic modeling and fine - scale analysis of craters and rocks at the Chang'E - 4 landing site. Earth and Planetary Science Letters, 553, 116666. https: //doi. org/10. 1016/j. epsl. 2020. 116666 .

[58] Wu, W. , Li, C. , Zuo, W. , Zhang, H. , Liu, J. , Wen, W. , et al. (2019). Lunar farside to be explored by Chang'e - 4. ature Geoscience, 12 (4), 222 - 223. https: //doi. org/10. 1038/s41561 - 019 - 0341 - 7 .

[59] Xiao, Z. , Ding, C. , Xie, M. , Cai, Y. , Cui, J. , Zhang, K. , & Wang, J. (2021) . Ejecta from the Orientale basin at the Chang'E - 4 landing site. Geophysical Research Letters, 47, e2020GL090935. https: //doi. org/

10. 1029/2020GL090935.

[60] Xie，M.，Zhu，M. - H.，Xiao，Z.，Wu，Y.，& Xu，A.（2017）. Effect of topography degradation on crater size - frequency distributions：Implications for populations of small craters and age dating. Geophysical Research Letters，44，10171 - 10179. https：//doi. org/10. 1002/2017GL075298.

A 2 - year locomotive exploration and scientific investigation of the lunar farside by the Yutu - 2 rover

L. Ding[1], R. Zhou[1], Y. Yuan[1], H. Yang[1], J. Li[2], T. Yu[2], C. Liu[2,3], J. Wang[2], S. Li[1], H. Gao[1], Z. Deng[1], N. Li[1], Z. Wang[1], Z. Gong[1], G. Liu[4], J. Xie[2], S. Wang[2], Z. Rong[2], D. Deng[2], X. Wang[2,3], S. Han[2], W. Wan[5], L. Richter[6], L. Huang[1], S. Gou[5], Z. Liu[1], H. Yu[1], Y. Jia[7], B. Chen[7], Z. Dang[7], K. Zhang[2], L. Li[2], X. He[2], S. Liu[2], K. Di[5]

1. State Key Laboratory of Roboticsand System, Harbin Institute of Technology, Harbin 150080, China
2. Beijing Aerospace Control Center, Beijing 100094, China
3. Key Laboratory of Science and Technology on Aerospace Flight Dynamics, Beijing 10009, China.
4. Department of Aerospace Engineering, Ryerson University, Toronto, ONM5B 2K3, Canada
5. State Key Laboratory of Remote Sensing Science, Aerospace Information Research Institute, Chinese Academy of Sciences, Beijing 100101, China
6. Large Space Structures GmbH, Hauptstrassel, D - 85386Eching, Germany
7. China Academy of Space Technology, Beijing 100094, China

The lunar nearside has been investigated by many uncrewed and crewed missions, but the farside of the Moon remains poorly known. Lunar farside exploration is challenging because maneuvering rovers with efficient locomotion in harsh extraterrestrial environment is necessary to explore geological characteristics of scientific interest. Chang'E - 4 mission successfully targeted the Moon's farside and deployed a teleoperated rover (Yutu - 2) to explore inside the Von Kármán crater, conveying rich information regarding regolith, craters, and rocks. Here, we report mobile exploration on the lunar farside with Yutu - 2 over the initial 2 years. During its journey, Yutu - 2 has experi enced varying degrees of mild slip and skid, indicating that the terrain is relatively flat at large scales but scattered with local gentle slopes. Cloddy soil sticking on its wheels implies a greater cohesion of the lunar soil than encountered at other lunar landing sites. Further identification results indicate that the regolith resembles dry sand and sandy loam on Earth in bearing properties, demonstrating greater bearing strength than that identified during the Apollo missions. In sharp contrast to the sparsity of rocks along the traverse route, small fresh craters with unilateral moldable ejecta are abundant, and some of them contain high-reflectance materials at the bottom, suggestive of secondary impact events. These findings hint at notable differences in the surface geology between the lunar farside and nearside. Experience gained with Yutu-2 improves the understanding of the farside of the Moon, which, in return, may lead to locomotion with improved efficiency and larger range.

Introduction

Planetary exploration using orbiters or rovers has continued to extend the boundaries of science and technology in the last few decades (1-3). The Moon, the closest celestial body to Earth with potentially exploitable chemical and mineralogical resources (4), is one of the most substantial destinations of exploration missions. The lunar exploration effort peaked after the successful Apollo program; since then, there have been 20 successful landers on the near side of the Moon as a result of crewed and rover-based missions. However, no astronauts or landers arrived on the lunar farside until 2019, mainly due to the blindness of the side with respect to communication and monitoring, although evidence collected through remote sensing had indicated the distinction of the two sides in gravitational field (5) and material composition (6), providing a strong impetus for in situ exploration of the farside. The Chang'E-4 (CE-4) lunar exploration mission with the Yutu-2 rover was the first mission attempting to land on the farside of the Moon and was expected to make valuable and important scientific discoveries through in situ exploration.

Since the successful landing of the CE-4 probe (fig. S1A) on 3 January 2019, at 177.6°E, 45.4°S on the lunar farside in the Von Kármán crater inside the South Pole-Aitken (SPA) basin, Yutu-2 has been continuing scientific exploration for more than two Earth years (25 lunar days) (7). Yutu-2, as the lightest-ever lunar rover (135kg), is a six-wheeled high-performance off-road robot equipped with four steering motors at the corner wheels, having a maximum speed capability of 200 m/hour (about 0.056 m/s). Benefiting from the well-designed rocker-bogie structure, Yutu-2 can climb up 20° slopes and surmount obstacles up to 200-mm high (8). The meshed surface of the wheels makes them light (735g; diameter of 300mm and width of 150mm) (fig. S1B), and the lugs attached to the wheelsenhance the rover's driving performance. Yutu-2 carries four scientific payloads (fig. S1C) to acquire high-resolution images and high-accuracy data, including a panoramic camera (Pancam), a visible and near-infrared image spectrometer (VNIS), a lunar penetrating radar (LPR) (9), and an Advanced Small Analyzer for Neutrals (ASAN) (10). Its high-reliability locomotion system enabled Yutu-2 to survive beyond the designed 3-month life span, expanding its exploration distance to 600.55 m at the end of the first 2 years, leading to richer scientific insights from the unexplored zone. Over the course of the first 2 years, about 16.46 gigabytes of scientific data have been return-transmitted and analyzed, thus filling the gaps of geological knowledge of the lunar farside and deepening the understanding of the Moon's formation and evolution.

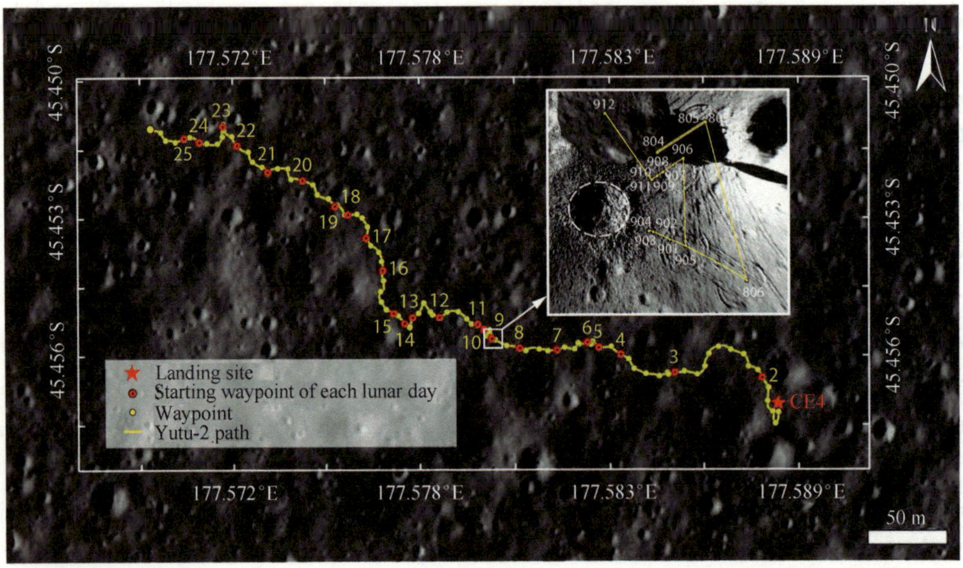

(a)

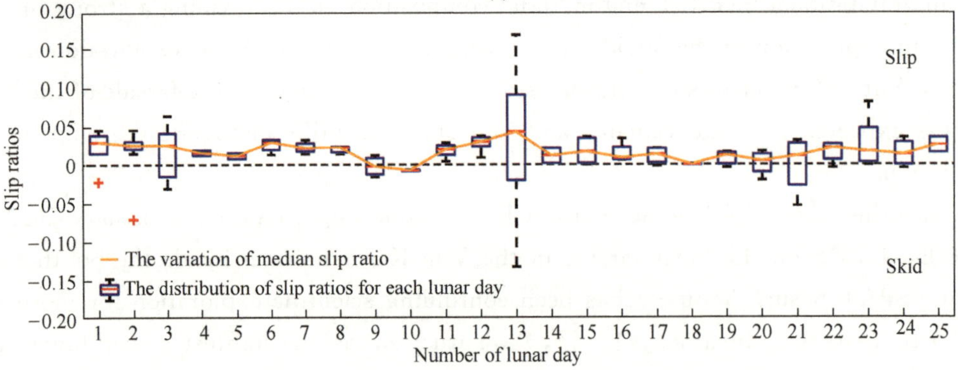

(b)

Fig. 1　Yutu-2's routing path and slippage on the farside of the Moon. (a) Routing path of the rover. The yellow dots represent waypoints. The red dots represent Yutu-2's starting waypoint of each lunar day; they are also the sleeping waypoint of the last lunar day. The rover passed through 132 waypoints in the first 25 lunar days. The exploration phase of the eighth and ninth lunar day is zoomed in to show detailed traverse routes. The base image is a high-resolution (0.9m per pixel) digital orthophoto map (DOM) (60) generated from LROC narrow angle camera (NAC) imagery (61). (b) Box plot of Yutu-2's wheel slip ratios along the routing path on each lunar day. Paths that are too short were dismissed in the statistics because they were usually without visual localization results or introduced too many errors. Paths controlled in driving time without planned target waypoints were also dismissed

Here, we present locomotive data and images collected by Yutu - 2 rover that reveal unique characteristics of the lunar farside. In addition, results of wheel slippage and sinkage, along with an analysis of the lunar regolith properties using the wheel - terrain interaction model, are presented. A comprehensive review of the scientific investigations carried out by Yutu - 2 in craters, rocks, and stratum was carried out with additional analyses. On the basis of the discovered distinct geological features on the lunar farside, more powerful rovers, as well as scientific payloads and theoretical conundrums on complex wheel - terrain interaction mechanisms, are exposed and realized to be necessary for future robotic lunar explorations.

Results

Travel route and wheel slippage

The Queqiao relay satellite (11), which is the first relay communication satellite outside Earth's orbit and features the largest satellite umbrella antenna diameter (4.2m), acts as a high - speed bridge to transmit data and commands between the ground station and Yutu - 2, with a time delay of 1 to 2 s in one - way transmission. Yutu - 2 has navigated toward the northwest (Fig. 1a) using a three - layered planning architecture, in which three - dimensional (3D) lunar surface data with 7 - m resolution obtained from the CE - 2 mission were used for long - term planning; the panoramic images were analyzed to generate roving trajectories for each lunar day, whereas the navigation images were used for replanning. Primarily, Yutu - 2 was directly controlled by uploaded commands generated from ground teleoperation; however, it can also work in autonomous mode on relatively flat terrains, planning traversable routes using its laser lattice or hazard avoidance camera (Hazcam). The autonomous mode of Yutu - 2 was used in a traverse of 2.7m on the fifth lunar day, in which autonomous planning was carried out five times, and each planning segment was about 0.5m. A vision - based decision - supporting method (12) consisting of obstacle map generation and path evaluation was used to ensure Yutu - 2's efficiency and safety. The physical properties of surface regolith strongly affected Yutu - 2's traverse trajectories, wheel slippages, and sinkage, which was, in turn, used to infer terrain properties in the tangential and normal directions.

The wheel slip ratios (hereafter referred to as slip ratios) for each lunar day (calculated using planned and actual travel distances derived from visual measurements between adjacent waypoints) (Fig. 1b) indicate that Yutu - 2 experienced relatively small wheel slipping and skidding. The slip ratios (defined as a function of the theoretical circumference velocity and the actual traveling longitu longitudinal velocity of the wheel) ranged from −0.15 to 0.15 and, in most cases, were within the range of 0 to 0.05, consistent with the observed small elevation variations along the traversed routes (fig. S2). On the ninth lunar day, the rover struggled to explore a localized fresh crater with shiny materials through several short approaching attempts without accurate visual localization on waypoints. Therefore, it was not possible to calculate the corresponding average slip ratios between any two adjacent waypoints, and they are missing from Fig. 1b. However, wheel trace imprints left by Yutu - 2 gave clues to the wheel slip ratios experienced during the complex movements (a to b, b to a, a to c, c to d, and d to c) as shown in Fig. 2. Fine - grained slip ratios based on clear and intact trace imprints (red frames in Fig. 2) were estimated using the slip ratio estimation model (13) by extracting the trace units as shown in fig. S3. When retracting from the crater (b to a and d to c), the estimated slip ratio was about -0.1, which was consistent with the decreased slope (3.10°) when departing from the crater rim. According to the symmetry mechanism of the slip ratio in the elastic phase ($-0.1<s<0.1$) (14), the slip ratio when maneuvering up the slope (a to b and c to d) was estimated as 0.1. The calculated slippage result (b to a; $s\approx-0.1$) was close to that when backing out from the other side of the crater (d to c; $s\approx-0.1$) but larger than that when roving on the flat surface (a to c; $s\approx0.0$), hinting that the stack ejecta was similar in terms of slope and properties on these two sides of the crater. Slip ratios derived from wheel trace imprints over short traverses helped to characterize the wheel slip experienced in local terrain, complementing the average slip ratios between two adjacent waypoints as shown in Fig. 1b.

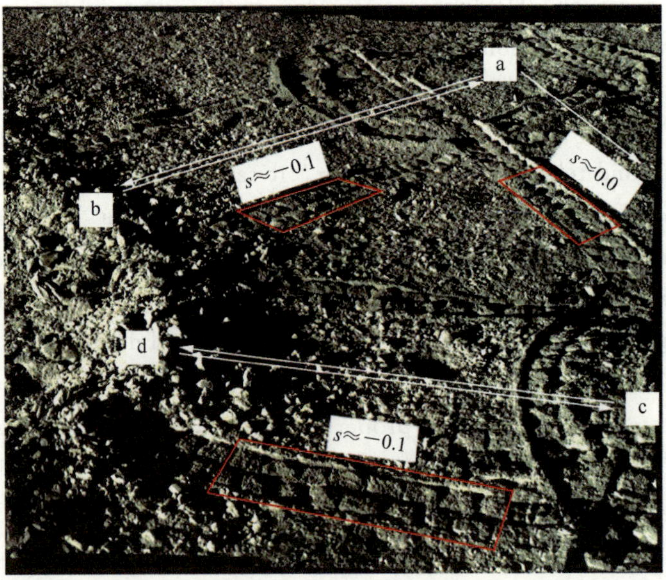

Fig. 2 Yutu - 2's wheel imprints and reflected slippage. Yutu - 2 first drove toward the target crater along trajectory a→b but failed in acquiring adequate data. Thus, it planned to approach the crater again from the northwest of the crater along the route c→d after coming down from the crater and turning around in circles to adjust direction. The wheel slip ratios estimated based on fragments of clear and intact trace imprints outlined by the red lines during the traverse of b→a, a→c, and d→c were estimated approximately as -0.1, 0, and -0.1, respectively. The image was taken by Pancam at the ninth lunar day at waypoint LE00912

Soil physical properties

After a 1.96 - m traverse from waypoints LE00804 to LE00805 (while backing from the crater), large agglomerates of cloddy soil were observed adhering onto Yutu - 2's wheels (Fig. 3a) . This can be compared with a previous CE - 3 mission during which the earlier Yutu rover experienced a similar traverse at the mission site of Mare Imbrium; only sticky and fine particles were observed on the wheels (Fig. 3b) . Although the wheels of Yutu were in the same configuration as that of Yutu - 2, the soil adhesion was negligible. According to the region division results (fig. S4), more than 46% of the wheel surface shown in the subfigure of Fig. 3a was covered with the cloddy soil, which is much larger than the coverage ratio observed on the wheels of CE - 3 Yutu (~2%; calculated using the subfigure of Fig. 3b) . Similar sharp contrasts between the Yutu - 2 and Yutu rovers, with respect to soil sticking to wheels, are also presented on their mutual shooting images (fig. S5) . The sticky fine particles on Yutu's wheels during CE - 3 were most likely due to electrostatic adhesion, but the larger pieces of soil on Yutu - 2's wheels during CE - 4 are unusual. Furthermore, Yutu - 2 observed similar phenomena on other lunar days, as portrayed by wheel images taken throughout the 4th, 5th, 6th, 19th, and 23rd lunar days of the entire 600.55 - m journey (fig. S6) . Thus, it is reasonable to assume that, at the CE - 4 landing site, the greater cohesion of the soil is regional rather than localized. The increased cohesion of the lunar regolith on the CE - 4 landing site is likely attributed to the higher percentage of agglutinates in the lunar regolith, which is in proportion with the soil maturity. This is supported by the findings that the Is/FeO maturity index of the CE - 4 landing site (82 ± 15) (15) is higher than that of the CE - 3 landing site (~53) (16), which indicates that the lunar soil at the CE - 4 landing site is more mature than that at the CE - 3 landing site (17) . According to the theory of plastic equilibrium, cohesion of materials is the bond that cements particles together, irrespective of the normal pressure exerted by one particle upon the others. On the other hand, particles of frictional masses can be held together only when a normal pressure exists among them (18) . With agglutinates, soil particles are more likely to agglomerate (hold together or build bond) when ground by the wheel, which changes the loose structure of aeolian sand soil, resultingin an increase in the solidification cohesion of the soil. Such blocky soil on the wheels would weaken the digging effect of the wheel lugs and consequently reduce the drawbar pull, which would be unfavorable to the rover's traction. Because most soil was observed to adhere on the wheel lugs rather than the high - porosity meshed surface, this situation may be improved in future missions by coating the lugs' surface with a special anti - adhesion material.

Although Yutu - 2 does not carry specific soil parameter identification instruments, the rover wheel can be used as a trenching device to identify soil parameters based on the wheel - terrain interaction. Regarding regolith properties, the normal bearing

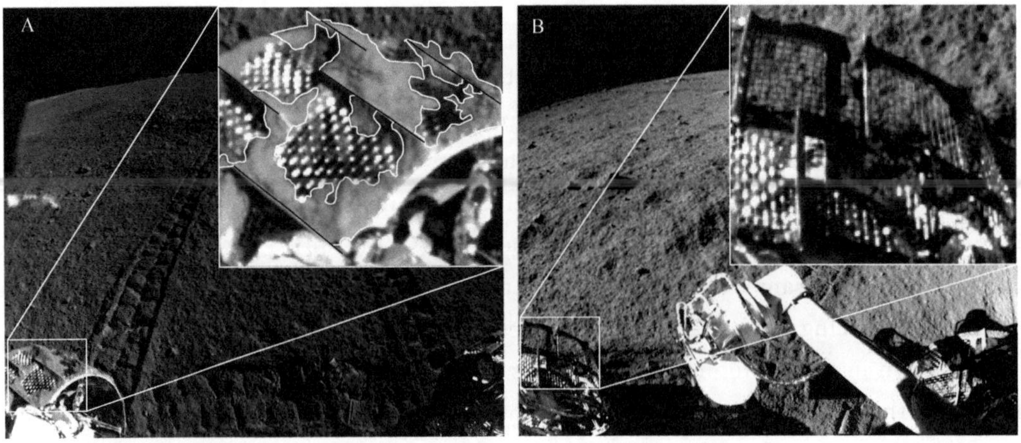

Fig. 3 Soil sticky phenomenon of Yutu - 2 compared with Yutu. (a) Large pieces of irregular - shaped lunar soil sticking to Yutu - 2's front left wheel. The image was taken by Hazcam of Yutu - 2 on the ninth lunar day on CE - 4 mission. (b) No soil adhesion on Yutu's front left wheel, which is clean with only fine particles. The image was taken by Hazcam after Yutu's arm dropped on CE - 3 mission

characteristics of the regolith around the landing site can be inferred based on wheel sinkage. During Yutu - 2's drive to point A (fig. S7) for mutual shooting, its wheels were almost supported only by lugs (the wheel rim was visibly in contact with the ground), but it sometimes sank slightly into the terrain and experienced moderate soil seepage into the wire mesh screen. By correcting the view to the orthographic angle of the wheel side images and calculating the proportion of the wheel sinkage to the wheel radius in the image (see Materials and Methods), the equivalent wheel sinkage was estimated to be 5 to 15mm. Compared with the wheel sinkage estimated at the CE - 3 landing site (5.6 to 11.5mm) (18), its range in CE - 4 (5 to 15mm) is a bit larger, under almost the same wheel load (22.5kg per wheel). The wheel sinkage of Lunokhod 1 in the Luna 17 mission was much larger (10.7 to 25.4mm) (19) under a heavier load (94.5kg per wheel). According to the terramechanics model for lugged wheels (20) (Fig. 4a) and wheel - terrain interaction parameters (table S1), we predicted the curves reflecting the regolith bearing property parameters under different sinkage conditions (Fig. 4b). Comparatively, the region of the estimated bearing parameters at CE - 3 (the region within two purple curves, Fig. 4a) mostly overlaps with that of the Luna 17 landing site (the region between two orange curves, Fig. 4a), whereas it has a larger equivalent sinkage modulus under the same sinkage exponent than that of CE - 4 on average. This indicates that the lunar regolith at the CE - 3 and Luna 17 sites is relatively stronger in bearing strength than that at the CE - 4 site. Assuming that the equivalent stiffness was 820kPa/m^N (21) (close to the typical value of the friction deformable modulus for lunar soil derived from Apollo missions; shown as the gray dashed line in Fig. 4b), the sinkage exponent (denoted by N) ranged from 0.83 to 1.0. When the sinkage exponent was assumed to be

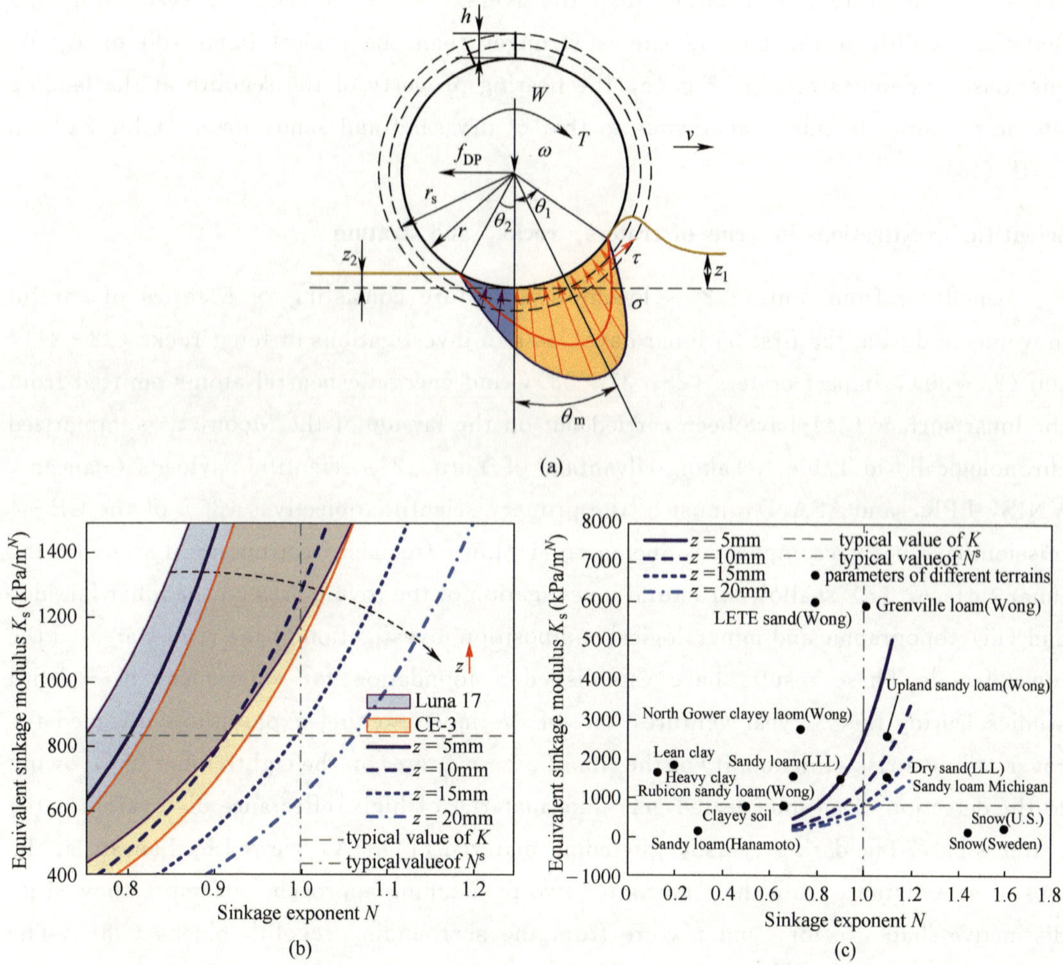

Fig. 4　Soil parameter identification. (a) Force diagram of wheel - terrain interaction of a planetary rover (20). The blue area is of active earth pressure, and the yellow area is of passive earth pressure. The blue curve circled area is where the normal stress occurred, whereas the red curve circled area is where the shear stress occurred. (b) Bearing characteristic curve under different wheel sinkage conditions. The blue region within the two purple lines corresponds to the bearing parameters under Lunokhod 1's wheel sinkage range in Luna 17 mission. The yellow region between the two orange lines corresponds to the bearing parameters under Yutu's wheel sinkage range in CE - 3 mission. (c) Bearing characteristic curve with bearing parameters of terrains on Earth. The bearing parameters of terrains on Earth are listed in table S2. In (b) and (c), N is the variable sinkage exponent of wheel - terrain interaction; for black scatter points representing parameters of terrains on Earth and the dashed - dotted lines representing typical parameters of terrains on the Moon, $N = n$ (intrinsic sinkage exponent of the terrain)

1.0 (22) (the typical value of lunar soil deduced from Apollo missions), the corresponding equivalent stiffness was 840 to 2800kPa/m^N. When the sinkage exponent and the equivalent stiffness modulus (denoted by K_s) were both assumed to be above the typical values of lunar regolith, the wheel sinkage (the intersection of the two gray dashed lines) was about 15.5mm, larger than the average wheel sinkage observed, implying that the regolith at the landing site is stronger than the typical lunar soil of Apollo missions. As demonstrated in Fig. 4c, the bearing property of the regolith at the landing site on the lunar farside is analogous to that of dry sand and sandy loam (table S2) on Earth (18).

Scientific investigations in terms of craters, rocks, and stratum

Benefiting from Yutu-2's locomotive venture consisting of a series of careful movements during the first 25 lunar days, in situ investigations of lunar rocks (22-24), soil (25-30), impact craters (28, 31-33), and energetic neutral atoms emitted from the lunar surface (34) have been carried out on the farside of the Moon, as summarized chronologically in Table 1. Taking advantage of Yutu-2's scientific payloads (Pancam, VNIS, LPR, and ASAN), most of the primary scientific objectives (35) of the CE-4 mission have been accomplished, including (i) low-frequency astronomical study on the lunar surface, (ii) shallow structural investigation of the roving area on the lunar farside, and (iii) topographic and mineralogical composition investigation of the roving area on the lunar farside. These results have established a foundation for subsequent in-depth studies. During this 2-year venture, one of the most exciting explorations involved the rover taking a risk of instability at the rim of a 2-m crater on the eighth lunar day, owing to the detection of an unexpected gel-like material of high reflectance discovered at the crater's base. The dark greenish, glistening material (Fig. 5A) viewed by Pancam at the closest safe distance from the crater after two painstaking approach-attempts showed its distinctive shape, color, and texture from the surrounding regolith blocks (28). The material's appearance, which resembles lunar samples 15466 and 70019 obtained by the Apollo 15 and 17 missions, suggests that it could be an impact melt breccia or a glass-coated agglutinitic regolith breccia with a possible complex formation of impact-generated welding, cementing, and agglutination of lunar regolith and breccia (28).

Table 1 Daily scientific exploration and key findings. The total working time of Yutu-2 was about 4019 hours for the first 25 lunar days. On most lunar days, Yutu-2 carried out routine navigation and detection using VNIS and LPR. It switched to rest mode at lunar night and noon due to the extreme temperatures. Key findings based on data collected on each lunar day are listed, beginning with asterisks and corresponding payloads, whereas special scientific exploration activities are listed without asterisks. LDay is an abbreviation for lunar day

Lunar day number	Accumulated distance (m)	Scientific exploration and key findings
1	44.19	* VNIS: Mental-derived materials in low-calcium pyroxene and olivine (30)
		Two-way mutual shooting
2	120.01	* LPR (LDay 1 and 2): The first high-resolution image of a lunar ejecta sequence and its thickness as well as internal architecture (41)

（续）

Lunar day number	Accumulated distance (m)	Scientific exploration and key findings
		* LPR：Lunar regolith source and a three - unit substructure at the landing site (25)
3	162.99	* VNIS and Pancam：An olivine - norite rock likely crystallized from the SPA - impact melt pool rich in plagioclase probably ejected from the Zhinyu crater (22)
4	178.97	Routine navigation and detection
5	190.77	* LPR：Dielectric properties and structure of lunar regolith (27)
		Detection of the first solar energetic particle (SEP) event with Lunar Lander Neutron and Dosimetry on the CE - 4 lander (34)
6	213.04	Imaging of wheel trace with Pancam
7	237.92	The farthest distance of a single walk (9.02m) on the lunar farside
		Neutral atom detection facing the sun and facing away from the sun for 4 hours，respectively，with ASAN
8	271.03	* Pancam：Discovery of high - reflection materials
9	284.66	* VNIS and Pancam：A dark greenish and glistening impact melt breccia in a crater and surrounding regolith in mixture of multiple sources (28)
		* LPR：The same physical properties between the beneath blocky ejecta deposits and the surrounding regolith，and observed fragments not originated from the subsurface (32)
		* LPR (LDay 1 to 9)：Deep subsurface structure of at least four layers of distinct lava flows and their thickness (43)
10	289.77	* VNIS：Ground - truth Hapke parameters provided for remote sensing observation (63)
11	318.62	Multipoint detection
12	345.06	* LPR：The shallow regolith structure and deep geological strata below the landing site (42)
13	357.70	Neutral atom detection for solar wind energy spectrum distribution
14	367.25	Neutral atom detection facing the sun and facing away from the sun for 4 hours，respectively，with ASAN
15	399.79	Detection of impact ejecta in front of a crater，and panoramic color imaging of the impact crater and mysterious substances in its center
16	424.45	Routine navigation and detection
17	447.68	* Pancam：Discovery of a crater with gel - like material
18	447.68	Due to the plan of launch Tianwen - 1，scientific payloads of Yutu - 2 stay switch off
19	463.26	Detection of a crater found on the 17th lunar day with VNIS and its ejecta on the west side of the crater
20	490.90	Routine navigation and detection
21	519.29	Routine navigation and detection
22	547.18	Routine navigation and detection
23	565.90	Panoramic color imaging of a large impact crater named Longtan，detection of small rocks on the southwest side of the impact crater，and updating of neutral atom detector parameters
24	589.64	Detection of surrounding solar wind particles with neutral atom detector
25	600.55	Six times of infrared spectrum detection and neutral atom detection

An inspection of the gel - like material，and even entering the crater，to conduct close - focused investigation using payloads would be of great scientific interest. However, because of the high probability of locomotor failures, the rover has been advised to limit any further movement. If Yutu - 2 drove down along the steep crater wall, it might be out of control due to the skid failure of the wheels. Furthermore, even if it enters the crater successfully, the rover will find maneuvering out of the crater highly difficult, according to antisymmetrical slip versus skid mechanics (14). For a wheel with skidding, the resistance force is in the same direction as the tractive force, both of which are resistance forces that increase with skid ratio. However, the resistance force counteracts the tractive force to decrease the drawbar pull when a wheel slips. As a result, even if the rover moves downward and upward along the same slope, its wheels may not be able to obtain enough drawbar pull for climbing back. Climbing attempts with severe wheel slippage is prone to cause large wheel sinkage, or even rover immobilization, similar to what was experienced by Spirit (36) and Opportunity (36), especially when traversing on soft and deformable terrain. One of the kernels of solving these problems is modeling fundamental principles of movement for the substrate of complex properties (like solidification or fluidization) during wheel - terrain interaction that relied on "robophysics" study (37, 38) with systematic tests. New theories based on simplified model of dynamic granular intrusion (39) and strategies of unconventional gaits of the rover (40) developed upon robophysics analysis are promising to cope with these dilemmas.

The CE - 4 landing site lacked surface boulders, but Yutu - 2 encountered several craters (fig. S8) (32). On the basis of the analysis of Lunar Reconnaissance Orbiter Camera (LROC) imagery from the Lunar Reconnaissance Orbiter launched in 2009, 88 impact craters can be quantifiably identified within a 50 - m range along Yutu - 2's routing path of the first 25 lunar days (Fig. 5B), ranging in diameter from 4.68 to 61.83m (mean value of 11.89m). Among them, more than 60% are less than 10m in diameter, with craters of diameter more than 20m being rare. From more detailed panoramic stereo images, 20 craters within a 10 - m distance from the rover can be measured, all with depths shallower than 0.6m and depth/diameter ratios (d/D) between 0.04 and 0.12, with the mean d/D value being 0.07 (33), suggesting that the craters are relatively fresh. Images acquired by the rover's visual systems along the traverse routes indicate broad morphologies of craters that can be mainly divided into three categories: first are heavily degraded craters (Fig. 5, C and D), with gentle slopes and flat edges bordering the ground, indicating that they have experienced a relatively long period of space weathering; in contrast, other craters (Fig. 5, E and F) have varied grain size ejecta (from fine particles to clods), with coarse walls and bottoms. These craters were further classified into two categories according to the different distributions of their ejecta, either evenly distributed or a skewed distribution to one side.

A high spatial density of fresh craters with unilateral ejecta can be observed at the CE - 4 landing site (fig. S8). High - resolution observations (fig. S9) show that the craters are almost all oriented to the northwest; thus, they are inferred to be sourced from the same set of impact events. Considering the impact cratering mechanism, the orientation of the ejecta is in line with the horizontal component of the impact force. Considering the fast overturn rate of surface regolith and small crater sizes, these craters are not assumed to be primary craters, but are most likely secondary craters formed by the Zhinyu crater, which is located to the west of the CE-4 landing site (32). Similar ejecta in skewed distribution were also observed at the landing site, as seen from a color image of the lander (fig. S10A). During the near-ground landing phase, the landing legs impacted the surface with oblique forces and dug out subsurface materials to form a one-sided deposit of fine-grained particles and relatively large clods on the summit. In contrast, the ejecta around the CE-3's lander legs (fig. S10B) resembles loose particles and are darker in color, compared with the surficial materials, and accompanied by small gravel.

Rocks along the traverse route (Fig. 6) are relatively rare in comparison with those scattered around the CE-3 landing site in meter size (fig. S11). Some rocks are perched on the surface as a result of projectiles and fragments of impact. As a comparison, unlike the densely clustered, various-sized fragments imaged by Pancam at the CE-3 landing site, rocks at the CE-4 landing site were relatively isolated and of a smaller scale (less than 40mm); some rocks were sparsely clustered, resembling pearls, whereas others were irregular or angular. In the CE-4 mission, rocks partially buried by regolith were scarcely observed; furthermore, relatively large rocks, such as those shown in fig. S11 (A to C) from the Mare Imbrium CE-3 site, are absent from observations of the first 25 lunar days. For CE-4, a distinct rock more than 200 mm in size measured by VNIS on the third lunar day was referred to as olivine norite from its spectrogram (similar in mineral composition as the regolith) and its fine-to-medium grain size texture, which was probably the result of relatively fast crystallization related to its origination from the impact melt pond, produced via the melting of the lunar lower crust and mantle materials by the SPA basin-forming event (22). A medium-sized (about 300 mm in diameter) angular rock isolated from its surroundings was observed in the onboard navigation camera (Navcam) image taken on the 23rd lunar day (Fig. 6H). Yutu-2 attempted to approach the rock by detouring around an extraordinary crater nicknamed "Longtan," but eventually ceased efforts due to the highly rugged and steeply sloped terrain (12°) and the presence of abundant rocks in the detection area.

The subsurface structure at the landing site, primarily consisting of low-loss, highly porous, granular materials, with embedded boulders of various sizes (41), was revealed on the basis of a dual-channel LPR data (CH-1, 60MHz; CH-2, 500MHz) (42) in both the shallow regolith and the deeper geological strata. The shallow regolith can be separated into three distinct layers according to the distribution, quantity, and dimension of inhomogeneities illustrated in the high-frequency LPR measurements (table

S3). At least five strata were detected to the depth of 328m (43), for which a uniform basal layer may be present between 38 and 52m, followed by further basal layers from different lunar geological periods. The buried ejecta is overlain by at least four layers of distinct lava flows that probably occurred during the Imbrium Epoch, with their thicknesses ranging from 12m to more than 100m, providing direct evidence of multiple lava - infilling events that occurred within the Von Kármán crater (43). To improve the estimation accuracy of the stratum depth, rectification and analysis of the LPR data are expected to be carried out with consideration of the rover's varying attitude. Furthermore, in situ sampling through drilling of great depth and contact measurements to be carried out in the future missions will accelerate the progress of stratigraphy research in the extraterrestrial environment.

Discussion

This work has described the lunar farside in situ exploration with the Yutu - 2 rover on a long - duration, multiyear CE - 4 mission. During the mission, Yutu - 2 experienced mild wheel slips and observed the surface characteristics and subsurface structure of the lunar farside. Studies into the lunar regolith have revealed increased cohesion in the regolith properties, as compared with that in the CE - 3 landing site, presumed as a result of higher percentage of agglutinate due to the longer exposure time and greater degree of space weathering. Summarization of Yutu - 2's in situ scientific investigations and discoveries with respect to the observed gel - like material, crater, and rock characteristics, as well as shallow subsurface structure, and additional analysis on differentiated fresh crater morphologies and ejecta distributions provide a comprehensive landscape overview of the lunar farside. Our study explicitly revealed the difference in the soil properties between the nearside and farside of the Moon with supportable in situ evidence. The scientific data and discoveries obtained by Yutu - 2 provide new resources on multiple aspects for further studies of the lunar farside and complement our understanding of the Moon.

Although the interesting soil cohesion phenomenon is analyzed to be regional with multiday observations, intensive study of the chemical composition and physical properties of the lunar regolith is necessary to be carried out for further analysis. More notable findings are expected to be discovered by taking full advantage of the rover's capability. For instance, trenching experiments can be conducted using one of the rover's wheels to expose the near - subsurface soil and estimate undisturbed terrain parameters more accurately.

On the basis of in - depth understanding of the lunar farside, more powerful locomotion, higher intelligence, and advanced scientific payloads are needed for future lunar exploration with space robots in larger scales and harsher environments. For example, legged robots, hybrid wheel - legged robots, or tethered rovers can be considered for future lunar crater or cave exploration. Physical intelligence (understanding the scenes surpassing purely geometric aspect), through active probing strategies for supportability or stability assessment, is necessary to be developed to cope with the unknown nature of the environment. Besides, targets of interest may be selected and targeted in a more efficient way, as implemented in the Curiosity rover (44), and explorations can be carried out in new discovery - oriented autonomous paradigm, with enhanced autonomous situated experimenting abilities. In terms of scientific payloads, diverse payloads for in situ sampling and chemical, physical, and biological analysis may be used with powerful executors and advanced data processing subsystems to break through the content, depth, and efficiency limitations and to satisfy the increasing requirement of comprehensive investigation. In general, multidisciplinary knowledge are expected to be incorporated to make an all - round improvement of rovers for future autonomous exploration to discover new insights into the Moon.

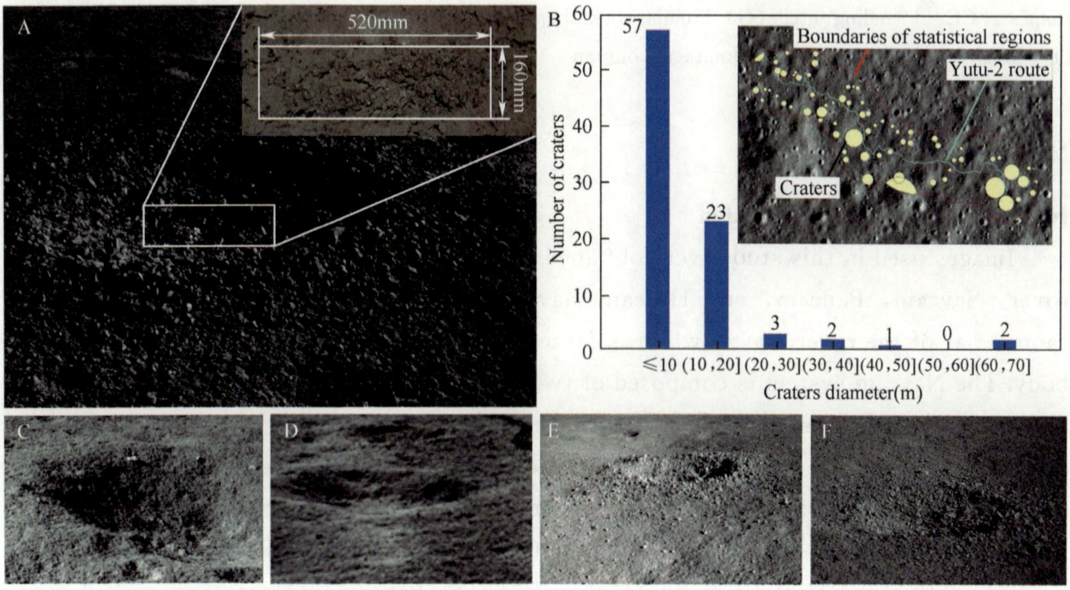

Fig. 5　Craters encountered by Yutu - 2 along the traversed route. (A) High - reflection materials encountered by Yutu - 2 on the ninth lunar day. (B) The statistical results on the diameter of craters are identified. The number of craters less than 10m in diameter is the highest. There are 88 craters identified within 50m of the routing path of Yutu - 2 on the first 25 lunar days. The base image is a LROC NAC image, whose resolution is 1m. The craters are visually identified and manually mapped based on the LROC NAC mosaic using the ArcGIS add - in tool of "CraterTools," which is independent of the map projection and has been widely used in planetary crater mapping (62) . Three evenly distributed points on the crater rim are digitized to derive the crater diameter with CraterTools. When digitizing the craters, a local tangent plane projection is adopted so that the craters appear circular, making crater mapping convenient and precise. (C and D) Old craters that have experienced a relatively long period of space weathering. (E and F) Fresh craters with ejecta on one side or around

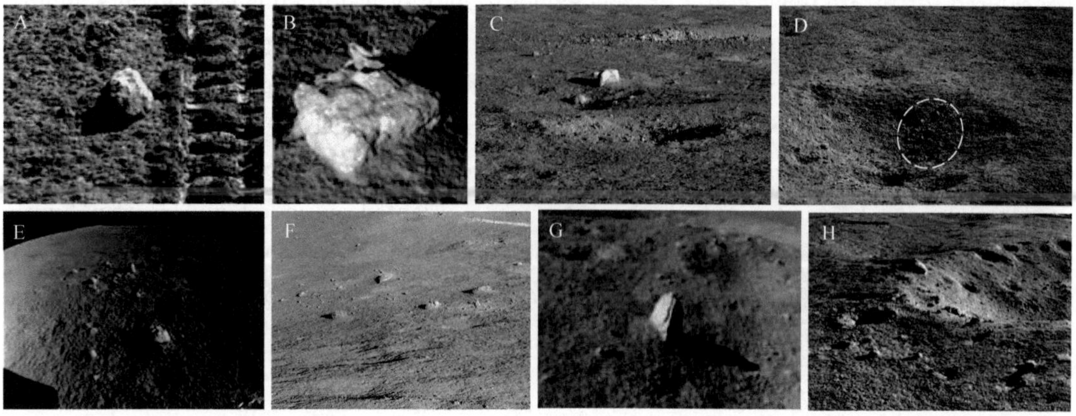

Fig. 6　Rocks encountered by Yutu - 2 along the traversed route. (A) Isolated rock. (B) Rock nicknamed "Qiyuan" discovered on the third lunar day from (15). (C) Rock perched on the surface. (D) Sparsely clustered gravels that are round, like pearls. (E and F) Sparsely clustered fragments in irregular or angular shapes at CE - 4 landing site. (G) Unique rock shaped like a sharp fin with a notable bulge in the edge. (H) Rock at the rim of crater nicknamed Longtan

Materials and methods

Instruments and data description

Images used in this study were obtained by three pairs of stereo cameras onboard the rover: Navcam, Pancam, and Hazcam. Navcam and Pancam are mounted on the same camera bar of the rover mast, whereas Hazcam is located at the lower front of the rover body. The Navcam system is composed of two identical cameras, and each camera has the ability to take gray - scale images with a field of view (FOV) of 46.6°. Navcam has a stereo base of 270mm, with a focal length of 17.7mm, and the effective image size is 1024 pixels by 1024 pixels (45). Each camera of the Pancam has the ability to take color (red: 640nm, green: 540nm, blue: 470 nm) and panchromatic (420 to 700 nm) images, with an FOV of 19.6° by 14.5° (45). The Pancam's stereo base is 270 mm, and the stereo images can be used to generate high - resolution topographic maps of the lunar surface using photogrammetric techniques. The effective image size is 2352 pixels by 1728 pixels in color mode and 1176 pixels by 864 pixels in panchromatic mode. The Pancam can acquire a 360° panoramic view by taking 56 pairs of images (28 different azimuth angles at two elevationangles) (45). In this study, we have used the images acquired by Hazcam to infer the property of the lunar regolith at the CE - 4 landing site. Each camera of Hazcam has a fisheye lens that offers a 170° FOV. The Hazcam acquires 1024 pixel - by - 1024 pixel stereo images, which are used to obtain 3D terrain information in front of the rover to enable hazard detection (e.g., large rocks and craters) along the rover's pathways (28).

Wheels of Yutu - 2 are used as a device to characterize the slippage experienced by the rover and identify soil parameters, based on the wheel - terrain interaction. The meshed wheel of Yutu - 2 is accompanied by lugs, and each wheel weighs 0.735 kg. Considering the gravity on the Moon, the vertical load on each wheel is estimated, based on the rover configuration and mass distribution at 36.5 N under a quasi - static state. The distance between the left and right wheels is 800mm. The radius of the wheel is 150 mm, and the width of the wheel is 150 mm. Each wheel is equipped with 24 lugs arranged in two staggered rows. The lugs in each row are evenly arranged on the wheel's outer cylinder surface, at a height of 10 mm.

Cross - site visual localization of Yutu - 2

The precise localization of the Yutu - 2 rover (hereafter referred to as Yutu - 2) was achieved using cross - site visual localization, based on the rover's Navcam images acquired at adjacent sites (waypoints) (45). In general, localization of rovers is achieved using the following key steps (46): (i) projection: cross - site images are projected to the ground plane to reduce distortions, if captured in opposite directions; (ii) area determination: initial localization results obtained by dead - reckoning are used to calculate the corresponding image area of the adjacent site (considered as the initial matching region) to limit the search range and ensure reliability of the subsequent matching; (iii) matching: images are enhanced for texture and grayscale consistency, and corresponding points are detected by the affine - scale invariant feature transform (ASIFT) matching algorithm (47) in the initial matching regions, and outlier detection is performed to obtain valid matching results; (iv) solving: accurate localization results are finally generated from the exterior orientation parameters derived from the bundle adjustment solution (48). Cross - site visual localization was conducted at the teleoperation center in near real time to provide localization results consisting of positions and attitudes in the landing site local (LSL) reference system. The LSL reference system is a north - east - down (NED) right - handed coordinate system, with the origin at the first waypoint; its z axis points down in the local normal direction, x axis points to the north pole, and y axis is orthogonal to the x and z axes. After a simple transformation, the coordinates of Yutu - 2's waypoints are provided in the landing site cartographic (LSC) coordinate system, which is an east - north - up (ENU) right - handed local system having the same origin as the LSL reference system, but its z axis points in the local normal direction, y axis points to the north pole, and x axis is orthogonal to the y and z axes (49). Previous research (50) has reported that the cross - site visual localization yields an accuracy of 4%, which is better than the nominal localization accuracy of 15% obtained from dead reckoning based on inertial measurement unit (IMU) and wheel odometer data. In practice, within a traverse distance of up to 10m between neighboring waypoints, the localization results of the cross - site visual localization and the dead reckoning from IMU and wheel odometry were generally consistent with differences in the levels of centimeters to decimeters, depending on the severity of rover slippage (51) .

Average slip ratio estimation between two adjacent waypoints

The slip ratio s of a lugged wheel at each moment t is defined as follows:

$$s(t)=\begin{cases}1-v(t)/(r_s(t)\omega(t))(r_s(t)\omega(t)\geqslant v(t),\ 0\leqslant s(t)\leqslant 1)\\ r_s(t)\omega(t)/v(t)-1(r_s(t)\omega(t)<v(t),\ -1\leqslant s(t)<0)\end{cases} \tag{1}$$

where ω (t) is the angular velocity function, r_s (t) is the shearing radius function, and $v(t)$ is the linear velocity function. When the value of s is larger than zero, it indicates that the wheel is slipping; when s is equal to 0, the wheel rolls without slipping or skidding; when s isless than 0, it indicates that the wheel is skidding, with $|s|$ being the values of the skid ratio. In general, a driving wheel slips when moving on a flat terrain or climbing up a slope, whereas it skids when moving down a slope (13) .

The average slip ratio s of a lugged wheel on Yutu - 2 over a traverse between two adjacent waypoints can be expressed using the average angular velocity $\overline{\omega}$, average linear velocity $\overline{v}$, and average shearing radius $\overline{r}_s$ as follows:

$$\overline{\omega}=\int_0^{\Delta t_p}\omega(\tau)d\tau/\Delta t_p \tag{2A}$$

$$\overline{v}=\int_0^{\Delta t_p}v(\tau)d\tau/\Delta t_p \tag{2B}$$

$$\overline{s}=\begin{cases}1-\overline{v}/(\overline{r}_s\overline{\omega})(\overline{r}_s\overline{\omega}\geqslant\overline{v},\ 0\leqslant s\leqslant 1)\\ \overline{r}_s\overline{\omega}/\overline{v}-1(\overline{r}_s\overline{\omega}<\overline{v},\ -1\leqslant s<0)\end{cases} \tag{2C}$$

where Δt_p is the travel time of Yutu - 2 between two adjacent waypoints. The average shearing radius, $\overline{r}_s$, can be computed as follows (13):

$$\overline{r}_s=r+\lambda_s h \tag{3}$$

where r is the wheel radius, h is the height of the lugs, and λ_s $(1\leqslant\lambda_s\leqslant 0)$ is the lug coefficient determined by the number of lugs and the internal friction angle of the soil.

Equation 2C can be rearranged as follows:

$$\overline{s}=\begin{cases}1-\overline{v}\Delta t_p/(\overline{r}_s\overline{\omega}\Delta t_p)(\overline{r}_s\overline{\omega}\Delta t_p\geqslant\overline{v}\Delta t_p,\ 0\leqslant s\leqslant 1)\\ \overline{r}_s\overline{\omega}\Delta t_p/(\overline{v}\Delta t_p)-1(\overline{r}_s\overline{\omega}\Delta t_p<\overline{v}\Delta t_p,\ -1\leqslant s<0)\end{cases} \tag{4}$$

and further simplified as

$$\overline{s}=\begin{cases}1-L_{gt}/L_p(L_p\geqslant L_{gt},\ 0\leqslant s\leqslant 1)\\ L_p/L_{gt}-1(L_p<L_{gt},\ -1\leqslant s<0)\end{cases} \tag{5}$$

where L_p is the planned travel mileage of Yutu - 2 derived from locomotion commands and L_{gt} is the actual travel mileage.

Note that Yutu - 2 takes images only at waypoints, and no images are captured during the traverse; hence, only the actual travel distance between two waypoints can be obtained. Thus, when Yutu - 2 travels in curves, the traverse mileage (arc length) is replaced with its corresponding chord length, which is the travel distance. Considering the locomotion states obtained during a traverse between two adjacent waypoints, Eq. 5 can be further rearranged as follows:

$$\bar{s}=\begin{cases}1-D_{gt}/D_p(D_p\geqslant D_{gt},\ 0\leqslant s\leqslant 1)\\ D_p/D_{gt}-1(D_p<D_{gt},\ -1\leqslant s<0)\end{cases}\quad(6)$$

where D_p is the planned travel distance of Yutu - 2 calculated using the coordinates of the starting and ending waypoints represented by x and y, respectively, and D_{gt} is the actual travel distance of Yutu - 2 estimated by visual localization.

In general, accurate wheel slip ratio at each moment can be estimated using Eq. 1 with the measured rover speed. Unfortunately, the speed of Yutu - 2 during movement was not available, and Eq. 6 was used for average slip ratio estimation over two adjacent waypoints using the available distance data. In practice, on the basis of the raw localization results of Yutu - 2 coordinates in the LSC reference system, the relative position between two adjacent waypoints can be computed as (Δx, Δy), and the actual travel distance can be calculated as $\sqrt{\Delta x^2+\Delta y^2}$. On the other hand, in the CE - 4 mission, the planned waypoints of Yutu - 2 are known and given in (x, y) coordinates in the LSC reference system. Thus, it is simple to calculate the planned distance between two adjacent waypoints according to their relative position in the x and y directions, according to the actual travel distance. Although the results calculated using Eq. 6 were obtained without knowing the vehicle speed, they are meaningful because the distance between two waypoints is short (less than 14. 28 m) and the terrain slope variation is small. Because, under these conditions, the fluctuation of the wheel slip ratio should not be large, implying that the fluctuation of rovers' velocity is small, together with the equal and constant angular velocities of all the wheels, the effect of speed variation of the vehicle during the movement should be small.

Slip ratio estimation based on wheel trace imprint

Rover trace contours can be used to infer the wheel - terrain interaction characteristics. A clear and intact wheel imprint with a longitudinal slip, as shown in fig. S3, consists of a series of trace units, and each trace unit can be divided into two areas: the lug - sheared area dug out by the lug itself and the hub - pressed area compressed by the wheel hub. The distance between two adjacent trace units is called the trace unit width and is denoted as Δx_p.

The lug - sheared and hub - pressed areas occur alternatively, resulting in a staggered pattern of the trace imprint. A larger slip ratio of the wheel is accompanied by a more compact pattern of lug imprint, and the lug imprint of the trace is condensed with an increase in the wheel slip ratio. Thus, the slip ratio can be estimated based on the structural features of the wheel trace imprints without considering the lateral slip ratio, as follows (52):

$$\bar{s}=1-\Delta x_p n/(2\pi\bar{r}_s)(0\leqslant\bar{s}\leqslant 1)\quad(7)$$

where n is the number of lugs on the wheel. Because n and $\bar{r}_s$ are constants determined by the wheel configuration, $\bar{s}$ can be estimated by measuring Δx_{p}.

By taking advantage of the same ratio of the trace unit width relative to the wheel width on both the orthograph and the actual size, we can estimate the distance between two adjacent trace units, Δx_{p}, for a measured length of the trace unit on the trace imprint orthograph, p_{u}, relative to the length of the wheel width on the trace imprint orthograph p_{w}:

$$\Delta x_{\mathrm{p}} = b \times \frac{p_{\mathrm{u}}}{p_{\mathrm{w}}} \tag{8}$$

where b is the wheel width.

An image processing method was introduced to extract Δx_{p} with two key steps: view correction and trace feature extraction. The original image was first corrected with the camera extrinsic matrix to a top - view image. Then, each trace imprint unit was manually divided into independent areas, and trace imprint information, such as the trace unit width, were acquired. The wheel slip ratio was then estimated from the trace imprint using Eq. 7. When a series of continuous trace units were almost of the same width, their average trace unit widths were calculated to reduce the measurement error, and the average wheel slip ratio was deduced to represent the wheel slippage experienced by Yutu - 2.

Soil parameter identification

Because Yutu - 2 does not carry specific soil parameter identification instruments, the rover wheel was used as a trenching device to identify soil parameters, based on the wheel - terrain interaction. To estimate the soil parameters, the wheel - terrain interaction should be modeled. For a lugged rover wheel moving on the soft and deformable terrain with an angular velocity ω, as shown in Fig. 4A, the wheel is applied with a vertical load W and a resistance force f_{DP} from the vehicle suspension, as well as a driving torque T at the wheel rotation axis by an actuator. The terrain interacts with the wheel circumference in the contact region, which corresponds to the angle divided into two parts: the entrance angle θ_1 from the vertical, at which the wheel begins to make contact with the soil, and the exit angle θ_2 from the vertical, where the wheel loses contact with the soil. In the wheel - terrain interaction region ($\theta_1 + \theta_2$), the continuous normal stress to support the wheel, and the shearing stress τ due to the relative movement are exerted on the wheel surface, as shown in Fig. 4A. The point of maximum stress is denoted as θ_{m}, according to which the stress region is divided into a forward part (σ_1, τ_1), corresponding to the angle from θ_1 to θ_{m}, and a rear part (σ_2, τ_2), corresponding to the angle from θ_{m} to θ_2.

Considering the shape of the wheel surface, the distributed normal stress along the wheel circumference is deduced from the Wong - Reece model (53) and improved by considering the slip sinkage and lug effects. The improved model is named Ding's model (54), which uses $n_0 + n_1 s$ to substitute for the conventional sinkage exponent n and can be used for small wheels with high slip sinkage. Ding's model can be expressed as follows:

$$\begin{cases}\sigma_1(\theta)=\left(\dfrac{k_c}{b}+k_\varphi\right)r^N(\cos\theta-\cos\theta_1)^N(\theta_m\leqslant\theta\leqslant\theta_1)\\ \sigma_2(\theta)=\left(\dfrac{k_c}{b}+k_\varphi\right)r^N\left\{\cos\left[\theta_1-\dfrac{\theta-\theta_2}{\theta_m-\theta_2}(\theta_1-\theta_m)\right]-\cos\theta_1\right\}^N(\theta_2\leqslant\theta<\theta_m)\end{cases}\tag{9}$$

where b is the width of the wheel, k_c is the cohesive modulus of the soil, k_φ is the frictional modulus of the soil, and N is the variable sinkage exponent of wheel - terrain interaction that can be represented by a linear function of slip ratio s, as follows:

$$N=n_0+n_1 s\tag{10}$$

where n_0 represents the static sinkage exponent and n_1 represents the dynamic sinkage resulting from wheel slippage. The values of n_0 and n_1 were determined on the basis of experiments, and the introduction of the linear soil sinkage exponent led to a high affinity for fitting the slip - sinkage phenomenon of lugged wheels (20). Previously, the dynamic sinkage caused by slip ratio was not considered; thus, it was deemed that $N=n$, where n is the intrinsic sinkage exponent of the terrain.

The entrance angle θ_1 is a function of wheel sinkage z and can be expressed as

$$\theta_1=\arccos[(r-z)/r]\tag{11}$$

and the exit angle θ_2 can be computed as

$$\theta_2=c_3\theta_1\tag{12}$$

where c_3 is the coefficient of the exit angle, characterizing the ratio of the exit angle θ_2 to the entrance angle θ_1. According to Rankine's theory of earth pressure (55), the active earth pressure (caused by the soil rebound) on deformable terrain (stress integration from θ_m to θ_2) is much smaller than the passive pressure (stress integration from θ_1 to θ_m). Therefore, the active earth pressure is generally negligible (56), and the corresponding exit angle θ_2 is assumed to be 0. Consequently, the coefficient c_3 for calculating the exit angle θ_2 is taken as zero in the calculation.

The maximum stress angle θ_m can be computed as

$$\theta_m=(c_1+c_2 s)\theta_1\tag{13}$$

where c_1 and c_2 are the coefficients of the wheel - terrain interaction angle. In the calculation, c_1 and c_2 are set to 0.5 and 0, respectively, and the value of θ_m is equal to 0.5 times of θ_1. As shown in Fig. 4A, the stress distribution curves (blue line: the normal distribution curve; red line: the shear stress distribution curve) are asymmetric with respect to the angular location of the maximum stress θ_m. However, in this case, θ_2 is assumed to be 0; thus, the angular location of maximum stress θ_m happens to occur midway between θ_1 and θ_2.

The tangential stress $\tau(\theta)$ can be computed as

$$\tau(\theta)=(c+\sigma(\theta)\tan\varphi)+\{1-\exp(-r_s[(\theta'_1-\theta)-(1-s)(\sin\theta'_1-\sin\theta)]/K)\}\tag{14}$$

where c is the cohesion of the soil, φ is the internal friction angle, K is the shearing deformation modulus, r_s is the equivalent shearing radius, and θ_1' is the equivalent entrance angle. With the lug effect, the shearing stress occurs on the surface of the soil sticking to the wheel circumference due to the wheel lugs instead of the wheel outer cylinder surface, and the equivalent shearing radius r_s can therefore be defined as follows (13):

$$r_s = r + \lambda_s h \, (0 \leqslant \lambda_s \leqslant 1) \tag{15}$$

where h and λ_s have aforementioned in Eq. 3. For Yutu - 2's wheels, which are evenly equipped with lugs at 10 - mm height, the value of λ_s is about 0.5, according to the experimental results in (13). The equivalent entrance angle θ_1' of a lugged wheel (20) can becomputed as follows:

$$\theta_1' = \arccos[(r - z)/(r + h)] \tag{16}$$

When θ approaches θ_m, the corresponding normal stress and tangential stress approach their maximum, expressed as follows:

$$\sigma_m = \left(\frac{k_c}{b} + k_\varphi\right) r^N (\cos\theta_m - \cos\theta_1)^N \tag{17}$$

$$\tau_m = (c + \sigma_m \tan\varphi) \times \{1 - \exp(-r_s[(\theta_1' - \theta_m) - (1 - s)(\sin\theta_1' - \sin\theta_m)]/K)\} \tag{18}$$

When the wheel is in a quasi - static state, the effect of the distributed stress (normal stress σ and tangential stress τ) can be simplified to the normal force F_N, drawbar pull F_{DP}, and driving resistance torque M_R by integrating along the wheel - terrain interaction area, which are balanced with the wheel load W, resistance force f_{DP}, and driving torque T, respectively. For wheels on Yutu - 2 having a maximumspeed of 200 m/hour, the quasi - static condition is valid because the dynamic effects are negligible at low speeds. Therefore, the force/torque balance equations for Yutu - 2's lugged wheels can be expressed as follows:

$$F_N = b\{\int_{\theta_2}^{\theta_m} [r\sigma_2(\theta)\cos\theta + r_s\tau_2(\theta)\sin\theta]d\theta + \int_{\theta_m}^{\theta_1} [r\sigma_1(\theta)\cos\theta + r_s\tau_1(\theta)\sin\theta]d\theta\} = W \tag{19A}$$

$$F_{DP} = b\{\int_{\theta_2}^{\theta_m} [r_s\tau_2(\theta)\cos\theta - r\sigma_2(\theta)\sin\theta]d\theta + \int_{\theta_m}^{\theta_1} [r_s\tau_1(\theta)\cos\theta - r\sigma_1(\theta)\sin\theta]d\theta\} = f_{DP} \tag{19B}$$

$$M_R = r_s^2 b[\int_{\theta_2}^{\theta_m} \tau_2(\theta)d\theta + \int_{\theta_m}^{\theta_1} \tau_1(\theta)d\theta] = T \tag{19C}$$

According to Shibly et al. (57), the wheel - terrain interaction stress can be linearized as follows:

$$\begin{cases} \sigma_1(\theta) = \sigma_m(\theta_1 - \theta)/(\theta_1 - \theta_m)(\theta_m \leqslant \theta \leqslant \theta_1) \\ \sigma_2(\theta) = \sigma_m(\theta - \theta_2)/(\theta_m - \theta_2)(\theta_2 \leqslant \theta \leqslant \theta_m) \end{cases} \tag{20}$$

$$\begin{cases} \tau_1(\theta) = \tau_m(\theta_1 - \theta)/(\theta_1 - \theta_m)(\theta_m \leqslant \theta \leqslant \theta_1) \\ \tau_2(\theta) = \tau_m(\theta - \theta_2)/(\theta_m - \theta_2)(\theta_2 \leqslant \theta \leqslant \theta_m) \end{cases} \tag{21}$$

Because the entrance angle is usually not large, and when θ approaches θ_1, the corresponding normal stress and tangential stressall approach zero; Ding (58) proposed that the product of $\cos\theta$ and stress can also be linearized, as shown in Eqs. 22 and 23. Let $\sigma'_m = \sigma_m \cos\theta_m$ and $\tau'_m = \tau_m \cos\theta_m$.

$$\begin{cases} \sigma_1(\theta)\cos\theta = \sigma'_m(\theta_1 - \theta)/(\theta_1 - \theta_m) \\ \sigma_2(\theta)\cos\theta = \sigma'_m(\theta - \theta_2)/(\theta_m - \theta_2) \end{cases} \tag{22}$$

$$\begin{cases} \tau_1(\theta)\cos\theta = \tau'_m(\theta_1 - \theta)/(\theta_1 - \theta_m) \\ \tau_2(\theta)\cos\theta = \tau'_m(\theta - \theta_2)/(\theta_m - \theta_2) \end{cases} \tag{23}$$

By applying Eqs. 20 to 23 in the wheel - terrain interaction model of Eq. 19 and ignoring the vertical component of the shearing stress, we obtained a simplified expression of F_N as follows

$$F_N \approx rb(\theta_1 - \theta_2)\sigma_m \cos\theta_m / 2 \tag{24}$$

whereas Eq. 24 can be rearranged as

$$\sigma_m = \frac{2F_N}{rb(\theta_1 - \theta_2)\cos\theta_m} \tag{25}$$

Note that Eqs. 17 and 25 represent two expressions for the maximum normal stress σ_m. σ_m can be calculated using Eq. 25 using the estimated F_N, wheel sinkage, and wheel configuration parameters. The three bearing characteristic parameters, namely, cohesive modulusof soil k_c, frictional modulus of soil k_φ, and sinkage exponent N, remained unknown in Eq. 17. Because Yutu - 2's wheels rolled in different contact states with the soil, various σ_m, θ_m, and θ_1 were obtained, which were then used with Eq. 17 to solve the unknown characteristic parameters in that equation. Because the wheel width b for Yutu - 2 is a constant and the bearing characteristic parameters k_c, k_φ and N for the same type of soil are also constants, the term $(k_c/b + k_\varphi)$ in Eq. 17 is a constant. Thus, under these conditions, $k_c$and k_φ cannot be solved using Eq. 17.

Therefore, the equivalent stiffness modulus K_s ($K_s = k_c/b + k_\varphi$) was estimated, instead of k_c and k_φ. The typical values of N obtained from the Apollo 15 mission were used with Eq. 17 to obtain the estimated soil stiffness K_s, which indicates the comparative stiffness of the lunar soil on the farside compared with that estimated for the soil sample obtained from the Apollo missions. The same method was used to estimate N of the soil. The process is illustrated in fig. S12.

In practice, the values of wheel radius r, wheel width b, wheel lug height h, and equivalent radius r_s were constant and determined by Yutu - 2's wheel configuration, as shown in table S1. The normal force F_N was estimated from the vertical load W, which can be computed from a quasi - static force analysis of the rover with the knowledge of rover configuration and mass distribution. The two key wheel motion state variables, slip ratio s and sinkage z, can be computed using vision - based techniques or by kinematic analysis of the rover suspension. The wheel sinkage in this study was estimated on the basis of images (fig. S6) that show the edges of the rover wheels. Through image processing (59), the circular contour of the wheel and the crooked wheel - interaction curve were extracted on the orthographic image of the rover wheel. Then, the curve was fitted into a straight line using the least squares method on the pixel plane, as shown in fig. S13. Furthermore, the radius of the wheel r_p and the distance of the wheel center to the wheel - terrain interaction line l_p on the pixel plane were acquired, and the wheel sinkage was then estimated using the following equation:

$$z = \frac{r_p - l_p}{r_p} r_m \tag{26}$$

where z is the wheel sinkage and r_m is the actual radius of the wheel. The soil - bearing characteristic parameters of the Apollo mission from (21) - including the cohesive modulus k_c, frictional modulus k_φ, and sinkage exponent N - are also shown in table S1. On the basis of the given parameters, for a wheel sinkage of 10mm, the equivalent stiffness modulus (denoted by K_s) was estimated to be 1280kPa/m^N, and the sinkage exponent N was estimated to be 0.91.

Acknowledgments

We thank all the scientists and engineers who contributed to the Chang'E - 4 mission - in particular those from the China Academy of Space Technology (CAST), Shanghai Academy of Spaceflight Technology (SAST), Harbin Institute of Technology (HIT), and Beijing Aerospace Control Center (BACC) - for their dedicated work on the Yutu - 2 lunar rover.

References and notes

[1] C. M. Pieters, J. N. Goswami, R. N. Clark, M. Annadurai, J. Boardman, B. Buratti, J. - P. Combe, M. D. Dyar, R. Green, J. W. Head, C. Hibbitts, M. Hicks, P. Isaacson, R. Klima, G. Kramer, S. Kumar, E. Livo, S. Lundeen, E. Malaret, T. McCord, J. Mustard, J. Nettles, N. Petro, C. Runyon, M. Staid, J. Sunshine, L. A. Taylor, S. Tompkins, P. Varanasi, Character and spatial distribution of OH/H_2O on the surface of the Moon seen by M3 on Chandrayaan - 1. Science 326, 568 - 572 (2009).

[2] C. Zeitlin, D. M. Hassler, F. A. Cucinotta, B. Ehresmann, R. F. Winner - Schweingruber, D. E. Brinza, S. Kang, G. Weigle, S. Bottcher, E. Bohm, S. Burmeister, J. Guo, J. Kohler, C. Martin, A. Posner, S. Rafkin, G. Reitz, Measurements of energetic particle radiation in transit to Mars on the Mars Science Laboratory. Science 340, 1080 - 1084 (2013).

[3] L. Xiao, P. Zhu, G. Fang, Z. Xiao, Y. Zou, J. Zhao, N. Zhao, Y. Yuan, L. Qiao, X. Zhang, H. Zhang, J. Wang, J. Huang, Q. Huang, Q. He, B. Zhou, Y. Ji,

Q. Zhang, S. Shen, Y. Li, Y. Gao, A young multilayered terrane of the northern Mare Imbrium revealed by Chang'e - 3 mission. Science 347, 1226 - 1229 (2015).

[4] M. Anand, I. A. Crawford, M. Balat - Pichelin, S. Abanades, W. van Westrenen, G. Péraudeau, R. Jaumann, W. Seboldt, A brief review of chemical and mineralogical resources on the Moon and likely initial in situ resource utilization (ISRU) applications. Planet. Space Sci. 74, 42 - 48 (2012).

[5] N. Namiki, T. Iwata, K. Matsumoto, H. Hanada, H. Noda, S. Goossens, M. Ogawa, N. Kawano, K. Asari, S. Tsuruta, Y. Ishihara, Q. Liu, F. Kikuchi, T. Ishikawa, S. Sasaki, C. Aoshima, K. Kurosawa, S. Sugita, T. Takano, Farside gravity field of the Moon from four - way doppler measurements of SELENE (Kaguya). Science 323, 900 - 905 (2009).

[6] D. J. Lawrence, W. C. Feldman, B. L. Barraclough, A. B. Binder, R. C. Elphic, S. Maurice, D. R. Thomsen, Global elemental maps of the Moon: The lunar prospector gamma - ray spectrometer. Science 281, 1484 - 1489 (1998).

[7] J. Liu, X. Ren, W. Yan, C. Li, H. Zhang, Y. Jia, X. Zeng, W. Chen, X. Gao, D. Liu, X. Tan, X. Zhang, T. Ni, H. Zhang, W. Zuo, Y. Su, W. Wen, Descent trajectory reconstruction and landing site positioning of Chang' e - 4 on the lunar farside. Nat. Commun. 10, 4229 (2019).

[8] P. Norris, Returning People to the Moon After Apollo: Will It Be Another Fifty Years (Springer, 2019).

[9] G. Fang, B. Zhou, Y. Ji, Q. Zhang, S. Shen, Y. Li, H. Guan, C. Tang, Y. Gao, W. Lu, S. Ye, H. Han, J. Zheng, S. Wang, Lunar Penetrating Radar onboard the Chang' e - 3 mission. Res. Astron. Astrophys. 14, 1607 - 1622 (2014).

[10] M. Wieser, S. Barabash, X. - D. Wang, A. Grigoriev, A. Zhang, C. Wang, W. Wang, The advanced small analyzer for neutrals (ASAN) on the Chang' e - 4 rover Yutu - 2. Space Sci. Rev. 216, 73 (2020).

[11] L. Zhang, L. Xiong, J. Sun, L. Chen, One year on - orbit operation of Queqiao lunar relay communications satellite. J. Aerosp. China 20, 5 - 13 (2019).

[12] Y. Wang, W. Wan, S. Gou, M. Peng, Z. Liu, K. Di, L. Li, T. Yu, J. Wang, X. Cheng, Vision - based decision support for rover path planning in the Chang' e - 4 mission. Remote Sens. 12, 624 (2020).

[13] L. Ding, H. Gao, Z. Deng, K. Yoshida, Slip ratio for lugged wheeled of planetary rover in deformable soil: Definition and estimation, in Proceedings of the 2009 IEEE/RSJ International Conference on Intelligent Robots and Systems (IEEE, 2009), pp. 3343 - 3348.

[14] L. Ding, H. Gao, Z. Deng, J. Guo, G. Liu, Longitudinal slip versus skid of planetary rovers' wheels traversing on deformable slopes, in Proceedings of the 2013 IEEE/RSJ International Conference on Intelligent Robots and Systems (IEEE, 2013), pp. 2842 - 2848.

[15] S. Gou, Z. Yue, K. Di, W. Wan, Z. Liu, B. Liu, M. Peng, Y. Wang, Z. He, R. Xu, In situ spectral measurements of space weathering by Chang' e - 4 rover. Earth Planet. Sci. Lett. 535, 116117 (2020).

[16] Z. Wang, Y. Wu, D. T. Blewett, E. A. Cloutis, Y. Zheng, J. Chen, Submicroscopic metallic iron in lunar soils estimated from the in situ spectra of the Chang' e-3 mission. Geophys. Res. Lett. 44, 3485-3492 (2017).

[17] Z. Tang, J. Liu, X. Wang, X. Ren, W. Chen, W. Yan, X. Zhang, X. Tan, X. Zeng, D. Liu, H. Zhang, W. Wen, W. Zuo, Y. Su, J. Yang, C. Li, Physical and mechanical characteristics of lunar soil at the Chang' e - 4 landing site. Geophys. Res. Lett. 47, e2020GL089499 (2020).

[18] J. Y. Wong, Theory of Ground Vehicles (John Wiley & Sons, 2008).

[19] Y. Gao, C. Spiteri, C. - L. Li, Y. - C. Zheng, Lunar soil strength estimation based on Chang'e-3 images. Adv. Space Res. 58, 1893-1899 (2016).

[20] L. Ding, Z. Deng, H. Gao, J. Tao, K. D. Iagnemma, G. Liu, Interaction mechanics model for rigid driving wheels of planetary rovers moving on sandy terrain with consideration of multiple physical effects. J. Field Robot. 32, 827 - 859 (2015).

[21] B. M. French, G. Heiken, D. Vaniman, Lunar Sourcebook: A User's Guide to the Moon (Cambridge Univ. Press, 1991).

[22] H. Lin, Z. He, W. Yang, Y. Lin, R. Xu, C. Zhang, M. Zhu, R. Chang, J. Zhang, C. Li, H. Lin, Y. Liu, S. Gou, Y. Wei, S. Hu, C. Xue, J. Yang, J. Zhong, X. Fu, W. Wan, Y. Zou, Olivine - norite rock detected by the lunar rover Yutu-2 likely crystallized from the SPA-impact melt pool. Natl. Sci. Rev. 7, 913-920 (2019).

[23] S. Gou, K. Di, Z. Yue, Z. Liu, Z. He, R. Xu, H. Lin, B. Liu, M. Peng, W. Wan, Y. Wang, J. Liu, Lunar deep materials observed by Chang' e - 4 rover. Earth Planet. Sci. Lett. 528, 115829 (2019).

[24] J. Huang, Z. Xiao, L. Xiao, B. Horgan, X. Hu, P. Lucey, X. Xiao, S. Zhao, Y. Qian, H. Zhang, C. Li, R. Xu, Z. He, J. Yang, B. Xue, Q. He, J. Zhong, H. Lin, C. Huang, J. Xie, Diverse rock types detected in the lunar South Pole - Aitken Basin by the Chang' e-4 lunar mission. Geology 48, 723-727 (2020).

[25] J. Zhang, B. Zhou, Y. Lin, M. Zhu, H. Song, Z. Dong, Y. Gao, K. Di, W. Yang, H. Lin, J. Yang, E. Liu, L. Wang, Y. Lin, C. Li, Z. Yue, Z. Yao, Z. Ouyang, Lunar regolith and substructure at Chang' e-4 landing site in South Pole-Aitken basin. Nat. Astron. 5, 25-30 (2021).

[26] H. Lin, R. Xu, W. Yang, Y. Lin, Y. Wei, S. Hu, Z. He, L. Qiao, W. Wan, In situ photometric experiment of lunar regolith with visible and near-infrared imaging spectrometer on board the Yutu - 2 lunar rover. J. Geophys. Res. - Planets 125, e2019JE006076 (2020).

[27] J. Lai, Y. Xu, X. Zhang, L. Xiao, Q. Yan, X. Meng, B. Zhou, Z. Dong, D. Zhao, Comparison of dielectric properties and structure of lunar regolith at Chang' e - 3 and Chang' e - 4 landing sites revealed by ground - penetrating radar. Geophys. Res. Lett. 46, 12783 - 12793 (2019).

[28] S. Gou, Z. Yue, K. Di, J. Wang, W. Wan, Z. Liu, B. Liu, M. Peng, Y. Wang, Z. He, R. Xu, Impact melt breccia and surrounding regolith measured by Chang'e - 4 rover. Earth Planet. Sci. Lett. 544, 116378 (2020).

[29] Z. Dong, X. Feng, H. Zhou, C. Liu, Z. Zeng, J. Li, W. Liang, Properties analysis of lunar regolith at Chang'E - 4 landing site based on 3D velocity spectrum of Lunar Penetrating Radar. Remote Sens. 12, 629 (2020).

[30] C. Li, D. Liu, B. Liu, X. Ren, J. Liu, Z. He, W. Zuo, X. Zeng, R. Xu, X. Tan, X. Zhang, W. Chen, R. Shu, W. Wen, Y. Su, H. Zhang, Z. Ouyang, Chang' e - 4 initial spectroscopic identification of lunar far - side mantle - derived materials. Nature 569, 378 - 382 (2019).

[31] X. - H. Fu, L. Qiao, J. Zhang, Z. - C. Ling, B. Li, The subsurface structure and stratigraphy of the Chang'E - 4 landing site: Orbital evidence from small craters on the Von Kármán crater floor. Res. Astron. Astrophys. 20, 008 (2020).

[32] C. Ding, Z. Xiao, B. Wu, Y. Li, N. C. Prieur, Y. Cai, Y. Su, J. Cui, Fragments delivered by secondary craters at the Chang ' e - 4 landing site. Geophys. Res. Lett. 47, e2020GL087361 (2020).

[33] M. Jia, K. Di, Z. Yue, B. Liu, W. Wan, S. Niu, J. Liu, W. Cheng, Y. Ling, Multi - scale morphologic investigation of craters in the Chang' e - 4 landing area. Icarus 355, 114164 (2021).

[34] Z. Xu, J. Guo, R. F. Wimmer - Schweingruber, J. L. Freiherr von Forstner, Y. Wang, N. Dresing, H. Lohf, S. Zhang, B. Heber, M. Yang, First solar energetic particles measured on the lunar far - side. Astrophys. J. Lett. 902, L30 (2020).

[35] Y. Jia, Y. Zou, J. Ping, C. Xue, J. Yan, Y. Ning, The scientific objectives and payloads of Chang' e - 4 mission. Planet. Space Sci. 162, 207 - 215 (2018).

[36] R. Gonzalez, K. Iagnemma, Slippage estimation and compensation for planetary exploration rovers: State of the art and future challenges. J. Field Robot. 35, 564 - 577 (2018).

[37] Y. O. Aydin, J. M. Rieser, C. M. Hubicki, W. Savoie, D. I. Goldman, Robotic Systems and Autonomous Platforms: Advances in Materials and Manufacturing (Woodhead Publishing, 2019).

[38] J. Aguilar, T. Zhang, F. Qian, M. Kingsbury, B. Mclnroe, N. Mazouchova, C. Li, R. Maladen, C. Gong, M. Travers, R. L. Hatton, H. Choset, P. B. Umbanhowar, D. I. Goldman, A review on locomotion robophysics: The study of movement at the intersection of robotics, soft matter and dynamical

systems. Rep. Prog. Phys. 79，110001 (2016).

[39] S. Agarwal，A. Karsai，D. I. Goldman，K. Kamrin，Surprising simplicity in the modeling of dynamic granular intrusion. Sci. Adv. 7，abe0631 (2021).

[40] S. Shrivastava，A. Karsai，Y. O. Aydin，R. Pettinger，W. Bluethmann，R. O. Ambrose，D. I. Goldman，Material remodeling and unconventional gaits facilitate locomotion of a robophysical rover over granular terrain. Sci. Robot. 5，aba3499 (2020).

[41] C. Li，Y. Su，E. Pettinelli，S. Xing，C. Ding，J. Liu，X. Ren，S. E. Lauro，F. Soldovieri，X. Zeng，X. Gao，W. Chen，S. Dai，D. Liu，G. Zhang，W. Zuo，W. Wen，Z. Zhang，X. Zhang，H. Zhang，The Moon's farside shallow subsurface structure unveiled by Chang'e-4 Lunar Penetrating Radar. Sci. Adv. 6，eaay6898 (2020).

[42] L. Zhang，J. Li，Z. Zeng，Y. Xu，C. Liu，S. Chen，Stratigraphy of the Von Kármán crater based on Chang'e-4 Lunar Penetrating Radar data. Geophys. Res. Lett. 47，e2020GL088680 (2020).

[43] J. Lai，Y. Xu，R. Bugiolacchi，X. Meng，L. Xiao，M. Xie，B. Liu，K. Di，X. Zhang，B. Zhou，S. Shen，L. Xu，First look by the Yutu-2 rover at the deep subsurface structure at the lunar farside. Nat. Commun. 11，3426 (2020).

[44] R. Francis，T. Estlin，G. Doran，S. Johnstone，D. Gaines，V. Verma，M. Burl，J. Frydenvang，S. Montano，R. C. Wiens，S. Schaffer，O. Gasnault，L. DeFlores，D. Blaney，B. Bornstein，AEGIS autonomous targeting for ChemCam on Mars Science Laboratory：Deployment and results of initial science team use. Sci. Robot. 2，aan4582 (2017).

[45] Z. Liu，K. Di，J. Li，J. Xie，X. Cui，L. Xi，W. Wan，M. Peng，B. Liu，Y. Wang，S. Gou，Z. Yue，T. Yu，L. Li，J. Wang，C. Liu，X. Xin，M. Jia，Z. Bo，J. Liu，R. Wang，S. Niu，K. Zhang，Y. You，B. Liu，J. Liu，Landing site topographic mapping and rover localization for Chang'e-4 mission. Sci. China Inf. Sci. 63，140901 (2020).

[46] W. Wan，Z. Liu，K. Di，B. Wang，J. Zhou，A cross-site visual localization method for Yutu rover. ISPRS J. Photogramm. Remote Sens. 40，279 (2014).

[47] J. M. Morel，G. Yu，ASIFT：A new framework for fully affine invariant image comparison. SIAM J. Imaging Sci. 2，438-469 (2009).

[48] B. Triggs，A. Zisserman，R. Szeliski，Vision Algorithms：Theory and Practice (Springer，1999).

[49] Z. Liu，K. Di，M. Peng，W. Wan，B. Liu，L. Li，T. Yu，B. Wang，J. Zhou，H. Chen，High precision landing site mapping and rover localization for Chang'e-3 mission. Sci. China Phys. Mech. Astron. 58，1-11 (2015).

[50] W. Wan，thesis，Chinese of Academy of Sciences (2012).

[51] J. Wang，J. Li，S. Wang，T. Yu，Z. Rong，X. He，Y. You，Q. Zou，W. Wan，Y. Wang，S. Gou，B. Liu，M. Peng，K. Di，Z. Liu，M. Jia，X. Xin，Y. Chen，

X. Cheng, X. Feng, C. Liu, S. Han, X. Liu, Computer vision in the teleoperation of the Yutu - 2 rover, in Proceedings of the Remote Sensing and Spatial Information Science ISPRS Geospatial Week (International Society for Photogrammetry and Remote Sensing, 2020), pp. 595 - 602.

[52] N. Li, thesis, Harbin Institute of Technology (2019).

[53] J. Y. Wong, A. R. Reece, Prediction of rigid wheel performance based on the analysis of soil - wheel stresses part I. Performance of driven rigid wheels. J. Terramech. 4, 81 - 98 (1967).

[54] L. Ding, H. Gao, Z. Liu, Z. Deng, G. Liu, Identifying mechanical property parameters of planetary soil using in - situ data obtained from exploration rovers. Planet. Space Sci. 119, 121 - 136 (2015).

[55] W. J. M. Rankine II, On the stability of loose earth. Philos. Trans. R. Soc. Lond. 147, 9 - 27 (1857).

[56] K. Terzaghi, R. B. Peck, G. Mesri, Soil Mechanics in Engineering Practice (John Wiley & Sons, 1996).

[57] H. Shibly, K. Iagnemma, S. Dubowsky, An equivalent soil mechanics for mulation for rigid wheels in deformable terrain, with application to planetary exploration rovers. J. Terramech. 42, 1 - 13 (2005).

[58] L. Ding, thesis, Harbin Institute of Technology (2009).

[59] H. Gao, F. Lv, B. Yuan, N. Li, L. Ding, N. Li, G. Liu, Z. Deng, Sinkage definition and visual detection for planetary rovers wheels on rough terrain based on wheel - soil interaction boundary. Robot. Auton. Syst. 98, 222 - 240 (2017).

[60] B. Liu, S. Niu, X. Xin, M. Jia, K. Di, Z. Liu, M. Peng, Z. Yue, High precision DTM and DOM generation using multi - source orbital data on Chang'E - 4 landing site, in Proceedings of the International Archives of the Photogrammetry, Remote Sensing and Spatial Information Sciences (International Society for Photogrammetry and Remote Sensing, 2019), pp. 1413 - 1417.

[61] M. S. Robinson, S. M. Brylow, M. Tschimmel, D. Humm, S. J. Lawrence, P. C. Thomas, B. W. Denevi, E. Bowman - Cisneros, J. Zerr, M. A. Ravine, M. A. Caplinger, F. T. Ghaemi, J. A. Schaffner, M. C. Malin, P. Mahanti, A. Bartels, J. Anderson, T. N. Tran, E. M. Eliason, A. S. McEwen, E. Turtle, B. L. Jolliff, H. Hiesinger, Lunar Reconnaissance Orbiter Camera (LROC) instrument overview. Space Sci. Rev. 150, 81 - 124 (2010).

[62] T. Kneissl, S. van Gasselt, G. Neukum, Map - projection - independent crater size - frequency determination in GIS environments - New software tool for ArcGIS. Planet. SpaceSci. 59, 1243 - 1254 (2011).

[63] Y. Yang, H. Lin, Y. Liu, Y. Lin, Y. Wei, S. Hu, W. Yang, R. Xu, Z. He, Y. Zou, The effects of viewing geometry on the spectral analysis of lunar regolith as in ferred by insitu spectrophotometric measurements of Chang'e - 4. Geophys. Res. Lett. 47, e2020GL087080 (2020).

Dielectric Properties of Lunar Subsurface Materials

Zehua Dong[1,2], Guangyou Fang[1,3,4], Di Zhao[1,3], Bin Zhou[1,4], Yunze Gao[1], and Yicai Ji[1]

1. Key Laboratory of Electromagnetic Radiation and Sensing Technology, Aerospace Information Research Institute (AIR), Chinese Academy of Sciences (CAS), Beijing, China

2. School of Information and Electronics, Beijing Institute of Technology, Beijing, China

3. School of Electronic, Electrical and Communication Engineering, University of Chinese Academy of Science, Beijing, China

4. Greater Bay Area Research Institute of AIR-CAS, Guangzhou, China

Abstract: The dielectric permittivity is one of the most significant properties of lunar material. For the past decades, researchers have assumed or used an indirect density versus depth relation to constrain the subsurface permittivity on the Moon. On 3 January 2019, Chang'E-4 (CE-4) rover-borne lunar penetrating radar (LPR) successfully landed on the Von Kármán crater on the Moon's farside. CE-4 LPR is the first surface radar on the Moon's farside for exploring regolith thickness and geological structure. Numerous parabolic-shaped diffractions (PSD), which indicate subsurface objects, are identified in LPR data. We estimate the permittivity of the first ~50m at the landing region using these PSDs based on a hyperbolic fitting method. A best-fit permittivity/density versus depth relation is obtained, and this relation could be used at different areas on the Moon.

Plain Language Summary: Currently, rover-deployed/spaceborne radar systems are frequently employed in lunar exploration. The depth and thickness of lunar subsurface structure observed from radar data are completely influenced by the dielectric properties of subsurface materials. The dielectric properties also provide vital clues on compositions and origin of subsurface materials. However, except the results measured with returned sample from lunar surface, we have no more information about dielectric properties of global and deeper lunar subsurface materials. In January, China's CE-4 spacecraft softly landed on the farside of the Moon for the first time and released the lunar penetrating radar (LPR). In this study, we estimate the dielectric permittivity of the first ~50m at the landing region with LPR data and propose a best-fit permittivity/density versus depth relation which could be used at different areas on the Moon.

Key Points:

- The permittivity of lunar subsurface materials at CE-4 landing site is estimated

- A best－fit permittivity/density versus depth relation is obtained
- The obtained best－fit relation can be used at different areas on the Moon

1 Introduction

The Earth－based radar has been used for mapping the Moon since 1960s (Campbell et al., 2014; Morgan et al., 2016). After Apollo Lunar Sounder Experiment (ALSE), subsurface radar sounding became another effective remote sensing technique for the Moon (Kofman et al., 2010), and sounder studies on Mars are also extensive (Seu et al., 2007). Compared with other observations (infrared or optical analysis), it could map the stratigraphy versus depth by detecting dielectric discontinuities, which can be of fundamental importance to better understand the dynamics and the history (Hiesinger, 2006). For radar methods, the dielectric property of subsurface materials is regarded as the most significant physical property. It determines the propagating velocity and signal attenuation of electromagnetic (EM) waves, which determine the correct time－to－depth inversion and the geological interpretation of radargrams. Furthermore, in the absence of moisture, the dominant parameters which affect the permittivity of subsurface materials are density and material composition (Strangway et al., 1972), so it is possible to retrieve some information on these petrophysical properties with estimated permittivity in an feasible way. Thus, the subsurface dielectric properties versus depth on the Moon are key problems to be resolved.

However, although lunar surface dielectric permittivity has been investigated by previous studies on the basis of different methods such as laboratory measurement on lunar samples (Chung et al., 1970; Frisillo et al., 1975; Troitsky & Shmulevich, 1973), empirical relation between permittivity, density, and FeO&TiO_2 contents (Fa et al., 2011; Ishiyama et al., 2013; Shkuratov & Bondarenko, 2001), and radar observations (Dong et al., 2017), the dielectric properties of lunar subsurface materials at various depths are not well constrained. So far, two compromised solutions are applied for the lunar subsurface dielectric permittivity when it is requisite. The first method is to directly assign assuming dielectric constant for the subsurface materials (Fa et al., 2015; Ono et al., 2009; Xiao et al., 2015), for instance, 2—4 for lunar regolith and 6—8 for mare basalts. This kind of simplified assumption is a compromise due to the lacking constraint of lunar subsurface dielectric permittivity, which may lead to inaccurate estimated depths and ambiguous interpretations. The other method is based on the empirical relations derived from lunar samples (Kobayashi et al., 2010; Zhang et al., 2015). The dielectric properties of lunar samples were measured as a function of density under controlled laboratory conditions (Carrier et al., 1991). Then, using a density versus depth profile determined from in situ core tube measurements and laboratory compression studies on lunar samples (Carrier et al., 1991), a model could be presented for dielectric properties versus depth in the lunar surface (Olhoeft & Strangway, 1975).

However, this method has some inherent drawbacks: (1) The dielectric permittivity versus depth relation is indirectly obtained from density, which may cause additional errors; (2) the results were carried out by laboratory sample measurement instead of in situ measurement, neglecting the influence from surroundings; (3) the limited quantity and depth of the core tubes are inadequate to well constrain the variation versus depth; and (4) these density versus depth relations are fixed for any areas, but the surface properties of lunar materials vary from place to place. As a result, existing approaches are deficient to constrain the dielectric properties of lunar subsurface materials.

China's Chang'E - 4 mission landed at 45.4446°S, 177.5991°E on the Moon on 3 January 2019 and subsequently released the Yutu - 2 rover (Wu et al., 2019). The CE - 4 LPR onboard Yutu - 2 is the first surface radar system on the Moon's farside, enabling an investigation of the subsurface with a finer time delay and horizontal sampling than that from orbital sounders (Xiao et al., 2015). The landing site is located on the southeast of the Von Kármán crater, within the South Pole - Aitken (SPA) basin. It is a key area to figure out internal structure, geological history, and thermal evolution of the Moon (Huang et al., 2018). The FeO&TiO_2 abundance of lunar regolith at CE - 4 landing site is estimated to be 12 - 15 wt.% (Ohtake et al., 2014; Sato et al., 2017), much less than the estimated result of ~25 - 27 wt.% from Chang'E - 3 (CE - 3) LPR on the Mare Imbrium (Dong et al., 2017; Ling et al., 2015). Therefore, the subsurface materials of CE - 4 landing region are fairly transparent to radio waves, facilitating the propagation of LPR signals. The LPR consists of two transducer channels with different frequencies and bandwidths, allowing subsurface structures at the landing site to be resolved at different depths and vertical resolutions. The dominant frequency and the bandwidth of Channel - Ⅰ are 60 and 40MHz, while for Channel - Ⅱ are 450 and 500MHz, respectively (Fang et al., 2014). The preliminary detected results demonstrate that the penetrating depths of Channel - Ⅰ is up to ~300m deep with a vertical resolution of ~3 m, while the vertical resolution of Channel - Ⅱ could be less than ~30cm within the upper ~50m depth. Allowing for the relatively constant dielectric permittivity of mare basalts, only the variations in lunar regolith and ejecta deposits are effective to constrain the dielectric permittivity versus depth relation. Local geological context suggests that unlike the infilled mare basalts below ~10m at CE - 3 landing site (Xiao et al., 2015), the shallow subsurface materials of CE - 4 landing region are composed of ejecta deposits up to more than ~50m, which covers the detection depth of Channel -Ⅱ. Therefore, the dielectric permittivity of lunar regolith and ejecta deposits versus depth was resolved with LPR Channel -Ⅱ data in this paper. Here we show the results obtained during the first 4 lunar days.

2 Materials and Methods

2.1 LPR Data and Processing Methods

The CE - 4 lunar penetrating radar (LPR) is a kind of carrier - free time - domain impulse pulse radar which is composed of transmitter, receiver, and antennas. The LPR data are saved as traces, and each trace of Channel - Ⅱ is composed of 2 048 points with a time window of 640ns (Fang et al., 2014). Yutu - 2 drove 178. 9 m before the restarting of its fourth working lunar day (29 April 2019) (Fig 1). The LPR restarts to working as soon as Yutu - 2 rover is awakened every lunar daytime and has collected a mass of data along the whole track of Yutu - 2. However, the majority of data were collected when the LPR stopped or moved slowly. According to velocity and telemetry information of the Yutu - 2 rover, we deleted the repetitive data and selected 4 844 traces of LPR data between waypoint X and LE402 (Fig. 1).

The raw LPR data are attenuated signals with low signal - to - noise ratio (SNR) due to the radar system noise, the out - of - band environment noise, and the absorption of radar wave in lunar subsurface (Dong et al., 2017). Thus, it is essential to improve the quality of radar signals that adopts a series of processing methods before the estimation of

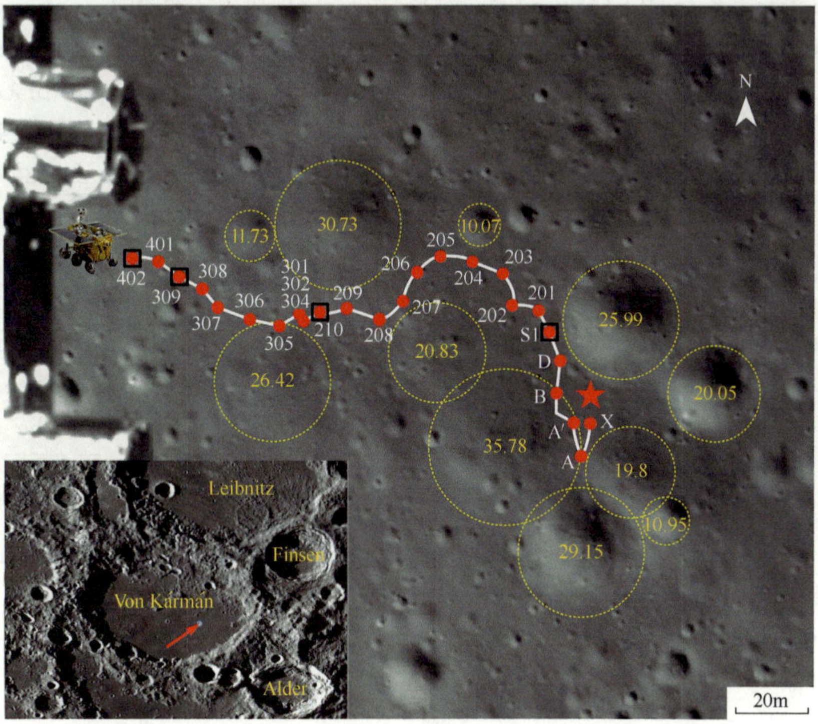

Fig. 1　The Yutu - 2's path on the Moon. Yutu - 2 has drove 175. 3m (white line) along 26 waypoints (red dots) from X to LE402. The black squares at S1, LE210, LE309, and LE402 mark where Yutu - 2 rested for lunar nights, which ended on 30 January 2019, 28 February 2019, 29 March 2019, and 29 April 2019, respectively. The yellow circles are small - sized craters (inside values indicate diameters) along Yutu - 2's track. The image was taken by the CE - 4 landing camera, and the left side was blocked by the CE - 4 lander. The inserted image shows the CE - 4 landing region. The red arrow points to the landing site, located on the southeast of the Von Kármán crater. The image was taken by the Lunar Reconnaissance Orbiter Camera Narrow Angle Camera (LROC NAC)

lunar regolith permittivity, as usually used for ground penetrating radar data on the Earth applications. Only amplitude compensation, direct current (DC) removing, background removing, and filtering (see supporting information in detail) are applied for LPR Channel - Ⅱ data in this study in order to not influence the features of subsurface objects. The SNR of the processed LPR data has been obviously improved compared with the raw data, facilitating the analysis and estimation of subsurface dielectric properties at the CE - 4 landing region.

2.2 Estimation Methods

LPR observes subsurface structures (such as boulders and interfaces) by detecting dielectric discontinuities. As the distance between the antenna and the subsurface object changes with the moving of Yutu - 2 rover, single target buried in the subsurface produces parabolic - shaped diffraction (PSD) in the radar cross section (Ristic et al., 2009). The reflection pattern can well be approximated by a parabola if the subsurface target has a size $<\lambda$ (where λ denotes the probing wavelength); differently, for objects larger than 3λ, the reflection pattern is more complicated, and targets having a size of tens of λ, generate semicontinuous, flat reflections. Here we identified plenty of well - defined PSDs in LPR Channel - Ⅱ data to estimate the dielectric permittivity versus depth on the basis of a hyperbolic fitting method.

Considering that rocks in the regolith mostly have irregular shapes and the effect of a rock's diameter will be reduced as the depth increases, we approximate the object as a point reflector in order to simplify the model. Fig. 2 demonstrates the mechanism of target detection for LPR. The propagating distance of EM wave from LPR to the subsurface object (L_1) and the minimum propagating distance when LPR is over the subsurface object (L) conform to Equation 1 based on the geometric relationship:

$$L_1^2 - L^2 = D^2 \tag{1}$$

where D is the horizontal distance from LPR to the subsurface object, which can be derived according to the location of Yutu - 2 rover. Moreover, L_1 and L could be described as Equation 2, respectively:

$$L_1 = \frac{vt_1}{2},\ L = \frac{vt}{2} \tag{2}$$

where t_1 and t are the two - way propagating times corresponding to L_1 and L and v is the average propagating velocity of EM wave in lunar subsurface materials.

The typical range of mass susceptibility for lunar regolith is around $100-10,000\times 10^{-9}\,m^3kg^{-1}$ (Rochette et al., 2010), which means the relative magnetic permeability approximately equals to 1. Besides, the lunar regolith also shows very small conductivity which was estimated as $\sim10^{-3}-10^{-4}$ S/m at 450MHz (Dong et al., 2017; Gold et al., 1972). Therefore, the lunar regolith material could be regarded as an ideal dielectric in which case the velocity of EM wave (v) is determined by the relative dielectric

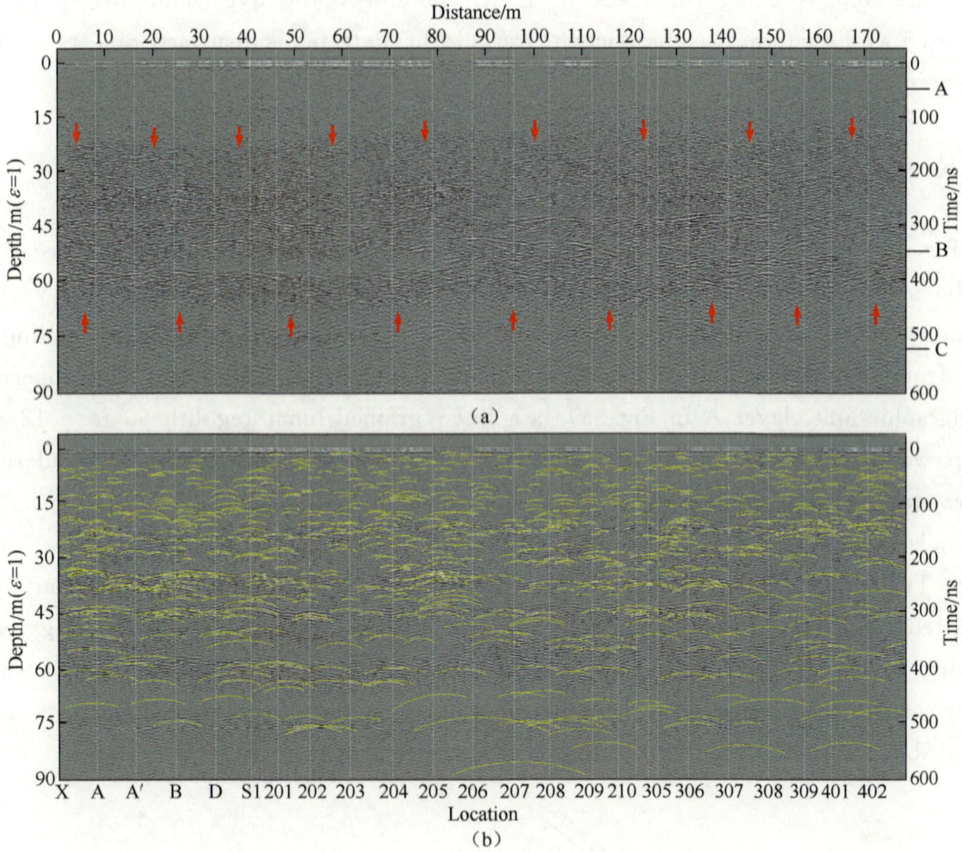

Fig. 2　Mechanism of the hyperbolic fitting method

(a) A subsurface object below LPR's track. (b) The LPR echo from the subsurface object

permittivity (ε_r) as follows:

$$v = \frac{c}{\sqrt{\varepsilon_r}} \tag{3}$$

where c is the speed of light in vacuum. Replacing the propagating distance in Equation 1 and the propagating velocity in Equation 2 with their respective expressions yields the following equation:

$$\varepsilon_r = \left(\left(c\sqrt{t_1^2 - t^2}\right)/2D\right)^2 \tag{4}$$

As a result, the average relative dielectric permittivity of the upper materials above the subsurface boulders can be calculated with LPR data. It should be noted that each estimated value indicates the average permittivity of lunar regolith above the subsurface boulder instead of the permittivity of subsurface boulder or lunar regolith at some depth. It is also worth mentioning that due to the spherical nature of the emitted waves, reflections do not only occur while LPR is directly over the object but also take place when LPR is moving nearby (Feng et al., 2017). However, Fig. 2 shows that the geometric relationship is applicable not just for the objects right under the LPR but for some lateral subsurface objects along the track of LPR. Therefore, the hyperbolic fitting method expressed as Equation 1 is practicable for all the detected subsurface objects in LPR radargram.

3 Results

Fig. 3 shows the processed LPR data with improved SNR. Several subsurface layers can be identified in the radargram. Visible and near infrared spectrometer (VNIS), another payload onboard Yutu - 2, proves that the surface materials at CE - 4 landing site came from the Finsen crater (Fig. 1) (Li et al., 2019). Therefore, the uppermost stratigraphic unit (layer A in Fig. 3a) is a fine - grained lunar regolith (0 to ~12m). (Here we roughly use a typical value of $\varepsilon = 3$ for all depths.) Below that, the radargram shows stacked and complicated reflections (layer B in Fig. 3a), which indicates that the materials (~12 to ~40m) are probably remnant part of ejecta deposits from the Finsen crater. The total thickness of overlying ejecta blanket from the Finsen crater is about 40m, consistent with the estimated depth of the surface lava deposits at the landing region (Huang et al., 2018; Qiao et al., 2019). The lowermost stratigraphic unit shows uniform radar echoes similar to the uppermost lunar regolith layer, which suggests that it is probably the paleo - regolith weathered from mare basalts.

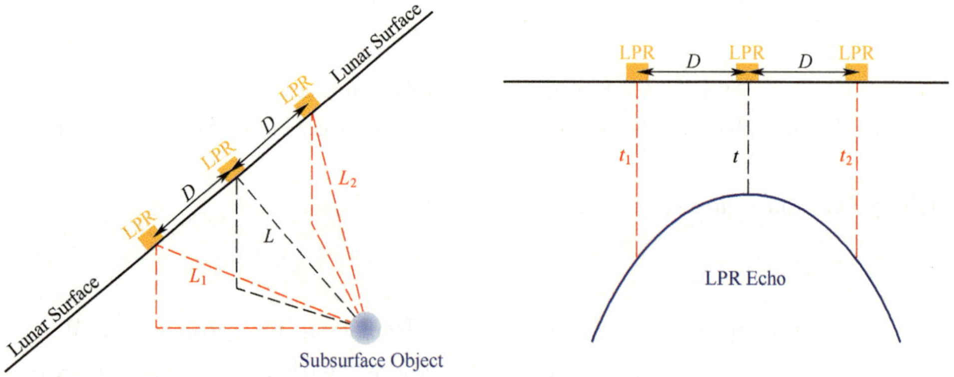

(a) (b)

Fig. 3 LPR radargrams. (a) Processed LPR Channel - Ⅱ data. The subsurface interfaces are marked with red arrows, and subsurface layers are labeled from A to C. (b) The subsurface objects in LPR data. The yellow arcs represent 823 identified PSDs. The white vertical lines indicate the waypoints corresponding to those marked in Fig. 1

Numerous PSD can be clearly recognized in LPR profile. The accurate identification of these PSDs in the radargram is prerequisite for the permittivity estimation of lunar subsurface materials. Various methods have been proposed to automatically detect the diffractions, for example, a fuzzy shell clustering approach (Delbo et al., 2000), a pattern recognition based on neural networks (Al - Nuaimy et al., 2000), and a support vector machine classifier (Pasolli et al., 2009). However, these methods were mostly employed for real - time work and involving processing of large volumes of data. Because there is extensive clutter and noise around the hyperbolas in our LPR images, it is difficult to achieve accurate, robust identification using these methods and the extracted results still required reconfirmation. With consideration of the accuracy and integrality of the identification, we manually identified all the PSDs and uncertain subsurface objects were not chosen for inclusion in our analysis.

After careful checking, 823 PSDs, which indicate subsurface boulders distributed at various depths, were identified in LPR radargram as shown in Fig. 3b. Deeper objects show wider and flatter diffractions than upper ones, because the lower objects would be "seen" by LPR for a longer period due to the fixed radiation angle of transmitting antenna. The hyperbolic fitting method was used for these PSDs, and the estimated dielectric constant from each subsurface boulder was obtained as shown in Fig. 4a.

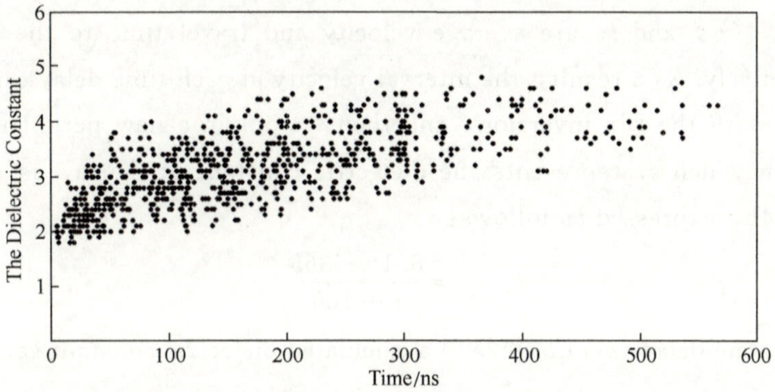

(a)

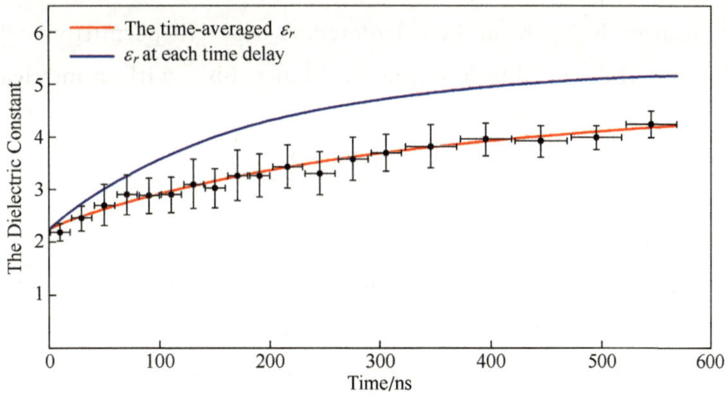

(b)

Fig. 4　Estimated results of the relative permittivity. (a) Estimated results of dielectric constant from each PSD in Fig. 3. The vertical lines indicate the divided sections. (b) Statistics of the estimate results. The spots indicate average permittivity within each section. The horizontal and vertical error bars denote the time range and 1σ error within each section. The red and blue curves indicate the best - fit relations of averaged permittivity and interval permittivity versus propagation time, respectively

Statistics and error analysis were carried out by dividing the estimated values into sections (20ns for 0—+200ns, 30ns for 200—320ns, and 50ns for 320—570ns). The average permittivity and one standard deviation (1σ) error within each subsection were identified as shown in Fig. 4b. The red curve in Fig. 4b is the best-fit rational relationship for the dielectric constant and propagation time of EM wave.

Note that these estimated results and the best-fit function represent the depth-averaged permittivity (mean values from depths of each subsurface boulder to the surface) instead of permittivity at each depth. This is because the dielectric constants are estimated from the velocity of EM wave in lunar materials in this study, the calculated values from our proposed method represent the depth-averaged velocity rather than interval velocity. Therefore, the interval velocity v_{int} should be derived using Dix inversion (Bradford & Harper, 2005; Dix, 1955):

$$v_{int}=\sqrt{\frac{v_{avg2}^2 t_2 - v_{avg1}^2 t_1}{t_2 - t_1}} \tag{5}$$

where v_{avg} (n) and t_n are average velocity and travel time to the nth point (or pixel), respectively. As a result, the interval velocity at each time delay can be calculated point by point with the Dix inversion, and then we obtain a new permittivity-versus-time relation in which ε_r represents the dielectric constant at depth, as the blue curve shown in Fig. 4b, expressed as follows:

$$\varepsilon_r=\frac{6.1t+365}{t+166} \tag{6}$$

A certain time delay (t(in ns)) and a calculated dielectric constant (ε_r) by Equation 6 lead to a certain corresponding depth (d (in m)) on the basis of $d=ct/2\sqrt{\varepsilon_r}$, so the dielectric constant at each depth can be calculated. A best-fit permittivity-versus-depth equation can be constrained (black curve in Figure 5b) with numerical method and given by

$$\varepsilon_r=\frac{6.3d+32.5}{d+14.8} \tag{7}$$

The estimated results fall into a range of 2.2—5.0, consistent with previous measurement of lunar regolith and ejecta deposits. The result shows that the estimated permittivity gets larger on the whole with increasing depth, indicating the permittivity (density) of in situ subsurface lunar materials suggest strong positive correlations with depth. Watters et al. (2017) suggested that grain size can have a major impact on the depth-averaged dielectric constant, providing possible new clues for us to explain how the CE-4 results are framed. The studied materials are composed of fine-grained regolith and remnant part of ejecta deposits. Therefore, the permittivity increases with depth probably due to larger grain size and higher compaction degree of the deeper lunar subsurface materials.

4 Discussions

The estimated dielectric constants show a little difference at similar depths. Theoretically, an ideal PSD in radar data can lead to an accurate calculated permittivity of lunar regolith using the proposed method, but several cases would affect the accuracy of the estimation in field application: (1) Yutu - 2 moves with a nonlinear track which would broaden the shape of PSD and, therefore, diminish the estimated permittivity; (2) the velocity of Yutu - 2 is not extremely uniform, leading to a rather large or small estimated value; (3) the regolith and the ejecta deposits are mixture of fine - grained soil and distributed boulders, so the propagation path of EM wave with accidently more boulders or less boulders would result in a larger or smaller dielectric constant, respectively; and (4) the dielectric permittivity varies due to the variety of regolith properties over small area on the Moon. Although all of these reasons impact the estimated results in this study, the influence of each case is fairly slight.

According to the local geological context (Huang et al., 2018; Qiao et al., 2019) and stratigraphy from LPR radargram (Fig. 3), the depth of the underlying lava basalts is constrained to be greater than ~45m, which means the lunar materials studied in this paper are regolith materials and ejecta deposits. The physical properties (e. g., density and permittivity) between mare basalts and lunar regolith or ejecta deposits are quite different. The permittivity would sharply increase to more than ~6.0 — 8.0 when encountering basaltic materials (Carrier et al., 1991). Therefore, the result of dielectric permittivity versus depth in this study is not applicable to infilled mare basalt layer.

In past decades, the only approach to estimate the subsurface dielectric constant at some depth is based on empirical relations of density versus depth determined from in situ core tube measurements and laboratory compression studies on lunar samples. The most common one is given by (Carrier et al., 1991)

$$\rho = 1.92 \frac{z + 12.2}{z + 18} \tag{8}$$

where ρ (in g/cm^3) is the density at depth z (incm). Besides, Hayne et al. (2017) mapped globally the thermophysical properties of lunar regolith using Diviner radiometer observations, and their results suggested an exponent density profile given by

$$\rho = 1.8 - (1.8 - 1.1) e^{-\frac{z}{0.07}} \tag{9}$$

The relationship of density to dielectric constant was studied in detail by Ulaby et al. (1988):

$$\varepsilon_r = 1.919^{\rho} \tag{10}$$

Then, we inverted the obtained relation of permittivity versus depth to a best - fit density versus depth on the basis of Equation 9 as shown in Fig. 5a. It can be seen that our density versus depth relation has similar growth trend with previous functions, but there are two obvious distinctions between them: One is that the previous functions almost

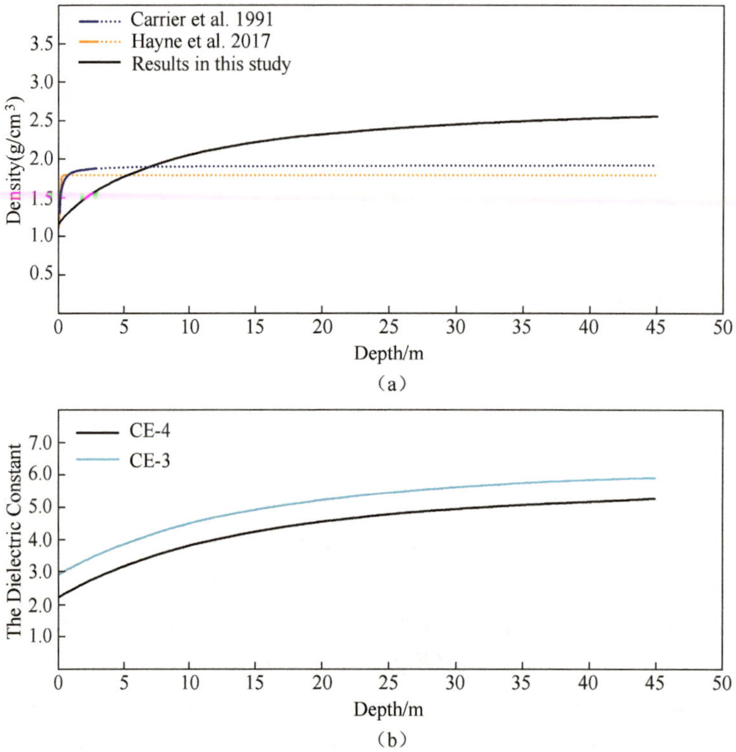

Fig. 5　(a) Comparison of density versus depth results in this study and by previous studies. The solid parts of the blue and orange lines are maximum depths in previous studies (<2.8 and <1 m, respectively). (b) The permittivity versus depth at CE-3 and CE-4 landing sites

become a constant deeper than ~1m; the other is that the previous functions sharply increase at depth <~1m, while our results indicate that the permittivity increases gradually with depth. Note that our results are derived from subsurface objects within ~50m, so the applicable depth range of our proposed permittivity (density) versus depth function is <50m in principle.

Compare with the permittivity versus depth results in this study and previous relations derived from laboratory lunar sample, our results show some obvious advantages. The quantity of PSDs used for obtaining the permittivity versus depth is up to 823, and the depth of these PSDs is more than ~40m. However, the maximum depths of lunar samples collected on the Apollo 15—17 missions are no more than 0.6 and 2.8 m for core tubes and drill cores, respectively. It is inadequate to estimate the subsurface permittivity (density) lower than ~3m using the previous relations.

Moreover, the previous density versus depth relations are fixed on lunar global surface, leading to a constant surface density (permittivity) and the same subsurface density at a certain depth at different places on the Moon. Actually, numerous studies have suggested that the properties of lunar surface materials, let alone of subsurface materials, vary from place to place (e.g., Fa et al., 2011). Thus, a fixed permittivity versus depth relation should not be used for different study region on the Moon. Allowing for the different local surface permittivity on the Moon, we describe the permittivity versus depth relation as a sum of the surface permittivity (ε_s) and the increment with depths on the basis of Equation 6, expressed as follows:

$$\varepsilon_r = \varepsilon_s + \frac{4.1d}{d + 14.8} \tag{11}$$

This function is to reduce the influence of various geological factors at different regions on the Moon as possible by taking the local surface permittivity into consideration. As a result, this permittivity - versus - depth result could be more helpful to constrain the global permittivity - depth relation on the Moon compared with using a fixed function as before. The determination of surface permittivity needs to be combined with other data and results such as radar or optical methods. Many previous studies researched the lunar global surface permittivity (e. g. , Fa & Wieczorek, 2012; Wang et al. , 2010), and the surface permittivity at some places has been more accurately measured by lunar missions. For example, the surface permittivity at CE - 3 landing site on the Mare Imbrium was estimated to be ~2. 9 (Dong et al. , 2017) . Consequently, the best - fit permittivity versus depth at CE - 3 landing site is obtained as shown in Fig. 5b. The lower surface permittivity at CE - 4 landing site is probably due to the lower TiO_2 & FeO content at CE - 4 landing region than that at CE - 3 landing region.

5 Conclusions

CE - 4 LPR is the first surface radar system on the Moon's farside. In this study, using the LPR Channel - Ⅱ data, we estimated the dielectric permittivity of lunar regolith on the basis of a hyperbolic fitting method. The 823 PSDs up to ~50m deep were extracted from processed LPR data, and the relative permittivity from each PSD was obtained. We identified a best - fit equation of permittivity versus depth at CE - 4 landing region with the estimated results. Furthermore, the obtained permittivity versus depth relation could be expressed as a sum of the surface permittivity and the increment with depths, which means that the relation is applicable to global regolith on the Moon when the local surface permittivity is given.

References

[1] Al - Nuaimy, W. , Huang, Y. , Nakhkash, M. , Fang, M. , Nguyen, V. , & Eriksen, A. (2000) . Automatic detection of buried utilities and solid objects with GPR using neural networks and pattern recognition. Journal of Applied Geophysics, 43 (2 - 4), 157 - 165. https://doi.org/10.1016/S0926 - 9851 (99) 00055 - 5.

[2] Bradford, J. H. , & Harper, J. T. (2005) . Wave field migration as a tool for estimating spatially continuous radar velocity and water content in glaciers. Geophysical

Research Letters, 32, L08502. https://doi.org/10.1029/2004GL021770.

[3] Campbell, B. A., Hawke, B. R., Morgan, G. A., Carter, L. M., Campbell, D. B., & Nolan, M. (2014). Improved discrimination of volcanic complexes, tectonic features, and regolith properties in Mare Serenitatis from Earth-based radar mapping. Journal of Geophysical Research: Planets, 119, 313-330. https://doi.org/10.1002/2013JE004486.

[4] Carrier, W. D., Olhoeft, G. R., & Mendell, W. (1991). Physical properties of the lunar surface. In G. H. Heiken, D. T. Vaniman, B. M. French (Eds.), Lunar source-book: A user's guide to the moon (pp. 475-552). New York: Cambridge University Press.

[5] Chung, D. H., Westphal, W. B., & Simmons, G. (1970). Dielectric properties of Apollo 11 lunar samples and their comparison with Earth materials. Journal of Geophysical Research, 75 (32), 6524-6531. https://doi.org/10.1029/JB075i032p06524.

[6] Delbo, S., Gamba, P., & Roccato, D. (2000). A fuzzy shell clustering approach to recognize hyperbolic signatures in subsurface radar images. IEEE Transactions on Geoscience and Remote Sensing, 38 (3), 1447-1451.

[7] Dix, C. H. (1955). Seismic velocities from surface measurements. Geophysics, 20, 68-86.

[8] Dong, Z., Fang, G., Ji, Y., Gao, Y., Wu, C., & Zhang, X. (2017). Parameters and structure of lunar regolith in Chang'E-3 landing area from lunar penetrating radar (LPR) data. Icarus, 282, 40-46.

[9] Fa, W., & Wieczorek, M. A. (2012). Regolith thickness over the lunar nearside: Results from Earth-based 70-cm Arecibo radar observations. Icarus, 218 (2), 771-787. https://doi.org/10.1016/j.icarus.2012.01.010.

[10] Fa, W., Wieczorek, M. A., & Heggy, E. (2011). Modeling polarimetric radar scattering from the lunar surface: Study on the effect of physical properties of the regolith layer. Journal of Geophysical Research, 116, E03005. https://doi.org/10.1029/2010JE003649.

[11] Fa, W., Zhu, M. H., Liu, T., & Plescia, J. B. (2015). Regolith stratigraphy at the Chang'E-3 landing site as seen by lunar penetrating radar. Geophysical Research Letters, 42, 10, 179-10, 187. https://doi.org/10.1002/2015GL066537.

[12] Fang, G. Y., Zhou, B., Ji, Y. C., Zhang, Q. Y., Shen, S. X., Li, Y. X., et al. (2014). Lunar penetrating radar onboard the Chang'e-3 mission. Research in Astronomy and Astrophysics, 14 (12), 1607.

[13] Feng, J., Su, Y., Ding, C., Xing, S., Dai, S., & Zou, Y. (2017). Dielectric properties estimation of the lunar regolith at CE-3 landing site using lunar penetrating radar data. Icarus, 284, 424-430. https://doi.org/10.1016/j.icarus.2016.12.005.

[14] Frisillo, A. L., Olhoeft, G. R., & Strangway, D. W. (1975). Effects of vertical

stress, temperature and density on the dielectric properties of lunar samples 72441, 12, 15301, 38 and a terrestrial basalt. Earth and Planetary Science Letters, 24 (3), 345 - 356. https://doi.org/10.1016/0012 - 821X (75) 90140 - 5.

[15] Gold, T., Bilson, E., & Yerbury, M. (1972). Grain size analysis, optical reflectivity measurements and determination of high frequency electrical properties for Apollo 14 lunar samples. Proceedings Lunar Science Conference 3rd, 3187 - 3193.

[16] Hayne, P., Bandfield, J., Siegler, M., Vasavada, A., Ghent, R., Williams, J., et al. (2017). Global regolith thermophysical properties of the Moon from the Diviner Lunar Radiometer Experiment. The Journal of Geophysical Research Planets, 122 (12), 2371 - 2400.

[17] Hiesinger, H. (2006). New views of lunar geoscience: An introduction and overview. Reviews in Mineralogy and Geochemistry, 60 (1), 1 - 81. https://doi.org/10.2138/rmg.2006.60.1.

[18] Huang, J., Xiao, Z., Flahaut, J., Martinot, M., Head, J., Xiao, X., et al. (2018). Geological characteristics of Von Kármán crater, northwestern South Pole - Aitken Basin: Chang'E - 4 landing site region. Journal of Geophysical Research: Planets, 123, 1684 - 1700. https://doi.org/10.1029/2018JE005577.

[19] Ishiyama, K., Kumamoto, A., Ono, T., Yamaguchi, Y., Haruyama, J., Ohtake, M., et al. (2013). Estimation of the permittivity and porosity of the lunar uppermost basalt layer based on observations of impact craters by SELENE. Journal of Geophysical Research: Planets, 118, 1453 - 1467. https://doi.org/10.1002/jgre.20102.

[20] Kobayashi, T., Kim, J. H., Lee, S. R., Araki, H., & Ono, T. (2010). Simultaneous observation of lunar radar sounder and laser altimeter of Kaguya for lunar regolith layer thickness estimate. IEEE Geoscience and Remote Sensing Letters, 7 (3), 435 - 439. https://doi.org/10.1109/LGRS.2009.2038499.

[21] Kofman, W., Orosei, R., & Pettinelli, E. (2010). Radar signal propagation and detection through ice. Space Science Reviews, 153 (1 - 4), 249 - 271.

[22] Li, C., Liu, D., Liu, B., Ren, X., Liu, J., He, Z., et al. (2019). Chang'E - 4 initial spectroscopic identification of lunar far - side mantle - derived materials. Nature, 569 (7756), 378 - 382. https://doi.org/10.1038/s41586 - 019 - 1189 - 0.

[23] Ling, Z., Jolliff, B. L., Wang, A., Li, C., Liu, J., Zhang, J., et al. (2015). Correlated compositional and mineralogical investigations at the Chang'E - 3 landing site. Nature Communications, 6 (1), 8880. https://doi.org/10.1038/ncomms9880.

[24] Morgan, G. A., Campbell, B. A., Campbell, D. B., & Hawke, B. R. (2016). Investigating the stratigraphy of Mare Imbrium flow emplacement with Earth -

based radar. Journal of Geophysical Research: Planets, 121, 1498 - 1513. https://doi. org/10. 1002/2016JE005041.

[25] Ohtake, M., Uemoto, K., Yokota, Y., Morota, T., Yamamoto, S., Nakamura, R., et al. (2014). Geologic structure generated by large - impact basin formation observed at the South Pole - Aitken basin on the Moon. Geophysical Research Letters, 41, 2738 - 2745. https://doi. org/10. 1002/2014GL059478.

[26] Olhoeft, G. R., & Strangway, D. W. (1975). Dielectric properties of the first 100 meters of the Moon. Earth and Planetary Science Letters, 24 (3), 394 - 404.

[27] Ono, T., Kumamoto, A., Nakagawa, H., Yamaguchi, Y., Oshigami, S., Yamaji, A., et al. (2009). Lunar radar sounder observations of subsurface layers under the nearside maria of the Moon. Science, 323 (5916), 909 - 912. https://doi. org/10. 1126/science. 1165988.

[28] Pasolli, E., Melgani, F., & Donelli, M. (2009). Automatic analysis of GPR images: A pattern - recognition approach. IEEE Transactions on Geoscience and Remote Sensing, 47 (7), 2206 - 2217. https://doi. org/10. 1109/TGRS. 2009. 2012701.

[29] Qiao, L., Ling, Z., Fu, X., & Li, B. (2019). Geological characterization of the Chang'E - 4 landing area on the lunar farside. Icarus, 333, 37 - 51. https://doi. org/10. 1016/j. icarus. 2019. 05. 029.

[30] Ristic, A. V., Petrovacki, D., & Govedarica, M. (2009). A new method to simultaneously estimate the radius of a cylindrical object and the wave propagation velocity from GPR data. Computational Geosciences, 35(8), 1620 - 1630. https://doi. org/10. 1016/j. cageo. 2009. 01. 003.

[31] Rochette, P., Gattacceca, J., Ivanov, A. V., Nazarov, M. A., & Bezaeva, N. S. (2010). Magnetic properties of lunar materials: Meteorites, Luna and Apollo returned samples. Earth and Planetary Science Letters, 292 (3 - 4), 383 - 391. https://doi. org/10. 1016/j. epsl. 2010. 02. 007.

[32] Sato, H., Robinson, M. S., Lawrence, S. J., Denevi, B. W., Hapke, B., Jolliff, B. L., & Hiesinger, H. (2017). Lunar mare TiO_2 abundances estimated from UV/Vis reflectance. Icarus, 296, 216 - 238. https://doi. org/10. 1016/j. icarus. 2017. 06. 013.

[33] Seu, R., Phillips, R. J., Biccari, D., Orosei, R., Masdea, A., Picardi, G., et al. (2007). SHARAD sounding radar on the Mars Reconnaissance Orbiter [J]. Journal of Geophysical Research, 112, E05S05. https://doi. org/10. 1029/2006JE002745.

[34] Shkuratov, Y. G., & Bondarenko, N. V. (2001). Regolith layer thickness mapping of the Moon by radar and optical data. Icarus, 149 (2), 329 - 338. https://doi. org/10. 1006/icar. 2000. 6545.

[35] Strangway, D. W., Chapman, W. B., Olhoeft, G. R., & Carnes, J. (1972). Electrical properties of lunar soil dependence on frequency, temperature and moisture. Earth and Planetary Science Letters, 16 (2), 275 - 281. https://doi. org/

10. 1016/0012 - 821X (72) 90203 - 8.

[36] Troitsky, V. S., & Shmulevich, S. A. (1973). Dependence of dielectric properties of rocks on their volume weight. Journal of Geophysical Research, 78 (29), 6933 - 6935. https://doi. org/10. 1029/JB078i029p06933.

[37] Ulaby, F. T., Bengal, T., East, J., Dobson, M. C., Garvin, J. & Evans, D. (1988). Microwave Dielectric Spectrum of Rocks, Univ. of Michigan, Ann Arbor, Mich.

[38] Wang, Z., Li, Y., Jiang, J., & Li, D. H. (2010). Lunar surface dielectric constant, regolith thickness, and He abundance distributions retrieved from the microwave brightness temperatures of CE - 1 Lunar Microwave Sounder. Science China Earth Sciences, 53 (9), 1365 - 1378. https://doi. org/10. 1007/s11430 - 010 - 4022 - z.

[39] Watters, T. R., Leuschen, C. J., Campbell, B. A., Morgan, G. A., Cicchetti, A., Grant, J. A., et al. (2017). Radar sounderevidence of thick, porous sediments inMeridiani Planum and implications forice - filled deposits on Mars. Geophysical Research Letters, 44, 9208 - 9215. https://doi. org/10. 1002/2017GL074431.

[40] Wu, W., Li, C., Zuo, W., Zhang, H., Liu, J., Wen, W., et al. (2019). Lunar farside to be explored by Chang'E - 4. Nature Geoscience, 12 (4), 222 - 223. https://doi. org/10. 1038/s41561 - 019 - 0341 - 7.

[41] Xiao, L., Zhu, P., Fang, G., Xiao, Z., Zou, Y., Zhao, J., et al. (2015). A young multilayered terrane of the northern Mare Imbrium revealed by Chang'E - 3 mission. Science, 347 (6227), 1226 - 1229. https://doi. org/10. 1126/science. 1259866.

[42] Zhang, J., Yang, W., Hu, S., Lin, Y., Fang, G., Li, C., et al. (2015). Volcanic history of the Imbrium basin: A close - up view from the lunar rover Yutu. Proceedings of the National Academy of Sciences, 112 (17), 5342 - 5347. https://doi. org/10. 1073/pnas. 1503082112.

Properties of Lunar Regolith on the Moon's Farside Unveiled by Chang'E-4 Lunar Penetrating Radar

Zehua Dong[1,2], Guangyou Fang[2,3,4], Bin Zhou[2,4], Di Zhao[2,3], Yunze Gao[2], and Yicai Ji[2]

1. School of Information and Electronics, Beijing Institute of Technology, Beijing, China
2. Key Laboratory of Electromagnetic Radiation and Sensing Technology, Aerospace Information Research Institute, Chinese Academy of Sciences, Beijing, China
3. University of Chinese Academy of Science, Beijing, China
4. Greater Bay Area Research Institute of AIR-CAS, Guangzhou, China

Abstract: The complex thermal history of the Moon leads to an unequal distribution of volcanic products between the lunar nearside and the farside. So far, no lunar materials have been sampled from the Moon's farside and no detailed properties of lunar regolith on the farside have been detected before. On January 3, 2019, Chang'E-4 (CE-4) touched down onto the Von Kármán crater on the Moon's farside. CE-4 Lunar Penetrating Radar (LPR) onboard the Yutu-2 rover is the first surface radar on the Moon's farside. Here we show the subsurface structure and properties of regolith materials at the landing region with LPR data during the first five lunar days. The thickness of lunar regolith is constrained as ~12m, much thicker than that at Chang'E-3 (CE-3) landing site, which is expected since CE-3 landed on lunar maria. The relative permittivity of lunar surface (<30cm) at CE-4 landing region is identified to be 2.35 ± 0.20. The loss tangent and $TiO_2 + FeO$ content of the regolith materials layer (0~12m) are constrained to be $(4.4 \pm 0.5) \times 10^{-3}$ and 11.6 ± 1.1 wt. %, respectively, much lower than those at CE-3 landing site. It indicates that local surface materials possess less attenuation for radiowave, in accordance with the greater penetrating depth of CE-4 LPR than that of CE-3 LPR. Furthermore, the results also prove that the growth rate of lunar weathered regolith successively declines over time and the growth rate of lunar regolith on the Moon's farside may well be higher than that on the nearside due to the more frequent meteorite impacts.

Plain Language Summary: Lunar regolith (informally called the lunar soil) is the uppermost material on the lunar surface. It preserves vital clues about geology and impact history of the Moon. Although numerous missions to the Moon have landed on the nearside, so far, no lunar materials have been sampled from the Moon's farside. In 2019, China's Chang'E-4 (CE-4) spacecraft softly landed on the farside of the Moon for the first time and deployed its rover, Yutu-2. The Lunar Penetrating Radar (LPR) onboard Yutu-2 can transmit electromagnetic pulses to detect the lunar subsurface structure and properties of the regolith. The relative permittivity, loss tangent and TiO_2 + FeO content of lunar regolith materials at landing site are constrained with LPR data in this paper. The results indicate that the farside may be bombarded more frequently, leading to different regolith accumulation rates on the lunar nearside versus farside.

Key Points:

• Detailed stratigraphy of lunar regolith on the Moon's farside is unveiled with Chang'E—4 Lunar Penetrating Radar data

• The dielectric properties and ilmenite content of regolith materials at the landing region are identified

• The thickness and accumulated rate of lunar regolith at the landing region are constrained and analyzed

1 Introduction

The lunar surface is covered by a layer of fine-grained material (lunar regolith) with different thickness range from 0 to 15m (e. g., Shkuratov & Bondarenku, 2001; Shoemaker et al., 1970). The properties of lunar regolith vary from place to place and are dependent on the mineral composition and the impact history. The shallow subsurface structure and mineral compositions of the lunar regolith offer the key to further understand the geology and the local history of basaltic volcanism of the Moon (Fa et al., 2015). The thickness of regolith layer can be used to constrain the Moon's impact history. Therefore, the properties of lunar regolith have important implications on unraveling the lunar surface age, lava eruption volumes and the dichotomy of volcanic products between the Earth-facing side and the far-side (Wilhelms et al., 1987). In detail, the properties of lunar regolith focused in this paper include the thickness, permittivity, density, loss tangent, and TiO_2 + FeO content.

The approximate thickness of lunar regolith has been estimated in large amounts by morphologic methods and remote sensing observations. During the Apollo and Luna era, that is, 1960s, in-situ experiments were initially conducted at several locations. In the following decades, modeling on the basis of the crater morphology and size-frequency distributions became the most prevailing methods (e. g., Bart et al., 2011; Gault et al., 1966; Oberbeck & Quaide, 1968; Quaide & Oberbeck, 1968; Wilcox et al., 2005). Gradually, Earth-based/space-borne radar systems were employed more frequently. For instance, the thickness of surface layer on the Moon's near-side was roughly estimated with Earth - based 70cm Arecibo radar data by Thompson (1987) and Shkuratov and Bondarenku (2001). Fa and Wieczorek (2012) took the scattering from the surface and subsurface rocks into consideration upon previous results and obtained an improved relation among the dielectric constant, bulk density and regolith composition. In addition, Fa and Jin (2007) studied the brightness temperature of the Moon and then mapped the thickness of lunar regolith. Based on Kaguya Lunar Radar Sounder and Laser altimeter data, Kobayashi et al. (2010) also obtained the global thickness map of lunar surface regolith. Moreover, Chang'E - 1 Lunar Microwave Sounder (CELMS) data were used on the basis of a back propagation neural network method to estimate the thickness of the surface layer on the Moon (Meng et al., 2014). In conclusion, the thickness of lunar regolith at large scale have been estimated with various methods.

The dielectric properties (permittivity and loss tangent) of the regolith layer are influenced by its ilmenite content and density (porosity) (Olhoeft & Strangway, 1975; Ulaby et al., 1990). Numerous previous studies measured relative dielectric permittivity with radar experiments performed in Surveyor and Luna programs at high - frequency, for example, 2.00 - 3.28 at ~13GHz (Muhleman et al., 1969) and 1.7 - 4.4 at ~1.8 - ~10GHz (Kroupenio, 1972, 1973, Kroupenio, et al., 1975). In addition, the results of the Apollo program and laboratory studies on lunar samples have confirmed that most lunar soils have dielectric constant about 3, while solid rocks have dielectric constant about 7.5 (Gold et al., 1970; Strangway & Olhoeft, 1977). More recently, a common method is to use spectral analysis to deduce the permittivity of lunar regolith (e.g., Shkuratov & Bondarenku, 2001): first, the ilmenite content of lunar regolith can be identified according to the spectrograph or gamma - ray spectrometer (GRS) data, and then the dielectric constant can be reversely obtained based on the existing fitting relationships between permittivity and TiO_2 + FeO abundance (e.g., Olhoeft & Strangway, 1975). Moreover, Ishiyama et al. (2013) improved the method by combining the thickness of regolith derived from spectral data with the Kaguya lunar radar sounder (LRS) data. The velocity of electromagnetic waves within the regolith layer is completely influenced by the average permittivity of regolith materials, which can be constrained by the propagating time between the surface and subsurface echoes in radargrams. Vijayan et al. (2015) proposed a method to estimate both the permittivity and thickness of the lunar regolith with simulated brightness temperature data from a dual frequency SAR of Chandrayaan - 2 mission. So far, the permittivity of lunar regolith has been roughly estimated using many indirect methods.

With regards to all these experiments, space - borne/rover - deployed ground penetrating radar (GPR) is the most direct and effective geophysical tool to characterize the physical properties of the lunar regolith (Seu et al., 2007). GPR could map the structure versus depth by detecting dielectric discontinuities, differs from other observations (e.g., infrared or optical analysis). Several such experiments have been conducted: the Apollo Lunar Sounder Experiment (ALSE) operated during NASA's Apollo 17 mission (Peeples et al., 1978; Porcello et al., 1974) and the LRS onboard Japan's Kaguya mission (Ono et al., 2009) profiled lunar surface topographic variations and inferred subsurface structure along the spacecraft ground track. However, satellite-borne radar sounders lead to coarse resolution at depth and low signal-to-noise ratio (SNR) due to the inherent bandwidths and environmental clutters, respectively, inadequate to identify the detailed fine structure and electrical properties of lunar surface regolith. The maximum penetrating depths of these two space-borne radar systems are greater than ~1km, but their vertical resolutions are only ~75-400m. The first attempt to conduct a rover-deployed GPR to detect the lunar subsurface was made during China's Chang'E-3 (CE-3) mission (Morgan et al., 2016). Although CE-3 LPR collected only 114m of scientific data before dormancy, many significant results with respect to the structure and properties of lunar regolith materials at the landing region on the Mare Imbrium were interpreted and revealed (e.g., Dong et al., 2017; Ling et al., 2015; Xiao et al., 2015; Zhang et al., 2015). However, so far, no detailed subsurface structure and properties of lunar regolith on the Moon's farside has ever been detected before Chang'E-4 (CE-4) mission due to surface clutter, coarse range resolution and problems in data interpretation, and no lunar samples on the Moon's farside has ever been returned, either.

On January 3, 2019, China's Chang'E-4 mission touched on the Moon at 45.4446°S, 177.5991°E and subsequently employed the Yutu-2 rover (Wu et al., 2019) (Figure. 1a). The CE-4 LPR onboard Yutu-2 is the first surface radar system on the Moon's farside, enabling an investigation of the subsurface with a higher range (i.e., vertical) and horizontal spatial resolution than that from orbital sounders. The landing site is located at the southeast of the Von Kármán crater (diameter = ~186km; central coordinates as 44.4°S, 176.2°E) (Figure 1b) which formed in the pre-Nectarian, within the largest known impact structure in the solar system: the South Pole-Aitken (SPA) basin. It is a key area to figure out internal structure, geological history and thermal evolution of the Moon (Hiesinger, 2006; Huang et al., 2018). Parts of the Von Kármán crater were flooded by mare basalts during the Imbrian Period (Huang et al., 2018; Morota et al., 2011; Spudis et al., 1994) (Figure 2a), and then several large Imbrian or post-Imbrian impact craters were formed around the Von Kármán crater, including the Von Kármán M crater, the Leibnitz crater, the Alder crater, the Finsen crater, and the Von Kármán L crater (Figure 1b). The ages of these craters are suggested to be: Von Kármán M > Von Kármán > Leibnitz > Alder > Finsen > Von Kármán L on the basis of Wilhelms Geological Maps (Fortezzo & Hare, 2013; Wilhelms et al., 1979) (Figure 2d). The CE-4 landing site is a generally flat region with an observed gravity anomaly, which probably suggests a distinctive subsurface geological framework (Andrews-Hanna et al., 2014) (Figure. 2b). This region also shows fairly high-level

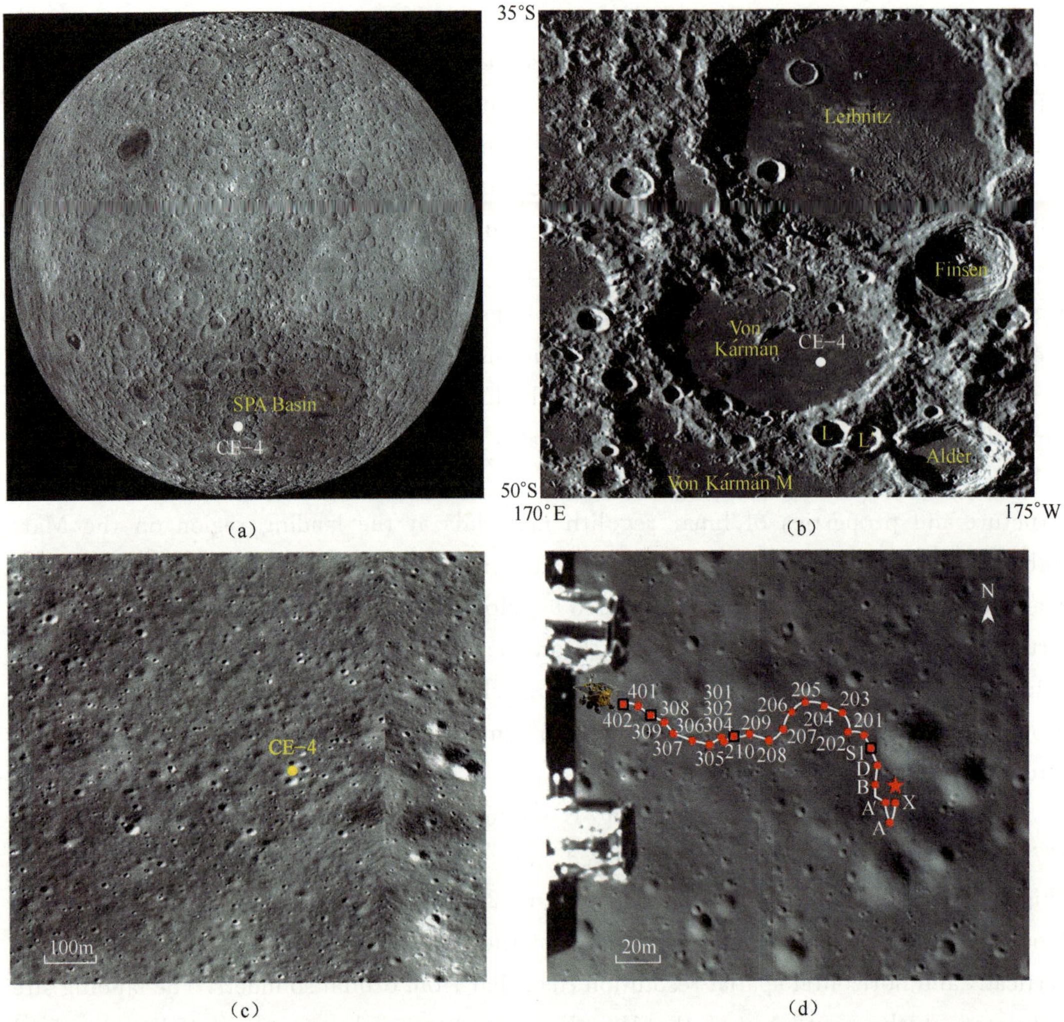

Figure 1　The location of Chang'E - 4 (CE - 4) landing site and the track of Yutu - 2 rover. (a) CE - 4 landing site on the Moon's farside. (b) Large - scale craters surrounding the CE - 4 landing site. (c) Local details at CE - 4 landing site obtained by the Lunar Resonant Orbiter (https://lroc. sese. asu. edu/). (d) Yutu - 2's track on the Moon. The red star marks the CE - 4 lander. Yutu - 2 has driven ~178. 9m (white line) along 26 waypoints (red dots) from X to LE402 during the first four lunar days. The black squares at S1, LE210, LE309, and LE402 indicate where Yutu - 2 rested for lunar nights, which ended on January 30, 2019, February 28, 2019, March 29, 2019, and April 29, 2019, respectively. The image was taken by the CE - 4 Landing Camera and the left side was blocked by the CE - 4 lander

optical maturity (Lemelin et al., 2016) (Figure 2c), which suggests the local surface regolith has been weathered for a long period and has probably accumulated a fairly thick regolith layer. The topography of the landing site is relatively flat (Figure 1c). The landing site lies with an area of secondary scour from Finsen crater and is surrounded by three craters whose diameters are 26, 22, and 28m, respectively (Figure 1d). The northwest trending Yutu - 2 traverse skirts around the north side of a degraded 40m crater and then continues along an approximately straight northwest trending traverse (208 - 402) (Figure 1d).

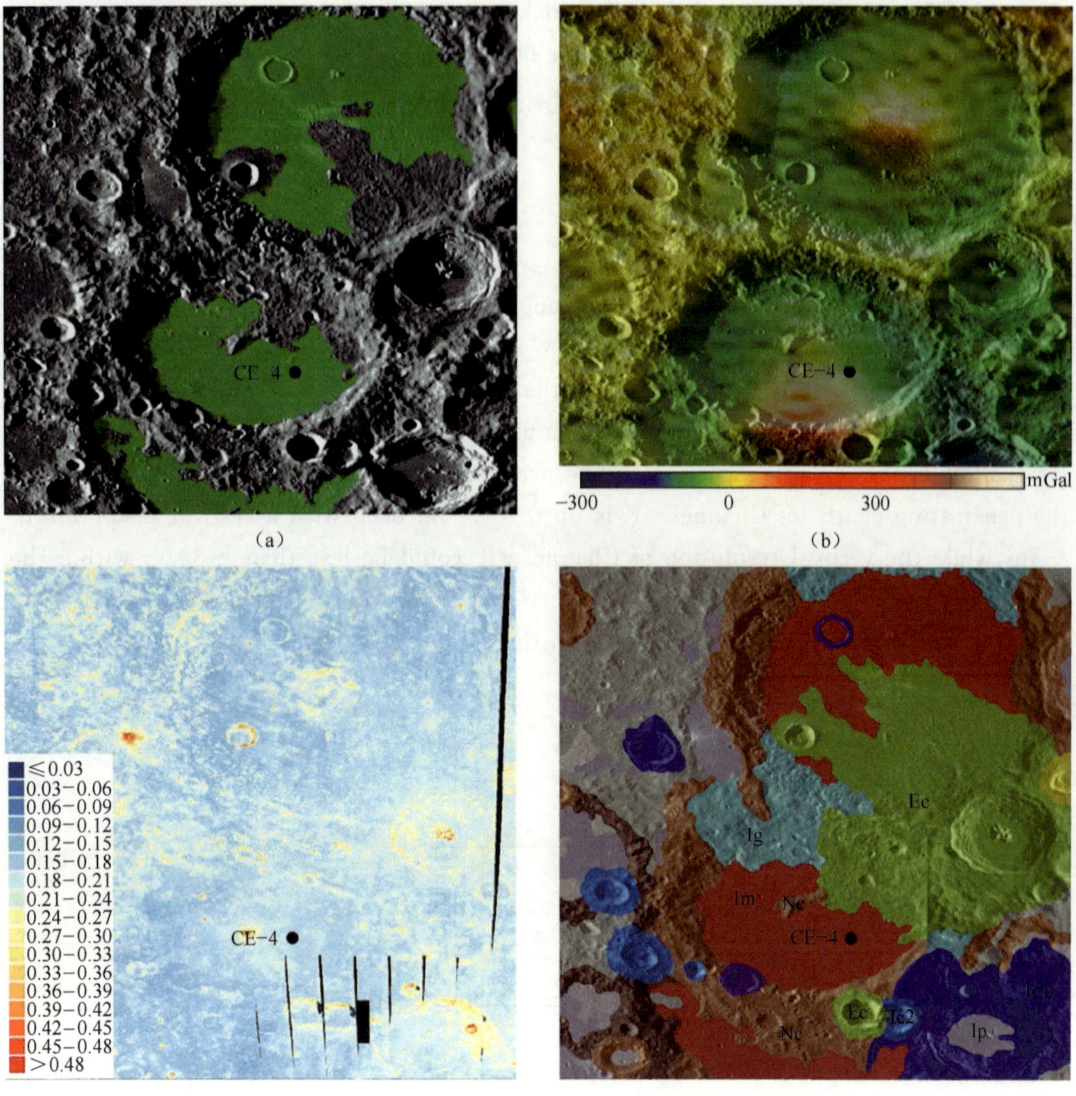

Figure 2　Geological background at Chang'E - 4 (CE - 4) landing region. (a) Mosaic of high - resolution LROC images of the lunar surface. The green areas indicate where was flooded by mare basalts. The southeast part of Leibnitz crater is presumably covered by ejecta from the Finsen crater. (b) Bouguer gravity filtered from degree 6 to 660 to emphasize regional anomalies, which are derived from the GRAIL GRGM900. C gravity mode. Colors indicate different intensity of gravity field in *m*Gal. (c) Derived Optical Maturity (OMAT) mosaic created from Mineral Mapper reflectance data acquired by the Kaguya mission. Colors indicate different local reflectivity. (d) Wilhelms Geologic Map at CE - 4 landing region. Symbols marked on various colors indicate different geological units. Im, Ip, Ec, Nc, Ic1, and Ic2 indicate mare material, plains material, crater material younger than most mare material, crater material younger than Nectaris basin but older than Imbrian basin, lower Imbrian crater materials, and upper Imbrian crater materials, respectively

In this study, we unveil the properties (including permittivity, loss tangent and TiO_2 + FeO abundance) of regolith materials at CE-4 landing region on the Moon's far-side with CE-4 LPR data, and then the thickness, shallow stratigraphy and weathering rate of local lunar regolith are constrained. The comparison of our results and previous studies at CE-3 landing site or other regions on the Moon's nearside is also analyzed.

2 Materials and Methods

CE-4 LPR adopts the same mode as CE-3 LPR which is composed of two independent channels with respective operating frequency (Channel-Ⅰ: 60MHz, Channel-Ⅱ: 450MHz) and bandwidth (Channel-Ⅰ: 40MHz, Channel-Ⅱ: 500MHz) (Fang et al., 2014), to balance the penetrating depths and range resolution of radar-grams. The Channel-Ⅰ is for detecting subsurface structure of lunar crust, while the Channel-Ⅱ is for shallow lunar regolith exploration. The preliminary detected results demonstrate that the penetrating depths of Channel-Ⅰ is up to ~300m deep with a vertical resolution of ~3m, while the vertical resolution of Channel-Ⅱ could be less than ~30cm within the upper ~50m depth. Therefore, the structure and properties of lunar regolith materials were revealed with LPR Channel-Ⅱ data in this study.

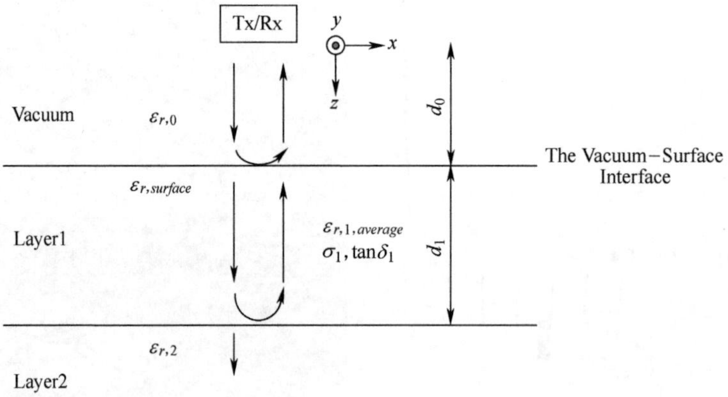

Figure 3　The two-dimensional model of lunar subsurface layers. $\varepsilon_{r,surface}$ is the relative permittivity of the vacuum-surface interface, $\varepsilon_{r,2}$ is the relative permittivity of layer2, $\varepsilon_{r,1,average}$ is the average relative permittivity of layer1, σ_1 and tan δ_1 are conductivity and loss tangent of layer1, respectively

2.1 Data Processing of LPR Data

A trace of the receiving signals of CE-4 LPR Channel-Ⅱ consists of 2, 048 sampling points within 640ns (Fang et al., 2014). Yutu-2 drove 178.9m before the restarting of its fifth working lunar day (May 29, 2019) (Figure 1d). The LPR begins to collect data as soon as Yutu-2 rover is awakened every lunar day and has collected a mass of data along the whole track of Yutu - 2. For the sake of the consistency of radargrams and in - situ subsurface structures, we removed the data received when Yutu - 2 rover cruised in very slow motion or stopped. Finally, 4, 844 high - quality traces of Channel - Ⅱ data were obtained between waypoints X to LE402 (Figure 1d).

The raw LPR data are attenuated signals with low signal - to - noise ratio (SNR) due to the inherent electrical noise, the spatial clutter and noise, and the attenuation of radar wave by lunar materials (Dong et al., 2017). Consequently, employing effective noise removal and enhancing methods is one of the prerequisites for further analysis. In this work, amplitude compensation, direct - current - removing, background - removing and band - pass filtering, as commonly applied for GPR data on the Earth, are carried out for Channel - Ⅱ data (Supporting Information). The quality and SNR of the processed results have been obviously enhanced compared with those of the raw data, facilitating the explanation and estimation of the structure and properties of regolith materials at the landing region. In addition, a modified trace correlation method (Dong et al., 2017) is also employed to extract subsurface layers and to identify in - situ shallow stratigraphy (Supporting Information).

2.2 Estimation of Properties of Lunar Regolith

Referring to the GPR applications in civil engineering and geophysical exploration (Dong et al., 2016), the stratigraphic model at CE - 4 landing region can be simplified as three two - dimensional layers that includes vacuum, regolith materials (Layer 1) and subsurface lunar materials (Layer 2) from top to bottom (Figure 3). Based on this model, the relative permittivity of the vacuum - surface interface ($\varepsilon_{r,surface}$) is given by Equation 1 (Supporting Information):

$$\varepsilon_{r,\ surface}=\varepsilon_{r,\ 0}\left(\frac{1-A_0}{1+A_0}\right)^2 \tag{1}$$

Where $\varepsilon_{r,0}$ is the relative permittivity of vacuum, equals to 1. A_0 is the relative reflected amplitude that indicates the ratio of amplitudes between the surface reflection and the incident pulse. Therefore, both the amplitudes of the surface echo and the incident signal are required to derive the value of A_0. However, only the amplitude of the surface reflected pulse can be identified with the LPR data. In order to obtain the amplitude of the incident pulse, a calibration experiment had been carried out on the Earth before the LPR was launched. We placed the LPR on an adequately large plate which is made of copper. Then, the reflection occurring at the surface of the plate can be approximated as total reflection, which means the amplitude of the surface reflected pulse is equal to that of the incident pulse (Dong et al., 2016). As a result, the value of A_0 can be obtained to calculate the relative permittivity of lunar regolith materials.

Thus, we can use the average relative permittivity ($\varepsilon_{r,1,average}$) and the two - way propagating time within the regolith materials (t_1) observed from LPR radargram to accurately estimate the thickness of lunar regolith layer (d_1):

$$d_1 = \frac{ct_1}{2\sqrt{\varepsilon_{r,\ 1,\ average}}} \tag{2}$$

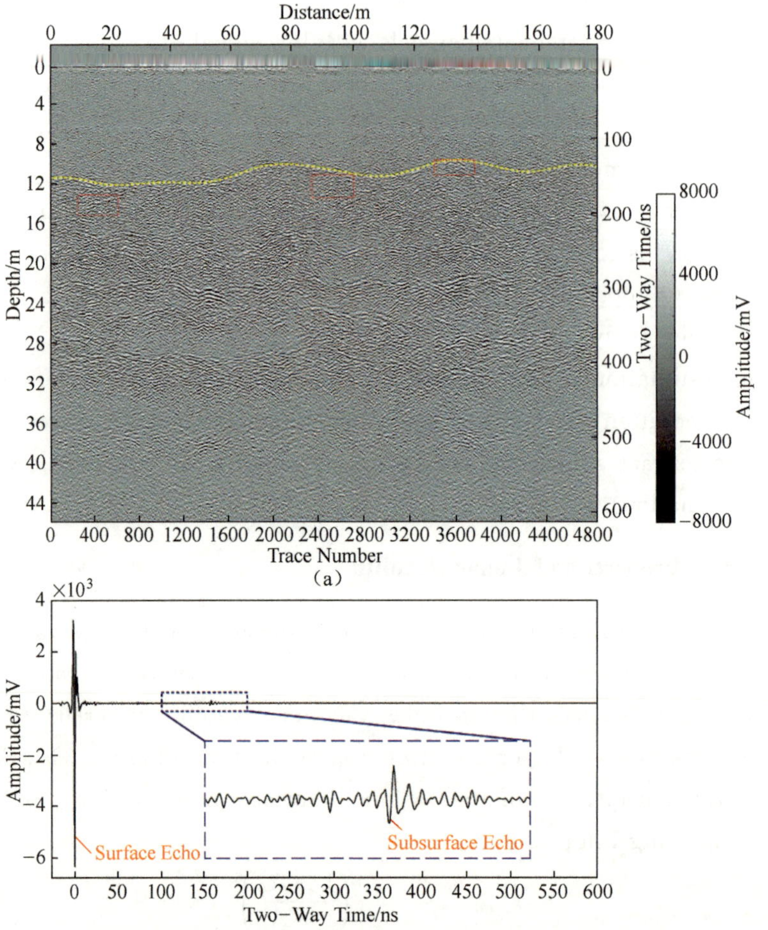

Figure 4　Chang'E - 4 (CE - 4) Lunar Penetrating Radar (LPR) Channel - Ⅱ data. (a) Processed CE - 4 LPR Channel - Ⅱ radargram. The yellow dashed line indicates the interface between regolith materials and the subsurface materials. The red boxes indicate three hyperbolic features. (b) One trace of Processed Channel - Ⅱ data and the surface and subsurface echoes were pointed out

The results measured by lunar regolith samples returned from Apollo missions suggest that the relative permittivity (ε_r) has a relationship with its bulk density (porosity) (Carrier et al., 1991). The bulk density (ρ (g/cm^3)) can be constrained as:

$$\varepsilon_r = 1.919^{\rho} \tag{3}$$

Furthermore, the conductivity (σ_1) of the regolith materials (layer1), is given by the following equation (Supporting Information):

$$\sigma_1 = \frac{\sqrt{\varepsilon_{r,1}}}{d_1}\frac{1}{\eta_0}\log\left[\frac{(1-A_0^2)(\sqrt{\varepsilon_{r,2}}-\sqrt{\varepsilon_{r,1}})}{-A_1(\sqrt{\varepsilon_{r,2}}+\sqrt{\varepsilon_{r,1}})}\right] \tag{4}$$

where, $\varepsilon_{r,1}$ and $\varepsilon_{r,2}$ are the relative permittivity of regolith materials and subsurface materials. A_0 and A_1 are relative amplitudes (signed) as referred in Equation 1. Then, the loss tangent ($\tan\delta_1$) of the regolith materials (layer1) can be determined as follows:

$$\tan\delta_1 = \frac{\sigma_1}{\omega\varepsilon_0\varepsilon_{r,1,average}} \tag{5}$$

The dielectric properties of Apollo lunar samples were measured in different lab conditions (e. g., in air, nitrogen, and vacuum) with different frequencies (1 - 450MHz) in previous studies. In addition, atmospheric moisture contamination also affects the measurement of lunar samples (Olhoeft et al., 1975). Based on lab experiments, the correlation between the bulk density, the TiO_2 + FeO content and the loss tangent of regolith materials has been studied. Therefore, the TiO_2 + FeO content ($\%TiO_2 + \%FeO$) of Layer1 can be derived from the loss tangent ($\tan\delta_1$) and bulk density ($\rho_{bulk,1}$) of Layer1 with a best - fit empirical relationship expressed as (Carrier et al., 1991):

$$\tan\delta = 10^{(0.038\times(\%TiO_2+\%FeO)+0.312\rho-3.26)} \tag{6}$$

As a result, the properties of lunar regolith materials (Layer1) including the relative permittivity, the thickness, the density, the loss tangent and the TiO_2+FeO abundance can be estimated as the procedures mentioned above in sequence.

3 Results

We calibrated and processed the CE - 4 LPR Channel - Ⅱ data with the methods in Section. 2. 1 and obtained a radargram with high resolution and improved SNR as shown in Figure 4a. The surface regolith layer was identified with the modified trace correlation method and the lower interface was marked with yellow line in Figure. 4a. The two - way time delay of this layer is ~150ns and varies with a range of ~20ns along Yutu - 2's track, which indicates a variation of regolith depth over small area on the Moon (Wilcox et al., 2005). The layer 1 (0—~12m) displays relatively uniform characteristics with weak internal reflections, which suggest homogenous and fine - grained materials with a small quantity of buried rocks, while the layer 2 (~12—~35m) shows abundant and cluttered reflections within which there are numerous obvious parabola - shaped diffractions that probably indicate various - sized subsurface boulders or fragments. Together, these observations suggest that the upper ~35m of lunar material were originally deposited as ejecta from surrounding craters, and that the uppermost ~12m were subsequently modified, mixed and overturned by the bombardment of numerous micrometeorites, forming the surface layer (0—~12m) with reworked ejecta and the subsurface layer (~12—~35m) with remnant ejecta deposits. The uppermost fine - grained regolith (0—~12m) and the underlying multi - layered ejecta (~12—

~35m)have been commonly accepted based on the obvious radar features, spectral data, and local geological context (Lai et al., 2019; Li, et al., 2019, 2020; Zhang et al., 2020), but the materials below ~35m is a controversial layer, which is speculated to be coarse ejecta from other craters (Li, et al., 2020; Zhang et al., 2020), or fragmented basalt (Lai et al., 2019). A trace of processed LPR signal was plotted in Figure 4b with easily recognized surface echo and subsurface echo of which amplitudes can be used to calculate the relative permittivity of lunar regolith.

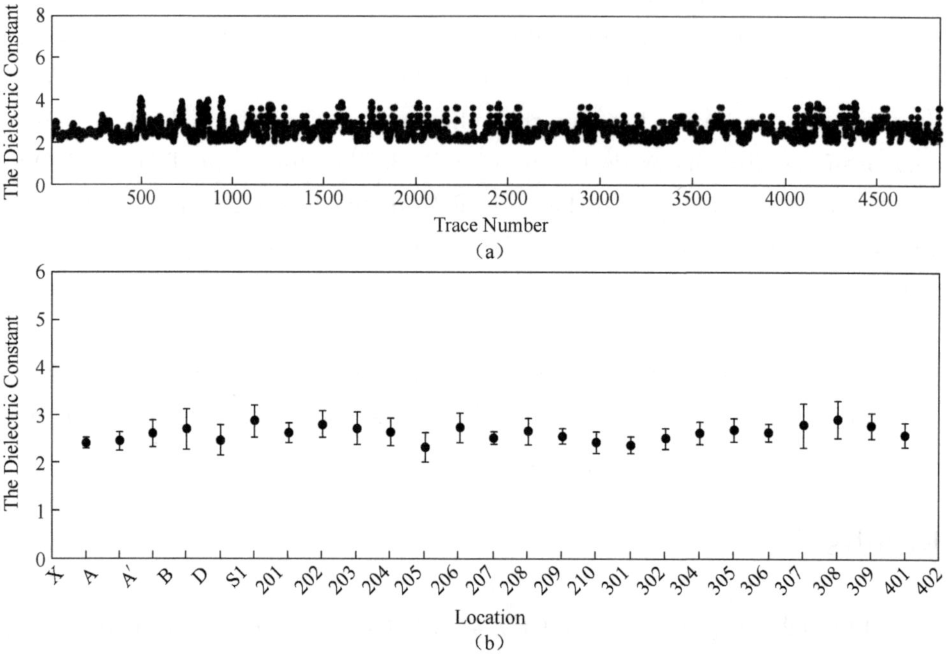

Figure 5 Estimated relative permittivity of the lunar surface. (a) Calculated values of the relative dielectric constant from the surface echo of each trace of ground penetrating radar (GPR) data along Yutu-2's track. (b) Statistics between each two waypoints as marked in Figure 1d. The spots indicate the mean values and the error bars indicate 1 σ error

The relative permittivity of the regolith layer was calculated trace by trace using the methods described in Section 2.2 and the estimated results were shown in Figure 5a. The calculated relative dielectric constants range from ~2 to 4 and vary along Yutu-2's track. It is probably due to the surrounding small craters and small-sized rocks exposed on the lunar surface. According to the waypoints marked in Figure 1d, we divided the estimated results into segments, and then statistics and error analysis were carried out. The average permittivity and one standard deviation (1σ) error within each subsection are identified as shown in Figure 5b.

The relative permittivity ($\varepsilon_{r,surface}$) of the lunar surface at CE-4 landing region is constrained to be 2.35 ± 0.20, consistent with the measurements from lunar regolith samples (Carrier et al., 1991), lower than the estimated results of 2.9 ± 0.4 at CE-3 landing region (Dong et al., 2017). Note that this result indicates the relative permittivity of the vacuum - surface interface (< ~30cm allowing for the vertical resolution of LPR) because it is calculated from the surface echo (Figure 3). In addition, the bulk density of surface regolith materials (< ~30cm) at CE - 4 landing region is estimated to be $1.31 \pm 0.20 g/cm^3$ with Equation 3.

The bulk density of the lunar regolith below ~6m at CE - 4 landing site has been estimated as ~ $1.9 g/cm^3$ (Li et al., 2020), which corresponds to the relative permittivity of the upper 6m thick regolith materials of ~3.5. Considering that the permittivity of lunar regolith increases rapidly at shallow depth, but increases more slowly at deeper depths (Carrier et al., 1991; Li et al., 2020), we assumed the average permittivity ($\varepsilon_{r,1,average}$) of regolith materials as ~3.6, consistent with that estimated from empirical functions of bulk density and regolith thickness (e.g., $\rho = 1.92 \frac{z+12.2}{z+18}$, where ρ is bulk density in g/cm^3, z is thickness in m) (Carrier et al., 1991). Figure 4a shows that the two - way time delay of the regolith layer is about 140—150ns. Then, the thickness of regolith layer is calculated to be ~11.5—~12.4 m along Yutu - 2's track. The time - to - depth conversion is marked in Figure 4a according to the derived relative permittivity, higher than the estimated results (2.5 — 7.5m) based on crater morphology by Huang et al. (2018).

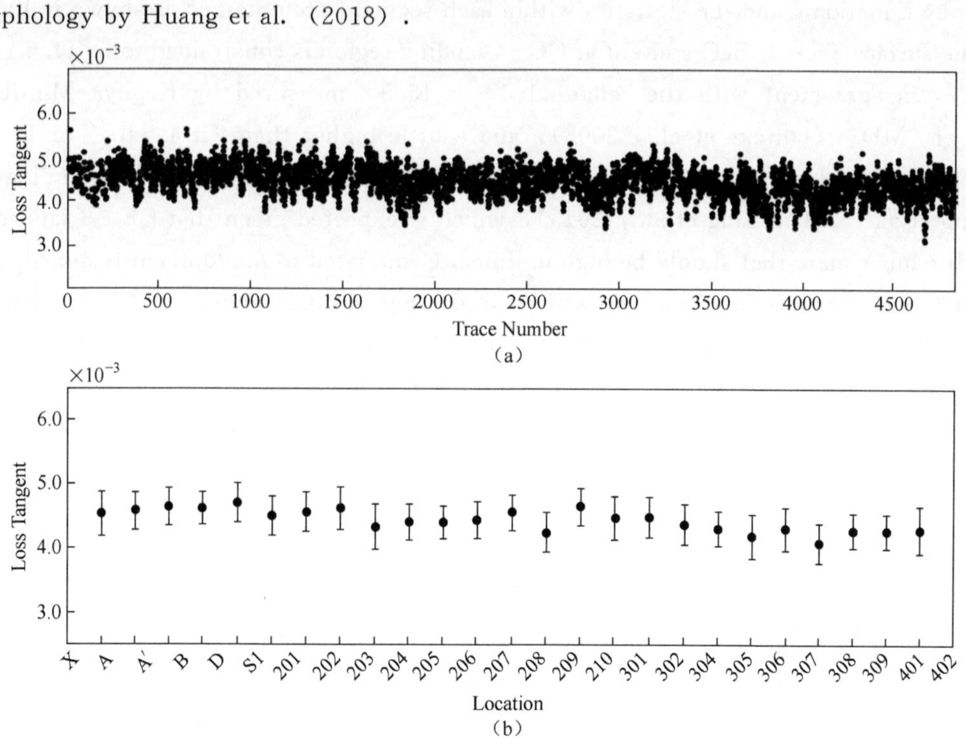

Figure 6　Estimated loss tangent of the regolith materials layer. (a) Calculated results of the loss tangent for each trace along Yutu - 2's track. (b) Statistics between each two waypoints as marked in Figure 1d. The spots indicate the mean values and the error bars indicate 1σ error

Furthermore, the relative permittivity of the subsurface layer (Layer 2) is estimated using a common hyperbolic fitting method (Supporting Information) (Jol, 2009). Three subsurface hyperbolic features are identified in subsurface layers, as marked in Figure 4a, give estimated relative permittivity of 4.0, 4.0, and 4.1, respectively. Therefore, the relative permittivity ($\varepsilon_{r,2}$) of the subsurface layer is assumed as 4.0, and then the loss tangent ($\tan\delta_1$) of the regolith layer is derived from each LPR trace as shown in Figure 6a. The average value and 1σ error are calculated within each segment as well (Figure 6b).

The results show that the loss tangent of the regolith layer at CE-4 landing region is constrained to be $(4.4\pm0.5)\times10^{-3}$ at the frequency of ~450MHz, consistent with $(5\pm2)\times10^{-3}$ found by Li et al. (2020), higher than $\sim1\times10^{-3}$ measured at the frequency of 450MHz from Apollo 14 lunar samples (Gold et al., 1972), but lower than ~0.01 at CE-3 landing site (Dong et al., 2017). Note that the LPR data are wideband signals (~200-700MHz), so here the conductivity and loss tangent of regolith materials are estimated with a central frequency (~450MHz), which may cause potential uncertainty with respect to the dependence of the loss tangent and conductivity on frequency.

Similarly, the TiO_2+FeO abundance of lunar regolith that estimated from each LPR trace by Equation 6 and the statistics within each segment are obtained as shown in Figure 7. The surface TiO_2+FeO content at CE-4 landing region is constrained to be 11.6 ± 1.1 wt.%, in agreement with the range 11.4%-15.9% measured by Kaguya Multiband Imager (MI) (Ohtake et al., 2008), and a little higher than 9 ± 4 wt.% or 11 ± 4 wt.% found by Li et al. (2020), lower than ~23 wt.%-30 wt.% at CE-3 landing site (Dong et al., 2017; Ling et al., 2015), which is expected given that CE-3 landed on basaltic lunar mare that should be high in ilmenite compared to predominantly anorthositic highlands material. It is consistent with that the higher loss tangent at CE-4 landing region than that at CE-3 landing site. Respectively, the variation of loss tangent ($\tan\delta_1$) and TiO_2+FeO abundance in this layer of regolith materials are constrained within a range of $\sim1.0\times10^{-3}$ and ~1% if the subsurface relative permittivity ($\varepsilon_{r,2}$) changes ±0.1.

4 Discussions

CE-4 LPR is an in-situ radar system working on the lunar surface, not like previous satellite-borne exploration systems. The heights of Channel-Ⅱ antennas are approximately 30cm, which indicates a radar cross section (RCS) less than 1×1m, according to its designed antenna beam angle (Fang et al., 2014). In addition, Yutu-2 is inclined to choose potential direction and track with a tendency to be flat and smooth using the real-time data of Panorama Camera (PC) onboard the rover, so the Channel-

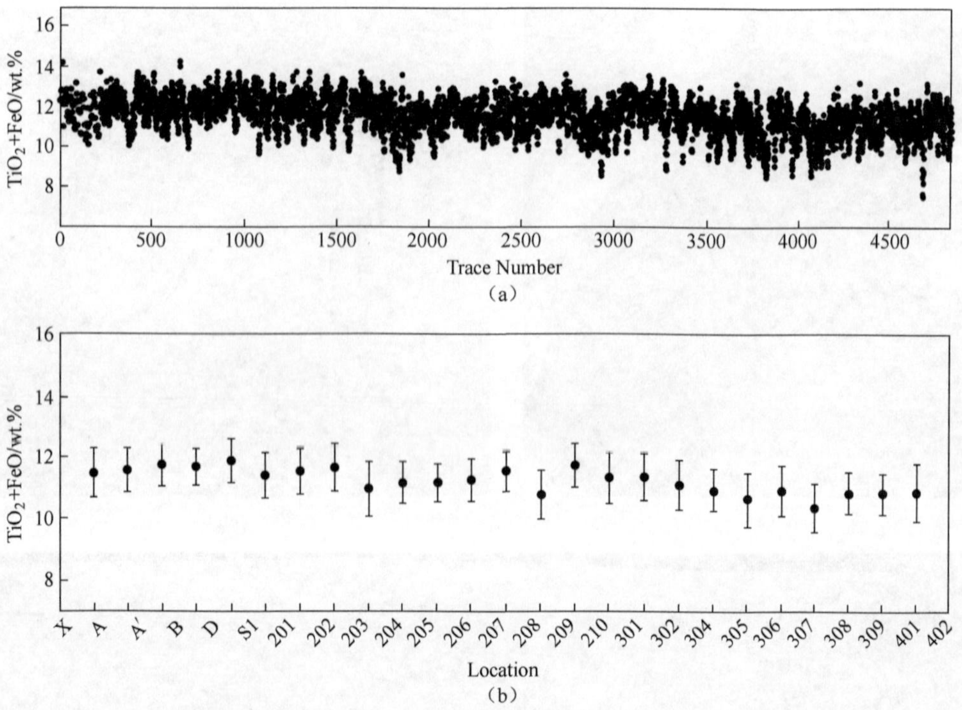

Figure 7　Estimated ilmenite content of the regolith materials layer. (a) Calculated results of the TiO_2 + FeO content of regolith materials (Layer 1) for each trace along the track. (b) Statistics between each two waypoints as marked in Figure. 1d. The spots indicate the mean values and the error bars indicate 1σ error

Ⅱ antenna almost keeps horizontally moving all the time. Therefore, only the small - sized ($< 1m^2$) local roughness and slope or some large - sized surface objects, for example, boulders and pits under the rover, would affect the estimated results of the relative permittivity. Figures. 8a and 8b show relatively flat and smooth lunar surface with no obvious rough terrain or large - sized rocks along Yutu - 2's track. As a result, we are inclined to attribute the varied estimated results of relative permittivity to the diversified dielectric properties of the regolith materials instead of the surrounding roughness.

It can be seen from Figure 4a that CE - 4 LPR Channel - Ⅱ received clear radar echoes at the depths more than ~40m, greatly exceeding the lowest radar echoes from CE - 3 LPR Channel - Ⅱ data of ~10m (Xiao et al., 2015). The improved penetrating depth at CE - 4 landing region suggests that local regolith materials are more transparent for electromagnetic waves, consistent with its lower loss tangent ($(4.4 \pm 0.5) \times 10^{-3}$) than that (~ 0.01) of the CE - 3 landing region. The radiowave attenuation of regolith materials depends primarily on the FeO + TiO_2 abundance: The lower the abundance of ilmenite, the lower the loss tangent and radiowave attenuation rate are. This is also shown by the lower FeO + TiO_2 abundance (12.1 ± 1.1wt. %) at (CE - 4 landing region compared with that (~23wt. %—30wt. %) at CE - 3 landing site (Dong et al., 2017;

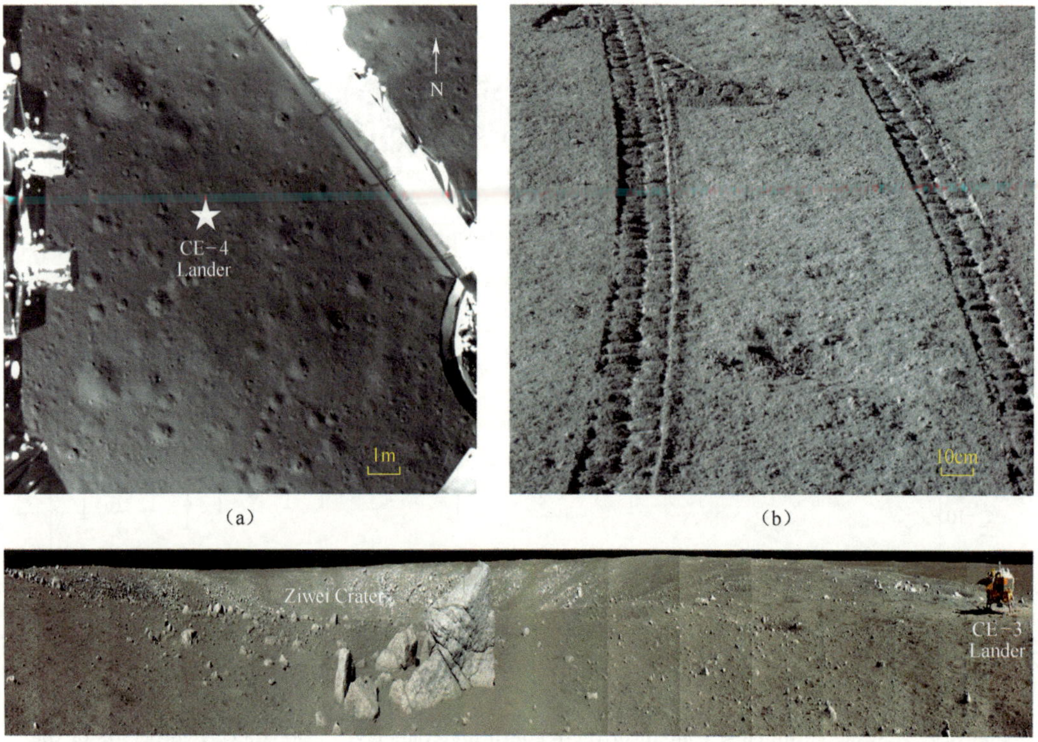

(a)　(b)

(c)

Figure 8　Lunar surface images. (a) Lunar surface image with high resolution at Chang'E - 4 (CE - 4) landing site taken by the CE - 4 Landing Camera. (b) Yutu - 2's track taken by the CE - 4 Panorama Camera (PC). (c) The image of Chang'E - 3 (CE - 3) landing region taken by the CE - 3 PC

Ling et al., 2015).

Additionally, the thickness (~12m) of the regolith materials at CE - 4 landing site is much thicker than the average level in lunar maria which was estimated as 4~5m by Mckay et al. (1991), also much thicker than the thickness of ~1m at CE - 3 landing region on the Mare Imbrium (Dong et al., 2017), consistent with the high maturity degree at CE - 4 landing region. Figures 8b and 8c show the lunar surface images at CE - 4 and CE - 3 landing region, respectively, which were taken by Panorama Camera (PC) onboard the rovers. It can be seen that there are numerous surface rocks (up to meters - sized) that widely distributed at CE - 3 landing region. Such immature local material deposition was produced by a Copernican crater named "Zi Wei" (diameter = ~420m) beside the CE - 3 landing site (Figure 8c). The model age of the "Zi Wei" crater was constrained to be only ~27—81Ma (Xiao et al., 2015), which leads to a thinner regolith layer of ~1m. On the contrary, there are seldom surface rocks at CE - 4 landing region and the grain size of local regolith is fairly tiny (Figure 8b), which indicate a fine - weathered process and thus a thicker regolith layer was accumulated.

Moreover, the estimated thickness of local surface regolith can help to constrain the lunar regolith growth rate. Visible and Near Infrared Spectrometer (VNIS), another payload onboard Yutu - 2, indicates that the uppermost materials at CE - 4 landing site came from the Finsen crater (Li et al., 2019). In addition, higher albedo linear features on the mare basalt of the Von Kármán crater were observed to converge toward the Finsen crater, also indicating that the local surface material were excavated from Finsen crater (Huang et al., 2018). Therefore, the surface regolith at CE - 4 landing site was probably weathered from the ejecta deposits from the Finsen crater. The age of the last basalt eruption of the Von Kármán crater was dated 3.15—3.6Ga (Haruyama et al., 2009; Huang et al., 2018), which provides a key reference to constrain the formation age of local surface regolith.

The average regolith growth rate at CE - 4 landing site was constrained as 1cm per ~3 million years, which is much less than that of 1cm per ~0.6—0.7million years at CE - 3 landing site (Dong et al., 2017; Fa et al., 2015). It indicates that the growth rate of lunar weathered regolith successively declines over time, which proves the previous conjectures about the regolith growth rate: Carrier et al. (1991) supposed three models of regolith accumulation rate, which assumed linear and nonlinear impact flux models at Apollo 11 and 12 landing sites. Ishiyama et al. (2013) modified the model as the regolith accumulation rate decreases with time: 5m/Ga in 4.0—3.5Ga ago, 2m/Ga in 3.5 - 3.0 Ga ago, and 1m/Ga in 3.0 - 0Ga ago, as shown in Figure 9.

In general, the obtained results of the accumulated thickness of lunar weathered regolith conform to the trend that proposed by Ishiyama et al. (2013). Our results suggest that the accumulated thickness of surface regolith materials tends to be a logarithmic function with time instead of linear relations such as the accumulated rate of 1.2m/Ga estimated at the Apollo 11 landing site (Melosh, 1989) or the accumulated rate of approximate 1cm per 2—3million years estimated by DiGiuseppe et al. (2009).

However, the regolith layer at CE - 4 landing region is a little thicker than that constrained by the previous models (e.g., ~2.5—7.5m by Huang et al., 2018). This is probably because the landing region is located on the farside of the Moon. Several factors affect the formation and weathering process of lunar regolith and the continuous impact of large and small meteoroids is the most significant one. As a result, the growth rate of lunar regolith on the Moon's farside may well be higher than that on the nearside due to the more frequent meteorite impacts.

So far, no lunar materials have been sampled from the Moon's farside; no detailed stratigraphy and properties of lunar regolith on the farside have been detected before, either. CE - 4 LPR which directly works on the lunar surface is the most effective tool to detect the structure and properties of regolith materials on the farside at present. It is expected that more detailed and important information will be returned from CE - 4 LPR with its on - going detection. Our results also help the selection of landing site for future exploration missions such as China's Chang'E - 6 mission, a sample return mission to the lunar farside for the first time.

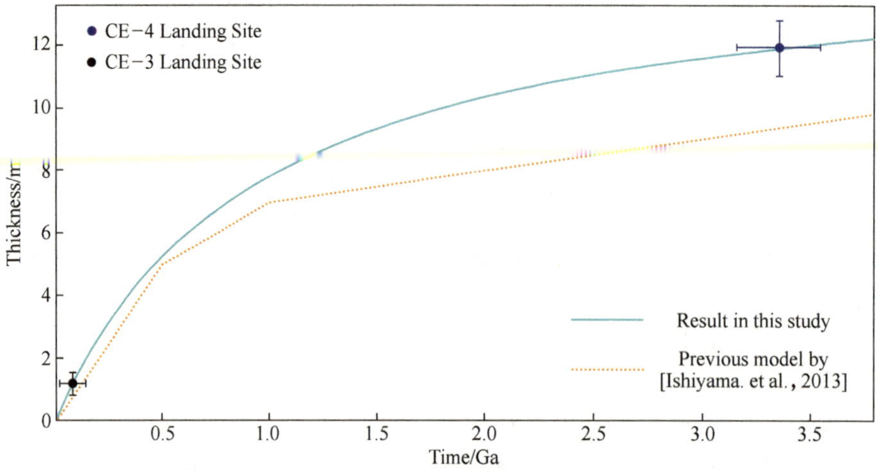

Figure 9　Accumulated thickness of lunar weathered regolith over time. The orange dashed line and the cyan line show the model of regolith accumulation rate raised by Ishiyama et al. (2013) and in this study, respectively. Two spots indicate the thickness of lunar regolith and the time of its formation at Chang'E-3 (CE-3) and Chang'E-4 (CE-4) landing sites, respectively

5　Conclusions

CE-4 LPR is the first surface radar system on the farside of the Moon. With the latest LPR data, we studied the structure and properties of lunar regolith at CE-4 landing region and draw the main conclusions as follows:

1. We processed the LPR Channel-Ⅱ data and obtained a high-quality radargram with improved SNR. The surface echo and subsurface echoes can be clearly recognized in both the radargram and a trace of LPR data.

2. The thickness of lunar regolith at CE-4 landing region was constrained to be ~11.5—~12.4m along Yutu-2's track. It is much thicker than that at CE-3 landing region on the Mare Imbrium and also thicker than the estimated average thickness in lunar maria.

3. The relative permittivity of surface regolith materials (<30cm) at CE-4 landing region was identified to be 2.35 ± 0.203 which is lower than that at CE-3 landing site, consistent with the difference of grain size, exposed rocks, and maturity of lunar surface between CE-3 and CE-4 landing regions. The bulk density of surface regolith materials (<30cm) at the landing region was estimated to be $1.31 \pm 0.20 g/cm^3$.

4. The loss tangent of the ~12m-thick layer of regolith materials at CE-4 landing region was constrained to be $(4.4 \pm 0.5) \times 10^{-3}$ at the frequency of ~450MHz, much lower than ~0.01 at CE-3 landing region, in accord with the greater penetrating depth of CE-4 LPR than that of CE-3 LPR.

5. The TiO_2 + FeO content of the ~12m - thick layer of regolith materials at CE - 4 landing region was constrained to be 11. 6±1. 1 wt. %, lower than that at CE - 3 landing region.

6. The growth rate of lunar weathered regolith successively declines over time and the accumulated thickness of lunar regolith tends to be a logarithmic function with time instead of linear relations. The growth rate of lunar regolith on the Moon's farside may well be higher than that on the nearside due to the more frequent meteorite impacts.

References

[1] Andrews - Hanna, J. C., Besserer, J., Head, III, J. W., Howett, C. J. A., Kiefer, W. S., Lucey, P. J., et al. (2014). Structure and evolution of the lunar Procellarum region as revealed by GRAIL gravity data. Nature, 514, 68 - 71. https://doi. org/10. 1038/nature13697.

[2] Bart, G. D., Nickerson, R. D., Lawder, M. T., & Melosh, H. J. (2011). Global survey of lunar regolith depths from lroc images. Icarus, 215 (2), 485 - 490. https://doi. org/10. 1016/j. icarus. 2011. 07. 017.

[3] Carrier, W. D., Olhoeft, G. R., & Mendell, W. (1991). Physical properties of the lunar surface. Lunar sourcebook, 475 - 594.

[4] DiGiuseppe, M., Pirich, R., & Kraut, V. (2009). Lunar regolith control and resource utilization. IEEE Long Island Systems, Applications and Technology Conference, 1 - 5. https://doi. org/10. 1109/LISAT. 2009. 5031560.

[5] Dong, Z. (2021). Properties of lunar regolith on the moon's farside unveiled by chang'E - 4 lunar penetrating radar [Data set]. Zenodo, https:// doi. org/10. 5281/zenodo. 4724190.

[6] Dong, Z., Fang, G., Ji, Y., Gao, Y., Wu, C., Zhang, X. (2017). Parameters and structure of lunar regolith in Chang'E - 3 landing area from lunar penetrating radar (LPR) data. Icarus, 282, 40 - 46. https://doi. org/10. 1016/j. icarus. 2016. 09. 010.

[7] Dong, Z., Ye, S., Gao, Y., Fang, G., Zhang, X., Xue, Z., & Zhang, T. (2016). Rapid detection methods for asphalt pavement thicknesses and defects by a vehicle - mounted ground penetrating radar (GPR) system. Sensors, 16 (12), 2067. https://doi. org/10. 3390/s16122067.

[8] Fa, W., & Jin, Y. (2007). Simulation of brightness temperature from lunar surface and inversion of regolith - layer thickness. Journal of Geophysical Research, 112, E05003. https://doi. org/10. 1029/2006je002751.

[9] Fa, W., & Wieczorek, M. A. (2012). Regolith thickness over the lunar nearside: Results from Earth - based 70 - cm Arecibo radar observations. Icarus, 218 (2), 771 - 787. https://doi. org/10. 1016/j. icarus. 2012. 01. 010.

[10] Fa, W., Zhu, M. H., Liu, T., & Plescia, J. B. (2015). Regolith stratigraphy at the Chang'E - 3 landing site as seen by lunar penetrating radar. Geophysical Research Letters, 42 (23), 10179 - 10187. https://doi.org/10.1002/2015gl066537.

[11] Fang, G. -Y., Zhou, B., Ji, Y. -C., Zhang, Q. -Y., Shen, S. -X., Li, Y. - X., et al. (2014). Lunar penetrating radar onboard the Chang'e - 3 mission. Research in Astronomy and Astrophysics, 14 (12), 1607 - 1622. https://doi.org/10.1088/1674-4527/14/12/009.

[12] Fortezzo, C. M., & Hare, T. M. (2013). Completed digital renovation of the 1 : 5, 000, 000 lunar geologic map series. Lunar and planetary science conference, 44, 2114.

[13] Gault, D. E., Quaide, W. L., Oberbeck, V. R., & Moore, H. J. (1966). Luna 9 photographs: Evidence for a fragmental surface layer. Science, 153, 985 - 988. https://doi.org/10.1126/science.153.3739.985

[14] Gold, T., Bilson, E., & Yerbury, M. (1972). Grain size analysis, optical reflectivity measurements and determination of high frequency electrical properties for Apollo 14 lunar samples. Proceedings of the Lunar Science Conference, 3, 3187 - 3193.

[15] Gold, T., Campbell, M. J., & O'Leary, B. T. (1970). Optical and high frequency electrical properties of lunar samples. Proceedings of the Apollo 11 Lunar Science Conference, 167, 707 - 709. https://doi.org/10.1126/science.167.3918.707

[16] Haruyama, J., Ohtake, M., Matsunaga, T., Morota, T., Honda, C., Yokota, Y., et al. (2009). Long - lived volcanism on the lunar farside revealed by SELENE Terrain Camera. Science, 323 (5916), 905 - 908. https://doi.org/10.1126/science.1163382.

[17] Hiesinger, H. (2006). New views of lunar geoscience: An introduction and overview. Reviews in Mineralogy and Geochemistry, 60, 1 - 81. https://doi.org/10.2138/rmg.2006.60.1.

[18] Huang, J., Xiao, Z., Flahaut, J., Martinot, M., Head, J., Xiao, X., et al. (2018). Geological characteristics of Von Kármán Crater, Northwestern South Pole - Aitken Basin: Chang'E - 4 landing site region. Journal of Geophysical Research: Planets 123 (7), 1684 - 1700. https://doi.org/10.1029/2018je005577.

[19] Ishiyama, K., Kumamoto, A., Ono, T., Yamaguchi, Y., Haruyama, J., Ohtake, M., et al. (2013). Estimation of the permittivity and porosity of the lunar uppermost basalt layer based on observations of impact craters by SELENE. Journal of Geophysical Research: Planets, 118, 1453 - 1467. https://doi.org/10.1002/jgre.20102.

[20] Jol, H. (2009). Ground penetrating radar: Theory and applications. Elsevier Science.

[21] Kobayashi, T., Kim, J. H., Lee, S. R., Araki, H., Ono, T. (2010). Simultaneous observation of lunar radar sounder and laser altimeter of Kaguya for lunar regolith layer thickness estimate. IEEE Geoscience and Remote Sensing Letters, 7 (3), 435 - 439. https://doi.org/10.1109/ lgrs. 2009. 2038499.

[22] Kroupenio, N. N. (1972). Some characteristics of the Venus surface. Icarus, 17, 692 - 698. https://doi.org/10.1016/0019 - 1035 (72) 90033 - 4.

[23] Kroupenio, N. N. (1973). Results of radar experiments performed on automatic stations Luna 16 and Luna 17. In COSPAR space Research XIII (pp. 969 - 973). Berlin: Akademie - Verlag.

[24] Kroupenio, N. N., Balo, A. G., Ruzskii, E. G., Ladyghin, V. A., Cherkasov, V. V., & Fomin, V. S. (1975). Results of radar experiments performed aboard the Luna 19 and 20 automatic stations. In COSPAR space Research XV (pp. 615 - 620). Berlin: Akademie - Verlag.

[25] Lai, J., Xu, Y., Zhang, X., Xiao, L., Yan, Q., Meng, X., et al. (2019). Comparison of dielectric properties and structure of lunar regolith at Chang'e - 3 and Chang'e - 4 landing sites revealed by ground - penetrating radar. Geophysical Research Letters, 46 (22), 12783 - 12793. https:// doi.org/10.1029/ 2019gl084458.

[26] Lemelin, M., Lucey, P. G., Gaddis, L. R., Hare, M. T., & Ohtake, M. (2016). Global map products from the Kaguya multiband imager at 512 ppd: Minerals, FeO, and OMAT. Lunar and Planetary Science Conference, 47, 2994.

[27] Li, C., Liu, D., Liu, B., Ren, X., Liu, J., He, Z., et al. (2019). Chang' E - 4 initial spectroscopic identification of lunar far - side mantle - derived materials. Nature, 569, 378 - 382. https://doi.org/10.1038/s41586 - 019 - 1189 - 0.

[28] Li, C., Su, Y., Pettinelli, E., Xing, S., Ding, C., Liu, J., et al. (2020) The moon's farside shallow subsurface structure unveiled by Chang'E - 4 Lunar penetrating radar. Science Advances, 6 (9), eaay6898. https://doi.org/10.1126/ sciadv. aay6898.

[29] Ling, Z., Jolliff, L. B., Wang, A., Li, C., Liu, J., Zhang, J., et al. (2015). Correlated compositional and mineralogical investigations at the Chang'e - 3 landing site. Nature Communications, 6, 8880. https://doi.org/10.1038/ ncomms9880.

[30] Mckay, D. S., Heiken, G., Basu, A., Blanford, G., Simon, S., Reedy, R., et al. (1991). The lunar regolith. In H. Heiken, T. Vaniman, & M. French (Eds.), Lunar source - book: A user's guide to the Moon. (pp. 285 - 356). York: Cambridge University Press.

[31] Melosh, H. J. (1989). Impact cratering: A geologic process. New York: Oxford University of Press.

[32] Meng, Z., Xu, Y., Zheng, Y., Zhu, Y., Jia, Y., & Chen, S. (2014). Inversion of lunar regolith layer thickness with CELMS data using BPNN method. Planetary and Space Science, 101 (2), 1 - 11. https://doi.org/10.1016/j.pss.2014.05.020.

[33] Morgan, G. A., Campbell, B. A., Campbell, D. B., & Hawke, B. R. (2016). Investigating the stratigraphy of mare imbrium flow emplacement with earth-based radar. Journal of Geophysical Research: Planets, 121, 1498 - 1513. https://doi.org/10.1002/2016je005041.

[34] Morota, T., Haruyama, J., Ohtake, M., Matsunaga, T., Honda, C., Yokota, Y., et al. (2011). Timing and characteristics of the latest mare eruption on the Moon. Earth and Planetary Science Letters, 302, 255 - 266. https://doi.org/10.1016/j.epsl.2010.12.028.

[35] Muhleman, D. O., Brown, W. E. Jr, Davids, L., Negus, J. & Peake, W. H. (1969). Lunar surface electromagnetic properties. In Surveyor program results (pp. 203 - 269). Surveyor Program Office.

[36] Oberbeck, V. R., & Quaide, W. L. (1968). Genetic implications of Lunar regolith thickness variations. Icarus, 9, 446 - 465. https://doi.org/10.1016/0019-1035(68)90039-0.

[37] Ohtake, M., Haruyama, J., Haruyama, J., Matsunaga, T., Yokota, Y., Morota, T., & Honda, C. (2008). Performance and scientific objectives of the SELENE (KAGUYA) multiband imager. Earth Planets and Space, 60, 257 - 264. https://doi.org/10.1186/bf03352789.

[38] Olhoeft, G., Strangway, D., & Pearce, G. (1975). Effects of water on electrical properties of lunar fines. Proc. Lunar Sci. Conf. 6th, p. 3333 - 3342.

[39] Olhoeft, G. R., & Strangway, D. W. (1975). Dielectric properties of the first 100 meters of the Moon. Earth and Planetary Science Letters, 24, 394 - 404. https://doi.org/10.1016/0012-821x(75)90146-6.

[40] Ono, T., Kumamoto, A., Nakagawa, H., Yamaguchi, Y., Oshigami, S., Yamaji, A., et al. (2009). Lunar radar sounder observations of subsurface layers under the nearside Maria of the Moon. Science, 323, 909 - 912. https://doi.org/10.1126/science.1165988.

[41] Peeples, W. J., William, R. S., Thomas, W. M., Stanley, H. W., Roger, J. P., Rolando, L. J., et al. (1978). Orbital radar evidence for lunar subsurface layering in Maria Serenitatis and Crisium. Journal of Geophysical Research, 83, 3459 - 3468. https://doi.org/10.1029/jb083ib07p03459.

[42] Porcello, L. J., Jordan, R. L., Zelenka, J. S., Adams, G. F., Phillips, R. J., Brown, W. E., et al. (1974). The Apollo lunar sounder radar system. Proceedings of the IEEE, 62, 769 - 783. https://doi.org/10.1109/proc.1974.9517.

[43] Quaide, W. L. , & Oberbeck, V. R. (1968) . Thickness determinations of the lunar surface layer from lunar impact craters. Journal of Geophysical Research, 73, 5247 - 5270. https://doi. org/10. 1029/jb073i016p05247.

[44] Seu, R. , Phillips, R. J. , Biccari, D. , Orosei, R. , Masdea, A. , Picardi, G. , et al. (2007) . SHARAD sounding radar on the Mars reconnaissance orbiter. Journal of Geophysical Research, 112 (E5), 37 - 55. https://doi. org/10. 1029/2006je002745.

[45] Shkuratov, Y. G. , & Bondarenku, N. V. (2001) . Regolith layer thickness mapping of the moon by radar and optical data. Icarus, 149: 329 - 338. https://doi. org/10. 1006/icar. 2000. 6545.

[46] Shoemaker, E. M. , Hait, M. H. , Swann, G. A. , Schleicher, D. L. , Dahlem, D. H. , Schaber, G. G. , & Sutton, R. L. (1970) . Lunar regolith at tranquillity base. Science, 167 (3918), 452 - 455. https://doi. org/10. 1126/science. 167. 3918. 452.

[47] Spudis, P. D. , Gillis, J. J. , & Reisse, A. R. (1994) . Ancient multiring basins on the moon revealed by clementine laser altimetry. Science, 266, 1848 - 1851. https://doi. org/10. 1126/science. 266. 5192. 1848.

[48] Strangway, D. W. , & Olhoeft, G. R. (1977) . Electrical properties of planetary surfaces. Philosophical Transactions of the Royal Society of London - Series A, 285, 441 - 450.

[49] Thompson, T. W. (1987) . High - resolution lunar radar at 70 - cm wave length. Earth, Moon, and Planets, 37, 59 - 70 https://doi. org/10. 1007/ bf00054324.

[50] Ulaby, F. T. , Bengal, T. H. , Dobson, M. C. , East, J. R. , Garvin, J. B. , Evans, D. L. , et al. (1990) . Microwave dielectric properties of dry rocks. IEEE Transactions on Geoscience and Remote Sensing, 28, 325 - 336. https://doi. org/10. 1109/36. 54359.

[51] Vijayan, S. , Mohan, S. , & Murty, S. S. (2015) . Lunar regolith thickness estimation using dual frequency microwave brightness temperature and influence of vertical variation of FeO + TiO_2. Planetary and Space Science, 105, 123 - 132. https://doi. org/10. 1016/j. pss. 2014. 11. 017.

[52] Wilcox, B. B. , Robinson, M. S. , Thomas, P. C. , & Hawke, B. R. (2005) . Constraints on the depth and variability of the lunar regolith. Meteoritics & Planetary science, 40, 695 - 710. https://doi. org/10. 1111/j. 1945 - 5100. 2005. tb00974. x.

[53] Wilhelms, D. E. , Howard, K. A. , & Wilshire, H. G. (1979) . Geologic map of the south side of the Moon. Denver: US Geological Survey.

[54] Wilhelms, D. E. , Mccauley, F. , & Trask, J. (1987) . The geologic history of the Moon. No. 1348. https://doi. org/10. 3133/pp1348.

[55] Wu, W. R. , Li, C. , Zuo, W. , Zhang, H. , Liu, J. , Wen, W. , et al. (2019) . Lunar farside to be explored by Chang'e - 4. Nature Geoscience, 12, 222 -

223. https://doi. org/10. 1038/s41561 – 019 – 0341 – 7.

[56] Xiao, L., Zhu, P., Guangyou, F., Xiao, Z., Zou, Y., Zhao, J., et al. (2015). A young multilayered terrane of the northern Mare Imbrium revealed by Chang' E – 3 mission. Science, 347, 1226 – 1229. https://doi. org/10. 1126/science. 1259866.

[57] Zhang, J., Yang, W., Hu, S., Lin, Y., Guangyou, F., Li, C., et al. (2015). Volcanic history of the Imbrium basin: A close – up view from the lunar rover Yutu. Proceedings of the National Academy of Sciences, 112, 5342 – 5347. https://doi. org/10. 1073/pnas. 1503082112.

[58] Zhang, J., Zhou, B., Lin, Y., Zhu, M. – H., Song, H., Dong, Z., et al. (2020). Lunar regolith and substructure at Chang'E – 4 landing site in South Pole – Aitken basin. Nature Astronomy, 5, 25 – 30. https://doi. org/10. 1038/s41550 – 020 – 1197 – x.

Properties Analysis of Lunar Regolith at Chang'E - 4 Landing Site Based on 3D Velocity Spectrum of Lunar Penetrating Radar

Zejun Dong[1,2], Xuan Feng[1,2,3], Haoqiu Zhou[1,2], Cai Liu[1,4], Zhaofa Zeng[1,4], Jing Li[1,3] and Wenjing Liang[1,2]

1. College of Geo - Exploration Science and Technology, Jilin University, No. 938 Xi MinZhu Street, Changchun 130026, China; dzj19@mails. jlu. edu. cn (Z. D.); zhouhq17@mails. jlu. edu. cn (H. Z.); liucai@jlu. edu. cn (C. L.); zengzf@jlu. edu. cn (Z. Z.); inter_lijing@jlu. edu. cn (J. L.); liangwj@jlu. edu. cn (W. L.)

2. Science and Technology on Near - Surface Detection Laboratory, Wuxi 214035, China

3. Institute of National Development and Security Studies, Jilin University, No. 2699 Qianjin Street, Changchun 130012, China

4. Key Laboratory of Geophysical Exploration Equipment, Ministry of Education (Jilin University), No. 938 Xi MinZhu Street, Changchun 130026, China

Abstract: The Chinese Chang'E - 4 mission for moon exploration has been successfully completed. The Chang'E - 4 probe achieved the first - ever soft landing on the floor of Von Kármán crater (177. 59°E, 45. 46°S) of the South Pole - Aitken (SPA) basin on January 3, 2019. Yutu - 2 rover is mounted with several scientific instruments including a lunar penetrating radar (LPR), which is an effective instrument to detect the lunar subsurface structure. During the interpretation of LPR data, subsurface velocity of electromagnetic waves is a vital parameter necessary for stratigraphic division and computing other properties. However, the methods in previous research on Chang'E - 3 cannot perform velocity analysis automatically and objectively. In this paper, the 3D velocity spectrum is applied to property analysis of LPR data from Chang'E - 4. The result shows that 3D velocity spectrum can automatically search for hyperbolas; the maximum value at velocity axis with a soft threshold function can provide the horizontal position, two - way reflected time and velocity of each hyperbola; the average maximum relative error of velocity is estimated to be 7. 99%. Based on the estimated velocities of 30 hyperbolas, the structures of subsurface properties are obtained, including velocity, relative permittivity, density, and content of FeO and TiO_2.

Keywords: Chang'E - 4; lunar penetrating radar (LPR); 3D velocity spectrum;

properties analysis

1 Introduction

The Chinese Chang' E – 4 mission for moon exploration has been successfully completed. The Chang' E – 4 probe achieved the first – ever soft landing on the floor of Von Kármán crater (177.59°E, 45.46°S) of the South Pole – Aitken (SPA) basin on January 3, 2019 [1 – 3]. SPA basin is the broadest basin on the Moon. This ancient basin was born from asteroid impacts 4 billion years ago, recording the evolutionary history of the far side of the Moon, and is of great significance for researching the internal materials and structures of the Moon [4 – 6]. Von Kármán crater is one of the primary craters in the SPA basin, with a diameter of 186 km. Recent studies have revealed that the ejecta from adjacent craters have various contributions to the subsurface material of Von Kármán crater, which results in the complex subsurface structure at the Chang' E – 4 landing site [1, 3, 4].

In order to patrol and investigate the lunar surface, a Yutu – 2 rover is carried by Chang' E – 4 probe. Yutu – 2 rover is mounted with several scientific instruments containing a lunar penetrating radar (LPR) which has been verified to be an effective device for detecting the lunar subsurface structure [7, 8]. The LPR is equipped with two types of channels (CH – 1 and CH – 2), the center frequencies of which are 60MHz and 500MHz [9, 10]. In addition, the CH – 2 possesses one transmitting antenna and two receiving antennas of different offsets which are also known as CH – 2A and CH – 2B [10 – 12]. With different frequencies, CH – 1 and CH – 2 have different detecting missions. The objective of CH – 1 is to detect the deep structure of Von Kármán crater [13]; CH – 2 is to map the detail of near surface layers, leading to the property analysis of lunar regolith [14, 15].

During the interpretation of LPR data, subsurface velocity of electromagnetic waves is a vital parameter necessary for stratigraphic division and computing other properties. However, the velocity analysis of field LPR data faces many difficulties. Complex subsurface structure and interference of noise always result in the incomplete, interlaced, and amplitude – varying hyperbolas. In previous studies of Chang' E – 3, Feng et al. proposed hyperbolic fitting method for velocity analysis of LPR data [10]; Lai et al. used two – way delay method to acquire the velocity [16]; Zhang et al. applied CH – 2A and CH – 2B data of different offsets to estimate the velocity [17]. However, these methods need humans to select the hyperbolas, which is highly subjective. Thus, in this article, we applied 3D normalized velocity spectrum to estimate the velocity [18]. This method can automatically and objectively select hyperbolas and analyze the velocities; the normalization processing can solve the error brought by different amplitudes of different positions on hyperbolas; subsequently, during the computation, we applied a variable horizontal computation window along longitudinal direction to satisfy the field situation that rock sizes increase vertically.

This paper is organized as follows. Section 2 introduces basic theory of 3D velocity spectrum and properties analysis. In Section 3, firstly a model testis performed to verify the feasibility of 3D velocity spectrum; subsequently the method is applied to the CH - 2B data analysis. Subsequently, based on the 3D velocity spectrum, we obtain the positions and velocities of the points where hyperbolas exist; then the property structures of lunar regolith are computed including velocity, relative permittivity, density, and content of FeO and TiO_2. In Section 4, we discuss the computation error of each hyperbola. Section 5 is the conclusion.

2 Methods

2.1 3D Velocity Spectrum

The velocity spectrum analysis method is originally based on common middle point (CMP) ground penetrating radar (GPR) data [19, 20], but it can also be applied to common offset data processing such as LPR data processing. In order to estimate velocity, a stacked amplitude is computed as follows:

$$S_{i,j,k} = \sum_{j=1}^{N_i} f(t_{i,j,k}, x_j) \tag{1}$$

$$i = 1, \cdots nt;\ j = 1, \cdots nx;\ k = 1, \cdots nv$$

where f is the LPR data in $t-x$ domain; N_i denotes the selected horizontal computation region size at ith time; to satisfy the field situation that rock sizes increase vertically, we make N_i variations along longitudinal direction; nt, nx, and nv are the number of sampling points of each trace, number of traces, and number of velocities used in computation; x_j represents the horizontal distance between the j th point and the extreme point of the hyperbola; $t_{i,j,k}$ is the two - way time of the j th points of the hyperbola [21, 22], which can be obtained using the formula below:

$$t_{i,j,k} = \left(t_i^2 + \frac{4x_j^2}{v_k^2}\right)^{1/2} \tag{2}$$

where t_i is the two - way time of extreme point of the hyperbola; v_k represents the velocity used in computation.

However, the method of (1) is not always effective in the field as the hyperbolas are incomplete, interlaced, with varying amplitude. To solve the error brought about by these situations, the normalization form of stacked amplitude is applied:

$$C_{i,j,k} = \frac{1}{N_i L} \sum_{l=i}^{L+i-1} \left[\frac{S_{i,j,k}^2}{\sum_{j=1}^{N} f^2(t_{i,j,k}, x_j)} \right] \tag{3}$$

where L is the time gate; the average in the time gate is adopted because the hyperbolas on radargram are not curves but regions with hyperbolic shapes. The normalized form can guarantee that the result $C_{i,j,k}$ is not affected by the varying amplitude. Importantly, $1/N_i$ is added in this form to compensate the energy differences caused by using different windows N_i along longitudinal direction. $C_{i,j,k}$ is 3D data, whose local maximum values indicate the time, horizontal position, and velocity of the points with hyperbolas.

2.2 Properties Computation

After we obtain the velocities from the 3D spectrum, the properties can be computed. Without considering magnetic permeability, the relative permittivity can be easily derived using the following formula:

$$\varepsilon = (c/v)^2 \tag{4}$$

where c equals 0.3m/ns.

According to the studies of lunar regolith samples of Apollo, there is a relation between the relative permittivity and density of lunar regolith [9, 17]:

$$\rho = \log_{1.919}(\varepsilon) \tag{5}$$

$$\tan\delta = 10^{0.440\rho-2.943} \tag{6}$$

The $\tan\delta$ is the loss tangent which represents the ratio of the imaginary part of the dielectric constant to its real part. The density of lunar regolith is shown in Figure 14. Subsequently, previous studies show that the loss tangent is also related to the TiO_2 and FeO weight percentage values [9, 17]:

$$\tan\delta = 10^{[0.038\times(\omega(TiO_2)+\omega(FeO))+0.312\rho-3.26]} \tag{7}$$

Combining (6) and (7), we can obtain the TiO_2 and FeO weight percentage of lunar regolith.

3 Results

3.1 Model Test

Before the processing of LPR data, a model test was performed to verify the feasibility of 3D velocity spectrum. The relative permittivity model used in test and its simulated GPR profile are shown in Figure 1a, b, respectively. In the simulation, the frequency of electromagnetic wave is 100MHz; the time window and the sampling interval are 100ns and 0.32ns; the trace interval is 0.1m; the transmitting and receiving antennas are placed on the ground surface. The radargram shows there are three pairs of hyperbolas (H1, H2, and H3); the double hyperbolas generate from the upper and bottom surface of the scatters. Subsequently, the 3D velocity spectrum is computed, whose result is shown in Figure 1c.

In fact, for a point (t_i, x_j), there is only one true velocity; this true velocity corresponds to the maximum value along velocity axis; the maximum value can help to find the position of hyperbolas on (t_i, x_j) profile. Thus, the maximum values $C_{i,j}$ along velocity axis are figured out using (8). Subsequently, in order to suppress the noise interference and obtain the accurate positions of hyperbolas, we apply a soft threshold function [23] as (9) shows. The value of ε is set to 0.3 and the result is shown in Figure 1d. The energy stacks in three rectangles indicate the positions of hyperbolas. The irregular energy stack in the ellipse is generated by the intersection of two hyperbolas.

$$C_{i,j} = \max_k C_{i,j,k} \tag{8}$$

$$C_{i,j} = \begin{cases} C_{i,j} - \varepsilon & C_{i,j} > \varepsilon \\ 0 & C_{i,j} < \varepsilon \end{cases} \tag{9}$$

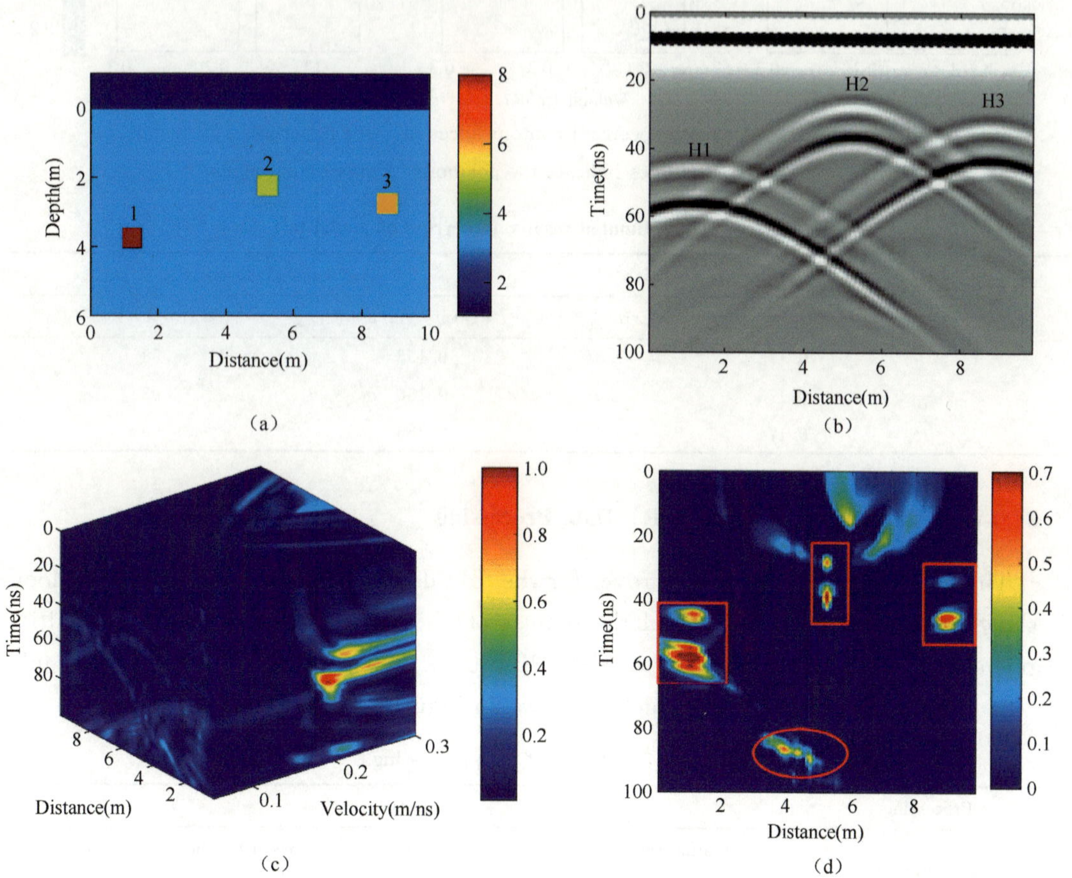

Figure 1　Model test results. (a) The model used in simulation; (b) the simulated GPR profile; (c) 3D velocity spectrum of (b); (d) maximum value along velocity axis with the soft threshold function processing.

After the x position is located, the velocity spectrum slices can be selected out to estimate the velocities. The spectrum contour slices of the three hyperbolas are shown in Figure 2. From Figure 2, the velocities can be read out. The estimated results and errors are shown in Table 1. The results show that the estimation errors of upper hyperbolas are about 1%, and the estimation errors of bottom hyperbolas are about 10%. The errors are in acceptable range; the relatively higher errors of bottom hyperbolas result from the effects of scatters; the low velocities of electromagnetic waves inside scatters reduce the estimated velocities.

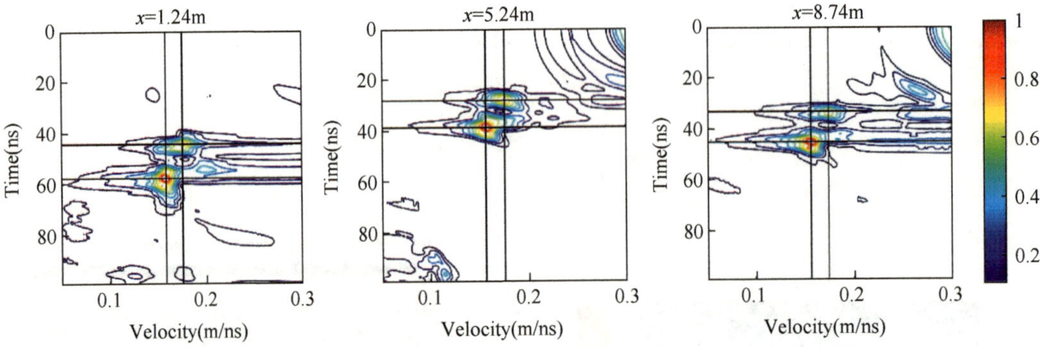

Figure 2　Velocity spectrum contour slices of three hyperbolas. The black crossed lines indicate the positions of maximum value

Table 1　The estimated results and errors of model test

Scatters	Upper Hyperbolas		Bottom Hyperbolas	
	Velocity (m/ns)	Error (%)	Velocity (m/ns)	Error (%)
1	0. 175	1. 04	0. 158	8. 78
2	0. 175	1. 04	0. 156	9. 93
3	0. 174	0. 46	0. 156	9. 93

3.2　Lunar Penetrating Radar (LPR) Data Processing

The walking route of Yutu-2 rover for the first day and the second day on the Moon is shown in Figure 3. The length of the path is about 105 m. Subsequently, the multi-segment data are spliced to achieve the CH-2B profile. After a series of processing (Table 2) the LPR profile is obtained, which is shown in Figure 4.

Table 2　Details of data processing.

Processing	Details
Traces amending	Adjusting the longitudinal displacement of traces, based on the phase of a strong reflection event
Traces selecting	The rover might stop at some points on the way to collect other scientific data but LPR never stops measurement, resulting in repeated acquisition of multiple traces at the same location. We average the repeated traces.
Time lag adjustment	There is a 28 ns delay for the start time of the transmitting antenna compared to the receiving antenna.
Useless data	

Continued Table

Processing	Details
deleting	Signals after 500ns are removed since these signals are not reliable.
Attenuation compensation	Conducting automatic gain control (AGC) to make deep information more visible
Background removal	Removing the average data along the rover path.
Band - pass filtering	Adopting band - pass filtering to suppress the low - frequency and high - frequency noise.

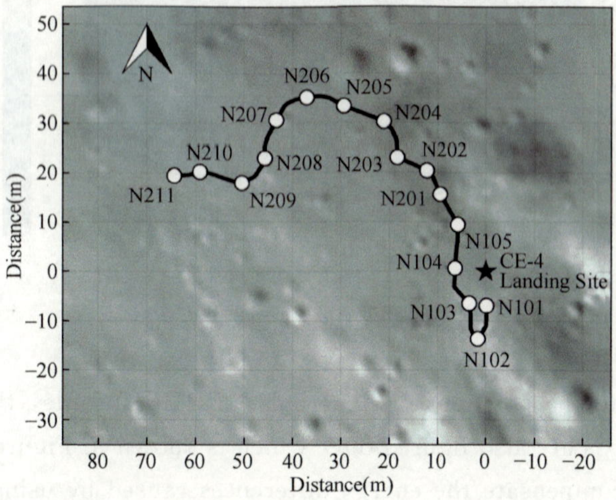

Figure 3　The walking route of Yutu - 2 rover for the first day and the second day on the Moon

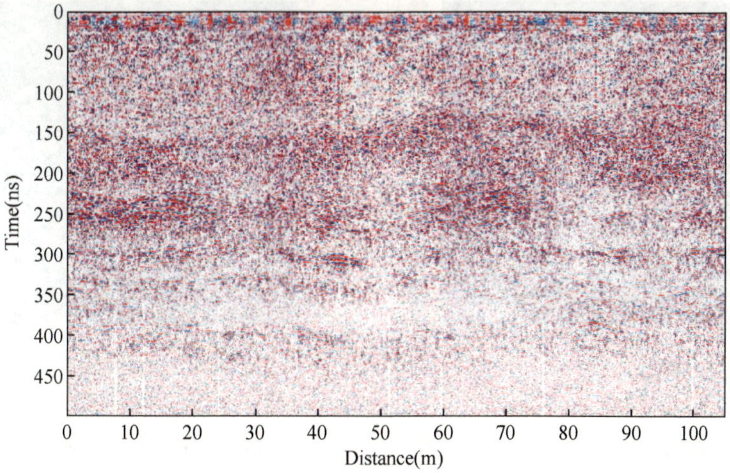

Figure 4　The CH - 2B profile of lunar penetrating radar (LPR) data

3.3　3D Velocity Spectrum

Based on (1) - (3), the 3D velocity spectrum is computed; the range of values are limited to 0 to 1; the 3D result without compensation of $1/N_i$ is shown in Figure5.

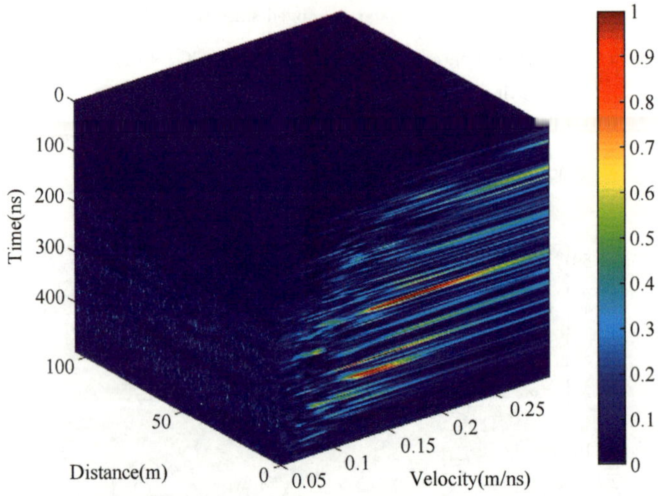

Figure 5 3D velocity spectrum

In order to locate the positions of hyperbolas on GPR profile, the maximum values $C_{i,j}$ along velocity axis are also figured out, which is shown in Figure 6a. Subsequently, $1/N_i$ is adopted to compensate the energy differences caused by using different windows N_i along longitudinal direction, the result is shown in Figure 6b. The result shows the energies in shallow regions are compensated.

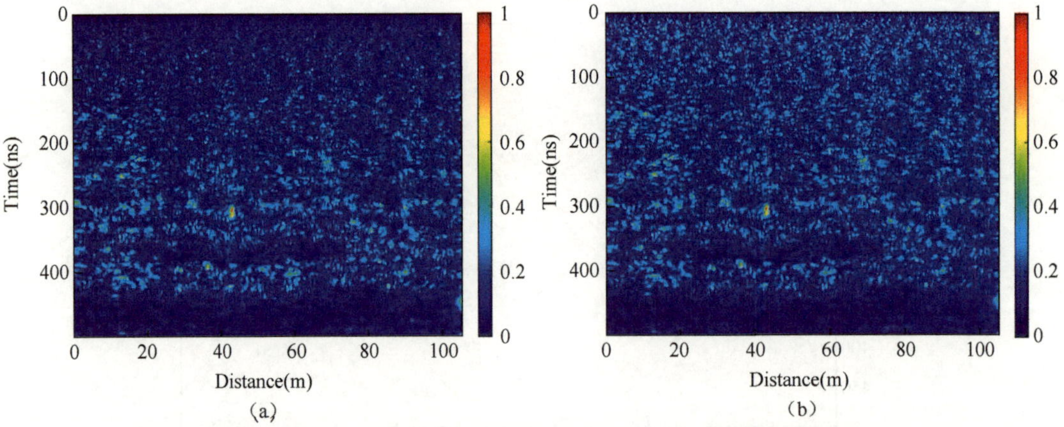

Figure 6 Maximum value along velocity axis. (a) Before compensation of $1/N_i$; (b) after compensation of $1/N_i$

Subsequently, the soft threshold function is also applied. The value of ε is set to 0.6. Subsequently, based on the result of soft threshold processing, the locations of local maximums are selected which are shown in Figure 7.

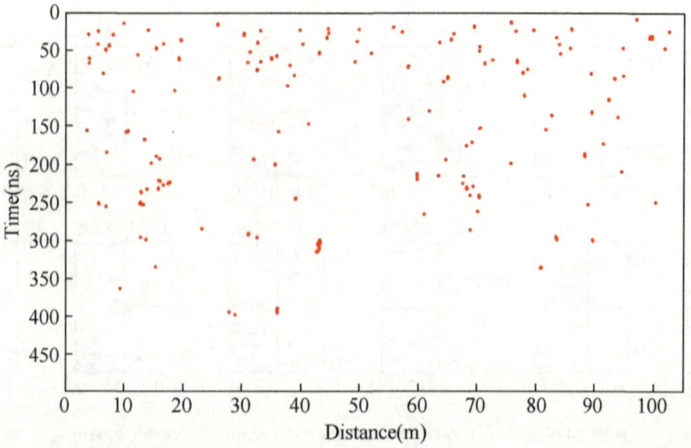

Figure 7　Initial positioning of hyperbolas based on soft threshold

Based on Figure 7, we can search for the velocities in 3D velocity spectrum in Figure 5. However, not all the points in Figure 7 are useful, there are three reasons:

(1) For a highly flat interface, the estimated velocity will reach to 0. 3m/ns, which is obviously wrong; the lunar regolith velocity is less than 0. 2m/ns [24], so we delete the points with velocities close to 0. 3m/ns.

(2) There may be a large stacked energy even if there are no hyperbolas but interlaced regions of several hyperbolas; for this situation, we should delete the non-hyperbolic points.

(3) The hyperbolas on radargram are not curves but regions with hyperbolic shapes, so there will be several excess points at one hyperbola, which should be deleted.

After filtering using the above method, the 30 preserved hyperbolas are noted in Figure 8a; the velocity spectrum slices of these hyperbolas are shown in Figure 8b.

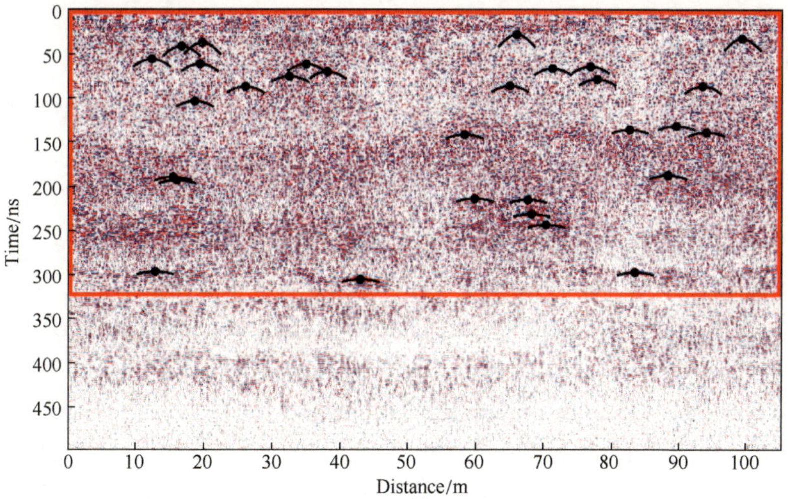

(a)

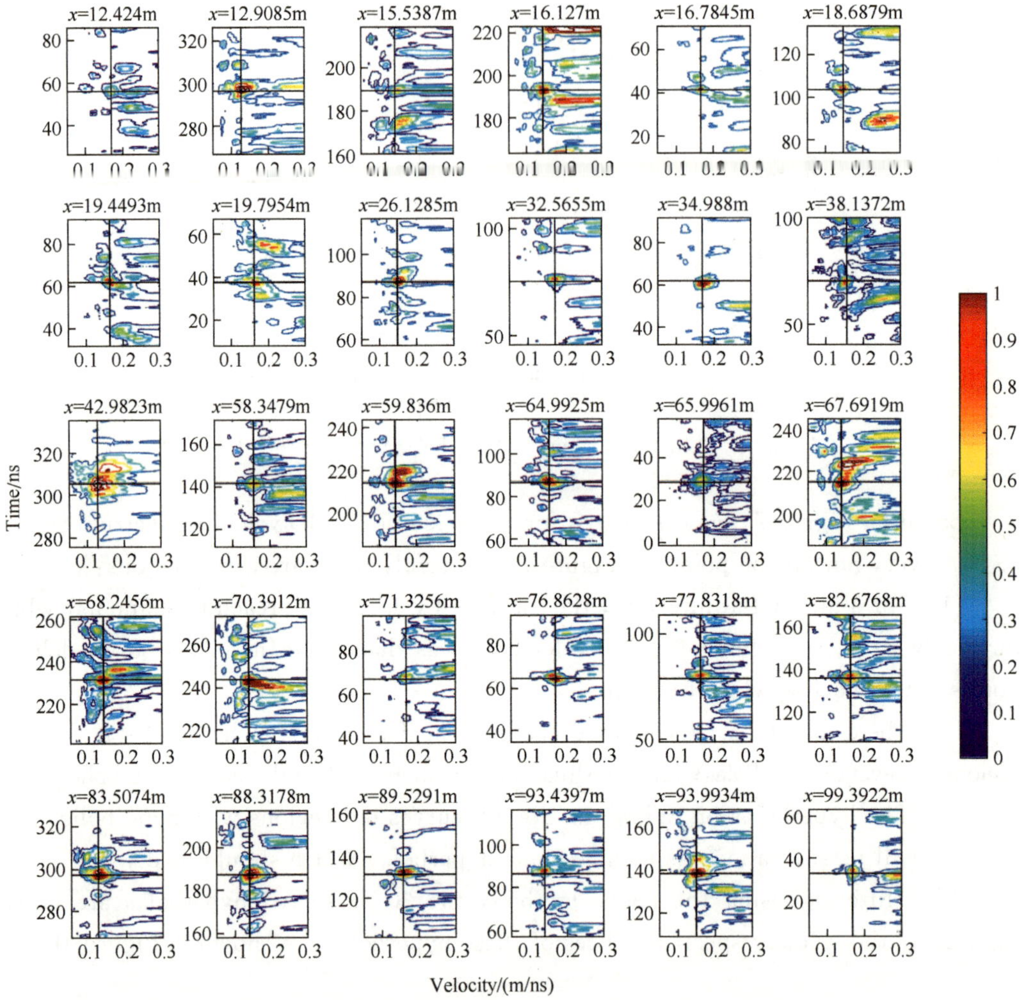

(b)

Figure 8　Result of hyperbolaselection. (a) Hyperbolas on radargram; (b) velocity spectrum slices of 30 hyperbolas - the black crossed lines indicate the positions of maximum values

3.4　Analysis of Radargram

For Figure 8a, the hyperbolas located inside of the red rectangle, the region of 320 - 500ns cannot be analyzed, so we focus on the region of 0 - 320ns. Based on the reflection energy, several distinct layers and special regions are noted on the radargram (Figure 9). Layer 1 is a surface layer; a distinct reflection energy difference can be seen at 25 ns. Layer 3 is the bottom layer; although its adjacent upper and lower areas show few reflections, the interface shows strong reflection energy. Layer 2 is a complex layer. The zone of 25 - 150ns is relatively more uniform than the deeper zone; it contains few reflections indicating its high weathering degree. The zone of 150 - 300ns is more complex possessing strong reflections; the complex signals are formed by the interlaced hyperbolas generated by the scattering of waves on rocks. Four obvious regions are selected out. The strong reflections within Regions 1 and 3 are distinct from the adjacent areas. The reflection energy in Region 2 is relatively lower than that of Regions 1 and 3, which means the material and horizontal structure changes between Regions 1 and 3. Region 4 also contains strong reflections, but its vertical size is shorter because the material and horizontal structure changes again at the position of 80 m; the material at about 250 ns in Region 3 does not extend to Region 4. Overall, the subsurface material and structure at the Chang' E - 4 landing site vary both vertically and horizontally, so the simple horizontal stratigraphic division is not appropriate within the zone of 25 - 300ns.

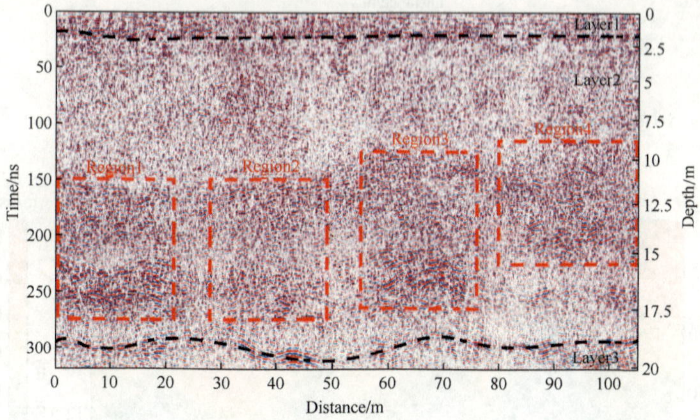

Figure 9　Analysis of radargram

3.5　Properties Analysis

Subsequently, properties analysis is applied to LPR data. Firstly, we combine the positions and velocities of selected hyperbolas in Figure 8 and obtain a velocity scatter figure (Figure 10). The region size is the same with Figure 9, the color indicates the velocity of each point.

Subsequently, the image interpolation is applied to obtain the subsurface velocity of electromagnetic wave structure, then the smoothing is performed to remove the effects of outliers. Finally, as the velocities calculated by hyperbolic fitting method are the root - mean - square velocity (RMS), the interval velocities should be derived using the Dix formula:

$$v_{int,\ n} = \sqrt{\frac{v_{rms,\ n}^2 t_n - v_{rms,\ n-1}^2 t_{n-1}}{t_n - t_{n-1}}} \tag{10}$$

where $v_{int,n}$ and $v_{rms,n}$ represent the interval velocity and root - mean - square velocity of n th layer respectively, t_n is the travel time to the nth layer. By using (10), the subsurface interval velocity structure is computed, and the profile is shown in Figure 11. Subsequently, based on the velocities, the relative permittivity, density, and content of

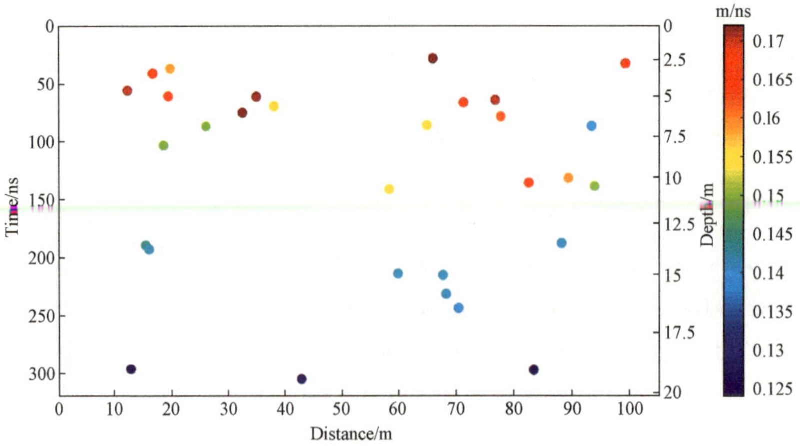

Figure 10　Velocity scatter figure

FeO and TiO_2 can be derived using (4) – (7), the results are shown in Figures 12 – 14, respectively.

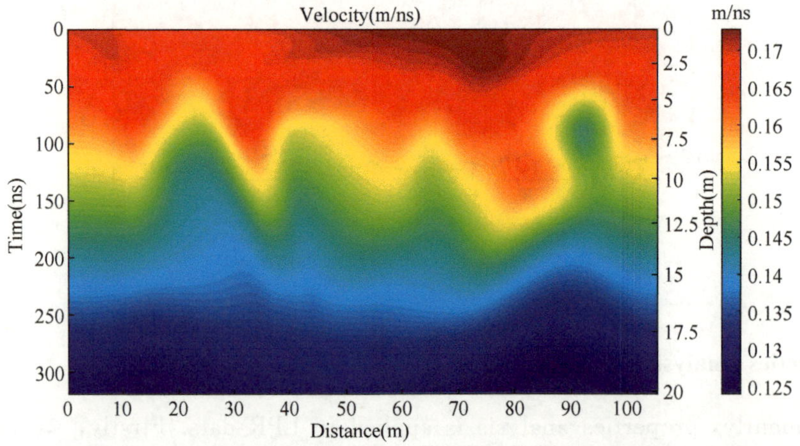

Figure 11　The subsurface electromagnetic wave velocity structure

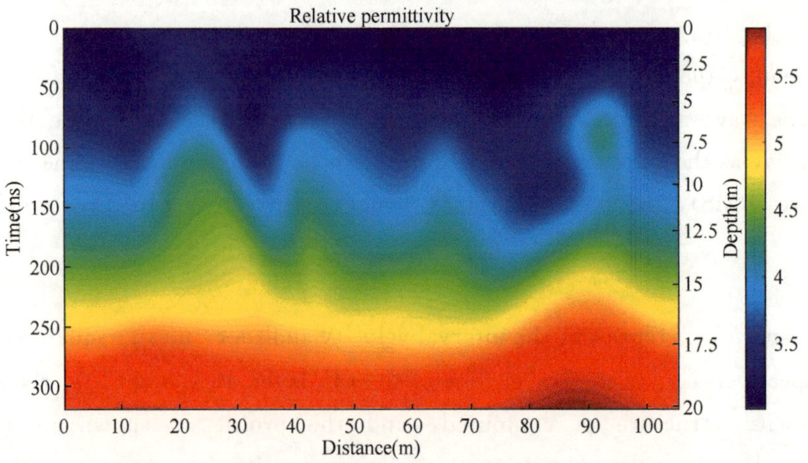

Figure 12　The subsurface relative permittivity structure

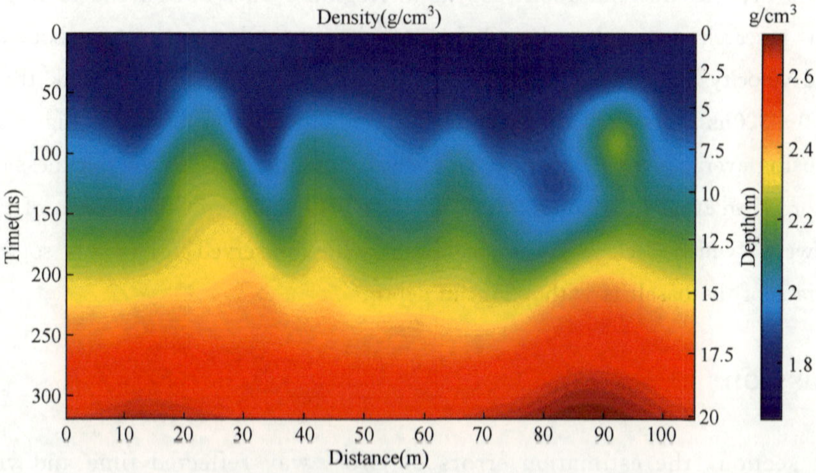

Figure 13　The density structure of lunar regolith

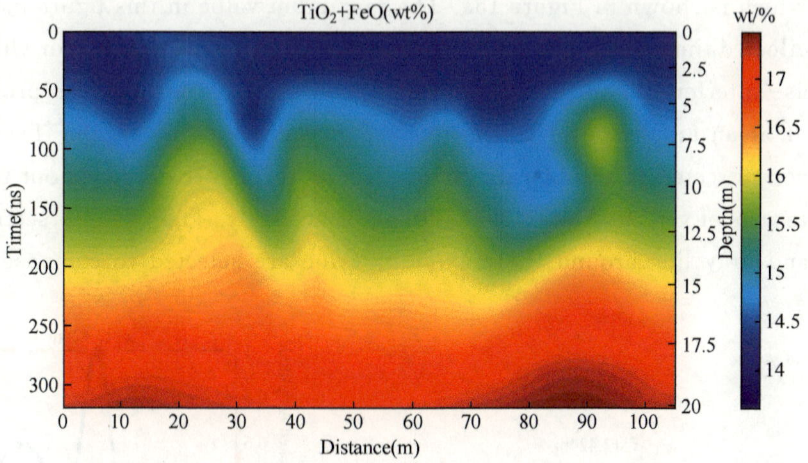

Figure 14　The TiO_2 and FeO content of lunar regolith

In Figure 10, the points with different velocities distribute in different layers: 0.15m/ns seem to be a watershed; the points whose velocities are bigger than 0.15m/ns are mainly distributed in 0 – 150 ns zone; the points whose velocities are smaller than 0.15m/ns are mainly distributed in 150 – 320 ns zone; this indicates the clear property difference between these two parts which can also be seen from Figures 11 – 14. These phenomena also correspond to the characteristic of radargram that 0 – 150ns zone is of high weathering degree with few rock reflections and 150 – 320ns zone is of low weathering degree with strong reflections.

Figures 11 – 14 show that the velocity descends as the depth increases, and the opposite varieties for velocity and relative permittivity are due to (4). In these figures, we can also find that there is a fluctuation at about 250 ns after position of 80 m, which also corresponds to the analysis in which the material and horizontal structure changes at a position of 80 m; the material about 250 ns in Region 3 does not extend to Region 4.

Previous research shows that the relative permittivity of lunar basalt is about 8 [24], so based on the velocity and relative permittivity structure we obtained, we think the region we researched (0 - 320ns) is mainly the lunar regolith. The depth coordinates in Figures 9 - 14 are determined using average root - mean - square velocity of all traces; so the thickness of the lunar regolithis more than 20m. In addition, the TiO_2 and FeO content in the bottom of Figure 14 is about 17.5 wt%, which is close to the shallow lava flow observed in [3, 4], so we speculate that the surface of the basalt is within 25m in depth.

4 Discussion

In this section, the estimation errors of two - way reflected time and velocity are discussed. Take the first slice in Figure 8b for instance, a 3D surface plot can be computed, which is shown in Figure 15a. The maximum value in this figure indicates the two - way reflected time and velocity of this hyperbola. However, some error will generate from the noise interference and incomplete hyperbola; the correct point is probably in a region; a flat at an amplitude of 0.8 is selected to split out this region. Two profiles, which are the red rectangle and black rectangle in Figure 15, are extracted out to compute the error intervals of velocity and reflected time, respectively. The two intersections of the blue curve and gray line around the maximum value are selected to compose the error limit.

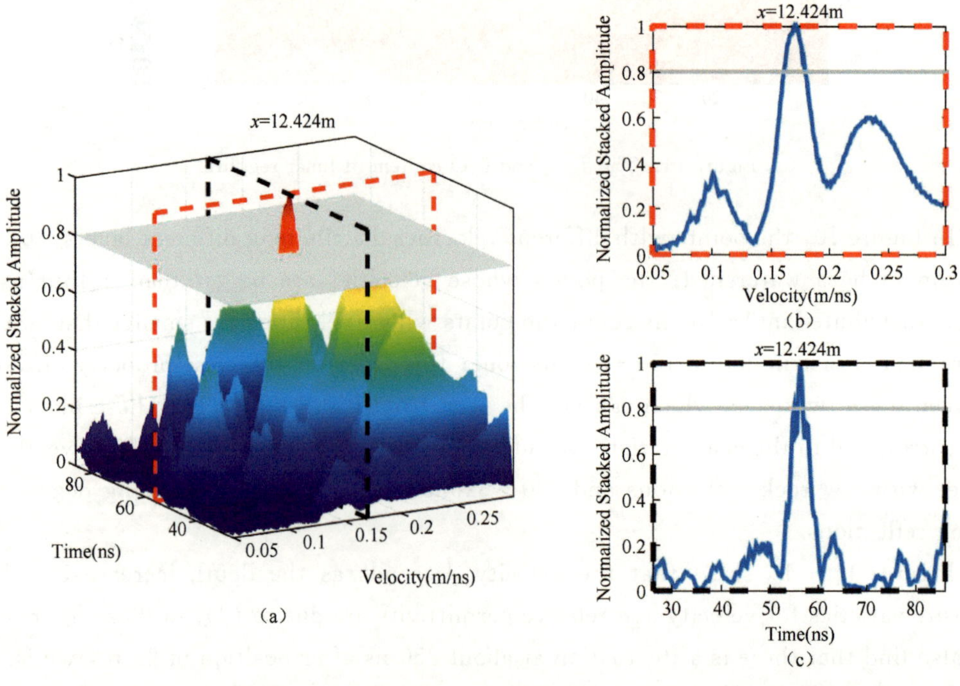

Figure 15　Error computation. (a) 3D surface plot of the first slice in Figure 8b. (b) The profile of in the red rectangle of (a). (c) The profile of in the black rectangle of (a)

In Figure 15, the error ranges of velocity and two - way reflected time are 0.161 - 0.179 m/ns and 55.8 - 56.84 ns, respectively. So, the real velocity and time may be within these two ranges. A maximum relative error (MRE) can be adopted to quantitatively describe the error:

$$\begin{cases} MRE_v = |\max(v - v_{ran})|/v \\ MRE_t = |\max(t - t_{ran})|/t \end{cases} \quad (11)$$

where v and t are the estimated values using 3D velocity spectrum, v_{ran} and t_{ran} represent the error ranges. So, the MRE_v and MRE_t in Figure 13 are 5.29% and 1.05%, respectively.

Subsequently, the errors of all hyperbolas are computed. The average MREs of time and velocity are 0.68% and 7.99%, respectively. It is clear that the error of two - way reflected time can be omitted, which is far less than the error of velocity. Furthermore, we computed the distributions of velocities and error ranges in horizontal and vertical directions which are shown in Figure 16.

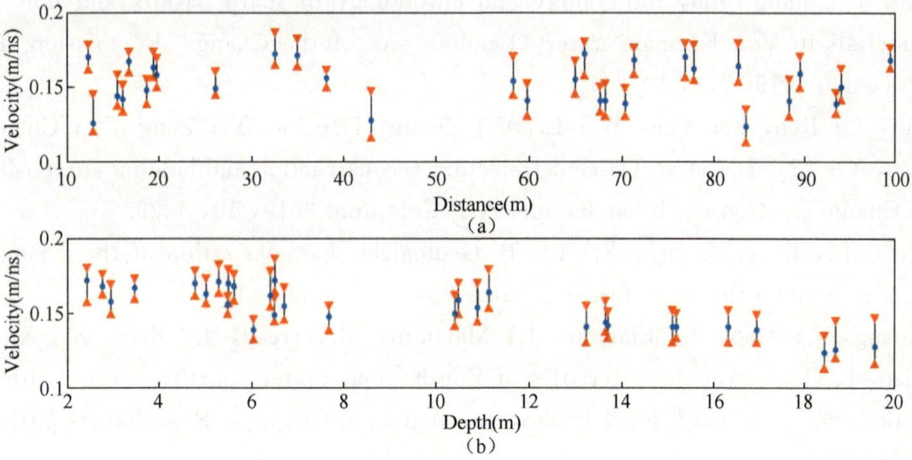

Figure 16　Distributions of velocities and error ranges. (a) Velocities and error ranges with distance. (b) Velocities and error ranges with depth. The triangular points represent the error ranges; the blue dots represent the estimated velocities

5　Conclusions

This paper applies the 3D velocity spectrum to property analysis of lunar penetrating radar data from Chang' E - 4. The result shows that 3D velocity spectrum can automatically search for hyperbolas; the maximum value along velocity axis with a soft threshold function can provide the horizontal position, two - way reflected time and velocity of each hyperbola. Based on the estimated velocities of 30 hyperbolas, the subsurface properties' structures are obtained, including velocity, relative permittivity, density, and content of FeO and TiO_2. Combining these properties and the radargram, we can conclude that 0 – 12m is the fine – grained regolith; 12 – 20m is the coarse regolith of different sources; the thickness of the lunar regolith is more than 20m; the surface of the shallow basalt is within 25m in the depth. Importantly, the subsurface material and structure at the Chang'E – 4 landing site vary both vertically and horizontally, so the simple horizontal stratigraphic division is not appropriate within the zone of 25 – 300ns. Finally, the estimation error is discussed, the errors of horizontal position and two – way reflected time are small which can be omitted; the average maximum relative error of velocity is 7.99%. the distributions of velocities and error ranges in horizontal and vertical directions are obtained.

References

[1] Ling, Z.; Qiao, L.; Liu, C.; Cao, H.; Bi, X.; Lu, X.; Zhang, J.; Fu, X.; Li, B.; Liu, J. Composition, mineralogy and chronology of mare basalts and non – mare materials in Von Kármán crater: Landing site of the Chang' E4 mission. Planet. Space Sci. 2019, 179, 104741.

[2] Liu, J.; Ren, X.; Yan, W.; Li, C.; Zhang, H.; Jia, Y.; Zeng, X.; Chen, W.; Gao, X.; Liu, D.; et al. Descent trajectory reconstruction and landing site positioning of Chang'E – 4 on the lunar farside. Nat. Commun. 2019, 10, 4229.

[3] Qiao, L.; Ling, Z.; Fu, X.; Li, B. Geological characterization of the Chang'e – 4 landing area on the lunar farside. Icarus 2019, 333, 37 – 51.

[4] Huang, J.; Xiao, Z.; Flahaut, J.; Martiont, M.; Head, J.; Xiao, X.; Xie, M.; Xiao, L. Geological characteristics of Von Kármán crater, northwestern South Pole – Aitken Basin: Chang' E – 4 landing site region. J. Geophys. Res. Planets 2018, 123, 1684 – 1700.

[5] Di, K.; Liu, Z.; Liu, B.; Wan, W.; Peng, M.; Li, J.; Xie, J.; Jia, M.; Niu, S.; Xin, X.; et al. Topographic Analysis of Chang' E – 4 Landing Site Using Orbital, Descent and Ground Data. In Proceedings of the International Archives of the Photogrammetry, Remote Sensing and Spatial Information Sciences, Enschede, The Netherlands, 10 – 14 June 2019; pp. 1383 – 1387.

[6] Hu, X.; Ma, P.; Yang, Y.; Zhu, M.; Jiang, T.; Lucey, P. G.; He, Z. Mineral abundances inferred from in situ reflectance measurements of Chang'E – 4 landing site in South Pole – Aitken basin. Geophys. Res. Lett. 2019, 46, 9439 – 9447.

[7] Xiao, L.; Zhu, P.; Fang, G.; Xiao, Z.; Zou, Y.; Zhao, J.; Zhao, N.; Yuan, Y.; Qiao, L.; Zhang, X.; et al. A young multilayered terrane of the northern Mare Imbrium revealed by Chang'E – 3 mission. Science 2015, 347, 1226 – 1229.

[8] Fa, W.; Zhu, M.; Liu, T.; Plescia, J. B. Regolith stratigraphy at the Chang'E – 3 landing site as seen by lunar penetrating radar. Geophys. Res. Lett. 2015, 42, 10,

179 - 10, 187.

[9] Dong, Z.; Fang, G.; Ji, Y.; Gao, Y.; Wu, C.; Zhang, X. Parameters and structure of lunar regolith in Chang' E - 3 landing area from lunar penetrating radar (LPR) data. Icarus 2017, 282, 40 - 46.

[10] Feng, J.; Su, Y.; Ding, C.; Xing, S.; Dai, S.; Zou, Y. Dielectric properties estimation of the lunar regolithat CE - 3 landing site using lunar penetrating radar data. Icarus 2017, 284, 424 - 430.

[11] Su, Y.; Fang, G.; Feng, J.; Xing, S.; Jing, Y.; Zhou, B. Data processing and initial results of Chang' E - 3 lunar penetrating radar. Res. Astron. Astrophys. 2014, 14, 1623 - 1632.

[12] Wang, K.; Zeng, Z.; Zhang, L.; Xia, S.; Li, J. A compressive - sensing based approach to reconstruct regolith structure from lunar penetrating radar data at the Chang' E - 3 landing site. Remote Sens. 2018, 10, 1925.

[13] Jia, Z.; Liu, S.; Zhang, L.; Hu, B.; Zhang, J. Weak Signal Extraction from Lunar Penetrating Radar Channel 1 Data Based on Local Correlation. Electronics 2019, 8, 573.

[14] Zhang, L.; Zeng, Z.; Li, J.; Lin, J.; Hu, Y.; Wang, X.; Sun, X. Simulation of the Lunar Regolith and Lunar - Penetrating Radar Data Processing. IEEE J. Sel. Top. Appl. Earth Obs. Remote Sens. 2018, 11, 655 - 663.

[15] Zhang, J.; Zeng, Z.; Zhang, L.; Lu, Q.; Wang, K. Application of Mathematical Morphological Filtering to Improve the Resolution of Chang' E - 3 Lunar Penetrating Radar Data. Remote Sens. 2019, 11, 524.

[16] Lai, J.; Xu, Y.; Zhang, X.; Tang, Z. Structural analysis of lunar subsurface with Chang'E - 3 lunar penetrating radar. Planet. Space Sci. 2016, 120, 96 - 102.

[17] Zhang, L.; Zeng, Z.; Li, J.; Huang, L.; Huo, Z.; Wang, K.; Zhang, J. Parameter Estimation of Lunar Regolith from Lunar Penetrating Radar Data. Sensors 2018, 18, 2907.

[18] Yilmaz, Ö. Seismic Data Analysis, 3rd ed.; Society of Exploration Geophysicists: Tulsa, Oklahoma, 2001.

[19] Feng, X.; Sato, M. Pre - stack migration applied to GPR for landmine detection. Inverse Probl. 2004, 20, S99 - S115.

[20] Fisher, E.; McMechan, G. A.; Annan, A. P. Acquisition and processing of wide - aperture ground - penetrating radar data. Geophysics 1992, 57, 495 - 504.

[21] Greaves, P. J.; Lesmes, D. P.; Lee, J. M.; Toksoz, M. N. Velocity variations and water content estimated from multi - offset, ground - penetrating radar. Geophysics 1996, 61, 683 - 695.

[22] Leparoux, D.; Gibert, D.; Ĉote, P. Adaptation of prestack migration to multi - offset ground - penetrating radar (GPR) data. Geophys. Prospect. 2001, 49, 374 - 386.

[23] Hu, B.; Wang, D.; Zhang, L.; Zeng, Z. Rock Location and Quantitative Analysis of Regolithat the Chang'e 3 Landing Site Based on Local Similarity Constraint. Remote Sens. 2019, 11, 530.

[24] Sen, P. N.; Scala, C.; Cohen, M. H. A self - similar model for sedmentary rocks with application to the dielectric constants for fused glass beads. Geophysics 1981, 46, 781 - 795.

Dielectric properties and stratigraphy of regolith in the lunar South Pole - Aitken basin: Observations from the Lunar Penetrating Radar

Jianqing Feng[1,2], Matthew. A. Siegler[1,2], and Mackenzie N. White[1,2]

1. Planetary Science Institute, 1700 East Fort Lowell, Tucson, AZ 85719, USA
2. Roy M. Huffington Department of Earth Sciences, Southern Methodist University, 6425 Boaz Lane, Dallas, TX 75205, USA

Abstract: Aims. We examine data obtained by the Lunar Penetrating Radar (LPR) onboard the Chang'E - 4 (CE - 4) mission to study the dielectric properties and stratigraphy of lunar regolith on the far side of the Moon.

Methods. The data collected from January 2019 to September 2020 were processed to generate a 540m radargram. The travel velocity of the radar signal and the permittivity of the regolith were deduced from hyperbolas in the radargram. As CE - 4 LPR detected distinct planar reflectors, we evaluated the dielectric loss from the maximum penetration depth based on the radar equation. The derived dielectric properties are compared with the measurements of Apollo samples and Chang'E - 2 microwave radiometer observations.

Results. The results suggest that regolith at the landing site has a permittivity of 2.64 - 3.85 and a loss tangent of 0.0032 - 0.0044, indicating that the local regolith is composed of a fine - grained, low - loss material that is much more homogeneous than that found at the Chang'E - 3 landing site. The total thickness of weathered material is ~ 40m, with several regolith layers and a buried crater identified in the reconstructed subsurface structure.

Conclusions. These layers clearly record a series of impact events from the adjacent regions. We suggest that the top layer is primarily made up of the ejecta from a large crater 140km away. In contrast, the material source of other thinner layers comes from nearby smaller craters.

Keywords: Moon - Planets and satellites; surfaces - space vehichles; instruments - methods; data analysis

1 Introduction

Impact cratering has shaped the surface of the Moon throughout the Moon's evolutionary history. Due to continuous impact gardening, nearly the entire lunar surface consists of a layer of regolith that completely covers the underlying bedrock (Heiken et al. 1991). Generally, the regolith has an average thickness of 4 - 5m in the maria and 10 - 15m in the highlands (McKay et al. 1991). Subsurface structures in the regolith record substantial evidence of impact cratering processes and volcanic eruptions. In 2014, Chang'E - 3 (CE - 3) mission surveyed the lunar subsurface at its landing site in northern Mare Imbrium using the Lunar Penetrating Radar (LPR)- the first ground penetrating radar (GPR) onboard a rover on the Moon (e. g., Su et al. 2014; Feng et al. 2017). In 2019, Chang'E - 4 (CE - 4) landed on the eastern floor of Von Kármán crater on the far side of the moon. The Yutu - 2 rover of CE - 4 mission carried an LPR with the same configuration as CE - 3 (e. g., Li et al. 2020; Lai et al. 2019; Ding et al. 2020). The LPR is an ultra - wideband carrier - free pulse radar with two channels. The first channel installed on the back of the rover operates at a center frequency of 60MHz with a bandwidth of 40MHz, while the second channel installed on the underside of the rover works at a center frequency of 500MHz with a bandwidth of 500MHz (Zhang et al. 2014; Li et al. 2020).

Studies have investigated the thickness of lunar regolith in various ways. From the Lunar Orbiter program in the 1960s to the Lunar Reconnaissance Orbiter (LRO) mission throughout the last 12 years, many researchers have used the morphology of small craters and crater counting techniques to constrain the regolith thickness (e. g., Bart et al. 2011; Bart 2014; Oberbeck & Quaide 1967, 1968; Shoemaker et al. 1969; Wilcox et al. 2005). During the Apollo 15, 16, and 17 missions, astronauts took in - situ measurements at the landing sites. The maximum penetration depth of the drills of Apollo 15, 16, and 17 are 2. 4m, 2. 25m, and 2. 95m, respectively (Allton & Waltz 1980). Although the drills did not touch the base rock, they provided an upper boundary constraint on the regolith thickness at the landing sites. Passive seismic experiments estimated the thickness from the shear - wave resonance at the Apollo 11, 12, and 15 sites to be 4. 4, 3. 7, and 4. 4m, respectively (Nakamura et al. 1975). As microwave and radio wave signals can penetrate the regolith much deeper than optical signals, orbital radar technology has been effectively applied to detect lunar subsurface structures since the 1970s (e. g., Phillips et al. 1973; Spudis et al. 2009; Ono et al. 2010). Earth - based radio telescopes are also used to observe the near side of the Moon and calculate the regolith thickness and associated dielectric properties (e. g., Thompson 1987; Shkuratov & Bondarenko 2001; Fa & Wieczorek 2012; Campbell et al. 2014).

The dielectric properties of the lunar regolith are primarily characterized by permittivity (the real component of the complex permittivity) and dielectric loss. The permittivity is used to quantify the electric polarizability of a dielectric medium. Dielectric loss (parameterized in terms of the loss tangent) quantifies a dielectric material' s inherent dissipation of electromagnetic (EM) energy (e. g. , heat) . The permittivity determines the propagation velocity of EM waves, while the dielectric loss constrains how deeply the EM waves can penetrate into a medium. The dielectric properties of Apollo samples have been measured in a laboratory setting (e. g. , Carrier et al. 1991; Olhoeft & Strangway 1975) and these experiments showed that the relative permittivity of lunar regolith is density - dependent, while the loss tangent is a function of both density and TiO_2+FeO content. Recently, Fa & Wieczorek (2012) reanalyzed the Apollo samples data and suggested that the loss tangent has a significant dependence on titanium but not iron. Using LRO Diviner surface temperatures and the microwave radiometer (MRM) on board the Chang' E - 2 (CE - 2) orbiter, Feng et al. (2020) and the companion Siegler et al. (2020) produced global maps of loss tangent at 3, 7. 8, 19. 35, and 37GHz. These maps not only showed a positive correlation between the loss tangent and titanium abundance, but also unveiled differences in highland and mare material as a function of frequency.

The CE - 4 landing site (45. 456°S, 177. 588°E) is located on the eastern floor of the Von Kármán crater, in the South Pole - Aitken (SPA) basin - the largest impact basin on the Moon. The Von Kármán crater formed in the Nectarian epoch with its floor flooded by low - albedo mare basalt in the late Imbrian age (Pasckert et al. 2018) . The subsurface structure of the CE - 4 landing site preserves a record of the sequence of major impact events of this region. Notably, there are several younger large (>25km) impact craters around the Von Kármán that may contribute ejecta materials to the local regolith in the Von Kármán: Leibnitz, Von Kármán L, Alder, and Finsen craters (Wilhelms et al. 1979) . The eastern part of the mare surface where CE - 4 landed is covered by bright ejecta. Mineralogy and morphology analysis suggest that the Finsen crater is the primary source of this bright ejecta (Huang et al. 2018) . Figure 1a shows the mosaic of the LRO Wide Angle Camera (WAC) for the regional vicinity of the Von Kármán crater. Figure 1b gives the WAC mosaic at 415 nm after photometric normalization, showing that bright ejecta from Finsen crater covers a large portion of Von Kármán crater and Leibnitz crater. Figure 1c shows an LRO Narrow Angle Cameras (NAC) image of the landing site captured in September, 2020. The CE - 4 lander and Yutu - 2 rover can be seen from the NAC image.

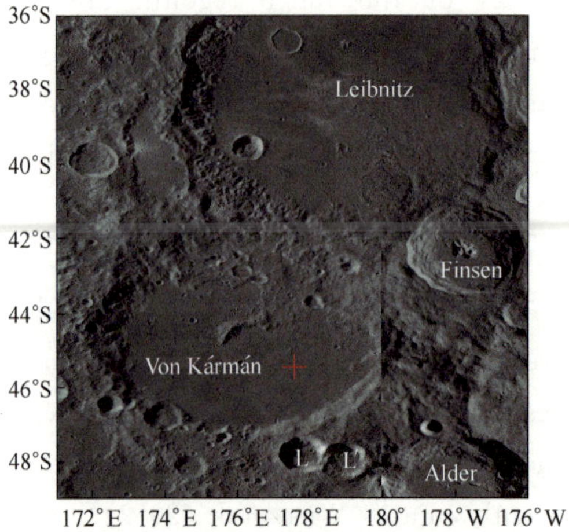

(a)

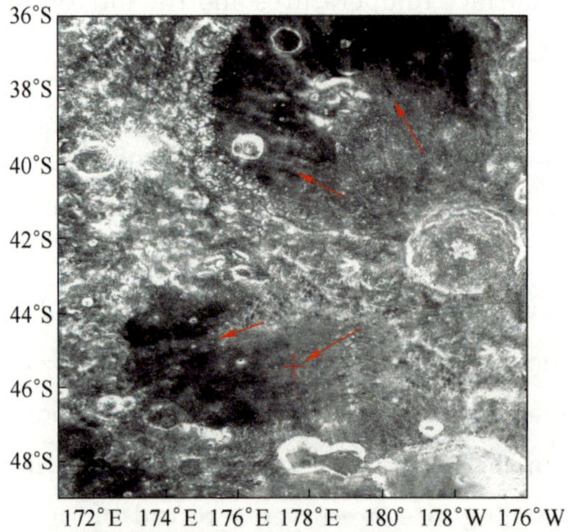

(b)

(c)

Fig. 1 (a) LRO WAC image mosaic of Von Kármán crater and adjacent region. The CE－4 landing site is marked by a red cross. (b) LRO WAC mosaic at 415 nm after photometric normalization. The red arrows indicate the direction of bright ejecta from Finsen crater. (c) The CE－4 lander and Yutu－2 rover captured by LRO NAC in September, 2020. Rover tracks are faintly visible between the lander and rover

2 Data and method

We processed the published LPR second channel data obtained from January 2019 to September 2020. The CE - 4 LPR is a common - offset radar and its data processing includes some regular steps widely used in GPR, such as time - zero correction, bandpass filtering, and background removal (Jol 2008) (similar to the data processing of CE - 3 LPR). The raw data is shown in Fig. 2a, with all the repeating traces deleted. In contrast to CE - 3 LPR data, a continuous subsurface interface is visible in the CE - 4 raw data without any processing (between 400 - 800 samples). A finite impulse response (FIR) filter with a pass range of 250 - 750MHz was used for the bandpass filtering. To remove the background, we take the mean of all traces and subtract it from each trace in each section. Strong "ringing" and reverberation in the shallow near - surface layer are removed through the background removal. A spherical exponential compensation (SEC) gain is then implemented to correct the signal amplitude for loss of energy due to the geometrical spreading effect. After the data processing, the subsurface structure in deep regolith with more details are revealed (Fig. 2b). Multiple interfaces are able to be distinguished, with the deepest one at the two - way travel time larger than 400ns.

The permittivity ε of lunar regolith can be computed from velocity v of EM waves traveling in the regolith as $\varepsilon = (c/v)^2$, where c is the light speed in the vacuum. We employ hyperbolic velocity analysis (Jol 2008) in our study because buried targets beneath the surface usually reflect as hyperbola patterns in GPR radargrams. The same method has been successfully used on CE - 3 LPR data (Feng et al. 2017; Lai et al. 2016). The equation relating the two - way travel time, t, of the reflection from a target and the radar position, x, can be expressed in the hyperbolic form:

$$\frac{t^2}{t_0^2} - \frac{(x - x_0)^2}{(vt_0/2)^2} = 1 \tag{1}$$

where x_0 and t_0 are the horizontal and vertical position of the subsurface target in the radargram ($x - t$ plane), respectively, and v and t_0 determine the shape of the hyperbolas (or the angles between their asymptotes). By using a least - squares method to fit the hyperbolic form of the observed data, the travel speed v and the target's position (x_0, t_0) can be computed. To avoid incorrect recognitions, we selected the twelve clearest hyperbolas from the LPR radargram. Figure 3 shows examples of these hyperbolas. These pronounced curves represent strong echoes from subsurface isolated rocks.

The decay of a radar signal through a medium is governed by geometric spreading loss, scattering loss, and dielectric loss. At the maximum penetration depth, the radar signal decays to an undetectable level and is lost in the system noise. Here, we use the radar equation and penetration depth of LPR in the landing site to estimate the average attenuation in the regolith. The basic radar range equation for GPR (e. g. , Annan &

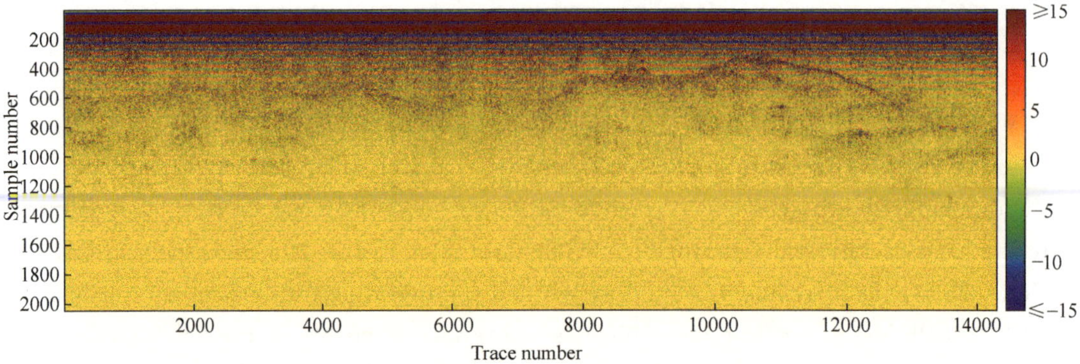

(a)

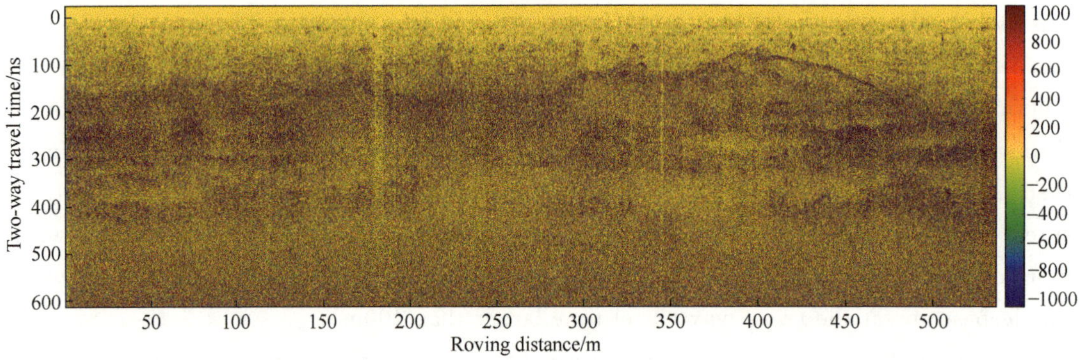

(b)

Fig. 2 (a) Raw data obtained by CE - 4 LPR second channel (500MHz) from January 2019 to September 2020. The sampling interval is 0.3125ns and the average trace interval is 3.75cm. The colors represent relative electric field strength of echoes received by antenna. The radargram is shown in a dynamic range of −15 to 15 to reveal more details in deeper regolith. (b) LPR data after bandpass filtering, background removal, time - zero correction, and spherical exponential compensation

Davis 1977; Hamran et al. 1995; Skolnik Merrill 1990) is: $\frac{P_r}{P_t}=\frac{G_r G_t \lambda^2 \sigma}{4(\pi)^3 R^4}e^{-4\alpha R}$, where P_r is the received power, P_t is the transmitted power, G_r is the gain of receiving antenna, G_t is the gain of transmitting antenna, and λ is the wavelength of EM signal in the medium. R is the distance between the target and radar, α is the attenuation constant in the medium, and σ is the backscatter radar cross - section (RCS) of the subsurface target or interface. According to Noon et al. (1998) we rewrite the radar equation for CE - 4 LPR as:

$$G_{sys}=\frac{P_t G_r G_t}{P_{min}}=\left[\frac{T^2 \lambda^2 \sigma}{(4\pi)^3 R_{max}^4}e^{-4\alpha R_{max}}\right]^{-1} \tag{2}$$

where G_{sys} is the system gain of LPR, which is 133.3dB (Zhang et al. 2014); P_{min} is the minimum detectable signal power; T is the transmissivity at the vacuum - regolith interface; and R_{max} is the maximum penetration depth. We assume the attenuation α is dominated by an effective dielectric loss, which includes the scattering effect. The laboratory measurements of Apollo samples at 450MHz indicate that lunar regolith has a loss tangent, $\tan\delta$, smaller than 0.03 (Carrier et al. 1991). When the $\tan\delta \ll 1$, $\alpha \approx \pi \frac{1}{\lambda} \tan\delta$. In this study, since subsurface layers at the CE－4 landing site are detected (shown in Fig. 2b), we consider σ as the RCS of a planar reflector. Given that the roughness of subsurface interfaces relative to the LPR center wavelength is greater than $0.5/\pi$, these interfaces are considered as rough planar reflectors, and their cross－section can be taken as the area of the "first Fresnel zone," with $\sigma = \Gamma\pi\lambda R/2$ (e.g., Annan & Davis 1977; Cook 1975; Grimm et al. 2006; Noon et al. 1998). Therefore, the average loss tangent of regolith could be calculated from the maximum signal penetration distance by:

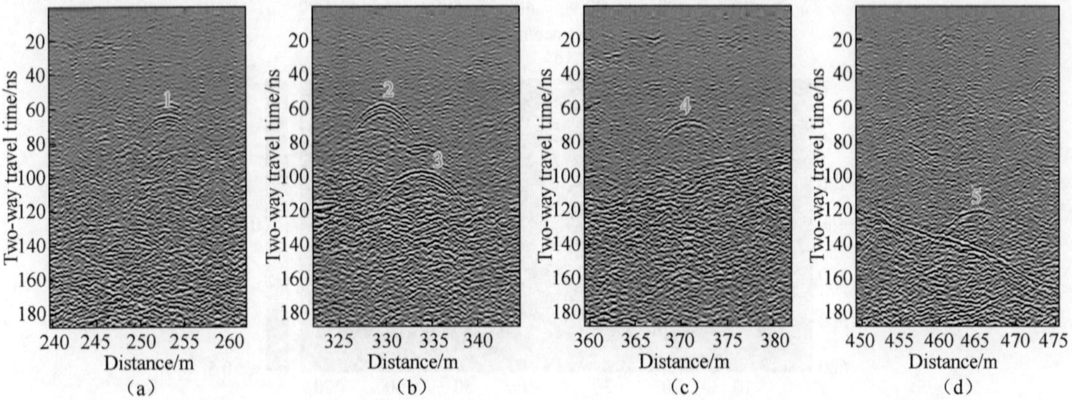

Fig. 3　Noticeable hyperbolas caused by subsurface rocks. The travel velocities of radar signal estimated by curve fitting from these hyperbolas are (1) 15.97 ± 0.48, (2) 15.63 ± 0.30, (3) 16.38 ± 0.30, (4) 17.08 ± 0.36, and (5) 16.50 ± 0.31cm/ns

$$\tan\delta = \frac{\lambda}{4\pi R_{\max}} \ln\left[\frac{T^2 \lambda^3 \Gamma G_{\mathrm{sys}}}{128\pi^2 R_{\max}^3}\right] \tag{3}$$

Here, $R_{\max}$ can be replaced by maximum travel time $t_{\max}$ in the form of $R_{\max} = f\lambda t_{\max}/2$, where f is the signal frequency, λ is the signal wavelength in regolith ($\lambda = \lambda_0/\sqrt{\varepsilon}$, where λ_0 is the signal wavelength in vacuum). Equation (3) then becomes:

$$\tan\delta = \frac{1}{2\pi f t_{\max}} \ln\left[\frac{T^2 \Gamma G_{\mathrm{sys}}}{16\pi^2 f^3 t_{\max}^3}\right] \tag{4}$$

For low－loss mediums, the reflectivity Γ of the substrate can be calculated from $\Gamma = \left|\frac{\sqrt{\varepsilon} - \sqrt{\varepsilon_1}}{\sqrt{\varepsilon} + \sqrt{\varepsilon_1}}\right|^2$, where ε and ε_1 are the permittivity of regolith and rocks (we assume $\varepsilon_1 = 7$ in this study (Carrier et al. 1991)). The transmissivity of the regolith surface is determined by $T = \frac{4\sqrt{\varepsilon}\sqrt{\varepsilon_0}}{\left|\sqrt{\varepsilon} + \sqrt{\varepsilon_0}\right|^2}$, where ε_0 is the permittivity of the vacuum. In Figure 4, we compare the radargram of CE－4 LPR with that of CE－3 LPR, both of which are shown in a very small dynamic range (−0.5 to 0.5) to display the boundary between signals and noise. It can be seen that the maximum penetration time at the CE-4 landing site is roughly four times longer than that at the CE-3 landing site.

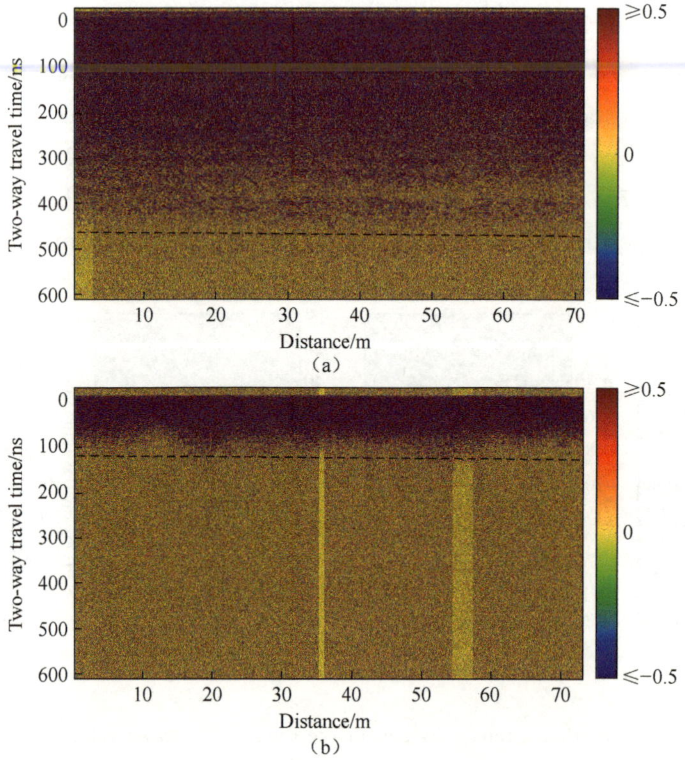

Fig. 4　Boundary between signals and noise in the LPR data.

(a) A subset of CE-4 LPR radargram shown in a small dynamic range (b) compared with that of CE-3 LPR. The dashed lines indicate the maximum penetration time of the radar signals.

To better interpret the subsurface features in the deposits and convert the travel time to penetration depth, an imaging method was applied to the radargram. The algorithm we used is the Kirchhoff migration. In migration, observed reflections are moved from the apparent locations back to their real spatial locations. A hyperbola in the radargram will be "refocused" to a target's position. Kirchhoff migration is based on an integral solution of the scalar wave equation. According to Schneider (1978) and Özdemir et al. (2014), in a 2-D situation the far-field approximation of Kirchhoff integral for LPR can be expressed as:

$$U(x_0, z_0)=\frac{1}{2\pi}\int\left[\frac{\cos\theta}{v_m R}\frac{\partial}{\partial t}U(x, z, t)\right]\mathrm{d}x \tag{5}$$

where $U(x_0, z_0)$ is the wave field at a subsurface scatter point (x_0, z_0), $U(x, z, t)$ is the wave field observed by LPR at surface position (x, z) at time two-way travel time t, θ is the refraction angle of wave field in the medium, R is the propagation distance of radar signal from the point to the observation position, and v_{m} is half of the travel speed in the regolith.

3　Result

3.1　Permittivity and dielectric loss

As shown in Fig. 5a, the average EM velocity in the regolith at the landing site ranges from 15.30 to 18.46cm/ns. The EM velocity at the CE-3 landing site (Feng et al. 2017) is also plotted in the figure for comparison. Accordingly, the depth-averaged permittivity of CE-4 regolith varies from 2.64 to 3.85 (Fig. 5b), consequently falling within the range

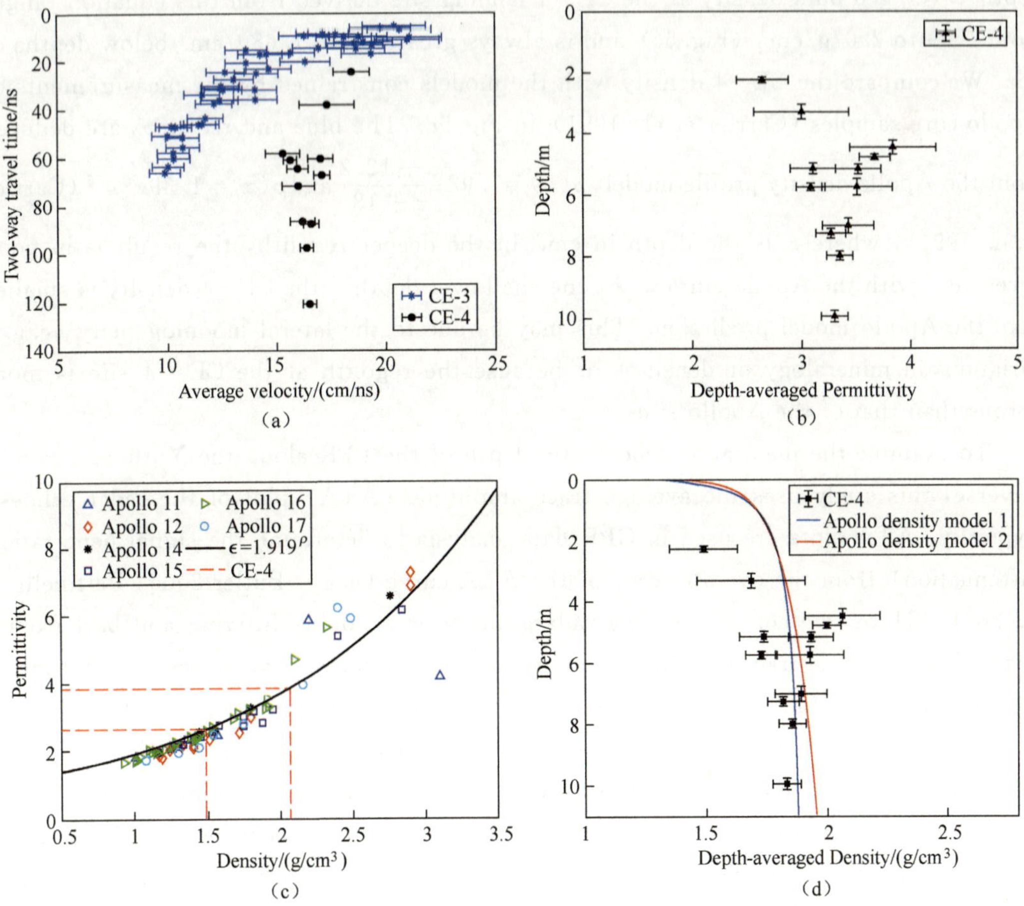

Fig. 5　Permittivity and density of lunar regolith estimated from LPR data.

(a) Average propagation velocity of radar signal estimated from hyperbolas as a function of two-way travel time. (b) Depth-averaged permittivity of regolith at CE-4 landing site versus depth. (c) Permittivity of Apollo soil samples at 450MHz as a function of density (Olhoeft & Strangway 1975). The dashed lines constrain the CE-4 permittivity range and the corresponding density range. (d) Inverted depth-averaged bulk density of regolith as a function of depth. The Apollo density models come from Carrier et al. (1991)

constrained by Apollo soil samples with a density lower than 2.1g/cm^3 (Olhoeft & Strang - way 1975) (Fig. 5c). Because we did not find any clear hyperbolas before 20ns in the radargram, the shallowest point in Fig. 5b for which we can calculate the dielectric constant is at a depth of 2m. Generally, the permittivity increases with depth in the upper ~4m but remains within the range between three and four for 4 to 10m. This implies that the local regolith (at least the first layer) consists of fine - grained soil and is much more homogeneous than the regolith at the CE - 3 landing site, where the weathered material contains more rocks and the permittivity reaches six at ~3m depth (Feng et al. 2017; Lai et al. 2016). Carrier et al. (1991) proposed that the permittivity of lunar regolith is density - dependent in the form of $\varepsilon = 1.919^{\rho}$, where ρ is bulk density. The depth - averaged bulk density at the CE - 4 landing site derived from this equation ranges from 1.49 to 2.07g/cm^3 (Fig. 5c) and is always greater than 1.68g/cm^3 below depths of 3m. We compare the CE - 4 density with the models constrained by the measurements of Apollo core samples (Carrier et al. 1991) in Fig. 5d. The blue and red lines are deduced from the Apollo density profile model, $\rho(z)=1.92\frac{z+12.2}{z+18}$ and $\rho(z)=1.39z^{0.056}$ (Carrier et al. 1991), where z is the depth in cm. In the deeper regolith, the result is in good agreement with the Apollo curves. At the shallower depths, the CE - 4 density is smaller than the Apollo model prediction. This may be due to the lateral inhomogeneity (e. g., variations in mineralogy or density) or because the regolith at the CE - 4 site is more porous than that of the Apollo sites.

To examine the maximum penetration depth of the LPR along the Yutu - 2 rover's traverse, this study uses the average trace amplitude (ATA) plot of the radar echoes. Typically, ATA plots are used in GPR data analysis to determine the signal penetration (attenuation) from the overall decay of the ATA curve (e. g., Butler 2005; Pettinelli et al. 2021). Here, we compressed the radargram after bandpass filtering and background removal into a single trace by taking the mean absolute value over all traces at each time to produce the ATA plot. Figure 6a gives the ATA curve of a 100 - m subsets of CE - 4 radar cross - section. Given that the radar systems and rover configurations of CE - 4 LPR are identical to CE - 3 LPR, we compare the ATA of data from both instruments. The CE - 4 LPR signal reaches the noise floor after 460ns. The small peaks on the ATA curve represent the reflections from subsurface interfaces, agreeing with the features shown in Fig. 2b. The last peak at ~510ns refers to the echoes from the deepest planar reflector LPR could detect, so 460 - 510ns is considered as the range of maximum travel time t_{max} of the LPR signal. The loss tangent of local regolith as a function of t_{max} is shown in Fig. 6b, wherein a penetration curve corresponds to a specific permittivity of local regolith. Penetration curves with permittivities of 2.64 and 3.85 are used as the upper and lower boundaries at the CE - 4 landing site, resulting in dielectric loss tangents ranging from 0.0031—0.0044, according to the t_{max} range.

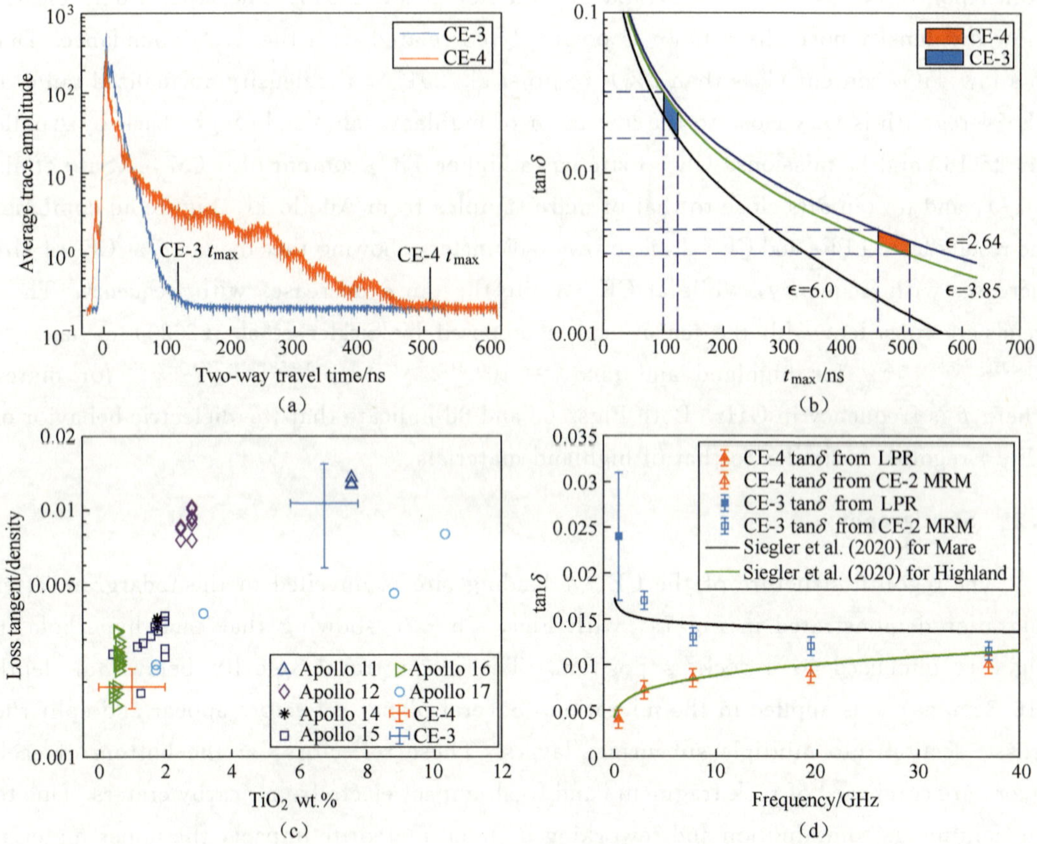

Fig. 6　Dielectric loss of lunar regolith estimated from LRP data.

(a) Average LPR trace amplitudes from 100m of CE－3 data and 100m of CE－4 data, used to assess signal decay in the lunar subsurface. The position of time－zero is corrected according to the largest amplitude. (b) The loss tangent of lunar regolith as a function of maximum penetration time. (c) The density normalized loss tangent as a function of TiO_2 weight percentage. The Apollo data points only represent measurements at 450MHz. (d) The loss tangent at CE－3 and CE－4 landing sites as a function of frequency. The data at 3, 7.8, 19.35, and 37GHz are taken from Siegler et al. (2020)

In contrast to the CE－4 landing site, the attenuation at the CE－3 landing site is much greater, with echoes reaching noise floor at 100－120ns. According to the calculation from Lai et al. (2016) and Feng et al. (2017), we used 3.85 and 6 as the upper and lower boundaries of CE－3 regolith permittivity in Fig. 6b. This is meant to lead to a loss tangent falling in the range of 0.016－0.032. However, because there is no apparent planar reflector or interface in the regolith at the CE－3 landing site (Feng et al. 2017; Lai et al. 2016), Eqs. (3) and (4) are not adequately applicable and the loss tangent may be inaccurate. The calculated value may just provide an upper boundary of CE－3 tan δ, since a regolith or bedrock interface is assumed to be at the maximum penetration depth. Even so, from these estimates we can infer that the lunar regolith is more highly attenuating beneath CE－3 than under CE－4.

Figure 6c compares tan δ of CE－4 regolith with that of soil samples ($\rho<2.1g/cm^3$)

from Apollo 11, 12, 14, 15, 16, and 17 (Carrier et al. 1991). The Apollo data clearly show the density normalized tan δ is positively correlated with the TiO_2 abundance. Due to a low TiO_2 content (less than 2%) (Sato et al. 2017), the density normalized tan δ of CE-4 regolith is very close to the low tan δ of highland samples brought back by Apollo 14, 15 16, and 17 missions. CE-3 site has a higher TiO_2 content of ~6.7% (Sato et al. 2017), and its tan δ is close to that of mare samples from Apollo 11. Figure 6d combines the results from LPRs and CE-2 microwave radiometer, showing that the tan δ at CE-4 site increases with frequency, while at CE-3 site the tan δ decreases with frequency. These trends are consistent with the tan δ models proposed by Siegler et al. (2020) as $\tan\delta = 10^{0.312\rho + f^{0.069} - 3.79}$ for highland and $\tan\delta = 10^{0.312\rho + (f^{-0.0025} - 0.958) \times \% TiO_2 - 2.65}$ for mare, where f is frequency in GHz. Both Figs. 6c and 6d indicate that the dielectric behavior of CE-4 regolith is similar to that of highland materials.

3.2　Stratigraphy

The regolith structure of the CE-4 landing site is unveiled in the radargram after migration demonstrated in Fig. 7a, with Figs. 7b-7e showing that each hyperbola in Fig. 3 is refocused to a rock's position. The mean travel velocity below 3m depth (16.3cm/ns) was applied in the migration. Several planar reflectors appear and split the cross-section into multiple subsurface layers. These reflectors, at the bottom of each layer, are composed of rock fragments and fresh impact ejecta from nearby craters. Due to the continuous comminution and reworking by micrometeorite impact, the upper material in each layer is overturned and pulverized much more frequently, thus taking on a smaller and more uniform grain size. The top layer thickness is ~7-16m with a relatively "transparent" background, indicating a mature fine grain material with low rock abundance. This is consistent with the smooth surface seen in the picture of the landing site captured by the panoramic camera on board the rover (Li et al. 2020). The permittivity result also shows that this fine-grained layer that lacks scatterers is relatively homogeneous along the rover's traverse. The other layers in the cross-section are thinner and have higher rock abundances, indicating a shorter period of gardening and evolution. These may be the ejecta delivered by craters smaller than the source of the first layer. The deepest reflector distinguishable is at a depth of ~40m. Its echo is weak, continuous, and flat. This may reflect the boundary between paleoregolith and basaltic base rock.

In the middle of the radargram (between roughly 200-300m along the traverse), subsurface reflectors or layers are disrupted by a symmetric bowl-shaped structure filled with point reflectors, as illustrated in Fig. 8. It penetrates three subsurface layers with a height of ~13-15m and a horizontal length of ~128m on its top. The echoes beneath this structure become weaker than those on other parts of the horizontal interfaces. Using LRO NAC Digital Terrain Models, Stopar et al. (2017) recently found that craters with diameters less than 400m have a depth-to-diameter (d/D) ratio ranging from 0.11 to

0.17. As the d/D ratio of this structure is ~0.10–0.12, we speculate it is the profile of an impact crater. This finding is also supported by the recent work from Zhang et al. (2021). This crater might form before the new material covered the paleoregolith surface. It is unlikely to be a secondary crater caused by the ejected debris because there is no topographic evidence on the surface of top layer (Fig. 1c) and the crater was. filled with fragments before it was buried. As the rock fragments filled in the crater scatter the EM signal in all directions, the echoes from deeper layers attenuate faster and the penetration depth of LPR is shallower under this carter.

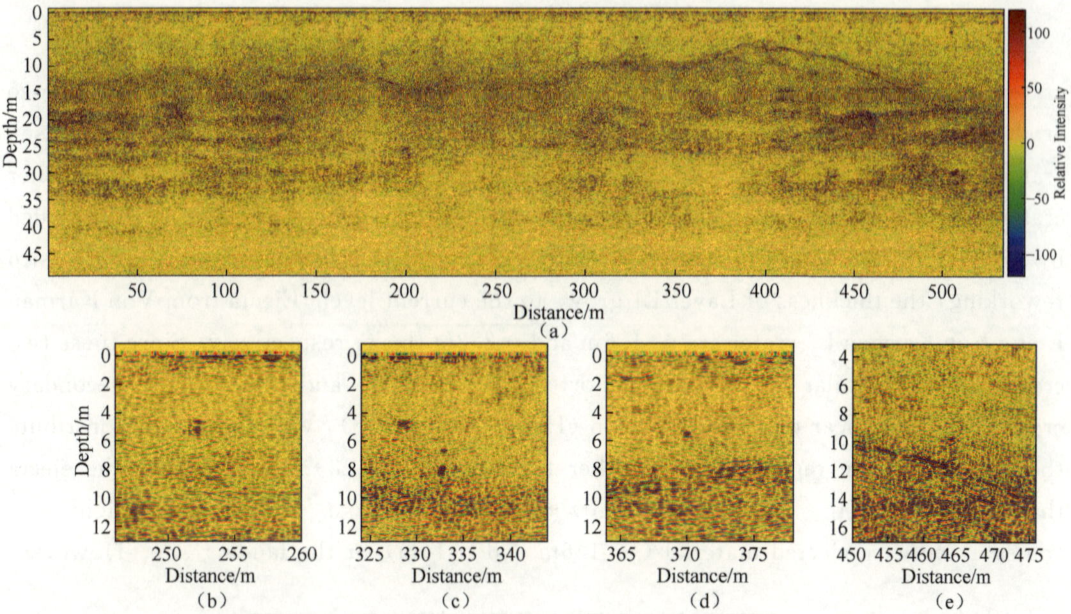

Fig. 7 Subsurface profile of lunar regolith at the CE-4 landing site.

(a) Reconstruction of lunar regolith subsurface structure using migration. The colors represent relative intensity of reflections from subsurface objects. (b)-(e) display the migration results of the hyperbolas given in Fig. 3

The stratigraphic diagram illustrated in Fig. 8 displays six layers from L1 at the top to L6 at the bottom. It is inferred from the local geological context that the soil at the landing site may consist of ejecta coming from the Finsen, Von Kármán L, Von Kármán L', Alder, and Leibnitz craters. To analyze the contributions of nearby impact events to the regolith at the landing site, we estimated the thickness of distal ejecta from each crater. As reported by McGetchin et al. (1973), the ejecta thickness model is described by $t=T(r/R)^b$, where t is the thickness at radial distance r from the crater center, T is the ejecta thickness at the rim of the crater, R is the radius of the crater, and b is the exponent of the power-law. Pike (1974) pointed out R should be the transient radius and defined T and b as: $T=0.033R$, $b=-3$.

Figure 9 demonstrates the computed ejecta thickness based on this model. The result suggests that ~8.1m thickness of material comes from Finsen. Finsen is an Eratosthenian

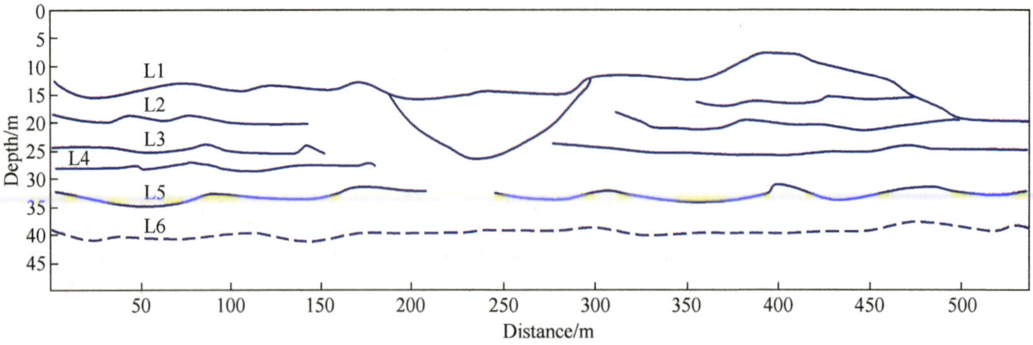

Fig. 8　Stratigraphic diagram of regolith at the CE-4 landing site

crater, and therefore younger than the Imbrian Mare in the Von Kármán crater (Fortezzo et al. 2020; Wilhelms et al. 1987). Figure 1b indicates the ejecta blanket of the Finsen crater lies on the top of the local regolith and is not overlapped by ejecta from other craters. Therefore, it is reasonable to propose that the majority of top layer L1 is provided by Finsen crater. After billions of years of quiescent soil development and in-situ reworking, the thickness of Layer L1 grows to the current level. Ejecta from Von Kármán L and Von Kármán L' crater are ~1.6m and ~2.2m thick, respectively. Since these two craters are of a similar age to Finsen (Fortezzo et al. 2020) and some of their secondary craters are found near the landing region (Huang et al. 2018), we believe they contribute the majority of the regolith of two other layers (L2 and L3). According to the ejecta thickness model, the Nectarian Leibnitz crater and Imbrian Alder crater might also transport massive ejected material (~146m and ~14m) at the landing site. However,

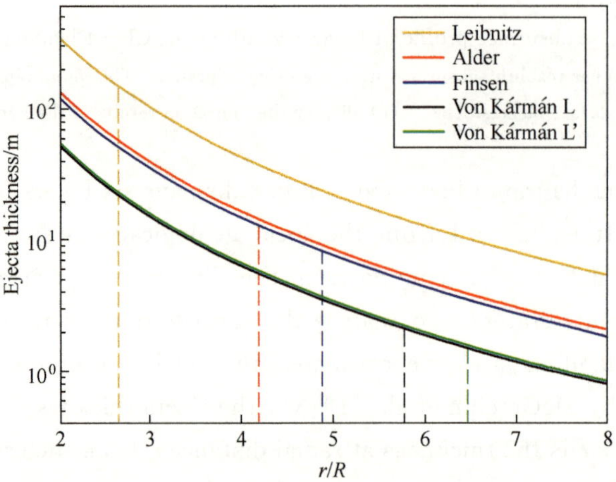

Fig. 9　Thickness of distal ejecta from each crater as a function of normalized distance (the ratio of distance to transient radius) from the crater's center. The dashed lines represent the normalized distance at the landing site. The transient radius is calculated according to the model described in (Melosh 2011)

the latest studies suggest that they are older than the Imbrian mare unit in Von Kármán (e. g. , Fortezzo et al. 2020; Lu et al. 2021), indicating that material from Leibnitz and Alder may be buried underneath the mare basalt layer (Huang et al. 2018). For layers L4 -L6, we did not find any potential source from outside the Von Kármán crater and, thus, their source might come from small craters close to the landing site inside this crater. These craters were formed after the emplacement of flood basalt lava flows and then buried by L3, L2, and L1; therefore it is hard to tell their exact locations from the optical images.

4 Conclusions

The LPR onboard the Yutu - 2 rover of the CE - 4 mission provides an effective method to survey the dielectric properties and subsurface structure of lunar regolith on the far side of the Moon. In this study, we estimate the permittivity, evaluate the dielectric loss, and reveal the shallow stratum of regolith at CE - 4 landing site using LPR data.

The average permittivity increases with depth in the upper 4m of regolith and remains constant in the deeper regolith. The depth - averaged bulk density profile inverted from permittivity is generally consistent with models from Apollo core samples. The dielectric loss is small, with the maximum penetration time of the LPR signal reaching ~40m depth. The comparison of our results with measurements of Apollo samples shows that the loss tangent of CE - 4 regolith is close to that of highland materials. This is also supported by the combination of LPR and CE - 2 MRM observations. Both the permittivity and dielectric loss are much lower at the CE - 4 landing site than at the CE - 3 landing site. This implies that the CE - 4 regolith is made up of highly - porous, low - loss, fine - grained materials, while the CE - 3 regolith has higher ilmenite content and is dominated by large rock debris ejected from the nearby Ziwei crater. The stratification structure in the post - processed radargram shows local deposit consisting of multiple layers preserves ejecta from several nearby large craters. We suggest that the bowl - shape structure with a d/D ratio of 0. 10～0. 12 identified in the cross - section is a buried paleo crater. The crater - ejecta thickness model and 415nm LRO WAC mosaic indicate that the majority of the top layer is ejecta from Finsen crater. The Von Kármán L crater and Von Kármán L' crater also contribute the two additional, thinner layers beneath the top layer.

Acknowledgements

The Chang'E - 3 and Chang'E - 4 Lunar Penetrating Radar data are downloaded from https://moon. bao. ac. cn/web/enmanager/home. The Lunar Reconnaissance Orbiter (LRO) Wide Angle Camera images and the LRO Narrow Angle Camera image are available at http://wms. lroc. asu. edu/lroc.

References

[1] Allton, J. & Waltz, S. 1980, in Lunar and Planetary Science Conference Proceedings, Vol. 11, 1463 - 1477.

[2] Annan, A. & Davis, J. 1977, Geological Survey of Canada, 77, 117.

[3] Bart, G. D. 2014, Icarus, 235, 130.

[4] Bart, G. D., Nickerson, R. D., Lawder, M. T., & Melosh, H. J. 2011, Icarus, 215, 485.

[5] Butler, D. K. 2005, Near - surface geophysics (Society of Exploration Geophysicists).

[6] Campbell, B. A., Ray Hawke, B., Morgan, G. A., et al. 2014, Journal of Geo - physical Research: Planets, 119, 313.

[7] Carrier, W. D., Olhoeft, G. R., & Mendell, W. 1991, Physical properties of the lunar surface, ed. G. Heiken, D. Vaniman, & B. French (Cambridge, UK: Cambridge University Press), 475 - 594.

[8] Cook, J. C. 1975, Geophysics, 40, 865.

[9] Ding, C., Xiao, Z., Wu, B., et al. 2020, Geophysical Research Letters, 47, e2020GL087361.

[10] Fa, W. & Wieczorek, M. A. 2012, Icarus, 218, 771.

[11] Feng, J., Siegler, M. A., & Hayne, P. O. 2020, Journal of Geophysical Research: Planets, 125, e2019JE006130.

[12] Feng, J., Su, Y., Ding, C., et al. 2017, Icarus, 284, 424.

[13] Fortezzo, C., Spudis, P., & Harrel, S. 2020, in Lunar and Planetary Science Con - ference, 2760.

[14] Grimm, R. E., Heggy, E., Clifford, S., et al. 2006, Journal of Geophysical Re - search: Planets, 111.

[15] Hamran, S. - E., Gjessing, D., Hjelmstad, J., & Aarholt, E. 1995, Journal of Applied Geophysics, 33, 7.

[16] Heiken, G., Vaniman, D., & French, B. 1991, Lunar sourcebook: A user's guide to the Moon (Cambridge, UK: Cambridge University Press).

[17] Huang, J., Xiao, Z., Flahaut, J., et al. 2018, Journal of Geophysical Research: Planets, 123, 1684.

[18] Jol, H. M. 2008, Ground penetrating radar theory and applications (Elsevier).

[19] Lai, J., Xu, Y., Zhang, X., & Tang, Z. 2016, Planetary and Space Science, 120, 96.

[20] Lai, J., Xu, Y., Zhang, X., et al. 2019, Geophysical Research Letters, 46, 12783.

[21] Li, C., Su, Y., Pettinelli, E., et al. 2020, Science advances, 6, eaay6898.

[22] Lu, Y., Wu, Y., Michael, G. G., et al. 2021, Icarus, 354, 114086.

[23] McGetchin, T. R., Settle, M., & Head, J. 1973, Earth and Planetary Science Letters, 20, 226.
[24] McKay, D. S., Heiken, G., Basu, A., et al. 1991, The lunar regolith, Vol. 567, 285-356.
[25] Melosh, H. J. 2011, Planetary surface processes, Vol. 13 (Cambridge University Press).
[26] Nakamura, Y., Dorman, J., Duennebier, F., Lammlein, D., & Latham, G. 1975, The Moon, 13, 57.
[27] Noon, D. A., Stickley, G. F., & Longstaff., D. 1998, Journal of Applied Geophysics, 40, 127.
[28] Oberbeck, V. R. & Quaide, W. L. 1967, Journal of Geophysical Research, 72, 4697.
[29] Oberbeck, V. R. & Quaide, W. L. 1968, Icarus, 9, 446.
[30] Olhoeft, G. & Strangway, D. 1975, Earth and Planetary Science Letters, 24, 394.
[31] Ono, T., Kumamoto, A., Kasahara, Y., et al. 2010, Space science reviews, 154, 145.
[32] Özdemir, C., Demirci, Ş., Yiğit, E., & Yilmaz, B. 2014, Mathematical Problems in Engineering, 2014.
[33] Pasckert, J. H., Hiesinger, H., & van der Bogert, C. H. 2018, Icarus, 299, 538.
[34] Pettinelli, E., Lauro, S. E., Mattei, E., Cosciotti, B., & Soldovieri, F. 2021, Nature Astronomy, 1.
[35] Phillips, R., Adams, G., Brown Jr, W., et al. 1973, in Lunar and Planetary Science Conference Proceedings, Vol. 4, 2821.
[36] Pike, R. J. 1974, Earth and Planetary Science Letters, 23, 265.
[37] Sato, H., Robinson, M. S., Lawrence, S. J., et al. 2017, Icarus, 296, 216.
[38] Schneider, W. A. 1978, Geophysics, 43, 49.
[39] Shkuratov, Y. G. & Bondarenko, N. V. 2001, Icarus, 149, 329.
[40] Shoemaker, E., Batson, R., Holt, H., et al. 1969, Journal of Geophysical Research, 74, 6081.
[41] Siegler, M. A., Feng, J., Lucey, P. G., et al. 2020, Journal of Geophysical Research: Planets, 125, e2020JE006405.
[42] Skolnik Merrill, I. 1990, New York st. louis: MCGRAW-Hill Publishing Company.
[43] Spudis, P., Nozette, S., Bussey, B., et al. 2009, Current Science, 533.
[44] Stopar, J. D., Robinson, M. S., Barnouin, O. S., et al. 2017, Icarus, 298, 34.
[45] Su, Y., Fang, G.-Y., Feng, J.-Q., et al. 2014, Research in Astronomy and Astrophysics, 14, 1623.
[46] Thompson, T. 1987, Earth, Moon, and Planets, 37, 59.
[47] Wilcox, B., Robinson, M., Thomas, P., & Hawke, B. 2005, Meteoritics &

Planetary Science, 40, 695.

[48] Wilhelms, D. E., Howard, K. A., & Wilshire, H. G. 1979, Geologic map of the south side of the Moon (Department of the Interior, US Geological Survey).

[49] Wilhelms, D. E., John, F., & Trask, N. J. 1987, The geologic history of the Moon, Report 2330-7102.

[50] Zhang, H.-B., Zheng, L., Su, Y., et al. 2014, Research in Astronomy and Astrophysics, 14, 1633.

[51] Zhang, L., Xu, Y., Bugiolacchi, R., et al. 2021, Earth and Planetary Science Letters, 564, 116912.

Exploring the Lunar Regolith's Thickness and Dielectric Properties Using Band-Limited Impedance at Chang'E-4 Landing Site

Yongjiu Feng[1,2], Shurui Chen[1,2], Xiaohua Tong[1,2], Chao Wang[1,2], Pengshuo Li[1,2], Mengrong Xi[1,2], and Changjiang Xiao[1,2]

1. College of Surveying & Geo-Informatics, Tongji University, Shanghai, China
2. The Shanghai Key Laboratory of Space Mapping and Remote Sensing for Planetary Exploration, Tongji University, Shanghai, China

Abstract: The Yutu-2 rover onboard China's Chang'E-4 was the first to land in the Von Kármán crater on the lunar farside, where structural and dielectric properties may provide previously unknown clues to lunar formation and evolution. Based on the Lunar Penetrating Radar (LPR) with 500MHz onboard the Yutu-2 rover, we proposed a band-limited impedance (BLIMP) inversion method to generate continuous dielectric 2-D profiles with the help of dielectric constants retrieved from point reflectors (e.g., buried rocks), which provide low-frequency information that may be lost or distorted in the LPR data pre-processing. The dielectric constants retrieved from point reflectors were transformed from dielectric constant fitting curves, where these discrete dielectric constants were calculated using the hyperbolic fitting method. We estimated that the top fine-grained regolith thickness along the rover path varies from ~5.3 to ~15.0m in the first 27 lunar days of operation. The thickness variation could most likely be attributed to changes in ancient surface topography buried underground and ejecta from nearby craters. Compared to methods based on a singular dielectric constant, the estimated 2-D dielectric profile in this study can reduce the uncertainties in lunar regolith thickness estimation, especially in the shallow layers. The BLIMP method and the estimated regolith thickness can improve our understanding of lunar subsurface structure and formation.

Plain Language Summary: The lunar regolith thickness is an important stratigraphic feature for scientific exploration such as temperature inversion and mineral content estimation on the Moon. The Lunar Penetrating Radar (LPR) onboard the Yutu - 2 rover (Chang'E - 4) provides a unique opportunity to probe the dielectric properties of lunar regolith and thereby estimate the regolith thickness on the farside of the Moon. We analyzed the variations in the dielectric constants of the Yutu - 2 LPR data using the band - limited impedance (BLIMP) method. The analysis reveals a two - layer subsurface structure with the top layer of fine - grained ranging from ~5.3 to ~15.0m. Analysis suggests that the variations in the regolith thickness were most likely attributed to changes in ancient surface topography buried underground and nearby craters. Compared to methods based on a singular dielectric constant, the estimated 2 - D dielectric profile in this study can reduce the uncertainties in lunar regolith thickness estimation, especially in the shallow layers.

Key Points:

· We integrated BLIMP and hyperbolic fitting to inverse continuous 2 - D dielectric profiles

· We estimated that the top fine - grained regolith thickness along the moving path varies from ~5.3 to ~15.0m in the first 27 lunar days

· The thickness variation could be attributed to the ejecta from nearby small craters, hence impacting the calculation of dielectric property

1 Introduction

The lunar surface is predominantly covered by a regolith layer with a few bedrocks exposed in lava channels and steep - sided crater walls (McKay et al., 1991). Although optical equipment can provide fine - resolution lunar surface photographs, topography maps and surface mineral composition maps (Hu et al., 2013; Sun et al., 2016; Tran et al., 2010), meteoritic impacts and solar weathering may bury critical clues retained on the surface. Radars with deep penetration ability are the optimal choice for subsurface exploration considering the desiccated environments of the Moon. In the last 10 years, the lunar subsurface and the global regolith thickness have been explored (Fa & Jin, 2010) using a microwave radiometer (MRM) onboard Chang'E - 1. Chang'E - 1 MRM operates on four channels including the center frequencies of 3, 7.8, 19.35, and 37GHz. It is noted that the maximum detection depth of Chang'E - 1 was estimated to be nearly 6 m on a global average considering the minimum detector's emission frequency (3GHz) and the dielectric constant (from 2.4 to 4.5) of known lunar regolith (Wang et al., 2010). This depth is roughly equal to the widely acknowledged average thickness of regolith in maria, far smaller than the average thickness of about 10 - 15m in the highland regions (McKay et al., 1991). Compared to the orbital radar, in situ radar detectors and suitable analysis methods can reveal a much more detailed profile of shallow subsurface structure and dielectric property distribution.

An in situ Lunar Penetrating Radar (LPR) detector onboard China's Chang'E - 4 provides vital scientific data for exploring the subsurface structure of the lunar farside. The landing site of Chang'E - 4, the South Pole - Aitken (SPA) basin, is one of the important scientific investigation areas of China's lunar exploration mission. The SPA basin, lying between the South Pole and the Aitken crater, is the largest known impact structure with ~2,500km diameter and centered at (53°S, 191°E) on the Moon (Stuart - Alexander, 1978) . Because of its huge size and diverse surface materials, the SPA forming impact is believed to excavate lower crust and/or upper mantle mate rials (Ohtake et al. , 2014) and has geophysically and petrologically influenced not only the SPA but also the entire Moon. The impact could have affected lunar nearside evolution by the offset - antipodal shocks (Schultz & Crawford, 2011), differentiated the impact melt sheet from shallow noritic materials (Hurwitz & Kring, 2014) and enriched the mafic minerals in the surrounding highlands (Lawrence et al. , 2002) . As a major lunar terrain (Jolliff et al. , 2000), SPA's surface materials are diverse and can be divided into four sub - zones according to their surface components (Moriarty III and Pieters, 2018) . Therefore, the SPA area is of great significance to reveal the subsurface structure and dielectric properties using datasets from radar detectors.

The first radar detection covering the SPA was a bistatic radar experiment carried out by Clementine (Nozette et al. , 1996) . After Clementine, the world's major countries launched different lunar orbiters including Lunar Prospector (Halekas et al. , 2009), SELENE (Kato et al. , 2008), Chandrayaan - 1 (Goswami & Annadurai, 2009), Lunar Reconnaissance Orbiter (Vondrak et al. , 2010), Chang'E - 1 (Li et al. , 2010), and Chang'E - 2 (Zuo et al. , 2014), providing optical and radar data for lunar exploration in various impact basins like SPA. Scientists have made great progress in vigorous investigations of SPA using orbital data (Cahill et al. , 2014; Huang et al. , 2018; Moriarty et al. , 2013) . Considering the large diameter of SPA and the abundance of FeO and TiO_2, which is consistent with the equivalent mixture of mantle rock and lower crustal material (Lucey et al. , 1998), as well as the main spectra signature of low - calcium pyroxene, scientists suggest that the SPA has excavated mantle materi als of the Moon's upper mantle, which is rich in low - calcium pyroxene (Melosh et al. , 2017) . Similar ultramafic compositions were also detected in the central peaks of the fresh craters of SPA, suggesting a homogeneous layer beneath the SPA surface (Nakamura et al. , 2009) . However, when studying the subsurface layer of SPA, most high frequency (> 2. 3GHz) orbital radar detectors have difficulty penetrating the lunar surface to reveal the fine subsurface structure and regolith dielectric properties. Although the Lunar Radar Sounder onboard KAGUYA operates at 4 - 6MHz, its coarse vertical resolution (~75m in vacuum) limits its ability to provide accurate subsurface structures (Kobayashi et al. , 2018) . In addition, previous in situ radar detectors (Fa et al. , 2015) and seismic experiments (Cooper et al. , 1974; Latham et al. , 1970) were all focused on the lunar nearside until the Yutu - 2 rover successfully landed at the Von Kármán crater in SPA on the lunar farside.

The LPR mounted on Yutu - 2 was aimed to acquire subsurface radar echoes at depths ranging from tens of meters (high frequency 500MHz) to hundreds of meters (low frequency 60MHz). The same type of LPR equipment onboard Yutu - 1 (on Chang'E - 3) has provided valuable information on the subsurface structure and regolith thick ness in the north Mare Imbrium. The maximum thickness of regolith over lunar mares varies from 1 to 3m as reported in recent studies using the Chang'E - 3 LPR data (Dong et al., 2017; Fa et al., 2015; Feng et al., 2017; Wang et al., 2018), providing estimates of regolith thickness in local areas of the lunar nearside. Recent studies indicated that the penetration depth of Chang'E - 4 is three times that of Chang'E - 3 (Lai et al., 2019). The radar echoes of Chang'E - 4 reflected from deep layers may be mixed with diffracted and reflected waves coming from different layers, and these have relatively low signal - to - noise ratios and require more refined processing than the Chang'e - 3 LPR data.

Because the dielectric constant determines the velocity of the radar wave and the LPR data are collected in a round - trip (two - way) time, different dielectric constants used in the time - depth conversion of LPR data may affect the strata thickness inversion. In this regard, when using the same LPR low frequency data (60MHz), some reported a penetration depth of ~360m (Zhang et al., 2020) while others reported ~500m (Zhang et al., 2021). In studies of thickness estimation, modelers have typically applied a singular dielectric constant for regolith, for example, $\varepsilon_r'=3$ in Fa et al. (2015) and $\varepsilon_r'=4.3$ in Lai et al. (2021), and calculated the thickness by multiplying half of the observed round - trip time at the stratigraphic interface and the propagation velocity of radar wave in a medium with the corresponding dielectric constant. If the round - trip time at the stratigraphic interface is very short, then an average dielectric constant ($\varepsilon_r'=3$) based on Apollo samples was considered appropriate (Krupenio & Kroupenio, 1973; Muhleman et al., 1969); otherwise, the influence of variation in the dielectric constant of the deep material should be investigated. These sampling sites were located at different regions in the low - and middle latitude areas on the lunar nearside as well as at different sampling depths from 0 to ~305cm. As such, a fixed dielectric constant summarized from these samples may not be most suitable to describe the dielectric properties of the Chang'E - 4 landing site since it is located at the farside of the Moon. Furthermore, the dielectric constant at the landing site should vary with location and depth because they are affected by many factors such as bulk density and material composition.

In this study, we focused on the analysis of the LPR data collected by the Yutu - 2 rover during the 1st to 27th lunar days, aiming to retrieve regolith dielectric properties and estimate the regolith thickness. We simulated LPR radar echoes based on a 2 - D relative dielectric random disturbance model and validated the band - limited impedance (BLIMP) method in the estimation of regolith dielectric constants. Using Chang'E - 4 LPR data, we

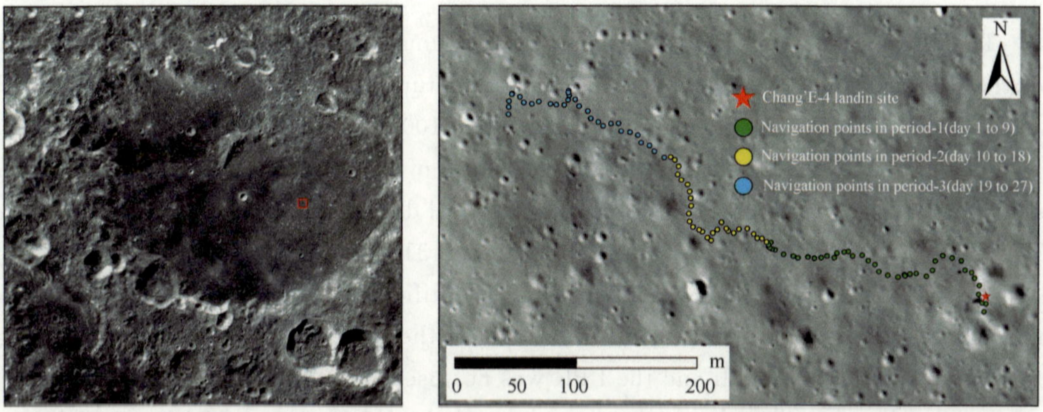

(a)　　(b)

Figure 1　The Chang'E-4 landing region and the Yutu-2 rover path. (a) A digital orthophoto map (DOM) of the Von Kármán crater produced from Chang'E-2 optical images; (b) A digital terrain map (DTM) of the rover path from day 1 to day 27 produced using the Lunar Reconnaissance Orbiter Camera (LROC) narrow-angle camera images

detected the boundary between the regolith layers and generated a 2-D dielectric profile for regolith dielectric characterization and regolith thickness calculation.

2　The Landing Area and Data Sets

2.1　The Landing Area

Chang'E-4 successfully landed at 177.5991°E/45.4446°S on 3 January 2019 in the northwestern region of the SPA basin (Figure 1). The landing site is in the eastern part of Von Kármán crater with an elevation of −5,935m based on the definition of a lunar spheroid with a radius of 1,737.4km. There are numerous small impact craters around the Von Kármán crater and earlier studies (Fu et al., 2020; Huang et al., 2020) suggested that some of them may provide clues about the formation of the Von Kármán crater. Recent geological investigations indicated that the Von Kármán crater is a pre-Nectarian crater (Yingst et al., 2017) near the SPA compositional anomaly (SPACA) with Mg-rich Pyroxenes. Furthermore, radially distributed higher albedo features observed on the mare basalt of the Von Kármán crater converge at the Finsen crater (located in SPACA), indicating that the upper material of the Von Kármán crater could be the ejecta derived from the Finsen crater (Huang et al., 2018). In the forthcoming sections, we analyzed the subsurface structure and dielectric properties based on three time-periods: the 1st-9th days, the 10th-18th days, and the 19th-27th days, which are labeled in green, yellow, and blue in Figure 1b, respectively.

2.2 The Data Sets

The Yutu - 2 rover works only during the lunar daytime, and its onboard LPR provides two channels to detect the subsurface structures. The first channel (CH1) with a central frequency of 60MHz can probe more than 100m in depth with 1m resolution and the second channel (CH2) with a central frequency of 500MHz can detect materi als 30m under the surface with 30cm resolution in the vertical direction (Fang et al., 2014). The CH2 data used in this paper were sampled with a 0.3125ns interval by two receivers. By 19 February 2021, Yutu - 2 had completed the scientific task in the first 27 lunar days and moved 652.56 m from the landing site. The data of the 18th lunar day was absent because the Yutu - 2 rover stood still and the LPR was not used.

To visualize the Chang'E - 4 landing site and the Yutu - 2 rover path, we collected the optical images from Chang'E - 2 and LROC. The DOM generated using the Chang'E - 27 - m resolution image (CE2_GRAS_DOM_07m_K136_45S175E_A) shows the overall morphology of the Von Kármán impact crater (Figure 1a). A high resolution (140cm) DTM produced using the LROC narrow - angle camera image (nac_dtm_change4_m1303640934_140cm) reveals both the detailed terrain information around the location site and along the Yutu - 2 rover path (Figure 1b).

3 Methodology

3.1 LPR Data Processing

The LPR CH2 transmitter sends energy pulses and then the receiver samples the reflected radar waveform in a limited time, that is, CH2 generates a radar trace composed of 2,048 timesteps (one - dimensional data) within 640ns. As the Yutu - 2 rover moves, the received LPR data can form a two - dimensional radar diagram. To explore the subsurface structure, it is necessary to conduct the following preprocessing based on the level 2B data: (a) removal of repetitive traces caused by the rover's stationary state, (b) time - zero correction to remove the signals in the first 28.203ns, which is caused by recording delay (Fang et al., 2014), (c) median filter to alleviate back ground noise, (d) bandpass filtering via a Finite Impulse Response filter with a cutoff range of 200 - 600MHz, to suppress the noise in other frequency and preserving the vital information around the central frequency (Feng et al., 2017), and (e) amplitude gain to compensate for signal attenuation, where an exponential gain function based on an average loss tangent of 0.0045 was used in this study by following recent work (Lai et al., 2019).

3.2 Hyperbolic Fitting Method

No reflective waveforms appear in the radar echo diagrams when the radar echoes propagate in materials with homogeneous dielectric properties, while it may cause strong reflective waveforms where there is measurable dielectric contrast. Figure 2 shows that, in the LPR data, hyperbolic curves caused by discrete targets are useful to determine the dielectric constant located on the vertex of the curve. It has been recognized that lunar regolith can be regarded as an ideal dielectric considering their weak or absent magnetic properties and their low electrical conductivity (Feng et al., 2017; Olhoeft et al., 1974; Rochette et al., 2010); thus, the relative magnetic permea - bility is commonly assigned a constant of 1 in the lunar research (Feng et al., 2017). Under this assumption, the average stacking velocity of electromagnetic wave propagation between the LPR and the reflector can be deter mined by the average dielectric constant computed at this vertex. We assume that a spherical target located at X_0 with T_0 time depth in the diagram and the rover moves from X_0 to X_1, the receiver may record the curve between (X_0, T_0) and (X_1, T_1). The average propagating velocity (V) can be derived by

$$\left(\frac{VT_1}{2}\right)^2 = (X_0 - X_1)^2 + \left(\frac{VT_0}{2}\right)^2 \tag{1}$$

The radar pulses meet the relationship that the average propagating velocity (V) in the material with average dielectric constant (ε_r') is equal to c、$\sqrt{\varepsilon_r'}$ ($\varepsilon_r' \geqslant 1$). Hence, ε can be inferred by Feng et al. (2017):

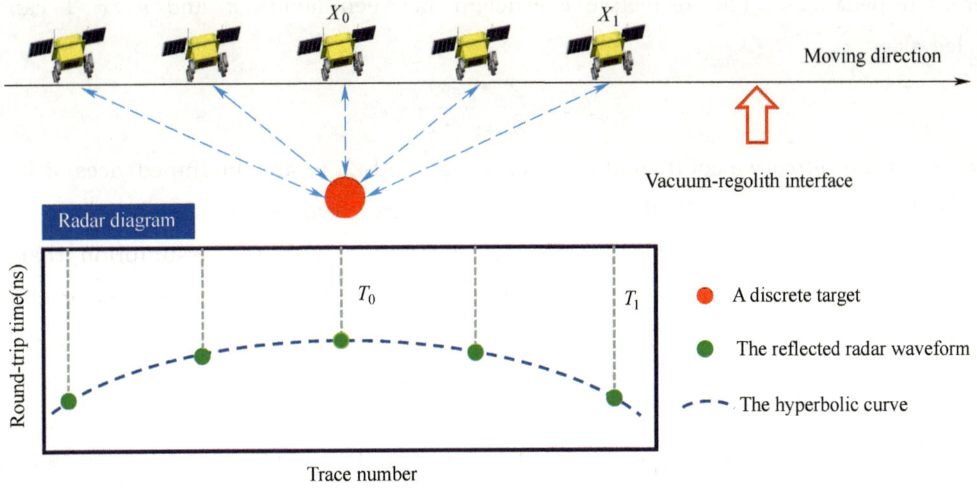

Figure 2 Diagrammatic sketch of the hyperbolic fitting method (Lai et al., 2016)

$$\varepsilon'_r = \frac{c^2 \times (T_0^2 - T_1^2)}{4 \times (X_0 - X_1)^2} \tag{2}$$

where c is the propagating velocity of the radar pulse in a vacuum. With the hyperbolic fitting method, we can first constrain the dielectric properties of the regolith at these particular sites.

3.3 BLIMP Inversion and Regolith Thickness Calculation

Impedance is an important property of a physical medium, which can be determined by a variety of basic param eters including density, velocity, and electrical and magnetic properties. In electromagnetics, the impedance represents the ratio of an electric field to a magnetic field, which can be calculated by the square root of the ratio of permeability to dielectric constant:

$$Z=\sqrt{\frac{\mu}{\varepsilon_r'}} \tag{3}$$

where Z is the impedance; μ is the permeability; ε_r' is the real part of dielectric constant. The BLIMP method was adapted for ground penetrating radar by Schmelzbach et al. (2012) from what was originally a seismic acoustic impedance inversion technique (Oldenburg et al., 1982). The acoustic impedance relates to the propagation of pressure waves through different mediums, depending on the density and elasticity of the medium, while electrical impedance is the analogous propagation of electromagnetic waves through different mediums, depending on the dielectric properties of these materials, which are a function of frequency, density, mineralogy and water content.

The received radar trace can be considered as a consequence of the incident radar waves undergoing reflection and transmission at each boundary between layers that have different impedances. The reflective coefficient between layers n and $n+1$ can be modeled as

$$R_n=\frac{Z_{n+1}-Z_n}{Z_{n+1}+Z_n} \tag{4}$$

where R_n is the reflective coefficient of layer n; Z_n and Z_{n+1} are the impedances of layers n and $n+1$, respectively. With Equation (4), we can recursively derive the impedance of each layer by the corresponding reflective coefficient with the assumption that the impedance of layer 1 is known, and their relationship can be defined as (Zeng et al., 2015):

$$Z_{n+1}=Z_1\prod_{k=1}^{n}\frac{1+R_k}{1-R_k} \tag{5}$$

To simplify the calculation in Equation5, Oldenburg et al. (1982) proposed and validated an approximation to the continuous impedance:

$$Z_{n+1}=Z_1\exp\left(\gamma\sum_{k=1}^{n}A_k\right) \tag{6}$$

where γ is the scaling factor between the reflected signal amplitude and the reflection coefficient and it is approx - imately 2 when the reflective coefficient is smaller than 0.3 (Oldenburg et al., 1982); A_n is the reflected signal amplitude in layer n. The relationship between A_n and R_n is that:

$$A_n = \frac{2R_n}{\gamma} \tag{7}$$

In BLIMP, constraints on the depth profile of impedance are needed to provide the low-frequency compo nents that might be lost or distorted when collecting, storing and pre-processing LPR data. In this study, the impedance - depth constraints ($\hat{Z}$) were transformed from fitting curves based on discrete dielectric constants, calculated from the hyperbolic fitting method [Equation 3], and Z_1 in Equations 5 and 6 was set as the surface value of $\hat{Z}$. Figure 3 shows the flow chart of the BLIMP method and how it connects to the hyperbolic fitting method. The cut-off frequency for the band-limited integration filter was set to 200 - 600MHz (Feng et al., 2017; Zhang et al., 2014). The linear trend of impedance ($\Delta\hat{Z}$) is subtracted from ($\Delta\tilde{Z}$) to reduce the edge effects (Chambers et al., 2013; Ferguson & Margrave, 1996; Leise, 1981) in the frequency domain (e. g., the bandpass filter). With the continuous record of impedance profile (Z_f), we derived the final dielectric constant profile through an inverse transformation of Equation 3. We then calculated the propagation distance in each pixel of the radargram and vertically accumulated them for the regolith thickness at any trace.

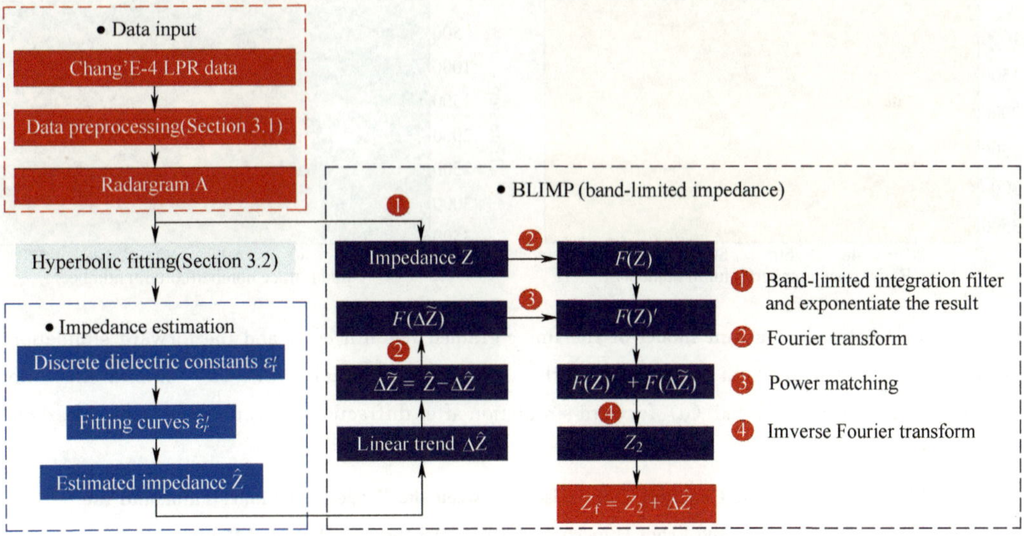

Figure 3　Diagrammatic sketch of the BLIMP and how it connects to the hyperbolic fitting method

3.4　Simulation and Validation of BLIMP

Given that the thickness of the fine-grained regolith layer is 12～13m (about 150ns in round-trip time) in the first 5 lunar days (Lai et al., 2019), we constructed a multi-layered 2-D dielectric model with 15m in height and 15m in width (Figure 4a). The first 2m in height was considered the vacuum and the remaining 13m in height was considered the regolith. The regolith part of the model was designed as vertically multi - layered and horizontally homogeneous with increasing dielectric constant. A clipped Gaussian random field was added to simulate the dielectric constant variation of the lunar regolith (Li et al. , 2017) . After conducting the radar forward simulation (Figure 4b), the first 500 lines (the vacuum) were removed to better show the reflected echo of the simulated regolith (Figure 4c) . We then implemented diffraction correction to the reflected echo map, where nine notable hyperbolae were found and marked using red circles (Figure 4d) . Table 1 shows the comparison between the predefined model dielectric constant and the dielectric constant estimated from hyperbolic fitting, suggesting only a minor difference of <8. 3% of the predefined value.

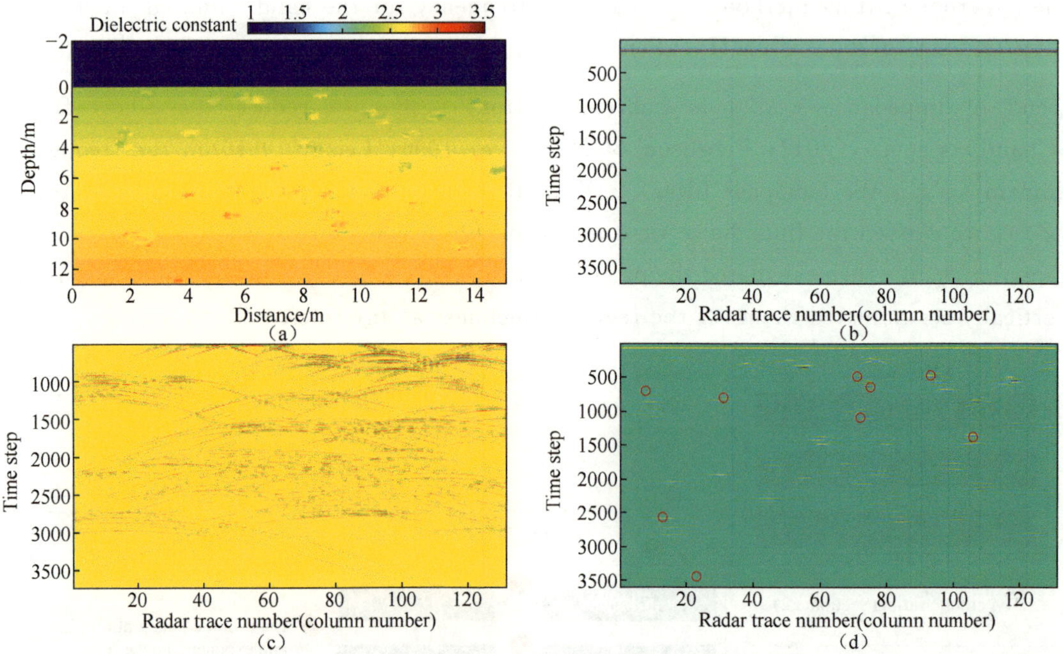

Figure 4 The dielectric constant model of the fine - grained regolith layer and the forward simulation. (a) Random equivalent medium model of regolith; (b) forward simulation; (c) forward simulation with the first 20ns removed; and (d) forward simulation after diffraction correction with nine hyperbolae marked

Table1 The Dielectric Constant Comparison Between the Predefined Model Value and the Estimated Value From Hyperbolic Fitting for the Nine Selected Points

Pointno.	Time (ns)	Model value	Estimated value	Error	Error (%)
1	25. 20	2. 36	2. 44	0. 08	3. 39
2	25. 52	2. 42	2. 22	−0. 20	8. 26
3	31. 84	2. 44	2. 46	0. 02	0. 81
4	34. 24	2. 45	2. 28	−0. 17	6. 94
5	38. 28	2. 55	2. 35	−0. 20	7. 84
6	49. 72	2. 48	2. 44	−0. 04	1. 61
7	61. 52	2. 52	2. 65	0. 13	5. 16
8	108. 8	2. 73	2. 88	0. 15	5. 49
9	143. 68	2. 75	2. 89	0. 14	5. 09

Figure 5 demonstrates the spatial differences between the predefined dielectric profile and the BLIMP estimated result by using the fitting line of the nine points as global constraints for BLIMP. The result shows a multilayer structure (Figure 5a) with a smaller time step interval compared to the structure of the predefined dielectric profile (Figure 5b). The result has a similar dielectric constant and no obvious disturbance in the horizontal direction, indicating the BLIMP's ability to capture the variation of the dielectric constant along the round-trip time. Although BLIMP cannot perfectly quantify this contrast (Figure 5c), the relative residual errors are within 15% of the predefined value. Figure 5d presents the relative estimation error of three radar traces (white dotted lines in Figure 5c), which are all within 9% of the predefined value. For the nine selected hyperbolae, Figure 5e shows the similarity of the fitting lines between the predefined dielectric profile and the BLIMP results, indicating the significant effect of hyperbolic fitting in BLIMP. Figure 5f demonstrates that most relative inversion errors are within 4% of the predefined value.

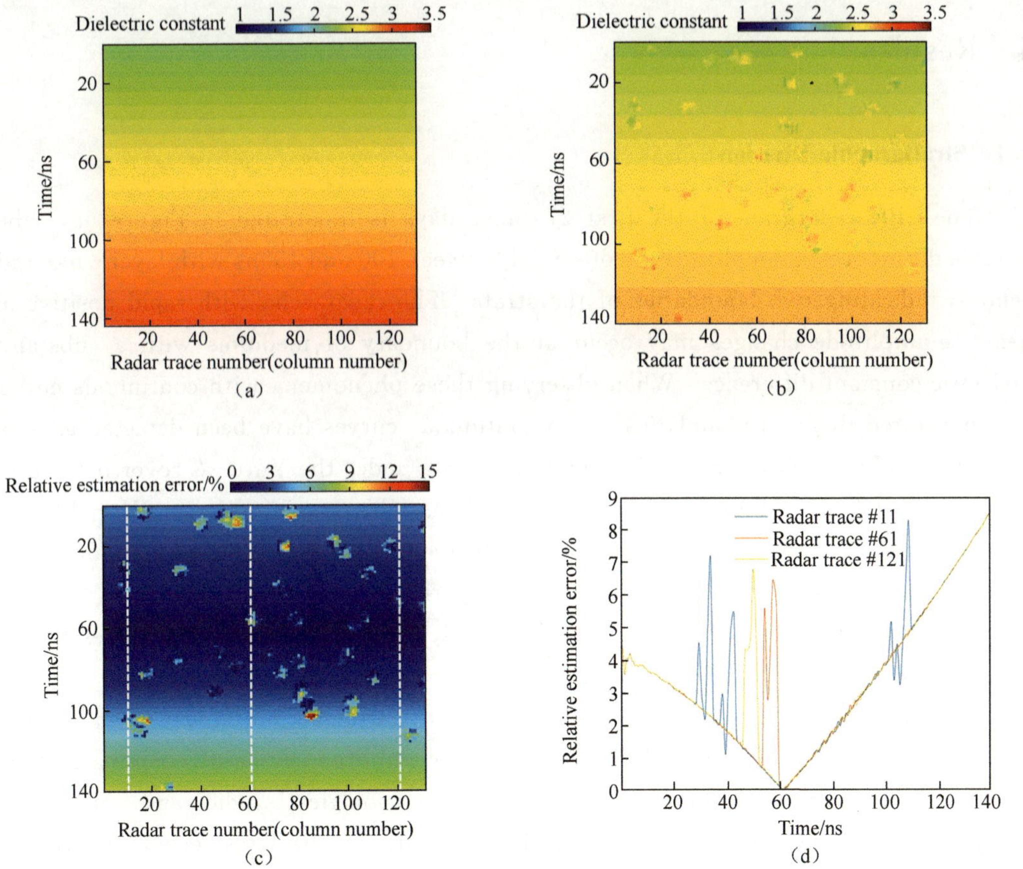

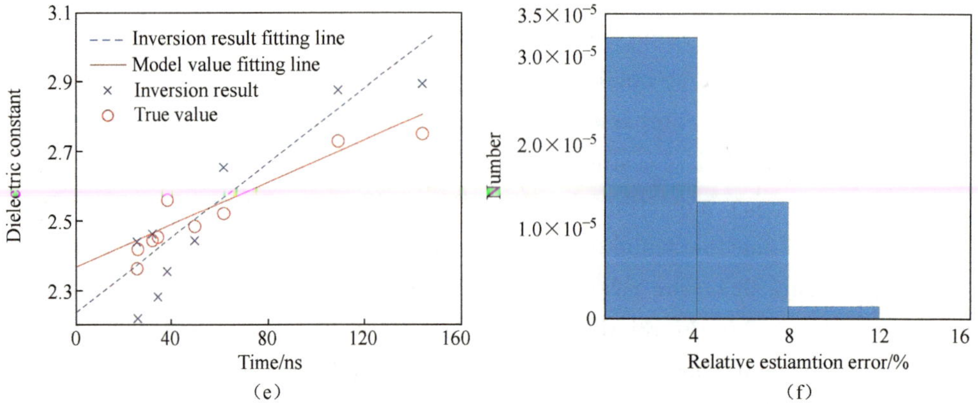

Figure 5　The BLIMP results and their statistics. (a) The BLIMP results; (b) the predefined model dielectric constant in the time and the sampling traces, which were transformed from Figure 6a; (c) the difference map between the estimated and predefined results; (d) the relative inversion errors in the three selected radar traces; (e) the fitting lines of the nine selected hyperbolae; (f) the histogram of the relative errors between the estimated and predefined results

4　Results

4.1　Stratigraphic Division

The LPR radargram of the first 27 lunar days is illustrated in Figure 6a, where notable differences in amplitude are observed between 100 and 450ns with layer-like radar echoes, indicating two boundaries of the strata. The radar echo with rapid positive and negative amplitude changes may occur at the boundary of mediums with a substantial dielectric constant difference. When observing these phenomena with continuous curves, we interpreted them as boundaries. Two continuous curves have been detected as strata boundaries (Figure 6a), and we interpret the strata under the Yutu-2 rover path as two layers (Layer I-II in Figure 6b). After the 450ns, the remaining Layer III yields weak radar signals and strong noise that make it difficult to distinguish the strata information.

1. Layer I (fine-grained regolith): The uppermost layer has a weak ampli tude and a dielectric constant of less than 5.5 (see Section 4.3 for details), which implies that the stratum is composed of fine-grained regolith.

2. Layer II (ejecta material): A stratum with stronger amplitudes than Layer I, which may be caused by rocks. They have a higher amplitude (~15dB) compared to the overlying fine-grained regolith of Layer I. Recent geological studies and multispectral analysis suggest that the ejecta from the Finsen crater is the latest ejecta covering the Chang'E-4 landing site (Huang et al., 2018; Li et al., 2019). Therefore, their ejecta material may come from the Finsen crater. We also observed clusters of weak echo features in Layer II, interpreting them as fine-grained material like Layer I. This might

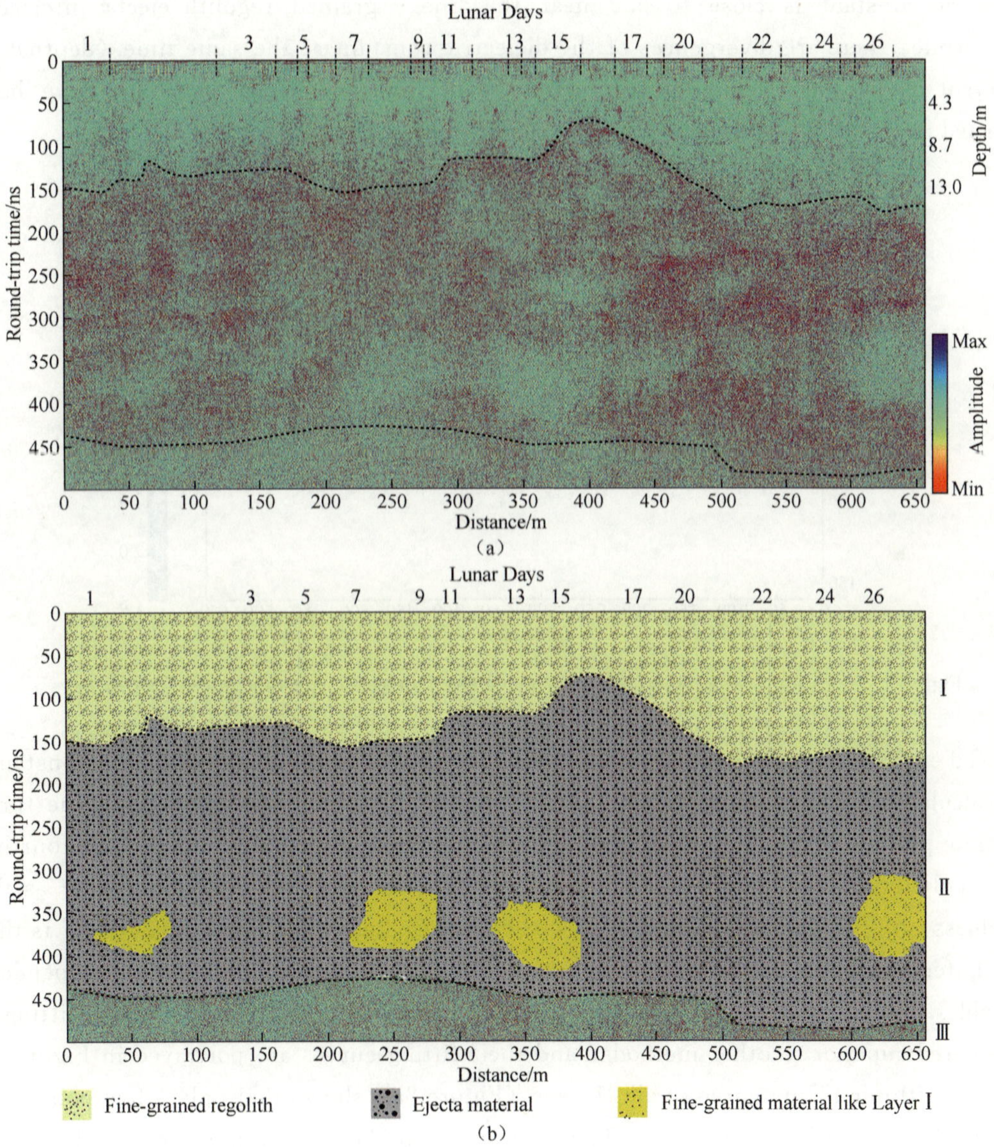

Figure 6　The LPR data in the first 27 lunar days after pre - processing (a), and the stratigraphic division under the Yutu - 2 rover path in the first 27 days (b)

be speculated as a paleoregolith, but more information is needed to support this determination.

4.2　Regolith Dielectric Properties

We applied the hyperbolic fitting method to estimate the regolith dielectric properties. Figure 7 demonstrates 78 discrete reflectors with a near - equidistant distribution in the horizontal direction, where most points are located in the upper 100ns (i. e. , Layer I) . After 20ns, the dielectric constant increases with the round - trip time and the maximum dielectric constant is close to 5.5 near the fine - grained regolith/ejecta interface. Meanwhile, there are divergences of the dielectric constant at the same time (depth) on different days. For example, more points with higher dielectric constant have been detected in the 10th - 17th days with the same round - trip time.

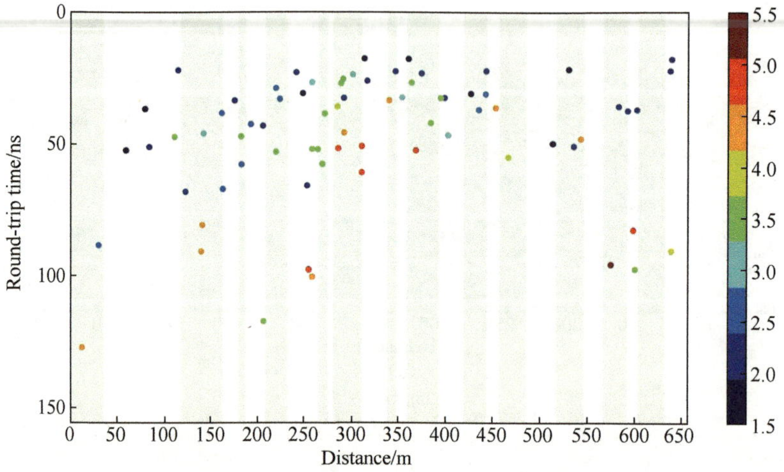

Figure 7　Location of the hyperbolic apexes and their dielectric constant in the first 27 days

To explore the relationship between the round - trip time and the dielectric constant, we calculated best - fit models (a) for the total 27 days and (b) for each of the three separate periods with a 9 - day interval. We tested first -, second -, third - and fourth - order polynomial and logarithmic functions and reported their R^2 and adjusted R^2 as the goodness - of - fit metrics (Table 2) . The difference between R^2 and adjusted R^2 is that the latter evaluates the fitting function accounting for the number of independent variables. We adopt adjusted R^2 as the metric, where the first - order polynomial fitting is considered superior to other methods, and their fitting curves are portrayed in Figure 8. The logarithmic fitting curve of all 27 days (Figure 8a) shows that only a few points are covered by the confidence interval. This indicates that the vertical dielectric profile is not globally consistent, and the regolith dielectric properties are changing along the Yutu - 2 rover path. Figure 8b shows the fitting result of each period, where more points are covered by the corresponding confidence interval. Only a minor discrepancy is observed in the shallow subsurface ($<$ 20ns) for all three periods, and the discrepancy becomes greater when the round - trip time increases, especially between period - 2 and the other two periods.

Table 2　The Fitting Performance of the Calculated Dielectric Constant Based on the Round - Trip Time

Fitting method	Period - 1		Period - 2		Period - 3		Periods 1 - 3	
	R^2	Adjusted R^2	R^2	Adjusted R^2	R^2	Adjusted R^2	R^2	Adjusted R^2
First - order polynomial	0.4597	0.4428	0.6600	0.6464	0.4856	0.4553	0.2969	0.2880

续表

Fitting method	Period - 1		Period - 2		Period - 3		Periods 1 - 3	
	R^2	Adjusted R^2	R^2	Adjusted R^2	R^2	Adjusted R^2	R^2	Adjusted R^2
Second - order polynomial	0.4621	0.4272	0.6602	0.6319	0.4877	0.4237	0.3181	0.3006
Third - order polynomial	0.4658	0.4124	0.6604	0.6161	0.5285	0.4342	0.3202	0.2938
Fourth - order polynomial	0.4839	0.4128	0.6784	0.6199	0.5286	0.3939	0.3264	0.2909
Logarithmic	0.4450	0.4277	0.6458	0.6316	0.4583	0.4264	0.3219	0.3133

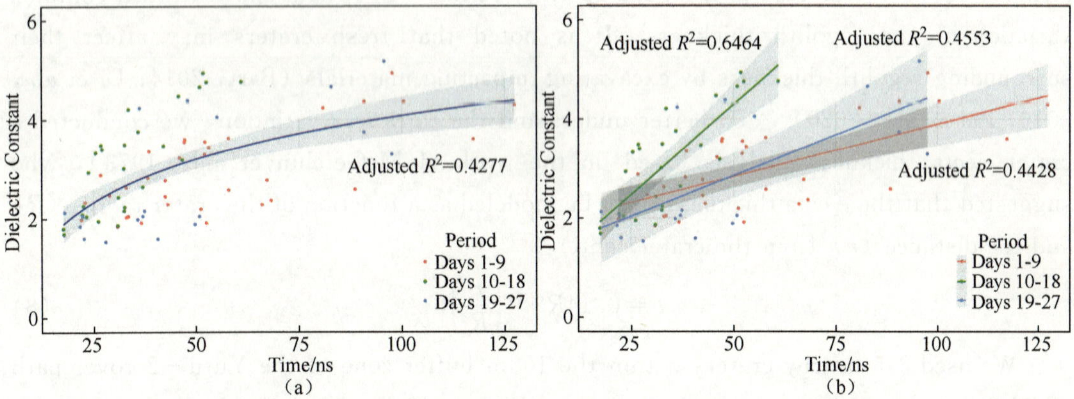

Figure 8 The fitting curves of the vertical dielectric profile. (a) Logarithmic function is the best - fit curves for the whole data set; (b) First - order polynomials are the best - fit curves for the data subset in each of the three periods

4.3 Variation of Regolith Thickness in the Chang'E - 4 Landing Site

We applied the validated BLIMP method to estimate the regolith thickness in the Chang'E - 4 landing site using the LPR data for the first 27 days. We produced the diagram of the bottom curve of Layer I within 150ns because the hyperbolic reflectors were distributed in the upper 150ns, and the range of the estimated dielectric constant was limited to the range of hyperbolic fitting results. Figure 9 presents the dielectric constant profile under the Yutu - 2 moving path, where major differences exist for the three periods. At the same depth, the dielectric constant estimated by BLIMP is the largest in period - 2, followed by period - 1 and period - 3. It is noted that a higher dielectric constant indicates a slower radar propagating velocity, and there is a significant variation in the regolith thickness along the Yutu - 2 moving path as shown by the vertical change of the diagram in Figure 9. Specifically, the 15th - 16th lunar days experienced the shortest round - trip time to the Layer I - II interface (75ns) thus the shallowest regolith, and the 22nd - 27th lunar days had the largest round - trip time (200ns) thus the thick est regolith. After the time - depth conversion, the shallowest regolith thickness was calculated as ~5.3m (the 15th - 16th days) and the deepest was calculated as ~15.0m (the 22nd－27th days).

5 Discussion

5.1 Regolith Thickness Variation

Regolith thickness is a vital parameter to estimate the He^3 reserves (Fa & Jin, 2007) and to understand the origin and evolutionary history of the lunar regolith (Langevin & Arnold, 1977). Our results from the recently released LPR data show a much complex variation of the regolith thickness. It is noted that fresh craters may affect their surrounding regolith thickness by excavating impacting materials (Bart, 2014; Di et al., 2016; Lai et al., 2020). To better understand the thickness variation, we conducted a crater ejecta thickness analysis based on the work of McGetchin et al. (1973), who suggested that the ejecta thickness t can be modeled as a function of the crater radius (R) and the distance (r) from the crater center:

$$t = 0.14R^{0.74}\left(\frac{r}{R}\right)^{-3} \tag{8}$$

We used 315 nearby craters within the 100m buffer zone of the Yutu－2 rover path and 5 distant craters outside the buffer zone (Figure 10) in the ejecta thickness analysis. The central point and the diameter of each crater were determined by Crater Helper Tools for ArcGIS 10 (Kneissl et al., 2011). The impact craters near the Chang'E－4 landing site and the Yutu－2 rover path are small in diameter (＜30m), and the histogram of impact crater diameter is shown in the top panel of Figure 11. The 5 distant craters have larger diameters (more than 200m). We calculated the ejecta thickness by averaging the ejecta layer thickness inferred at each rover position (i.e., Layer II in Figure 6b) with the thickness calculated at each nearby crater from their diameter. These calculated ejecta thicknesses were grouped into lunar days (Figure 11) for subsequent comparison and analysis. The bottom panel of Figure 11 presents the estimated ejecta thickness results, where the ejecta thickness contributed from these craters ranges from 0.45 to 3.06m and the contribution of distant craters fluctuates slightly (0.21－0.35m). Hence, although this ejecta from distant craters may affect the regolith thickness, the thickness difference is small and cannot be the dominant factor in thickness variation. For nearby craters, their ejecta play a more vital role in regolith thickness variation (0.18－2.83m) than that of the distant crater, but this cannot well explain the great thickness variation during the first 27 lunar days. A recent investigation detected a buried crater (～270 diameters) located west of the landing site, and the rover path between the 15th and 20th lunar days passed through the crater edge (Lai et al., 2021). This impact and its ejecta may substantially affect the regolith thickness variation along the Yutu－2 rover path, and hence may explain the variations detected on the 15th and 16th lunar days.

5.2 Further Analysis of the Dielectric Constant - Depth Relationship

Our results confirm that depth is positively related to the dielectric constant, that is, the dielectric constant increases with increasing depth. Based on lunar regolith samples from Apollo 15 - 17 missions, Olhoeft and Strangway (1975) indicated a power law of the dielectric constant - density relationship and then reported the dielectric constant - depth relationship in the "Lunar Sourcebook" as (Carrier et al., 1991) follows:

$$\varepsilon_r' = 1.919^{1.92\frac{z_{cm}+12.2}{z_{cm}+18}} \tag{9}$$

where z_{cm} is the depth (cm) with a dielectric constant range between 2.33 and 3.50.

The largest penetration depths were 236, 224 and 305cm in Apollo 15, 16, and 17, respectively (Carrier, 1974). To match the largest depth in Apollo Missions, the dielectric - depth relationships of this study were compared with those in Equation 9 within a depth of 305cm (Figure 12). The dielectric constant in the best - fit logarithmic and first - order polynomial curves (Figure 8) was derived from the hyperbolic fitting method and is typically referred to as the "average dielectric constant" (Dix, 1955). It represents the average dielectric constant of the radar wave propagation path between the corresponding depth (round - trip time) and the LPR antennas. To compare the average dielectric constant with that of the Apollo samples, we converted the former to an interlayer dielectric constant (dielectric constant of the corresponding layer/depth) by following (Feng et al., 2017):

$$\varepsilon_{\text{interlayer}} = \frac{(t_m - t_n)(\varepsilon_{ave_m}\varepsilon_{ave_n})}{t_m\varepsilon_{ave_n} - t_n\varepsilon_{ave_m}} \tag{10}$$

where $\varepsilon_{\text{interlayer}}$ is the transformed interlayer dielectric constant between layers m and n; t_x and ε_{aue_x} are the round - trip time and average dielectric constant to layer x, respectively.

The original fitting of Figures 12a and 12b is successively in logarithmic and first - order polynomial forms. The similarity of the two curves in Figure 12a suggests that if we do not account for the location and spatial

heterogeneity, the dielectric constant - depth relationship in Chang'E - 4 landing site is similar to the Apollo samples. However, the sharp contrast in Figure 12b and the computed dielectric constant in Figure 7, as well as their fitted curves in Figure 8b, indicate a substantial difference in the dielectric constant across space, which is consistent with the difference in permittivity between measured values of Apollo samples and the best - fit power law (Equation 9), as plotted in Figure 9.52 in Carrier et al. (1991). In future studies of lunar regolith that rely on dielectric constant modeling (e.g., experiments with radar wave processing methods), more attention should be paid to the spatial heterogeneity of dielectric constants.

5.3 Impact of Dielectric Constant on Regolith Thickness Inversion

The dielectric constant determines the propagation velocity of radar waves, thus

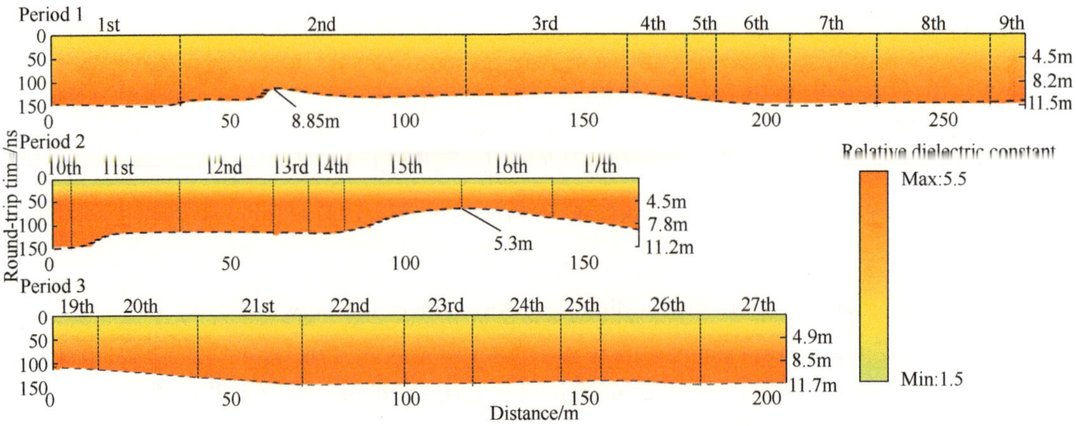

Figure 9　Distribution of the dielectric constant estimated by BLIMP under the Yutu－2 rover path using the best－fit dielectric－depth curve for each period as given in Section 4. 2

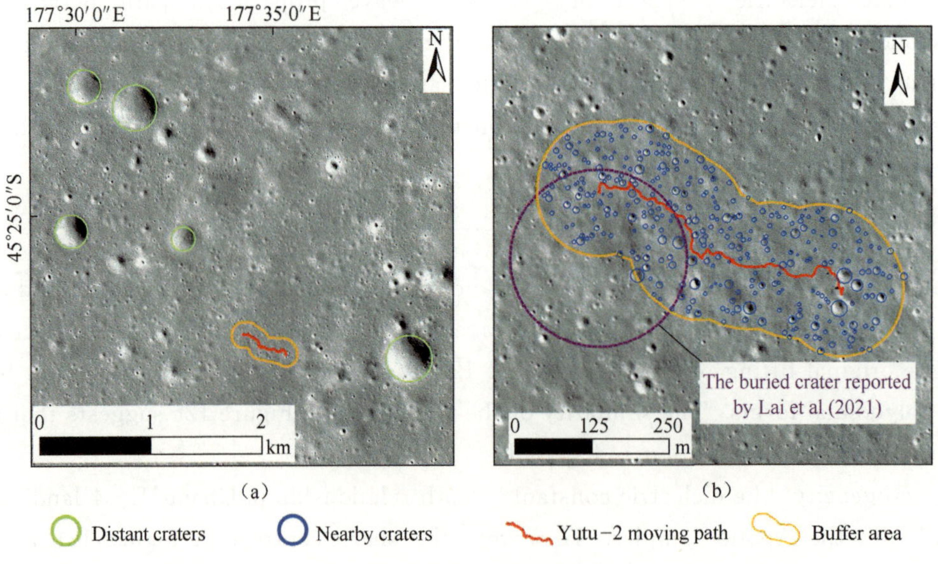

Figure 10　Locations of craters used in ejecta thickness analysis

playing a vital role in the strata thickness inversion. The accurate dielectric data from lunar regolith samples are only available on the lunar nearside, and the samples showed that the dielectric constant increases with the regolith depth due to increasing regolith density (Carrier et al., 1991). From Figures 7 and 8, we observed strong spatial heterogeneity, that is, the relationship between round－trip time and dielectric constant is spatially different, a phenomenon that could not be detected if all data were used to fit at once. For example, the first－order polynomial fitting curves (c. f. Figure 8b) show rapid growth of the green curve (period－2), which is related to the shallow regolith

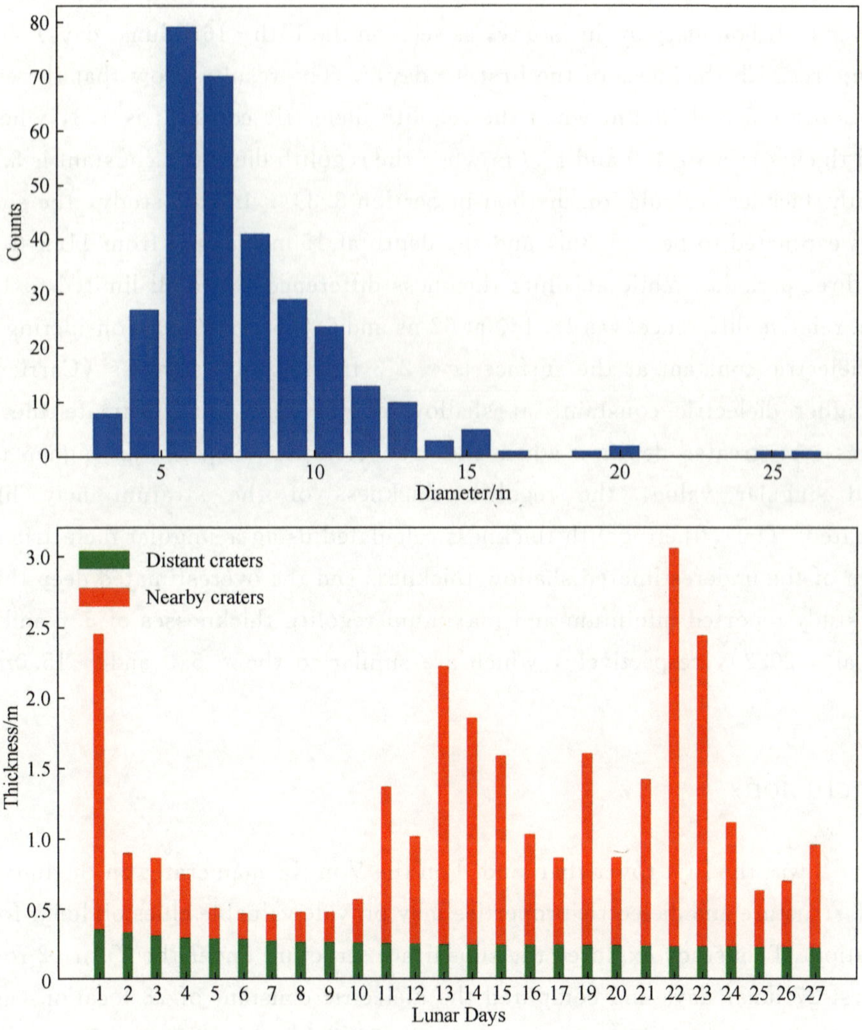

Figure 11　Top panel: the histogram of diameters of nearby craters; Bottom panel: the estimated average ejecta thickness on each lunar day

thickness. This indicates that the regolith density in period - 2 (the second 9 - day) increases with depth more than in the other two periods. The dielectric constant estimated at the same echo time for the second 9 - day is also higher than the first and third 9 - day two periods, which also proves that the regolith thickness is heterogeneous along the 27 - day profile. Hence, the average dielectric constant derived from the hyperbolic fitting method or the analytical results of Apollo samples do not fully characterize the spatial characteristics of the dielectric constant, leading to a possible overestimation or underestimation of the lunar regolith thickness.

To compare the influences of diverse dielectric constants, two of them (i. e. , $\varepsilon_r' = 4.3$ in Lai et al. (2021) and $\varepsilon_r' = 3.52$ in Li et al. (2020)) were used to estimate the regolith thickness. We calculated the regolith thickness at 62ns (the shallowest position of the Layer I - II boundary in this study, as seen on the 15th - 16th lunar days) and 150ns (an average regolith thickness of the first 9 - day) . The results show that the estimated thicknesses are 4.5 and 10.9m when the regolith dielectric constant is 4.3, whereas the estimated thicknesses are 4.9 and 12.0m when the regolith dielectric constant is 3.52 (see the regolith thickness calculation method in Section 3.3) . In this study, the shallowest depth was estimated to be ~5.3m, and the depth at 150ns ranged from 11.2 to 11.7 m over the three periods. While absolute thickness differences were all limited to 1 m, the maximum relative difference was 15.1% at 62 ns and 6.8% at 150ns. Considering that the average dielectric constant at the surface is ~2.3 from Apollo samples (Carrier et al., 1991), higher dielectric constants at shallow depths may underestimate the regolith thicknesses. At greater depths, when the true dielectric constant is higher than the predefined singular value, the regolith thickness of the stratum may likely be overestimated. Thus, the regolith thickness calculated using a singular dielectric constant is the sum of the underestimated shallow thickness and the overestimated deep thickness. A recent study reported minimum and maximum regolith thicknesses of 5.7 and 15.6 m (Chen et al., 2022), respectively, which are similar to the ~5.3 and ~15.0m in our study.

6 Conclusions

Yutu - 2 was the first rover that landed on the Von Kármán crater on the lunar farside where the structure and dielectric properties may provide valuable clues on lunar formation and evolution. This study explored the subsurface structure under the Yutu - 2 rover path for the first 27 lunar days and computed the dielectric constant of 78 locations using the hyperbolic fitting method. The fitting line of the 78 dielectric constant values was used to provide the low - frequency component of the LPR data, which was input into the BLIMP method to compute the 2 - D dielectric properties along the Yutu - 2 rover path.

The results indicated that the subsurface structure of the Chang'E - 4 landing site can be divided into fine - grained regolith and ejecta material layers, and the dielectric constants in the upper fine - grained layer vary spatially. The regolith thickness varies from ~5.3 to ~15.0m along the Yutu - 2 rover path. Ejecta thickness from distant craters fluctuates slightly along the rover path, and nearby craters within the 100 - m buffer may contribute ~2.5m thick ness variation. Considering that the regolith thickness fluctuates in the range of ~10m, a plausible explanation is that the ancient impact near the rover path may have contributed to most of the observed regolith thickness variation. Using a pre - defined 2D dielectric model of the lunar near - surface structure for validation, we found that the uncertainty of the dielectric constant obtained by BLIMP is within 4%, and then applied BLIMP to the LPR radar data to retrieve the 2D dielectric

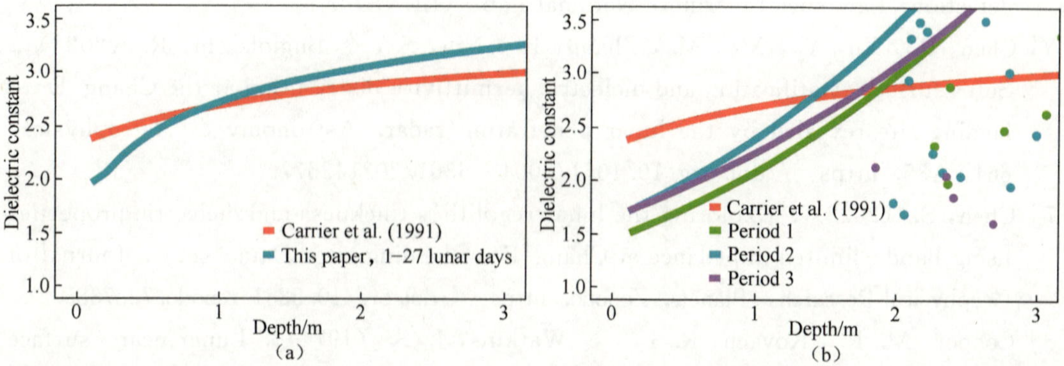

Figure 12 Comparison of the dielectric constant - depth relationship between Carrier et al. (1991) and 1 - 27 lunar days (a), and between Carrier et al. (1991) and those of the three periods (b). The scatter points in (b) are the time - depth conversion data points from Figure 8b, which were used for comparison in the interlayer dielectric constant curves

properties of the subsurface. These dielectric properties collected at the Chang'E - 4 landing site suggest that the dielectric constant - time relationship is varied across space and a detailed preliminary investigation of the dielectric constant is valuable for regolith thickness inversion and further exploration.

Acknowledgements:

The authors would like to thank the two anonymous reviewers for their helpful comments, which greatly improved the quality of our paper.

References

[1] Bart, G. D. (2014). The quantitative relationship between small impact crater morphology and regolith depth. Icarus, 235, 130 - 135. https://doi.org/10.1016/j.icarus.2014.03.020.

[2] Cahill, J. T. S., Thomson, B. J., Patterson, G. W., Bussey, D. B. J., Neish, C. D., Lopez, N. R., et al. (2014). The miniature radio frequency instrument's (Mini - RF) global observations of Earth's Moon. Icarus, 243, 173 - 190. https://doi.org/10.1016/j.icarus.2014.07.018.

[3] Carrier, W. D. (1974). Apollo drill core depth relationships. The Moon, 10 (2), 183 - 194. https://doi.org/10.1007/bf00655719.

[4] Carrier, W. D., Olhoeft, G. R., & Mendell, W. (1991). Physical properties of the lunar surface: Lunar sourcebook: A user's guide to the Moon (Vol. 475).

[5] Chambers, D., Paglieroni, D., Mast, J., & Beer, N. (2013). Real - time vehicle - mounted multistatic ground penetrating radar imaging system for buried object

detection. Lawrence Livermore National Lab. (LLNL).

[6] Chen, R., Xu, Y., Xie, M., Zhang, L., Niu, S., & Bugiolacchi, R. (2022). Sub-surface stratification and dielectric permittivity distribution at the Chang'E-4 landing site revealed by the lunar penetrating radar. Astronomy & Astrophysics, 664, A35. https://doi.org/10.1051/0004-6361/202142677.

[7] Chen, S. (2022). Exploring the lunar regolith's thickness and dielectric properties using band-limited impedance at Chang'E-4 landing site [Data-set]. Journal of Geophysical Research-Planets. Zenodo. https://doi.org/10.5281/zenodo.7237385.

[8] Cooper, M. R., Kovach, R. L., & Watkins, J. S. (1974). Lunar near-surface structure. Reviews of Geophysics, 12 (3), 291-308. https://doi.org/10.1029/rg012i003p00291.

[9] Di, K., Sun, S., Yue, Z., & Liu, B. (2016). Lunar regolith thickness determination from 3D morphology of small fresh craters. Icarus, 267, 12-23. https://doi.org/10.1016/j.icarus.2015.12.013.

[10] Dix, C. H. (1955). Seismic velocities from surface measurements. Geophysics, 20 (1), 68-86. https://doi.org/10.1190/1.1438126.

[11] Dong, Z., Fang, G., Ji, Y., Gao, Y., Wu, C., & Zhang, X. (2017). Parameters and structure of lunar regolith in Chang'E-3 landing area from lunar penetrating radar (LPR) data. Icarus, 282, 40-46. https://doi.org/10.1016/j.icarus.2016.09.010.

[12] Fa, W., & Jin, Y.-Q. (2007). Quantitative estimation of helium-3 spatial distribution in the lunar regolith layer. Icarus, 190 (1), 15-23. https://doi.org/10.1016/j.icarus.2007.03.014.

[13] Fa, W., & Jin, Y.-Q. (2010). A primary analysis of microwave brightness temperature of lunar surface from Chang-E 1 multi-channel radiometer observation and inversion of regolith layer thickness. Icarus, 207 (2), 605-615. https://doi.org/10.1016/j.icarus.2009.11.034.

[14] Fa, W., Zhu, M. H., Liu, T., & Plescia, J. B. (2015). Regolith stratigraphy at the Chang'E-3 landing site as seen by lunar penetrating radar. Geophysical Research Letters, 42 (23), 10179-10187. https://doi.org/10.1002/2015gl066537.

[15] Fang, G.-Y., Zhou, B., Ji, Y.-C., Zhang, Q.-Y., Shen, S.-X., Li, Y.-X., et al. (2014). Lunar penetrating radar onboard the Chang'E-3 mission. Research in Astronomy and Astrophysics, 14 (12), 1607-1622. https://doi.org/10.1088/1674-4527/14/12/009.

[16] Feng, J., Su, Y., Ding, C., Xing, S., Dai, S., & Zou, Y. (2017). Dielectric properties estimation of the lunar regolith at CE-3 landing site using lunar penetrating radar data. Icarus, 284, 424-430. https://doi.org/10.1016/j.icarus.2016.12.005.

[17] Ferguson, R. J., & Margrave, G. F. (1996). A simple algorithm for band-limited impedance inversion (Vol. 8, p. 10).

[18] Fu, X.-H., Qiao, L., Zhang, J., Ling, Z.-C., & Li, B. (2020). The subsurface structure and stratigraphy of the Chang'E-4 landing site: Orbital evidence from small craters on the Von Kármán crater floor. Research in Astronomy and Astrophysics, 20 (1), 008. https://doi.org/10.1088/1674-4527/20/1/8.

[19] Goswami, J. N., & Annadurai, M. (2009). Chandrayaan-1: India's first planetary science mission to the Moon. Current Science, 486-491.

[20] Halekas, J., Delory, G., Lin, R., Stubbs, T., & Farrell, W. (2009). Lunar Prospector measurements of secondary electron emission from lunar regolith. Planetary and Space Science, 57 (1), 78-82. https://doi.org/10.1016/j.pss.2008.11.009.

[21] Hu, W., Di, K., Liu, Z., & Ping, J. (2013). A new lunar global DEM derived from Chang'E-1 Laser Altimeter data based on crossover adjustment with local topographic constraint. Planetary and Space Science, 87, 173-182. https://doi.org/10.1016/j.pss.2013.08.004.

[22] Huang, J., Xiao, Z., Flahaut, J., Martinot, M., Head, J., Xiao, X., et al. (2018). Geological characteristics of Von Kármán crater, northwest-ern South Pole-Aitken basin: Chang'E-4 landing site region. Journal of Geophysical Research: Planets, 123 (7), 1684-1700. https://doi.org/10.1029/2018je005577.

[23] Huang, J., Xiao, Z., Xiao, L., Horgan, B., Hu, X., Lucey, P., et al. (2020). Diverse rock types detected in the lunar South Pole-Aitken Basin by the Chang'E-4 lunar mission. Geology, 48 (7), 723-727. https://doi.org/10.1130/g47280.1.

[24] Hurwitz, D. M., & Kring, D. A. (2014). Differentiation of the South Pole-Aitken basin impact melt sheet: Implications for lunar exploration. Journal of Geophysical Research: Planets, 119 (6), 1110-1133. https://doi.org/10.1002/2013je004530.

[25] Jolliff, B. L., Gillis, J. J., Haskin, L. A., Korotev, R. L., & Wieczorek, M. A. (2000). Major lunar crustal terranes: Surface expressions and crust-mantle origins. Journal of Geophysical Research, 105 (E2), 4197-4216. https://doi.org/10.1029/1999je001103.

[26] Kato, M., Sasaki, S., Tanaka, K., Iijima, Y., & Takizawa, Y. (2008). The Japanese lunar mission SELENE: Science goals and present status. Advances in Space Research, 42 (2), 294-300. https://doi.org/10.1016/j.asr.2007.03.049.

[27] Kneissl, T., van Gasselt, S., & Neukum, G. (2011). Map-projection-independent crater size-frequency determination in GIS environments—New software tool for ArcGIS. Planetary and Space Science, 59 (11), 1243-1254.

https：//doi. org/10. 1016/j. pss. 2010. 03. 015.

[28] Kobayashi，T.，Kim，J. -H.，& Lee，S. R.（2018）. HF（5MHz）imaging of the Moon by Kaguya lunar radar sounder off nadir echo data. IEEE Transactions on Geoscience and Remote Sensing，56（7），3709 - 3714. https：//doi. org/10. 1109/tgrs. 2018. 2806910.

[29] Krupenio，N.，& Kroupenio，N.（1973）. Results of radar experiments performed on automatic stations Luna 16 and Luna 17. Proceedings space research conference（Vol. 2，pp. 969 - 973）.

[30] Lai，J.，Xu，Y.，Bugiolacchi，R.，Meng，X.，Xiao，L.，Xie，M.，et al.（2020）. First look by the Yutu - 2 rover at the deep subsurface structure at the lunar farside. Nature Communications，11（1），3426. https：//doi. org/10. 1038/s41467 - 020 - 17262 - w.

[31] Lai，J.，Xu，Y.，Bugiolacchi，R.，Wong，H. K.，Xu，L.，Zhang，X.，et al.（2021）. A complex paleo - surface revealed by the Yutu - 2 rover at the lunar farside. Geophysical Research Letters，48（20）. https：//doi. org/10. 1029/2021gl095133.

[32] Lai，J.，Xu，Y.，Zhang，X.，& Tang，Z.（2016）. Structural analysis of lunar subsurface with Chang'E - 3 lunar penetrating radar. Planetary and Space Science，120，96 - 102. https：//doi. org/10. 1016/j. pss. 2015. 10. 014.

[33] Lai，J.，Xu，Y.，Zhang，X.，Xiao，L.，Yan，Q.，Meng，X.，et al.（2019）. Comparison of dielectric properties and structure of lunar regolith at Chang'E - 3 and Chang'E - 4 landing sites revealed by ground - penetrating radar. Geophysical Research Letters，46（22），12783 - 12793. https：// doi. org/10. 1029/2019gl084458.

[34] Langevin，Y.，& Arnold，J.（1977）. The evolution of the lunar regolith. Annual Review of Earth and Planetary Sciences，5（1），449 - 489. https：// doi. org/10. 1146/annurev. ea. 05. 050177. 002313.

[35] Latham，G. V.，Ewing，M.，Press，F.，Sutton，G.，Dorman，J.，Nakamura，Y.，et al.（1970）. Passive seismic experiment. Passive Seismic Experiment：Science，167（3918），455 - 457. https：//doi. org/10. 1126/science. 167. 3918. 455.

[36] Lawrence，D. J.，Feldman，W. C.，Elphic，R. C.，Little，R. C.，Prettyman，T. H.，Maurice，S.，et al.（2002）. Iron abundances on the lunar surface as measured by the Lunar Prospector gamma - ray and neutron spectrometers. Journal of Geophysical Research，107（E12），13 - 11 - 13 - 26. http s：//doi. org/10. 1029/2001je001530.

[37] Leise，J. A.（1981）. A segmented editing and detrending method for discrete Fourier transforms.

[38] Li，C.，Liu，D.，Liu，B.，Ren，X.，Liu，J.，He，Z.，et al.（2019）. Chang'E - 4 initial spectroscopic identification of lunar far - side mantle - derived materials. Nature，569（7756），378 - 382. https：//doi. org/10. 1038/s41586 - 019 - 1189 - 0.

[39] Li, C. , Liu, J. , Ren, X. , Mou, L. , Zou, Y. , Zhang, H. , et al. (2010) . The global image of the Moon obtained by the Chang'E - 1: Data processing and lunar cartography. Science China Earth Sciences, 53 (8), 1091 - 1102. https://doi.org/10.1007/s11430 - 010 - 4016 - x.

[40] Li, C. , Su, Y. , Pettinelli, E. , Xing, S. , Ding, C. , Liu, J. , et al. (2020) . The Moon's farside shallow subsurface structure unveiled by Chang'E - 4 Lunar Penetrating Radar. Science Advances, 6 (9), eaay6898. https://doi.org/10.1126/sciadv.aay6898.

[41] Li, J. , Zeng, Z. , Liu, C. , Huai, N. , & Wang, K. (2017) . A study on lunar regolith quantitative random model and lunar penetrating radar parameter inversion. IEEE Geoscience and Remote Sensing Letters, 14 (11), 1953 - 1957. https://doi.org/10.1109/lgrs.2017.2743618.

[42] Lucey, P. G. , Taylor, G. J. , Hawke, B. R. , & Spudis, P. D. (1998) . FeO and TiO_2 concentrations in the South Pole - Aitken basin: Implications for mantle composition and basin formation. Journal of Geophysical Research, 103 (E2), 3701 - 3708. https://doi.org/10.1029/97je03146.

[43] McGetchin, T. R. , Settle, M. , & Head, J. W. (1973) . Radial thickness variation in impact crater ejecta: Implications for lunar basin deposits. Earth and Planetary Science Letters, 20 (2), 226 - 236. https://doi.org/10.1016/0012 - 821x (73) 90162 - 3.

[44] McKay, D. S. , Heiken, G. , Basu, A. , Blanford, G. , Simon, S. , Reedy, R. , et al. (1991) . The lunar regolith. Lunar Sourcebook, 567, 285 - 356.

[45] Melosh, H. J. , Kendall, J. , Horgan, B. , Johnson, B. C. , Bowling, T. , Lucey, P. G. , & Taylor, G. J. (2017) . South Pole - Aitken basin ejecta reveal the Moon's upper mantle. Geology, 45 (12), 1063 - 1066. https://doi.org/10.1130/g39375.1.

[46] Moriarty, D. P. , III, & Pieters, C. M. (2018) . The character of South Pole - Aitken Basin: Patterns of surface and subsurface composition. Journal of Geophysical Research: Planets, 123 (3), 729 - 747. https://doi.org/10.1002/2017je005364.

[47] Moriarty, D. P. , Pieters, C. M. , & Isaacson, P. J. (2013) . Compositional heterogeneity of central peaks within the South Pole - Aitken Basin. Journal of Geophysical Research: Planets, 118 (11), 2310 - 2322. https://doi.org/10.1002/2013je004376.

[48] Muhleman, D. O. , Brown, W. E. , Jr. , Davids, L. , Negus de Wys, J. , & Peake, W. H. (1969) . Lunar surface electromagnetic properties (Vol. 184, p. 203) . NASA Special Publication.

[49] Nakamura, R. , Matsunaga, T. , Ogawa, Y. , Yamamoto, S. , Hiroi, T. , Saiki, K. , et al. (2009) . Ultramafic impact melt sheet beneath the South Pole - Aitken

basin on the Moon. Geophysical Research Letters, 36 (22), L22202. https: //doi. org/10. 1029/2009gl040765.

[50] Nozette, S. , Lichtenberg, C. L. , Spudis, P. , Bonner, R. , Ort, W. , Malaret, E. , et al. (1996) . The clementine bistatic radar experiment. Science, 274 (5292), 1495 - 1498. https: //doi. org/10. 1126/science. 274. 5292. 1495.

[51] Ohtake, M. , Uemoto, K. , Yokota, Y. , Morota, T. , Yamamoto, S. , Nakamura, R. , et al. (2014) . Geologic structure generated by large - impact basin formation observed at the South Pole - Aitken basin on the Moon: Rock and melt pool within the SPA Basin. Geophysical Research Letters, 41 (8), 2738 - 2745. https: //doi. org/10. 1002/2014gl059478.

[52] Oldenburg, D. , Scheuer, T. , & Levy, S. (1982) . Recovery of acoustic impedance from reflection seismograms (pp. 202 - 204) . SEG Technical Program Expanded Abstracts 1982, Society of Exploration Geophysicists.

[53] Olhoeft, G. R. , Frisillo, A. L. , Strangway, D. W. , & Sharpe, H. (1974) . Temperature dependence of electrical conductivity and lunar temperatures. The Moon, 9 (1), 79 - 87. https: //doi. org/10. 1007/bf00565394.

[54] Olhoeft, G. R. , & Strangway, D. W. (1975) . Dielectric properties of the first 100 meters of the Moon. Earth and Planetary Science Letters, 24 (3), 394 - 404. https: //doi. org/10. 1016/0012 - 821x (75) 90146 - 6.

[55] Rochette, P. ,Gattacceca, J. , Ivanov, A. , Nazarov, M. , & Bezaeva, N. (2010). Magnetic properties of lunar materials: Meteorites, Luna and Apollo returned samples. Earth and Planetary Science Letters, 292 (3 - 4), 383 - 391. https: //doi. org/10. 1016/j. epsl. 2010. 02. 007.

[56] Schmelzbach, C. , Tronicke, J. , & Dietrich, P. (2012) . High - resolution water content estimation from surface - based ground - penetrating radar reflection data by impedance inversion. Water Resources Research, 48 (8), W08505. https: //doi. org/10. 1029/2012wr011955.

[57] Schultz, P. H. , & Crawford, D. A. (2011) . Origin of nearside structural and geochemical anomalies on the Moon, recent advances and current research issues in lunar stratigraphy. Geological Society of America.

[58] Stuart - Alexander, D. E. (1978) . Geologic map of the central far side of the Moon (p. 1047) . US Geological Survey Report.

[59] Sun, L. , Ling, Z. , Zhang, J. , Li, B. , Chen, J. , Wu, Z. , & Liu, J. (2016) . Lunar iron and optical maturity mapping: Results from partial least squares modeling of Chang'E - 1 IIM data. Icarus, 280, 183 - 198. https: //doi. org/10. 1016/j. icarus. 2016. 07. 010.

[60] Tran, T. , Rosiek, M. , Beyer, R. A. , Mattson, S. , Howington - Kraus, E. , Robinson, M. , et al. (2010) . Generating digital terrain models using LROC NAC images. In Proceedings at ASPRS/CaGIS 2010 fall specialty conference.

[61] Vondrak, R., Keller, J., Chin, G., & Garvin, J. (2010). Lunar reconnaissance orbiter (LRO): Observations for lunar exploration and science. Space Science Reviews, 150 (1-4), 7-22. https://doi.org/10.1007/s11214-010-9631-5.

[62] Wang, K., Zeng, Z., Zhang, L., Xia, S., & Li, J. (2018). A Compressive sensing-based approach to reconstructing regolith structure from lunar penetrating radar data at the Chang'E-3 landing site. Remote Sensing, 10 (12), 1925. https://doi.org/10.3390/rs10121925.

[63] Wang, Z., Li, Y., Jiang, J., & Li, D. (2010). Lunar surface dielectric constant, regolith thickness, and 3He abundance distributions retrieved from the microwave brightness temperatures of CE-1 Lunar Microwave Sounder. Science China Earth Sciences, 53 (9), 1365-1378. https://doi.org/10.1007/s11430-010-4022-z.

[64] Yingst, R., Chuang, F., Berman, D., & Mest, S. (2017). Geologic mapping of the Planck Quadrangle of the Moon (LQ-29). In Proceedings lunar and planetary science conference (pp. 1680).

[65] Zeng, Z.-F., Xiong, C., Jing, L., Chen, L.-N., Qi, L., & Liu, F.-S. (2015). Recursive impedance inversion of ground-penetrating radar data in stochastic media. Applied Geophysics, 12 (4), 615-625. https://doi.org/10.1007/s11770-015-0514-0.

[66] Zhang, H.-B., Zheng, L., Su, Y., Fang, G.-Y., Zhou, B., Feng, J.-Q., et al. (2014). Performance evaluation of lunar penetrating radar onboard the rover of CE-3 probe based on results from ground experiments. Research in Astronomy and Astrophysics, 14 (12), 1633-1641. https://doi.org/10.1088/1674-4527/14/12/011.

[67] Zhang, J., Zhou, B., Lin, Y., Zhu, M.-H., Song, H., Dong, Z., et al. (2021). Lunar regolith and substructure at Chang'E-4 landing site in South Pole-Aitken basin. Nature Astronomy, 5 (1), 25-30. https://doi.org/10.1038/s41550-020-1197-x.

[68] Zhang, L., Li, J., Zeng, Z., Xu, Y., Liu, C., & Chen, S. (2020). Stratigraphy of the Von Kármán crater based on Chang'E-4 lunar penetrating radar data. Geophysical Research Letters, 47 (15), e2020GL088680. https://doi.org/10.1029/2020gl088680.

[69] Zuo, W., Li, C., & Zhang, Z. (2014). Scientific data and their release of Chang'E-1 and Chang'E-2. Chinese Journal of Geochemistry, 33 (1), 24-44. https://doi.org/10.1007/s11631-014-0657-3.

The subsurface structure and stratigraphy of the Chang'E – 4 landing site: Orbital evidence from small craters on the Von Kármán crater floor

Xiao – Hui Fu[1,2], Le Qiao[1], Jiang Zhang[1], Zong Cheng Ling[1], Bo Li[1,3]

1. Shandong Provincial Key Laboratory of Optical Astronomy and Solar – Terrestrial Environment, Institute of Space Sciences, Shandong University, Weihai 264209, China

2. CAS Center for Excellence in Comparative Planetology, China

3. State Key Laboratory of Remote Sensing Science, Institute of Remote Sensing and Digital Earth, Chinese Academy of Sciences, Beijing 100101, China

Abstract: Chang'E – 4 (CE – 4) successfully landed on the floor of Von Kármán crater within the South Pole – Aitken basin (SPA). One of its scientific objectives is to determine the sub – surface structure and thickness of lunar regolith at the landing site and along the traverse route of the Yutu – 2 rover. Using orbital data, we employed small craters (diameters <1km) on the floor of Von Kármán crater as probes to investigate the subsurface struc – ture and stratigraphy of the CE – 4 landing site. In this study, 40 dark – haloed craters that penetrate through the surface Finsen ejecta and excavate underlying mare deposits were identified, and 77 bright ray craters that expose only the underlying fresh materials but do not penetrate through the surface Finsen ejecta were found. The excavation depths of these craters and their distances from the Finsen crater center were calculated, and the thickness distribution of Finsen ejecta on the Von Kármán floor was systematically investigated. The boundary between Finsen ejecta and underlying mare basalt at the CE – 4 landing site is constrained to a depth of 18m. We proposed the stratigraphy for the CE – 4 site and interpreted the origins of different layers and the geologic history of Von Kármán crater. These results provide valuable geological background for interpreting data from the Lunar Penetrating Radar (LPR) and Visible and Near – infrared Imaging Spectrometer (VNIS) on the Yutu – 2 rover. The CE – 4 landing site could provide a reference point for crater ejecta distribution and mixing with local materials to test and improve the ejecta thickness models according to the in situ measurements of the CE – 4 LPR.

Keywords: Chang'E – 4; dark – haloed crater; ejecta thickness; Moon

1 Introduction

On the 3rd January 2019, Chang'E – 4 (CE – 4) safely landed in the eastern Von Kármán crater (186km in diameter; central coordinates 44.4°S, 176.2°E) within the 2500 – km – diameter South Pole – Aitken basin (SPA) (Li et al. 2019; Di et al. 2019). CE – 4 is the first in situ exploration mission on the lunar farside as well as in the SPA basin. One scientific object of this mission is to determine the mineral compositions of surface materials and perform geological characterization of the landing site (Jia et al. 2018). The Visible and Near – infrared Imaging Spectrometer (VNIS) and Lunar Penetrating Radar (LPR) aboard the Yutu – 2 rover are the two key instruments to reveal the surface composition and subsurface structure of the landing site and the geological history of this region (Jia et al. 2018).

Von Kármán crater is a perfect location to investigate both lunar farside volcanism and deep – seated materials possibly from the lunar upper mantle. The SPA is an ancient impact structure on the lunar farside (Wilhelms et al. 1979), which shows distinctive geochemical characteristics and has a unique geological history (e.g., Pieters et al. 1997; Lucey et al. 1998; Jolliff et al. 2000). Except for ultramafic mantle materials exposure, previous studies also suggest the presence of differentiated impact melts at the bottom of the SPA basin (e.g., Morrison 1998; Hurwitz & Kring 2014; Vaughan & Head 2014). Von Kármán crater lies within the Mg pyroxene annulus defined by Moriarty Ⅲ & Pieters (2018). Its floor has been filled by mare basalts in the Late Imbrian Period (Pasckert et al. 2018). The basalt plains were subsequently reshaped by ejected materials from surrounding young craters. Both mare basalts and mafic materials with abundant Mg rich pyroxene are exposed on the floor of Von Kármán crater. The morphology, mineralogy, and geochemistry of Von Kármán crater were investigated (Huang et al. 2018; Ling et al. 2018, 2019; Qiao et al. 2019) using orbital data because it had been listed as the landing site of CE – 4 mission (Jia et al. 2018).

The subsurface structure of the CE – 4 landing site contains a direct record of the sequence of major geological events and the geological history of this area. The impact structures in the regional vicinity of Von Kármán crater span the vast expanse of geological time, from the Pre – Nectarian SPA basin (Wilhelms et al. 1979) to the Nectarian Von Kármán crater (Stuart – Alexander 1978), the Imbrian Alder crater (82km in diameter) (Wilhelms et al. 1987) and the Copernican Finsen crater (73km in diameter) (Ling et al. 2019). Superpositions among these crater materials and mare basalt units establish local stratigraphic sequences. From lunar surface mineralogy and morphology analysis (Pasckert et al. 2018; Huang et al. 2018; Qiao et al. 2019), it has been widely accepted that the Finsen crater is the main source of surface ejecta in the eastern portion of Von Kármán crater floor as well as at the CE – 4 landing site (Pasckert et al. 2018; Li et al.

2019). However, several key questions about the subsurface structure and stratigraphy are still open. What is the thickness of Finsen ejecta? How much ejected materials did these craters deliver to the Von Kármán crater floor? Are the Alder ejecta below or above the mare basalt unit? The stratigraphic investigations of the CE-4 landing site allow reconstruction of the general sequence of geological events in the region.

Small craters can penetrate through lunar surface geological units and act as probes into the under-lying materials. Such craters have been widely used to determine the thickness of the surface unit and the compositions of lava flows, impact melt sheets beneath them and other hidden layers (Budney & Lucey 1998; Chen et al. 2018). Dozens of small craters (less than 2km) are found on the floor of Von Kármán crater. Based on these small crater studies, Huang et al. (2018) reported that the thickness of the 'fine grain' regolith varies from ~2.5 to 7.5m by the method of Quaide & Oberbeck (1968) in the landing region. In the present study, we attempt to use these craters to derive the thickness of Finsen ejecta and constrain the subsurface structure of the CE-4 landing site.

The small craters on the Von Kármán floor can be categorized into two distinctive groups by the ejected material albedo around individual craters. One group is the craters surrounded by bright rays or ejecta patterns. These bright ray craters (BRCs) are relatively young. The bright continuous ejecta blanket is likely formed due to fresh material excavated by the primary crater. The other group comprises craters surrounded by low-albedo deposits known as dark-haloed craters (DHCs) (Head & Wilson 1979; Salisbury et al. 1968). The DHCs are interpreted as resulting from the excavation of low-albedo material from subsurface, such mare basalts, glass materials or impact melts, although the nature and origin of DHCs are not completely understood (Head & Wilson 1979; Kaydash et al. 2014). The presence of DHCs has been taken as an indicator of ancient lunar volcanic deposits or cryptomaria (Bell & Hawke 1984; Whitten & Head 2015).

The crater excavation method can assess the thickness of Finsen ejecta and determine the vertical boundary between Finsen ejecta and mare basalts. On the Von Kármán floor, DHCs are craters that punch through overlying ejecta to excavate underlying mare material. BRCs expose only the underlying fresh materials but do not exhume the mare basalt layer below Finsen ejecta. Coordinated analyses of the crater excavation depths and their distances from the Finsen crater center can constrain the ejecta thickness.

To test this hypothesis, we searched the small craters in this area and identified the DHCs and BRCs. Their diameters (as well as estimated excavation depths) and distances from Finsen crater were collected. Using these orbital datasets, we investigated the thickness and distribution of the Finsen ejecta on the Von Kármán floor. This study provides substantial geological background for CE-4 exploration, such as the stratigraphic sequences and major geological events in the SPA. The thickness estimates for different geological units at the CE-4 landing site are robust. Future measurements by the Yutu - 2 rover will provide further insights into the subsurface setting of this area.

2 Data and Methods

2.1 Data

We generated regional data mosaics of Lunar Reconnaissance Orbiter (LRO) Wide - Angle Camera (LROC WAC, 100m/pixel, Robinson et al. 2010), Narrow - Angle Camera (NAC, up to 0.60m/pixel, Robinson et al. 2010), and Kaguya Terrain Camera (TC) Ortho mosaic (10m pixel size; Haruyama et al. 2008) to perform the identification and topographic analysis of small craters on the Von Kármán floor. More details about WAC and NAC image processing are described in Qiao et al. (2019). LROC NAC images, the highest spatial resolutions (0.5—2m) of the lunar surface among all available orbital data, were acquired at different incidence angles. The NAC images covering the Von Kármán crater are all at moderate/high solar incidence angles (> 40°).

The global mineral abundance maps (60m/pixel) produced by Lemelin et al. (2016) from Kaguya MI spectrometer data were also used here. Quantitative mineralogical analysis for major rock - forming minerals (olivine, orthopyroxene, clinopyroxene and plagioclase) is obtained through radiative transfer modeling calculations. The Kaguya MI - extracted mineral abundance has a $1\sigma = 7$ wt% uncertainty from the Lunar Soil Characterization Consortium (LSCC) data (Lemelin et al. 2019). Kaguya MI - derived silicate abundance maps allow detailed mineralogical characterization of the Von Kármán floor and identification of small craters.

The identification of the DHCs and BRCs on the Von Kármán crater floor included three steps. First, the DHCs and BRCs were identified in the Kaguya TC mosaics, which provide nearly 100% coverage of the Von Kármán floor. Second, the DHCs and BRCs were confirmed on the NAC images with solar incidence angles less than 55°. In the third step, the identified DHCs and BRCs were re - checked on the Kaguya MI - extracted mineral abundance maps. DHCs on the Von Kármán floor should expose basaltic materials that are obviously different from the surrounding Finsen ejecta. The materials excavated by BRCs usually show mineral compositions similar to those of the surrounding terrains. This similarity occurs because these small craters do not penetrate the uppermost Finsen ejecta. Accurate measurements of small crater sizes were also performed based on Kaguya TC images and LROC NAC images.

In the present study, crater identifications and diameter measurements were performed manually using the CraterTools extension in ArcMap (Kneissl et al. 2011). This tool has been extensively used in crater counting and diameter measurements in planetary science (e.g., Head et al. 2010). Crater diameter measurements are defined on the basis of a best - fit circle, fit to the crater rim based on three points selected along the rim. The center of this circle was taken as the crater position.

2.2 Ejecta thickness models and crater excavation depth

The radial variations in Finsen ejecta thickness have also been estimated by ejecta thickness models. Based on observed ejecta thicknesses from explosion craters, terrestrial craters, and laboratory craters and on observations of ejecta from Copernicus and the Imbrium basin, several models have been developed to estimate the thickness of ejecta at a given distance from a crater. All these models are summarized in Tab. 1. In these equations, t is ejecta thickness, r is the distance from the center of the crater to a given location, and R_t is the radius of the transient crater. All these variables are in meters. For the present study, we utilized all these models to estimate the Finsen ejecta distributions on the Von Kármán crater floor. More details about these models are discussed in Section 4.

In the present study, the excavation depth (H_{exc}) of each crater was calculated with the following equations in Melosh (1989): $H_{exc} = 0.1D_t$, $D_t = 0.84D$, where D_t and D are the diameter of the transient crater and the final diameter, respectively. The final diameter of a small crater was obtained by the CraterTools extension in ArcMap. For the Finsen crater, the diameter (D) is 73km, and the R_t is 30.66km.

Tab. 1　Multiple models to estimate the thickness of ejecta

	Equations	References
1	$t = 0.14R_t^{0.74}(r/R_t)^{-3.0}$	McGetchin et al. (1973)
2	$t = 0.033R_t(r/R_t)^{-3.0}$	Pike (1974)
3	$t = 0.0078R_t(r/R_t)^{-2.61}$	Housen et al. (1983); Petro & Pieters (2008)
4	$t = 3.95(\pm 1.19)R_t^{0.399}(r/R_t)^{-3.0}$	Sharpton (2014)

2.3 Description of the study area

Von Kármán crater is a Nectarian crater (Stuart - Alexander 1978) with degraded topographic and morphologic features (Fig. 1). This crater is surround by several other large impact craters: Leibnitz (~3.88Ga, D= 236km, Yingst et al. 2017) in the north, Finsen in the northeast, Von Kármán M (D=225km) /L/L' in the south, Alder in the southeast, and Oresme (Pre - Nectarian, D=73km, Wilhelms et al. 1979) in the west (Fig. 1). The Oresme crater, characterized by a damaged outer rim, is possibly older than the Von Kármán crater based on the geological map by Wilhelms et al. (1979). The Von Kármán M crater is heavily destroyed by Von Kármán and thus relatively older. The ejecta of the Leibnitz crater bury the north rim crest of the Von Kármán crater, indicating its relatively young age. Clearly, Leibnitz and other young craters all delivered significant amounts of ejected materials onto the floor of Von Kármán crater (Fig. 1).

The Finsen ejecta and mare basalts that filled in the Von Kármán crater exhibit

Fig. 1　LROC WAC image mosaic of the Von Kármán crater within the South Pole - Aitken (SPA) basin. The bold white circle marks the rim location of the Von Kármán crater, and the dotted white circles mark the rim locations of the Von Kármán M, Leibnitz, Oresme and Finsen craters. The yellow square marks the position of Figure 2. The projection is a Lambert conformal projection centered at 176° E, 44.5° S, and north is up

different mineral and chemical compositions. The eastern portion of the Von Kármán crater floor is homogeneous in surface FeO (11 — 13 wt%) and TiO_2 (1 — 2 wt%) contents using Kaguya MI - derived elemental maps (Qiao et al. 2019). The mare deposits exposed by the Zhinyu crater show relatively elevated FeO abun dances (13—18 wt%) and broader TiO_2 (1 — 3 wt%) range. Nearby mare materials show relatively higher average olivine (9± 3wt %) and clinopyroxene (16±4 wt%) abundances (Fig. 2). Finsen ejecta contain more feldspathic components, with an estimated average plagioclase abundance of 45±4 wt% (Qiao et al. 2019) (Fig. 2).

We define the study area for the small crater search using the Kaguya MI - derived plagioclase abun dance map. Finsen ejecta are characterized by relatively elevated plagioclase abundance. On the plagioclase abundance map (Fig. 2B), Finsen ejecta show an unique yellow tone. We define this area as the continuous ejecta blanket of the Finsen

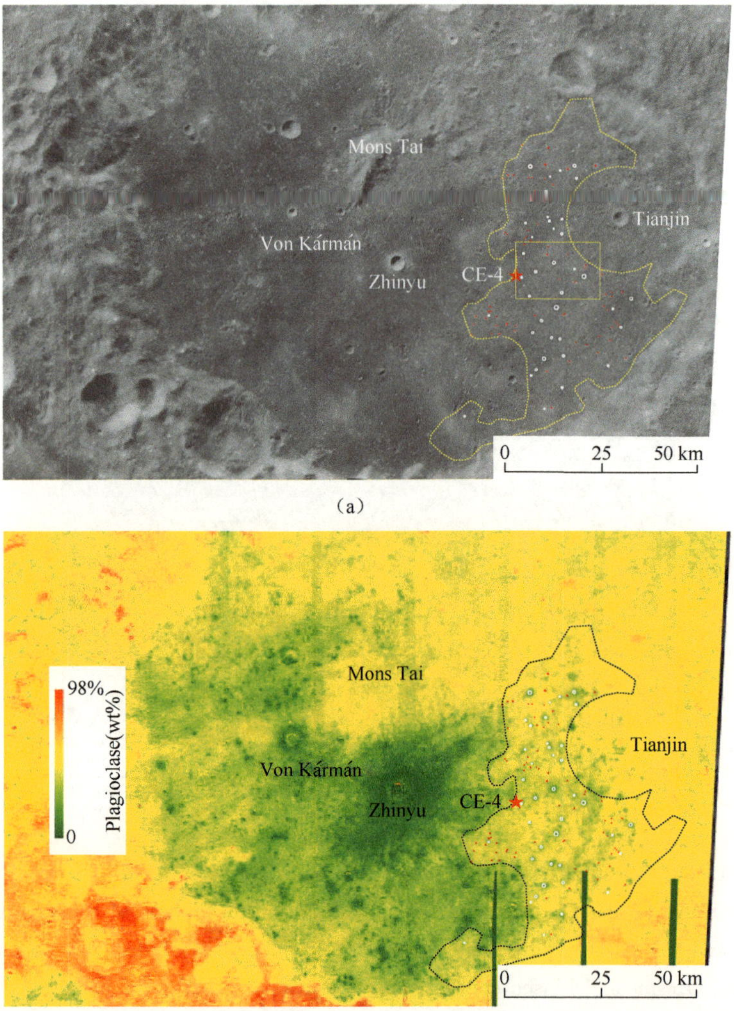

(a)

(b)

Fig. 2　Geological context of the study area and the identified small craters. Panel A is the Kaugya TC mosaic; Panel B shows the Kaguya MI - retrieved abundance mapping of plagio - clase in the study area (Lemelin et al. 2016) . The dotted line indicates the study area. The radial ejecta of the Zhinyu crater and 15. 2 - km (4x its diameter) regions around the Tianjin crater are excluded for the following small crater mapping to minimize the effects on the es - timation of Finsen ejecta thickness. The white and red circles represent the DHCs and BRCs identified in the present study. The yellow square marks the position of Fig. 3. All the maps in this study are projected into the Lambert conformal 225 projection, centered at 176° E, 44. 5° S, and north is up

crater (dotted line polygon in Fig. 2) . The boundary between the Von Kármán crater floor and wall is adapted from the mare boundary defined by Nelson et al. (2014) . The CE - 4 landing site (red star in Fig. 2) is located on the western edge of the study area.

At the edges of the study area, we observe two craters with diameters > 3. 5km

(Fig. 2) . The Zhinyu (3. 6km in diameter) crater is fresh and has a sharp and raised rim. The Tianjin crater (3. 6km in diameter) lies at the base of the Von Kármán inner wall and is moderately degraded. On the plagioclase abundance map, the Zhinyu crater produces radial mare deposits with lower plagioclase and delivers them to the adjacent regions. The regions that may contain ejecta derived from this crater are all excluded here. The Tianjin crater does not show an obvious compositional contrast with the surrounding terrains. Nevertheless, any craters within the 15. 2 - km region (4 times its diameter) around the Tianjin crater are ruled out for the small crater search.

3 RESULTS

3. 1 The identification of dark - haloed craters

Dark - haloed craters were identified in the study area following the three steps in Section 2. 1. The DHC candidates firstly identified on the Kaguya TC images were then reexamined on these NAC images (see examples in Fig. 3) . Previous studies proposed that the solar incidence angles can affect small craters' detection and measurements (Ostrach et al. 2011; Antonenko et al. 2013) . To examine this, we compared numerous NAC images of the same area taken with different incidence angles. Small craters and their surrounding ejecta show significant albedo variations and no shadows in the NAC images with low incidence angles. The images with incidence angles less than 20° are the idea NAC dataset for the small crater search. However, the NAC images that cover the study area were all acquired at the incidence angles larger than 40°. Therefore, the NAC images with incidence angles 40—55° were used for small craters' identification and validation. Due to the incomplete coverage of the NAC images with incidence angles 40—55° in the study area, some BRCs candidates can not be verified. And these craters were not involved in the following ejecta estimation.

The abundance maps of silicates obtained by Kaguya MI (Lemelin et al. 2016) were also applied for verification of DHCs. As mentioned above, mare basalts are characterized by lower abundance of plagioclase (Pl) and higher abundances of olivine (Ol) and clinopyroxene (CPX) than Finsen ejecta. On the abundance map of Pl, we can easily recognize the isolated green spots on the yellow - toned Finsen ejecta (Fig. 2B and 4C) . The green spots, consisting of green - toned ejecta from the Zhinyu crater, represent mare materials from the mare basalt layer beneath Finsen ejecta. Therefore, Pl abundance variations between exhumed mare deposits and Finsen ejecta distinguish them and enable us to check the DHCs. Fig. 4 shows 8 typical DHCs found near the CE - 4 landing site. On the abundance maps of Pl, Ol and CPX (Fig. 4 B - D), we note that the locations and sizes of the DHCs measured on the Kaguya TC and LROC NAC images match well with the crater shapes outlined by mineral distributions. All the DHCs identified on Kaguya TC

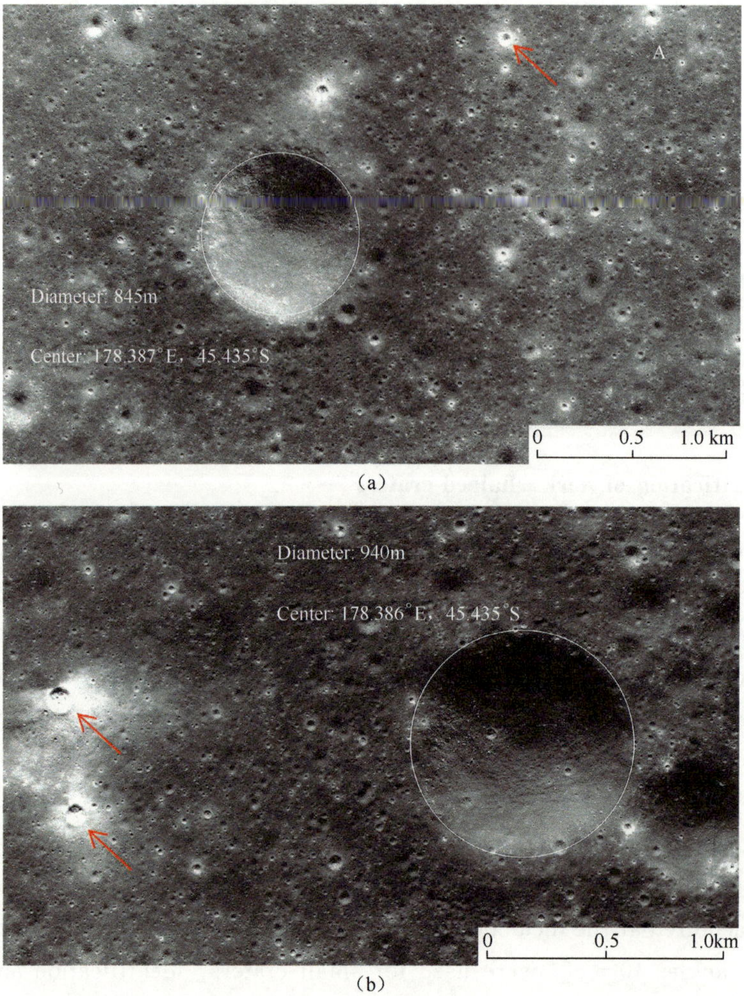

Fig. 3　Typical DHCs and BRCs on the eastern floor of the Von Kármán crater. White circles represent the best-fit circles of two DHC rims. BRCs with high-albedo rays are also noted (red arrows). The NAC image IDs are M141099568R and M1148327111L

and LROC NAC images are confirmed on the abundance maps of silicates.

3.2　The identification of bright ray craters

Bright ray craters are characterized by high-albedo deposits occurring around the crater (Hawke et al. 2004). Most small craters with typical bright rays can be identified in Kaguya TC images. However, due to the low spatial resolution of Kaguya TC images, it is difficult to accurately measure the diameters of these craters. The high spatial resolution images acquired by the LROC NAC were used. In the NAC images with incidence angles 40—55°, fresh ejecta of these craters show enhanced contrast with the background, which makes BRCs easily recognized (Fig. 3). We identified 77 BRCs and determined their diameters and center locations on the NAC images. On Kaguya MI-retrieved abundance maps of silicates, these craters usually show no obvious compositional contrast with the surrounding terrains, indicating that they did not penetrate through the Finsen ejecta (Fig. 4).

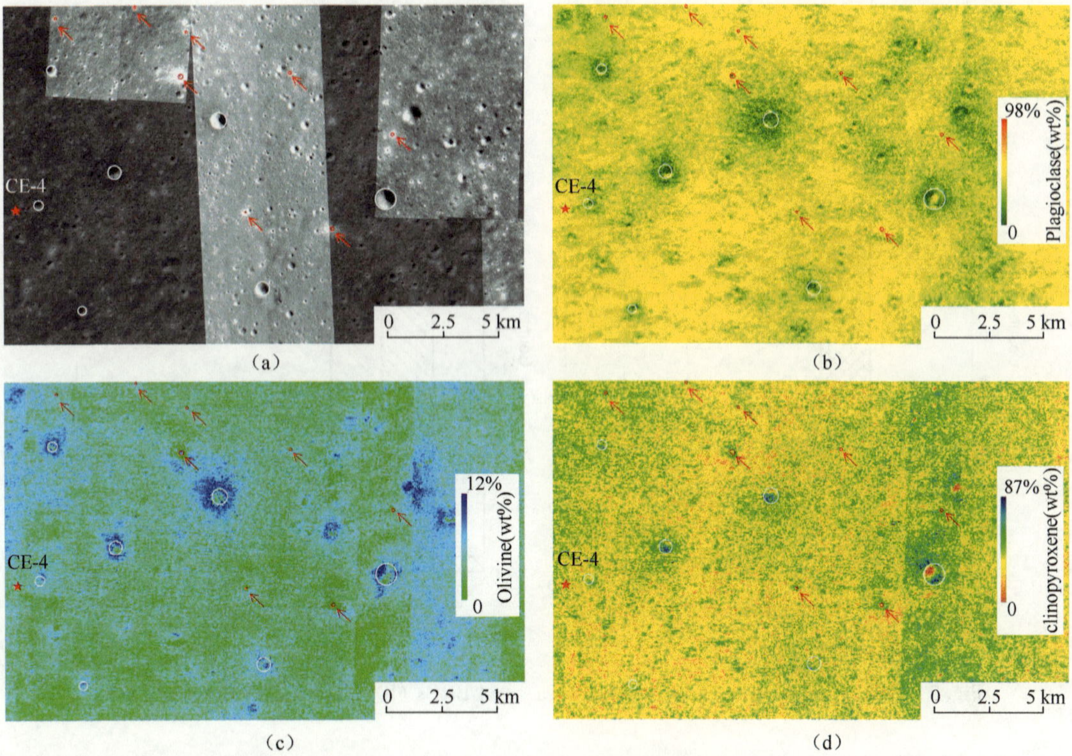

Fig. 4 DHCs and BRCs identified on the orbital images and Kaguya MI - retrieved abundance maps of silicates. Panel A presents LROC NAC images overlaid on the Kaguya TC mosaic. The IDs of the NAC images used here for small crater identifications are M1145979809L, M1145979809R, M143453659L, M143453659R, M1229577647L, and M1229577647R. Panels B, C, and D are Kaguya MI - retrieved abundance maps of plagioclase, olivine, and clinopyroxene, respectively. The white and red circles in the 4 panels represent the best - fit circles of DHCs and BRCs, respectively. DHCs exhume basaltic materials with lower Pl and higher Ol and CPX contents. Red arrows indicate the identified BRCs, which show no compositional contrast with the background. The red star shows the location of the CE - 4 landing site

3.3 Crater excavation depth and ejecta thickness

We estimated the thickness of Finsen ejecta on the floor of the Von Kármán crater using the small craters that completely penetrated (DHCs) and failed to penetrate (BRCs) through this ejecta layer. In the study area, 40 DHCs and 77 BRCs were identified. Their diameters, crater centers, distances from the Finsen crater center, and estimated excavation depths are listed in Tab. 2 and 3. The excavation depth of each crater was estimated with the crater excavation equation of Melosh (1989) (see the equation in section 2. 2) . All of the craters are simple impact structures with bowl - shaped interiors and smooth walls (see the examples in Fig. 3) . Their distances from the center of the Finsen crater vary from 110 to 180km. Most of the DHCs range from 400 to 800m in diameter (Fig. 5) . The BRCs are relatively small, with diameters less than 250m (Fig. 5) .

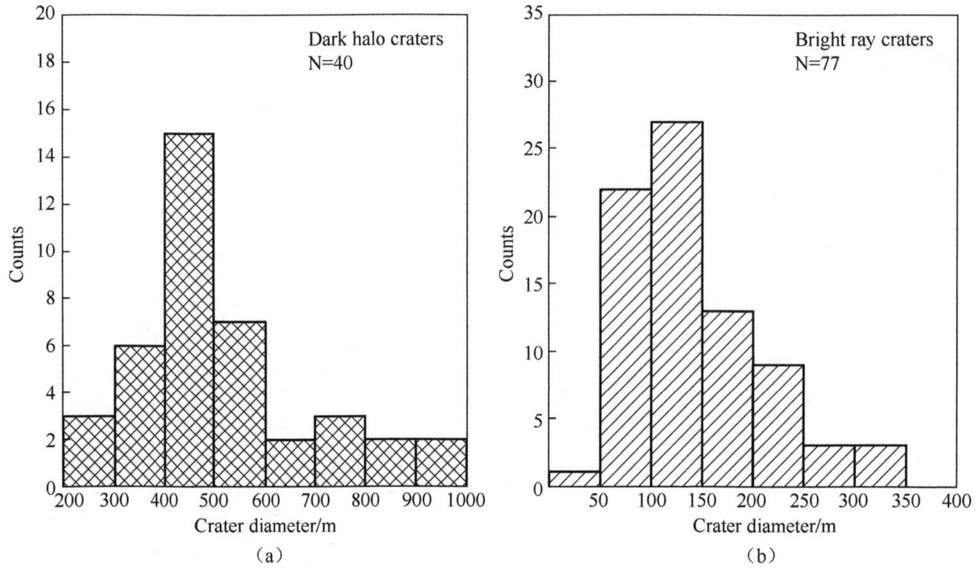

Fig. 5　Histogram of diameters for DHCs (a) and BRCs (b) identified in the study area

Fig. 6 shows the correlation of crater excavation depths with their distances from the Finsen crater center. Note that the DHCs (red circles) and BRCs (green diamonds) clearly plot in two clusters. An obvious boundary is found between these clusters. The DHCs are concentrated above this boundary, while the BRCs are located below the boundary. This result is reasonable because the DHCs have larger diameters and excavation depths than the BRCs. The boundary between the 'penetrating' DHCs and 'non - penetrating' BRCs (Fig. 6) approaches the true thickness of Finsen ejecta distributed on the eastern portion of the Von Kármán crater floor.

Finsen ejecta thickness was also estimated with the ejecta thickness models (Tab. 1). The Finsen ejecta distributions derived from these models follow the same trend (Fig. 6): the ejecta thickness decays exponentially with the radial distance. However, the curves built based on these models are all lower than the boundary constrained by small craters (Fig. 6) . This result suggests that all these models might underestimate the thickness of the Finsen ejecta. Among the five models, the curve built based on the equation of Pike (1974) is closest to the boundary in Fig. 6. We followed the approach of Pike (1974) and revised their equation as $t = 0.053R_t \ (r/R_t)^{-3}$. This revised equation (black curve in Fig. 6) gives the best fit to the actual ejecta thickness in the normalized distance range in

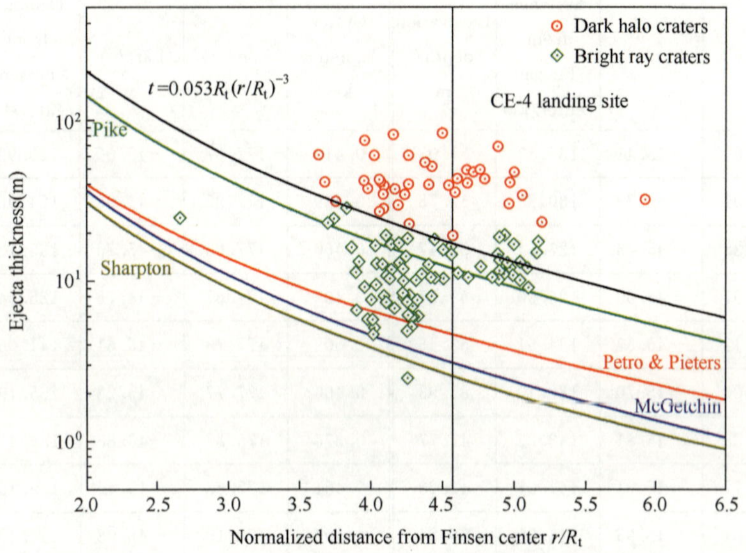

Fig. 6　Thickness estimates for the Finsen ejecta plotted as a function of the distance from Finsen center. The red circles and green diamonds represent the DHCs and BRCs identified in the study area. Their distance from Finsen crater center (Tab. 2 and 3) have been nor-malized to the transient radius of Finsen crater (R_t = 30.66km). The solid curves present the thickness of Finsen ejecta calculated using the different models (see equations in Tab. 1). The vertical gray line represents the CE-4 landing site (r = 140.15km, r/R_t = 4.57). Its distance from the Finsen crater center (r = 140.15km) is calculated based on the landing location (177.588°E, 45.457°S) reported by Di et al. (2019)

3.5—4.7. In section 4.1, we discussed the possible reasons for the deviations of these ejecta thickness models.

4　Discussion

4.1　Possible reasons for the underestimation of ejecta thickness models

Several ejecta thickness models have been proposed to estimate the ejecta thickness and spatial distribution around a crater. The ejecta thickness can be expressed as a function of the crater size and the distance between the crater center and a location of interest. It is generally described as $t = T(r/R_t)_b$, where t is the thickness at radial distance r from the crater center, T is the ejecta thickness at the rim of the crater, R_t is the radius of the transient crater, and b is the exponent of the power law.

The models of McGetchin et al. (1973) and Pike (1974) are the most cited models

Table2　DHCs identified in the study area

Crater diameter/km	Lon/(°)	Lat/(°)	Distance from Finsen center/km	Excavation depth/m	Crater diameter/km	Lon/(°)	Lat/(°)	Distance from Finsen center/km	Excavation depth /m
0.607	177.80	45.41	135.82	51.03	0.410	177.87	−45.00	129.99	34.48
0.712	178.01	−45.33	130.78	59.78	0.489	177.95	−44.98	124.90	41.08
0.958	178.38	−45.43	127.54	80.47	0.460	177.65	−45.26	135.12	38.63
0.709	178.97	−45.60	123.50	59.57	0.407	177.67	−44.76	125.64	34.16
0.636	178.12	−45.57	134.51	53.46	0.502	177.73	−45.61	141.06	42.13
0.975	178.06	−45.70	138.10	81.93	0.400	177.77	−46.28	155.19	33.62
0.569	177.88	−45.81	143.23	47.77	0.376	176.87	−46.88	181.47	31.56
0.503	177.77	−45.91	147.01	42.23	0.461	177.63	−45.46	139.43	38.75
0.558	178.01	−45.95	144.55	56.84	0.506	178.10	−45.08	124.47	42.53
0.473	178.11	−45.96	143.30	39.77	0.328	178.00	−45.04	125.33	27.58
0.555	177.83	−46.23	153.26	46.59	0.442	178.09	−44.96	122.11	37.13
0.805	177.92	−46.13	149.85	67.58	0.318	177.92	−44.95	124.69	26.71
0.574	178.16	−46.10	145.99	48.19	0.222	177.96	−45.73	140.24	18.68
0.713	178.23	−44.51	111.50	59.88	0.415	178.71	−46.17	140.73	34.84
0.880	177.69	−44.53	121.30	83.89	0.480	178.15	−46.37	152.41	40.34
0.454	177.72	−44.80	125.51	38.17	0.484	178.26	−44.61	112.85	40.61
0.439	178.26	−45.37	128.00	36.84	0.352	177.24	−45.78	152.20	29.57
0.368	178.07	−44.55	115.09	30.90	0.272	177.94	−46.55	159.44	22.83
0.473	178.85	−45.84	130.85	39.76	0.500	177.90	−44.73	121.20	41.98
0.346	178.75	−45.69	128.54	29.07	0.265	177.75	−45.12	130.91	22.22

for eject thickness estimation (e.g., Xie & Zhu 2016; Zhu et al. 2015). McGetchin et al. (1973) developed the power law equation based on the results of nuclear and high explosive cratering events, small - scale laboratory cra - tering experiments, natural meteorite impact craters, and estimates for lunar craters, including numerous cratering events. They defined T and b as $T = 0.14R_t^{0.74}$, $b = -3$. Pike (1974) modified this equation in consideration of the thickness of ejecta at the rim of terrestrial analogs and experimental explosion craters. Pike (1974) revised the equation for T: $T = 0.033R_t$. Generally, the Pike equation predicts approximately an order of magnitude greater amount of material introduced to any given location on the surface than the McGetchin model.

Housen et al. (1983) developed another model based on the dimension analysis of impact cratering. This model is applicable to different planetary surfaces and contains several variables depending on the target properties. Petro & Pieters (2004) simplified the Housen et al. (1983) scaling equation and suggested $T = 0.0078R_t$, $b = -2.61$. Sharpton (2014) investigated lunar craters with diameters from 2.2 to 45km. They followed the model of McGetchin et al. (1973) and proposed that the ejecta thickness at the rim of the complex crater is $T = 3.95(\pm 1.19)R_t^{0.399}$.

The underestimation of these models, relative to the ejecta thicknesses indicated by the DHCs and BRCs identified in this study, may be related to their own assumptions and limitations. The McGetchin model was noted to have been established from inadequate lunar data and to be inconsistent with the general form of either empirical or theoretical studies of craters (Pike 1974; Xie & Zhu 2016). The Housen et al. (1983) model was corrected to account for the curvature of the Moon (Petro & Pieters 2004) to investigate the mixing of materials excavated by impact basins on a global scale. This is clearly not the case for the Finsen ejecta distribution on the Von Kármán crater floor. The Sharpton (2014) model was built on investigations of 21 small - scale impact craters on the Moon. This equation should not be extrapolated to large complex craters such as the Finsen crater (73km in diameter, with a central peak) because the formation conditions (such as the size of projectiles and transient cavity diameter; Xie & Zhu 2016) of the Finsen crater were different from those of craters with diameters of 2—45km.

The underestimation of the ejecta models may be due to the contribution of other craters in addition to Finsen. Based on the surface morphology and mineralogical features of the study area, Finsen crater is the main source of the surficial ejecta layer. However, other craters, older than Finsen but younger than Von Kármán volcanism, might also have contributed. Such an ejecta layer should lie between the surface Finsen materials and the mare basalts. Alder crater is one candidate to transport ejected materials to the study area (Fig. 1). Alder is an Imbrian crater that is clearly older than Finsen crater. However, the sequence of the Alder impact event and the Von Kármán volcanism is quite ambiguous. In the western portion of Von Kármán, Finsen ejecta becomes thin to discontinuous with increasing distance from the Finsen center. If Alder ejecta lie on mare basalt units, there should be unique morphological features for tracking their source region, such as linear features and secondary impact craters/crater - chains. However, we found no morphological evidence in the western portion of the Von Kármán floor. This observation indicates that the materials in the western portion of the Von Kármán floor should not have been derived from Alder crater. We suggest that the formation of Alder crater should have been earlier than the emplacement of mare basalt; thus, Alder ejecta should lie beneath the basalt layer. There are two craters on the rim of the Von Kármán crater: Von Kármán L (Eratosthenian, D=29km) and L' (Imbrian, D=29km). Their contributions can also be ignored because of their young ages and relatively small sizes. In summary, the superposition relationships rule out the possible contribution of ejecta materials from other craters.

Table 3　BRCs identified in the study area

Crater diameter /km	Lon/(°)	Lat/(°)	Distance from Finsen center/km	Excavation depth /m	Crater diameter /km	Lon/(°)	Lat/(°)	Distance from Finsen center/km	Excavation depth /m
0.001	177.40	44.83	130.71	7.90	0.124	177.80	−45.77	143.49	10.43
0.123	177.57	−45.90	149.69	10.35	0.192	178.58	−46.07	139.68	16.13
0.124	177.59	−45.87	148.72	10.41	0.176	178.44	−46.00	139.90	14.76
0.138	177.35	−45.87	152.32	11.62	0.059	177.76	−44.73	123.49	4.98
0.173	177.48	−45.84	149.81	14.52	0.090	177.69	−44.62	122.96	7.54
0.155	177.35	−45.79	150.65	13.01	0.068	177.73	−44.62	122.16	5.69
0.127	177.22	−45.81	153.05	10.65	0.108	177.93	−44.77	121.28	9.08
0.107	177.08	−45.86	156.31	8.99	0.150	177.80	−44.80	124.04	12.57
0.155	177.10	−45.79	154.60	13.00	0.067	177.88	−44.81	122.92	5.65
0.228	177.25	−45.75	151.37	19.15	0.132	178.05	−44.78	119.44	11.07
0.132	177.28	−45.74	150.82	11.05	0.142	177.74	−46.31	156.34	11.94
0.220	177.27	−45.71	150.30	18.51	0.129	177.66	−45.18	133.50	10.86
0.212	178.02	−46.10	136.45	17.83	0.076	177.88	−44.63	119.50	6.42
0.119	178.74	−45.85	132.34	10.04	0.270	178.14	−44.54	113.50	22.71
0.123	178.59	−45.93	136.26	10.37	0.149	178.72	−45.95	135.18	12.51
0.087	178.83	−45.83	130.87	7.27	0.200	177.38	−45.93	153.29	16.81
0.057	178.79	−45.80	130.64	4.81	0.153	177.96	−46.24	151.80	12.83
0.073	178.80	−45.82	130.76	6.11	0.131	178.30	−45.99	141.32	11.01
0.099	178.68	−45.70	129.59	8.30	0.123	178.17	−45.25	126.81	10.30
0.118	178.63	−45.67	129.59	9.90	0.150	178.40	−45.34	125.27	12.57
0.194	178.86	−45.55	123.72	16.33	0.155	177.61	−44.35	120.15	13.00
0.096	178.73	−45.56	125.53	8.04	0.204	177.38	−44.63	128.78	17.09
0.138	178.65	−45.57	127.02	11.62	0.055	178.59	−45.38	123.32	4.66
0.124	178.73	−45.59	126.36	10.41	0.112	178.60	−45.41	123.95	9.43
0.167	178.27	−45.48	130.24	14.04	0.088	178.39	−45.36	125.92	7.40
0.124	177.42	−45.08	135.78	10.40	0.135	177.47	−45.86	150.20	11.33
0.094	177.44	−45.10	135.63	7.90	0.088	178.08	−45.46	132.57	7.38
0.123	178.15	−45.76	138.20	10.30	0.142	177.41	−44.71	129.54	11.95
0.181	178.16	−45.75	137.84	15.22					

4.2　The stratigraphy of the CE - 4 landing site

The SPA is the largest and oldest recognizable impact basin on the Moon. It is a high - priority landing site for scientific human and robotic exploration missions (e. g. , Jolliff et al. 2012) . The nature and formation of the SPA basin have implications for several fundamental questions about lunar early evolution (Jolliff et al. 2017; Shearer et al. 2011) including lunar impact chronology, large impact basin formation, and lunar farside volcanism. Lunar stratigraphy is significant for understanding the geological history and geological event sequence of the study area (Fig. 1) . Von Kármán crater is located on the impact breccia/melt sheet produced by the SPA impact event. After the formation of Von Kármán crater (Nectarian Period), its floor was resurfaced by multiple ejecta blankets from the surrounding craters and filled with low - Ti mare basalts. In this section, we investigated the superposition of materials on the floor of Von Kármán crater and constructed the stratigraphy of the study area as well as the CE - 4 landing site.

Finsen ejecta compose the uppermost layer in the study area (Fig. 7) . Finsen ejecta, the yellow tones in the Pl abundance map (Fig. 2B), widely cover the study area. These OPX - rich (and CPX - poor) materials may have originated from the lunar mantle (Li et al. 2019), which will be continuously investigated by the Yutu - 2 rover. According to the mineralogical characterization of materials exhumed by craters, low - Ti mare basalt represents the layer below the Finsen ejecta. The low - Ti basalt emplace - ment in the Von Kármán crater occurred at 3. 6Ga (Huang et al. 2018) . This basalt layer was finally covered by Finsen ejecta at 969± 130Ma (Ling et al. 2019) . Based on the DHC and BRC excavation depth calculations, we conclude that the vertical depth of the boundary between the overlying ejecta layer and the mare basalt layer is approximately 18m at the CE - 4 landing site.

Both the Finsen ejecta and mare basalt layers have long exposure histories on the lunar surface. Regolith layers should have developed on the top surfaces as a result of long continued bombardment. We proposed that, prior to the superposition of the Finsen ejecta, a basaltic regolith layer was developed on the surface of mare basalt plains. The surface of the mare deposits were affected by subsequent impact events since their emplacement at 3. 6Ga and gradually reworked by micrometeorites and solar wind ions until the surface was finally buried by Finsen ejecta at 969± 130Ma (Ling et al. 2019) . We estimated that the thickness of the ancient basaltic regolith layer is approximately 4—5m, the same as the average regolith thickness in lunar mare regions (McKay et al. 1991). The ancient basaltic regolith layer was very likely disturbed by the impact transport of Finsen ejecta and mixed with these foreign materials. This mixing occurred because the material in the primary ejecta curtain interacted with and incorporated local material (Oberbeck 1975) . Thus, there should be a transitional and mixed zone developed near the boundary between Finsen ejecta and mare basalt (shaded regions in Fig. 7) . However, it is difficult to assess the thickness of the mixed zone.

The Finsen ejecta has also been continually reworked by small young craters since ejecta formation. We suggested that a regolith layer exists on the top surface of the Finsen ejecta. The regolith layer should be a fine - grained, homogeneous mixture of local materials, compared with the underlying rocks. Huang et al. (2018) assessed the regolith thickness in a rectangular area (average 5.8km in the E-W direction and 2.4km in the N-S direction, including the CE-4 landing site) using NAC images with incidence angles less than 55°. They estimated that the regolith thickness in this area varies from ~2.5 to 7.5m. It is not possible to estimate the regolith thickness at the CE-4 landing site using the same method because the CE-4 site is not covered by these NAC images with incidence angles less than 55°. However, the rectangular area in the eastern portion of the Von Kármán crater is a relatively small and smooth plain, which clearly has the same exposure history and impact flux. Therefore, we assume that the local regolith layer of the CE-4 landing site has the same thickness as the rectangular area of Huang et al. (2018).

To reveal the deeper layers, relatively large craters on the floor of Von Kármán crater are needed. The largest crater near the study area is the Zhinyu crater. It is approximately 3.6km in diameter with an estimated excavation depth of 310m. This crater exposed massive mare deposits (Fig. 2B), indicating that it did not penetrate through the mare basalt layer. The bottom of the mare basalt layer must be deeper than 310m. There are no craters on the Von Kármán floor that exhume materials beneath the mare basalts. Therefore, the crater excavation method cannot provide further stratigraphic information at greater depths.

The superposed relationships of these 'hidden' layers under the mare basalt layer have been interpreted according to the ages of the craters (Huang et al. 2018). The Nectarian Leibnitz crater and Imbrian Alder crater both delivered ejecta to the Von Kármán floor. The CE - landing site is 224 and 134km from the Leibnitz and Alder crater centers, respectively. Using the proposed model (the equation in Fig. 6), we estimate that the thicknesses of Leibnitz and Alder ejecta at the CE-4 landing site are approximately 520 and 29m, respectively. The materials below Leibnitz ejecta should be impact breccias/melts, which usually occur on the fractured floor of a crater according to terrestrial impact structures studies. We assume that the SPA basin is the oldest lunar basin; thus, its impact breccia/melts form the base of the stratigraphy of the study area.

The stratigraphy of the CE-4 landing site is determined based on the results of small crater mapping and geological investigations of the study area (Fig. 7). A brief geological history of the study area is proposed as follows: A Nectarian impact event formed the Von Kármán crater with a central peak, and impact melts/breccia spread on the crater floor; The Leibnitz impact event produced mass materials that overlay the northern rim and floor of the Von Kármán crater; The formation of Alder crater also transported a thin layer of exhumed materials; Basaltic volcanism was active at 3.6Ga, and low-Ti basalt flows filled most of the floor; A basaltic regolith layer formed on the surface of the mare basalt plain; In the long Eratosthenian Period, no large crater formed in the adjacent area, and basaltic regolith formed on the surface of mare basalt; The Copernican Finsen impact ejecta overlay the basalt unit and covered the study area; the following small impact events struck the surface, and some of them excavated the underlying basaltic materials and formed the craters with dark halos.

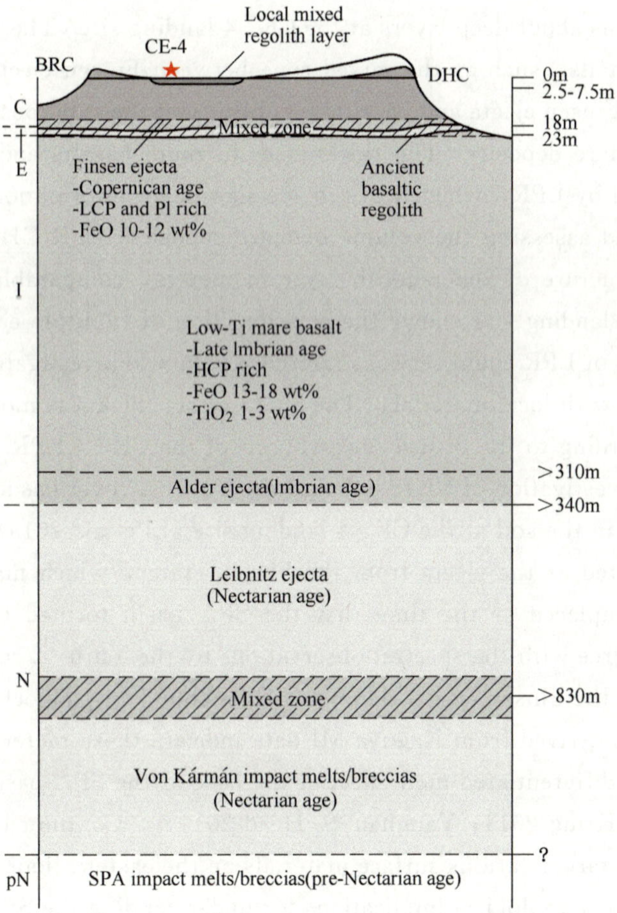

Fig. 7　The stratigraphy and subsurface structure proposed in the present study. The depth and layer thickness are not scaled in this profile. The red star indicates the CE－4 landing site. Typical DHCs and BRCs are shown here. The chemical and mineralogical compositions of Finsen ejecta and mare basalts are from Qiao et al. (2019) and Ling et al. (2019). The shaded regions indicate the mixed zone between the impact ejecta and the underlying layer. The letters on the left represent lunar geological periods: pN for Pre－Nectarian, N for Nectarian, I for Imbrian, E for Eratosthenian, and C for Copernican

4.3　Implications for CE－4 mission

The LPR aboard the Yutu－2 rover is planned to determine the subsurface structure and the depth of the subsurface layers. It covers two channels with different penetration depths: a low－frequency channel (CH－1, 60MHz) and a high－frequency channel (CH－2, 500MHz) (Guo et al. 2019; Jia et al. 2018). CH－1 is designed to determine the subsurface structure to a depth of several hundreds of meters. CH – 2 is intended to detect the fine structure of the lunar regolith with a penetration depth of 10m (Guo et al. 2019; Jia et al. 2018).

The results of LPR CH – 1 can refine the established stratigraphic column and answer several key questions about deep layers at the CE – 4 landing site. These results will reveal more structural details, such as the mixed zone between Finsen ejecta and mare basalt, the thicknesses of Finsen ejecta and basaltic regolith, and the superposition relationship of Alder ejecta and mare deposits. The thicknesses of mare basalts and possible sublayers will also be yielded by LPR, which will provide significant information for understanding lunar volcanism and assessing the volume of mare deposits. LPR CH – 2 will reveal the fine subsurface structure of the regolith layer in meters, comparable with Apollo drill cores. The CE – 4 landing site shows the superposition of multiple ejecta layers. The in situ measurements of LPR could serve as reference points to investigate the ejecta thickess and mixing effects with local materials. The above ejecta thickness models could be tested and improved according to the *in situ* observations of the CE – 4 LPR.

The in *situ* investigation of VNIS onboard the Yutu – 2 rover has identified olivine and low – Ca pyroxene in the soil at the CE – 4 landing site (Li et al. 2019). These materials have been interpreted as the ejecta from the Finsen crater, which may represent deep – seated materials emplaced at the time that the SPA basin formed (Li et al. 2019). Kaguya MI data agree with the spectral observations by the Yutu – 2 rover on the presence of OL and OPX in the Finsen ejecta. However, the abundant plagioclase (average 45± 4 wt%) components derived from Kaguya MI data indicate these materials are more likely to origin from the differentiated melt sheet at the base of the SPA basin (e. g., Morrison 1998; Hurwitz & Kring 2014; Vaughan & Head 2014). Continued exploration by the Yutu – 2 rover will target various surface materials on the eastern floor of the Von Kármán crater and reveal their geological implications for understanding the SPA evolution. Using the orbital dataset, this study identified the craters that may have delivered materials to the CE – 4 landing site and specifies their ejecta thicknesses and superposition relationships. This information will contribute to the CE – 4 mission and help us understand the geologic context and origin of various materials.

5 Conclusions

We identified 40 dark – haloed craters on the Von Kármán crater floor that penetrated through the surface Finsen ejecta and excavated underlying mare deposits and 77 bright ray craters that expose only the underlying fresh materials. Based on the small cater mapping, we constrained the Finsen ejecta thickness at the CE – 4 landing site. We provided a proposed stratigraphy for the CE – 4 site and interpreted the origins of different layers and the geologic history of the Von Kármán crater. These results provide valuable geological background for interpreting data from LPR and VNIS on the Yutu－2 rover. We suggest that the CE－4 landing site could serve as a reference point for crater ejecta distributions and mixing effects with local materials, which could be used to test and improve the ejecta thickness model according to the results of the CE－4 LPR.

Acknowledgements

The authors thank Professor Meng－Hua Zhu, Professor Feng Zhang, and Dr. Ming Gang－Xie for their helpful discussions. We gratefully acknowledge very constructive and helpful reviews provided by Dr. Joshua Snape and an anonymous reviewer that significantly improved the original version of the manuscript.

Fu, X. H., was supported by the National Natural Science Foundation of China (Grant Nos. 41490633 and 41590851), the open fund of the State Key Laboratory of Lunar and Planetary Sciences (Macau University of Science and Technology) (Macau FDCT grant No. 119/2017/A3), and the open fund of the Key Laboratory of Lunar and Deep Space Exploration, Chinese Academy of Sciences. Ling, Z. C., was supported by the Key Research Program of the Chinese Academy of Sciences (Grant No. XDPB11) and the Foundation of the State Key Laboratory of Transient Optics and Photonics (Chinese Academy of Sciences) (SKLST201504). Li, B., was supported by the Program for JLU Science and Technology Innovative Research Team (JLUSTIRT, 2017TD－26), the Focus on Research and Development Plan in Shandong Province (2018GGX101028), and the Shandong Provincial Natural Science Foundation (ZR2019MD015).

All original LROC images are available at NASA's Planetary Data System (http://pds.nasa.gov/), Kaguya TC images are archived at the SELENE Data Archive (https://darts.isas.jaxa.jp/planet/pdap/selene/), and Kaguya MI mineral maps are available at the Imaging Annex of the Planetary Data System (http://astrogeology.usgs.gov/pds/annex).

References

[1] Antonenko, I., Robbins, S. J., Gay, P. L., Lehan, C., & Moore, J. 2013, in Lunar and Planetary Science Conference, vol. 44, 2705 5.

[2] Bell, J. F., & Hawke, B. R. 1984, Journal of Geophysical Research: Planets, 89, 6899 2.

[3] Budney, C. J., & Lucey, P. G. 1998, Journal of Geophysical Research: Planets, 103, 16855 2.

[4] Chen, Y., Li, C., Ren, X., et al. 2018, Journal of Geophysical Research: Planets, 123, 630 2.

[5] Di, K., Liu, Z., Liu, B., et al. 2019, Journal of Remote Sensing, 23, 177

1，10.

[6] Guo，D.，Fa，W.，Zeng，X.，et al. 2019，in Lunar and Planetary Science Conference 14.

[7] Haruyama，J.，Matsunaga，T.，Ohtake，M.，et al. 2008，Earth，Planets and Space，60，243 3.

[8] Hawke，B.，Blewett，D.，Lucey，P.，et al. 2004，Icarus，170，1 7.

[9] Head,J. W.，Fassett，C. I.，Kadish，S. J.，et al. 2010，Science，329，1504 3.

[10] Head，J. W.，& Wilson，L. 1979，in Lunar and Planetary Science Conference，vol. 10，2861—2897 2.

[11] Housen,K. R.，Schmidt，R. M.，& Holsapple，K. A. 1983，Journal of Geophysical Research：Planets，88，2485 4，10.

[12] Huang,J.，Xiao，Z.，Flahaut，J.，et al. 2018，Journal of Geophysical Research：Planets 2，13.

[13] Hurwitz，D. M.，& Kring，D. A. 2014，Journal of Geophysical Research：Planets，119，1110 2，14.

[14] Jia，Y.，Zou，Y.，Ping，J.，et al. 2018，Planetary and Space Science，162，207 1，2，14.

[15] Jolliff，B. L.，Gillis，J. J.，Haskin，L. A.，Korotev，R. L.，& Wieczorek，M. A. 2000，Journal of Geophysical Research：Planets，105，4197 2.

[16] Jolliff，B. L.，Shearer，C. K.，& Cohen，B. A. 2012，in Annual Meeting of the Lunar Exploration Analysis Group，vol. 1685，3047 12.

[17] Jolliff，B. L.，Shearer，C. K.，Papanastassiou，D. A.，Liu，Y.，& Team，M. S. 2017，in Lunar and Planetary Science Conference，vol. 48，1300 12.

[18] Kaydash，V.，Shkuratov，Y.，& Videen，G. 2014，Icarus，231，22 2.

[19] Kneissl，T.，van Gasselt，S.，& Neukum，G. 2011，Planetary and Space Science，59，1243 3.

[20] Lemelin，M.，Lucey，P. G.，Gaddis，L. R.，Hare，T.，& Ohtake，M. 2016，in Lunar and Planetary Science Conference，vol. 47，2994 3，6.

[21] Lemelin，M.，Lucey，P. G.，Miljkovic，K.，et al. 2019，Planetary and Space Science，165，230 3.

[22] Li，C.，Liu，D.，Liu，B.，et al. 2019，Nature，569，378 1，2，13，14.

[23] Ling，Z.，Qiao，L.，Liu，C.，et al. 2019，Planetary and Space science 2，13，15.

[24] Ling，Z. C.，Jolliff，B. L.，Liu，C. Q.，et al. 2018，in Lunar and Planetary Science Conference，vol. 49 2.

[25] Lucey，P. G.，Taylor，G. J.，Hawke，B . R .，& Spudis，P. D. 1998，Journal of Geophysical Research：Planets，103，3701 2.

[26] McGetchin，T. R.，Settle，M.，& Head，J. W. 1973，Earth and Planetary Science Letters，20，226 4，9，10.

[27] McKay，D. S.，Heiken，G.，Basu，A.，et al. 1991，Lunar sourcebook，285—

356 13.

[28] Melosh, H. J. 1989, Impact cratering: A geologic process, vol. 11 (Oxford University Press) 4, 8.

[29] Moriarty III, D. P., & Pieters, C. M. 2018, Journal of Geophysical Research: Planets, 123, 729 2.

[30] Morrison, D. A. 1998, in Lunar and Planetary Science Conference, vol. 29 2, 14.

[31] Nelson, D. M., Koeber, S. D., Daud, K., et al. 2014, in Lunar and Planetary Science Conference, vol. 45, 2861 4.

[32] Oberbeck, V. R. 1975, Reviews of Geophysics, 13, 337 13.

[33] Ostrach, L. R., Robinson, M. S., Denevi, B. W., & Thomas, P. C. 2011, in Lunar and Planetary Science Conference, vol. 42, 1202 5.

[34] Pasckert, J. H., Hiesinger, H., & van der Bogert, C. H. 2018, Icarus, 299, 538 2.

[35] Petro, N. E., & Pieters, C. M. 2004, Journal of Geophysical Research: Planets, 109 10.

[36] Petro, N. E., & Pieters, C. M. 2008, Meteoritics & Planetary Science, 43, 1517 4.

[37] Pieters, C. M., Tompkins, S., Head, J. W., & Hess, P. C. 1997, Geophysical Research Letters, 24, 1903-2.

[38] Pike, R. J. 1974, Earth and Planetary Science Letters, 23, 265 4, 9, 10.

[39] Qiao, L., Ling, Z., & Fu, X. 2019, Icarus 2, 3, 4, 15.

[40] Quaide, W. L., & Oberbeck, V. R. 1968, Journal of Geophysical Research, 73, 5247 2.

[41] Robinson, M. S., Brylow, S. M., Tschimmel, M., et al. 2010, Space science reviews, 150, 81 3.

[42] Salisbury, J. W., Adler, J. E. M., Smalley, V. G., & Ring, J. 1968, Monthly Notices of the Royal Astronomical Society, 138, 245 2.

[43] Sharpton, V. L. 2014, Journal of Geophysical Research: Planets, 119, 154 4, 10.

[44] Shearer, C. K., Jolliff, B. L., Alkalai, L., et al. 2011, in Solar System Sample Return Missions, vol. 1611, 5018 12.

[45] Stuart-Alexander, D. E. 1978, Geologic map of the central far side of the Moon (United States Geological Survey) 2, 4.

[46] Vaughan, W. M., & Head, J. W. 2014, Planetary and Space Science, 91, 101 2, 14.

[47] Whitten, J., & Head, J. W. 2015, Planetary and Space Science, 106, 67 2.

[48] Wilhelms, D. E., Howard, K. A., & Wilshire, H. G. 1979, Geologic map of the south side of the Moon (United States Geological Survey) 2, 4.

[49] Wilhelms, D. E., John, F., & Trask, N. J. 1987, The geologic history of the

Moon (USGS Professional Paper) 2.
[50] Xie, M., & Zhu, M. -H. 2016, Earth and Planetary Science Letters, 440, 71 9, 10, 11.
[51] Yingst, R. A., Chuang, F. C., Berman, D. C., & Mest, S. C. 2017, in Lunar and Planetary Science Conference, vol. 48 4.
[52] Zhu, M. - H., Wunnemann, K., & Potter, R. W. K. 2015, Journal of Geophysical Research: Planets, 120, 2118 9.

Inferring the Shallow Layered Structure at the Chang'E – 4 Landing Site: A Novel Interpretation Approach Using Lunar Penetrating Radar

Iraklis Giannakis[1], Feng Zhou[2], Craig Warren[3], and Antonios Giannopoulos[4]

1. School of Geosciences, University of Aberdeen, Kings College, Aberdeen, UK
2. School of Mechanical Engineering and Electronic Information, China University of Geosciences (Wuhan), Wuhan, China
3. Department of Mechanical and Construction Engineering, Northumbria University, Newcastle, UK
4. School of Engineering, The University of Edinburgh, Edinburgh, UK

Abstract: The current paper investigates the shallow layers of the lunar regolith at the Chang'E – 4 landing site. Four layers between 0 and 10m were identified using lunar penetrating radar. Based on these outputs, a revised stratigraphic model is suggested for the post – Imbrian ejecta at the Von Kármán crater. The layers were previously unseen due to the smooth boundaries between them. The revised model was inferred using an advanced hyperbola – fitting scheme. Applying conventional hyperbola – fitting to non – homogeneous media results in errors and inaccuracies that are often wrongly assumed to be negligible. We propose a novel hyperbola – fitting scheme that is, not constrained to homogeneous media and can be applied subject to any arbitrary one – dimensional permittivity distribution. Via this approach, we can estimate the permittivity profile of an investigated area and detect layered structures that were previously transparent to electromagnetic waves due to the gradational dielectric properties at their interfaces.

Plain Language Summary: The landing site of Chang'E – 4 is at the Von Kármán (VK) crater at the South Pole – Aitken (SPA) basin. The SPA basin is the oldest and biggest basin on the Moon created at the early stages of its evolution by an impact that is, believed to have penetrated the lunar crust and uplifted materials from the top mantle. Understanding the geology and stratigraphy of the SPA basin can help us understand cratering processes and shed light on the evolution of the Moon. To that extent, ground penetrating radar (GPR) has been added to the payload of Yutu – 2 to investigate the stratigraphy of the Chang'E – 4 landing site. One drawback of GPR is that conventional processing methods fail to detect layers with smooth boundaries between them. Consequently, GPR data might give the false impression that an area is relatively homogenous, while in fact it might consist of numerous gradational layers representing a much more complex geological history. In the current paper, a novel interpretation tool is described, capable of detecting smooth transitions between layers. The proposed methodology revealed a previously unseen layered structure for the first ~10m of the VK crater.

Key Points:

• We suggest a novel hyperbola-fitting technique that assumes an arbitrary permittivity distribution with respect to depth

• The proposed method is used to map the lunar regolith at the Chang'E-4 landing site

• A layered structure is revealed at the first 10 meters. A new stratigraphic model is suggested for the Von Kármán crater

1 Introduction

Ground penetrating radar (GPR) is a mature geophysical technique (Daniels, 2004) with a unique span of applications ranging from landmine detection (Feng et al., 2012; Giannakis et al., 2016) and concrete inspection (Giannakis et al., 2020; Wai-Lok Lai et al., 2018), to glaciology (Williams et al., 2014) and archeology (Conyers, 2004). In planetary sciences, GPR has been applied both for satellite (Lauro et al., 2020) and in-situ measurements (Li et al., 2020), with promising results for mapping sub-glacial water bodies in Mars (Lauro et al., 2020), and for inferring the layered structure of the lunar regolith (Lai et al., 2020; Li et al., 2020; Zhang et al., 2020).

Subject to the application and the employed measurement configuration, various GPR processing and in-terpretation techniques have been suggested over the years (Daniels, 2004). From typical signal processing (Cassidy, 2009; Li et al., 2015) linear Born approximations (Boero et al., 2018) and machine learning (Giannakis et al., 2019) to full-waveform inversion (Meles et al., 2010). Within that context, hyperbola-fitting is considered one of the most mainstream techniques for the interpretation of common-offset GPR data (Mertens et al., 2016). The simplicity and computational efficiency of hyperbola-fitting make it an appealing choice for mapping the dielectric properties of an investigated medium, and for estimating the coordinates of subsurface targets (Mertens et al., 2016).

Hyperbola-fitting has been used in both Chang'E-3 and Chang'E-4 missions (Dong, Feng, Zhao, et al., 2020; Dong, Feng, Zhou, et al., 2020; Fa, 2020; Lai et al., 2019; Li et al., 2020) for estimating the electric permittivity of the lunar regolith and subsequently inferring its density and mineralogical composition (Dong, Feng, Zhou, et al., 2020; Li et al., 2020). Nonetheless, the underlying assumptions of hyperbola-fitting constrain its applicability, especially in complex environments where permittivity varies with depth. To mitigate that, conventional hyperbola-fitting is often complemented with the Dix conversion (Dix, 1955; Dong, Feng, Zhao, et al., 2020) in order to transform the estimated bulk velocity to actual velocity. Through a series of numerical examples presented in this paper, it is illustrated that the conventional approach (Dong, Feng, Zhao, et al., 2020) has limited applicability for the lunar regolith and should be used with caution. To that extent, we present a novel hyperbola-fitting that tackles this problem by simultaneously fitting multiple hyperbolas subject to any arbitrary 1D permittivity distribution.

The proposed scheme is applied to the lunar penetrating radar (LPR) data collected by the Yutu-2 rover during the first two lunar days of the Chang'E-4 mission at the Von Kármán (VK) crater (Li et al., 2020). Four distinct layers-that were previously not visible due to the smooth boundaries between them-were identified within the first 10m. This outcome differs significantly from previous theories suggesting that the first 12m of the landing site are fairly homogeneous, part of the weathered fine-grained regolith that lies on top of the ejecta from the Finsen crater (Zhang et al., 2020). Based on the revised permittivity profile and the available literature on the geology of the Chang'E-4 landing site, we suggest a new post-Imbrian strati-graphic model for the VK crater, in which an approximately ~3m weathered fine-grained layer is followed by ~8m of ejecta from the VK L and L' craters overlying the ejecta from the Finsen crater.

2 Methodology

2.1 Advanced Hyperbola-Fitting

In this section, a novel hyperbola-fitting framework is described, capable of dealing with half-spaces with arbitrary 1D relative permittivity distributions $\epsilon(y)$ (see Figure S1). A detailed description of conventional hyperbola fitting using the Dix convention is given in Text S1. Similar to typical hyperbola-fitting, in order to avoid non-uniqueness (Giannakis et al., 2019; Mertens et al., 2016), the proposed scheme assumes that the radius (R) of the investigated target equals zero. Subject to a varying velocity with depth, the two way travel time t that it takes for the wave to travel from the point $\vec{B}=\langle x, y\rangle$ to the point $\vec{A}=\langle x_0, d\rangle$ via the parametric curve $\vec{q}(m)=\langle q_x(m), q_y(m)\rangle$ (where $m\epsilon\,[0-1]$ and d is the depth of the target) can be calculated using a scalar line integral over $\vec{q}(m)$

$$t=\frac{2}{c_0}\int_0^1 \sqrt{\epsilon(y)}\left\|\frac{\partial \vec{q}(m)}{\partial m}\right\| \mathrm{d}m \tag{1}$$

Given a specific velocity structure, the path $\vec{q}(m)$ can be calculated using Fermat's principle (Aldo, 1996). The notation $\frac{\partial \vec{q}(m)}{\partial m}$ is used to denote the norm of the first derivative of the parametric curve $\vec{q}(m)$ with respect to the parameter $m\epsilon[0-1]$. It is shown that if we simplify Equation 1 and make the assumption that the path $\vec{q}(m)$ is the straight line that connects the antenna to the center of the target, it leads to an elegant and computationally efficient formulation without compromising accuracy (more details are given in Section 2. 2) . The straight line that connects the antenna to the target can be expressed via the parametric curve $\vec{q}(m)=\vec{A}+(\vec{B}-\vec{A})m$. Substituting this into Equation 1 results in

$$t=\frac{2\|\vec{B}-\vec{A}\|}{c_0}\int_0^1\sqrt{\epsilon(y)}\,\mathrm{d}m. \tag{2}$$

The linear path of the integral in Equation 2 can be written as $\vec{q}(m)=\langle x_i+m(x_0-x_i),\ y_i+m(d-y_i)\rangle$, where x_i, y_i are the coordinates of the antenna at the ith position. Consequently, the yvariable in Equation 2 can be substituted by $y=y_i+m(d-y_i)$, which implies that, for $y_i=0$ (antenna placed at the surface) $\partial m=\frac{\partial y}{d}$ and that $y=d$ for $m=1$. Therefore, Equation 2 can be re - written as

$$t=\frac{2\|\vec{B}-\vec{A}\|}{c_0 d}\int_0^d\sqrt{\epsilon(y)}\,\mathrm{d}y \tag{3}$$

Solving the integral numerically yields

$$t\approx\frac{2\|\vec{B}-\vec{A}\|}{c_0 d}\sum_{s=0}^{Q}\sqrt{\epsilon(s\cdot\Delta y)}\,\Delta y \tag{4}$$

where Δy is the discretization step and $Q=d/\Delta y$. Notice that the summation term $N(d,\epsilon)=\sum_{s=0}^{Q}\sqrt{\epsilon(s\cdot\Delta y)}\,\Delta y$ is independent of the position of the antenna and needs to be calculated just once. The final formulation for the proposed scheme is given by

$$t\approx\frac{2\|\vec{B}-\vec{A}\|}{c_0 d}N(d,\epsilon) \tag{5}$$

where the only unknowns are the relative permittivity function $\epsilon(y)$ and the depth of the target d. The parameter x_0 can be easily derived from the apex of the hyperbola at the measured B—Scan. Subject to a given $\epsilon(y)$, the depth of the target is calculated using the apex of the hyperbola (x_0, t_0), where $\|\vec{B}-\vec{A}\|=d$

$$t_0\approx\frac{2}{c_0}N(d,\epsilon) \tag{6}$$

For a given $\epsilon(y)$, the only unknown in Equation 6 is the depth d that is, estimated numerically using the bisection method. Notice that both Equation 6 and the summation $N(d,\epsilon)$ in Equation 4 need to be evaluated just once. The only term in Equation 5 that needs to be updated as the scan progresses is the distance $\|\vec{B}-\vec{A}\|$. To summarize, given a permittivity distribution $\epsilon(y)$ and the apex of a hyperbola (x_0, t_0), the depth d of the target is estimated by numerically solving Equation 6 using the bisection method. Subsequently, $N(d,(\epsilon))$ is evaluated and furthermore used in Equation 5 to calculate the arrival times.

The proposed scheme utilizes numerous hyperbolas and tries to find the optimum $\epsilon(y)$ that simultaneously minimizes $\min\limits_{\epsilon(y)}\sum_{i=1}^{Z}\|\boldsymbol{t}_i-\boldsymbol{T}_i\|$, where $\boldsymbol{T}_i\in\mathbb{R}^{n_i}$ and $\boldsymbol{t}_i\in\mathbb{R}^{n_i}$ represent the measured and predicted arrival times for the i th hyperbola, Z is the total number of the employed hyperbolas and n_i is the number of discretization points for the i th hyperbola. To further simplify the problem, the permittivity is discretized with K equidistant points and subsequently a cubic interpolation is applied to map ϵ with respect to y in a continuous manner. Therefore, the minimization is re - written as $\min_{\epsilon(k)k\in\mathbf{R}^K} \sum_{i=1}^{Z} \| \boldsymbol{t}_i - \boldsymbol{T}_i \|$ with only K unknowns. This is a non - linear and non - convex problem that can be solved using global optimizers. A Particle Swarm Optimisation (PSO) (Kennedy & Eberhart, 1995), with 50 particles and uniform PSO parameters, was proven to be very efficient for reconstructing $\epsilon(y)$, given a sufficient number of measured hyperbolas. The number of equidistant points K is estimated by plotting the Error - K curve. This approach is based on the L - curve method (Hansen, 1992) that tries to balance accuracy and constraints. Within that context, we choose the K value for which the solution balances accuracy and simplicity. In particular, the minimization is executed multiple times with increasing K until the error starts to converge. The K value is chosen at the earliest point of convergence. Greater K values can potentially result (if a sufficient number of hyperbolas is not present) in unnecessary complicated permittivity structures without increasing the fitting accuracy.

2.2 Numerical Experiments

Two numerical 2D case studies (illustrated in Fig. 1a) are used in this section to evaluate the performance of the proposed scheme. Both models are non - dispersive, non - conductive, and non - magnetic, with a varying relative permittivity with respect to depth $\epsilon(y)$. In both scenarios, the relative permittivity varies with respect to depth and ranges from 2 to 10. Nine perfect electric conductors are distributed randomly within a $2\times 1m^2$ domain. The targets have a cylindrical shape with 5cm diameter and their main axis is perpendicular to the acquisition line. Measurements are taken every 2cm along the x - axis using a line source with 1GHz central frequency. The offset between the transmitter and the receiver is 1cm. The numerical simu - lations were executed using gprMax (Warren et al., 2016, 2019), an open source electromagnetic solver that uses a second order (in both space and time) finite - difference time domain (FDTD) method (Yee, 1966). The spatial discretization step of the FDTD grid is $\Delta x = \Delta y$ = 5mm, and the time step Δt is calculated using the Courant limit (Taflove & Hagness, 2000). The boundaries of the domain are truncated using the recursive integration perfectly matched layer (Giannopoulos, 2008). The gprMax input files used to generate both models are given in Texts S2 and S3.

The results are shown in Fig. 1b where most of the hyperbolas (and their multiples) are clearly visible, as expected in this clutter - free environment. From Figure 1b, it is apparent that even in these clinical clut - ter - free numerical experiments, the reflections from the layers are very weak and not visible in the measured radargrams. This is due to the smooth transition between the layers that can greatly decrease their reflection coefficient (Bano, 2006; Diamanti et al., 2014). This gives the false impression that a medium is homogeneous when in fact it can be as complex as the investigated case studies

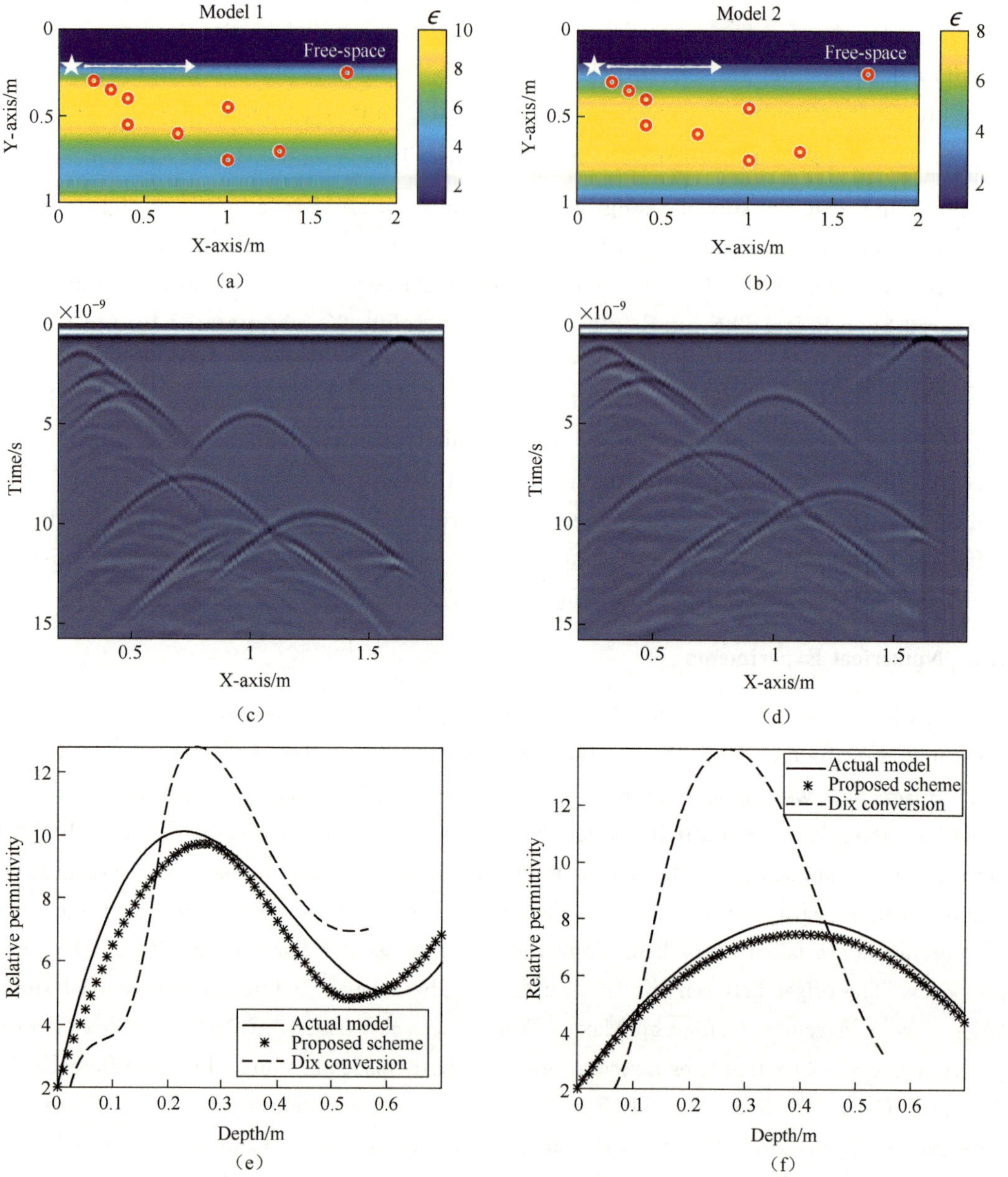

Fig. 1 (a) The investigated numerical experiments. Nine cylindrical targets are buried in two media with varying permittivity with respect to depth. Measurements are taken every 2cm (from left to right) using a ground - coupled line source (white star) with 1GHz central frequency. (b) Resulting B - Scans. It is evident that due to the smooth boundaries between the layers, no reflections are visible on the resulting radargrams. The shapes of the hyperbolas are the only features that can be used to infer the permittivity profile. (c) The resulting permittivity profile for the models shown in panel (a)

(see Fig. 1a), with four clear and distinct layers. This is very important when interpreting radargrams from the lunar regolith, where smooth transitions between layers are expected due to impact gardening and the reworking of the materials during crater formation (Ding et al. , 2020) . The term "impact gardening" used throughout the paper refers to the alteration of lunar surface due to the gradual gardening of lunar materials via small meteorite impacts (Horz et al. , 1991; Pieters & Noble, 2016) .

The proposed scheme and the typical hyperbola - fitting with Dix conversion (Dong, Feng, Zhao, et al. , 2020) (see Text S1) were applied to the radargrams shown in Fig. 1b. In both models, the proposed methodology outperforms conventional hyperbola - fitting, and manages to sufficiently estimate the permittivity profile and the underlying layered structure in an efficient manner (see Fig. 1c) . Small errors observed in Fig. 1c can be due to: the linear - path simplification; manual picking of the hyperbolas (Ding et al. , 2020); non - accurate time - zero correction (Yelf, 2004); and/or non - ideal targets that is, $R \neq 0$. From Fig. 1c, it is apparent that conventional hyperbola fitting using the Dix conversion overestimates the resulting permittivity and fails to reconstruct complex velocity profiles despite the dense velocity sampling (nine hyperbolas) used in the numerical case studies.

Typical hyperbola fitting using the Dix conversion (see Text S1) is based on the assumption that the investigated medium consists of homogeneous horizontal layers (Dix, 1955) . The number of layers equals the number of apparent velocities used in the conversion. For example, if five hyperbolas are used, then the underlying assumption is that the investigated medium consists of five homogeneous layers. In the presence of numerous layers or smooth variations of velocity, the Dix conversion requires a large number of hyperbolas in order to accurately reconstruct the 1D velocity profile. The proposed methodology overcomes this limitation by assuming a smooth velocity profile. The latter is independent to the number of hyperbolas used. The resulting model using our approach is the optimized 1D velocity profile that satisfies the measured hyperbolas.

2.3　Results From Chang'E - 4 Channel 2 Data

The proposed methodology is applied to the high frequency data collected by the Yutu - 2 rover at the VK crater during the first two lunar days of the Chang'E - 4 mission (Li et al. , 2020) . During the first two lunar days, the rover followed an irregular path and managed to cover ~106m (Li et al. , 2020) . The current paper focuses on the first 150ns of the scan in order to effectively map the shallow layers (~10 — 12m) of the rego - lith. Based on the results, a revised stratigraphy for the VK crater is proposed (in the next chapter) that takes into account a previously unseen layered structure within the first ~10m of lunar regolith.

The radargram was processed using a typical GPR processing pipeline that involves zero - time correction, dewow, time - gain (exponential gain), and background removal (Cassidy, 2009) . The resulting B - Scan for the first 150ns is illustrated in Fig. 2 (Li et al. , 2020) . The overall signal to clutter ratio is substantially higher compared to Chang'

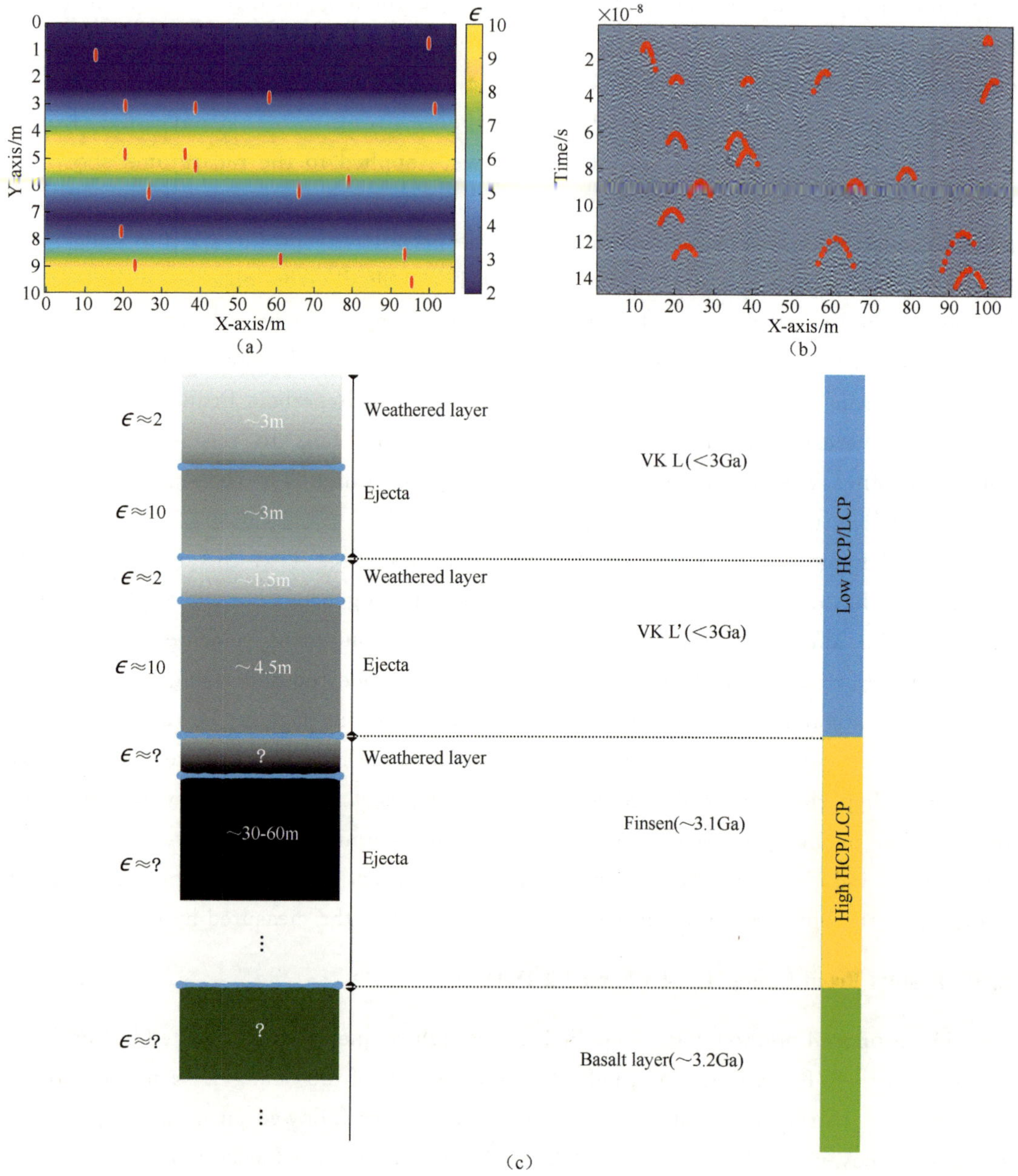

Fig. 2 (a) The resulting relative permittivity profile $\epsilon(y)$ at the landing site of Chang'E－4 mission using the advanced hyperbola fitting. The coordinates of the investigated targets are illustrated with red dots. (b) The fitted hyperbolas subject to the permittivity profile shown in panel (a). More details regarding the fitted hyperbolas can be found in Fig. S6. (c) The proposed stratigraphy model for the Chang'E－4 landing site. The first ~6m consists of a top weathered layer overlaying the ejecta from VK L crater. Below that, is a low permittivity layer that corresponds to the weathered ejecta of the VK L'crater. The VK L'ejecta extends to ~12m depth, where the Finsen and Alder ejecta lay on top of the Imbrian basaltic layer. Dates are based on Lu et al. (2021) and the chemical composition on Huang et al. (2018)

E-3 mission (Lai et al., 2019; Li et al., 2020) (potentially due to lack of ilmenite) which results in clear hyperbolic features that can be utilized to deduce the shallow layered structure at the first 10-12m of the landing site.

Fig. 2a illustrates the resulting relative permittivity profile using the proposed advanced hyperbola-fit-ting, subject to the hyperbolas shown in Fig. 2b. It is evident that there is a layered structure with four layers in the first 10m of the regolith. The first and the third layers have low permittivity values while the second and the fourth layers have permittivity up to $\epsilon \approx 10$ (see Fig. 2a). Typical lunar soils have low permittivity values although there are reported high-density lunar samples with relative permittivity up to $\epsilon \approx 10$ (Chung et al., 1970; Olhoeft & Strangway, 1975). To further evaluate the validity of these results a sensitivity analysis is presented in Fig. S3 and S4 where it is shown that the permittivity value for the low permittivity layers can range between $\epsilon \sim 1.7-2.3$ without compromising accuracy. For the high permittivity layers, there is an optimisation plateau at $\epsilon \sim 8-10$ where there is no significant error improvement.

We would like to highlight that current knowledge regarding the permittivity of lunar soils is based primarily on shallow samples brought back to Earth during the Apollo missions. Superficial lunar samples are not representative of deeper layers since they are exposed to impact gardening. The latter, results in an increased porosity (Horz et al., 1991) and vitrification (Nash & Conel, 1973). Increased porosity decreases the overall permittivity of multi-phase media (Dobson et al., 1985; Hallikainen et al., 1985; Peplinski et al., 1995), which results in a well-known causal relationship between permittivity and density (Chung et al., 1970; Olhoeft & Strangway, 1975). In addition, vitrification transforms lunar materials to glassy fragments that have lower permittivity values compared to typical lunar minerals (pyroxenes, plagioclase, olivine, ilmenite etc.) (Campbell & Ulrichs, 1969). The lunar regolith is, therefore, a low permittivity medium, established through laboratory (Carrier et al., 1991), satellite (Bhattacharya et al., 2015; Gong et al., 2015), and in-situ measurements (Dong et al., 2021). The semi-empirical models tuned for lunar soils are primarily based on shallow regolith samples (Carrier et al., 1991; Chung et al., 1970; Frisillo et al., 1975; Shkuratov & Bondarenko, 2001). The latter are affected by impact gardening making them unreliable for estimating the dielectric properties of deeper ejecta. Estimation of the dielectric properties of deeper layers is still an ongoing research area that is, primarily based on LPR measurements and typical hyperbola-fitting (Dong, Feng, Zhao, et al., 2020). An indirect method to infer the permittivity is using the Complex Refractive Index Model (CRIM) which is a mainstream formula for computing the bulk permittivity of multi-phase media (Birchak et al., 1974) based on the permittivity of their components. Typical minerals in lunar soils are pyroxene, plagioclase, olivine and ilmenite (Zongcheng et al., 2015). The relative electric permittivity of plagioclase is $\epsilon_{pl} \approx 6.5$ (Zheng et al., 2005). The relative permittivity of pyroxene varies from $\epsilon_{py} \approx 7$

−10 (Campbell & Ulrichs, 1969; Zheng et al., 2005), in particular for augite and for pyroxenes in anorthosites the relative permittivity can be as high as $\epsilon_{py} \approx 11$ (Zheng et al., 2005). The relative permittivity of olivine is $\epsilon_{ol} = 8$ (Zheng et al., 2005) and for ilmenite $\epsilon_{il} \approx 50$ (Zheng et al., 2005). Using CRIM, it can be proven, that some realistic mineralogical composition can give rise to high permittivity values for example, $V_{air} = 10\%$, $V_{il} = 8\%$, $V_{pl} = 40\%$, $V_{py} = 42\%$. The aforementioned soil sample has bulk relative permittivity (estimated using CRIM) $\epsilon \approx 9$. High relative permittivity values as high as $\epsilon = 8.9$ have also been calculated via hyperbola fitting at Chang'E-3 landing site (Lai et al., 2016).

3 Stratigraphy of the Landing Site

3.1 Geological Background

The Chinese lunar probe Chang'E-4, carrying the Yutu-2 rover, was the first human-made object that land ed on the far-side of the Moon on 3rd of January 2019 (Li et al., 2019; Tang et al., 2020). The landing site is located at the South Pole-Aitkens (SPA) basin-the oldest and biggest crater on the Moon (Huang et al., 2018; Hu et al., 2019; James et al., 2019). The SPA basin is pre-Nectarian in age and has an elliptical shape (Garrick-Bethell & Zuber, 2009) with an approximate diameter of 2,100−2,500km (Moriarty et al., 2013). The transient cavity of the SPA basin has been estimated between 840 and 1,400km (Moriarty et al., 2013; Potter et al., 2012). The maximum excavation depth of lunar craters with a diameter below < 10km is approximately 10% of their diameter (Stopar et al., 2017). There is an ongoing debate whether or not this ratio holds true for larger craters (Potter et al., 2012; Wieczorek & Phillips, 1999). Simulations suggest that this ratio can be extrapolated for craters with diameter up to 1,100km (Potter et al., 2012) while gravity measurements indicate that this ratio decreases for craters with diameter >500km (Wieczorek & Phillips, 1999). The above, suggest that the SPA basin might have excavated through lunar crust and into the mantle (Hammond et al., 2009; Hurwitz & Kring, 2014; Melosh et al., 2017; Moriarty et al., 2013, 2021; Uemoto et al., 2017; Vaughan & Head, 2014). This Premise is based on the maximum depth of lunar crust ~60km (~45km on the far-side [Moriarty et al., 2021]), as estimated by the Gravity Recovery and Interior Laboratory mission (Wieczorek et al., 2013), which is in good agreement with seismic data from the Apollo missions (Khan, 2002). The shallow mantle layer was most likely melted during the impact (Hurwitz & Kring, 2014; Melosh et al., 2017; Moriarty et al., 2013, 2021; Potter et al., 2012; Vaughan & Head, 2014) and parts of it are expected to occur within the SPA basin, forming an underlying sheet of non-crustal materials (Moriarty et al., 2013, 2021; Potter et al., 2012). These materials are of paramount importance since they can constrain the composition of the upper mantle and provide an insight into the early evolution of the Moon (Moriarty et al. , 2013; Yamamoto et al. , 2010) .

Based on previous models of lunar evolution - that suggest an upper mantle predominantly composed of olivine (Charlier et al. , 2018; Y. Lin et al. , 2017; Moriarty et al. , 2021; Snyder et al. , 1992; Yamamo to et al. , 2010) - strong spectral signatures of olivine were expected to be present within SPA (Ivanov et al. , 2018; Lucey, 2004; Moriarty et al. , 2021) . Nonetheless, data from CLEMENTINE and SELENE did not support this premise (Lemelin et al. , 2019; Matsunaga et al. , 2008; Tompkins & Pieters, 1999; Yamamoto et al. , 2010), apart from small occurrences of olivine clusters (Lemelin et al. , 2019; Yamamoto et al. , 2010) most likely originated from crustal materials, due to their location (the exterior of the SPA) and the high content of feldspar in their near proximity (Moriarty & Pieters, 2018) . Remote sensing data from the peak of craters indicate that SPA is dominated by Mg rich orthopyroxene (Moriarty & Pieters, 2018; Nakamura et al. , 2009) . Many different theories have been proposed to explain the unexpected lack of olivine and the abundance of pyroxene (Ivanov et al. , 2018) . Low olivine content supports the premise that SPA did not penetrate through the crust (Lucey, 2004) due to an oblique impact (Garrick - Bethell & Zuber, 2009; Schultz & Crawford, 2011) . Lack of olivine can also be explained by a pyroxene - rich composition of the upper mantle (Kuskov O. L. , 2015; Melosh et al. , 2017) . Moreover, the large impact that created SPA could potentially cause large lava bodies that undergo fractional crystallization which can lead to the separation and sinking of olivine and the accumulation of pyroxene at the top (Hurwitz & Kring, 2014; Vaughan & Head, 2014) . The SPA is dominated by mafic materials and in particular by Mg rich and low - Ca pyroxene (Huang et al. , 2018; Moriarty & Pieters, 2015, 2018; Nakamura et al. , 2009) . CLEMENTINE measurements reveal an inner zone with Fe abundance and an outer zone with lower Fe content (Jolliff et al. , 2000; Lucey et al. , 1998; Moriarty & Pieters, 2018) . Furthermore, using data from the Moon Minerology Mapper (M^3), Moriarty and Pieters (2018) have divided the SPA into four zones. The first zone is the inner SPA area called SPACA, with characteristic Ca -pyroxene abundance that lies at the center of the SPA. The second zone sur - rounds SPACA, and it is an area with Mg rich pyroxenes. Based on spectral analysis of the central peaks of the craters within SPACA, strong indications were given to support the premise that SPACA lies on top of the Mg rich area (Moriarty & Pieters, 2015, 2018) . The third zone is a heterogeneous annulus that consists of pyroxene and feldspar, and acts as the intermediate stage between the SPA and its exterior. The latter is the fourth zone, a mafic - free area with high content of feldspar, similar to lunar highlands (Moriarty & Pieters, 2015, 2018) .

The landing site of Chang'E - 4 is within the Mg - rich annulus (Moriarty & Pieters, 2018) and in particular in the interior of the VK crater (177.588°E, 45.4578°S, see Figure S5) (Di et al. , 2019; Li, 2021; Li et al. , 2020; Qiao et al. , 2019; Zhang et al., 2020). VK is an elliptical crater (Zhang et al., 2020) with approximately ~186km diameter (Guo et al., 2021; Huang et al., 2018). The age of VK was estimated pre-Nectarian (Huang et al., 2018; Losiak et al., 2009) and recent studies have placed it at ~4.2Ga (Lu et al., 2021), very close to the formation of SPA (Garrick-Bethell & Miljković, 2018; Hiesinger et al., 2012; Lu et al., 2021). The creation of Leibnitz crater affected the north part of VK and contributed to the ejecta layer prior to the Imbrian basaltic flood (Huang et al., 2018; Yingst et al., 2017). Ejecta from Alder crater (widely accepted as an Imbrian crater (Losiak et al., 2009) and recently dated at~3.5Ga [Lu et al., 2021]) are also expected to the pre-basaltic layers (Huang et al., 2018; Lu et al., 2021). The VK crater was flooded with basalts during the Imbrian period (Huang et al., 2018; Paskert et al., 2018) around ~3.2—3.3Ga (Haruyama et al., 2009; Lu et al., 2021). The ejecta from Finsen crater are expected to have contributed to the post-Imbrian VK layers (Huang et al., 2018). Finsen is an Eratosthenian crater (Losiak et al., 2009; Wilhelms & McCauley, 1987) and its ejecta and floor are dated at ~3.1 and 3Ga respectively (Lu et al., 2021). Recent studies suggest that Orientale crater might have added to the post-Imbrian VK layers as well (Xiao et al., 2021). Subsequently, the Eratosthenian craters VK L and L' were formed (Paskert et al., 2018; Zhang et al., 2020). The VK, Leibnitz, Alder, VK L, and L' lie within the Mg-pyroxene annulus while Finsen is within the SPACA region (Moriarty & Pieters, 2018).

Geological context suggests that the craters VK L, L', Finsen and Orientale have contributed to most of the post-Imbrian ejecta layers of the VK crater (Huang et al., 2018; Di et al., 2019; Paskert et al., 2018; Xiao et al., 2021). The thickness of the Finsen ejecta at the area of the Chang'E-4 landing site is estimated—via numerical simulations (Di et al., 2019)—at ~30m. This is not in good agreement with the results obtained using dark-halo and non-dark halo craters (Guo et al., 2021; Li et al., 2020) that suggest a thicker post-ba-saltic layer, probably due to the presence of Orientale ejecta (Xiao et al., 2021). Nonetheless, contradicting data (Yue et al., 2020) place the date of Orientale to be older than the Imbrian basaltic flood, which implies that there might be another source that contributed to the post-basaltic VK layers, probably additional ejecta from the Eratosthenian craters (VK L and L') at the rim of VK (Paskert et al., 2018).

The surface of the landing site is smooth with a small number of boulders (Ding et al., 2020; Weiren et al., 2019), most of them being glassy fragments and breccias from secondary craters (H. Lin et al., 2020). From in situ reflectance data, the visible surface at the landing site is not olivine-pyroxene rich and consists of 56%—72% plagioclase, similar to lunar highlands (Hu et al., 2019; Li et al., 2019) with Mg rich orthopy-roxene (Gou et al., 2020; Huang et al., 2018). The thickness of the regolith (weathered top soil) is estimated using LROC NAC images at ~2.5—7.5m (Huang et al., 2018) using the technique described by (Quaide Wil liam & Oberbeck, 1968).

Based on the M^3 reflectance data, it is estimated that below the top weathered soil, lays a low - calcium pyroxene (LCP) layer (Gou et al., 2019; Li et al., 2019) ranging from ~8—13m followed by a high - calcium pyroxene (HCP) layer from ~13—53m (Huang et al., 2018).

3.2 Previous Studies Using LPR Data

Further insights on the ejecta at the VK crater are provided by the LPR mounted to the Yutu - 2 rover of the Chang'E - 4 mission (Li et al., 2020). The first attempt to examine the lunar surface with in - situ LPR equipment occurred during the Chang'E - 3 mission on the near side of the moon (Ding et al., 2020; Fa et al., 2015; Lai et al., 2019; Yuan et al., 2017). Similar antenna configurations were employed for both Chang'E - 3 and Chang'E - 4 missions (Li et al., 2020). In particular, two antennas with 500MHz central frequency (at the bottom of the rover), and one low frequency antenna (mounted at the back of the rover) with 60MHz central frequency (Li et al., 2020). The low frequency antenna from the Chang'E - 4 mission gave thin indications of four different lava flows that probably occurred during the Imbrian period (Lai et al., 2020). Unfortunately, the low frequency data in both missions suffer from ringing noise due to the coupling between the antenna and the rover, which resulted in erroneous reflections and noisy data (Li et al., 2018; Zhang et al., 2020). In contrast to the Chang'E - 3 landing site (Lai et al., 2019), in the VK crater, the ilmenite content is fairly low, making the ejecta layers transparent to LPR (Dong, Feng, Zhao, et al., 2020). This resulted in good quality data that clearly demonstrated a complex layered structure for the first 50m of the VK crater (Li et al., 2020; Zhang et al., 2020). In addition, using a conventional hyperbola - fitting (assuming a homogeneous medium) with Dix conversion, the electric permittivity of the ejecta layers was estimated, and furthermore used to infer the mineralogical (Fe and Ti content) (Li et al., 2020) and the mechanical (density) properties of the lunar regolith (Dong, Feng, Zhao, et al., 2020; Dong, Feng, Zhou, et al., 2020), based on semi - empirical formulas fine - tuned for lunar soils (Carrier et al., 1991; Hickson et al., 2018; Olhoeft & Strangway, 1975), The relative electric permittivity at the landing site monotonically increases from ~3—6 with respect to depth, as estimated using typical hyperbola fitting (Dong, Feng, Zhao, et al., 2020). This corresponds to a density that starts from ~1gr/cm^3 at the surface and reaches 2.5gr/cm^3 at 50m depth (Dong, Feng, Zhao, et al., 2020).

3.3 Revised Stratigraphic Column Using Advanced Hyperbola Fitting

The suggested stratigraphy model is based on the LPR results shown in Fig. 2a and the following premises:

1. Finsen, VK L and L' craters are the predominant sources of the post - Imbrian ejecta in the VK crater (Huang et al., 2018; Paskert et al., 2018; Zhang et al.,

2020).

2. Finsen crater was developed before VK L and L' craters (Huang et al., 2018; Zhang et al., 2020).

3. Finsen ejecta at the landing site are estimated via numerical simulations at ~30m (Di et al., 2019).

4. The thickness of the weathered top soil was calculated at ~2.5—7.5m (Huang et al., 2018) using the geometrical method described in (Quaide William & Oberbeck, 1968). This is consistent with the average regolith growth rate (~1.5m/Ga) from ~3.5 Ga to the present day (Gou et al., 2021; Horz et al., 1991; Oberbeck & Quaide, 1968; Quaide & Oberbeck, 1975).

5. Below the weathered top soil there is an LCP layer down to~13m (Gou et al., 2019; Huang et al., 2018; Li et al., 2019).

6. Below the LCP layer there is a thick layer (>30m) with high HCP/LCP ratio (Huang et al., 2018).

7. Finsen crater is within the SPACA region and therefore it is expected that its excavated materials have an increased HCP/LCP ratio (Moriarty & Pieters, 2018).

8. The ejecta materials from VK L and L' craters have low HCP/LCP ratio (Huang et al., 2018; Hu et al., 2019; Ling et al., 2019).

9. There is a clear sharp boundary observed on LPR data (Dong, Feng, Zhou, et al., 2020; Li et al., 2020; Zhang et al., 2020) at~13m, most-likely between the LCP and the HCP layer.

The proposed stratigraphy model suggests that the HCP layer overlying the Imbrian basalts is the ejecta from the Finsen crater (Huang et al., 2018); and maybe Orientale crater too (Xiao et al., 2021). This premise is consistent both with the size of this layer (as predicted by numerical simulations [Di et al., 2019]) and with the chemical composition of the Finsen crater (Moriarty & Pieters, 2018; Moriarty et al., 2011). On top of the Finsen ejecta, it is expected to encounter ejecta from Eratosthenian post-Finsen craters. A homogeneous weathered layer with 12m depth as suggested by Zhang et al. (2020) is not consistent with LROC NAC images that points to a ~2.5—7.5m regolith (Huang et al., 2018) and by the layered structure revealed by the proposed hyperbola-fitting Scheme lsee Figure 2a). In addition, a 12-13 m regolith indicates a regolith growth rate of~3—4m/Ga which is twice as fast compared to the growth rate expected from ~3.5Ga to the present day (Gou et al., 2021; Horz et al., 1991; Oberbeck & Quaide, 1968; Quaide & Oberbeck, 1975). In general, the average regolith depth in mare areas is expected to be around ~4—5m while a ~10—15m regolith is more common in old lunar highlands (McKay et al., 1991).

We propose that the top ~10—12m of the landing site consists of ejecta from the VK L and L' craters. This is in good agreement with the LCP content of the VK L and L' craters (Huang et al., 2018; Hu et al., 2019; Ling et al., 2019) and with the layered structure illustrated in Fig. 2a. The suggested stratigraphic model of the post - basaltic flood ejecta of VK crater is shown in Figure 2c. The ejecta of VK L’ (≈5.5m) were deposited on top of the Finsen ejecta at early Eratosthenian. Impact gardening degraded the first ~1.5m of the ejecta decreasing its density (Horz et al., 1991) and consequently its electric permittivity (due to the causal rela - tionship between permittivity and density [Chung et al., 1970; Olhoeft & Strangway, 1975]). The width of the VK L’ regolith is in good agreement with the literature which suggests that rapid gardening is expected at young ejecta, a phenomenon that has also been observed at the Chang’E - 3 landing site (Fa et al., 2015; Gou et al., 2021). The ejecta from VK L is subsequently deposited on top of the weathered layer creating a top layer with ~6m width. The long gardening process, from early Eratosthenian till now, gave rise to a ~3m of loose lunar soil with low electric permittivity as predicted by Fig. 2a. This is in good agreement with the regolith thickness as estimated using LROC NAC images (Huang et al., 2018); and also with the average regolith growth rate (~1.5m/Ga) expected from ~3.5 Ga to the present day (Gou et al., 2021; Horz et al., 1991; Oberbeck & Quaide, 1968; Quaide & Oberbeck, 1975).

4 Conclusions

A novel interpretation tool was described capable of estimating the permittivity profile of the shallow lunar surface using LPR. The validity and the superiority of the suggested scheme compared to typical hyperbola - fitting was demonstrated via a set of numerical experiments that have clearly shown that the proposed scheme is capable of reconstructing complicated permittivity profiles using the shape of multiple hyperbolas as the only inputs. The proposed methodology is suitable for any arbitrary one - dimensional permittivity distribution, which makes it an appealing choice for inferring the dielectric properties of the lunar regolith. The advanced hyperbola - fitting was applied to the high frequency data collected during the first two lunar days of the Chang’E - 4 mission. The resulting permittivity profile indicates a layered structure within the first 10m of the regolith. These shallow layers are not visible in the measured radargram due to the smooth boundaries between them, making them undetectable using traditional signal processing approaches. It is argued that the multiple layers detected within the shallow lunar regolith can be the ejecta of the Eratosthenian craters VK L and L’, laying on top of the late - Imbrian ejecta of Finsen crater.

References

[1] Aldo, V. (1996). Ray tracing based on Fermat’s principle in irregular grids. Geophysical Prospecting, 44 (5), 741—760.

[2] Bano, M. (2006). Effects of the transition zone above a water table on the reflection of GPR waves. Geophysical Research Letters, 33 (13). L13309. https://doi.org/10.1029/2006gl026158.

[3] Bhattacharya, A., Porwal, A., Dhingra, S., De, S., & Venkataraman, G. (2015). Remote estimation of dielectric permittivity of lunar surface regolith using compact polarimetric synthetic aperture radar data. Advances in Space Research, 56 (11), 2439-2448. https://doi.org/10.1016/j.asr.2015.10.007.

[4] Birchak, J. R., Gardner, C. G., Hipp, J. E., & Victor, J. M. (1974). High dielectric constant microwave probes for sensing soil moisture. Proceedings of the IEEE, 62 (1), 93-98. https://doi.org/10.1109/PROC.1974.9388.

[5] Boero, F., Fedeli, A., Lanini, M., Maffongelli, M., Monleone, R., Pastorino, M., et al. (2018). Microwave tomography for the inspection of wood materials: Imaging system and experimental results. IEEE Transactions on Microwave Theory and Techniques, 66 (7), 3497-3510. https://doi.org/10.1109/tmtt.2018.2804905.

[6] Campbell, M. J., & Ulrichs, J. (1969). Electrical properties of rocks and their significance for lunar radar observations. Journal of Geophysical Research, 74 (25), 5867-5881. https://doi.org/10.1029/jb074i025p05867.

[7] Carrier, W. D., Olhoeft, G. R., & Mendell, W. (1991). Physical properties of the lunar surface. In Lunar sourcebook (pp. 522-530).

[8] Cassidy, N. J. (2009). Chapter 5 - Ground penetrating radar data processing, modelling and analysis. In H. M. Jol (Ed.), Ground penetrating radar theory and applications (pp. 141—176). Amsterdam: Elsevier. https://doi.org/10.1016/b978—0—444—53348—7.00005—3.

[9] Charlier, B., Grove, T. L., Namur, O., & Holtz, F. (2018). Crystallization of the lunar magma ocean and the primordial mantle - crust differentiation of the moon. Geochimica et Cosmochimica Acta, 234, 50—69. https://doi.org/10.1016/j.gca.2018.05.006.

[10] Chung, D. H., Westphal, W. B., & Simmons, G. (1970). Dielectric properties of Apollo 11 lunar samples and their comparison with Earth materials. Journal of Geophysical Research, 75 (32), 6524-6531. https://doi.org/10.1029/jb075i032p06524.

[11] Conyers, L. B. (2004). Ground penetrating radar for archaeology. Walnut Creek, CA, USA: AltaMira Press.

[12] Daniels, D. J. (2004). Ground penetrating radar. U.K: Institution of Engineering and Technology.

[13] Di, K., Zhu, M. H., Yue, Z., Lin, Y., Wan, W., Liu, Z., & Xue, B. (2019). Topographic evolution of Von Kármán crater revealed by the lunar rover Yutu - 2. Geophysical Research Letters, 46, 12764-12770. https://doi.org/10.1029/2019gl085252.

[14] Diamanti, N., Annan, A. P., & Redman, J. D. (2014). Impact of gradational electrical properties on GPR detection of interfaces. In Proceedings of the 15th international conference on ground penetrating radar (pp. 529 - 534). https://doi.org/10.1109/icgpr.2014.6970480.

[15] Ding, C., Xiao, Z., Su, Y., & Cui, J. (2020). Hyperbolic reflectors determined from peak echoes of ground penetrating radar. Icarus, 358, 114280.

[16] Dix, C. H. (1955). Seismic velocities from surface measurements. Geophysics, 20, 68 - 86. https://doi.org/10.1190/1.1438126.

[17] Dobson, M. C., Ulaby, F. T., Hallikainen, M. T., & El - rayes, M. A. (1985). Microwave dielectric behavior of wet soil - part ii: Dielectric mixing models. IEEE Transactions on Geoscience and Remote Sensing, 23 (1), 35 - 46. https://doi.org/10.1109/tgrs.1985.289498.

[18] Dong, Z., Fang, G., Zhou, B., Zhao, D., Gao, Y., & Ji, Y. (2021). Properties of lunar regolith on the moon's farside unveiled by chang'E - 4 lunar penetrating radar. Journal of Geophysical Research: Planets, 126 (6), e2020JE006564. https://doi.org/10.1029/2020je006564.

[19] Dong, Z., Feng, G., Zhao, D., Zhou, B., Gao, Y., & Ji, Y. (2020). Dielectric properties of lunar subsurface materials. Geophysical Research Letters, 47, e2020GL089264. https://doi.org/10.1029/2020gl089264.

[20] Dong, Z., Feng, X., Zhou, H., Liu, C., Zeng, Z., Li, J., & Liang, W. (2020). Properties analysis of lunar regolith at Chang'E - 4 landing site based on 3d velocity spectrum of lunar penetrating radar. Remote Sensing, 12, 629. https://doi.org/10.3390/rs12040629.

[21] Fa, W. (2020). Bulk density of the lunar regolith at the Chang'E - 3 site as estimated from lunar penetrating radar. Earth and Space Science, 7 (2), e2019EA000801. https://doi.org/10.1029/2019ea000801.

[22] Fa, W., Zhu, M. - H., Liu, T., & Plescia, J. B. (2015). Regolith stratigraphy at the Chang'E - 3 landing site as seen by lunar penetrating radar. Geophysical Research Letters, 42 (23), 10179 - 10187. https://doi.org/10.1002/2015gl066537.

[23] Feng, X., Sato, M., & Liu, C. (2012). Subsurface imaging using a handheld GPR md system. IEEE Geoscience and Remote Sensing Letters, 9 (4), 659 - 662. https://doi.org/10.1109/lgrs.2011.2177514.

[24] Frisillo, A. L., Olhoeft, G. R., & Strangway, D. W. (1975). Effects of vertical stress, temperature and density on the dielectric properties of lunar samples 72441, 12, 15301, 38 and a terrestrial basalt. Earth and Planetary Science Letters, 24, 345 - 356. https://doi.org/10.1016/0012 - 821x (75) 90140 - 5.

[25] Garrick - Bethell, I., & Miljković, K. (2018). Age of the lunar South Pole Aitken basin. In Lunar and planetary science conference.

[26] Garrick-Bethell, I., & Zuber, M. T. (2009). Elliptical structur e of the lunar South Pole-Aitken basin. Icarus, 204 (2), 399-408. https://doi.org/10.1016/j.icarus.2009.05.032.

[27] Giannakis, I., Giannopoulos, A., & Warren, C. (2016). A realistic FDTD numerical modeling framework of ground penetrating radar for landmine detection. IEEE Journal of Selected Topics in Applied Earth Observations and Remote Sensing, 9 (1), 37-51. https://doi.org/10.1109/jstars.2015.2468597.

[28] Giannakis, I., Giannopoulos, A., & Warren, C. (2019). A machine learning based fast-forward solver for ground penetrating radar with application to full-waveform inversion. IEEE Transactions on Geoscience and Remote Sensing, 57 (7), 4417-4426. https://doi.org/10.1109/tgrs.2019.2891206.

[29] Giannakis, I., Giannopoulos, A., & Warren, C. (2020). A machine learning scheme for estimating the diameter of reinforcing bars using ground penetrating radar. IEEE Geoscience and Remote Sensing Letters, 18, 461-465.

[30] Giannopoulos, A. (2008). An improved new implementation of complex frequency shifted PML for the FDTD method. IEEE Transactions on Antennas and Propagation, 56 (9), 2995-3000. https://doi.org/10.1109/tap.2008.928789.

[31] Gong, X., Paige, D. A., Siegler, M. A., & Jin, Y.-Q. (2015). Inversion of dielectric properties of the lunar regolith media with temperature profiles using chang'E microwave radiometer observations. IEEE Geoscience and Remote Sensing Letters, 12 (2), 384-388. https://doi.org/10.1109/LGRS.2014.2343617.

[32] Gou, S., Di, K., Yue, Z., Liu, Z., He, Z., Xu, R., et al. (2019). Lunar deep materials observed by chang'E-4 rover. Earth and Planetary Science Letters, 528, 115829. https://doi.org/10.1016/j.epsl.2019.115829.

[33] Gou, S., Di, K., Yue, Z., Liu, Z., He, Z., Xu, R., et al. (2020). Forsteritic olivine and magnesium-rich orthopyroxene materials measured by Chang'E-4 rover. Icarus, 345, 113776. https://doi.org/10.1016/j.icarus.2020.113776.

[34] Gou, S., Yue, Z., Di, K., Cai, Z., Liu, Z., & Niu, S. (2021). Absolute model age of lunar finsen crater and geologic implications. Icarus, 354, 114046. https://doi.org/10.1016/j.icarus.2020.114046.

[35] Guo, D., Fa, W., Zeng, X., Du, J., & Liu, J. (2021). Geochemistry of the von Kármán crater floor and thickness of the non-mare ejecta over the chang'E-4 landing area. Icarus, 359, 114327. https://doi.org/10.1016/j.icarus.2021.114327.

[36] Hallikainen, M. T., Ulaby, F. T., Dobson, M. C., El-rayes, M. A., & Wu, L. (1985). Microwave dielectric behavior of wet soil-part 1: Empirical models and experimental observations. IEEE Transactions on Geoscience and Remote Sensing, 23 (1), 25-34. https://doi.org/10.1109/tgrs.1985.289497.

[37] Hammond, N. , Nimmo, F. , & Korycansky, D. (2009) . Hydrocode modeling of the South Pole Aitken basin - forming impact. In Lunar and planetary sciences conference (LPSC) .

[38] Hansen, P. C. (1992) . Analysis of discrete ill - posed problems by means of the l - curve. SIAM Review, 34, 561 - 580. https://doi.org/10.1137/1034115.

[39] Haruyama, J. , Ohtake, M. , Matsunaga, T. , Morota, T. , Honda, C. , Yokota, Y. , et al. (2009) . Long lived volcanism on the lunar farside revealed by selene terrain camera. Science, 323 (5916), 905 - 908. https://doi.org/10.1126/science.1163382.

[40] Hickson, D. , Boivin, A. , Daly, M. G. , Ghent, R. , Nolan, M. C. , Tait, K. , & Tsai, C. A. (2018) . Near surface bulk density estimates of NEAs from radar observations and permittivity measurements of powdered geologic material. Icarus, 306. https://doi.org/10.1016/j.icarus.2018.01.018.

[41] Hiesinger, H. , vander Bogert, C. H. , Pascker, J. H. , Schmedemann, N. , Robinson, M. S. , Jolliff, B. , & Petro, N. (2012) . New crater size - fre - quency distribution measurements of the South Pole Aitken basin. In Lunar and planetary science conference.

[42] Horz, F. , Grive, R. A. F. , Heiken, G. , Spudis, P. D. , & Binder, A. B. (1991) . Lunar surface processes. In G. Heiken, D. Vaniman, & B. M. French (Eds.), Lunar sourcebook (Vol. 121, pp. 61 - 120) . New York: Cambridge University Press.

[43] Hu, X. , Ma, P. , Yang, Y. , Zhu, M. , Jiang, T. , Lucey, P. G. , et al. (2019) . Mineral abundances inferred from insitu reflectance measurements of Chang'E - 4 landing site in South - Pole Atkins. Geophysical Research Letters, 46, 9439 - 9447. https://doi.org/10.1029/2019gl084531.

[44] Huang, J. , Xiao, Z. , Flahaut, J. , Martinot, M. , Head, J. , Xiao, X. , et al. (2018) . Geological characteristics of Von Kármán crater, north - western South Pole - Aitken basin: Chang'E - 4 landing site region. Journal of Geophysical Research: Planets, 123, 1684 - 1700. https://doi.org/10.1029/2018je005577.

[45] Hurwitz, D. M. , & Kring, D. A. (2014) . Differentiation of the South Pole - Aitken basin impact melt sheet: Implications for lunar exploration. Journal of Geophysical Research: Planets, 119 (6), 1110 - 1133. https://doi.org/10.1002/2013JE004530.

[46] Ivanov, M. A. , Hiesinger, H. , van derBogert, C. H. , Orgel, C. , Pasckert, J. H. , & Head, J. W. (2018) . Geologic history of the northern portion of the South Pole - Aitken basin on the Moon. Journal of Geophysical Research: Planets, 123, 2585 - 2612. https://doi.org/10.1029/2018je005590.

[47] James, P. B. , Smith, D. E. , Byrne, P. K. , Kendall, J. D. , Melosh, H. J. , & Zuber, M. T. (2019) . Deep structure of the lunar South Pole - Aitken basin.

Geophysical Research Letters, 46, 5100 - 5106. https://doi.org/10.1029/2019gl082252.

[49] Jolliff, B. L., Gillis, J. J., Haskin, L. A., Korotev, R. L., & Wieczorek, M. A. (2000). Major lunar crustal terranes: Surface expressions and crust - mantle origin. Journal of Geophysical Research, 105, 4197 - 4216. https://doi.org/10.1029/1999je001103.

[49] Kennedy, J., & Eberhart, R. (1995). Particle swarm optimization. In Proceedings of ICNN'95 - International Conference on Neural Networks, 4, 1942 - 1948.

[50] Khan, A. (2002). An inquiry into the lunar interior: A nonlinear inversion of the Apollo lunar seismic data. Journal of Geophysical Research, 107, 3 - 1 - 3 - 23. https://doi.org/10.1029/2001je001658.

[51] Kuskov, O. L., Kronrod, E. V., & Kronrod, V. A. (2015). Thermochemical constraints on the thermal state, composition, and mineralogy of the upper mantle of the moon: Evidence from the seismic models. Solar System Research, 49, 75 - 91. https://doi.org/10.1134/s0038094615010049.

[52] Lai, J., Xu, Y., Bugiolacchi, R., Meng, X., Xiao, L., Xie, M., et al. (2020). First look by the Yutu - 2 rover at the deep subsurface structure at the lunar farside. Nature Communications, 11, 3426. https://doi.org/10.1038/s41467 - 020 - 17262 - w.

[53] Lai, J., Xu, Y., Zhang, X., & Tang, Z. (2016). Structural analysis of lunar subsurface with chang e - 3 lunar penetrating radar. Planetary and Space Science, 120, 96 - 102. https://doi.org/10.1016/j.pss.2015.10.014.

[54] Lai, J., Xu, Y., Zhang, X., Xiao, L., Yan, Q., Meng, X., et al. (2019). Comparison of dielectric properties and structure of lunar regolith at Chang'E - 3 and Chang'E - 4 landing sites revealed by ground - penetrating radar. Geophysical Research Letters, 46, 12783 - 12793. https://doi.org/10.1029/2019gl084458.

[55] Lauro, S. E., Pettinelli, E., Caprarelli, G., Guallini, L., Rossi, A. P., Mattei, E., & R, O. (2020). Multiple subglacial water bodies below the south pole of Mars unveiled by new MARSIS data. Nature Astronomy, 5. https://doi.org/10.1038/s41550 - 020 - 1200 - 6.

[56] Lemelin, M., Lucey, P. G., Miljković, K., Gaddis, L. R., Hare, T., & Ohtake, M. (2019). The compositions of the lunar crust and upper mantle: Spectral analysis of the inner rings of lunar impact basins. Planetary and Space Science, 165, 230 - 243. https://doi.org/10.1016/j.pss.2018.10.003.

[57] Li, C., Liu, D., Liu, B., Ren, X., Liu, J., He, Z., et al. (2019). Chang'e - 4 initial spectroscopic identification of lunar far - side mantle - derived materials. Nature, 569, 378 - 382. https://doi.org/10.1038/s41586 - 019 - 1189 - 0.

[58] Li, C., Su, Y., Pettinelli, E., Xing, S., Ding, C., Liu, J., & Zhang, H. (2020). The moon's farside shallow subsurface structure unveiled by Chang'E - 4

lunar penetrating radar. Science Advances, 6 (9) . https://doi.org/10.1126/sciadv.aay6898.

[59] Li, C., Xing, S., Lauro, S. E., Su, Y., Dai, S., Feng, J., et al. (2018) . Pitfalls in GPR data interpretation: False reflectors detected in lunar radar cross sections by Chang'E - 3. IEEE Transactions on Geoscience and Remote Sensing, 56 (3), 1325 - 1335. https://doi.org/10.1109/tgrs.2017.2761881.

[60] Li, C., Zuo, W., Wen, W., Zeng, X., Gao, X., Liu, Y., et al. (2021) . Overview of the chang'E - 4 mission: Opening the fr ontier of scientific exploration of the lunar far side. Space Science Reviews, 217, 35. https://doi.org/10.1007/s11214 - 021 - 00793 - z.

[61] Li, J., Liu, C., Zeng, Z., & Chen, L. (2015) . GPR signal denoising and target extraction with the CEEMD method. IEEE Geoscience and Remote Sensing Letters, 12 (8), 1615 - 1619. https://doi.org/10.1109/lgrs.2015.2415736.

[62] Lin, H., Lin, Y., Yang, W., He, Z., Hu, S., Wei, Y., & Zou, Y. (2020) . New insight into lunar regolith forming processes by the lunar rover Yutu - 2. Geophysical Research Letters, 47, e2020GL087949. https://doi.org/10.1029/2020gl087949.

[63] Lin, Y., Tronche, E. J., Steenstra, E. S., & van Westrenen, W. (2017) . Evidence for an early wet moon from experimental crystallization of the lunar magma ocean. Nature Geosciences, 10, 14 - 18. https://doi.org/10.1038/ngeo2845.

[64] Ling, Z., Qiao, L., Liu, C., Cao, H., Bi, X., Lu, X., et al. (2019) . Composition, mineralogy and chronology of mare basalts and non - mare mate - rials in Von Kármán crater: Landing site of the Chang'E - 4 mission. Planetary and Space Science, 179, 104741. https://doi.org/10.1016/j.pss.2019.104741.

[65] Losiak, A., Wilhelms, D. E., Byrne, C. J., Thaisen, K., Weider, S. Z., Kohout, T., & Kring, D. A. (2009) . A new lunar impact crater database. In Lunar and planetary science conference.

[66] Lu, Y., Wu, Y., Michael, G. G., Ma, J., Cai, W., & Qin, N. (2021) . Chronological sequence of chang'E - 4 landing zone within von Kármán crater. Icarus, 354, 114086. https://doi.org/10.1016/j.icarus.2020.114086.

[67] Lucey, P. G. (2004) . Mineral maps of the moon. Geophysical Research Letters, 31 (8), L08701. https://doi.org/10.1029/2003gl019406.

[68] Lucey, P. G., Taylor, G. J., Hawke, B. R., & Spudis, P. D. (1998) . Feo and tio2 concentrations in the South Pole - Aitken basin: Implications for mantle composition and basin formation. Journal of Geophysical Research, 103 (E2), 3701 - 3708. https://doi.org/10.1029/97je03146.

[69] Matsunaga, T., Ohtake, M., Haruyama, J., Ogawa, Y., Nakamura, R., Yokota, Y., et al. (2008) . Discoveries on the lithology of lunar crater central peaks. Geophysical Research Letters, 35, L23201. https://doi.org/10.1029/

2008gl035868.

[70] McKay, D. S., Heiken, G., Basu, A., Blanford, G., Simon, S., Reedy, R., & Papike, J. (1991). Physical properties of the lunar surface. In Lunar sourcebook (pp. 522 - 530).

[71] Meles, G. A., Van der Kruk, J., Greenhalgh, S. A., Ernst, J. R., Maurer, H., & Green, A. G. (2010). A new vector waveform inversion algo - rithm for simultaneous updating of conductivity and permittivity parameters from combination crosshole/borehole - to - surface GPR data. IEEE Transactions on Geoscience and Remote Sensing, 48 (9), 3391 - 3407. https://doi.org/10. 1109/tgrs. 2010. 2046670.

[72] Melosh, H., Kendall, J., Horgan, B., Johnson, B., Bowling, T., Lucey, P., & Taylor, G. (2017). South Pole - Aitken basin ejecta reveal the Moon's upper mantle. Geology, 45 (12), 1063 - 1066. https://doi.org/10.1130/g39375.1.

[73] Mertens, L., Persico, R., Matera, L., & Lambot, S. (2016). Automated detection of reflection hyperbolas in complex GPR images with no a priori knowledge on the medium. IEEE Transactions on Geoscience and Remote Sensing, 54 (1), 580 - 596. https://doi.org/10.1109/tgrs.2015.2462727.

[74] Moriarty, D. P., Ⅲ, & Pieters, C. M. (2015). The nature and origin of mafic mound in the South Pole - Aitken basin. Geophysical Research Letters, 42 (19), 7907 - 7915. https://doi.org/10.1002/2015gl065718.

[75] Moriarty, D. P., Ⅲ, Watkins, R. N., Valencia, S. N., Kendall, J. D., Evans, A. J., Dygert, N., & Petro, N. E. (2021). Evidence for a stratified upper mantle preserved within the South Pole - Aitken basin. Journal of Geophysical Research: Planets, 126 (1), e2020JE006589. https://doi.org /10.1029/2020je006589.

[76] Moriarty, D. P., & Pieters, C. M. (2018). The character of South Pole - Aitken Basin: Patterns of surface and subsurface composition. Journal of Geophysical Research: Planets, 123, 729 - 747. https://doi.org/10.1002/2017je005364.

[77] Moriarty, D. P., Pieters, C. M., & Isaacson, P. J. (2013). Compositional heterogeneity of central peaks within the South Pole - Aitken Basin. Journal of Geophysical Research: Planets, 118, 2310 - 2322. https://doi.org/10.1002/2013je004376.

[78] Moriarty, D. P., Pieters, C. M., Nettles, J., Isaacson, P. J., Cheek, L., Head, J. W., & Petro, N. (2011). Finsen and alder: A compositional study of lunar central peak craters in the South Pole - Aitken basin. In Lunar and planetary science conference.

[79] Nakamura, R., Matsunaga, T., Ogawa, Y., Yamamoto, S., Hiroi, T., Saiki,

K., & Yokota, Y. (2009). Ultramafic impact melt sheet beneath the South Pole - Aitken basin on the moon. Geophysical Research Letters, 36 (22), L22202. https://doi.org/10.1029/2009gl040765.

[80] Nash, D. B., & Conel, J. E. (1973). Vitrification darkening of rock powders: Implications for optical properties of the lunar surface. The Moon, 8 (3), 346 - 364. https://doi.org/10.1007/bf00581729.

[81] Oberbeck, V. R., & Quaide, W. L. (1968). Genetic implications of lunar regolith thickness variations. Icarus, 9 (1), 446 - 465. https://doi.org/10.1016/0019 - 1035 (68) 90039 - 0.

[82] Olhoeft, G. R., & Strangway, D. W. (1975). Dielectric properties of the first 100 meters of the moon. In Lunar sourcebook (Vol. 24). https://doi.org/10.1016/0012 - 821x (75) 90146 - 6.

[83] Paskert, J. H., Hiesinger, H., & van der Bogert, C. H. (2018). Lunar far side volcanism in and around the South Pole - Aitken basin. Icarus, 299, 538 - 562.

[84] Peplinski, N. R., Ulaby, F. T., & Dobson, M. C. (1995). Dielectric properties of soils in the 0.3 - 1.3 - ghz range. IEEE Transactions on Geoscience and Remote Sensing, 33 (3), 803 - 807. https://doi.org/10.1109/36.387598.

[85] Pieters, C. M., & Noble, S. K. (2016). Space weathering on airless bodies. Journal of Geophysical Research: Planets, 121 (10), 1865 - 1884. https://doi.org/10.1002/2016je005128.

[86] Potter, R. W. K., Collins, G. S., Kiefer, W. S., McGovern, P. J., & Kring, D. A. (2012). Constraining the size of the South Pole - Aitken basin impact. Icarus, 220, 730 -743. https://doi.org/10.1016/j.icarus.2012.05.032.

[87] Qiao, L., Ling, Z., Fu, X., & Li, B. (2019). Geological characterization of the chang'E - 4 landing area on the lunar farside. Icarus, 333, 37 - 51. https://doi.org/10.1016/j.icarus.2019.05.029.

[88] Quaide, W., & Oberbeck, V. (1975). Development of the mare regolith: Some model considerations. The Moon, 13 (1), 27 - 55. https://doi.org/10.1007/bf00567506.

[89] Quaide William, L., & Oberbeck, V. R. (1968). Thickness determinations of the lunar surface layer from lunar impact craters. Journal of Geophysical Research, 73 (16), 5247 - 5270. https://doi.org/10.1029/jb073i016p05247.

[90] Schultz, P. H., & Crawford, D. A. (2011). Origin of nearside structural and geochemical anomalies on the Moon. In Recent Advances and Current Research Issues in Lunar Stratigraphy.

[91] Shkuratov, Y. G., & Bondarenko, N. Y. (2001). Regolith layer thickness mapping of the Moon by radar and optical data. Icarus, 149, 329 - 338. https://doi.org/10.1006/icar.2000.6545.

[92] Snyder, G. A., Taylor, L. A., & Neal, C. R. (1992). A chemical model for

generating the sources of mare basalts: Combined equilibrium and fractional crystallization of the lunar magmasphere. Geochimica et Cosmochimica Acta, 56 (10), 3809 - 3823. https://doi.org/10.1016/0016 - 7037 (92) 90172 - f.

[93] Stopar, J. D., Robinson, M. S., Barnouin, O. S., McEwen, A. S., Speyerer, E. J., Henriksen, M. R., & Sutton, S. S. (2017). Relative depths of simple craters and the nature of the lunar regolith. Icarus, 298, 34 - 48. https://doi.org/10.1016/j.icarus.2017.05.022.

[94] Taflove, A., & Hagness, S. C. (2000). Computational electrodynamics: The finite - difference time - domain method. Norwood, MA, USA: Artech House.

[95] Tang, Z., Liu, J., Wang, X., Ren, X., Chen, W., Yan, W., et al. (2020). Physical and mechanical characteristics of lunar soil at the Chang'E - 4 landing site. Geophysical Research Letters, 47, e2020GL089499. https://doi.org/10.1029/2020gl089499.

[96] Tompkins, S., & Pieters, C. M. (1999). Mineralogy of thelunar crust: Results from Clementine. Meteoritics & Planetary Science, 34, 25 - 41. https://doi.org/10.1111/j.1945 - 5100.1999.tb01729.x.

[97] Uemoto, K., Ohtake, M., Haruyama, J., Matsunaga, T., Yamamoto, S., Nakamura, R., et al. (2017). Evidence of impact melt sheet differentiation of the lunar South Pole - Aitken basin. Journal of Geophysical Research: Planets, 122 (8), 1672 - 1686. https://doi.org/10.1002/2016je005209.

[98] Vaughan, W. M., & Head, J. W. (2014). Impact melt differentiation in the South Pole - Aitken basin: Some observations and speculations. Planetary and Space Science, 91, 101 - 106. https://doi.org/10.1016/j.pss.2013.11.010.

[99] Wai - Lok Lai, W., Dérobert, X., & Annan, P. (2018). A review of ground penetrating radar application in civil engineering: A 30 - year journey from locating and testing to imaging and diagnosis. NDT & E International, 96, 58 - 78. https://doi.org/10.1016/j.ndteint.2017.04.002.

[100] arren, C., Giannopoulos, A., & Giannakis, I. (2016). gprMax: Open source software to simulate electromagnetic wave propagation for ground penetrating radar. Computer Physics Communications, 209, 163 - 170. https://doi.org/10.1016/j.cpc.2016.08.020.

[101] Warren, C., Giannopoulos, A., Gray, A., Giannakis, I., Patterson, A., Wetter, L., & Hamrah, A. (2019). A CUDA - based GPU engine for gprMax: Open source FDTD electromagnetic simulation software. Computer Physics Communications, 237, 208 - 218. https://doi.org/10.1016/j.cpc.2018.11.007.

[102] Weiren, W., Chunlai, L., Wei, Z., Hongbo, Z., Jianjun, L., Weibin, W., & Ziyuan, O. (2019). Lunar farside to be explored by chang'e - 4. Nature Geoscience, 12 (7), 222 - 223.

[103] Wieczorek, M. A., Neumann, G. A., Nimmo, F., Kiefer, W. S., Taylor, G. J., Melosh, H. J., et al. (2013). The crust of the Moon as seen by GRAIL. Science, 339, 671-675. https://doi.org/10.1126/science.1231530.

[104] Wieczorek, M. A., & Phillips, R. J. (1999). Lunar multiring basins and the cratering process. Icarus, 139 (2), 246-259. https://doi.org/10.1006/icar.1999.6102.

[105] Wilhelms, D. E., & McCauley, J. F. (1987). The geologic history of the moon.

[106] Williams, R. M., Ray, L. E., Lever, J. H., & Burzynski, A. M. (2014). Crevasse detection in ice sheets using ground penetrating radar and machine learning. IEEE Journal of Selected Topics in Applied Earth Observations and Remote Sensing, 7 (12), 4836-4848. https://doi.org/10.1109 /jstars.2014.2332872.

[107] Xiao, Z., Ding, C., Xie, M., Cai, Y., Cui, J., Zhang, K., & Wang, J. (2021). Ejecta from the orientale basin at the chang'E-4 landing site. Geophysical Research Letters, 48, e2020GL090935. https://doi.org/10.1029/2020GL090935.

[108] Yamamoto, S. R., Nakamura, R., Matsunaga, T., Ogawa, Y., Ishihara, Y., Morota, T., et al. (2010). Possible mantle origin of olivine around lunar impact basins detected by SELENE. Nature Geosciences, 3, 533-536. https://doi.org/10.1038/ngeo897.

[109] Yee, K. (1966). Numerical solution of initial boundary value problems involving Maxwell's equations in isotropic media. IEEE Transactions on Antennas and Propagation, 14 (3), 302-307. https://doi.org/10.1109/tap.1966.1138693.

[110] Yelf, R. (2004). Where is true time zero? In Proceedings of the tenth international conference on grounds penetrating radar, 2004. GPR 2004 (Vol. 1, pp. 279-282).

[111] Yingst, R. A., Chuang, F. C., Berman, D. C., & Mest, S. C. (2017). Geologic mapping of the planck quadrangle of the moon. In Lunar and planetary science conference.

[112] Yuan, Y., Zhu, P., Zhao, N., Xiao, L., Garnero, E., Xiao, Z., et al. (2017). The 3-d geological model around chang'E-3 landing site based on lunar penetrating radar channel 1 data. Geophysical Research Letters, 44 (13), 6553-6561. https://doi.org/10.1002/2017gl073589.

[113] Yue, Z., Yang, M., Jia, M., Michael, G., Di, K., Gou, S., & Liu, J. (2020). Refined model age for orientale basin derived from zonal crater dating of its ejecta. Icarus, 346, 113804. https://doi.org/10.1016/j.icarus.2020.113804.

[114] Zhang, L., Li, J., Zeng, Z., Xu, Y., Liu, C., & Chen, S. (2020). Stratigraphy of the Von Kármáncrater based on Chang'E-4 lunar penetrating

radar data. Geophysical Research Letters，47，e2020GL088680. https://doi. org/10. 1029/2020gl088680.

[115] Zheng，Y.，Wang，S.，Feng，J.，Ouyang，Z.，& Li，X. (2005) . Measurement of the complex permittivity of dry rocks and minerals：Application of polythene dilution method and Licht enecker's mixture formulae. Geophysical Journal International，163 (3)，1195 - 1202. https://doi. org/10. 1111/j. 1365-246x. 2005. 02718. x.

[127] Zongcheng，L.，Bradley，L. J.，Alian，W.，Chunlai，L.，Jianzhong，L.，Jiang，Z.，et al. (2015) . Correlated compositional and mineralogical investigations at the Chang'E-3 landing site. Nature Communication，6，880.

Geologically Old but Freshly Exposed Rock Fragments Encountered by Yutu – 2 Rover

Sheng Gou[1,2,3], Zongyu Yue[1,4], Kaichang Di[1,4], Roberto Bugiolacchi[2,3,5], Wenhui Wan[1], Meiping Yang[1,6], and Lejia Ye[1,6]

1. State Key Laboratory of Remote Sensing Science, Aerospace Information Research Institute, Chinese Academy of Sciences, Beijing, China

2. State Key Laboratory of Lunar and Planetary Sciences, Macau University of Science and Technology, Macau, China

3. CNSA Macau Center for Space Exploration and Science, Macau, China

4. Center for Excellence in Comparative Planetology, Chinese Academy of Sciences, Hefei, China

5. Earth Sciences, University College London, London, UK

6. University of Chinese Academy of Sciences, Beijing, China

Abstract: During the exploration traverse of the Von Kármán crater on the lunar farside, the Yutu – 2 rover, part of China's Chang'e – 4 mission, come across rock fragments with higher albedo than the surrounding surface. This study investigates their origin, emplacement, and composition based on in situ spectral analysis and geological considerations. The spectral signatures of two fragments show a clear dominance of low – calcium pyroxenes against the high – calcium phases characteristic of the basalts infilling the Von Kármán crater. Based on spectral considerations, this suggests an allochthonous origin for the ejecta fragments, with the most likely source attributable to the Imbrian 73 – km Finsen crater. We propose that these rock fragments might have had a convoluted history, born as ejected target materials from the Imbrian Finsen impact, possibly buried and resurfaced several times within Von Kármán by the churning mechanisms that continuously reshape the lunar surface, and finally exhumed by a secondary impact from the Copernican 3. 8 – km Zhinyu crater.

Plain Language Summary: Due to widespread micro to macro impacts and diurnal temperature cycling, most exposed rocks on the lunar surface turn into small fragments and dust within a few millions of years, leaving the surface covered by a type of soil, called regolith, with a relatively uniform color and texture. The Yutu – 2 rover from China is exploring the floor of the Von Kármán crater on the relatively unknown lunar farside. On 2 days of its journey, a few light – colored rock fragments stood out from the predominantly dark materials making up most of its surface. By analyzing their color and composition through dedicated instruments, it becomes apparent that these rocks are out of place and very similar in composition, at least within one mineral component called pyroxene, to the Imbrian Finsen crater materials. Some fragments string out in long chains around a 15 - m elliptical fresh crater, pointing to their origin. Further analysis suggests that these rock fragments might have been buried in the regolith, and then exposed again by a secondary crater formed by ejecta from the "recent" Copernican Zhinyu crater.

Key Points:

- Yutu—2 rover observed two rock fragments with different physical attributes from the surrounding surface
- Spectral evidence suggests they might originate from the 73—km Finsen crater, about 140 km away from the finding site
- The fragments might have been brought to the surface about 16 Myr by the ejecta from the 3. 8—km Zhinyu crater

1 Introduction

In January 2019, China's Chang'e - 4 lander touched down on the floor of Von Kármán crater at (177. 588°E, 45. 457°S) within the South Pole - Aitken (SPA) basin on the lunar farside. It then deployed a rover called Yutu - 2 to conduct measurements along its traverse path (Fig. 1) . The Von Kármán is a complex crater with an estimated model age of 3. 97Ga (Yingst et al. , 2017), a diameter of 186 and a maximum depth of about 2. 4km. The crater was infilled with basalts by several extrusive events during the Imbrian period (Gou et al. , 2021a; Huang et al. , 2018), each time resetting the low - to - mid - size crater populations. Based on orbital and in situ measured data, the surface materials of the Chang'e - 4 landing site are generally thought to represent Imbrian Finsen crater ejecta (e. g. , Di et al. , 2019; Gou et al. , 2019, 2020b; Huang et al. , 2018; Lai et al. , 2020; Li et al. , 2019b; Lin et al. , 2020), which has been gardened into mature lunar regolith over billions of years (Gou et al. , 2020b) .

As of December 21, 2020, the Yutu - 2 rover had already operated for 25 lunar days and traversed about 601m within the Von Kármán crater. During the 3rd (February 28 to March 13, 2019) and 13th (December 20, 2019 to January 2, 2020) lunar days, the Yutu - 2 rover encountered unusual rock fragments of various sizes, characterized by higher albedo than the background. Due to the combined entropic action of micro to macro impacts and the extreme thermal excursions in the airless lunar environment, most exposed rock fragments on the floor of Von Kármán crater should have been fragmented over millions of years (Basilevsky et al. , 2013; Hörz et al. , 1975; Molaro & Byrne,

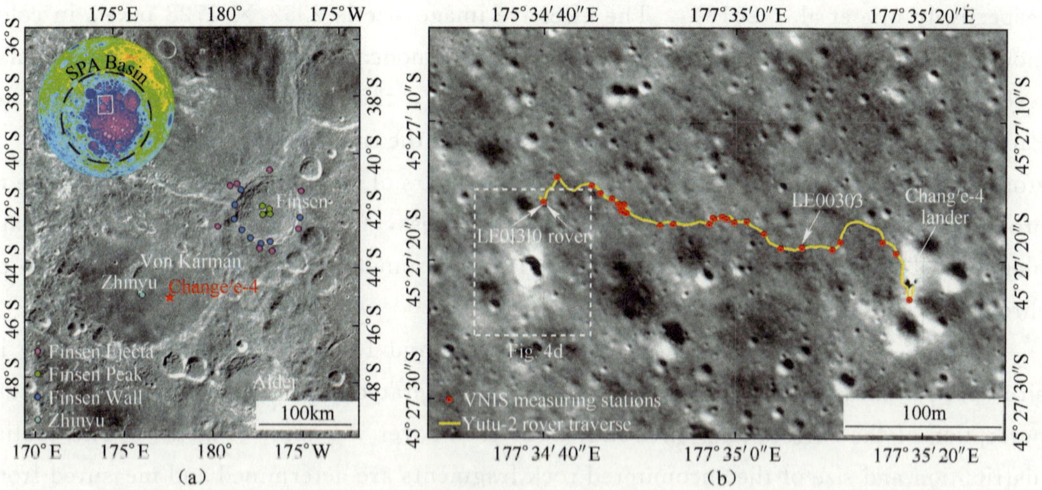

Fig. 1　(a) Geologic context of Chang'e－4 landing site, marked by a red star. The base map is a digital orthophoto map mosaic (～7m/pixel) generated from Chang'e－2 images (Ren et al., 2014). The Lunar Orbiter Laser Altimeter (LOLA) topography elevation (Smith et al., 2010) inset shows the extent of the figure (white box) within the SPA basin. Colorsr epresent different heights in the inset, with red being the highest and purple the lowest. Dots with different colors correspond to locations of Moon Mineralogy Mapper (M^3) spectra, which are used for spectral comparison (see Section 2.3). This figure is modified after Fig. 1 in Gou et al. (2020c); (b) Traverse map of Yutu－2 rover over the past 13 lunar days. The base map is a Lunar Reconnaissance Orbiter Camera (LROC) Narrow Angle Camera (NAC; Robinson et al., 2010) image (～1m/pixel), which was acquired on December 24, 2019 (ID: M1331814485LE). SPA, South Pole－Aitken

2012), leaving the surface covered by a compacted layer of regolith (agglutinated fragments) and dust, with a largely uniform color and texture.

To unravel the origin and composition of the rock fragments, we analyzed images and spectra acquired by the panoramic camera (Pancam) and visible/near－infrared spectrometer (VNIS) instruments onboard the Yutu－2 rover. This work aims at improving our understanding of the evolution of regolith in the vicinity of the Chang'e－four landing site within the Von Kármán crater.

2　Data

2.1　Pancam Images

The Pancam instrument, mounted on the mast of Yutu－2 rover, is composed of two cameras with identical functions, performances, and interfaces (Jia et al., 2018). Each camera has a stereo baseline of 270mm and acquires an image in a spectral range of 420—700nm, with a horizontal and vertical field of views (FOVs) being 19.7° and 14.6°, respectively (Jia et al. , 2018) . The effective image size is 2352 × 1728 pixels in color mode and 1, 176 × 864 pixels in panchromatic mode. The Pancam has an angular resolution of 0. 15 mrad (about 120 pixels/degree), thus the spatial resolution of the Pancam image depends strongly on the distance from the camera (rover) . When the rover stops at a certain waypoint, the Pancam takes 28 pairs of stereo images at a pitch angle to produce a 360° panoramic view for location - awareness, terrain reconstruction, and path planning. To cover a larger vertical FOV, Pancam sometimes acquires images at two or three pitch angles (Liu et al. , 2020) .

Pancam acquired stereo images during the 3rd and 13th lunar day before making in situ spectral measurements. Pancam mosaic is generated from these images for visual perception of the surrounding environment at each rover waypoint (Fig. 2) . The spatial distribution and size of the encountered rock fragments are determined and measured from the Pancam stereo images using an in - house developed photogrammetric mapping software (Di et al. , 2016) . Note that the size discussed in this study is the apparent width, which is calculated as the distance between the leftmost and rightmost ends of the rock fragment's visible facies.

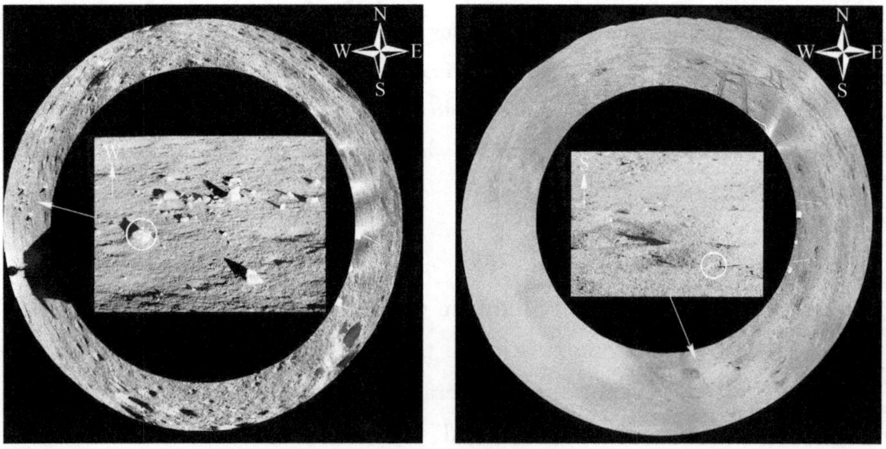

(a)　(b)

Fig. 2 Rock fragments encountered by the Yutu - 2 rover. (a and b) are Pancam mosaics generated from Pancam's left - eye images acquired during the 3rd and 13th lunar day, respectively. The targeted fragments at waypoints LE00303 and LE01310 for spectral measurements are indicated by white circles in each Pancam image. The width of each rover rut is 15cm, and the tread of the rover is about 85cm. They can be used as scale references in both mosaics

2. 2 In Situ Measured Spectra

The VNIS instrument includes a visible/near - infrared (VIS/NIR) imaging spectrometer in the range 450—945nm with a spectral resolution of 2—7nm, a shortwave infrared (SWIR) detector that works from 900 to 2,395nm with a spectral resolution of 3—12nm, and a white panel for calibration and dust-proofing (Li et al., 2019b). The effective image size of the VIS/NIR imaging spectrometer is 256 × 256 pixels, and the spatial resolution is ~1mm/pixel typically from 1m distance. The FOV of the SWIR detector corresponds to a circular region in the VIS/NIR image with the center located at (96, 128) and a radius of 54 pixels (Li et al., 2019b).

To ensure the safety of the Yutu-2 rover, the selection of the observation target is determined by the teleoperation team. Therefore, as of the end of the 13th lunar day, the VNIS only made in situ spectral measurements on two rock fragments at waypoints LE00303 and LE01310, respectively. The key observation target at the rest VNIS observation waypoints was the regolith, in particular, the wheel-trenched regolith compared to the background (a VNIS observation summary is provided as Table S1). The measured raw data were preprocessed and released to the public as level 2B radiance spectra. The radiance spectra were then processed to derive the full-range reflectance spectra (450—2,395nm) by the solar irradiance calibration method (Gou et al., 2019) for further analysis. The spectra of the two rock fragments (i.e., LE00303 and LE01310) and examples of their surrounding regolith (i.e., LE00304 and LE01305) are shown in Fig. 3. The average albedos of the rock fragments at waypoints LE00303 and

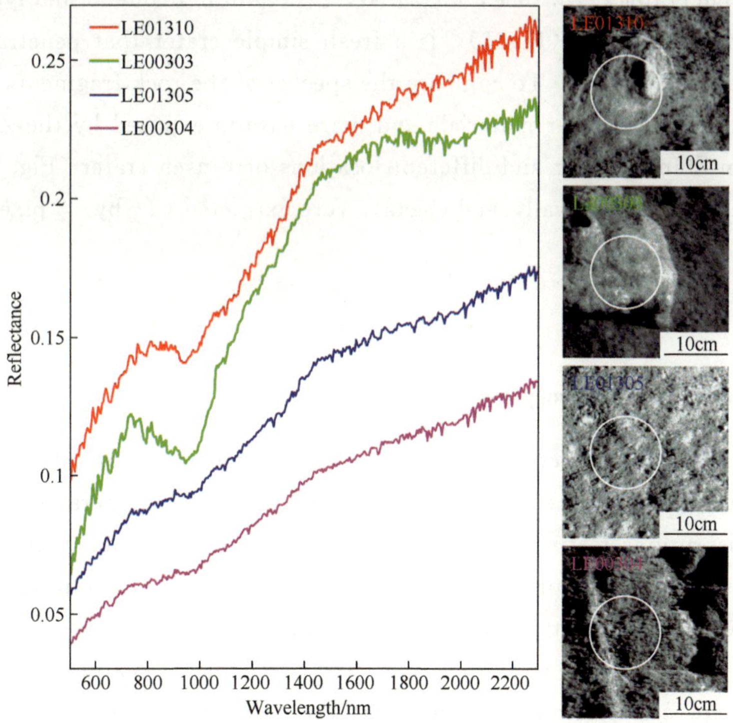

Fig. 3　Spectra of rock fragments and examples of their surrounding regolith measured by the VNIS. The white circle in each image indicates the FOV of the SWIR detector. FOV, field of view; SWIR, shortwave infrared; VNIS, visible/near-infrared spectrometer

LE01310 are 0.1747 and 0.1966, respectively, while the average albedos of the regolith at waypoints LE00304 and LE01305 are 0.0926 and 0.1265, respectively. These values are consistent with the observation that rock fragments typically have higher albedos than the regolith of similar composition (Pieters & Englert, 1993).

2.3 Moon Mineralogy Mapper Spectra

The Moon Mineralogy Mapper (M^3) that flew onboard India's Chandrayaan-1 lunar orbiter, was a pushbroom imaging spectrometer operating from the visible into the NIR (0.42—3.0μm). M^3 observed the lunar surface either in a high-resolution target mode (80m/pixel; 260 spectral channels) or in a low-resolution global mode (140m/pixel; 85 spectral channels) from a 100km orbit (Pieters et al., 2009). The M^3 Level 2 spectral reflectance images (REFIMG), which have been photometrically calibrated and thermal emission corrected (Besse et al., 2013; Clark et al., 2011), are used in this study.

The surface materials observed by the Yutu-2 rover are generally thought to be Finsen crater ejecta, not the nderlying mare basalts (e.g., Di et al., 2019; Gou et al., 2019, 2020a; Huang et al., 2018; Lai et al., 2020; Li et al., 2019a; Lin et al., 2020). Finsen crater (42.29°S, 177.72°W), ~73km in diameter and lying northeast of Von Kármán crater, is a well-preserved complex crater that has a prominent central peak (Fig. 1). Zhinyu crater (176.15°E, 45.34°S), ~3.8km in diameter and lying northwest of Chang'e-4 landing site (Fig. 1), is a fresh simple crater that penetrates the mare basalts inside the host crater. To compare the spectra of the rock fragments with those of attributed to the Finsen crater materials and mare basalts exposed by the Zhinyu crater, M^3 spectra from Zhinyu crater and different locations of Finsen crater (Fig. 1), including Finsen crater central peak, wall, and ejecta, were extracted (3-by-3 pixel averages).

3 Methods

3.1 Analysis of Spectral Absorption Features

Both rock fragments spectra show absorptions around 1,000 and 2,000nm (Fig. 3), which are the indicators of the presence of pyroxene minerals. Generally, orthopyroxene (OPX) has absorptions features centered at ~900 and ~1,900nm while clinopyroxene (CPX) has absorptions at shorter frequencies, ~1,000 and ~2,200nm (Adams, 1974). Therefore, the exact absorption positions, also known as band centers, around 1,000 and 2,000nm, often referred to as the "one-and two-micron features," of both M^3 and VNIS spectra are calculated based following the removal of the "continuum." This technique was developed to amplify weak absorption features, often due to the presence of opaque fractions within the sample, such as nanometer-scale particles of metallic iron ($npFe^0$) within the regolith. Their presence, which accumulates in time with exposure to cosmic weathering, tends to "flatten the curve" (i. e., the absorption features strength) and "redden the continuum slope" (i. e., increase reflectance in the NIR; Hapke et al., 1975; Morris, 1980). Using this approach, the spectra are smoothed by a Savitzky - Golay filter (Savitzky & Golay, 1964) with a 15 - point quadratic polynomial (Fig 4). Next, a straight - line continuum is applied in this study by selecting the reflectance maxima on either side of each absorption region, which removes any spectral slope effects due to phase angle differences (Mitchell et al., 2020). The three reflectance maxima of a rock fragment spectrum are generally around 750, 1,500 and 2,350nm (Fig. 4). Third, a second - order polynomial fit is then made to the continuum - removed spectral values around each absorption region, and the minimum of the polynomial is determined to be the band center.

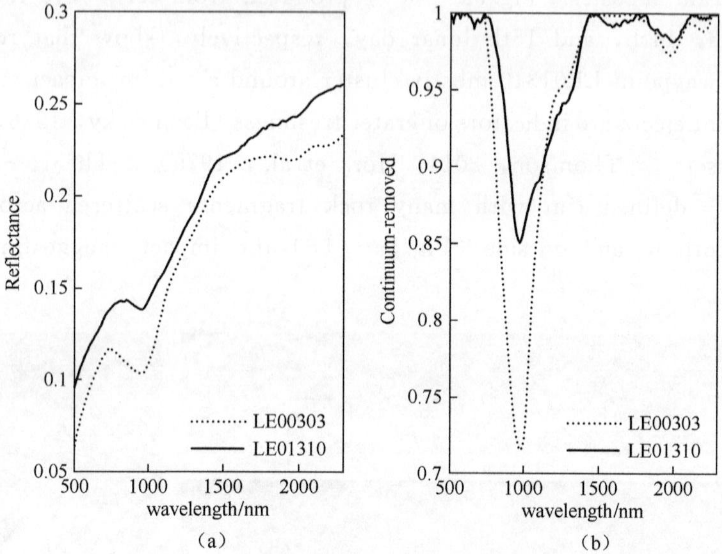

Fig. 4 Spectra of rock fragments measured at waypoints LE00303 and LE01310, respectively

(a) Smoothed. (b) Contiuum - removed

3.2 Determination of Absolute Model Age

Based on our interpretations of the data as discussed in sections 4.3 and 4.4, the spectral features of the two rock fragments are comparable to those of surface materials of Imbrian associated to the Finsen crater. They might represent freshly exposed rock fragments by a secondary impact from Zhinyu crater. To estimate the exposure time of these rock fragments, an absolute model age (AMA) of Zhinyu crater was derived us - ing crater size - frequency distribution (CSFD) analysis, which combines information from crater formation rates and radiometric ages of lunar rock samples (Neukum, 1983). The CSFD data were obtained following these steps: (1) outline a representative area of the Zhinyu crater ejecta blanket; (2) map craters using CraterTools (Kneissl et al., 2011), an application written for the ArcGIS software, within the outlined represent - ative area. Due to the inherent uncertainties in diameter measurements (~10%; Robbins et al.,

2014; Wang et al. , 2020), only craters with diameters larger than 5m were utilized; (3) compare the CSFD of the area to lunar chronology models (Neukum, 1983) after randomness analysis (Michael et al. , 2012) . This step was carried out utilizing Craterstats (Michael & Neukum, 2010; Michael et al. , 2012) . Further information on the application of SCFD measurements to determine the AMA of Zhinyu crater is found in the supplementary material.

4 Results and Discussions

4.1 Distribution of Rock Fragments

Three Pancam mosaics (Fig. 5a — 5c), produced from left - eye images acquired during the 13th, 14th, and 15th lunar day, respectively, show that rock fragments encountered at waypoint LE01310 mostly cluster around a ~15m impact crater. A sharp rim and abundant ejecta are indicators of crater freshness (Basilevsky, 1976; Basilevsky et al. , 2013; Fassett & Thomson, 2014; Hörz et al. , 1975) . The 15 - m crater has indeed a well - defined rim with many rock fragments scattered across the whole morphology, both in and outside (Fig. 5a — 5c) the impact, suggesting a relatively young age.

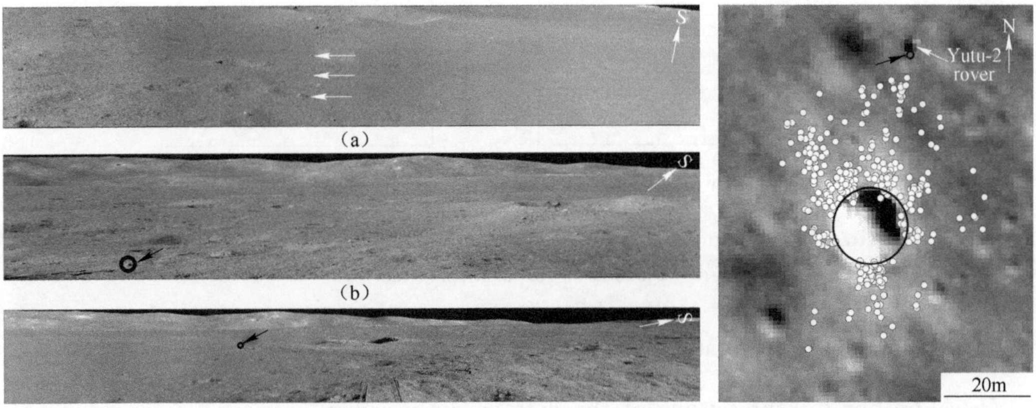

Fig. 5　(a - c) Pancam mosaics of the 15 - m fresh crater with a blocky rim. These mosaics were produced from Pancam's left - eye images acquired during the 13th, 14th, and 15th lunar day, respectively. The white arrows in (a) mark one of the rock fragments chains. (d) The spatial distribution of rock fragments around the 15 - m crater. Each white dot represents a rock fragment. The base map is the same as that of Fig. 1b. The Yutu - 2 rover approached the rock fragment for a close VNIS observation at waypoint LE01310 and then withdrew during the 13th lunar day. Due to the Pancam's viewing geometry (the rock fragment is very close to the Pancam/rover), the VNIS measured rock fragment cannot be seen on the Pancam mosaic shown in (a) . The location of the VNIS observed rock fragment is marked by the black circle with an arrow in (b—d) . VNIS, visible/near - infrared spectrometer

Due to the shooting angle of the Pancam, most of the area south of the 15 - m impact crater is within a blind area. Nevertheless, rock fragments are distributed radially around the 15 - m impact crater, with density decreasing with distance from the crater rim (Fig. 5d). Craters devoid of rock fragments form solely within the loose and unconsolidated regolith layer, and craters with blocks are an indicator of (partially) penetration of a coarse - grained or coherent bedrock layer (Wilcox et al., 2005). Rock fragments at waypoint LE01310 string out in long chains in multiple directions and extend more than two crater radii from the rim crest (Fig. 5d), indicating a common origin. A Pancam mosaic shows the orientation of one of the chains extends northwest - southeastward (Fig. 5a). The observed rock fragments radiating in different directions are thus essentially chain clusters of ejecta from the 15 - m impact crater. The evidence points to a crater that was excavated on a target consisting of a loose and unconsolidated regolith layer over a coarse - grained or possibly a coherent layer (bedrock) which was (partially) penetrated during crater formation.

Unlike what is observed for the rock fragments at waypoint LE01310 (Fig. 1b), the distribution of rock fragments at waypoint LE00303 cannot be related to their surroundings based on the Pancam mosaic (Fig. 2a). Therefore, it is impossible to determine their spatial distribution.

4.2 Implications for Local Regolith

Much of the evidence proposes that the Chang'e - 4 landing site was covered by ejecta from Finsen crater (e. g., Di et al., 2019; Gou et al., 2019, 2020a; Huang et al., 2018; Lai et al., 2020; Li et al., 2019a; Lin et al., 2020). After deposition of Finsen ejecta, the surface of the landing site has been impacted and gardened by meteorites for billions of years. As a result, a fine - grained unconsolidated layer termed "regolith" accumulated gradually above a coarse - grained or even coherent (e. g., bedrock) layer. If one impact penetrates through the fine - grained regolith layer, it will excavate materials in the underlying layer and deposit them around the crater. Therefore, the thickness of regolith could be estimated from the excavation depth of the smallest crater with a blocky rim (Wilcox et al., 2005), like the one encountered at waypoint LE01310.

For a simple crater on the lunar surface, its maximum excavation depth (H_{exc}) is approximately equal to 1/10 the transient crater diameter (D_t), which is about 0.84 times of the crater rim - to - rim diameter (D), that is, $H_{exc} = 0.1D_t = 0.084D$ (Melosh, 1989). After checking all the available Pancam images, no other crater with obvious rock fragments around its rim was observed. The 15 - m crater becomes the only example in this area. Because rock fragments are exposed during the impact - cratering process, it is likely that the thickness of local fine - grained loose regolith at waypoint LE01310 should be no more than 1.3m, which is calculated from maximum excavation depth of the 15 - m crater. However, previous works reported that the thickness of a regolith layer at Chang'e－4 landing site is about 12m, which is derived from the 500MHz high－frequency lunar penetrating radar (LPR) data (Lai et al., 2019; Li et al., 2020). It is unlikely that the local thickness of the lunar regolith would differ so drastically from the average. One possible scenario to explain the difference of nearly an order of magnitude is that there might be abundant rock fragments admixed in the regolith, which would have been exposed by the impact event. No rock fragments smaller than 1 cm are measured in this study, due to limited resolution of the Pancam images. Measurements reveal that the mean and variance of the sizes of rock fragments are about 7.3 and 5.9cm, respectively. However, the resolution of the high－frequency LPR data is about 17cm (Li et al., 2018). Therefore, these buried rock fragments may not be visible in the LPR radargram. Another possible explanation might be rooted in semantics, in terms of the definition of regolith between this study and the works of Lai et al. (2019) and Li et al. (2020), that is, fine－grained regolith versus regolith with coarse rock fragments embedded inside.

4.3 Source of Rock Fragments

Spectral analysis reveals the absorption features of the two rock fragments are similar to that of Finsen crater materials (rich in low－calcium pyroxene) and are clearly compositionally different from that of the underlying mare basalts (rich in high－calcium pyroxene) excavated by the Zhinyu crater (Fig. 6). It isreasonable to deduce that these

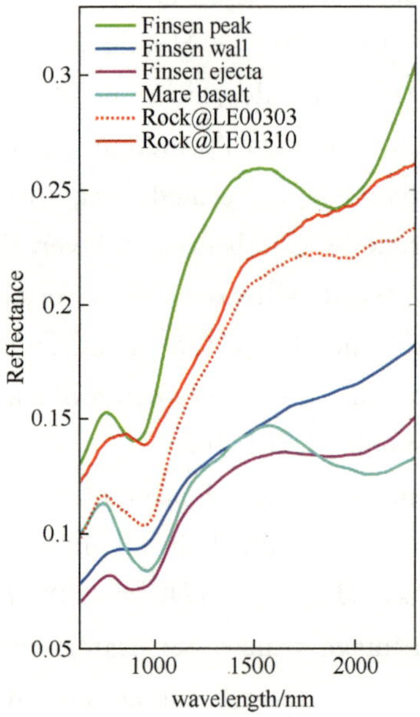

(a)

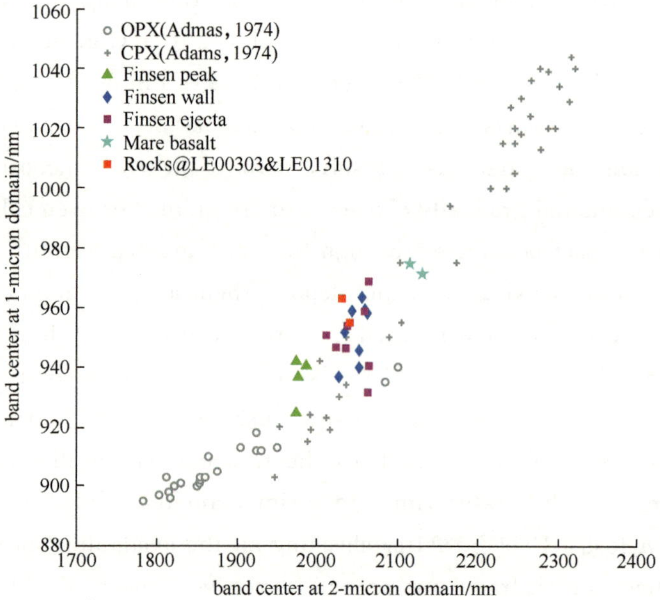

(b)

Fig 6　Spectral features of the Finsen crater materials, basalts exposed by the Zhinyu crater, and the two rock fragments measured by the Yutu－2 rover

(a) Smoothed spectra. Only 1 M^3 spectrum of each type is shown here, and the rest are available from the data repository, (b) absorption band centers comparison buried rock fragments might also be ejecta from the Finsen crater, which is Imbrian (Ivanov et al., 2018; Gou et al., 2021b). The sizes of the two rock fragments are about 25 and 20cm, respectively. Survival times of meter-sized rocks on the lunar surface is usually no more than 300Ma (Basilevsky et al., 2013). Very few examples of micrometeorite impacts are observed in the close-up images (Fig. 3), which leads to the inference that the rocks fragments have been exposed to space weathering for a relatively short geological time. We conclude that the Yutu-2 rover might have stumbled across rock fragments that were buried within the regolith with time, maybe several times, before eventually being exhumed and redeposited on the surface. This is also supported by the observation that both rock fragments show limited space weathering effects with strong mafic absorption features (Fig. 3; Pieters & Noble, 2016): A heavily space weathered rock fragment would have developed a thick patina coating, whose spectral reflectance properties would differ from that of the underlying rock (Frushour et al., 2014; Morrison et al., 1973; Wentworth et al., 1999).

4.4 The Exposure Time of Rock Fragments

The 15-m fresh crater is not the only crater visited by the rover during its traverse inside the Von Kármán crater. Along with heavily degraded craters, there are many fresh craters with blocky rims and interiors in this area (Fig. 7). Some of the small blocks around these fresh craters were pulverized by the rover's wheels and their derived composition show similarities with the surrounding Finsen ejecta-sourced regolith, indicating these blocks to be regolith conglomerates (not cohesive rock fragments) possibly belonging to the Finsen ejecta (Gou et al., 2020c). Considering the fast regolith growth rate at the Chang'e-4 landing site (Gou et al., 2021b; Lai et al., 2019) and quick breakdown of boulders on the lunar surface (Basilevsky et al., 2013), the high spatial density of these fresh craters in such a small region is not typical of primary craters (Ding et al., 2020).

Zhinyu crater (176.15°E, 45.34°S), ~3.8km in diameter and ~800m in depth, is a fresh simple crater that has a faint impact ray system extending to the Chang'e-4 landing site (Ding et al., 2020). Ejecta from craters such as Zhinyu would cause secondary craters, which are usually elongated radially with respect to the primary crater and form shallow depressions, crater chains, and crater clusters (e.g., McEwen & Bierhaus, 2006). The axis lengths of northwest-southeast and southwest—northeast directions of the 15-m crater under investigation are ~18 and ~15m, respectively, indicating the impact direction might be north-west-southeastward. These considerations point to the most likely origin as an elliptical secondary crater by ejecta from the Zhinyu crater. CSFD measurement on ejecta blanket of the Zhinyu crater yields an AMA of μ16 (+5, −4) Ma (Fig. 8). Therefore, the rock fragments encountered by the Yutu-2 rover during the 13th lunar day might be buried Imbrian Finsen crater materials that were freshly exposed by a secondary crater formed by ejecta from the 10 - 20 - Ma - old Zhinyu crater.

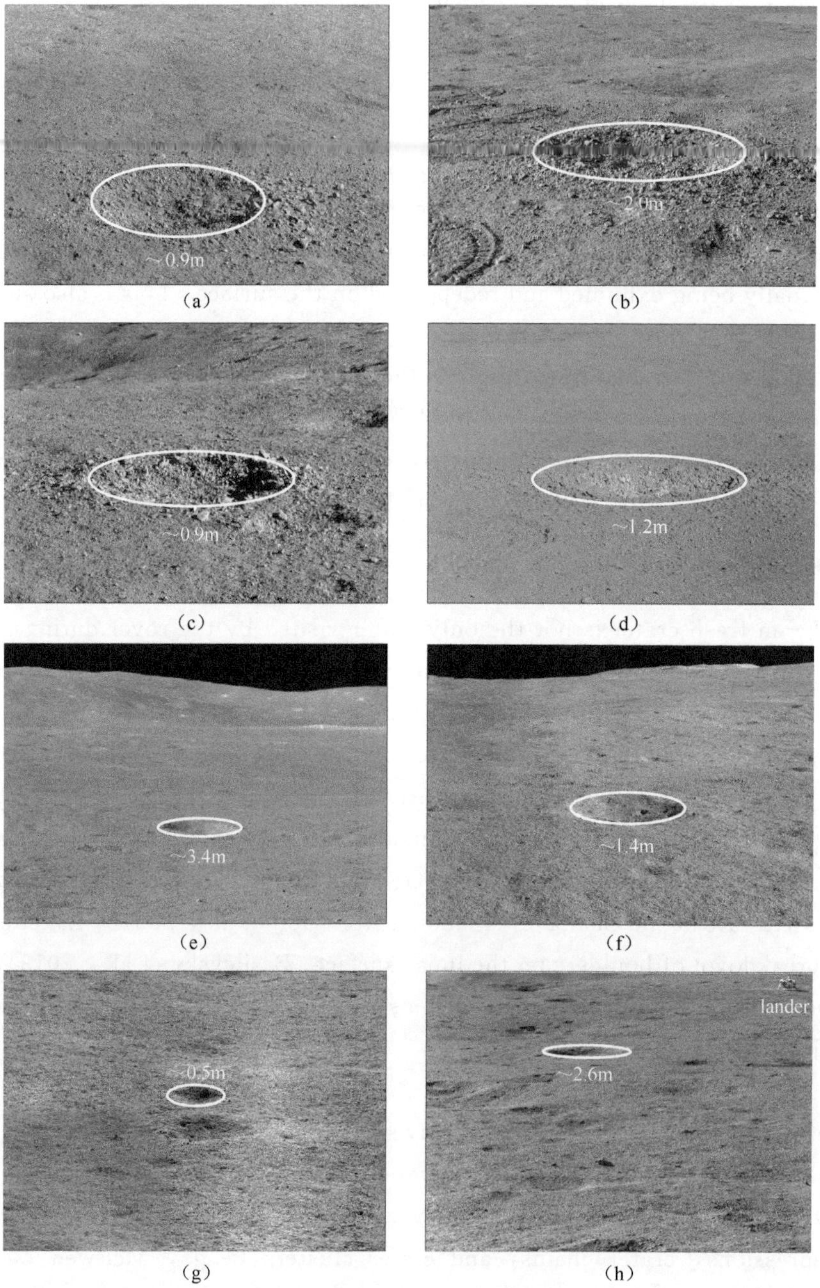

Fig. 7　Pancam images of fresh craters (a—d) and heavily degraded craters (e - h) in the Chang'e - 4 landing area. Because the spatial resolution of the Pancam image depends strongly on distance f rom the camera (rover), the size of each crater is marked as an alternative for scale bar. The rover wheel rut seen in (b and d) is 15cm wide

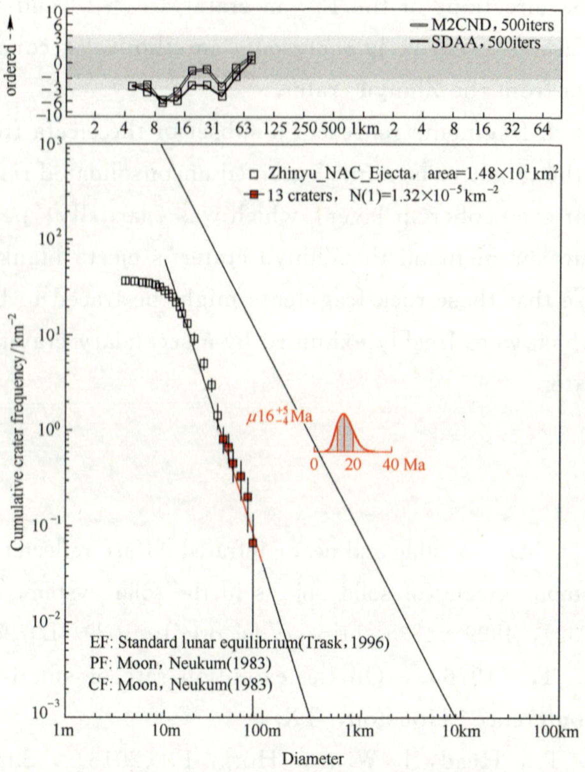

Fig 8　Absolute model age of Zhinyu crater. According to randomness analysis，craters with a diameter ⩾40m are used for fitting and are marked as red filled squares

5　Conclusion

The Yutu－2 rover，part of the Chang'e－4 lunar mission to the lunar farside，come across rock fragments during its traverse of the Von Kármán crater floor. Two rock fragments were studied using spectral analysis at waypoints LE00303 and LE01310 (Fig. 1b) . Their spectral absorption characteristics were found to differ from the surrounding mare basalts，but rather to resemble those from the Finsen crater materials derived through remote sensing (rich in low－calcium pyroxene)，indicating these rock fragments might be related to the impact.

The rock fragments encountered at waypoint LE01310 are distributed around ～15m impact crater，which appears relatively fresh. Some of the rock fragments line up in long chains radiating on multiple directions extending more than two crater radii from the rim crest，indicating a common origin. The Zhinyu crater is a fresh simple crater that has an optically faint impact ray system that can be traced up to the Chang'e－4 landing site. Ejecta from Zhinyu crater also excavated secondary craters with an asymmetric axis with the longer pointing back at the source. The axis lengths of northwest－southeast and southwest - northeast directions of the 15 - m crater are ~18 and ~15m, respectively, suggesting that the crater most likely represents an elliptical secondary crater formed by the low - angle ejecta from the Zhinyu crater.

The observed rock fragments represent a subset of the ejecta from the 15 - m impact crater, suggesting the target to be a fine - grained unconsolidated regolith layer overlaying a coarser - grained or even coherent layer, which was (partially) penetrated during crater formation. CSFD measurement on the Zhinyu crater's ejecta blanket yields an AMA of μ16 Ma. We propose that these rock fragments might be traced to buried Imbrian Finsen crater materials, which were freshly exhumed by a secondary crater formed by the ejecta from the Zhinyu crater.

References

[1] Adams, J. B. (1974). Visible and near - infrared diffuse reflectance spectra of pyroxenes as applied to remote sensing of solid objects in the solar system. Journal of Geophysical Research, 79 (32), 4829—4836. https://doi.org/10.1029/JB079i032p04829.

[2] Basilevsky, A. T. (1976). On the evolution rate of small lunar craters. In 7th lunar science conference. Houston, TX.

[3] Basilevsky, A. T., Head, J. W., & Horz, F. (2013). Survival times of meter - sized boulders on the surface of the Moon. Planetary and Space Science, 89, 118—126. https://doi.org/10.1016/j.pss.2013.07.011.

[4] Besse, S., Yokota, Y., Boardman, J., Green, R., Haruyama, J., Isaacson, P., et al. (2013). One Moon, many measurements 2: Photometric corrections. Icarus, 226 (1), 127—139. https://doi.org/10.1016/j.icarus.2013.05.009.

[5] Clark, R. N., Pieters, C. M., Green, R. O., Boardman, J. W., & Petro, N. E. (2011). Thermal removal from near - infrared imaging spectroscopy data of the Moon. Journal of Geophysical Research, 116 (E6), E00G16. https://doi.org/10.1029/2010JE003751.

[6] Di, K., Xu, B., Peng, M., Yue, Z., Liu, Z., Wan, W., et al. (2016). Rock size - frequency distribution analysis at the Chang'E - 3 landing site. Planetary and Space Science, 120, 103—112. https://doi.org/10.1016/j.pss.2015.11.012.

[7] Di, K., Zhu, M. — H., Yue, Z., Lin, Y., Wan, W., Liu, Z., et al. (2019). Topographic evolution of Von Kármán crater revealed by the lunar rover Yutu - 2. Geophysical Research Letters, 46 (22), 12764—12770. https://doi.org/10.1029/2019GL085252.

[8] Ding, C., Xiao, Z., Wu, B., Li, Y., Prieur, N. C., Cai, Y., et al. (2020). Fragments Delivered by Secondary Craters at the Chang'E - 4 Landing Site. Geophysical Research Letters, 47 (7), e2020GL087361. https://doi.org/10.1029/2020GL087361.

[9] Fassett, C. I., & Thomson, B. J. (2014). Crater degradation on the lunar maria: Topographic diffusion and the rate of erosion on the Moon. Journal of Geophysical Research: Planets, 119 (10), 2255 — 2271. https://doi.org/10.1002/2014JE004698.

[10] Frushour, A., Noble, S., Christoffersen, R., & Keller, L. P. (2014). Alteration of lunar rock surfaces through interaction with the space environment. In 45th lunar and planetary science conference. Houston, the Woodlands.

[11] Gou, S., Di, K., Yue, Z., Liu, Z., He, Z., Xu, R., et al. (2019). Lunar deep materials observed by Chang'e - 4 rover. Earth and Planetary Science Letters, 528, 115829. https://doi.org/10.1016/j.epsl.2019.115829.

[12] Gou, S., Di, K., Yue, Z., Liu, Z., He, Z., Xu, R., et al. (2020a). Forsteritic olivine and magnesium - rich orthopyroxene materials measured by Chang'e - 4 rover. Icarus, 345, 113776. https://doi.org/10.1016/j.icarus.2020.113776.

[13] Gou, S., Yue, Z., Di, K., Bugiolacchi, R., Zhu, M. — H., Pinet, P. C., & Cai, Z. (2021a). Mare basalt flooding events surrounding Chang'e - 4 landing site as revealed by Zhinyu crater ejecta. Icarus, 360, 114370. https://doi.org/10.1016/j.icarus.2021.114370.

[14] Gou, S., Yue, Z., Di, K., Cai, Z., Liu, Z., & Niu, S. (2021b). Absolute model age of lunar Finsen crater and geologic implications. Icarus, 354, 114046. https://doi.org/10.1016/j.icarus.2020.114046.

[15] Gou, S., Yue, Z., Di, K., Wan, W., Liu, Z., Liu, B., et al. (2020b). In situ spectral measurements of space weathering by Chang'e - 4 rover. Earth and Planetary Science Letters, 535, 116117. https://doi.org/10.1016/j.epsl.2020.116117.

[16] Gou, S., Yue, Z., Di, K., Wang, J., Wan, W., Liu, Z., et al. (2020c). Impact melt breccia and surrounding regolith measured by Chang'e - 4 rover. Earth and Planetary Science Letters, 544, 116378. https://doi.org/10.1016/j.epsl.2020.116378.

[17] Hapke, B., Cassidy, W., & Wells, E. (1975). Effects of vapor - phase deposition processes on the optical, chemical, and magnetic properties OE the lunar regolith. The Moon, 13 (1), 339 — 353. https://doi.org/10.1007/BF00567517.

[18] Hörz, F., Schneider, E., Gault, D. E., Hartung, J. B., & Brownlee, D. E. (1975). Catastrophic rupture of lunar rocks: A Monte Carlo simulation. The Moon, 13 (1), 235—258.

[19] Huang, J., Xiao, Z., Flahaut, J., Martinot, M., Head, J., Xiao, X., et al. (2018). Geological Characteristics of Von Kármán Crater, Northwestern South Pole - Aitken Basin: Chang'E - 4 Landing Site Region. Journal of

Geophysical Research: Planets, 123 (7), 1684 — 1700. https://doi.org/10.1029/2018JE005577.

[20] Ivanov, M. A., Hiesinger, H., van der Bogert, C. H., Orgel, C., Pasckert, J. H., & Head, J. W. (2018). Geologic History of the Northern Portion of the South Pole - Aitken Basin on the Moon. Journal of Geophysical Research: Planets, 123 (10), 2585—2612. https://doi.org/10.1029/2018JE005590.

[21] Jia, Y., Zou, Y., Ping, J., Xue, C., Yan, J., & Ning, Y. (2018). The scientific objectives and payloads of Chang'E - 4 mission. Planetary and Space Science, 162, 207—215. https://doi.org/10.1016/j.pss.2018.02.011.

[22] Kneissl, T., Van Gasselt, S., & Neukum, G. (2011). Map - projection - independent crater size - frequency determination in GIS environ - ments - New software tool for ArcGIS. Planetary and Space Science, 59 (11), 1243 — 1254. https://doi.org/10.1016/j.pss.2010.03.015.

[23] Lai, J., Xu, Y., Bugiolacchi, R., Meng, X., Xiao, L., Xie, M., et al. (2020). First look by the Yutu - 2 rover at the deep subsurface structure at the lunar farside. Nature Communications, 11 (1), 3426. https://doi.org/10.1038/s41467—020—17262—w.

[24] Lai, J., Xu, Y., Zhang, X., Xiao, L., Yan, Q., Meng, X., et al. (2019). Comparison of Dielectric Properties and Structure of Lunar Regolith at Chang'e - 3 and Chang'e - 4 Landing Sites Revealed by Ground Penetrating Radar. Geophysical Research Letters, 46, 12783—12793. https://doi.org/10.1029/2019GL084458.

[25] Li, C., Liu, D., Liu, B., Ren, X., Liu, J., He, Z., et al. (2019a). Chang'E - 4 initial spectroscopic identification of lunar far—side mantle - derived materials. Nature, 569 (7756), 378 — 382. https://doi.org/10.1038/s41586 — 019—1189—0.

[26] Li, C., Shuguo, X., Lauro, S., Su, Y., Dai, S., Feng, J., et al. (2018). Pitfalls in GPR Data Interpretation: False Reflectors Detected in Lunar Radar Cross Sections by Chang'e - 3. IEEE Transactions on Geoscience and Remote Sensing, 56 (3), 1325—1335. https://doi.org/10.1109/TGRS.2017.2761881.

[27] Li, C., Su, Y., Pettinelli, E., Xing, S., Ding, C., Liu, J., et al. (2020). The Moon's farside shallow subsurface structure unveiled by Chang'E - 4 Lunar Penetrating Radar. Science Advances, 6 (9), eaay6898. https://doi.org/10.1126/sciadv.aay6898.

[28] Li, C., Wang, Z., Xu, R., Lv, G., Yuan, L., He, Z., & Wang, J. (2019b). The Scientific Information Model of Chang'e - 4 Visible and Near - IR Imaging Spectrometer (VNIS) and In - Flight Verification. Sensors, 19 (12), 2806. https://doi.org/10.3390%2Fs19122806.

[29] Lin, H., He, Z., Yang, W., Lin, Y., Xu, R., Zhang, C., et al. (2020). Olivine - norite rock detected by the lunar rover Yutu - 2 likely crystallized from the

SPA impact melt pool. National Science Review, 7, 913－920. https://doi.org/10.1093/nsr/nwz183.

[30] Liu, Z., Di, K., Li, J., Xie, J., Cui, X., Xi, L., et al. (2020). Landing site topographic mapping and rover localization for Chang'e - 4 mission. Science China Information Sciences, 63 (4), 140901. https://doi.org/10.1007/s11432－019－2796－1.

[31] McEwen, A. S., & Bierhaus, E. B. (2006). The importance of secondary cratering to age constraints on planetary surfaces. Annual Review of Earth and Planetary Sciences, 34 (1), 535 － 567. https://doi.org/10.1146/annurev.earth.34.031405.125018.

[32] Melosh, H. J. (1989). Impact cratering: A geologic process. New York, NY: Oxford University Press.

[33] Michael, G. G., & Neukum, G. (2010). Planetary surface dating from crater size - frequency distribution measurements: Partial resurfacing events and statistical age uncertainty. Earth and Planetary Science Letters, 294 (3), 223－229. https://doi.org/10.1016/j.epsl.2009.12.041.

[34] Michael, G. G., Platz, T., Kneissl, T., & Schmedemann, N. (2012). Planetary surface dating from crater size - frequency distribution measurements: Spatial randomness and clustering. Icarus, 218 (1), 169－177. https://doi.org/10.1016/j.icarus.2011.11.033.

[35] Mitchell, A. M., Reddy, V., Sharkey, B. N. L., Sanchez, J. A., Burbine, T. H., Le Corre, L., & Thomas, C. A. (2020). Constr aining ordinary chondrite composition via near - infrared spectroscopy. Icarus, 336, 113426. https://doi.org/10.1016/j.icarus.2019.113426.

[36] Molaro, J., & Byrne, S. (2012). Rates of temperature change of airless landscapes and implications for thermal stress weathering. Journal of Geophysical Research, 117 (E10), E10011. https://doi.org/10.1029/2012JE004138.

[37] Morris, R. V. (1980). Origins and size distribution of metallic iron particles in the lunar regolith. In 11th lunar and planetary science conference. Houston, TX.

[38] Morrison, D., McKay, D., Fruland, R., & Moore, H. (1973). Microcraters on Apollo 15 and 16 rocks. In 4th lunar science conference. Houston, TX.

[39] Neukum, G. (1983). Meteoriten bombardement und Datierung planetarer Oberflächen (German original). University of Munich.

[40] Pieters, C. M., Boardman, J., Buratti, B., Chatterjee, A., Clark, R., Glavich, T., et al. (2009). The Moon mineralogy mapper (M3) on Chan－drayaan - 1. Current Science, 96 (4), 500－505.

[41] Pieters, C. M., & Englert, P. A. J. (1993). Remote geochemical analysis: Elemental, and mineralogical compostion. New York, NY: Cambridge University Press.

[42] Pieters, C. M., & Noble, S. K. (2016). Space weathering on airless bodies. Journal of Geophysical Research: Planets, 121 (10), 1865 — 1884. https://doi.org/10.1002/2016JE005128.

[43] Ren, X., Liu, J. J., Wang, F. F., Wang, W. R., Mu, L. L., & Li, H. H. (2014). A new lunar global topographic map products from Chang'E - 2 stereo camera image data. Cascais, Portugal. European Planetary Science Congress.

[44] Robbins, S. J., Antonenko, I., Kirchoff, M. R., Chapman, C. R., Fassett, C. I., Herrick, R. R., et al. (2014). The variability of crater identification among expert and community crater analysts. Icarus, 234, 109 — 131. https://doi.org/10.1016/j.icarus.2014.02.022.

[45] Robinson, M. S., Brylow, S. M., Tschimmel, M., Humm, D., Lawrence, S. J., Thomas, P. C., et al. (2010). Lunar Reconnaissance Orbiter Camera (LROC) instrument overview. Space Science Reviews, 150, 81 — 124. https://doi.org/10.1007/s11214—010—9634—2.

[46] Savitzky, A., & Golay, M. J. E. (1964). Smoothing and differentiation of data by simplified least squares procedures. Analytical Chemistry, 36, 1627 — 1639. https://doi.org/10.1021/ac60214a047.

[47] Smith, D. E., Zuber, M. T., Neumann, G. A., Lemoine, F. G., Mazarico, E., Torrence, M. H., et al. (2010). Initial observations from the Lunar Orbiter Laser Altimeter (LOLA). Geophysical Research Letters, 37 (18), L18204. https://doi.org/10.1029/2010GL043751.

[48] Wang, Y., Xie, M., Xiao, Z., & Cui, J. (2020). The minimum confidence limit for diameters in crater counts. Icarus, 341, 113645. https://doi.org/10.1016/j.icarus.2020.113645.

[49] Wentworth, S. J., Keller, L. P., McKAY, D. S., & Morris, R. V. (1999). Space weathering on the Moon: Patina on Apollo 17 samples 75075 and 76015. Meteoritics & Planetary Sciences, 34 (4), 593 — 603. https://doi.org/10.1111/j.1945—5100.1999.tb01366.x.

[50] Wilcox, B., Robinson, M., Thomas, P., & Hawke, B. (2005). Constraints on the depth and variability of the lunar regolith. Meteoritics & Planetary Sciences, 40, 695—710. https://doi.org/10.1111/j.1945—5100.2005.tb00974.x.

[51] Yingst, R. A., Chuang, F. C., Berman, D. C., & Mest, S. C. (2017). Geologic Mapping of the Planck Quadrangle of the Moon (LQ—29). In 48th lunar and planetary science conference, the Woodlands, TX.

References From the Supporting Information

[1] Clark, J. D., van der Bogert, C. H., & Hiesinger, H. (2018). Investigating

the Mandel'shtam lobate scarp complex. European Planetary Science Congress, 12.

[2] McGetchin, T. R., Settle, M., & Head, J. W. (1973). Radial thickness variation in impact crater ejecta: implications for lunar basin deposits. Earth and Planetary Science Letters, 20 (2), 226 − 236. https://doi.org/10.1016/0012−821X (73) 90162−3.

[3] Moore, H. J., Hodges, C. A., & Scott, D. H. (1974). Multiringed basins − illustrated by Orientale and associated features. Houston, TX. 5th Lunar Science Conference. (pp. 71−100).

Geochemistry of the Von Kármán crater floor and thickness of the non – mare ejecta over the Chang'e – 4 landing area

Dijun Guoa[1], Wenzhe Fa[1,2,3], Xiaojia Zeng[4], Jun Du[1], Jianzhong Liu[3,4]

1. Institute of Remote Sensing and Geographical Information System, School of Earth and Space Sciences, Peking University, Beijing, China
2. State Key Laboratory of Lunar and Planetary Sciences, Macau University of Science and Technology, Macau, China
3. CAS Center for Excellence in Comparative Planetology, Hefei, China
4. Center for Lunar and Planetary Sciences, Institute of Geochemistry, Chinese Academy of Sciences, Guiyang, China

Abstract: China's Chang'e – 4 mission has carried out the first ever lunar farside landing exploration on the floor of the Von Kármán crater, a geologically complex region located in the most ancient and deepest South Pole – Aitken (SPA) basin. In order to demonstrate the characteristics of materials in the landing area, we investigated the regional geochemistry and thickness of non – mare ejecta overlaying the mare basalts. Comparative analyses of FeO, TiO_2 and Th concentrations suggest that the landing site surface is dominated by non – mare ejecta from nearby craters (e. g., Finsen crater) with part of basaltic materials. The ejecta thickness is estimated based on the excavation depth of dark – haloed and non – dark – haloed craters by using support – vector machine, a supervised machine learning method for classification. The results show that the ejecta thickness in the region of 40km across the landing site varies from near zero to ~80m with a mean value of ~41m. The ejecta at the Chang'e – 4 landing site is ~40m thick, which is comparable to the in situ observations by the Lunar Penetrating Radar onboard the Yutu – 2 rover. Our results provide valuable information for interpretation of the on – going returned data and geologic analysis of the Chang'e – 4 exploration region.

Keywords: Moon; Chang'e – 4; Von Kármán crater; Ejecta deposits; Dark – haloed crater

1 Introduction

1.1 Geologic context

The Earth's Moon is characterized by its asymmetry of the nearside and the farside (e.g., Wieczorek et al., 2013). Since the 1960s, the nearside has been visited more than 10 times during previous landed missions, based on which, fundamental concepts of lunar science have been established. On the other hand, the farside has remained as an untouched territory until 3 January 2019, the lander of China's Chang'e - 4 (CE - 4) probe touched down at the Statio Tianhe (177.5991°E, 45.4446°S) in the southern floor of the Von Kármán crater (~186km in diameter) (Fig. 1; Li et al., 2019). Assembled with instruments including Panoramic Camera (PCAM), Visible and Near Infrared Spectrometer (VNIS) and Lunar Penetrating Radar (LPR) (Di et al., 2019; Li et al., 2019; Li et al., 2020), the CE - 4 mission provides an unprecedented opportunity to investigate the Moon's farside. The mission is anticipated to improve our knowledge of the composition, subsurface structure and geologic evolution of the unique South Pole - Aitken (SPA) terrane (Jia et al., 2018; Jolliff et al., 2000).

The exploration target (the Von Kármán crater), was carefully chosen to be inside the giant SPA basin, one of the Moon's most scientifically rich regions (e.g., Jolliff et al., 2017). About 2400km in diameter (Garrick - Bethell and Zuber, 2009), the SPA basin is the largest impact depression on the Moon and is thought to be formed from an impact that penetrated through the ancient lunar crust and excavated the mantle materials whose composition is highly debatable (e.g., Charlier et al., 2018; Elkins - Tanton et al., 2011). As the oldest impact basin on the Moon, the SPA basin was formed ~4.2 —4.3Ga based on crater statistics (Garrick - Bethell et al., 2020; Orgel et al., 2018), which might correspond to the termination of the magma ocean solidification (Elkins - Tanton et al., 2011). In such a scenario, the SPA impact excavated the primordial lunar crust that was in special thermal condition compared with the subsequent great impacts. Considering the extremely large scale, the SPA basin could produce significant volume of impact melts that might undergo differentiation (e.g., Uemoto et al., 2017; Vaughan and Head, 2014). Exploring at the surface of the interior SPA basin, the scientific data returned from the CE - 4 mission are critical for deciphering the formation and evolution of the SPA basin.

As shown in Fig. 1b, the floor of the Von Kármán crater is ~ —6km in elevation. Based on the Gravity Recovery and Interior Laboratory (GRAIL) data, the crustal thickness in the Von Kármán region is estimated to be less than 10km (Wieczorek et al., 2013). Scaling law suggests that a crater of 180km diameter can excavate 12km in depth (Melosh, 1989), thus it is conceivable that the Von Kármán crater could expose materials from the lower crust or the upper mantle. The model age of the Von Kármán

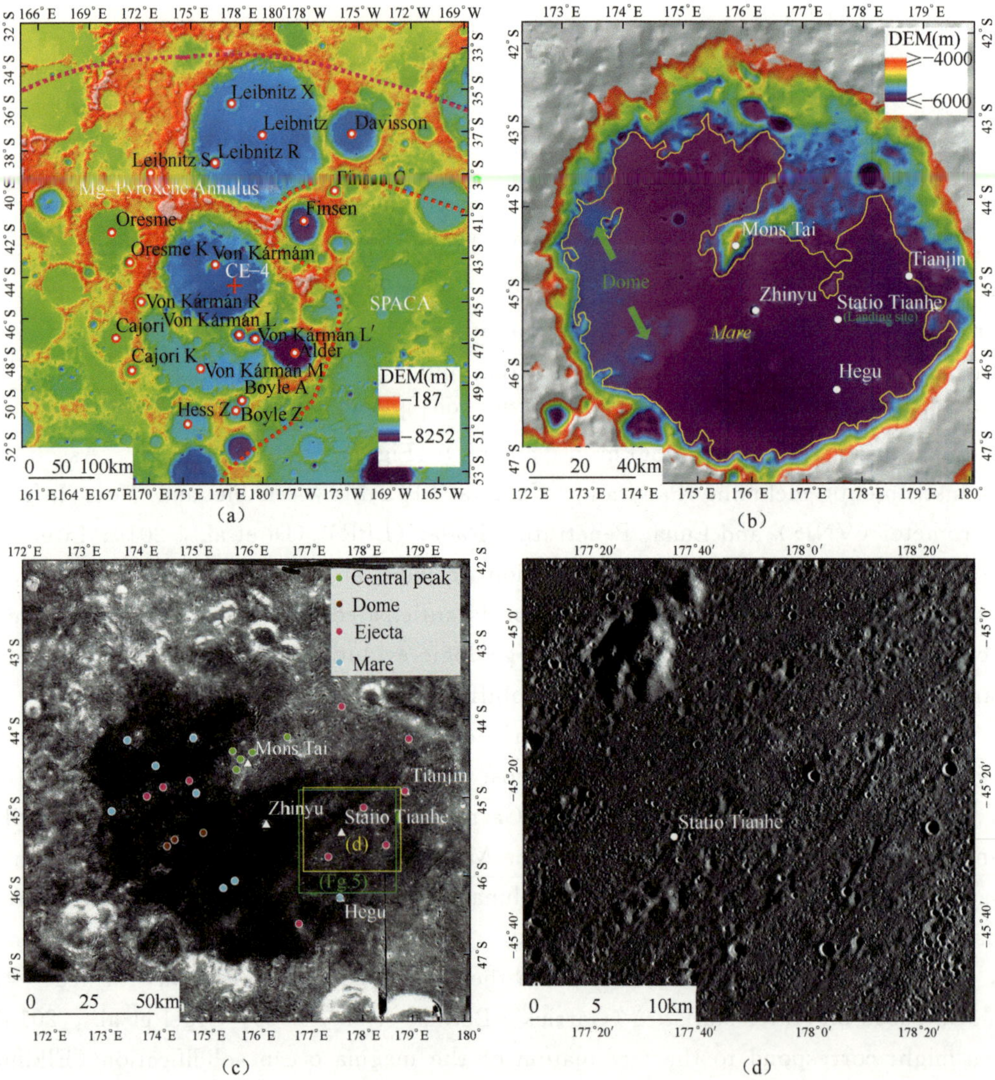

Fig. 1　(a) Topography and geologic features around the Von Kármán crater. SPA compositional anomaly (SPACA) and Mg Pyroxene represent the annuli containing elevated Ca, Fe - rich pyroxene and Mg pyroxene, respectively (Moriarty and Pieters, 2018). Von Kármán L' is an unofficially named crater given by Huang et al. (2018). (b) Topography of the interior of the Von Kármán crater. he yellow line represents the boundary of mare basalts identified from Lunar Reconnaissance Orbiter Camera (LROC) images (Nelson et al., 2014). White dots designate the new nomenclatures accepted by IAU after the landing of the CE - 4 mission, where Statio Tianhe is the CE - 4 landing site and Mons Tai is the central peak of the Von Kármán crater. Two dome - like features are marked with green arrows. The elevation is obtained from SELENE and LRO DEM 2015 (SLDEM2015) (Barker et al., 2016). (c) Kaguya Multiband Imager (MI) 1000nm band image. The dots represent locations of different geologic units for geochemical data extraction in Section 2. 2. Green square in (c) marks the extent of Fig. 5 where the DHCs and non -DHCs are identified. Yellow square in (c) marks the extent of (d), which shows the linear features from northeast to southwest in Kaguya Terrain Camera (TC) images (～7. 4m/pixel). Based on the texture and alignment, the linear features may represent the ejecta deposits from Finsen crater. (For interpretation of the references to colour in this figure legend, the reader is referred to the web version of this article)

crater is ～4.2Ga (Lu et al., 2021), and the floor of this Pre - Nectarian crater was flooded by mare basalts during Imbrian period (～3.2Ga) (Lu et al., 2021), resulting in flat topography (Fig. 1b) and low albedo (Fig. 1c). After emplaced, large areas of the mare have been modified and covered by feldspathic materials that appear brighter than mare basalts (Fig. 1c). As shown in Fig. 1d, the bright materials are distributed with secondary impacting linear features on the surface, indicating that the ejecta approached from the northeast and that the parent crater could be the Eratosthenian - aged Finsen crater (～73km in diameter) (Di et al., 2019; Huang et al., 2018; Lu et al., 2021). Young craters that are large enough to penetrate through the ejecta deposits can excavate substratum low - albedo mare basalts (Supplementary Fig. 1a), appearing as dark - haloed craters (DHCs) (Antonenko et al., 1995; Bell and Hawke, 1984). Such craters can be used to estimate the thickness of the non - mare ejecta emplaced above mare basalts, which will be discussed in Section 3.

1.2 Motivation of this study

On the surface of the Von Kármán floor, the composition is nonuniform due to modifications of subsequent evolution processes: eruption of the mare basalts, emplacements of extraneous non - mare ejecta, mixture through local cratering events (Huang et al., 2018). Compositional signatures of the material and thickness of accumulated extraneous ejecta are key information to understand the geologic evolution and interpret the data returned from the CE - 4 mission. At the CE - 4 landing site, the albedo is obviously higher than mare basalts (Fig. 1c) while their optical maturity (OMAT) does not show any significant variations (Fig. 2), indicating that feldspathic component is rich in materials of the landing site as suggested by previous studies (e.g., Gou et al., 2019; Lin et al., 2019). Ejecta deposits are also evident from the linear features where the CE - 4 is landed (Fig. 1d). However, it is debated about the origin of regolith minerals. Initial analysis of the CE - 4 VNIS spectroscopic data suggests that the regolith at the landing region is rich in low - calcium pyroxene (LCP) and olivine that originated from the mantle (Li et al., 2019), which is supported by Gou et al. (2019). However, other studies argued that the materials were originated from impact melt crystallization inside the SPA basin based on their enriched plagioclase (Hu et al., 2019; Lin et al., 2019). Remote sensing investigation of surface compositions over a wide area is critical for unveiling the sources of the materials. Several studies have estimated the thickness of extraneous ejecta at the landing site using different approaches and data (e.g., Di et al., 2019; Huang et al., 2018; Lai et al., 2019; Li et al., 2020), but their results are inconsistent and vary from ～30m to more than 70m. Using ejecta thickness decay models, Huang et al. (2018) estimated that the thickness at the landing area is ～7.2 — 15.5m. However, this model prediction is not supported by direct measurements. Based on the CE - 4 LPR observations, Lai et al. (2019) proposed that the ejecta thickness is ～26m by assuming that the real parts of the dielectric permittivity values before and after 154 ns are 4. 3 and 6, respectively. Through topographic difference and impact cratering simulations of Finsen and Alder craters, Di et al. (2019) suggested that the thickness of ejecta from these two craters accumulated to ～70m at the CE－4 landing site. But the ejecta thickness model was not tested through observations of craters at comparative scale, and the study area is located in the discontinuous ejecta field where the thickness could vary substantially. Using LPR data of the first two lunar days, Li et al. (2020) reported that the ejecta are ～40m thick at the landing site, which is in the range of depth estimated from DHCs and non－DHCs. However, their estimation has a wide range from 34m to 78m, and therefore the constraint to the radar observations is weak. To get reliable results, the in situ radar detections should be well constrained by geological evidences.

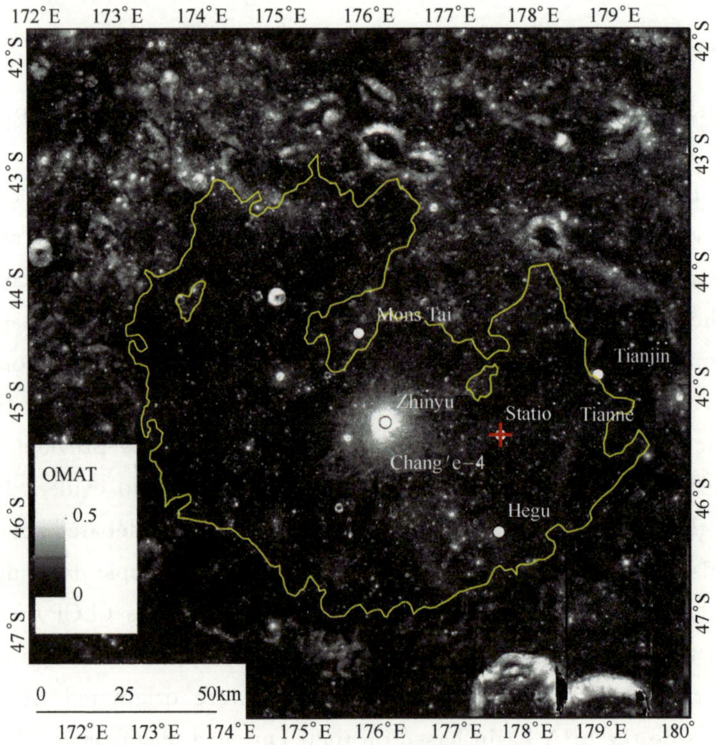

Fig. 2 OMAT of the Von Kármán crater derived from MI data using the method in Lemelin et al. (2016). The yellow line represents the boundary of mare basalts in Fig. 1b. The CE－4 landing site is marked by the red cross at the Statio Tianhe (For interpretation of the references to colour in this figure legend, the reader is referred to the web version of this article)

With a systematic study of compositions and sources of surface materials in the CE－4 landing area, new results can be used to resolve the controversial results in the aforementioned studies. The VNIS is able to detect the surface compositions, but the observation targets are limited on the traverse trail. By combining the geologic context, geochemical analysis (especially TiO_2, FeO and Th distribution) in a broader view is an important complement to interpret the data and analyze the geologic processes and evolution of the exploration region. TiO_2 and FeO concentrations are essential to estimate the complex dielectric permittivity, which is critical in processing the LPR data (Fa and Wieczorek, 2012). On the other hand, even though the LPR can reflect the subsurface structures, additional geologic information is necessary in interpreting the observed radargram. Overlaying on the mare basalts, the non-mare ejecta are the most important component of local regolith and their thickness range is a key prior knowledge for LPR data interpretation. Properties of ejecta deposits are also fundamental in understanding cratering process and lunar surface evolution. This work carries out comparative analysis of the geochemistry in different units of the Von Kármán crater, which is used to analyze the compositional characteristics and evolution history in the landing area (Section 2). By employing both DHCs and non-DHCs on the Von Kármán crater floor, we proposed an innovative method to estimate the regional thickness of accumulated non-mare ejecta above mare basalts (Section 3). The ejecta thickness results are compared with the CE-4 LPR in situ measurements in Section 4. In Section 5, we discussed the ejecta thickness at the CE-4 landing site, the possible ejecta sources, and the regional evolution.

2 Geochemistry in the Von Kármán crater floor

Remote sensing observations and sample analyses reveal that lunar surface materials show diverse chemical compositions in FeO and Th concentrations (e.g., Korotev, 2005): the feldspathic-rich highland materials are characterized by relatively low concentrations of FeO and Th; basaltic rocks are mafic-rich with relatively high concentration of FeO and low concentration of Th; and the KREEP-rich samples are distinctly rich in Th (i.e., >4ppm). This indicates that the FeO and Th concentrations could be used to distinguish different lithologies on the Moon. As mare basalts exhibit a wide range of TiO_2 concentration (i.e., from <1 wt.% to up to >14 wt.%; Neal and Taylor, 1992), different types of lunar basaltic lithologies can be well distinguished by comparing the FeO and TiO_2 contents. In the case of surface materials at the Von Kármán crater, their chemical features (i.e., FeO, TiO_2 and Th) can help us to understand the types and characteristics of regolith in this area. This study used multiple datasets to address the chemical composition of the Von Kármán region and the CE-4 landing site. Kaguya Multiband Imager (MI) data (~20m/pixel) are used to derive the FeO and TiO_2 concentrations based on the algorithms in Lemelin et al. (2015) and Otake et al. (2012), respectively. Comparing with other multispectral data such as Clementine Ultraviolet/Visible camera (UVVIS) and LRO Wide Angle Camera (WAC) data, MI data have a higher spatial resolution and a good coverage. The algorithms have a standard deviations of 0. 43 wt. % for TiO_2 and a root - mean - square error of 1 wt. % for FeO (Lemelin et al. , 2015; Otake et al. , 2012) . The Th concentration is ob - tained from the Lunar Prospector (LP) Gamma Ray Spectrometer (GRS) data with a spatial resolution of 0. 5°/pixel and an uncertainty of 0. 5ppm (Lawrence et al. , 2002)

2. 1 Overall distribution

Fig. 3 shows the FeO, TiO_2 and Th concentrations of the Von Kármán crater. It can be seen that iron and titanium contents are higher on the floor than those of the wall and exterior parts, and their distributions are spatially highly inhomogeneous (Fig. 3) . Measurements of lunar samples show that mare basalts have Th concentration of ~0. 2—3. 2 ppm and FeO concentration of ~16—23 wt. % (Lucey et al. , 2006) . To address the possible occurrences of the purest mare basalts on the Von Kármán floor and assess the basaltic materials at the CE - 4 landing region, the FeO concentration is divided into three groups: the high - FeO group, the intermediate - FeO group, and the low - FeO group (Fig. 3b) . The high - FeO group represents the purest basalts in the investigation region. Considering the uncertainty in the FeO calculation algorithm (Lemelin et al. , 2015), the threshold of the high - FeO group is 15 wt. % (Fig. 3b) . The low - FeO group represents the regions where the composition is dominated by non - mare ejecta transported from craters outside of the Von Kármán. Their FeO concentrations are generally less than 14 wt. % (Fig. 3b) . The group of intermediate FeO is 14 — 15 wt. % in concentration, which generally encompasses the high - FeO units (Fig. 3b) and could represent high - FeO margins that are slightly contaminated by non - mare materials. Based on Fig. 3b, the purest mare basalts cover very limited regions of the Von Kármán floor and the largest patch of the high - FeO unit is around the Zhinyu crater (176. 15°E, 45. 34°S) . With a rim diameter of ~3. 8km, Zhinyu crater could excavate to the maximum depth of ~320m based on scaling laws (Melosh, 1989) . However, the FeO and TiO_2 concentrations of the ejecta near the Zhinyu crater rim where the materials are from the deepest region are relatively lower than those of the Zhinyu ejecta that deposited in further distance (Fig. 3a - c) . It indicates that Zhinyu crater penetrated through the basaltic layer and excavated the pre - mare substrate, therefore the local mare basalts can not extend to more than 320m in depth. The TiO_2 concentration of mare basalts on the Von Kármán crater floor is less than 2. 5 wt. % (Fig. 3c and Fig. 4b) and belongs to low - titanium basalts (Lucey et al. , 2006) . Th is enhanced in the east crater floor where more ejecta from Finsen crater were emplaced. Garrick - Bethell and Zuber (2005) suggests the Th anomaly in the SPA basin is an indigenous unit and likely emplaced when the basin was formed. Some other studies ascribe the Th - enriched material to the antipodal ejecta convergence of Imbrium or Serenitatis basin (Haskin et al. , 2004; Wieczorek and Zuber, 2001) .

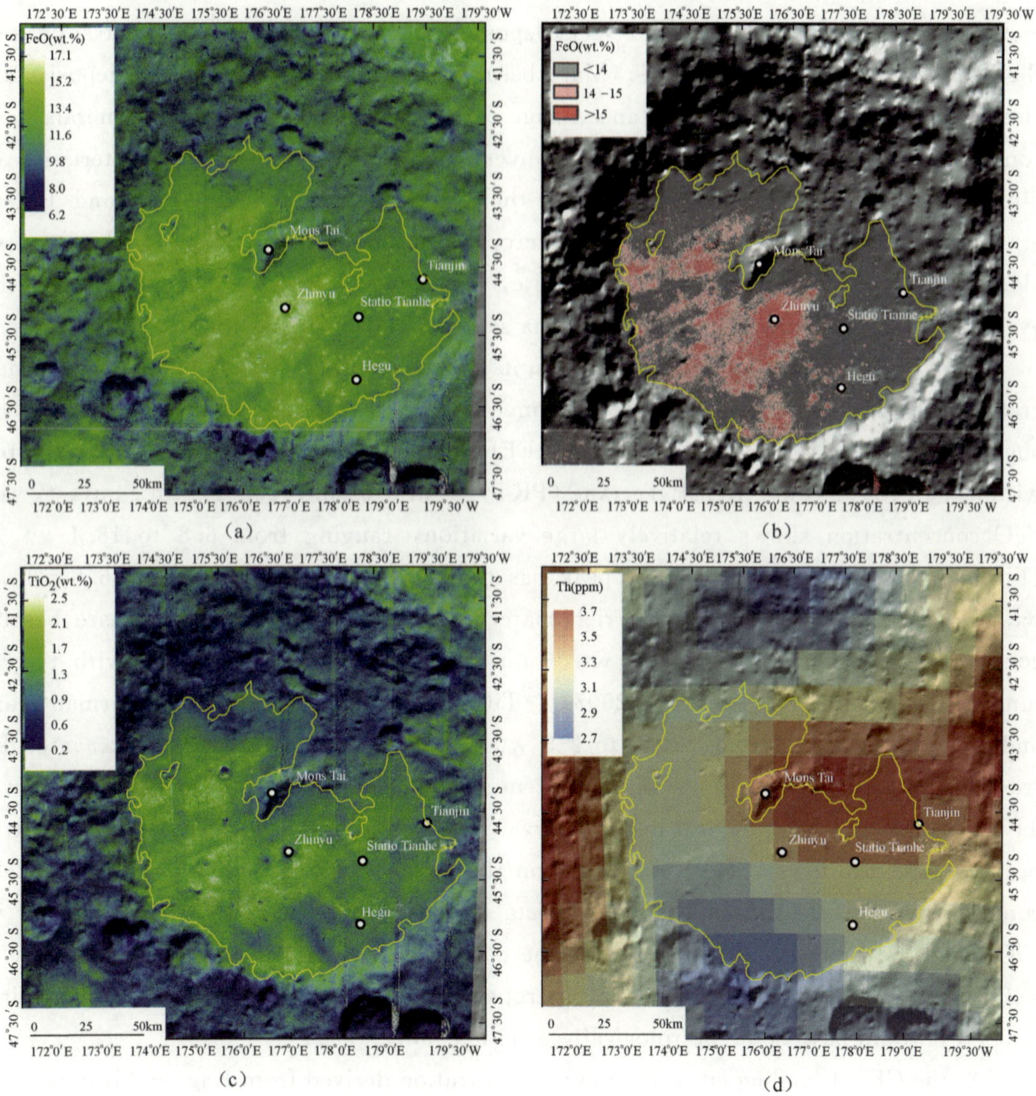

Fig. 3 (a), (c) and (d) are FeO, TiO_2 and Th concentrations of the Von Kármán crater. (b) shows the three FeO concentration groups. The yellow line represents the boundary of mare basalts as Fig. 1b. The standard deviation of TiO_2 concentration is 0.43 wt. % (Otake et al., 2012); the root-mean-square error of FeO concentration is 1 wt. % (Lemelin et al., 2015). The uncertainty of the thorium concentration is <0.5ppm (Lawrence et al., 2002; Lawrence et al., 2000) (For interpretation of the references to colour in this figure legend, the reader is referred to the web version of this article)

2.2 Geochemistry in different units

The latest geologic maps show that the floor of the Von Kármán crater consists of several individual geologic units (Fortezzo et al., 2020; Ling et al., 2019), which suggests various geologic processes in the regional evolution history. Based on the geologic maps and topographic and optical data, we identified outcrops of several prominent units including mare basalts, extraneous ejecta deposits, dome-like areas, and central peak of the Von Kármán crater (Fig. 1c). Mare basalts and the superposed ejecta deposits are clearly separated on the MI reflectance image of 1000nm band where mafic minerals have conspicuous absorption (Fig. 1c). To investigate the geochemical characteristics of different units in the Von Kármán crater, the concentrations of FeO, TiO_2 and Th are extracted from the exact pixels of the outcrop locations shown in Fig. 1c. The results, along with the data of the CE-4 landing site, are shown in Fig. 4 and compared with the available bulk chemistry of the Apollo/Luna samples and lunar meteorites.

As shown in Fig. 4a, Th concentration in the four geologic units of the Von Kármán crater shows a narrow range (2.9—3.7ppm). The relatively low concentration of Th suggests that this region is deficit of KREEP materials, which are widely distributed within the Procellarum KREEP Terrane (PKT) (Jolliff et al., 2000). In contrast, the FeO concentration shows relatively large variation, ranging from 6.8 to 18.4 wt.% (Fig. 4). Specifically, the central peak has the lowest FeO concentration (6.8—11.3 wt.%), suggesting a feldspathic-rich material (Korotev, 2005), while mare basalts contain FeO concentration of ~15 wt.% on average, which is consistent with typical lunar basaltic materials (Korotev, 2005). TiO_2 concentration of the Von Kármán mare and dome units varies from ~0.8 to 2.3 wt.%, and the mare materials excavated by fresh DHCs are ~0.9—1.8 wt.%, consistent with the very low-Ti (<1 wt.%) to low-Ti (1—6 wt.%) basalts (Neal and Taylor, 1992). While the feldspathic-rich central peak contains the lowest TiO_2 concentration that is generally less than 1 wt.%. Likely contaminated by basaltic material, the ejecta unit including the landing site is richer in TiO_2 than the central peak (Fig. 4b). The dome-like areas are compositionally (i.e., FeO, Th, and TiO_2) similar to the outcrops of mare basalts (Fig. 3a, b), indicating that they would have some petrologically genetic connections.

At the CE-4 landing site, the FeO concentration derived from Kaguya MI data is ~13.1 wt.% (Fig. 4), which is lower than the typical values of mare basalt samples (i.e., FeO >15 wt.%) and is higher than that of highland rocks (i.e., FeO <7 wt.%) (Korotev, 2005). Comparing with different geologic units, the landing site has a TiO_2 concentration (~1.8 wt.%) higher than most of the ejecta material and a medium FeO concentration between mare basalts and ejecta (Fig. 4b). Such intermediate FeO concentration suggests that the regolith at the CE-4 landing site are most likely a mixture of iron-rich mare components and iron-poor highland lithologies at or around the Von Kármán crater. LP GRS data show that the Th concentration at the CE-4 landing site is 3.55ppm, which is distinctly lower than that of KREEP-bearing lithologies from Procellarum (>5ppm) (Korotev, 1997). Compared with the Apollo and Luna regolith samples, the bulk chemical composition of the regolith at the CE-4 landing site (i.e., FeO and Th) is similar to the Apollo 15 soils (Fig. 4a).

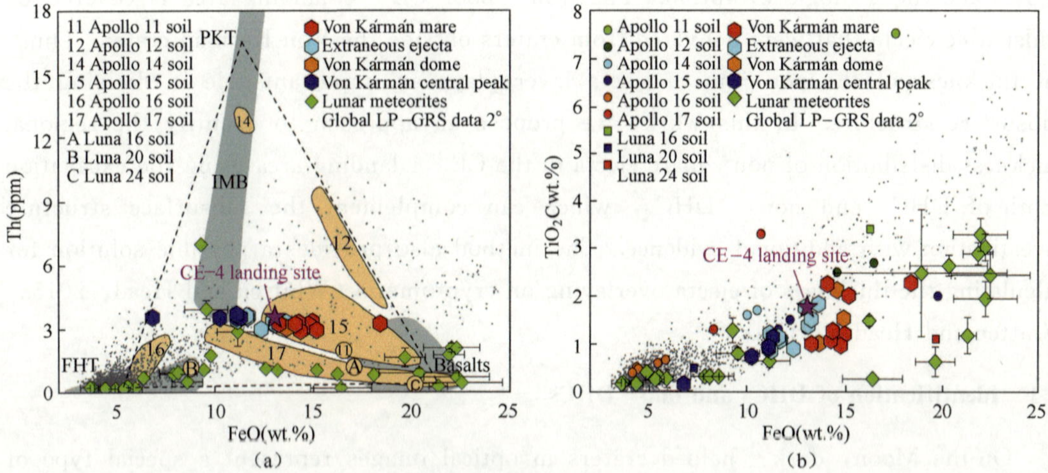

Fig. 4　(a) Th - FeO and (b) TiO_2 - FeO concentrations of the locations marked in Fig. 1c. The bulk chemistry of the Apollo/Luna samples and lunar meteorites are also plotted for comparison (Korotev, 2005). The geochemical data of the CE - 4 landing site is marked by the pink star. Gray dots are the global data from LP GRS observations with a spatial resolution of 2 degrees (Prettyman et al., 2006). PKT: Procellarum KREEP Terrane; FHT: Feldspathic Highlands Terrane; IMB: impact - melt breccia. (For interpretation of the references to colour in this figure legend, the reader is referred to the web version of this article.)

Based on stratigraphic relationship and geologic context (in Section 1), a timeline for the formation of the geochemical units can be recog nized. The central peak, low in iron and titanium concentrations, is the oldest unit that was formed simultaneously with the Von Kármán crater. The most significant modification to the Von Kármán crater floor was the eruption of the iron - rich mare basalts at ~3.2Ga (Lu et al., 2021). Endogenic processes also created two domes which have similar geochemistry with erupted mare basalts but might be formed in a later period (Head and Wilson, 2017). After the eruptions of mare basalts, emplacements of feldspathic ejecta from external craters (e.g., Finsen) covered a significant portion of the mare. In the landing region, geologic evidences indicate that the top layer is dominated by Finsen crater ejecta of low FeO concentration. Due to the processes in the following evolution, basaltic material might be mixed into the feldspathic ejecta and enhanced the local iron abundance.

3　Ejecta thickness estimation based on DHCs and non - DHCs

Our above geochemical analyses suggest that the floor of the Von Kármán crater is laterally inhomogeneous in composition and consists of two major components: the iron - rich mare basalts and the iron - poor non - mare ejecta. Subsurface structure, which can present the vertical extent of stratigraphic units, is another important aspect to understand the geologic evolution. The iron - poor CE - 4 landing area is covered by feldspathic ejecta that were delivered from craters outside the Von Kármán crater. Thus, the thickness of the non - mare ejecta layer plays an important role in deriving the subsurface structure. In this work, we propose an approach to quantify the regional thickness distribution of non - mare ejecta in the CE - 4 landing area using the excavation depth of DHCs and non - DHCs, which can complement the subsurface structure investigation with additional evidence. The method also provides a possible solution for calculating the thickness of ejecta overlaying on cryptomaria (Whitten and Head, 2015a; Whitten and Head, 2015b).

3.1 Identification of DHCs and non - DHCs

On the Moon, dark - haloed craters in optical images represent a special type of craters that are surrounded by a dark halo consisting of low - albedo ejecta (Antonenko et al., 1995; Bell and Hawke, 1984). Dark halo is especially prominent in the continuous ejecta facies that are about one radius from crater rim (Supplementary Fig. 1a), where crater ejecta are the most abundant. The low - albedo material is made of mare basalts that were buried by high - albedo non - mare ejecta and excavated during the cratering process of a DHC (Bell and Hawke, 1984; Schultz and Spudis, 1979). If the crater is not large enough to penetrate through the superposed ejecta deposits, the dark halo can not be formed and it appears as a non - DHC (Supplementary Fig. 1b). Based on the scaling relationship of crater diameter and its maximum excavation depth (Melosh, 1989), DHCs and non - DHCs can constrain the burial depth of mare basalts, i. e., the thickness of overlaying non - mare ejecta materials.

As the dark materials associated with DHCs are mare basalts that have a strong absorption at 1μm band and thus is low in reflectance, optical images are effective data to distinguish DHCs and non - DHCs after eliminating the influence of maturity (Guo et al., 2019). In this study, we use Kaguya MI images (~20m/pixel) of 1μm band to identify DHCs and non - DHCs. As shown in Fig. 2, the OMAT does not display significant variations among different areas. In the CE - 4 landing area, low albedo materials of DHCs can extend to 1—3 radii from crater rims. Examples of a typical DHC and a non - DHC in the investigation area can be seen in Supplementary Fig. 1. As observed at a low altitude by the CE - 4 landing camera (Supplementary Fig. 2), DHCs are widely distributed on the Von Kármán crater floor. In the ~40km × 40km region around the CE - 4 landing site (i. e., Statio Tianhe), 35 DHCs (with diameters of ~277—1422m) and 50 non - DHCs (with diameters of ~181—531m) were identified and mapped out (Fig. 5, Supplementary Tab. 3). Both DHCs and non - DHCs have similar optical maturity (OMAT) distribution (Supplementary Fig. 3), and therefore the dark halos are not caused by regolith maturity. Spatial distribution in Fig. 5 shows that more DHCs were identified in the southwest than in the northeast, indicating that ejecta in the northeast region are probably relatively thicker. At the CE－4 landing site, the closest DHC to the CE－4 landing site is much further than the closest non－DHC, which may suggest that the landing site locates in a region with relatively thick ejecta deposits.

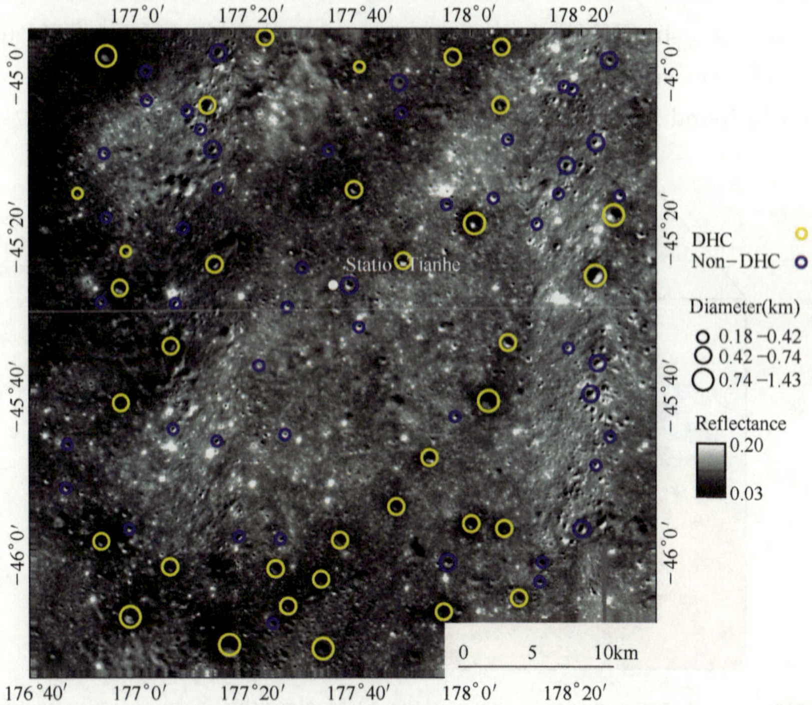

Fig. 5　The DHCs (yellow) and non－DHCs (blue) identified in the area around the CE－4 landing site. The circle size represents the relative size of three diameter groups instead of the real diameters. Background image is the MI reflectance at 1000nm band. Statio Tianhe (white dot) is the landing site of the CE－4 mission. (For interpretation of the references to colour in this figure legend, the reader is referred to the web version of this article.)

3.2　Method for ejecta thickness estimation

A direct way to obtain the thickness of extraneous ejecta overlaying the mare basalt is to identify the ejecta－mare interface, and the ejecta thickness can be easily obtained by subtracting the elevation of the interface from the surface elevation. As the ejecta－mare interface is spatially continuous and lays between the deepest excavations of DHCs and non－DHCs, these two types of craters can be used to constrain locations of the ejecta－mare interface. From another perspective, the ejecta－mare interface can be regarded as a classifier that can ideally separate the DHCs and non－DHCs by their maximum excavate depths, as demonstrated in Fig. 6. In this study, a classifier is generated from DHCs and non－DHCs data using the support－vector machine (SVM) model (Cortes and Vapnik, 1995). The hyperplane of the derived classifier represents the ejecta－mare interface

(Fig. 6) . As a widely - used supervised machine learning model, SVM can make good use of the training data by calculating the data in other dimensions in which different classes can be separated more accurate. The hyperplane of SVM classifier is defined as the plane that has the largest distance to different classes in multi - dimensional space (Cortes and Vapnik, 1995) . In SVM, the kernel function is flexible to be defined to meet the requirements of the specified application scenarios. The theoretical basis of SVM algorithm can be found in Appendix A.

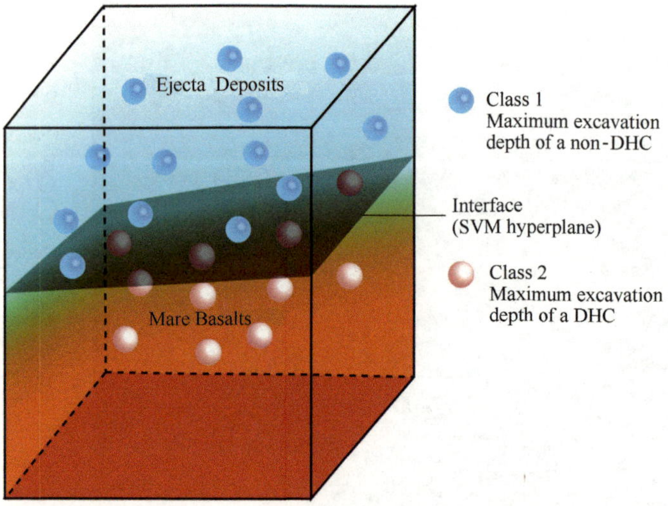

Fig. 6　A schematic diagram showing the SVM algorithm used to estimate the thickness of ejecta deposits. Each ball represents the maximum excavation depth of a non - DHC (blue) or DHC (red) (For interpretation of the references to colour in this figure legend, the reader is referred to the web version of this article)

Our estimate of the ejecta thickness is implemented through several steps. First, calculating the maximum excavation depths (h_e) of DHCs and non - DHCs. For simple craters (usually <20km in diameter), cratering mechanism analysis suggests that the maximum excavation depth is about 1/10 of the transient cavity diameter, which is about 0.84 of the rim - to - rim diameter (Melosh, 1989) . Thereafter, the background elevation (h) of each crater is extracted from SELENE and LRO DEM 2015 (SLDEM2015) lunar DEM model (~60m/pixel), which has the vertical accuracy of ~3—4m (Barker et al. , 2016) . To exclude the ejecta from the crater itself, h is defined as the mean elevation of the annulus 2.5—3 radii from the crater center. With the results of the above two steps, we can get the elevation at the maximum cratering excavation (z) through $h - h_e$. The three - dimensional coordinates of crater maximum excavation make up the training vectors ($[x_i, y_i, z_i]$, where x_i and y_i are the horizontal coordinates of a crater center, i represents the crater index) . Two target values for classification ($[k_i]$) are self - defined (such as 1 and - 1) to represent DHC and non - DHC, respectively. Through SVM training, we can obtain the classification hyperplane, which shows the location of the ejecta - mare interface. By subtracting hyperplane from the regional elevation, thickness of the extraneous ejecta can be obtained.

In this study, the SVM process is implemented using scikit - learn package (https://scikit - learn. org/), a machine learning module for Python. Penalty parameter C and kernel parameter γ are the most important parameters in our SVM model with radial basis function (RBF) kernel. As the parameter γ increases, the hyperplane becomes more complex and has higher accuracy, but may result in overfitting at large γ. Meanwhile, a larger C value also gives higher accuracy with the risk of overfitting. On the contrary, smaller γ and C result in lower accuracy and may be underfitted.

We tested parameter C from 5 to 100 and parameter γ from 0. 1 to 10 (Appendix A). Part of the results is shown in Supplementary Fig. 4. To avoid either underfitting or overfitting, we finally chose the results of $C = 30$ and $\gamma =$ 'scale' (≈ 4) (Supplementary Fig. 4; Supplementary Tab. 1 and 3; Supplementary Movie 1). With 2 DHCs and 4 non - DHCs outliers in the 85 training craters, the accuracy score of our SVM classifier is 0. 93 (79/85, Supplementary Tab. 1 and 3), which means that 93% of the DHCs and non - DHCs can be correctly separated by the classifier hyperplane. It indicates that the mare - ejecta interface is well - modeled by the SVM hyperplane.

3. 3　Ejecta thickness estimation results

Using the SVM model, we obtained the three - dimensional coordinates of ejecta - mare interface, i. e. , the hyperplane of the SVM classifier (Fig. 6). By subtracting the elevation of the interface from SLDEM2015, thickness distribution of the non - mare ejecta in the CE - 4 landing area is calculated and is shown in Fig. 7. Instead of only get - ting the range of areal ejecta thickness (e. g. , Li et al. , 2020), we obtained the thickness value at every location of the investigation area, which provides important information for understanding geology and materials of the CE - 4 landing area. Spatial resolution of the thickness data is 60m/pixel, which is the same as the DEM model SLDEM2015 (Barker et al. , 2016). As shown in Fig. 7a, the ejecta show a southwest - northeast distribution pattern and are thicker in the northeast, indicating that Finsen crater is one of the important ejecta sources. The distance (117 — 166km) of the investigation area is more than one radius away from the rim of the Finsen crater, therefore the ejecta are discontinuous and appear as linear ridges up to tens of meters high (Fig. 1d, Fig. 7a, Supplementary Fig. 5). The thickest ejecta in the investigation region is ~80m whereas some areas are nearly free of non - mare ejecta (Fig. 7a), and the mean thickness is 41m. At the CE - 4 landing site, the ejecta are ~40m thick (Fig. 7b), and the uncertainty is less than 2m according to LPR detections (Section 4). Due to the nonuniform distribution of the ejecta, there is a substantial variation of ~30m even within a region of 100s m across (Fig. 7b).

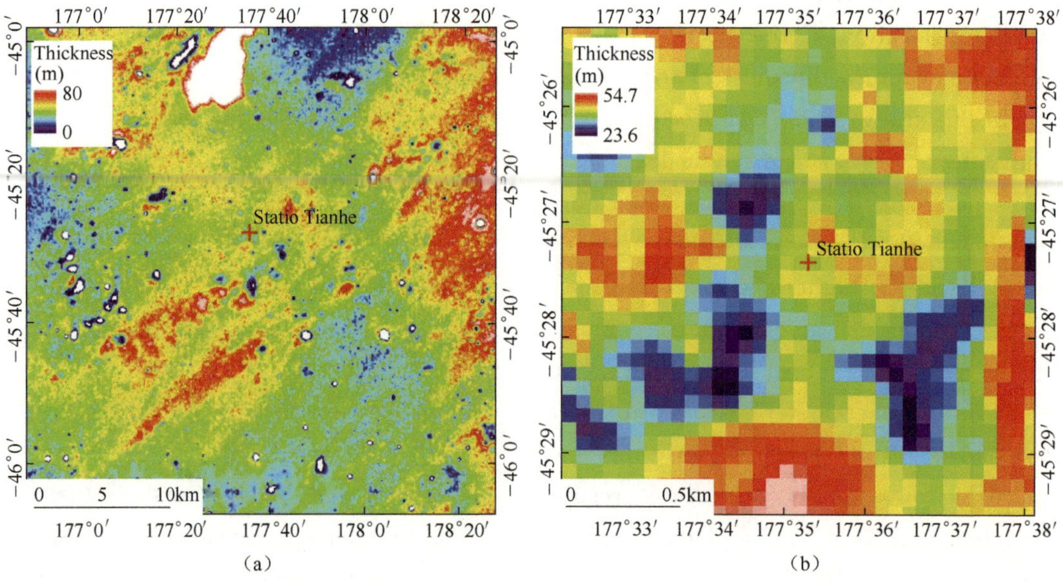

Fig. 7　(a) Ejecta thickness map for the region 40km across the CE－4 landing site (Statio Tianhe). (b) Ejecta thickness in local area of the CE－4 landing site. The spatial resolution is 60m/pixel. To get rid of the invalid ejecta thickness estimate in the region, the pixels are obliterated and shown in white if the thickness value is negative (e. g., floor of large craters) or greater than 80m (e. g., the central peak), which is almost the deepest excavation of the identified craters

There are several factors that may affect the accuracy of the SVM classifier and result in errors in the ejecta thickness estimate. (1) Impact experiments suggest that a dark－haloed crater can not be created until the excavation penetrates in the mare basalts deeper than just touching the pre－existing mare surface (Antonenko, 1999). Thus, the calculated ejecta thickness is likely overestimated to some extent. However, no reliable correction methods can be applied at present. (2) The rough terrain surface around the crater causes uncertainty in the background elevation estimation, which can be seen from the standard deviation (2.1—23.5m) of crater background elevation in Supplementary Tab. 3. This effect can be minimized as SVM classifier is produced from the entire crater database. After excluding the three craters with unusually large DEM standard deviations (>10m) (Supplementary Tab. 3), we got almost the same ejecta thickness of ～40m at the landing site. (3) The nonuniform distribution of craters, especially the absence of either DHCs or non－DHCs in sub－areas can lead to poor constraints and increase the uncertainties at local scale, such as the area southwest to the CE－4 landing site where no DHCs are observed. Nevertheless, such problems can not be removed even using other methods, and the results of this study can be compared with in situ observations of the CE－4 LPR.

4 Ejecta thickness based on the CE-4 LPR observations

CE-4 LPR is a dual-frequency ground penetrating radar operating at center frequencies of 60 and 500MHz (Fang et al., 2014). According to the Kaguya MI observations, the FeO and TiO_2 concentrations of the regolith at the CE-4 landing site are 12.7 wt.% and 1.3 wt.%, respectively. Using the relation among dielectric permittivity, bulk density, and composition of regolith (Fa and Wieczorek, 2012) and a typical porosity of 0.45, dielectric permittivity of the lunar surface materials at the landing site is estimated to be 3.07 +0.020i. Such extremely low loss tangent (0.0064) of the surface material allows the radar waves penetrate as deep as several tens of meters, providing a chance to constrain the ejecta thickness. Here we compare the estimated ejecta thickness with the in situ LPR high-frequency observations during the first lunar day.

During the first lunar day (Fig. 8a), the rover traveled from point X to S1, with a surface traverse of ~37.8m, which is in a pixel size of the non-mare ejecta thickness data (Fig. 7). After removing the repetitive observations at the navigation points (e.g., points X, A, B, C, D, and S1), there are in total 1038 effective tracks of raw data for the high frequency channel. From point X to A, it is the parameters adjustment period. Cumulative number of the LPR was adjusted until to point A, and a constant value of 12 was set. Therefore, we selected 880 tracks of LPR data from point A to S1, and the surface traverse is ~29m. The raw data were then processed following the standard ground penetrating radar (GPR) data processing procedure, including horizontal band removal, band-pass filtering, and compensation of geometrical spreading and dielectric attenuation (e.g., Fa et al., 2020; Fa et al., 2015). The processed LPR image is shown in B-scan format as a function of lateral distance (horizontal) and one-way travel time (vertical), or equivalently, depth (Fig. 8b). To convert the time delay to depth, a real dielectric permittivity of 3.1 is used. Absolute signal amplitude calibration is not conducted at the moment and the tone in Fig. 8b shows the relative amplitude with white representing strong echoes and gray denoting weak echoes.

In Fig. 8b, a homogenous region is obvious from the surface to a depth of ~11–14m. This region contains many random irregular layers with lateral continuity only a few meters, and several hyperbolic curves, which are most probably caused by subsurface rocks of sizes larger than ~0.35m (the wavelength of the LPR waves) (Fa et al., 2015). Below this layer to a depth of ~36–38m is a region that appears bright and dark alternatively from the top to the bottom. Within this region, two echoes at a depth of ~25m and ~27.5m are very obvious. The deepest echo can be detected is at ~42m, though very weak. Below 42m is a region with random noises that is beyond the penetration ability of the high-frequency LPR. Here the weak echo at a depth of ~42m could be the base of the ejecta, and this value is also consistent with the estimate from

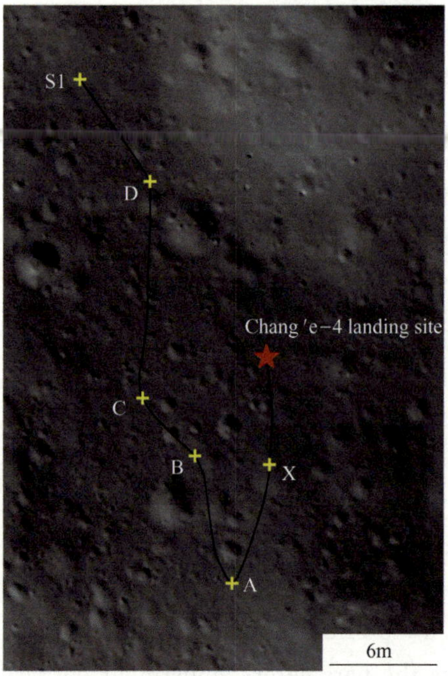

(a)

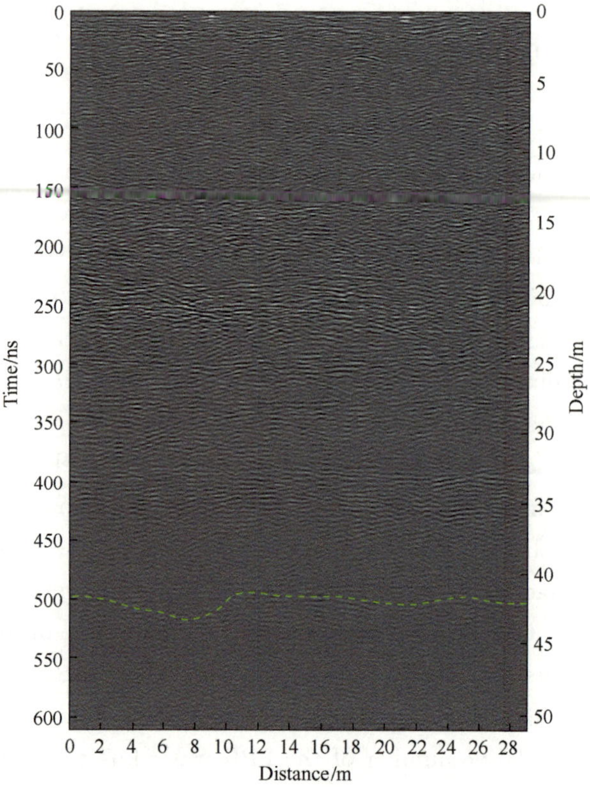

(b)

Fig. 8　(a) An optical image from the CE-4 landing camera (LCAM, image ID: CE4_GRAS_LCAM-1-3050_SCI_N_20190103022520_20190103022520_0001_A) showing the geologic context of the landing site; the red star is the CE-4 landing site; the black line shows the traverse line of the Yutu-2 rover; and X, A, B, C, D and S1 are surface navigation points during the first lunar day. (b) The processed LPR image from Point A to S1 along the Yutu-2 survey line. The green dashed line shows the deepest echoes at a depth of ~42m. The one-way travel time between the antenna and the surface is offset so that zero time corresponds to the lunar surface, and the depth is obtained with a real dielectric permittivity of 3.1 (For interpretation of the references to colour in this figure legend, the reader is referred to the web version of this article)

DHCs and non-DHCs (Section 3). The ejecta thickness constrained from DHCs is 40m, and the LPR observations show that subsurface to a depth of 42m is inhomogeneous at the CE-4 landing site. This implies that the ejecta at the CE-4 landing site are mostly probably accumulated from multiple impact events, as evidenced by the secondary crater chains in different directions (Section 1).

5　Discussion

Based on our analyses, the ejecta thickness of the CE-4 landing region is highly nonuniform and ranges from near zero to ~80m, and the thickness at the landing site is

~40m. Comparing with previous studies (e. g., Di et al., 2019; Huang et al., 2018; Lai et al., 2019; Li et al., 2020), our study used both geologic analysis (DHCs and non-DHCs) and in situ LPR observations, and the results are consistent with each other. In the local region around the CE-4 landing site, we identified six DHCs and non-DHCs (denoted by C1 to C6, Fig. 9), which can provide close constraints to the ejecta thickness at local scale. As can be seen in Fig. 9, the nearest DHC is crater C2, ~565m in diameter situated in the northeast and is ~4.7km from the landing site, and the nearest non-DHC is crater C5, ~452m in diameter situated at ~1km from the landing site in the east. Based on scaling relation estimation (Melosh, 1989), the maximum excavation depths of the DHC (C2) and the non-DHC (C5) are about 47m and 38m, respectively. The elevation of the CE-4 landing site is almost the same as the background elevation of crater C5, but is ~7m lower than that of crater C2. Based on these craters, the ejecta thickness at the CE-4 landing site is ~40m, which is in agreement with the values based on SVM estimate and LPR observations.

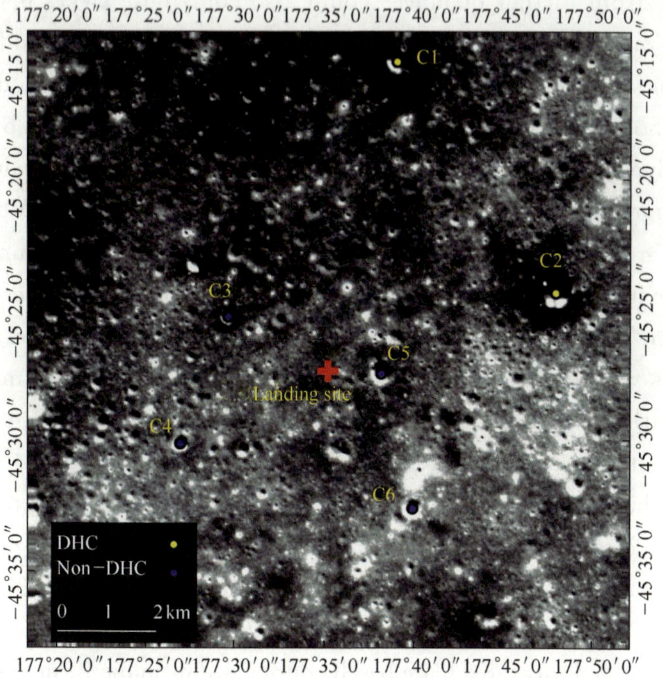

Fig. 9　DHCs (yellow) and non-DHCs (blue) near the CE-4 landing site. The diameters of craters C1, C2, C3, C4, C5 and C6 are 480, 565, 319, 316, 452 and 363m, respectively. Background is MI reflectance image of 1000nm band (For interpretation of the references to colour in this figure legend, the reader is referred to the web version of this article)

There are several craters located around the Von Kármán crater (Fig. 1a), three of them might deliver non-mare ejecta to the CE-4 landing site, including Finsen, Von Kármán L, Von Kármán L' (Supplementary Tab. 2). The Alder crater was also considered as an ejecta source (e. g. , Di et al. , 2019), but a latest study suggests that its age is older than the mare units (Lu et al. , 2021) . As Zhinyu crater is small (3. 8km in diamter) and is far (>15km) from the study area, its influence to the ejecta thickness of the CE - 4 landing area is negligible. In order to evaluate the major ejecta sources, we modeled the accumulated ejecta thickness based on crater ejecta thickness decay model of Pike (1974); and the mixture process with local material was considered in modeling (Oberbeck et al. , 1975) . Model equations are given in Supplementary Text 1. The results suggest that the accumulated ejecta are ~9. 5m thick at the CE - 4 landing site (Supplementary Fig. 7) . The value is significantly less than SVM estimates and LPR observations. One reason is that the landing site is located outside of the continuous ejecta of all the three craters, where the ejecta are sedimented alternately and the thickness variation can not be predicted by ejecta thickness decay laws. The original data from which the model was generated also show great variety in discontinuous ejecta facies (McGetchin et al. , 1973; Pike, 1974) . Moreover, we only considered three craters in the modeling, but young craters from further distances could also contribute ejecta materials. Based on cratering simulation of Di et al. (2019), ejecta from Finsen crater could be ~30m thick at the CE - 4 landing site, which is also inconsistent with ejecta decay model but is close to our estimates. Though absolute ejecta thickness was not well predicted by the ejecta decay model, the relative thickness can be concluded. At the landing site, calculation from the Pike's model suggests that ejecta from Finsen crater are ~4m, from Von Kármán L and Von Kármán L' are both less than 1m, and that about half of the accumulated ejecta are excavated local material. Therefore, the superposed non - mare materials are mainly a mixture of Finsen crater ejecta and excavated mare basalts, and the fraction of local material increases with depth (Xie et al. , 2020) . The ejecta - mare mixture scenario is also consistent with the geochemical characteristics (Section 2) .

Located in the region of ~4. 2Ga old Von Kármán crater, the CE - 4 landing area has a complex evolution history. Geochemical analysis indicates that the surface of the Von Kármán crater floor consists of two major compositional components: the iron - rich mare basalts and the iron - poor non - mare ejecta. The youngest of the major ejecta suppliers is Finsen crater. At the CE - 4 landing site, the non - mare ejecta are ~40m thick and the materials are mainly a mixture of Finsen crater ejecta and excavated mare basalts during sedimentation, in which the fractions of mare materials increase with depth. Geochemistry of the landing site also shows an intermediate FeO concentration between mare basalts and feldspathic highland material. After the sedimentations of extraneous ejecta, due to space weathering processes such as micrometeorite bombardments and solar wind implantation, the primary ejecta have been shattered and overturned to some depth. On the other hand, postdated impact events could excavate and transport basaltic materials to the surface, as shown in Supplementary Fig. 6. Therefore, it is expected that the materials observed by VNIS on the Yutu - 2 rover are predominated by the feldspathic ejecta from the Finsen crater with a significant portion of basaltic materials (e. g. ,

Hu et al., 2019; Lin et al., 2019).

Our study quantified the compositional variations in different geologic units of the Von Kármán crater floor. The CE-4 spacecraft landed in an area dominated by ejecta from Finsen crater and its geochemical characteristics are distinct from typical mare basalts, which is consistent with previous studies (e.g., Huang et al., 2018; Lin et al., 2019; Ling et al., 2019). The SVM algorithm employed in our study revealed regional thickness of extraneous ejecta of the CE-4 landing area, shedding new light on the regional geologic evolution. Supported by LPR observations and local geologic features, we estimated that the ejecta thickness at the CE-4 landing site is ~40m, which is consistent with Li et al. (2020) but inconsistent with other studies (e.g., Di et al., 2019; Huang et al., 2018; Lai et al., 2019).

6 Conclusions

In this work, we explored the geochemical characteristics and non-mare ejecta distribution of the Von Kármán crater with special emphasis on the landing area of the Chang'e-4 probe. Geochemistry of the Von Kármán crater indicates that the regolith in the CE-4 landing area is distinct from typical mare basalts. Geologic characteristics and the concentrations of FeO, TiO_2 and Th indicate that the landing site materials are mixed from feldspathic ejecta and mare basalts, with Finsen crater ejecta material as a major component. To address the thickness of superposed ejecta above mare basalts, we identified and analyzed 35 DHCs and 50 non-DHCs in the landing area. By adopting the SVM algorithm, we obtained the equivalent ejecta-mare interface and calculated the thickness of non-mare ejecta over the CE-4 landing region. The results suggest that the ejecta thickness is substantially varied and is ~40m at the landing site, in agreement with the LPR measurements of the reflectors boundary at ~42m depth. Based on ejecta distribution models, Finsen crater is the major source of foreign ejecta accumulated at the CE-4 landing site and a significant portion of local materials are mixed to the non-mare ejecta. Results from this work demonstrated important properties of the materials in the Chang'e-4 landing area in a broad view and at specified locations, which can be used in the Yutu-2 data interpretation and rover traverse design, e.g., searching for mare basalts outcrops where non-mare ejecta are thin.

Acknowledgments

This work was supported partly by B-type Strategic Priority Program of the Chinese Academy of Sciences (No. XDB41000000), National Natural Science Foundation of China (No. 41902317, 41941002, 41773065), National Key Research and Development project (2019YFE0123300) and China Postdoctoral Science Foundation (No. 2019M650319).

SLDEM2015 and Th concentration data can be found in the NASA's Planetary Data System (http://pds.nasa.gov/). The Kaguya MI data are archived at SELENE Data Archive (https://darts.isas.jaxa.jp/planet/pdap/selene/). The CE-4 data are provided by China National Space Administration, and are available at Data Publishing and Information Service System of China's Lunar Exploration Program (http://moon.bao.ac.cn/). The authors acknowledge editor Brandon Johnson for handling the manuscript. We thank Noah Petro, James W. Head and an anonymous reviewer for helpful comments that improved the manuscript. This is PKU PRSL contribution 15.

Supplementary data to this article can be found online at https://doi.org/10.1016/j.icarus.2021.114327.

Appendix A. Support-vector machine (SVM)

A support-vector machine is a learning machine originally for two-group classification problems (Cortes and Vapnik, 1995). Given the training data:

$$(\overrightarrow{x_1}, y_1), \cdots, (\overrightarrow{x_n}, y_n) \tag{A1}$$

where $\overrightarrow{x_i}$ is a p-dimensional vector, y_i is either -1 or 1, representing which class the point $\overrightarrow{x_i}$ belongs to. The SVM classifier is defined by finding the optimal $(p-1)$-dimensional hyperplane, which has the maximum distance from the nearest point of either $y_i=-1$ class or $y_i=1$ class. The hyperplane can be expressed as

$$\vec{w} \cdot \vec{x} + b = 0 \tag{A2}$$

where $\vec{w}$ is the normal vector to the hyperplane, and b is the bias (Fig. A1). If the training data are linearly separable, the margins of the two groups can be defined as:

$$\vec{w} \cdot \vec{x} + b = -1 \tag{A3}$$

$$\vec{w} \cdot \vec{x} + b = 1 \tag{A4}$$

The points located on the margins are called support vectors. For any data $\overrightarrow{x_i}$ in the class of $y_i=1$,

$$\vec{w} \cdot \overrightarrow{x_i} + b \geqslant 1 \tag{A5}$$

For $\overrightarrow{x_i}$ in the class of $y_i=-1$,

$$\vec{w} \cdot \overrightarrow{x_i} + b \leqslant -1 \tag{A6}$$

SVM works when the training data are linearly separable, which, however, is hard to be satisfied in most actual scenarios. Therefore, kernel functions are usually employed to map $\overrightarrow{x_i}$ to $\phi(\overrightarrow{x_i})$, which is more separable than original data (Cortes and Vapnik, 1995). The kernel function can be self-defined or taken from the generally used ones such as linear function, polynomial function, or radial basis function (RBF). RBF has the ability to project the original data space into infinite dimensional space by adjusting the parameters and is very flexible (Shawe-Taylor and Cristianini, 2004). In RBF kernel, the value of a given point depends on the distance of the other points, the closer ones have

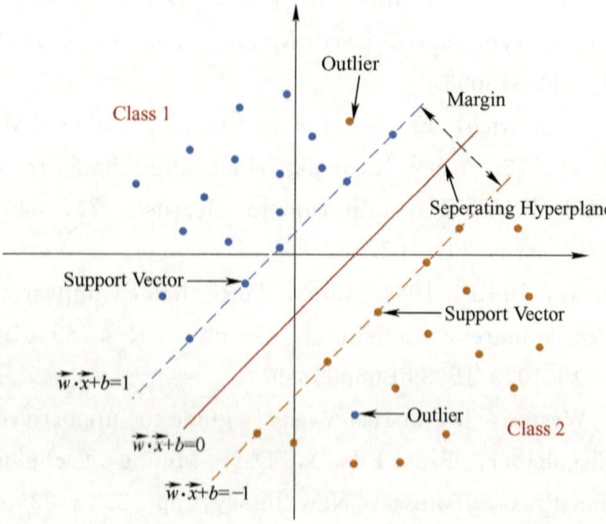

Fig. A1　Schematic diagram of SVM classifier. The two classes are represented by points in orange ($y=-1$) and blue ($y=1$), respectively (For interpretation of the references to colour in this figure legend, the reader is referred to the web version of this article)

stronger influence, which meets the Tobler's first law of geography (Tobler, 1970), and thus used in our process.

SVM training with RBF kernel uses two parameters to control the fitting results, C and γ. Their effects to the classifier can be seen in Supplementary Fig. 4. The parameter C is the penalty parameter and is common to all SVM kernels. A low C makes the decision surface smooth and has the risk of underfitting, while a high C aims at classifying all training examples correctly and has the risk of overfitting (Ben - Hur and Weston, 2010). The decision of parameter C is a trade - off between the margin and the size of the slack variables in specified cases (Shawe - Taylor and Cristianini, 2004) . The parameter γ, $\gamma=1/(2\sigma^2)$ (where σ is the standard deviation), decides how far the influence of a single training example reaches, and the influence distance is further with smaller value and vice versa (Ben - Hur and Weston, 2010) . If γ is too large, the influence of support vectors only reaches the support vector itself; while if γ is too small, the support vectors would influence the whole training set and the model cannot capture the complexity or "shape" of the data.

References

[1] Antonenko, I. , 1999. Volumes of Cryptomafic Deposits on the Western Limb of the Moon: Implications for Lunar Volcanism. Department of Geological Sciences, Brown University.

[2] Antonenko, I., Head, J. W., Mustard, J. F., Hawke, B. R., 1995. Criteria for the detection of lunar cryptomaria. Earth Moon Planet. 69 (2), 141—172. https://doi. org/10. 1007/ Bf00613096.

[3] Barker, M. K., Mazarico, E., Neumann, G. A., Zuber, M. T., Haruyama, J., Smith, D. E., 2016. A new lunar digital elevation model from the lunar orbiter laser altimeter and SELENE terrain camera. Icarus. 273, 346 — 355. https://doi. org/10. 1016/j. icarus. 2015. 07. 039.

[4] Bell, J. F., Hawke, B. R., 1984. Lunar dark - haloed impact craters: origin and implications for early mare volcanism. J. Geophys. Res. 89 (B8), 6899 — 6910. https://doi. org/ 10. 1029/JB089iB08p06899.

[5] Ben - Hur, A., Weston, J., 2010. A user's guide to support vector machines. In: Carugo, O., Eisenhaber, F. (Eds.), Data Mining Techniques for the Life Sciences. Humana Press, Totowa, New Jersey, pp. 223—239. https://doi. org/10. 1007/978—1—60327—241—4 _13.

[6] Charlier, B., Grove, T. L., Namur, O., Holtz, F., 2018. Crystallization of the lunar magma ocean and the primordial mantle - crust differentiation of the moon. Geochim. Cosmochim. Acta 234, 50 — 69. https://doi. org/10. 1016/j. gca. 2018. 05. 006.

[7] Cortes, C., Vapnik, V., 1995. Support - vector networks. Mach. Learn. 20 (3), 273—297. https://doi. org/10. 1007/BF00994018.

[8] Di, K., Zhu, M. H., Yue, Z., Lin, Y., Wan, W., Liu, Z., Gou, S., Liu, B., Peng, M., Wang, Y., Niu, S., Zhang, J., Li, J., Xie, J., Xi, L., Yang, J., Xue, B., 2019. Topographic evolution of Von Kármán crater revealed by the lunar rover Yutu - 2. Geophys. Res. Lett. 46 (22), 12764 — 12770. https://doi. org/10. 1029/2019gl085252.

[9] Elkins - Tanton, L. T., Burgess, S., Yin, Q. Z., 2011. The lunar magma ocean: reconciling the solidification process with lunar petrology and geochronology. Earth Planet. Sci. Lett. 304 (3—4), 326—336. https://doi. org/10. 1016/j. epsl. 2011. 02. 004.

[10] Fa, W., Wieczorek, M. A., 2012. Regolith thickness over the lunar nearside: results from earth - based 70 - cm Arecibo radar observations. Icarus. 218 (2), 771—787. https://doi. org/10. 1016/j. icarus. 2012. 01. 010.

[11] Fa, W., Zhu, M. —H., Liu, T., Plescia, J. B., 2015. Regolith stratigraphy at the Chang' E - 3 landing site as seen by lunar penetrating radar. Geophys. Res. Lett. 42 (23), 10179—10187. https://doi. org/10. 1002/2015gl066537.

[12] Fa, W., Guo, D., Liu, X., Du, J., 2020. Shallow subsurface structure of the moon at China's Chang'E 4 landing site observed by the lunar penetrating radar. In: The 51st Lunar and Planetary Sciecne Conference (p. #1799).

[13] Fang, G. - Y., Zhou, B., Ji, Y. —C., Zhang, Q. - Y., Shen, S. —X., Li, Y. —X., Guan, H. —F., Tang, C. — J., Gao, Y. —Z., Lu, W., Ye, S. —

B. , Han, H. - D. , Zheng, J. , Wang, S. —Z. , 2014. Lunar penetrating radar onboard the Chang'e - 3 mission. Res. Astron. Astrophys. 14 (12), 1607—1622. https://doi.org/10.1088/1674—4527/14/12/009.

[14] Fortezzo, C. M. , Spudis, P. D. , Harrel, S. L. , 2020. Release of the digital unified global geologic map of the Moon at 1 : 5,000,000 - scale. In: Lunar and Planetary Science Conference (p. #2760) .

[15] Garrick - Bethell, I. , Zuber, M. T. , 2005. An indigenous origin for the south pole Aitken basin thorium anomaly. Geophys. Res. Lett. 32, L13203 https://doi.org/10.1029/2005gl023142.

[16] Garrick - Bethell, I. , Zuber, M. T. , 2009. Elliptical structure of the lunar south pole - Aitken basin. Icarus 204 (2), 399 — 408. https://doi.org/10.1016/j.icarus.2009.05.032.

[17] Garrick - Bethell, I. , Miljković, K. , Hiesinger, H. , van der Bogert, C. H. , Laneuville, M. , Shuster, D. L. , Korycansky, D. G. , 2020. Troctolite 76535: a sample of the moon's south pole - Aitken basin? Icarus 338, 113430. https://doi.org/10.1016/j.icarus.2019.113430.

[18] Gou, S. , Di, K. , Yue, Z. , Liu, Z. , He, Z. , Xu, R. , Lin, H. , Liu, B. , Peng, M. , Wan, W. , Wang, Y. , Liu, J. , 2019. Lunar deep materials observed by Chang'e - 4 rover. Earth Planet. Sci. Lett. 528, 115829. https://doi.org/10.1016/j.epsl.2019.115829.

[19] Guo, D. , Fa, W. , Zeng, X. , Cai, Y. , Du, J. , Zhao, M. , 2019. Geological investigation of the Chang'e - 4 landing site and the expected scientific return from the lunar penetrating radar. In: The 50th Lunar and Planetary Science Conference (#1844) .

[20] Haskin, L. A. , McKinnon, W. B. , Benner, L. A. M. , Jolliff, B. L. , 2004. Thorium anomalies in the NW quadrant of the South Pole - Aitken basin. In: The 35th Lunar and Planetary Science Conference (#1461) .

[21] Head, J. W. , Wilson, L. , 2017. Generation, ascent and eruption of magma on the moon: new insights into source depths, magma supply, intrusions and effusive/explosive eruptions (part 2: predicted emplacement processes and observations) . Icarus 283, 176—223. https://doi.org/10.1016/j.icarus.2016.05.031.

[22] Hu, X. , Ma, P. , Yang, Y. , Zhu, M. —H. , Jiang, T. , Lucey, P. G. , Sun, L. , Zhang, H. , Li, C. , Xu, R. , He, Z. , Lin, H. , Huang, C. , Sun, Y. , 2019. Mineral abundances inferred from in situ reflectance measurements of Chang'e - 4 landing site in south pole - Aitken basin. Geophys. Res. Lett. 46 (16), 9439—9447. https://doi.org/10.1029/2019GL084531.

[23] Huang, J. , Xiao, Z. , Flahaut, J. , Martinot, M. , Head, J. , Xiao, X. , Xie, M. , Xiao, L. , 2018. Geological characteristics of Von Kármán crater, northwestern south pole - Aitken basin: Chang'e - 4 landing site region. J.

Geophys. Res. Planet. 123 (7), 1684 — 1700. https://doi.org/10.1029/2018JE005577.

[24] Jia, Y., Zou, Y., Ping, J., Xue, C., Yan, J., Ning, Y., 2018. The scientific objectives and payloads of Chang'E-4 mission. Planet. Space Sci. 162, 207—215. https://doi.org/10.1016/j.pss.2018.02.011.

[25] Jolliff, B.L., Gillis, J.J., Haskin, L.A., Korotev, R.L., Wieczorek, M.A., 2000. Major lunar crustal terranes: surface expressions and crust-mantle origins. J. Geophys. Res. Planet. 105 (E2), 4197 — 4216. https://doi.org/10.1029/1999je001103.

[26] Jolliff, B.L., Shearer, C.K., Papanastassiou, D.A., Liu, Y., 2017. Why do we need samples from the Moon's South Pole-Aitken basin and what would we do with them? In: The 48th Lunar and Planetary Science Conference (#1300).

[27] Korotev, R.L., 1997. Some things we can infer about the moon from the composition of the Apollo 16 regolith. Meteorit. Planet. Sci. 32 (4), 447—478. https://doi.org/10.1111/j.1945—5100.1997.tb01291.x.

[28] Korotev, R.L., 2005. Lunar geochemistry as told by lunar meteorites. Geochemistry 65 (4), 297—346. https://doi.org/10.1016/j.chemer.2005.07.001.

[29] Lai, J., Xu, Y., Zhang, X., Xiao, L., Yan, Q., Meng, X., Zhou, B., Dong, Z., Zhao, D., 2019. Comparison of dielectric properties and structure of lunar regolith at Chang'e-3 and Chang'e-4 landing sites revealed by ground-penetrating radar. Geophys. Res. Lett. 46 (22), 12783—12793. https://doi.org/10.1029/2019gl084458.

[30] Lawrence, D.J., Feldman, W.C., Barraclough, B.L., Binder, A.B., Elphic, R.C., Maurice, S., Miller, M.C., Prettyman, T.H., 2000. Thorium abundances on the lunar surface. J. Geophys. Res. —Planet. 105 (E8), 20307—20331. https://doi.org/10.1029/1999je001177.

[31] Lawrence, D.J., Elphic, R.C., Feldman, W.C., Gasnault, O., Genetay, I., Maurice, S., Prettyman, T.H., 2002. Small-area thorium enhancements on the lunar surface. In: The 33rd Lunar and Planetary Science Conference (#1970).

[32] Lemelin, M., Lucey, P.G., Song, E., Taylor, G.J., 2015. Lunar central peak mineralogy and iron content using the Kaguya multiband imager: reassessment of the compositional structure of the lunar crust. J. Geophys. Res. Planet. 120 (5), 869—887. https://doi.org/10.1002/2014JE004778.

[33] Lemelin, M., Lucey, P.G., Gaddis, L.R., Hare, T., Ohtake, M., 2016. Global map products from the Kaguya Multiband Imager at 512 ppd: Minerals, FeO, and OMAT. In: The 47th Lunar and Planetary Science Conference (#2994).

[34] Li, C., Liu, D., Liu, B., Ren, X., Liu, J., He, Z., Zuo, W., Zeng, X., Xu, R., Tan, X., Zhang, X., Chen, W., Shu, R., Wen, W., Su,

Y., Zhang, H., Ouyang, Z., 2019. Chang'E - 4 initial spectroscopic identification of lunar far - side mantle - derived materials. Nature 569 (7756), 378—382. https://doi.org/10.1038/s41586—019—1189—0.

[35] Li, C., Su, Y., Pettinelli, E., Xing, S., Ding, C., Liu, J., Ren, X., Lauro, S.E., Soldovieri, F., Zeng, X., Gao, X., Chen, W., Dai, S., Liu, D., Zhang, G., Zuo, W., Wen, W., Zhang, Z., Zhang, X., Zhang, H., 2020. The Moon's farside shallow subsurface structure unveiled by Chang'E - 4 lunar penetrating radar. Sci. Adv. 6 (9) https://doi.org/10.1126/sciadv.aay6898 eaay6898.

[36] Lin, H., He, Z., Yang, W., Lin, Y., Xu, R., Zhang, C., Zhu, M.—H., Chang, R., Zhang, J., Li, C., Lin, H., Liu, Y., Gou, S., Wei, Y., Hu, S., Xue, C., Yang, J., Zhong, J., Fu, X., Wan, W., Zou, Y., 2019. Olivine - norite rock detected by the lunar rover Yutu - 2 likely crystallized from the SPA impact melt pool. Natl. Sci. Rev. https://doi.org/10.1093/nsr/nwz183 nwz183.

[37] Ling, Z., Qiao, L., Liu, C., Cao, H., Bi, X., Lu, X., Zhang, J., Fu, X., Li, B., Liu, J., 2019. Composition, mineralogy and chronology of mare basalts and non - mare materials in Von Kármán crater: landing site of the Chang'E - 4 mission. Planet. Space Sci. 179, 104741. https://doi.org /10.1016/j.pss.2019.104741.

[38] Lu, Y., Wu, Y., Michael, G.G., Ma, J., Cai, W., Qin, N., 2021. Chronological sequence of Chang'E - 4 landing zone within Von Kármán crater. Icarus 354. https://doi.org/10.1016/j.icarus.2020.114086.

[39] Lucey, P.G., Korotev, R.L., Gillis, J.J., Taylor, L.A., Lawrence, D., Campbell, B.A., Elphic, R., Feldman, B., Hood, L.L., Hunten, D., Mendillo, M., Noble, S., Papike, J.J., Reedy, R.C., Lawson, S., Prettyman, T., Gasnault, O., Maurice, S., 2006.

[40] Understanding the lunar surface and space - moon interactions. In: Jolliff, B., Wieczorek, M., Shearer, C., Neal, C.R. (Eds.), New Views of the Moon. Mineralogical Society of America, Chantilly, Virginia, pp. 83—220. https://doi.org/10.2138/rmg.2006.60.2.

[41] McGetchin, T.R., Settle, M., Head, J.W., 1973. Radial thickness variation in impact crater ejecta: implications for lunar basin deposits. Earth Planet. Sci. Lett. 20 (2), 226—236. https://doi.org/10.1016/0012—821X (73) 90162—3.

[42] Melosh, H.J., 1989. Impact Cratering: A Geologic Process. Oxford University Press New York.

[43] Moriarty, D.P., Pieters, C.M., 2018. The character of south pole - Aitken basin: patterns of surface and sub - surface composition. J. Geophys. Res. Planet. 123 (3), 729—747. https://doi.org/10.1002/2017JE005364.

[44] Neal, C. R., Taylor, L. A., 1992. Petrogenesis of mare basalts: a record of lunar volcanism. Geochim. Cosmochim. Acta 56 (6), 2177－2211. https://doi.org/10.1016/0016－7037 (92) 90184－K.

[45] Nelson, D. M., Koeber, S. D., Daud, K., Robinson, M. S., Watters, T. R., Banks, M. E., Williams, N. R., 2014. Mapping lunar maria extents and lobate scarps using LROC image products. In: The 45th Lunar and Planetary Science Conference (#2861).

[46] Oberbeck, V. R., Hörz, F., Morrison, R. H., Quaide, W. L., Gault, D. E., 1975. On the origin of the lunar smooth-plains. Moon 12 (1), 19－54.

[47] Orgel, C., Michael, G., Fassett, C. I., van der Bogert, C. H., Riedel, C., Kneissl, T., Hiesinger, H., 2018. Ancient bombardment of the inner solar system: reinvestigation of the "fingerprints" of different impactor populations on the lunar surface. J. Geophys. Res. Planet. 123 (3), 748－762. https://doi.org/10.1002/2017je005451.

[48] Otake, H., Ohtake, M., Hirata, N., 2012. Lunar iron and titanium abundance algorithms based on SELENE (Kaguya) Multiband Imager data. In: The 43rd Lunar and Planetary Science Conference (#1905).

[49] Pike, R. J., 1974. Ejecta from large craters on the moon: comments on the geometric model of McGetchin et al. Earth Planet. Sci. Lett. 23 (3), 265－271. https://doi.org/10.1016/0012－821X (74) 90114－9.

[50] Prettyman, T. H., Hagerty, J. J., Elphic, R. C., Feldman, W. C., Lawrence, D. J., McKinney, G. W., Vaniman, D. T., 2006. Elemental composition of the lunar surface: analysis of gamma ray spectroscopy data from lunar prospector. J. Geophys. Res. Planet. 111, E12007 https://doi.org/10.1029/2005je002656.

[51] Schultz, P. H., Spudis, P. D., 1979. Evidence for ancient mare volcanism. In: The 10th Lunar and Planetary Science Conference, pp. 2899－2918.

[52] Shawe-Taylor, J., Cristianini, N., 2004. Kernel Methods for Pattern Analysis. Cambridge University Press.

[53] Tobler, W. R., 1970. A computer movie simulating urban growth in the Detroit region. Econ. Geogr. 46 (supl), 234－240. https://doi.org/10.2307/143141.

[54] Uemoto, K., Ohtake, M., Haruyama, J., Matsunaga, T., Yamamoto, S., Nakamura, R., Yokota, Y., Ishihara, Y., Iwata, T., 2017. Evidence of impact melt sheet differentiation of the lunar south pole-Aitken basin. J. Geophys. Res. Planet. 122, 1672－1686. https://doi.org /10.1002/2016JE005209.

[55] Vaughan, W. M., Head, J. W., 2014. Impact melt differentiation in the south pole-Aitken basin: some observations and speculations. Planet. Space Sci. 91, 101－106. https://doi.org/10.1016/j.pss.2013.11.010.

[56] Whitten, J., Head, J. W., 2015a. Lunar cryptomaria: mineralogy and composition of ancient volcanic deposits. Planet. Space Sci. 106, 67－81. https://

doi. org/10. 1016/j. pss. 2014. 11. 027.

[57] Whitten,J. L. , Head, J. W. , 2015b. Lunar cryptomaria: physical characteristics, distribution, and implications for ancient volcanism. Icarus 247, 150 — 171. https:// doi. org/10. 1016/j. icarus. 2014. 09. 031.

[58] Wieczorek, M. A. , Zuber, M. T. , 2001. A Serenitatis origin for the Imbrian grooves and south pole - Aitken thorium anomaly. J. Geophys. Res. 106 (E11), 27853—27864. https://doi. org/10. 1029/2000je001384.

[59] Wieczorek, M. A. , Neumann, G. A. , Nimmo, F. , Kiefer, W. S. , Taylor, G. J. , Melosh, H. J. , Phillips, R. J. , Solomon, S. C. , Andrews - Hanna, J. C. , Asmar, S. W. , Konopliv, A. S. , Lemoine, F. G. , Smith, D. E. , Watkins, M. M. , Williams, J. G. , Zuber, M. T. , 2013. The crust of the moon as seen by GRAIL. Science 339 (6120), 671 — 675. https://doi. org/10. 1126/science. 1231530.

[60] Xie,M. , Liu, T. , Xu, A. , 2020. Ballistic sedimentation of impact crater ejecta: implications for the provenance of lunar samples and the resurfacing effect of ejecta on the lunar surface. J. Geophys. Res. Planet. 125 (5) https://doi. org/10. 1029/2019je006113.

Millimeter - to Decimeter - Scale Surface Roughness of the Moon at the Chang'e - 4 Exploration Region

Dijun Guo[1], Wenzhe Fa[1,2,3], Bo Wu[4], Yuan Li[5], and Yang Liu[6]

1. Institute of Remote Sensing and Geographical Information System, School of Earth and Space Sciences, Peking University, Beijing, China
2. State Key Laboratory of Lunar and Planetary Sciences, Macau University of Science and Technology, Macau, China
3. Center for Excellence in Comparative Planetology, Chinese Academy of Sciences, Hefei, China
4. Planetary Remote Sensing Laboratory, Department of Land Surveying and Geo - Informatics, The Hong Kong Polytechnic University, Hong Kong, China
5. School of Geospatial Engineering and Science, Sun Yat - Sen University, Zhuhai, China
6. State Key Laboratory of Space Weather, National Space Science Center, Chinese Academy of Sciences, Beijing, China

Abstract: The surface slope and roughness of the Moon have been investigated extensively over a wide baseline range except millimeter to decimeter scales. In this study, we present for the first time millimeterto decimeter - scale surface slope and roughness of the Moon at China's Chang'e - 4 landing regions (~20m across) using the Digital Terrain Model (DTM) with a resolution of 5mm/pixel. The bidirectional slope at the 7mm scale can be larger than 40° with a median value of ~10°. The root - mean - square (RMS) height within a window size of 125mm varies from ~1mm to ~18mm with a median value of ~4mm. Both the bidirectional slope and RMS height show scale - dependent behaviors and the parameter of scale dependence, the Hurst exponent, is ~0.6 - 0.85. We also synthesized the bidirectional slope at baseline from micrometer to kilometer, showing that bidirectional slope decreases from ~60° at micrometer to ~1° at a kilometer. At millimeter - scale, surface roughness is mainly controlled by small impact craters, rocks, and regolith properties. Our roughness results not only bridge the gap in understanding surface roughness from traditional topographic data sets to radar and thermal observations, but also provide valuable information about lunar regolith characteristics, and small - scale geological processes.

Plain Language Summary: The roughness of a planetary body quantitatively describes terrain surface topographic undulation at a horizontal scale and is widely used in geological and geomorphological studies. Roughness is an important parameter in radar observations due to its modulation of radar wave behaviors. Lunar surface roughness at a wide baseline range has been investigated before, but not in the baseline range of millimeter to decimeter. The Panoramic Camera onboard the Yutu－2 rover collected millimeter－scale resolution stereo images over China's Chang'e－4 landing region. We generated a 5mm/pixel resolution DTM covering an area of ～20m wide and then calculated three roughness parameters. We found that small impact craters, rocks, and regolith properties are the controlling factors affecting the millimeter － scale roughness of the lunar surface. Combining with previous roughness results, we obtained lunar surface roughness versus baseline at scales ranging from micrometer to kilometer, which can be applied in a variety of studies.

1　Introduction

Surface roughness is a geomorphologic variable describing topography variation with respect to horizontal scale (known as baseline). The topographic roughness of a planetary body is directly related to its surface geological features of various scales and hence provides clues about surface processes and evolutionary history (e. g., Cai & Fa, . 2020; Kreslavsky & Head, 2000; Kreslavsky et al., 2013; Kreslavsky et al., 2014). In remote sensing observations, roughness is one of the few predominant parameters determining optical reflectance (e. g., Hapke, 1984), thermal emissivity (e. g., Bandfield et al., 2015), and radar albedo (e. g., Fa et al., 2011).

During the past decade, the surface roughness of the Moon has been studied extensively over a wide baseline range using the topography data acquired from orbital missions (e. g., Cai & Fa, 2020; Kreslavsky et al., 2013; Rosenburg et al., 2011). The results show that the topographic roughness of the Moon is scale－dependent and that its scale－dependent behaviors are related to different geological processes. At the kilometer scale, lunar surface roughness is controlled by large－scale impact cratering events, volcanic eruptions, and tectonic features (Kreslavsky et al., 2013; Rosenburg et al., 2011). At the hectometer scale, the controlling factors are small impact events and regolith accumulation processes (Kreslavsky et al., 2013). At the meter scale, the difference in surface roughness between the maria and highlands is noticeable but not as large as those at hectometer and kilometer scales, and small degraded impact craters (<1km) are the dominant factor affecting surface roughness (Cai & Fa, 2020). At the submillimeter scales, the root－mean－square (RMS) slope of the Fra Mauro region (a typical highland terrain) is significantly higher than those at the Apollo 11 and 12 landing regions (typical maria) (Helfenstein & Shepard, 1999). Individual impact craters generally possess noticeable roughness characteristics at different subunits (i. e., floor, wall, and continuous ejecta blanket) depending on crater age and terrain type (Wang et al., 2020). On the other hand, the scale-dependent roughness characteristics show fractal-like behaviors that can be quantitively described by the Hurst exponent (e. g., Helfenstein & Shepard, 1999; Shepard et al., 2001). Based on the roughness at hectometer to kilometer scales, the Hurst exponents of highlands are significantly greater than those of the maria (Rosenburg et al., 2011), whereas, at the meter to hectometer scales, the Hurst exponent distributions of the maria and highlands are overlapping (Cai & Fa, 2020).

On the other hand, lunar surface roughness has been investigated indirectly from radar and thermal infrared observations through its modulation on surface reflectivity and emissivity. The estimated roughness is an effective value at the wavelength scale within the spatial resolution of the observations. For example, Arecibo P-band radar observations suggested that the upper limits of the RMS slope at the Apollo and Surveyor landing sites are from 3° to 7° (Fa & Wieczorek, 2012). Bandfield et al. (2015) found that the Moon is highly anisothermal because of surface roughness and proposed two ways to derive surface roughness from LRO Diviner radiometer measurements. Analyses of Diviner nadir observations at four units suggest an RMS slope of 20° at scales larger than 0.5-5mm, whereas multiple emission angle measurements at several sites show 20-35° RMS slope distributions (Bandfield et al., 2015). These results provide complementary information to large-scale surface roughness calculated directly from topography data sets.

Despite previous studies, lunar roughness at the decimeter-scale has not yet been reported due to the lack of available topography data. Though millimeter-scale roughness at spatial scales from 8.5μm to 8.5cm was previously investigated by Helfenstein and Shepard. (1999), the DTMs used are only ~8cm in coverage at three Apollo landing sites. However, rich information about small-scale lunar regolith characteristics and processes (e. g., micrometeorite impacts) can be revealed by millimeter-scale roughness and is thus worthwhile to investigate. In addition, the characteristics of meter-scale craters and their impact fragments can be investigated by using millimeter-scale DTMs. Understanding the relation between surface roughness calculated directly from traditional topographic data sets to those estimated indirectly from radar and thermal observations also requires a thorough investigation of surface roughness at decimeter scales and smaller. In China's Chang'e-4 (CE-4) lunar farside exploration mission, the Yutu-2 rover carries a Panoramic Camera (PCAM), which can collect stereo images of the lunar surface at an unprecedented resolution of millimeter (Jia et al., 2018). This provides an opportunity to create high resolution DTM, allowing an investigation of lunar surface roughness at millimeter to decimeter scales.

2 Topographical Data of the Study Region and Roughness Measures

The CE-4 spacecraft landed at an area dominated by non-mare ejecta mainly from the 3. 1 Ga aged Finsen crater (e. g. , Guo et al. , 2021; Lu et al. , 2021) . As seen from the Lunar Reconnaissance Orbiter (LRO) Narrow Angle Camera (NAC) image (Figure. 1a), the CE - 4 landing area is generally flat and the dominant surface features are impact craters with diameters smaller than 100m. These craters are all degraded to some extent based on their morphologic characteristics. With the best resolution of 3mm/pixel, the PCAM has made detailed in situ observations of the CE - 4 exploration area (Jia et al. , 2018; Wu et al. , 2021) . On the 9th and 10th lunar days (Figure 1a), the Yutu - 2 rover moved to a location ~185m west to the landing site and collected stereo images covering an enclosed region of ~20m in width, which is used to generate a DTM for our roughness study.

In total, 66 pairs of PCAM stereo images collected on the ninth and tenth lunar days are used, and their ground sampling distances (GSD) in the DTM region vary from 1. 5 to 3. 5mm/pixel depending on the distance from the rover. These images are processed using the photogrammetry and photoclinometry techniques as discussed in Wu et al. (2021) . The photogrammetric process included tie point matching, bundle adjustment, coregistration of image blocks, dense image matching, and 3D point cloud generation, from which a photogrammetric DTM with a spatial resolution of 1cm/pixel was generated. The photogrammetric DTM was further refined using photoclinometry (Wu et al. , 2018) based on the illumination and observation information of the PCAM images when they were collected, and a refined DTM with a resolution of 5mm/pixel was generated (Figure. 1b) . As the resolution of PCAM images decreases with increasing distance from the rover, and hence that of the DTM behaves in a similar manner. Considering the lowest GSD of the images, the effective resolution of the entire DTM can be regarded as 5mm/pixel. The elevation accuracy is better than ~1cm near the inner edge of the study region (Figure 1b) and decreases with increasing range. Due to pixel - locking effects and the scale limitation of the images (Shimizu & Okutomi, 2005), the DTM has ring - shaped noise artifacts (Figures 1b and S1) . Because of the shadows and solar glares in PCAM images and the decrease of effective resolution with increasing range (Wu et al. , 2021), the quality of the DTM varies with locations. Most of the PCAM images (54 out of 66 pairs) used for DTM generation were taken on the 10th lunar day when the rover was at the center of the region. Because of the viewing geometry, a crater wall near the camera can be blocked out and hence resulted in the asymmetric effective resolution of the crater. In our analysis below, the areas where effective resolutions are notably low or the data quality is severely poor (referred as "null area", see Figure. S1) will be excluded in calculating statistics. Uncertainty analysis shows that the influences of such noise artifacts on the bidirectional slope and RMS height are less than 14% and 6%, respectively (Text. S1) .

The produced DTM covers an area of ~18. 5×20. 7m and the topography decreases from north to south with an elevation difference of ~1. 5m (Figure 1b) . The region

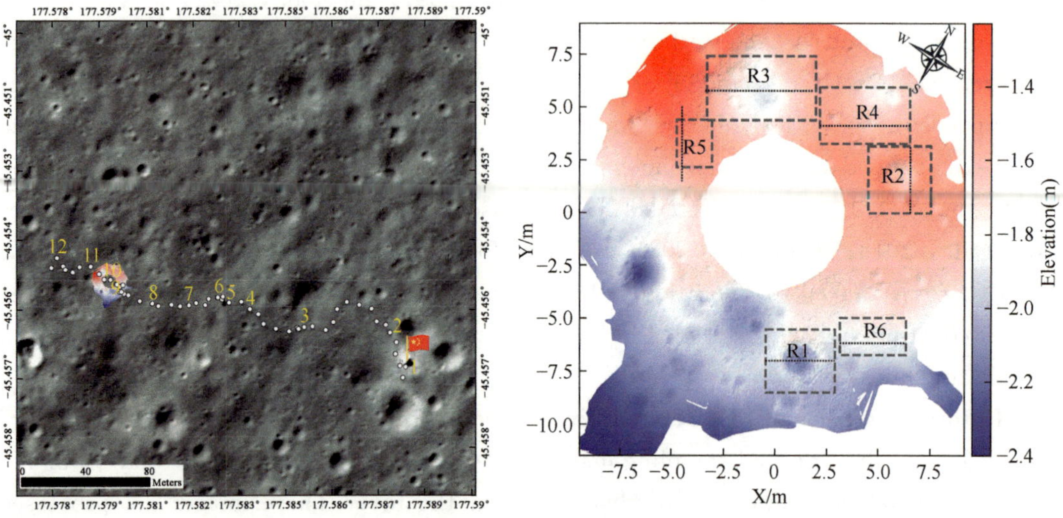

Figure 1　(a) Yutu－2 traverse route during the first 12 lunar days overlaying on an lunar reconnaissance orbiter narrow angle camera image (M1311886645RE). The white dots are the navigation points of the rover, and numbers indicate the first navigation point of each lunar day. The Chang'e－4 lander is marked by the red flag. (b) Hillshaded digital terrain model (DTM) image, where the white regions represent those with no topography data. The elevation values are given in a local camera－based coordinate system. Note that the DTM is ungeoreferenced and has a rotation of 29° in the clockwise direction. The squares (R1－R6) indicate 6 subregions for detailed roughness analysis. The dotted lines in panel (b) denote locations where elevation profiles are extracted for Hurst exponent calculation (Figure S7)

shows diverse morphologic types. Both fresh and degraded craters of meter size are widely distributed in this region, along with abundant craters and rocks at centimeter to decimeter scales. The southeast region is heavily churned by the Yutu－2 rover as indicated by the surface tracks. In order to demonstrate the spatial variation of roughness, we selected six subregions based on crater degradation state (depth/diameter ratio, d/D), geomorphologic diversity, and the quality of the DTM data (Figure 1b). These six subregions are a fresh crater (D: ~2m and d/D: ~0.15; R1), a moderately degraded crater (D: ~2m and d/D: ~0.11; R2), a highly degraded crater (D: ~4.5m and d/D: ~0.07; R3), an area with large craters (D: 0.4－1m; R4), an area with small craters (D<0.2m; R5) and rover tracks (R6).

In order to compare roughness at different scales, we also use the LRO NAC DTM with 5m/pixel resolution and SELENE and LRO merged DEM (SLDEM2015) with 60m/pixel resolution (Barker et al., 2016) to calculate the roughness in the CE－4 landing region at meter to hectometer scales. For the comparison purpose, we also calculated the decimeter to meter scale roughness at the CE－3 landing site using the DTM derived from the CE－3 landing camera (Liu et al., 2015).

In this study, we use the three most common parameters in surface slope and roughness analysis: bidirectional slope, RMS height, and the Hurst exponent (e. g. , Kreslavsky et al. , 2013; Shepard et al. , 2001) . The bidirectional slope is selected because it is investigated extensively at kilometer and submillimeter scales, making it possible for comparison over a wide scale range. We calculated the bidirectional slope using the method in Cai and Fa (2020) . RMS height is calculated as the standard deviation of surface elevations within a two - dimensional window (Cai & Fa, 2020; Shepard et al. , 2001) . RMS height within centimeter scale windows serves as a key parameter in analyzing imaging radar data of the Moon, and our DTM provides the first available data set to calculate it at radar wavelength scale. The Hurst exponent ($0 \leqslant H \leqslant 1$) is a measure of scale dependent behaviors for vertical roughness variation with respect to horizontal scale (Cai & Fa, 2020; Shepard et al. , 2001) . The Hurst exponent is estimated as the linear slope in the *log - log* plot of RMS deviation versus baseline (e. g. , Cai & Fa, 2020; Shepard et al. , 2001) . In the calculation, a raw topographic profile is first detrended in order to remove large - scale slopes, and a threshold value of 0. 999 is used in linear regression (see Cai & Fa, 2020 for detail) .

3 Topographic Slope and Roughness

3. 1 Bidirectional Slope and RMS Height

Using the 5mm/pixel resolution DTM, we calculated the bidirectional slope at baselines from 7 to 283mm and the RMS height at the window size from 15 to 1, 255mm (Figures S2 and S3) . As examples, Figures 2a and 2b show the bidirectional slope at 7mm baseline and RMS height at 125mm window size, respectively. For comparison purpose, we also calculated the bidirectional slope of the Von Kármán crater at the decameter scale and that of the CE - 4 landing region at the meter scale (Figure. S4) . The bidirectional slope and RMS height of the six subregions versus scale are shown in Figures 2c and 2d.

The large slopes of the studying area are mainly associated with craters and tracks of the Yutu - 2 rover (Figures 2a and S1) . As for craters, the steepest slopes occur on the walls whereas the floor of a degraded crater is relatively smooth. The walls of most craters show asymmetric slopes that the side close to the outer edge of the DTM is rougher than the opposite side, which is in the shadow region of the images where the resolution of the DTM is relatively lower. The 7mm scale bidirectional slopes at the roughest areas can be greater than 40° (Figures 2a and S5c) . The median bidirectional slope at millimeter scale is ~10°, which is significantly larger than those at the meter to decameter scale (2°-2. 6°; Figure S4) . The meter to decameter scale slopes (Figure S4) is comparable to those of the lunar maria (2. 0°- 3. 4°) yet notably smaller than those of the highlands (7. 5°- 8. 2°) (Cai & Fa, 2020; Rosenburg et al. , 2011) . The six subregions are quite different in the millimeter - scale bidirectional slope map (Figure. 2a) . The fresh crater area (R1) is significantly rougher than the other subregions. The bidirectional slopes at the roughest locations are even greater than 40° and the crater wall has larger bidirectional slopes than the ejecta deposits (Figures 2a and S5c) . In the ejecta deposits, the bidirectional slopes of the north - south direction are smaller than those of the east - west direction (Figures 2a and S5c) . As the millimeter - scale bidirectional slope in this region is associated with fragments in the ejecta, the roughness planform suggests that this fresh crater may be formed by an oblique impact (Guo et al. , 2018; Herrick & Hessen, 2006) . The moderately and highly degraded regions (R2 and R3) both show greater bidirectional slopes on the wall than the other parts. The subregions with large and small craters (R4 and R5) show increased bidirectional slopes associated with craters. In the rover tracks region (R6), the increase in bidirectional slope is associated with the regolith being churned by the rover wheels.

Figure 2c shows the median bidirectional slopes of the six subregions at baselines of ~7 - 300mm (the relative values to the fresh crater subregion are shown in Figure. S6a) . At the 7mm scale, the median value of the fresh crater subregion is ~17°, which is significantly larger than the 9 - 11° of of the other five subregions. The other five subregions (R2 - R5) have similar slopes at the millimeter scale, which may suggest that the small - scale morphology of the mature regolith is relatively steady. The median slope similarity of rover track subregion (R6) and the degraded craters (R2 and R3) at the millimeter scale indicates that surface disturbance from the rover does not significantly change its millimeter - scale morphology of regolith. With the increase of baseline, the bidirectional slope decreases, and the effects of impact fragments and rover tracks are overtaken by large scale features such as craters (Figure. S2) . The variation behaviors of bidirectional slope versus baseline for the six subregions are different. The slope of the fresh crater subregion quickly declines and becomes comparable to the highly degraded crater at ~300mm baseline (Figures 2c and S6a) . When the baseline is less than 30mm, the bidirectional slopes of the rover track region are slightly larger than the degraded subregions, but the slope decreases with increasing baseline when greater than 30mm (Figures 2c and S6a) .

The RMS height of the study region at the 125mm scale is shown in Figure 2b, which has a median value of 4mm and the largest RMS height can be greater than 15mm. Large RMS heights are mainly associated with craters, especially on the walls where the greatest elevation variations exist. In the intercrater areas, the surface is relatively smooth with RMS height generally smaller than 2mm (Figure 2b) . The fresh crater (R1) is the most prominent feature with high RMS height, but it is mainly caused by the morphologic changes and groups of impact fragments, indicated by the RMS height distributions on the wall. The rover tracks (R6) are shown with increased RMS height (Figure. 2b), though not as prominent as they are in the bidirectional slope map (Figure. 2a) . As the window

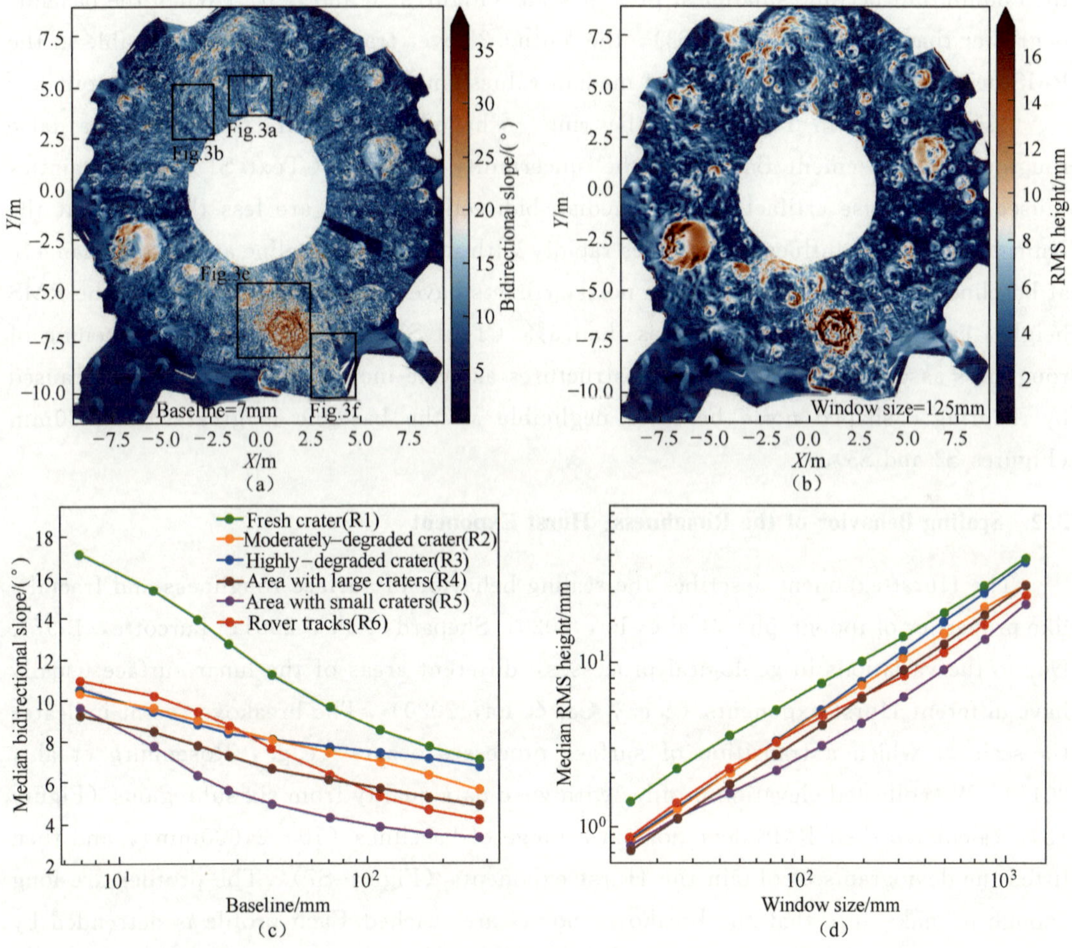

Figure 2　(a) Bidirectional slope at 7mm baseline. (b) Root mean square (RMS) height at a window size of 125mm. (c) Median bidirectional slopes of the six subregions versus baseline. (d) Median RMS heights for the six subregions versus window size. Boxes in panel (a) denote the locations of panoramic camera (PCAM) images in Figure 3

size increases, the effects of small objects and structures (e.g., rocks, and small craters) that are evident at small scales become weak to unrecognizable (Figures 2d and S3). The median RMS heights of the six subregions in the window size from 15 to 1, 255mm are shown in Figure 2d and the relative values to the fresh crater are shown in Figure S6b. With increasing scale, the RMS height increases in general (Figures 2d and S6). On the other hand, the six subregions show different behaviors as scale changes, which may be caused by different controlling geological features and their size variations. The fresh crater (R1) is rougher than the other regions at the investigated scales, and the subregion with small craters (R5) has the minimum RMS height except at 15mm window size. Except for the fresh crater subregion (R1), the region of the

Yutu - 2 track has a slightly larger RMS height than the other subregions at scales smaller than 80mm but become smaller at greater scale (Figures 2d and S6b) . When the baseline is greater than 200mm (Figure S3), the Yutu - 2 rover tracks are almost invisible in the RMS height map with differences of median values smaller than 5% to the background.

As illustrated in Figure 2a, the ring - shaped noise artifacts result in a false roughness increasement. Based on the uncertainty analysis (Text. S1), uncertainties caused by the noise artifacts to the median bidirectional slope are less than 14% at the 7mm baseline. The influence decreases rapidly with increasing baseline and is less than 4% at baselines greater than 40mm. The noise artifacts have much smaller effects on the RMS height, causing uncertainties of less than 6% (Text S1) . Because of the weakening of roughness associated with small size structures as scale increases, the roughness caused by the ring - shaped noise becomes negligible as the baseline is greater than 30mm (Figures S2 and S3).

3.2 Scaling Behavior of the Roughness: Hurst Exponent

The Hurst exponent describes the scaling behavior of surface roughness and fractal - like properties of topography (Cai & Fa, 2020; Shepard et al. , 2001; Turcotte, 1997). Due to the variations in geological processes, different areas of the lunar surface usually have different Hurst exponents (e. g. , Cai & Fa, 2020) . The breakover point indicates the scale at which a transition of surface processes occurs (e. g. , Rosenburg et al. , 2011) . We collected elevation profiles with good data quality from six subregions (Figure 1b), calculated their RMS deviations at a range of baselines (10 - 2,500mm), and then fitted the deviograms to obtain the Hurst exponents (Figure S7) . The profiles are long enough to make sure that the breakover points are reached. Each profile is detrended by subtracting a linear fit of the profile to remove the regional tilt. We define the breakover point with the Pearson correlation coefficient threshold of 0. 999 for the deviogram fit. The Hurst exponents of the six subregions (in order of R1, R2, ..., R6) are 0. 78, 0. 85, 0. 85, 0. 71, 0. 51, and 0. 61, and the breakover points are at 1, 180, 970, 1, 950, 1,670, 500, and 1240mm, respectively (Table S1) . The Hurst exponents are representative values in the corresponding subregions (Figure S8) . Because of the ring - shaped noise artifacts, the RMS deviation shows weak fluctuations against the baseline at scales smaller than 50mm (Figure S7) . The analysis reveals that the uncertainty of the Hurst exponent caused by the noise is less than ~2. 0% (Tex S1) .

As shown in Figure 2c, the bidirectional slopes of the moderately degraded crater (R2) and the highly degraded crater (R3) are very close at baselines smaller than 80mm, which is consistent with the fact that their Hurst exponents are the same. However, their breakover points are very different (i. e. , 970±250mm vs. 1,950±200mm), which may be related to the size differences of the two craters (i. e. , 2vs. 4. 5m) . Hurst exponent of the fresh crater (R1) is 0. 78, which is smaller than the degraded craters. The subregion with small craters (R5) has a Hurst exponent of ~0.5, which is a typical "Brownian motion" behavior (Shepard et al., 2001; Turcotte, 1997). The Hurst exponent of the Yutu-2 rover track region (R6) is 0.61, a relatively small value compared with other regions. The low Hurst exponent indicates that the surface is rough at a small scale and smooth rapidly as the scale increases, which is consistent with the bidirectional slope change with baseline (Figure 2c). The different scaling behaviors suggest the differences between artificial disturbances and natural evolutions. In general, the Hurst exponent of the studying area is smaller than that of the highlands and maria at meter scale (Cai & Fa, 2020), which suggests that the surface is less likely to maintain its roughness at the submeter scale than at the meter scale (Shepard et al., 2001). One possible reason is that the submeter-scale roughness is mainly related to geological features and characteristics of lunar regolith, which has a large porosity of 83±3% (Hapke & Sato, 2016) and the material has high mobility under external force (Melosh, 2011). However, the meter-scale roughness is mainly related to craters of hundreds of meters to kilometers in diameter (Cai & Fa, 2020), which have dramatic topographic changes. Further investigations on terrains of different ages and types (e.g., mare and highland) are needed to better understand the observations.

4 Discussion

4.1 Factors Affecting Millimeter-Scale Roughness

As suggested by previous studies, the topographic roughness of the Moon is scale-dependent and such behaviors are related to surface geologic processes (e g., Cai & Fa, 2020; Fa et al., 2016; Kreslavsky et al., 2013). Analyzing factors controlling the millimeter scale roughness can help understand processes shaping small scale surface features. As shown in Figures 2a and 3a, several surface features can increase the roughness with respect to undisturbed, smooth background regions. The most prominent feature is the crater of ~70cm in diameter, where impact fragments and breccia of centimeter size and smaller result in slopes of 20° to more than 30°. Smaller craters <10cm in diameter can cause bidirectional slopes as large as 20°, independent of the presence of fragments (Figure 3a). Four rocks with the size of 5-20cm significantly increase the roughness with a similar magnitude as the large fresh crater (R1), where the surface is covered with impact fragments of centimeter-to-decimeter size (Figures 3e and S5b). Figure 3b shows a relatively flat surface, in which, however, the regolith is very rough at small scales (Figures 3c and 3d) as is shown in the bidirectional slope map (Figure 2a). Figure. 3c shows the appearance of craters with a diameter of 10-20cm that result in a substantial increase (to ~20°) in the bidirectional slope. The region in Figure 3d is devoid of impact craters, fragments, and rocks, but the surface is nonuniform and hummocky at small scales and exhibits bidirectional slopes at the 7mm scale in rough areas from 10° to 20°. Similar roughness characteristics of the surface can also be observed in the photograph taken by Apollo astronaut on the lunar surface (Figure. S9) . We attribute such millimeter - scale roughness variations to the evolution of regolith which may be affected by several factors. (a) The ingredients, grain size, and cohesive strength of regolith materials, which mainly affect the geomechanics of regolith (Colwell et al. , 2007; Langevin, 1997; Mckay et al. , 1991; Spray, 2016) . (b) Micrometeorite impacts, which crush rocks and soil aggregates to smaller grains and stir up the regolith materials (e. g. , Hörz et al. , 2020; Spray, 2016) . On the other hand, impact shock sintering and agglutination can create aggregates, increasing the size and bearing capacity of regolith particles (e. g. , Spray, 2016) . (3) Soil movements, such as soil creep that develops under impact and seismic shaking and transports surface loose material downhill, forming “elephant hide” slope textures (Melosh, 2011; Xiao et al. , 2013) . Figures 3a - 3d clearly show the differences in regolith grain size, and the areas with coarse grains (e. g. , undegraded craters) are rougher than the fine - grained areas (Figure 3a) . The PCAM image in Figure 3d does not show evidence of soil creep, due to that the background topography is very smooth (e. g. , the absolute slopes are smaller than 3° at the 10cm scale) . Lunar aggregates are usually smaller than 1mm, and a portion of them can be as large as 10mm (Heiken et al. , 1991) . Apollo 17 samples reveal that mature soils have a smaller mean grain size than immature soils though the former have higher aggregate contents (McKay et al. , 1974) . Comminution and agglutination of lunar regolith are dynamic processes caused by micrometeorite impacts and the first 1mm of the lunar surface becomes dynamic equilibrium in 1. 5 Myr (Langevin, 1997) . However, how these regolith erosion processes influence surface roughness is unknown. These factors should be investigated in the future with additional analysis techniques and in regions with various geologic contexts. Based on the above analyses from the high resolution PCAM images and the roughness parameters, we conclude that impact craters, rocks, and regolith properties are important factors controlling the roughness at millimeter scale.

Using the derived DTM, we can investigate the roughness of a meter - scale fresh crater, which has not been reported yet. As shown in Figure 2a, the 2 - m - in - diameter fresh crater is the most pronounced feature in the studying area and is the roughest structure. PCAM images in Figures 3e and S5b show the details of this crater. The crater depth is ~0. 3m and thus the depth - diameter - ratio is 0. 14±0. 004, which falls in the range of the fresh simple lunar craters (Stopar et al. , 2017) . Abundant centimeter - to - decimeter scale fragments are found in the crater interior and ejecta (Figures 3e and S5b), causing an increase in the bidirectional slope at the millimeter scale (>40°, Figure S5c). However, the crater interior has more fragments than the exterior, resulting in a rougher surface in the interior than the exterior at the millimeter scale, and the bidirectional slope decreases with increasing distance until to ~2 radii, which is the boundary of continuous ejecta (Figure S5d). Such phenomenon can also be observed for larger craters, such as the South Ray crater, which is about 700m in diameter and 2Ma in age (Watkins et al., 2019). As shown in Figures S10a and S10b, the South Ray crater is rich in fragments of <14m in size (Watkins et al., 2019) and the rock abundance (RA) is ~0.07 inside the crater and decreases to ~0.02 at ~2 radii (Bandfield et al., 2011). The bidirectional slope at the meter scale shows similar spatial distributions, that is, the roughness is greater in the crater interior than the exterior, and the slopes decrease from ~35° inside the crater to ~2° at ~2 radii (Figures S10c and S10d). The roughness increasements caused by impact fragments are also shown in young craters of tens of kilometers in diameter (Wang et al., 2020). Crater degradation can significantly affect surface roughness as well. Wang et al. (2020) revealed that the difference in roughness between crater interior and exterior regions decreases as a crater degrades. It is supported by the differences between the interior and exterior of the craters in subregions R1, R2, and R3, which are 66±3%, 61±1%, and 44±1%, respectively. For craters smaller than 1km in diameter, Cai and Fa (2020) suggest that their meter-scale roughness (indicated by RMS deviation) decreases with aging due to diffusive degradation. A similar roughness-age relationship was also found in hectometer-scale roughness of large craters (diameter > 70km), which was attributed to the thicker regolith of older craters than the young craters (Kreslavsky et al., 2013). With additional high-resolution topography data from future missions, the relationship between roughness and age for meter-size craters can be further investigated. In the discontinuous ejecta facies of the fresh crater, the surface materials are dominated by sparsely distributed fine-grained ejecta (Figure 3e) and the roughness statistics are not obviously affected (Figure S5d). As discussed in Section 3.1, the influence of the fragments on roughness decreases with increasing baseline and becomes unrecognizable as the baseline is greater than 200mm (Figure 2c).

The influence of the artificial process on the surface roughness can be demonstrated by the Yutu-2 rover moving tracks (Figures 2 and 3f). The Yutu-2 rover is weighted 135kg and its wheels are 30cm in diameter and 15cm in width (Tang et al., 2020). After being churned by the rover, an elevation difference of ~10-20mm was formed on the surface, which results in a locally intense slope increasement that is even larger than the ejecta deposits of the fresh crater (R1), as illustrated by the triangle area in Figure 3f. The response of regolith to rover churning depends on the strength of lunar regolith, as reported by Tang et al. (2020). An important factor affecting regolith mechanical properties is the content of agglutinates, which show different structures from regolith grains and are stronger in their bearing capacity (e.g., Slyuta, 2014). The formation of agglutinate is age-dependent and is more abundant in mature regolith (e.g., Pidgeon et al., 2007; Pieters & Noble, 2016). As an indicator of regolith modifications by rover churning, the millimeter-scale roughness can be related to regolith bearing capacity,

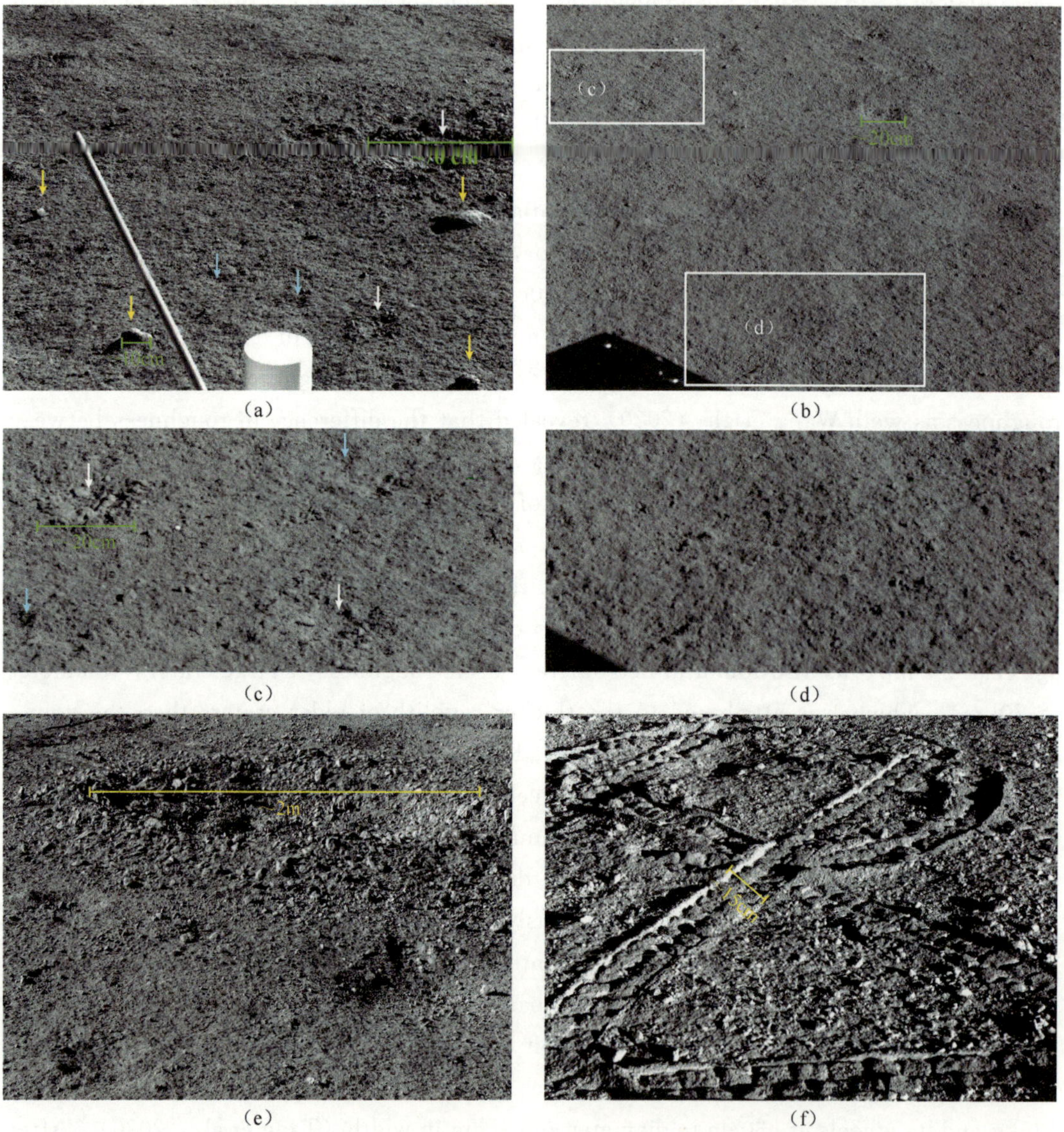

Figure 3　Yutu - 2 panoramic camera images showing the details of specific locations in the roughness study region. (a) A relatively rough region with craters and rocks. (b) A generally smooth area. Panels (c) and (d) show the subareas marked in panel (b). (e) A fresh crater of ～2m in diameter. (f) The area was churned by the Yutu - 2 rover. The image coverages are shown by black boxes in Figure 2a. The yellow, white, and cyan arrows represent rocks, craters with fragments, and degraded craters, respectively. Scale bars are given by green and yellow lines (note the scale changes with distance). Image IDs are listed in Table S2

fraction angle, and repose angle. Such effects can be studied with the returned samples from future missions, such as Chang'e - 5/6.

4.2 Bidirectional Slope at Micrometer to Kilometer Scales

By incorporating previous results (Helfenstein & Shepard, 1999; Rubanenko et al., 2020), we obtained for the first time the bidirectional slope of the Moon at scales ranging from micrometer to kilometer (Figure 4). In the centimeter to meter scales, the slopes are calculated using the CE-3 LCAM DTM with a resolution of 5cm/pixel (Liu et al., 2015). LRO NAC DTM (5m/pixel) is used to calculate the roughness at meter to hectometer scales over the CE-4 landing region. Though calculated from different DTMs over different regions, the bidirectional slopes generally show good consistency. The slope based on the CE-4 DTM is consistent with the results calculated from the Apollo Lunar Surface Closeup Camera (ALSCC) DTM (Helfenstein & Shepard, 1999). The bidirectional slope of the CE-3 landing area at the small scale is consistent with that of the Fra Mauro region (Helfenstein & Shepard, 1999). At decameter to hectometer scales, the bidirectional slope of the CE-4 area based on LRO NAC DTM is in good accordance with that of the maria region calculated from LOLA elevation data (Rubanenko et al., 2020).

The bidirectional slope at the full range from micrometer to kilometer scales reveals important topographic properties of the Moon. Figure 4 shows two major characteristics of the lunar surface: (a) The bidirectional slope decreases with increasing baseline in general; (b) The slopes at the same scale vary among different terrains. Due to the correlation between roughness and geological features, the slopes at different scales can reflect different geological processes and degradation effects. At millimeter to centimeter scales, our results reveal additional controlling factors including the occurrences of submeter size craters and centimeter-to-decimeter-size rocks. At subcentimeter scales, surface roughness reveals regolith gardening processes that modify the top centimeters of regolith (Helfenstein & Shepard, 1999). As inferred by Helfenstein and Shepard (1999), the types of geological processes (mare eruption, impact cratering) from which local materials are emplaced can influence surface roughness at the centimeter and larger scales. Ejecta deposits and the degradation state of craters can significantly affect the surface roughness. The CE-3 landing area is on top of the rocky continuous ejecta deposits of the relatively young Zi Wei crater (Fa, 2020), whereas the CE-4 area is located in the discontinuous ejecta field of the 3.1Ga old Finsen crater (Guo et al., 2021). LRO Diviner RA (Bandfield et al., 2011) shows that RA at the CE-3 landing area is ~4 times greater than the CE-4 landing area (0.009 vs. 0.002, Figure S12). In turn, the roughness results show that the CE-3 area is rougher than the CE-4 area at the submeter scale (Figure 4). At meter to hectometer scales, young (1-2Ga) craters are the most important factor affecting surface roughness, and the surface roughness increases with age until reaching a steady-state (e.g., Cai & Fa, 2020; Kreslavsky et al. 2013). As shown in Figure 4, the bidirectional slope of the CE-3 area is smaller than the CE-4 area at the scale of ~7m, which may be caused by the fact that the older CE-4 region accumulated more smaller craters than the younger CE-3 region. At kilometer-scale, roughness is controlled by early major geological events such as large impacts, volcanic eruption, and tectonism (Kreslavsky et al., 2013). Thus, the roughness shows a dichotomy of lunar maria and highlands at decameter to kilometer scale (Figure 4). The scale-dependent bidirectional slopes in Figure 4 also imply that at meter scale and larger, lunar surface topography and roughness are mainly dominated by the gravity of the Moon and the strength of materials, whereas at millimeter scale and smaller surface roughness depends mostly on the size and shape of regolith particles that are majorly controlled by strength. On the other hand, the two poles of the Moon show significantly higher roughness than the nonpolar regions (Figure 4), which may be related to the different morphologic characteristics associated with the possible presence of water ice (e.g., Deutsch et al., 2020; Rubanenko et al., 2019), but needs to be further investigated.

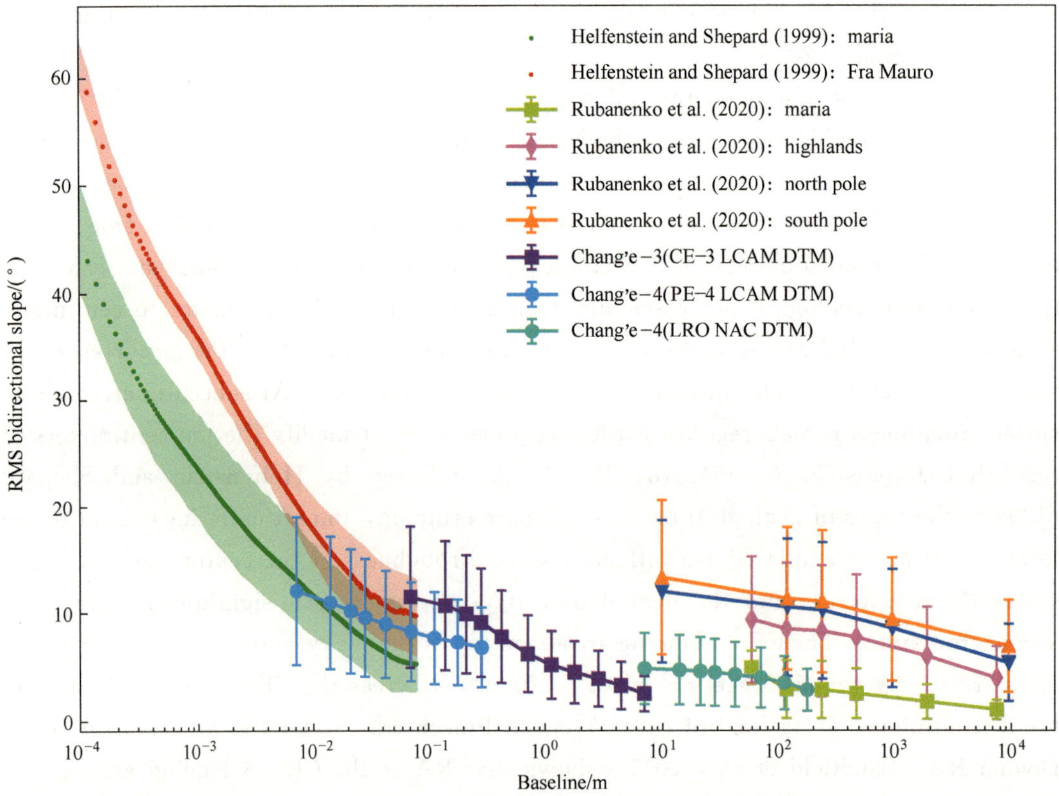

Figure 4　Bidirectional slopes of the Moon from micrometer to kilometer scale. The micrometer scale slope is converted from the unidirectional data in Helfenstein and Shepard (1999) by a factor of $\sqrt{2}$ and the shadow ranges indicate one standard deviation. Bidirectional slopes from decameter to kilometer scales are from Rubanenko et al. (2020). The slopes of the CE-3 area are calculated from LCAM digital terrain model (DTM) (5cm/pixel) (Figure S11; Liu et al., 2015). Lunar reconnaissance orbiter narrow angle camera DTM (5m/pixel) covering the CE-4 exploration area is used to calculate the slope at meter to hectometer scales. Error bars represent one standard deviation.

The results in Figure 4 shows that the scale - dependent roughness behaviors are related to geological processes, which can improve our understanding of the geological evolution of the lunar surface. In addition, our results not only provide constraints for lunar morphologic simulations (e. g. , Richardson, 2009), but also serves as a reference for roughness parameter in remote sensing applications, such as photometric models (e. g. , Hapke, 1984, 2012), radar scattering models (e. g. , Fa et al. , 2011; Ulaby et al. , 2014), and thermal infrared emission models (Bandfield et al. , 2015) . With a full - scale coverage, the data shown in Figure 4 will be of great importance for comparing the topographic contrast at different scales between different planetary bodies in a comparative planetary study.

4.3 Application to Radar Observations

During the past two decades, several imaging radars, such as Arecibo P - and S - band radar, LRO, and Chandrayaan - 1 Miniature Radio Frequency (Mini - RF) radars, have been used to study lunar regolith properties. Both radar observations and modeling results show that radar echo from the lunar surface is dominated by scattering from the rough surface and rocks (either perched on the surface or buried in the regolith) (Thompson et al. , 2011) . In radar scattering, RMS height at the radar wavelength scale is usually the key indicator for surface roughness. As other studies focused on larger scales (e. g. , Cai & Fa, 2020; Rosenburg et al. , 2011), the RMS heights in Figures 2b and 2d are the only roughness information directly calculated from topography considering the wavelengths of lunar observation radar. Here we use the Fraunhofer criterion (Ulaby et al. , 2014) to assess whether or not a surface is radar smooth, which is,

$$s < \frac{\lambda}{32\cos\theta} \tag{1}$$

where s is the RMS height, λ is the wavelength, and θ is the radar incidence angle (Ulaby et al. , 2014) . If s is smaller than the right - hand side (RHS), a surface can be treated as radar smooth.

For LRO Mini - RF, the wavelengths are 12. 6cm (S - band) and 4. 2cm (C - band) and the nominal incidence angle is 51°. As shown in Figure 2b, the median RMS height at the window size of 125mm is 3. 9mm, which is smaller than 6mm of the RHS, indicating that the surface is radar smooth at the S band. Such a conclusion is also true for Chandrayaan - 1 Mini - SAR with an incidence angle of 35°. At the baseline of 45mm, the median RMS height is 1. 7mm, which is smaller than 2. 0mm of the RHS of Equation 1, and hence lunar surface can be regarded as smooth for the C band radar as well. The analyses here show that for both LRO and Chandryaann - 1 Mini - RF radar configurations, surface scattering is not the dominant scattering mechanism. This is also consistent with both observations and modeling results that at incidence angle larger than 35°, volume scattering from rocks is the most dominant scattering mechanism (Fa et al. ,

2011; Thompson et al., 2011).

Fa and Wieczorek (2012) analyzed Arecibo P - band radar observations (wavelength: 70cm) at the Apollo and Surveyor landing sites and found that upper limits of the RMS slope are from 3° to 7°. These values are comparable to the RMS slope of 6.2 ±3.6° at 71cm baseline in the CE - 3 landing site (Figure 4). All these results suggest that the lunar surface is also smooth at the P - band radar wavelength scale.

4.4 Comparison to Thermal Surface Roughness

Bandfield et al. (2015) quantitatively investigated the relation between anisothermality and surface roughness by assuming that the distribution of lunar surface slope is Gaussian. Their results show that Diviner radiometer nadir observations at four geological units can be well fit by a 20° RMS slope distribution and that multiple emission angle measurements at several sites can be mostly modeled using 20 - 35° RMS slope distributions. They found similar surface roughness values among different geological units and attributed this to the impact gardening process on a rapid timescale.

The thermal roughness derived from the Diviner radiometer is an effective value at scales larger than 0.5 - 5mm within a field view of ~160m×320m. For comparison, we also calculated the RMS slope at a 5mm baseline in the CE - 4 exploration region (Figure S14). The RMS slope varies from ~1° to more than 30° with a mean value of 7±5.6°, which is generally smaller than Diviner thermal surface roughness. Several possible factors could result in the difference between our results and Diviner estimations, such as the scale of sensitivity, region of study, and model assumptions. For example, our results show that the distribution of surface roughness (e.g., bidirectional slopes and RMS height) is generally non - Gaussian with a long tail (Figures S15 and S16), whereas Bandfield et al. (2015) assumed a Gaussian surface slope distribution. We noticed that RMS slopes of several rough features (e.g., impact craters, rocks) are comparable to Diviner thermal surface roughness. We attributed the large roughness of these features to surface rocks and regolith properties, which is consistent with the explanation of the thermal roughness as given by Bandfield et al. (2015).

5 Conclusions

In this study, we analyzed the roughness characteristics of the lunar surface at millimeter to decimeter scales over China's Chang'e - 4 exploration region using the high - resolution DTM data (5mm/pixel) obtained by the PCAM stereo images. The major conclusions are:

1. The median bidirectional slope at a 7mm scale over the CE - 4 region is 10°, which is much larger than those at the meter and decameter scales (2° - 2.6°). By integrating previous results, we constructed the bidirectional slope versus baseline at scales from

micrometer to kilometer for the first time. The results show that bidirectional slope decreases from ~45 - 60° at micrometer to ~1 - 4° at kilometer scale.

2. The RMS height at a 125mm scale varies from smaller than 2mm to more than 15mm with a median value of ~4mm. This suggests that the lunar surface is smooth at Mini - RF S - and X - band, providing valuable information on deciphering radar data.

3. At small scales, the Hurst exponents (0.6 - 0.85) are generally less than those (~0.75 - 0.95) derived from meter - scale roughness of highlands and maria.

4. Even for a study region as small as ~20m across, the bidirectional slope and RMS height show substantial variation depending on geological context and degradation state of surface features. By comparison with high - resolution PCAM images, we found that small impact craters and fragments, rock and regolith properties are the controlling factors affecting millimeter - scale lunar roughness.

As millimeter - to decimeter - scale lunar surface roughness has not been reported before, our results can expand the knowledge on the small - scale features and processes of the Moon. The full - scale roughness constructed in this study provides important information to understand lunar geological processes and evolution history, and can also be applied in wide study scenarios.

Acknowledgments

The authors thank Hannah Susorney and Paul Helfenstein for their constructive and helpful comments. This work was supported partly by the National Natural Science Foundation of China (No. 41941002, 41902317, 41773065, 11941001), B—type Strategic Priority Program of the Chinese Academy of Sciences (No. XDB41000000), China Postdoctoral Science Foundation (No. 2019M650319), and the Research Grants Council of Hong Kong (RIF Project No: R5043—19).

References

[1] Bandfield, J. L., Ghent, R. R., Vasavada, A. R., Paige, D. A., Lawrence, S. J., & Robinson, M. S. (2011). Lunar surface rock abundance and regolith fines temperatures derived from LRO Diviner Radiometer data. Journal of Geophysical Research: Planets, 116 (E12), E00H02. https://doi.org/10.1029/2011JE003866.

[2] Bandfield, J. L., Hayne, P. O., Williams, J. -P., Greenhagen, B. T., & Paige, D. A. (2015). Lunar surface roughness derived from LRO Diviner Radiometer observations. Icarus, 248, 357 - 372. https://doi.org/10.1016/j.icarus.2014.11.009.

[3] Barker, M. K., Mazarico, E., Neumann, G. A., Zuber, M. T., Haruyama, J., & Smith, D. E. (2016). A new lunar digital elevation model from the Lunar Orbiter Laser Altimeter and SELENE Terrain Camera. Icarus, 273, 346 -

355. https://doi. org/10. 1016/j. icarus. 2015. 07. 039.

[4] Cai, Y., & Fa, W. (2020). Meter - scale topographic roughness of the Moon: The effect of small impact craters. Journal of Geophysical Research: Planets, 125 (8), e2020JE006429. https://doi. org/10. 1029/2020je006429.

[5] Colwell, J. E., Batiste, S., Horányi, M., Robertson, S., & Sture, S. (2007). Lunar surface: Dust dynamics and regolith mechanics. Reviews in Geophysics, 45 (2), 2005RG000184. https://doi. org/10. 1029/2005RG000184.

[6] Deutsch, A. N., Head, J. W., & Neumann, G. A. (2020). Analyzing the ages of south polar craters on the Moon: Implications for the sources and evolution of surface water ice. Icarus, 336, 113455. https://doi. org/10. 1016/j. icarus. 2019. 113455.

[7] Fa, W. (2020). Bulk density of the lunar regolith at the Chang'e - 3 landing site as estimated from lunar penetrating radar. Earth Space and Sciences, 7 (2), e2019EA000801. https://doi. org/10. 1029/2019ea000801.

[8] Fa, W., Cai, Y., Xiao, Z. Y., & Tian, W. (2016). Topographic roughness of the northern high latitudes of Mercury from MESSENGER Laser Altimeter data. Geophysical Research Letters, 43 (7), 3078 - 3087. https://doi. org/10. 1002/2016gl068120.

[9] Fa, W., & Wieczorek, M. A. (2012). Regolith thickness over the lunar nearside: Results from Earth - based 70 - cm Arecibo radar observations. Icarus, 218 (2), 771 - 787. https://doi. org/10. 1016/j. icarus. 2012. 01. 010.

[10] Fa, W., Wieczorek, M. A., & Heggy, E. (2011). Modeling polarimetric radar scattering from the lunar surface: Study on the effect of physical properties of the regolith layer. Journal of Geophysical Research: Planets, 116, E03005. https://doi. org/10. 1029/2010je003649.

[11] Guo, D., Fa, W., Zeng, X., Du, J., & Liu, J. (2021). Geochemistry of the Von Kármán crater floor and thickness of the non - mare ejecta over the Chang'e - 4 landing area. Icarus, 359, 114327. https://doi. org/10. 1016/j. icarus. 2021. 114327 .

[12] Guo, D., Liu, J., Head, J. W., & Kreslavsky, M. A. (2018). Lunar Orientale impact basin secondary craters: Spatial distribution, size - frequency distribution, and estimation of fragment size. Journal of Geophysical Research: Planets, 123 (6), 1344 - 1367. https://doi. org/10. 1029/2017je005446.

[13] Hapke, B. (1984). Bidirectional reflectance spectroscopy: 3. Correction for macroscopic roughness. Icarus, 59 (1), 41 - 59. https://doi. org/10. 1016/0019 - 1035 (84) 90054 - X.

[14] Hapke, B. (2012). Theory of reflectance and emittance spectroscopy. Cambridge university press.

[15] Hapke, B., & Sato, H. (2016). The porosity of the upper lunar regolith. Icarus, 273, 75 - 83. https://doi. org/10. 1016/j. icarus. 2015. 10. 031

[16] Heiken, G. H., Vaniman, D. T., & French, B. M. (1991). Lunar sourcebook:

A user's guide to the Moon. Cambridge University Press.

[17] Helfenstein, P. , & Shepard, M. K. (1999) . Submillimeter - scale topography of the lunar regolith. Icarus, 141 (1), 107 - 131. https://doi. org/10. 1006/icar. 1999. 6160.

[18] Herrick, R. R. , & Hessen, K. K. (2006) . The planforms of low - angle impact craters in the northern hemisphere of Mars. Meteoritics and Planetary Sciences, 41 (10), 1483 - 1495. https://doi. org/10. 1111/j. 1945 - 5100. 2006. tb00431. x .

[19] Hörz, F. , Basilevsky, A. T. , Head, J. W. , & Cintala, M. J. (2020) . Erosion of lunar surface rocks by impact processes: A synthesis. Planetary and Space Sciences, 194, 105105. https://doi. org/10. 1016/j. pss. 2020. 105105.

[20] Jia, Y. , Zou, Y. , Ping, J. , Xue, C. , Yan, J. , & Ning, Y. (2018) . The scientific objectives and payloads of Chang ' E - 4 mission. Planetary and Space Sciences, 162, 207 - 215. https://doi. org/10. 1016/j. pss. 2018. 02. 011.

[21] Kreslavsky, M. A. , & Head, J. W. (2000) . Kilometer - scale roughness of Mars: Results from MOLA data analysis. Journal of Geophysical Research: Planets, 105 (E11), 26695 - 26711. https://doi. org/10. 1029/2000je001259.

[22] Kreslavsky, M. A. , Head, J. W. , Neumann, G. A. , Rosenburg, M. A. , Aharonson, O. , Smith, D. E. , & Zuber, M. T. (2013) . Lunar topographic roughness maps from Lunar Orbiter Laser Altimeter (LOLA) data: Scale dependence and correlation with geologic features and units. Icarus, 226 (1), 52 - 66. https://doi. org/10. 1016/j. icarus. 2013. 04. 027.

[23] Kreslavsky, M. A. , Head, J. W. , Neumann, G. A. , Zuber, M. T. , & Smith, D. E. (2014) . Kilometer - scale topographic roughness of Mercury: Correlation with geologic features and units. Geophysical Research Letters, 41 (23), 8245 - 8251. https://doi. org/10. 1002/2014gl062162.

[24] Langevin, Y. (1997) . The regolith of Mercury: Present knowledge and implications for the Mercury Orbiter mission. Planet and Space Sciences, 45 (1), 31 - 37. https://doi. org/10. 1016/s0032 - 0633 (96) 00098 - 0.

[25] Liu, Z. , Di, K. , Peng, M. , Wan, W. , Liu, B. , Li, L. , et al. (2015). High precision landing site mapping and rover localization for Chang ' e - 3 mission. Science China Physics, Mechanics & Astronomy, 58 (1), 1 - 11. https://doi. org/10. 1007/s11433 - 014 - 5612 - 0.

[26] Lu, Y. , Wu, Y. , Michael, G. G. , Ma, J. , Cai, W. , & Qin, N. (2021). Chronological sequence of Chang ' E - 4 landing zone within Von Kármán crater. Icarus, 354, 114086. https://doi. org/10. 1016/j. icarus. 2020. 114086.

[27] McKay, D. S. , Fruland, R. M. , & Heiken, G. H. (1974) . Grain size and the evolution of lunar soils. Paper presented at 5th Lunar Conference. (pp. 887 - 906) .

[28] Mckay, D. S. , Heiken, G. , Basu, A. , Blanford, G. , Simon, S. , Reedy, R. , et al. (1991) . The lunar regolith. In G. H. Heiken, D. T. Vaniman, &

B. M. French (Eds.), Lunar Sourcebook: A user's guide to the Moon. Cambridge University Press.

[29] Melosh, H. J. (2011). Planetary surface processes. Cambridge University Press.

[30] Pidgeon, R. T., Nemchin, A. A., van Bronswijk, W., Geisler, T., Meyer, C., Compston, W., & Williams, I. S. (2007). Complex history of a zircon aggregate from lunar breccia 73235. Geochimica Cosmochimica Acta, 71 (5), 1370 - 1381. https://doi. org/10. 1016/j. gca. 2006. 11. 021.

[31] Pieters, C. M., & Noble, S. K. (2016). Space weathering on airless bodies. Journal of Geophysical Research Planets, 121 (10), 1865 - 1884. https://doi. org/10. 1002/2016JE005128.

[32] Richardson, J. E. (2009). Cratering saturation and equilibrium: A new model looks at an old problem. Icarus, 204 (2), 697 - 715. https://doi. org/10. 1016/j. icarus. 2009. 07. 029.

[33] Rosenburg, M. A., Aharonson, O., Head, J. W., Kreslavsky, M. A., Mazarico, E., Neumann, G. A., et al. (2011). Global surface slopes and roughness of the Moon from the Lunar Orbiter Laser Altimeter. Journal of Geophysical Research Planets, 116, E02001. https://doi. org/10. 1029/2010je003716.

[34] Rubanenko, L., Schorghofer, N., Greenhagen, B. T., & Paige, D. A. (2020). Equilibrium temperatures and directional emissivity of sunlit airless surfaces with applications to the Moon. Journal of Geophysical Research: Planets, 125 (6), e2020JE006377. https://doi. org/10. 1029/2020je006377.

[35] Rubanenko, L., Venkatraman, J., & Paige, D. A. (2019). Thick ice deposits in shallow simple craters on the Moon and Mercury. Nature Geoscience, 12 (8), 597 - 601. https://doi. org/10. 1038/s41561 - 019 - 0405 - 8.

[36] Shepard, M. K., Campbell, B. A., Bulmer, M. H., Farr, T. G., Gaddis, L. R., & Plaut, J. J. (2001). The roughness of natural terrain: A planetary and remote sensing perspective. Journal of Geophysical Research Planets, 106 (E12), 32777 - 32795. https://doi. org/10. 1029/2000je001429.

[37] Shimizu, M., & Okutomi, M. (2005). Sub - pixel estimation error cancellation on area - based matching. International Journal of Computer Vision, 63 (3), 207 - 224. https://doi. org/10. 1007/s11263 - 005 - 6878 - 5.

[38] Slyuta, E. N. (2014). Physical and mechanical properties of the lunar soil (a review). Solar System Research, 48 (5), 330 - 353. https://doi. org/10. 1134/S0038094614050050.

[39] Spray, J. G. (2016). Lithification mechanisms for planetary regoliths: The glue that binds. Annual Review of Earth and Planetary Sciences, 44 (1), 139 - 174. https://doi. org/10. 1146/annurev - earth - 060115 - 012203.

[40] Stopar, J. D., Robinson, M. S., Barnouin, O. S., McEwen, A. S., Speyerer, E. J., Henriksen, M. R., & Sutton, S. S. (2017). Relative depths of simple

craters and the nature of the lunar regolith. Icarus，298，34 - 48. https://doi. org/10. 1016/j. icarus. 2017. 05. 022.

[41] Tang，Z.，Liu，J.，Wang，X.，Ren，X.，Chen，W.，Yan，W.，et. al. (2020). Physical and mechanical characteristics of lunar soil at the Chang'e - 4 landing site. Geophysical Research Letters，47 (22)，e2020GL089499. https://doi. org/10. 1029/2020gl089499.

[42] Thompson，T. W.，Ustinov，E. A.，& Heggy，E. (2011). Modeling radar scattering from icy lunar regoliths at 13 cm and 4 cm wavelengths. Journal of Geophysical Research Planets，116，E01006. https://doi. org/10. 1029/2009je003368.

[43] Turcotte，D. L. (1997). Fractals and Chaos in geology and geophysics (2 ed.). Cambridge University Press.

[44] Ulaby，F. T.，Long，D. G.，Blackwell，W. J.，Elachi，C.，Fung，A. K.，Ruf，C.，et al. (2014). Microwave radar and Radiometric remote sensing. The University of Michigan Press.

[45] Wang，J.，Kreslavsky，M. A.，Liu，J.，Head，J. W.，Zhang，K.，Kolenkina，M. M.，& Zhang，L. (2020). Quantitative characterization of impact crater materials on the Moon: Changes in topographic roughness and thermophysical properties with age. Journal of Geophysical Research: Planets，125 (10)，e2019JE006091. https://doi. org/10. 1029/2019je006091.

[46] Watkins，R. N.，Jolliff，B. L.，Mistick，K.，Fogerty，C.，Lawrence，S. J.，Singer，K. N.，& Ghent，R. R. (2019). Boulder distributions around young，small lunar impact craters and implications for regolith production rates and landing site safety. Journal of Geophysical Research: Planets，124 (11)，2754 - 2771. https://doi. org/10. 1029/2019JE005963.

[47] Wu，B.，Li，Y.，Liu，W. C.，Wang，Y.，Li，F.，Zhao，Y.，& Zhang，H. (2021). Centimeter - resolution topographic modeling and fine - scale analysis of craters and rocks at the Chang'E - 4 landing site. Earth and Planetary Science Letters，553，116666. https://doi. org/10. 1016/j. epsl. 2020. 116666.

[48] Wu，B.，Liu，W. C.，Grumpe，A.，& Wöhler，C. (2018). Construction of pixel - level resolution DEMs from monocular images by shape and albedo from shading constrained with low - resolution DEM. ISPRS Journal of Photogrammetry and Remote Sensing，140，3 - 19. https://doi. org/10. 1016/j. isprsjprs. 2017. 03. 007.

[49] Xiao，Z.，Zeng，Z.，Ding，N.，& Molaro，J. (2013). Mass wasting features on the Moon - How active is the lunar surface? Earth and Planetary Science Letters，376，1 - 11. https://doi. org/10. 1016/j. epsl. 2013. 06. 015.

Population of Degrading Small Impact Craters in the Chang'E – 4 Landing Area Using Descent and Ground Images

Teng Hu[1,2], Ze Yang[1,2], Zhizhong Kang[1,2], Hongyu Lin[3], Jie Zhong[4], Dongya Zhang[1,2], Yameng Cao[1,2] and Haomin Geng[1,2]

1. Department of Remote Sensing and Geo – Information Engineering, School of Land Science and Technology, China University of Geosciences, Xueyuan Road, Beijing 100083, China; huteng@cugb.edu.cn (T. H.); yangze1204@cugb.edu.cn (Z. Y.); 2012170008@email.cugb.edu.cn (D. Z.); 2012170007@email.cugb.edu.cn (Y. C.); 2012170009@email.cugb.edu.cn (H. G.)
2. Subcenter of International Cooperation and Research on Lunar and Planetary Exploration, Center of Space Exploration, Ministry of Education of The People's Republic of China, No. 29 Xueyuan Road, Haidian District, Beijing 100083, China
3. Beijing Institute of Space Mechanics and Electricity, 104 Youyi Road, Beijing 100094, China; linhy_bisme@spacechina.com
4. Institute of Optics and Electronics, Chinese Academy of Sciences, Chengdu 610209, China; zhongjie@ioe.ac.cn

Abstract: The landing camera (LCAM) of Chang'e – 4 lander provides a series of low (46cm/pixel) to high (2.3cm/pixel) resolution images, which are suitable for centimeter – scale craters. In this paper, we analyze the degradation of those small – sized craters to provide detailed information on the local geological evolution of the lunar surface. From the mosaicked descent image, 6316 craters were extracted and classified into four degradation levels based on their morphology on the image: fresh, slightly degraded, moderately degraded, and severely degraded. The ground terrain camera (TCAM) image and the DEM of the Yutu – 2 panoramic camera (PCAM) validate the crater degradation levels from a qualitative and quantitative perspective, respectively. The results show that the smaller the size of the craters, the more easily they are degraded. The crater populations in equilibrium in the four study areas indicate that the cumulative size – frequency distribution (SFD) slope is different from previous research results, and the smaller the craters, the more difficult to reach an equilibrium state (for craters smaller than a given size, the production rate is exactly balanced by the removal rate), which may be due to secondary cratering and surface resurfacing caused by the burial of ejecta from neighboring craters.

Keywords: meter to centimeter size impact craters; Yutu - 2 panoramic camera; landing camera; terrain camera; crater populations in equilibrium; surface resurfacing

1 Introduction

The Chang'e - 4 mission was the world's first soft landing on the far side of the Moon and is also China's second lunar surface soft landing mission [1 - 3]. The lander is equipped with a variety of cameras, capturing in situ field images at different resolutions and positions and providing imagery of impact craters down to centimeter scale. These high - resolution data from the Chang'e - 4 landing area provide the possibilities for smallscale (especially centimeter - scale) geomorphological features, such as impact craters and boulders [4, 5], enabling a more detailed investigation of the morphological features associated with them in the landing area [5, 6]. Therefore, the analysis of the centimeter - scale geomorphological features of the Chang'e - 4 landing area on the lunar far side, especially the impact crater population, is valuable for obtaining statistical characteristics of the centimeterscale impact crater distribution, and detailed information on their geomorphology.

Craters are among the most common features on the lunar surface and reveal significant information primarily regarding relative and absolute surface chronologies and erosion processes [7 - 9]. Craters after the formation of impacts are affected by the micrometeorite bombardment, emplacement of ejecta from distal larger craters, and stresses associated with geological deformation [10 - 13]. Consequently, the spatial morphology changes and gradually degenerates [14]. The first lunar image captured by Ranger 7 confirmed the degradation of the lunar surface. Among the factors influencing the degradation of impact craters, direct micrometeorite bombardment and ejecta from other impact craters result in incomplete craters, disappearance of the crater rims, and the infilling of the crater cavities. Additionally, internal geological processes such as geomorphological processes (avalanches, talus slides, landslides, and creep) influence the morphology of impact craters, especially small craters [10, 13].

From formation to obliteration, impact craters experience different stages of degradation with various morphological characteristics. The degradation of a population of craters reflects the age of the local geological units, providing insights into the processes of lunar geological evolution. The relationship between the degradation of impact crater populations and the age of geological units has been the focus of study. Craddock and Howard [11] studied the crater degradation process and described the relative age of geological units with the concept of degradation parameters; the degradation pattern of impact craters and the impact history of the region have also been studied in much detail in recent years [15 - 17]. Meanwhile, the evaluation of the impact degradation level is also of interest in the studies. Basilevsky [10], Wood et al. [18], and Ivanov and Basilevsky

[19] distinguished the degradation level according to crater morphology; Li [20] used craters of typical lunar maria acquired from Chang'e - 2 digital orthophoto map (DOM) and digital elevation model (DEM) to study crater shape and degradation, and crater depth was used as the critical parameter to evaluate crater degradation. Wood, Head, and Cintala [19] Ivanov and Basilevsky [20] Li [16 - 18, 21, 22]. The image - based ray brightness, rim, and terrace sharpness, texture, and digital elevation model (DEM) data - based impact crater depth, profile, and slope are the key elements of impact crater degradation analysis in existing studies. However, considering the high resolution of the Landing Camera (LCAM, cm - level resolution at the center of the landing area), in the absence of a high - resolution DEM, we analyze the degradation of small impact craters mainly using the features obtained from the LCAM images [13, 15 - 17, 21 - 23].

Some craters, which we call fresh impact craters, have sharp edges and bright ejecta in optical images, while in the case of other craters, the edges are softened or incomplete, and their walls are even collapsed and discontinuous. The bottoms of these craters, which we call degraded impact craters, are filled, and their reflectance is consistent with the surrounding background [11, 13, 24]. To better understand crater degradation, an image - based statistical method based on the degradation classification of impact craters is developed. Chapman [25] defined four levels of lunar impact crater degradation. Based on this work, we classify the degradation of the crater population into four classes based on the features shown in the images, namely (1) the presence of ray patterns around the craters, (2) the clarity of the crater rims, and (3) the contrast between the bright and shadowed regions of the crater interiors. Impact craters that are clear and complete, with sharp contrasting bright and shadowed interiors, are considered fresh impact craters. The impact craters are classified into the following types according to the degree of degradation: fresh craters (C1), slightly degraded craters (C2), moderately degraded craters (C3), and severely degraded craters (C4). Overall, empirical measures of crater degradation levels based on orbiter image data are highly dependent on the appearance of impact craters, and studies of small, degraded craters are subject to the image resolution of the orbital data.

Existing impact crater degradation studies have generally concluded that the smaller the crater is, the more likely it is to be filled or destroyed by subsequent impacts, while large craters are relatively stable and less prone to degradation [11, 13, 26 - 28]. However, few studies have focused on the population degradation of small craters, especially those smaller than one meter. Therefore, in this study, we focus on the analysis of population degradation statistics for impact craters with sizes of centimeters to one meter, which has not been addressed in previous studies. In addition, we do not analyze the effect of secondary impact craters on the degradation of impact craters because the sizes of the craters involved in this study are small and the distribution is dense; thus, it is currently not possible to distinguish primary and secondary craters by their spatial distribution patterns. The main contributions include the following:

(1) Impact craters of C1, C2, C3, and C4 degradation levels at the centimeter to meter scale in the Chang'e－4 landing area were acquired based on LCAM mosaic imagery.

(2) The impact crater degradation level classification was qualitatively verified from a side－view perspective using TCAM images, and the crater depth－to－diameter ratio calculated by DEM using PCAM of Yutu－2 was used to quantitatively verify the impact crater degradation level.

(3) The equilibrium state of the centimeter to meter impact crater population in theChang'e－4 landing area was evaluated. Moreover, the local surface resurfacing was analyzed.

Degradation analysis of lunar impact craters extends and deepens our understanding of crater morphology, crater identification, lunar cratering history, and lunar chronology [12]. Degradation analysis of small－sized impact crater populations with sizes on the meter to centimeter scale can provide detailed information on the local geological evolution of the lunar surface relative to larger－sized impact craters (hundreds of meters), providing the possibility of obtaining localized surface resurfacing of the lunar surface.

2 Methodology

In this paper, high－resolution descent images are used to explore the distribution and degradation of small impact crater populations in the landing area. The descent images, however, have no geolocation information. Thus, we needed to mosaic and register them based on existing georeferenced data. All the craters in the mosaic images were manually identified based on the crater textures, as the accuracy and completeness of the automatic identification techniques are limited. Craters were manually classified to different degradation levels to permit a quantitative analysis of degradation in the crater population, a multi－vote process to compensate for human subjective factors. The impact crater degradation level classification was verified qualitatively and quantitatively using the side－view morphology of the impact crater from the images of the ground terrain camera (TCAM) and the depth－to－diameter ratio of the DEM acquired by the Yutu－2 panoramic camera (PCAM). The specific flow is shown in Figure 1. A detailed description of the method is described below.

2.1 The Mosaic and Registration of the Images from the LCAM

Typically, we used the structure from motion (SfM) algorithm [29] to estimate the photo position and attitude to obtain the normalized image; however, the final stage of the Chang'e－4 lander was the vertical landing, and the images were acquired along the main optical axis direction, leading to large errors in the position and attitude calculated by the

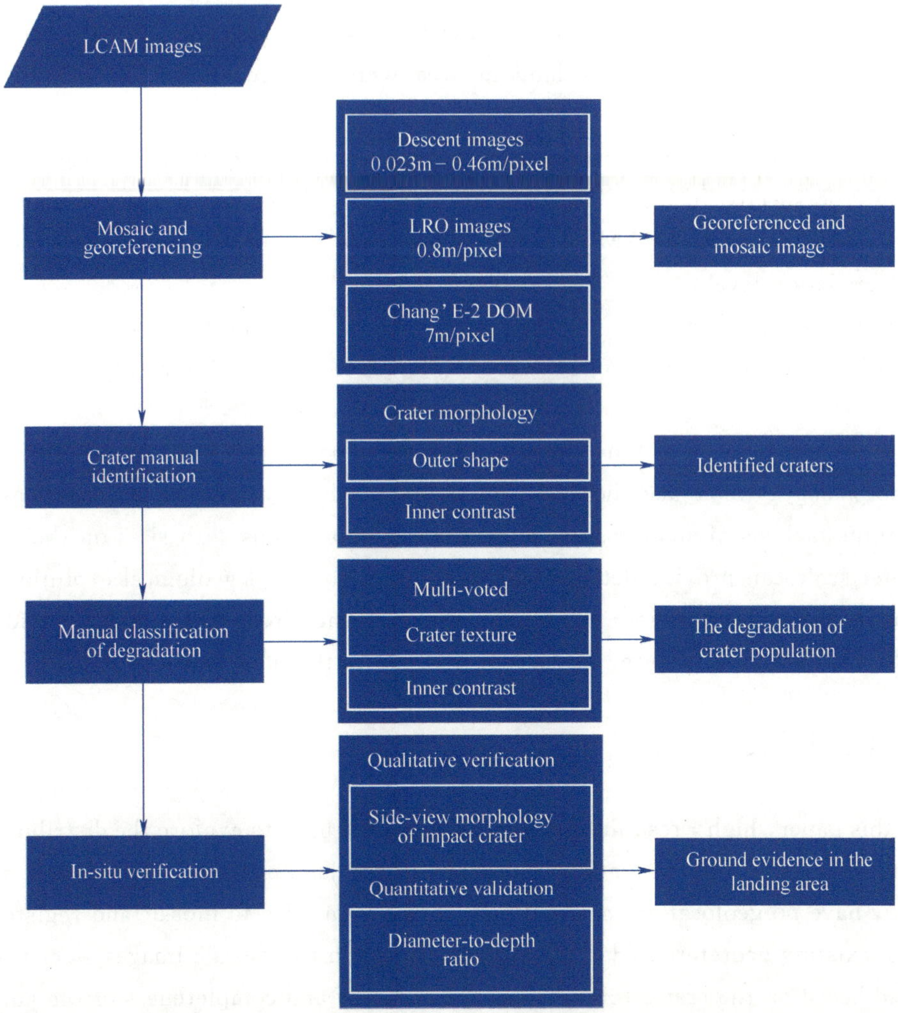

Figure 1　Flow chart of the analysis of impact crater population degradation

SfM algorithm [30, 31]. Therefore, this paper adopted the mosaic method.

Since the descent images lacked geo-referencing, we registered the descent images before mosaicking, making them have the same coordinate reference as Chang'e-2 DOM. However, the resolution of Chang'e-2 DOM is 7m/pixel, which is quite different from the first descent image (46 cm/pixel). Therefore, we considered using the lunar reconnaissance orbiter camera (LROC) Narrow-Angle Camera (NAC) image to register with the Chang'e-2 DOM image first and then registered the descent images based on the georeferenced LROC NAC image.

We used ArcGIS for the image registration and applied the spline transformation model to transfer the images. For the registration of the LROC NAC image and the Chang'e-2 DOM, we selected 300 pairs of corresponding points and ensured that they were evenly distributed. For the registration of the descent images and the LROC NAC image, we needed to register 25 images. For each image pair, 100 pairs of corresponding points were selected. Experimentally, 100 pairs of random points were selected to test the registration error and the results showed that the error was less than 2 pixels. Furthermore, the experiments selected larger impact craters (with diameters larger than 50 pixels) and compared their diameters to the LROC NAC image to determine the difference in crater diameter. The maximum diameter difference was no more than 2 pixels. This finding suggested that image distortion due to image registration had little effect on subsequent crater size acquisition. The resolution of the georeferenced LCAM mosaic image is 0.023m/pixel (center) to 0.46m/pixel (edge).

2.2 Identification of Impact Craters

As impact craters are the most typical geomorphic structure on the lunar surface, they provide important information on past and present geological processes and have also been applied to age dating of geologic surfaces [32, 33]. In this paper, impact craters were extracted around the landing site of Chang'e-4 using manual visual identification, as shown in Figure 2a, b. Figure 2a illustrates the contrast between the shadowed areas (dark areas, blue outline) and the illuminated areas (light areas, red outline) presented internally along the illumination direction of the impact crater in the image. Figure 2b shows the circular shape of craters (yellow arrows mark the impact crater boundaries); thus, they can be distinguished from the background.

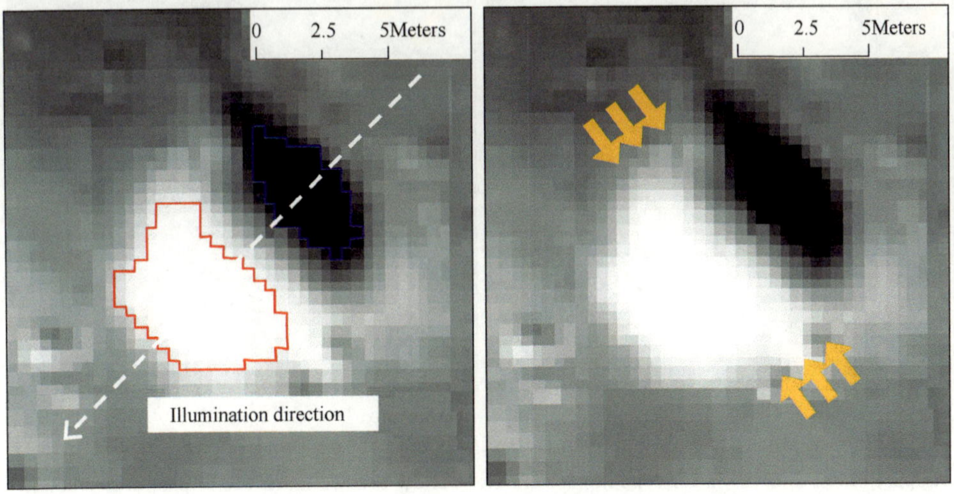

(a)　　(b)

Figure 2　Visual identification criteria of impact crater features. (a) Crescent-like bright and dark regions. (b) Circular shape, yellow arrows mark the impact crater boundaries

Therefore, craters can be identified through the combination of contrast and shape attributes. Identifying craters manually means craters can be detected, marked, and verified through human eyes, decreasing the likelihood of incomplete and mistaken identification evident in currently deployed automatic methods.

For the identification of impact craters, a minimum crater diameter of 5 pixels (e.g., Figure 3a, b) or more is needed, but for the degradation analysis of impact craters , a minimum crater diameter of 10 pixels (e.g., Figure 3c, d) is required due to the detailed information on the shape of the crater, the contrast between bright and dark and the rays around the crater.

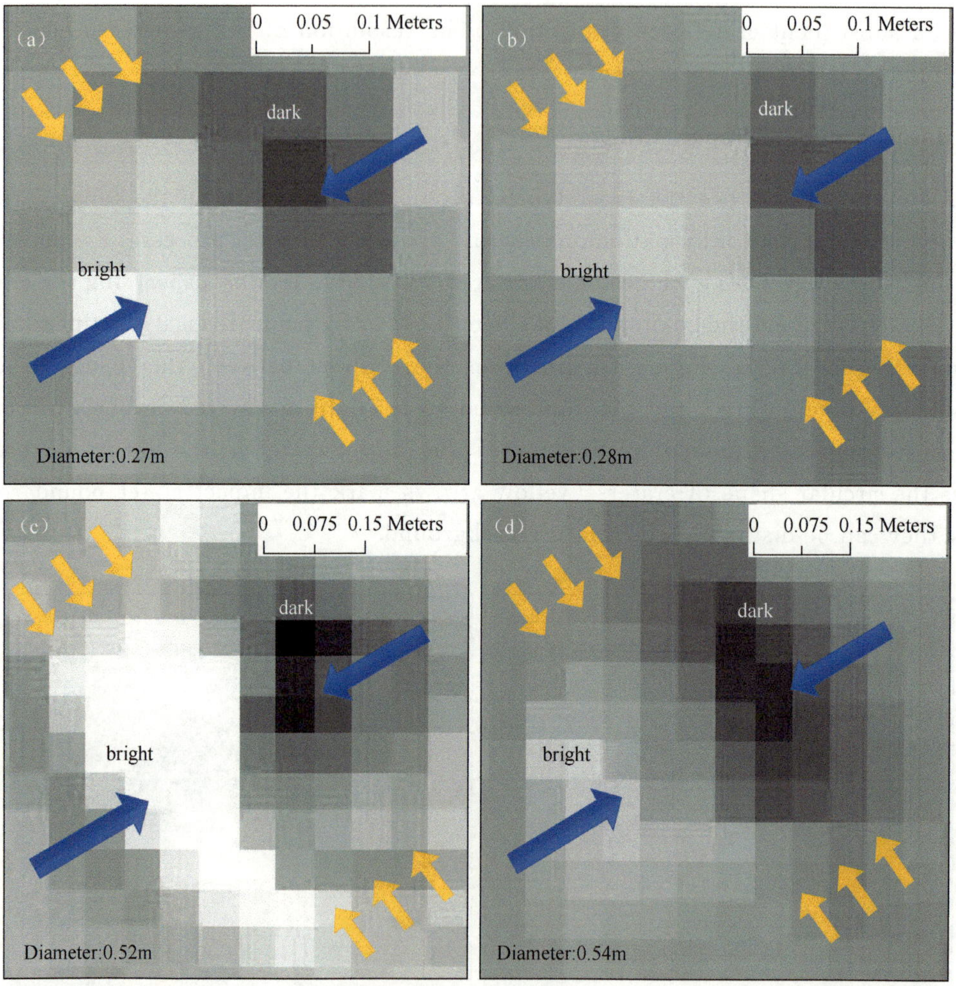

Figure 3 Diagram of impact craters with diameters of 5 pixels (a, b) and 10 pixels (c, d), with (c) a C2 type impact crater and (d) a C3 type impact crater. Yellow arrows mark the impact crater boundaries

2.3 Classification of Degradation Levels of Impact Craters

Once formed, impact craters begin to degrade gradually. As the degradation of the impact crater becomes more severe, the distal ejecta rays (if any) around the fresh impact crater gradually disappear, and the contrast between the bright and dark areas of the impact crater is no longer sharp, while the boundary gradually becomes indistinguishable and nearly blends in the surrounding lunar surface [2, 13, 28, 34, 35]. Statistical studies of morphological and other related characteristics of their degradation level are important to understand the correlation between the evolution of crater morphology and the age of the corresponding geological units. Unlike existing studies that have targeted larger impact craters (meters and even hundreds of meters), most craters in our study are smaller than 2m in diameter. Unfortunately, there is no DEM with a resolution higher than 0.5m/pixel to calculate the exact value of the quantitative classification indicators for craters. Thus, a qualitative classification was applied in this paper based on the traditional method [25]. As shown in Figure 4, we used a qualitative identification method to define and determine the degradation levels of impact craters. The latter were divided into four types (C1 - C4):

(1) C1 type: fresh craters, with obvious rays or clear edges, and a sharp contrast between bright and dark.

(2) C2 type: slightly degraded craters, with clear edges, and a relatively clear contrast between bright and dark.

(3) C3 type: moderately degraded craters, which show clear signs of degradation with small differences between bright and dark, raised crater floors due to filling, and fuzzy circular outlines.

(4) C4 type: severely degraded craters with unclear outlines, shallow crater floors, and a circular outline that can be distinguished when the image is stretched.

The manual qualitative classification of craters usually introduces human subjective factors, so judgment errors are inevitable. For example, the C3 - type craters in Figure 4 are like C4 - type craters and may be misclassified; thus, a multi - voting process for all craters was adopted. In the voting process, first, all impact craters were individually classified by three researchers into degradation levels. Then, we summarize the obtained impact crater degradation levels. For a given impact crater, the degradation level of the is crater is confirmed only when the number of votes is greater than or equal to 2. Otherwise, the crater is considered to have an uncertain degradation level and needs to be reclassified. Finally, the impact crater degradation level will not be determined until all impact crater degradation levels have obtained 2 or more equal votes. In this process, the number of impact craters with uncertain degradation levels was less than 15% when researchers first independently classified the crater degradation levels, and after two rounds of reclassifying craters with uncertain degradation levels, the degradation levels of all impact craters were determined. The specific crater degradation levels classification results can be downloaded from the Science Data Bank of Beijing, China.

Nevertheless, we are concerned that the very small number of C3 and C4 impact craters may still be misclassified. Therefore, in the subsequent analysis, we also combined the statistical analysis of C3 and C4 impact craters. In addition, the qualitative and quantitative methods for classifying impact crater degradation levels are described in the next section.

2.4 Qualitative and Quantification Analysis of the Degradation Level of Impact Craters Using Ground TCAM Images and DEMs Acquired by Yutu - 2 PCAM

We compared craters in images from both the TCAM and LCAM. We manually identified the craters in both types of images and analyzed the degradation of the crater population in the landing area. Then, we verified our results by comparing the identified craters on the LCAM mosaic images and the TCAM images. Figure 5 shows two impact craters with different degradation levels in the LCAM mosaic image and TCAM image (red circles are C3 type impact craters and blue circles are C4 type impact craters).

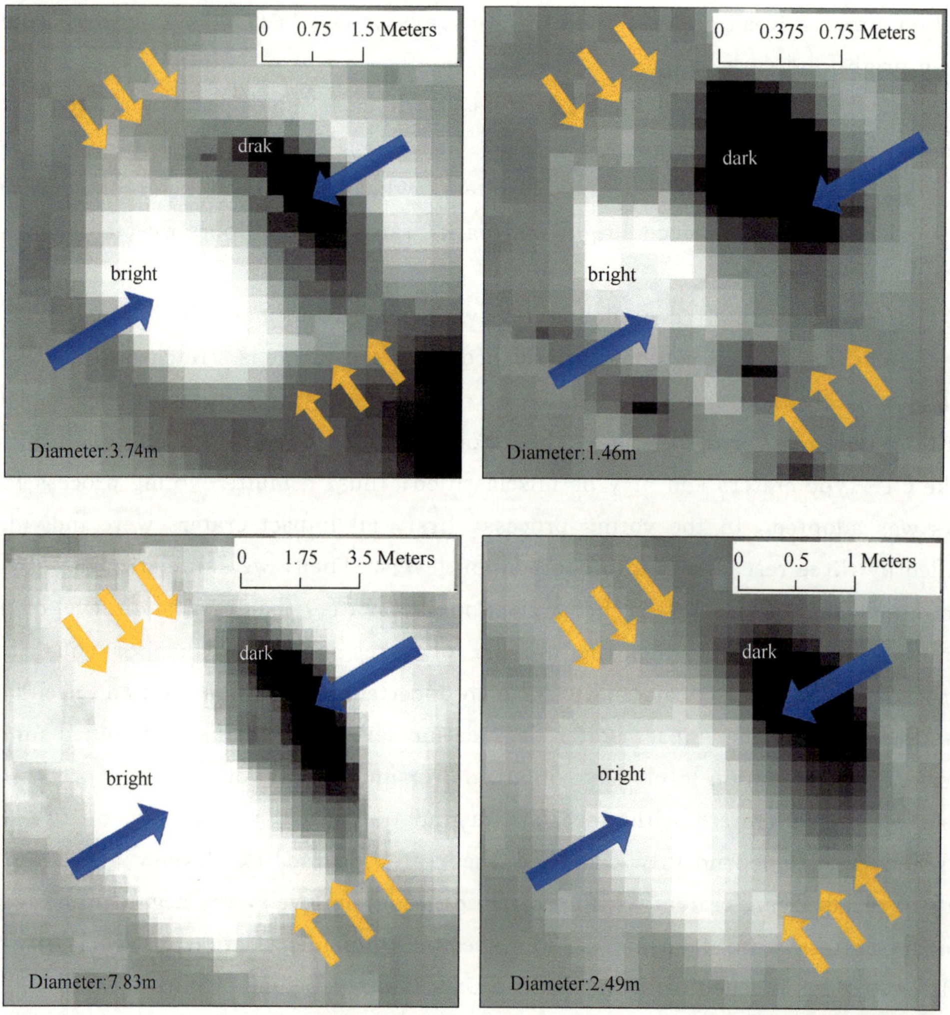

(a)　　(b)

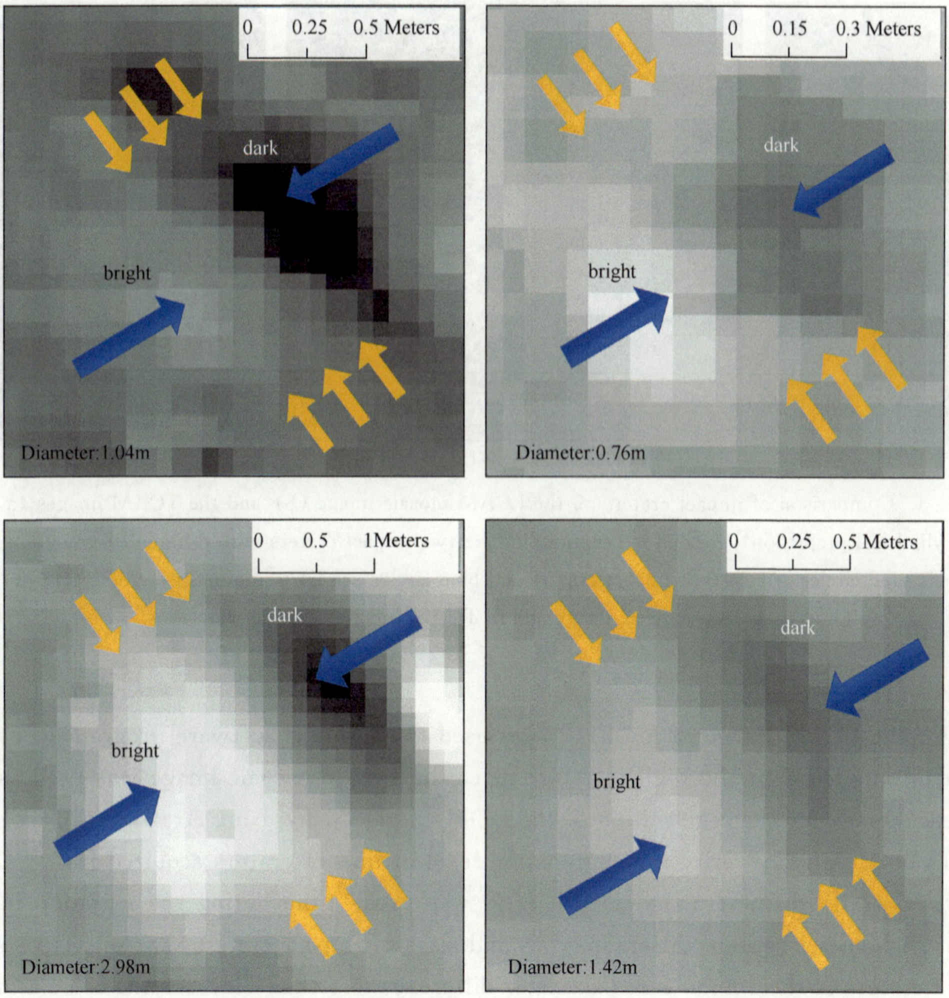

(c)　　(d)

Figure 4　Examples of craters at four degradation levels, where (a－d) correspond to impact craters of types C1, C2, C3, and C4, respectively. Yellow arrows mark the impact crater boundaries. The diameters of the impact craters are written in the figures

As shown in Figure 5, we show two pairs of corresponding impact craters on TCAM images and LCAM mosaic images, where the blue circles are C4 level impact craters, and the red circles are C3 level impact craters. The TCAM images are obtained from a side view, for the C3 impact crater compared to the C4 impact crater, part of the crater rim can be identified, while the concave interior of the C3 impact crater is also more obvious. We carefully compared the LCAM mosaic image and the corresponding impact crater on the TCAM images from different angles for close－up verification. In the central part of the landing area, the degradation levels and actual spatial forms of craters were checked one by one at close range. This comparison helped us to understand the actual morphology of

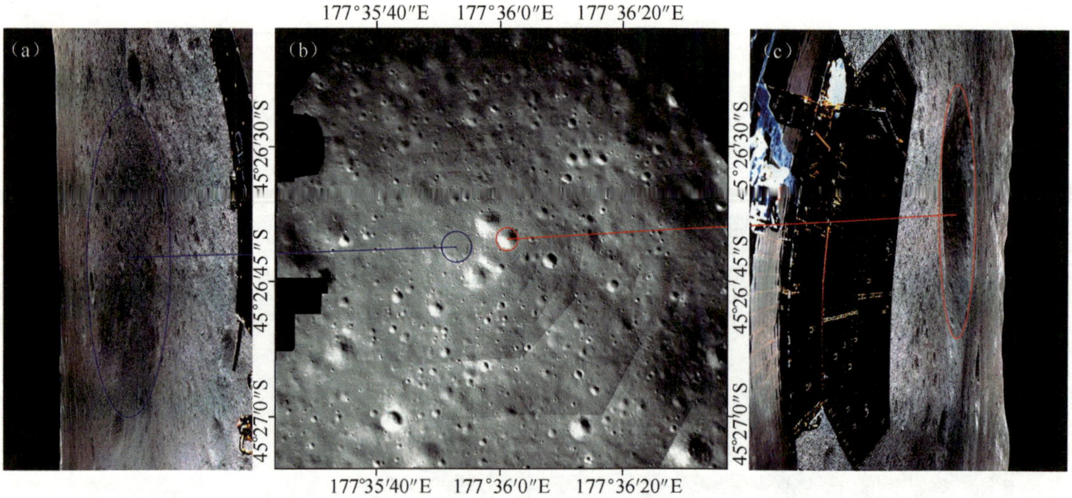

Figure 5　Comparison of impact craters on the LCAM mosaic image (b) and the TCAM images (a, c: cylindrical projection). Circled in red and blue are two impact craters with different degradation levels (red circles are C3 type impact craters and blue circles are C4 type impact craters) shown in LCAM mosaic image and in TCAM images, respectively

craters in the descent image.

For the PCAM image of Yutu - 2, we used the Context software program to obtain the DOM and DEM of the region. After that, the LCAM mosaic image was used as the reference image for the registration of the DOM and DEM by using ArcGIS.

To quantitatively verify the degradation level of impact craters acquired from LCAM images, a crater diameter - to - depth ratio was used. Considering the variation in the depthto - diameter ratio of different types and sizes of impact craters, we selected simple impact craters of the similar size (diameter difference less than 1m) for the calculation of the depth - to - diameter ratio. For the calculation of the impact crater depth, we used the average of the elevations of the eight directions at the edge of the crater to represent the topographic elevation and obtained the crater depth by subtracting this value from the depth at the center of the DEM impact crater.

3　Results and Analysis

The difference in the resolution of different regions of the mosaic image makes the minimum diameter of the impact crater that can be extracted different for each part. Therefore, we divided the study area into four study areas based on the minimum diameter of the impact crater that can be extracted from each area. In addition, the degradation statistics of crater populations in each study area were analyzed. Finally, TCAM and LCAM images were used to analyze the impact crater degradation classification qualitatively and quantitatively.

3.1 Study Area and Data

Chang'e - 4 landed on the floor of the Von Kármán crater (186km in diameter), located in the central part of the northwestern South Polar Aitken (SPA) basin. Recent geological mapping surveys show that this crater is approximately 3.6 - 3.7 Gyr old (pre - Nectarian) [36]. The landing site (45.457°S, 177.588°E) [37] is approximately 350km northwest of the SPA ellipse center (53°S, 169°W) [38]. During the landing process, the images captured by the LCAM of Chang'e - 4 provided an unprecedented opportunity to study the degradation of the population of small impact craters with diameters of 10m or less in the area.

We selected 25 images captured at different landing altitudes (600 to 30m). The ground resolution of the captured images increased gradually as the camera descended, and impact craters of sub - meter size can gradually be identified. Because of their high spatial resolution, descent images are ideal for interpreting the small geological structures in landing areas and analyzing the geological evolution of planetary surfaces. An accurate geographic reference is the foundation for mapping and interpretation of planetary geology using descent images. The images from the LCAM, however, do not have geographic coordinates. Therefore, we must introduce images with geographic references to register descent images.

The Chang'e - 2 DOM (7m/pixel) is the lunar global DOM with the highest ground resolution and is thus used as the reference image when registering the descent images. The great difference in spatial resolution between the Chang'e - 2 DOM and the descent images, however, presented a serious challenge when selecting corresponding pixels. Thus, we used the LROC NAC image (M1298916428LR) to bridge the gap in resolutions [39].

We used full - color mosaic images, captured by the lander, and published by China's Lunar and Deep Space Exploration (CLEP), containing a high - resolution TCAM image of the landing zone displayed by azimuth and cylindrical projections, for the qualitative comparison of craters found in LCAM and TCAM images. To quantitatively compare the results of the impact crater degradation classification, a DEM was acquired from the Yutu -2 rover's PCAM image to calculate the diameter - to - depth ratio. Considering that the data around the landing site might be affected by the impact of the lander during the landing process resulting in inaccurate DEM results, we selected the PCAM images of the third lunar day of Yutu - 2.

3.2 The Mosaic of the Descent Images

We registered and mosaicked the descent images using the method in Section 2.1. For the acquired LCAM mosaic image, we selected 100 pairs of homonymous points with a uniform distribution between the LCAM mosaic image and the georeferenced LROC NAC images and 50 pairs of corresponding impact craters with diameters larger than 50× pixels to calculate the registration error of the LCAM mosaic image and the diameter deviation of the corresponding impact craters, respectively. The mean and root mean square error (RMSE) of the homonymous points deviation, as well as the RMSE and mean value of the diameter difference of the corresponding impact craters were calculated separately. The results show that the mean value of the homonymous point deviation is 2 pixels and the RMSE is 1 pixel. Moreover, the mean value of impact crater diameter deviation is 2 pixels and the RMSE is 1 pixel, which satisfies the need for subsequent degradation statistical analysis of the crater population.

Based on the obtained LCAM mosaic image, the study area centered on the landing point was defined, covering an area of 146,687m^2 (408m from east to west and 386m from north to south), as shown in Figure 5b. The resolution is 0.023m/pixel in the central area and 0.46m/pixel on the edge. The descent images used in this paper were captured along the optical axis during landing; thus, the size of the extractable impact craters gradually decreased as the ground resolution of the descent images increased, especially in the central landing area where the extractable size could reach the decimeter level. Due to the improved spatial resolution, impact craters as small as 15cm in diameter could be identified over the central landing area. Therefore, the descent images provided a reliable data source for analyzing impact craters around the landing site.

3.3 Impact Crater Distribution Results

In total, 6316 impact craters were identified and extracted manually. The crater distribution map and the boundaries of craters with different sizes are circled in Figure 6. As shown in Figure 6, influenced by the high central resolution and low edge resolution of the LCAM mosaic image, the impact craters are very dense near the center of the landing site and gradually become sparse as they approach the boundary (the pink line of Figure 6) of the study area. At the same time, smaller size (0.15 – 0.5m) impact craters are also distributed around the landing site, while the impact craters become larger at the boundary of the study area. This is because the resolution of the mosaic image obtained from LCAM images is highest around the landing site at 0.023m/pixel, where the largest number of small impact craters can be identified and decreases to 0.46m/pixel closer to the boundary of the study area, where only relatively large (>2m) impact craters can be identified due to resolution limitations. The study data are a mosaic of LCAM images from higher to lower resolution, and the difference in resolution leads to different minimum sizes of impact craters that can be identified in various parts of the study area. As shown in Figure 6, the study area is divided into four parts, where the area within the red line, the area within the yellow line, the area within the aquamarine line, and the entire mosaic image range corresponds to the minimum diameter of identifiable impact craters of 0.15, 0.5, 1, and 2m, accordingly. The minimum diameter of identifiable impact craters corresponding to each of the study areas is divided in equal proportion to the total number of impact crater populations, making about 2000 impact craters in each of the four areas. Therefore, the red outlined area, yellow outlined area, aquamarine outlined area, and pink outlined area (the entire area) contain 2341, 2160, 2926, and 2018 craters with diameters of 0.15m or more, 0.5m or more, 1m or more, and 2m or more, respectively.

The completeness of the impact crater population statistics is mainly limited by the resolution of the images, with impact crater diameters greater than 10× pixels having a high confidence level [40]. Therefore, to make the statistical analysis of the crater degradation significant, we calculated histograms of the size distribution of the craters above the minimum identified size for the above four study areas and considered the middle 95% of the craters in each area to have a high confidence level. For the four study areas, the minimum statistical diameter of the impact crater is 0.268m, which is greater than 10×pixels. As shown in Figure 7, for the red, yellow, aquamarine, and pink outlined areas obtained above, the corresponding diameter intervals of 95% craters in each area are 0.268-1.725m (contains 2223 impact craters), 0.512 - 2.612m (contains 2050 impact craters), 1.026 - 8.963m (contains 2781 impact craters), and 2.040 - 15.449m, (contains 1916 impact craters). This allows us to perform a degradation analysis of the crater population with crater diameters in the range of 0.268 - 15.449m.

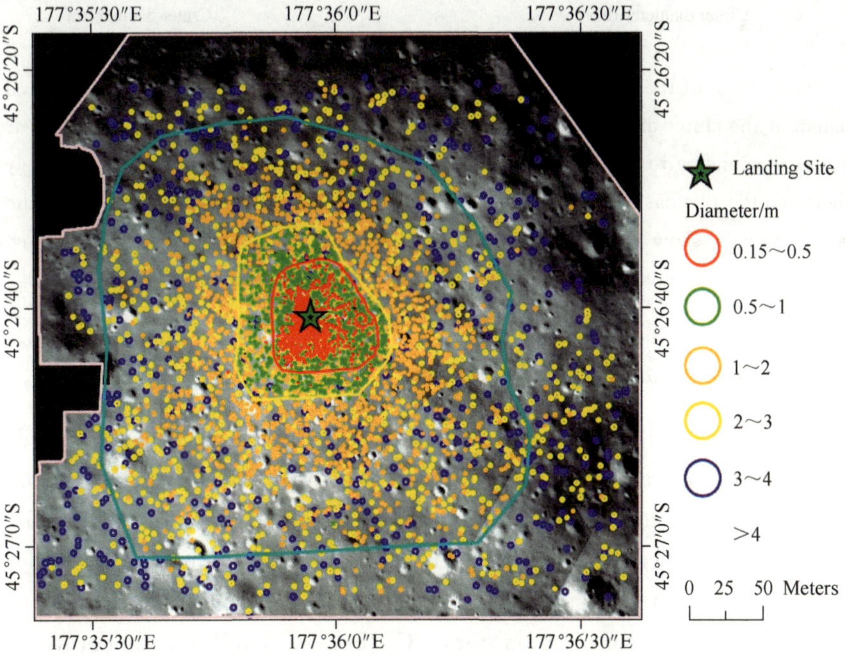

Figure 6　The mosaicked LCAM image with boundaries of areas defined by different sizes of visible craters. The red, the yellow, the aquamarine, and the pink outlined areas correspond to impact craters with a minimum diameter of 0.15, 0.5, 1, and 2m that can be identified, respectively, as constrained by the limitations of the resolution of descent images

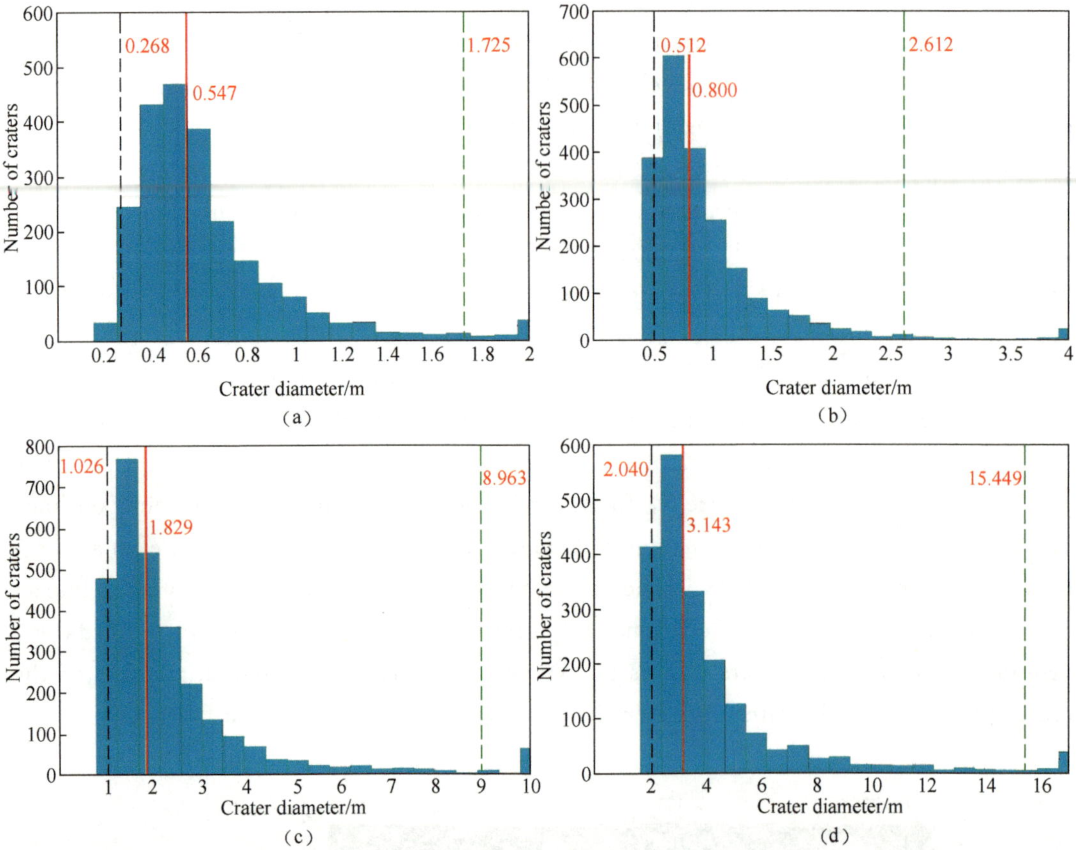

Figure 7　Histogram of the crater diameter distribution in each study region, where the red line is the median of the crater diameter distribution, and the black and the green dashed lines are the diameters corresponding to 2.5% and 97.5% of the number of impact craters, respectively. The interval between the two dashed lines corresponds to the diameter distribution of 95% of the impact craters. (a - d) show histograms of the crater diameter distribution of the red, yellow, aquamarine, and pink outlined areas, respectively

3.4　Impact Crater Degradation and Crater Size

All craters in the landing area were classified into degradation levels according to the morphology of the impact craters. The study areas with the classified craters are visualized in Figure 8. Detailed degradation results for the crater population are provided in Table 1.

As shown in Figure 8, the C1 craters (red, fresh craters) in all four study areas are rare, and the distribution is sparse. C2 impact craters (aquamarine, slightly degraded craters) cover more area than C1 craters. C3 craters (yellow, moderately degraded craters) and C4 craters (green, severely degraded craters) extend over a large range of the landing area. Therefore, the craters in the landing area are dominated by C3 and C4 craters, which occupy more than 90% of the craters in the landing area.

In the above, we obtained a 95% confidence interval for the diameter range of impact craters in each study area, and the following statistical analysis of the degradation of the impact crater population was performed using the craters within the diameter range of each study area as a statistical sample. We then calculated the proportion of C1, C2, C3, and C4 in each impact crater statistical diameter interval. The percentage of each degradation level was also calculated for the total number of impact craters in each study area. Additionally, considering that the C3 and C4 types are more likely to be confused and that misclassification may still occur despite the multiple voting method used in the previous section, we also used the C3 + C4 type of impact crater count to represent the more severely degraded craters.

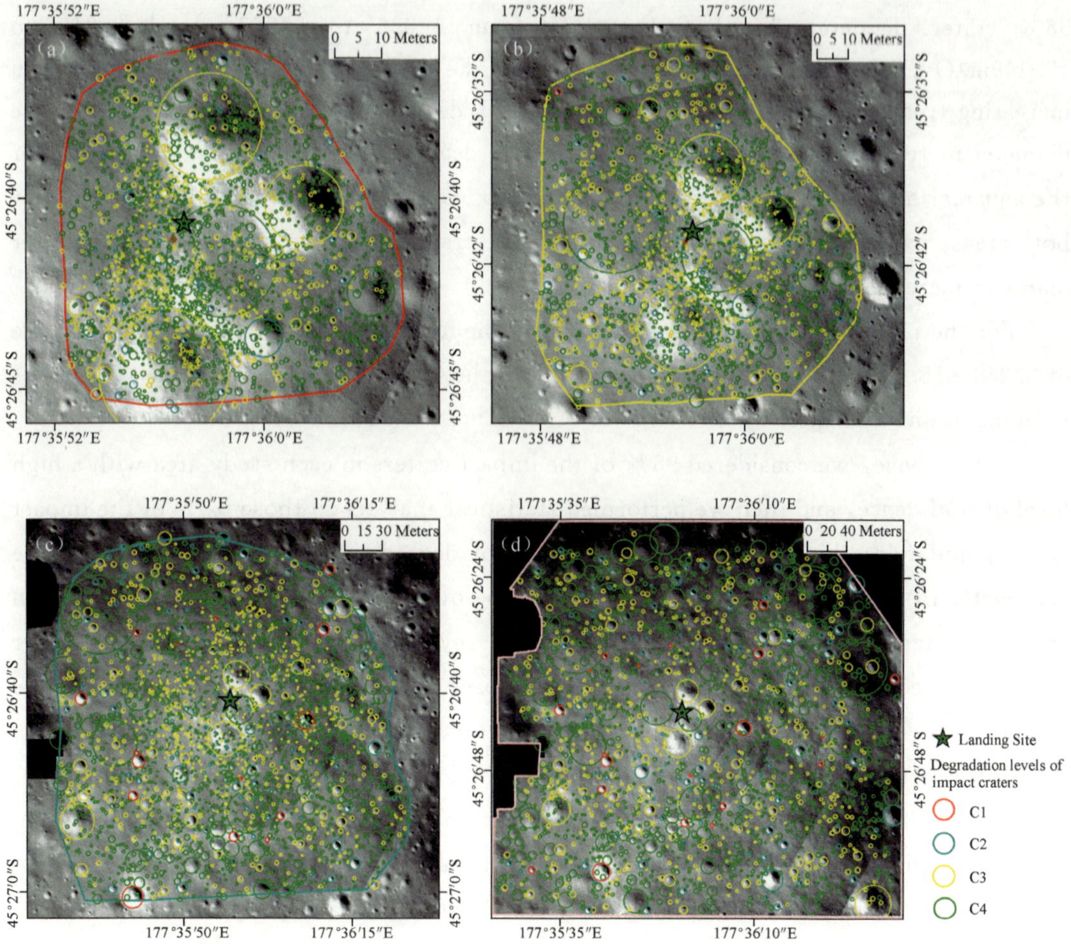

Figure 8　Crater distribution map with different degradation levels in different study areas. (a - d) show the red, yellow, aquamarine, and pink outlined study areas. Red, aquamarine, yellow, and green circles are C1, C2, C3, and C4 impact craters

As shown in Table 1, for the total number of impact craters counted in each study area, C1 craters are the lowest proportion (less than 1%) and C4 craters are the highest (greater than 64%), with C3 + C4 craters accounting for more than 93% of the craters, indicating that the study area is heavily degraded. Furthermore, in those areas, the smaller the impact crater diameter is, the fewer C1 - C3 impact craters and the more C4 impact craters there are. In addition, as the crater diameter increases, the proportion of C1 - C3 impact craters increases, while the proportion of C4 impact craters decreases (the decrease is 79.46% to 63.67% for the red outlined area and 79.53% to 57.62% for the yellow outlined area). In addition, the proportion of C3 + C4 type impact craters, i.e., more heavily degraded impact craters, decreases as the diameter of the impact craters increases. This finding suggests that the smaller the craters are, the easier they degrade.

For the aquamarine outlined area and the full study area - wide crater population, the 95% crater diameter distribution ranges from 1.026 to 8.963m and 2.040 to 15.449m. Overall, the number of craters within the aquamarine outlined area shows an increasing trend for C1, C2, and C3 craters and a decreasing trend for C4 craters as the diameter increases. The percentage of C4 craters decreases from 72.41% to 57.18% for the aquamarine outlined area and from 72.65% to 56.37% for the overall study area. For both areas, overall, the proportion of C3 + C4 craters tends to decrease as the crater diameter increases.

For the population of impact craters within the four study areas in the diameter range of 0.268 - 15.449m, the proportion of craters in the C3 + C4 degradation class decreases with increasing diameter.

In the above, we considered 95% of the impact craters in each study area with a high level of confidence, and then we performed statistical analysis on those parts of the impact crater population. The analysis shows a trend of degradation of impact craters on the centimeter to meter scale (0.268 - 15.449m). The overall population of impact craters in this size range also shows a trend toward more severe degradation with smaller crater sizes. The specific degradation of these small impact crater populations is summarized as follows:

(1) For each study area, the proportion of C4 type impact craters is the highest, exceeding 50% for most areas, while the proportion of fresh impact craters is less than 10%, indicating that impact crater degradation in the Chang'e - 4 landing area is severe. The sizes of the craters in the landing area were relatively small (the maximum diameter was 67m).

(2) As the size of the impact craters increased, the proportions of C1, C2, and C3 craters all gradually increased, while the proportion of C4 craters gradually decreased. The C3 + C4 impact craters show the same degradation pattern, which is consistent in all four study areas (for the effective statistical interval, i.e., diameter range corresponding to 95% of the impact crater samples in the study area).

Table 1　Impact crater population diameter and degradation level statistics for the four study areas

	Data Range	0 - 25%	25 - 50%	50 - 75%	75 - 100%	Total
Red outlined area (4334m²)	Diameter (m)	0.268 - 0.430	0.430 - 0.547	0.547 - 0.714	0.714 - 1.725	0.268 - 1.725
	C1	0.00%	0.00%	0.00%	0.18%	0.84%
	C2	1.26%	1.08%	1.26%	4.68%	6.00%
	C3	19.28%	20.68%	23.38%	31.47%	28.91%
	C4	79.46%	78.24%	75.36%	63.67%	64.25%
	C3 + C4	98.74%	98.92%	98.74%	95.14%	93.16%
	Data range	0 - 25%	25 - 50%	50 - 75%	75 - 100%	Total
	Diameter (m)	0.512 - 0.627	0.627 - 0.800	0.800 - 1.059	1.059 - 2.612	0.512 - 2.612
Yellow outlined area (9569m²)	C1	0.00%	0.00%	0.00%	0.39%	0.40%
	C2	0.97%	1.56%	1.95%	5.86%	4.06%
	C3	19.49%	25.00%	29.63%	36.13%	28.30%
	C4	79.53%	73.44%	68.42%	57.62%	67.24%
	C3 + C4	99.03%	98.44%	98.05%	93.75%	95.54%
	Data range	0 - 25%	25 - 50%	50 - 75%	75 - 100%	Total
	Diameter (m)	1.026 - 1.381	1.381 - 1.829	1.829 - 2.597	2.597 - 8.963	1.026 - 8.963
Aquamarine outlined area (71,763m²)	C1	0.14%	0.00%	0.14%	1.29%	0.10%
	C2	1.44%	3.31%	2.88%	8.62%	2.59%
	C3	26.01%	25.90%	28.39%	32.90%	27.56%
	C4	72.41%	70.79%	68.59%	57.18%	69.76%
	C3 + C4	98.42%	96.69%	96.97%	90.09%	97.32%
	Data range	0 - 25%	25 - 50%	50 - 75%	75 - 100%	Total
	Diameter (m)	2.040 - 2.492	2.492 - 3.141	3.141 - 4.408	4.408 - 15.449	2.040 - 15.449
Pink outlined area (146,687m²)	C1	0.21%	0.21%	0.63%	2.30%	0.04%
	C2	2.51%	4.80%	5.43%	11.27%	2.07%
	C3	24.63%	31.11%	29.85%	30.06%	23.71%
	C4	72.65%	63.88%	64.09%	56.37%	74.18%
	C3 + C4	97.29%	94.99%	93.95%	86.43%	97.89%

3.5　In Situ Qualitative and Quantitative Validation

We used TCAMimages from the Chang'e - 4 lander and panoramic camera (PCAM) images from the Yutu - 2 Rover for the in - situ cross - check; the results are shown in Figures 9 and 10. The TCAM images were used to qualitatively analyze impact craters of different degradation levels from a side - view perspective. The PCAM images are used to obtain the DOM and DEM, where the DEM can be used to calculate the crater depth - to -

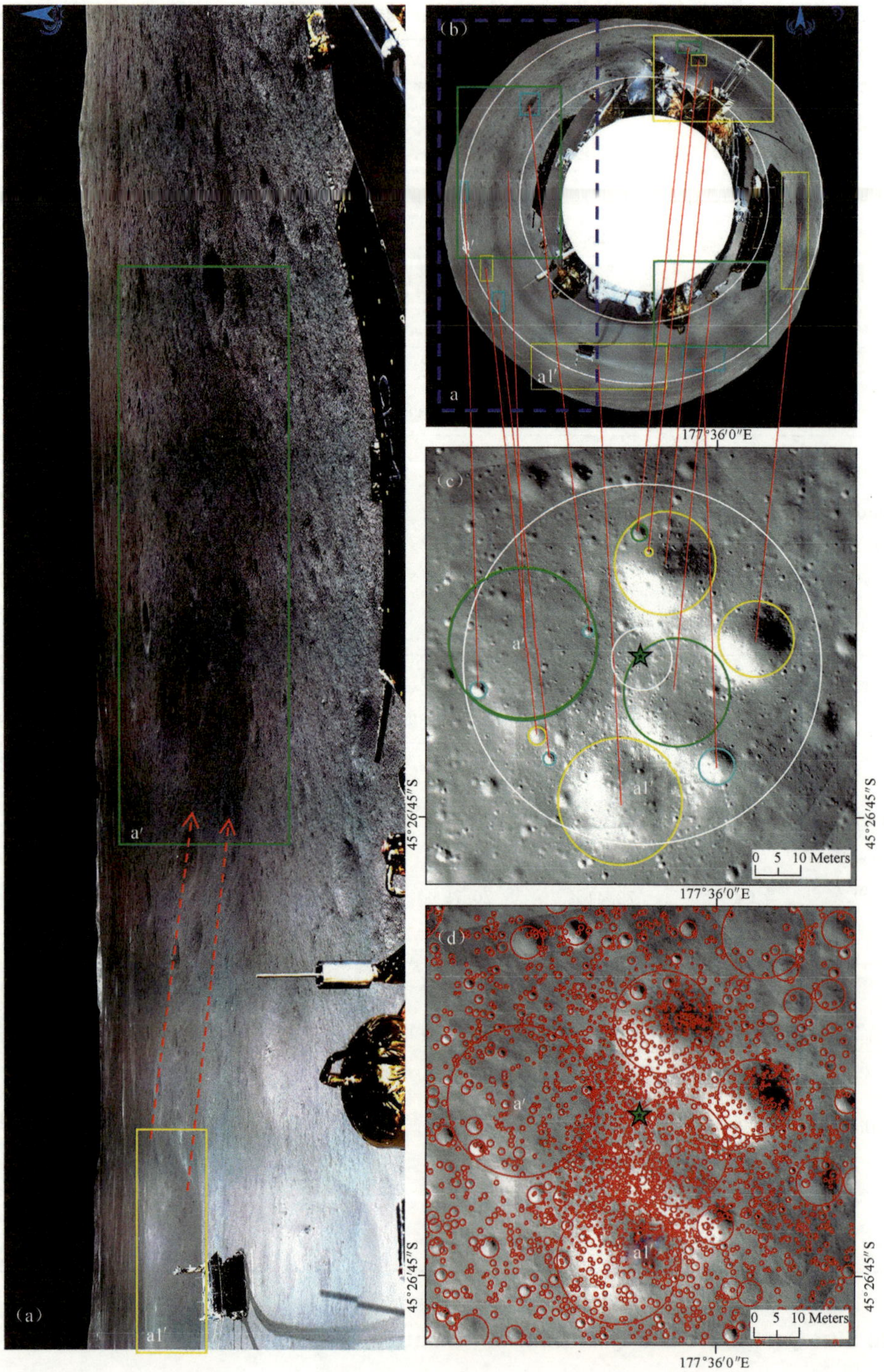

Figure 9 The corresponding impact craters in the TCAM and LCAM mosaic images. (a) Part of the enhanced ground TCAM image (cylindrical projection), corresponding to the area shown in the blue dashed box in (b). The red dashed arrow shows the direction of ejection from crater a1' to crater a'. (b) Ground TCAM image (azimuthal projection). The red solid lines show 12 corresponding craters with different degradation levels. Aquamarine, yellow, and green outlined areas in (b, c) are C2, C3, and C4 craters. The areas within the white circles in (b, c) are the study areas where TCAM images were used to verify the degradation level of impact craters. (c) The LCAM mosaic image. (d) Distribution of LCAM impact craters corresponding to the TCAM image. The green pentagram is the Chang'e-4 landing site

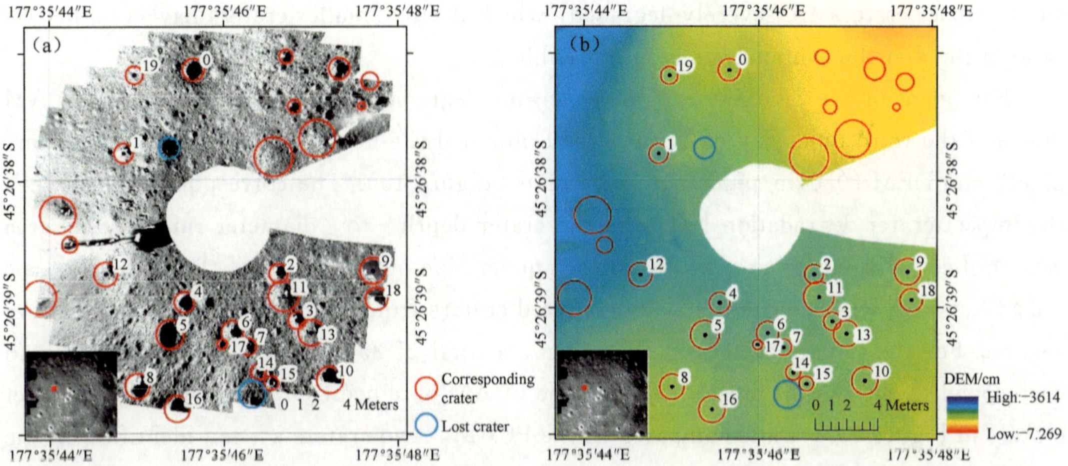

Figure 10 DOM (a) and DEM (b) of the third lunar day PCAM of Yutu - 2, where the red impact craters are the corresponding craters of the PCAM and LCAM, and the blue represents impact craters not identified by the LCAM. The areas shown in (a, b) correspond to the red area of the LCAM mosaic image thumbnail in the lower left corner of the figure

diameter ratios of impact craters and to quantitatively analyze impact craters of different degradation levels.

For TCAM, we performed a degradation analysis of the impact crater within part of the TCAM panoramic surround image (the region bounded by two white circles in Figure 9b, corresponding to the area between the two white circles of the LCAM mosaic image in Figure 9c), which was used to support our classification of the degradation of impact craters. The circular area contains 2371 impact craters in the LCAM mosaic image, representing approximately 36% of all impact craters. Considering the large distortion of the terrain camera data, we compared the morphologies of 100 pairs of more discernible craters and input the results into the classification of the crater degradation levels. In Figure 9b, we show 12 crater pairs that have been labeled in the image to provide a side-view morphology of the impact craters at different degradation levels from the TCAM.

As shown in Figure 9b, c, 12 pairs of corresponding craters (four C2, five C3, and three C4 craters) on the ground TCAM image and LCAM mosaic image are linked by red lines. C2 craters (aquamarine) have clear circular shapes, and the inner contrast between bright and dark is clear on the ground TCAM image. C3 craters (yellow) have circular shapes but no discernible rim. C4 craters (green) have hidden circular shapes, the bottoms are infilled, and the crater rims disappear. The classified degradation levels on the LCAM mosaic image are consistent with the TCAM camera images.

Moreover, there are many impact craters of different sizes around the landing point, with varying degrees of degradation. Most craters on the surface are small craters, and their distribution is dispersed across the surface and consistent with their distribution in the descent images. However, only a few small craters have a clear circular shape. Most small impact craters are severely degraded, which is why smaller craters have a higher C4 ratio in the distribution analysis results (Table 1).

For quantitative validation of impact crater degradation levels, we used the PCAM image of the third lunar day of Yutu - 2 and obtained the georeferenced DOM (0.0014m/pixel) and DEM (0.01m/pixel) of the region (Figure 10). The correspondence between the impact crater degradation level and the crater depth - to - diameter ratio in this area was verified. This study area is within the aquamarine outlined area of the LCAM mosaic image, an area where we statistically analyzed craters from 1.026 to 8.963m in the above section. For the impact craters in this range, a total of 25 craters can be established to correspond to the LCAM mosaic data and the PCAM obtained DOM and DEM (red impact craters in Figure 10), while compared to the PCAM, two craters are not identified on the LCAM mosaic image in this diameter range. These two impact craters are relatively small, highly degraded, and not identified due to the resolution limitations of this area in the LCAM mosaic image. Thus, the total number of impact craters obtained from the above content impact crater count may be underestimated, but our method identifies at least 92% of the impact craters in the study area and still has a high degree of reliability.

For the relationship between the level of crater degradation and the crater depth - to - diameter ratio, we selected the craters numbered 0 - 19 in Figure 10b for the analysis because these craters are in a region with low elevation change and the topography has less influence. However, for the craters numbered 15 and 17, we did not use these two craters for the calculation of the depth - to - diameter ratio because the difference between their sizes and the size of the largest crater in the region exceeded 1m, considering that the depth - to - diameter ratio was calculated in the case of similar crater sizes. Then, we sorted the diameter - to - depth ratios of the above 18 craters from lowest to highest diameter - todepth ratio and show the results in Table 2. Since there are only craters with degradation levels of C3 and C4 in the study area, for these 18 craters, the diameter - to - depth ratios of C4 craters tend to be smaller, while those of C3 craters tend to be larger (Table 2). Overall, both C3 and C4 impact craters are very shallow, with maximum depths of less than 0.2cm and impact crater diameter - to - depth ratios of less than 1.121×10^{-3}. Consequently, the DEM - calculated crater depth - to - diameter ratios validate our classification of impact crater degradation levels C3 and C4 as relatively accurate.

Table 2 Diameter - to - depth ratio of the impact crater in the third lunar day experiment area from the Yutu - 2 PCAM

Depth - to - Diameter Ratio ($\times10^{-3}$)	Crater Number	Diameter (m)	Depth (cm)	Crater Degradation Levels
0.023	19	2.015	0.005	C4
0.043	5	1.907	0.008	C4

Continued Table 2

Depth - to - Diameter Ratio (×10^{-3})	Crater Number	Diameter (m)	Depth (cm)	Crater Degradation Levels
0.288	14	1.017	0.029	C4
0.321	12	1.592	0.051	C4
0.363	13	1.744	0.063	C4
0.385	7	1.123	0.043	C4
0.390	0	1.466	0.057	C3
0.432	10	1.798	0.078	C3
0.450	6	1.531	0.069	C3
0.465	18	1.499	0.070	C3
0.495	9	1.734	0.086	C3
0.518	8	1.673	0.087	C3
0.557	3	1.176	0.065	C3
0.601	4	1.374	0.083	C3
0.700	11	2.040	0.143	C3
0.794	2	1.256	0.100	C3
0.804	1	1.295	0.104	C3
1.121	16	1.752	0.196	C3

4 Discussion

With continued bombardment, the impact crater density of any diameter will eventually reach an upper limit, especially for smaller impact craters [41, 42]. A crater population smaller than a given diameter is considered to be in equilibrium when the production rate of craters is balanced by the removal rate [15]. Impact - crater - based lunar surface age determinations should always be performed on impact crater populations larger than the equilibrium onset diameter (D_{eq}, for a given crater population, the diameter of the largest crater at which the population is in equilibrium), otherwise, misleading dating results will be obtained [15, 43].

The methods of Gault [43] and Hartmann [44] are the widely used methods for evaluating the equilibrium state of impact crater populations, both of which regard certain empirical crater densities as the equilibrium level. Therefore, cumulative size - frequency distribution (SFD) slopes around - 2 are often used as empirical equilibrium levels for different impact crater populations. However, reference to the method of Gault and Hartmann can lead to different interpretations of impact crater populations, especially for small - sized impact crater populations in older geological units [36]. In particular, the age of the geological units in our study area is approximately 3.6 – 3.7 Gyr old [36], while the impact craters are mostly below 2m in diameter and there may also be many secondary impact craters. We used the Craterstats 2 [45] program to plot Hartmann's equilibrium lines [44] (black lines in Figure 11) for the craters in the four study areas and found that the empirical equilibrium levels of Hartmann were not appropriate for our four study areas [44]. Alternatively, we used a power – law index to obtain the cumulative SFD slope according to Xiao and Werner [15]. The details are as follows.

In the above, we divided the LCAM mosaic image into four study areas (Figure 6). In the statistical analysis of impact craters at each degradation level, the small number of impact craters outside the 95% impact crater population, limited by the resolution and size of the study area, will lead to low confidence in the results of the obtained impact crater degradation statistics. Therefore, we performed degradation statistics for 95% of the impact craters in each study area in the above section. However, when analyzing the crater accumulation process, we analyzed the full population of craters to obtain a localized geographical evolution of the four study areas.

For each of the four study areas, the equilibrium state of their impact crater populations was evaluated. We fitted power laws to craters in each study area using the technique by Clausi et al. [46] to obtain the cumulative SFD slope (the power law exponent is less than the cumulative SFD slope by 1) and the equilibrium onset diameter D_{eq} (as shown in Table 3) [10, 15, 16, 47]. The results show that the crater populations in equilibrium have a cumulative SFD slope of <-2 for both the red and yellow outlined areas (−2.696 and −2.392 for the red and yellow regions, respectively), while the crater populations in equilibrium have a cumulative SFD slope of about −2 for both the aquamarine and pink outlined areas (−1.952 and −1.917 for the red and yellow regions, respectively).

Table 3 Summary of the equilibrium states of the crater populations on the counting areas

Counting Areas	Cumulative SFD Slopes of Crater Population in Equilibrium	D_{eq} (m)
Red outlined area	−2.696	0.779
Yellow outlined area	−2.392	0.810
Aquamarine outlined area	−1.952	2.090
Pink outlined area	−1.917	3.037

In previous studies, the cumulative SFD slope of decimeters to hundreds of meters in equilibrium for the crater population was −2, while the production SFD slope was −3 [15, 43, 49]. In contrast, only the cumulative SFD slopes of crater populations in the equilibrium of the pink outline area and aquamarine outline area were approximately −2, a parameter that has been considered to represent equilibrium in previous studies, while the equilibrium onset diameters of the craters in those areas are 3 and 2m. For impact crater populations with higher resolution and smaller diameter ranges, the red outlined area and the yellow outlined area are found to have the highest resolution areas with steeper cumulative SFD slopes (of −2.696 and −2.392, respectively). In previous studies, smaller diameter impact crater populations were more likely to reach equilibrium [15]; thus, the red outlined area should be the most likely of the study areas to be in equilibrium. However, this area has the highest cumulative SFD slopes; it seems that the statistical area is far from equilibrium, while the large areas easily reach equilibrium.

Regarding the inconsistency of our findings relative to previous impact crater population equilibrium studies, we also suggest that it is possible that the Chang'e - 4 landing area is old with a thick regolith layer [24], and thus may have a large distribution of secondary impact craters. Such secondary impact craters may be more widely distributed in the centimeter diameter range. A large distribution of secondary impact craters would result in a steeper cumulative SFD slope [50], resulting in a higher resolution area with a steeper cumulative SFD slope.

Surface resurfacing may also be causing the red and the yellow outlined areas to not be in equilibrium. Figure 11 shows the cumulative SFDs for four statistical areas. Figure 11a, b show inflection points, it seems that some magma resurfacing events occurred in the landing area, resulting in some small craters being infilled and thus disappearing in the crater counting results. However, recent research [51] shows that the landing area is 70m higher than the surrounding lunar basalt and that the surface materials were mainly ejecta from the Finsen and Alder craters. Therefore, the surface may have not been resurfaced by lava flows.

As shown in Figure 9a, the ejecta from C3 type crater a1' infilled and partly resurfaced the neighboring C4 type crater a' and caused a lower crater density in Figure 9d. The corresponding crater diameters at the inflection points of Figure 11a, b are 3 and 8m, respectively, and the crater excavation depths can be scaled by diameters (0.084 : 1 for a simple lunar crater [41, 52]); thus, the initial depth for craters with diameters of 3 and 8m are 25.3 and 67.2cm. The depth results show that the smaller the craters, the easier they are infilled by ejecta, while the larger craters show bigger depths to resist this influence.

Considering the existence of five craters with diameters of 15 - 22m in the red statistical area and their different degradation levels, these five impact craters formed at different times due to the different degradation statuses. During the impact formation process, the materials were ejected into the surrounding area, thereby covering the neighboring tiny craters and producing an effect like magma resurfacing events.

This ejecta reshaping process may have caused the reduction in the number of craters smaller than 3m in diameter in the red statistical area. Then, the cumulative SFD of the crater population was reset, and the process of crater accumulation restarted. As the

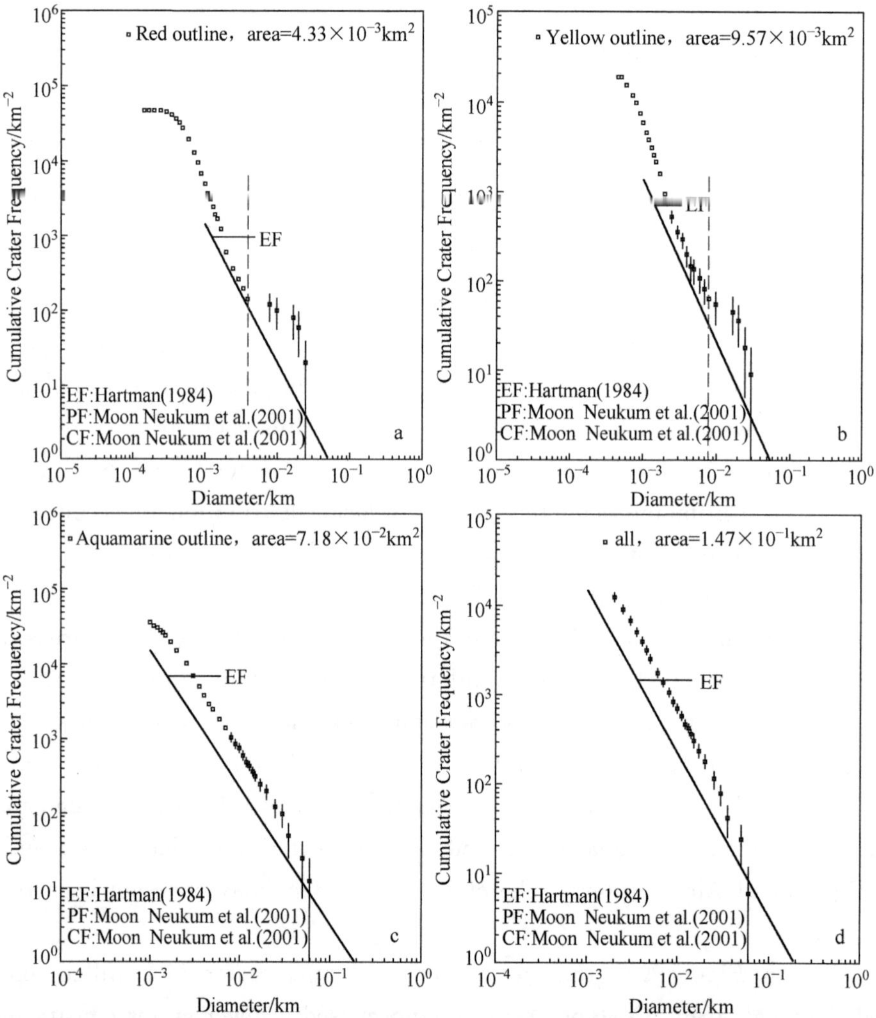

Figure 11　Craterstats 2 program calculated cumulative SFD and equilibrium lines in the red, yellow, aquamarine, pink outlined areas. EF refer to the equilibrium function of Hartmann [44]. PF and CF refer to the production function and the chronology function of Neukum et al. [48], respectively. Red dashed lines show the inflection points. The equilibrium lines (criterions from Hartmann [44]) are displayed as black lines in (a－d). (a) Red outlined area. (b) Yellow outlined area. (c) Aquamarine outlined area. (d) Pink outlined area.

statistical area gradually becomes larger, the crater size corresponding to the inflection points gradually becomes larger (8m shown in Figure 11b) and nearly disappears in Figure 11c, d. This shows that the surface is increasingly difficult to resurfaced by ejecta materials as the craters become larger, and this is likely the reason that the inflection points of the statistical curve of impact craters gradually increase until it disappears.

Unlike previous works [15, 42, 43], small craters (diameter smaller than 3m in this work) show an unusual cumulative SFD slope in equilibrium that smaller craters could be resurfaced by the ejecta from neighboring craters and uneasily reach equilibrium [16].

Considering that, the smaller the impact crater mentioned above is, the more severe the degradation is, the smaller the size of the impact crater is, and the more difficult it is for them to achieve crater equilibrium due to secondary craters and the local resurfacing events caused by ejecta from neighboring craters.

5 Future Research Considerations

Considering that Chang'e - 4 is the first exploration mission to the far side of the Moon, its research data are of high scientific value. Compared with the lunar orbiter images, the Chang'e - 4 lander and Yutu - 2 rover provide a large number of high - resolution images in situ, such as the lander's LCAM images, Yutu - 2 PCAM images, navigation camera images, etc. Combining these data and obtaining possible DEMs will help subsequent detailed population degradation analyses of centimeter - to meter - scale impact craters around the landing site and rover route. Furthermore, statistical analysis of the small - sized impact crater population degradation can also be extended to other landing areas for future lunar landing exploration and even for other celestial bodies, providing more detailed information for analyzing local geological evolution (e. g. , potential surface resurfacing).

6 Conclusions

Small - sized (centimeter to meter scale) craters are the focus of our study. In contrast to the existing studies, which have mainly concentrated on large - sized craters (hundreds of meters), our study considers the degradation statistics of these small - sized crater populations to provide detailed information on the analysis of the surface resurfacing of local areas. Based on this, 6316 craters with diameters from 0. 15 to 67m were identified and classified into different degradation levels using their morphological differences on the high - resolution Chang'e - 4 LCAM mosaic images. The ground TCAM image and the crater diameter - to - depth ratio obtained from the DEM of the Yutu -2 PCAM validate the classification of crater degradation levels from a qualitative and quantitative perspective, respectively. The results of the study indicated that:

(1) The crater population is severely degraded, with over 90% of the craters being moderately degraded (C3) and severely degraded (C4). Moreover, like the degradation pattern of the large crater population, the smaller the crater size, the more severe the degradation.

(2) The red and yellow outlined areas cover more small - sized craters (with minimum crater diameters of 0. 15 and 0. 5m, respectively) than the other two study areas in this paper; unlike previous findings, they have more difficulty reaching equilibrium (cumulative SFD slopes of - 2. 696 and - 2. 392 for these two areas). This may be due to

secondary craters and/or resurfacing events caused by ejecta from adjacent craters.

Acknowledgments

The Chang'e-4 mission was carried out by the Chinese Lunar Exploration Program, and the data can be downloaded from the Data Release and Information Service System of China's Lunar Exploration Program (https://moon.bao.ac.cn/ce5web/searchOrder_dataSearchData.search, accessed on 4 January 2022).

References

[1] Li, C.; Zuo, W.; Wen, W.; Zeng, X.; Gao, X.; Liu, Y.; Fu, Q.; Zhang, Z.; Su, Y.; Ren, X. Overview of the Chang'e-4 mission: Opening the frontier of scientific exploration of the lunar far side. Space Sci. Rev. 2021, 217, 1-32.

[2] Liu, Z.; Di, K.; Li, J.; Xie, J.; Cui, X.; Xi, L.; Wan, W.; Peng, M.; Liu, B.; Wang, Y. Landing site topographic mapping and rover localization for Chang'e-4 mission. Sci. China Inf. Sci. 2020, 63, 1-12.

[3] Yang, C.; Zhao, H.; Bruzzone, L.; Benediktsson, J.A.; Liang, Y.; Liu, B.; Zeng, X.; Guan, R.; Li, C.; Ouyang, Z. Lunar impact crater identification and age estimation with Chang'E data by deep and transfer learning. Nat. Commun. 2020, 11, 1-15.

[4] Guo, D.; Fa, W.; Wu, B.; Li, Y.; Liu, Y. Millimeter-to Decimeter-Scale Surface Roughness of the Moon at the Chang'e-4 Exploration Region. Geophys. Res. Lett. 2021, 48, e2021GL094931.

[5] Wu, B.; Li, Y.; Liu, W.C.; Wang, Y.; Li, F.; Zhao, Y.; Zhang, H. Centimeter-resolution topographic modeling and fine-scale analysis of craters and rocks at the Chang'E-4 landing site. Earth Planet. Sci. Lett. 2021, 553, 116666.

[6] Jia, M.; Di, K.; Yue, Z.; Liu, B.; Wan, W.; Niu, S.; Liu, J.; Cheng, W.; Lin, Y. Multi-scale morphologic investigation of craters in the Chang'e-4 landing area. Icarus 2021, 355, 114164.

[7] Hiesinger, H.V.; Van Der Bogert, C.; Pasckert, J.; Funcke, L.; Giacomini, L.; Ostrach, L.; Robinson, M. How old are young lunar craters? J. Geophys. Res. Planets 2012, 117, E00H10.

[8] Geiss, J.; Rossi, A.P. On the chronology of lunar origin and evolution. Astron. Astrophys. Rev. 2013, 21, 1-54.

[9] Massironi, M.; Cremonese, G.; Marchi, S.; Martellato, E.; Mottola, S.; Wagner, R.J. Mercury's geochronology revised by applying Model Production Function to Mariner 10 data: Geological implications. Geophys. Res. Lett. 2009, 36, L21204.

[10] Basilevsky, A. T. On the evolution rate of small lunar craters. In Proceedings of the 7th Lunar Planetary Science Conference, Houston, TX, USA, 15 - 19 March 1976; Volume 1, pp. 1005 - 1019.

[11] Craddock, R. A. ; Howard, A. D. Simulated degradation of lunar impact craters and a new method for age dating farside mare deposits. J. Geophys. Res. Planets 2000, 105, 20387 - 20401.

[12] Molaro, J. ; Byrne, S. Rates of temperature change of airless landscapes and implications for thermal stress weathering. J. Geophys. Res. Planets 2012, 117, E10011.

[13] Fassett, C. I. ; Thomson, B. J. Crater degradation on the lunar maria: Topographic diffusion and the rate of erosion on the Moon. J. Geophys. Res. Planets 2014, 119, 2255 - 2271.

[14] Wilhelms, D. E. ; John, F. ; Trask, N. J. The Geologic History of the Moon; USGS Professional Paper 1348; Geological Survey: Reston, VA, USA, 1987; pp. 2330 - 7102.

[15] Xiao, Z. ; Werner, S. C. Size - frequency distribution of crater populations in equilibrium on the Moon. J. Geophys. Res. Planets 2015, 120, 2277 - 2292.

[16] Xie, M. ; Xiao, Z. ; Xu, A. Time - dependent production functions of lunar simple craters on layered targets with consideration of topographic degradation. Geophys. Res. Lett. 2019, 46, 10987 - 10996.

[17] Xie, M. ; Zhu, M. H. ; Xiao, Z. ; Wu, Y. ; Xu, A. Effect of topography degradation on crater size - frequency distributions: Implications for populations of small craters and age dating. Geophys. Res. Lett. 2017, 44, 10171 - 10179.

[18] Wood, C. ; Head, J. ; Cintala, M. Crater degradation on Mercury and the moon - Clues to surface evolution. In Proceedings of the 8th Lunar and Planetary Science Conference Proceedings, Houston, TX, USA, 14 - 18 March 1977; Volume 3, pp. 3503 - 3520.

[19] Ivanov, M. ; Basilevsky, A. Morphology and size - frequency distribution of kilometer - scale impact craters on Callisto and Ganymede derived from Galileo data. Sol. Syst. Res. 2002, 36, 447 - 457.

[20] Li, K. Study on Small - Scale Lunar Craters' Morphology and Degradation; Wuhan University: Wuhan, China, 2013.

[21] Howard, A. D. Simulating the development of Martian highland landscapes through the interaction of impact cratering, fluvial erosion, and variable hydrologic forcing. Geomorphology 2007, 91, 332 - 363.

[22] Kreslavsky, M. A. ; Head, J. W. ; Neumann, G. A. ; Rosenburg, M. A. ; Aharonson, O. ; Smith, D. E. ; Zuber, M. T. Lunar topographic roughness maps from Lunar Orbiter Laser Altimeter (LOLA) data: Scale dependence and correlation with geologic features and units. Icarus 2013, 226, 52 - 66.

[23] Xie, M. ; Liu, T. ; Xu, A. Ballistic Sedimentation of Impact Crater Ejecta: Implications for the Provenance of Lunar Samples and the Resurfacing Effect of Ejecta on the Lunar Surface. J. Geophys. Res. Planets 2020, 125, e06113.

[24] Qiao, L. ; Ling, Z. ; Fu, X. ; Li, B. Geological characterization of the Chang'e - 4 landing area on the lunar farside. Icarus 2019, 333, 37 - 51.

[25] Chapman, C. The system of lunar craters. Quadr. I Comm. 1963, 2, 71 - 78.

[26] Hirabayashi, M. ; Minton, D. A. ; Fassett, C. I. An analytical model of crater count equilibrium. Icarus 2017, 289, 134 - 143.

[27] Povilaitis, R. ; Robinson, M. ; Van der Bogert, C. ; Hiesinger, H. ; Meyer, H. ; Ostrach, L. Crater density differences: Exploring regional resurfacing, secondary crater populations, and crater saturation equilibrium on the moon. Planet. Space Sci. 2018, 162, 41 - 51.

[28] Minton, D. A. ; Fassett, C. I. ; Hirabayashi, M. ; Howl, B. A. ; Richardson, J. E. The equilibrium size - frequency distribution of small craters reveals the effects of distal ejecta on lunar landscape morphology. Icarus 2019, 326, 63 - 87.

[29] Salaün, Y. ; Marlet, R. ; Monasse, P. Line - based robust SfM with little image overlap. In Proceedings of the 2017 International Conference on 3D Vision (3DV), Qingdao, China, 10 - 12 October 2017; pp. 195 - 204.

[30] Li, R. ; Fei, M. ; Xu, F. ; Matthies, L. H. ; Olson, C. F. ; Arvidson, R. E. Localization of Mars rovers using descent and surface - based image data. J. Geophys. Res. Planets 2002, 107, FIDO 4 - 1 - FIDO 4 - 8.

[31] Kang, Z. Z. ; Zhang, Z. X. ; Yang, F. L. Relative Orientation and Epipolar Arrangement based on Forward Moving Image Pairs along the Optical Axis. Acta Geod. Cartogr. Sin. 2007, 36, 56 - 61.

[32] Barlow, N. G. Constraining geologic properties and processes through the use of impact craters. Geomorphology 2015, 240, 18 - 33.

[33] Wang, J. ; Ming, D. ; Cheng, W. ; Sensing, R. Identification of Multiscale Spatial Structure of Lunar Impact Crater: A Semivariogram Approach. IEEE J. Sel. Top. Appl. Earth Obs. 2021, 14, 2731 - 2739.

[34] Huang, Y. H. ; Minton, D. A. ; Hirabayashi, M. ; Elliott, J. R. ; Richardson, J. E. ; Fassett, C. I. ; Zellner, N. E. Heterogeneous impact transport on the Moon. J. Geophys. Res. Planets 2017, 122, 1158 - 1180.

[35] Kinczyk, M. J. ; Prockter, L. M. ; Byrne, P. K. ; Susorney, H. C. ; Chapman, C. R. A morphological evaluation of crater degradation on Mercury: Revisiting crater classification with MESSENGER data. Icarus 2020, 341, 113637.

[36] Ling, Z. ; Qiao, L. ; Liu, C. ; Cao, H. ; Bi, X. ; Lu, X. ; Zhang, J. ; Fu, X. ; Li, B. ; Liu, J. Composition, mineralogy and chronology of mare basalts and non - mare materials in Von Kármán crater: Landing site of the Chang'E - 4 mission. Planet. Space Sci. 2019, 179, 104741.

[37] Di, K.; Zhu, M. H.; Yue, Z.; Lin, Y.; Wan, W.; Liu, Z.; Gou, S.; Liu, B.; Peng, M.; Wang, Y. Topographic evolution of Von Kármán crater revealed by the lunar rover Yutu-2. Geophys. Res. Lett. 2019, 46, 12764-12770.

[38] Garrick-Bethell, I.; Zuber, M. T. Elliptical structure of the lunar South Pole-Aitken basin. Icarus 2009, 204, 399-408.

[39] Wu, B.; Li, F.; Hu, H.; Zhao, Y.; Wang, Y.; Xiao, P.; Li, Y.; Liu, W. C.; Chen, L.; Ge, X. Topographic and Geomorphological Mapping and Analysis of the Chang'E-4 Landing Site on the Far Side of the Moon. Photogramm. Eng. Remote Sens. 2020, 86, 247-258.

[40] Wang, Y.; Xie, M.; Xiao, Z.; Cui, J. The minimum confidence limit for diameters in crater counts. Icarus 2020, 341, 113645.

[41] Melosh, H. J. Impact Cratering: A Geologic Process; Oxford University Press: New York, NY, USA; Clarendon Press: Oxford, UK, 1989; pp. 1-255.

[42] Richardson, J. E. Cratering saturation and equilibrium: A new model looks at an old problem. Icarus 2009, 204, 697-715.

[43] Gault, D. E. Saturation and equilibrium conditions for impact cratering on the lunar surface: Criteria and implications. Radio Sci. 1970, 5, 273-291.

[44] Hartmann, W. K. Does crater 'saturation equilibrium' occur in the solar system? Icarus 1984, 60, 56-74.

[45] Michael, G.; Neukum, G. Planetary surface dating from crater size-frequency distribution measurements: Partial resurfacingevents and statistical age uncertainty. Earth Planet. Sci. Lett. 2010, 294, 223-229.

[46] Clauset, A.; Shalizi, C. R.; Newman, M. E. Power-law distributions in empirical data. SIAM Rev. 2009, 51, 661-703.

[47] Crater Analysis Techniques Working Group. Standard techniques for presentation and analysis of crater size-frequency data. Icarus 1979, 37, 467-474.

[48] Neukum, G.; Ivanov, B.; Hartmann, W. K. Cratering records in the Inner solar system in relation to the lunar reference system. Space Sci. Rev. 2001, 96, 55-86.

[49] Moore, H. J. Density of small craters on the lunar surface. In United States Geological Survey Astrogeology Studies; Annual Program Report: Washington, DC, USA, 1964; pp. 34-51.

[50] Xiao, Z.; Strom, R. G. Problems determining relative and absolute ages using the small crater population. Icarus 2012, 220, 254-267.

[51] Di, K.; Liu, Z.; Liu, B.; Wan, W.; Peng, M.; Wang, Y.; Gou, S.; Yue, Z.; Xin, X.; Jia, M. Chang'e-4 lander localization based on multi-source data. J. Remote Sens. 2019, 23, 177-184.

[52] Thomson, B. J.; Grosfils, E. B.; Bussey, D. B. J.; Spudis, P. D. A new technique for estimating the thickness of mare basalts in Imbrium Basin. Geophys. Res. Lett. 2009, 36, L12201.

Multi – scale morphologic investigation of craters in the Chang'e – 4 landing area

Mengna Jia[1,2], Kaichang Di[1,3,6], Zongyu Yue[1,6], Bin Liu[1], Wenhui Wan[1], Shengli Niu[3], Jianzhong Liu[4,6], Weiming Cheng[5,6], Yangting Lin[7]

1. State Key Laboratory of Remote Sensing Science, Aerospace Information Research Institute, Chinese Academy of Sciences, Beijing 100101, China
2. China Academy of Space Technology (Xi'an), Xi'an 710100, China
3. State Key Laboratory of Lunar and Planetary Sciences, Macau University of Science and Technology, Macau 999078, China
4. Institute of Geochemistry Chinese Academy of Sciences, Guiyang 550002, China
5. Institute of Geographic Sciences and Natural Resources Research, Chinese Academy of Sciences, Beijing 100101, China
6. CAS Center for Excellence in Comparative Planetology, Hefei 230026, China
7. Key Laboratory of Earth and Planetary Physics, Institute of Geology and Geophysics, Chinese Academy of Sciences, Beijing 100029, China

Abstract: China's Chang'e – 4 (CE – 4) rover landed in Von Kármán crater on January 3, 2019, and has acquired a large amount of data of the landing site. Craters are an important landform in lunar surface and the analysis of craters at the landing site with high – resolution images is of great significance for both scientific study and engineering safety. In this study, multi – sources data, including Lunar Reconnaissance Orbiter Camera Narrow Angle Camera (NAC) images, NAC digital elevation model (DEM), Chang'e – 2 (CE – 2) digital orthophoto map (DOM), SLDEM2015 and stereo images from panoramic cameras onboard the CE – 4 rover, are involved to reveal the crater morphologic features at the landing site. The diameter, depth, spatial density and depth/diameter ratio (d/D) of the mapped craters have been investigated. Randomness analysis and the size – frequency distributions of craters in two scales are conducted to thoroughly elaborate the crater distribution in the landing area. The multi – scale crater investigation establishes a foundation for subsequent in – depth studies, e.g., crater evolution and degra – dation of lunar farside and CE – 4 landing area dating.

Keywords: Craters; Chang'e – 4; Panoramic image

1 Introduction

Chang'e - 4 (CE - 4) mission is China's second lunar soft landing mission and it was initially designed as the back - up of Chang'e - 3 (CE - 3) mission, both of which constitute the second phase of China Lunar Exploration Project (CLEP, 2018; Jia et al., 2018). CE - 4 mission includes a communication relay satellite, a lander, and a rover (Yutu - 2). The relay satellite (Magpie Bridge), which was launched on May 5, 2018, provides Earth - Moon relay communications between the CE - 4 lunar probe and the control center in the Earth (Gao et al., 2019; Qin et al., 2019). The CE - 4 probe was launched on December 8, 2018 and successfully landed in the Von Kármán crater at (177.588°E, 45.457°S) on January 3, 2019 (Di et al., 2019a).

The Von Kármán crater was formed in pre - Nectarian period and it is inside the oldest, deepest, and largest lunar basin, i.e., the South - Pole Aiken (SPA) basin. The SPA basin was thought to have exposed the lunar lower crust and probably upper mantle material according to the spectral analysis of remote sensing observations and numerical impact cratering simulations (e.g., Head et al., 1993; Lucey et al., 1998; Ohtake et al., 2014; Melosh et al., 2017). The in - situ investigation by CE - 4 provides a unique opportunity for in - depth research. CE - 4 mission has returned a large amount of data with the onboard payloads including a landing camera, a terrain camera, a pair of panoramic cameras, a visible and near - infrared imaging spectrometer, and ground penetrating radar (Wu et al., 2019). Much information on the lunar farside, such as the composition of the deep material (Gou et al., 2019; Li et al., 2019), local mineral abundances (Hu et al., 2019), and topographic evolution (Di et al., 2019b), have been obtained and more findings are expected with the ongoing interpretation of the returned data.

As one of the most important landforms in lunar surface, craters are significant in understanding the geologic history (Wilhelms et al., 1987), subsurface composition and structure (Soderblom et al., 1974; Bland et al., 2016), surface gardening process (Melosh, 1989), etc. Therefore, much work has been carried out in lunar crater study, especially creating catalogues of global or regional craters and making related analyses. However, the existing lunar global crater catalogues are generally established using low - resolution images. For example, the most recently created lunar global crater database is mainly based on Lunar Recon - naissance Orbiter Camera (LROC) Wide Angle Camera (WAC) mosaic at 100m/pixel and the minimum diameter of the craters is ~1.0km (Robbins, 2019). On the other hand, the near global covering of LROC Narrow Angle Camera (NAC) images at the resolution of ~0.5 — 2.0m/ pixel and the returned panoramic camera (PanCam) images by Yutu - 2 rover have offered the opportunity for creating a much finer regional crater catalogue. It is significant for understanding of impact physics at this scale, estimating the gardening process in lunar shallow surface, interpreting the returned data especially for the ground penetrating radar, and even for evaluating the engineer safety in future rover traveling.

In this research, a 0.9m - resolution DOM mosaic is produced from the LROC NAC images, based on which two different - scaled crater catalogues are established. Besides, smaller scaled craters are also digitalized from stereo images obtained by the PanCam mounted on the CE - 4 rover. According to these catalogues, the crater - related morphological features in the landing area, including multi - scale spatial density, depth/diameter ratio (d/D) and size - frequency distribution (SFD), etc., are investi - gated in detail.

2 Data and method

2.1 Data

The datasets used in this study include images from LROC NAC (Robinson et al., 2010), Chang'e - 2 (CE - 2) CCD camera (CLEP, 2018), and Yutu - 2 rover PanCam. Topographic data from SLDEM2015 (Barker et al., 2016) and LROC NAC image generated DEM (Henriksen et al., 2017) are used for depth measurements. The LROC NAC images and DEM products, and SLDEM2015 can be downloaded from the Planetary Data System (PDS) Geosciences Node (https://ode.rsl.wustl. edu/moon/), while the CE - 2 CCD digital orthophoto map (DOM) can be obtained from the website of Data Publishing and Information Service System (DPISS) of CLEP (http://moon.bao.ac.cn/).

The LROC NAC images are generally at the resolutions ~0.5—2.0m/ pixel, which are the highest resolution lunar orbital images available at present (Robinson et al., 2010). LROC NAC DEMs are generated by two NAC images covering the same area on the ground while taken from different view angles (Henriksen et al., 2017). The NAC DEM used in this study has a ground sample distance of 5.0m and was created from stereo images of M1303619844 and M1303640934 (Robinson, 2019).

The two - line push - broom time delay and integration CCD camera is one of the important payloads of CE - 2 mission. It is composed of a forward - looking and a backward - looking line camera to obtain orbital stereo images. The CE - 2 stereo cameras acquired stereo images of global coverage at a resolution of 7m/pixel in a 100km ×100km polar circular orbit (Zhao et al., 2011). The global 7m DOM product of CE - 2 has been produced and released on DPISS website (http://moon.bao.ac.cn/; CLEP, 2018).

The SLDEM2015 was generated by a combination of LRO Lunar Orbiter Laser Altimeter and DEMs from stereo images obtained by Kaguya Terrain Camera. SLDEM2015 covers latitudes within ±60° and has an effective resolution of approximately 60m/pixel at the equator and a typical vertical accuracy of 3m - 4m (Barker et al., 2016).

The selected landing area for CE - 4 is within 176.4°E - 178.8°E, 45°S - 46°S (Fig. 1a). To generate the high - resolution DOM in the CE - 4 landing area, 100 LROC NAC Experimental Data Record (EDR) level images with similar illumination condition and good imaging quality were selected. The radiometrical and geometric corrections are subsequently made to these EDR level NAC images with the software of Integrated System for Imagers and Spectrometers (ISIS) (Anderson et al., 2014). Then, a program is developed to process the geometric inconsistencies between NAC images (Liu et al., 2019). Unfortunately, there are some gaps between some LROC NAC images, and then one CE - 2 DOM image (CE2 _GRAS _DOM _07m _K136 _45S175E _A) is used for these areas. The final DOM mosaic of the landing area is shown in Fig. 1b in a Lambert conformal conic projection, which has a ground sample distance of ~0.9m and an image size of 57483 columns ×34084 rows.

The PanCam on the CE - 4 rover comprises two identical cameras with 19.7°×14.5° field of view, which can obtain stereo image pairs around the rover with a baseline of 270mm (Jia et al., 2018). The PanCam has the ability of 360° imaging with the assistance of the rotary stage and the acquired images have the size of 2352 pixels ×1728 pixels. The PanCam images used in this study were obtained during the first lunar day of the mission (Di et al., 2019b) and 58 stereo pairs were adopted for crater measurement. The data can also be downloaded from the DPISS website (http://moon.bao.ac.cn/). It is worth noting that very small craters surrounding the landing site are affected by rocket engine plume during soft landing. The affected region (Gou et al., 2020) is mainly distributed on the western direction of the lander and PanCam images in this study is obtained at the northwest of the lander with a straight - line distance of ~16m. Therefore, the rocket engine plume effect is very limited in the research area of this study.

2.2 Crater mapping and analysis in CE - 4 landing area

The craters in the CE - 4 landing area are mapped in three scales based on the produced DOM, the DEMs (entire landing area or local area surrounding the landing site) or the PanCam stereo images. Within 1° in latitude and 2.4° in longitude (30km ×51km) LROC NAC mosaic in the landing area (Fig. 1b), craters with the minimum diameter of ~100m are mapped, and their depth are measured with SLDEM2015. For a small area of 3.2km ×3.0km surrounding the landing site (red rectangular in Fig. 1a), where a very high - resolution DEM of ~5m/pixel is available, craters larger than several meters are mapped and measured. It can be inferred from previous researches (Fig. 2b of Huang et al., 2018 and Fig. 5c of Qiao et al., 2019) that the local slope for the CE - 4 landing area is very small and the slope effect could be ignored in the following investigation.

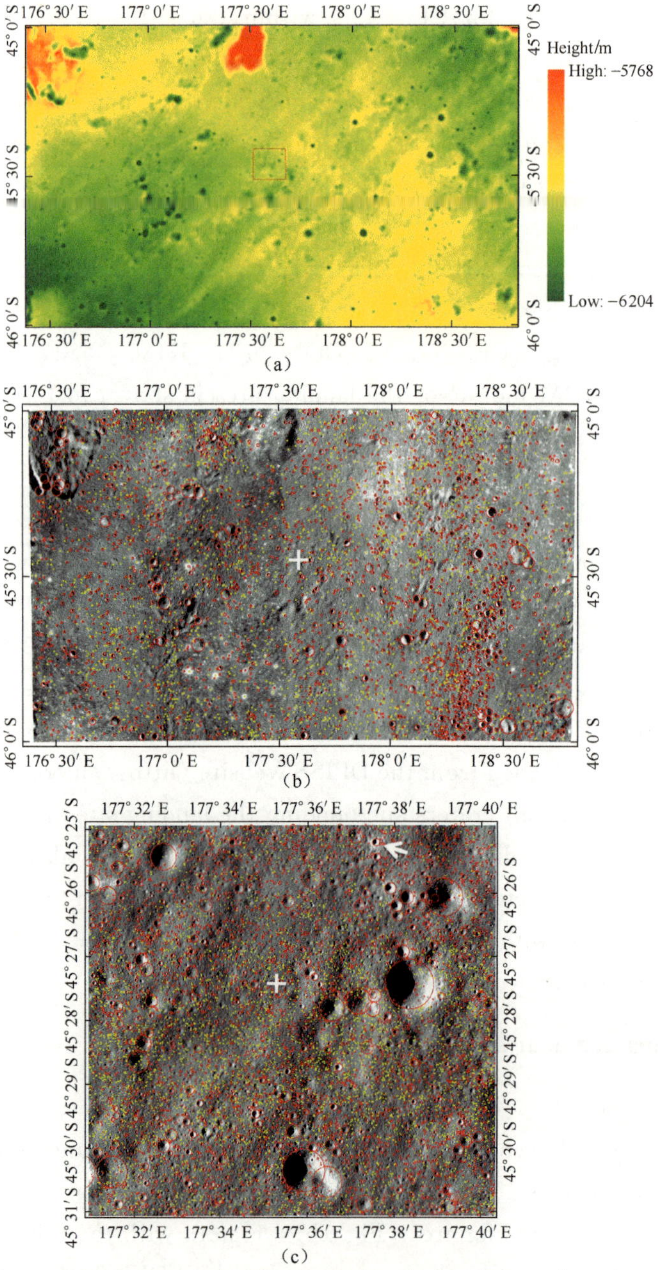

Fig. 1 (a) Crater mapping areas in the research. Craters larger than 100m in diameter are mapped in the above whole area, and their depths are measured with SLDEM2015. The craters larger than several meters in diameter are mapped within the red rectangle, and the depths are measured with an available high resolution DEM from LROC NAC stereo images. The basemap is SLDEM2015 (Barker et al., 2016). (b) The mapped craters larger than 100m in the CE - 4 landing area overlaid on the LROC NAC DOM mosaic. (c) The distribution of the mapped craters in the 3.2km ×3.0km area surrounding the CE - 4 landing site. The white arrow in (c) points to a relatively fresh crater, whose detailed information can be found in Table 1 (Crater ID: C1). The craters larger than the completeness diameters (127m and 9m for (b) and (c) respectively) are plotted in red while the smaller ones are in yellow. The CE - 4 landing site is marked as the white crosses in (b) and (c) (For interpretation of the references to colour in this figure legend, the reader is referred to the web version of this article)

When mapping the craters with the NAC mosaic, crater rims are first manually extracted with an ArcGIS add - in tool of "CraterTools", which is independent of the projection and has been widely used in planetary crater mapping (Kneissl et al., 2011; Povilaitis et al., 2018). In addition, to digitalize crater rims more precisely, a local tangent plane projection is adopted and the projection center is set at the center of the research area. A strategy of fitting the crater with three evenly distributed points on their rims is always used. When applicable, the mapped crater rims are further optimized with another ArcGIS add - in tool developed by Liu et al. (2018) with the assistant of SLDEM2015 and NAC DEM for both the larger and smaller classes of craters, respectively. The rim - to - floor depths of these craters are also automatically calculated with the Arc - GIS add - in tool by Liu et al. (2018).

Much smaller craters, in sub - meter or meter scale in diameter, are mapped in the PanCam images from CE - 4 rover. A program is developed for crater mapping and measurements in the PanCam stereo images. Fig. 2 shows a pair of stereo images from the PanCam. Two rim end - points of one diameter line of the crater are manually identified in the left image, and the corresponding points in the right image are automatically matched. Consequently, the 3 - D ground coordinates of these endpoints are calculated by triangulation with the assistance of the image interior and exterior orientation parameters. As a result, the crater diameter is derived as the distance between the two rim points.

(a) (b)

Fig. 2 Crater measurement from a pair of PanCam images. The crater rim points at the ends of one diameter line are manually identified in the left image (a), and the corresponding homologous points are automatically obtained in the right image (b). The diameter of the crater is finally derived as the distance of the two end points

The rim - to - floor depth of a crater from PanCam stereo images is calculated as the distance between crater floor and the fitted rim plane. To be specific, a series points distributed evenly on the crater rim (red crosses in Fig. 3) and a profile crossing the crater floor (green crosses in Fig. 3) are identified, respectively. Their ground coordinates are calculated by photogrammetric method. A plane is fitted with the rim points and the distance from each green point on the profile to the fitted plane is calculated. A quadratic polynomial is adopted to fit the resultant distances, and the final crater depth is searched as the distance at the fitted vertex. By this way, the influence of topographic undulation on depth measurement can be alleviated.

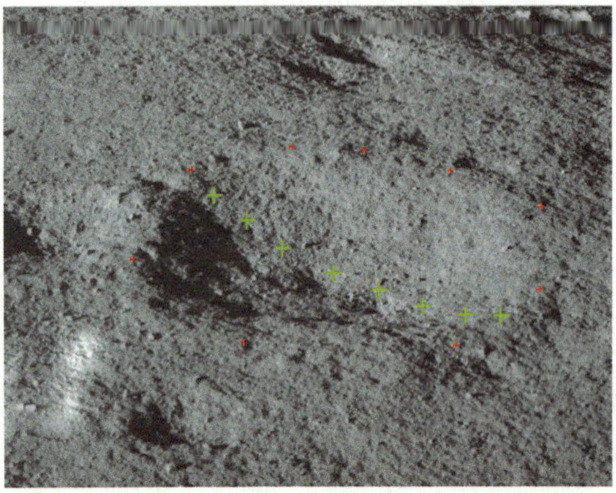

Fig. 3 Points used to determine the crater depth from PanCam stereo images. The crater rim plane is fitted by points labeled with red small crosses, and some green crosses passing through the crater center as much as possible are also identified. The crater depth is defined as the distance from the vertex of the fitted profile to the fitted crater rim plain. See text for details (For interpretation of the references to colour in this figure legend, the reader is referred to the web version of this article)

To further reveal the crater population, the randomness analysis function in the CraterStats2 tool (Michael et al., 2012) is used, which can quantitatively describe the crater clustering condition at different size ranges relative to a series of randomly distributed populations (Michael et al., 2012). In the process, craters are separated into several groups according to diameters in a $\sqrt{2}$ - spaced interval. A Monte Carlo simulation is then used to give the random distribution of samples in each group and meanwhile the crater size and the counting area are all considered. The measured distance between craters is taken as the parameter to evaluate the randomness by comparing the real measurement and the simulated result. If the former is far below the expected value, it proves that the population is non - random and more clustered, while if the distance measurement lies within the bulk of the histogram, it indicates that the test does not reject randomness and craters are more likely in random distribution. In this study, the selected parameter is "mean 2nd - closest neighbor distance (M2CND)" and the iteration times of the Monte Carlo simulation is set to be 10000.

3 Catalogue of craters in CE - 4 landing area

Fig. 1b shows the mapped craters in the landing area with DOM mosaic from LROC

NAC and the corresponding incremental SFD is shown in Fig. 4 with blue stars. The mapping result shows that the diameter（*D*）corresponding to the peak number is ~100m, and the completeness diameter is determined as 127m with the method proposed by Robbins et al.（2018）. There are totally 4333 craters mapped in the landing area, within which 3760 and 2298 craters are larger than 100m（yellow circle in Fig. 1b）and 127m（red circle in Fig. 1b）, respectively. Besides, 97 and 13 craters have diameters larger than 500m and 1km, respectively.

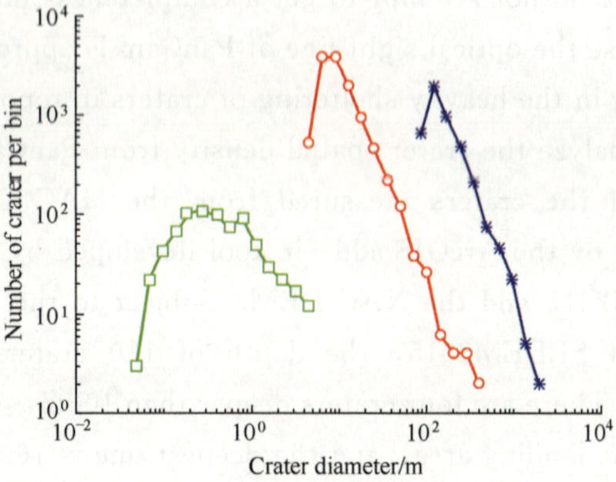

Fig. 4　The incremental size-frequency distribution of craters shown in a log　log plot. The diameter interval is $\sqrt{2}D$m and the crater diameter refers to the middle value of each bin. The blue "*"s represent the results of craters mapped from the 2.4°×1° landing area. Craters from the 3.2km ×3.0km area surrounding the landing site are indicated by red circles. The green squares show craters from PanCam images（For interpretation of the references to colour in this figure legend, the reader is referred to the web version of this article）

In the 3.2km ×3.0km area surrounding the landing site, we tried to digitalize all measureable craters and finally 11,787 craters from several meters to more than hundred meters in diameter are mapped. The completeness diameter is found to be 9m and all craters are depicted in Fig. 1c with the red circles indicating the 5108 craters larger than 9m and yellow for the residual craters. There are 27 craters larger than 100m with the largest being 393m, while the smallest crater measured in the area is ~3.6m subjected to limitation of the image resolution. The incremental SFD of craters in this scale is shown in Fig. 4 with red circles.

For mapping craters from PanCam images, only those craters within the distance to the rover less than 60m are mapped in this study. This is because the measurement precision with PanCam images will decrease greatly in the far distance. Totally, 738 craters surrounding the rover are mapped and measured from the PanCam stereo images（Fig. 5）, and most of the mapped craters（95%）are within the distance of 30m to the rover. There are several large craters with the diameters ~25m surrounding the lander, and because they can also be observed from the NAC mosaic, they are not included in the

PanCam image crater analysis.

The largest crater measured from PanCam images has a diameter of 4.66m (labeled by No. 1 in Fig. 5b) and it is ~27m far from the rover. Totally, there are 21 mapped craters larger than 3m, ~85% of the craters are smaller than 1 m, and ~7% of them are in centimeter level. The mean diameter of the whole catalogue is ~0.62m. The size - frequency distribution of the mapped craters is shown in Fig. 4 with green square. It seems that the decrease of frequency as diameter in - creases for craters larger than 0.8m tends to be stable. However, we do not attempt to get a completeness diameter for craters from PanCam images, because the optical sight line of PanCam is approximately parallel to the lunar surface, resulting in the heavily sheltering of craters in topographic undulated area. Similarly, we do not analyze the crater spatial density from PanCam images either.

The depth (d) of the craters measured from the NAC DOM mosaic is further automatically calculated by the ArcGIS add - in tool developed by Liu et al. (2018) with the assistant of SLDEM2015 and the NAC DEM. Subject to the ground sample distance and height precision of SLDEM2015, the depth of 110 craters larger than 480m in diameter are measured. There are four craters deeper than 100m, which are all distributed in the eastern part of the landing area, and the deepest one is 169m with the diameter of 956m. In addition, approximately 33% of crater depths are less than 25m. The relationship between the diameter and the depth of these 110 craters are displayed in Fig. 6a with red triangles.

The height precision of the released LROC NAC DEM adopted in this research is ~1.1m (Robinson, 2019; http://lroc.sese.asu.edu/). Thus, the depths of 28 craters deeper than 5m and with the size range of ~60m - 393m are obtained from the NAC DEM and the results also demonstrated in Fig. 6 (blue diamonds). The maximum depth is 42m corresponding to the crater with the diameter of ~393m. Twenty - three craters, or 82% of the total measured craters, have the depth within 5m - 14m, and their diameters ranging from 60m to 320m.

Subjected to the variation of the resolution and illumination conditions of PanCam images, the depths of only 20 craters (labeled with green in Fig. 5) are measured and ~80% of them are less than 10m to the rover. The depth of the 20 craters is shown in Fig. 6a with green crosses, and all of them are shallower than 0.6m. Among the 20 craters, the diameters of the largest two are ~4.3m, both of which are more than 30m far from the rover, and they also have the largest depths (0.36m and 0.26m).

The relationship between diameter and depth of all depth - attainable craters in this research is fitted with a power law as done in previous studies (e.g., Wood and Anderson, 1978; Daubar et al., 2014; Sun et al., 2018) and the result is $d = 0.0726D^{0.9493}$. The exponent is close to 1.0, indicating the depth and the diameter of the craters are almost linear. However, a linear fitting of the diameter and depth get unsatisfactory results in this study ($R^2 = 0.37$), which greatly diverge from smaller craters.

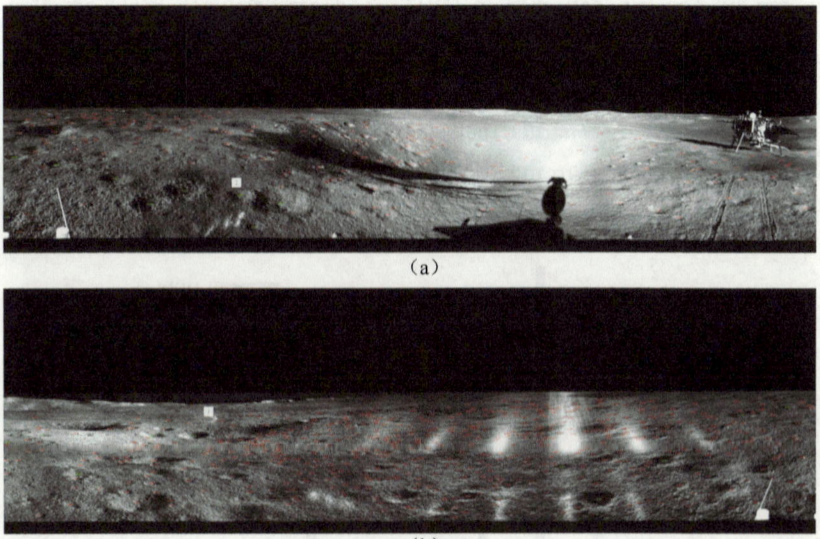

(a)

(b)

Fig. 5　The mapped craters from the PanCam images with diameters annotated. The sub－figure (a) and (b) together make up the full 360°PanCam mosaic

Fig. 6b shows the variation of the d/D against the diameter of the mapped craters in CE－4 landing area. The crater d/D in the research area has a wide distribution especially for large craters measured from the SLDEM2015, ranging from 0.02 to 0.18 with the mean value of 0.06. For craters with the depth measured from SLDEM2015, the top several craters in Fig. 6b with d/D less than 0.04 have relatively shallower depth and are possibly secondary craters or eroded craters. There are 18 craters with d/D larger than 0.1 and the largest one corresponds to the crater of 956m in diameter, most of which are fresh primary craters by visual interpretation. From Fig. 7, the bin with d/D centered in 0.04 contains the largest crater numbers for craters with depth measured from SLDEM2015,

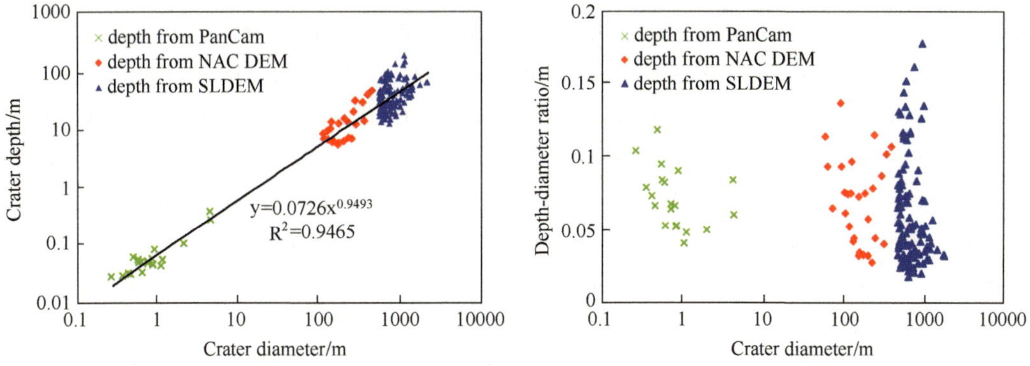

(a)　　(b)

Fig. 6　The relationship between the diameter and (a) depth, (b) depth/diameter ratio for craters in the landing area. The samples with different colors and shapes indicate craters measured from PanCam, NAC DEM and SLDEM2015, respectively closely followed by the bin centered in 0.02, and craters with d/D smaller than 0.06 accounting for 61% of the group.

The d/D of craters measured from NAC DEM is all less than ~0.14, with the mean value of ~0.07. Six craters have d/D larger than 0.1, spreading in all size ranges (60m to 393m). Compared with craters measured from SLDEM, the crater number distribution of d/D for this group is more uniform especially between 0.01 and 0.09 (Fig. 7).

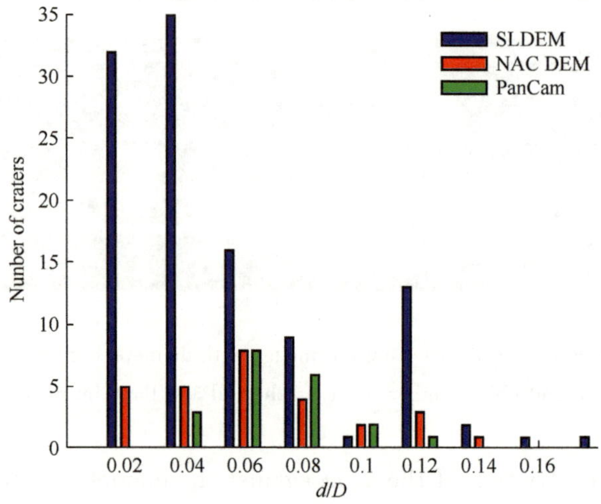

Fig. 7 Histogram of d/D of craters with depth measured from SLDEM2015 (blue), NAC DEM (red) and PanCam image (green), the bin interval is set to be 0.02 (For interpretation of the references to colour in this figure legend, the reader is referred to the web version of this article)

As demonstrated in Fig. 6b, the d/D of craters measured from the PanCam images ranges from 0.04 to 0.12 with the mean value of 0.07, only two of which are larger than 0.1. The largest d/D corresponds to a crater diameter of 0.5m, and this crater is on the east of the rover with a distance of about 4.1m (labeled by No. 2 in Fig. 5a). The d/D for this group of craters mainly concentrates between 0.05 and 0.09, up to 70% of whole group (Fig. 7). The detailed comparison of the d/D can be found in Section 4.4.

The spatial density of craters larger than 127m in the landing area is analyzed in a 500m ×500m grid with a search radius of 1km (Fig. 8a). The high-density areas are mainly between 178.0°E and 178.5°E and several other high-density clusters scattered elsewhere. By comparing with Fig. 1b, all these high-density areas correspond to secondary chains or clusters, indicating that secondary craters are very widespread in the landing area. After removing possible secondary craters (Fig. 8b), the spatial density of craters tends to be more evenly distributed, although the eastern part still has higher spatial density than that in the west.

The spatial density for smaller craters surrounding the landing site is calculated in the 50m ×50m grid with the search radius of 100m and the results for craters larger than 9m

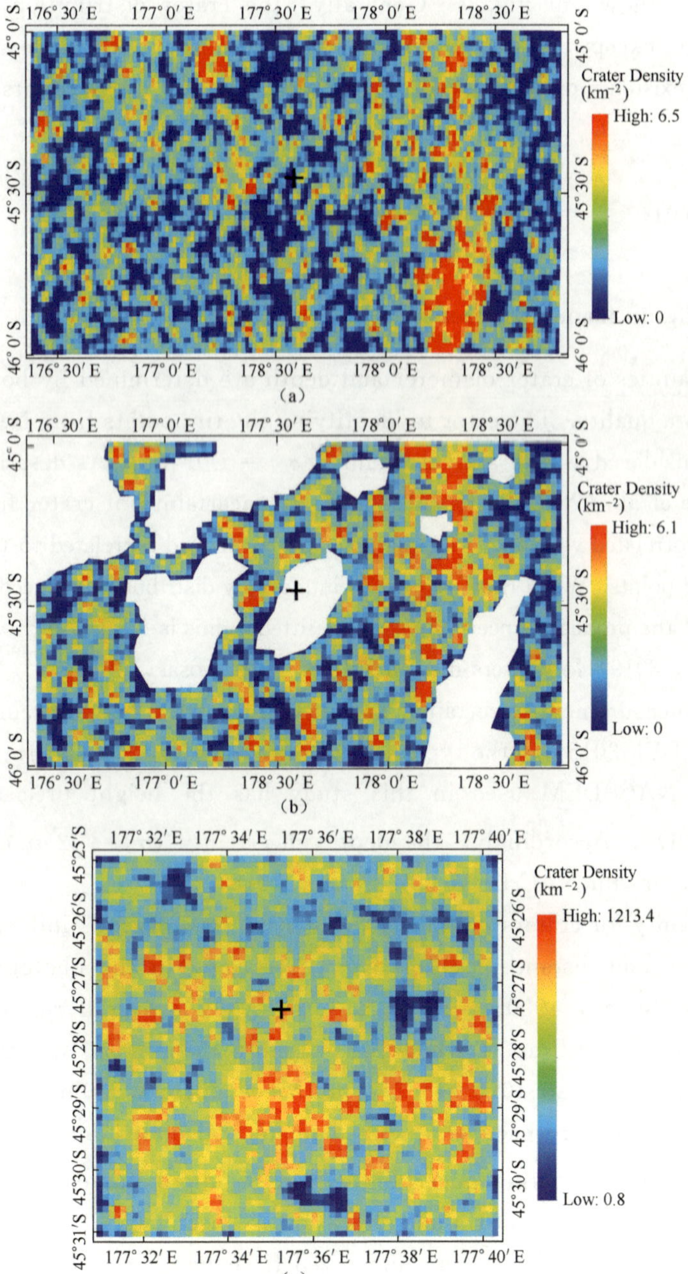

Fig. 8 (a) The spatial density map of all craters larger than 127m. (b) The spatial density map of craters larger than 127m excluding possibly secondaries in the landing area. The white areas in (b) indicate the removed secondary chains or clusters. (c) The spatial density map of craters larger than 9m in diameter in a local 3. 2km × 3. 0km area surrounding the landing site. The CE-4 landing site is marked by black crosses. The search radii (1km for (a), (b) and 100m for (c)) are set to make the density maps more smoothly and more visually friendly. Consequently, some craters have been counted more than one time and it is not recommended to deduce the crater number directly from the density maps in diameter are shown in Fig. 8c. Generally, the crater distribution in this scale is relatively uniform except that craters on the northern border are a little sparse. It can be attributed to the existence of several very large craters and this also accounts for other low - density areas.

4 Discussion

4.1 Uncertainties in the measurements

The uncertainties of crater diameter and depth are determined by both measurement precision and data quality. The error in identifying the rim points from NAC DOM mosaic (σ_N) can be considered as half a pixel, namely $\sigma_N = \pm 0.45$m. As described in Jia et al. (2019) and Yue et al. (2019), the measurement uncertainty of crater fitting with three rim points is uncorrelated with the circle diameter but is instead correlated to the distribution of the three selected points. When the three rim points evenly distributed and according to the error propagation law, the produced uncertainty for the fitted radius is $1.16\sigma_N = \pm 0.52$m (Jia et al., 2019; Yue et al., 2019) for craters measured from NAC mosaic.

The depth measurement uncertainty depends on the DEM height accuracy (Liu et al., 2018). The SLDEM2015 has the typical height accuracy (σ_h) of 3－4m (Barker et al., 2016) and the NAC DEM used in this study has the height precision of ～1.1m (Robinson, 2019). Accordingly, the depth uncertainty ($\sigma_d = \sqrt{2}\,\sigma_h$) is ± 5.7m for SLDEM and ± 1.6m for NAC DEM (Liu et al., 2018).

The uncertainty for craters digitalized from PanCam images is influenced by the line of sight direction and distance to the rover. According to the photogrammetric error propagation law (Wang, 1990; Peng et al., 2014), the overall rim point calculation uncertainties (σ_p) are estimated which are no more than ± 0.04m, ± 0.36m and ± 0.99m for craters less than 10m (48.2% of the catalogue), 30m (95.1%) and 50m (99.5%) to the rover, respectively. For diameter calculated by the distance of two ground points, the propagating uncertainties can be approximately $\sqrt{2}\,\sigma_p$ (Liu et al., 2018), namely ± 0.06m, ± 0.51m and ± 1.40m, respectively.

The depth uncertainty for craters digitalized from PanCam images is similar to that measured from SLDEM or NAC DEM. The topographic influence on depth measurement has been eliminated similar as that in Liu et al. (2018). Due to the closer ranges of the craters for depth measurement, the uncertainties also are smaller and for the smallest crater with a depth of 0.028 and distance of 3.4m, the depth uncertainty is ± 0.006m.

4.2 Crater randomness analysis of the multi - scale crater catalogues

The randomness analysis function in CraterStats2 tools is utilized to analyze the crater distribution characteristics. Craters larger than 100m in the landing area are divided into 9 bins by diameters and the clustering analysis result is displayed in Fig. 9. It can be observed that craters in most ranges tend to be clustered.

For the first and second diameter ranges, i. e. , 88m – 120m (Fig. 9a) and 120m – 180m (Fig. 9b), the M2CND test rejects randomness with 100% confidence, indicating that craters within these two scales are clustered. In both scales, the denser and sparser areas are substantially consistent with these in the spatial density map (Fig. 8) . The incomplete statistic of craters between 88m – 120m may also be partially accounted for the cluster result. Craters within the diameter range of 120m – 180m with the amount of 1238 are denser in the eastern part of the landing area, corresponding to the high – density area between 178. 0°E and 178. 5°E in Fig. 8a, and are likely secondary crater chains. In the next crater bin with the diameter range of 180m – 250m (Fig. 9c), the M2CND is 1. 17km and it is towards the clustered end of the simulated histogram with a low confidence level of 1. 78% to accept the randomness distribution assumption. The crater concentration in the low – right corner of the region mainly accounts for the non – randomness result and it is also applicable for the following two bins. In the crater bin of 500m – 710m (Fig. 9f), it has a relatively high chance (28. 9%) of not rejecting randomness even though the craters in the low – right seem still denser than other areas. The larger crater bins, e. g. , the bins from 710m – 1. 4km (Fig. 9g h), also tend to reject the randomness assumption. The distribution estimation method for the last bin (Fig. 9i) is mean closest neighbour distance since only two craters are involved. The above analysis also demonstrates that the crater chain in the southeast corner contains secondaries in very large size ranges (~180m – 700m), together with their diversely degradation condition from NAC DOM mosaic, likely indicating that they are not formed at the same time period.

The cluster condition of small – sized craters in the local area surrounding the landing site is also investigated in the same way (Fig. 10) and there are 13 crater groups involved. The first two bins are below completeness diameter and they are ignored. Within the diameter range of 7. 8m – 11m (Fig. 10c), the crater population covers almost the whole area and it tends to be randomly distributed by visual interpretation except some areas influenced by the large craters. The M2CND is at the cluster end of the simulated histogram and the low confidence level (1. 98%) of accepting the randomness is likely caused by the areas including large craters and several small – scale crater clusters. The next crater bin (11m – 16m, Fig. 10d) gives a M2CND of 59. 4m, which is larger than the simulated median distance and indicates that craters in this diameter range appear consistent with a random population. Holes still exist due to the presence of large craters. The following two groups (16m – 31m, Fig. 10e – f) have analogical crater distributions and confidence levels, demonstrating that the populations are consistent with the randomness simulation. For craters within the diameter interval of 31m – 44m (Fig. 10g), the degree of confidence is relatively low compared with the adjacent bins but still within the simulated histogram. The following four M2CND values (Fig. 10h – k) are nearly

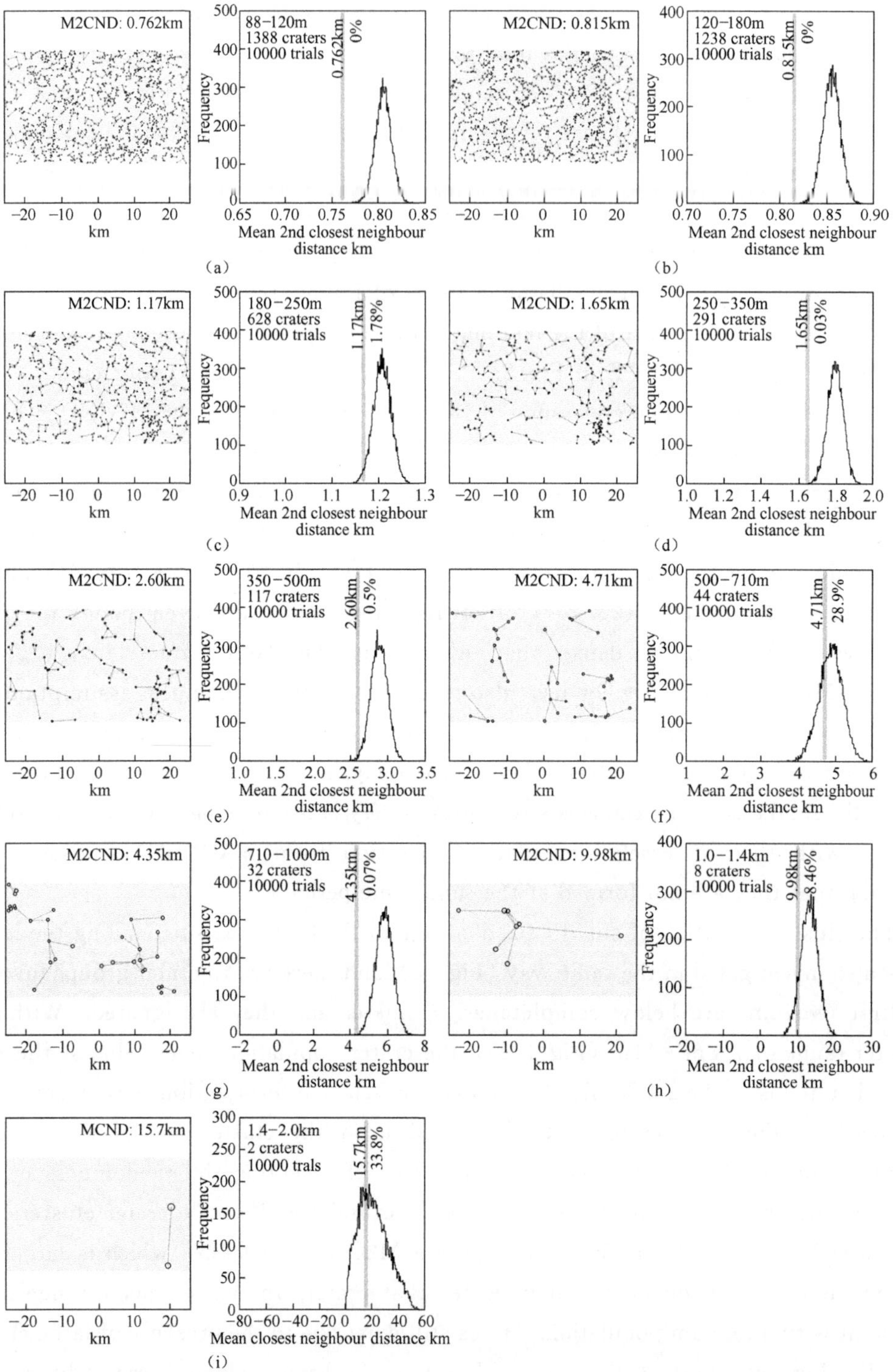

Fig. 9　Randomness analysis of craters larger than ~100m in diameter (divided into nine groups by diameter) in the landing area. The left plot of each group shows the crater distribution in this diameter range and the right plot compares the M2CND of the measurement with that of the simulated results. The lines connecting craters in the left plots of each group show the 2nd closest neighbour for each crater (Michael et al., 2012)

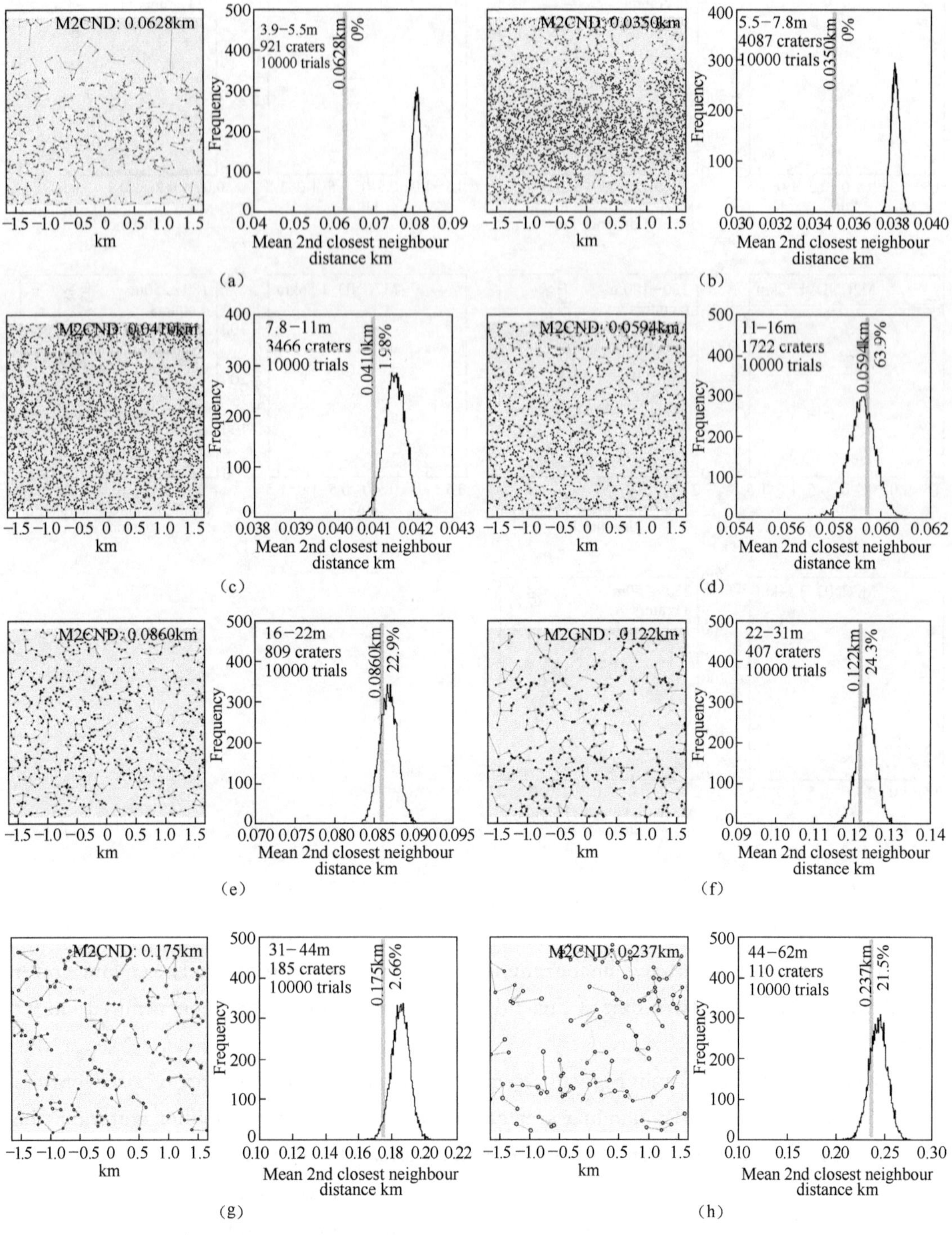

Fig. 10　Randomness analysis of small-sized craters within the 3.2km × 3.0km area around the CE-4 landing site. Craters are separated into 13 groups and the configuration of each group is same as that of Fig. 9

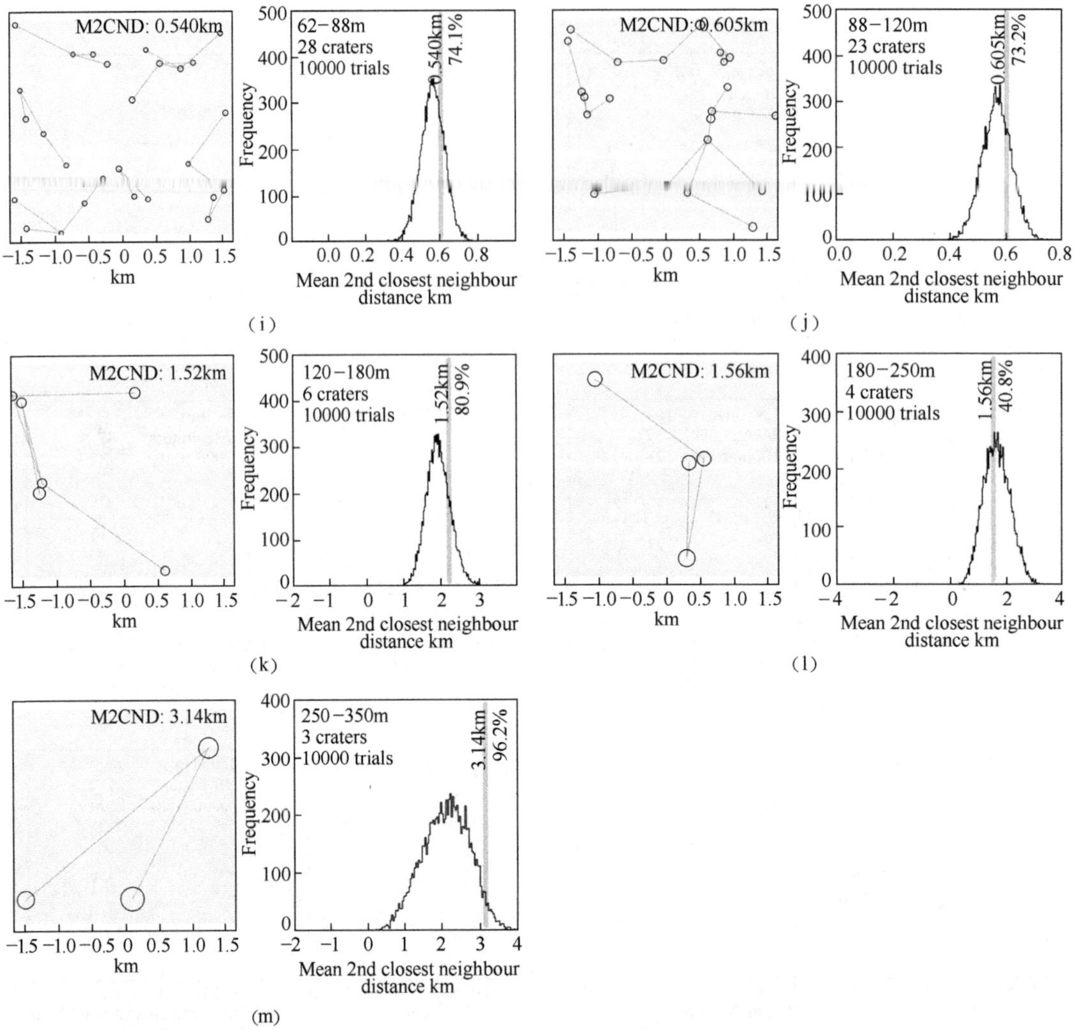

Fig. 10 (continued).

centered within the simulated histograms, suggesting the approximately random crater distribution. The last two bins (Fig. 10l − m) are ignored for their limited crater numbers.

By comparing Fig. 9 and Fig. 10, it can be deduced that the large − sized craters (>100m in diameter) in the landing area are more clus − tered than smaller craters in the local 3.2km×3.0km area surrounding the landing site. One of the dominating reasons is that numerous secondary craters larger than 100m are distributed in the large − scale landing area. There has always been controversial about whether and what scales of small craters on lunar surface is dominant by secondaries. The uniform distribution of small craters in the area surrounding the CE − 4 landing site seems to indicate that most of these small craters are primaries and as the continuous observation of CE − 4 rover, more evi −

dence may be discovered to verify such deduction.

For the PanCam images are obtained with the optical sight line approximately parallel to the lunar surface, craters could not be measured completely by the occlusion in undulated topography. So craters derived from PanCam images are not involved in randomness analysis and size - frequency distribution analysis.

4.3 Crater size - frequency distribution in the landing area

Crater size - frequency distribution analysis is a powerful tool for studies of planetary surface age (Neukum et al., 1975; Neukum, 1983), bombardment rate (Ivanov, 2006; Ivanov, 2008), surface modification processes (Michael and Neukum, 2010), etc. Thus, even indicated in Section 4.2 that craters larger than 100m tend to be clustered in many diameter ranges, we attempted to remove obvious secondary craters by visual interpretation (e.g., Shoemaker et al., 1962; Oberbeck and Morrison, 1973) and to investigate the SFDs of other craters measured from NAC images in two - scales. The removed secondary - crater - related area, which takes ~1/3 of the whole landing area, can be inferred from Fig. 8b.

The CraterStats2 tool (Michael and Neukum, 2010; http://www.geo.fu — berlin.de/) is used to analyze the SFD in differential (Fig. 11a) and relative (Fig. 11b) modes, respectively. The relative SFD (short as R - plot hereafter) is advantageous in revealing subtle changes (Crater Analysis Techniques Working Group, 1979; Robbins et al., 2018) and the differential plot (D - plot) is useful for identifying possible partial resurfacing events (Michael and Neukum, 2010). The production function (PF) and chronology function (CF) adopted are from Neukum (1983).

For craters obtained from the 3.2km × 3.0km area around the landing site (black squares in Fig. 11), it can be discerned from the D - plot that craters smaller than ~90m deviate from the aging curves. It has been demonstrated that the craters smaller than the equilibrium diameters have lower density than their normal production crater pop - ulations (Gault, 1970; Richardson, 2009; Xiao and Werner, 2015). The equilibrium diameter is a demarcation, smaller than which the formation of new craters is balanced by obliteration of old craters due to degradation. Thus, this deviation of crater distribution slope can be attributed to the saturation equilibrium.

The equilibrium phenomenon can also be judged from the R - plot. Several empirical equilibrium levels are included in Fig. 11b with the orange solid line from Hartmann (1984). The other three light - blue dashed horizontal lines are 1%, 5% and 10% of the geometric saturation (N_{gs}), respectively, which was pointed out by Gault (1970) that crater population reaches equilibrium within 1—10% of the geometric saturation level. The cumulative SFD can be expressed as $N = aD^b$, where b indicates the log - log slope of cumulative SFD and the power law index of R - plot is less than the cumulative slope by 1 (Crater Analysis Techniques Working Group, 1979). Another extensively used criterion for crater population equilibrium judgment is the cumulative log - log slope of ~2

(Gault, 1970; Hartmann, 1984; Xiao and Werner, 2015). It can be inferred that all craters in the 3.2km ×3.0km area seem to have a cumulative slope of ~2 except some fluctuation between ~90m and 260m. The method of Hartmann (1984) gives similar estimation result. Consequently, it can be concluded that craters smaller than 90m in the 3.2km ×3.0km area have reached saturation equilibrium.

For craters larger than 100m (red stars in Fig. 11), a slope of ~2 can be interpreted for craters smaller than ~250m from R - plot, and a slight crater distribution density decline can also be observed from Fig. 11a, indicating that craters smaller than 250m seems in saturation equilibrium. The law of Hartmann (1984) fails to give equilibrium judgment due to craters in this area having holistic low crater distribution density.

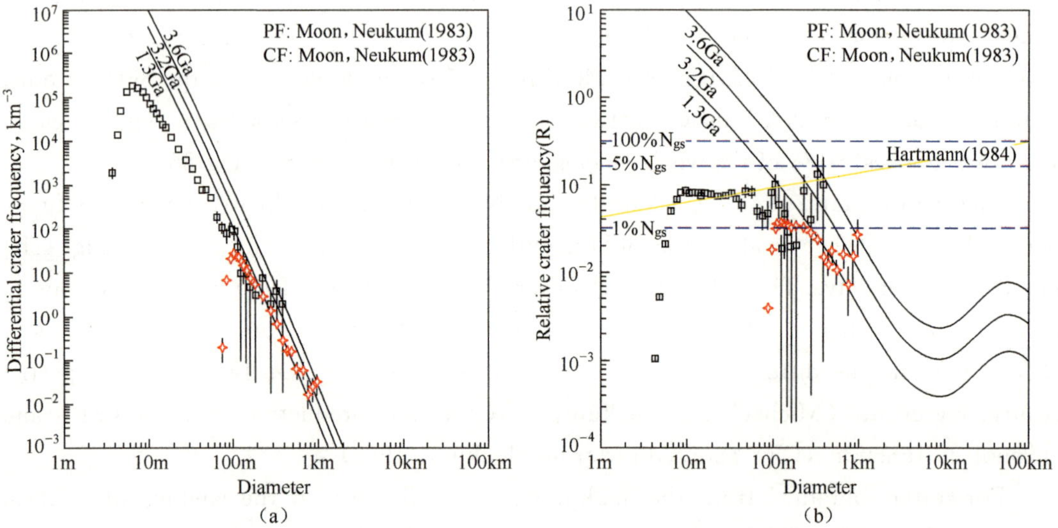

Fig. 11 The size - frequency distributions of craters in differential (a) and relative (b) modes in the CE - 4 landing area. Four equilibrium lines (10%, 5%, 1% of the geometric saturation (N_{gs}) and criterion from Hartmann (1984)) are included in the R - plot. Three isochrons (1.3Ga, 3.2Ga, 3.6Ga) are also displayed in two plots. The craters within the 3.2km ×3.0km area around the landing site are represented by black squares and larger craters from the whole landing area are shown with red stars (For interpretation of the references to colour in this figure legend, the reader is referred to the web version of this article)

Even though the apparent secondary craters and related areas have been removed, it is still hard to determine the model age from the differential SFD distribution of residual craters. One of the reasons is likely to be the remaining of inconspicuous secondary craters, as described in Qiao et al. (2019) that secondaries are pervasive in the whole landing area. The inconformity formation age of craters can also cause the scattered SFD, which means that a part of craters within the landing area may be formed superposed on the ejecta of other craters, mainly from Finsen crater (Huang et al., 2018; Qiao et al., 2019). Three isochrons are displayed in Fig. 11 (1.3Ga, 3.2Ga, 3.6Ga) and some craters bins between ~250m and 800m tend to be closed to the ~1.3Ga isochron. Age of

Finsen crater has been considered as Copernican (Qiao et al., 2019), Eratosthenian, (Wilhelms et al., 1979; Pasckert et al., 2018), or late Imbrium (Ivanov et al., 2018). Thus, the interpretation of craters lying near the ~1.3Ga isochron partially relies on further careful determination of the age of Finsen crater or other nearby craters whose ejecta covers the CE-4 landing area.

4.4 Depth/diameter ratios of the multi-scale crater catalogues

Continuous observation of lunar surface with high-resolution orbital images and the implementation of lunar landing missions provides the opportunity for the investigation of d/D of very small craters, e.g. in meter scale or even smaller. In this study, the characteristics of d/D in multi-scales, from sub-meter level to kilometer level, are compared and discussed.

From Fig. 6, the distribution of d/D against diameter for all three-scale craters is very scattered especially for those craters with the depth larger than 480m. Besides, compared with the other two groups, even though more than a half of craters have d/D lower than 0.05, the group with depth obtained from SLDEM2015 contains ten largest d/D (>0.12) in the whole research area, indicating very scattering d/D distribution and the variety of degradation states of the craters. The d/D for craters measured from NAC DEM and PanCam images has similar distribution range, except those from NAC DEM having a few smaller d/D. One of the reasons for the low d/D values of craters measured from PanCam images is that only those craters close to CE-4 rover or shallow craters in a distance can be measured.

The largest three d/D measured from SLDEM (D>480m) are 0.18, 0.16 and 0.15, respectively, which have relatively clear rim border but their ejecta blanket is almost indistinguishable. According to historical degradation state grouping criteria, they can be catalogued as B-class craters (Basilevsky et al., 2014; Stopar et al., 2017), and these d/D are roughly consistent with that of Stopar et al. (2017), e.g. 0.192 ±0.005 for B-class craters with diameter between 400m-10km. It is short of fresher large craters, namely A or AB-classes craters, in CE-4 landing area.

For smaller craters measured from NAC DEM (D<400m), the d/D is obviously smaller than that of larger craters from SLDEM from Fig. 6. The observation in this study is in accord with other exsiting researches focused on d/D of small craters (e.g., Basilevsky et al., 2014; Daubar et al., 2014; Mahanti et al., 2014; Stopar et al., 2017; Sun et al., 2018), that vast majority smaller craters possess d/D lower than that of larger craters (Pike, 1977; Stopar et al., 2017). Besides, there are very limited fresh craters within the 3.2km ×3.0km area with only one crater possessing bright ejecta and reliable depth measurement, which can be found from Fig. 1c marked by white arrow and has the d/D of 0.11 (C1 in Tab. 1). To avoid this low d/D and lack of fresh crater is caused by the limited study area, several small fresh craters with obvious bright ejecta blanket and clear rim border were found from NAC image m1303640934 within the landing area, whose parameters are listed in Table 1. Their d/D values are varied from 0.11 to 0.18 with the mean of 0.14 ± 0.02. The maximum d/D (0.18) is similar to that of larger craters measured from SLDEM, while the larger craters have larger d/D than the craters measured here as a whole.

Tab. 1　Information about some relatively fresh craters surrounding the CE－4 landing site. The C1 crater can be seen from Fig. 1c pointed by a white arrow

Crater ID	Longitude/(°)	Latitude/(°)	Diameter/m	Depth/m	d/D
C1	177.626	−45.420	59.7	6.3	0.11
C2	177.584	−45.866	113.19	20.33	0.18
C3	177.347	−45.868	147.16	22.44	0.15
C4	177.670	−45.526	118.66	17.70	0.15
C5	177.735	−45.499	116.17	17.24	0.15
C6	177.486	−45.777	107.26	15.37	0.14
C7	177.517	−45.619	92.33	12.51	0.14
C8	177.625	−45.744	170.89	21.92	0.13
C9	177.572	−45.903	126.86	16.15	0.13
C10	177.576	−45.577	103.96	12.88	0.12
C11	177.619	−45.569	63.60	7.13	0.11

The reasons for lower d/D for smaller craters can mainly be attributed to that smaller craters essentially have lower d/D values or smaller craters tend to degrade more easily (Fassett and Thomson, 2014), given that they are formed contemporarily with the larger ones. Previous studies (e.g., Melosh, 1989; Stopar et al., 2017) pointed out that the formation of smaller craters are strength－dominated versus gravity－dominated for larger craters. In Sun et al. (2018), it was thought that the difference of strength properties of the surface and shallow subsurface causes the d/D inconsistency of smaller and larger crater. Though craters in all possible degradation levels and even more smaller craters (sub－meter in diameter) are included in this research, the conclusion resembles that craters larger than 480m on the whole contains larger d/D than the other two group craters. From this point of view, it can be deduced that the d/D of craters seems to increase from sub－meter craters until to those craters on the order of 1km in diameter since the variation of target property. On the other hand, with similar diameter and apparent morphology, small fresh craters listing in Table 1 have very scattered d/D and it is likely to indicate that smaller craters tend to be more likely to degrade or to be influenced by the target properties, even if their bright ejecta keeps not to be eroded.

Compared with existing results (e.g., Basilevsky et al., 2014; Mahanti et al., 2015; Stopar et al., 2017; Sun et al., 2018), the most distinct feature of d/D in CE－4 landing area is the scarcity of craters with large d/D (>0.18), no matter for crater measures from SLDEM or NAC DEM. This can be partially ascribed to old geologic age of the area. In addition, the subsequent post - Von Kármán impacting events, e. g. , Alder and Finsen cratering, may also enhance the degradation of craters, causing the scarcity of large fresh craters in the research area.

5 Conclusion

This research mapped three crater catalogues with different scales in the CE - 4 landing area, and the spatial distribution characteristics and morphological properties are analyzed. The analysis focuses on three scales: craters larger than 100m covering the whole landing area (176. 4°E - 178. 8°E, 45°S—46°S), all craters in local scale (3. 2km × 3. 0km) surrounding the landing site, and craters surrounding the rover measured from one station of the Yutu - 2 PanCam images. Completeness analysis demonstrates that the first crater catalogue completed in ~127m and the second one has a complete diameter of ~9m.

The spatial density investigation of the catalogue with crater larger than 127m indicates that distinct secondary craters cover at least 1/3 of the landing area and mainly influence the eastern part of the landing area except several scattered secondary clusters in the west. The spatial density for small - scaled craters within the 3. 2km ×3. 0km area is relatively uniform. The randomness analysis reveals more elaborate crater spatial distribution and results show that craters larger than 100m are almost in non - random distribution in all the size ranges.

The size - frequency distribution reveals that craters smaller than 90m within 3. 2km × 3. 0km area and smaller than 250m of the whole landing area tend to reach saturation equilibrium. The comparison of d/D of the three - scale crater catalogues indicates that larger craters have relatively larger d/D. The surface and subsurface material difference may partially account for this phenomenon. One - hundred - meter - scale fresh craters with bright ejecta and clear rim border have varied d/D values, indicating that small craters seem to be more easily to degrade or or to be influenced by the target properties, even their bright ejecta keeps uneroded. The scarcity of craters with large d/D (>0. 18) in the research areas can be partially ascribed to old geologic age of the area as well as the subsequent post - Von Kármán impacting events. Further effort, like physical impact experiment or numerical simulation can devote to uncover the mechanism of the observation.

Acknowledgements

The authors gratefully acknowledge all those who worked on the Planetary Data System and Data Publishing and Information Service System (DPISS) of China's Lunar Exploration Project archives to make the LROC NAC imagery, NAC DEM, SLDEM2015 and Chang'e－2 DOM mosaic publicly available. The authors also appreciate the Editor, Brandon Johnson, the reviewer, Mikhail Kreslavsky and two anonymous reviewers for their constructive criticism and thoughtful reviews. Finally, thanks Thomas Kneissl for the software CraterTools. This research is supported by the B－type Strategic Priority Program of the Chinese Academy of Sciences (grant No. XDB41000000), the National Natural Science Foundation of China (grant Nos. 41941003, 41590851, 41671458).

References

[1] Neukum, G., 1983. Meteoritenbombardement und Datierung planetarer Oberflächen, Habilitationsschrift (habilitation thesis), Univ. München, Munich, Germany. Meteorite bombardment and dating of planetary surfaces (English translation, 1984). Habilitation Thesis for Faculty Membership. University of Munich, p. 186.

[2] Anderson, J. A., Sides, S. C., Soltesz, D. L., et al., 2014. Modernization of the integrated software for imagers and spectrometers. In: Lunar and planetary science conference. Abstract #2039.

[3] Barker, M. K., Mazarico, E., Neumann, G. A., et al., 2016. A new lunar digital elevation model from the lunar orbiter laser altimeter and SELENE terrain camera. Icarus 273, 346—355.

[4] Basilevsky, A. T., Kreslavsky, M. A., Karachevtseva, I. P., et al., 2014. Morphometry of small impact craters in the Lunokhod－1 and Lunokhod－2 study areas. Planet. Space Sci. 92, 77—87.

[5] Bland, M. T., Raymond, C. A., Schenk, P. M., et al., 2016. Composition and structure of the shallow subsurface of Ceres revealed by crater morphology. Nat. Geosci. 9 (7), 538.

[6] CLEP, 2018. The data release of CE－2 Global digital orthophoto model (7m) [2019—4—15]. http://moon. bao. ac. cn/pubMsg/detail－dom7EN. jsp.

[7] Crater Analysis Techniques Working Group, 1979. Standard techniques for presentation and analysis of crater size－frequency data. Icarus 37, 467—474. https://doi. org/10. 1016/0019—1035(79)90009—5.

[8] Daubar, I. J., Atwood－Stone, C., Byrne, S., McEwen, A. S., Russell, P. S., 2014. The morphology of small fresh craters on Mars and the Moon. J. Geophys. Res. 119 (12), 2620—2639. https://doi. org/10. 1002/2014JE0 04671.

[9] Di, K., Liu, Z., Liu, B., et al., 2019a. Chang'e－4 lander localization based on multi－source data. J. Remote Sensing 23 (1), 177—184.

[10] Di, K., Zhu, M., Yue, Z., et al., 2019b. Topographic evolution of Von Kármán crater revealed by the lunar rover Yutu－2. Geophys. Res. Lett. 46, 12764—12770.

[11] Fassett, C. I., Thomson, B. J., 2014. Crater degradation on the lunar maria: topographic diffusion and the rate of erosion on the Moon. J. Geophys. Res. 119 (10), 2255—2271.

[12] Gao, Y., Ge, Y., Ma, L., et al., 2019. Optimization design of configuration and layout for Queqiao relay satellite. Adv. Astronaut. Sci. Technol. 1—6.

[13] Gault, D. E., 1970. Saturation and equilibrium conditions for impact cratering on the lunar surface: criteria and implications. Radio Sci. 5, 273—291.

[14] Gou, S., Di, K., Yue, Z., et al., 2019. Lunar deep materials observed by Chang'e-4 rover. Earth Planet. Sci. Lett. 528 (115829), 1—9.

[15] Gou, S., Yue, Z., Di, K., et al., 2020. In situ spectral measurements of space weathering by Chang'e-4 rover. Earth Planet. Sci. Lett. 535, 116—117.

[16] Hartmann, W. K., 1984. Does crater "saturation equilibrium" occur in the solar system? Icarus 60 (1), 56—74.

[17] Head, J. W., Murchie, S., Mustard, J. F., et al., 1993. Lunar impact basins: new data for the western limb and far side (Orientale and south pole-Aitken basins) from the first Galileo flyby. J. Geophys. Res. 98 (E9), 17149—17181.

[18] Henriksen, M. R., Manheim, M. R., Burns, K. N., et al., 2017. Extracting accurate and precise topography from LROC narrow angle camera stereo observations. Icarus 283, 122—137.

[19] Hu, X., Ma, P., Yang, Y., et al., 2019. Mineral abundances inferred from in situ reflectance measurements of Chang'E-4 landing site in South Pole-Aitken Basin. Geophys. Res. Lett. 46 (16), 9439—9447.

[20] Huang, J., Xiao, Z., Flahaut, J., et al., 2018. Geological characteristics of Von Kármán crater, northwestern south pole-Aitken Basin: Chang' E-4 landing site region. J. Geophys. Res. 123 (7), 1684—1700.

[21] Ivanov, B. A., 2006. Earth/Moon impact rate comparison: searching constraints for lunar secondary/primary cratering proportion. Icarus 183 (2), 504—507.

[22] Ivanov, B. A., 2008. Size-Frequency Distribution of Asteroids and Impact Craters: Estimates of Impact Rate. Catastrophic events caused by cosmic objects. Springer, Dordrecht, pp. 91—116.

[23] Ivanov, M. A., Hiesinger, H., van der Bogert, C. H., et al., 2018. Geologic history of the northern portion of the South Pole-Aitken basin on the Moon. J. Geophys. Res. 123 (10), 2585—2612.

[24] Jia, Y., Zou, Y., Ping, J., et al., 2018. The scientific objectives and payloads of Chang'E-4 mission. Planet. Space Sci>162, 207—215.

[25] Jia, M., Yue, Z., Di, K., et al., 2019. A reanalysis of the relationship between the size of boulders and craters in lunar surface. Icarus 331, 116—126.

[26] Kneissl, T., Van Gasselt, S., Neukum, G., 2011. Map-projection-independent crater size-frequency determination in GIS environments-new

software tool for ArcGIS. Planet. Space Sci. 59, 1243—1254.

[27] Li, C., Liu, D., Liu, B., et al., 2019. Chang'E-4 initial spectroscopic identification of lunar far-side mantle-derived materials. Nature 569, 378—382.

[28] Liu, Z., Yue, Z., Michael, G., Gou, S., Di, K., Sun, S., Liu, J., 2018. A global database and statistical analyses of (4) Vesta craters. Icarus 311, 242—257.

[29] Liu, B., Niu, S., Xin, X., et al., 2019. High precision DTM and DOM generating using multi-source orbital data on Chang'e-4 landing site. In: ISPRS Geospatial Week 2019— GSW2019, "International workshop on Planetary Remote Sensing and Mapping", 10—14 June, Enschede, The Netherlands, pp. 1413—1417.

[30] Lucey, P.G., Taylor, G.J., Hawke, B.R., Spudis, P.D., 1998. FeO and TiO_2 concentrations in the South Pole-Aitken basin: implications for mantle composition and basin formation. J. Geophys. Res. 103 (E2), 3701—3708.

[31] Mahanti, P., Robinson, M.S., Stelling, R., 2014. How deep and steep are small lunar craters? New insights from LROC NAC DEMs. In: 45th Lunar and Planetary Science Conference. Abstract #1584.

[32] Mahanti, P., Robinson, M.S., Stelling, R., 2015. How old are small lunar craters? A depth-to-diameter ratio based Analyis. In: 46th lunar and planetary science conference. Abstract #1615.

[33] Melosh, H.J., 1989. Impact cratering: A geologic process. Oxford University Press, New York, p. 253.

[34] Melosh, H.J., Kendall, J., Horgan, B., Johnson, B.C., Bowling, T., Lucey, P.G., Taylor, G. J., 2017. South Pole-Aitken basin ejecta reveal the Moon's upper mantle. Geology 45 (12), 1063—1066.

[35] Michael, G.G., Neukum, G., 2010. Planetary surface dating from crater size-frequency distribution measurements: partial resurfacing events and statistical age uncertainty. Earth Planet. Sci. Lett. 294, 223—229.

[36] Michael, G.G., Platz, T., Kneissl, T., et al., 2012. Planetary surface dating from crater size-frequency distribution measurements: spatial randomness and clustering. Icarus 218 (1), 169—177.

[37] Neukum, G., Koenig, B., Arkani-Hamed, J., 1975. A study of lunar impact crater size-distributions. Moon 12 (2), 201—229.

[38] Oberbeck, V.R., Morrison, R.H., 1973. On the formation of the lunar herringbone pattern. In: 4th Lunar Planetary Science Conference, 4, pp. 107—123.

[39] Ohtake, M., Uemoto, K., Yokota, Y., et al., 2014. Geologic structure generated by large-impact basin formation observed at the South Pole-Aitken basin on the Moon. Geophys. Res. Lett. 41 (8), 2738—2745.

[40] Pasckert, J. H., Hiesinger, H., van der Bogert, C. H., 2018. Lunar farside volcanism in and around the South Pole - Aitken basin. Icarus 299, 538—562.

[41] Peng, M., Wan, W., Wu, K., et al., 2014. Topographic mapping capability analysis of Chang'e - 3 Navcam stereo images and three - dimensional terrain reconstruction for mission operations. J. Remote Sensing 18, 995—1002.

[42] Pike, R. J., 1977. Apparent depth/apparent diameter relation for lunar craters. In: 8th Lunar and planetary science conference, pp. 3427—3436.

[43] Povilaitis, R. Z., Robinson, M. S., van der Bogert, C. H., Hiesinger, H., Meyer, H. M., Ostrach, L. R., 2018. Crater density differences: exploring regional resurfacing, secondary crater populations, and crater saturation equilibrium on the moon. Planet. Space Sci. 162, 41—51.

[44] Qiao, L., Ling, Z., Fu, X., et al., 2019. Geological characterization of the Chang'e - 4 landing area on the lunar farside. Icarus 333, 37—51.

[45] Qin, S. H., Huang, Y., Li, P. J., et al., 2019. Orbit and tracking data evaluation of Chang'E - 4 relay satellite. Adv. Space Res. 64, 836—846.

[46] Richardson, J. E., 2009. Cratering saturation and equilibrium: a new model looks at an old problem. Icarus 204, 697—715.

[47] Robbins, S. J., 2019. A new global database of lunar impact craters >1—2km: 1. Crater locations and sizes, comparisons with published databases, and global analysis. J. Geophys. Res. 124.

[48] Robbins, S. J., Riggs, J. D., Weaver, B. P., et al., 2018. Revised recommended methods for analyzing crater size - frequency distributions. Meteorit. Planet. Sci. 53, 891—931.

[49] Robinson, M., 2019. Topographic map of the Chang'e - 4 Site [2019—4—30]. http://lroc. se se. asu. edu/posts/1100.

[50] Robinson, M. S., Brylow, S. M., Tschimmel, M., et al., 2010. Lunar reconnaissance orbiter camera (LROC) instrument overview. Space Sci. Rev. 150, 81—124.

[51] Shoemaker, E. M., Hackman, R. J., Eggleton, R. E., 1962. Interplanetary correlation of geologic time. Adv. Astronaut. Sci. 8, 70—79.

[52] Soderblom, L. A., Condit, C. D., West, R. A., et al., 1974. Martian planetwide crater distributions: implications for geologic history and surface processes. Icarus 22 (3), 239—263.

[53] Stopar, J. D., Robinson, M. S., Barnouin, O., et al., 2017. Relative depths of simple craters and the nature of the lunar regolith. Icarus 298, 34—48.

[54] Sun, S., Yue, Z., Di, K., 2018. Investigation of the depth and diameter relations of subkilometer - diameter lunar craters. Icarus 309, 61—68.

[55] Wang, Z. Z., 1990. Principles of Photogrammetry (With Remote Sensing). Press of Wuhan Technical University of Surveying and Mapping and Publishing House of

Surveying and Mapping, Beijing, pp. 327—330.

[56] Wilhelms, D. E., Howard, K. A., Wilshire, H. G., Kozak, R., 1979. Geologic map of the south side of the moon. IMAP 1162.

[57] Wilhelms, D. E., McCauley, J. F., Trask, N. J., 1987. The geologic history of the Moon. USGS Professional Paper 1348. US Government Printing Office, Washington DC, pp. 1—281.

[58] Wood, C. A., Anderson, L., 1978. New morphometric data for fresh lunar craters. In: Lunar and planetary science conference proceedings, pp. 3669—3689.

[59] Wu, W., Li, C., Zuo, W., et al., 2019. Lunar farside to be explored by Chang'e-4. Nat. Geosci. 12 (4), 222—223.

[60] Xiao, Z., Werner, S. C., 2015. Size-frequency distribution of crater populations in equilibrium on the Moon. J. Geophys. Res. Planets 120 (12), 2277—2292.

[61] Yue, Z., Di, K., Liu, Z., et al., 2019. Lunar regolith thickness deduced from concentric craters in the CE-5 landing area. Icarus 329, 46—54.

[62] Zhao, B. C., Yang, J. F., Wen, D. S., et al., 2011. Overall scheme and on-orbit images of Chang'E-2 lunar satellite CCD stereo camera. SCIENCE CHINA Technol. Sci. 54 (9), 2237.

Dielectric Properties of Lunar Materials at the Chang'e-4 Landing Site

Jialong Lai[1], Feifei Cui[1], Yi Xu [2,*], Chaofei Liu[1] and Ling Zhang[2]

1. School of Science, Jiangxi University of Science and Technology, Ganzhou 341000, China; laijialong@jxust.edu.cn (J. L.); 9120170119@jxust.edu.cn (F. C.); liuchaofei@jxust.edu.cn (C. L.)
2. State Key Laboratory of Lunar and Planetary Sciences, Macau University of Science and Technology, Macau 999078, China; linzhang@must.edu.mo

Abstract: On January 3rd 2019, the Chang'e-4 mission successfully landed in the Von Kármán Crater inside the South Pole-Aitken (SPA) basin and achieved the first soft landing on the farside of the Moon. Lunar penetrating radar (LPR) equipped on the rover measured the shallow subsurface structure along the motion path for more than 700m. LPR data could be used to obtain the dielectric properties of the materials beneath the exploration area, providing important clues as to the composition and source of the materials. Although the properties of the upper fine-grained regolith have been studied using various methods, the underlying coarse-grained materials still lack investigation. Therefore, this paper intends to estimate the loss tangent of the coarse-grained materials at depth ranges of ~12 and ~28m. Stochastic media models with different rock distributions for the LPR finite-difference time-domain (FDTD) simulation are built to evaluate the feasibility of the estimation method. Our results show that the average loss tangent value of coarse-grained materials is 0.0104±0.0027, and the abundance of $FeO_T + TiO_2$ is 20.08 wt.%, which is much higher than the overlying fine-grained regolith, indicating different sources.

Keywords: Chang'e-4; lunar penetrating radar; loss tangent

1 Introduction

In January 2019, the Chang'e-4 lander successfully landed in the Von Kármán Crater on the lunar farside, becoming the first probe to land on the farside of the moon. The Yutu-2 rover is equipped with instruments including a multi-antenna, two channels (channel one and two) lunar penetrating radar (LPR) system. The central frequency of channel one is 60MHz, which is utilized to detect the relatively deep part of the subsurface structure, specifically the basalt layered structure, basalt thickness, etc. The central frequency of channel two is 500MHz, which is used to detect the shallow subsurface of the lunar regolith, including the thickness of the lunar regolith and the depth of the basalt layer and the physical parameters of the lunar regolith, such as the electrical constant, loss tangent and density[1,2].

Up to now, most lunar missions have been flyby or orbital explorations, and the more than 20 landing sites are all on the nearside of the Moon. Therefore, the successful implementation of the Chang'e - 4 exploration mission provides an unprecedented opportunity to observe the subsurface structure of the lunar farside[3].

The landing site of the Chang'e - 4 is located at 177. 588°E, 45. 457°S in the Von Kármán Crater, marked by a red triangle in Fig. 1[4,5]. The Von Kármán Crater is inside the South Pole - Aitken (SPA) basin, the largest and oldest impact crater on the Moon. According to the spectrum results and impact numerical simulation, it is generally believed that the Aitken Basin at the Lunar South Pole sputters the lower lunar crust and even the upper lunar mantle material on the lunar surface, so SPA Basin has important research value to understand the lunar interior and history[6]. The Von Kármán Crater has experienced multiple impact ejecta filling events and multiple stages of lava flow filling[7-9]. This complex geological history has brought uncertainty to the interpretation of the Chang'e - 4 LPR data. Huang et al. (2018) utilized the Lunar Reconnaissance Orbiter Camera (LROC) Wide - Angle Camera (WAC) data to obtain the global normalized reflectance image, and found that crater rays and secondary crater chains around the Chang'e - 4 landing site pointed to the Finsen Crater[7]. In situ spectral data collected by the Yutu - 2 rover showed that the characteristics of surface materials around the motion path are similar to those of Finsen's ejecta, which were analyzed with the Moon Mineralogy Mapper spectral data. This may indicate that the deposits in these two locations share the same source[10,11]. Furthermore, Lai et al. (2020) used the impact crater ejecta empirical model to estimate the ejecta thickness of the Finsen Crater in the Chang'e - 4 landing site to be 7. 9meters, close to the thickness of the lunar regolith based on the radar observation[12]. In summary, recent research showed that there is a 12m - thick fine regolith layer under the Yutu - 2 rover's motion path, and most of surficial materials came from the Finsen Crater[13-17].

Chang'e - 4 LPR observed an important feature of the subsurface structure of the landing site - at a depth of about 12m, there is a clear boundary between fine and coarse materials (Section 2. 1) . The fine materials are interpreted as fine - grained regolith, mainly formed by the deposits from Finsen[13]. The coarse materials are recognized as impact ejecta resulting from multiple impact events[16].

The composition and the properties of the fine - grained regolith were analyzed carefully using the in situ LPR data and spectral data. Lai et al. (2019), Li et al. (2020)

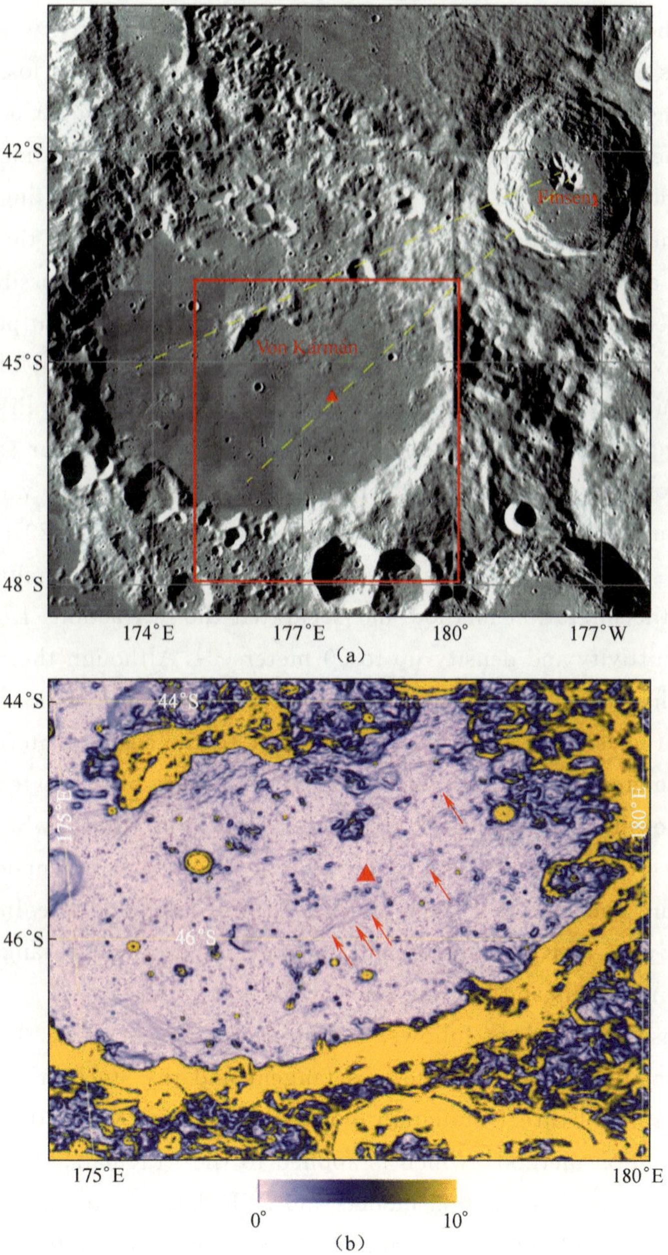

Fig. 1　The landing site of Chang'e - 4. (a) The base map obtained from Lunar Reconnaissance Orbiter Camera (LROC) Wide - Angle Camera (WAC) images[4]. The landing site of Chang'e - 4 is marked by a red triangle; the yellow dashed lines show the directions of Finsen ejecta. (b) Enlarged view of the red rectangle in (a). The landing site of Chang'e - 4 is marked by a red triangle. Red arrows mark the linear features in the northeast - southwest direction created by Finsen ejecta with elevated topography. The base map is of the SLDEM slope, obtained from Barker et al. (2016)[5]. The color bar is the degree of the SLDEM slope

and Dong et al. (2020) utilized the hyperbola fitting method to estimate the permittivity and density of the subsurface fine regolith[13,14,18]. The permittivity value of the fine regolith layer was estimated to be ~4. Lai et al. (2019) used signal loss to calculate the loss tangent of the fine - grained regolith layer, which was about 0.0039[13]. Li et al. (2020) determined the average loss tangent of the fine - grained material under the Chang'e - 4 landing site to be $(5 \pm 2) \times 10^{-3}$, based on the shifting of the centroid frequency with the depth. Since the loss tangent value is smaller than the normal value of the mare basalt, an intact, dense lava flow layer is excluded from possibility[14]. Dong et al. (2021) utilized the surface, subsurface reflections and the incident pulse amplitude to infer the permittivity and density of the fine - grained regolith layer[19]. They also obtained the surface permittivity of the Chang'e - 4 landing site for the first time. Dong et al. utilized the empirical model developed with measurements of lunar samples to obtain the loss tangent of the fine - grained regolith layer materials, which was $(4.4 \pm 0.5) \times 10^{-3}$[19]. Based on the 3D velocity spectrum method, Dong et al. (2020) calculated the properties of the subsurface materials, including velocity, relative permittivity, etc.[20]. Li and Zhang used the migration method and separated the diffractions from reflections to estimate the permittivity and density up to 50 meters[21]. Although the properties of fine materials under the Chang'e - 4 landing site have been inversed by various methods, there is still a lack of studies on the properties and source of the coarse materials, such as the loss tangent parameter. The loss tangent and the permittivity of the materials are the most important indicators for the composition of materials on the Moon, which can be further used to speculate on the source of the materials and acquire a better understanding of the geological structure and history around the Chang'e - 4 landing site. In this paper, we utilize the Chang'e - 4 LPR data of channel two to estimate the loss tangent of the coarse and sub - surface materials and indicate $FeO_T + TiO_2$ content.

This paper is organized as follows. Section 2 introduces the LPR data processing method, the estimation method of the loss tangent and the stochastic media models. In Section 3, the results of three simulation models are given to verify the feasibility of the loss tangent estimation method, which is applied to the LPR data. A detailed discussion on the results of the three simulation models and LPR data, including the estimated loss tangent and the total FeO and TiO_2 contents of coarse materials inferred using empirical formulas, is given in Section 4. Section 5 contains the conclusion.

2 Materials and Methods

2.1 LPR Data

Chnnel two of LPR has one transmitting and two receiving antennas, which are deployed on the bottom of the Yutu - 2 rover, named as 2A and 2B, respectively. In this study, we only analyzed LPR 2B data. The bandwidth of LPR channel 2B is 450MHz, which indicates that the resolution is 0.3m in a vacuum[22]. The Yutu－2 rover has been working on the Moon for more than two years, and the LPR has detected a more than 700m long subsurface structures along its motion path. The first 15 days of LPR data were utilized in this work (Fig. 2). The Data File IDs were listed in Tab. S2.

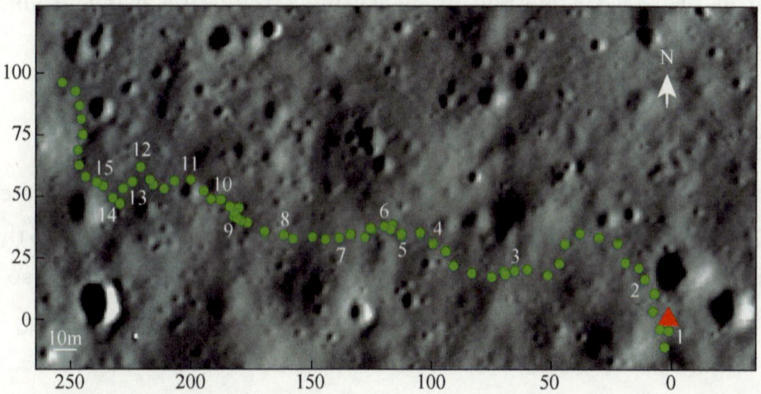

Fig. 2　Motion path of the Yutu－2 rover during the first fifteen lunar days. Background image is the LROC (Robinson et al., 2010) Narrow－Angle Camera (NAC) image of the Yutu－2 rover's surveying area (NAC image ID: M134022629LE, 1.27m/pixel, 77.55° incidence angle)[4]. The white arrow indicates north. The landing site of the Chang'e－4 mission is marked by a red triangle and the waypoints of the rover's path are marked as green dots. The number near the waypoint is the starting position of the corresponding lunar day

Data preprocessing is necessary before interpreting radar data. Based on the characteristics of the Chang'e－4 LPR data, a set of preprocessing methods is proposed[13]. In this paper, the data processing scheme is modified to retain the original amplitude information.

1. Data reading: 2B processing level data are utilized in the paper, which are generated with raw data after integration, conversion from unsigned integers to signed integers, normalization, removing direct current offset, and adding geometric coordinates.

2. Redundant data removal: when LPR is operating, the Yutu－2 rover may stop to perform the operations of other instruments, which generated duplicate data. So, data collected during static sampling should be removed.

3. Data file stitching: the files of different lunar days are stored separately. In this step, these data need to be stitched together.

4. Adjust the time delay: the receiving antenna is turned on 28.203ns earlier than the transmitting antenna, so the initial 28.203ns of all the data traces were removed.

5. Background removal: subtracting the average value from the LPR data to reduce

the 'ringing' effects and periodic noises caused by the antenna - ground coupling.

6. Band - pass filter: the finite impulse response (FIR) filter is used to reduce noise. Its cut - off frequencies are set as 150250750, and 850MHz.

2.2 Loss Tangent Estimation Method

Grimm et al. proposed a method of measuring attenuation using ground penetrating radar[23]. The radar range equation, given in Skolnik[24], is:

$$\frac{P_r}{P_t}=\frac{G^2\lambda^2\xi}{(4\pi)^3R^4}e^{-4\alpha R} \tag{1}$$

where P_r is the received power, P_t is the transmitted power, G is the system gain, λ is the wavelength in the medium, ξ is the backscatter cross section, $e^{-4\alpha R}$ is the attenuation in the medium, and α is the spatial attenuation coefficient. Later, α can also be expressed as η (α in dB) in unit dB/m, $\eta = 20\log_{10}(e^{\alpha}) = 8.686\alpha$.

In general, α can be expressed as the following formula[25]:

$$\alpha=\omega\sqrt{\frac{\mu\varepsilon}{2}\left(\sqrt{1+tan^2\delta}-1\right)} \tag{2}$$

where ω is the angular frequency, μ is the magnetic permeability, ε is the dielectric permittivity, and $\tan\delta$ is the loss tangent. For a lunar sample, $\tan\delta \ll 1$, and is assumed to be the free space magnetic permeability. Then,

$$\eta=0.1\times10^{-8}\sqrt{\varepsilon'}f\tan\delta \tag{3}$$

where ε' is the real part of the relative permittivity.

Formula (1) indicates that the attenuation loss can be divided into two parts—the scattering loss and the intrinsic (or absorption) loss. If the scattering loss can be effectively compensated for, then the absorption loss can be estimated.

For different reflected targets, the ratio between the received power and the transmitted power is proportional to the reciprocal power of the propagation distance:

$$P_R/P_t \propto 1/R^n \tag{4}$$

In this case, three different models of reflected targets are considered: $n=2$ for a planar and smooth reflected target; $n=3$ for the Fresnel zone case; and $n=4$ for discrete scatters. After compensating for the scatter signals, the attenuation can be estimated by least squares fitting of the two - way distance versus the ratio of the transmitted power to received power in unit dB/m. After obtaining the attenuation η, $\tan\delta$ can be estimated by Formula (3).

2.3 Simulation Model

2.3.1 Stochastic Media Model

In order to verify the feasibility of the loss tangent estimation method, a simulation model was utilized. Previous studies have shown that the internal structure of the lunar regolith is close to the stochastic media model[26]. Hence, the stochastic media model with an ellipsoidal autocorrelation function was used in this study to simulate the lunar regolith structure[27]. The method is described as follows:

1. A two - dimensional stochastic equivalent medium model is established by using ellipsoidal autocorrelation function:

$$f(x, y)=\exp\left[-\left(\frac{x^2}{a^2}+\frac{y^2}{b^2}\right)\right]^{\frac{1}{1+r}} \tag{5}$$

where r is the fuzzy factor of the stochastic media model, and a and b represent the autocorrelation length in the X and Y directions, respectively.

2. The $f(x, y)$ is transformed into the wavenumber domain ($F(k_x, k_y)$) by the twodimensional Fourier transform equation, and then the power spectrum function $R(k_x, k_y)$ is calculated.

$$R(k)=\| F(k_x, k_y) \|^2 \tag{6}$$

3. Generate the independent and uniform random numbers $\phi(k_x, k_y)$ in the range of $[0, 2\pi)$.

4. By adding the random number $\phi(k_x, k_y)$ into Formula (7), a new energy spectral density function is obtained:

$$F'(k_x, k_y)=\sqrt{R(k)}\exp[+i\phi(k_y, k_y)] \tag{7}$$

5. The inversed two - dimensional Fourier transform is performed to transform $F'(k_x, k_y)$ from the wavenumber domain to the spatial domain to obtain a new $f'(x, y)$.

2.3.2　Simulation Model

In order to model different scenarios of lunar regolith structure, three stochastic media models were used. Model One is shown in Fig. 3a. In the models (e.g., Fig. 3a, b), the material of the first layer is free space. Model Two and Model Three have the same layered structure, but with different abundances and distribution of rocks. Five rocks are placed in the upper part of Model Two, while in Model Three, they are in the lower part. The autocorrelation lengths in the X and Y directions are both 2. The fuzzy factor r is 0 in this model. The model is 10m in length, and 11m in depth, and the spatial steps in the X (length) and Y (depth) direction are both 0.005m. The time window is 150ns, the time step is 0.0118ns. The moving steps of both transmitter and receiver antenna are 0.05m. The average permittivity of background material is 3.5, which is consistent with the result of Lai et al. [13] andWang et al. [15]. The relative loss tangent is set as 0.02, which indicates the average conductivity of the model is about 2×10^{-3} (e.g., Fig. 3b) . The transmitter antenna and receiver antenna are 0.3m higher than the zero depth. The range between the transmitter antenna and the receiver antenna is 0.32m, the same as the settings of the Chang'e 4 LPR.

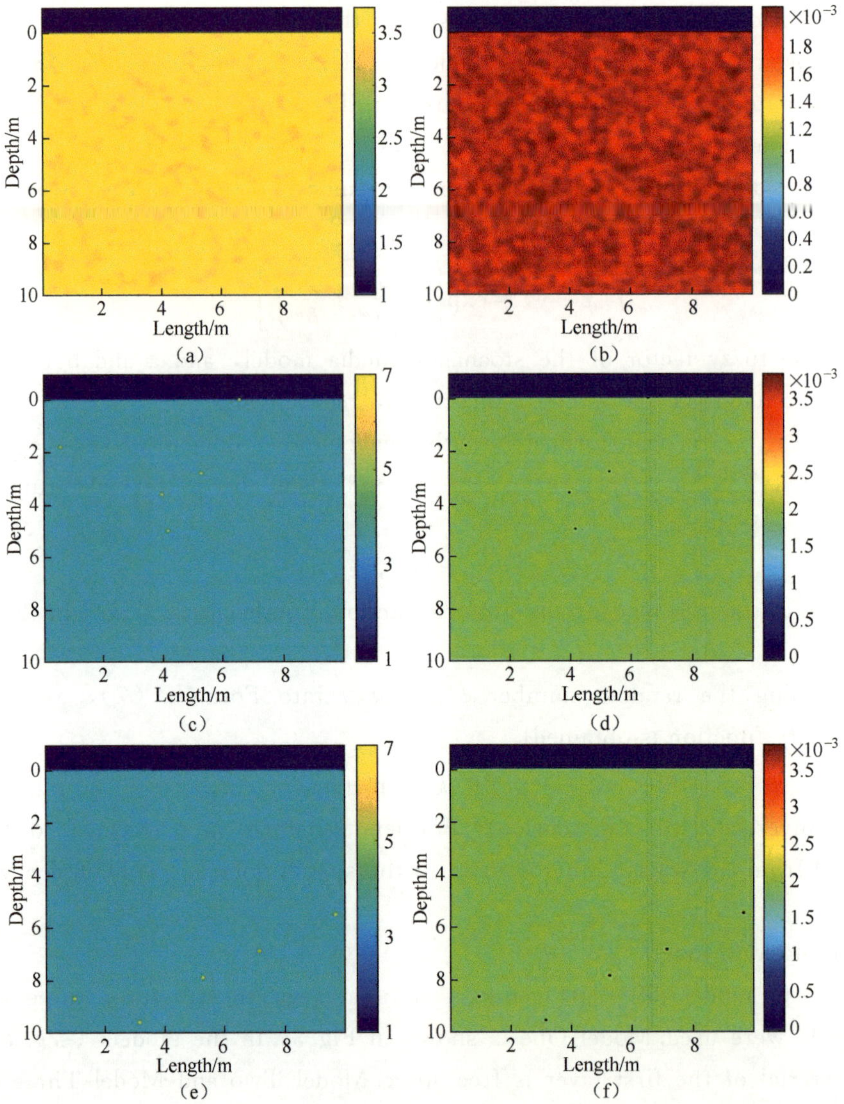

Fig. 3　Three simulation models. (a) The permittivity of the simulation in Model One. The color bar is the relative permittivity. (b) The conductivity of the simulation in Model One. The color bar is the conductivity, and the unit of conductivity is in Siemens/meter. (c, d) are the permittivity and the conductivity of the simulation in Model Two, respectively. (e, f) are the permittivity and the conductivity of the simulation in Model Three, respectively

3　Results

3.1　Synthetic Data Results

The finite-difference time-domain (FDTD) simulation method is utilized to verify the correctness of the loss tangent estimation method, which is mentioned in Section

2.2. gprMax software, an open source software package for numerical modeling of ground penetrating radar (GPR), is used for simulation[28]. It is foreseeable that the reflection signal of the buried rock and the multiple scattering signals among the rocks will affect the results of the loss tangent estimation, so three models are used in this experiment to evaluate the influence of rocks. The base stochastic media model is the same for these three models, except for locations of the rocks. Fig. 4 shows the comparisons between the simulation results of these three models.

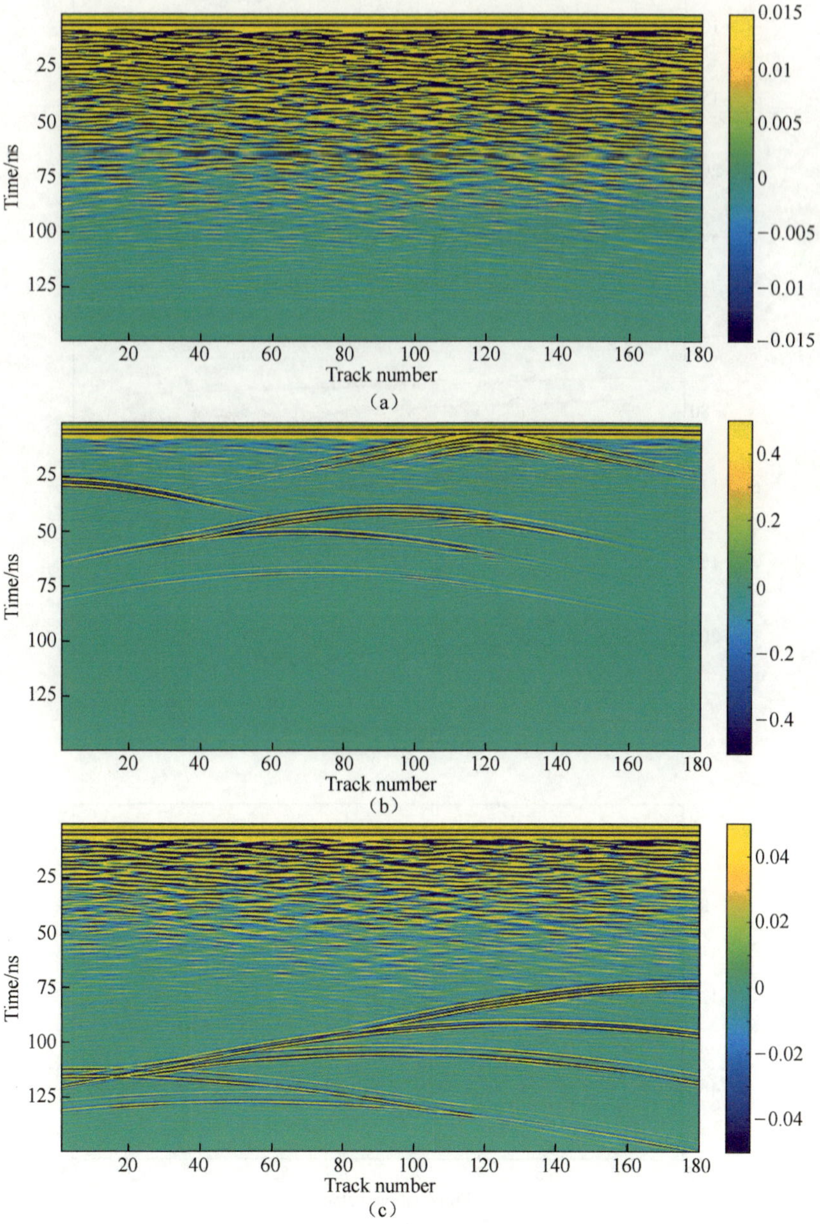

Fig. 4　Simulation results of three stochastic media models. (a), (b) and (c) are the simulation results of Models One, Two and Three in Fig. 3, respectively

The method mentioned in Section 2. 2 is utilized to estimate the loss tangent of the simulation model. In order to reduce the influence of noises, the mean square of each radar trace is calculated before estimating the loss tangent. Fig. 5 shows the estimated attenuation result.

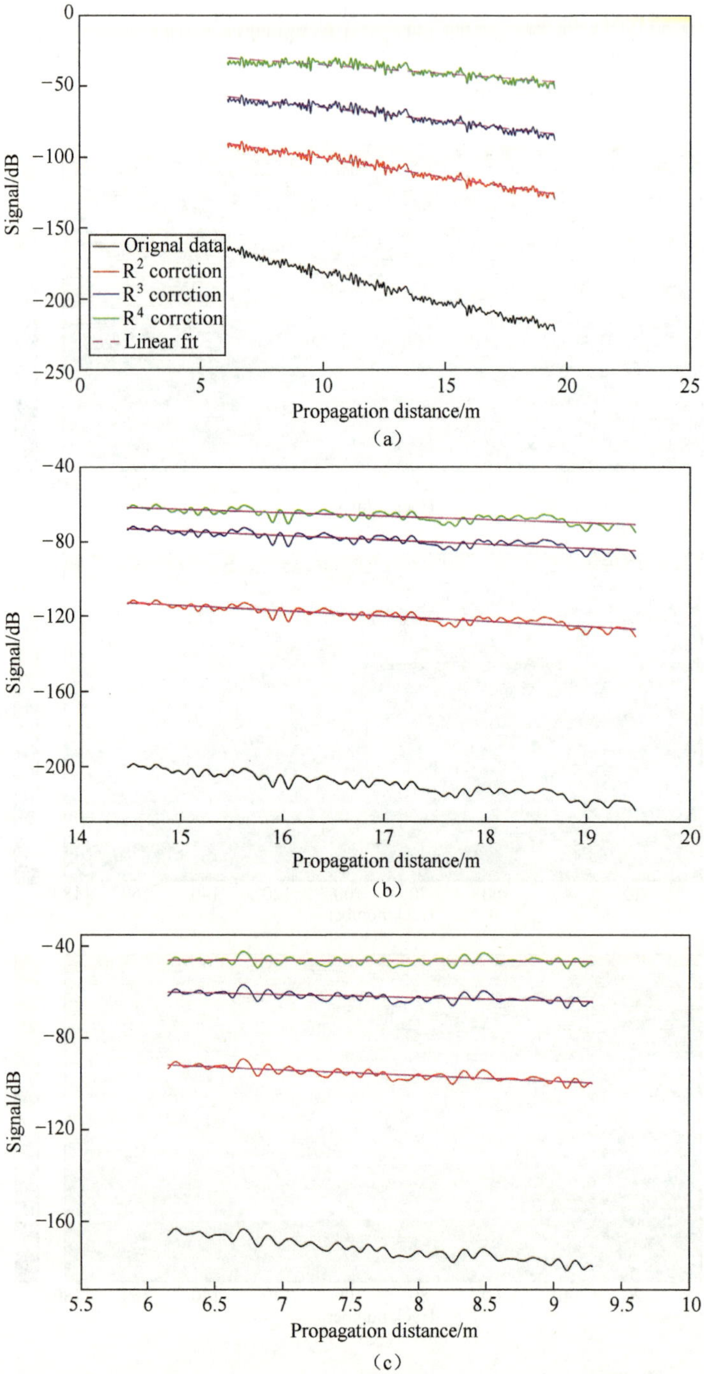

Fig. 5　Estimated attenuation results. (a), (b) and (c) present the attenuation estimated result of the Model one, two and three, respectively. ε= 3. 5 is used for time - to - propagation distance conversion

In Fig. 5, the red, blue and green lines represent the synthetic data after R^2, R^3, and R^4 backscatter/spreading correction, respectively. The dashed purple line indicates the least linear fitting results. The slope of the dashed purple line is equal to the attenuation η. Then, Formula (3) is used to calculate loss tangent parameters. Tab. 1 shows the attenuation's estimated result.

Tab. 1　The attenuation's estimated result of three models

Model	Correction Method	Slope	Loss Tangent	Error Bar	R - Squared
Model one	R^2	−2.6730	0.0314	0.0001	0.9665
Model one	R^3	−1.9526	0.0229	0.0001	0.9293
Model one	R^4	−1.2322	0.0145	0.0002	0.8103
Model two	R^2	−2.8368	0.0333	0.0006	0.8332
Model two	R^3	−2.3228	0.0273	0.0006	0.7699
Model two	R^4	−1.8087	0.0212	0.0006	0.6695
Model three	R^2	−2.4643	0.0290	0.0009	0.7055
Model three	R^3	−1.3300	0.0156	0.0009	0.4133
Model three	R^4	−0.4246	0.0023	0.0009	0.0151

The average loss tangent of Model One is 0.0229 ± 0.0084, which is consistent with the simulation model (0.02). The average loss tangents of Model Two and Model Three are 0.0273 ± 0.0061 and 0.0156 ± 0.0134, respectively. Thus, the presence of rocks introduces some errors to the results, but within a tolerable range. The error of results of Model Two and Model Three is 36.5% and 22%, respectively. The distribution of rocks also affects the inferred loss tangent. In this example, the rocks at the bottom result in fewer errors of the estimated result than that of rocks in the upper part. Next, the algorithm will be used in the Chang'e - 4 LPR data.

3.2　LPR Results

The loss tangent estimation method introduced in Section 2.2 is used to calculate the loss tangent of the coarse material. In order to mitigate the influence of the rocks, we only select the depth range of relatively uniform materials (the sampling time exceeds 150ns for the 6th to the 12th lunar days in Fig. 6 have been selected. After averaging the energy of adjacent 437 tracks of LPR data, which is obtained after data preprocessing described in Section 2.1, the original root mean square data are obtained (black line in Fig. 7). Then, the spreading and backscatter correction factor (R^2, R^3, R^4) are utilized to compensate for the spreading and backscatter loss. Then, the least squares fitting method is used to fit the data after exponential function compensation (dashed purple lines in Fig. 7). The estimated results of signal attenuation of LPR data are shown in Fig. 7. The corresponding loss tangent results are given in Tab. 2.

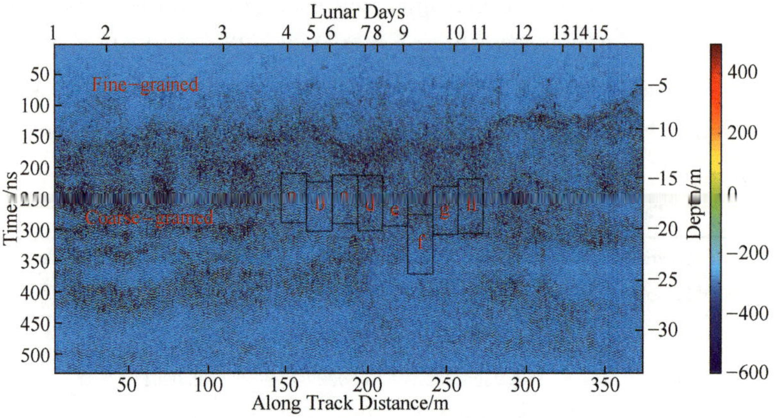

Fig. 6 The first fifteen lunar days' LPR data image after processing. The color bar is the amplitude of the LPR signal after processing. Groups a—g indicate the areas used to calculate the loss tangent

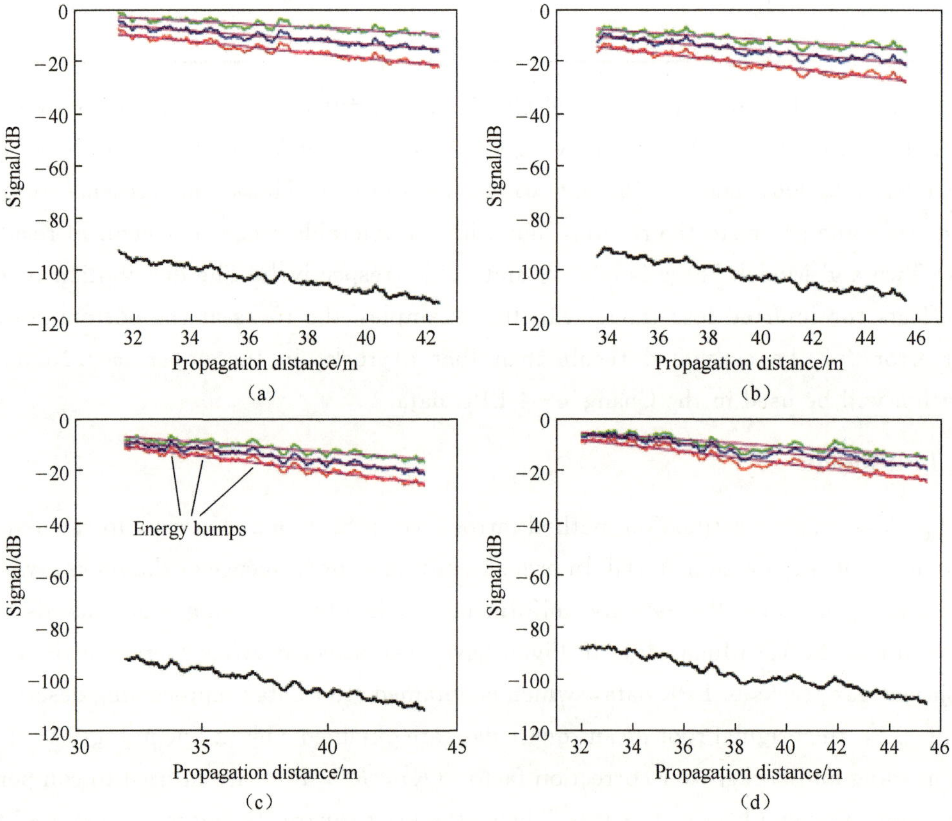

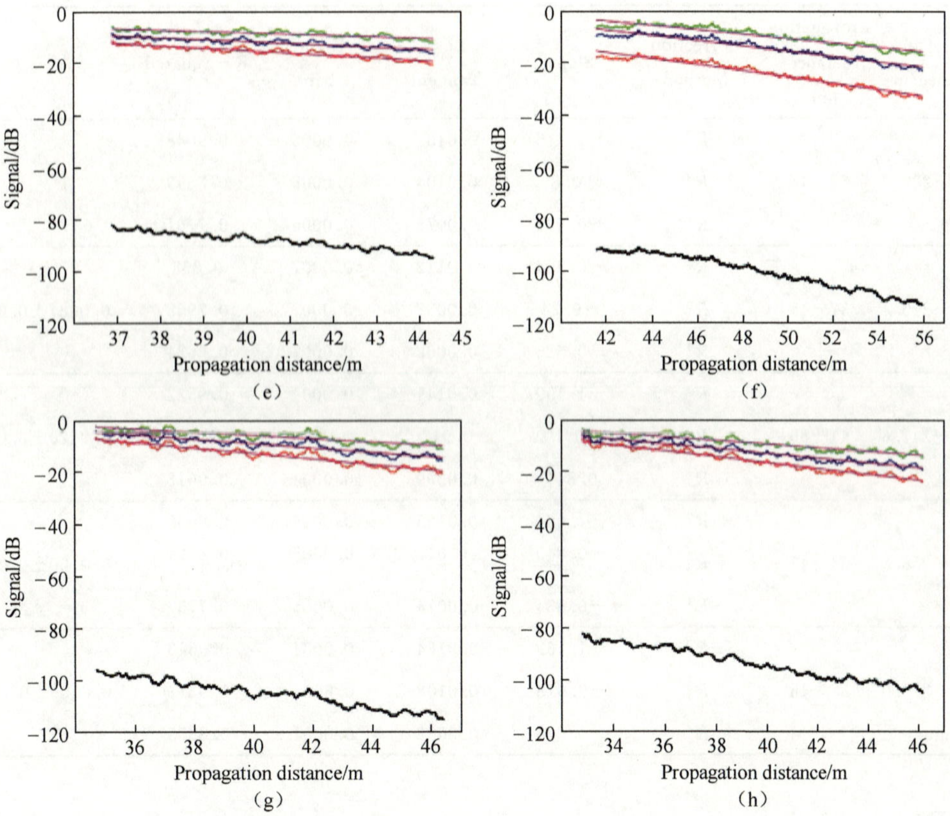

Fig. 7　The estimated result of signal attenuation of LPR

ε= 3.5 is used for time - to - propagation distance conversion. (a—h) present the attenuation estimated result of correspond to the eight calculation regions in Fig. 6, respectively

Tab. 2　The loss tangent's estimated results

Along Track Distance/m	Propagation Distance/m	Correction Method	Slope	Loss Tangent	Error bar	R - Squared	Average
146.8—162.9	31—43	R^2	−1.098	0.0129	0.0005	0.9233	0.0101±0.0028
		R^3	−0.862	0.0101	0.0005	0.8816	
		R^4	−0.625	0.0073	0.0005	0.7972	
162.9—178.3	33—46	R^2	−1.082	0.0127	0.0005	0.9105	0.0101±0.0027
		R^3	−0.862	0.0101	0.0005	0.8676	
		R^4	−0.641	0.0075	0.0005	0.7863	
178.3—194.0	32—44	R^2	−1.181	0.0139	0.0004	0.9437	
		R^3	−0.95	0.0112	0.0004	0.9144	
		R^4	−0.72	0.0085	0.0004	0.8574	

Continued Table

Along Track Distance/m	Propagation Distance/m	Correction Method	Slope	Loss Tangent	Error bar	R - Squared	Average
194.0—209.8	32—46	R^2	−1.115	0.0131	0.0005	0.9048	
		R^3	−0.89	0.0104	0.0005	0.8599	
		R^4	−0.664	0.0078	0.0005	0.7761	
209.8—225.4	37—44	R^2	−0.954	0.0112	0.0007	0.868	0.0087±0.0025
		R^3	−0.74	0.0087	0.0007	0.7968	
		R^4	−0.526	0.0062	0.0007	0.6624	
225.4—241.6	41—56	R^2	−1.199	0.0141	0.0004	0.9522	0.0120±0.0021
		R^3	−1.02	0.012	0.0004	0.9329	
		R^4	0.841	0.0099	0.0004	0.9012	
241.6—258.3	34—47	R^2	−1.064	0.0125	0.0005	0.9086	0.0100±0.0025
		R^3	−0.849	0.01	0.0005	0.8624	
		R^4	−0.634	0.0074	0.0005	0.7757	
258.3—273.0	33—46	R^2	−1.139	0.0134	0.0004	0.9543	0.0109±0.0026
		R^3	−0.918	0.0108	0.0004	0.9311	
		R^4	−0.696	0.0082	0.0004	0.8855	

4 Discussion

4.1 Loss Tangent of the Simulation Model

Fig. 5 and Tab. 1 show estimated results of the loss tangent of the simulation models. For the simulation model, the R^3 correction result, which indicates that reflections come from a Fresnel-zone sized area, is plausible. The loss tangent of Model One after R^3 correction is 0.0229, compared to the parameter ~0.02 used in the stochastic media model, and the error is less than 15%. Taking into account the result after over-compensation (R^4 correction) and under-compensation (R^2 correction), the real value is within the range of estimation values (0.0229±0.0084). Therefore, the simulation results show that the estimation method can effectively obtain the loss tangent when little rock is present.

Model Two includes five stones in the upper part of layer two. It is expected that the reflection signal from the rocks and the scattering signal among the rocks will affect the radar signals. The locations of rocks are known in the simulation model. The deepest rock is placed at the propagation distance of about 10m, so we selected the echo signal with a propagation distance between 14 and 19 meters to calculate the energy attenuation curve. The average loss tangent of Model Two is 0.0273±0.0061, which is higher than the actual setting. Judging from the results, rocks cause scattering of the EM wave and result in a slightly larger loss tangent than the simulation parameter, as the interference of rocks increases the signal attentions. Therefore, avoiding rocks in the calculation area could improve the accuracy of the loss tangent. However, if rocks are ubiquitous, the result can also be considered as the upper limit of the loss tangent.

Model Three has five stones distributed on the bottom. The shallowest rock is about ~10m in propagation distance, so we selected the echo signal with a propagation distance between 6 and 8.5m to calculate the energy attenuation curve. The average loss tangent of Model Three is 0.0231±0.0144, which is consistent with the real setting and better than the case in which rocks are above the study region (Model Two). Hence, the result of the loss tangent can be improved by selecting the area of calculation appropriately. It is worth noting that although the results of Model One and Model Two are similar, the confidences of these two models (0.0084 vs. 0.0144) are different due to the different horizontal length of the calculation area. Increasing the calculation area can reduce the confidence interval. In addition, by comparing Model One and Model Three, it can be concluded that a longer calculation interval in depth could reduce the error bar.

In actual situations, if the selected area contains rocks in the lower part, the tail of the energy attenuation curve may go up, resulting in the estimated loss tangent being smaller than the actual situation. For example, in Model Three, the calculation area is expanded to 9.5m, so the rocks are included, then the average loss tangent is changed to 0.0156±0.0133. In this case, it can be considered that the calculated loss tangent is the lower limit of the actual situation.

4.2 Loss Tangent of the Coarse Material

Fig. 6 shows the LPR data after processing with the steps proposed by Lai et al.[13]. The significant feature of the LPR image is that there is a continuous horizontal reflection around the depth of ~12m. In order to calculate the loss tangent of coarse materials, a study area in the depth range of ~146 to ~273m on the Yutu-2 rover's motion path was selected in this study. LPR data were divided into a set of groups along the horizontal direction, with each group being 15m long. The total number of groups is eight, which are named as group a, b, …, h, respectively (Fig. 6). The attenuation of these eight groups was calculated separately, and the corresponding results are shown in Fig. 7 and Tab. 2. Then, we utilized Formula (3) to estimate the loss tangent. From Fig. 7 and Tab. 2, it can be seen that each data group used different depth ranges to calculate the loss tangent to reduce the noises caused by rocks. Tab. 2 shows that the estimated loss tangent of groups a, b, c, d, g and h are similar, the average value of which is 0.0105. The estimated loss tangent of group e is larger than the average result, while group f's value is smaller.

The main reason for deviations in the estimation results is that there are several energy bumps (Fig. 7e) at bottom (40—43m) of the group e, probably due to rock scattering, which causes the result of group e to be relatively small. The group e can be regarded as the situation that the tail of the energy attention curve is affected by the rock reflection signal, which is similar to the situation of Model three, and it can be considered as the lower limit of the loss tangent.

From Fig. 7, the energy attenuation curve of group f changes relatively smoothly compared to other groups, and the error bar is smaller than those of other groups. However, the loss tangent of group f is larger than the average value, which may indicate that some rocks might be distributed above the selected area. Similar to the results of Model Two, scattering caused by rocks affects the estimation of loss tangent. In this situation, the upper limit of the loss of coarse materials can be obtained.

In groups a, b, c and d, the maximum value of the estimated loss tangent after R^3 correction is 0.0112 and the minimum value is 0.0101. The difference between these two values is about 10%. To improve the confidence of the results, the sampled data on a longer motion path are used to calculate the energy attenuation curve and then estimate the loss tangent. In this study, we use the energy attenuation curve of adjacent 1700 tracks of LPR data to re-estimate the loss tangent, and obtain Fig. 8.

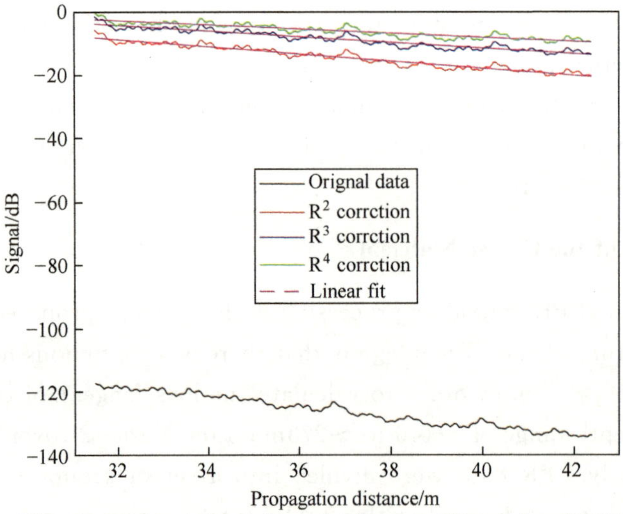

Fig. 8　The attenuation estimated result adjacent 1700 tracks of LPR data. $\varepsilon = 3.5$ is used for time-to-propagation distance conversion

The corresponding result of Fig. 7 is calculated in Tab. 3.

Tab. 3　The attenuation estimated result of adjacent 1700 tracks of LPR data

Along Track Distance/m	Propagation Distance/m	Correction Method	Slope	Loss Tangent	Error Bar	R-Squared	Average
146.8−208.1	31.5−42.5	R^2	−1.1196	0.0132	0.0004	0.9507	0.0104±0.0027
		R^3	−0.8868	0.0104	0.0004	0.9246	
		R^4	−0.6540	0.0077	0.0004	0.8709	

From Fig. 8, we can conclude that increasing the size of the calculation area can effectively improve the R - square value, and can fit the curve better to reduce the error bars. The average value of the loss tangent is 0.0104, which is consistent with the average value of the loss tangent of group a, b, c, d, g and h (0.0105).

Based on the previous analysis, it is concluded that the estimated average loss tangent of the coarse materials in the calculated area is 0.0104 ± 0.0027. If empirical results of the simulation model are used, the upper limit of the loss tangent is 0.0120, and the lower limit is 0.087.

4.3. The Geological History of the Chang'e - 4 Landing Site

In radargrams, there are continuous horizontal reflectors inside the coarse - grained materials[13-17]. Zhang et al. (2020) argued that there are 4 - 5 layers of impact ejecta blanket coverage between 12 and 40 meters under the motion path of the Yutu - 2 rover[16]. Xu et al. (2021) found four impact craters (the thickness of the local materials and ejecta deposits are more than 1 meter) that can have a greater impact on the landing site through the model[29]. In order to better analyze the subsurface structure of the landing site, the migration method and topographic correction are applied to the radargram in Fig. 2. Fig. 9a shows the radar image after processing. There are some horizontal reflectors around ~210, ~270 and ~360 ns, which are marked by red lines in Fig. 9b. These continuous interface reflections appear on both sides of the loss tangent estimated area (Fig. 6 and 9). The interruption of the continuity is because of a bowl - shaped area marked by the black dotted line in Fig. 9b. Zhang et al. (2021) and Zhou et al. (2021) believed that this area is a buried impact crater, and the continuous reflective interfaces were destroyed by the subsequent impact event[30,31]. The materials under the impact crater were mixed and are relatively uniform, so the loss tangent estimation model can be applied. The estimated loss tangent is the loss tangent of the mixed multi - layered coarse materials.

Considering that the central frequency of the Yutu - 2 high - frequency radar is 500MHz, the empirical formula based on the data points only representing measurements at 450MHz is utilized[32].

$$\tan\delta = 10^{[0.038\times(\%TiO_2+\%FeO)-2.746]} \tag{8}$$

Substitute the average loss obtained by the previous calculation (0.0104) into Formula (8), and then we can obtain the total FeO_T and TiO_2 concentrations of the coarse materials at the CE - 4 landing site as 20.08 wt. %. Additionally, with the upper and lower loss tangent bounds (0.0120 and 0.0087), the total $FeO_T + TiO_2$ concentrations will be 21.7 and 18.0 wt. %, respectively. The FeOT + TiO_2 concentration of the surface fine materials around the Chang'e - 4 landing site is lower than 15 wt. %, and the TiO_2 content of which is between 1 and 2 wt. %[9,33-36]. This value is below even the lowest boundary estimates based on Equation (8) of the coarse materials. This may indicate that the surface fine material has much lower FeO_T content than that of the underlying coarse substance, and may share a different source. Qiao et al. (2019) proposed that dozens of craters (about 330m - 1. 1km in diameter) have relatively high FeO_T in the crater interiors and ejecta blankets[36]. The high $FeO_T + TiO_2$ concentrations of the coarse materials may be causedby the iron - rich nature of the ejecta blankets.

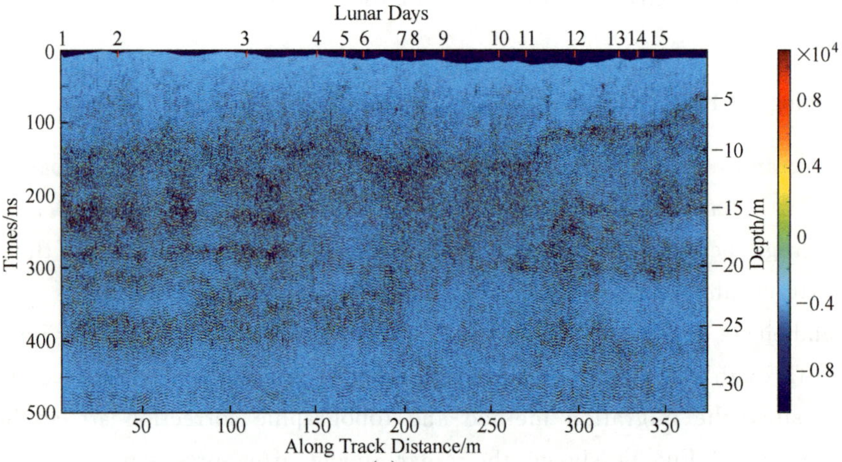

(a)

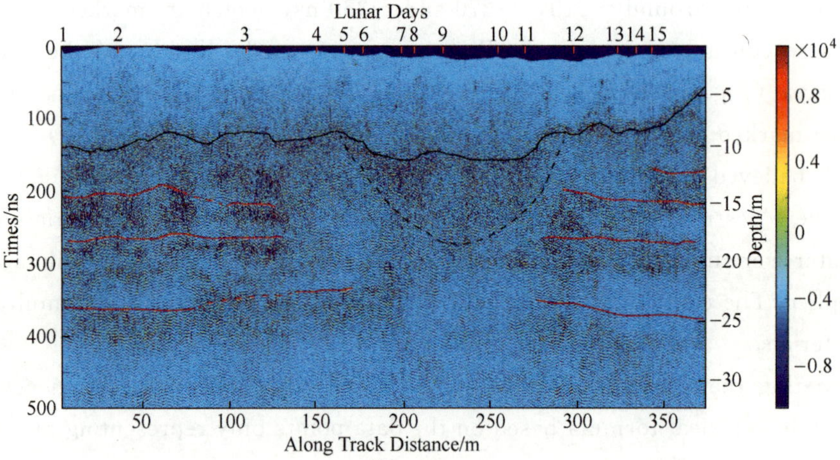

(b)

Fig. 9 LPR data after migration and topographic correction. (a) The processed LPR image. (b) The LPR image with interpretation. The depth of the y - axis is converted with $\bar{\varepsilon} = 4.3$ ($\tau \leqslant 154ns$), $\bar{\varepsilon} = 6$ ($\tau > 154ns$). The black solid line marks the interface reflector between the fine and coarse materials. Red solid lines show the internal layered interface of coarse - grained materials, while shows the possible internal layered interface of coarse - grained materials

5 Conclusions

In this study, we use signal attenuation to estimate the loss tangent value of the coarse material in lunar regolith beneath the motion path of the Chang'e-4 rover. Three stochastic media models are utilized to verify the estimation method. The simulation results demonstrate that the estimation is close to the real setting when the propagation medium is uniform with little scattering, and the error is about 15%. If, however, scattering targets are present in the study area, the results will be affected. For example, the result becomes larger when rocks appear above the calculation area. In this case, the estimated result can be considered as the upper limit of the loss tangent. When the rocks appear below the calculation area, the estimated result is smaller than the actual value. In addition, increasing the calculation area could improve the confidence of the estimation results.

The average loss tangent value of the coarse materials in the Chang'e-4 landing site is about 0.0104; considering the influence of the rocks, the upper limit of the loss tangent is 0.0120, and the lower limit is 0.087. Based on the empirical model of Apollo samples, the total FeO_T and TiO_2 content of coarse-grained materials was estimated to be 20.08 wt.%, higher than that of the upper fine-grained materials, which indicates different origins of these two parts.

Supplementary Materials: The following are available online at https://www.mdpi.com/article/10.3390/rs13204056/s1, Table S1: IDs of LPR CH-2 data.

Data Availability Statement: CE-4 LPR data are available at Data Publishing and Information Service System of China's Lunar Exploration Program (http://moon.bao.ac.cn). All the LPR data IDs are listed in Supplementary Table S1. Data sources of Figures 1 and 2 are given in the captions.

Acknowledgments

Scientific data are provided by the China National Space Administration (CNSA). We are grateful for the support from the team members of the Ground Application and Research System (GRAS), who contributed to data receiving and preprocessing.

References

[1] Fang, G.-Y.; Zhou, B.; Ji, Y.-C.; Zhang, Q.-Y.; Shen, S.-X.; Li, Y.-X.; Guan, H.-F.; Tang, C.-J.; Gao, Y.-Z.; Lu, W.; et al. Lunar Penetrating Radar onboard the Chang'e-3 mission. Res. Astron. Astrophys. 2014, 14, 1607—1622.

[2] Jia, Y. Z. ; Zou, Y. L. ; Ping, J. S. ; Xue, C. B. ; Yan, J. ; Ning, Y. M. The scientific objectives and payloads of Chang'E-4 mission. Planet. Space Sci. 2018, 162, 207—215.

[3] Wu, W. R. ; Liu, J. Z. ; Tang, Y. H. ; Yu, D. Y. ; Yu, G. B. ; Zhang, Z. China Lunar Exploration Program. J. Deep Space Explor. 2019, 6, 405—416.

[4] Robinson, M. S. ; Brylow, S. M. ; Tschimmel, M. ; Humm, D. ; Lawrence, S. J. ; Thomas, P. C. ; Denevi, B. ; Bowman-Cisneros, E. ; Zerr, J. ; Ravine, M. A. ; et al. Lunar Reconnaissance Orbiter Camera (LROC) Instrument Overview. Space Sci. Rev. 2010, 150, 81—124.

[5] Barker, M. K. ; Mazarico, E. ; Neumann, G. A. ; Zuber, M. T. ; Haruyama, J. ; Smith, D. E. A new lunar digital elevation model from the Lunar Orbiter Laser Altimeter and SELENE Terrain Camera. Icarus 2016, 273, 346—355.

[6] Li, C. ; Liu, D. ; Liu, B. ; Ren, X. ; Liu, J. ; He, Z. ; Zuo, W. ; Zeng, X. ; Xu, R. ; Tan, X. ; et al. Chang'E-4 initial spectroscopic identification of lunar far-side mantle-derived materials. Nat. Cell Biol. 2019, 569, 378—382.

[7] Huang, J. ; Xiao, Z. Y. ; Flahaut, J. ; Martinot, M. ; Head, J. ; Xiao, X. ; Xie, M. G. ; Xiao, L. Geological characteristics of Von Kármán crater, northwestern south pole-Aitken Basin: Chang'E-4 landing site region. J. Geophys. Res. Planet. 2018, 123, 1684—1700.

[8] Pasckert, J. H. ; Hiesinger, H. ; van der Bogert, C. H. Lunar farside volcanism in and around the South Pole-Aitken basin. Icarus 2018, 299, 538—562.

[9] Ling, Z. C. ; Qiao, L. ; Liu, C. Q. ; Cao, H. J. ; Bi, X. Y. ; Lu, X. J. ; Zhang, J. ; Fu, X. H. ; Li, B. ; Liu, J. Composition, mineralogy and chronology of mare basalts and non-mare materials in Von Kármán crater: Landing site of the Chang'E-4 mission. Planet. Space Sci. 2019, 179, 104741.

[10] Gou, S. ; Yue, Z. ; Di, K. ; Bugiolacchi, R. ; Wan, W. ; Yang, M. ; Ye, L. Geologically Old but Freshly Exposed Rock Fragments En-countered by Yutu-2 Rover. J. Geophys. Res. Planet. 2021, 126, e2020JE006565.

[11] Lin,H. ; He, Z. ; Yang, W. ; Lin, Y. ; Xu, R. ; Zhang, C. ; Zhu, M. —H. ; Chang, R. ; Zhang, J. ; Li, C. ; et al. Olivine-norite rock detected by the lunar rover Yutu-2 likely crystallized from the SPA-impact melt pool. Natl. Sci. Rev. 2020, 7, 913—920.

[12] Lai, J. ; Xu, Y. ; Bugiolacchi, R. ; Meng, X. ; Xiao, L. ; Xie, M. ; Liu, B. ; Di, K. ; Zhang, X. ; Zhou, B. ; et al. First look by the Yutu-2 rover at the deep subsurface structure at the lunar farside. Nat. Commun. 2020, 11, 1—9.

[13] Lai, J. ; Xu, Y. ; Zhang, X. ; Xiao, L. ; Yan, Q. ; Meng, X. ; Zhou, B. ; Dong, Z. ; Zhao, D. Comparison of Dielectric Properties and Structure of Lunar Regolith at Chang'e-3 and Chang'e-4 Landing Sites Revealed by Ground-Penetrating Radar. Geophys. Res. Lett. 2019, 46, 12783—12793.

[14] Li, C. ; Su, Y. ; Pettinelli, E. ; Xing, S. ; Ding, C. ; Liu, J. ; Ren, X. ; Lauro, S. E. ; Soldovieri, F. ; Zeng, X. ; et al. The Moon's farside shallow subsurface structure unveiled by Chang'E - 4 Lunar Penetrating Radar. Sci. Adv. 2020, 6, eaay6898.

[15] Wang, R. ; Su, Y. ; Ding, C. ; Dai, S. ; Liu, C. ; Zhang, Z. ; Hong, T. ; Zhang, Q. ; Li, C. A Novel Approach for Permittivity Estimation of Lunar Regolith Using the Lunar Penetrating Radar Onboard Chang'E - 4 Rover. Remote. Sens. 2021, 13, 3679.

[16] Zhang, L. ; Li, J. ; Zeng, Z. ; Xu, Y. ; Liu, C. ; Chen, S. Stratigraphy of the Von Kármán Crater Based on Chang'E - 4 Lunar Pene - trating Radar Data. Geophys. Res. Lett. 2020, 47, e2020GL088680.

[17] Zhang, J. ; Zhou, B. ; Lin, Y. ; Zhu, M. H. ; Song, H. ; Dong, Z. ; Ouyang, Z. Lunar regolith and substructure at Chang'E - 4 landing site in South Pole - Aitken basin. Nat. Astron. 2021, 5, 25 - 30.

[18] Dong, Z. ; Fang, G. ; Zhao, D. ; Zhou, B. ; Gao, Y. ; Ji, Y. Dielectric Properties of Lunar Subsurface Materials. Geophys. Res. Lett. 2020, 47, e2020GL089264.

[19] Dong, Z. ; Fang, G. ; Zhou, B. ; Zhao, D. ; Gao, Y. ; Ji, Y. Properties of Lunar Regolith on the Moon's Farside unveiled by Chang'E - 4 Lunar Penetrating Radar. J. Geophys. Res. Planet. 2021, 126, e2020JE006564.

[20] Dong, Z. ; Feng, X. ; Zhou, H. ; Liu, C. ; Zeng, Z. ; Li, J. ; Liang, W. Properties Analysis of Lunar Regolith at Chang'E - 4 Landing Site Based on 3D Velocity Spectrum of Lunar Penetrating Radar. Remote Sens. 2020, 12, 629.

[21] Li, C. ; Zhang, J. Velocity Analysis Using Separated Diffractions for Lunar Penetrating Radar Obtained by Yutu - 2 Rover. Remote Sens. 2021, 13, 1387.

[22] Su, Y. ; Fang, G. - Y. ; Feng, J. - Q. ; Xing, S. —G. ; Ji, Y. —C. ; Zhou, B. ; Gao, Y. — Z. ; Li, H. ; Dai, S. ; Xiao, Y. ; et al. Data processing and initial results of Chang'e - 3 lunar penetrating radar. Res. Astron. Astrophys. 2014, 14, 1623—1632.

[23] Grimm, R. E. ; Heggy, E. ; Clifford, S. ; Dinwiddie, C. ; McGinnis, R. ; Farrell, D. Absorption and scattering in ground - penetrating radar: Analysis of the Bishop Tuff. J. Geophys. Res. Space Phys. 2006, 111.

[24] O'Donnell, R. M. Introduction to Radar Systems; MIT Open Course Ware, Primavera; Massachusetts Institute of Technology: Cambridge, MA, USA, 2007.

[25] Ward, S. H. ; Hohmann, G. W. Electromagnetic Theory for Geophysical Applications. In Electromagnetic Methods in Applied Geophysics; Society of Exploration Geophysicists: Beijing, China, 1988; pp. 130—311.

[26] Zhang, X. ; Lv, W. ; Zhang, L. ; Zhang, J. ; Lin, Y. ; Yao, Z. Self - Organization Characteristics of Lunar Regolith Inferred by Yutu - 2 Lunar

Penetrating Radar. Remote Sens. 2021, 13, 3017.
[27] Jiang, Z.; Zeng, Z.; Li, J.; Liu, F.; Li, W. Simulation and analysis of GPR signal based on stochastic media model with an ellip - soidal autocorrelation function. J. Appl. Geophys. 2013, 99, 91—97.
[28] arren, C.; Giannopoulos, A.; Giannakis, I. gprMax: Open source software to simulate electromagnetic wave propagation for Ground Penetrating Radar. Comput. Phys. Commun. 2016, 209, 163—170.
[29] Xu, L.; Zhang, X.; Qiao, L.; Lai, J. Evaluating the Thickness and Stratigraphy of Ejecta Materials at the Chang'e-4 Landing Site. Astron. J. 2021, 162, 29.
[30] Zhang, L.; Xu, Y.; Bugiolacchi, R.; Hu, B.; Liu, C.; Lai, J.; Zeng, Z.; Huo, Z. Rock abundance and evolution of the shallow stratum on Chang'e-4 landing site unveiled by lunar penetrating radar data. Earth Planet. Sci. Lett. 2021, 564, 116912.
[31] Zhou, H.; Feng, X.; Ding, C.; Dong, Z.; Liu, C.; Zhang, Y.; Meng, Z. Yutu-2 Radar Sounding Evidence of a Buried Crater at Chang'E-4 Landing Site. IEEE Trans. Geosci. Remote Sens. 2021, PP, 1—19.
[32] Carrier, E. D., Ⅲ; Olhoeft, G. R.; Mendell, W. Physical properties of the lunar surface. In The Lunar Sourcebook; Cambridge University Press: Cambridge, UK, 1991; p. 543.
[33] Guo, D.; Fa, W.; Zeng, X.; Du, J.; Liu, J. Geochemistry of the Von Kármán crater floor and thickness of the non - mare ejecta over the Chang'e-4 landing area. Icarus 2021, 359, 114327.
[34] Zeng, Q.; Chen, S.; Zhang, Y.; Mu, Y.; Dai, R.; Yang, C.; Li, A.; Lu, P. Mineralogical and chemical properties inversed from 21 - lunar - day VNIS observations taken during the Chang'E-4 mission. Sci. Rep. 2021, 11, 1—10.
[35] Ohtake, M.; Haruyama, J.; Matsunaga, T.; Yokota, Y.; Morota, T.; Honda, C. Performance and scientific objectives of the SE - LENE (KAGUYA) Multiband Imager. Earth Planet. Space 2008, 60, 257—264.
[36] Qiao, L.; Ling, Z.; Fu, X.; Li, B. Geological characterization of the Chang'e-4 landing area on the lunar farside. Icarus 2019, 333, 37—51.

A Complex Paleo – Surface Revealed by the Yutu – 2 Rover at the Lunar Farside

Jialong Lai[1,2], Yi Xu[1], Roberto Bugiolacchi[1,3], Hon Kuan Wong[1], Luyuan Xu[1], Xunyu Zhang[1], Ling Zhang[1], Xiaoping Zhang[1], Long Xiao[4], Bin Liu[5], and Kaichang Di[5]

1. State Key Laboratory of Lunar and Planetary Sciences, Macau University of Science and Technology, Macau
2. School of Science, Jiangxi University of Science and Technology, Ganzhou, China
3. University College London, Earth Sciences, London, UK
4. School of Earth Sciences, Planetary Science Institute, China University of Geosciences, Wuhan, China
5. State Key Laboratory of Remote Sensing Science, Institute of Remote Sensing and Digital Earth, Chinese Academy of Sciences, Beijing, China

Abstract: The observable lunar surface is represented by a ubiquitous layer of fine – grained materials produced by billions of years of hypervelocity pounding of its crustal layer. The data from the Lunar Penetrating Radar onboard Chang'e – 4 (CE – 4) rover (Yutu – 2), which is exploring the Von Kármán Crater on the lunar farside, are helping to peel back the upper layer of finely comminuted materials, interpreted as a thick layer of ejecta from the neighboring Finsen crater, to reveal a complex paleo – surface morphology. During the rover's 560m journey, from depths of 7 — 20m, distinct variations in the returned signal characteristics reveal a possible 270 ± 10m buried crater with an estimated age of less than 100m. y. A smooth surface depression that lies to the southwest of the Yutu – 2 rover's travel path might not be the remnant of a degraded crater but a related surface expression of the hidden structure.

Plain Language Summary: The absence of an atmosphere and substantial resurfacing events in the last couple of billion years means that the Moon's surface has journeyed through space exposed to anything that travels through it and, unlike the Earth, most of these "encounters" have left a mark. Thus, the good news for scientists is that the lunar surface represents a record of astrophysical phenomena within the Solar System that is no longer available on its larger companion: our planet. However, the larger impacts have indeed contributed to modify its surface hiding some of its old terrains. The Lunar Penetrating Radar (LPR) onboard Chang'e – 4 (CE – 4) rover (Yutu – 2) can see below the surface as it makes its way across the lunar ground, revealing the ancient lunar surface (paleo-surface) at the depth of ~12m. It has mapped its elevation profile, which now we can see that it has been substantially modified following its formation some 3.1 billion years ago. Notably, the LPR has discovered a 270m sized buried crater corresponding to a large ground depression seen on the surface, a first achievement for lunar radar sounding.

Key Points:

· The CE-4 LPR reveals the paleosurface of the lunar farside at a depth between 7 and 20 m

· The LPR has mapped a 270 m sized buried crater probably related to a larger circular depression on the lunar surface

· The geomorphology, the subsurface structure, the impact melt deposits, and the exposure time of the buried crater, are analyzed

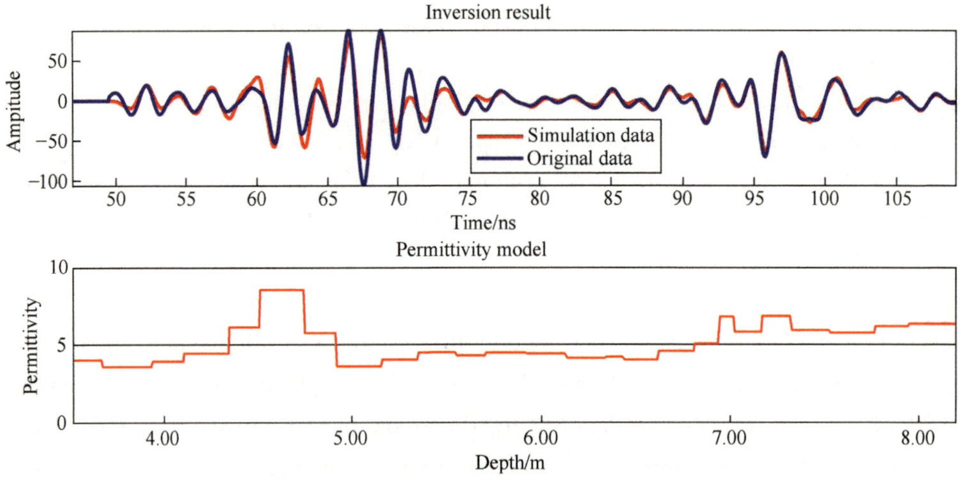

Fig. 1　Permittivity profile estimated with deconvolution method

1　Introduction

Impact craters represent the most salient geomorphological characteristic of the lunar surface and are a key to a better understanding of the geology of our satellite. Crater chronology, a dating methodology derived from returned lunar samples, remains the sole remote sensing methodology for estimating absolute surface ages across the solar system. The geomorphology of craters can constrain their age and shed light on the regional stratigraphy, set constraints on the thickness of lunar regolith, and assist in estimating the depth of the basalt layers. The impact events also exhume materials from depth granting access to analysis. However, the relentless reworking of the surface, from the micro to the macro level, remodels the lunar surface through a process of destruction and burial (Carrier et al., 1991). Gravity field variations interpreted from the GRAIL data (Evans et al., 2016; Sood et al., 2017) have revealed tens to hundreds of km sized craters hidden below the lunar maria. Ground Penetrating Radars (GPR) can reveal the shallow subsurface stratigraphy, including buried craters, and they have been deployed on Earth, Mars, and the Moon. On our planet, the scientific yield has been modest probably due to the high attenuation of GPR signals in moist environments and interferences caused by vegetation and human activities (Heggy & Paillou, 2006). On Mars, the dense data coverage of the Shallow Radar (SHARAD) sounder in the polar region enables a three-dimensional view of hidden structures and reveals a couple of km-scale buried craters (Putzig et al., 2018). However, due to their limited spatial resolution, SHARAD and GRAIL are not suitable for detecting small-scale and close-to-surface geological discontinuities.

The Lunar Penetrating Radar (LPR) onboard Chang'e-4 (CE-4) rover (Yutu-2) has been surveying the shallow subsurface around its landing site on the lunar farside for over 2 years. It shows the stratigraphic structure up to 330—500m deep (Lai et al., 2020; J. Zhang, Xu, et al., 2021), including a ~12-m thick regolith layer (Guo et al., 2021; Lai et al., 2019; Li et al., 2020; L. Zhang, Xu, et al., 2021), the basalt layers deposited by multiple lava eruption events (Lai et al., 2020; Yuan et al., 2021; L. Zhang et al., 2020), and evidence for surface modification events that occurred in the Von Kármán crater (VK) inside the South Pole-Aitken (SPA) basin (Lu et al., 2021; Qiao et al., 2019; Xiao et al., 2021; Xu et al., 2021) (Figure S1). Here, we report the discovery of a 270-m sized buried crater with clear diagnostic geomorphological features on the lunar farside using LPR data from the first 25 lunar days of the CE-4 mission. The LPR also reveals the paleo-surface below the area surveyed by the Yutu-2 rover at the depth of about 12m that was exposed at a time before the Finsen ejecta deposition (Huang et al., 2018).

2 Methods

The LPR detected a continuous subsurface reflector at the depth of ~12m, which has been subsequently interpreted as the lower interface of the fine-grained lunar regolith layer (Guo et al., 2021; Lai et al., 2019, 2020). As the Yutu-2 rover exploration journey progressed (Fig. 2), the reflector's signal has increased in depth for then reaching a plateau (details in Section 3), drawing what appears to be the outline of a small basin, probably an impact crater. Further, the radar signal strength and pulse width changed when the rover moved from the outside of the alleged crater to the rim (Fig. 3b), indicating that the dielectric properties of materials outside the crater are different from those on the crater walls. To test this hypothesis, we applied two methods to estimate the permittivity value of the materials.

2.1 Permittivity Value Estimation With Deconvolution Method

The normalized waveform of the output pulse of LPR CH-2 antenna $w(t)$ is obtained based on the permittivity value of rock samples of Apollo missions and received waveform given in Fang et al. (2014). We assume that the received radar signal $x(t)$ is the convolution results of $w(t)$ and reflection coefficients $r(t)$ (Robinson, 1985), if noises are ignored.

$$x(t)=w(t)*r(t) \tag{1}$$

$$r(t)=\{R_1, R_2, \cdots, R_n\} \tag{2}$$

R_n depends on the permittivity values of neighboring layers, as follows:

$$R_n=\frac{\sqrt{\varepsilon_n}-\sqrt{\varepsilon_{n+1}}}{\sqrt{\varepsilon_n}+\sqrt{\varepsilon_{n+1}}} \tag{3}$$

Xia et al. (2004) employed the deconvolution method (Yilmaz, 2001) in inverting a series of reflection coefficients of low - loss materials using GPR data (Xia et al., 2004); however, they obtained the non - discrete reflection coefficient. In this work, we applied a genetic algorithm (Holland, 1992) to infer the discrete reflection coefficients and corresponding permittivity values. Real - coded genetic algorithms (Goldberg, 1991) and adaptive genetic algorithm (Yang et al., 2016) are adopted here. The target function is set as

$$\psi=\min(\Sigma \mid x_r(t)-x_s(t) \mid) \tag{4}$$

where $x_r(t)$ is the measured value obtained from LPR data and $x_s(t)$ is the inversed waveform. The target function aims to minimize the differences between the measured values and the synthetic waveform. $x_s(t)$ can be obtained from quation 1, $r(t)$ is obtained from Equation 3, and model m of subsurface materials:

$$m=\{\varepsilon_1, \varepsilon_2, \cdots, \varepsilon_n, h_1, h_2, \cdots, h_n\} \tag{5}$$

where ε_n is the permittivity value of n_{th} layer and h_n is the thickness of n_{th} layer.

Here, we used radar data track No. 10976 as an example for permittivity inversion. Both the declining reflector (black arrows in Fig. 3b) caused by the suspected buried crater and hyperbolic curve caused by the rock in the regolith appears in the selected radar data. The geometric spreading correction of radar signal should be considered and performed as follows:

$$A_0=A\bar{v}t\,e^{\alpha t} \tag{6}$$

where A_0 is amplitude after compensation, A is the amplitude of receiving signals, t is the two - way timing delay, $\bar{v}$ is the average propagation speed of an electromagnetic wave in the subsurface materials, and α is the absorption factor, which can be expressed by 7.

$$\alpha=\pi f \tan\delta \sqrt{\varepsilon'}/c_0 \tag{7}$$

where f is the frequency of radar, $\tan\delta$ is loss tangent, c_0 is the light speed in vacuum, and ε' is the average permittivity value of materials. We assume that $\tan\delta=0.004$, $\varepsilon'=4$ (Lai et al., 2019) to compensate for the amplitude of the radar signal. The inversed permittivity profile is shown in Fig. 1.

The upper figure shows the synthetic waveform (red) and LPR data (blue). The figure below shows the inversed permittivity profile with depth.

The inferred permittivity of the rock that results in the hyperbolic shaped signal is

8. 6, while the ε' of the surface of the buried crater is 6. 8, the ε' of the material below the crater is 6.

2. 2 Permittivity Estimation Based on the Continuity of the Subsurface Reflector

Two continuous reflectors appear at the depth of ~250ns in the radar image, indicated by the black solid line and red dash line in Fig. S3a, when the aggregated traveling distance of the Yutu - 2 rover is 420 — 480m and 480 — 520m, respectively. Although the two reflectors connect at ~480m and probably are caused by the same interface, they do not follow the same trend. The main reason could be the different permittivity values of materials overlying the reflectors change the apparent depth of two reflectors in the radar image.

Therefore, we could infer the average permittivity value of materials in the area between the red dash line and the black solid line by assuming the two reflectors follow the same trend.

$$y(x) = y_1(x) + (y_2(x) - y_1(x)) * \mathrm{sqrt}(\bar{\varepsilon}_{up} / \bar{\varepsilon}_{down})$$

where x is the horizontal distance, $y_1(x)$ is the delay of the reflector indicated by the red dash line, $y_2(x)$ is the delay of the reflector indicated by the black solid line, $y(x)$ is the modified delay based on the unknown permittivity value. $\bar{\varepsilon}_{up}$ and $\bar{\varepsilon}_{down}$ are average permittivity values of the materials above and below the red dash line (the surface of the buried crater), respectively. Assuming $\bar{\varepsilon}_{up} = 4$, we obtaine that $\bar{\varepsilon}_{down} = 5.8$ (green dash line in Fig. S3b) and $\bar{\varepsilon}_{down} = 6.2$ (black dash line in Fig. S3b). From Fig. S3b, we observe the aforementioned reflector becomes straight in the region of 420—530m when $\bar{\varepsilon}_{down} = 5.8 - 6.2$. The estimated permittivity value also agrees with the result ($\varepsilon = 6$) inferred with the deconvolution method. Therefore, the subsurface layer (black/green dash line) shown in Fig. S3b might be a plausible geologic structure.

2. 3 Pearson Correlation

We investigated whether the paleo - surface revealed by LPR correlates to the observed surface topography by using the Pearson correlation parameter. The correlation value r is obtained with the following equation:

$$r = \frac{\sum XY - \frac{\sum X \sum Y}{N}}{\sqrt{\left(\sum X^2 - \frac{(\sum X)^2}{N}\right)\left(\sum Y^2 - \frac{(\sum Y)^2}{N}\right)}}$$

X, Y are input data. For example, X, Y to calculate the correlation between the paleo - surface and surface is the elevation data of paleo - surface and surface, respectively.

3　Results and Discussions

3.1　LPR Results

The LPR employs two sets of antennae each operating at different frequencies (centered at 500 and 60MHz) . The LPR data from the higher frequency channel are used in this work, with a resolution of 0.3m in vacuum (Fang et al. , 2014), and were collected during a ~563m long surveying path of the Yutu-2 rover. The current location of the Yutu-2 rover is close to a ~180m diameter rounded depression, resembling a highly degraded crater (Figure 2a) .

The LPR raw data underwent a series of steps, including repetitive data conditioning, noise filtering, background removal, and amplitude compensation. Details can be found in the Text S1 in Supporting Information S1.

Fig. 3 shows the processed radar image and the LPR signal in the green, uppermost subsurface area (Fig. 3c) to be relatively weak, indicating the materials within the region are rather homogenous and interpreted as representing the lunar regolith layer by previous studies (Lai et al. , 2020; J. Zhang, Xu, et al. , 2021; L. Zhang et al. , 2020) . The surface materials are interpreted as Eratosthenian ejecta materials, mostly originating from the nearby Finsen crater in the northeast of VK based on in-situ visible and near-infrared spectral observations of the Yutu-2 rover (Gou et al. , 2020; Lin et al. , 2020) (Fig. S1c) . The empirical model also predicts that the thickness of ejecta deposits from Finsen impact events is comparable to the lunar regolith stratum (the green layer in Fig. 3c) (Lai et al. , 2020; Yuan et al. , 2021) . Below the regolith layer, various scattering signals are present, along with several continuous LPR reflectors (red dash lines in Fig. 3c): these horizontal features are interpreted as the base of ejecta delivered by multiple impact events (Lai et al. , 2019) . From the sixteenth lunar day [indicated by the red triangle in Fig. 3a, the subsurface reflector (black solid line in Fig. 3c)] caused by the interface between regolith and ejecta appears to rise (black triangle in Fig. 3a and 3c) for a short distance at the traveling distance of 340m then declining sharply between 380 and 480m (between the red and green triangles in Fig. 3a and 3c) . After 480m, the reflector flattens and then gradually rises again at the end of the travel path. The shape of this part of the subsurface reflector evokes the cross-section of an impact crater.

Please note that although piece-wise permittivity is adopted here for delay-depth conversion, the real permittivity varies with location and will introduce some distortion of the subsurface geometry. In Fig. S3b, the boundary between the two values of permittivity is set as the subsurface reflector and marked by the black line in Fig. 3c.

The geometry correction for the varying permittivity does not alter the crater-like shape but may cause some errors in estimating the slope of the crater wall.

Overall, this strong reflection boundary might represent the paleo - surface beneath the surveyed path (Fig. 3c) . Below this depth, several linear discontinuities suggest structural geology features, such as the "funnel" - like structure reminiscent of a flattened horst and graben region, with associated normal faults (yellow dash lines in Fig. 3c) . The layered structure of ejecta deposits, still visible and horizontal on the left of the image, might have been dislodged and plastically deformed by the impact event (red dashed lines in Fig. 3c) . It may represent the first direct evidence of the consequences of the shock effects in the lunar subsurface, i. e. , faulting, fractures, and chaotic bedrock (Heggy & Paillou, 2006) .

3. 2 Buried Crater

The visible surface depression that lies to the southwest of the Yutu - 2 rover's travel path outlined in Fig. 2 might represent the surface expression of the potential buried crater detected by the LPR onboard the rover. The contour lines and 3D view (Fig. 2b and 2c) shows the shape of the depression and the relative positions of the LPR scanning line to the center of the depression. When the rover crossed the edge of the crater (black dash line in Fig. 2b), the observed subsurface reflector began to fall, indicated by the red dots in Fig. 2b. The distance from the rover to the depression center has a strong correlation with the topography of the subsurface reflector, illustrated by the purple line and red dash line in Figure S3. The Pearson correlation value between the two sets of data is 0. 96. The correlation is relatively weak toward the end of the path because of the possible effect of another depression to the north (Fig. 2c) or resulting from the modification stage of the underlying crater. Fig. 2b shows that the rover only traveled along a short yet non - straight path inside the buried crater and did not cross the center, so the flat part of the subsurface reflector probably does not represent the crater floor (Fig. S4) . The slope of the subsurface reflector is 10°. Assuming the depression (circled by the white dash line in Fig. 2a) is caused by the detected buried crater and, given its deduced size, it represents a simple bowl - shaped crater, the diameter of the buried crater can be estimated at 270 ± 10m and the radial gradient of the slope at 29°. Considering the resolution of the surface DEM is 5 m, the range of the radial gradient is 24°—36. 5°. Craters can be classified as A, AB, B, BC, C types, where A type represents the youngest impacts with walls having the steepest angle of slope and the most pristine rim morphology, while C is the most degraded (Basilevskii, 1976) . Accordingly, the estimated age of the crater can be inferred based on these considerations and relating to their diameter (Basilevskii, 1976) . Within this model, the buried crater would belong to either the A or AB type, and its age could be no older than 100m. y. according to the current models of destruction rates on the lunar surface (Basilevskii, 1976; Fassett & Thomson, 2014) .

Multiple scattering signals draw a detectable but fuzzy upper subsurface boundary for most of the Yutu - 2 journey (red arrows in Fig. 3b), but between 390 and 440m of the

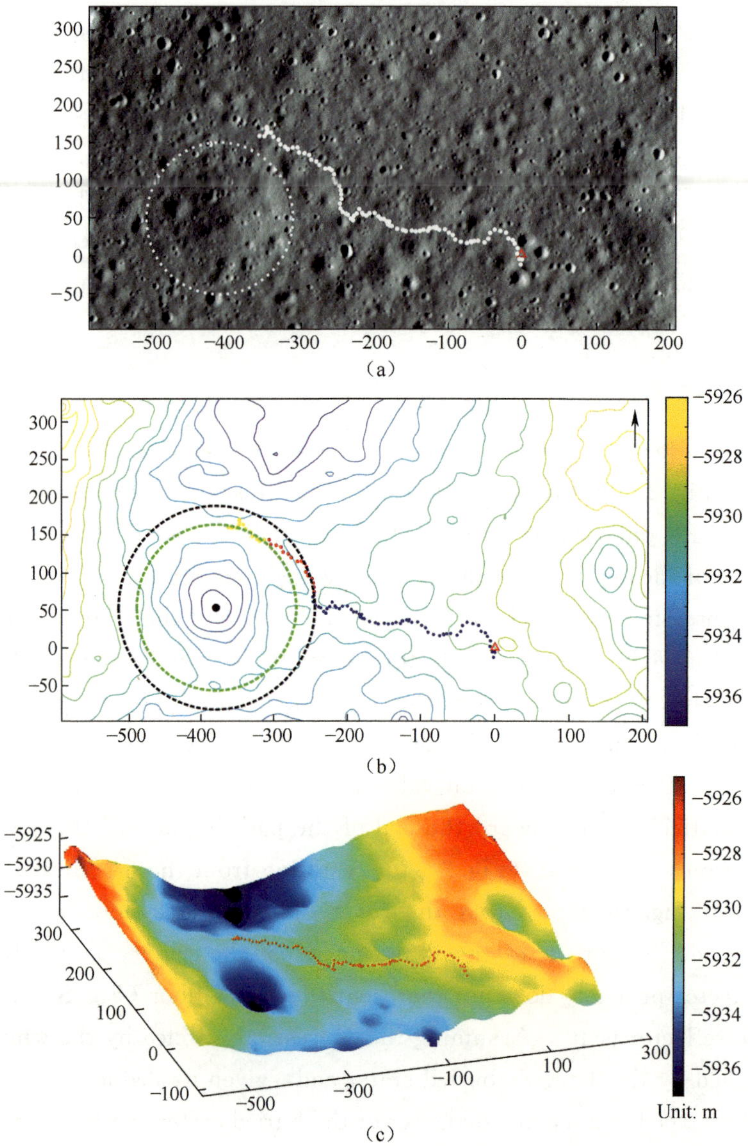

Fig. 2　The first 25 lunar days traversing path of the Yutu - 2 rover. (a) The LROC (Robinson et al., 2010) Narrow - Angle Camera (NAC) image of Yutu - 2 rover surveying area. The black arrow indicates North. The landing site of the Chang'e - 4 mission is marked by a red triangle and the waypoints of the rover's path are marked as white dots. The current cumulative length of LPR data collection is ～563m. The circle indicated by the white dash line marks the outline of the depression close to the traveling path of the Yutu - 2 rover (NAC image ID: M134022629LE, 1.27m/pixel, 77.55° incidence angle). (b) In the contour plot of the CE - 4 landing site, the color bar refers to the elevation in meters. The blue, yellow, and red dots represent the traveling path, and the color differences help tracing the underlying paleo - surface revealed by radar image by turning red above the descending and yellow when flat or rising. The starting and ending positions of the declining part are also marked as red and green triangles in the radar image shown in Fig. 3a and 3c, respectively. (c) The 3D view of the surveying area. Surface DEM data were derived from Lunar Reconnaissance Orbiter (LRO) Narrow Angle Camera (NAC) stereo pair image (Robinson, 2019). The unit of X, Y axle and color bar are in meters

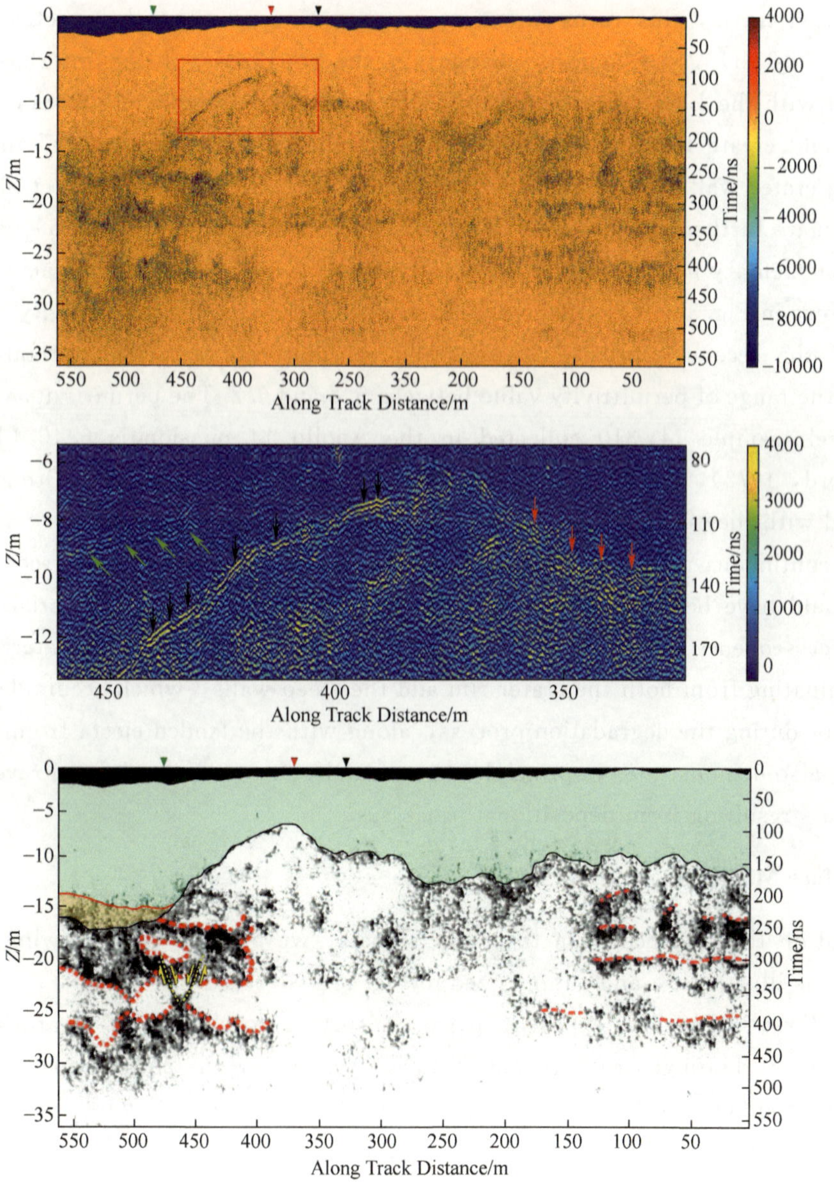

Fig. 3　Processed Lunar Penetrating Radar results of the Chang'e - 4 surveyed area. (a) Lunar Penetrating Radar results of CE - 4 site of the first 25 lunar days after processing. The color bar refers to the amplitude of the radar signal. The Z position of the subsurface reflectors is converted with $\bar{\varepsilon}=4.3(\leqslant 154\text{ns})/6(>154\text{ns})$. The black marker above the figures shows the starting position of the rim of the buried crater, the red triangle marks the position where the subsurface reflector starts to decline, and the green triangle marks the position where the subsurface reflector flattens. (b) Enlarged radar image of the red rectangle as in Figure 3a. The color bar in (a) and (b) refers to the amplitude of the radar signal in mV. Red arrows point to a fuzzy subsurface boundary containing multiple scattering signals. Black arrows indicate that the boundary becomes clear and sharp when the traverse distance is between 390 and 440m. Green arrows denote the continuous reflectors within the buried crater that could have been caused by the depositing events. (c) Enhanced radar image. The Green region highlights the relatively homogeneous radar reflections overlying the alleged paleo - surface. Red dashed lines mark possible disruption boundaries of plastic deformations of the rocks at depth due to the force of the impact. The yellow dashed lines represent a fault - like morphology. The brown "wedge" represents the probable accumulation of debris (scree) at the bottom of the crater. Note that the scanning sequence of the LPR shown here runs from right to left

accumulated travel, this turns into a strong, continuous, and sharp reflector (black arrows in Fig. 3b). It suggests changes in the roughness of the interface that are consistent with the LPR scanning the parts of the buried crater wall. Rocks or a rough surface could create hyperbolic - shaped radar signals, while "smoother" impact melt plastering crater walls would produce sharp and clear reflectors (Stopar et al., 2014). To investigate further the presence of impact melt - rich deposits (Fig. S5), we applied two complementary methods to obtain the permittivity values of the materials beneath the black arrows in Fig. 3b. The deconvolution method estimates the permittivity value to be 6.8, and the second approach assumes the continuity of a subsurface interface and provides the range of permittivity value between 5.8 and 6.2. The permittivity value of the impact melt sample 14, 310 collected in the Apollo 14 mission is 6.46 (Bassett & Shackelford, 1972), which is compatible with our estimation results. Other materials associated with the permittivity value in the range of 5.8—6.5 are breccia.

The continuous reflectors within the buried crater indicated by the green arrows in Fig. 3b could have been caused by the depositing events of infilling materials or a thin ejecta layer sequence of a series of small impact events. The infilling materials include scree originating from both the crater rim and the steep walls, which accumulated on the crater floor during the degradation process, along with the landed ejecta from the Finsen crater. It also demonstrates in principle that ground - penetrating radar can reveal level or tilted layers resulting from depositional processes.

3.3 Surface Modification

Based on the evidence from the radar signals, we propose the following geological scenario, as illustrated, and summarized in Fig. 4:

(A) First stage: The exposed (paleo -) surface post - impact, featuring a fresh ~300m crater. Through erosional and diffusive processes, the post - impact topography gradually smoothens; (B) Second stage: this process continues while scree materials accumulate, regolith forms, and impact ejecta from distal events are deposited; (C) Third stage: ejecta from the Finsen impact event arrives burying the crater, betraying its presence on the surface by a subdued ghost topogravaphy; Calculations suggest less than 100m. y. passed between the excavation of the buried crater and the Finsen impact; and (D) Fourth stage: After surface modification of about 3.1Ga (Lu et al., 2021), only a shallow surface depression remains visible, barely visible to the naked eye as a highly degraded crater Gravity - led wasting, impacts and space weathering have reshaped the lunar surface profoundly. The elevation variation between the current surface and the alleged paleo - surface is 11.5m on average, ranging from 5.8 to 17.6m. It should be noted that the Yutu - 2 rover traveling path is designed to minimize hazards, principally surface roughness and steep gradients: consequently, although radar image shows that the paleo - surface changed to a much flatter surface with average topographical variations

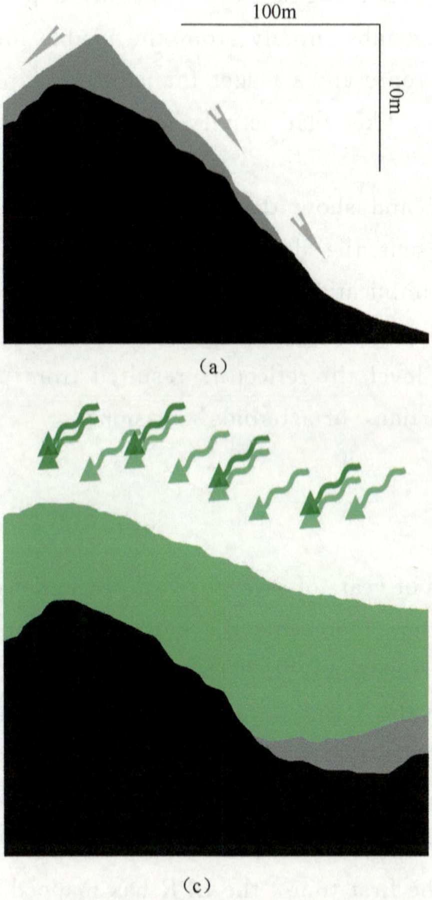

Fig. 4　Conceptual model of the surface modification processes occurred at the Chang'e - 4 landing site. (a) First stage: early erosional and diffusional phase. (b) Second stage: the (paleo -) surface undergoes further erosion, the effects of space weathering, and the accumulation of random ejecta material. (c) Third stage: ejecta from the Finsen impact buries the crater. (d) Fourth stage: gardening processes endure. The black color indicates the detected paleo - surface boundary; the gray color represents the gravitationally mobilized materials produced during the erosional and diffusional phase; the green parts are delivered ejecta materials

of 1.1m, it is not sufficient to prove that the VK crater floor before Finsen impact is much rougher than the current surface is.

Although the lunar surface was reshaped by the events described above, the paleo - surface still affects the elevation of the surface. The Pearson correlation value between the paleo - surface and the current surface of the entire traveling path is 0.89, and the value increases to 0.95 for the buried crater part. The alignment between the two surface profiles might be explained by the absence of other significant ejecta/impact events post - Finsen.

CE - 4 LPR provides a unique opportunity to observe ancient lunar surfaces ("paleo -

surfaces") hidden by the ubiquitous regolith layer (ranging from 7 to 20m, depending on location) which could, as in this case, differ substantially from the visible interface. Also, the location of a paleo - surface could represent a target for a future exploration mission trying to study the space environment, like solar activity levels, at the time of burial (Crawford et al. , 2021) .

A 270m sized buried crater is identified and show that the corresponding highly degraded crater on the surface might represent the depression originating from the underlying structure, which has important implications for future crater geomorphic studies and crater statistical analysis. LPR also demonstrates its capability to capture the features of middle - small scale craters, and level/tilt reflectors resulted from previous depositions as references for future lunar, Martian, or asteroids' missions.

4 Conclusions

The lunar surface bears witness to billions of years of astrophysical phenomena within the solar system and the study of impact craters represents the most important tool in decoding its complex geologic history (e. g. , Hartmann, 1970) .

The ground - penetrating radar (LPR) on the Yutu - 2 rover of Change - 4 (CE - 4) mission has been surveying the lunar farside since the beginning of 2019. In this paper, we report the first 25 lunar days of LPR data revealing a complex ancient lunar surface (paleo - surface), the elevation profile of which has been substantially modified after about 3. 1 billion years since formation. For the first time, the LPR has mapped a 270m sized buried crater resembling a highly degraded crater as seen on the lunar surface. The geomorphology, the subsurface structure affected by the shock wave, the impact melt deposits, and the derived exposed time of the crater, are analyzed in the paper.

Previous publications on CE - 4 LPR results mainly focus on the layered subsurface structure of the CE - 4 surveying area and corresponding geologic history. However, this paper offers a different perspective from prior LPR results by showing the paleo - surface of the lunar farside at the depth of 12m and the first mapping of the clear profile of a buried crater that had not been mentioned or discussed in other related work. The significance of the findings is high: although other missions have found large - scale buried craters on both Mars and the Moon, our results are outstanding in terms of data resolution and accuracy.

Acknowledgments

Scientific data are provided by the China National Space Administration (CNSA). We are grateful for the support from the team members of the Ground Application and

Research System (GRAS), who contributed to data receiving and preprocessing. We are grateful for the constructive suggestions of two anonymous reviewers. This study is supported by the Science and Technology Development Fund (FDCT) of Macau (Grants 0042/2018/A2, 0089/2018/A3, 0079/2019/A2, and 0049/2020/A1), The National Natural Science Foundation of China, Grant number12103020, the Pre－research Project on Civil Aerospace Technologies of CNSA (D020101), the Science and technology project of Jiangxi education department (Grant GJJ200821) and the Scientific Research Starting Foundation for scholars from Jiangxi University of Science andTechnology (Grant jxxjbs18017)

References

[1] Basilevskii, A. T. (1976). On the evolution rate of small lunar craters. The Lunar and Planetary Science Conference, 1, 1005－1020.

[2] Bassett, H. L., & Shackelford, R. G. (1972). Dielectric properties of Apollo 14 lunar samples at microwave and millimeter wavelengths. Lunar and Planetary Science Conference Proceedings (Vol. 3, p. 3157).

[3] Carrier, W. D., Olhoeft, G. R., & Mendell, W. (1991). Physical Properties of the Lunar Surface. Lunar Sourceb, 522－530.

[4] Crawford, I. A., Joy, K. H., Pasckert, J. H., & Hiesinger, H. (2021). The lunar surface as a recorder of astrophysical processes. Philosophical Transactions of the Royal Society A: Mathematical Physical and Engineering Sciences, 379, 20190562. https://doi.org/10.1098/rsta.2019.0562

[5] Evans, A. J., Soderblom, J. M., Andrews - Hanna, J. C., Solomon, S. C., & Zuber, M. T. (2016). Identification of buried lunar impact craters from GRAIL data and implications for the nearside maria. Geophysical Research Letters, 43, 2445－2455. https://doi.org/10.1002/2015gl067394

[6] Fang, G. Y., Zhou, B., Ji, Y. C., Zhang, Q. Y., Shen, S. X., Li, Y. X., et al. (2014). Lunar Penetrating Radar onboard the Chang'e - 3 mission. Research in Astronomy and Astrophysics, 14 (12), 1607－1622. https://doi.org/10.1088/1674－4527/14/12/009

[7] Fassett, C. I., & Thomson, B. J. (2014). Crater degradation on the lunar maria: Topographic diffusion and the rate of erosion on the Moon. Journal of Geophysical Research: Planets, 119 (10), 2255 － 2271. https://doi.org/10.1002/2014je004698

[8] Goldberg, D. (1991). Real - coded Genetic Algorithms, Virtual Alphabets, and Blocking. Complex Systems, 5.

[9] Gou, S., Yue, Z., Di, K., Wang, J., Wan, W., Liu, Z. et al. (2020).

Impact melt breccia and surrounding regolith measured by Chang'e - 4 rover. Earth and Planetary Science Letters, 544, 116378. https://doi.org/10.1016/j.epsl.2020.116378

[10] Guo, D., Fa, W., Zeng, X., Du, J., & Liu, J. (2021). Geochemistry of the Von Kármán crater floor and thickness of the non - mare ejecta over the Chang'e -4 landing area. Icarus, 359, 114327. https://doi.org/10. 1016/j. icarus. 2021. 114327.

[11] Hartmann, W. K. (1970). Lunar cratering chronology. Icarus, 13 (2), 299—301. https://doi.org/10.1016/0019—1035 (70) 90059—x.

[12] Heggy, E., & Paillou, P. (2006). Probing structural elements of small buried craters using ground - penetrating radar in the southwestern Egyptian desert: Implications for Mars shallow sounding. Geophysical Research Letters, 33 (5). https://doi.org/10.1029/2005gl024263.

[13] Holland, J. H. (1992). Adaptation in natural and artificial systems: An introductory analysis with applications to biology, control, and artificial intelligence. MIT press.

[14] Huang, J., Xiao, Z., Flahaut, J., Martinot, M., Head, J., Xiao, X., et al. (2018). Geological characteristics of Von Kármán crater, northwestern south pole - Aitken Basin: Chang'E - 4 landing site region. Journal of Geophysical Research: Planets, 123 (7), 1684—1700. https://doi.org/10.1029/2018je005577.

[15] Lai, J., Xu, Y., Bugiolacchi, R., Meng, X., Xiao, L., Xie, M., & Xu, L. (2020). First look by the Yutu - 2 rover at the deep subsurface structure at the lunar farside. Nature Communications, 11 (1), 1—9. https://doi.org/10.1038/s41467—020—17262—w.

[16] Lai, J., Xu, Y., Zhang, X., Xiao, L., Yan, Q., Meng, X., et al. (2019). Comparison of dielectric properties and structure of lunar regolith at Chang'e - 3 and Chang'e - 4 landing sites revealed by ground - penetrating radar. Geophysical Research Letters, 46 (22), 12783 — 12793. https://doi.org/10.1029/2019GL084458.

[17] Li, C., Su, Y., Pettinelli, E., Xing, S., Ding, C., Liu, J., & Zhang, H. (2020). The Moon's farside shallow subsurface structure was unveiled by Chang'E - 4 Lunar Penetrating Radar. Science Advances, 6 (9), eaay6898. https://doi.org/10.1126/sciadv.aay6898.

[18] Lin, H., Lin, Y., Yang, W., He, Z., Hu, S., Wei, Y., & Zou, Y. (2020). New insight into lunar regolith - forming processes by the lunar rover Yutu - 2. Geophysical Research Letters, 47 (14), e2020GL087949. https://doi.org/10.1029/2020gl087949.

[19] Lu, Y., Wu, Y., Michael, G. G., Ma, J., Cai, W., & Qin, N. (2021). Chronological sequence of Chang'E - 4 landing zone within Von Kármán

crater. Icarus，354，114086. https://doi. org/10. 1016/j. icarus. 2020. 114086.

[20] Putzig,N. E.，Smith，I. B.，Perry，M. R.，Foss，F. J.，Ⅱ，Campbell，B. A.，Phillips，R. J.，& Seu，R. (2018). Three - dimensional radar imaging of structures and craters in the Martian polar caps. Icarus，308，138－147. https://doi. org/10. 1016/j. icarus. 2017. 09. 023.

[21] Qiao，L.，Ling，Z.，Fu，X.，& Li，B. (2019). Geological characterization of the Chang'e - 4 landing area on the lunar farside. Icarus，333，37 - 51. https://doi. org/10. 1016/j. icarus. 2019. 05. 029.

[22] Robinson,E. A. (1985). Seismic time - invariant convolutional model. Geophysics，50 (12)，2742－2751. https://doi. org/10. 1190/1. 1441894.

[23] Robinson，M. (2019). Topographic Map of the Chang'e 4 Site. Retrieved from http: //www. lroc. asu. edu/posts/1100.

[24] Robinson，M. S.，Brylow，S. M.，Tschimmel，M.，Humm，D.，Lawrence，S. J.，Thomas，P. C.，et al. (2010). Lunar reconnaissance orbiter camera (LROC) instrument overview. Space Science Reviews，150 (1－4)，81－124. https://doi. org/10. 1007/978－1－4419－6391－8 _ 6.

[25] Sood，R.，Chappaz，L.，Melosh，H. J.，Howell，K. C.，Milbury，C.，Blair，D. M.，& Zuber，M. T. (2017). Detection and characterization of buried lunar craters with GRAIL data. Icarus，289，157－172. https://doi. org/10. 1016/j. icarus. 2017. 02. 013.

[26] Stopar，J. D.，Hawke，B. R.，Robinson，M. S.，Denevi，B. W.，Giguere，T. A.，& Koeber，S. D. (2014). Occurrence and mechanisms of impact melt emplacement at small lunar craters. Icarus，243，337－357. https://doi. org/10. 1016/j. icarus. 2014. 08. 011.

[27] Xia，J.，Franseen，E. K.，Miller，R. D.，& Weis，T. V. (2004). Application of deterministic deconvolution of ground - penetrating radar data in a study of carbonate strata. Journal of Applied Geophysics，56 (3)，213－229. https://doi. org/10. 1016/s0926－9851 (04) 00054－0.

[28] Xiao，Z.，Ding，C.，Xie，M.，Cai，Y.，Cui，J.，Zhang，K.，& Wang，J. (2021). Ejecta From the Orientale Basin at the Chang'E - 4 Landing Site. Geophysical Research Letters，48 (3)，e2020GL090935. https://doi. org/10. 1029/2020gl090935.

[29] Xu，L.，Zhang，X.，Qiao，L.，& Lai，J. (2021). Evaluating the thickness and stratigraphy of Ejecta materials at the Chang'e - 4 landing site. The Astronomical Journal，162 (1)，29. https://doi. org/10. 3847/1538－3881/abf8b0.

[30] Yang，C.，Qian，Q.，Wang，F.，& Sun，M. (2016). An improved adaptive genetic algorithm for function optimization. In 2016 IEEE International Conference on Information and Automation (ICIA) (pp. 675－680). IEEE. https://doi. org/

10. 1109/icinfa. 2016. 7831905.

[31] Yilmaz, Ö. (2001) . Processing, inversion, and interpretation of seismic data. Seismic Data Analysis, 231—233.

[32] Yuan, Y. , Zhu, P. , Xiao, L. , Huang, J. , Garnero, E. J. , Deng, J. , & Li, W. (2021) . Intermittent volcanic activity detected in the Von Kármán crater on the farside of the Moon. Earth and Planetary Science Letters, 569, 117062. https://doi. org/10. 1016/j. epsl. 2021. 117062.

[33] Zhang, J. , Zhou, B. , Lin, Y. , Zhu, M. H. , Song, H. , Dong, Z. , et al. (2021) . Lunar regolith and substructure at Chang'E - 4 landing site in the South Pole - Aitken basin. Nature Astronomy, 5 (1), 25—30. https://doi. org/10. 1038/s41550—020—1197—x.

[34] Zhang, L. , Li, J. , Zeng, Z. , Xu, Y. , Liu, C. , & Chen, S. (2020) . Stratigraphy of the Von Kármán crater based on Chang' E - 4 lunar penetrating radar data. Geophysical Research Letters, 47 (15), e2020GL088680. https://doi. org/10. 1029/2020gl088680.

[35] Zhang, L. , Xu, Y. , Bugiolacchi, R. , Hu, B. , Liu, C. , Lai, J. , & Huo, Z. (2021) . Rock abundance and evolution of the shallow stratum on Chang'e - 4 landing site unveiled by lunar penetrating radar data. Earth and Planetary Science Letters, 564, 116912. https://doi. org/10. 1016/j. epsl. 2021. 116912.

First look by the Yutu – 2 rover at the deep subsurface structure at the lunar farside

Jialong Lai[1,2], Yi Xu[1], Roberto Bugiolacchi[1,3], Xu Meng[1,4], Long Xiao[1,5], Minggang Xie[1,6], Bin Liu[7], Kaichang Di[7], Xiaoping Zhang[1], Bin Zhou[8,9], Shaoxiang Shen[8,9] & Luyuan Xu[1]

1. State Key Laboratory of Lunar and Planetary Sciences, Macau University of Science and Technology, Macau, China

2. School of Science, Jiangxi University of Science and Technology, Ganzhou, China

3. University College London, Earth Sciences, London, UK

4. School of Civil Engineering, Guangzhou University, Guangzhou, China

5. Planetary Science Institute, School of Earth Sciences, China University of Geosciences, Wuhan, China

6. College of Science, Guilin University of Technology, Guilin, China

7. State Key Laboratory of Remote Sensing Science, Aerospace Information Research Institute, Chinese Academy of Science, Beijing, China

8. Key Laboratory of Electromagnetic Radiation and Detection Techonology, Chinese Academy of Sceience, Beijing, China

9. Aerospace Information Research Institute, Chinese Academy of Science, Beijing, China

The unequal distribution of volcanic products between the Earth – facing lunar side and the farside is the result of a complex thermal history. To help unravel the dichotomy, for the first time a lunar landing mission (Chang'e – 4, CE – 4) has targeted the Moon's farside landing on the floor of Von Kármán crater (VK) inside the South Pole – Aitken (SPA). We present the first deep subsurface stratigraphic structure based on data collected by the ground – penetrating radar (GPR) onboard the Yutu – 2 rover during the initial nine months exploration phase. The radargram reveals several strata interfaces beneath the surveying path: buried ejecta is overlaid by at least four layers of distinct lava flows that probably occurred during the Imbrium Epoch, with thicknesses ranging from 12m up to about 100m, providing direct evidence of multiple lava – infilling events that occurred within the VK crater. The average loss tangent of mare basalts is estimated at 0. 0040—0. 0061.

Unraveling the shallow subsurface structure of the lunar mare offers the key to a better understanding of the local history of basaltic volcanism, an important process coupled to the Moon's thermal evolution[1]. The thickness and surface area of basalt layers can be used to constrain lava eruption volumes. A range of remote – sensing data including the study of impact craters morphology[2,3], the analysis of high - resolution gravity data[4], and the reflectance spectra of crater ejecta deposits[5-8] have contributed to developing the current model of lunar evolution.

Ground - penetrating radars on the lunar surface and radar sounders onboard orbiting spacecraft have helped to investigate the physical properties of the subsurface materials and their possible stratigraphy. The Apollo Lunar Sounder Experiment, part of the Apollo 17 mission, was the first instrument to detect deep subsurface reflectors corresponding to the interface between mare and bedrock[9] at average apparent depths of 1—1.6km in Mare Serenitatis, Mare Crisium, and Oceanus Procellarum[9-11]. The apparent depth is defined as the propagation depth of a radar signal with the speed of light in the vacuum. Later, the Lunar Radar Sounder onboard the Kaguya spacecraft (SELENE) observed relatively shallow reflectors interpreted as subsurface boundaries between distinct basaltic rock layers in the nearside maria at apparent depths in the range of hundreds of meters[12-14].

Compared with the spaceborne radar experiment, the lunarpenetrating radar (LPR) onboard Chang'e - 3 (CE - 3) and Chang'e - 4 (CE - 4) rover have a much higher range resolution (1 — 2m in the mare basalt for the 60MHz channel), thus offering a unique opportunity to survey in greater detail the shallow subsurface of both the lunar nearside and farside[15-17]. CE - 4 landed in the South Pole - Aitken (SPA) Basin, the largest known impact structure on the Moon and a key region ideally suited to address several outstanding geological questions as the impact might have even penetrated the entire lunar crust[18,19]. The radargrams produced from the data acquired by the CE - 4 instrument reveal the basalt layer thickness of each lava eruption and the time sequence of surface modification events that occurred in the Von Kármán (VK) crater (Supplementary Fig. 1). In a broader context, this new information adds to our limited understanding of the igneous history of the SPA Basin, which is thought to have been significantly shorter and less extensive than its equivalent on the nearside[1,20]. The reason for the asymmetric distribution between the lunar sides is understood to relate either to differences in crustal thickness, to the abundance of radioactive elements, or to the geological consequences of the large SPA - forming impact itself[21-27].

In this work, we report the LPR results for the first 9 months derived from channel one (CH - 1, 60MHz) data and test our interpretations using LPR simulation. A stratigraphic model of the surveying area (landing coordinates based on Lunar Reconnaissance Orbiter terrain data: 177.5885° E, 45.4561° S, — 5927m[28,29]) was generated from the extracted reflectors profile, which suggests possible lava flows sources and a potentially complex buried topography. The local geological history of the CE - 4 landing site is inferred based on the revealed stratigraphy.

Results

Geologic settings. VK crater (171km) lies within the SPA basin, an impact crater about 2500km in diameter. The thermal history of the crater and its neighborhood thus should be interpreted within an atypical geological context[19-25]. During the Late Heavy Bombardment (LHB[30]) period, several giant impacts including Imbrium on the nearside and VK's northern neighbor, crater Leibnitz (245km in diameter) were produced. Post LHB, the region underwent a relatively prolonged phase of lava infill, which lasted about 200 — 600Ma[21,31], with the youngest flows estimated between 3.15 and 3.6Ga[19,21]. However, currently, no direct evidence of the volcanic history of VK crater indicates whether the mare deposits were formed by one episode of basaltic volcanism based on the uniform reflectance spectral characteristics or multiple lava - infilling events[19]. LPR can provide first - hand data to disclose the subsurface stratigraphy and constraint the thermal history.

VK's neighboring region is geologically highly complex: the map I - 1047[31] and the inset[32] (Supplementary Fig. 1) show a superposition of impact morphologies spanning from the pre - Nectarian to the Copernican epochs. The neighboring impacts produced ejecta materials that punctuated the infill and post - infill phase of the VK crater. The time sequence of these craters is relevant to the interpretation of the stratigraphy at the CE - 4 exploration path, which is analyzed in Supplementary Note 1.

Lunar—penetrating radar results. The penetrating depth of LPR CH - 1 can reach up to ~330m (Supplementary Note 2 and Supplementary Fig. 2), although the top section of the radar signals becomes saturated due to the strong coupling effects from the electromagnetic interaction with the metal in the rover. However, channel two (CH - 2) of the LPR data, the center frequency of which is 500MHz and can penetrate up to ~35m[17], can be employed to complement the profile of the close - to - surface section[17] (Fig. 1c). Here we focus on the LPR data analysis between 52 and 328m.

The prominent and continuous subsurface reflectors A - E at depths of (A) 51.8 ± 1.1m, (B) 63.2 ± 1.2m, (C) 96.2 ± 3.2m, (D) 130.2 ± 3.7m, and (E) 225.8 ± 5.5m can be observed both in the processed radar image and aggregated data traces displayed in terms of signal strength (dB, yellow line) (Fig. 1a). The horizontal reflectors appear relatively constant running parallel to the surface (see Fig. 1), except for the horizontal reflectors D that shows a gradual rise of 7.1m in the right end. This is probably due to the change in subsurface topography, e.g., crater at depth of 130m (see simulation results in Supplementary Fig. 4). From around waypoint 42, the reflector D becomes flat, because the rover conducted a local exploration mission to collect other scientific data at the end of the ninth month exploration with consequent little variation of the subsurface topography. Nonetheless, this localized and repeated sampling phase helps to constrain the consistency and reliability of the data gathering process.

The materials between the most prominent horizontals are rather uniform and strong radar echo are rare (e.g., A—B, B - C, C - D in Fig. 1b); however, a couple of subtle features stand out at the bottom part region of the radargram using image enhancement technique. Some relatively short lines appear in the D - E strata and more continuous ones occur below reflector E, which are interpreted as ejecta at a different scale. As the

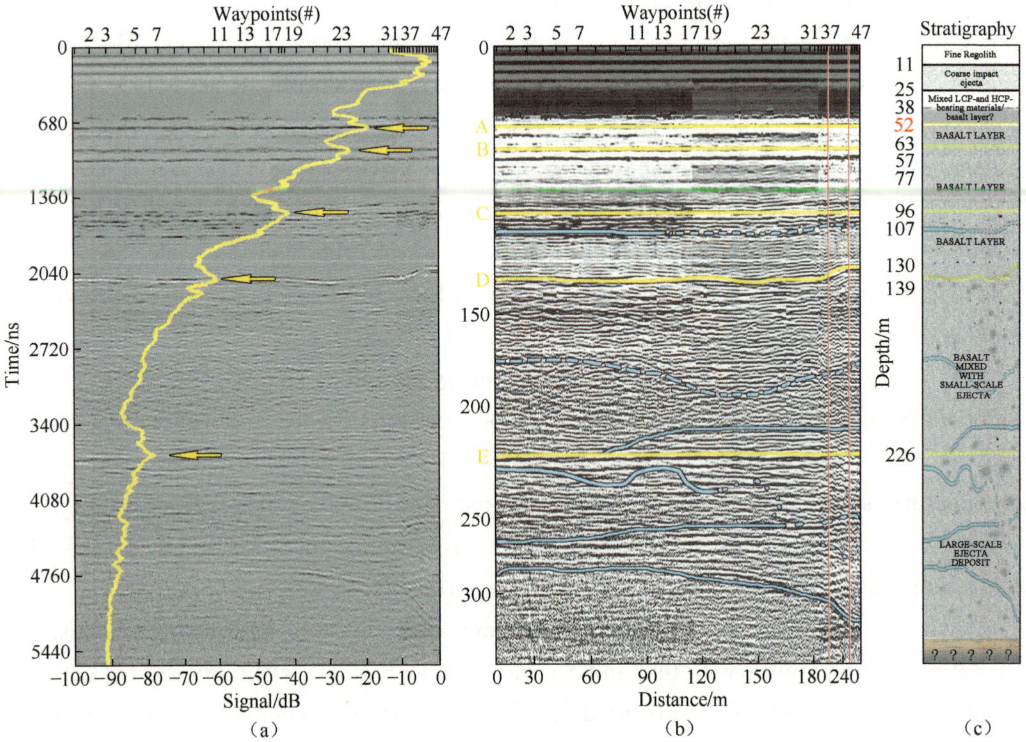

Fig. 1　Chang'e－4 radargram for the 60MHz channel. a LPR CH－1 radargram of CE－4 landing site using an Automatic Gain Control (AGC) method[49] for amplitude compensation. The yellow line represents the aggregated data traces to show the enhanced subsurface echoes A－E. This approach also minimizes "anomalous" signal points along the travel path that might arise from the random distribution of rocks and debris among more heterogeneous layers. (b) Interpreted LPR CH－1 radargram of CE－4 landing site using image enhancement techniques. The cumulative length of the traversed path is 284.6m. Yellow lines represent enhanced subsurface echoes A－E (a); light blue lines are subtle boundaries denoting differences in "stripe" directions and sharpness; dashed lines denote higher uncertainty in location. The two red vertical lines indicate waypoint 37 and 42, respectively. Please note that LPR data were not collected at a fixed speed. For example, the jump from 180 to 240m at the end of X axis is because LPR CH－1 collected much fewer data at the end of the traverse path than at the beginning stage when rover traveled around a small crater. The enlarged images of the end of the traverse are given in Supplementary Fig. 3. (c) The interpreted stratigraphy structure inferred from the LPR results. $\varepsilon = 4.5$ (≤52m) and 6.5 (>52m) is used for time－to－depth conversion

thickness of stratum D－E is about 100m, it is possible that it was formed by multiple lava eruption events interposed by small－scale ejecta deposits or thin regolith that was formed in the lull period: this geologically complex admix may reflect in the scattered features in the radar results as evident in Fig. 1b. Alternatively, the largescale ejecta layers at depths of over 200m may produce relatively continuous signal discontinuities, but with more pronounced fluctuations than a well－defined interface.

Simulation results. To test our geological interpretation of the radar data, several

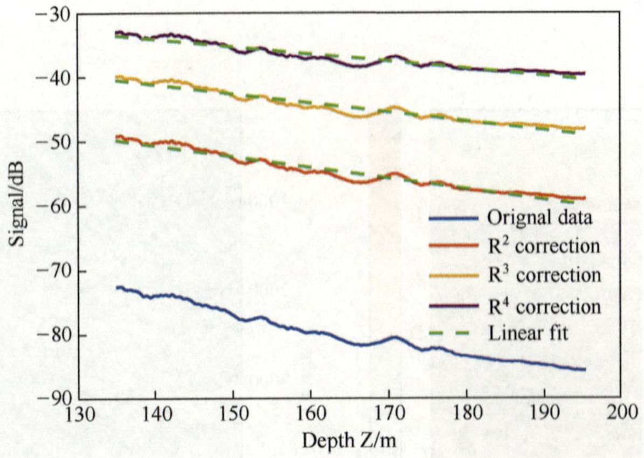

Fig. 2　Depth vs. signal power profile. R^2, R^3, R^4 backscatter/spreading corrections are applied, respectively. The best - fit lines are used to calculate the attenuation η (dB/m). $\varepsilon = 6.5$ is used for time - to - depth conversion

subsurface models were designed for LPR simulation with various sets of loss tangent and permittivity values.

The average loss tangent (the ratio of the imaginary and real part of permittivity, tanδ) in CE - 4 site is inverted with three types of geometric spreading corrections, as shown in Fig. 2. For R^2 correction, $\tan\delta = 0.0060 \pm 0.0001$; for R^3 correction, $\tan\delta = 0.0051 \pm 0.0001$; for R^4 correction, $\tan\delta = 0.0041 \pm 0.0001$. In the case of rough interface (R^3 correction), we confirm the estimation value with the result inferred from the penetrating depth of LPR CH - 1, which is 0.005 (see Supplementary Note 3). The first model (Fig. 3a) adopts the derived loss tangent value and models the subsurface structure underneath a short path, including regolith, ejecta from nearby craters, basalt layers admixed with small - scale ejecta, and large - scale ejecta formed in a time sequence from recent to remote. The simulation results show the clear boundary between the regolith and mare basalts (Fig. 3b). Also, both smalland large - scale ejecta could produce clear radar echoes, especially when larger debris is present at depths of over 200m, which would produce flatter horizontal lines instead of large hyperbola shaped signals. The echoes of small - scale ejecta appear shorter and less continuous, comparable to the observations of the LPR results in D - E section. The simulation results show similar characteristics to the CE - 4 LPR radargram, confirming the plausibility of the subsurface model. The other two models (Fig. 3c, d) illustrate how different loss tangent values affect the radargram: Fig. 3c shows that the penetrating depth of the case of $\tan\delta = 0.001$ is 6400ns, much deeper than 4900ns in the case of $\tan\delta = 0.005$, and reflections from the ejecta are clearly visible, while LPR attenuates faster with $\tan\delta = 0.009$ (Fig. 3d), the reflected signals become weak beneath 3000ns. Simulation results with different permittivity values of mare basalt are given in the Supplementary Note 4 and Supplementary Fig. 5.

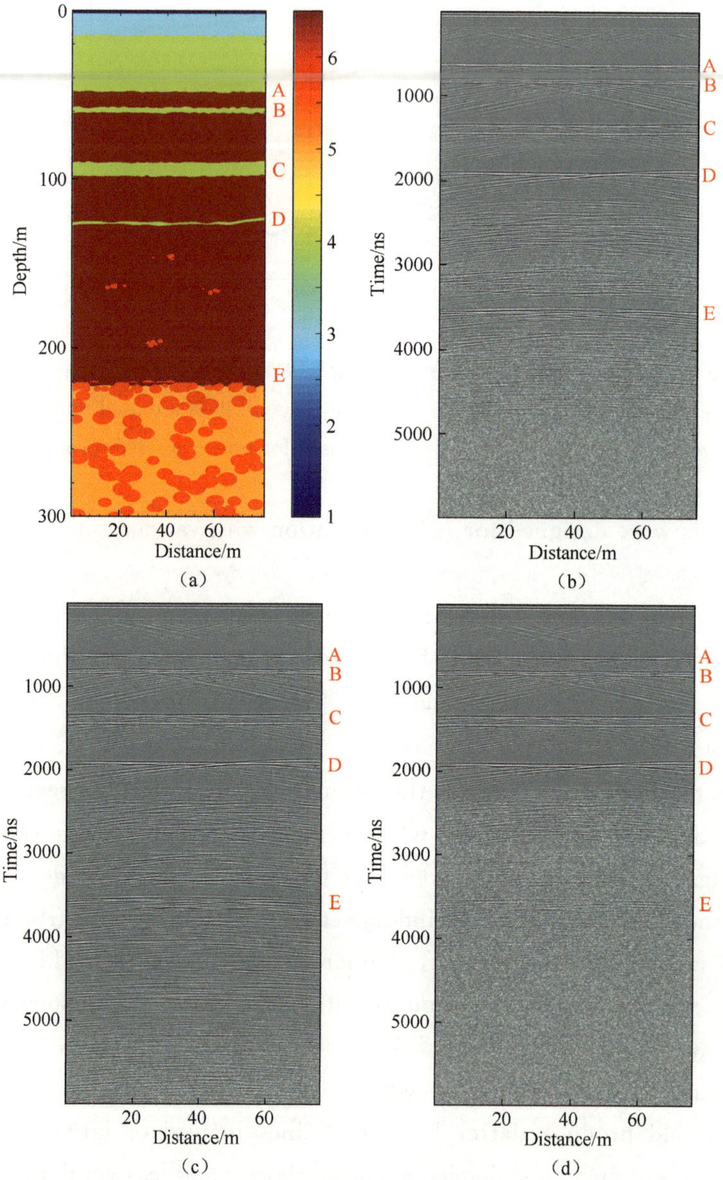

Fig. 3　Radar simulation results. (a) An 80m - long path simulation model. (b) Simulation LPR result with loss tangent tanδ=0.005. (c) Loss tangent tanδ=0.001. (d) Loss tangent tanδ= 0.009. The color bar in a shows the permittivity value used for each layer. From top to bottom, the simulation model contains (a) vacuum, regolith layer, ejecta, and multiple basalt layers with interlayered regolith. In (a), both small and large amount of ejecta is modeled at depth ranges of 120—220m and 220—300m, respectively

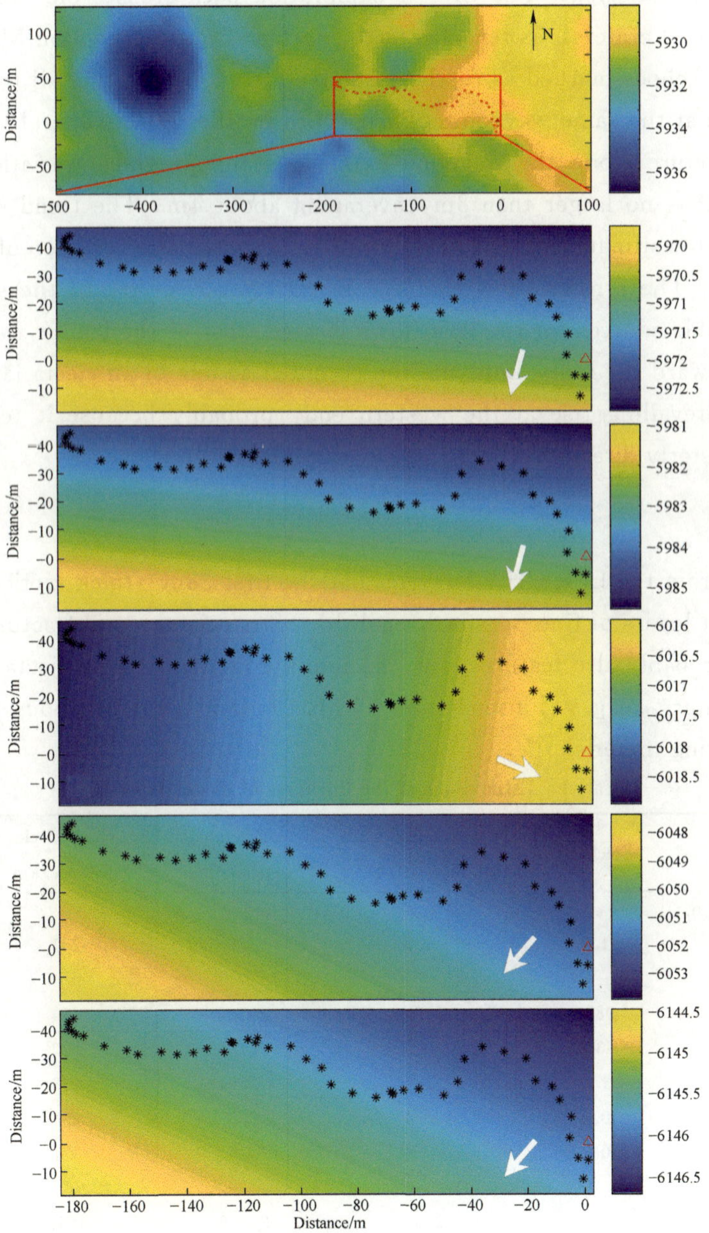

Fig. 4　Trend surface analysis results. Top image shows the surface DEM. A – E are the trend surface analysis of the subsurface structure of the CE – 4 surveying area, corresponding to Fig. 1. The color bars indicate the elevation variation within each layer and the arrows show the rising direction of the layer. The coordinate system is based on the lander location marked by the red triangle. The black dots represent the rover's path during the initial nine months. Surface DEM data were derived from Lunar Reconnaissance Orbiter (LRO) Narrow Angle Camera (NAC) stereo pair image[28]. The positive direction of Y axis points to the north and the positive direction of X axis points to the East

Trend surface analysis. The CE－4 exploration path crosses old surfaces scattered with different sized craters, as revealed in the surface Digital Elevation Model (Fig. 4). The subsurface echoes marked as A－E are derived from aggregating all the repeated data tracks collected at the same waypoint. Their depth profile reflects lateral variations of the subsurface structure along the traveling path. The topographical variation within each subsurface layer is no larger than 8m, averaging about 4m. The trend surface analysis method is used to estimate the relatively large－scale systematic variation of the subsurface layer (Fig. 4). The arrows show that A, B, D layer rise toward the same direction, indicating possible sources for the lava flows located west of the lander, whereas layer C tends to rise toward the east, implying the potential source of an ejecta layer. E stratum also shows a prevailing rise at the western end, probably because it tends to become thicker in a westerly direction.

Discussions

The data from the LPR CH－2 reveal that the first 38m－thick section could consist of three distinct layers of fine－grained regolith, coarse ejecta, and fractured basalt. The bottom layer probably also features low－calcium pyroxene－bearing ejecta materials from neighboring craters, e. g., Finsen crater (Tab. 1), and autochthonous high－calcium pyroxene－bearing materials[19].

Tab. 1　Estimations of the thickness of ejecta deposits

Source	Ejecta/m	Local material/m	Ejecta deposits/m	Great circle distance/km	Age (Ga)
Pre－Nectarian epoch					
Hess	1.2	1.5	2.7	281	
Nectarian epoch					
Ingenii	98.2	195.4	293.6	484	3.91[36]
Leibnitz	63.2	63.7	126.9	222	3.88[36]
Davisson	1.3	1.6	2.9	283	
Boyle	0.4	0.4	0.8	238	
Abbe	0.2	0.3	0.5	371	
Imbrium epoch					
Imbrium Basin	18.3	276.3	294.6	4996	－3.85
Orientale Basin	15.5	118.2	133.7	2269	<3.85
Maksutov	0.5	0.8	1.3	336	
Alder	8.0	5.2	13.2	134	
Eratosthenian and Copernican epochs					
Finsen	4.7	3.2	7.9	140	
Von Kàrmàn L	1.5	0.6	2.1	72	

The ejecta deposits are the mixture of ejecta delivered by impact events and excavated local material. Great circle distance means the shortest distance on the lunar surface between the center of the crater and CE－4 landing site. Derivation formulas are described in the "Methods" section.

The first recognized reflector A in LPR CH－1 radargram (Fig. 1a) is found at the depth of 52m; a uniform basalt layer may be present between 38 and 52m. Alternatively, the possible boundary within this range is narrower than the CH－1 resolution, thus preventing detection.

Reflectors B, C, D probably represent interfaces between basalt layers from different periods, caused by the high permittivity contrast between solid mare basalt and high－porosity deposits formed during latent periods of lava activity. In this scenario, a stratum of regolith would have sufficient time to develop through a process of surface weathering, admixed with random ejecta deposits[12]. The trend surface analysis suggests that layer C rises towards the East, unlike layers A, B, D. This could be due to the presence of ejecta deposits delivered from the East, plausibly originating from crater Alder.

The 100m－thick D－E stratum is interpreted as a layer formed by an undefined number of intermittent lava flows and smallscale ejecta deposits, possibly interposed by shallow regolith layers. This interpretation agrees with a prior study using small craters close to the CE－4 landing site, which found evidences of mare basalts at a depth between ～30 and 90m[32].

The occurrence of large－scale, multiple lava flooding events within the VK crater is also revealed by several geomorphological features relating to a prominent dome located west of the crater. The elevation cross－sections of the 40km mound/dome structure (Fig. 5b, c) suggest that it represents the last, less voluminous, and possibly more viscous lava flows that accumulated relatively close to the vent/s, probably located at the foot of the crater terrace. Three finger－shaped flow lobes have heights of around 110m (section a—b) with a slope of about 4.6°, whereas to the south and close to the rim, the flow also drops relatively abruptly by 100m (section 1) with a 2.6° slope, which is comparable to the thickness estimation of the D－E stratum. These are substantial even in comparison with the well－studied mare Imbrium lobes, which range between 40 and 65m[33,34]. This suggests that the infill history of the basin was punctuated and probably prolonged in time. However, much of VK's crater floor has been considerably scarred by countless secondary impact events since its formation, some very recently judging by the secondaries' size and freshness, rendering any attempt to estimate ages based on craters sizefrequency distribution surveys arduous at best. However, given that these types of sharp geological boundaries have a long but limited lifespan on the lunar surface, it offers the intriguing possibility that the erupting activity might have continued into the Eratosthenian epoch, comparable to the Mare Imbrium 'young' flows[35]. Pasckert et al.[21] based on Neukum fits of the sizefrequency distribution of craters within the VK crater floor derived a temporal interval between 3.75 and 3.15Ga as the last eruptive phase. Using the same technique, extrusive events within the SPA basin are estimated to have peaked in the Late Imbrian period, ～3.74－3.71Ga[36] ending about $3.6^{+0.09}_{-0.2}$ Ga[19].

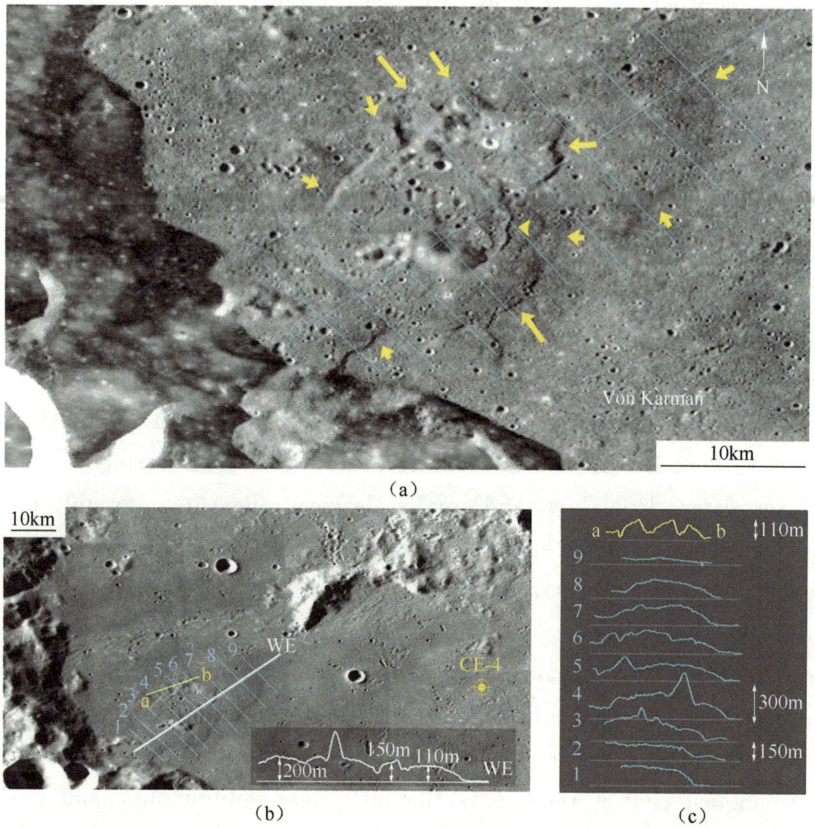

Fig. 5　Western Von Kármán flow structure. Lobate fronts and other morphologic details of the flow structure are shown in the figure. Cross - sections altimetry data were generated from SLDEM2015 + LOLA data maps[50]. Centre lat - lon coordinates for image (a) are: 45. 45°S, 174. 10°E, respectively. Arrows point at prominent flow fronts. LROC Wide Angle Camera (WAC)[51] image mosaic

The ejecta of the nearby fresh crater Zhinyu with a diameter of 3. 8km represent target materials down to about 300m and the radial variation in olivine estimation content suggests the existence of at least three distinct layers with different olivine abundance (Fig. 6). The excavation process from impacts produces an inverse stratigraphy of ejected materials, with those closest to the crater rim representing the deepest part of the excavation. The spectrally derived olivine %wt abundances[37] reveal at least three types of potentially heterogeneous composition as shown by concentric circles marked in Fig. 6. Spectrally derived compositional data of shocked materials should be interpreted with a caution of course; however, the heterogeneous concentric pattern associated with Zhinyu is uncommon, thus making it more likely of reflecting actual compositional/petrological differences with depth.

The deepest strata seen by the LPR represent the large - scale ejecta deposits located at the depth of about 230m with thickness more than 116m, which is within the range of

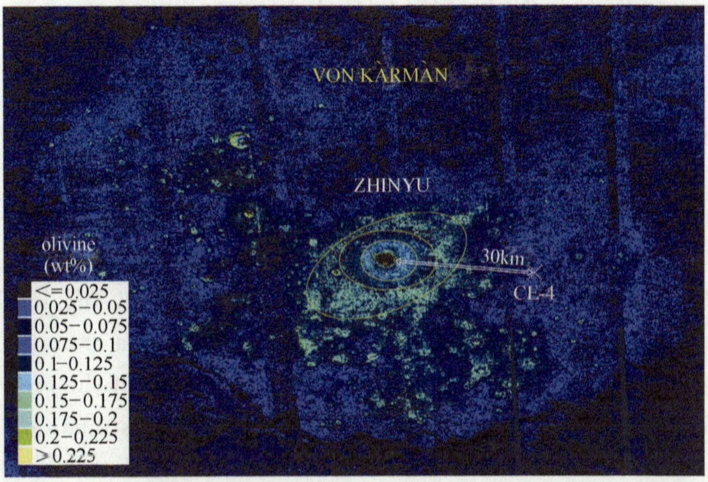

Fig. 6　Olivine abundance (wt%) map mosaic of the VK crater. Crater Zhinyu (3.8km) displays an extended ejecta apron characterized by three different spectral signatures interpreted as relating to different olivine content of the radially distributed materials[37]

estimated ejecta deposits from several source candidates (Tab. 1). The subsurface structure indicates the ejecta might come from the western direction, as it becomes thicker toward the right end of the radargram series (Fig. 1). Based on the analysis of the timing sequence of impacts, estimated thickness values, and the direction of the ejecta source, one possibility is that the strata E were emplaced by the Imbrium impact event or a mix of both Imbrium and Orientale. Based on broader geological considerations[18], the fact that the CE-4 landing site coincides to the antipodal position of the Imbrium impact suggests that the geologic unit Ig (Supplementary Fig. 1d) might be the product of its ejecta surge. However, VK was heavily impacted by many later events and ejecta distribution is notoriously difficult to associate to a distal impact unless it is deposited within a clear ballistic path. Therefore, we cannot exclude other possibilities as sources of large-scale ejecta.

Based on the surface age derivations from size-frequency distribution of impact craters in the SPA basin, the volcanic activity appears to have peaked in the Late Imbrian. However, the start time of the volcanism in the region is not well constrained. The eruptive activities may have started as early as several million years later after the VK impact event (3.97Ga). Later, large impact events such as Ingenii (~3.91Ga), Leibnitz (~3.88Ga), Imbrium, and Orientale, etc. together produced up to over 200m of ejecta at the CE-4 landing site region (Tab. 1).

The mare infill of the basin probably followed this main deposition phase. This stage was punctuated by the arrival of small-scale ejecta from other distant impact craters or nearby relatively small craters. For example, the ejecta from Alder crater (Tab. 1) in the east might be buried by mare basalts at the depth of 96m. In this scenario, the thickness of mare basalt would lie beyond the LPR CH - 1 detection limit.

Using the excavation depths of the largest impacts that did not penetrate to the crater basement, the maximum thickness of the mare infill has been estimated to about 200m[21] and possibly over 300m. A higher estimate of 310m was derived looking at the spectral characteristics of crater Zhinyv[19] ejecta, which is 32km west of the CE - 4 landing site. Our findings based on LPR observations also align with an overall basalt layer thickness larger than ~300m.

Overall, the LPR data lead to an interpretative model of the local stratigraphy, which is comparable to that inferred from reflectance spectra data of crater ejecta[31]. The main difference between the methodologies relates to the depth of the proposed layers, in the case of the LPR results, consistently deeper than previous estimations[19,21]. Another new insight is that the volcanism within VK was punctuated and prolonged, with at least four major infill events that can be interpreted from both the radargrams and geological considerations. The radargram provides direct evidence of multiple lava - infilling events having occurred within the VK crater, resulting in 12, 33, 34, and 96 mthick lava layers at the CE - 4 site. The radargram also shows that large - scale, multiple lava flooding occurrences were punctuated by the arrival of ejecta from impacts of different sizes and origin. In this work, we also derive an estimate for the average loss tangent of mare basalts on farside is inferred as 0.0040—0.0061.

Methods

Lunar - penetrating radar data. CH - 1 of LPR operates at the center frequency of 60 MHz with a 40 MHz bandwidth. The monopole antennas (12mm in diameter and 1150mm in length) are located at the back of the rover standing about 60cm above the ground. In this work, we analyze LPR CH - 1 data (file name list is given in Supplementary Tab. 1) collected in the first 9 months by the Yutu - 2 rover along its 284.6m - long exploration journey (see exploration path in Fig. 7).

Data processing. The radargram from the LPR CH - 1 data was derived after removing repetitive data and background noise, applying filtering and amplitude compensation. Further processing details can be found in Supplementary Note 5 and Supplementary Fig. 6 and 7. The actual depth of the subsurface reflector, d, is converted from the two - way traveling time t at the reflector and the relative permittivity of lunar basalt, ε, usingd

$$d=\frac{ct}{2\sqrt{\varepsilon}} \tag{1}$$

The permittivity of the Apollo regolith and basalt samples[17,38], $\varepsilon = 4.5$ ($\leqslant$52m) and 6.5 (>52m) asured at 60MHz is adopted in this work.

To identify the subsurface reflectors, not only the radargrams showing LPR data collected in the motion state were used but also the data trace from each waypoint (total 49 waypoints in this work, shown as yellow dots in Fig. 7) was generated by aggregating all the acquired repetitive data (～400—1000 tracks) at the same location to further reduce random noise and increase the signal - to - noise ratio. Furthermore, the identified reflectors using CE - 4 LPR CH - 1 data were compared with those derived from CE - 3 data (Supplementary Note 6 and Supplementary Fig. 8—10) to avoid signal artifacts caused by the inherent system noise[39].

The trend surface analysis was performed with the reflector location of each waypoint and low - order polynomial fitting[40].

Radar signal simulation. To evaluate our geological interpretations based on the LPR radargram, simulation of the proposed subsurface stratigraphy model was conducted in the transverse electric mode with a two - dimensional finite - difference time - domain method using gprMax[41] and Gaussian noise set as the average signal level below the LPR CH - 1 detection limit was included in the simulation. The detailed model and permittivity value of each layer are shown in Fig. 3 and Supplementary Note 4.

Ejecta deposition estimation. Large volumes of ejecta were delivered to the CE - 4 landing site by several impacts[42]. The ejecta thickness was estimated using

$$T = 0.068 R_t (r/R_{at})^{-3} \tag{2}$$

where r is the distance from crater center to the landing site with the consideration of the curvature of the Moon and R_{at} is the radius of a transient cavity at the preimpact surface in meters[43,44]. For the complex craters,

$$R_{at} = 0.4906(2R)^{0.85},\ R > 9.5\text{km} \tag{3}$$

R is rim - to - rim radius of a final crater[45]. For the Imbrium and Orientale basins, which were formed after the VK crater, R_{at} were obtained from Miljković et al. [46]. The apparent radius of Ingenii crater is 114km[47]. The thickness of ejecta delivered to landing site is listed in Tab. 1.

The cratering efficiency (μ) is the ratio between the thickness of local material excavated by the impact of ejecta and the ejecta.

$$\mu = 0.0092 r_{gc}^{0.87} \tag{4}$$

is adopted from Petro and Pieters[48], where r_{gc} is the great circle distance. The thickness of ejecta deposits including ejecta and local excavated materials can be obtained from

$$h = T \times (1 + \mu) \tag{5}$$

Data availability

CE - 3 LPR data and CE - 4 LPR data are available at Data Publishing and Information Service System of China's Lunar Exploration Program (http://moon.bao.ac.cn/). All the LPR data IDs are listed in Supplementary Tab. 1. Data for Fig. 1 and 2 are available at https://doi.org/10.5281/zenodo.3763355. Data sources of Fig. 4—6 are given in the captions. Additional data related to this paper are available from the corresponding author upon reasonable request.

Acknowledgements

Scientific data of Chang' e missions are provided by the China National Space Administration (CNSA) . We are grateful for the support from the team members of the Ground Application and Research System (GRAS), who contributed to data receiving and preprocessing. This study is supported by the Science and Technology Development Fund (FDCT) of Macau (Grants 0042/2018/A2, 0089/2018/A3, 005/2017/A1, and 0079/2019/A2), the Pre - research Project on Civil Aerospace Technologies of CNSA (D020101), the Science and technology project of Jiangxi education department (Grant GJJ180489), and the Scientific Research Starting Foundation for scholars from Jiangxi University of Science and Technology (Grant jxxjbs18017) .

References

[1] Yingst, R. A. & Head, J. W. Volumes of lunar lava ponds in South Pole - Aitken and Orientale Basins: implications for eruption conditions, transport mechanisms, and magma source regions. J. Geophys. Res. Planets 102, 10909—10931 (1997) .

[2] De Hon, R. A. Thickness of mare material in the Tranquillitatis and Nectaris basins. In 5th Lunar Sci. Conf. Proc. Houston, TX, 18—22 March 1974. Vol. 1 (A75—39540 19—91) 53—59 (Pergamon Press. Inc. , New York, 1974) .

[3] Hon De, R. A. Thickness of the western mare basalts. In 10th Lunar Planet. Sci. Conf. Houston, TX, 19—23 March 1979. Vol. 3 (A80—23677 08—91) 2935—2955 (Pergamon Press, Inc. , New York, 1979) .

[4] Gong, S. et al. Thicknesses of mare basalts on the Moon from gravity and topography. J. Geophys. Res. Planets 121, 854—870 (2016) .

[5] Budney, C. J. & Lucey, P. G. Basalt thickness in Mare Humorum: the crater excavation method. J. Geophys. Res. Planets 103, 16855—16870 (1998) .

[6] Gillis, J. J. & Spudis, P. D. Geology of the Smythii and Marginis region of the Moon: using integrated remotely sensed data. J. Geophys. Res. Planets 105, 4217—4233 (2000) .

[7] Heather, D. & Dunkin, S. A stratigraphic study of southern Oceanus Procellarum using Clementine multispectral data. Planet. Space Sci. 50, 1299—1309 (2002) .

[8] Thomson, B. J. , Grosfils, E. B. , Bussey, D. B. J. & Spudis, P. D. A new technique for estimating the thickness of mare basalts in Imbrium Basin. Geophys. Res. Lett. 36, 1—5 (2009) .

[9] Cooper, B. L. , Carter, J. L. & Sapp, C. A. New evidence for graben origin of Oceanus Procellarum from lunar sounder optical imagery. J. Geophys. Res. 99, 3799 (1994) .

[10] Peeples, W. J. et al. Orbital radar evidence for lunar subsurface layering in Maria Serenitatis and Crisium. J. Geophys. Res. Solid Earth 83, 3459—3468 (1978).

[11] Phillips, R. J. et al. in Apollo 17 Preliminary Science Report (NASA SP—330) 22—1:22—26 (National Aeronautics and Space Administration, 1973).

[12] Ono, T. et al. Lunar radar sounder observations of subsurface layers under the nearside maria of the Moon. Science 323, 909—912 (2009).

[13] Oshigami, S. et al. Distribution of the subsurface reflectors of the western nearside maria observed from Kaguya with Lunar Radar Sounder. Geophys. Res. Lett. 36, L18202 (2009).

[14] Oshigami, S. et al. Mare volcanism: reinterpretation based on Kaguya Lunar Radar Sounder data. J. Geophys. Res. Planets 119, 1037—1045 (2014).

[15] Xiao, L. et al. A young multilayered terrane of the northern Mare Imbrium revealed by Chang'E-3 mission. Science (80) 347, 1226 (2015).

[16] Lai, J., Xu, Y., Zhang, X. & Tang, Z. Structural analysis of lunar subsurface with Chang' E - 3 lunar penetrating radar. Planet. Space Sci. 120, 96 — 102 (2016).

[17] Lai, J. et al. Comparison of dielectric properties and structure of lunar Regolith at Chang'e - 3 and Chang'e - 4 landing sites revealed by ground penetrating radar. Geophys. Res. Lett. 46, 12783—12793 (2019).

[18] Stuart - Alexander, D. E. Geologic Map of the Central Far Side of the Moon I - 1047, https://doi.org/10.3133/i1047 (US Geological Survey, 1978).

[19] Huang, J. et al. Geological characteristics of Von Kármán Crater, Northwestern South Pole - Aitken Basin: Chang'E - 4 landing site region. J. Geophys. Res. Planets 123, 1684—1700 (2018).

[20] Head, J. W. Lunar volcanism in space and time. Rev. Geophys. 14, 265 (1976).

[21] Pasckert, J. H., Hiesinger, H. & van der Bogert, C. H. Lunar farside volcanism in and around the South Pole - Aitken basin. Icarus 299, 538 — 562 (2018).

[22] Head, J. W. & Wilson, L. Generation, ascent and eruption of magma on the Moon: New insights into source depths, magma supply, intrusions and effusive/explosive eruptions (Part 2: Predicted emplacement processes and observations). Icarus 283, 176—223 (2017).

[23] Wieczorek, M. A. & Phillips, R. J. The "Procellarum KREEP Terrane": implications for mare volcanism and lunar evolution. J. Geophys. Res. Planets 105, 20417—20430 (2000).

[24] Wieczorek, M. A., Zuber, M. T. & Phillips, R. J. The role of magma buoyancy on the eruption of lunar basalts. Earth Planet. Sci. Lett. 185, 71—83 (2001).

[25] Wieczorek, M. A. et al. The crust of the Moon as seen by GRAIL. Science (80) 339, 671—675 (2013).

[26] Wilson, L. & Head, J. W. Generation, ascent and eruption of magma on the Moon: new insights into source depths, magma supply, intrusions and effusive/explosive eruptions (Part 1: theory) . Icarus 283, 146—175 (2017) .

[27] Laneuville, M. , Wieczorek, M. A. , Breuer, D. & Tosi, N. Asymmetric thermal evolution of the Moon. J. Geophys. Res. Planets 118, 1435—1452 (2013) .

[28] Topographic Map of the Chang'e 4 Site. Available at: http://www.lroc.asu.edu/posts/1100. Accessed date: June 20, 2020.

[29] Liu, J. et al. Descent trajectory reconstruction and landing site positioning of Chang'E-4 on the lunar farside. Nat. Commun. 10, 4229 (2019) .

[30] Cohen, B. A. Support for the Lunar Cataclysm hypothesis from Lunar Meteorite Impact Melt Ages. Science (80) 290, 1754—1756 (2000) .

[31] Ling, Z. et al. Composition, mineralogy and chronology of mare basalts and non-mare materials in Von Kármán crater: landing site of the Chang'E-4 mission. Planet. Space Sci. 179, 104741 (2019) .

[32] Qiao, L. , Ling, Z. , Fu, X. & Li, B. Geological characterization of the Chang'e-4 landing area on the lunar farside. Icarus 333, 37—51 (2019) .

[33] Schaber, G. G. , Boyce, J. M. & Moore, H. J. The scarcity of mappable flow lobes on the lunar maria: unique morphology of the Imbrium flows. In 7th Lunar Planet. Sci. Conf. Proc. Houston, TX, 15—19 March 1976. Vol. 3 (A77—34651 15—91) 2783—2800 (Pergamon Press, Inc. , New York, 1976) .

[34] Garry, W. B. , Robinson, M. S. & Team, L. Observations of flow lobes in the phase I Lavas, Mare Imbrium, the Moon. In 41st Lunar Planet. Sci. Conf. 1—5 March 2010, Woodlands, TX. LPI Contribution No. 1533, 2278 (2010) .

[35] Schaber, G. G. Lava flows in Mare Imbrium: geologic evaluation from Apollo orbital photography. In 4th Lunar Sci. Conf. 5—8 March 1973, Houston, TX. 4, 73 (1973) .

[36] Yingst, R. A. , Chuang, F. C. , Berman, D. C. & Mest, S. C. Geologic Mapping of the Planck Quadrangle of the Moon (LQ—29) . In 48th Lunar Planet. Sci. Conf. 20—24 March 2017, Woodlands, TX. LPI Contribution No. 1964, 1680 (2017) .

[37] Lemelin, M. , Lucey, P. G. , Gaddis, L. R. , Hare, T. & Ohtake, M. Global map products from the Kaguya Multiband Imager at 512 ppd: Minerals, FeO and OMAT. In 47th Lunar Planet. Sci. Conf. 21—25 March 2016, Woodlands, TX. LPI Contribution No. 1903, 2994 (2016) .

[38] Carrier, W. D. , Olhoeft, G. R. & Mendell, W. Physical properties of the lunar surface. Lunar Sourcebook 522—530 (Lunar and Planetary Institute, 1991) .

[39] Li, C. et al. Pitfalls in GPR data interpretation: false reflectors detected in lunar radar cross sections by Chang'e-3. IEEE Trans. Geosci. Remote Sens. 56, 1325—1335 (2018) .

［40］Davis，J. C. Statistics and Data Analysis in Geology（Wiley，1986）.

［41］Warren，C.，Giannopoulos，A. & Giannakis，I. gprMax：open source software to simulate electromagnetic wave propagation for Ground Penetrating Radar. Comput. Phys. Commun. 209，163—170（2016）.

［42］Lin，H. et al. Olivine - norite rock detected by the lunar rover Yutu - 2 likely crystallized from the SPA impact melt pool. Natl. Sci. Rev. 7，913—920（2020）.

［43］Pike，R. J. Ejecta from large craters on the Moon：Comments on the geometric model of McGetchin et al. Earth Planet. Sci. Lett. 23，265—271（1974）.

［44］Xie，M. & Zhu，M. — H. Estimates of primary ejecta and local material for the Orientale basin：Implications for the formation and ballistic sedimentation of multi - ring basins. Earth Planet. Sci. Lett. 440，71—80（2016）.

［45］Xie，M.，T. Liu，and A. Xu Ballistic sedimentation of impact crater ejecta：Implication for resurfacing and provenance of lunar samples. J. Geophys. Res. Planets 125，e2019JE006113（2020）.

［46］Miljković, K. et al. Asymmetric distribution of lunar impact basins caused by variations in target properties. Science 342，724—726（2013）.

［47］Miljković, K. et al. Subsurface morphology and scaling of lunar impact basins. J. Geophys. Res.：Planets 121，1695—1712，https://doi. org/10. 1002/2016je005038（2016）.

［48］Petro，N. E. & Pieters，C. M. Modeling the provenance of the Apollo 16 regolith. J. Geophys. Res. 111，E09005（2006）.

［49］Rabiner，L. R. & Gold，B. Theory and Application of Digital Signal Processing（Prentice - Hall，1975）.

［50］Barker，M. K. et al. A new lunar digital elevation model from the Lunar Orbiter Laser Altimeter and SELENE Terrain Camera. Icarus 273，346—355（2016）.

［51］Robinson，M. S. et al. Lunar Reconnaissance Orbiter Camera（LROC）instrument overview. Space Sci. Rev. 150，81—124（2010）.

Comparison of Dielectric Properties and Structure of Lunar Regolith at Chang'e - 3 and Chang'e - 4 Landing Sites Revealed by Ground - Penetrating Radar

Jialong Lai[1,2], Yi Xu[1], Xiaoping Zhang[1], Long Xiao[1,3], Qi Yan[1], Xu Meng[1], Bin Zhou[4], Zehua Dong[4], and Di Zhao[4]

1. State Key Laboratory of Lunar and Planetary Sciences, Macau University of Science and Technology, Macau, China
2. School of Science, Jiangxi University of Science and Technology, Ganzhou, China
3. Planetary Science Institute, School of Earth Sciences, China University of Geosciences, Wuhan, China
4. Key Laboratory of Electromagnetic Radiation and Detection Technology, Institute of Electronics, Chinese Academy of Science, Beijing, China

Abstract: On 3 January 2019, the Chang'e - 4 (CE - 4) touched down on the Von Karman crater located inside the South Pole - Aitken Basin, providing for the first time the opportunity for in situ measurements of the lunar regolith at the farside of the Moon. The CE - 4 ground penetrating radar reveals that fine - grained regolith, coarse impact ejecta, and fractured bedrocks lie beneath the exploration path of the Yutu - 2 rover. The variations of regolith permittivity with depth and the radargrams indicate that the CE - 4 site has a fine - grained regolith layer thickness of 11. 1m, which is about 1. 3—3 times higher than the in situ measurement results at the Apollo and Chang'e - 3 (CE - 3) sites except for Apollo 16, possibly due to a faster weathering rate of ejecta deposits compared with coherent basalt substrates. The penetration depth of CE - 4 is about 2. 85 times (in terms of round - way delay) deeper than CE - 3, probably due to the differences in abundances of ilmenite and rocks in the regolith.

Plain Language Summary: CE - 4 is the first craft in history to land on the lunar farside. Its rover is equipped with a ground - penetrating radar (GPR), the same as the one mounted on the CE - 3 rover, which uses pulses of electromagnetic energy to reveal the subsurface structure and properties, especially useful to study the lunar soil layer ("regolith") that mantles most of the lunar surface. The results from the GPR show that the thickness of the regolith at the CE - 4 site is 1. 3 - 3 times higher than the in situ measurements at the

Apollo 11, 12, 14, 15, 17, and CE－3 sites even though the ages of the surfaces of Apollo 11 and Apollo 17 are thought to be comparable with that of the CE－4 site. It implies a faster regolith growth speed at the CE－4 site. The mineral mix of the surface materials at the CE－4 site is different from that at the CE－3: this results in less radar signal attenuation, thus increasing the detection limit up to the depth of 35m, 2.85 times deeper than the CE－3.

Key Points:

• The LPR measurements at the CE－4 landing site reveal that the shallow structure comprises of fine－grained regolith, coarse impact ejecta, and fractured bedrocks

• The CE－4 landing site has a thicker regolith layer (~11 m) compared with the CE－3, hinting at a longer weathering history

• A lower abundance of ilmenite and rocks in the regolith at the CE－4 site allows 2.85 times the penetration depth compared with CE－3 site

1 Introduction

The lunar surface is mantled by a layer of regolith made of broken, melted, and altered bedrock that resulted from countless impacts and continuous exposure to space weathering events (Shoemaker et al., 1968, 1970). A better understanding of the properties of the regolith layer at depth is fundamental in improving our evolutionary model of the Moon. Our present knowledge of the lunar surface comes from experiments carried out in situ by the Apollo and CE－3 missions (US and China, respectively), returned samples (from US and Russia missions), and remote sensing from a wide range of multinational instruments and techniques. Due to the absence of an atmosphere, the regolith interfaces directly with the outer space, resulting in a range of modifying factors scaling from macro (asteroids/comet impacts) to micro (microasteroids, solar－wind particles, cosmic rays, etc.). However, the destructive processes also help to preserve vital information on the composition and the early history of the Sun, including the nature and history of cosmic rays (Mckay et al., 1991). Characterizing the lunar regolith has important implications for the understanding of the lunar geologic history, identifying lunar resources and suitable areas for future Moon bases (Jaumann et al., 2012; Neal, 2009; Sanders & Larson, 2011; Wilhelms et al., 1987).

Samples of lunar regolith were collected by scooping or drilling to a depth between about 10 (Apollo 11) and 300cm (Apollo 17) at the Apollo and Luna landing sites; however, drill core failed to reach the rocky substrate. The core tube samples were examined carefully in the laboratory, and several discrete thin layers could be identified showing different petrographic, chemical, and maturity profiles, each the result of a com－plex distinct mixing and emplacement history (Mckay et al., 1991; Taylor et al., 1979; Vaniman et al., 1979; Vaniman & Papike, 1977). The relative permittivity values of the Surveyor regolith samples range from 2.0 to 3.7 and those of Luna regolith samples from 1.7 to 4.4, which are strongly dependent on density (Carrier et al., 1991; Olhoeft & Strangway, 1975). Passive seismic experiments and multifrequency electromagnetic probing experiments were conducted at the Apollo stations, and regolith thickness could be determined from changes of seismic velocities (Cooper et al., 1974; Duennebier et al., 1974; Grimm, 2018; Mendell, 1985; Nakamura et al., 1975; Simmons et al., 1973). On average, the regolith layer is about 4— to 5-m thick in the mare regions and about 10-15m in older highland terrains (e.g., Mckay et al., 1991).

S and P band Earth-based radar data have been used to infer the regolith layer thicknesses over the near side of the Moon (Campbell et al., 2014; Fa & Wieczorek, 2012; Shkuratov & Bondarenko, Thompson, 1987). The thickness and dielectric properties of the regolith can be constrained with high-resolution optical images using impact crater morphology and size-frequency distribution (Bart et al., 2011; Oberbeck & Quaide, 1967, 1968; Quaide & Oberbeck, 1968), Chang'e microwave brightness temperature data (Fa & Jin, 2010; Meng et al., 2014), and observations of the Lunar Radar Sounder of Kaguya (Ishiyama et al., 2013; Kobayashi et al., 2010). The Earth-based and satellite-borne approaches have revealed large-scale or even global-scale variations of lunar regolith at a coarse resolution.

On 14 December 2013, Chang'e-3 (CE-3) landed at 340.49°E, 44.12°N in northern Mare Imbrium and released the Yutu rover. On 3 January 2019, Chang'e-4 (CE-4) landed at 177.588° E, 45.457° S (Statio Tianhe) in the Von Kármán crater of the South Pole-Aitken (SPA) basin and released the Yutu-2 rover (Di et al., 2019). The two rovers both equipped with ground-penetrating radars (hereafter referred to as Lunar Penetrating Radar, LPR) provide unique data sets of in situ measurements of the lunar regolith (Fang et al., 2014). The aim of this work is to report the first five lunar days of CE-4 LPR results at 500MHz and compare the results between the two sites.

2 Chang'e-3 and Chang'e-4 Landing Sites

The CE-3 probe landed in the Imbrium basin on the basalts inferred as Eratosthenian in age. These represent some of the youngest units on the Moon at about 2.35—2.5 billion years (Gy) old (Zhao et al., 2014, Wu et al., 2018), for instance, compared with samples returned by the Apollo and Luna missions. Estimations of mafic components range from $TiO_2 = 5 \pm 1$ weight percent (wt%) and $FeO = 20 \pm 2$ wt% (derived from the gamma-ray spectrometer data onboard Lunar Prospector; Prettyman et al., 2006) to $TiO_2 = 4.0$ wt% and $FeO = 20.7$wt% (data from the Active Particle X-ray Spectrometer (APXS) instruments on the Chang'e-3 rover; X. Zhang et al., 2015). The titanium dioxide abundance of the geological unit Im1, ~15km to the north, is lower at ~1.4wt%, also suggested by the low albedo (see Fig. 1a; Zhao et al., 2014).

The Yutu rover surveyed in total about a 114-m-long track within the continuous ejecta deposits of the ～450m sized Ziwei crater. Using crater size - frequency distribution measurements of neighboring terrains, crater morphology considerations, and the distribution of meter - sized boulders observed on the crater rim, the age of the Ziwei crater has been estimated to be between 27 and 80 million years (Xiao et al., 2015).

At the time of writing, the CE - 4 Yutu - 2 rover has travelled almost 200m within the Von Kármán crater lying in the northwestern of the SPA basin, as shown in Fig. 1b and 1c, which is the largest and oldest impact basin on the Moon and may contain materials originating from deep in the crust or the upper mantle that would provide insights into early planetary differentiation (Spudis et al., 1994; Stuart - Alexander,

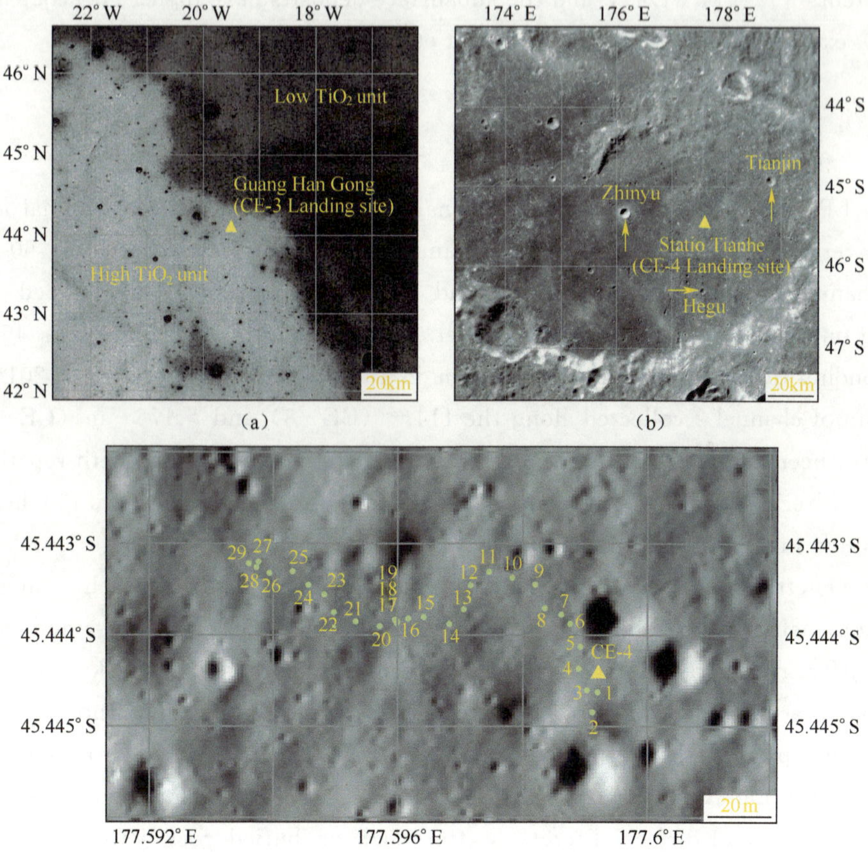

(c)

Fig. 1 (a) The CE - 3 landing site on TiO_2 map derived from the Lunar Reconnaissance Orbiter Camera (LROC) Wide - Angle Camera (WAC) images (Sato et al., 2017). The bright color indicates the high - titanium area. (b) The CE - 4 landing site on CE - 2 Digital Orthophoto Map (DOM) with a resolution of 7m/pixel. File ID: CE2 _GRAS _DOM _07m _K136 _45S175E _A. tif. (c) The motion path of Yutu - 2 rover. The base image was taken by the Lunar Reconnaissance Orbiter (LRO) Narrow Angle Camera (NAC) Camera. File ID: M134022629LE. The green circles indicate the waypoints. (a) — (c) are shown in Mercator projection centered at CE - 3 and CE - 4 landing sites, respectively

1978; Jolliff et al. , 2003) . The Von Kármán crater formed in the pre - Nectarian era, and it was infilled with mare basalts during the Imbrian period. Subsequently, countless impact events both within and from outside the area infilled the crater with ejecta (Huang et al. , 2018) . In particular, large portions of the northeastern crater rim and crater floor were covered by ejecta from the nearby 72 - km - diameter Finsen crater (Pasckert et al. , 2018) . Based on crater size - frequency distribution measurements, Huang et al. (2018) derived an absolute model age for the floor of Von Kármán crater of $3.6^{+0.09}_{-0.2}$ Ga. Pasckert et al. (2018) derived two different ages of mare deposits (3.15 and 3.75Ga) on the floor, indicating possible multiple lava eruption events. The surface materials have lower FeO contents (12—14 wt%), and the subsurface deposits have higher FeO contents (17—18 wt%) excavated by fresh craters (Pasckert et al. , 2018) .

3 Data and Methods

The LPR transmits a pulsed radar signal and receives the reflected signal along the rover traverse path. It operates at two channels with center frequencies at 60 and 500 MHz. Channel 2 has one transmitting and two receiving bow - tie shaped antennas deployed on the bottom board of the rover. The bandwidth of channel 2 is 450 MHz, corresponding to range resolutions of 0.3m in the vacuum (Fang et al. , 2014) . The radar data of channel 2 collected along the 114— (CE - 3) and ~176 - m (CE - 4) long paths have been compared in this work. The LPR data were processed with repetitive data removal, filtering, background removal, and amplitude compensation. Further details can be found in supporting information S1.

The penetration depth of the LPR is calculated using an approach similar to the correlation coefficient method proposed by Xing et al. (2017) . The noise and radar echo can be differentiated by the corrections between neighboring data tracks. The detection limit is approached when the correlation reaches a low level due to high noise.

The real part of the permittivity of lunar regolith is inverted with the hyperbola method used in Lai et al. (2016) and Feng et al. (2017) . The hyperbolic shaped signal is caused by the subsurface discrete scatter such as buried rock with about 10 - cm diameter within the regolith. The average value of permittivity can be inferred with the following:

$$\bar{\varepsilon}=\frac{c^2(t^2-t_0^2)}{4(x-x_0)^2} \tag{1}$$

where c in the formula represents the speed of light in vacuum and $\bar{\varepsilon}$ is the average value of the real part of permittivity to a certain depth, x_0 and t_0 are the horizontal location of the vertex position of the hyperbola and the two - way traveling time of the vertex position of the hyperbola, respectively, and x and t are the horizontal location of a point in the hyperbola and its two - way traveling time, respectively. The hyperbola - shaped signals are marked in Fig. 3b, and the corresponding $\overline{\varepsilon}$, t_0, and errors are given in Tab. S2. The instantaneous permittivity ε is the permittivity at a certain depth, which is obtained with the fitting line of data points of $\overline{\varepsilon}$ using the minimum χ^2 method. The bulk density ρ can be inferred with equation (2) (Carrier et al., 1991) and the instantaneous permittivity ε.

$$\varepsilon = 1.919\rho \tag{2}$$

The average loss tangent (the ratio of the imaginary and real part of permittivity, tanδ) is inverted with the radar equation given in Skolnik (2001).

$$\frac{P_r}{P_t} = \frac{G^2\lambda^2\xi}{(4\pi)^3R^4}e^{-4\alpha R} \tag{3}$$

where P_r is the received power, P_t is the transmitted power, G is the system gain, λ is the wavelength in the medium, $e^{-4\alpha R}$ is the attenuation in the medium, and ξ is the backscatter cross section. Three models of the reflection targets should be considered: for a smooth and planar reflector, $P_t/P_t \propto 1/R^2$; for discrete scatters, $P_r/P_t \propto 1/R^4$; for the Fresnel zone case, $P_t/P_t \propto 1/R^3$ (Grimm et al., 2006).

The mean square of the radar traces within a horizontal distance (10—20m in this paper) is obtained, then R^2, R^3, and R^4 are used for geometric spreading or backscatter correction. The attenuation η (dB/m) is measured with linear least squares fitting of the two-way distance versus dB power. The tanδ is calculated by the following:

$$\tan\delta = \eta/(9.1\times10^{-8}\sqrt{\varepsilon}f) \tag{5}$$

where f is the frequency.

The relationship between loss tangent and chemical components and density is given as follows (Carrier et al., 1991):

$$\tan\delta = 10^{[(0.038(\%TiO_2+\%FeO)+0.312\rho-3.260)]} \tag{6}$$

4 Results and Discussions

4.1 Penetration Depth

The penetration depth of the CE-3 is 131.0—176.5ns in terms of round-way transmission time (Xing et al., 2017). The CE-4 LPR can reach deeper up to 426.0—449.7ns, 2.85 times the detection limit of the CE-3 LPR, as shown in Fig. S3. The detection ability of the LPR is closely related to tanδ, considering that the CE-3 and the CE-4 LPRs share the same set of radar parameters. Based on equation (6), the tanδ at the CE-3 site is estimated as 0.13 with FeO and TiO_2 abundances derived from the APXS experiment (Zhang et al., 2015; Zhang, Xie, et al., 2015), while tanδ at the CE-4 site is 0.004—0.005, based on the elemental map derived from the Kaguya Multiband Imager (MI) data (Qiao et al., 2019). The different loss tangent values and nonneglectable geometric spread produces the 2.1—2.4 times difference in the detection depths between the two sites (details can be found in Text S2), close to the observations of the LPR results. One factor contributing further to the increase of the penetration depth at CE-4 site could be the lower abundance of near-surface large rocks than at the CE-3 site (see supporting information text S3). The scattering from rocks increases the attenuations of the radar signal. Moreover, the thin fine-grained regolith layer at CE-3 site results in the LPR signals to decay rapidly in the underlying coarse-grained and blocky megaregolith layer, for example, $\tan\delta$ of Apllo17 site at the depth of 8m can reach 0.08 ± 0.04 (Carrier et al., 1991), which may also explain the large difference of penetration depths at CE-3 and CE-4.

4.2 Permittivity, Bulk Density, and Loss Tangent

Fig. 2a shows the comparisons between the average value of the real part of permittivity at CE-3 and CE-4 sites using the hyperbola method used in Lai et al. (2016). At the CE-4 site, $\bar{\varepsilon}$ increases slightly with depth and reaches 4.3 at 130ns, while that of CE-3 is 3.5 of the top 40ns (in terms of round-way delay) and then changes abruptly to 6 in the underlying layer, suggesting that the fine-grained lunar regolith layer at the CE-4 site is probably much thicker than at the CE-3 site given that the permittivity of Apollo rock samples is 6—14 (Carrier et al., 1991). The permittivity values of the regolith layer at the depth of 20—40ns in the CE-4 site are within a similar range as those in the CE-3 site. The instantaneous permittivity profile of CE-4 is shown in Fig. 2b. It increases very rapidly at the near-surface section of the regolith (<2m) and then grows slowly to the maximum value of about 4.5 at the depth of 11m. The deeper part of permittivity cannot be inferred due to the absence of hyperbolic data.

By applying equation (2), we derive a bulk density for the regolith at CE-4 of ρ= 1.22—2.23g/cm^3, increasing with depth as shown in Fig. 2c. This trend is comparable with observations at the Apollo sites, and the estimated bulk density is in the higher end of the range measured from the Luna and Apoolo samples, similar to that of Apollo 17 samples, which is 1.57—2.29g/cm^3. The surface regolith density value is extrapolated by a line fit: the low confidence is due to the paucity of data points for the top surface section (< 20ns), and the errors within the hyperbola method increases with depth, as illustrated in Fig. 2a and 2b. The average density of the surface regolith at the CE-3 site was estimated to be ρ=1.63±0.19g/cm^3 (Carrier et al., 1991; Dong et al., 2017).

Fig. 2d shows an example of the horizontally averaged power versus depth curve and those after R^2, R^3, and R^4 backscatter/spreading correction at the CE-4 site. The dashed line indicates the linear fitting results. The corresponding standard deviations of the fit for R^2 corrections are given in Tab. S4. The range of the fitting line begins at about 3.5m where the interference of the surface wave is weak, and it ends where the first subsurface reflector appears. The curve of the R^3 correction representing the case of a rough

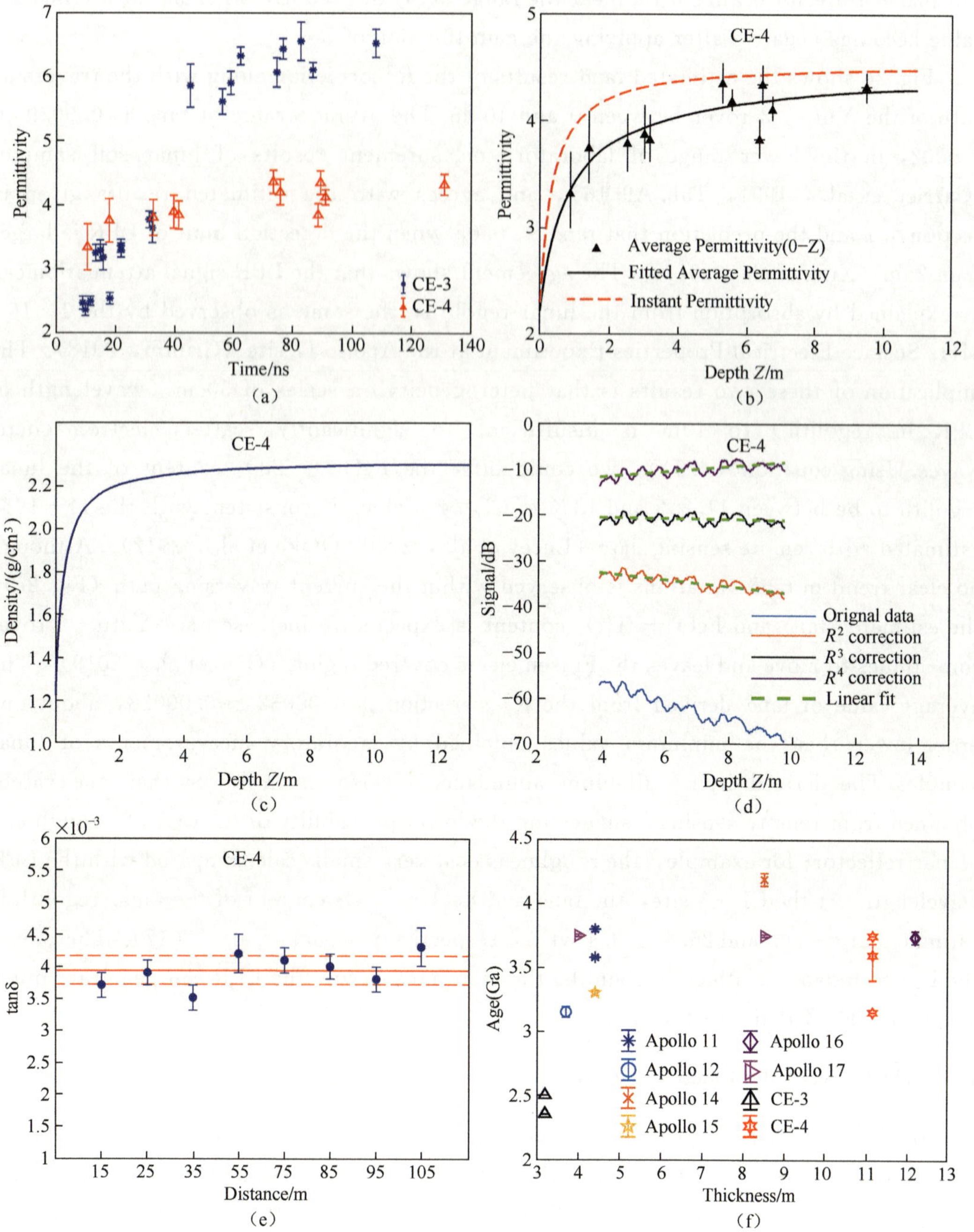

Fig. 2　(a) Average permittivity estimated with the hyperbola method versus two - way delay at the CE - 3 (Lai et al., 2016) and the CE - 4 sites. (b) The depth profile of the fitted average permittivity and instantaneous permittivity at the CE - 4 site. (c) The derived density—depth curve at the CE - 4 site. (d) At the CE - 4 site, depth versus signal power profile and after R^2, R^3, R^4 backscatter/spreading correction. The best - fitted line of R^2 correction is used to calculate the attenuation η (dB/m). $\varepsilon = 4$ is used for time - to - depth conversion. (e) The $\tan\delta$ calculated by equation (5) along the rover path between 0 and 105m. The solid and dotted lines indicate the average value and uncertainties of $\tan\delta$. (f) The comparisons of the thickness of the lunar regolith estimated by the in situ measurements in the Apollo, CE - 3, and CE - 4 sites versus thesurface age and planar reflector is almost flat near the range decay of −20dB, whereas the attenuation value becomes negative after applying the gain function of R^4.

Fig. 2e shows the estimated $\tan\delta$ results of the R^2 correction along with the traversing path of the Yutu - 2 rover between 0 and 105m. The average value of $\tan\delta$ is 0.0039 ± 0.0002, in the lower range of laboratory measurement results of lunar soil samples (Carrier et al., 1991, Tab. A9.16), and agrees with the estimated results given in section 4.1 and the prediction that $\tan\delta < 0.05$ when the detection limit of LPR is larger than 25m (Xing et al., 2017). The agreement shows that the LPR signal attenuation can be explained by absorption from the lunar regolith, the same as observed by the 1−16 - MHz Surface Electrical Properties Experiment at the Apollo 17 site (Grimm, 2018). The implication of these two results is that heterogeneity on scales of 30cm (wavelength of LPR in regolith) to 10m is insufficient to significantly scatter electromagnetic waves. Using equation ((6)), we could infer the $FeO + TiO_2$ content of the lunar regolith to be between 11.8% and 13% at CE - 4, which is consistent with the 11−19% estimated with remote sensing data (Lucey et al., 2000; Otake et al., 2012). Although no clear trend in $\tan\delta$ variations is observed within the current traversing path (Fig. 2e), the estimated $\tan\delta$ and $FeO + TiO_2$ content is expected to increase once Yutu - 2 rover con - tinues to move and leaves the Finsen ejecta covered region (Qiao et al., 2019). The average value of $\tan\delta$ derived from the R^3 correction is 0.00032 ± 0.00028, about one order lower than the minimum values obtained by laboratory measurements of lunar samples. The derived iron - titanium abundance is also much lower than the values obtained from remote sensing, suggesting the lower probability of the case of a rough and planar reflector; for example, the roughness is at very small scale compared with the LPR wavelength. At the CE - 3 site, the $\tan\delta$ and $FeO + TiO_2$ content of the lunar regolith is estimated at ~0.02 and 26.4 ± 3.4 wt%, respectively (Dong et al., 2017). Therefore, the low abundance of ilmenite could be the main reason why the LPR can penetrate much deeper at CE - 4 than at CE - 3.

4.3 Thickness of the Lunar Regolith

Based on the derived permittivity, a piecewise function $\bar{\varepsilon}=4.3$ ($\tau \leqslant 154$ns) and $\bar{\varepsilon}=6$ ($\tau > 154$ns) is used to carry out the time - to - depth conversion in radargram, which gives an estimate for the thickness of fine - grained lunar regolith at the CE - 4 location of 11.1 ± 0.4m. The thickness of the lunar regolith at the CE - 4, CE - 3, and Apollo sites are compared in Fig. 2f and summarized in Tab. S3.

The observation that the fine regolith layer would be thicker at CE - 4 agrees well with the general consensus that the regolith layer grows thicker with age, especially in consideration that the rate of regolith growth would have been higher earlier than 3.5 billion years ago when the impact flux was much higher (e.g., Shkuratov & Bondarenko). The CE - 4 site is older (>3.5Ga) compared with the CE - 3 site

(<3. 0Ga) . In addition, comparing the regolith thickness estimated by the nearside in situ measurements, the CE - 4 site is thicker compared with the Apollo sites, about 1. 3 —3 times even though the surface age at the CE - 4 site is similar to the Apollo 11 and 17 sites as shown in Fig. 2f. Hence, the formation speed of regolith at the CE - 4 site (3. 1m/Gyr) is faster than the rate of 1. 2m/Gyr at the Apollo 11 landing site (Melosh, 1989) but close to the growth rate of about 1cm per 2—3 million years given in DiGiuseppe et al. (2009) . The exception is for the Apollo 16 site that has a regolith layer that is 1m thicker than that of the CE - 4 because in the high lands the regolith thickness is generally greater than mare due to different strengths of the substrate materials (Fa & Wieczorek, 2012; Mckay et al. , 1991) . The results are consistent with remote sensing obser - vations that the regolith layer thickness on the farside of the Moon is larger than that on the nearside (Bart et al. , 2011; Fa & Jin, 2010; Mckay et al. , 1991) . The main reason of the thick fine - grain regolith could be the faster regolith growth rate on the impact ejecta deposits than on the solid basalt substrates as substantial ejecta were delivered from surrounding large impact craters to the location of CE - 4 site, whereas the CE - 3 and Apollo 11, 12, 15, 17 surfaces are not mantled by such a large amount of ejecta. The arrival of ejecta also speeds up the development of the fine - grain regolith by pulverizing, mixing and overturning the surface materials. The regolith accumulation speed is mainly affected by the impact flux of meteorite bombardment, which shows an asymmetric distribution between the leading and trailing side of the lunar surface intro - duced by the synchronous rotation of the Moon (Ito & Malhotra, 2010; Morota & Furumoto, 2003; Werner & Medvedev, 2010) . However, based on the analytical prediction of crater density spatial distribution, the impact flux in CE - 4 site is relatively low compared to other in situ measurement sites (Werner & Medvedev, 2010). Nevertheless, the hypothesis that the orientation of the Moon has changed in the past may result in an opposite interpretation on the impact flux distribution in ancient time (Wieczorek & Le Feuvre, 2009) .

4. 4　Stratigraphy Structure of the Lunar regolith

Figure 3a shows the LPR image of the CE - 3 traverse, where several horizontal and relatively short lines appear within the first meter, which are probably reflectors from the ejecta of small craters around the rover. Below that, the most prominent feature is the presence of several hyperbolic curves probably caused by the buried rocks resulted from Ziwei impact event that produce random signals and dominate the whole image. Further down, the scattering signals become faint bordering to noise beneath the depth of ~120ns. We observe that the expected reflectors from the interfaces between the reworked zone and ejecta (~1m) and between ejecta and paleoregolith (~2. 3m), or the deeper one between paleoregolith and Eratosthenian basalts (5~6m) are difficult to recognize from the radar image (the geologic analysis of the stratigraphy structure at the CE - 3 site is given in Text S4) . With the similar CE - 3 data set, researchers have reached different interpretations because of these uncertainties (Fa et al., 2015; Feng et al., 2017; Lai et al., 2016; Su et al., 2014; Xiao et al., 2015; Zhang, Yang, et al., 2015). Lai et al. (2016) and Feng et al. (2017) identified the interfaces based on drastic changes of permittivity profiles inferred from the radar data. Some researchers draw the boundaries of the base of the regolith at the locations where signals die out; however, these depths coincide with the attenuation coefficients set at different traverse sections (Su et al., 2014). Additionally, the radar signal simulation shows that the scattering signals from rocks at shallower depths could appear as if they were at deeper depths due to multiple reflections (Lai et al., 2017). Hence, the location where signals die out may not represent the interface between regolith and bedrock (Feng et al., 2017). Although people select boundaries based on different criteria, they all agree that the thickness of layers along the survey line varies dramatically; for example, the ejecta layer of Ziwei crater with an average thickness of 3m could vary by 4m within the short distance of ~20m. The observation suggests that the local stratigraphic profile must be highly heterogeneous at the CE-3 site.

We applied automatic gain control and exponential gain function (Rabiner & Gold, 1975) to the CE-4 radargram to show the texture (Fig. 3b) and relative strength of the radar echo (Fig. 3c), respectively. At the top 2m of the CE-4 radargram (the dotted lines in Fig. 3b), multiple lateral and continuous reflectors caused by the ejecta of adjacent small craters or secondary craters can be seen. Beneath that is a smooth and uniform fine-grain regolith layer based on previous permittivity analysis and radargram observations shown in Fig. 3c.

Below the regolith layer, the radar signals become stronger upon the reflector as indicated by "hr1" in Fig. 3c, which probably represents coarse ejecta of large impact events. The lateral and intermittent reflectors are at depths of 10.8—11.5, 16.2—17.3, and 20.2—20.9m as shown in "hr1", "hr2", and "hr3" in Fig. 3c. The hyperbolic shaped signals (e.g., marked by rectangles in Fig. 3c) between reflectors may be caused by rocks of ejecta deposits that survived the impact and weathering processes as most impacts cannot penetrate to such depths. Except for the hyperbola, the region between reflectors shows a weak echo, which indicates that the materials are rather homogenous and could represent paleoregolith mixed with the ejecta. The local geological context suggests that the Finsen, Von Kármán L, and Von Kármán L′ craters have contributed most of the ejecta deposits within the CE-4 landing region (Huang et al., 2018; Petro & Pieters, 2006; Sharpton, 2014). In Fig. 3c, the surface of the traversing path goes down from left to right, indicating that the regolith layer becomes thinner by about 1m as the Yutu-2 rover moved mainly westward and the distance from the Finsen crater increases. The crater morphology method (Quaide @ Oberbeck, 1968) estimates that the low-calcium pyroxene-bearing materials in ejecta deposits of the aforementioned three craters are at the depth of at least 8—13m. The model using both crater diameter and

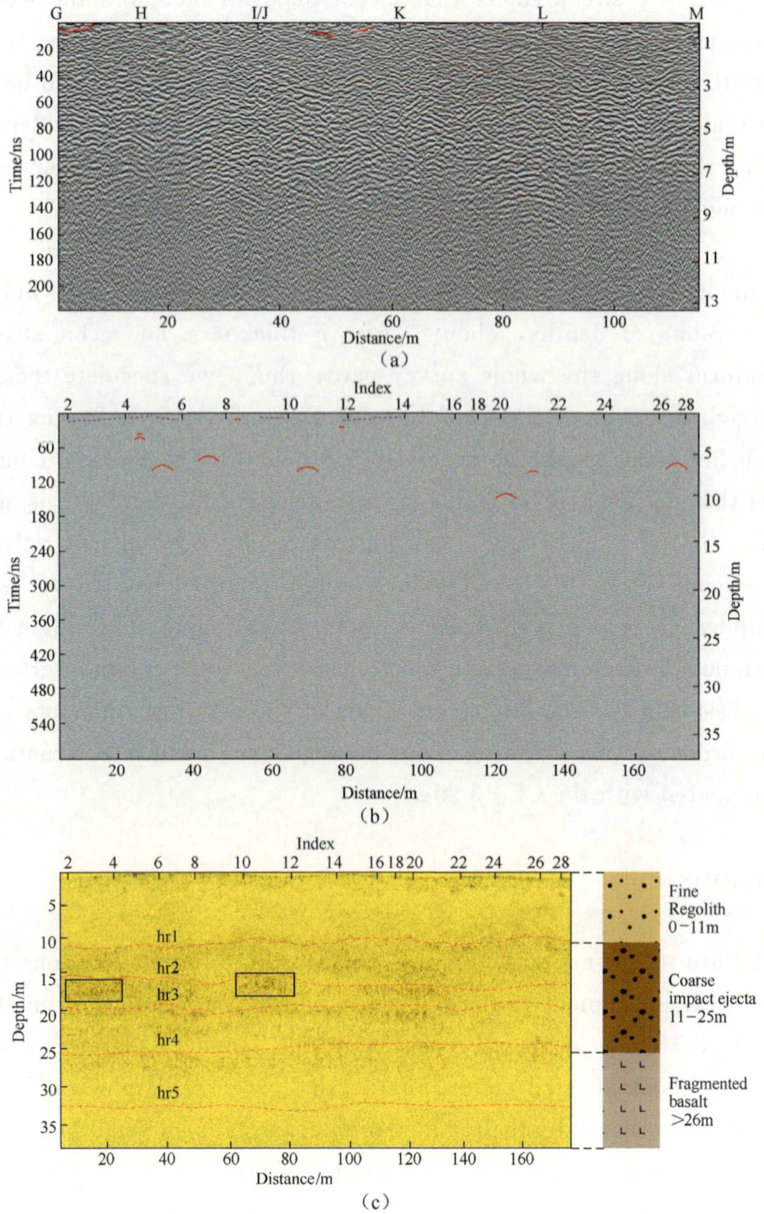

Fig. 3　(a) LPR radargram of CE－3 using the automatic gain control (AGC) method (Rabiner & Gold, 1975). The total length of the traversed path is 114m. The red lines indicate the ejecta layers of local small craters. (b) The first 5 days of LPR results of the CE－4 site with the AGC method. Current cumulative length of LPR data collection is ～176m. The dotted red lines indicate the small ejecta of local craters and buried small crater. The red arched lines represent the selected hyperbolic signal for permittivity estimation. (c) LPR radargram of CE－4 site using an exponential equation gain function (Rabiner @ Gold, 1975) for amplitued compensation. The geologic interpretation of LPR results is given on the panel on the right. "hr" stands for the horizontal reflector, also indicated by the dotted line. The rectangles circled the hyperbolic signals caused by the discrete scatters within the ejecta layers distance to the CE - 4 site predicts that ejecta deposits including the excavated local material and the total accumulated ejecta are up to 15.5 - m thick (Huang et al., 2018), which suggests that the reflectors deeper than 16m in the radargram could be the result of impact events that occurred before the forming of the Finsen crater. Considering the depth of the ejecta and the age of the impact event, the nearby Alder crater represents a possible source candidate (Pasckert et al., 2018; stuart - Alexander, 1978; Wilhelms et al., 1979).

The continuous reflector at the depth of 25—26m indicated by "hr4" in Fig. 3c spreads over a range of depths, about 3.6m in thickness. The echo strength of the reflector is uniform along the whole survey path; thus, we speculate that it might be caused by a rough interface of fractured bedrock. Beneath that, no strong radar echo or obvious hyperbola - shaped signal is observed, probably due to uniform composition and ejecta deposits that do not penetrate this layer. Furthermore, the depth is similar to the prediction about the position of high - calcium pyroxene - bearing materials in the local basalt layer (Huang et al., 2018). The interpreted fractured basalt layer is at 26m and goes slightly up by ~1.2m along the traveling path. At the depth of 32—34m (Fig. 3b), a weak but continuous reflector appears, which may indicate the rough interface of another fractured paleobasalt layer. The lateral variations of the depth of reflectors at the CE - 4 site are in the order of 1m, showing more homogenous subsurface structure along the survey path compared with the CE - 3 site.

5 Conclusions

Yutu and Yutu - 2 rovers surveyed two geologically different locations on the Moon with the same LPR instrument payload. Yutu surveyed Mare Imbrium on the nearside, and Yutu - 2 is currently in the SPA basin on the farside. Radargrams of both LPR experiments show a complicated picture of the shallow stratigraphic profile: numerous short lateral reflectors caused by the accumulation of ejecta from small local craters appear in the near - surface region. Instead of an expected sharp and flat interface between the lunar regolith and the bedrock, we have found a chaotic scattering of signals caused by ejecta from a young nearby crater, which dominate the CE - 3 LPR results. Several rough and intermittent interfaces with rocks and paleoregolith admixed are identified in the CE - 4 LPR results, which may correspond to more that one large - scale ejecta layers or the transition zones between regolith and bedrock.

The LPR results reveal a profoundly different lunar regolith structure and properties at the CE - 3 and CE - 4 landing sites. The high ilmenite abundance in CE - 3 leads to high loss tangent in the regolith; combined with a thin regolith layer and an large buried rock population, these combining factors cause the penetration depth to be 2.85 times (in terms of two - way delay) smaller than at the CE - 4 site. The 11.1 - m thick fine rego -

lith layer derived from an extracted average permittivity of 4.3 at the CE - 4 site is much thicker than that of the CE - 3 and other nearside measurements in maria. At the CE - 4 site, beneath the regolith, we detected the subsurface echoes from the early ejecta layers delivered by local and distal impact events preceding the near Finsen crater. The rough interface above the fractured bedrock at the depth of 25.8 and 33—35m are also revealed by the LPR. Due to the detection limit, CE - 3 LPR does not provide clear boundaries of the underlying layers.

Acknowledgments

We deeply appreciate the support from the team members of the Ground Application and Research System (GRAS), who contributed to datareceiving and preprocessing. The authors thank Robert E. Grimm from Southwest Research Institute and an anonymous reviewer for their constructive comments and suggestions that improved the quality of the paper. We thank Roberto Bugiolacchi and Luyuan Xu for useful discussions and editing the paper. This study was supported by the Science and Technology Development Fund (FDCT) of Macau (grants 2018/2018/A2, 121/2017/A3, and 0089/2018/A3), the Science and Technology Project of Jiangxi Education Department (grant GJJ180489), and the Scientific Research Starting Foundation for scholars from Jiangxi University of Science and Technology (grant jxxjbs18017). Chang'e - 3 (CE - 3) Lunar Penetrating Radar (LPR) data, Panorama Camera images, and CE - 2 Digital Orthophoto Map (DOM) were provided by the teamof Ground Application and Research System (GRAS) and are available from the Data Publishing and Information Service System of China's Lunar Exploration Program (http://moon.bao.ac.cn). File IDs are given in the caption. CE - 4 LPR data and panorama camera images will be released in http://moon.bao.ac.cn. File IDs are listed in Table S1. Data used in Figures 2a, 2b, 2e, and 2f are provided in Tables S2 - S4. LPR data as shown in Figures 2d, 3b, and 3c are available at http://doi.org/10.5281/zenodo.3475664. The Lunar Reconnaissance Orbiter Wide Angle Camera image for deriving TiO_2 map and Lunar Reconnaissance Orbiter (LRO) Narrow Angle Camera (NAC) image are available from the LROC team (http://wms.lroc.asu.edu/lroc).

References

[1] Bart, G. D., Nickerson, R. D., Lawder, M. T., & Melosh, H. J. (2011). Global survey of lunar regolith depths from LROC images. Icarus, 215 (2), 485—490. https://doi.org/10.1016/J.ICARUS.2011.07.017.

[2] Campbell, B. A., Ray Hawke, B., Morgan, G. A., Carter, L. M., Campbell, D. B., & Nolan, M. (2014). Improved discrimination of volcanic complexes, tectonic features,

and regolith properties in Mare Serenitatis from Earth - based radar mapping. Journal of Geophysical Research: Planets, 119, 313—330. https://doi. org/10. 1002/2013JE004486.

[3] Carrier, W. D. , Olhoeft, G. R. , & Mendell, W. (1991) . Physical properties of the lunar surface. Lunar Sourcebook, 522—530.

[4] Cooper, M. R. , Kovach, R. L. , & Watkins, J. S. (1974) . Lunar near - surface structure. Reviews of Geophysics, 12 (3), 291. https://doi. org/10. 1029/RG012i003p00291.

[5] Di, K. , Liu, Z. Liu, B. , Wan, W. , Peng, M. , Li, J. , ... Miao, Y. (2019). TOPOGRAPHIC ANALYSIS OF CHANG'E - 4 LANDING SITE USING ORBITAL, DESCENT AND GROUND DATA. ISPRS - International Archives of the Photogrammetry, Remote Sensing and Spatial Information Sciences, XLII - 2/W13, 1383—1387. https://doi. org/10. 5194/isprs - archives - xlii - 2 - w13—1383—2019.

[6] DiGiuseppe, M. , Pirich, R. , & Kraut, V. (2009) . Lunar regolith control and resource utilization. In 2009 IEEE Long Island Systems, Applications and Technology Conference (pp. 1—5) . IEEE. https://doi. org/10. 1109 /LISAT. 2009. 5031560.

[7] Dong, Z. , Fang, G. , Ji, Y. , Gao, Y. , Wu, C. , & Zhang, X. (2017). Parameters and structure of lunar regolith in Chang'e - 3 landing area from Lunar Penetrating Radar (LPR) data. Icarus, 282, 40 — 46. https://doi. org/10. 1016/j. icarus. 2016. 09. 010.

[8] Duennebier, F. K. , Watkins, J. , & Kovach, R. (1974) . Results from the lunar surface profiling experiment. In the Lunar and Planetary Science Conference (p. 183) .

[9] Fa, W. , & Jin, Y. - Q. (2010) . A primary analysis of microwave brightness temperature of lunar surface from Chang e 1 multi - channel radiometer observation and inversion of regolith layer thickness. Icarus, 207 (2), 605 — 615. https://doi. org/10. 1016/j. icarus. 2009. 11. 034.

[10] Fa,W. , & Wieczorek, M. A. (2012) . Regolith thickness over the lunar nearside: Results from Earth - based 70 - cm Arecibo radar observations. Icarus, 218 (2), 771—787. https://doi. org/10. 1016/J. ICARUS. 2012. 01. 010.

[11] Fa, W. , Zhu, M. , Liu, T. , & Plescia, J. B. (2015) . Regolith stratigraphy at the Chang'e - 3 landing site as seen by lunar penetrating radar. Geophysical Research Letters, 42, 10, 179—10, 187. https://doi. org /10. 1002/2015GL066537.

[12] Fang, G. - Y. , Zhou, B. , Ji, Y. —C. , Zhang, Q. - Y. , Shen, S. —X. , Li, Y. — X. , et al. (2014) . Lunar Penetrating Radar onboard the Chang ' e - 3 mission. Research in Astronomy and Astrophysics, 14 (12), 1607—1622. https://doi. org/10. 1088/1674—4527/14/12/009.

[13] Feng, J. , Su, Y. , Ding, C. , Xing, S. , Dai, S. , & Zou, Y. (2017).

Dielectric properties estimation of the lunar regolith at CE－3 landing site using lunar penetrating radar data. Icarus，284，424－430. https://doi.org/10.1016/j.icarus.2016.12.005.

[14] Grimm，R. E.（2018）. New analysis of the Apollo 17 surface electrical properties experiment. Icarus，314，389－399. https://doi.org/10.1016/J.ICARUS.2018.06.007.

[15] Grimm，R. E.，Heggy，E.，Clifford，S.，Dinwiddie，C.，McGinnis，R.，& Farrell，D.（2006）. Absorption and scattering in ground－penetrating radar：Analysis of the Bishop Tuff. Journal of Geophysical Research，111，E06S02. https://doi.org/10.1029/2005JE002619.

[16] Huang，J.，Xiao，Z.，Flahaut，J.，Martinot，M.，Head，J.，Xiao，X.，et al.（2018）. Geological characteristics of Von Kármán crater，northwestern South Pole－Aitken Basin：Chang'E－4 landing site region. Journal of Geophysical Research：Planets，123，1684－1700. https://doi.org/10.1029/2018JE005577.

[17] Ishiyama，K.，Kumamoto，A.，Ono，T.，Yamaguchi，Y.，Haruyama，J.，Ohtake，M.，et al.（2013）. Estimation of the permittivity and porosity of the lunar uppermost basalt layer based on observations of impact craters by SELENE. Journal of Geophysical Research：Planets，118，1453－1467. https://doi.org/10.1002/jgre.20102.

[18] Ito，T.，& Malhotra，R.（2010）. Asymmetric impacts of near－Earth asteroids on the Moon. Astronomy and Astrophysics，519，A63. https://doi.org/10.1051/0004－6361/200912901.

[19] Jaumann，R.，Hiesinger，H.，Anand，M.，Crawford，I. A.，Wagner，R.，Sohl，F.，et al.（2012）. Geology，geochemistry，and geophysics of the Moon：Status of current understanding. Planetary and Space Science，74（1），15－41. https://doi.org/10.1016/j.pss.2012.08.019.

[20] Jolliff，B. L.，Haskin，L. A.，Korotev，R. L.，Papike，J. J.，Shearer，C. K.，Pieters，C. M.，& Cohen，B. A.（2003，March）. Scientific expectations from a sample of regolith and rock fragments from the interior of the lunar South Pole－Aitken basin. In Lunar and Planetary Science Conference（Vol. 34）.

[21] Kobayashi，T.，Kim，J. H.，Lee，S. R.，Araki，H.，& Ono，T.（2010）. Simultaneous observation of lunar radar sounder and laser altimeter of Kaguya for lunar regolith layer thickness estimate. IEEE Geoscience and Remote Sensing Letters，7（3），435－439. https://doi.org/10.1109/LGRS.2009.2038499.

[22] Lai，J.，Xu，Y.，Zhang，X.，& Tang，Z.（2016）. Structural analysis of lunar subsurface with Chang'e－3 lunar penetrating radar. Planetary and Space Science，120，96－102. https://doi.org/10.1016/j.pss.2015.10.014.

[23] Lai，J.，Xu，Y.，Zhang，X.，& Tang，Z.（2017）. Lunar regolith stratigraphy analysis based on the simulation of lunar penetrating radar signals. Advances in Space Research，60（9），2099－2107. https://doi.org/10.1016/J.ASR.

2017. 07. 032.

[24] Lucey, P. G., Blewett, D. T., & Jolliff, B. L. (2000). Lunar iron and titanium abundance algorithms based on final processing of Clementine ultraviolet - visible images. Journal of Geophysical Research, 105 (E8), 20297—20305. https://doi.org/10.1029/1999JE001117.

[25] Mckay, D. S., Heiken, G., Basu, A., Blanford, G., Simon, S., Reedy, R., et al. (1991). The lunar regolith. In The Lunar Sourcebook, (pp. 285—356). Cambridge, England: Cambridge University Press.

[26] Melosh, H. J. (1989). Impact cratering: A geologic process. Research Supported by NASA. New York, Oxford University Press (Oxford Monographs on Geology and Geophysics, No. 11), 1989, 253 P., 11. Retrieved from http://adsabs.harvard.edu/abs/1989QB603.C7M45....

[27] Mendell, W. W. (1985). Lunar bases and space activities of the 21st century. Lunar and Planetary Institute. Retrieved from http://ads.harvard.edu/books/lbsa/.

[28] Meng, Z., Xu, Y., Zheng, Y., Zhu, Y., Jia, Y., & Chen, S. (2014). Inversion of lunar regolith layer thickness with CELMS data using BPNN method. Planetary and Space Science, 101, 1 — 11. https://doi.org/10.1016/J.PSS.2014.05.020.

[29] Morota, T., & Furumoto, M. (2003). Asymmetrical distribution of rayed craters on the Moon. Earth and Planetary Science Letters, 206 (3—4), 315—323. https://doi.org/10.1016/S0012—821X (02) 01111—1.

[30] Nakamura, Y., Dorman, J., Duennebier, F., Lammlein, D., & Latham, G. (1975). Shallow lunar structure determined from the passive seismic experiment. The Moon, 13 (1—3), 57—66. https://doi.org/10.1007/BF00567507.

[31] Neal, C. R. (2009). The Moon 35 years after Apollo: What's left to learn? Geochemistry, 69 (1), 3 — 43. https://doi.org/10.1016/J.CHEMER.2008.07.002.

[32] Oberbeck, V. R., & Quaide, W. L. (1967). Estimated thickness of a fragmental surface layer of Oceanus Procellarum. Journal of Geophysical Research, 72 (18), 4697—4704. https://doi.org/10.1029/JZ072i018p04697.

[33] Oberbeck, V. R., & Quaide, W. L. (1968). Genetic implications of Lunar regolith thickness variations. Icarus, 9 (1 — 3), 446 — 465. https://doi.org/10.1016/0019—1035 (68) 90039—0.

[34] Olhoeft, G. R., & Strangway, D. W. (1975). Dielectric properties of the first 100 meters of the Moon. Earth and Planetary Science Letters, 24 (3), 394—404. https://doi.org/10.1016/0012—821X (75) 90146—6.

[35] Otake, H., Ohtake, M., & Hirata, N. (2012). Lunar iron and titanium abundance algorithms based on SELENE (Kaguya) multiband imager data. In 43rd

Lunar and Planetary Science Conference (Vol. 43, p. 1905) . Woodlands, Texas. Retrieved from http://adsabs. harvard. edu/ abs/2012LPI...43. 1905O.

[36] Pasckert, J. H. , Hiesinger, H. , & van der Bogert, C. H. (2018) . Lunar farside volcanism in and around the South Pole - Aitken basin. Icarus, 299, 538—562. https://doi. org/10. 1016/j. icarus. 2017. 07. 023.

[37] Prettyman, T. H. , Hagerty, J. J. , Elphic, R. C. , Feldman, W. C. , Lawrence, D. J. , McKinney, G. W. , & Vaniman, D. T. (2006) . Elemental composition of the lunar surface: Analysis of gamma ray spectroscopy data from Lunar Prospector. Journal of Geophysical Research, 111 (E12) . https://doi. org/10. 1029/2005je002656.

[38] Petro, N. E. , & Pieters, C. M. (2006) . Modeling the provenance of the Apollo 16 regolith. Journal of Geophysical Research, 111, E09005. https://doi. org/10. 1029/2005JE002559.

[39] Qiao, L. , Ling, Z. , Fu, X. , & Li, B. (2019) . Geological characterization of the Chang'e - 4 landing area on the lunar farside. Icarus, 333, 37—51. https://doi. org/10. 1016/j. icarus. 2019. 05. 029.

[40] Quaide, W. L. , & Oberbeck, V. R. (1968) . Thickness determinations of the lunar surface layer from lunar impact craters. Journal of Geophysical Research, 73 (16), 5247—5270. https://doi. org/10. 1029/JB073i016p05247.

[41] Rabiner, L. R. , & Gold, B. (1975) . Theory and application of digital signal processing. Englewood Cliffs, NJ: Prentice - Hall Inc.

[42] Sanders, G. B. , & Larson, W. E. (2011) . Integration of in - situ resource utilization into lunar/Mars exploration through field analogs. Advances in Space Research, 47 (1), 20—29. https://doi. org/10. 1016/j. asr. 2010. 08. 020.

[43] Sato, H. , Robinson, M. S. , Lawrence, S. J. , Denevi, B. W. , Hapke, B. , Jolliff, B. L. , & Hiesinger, H. (2017) . Lunar mare TiO_2 abundances estimated from UV/Vis reflectance. Icarus, 296, 216 — 238. https://doi. org/10. 1016/J. ICARUS. 2017. 06. 013.

[44] Sharpton, V. L. (2014) . Outcrops on lunar crater rims: Implications for rim construction mechanisms, ejecta volumes and excavation depths. Journal of Geophysical Research: Planets, 119, 154 — 168. https://doi. org/10. 1002/2013JE004523.

[45] Shkuratov, Y. G. , & Bondarenko, N. V. (2001) . Regolith layer thickness mapping of the Moon by radar and optical data. Icarus, 149 (2), 329 — 338. https://doi. org/10. 1006/ICAR. 2000. 6545.

[46] Shoemaker, E. M. , Batson, R. M. , Holt, H. E. , Morris, E. C. , Rennilson, J. J. , & Whitaker, E. A. (1968) . Television observations from Surveyor 3. Journal of Geophysical Research, 73 (12), 3989—4043. https://doi. org/10. 1029/JB073i012p03989.

[47] Shoemaker, E. M. , Hait, M. H. , Swann, G. A. , Schleicher, D. L. , Dahlem,

D. H. , Schaber, G. G. , & Sutton, R. L. (1970) . Lunar Regolith at tranquillity base. Science, 167 (3918), 452 — 455. Retrieved from. https://www.jstor.org/stable/1728711? seq = 1 # metadata _info _tab _con - tents, https://doi.org/10.1126/science.167.3918.452.

[48] Simmons, G. , Strangway, D. W. , Annan, A. P. , Baker, R. , Bannister, L. , Brown, R. , et al. (1973) . The surface electrical properties experiment, Apollo 17, Preliminary Science Report. In NASA SP—330 (pp. 15—1—15—14) .

[49] Skolnik, M. I. (2001) . Introduction to radar systems. New York: McGraw Hill.

[50] Spudis,P. D. , Gillis, J. J. , & Reisse, R. A. (1994) . Ancient multiring basins on the Moon revealed by Clementine Laser Altimetry. Science, 266 (5192), 1848—1851. https://doi.org/10.1126/science.266.5192.1848.

[51] Stuart -Alexander, D. E. (1978) . Geologic map of the central far side of the Moon. US Geological Survey, I - 1047. https://doi.org/10.3133/i1047.

[52] Su, Y. ,Fang, G. - Y. , Feng, J. - Q. , Xing, S. —G. , Ji, Y. —C. , Zhou, B. , et al. (2014) . Data processing and initial results of Chang'e - 3 lunar pene - trating radar. Research in Astronomy and Astrophysics, 14 (12), 1623 — 1632. https://doi.org/10.1088/1674—4527/14/12/010.

[53] Taylor,G. J. , Warner, R. D. , & Keil, K. (1979) . Stratigraphy and depositional history of the Apollo 17 drill core. In: Lunar and Planetary Science Conference, 10th, Houston, Tex. , March 19 — 23, 1979, Proceedings. Volume 2. (A80 — 23617 08 — 91) New York, Pergamon Press, Inc. , 1979, p. 1159 — 1184. , 10, 1159—1184. Retrieved from http://adsabs.harvard.edu/abs/1979LPSC...10.1159T.

[54] Thompson, T. W. (1987) . High - resolution lunar radar map at 70 - cm wavelength. Earth, Moon and Planets, 37 (1), 59 — 70. https://doi.org/10.1007/BF00054324.

[55] Vaniman, D. T. , Labotka, T. C. , Papike, J. J. , Simon, S. B. , & Laul, J. C. (1979) . The Apollo 17 drill core - Petrologic systematics and the identification of a possible TYCHO component. In: Lunar and Planetary Science Conference, 10th, Houston, Tex. , March 19 — 23, 1979, Proceedings. Volume 2. (A80 — 23617 08—91) New York, Pergamon Press, Inc. , 1979, p. 1185 — 1227. , 10, 1185 — 1227. Retrieved from http://adsabs.harvard.edu/abs/1979LPSC...10.1185V.

[56] Vaniman, D. T. , & Papike, J. J. (1977) . The Apollo 17 drill core - Characterization of the mineral and lithic component/sections 70007, 70008, 70009. In 8th Lunar Science Conference, (pp. 3123—3159) . Houston, Texus: New York, Pergamon Press, Inc.

[57] Werner, S. C. , & Medvedev, S. (2010) . The Lunar rayed - crater population - Characteristics of the spatial distribution and ray retention. Earth and Planetary Science Letters, 295 (1—2), 147—158. https://doi.org/10.1016/J.EPSL.2010.03.036.

[58] Wieczorek, M. A. , & Le Feuvre, M. (2009) . Did a large impact reorient the Moon?

Icarus, 200 (2), 358—366. https://doi. org/10. 1016/J. ICARUS. 2008. 12. 017.

[59] Wilhelms, D. E. , Howard, K. A. , & Wilshire, H. G. (1979) . Geologic map of the south side of the Moon. IMAP. https://doi. org/10. 3133/i1162.

[60] Wilhelms, D. E. , McCauley, J. F. , & Trask, N. J. (1987) . The geologic history of the Moon. Professional Paper. https://doi. org/10. 3133/PP1348.

[61] Wu, Y. , Li, L. , Luo, X. , Lu, Y. , Chen, Y. , Pieters, C. M. , et al. (2018). Geology, tectonism and composition of the northwest Imbrium region. Icarus, 303, 67—90. https://doi. org/10. 1016/j. icarus. 2017. 12. 029.

[62] Xiao, L. , Zhu, P. , Fang, G. , Xiao, Z. , Zou, Y. , Zhao, J. , et al. (2015) . A young multilayered terrane of the northern Mare Imbrium revealed by Chang'e - 3 mission. Science, 347 (6227), 1226—1229. https://doi. org/10. 1126/science. 1259866.

[63] Xing, S. G. , Su, Y. , Feng, J. Q. , Dai, S. , Xiao, Y. , Ding, C. Y. , & Li, C. L. (2017) . The penetrating depth analysis of Lunar Penetrating Radar onboard Chang'e - 3 rover. Research in Astronomy and Astrophysics, 17 (5) . https://doi. org/10. 1088/1674—4527/17/5/46.

[64] Zhang, J. , Yang, W. , Hu, S. , Lin, Y. , Fang, G. , Li, C. , et al. (2015). Volcanic history of the Imbrium basin: A close - up view from the lunar rover Yutu. Proceedings of the National Academy of Sciences of the United States of America, 112 (17), 5342—5347. https://doi. org/10. 1073/pnas. 1503082112.

[65] Zhang, X. , Xie, M. , Zhu, M. — H. , Dong, W. , Tang, Z. , & Xu, A. (2015) . Major elements abundances in Chang'E - 3 landing site from active particle - induced X - ray spectrometer. EGU General Assembly, Held 12 — 17 April, 2015 in Vienna, Austria. Id. 726, 17. Retrieved from http://adsabs. harvard. edu/abs/2015 EGUGA. . 17. . 726Z.

[66] Zhao, J. , Huang, J. , Qiao, L. , Xiao, Z. , Huang, Q. , Wang, J. , et al. (2014) . Geologic characteristics of the Chang'E - 3 exploration region. Science China Physics, Mechanics and Astronomy, 57 (3), 569 — 576. https://doi. org/10. 1007/s11433—014—5399—z.

High – Resolution Terrain Analysis for Lander Safety Landing and Rover Path Planning Based on Lunar Reconnaissance Orbiter Narrow Angle Camera Images: A Case Study of China's Chang'e – 4 Probe

Bo Li[1,2,3], Zongyu Yue[2], Jiang Zhang[1], Xiaohui Fu[1], Zongcheng Ling[1], Shengbo Chen[3], Jian Chen[1], and Peiwen Yao[1]

1. Shandong Provincial Key Laboratory of Optical Astronomy and Solar – Terrestrial Environment; Institute of Space Sciences, Shandong University, Weihai, China
2. State Key Laboratory of Remote Sensing Science, Institute of Remote Sensing and Digital Earth, Chinese Academy of Sciences, Beijing, China
3. College of Geoexploration Science and Technology, Jilin University, Changchun, China

Abstract: China's Chang'e – 4 (CE – 4) probe will explore the South Pole – Aitken basin in late 2018. Its preselected landing area is located on the southeastern floor of the Von Kármán crater. Landing experience of China's Chang'e – 3 probe may not be applied to CE – 4 mission directly because the topography of lunar farside is more rugged. Moreover, due to scale dependence and smooth effects, the previous topographic studies derived from digital elevation model data with lower resolution cannot represent the meter – scale topographic features. Because of low time and space complexities in lunar images, image segmentation algorithm is especially suitable for recognizing lunar features. We divided the narrow angle camera (NAC) mosaic image into three parts: dark area, bright area, and flat area, based on a double – threshold Otsu method. The first two parts corresponded to the undulating areas (positive and negative terrains). And then we calculated the flat area percentage (*Fap*) of the previous successful lunar landing missions (including Chang'e – 3, Apollo, Surveyor, and Luna series) and obtained the *Fap* threshold (>0.6) for lunar safe landing. According to *Fap* map (~ 1.5m per pixel) of CE – 4 preselected landing area, the divided square grids with a size of $0.01°$ can be classified into safe grids ($Fap > 0.6$) and unsafe grids ($Fap \leqslant 0.6$). The CE – 4 preselected landing areas can be greatly reduced to five potential landing areas. In the last, we proposed a route planning method, which took both the distance and security into account, for CE - 4 rover to dynamically generate a safe and short route between current grid and target grid.

Plain Language Summary: China's Chang'e - 4 (CE - 4) probe will land on the South Pole - Aitken basin at the end of 2018. Its preselected landing area (45 - 46°S, 176.4 - 178.8°E) is located on the southeastern floor of the Von Kármán crater. It is a very large area, and there are many small (meter - scale) topographic features such as rocks, craters, ridges, and troughs. Therefore, it is necessary to narrow the scope and choose the safe landing areas from the preselected landing area. The flat area percentage (*Fap*) isan important factor to measure the safety of landing area. Based on an image segmentation method, we extracted *Faps* in square grids with a size of 0.01°and then the CE - 4 landing areas were reduced to five safe areas. After CE - 4 lands at the safe site, a rover will be released to travel around. We proposed a path planning method for CE - 4 rover moving from current grid to target grid. This method took both the distance and security into account to find out the best path with safety and short distance.

1 Introduction

Since the Luna - 1 mission in 1959, there have been many successful lunar explorations by orbiters and landers, such as Luna, Surveyor, Apollo, and Chang'e missions. However, the early lunar landing sites were restricted to the midlatitudes because of technological limitations. Today, the recent lunar missions (such as Chang'e - 4 [CE - 4] and Luna - 25) are interested in the lunar subpolar areas (±65 - 85° latitudes; Kokhanov et al., 2017; Ye et al., 2017).

Luna - 25 is designed for launching and landing on a lunar subpolar area in 2019. The objective of Luna - 25 is to explore the lunar polar resources and the lunar exosphere to study the content of polar water ice. For Luan - 25, 12 potential landing sites (PLSs) were selected according to several criteria. For example, surface slopes at the landing site must be less than 7° at a baseline of 2.5 m, and the illumination level within the landing area must be more 40% over the day, and so on (Mitrofanov et al., 2016). And then a topographical and photogeological study of three PLSs (1, 4, and 6) was carried out to characterize their safety conditions and possible sources of materials (Ivanov et al., 2018). And the hilly unit in PLS 6 is among the safest terrains for the Luna - 25 landing.

China's CE - 4, the first in situ exploration of lunar farside, will explore the South Pole - Aitken basin in late 2018 (Wu et al., 2017). The preselected landing area (45 - 46°S, 176.4 - 178.8°E) for CE - 4 is located on the southeastern floor of the Von Kármán crater (Figure 1a). The diameter of Von Kármán crater is ~186km, and its central coordinates is 44.4°S, 176.2°E. Von Kármán crater is a degraded pre - Nectarian impact crater, flooded with mare basalts in its floor. CE - 4 will choose a safe landing site to land, which is orbited by a rover to execute some geologic investigation tasks. China's Chang'e -

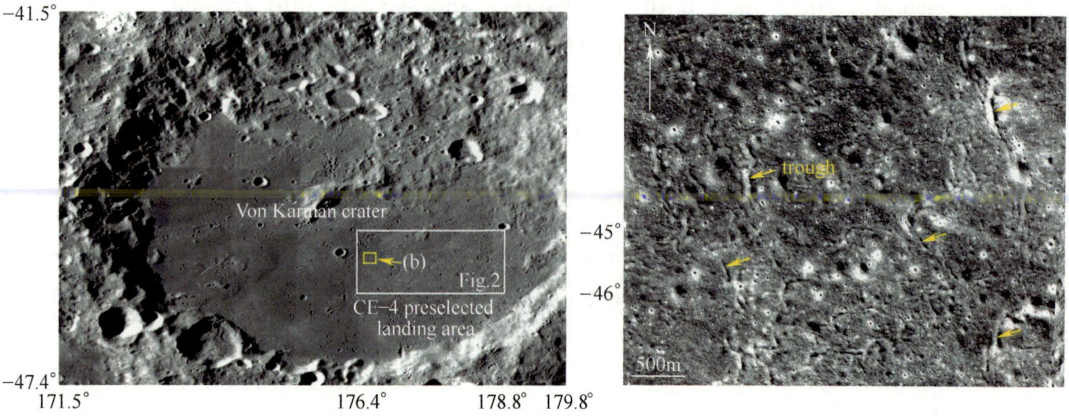

(a)　　(b)

Figure 1　(a) Lunar Reconnaissance Orbiter wide angle camera mosaic of Von Kármán crater The white box shows the Chang'e - 4 (CE - 4) preselectedlandingarea. (b) The troughs (yellow arrows) in the floor of Von Kármán crater

3 (CE - 3) lunar probe has successfully landed on a young lava plain in northern Mare Imbrium, in December 2013. And Yutu rover explored the lunar surface targets and subsurface near a young crater using its main scientific instruments (Di et al., 2016; Ling et al., 2015; Xiao et al., 2015). But the successful experience of CE - 3 landing and roving may not be applied to CE - 4 mission directly, because the topographies of lunar farside are more rugged and difficult for the safe landing and traversing objective. (1) The preselected landing region for CE - 4 is older than the landing site of CE - 3 mission. The landing area of CE - 3 can be divided into two basaltic units, and the absolute model ages of these two units are ~3.17 and ~3.48 Ga (Li, Ling, et al., 2016). Based on different TiO_2 contents of basaltic units, Ling et al. (2018) classified the mare region on Von Kármán crater's floor into two units, whose absolute model ages were ~3.70 (Haruyama et al., 2009) and ~3.65 Ga. Therefore, the CE - 4 preselected landing area has more large impact craters and a higher crater density than CE - 3 landing site. Large craters within a landing site represent potential hazards to landers as well as navigational threats to rovers. (2) There are more small geomorphic features such as degraded secondary crater chains, sinuous ridges, and troughs (Figure 1b) in CE - 4 preselected landing area than the CE - 3 landing site. These ridges are asymmetric in shape and extend in a sinusoidal - like shape, whose widths are tens of meters (Huang et al., 2018). While sinuous troughs are formed between the ridges, both the troughs and ridges are concentrated in the west of the landing region.

For early lunar landing missions (such as Luna and Surveyor series), lunar probes did not have obstacle identification and avoidance capabilities due to limited techniques at that time. While in Apollo missions, the success rate of safe landing was improved because the astronauts observed the landing area and manipulated the control system to avoid the obstacles (Sears, 1964). In the future unmanned Martian and lunar landing missions, the autonomous obstacle - avoiding technology using laser imaging radar is proposed (Johnson et al., 2002; Wong & Singh, 2002). When the probe drops to a certain height, the lunar surface is scanned by laser imaging radar. Obstacles are identified based on the slope and roughness maps derived from the three - dimensional DEM data. But this new technology has not been achieved and applied. Several studies have analyzed the terrains and landforms of the CE - 4 and CE - 5 preselected landing area based on digital elevation model (DEM) data. For example, Wu et al. (2015) proposed an autonomous hazard detection and rough - fine hazard avoidance guidance method for soft lunar landing. This method mitigates the landing risks due to the uncertainty and hazards of the lunar terrain, such as surface roughness and slope. Huang et al. (2018) thought that the proposed CE - 4 landing area was usually flat, based on the fact that almost 99% of the area has a slope less than 15° and 94% of the area has a slope less than 5° at a ~59 - m scale length. Zhao et al. (2017) found that Rümker plateau (a candidate landing site for the CE - 5 mission) stands 200 - 1, 300m above the surrounding mare surface and 75% of the plateau has a slope of less than 3° at a baseline length of ~30m. However, these results were not suitable for guiding CE - 4 landing and rover route planning. Because these topographic analyses were performed based on the merged DEM and the resolutions of the produced slope map was ~59 - and ~30 - m scale. According to the previous studies, the lunar topographic textures (such as slope, roughness, and entropy values) show scale dependence (Cao et al., 2015; Kreslavsky et al., 2013; Li et al., 2015; Li, Wang, et al., 2016; Robbins, 2018). Kreslavsky et al. (2013) found that hectometer - scale lunar roughness poorly correlated with kilometer - scale lunar roughness, because they reflected different sets of geologic processes and time scales. Li, Wang, et al. (2016) used the statistical moments of gray level histograms in different - sized neighborhoods (e. g., 3, 5, 10, 20, 40, and 80 pixels) to compute the kilometer - scale roughness and entropy values based on a WAC mosaic image. They found that large roughness and entropy signatures were only shown in larger - scale maps, while the smallest three - pixel scale map had more disorderly and unsystematic textures. Perhaps, there is a smooth effect in the process of generating larger - scale textures. In this process smaller textures resulting from tiny topographical and compositional changes may be removed. Thus, the resolutions of the slope maps derived from the DEM data with ~59 - and ~30 - m scale are not high enough to represent the smaller topographic features (such as rocks, ridges, and troughs) with a meter scale.

In this paper, we used Lunar Reconnaissance Orbiter (LRO) narrow angle camera (NAC) images with a decimeter resolution to (1) divide NACs covering the CE - 4 preselected landing site into three parts: dark area, bright area, and flat area, based on an image segmentation algorithm. The first two parts correspond to the undulating area (positive and negative terrains, PNTs) of the lunar surface. (2) Calculate the flat area percentage (*Fap*) of the previous successful lunar landing missions (including CE - 3, Apollo, Surveyor and Luna series) and obtain the *Fap* threshold for lunar safety landing. (3) Classify the CE - 4 preselected landing area into regular grids and evaluate the safety of each grid to choose potential landing areas (PLAs). (4) Traverse out all the paths between the current grid and target grid of CE - 4 rover to find out the best path with safety and short distance.

2 NAC Image Segmentation Using a Double - Threshold Otsu Method

2.1 LRO NAC Images

LRO was launched on 18 June 2009 and entered lunar orbit on 23 June 2009. The LRO Camera consists of two NACs with about 50cm per pixel resolution from an altitude of 50km (Robinson et al., 2010). LRO is an exploration mission with a primary goal of identifying desirable safe landing sites on the Moon. The LRO NAC images provide information on the distribution of craters and slopes over small baselines, with the technique of shadowed - area analysis (Abdrakhimov et al., 2015). Here we selected 27 single NAC images to cover the CE - 4 preselected landing area. The details of these NAC images were listed in the Table 1. As far as possible, the NAC images with the incidence angles between 60° and 80° were selected to ensure that the image contrast is large, and the PNTs are obvious in these images. In order to ensure the accuracy of *Fap* measurements, Lambert Conformal Conic Projection was used to reduce the projective deformation. This projection is based on the GCS _ MOON _ 2000 (a geographic coordinate system) whose datum is the D _ Moon _ 2000 in GIS software. In this projection coordinate system, all NAC images were registered and then mosaiced into one image with a resolution ~1.5m per pixel (Figure 2a), but with some small residual gaps (white areas marked by yellow arrows in Figure 2a).

Table 1 Details of NAC Images Covering the CE - 4 Preselected Landing Area

NAC image ID	Resolution (meters per pixel)	Incidence angle/(°)	Solar azimuth angle/(°)
M1220160190L	0.578	65.75	150.73
M143453659L	0.629	53.36	225.41
M143453659R	0.628	53.43	224.01
M161149639L	1.155	71.27	201.91
M161149639R	1.153	71.35	200.83
M161142858L	1.155	71.08	201.89
M161142858R	1.154	71.17	200.82
M134022629L	1.269	77.55	167.75
M134022629R	1.268	77.64	167.94
M180000082L	1.265	83.55	173.82

续表

NAC image ID	Resolution (meters per pixel)	Incidence angle/ (°)	Solar azimuth angle/ (°)
M180000082R	1.263	83.65	173.8
M115143943L	1.267	76.24	194.77
M115143943R	1.265	76.33	193.9
M128121241R	0.564	58.91	219.08
M1242491605L	0.82	55.89	229.45
M1242491605R	0.819	55.98	227.9
M180007222R	1.251	83.62	173.67
M178833263R	1.398	86.79	201.82
M1095330314L	1.319	84.73	173.98
M1095330314R	1.317	84.83	173.96
M189435642L	0.622	54.31	232.19
M1128338348L	0.658	67.75	155.39
M1095344559R	1.356	84.51	174.1
M180007222L	1.252	83.52	173.68
M171761396R	0.648	43.19	266.16
M1110639845R	1.234	86.39	176.84
M1145986900R	0.617	53.56	128.73

Note. NAC = narrow - angle camera; CE - 4 = Chang'e - 4.

2.2　Double - Threshold Otsu Method

Lunar and Martian landers are sensitive to PNTs (including rocks, craters, ridges, troughs, and secondary crater chains). PNTs can lead lander to tilt or tip over with remaining horizontal velocity, impact the under - side and damage the lander, and can prevent complete opening of the solar arrays after landing (Golombek et al., 2012). In addition, when the rover patrols, the area below the rover must be free of PNTs likely damaging the rover's lower structure. Therefore, PNTs on the surface of a planet represent an obvious hazard to landing spacecraft and a potential impediment to traveling. The more exactly PNTs on a landing site can be recognized, the better the potential hazards to the lander and rover can be described and avoided. Identifying the PNTs at PLSs through remote sensing data is thus critical for the success of landing tasks.

On lunar and other planetary surfaces, shadows and bright areas in remote sensing images or in situ photographs are caused by the presence of raised or depressed features, which are interpreted as PNTs, that is, uneven areas (Golombek et al., 2008). However, the brightness changes on lunar images may also be caused by different component distributions. Von Kármán crater is a degraded pre - Nectarian impact crater, flooded with mare basalts on its floor. Due to longterm and continuous impact events from the pre - Nectarian period to the present, the regolith covering Von Kármán crater's floor was fully mixed. Thus, we thought that the brightness changes on lunar images caused by compositional difference were subtle and even negligible. Several studies have applied this criterion to identify individual rocks on lunar and Martian surfaces using remote sensing or in situ images (Cintala & Mcbride, 1995; Di et al., 2016; Li et al., 2017; Liasis & Stavrou, 2016; Wang et al., 2017).

Each PNT consists of two parts (dark and bright areas) in lunar images under the sunlight. Here we used an image segmentation algorithm to identify PNTs and extract flat areas on CE-4 preselected landing area based on the mosaic NAC image. (1) Nowadays, there is almost no internal geological process and few external impact processes on the Moon. Compared with the weathering on Earth, the only space weathering process on the Moon is too weak to modify the lunar surface morphology. The topographic and structural features on lunar surface almost do not vary with time at present. Therefore, lunar remote sensing images have much lower time complexity than Earth remote sensing images. (2) Compared with the abundant activities on Earth, there is no production, life and modification activity caused by human, animals, and plants on the Moon. And the ground objects are rich in color, shape, and texture on Earth, while there are only monochrome and simple objects on the Moon. Therefore, lunar remote sensing images have much lower space complexity than Earth remote sensing images. Because of low complexity in time and space of lunar images, image segmentation algorithm is especially suitable for recognizing and extracting lunar features.

Otsu's method (Otsu, 1979) is optimum in the sense that it maximizes the between-class variance, a wellknown measure used in statistical discriminant analysis (Gonzalez & Woods, 2007). The basic idea is that well-thresholded classes should be distinct with respect to the intensity values of their pixels and, conversely, that a threshold giving the best separation between classes in terms of their intensity values would be the best (optimum) threshold. The Otsu method can be extended to an arbitrary number of thresholds, because the separability measure on which it is based also extends to an arbitrary number of classes (Fukunaga, 1990). In this paper, the NAC images were classified into three parts, which were dark areas, bright areas, and flat areas. Therefore, we focused attention on image segmentation using two optimum thresholds.

Let $\{0, 1, 2, \ldots, L-1\}$ denote the L distinctintensity levels in a digital image of size $M \times N$ pixels, and let n_i denote the number of pixels with intensity i. The total number, MN, of pixels in the image is $MN = n_0 + n_1 + n_2 + \ldots + n_{L-1}$, where $p_i = n_i / MN$. For three classes consisting of three intensity intervals (which are separated by two thresholds) the between-class variance (σ_B) is given by

$$\sigma_B^2 = P_1(m_1 - m_G)^2 + P_2(m_2 - m_G)^2 + P_3(m_3 - m_G)^2 \tag{1}$$

where

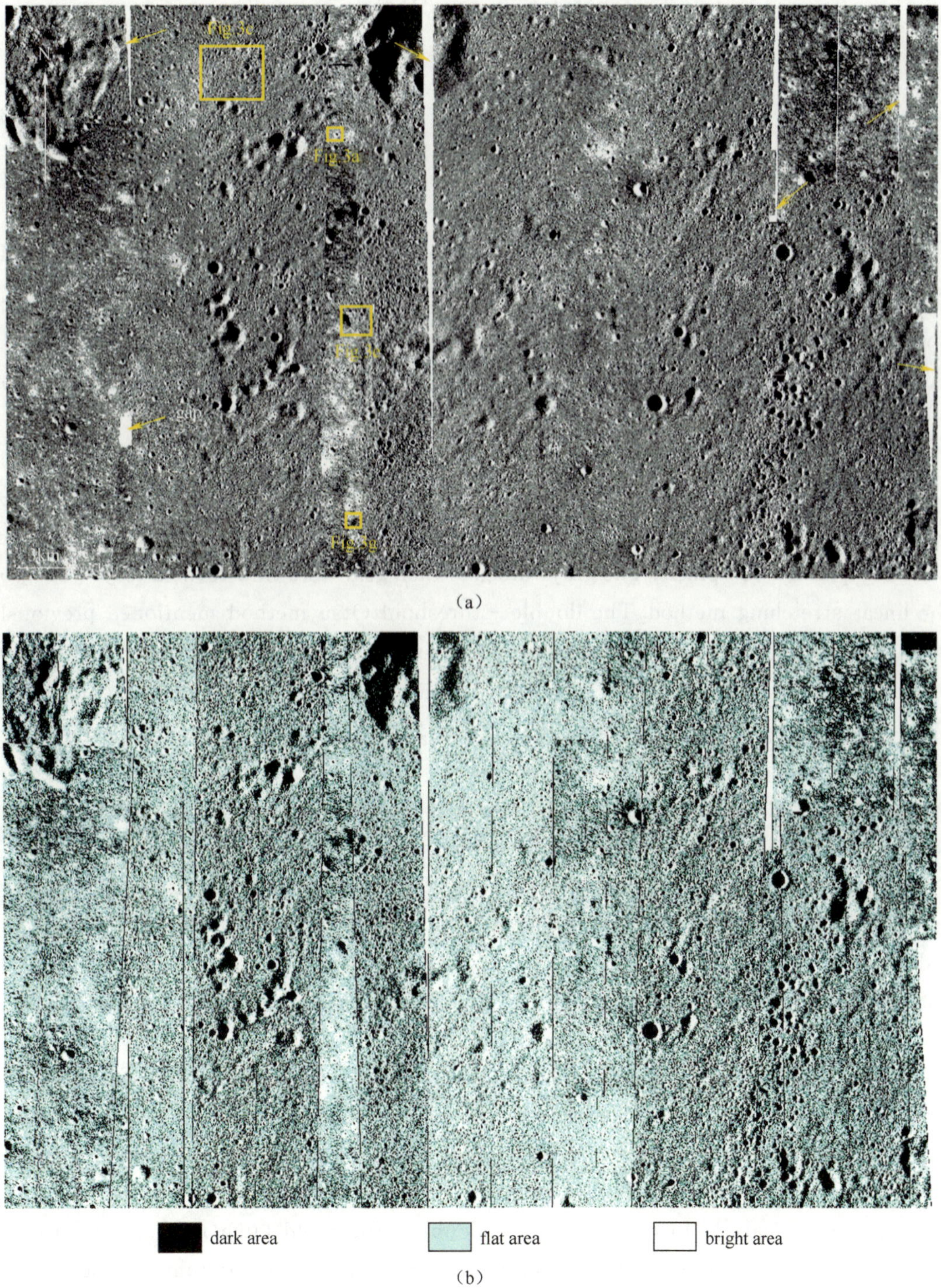

Figure 2　(a) Lunar Reconnaissance Orbiter narrow - angle camera mosaic of CE - 4 preselected landing area marked with a white box in Figure 1. The Lambert Conformal Conic Projection was used and some small gaps (white areas pointed out by yellow arrows) remained.　(b) The image segmentation results of CE - 4 preselected landing area through a doublethreshold Otsu method. CE - 4 = Chang'e - 4

$$p_1 = \sum_{i=0}^{k_1} p_i, \; P_2 = \sum_{i=k_1+1}^{k_2} p_i, \; P_3 = \sum_{i=k_2+1}^{L-1} p_i \tag{2}$$

and

$$m_1 = \frac{1}{P_1} \sum_{i=0}^{k_1} i p_i, \; m_2 = \frac{1}{P_2} \sum_{i=k_1+1}^{k_2} i p_i, \; m_3 = \frac{1}{p_3} \sum_{i=k_2+1}^{L-1} i p_i \tag{3}$$

The two optimum threshold values, k_1^* and k_2^*, are the values that maximize $\sigma_B^2 = (k_1, k_2)$. In other words, we find the two optimum thresholds by finding:

$$\sigma_B^2(k_1^*, k_2^*) = \max_{0<k_1<k_2<L-1} \sigma_B^2(k_1, k_2) \tag{4}$$

The thresholded image is then given by $g(x, y) = a$, if $f(x, y) <= k_1^*$; $g(x, y) = b$, if $k_2^* < f(x, y) <= k_2^*$; $g(x, y) = c$, if $f(x, y) > k_2^*$, where a, b, and c are the values standing for the dark area, flat area, and bright area separately.

2.3 Image Segmentation Results

In order to ensure the consistency of illumination conditions in the segmentation region, the mosaic NAC image was divided into regular grids (or subimages) whose sizes were 500 pixels×500 pixels. The gray values of each grid were transformed to [0 - 255] by the linear stretching method. The double - threshold Otsu method mentioned previously was used to every grid to calculate the two optimum thresholds, k_1^* and k_2^*, and the original value (v) in each grid can be divided into the dark part ($v < k_1^*$), the bright part ($k_1^* < v < k_2^*$) and the flat part ($v > k_2^*$). The image segmentation results of CE - 4 preselected landing area were shown in Figure 2b. The black and white colors represented the dark and bright areas, which were uneven areas, while the light green color stood for the flat areas. There were several observations:

(1) PNTs identification. In Figures 3a and 3b, all the impact craters can be identified as dark and bright areas, and rocks (bigger than ~3.0m, yellow arrows in Figure 3a and red arrows in Figure 3b) around impact craters can also be distinguished. In Figures 3c and 3d, there are many small ridges and troughs (yellow arrows in Figure 3c), which can be seen as risks of landing and patrolling for CE - 4 mission. Fortunately, these tiny undulating objects can be identified as dark and bright areas (red arrows in Figure 3d) using our method. Therefore, we can see that there are good classification results for CE - 4 preselected landing area by a double - threshold Otsu method. The PNTs can be identified accurately and completely, and the left parts are the flat regions.

(2) Error analysis. In our paper, the double - threshold Otsu method was used in each grid of CE - 4 preselected landing area, and there were two optimum thresholds (k_1^* and k_2^*) in each grid. The k_1^* and k_2^* of the adjacent grids may be inconsistent; thus, the segmentation results may be of discontinuity in the same object, which is located in several grids. In Figure 3f, the discontinuity (the horizontal yellow dotted line on the left) appears, on the edge between parts A and B coming from the same NAC image,

due to the different segmentation thresholds of the two parts. In addition, parts A, B, and part C come from two different NAC images. Because of the different illumination conditions and segmentation thresholds, discontinuities (the yellow vertical dotted line on the right) also occur in their adjacent boundaries. Moreover, in the study area, there are several dark and bright areas due to compositional inhomogeneity, which can cause the flat areas to be mistakenly classified into undulating areas. As shown in Figures 3g and 3h, there is only the dark area (the yellow dotted polygon) without a corresponding bright area, indicating that the dark area maybe result from the compositional heterogeneity on lunar surface. Luckily, this type of brightness changes caused by compositional heterogeneity is very rare in the study area and can be ignored.

(3) Result verification. In order to verify the accuracy of the image segmentation results, we evaluated the consistency between the traverse route of Yutu rover and the flat areas identified by our method at CE - 3 landing site. In Figures 3i - 3k, we can see that the route (blue line) of Yutu rover fall into the flat area extracted by our method and is kept away from the dark and bright areas, which are inferred as PNTs. This comparison result shows a good consistency between a real route and our experimental segmentation results. Therefore, the double - threshold Otsu method can be used to identify obstacles on lunar surface and ensure the safety of CE - 4 landing and patrolling tasks.

3 Safety Assessment for CE - 4 Preselected Landing Area

3.1 Calculating the *Faps* of Square Grids

In order to assess the safety of different locations, we divided the CE - 4 preselected landing area into regular square grids. Both the lander and the rover of CE - 4 were designed as a backup for the CE - 3 mission (Ye et al., 2017), and the flight process and working modes of CE - 4 were fully inherited from those of CE - 3 probe (Jia et al., 2018). Therefore, according to the CE - 3 landing process, the divided square grid size in the CE - 4 preselected landing area can be determined. The powered descent process of the CE - 3 mission includes stages of major deceleration, rapid adjustment, approach, hovering, obstacle avoidance, and low - speed descent (Liu et al., 2014). The whole process lasted about 720s. When CE - 3 probe descended to a height of about 100m, the probe entered the hovering stage. Because the time taken by this stage was just 25s, the motion of CE - 3 probe is small (~0.001° in longitude, from - 19.5134° to - 19.5125°). While in the whole descent process, the entire motion of CE - 3 probe is ~0.1097° in longitude, from - 19.4027° to - 19.5124°. According to the above analysis, the size of divided square grids used to evaluate safety was set to 0.01°. Based on the segmentation results described on Section 2, the *Fap* of each divided square grid with a size of 0.01° was calculated and was shown in Figure 4.

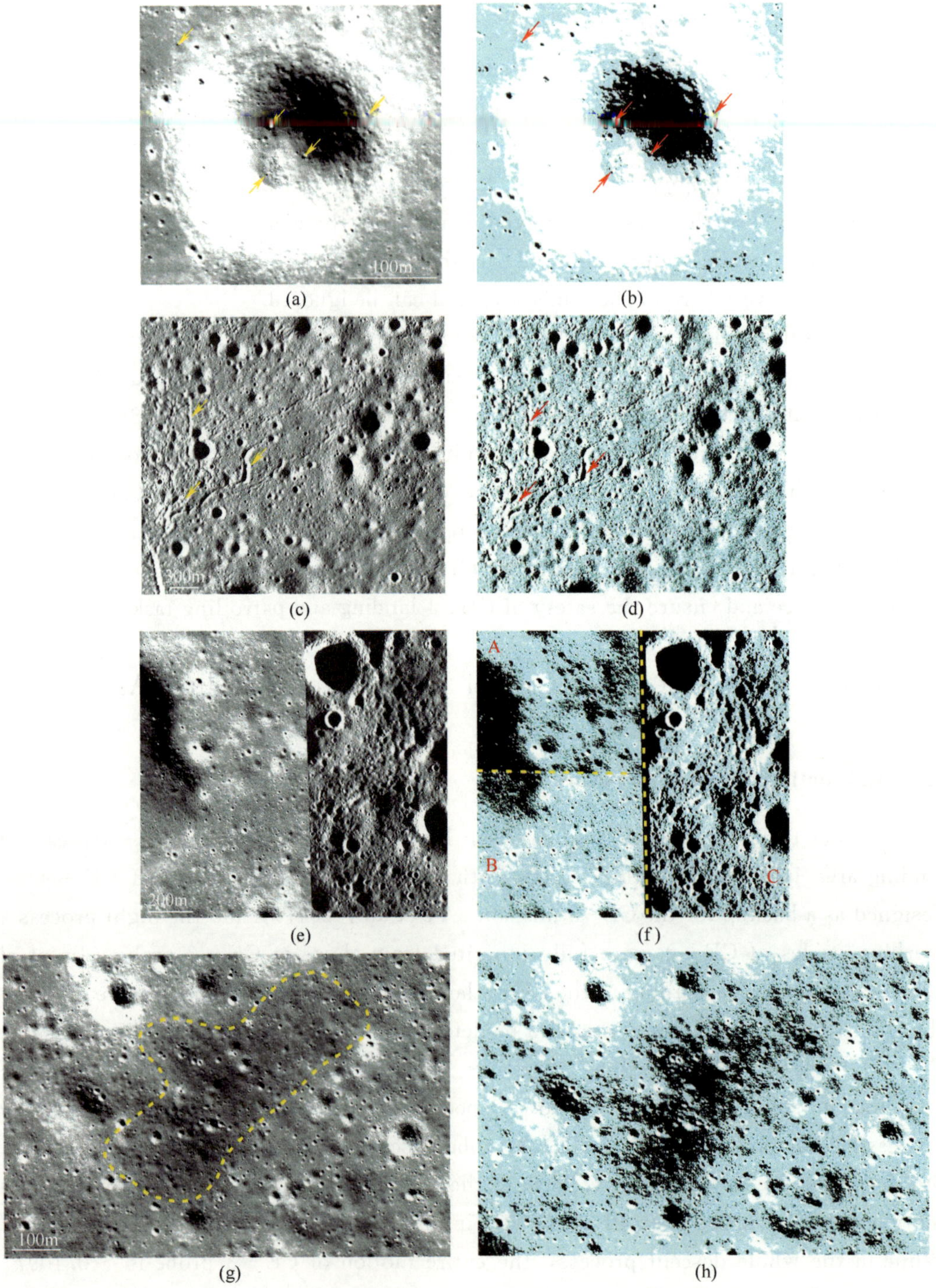
100m
(a)
(b)
300m
(c)
(d)
200m
(e)
A
B
C
(f)
100m
(g)
(h)

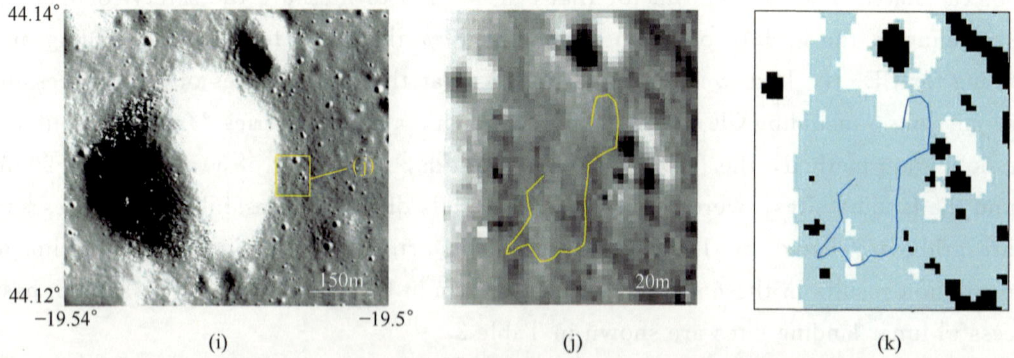

Figure 3　(a and b) The NAC image and the segmentation results of a crater and several surrounding rocks (the yellow arrows). (c and d) The NAC image and the segmentation results of the small ridges and troughs (yellow and red arrows). (e and f) The discontinuities (the two yellow dotted lines) due to the different illumination conditions and segmentation thresholds in the same NAC image and two adjacent NAC images. (g and h) The brightness changes (the yellow dotted polygon) because of the compositional heterogeneity on lunar surface. (i - k) The CE - 3 landing site in the NAC image M181302794LC with a resolution of ~1.57m, the route of Yutu rover, and the image segmentation results of the CE - 3 landing site. The boundaries of (a), (c), (e), and (g) are marked in the yellow boxes in Figure 2. NAC = narrow - anglecamera; CE - 3 = Chang'e - 3

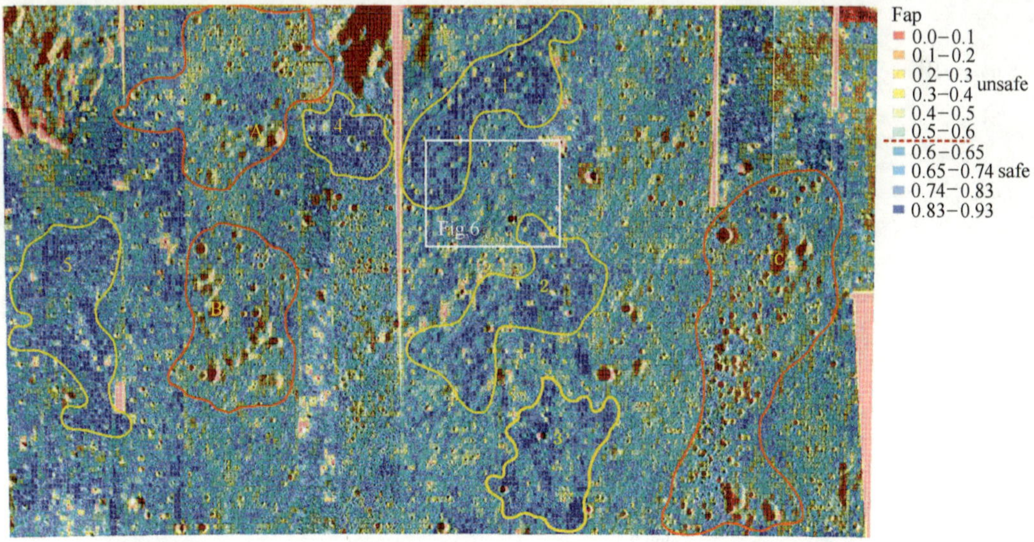

Figure 4　The *Fap* of each divided square grid with a size of 0.01° in Chang'e - 4 preselected landing area, superposing on the narrow - angle camera mosaic image. The five potential landing areas (yellow polygons) are marked with numbers 1 - 5, while the three least safe landing areas (red polygons) are marked with A - C

3.2　Determining the *Fap* Threshold for Lunar Safe Landing

After both flat areas and uneven areas were identified in a grid, the *Fap* of this grid can be calculated. The *Fap* is a factor that can be used to measure the safety for landing and traveling in the grid. In order to obtain the *Fap* threshold for the safe landing and traveling for CE－4, here we calculated the *Faps* at the landing sites of past successful lunar missions, including CE－3, Apollo, Surveyor, and Luna series. Using the double－threshold Otsu method, the *Faps* of the square grids, with sizes of 50, 100, and 200m around the landing sites, were computed. The details of past successful landing sites and square grids are listed in Table S1 in the supporting information, and the image segmentation results in these square grids are shown in Figures S1－S13. The Faps in past successful lunar landing sites are shown in Table 2.

Table 2　The Faps of Square Grids Around the Past Successful Lunar Landing Sites

Landing Site	Square size (m)	*Fap* (%)	Landing site	Square size (m)	*Fap* (%)
Luna 17	50	53.6	Apollo11	50	49.2
	100	57.3		100	47.4
	200	55.1		200	53.6
Luna 21	50	56.5	Apollo12	50	57.4
	100	62.3		100	47.0
	200	64.0		200	56.0
CE－3	50	37.8	Apollo14	50	61.2
	100	39.4		100	62.3
	200	35.0		200	69.0
Surveyor1	50	65.3	Apollo15	50	44.4
	100	74.3		100	63.2
	200	71.2		200	62.8
Surveyor5	50	51.4	Apollo16	50	65.0
	100	48.2		100	70.8
	200	58.3		200	68.7
Surveyor6	50	59.7	Apollo17	50	70.3
	100	54.2		100	66.9
	200	52.8		200	71.7
Surveyor7	50	70.1			
	100	73.5			
	200	68.6			

We can see that for the past successful landing sites, the *Faps* in square grids with a size of 200m are all bigger than ～53.6%, except for CE－3 landing site with a *Fap* of only ～35%. The CE－3 probe landed on the rim of Zi Wei crater (～27－80 Myr old; Xiao et al., 2015). The diameter of this crater is ～450m, which would have excavated ～40－50m beneath the surface. The entire CE－3 landing site was blocky and covered by the ejecta and big boulders of Zi Wei crater (Ling et al., 2015). Although CE－3 landing area has the smallest *Fap*, the uneven areas in it do not occur randomly but tend to be clustered together. In Figure 3j, almost all the uneven areas were gathered near the left of CE - 3 landing site, and CE - 3 probe landed in the flat area on the right of the Zi Wei crater. Therefore, the distribution pattern of the uneven features in the landing area is also an important factor for landing safety assessment. In this work, we only considered the *Fap* factor but ignored the distribution pattern of the uneven features in the landing area. According to Table 2, the average *Fap* of all successful landing sites was 58.7%. If CE - 3 landing site was excluded, the average *Fap* increased to 60.6%. Therefore, we thought it was safe for CE - 4 landing in a grid if its *Fap* was bigger than 0.6.

3.3 Safety Assessment for CE - 4 Preselected Landing Area

According to the *Fap* of CE - 4 preselected landing area, the divided square grids can be classified into safe grid ($Fap > 0.6$) and unsafe grid ($Fap \leqslant 0.6$). In Figure 4, the grids with blue and light blue colors were safe for CE - 4 landing, while the grids with red and light red colors were unsafe for CE - 4 landing. And the adjacent safe grids were merged, so that the range of CE - 4 preselected landing area can be greatly reduced. As shown in Figure 4, five PLAs (yellow polygons marked with numbers 1 - 5) were labeled. PLAs 1 - 3 were located on the right side of the preselected landing area, while PLAs 4 and 5 were located on the left of the preselected landing area. The details of the five PLAs were shown in Table 3. The average Faps of the five PLAs were all larger than 0.7 except for PLA 2. PLA 1 and PLA 5 were the best choices for CE - 4 landing because of their bigger areas and higher average *Faps*. In addition, three least safe landing areas (LLAs, red polygons marked with letters A - C) were labeled in Figure 4. These three LLAs with dense PNTs are not suitable for CE - 4 lander to land because of the lowest average *Fap* values. The details of the three LLAs were shown in Table 4.

Table 3　The Details of Five PLAs in CE - 4 Preselected Landing Areas

PLA	Area (km^2)	Number of divided square grid	Average *Fap*
1	48	801	0.712
2	53	873	0.687
3	39.2	656	0.701
4	16.8	298	0.745
5	51.6	884	0.716

Note. PLA = potential landing area; CE - 4 = Chang'e - 4.

Table 4　The Details of Five LLAs in CE - 4 Preselected Landing Areas

LLA	Area (km^2)	Number of divided square grid	Average *Fap*
A	78.1	1317	0.558
B	61.9	1036	0.543
C	125.6	2110	0.522

Note. PLA=least safe landing area; CE - 4 = Chang'e - 4.

4 Route Planning for CE-4 Rover Moving From Current Grid to Target Grid

The CE-4 Lunar Probe is comprised of a lander, a rover, and a relay satellite. After CE-4 lands on the floor of Von Kármán crater, a rover will be released to travel around. The payload elements of CE-4 rover contain a visible/near-infrared imaging spectrometer, a ground penetrating radar and a panorama camera, etc. The rover should be able to carry out imaging, spectral and radar measurements. However, due to the limited field of view of the panorama camera, it is difficult for rover to select a long-distance (several or tens of kilometers) and safe patrol route in real time. In this section, we proposed a route planning method, which took both the distance and security into account, for CE-4 rover to dynamically generate a safe and short route from current grid A to target grid B. This method has two steps including route traversal and route planning.

4.1 Route Traversal

In this step, all the paths between grid A and grid B are traversed in a computer, and all passing grids in each path are saved in a file. First, a rectangle with a diagonal line AB is generated. The default is that grid A is in the upper left corner, otherwise the rectangle is turned over. We assume that the routes from grid A to grid B are in the rectangle AB and cannot be out of the range. And then taking grid A as the starting grid, a path set *P* containing all the paths is generated using a path traversal algorithm, which travels the whole rectangle AB in an order of the rightward, downward, and diagonal directions. The sketch map is shown in Figure 5 and pseudocode of path traversal algorithm is shown as below:

```
Pseudo code
    create a stack S
    create an array status [grid (m, n)]
dir [3] [2] = { {0, 1}, {1, 0}, {1, 1} }; //rightward, downward and diagonal
direction
    //m and n are the row and column number of the grid, status [grid (m, n) ]
==0means this grid is //not
    be visited, while status [grid (m, n)] ==1 indicates it has been traveled
    procedure PathTraverse (grid (i, j)):
      if (i == m-1 && j == n-1) // a path has been found
        pop all grids from S and record this path;
        return;
      if (i>= m || j >= n) //out of range of the rectangle
```

```
• return;
```

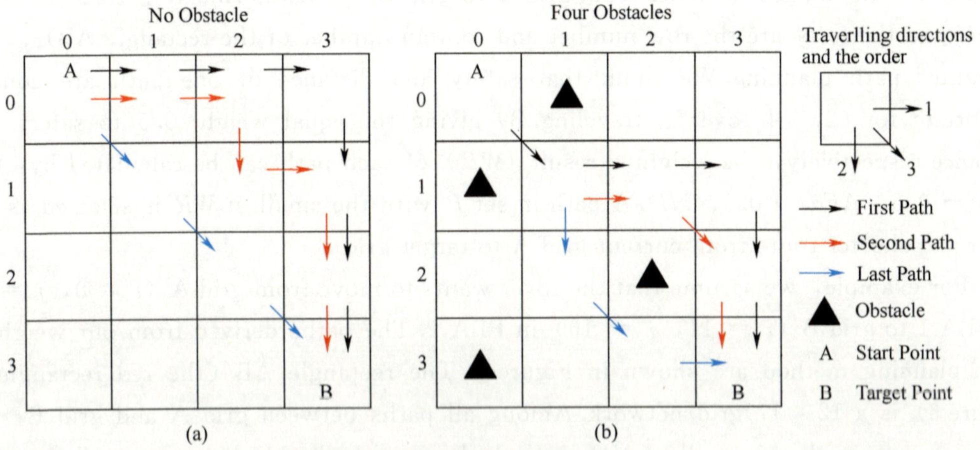

Figure 5 The sketch map of our path traversal algorithm. Panels (a) and (b) show the results of traveling in a 4×4 network with no obstacle and four obstacles. The black, red, and blue arrows show the first, second, and last path from grid A to grid B using our algorithm

```
for (int k=0; k < 3; k++) //travel in three directions
•     ii = i + dir [k] [0];
•     ij = j + dir [k] [1]; // the row and column number of the next
                                travelled grid
•       if (ii < m &&ij< n && status [grid (ii, ij) ] ==0)
•         status [grid (ii, ij) ] =1; //this grid has been visited
•         S. push (grid (ii, ij) ); •//push this grid into the stack
•         PathTraverse (grid (ii, ij) ); // recursion
•           status [grid (ii, ij)] = 0; // backtrack
•         S. pop (); // three directions have been travelled and this grid pop from
                                the stack
end procedure
```

4.2 Route Planning

In this step, we selected the safer and shorter route from the path set P. (1) Removing the inappropriate paths. The paths including the grid with smaller Fap should be removed from the path set P. (2) Calculating the average Fap ($AFap$) of each path. The $Faps$ of all grids on a path are summed up and then divided into the number of the passing grids in the path to obtain the average Fap. (3) Computing the normalized distance (ND) of each path. If traveling in the rightward and downward directions, the distance between current grid and next grid is the length of the grid. While if traveling in the diagonal direction, the distance between the two grids is $\sqrt{2}$ times of the square grid's length. The distance of each path is normalized to a range of 0 - 1.0 after divided into the longest distance from grid A to grid B, which is $rowN + colN - 2$ (the $rowN$ and the $colN$ are the row number and column number of the rectangle AB). (4) Weighted path planning. We think that safety and distance of one path are equally important for CE - 4 rover's traveling. By giving the equal weight 0.5 to safety and distance respectively, the weighted result (WR) of each path can be calculated by: $WR = 1 - 0.5 * AFap + 0.5 * ND$. A path in set P with the smallest WR is selected as the safer and shorter route from current grid A to target grid B.

For example, we assume that the rover wants to move from grid A ($i = 0$, $j = 0$) in PLA 1 to grid B ($i = 11$, $j = 16$) in PLA 2. The paths derived from our weighted path planning method are shown in Figure 6. The rectangle AB (the red rectangle in Figure 6) is a 12×17 grid network. Among all paths between grid A and grid B, the shortest route with the smallest ND is marked out with the black arrows in Figure 6b, the safest route with the biggest $AFap$ is marked out with the red arrows in Figure 6b, and the best weighted path with the smallest WR is shown with blue arrows in Figure 6b. In addition, if the rover wants to add an intermediate grid C ($i = 3$, $j = 10$, Figures 6a and 6b) in its route, that is to say, the rover travels from grid A to grid C and then moves to grid B. In such a situation, the all paths from grid A to grid B are found out through our route traversal method. The paths containing grid C are selected and their WRs are calculated. The path with the smallest WR is the best path from grid A to B going through grid C, which is shown with green circles in Figure 6b. The details of the four paths derived from our method are shown in Table 5.

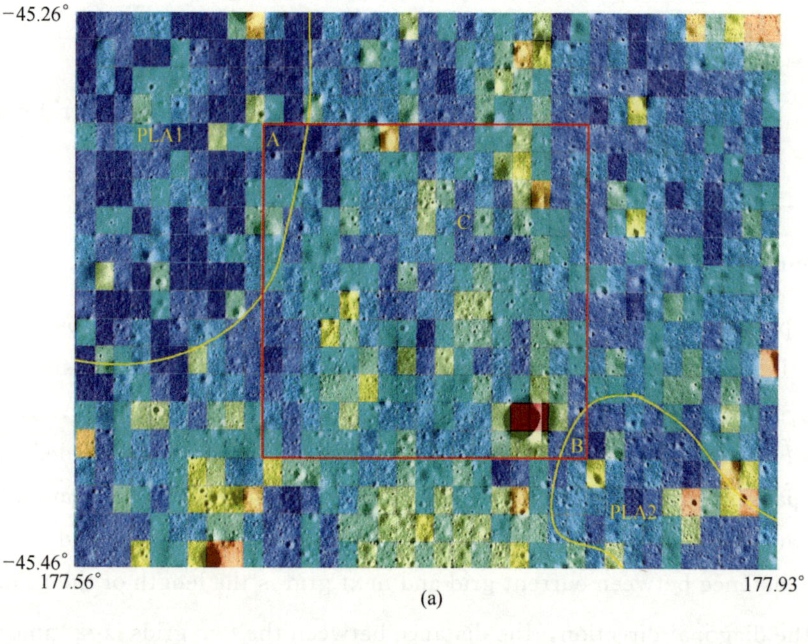

(a)

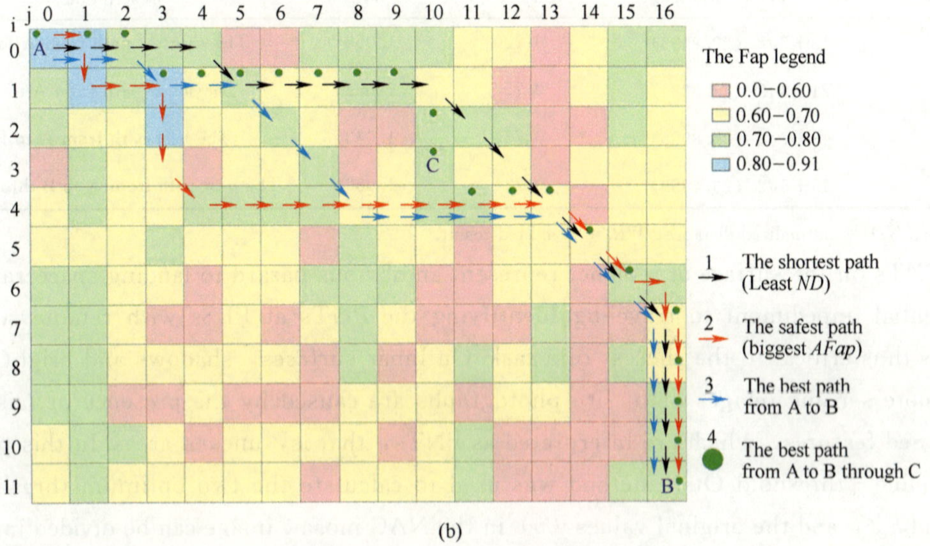

Figure 6　(a) The *Fap* map of the study area, marked in the white rectangle in Figure 4, superposing on the narrow - angle camera mosaic image. Grid A in PLA 1 and grid B in PLA 2 are current grid and target grid. Grid C is an intermediate grid. (b) Route planning results. The black arrows show the shortest path, while the red and blue arrows show the safest path and the best path from grid A to grid B. The green circles show the best path from grid A to grid B going through grid C. Number *i* and *j* are the row and column numbers of the grid in rectangle AB. PLA = potential landing area; ND = normalized distance

5　Summary and Future Work

China's CE - 4 probe will explore the South Pole - Aitken basin at the end of 2018 to achieve the first on - site exploration of the lunar farside. The south of the Von Kármán crater is chosen as its PLA. However, the successful experience of CE - 3 landing and patrolling in Mare Imbrium in 2013 could not be directly applied to the CE - 4 mission, because of the more rugged far side of Moon, which leads to the difficulties of safe landing and traversing. Moreover, due to scale dependence and smoothing effects, smaller topographic features in the CE - 4 preselected landing zone could not be displayed through the topographical studies previously derived from DEM data with low - resolution (~59 m per pixel). In this paper, we proposed a method to identify PNTs and choose PLAs for CE - 4 probe using LRO NAC images with a decimeter resolution. This method can also be applied to select the safe landing areas for China's CE - 5 and Mars exploration missions in the future.

Table 5　The Details of Four Paths in Figure 6

Path	*ND* (times of grid length)	*AFap* (%)	*WR*	Description
1	$13+7\sqrt{2}$ (22.8995)	0.730	1.058	The shortest path from A to B
2	$21+3\sqrt{2}$ (25.8284)	0.757	1.082	The safest path from A to B
3	$13+7\sqrt{2}$ (22.8995)	0.734	1.043	The best path from A to B
4	$15+6\sqrt{2}$ (23.4853)	0.734	1.060	The best path from A to B through C

Note. *ND* = normalized distance; *WR* = weighted result.

PNTs on the surface of a planet represent an obvious hazard to landing spacecraft and a potential impediment to traveling. Identifying the PNTs at PLSs with remote sensing data is thus critical to the success of a task. On lunar surfaces, shadows and bright areas in remote sensing images or in situ photographs are caused by the presence of raised or depressed features, which are interpreted as PNTs, that is, uneven areas. In this paper, the double - threshold Otsu method was used to calculate the two optimum thresholds, k_1^* and k_2^*, and the original values (v) in the NAC mosaic image can be divided into the dark parts ($v<k_1^*$), the bright parts ($k_1^*<v<k_2^*$), and the flat parts ($v>k_2^*$). The image segmentation results of CE - 4 preselected landing area were shown in Figure 2b. The black and white colors represented the dark and bright areas, which were uneven areas, while the light green color stood for the flat areas.

In order to obtain the *Fap* threshold for the safe landing and traveling, we calculated the *Faps* at the landing sites of past successful lunar missions, including CE - 3, Apollo, Surveyor, and Luna series. The average *Fap* of all successful landing sites was 58.7%. If CE - 3 landing site was excluded, the average *Fap* increased to 60.6%. Therefore, it is safe for CE - 4 landing in a grid if its *Fap* is bigger than 0.6. According to the *Fap* of CE - 4 preselected landing area, the divided square grids can be divided into safe grid ($Fap > 0.6$) and unsafe grid ($Fap \leqslant 0.6$). The adjacent safe grids were merged, and the range of CE - 4 preselected landing area can be greatly reduced to five PLAs. The PLA 1 and PLA 5 are the best choices for CE - 4 landing because of their bigger areas and higher average *Faps*.

After CE - 4 lands on the floor of Von Kármán crater, a rover will be released to travel around. Due to the limited field of view of the panorama cameras, it is difficult for rover to select the long - distance and safe route in real time. In this paper, we proposed a route planning method, which took both the distance and security into account, for CE - 4 rover to dynamically generate a safer and shorter route between current grid A and target grid B. First, all the paths between grid A and grid B are traversed in a computer, and all passing grids in each path are saved in a file. By giving the equal weight 0.5 to safety and distance respectively, the weighted result of each path can be calculated by $WR = 1 - 0.5 * AFap + 0.5 * ND$. A path with the smallest *WR* is selected as the safer and shorter route from current grid A and target grid B.

In the future, we plan to continue the following work: (1) In this paper, we only used the *Fap* factor to evaluate the grid safety without considering the concentration degree of the uneven areas. If the grid has a smaller *Fap* but clustered uneven objects, just like the CE - 3 landing site, this grid is still safe for lander landing and rover traveling. We will continue to improve the grid safety evaluation method by taking both the *Fap* and concentration degree into account. (2) Due to time and energy constraints, we performed the path traversal in only three directions; in fact, there are eight directions when the rover moves from one grid to another next grid. Moreover, because of the huge computation, the size of the divided square gird in CE - 4 preselected landing area was only 0.01°. In the future work, we will further divide the grid into 0.005° and 0.001° and obtain more accurate security evaluation results.

Acknowledgments

LRO NAC images were downloaded from the PDS (https://pds.nasa.gov/). This work is supported by The Program for JLU Science and Technology Innovative Research Team (JLUSTIRT, 2017TD - 26), the Focus on research and development plan in Shandong province (2018GGX101028), the Open Fund of State Key Laboratory of Remote Sensing Science (OFSLRSS201802), the Key Research Program of the Chinese Academy of Sciences (XDPB11), Chinese Academy of Sciences Frontier Science Research Project (QYZDY - SSW - DQC028), the National Science And Technology Infrastructure Work Projects (2015FY210500), and the National Natural Science Foundation of China (41772346 and 41490633). The efforts of the science and engineering teams behind all the data sets used in this study, particularly the LRO mission, NAC, and WAC instruments, are gratefully acknowledged. Thanks very much for the suggestions of the Editor Jonathan Jiang to improve our manuscript. We would like to thank the three anonymous reviewers who provided very helpful and useful suggestions of the manuscript.

References

[1] Abdrakhimov, A. M., Basilevsky, A. T., Ivanov, M. A., Kokhanov, A. A., Karachevtseva, I. P., & Head, J. W. (2015). Occurrence probability of slopes on the lunar surface: Estimate by the shaded area percentage in the LROC NAC images. Solar System Research, 49(5), 285 - 294. https://doi.org/10.1134/S0038094615050019

[2] Cao, W., Cai, Z., & Tang, Z. (2015). Fractal structure of lunar topography: An interpretation of topographic characteristics. Geomorphology, 238, 112 - 118. https://doi.org/10.1016/j.geomorph.2015.03.002

[3] Cintala, M. J., Mcbride, K. M., 1995. Block distribution on the lunar surface: A

comparison between measurements obtained from surface and orbital photography. Lunar and Planetary Science Conference (Vol. 25) . Lunar and Planetary Science Conference.

[4] Di, K., Xu, B., Peng, M., Yue, Z., Liu, Z., Wan, W., et al. (2016). Rock size - frequency distribution analysis at the chang'e - 3 landing site. Planetary and Space Science, 120, 103 - 112. https://doi.org/10.1016/j.pss.2015.11.012

[5] Fukunaga, K. (1990). Introduction to statistical pattern recognition (2nd ed.). New York: Academic Press.

[6] Golombek, M., Grant, J., Kipp, D., Vasavada, A., Kirk, R., Fergason, R., et al. (2012). Selection of the mars science laboratory landing site. Space Science Reviews, 170 (1 - 4), 641 - 737. https://doi.org/10.1007/s11214 - 012 - 9916 - y

[7] Golombek, M. P., Huertas, A., Marlow, J., Mcgrane, B., Klein, C., Martinez, M., et al. (2008). Size - frequency distributions of rocks on the northern plains of Mars with special reference to Phoenix landing surfaces. Journal of Geophysical Research, 113, E00A09. https://doi.org/10.1029/2007JE003065

[8] Gonzalez, R. C., & Woods, R. E. (2007). Digital image processing (3rd ed.). Upper Saddle River, NJ: Prentice - Hall, Inc.

[9] Haruyama, J., Ohtake, M., Matsunaga, T., Morota, T., Honda, C., Yokota, Y., et al. (2009). Long - lived volcanism on the lunar farside revealed by SELENE Terrain camera. Science, 323 (5916), 905 - 908. https://doi.org/10.1126/science.1163382.

[10] Huang, J., Xiao, Z. Y., Flahaut, J., Martinot, M., Head, J., Xiao, X., et al. (2018). Geological characteristics of Von Kármán crater, northwestern South Pole - Aitken Basin: Chang'E - 4 landing site region. Journal of Geophysical Research: Planets, 123, 1684 - 1700. https://doi.org/10.1029/2018JE005577.

[11] Ivanov, M. A., Abdrakhimov, A. M., Basilevsky, A. T., Demidov, N. E., Guseva, E. N., Head, J. W., et al. (2018). Geological characterization of the three high - priority landing sites for the Luna - Glob mission. Planetary and Space Science, 162, 190 - 206.

[12] Jia, Y. Z., Zou, Y. L., Xue, C. B., Ping, J. S., Yan, J., & Ning, Y. M. (2018). Scientific objectives and payloads of Chang'E - 4 Mission (in Chinese). Chinese Journal of Space Science, 38 (1), 118 - 130.

[13] Johnson, A. E., Klumpp, A. R., Collier, J. B., & Wolf, A. A. (2002). Lidar - based hazard avoidance for safe landing on Mars. Journal of Guidance, Control, and Dynamics, 25 (6), 1091 - 1099. https://doi.org/10.2514/2.4988

[14] Kokhanov, A. A., Karachevtseva, I. P., Zubarev, A. E., Patraty, V., Zh, F. R., & Oberst, J. (2017). Mapping of potential lunar landing areas using LRO and SELENE data. Planetary and Space Science, 162, 179 - 189.

[15] Kreslavsky, M. A., Head, J. W., Neumann, G. A., Rosenburg, M. A., Aharonson, O., Smith, D. E., & Zuber, M. T. (2013). Lunar topographic roughness maps from Lunar Orbiter Laser Altimeter (LOLA) data: Scale dependence and correlation with geologic features and units. Icarus, 226 (1), 52 - 66. https://doi.org/10.1016/j.icarus.2013.04.027.

[16] Li, B., Ling, Z., Zhang, J., & Chen, J. (2017). Rock size - frequency distributions analysis at lunar landing sites based on remote sensing and in - situ imagery. Planetary and Space Science, 146, 30 - 39. https://doi.org/10.1016/j.pss.2017.08.008.

[17] Li, B., Ling, Z., Zhang, J., Chen, J., Wu, Z., Ni, Y., & Zhao, H. (2015). Texture descriptions of lunar surface derived from LOLA data: Kilometer - scale roughness and entropy maps. Planetary and Space Science, 117, 303 - 311. https://doi.org/10.1016/j.pss.2015.07.004.

[18] Li, B., Ling, Z. C., Zhang, J., Chen, J., Sun, L. Z., & Zhao, H. W. (2016). Geochronology, petrogenesis and geological significance of the lunar basalts around CE - 3 landing site. Acta Petrologica Sinica, 32, 19 - 28.

[19] Li, B., Wang, X. Q., Zhang, J., Chen, J., & Ling, Z. (2016). Lunar textural analysis based on WAC - derived kilometer - scale roughness and entropy maps. Planetary and Space Science, 125, 62 - 71. https://doi.org/10.1016/j.pss.2016.03.004.

[20] Liasis, G., & Stavrou, S. (2016). Satellite images analysis for shadow detection and building height estimation. Isprs Journal of Photogrammetry & Remote Sensing, 119, 437 - 450. https://doi.org/10.1016/j.isprsjprs.2016.07.006.

[21] Ling, Z., Jolliff, B. L., Liu, C., Bi, X., Qiao, L., Li, B., Zhang, J., et al. (2018). Composition, mineralogy and chronology of mare basalts in Von Kármán crater: A candidate landing site of Chang'e - 4. 49th Lunar and Planetary Science Conference.

[22] Ling, Z., Jolliff, B. L., Wang, A., Li, C., Liu, J., Zhang, J., et al. (2015). Correlated compositional and mineralogical investigations at the Chang'e - 3 landing site. Nature Communications, 6 (1), 8880. https://doi.org/10.1038/ncomms9880

[23] Liu, J. J., Yan, W., Li, C. L., Tan, X., Ren, X., & Mu, L. L. (2014). Reconstructing the landing trajectory of the CE - 3 lunar probe by using images from the landing camera. Research in Astronomy and Astrophysics, 14 (12), 1530 - 1542. https://doi.org/10.1088/1674 - 4527/14/12/003

[24] Mitrofanov, I., Djachkova, M., Litvak, M., & Sanin, A. (2016). The method of landing sites selection for Russian lunar lander missions. Geophysical Research Abstracts, 18, EGU2016 - EGU10018.

[25] Otsu, N. N. (1979). A threshold selection method from gray - level. IEEE

Transactions on Systems, Man, and Cybernetics, 9 (1), 62 - 66. https://doi. org/10. 1109/TSMC. 1979. 4310076.

[26] Robbins, S. J. (2018) . The fractal nature of planetary landforms and implications to geologic mapping. Earth & Space Science, 5 (5), 211 - 220. https://doi. org/10. 1002/2018EA000372.

[27] Robinson, M. S. , Brylow, S. M. , Tschimmel, M. , Humm, D. , Lawrence, S. J. , Thomas, P. C. , et al. (2010) . Lunar Reconnaissance Orbiter camera (LROC) instrument overview. Space Science Reviews, 150 (1 - 4), 81 - 124. https://doi. org/10. 1007/s11214 - 010 - 9634 - 2.

[28] Sears, N. E. (1964) . Apollo guidance and navigation. Volume 1 Primary G&N System Lunar Orbit Operations, NASA - CR - 117537.

[29] Wang, Y. , Di, K. , Xin, X. , &Wan, W. (2017) . Automatic detection of Martian dark slope streaks by machine learning using HiRISE images. Isprs Journal of Photogrammetry & Remote Sensing, 129, 12 - 20. https://doi. org/10. 1016/j. isprsjprs. 2017. 04. 014.

[30] Wong, E. C. , & Singh, G. (2002) . Autonomous guidance and control design for hazard avoidance and safe landing on Mars. In: Proceedings of AIAA/AAS Astrodynamics Specialist Conference and Exhibit. Monterey: AIAA, 2002. 2002 - 4619.

[31] Wu, W. R. , Wang, D. Y. , Huang, X. Y. , Ji, L. , & Zhe, Z. (2015). Autonomous hazard detection and avoidance guidance method for soft lunar landing (in Chinese). SCIENTIA SINICA Informationis, 45, 1046 - 1059.

[32] Wu, W. R. , Wang, Q. , Tang, Y. H. , Yu, G. B. , Liu, J. Z. , & Zhang, W. (2017) . Design of Chang'E - 4 lunar farside soft - landing mission (in Chinese). Journal of Deep Space Exploration, 4 (2), 111 - 117.

[33] Xiao, L. , Zhu, P. , Fang, G. , Xiao, Z. , Zou, Y. , Zhao, J. , Zhao, N. , et al. (2015) . A young multilayered terrane of the northern Mare Imbrium revealed by Chang'e - 3 mission. Science, 347 (6227), 1226 - 1229. https://doi. org/10. 1126/science. 1259866.

[34] Ye, P. , Sun, Z. Z. , Zhang, H. , & Li, F. (2017) . An overview of the mission and technical characteristics of Change' 4 lunar probe. Science china Technological Sciences, 60 (5), 658 - 667. https://doi. org/10. 1007/s11431 - 016 - 9034 - 6.

[35] Zhao, J. , Xiao, L. , Qiao, L. , Glotch, T. D. , & Huang, Q. (2017) . The Mons Rümker volcanic complex of the Moon: A candidate landing site for the Chang'e - 5 mission. Journal of Geophysical Research: Planets, 122, 1419 - 1442. https://doi. org/10. 1002/2016JE005247.

The Moon's farside shallow subsurface structure unveiled by Chang'E - 4 Lunar Penetrating Radar

Chunlai Li[1,2], Yan Su[1,2], Elena Pettinelli[3], Shuguo Xing[1], Chunyu Ding[4], Jianjun Liu[1,2], Xin Ren[1], Sebastian E. Lauro[3], Francesco Soldovieri[5], Xingguo Zeng[1], Xingye Gao[1], Wangli Chen[1], Shun Dai[1], Dawei Liu[1], Guangliang Zhang[1], Wei Zuo[1,2], Weibin Wen[1], Zhoubin Zhang[1], Xiaoxia Zhang[1], Hongbo Zhang[1]

1. Key Laboratory of Lunar and Deep Space Exploration, National Astronomical Observatories, Chinese Academy of Sciences, Beijing 100101, China

2. University of Chinese Academy of Sciences, Beijing 100049, China

3. Mathematics and Physics Department of Roma Tre University, Roma 00146, Italy

4. School of Atmospheric Sciences, Sun Yat—sen University, Zhuhai 519000, China

5. Istituto per il Rilevamento Elettromagnetico dell'Ambiente IREA - CNR, I - 80124 Naples, Italy

On 3 January 2019, China's Chang'E - 4 (CE - 4) successfully landed on the eastern floor of Von Kármán crater within the South Pole - Aitken Basin, becoming the first spacecraft in history to land on the Moon's farside. Here, we report the observations made by the Lunar Penetrating Radar (LPR) onboard the Yutu - 2 rover during the first two lunar days. We found a signal penetration at the CE - 4 landing site that is much greater than that at the CE - 3 site. The CE - 4 LPR images provide clear information about the structure of the subsurface, which is primarily made of lowloss, highly porous, granular materials with embedded boulders of different sizes; the images also indicate that the top of the mare basal layer should be deeper than 40m. These results represent the first high - resolution image of a lunar ejecta sequence ever produced and the first direct measurement of its thickness and internal architecture.

Introduction

The stratigraphic architecture of the lunar subsurface is the result of a complex and long process of emplacement and modification of the deposited materials (1). Such architecture can be inferred from remote sensing observations (2) or can be directly imaged using geophysical techniques like seismic or electromagnetic wave prospections (3). Both methods were successfully tested during the Apollo missions, providing ground - breaking information about the Lunar interior. In the past 20 years, spaceborne/rover - deployed Ground Penetrating Radar (GPR) has progressively become the most suitable geophysical technique to investigate planetary subsurface stratigraphy. Two orbiting radars were hosted on NASA's Apollo 17 and Japan Aerospace Exploration Agency's Kaguya missions, the Apollo Lunar Sounder Experiment and Kaguya Lunar Radar Sounder (4, 5), which were able to survey the subsurface structure down to a depth of 1 to 2km. The first attempt to survey the Moon's subsurface using a GPR onboard a rover was made during the Chang'E - 3 (CE - 3) mission, which was equipped with low - and high - frequency antennas (6). The high - frequency GPR observations revealed a detailed subsurface structure up to a depth of about 10m (7). Conversely, interpretation of the low - frequency data remains quite controversial due to antenna coupling with the rover and the lunar surface, combined with the limited length of the radar profile (8 - 10).

On 3 January 2019, the Chang'E - 4 (CE - 4) lander touched down on the eastern floor of Von Kármán crater at 45.4446°S, 177.5991°E (Fig. 1A) (11, 12), becoming the first spacecraft to land safely on the Moon's farside. The Yutu - 2 rover, equipped with the same dual - frequency Lunar Penetrating Radar (LPR) as CE - 3, was successfully deployed to the surface about 12 hours later. The pre - Nectarian Von Kármán crater (44. 45°S, 176. 3°E; diameter equal to about 186. 3km) is located in the northwestern sector of the South Pole - Aitken basin, the oldest and largest impact structure on the Moon (13). The overall terrain of Von Kármán crater is relatively flat, and its floor was flooded by Imbrian - aged mare basalts that appear dark in optical images (14). The absolute model age of the mare basalts was dated to be ~3. 6 billion years (Ga) based on the size - frequency distribution of superposed impact craters (2, 14). Judging from the remote optical and multispectral images, a large portion of mare basalt was later overprinted by distal ejecta that appear bright, which were most probably delivered from surrounding large craters such as Finsen and Von Kármán L (2). As a result, secondary craters are widespread over the floor of Von Kármán, as indicated by their spatial patterns (e. g., chains or clusters with herringbone - shaped morphology) (2, 15). The CE - 4 spacecraft landed on the bright distal ejecta from surrounding large craters (Fig. 1A and fig. S8), in an area with a low abundance of rocks and boulders on the surface (fig. S1).

The CE - 4 LPR is a dual - frequency GPR system, operating at 60MHz (low frequency) and 500MHz (high frequency), with a frequency band of 40 to 80MHz and 250 to 750MHz, respectively (16, 17). The radar data were collected during the first two lunar days along the Yutu - 2 route (Fig. 1B). In this work, we processed and interpreted only the high - frequency LPR data. As in the CE - 3 mission, the low -

frequency antennas are installed on the back of the CE－4 rover; unfortunately, this leads to electromagnetic coupling with the rover's metallic body (18), resulting in strong disturbances that largely overlap the signals coming from the subsurface. These disturbances could easily be misinterpreted as subsurface reflectors (10), and thus, more data should be acquired to effectively mitigate/remove these artifacts from the radargram.

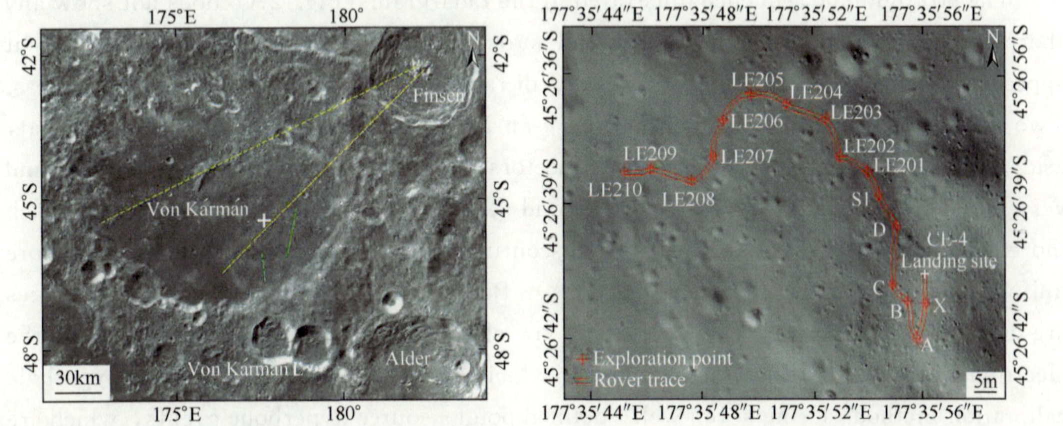

(a)　　(b)

Fig. 1　The CE－4 landing region and the Yutu－2 rover route. (a) CE－4 landed in the eastern floor of Von Kármán crater (44.45° S, 176.3° E; diameter ～186.3km), as indicated by the white cross (177.5991°E, 45.4446°S) on a bright ejecta blanket. The yellow and green lines show the ejecta direction from Finsen (12) and Von Kármán L, respectively. The image is CE－27－m resolution Digital Orthophoto Map (DOM). (b) Yutu－2 rover route during the first two lunar days. Two red lines show the tracks of the left and right wheels on the Yutu－2 rover. The LPR performed observation along the ～106－m route from exploration points A to LE210. The background image is mosaicked from images obtained during the landing process, where the spatial resolution is 5cm

Results

LPR image and physical parameter estimation

The collected high－frequency radar data from point A to point LE210 (Fig. 1B) were processed using standard GPR signal processing methods (fig. S2) [e. g., (19)] and are represented as a seismic color image (radargram) (Fig. 2A). The radar section indicates a very good signal penetration, and radar features are well detectable down to a two－way travel time of about 500 ns. This penetration is much larger (over three times in terms of two－way travel time) than that reached by the CE－3 LPR at the Zi Wei crater landing site (for comparison, see fig. S3). As a result, the subsurface at the CE－4 landing site is much more transparent to radio waves, and this qualitative observation suggests a totally different geological context for the two landing sites. A quantitative estimation of the signal attenuation in the subsurface material is given by the loss tangent value (20), which, for the CE-4 landing site, was estimated to be $\tan\delta = (5 \pm 2) \times 10^{-3}$; this value appears to preclude an intact, dense lava flow layer in this sequence, because such a material would increase the loss tangent ($\tan\delta > 10^{-2}$) and prevent deep penetration (21).

The stratigraphic sequence illustrated in the radargram (Fig. 2A) does not show any sharp and continuous reflectors that extend across the entire radar profile, but rather, it appears to be made of a complex distribution of radar features that cluster in specific areas. Two main depth sections can be identified: an initial section (about 0 to 150ns) that, aside from a few scattered hyperbolic reflectors (fig. S4), appears fairly uniform and very transparent to radio waves, and a second section (about 150 to 500ns) that is much more irregular, with areas of strong discontinuous reflectors alternating with more uniform and transparent intervals (Fig. 2A). Below ~500ns, no evident radar features are detectable. To convert the two-way travel time into depth, an estimation of the electromagnetic wave velocity is necessary, which we calculated by applying the hyperbola calibration method. Using seven well-defined point-source hyperbolic events, which are only present within the first section (between 30 and 90ns), we computed the average wave velocity at different time delays (fig. S4). Such velocity progressively decreases and reaches a constant value, $v = 0.16\text{m/ns}$, at about 70ns. Given the lack of any other information, we assumed that this velocity does not appreciably change with depth and that it does not vary across the entire radar section. On the basis of this value, the thickness of the homogeneous material at the top of the radar section (0 to 150ns) is about 12m, whereas the underlying material (150 to 500ns) is about 28m thick. We used the set of retrieved velocity values to estimate the material bulk density in the first 7 m (fig. S5) of the radar section. Applying two different models (22, 23), we obtained two slightly different trends reaching the asymptotic values $1.90 \pm 0.08\text{g/cm}^3$ (22) and $1.67 \pm 0.07\text{g/cm}^3$ (23) at about 5.8m depth. Last, applying the relationship between loss tangent, density, and oxide content defined for the lunar nearside (21), we compute the ($FeO + TiO_2$) abundance at the CE-4 landing site (fig. S6). We found a ($FeO + TiO_2$) content of $9 \pm 4\%$ (22) and $11 \pm 4\%$ (23), in good agreement with the range, 11.4 to 15.9%, measured by Kaguya Multiband Imager (MI) (24) at this site. This compatibility is quite notable, as the two techniques probe different portions of the lunar regolith: Kaguya data refer to the surface material, whereas LPR data refer to the large volume of material located to a depth of 6m.

Tomographic reconstruction of the stratigraphic sequence

Despite the good quality of the radar image, the complexity of the spatial distribution and shape of the radar features makes identification of the geological structures/events that generated such features quite difficult (25). To overcome this problem, we applied a tomographic inversion algorithm (26) that is capable of gaining information about the size

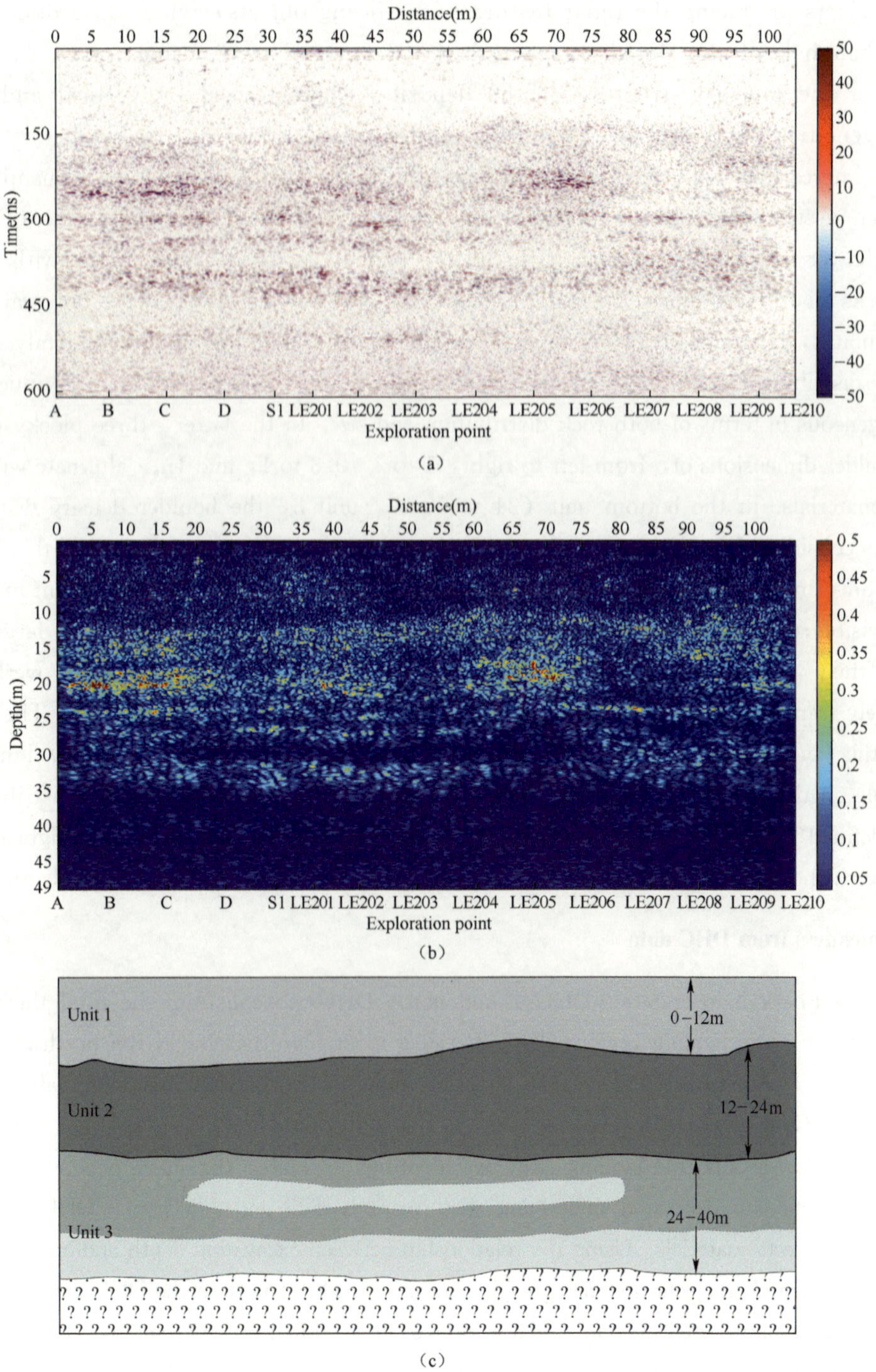

Fig. 2　LPR data at 500MHz.　(a) LPR 500 - MHz radargram represented in standard seismic colors after applying Dewow, background subtraction, and spherical and exponential compensation (SEC) gain and migration. The x axis is the rover distance (top, starting point on the left) and the exploration points (bottom), and the y axis indicates the two - way travel time and depth; the depth is calculated on the basis of the average electromagnetic wave velocity of 0.16m/ns.　(b) Tomographic reconstruction of the radar data, where red represents high reflectivity (large electromagnetic contrast) and blue is low reflectivity (small electromagnetic contrast).　(c) Schematic of the stratigraphic sequence highlighting the contacts between units and the relevant thicknesses based on the radargram (a) and the tomographic reconstruction (b). Gray tone indicates finer (light gray) or coarser (dark gray) materials

of the objects producing the radar features (scattering objects) when their dimension is larger than half of the signal wavelength (~0.2m). The resulting image (Fig. 2B) illustrates the internal structure of the deposits, characterized by vertical and lateral inhomogeneities, with groups of boulders and rocks (of various sizes and located at different depths) intercalated with finer materials. On the basis of the distribution, quantity, and dimension of such inhomogeneities, the image can be separated in three main units (Fig. 2c). The shallower unit (depth interval, 0 to 12m), unit 1, looks rather uniform, with sporadic large rocks distributed along the radar section. The underlying unit (depth interval, 12 to 24m), unit 2, is composed of an upper part containing a large amount of randomly (almost evenly) distributed 0.2 - to 1 - m - wide rocks and a deeper part that is much more inhomogeneous in terms of both rock distribution and size. In the latter, three blocky regions, with boulder dimensions of, from left to right, 1 to 3, 0.3 to 1, and 1m, alternate with zones of fine materials. In the bottom unit (24 to 40m), unit 3, the boulder density distribution decreases considerably, and the rocky material is essentially localized on the top of this interval. Below 28m, in the central part of the image, a lens of rather uniform and transparent material is clearly visible; this lens is probably made of very fine grained material with no large rocks. Beneath this, there is a zone with relative uniform distribution of small rocks overlaying a completely transparent layer and thus again probably made of very fine materials. The bottom part of this layer (40m) is not clearly visible in the tomographic image (Fig. 2b) but is well detectable on the radargram (Fig. 2a), and it is delimited by a discontinuous distribution of poorly defined scatterers. The sequence of interposed coarser and finer materials might continue below 40m; however, it is impossible to discriminate signal from noise at such a depth.

Ejecta thickness from DHC data

We used dark halo craters (DHCs) and non - DHCs to constrain the total thickness of distal ejecta over the landing region (27). Because most mare basalts in the northwest of the landing area are undisturbed and no DHCs can be found in that region, we only analyzed DHCs and non - DHCs in the area covered by ejecta in the southeast (fig. S7, A and B). Considering a radius of 20km around the landing site, we identified 25 DHCs that have penetrated through the bright ejecta and excavated underlying dark mare basalts and 163 non - DHCs that only excavated the ejecta materials. Using the relationship between excavation depth and crater diameter (28), we estimated an ejecta thickness ranging from 34 to 78m based on the DHC data (fig. S7C) and greater than 39m based on non - DHCs located about 2km from the landing site (fig. S7C). The observed deepest reflectors at 40 m are consistent with the lower end of this range, but the ejecta thickness may extend below this detectability limit because the ejecta thickness is estimated to be 30 to 90m based on the excavation depth of the iron - rich craters (29).

Discussion

Combining the information provided by the radargram, the tomographic image, and the quantitative analysis (estimated wave velocity and loss tangent), we can conclude that the subsurface internal structure at the landing site is essentially made by low-loss, highly porous granular materials embedding boulders of different sizes. Given such a strong geological constraint, the most plausible interpretation is that the sequence is made of a layer of regolith overlaying a sequence of ejecta deposits from various craters (Fig. 3), which progressively accumulated after the emplacement of the mare basalts on the floor of Von Kármán crater. The layer of regolith (unit 1) is quite thick (up to 12m), is rather homogenous both laterally and vertically, and is mostly composed of fine materials. It developed from the uppermost portion of the ejecta deposits, which were thicker than 12 m and were delivered to this area by multiple impact craters, mostly Finsen, Von Kármán L, and Von Kármán L' craters (table S1) (2). Unit 2 (depth, 12 to 24m) is characterized by large rocks and boulders that are interbedded with thin layers of fine materials. Ejecta deposition never occurs via simple carpeting, but it is accompanied by horizontal shearing and mixing, excavation, and subsurface structural disturbances (28), especially in areas beyond the continuous ejecta deposits of the source crater. Fine materials can be produced by the internal shearing of ejecta deposits, which reduces particle sizes, and the fine layers can also be formed due to regolith development during the intervals between different impact events. On the other hand, the CE-4 landing site is located within a secondary crater chain linked to the Finsen crater (fig. S8C), and the depth of the secondary crater is about 16m (fig. S8D); this is the minimum depth of the structural disturbances formed during the landing of the ejecta, as the secondary crater is highly degraded. Therefore, unit 2 is likely formed due to a combination of (i) coarse ejecta that were not mobilized during the landing of impact ejecta (28), (ii) structural disturbances in local materials caused by the landing ejecta, and (iii) fine materials generated during or after the ejecta deposition. Unit 3 (24 to 40m) contains alternating layers of coarse and fine materials. This unit can be interpreted as a combination of ejecta deposits, which were delivered by various craters older than Finsen and regolith developed during the impact intervals. Finsen is an Eratothenian-aged crater (<3.2Ga), and many older craters have contributed to deliver ejecta deposits on the crater floor of Von Kármán before that age (30). However, tracing the source crater for each individual layer of ejecta deposits is not possible. Last, the lack of a detectable radar signal below 40m does not allow us to make any definite conclusion about the properties of the materials at the base of the ejecta, although we can speculate that the granular materials extend beneath such a depth; accordingly, the upper contact of the mare basal layer must be deeper than 40m.

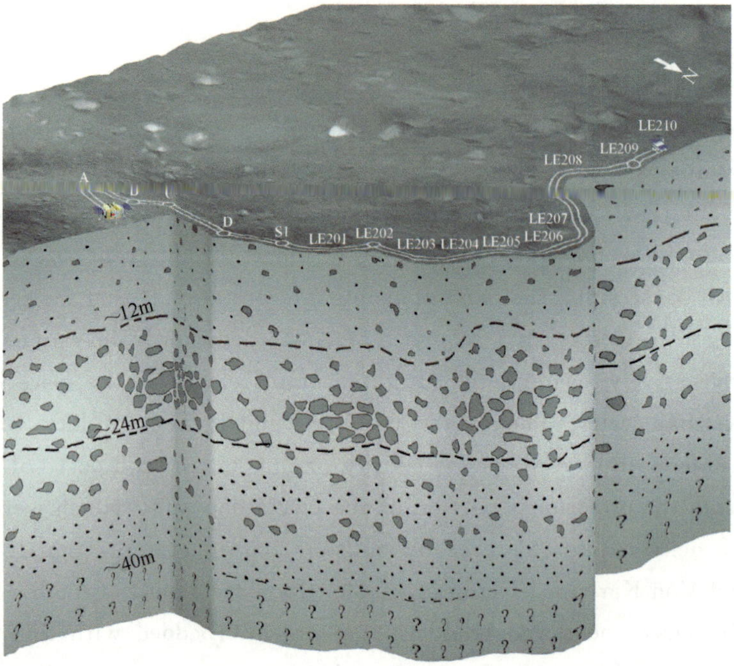

Fig. 3　Schematic representation of the subsurface geological structure at the CE－4 landing site inferred from LPR observations. The subsurface can be divided into three units: Unit 1 (up to 12m) consists of lunar regolith, unit 2 (depth range, 12 to 24 m) consists of coarser materials with embedded rocks, and unit 3 (depth range, 24 to 40 m) contains alternating layers of coarse and fine materials

The results of the radar data collected by the LPR during the first 2 days of lunar operation provide the first electromagnetic image of the farside subsurface structure and the first "ground truth" of the stratigraphic architecture of an ejecta deposit. The results illustrate, in an unprecedented way, the spatial distribution of the different products that contribute to form the ejecta sequence and their geometrical characteristics. The data also permitted the first unambiguous measurement of the lunar regolith layer thickness made with a radar, as previous knowledge regarding lunar soil thickness and ejecta deposition was based on impact models and indirect measurements. This work shows that the extensive use of the LPR could greatly improve our understanding of the history of lunar impact and volcanism and could shed new light on the comprehension of the geological evolution of the Moon's farside.

Materials and Methods

LPR operation principle, data acquisition, and processing

CE－4 LPR is equipped with two distinct GPR systems. The low－frequency system is equipped with two monopole antennas having a nominal central frequency of 60MHz and 40 - to 80 - MHz bandwidth. The monopoles have a length of 1. 15m and are spaced 0. 8 m from each other. These antennas are installed in the back of the rover and suspended 0. 6m above the ground. The high - frequency system is equipped with one transmitting and two receiving bowtie antennas (CH2A and CH2B), with a nominal central frequency of 500MHz and 250 - to 750 - MHz bandwidth. The antennas are located at the bottom of the Yutu - 2 rover, ~0. 3m above the ground (contactless), and are separated 0. 16m from each other (16). LPR data were collected along the radar route, while the rover was traveling. The time intervals between two adjacent traces are 1. 536 s at 60MHz and 0. 6636s at 500MHz. The velocity of the rover is ~5. 5cm/s, which gives an approximate step size of ~3. 6cm for the 500MHz and ~8. 5cm for the 60MHz. During the first two lunar days from point X to LE210, the LPR worked 6 hours 28min 53s and collected in total 13, 400 and 31, 749 traces at low and high frequency, respectively. In the LPR commissioning phase from point X to A, several preliminary tests were performed, and therefore, we only used LPR data collected along the route from point A to LE210 whose length is ~106m.

Because of the strong coupling of the antennas with both the metallic rover and the lunar surface, which generates severe disturbanceson the transmitted and received signals, low - frequency data have not been analyzed at this stage, as they need to be properly calibrated and accurately checked before confirming their reliability (10). Furthermore, because the first receiving antenna (CH2A) in the high - frequency system is affected by a strong cross—talk, we have analyzed only the data collected by the second antenna (CH2B).

The high - frequency raw data are of good quality and were processed according to the standard GPR procedure. We used a commercial software (EkkoProject V. 5, Sensors and Software Inc., Canada), and we applied the following processing (19): (i) high - loss temporal filter, known as Dewow, to reduce the low - frequency signal due to the coupling between the receiver and the transmitter; (ii) background subtraction to suppress ring - down response; and (iii) Stolt - FK migration (31) using a wave velocity v=0. 16m/ns. Moreover, we plotted the radargram using a SEC (spherical and exponential compensation) (19) gain in seismic standard color mode. Raw and processed data for the central part of the radar section (distance between 40 and 80m along the profile) are shown in fig. S2. Note that the migration improves the signal to noise ratio and allows us to better detect the deeper reflectors.

Comparison between CE - 3 and CE - 4 data

We compared 500MHz data collected by the LPR on board CE - 3 and CE - 4 (fig. S3) in terms of the maximum two - way travel time (penetration depth) at which the radar signal is still detectable. Such a comparison has been made using the same attenuation and maximum gain values applied for the radargrams shown in fig. S2C but using a different starting point for CE－3 (0.7) to better show the shallower reflectors. At the CE－4 landing site, the radar signal is detected up to a maximum two－way travel time of ~500ns, whereas that at the CE－3 landing site is only ~150ns. The difference in signal penetration is due to the difference in electromagnetic properties of the subsurface at the two sites, as highlighted by the loss tangent values (dictating the attenuation of the electromagnetic wave), which is $\tan\delta = (5 \pm 2) \times 10^{-3}$ for the CE－4 landing site and $\tan \geqslant 10^{-2}$ for the CE－3 landing site (32) (see below).

Physical parameter estimation

Wave velocity and density

Individual targets buried in the subsurface produce different radar reflection features in the radar cross section. If the target has a size < 0.1 (where λ denotes the probing wavelength), the reflection pattern is well approximated by a hyperbola; in contrast, for objects larger than 1λ to 3λ, the reflection pattern is more complicated, while targets having a size of tens of λ generate semicontinuous, flat reflections (33). To convert the two－way travel time into depth, we identified seven well－defined point targets in the first interval of the section (0 to 150ns) to estimate the radar wave velocity using the common well－established hyperbolic fitting method (fig. S4) (19). For each hyperbolic event, the computed value represents the average velocity in the volume of material between the surface and the point target; the estimated value decreases with depth, reaching an asymptotic limit of 0.16m/ns at about 70ns (5.8m). Laboratory measurements have shown that the wave velocity (i. e., the real part of the dielectric permittivity) depends on material bulk density. Different equations relating such quantities have been proposed; here, we exploited the Olhoeft and Strangway formula (22)

$$\rho = 2\frac{\log\left(\frac{c}{v}\right)}{\log(1.92)} \tag{1}$$

and the equation proposed by Hickson et al. (23)

$$\rho = \frac{\left(\frac{c}{v}\right)^{\frac{2}{3}} - 1}{0.307} \tag{2}$$

to compute the average density at different depths (fig. S5). In general, the density increases with depth, although the values computed using Eq. 2 (red dots in fig. S5) are systematically lower than those estimated through Eq. 1 (black dots in fig. S5), which are in good agreement with the trend (blue dashed line in fig. S5) estimated for the lunar nearside (21). Note that the first hyperbolic event is detectable at about 30ns (~3m depth), therefore far below the maximum excavation depth reached during the Apollo missions (21). Below 5.8m, the density approaches a constant value, $1.90 \pm 0.08\text{g/cm}^3$ according to Eq. 1 and $1.67 \pm 0.07\text{g/cm}^3$ according to Eq. 2, where the uncertainties have been computed using the standard propagation formula (34).

Loss tangent and $FeO + TiO_2$ abundances

The material loss factor can be computed using a frequency shift method; in the literature, there are several techniques proposed to extract such a parameter from radar data (35, 36). In this work, we applied a similar approach presented in (20), which estimates the decrease of the centroid frequency as a function of depth in the spectrum of the LPR signal. The frequency decay has been computedfor the two-way travel time from 0 to about 450ns; thus, the average loss tangent, $\tan\delta = (5 \pm 2) \times 10^{-3}$, represents an average value for the entire radar section over that interval.

Laboratory measurements show that the loss tangent depends on bulk density and ($FeO + TiO_2$) abundance as follows (21)

$$\tan\delta = 10^{0.038(\%FeO + \%TiO_2) + 0.312\rho - 3.26} \quad (3)$$

Assuming a specific value for the loss tangent, we can use Eq. 3 to plot the joint probability density (37) as a function of bulk density and oxide content. Such distribution (fig. S6) allows us to estimate the values of the bulk density if the ($FeO + TiO_2$) content is known and vice versa. Assuming for the CE-4 landing site a loss tangent value $\tan\delta = (5 \pm 2) \times 10^{-3}$, we can use Kaguya MI (23) measurement (range, 11.4 to 15.9%) to estimate the bulk density (i. e., $1.4 \pm 0.3 g/cm^3$). On the other hand, using the bulk density values computed applying Eqs. 1 and 2, it is possible to determine the oxide content, that is, $9 \pm 4\%$ and $11 \pm 4\%$, respectively.

Inversion tomography

The method, used to obtain the tomographic image shown in Fig. 3B, belongs to the set of migration approaches (26, 31, 38, 39, 40), which are based on a simplified model of electromagnetic scattering. In particular, the model underlying the migration approaches neglects the mutual interactions between the targets and allows us to estimate the properties of the buried scatterers, in terms of location and geometrical features (26, 39); conversely, the quantitative reconstruction of the electromagnetic properties is possible only in the case of weak scatterers (41).

If we consider a point-like scatterer located in the object space at $\bar{r} = (x, y, z)$, and denote as $\bar{r}_m$, the measurement point belonging to the measurement domain Σ, and let $i(t)$ be the transmitted signal, the corresponding backscattered field is given by

$$e_s(\bar{r}_m, t) = i\left(t - 2\frac{|\bar{r}_m - \bar{r}|}{v}\right) \quad (4)$$

where the amplitude factor due to the propagation spreading has been neglected, $i(t)$ is a short-duration pulse or the pulse after range compression has occurred, and v is the velocity of the electromagnetic wave in the investigated medium. Actually, due to the finite directivity of the antennas, a point scatterer appears as a diffraction hyperboloid whose apex is located at (x, y, $\frac{2z}{v}$), if the measurement domain Σ is located at $z=0$. The aim of migration is to compensate such a spreading by refocusing each segment of the hyperbola to its apex. A common way to implement the migration is through the so-called diffraction summation (38). In this method, the object space is divided in pixels, and for each of them, a diffraction hyperboloid is built in the data space. After, the reconstruction at each pixel is achieved by summing up contributions, where the synthetic hyperboloid intersects data. From a mathematical point of view, the reconstruction is provided by $R(\bar{r})$

$$R(\bar{r})=\int_{\Sigma}\int_{T} e_s(\bar{r}_m, t)\delta\left(t-2\frac{|\bar{r}_m-\bar{r}|}{v}\right)dtd\bar{r}_m \tag{5}$$

where T is the time interval during which data are collected and Σ is the measurement domain over which the measurements are collected. After Fourier transforming with respect to time, Eq. 5 can be recast (apart an unessential factor) as

$$R(\bar{r})=\int_{\Sigma}\int_{B_f} E_s(\bar{r}_m, f)e^{j2k|\bar{r}_m-\bar{r}|}dfd\bar{r}_m \tag{6}$$

where $E_s(\bar{r}_m, f)$ is the backscattered field collected at the generic measurement point $\bar{r}_m$ and frequency f, k is the wave number in the background medium, and B_f is the working frequency band. E_s ($\bar{r}_m$, f) is achieved by means of the Fourier transform of the GPR trace collected at the generic measurement point $\bar{r}_m$.

By sampling both the measurement domain and the working frequency band, the integral is replaced by the summation

$$R(\bar{r})=\sum_m\sum_i E_s(\bar{r}_m, f_i)e^{j2k_i|\bar{r}_m-\bar{r}|}\Delta f\Delta\bar{r}_m \tag{7}$$

In particular, we have considered a working frequency band $B_f=[250, 750]$ MHz sampled with a frequency step equal to $\Delta f=1.56$MHz (320 working frequencies). The measurement domain has an extent equal to 10m, centered with respect to the point to be imaged and sampled with a step equal to 0.0355m. The wave number is evaluated by assuming the relative dielectric permittivity of the subsurface equal to 3.52 (corresponding to a velocity of the electromagnetic wave equal to 0.16m/ns).

We present the tomographic image as the modulus of the radar reflectivity map $|R(\bar{r})|$ that accounts for the location and geometry of the scattering objects inside the investigation domain. In particular, the investigation domain has a horizontal extent of 106m anda depth ranging from 0 to 48.9m. The dimensions of the objects were computed by evaluating the transverse extent (along x axis) of the focused spots on the tomographic image.

Ejecta thickness estimation using impact craters

We estimated the thickness of ejecta using the excavation depth of DHCs and non-DHCs. Because most mare basalts in the northwest of the landing area are undisturbed and no DHCs can be found in this region, we only analyzed DHCs and non - DHCs in ejecta blanket southeast of the landing site. In fig. S7 (A and B), we identified a total of 25 DHCs (red circles) with diameters of 327 to 1064m and 163 non - DHCs (yellow circles) with diameters of 34 to 777m within a distance of 20km from the CE - 4 landing site (the white cross) in the CE - 2 Digital Orthophoto Map (DOM) and the Kaguya Multiband Imager (MI) map. DHCs are thought to have penetrated the ejecta layer, so an upper limit of the ejecta thickness can be estimated, whereas non - DHCs did not penetrate the ejecta, and hence, a lower limit of the ejecta thickness can be obtained (27). Using the 1/10 relation between excavation depth and the transient crater diameter (which is 0.84 of the final diameter), the excavation depth (t) of an impact crater is calculated as $t = 0.084D$, *where* D is the final rimto - rim diameter (28). Figure S7C shows the excavation depth as a function of distance from the CE - 4 landing site. From this figure, excavation depths from DHCs are generally larger than those from non - DHCs, as expected. Using the 5 and 95% percentiles of DHC diameters (405 and 924 m), 90% of the surface have an ejecta thickness of 34 to 78m. However, there are several cases where excavation depths from DHCs are less than those from non - DHCs, indicating lateral variations of ejecta thickness. Using the non - DHCs 2km within the landing site, the distal eject thickness should be at least larger than 39m. All these numbers are consistent with the subsurface reflectors at 40m, indicating that ejecta should be thicker than this value.

Ejecta thickness model

We used a power function of ejecta thickness (42) to estimate the thickness of distal ejecta at the CE - 4 landing site. The ejecta thickness t can be modeled as

$$t = T\left(\frac{r}{R}\right)^{-3} \tag{8}$$

where r is the distance from the crater center, R is crater radius, and T is the ejecta thickness at the rim of the crater. The relation between T and crater radius can be expressed as

$$T = 0.14\, R^{0.74} \tag{9}$$

As the distance from the crater center is known, the ejecta thickness can be estimated using the above two equations.

By analyzing the trend of the rays in the CE - 2 DOM image, the distal ejecta over the CE - 4 landing regions are most probably from Finsen, Von Kármán L, and several other craters (2). Table S1 shows the estimated thickness for the major ejecta - contributors at the CE - 4 landing site using the ejecta thickness model, showing that ejecta thickness at the CE - 4 landing site is ~8.1m, which is comparable with that estimated by Huang et al. (2).

References and notes

[1] H. Hiesinger, J. W. Head, New views of lunar geoscience: An introduction and overview. Rev. Mineral. Geochem. 60, 1-81 (2006).

[2] J. Huang, Z. -Y. Xiao, J. Flahaut, M. Martinot, J. W. Head, X. Xiao, M. -G. Xie, L. Xiao, Geological characteristic of Von Kármán Crater, Northwestern South Pole Aitken Basin: Chang'E 4 Landing Site Region. J. Geophys. Res. 123, 1684-1700 (2018).

[3] J. M. Reynolds, An Introduction to Applied and Environmental Geophysics (John Wiley & Sons, 2011).

[4] W. J. Peeples, W. R. Sill, T. W. May, S. H. Ward, R. J. Phillips, R. L. Jordan, E. A. Abbott, T. J. Killpack, Orbital radar evidence for lunar subsurface layering in Maria Serenitatis and Crisium. J. Geophys. Res. 83, 3459-3468 (1978).

[5] T. Ono, A. Kumamoto, H. Nakagawa, Y. Yamaguchi, S. Oshigami, A. Yamaji, T. Kobayashi, Y. Kasahara, H. Oya, Lunar radar sounder observations of subsurface layers under the nearside maria of the Moon. Science 323, 909-912 (2009).

[6] W. H. Ip, Y. Jun, C. -L. Li, Z. -Y. Ouyang, Preface: The Chang'e-3 lander and rover mission to the Moon. Res. Astron. Astrophys. 14, 1511-1513 (2014).

[7] W. -Z. Fa, M. -H. Zhu, T. -T. Liu, J. -B. Plescia, Regolith stratigraphy at the Chang'E-3 landing site as seen by lunar penetrating radar. Geophys. Res. Lett. 42, 10179-10187 (2015).

[8] L. Xiao, P. -M. Zhu, G. -Y. Fang, Z. -Y. Xiao, Y. -L. Zou, J. -N. Zhao, N. Zhao, Y. - F. Yuan, L. Qiao, X. - P. Zhang, H. Zhang, J. Wang, J. Huang, Q. Huang, Q. He, B. Zhou, Y. -C. Ji, Q. -Y. Zhang, S. -X. Shen, Y. - X. Li, Y. - Z. Gao, A young multilayered terrane of the northern Mare Imbrium revealed by Chang'E-3 mission. Science 347, 1226-1229 (2015).

[9] J. Zhang, W. Yang, S. Hu, Y. Lin, G. Fang, C. Li, W. Peng, S. Zhu, Z. He, B. Zhou, H. Lin, J. Yang, E. Liu, Y. Xu, J. Wang, Z. Yao, Y. Zou, J. Yan, Z. Ouyang, Volcanic history of the Imbrium basin: A close-up view from the lunar rover Yutu. Proc. Natl. Acad. Sci. U. S. A. 112, 5342-5347 (2015).

[10] C. Li, S. Xing, S. E. Lauro, Y. Su, S. Dai, J. Feng, B. Cosciotti, F. D. Paolo, E. Mattei, Y. Xiao, C. Ding, E. Pettinelli, Pitfalls in GPR data interpretation: False reflectors detected in Lunar radar cross sections by Chang'e-3. IEEE Trans. Geosci. Remote Sens. 56, 1325-1335 (2018).

[11] W. Wu, C. Li, W. Zuo, H. Zhang, J. Liu, W. Wen, Y. Su, X. Ren, J. Yan, D. Yu, G. Dong, C. Wang, Z. Sun, E. Liu, J. Yang, Z. Ouyang, Lunar farside to be explored by Chang'e-4. Nat. Geosci. 12, 222-223 (2019).

［12］ C. Li，D. Liu，B. Liu，X. Ren，J. Liu，Z. He，W. Zuo，X. Zeng，R. Xu，X. Tan，X. Zhang，W. Chen，R. Shu，W. Wen，Y. Su，H. Zhang，Z. Ouyang，Chang'E－4 initial spectroscopic identification of lunar far－side mantle－derived materials. Nature 569，378 (2019).

［13］ D. E. Stuart－Alexander，"Geologic map of the central far side of the Moon" (IMAP 1047，U. S. Geological Survey，1978).

［14］ J. H. Pasckert，H. Hiesinger，C. H. van der Bogert，Lunar farside volcanism in and around the South Pole－Aitken basin. Icarus 299，538－562 (2018).

［15］ Z. Xiao，Size－frequency distribution of different secondary crater populations：1. Equilibrium caused by secondary impacts. J. Geophys. Res. 121，2404－2425 (2016).

［16］ G.－Y. Fang，B. Zhou，Y.－C. Ji，Q.－Y. Zhang，S.－X. Shen，Y.－X. Li，H.－F. Guan，C.－J. Tang，Y.－Z. Gao，W. Lu，S.－B. Ye，H.－D. Han，J. Zheng，S.－Z. Wang，Lunar penetrating radar onboard the Chang'e－3 mission. Res. Astron. Astrophys. 14，1607－1622 (2014).

［17］ Y. Su，G.－Y. Fang，J.－Q. Feng，S.－G. Xing，Y.－C. Ji，B. Zhou，Y.－Z. Gao，H. Li，S. Dai，Y. Xiao，C.－L. Li，Data processing and initial results of Chang'E－3 lunar penetrating radar. Res. Astron. Astrophys. 14，1623－1632 (2014).

［18］ M. Angelopoulos，D. Redman，W. H. Pollard，T. W. Haltigin，P. Dietrich，Lunar ground penetrating radar：Minimizing potential data artifacts caused by signal interaction with a rover body. Adv. Space Res. 54，2059－2072 (2014).

［19］ H.M. Jol，Ground Penetrating Radar：Theory and Applications (Elsevier Science，2009).

［20］ S. E. Lauro，E. Mattei，B. Cosciotti，F. Di Paolo，S. A. Arcone，M. Viccaro，E. Pettinelli，Electromagnetic signal penetration in a planetary soil simulant：Estimated attenuation rates using GPR and TDR in volcanic deposits on Mount Etna. J. Geophys. Res. 122，1392－1404 (2017).

［21］ W. D. Carrier，G. R. Olhoeft，W. Mendell，Physical properties of the lunar surface，in Lunar Source－Book：A User's Guide to the Moon，G. H. Heiken，D. T. Vaniman，B. M. French，Eds. (Cambridge Univ. Press，1991)，chap. 9.

［22］ G. R. Olhoeft，D. W. Strangway，Dielectric properties of the first 100 meters of the Moon. Earth Planet. Sci. Lett. 24，394－404 (1975).

［23］ D. Hickson，A. Boivin，M. G. Daly，R. Ghent，M. C. Nolan，K. Tait，A. Cunje，C. A. Tsai，Near surface bulk density estimates of NEAs from radar observations and permittivity measurements of powdered geologic material. Icarus 306，16－24 (2018).

［24］ M. Ohtake，J. Haruyama，T. Matsunaga，Y. Yokota，T. Morota，C. Honda；LISM Team，Performance and scientific objectives of the SELENE (KAGUYA)

Multiband Imager. Earth Planets Space 60, 257 – 264 (2008).

[25] P. S. Russell, J. A. Grant, K. K. Williams, L. M. Carter, W. B. Garry, I. J. Daubar, Ground penetrating radar geologic field studies of the ejecta of Barringer Meteorite Crater, Arizona, as a planetary analog. J. Geophys. Res. Planets 118, 1915 – 1933 (2013).

[26] I. Catapano, F. Soldovieri, G. Alli, G. Mollo, L. A. Forte, On the reconstruction capabilities of beamforming and a microwave tomographic approach. IEEE Geosci. Remote Sens. Lett. 12, 2369 – 2373 (2015).

[27] J. L. Whitten, J. W. Head, Lunar cryptomaria: Physical characteristics, distribution, and implications for ancient volcanism. Icarus 247, 150 – 171 (2015).

[28] H. J. Melosh, Impact Cratering: A Geologic Process (Oxford Univ. Press, 1989).

[29] L. Qiao, Z. Ling, X. Fu, B. Li, Geological characterization of the Chang'e – 4 landing area on the lunar farside. Icarus 333, 37 – 51 (2019).

[30] R. A. Yingst, F. C. Chuang, D. C. Berman, S. C. Mest, Geologic mapping of the Planck Quadrangle of the Moon (LQ – 29), in *48*th Lunar and Planetary Science Conference, The Woodlands, TX, 20 to 24 March 2017.

[31] R. H. Stolt, Migration by Fourier transform. Geophysics 43, 23 – 48 (1978).

[32] S. – G. Xing, Y. Su, J. – Q. Feng, S. Dai, Y. Xiao, C. – Y. Ding, C. – L. Li, The penetrating depth analysis of Lunar Penetrating Radar onboard Chang'e – 3 rover. Res. Astron. Astrophys. 17, 046 (2017).

[33] N. J. Cassidy, E. S. Calder, A. Pavez, L. Wooller, GPR – derived facies architectures: A new perspective on mapping pyroclastic flow deposits, in Studies in Volcanology: The Legacyof George Walker (The Geological Society for IAVCEI, 2009), pp. 181 – 210.

[34] J. Taylor, Introduction to Error Analysis, the Study of Uncertainties in Physical Measurements (University Science Books, 1997).

[35] J. D. Irving, R. J. Knight, Removal of wavelet dispersion from ground – penetrating radar data. Geophysics 68, 960 – 970 (2003).

[36] Y. Quan, J. M. Harris, Seismic attenuation tomography using frequency shift method. Geophysics 62, 895 – 905 (1997).

[37] A. Tarantola, Inverse Problem Theory and Methods for Model Parameter Estimation (Siam, 2005).

[38] J. Gazdag, P. Sguazzero, Migration of seismic data. Proc. IEEE 72, 1302 – 1315 (1984).

[39] R. Solimene, I. Catapano, G. Gennarelli, A. Cuccaro, A. Dell'Aversano, F. Soldovieri, SAR imaging algorithms and some unconventional applications: A unified mathematical overview. IEEE Signal Process. Mag. 31, 90 – 98 (2014).

[40] R. Markelein, K. Mayer, R. Hannemann, T. Krylow, K. Balasubramanian, K.

J. Langenberg, V. Schmitz, Linear and nonlinear algorithms applied in nondestructive evaluation. Inv. Probl. 18, 1733 - 1757 (2002).

[41] M. Slaney, A. C. Kak, L. E. Larsen, Limitations of imaging with first - order diffraction tomography. IEEE Trans. Microw. Theory Tech. 32, 860 - 874 (1984).

[42] T. R. McGetchin, M. Settle, J. W. Head, Radial thickness variation in impact crater ejecta: Implications for lunar basin deposits. Earth Planet. Sci. Lett. 20, 226 - 236 (1973).

Acknowledgments

This research was supported by the CE - 4 mission of the Chinese Lunar Exploration Program and funded by the National Natural Science Foundation of China (grants nos. 11941002 and 41601374). We thank the team members of the Ground Research and Application System, who have contributed to data receiving and release. We thank Dr. Brad Thomson and the other two anonymous reviewers for their very helpful comments.

Preserving signal during random noise attenuation through migration enhancement and local orthogonalization

Chao Li[1] and Jinhai Zhang[2]

1. Chinese Academy of Sciences, Key Laboratory of Earth and Planetary Physics, Institute of Geology and Geophysics, Beijing, China; Chinese Academy of Sciences, Innovation Academy of Earth Science, Beijing, China; and University of Chinese Academy of Sciences, College of Earth and Planetary Sciences, Beijing, China. E-mail: superlee@mail.iggcas.ac.cn.
2. Chinese Academy of Sciences, Key Laboratory of Earth and Planetary Physics, Institute of Geology and Geophysics, Beijing, China; Chinese Academy of Sciences, Innovation Academy of Earth Science, Beijing, China. E-mail: zjh@mail.iggcas.ac.cn (corresponding author).

Abstract: Denoising methods are usually imperfect, causing some signal leakage in the residual data. The local orthogonalization method is powerful for retrieving the leaked signal; however, its original implementation could not effectively retrieve weak residual because the residual data have weak similarity. We develop a process for extracting weak signal by the local orthogonalization method in the migrated domain and then eliminating artifacts and compensating amplitude with two more rounds of local orthogonalization in the original domain. In the migrated domain, the weak signal can be enhanced by energy focusing, but the random noise cannot be systematically enhanced, which allows us to retrieve most of the residual signal using local orthogonalization. After demigrating the retrieved weak signal back to the original domain, we may encounter artifacts caused by the false focus of random noise in the migrated domain. Therefore, we perform the second round of local orthogonalization between the current and initial denoised results to eliminate these artifacts. We further compensate for the amplitude loss caused by migration and demigration using the third round of local orthogonalization between the updated signal and noise sections. Numerical experiments illustrate the superiority of our method over existing methods for retrieving the weak signal for pre- and poststack seismic profiles and ground-penetrating radar data.

Introduction

Random noise in seismic data and ground-penetrating radar (GPR) data originates from various sources, contaminating the entire data profile. Random noise attenuation is a persistent task in seismic exploration and GPR data processing (Yilmaz, 2001; Bonar and Sacchi, 2012; Chen and Ma, 2014; Liu et al., 2014; Bi et al., 2018; Liu and Li, 2018; Lin and Zhang, 2020, 2021), and many methods have been developed. In general, there are five categories of commonly used denoising methods (Zu et al., 2019). The first category is the prediction - based methods, such as the $f-x$ deconvolution method (Canales, 1984; Soubaras, 1995; Chen and Ma, 2014), $t-x$ predictive filtering method (Abma and Claerbout, 1995; Liu and Li, 2018), and nonstationary autoregression method (Yang et al., 2014; Wu et al., 2016). The second category is the median filtering (Bednar, 1983; Huo et al., 2012) and its variants (Bonar and Sacchi, 2012), which is proper for dealing with spike - like noise, such as a deblending technique for simultaneous - source seismic data (e.g., Chen et al., 2014; Chen, 2015) based on the difference between signal and noise. The third category is the decomposition - based methods, such as singular - value decomposition (SVD) - based methods (Bekara and Van der Baan, 2007; Bai et al., 2018), empirical - mode decomposition (Huang et al., 1998; Chen and Ma, 2014; Chen et al., 2016a, 2016b), morphological decomposition (Huang et al., 2017b), and continuous wavelet transform (Mousavi et al., 2016). The fourth category is the rank - reduction based methods, which assume that the data have a low - rank structure and attenuate noise by keeping the low - rank components. The commonly used rank - reduction - based methods include Cadzow filtering (Cadzow, 1988; Zu et al., 2017) and singular spectrum analysis (Chen, 2017). The fifth category is the sparse - transform - based methods (Rubinstein et al., 2009; Chen et al., 2016c), including the Fourier transform (Naghizadeh and Sacchi, 2012), the Radon transform (Xue et al., 2014), wavelet transform (Zhang and Ulrych, 2003; Anvari et al., 2017), curvelet transform (Hennenfent and Herrmann, 2006; Neelamani et al., 2008; Ma and Plonka, 2010; Bao et al., 2014), shearlet transform (Kong et al., 2016), and seislet transform (Fomel and Liu, 2010). The machine learning - based method also has been applied for noise attenuation in recent years (e.g., Zu et al., 2018, 2019).

Despite many denoising methods developed previously, there is still strong (or at least visible) signal leakage in the residual data (i.e., the noise section), which could not be effectively separated. Chen and Fomel (2015) propose a local signal - and - noise orthogonalization method to retrieve the leaked signal from the noise section by applying a weighting factor to the estimated signal section. This method can retrieve some leaked signal from the noise section and can be combined with other denoising methods to strengthen the final output. Recently, the local orthogonalization method has been widely used in elastic wave - vector decomposition (Sripanich et al., 2017), detecting a microseismic event (Huang et al., 2017a), removing internal multiple crosstalks (Zhang et al., 2021a), and blending noise attenuation (Chen and Fomel, 2021; Huang et al., 2021). However, there is still some signal leakage for the local orthogonalization method when the local similarity between the denoised and noise sections is too small. Thus, further efforts are necessary to retrieve more portions of weak signal from strong noise.

In this paper, we propose a novel scheme of weak - signal extraction based on the migration enhancement (Huo et al., 2019; Li et al., 2021) and the local orthogonalization method (Chen and Fomel, 2015). The basic idea of our scheme is to collect weak signal by focusing migration, which can enhance the local similarity and allow a better retrieval of the weak signal using local orthogonalization; meanwhile, the random noise is overmigrated and thus is not systematically enhanced. Thus, we can better retrieve weak signal in the migrated domain. However, the demigration result of the retrieved weak signal in the migrated domain would have visible artifacts associated with migration and demigration. These artifacts usually have hyperbolic shapes (such as scattering waves), as the demigration results from some local speckle noise associated with a false focus of random noise in the migrated domain. Fortunately, these artifacts have low similarity with the initial denoising result and thus can be greatly reduced by a new local orthogonalization. Next, we perform an additional local orthogonalization to compensate for the amplitude. Numerical experiments show the great superiority of the proposed method over traditional methods in extracting more portions of weak signal, leaving almost invisible signal in the final noise section in the migrated and original domains.

Method

Local signal - and - noise orthogonalization

Given an initial denoising method (as mentioned in the "Introduction" section), the input noisy data d can be divided as

$$d = s_0 + n_0 \tag{1}$$

where s_0 and n_0 are the estimated signal and noise sections, respectively. However, the denoising result s_0 is usually imperfect, with some visible or invisible weak signal left in the noise section n_0. The local signal - and - noise orthogonalization method (Chen and Fomel, 2015) can further retrieve some portions of weak signal, if it has an evident similarity with the estimated signal s_0 as follows:

$$\bar{s}_r = w \circ s_0 \tag{2}$$

where $\bar{s}_r$ is the newly retrieved weak signal from the noise section n_0, $\circ$ represents the Hadamard product, $w \circ s_0 = \text{diag}(w)\ s_0 = \text{diag}(s_0)\ w$, diag (·) is the diagonal matrix operator, and w is the local orthogonalization weight (LOW). After the LOW is estimated, the signal s_r and noise n_r processed by the local orthogonalization method can be generated by

$$s_r = s_0 + \bar{s}_r = s_0 + w \circ s_0 = (I + \text{diag}[w])s_0 \tag{3}$$

$$n_r = n_0 - \bar{s}_r = d - s_0 - \bar{s} = d - (I + \text{diag}[w])s_0 \tag{4}$$

The local orthogonalization method assumes that the noise s_r and signal n_r should be orthogonal to each other and can be solved by

$$\min_{w} \| n_0 - \bar{s}_r \|_2^2 \tag{5}$$

To ensure continuity and smoothness, the LOW can be derived from a least - squares problem with added regularization operator (Fomel, 2007a):

$$w = \arg\min_{w} \| n_0 - \mathrm{diag}(s_0) w \|_2^2 + R(w) \tag{6}$$

where diag (s_0) is the diagonal matrix composed of s_0 and $R(w)$ is the smoothing regularization operator. Equation 6 can be solved by shaping regularization with a triangle smoothing operator Γ (Fomel, 2007b):

$$w = \{\lambda^2 I + \Gamma[\mathrm{diag}(s_0)^T \mathrm{diag}(s_0) - \lambda^2 I]\}^{-1} \Gamma \mathrm{diag}(s_0)^T n_0 \tag{7}$$

where $\lambda = \| \mathrm{diag}(s_0)^T \mathrm{diag}(s_0) \|_2$ represents a scaling parameter to enable fast convergence of equation 7. The width of the triangle smoother (smoothing radius) is the controlling parameter for calculating the LOW, and LOW is shown to be robust using different smoothing radiuses (Chen and Fomel, 2015).

Signal - enhancement - based orthogonalization

Fomel (2007a) defines the local similarity between matrices M and N as

$$c = \sqrt{c_1^T c_2} \tag{8}$$

where c_1 and c_2 can be solved by

$$c_1 = \{\lambda_1^2 I + \Gamma[\mathrm{diag}(M)^T \mathrm{diag}(M) - \lambda_1^2 I]\}^{-1} \Gamma \mathrm{diag}(M)^T N \tag{9}$$

$$c_2 = \{\lambda_2^2 I + \Gamma[\mathrm{diag}(N)^T \mathrm{diag}(N) - \lambda_2^2 I]\}^{-1} \Gamma \mathrm{diag}(N)^T M \tag{10}$$

where $\lambda_1 = \| \mathrm{diag}(M)^T \mathrm{diag}(M) \|_2$ and $\lambda_2 = \| \mathrm{diag}(N)^T \mathrm{diag}(N) \|_2$ are the scaling parameters, which are consistent with equation 7. Therefore, the local similarity c_{sn} between the initial denoised signal section s_0 and the initial noise section n_0 can be written as

$$c_{sn} = \sqrt{w_n^T w} \tag{11}$$

with $w_n = \{\lambda_n^2 I + \Gamma[\mathrm{diag}(n_0)^T \mathrm{diag}(n_0) - \lambda_n^2 I]\}^{-1} \Gamma \ \mathrm{diag}(n_0)^T s_0$. Consequently, if the initial signal s_0 and noise n_0 have a high local similarity value (Figure 1f), the leaked signal can be estimated by equation 2 with a large LOW; in contrast, if the signal s_r and noise n_r have a low local similarity value (Figure 1i), especially after applying the local orthogonalization method, the leaked signal cannot be predicted by equation 2. There are two possibilities for causing a low local similarity value between signal s_r and noise n_r: (1) the leaked signal is completely removed from the initial noise section and (2) the leaked signal energy is comparable to or smaller than the background - noise energy (Figure 1h). In the second case, the leaked signal cannot be retrieved from the initial noise section by the local orthogonalization method again.

To retrieve the weak signal from strong background noise (Figure 1h), we use focusing migration (e. g., Stolt, 1978) to enhance the weak signal in the migrated

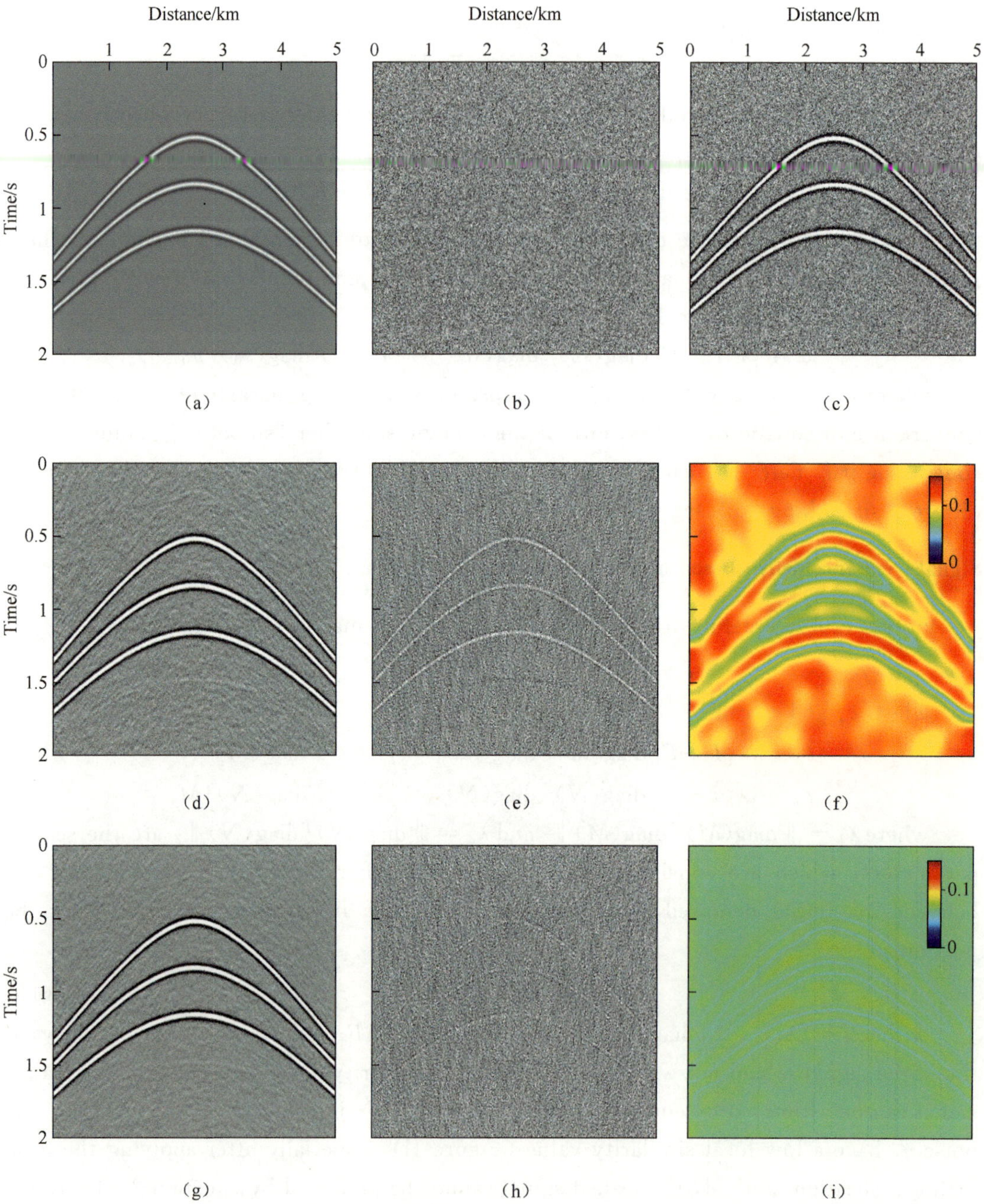

Figure 1　Denoising result of synthetic shot gather using the $f-x$ deconvolution and the local orthogonalization method.

(a) Clean shot gather, (b) random noise, (c) noisy shot gather, (d) denoised result by the $f-x$ deconvolution method, (e) noise removed by the $f-x$ deconvolution method, (f) local similarity map between (d and e), (g) denoised result by the local orthogonalization method, (h) noise removed by the local orthogonalization method, and (i) local similarity map between (g and h) domain, improving the local similarity between the processed signal s_r and noise n_r thus allowing the retrieval of more portions of the weak leaked signal by applying the local orthogonalization method. As shown in Figure 2, in the migrated domain, the denoised section (Figure 2a) and the leaked signal (Figure 2b) are concentrated in some local regions and become more evident. This indicates that the local similarity (Figure 2c) is greatly enhanced after applying the migration, which allows us to apply the local orthogonalization method to retrieve more portions of the leaked signal (Figure 3a) that cannot be retrieved in the original domain.

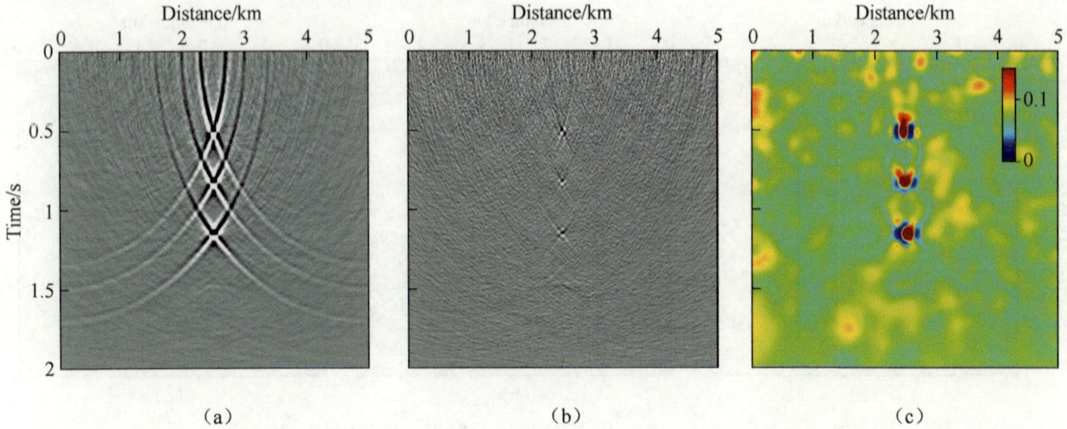

Figure 2　Signal enhancement by Stolt migration process.

(a) The migrated signal section after using the local orthogonalization method. (b) The migrated noise section after using the local orthogonalization method. (c) Local similarity map between (a and b), with the same scale as shown in Figure 1f and 1i

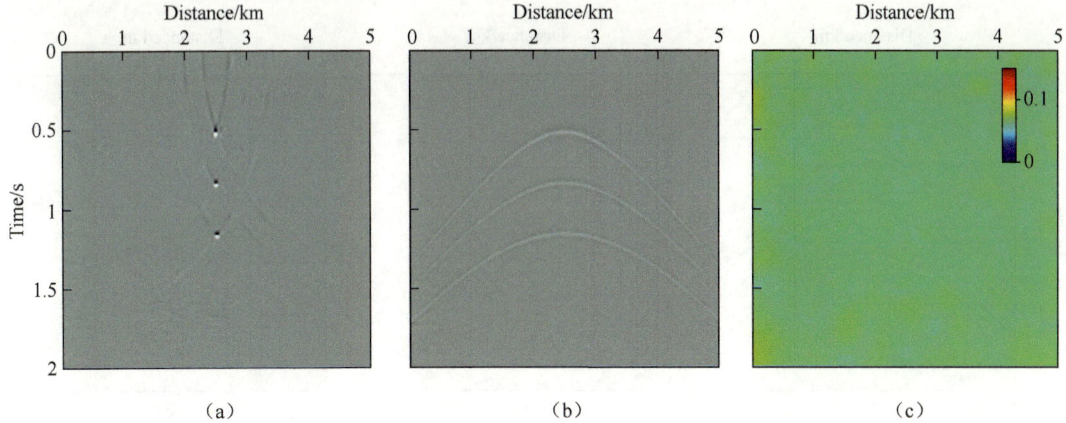

Figure 3　Leaked signal retrieved by the local orthogonalization method. (a) The retrieved signal in the migrated domain. (b) The retrieved signal in the original domain (the demigrated result of [a]). (c) Local similarity map between Figure 2a and 2b after adding and subtracting (a), respectively

As shown in Figure 4a – 4c, the result of applying local orthogonalization in the migrated domain has much less random noise than that of directly applying the local orthogonalization in the original domain (Figure 4a and 4b) because the migration could reduce random noise while enhancing the weak signal. Thus, we directly apply the local orthogonalization method in the migrated domain and then demigrate the retrieved signal back to the original domain. However, as shown in Figure 4b, the demigrated result still brings some artifacts and distorts the amplitude of the retrieved signal. We can apply two more rounds of local orthogonalization in the original domain to eliminate these artifacts and compensate for the amplitude (Figures 5, 6a, and 6b).

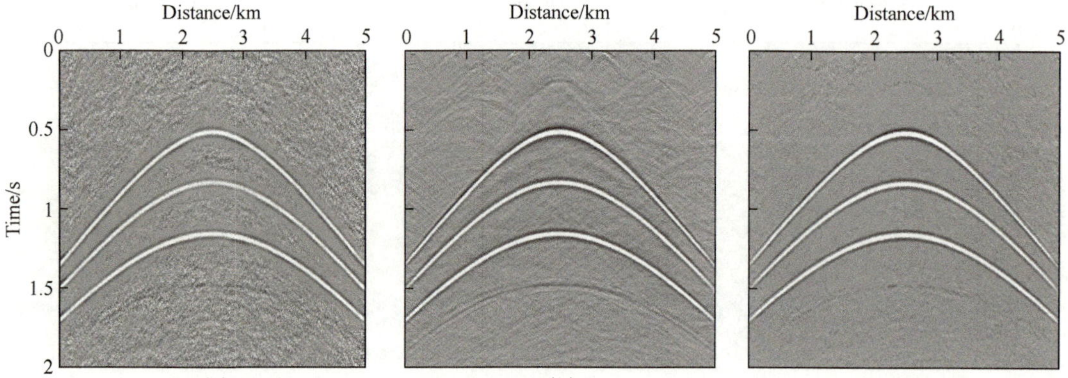

Figure 4　Comparison of retrieved signal using the local orthogonalization method in the original domain and migrated domain.

(a) The retrieved signal using the local orthogonalization method between Figure 1d and 1e in the original domain. (b) The demigrated retrieved signal using the local orthogonalization method between Figure 1d and 1e in the migrated domain. (c) The retrieved signal using local orthogonalization between (b) and Figure 1d

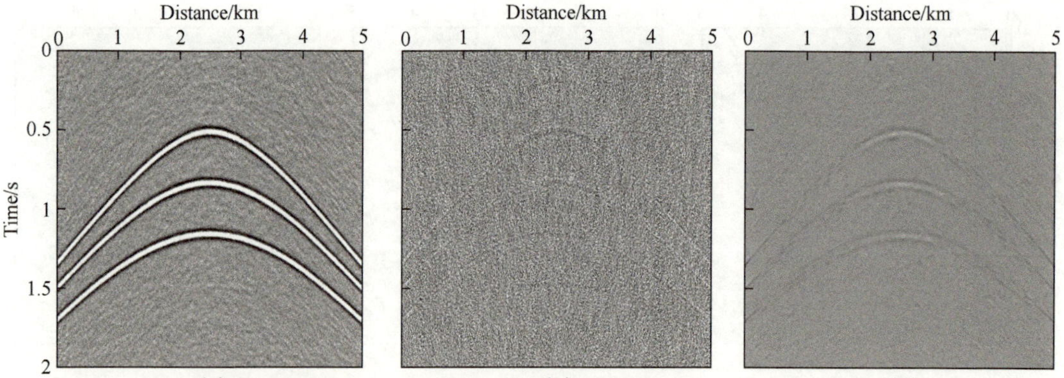

Figure 5　The results generated by the second and third rounds of local orthogonalization method.

(a) The denoised section after the second round of local orthogonalization method. (b) The noise section after the second round of local orthogonalization method. (c) The retrieved signal using the third round of local orthogonalization method

In summary, the migration enhancement and local orthogonalization - based method for weak - signal extraction consists of the following steps (Figure 7):

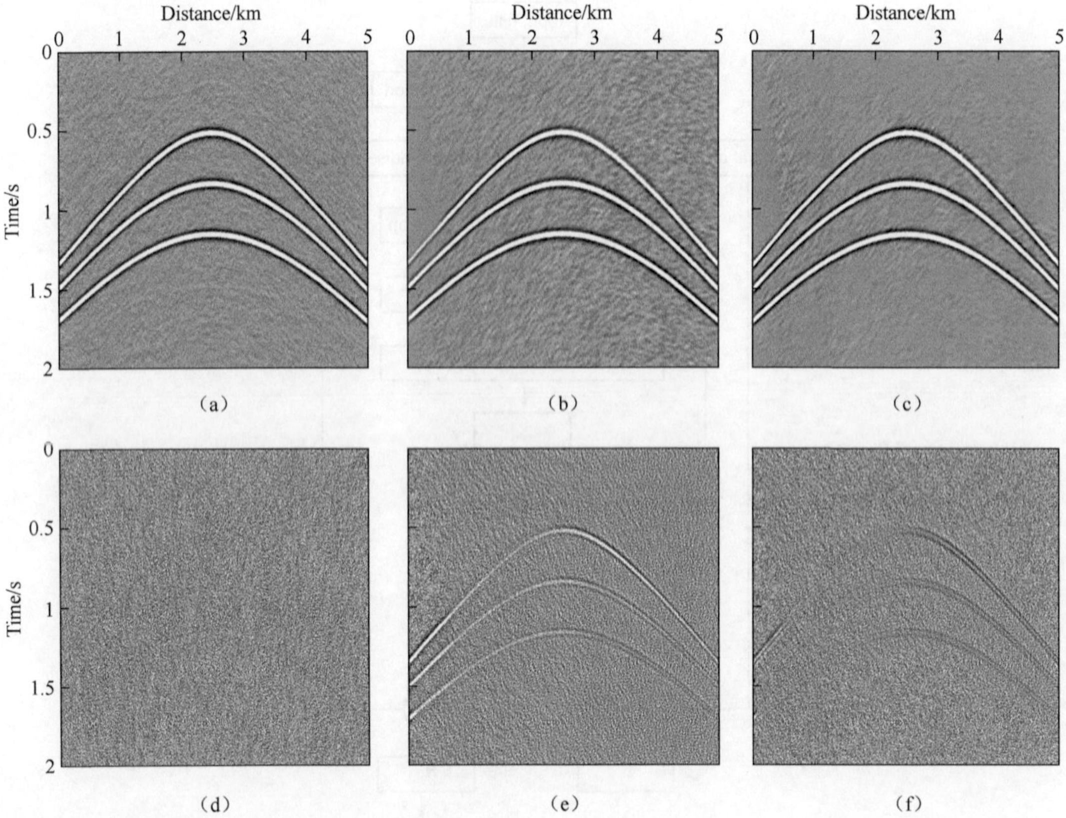

Figure 6　Comparison of denoising performance using different methods.

(a) The signal section estimated by the proposed method using the result of the $f-x$ deconvolution method as input. (b) The signal section estimated by the $t-x$ SOPF method. (c) The signal section estimated by the proposed method using the result of the $t-x$ SOPF method as input. (d) The noise section estimated by the proposed method using the result of the $f-x$ deconvolution method as input. (e) The noise section estimated by the $t-x$ SOPF method. (f) The noise section estimated by the proposed method using the result of the $t-x$ SOPF method as input

1) using the initial denoising method to generate the initial denoised section s_0 and noise section n_0,

2) migrating s_0 and n_0 simultaneously to enhance the local similarity between them in the migrated domain,

3) applying the first round of local orthogonalization (in the migrated domain) between S_0' and N_0' (migrated s_0 and n_0) to retrieve the leaked signal $\overline{S}_0$.

4) demigrating $\overline{S}_0$ back to the original domain ($\overline{s}_0$) and applying the second round of local orthogonalization (in the original domain) between the $\overline{s}_0$ and initial denoised section s_0 to retain the similar parts of $\overline{s}_1$ and eliminate artifacts caused by the migration and demigration,

5) adding the retrieved signal $\overline{s}_1$ to the initial denoised section s_0 and subtracting it from the noise section n_0 to obtain an updated denoised section s_1 and noise section n_1,

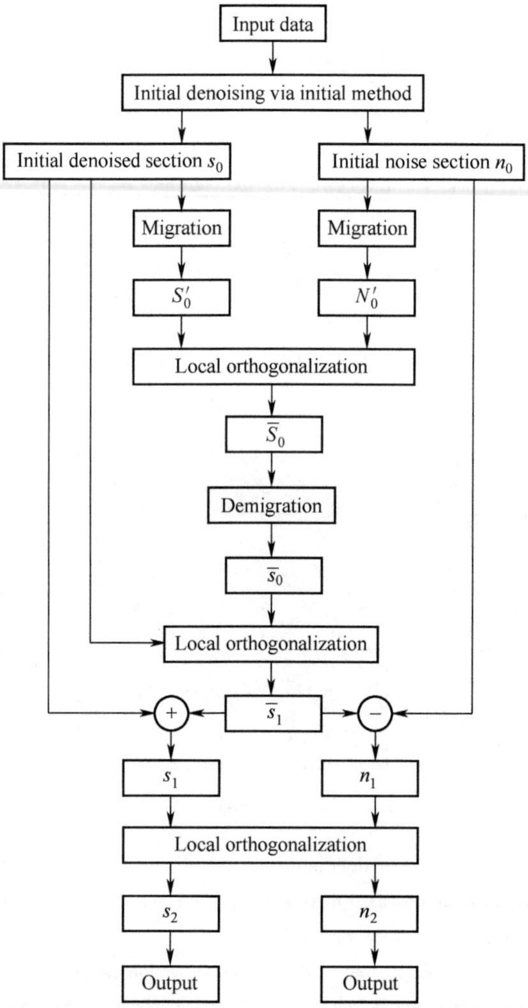

Figure 7　The workflow of the proposed method

respectively, and

6) applying the third round of local orthogonalization (in the original domain) between s_1 and n_1 to compensate for the amplitude and obtain the final denoised section s_2 and noise section n_2.

Table 1　Comparison of S/N of different denoising methods

Method/data	S/N
Input data	−2.07
$f-x$ deconvolution method	4.56
$f-x$ deconvolution + local orthogonalization method	6.62
$f-x$ deconvolution + the proposed method	7.08
$t-x$ SOPF method	8.57
$t-x$ SOPF method + the proposed method	12.91

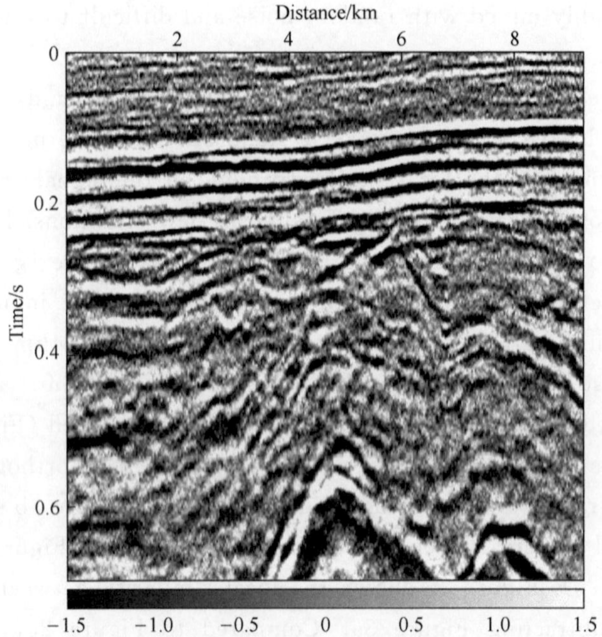

Figure 8　Poststack seismic data

Synthetic data example

We use a synthetic shot gather (Figure 1a) similar to Liu et al. (2012) to test the effectiveness of the proposed method. The synthetic data consist of three hyperbolic events and 501 traces. Three typical methods are compared – the $f-x$ deconvolution method (Canales, 1984), the local orthogonalization method (Chen and Fomel, 2015), and the $t-x$ domain streaming orthogonal prediction filter (SOPF) method (Liu and Li, 2018). The SOPF method is a streaming and orthogonalization – based prediction filter, which can generate a robust result for random noise attenuation. To accurately compare the denoising results of different methods, we use the same scale for the gray images shown in Figures 1 – 3, 5, and 6. We add some random noise (Figure 1b), which is normally distributed noise with a variance of 0.001, to the clean data to generate a noisy shot gather (Figure 1c). The denoised result (Figure 1d) using the $f-x$ deconvolution method (using 10 traces for filtering with a window size of 20 traces) contains considerable random noise, and there is obvious signal leakage in the corresponding noise section (Figure 1e). The local similarity map (Figure 1f) between Figure 1d and 1e shows three visible strong energy hyperbolic events, which is the basis for the local orthogonalization method. Figure 1g and 1h shows the denoised result and noise section, respectively, using the local orthogonalization method (using a smoothing radius of 15). The energy of the signal is enhanced, and the leaked signal is partially retrieved from the noise section compared to Figure 1e. Meanwhile, Figure 1i shows weak local similarity between Figure 1g and 1h. However, we can still obse rve some signal leakage in the noise section, which is highly mixed with random noise and difficult to retrieve using existing methods.

To retrieve a weak leaked signal from strong noise background, we need to increase the intensity of the leaked signal to make it stand out from the noise section. In other words, we should increase the local similarity between the signal and noise sections to make the local orthogonalization method feasible in new situations. Therefore, we use a series of velocities to implement Stolt migration to signal (Figure 1g) and noise sections (Figure 1h) and select the most focused image result as the input of the following process. As shown in Figure 2, after migration with a velocity of 4km/s, the leaked signal in the migrated noise section (Figure 2b) becomes more evident. The local similarity between the migrated signal and noise sections is further enhanced (Figure 2c).

Figure 3a shows the retrieved signal after applying the local orthogonalization method (using a smoothing radius of 15) in the migrated domain. Figure 3b shows the retrieved signal in the original domain, which is the demigration result of Figure 3a using the same velocity (i. e. , 4km/s). Figure 3c shows the local similarity between Figure 2a and 2b after adding and subtracting Figure 3a. Compared to Figure 2c, the value of local similarity in Figure 3c is significantly reduced, which means more portions of the leaked signal had been retrieved from the noise section, suggesting that the local orthogonalization method can retrieve more leaked signals after enhancing the local similarity by migration.

In addition, the retrieved signal (Figure 4a), by directly applying the local orthogonalization method (using a smoothing radius of 15) between the initial denoised section (Figure 1d) and noise section (Figure 1e), contains considerable noise. However, the demigrated retrieved signal (Figure 4b), by applying the local orthogonalization method (using a smoothing radius of 15) between the migrated initial denoised section and noise section, has less random noise but extra artifacts. To eliminate these artifacts, we apply the second round of local orthogonalization to Figures 1d and 4b, and the retrieved signal (Figure 4c) has less noise retained compared to Figure 4a. Subsequently, after adding the retrieved signal (Figure 4c) to the initial denoised section (Figure 1d) and subtracting the retrieved signal (Figure 4c) from the initial noise section (Figure 1e), we obtain the updated denoised section (Figure 5a) and noise section (Figure 5b). Finally, we use the third round of local orthogonalization to compensate for the amplitude loss and obtain the final signal section (Figure 6a) and the final noise section (Figure 6d), respectively. Compared to the denoised result of the $f-x$ deconvolution method (Figure 1d and 1e) and the local orthogonalization method (Figure 1g and 1h) applied in the original domain, the denoised section of the proposed method has been enhanced, and the leaked signal in the noise section is almost invisible. Compared to the $t-x$ SOPF method (Figure 6b and 6e, using the same parameters as described in Liu and Li, 2018), the leaked signal in the noise section is significantly reduced by the proposed method (using a migration velocity of 4km/s and a smoothing radius of 15). Although the final signal section (Figure 6a) has more random noise than the $t-x$ SOPF method (Figure 6b), it is caused by the performance difference between the two methods ($f-x$ deconvolution and $t-x$ SOPF methods). In other words, the performance of the proposed method relies on the original denoising method, and we can retrieve the leaked signal in the noise section rather than eliminating the leaked noise in the signal section. To compare the performance of the proposed method and the $t-x$ SOPF method more appropriately, we take Figure 6b and 6e as our input, and we obtain the corresponding final signal section (Figure 6c) and noise section (Figure 6f). The level of noise seems similar in Figure 6b and 6c, but the signal is strengthened in Figure 6c because the amplitude of the leaked signal is further reduced in the noise section as shown in Figure 6f.

To quantitatively compare the denoising result of different methods, we use the signal-to-noise ratio (S/N) as a criterion. As shown in Table 1, the S/N of the noisy data is −2.07. After applying the local orthogonalization method to the result obtained by the $f-x$ deconvolution method, the S/N is increased from 4.56 to 6.62. After applying the proposed method to the result obtained by the $f-x$ deconvolution method, the S/N reaches 7.08. Similarly, after applying the proposed method to the result of the $t-x$ SOPF method, the S/N is enhanced from 8.57 to 12.91. The proposed method has the highest S/N among all the methods listed, which quantitatively verifies the effectiveness of the proposed method in improving the S/N.

Field data example

In this section, we use three field data sets to evaluate the performance of the proposed method. The first one is poststack seismic data from Liu and Chen (2013), the second one is lunar penetrating radar (LPR) data from the Chang'E-4 (CE4) mission, (Li and Zhang, 2021; Zhang et al., 2021b), and the third one is prestack seismic data provided by Geofizyka Torun Sp. Z. o. o, Poland from FreeUSP (2015). In addition to the $f-x$ deconvolution method, the local orthogonalization method, and the $t-x$ SOPF method for processing the poststack seismic data, we also use the curvelet domain thresholding method and local SVD method for processing the LPR and prestack seismic data. To better show the result of different methods, we use local similarity (Fomel, 2007a) to evaluate the denoising performance and show all gray images on the same scale for the same data set.

Figure 8 shows a poststack seismic data set, which has 310 traces and 700 temporal samples with a sampling interval of 1ms. Figure 9a and 9b shows the denoised signal section and obtained noise section using the $f-x$ deconvolution method (using four traces for filtering with a window size of 20 traces). There is an obvious layered signal and anticline signal remaining in the noise section, and the local similarity map (Figure 9c) shows a relatively large correlation between Figure 9a and 9b. After using the local orthogonalization method (using a smoothing radius of 20) in the original domain, the amplitude of the denoised signal section (Figure 9d) is enhanced, and the leaked signal is partially retrieved from the noise section (Figure 9e). The local similarity map (Figure 9f) shows a smaller correlation between Figure 9d and 9e . However, some leaked signal

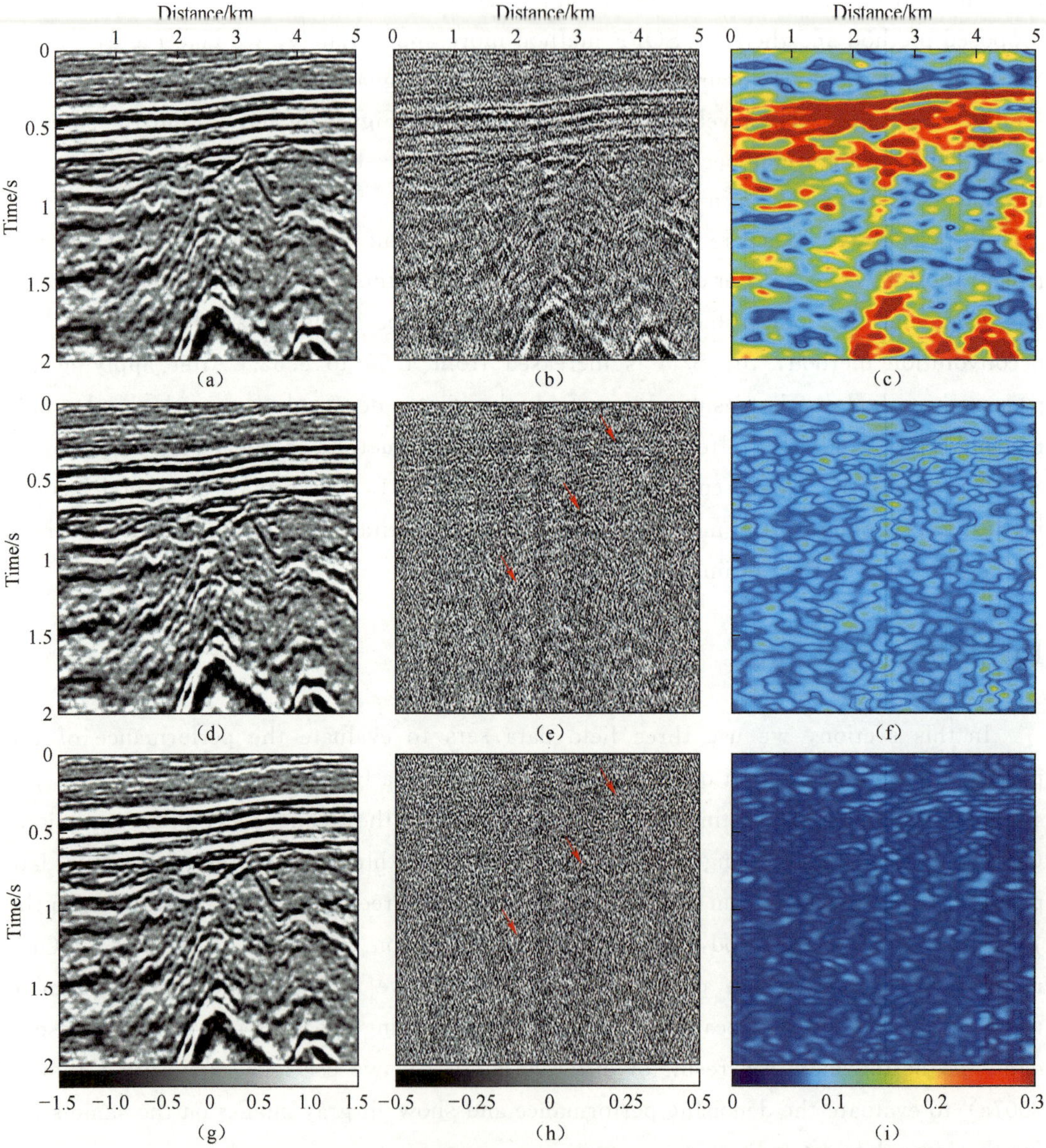

Figure 9　Denoising result of field data using the $f-x$ deconvolution, the local orthogonalization method, and the proposed method.

(a) Denoised result by the $f-x$ deconvolution method. (b) Noise removed by the $f-x$ deconvolution method. (c) The local similarity map between (a and b). (d) Denoised result by the local orthogonalization method. (e) Noise removed by the local orthogonalization method. (f) The local similarity map between (d and e). (g) Denoised result by the proposed method using the result of $f-x$ deconvolution as input. (h) Noise removed by the proposed method. (i) The local similarity between (g and h) still remains in the noise section, especially for some weak anticline signal, which is interfered by strong noise and difficult to retrieve. Therefore, we perform migration (with a velocity of 5km/s) on the signal (Figure 9a) and the noise (Figure 9b) sections to make the anticline signal focused in some local regions and enhance the leaked signal's energy simultaneously. Figure 9g and 9h shows the denoised and noise sections using the proposed method (using a smoothing radius of 20). It is clear that the denoised signal is further enhanced and that the leaked signal in the final noise section (Figure 9h) is reduced. The local similarity map (Figure 9i) of Figure 9g and 9h has the minimum value compared with Figure 9c and 9f, which shows that more portions of the signal are retrieved by the proposed method. For comparison, Figure 10a and 10b shows the denoised signal section and noise section using the $t-x$ SOPF method (using the same parameters as described in Liu and Li, 2018), whose local similarity map (Figure 10c) shows a small value compared to that of the $f-x$ deconvolution method (Figure 9c), suggesting that the $t-x$ SOPF method is more effective than the $f-x$ deconvolution method. Even if the visual denoising performance of the $t-x$ SOPF method is excellent,

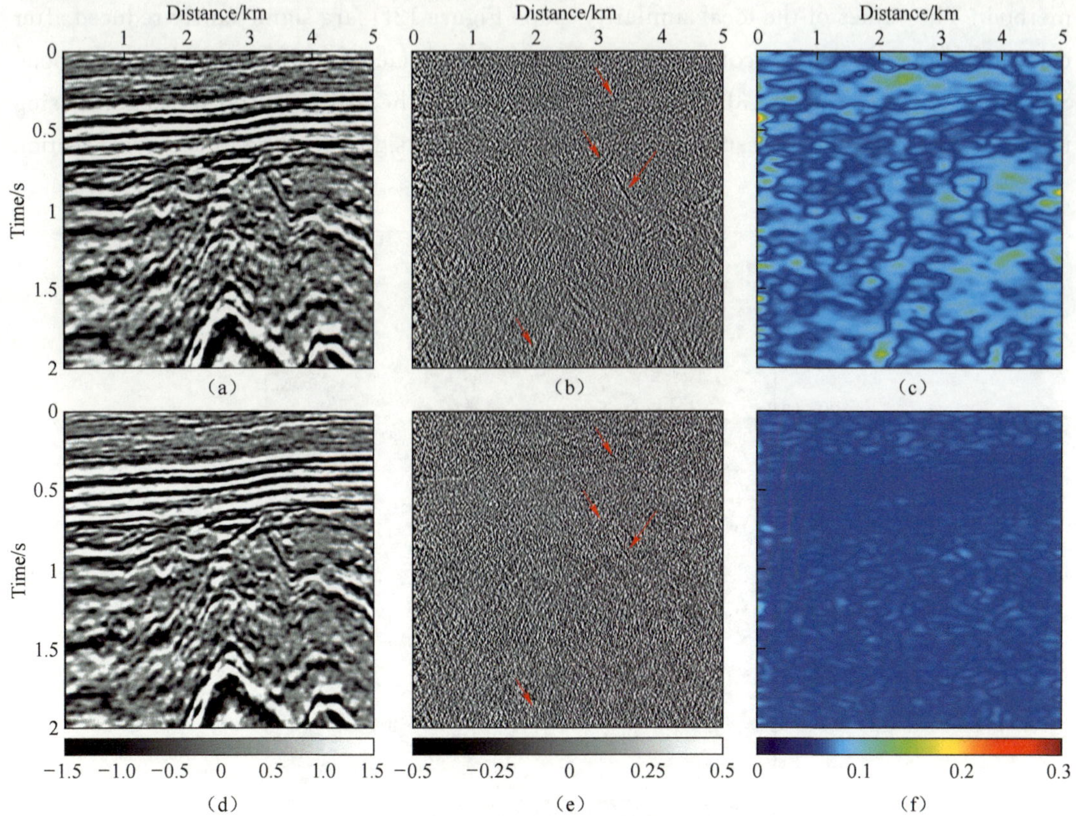

Figure 10　Denoising result of field data using the $t-x$ SOPF method and the proposed method.
(a) Denoised result by the $t-x$ SOPF method. (b) Noise removed by the $t-x$ SOPF method. (c) The local similarity map between (a and b). (d) Denoised result by the proposed method. (e) Noise removed by the proposed method. (f) The local similarity between (d and e)

our method (using a migration velocity of 5km/s and a smoothing radius of 10) can still extract more portions of the leaked signal to gain a better denoised section (Figure 10d) and noise section (Figure 10e). The values of the local similarity map (Figure 10f) are reduced compared with Figure 10c, which indicates that the proposed method can even compensate for signal leakage for an advanced denoising method.

As shown in Figure 11, the CE4 LPR data were recorded in the first three lunar days. The central frequency is 500MHz, and the traveling distance is 163m (Zhang et al., 2021b). There are 2048 temporal samples in each trace with a sampling interval of 0.3125ns. Figure 12a and 12b shows the denoised and the estimated noise sections using the curvelet domain thresholding method, respectively. The local similarity map between them is relatively high, which indicates that the noise section (Figure 12b) has evident signal leakage. After using the proposed method (using a migration velocity of 2km/s and a smoothing radius of 15), the energy of the denoised result (Figure 12d) is enhanced, especially in the shallow region (from 100 to 150ns), and the leaked signal is almost completely retrieved from the noise section (Figure 12e) after applying the proposed method. The values of the local similarity map (Figure 12f) are significantly reduced after using the proposed method compared with using the curvelet domain thresholding method. Similarly, Figure 13a and 13b shows the denoised and the estimated noise sections using the local SVD method, respectively. There is obvious signal leakage in the noise section

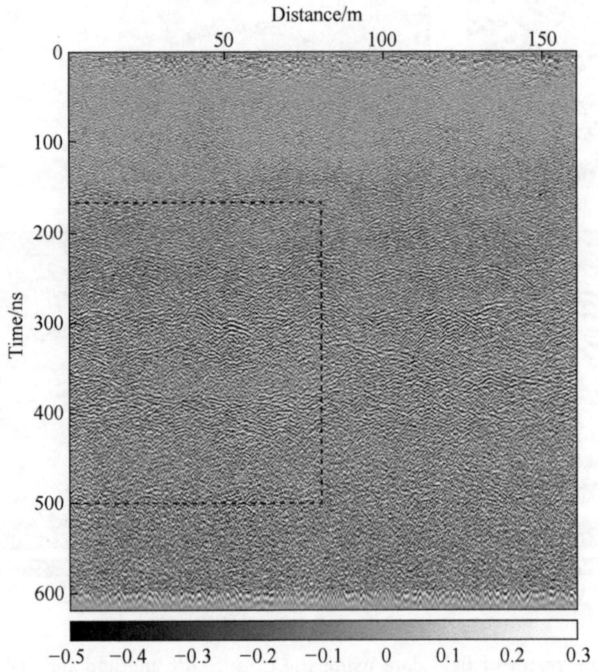

Figure 11　The LPR data set. The dashed black box denotes an enlarged area for detailed comparison

(Figure 13b), which also is indicated by the local similarity map (Figure 13c). The denoised result (Figure 13d) is enhanced after applying the proposed method (using a migration velocity of 2km/s and a smoothing radius of 15). Most portions of the leaked signal are retrieved from the noise section (Figure 13e). In addition, the local similarity map (Figure 13f) shows less correlation between Figure 13d and 13e, verifying the validity of the proposed method for retrieving the weak signal. Figure 14 shows the enlarged result of Figures 12 and 13, as shown with the dashed black box in Figure 11, which clearly shows that the proposed method can retrieve most leaking signals caused by the initial denoising method.

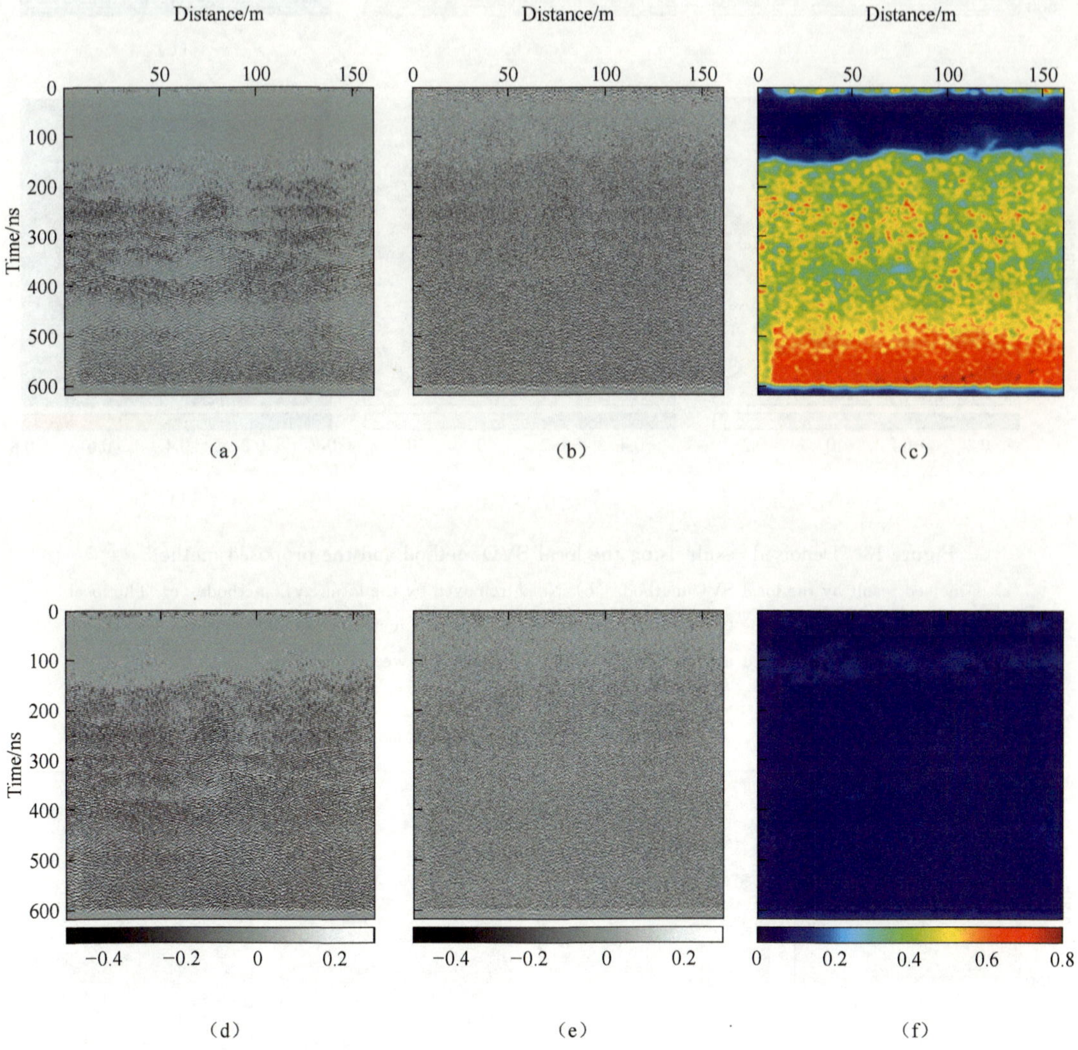

Figure 12 Denoised result using the curvelet domain thresholding method and the proposed method. (a) Denoised result by the curvelet domain thresholding method. (b) Noise removed by the curvelet domain thresholding method. (c) The local similarity between (a and b). (d) Denoised result by the proposed method. (e) Noise removed by the proposed method. (f) The local similarity between (d and e)

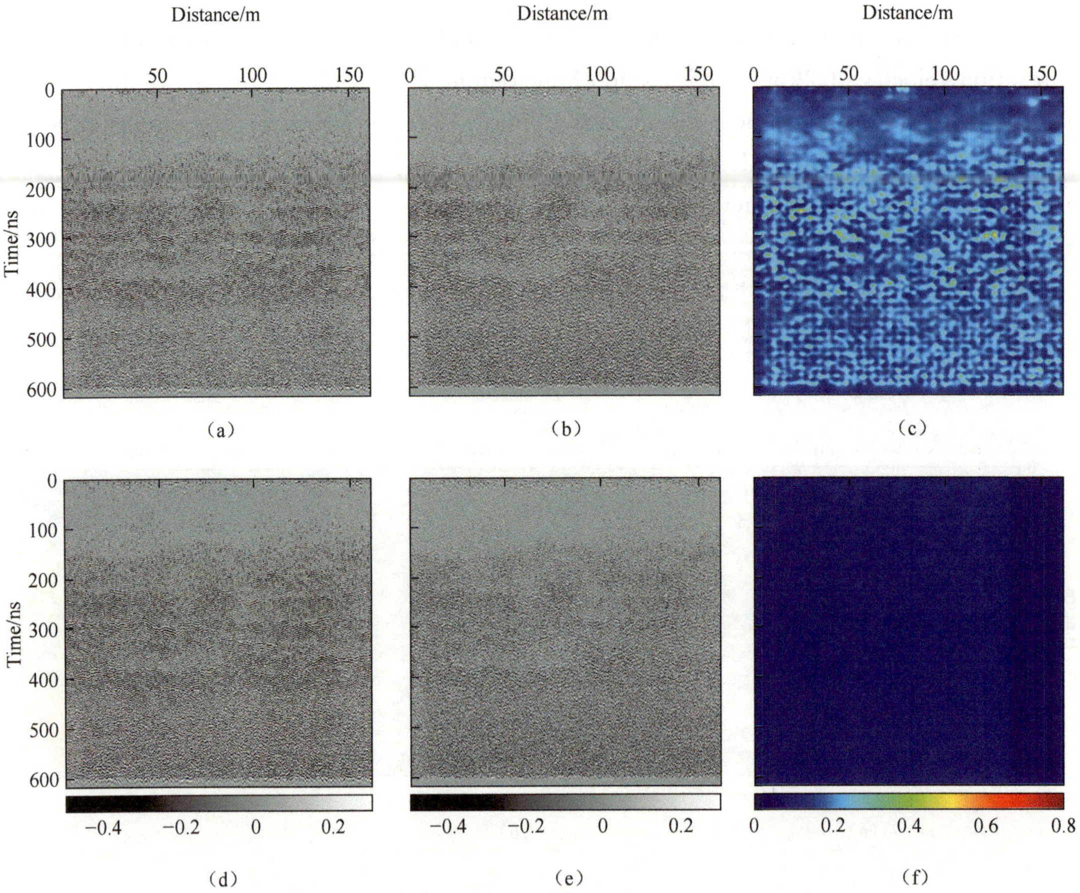

Figure 13　Denoised result using the local SVD method and the proposed method.
(a) Denoised result by the local SVD method. (b) Noise removed by the local SVD method. (c) The local similarity between (a and b). (d) Denoised result by the proposed method. (e) Noise removed by the proposed method. (f) The local similarity between (d and e)

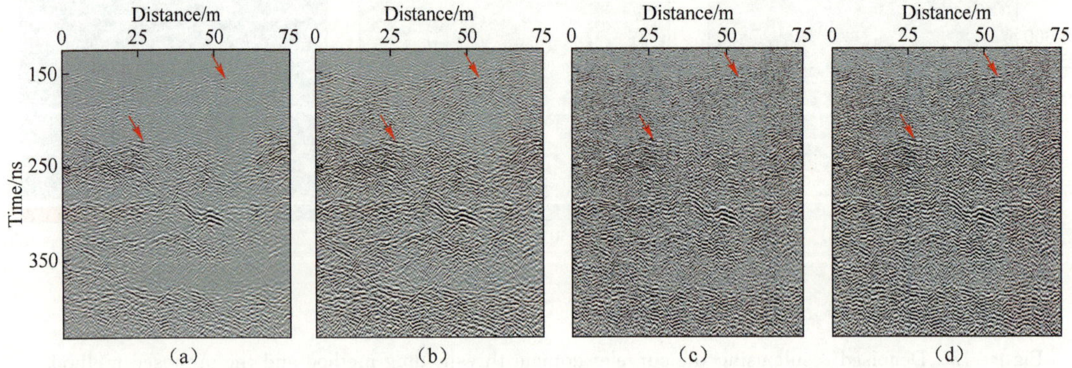

Figure 14　Local details of the denoised results and noise sections for the dashed black box area in Figure 11. The first row denotes the denoised results and the second row denotes the separated noise. The method used is shown under each column

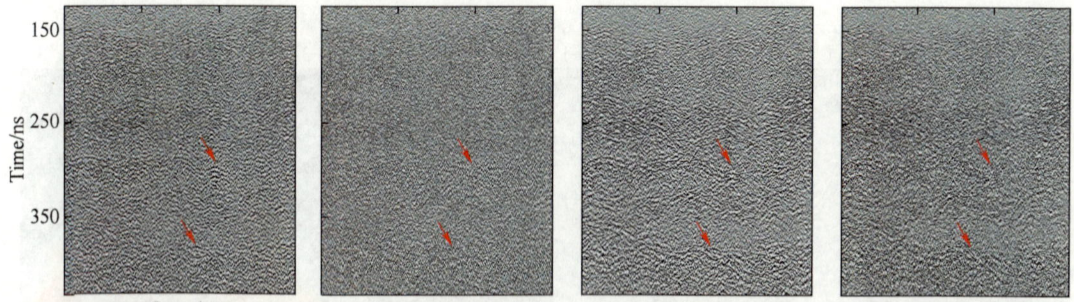

Curvelet (e)　　Curvelet+proposed method (f)　　SVD (g)　　SVD+proposed method (h)

Figure 14　Local details of the denoised results and noise sections for the dashed black box area in Figure 11. The first row denotes the denoised results and the second row denotes the separated noise. The method used is shown under each column. (Continued)

(a) Enlarged area of Figure 12a, (b) enlarged area of Figure 12d, (c) enlarged area of Figure 13a, (d) enlarged area of Figure 13d, (e) enlarged area of Figure 12b, (f) enlarged area of Figure 12e, (g) enlarged area of Figure 13b, and (h) enlarged area of Figure 13e

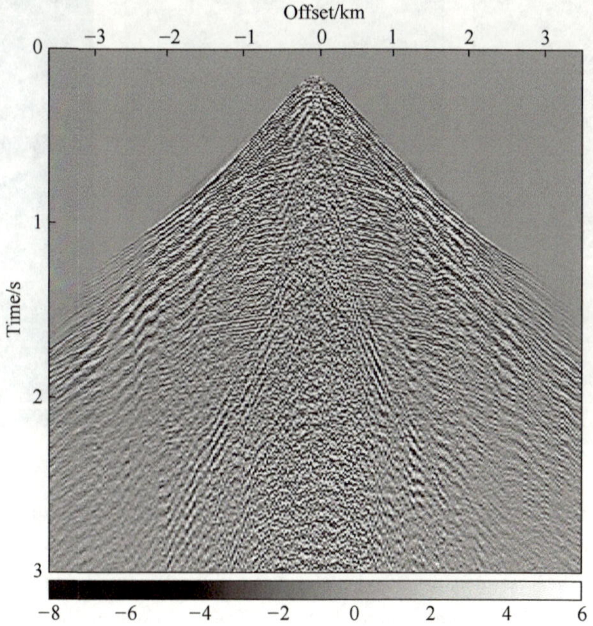

Figure 15　Prestack seismic data

Figure 15 shows a prestack seismic profile, which has 282 traces and 751 temporal samples with a sampling interval of 4ms. Figure 16a and 16b shows the denoised and the estimated noise sections using the curvelet domain thresholding method, respectively. The amplitudes of the denoised section (Figure 16d) using the proposed method (using a migration velocity of 1.5km/s and a smoothing radius of 15) have been compensated, especially as indicated by the red arrows, and the noise section (Figure 16e) has less leaked signal. Compared to the local similarity map using the proposed method (Figur 16f) and the

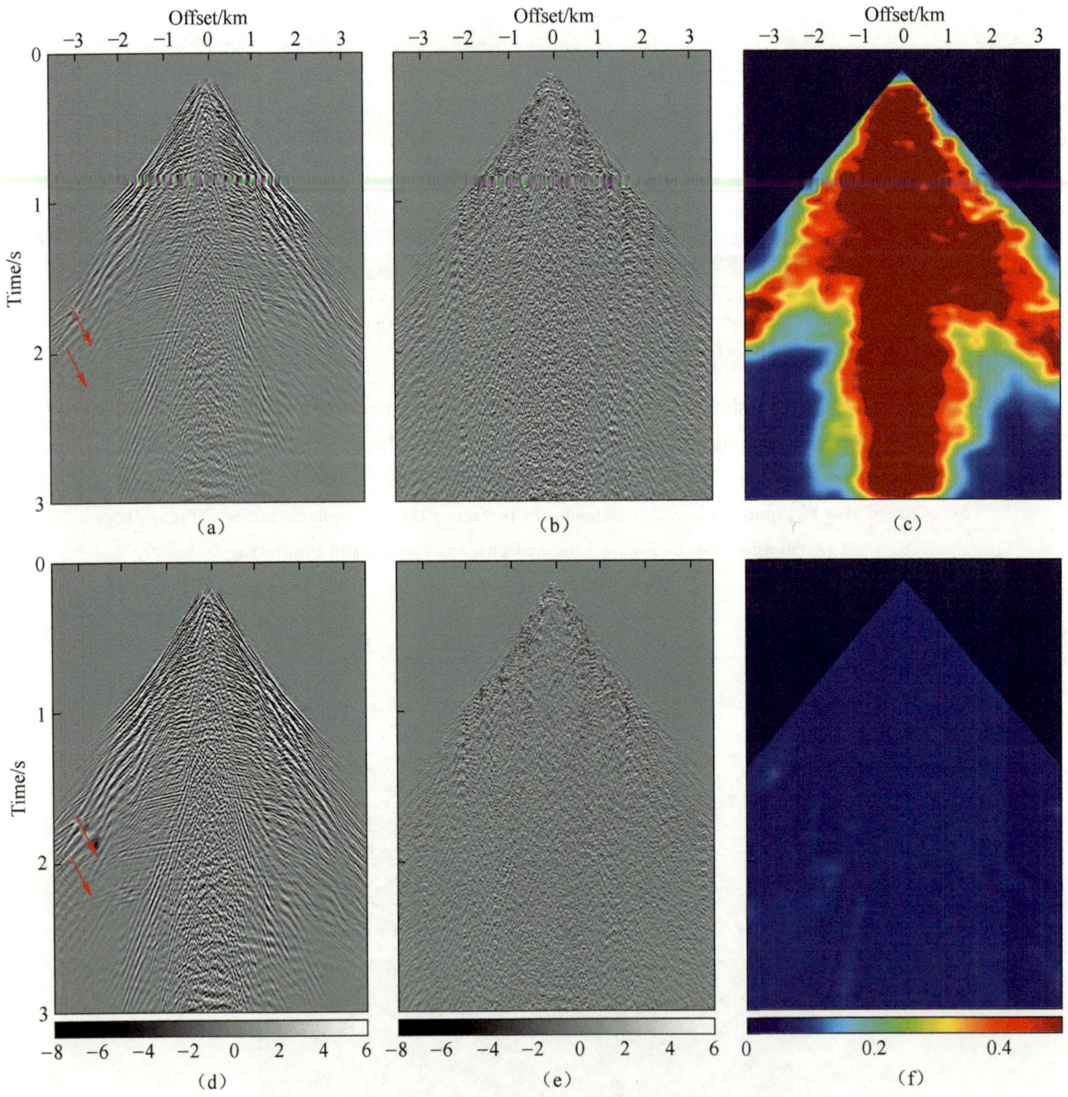

Figure 16　Denoised result using the curvelet domain thresholding method and the proposed method

(a) Denoised result by the curvelet domain thresholding method. (b) Noise removed by the curvelet domain thresholding method. (c) The local similarity map between (a and b). (d) Denoised result by the proposed method. (e) Noise removed by the proposed method. (f) The local similarity between (d and e)

curvelet domain thresholding method (Figure 16c), the proposed method can retrieve most portions of the leaked signal caused by the curvelet domain thresholding method. Figure 17a and 17b shows the denoised and noise sections using the local SVD method, respectively. The local similarity map (Figure 17c) shows a relatively high similarity between them. The reflections (Figure 17d) are enhanced after applying the proposed method (using a migration velocity of 1.5km/s and a smoothing radius of 15). The noise section (Figure 17e) contains less leaked signal (as indicated by the red arrows). The values of the local similarity (Figure 17f) are smaller than that in Figure 17e, which shows that most portions of the leaked signal are retrieved

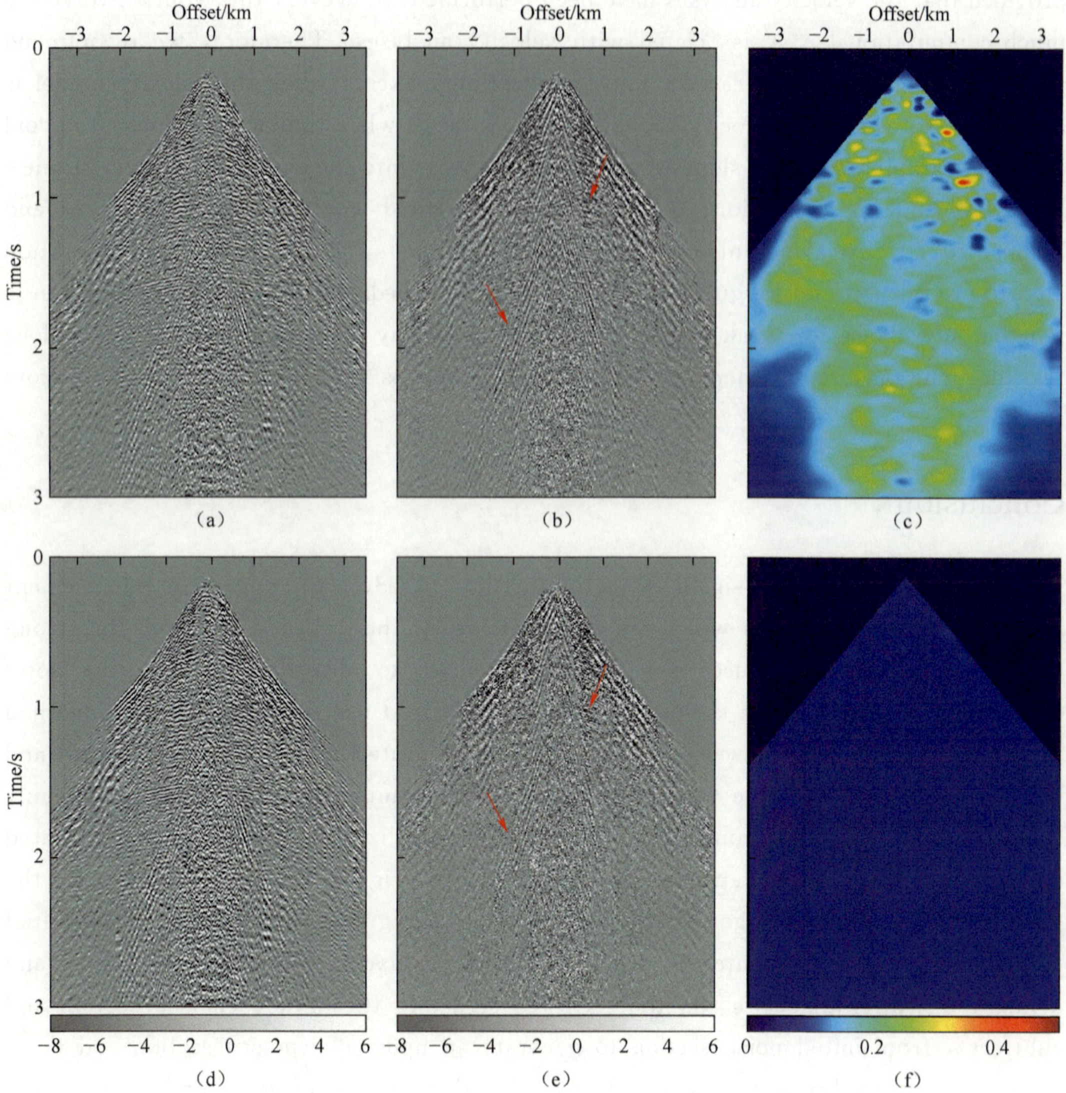

Figure 17　Denoised result using the local SVD method and the proposed method.
(a) Denoised result by the local SVD method. (b) Noise removed by the local SVD method. (c) The local similarity between (a and b). (d) Denoised result by the proposed method. (e) Noise removed by the proposed method. (f) The local similarity between (d and e)

by the proposed method.

Discussion

In all numerical experiments listed in the text, we use a constant velocity for migration and extract most residual signals in the migrated domain. This indicates that our method is not sensitive to the velocity used. A better velocity model would be helpful to concentrate more signals to the apices within smaller regions in the migrated domain, provided that the velocity analyses have been performed; however, this would involve too much computational cost associated with velocity analyses. Therefore, we recommend using constant velocity instead for most applications, except when the velocity model is available. Note that the velocity should not be too large when tuning the velocity to avoid potential damping of high - slope wavefield components into the evanescent region of one - way wave equation migration; otherwise, the wavefield would be irreversible (Li and Zhang, 2022). To avoid introducing numerical artifacts, a reversible Stolt migration algorithm (Burnett and Ferguson, 2008) is recommended. In addition, the migration is not the only choice for weak signal retrieval because any invertible transform would be feasible. The only requirement is that the weak signal is recognizable or separable from the noise.

Conclusion

We proposed a weak - signal extraction method based on the migration enhancement and local orthogonalization, which can retrieve more of the leaked signal from the strong noise section and introduce less noise compared to directly applying the local orthogonalization method in the original domain. We first use an initial denoising method to generate the initial denoised result; second, we migrate the initial denoised result and noise section using the same velocity to enhance their similarity in the migrated domain; then, we apply the first round of the local orthogonalization method in the migrated domain and demigrate the retrieved signal back to the original domain; next, we use the second round of local orthogonalization between the demigrated retrieved signal and initial denoised signal to eliminate hyperbolic - shape artifacts caused by migration and demigration; after that, we add the retrieved signal to the initial denoised section and subtract it from initial noise section to generate an updated denoised section and noise section, respectively; and finally, we use the third round of local orthogonalization between the updated denoised section and noise section to compensate amplitude loss caused by demigration. Numerical experiments on the synthetic and field data verify the performance of the proposed method for random noise attenuation and weak signal extraction.

Acknowledgments

We thank four anonymous reviewers for their helpful suggestions, which greatly improved the clarity and readability of the manuscript. This work was supported by the Key Research Program of the Chinese Academy of Sciences (ZDBS - SSW - TLC00104) and the National Natural Science Foundation of China (41941002). We thank the Supercomputing Laboratory of IGGCAS for providing computing resources.

References

[1] Abma, R. , and J. Claerbout, 1995, Lateral prediction for noise attenuation by $t-x$ and $f-x$ techniques: Geophysics, 60, 1887 - 1896, doi: 10.1190/1.1443920.

[2] Anvari, R. , M. A. N. Siahsar, S. Gholtashi, A. R. Kahoo, and M. Mohammadi, 2017, Seismic random noise attenuation using synchrosqueezed wavelet transform and low - rank signal matrix approximation: IEEE Transactions on Geoscience and Remote Sensing, 55, 6574 - 6581, doi: 10.1109/TGRS.2017.2730228.

[3] Bai, M. , J. Wu, S. Zu, and W. Chen, 2018, A structural rank reduction operator for removing artifacts in least - squares reverse time migration: Computers & Geosciences, 117, 9 - 20, doi: 10.1016/j.cageo.2018.04.003.

[4] Bao, Q. Z. , Q. C. Li, and W. C. Chen, 2014, GPR data noise attenuation on the curvelet transform: Applied Geophysics, 11, 301 - 310, doi: 10.1007/s11770 - 014 - 0444 - 2.

[5] Bednar, J. B. , 1983, Applications of median filtering to deconvolution, pulse estimation, and statistical editing of seismic data: Geophysics, 48, 1598 - 1610, doi: 10.1190/1. 1441442.

[6] Bekara, M. , and M. Van der Baan, 2007, Local singular value decomposition for signal enhancement of seismic data: Geophysics, 72, no. 2, V59 - V65, doi: 10.1190/1. 2435967.

[7] Bi, W. , Y. Zhao, C. An, and S. Hu, 2018, Clutter elimination and randomnoise denoising of GPR signals using an SVD method based on the Hankel matrix in the local frequency domain: Sensors, 18, 3422, doi: 10.3390/s18103422.

[8] Bonar, D. , and M. Sacchi, 2012, Denoising seismic data using the nonlocal means algorithm: Geophysics, 77, no. 1, A5 - A8, doi: 10.1190/geo2011 - 0235.1.

[9] Burnett, W. A. , and R. J. Ferguson, 2008, Reversible Stolt migration: CREWES Research, 20, 1 - 8.

[10] Cadzow, J. A. , 1988, Signal enhancement - A composite property mapping algorithm: IEEE Transactions on Acoustics, Speech, and Signal Processing, 36, 49 - 62, doi: 10.1109/29.1488.

[11] Canales, L. L. , 1984, Random noise reduction: 54th Annual International Meeting, SEG, Expanded Abstracts, 525 - 527, doi: 10.1190/1.1894168.

[12] Chen, Y. , 2015, Deblending using a space - varying median filter: Exploration Geophysics, 46, 332 - 341, doi: 10.1071/EG14051.

[13] Chen, Y. , 2017, Fast dictionary learning for noise attenuation of multidimensional seismic data: Geophysical Journal International, 209, 21 - 31, doi: 10.1093/gji/ggw492.

[14] Chen, Y. , and S. Fomel, 2015, Random noise attenuation using local signaland -

noise orthogonalization: Geophysics, 80, no. 6, WD1 - WD9, doi: 10.1190/geo2014 - 0227.1.

[15] Chen, Y., and S. Fomel, 2021, Non - stationary local signal - and - noise orthogonalization: Geophysics, 86, no. 5, V409 - V418, doi: 10.1190/geo2020 - 0151.1.

[16] Chen, Y., and J. Ma, 2014, Random noise attenuation by $f-x$ empirical - mode decomposition predictive filtering: Geophysics, 79, no. 3, V81 - V91, doi: 10.1190/geo2013 - 0080.1.

[17] Chen, Y., J. Yuan, Z. Jin, K. Chen, and L. Zhang, 2014, Deblending using normal moveout and median filtering in common - midpoint gathers: Journal of Geophysics and Engineering, 11, 045012, doi: 10.1088/1742 - 2132/11/4/045012.

[18] Chen, W., Y. Chen, W. Liu, and Z. Cheng, 2016a, Nonstationary signal decomposition via improved complete ensemble empirical mode decomposition and its application in ground roll noise attenuation: 86th Annual International Meeting, SEG, Expanded Abstracts, 4139 - 4144, doi: 10.1190/segam2016 - 13531013.1.

[19] Chen, W., Y. Chen, J. Xie, S. Zu, and Y. Zhang, 2016b, Multiples attenuation using trace randomization and empirical mode decomposition: 86th Annual International Meeting, SEG, Expanded Abstracts, 4139 - 4144, doi: 10.1190/segam2016 - 13771915.1.

[20] Chen, Y., J. Ma, and S. Fomel, 2016c, Double - sparsity dictionary for seismic noise attenuation: Geophysics, 81, no. 2, V103 - V116, doi: 10.1190/geo2014 - 0525.1.

[21] Fomel, S., 2007a, Local seismic attributes: Geophysics, 72, no. 3, A29 - A33, doi: 10.1190/1.2437573.

[22] Fomel, S., 2007b, Shaping regularization in geophysical - estimation problems: Geophysics, 72, no. 2, R29 - R36, doi: 10.1190/1.2433716.

[23] Fomel, S., and Y. Liu, 2010, Seislet transform and seislet frame: Geophysics, 75, no. 3, V25 - V38, doi: 10.1190/1.3380591.

[24] FreeUSP (Free Unix Seismic Processing), 2015, http://www.freeusp.org/RaceCarWebsite/TechTransfer/Tutorials/Processing_2D, accessed September 2015.

[25] Hennenfent, G., and F. J. Herrmann, 2006, Seismic denoising with nonuniformly sampled curvelets: Computing in Science & Engineering, 8, 16 - 25, doi: 10.1109/MCSE.2006.49.

[26] Huang, N. E., Z. Shen, S. R. Long, M. C. Wu, H. H. Shih, Q. Zheng, N. C. Yen, C. C. Tung, and H. H. Liu, 1998, The empirical mode decomposition and the Hilbert spectrum for nonlinear and non - stationary time series analysis: Proceedings of the Royal Society of London Series A: Mathematical, Physical and Engineering Sciences, 454, 903 - 995, doi: 10.1098/rspa.1998.0193.

[27] Huang, W., R. Wang, H. Li, and Y. Chen, 2017a, Unveiling the signals from

extremely noisy microseismic data for high - resolution hydraulic fracturing monitoring: Scientific Reports, 7, 1 - 16, doi: 10.1038/s41598 - 017 - 09711 - 2.

[28] Huang, W., R. Wang, X. Chen, Y. Zhou, Y. Chen, and J. You, 2017b, Low - frequency noise attenuation of seismic data using mathematical morphological filtering: 87th Annual International Meeting, SEG, Expanded Abstracts, 5011 - 5016, doi: 10.1190/segam2017 - 17665921.1.

[29] Huang, G., D. Zhang, W. Chen, and Y. Chen, 2021, Accelerated signal - and - noise orthogonalization: IEEE Transactions on Geoscience and Remote Sensing, 60, 1 - 9, doi: 10.1109/TGRS.2021.3054839.

[30] Huo, S., Y. Luo, and P. G. Kelamis, 2012, Simultaneous sources separation via multidirectional vector - median filtering: Geophysics, 77, no. 4, V123 - V131, doi: 10.1190/geo2011 - 0254.1.

[31] Huo, J., B. Zhou, Q. Zhao, I. M. Mason, and Y. Rao, 2019, Migration - based filtering: Applications to geophysical imaging data: Geophysics, 84, no. 4, S219 - S228, doi: 10.1190/geo2018 - 0703.1.

[32] Kong, D., Z. Peng, H. Fan, and Y. He, 2016, Seismic random noise attenuation using directional total variation in the shearlet domain: Journal of Seismic Exploration, 25, 321 - 338.

[33] Li, C., and J. Zhang, 2021, Velocity analysis using separated diffractions for lunar penetrating radar obtained by Yutu - 2 rover: Remote Sensing, 13, 1387, doi: 10.3390/rs13071387.

[34] Li, C., and J. Zhang, 2022, Wavefield separation using irreversible - migration filtering: Geophysics, 87, no. 3, A43 - A48, doi: 10.1190/geo2021 - 0607.1.

[35] Li, C., Y. Lin, W. Lv, and J. Zhang, 2021, Eliminating above - surface diffractions from ground - penetrating radar data using iterative Stolt migration: Geophysics, 86, no. 1, H1 - H11, doi: 10.1190/geo2019 - 0796.1.

[36] Lin, Y., and J. Zhang, 2020, Progressive denoising of seismic data via robust noise estimation in dual domains: Geophysics, 85, no. 1, V99 - V118, doi: 10.1190/geo2019 - 0010.1.

[37] Lin, Y., and J. Zhang, 2021, A multispectral denoising framework for seismic random noise attenuation: IEEE Transactions on Geoscience and Remote Sensing, 60, 1 - 17, doi: 10.1109/TGRS.2020.3047633.

[38] Liu, G., and X. Chen, 2013, Noncausal $f - x - y$ regularized nonstationary prediction filtering for random noise attenuation on 3D seismic data: Journal of Applied Geophysics, 93, 60 - 66, doi: 10.1016/j.jappgeo.2013.03.007.

[39] Liu, Y., and B. Li, 2018, Streaming orthogonal prediction filter in the $t - x$ domain for random noise attenuation: Geophysics, 83, no. 4, F41 - F48, doi: 10.1190/geo2017 - 0322.1.

[40] Liu, G., X. Chen, J. Du, and K. Wu, 2012, Random noise attenuation using $f - x$

regularized nonstationary autoregression: Geophysics, 77, no. 2, V61 - V69, doi: 10.1190/geo2011 - 0117.1.

[41] Liu, Y., N. Liu, and C. Liu, 2014, Adaptive prediction filtering in $t-x-y$ domain for random noise attenuation using regularized nonstationary autoregression: Geophysics, 80, no. 1, V13 - V21, doi: 10.1190/geo2014 - 0011.1.

[42] Ma, J., and G. Plonka, 2010, The curvelet transform: IEEE Signal Processing Magazine, 27, 118 - 133, doi: 10.1109/MSP.2009.935453.

[43] Mousavi, S. M., C. A. Langston, and S. P. Horton, 2016, Automatic microseismic denoising and onset detection using the synchrosqueezed continuous wavelet transform: Geophysics, 81, no. 4, V341 - V355, doi: 10.1190/geo2015 - 0598.1.

[44] Naghizadeh, M., and M. Sacchi, 2012, Multicomponent $f-x$ seismic random noise attenuation via vector autoregressive operators: Geophysics, 77, no. 2, V91 - V99, doi: 10.1190/geo2011 - 0198.1.

[45] Neelamani, R., A. I. Baumstein, D. G. Gillard, M. T. Hadidi, and W. L. Soroka, 2008, Coherent and random noise attenuation using the curvelet transform: The Leading Edge, 27, 240 - 248, doi: 10.1190/1.2840373.

[46] Rubinstein, R., M. Zibulevsky, and M. Elad, 2009, Double sparsity: Learning sparse dictionaries for sparse signal approximation: IEEE Transactions on Signal Processing, 58, 1553 - 1564, doi: 10.1109/TSP.2009.2036477.

[47] Soubaras, R., 1995, Prestack random and impulsive noise attenuation by $f-x$ projection filtering: 65th Annual International Meeting, SEG, Expanded Abstracts, 711 - 714, doi: 10.1190/1.1887383.

[48] Sripanich, Y., S. Fomel, J. Sun, and J. Cheng, 2017, Elastic wave - vector decomposition in heterogeneous anisotropic media: Geophysical Prospecting, 65, 1231 -1245, doi: 10.1111/1365 - 2478.12482.

[49] Stolt, R. H., 1978, Migration by Fourier transform: Geophysics, 43, 23 - 48, doi: 10.1190/1.1440826.

[50] Wu, G., S. Fomel, and Y. Chen, 2016, Data - driven time - frequency analysis of seismic data using regularized nonstationary autoregression: 86th Annual International Meeting, SEG, Expanded Abstracts, 1700 - 1705, doi: 10.1190/segam2016 - 13865416.1.

[51] Xue, Y., J. Ma, and X. Chen, 2014, High - order sparse Radon transform for AVO - preserving data reconstruction: Geophysics, 79, no. 2, V13 - V22, doi: 10.1190/geo2013 - 0002.1.

[52] Yang, W., R. Wang, J. Wu, and Y. Chen, 2014, Random noise attenuation using a new spectral decomposition method: 84th Annual International Meeting, SEG, Expanded Abstracts, 4366 - 4370, 10.1190/segam2014 - 0175.1.

[53] Yilmaz, Ö., 2001, Seismic data analysis: Processing, inversion, and interpretation

of seismic data：SEG.

[54] Zhang，R.，and T. J. Ulrych，2003，Physical wavelet frame denoising：Geophysics，68，225－231，doi：10.1190/1.1543209.

[55] Zhang，D.，D. Verschuur，M. Davydenko，Y. Chen，A. M. Alfaraj，and S. Qu，2021a，Local primary－and－multiple orthogonalization for leaked internal multiple crosstalk estimation and attenuation on full－wavefield migrated images：Geophysics，86，no. 1，A7－A13，doi：10.1190/geo2020－0448.1.

[56] Zhang，J.，B. Zhou，Y. Lin，M. H. Zhu，H. Song，Z. Dong，Y. Gao，K. Di，W. Yang，H. Lin，J. Yang，E. Liu，L. Wang，Y. Lin，C. Li，Z. Yue，Z. Yao，and Z. Ouyang，2021b，Lunar regolith and substructure at Chang'E－4 landing site in South Pole－Aitken basin：Nature Astronomy，5，25－30，doi：10.1038/s41550－020－1197－x.

[57] Zu，S.，H. Zhou，W. Mao，D. Zhang，C. Li，X. Pan，and Y. Chen，2017，Iterative deblending of simultaneous－source data using a coherency－pass shaping operator：Geophysical Journal International，211，541－557，doi：10.1093/gji/ggx324.

[58] Zu，S.，H. Zhou，R. Wu，W. Mao，and Y. Chen，2018，Hybrid－sparsity constrained dictionary learning for iterative deblending of extremely noisy simultaneous－source data：IEEE Transactions on Geoscience and Remote Sensing，57，2249－2262，doi：10.1109/TGRS.2018.2872416.

[59] Zu，S.，H. Zhou，R. Wu，M. Jiang，and Y. Chen，2019，Dictionary learning based on dip patch selection training for random noise attenuation：Geophysics，84，no. 3，V169－V183，doi：10.1190/geo2018－0596.1.

Velocity Analysis Using Separated Diffractions for Lunar Penetrating Radar Obtained by Yutu－2 Rover

Chao Li[1,2,3] and Jinhai Zhang[1,2]

1. Key Laboratory of Earth and Planetary Physics, Institute of Geology and Geophysics, Chinese Academy of Sciences, Beijing 100029, China; superlee@mail. iggcas. ac. cn

2. Innovation Academy of Earth Science, Chinese Academy of Sciences, Beijing 100029, China

3. College of Earth and Planetary Sciences, University of Chinese Academy of Sciences, Beijing 100029, China

Abstract: The high－frequency channel of lunar penetrating radar (LPR) onboard Yutu－2 rover successfully collected high quality data on the far side of the Moon, which provide a chance for us to detect the shallow subsurface structures and thickness of lunar regolith. However, traditional methods cannot obtain reliable dielectric permittivity model, especially in the presence of high mix between diffractions and reflections, which is essential for understanding and interpreting the composition of lunar subsurface materials. In this paper, we introduce an effective method to construct a reliable velocity model by separating diffractions from reflections and perform focusing analysis using separated diffractions. We first used the plane－wave destruction method to extract weak—energy diffractions interfered by strong reflections, and the LPR data are separated into two parts: diffractions and reflections. Then, we construct a macro－velocity model of lunar subsurface by focusing analysis on separated diffractions. Both the synthetic ground penetrating radar (GPR) and LPR data shows that the migration results of separated reflections have much clearer subsurface structures, compared with the migration results of un－separated data. Our results produce accurate velocity estimation, which is vital for high－precision migration; additionally, the accurate velocity estimation directly provides solid constraints on the dielectric permittivity at different depth.

Keywords: lunar penetrating radar; plane－wave destruction; focusing analysis; diffractions; velocity analysis

1 Introduction

The lunar subsurface material and geological structure recorded the early evolution history of the Moon. In the past half century, several radar detections have been carried out to study subsurface structures of the Moon, such as Earth based radar[1,2], orbital radar[3-5], and lunar penetrating radar (LPR)[6-8]. The orbital radar can penetrate deep into the ground because of using low - frequency radar waves and can obtain a global coverage; however, it has relatively low spatial resolution on the local details of stratifications[9]. In contrast, the LPR emits and receives the reflected high - frequency radar waves directly from the lunar subsurface anomalies, such as reflection layers and small diffractors, thus can detect local details of subsurface structures (Fig. 1)[6]. Chang'E - 3 landed in the Imbrium basin on the near side of the Moon in 2013, and the Yutu rover performed the first roving detection by LPR on an extraterrestrial planet[7]. There are two channels of the LPR payload: Channel 1 works with a central frequency of 60MHz, which is to detect the deep subsurface structures with a resolution of 10m within the depth of 500m, and Channel 2 works with a central frequency of 500MHz, which is to detect the shallow subsurface structures and lunar regolith with a resolution of 0. 3m within the depth of 50m[6]. On 3 January 2019, Chang'E - 4 landed in the Von Kármán crater on the farside of the Moon[8]. The Yutu - 2 rover was equipped with similar LPR payloads as the Yutu rover. The LPR onboard Yutu - 2 rover had accumulated a lot of data within its current traveling distance of ~600m, after working 26 lunar days. The LPR data revealed that the geological sediments consist of several ejecta layers that are from adjacent craters and multi - episode volcanic eruptions[8], which clearly exhibits the transportation history of lunar shallow materials.

The model of dielectric permittivity is important for high accuracy imaging and for estimating the material composition of lunar subsurface. The velocity of radar propagation within the lunar rocks and regolith can be converted from the dielectric permittivity as follows,

$$v = \frac{c}{\sqrt{\varepsilon_r}}, \tag{1}$$

where v is the velocity of electromagnetic waves, ε_r is the relative dielectric permittivity and c is the speed of light in vacuum[10]. With this approximation, we can directly extend many sophisticated methods developed in the seismic exploration fields, such as depth migration[7] and velocity analysis, to image and identify complex structures detected by the LPR. Currently, there are two approaches for building up lunar velocity models: (1) onedimensional empirical relations derived from statistical method based on measurements on Apollo lunar samples[7,8,11,12]; (2) hyperbola fitting on identified diffractions[13-16]. The 1D layered model assumption was only valid for the situation that the subsurface velocity variations are not changed dramatically, such as horizontal layered structure, but would have an evident error in the presence of lateral velocity variations or dipping structures. In contrast, the hyperbola fitting is effective for determining the single - point velocity at the apex of diffractions; however, this method has two inherent drawbacks: (1) the effectiveness of this method highly depends on the human subjectivity and it is time consuming for processing a large number of diffractions; (2) it is highly dependent on the signal to noise ratio of diffractions and are usually influenced by strong energy of reflections; whereas, it is usually difficult to separate the diffractions from noisy background and strong reflections, since the diffractions have much weak energy compared with the reflections. Therefore, the hyperbola fitting method can only obtain good results when the diffractions are clear and identifiable; otherwise, it will fail to handle weak diffractions that are fully mixed with strong energy reflections. Unfortunately, typical lunar models (ejecta and brecciated basalt) are shown to be rich in diffractions from small rocks or caves[9], which are buried under strong energy reflections from boulders or layer boundaries that are nearly horizontal.

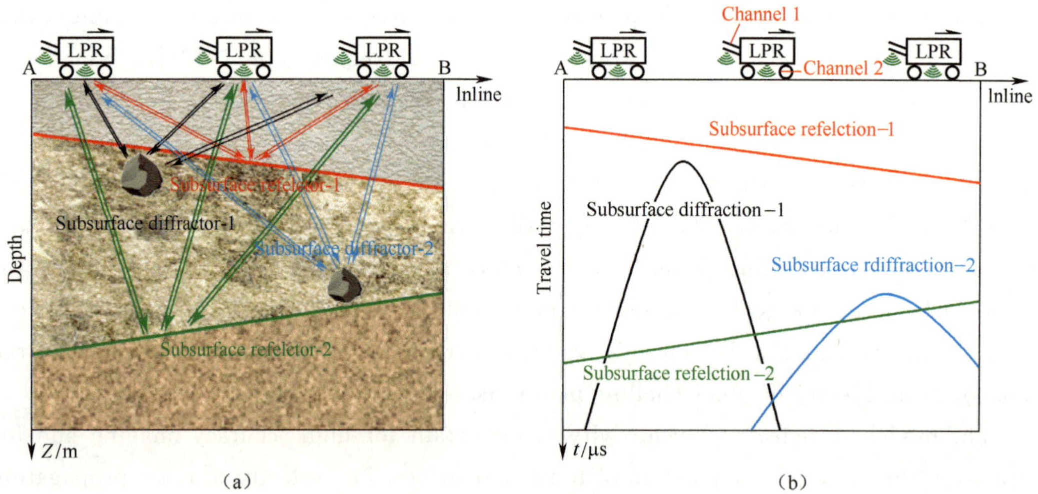

Fig. 1　A schematic diagram of lunar penetrating radar (LPR) observation (a) and the recorded (b)

In the field of seismic exploration, many methods have been proposed to separate diffractions and reflections. Harlan et al.[17] proposed to separate diffractions from reflections and conducted velocity analysis by diffraction focusing. Based on local slopes, Fomel et al.[18] proposed a two - step workflow for diffraction separation and velocity analysis by plane - wave destruction (PWD) and focusing analysis. Berkovitch et al.[19] used a new time - correction formula to estimate and separate diffractions, which focused the energy of diffractions and scattered the energy of reflections. Klokov and Fomel[20] derived analytical equation for diffractions in dip - angle domain and separated diffractions and reflections by their difference in shape using Radon transform. Asgedom et al.[21] enhanced the diffractions by common - reflection - surface method and replacement - media method. Zhang and Zhang[22] used shot and opening angle gathers to image weak diffractions. Merzlikin and Fomel[23] imaged diffractions by path - summation method and significantly reduced the computational cost by two fast Fourier transforms. Merzlikin et al. [24] divided the full wavefield into reflections, diffractions and noise by a framework consisting of Kirchhoff modeling, plane - wave destruction and path - summation integral filter. Tschannen et al. [25] used machine learning method to separate diffractions and reflections. Wang et al. [26] separated diffractions by localized rank - reduction method.

Due to the large differences between seismic and ground penetrating radar (GPR) data in the frequency components and velocity ranges, few studies have been applied to separate diffractions to GPR data. Yuan et al. [27] applied the method of Fomel et al. [18] to GPR data and the estimated velocity was shown to be consistent with the crosshole GPR velocity estimation. The framework in Fomel et al. [18] converts the diffraction separation problem into a pure signal processing task, where the GPR data are regarded as dimensionless[27], which greatly facilitates the extension of sophisticated methods that are developed for seismic data to GPR data. Economou et al. [28] used local slopes to build up a new summation weight and applied the weighted summation of constant - velocity migrated GPR profiles to achieve diffraction focusing without a velocity model.

In this paper, we first review the principle and process of Fomel et al. [18] Secondly, we use synthetic GPR data to verify the effectiveness of this method. Thirdly, we use the PWD method to decompose Chang'E - 4 LPR data into diffractions and reflections. Then, we use the local focusing analysis on diffractions to obtain a two - dimensional velocity model. Next, we convert this model into relative dielectric permittivity. Finally, we compare the imaging results after migration and draw a conclusion.

2 Methods

There are two key steps for the method of Fomel et al. [18]: (1) separate diffractions and reflections using the PWD method[18,29,30], which maps each trace onto neighboring traces and find out the vertical shift to minimize residual energy among the mapped trace and its adjacent traces; (2) estimate velocity using separated diffractions by focusing analysis method.

2.1 Plane - Wave Destruction Method

The PWD filters can be defined as the solution of local plane differential equation[29,30].

$$\frac{\partial P}{\partial x}+\sigma\frac{\partial P}{\partial t}=0 \tag{2}$$

where $P\ (t,\ x)$ represents the wavefield, x and t are distance and time, $\sigma\ (t,\ x)$ represents the dominant local slopes, which describes the local variations of signals within the local window. The PWD filters formed a prediction of each trace from its adjacent traces to estimate local dominant slopes. Fomel[30] constructed a least－squares equation to estimate the local slopes

$$\boldsymbol{C}(\sigma)\boldsymbol{d} \approx 0 \tag{3}$$

where $\boldsymbol{d}$ represents the data, and $\boldsymbol{C}$ (σ) denotes the operator of convolving the data with the 2D filter, which represents a transformed solution of Equation (2)[30]. We can get the local slopes by solving Equation (3) through linear iterative optimization method[30]. After local slopes are estimated, the PWD filter can remove all events with the same planer local slopes, such as reflections. Diffractions and noises are then leaved in the data, because they have non－planer local slopes. In other words, planer events mean that values of local slopes do not change within the local window and non－planer events mean that values of local slopes vary within the local window. The PWD method assumes that the reflections have similar local slopes within the local window while the diffractions have rapidly changing or even contrary local slopes within the local window[31].

2.2 Focusing Analysis Method

After the diffractions are separated, the focusing analysis method can be applied for velocity estimation. The varimax norm is a good measure of focusing level for separated diffractions[18,32,33]. The varimax norm is defined as

$$\phi = N\sum_{i=1}^{N} s_i^4 / \Big(\sum_{i=1}^{N} s_i^2\Big)^2 \tag{4}$$

where s_i are signal amplitudes inside a window of size N. Fomel[34] uses two continuously variable quantities to replace the varimax norm, which are solutions of the following regularized optimization problems

$$\min_{p_i}\Big(\sum_{i=1}^{N}(s_i^2 - p_i)^2 + R[p_i]\Big) \tag{5}$$

$$\min_{q_i}\Big(\sum_{i=1}^{N}(1 - q_i s_i^2)^2 + R[q_i]\Big) \tag{6}$$

where R represents a regularization operator, p_i and q_i are continuously variable quantities. By solving Equations (5) and (6), the most focused image can be found.

In summary, we first estimate the dominant local slopes of input data and use the PWD filter to separate reflections and diffractions. Second, we use velocity continuation[35,36] to get a series of migrated images of the separated diffractions. Then, we use focusing analysis method to find out the most focused image and corresponding migration velocity (i. e. , rootmean－square velocity) . Furthermore, we transform the root－mean－square velocity into Dix velocity and further convert it from time domain to depth domain. Next, we perform migration on the full－wavefield LPR data, the separated reflections, and the separated diffractions, respectively. Finally, we obtain clearer and focused imaging results of the separated reflections that had been migrated, compared with the migrated full－wavefield LPR data using the same velocity model.

3 Results

3.1 Synthetic Data Results

The gprMax is open source software, which uses finite - difference time - domain (FDTD) method to simulate electromagnetic wave propagation[37,38]. To verify the effectives of the PWD method and focusing analysis method, we use the gprMax to simulate a 2D synthetic GPR profile. The model consists of three dipping layers and many point - like diffractors with different relative permittivity in each layer, as shown in Fig. 2. The model is 40m in length, 20m in depth, and the grid spacing is 0.05m. From the top to the bottom, the background relative permittivity of each layer is 1, 3, 4 and 5, respectively. To simulate the phenomenon of weak diffractions, we set the relative permittivity of point - like diffractors and reflectors within each layer 0.5 higher than that of the background. Additionally, for consistency with the high - frequency LPR, the central frequency of simulated antenna is 500MHz, the sampling interval is 0.5m, and the temporal step is 1.179ns. Fig. 2b shows the simulated GPR profile, which consists of two strong - energy reflections and lots of diffractions. For diffractions, only those in the first subsurface layer can be well identified, while those in the deeper parts are difficult to identify. If we use the original GPR profile to conduct velocity analysis by hyperbola fitting method, we can only get accurate velocity estimation for shallow parts and cannot get reliable velocity estimation for deep parts, especially in the presence of strong random noise. Therefore, we need to separate diffractions from strong - energy reflections to enhance the recognizability of weak - energy diffractions, especially for diffractions from deep depth.

Fig. 3 shows the results after applying the PWD method. In Fig. 3a, the two wings of the diffractions have different local slopes, which are the foundation to identify and separate reflections and diffractions. The separated reflections (Fig. 3b) are clearer and more continuous compared with the original GPR profile, and only small part of energy remains around the positions of diffractors. Additionally, the diffractions are effectively separated from the strong - energy reflections, and some weak - energy diffractions that are hard to identify in the original GPR profile are also successfully identified in the separated results (see the red dashed box in Fig. 3c). By comparing the original model with the separated diffractions, we find that the apexes of separated diffractions and small anomalies in each layer correspond to each other, which proves that the PWD method is powerful on separating the reflections and diffractions.

We use the separated diffractions to implement focusing analysis method. Fig. 4a shows the results of the generated velocity model, which are transformed from the time domain to depth domain. The two dipping black dash lines indicate the two layer boundaries in original model. From the generated velocity model, we can observe that the focusing analysis method can obtain accurate velocity estimation from separated diffractions. Meanwhile, the migrated diffractions (Fig. 4d) show that all diffractions are well focused, and the migrated reflections (Fig. 4c) are much clearer than the results of migrated input GPR data (Fig. 4b). Consequently, we verify by synthetic GPR data that the focusing analysis method can get accurate velocity estimation using separated diffractions.

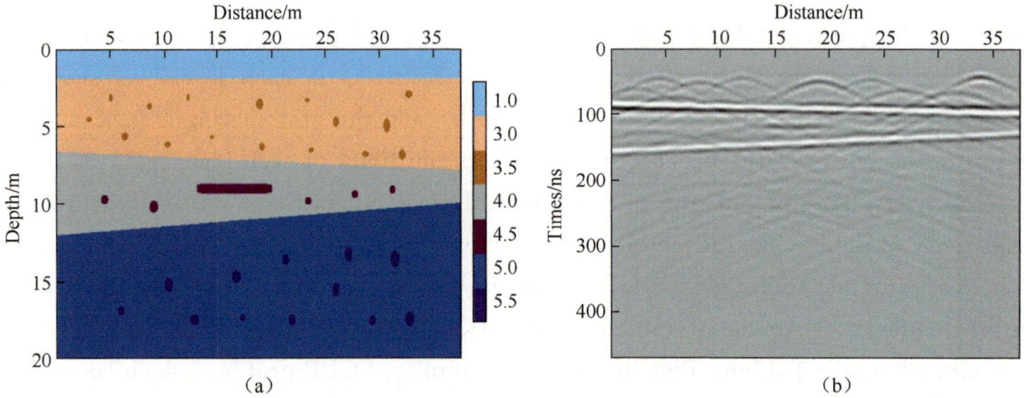

Fig. 2　The relative dielectric permittivity model (a) and synthetic ground penetrating radar (GPR) data (b)

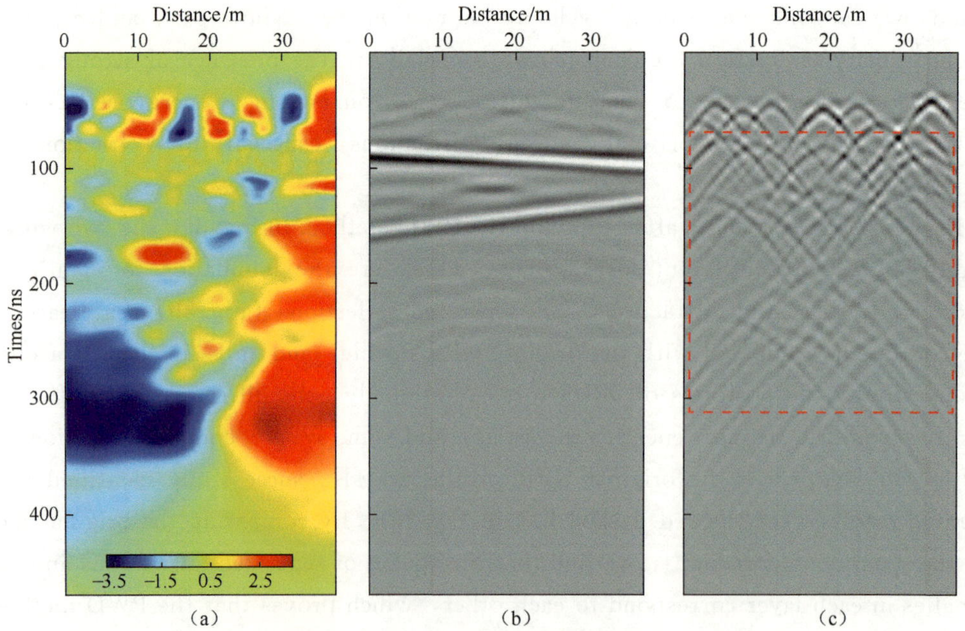

Fig. 3　The separation results of synthetic GPR data after applying the plane-wave destruction (PWD) method. (a) The local slopes. (b) the separated reflections. (c) the separated diffractions. The red dashed box indicates weak-energy diffractions that are originally mixed with strong energy reflections in the original GPR profile shown in Fig. 2b

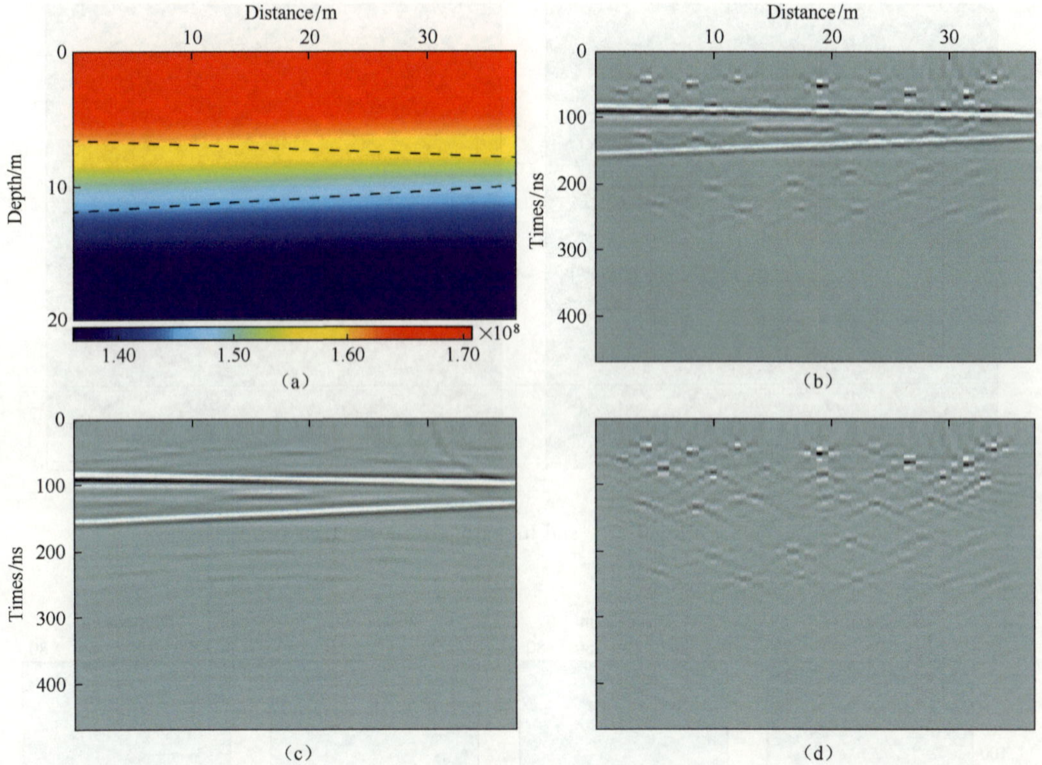

Fig. 4　The results of focusing analysis method and time migration. (a) The velocity structure in the depth domain (the black dash line indicates the layer boundary) . (b) the migrated input GPR data. (c) the migration results of separated reflections. (d) the migration results of separated diffractions

3. 2　LPR Data Results

The high - frequency LPR onboard Yutu - 2 rover was used to detect the near - surface structure. It consists of two receiving antennas (Channel 2A and Channel 2B) . The channel 2A is closer to the transmitting antenna thus has a lower signal - to - noise ratio than channel 2B. Therefore, we use the channel 2B LPR data in our experiments. The data are recorded in the first two lunar days, until 9 February 2019, with a total length of 88m (Fig. 5) . The time step of channel 2B is 0. 3125ns and spatial interval is 3. 65cm. Before we apply the PWD method, the channel 2B data were processed by decoding, removing duplicative traces, removing time delay, band - pass filter, and removing background (Fig. 6a, b)[7,8].

After applying the PWD method, the preliminary processed LPR data (Fig. 6b) are divided into two parts: reflections (Fig. 6c) and diffractions (Fig. 6d) . Obviously, the separated reflections show more continuous hierarchical structures, because most hyperbola - shaped diffractions are eliminated and only small part of energy remained at the apexes of diffractions. We can clearly observe the changes of subsurface layers

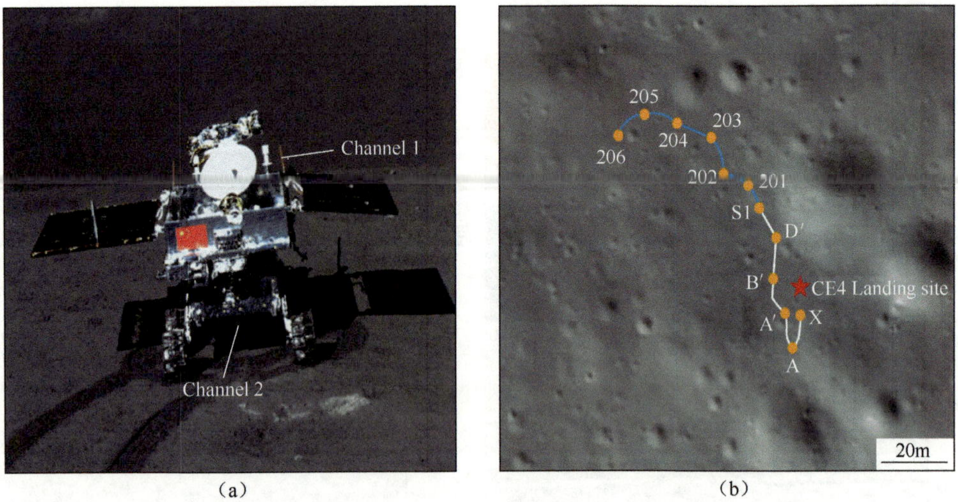

Fig. 5　The Yutu-2 rover (a) and its path (b) in the first two lunar days

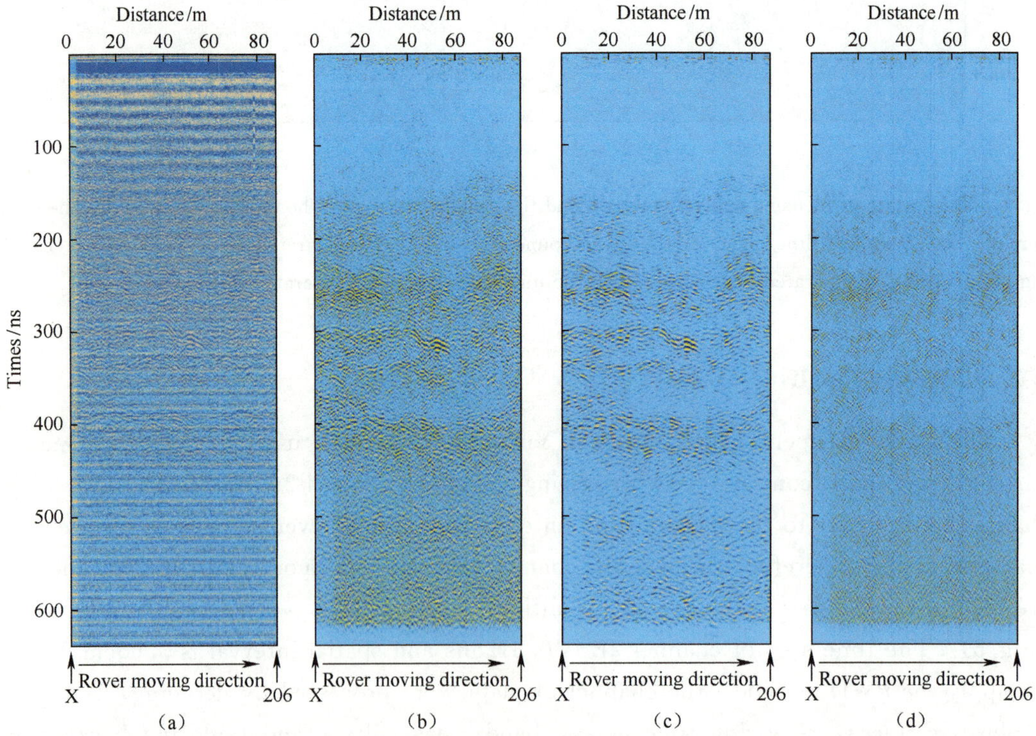

Fig. 6　The separation results of LPR data after applying the PWD method. (a) The original LPR data (after decoding and removing duplicative traces). (b) the preliminary processed LPR data (after removing time delay, band-pass filter, and removing background). (c) the separated reflections. (d) the separated diffractions, where symbols "x" and "206" represent the starting and ending point of Yutu-2 rover shown in Fig. 5, respectively

compared with the preliminary processed LPR data. Meanwhile, after separating from the strong energy reflections, some weak - energy diffractions are further enhanced, which provide more available data for the velocity analysis.

Afterwards, we use separated diffractions for focusing analysis and derive a 2D velocity model, which is further converted into the relative permittivity model (Fig. 7) . Finally, we perform migration on the preliminary processed LPR data, the separated reflections, and the separated diffractions, respectively (Fig. 8) . From the estimated velocity model, we find that the lateral velocity variations are gentle and velocity interfaces are generally horizontal, which means that there are no significant kinetic effects in this region. Furthermore, the relative permittivity shows that the lunar subsurface can be roughly divided into three layers from 0 mto 50m, according to the variations of relative permittivity. This is consistent with the results in Zhang et al.[8] The first layer is interpreted as a lunar regolith layer from 0m to ~15m, whose relative permittivity is about 3~4. The second layer is a material transition zone from ~15 m to ~25 m, which gradual changes from lunar regolith to ejecta that has a relative permittivity about 4~5. The third layer is a basalt layer from ~25m to 50m, whose relative permittivity is about 5~6. 6.

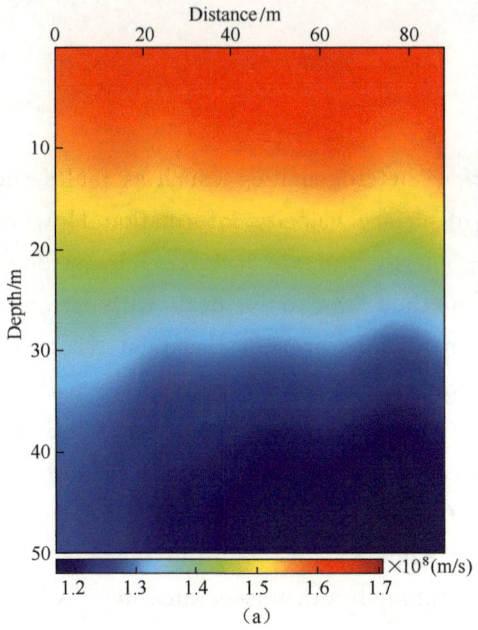

Fig. 7　The estimated velocity model (a) and relative permittivity of lunar subsurface (b) for the LPR profile in the first two lunar days

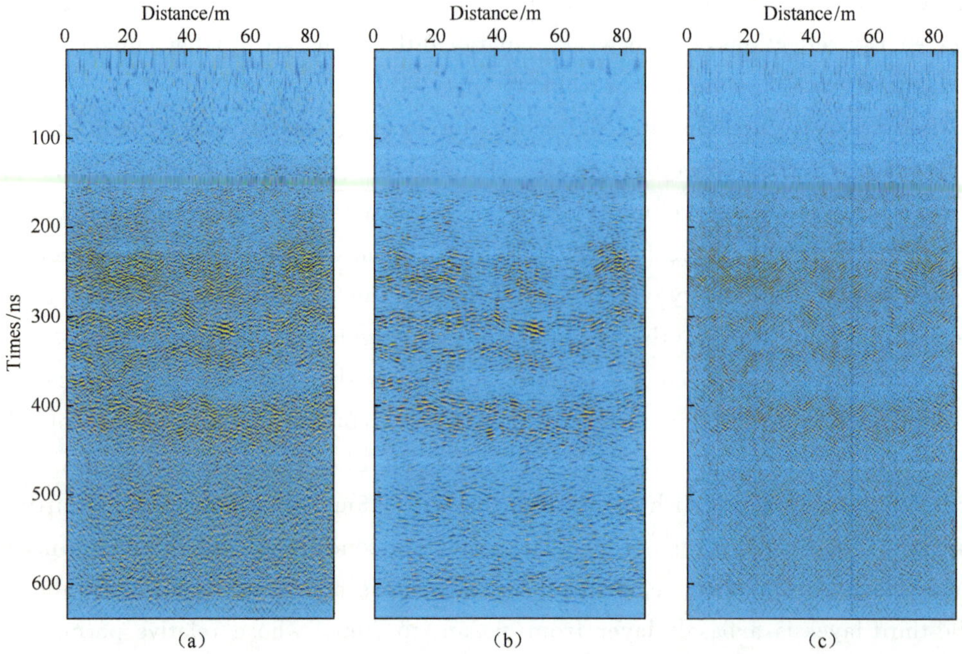

Fig. 8　Comparison between the migrated results. (a) The migrated preliminary processed LPR data. (b) the migrated reflections after separation. (c) the migrated diffractions after separation

4　Discussion

Diffractions from small - scale subsurface heterogeneities (such as faults and small rocks) of lunar subsurface carried rich subsurface velocity information. However, the diffractions usually have relatively weak - energy and are interfered by strong energy reflections. Traditional hyperbola fitting method can only deal with strong energy diffractions and does not work when the diffractions are not clear and recognizable. To perform velocity estimation from top to bottom for the LPR data, we use the PWD method to separate weak - energy diffractions from reflections, which can evidently enhance the identifiability of diffractions and avoid the influence from reflections.

In the separated reflections, we can still see some residual diffractions remained, which can influence the imaging results to some extent, but the remained energy of diffractions is relatively extraordinary low compared with unseparated data. Additionally, the residual diffractions mainly located around diffractors, where only a small part of reflections can be interfered. Certainly, in future works, we should find a more powerful method to further separate the residual diffractions, which would gain a better imaging result of reflections.

5 Conclusions

We introduce an effective velocity analysis method by separating reflections and diffractions of the LPR data. We illustrate the effectiveness of this method first by the synthetic GPR data. The estimated velocity model from the separated synthetic diffractions showed a good agreement with the true velocity model. We separate the reflections and diffractions of the Chang'E - 4 LPR data, and the separated diffractions are used to perform velocity analysis. The method of separating diffractions has two advantages: (1) the weakenergy diffractions can be enhanced and we have more available data of diffractions for accurate velocity analysis; (2) the separated reflections can eliminate the interference of cluttered diffractions and the lunar subsurface structures are clearer after imaging. These are important for reliable geological interpretations, especially when the reflections are highly mixed with the diffractions.

The velocity model is essential for obtaining well - focused migration results. Meanwhile, the relative permittivity converted from the velocity can provide a direct constraint on identifying the material composition[11]. We construct an accurate 2D velocity model for the lunar subsurface using the separated diffractions. Based on this model, we perform migration on the LPR data using the velocity model. The results show that lunar subsurface can be divided into three zones, which are regolith zone, the transition zone and the basalt zone, respectively, according to values of relative permittivity. We can see a more detailed structure of the lunar subsurface from the separated reflections, and such a continuous distribution of lunar subsurface layer has never been obtained by previous studies in the same research area. This method could be a powerful tool on imaging the detailed structures detected by ground penetrating radar on both lunar surface and other extraterrestrial planets.

Acknowledgments

We thank the Supercomputing Laboratory of Institute of Geology and Geophysics, Chinese Academy of Sciences (IGGCAS) for providing computing resources and Madagascar software package for providing open - source codes (http://www.ahay.org, accessed on 3 April 2021).

References

[1] Campbell, B. A.; Ray Hawke, B.; Morgan, G. A.; Carter, L. M.; Campbell, D. B.; Nolan, M. Improved Discrimination of Volcanic Complexes, Tectonic Features, and Regolith Properties in Mare Serenitatis from Earth - Based Radar

Mapping. J. Geophys. Res. Planets 2014, 119, 313—330.

[2] Morgan, G. A.; Campbell, B. A.; Campbell, D.; Hawke, B. Investigating the Stratigraphy of Mare Imbrium Flow Emplacement with Earth - Based Radar. J. Geophys. Res. Planets 2016, 121, 1498—1513.

[3] Peeples, W. J.; Sill, W. R.; May, T. W.; Ward, S. H.; Phillips, R. J.; Jordan, R. L.; Abbott, E. A.; Killpack, T. J. Orbital Radar Evidence for Lunar Subsurface Layering in Maria Serenitatis and Crisium. J. Geophys. Res. Solid Earth 1978, 83, 3459—3468.

[4] Sharpton, V. L.; Head, J. W., III. Stratigraphy and Structural Evolution of Southern Mare Serenitatis: A Reinterpretation Based on Apollo Lunar Sounder Experiment Data. J. Geophys. Res. Solid Earth 1982, 87, 10983—10998.

[5] Ono, T.; Kumamoto, A.; Nakagawa, H.; Yamaguchi, Y.; Oshigami, S.; Yamaji, A.; Kobayashi, T.; Kasahara, Y.; Oya, H. Lunar Radar Sounder Observations of Subsurface Layers under the Nearside Maria of the Moon. Science 2009, 323, 909.

[6] Fang, G. Y.; Zhou, B.; Ji, Y. C.; Zhang, Q. Y.; Shen, S. X.; Li, Y. X.; Guan, H. F.; Tang, C. J.; Gao, Y. Z.; Lu, W.; et al. Lunar Penetrating Radar Onboard the Chang'e - 3 Mission. Res. Astron. Astrophys. 2014, 14, 1607—1622.

[7] Zhang, J.; Yang, W.; Hu, S.; Lin, Y.; Fang, G.; Li, C.; Peng, W.; Zhu, S.; He, Z.; Zhou, B.; et al. Volcanic History of the Imbrium Basin: A Close - up View from the Lunar Rover Yutu. Proc. Natl. Acad. Sci. USA 2015, 112, 5342—5347.

[8] Zhang, J.; Zhou, B.; Lin, Y.; Zhu, M. — H.; Song, H.; Dong, Z.; Gao, Y.; Di, K.; Yang, W.; Lin, H.; et al. Lunar Regolith and Substructure at Chang'E - 4 Landing Site in South Pole - Aitken Basin. Nat. Astron. 2021, 5, 25—30.

[9] Lv, W.; Li, C.; Song, H.; Zhang, J.; Lin, Y. Comparative Analysis of Reflection Characteristics of Lunar Penetrating Radar Data Using Numerical Simulations. Icarus 2020, 350, 113896.

[10] Conyers, L. B.; Goodman, D. Ground - Penetrating Radar: An Introduction for Archaeologists; AltaMira Press: Walnut Creek, CA, USA, 1997.

[11] Heiken, G. H.; Vaniman, D. T.; French, B. M. Lunar Sourcebook, a User's Guide to the Moon; CUP Archive: Cambridge, UK, 1991.

[12] Kobayashi, T.; Kim, J. H.; Lee, S. R.; Araki, H.; Ono, T. Simultaneous Observation of Lunar Radar Sounder and Laser Altimeter of Kaguya for Lunar Regolith Layer Thickness Estimate. IEEE Geosci. Remote Sens. Lett. 2010, 7, 435—439.

[13] Feng, J.; Su, Y.; Ding, C.; Xing, S.; Dai, S.; Zou, Y. Dielectric Properties Estimation of the Lunar Regolith at CE - 3 Landing Site Using Lunar Penetrating

Radar Data. Icarus 2017, 284, 424—430.

[14] Dong, Z.; Fang, G.; Zhao, D.; Zhou, B.; Gao, Y.; Ji, Y. Dielectric Properties of Lunar Subsurface Materials. Geophys. Res. Lett. 2020, 47, e2020GL089264.

[15] Dong, Z.; Feng, X.; Zhou, H.; Liu, C.; Zeng, Z.; Li, J.; Liang, W. Properties Analysis of Lunar Regolith at Chang'E-4 Landing Site Based on 3D Velocity Spectrum of Lunar Penetrating Radar. Remote Sens. 2020, 12, 629.

[16] Song, H.; Li, C.; Zhang, J.; Wu, X.; Liu, Y.; Zou, Y. Rock Location and Property Analysis of Lunar Regolith at Chang'E-4 Landing Site Based on Local Correlation and Semblance Analysis. Remote Sens. 2020, 13, 48.

[17] Harlan, W. S.; Claerbout, J. F.; Rocca, F. Signal/Noise Separation and Velocity Estimation. Geophysics 1984, 49, 1869—1880.

[18] Fomel, S.; Landa, E.; Taner, M. T. Poststack Velocity Analysis by Separation and Imaging of Seismic Diffractions. Geophysics 2007, 72, U89—U94.

[19] Berkovitch, A.; Belfer, I.; Hassin, Y.; Landa, E. Diffraction Imaging by Multifocusing. Geophysics 2009, 74, WCA75—WCA81.

[20] Klokov, A.; Fomel, S. Separation and Imaging of Seismic Diffractions Using Migrated Dip-Angle Gathers. Geophysics 2012, 77, S131—S143.

[21] Asgedom, E. G.; Gelius, L. J.; Tygel, M. 2D Common-Offset Traveltime Based Diffraction Enhancement and Imaging. Geophys. Prospect. 2013, 61, 1178—1193.

[22] Zhang, J.; Zhang, J. Diffraction Imaging Using Shot and Opening Angle Gathers: A Prestack Time Migration Approach. Geophysics 2014, 79, S23—S33.

[23] Merzlikin, D.; Fomel, S. Analytical Path-Summation Imaging of Seismic Diffractions. Geophysics 2017, 82, S51—S59.

[24] Merzlikin, D.; Fomel, S.; Sen, M. K. Least-Squares Path-Summation Diffraction Imaging Using Sparsity Constraints. Geophysics 2019, 84, S187—S200.

[25] Tschannen, V.; Ettrich, N.; Delescluse, M.; Keuper, J. Detection of Point Scatterers Using Diffraction Imaging and Deep Learning. Geophys. Prospect. 2020, 68, 830—844.

[26] Wang, H.; Liu, X.; Chen, Y. Separation and Imaging of Seismic Diffractions Using a Localized Rank-Reduction Method with Adaptively Selected Ranks. Geophysics 2020, 85, V497—V506.

[27] Yuan, H.; Montazeri, M.; Looms, M. C.; Nielsen, L. Diffraction Imaging of Ground-Penetrating Radar Data. Geophysics 2019, 84, H1—H12.

[28] Economou, N.; Vafidis, A.; Bano, M.; Hamdan, H.; Ortega-Ramirez, J. Ground-Penetrating Radar Data Diffraction Focusing without a Velocity Model. Geophysics 2020, 85, H13—H24.

[29] Claerbout, J. F.; Abma, R. Earth Soundings Analysis: Processing Versus Inversion; Blackwell Scientific Publications: London, UK, 1992; Volume 6.

[30] Fomel, S. Applications of Plane - Wave Destruction Filters. Geophysics 2002, 67, 1946—1960.
[31] Decker, L. A. Seismic Diffraction Imaging Methods and Applications. Ph. D. Thesis, The University of Texas at Austin, Austin, TX, USA, 2014.
[32] Wiggins, R. A. Minimum Entropy Deconvolution. Geoexploration 1978, 16, 21—35.
[33] Levy, S.; Oldenburg, D. Automatic Phase Correction of Common - Midpoint Stacked Data. Geophysics 1987, 52, 51—59.
[34] Fomel, S. Local Seismic Attributes. Geophysics 2007, 72, A29—A33.
[35] Fomel, S. Time - Migration Velocity Analysis by Velocity Continuation. Geophysics 2003, 68, 1662—1672.
[36] Fomel, S. Velocity Continuation and the Anatomy of Residual Prestack Time Migration. Geophysics 2003, 68, 1650—1661.
[37] Giannopoulos, A. Modelling Ground Penetrating Radar by GprMax. Constr. Build. Mater. 2005, 19, 755—762.
[38] Warren, C.; Giannopoulos, A.; Giannakis, I. gprMax: Open Source Software to Simulate Electromagnetic Wave Propagation for Ground Penetrating Radar. Comput. Phys. Commun. 2016 209, 163—170.

An integrated photogrammetric and photoclinometric approach for illumination – invariant pixel – resolution 3D mapping of the lunar surface

Wai Chung Liu, Bo Wu

Abstract: High – resolution 3D information on lunar and planetary surfaces is crucial for planetary exploration missions and science. Photogrammetry is a state – of – the – art technology for generating 3D topographic models of surfaces such as digital elevation models (DEMs) . The performance of the photogrammetry and the resulting DEMs is affected by image matching. However, most matching algorithms fail when the images have large differences in illumination and subtle textures. This problem can be addressed by integrating photoclinometry into the photogrammetric process. This paper presents an integrated photogrammetric and photoclinometric approach that is able to generate pixel – resolution DEMs of the lunar surface and is illumination invariant. The incorporation of photoclinometry into photogrammetry involves two main steps. First, a photoclinometry assisted image matching (PAM) approach is developed by integrating photometric stereo analysis in the image matching to create pixel – wise matches, even for images with large illumination differences. Second, the DEM derived from photogrammetry using the matching results is further refined to pixel – wise resolution using photoclinometry with a shadow constraint. The proposed approach has been used for high – resolution topographic mapping at the Chang'E – 4 and Chang'E – 5 landing sites using Lunar Reconnaissance Orbiter Camera (LROC) Narrow Angle Camera (NAC) images acquired under different illumination conditions. The results indicate that the proposed approach is robust to severe inconsistencies in illumination and subtle textures in cases where the conventional approaches fail. The approach is able to achieve geometric accuracies comparable to photogrammetry but more small – scale topographic details. The proposed approach can also be used for high – resolution topographic mapping of other planetary bodies such as Mercury or asteroids, and provides a useful reference for similar topographicmapping on Earth.

Keywords: Photogrammetry; Photoclinometry; Shape – from – shading; Image matching; Moon

1 Introduction

Photogrammetry is a technology that allows 3D measurements of objects from photographs (Konecny, 1985; Heipke et al., 2007). Combining a variety of techniques, such as camera calibration, image matching, and bundle adjustment, photogrammetry is able to derive 3D information, e.g., point clouds and digital elevation models (DEMs), from a pair or set of overlapping images (Wu, 2017). In these techniques, image matching plays a vital role throughout the photogrammetric process. Dense and reliable image matching is a prerequisite for generating DEMs with favorable geometric accuracy and spatial resolution (Wu et al., 2011, 2012). However, this requires sufficient image texture and adequate similarities between the images. In the case of lunar and other planetary surfaces, images are often acquired under different illumination conditions. Significant differences in the illumination of remote sensing images can result in dissimilar image patterns of the same terrain surface, which in turn limits the use of these images and poses challenges to creating high - quality DEMs (Wu et al., 2012; Preusker et al., 2017).

Photoclinometry, or shape - from - shading (SfS), is another technology that reconstructs the 3D geometry (shape) based on the relationships between the image intensity (shading) and illumination and the viewing directions (Horn, 1977; Kirk, 1987). The intensity of each pixel contains information on the interactions between the incoming radiation (e.g., sunlight) and the surface. Photoclinometry then derives the surface gradients for each image pixel and estimates the underlying surface from its gradients. The technology can recover pixel - wise surface shapes from a single image (Grumpe et al., 2014; Wu et al., 2018a) or multiple images (Woodham, 1980; Liu et al., 2018), with the latter called photometric stereo.

However, although not favored in photogrammetry, varying illumination can provide the advantage of using photometric stereo for 3D surface reconstruction. Photometric stereo (Woodham, 1980) is an extension of photoclinometry in which images are simultaneously analyzed under different illumination conditions (e.g., images with sunlight coming from different directions) and the surface gradients are derived for each overlapping pixel. In contrast with the conventional image matching in photogrammetry, the overall reconstruction robustness of photometric stereo increases with the availability of inconsistent illuminations. Hence, this approach has the potential to be incorporated into image matching to handle severe differences in illumination and thus generate better matching results, which is impossible to achieve using the conventional image matching methods. In addition, photogrammetry can normally generate DEMs with resolutions only several times (e.g., 3 - 5 times, or greater) that of the original image resolution, depending on the textural condition of the image (Heipke et al., 2007; Wu et al., 2012).

Photoclinometry can then be used to refine the photogrammetric DEMs such that they have resolutions comparable to the image resolution (Grumpe et al., 2014; Lohse and Heipke, 2004; Wu et al., 2018a).

This paper presents a novel approach to integrating photogrammetry and photoclinometry for the illumination - invariant 3D mapping of the lunar surface, which is able to generate pixel - wise resolution DEMs from images with large illumination differences. First, a photoclinometry assisted image matching (PAM) algorithm that is invariant to changes in illumination is presented. PAM is able to generate surface gradients from the candidate matching patches using photometric stereo reconstruction. Given that correct matches usually generate a reasonable surface while incorrect matches do not, PAM finds matches by evaluating the likelihood that they will generate a reasonable surface. The output of PAM is disparity maps that can be directly used to generate DEMs through photogrammetric intersection based on the image interior and exterior orientation parameters. Second, the photogrammetric DEM from the previous step is further refined by aphotoclinometric technique (Wu et al., 2018a) such that a pixel - wise resolution of the DEM can be achieved. A new shadow constraint is also developed and incorporated into our existing photoclinometric framework (Wu et al., 2018a) for better performance.

The rest of the paper is organized as follows. Section 2 reviews the related literature. In Section 3, we provide detailed descriptions of the integrated photogrammetric and photoclinometric approach, PAM, and photoclinometric refinement. Section 4 presents experimental validations of the proposed approach using various Lunar Reconnaissance Orbiter Camera (LROC) Narrow Angle Camera (NAC) images covering the Chang'E - 4 landing site and in the candidate Chang'E - 5 landing region. Finally, concluding remarks are made in Section 5.

2 Related work

Image matching is a critical step in photogrammetry, and successful photogrammetric processing of remote sensing images requires accurate and sufficient matching points. The scale invariant feature transform (SIFT) algorithm (Lowe, 2004) is one of the most popular image matching methods used in the photogrammetry and computer vision communities. SIFT extracts the distinctive features of images and matches them according to descriptors constructed based on the intensities in the scale space. In images of the lunar and planetary surfaces, these descriptors usually correspond to topographic features such as boulders and crater rims. With SIFT and other feature - based algorithms, matching is only performed for the extracted feature points. For pixel - wise matching algorithms, semi - global matching (SGM) is a wellknown method used for dense image matching (Hirschmüller, 2008). SGM requires images with epipolar geometry such that only the translational parameter (i. e., disparity) needs to be solved by matching. A number of matching cost functions are commonly used in SGM, such as the Birchfield and Tomasi measure (Hirschmüller, 2008; Wu et al., 2018b) and the rank transform and census transform order - based methods (Banz et al., 2012). However, most intensity - based matching methods fail when the images have obvious illumination differences (Grumpe et al., 2014), because the changes in illumination lead to complex changes in the intensities of the pixels and the subsequent image patterns (Wu et al., 2018b). To overcome this problem, several methods have been proposed for matching images with illumination differences. For example, the phase - correlation and phase - congruency methods were shown to be illumination invariant (Foroosh et al., 2002). Wu et al., 2018b presented an illumination - invariant SIFT by suppressing the descriptor based on the distribution of the dominant orientation of the feature points. However, this approach can only be used for feature matching. Other studies have addressed this problem by estimating the underlying surface gradients of each image using photoclinometry (Grumpe and Wöhler, 2011; Grumpe et al., 2014; Liuand Wu, 2018). As the surface gradients are supposed to be illumination - invariant representations of the image content, matching can be performed without encountering apparent difficulties related to the differences in illumination. The proposed PAM approach belongs to this group of methods, in that the matching is performed based on the surface gradients. However, whereas most other methods rely on the photoclinometric processing of individual monocular images, PAM jointly analyzes both images to directly create robust matches. As photoclinometry and photometric stereo reconstruction are able to perform pixel - wise estimation, PAM is able to perform pixel - wise matching and hence provides a good complement to the popular SGM framework, which is also a pixel - wise matching algorithm.

In regard to photoclinometry, Horn (1977, 1990) and Kirk (1987) proposed early algorithms for implementing photoclinometric surface reconstruction. Using a reflectance model, photoclinometry estimates the gradient of a surface by the ratio of the energy reflection embedded in the intensity value of each pixel. Reflectance models describe how a surface reflects energy from an illumination source to the observer. Popular reflectance models include the Lambert model, the Lunar - Lambert model (McEwen, 1991), the Hapke model (Hapke, 1981, 1986, 2002, 2012), and variants of the aforementioned models (Oren and Nayar, 1994; Labarre et al., 2017). Recent developments have also used low - order polynomial functions with in - situ data to simplify the complexity of the reflectance model (Wohlfarth et al., 2018). Photoclinometry produces DEMs and other types of 3D models with resolutions comparable to those of the input images, which are apparently higher than the models generated using methods such as photogrammetry and laser altimetry. The approach was further developed to be applied in planetary remote sensing (Kirk et al., 2003a; O'Hara and Barnes, 2012). For example, photoclinometrically derived products such as slope maps and DEMs were used for landing site assessment for Mars missions such as the Mars Exploration Rover missions (Beyer et al. , 2003; Kirk et al. , 2003b), Mars Science Laboratory mission (Beyerand Kirk, 2012), and the InSight mission (Beyer, 2017) . Photoclinometry has also been used in lunar surface spectral analysis (Wöhler et al. , 2014), lunar crater analysis (Salamunić-car et al. , 2014), and mapping of geologic features (Wöhler et al. , 2017) . Although photoclinometry performs well on a local scale, it has poor accuracy on large scales. Hence, in recent years, methods have been developed to incorporate existing low - resolution 3D information (e. g. , low - resolution DEMs from photogrammetry or laser altimetry) into photoclinometry (Grumpe et al. , 2014; Wu et al. , 2018a) and satisfactory results have been achieved. However, most of these approaches do not explicitly consider shadows. The shadows hide all of the intensity information needed for topographic reconstruction, thus leading to defective DEMs. This is particularly important for airless planetary bodies such as the Moon, because there is no atmosphere to scatter the light in other directions and illuminate the shadowed regions. Our proposed approach takes shadows into account in the photoclinometry to facilitate robust reconstruction around the shadows.

For photometric stereo, Woodham (1980) first proposed and demonstrated the use of multiple images, each under a different illumination condition, to estimate the surface gradients, which is referred to as photometric stereo reconstruction. Multiple illuminations provide redundancy and hence allow more reliable determination of the surface. Photometric stereo reconstruction was applied to close - range and small - scale scene reconstruction (Ackermann et al. , 2012; Jung et al. , 2015) . Heipke (1992) introduced the use of photometric stereo in topographic mapping, which is referred to as multiple - image shape - fromshading (MI - SfS) and applied the technique to earth observation images (Piechullek and Heipke, 1996; Piechullek et al. , 1998) . This approach was further developed and investigated for planetary mapping and surface reconstruction (Lohse and Heipke, 2004; Wöhler, 2004; Lohse et al. , 2006; Liu and Wu, 2017; Liu et al. , 2018) . As photometric stereo requires corresponding pixels on all of the images, co - registration of the images is needed. As it is difficult to match and register images under different illumination conditions, image co - registration for photometric stereo usually requires different levels of manual work. In contrast, the proposed approach is able to automatically obtain pixel - synchronous images from the photogrammetric processing, which facilitates automatic photometric stereo reconstruction.

3　The integrated photogrammetric and photoclinometric approach

3.1　Overview of the approach

The workflow of the integrated photogrammetric and photoclinometric approach is illustrated in Fig. 1. The approach starts with a pairof lunar surface images with their corresponding exterior orientation (EO) parameters. The images may contain significant illumination differences. After epipolarly rectifying the images based on their EO parameters, the images are forwarded to the PAM algorithm to obtain the disparity maps. PAM is illumination - invariant and is able to match images with large illumination differences. A photogrammetric DEM is then generated using the disparity maps and the EO parameters of the images. Finally, the photogrammetric DEM is refined to pixel - resolution based on a photoclinometric method incorporating shadow constraint.

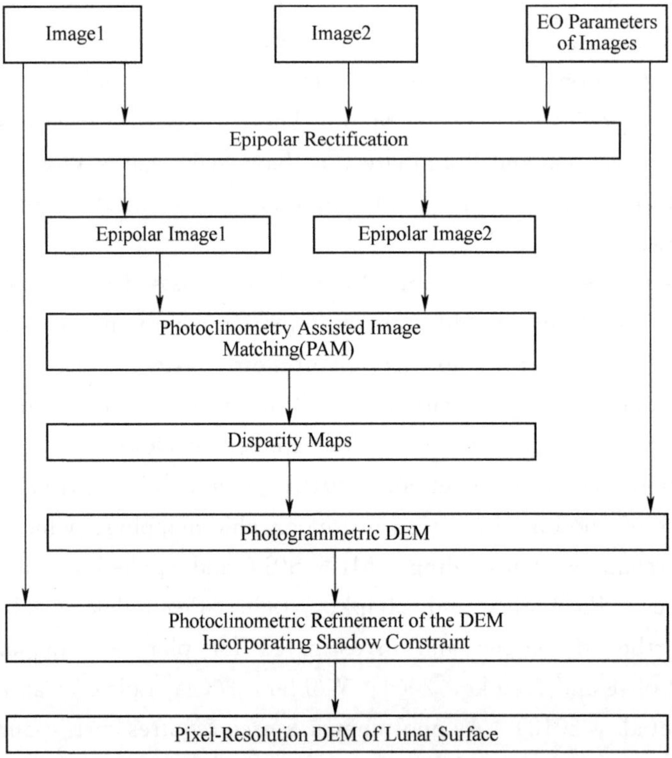

Fig. 1　Workflow of the integrated photogrammetric and photoclinometric approach

3.2　Photoclinometry assisted image matching (PAM) invariant to illumination changes

3.2.1　Photometric stereo for image matching

Fig. 2 illustrates the photometric relationships for image acquisition on a surface. Assuming that a single light source from a certain direction illuminates the surface, the incidence angle (i) is formed between the vector pointing to the light source and the normal vector of the surface. The surface absorbs some of the light energy and reflects/scatters the remaining light in different directions. When an observer (e.g., a camera) is viewing from a certain direction, the emission angle (e) is formed between the vector pointing to the observer and the normal vector of the surface. Then, the

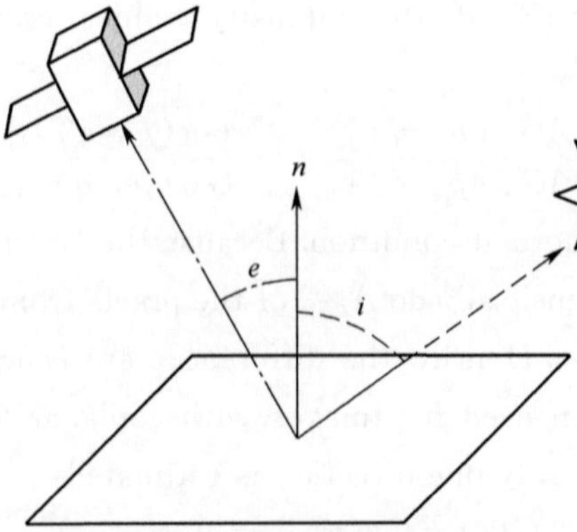

Fig. 2　Conceptual illustration of the relationships between the surface, the sun, and the observer (camera) for image acquisition

amount of light received by the observer depends on (i) the incidence angle and the emission angle; and (ii) the surface reflecting properties (i. e. , the intrinsic albedo) .

Mathematically, the photometric relationships can be modeled as:

$$I = AG(p, q) \tag{1}$$

where the image intensity I is the product of the surface reflecting properties (i. e. , albedo A) and the reflectance $G(p, q)$ generated from the surface $Z(x, y)$, and p and q denote the normal vector of the surface along the x - direction and y - direction, respectively. The normal vector of the surface is the negative of the surface gradients $\left(\text{i. e. , } p = \frac{\partial Z}{\partial x};\ q = -\frac{\partial Z}{\partial y}\right)$. The function $G(p, q)$ is also known as the reflectance model, which geometrically describes how a surface reflects incoming light from a source to an observer. In this paper, the Lunar - lambert model (McEwen, 1991) is used as follows:

$$G(p, q) = (1-\lambda)\mu_0(p, q) + 2\lambda \frac{\mu_0(p, q)}{\mu_0(p, q) + \mu(p, q)} \tag{2}$$

where μ_0 is the cosine of the incidence angle (i), μ is the cosine of the emission angle (e), and λ is the phase function which describes how the surface particles reflect light depending on the phase angle (i. e. , the angle between the vector pointing to the light source and the vector pointing to the observer) . Common models for the phase function in lunar and planetary photometry include the polynomial function (Lohse et al. , 2006) and the exponential function (Gaskell, 2008) .

If two images have been acquired of the same surface, they may or may not have consistent directions of illumination. It is quite common for images of lunar and planetary surfaces to have different illumination directions. Although matching images with different directions of illumination is difficult and not favored in photogrammetry, it can be achieved through photometric stereo as follows. Given the pixel intensity value of a point (x, y) on a base image ($I_B|_{xy}$) and the value of its corresponding pixel on the matching image (I_M), the ratio of the intensity values creates an albedo - free observation as follows:

$$\left.\frac{I_B}{I_M}\right|_{x,y}=\left.\frac{AG_B(p, q)}{AG_M(p, q)}\right|_{x,y}=\left.\frac{G_B(p, q)}{G_M(p, q)}\right|_{x,y} \tag{3}$$

This is referred to as the quotient condition. Because the two pixels correspond to the same physical location, the intrinsic albedo $A_{x,y}$ of the pixels from the different images in this location should be the same. Hence, the differences are canceled out by division in Eq. (3) . This method has been used for images with coplanar light sources (Wöhler, 2004) where the gradient along only one direction is estimated:

$$G(s)=G(sp_u, sq_u) \tag{4}$$

where s refers to the gradient along a predefined direction, and $\begin{bmatrix} p_u \\ q_u \end{bmatrix}$ is a unit vector denoting the predefined direction. Liu et al. (2018) used this approach to estimate the gradient along a predefined direction using the following quotient condition:

$$\left.\frac{G_B(s)}{G_M(s)}\right|_{x,y}=\left.\frac{I_B}{I_M}\right|_{x,y} \tag{5}$$

where G_B and G_M are the reflectance functions (Eq. (4)) under the illumination and viewing conditions of the base image and matching image, respectively. The predefined direction is named the principal direction (Liu et al., 2018) and is determined according to Fig. 3. L1 and L2 are horizontal (azimuthal) vectors pointing toward each of the illumination sources, and v is a horizontal vector bisecting the angle L1 - O - L2 such that $\angle\alpha 1 = \angle\alpha 2$. The principal direction u is determined by rotating vector v 90° clockwise. Our previous study (Liu et al., 2018) indicated that the ratio of reflectance (Eq. (5)) can be best represented along the principal direction because G_B and G_M behave in reverse manner as the gradient along this direction changes. Specifically, when the surface is inclined toward L1 (i. e., brighter G_B), its reflectance with respect to L2 will be lower (i. e., darker G_M), and vice versa, resulting in a wider spectrum of the ratio of reflectance as the gradient changes. However, for the gradient along the perpendicular of the principal direction (i. e., v), G_B and G_M change similarly as the gradient along v changes (i. e., both brighter or both darker) . Thus, in this case, the ratio of reflectance will be close to constant as the gradient changes. As a result, the photoclinometric processing of both images along the principal direction will be more robust and less biased to any side, which will facilitate the subsequent photoclinometry assisted image matching.

A gradient map can be computed by solving for s in Eq. (5) at each pixel (x, y). Because the map presents the gradient along the principal direction (i. e., u in Fig. 3), it provides more topographic information along this direction and less information along its perpendicular (i. e., v in Fig. 3) . Hence, the gradient changes more frequently along direction u and less frequently along direction v. This is similar to the situation in which an image shows more changes in intensity along the direction of illumination and less along its perpendicular.

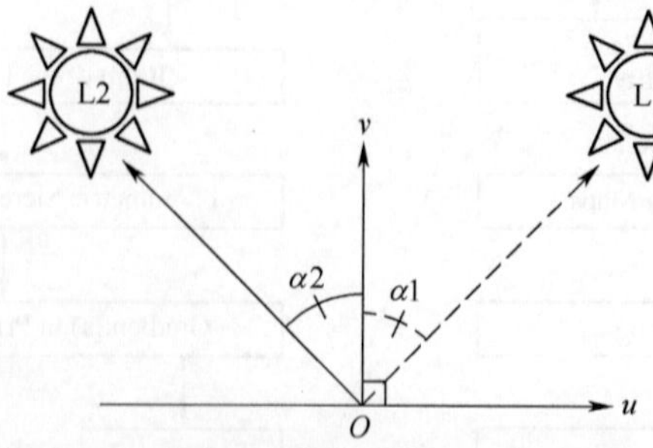

Fig. 3　Determination of the principal direction in photometric stereo. All vectors are on a 2D $x - y$ plane

Based on the aforementioned principles, we develop a PAM algorithm that is able to handle images with large differences in illumination. The concept of photometric stereo is used to calculate the matching score and evaluate the similarity of the candidate matches. In the process, the PAM algorithm finds matches by evaluating the underlying terrain instead of the direct image intensities. As a result, the differences in image intensity caused by the illumination differences do not significantly influence the matching.

The workflow of the PAM algorithm is illustrated in Fig. 4. Here, the design of PAM follows the general architecture of SGM (Hirschmüller, 2008), in which the base image (I_b) and matching image (I_m) are epipolarly rectified according to their EO parameters. The objective is to find the horizontal disparity d_x and vertical disparity dy such that a pixel on the matching image I_m $(x + dx, y + dy)$ correctly corresponds to the pixel on the base image I_b (x, y). Using disparities in two directions (x and y) ensures the matching is robust when the epipolar images are not perfectly aligned. For each pixel within a predefined disparity range (dx, dy), PAM extracts a local patch centered at I_b (x, y) and I_m $(x + dx, y + dy)$. Using photometric stereo, PAM then analyzes the underlying terrain produced by the pair of image patches and generates a matching score representing the likelihood of the resulting terrain. The algorithm chooses the one with the highest matching score as the best match. PAM also considers the local smoothness and conducts multiple - scale evaluations when computing the matching score. Matches with significant changes in disparity are treated as mismatches and are filtered out, and the remaining matches are forwarded to the sub - pixel refinement. As the PAM algorithm incorporates the principle of photometric stereo, it requires the illumination directions with respect to the images to be known. This information can be obtained from the image metadata or derived by manual observation.

3. 2. 2　Computation of the matching score

For any local patch on the base image centered at pixel (x, y), a patch of the same dimension is extracted from the matching image with its center pixel translated by the disparities (dx, dy) under evaluation. A ratio of the two local patches is computed by

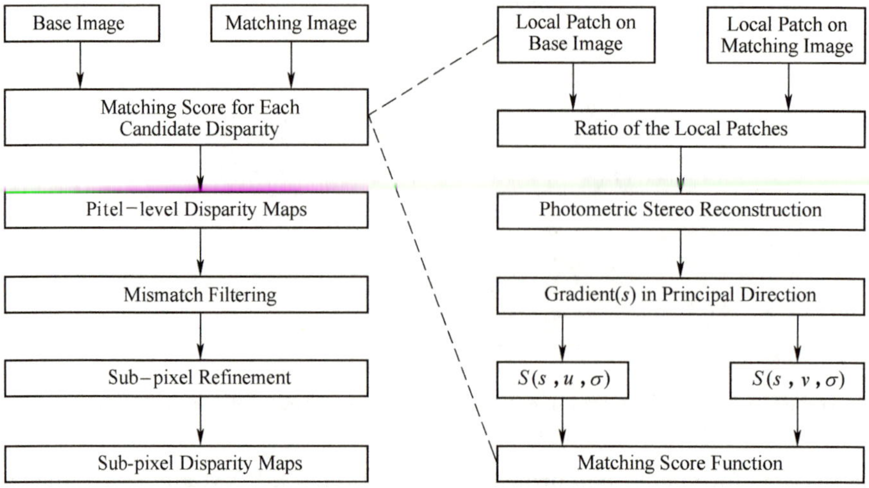

Fig. 4　The workflow of photoclinometry assisted image matching (PAM)

dividing the patch on the base image by the one on the matching image. Using the ratio minimizes the effects of varying albedo in the photometric stereo analysis, as can be seen in Eq. (5). The photometric stereo technique is then used to determine the gradient map of the pair along the principal direction. The rationale behind PAM is that when the two image patches are matched correctly, they produce a highly reasonable gradient map due to the removal of the albedo variations by Eq. (5) and the accurate alignment of the corresponding pixels. However, when the two patches are matched incorrectly, they produce a biased gradient map due to misalignment. Biased gradient maps usually show similar roughness along multiple directions and have similar patterns of input images (influenced by the illumination), while reasonable gradient maps usually show high roughness only along the principal direction and low roughness along its perpendicular. Hence, the roughness of the gradient map can be evaluated along two orthogonal directions. The two directions are then combined using a mathematical function representing the matching score.

When a patch on the base image is matched with a candidate patch on the matching image, a gradient map can be generated using the aforementioned photometric stereo analysis. If the two patches are correctly matched, the gradient map is likely to have more frequent changes along the principal direction and less along its perpendicular. Otherwise, the map will have frequent changes along different directions due to "leakage" of the topographic and albedo information as a result of misalignment. PAM uses the gradient roughness to measure the changes. A reasonable gradient map is evaluated as having high gradient roughness along the principal direction and low gradient roughness orthogonal to the principal direction. The roughness of the gradient (s) map along direction t is evaluated using the following function:

$$S(s, t, \sigma)_{x, y, dx, dy} = g(s_{tt}^2, \sigma)_{x, y, dx, dy} \tag{6}$$

where $S(s, t, \sigma)_{x,y,dx,dy}$ is the mean squared deviation of s_{tt}^2, the squared second derivative of s along the direction t, within a local region centered at pixel (x, y) with disparity (dx, dy). $g(*, \sigma)$ is a Gaussian function with the parameter σ controlling the size of the local region in the multiple - scale analysis and the relative contribution of each component within the region. s_{tt}^2 can be approximated by finite differences:

$$s_{tt}^2 = (s_{t-1} - 2s_t - s_{t+1})^2 \tag{7}$$

A correct match is then defined as a high $S(s, u, \sigma)$ and a low (s, v, σ), of which u is the principal direction and v is its perpendicular. To achieve this, $S(s, u, \sigma)$ and $S(s, v, \sigma)$ are connected using a score function:

$$M(x, y, d_x, d_y, \sigma) = 1 - e^{-w_x S(s, u, \sigma)} + e^{-w_y S(s, v, \sigma)}|_{x, y, dx, dy} \tag{8}$$

In the above function, the first component $1-e^{-w_x S(s,u,\sigma)}$ increases when $S(s, u, \sigma)$ increases, while the second component $e^{-w_y S(s,v,\sigma)}$ penalizes a high $S(s, v, \sigma)$. w_x and w_y are weighting factors used to control the relative contribution of each component. A larger w_x and w_y mean the score is more easily affected by a small change in $S(s, u, \sigma)$ and $S(s, v, \sigma)$, respectively, and vice versa. M ranges from 0 to 2, with the best match approaching 2.

To ensure robustness of the matching score, PAM takes the multiplescale evaluation into consideration by aggregating the matching scores computed from multiple image scales:

$$M_{final}(x, y, d_x, d_y) = \sum_{k \in K} M_k(x, y, d_x, d_y, \sigma_k) \tag{9}$$

where $M_k(x, y, d_x, d_y, \sigma_k)$ is the matching score at a scale denoted by σ_k within a predefined number K of scales. For computational consideration, this can be approximated by down - sampling the image by a factor (2 in this research) and computing the matching score for each downsampled image. For each pixel (x, y), the disparities (dx, dy) with the highest matching score $M_{final}(x, y, d_x, d_y)$ are chosen as the best match.

3.2.3　Filtering of mismatches

Mismatches are unavoidable in most cases. In this paper, mismatches are defined as pixels with a disparity significantly different from their adjacent neighbors:

$$D_x(x, y) = | d_x(x, y) - d_x \in n | \geqslant \varepsilon$$
$$D_y(x, y) = | d_y(x, y) - d_y \in n | \geqslant \varepsilon \tag{10}$$

where n is the vicinity of any pixel (x, y), and is defined as 1 pixel adjacent to the center pixel. ε is a predefined threshold default at 1. Eq. (10) is implemented by using morphological operators on the disparity maps:

$$D_x(x, y) = | morph_{dil}(d_x(x, y), n) - morph_{ero}(d_x(x, y), n) |$$
$$D_y(x, y) = | morph_{dil}(d_y(x, y), n) - morph_{ero}(dy(x, y), n) | \tag{11}$$

where $morph_{dil}(*, n)$ is a dilation operator whereby the value of a pixel is replaced by the highest value within a neighborhood defined by n, and $morph_{ero}(*, n)$

is an erosion operator whereby the value of a pixel is replaced by the smallest value within the neighborhood. The neighborhood is defaulted as a 3x3 kernel, which means that only the disparities of the adjacent pixels are compared. A pixel is labelled as a mismatch when $D_x(x, y) \geqslant \varepsilon$ or $D_y(x, y) \geqslant \varepsilon$. Pixels enclosed by mismatches are also labelled as mismatches, which can be achieved by using an algorithm such as flood fill or watershed analysis. After filtering the mismatches, the sub-pixel values of the matches are estimated by fitting a quadratic curve over the matching score of the neighboring disparities and finding the minima (Hirschmüller, 2008).

3.3 Photoclinometric refinement of photogrammetric DEM incorporating shadow constraint

3.3.1 Surface gradients from image intensities constrained by photogrammetric DEM

Using the above matching results from PAM, a photogrammetric DEM can be generated based on the EO parameters. Single-image SfS is then used to recover details that exist on the image but not in the photogrammetric DEM. Only one image of the stereo pair is needed for this process. Hence, the image with better illumination conditions, such as fewer shadows and adequate topographic shading, is chosen. Similar to Grumpe et al., 2014; Wu et al., 2018a, the SfS algorithm introduces a low-resolution DEM (i.e., the photogrammetric DEM) as a constraint and generates a high-resolution DEM with comparable resolution and details to the image. The image used in the SfS is co-registered to the photogrammetric DEM by ortho-rectification based on the image EO parameters. In Wu et al., 2018a, the reflectance of a pixel is estimated from its neighboring pixels by assuming locally constant albedo, yielding an albedo-free equation:

$$\left.\frac{A_{x,y}G_{x,y}}{A_jG_j} = \frac{I_{x,y}}{I_j}\right|_{j\in N}$$

$$G_{x,y} = \frac{I_{x,y}}{n}\sum_{j\in N_{x,y}}\frac{G_j}{I_j} \tag{12}$$

where each I_j is an image pixel within a predefined local neighborhood $N_{x,y}$ of the center pixel $I_{x,y}$. It is assumed that the albedo at the center $A_{x,y}$ equals the A_j of its vicinity, and hence that the ratio of the reflectance value equals that of the image intensity. The estimated reflectance $G_{x,y}$ is then computed as the mean of all of the valid candidate reflectance values $\frac{I_{x,y}}{I_j}G_j$ within $N_{x,y}$. The shadowed regions are not considered as valid pixels in this process. The reflectance value of each pixel is initialized from the initial DEM, with the values being derived from the photogrammetric DEM in this case. The normal vectors along the x-direction (p) and y-direction (q) are then obtained by minimizing:

$$E_{sfs} = G(p_0, q_0)|_{x,y} + \frac{\partial G}{\partial p}\Delta p + \frac{\partial G}{\partial q}\Delta q - G_{x,y} \tag{13}$$

where p_0 and q_0 are the surface gradients of the DEM at the current iteration and Δp, Δq are the update values for the surface gradients. The photogrammetric DEM serves as the initial DEM for the initial conditions and constraints during the refinement. Wu et al., 2018a incorporate the initial DEM constraint by a cost function:

$$F(p, q) = (p - p_0)^2 + (q - q_0)^2 + w_{ini}[(g(p, \sigma) - p_{ini})^2 + (g(q, \sigma) - q_{ini})^2] \tag{14}$$

The first two components, which are referred to as the integrability constraints (Horn, 1990), encourage the surface normal vectors to approach the closest integrable solutions p_0 and q_0 of the current iteration. The last two components, which are referred to as the initial DEM constraints, encourage the low-resolution versions of the normal vectors (i. e., $g(p, \sigma)$ and $g(q, \sigma)$) to approach those of the initial DEM (i. e., p_{ini} and q_{ini}). w_{ini} controls the contribution of the initial DEM constraint. The reflectance constraint E_{sfs} is treated as an absolute constraint, which separates the more important E_{sfs} from the other geometrical constraints $F(p, q)$ and ensures its effect. p and q are then updated using the Lagrange multiplier (Wu et al., 2018a) by minimizing $F(p, q)$ subject to ($E_{sfs}=0$).

Table 1　Comparative statistics of the DEMs of the Chang'E-4 landing site

	Photogrammetric DEM	Integrated DEM
RMSE (m)	4.32	3.47
Maximum absolute deviation (m)	38.49	19.84
Absolute deviation at 99.5 percentile (m)	21.66	11.17

3.3.2　Incorporating the shadow constraint into photoclinometry

In shadowed regions, it is not appropriate to use photoclinometry because there is no shading information. However, shadows also provide information about the surface normal vectors under the shadows. The magnitude limit of the normal vectors within the shadows can be obtained by:

$$s_{shw} = -\frac{\cos i}{\sin i} = \frac{1}{\tan i} \tag{15}$$

where i is the incidence angle and s_{shw} is the signed magnitude of the vector of a shadow along the illumination direction, which provides an upper limit of the normal vectors of a shadowed surface. When the normal vectors p and q are projected in the direction of the illumination, the magnitude cannot be higher than s_{shw} because, in that case, the surface would be illuminated and no longer shadowed. A straightforward way to achieve this is to force the normal vectors of each shadowed pixel to comply to s_{shw} when they are evaluated along the direction of the illumination. This constraint can be formulated as:

$$E_{shw} = \max\left(L^T \begin{bmatrix} p \\ q \end{bmatrix} - s_{shw}, 0\right) \tag{16}$$

where $L=\begin{bmatrix} p_L \\ q_L \end{bmatrix}$ is the unit vector pointing toward the illumination source, and $L^T \begin{bmatrix} p \\ q \end{bmatrix}$ projects the normal vector (p, q) to L. For a shadowed pixel, the projected magnitude is considered to be equal to or less than ε_{shw} and, hence, E_{shw} equals zero. Similar to E_{sfs}, E_{shw} is treated as an absolute constraint for the shadowed pixels. As E_{sfs} is switched off for the shadowed pixels, the optimization is to minimize F (p, q) subject to ($E_{shw}=0$). To summarize, each pixel contains two solutions of normal vectors, one for the illuminated pixels and the other for the shadowed pixels:

$$\begin{bmatrix} p_{sfs} \\ q_{sfs} \end{bmatrix} = \arg min_{p,\,q} F(p,\ q)\ s.t.\ E_{sfs}=0$$

$$\begin{bmatrix} p_{shw} \\ q_{shw} \end{bmatrix} = \arg min_{p,\,q} F(p,\ q)\ s.t.\ E_{shw}=0 \tag{17}$$

where $\begin{bmatrix} p_{sfs} \\ q_{sfs} \end{bmatrix}$ is the solution derived based on the reflectance constraint E_{sfs}, and $\begin{bmatrix} p_{shw} \\ q_{shw} \end{bmatrix}$ is the solution derived based on the shadow constraint E_{shw}. We assume that a shadow map of the image is available that contains the weights of the shadows for each pixel, which range from 0 to 1. The weights represent the likelihood of a pixel being shadowed based on certain shadow detection algorithm, or the fraction of a pixel that is shadowed. The final estimated normal vector is then a linear combination of $\begin{bmatrix} p_{sfs} \\ q_{sfs} \end{bmatrix}$ and $\begin{bmatrix} p_{shw} \\ q_{shw} \end{bmatrix}$ based on the weight values of the shadow map:

$$\begin{bmatrix} p_{final} \\ q_{final} \end{bmatrix} = (1-w_{shw}) \begin{bmatrix} p_{sfs} \\ q_{sfs} \end{bmatrix} + w_{shw} \begin{bmatrix} p_{shw} \\ q_{shw} \end{bmatrix} \tag{18}$$

where w_{shw} is the weight value of each pixel on the shadow map and $\begin{bmatrix} p_{final} \\ q_{final} \end{bmatrix}$ is the final estimated normal vector of the surface. In cases where the shadow map is a mask that has only ones for all of the shadowed pixels and zeros otherwise, $\begin{bmatrix} p_{final} \\ q_{final} \end{bmatrix}$ then solely depends on $\begin{bmatrix} p_{shw} \\ q_{shw} \end{bmatrix}$ for the shadowed regions and $\begin{bmatrix} p_{sfs} \\ q_{sfs} \end{bmatrix}$ otherwise.

3.3.3 Refining photogrammetric DEMs from photoclinometrically estimated gradients

Following the algorithm developed by Wu et al., 2018a, the elevation of each individual node on the DEM is optimized based on the photoclinometrically estimated surface gradients computed from the previous steps. The surface gradients of the DEM are obtained by finite differences:

$$p=-\frac{\partial Z}{\partial x}=\frac{1}{2h_x}\sum\begin{bmatrix}1 & -1\\1 & -1\end{bmatrix}\circ\begin{bmatrix}Z_{x,\ y+1} & Z_{x+1,\ y+1}\\Z_{x,\ y} & Z_{x+1,\ y}\end{bmatrix}$$

$$q=-\frac{\partial Z}{\partial y}=\frac{1}{2h_y}\sum\begin{bmatrix}-1 & -1\\1 & 1\end{bmatrix}\circ\begin{bmatrix}Z_{x,\ y+1} & Z_{x+1,\ y+1}\\Z_{x,\ y} & Z_{x+1,\ y}\end{bmatrix} \tag{19}$$

where h_x and h_y are the cell sizes of the DEM in the x - and y - direction, respectively, and [*]° [*] is an element - wise multiplication operator. As depicted in Fig. 5, the surface normal vectors of each cell $\begin{bmatrix}p_{DEM}\\q_{DEM}\end{bmatrix}_{i,j}$ are computed by the four neighboring grid nodes. Hence, the optimization of the center node $Z_{x,y}$ (the middle square in Fig. 5) is optimized in a least - squares solution according to the adjacent surface normal vectors estimated by photoclinometry. The optimization uses a relaxation strategy whereby only one DEM node is optimized and updated at a time. The refined DEM is then forwarded to the reflect ance estimation, as described in Section 3. 3. 1, and the process iterates until the changes per iteration are negligible. Creating a DEM from the estimated

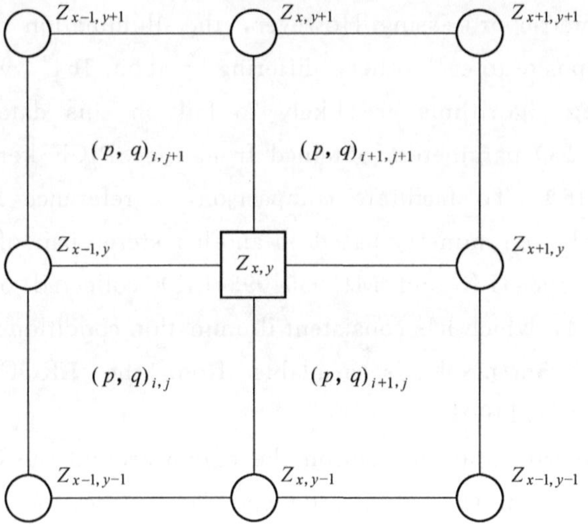

Fig. 5　The layout of a DEM node during height optimization.
The center node is optimized according to the surrounding surface gradients

normal vectors is important for the reconstruction process because it can strictly enforce integrability on the possibly non - integrable surface normal vector field and hence provide better initial estimates for the next iteration. A hierarchical approach is used to ensure that the reconstruction performance is robust. The image is first down - sampled to a predefined resolution (e. g. , the resolution of the initial DEM) and the refinement process is performed accordingly. The refined DEM is then up - sampled by a factor (e. g. , two) and the refinement process repeats until it reaches image resolution.

4 Experimental analysis

4.1 Experiment with LROC NAC images of the Chang'E－4 landing site

The LROC NAC images feature a high resolution, 0.5 – 2m/pixel, which is favorable for the detailed 3D mapping and analysis of the landing sites for lunar missions (Speyerer et al., 2016; Wu and Liu, 2017; Hu and Wu, 2018). Fig. 6 shows a pair of LROC NAC images covering the landing site of the Chinese Chang'E－4 mission, which successfully landed on the far side of the Moon on January 3, 2019 (Wu et al., 2019). To study the topographic conditions of the landing site and the nearby environment, it is critical to have high－resolution DEMs of the region to support the topographic analysis. The stereo pair of LROC NAC images shown in Fig. 6 are those only available before and immediately after the landing. As can be seen in Fig. 6, the difference in the slew angles (convergence angles) between the images is approximately 22°, which is ideal for photogrammetric processing. However, the illumination directions of the two images are almost opposite to each other, differing by about 164°, which implies that the conventional matching algorithms are likely to fail on this dataset. The images are associated with their EO parameters obtained from the SPICE kernels (Acton, 1996; Speyerer et al., 2016). To facilitate comparison, a reference DEM was generated independently using photogrammetry based on another stereo pair of LROC NAC images (image IDs: M1303619844LR and M1303640934LR) collected one month after the landing of Chang'E－4, which has consistent illumination conditions. The reference DEM has a resolution of 5m/pixel, is available from the LROC Archive (http://www.lroc.asu.edu/posts/1100).

We focus our experimental analysis on the region around the Chang'E－4 landing site. Fig. 7 (a) shows the image subset after epipolar rectification. The resolution of the epipolar image is resampled to 4m/pixel, which is considered optimal for photogrammetric processing, given the resolution of the original images. The resolution of 4m/pixel is also suitable for comparison with the reference DEM. The dimension of the image is 2000 × 1320 pixels. Due to the illumination differences, the conventional matching algorithms fail to produce any reasonable DEMs. The disparity maps are generated by matching the images using PAM and are shown in Fig. 7 (b). Although the disparity maps correspond to the general topography, the detailed topographic features (e.g., small craters) are missing because they are severely shadowed in both images; these regions are outlined in blue in Fig. 7 (b). In the left image, the western portions of the craters are shadowed; in the right image, the eastern portions are shadowed. Hence, when the images are analyzed during PAM matching, most of the craters are either shadowed by the left image or by the right image, and thus there is insufficient

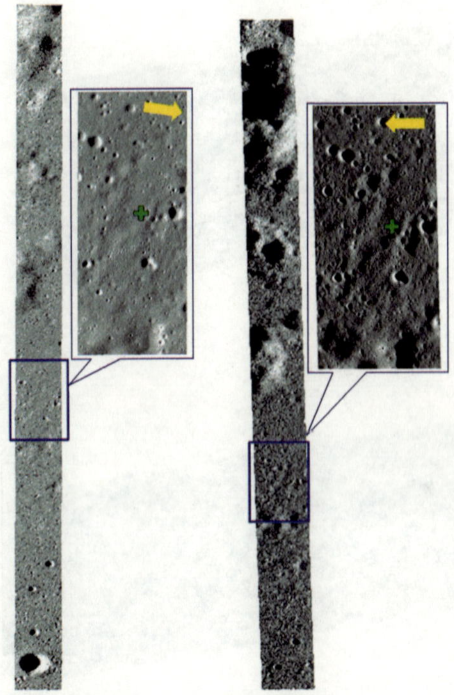

M134022629LC　　M178833263LC

Image ID	M134022629LE	M1145135367L
Slew angel (°)	0.0	−22.1
Illumination azimuth(°)	282.4	86.2
Incidence angle(°)	77.6	86.7
Emission angle(°)	1.7	23.0
Resolution (m/pixel)	1.3	1.4

Fig. 6　The LROC NAC image pair covering the Chang’E－4 landing site. Blue boxes show the enlarged views of the landing area, and green crosses indicate the landing site on each of the images. Yellow arrows indicate the illumination direction of each image. (For interpretation of the references to colour in this figure legend, the reader is referred to the web version of this article.)

information inside the craters for matching. In general, the image pixels are required to be illuminated in both images for the best performance of PAM. In situations like this, PAM matches the portions of the craters by their illuminated surroundings such as the crater rim, and hence the disparities of the inner parts are replaced by those of the surroundings, leading to a smooth surface. If the craters are shaded instead of shadowed, there will be information within the craters, and their disparities can be more accurately determined by PAM. For the regions not shadowed, the disparity maps successfully retrieve their shapes.

The 3D view of the photogrammetric DEM from PAM (Fig. 8b) closely corresponds with the reference DEM (Fig. 8a) in most of the regions except for the shadowed regions. This problem can be overcome by choosing the image with the least shadows from the image pair and performing photoclinometric refinement. Fig. 8 (d) shows the orthoimage generated using the photogrammetric DEM with the least shadows. The ortho－image (1.5m/pixel) and the photogrammetric DEM (4m/pixel) are used as inputs in the photoclinometric refinement, and an integrated DEM (1.5m/pixel) is generated using the method described in Section 3.3, as shown in Fig. 8 (c). Since the reference DEM

(a) The base image (left) and the matching image (right)

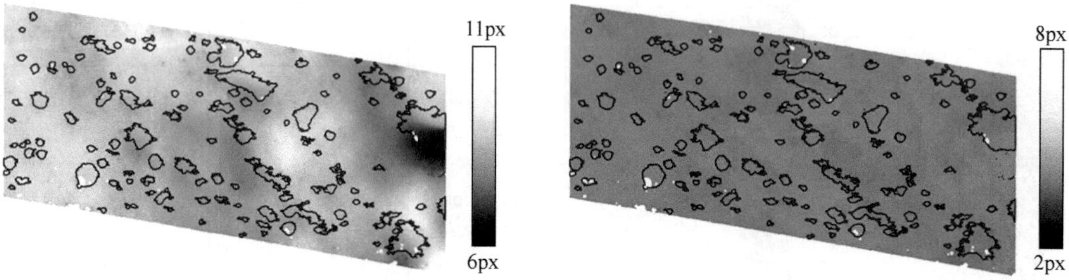

(b) The x-disparity (left) and y-disparity (right) map.

Fig. 7　The epipolar images used for matching and the disparity maps generated by PAM for the Chang'E - 4 landing site. Poor matches are outlined in blue. (For interpretation of the references to colour in this figure legend, the reader is referred to the web version of this article.)

(Fig. 8a) and the photo grammetric DEM (Fig. 8b) are produced independently using different sets of stereo images, small low - frequency deviations (about 1 - 3 m) in elevation, with respect to the reference DEM, can be noted on the photogrammetric DEM (Fig. 8b) and the integrated DEM (Fig. 8c) . The 3D views of the DEMs (Fig. 8) and their corresponding shaded relief (Fig. 9) indicate that the integrated DEM (Fig. 8c and 9b) recovers the missing details in the photogrammetric DEM and the reference DEM.

The topography of the Chang'E - 4 landing site derived from different DEMs is also compared. The integrated DEM provides the necessary information about the topography around the landing site, while no useful information can be derived from the other DEMs. Two profiles are derived from the DEMs and presented in Fig. 10. The first profile reveals two of the crater triplets surrounding the Chang'E - 4 landing site. It is apparent that only the integrated DEM is able to retrieve the shape of these small craters. The relatively larger craters in Fig. 10 (a) and (b) show that incorporating the shadow geometry allows precise reconstruction of the shadowed terrain where photogrammetric DEM fails to do so. There are small horizontal offsets in the shadowed regions, which likely stem from possible inaccuracies in the shadow detection.

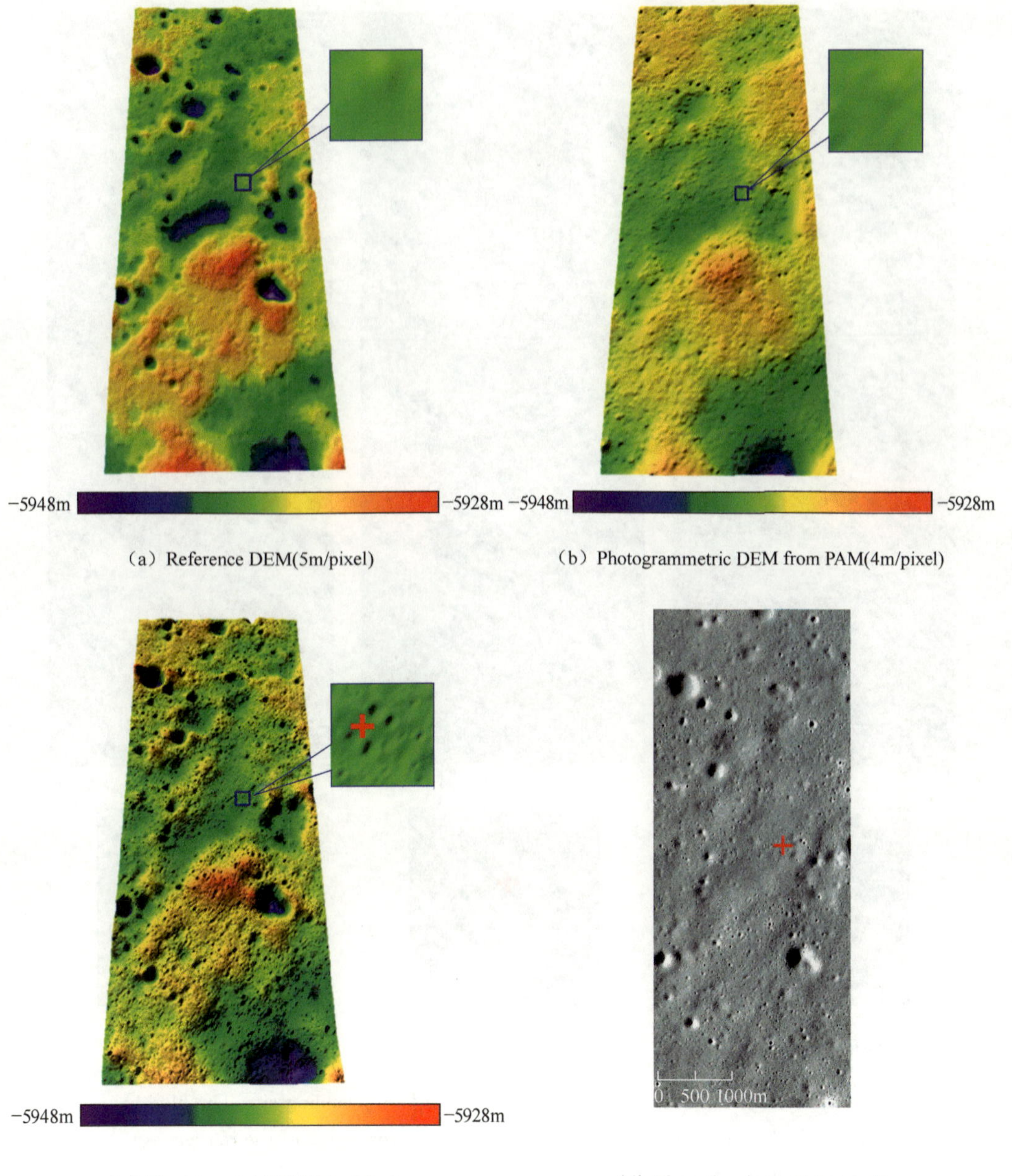

(a) Reference DEM(5m/pixel)

(b) Photogrammetric DEM from PAM(4m/pixel)

(c) The integrated DEM(1.5m/pixel)

(d) The ortho-image(1.5m/pixel)

Fig. 8　3D views of the DEMs and the corresponding ortho - image covering the Chang'E - 4 landing site. The blue boxes show close up views of the landing site (the red cross). (For interpretation of the references to colour in this figure legend, the reader is referred to the web version of this article)

The performance of the DEMs generated using the proposed approach is quantitatively analyzed using the root mean square error (RMSE) and the maximum absolute differences in elevations with respect to the reference DEM. The absolute deviation at the 99.5 percentile is also provided to enable a more comprehensive analysis of the

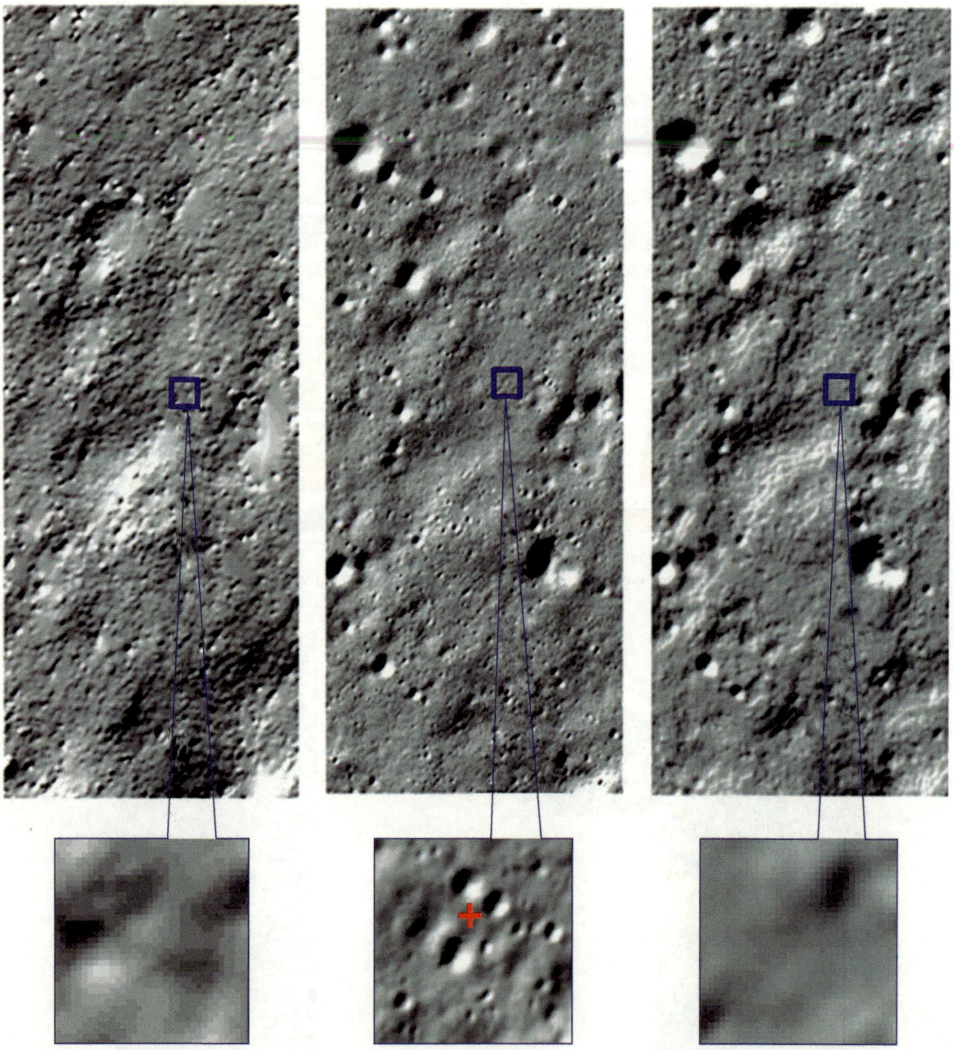

Fig. 9　The shaded relief of the DEMs covering the Chang'E－4 landing site. The blue boxes show close up views of the landing site (the red cross). (For interpretation of the references to colour in this figure legend, the reader is referred to the web version of this article.)

performance of the approach. The indicators are summarized in Table 1. The photogrammetric DEM has an RMSE of approximately 4.3m, an absolute maximum deviation of 38.5m, and an absolute deviation at the 99.5 percentile of 21.7m. The major contributor to the aforementioned error is that the image pair shadowed a significant amount of the topographic features, especially those corresponding to steeper slopes such as the crater walls. This problem is significantly improved in the integrated DEM, which has a slightly lower RMSE of 3.7m and a significantly reduced absolute maximum

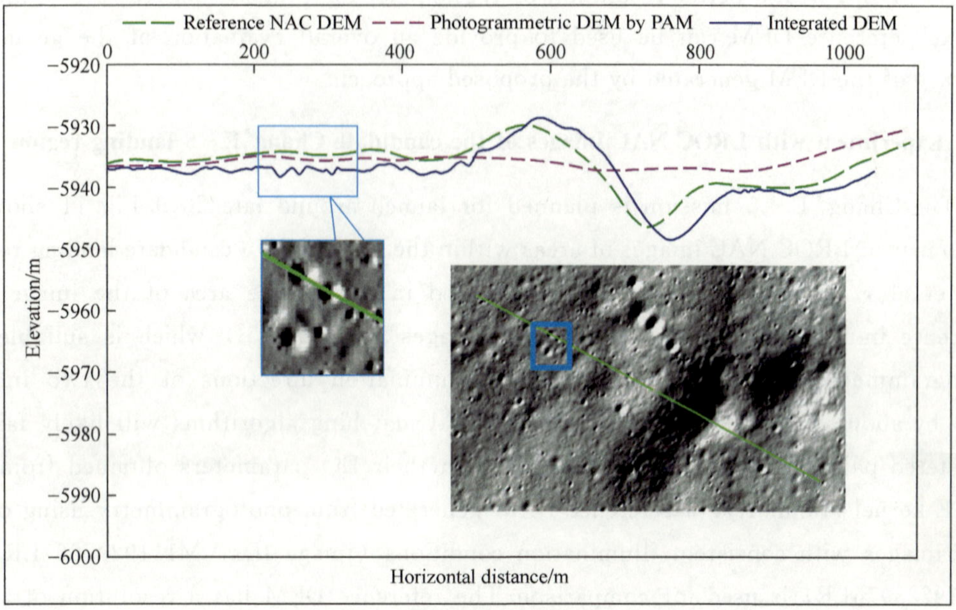

(a) Profile1

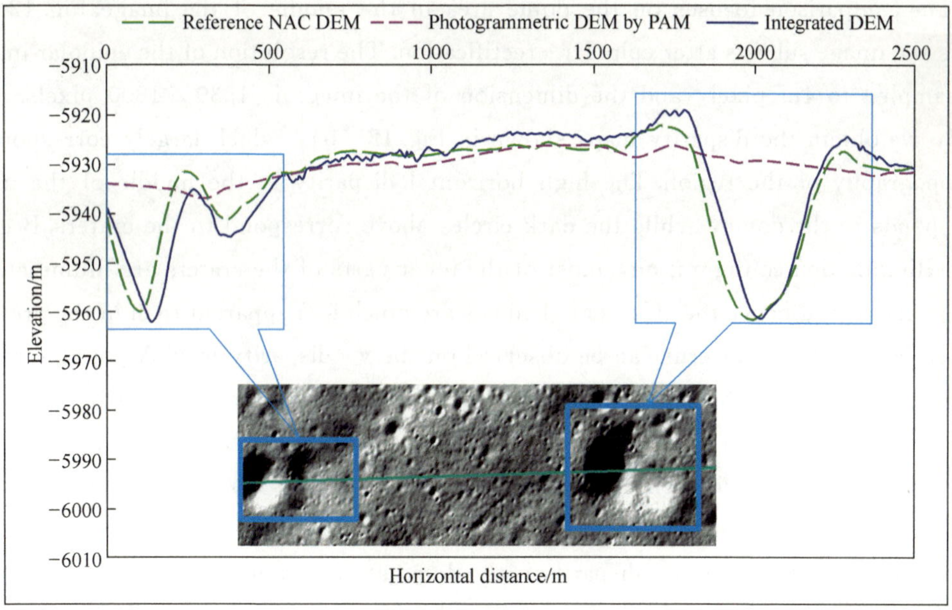

(b) Profile2

Fig. 10　Profile comparison of the DEMs of the Chang'E－4 landing site. The blue boxes indicate the topographic details revealed by the integrated DEM(For interpretation of the references to colour in this figure legend, the reader is referred to the web version of this article)

difference of 19. 8m, an approximately 19m reduction compared to the photo grammetric DEM. The 99. 5 percentile also drops to approximately 11. 2m, a 10m reduction with respect to the photogrammetric DEM. It should be noted that it is impossible to obtain a reference DEM with a better resolution of 1.5m/pixel for comparison. However, the 5-m/pixel reference DEM can be used to provide an overall evaluation of the geometric accuracy of the DEM generated by the proposed approach.

4.2 Experiment with LROC NAC Images of the candidate Chang'E-5 landing region

The Chang'E-5 mission is planned for launch around late 2019. Fig. 11 shows a stereo pair of LROC NAC images of areas within the Chang'E-5 candidate landing region (Wu et al., 2018c). There is a dome located in the middle area of the image. The difference in the slew angles between the images is about 15°, which is suitable for photogrammetric processing. However, the illumination directions of the two images differ by about 80°, indicating that conventional matching algorithms will likely fail on this stereo pair. The images are associated with their EO parameters obtained from the SPICE kernels. Similarly, a reference DEM generated from photogrammetry using other NAC images with consistent illumination conditions (image IDs: M1119207667LR and M1119228976LR) is used for comparison. The reference DEM has a resolution of 5 m/pixel.

The comparison focuses on the dome area in the middle of the image. Fig. 12 (a) shows the image subsets after epipolar-rectification. The resolution of the epipolar images is resampled to 4m/pixel, and the dimension of the image is 1330 × 1800 pixels. After PAM, we obtain the disparity maps shown in Fig. 12 (b), which largely correspond to the topography of the region. The high horizontal disparity in the middle of the image corresponds to the dome, while the dark circles above correspond to the craters. Because of the illumination configuration, most of the inner parts of the craters are illuminated in both images, and hence the effects of shadows are much less apparent than in the previous dataset. Subtle stripe patterns can be observed on the y-disparity map. A possible reason for their occurrence is that the proposed photometric stereo method (Eq. (4)) performs best when the horizontal illumination differences are close to 180°, whereas in cases with illumination horizontal differences significantly less than 180° (such as about 80° in this dataset), the performance declines and becomes less sensitive to subtle changes. Nevertheless, the x-disparity is the dominating parameter for 3D reconstruction in this dataset, and most of the topographic features are revealed by the x-disparities. Hence, with this dataset, the y-disparity has a subtle effect and may contain very slight discrepancies.

Fig. 13 shows the 3D views of the reference DEM (Fig. 13a), the photogrammetric DEM (4m/pixel) from PAM (Fig. 13b), the DEM (1.5m/pixel) from the integrated photogrammetric and photoclinometric approach (Fig. 13c), and the ortho-image based on the photogrammetric DEM and the image EO parameters (Fig. 13d). The

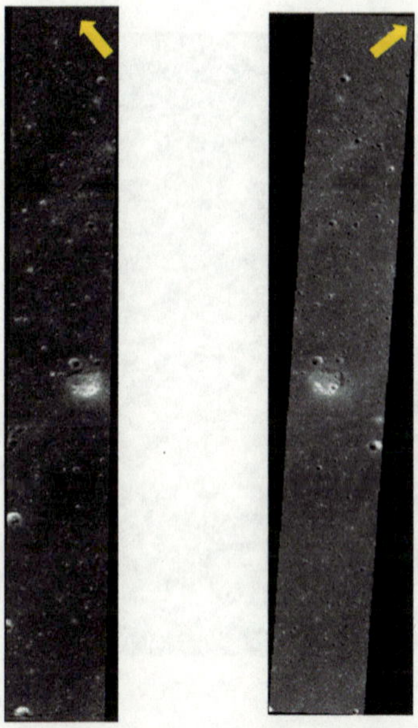

M1119207667RC　　M1145135367LC

Image ID	M1119207667R	M1145135367L
Slew angle (°)	14.8	0.0
Illumination azimuth(°)	154.4	233.0
Incidence angle(°)	45.7	56.0
Emission angle(°)	17.2	1.8
Resolution (m/pixel)	1.5	1.5

Fig. 11　The LROC NAC image pair within the Chang'E－5 candidate landing region. Yellow arrows indicate the illumination direction of each image. (For interpretation of the references to colour in this figure legend, the reader is referred to the web version of this article.)

conventional photogrammetric routines based on SGM fail to generate any reasonable DEMs for this dataset. In contrast, although the effective resolution is not as high as the reference DEM, photogrammetric processing with PAM successfully generates a plausible DEM. Moreover, the integrated DEM successfully recovers the details better than the other DEMs, as compared with the details shown in the ortho－image, which can also be observed from the shaded relief of the DEMs as shown in Fig. 14.

Two profiles are extracted from the DEMs and presented in Fig. 15. The first profile (Fig. 15a), which shows the larger crater near to the dome, demonstrates that the photogrammetric DEM is able to retrieve a precise topography while the integrated DEM is able to reconstruct the fine details. The profile of the dome (Fig. 15b) shows that the steep edges and artifacts from the photogrammetric DEM are corrected by photoclinometry. The profiles also show small craters and structures that are preserved on the integrated DEM but not on the photogrammetric DEM or the reference DEM.

The quantitative indicators of the performance of the proposed approach are summarized in Table 2. The RMSE of the photogrammetric DEM is about 5.8m with respect to the reference DEM, while the absolute maximum is about 54.5m and the absolute deviation at the 99.5 percentile is about 18.7 m. The pixels with large errors are usually located in the shadows and can be filtered out using operators such as a median

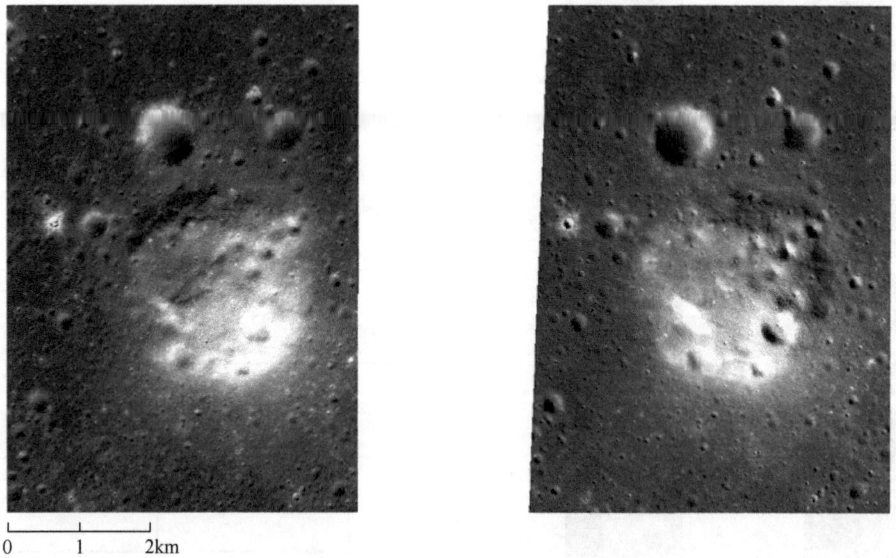

(a) The base image (left) and the matching image (right)

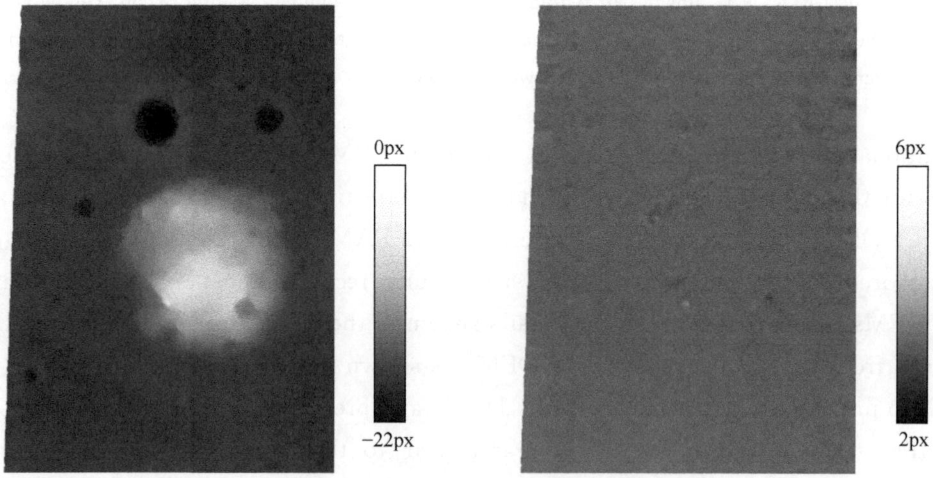

(b) The x-disparity (left) and y-disparity (right) maps

Fig. 12　The epipolar images used for matching and the disparity maps generated by PAM within the Chang'E－5 candidate landing region

filter. The integrated DEM has an RMSE of approximately 6. 6 m, with a maximum of 45 m and an absolute deviation at the 99. 5 percentile of 24. 8m. In general, the indicators are slightly higher than those of the photogrammetric DEM, except for the absolute maximum. In this case, there are slight gradual deviations in terms of very low－resolution surface gradients on the integrated DEM. This is possibly due to the spatially

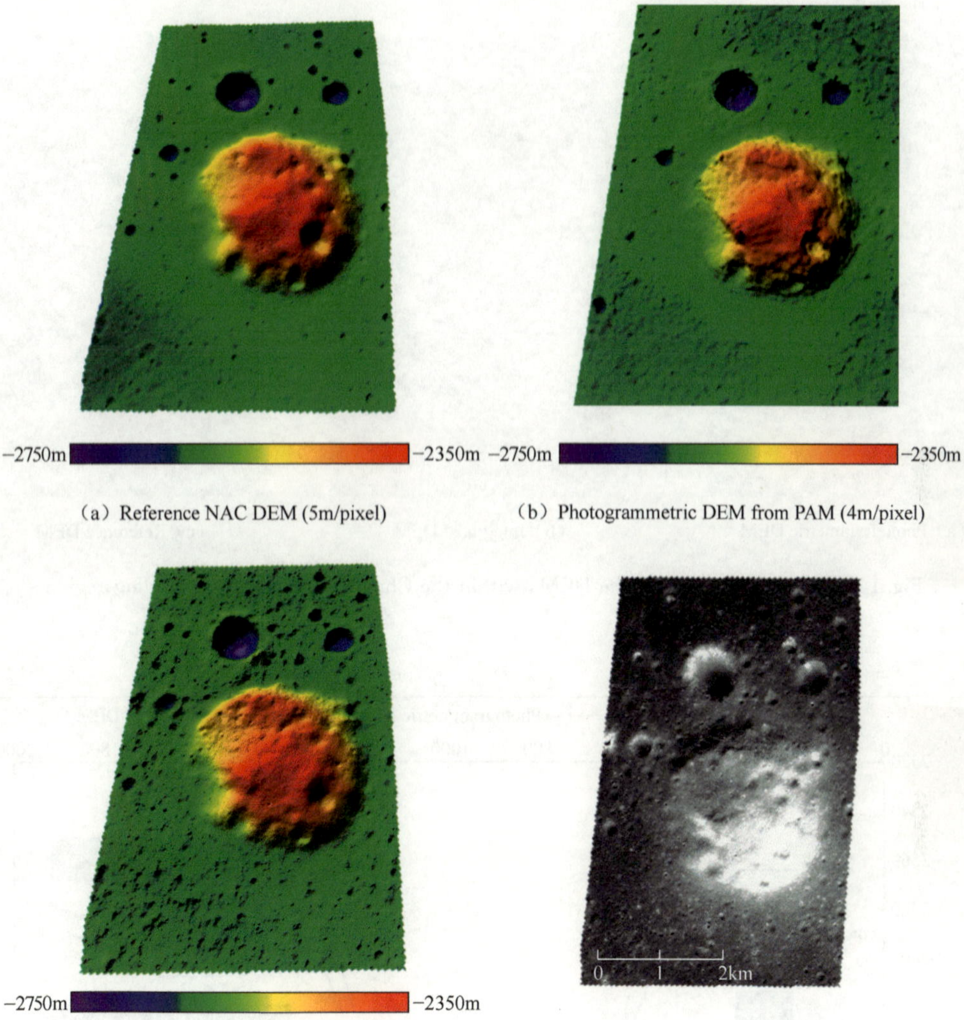

(a) Reference NAC DEM (5m/pixel)

(b) Photogrammetric DEM from PAM (4m/pixel)

(c) The integrated DEM (1.5m/pixel)

(d) The ortho-image (1.5m/pixel)

Fig. 13 3D views of the DEMs and the corresponding ortho - image within the Chang'E - 5 candidate landing region

varying albedo, especially across the lunar dome. This can be improved by tuning the parameters or incorporating photometric stereo reconstruction for a more robust estimation of the surface.

Table 2 Comparative statistics of the DEMs within the Chang'E - 5 candidate landing region

	Photogrammetric DEM	Integrated DEM
RMSE (m)	5.76	6.55
Maximum absolute deviation (m)	54.52	45.13
Absolute deviation at 99.5 percentile (m)	18.70	24.80

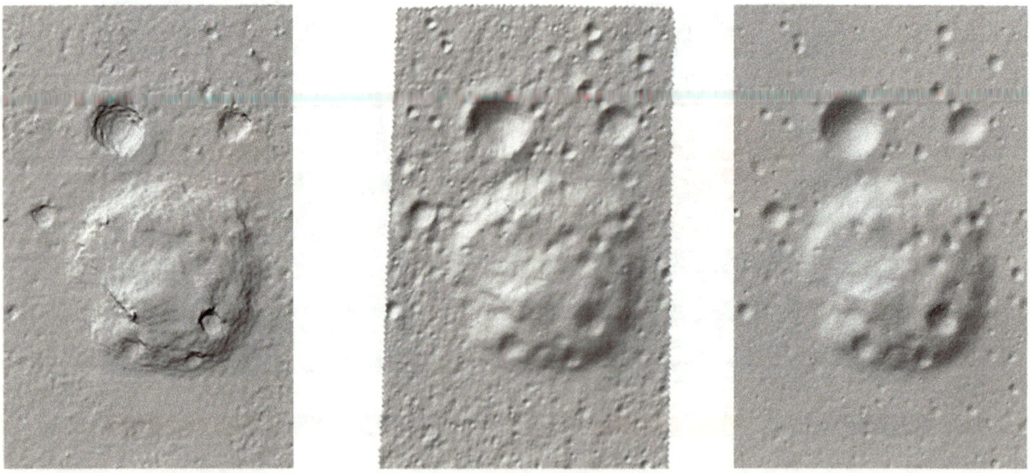

(a) Photogrammetric DEM　　(b) Integrated DEM　　(c) Reference DEM

Fig. 14　The shaded relief of the DEMs within the Chang'E－5 candidate landing region

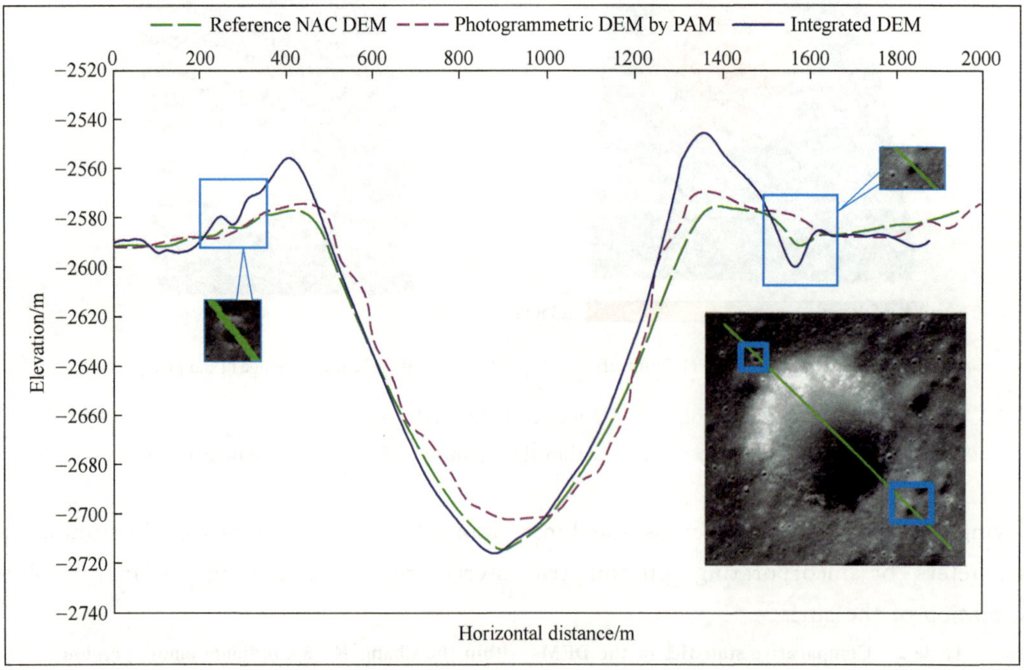

(a) Profile1

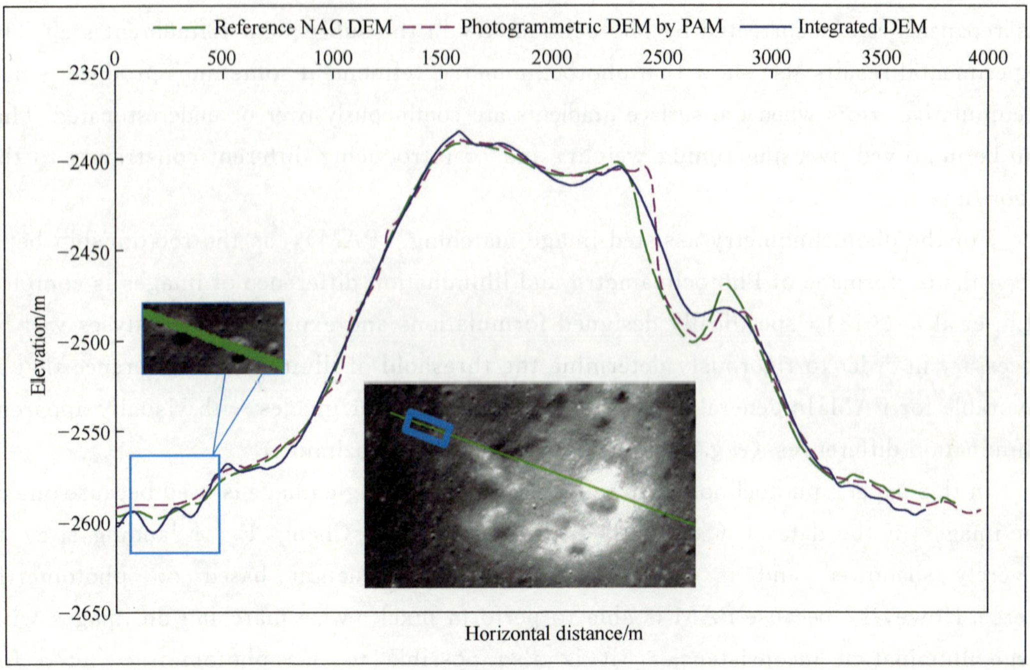

(b) Profile2

Fig. 15 Profile comparison of the DEMs within the Chang'E - 5 candidate landing site. The blue boxes indicate the small topographic details revealed by the integrated DEM

5 Conclusions and discussion

This paper presents an integrated photogrammetric and photoclinometric approach for illumination - invariant and pixel - resolution 3D mapping of the lunar surface. The approach fuses photoclinometry and photometric stereo techniques in the photogrammetric processing in a synergistic manner. Experimental analyses of the proposed framework using actual LROC NAC images with different illumination conditions show that the proposed approach outperforms the conventional approaches. The PAM algorithm used in the proposed approach is the key to handling images with complex illumination inconsistencies, such as orthogonal illumination and opposite illumination, and images with subtle textures. Based on the resulting photogrammetric DEM, photoclinometric refinement enables pixel - wise reconstruction of the terrain surface. Photoclinometry is able to recover the topographic details missing during matching and due to shadows. The RMSE of the generated DEMs ranges from 3 to 7m compared to the reference DEMs (i. e. , DEMs generated from photogrammetry using images with consistent illumination conditions) . There are occasions where the absolute error exceeds 40m, which is mostly due to shadows or overly bright regions. These discrepancies arise because the images of these regions contain no useful information for the approach to work on, and most of the discrepancies can be corrected by photoclinometry in the subsequent refinement step. The experimental results also show that photoclinometric refinement sometimes produces small accumulative errors when the surface gradients are continuously over or underestimated. This can be improved by using tuning weights and/or introducing different constraints to the algorithm.

For the photoclinometry assisted image matching (PAM), as the relationship bet－ween the performane of Photoclinametrg and illumination difference of images is complex (Liu et al., 2018), specifically designed formulations and experimental analyses will be necessary in order to rigorously determine the threshold of illumination difference that is favorable for PAM. In general, PAM will be applicable for images with visually apparent illumination differences (e. g., 30° or above in incidence azimuth).

In this paper, photoclinometric refinement using a single image is used because one of the images in the dataset (e. g., the stereo pair of the Chang'E－4 landing site) is severely shadowed and is not suitable for reconstruction based on photometric stereo. However, because PAM is able to perform pixel－wise matching on images with large illumination inconsistencies, it is also possible to use photometric stereo for pixelwise refinement as a final stage. In that case, pixel－synchronous images are needed, which can be obtained using PAM. Whether to use a single image or photometric stereo as the final refinement stage depends on the illumination conditions of the images and the user requirements.

The proposed approach has the advantage of being robust to illumination inconsistencies and subtle textures and is able to produce pixel－level resolution DEMs of the lunar surface. The approach can be used for high－resolution topographic mapping of other similar planetary bodies such as Mercury or asteroids, where substantial variations in illumination pose challenges (Preusker et al., 2017). However, for Mars, the atmospheric influence on the reflectance model needs to be investigated before it can be applied. Elements of the proposed approach may also be helpful in planetary photometric and spectral analysis. The proposed integrated photogrammetric and photoclinometric approach can also serve as a reference for high－resolution topographic mapping in Earth applications.

Acknowledgments

The work described in this paper was funded by a grant from the Research Grants Council of Hong Kong (Project No: PolyU 152086/15E) and grants from the National Natural Science Foundation of China (Project No: 41471345 and Project No: 41671426). The work was also supported by a grant from the Hong Kong Polytechnic University

(Project No: G－YBN8). The authors would also like to thank all those who worked on the archive of the LROC datasets to make them publicly available.

References

[1] Ackermann, J., Langguth, F., Fuhrmann, S., and Goesele, M., 2012. Photometric stereo for outdoor webcams. In: IEEE Conference on Computer Vision and Pattern Recognition, Providence, Rhode Island, pp. 262－269.

[2] Acton, C., 1996. Ancillary data services of NASA's navigation and ancillary information facility. Planet. Space Sci. 44 (1), 65－70.

[3] Banz, C., Pirsch, P., Blune, H., 2012. Evaluation of penalty functions for semi－global matching cost aggregation. Int. Arch. Photogramm. Rem. Sens. Spatial Inf. Sci. XXXIX－B3, 1－6.

[4] Beyer, R. A., McEwen, A. S., Kirk, R. L., 2003. Meter－scale slopes of candidate MER landing sites from point photoclinometry. J. Geophys. Res. 108 (E12) ROV 26－1.

[5] Beyer, R. A., Kirk, R. L., 2012. Meter－scale slopes of candidate MSL landing sites from point photoclinometry. Space Sci. Rev. 170 (1－4), 775－791.

[6] Beyer, R. A., 2017. Meter－scale slopes of candidate InSight landing sites from point photoclinometry. Space Sci. Rev. 211 (1－4), 97－107.

[7] Foroosh, H., Zerubia, J. B., Berthod, M., 2002. Extension of phase correlation to subpixel registration. IEEE Trans. Image Process. 11 (3), 188－200.

[8] Gaskell, R. W., Barnouin－Jha, O. S., Scheeres, D. J., Konopliv, A. S., Mukai, T., Abe, S., Saito, J., Ishiguro, M., Kubota, T., Hashimoto, T., Kawaguchi, J., Yoshikawa, M., Shirakawa, K., Kominato, T., Hirata, N., Demura, H., 2008. Characterizing and navigating small bodies with imaging data. Meteorit. Planet. Sci. 43 (6), 1049－1061.

[9] Grumpe, A., and Wöhler, C., 2011. DEM construction and calibration of hyperspectral image data using pairs of radiance images. In: 7th International Symposium on Image and Signal Processing and Analysis (ISPA), Dubrovnik, pp. 609－614.

[10] Grumpe, A., Belkhir, F., Wöhler, C., 2014. Construction of lunar DEMs based on reflectance modelling. Adv. Space Res. 53, 1735－1767.

[11] Hapke, B., 1981. Bidirectional reflectance spectroscopy: 1. Theory. J. Geophys. Res. 86 (B4), 3039－3054.

[12] Hapke, B., 1986. Bidirectional reflectance spectroscopy: 4. The extinction coefficient and the opposition effect. Icarus 67 (2), 264－280.

[13] Hapke, B., 2002. Bidirectional reflectance spectroscopy: 5. The coherent backscatter opposition effect and anisotropic scattering. Icarus 157 (2), 523－534.

[14] Hapke, B., 2012. Theory of Reflectance and Emittance Spectroscopy. Cambridge University Press, Cambridge, UK.

[15] Heipke, C., 1992. Integration of digital image matching and multiple image shape from shading. Int. Arch. Photogramm. Rem. Sens. XXIX – B3, 832 – 841.

[16] Heipke, C., Oberst, J., Albertz, J., Attwenger, M., 2007. Evaluating planetary digital terrain models: the HRSC DTM test. Planet. Space Sci. 55 (14), 2173 – 2191.

[17] Hirschmüller, H., 2008. Stereo processing by semi – global matching and mutual information. IEEE Trans. Pattern Anal. Mach. Intell. 30 (2), 328 – 341.

[18] Horn,B. P., 1977. Understanding image intensities. Artif. Intell. 8 (2), 201 – 231.

[19] Horn, B. K. P., 1990. Height and gradient from shading. Int. J. Comput. Vision 5 (1), 37 – 75.

[20] Hu, H., Wu, B., 2018. Block adjustment and coupled epipolar rectification of LROC NAC images for precision lunar topographic mapping. Planet. Space Sci. 160 (1), 26 – 38.

[21] Jung, J., Lee, J. – Y., Kweon, I. S., 2015. One – day outdoor photometric stereo via skylight estimation. In: IEEE Conference on Computer Vision and Pattern Recognition (CVPR), Boston, MA, pp. 4521 – 4529.

[22] Kirk, R. L., 1987. III. A Fast Finite – element Algorithm for Two – dimensional Photoclinometry. Ph. D. Thesis, Caltech, pp. 165 – 258.

[23] Kirk, R. L., Barrett, J. M., Soderblom, L. A., 2003a. Photoclinometry made simple? ISPRS Working Group Ⅳ/9 “Advances in Planetary Mapping” Workshop, Houston, TX.

[24] Kirk, R. L., Howington – Kraus, E., Redding, B., Galuszka, D., Hare, T. M., Archinal, B. A., Soderblom, L. A., Barrett, J. M., 2003b. High – resolution topomapping of candidate MER landing sites with Mars orbiter camera narrow – angle images. J. Geophys. Res. 108 (E12), 8088. https://doi.org/10.1029/2003JE002131.

[25] Konecny, G., 1985. The international society for photogrammetry and remote sensing: 75 years old, or 75 years young? Photogramm. Eng. Remote Sens. 51 (7), 919 – 933.

[26] Labarre, S., Ferrari, C., Jacquemoud, S., 2017. Surface roughness retrieval by inversion of the Hapke model: A multiscale approach. Icarus 290, 63 – 80.

[27] Liu, W. C., Wu, B., 2017. Photometric stereo shape – and – albedo – from – shading for pixellevel resolution lunar surface reconstruction. Int. Arch. Photogramm. Rem. Sens. Spatial Inform. Sci. XLII – 3 (W1), 91 – 97.

[28] Liu, W. C., Wu, B., 2018. An integrated photogrammetric and photoclinometric approach for pixel – resolution 3D modelling of lunar surface. International Archives of the Photogrammetry. Rem. Sens. Spatial Inform. Sci. XLII – 3, 1117 – 1121.

[29] Liu, W. C., Wu, B., Wöhler, C., 2018. Effects of illumination differences on photometric stereo shape - and - albedo - from - shading for precision lunar surface reconstruction. ISPRS J. Photogramm. Rem. Sens. 136, 58 - 72.

[30] Lohse, V., Heipke, C., 2004. Multi - image shape - from - shading: derivation of planetary digital terrain models using clementine images. Int. Arch. Photogramm. Rem. Sens. XXXV (B4), 828 - 833.

[31] Lohse, V., Heipke, C., Kirk, R. L., 2006. Derivation of planetary topography using multiimage shape - from - shading. Planet. Space Sci. 54 (7), 661 - 674.

[32] Lowe, D., 2004. Distinctive image features from scale - invariant keypoints. Int. J. Comput. Vision 60 (2), 91 - 110.

[33] McEwen, A. S., 1991. Photometric functions for photoclinometry and other applications. Icarus 92 (2), 298 - 311.

[34] O' Hara, R., Barnes, D., 2012. A new shape from shading technique with application to Mars Express HRSC images. ISPRS J. Photogramm. Remote Sens. 67, 27 - 34.

[35] Oren, M., Nayar, S. K., 1994. Generalization of Lambert's reflectance model. In: Proceedings of the 21st Conference on Computer Graphics and Interactive Techniques, Orlando, FL, USA, pp. 239 - 246.

[36] Piechullek, C., Heipke, C., 1996. DTM refinement using multi image shape from shading. Int. Arch. Photogramm. Rem. Sens. XXXI (B3), 644 - 651.

[37] Piechullek, C., Heipke, C., Ebner, H., 1998. Multi image shape from shading: Results using real aerial imagery. Int. Arch. Photogramm. Rem. Sens. XXXII (3/1), 160 - 167.

[38] Preusker, F., Stark, A., Oberst, J., Matz, K. - D., Gwinner, K., Roatsch, T., Watters, T. R., 2017. Toward high - resolution global topography of Mercury from MESSENGER orbital stereo imaging: a prototype model for the H6 (Kuiper) quadrangle. Planet. Space Sci. 142, 26 - 37.

[39] Salamuničar, G., Lončarić, S., Grumpe, A., Wöhler, C., 2014. Hybrid method for crater detection based on topography reconstruction from optical images and the new LU78287GT catalogue of lunar impact craters. Adv. Space Res. 53 (12), 1783 - 1797.

[40] Speyerer, E. J., Wagner, R. V., Robinson, M. S., Licht, A., Thomas, P. C., Becker, K., Anderson, J., Brylow, S. M., Humm, D. C., Tschimmel, M., 2016. Pre - flight and onorbit geometric calibration of the Lunar Reconnaissance Orbiter Camera. Space Sci. Rev. 200 (1 - 4), 357 - 392.

[41] Wohlfarth, K. S., Liu, W. C., Wu, B., Grumpe, A., and Wöhler, C., 2018. High resolution digital terrain models of the Martian surface: compensation of the atmosphere on CTX imagery. In: 49th Lunar and Planetary Science Conference, March 19 - 23, The Woodlands, Texas, Abstract No. 2498.

[42] Wöhler, C., 2004. Shape from shading under coplanar light sources. pattern recognition. Lect. Notes Comput. Sci. 3175, 278 - 285.

[43] Wöhler, C., Grumpe, A., Berezhnoy, A., Bhatt, M. U., Mall, U., 2014. Integrated topographic, photometric and spectral analysis of the lunar surface: application to impact melt flows and ponds. Icarus 235, 86 - 122.

[44] Wöhler, C., Grumpe, A., Rommel, D., Bhatt, M., and Mall, U., 2017. Elemental and topographic mapping of lava flow structures in Mare Serenitatis on the Moon. In: Proc. ISPRS Int. Symp. on Planetary Remote Sensing and Mapping, Hong Kong, 2017. The International Archives of the Photogrammetry, Remote Sensing and Spatial Information Sciences, XLII - 3 (W1) . pp. 163 - 170.

[45] Woodham, R., 1980. Photometric method for determining surface orientation from multiple images. Opt. Eng. 19 (1), 139 - 144.

[46] Wu, B., Liu, W. C., Grumpe, A., Wöhler, C., 2018a. Construction of pixel - level resolution DEMs from monocular images by shape and albedo from shading constrained with low - resolution DEM. ISPRS J. Photogramm. Remote Sens. 140, 3 - 19.

[47] Wu, B., Zeng, H., Hu, H., 2018b. Illumination invariant feature point matching for highresolution planetary remote sensing images. Planet. Space Sci. 152, 45 - 54.

[48] Wu, B., Zhang, Y., Zhu, Q., 2011. A triangulation - based hierarchical image matching method for wide - baseline images. Photogramm. Eng. Remote Sens. 77 (7), 695 - 708.

[49] Wu, B., Zhang, Y., Zhu, Q., 2012. Integrated point and edge matching on poor textural images constrained by self - adaptive triangulations. ISPRS J. Photogramm. Remote Sens. 68, 40 - 55.

[50] Wu, B., 2017. Photogrammetry: 3D from imagery. In: Richardson, D., Castree, N., Goodchild, M. F., Kobayashi, A., Liu, W., Marston, R. A. (Eds.), The International Encyclopedia of Geography. John Wiley & Sons Ltd, New York, NY, pp. 1 - 13.

[51] Wu, B., Liu, W. C., 2017. Calibration of boresight offset of LROC NAC imagery for precision lunar topographic mapping. ISPRS J. Photogramm. Remote Sens. 128, 372 - 387.

[52] Wu, B., Huang, J., Li, Y., Wang, Y., Peng, J., 2018c. Rock abundance and crater density in the candidate Chang' E - 5 landing region on the Moon. J. Geophys. Res. Planets 123 (12), 3256 - 3272.

[53] Wu, B., Li, F., Hu, H., Zhao, Y., Wang, Y., Xiao, P., Li, Y., Liu, W., Chen, L., Ge, X., Yang, M., Xu, Y., Ye, Q., Wu, X., Zhang, H., 2019. Topographic and geomorphological analysis of the Chang'E - 4 landing site on the far side of the Moon. Earth Space Sci.

Landing site topographic mapping and rover localization for Chang'e - 4 mission

Zhaoqin LIU[1], Kaichang DI[1], Jian LI[2], Jianfeng XIE[2], Xiaofeng CUI[2], Luhua XI[2], Wenhui WAN[1], Man PENG[1], Bin LIU[1], Yexin WANG[1], Sheng GOU[1], Zongyu YUE[1], Tianyi YU[2], Lichun LI[2], Jia WANG[2], Chuankai LIU[2], Xin XIN[1], Mengna JIA[1], Zheng BO[1], Jia LIU[1], Runzhi WANG[1], Shengli NIU[1], Kuan ZHANG[2], Yi YOU[2], Bing LIU[2] & Jiangang LIU[2]

1. State Key Laboratory of Remote Sensing Science, Aerospace Information Research Institute, Chinese Academy of Sciences, Beijing 100101, China
2. Beijing Aerospace Control Center (BACC), Beijing 100094, China

Abstract: This paper presents the techniques and results of landing - site topographic mapping and rover localization using orbital, descent and rover images in the Chang'e - 4 mission. High - resolution maps of the landing site are generated from orbital and descent images. Local digital elevation models and digital orthophoto maps with 0.02m resolution are generated at each waypoint. The location of the lander is determined as (177.588°E, 45.457°S) using festure - matching techniques. The cross - site visual localization method is routinely used to localize the rover at each waypoint to reduce error accumulation from wheel slippage and IMU drift in dead reckoning. After the first five lunar days, the rover travels 186.66m from the lander, according to the cross - site visual localization. The developed methods and results have been directly utilized to support the mission's operations. The maps and localization information are also valuable for supporting multiple scientific explorations of the landing site.

Keywords: Chang'e - 4; Yutu - 2 rover; landing site mapping; rover localization; descent images; rover images

1 Introduction

The Chang'e - 4 (CE - 4) probe, comprising a lander and a rover, was successfully landed in Von Kármán crater in the South Pole - Aitken (SPA) basin on January 3, 2019. The rover Yutu - 2 was separated from the lander and started surface exploration on the same day, marking the first soft landing of a human spacecraft on the far side of the Moon [1]. The SPA basin is the largest, deepest, and oldest impact basinon the Moon [2], considered as having exposed the lunar mantle materials [3]. In situ explorations at the landing site by the Yutu - 2 rover can reveal the surface materials and evolution history of SPA [4 — 6], thereby contributing in understanding the early history of the solar system.

High - precision topographic mapping of the landing site and localization of the rover are important for the safe navigation and accomplishment of science and engineering objectives [7, 8]. Before the CE - 4 mission, seven unmanned rovers had landed on and explored the surfaces of the Moon and Mars [9—15].

(a)　(b)　(c)

Figure 1　Descent images acquired at ~4.0km (a), ~100m (b) and ~8m (c) above the lunar surface

Orbital and rover images were used to map the landing sites and areas traversed based on photogrammetric mapping techniques. Many rover localization techniques, including onboard dead reckoning, visual odometry, bundle adjustment, and feature matching between rover images and an orbital basemap, were employed during different missions in varying scenarios [11—14, 16—18]. The mapping and localization methods and products are critical for supporting teleoperation tasks such as obstacle avoidance and rover path planning, as well as various scientific investigations.

To support the CE - 4 mission operations and scientific explorations, landingsite topographic mapping and rover localization were conducted at different scales and at different turnaround times using orbital, descent, and rover images. This paper presents the techniques and results of landing site mapping and rover localization for the CE - 4 mission. The mapping and localization products directly guarantee the success of surface operations, including safe traversing, efficient approach to science targets, and science data acquisition, as well as supported topographic, geologic, and geochemical studies of the landing site and Von Kármán crater.

2 Data and methods

2.1 Data

Landing site mapping and rover localization are based on orbital data, descent images, and rover images (i. e., images taken by the rover's navigation and panoramic cameras). The utilized reference systems are similar to those defined in the Chang'e-3 mission, including the lunar body-fixed (LBF), landing site local (LSL) and landing site cartographic coordinate (LSC) [14].

The orbital images and products, including the Chang'e-2 images and digital elevation model (DEM), digital orthophoto map (DOM), lunar reconnaissance orbiter camera (LROC), narrow angle camera (NAC) images, and SLDEM2015, were used for general analysis of the landing site and as basemaps for lander and rover localization. Before the CE-4 landing, 100 LROC NAC images with a resolution of ~1.0m were used to generate a high-resolution seamless DOM with the assistance of SLDEM2015 [19].

As a back-up for the Chang'e-3, the Chang'e-4 lander carries a descent camera, with a field of view (FOV) of 45.4° and the image size of 1024 × 1024 pixels [14, 20]. To transfer the data promptly, 59 selected descent images with a compression ratio of 1 : 64 were downlinked before the release of the rover from the lander. After the rover-lander separation, ~5300 descent images, compressed with a ratio of 1 : 8, were transferred. Figure 1 shows some descent images acquired at different altitudes above the lunar surface.

The Yutu-2 rover hosts three pairs of stereo cameras, i. e., the navigation camera (Navcam), panoramic camera (Pancam), and hazard avoidance camera (Hazcam). Navcam and Pancam are mounted on the same camera bar of the rover mast, while Hazcam is fixed on the lower front of the rover body. Navcam includes a stereo base of 270mm, a focal length of 17.7mm, an FOV of 46.6°, and image size of 1024 × 1024 pixels [14]. The rover stops at certain points (termed waypoints) taking images for in situ exploration and path planning. Figure 2 shows stereo Navcam images taken at adjacent waypoints A' and B'. Pancam also possesses a stereo base of 270mm, with horizontal and vertical FOVs of 19.8° and 14.6°, respectively. Twenty-eight pairs of stereo images are required to form a 360° panorama at a pitch angle. To cover a larger vertical FOV, Pancam images have been acquired at two or three pitch angles. Figure 3 shows a Pancam image mosaic generated from 56 left Pancam images (two pitch angles) taken at waypoint S1 on the first lunar day.

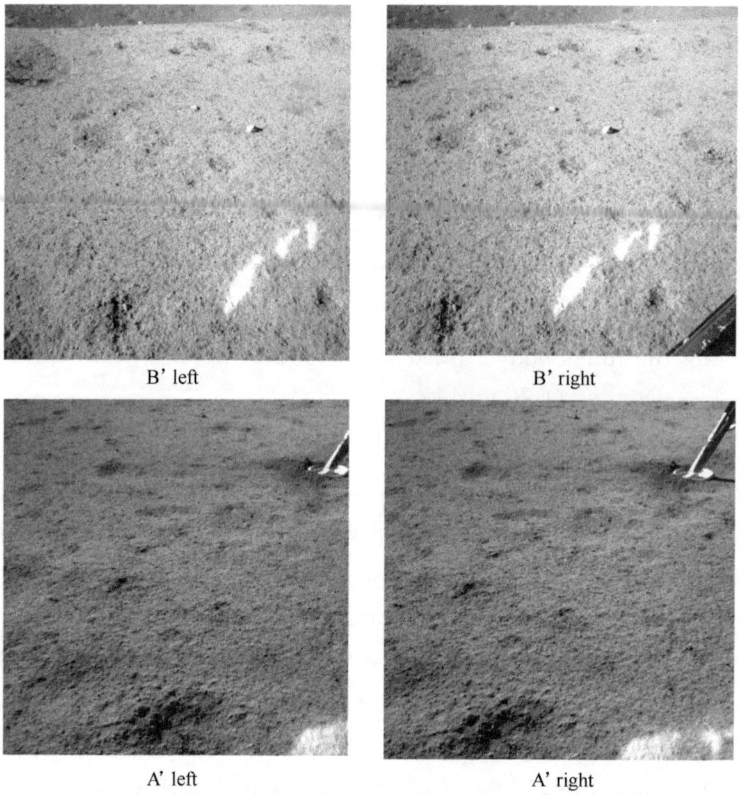

Figure 2　Stereo Navcam images taken at adjacent waypoints A' and B'

Figure 3　Pancam image mosaic generated from images taken at waypoint S1 on the first lunar day

2.2　Methods

Figure 4 shows the procedure for the landing site topographic mapping and rover localization, with details of some key techniques given below.

The EDR - level LROC NAC images were pre - processed with the integrated system for imagers and spectrometers (ISIS) for the SPICE kernels attached, radiometric correction, and echo effects removed with the "spiceinit", "lronaccal", "lronacecho" commands sequentially [21, 22]. The rational function model (RFM) of the images was established by fitting with a rigorous sensor model, which was built based on collinearity equations [23 - 25]. The LROC NAC images are ortho - rectified using RFM and SLDEM2015. To remove the geometric deviations among the LROC NAC images, a block adjustment based on the rectified images is performed. Finally, a geometrically seamless DOM of the landing area is produced via mosaicking.

Automated photogrammetric processing approaches are applied for the landing site mapping using descent images immediately after the rover - lander separation, and all descent images are downlinked. One hundred descent images taken at 400—8m altitudes were selected to generate the DOM using a method similar to that used in CE - 3 [14]. The DOM is used as a basemaps for rover localization and path planning.

Along the traverse, the Yutu - 2 rover acquired stereo images of full or partial (e.g., 60°, 120°) panoramic views, using its Navcam at each waypoint, and full panoramic views, using Pancam at some waypoints. The block bundle adjustment was implemented to refine the exterior orientation parameters of the Navcam and Pancam images. The topographic products, including DEM and DOM, were generated using photogrammetric processing techniques {14, 26} to support rover localization and path planning.

Three localization methods are employed to localize the rover, including dead - reckoning based on the inertial measurement unit (IMU) and wheel odometer data, cross - site visual localization based on Navcam images acquired at adjacent sites (waypoints), and the DOM matching method predicated on DOMs generated from Navcam stereo and descent images [14].

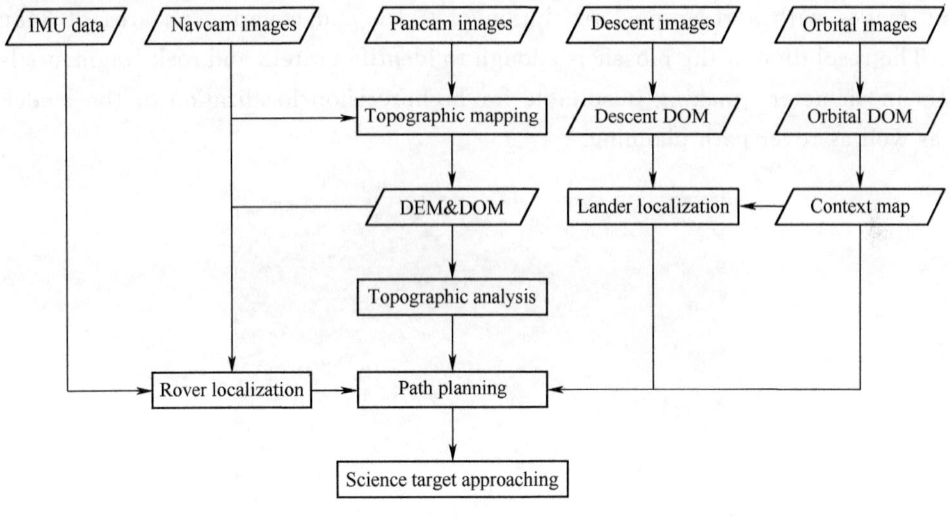

Figure 4　Procedure of landingsite topographic mapping and rover localization

3　Results

3.1　Landing site mapping

Figure 5 shows a seamless DOM generated from 100 LROC NAC images. The image size includes 57483 columns and 34084 rows, with a resolution of 0.9m. The map covers a large area of ~52km × 30km and is suitable for context analysis, such as science exploration area definition. The DOM is also used as a basemap complementary to the

Chang'e - 2 DOM[1] in lander localization and long - term rover traverse planning.

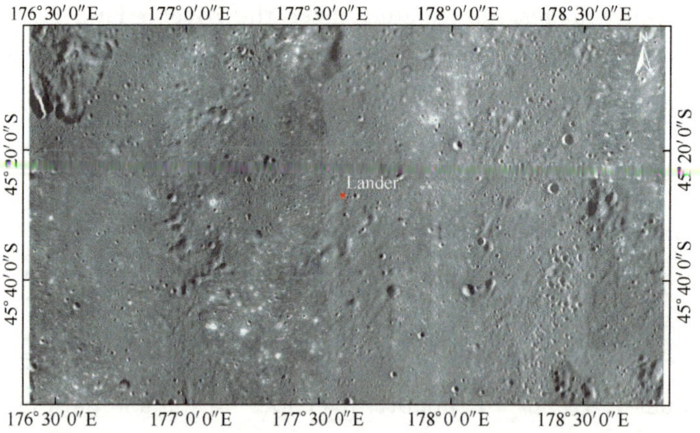

Figure 5　(Color online) Mosaiced DOM from LROC NAC with Lambert conformal conic projection in the CE - 4 landing site region

The resolutions of the descent images used in generating the DOM vary from 0.4 (at 400m altitude) to 0.01m (at 10m altitude). Figure 6 shows a DOM generated from descent images with a resampled resolution of 0.1m and covering an area of 348m × 315m. The resolution of the mosaic is enough to identify craters and rock fragments larger than 1m in diameter, making it suitable for highprecision localization of the lander and rover as well as rover path planning.

Figure 6　(Color online) Descent image - generated DOM at the landing site (red point is the landing point)

3.2　Results of waypoint mapping

The Yutu - 2 rover traversed 33 waypoints in five lunar days after its release from the

1) http://moon.bao.ac.cn/.

lander. Topographic products (DEM and DOM) were routinely generated quickly at each waypoint using Navcam images.

The resolutions of the Navcam images range from 0.001 to 0.02m at 1.0 to 20m from rover. Figure 7 shows the DEM and DOM with 0.02m resampled resolution generated from 18 Navcam stereo images at waypoint D', covering an area of 50m × 32 m with elevation ranging from −2.34 to 0.56m. The DEM is used to produce slope, aspect, and obstacle maps to support rover path planning and scientific investigations.

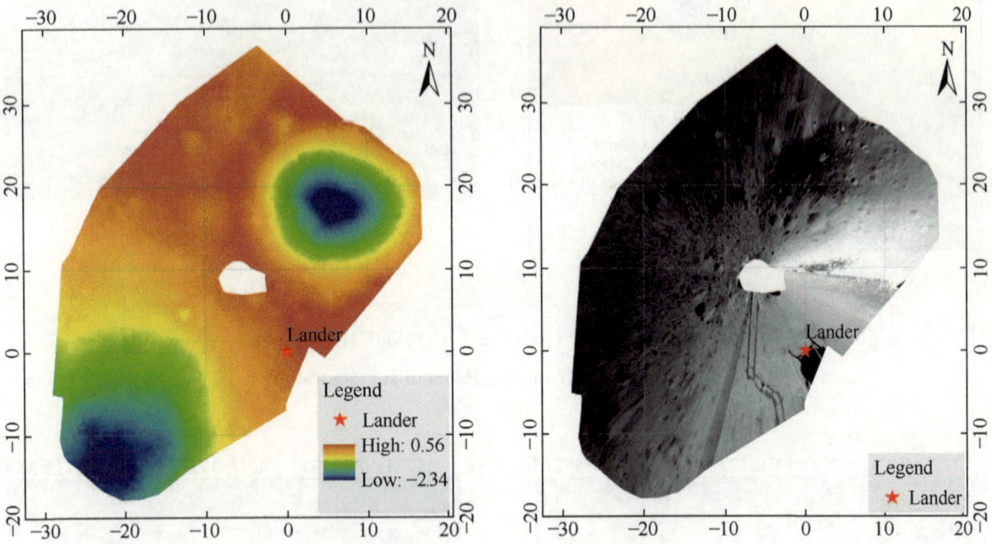

(a)　　(b)

Figure 7　(Color online) DEM (a) and DOM (b) generated from Navcam images taken at waypoint D' (unit: m)

A DEM and a DOM are generated using 58 Pancam stereo images, covering an area of 50m × 50m with a grid spacing of 0.02m. Figure 8 displays a DEM and a perspective view of the DOM overlapped on it. It illustrates that the rover is on a slope and near the rim of the 27-m-diameter crater (to the north-east of waypoint D'), with many other small craters along the traverse.

3.3 Lander and rover localization results

Immediately after landing, the lander location is first determined by a radio tracking technique and then refined by image feature matching and combined measurements from the descent camera, monitoring camera images of the Chang'e-4 lander and DOMs from Chang'e-2 and LROC NAC images. Consequently, the coordinates of the lander in the LROC NAC DOM are (177.588°E, 45.457°S) (Figure 9 (a)) [27]. Figure 9 (b) shows the lander (pointed by the larger arrow) and rover (pointed by the smaller arrow) location observed directly in the LRO NAC image (M1303619844) acquired on February

1,2019. According to Figure 9 (b), our lander and rover localization results are almost identical to the actual locations (green points are waypoints, yellow line is rover traverse), indicating that our localization results are very precise.

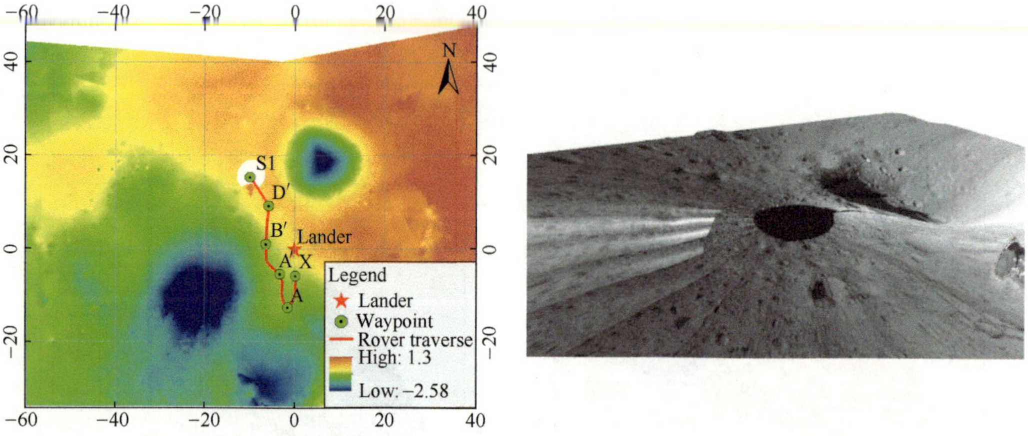

(a)　　(b)

Figure 8　(Color online) DEM (unit: m) (a) and 3D perspective view of the DOM (b) generated from the Pancam images taken at S1

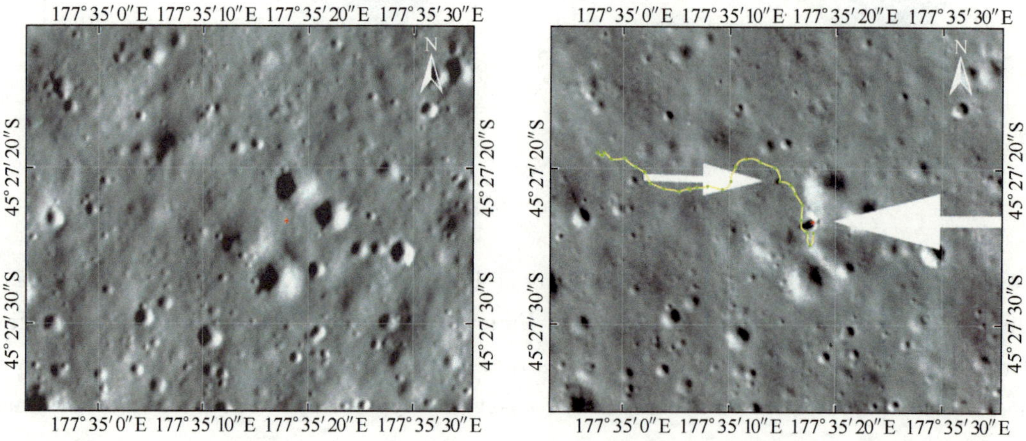

(a)　　(b)

Figure 9　(Color online) Lander localization and comparison with direct observation on LROC NAC image acquired CE－4 landed before (a) and after (b)

Figure 10 shows the rover traverse map derived from the cross－site visual localization. The basemap is a descent image mosaic of the landing site. All coordinates are converted to the LSC reference system with origin at the lander. Along the traverse direction, X is defined as the first waypoint after the rover departed from the lander, with S1, LE00210, LE00309, LE00402, and LE00506 being the waypoints where the rover stayed dormant during lunar nights.

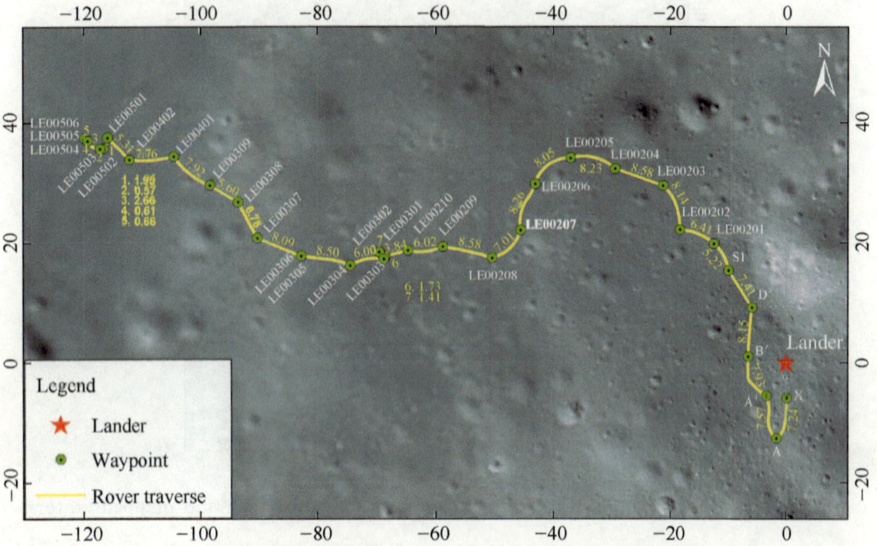

Figure 10　(Color online) Yutu - 2 rover traverse map for the first five lunar days derived from cross - site visual localization (unit: m)

As of May 11, 2019, the Yutu - 2 rover had traveled 190.66m from the lander according to the deadreckoning method [28]. According to the cross - site visual localization method, the rover traveled 182.26m from waypoint X excluding the 4.4m distance traveled from the lander to waypoint X. Therefore, the overall distance traversed from the lander is 186.66m after the first five lunar days. Previous research [29] has reported that the cross - site visual localization yields an accuracy of 4%, which is much better than the nominal localization accuracy of 15% from dead reckoning.

4　Support of mission operations

4.1　Separation of the rover from the lander

After successful landing, the separation of the rover from the lander is a critical step, ensuring that the rover safely lands on the lunar and starts the in situ exploration. Based on the determined lander location, the topography and surface features are analyzed using the descent image mosaic. Figure 11 shows the context map of the landing site. Over 600 craters with diameters greater than 1.0m are mapped and measured. The context map is helpful for general planning of the rover - lander separation and rover traverse.

The lander acquired a front view image using a monitoring camera, to support the rover - lander separation. Before descending, the rover acquired stereo images atop the lander by using Navcam. However, owing to the time constraint and communication

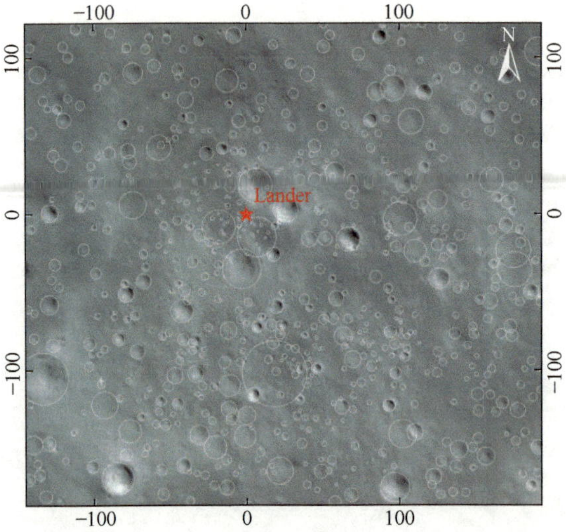

Figure 11　(Color online) Descent image mosaic (unit: m) and mapped craters

bandwidth constraints, only the monitoring camera image (see Figure 12 (a)) is downlinked to the Earth control center before the command of separation is made. Based on the monoscopic measurement technique, we measure the distance and size of the small crater in front of the rover and analyze the terrain and the communication blackout area. Figure 12 (a) shows the measurement and analysis results, including the distance to the front panel of the lander (white dotted line), predicted trajectories of the rover wheels (red line), predicted shadow area on the rover's top panel (yellow area), and communication blackout area (blue area) at 18: 00 on January 3, 2019. These measurements directly support decision - making for rover - lander separation. Figure 12 (b) shows another monitoring camera image taken after the rover landed on the lunar surface.

4.2 Topographic analysis and path planning

Along the rover traverse, the Yutu - 2 rover acquired stereo images at every waypoint using Navcam. Local DEMs and DOMs with 0. 02m grid spacing are routinely generated at each waypoint using the Navcam images. Based on the products, detailed topographic analyses, including slope calculation, and hazard recognition, are conducted to support waypoint - to - waypoint path planning. Figure 13 shows the slope and obstacle maps derived from the Navcam DEM. To match the size of the rover, the resolutions of the slope and obstacle maps are designed as 0. 3m. Based on these maps, the rover path is generated interactively, considering the nature of the terrain and mechanical characteristics of the rover.

The rover path is planned on the DEM - derived maps with an overhead view. To cross - check the planned path on the horizontal view, a 3D path is projected to the actual

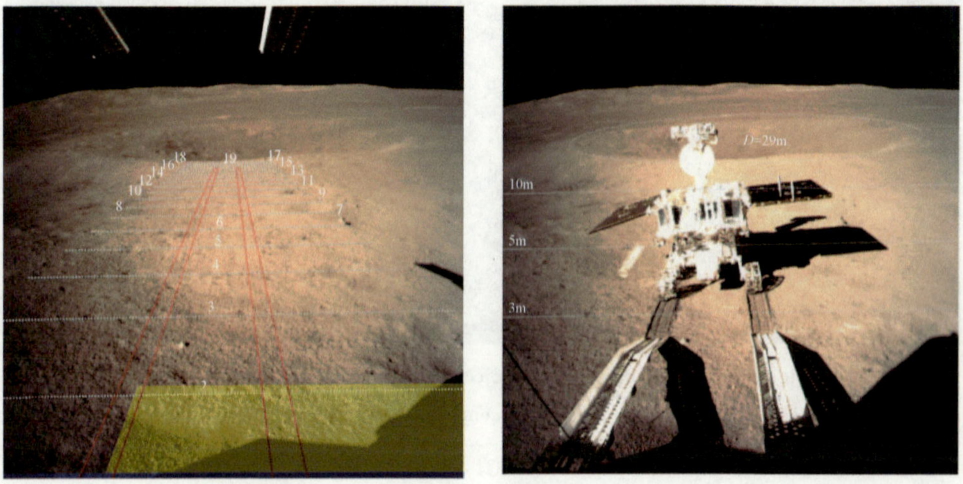

(a)　　(b)

Figure 12　(Color online) Monoscopic measurement supporting the separation of the rover from the lander (the measurement and analysis results of rover landed before (a) and after (b))

rover image based on the imaging model of the camera. Figure 14 shows the projection of a planned path from A' to B' on the Navcam image mosaic at waypoint A'. This horizontal view exhibits the planned path more intuitively and can aid in adjusting the path if necessary. The rover path projection technique is valuable in rover path planning during mission operations, and in evaluating effectiveness of the rover traverse, by projecting the planned path onto the image taken after traversing.

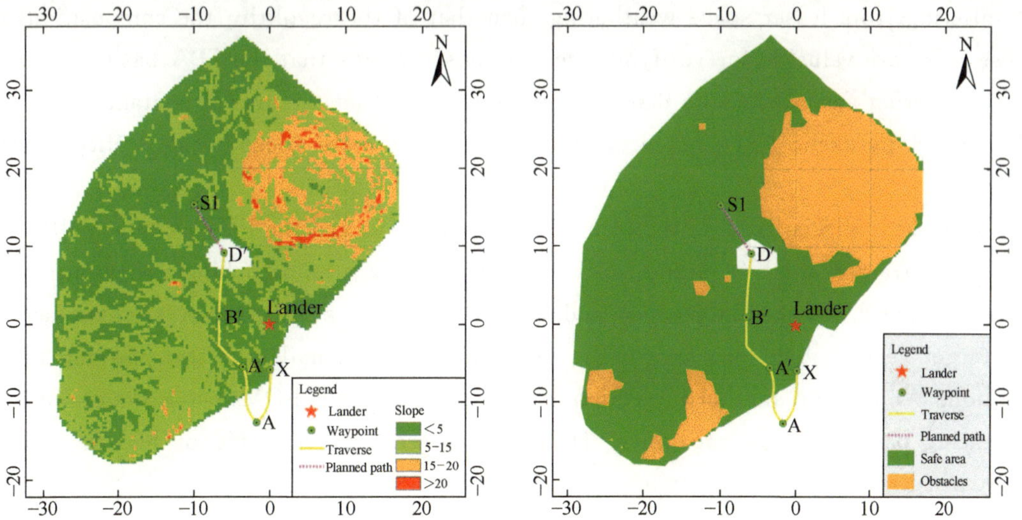

(a)　　(b)

Figure 13　(Color online) Slope (a) and obstacle (b) maps generated from Navcam DEM (unit: m)

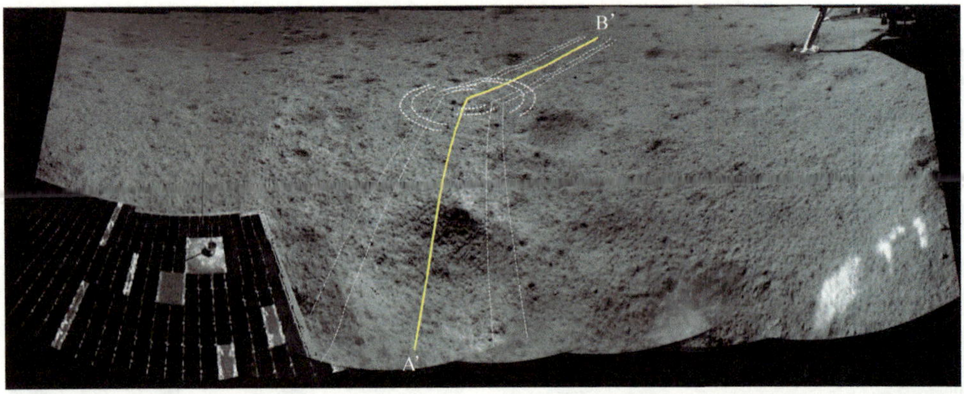

Figure 14　(Color online) Projection of the planned path on the Navcam image mosaic. The yellow line is rover center line and the white lines are wheel trajectories

4.3　Science target approaching

Besides safe and efficient traversing, it is extremely important for the rover to approach the designated science target and acquire scientific data using its science payloads. The visible and near-infrared imaging spectrometer (VNIS), comprising of a visible and near-infrared (VNIR) imaging spectrometer and a shortwave infrared (SWIR) spectrometer, is a significant instrument for investigating the mineralogical compositions of lunar surface materials [30]. The mineral composition and abundance at the CE-4 landing site represent substantial information on the evolution of the lunar far side, which can be partially deciphered by the full spectrum acquired by VNIS. Rock fragments of various sizes and tones are observed along the rover's traverse (Figure 15 (a)). Because of suffering from lesser space weathering than that of the regolith, the compositions of these rocks are valuable for verifying previous observations that the SPA basin penetrats the lunar crust and exposes its mantle material [3] and that the mantle contains abundant olivine and pyroxene [31]. Hence, examining appropriate rocks using VNIS is very important.

Because VNIS is fixed on the front panel of the rover and the FOVs of the VNIR and SWIR spectra are small (8.5°and 3.6° respectively) [32], approaching and pointing the science target depend on the position and orientation of the rover. This makes approaching the target very challenging, especially for approaching a small rock and pointing to its location. We develop another projection tool for predicting the FOVs on the rover images, enhancing path planning capability. Here, we present an example of how a rock is approached for in situ exploration.

At waypoint LE00301 (Figure 15 (a)), rock fragments of different sizes are observed, with scientists interested in obtaining VNIR and SWIR data for a representative rock. An initial path is planned for the rover to approach the designated rock, by defining

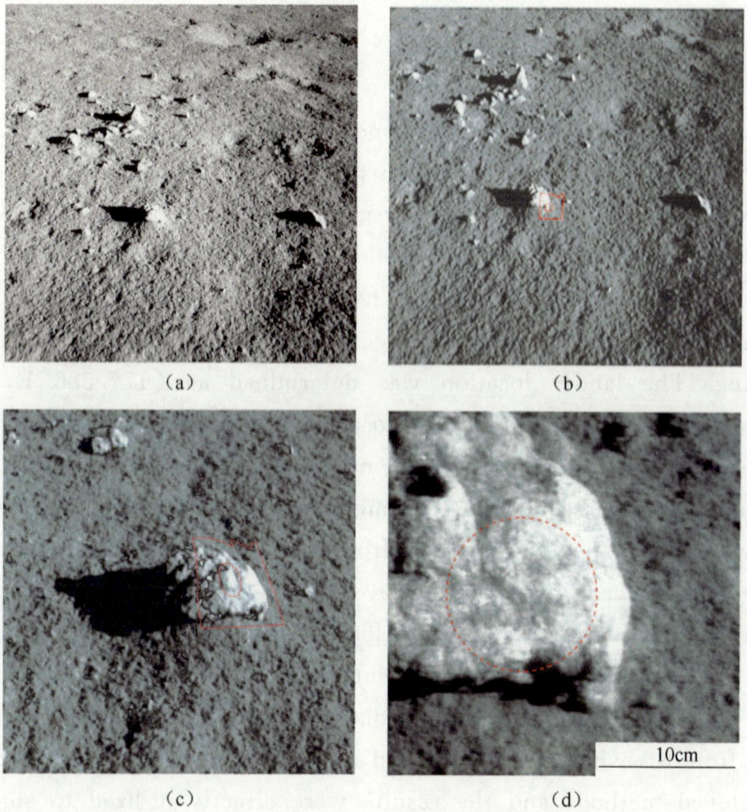

Figure 15　(Color online)　(a) Rocks identified on the Navcam image at LE00301; (b) predicted VNIS FOVs from the Navcam image at LE00301; (c) predicted VNIS FOVs from the Hascam image at LE00302; (d) image obtained by VNIS at LE00303 (the circle corresponds to the SWIR FOV)

LE00302 as the next waypoint. The VNIS FOVs are predicted on the Navcam image at LE00301 to reveal what the VNIS directs observation when the rover moves to LE00302 (Figure 15 (b)). This helps in adjusting and finalizing the path planning between LE00301 and LE00302. After the approaching drive, the rock is in the FOV of the VNIR, but the FOV of SWIR is not optimal. Thus, a short drive (a few centimeters) is planned and the FOVs are predicted from the Hazcam image at LE00302, showing the part of the rock exposed to the VNIS after the short drive (Figure 15 (c)). Finally, the rover achieves the final drive of this target-approaching task, and subsequently, acquires VNIR and SWIR data. Figure 15 (d) shows the acquired VNIR image and the FOV of SWIR, indicating that the science target approaching is implemented. The VNIR and SWIR data of the rock contribute significantly to the mineralogical study of the landing site.

5 Conclusion

This paper summarizes the techniques and results of landing - site mapping and rover localization in the CE - 4 mission based on the orbital, descent, and rover images. A seamless DOM mosaic of the landingsite region with a resolution 0.9m was produced before the CE - 4 launch. The DOM was used as a basemap, complementing the Chang'e - 2 DOM and DEM in support of the mission. The lander was localized using an image feature - matching method, involving descent images and orbital basemaps, immediately after CE - 4 landing. The lander location was determined as (177.588° E, 45.457° S). Meanwhile, a 0.1 - m resolution image mosaic of the landing site was produced using descent images. This high - precision image mosaic was used for contextual analyses and supporting the rover traverse planning. Along the rover traverse, 0.02 - m - resolution local DEMs were automatically generated from Navcam stereo images, with slope and obstacle maps were derived to support the rover path planning at each waypoint. Crosssite visual localization was routinely applied to localize the rover at each waypoint based on image feature matching and bundle adjustment of Navcam images for adjacent waypoints. By the end of the first five lunar days, the distance traveled by the rover from the lander was 190.66m from the wheel odometer, and 188.06m from visual localization.

The developed methods and the results were directly utilized to support mission operations at the Beijing Aerospace Control Center. As the Yutu - 2 continues exploring the far side of the Moon, the mapping and localization techniques are routinely employed to ensure safe traversing and efficient data acquisition. The maps and localization information also support scientific exploration of the landing site and Von Kármán crater.

Acknowledgements

This work was supported in part by Key Research Program of the Chinese Academy of Sciences (Grant No. XDPB11), National Natural Science Foundation of China (Grant Nos. 41671458, 41590851, 41941003). We thank the Lunar and Deep Space Exploration Science Applications Center of the National Astronomical Observatory for providing the Pancam images and VNIS data.

References

[1] Wu W R, Li C L, Zuo W, et al. Lunar farside to be explored by Chang'e - 4. Nat Geosci, 2019, 12: 222—223.

[2] Petro N E, Pieters C M. Surviving the heavy bombardment: ancient material at the surface of South Pole - Aitken basin. J Geophys Res, 2004, 109: 4.

[3] Melosh H J, Kendall J, Horgan B, et al. South Pole - Aitken basin ejecta reveal the Moon's upper mantle. Geology, 2017, 45: 1063—1066.

[4] Li C L, Liu D W, Liu B, et al. Chang'e - 4 initial spectroscopic identification of lunar far - side mantle - derived materials. Nature, 2019, 569: 378—382.

[5] Hu X, Ma P, Yang Y, et al. Mineral abundances inferred from in - situ reflectance measurements of Chang'e - 4 landing site in South Pole - Aitken basin. Geophys Res Lett, 2019, 46: 9439—9447.

[6] Gou S, Di K, Yue Z, et al. Lunar deep materials observed by Chang'e - 4 rover. Earth Planet Sci Lett, 2019, 528: 115829.

[7] Li R X, Squyres S W, Arvidson R E, et al. Initial results of rover localization and topographic mapping for the 2003 Mars Exploration Rover mission. Photogramm Eng Remote Sens, 2005, 71: 1129—1142.

[8] Arvidson R E, Anderson R, Bartlett P, et al. Localization and physical properties experiments conducted by spirit at Gusev Crater. Science, 2004, 305: 821—824.

[9] Zakrajsek J, McKissock D, Woytach J, et al. Exploration rover concepts and development challenges. In: Proceedings of the 1st Space Exploration Conference: Continuing the Voyage of Discovery, Orlando, 2005. 1—23.

[10] Golombek M P, Anderson R C, Barnes J R, et al. Overview of the Mars Pathfinder Mission: launch through landing, surface operations, data sets, and science results. J Geophys Res, 1999, 104: 8523—8553.

[11] Li R X, Archinal B A, Arvidson R E, et al. Spirit rover localization and topographic mapping at the landing site of Gusev crater, Mars. J Geophys Res, 2006, 111: 6.

[12] Li R X, Arvidson R E, Di K C, et al. Opportunity rover localization and topographic mapping at the landing site of Meridiani Planum, Mars. J Geophys Res, 2007, 112: 90.

[13] Di K C, Xu F L, Wang J, et al. Photogrammetric processing of rover imagery of the 2003 Mars Exploration Rover mission. ISPRS J Photogrammetry Remote Sens, 2008, 63: 181—201.

[14] Liu Z Q, Di K C, Peng M, et al. High precision landing site mapping and rover localization for Chang'e - 3 mission. Sci China Phys Mech Astron, 2015, 58: 019601.

[15] NASA. Where is Curiosity? 2019. https://mars. nasa. gov/msl/mission/whereistherovernow/.

[16] Cheng Y, Maimone M W, Matthies L. Visual odometry on the Mars Exploration Rovers - a tool to ensure accurate driving and science imaging. IEEE Robot Automat Mag, 2006, 13: 54—62.

[17] Maimone M, Cheng Y, Matthies L. Two years of visual odometry on the Mars Exploration Rovers. J Field Robotics, 2007, 24: 169—186.

[18] Di K C, Liu Z Q, Yue Z Y. Mars rover localization based on feature matching between ground and orbital imagery. Photogramm Eng Remote Sens, 2011, 77: 781—791.

[19] Barker M K, Mazarico E, Neumann G A, et al. A new lunar digital elevation model from the lunar orbiter laser altimeter and SELENE terrain camera. Icarus, 2016, 273: 346—355.

[20] Wan W, Liu Z, Di K, et al. A cross - site visual localization method for Yutu rover. In: Proceedings of ISPRS 2014 Technical Commission IV Symposium, Suzhou, 2014. 279—284.

[21] NAIF. Lunar reconnaissance orbiter camera (LROC) instrument kernel v18. 2014. http://naif. jpl. nasa. gov/pub/naif/pds/data/lro - l - spice - 6 - v1. 0/lrosp - 1000.

[22] Henriksen M R, Manheim M R, Speyerer E J, et al. Extracting accurate and precise topography from LROC narrow angle camera stereo observations. Int Arch Photogramm Remote Sens Spatial Inf Sci, 2016, XLI - B4: 397—403.

[23] Di K C, Xu B, Liu B, et al. Geopositioning precision analysis of multiple image triangulation using LRO NAC lunar images. Int Arch Photogramm Remote Sens Spatial Inf Sci, 2016, XLI - B4: 369—374.

[24] Liu B, Xu B, Di K C, et al. A solution to low RFM fitting precision of planetary orbiter images caused by exposure time changing. Int Arch Photogramm Remote Sens Spatial Inf Sci, 2016, XLI - B4: 441—448.

[25] Liu B, Jia M N, Di K N, et al. Geopositioning precision analysis of multiple image triangulation using LROC NAC lunar images. Planet Space Sci, 2018, 162: 20—30.

[26] Peng M, Wan W H, Wu K, et al. Topographic mapping capbility analysis of Chang' e - 3 Navcam stereo images and 3D terrain reconstruction for mission operations (in Chinese). J Remote Sens, 2014, 18: 995—1002.

[27] Di K C, Liu Z Q, Liu B, et al. Chang'e - 4 lander localization based on multi - source data. J Remote Sens, 2019, 23: 177—180.

[28] CLEP. Rover and lander of Chang'e - 4 have finished the work of the first five lunar days. 2019. http://www. clep. org. cn/n5982341/c6806279/content. html.

[29] Wan W H. Theory and Methods of Stereo Vision Based Autonomous Rover Localization in Deep Space Exploration. Dissertation for Ph. D. Degree. Beijing: Chinese Academy of Sciences, 2012.

[30] Jia Y, Zou Y, Ping J, et al. The scientific objectives and payloads of Chang'e - 4 mission. Planet Space Sci, 2018, 162: 207—215.

[31] Wieczorek M A, Jolliff B, Khan A, et al. The constitution and structure of the lunar interior. Rev Mineral Geochem, 2006, 60: 221—364.

[32] He Z P, Wang B Y, Lv G, et al. Visible and near - infrared imaging spectrometer and its preliminary results from the Chang'e 3 project. Rev Sci Instruments, 2014, 85: 083104.

Chronological sequence of Chang'E-4 landing zone within Von Kármán crater

Yu Lu[1,2,3], Yunzhao Wu[1,4], Gregory G. Michael[5], Jinsong Ma[6,7], Wei Cai[1], Nannan Qin[1]

1. Key Laboratory of Planetary Sciences, Purple Mountain Observatory, Chinese Academy of Sciences, Nanjing 210034, China
2. School of Astronomy and Space Science, University of Science and Technology of China, Hefei 230026, China
3. CAS Center for Excellence in Comparative Planetology, Hefei 230026, China
4. Space Science Institute, Macau University of Science and Technology, Macau, China
5. Planetary Sciences and Remote Sensing, Institute of Geological Sciences, Freie Unversitaet Berlin, Berlin 12249, Germany
6. School of Geography and Ocean Science, Nanjing University, Nanjing 210023, China
7. Jiangsu Center for Collaborative Innovation in Geographical Information Resource Development and Application, Nanjing 210023, China

Abstract: Sequence and timing are critical for understanding the geologic evolution. Here we performed detailed crater size-frequency distribution measurements within the Von Kármán crater, where China's Chang'E-4 (CE-4) mission landed, and its surroundings including Alder and Finsen craters, to understand the geologic evolution of the landing zone. The formation time of Von Kármán crater was estimated to be ~$4.2^{+0.03}_{-0.03}$ Ga, approaching the age of South Pole-Aitken basin. Three major layers under the landing site were dated: The ejecta from Alder crater was considered to be ~$3.5^{+0.01}_{-0.2}$ Ga old, belonging to Imbrian. The floor of Von Kármán crater was divided into mare basalt units and Finsen ejecta unit based on spectral differences. Two mare basalt units were estimated to be ~$3.2^{+0.07}_{-0.1}$ and $3.3^{+0.08}_{-0.2}$ Ga old (Imbrian), respectively, while Finsen crater and its ejecta, where CE-4 landed, were estimated to be ~$3.1^{+0.3}_{-0.5}$ Ga (Eratosthenian). The chronological sequence established above indicates that the CE-4 landing zone has experienced a very complex evolutionary history.

Keywords: Chang'E-4; Von Kármán crater; CSFD; Geologic evolution

1 Introduction

The enormous South Pole-Aitken (SPA) basin (diameter: ~2500km) is recognized as the largest and oldest impact basin on the Moon (Stuart - Alexander, 1978; Wilhelms, 1987). The large impact may have penetrated the lunar crust and exposed upper mantle materials (Potter et al., 2012; Melosh et al., 2017), and thus is of great importance in lunar exploration. On January 3, 2019, China's Chang'E - 4 (CE - 4) mission, including a lander and the Yutu - 2 rover, successfully landed on the floor of Von Kármán crater within the SPA, making it the first lunar farside soft landing mission in human history (Wu et al., 2019) (Fig. 1a). Until now the Yutu - 2 rover, with the aim of exploring the formation and evolution history of SPA, is still traveling on the lunar farside and making in situ measurements along its route.

Von Kármán crater (centered at 44.45° S, 176.25° E; diameter: ~186km) was believed to be a degraded Nectarian (e.g., Stuart - Alexander, 1978) or pre - Nectarian crater (e.g., Losiak et al., 2009; Yingst et al., 2017). It is superposed on the older Von Kármán M crater, and is superimposed by Leibnitz, Alder and Finsen craters (Fig. 1b). The northern and eastern parts of Von Kármán crater are covered by ejecta from Leibnitz and Finsen crater respectively, while the western part is occupied by mare basalts (Huang et al., 2018; Di et al., 2019).

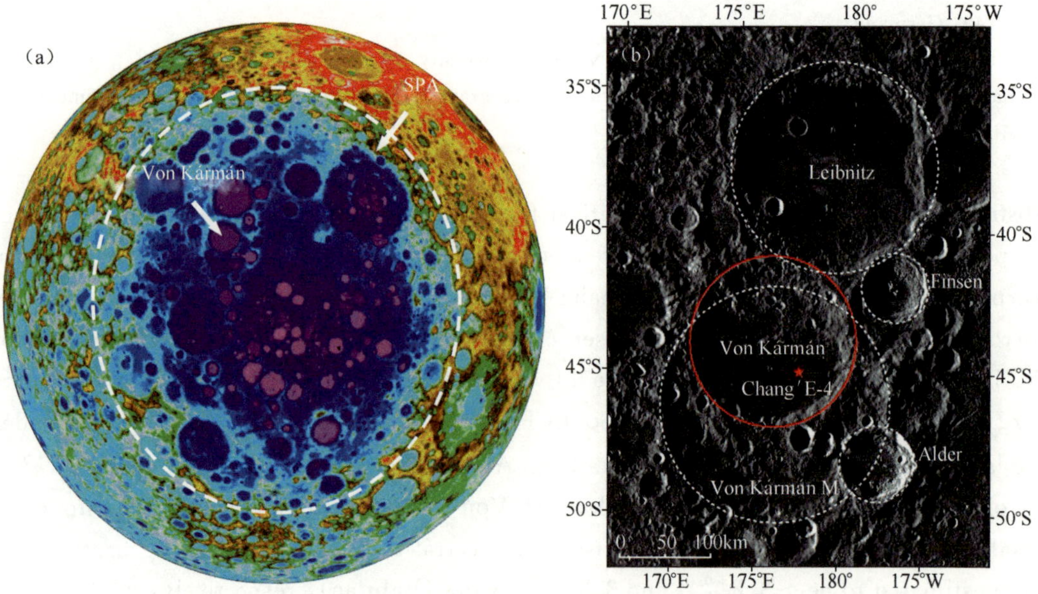

Fig. 1 (a) The topography of Von Kármán crater within the SPA basin, Lunar Orbiter Laser Altimeter DEM data. White arrows indicate the locations of SPA (dashed circle) and the Von Kármán crater. (b) The regional vicinity of Von Kármán crater, LROC WAC low - Sun mosaic. The red circle marks the rim of Von Kármán crater, and the red star marks the landing site of CE - 4. (For interpretation of the references to color in this figure legend, the reader is referred to the web version of this article.)

Previous studies have provided detailed reports on the geologic context, composition, chronology and stratigraphy of the CE - 4 landing zone within Von Kármán crater (e.g.,

Huang et al., 2018; Di et al., 2019; Fu et al., 2019). The stratigraphy of the landing zone was established by different authors with various methods, such as crater excavation depth method (Huang et al., 2018; Fu et al., 2019), results from the Lunar Penetrating Radar (LPR) onboard the Yutu-2 rover (Lai et al., 2019, 2020; Li et al., 2020), and numerical simulation (Di et al., 2019). The stratigraphy has been generally proposed to consist of three/four major layers: the top layer is the fine-grained regolith, with a thickness of ~2.5-7.5 m (Huang et al., 2018), ~11m (Lai et al., 2019, 2020), or ~12m (Li et al., 2020; Zhang et al., 2020). The next layer is considered to be the impact ejecta from Finsen crater, with a thickness of ~5.5-10.5m (Huang et al., 2018), ~14m (Lai et al., 2019, 2020), or ~32m (Di et al. 2019). Beneath the ejecta is the mare basalt layer, probably around 300m (Huang et al., 2018; Fu et al., 2019), or ~58m (Lai et al., 2020). The ejecta from Alder crater was considered to be below the mare basalts (Huang et al., 2018; Fu et al., 2019; Lai et al., 2020). The numerical simulation of impact ejecta shows that Alder crater deposited ~35-m-thick ejecta at the CE-4 landing site (Di et al., 2019). Recent studies using the LPR data from Yutu-2 rover showed that the ejecta from Alder crater might be ~11m (Lai et al., 2020), or ~17m (Zhang et al., 2020) thick, and below the Alder ejecta lied several other mare basalt layers and large-scale ejecta deposits from distant craters or basins (Lai et al., 2020).

The chronological sequence is very important to understanding the geologic evolution of the CE-4 landing zone, however, it has not yet been well constrained. The formation time of Von Kármán crater has been heavily debated to belong to either the Nectarian or pre-Nectarian. The ejecta of Finsen crater where CE-4 landed has not yet been dated, and the age of Finsen crater, equal to the age of the ejecta, also stayed controversial: whether Imbrian, Eratosthenian, or Copernican (e.g., Fu et al., 2019). The age of Alder crater was still unknown. Besides, Di et al. (2019) suggested that the CE-4 landing site is ~70m higher than the mare basalt, and the ejecta from both Alder (thickness: ~35m) and Finsen crater (~32m) are superposed on the mare basalt. The major difference is the stratigraphy of the mare basalt and the ejecta from Alder crater. Huang et al. (2018) and Fu et al. (2019) support that the ejecta from Alder crater is older than the mare basalt, and thus is below the basalt. Di et al. (2019) believed that the ejecta from Alder crater is above the mare basalt. Dating Alder crater and the mare basalts could provide critical clues to figure it out.

In this study, we first derived the formation time of Von Kármán crater, and then classified the geologic units within Von Kármán crater based on spectral data, and performed crater counting measurements on each unit and the surroundings of Von Kármán crater, including Alder and Finsen craters, to establish the chronological sequence of the CE-4 landing zone, and further discussed its geologic evolution.

2 Data and method

We employed Lunar Reconnaissance Orbiter Camera (LROC) images (Robinson et al., 2010) and Selenological and Engineering Explorer (SELENE/Kaguya) data (Haruyama et al., 2008) to distinguish geologic units and perform crater statistics analysis.

The crater size - frequency distribution (CSFD) method has been widely used to derive the absolute model ages (AMAs) of planetary surfaces, and the technique has been described in great detail by various authors (Arvidson et al., 1979; Neukum, 1983; Hiesinger et al., 2000; Hartmann and Neukum, 2001; Neukum et al., 2001). In general, these chronology models include two functions: the crater production function and the chronology function. The production function describes the size - frequency distribution of craters on planetary surfaces, and the chronology function calibrates the crater population accumulation to time. The absolute ages of planetary surfaces are obtained by linking the crater size - frequency distribution with the absolute ages of returned Apollo and Luna samples (Neukum, 1983; Neukum et al., 2001).

The accurate separation of geologic units is crucial to establishing ages and investigating the geologic evolution. It only makes sense to perform crater counting on a geologically homogeneous surface (Michael and Neukum, 2010). The false color images of multiband spectral data, such as Clementine, LROC Wide Angle Camera (WAC) 7 - band, Moon Mineralogy Mapper (M^3) integrated band depth (IBD), and element abundance maps, such as FeO, TiO_2 maps, aid in distinguishing different geologic units, and thus are widely used to separate them out (Hiesinger et al., 2000, 2011; Wu et al., 2018). In this study, the LROC empirically normalized WAC 7 - band mosaic, the global map of plagioclase abundance and FeO abundance derived from Kaguya Multiband Imager data (~59m/pixel) (Lemelin et al., 2016) clearly show the distribution of the ejecta from Finsen crater, and thus were used to distinguish the ejecta deposits from the mare basalt within Von Kármán crater.

After classifying the geologic units, several homogenous areas were identified to perform CSFD measurements. Note that areas with obvious secondary craters, such as crater chains or clusters, were excluded. The low - Sun illumination LROC WAC mosaic (~100m/pixel), with an incidence angle range of 69°- 84° (Wagner et al., 2015), was used to show the geomorphology of the study area and perform CSFD measurements on the ejecta blankets of the Von Kármán and Finsen crater. The Kaguya Terrain Camera morning map, with a resolution of ~10m/pixel (Haruyama et al., 2008), was used to perform CSFD analysis on the mare basalt units. All the images were imported into ArcMap for subsequent analysis, and craters were mapped with CraterTools extension tool within ArcMap (Kneissl et al., 2011). We also performed CSFD measurements on the floor of Finsen crater and the terrace of Alder crater in JMARS software (www.jmars.asu.edu) using LROC Narrow Angle Camera (NAC) high - resolution images (~0.5 - 2m/pixel) (Robinson et al., 2010). The derived crater size and numbers were imported into Craterstats2, developed by Michael and Neukum (2010), for subsequently analysis. The final ages were derived based on the chronology and production functions from Neukum et al. (2001) with a Poisson timing analysis (Michael et al., 2016).

3 Results

3.1 Division of the geologic units

The floor of Von Kármán crater was classified into two mare basalt units and an ejecta deposit unit based on the spectral data (Fig. 2). The CE - 4 landing zone is mainly covered by ejecta deposits from Finsen crater. The criteria distinguishing the ejecta unit from the mare basalt units are as follows: In the WAC false color image, the ejecta deposit unit is much brighter than the basalt units (Fig. 2a). In the plagioclase abundance image (Lemelin et al., 2016), the ejecta unit contains more plagioclase than the basalt units (Fig. 2b). In the FeO abundance image, the basalt units contain more iron than the ejecta unit (Fig. 2c).

3.2 Age of Von Kármán crater

Von Kármán crater is superposed on the older Von Kármán M crater (Fig. 1b). The rim between Von Kármán and Von Kármán M crater provided an ideal area for dating Von Kármán crater (Fig. 3a). Note that the eastern part of the rim was excluded to get rid of possible effects caused by ejecta blanketing of Alder crater. The CSFD result shows that Von Kármán crater formed at $\sim 4.2^{+0.03}_{-0.03}$ Ga (Fig. 3b), which is very close to the formation time of SPA basin (e.g., >4.29Ga, Fernandes et al., 2013; ~4.25Ga, Garrick - Bethell et al., 2020).

3.3 Age of Alder crater

The floor of Alder crater was considered as a possible counting area to date the crater: however, the CSFD showed a very young age (~2.7Ga) which is not plausible (Alder crater is widely accepted as a Imbrian crater, e.g., Stuart - Alexander, 1978; Wilhelms, 1987; Losiak et al., 2009). Geological processes commonly lead to crater - floor infill, which means that the age of the floor only places a lower limit on the formation time of Alder crater. As an alternative, the terrace to the southwest of Alder crater was considered to have formed together with Alder crater, and thus was used to date it. The selected area for dating is shown in Fig. 3a, and the derived absolute model age (AMA) is $\sim 3.5^{+0.01}_{-0.2}$ Ga (Fig. 3c), which means that the ejecta from Alder crater is also ~3.5Ga old.

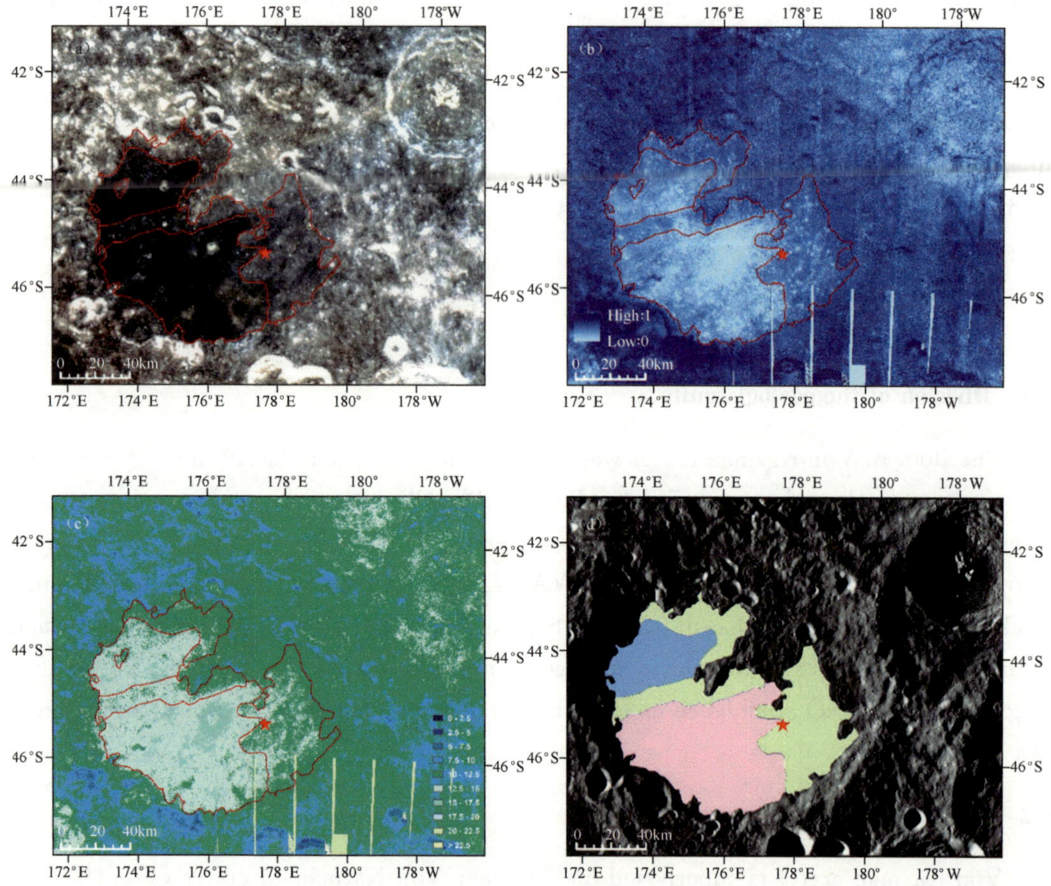

Fig. 2　The geologic map within Von Kármán crater. Red lines indicate the boundaries of the geologic units. Mare boundaries come from Nelson et al. (2014), and the red star marks the landing site of CE-4. (a) LROC WAC Empirically Normalized color mosaic (R: 689nm, G: 415nm, B: 321nm). (b) The plagioclase abundance map derived from Kaguya Multiband Imager data. (c) The FeO abundance map derived from Kaguya Multiband Imager data. (d) LROC WAC low-Sun image. The green color indicates ejecta from Finsen crater, the blue color represents mare basalt unit1 and the pink color represents mare basalt unit2. (For interpretation of the references to color in this figure legend, the reader is referred to the web version of this article.)

3.4　Age of mare basalt units

Many secondary craters were identified on the floor of Von Kármán crater (Huang et al., 2018). To avoid the effects of these secondary craters, two relatively homogeneous areas within the mare basalt units were identified and mapped at the western part of Von Kármán crater to derive the ages of them (Fig. 3a). Areas with obvious secondaries were excluded. Our CSFD measurements of these two mare basalt units show AMAs of $\sim 3.2^{+0.07}_{-0.1}$ and $3.3^{+0.08}_{-0.2}$ Ga (Fig. 3d, e).

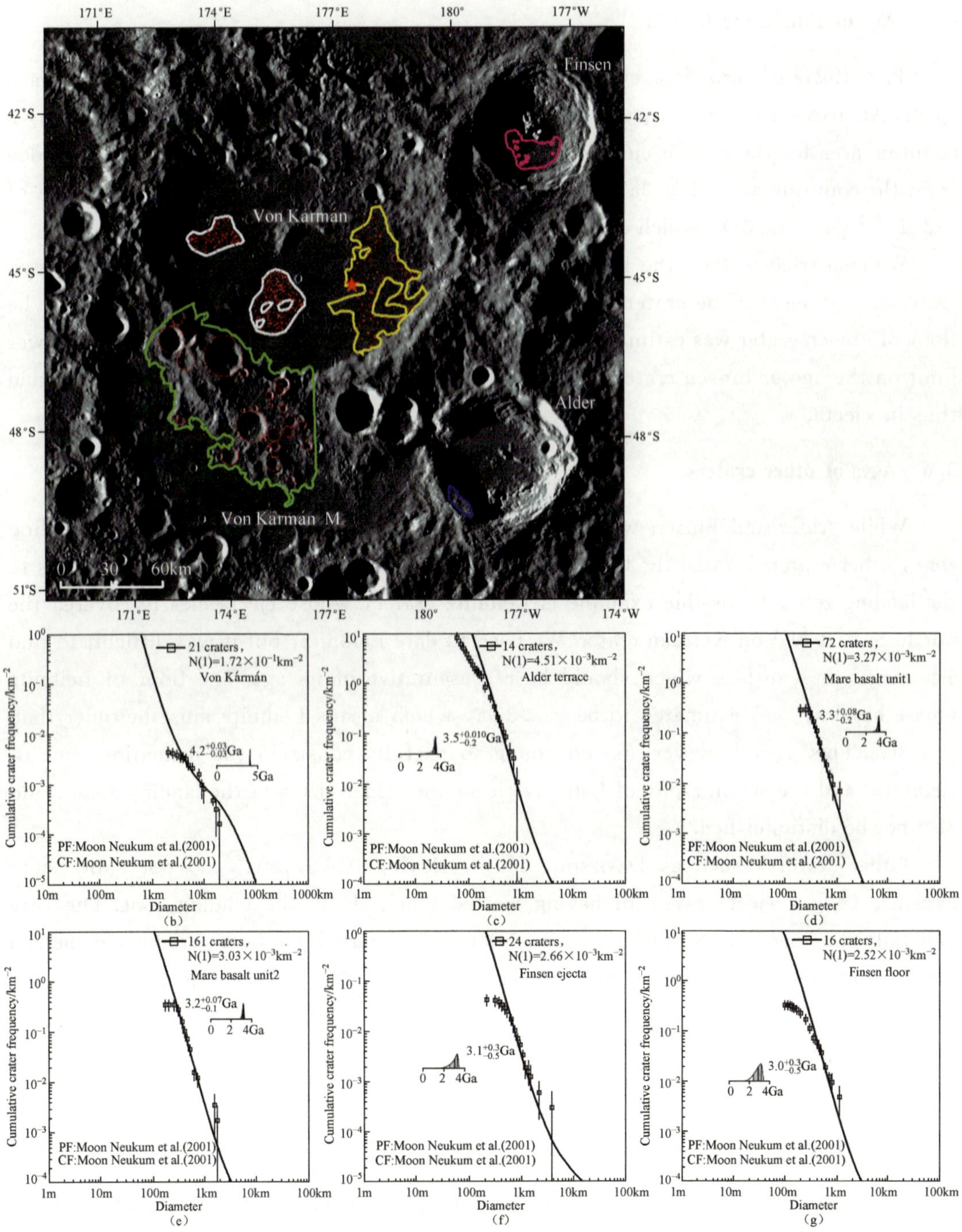

Fig. 3 (a) Locations of crater－counting areas of Von Kármán ejecta (green boundary), Alder terrace (blue boundary), mare basalt units (white boundaries, the left one is mare basalt unit1), Finsen ejecta (yellow boundary) and Finsen floor (pink boundary). Craters are outlined with red circles. The CE－4 landing site is marked with a red star. (b)－(g) CSFD plots of these units respectively. (For interpretation of the references to color in this figure legend, the reader is referred to the web version of this article.)

3.5 Age of Finsen ejecta unit

The effects of secondary craters and topography make it challenging to date Finsen ejecta. An area on the eastern part of the floor of Von Kármán crater was selected as a counting area for the Finsen ejecta deposit unit. Obvious secondary craters were excluded from the counting area (Fig. 3a). The CSFD measurement of the ejecta shows an AMA of $\sim 3.1^{+0.3}_{-0.5}$ Ga (Fig. 3f), which corresponds to the age of Finsen crater.

We also tried to date the floor of Finsen to constrain the formation time of Finsen crater and its ejecta. The crater - counting area and CSFD plot are shown in Fig. 3. The floor of Finsen crater was estimated to be $\sim 3.0^{+0.3}_{-0.5}$Ga old. As this value should be a lower limit on the age of Finsen crater, we believe ~3.1Ga is reasonable for Finsen crater and thus its ejecta.

3.6 Ages of other craters

While Alder and Finsen were considered to be the most influential on the landing zone, other craters around the CE-4 landing site could also contribute ejecta materials to the landing zone. A possible example is Leibnitz crater, whose ejecta clearly covered the northern part of Von Kármán crater. We tried to date Leibnitz, but it was difficult to find a homogeneous surface which should be representative of its age. The floor of Leibnitz (mare basalts) was estimated to be ~3.3Ga, which means Leibnitz must be older than ~3.3Ga. This age, however, is too young to usefully constrain the formation time of Leibnitz, and the stratigraphy of Leibnitz ejecta and Alder ejecta at the landing zone could also not be distinguished.

Other craters, such as Davisson, Hess, Boyle, Abbe, etc., do not show clear evidence (e.g., ejecta rays) of having ejected materials to the landing zone. They are generally considered as Nectarian or pre-Nectarian craters. If these craters have dispersed material into the landing zone, this material should underlie the Alder ejecta.

4 Discussion

4.1 Age results

Table 1 shows the comparison between some previous studies and the results derived in this study. The Von Kármán crater was determined to be $\sim 4.2^{+0.03}_{-0.03}$ Ga, close to the formation time of SPA basin (Fernandes et al., 2013; Garrick-Bethell et al., 2020). It belongs to pre-Nectarian, consistent with Wilhelms (1987), Losiak et al. (2009), and comparable with Yingst et al. (2017) (~3.97Ga). Alder crater was considered to belonging to Imbrian, consistent with Stuart-Alexander (1978) and Wilhelms (1987). The ages of the mare basalts within Von Kármán crater (~3.2-3.3Ga) are comparable

Table 1　Comparison of the age results in this study and those of previous studies

	Stuart - Alexander (1978)	Wilhelms (1987)	Haruyama et al. (2009)	Losiak et al. (2009)	Yingst et al. (2017)	Huang et al. (2018)	Ivanov et al. (2018)	Pasckert et al. (2018)	Ling et al. (2019)	This study
Von Kármán	Nectarian	Pre - Nectarian		Pre - Nectarian	3. 97Ga	Pre - Nectarian				4. 2Ga
Alder	Imbrian	Imbrian		Imbrian						3. 5Ga
Mare basalt	Imbrian	Imbrian	3. 35Ga		Imbrian	3. 6Ga		3. 15/3. 75Ga	3. 38/ 3. 57Ga	3. 2～3. 3Ga
Finsen	Eratosthenian	Eratosthenian		Eratosthenian	Imbrian	Imbrian	3. 61Ga			3. 1Ga

with the findings of other authors, falling in Imbrian period. For example, Haruyama et al. (2009) derived an age of ~3.35Ga. Ling et al. (2019) estimated that the younger lava flows were ~ 3.38Ga. We found that Finsen crater was formed during the Eratosthenian period, unlike Yingst et al. (2017), but in agreement with Stuart - Alexander (1978), Wilhelms (1987) and Losiak et al. (2009).

4.2 The chronological sequence

According to the age results derived above, there probably exists four major layers in the CE - 4 landing zone. From bottom to top are the ejecta from Alder crater, mare basalts, the ejecta from Finsen crater and the regolith (Fig. 4). The results show that Von Kármán crater formed at $\sim 4.2^{+0.03}_{-0.03}$ Ga, and by $\sim 3.5^{+0.01}_{-0.2}$ Ga, Alder crater had formed and ejected material into the Von Kármán crater. After that, mare basalts flooded the crater at $\sim 3.2^{+0.07}_{-0.1}$—$3.3^{+0.08}_{-0.2}$ Ga. Then Finsen crater formed at $\sim 3.1^{+0.3}_{-0.5}$ Ga, so the mare basalts were covered with Finsen ejecta at that time. After a long period of space weathering, the surface gradually accumulated a thick fine - grained regolith. This timing sequence is also consistent with the stratigraphy of the CE - 4 landing zone (Huang et al., 2018; Fu et al., 2019; Lai et al., 2019, 2020; Zhang et al., 2020).

4.3 Implications for rover data interpretation

Until now the Yutu - 2 rover is still working on the floor of Von Kármán crater, and much in situ data are being returned to the Earth continuously. The LPR on board the Yutu - 2 rover revealed the thickness and physical properties of the subsurface structures (Lai et al., 2019, 2020; Li et al., 2020; Zhang et al., 2020). However, where these materials came from and how they were formed cannot be determined by the LPR data only. Geological background knowledge is very important in interpreting these data. The chronological sequence established in this study showed that the shallow subsurface stratigraphy of the Chang'E - 4 landing zone is the regolith, ejecta from Finsen, mare basalts and ejecta from Alder respectively, from top to bottom, which should be considered while interpreting the Yutu - 2 data.

5 Conclusion

Sequence and timing are critical to understanding the geologic evolution. The successful landing of the Chang'E - 4 mission on the floor of Von Kármán crater within SPA provides a very unique opportunity to learn about the evolutionary history of the farside of the Moon. Based on the detailed CSFD measurements performed within Von Kármán crater and its surroundings including Alder and Finsen craters, we established a chronological sequence of CE - 4 landing zone to understand its geologic evolution. The results show that Von Kármán crater formed at around ~ 4.2Ga, shortly after the

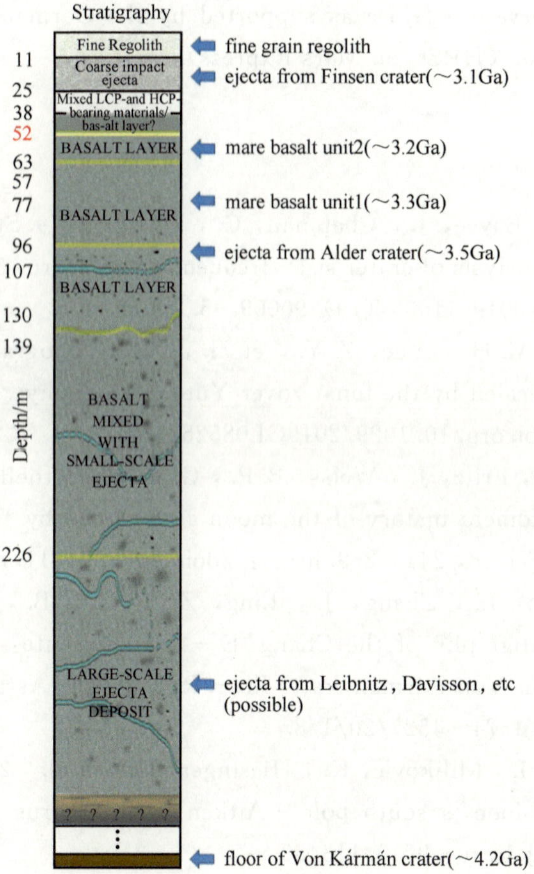

Fig. 4　The sketch map of the chronological sequence of the CE－4 landing zone (modified from Lai et al., 2020)

formation of the SPA basin. Alder crater formed at ~3.5Ga, and ejected material onto the landing zone at that time. About 200－300Ma later, mare basalts flooded the Von Kármán crater. After that, Finsen crater formed at ~3.1Ga, covering the landing zone with its ejecta. The age results above showed that the CE－4 landing zone has experienced a very complex evolutionary history. The timing sequence established in this study could also provide geologic background knowledge for interpretations of the spectral and radar data derived by Yutu－2 rover in the future.

Acknowledgments

We appreciate the anonymous reviewer for helpful comments and suggestions that greatly improved the manuscript. This work was supported by the Strategic Priority Research Program of Chinese Academy of Sciences (XDB 41000000), the Technical Advanced Research Project of Civil Space (D020201), the Macau Science and Technology Development Fund (0042/2018/A2 and 0090/2020/A), and Minor Planet Foundation of

Purple Mountain Observatory. GM was supported by the German Space Agency (DLR Bonn), grant 50QM1702 (HRSC on Mars Express).

References

[1] Arvidson, R. E., Boyce, J., Chapman, C., et al., 1979. Standard techniques for presentation and analysis of crater size-frequency data. Icarus 37, 467-474. https://doi.org/10.1016/0019-1035(79)90009-5.

[2] Di, K. C., Zhu, M. H., Yue, Z. Y., et al., 2019. Topographic evolution of Von Kármán crater revealed by the lunar rover Yutu-2. Geophys. Res. Lett. 46, 12764-12770. https://doi.org/10.1029/2019GL085252.

[3] Fernandes, V. A., Fritz, J., Weiss, B. P., Garrick-Bethell, I., Shuster, D. L., 2013. The bombardment history of the moon as recorded by $^{40}Ar-^{39}Ar$ chronology. Meteorit. Planet. Sci. 48, 241-269. https://doi.org/10.1111/maps.12054.

[4] Fu, X. H., Qiao, L., Zhang, J., Ling, Z. C., Li, B., 2019. The subsurface structure and stratigraphy of the Chang'E-4 landing site: orbital evidence from small craters on the Von Kármán crater floor. Res. Astron. Astrophys. 20, 8. https://doi.org/10.1088/1674-4527/20/1/8.

[5] Garrick-Bethell, I., Miljkovic, K., Hiesinger, H., et al., 2020. Troctolite 76535: a sample of the Moon's south pole-Aitken basin? Icarus 338, 113430. https://doi.org/10.1016/j.icarus.2019.113430.

[6] Hartmann, W. K., Neukum, G., 2001. Cratering chronology and the evolution of Mars. Space Sci. Rev. 96, 165-194. https://doi.org/10.1023/A:1011945222010.

[7] Haruyama, J., Matsunaga, T., Ohtake, M., et al., 2008. Global lunar-surface mapping experiment using the lunar imager/spectrometer on SELENE. Earth Planet. Space 60, 243-256. https://doi.org/10.1186/bf03352788.

[8] Haruyama, J., Ohtake, M., Matsunaga, T., et al., 2009. Long-lived volcanism on the lunar farside revealed by SELENE terrain camera. Science 323 (5916), 905-908. https://doi.org/10.1126/science.1163382.

[9] Hiesinger, H., Jaumann, R., Neukum, G., Head, J. W., 2000. Ages of mare basalts on the lunar nearside. J. Geophys. Res. Planets 105, 29239-29275. https://doi.org/10.1029/2000JE001244.

[10] Hiesinger, H., Head, J. W., Wolf, U., Jaumann, R., Neukum, G., 2011. Ages and stratigraphy of lunar mare basalts: a synthesis. Spec. Pap. Geol. Soc. Am. 477, 1-51. https://doi.org/10.1130/2011.2477(01).

[11] Huang, J., Xiao, Z., Flahaut, J., et al., 2018. Geological characteristics of Von Kármán crater, northwestern south pole-Aitken basin: Chang'E-4 landing site region. J. Geophys. Res. Planets 123, 1684-1700. https://doi.org/10.1029/2018JE005577.

[12] Ivanov, M. A., Hiesinger, H., van der Bogert, C. H., Orgel, C., Pasckert,

J. H. , Head, J. W. , 2018. Geologic history of the northern portion of the south pole – Aitken basin on the moon. J. Geophys. Res. Planets 123, 2585 – 2612. https://doi. org/10. 1029/ 2018JE005590.

[13] Kneissl, T. , van Gasselt, S. , Neukum, G. , 2011. Map – projection – independent crater size – frequency determination in GIS environments – new software tool for ArcGIS. Planet. Space Sci. 59, 1243 – 1254. https://doi. org/10. 1016/j. pss. 2010. 03. 015.

[14] Lai, J. L. , Xu, Y. , Zhang, X. P. , Xiao, L. , Yan, Q. , 2019. Comparison of dielectric properties and structure of lunar regolith at Chang'E – 3 and Chang'E – 4 landing sites revealed by ground penetrating radar. Geophys. Res. Lett. 46, 12783 – 12793. https://doi. org/ 10. 1029/2019GL084458.

[15] Lai, J. L. , Xu, Y. , Bugiolacchi, R. , et al. , 2020. First look by the Yutu – 2 rover at the deep subsurface structure at the lunar farside. Nat. Commun. 11, 3426. https://doi. org/ 10. 1038/s41467 – 020 – 17262 – w.

[16] Lemelin, M. , Lucey, P. G. , Gaddis, L. R. , Hare, T. , Ohtake, M. , 2016. Global map products from the Kaguya multiband imager at 512 ppd: minerals, FeO, and OMAT. In: 47th Lunar Planet. Sci. Conf. Abstract #2994.

[17] Li, C. L. , Su, Y. , Pettinelli, E. , et al. , 2020. The moon's farside shallow subsurface structure unveiled by Chang'E – 4 lunar penetrating radar. Sci. Adv. 6, eaay6898. https://doi. org/10. 1126/sciadv. aay6898.

[18] Ling, Z. C. , Qiao, L. , Liu, C. Q. , et al. , 2019. Composition, mineralogy and chronology of mare basalts and non – mare materials in Von Kármán crater: landing site of the Chang'E – 4 mission. Planet. Space Sci. 179, 104741. https://doi. org/ 10. 1016/j. pss. 2019. 104741.

[19] Losiak, A. , Wilhelms, D. E. , Byrne, C. J. , et al. , 2009. A new lunar impact crater database. In: 40th Lunar Planet. Sci. Conf. Abstract #1532.

[20] Melosh, H. J. , Kendall, J. , Horgan, B. , et al. , 2017. South pole – Aitken basin ejecta reveal the Moon's upper mantle. Geology 45, 1063 – 1066. https://doi. org/ 10. 1130/ G39375. 1.

[21] Michael, G. G. , Neukum, G. , 2010. Planetary surface dating from crater size frequency distribution measurements: partial resurfacing events and statistical age uncertainty. Earth Planet. Sci. Lett. 294, 223 – 229. https://doi. org/10. 1016/j. epsl. 2009. 12. 041.

[22] Michael, G. G. , Kneissl, T. , Neesemann, A. , 2016. Planetary surface dating from crater size – frequency distribution measurements: Poisson timing analysis. Icarus 277, 279 – 285. https://doi. org/10. 1016/j. icarus. 2016. 05. 019.

[23] Nelson, D. M. , Koeber, S. D. , Daud, K. , Robinson, M. S. , Watters, T. R. , Banks, M. E. , 2014. Mapping lunar maria extents and lobate scarps using LROC image products. In: 45th Lunar Planet. Sci. Conf. Abstract #2861.

[24] Neukum, G. , 1983. Meteoritenbombardement und datierung planetarer oberflachen.

In: Habilitation Thesis for Faculty Membership. University of Munich, p. 186 (English translation, 1984: Meteorite Bombardment and Dating of Planetary Surfaces). http://ntrs. nasa. gov/search. jsp? R=19840027189.

[25] Neukum, G., Ivanov, B. A., Hartmann, W. K., 2001. Cratering records in the inner solar system in relation to the lunar reference system. Space Sci. Rev. 12, 55 - 86. https:// doi. org/10. 1007/978 - 94 - 017 - 1035 - 0 _ 3.

[26] Pasckert, J. H., Hiesinger, H., van der Bogert, C. H., 2018. Lunar farside volcanism in and around the south pole - Aitken basin. Icarus 299, 538 - 562. https://doi. org/10. 1016/ j. icarus. 2017. 07. 023.

[27] Potter, R. W. K., Collins, G. S., Kiefer, W. S., McGovern, P. J., Kring, D. A., 2012. Constraining the size of the south pole - Aitken basin impact. Icarus 220, 730 - 743. https://doi. org/10. 1016/j. icarus. 2012. 05. 032.

[28] Robinson, M. S., Brylow, S. M., Tschimmel, M., et al., 2010. Lunar reconnaissance orbiter camera (LROC) instrument overview. Space Sci. Rev. 150 (124), 81. https:// doi. org/ 10. 1007/s11214 - 010 - 9634 - 2 (1 - 4).

[29] Stuart - Alexander, D. E., 1978. Geologic map of the central far side of the moon. In: U. S. Geol. Surv. Map I - 1047, Reston, VA.

[30] Wagner, R. V., Speyerer, E. J., Robinson, M. S., Team, L., 2015. New mosaicked data products from the LROC team. In: 44th Lunar Planet. Sci. Conf. Abstract #1473.

[31] Wilhelms, D. E., 1987. The geologic history of the moon. In: U. S. Geol. Surv. Prof. Pap. 1348.

[32] Wu, Y. Z., Li, L., Luo, X. X., et al., 2018. Geology, tectonism and composition of the northwest Imbrium region. Icarus 303, 67 - 90. https://doi. org/10. 1016/ j. icarus. 2017. 12. 029.

[33] Wu, W. R., Li, C. L., Zuo, W., et al., 2019. Lunar farside to be explored by Chang'E - 4. Nat. Geosci. 12, 222 - 223. https://doi. org/10. 1038/s41561 - 019 - 0341 - 7.

[34] Yingst, R. A., Chuang, F. C., Berman, D. C., Mest, S. C., 2017. Geologic mapping of the planck quadrangle of the Moon (LQ - 29). In: 48th Lunar Planet. Sci. Conf. Abstract #1680.

[35] Zhang, L., Li, J., Zeng, Z. F., et al., 2020. Stratigraphy of the Von Kármán crater based on Chang'E - 4 lunar penetrating radar data. Geophys. Res. Lett. Accepted. https:// doi. org/10. 1029/2020GL088680.

Comparative analysis of reflection characteristics of lunar penetrating radar data using numerical simulations

Wenmin Lv[1,2,3], Chao Li[1,2,3], Hanjie Song[4], Jinhai Zhang[1,2], Yangting Lin[1,2]

1. Key Laboratory of Earth and Planetary Physics, Institute of Geology and Geophysics, Chinese Academy of Sciences, Beijing, China
2. Innovation Academy for Earth Science, Chinese Academy of Sciences, Beijing, China
3. University of Chinese Academy of Sciences, Beijing, China
4. National Space Science Center, Chinese Academy of Sciences, Beijing, China

Abstract: The Moon and most extraterrestrial objects are lacking in water; thus, electromagnetic waves can penetrate deep into the ground and bring some structural information about the regolith and surrounding rocks. As an important assistant to the existing payloads, ground penetrating radar (GPR) is particularly suitable for deep space exploration, which is feasible and necessary for detecting the subsurface structures (mainly associated with the dielectric permittivity variations) of extraterrestrial objects. Unfortunately, there are not enough reference GPR profiles or drills on the extraterrestrial objects to constrain the acquired data. Here, we construct some typical models and perform numerical simulations. To make the results reasonable, various factors are considered, including the type, scale, and location of the abnormal bodies. Then, we review the reflecting phenomenon under the ground and perform comparative analyses on the reflection characteristics of different models. Finally, we present some criteria for identifying different reflectors using critical patterns of reflections. The configuration of our numerical experiments is comparable to that of the lunar penetrating radar onboard the Yutu-1 and -2; thus, the results would be reliable to support the geological interpretation of detected records as well as possible GPR detection on extraterrestrial objects in the future. This work is helpful for recognizing the reflection characteristics of the observed GPR data and would be beneficial to uncovering more critical details of subsurface structures.

Keywords: Lunar penetrating radar; Relative dielectric permittivity; Numerical simulation; Reflection characteristics; Chang'E-4

1 Introduction

Deep space exploration plays an important role in understanding the evaluation history of the solar system and is essential for the potential utilization of mineral resources from the Moon and asteroids. The commonly used telescope, camera, and spectrometer can provide key information on the topography, element distribution, and impact history of the objects; in contrast, radar exploration can detect key information of subsurface structures and provide powerful constraints for geological interpretations. Orbital radar exploration was popular before 2013 (e. g., Cooper et al., 1974; Phillips et al., 1974; Shkuratov and Bondarenko, 2001; Ono et al., 2010; Ishiyama et al., 2013), since it can provide a global map of the subsurface; however, in earlier studies, we were only able to identify one or two rough layers within the >1-km depth from the orbital radar profile, because its spatial resolution is limited by the effective dominant frequency (Fa and Jin, 2007; Ono et al., 2008; Nozette et al., 2010; Spudis et al., 2010); in addition, the energy of reflection is particularly weak due to absorption attenuation under the ground and spherical spreading attenuation at high orbital altitude (~100km) over the lunar surface. Thus, it is difficult to identify detailed local structures from the orbital radar profile.

In 2013, the Chang'E-3 mission successfully performed the first ground-based radar exploration in the Imbrium basin on the Moon (Fang et al., 2014; Fa et al., 2015; Xiao et al., 2015; Zhang et al., 2015). Although the rover Yutu only moved about 114m in distance, the lunar penetrating radar (LPR) revealed impressive local details of the lunar regolith (mixed with ejecta) within 5m depth and several basalt layers within 350m depth (Zhang et al., 2015; Hu et al., 2016; Ding et al., 2017; Dong et al., 2017; Lai et al., 2017). These details of the lunar subsurface provide the foundation and reference for our understanding of the formation of lunar regolith and the eruption history of lunar volcanoes at the landing site (Dai et al., 2014; Fang et al., 2014; Su et al., 2014; Zhang et al., 2014; Fa et al., 2015; Xiao et al., 2015; Zhang et al., 2015; Gao et al., 2016; Lai et al., 2017; Xing et al., 2017; Yuan et al., 2017; Zhang et al., 2018a, 2018b, 2019; Jia et al., 2019). At the beginning of 2019, the Chang'E-4 mission carried out the first landing detection on the far side of the Moon (Di et al., 2019; Li et al., 2019, 2020c; Lin et al., 2020; Zhang et al., 2020). The LPR, similar to that onboard the Yutu-1 rover, had obtained an approximately 424-m length profile along the track of the Yutu-2 rover within sixteen lunar days in the Von Karman crater, which revealed a much more complicated impact history on the far side, compared with the near side.

As a type of electromagnetic equipment, radar has many potential advantages, for example, it can detect the permanent shadow area (e. g. , lunar poles and impact craters) without the restriction of light (Ding et al. , 2015); in addition, the electromagnetic wave can propagate in vacuum freely and brings some high - resolution information of the dielectric permittivity variations under the ground surface. Thus, the GPR is a convenient, feasible and necessary tool for detecting the subsurface structures of extraterrestrial objects, as an important auxiliary tool to the existing payloads (Boisson et al. , 2011; Ciarletti et al. , 2011; Fa and Wieczorek, 2012).

Unfortunately, the GPR techniques (e. g. , data processing, imaging, and inversion) are not fully developed even in engineering applications on the Earth, since the physical properties and buried depths of our objects (e. g. , electric cables and pipelines) are usually well known before detection and excavation. In addition, the electromagnetic wave would encounter strong attenuation in the case of water, especially salty water, which limits the penetrating depth severely (Goodman, 1994) . In contrast, the Moon and most extraterrestrial objects are lacking in water; if any, it usually exists in the form of icy water under the ultra - low temperatures of space, which would not influence the propagation of the electromagnetic wave that much (similar to the rocks and regolith). Thus, the GPR detection is particularly suitable to and would be more and more popular for deep space exploration in the future.

As a powerful technique, there are still some obstacles existing for GPR detection. First, the GPR equipment onboard the aircraft or rover can only observe a limited quantity of reflected data, due to the limited spatial coverage and limited data - transmitting bandwidth. Besides, the equipment in space is greatly simplified compared to that on the Earth, due to the strict limitation on size, weight, and power, which means a poorer quality of the original observations. Furthermore, there are not enough reference GPR profiles or drills on the extraterrestrial objects to constrain the acquired data. Therefore, it is necessary to construct some typical models and perform numerical simulations at this stage, which would be essential for recognizing the reflection characteristics from the observed data and making a reasonable geological interpretation, even uncovering more critical details in the subsurface structures.

Fa (2013) performs echo simulation for stratified lunar subsurface, based on geometrical optics and ray tracing technique, which can take many factors into accounts, such as the reflection, transmission, attenuation, geometrical spreading, and the antenna directivity. However, this kind of method can only handle layered structures that are relatively flat. Many numerical simulation methods have been developed based on the finite - difference method (e. g. , Kunz and Luebbers, 1993; Bourgeois and Smith, 1996; Carcione, 1996; Roberts and Daniels, 1997; Daniels, 2005; Taflove and Hagness, 2005; Liu, 2006; Feng et al. , 2006; Huang et al. , 2010; Zhu et al. , 2016), which can well handle random media; unfortunately, its computational cost would be unusually heavy for large - scale models, especially when simulating high -

frequency components and high - permeability materials, due to numerical artifacts (e. g. , Zhang and Yao, 2013) . The GprMax is open - source software that can simulate electromagnetic wave propagation for complex models (Giannopoulos, 2005) . It can also handle dispersive media (Giannakis and Giannopoulos, 2014), rough surfaces (Giannakis et al. , 2016) and GPR antenna models (Giannakis et al. , 2018) . However, the input of the GprMax is only convenient for simple models but extremely complicated for random media, which have a lot of abnormal bodies buried in inhomogeneous permittivity background. The matGPR is another free GPR analysis and interpretation package, based on MATLAB (Tzanis, 2010) .

Most early works on the numerical simulation of LPR are based on layered models (e. g. , Fa, 2013; Dai et al. , 2014; Su et al. , 2014; Zhao et al. , 2018), which could not well present one of the most important properties of the lunar regolith: high heterogeneity caused by long - term meteoroid impacts. Recently, many works take the heterogeneity into account (e. g. , Hu et al. , 2016; Ding et al. , 2017; Feng et al. , 2017; Lai et al. , 2017; Li et al. , 2017; Zhang et al. , 2018b, 2019), which can better simulate the real cases; however, these works could not take some typical models into account (e. g. , cracks and ejecta). We should construct some new types of reasonable models to better simulate the wave propagation in the lunar regolith.

In this paper, we constructed four typical models (i. e. layer model, rocky abnormality model, cracked basalt model, and ejecta model) and then performed numerical simulations using the time - domain finite - difference method (Yee, 1966; Taflove and Hagness, 2005) . Furthermore, we delineated the reflection characteristics caused by different subsurface structures and performed comparative analyses. Finally, we presented some criteria for identifying different reflectors using critical patterns of reflections. The configuration of our numerical experiments is similar to that of the LPR onboard the Yutu - 1 and - 2 (Fang et al. , 2014; Zhang et al. , 2015; Lai et al. , 2016); thus, the acquired results are comparable to practical observations and would be reliable for supporting the interpretation of LPR records as well as further GPR detection on extraterrestrial objects in the future.

2 Construction of permittivity models

Regolith, rocks, ejecta, and basalt are typical geological elements that appear on the Moon and extraterrestrial objects (Jolliff et al. , 2006) . The GPR detection is sensitive to the contrast of relative dielectric permittivity upon and below the reflectors; meanwhile, the shapes of reflectors and the degree of heterogeneity of geological bodies can also introduce apparent pattern variations of reflected wavefields (Fa et al. , 2015). To build up a quantitative reference for distinguishing different types of geological units using the GPR detections, we constructed four typical relative dielectric permittivity models, i. e. , the layered homogeneous model, the rock abnormal body model, the cracked basalt model, and the ejecta model, which are taken as the inputs of our later numerical simulations. On the top of each model, a vacuum layer (0. 3m in thickness over the ground) is presented to simulate the distance from the LPR antennas to the lunar surface. Under the vacuum layer, we set several possible combinations of the lunar regolith, rocks, cracked basalt, and ejecta, either with a shallow or deep depth. The model size is 6. 2m×10. 2m for small models (or 12. 4m×10. 2m for buried deep layers), and the grid spacing is 0. 01m. Below is a detailed description of each model.

2. 1　Heterogeneous permittivity background

The lunar regolith is widely spread on the surface of the Moon, which is usually mixed with some rocks and ejecta (Heiken et al. , 1991) . Many works have been performed to study shallow structures using LPR detections (e. g. , Zhang et al. , 2015; Feng et al. , 2017; Lai et al. , 2017; Li et al. , 2017; Lai et al. , 2019; Li et al. , 2020c; Dong et al. , 2020; Zhang et al. , 2020) . Here, we consider the lunar regolith as a heterogeneous background rather than a homogeneous one to better represent the actual effect on the reflection pattern. First, we generate a gradually increasing dielectric permittivity background according to the typical 1D lunar regolith model (Olhoeft and Strangway, 1975; Heiken et al. , 1991).

$$\varepsilon = 1.919^{\rho} \tag{1}$$

with

$$\rho = 1.92 \frac{z + 12.2}{z + 18} \tag{2}$$

where ε is relative dielectric permittivity, ρ (g/cm^3) is density, and z (cm) is the depth. To generate a heterogeneous background, we add a perturbation, a low - pass filtered random sequence, into the 1D model on each trace, independently. The heterogeneous background model is shown in Fig. 1.

2. 2　Layered homogeneous model

The layered model is used to illustrate the formation of reflections from the subsurface structures. Besides the vacuum layer, we set four horizontal layers with suddenly increasing permittivity and add a suddenly decreasing permittivity layer at the depth of 3. 5m, as shown in Fig. 2. This model is ideal rather than practical but it is helpful for our understanding of the wave propagation in layered media, especially on the quantitative reflection coefficients. Fig. 2b shows the comparison of waveforms generated by our code and the GprMax (Giannopoulos, 2005) .

The first arrival of GPR data behaves as a large - amplitude plane wave, which is the reflection from the boundary between the vacuum and lunar surface. We muted the first arrival in the waveforms shown in Fig. 2b to better exhibit the weak reflections from the

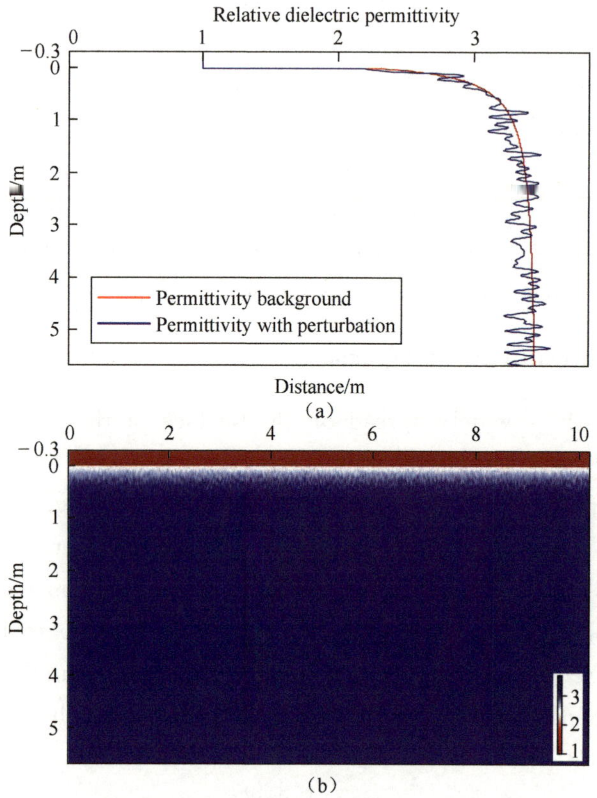

Fig. 1　Heterogeneous permittivity model. (a) The disturbance superimposed to the 1D background is a wavy rather than spiky shape, where the maximal deviation from the 1D model is ~0. 8; (b) the 2D profile of permittivity background.

subsurfaces. We can observe slightly waveform difference close to the first arrival. This could be caused by the slightly different operation on the source introduction (e. g. , using different smooth functions around the source location), which would cause different radiating amplitudes around the source location; fortunately, this is only affected at small travel times. As shown in Fig. 2b, we could only observe some differences close to the first arrival but no apparent difference for all reflections from the subsurface. The high consistency between these two results verifies that our method is correct in technique and has high numerical accuracy.

2. 3　Rocky block model

We set several standalone rocks with different sizes and shapes at different buried depths, to verify the influence of rocks, as shown in Fig. 3. The actual relative dielectric permittivity of the rock is usually considered to be ~7; however, a higher relative dielectric permittivity encounters stronger artifacts due to the numerical dispersion, which

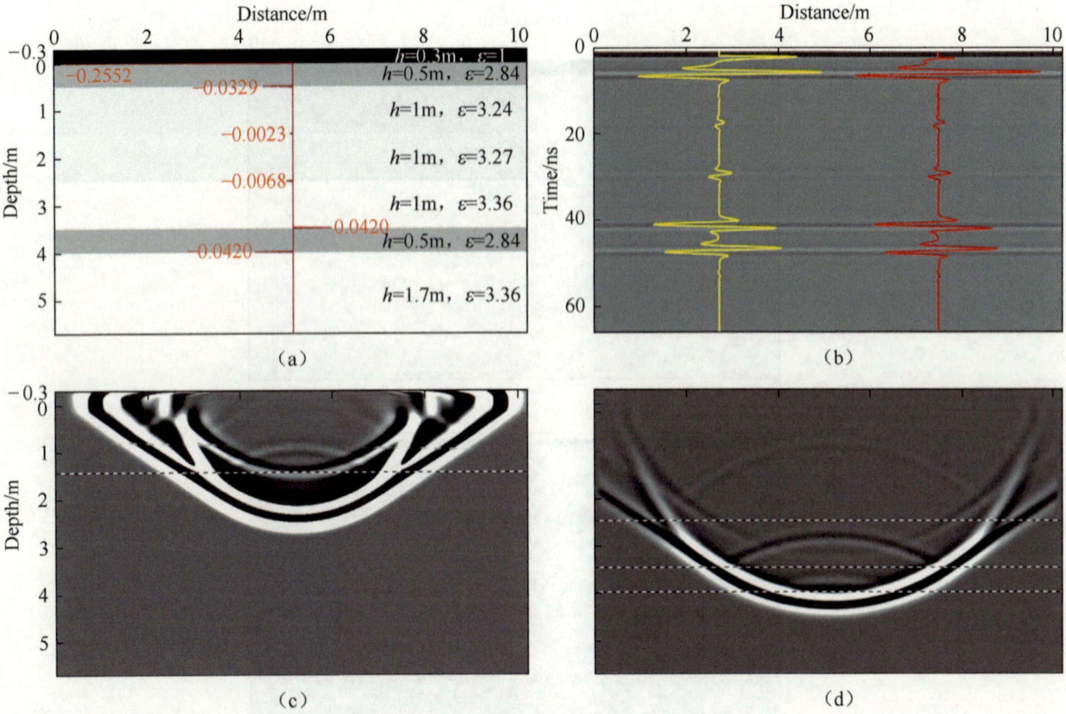

Fig. 2 Layered homogeneous model. (a) The permittivity and corresponding reflection coefficients (orange peaks) of the model; (b) simulation profile and one - channel record, where yellow wiggle is generated by the GprMax (Giannopoulos, 2005) and red wiggle is by our code; (c) wavefield snapshot at 18. 4ns; (d) wavefield snapshot at 27. 6ns. The dashed lines in (c) and (d) indicate the reflective interfaces in the model. (For interpretation of the references to color in this figure legend, the reader is referred to the web version of this article.)

may distort the waveform amplitude and reflection pattern, with the dominant frequency of 400MHz and grid spacing of 1cm. To reduce this kind of artifact, we usually need to employ smaller grid spacing at the cost of computational efficiency (Zhang and Yao, 2017) . Therefore, we intentionally apply a relatively smaller relative dielectric permittivity $\varepsilon = 5$ for the rocks (also for the basalt layer below) to speed up our simulation. The main side effect is that the amplitude of the reflections would be systematically smaller than the practical ones, but the general pattern of the reflections is almost the same. One may think about avoiding the numerical dispersion by intentionally tuning down the dominant frequency to 200— 300MHz, but this would lead to some miss of tiny reflection patterns of high - frequency components, which would be important for high - resolution imaging and inversion. In contrast, our method, intentionally tuning down the permittivity of the rock, would well keep the reflection pattern of high - frequency components.

We set six layers of abnormal bodies (including vacuum cavities and rocky blocks) in total (Fig. 3), where the permittivity is 1 for the vacuum cavity and 5 for rocky

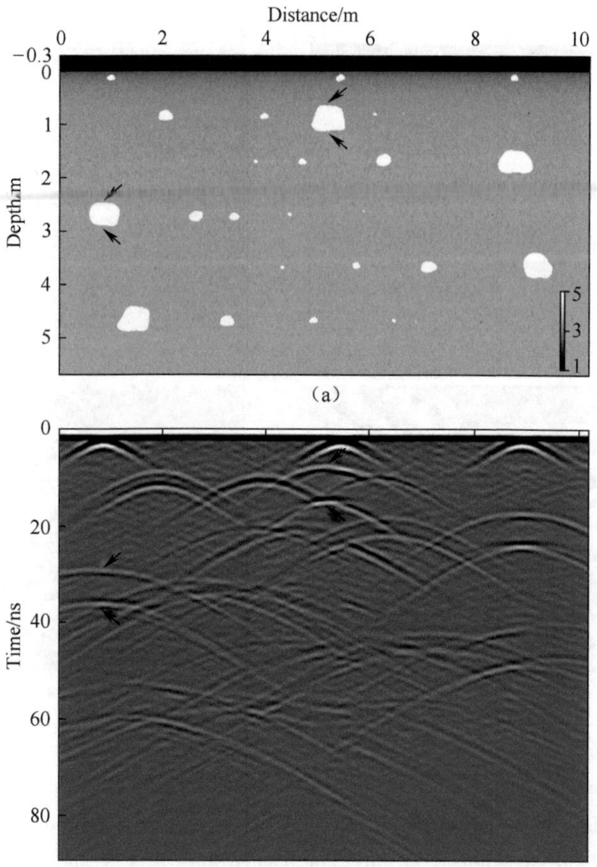

(a)

(b)

Fig. 3　Rocky block model (a) and its simulation result (b). The two blocks indicated by arrows in (a) are in similar size but with different depths. The arrows in (b) indicate the corresponding hyperbolas diffracted from the top and bottom of the two blocks

blocks. The upper - most three rocks are just buried under the lunar surface, with a diameter of ~10cm. The other five layers of rocks consist of a vacuum cavity and six rocky blocks at depths of ~0.8m, 1.5m, 2.5m, 3.5m, and 4.5m, with a diameter of 1cm for the cavity, and 1cm, 2cm, 5cm, 10cm, 20cm, and 50cm for rocky blocks, respectively. The cavities and rocky blocks are generally separated in distance and depth, to suppress the shielding effect; thus, we could clearly identify their reflections and perform comparative analyses on the effects of block size and depth.

2.4　Cracked basalt model

Meteorite impacts are the most important process on the Moon, which may shatter the basalt layers and cause both lateral and vertical transmissions of materials. Post - impact isostatic adjustments of large impact structures might have occurred but the rigidity of the lunar crust will inhibit this process and lead to a state of disequilibrium, and the remnant of cracks is one of the results (Jolliff et al., 2006). Thus, we build up the cracked basalt model with a basalt-dominated permittivity background and a series of tiny cracks (Fig. 4). The permittivity of the basalt background is again set to be 5, and the permittivity of cracks varies between 1 and 3.5 randomly assuming that the cracks are sometimes vacuum but partially filled with regolith depositions. The tree-root shape of cracks is to simulate a single impact event. The width of each branch of the cracks is at least 1cm but no >4cm. In addition, to simulate the buried basalt layer with cracks, we overlay a new layer of rocky blocks (with a thickness of ~5m) on top of the cracked basalt layer (Fig. 5).

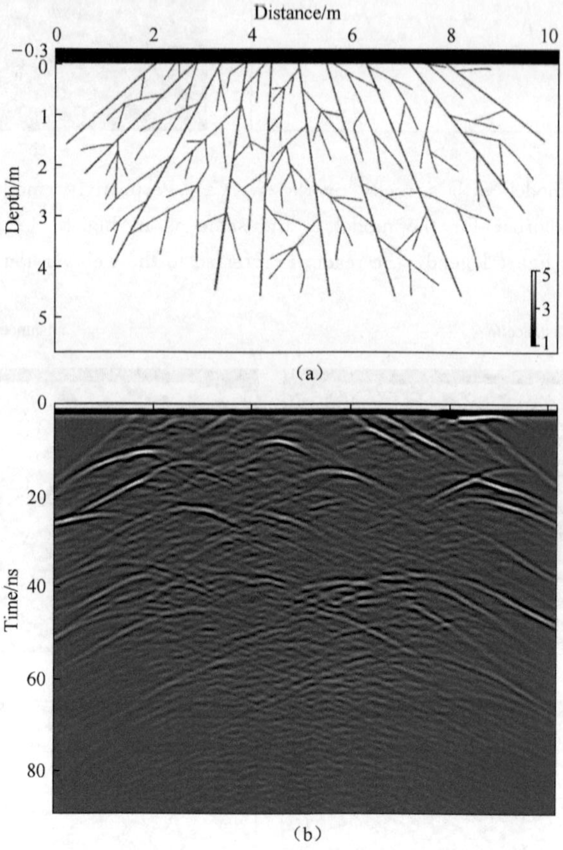

Fig. 4　Cracked basalt model (a) and its corresponding simulation profile (b)

2.5　Ejecta model

After the meteorite impacts, the rocky blocks and regolith are transported, mixed, and deposited in a wide range, layer by layer (Jolliff et al., 2006). Taking the sorting and distributing characteristics of blocks into account, we build up the ejecta model as some stacking of horizontally dominated blocks, as shown in Fig. 6. This model is similar

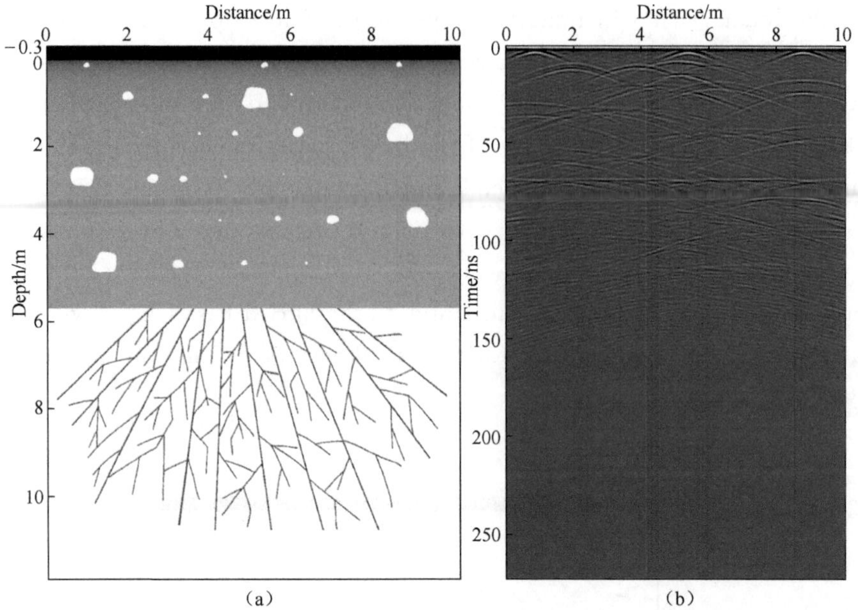

Fig. 5　Cracked basalt model with a rocky overlayer. (a) Permittivity model; (b) corresponding simulation profile. The colorbar for the model is the same as in Fig. 4. (For interpretation of the references to color in this figure legend, the reader is referred to the web version of this article.)

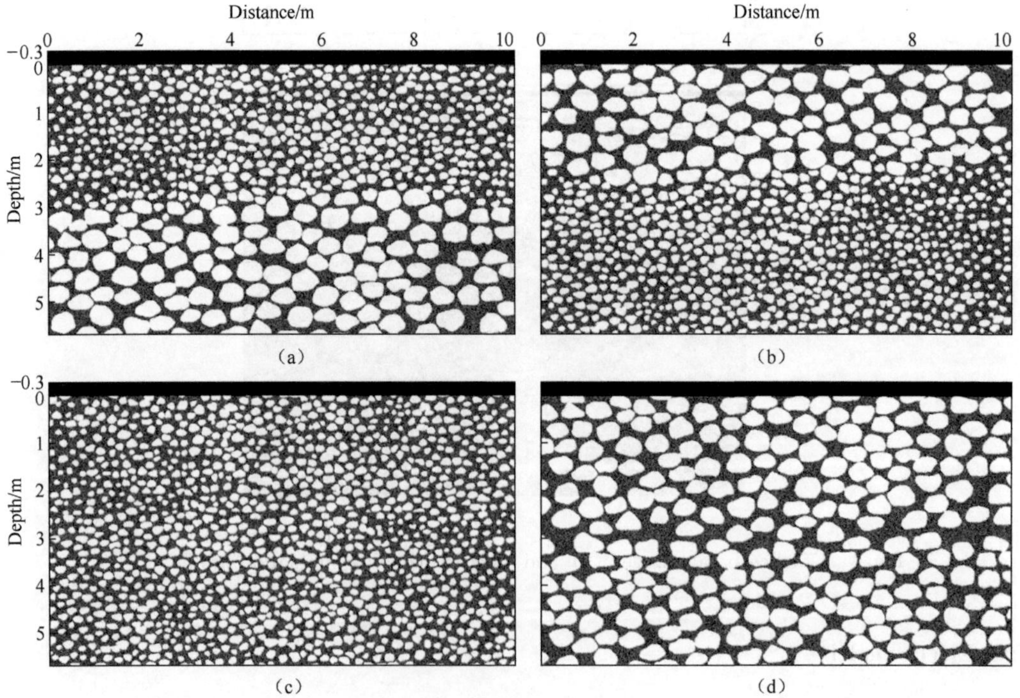

Fig. 6　Various ejecta permittivity model. (a) Normal ejecta model; (b) inverse ejecta model; (c) small- block ejecta model; (d) big - block ejecta model. The colorbar for the model is the same as in Fig. 4. (For interpretation of the references to color in this figure legend, the reader is referred to the web version of thisarticle.)

to but more complicated than the rocky block model and the cracked basalt model. The vacuum voids are partially filled with regolith so the permittivity there ranges from 1 to 3.5 randomly. Different from the crack model, this model is formed with a number of rocky blocks instead of an invariable background. The permittivity of blocks is set to be 5 again. Taking the sorting effect into account, we set the diameter to be ~15cm for the small blocks and ~ 50cm for the big blocks, respectively. Coherently, we also overlay a new rocky block layer (with a thickness of ~5m) on top of the normal ejecta model to simulate the buried ejecta layer (Fig. 7).

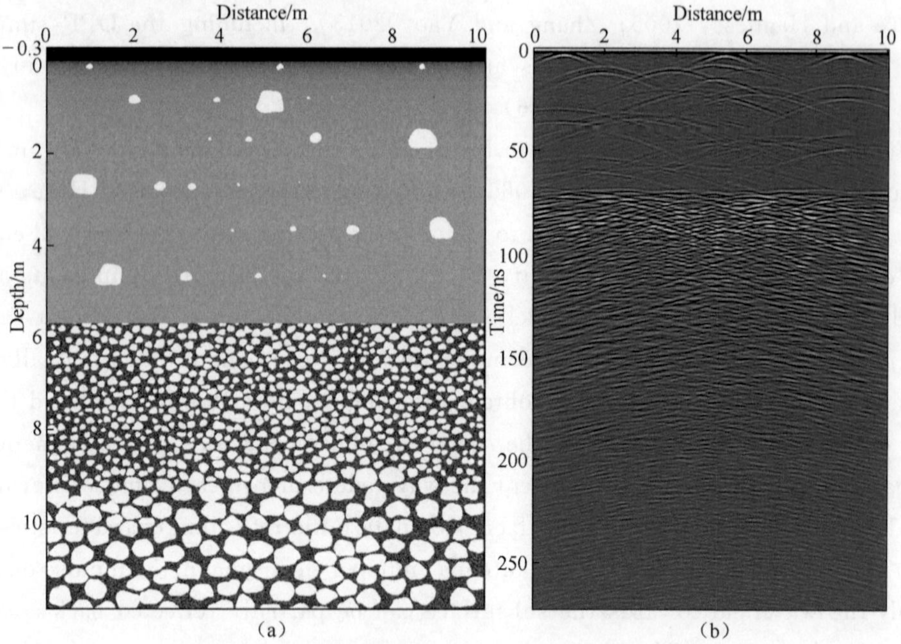

Fig. 7　Ejecta model with a rocky overlayer. (a) Permittivity model; (b) corresponding simulation profile. The colorbar for the model is the same as in Fig. 4. (For interpretation of the references to color in this figure legend, the reader is referred to the web version of this article.)

3　Numerical simulation using the finite - difference method

Numerical simulations can represent the wavefield propagation in complex media and are helpful for our understanding of observed reflections; furthermore, they are also the basis of complex - structure imaging and full waveform inversion (Schuster, 2017). There are mainly three kinds of methods to perform numerical simulations, such as the ray tracing method, spectral - element (or finite - element) method, and finite - difference method. The ray tracing method only takes kinematic characteristics of electromagnetic wave propagation into consideration but lacks dynamic information; thus, it can be well applied to massive - block models but is unfeasible in random or complex media (Fa, 2013), since it is difficult to handle multiple reflections. The spectral -

element method can well handle various complex models, due to its great freedom in describing the topography and shapes of reflectors, but its computational cost and memory demand are fairly huge, especially for random media; in addition, its model construction relies on professional tools so that it is relatively inflexible to update the input models. In contrast, the finite - difference method is the most suitable method overall, since it can acquire a fairly nice result with a simple numerical implementation; in addition, its computational cost and memory demand are much smaller than that of the spectral element method. The finite - difference method is dominant in many applications (Taflove and Hagness, 2005; Zhang and Yao, 2013), including the GPR simulations (Kunz and Luebbers, 1993; Roberts and Daniels, 1997; Daniels, 2005; Liu, 2006; Huang et al. , 2010; Zhu et al. , 2016) .

We discretized spatial derivatives in Maxwell' s curl equations (in a vacuum) using the finite - difference method (Yee, 1966; Taflove and Hagness, 2005) . Here, we only consider the transverse electric pattern to simplify the solving of the 2D Maxwell equations (Feng et al. , 2006; Irving and Knight, 2006) . First, the discrete components are only defined on the spatial grid of the model; then, the components at the current time step can be explicitly expressed by the components at two previous steps; finally, after iterating over all time steps, we can obtain the components at both spatial and temporal grids, which can clearly exhibit the detailed phenomena of electromagnetic wave propagation in the media. We can present the wavefield either in each time step or in terms of the LPR profile along the lunar surface, to analyze the reflection characteristics.

In our numerical simulation, the antenna emits an electromagnetic impulse outwards, but only the wavefield towards the subsurface can be partially reflected back and finally recorded by the antenna. The electromagnetic wave would propagate through the vacuum first and then deep into the lunar surface, where some energy is reflected on the interface between the vacuum and the lunar surface (called the surface reflection), while the other energy is transmitted into the lunar regolith and rocks, until part of the energy is reflected (called the primary reflection) at some new abnormality of relative dielectric permittivity, such as the interfaces in Fig. 2. In fact, the reflected waves would also experience the same procedure when they are transmitting upwards so that multiple reflections are produced. The multiple reflections are much weaker than the primaries, as we can see from the snapshots of the layered model (Fig. 2); thus, we only consider the primary reflections in most cases. We would obtain a trace of the radargram, which contains the surface reflection, primary reflection, and multiple reflection, for a given pair of source and receiver.

The LPR has two channels, with the dominant frequency of 60MHz (channel - 1) and 500MHz (channel - 2), respectively. We only consider the high - frequency channel, which is for detecting shallow structures. The channel - 2 antennas are 30cm over the surface, in Bowtie shape. The receivers (A and B) are close to the emitter with a distance of 15. 4cm and 31. 7 cm, respectively (Fang et al. , 2014) . Compared to the model scale, it is reasonable to set the offset to be zero. After navigating the source and receiver together from the upper left corner to the upper right corner of the model, 30cm over the lunar surface, we finally obtain a simulated 2D LPR profile (e. g. , Su et al. , 2014; Zhang et al. , 2018a, b), as shown in Fig. 8.

3. 1 The principle of diffraction and reflection

High - frequency electromagnetic waves will be reflected, diffracted, and refracted simultaneously when it meets the interface with discontinuous dielectric properties, such as blocks, voids, and cracks (Soldovieri and Solimene, 2010) . In other words, during the propagation of an electromagnetic wave, the abnormal bodies could be regarded as a new exciting source and generate disturbances, which can continuously spread around weak signals, as the diffraction waves in a radargram. As shown in Fig. 8, we assume the LPR move forward from x_0 to x over a point - like abnormal body, where the travel time of the diffractions follows.

$$(x-x_0)^2+h_0^2=v^2t^2 \tag{3}$$

It can be transformed into a hyperbolic form as

$$\frac{t^2}{\left(\frac{2h_0}{v}\right)^2}-\frac{(x-x_0)^2}{h_0^2}=1 \tag{4}$$

where h_0 is the depth of the reflector. The propagating velocity v can be approximated by the permittivity as (Zhang et al. , 2015)

$$v=\frac{c}{\sqrt{\varepsilon_r}} \tag{5}$$

where c is the light speed in a vacuum (Ristic et al. , 2009) .

When the frequency is high enough (e. g. , over 1MHz when the resistance is 10,000Ω • m), the electromagnetic wave mainly exhibits a propagation property, and the contribution of conductivity is negligible during propagation, which happens in lower frequency components. The magnetic permeability is assumed to be equal to the one in a vacuum, and the relative dielectric permittivity is shown to be independent of chemical or mineralogical composition (Olhoeft et al. , 1974; Heiken et al. , 1991; Rochette et al. , 2010) . Thus, we can see that the travel time of diffraction waves from a scattering point behaves as a hyperbola, as shown in Fig. 8, where the travel time is normalized by assuming that $v\equiv 1\text{m/s}$ (Claerbout, 1985) .

Obviously, the diffraction is the basic procedure of wave propagation, which can generate reflections from reflectors in any shape, as shown in Fig. 9. In fact, a continuous segment of reflectors can be considered as many adjacent scattering points; thus, when the scattering points are close enough to each other (as shown in Fig. 9a), the hyperbolic diffraction wave would be nearly canceled out and only those at the corner or endpoints could be presented. If we widen the interval of reflectors, the diffractions of every point would appear as a cluster of many independent hyperbolic shapes (Fig. 9c). Therefore, the diffraction wave is a general wave phenomenon while the reflection wave is a particular form of it when the scattering points are continuous in space. This idea (Claerbout, 1985) had been applied to perform the migration on the LPR data (Zhang et al., 2015).

3.2 The setting of modeling parameters

The spatial sampling interval of the LPR detection is ~5cm (Zhang et al., 2015), so we only emit and receive an electromagnetic wave every 5cm over the lunar surface (Zhang et al., 2015; Feng et al., 2017). The resulting profile is basically a zero-offset profile since the emitter and receiver are close to each other. To truthfully represent the shape of the cracks and small-size of the rocks, we set the grid spacing to be 1cm; meanwhile, such fine grid spacing is helpful to avoid strong numerical dispersion (Zhang and Yao, 2017). Besides, we intentionally reduced the dielectric permittivity from 7 to 5 to avoid strong numerical dispersion. The Ricker wavelet is taken as the source time function, whose dominant frequency is set to 400MHz, which is slightly lower than the actual dominant frequency (500MHz) to avoid apparent artifacts due to numerical dispersions. Without these configurations, the computational cost would be greatly aggravated and is unaffordable to us. The maximal travel time is ~100ns for the 6-m model or ~300ns for the 12-m model, respectively. The temporal interval is 0.02ns.

Restricted by the CPU storage and computational cost, the propagation of the electromagnetic wave has to be truncated in the numerical simulation even though it is practically infinite in free space. Therefore, an absorbing boundary condition is needed to suppress fake reflections from artificial boundaries. We applied the perfectly matched layers (20 layers on each side) to reduce the artifacts from the four boundaries (Berenger, 1994; Collino and Tsogka, 2001; Gao et al., 2017). The simulation results show that the influence of truncated boundaries is well handled; thus, our results are free of artificial reflections from model boundaries.

3.3 Distinguishing different models by the critical pattern of reflections

For the layered model (Fig. 2), we can observe clear and continuous reflections that are reflected from the sudden changes of permittivity, whose amplitude is proportional to the contrast of the permittivity above and below the reflectors; meanwhile, the polarity of the amplitude is consistent with the reflection coefficient (for the vertical incidence)

$$R = \frac{\sqrt{\varepsilon_1} - \sqrt{\varepsilon_2}}{\sqrt{\varepsilon_1} + \sqrt{\varepsilon_2}} \tag{6}$$

where ε_1 and ε_2 are the relative dielectric permittivity above and below the interface, respectively. Fig. 2b shows that the wavefronts are in opposite polarity when the permittivity variation is contrary. Besides, the amplitude is also associated with the incident angle (Zhang et al., 2018a, b); whereas, the LPR is almost zero-offset (the two receivers are 16 and 32cm away from the source, respectively) and the equipment is designed to concentrate on the downward 3D cone; therefore, we can only consider the vertical incidence case for the LPR reflections.

The reflections obtained from the rocky block model are shown in Fig. 3. Firstly, we use a simple block model (Fig. 10) to directly observe the propagation of electromagnetic waves in abnormal bodies. Obviously, the electromagnetic waves can be re-reflected for multiple times inside the bodies, so that we cannot simply take the abnormal bodies to be point-like reflectors for granted (Shihab and Al-Nuaimy, 2005). The reflections are extremely complex due to the multiple reflections inside the body and between the bodies. Nevertheless, we can still find out the relationship between each hyperbola and small scattering point (or from the top and bottom of the blocks), as indicated by the arrows in Fig. 3. The rocks close to the surface contribute to the strongest energy of the distinguishable reflections in the LPR profile. The reflections of deeper buried rocks are much weaker, which are contaminated by the coda waves of shallower reflectors. As pointed out by Zhang et al. (2015), the LPR could not observe the reflections from the steeply dipping reflectors just under the rover, since their reflections eventually propagate into other directions rather than back to the receiver location (e.g., downward, horizontal, or towards the ground surface but out of the receiver aperture). Therefore, we could only observe two nearly parallel hyperbolas for each large block, whose phases (or polarity) are on the contrary. In addition, the amplitude of the reflection from the top of the reflector is apparently larger than that from the bottom. Of course, for those irregularly-shaped rocks, the reflection is much more complicated, since it is difficult to find out their corresponding hyperbolas.

In terms of signal intensity, it is easy to notice that the signal of the shallow area can be preserved better than that of the deep area, and the small abnormal bodies (with diameter <2cm) could barely be detected. Besides, the cavity and blocks can be easily separated in the GPR profile (Fig. 3) since their diffractions can be identified. Of course, this is valid only for the reflectors in the shallow area (<2.5m). Specifically, the abnormalities at 4.5 m could be recognized but are buried in strong noises (i.e., the multiple reflections and coda waves associated with the shallower objects). For the same size reflector, a deeper buried depth would lead to hyperbola flattening (i.e., becoming wider in the horizontal) in the LPR profile, as indicated by the arrows in Fig. 3.

Fig. 4b shows the simulation result of the cracked basalt model. We can see that the diffractions are generally weak and chaotic, only nearly horizontal cracks can be detected while many other structures are hidden. This result is consistent with the principle that only when the normal direction of the reflector is towards the receiver aperture can it be detected (Claerbout, 1985; Zhang et al., 2015). Clearly, their reflections seldom behave as hyperbolas, since these reflections are mainly reflected from a segment of cracks towards the receiver aperture, which are more similar to the reflections from a segment of the dipping reflector, as shown in Fig. 9. In addition, the reflections have almost no nearly parallel reflections as that exhibited in the GPR profile of the rocky block model, which is mainly because the directions of cracks are generally random and too thin in size. A cracked basalt model is more likely a group of tightly arranged huge rocks, which have twists and turns on the body surface.

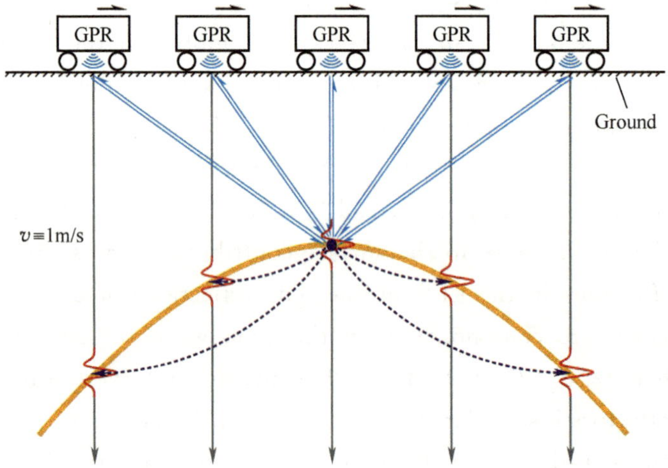

Fig. 8 Cartoon of the GPR observation and the diffraction in the profile. The dot is a point-like diffractor, the blue lines indicate the tracks of electromag-netic wave propagation, the red curve is the diffraction wave observed, and the orange hyperbola is the travel time of diffractions at different distances (assuming v≡1m/s). (For interpretation of the references to color in this figure legend, the reader is referred to the web version of this article.)

The reflections of the ejecta model are much more intricate and chaotic than that of the other models, as shown in Fig. 11. We notice that the diffractions of bigger blocks are more distinguishable than the smaller ones. The small blocks in the shallow areas can induce stronger diffractions than the big blocks in deep areas, which is due to the stronger interactions between small blocks than big ones. In other words, the diffractive wave born in the small blocks can propagate back and forth further and longer than that in the big blocks, and this phenomenon can be further illustrated by waveforms in Fig. 12. It is difficult to find a wide hyperbola since the hyperbolas are greatly compressed by the wavefield superposition and interaction between blocks. Generally, the GPR profile is formed with a pile of short earthworm-shaped reflections, with few clear hyperbolas even for the reflections from deep objects.

When the cracked basalt model and the ejecta model are buried underlying thick strata of rocky block layer, as shown in Figs. 5 and 7, we can still distinguish the overlying strata and the buried layer easily, since there is a sudden change of reflection pattern between them. Besides, there are still many multiple diffractions after we deepen the

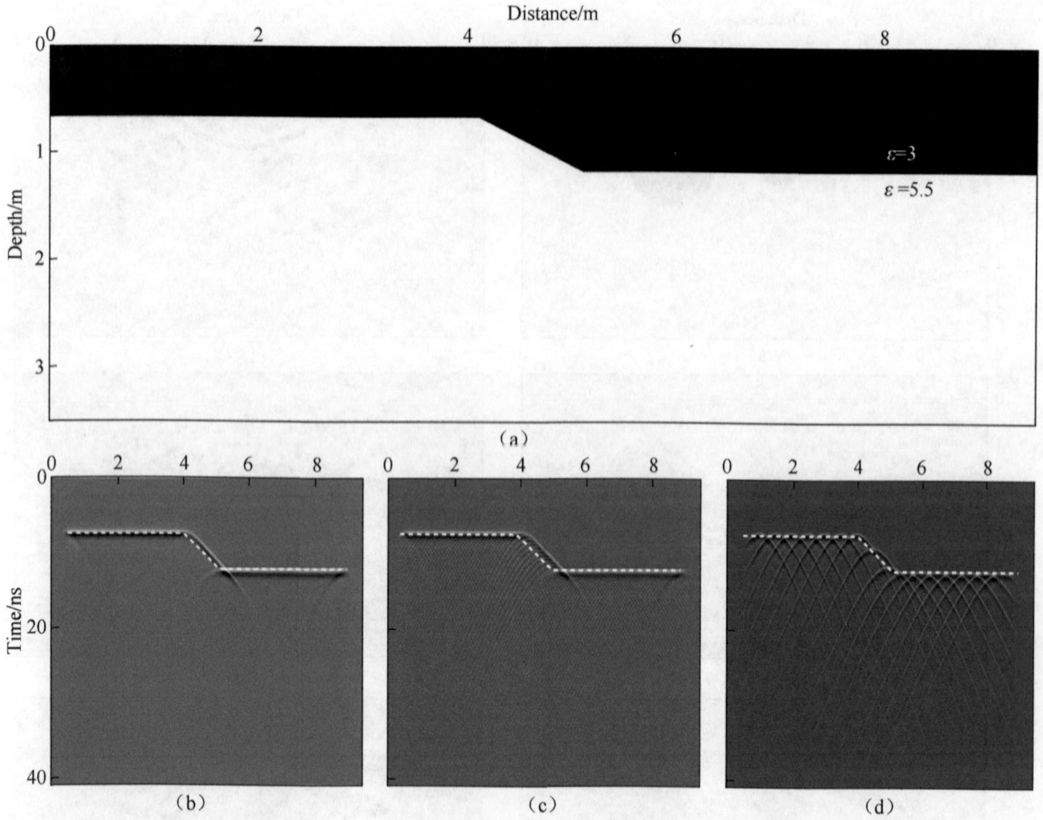

Fig. 9　Illustration of the diffractions and reflections. (a) The two - layered slope model of permittivity; (b - d) the simulation results of the slope model when the interval of the reflector is 0. 01 m, 0. 1 m, and 0. 5 m, respectively. The dashed line indicates the reflection interface

ejecta model (Fig. 7), while the profile of the cracked basalt model is almost blank and flat (Fig. 5) . However, it is difficult to distinguish the four ejecta models (Fig. 6a - d), since their reflections are almost the same and extremely chaotic (Fig. 11a - d) .

Fig. 12 shows the waveform of reflections for different models (at No. 100 trace). We abandon the first 23 temporal samples, since the direct wave and the reflections from the surface are too strong but the same for each profile. Obviously, the signal intensity of the ejecta model is stronger than that of the cracked basalt model. This is mainly caused by the strong reflectivity of the ejecta model, where most of the total energy is reflected back to the receiver, after undergoing an extremely complicated reflection - transmission process; in contrast, most of the total energy behaves as transmitted waves and is finally absorbed by the damping boundaries (Fig. 2c and d) . This reminds us that a too complicated shallow layer with strong reflections or diffractions may prevent the electromagnetic wave from penetrating down into subsurface structures, which would be unfavorable for detecting deep structures, as shown in Fig. 13a.

Fig. 13 shows the LPR profiles obtained by lunar rover Yutu - 1 and - 2. Obviously,

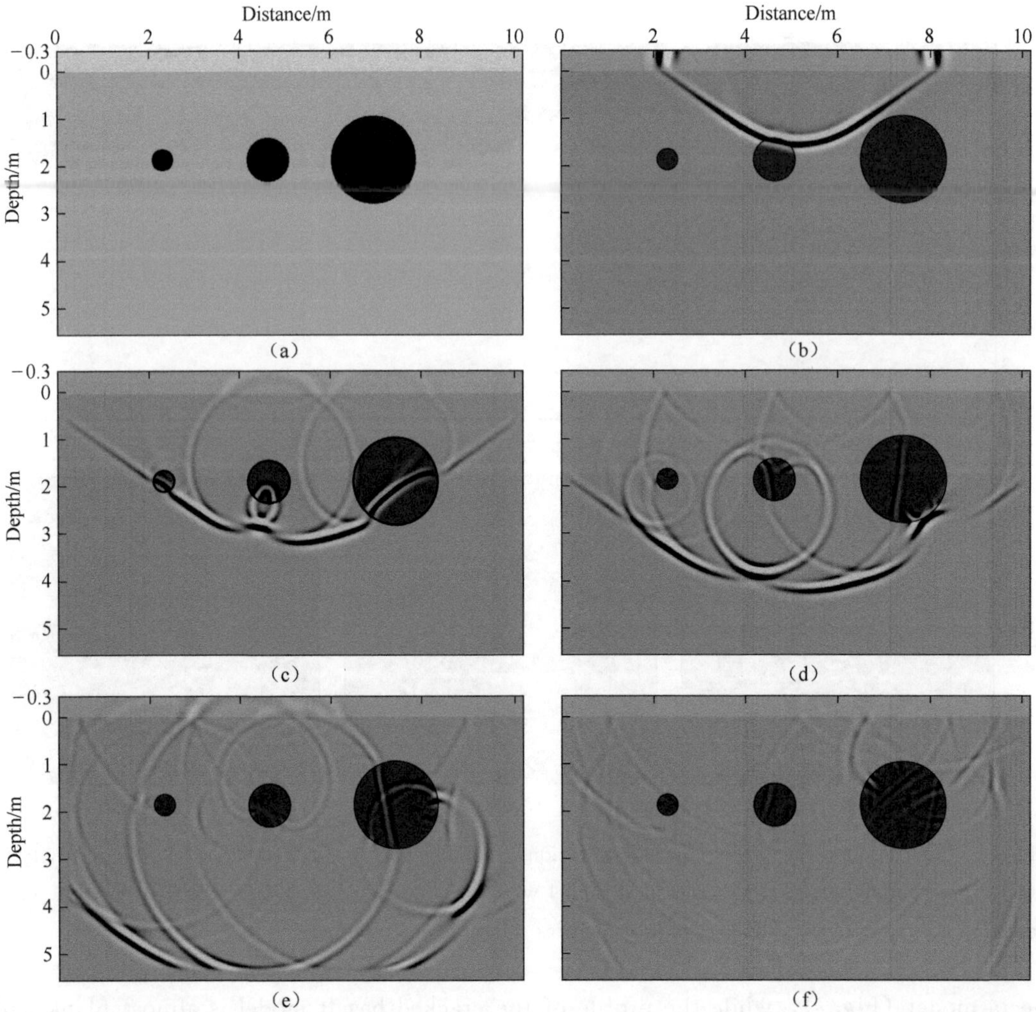

Fig. 10 The sphere model and corresponding wavefield snapshots at different times

we can observe that both profiles can be roughly divided into some layered structures, according to the rapid variation of reflection patterns, as shown by synthetic results in Figs. 5 and 7. This means that there are apparent changes in material properties, particularly in porosity, rock size and ejecta grain - size sorting. In addition, we can see that there are apparent differences between the regolith structures of the landing sites on the near side and far side. The LPR profile of the near side (Zhang et al., 2015) exhibits strong reflections at shallow depth, mainly within the travel time of 80ns; in contrast, the LPR profile of the far side (Zhang et al., 2020) exhibits strong reflections at middle depth, mainly after the travel time of 90ns. This indicates that the far side experienced a much stronger bombardment history, compared with the near side, followed by a relatively weak bombardment history by smaller meteorites. Furthermore, there are some apparent sublayers in the LPR profile of the far side, which could be multi - stage ejecta

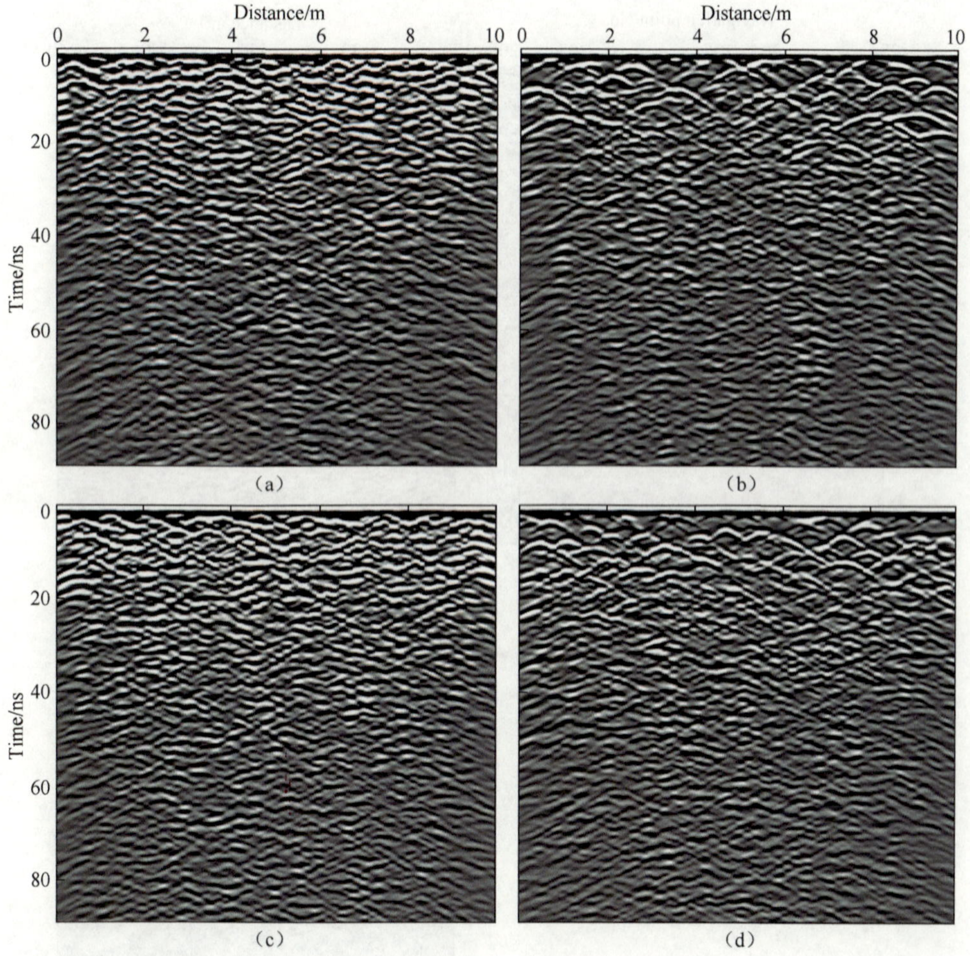

Fig. 11　Simulation results of various ejecta permittivity models. (a) Normal ejecta model; (b) inverse ejecta model; (c) small-block ejecta model; (d) big-block ejecta model

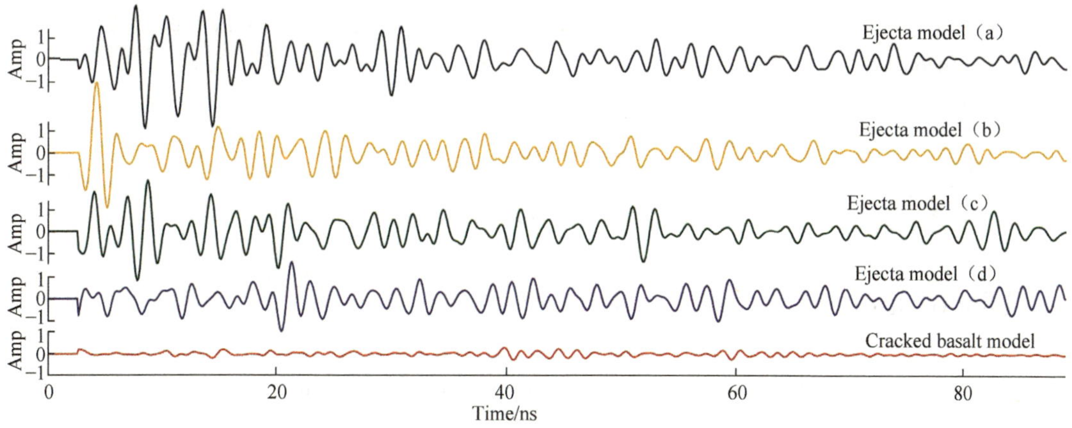

Fig. 12　Comparison of waveforms between different models

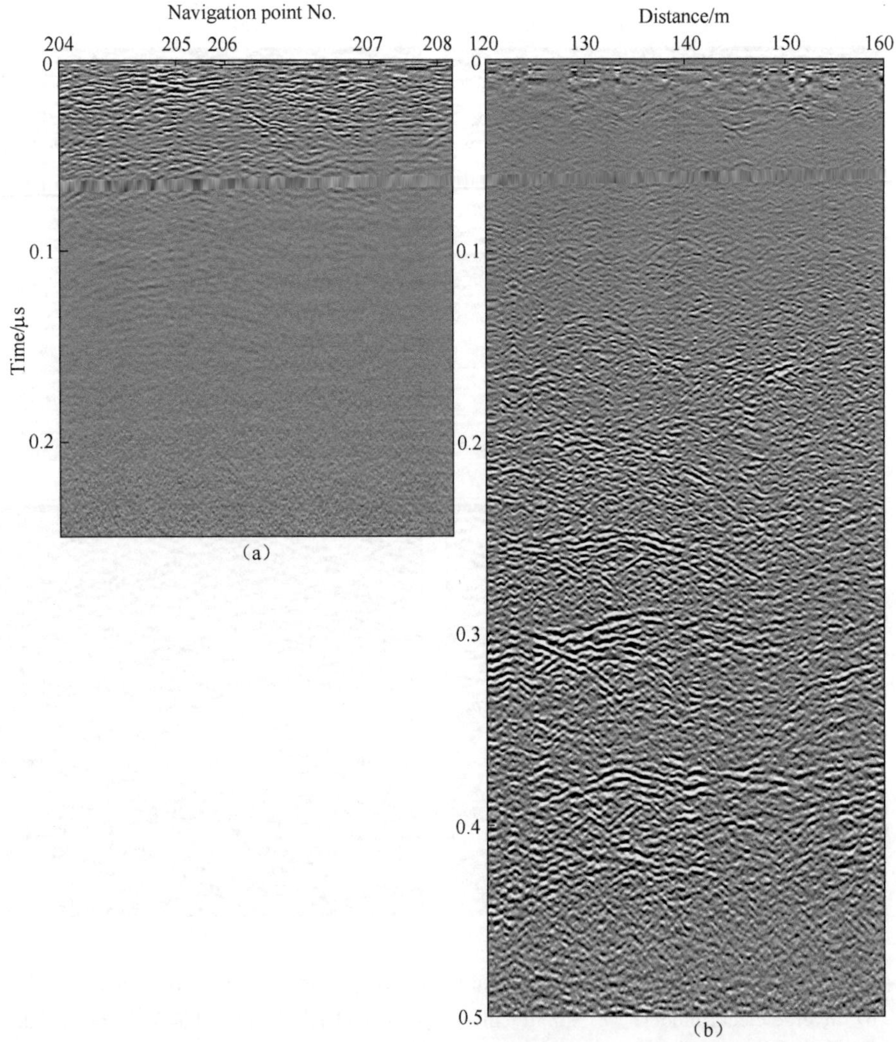

Fig. 13　Partial profiles of the LPR data obtained by Yutu－1 and－2 lunar rovers. (a) By Chang' E－3 mission, as part of Fig. S18 (a) in Zhang et al. (2015); (b) by Chang' E－4 mission, as part of Fig. S8 (a) in Zhang et al. (2020). These two profiles have a similar distance of ～40m

from different impact craters. The approximately symmetric hyperbolas in Fig. 13b indicate that the reflector should be big boulders, rather than cracks.

4　Discussion

The actual ejecta depth is about 10 to 100m (much deeper than the 5－m buried depth used here), which is difficult to be faithfully simulated due to heavy computational cost. This drawback is mainly because of the high－frequency source, fine－grid model and long－term record, and all three aspects make it unfeasible to obtain highly accurate reflections without visible numerical dispersions, even for a small-scale 2D model. To handle the task, we apply a small-scale 2D model (~5m or 10m in depth); then, we reduce the dominant frequency from 500MHz to 400MHz; finally, we apply a 1cm grid spacing even though the crack thickness and size of dust particles could be much smaller. In addition, to simplify the numerical simulation, we ignore the attenuation of electromagnetic waves during propagation, which is inevitable in practical records.

The amplitude of reflection is affected by many factors, including the geometrical spreading, shapes of the reflectors, velocity contrasts, and physical absorptions. The absorption is particularly important for evaluating the chemical components of the lunar regolith; however, a relatively accurate consideration of this factor highly depends on the accuracy of the compensation on geometrical spreading and the inversion accuracy of macro models. Obviously, these aspects are not well solved at the current stage; thus, we prefer to concentrate on considering the structure of models first. In addition, the physical absorptions would lead to a physical dispersion of waveforms, which requires a much higher accuracy of the simulation method to completely avoid the numerical dispersion. Unfortunately, most existing methods could not meet such a strict requirement at the current stage (e. g., Carcione, 1996; Zhu et al., 2016). We would be able to well consider physical absorptions in future works after we can handle 3D simulations more accurately and more efficiently.

A 2D model basically assumes that there is no variation along the y-direction (perpendicular to the page). However, practical LPR detection would be associated with all azimuthal angles within the 3D cone downward, rather than only within the $x-z$ plane. In other words, practical data would be much more complicated, even though their structures may be relatively simple compared with the 2D model here. Therefore, future works are still needed to perform similar comparable analyses but with larger-scale 3D models, which would be essential for detailed geological interpretations on critical structures.

As shown in Fig. 9, each scattering point in the model would be presented as a hyperbola in the observed LPR profile; thus, the LPR profile should be much more complicated compared with the corresponding model, as shown in Fig. 13. Consequently, it is difficult to distinguish details, such as the scale, shape, and density of rocks or cracks. Proper processing of reversely mapping the hyperbola onto the scattering point in the model domain is necessary, to make the details of reflectors much clearer and more distinguishable (Li et al., 2020b). The depth migration is such a powerful tool that can better image the depth, shape, and scale of the reflector (Claerbout, 1985; Sava and Asphaug, 2018), such as the depth migration performed on the LPR data from Chang'E-3 mission (Zhang et al., 2015; Lai et al., 2017; Li et al., 2020a) and Chang'E-4 mission (Li et al., 2020c; Zhang et al., 2020).

The field data obtained by the LPR have strong reverberations (or multiples) between the lunar surface and the bottom of the lunar rover, because the reflection coefficients between the vacuum and the metallic rover (or ground surface) are stronger than that between the lunar substructures. This is a big issue for practical data processing to reveal weak reflections (Zhang et al., 2015). Fortunately, numerical simulations can completely avoid this kind of strong energy artifact.

5 Conclusion

Ground penetrating radar is a promising tool in future deep-space explorations, but its technological accumulation is relatively weak on the Earth. Here, we present a series of comparative analyses on the reflection characteristics of the GPR data using 2D numerical simulations. Several possible typical models are fully considered with a comparable configuration as the LPR performed on the far side and near side of the Moon by Chang' E-3 and -4 missions. During the numerical simulation, many factors are taken into account to make the analyses reasonable, including the type, scale, and location of the permittivity of abnormal bodies. The reflections from small rocks appear as hyperbolas in the GPR profile; in contrast, the reflections from the top and bottom of large rocks appear as two nearly parallel hyperbolas, whose phases (or polarity) are on the contrary. The rocks close to the surface contribute to the strongest energy of the distinguishable reflections in the LPR profile. Deeply buried rocks have much weaker reflections, which are usually contaminated by the coda waves and multi-reflections of shallower reflectors. The reflections from the cracked basalt have almost no nearly parallel reflections as that exhibited in the GPR profile of the rocky block model. The reflections of the ejecta model are much more intricate and chaotic than the other models, without wide and clear hyperbolas but a pile of short earthworm-shaped reflections. These comparative analyses and empirical knowledge are helpful for us to better understand the LPR records and to distinguish different types of reflectors (e.g., rocky blocks, cracked basalt, and ejecta), which are essential for solid geological interpretations of the evolutionary history of the Moon and extraterrestrial objects.

Acknowledgement

This work was supported by the National Natural Science Foundation of China (41941002) and by the National Major Project of China (2017ZX05008-007). HS was supported by the Beijing Municipal Science and Technology Commission (Z181100002918003) and the pre-research project on Civil Aerospace Technologies No. D020201 funded by China National Space Administration (CNSA). JZ thanks the Foundation for Excellent Member of the Youth Innovation Promotion Association, Chinese Academy of Sciences (2016). We thank the Supercomputing Laboratory of

IGGCAS for providing computing resources.

References

[1] Berenger, J. P., 1994. A perfectly matched layer for the absorption of electromagnetic waves. J. Comput. Phys. 114 (2), 185 - 200, 10.106 - 3/1.4933045.

[2] Boisson, J., Heggy, E., Clifford, S. M., Yoshikawa, K., Anglade, A., Lognonné, P., 2011. Radar sounding of temperate permafrost in Alaska: analogy to the Martian midlatitude to high - latitude ice - rich terrains. J. Geophys. Res. Planets 116 (E11). https://doi.org/10.1029/2010JE003768.

[3] Bourgeois, J. M., Smith, G. S., 1996. A fully three - dimensional simulation of a ground with experiment. IEEE Trans. Geosci. Remote Sens. 34 (1), 36 - 44. https://doi.org/ 10.1109/36.481890.

[4] Carcione, J. M., 1996. Ground penetrating radar: wave theory and numerical simulation in lossy anisotropic media. Geophysics 61 (6), 1664 - 1677. https://doi.org/10.1190/ 1.1444085.

[5] Ciarletti, V., Corbel, C., Plettemeier, D., Cais, P., Clifford, S. M., Hamran, S. E., 2011. WISDOM GPR designed for shallow and high - resolution sounding of the Martian subsurface. Proc. IEEE 99 (5), 824 - 836. https://doi.org/10.1109/JPROC.2010.2100790.

[6] Claerbout, J. F., 1985. Imaging the Earth's Interior. Blackwell Scientific Publications Oxford.

[7] Collino, F., Tsogka, C., 2001. Application of the perfectly matched absorbing layer model to the linear elastodynamic problem in anisotropic heterogeneous media. Geophysics 66 (1), 294 - 307. https://doi.org/10.1190/1.1444908.

[8] Cooper, M. R., Kovach, R. L., Watkins, J. S., 1974. Lunar near - surface structure. Rev. Geophys. 12 (3), 291 - 308. https://doi.org/10.1002/9781118782118.ch3.

[9] Dai, S., Su, Y., Xiao, Y., Feng, J. Q., Xing, S. G., Ding, C. Y., 2014. Echo simulation of lunar penetrating radar: based on a model of inhomogeneous multilayer lunar regolith structure. Res. Astron. Astrophys. 14 (12), 1642 - 1653. https://doi.org/10.1088/ 1674 - 4527/14/12/012.

[10] Daniels, D. J., 2005. Ground penetrating radar. In: Encyclopedia of RF and Microwave Engineering. https://doi.org/10.1002/0471654507.eme152.

[11] Di, K., Liu, Z., Liu, B., Wan, W., Peng, M., Wang, Y., Gou, S., Yue, Z., Xin, X., Jia, M., Niu, S., 2019. Chang'e - 4 lander localization based on multi - source data. J. Remote Sens. 23 (1), 177 - 184, 10.07 - 4619 (2019) 01 - 0177 - 08.

[12] Ding, C., Jianqing, F., Lei, Z., 2015. A review of application of radar -

detection techniques in lunar exploration. Astron. Res. Technol. 12 (2), 228 - 242.

[13] Ding, C., Su, Y., Xing, S., Dai, S., Xiao, Y., Feng, J., Liu, D., Li, C., 2017. Numerical simulations of the lunar penetrating radar and investigations of the geological structures of the lunar regolith layer at the Chang'E 3 landing site. Int J. Antennas Propag. https://doi.org/10.1155/2017/3013249.

[14] Dong, Z., Fang, G., Ji, Y., Gao, Y., Wu, C., Zhang, X., 2017. Parameters and structure of lunar regolith in Chang'E - 3 landing area from lunar penetrating radar (LPR) data. Icarus 282, 40 - 46. https://doi.org/10.1016/j.icarus.2016.09.010.

[15] Dong, Z., Feng, X., Zhou, H., Liu, Cai, Zeng, Z., Li, Jing, Liang, W., 2020. Properties analysis of lunar regolith at Chang'E - 4 landing site based on 3d velocity spectrum of lunar penetrating radar. Remote Sens. 12 (4) https://doi.org/10.3390/rs12040629.

[16] Fa, W., 2013. Simulation for ground penetrating radar (GPR) study of the subsurface structure of the Moon. J. Appl. Geophys. 99, 98 - 108. https://doi.org/10.1016/j.jappgeo.2013.08.002.

[17] Fa, W., Jin, Y. Q., 2007. Simulation of brightness temperature from lunar surface and inversion of regolith - layer thickness. J. Geophys. Res. Planets 112, E05003. https:// doi.org/10.1029/2006JE002751.

[18] Fa, W., Wieczorek, M. A., 2012. Regolith thickness over the lunar nearside: results from Earth - based 70 - cm Arecibo radar observations. Icarus 218 (2), 771 - 787. https://doi.org/10.1016/j.icarus.2012.01.010.

[19] Fa, W., Zhu, M., Liu, T., Plescia, J. B., 2015. Regolith stratigraphy at the Chang'E - 3 landing site as seen by lunar penetrating radar. Geophys. Res. Lett. 42 (23), 10, 179 - 110, 187. https://doi.org/10.1002/2015gl066537.

[20] Fang, G., Zhou, B., Ji, Y., Zhang, Q., Shen, S., Li, Y., Guan, H., Tang, C., Gao, Y., Lu, W., 2014. Lunar penetrating radar onboard the Chang'e - 3 mission. Res. Astron. Astrophys. 14 (12), 1607 - 1622. https://doi.org/10.1088/16744527/1 - 4/12/009.

[21] Feng, D., Dai, Q., He, J., He, G., 2006. Finite difference time domain method of GPR forward simulation. Prog. Geophys. 2 (21), 630 - 636, 10.101 - 6/S1001 - 8042 (06) 60011 - 0.

[22] Feng, J., Su, Y., Ding, C., Xing, S., Dai, S., Zou, Y., 2017. Dielectric properties estimation of the lunar regolith at CE - 3 landing site using lunar penetrating radar data. Icarus 284, 424 - 430. https://doi.org/10.1016/j.icarus.2016.12.005.

[23] Gao, Y., Dong, Z., Fang, G. Y., Ji, Y., Zhou, B., 2016. The processing and analysis of lunar penetrating radar channel - 1 data from Chang'E - 3. J. Radars 4, 518 - 526. https://doi.org/10.12000/JR15030.

[24] Gao, Y., Song, H., Zhang, J., Yao, Z., 2017. Comparison of artificial absorbing boundaries for acoustic wave equation modelling. Explor. Geophys. 48 (1), 76-93. https://doi.org/10.1071/EG15068.

[25] Giannakis, I., Giannopoulos, A., 2014. A novel piecewise linear recursive convolution approach for dispersive media using the finite-difference time-domain method. IEEE Trans. Antennas Propag. 62 (5), 2669-2678. https://doi.org/10.1109/TAP.2014.2308549.

[26] Giannakis, I., Giannopoulos, A., Warren, C., 2016. A realistic FDTD numerical modeling framework of ground penetrating radar for landmine detection. IEEE J. Sel. Top. Appl. Earth Obs. Remote Sens. 9 (1), 37-51. https://doi.org/10.1109/JSTARS.2015.2468597.

[27] Giannakis, I., Giannopoulos, A., Warren, C., 2018. Realistic FDTD GPR antenna models optimised using a novel linear/non-linear full waveform inversion. IEEE Trans. Geosci. Remote Sens. 207 (3), 1768-1778. https://doi.org/10.1109/TGRS.2018.2869027.

[28] Giannopoulos, A., 2005. Modelling ground penetrating radar by GprMax. Constr. Build. Mater. 19 (10), 755-762. https://doi.org/10.1016/j.conbuildmat.2005.06.007.

[29] Goodman, D., 1994. Ground-penetrating radar simulation in engineering and archaeology. Geophysics 59 (2), 224-232. https://doi.org/10.1016/01489062(94)9-11045.

[30] Heiken, G. H., Vaniman, D. T., French, B. M., 1991. Lunar Sourcebook: A user's Guide to the Moon. The University of Cambridge.

[31] Hu, Y., Zeng, Z., Li, J., Liu, F., 2016. Simulation and processing of LPR onboard the rover of Chang'E-3 mission: based on multilayer lunar regolith structure stochastic media model. In: Paper Presented at the 2016 16th International Conference on Ground Penetrating Radar (GPR). https://doi.org/10.1109/ICGPR.2016.7572611.

[32] Huang, Q., Li, Z., Wang, Y., 2010. A parallel 3-D staggered grid pseudospectral time domain method for ground-penetrating radar wave simulation. J. Geophys. Res. 115 (B12) https://doi.org/10.1029/2010jb007711.

[33] Irving, J., Knight, R., 2006. Numerical modeling of ground-penetrating radar in 2-D using MATLAB. Comput. Geosci. 32 (9), 1247-1258. https://doi.org/10.1016/j.cageo.2005.11.006.

[34] Ishiyama, K., Kumamoto, A., Ono, T., Yamaguchi, Y., Haruyama, J., Ohtake, M., Yuto, K., Naoki, T., Oshigami, S., 2013. Estimation of the permittivity and porosity of the lunar uppermost basalt layer based on observations of impact craters by SELENE. J. Geophys. Res. Planets 118 (7), 1453-1467. https://doi.org/10.1002/jgre.20102.

[35] Jia, Z., Liu, S., Zhang, L., Hu, B., Zhang, J., 2019. Weak signal

extraction from lunar penetrating radar channel 1 data based on local correlation. Electronics 8, 573. https://doi. org/10. 3390/electronics8050573.

[36] Jolliff, B. L., Wieczorek, M. A., Shearer, C. K., Neal, C. R., 2006. New Views of the Moon, Reviews in Mineralogy and Geochemistry, 60. Mineralogical Society of America, Chantilly, VA.

[37] Kunz, K. S., Luebbers, R. J., 1993. The Finite Difference Time Domain Method for Electromagnetics. CRC press.

[38] Lai, J., Xu, Y., Zhang, X., Tang, Z., 2016. Structural analysis of lunar subsurface with Chang' E - 3 lunar penetrating radar. Planet. Space Sci. 120, 96 - 102. https://doi. org/ 10. 1016/j. pss. 2015. 10. 014.

[39] Lai, J., Xu, Y., Zhang, X., Tang, Z., 2017. Lunar regolith stratigraphy analysis based on the simulation of lunar penetrating radar signals. Adv. Space Res. 60 (9), 2099 - 2107. https://doi. org/10. 1016/j. asr. 2017. 07. 032.

[40] Lai, J., Xu, Y., Zhang, L., Xiao, Q., Yan, X., Meng, X., Zhou, B., Dong, D., Zhao, D., 2019. Comparison of dielectric properties and structure of lunar regolith at Chang'e - 3 and Chang'e - 4 landing sites revealed by ground - penetrating radar. Geophys. Res. Lett. 46, 12783 - 12793. https://doi. org/ 10. 1029/2019gl084458.

[41] Li, J., Chen, Y., Zeng, Z., 2020a. Estimated lunar regolith structure based on the least - squares Kirchhoff migration of CE - 3 lunar penetrating radar data. IEEE Geosci. Remote S. (in press). https://doi. org/10. 1109/LGRS. 2020. 2986033.

[42] Li, C., Lin, Y., Lv, W., Zhang, J., 2020b. Eliminating above - surface diffractions in GPR data using iterative Stolt migration. Geophysics (in revision).

[43] Li, J., Zeng, Z., Liu, C., Huai, N., Wang, K., 2017. A study on lunar regolith quantitative random model and lunar penetrating radar parameter inversion. IEEE Geosci. Remote Sens. Lett. 14 (11), 1953 - 1957. https://doi. org/ 10. 1109/ LGRS. 2017. 2743618.

[44] Li, C., Liu, D., Liu, B., Ren, X., Liu, J., He, Z., Zuo, W., Zeng, X., Xu, R., Tan, X., 2019. Chang'E - 4 initial spectroscopic identification of lunar far - side mantle - derived materials. Nature 569 (7756), 378. https://doi. org/ 10. 1038/s41586 - 019 - 1189 - 0.

[45] Li, C., Su, Y., Pettinelli, E., Xing, S., Ding, C., Liu, J., Ren, X., Lauro, S. E., Soldovieri, F., Zeng, X., Gao, X., Chen, W., Dai, S., Liu, D., Zhang, G., Zuo, W., Wen, W., Zhang, Z., Zhang, X., Zhang, H., 2020c. The Moon' s farside shallow subsurface structure unveiled by Chang'E - 4 lunar penetrating radar. Sci. Adv. 6 https://doi. org/10. 1126/ sciadv. aay6898 eaay6898.

[46] Lin, H., He, Z., Yang, W., Lin, Y., Xu, R., Zhang, C., Zhu, M., Chang, R., Zhang, J., Li, C., Lin, H., Liu, Y., Gou, S., Hu, S., Xue,

C. , Yang, J. , Zhong, J. , Fu, X. , Wan, W. , Zou, Y. , 2020. Olivine - norite rock detected by the lunar rover Yutu - 2 likely crystallized from the SPA - impact melt pool. Natl. Sci. Rev. 7 (5), 913 - 920. https://doi. org/10. 1093/nsr/nwz183.

[47] Liu, L. , 2006. Fracture characterization using borehole radar: numerical modeling. Water Air Soil Pollut. Focus 6 (1 - 2), 17 - 34. https://doi. org/10. 1007/s11267 - 005 - 9010 - 5.

[48] Nozette,S. , Spudis, P. , Bussey, B. , Jensen, R. , Raney, K. , Winters, H. , Lichtenberg, C. L. , Marinelli, W. , Crusan, J. , Gates, M. , Robinson, M. , 2010. The lunar reconnaissance orbiter miniature radio frequency (mini - RF) technology demonstration. Space Sci. Rev. 150, 285 - 302. https://doi. org/10. 1007/s11214 - 009 - 9607 - 5.

[49] Olhoeft, G. R. , Strangway, D. , 1975. Dielectric properties of the first 100 meters of the Moon. Earth Planet. Sci. Lett. 24 (3), 394 - 404. https://doi. org/10. 1016/0012 - 821X (75) 90146 - 6.

[50] Olhoeft, G. , Frisillo, A. , Strangway, D. , Sharpe, H. , 1974. Temperature dependence of electrical conductivity and lunar temperatures. Earth Moon Planet. 9 (1 - 2), 79 - 87. https://doi. org/10. 1007/BF00565394.

[51] Ono, T. , Kumamoto, A. , Yamaguchi, Y. , Yamaji, A. , Kobayashi, T. , Kasahara, Y. , Oya, H. , 2008. Instrumentation and observation target of the lunar radar sounder (LRS) experiment on - board the SELENE spacecraft. Earth Planet. Space 60 (4), 321 - 332. https://doi. org/10. 1186/BF03352797.

[52] Ono, T. , Kumamoto, A. , Kasahara, Y. , Yamaguchi, Y. , Yamaji, A. , Kobayashi, T. , Oshigami, S. , Nakagawa, H. , Goto, Y. , Hashimoto, K. , 2010. The lunar radar sounder (LRS) onboard the KAGUYA (SELENE) spacecraft. Space Sci. Rev. 154 (1 - 4), 145 - 192. https://doi. org/10. 1007/s11214 - 010 -9673 - 8.

[53] Phillips, R. , Adams, G. , Brown Jr. , W. , Eggleton, R. , Jackson, P. , Jordan, R. , Peeples, W. , Porcello, L. , Schaber, G. , Sill, W. , 1974. The Apollo 17 lunar sounder experiment: A progress report. In: Lunar & Planetary Science Conference.

[54] Ristic, A. V. , Petrovacki, D. , Govedarica, M. , 2009. A new method to simultaneously estimate the radius of a cylindrical object and the wave propagation velocity from GPR data. Comput. Geosci. 35 (8), 1620 - 1630. https://doi. org/10. 1016/j. cageo. 2009. 01. 003.

[55] Roberts,R. L. , Daniels, J. J. , 1997. Modeling near - field GPR in three dimensions using the FDTD method. Geophysics 62 (4), 1114 - 1126. https://doi. org/10. 1190/- 1. 1444212.

[56] Rochette, P. , Gattacceca, J. , Ivanov, A. , Nazarov, M. , Bezaeva, N. , 2010. Magnetic properties of lunar materials: meteorites, Luna and Apollo returned

samples. Earth Planet. Sci. Lett. 292 (3 - 4), 383 - 391. https://doi. org/10. 1016/j. epsl. 2010. 02. 007.

[57] Sava, P., Asphaug, E., 2018. 3D radar wavefield migration of comet interiors. Adv. Space Res. 62 (5), 1146 - 1164. https://doi. org/10. 1016/j. asr. 2018. 06. 009.

[58] Schuster, G. T., 2017. Seismic Inversion. Society of Exploration Geophysicists.

[59] Shihab, S., Al - Nuaimy, W., 2005. Radius estimation for cylindrical objects detected by ground penetrating radar. Subsurf. Sens. Technol. Appl. 6 (2), 151 - 166. https://doi. org/10. 1007/s11220 - 005 - 0004 - 1.

[60] Shkuratov, Y. G., Bondarenko, N. V., 2001. Regolith layer thickness mapping of the Moon by radar and optical data. Icarus 149 (2), 329 - 338. https://doi. org/10. 1006/ icar. 2000. 6545.

[61] Soldovieri, F., Solimene, R., 2010. Ground Penetrating Radar Subsurface Imaging of Buried Objects. In - Tech.

[62] Spudis, P. D., Bussey, D. B. J., Baloga, S. M., Butler, B. J., Carl, D., Carter, L. M., Chakraborty, M., Elphic, R. C., Gillis - Davis, J. J., Goswami, J. N., Heggy, E., Hillyard, M., Jensen, R., Kirk, R. L., LaVallee, D., McKerracher, P., Neish, C. D., Nozette, S., Nylund, S., Palsetia, M., Patterson, W., Robinson, M. S., Raney, R. K., Schulze, R. C., Sequeira, H., Skura, J., Thompson, T. W., Thomson, B. J., Ustinov, E. A., Winters, H. L., 2010. Initial results for the north pole of the Moon from mini - SAR, Chandrayaan - 1 mission. Geophys. Res. Lett. 37, L06204. https://doi. org/10. 1029/2009GL042259.

[63] Su, Y., Fang, G., Feng, J., Xing, S., Ji, Y., Zhou, B., Gao, Y., Li, H., Dai, S., Xiao, Y., 2014. Data processing and initial results of Chang'e - 3 lunar penetrating radar. Res. Astron. Astrophys. 14 (12), 1623 - 1632. https://doi. org/10. 1088/1674 - 4527/14/12/010.

[64] Taflove, A., Hagness, S. C., 2005. Computational Electrodynamics: The Finite - Difference Time - Domain Method. Artech House.

[65] Tzanis, A., 2010. matGPR release 2: a freeware MATLAB® package for the analysis & interpretation of common and single offset GPR data. Fast Times 15 (1), 17 -43.

[66] Xiao, L., Zhu, P., Fang, G., Xiao, Z., Zou, Y., Zhao, J., Zhao, N., Yuan, Y., Qiao, L., Zhang, X., 2015. A young multilayered terrane of the northern Mare Imbrium revealed by Chang'E - 3 mission. Science 347 (6227), 1226 -1229, 10. 11 -26/science. 1259866.

[67] Xing, S. G., Su, Y., Feng, J. Q., Dai, S., Xiao, Y., Ding, C. Y., Li, C. L., 2017. The penetrating depth analysis of lunar penetrating radar onboard Chang'e - 3 rover. Res. Astron. Astrophys. 17 (5), 79 - 90. https://doi. org/

10. 1088/1674 - 4527/17/5/46.

[68] Yee, K. , 1966. Numerical solution of initial boundary value problems involving Maxwell' s equations in isotropic media. IEEE Trans. Antennas Propag. 14 (3), 302 - 307. https://doi. org/10. 1109/TAP. 1966. 1138693.

[69] Yuan, Y. , Zhu, P. , Zhao, N. , Xiao, L. , Garnero, E. , Xiao, Z. , Zhao, N. , Qiao, L. , 2017. The 3 - D geological model around Chang'E - 3 landing site based on lunar penetrating radar channel 1 data. Geophys. Res. Lett. 44 (13), 6553 - 6561. https://doi. org/10. 1002/2017GL073589.

[70] Zhang, J. , Yao, Z. , 2013. Optimized finite - difference operator for broadband seismic wave modeling. Geophysics 78 (1), A13 - A18. https://doi. org/10. 1190/geo2012 - 0277. 1.

[71] Zhang, J. , Yao, Z. , 2017. Exact local refinement using Fourier interpolation for nonuniform - grid modeling. Earth Planet. Phys. 1 (1), 58 - 62. https://doi. org/10. 26464/epp2017008.

[72] Zhang, H. B. , Zheng, L. , Su, Y. , Fang, G. Y. , Zhou, B. , Feng, J. Q. , Xing, S. G. , Dai, S. , Li, J. D. , Ji, Y. C. , Gao, Y. Z. , Xiao, Y. , Li, C. L. , 2014. Performance evaluation of lunar penetrating radar onboard the rover of CE - 3 probe based on results from ground experiments. Res. Astron. Astrophys. 14 (12), 1633 - 1641. https://doi. org/10. 1088/1674 - 4527/14/12/011.

[73] Zhang, J. , Yang, W. , Hu, S. , Lin, Y. , Fang, G. , Li, C. , Peng, W. , Zhu, S. , He, Z. , Zhou, B. , 2015. Volcanic history of the Imbrium basin: a close - up view from the lunar rover Yutu. Proc. Natl. Acad. Sci. 112 (17), 5342 - 5347. https://doi. org/10. 1073/pnas. 1503082112.

[74] Zhang, L. , Zeng, Z. , Li, J. , Huang, L. , Huo, Z. , Wang, K. , Zhang, J. , 2018a. Parameter estimation of lunar regolith from lunar penetrating radar data. Sensors 18 (9), 2907. https://doi. org/10. 3390/s18092907.

[75] Zhang, L. , Zeng, Z. , Li, J. , Lin, J. , Hu, Y. , Wang, X. , Sun, X. , 2018b. Simulation of the lunar regolith and lunar - penetrating radar data processing. IEEE J. Sel. Top. Appl. Earth Obs. Remote Sens. 11 (2), 655 - 663. https://doi. org/10. 1109/JSTARS. 2017. 2786476.

[76] Zhang, L. , Zeng, Z. , Li, J. , Huang, L. , Huo, Z. , Zhang, J. , Huai, N. , 2019. A story of regolith told by lunar penetrating radar. Icarus 321, 148 - 160. https://doi. org/ 10. 1016/j. icarus. 2018. 11. 006.

[77] Zhang, J. , Zhou, B. , Lin, Y. , Zhu, M. , Song, H. , Dong, Z. , Gao, Y. , Di, K. , Yang, W. , Lin, H. , Yang, J. , Liu, E. , Wang, L. , Lin, Y. , Li, C. , Yue, Z. , Yao, Z. , Ouyang, Z. , 2020. Lunar regolith and substructure at Chang'E - 4 landing site in south pole - Aitken basin. Nat. Astron. (in revision) .

[78] Zhao, N. , Zhu, P. , Yuan, Y. , Zhang, B. Z. , Deng, J. , 2018. The shallow subsurface structures of Chang ' E - 3 landing site based on the wavefield

characteristics of LPR Channel - 2B data. Adv. Space Res. 62 (4), 884 - 889. https://doi.org/10.1016/j.asr.2018.05.023.

[79] Zhu, T., Carcione, J. M., Botelho, M. A. B., 2016. Reverse time imaging of ground - penetrating radar and SH - seismic data including the effects of wave loss. Geophysics 81 (4), H21 - H32. https://doi.org/10.1190/geo2015 - 0397.1.

A PRECISE VISUAL LOCALISATION METHOD FOR THE CHINESE CHANG'E – 4 YUTU – 2 ROVER

YouQing Ma

ShaoChuang Liu

Bing Sima

1. Institute of Remote Sensing and Digital Earth, Chinese Academy of Sciences, Beijing, China

Bo Wen

Song Peng

Yang Jia

2. Beijing Institute of Spacecraft System Engineering, Beijing, China

Abstract: Precise localisation of the Yutu – 2 moon rover plays an important role in path planning, obstacle avoidance and navigating to target features. To provide highprecision localisation information, a stereo bundle adjustment method using the theory of the unit quaternions is presented for the first time. To improve the precision and robustness of the proposed method, the rover's pose, from a visual odometry technique assisted by an inertial measurement unit and the rotation angles of the mast mechanism, is viewed as a pseudo – observation. A reasonable weighting strategy and a rational geometric constraint condition of the stereo cameras is also invoked. Experimental results demonstrate that the proposed method provides more accurate localisation results than either a bundle adjustment alone or a weighted total least – squares method. The proposed method has been successfully used in Chang'e – 4 mission operations.

Keywords: bundle adjustment, Chang'e – 4, pseudo – observation – SLAM, unit quaternions, visual localisation, Yutu – 2 rover

Introduction

Background

CHANG'E – 4, THE SECOND CHINESE lunar lander and rover, successfully landed at 177. 587°E, 45. 456°S in the von Karman crater of the moon on 3rd January 2019. Soon after landing, a small rover named Yutu – 2 (Jade Rabbit – 2 in English) was deployed and took its first tentative drive onto the airless regolith, as shown in Fig. 1. The service design life cycle of the Yutu – 2 rover is scheduled for three months. Due to the uncertainties on the lunar surface, the Yutu – 2 rover travels at a low speed between adjacent waypoints, which are referred to as stations. The Yutu – 2 rover's autonomous navigation capability involves visual simultaneous localisation and mapping (SLAM), path planning and rover control. When the Yutu – 2 rover moves within an unknown scene, visual SLAM is performed by finding distinctive parts of an image, then tracking these parts as the rover moves, which has been well studied in the photogrammetric computer vision and robotics communities (Milford, 2013; Kim and Eustice, 2015). Recently, more attention has been given to the rover's precise localisation. There are two primary motivations for the onboard localisation of the Yutu – 2 rover:

(a) Chang'e–4 lander　　(b) Yutu–2 rover

Fig. 1　The Chang'e – 4 probe. (a) An image taken by the Yutu – 2 rover's Pancam camera of the lander. (b) An image taken by the lander's topographic camera of the rover

(1) Accurate visual localisation of the rover plays an important role in 3D terrain reconstruction and path planning, offering safe guidance for obstacle avoidance and approaching targets.

(2) Visual SLAM technology is more useful than a star sensor (star tracker) (He et al., 2014; Lu et al., 2017) and very long baseline interferometry (VLBI) (Huang et al., 2012, 2014; Zheng et al., 2017). The onboard star sensor is only used in the initial orientation of the rover and does not work when the rover moves continuously. The measured lunar lander position accuracy is approximately 100m by VLBI, and the earth – based Chinese VLBI Network (CVN) is required (Zheng et al., 2017). Hence, the visual SLAM technology can continuously provide a more precise localisation for the lunar rover in orbit.

The Yutu – 2 rover, carrying science and engineering instrument payloads, has explored the landing site. Its onboard navigation system consists mainly of an inertial measurement unit (IMU), an odometer and several stereo cameras, including a Navcam, Pancam and Hazcam, in order to satisfy exploration requirements. Navigation cameras are the most important stereo - imaging systems for the Yutu - 2 rover's localisation and mapping research activities, which have an effective depth of field from 0.5m to infinity. The pixel count of the stereo navigation cameras is 1024×1024 pixels with a field of view of 46.9°, which has a stereo base of 27cm and a focal length of 8.72mm. The intrinsic and extrinsic (inner and exterior orientation) parameters of the navigation cameras can be acquired by using a selfcalibrating bundle adjustment (SCBA) model with control points conducted on the earth (Wang, 1990), which are listed in Tables Ⅰ and Ⅱ.

Table Ⅰ　Intrinsic parameters of the Yutu - 2 rover's navigation stereo cameras

Camera	Principal distance (pixels)	Principal point (x_0, y_0 in pixels)	Radial distortion k_1	Radial distortion k_2	Tangential distortion p_1	Tangential distortion p_2
Left	1179.59	502.33, 525.14	−0.024864	0.036222	−0.000249	0.001243
Right	1183.53	501.56, 544.62	−0.010946	0.002646	−0.000616	0.001279

Table Ⅱ　Extrinsic parameters of the Yutu - 2 rover's navigation stereo cameras

Translation vector (mm)	Rotation matrix		
−270.06	0.999994	−0.00173	0.002841
0.72	0.001744	0.999993	−0.0033
4.02	−0.00284	0.003309	0.999991

The following notation is employed throughout this work: F_M denotes a reference frame M; θ_J denotes the rotation angle of Yutu - 2's mast mechanism; point A represented in frame F_M is written as a 3×1 position vector $\mathbf{r}_A^M$; a transformation between two frames is represented by a homogenous transformation matrix $\mathbf{T}_N^M$ that transforms the coordinate representation of homogeneous points from frame F_N to frame F_M; and $\mathbf{R}_N^M$ denotes the 3×3 rotation matrix of the transformation. In Fig. 2, the coordinate frames listed in Table Ⅲ are depicted.

The coordinate system transformation framework, which includes the pitch frame F_p, the yaw frame F_y and the opening frame F_o, connects the frames F_{Rover} and F_C. Before the stereo navigation cameras are operated, the mast mechanism should first

Table Ⅲ　The coordinate system transformation framework in the Yutu - 2 rover

F_{xy}	Image frame
F_l and F_r	Left and right camera frames, respectively
F_C	Corner cube frame, which is mounted on top of the left camera
F_{Rover}	Rover body frame
F_{IMU}	Inertial frame
F_p	Pitch frame
F_y	Yaw frame
F_o	Opening frame
F_{NED}	World frame (north - east - down)

move in the vertical direction with an opening angle θ_o. Second, the camera platform on the mast mechanism has a pitching motion with the pitch angle θ_p. Third, the mast mechanism turns in a horizontal plane with the yaw angle θ_y, as shown in Fig. 2. The measurement precision of the angles θ_o, θ_y, θ_p of the mast mechanism is 0.3°. During the SCBA test, $\mathbf{r}_A^l$ can be obtained via the photogrammetric forward intersection method (Wang, 1990). Then, $\mathbf{r}_A^l$ can be transformed into $\mathbf{r}_A^{Rover}$ in one of two ways. The first way is a direct transformation between the frame F_l and the frame F_{Rover}, which constitutes the extrinsic parameters of the left navigation camera in F_{Rover}, where the mean absolute localisation precision σ_{XYZ} is 6.6mm in the range 0.5 to 5.9m. The second way is an indirect transformation procedure using frames F_l, F_C, F_p, F_y, F_o and F_{Rover}, where the precision σ_{XYZ} is 20.4mm. When visual localisation methods are applied, the measurement errors of the rotation angles of the mast mechanism will decrease the accuracy of the rover's localisation and mapping, which is noteworthy.

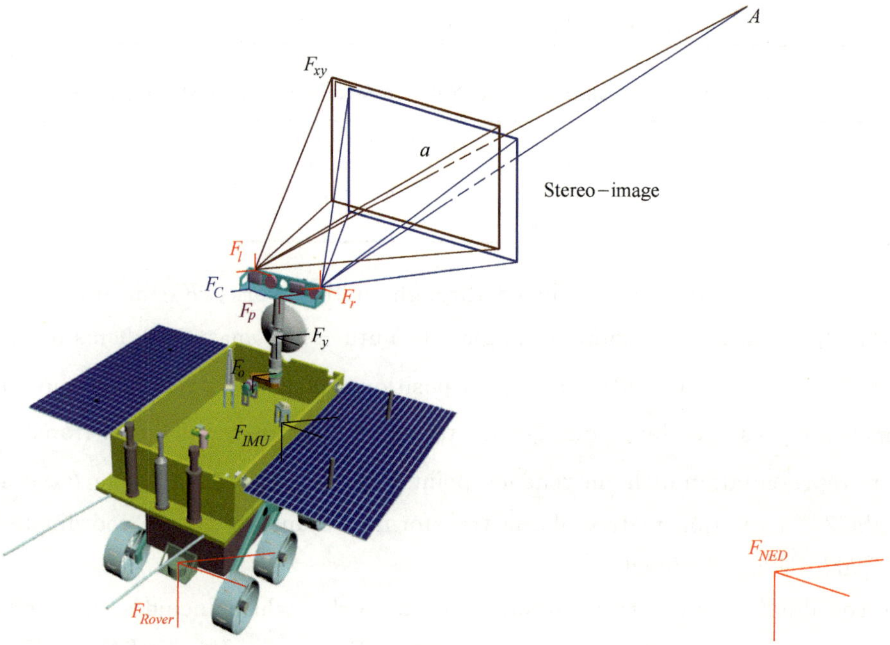

Fig. 2 Coordinate frames involved in the stereo camera system in the simulated lunar rover

Related Work

Current visual localisation methods for rovers, which have been widely used in the photogrammetric and image processing fields, can be mainly divided into one of the following three methods.

IMU-assisted Visual Odometry. The onboard IMU and wheel odometry-based localisation has been regularly performed with infrequent support from sun-finding techniques that improve the quality of the azimuth angle. In cases where the rover experiences slippage caused by traversing on loose soil terrain or against steep slopes (particularly when in craters), the onboard IMU and wheel odometry technique cannot provide accurate location results for the rover. To reduce this problem, visual odometry (VO) can be used to estimate camera motion from the analysis of feature correspondences within a sequence of images over time (Kitt et al., 2010). In recent years, improvements, assistance or a fusion to visual processing techniques using such sensors as an IMU, gravity-aided odometry (Liu et al., 2018) or a global navigation satellite system (GNSS) have positively contributed to VO solutions (Chermak et al., 2017; Schubert et al., 2018). Historically, there have been two main concepts towards approaching the IMU-assisted VO problem: (1) batch non-linear optimisation methods; and (2) recursive filtering methods. While the former jointly minimises the error originating from integrated IMU measurements and the reprojection errors from visual terms, recursive algorithms commonly use the IMU measurements for state propagation while the updates originate from the visual observations (Leutenegger et al., 2015). Accurate estimation of the camera motion during navigation depends on the camera's ability to successfully track features over successive images. When the rover is operating, lack of computational resources will make some IMU-assisted VO algorithms an unfavourable choice for online estimation. If the Chinese Chang'e-3 lunar rover and US (NASA) Mars rover are used as examples, due to computer processing and storage capacity issues, the rovers' vision systems acquire stereo-images over a distance of 0.5m as the rovers move, and store those images in the cache. As the rover moves, the images that were acquired at an earlier moment will be erased from the cache as required. Thus, the cumulative errors of the rover pose estimation problem, based on IMU-assisted VO methods, cannot be avoided. In practice, stereo-images t obtained over a distance of approximately 7.0m can be stored in the memory of the onboard computer in the rover. The IMU measurement data, the rotation angles of the mast mechanism and the stereoimages are compressed into telemetry data without loss of quality, which are then transformed into the rover's control system on the ground (earth).

Bundle Adjustment. One important task in photogrammetry is to derive 3D information from 2D measurements, which is often performed using a bundle adjustment (BA) process. Borlin and Grussenmeyer (2013) compare the convergence properties of different least squares adjustment algorithms, especially with respect to their "pull-in" radius applied to the BA problem. Then, the least squares adjustment method was studied as an optimisation problem and shown to be equivalent to the undamped Gauss-Newton (GN) optimisation method. In Di et al. (2008), Li et al. (2005) and Liu et al. (2015), BA is used to build an image network containing all images (along with any available IMU and VO results) to achieve a highly accurate localisation of the rover's position along the entire traverse, which has been successfully applied in the American Mars Exploration Rover (MER) mission and the Chinese Chang'e-3 mission, which can be conducted on the ground (earth) . In the MER mission, an incremental BA technique, developed for processing the MER rover images and applied for rover localisation and topographic mapping of the MER landing site, is presented (Li et al. , 2005; Di et al. , 2008) . In Liu et al. (2015), in order to reduce the error accumulation in the VO method, two types of image - based methods, namely cross - site visual localisation and digital orthophotomap (DOM) matching, are applied in the Chang'e - 3 mission. The steps of cross - site visual localisation, based on Navcam images, are as follows:

(1) Corresponding areas between the Navcam images acquired at adjacent sites were first extracted via initial localisation results from the VO system.

(2) Image matching between corresponding areas was performed with affine scaleinvariant feature transform (SIFT) matching to extract tie points, during which process gross errors were eliminated by a random sample consensus (RANSAC) algorithm.

(3) The image network was subsequently constructed by the tie points and BA was performed to obtain accurate localisation results of the present site with respect to the previous site.

This stepwise localisation method was routinely applied to support the teleoperation of the rover. Hence, the key feature of the rover's visual localisation methodology is the precise camera - pose estimation problem. To accomplish this, improved algorithms based on BA have been proposed, such as sparse BA (Lourakis and Argyros, 2009), sparse sparse BA (Konolige, 2010), G2o (Kümmerle et al. , 2011), relative BA (Blanco et al. , 2013), incremental BA (Indelman et al. , 2013) and parallaxBA (Zhao et al. , 2015), which are based on the ordinary least - squares solution. In Strasdat et al. (2010), the authors prove that SLAM based on BA outperforms the extended Kalman filter (EKF) in terms of consistency and precision. In these improved BA methods, an improper selection of the initial values will lead to the need for more iterations and convergence to a local optimal solution. To treat this initial value problem, the current authors, in Qing et al. (2018), proposed a weighted total least - squares solution to the rover's visual localisation model. This proposed method can quickly provide the mean relative localisation results (4. 59%) . In both the IMU assisted VO and BA methods, the rover's relative localisation result can be obtained. However, the cumulate errors of the rover's pose means an estimation problem based on those methods that cannot be truly avoided, especially with long - distance rover movement. Hence, a method for the rover's absolute localisation should be studied.

Absolute Localisation. As well as the lander and rover, missions will also involve an orbiter circling the moon or planet. Some a priori information, such as land markers formed by meteorite craters or rocks, can be obtained from the cameras on the Lunar or Mars Orbiter, which constitute the lander's vision system. Meteorite craters and stones are common features on the surface of the moon or on Mars. When the rover's vision system is operating, the rover's motion and orientation information can be obtained via image matching, image registration and space resection with the prior information. Matching or registration precision is the key issue for the rover's motion and orientation determination. Compared with the various BA methods, the superiority of the absolute localisation method is that it can adjust the cumulated errors (Di et al., 2008). Hence, the absolute localisation method is used for the rover's absolute localisation after long-distance movement. Taking the Yutu-2 rover as an example; ground tests indicate that the accumulated error caused by VO and BA is up to 15% and 3% to 5%, respectively. In the visual localisation of two adjacent stations, or during the initial movement, the rover's movement distance is less than 7m. Thus, the error in the rover localisation based on VO and BA is approximately 1.05m and 0.21 to 0.35m, respectively. The precision will decrease with an increasing distance: hence the VO and BA methods are only used for short-distance movement; the absolute localisation method will be used for long-distance movement. It is noted that the visual localisation precision based on the absolute localisation method depends on the differences in perspective and the resolution of the images from the rover, orbiter and lander. In other words, the absolute localisation method has a slightly lower precision than the spatial resolution of the orbiter's images. Meanwhile, due to the limits of the image coverage and the visit circle of the orbiter, the absolute localisation method cannot be readily available. The descent image also has limitations in terms of image coverage.

Aim

The classical rover localisation and landing-site mapping technology is based on a BA of an image network formed by navigation and Pancam stereo-images. When BA or other visual localisation methods are applied, the coordinate system transformation framework should be used twice. The camera pose is initialised with the value from the rover pose, so the procedure can be denoted as "forward transformation". When the camera pose is refined, a higher-precision rover pose can be obtained from the optimised camera pose. Accordingly, this procedure can be denoted as "backward transformation". Because of the complicated non-linear relationship between F_R (right camera frame) and F_C (corner cube frame of the left camera), the measurement errors of the rotation angles of the mast mechanism could not be cancelled out through the forward and backward transformations procedure. In addition, those IMU-assisted VO, BA and other SLAM methods, which are based on Euler angles, cannot suit cases with a large rotation due to the phenomena of coupling and singularity. Thus, a new stereo BA method based on unit quaternions is first proposed for the precise localisation of a planetary rover. Meanwhile, the rover pose from the IMU-assisted VO technique and the rotation angles of the mast mechanism have not been fully utilised. In contrast to Zhang et al. (2017) and Qing et al. (2018), the proposed method views the rotation angles of the mast mechanism and the rover pose from the IMU-assisted VO technique as part of the observations (pseudoobservations) for the stereo BA model. To improve the rover pose estimation accuracy, a more rational geometric constraint condition of the stereo cameras and the weighted matrix of the rover pose, as well as the matching image point, are involved. In contrast to Di et al. (2008) and Liu et al. (2015), the proposed method has the advantage of adding to the number of redundant observations.

The proposed method provides the following two contributions:

(1) A more reasonable solution is derived for the rover's visual localisation problem, which uses the stereo BA method based on the theory of the unit quaternion. The proposed method avoids the theoretical limitation of the rover's traditional visual localisation methods based on Euler angles.

(2) The stereo BA method for combining the rotation angles of the mast mechanism and the rover pose from the IMU-assisted VO technique is established with the geometric constraint condition of the stereo cameras. Additionally, a more reasonable weighting strategy is invoked. Thus, the proposed method is more robust compared with traditional visual BA methods. It should be noted that the aim is to incorporate a stereo BA model with the telemetry data and a rational geometric constraint condition of the stereo cameras, together with the weighted matrix in a visual localisation process to estimate the following quantities:

(a) the exterior orientation (extrinsic) parameters (EOPs) of the imagery involved;

(b) the "ground" coordinates of the tie points; and

(c) the pose information of the rover.

Methodology

Workflow of the Proposed Visual Localisation Method

This paper presents the proposed visual localisation method using tie-point selection based on speeded-up robust features (SURF) matching results (Bay et al., 2008). A tie point indicates that a feature point is reflected in stereo-images at adjacent locations. In practice, tie points are selected from natural or non-natural feature points. Examples of natural feature points are: the edge and corner of a small, angular rock; the start and end points of a line or curve; or a sharp change (discontinuity) in greyscale values in the image. Meanwhile, the corner of the Chang'e-4 lander and points in the rover's tracks are representative of nonnatural feature points. In most cases natural tie points are used, but in areas of weak texture or overexposure, non-natural feature points are an important supplement for the rover's visual localisation. Fig. 3 shows the workflow of the proposed visual localisation method.

The initial localisation results of the Yutu-2 rover can be obtained by the IMU-assisted VO method, which will be used to obtain the overlapping image area from adjacent stations. When the SURF method is applied, the overlapping image area can be considered as the initial region to improve the reliability and efficiency of the image - matching procedure. Next, tie points are detected and extracted via the SURF matching method in the above initial matching regions. The camera pose is initialised with the rover pose in the "forward transformation" procedure. Finally, the stereo BA model with the IMU - assisted VO measurements, the rotation angles of the mast mechanism and the geometry constraint condition of the stereo cameras is established. The least - squares solution is used to calculate the optimised camera pose. Through the " backward transformation" procedure, the rover localisation results are generated.

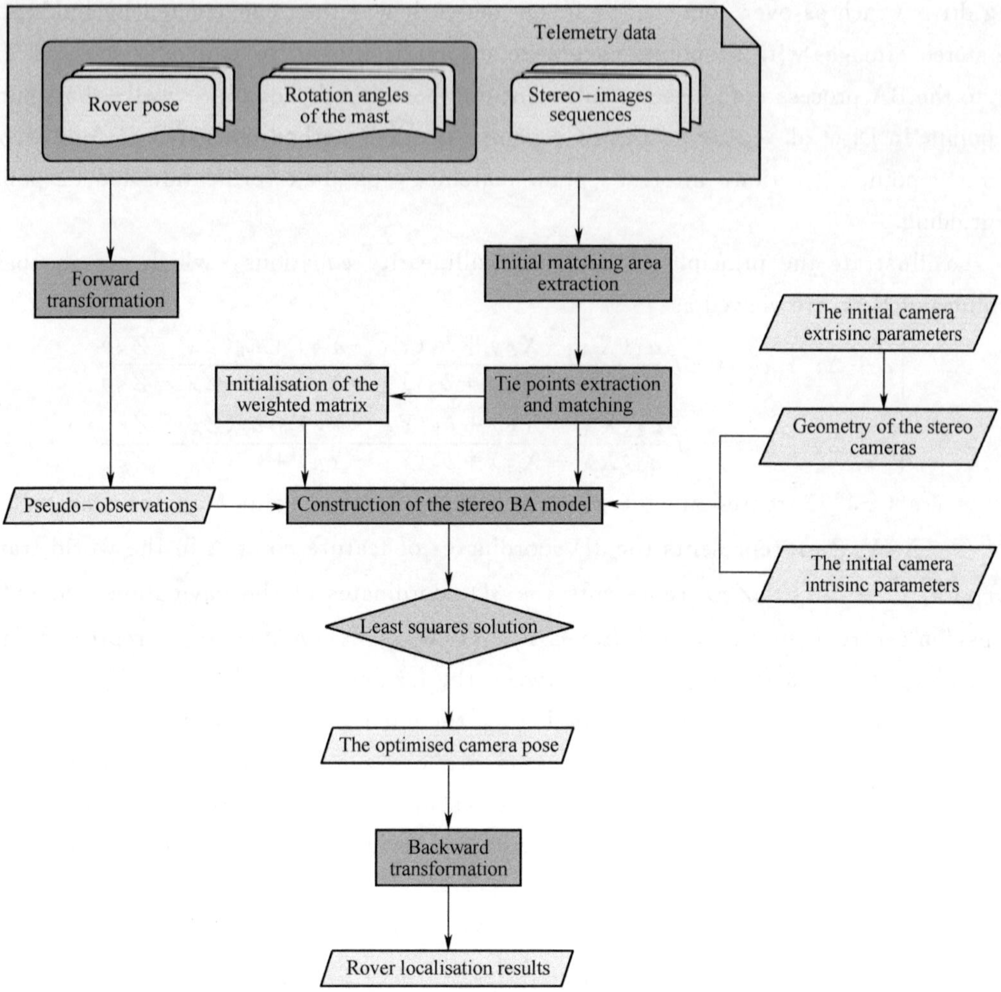

Fig. 3　Workflow of the proposed visual localisation method

Geometric constraints formed by tie points between the cross - station (cross - site) images are the basis for the rover pose estimation solution. An image network can be formed by linking all the images of two stations. Compared with the traditional VO and visual BA methods, this paper presents an approach which incorporates the IMU -

assisted VO measurements and the rotation angles of the mast mechanism into the stereo BA model.

Stereovision Imaging Model and Strict Geometric Relationships

Basic Structure of the Stereovision System. As Fig. 2 shows, panoramas and traversing stereo - images of the navigation cameras were taken at different stations. Pancam panoramas were only acquired at stations where substantial scientific exploration activities took place, and Navcam panoramas were more frequently used for navigation, obstacle avoidance and target approaching. To acquire the rover's precise location, stereo - images from adjacent stations were often obtained at the midpoint of a long drive (such as over 30m). The image network was then constructed by linking all the stereo - images with tie points that were automatically and/or manually selected. The key to the BA process is to select a sufficient number of high - quality, well - distributed tie points. In Di et al. (2008), the tie - point selection method consisted of four steps: interest - point extraction, interest - point matching, parallax verification and tie points by gridding.

To illustrate the principle of BA, the collinearity equations, which are the basic formulas of BA, are viewed as:

$$\begin{cases} x + \Delta x + x_0 = -f \dfrac{a_1(X_A - X_S) + b_1(Y_A - Y_S) + c_1(Z_A - Z_S)}{a_3(X_A - X_S) + b_3(Y_A - Y_S) + c_3(Z_A - Z_S)} \\ y + \Delta y + y_0 = -f \dfrac{a_2(X_A - X_S) + b_2(Y_A - Y_S) + c_2(Z_A - Z_S)}{a_3(X_A - X_S) + b_3(Y_A - Y_S) + c_3(Z_A - Z_S)} \end{cases} \tag{1}$$

where (x, y) represents the 2D coordinates of image point a in the frame F_{xy}; $\mathbf{r}_A^{NED} = (X_A Y_A Z_A)$ represents the 3D coordinates of feature point A in the world frame F_{NED}; $\mathbf{r}_l^{NED} = (X_S Y_S Z_S)$ represents the 3D coordinates of the navigation camera's projection centre l in the world frame F_{NED}; $\mathbf{R}_{NED}^l = (a_1 a_2 \cdots c_3)$ represents the elements of the 3×3 rotation matrix between the frames F_l and F_{NED}, which contains three independent elements (the roll angle φ_S, the pitch angle ω_S and the yaw angle k_S); f represents the principal distance of the camera; $(x_0\ y_0)$ represents the principal point of the camera in the frame F_{xy}; $(\Delta x \Delta y)$ represents the lens distortion. In Table I, the lens distortion coefficients (k_1, k_2, p_1, p_2) are derived from the usual Brown (1966, 1971) equations:

$$\Delta x = (x - x_0)(k_1 r^2 + k_2 r^4) + p_1(r^2 + 2(x - x_0)^2) + 2p_2(x - x_0)(y - y_0)$$
$$\Delta y = (y - y_0)(k_1 r^2 + k_2 r^4) + p_2(r^2 + 2(y - y_0)^2) + 2p_1(x - x_0)(y - y_0)$$
$$r^2 = (x - x_0)^2 + (y - y_0)^2$$

Exterior Orientation Parameters (EOPs) of the Stereo Cameras. The unknowns of the BA include the EOPs $(X_S, Y_S, Z_S, \varphi_S, \omega_S, k_S)$ of all the images (Navcam and/or Pancam) involved in the image network and the 3D coordinates of all tie points. Meanwhile, the BA will converge to an optimal solution with the appropriate initial values of the unknowns. Hence, the rotation matrix $\mathbf{R}_{NED}^l$ and the vector $\mathbf{r}_l^{NED}$, which contain the EOPs, should be studied.

The IMU - assisted VO method can provide the initial localisation results of the Yutu - 2 rover, and the telemetry data includes the rotation angles of the mast mechanism. In that case, the rotation matrix $\mathbf{R}_{NED}^{l}$ and the vector $\mathbf{r}_{l}^{NED}$ can be obtained by using the geometric relationship between the frames F_l, F_{Rover} and F_{NED}, as shown in equations (2) to (5).

The numerical relationship between the rotation matrix $\mathbf{R}_{NED}^{l}$ and $\mathbf{R}_{NED}^{Rover}$ can be expressed as:

$$\mathbf{R}_{NED}^{Rover} = \mathbf{R}_{NED}^{l}\mathbf{R}_{l}^{Rover} \tag{2}$$

where $\mathbf{R}_{NED}^{Rover}$ represents the rotation matrix between the frames F_{Rover} and F_{NED}; $\mathbf{R}_{l}^{Rover}$ represents the rotation matrix between the frame F_l and F_{Rover}. When the rover moves, the IMU will provide the measurement data of the roll angle φ_R, the pitch angle ω_R and the yaw angle k_R in the frame F_{NED}. The rotation matrix $\mathbf{R}_{NED}^{Rover}$ can be viewed as:

$$\mathbf{R}_{NED}^{Rover} = \boldsymbol{R}_X(\varphi_R)\mathbf{R}_Y(\omega_R)\boldsymbol{R}_Z(k_R) \tag{3}$$

where $\mathbf{R}_X(\varphi_R)$ represents the rotation matrix by φ_R around the X axis of the frame F_{NED}; $\mathbf{R}_Y(\omega_R)$ represents the rotation matrix by ω_R around the Y axis of the frame F_{NED}; $\mathbf{R}_Z(k_R)$ represents the rotation matrix by k_R around the Z axis of the frame F_{NED}. In equation (2), $\mathbf{R}_{l}^{Rover}$ can also be viewed as:

$$\mathbf{R}_{l}^{Rover} = \mathbf{R}_{O}^{Rover}\mathbf{R}_{opening}\mathbf{R}_{Y}^{O}\mathbf{R}_{yaw}\mathbf{R}_{P}^{Y}\mathbf{R}_{pitch}\boldsymbol{R}_{C}^{P}\mathbf{R}_{l}^{C} \tag{4}$$

where

$$\mathbf{R}_{opening} = \begin{bmatrix} 1 & 0 & 0 \\ 0 & \cos\theta_o & -\sin\theta_o \\ 0 & \sin\theta_o & \cos\theta_o \end{bmatrix}$$

is the rotation between frames F_{Rover} and F_o;

$$\mathbf{R}_{yaw} = \begin{bmatrix} \cos\theta_y & 0 & -\sin\theta_y \\ 0 & 1 & 0 \\ \sin\theta_y & 0 & \cos\theta_y \end{bmatrix}$$

is the rotation between frames F_o and F_y;

$$\mathbf{R}_{pitch} = \begin{bmatrix} 1 & 0 & 0 \\ 0 & \cos\theta_p & -\sin\theta_p \\ 0 & \sin\theta_p & \cos\theta_p \end{bmatrix}$$

is the rotation between frames F_y and F_p; and $\mathbf{R}_{O}^{Rover}$, $\mathbf{R}_{Y}^{O}$, $\mathbf{R}_{P}^{Y}$, $\mathbf{R}_{C}^{P}$ *and* $\mathbf{R}_{l}^{C}$ denote the rotation matrices involved in the coordinate system transformation framework, where angle measurement accuracy is 10.

The 3D coordinates of the navigation camera's projective centre l in the frame F_{NED} can be presented as:

$$\mathbf{r}_{l}^{NED} = \mathbf{R}_{NED}^{Rover}(\mathbf{R}_{O}^{Rover}\mathbf{R}_{opening}(\mathbf{R}_{Y}^{O}\mathbf{R}_{yaw}(\mathbf{R}_{P}^{Y}\mathbf{R}_{pitch}(\mathbf{R}_{C}^{P}\mathbf{t}_{l}^{C} + \mathbf{t}_{C}^{P}) + \mathbf{t}_{P}^{Y}) + \mathbf{t}_{Y}^{O}) + \mathbf{t}_{O}^{Rover}) + \mathbf{r}_{Rover}^{NED} \tag{5}$$

where $\mathbf{t}_l^C$, $\mathbf{t}_C^P$, $\mathbf{t}_P^Y$, $\mathbf{t}_Y^O$ *and* $\mathbf{t}_O^{Rover}$ denote the transformation vectors involved in the coordinate system transformation framework.

Camera's Fixed Relative Relationship. When the rover's visual localisation methods are applied, the multiviews at the adjacent stations should be involved. To enhance the robustness of the method, the geometry constraint condition of the stereo cameras in a single location should be invoked:

$$\begin{cases}\mathbf{R}_{NED}^r = \mathbf{R}_{NED}^l \mathbf{R}_l^r \\ \mathbf{r}_r^{NED} = \mathbf{R}_{NED}^l \mathbf{r}_r^l + \mathbf{r}_l^{NED}\end{cases} \tag{6}$$

where $\mathbf{r}_l^{NED}$, $\mathbf{r}_r^{NED}$ represent the 3D coordinates of the stereo camera's projective centre in the frame $\mathbf{F}_{NED}$; $\mathbf{r}_r^l$, $\mathbf{R}_l^r$ represent the extrinsic parameters of the rover's navigation stereo cameras, as illustrated in Table Ⅱ. Equations (1), (2), (5) and (6), which respectively denote the strict geometric relationships in the rover's stereovision system, are the foundations of the stereo BA model.

Construction of the Stereo BA Model

Error Equation of the Stereovision System. The basic idea of a BA is linearising each image-point collinear equation (equation (1)) to obtain the error equation where the correction variable is unknown. Then, from a sequence of images, the normal equations are obtained, which together form the entire imaging-network normal equations. Finally, the iterative solution of the normal equations is solved via a least-squares method.

Using Taylor's formula, the linearisation equation of formula (1) can be derived in the following Gauss-Markov model (GMM) mode:

$$\mathbf{v}_\mathbf{x} = \mathbf{A}_\mathbf{x} t + \mathbf{B}_\mathbf{x}\mathbf{x} - \mathbf{l}_\mathbf{x}; \ \mathbf{P}_\mathbf{x} \tag{7}$$

where $\mathbf{t} = (\Delta X_s, \Delta Y_S, \Delta Z_S, \Delta\varphi_S, \Delta\omega_S, \Delta k_S)^T$ represents the correction of the extrinsic (exterior orientation) parameters of the camera in the frame F_{NED}; $\mathbf{x} = (\Delta X, \Delta Y, \Delta Z)^T$ represents the correction of the 3D coordinates of the feature points A in the frame F_{NED}; $\mathbf{l}_\mathbf{x}$ represents the residual; $\mathbf{v}_\mathbf{x}$ represents the correction of the 2D image points in the frame F_{xy}; $\mathbf{A}_\mathbf{x}$; $\mathbf{B}_\mathbf{x}$ represent coefficient matrices; $\mathbf{P}_\mathbf{x}$ denotes the diagonal weight matrix.

EOP Error Equations using Unit Quaternions. In equations (3) and (4), the rover pose (including φ_R, ω_R, k_R) and the rotation angles of the mast mechanism (θ_o, θ_y, θ_p) can be viewed as the observed values, which should be considered with regard to their errors. Equation (2) can then be expressed as:

$$\mathbf{R}_{NED}^l = \mathbf{R}_{NED}^{Rover}(e_{\varphi R}, e_{\omega R}, e_{k R})(\mathbf{R}_l^{Rover}(e_{\theta_o}, e_{\theta_y}, e_{\theta_p}))^{-1} \tag{8}$$

where e: denotes the measurement error of the rotation angles and $(\cdot)^{-1}$ denotes the inverse matrix.

In comparison to the lower-accuracy measurements of the rotation angles (θ_o, θ_y, θ_p) of the mast mechanism, the angles in the matrices $\mathbf{R}_O^{Rover}$, $\mathbf{R}_Y^O$, $\mathbf{R}_P^Y$, $\mathbf{R}_C^P$ and $\mathbf{R}_l^C$ can be viewed as the observed values without error. In a practical application, these five **R** matrices are each close to being the identity matrix **I**. Representing the rotations in matrix form with orthonormality constraints often makes the linearisation of equations (1) and (8) challenging to solve as they are very nonlinear. In contrast, unit quaternions provide a singularity - free representation of rotations, and by restricting the first element to non - negative numbers, there is a one - to - one correspondence between the rotation matrices and quaternions (Fathian et al. , 2018) . Instead of nonlinear rotations using Euler angles, unit quaternions are used, which represent a rotation by four elements ($q_0 q_x q_y q_z$) such that $q_0^2 + q_x^2 + q_y^2 + q_z^2 = 1$. Note that the traditional control method of using Euler angles, such as the roll angle φ_R, the pitch angle ω_R and the yaw angle k_R, cannot suit the case of a large rotation because of the phenomena of coupling and singularity.

In equation (1), the associated rotation matrix $\mathbf{R}_{NED}^{l}$ contains three independent elements φ_S, ω_S, k_S. Compared with φ_S and ω_S, the value of the yaw angle k_S is relatively large, in the range $-180°$ to $+180°$ degrees. In the case of forward stereo - images from the previous station and backward stereo - images at the current station, the difference between the yaw angle k_S at such adjacent stations is large. Hence, there are some problems in the rover's visual localisation methods based on the BA. The rotation matrix $\mathbf{R}_{NED}^{l}$ is calculated either by equation (2) or equation (8) . At the initial stage of BA, the unknowns φ_S, ω_S, k_S can be obtained by equation (14) . However, the phenomena of coupling and singularity will appear in some special situations: for example, there are no unique solutions for the Euler angles in the case of a right angle. In addition, the three independent elements φ_S, ω_S, k_S will also be coupled. In the method proposed here, the associated rotation matrix $\mathbf{R}_{NED}^{l}$ using the theory of unit quaternions can be viewed as:

$$\mathbf{R}_{NED}^{l} = \begin{bmatrix} q_0^2 + q_x^2 - q_y^2 - q_z^2 & 2(q_x q_y - q_0 q_z) & 2(q_x q_z + q_0 q_y) \\ 2(q_y q_x + q_0 q_z) & q_0^2 - q_x^2 + q_y^2 - q_z^2 & 2(q_y q_z - q_0 q_x) \\ 2(q_z q_x - q_0 q_y) & 2(q_z q_y + q_0 q_x) & q_0^2 - q_x^2 - q_y^2 + q_z^2 \end{bmatrix} \tag{9}$$

Then, equation (7) can be reinterpreted as:

$$\mathbf{v_x} = \mathbf{A_x t} + \mathbf{B_x x} - \mathbf{l_x};\ \mathbf{P_x} \tag{10}$$

where $\mathbf{t} =$ (ΔX_S, ΔY_S, ΔZ_S, Δq_0, Δq_x, Δq_y, Δq_z)^{T} is the augmented vector; and $\mathbf{A_x}$ represents the coefficient matrix. The linearity of the constraint condition is given (Mercan et al. , 2018) as:

$$q_0 \Delta q_0 + q_x \Delta q_x + q_y \Delta q_y + q_z \Delta q_z + w = 0 \tag{11}$$

where $w = (q_0^2 + q_x^2 + q_y^2 + q_z^2 - 1)/2$. Then, equation (11) can be expressed as:

$$\mathbf{Ct} + \mathbf{w} = \mathbf{0} \tag{12}$$

where $\mathbf{C} = [0\ 0\ 0\ q_0\ q_x\ q_y\ q_z]$ represents the coefficient matrix of the condition equation; and $\mathbf{w} = [w]$ denotes the constant vector.

The initial values of the camera pose can be obtained by equation (8) . When the camera pose is expressed in unit quaternions, the augmented vector **t** will be initialised as:

$$\begin{bmatrix} q_0 \\ q_x \\ q_y \\ q_z \end{bmatrix} \begin{bmatrix} \cos\frac{\varphi_S}{2}\cos\frac{\omega_S}{2}\cos\frac{k_S}{2} - \sin\frac{\varphi_S}{2}\sin\frac{\omega_S}{2}\sin\frac{k_S}{2} \\ \sin\frac{\varphi_S}{2}\cos\frac{\omega_S}{2}\sin\frac{k_S}{2} - \cos\frac{\varphi_S}{2}\sin\frac{\omega_S}{2}\cos\frac{k_S}{2} \\ \sin\frac{\varphi_S}{2}\cos\frac{\omega_S}{2}\cos\frac{k_S}{2} + \cos\frac{\varphi_S}{2}\sin\frac{\omega_S}{2}\sin\frac{k_S}{2} \\ -\cos\frac{\varphi_S}{2}\cos\frac{\omega_S}{2}\sin\frac{k_S}{2} - \sin\frac{\varphi_S}{2}\sin\frac{\omega_S}{2}\cos\frac{k_S}{2} \end{bmatrix} \tag{13}$$

where φ_S, ω_S, k_S can be obtained as follows:

$$\begin{cases} \tan\varphi_S = \frac{a_3}{c_3} \\ \sin\omega_S = -b_3 \\ \tan k_S = \frac{b_1}{b_2} \end{cases} \tag{14}$$

As described before, the initial rover pose is obtained by the IMU - assisted VO method. Then, the EOPs (X_S, Y_S, Z_S, φ_S, ω_S; k_S) of all images (Navcam and/or Pancam) can be acquired before the BA is applied. Hence, the EOPs are not only the initial values, but they provide more observations for BA. In the proposed method, the extrinsic (exterior) orientation parameters of the camera in F_{NED} are viewed as pseudo - observations (Yuan, 2009). In this case, the number of observation values will increase, which improves the estimation accuracy of the EOPs in the BA process. Equations (5) and (13) can then be expressed as:

$$\mathbf{v_f} = \mathbf{It};\ \mathbf{P_f} \tag{15}$$

where **I** denotes the 7×7 identity matrix; $\mathbf{P_f}$ denotes the 7×7 diagonal weight matrix; and $\mathbf{v_f}$ represents the corrections to the extrinsic parameters of the camera.

Constraint Equation for the Relative Relationship of the Stereo Camera

The geometry described by equation (6) is used as a mandatory constraint and added to the stereo BA model. Previous BA methods usually design a local optimisation function to obtain the external parameters. However, the optimisation process does not fully consider the fact that the external parameters of the cameras are kept constant or treated as known. Based on this circumstance, the constraint is introduced that the external parameters between the cameras remain unchanged and known, which can reduce the amount of computation needed to a certain extent. Additionally, this approach can improve the estimation accuracy and robustness of the stereo BA model.

Equation (6) indicates the geometry constraint condition of the stereo cameras at a single location, which can be expressed as:

$$\begin{cases} \mathbf{R}_l^r = (\mathbf{R}_{NED}^l)^{\mathrm{T}}\mathbf{R}_{NED}^r \\ \mathbf{r}_r^l = (\mathbf{R}_{NED}^l)^{\mathrm{T}}(\mathbf{r}_r^{NED} - \mathbf{r}_l^{NED}) \end{cases} \tag{16}$$

There are two ways to use the geometry constraint condition in formula (16) as follows, depending on whether the extrinsic (exterior orientation) parameters of the navigation cameras are obtained in advance or not.

Extrinsic Parameters Not Obtained in Advance In this case, the geometry constraint condition of the stereo cameras remains unchanged in the jth or jth location. Thus, the following equation can be given as:

$$\begin{cases}(\mathbf{R}_l^r)_i = (\mathbf{R}_l^r)_j \\ (\mathbf{r}_r^l)_i = (\mathbf{r}_r^l)_j\end{cases} \tag{17}$$

where $(\mathbf{R})_i$ represents the rotation matrix $\mathbf{R}$ at the ith location. When the BA method is applied, there are $24=4\times 6$ variables, which are the extrinsic parameters of the stereo camera at the two locations. In equation (17), the number of constraints is $12=(3\times 3)+3$. However, there are only three independent elements in the rotation matrix $\mathbf{R}$. Hence, the number of independent constraint conditions in equation (17) is six. Based on the orthogonality of the rotation matrix, $(\mathbf{R}_l^r)_{21}$, $(\mathbf{R}_l^r)_{31}$, $(\mathbf{R}_l^r)_{32}$ are used to reduce the amount of calculation required and avoid data redundancy and improve practical relevance (Cui et al., 2016). Then, equation (17) can be viewed as:

$$H_{6\times 1} = \begin{bmatrix}(((\mathbf{R}_{NED}^l)^T\mathbf{R}_{NED}^r)_{21})_i - (((\mathbf{R}_{NED}^l)^T\mathbf{R}_{NED}^r)_{21})_j \\ (\mathbf{R}_{NED}^l)^T\mathbf{R}_{NED}^r)_{31})_i - (((\mathbf{R}_{NED}^l)^T\mathbf{R}_{NED}^r)_{31})_j \\ (((\mathbf{R}_{NED}^l)^T\mathbf{R}_{NED}^r)_{32})_i - (((\mathbf{R}_{NED}^l)^T\mathbf{R}_{NED}^r)_{32})_j \\ ((\mathbf{R}_{NED}^l)^T(\mathbf{r}_r^{NED}-\mathbf{r}_l^{NED}))_i - ((\mathbf{R}_{NED}^l)^T(\mathbf{r}_r^{NED}-\mathbf{r}_l^{NED}))_j\end{bmatrix} = \mathbf{0} \tag{18}$$

where the vector $\mathbf{h}$ represents the constant term.

Using Taylor's formula, the linearisation of equation (18) can be derived as:

$$\mathbf{D}_i^l\mathbf{t}_i^l + \mathbf{D}_i^r\mathbf{t}_i^r - \mathbf{D}_j^l\mathbf{t}_j^l - \mathbf{D}_j^r\mathbf{t}_j^r - \mathbf{h}^0 = \mathbf{0} \tag{19}$$

where $\mathbf{D}_i^l = \boldsymbol{\delta}\mathbf{h}/\boldsymbol{\delta}\mathbf{t}_i^l$, $\mathbf{D}_i^r = \boldsymbol{\delta}\mathbf{h}/\boldsymbol{\delta}\mathbf{t}_i^r$, $\mathbf{D}_j^l = \boldsymbol{\delta}\mathbf{h}/\boldsymbol{\delta}t_j^l$, $\mathbf{D}_j^r = \boldsymbol{\delta}\mathbf{h}/\boldsymbol{\delta}\mathbf{t}_j^r$ denote the coefficient matrices; $\mathbf{t}_i^l$, $\mathbf{t}_i^r$, $\mathbf{t}_j^l$, $\mathbf{t}_j^r$ are the corrections to the extrinsic parameters of the left camera l or right camera r at locations i or j; and $\mathbf{h}^0$ represents the constant vector when the initial values of the extrinsic parameters are all set as zero.

Extrinsic Parameters Obtained in Advance. This case means that $(\mathbf{R}_l^r)_i$ and $(\mathbf{R}_l^r)_j$ are the same constant matrices, and $(\mathbf{r}_\mathrm{r}^\mathrm{l})_i$ are $(\mathbf{r}_\mathrm{r}^\mathrm{l})_j$) are the same constant vectors. Then, the following equation can be given:

$$\mathbf{f}_{12\times 1} = \begin{bmatrix}(((\mathbf{R}_{NED}^l)^T\mathbf{R}_{NED}^r)_{21})_i \\ (((\mathbf{R}_{NED}^l)^T\mathbf{R}_{NED}^r)_{31})_i \\ (((\mathbf{R}_{NED}^l)^T\mathbf{R}_{NED}^r)_{32})_i \\ (((\mathbf{R}_{NED}^l)^T\boldsymbol{R}_{NED}^r)_{21})_j \\ (((\mathbf{R}_{NED}^l)^T\mathbf{R}_{NED}^r)_{31})_j \\ (((\mathbf{R}_{NED}^l)^T\mathbf{R}_{NED}^r)_{32})_j \\ ((\mathbf{R}_{NED}^l)^T(\mathbf{r}_r^{NED}-\mathbf{r}_l^{NED}))_i \\ ((\mathbf{R}_{NED}^l)^T(\mathbf{r}_r^{NED}-\mathbf{r}_l^{NED}))_j\end{bmatrix} = \mathbf{e} \tag{20}$$

where the vector **e** represents the constant term.

Using Taylor's formula, the linearisation of equation (20) can be derived as:

$$\mathbf{G}_i^l \mathbf{t}_i^l + \mathbf{G}_i^r \mathbf{t}_i^r + \mathbf{G}_j^l \mathbf{t}_j^l + \mathbf{G}_j^r \mathbf{t}_j^r - \mathbf{f}^0 - \mathbf{e} = \mathbf{0} \tag{21}$$

where $\mathbf{G}_i^l = \boldsymbol{\delta}\mathbf{f}/\boldsymbol{\delta}\mathbf{t}_i^l$, $\mathbf{G}_i^r = \boldsymbol{\delta}\mathbf{f}/\boldsymbol{\delta}\mathbf{T}\mathbf{t}_i^r$, $\mathbf{G}_j^l = \boldsymbol{\delta}\mathbf{f}/\boldsymbol{\delta}\mathbf{t}_j^l$, $\mathbf{G}_j^r = \boldsymbol{\delta}\mathbf{f}/\boldsymbol{\delta}\mathbf{t}_j^r$ are the coefficient matrices; and $\mathbf{f}^0$ represents the constant vector when the initial values of the extrinsic parameters are all set as zero.

In equation (20), the number of independent constraint conditions is 12. Compared with the first method, the second method (where the extrinsic parameters are obtained in advance) has the advantage of adding to the number of redundant observations when adopted in the stereo BA model. However, it requires prior information about the extrinsic parameters of the stereo cameras. When the stereo cameras are calibrated on the ground, the extrinsic parameters cannot remain unchanged due to the violent vibration during the flight of the rover. A self-calibration method can obtain the accurate extrinsic parameters of the stereo cameras in orbit using, as control points, the solar cells and other features on the rover (Zhang et al., 2017).

Least-squares Solution of the Stereo BA Model

According to the different combinations of the above observations, there are certain solutions to the stereo BA model for the rover's visual localisation problem. The extrinsic parameters of the stereo cameras and the 3D coordinates of the feature points are viewed as the unknowns; the 2D coordinates of the matched image points are viewed as the observation values. Thus, equation (10) is the core equation. At the same time, equation (12) is also a necessary constraint condition.

Suppose that the stereo camera operates at two locations, i and j, and it obtains four images $image_i^{left}$, $image_i^{right}$, $image_j^{left}$, $image_j^{right}$. The number of tie points is n, as shown in Fig. 4.

In equation (10), the number of observation values is $(2 \times 2 \times k \times n)$ and the number of unknowns is $(3 \times n + 2 \times k \times 7)$, where k denotes the number of locations. In equation (12), the number of constraint conditions is $2 \times k$.

The combination of the above equations can be resolved by making the equivalent Lagrange target function ϕ stationary. Then, the first-order condition equations can be set as zero to obtain the estimation value of the random errors and the unknowns. The sufficient condition for a multivariate minimum can be fulfilled because the second-order condition is positive definite.

Solution with the Geometry Constraint Condition of the Stereo Cameras

The initial values of the unknowns **t** are obtained by using equations (5) and (13). In a single location, the initial values of the unknowns **t** of the stereo cameras can be set to the same values. However, during the iterative process of the least-squares solution based on equations (10) and (12), equations (5) and (13) are used once to obtain the initial values. $\mathbf{z}=[\mathbf{tx}]$ denotes the vector of all the unknowns and $\mathbf{U}=[\mathbf{A_x B_x}]$ is the coefficient matrix. The expression $\mathbf{Ct}+\mathbf{w}=\mathbf{0}$ (equation (12)) should be rewritten as $\mathbf{C_0 z}+\mathbf{w}=\mathbf{0}$. It is important to note that $\mathbf{C}_0=[\mathbf{C}\ 0\ 0\ 0]=[0\ 0\ 0\ q_0\ q_x\ q_y\ q_z\ 0\ 0\ 0]$.

When the initial values of the unknowns $\mathbf{t}$ are obtained by using equations (5) and (13), the expression $\mathbf{v_f}=\mathbf{It}$ (equation (15)) should be rewritten as $\mathbf{v_f}=\mathbf{I_0 z}$. Then, formula (19) should also be invoked. Thus, the error equations are as follows:

$$\begin{cases}\mathbf{v_x}=\mathbf{Uz}-\mathbf{l_x};\ \mathbf{P_x}\\ \mathbf{v_f}=\mathbf{I_0 z};\quad \mathbf{P_f}\\ \mathbf{C_0 z}+\mathbf{w}=\mathbf{0}.\end{cases} \tag{22}$$

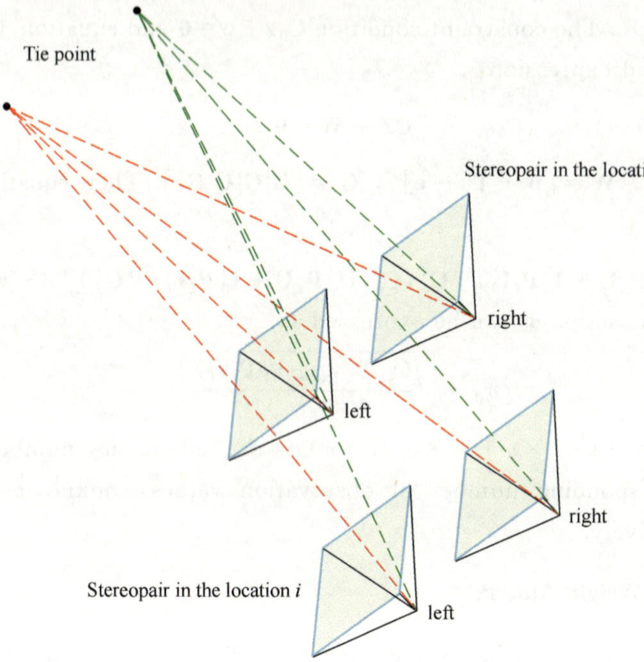

Fig. 4　Illustration of the geometric relationship involved in the stereo camera system in the simulated rover.

In equation (22), the number of observation values will increase to $(2\times2\times k\times n+2\times k\times7)$. The number of unknowns is $(3\times n+2\times k\times7)$ and the number of constraint conditions is $2\times k$.

The Lagrange target function ϕ can also be written as:

$$\phi(\mathbf{z},\ \boldsymbol{\theta})=\mathbf{v_x^T P_x v_x}+\mathbf{v_f^T P_f v_f}+2\boldsymbol{\theta}^T(\mathbf{C_0 z}+\mathbf{w})$$

where $\boldsymbol{\theta}$ denotes the vector of the unknown Lagrange multipliers. When the first - order condition equations $\partial\phi/\partial\mathbf{z}$, $\partial\phi/\partial\boldsymbol{\theta}$ are set to zero, the unknowns can be viewed as:

$$\mathbf{z}=\mathbf{u}+(\mathbf{U^T P_x U}+\mathbf{I_0^T P_f I_0})^{-1}\mathbf{C_0^T}(\mathbf{C_0}(\mathbf{U^T P_x U}+\mathbf{I_0^T P_f I_0})^{-1}\mathbf{C_0^T})^{-1}(-\mathbf{w}-\mathbf{C_0 u}) \tag{23}$$

where $\mathbf{u}=(\mathbf{U^T P_x U}+\mathbf{I_0^T P_f I_0})^{-1}\mathbf{U^T P_x L_x}$; and in a single image,

$$\mathbf{I}_0 = \begin{bmatrix} \mathbf{I}_{7\times7} & \mathbf{0} \\ \mathbf{0} & \mathbf{0}_{3\times3} \end{bmatrix}$$

Because the camera pose is expressed in unit quaternions, the initialisation of the unknowns **t** can be directly set to the default value (0, 0, 0, 1, 0, 0, 0)^{T}. Meanwhile, it should be noted that the initial 3D coordinates of the tie points have a lower precision by using the multiview forward intersection method in adjacent stations, which will slow the convergence of the least - squares solution. For the initial value of the camera pose with a certain precision from equation (13), the stereo BA method has a faster convergence ability than the existing BA algorithms with the default value of the unknowns **t**.

When the geometry constraint condition of the stereo cameras is invoked in the stereo BA model, equation (20) indicates that the number of constraint conditions will increase to $(2\times k+2\times 6\times k)$. The constraint condition $\mathbf{C_0 z}+\mathbf{w}=\mathbf{0}$ and equation (21) can also be rewritten in a general expression:

$$\ddot{\mathbf{C}}\mathbf{Z} + \ddot{\mathbf{W}} = \mathbf{0} \tag{24}$$

where $\ddot{\mathbf{C}} = [\mathbf{C_0}\ \ \mathbf{G}]^{T}$, $\ddot{\mathbf{W}} = [\mathbf{w} - \mathbf{f}^0 - \mathbf{e}]^{T}$, $\mathbf{G} = [\mathbf{G}_i^l \mathbf{G}_i^r \mathbf{G}_j^l \mathbf{G}_j^r]$. Then equation (23) can be written as:

$$\mathbf{z} = \mathbf{u} + (\mathbf{U}^{T}\mathbf{P_x}\mathbf{U} + \mathbf{I}_0^{T}\mathbf{P_f}\mathbf{I_0})^{-1}\mathbf{C}_0^{T}(\mathbf{C_0}(\mathbf{U}^{T}\mathbf{P_x}\mathbf{U} + \mathbf{I_0}^{T}\mathbf{P_f}\mathbf{I_0})^{-1}\mathbf{C}_0^{T})^{-1}(-\ddot{\mathbf{w}} - \ddot{\mathbf{C}}\mathbf{u}). \tag{25}$$

The standard deviation σ_0 can be expressed as:

$$\sigma_0 = \sqrt{\frac{(\mathbf{v_x}^{T}\mathbf{P_x}\mathbf{v_x} + \mathbf{v_f}^{T}\mathbf{P_f}\mathbf{v_f})}{t}} \tag{26}$$

where $t=a-b+c=(4\times k\times n+14\times k-3\times n)$ is the redundancy number, and a, b, c represent the corresponding number of observation values, unknowns and constraint conditions, respectively.

Initialisation of the Weight Matrix

In the initialisation of the weight matrices $\mathbf{P_f}$ and $\mathbf{P_x}$, the precision of the measurement value of the rover' s pose and the matching image points should be considered. Σ_0 denotes the precision value of the matching image points. In the traditional BA model, the resolution of each image pixel is considered the same, thus the weighted matrix $\mathbf{P_x}$ can be set equal to the identity matrix. However, due to the short baseline of the stereo cameras, the intersection angle of feature points over longer distances (over, say, 4.0m) is small and the localisation precision is lower, which will make the BA procedure fail. Hence, a more reasonable weighting strategy should be used. The weight matrix $\mathbf{P_x}$ can be set according to the camera - to - object distance of the feature point (Qing et al., 2018):

$$\begin{cases} P_x = 1.0, & d_j \leqslant 2.0\text{m} \\ P_x = 1.0 - 0.3\times(Z_j - 2.0), & 2.0\text{m} < d_j \leqslant 4.0\text{m} \\ P_x = 0.4\times(Z_j - 3.0)^2, & d_j > 4.0\text{m} \end{cases} \tag{27}$$

where d_j represents the distance between feature points j and the rover, and Z_j denotes the Z coordinate (camera axis direction) in the frame F_{Rover}. It is important to note that the general form of the weight matrix $\mathbf{P_x}$ depends on experience and is not the only form. $\mathbf{P_f}=\sigma_0^2/\sigma_f^2\ \mathbf{I}$ and σ_f represent the precision of the pseudo - observations. In practice, σ_f is set to $\sqrt{\sigma_R^2+\sigma_r^2}$ where σ_r denotes the precision of the rotation angles of the mast mechanism and σ_R represents the precision of the measurement data of the roll angle φ_R, pitch angle ω_R and yaw angle k_R. When equation (26) is invoked in the above solutions of the stereo BA model, the weight matrices $\mathbf{P_x}$ and $\mathbf{P_f}$ will be recalculated by the varying weight method (Wang, 1990).

In the above solutions with the pseudo - observation and the geometry constraint conditions of the stereo cameras, the number of the redundant observation components increases, and the stereo BA model can improve the estimation accuracy of the rover's visual localisation. However, the degree of freedom of the unknowns in the stereo BA model decreases, which will make the correction of the unknowns less pronounced.

Proposed Algorithm

The proposed algorithm for estimating the rover localisation is given as follows:

Algorithm. The stereo BA model with telemetry data and a rational geometric constraint condition of the stereo cameras and weight matrix.

(a) Set the initial value of $\mathbf{t}^0$ using formulas (5) and (13).

(b) Set the initial value of $\mathbf{x}^0$ using formula (28) (page 31).

(c) Set the initial value of $\mathbf{z}^0=[\mathbf{t}^0\mathbf{x}^0]$.

(d) Initialisation of the weight matrix $\mathbf{P_x}$ via equation (27), and $\mathbf{P_f}=\sigma_0^2/\sigma_f^2\mathbf{I}$.

(e) Construct the constraint condition $\ddot{\mathbf{C}}z+\ddot{\mathbf{w}}=\mathbf{0}$ through equation (24).

(f) Set the iteration count $m=0$.

(g) Calculate: $\mathbf{l_x}$ using equation (1).

(h) Update matrices $\mathbf{U}=[\mathbf{A_x B_x}]$ using equation (10).

(i) Estimate $\mathbf{z}^m$ using equation (25):

$$\delta\mathbf{z}^0=\mathbf{u}+(\mathbf{U}^T\mathbf{P_x}\mathbf{U}+\mathbf{I}_0^T\mathbf{P_f}\mathbf{I_0})^{-1}\mathbf{C}_0^T(\mathbf{C_0}(\mathbf{U}^T\mathbf{P_x}\mathbf{U}+\mathbf{I}_0^T\mathbf{P_f}\mathbf{I_0})^{-1}\mathbf{C}_0^T)^{-1}(-\ddot{\mathbf{w}}-\ddot{\mathbf{C}}u)$$

$$\mathbf{u}=(\mathbf{U}^T\mathbf{P_X}\mathbf{U}+\mathbf{I}^T\mathbf{P_F}\mathbf{I_0})^{-1}\mathbf{U}^T\mathbf{P_X}\mathbf{L_X}$$

$$\mathbf{z}^m=\delta\mathbf{z}^0+\mathbf{z}^0$$

(j) Obtain: $\mathbf{v_x}=\mathbf{Uz}-\mathbf{l_x}$, $\mathbf{v_f}=\mathbf{I_0 z}$.

(k) Update: $\mathbf{z}^0=\mathbf{z}^m$, $m=m+1$.

(l) $\mathbf{P_x}$ and $\mathbf{P_f}$ will be recalculated by the varying weight method (Wang, 1990).

(m) Repeat (g) to (l) while infinite norms $\|\delta\mathbf{z}^0\|_\infty>\epsilon_0=0.001$ or $m\leqslant k_{num}$, where $k_{num}=6$ denotes the number of iterations and ϵ_0 denotes the threshold.

(n) Obtain the standard deviation σ_0 using equation (26).

Results

The experimental results are presented in two sections. First, the proposed method (referred to as BA+ pseudo) is compared with two other methods. One is the BA method based on the ordinary least squares (Di et al., 2008) which is referred to as BA + geometry, which invokes equations (7) and (19). The other is a weighted total least - squares (referred to as WTLS) method (Qing et al., 2018). In contrast to BA + geometry and WTLS, the proposed BA+ pseudo method uses equation (25) to obtain the camera pose. Then, the rover localisation results are generated through the "backward transformations" procedure. These three methods are all self - programmed in C++. Second, the Yutu - 2 rover's visual localisation results are given for the Chang'e - 4 mission.

Experimental Results of the Yutu - 2 Rover's Visual Localisation

Before Chang'e - 4's mission began, the proposed localisation method was tested with a model rover, which has the same size and functions as Yutu - 2, in a simulated field with artificial craters and rocks. The 3D coordinates of all control points in the control reference frame $F_{control}$ ($OXYZ$) were measured by two electronic theodolites whose angular precision was 2.0 seconds and whose coordinate measurement precision was 0.12mm. The feature points on the rover and eight of the control points were then measured using so - called indoor GPS (IGPS, using the infrared laser signals emitted by two or more emitters) in a reference frame F_{IGPS}, whose angular measurement precision is 1.0 seconds and whose resulting coordinate precision is 0.1mm. Through common control points, the numerical conversion parameters between $F_{control}$ and F_{IGPS} can be obtained. Immediately following this, further numerical conversion parameters between F_{Rover} and F_{IGPS} can also be obtained through the common feature points. When the 3D coordinates of the feature points are measured using IGPS, the numerical conversion parameter between $F_{control}$ and F_{Rover} can be immediately calculated, which will be viewed as the true data of the rover's pose. The movement path of the model rover can be viewed in Fig. 5. At stations 1 to 6, seven or eight pairs (a backward pair was added) of stereo - images were captured by the stereo navigation cameras, which validated the robustness and performance of the developed software.

The precision of the image - point extraction, the distribution of the tie points and the distance between the various stations are the main factors that affect the rover's pose estimation precision. It is important to note that during the selection process the tie points should be evenly distributed in the overlapping areas between the stereo field of view (FOV) in the previous and current stations, and manual extraction should be utilised. The tie points are selected by the SURF method and the manual matching process; the precision of the extracted image points is nearly 1.0 pixel. If one tie point is perfectly multi-collinear or coplanar with its adjacent tie points, then those BA methods cannot obtain a high-precision pose estimation and may even fail (Qing et al., 2018).

The distribution of the selected tie points for station 2 (Fig. 5) can be viewed in Fig. 6, where the rover's pose (φ_R, ω_R, k_R and NED $\mathbf{r}_{Rover}$) and the rotation angles of the mast mechanism at the previous station (station 1) are listed in the second row of Table Ⅳ, and the telemetry pose from the IMU-assisted VO measurements and the rotation angles of the mast mechanism at the current station (station 2) are all listed in the third row of Table Ⅳ. The distance between stations 1 and 2 is 7.2m. For both BA+ pseudo and BA+ geometry there are six iterations. To test the relationships between the number of tie points and the rover's pose estimation based on different methods, various experiments were performed. The following four cases were employed to calculate the rover's pose at the current station 2 with different numbers of tie points:

Case 0: ten tie points (1, 2, 3, 4, 5, 6, 7, 8, 9, 10).

Case 1: nine tie points (1, 2, 4, 5, 6, 7, 8, 9, 10).

Case 2: eight tie points (1, 2, 3, 4, 5, 6, 9, 10).

Case 3: five tie points (1, 2, 3, 4, 5).

The rover's pose estimated results with different tie points are listed in Table Ⅴ. It should be noted that the rover's pose parameters at the previous station are viewed as true data.

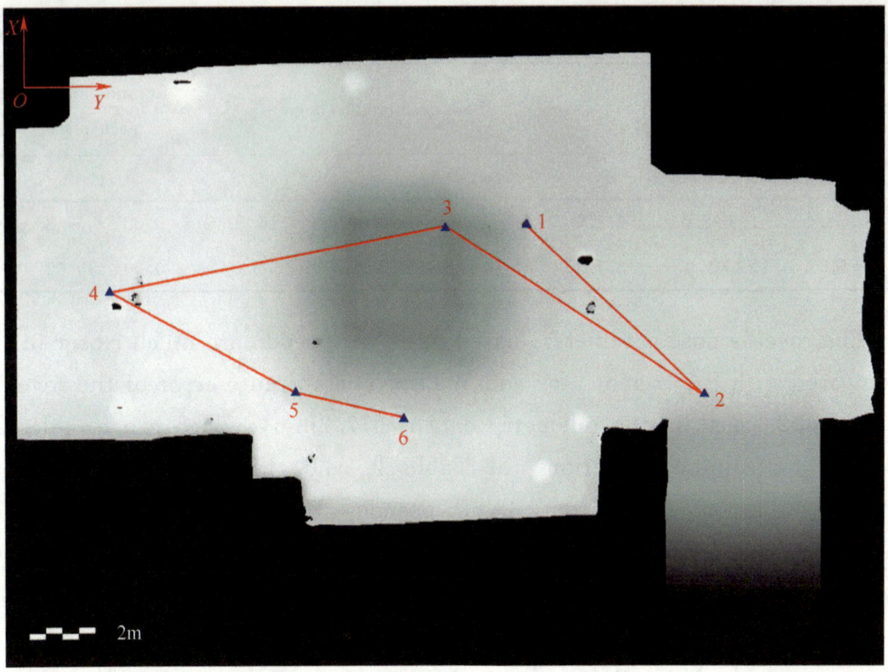

Fig. 5　The movement path of the model rover in a simulated field between stations 1 and 6

It is noted that the variance σ_0^2 in the two BA and WTLS methods are different, which respectively indicate the mean square error (MSE) of the reprojections and the rover pose (Qing et al., 2018). In the sixth column of Table Ⅴ, the symbol $\Delta_{XYZ} = \sqrt{(\Delta X_{true} - \Delta X')^2 + (\Delta Y_{true} - \Delta Y')^2 + (\Delta Z_{true} - \Delta Z')^2}$ denotes the absolute error value of the rover's pose estimation. The symbols $\Delta X'$, $\Delta Y'$ and $\Delta Z'$ (columns 3 to 5) represent the rover's pose estimation based on the above three methods or the telemetry data. ΔX_{true}, ΔY_{true} and ΔZ_{true} correspond to the true data of the rover's pose, which are −1704.1, 22 640.8 and −310.9mm, respectively. The results in Table Ⅴ indicate the following five points:

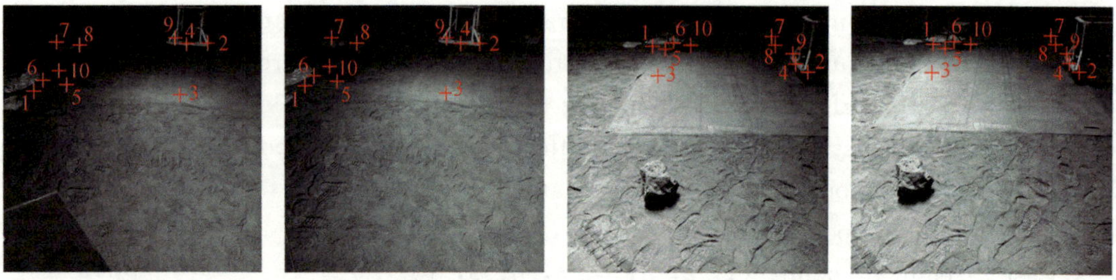

(a) (b) (c) (d)

Fig. 6 Ten tie points of the stereo-images in the two adjacent stations in the testfield. (a), (b) The seventh stereopair of images of the model rover taken at the previous station 1. (c), (d) The first pair of images of the model rover taken at the current station 2. Red crosses and numbers indicate tie points

Table Ⅳ The model rover's pose and the rotation angles of the mast mechanism at different stations

Station number	Location (mm)			Pose orientation (degrees)			Rotation angles of the mast mechanism (degrees)		
	X	Y	Z	φ_R	ω_R	k_R	θ_o	θ_y	θ_p
1	956.1	15 898.0	−371.0	−0.53	0.80	115.57	0.00	−59.88	−29.95
2	−1480.4	22001.9	−331.1	0.69	0.39	−62.07	0.00	40.00	−29.95

(1) The rover's pose parameters based on BA+ pseudo are much closer to the true data, compared with BA+ geometry and WTLS. The absolute error of the rover's pose Δ_{XYZ} at station 2 based on the telemetry data is 677.2mm, which can be calculated by using the second column (location) of Table Ⅳ and the true data. In case 0, the corresponding values of Δ_{XYZ} from BA + pseudo, BA + geometry and WTLS are 67.40%, 51.86% and 50.24% of those obtained by the telemetry data, respectively. In case 1, the corresponding values of Δ_{XYZ} are 73.91%, 58.71% and 58.62%, respectively. In case 2, the corresponding values of Δ_{XYZ} are 62.11%, 44.73% and 42.50%, respectively. Meanwhile, the case 3 rover pose parameters based on BA + pseudo, BA+ geometry and WTLS are much farther from the true data, compared with the telemetry data. Thus, the above three visual localisation methods can correct the rover's pose in cases 0, 1 and 2, but not in case 3.

(2) The rover's pose estimation precision based on BA+ pseudo is higher than BA+ geometry. Through the seventh column σ_0 of Table V, the precision σ_0 based on BA+ pseudo is 0.393, 0.338, 0.512 and 1.528 pixels in cases 0, 1, 2 and 3, respectively. Meanwhile, the corresponding σ_0 figure based on BA+ geometry is 0.793, 0.344, 0.859 and 1.267 pixels for cases 0 to 3, respectively.

(3) BA+ pseudo's computation time is longer than for either BA+ geometry and WTLS. To access the calculating efficiency, the calculation times of the above three methods, from the input of the matching image points to the calculation of the Yutu-2 rover's pose, are listed in the final column of Table V. Note that, other than case 0, the computation times of the two BA methods (BA + pseudo and BA+ geometry) are pretty comparable, but much lengthier than WTLS.

It should be noted that improper initialisation of the camera pose will lead the BA+ geometry method to converge to a local minimum, and is thus highly sensitive to the initialisation. In Table Ⅳ, the rover's yaw angle k_R, plus the pitch and yaw angles of the mast (θ_p and θ_y) are far from zero, whereas the other angles from the telemetry data are nearly zero, which will easily cause an improper initialisation of the camera pose without the telemetry data.

Table V　The model rover's pose results using different methods at station 2 for four different cases

		$\Delta X'$ (mm)	$\Delta Y'$ (mm)	$\Delta Z'$ (mm)	Δ_{XYZ} (mm)	σ_0	Times (ms)[a]
BA+pseudo	Case 0	−1838.6	22749.9	−447.9	220.8	0.393 pixels	202
	Case 1	−1714.9	22741.1	−455.9	176.7	0.338 pixels	130
	Case 2	−1833.4	22862.4	−312.5	256.6	0.512 pixels	141
	Case 3	−2462.0	22851.4	−271.6	787.6	1.528 pixels	118
BA+ geometry	Case 0	−1966.0	22 799.7	−422.5	326.0	0.793 pixels	138
	Case 1	−1869.0	22762.8	−500.8	279.6	0.344 pixels	131
	Case 2	−1958.0	22913.9	−278.8	374.3	0.859 pixels	117
	Case 3	−2463.8	22884.8	−274.6	798.8	1.267 pixels	110
WTLS	Case 0	−1978.0	22801.4	−423.7	337.0	3.2mm	2
	Case 1	−1869.8	22762.3	−501.3	280.2	3.0mm	2
	Case 2	−1984.1	22909.6	−279.2	389.4	4.5mm	2
	Case 3	−2463.1	22883.0	−276.8	797.4	10.3mm	2

[a] These computing times for the test environment uses an Intel Core™ i5@ 2.50GHz, 2GB Windows system.

(4) The greater the number of tie points, the higher the rover's pose estimation with the same level of extracted image points in cases 1, 2 and 3. This phenomenon does not occur in case 0 where the 3D coordinates of all ten tie points can be calculated by using the space intersection method with four images. The precision of reconstruction of case 0 is listed in Table Ⅵ.

In Table Ⅵ, σ_X; σ_Y, σ_Z represent the reconstruction precision of the coordinates of the tie points, where $\sigma_{XYZ}=\sqrt{\sigma_x^2+\sigma_Y^2+\sigma_Z^2}$. The rover's pose at station 1 is set as the true data, and the rover's pose at station 2 is selected from the pose derived from the IMU-assisted VO measurements and the rotation angles of the mast mechanism. In the space intersection method with four images, the error equation can be written as:

$$\mathbf{v}=\mathbf{A}\mathbf{x}-\mathbf{l};\ \mathbf{P}_{\mathbf{x}} \quad (28)$$

where $\mathbf{x}=(\Delta X, \Delta Y, \Delta Z)^{\mathrm{T}}$ represents the corrections to the 3D coordinates of the tie points in the frame F_{NED}; $\mathbf{l}$ represents the residual; $\mathbf{v}$ represents the corrections to the 2D image points in the frame F_{xy}; and $\mathbf{A}$ represents the coefficient matrix. The initial values of $\mathbf{x}$ can be obtained by the stereo reconstruction method with the stereoimages at station 1. Then, the reconstruction error σ_X; σ_Y; σ_Z can be calculated as:

$$\sqrt{(\mathbf{A}^{\mathrm{T}}\mathbf{P}_{\mathbf{x}}\mathbf{A})^{-1}} \times \sqrt{(\mathbf{v}^{\mathrm{T}}\mathbf{P}_{\mathbf{x}}\mathbf{v})/(4\times k\times n-3\times n)}$$

In Table Ⅵ, the precision σ_{XYZ} of tie point 3 is 68.9mm, and the precisions σ_{XYZ} of the other nine tie points have a range from 30.4 to 43.1mm. Thus, tie point 3 has a lower precision, which can result in a low localisation precision for the rover. The other tie points have similar precisions, which further proves that the greater the number of the tie points, the higher the visual localisation precision. In case 1, all tie points except point 3 are used to estimate the rover pose, and the rover's poses are much closer to the true data compared with cases 0, 2 and 3.

Table Ⅵ The reconstruction precision of all selected tie points (mm) in case 0

Tie point number	σ_X	σ_Y	σ_Z	σ_{XYZ}
1	20.8	20.0	21.6	36.0
2	28.6	22.7	22.8	43.1
3	35.4	38.3	45.1	68.9
4	26.9	21.4	22.3	41.0
5	21.5	19.8	20.4	35.6
6	18.7	17.8	29.2	39.0
7	14.0	12.7	25.9	32.1
8	14.1	12.5	23.8	30.4
9	24.2	19.3	20.5	37.1
10	15.7	14.7	24.3	32.4

(5) An even distribution of tie points has a key role in improving the visual localisation precision for the rover. In cases 0 and 2, the error σ_0 based on BA+ pseudo with ten and eight tie points is 0.393 and 0.512 pixels, respectively. Thus, the results reveal that an even distribution of tie points has a key role in improving the visual localisation precision because of the inclusion of tie points 7 and 8. After a comparison between the different number of the tie points based on BA+ geometry and WTLS, the same conclusion is arrived at.

In general, a variety of comparison tests and analyses between BA+ pseudo, BA+ geometry and WTLS in terms of precision should be performed. The stereo-images of the lunar rover cameras were obtained in six positions, with the results listed in Table Ⅶ.

In Table Ⅶ, the first column represents the numbers of the previous and current stations, with the 3D distance of the lunar rover between the two adjacent stations in the second column. The localisation "Absolute precision Δ_{XYZ}" represents the 3D distance between the calculated position at the current station and the measurement positions in $F_{control}$, whereas the "Relative precision" represents the ratio of the absolute precision to the relative distance. The three methods use the same data, including the matching image points and the calibration parameters of the rover's cameras. In the case of stations 1 and 2, the data of the rover's precision are from case 1 in Table Ⅴ.

In Table Ⅶ, the research comes to a similar conclusion that the mean localisation precision using BA+ pseudo is higher than either BA+ geometry or WTLS. From the "Relative precision" results, the mean relative localisation precisions of BA+ pseudo, BA+ geometry and WTLS are 3.01%, 3.73%, 3.70%, respectively, which satisfies the designed precision of 5%.

With BA + pseudo, equation (21) is used as the geometry constraint condition. Another similar visual localisation method (named BA + pseudo*) was studied, which is based on equations (22) and (19). The difference between BA + pseudo* and BA+ pseudo is that the former uses equation (19), while the latter uses equation (21) as the geometry constraint condition. Because the ordinary least - squares estimates are chosen to minimise the sum of the squared residuals (RSS) in BA, a phenomenon was found where the sum of the squared residuals for BA+ pseudo* and BA + pseudo are nearly the same. Due to the formula $RSS=\sigma_0^2 \times t$, the larger the redundancy number t, the smaller the value of the rover's pose estimation precision σ_0, so the redundancy number of BA+ pseudo is larger than that of BA+ pseudo*. Thus, the rover's pose estimation precision of BA+ pseudo is better, and adoption of equation (21) as the geometry constraint condition is more rational.

Table Ⅶ　Summary localisation results of the model rover with the indoor testfield

Stereo - image station number	Relative distance (m)	Absolute precision Δ_{XYZ} (mm)			Relative precision		
		BA + pseudo	BA + geometry	WTLS	BA + pseudo	BA + geometry	WTLS
1 - 2	7.2	176.7	279.6	280.2	2.45%	3.88%	3.89%
2 - 3	9.0	204.6	224.8	225.3	2.27%	2.50%	2.50%
3 - 4	11.3	293.2	388.7	368.3	2.59%	3.44%	3.26%
4 - 5	8.6	329.7	347.1	351.4	3.83%	4.04%	4.09%
5 - 6	5.6	218.0	269.4	267.4	3.89%	4.81%	4.78%

Applications of Visual Localisation to the Yutu - 2 Rover

In the Chinese Chang'e - 4 mission, the teleoperation system is responsible for visual localisation, topographic mapping, path planning and other operations with respect to the environmental perception, movement decision and scientific exploration. The stereo navigation cameras play an important role in the visual navigation of Yutu - 2, assisting in correcting the accumulated error caused by VO, whose expected error is predicted, by ground tests, to be up to 15% . The proposed method achieves autonomous visual localisation by means of corresponding area detection and feature matching based on SURF. With very difficult surfaces and poor texture conditions, manual visual localisation should also be utilised. The proposed visual localisation method and software has been deployed in a 64 - core image workstation in the ground teleoperation system and fully applied in the Chang'e - 4 mission operations. On 13th March 2019, the Yutu - 2 rover entered a sleeping state. According to the proposed visual localisation method, Yutu - 2 travelled 163. 2m, passing by 27 stations. The Yutu - 2 rover re - entered a work state on 30th March 2019. Thus, the service design goal (three months) of the Yutu - 2 rover was reached. The overall traverse of the Yutu - 2 rover is shown in Fig. 7. The base map was selected from the digital ortho - image map acquired by NASA' s Lunar Reconnaissance Orbiter (LRO) (Keller et al. , 2016) . The purple circles are the rover locations generated from the proposed visual localisation software. The double red lines are the recovered rover traverse based on the rover visual localisation results and the designed path. The rover' s traverse maps, generated at any time the rover stopped and took stereo -images, have been greatly supported by the surface operations of the rover and are also valuable in various scientific investigations of the landing site.

Due to the service design goal (three months) of the Yutu - 2 rover, this work provided the rover's visual localisation information until 13th March 2019. The precision and robustness of the localisation method proposed in this paper was verified by using LRO images. This is shown as in Fig. 8, where Fig. 8 (a) (M1303640934L) was taken on 1st February 2019 and Yutu - 2 was at a station labelled LE00202. The view in Fig. 8 (b) (M1305967189R) shows Yutu - 2 at station LE00210 on 28th February 2019, meaning that the rover had traversed approximately 46m to the west of LE002 during the month of February. All the images were first sampled to the same pixel scale (80cm) and were then expanded by a factor of two. The bright spot at the centre of the two images in Fig. 8 indicates the lander and the arrows indicate the rover. For more details, see the website http://lroc. sese. asu. edu/posts/1096. Just after midnight (UTC) on 1st February 2019, the LRO passed nearly overhead the Chang'e - 4 landing site. From an altitude of 82km, the LRO Narrow Angle Camera pixel ground sample distance was 0. 85m, allowing a sharper view of the lander and the Yutu - 2 rover (six hours later the left image in Fig. 8 (a) was taken, with a slew angle of 40°) . The LRO passes over any given place on the moon at least once every month (in daylight), allowing the westward progress of the Yutu - 2 rover to be seen. At the end of February, Yutu - 2 was a total of 69m from its home base, which is the Chang'e - 4 lander.

Yutu - 2 only occupies 1 pixel in the LRO images. The precise pixel coordinates of the

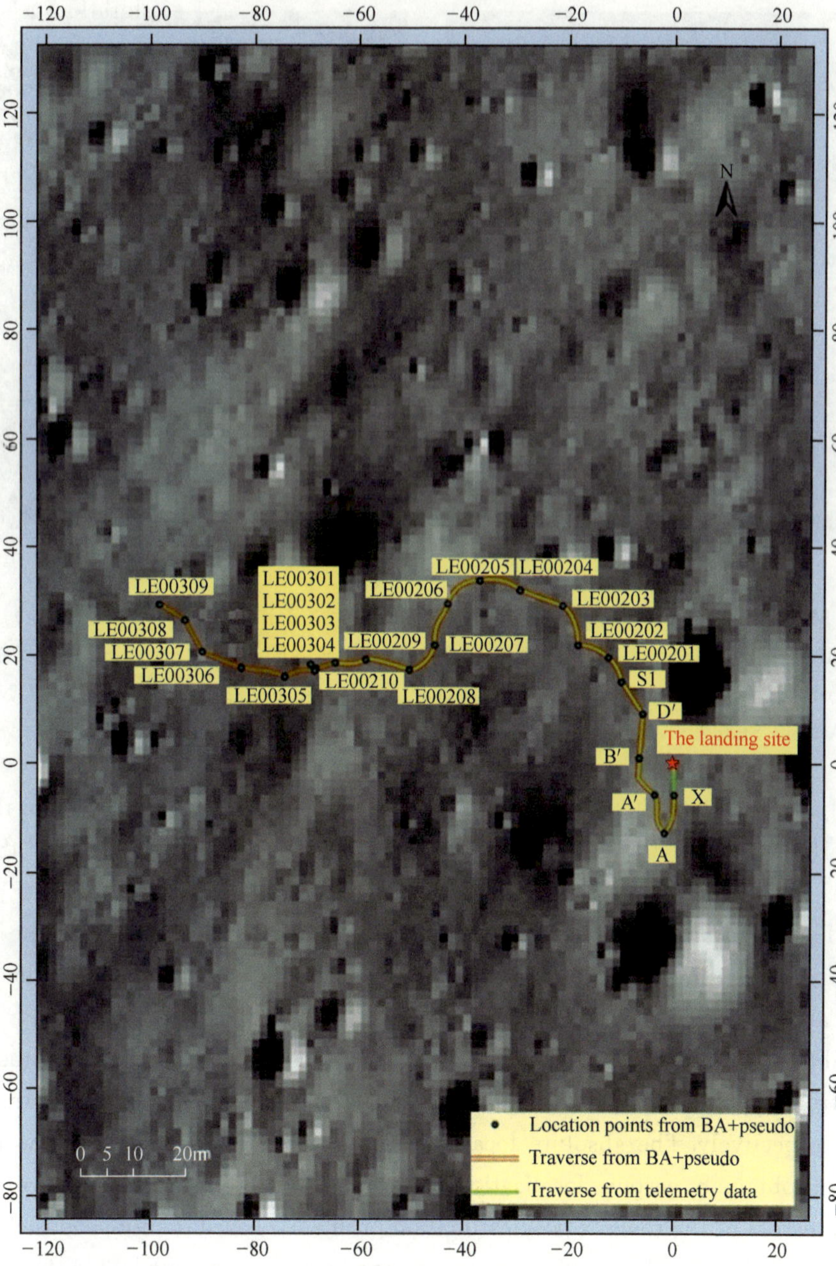

Fig. 7　The lunar rover's pose estimation results using the proposed visual localisation method (BA + pseudo). The green line represents the IMU-assisted VO data of the rover in real time. Base image from NASA's Lunar Reconnaissance Orbiter

lander (the Chang'e-4 landing site) is 177° 35′11.745″E, 45° 27′ 22.019″ S; those of the Yutu-2 rover at station LE00202 are 177° 35′ 08.137″ E, 45° 27′ 19.331″S, which are extracted from the image M1303640934L (Fig. 8 (a)) based on a corner detection method. The corresponding scaled pixel width and pixel height are 0.82 and 0.85m,

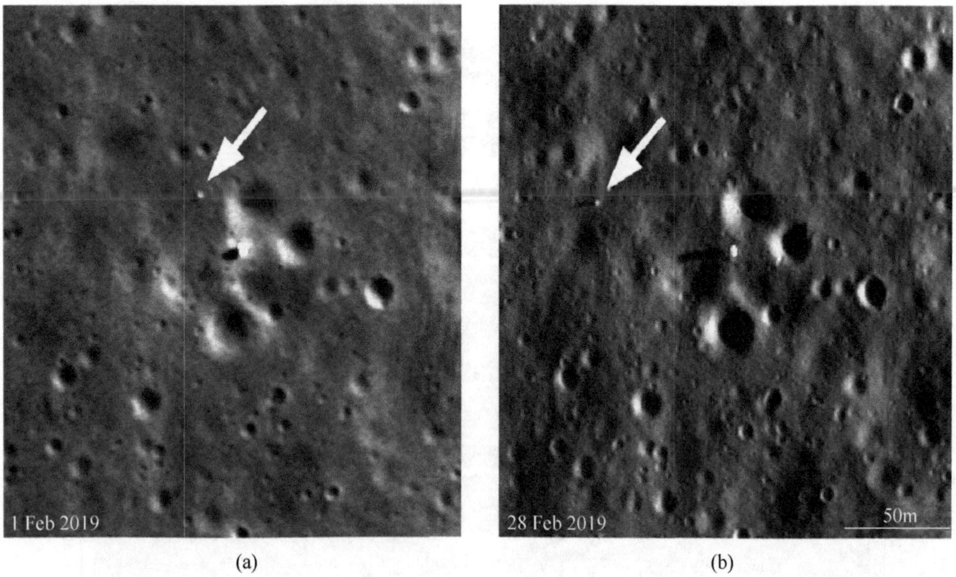

Fig. 8　Movement of the Chang'e-4 rover, Yutu-2, during February 2019. (a) Image M1303640934L taken on 1st February 2019. (b) Image M1305967189R taken on 28th February 2019. North at top. Images from NASA's Lunar Reconnaissance Orbiter (courtesy of NASA/GSFC/Arizona State University)

respectively. At station LE00210, the precise pixel coordinates of Yutu-2 from the image M1305967189R (Fig. 8 (b)) are 177° 34' 59.981" E, 45° 27' 19.834" S). The coordinates of the lander and Yutu-2 can also be expressed in the NED frame F_{NED}. The origin of the NED frame is set at the landing site. The localisation results of the Yutu-2 rover in Chang'e-4 mission operations, based on BA+ pseudo and the LRO images, are listed in Table Ⅷ.

In Table Ⅷ, the 2D Euclidean distances between the calculated and measurement positions from the LRO images at stations LE00202 and LE00210 are 1.8±1.1m and 6.2±1.1m, respectively. The absolute location results not only indicate that the proposed method can obtain a certain localisation precision in orbit, but also prove that the cumulative errors of the rover's pose estimation problem, based on the proposed method, cannot be avoided during long-distance movement by the rover. The larger the movement distance, the lower the rover's absolute pose estimation in the proposed method. Hence, an absolute localisation method with some prior information, such as the land marker in the images from the Orbiter or the lander's visual system, should be used periodically.

Discussion and conclusion

In this paper a precise visual localisation method for the Yutu-2 rover was developed. In the proposed method, the initial localisation results from the telemetry data were used to estimate the corresponding area in cross - station (cross - site) images. The bundle adjustment (BA) was based on the theory of unit quaternions for combining the rotation angles of the mast mechanism and the rover's pose from the IMU - assisted VO technique. The geometric constraint condition of the stereo cameras was then performed to localise the rover with high precision. In practice, the proposed method viewed the rotation angles of the mast mechanism and the rover pose from the IMU - assisted VO technique as pseudo - observations. Meanwhile, a more reasonable weighting strategy was invoked. Finally, the least - squares solution of the stereo BA model was given. The experimental results in the simulation field demonstrated that:

Table Ⅷ　Summary localisation results of the Yutu - 2 rover in Chang'e - 4 mission operations

Station	BA+pseudo (mm)		LRO images (mm)	
	X	*Y*	*X*	*Y*
LE00202	22.0	−18.0	19.7	−21.3
LE00210	19.1	−63.6	22.4	−68.8

(1) The proposed method can obtain a mean relative localisation precision of 3.01%, which will decrease the accumulated errors of the rover localisation that were derived from the IMU drift and wheel slippage. The 2D absolute precisions of Yutu - 2 at stations LE00202 and LE00210 are 1.8 ± 1.1m and 6.2 ± 1.1m, respectively, from the LRO images.

(2) The rover's pose parameters, based on the proposed BA+ pseudo method, are much closer to the true data compared with the BA + geometry and WTLS methods. Meanwhile, the rover pose estimation based on the proposed method is more precise than what was achieved using BA+ geometry.

(3) The proposed method takes a longer computation time than either BA + geometry or WTLS.

(4) The greater the number of tie points, the higher the rover's pose estimation precision in BA+ geometry, WTLS and the proposed BA+ pseudo method.

(5) An even distribution of tie points has the key role in improving the visual localisation precision for the rover.

The proposed visual localisation method and software have been successfully used in the teleoperations of the Yutu - 2 rover.

Acknowledgements

This work was supported by the National Natural Science Foundation of China (Grant No. 41601494).

References

[1] BAY, H., Ess, A., TUYTELAARS, T. and VAN Gool, L., 2008. Speeded - up robust features (SURF). Computer Vision and Image Understanding, 110 (3): 346 - 359.

[2] Blanco, J. - L., GonzÁlez - JimÉnez, J. and FernÁndez - Madrigal, J. - A., 2013. Sparser relative bundle adjustment (SRBA): constant - time maintenance and local optimization of arbitrarily large maps. IEEE International Conference on Robotics and Automation, Karlsruhe, Germany. Pages 70 - 77.

[3] Börlin, N. and Grussenmeyer, P., 2013. Bundle adjustment with and without damping. Photogrammetric Record, 28 (144): 396 - 415.

[4] Brown, D. C., 1966. Decentering distortion of lenses. Photogrammetric Engineering, 32 (3): 444 - 462.

[5] Brown, D.C., 1971. Close - range camera calibration. Photogrammetric Engineering, 37 (8): 855 - 866.

[6] Chermak, L., Aouf, N. and Richardson, M. A., 2017. Scale robust IMU - assisted KLT for stereo visual odometry solution. Robotica, 35 (9): 1864 - 1887.

[7] Cui, J., Huo, J. and Yang, M., 2016. Novel method of calibration with restrictive constraints for stereo - vision system. Journal of Modern Optics, 63 (9): 835 - 846.

[8] Di, K., Xu, F., Wang, J., Agarwal, S., Brodyagina, E., Li, R. and Matthies, L., 2008. Photogrammetric processing of rover imagery of the 2003 Mars Exploration Rover mission. ISPRS Journal of Photogrammetry and Remote Sensing, 63 (2): 181 - 201.

[9] Fathian, K., Ramirez - Paredes, J. P., Doucette, E. A., Curtis, J. W. and Gans, N. R., 2018. Quest: a quaternion - based approach for camera motion estimation from minimal feature points. IEEE Robotics and Automation Letters, 3 (2): 857 - 864.

[10] He, Z., Wang, X. and Fang, J., 2014. An innovative high - precision SINS/CNS deep integrated navigation scheme for the Mars rover. Aerospace Science and Technology, 39: 559 - 566.

[11] Huang, Y., Hu, X., Li, P., Cao, J., Jiang, D., Zheng, W. and Fan, M., 2012. Precise positioning of the Chang'E - 3 lunar lander using a kinematic statistical method. Chinese Science Bulletin, 57 (35): 4545 - 4551.

[12] Huang, Y., Chang, S., Li, P., Hu, X., Wang, G., Liu, Q., Zheng, W. and Fan, M., 2014. Orbit determination of Chang'E - 3 and positioning of the lander and the rover. Chinese Science Bulletin, 59 (29 - 30): 3858 - 3867.

[13] Indelman, V., Melim, A. and Dellaert, F., 2013. Incremental light bundle adjustment for robotics navigation. IEEE/RSJ International Conference on Intelligent

Robots and Systems, Tokyo, Japan. Pages 1952 - 1959.

[14] Keller, J. W., Petro, N. E. and Vondrak, R. R., 2016. The Lunar Reconnaissance Orbiter Mission - six years of science and exploration at the moon. Icarus, 273: 2 - 24.

[15] Kim, A. and Eustice, R. M., 2015. Active visual SLAM for robotic area coverage: theory and experiment. International Journal of Robotics Research, 34 (4 - 5): 457 - 475.

[16] Kitt, B., Geiger, A. and Lategahn, H., 2010. Visual odometry based on stereo image sequences with RANSAC - based outlier rejection scheme. 2010 IEEE Intelligent Vehicles Symposium, La Jolla, California, USA. Pages 486 - 492.

[17] Konolige, K., 2010. Sparse sparse bundle adjustment. British Machine Vision Conference, Aberystwyth, UK. Pages 1021 - 10211.

[18] Kümmerle, R., Grisetti, G., Strasdat, H., Konolige, K. and Burgard, W., 2011. G2o: a general framework for graph optimization. IEEE International Conference on Robotics and Automation, Shanghai, China. Pages 3607 - 3613.

[19] Leutenegger, S., Lynen, S., Bosse, M., Siegwart, R. and Furgale, P., 2015. Keyframe - based visual - inertial odometry using nonlinear optimization. International Journal of Robotics Research, 34 (3): 314 - 334.

[20] Li, R., Squyres, S. W., Arvidson, R. E., Archinal, B. A., Bell, J., Cheng, Y., Crumpler, L., Des Marais, D. J., Di, K., Ely, T. A., Golombek, M., Graat, E., Grant, J., Guinn, J., Johnson, A., Greeley, R., Kirk, R. L., Maimone, M., Matthies, L. H., Malin, M., Parker, T., Sims, M., Soderblom, L. A., Thompson, S., Wang, J., Whelley, P. and Xu, F., 2005. Initial results of rover localization and topographic mapping for the 2003 Mars Exploration Rover mission. Photogrammetric Engineering & Remote Sensing, 71 (10): 1129 - 1142.

[21] Liu, Z., Di, K., Peng, M., Wan, W., Liu, B., Li, L., Yu, T., Wang, B., Zhou, J. and Chen, H., 2015. High precision landing site mapping and rover localization for Chang'e - 3 mission. Science China Physics, Mechanics & Astronomy, 58 (1): 1 - 11.

[22] Liu, J., Wei, E., Jin, S. and Liu, J., 2018. Absolute navigation and positioning of Mars rover using gravityaided odometry. Journal of Navigation, 71 (3): 530 - 546.

[23] Lourakis, M. I. A. and Argyros, A. A., 2009. SBA: a software package for generic sparse bundle adjustment. ACM Transactions on Mathematical Software, 36 (1): article 2: 21 - 230.

[24] Lu, J., Lei, C. and Yang, Y., 2017. A dynamic precision evaluation method for the star sensor in the stellarinertial navigation system. Scientific Reports, 7 (1): article 4356. 12 pages.

[25] Mercan, H., Akyilmaz, O. and Aydin, C., 2018. Solution of the weighted symmetric similarity transformations based on quaternions. Journal of Geodesy, 92 (10): 1113 - 1130.

[26] Milford, M., 2013. Vision - based place recognition: how low can you go? International Journal of Robotics Research, 32 (7): 766 - 789.

[27] Qing, M. Y., Chuang, L. S., Bo, W., Shuo, Z., Lei, L. M. and Song, P., 2018. Weighted total least squares for the visual localization of a planetary rover. Photogrammetric Engineering & Remote Sensing, 84 (10): 605 - 618.

[28] Schubert, D., Goll, T., Demmel, N., Usenko, V., Stückler, J. and Cremers, D., 2018. The TUM VI benchmark for evaluating visual - inertial odometry. IEEE/RSJ International Conference on Intelligent Robots and Systems, Madrid, Spain. Pages 1680 - 1687.

[29] Strasdat, H., Montiel, J. M. M. and Davison, A. J., 2010. Real - time monocular SLAM: why filter? IEEE International Conference on Robotics and Automation, Anchorage, Alaska, USA. Pages 2657 - 2664.

[30] Wang, Z. Z., 1990. Principles of Photogrammetry (with Remote Sensing). Press of Wuhan Technical University of Surveying and Mapping, Wuhan, China. 574 pages.

[31] Yuan, X., 2009. Quality assessment for GPS - supported bundle block adjustment based on aerial digital frame imagery. Photogrammetric Record, 24 (126): 139 - 156.

[32] Zhang, S., Liu, S., Ma, Y., Qi, C., Ma, H. and Yang, H., 2017. Self calibration of the stereo vision system of the Chang'e - 3 lunar rover based on the bundle block adjustment. ISPRS Journal of Photogrammetry and Remote Sensing, 128: 287 - 297.

[33] Zhao, L., Huang, S., Sun, Y., Yan, L. and Dissanayake, G., 2015. ParallaxBA: bundle adjustment using parallax angle feature parametrization. International Journal of Robotics Research, 34 (4 - 5): 493 - 516.

[34] Zheng, W., Tong, F., Zhang, J., Liu, L. and Shu, F., 2017. Interferometry imaging technique for accurate deep - space probe positioning. Advances in Space Research, 60 (12): 2847 - 2854.

Résumé

La localisation précise du rover lunaire Yutu - 2 joue un rôle important dans la planification de trajectoires, l'évitement d'obstacles et la navigation jusqu'à des cibles. Pour fournir des informations de localisation de haute précision, une méthode de compensation par faisceaux stéréoscopique utilisant la théorie des quaternions unitaires est présentée pour la première fois. Pour améliorer la précision et la robustesse de la méthode proposée, la position et l'orientation du mobile, issues d'une technique d'odométrie visuelle assistée par une centrale inertielle et des angles de rotation du mécanisme du mât, sont considérées comme pseudoobservations. Une stratégie de pondération raisonnable et une condition rationnelle de contrainte géometrique des caméras stéréo sont également utilisées. Les résultats expérimentaux démontrent que la méthode proposée fournit des résultats de localisation plus précis qu' une compensation par faisceaux seule ou une méthode des moindres carrés pondérés. La méthode proposée a été utilisée avec succès dans les opérations de la mission Chang'e－4.

Zusammenfassung

Die präzise Ortsbestimmung des Yutu－2 Mondfahrzeugs spielt eine wichtige Rolle für die Pfadplanung, die Umgehung von Hindernissen und die Navigation zu den Zielobjekten. Hierzu wird zum ersten Mal eine Stereo－Bündelausgleichung, auf der Basis von Quaternionen der Norm Eins, zur Bereitstellung von hochpräziser Lokalisierungsinformation vorgestellt. Zur Verbesserung der Genauigkeit und Robustheit der vorgeschlagenen Methode, wird die Stellung des Rovers als Pseudo－Beobachtung betrachtet, die von einer visuellen Wegstreckentechnik, unterstützt durch eine inertiale Messeinheit und die Rotationswinkel des Mastenmechanismus abgeleitet wird. Zudem wird eine vernünftige Gewichtungsstrategie und eine rationale Zwangsbedingung der Geometrie der Stereokameras aktiviert. Die experimentellen Ergebnisse belegen, dass die vorgeschlagene Methode bessere Lokalisierungsergebnisse liefert, als eine Bündelausgleichung allein, oder eine gewichtete, Totale Kleinste Quadrate Methode. Die Methode wurde erfolgreich in Einsätzen der Chang'e－4 Mission angewandt.

Resumen

La localizacion precisa del rover lunar Yutu－2 juega un papel importante en la planificacion de la ruta, evitar obstaculos y la navegación hacia los objetivos. Para proporcionar información de localizacion de alta precisión, se presenta por primera vez un método de ajuste de haces estereo que utiliza la teor a de los cuaterniones unitarios. Para mejorar la precision y robustez del metodo propuesto, se introduce como una pseudo－observacion la actitud del vehículo explorador, obtenida con técnicas de odometría visual asistida por una unidad de medición inercial y los ángulos de rotación del mecanismo del mástil. En el ajuste también se contemplan una estrategia de ponderacion de pesos razonable y condiciones de restricción geométrica de las cámaras estéreo. Los resultados experimentales demuestran que el método propuesto proporciona resultados de localización más precisos que solo un ajuste de haces o un método de mnimos cuadrados ponderados. El método propuesto se ha utilizado con éxito en las operaciones de la misión Chang'e－4.

摘　　要

嫦娥四号玉兔-2号月球车在轨移动时，准确的位置将为月球车的路径规划、避障和科学探测提供基础信息。本文提出一种基于单位四元数理论的立体图像光束法平差模型，首次应用于嫦娥四号玉兔-2号巡视器的高精度定位问题。为增强该模型的算法鲁棒性和精度，将基于惯性导航设备和视觉里程计的巡视器遥测位置与姿态角信息、桅杆机构转角视为伪观测量，并引入一种更合理的立体相机几何约束条件。与基于欧拉角的光束法平差算法和加权总体最小二乘算法相比较，新算法在巡视器验证器的地面实验中具有更高的巡视器定位精度（平均相对定位精度为3.01％），且成功应用在嫦娥四号玉兔二号巡视器在轨任务中。

Prediction of terrain occlusion in Chang'e – 4 mission

YouQing Ma[1], Song Peng[2], Yang Jia[2], ShaoChuang Liu[1]

1. Institute of Remote Sensing and Digital Earth, Chinese Academy of Sciences, Beijing 100101, China
2. Beijing Institute of Spacecraft System Engineering, Beijing 100094, China

Abstract: This paper proposes a precise terrain occlusion prediction algorithm based on stereo image sequences from the China's Chang'e – 4 Yutu – 2 rover' visual system, which includes an on orbit stereo camera calibration method and a camera pose estimation method. The main advantage of the proposed algorithm is that the above two methods consider the problem that the coefficient matrix and observation vector should also be considered with regard to the error during optimization. In this case, the total least squares solution is applied instead of the ordinary least squares solution. Experimental results show that the proposed algorithm has higher precision (with average differences in the altitude angles and azimuthal angles of –0.37 degrees and –0.29 degrees) and application stability than the ordinary least squares method. The proposed algorithm has been successfully used in Chang'e – 4 mission operations.

Keywords: Yutu – 2 lunar; rover; Terrain occlusion angle; Camera self – calibration; Loop – closed block bundle adjustment; Skyline automatic extraction

1 Introduction

In China's Chang'e – 4 (CE – 4) lunar detection project, the CE – 4 probe consists of the "Queqiao" satellite, a lander and a lunar rover. The "Queqiao" satellite was launched into space on March 21, 2018, and the lander and rover set down within the Von Karman crater in on January 3, 2019. Soon after landing, a rover named Yutu – 2 [1, 2] was deployed and took its first tentative drive onto the airless regolith. CE – 4 mission has been planned to install six kinds of scientific payloads to complete the corresponding tasks, three kinds of payloads on the lander are the Landing Camera, the Terrain Camera, and the Low Frequency Spectrometer, and three kinds of payloads on the rover are the Panoramic Camera, the Lunar Penetrating Radar, and the Visible and Near – Infrared Imaging Spectrometer [2, 3]. The South Pole – Aitken (SPA) basin, approximately 2500km in diameter and 13km deep, is thought to have formed from an impact that penetrated through the Moon's distinctive plagioclaserich crust. Hence, the basin may expose fragments of the lunar mantle [1]. The Von Kármán crater, which is located at 45.456°S and 177.587°E within the SPA basin, is the best landing area for the lander. In the Von Kármán crater, there are several old impact craters and basins from the Nectrian Period with unique scientific value. The average elevation of this area is about 5,926m, with a standard deviation of 20.4m. The elevation ranges over about 321m. The highest geological feature in. the region is the mound located near the north boundary. The landing region is generally flat at a scale length of 59m, as nearly 99% of the area has a slope less than 15degrees, and the slopes of about 94% of the area are less than 5 degrees at a 59m scale length. Local steep slopes (>15degrees) are mostly associated with craters larger than 1km diameter [2]. Thus, due to the rugged terrain, the Yutu-2 rover will work in the lower-lying area. The altitude and azimuthal angles of the sun and the "Queqiao" satellite in the northeast-down (NED) frame can be acquired using an ephemeris. Situations in which the rover is in the shadow region or the "Que-qiao" satellite is outside the view of the rover's antenna, will limit the energy extraction and data communication of the Yutu-2 rover. To obtain terrain occlusion information, the terrain occlusion prediction method should be studied.

Due to the area of interest, research on a visual terrain occlusion prediction method for the planetary rover has not been extensively developed. Several studies have focused on the precise rover localization method using the skyline information, which de. nes the separation between sky and non-sky areas. Planetary rover localization is a challenging problem since no conventional methods such as Global Positioning System (GPS), structural landmarks etc. are available [4-6]. Horizon line is a promising visual cue which can be exploited for estimating the rover pose during planetary missions. By matching the horizon line detected in 2D images captured by the rover's visual system with virtually generated horizon lines from 3D terrain model (e.g., digital elevation map (DEM)), the localization problem can be solved in principle. In Ref. [7], position estimates are generated from two sources of information: visual imagery and attitude data collected and relayed by the rover, and maps of the rover's environment. Visual data about the skyline surrounding the rover is matched to data extracted from the maps, which is based on a probabilistic formulation: prior densities encode knowledge about the rover's location, and likelihood functions encode knowledge about the imaging device. Results indicate that the system can determine position estimates with accuracy from 84 to 200m (with 7.5min maps in USGS format) and accuracy from 186 to 564m (with 1 degree maps in USGS format). A7.5min map is a uniformly spaced grid with 30m resolution, while a 1 degree map is a uniformly spaced grid with 3min resolution. Both formats specify rectangular arrays of altitude values. Oiri et al. [8] presented a position estimation method for planetary rovers based on panoramic skyline images. The estimation is conducted by a comparison between two skyline images. One is extracted by image processing of an image from an omni – directional camera. The other is obtained by a geometry – based calculation from a DEM that was obtained by satellites. In Ref. [9], the authors introduced a system applicable to autonomous planetary exploration rovers that provides localization in the inevitable absence of a GPS. Several vision – based techniques (visual odometry, terrain matching and horizon visibility refinement) are applied using data from both live local rover and existing global orbital views to accurately determine the rover location within an orbital map. In Ref. [10], the authors proposed a horizon solution for precise rover orientation estimation, which is suitable for non – GPS or land marker information. The detected horizon in imagery provided by the on board navigation camera is matched with the horizon rendered over the existing terrain model. The set of rotation parameters (roll, pitch and yaw) that minimize the cost function between the two horizon curves corre – sponds to the rover estimated pose. In Ref. [11], the authors presented an efficient terrain – aided navigation method of generating position measurements from the visual appearance of the horizon profile (and hence terrain) surrounding the platform. An optimization process is used to align the measured horizon profile to an off – line pre – generated terrain – aided reference profile which allows for efficient position and attitude estimation. In those methods, sky – line extraction (horizon line detection) is one of the key steps.

Horizon line detection is a segmentation problem where a boundary between a sky and non – sky region is searched. Conventionally edge detection is performed as the first step followed by dynamic programming (DP) to find the shortest path which conforms to the detected horizon line. Recent work has proposed the use of machine learning to reduce the number of non – horizon edges to accurately detect the horizon line. Ahmad et al. [12] proposed a machine learning approach for horizon line detection using edge images and local features (i. e. , SIFT [13]). Given an image, first edge detection is applied. The authors refer to these edges as Maximally Stable Extremal Edges (MSEEs). Then a Support Vector Machine (SVM) classifier is used with ground truth information to classify MSEE pixels into two categories: horizon and non – horizon. Ahmad et al. [14] investigated the suitablity of various local texture features and their combinations to reduce the number of false classifications for a recently proposed horizon detection approach. The authors explored SIFT, LBP, HOG and their combinations SIFT – LBP, SIFTHOG, LBP – HOG and SIFT – LBP – HOG as features to train the SVM classifier and proposed various nodal costs which could result in enhanced accuracy of the detected horizon line. Ahmad et al. [15] presented a novel horizon line detection approach which uses machine learning and DP to extract the horizon line from a classification map instead of an edge map. The key idea is assigning a classification score to each pixel, which can be interpreted as the likelihood of the pixel belonging to the horizon line, and representing the classification map as a multi – stage graph. Using DP, the horizon line can be extracted by finding the path that maximizes the sum of classification scores. In contrast to edge maps which are typically binary (edge vs no - edge) and contain gaps, classification maps are continuous and contain no gaps, yielding significantly better solutions. In Ref. [16], a new approach to horizon line detection by coupling machine learning with DP is proposed. Increasing the stability and reliability of navigation for mobile objects of different classes is becoming increasingly significant. Satellite navigation systems have a fundamental defect: a vulnerability to hacking and spoofing. In the caseof stellar inertial navigation systems and correlation - extremal navigation, instability of the horizon line produces navigation errors. Grishin and Maslov [17] proposed a measurement procedure for horizon line position estimation. Processing of the results shows a horizon line direction instability of about 5 - 7 angular minutes during the day time. For some images with difficult environmental conditions like a foggy or cloudy sky, the conventional image processing methods are inherently inaccurate in identifying the correct horizon. Fefilatyev et al. [18] proposed an automatic skyline extraction method based on machine learning technology.

In other applications, such as unmanned aerial vehicle (UAV), small aircraft and unmanned surface vehicle, automatic or semiautomatic skyline extraction can be achieved by edge detection, classification, and segmentation methods. A vision - based obstacle detection system for small UAV is presented in Ref. [19]. Obstacles are detected by segmenting the image into sky and non - sky regions and treating the non - sky regions as obstacles. The feasibility of this approach is demonstrated by using the vision output to steer a small unmanned aircraft to fly towards an obstacle. Thurrowgood et al. [20] described a technique for estimating the attitude of a UAV by monitoring the visual horizon. An algorithm is developed that makes the best use of color and intensity informa - tion in an image to determine the position and orientation of the horizon, and infer the aircraft's attitude. In order to achieve robust image segmentation, machine learning is applied to a large data - base of images with many different types of skies [21]. In the GPS - free navigation problem of airborne vehicles, Grelsson et al. [22] proposed a highly accurate method to estimate the attitude (pitch and roll angle) by horizon detection in fisheye images. A edge detector and a probabilistic Hough voting scheme are used to compute an approximate attitude and the corresponding horizon line in the image. Horizon edge pixels are extracted in a band close to the approximate horizon line. The attitude estimates are refined through registration of the extracted edge pixels with the geometrical horizon from a DEM.

Above all, the skyline information is mainly usedto estimate the rover pose. The combination of the sequence images from ther over's visual system, a DEM and the rover pose can provide terrain occlusion information with certain accuracy. However, the estimation precision of the terrain angles is reduced for some realities, such as the low - precision DEM, the low image matching accuracy between the DEM and the sequence images and the low - precision rover pose. In the other hand, the estimation precision of the rover pose will be also reduced in the case of the low - precision DEM and the low image matching accuracy. Due to the differences in perspective and the resolution of the images from the rover, orbiter and lander, the rover pose does not have high precision. Thus, as a kind of rude localization method, the above rover pose estimation method will be only used to reduce the inuence of cumulative (visual) odometer localization error in the long - distance positioning.

Inspired by this finding, this paper proposes a new solution for the visual terrain occlusion prediction algorithm based on stereo image sequences and control points on the rover. Specially, in this work:

• Aself - calibration bundle adjustment (BA) method is adopted to obtain the intrinsic parameters of the stereo cameras using the image sequence and the feature points of the structure.

• Multi - degree (e. g. , two - degree, three - degree, four - degree or six - degree) overlapping image points in the stereo image sequence from one station are extracted to construct a loop - closed block BA model. In addition, the constraint conditions, including the 3D coordinates of the feature points and the length between the corners of the solar cells, should be invoked in the loop - closed block BA model. We can then obtain the rotation matrix between the camera frame and the lunar rover frame.

• Atotal least squares (TLS) solution is used in the self - calibration BA model and the loop - closed block BA model instead of the ordinary least squares (OLS) solution. The comparison between the TLS solution and the OLS solution is carried out.

• The lunar rover pose can be obtained with high precision using our previous work (weighted total least squares, abbreviated as WTLS) [4], which will limit the accumulated error of the rover pose.

• A visual terrain occlusion prediction algorithm is firstly presented to automatically recognize the skyline and calculate the terrain occlusion angles (the azimuthal and the altitude angles) with high precision. At the same time, a workflow is given.

In comparison to the matching method based on a lunar DEM and an image of the lunar rover's visual system, the proposed algorithm has three advantages. (1) We derive a more reasonable solution to the camera calibration problem, which uses control points on the structural part of the lunar rover. During the fight of the CE - 4 probe, the intrinsic parameters of the stereo camera system will change. Therefore, the precision of the terrain occlusion angles prediction decreases when we use the intrinsic parameters from the ground experiment. (2) The proposed algorithm adopts a loop - closed block BA method, which can reduce the rotation angle error (0. 3degrees, 3σ) of the mast. (3) The proposed algorithm adopts the TLS solution of the camera calibration model and the loop - closed block BA model instead of the OLS solution. Hence, there are several motivations for the terrain occlusion prediction on orbit.

(a) Accurate terrain occlusion information will help create instructions for deployment motion, antenna rotation and rover operation, which will save the Yutu - 2 rover's energy, improve the work efficiency and reduce the risk of the lunar exploration project.

(b) The proposed algorithm is more useful than the previous localization methods [7 - 11]. [7 - 11] only study the rover's visual pose estimation problem, and cannot provide the solution of the terrain occlusion information prediction problem. It is noted that one flaw of those methods [7 - 11] is that the rover is viewed as a single point, the actual angles (roll, pitch and yaw) are ignored. Hence, even though the rover is under illumination, the "- Y" solar wing is not toward the sun at some angles. For the same reason, the antenna cannot also communicate with the "Queqiao" satellite at some angles. In order to obtain the accurate pose of the rover, the "- Y" solar wing and the antenna, the telemetry data from the rover's measurement equipments should be used, which include a sequence of images, the rover pose and the rotation angles of the mast [4]. Hence, the proposed algorithm can continuously provide precise terrain occlusion information for the lunar rover on orbit.

The remainder of this paper is structured as follows. Section 22 describes the proposed algorithm. In Section 33, experimental results are presented to demonstrate the effectiveness and precision of the proposed algorithm compared to OLS. Section 44 concludes the paper.

2 Methodology

2.1 Principal of terrain occlusion prediction

Similar to China's CE - 3 Yutu lunar rover, the visual system of the Yutu - 2 rover includes navigation cameras, panoramic cameras and hazard cameras, the first two of which are mounted on the mast of the rover. Thus, the navigation and panoramic camera system can obtain a panoramic image sequence, as shown in Fig. 1. However, the information about the surrounding terrain exists only in the sequence of images, and this information can provide a solution to the terrain occlusion prediction problem.

To improve the working efficiency, this paper first proposes an exact terrain occlusion prediction algorithm based on stereo image sequences and control points. Taking the navigation camera as an example, the skyline can be obtained automatically in the naviga - tion images using a feature extraction algorithm. Edge detection can be completed by detecting all of the points whose grey values change substantially and connecting them to form several lines. Because there is no air in the lunar environment, sunlight is directly irradiated, and diffuse reflection and scattering phenomena on the surfaces of the Earth and Mars are not common. Therefore, the demarcation of the sky and non - sky areas is more obvious. First, a camera image is denoised using the 3×3 template algorithm, and the denoised image is then binarized with a 10～30 threshold. Second, the Canny operator

[23] is used for edge detection, and the contour threshold is set to 3. The edge points are then selected using the Sobel operator, and the core is set to a 3×3 matrix. Finally, the pixels in the image are judged from top to bottom. If a change occurs, the image point is considered to be a feature point of the skyline. Because the skyline can be considered to be a continuous curve, theyimage coordinates of the adjacent feature points of the skyline should be less than 2 pixels in practice.

In Fig. 1, a feature point A in the line of the mountain will be reflected by the image point a, and the 2D coordinates are x, y. In the camera calibration procedure, the principal distance fand principal point x_0, y_0 of the camera in the camera frame and the lens distortion (k_1, k_2, p_1, p_2) can be obtained, which will compensate for the system error ($\Delta x \Delta y$) of the camera lens. The corrected image points can be represented as $\underline{x} = x - \Delta x$, $\underline{y} = y - \Delta y$, and the vector $\vec{\gamma}_S$ of the terrain occlusion angle in the camera frame is:

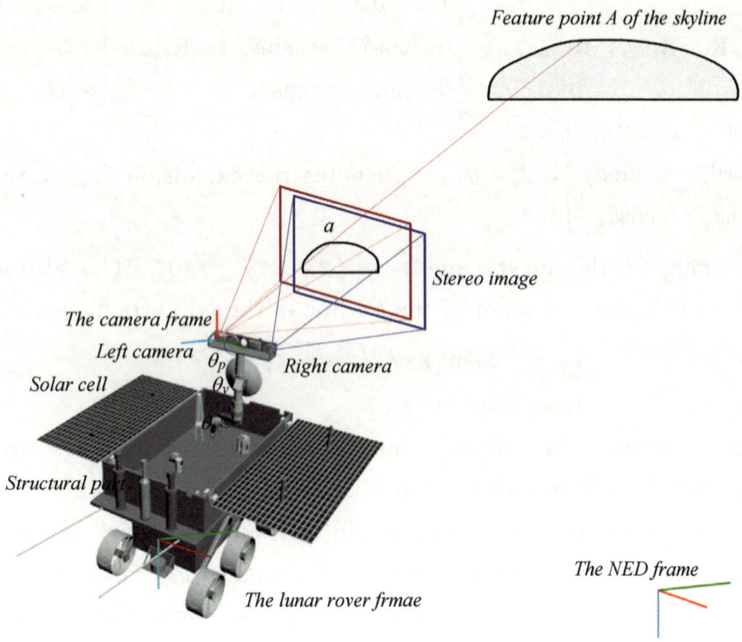

Fig. 1 Illustration of the Yutu - 2 rover's stereo camera system

$$\vec{\gamma}_S = [\cos\eta_S \sin\tau_S \quad \sin\eta_S \quad -\cos\eta_S \cos\tau_S]^T \tag{1}$$

where $\eta_S = a\tan((\underline{y} - y_0)/\sqrt{f^2 + (\underline{x} - x_0)^2})$ denotes the altitude angle of the feature point in the line of the mountain; $\tau_S = a\tan(\underline{x} - x_0)/f)$ denotes the azimuthal angle; and sin, cos and atan represent the sine, cosine, and arctangent functions, respectively.

At the beginning of one task period, the work frame (the NED frame) will be established, and the Yutu - 2 rover moves several times by a distance of approximately

20m to 30m. Compared with the distance between the Yutu - 2 rover and the feature point of the mountain (approximately 10km), the distance between the camera and the Yutu - 2 rover (1.2m) and the distance between the rover and the work frame origin can be ignored. When the Yutu - 2 rover and the mast are in their initial states, the vector from the camera frame origin to the feature point can be approximated as the vector from the rover frame origin to the feature point and the vector from the work frame origin to the feature point. In this case, the vector $\vec{\gamma}$ of the terrain occlusion angle in the NED frame can be obtained through a transformation of the rover's coordinate system [4] and the rover pose (the roll angle φ_R, the pitch angle ω_R and the yaw angle κ_R) via telemetry. Hence, $\vec{\gamma}$ can be written as:

$$\vec{\gamma} = \mathbf{R}_{NED-R}\mathbf{R}_{S-R}\overrightarrow{\gamma_S} \tag{2}$$

where $\mathbf{R}_{NED-R}$ denotes the rotation matrix between the rover frame and the NED frame, $\mathbf{R}_{S-R}$ denotes the rotation matrix between the camera frame and the rover frame, $\mathbf{R}_{S-R} = \mathbf{R}_{expansion}\mathbf{R}_{yaw}\mathbf{R}_{pitch}$, $\mathbf{R}_{expansion} = \begin{bmatrix} 1 & 0 & 0 \\ 0 & \cos\theta_e & -\sin\theta_e \\ 0 & \sin\theta_e & \cos\theta_e \end{bmatrix}$, $\mathbf{R}_{yaw} = \begin{bmatrix} \cos\theta_y & 0 & -\sin\theta_y \\ 0 & 1 & 0 \\ \sin\theta_y & 0 & \cos\theta_y \end{bmatrix}$, $\mathbf{R}_{pitch} = \begin{bmatrix} 1 & 0 & 0 \\ 0 & \cos\theta_p & -\sin\theta_p \\ 0 & \sin\theta_p & \cos\theta_p \end{bmatrix}$, θ_e, θ_y, θ_p denotes the expansion angle, the yaw angle, and the pitch angle of the mast, and $\vec{\gamma} = (\gamma_X \quad \gamma_Y \quad \gamma_Z)$. The altitude anglegand azimuthal angleτ of the feature point of the skyline in the NED frame can be written as

$$\begin{cases} \eta = -a\sin(\gamma_Z / \sqrt{\gamma_X^2 + \gamma_Y^2 + \gamma_Z^2}) \\ \tau = a\tan(\gamma_Y / \gamma_X) \end{cases} \tag{3}$$

where asin denotes the arcsine function, and atan denotes the arctangent function. Equation (3) indicates that the precisions of η and τ are affected by the image point coordinates, the intrinsic parameters, and the rotation angles of the mast and the rover pose. The measurable precision of the rotation angles (θ_e, θ_y, θ_p) is 0.3 degrees (3σ), and the precision of the rover pose is 1.0 degrees (3σ), which will havea greater impact on the terrain occlusion than the intrinsic parameters and the image point coordinates. In Ref. [6], the misalignment between the camera optical frame and the camera body frame, as tracked by a motion capture system, leads to a drift between the reconstructed trajectory and the reference one. The authors presented a calibration method which estimates the relative orientation and position between this two reference frames.

As shown in Fig. 1, several structures on the back of the Yutu - 2 rover are in the field of view of the stereo cameras, and the 3D coordinates of the feature points in the rover frame were measured before the flight. In practice, those 3D coordinates can be viewed as being unchanged when the Yutu - 2 rover works on orbit. When the pair of wings is in the field of view of the cameras, the solar cells on the wings can provide distance information, such as the length between the corners of the solar cells.

2.2 Precise camera self-calibration on orbit

Camera calibration is the first step in computer vision and photogrammetry. Over the past few decades, researchers have performed several studies on calibration methods for deep space rover stereo systems. Gennery [24] proposed a *CAHV* model, which was applied to the navigation cameras of the American "Spirit" and "Opportunity" Mars rovers. The *CAHV* model can provide the parameters to transform from the camera frame to the NED frame. Based on the *CAHV* model, Gennery [24] added a vector to describe the distortion parameters, and the *CAHVOR* model was applied to the panoramic cameras of the "Spirit" and "Opportunity" rovers. In the case of hazard cameras, the *CAHVORE* model was redesigned. Zhang et al. [25] proposed an integrated self-calibration method with two additional constraints. The first constraint was an error model of the mast mechanism, and the second constraint was an error model of the Inertial Measurement Unit. In this method, the actual value of the twist must be given and measured repeatedly. However, the onboard navigation system of the rover cannot provide the actual rotation angles of the mast mechanism when the rover works on orbit. Qing et al. [4] proposed a new solution to the relative orientation problem of the stereo cameras. However, the intrinsic parameters of the camera cannot be recalculated.

To use the control information of the structure of the Yutu-2 rover, this paper proposes a self-calibration BA model to obtain the camera parameters with high precision. Based on the collinear conditional equation (Eq. (4)), the model considers the 2D image point coordinates, the 3D coordinates of the points of the structure and the initial values of the camera pose as inputs:

$$\begin{cases} x - x_0 - \Delta x = -f\dfrac{a_1(X - X_S) + b_1(Y - Y_S) + c_1(Z - Z_S)}{a_3(X - X_S) + b_3(Y - Y_S) + c_3(Z - Z_S)} \\ \\ y - y_0 - \Delta y = -f\dfrac{a_2(X - X_S) + b_2(Y - Y_S) + c_2(Z - Z_S)}{a_3(X - X_S) + b_3(Y - Y_S) + c_3(Z - Z_S)} \end{cases} \tag{4}$$

where X_S, Y_S, Z_S denote the camera location in the rover frame; a_1, a_2, a_3, b_1, b_2, b_3, c_1, c_2, c_3. represent the elements of $\mathbf{R}_{S\text{-}R}$ between the camera frame and the rover frame, which contains three independent elements (φ, ω, κ); and X, Y, Z denote the 3D coordinates of the control points in the structure, which can be viewed as actual values. All of the image points have been corrected for distortion with (k_1, k_2, p_1, p_2), which will compensate for the system error (Δx, Δy). The linearization of Eq. (4) is as follows:

$$\begin{cases} v_x = \dfrac{\partial x}{\partial X_S}\Delta X_S + \dfrac{\partial x}{\partial Y_S}\Delta Y_S + \dfrac{\partial x}{\partial Zz}\Delta Z_s + \dfrac{\partial x}{\partial \varphi}\Delta\varphi + \dfrac{\partial x}{\partial \omega}\Delta\omega + \dfrac{\partial x}{\partial \kappa}\Delta\kappa \\ \quad + \dfrac{\partial x}{\partial f}\Delta f + \dfrac{\partial x}{\partial x_0}\Delta x_0 + \dfrac{\partial x}{\partial y_0}\Delta y_0 - l_x \\ v_Y = \dfrac{\partial y}{\partial X_S}\Delta X_S + \dfrac{\partial y}{\partial Y_S}\Delta Y_S + \dfrac{\partial y}{\partial Z_S}\Delta Z_S + \dfrac{\partial y}{\partial \varphi}\Delta\varphi + \dfrac{\partial y}{\partial \omega}\Delta\omega + \dfrac{\partial y}{\partial \kappa}\Delta\kappa + \dfrac{\partial y}{\partial f}\Delta f \\ \quad + \dfrac{\partial y}{\partial x_0}\Delta x_0 + \dfrac{\partial y}{\partial y_0}\Delta y_0 - l_y \end{cases} \tag{5}$$

where v_x, v_y represent the correction of the 2D image points; ΔX_S, ΔY_S, ΔZ_S, $\Delta\varphi$, $\Delta\omega$, $\Delta\kappa$ represent the correction of the camera pose; Δf, Δx_0, Δy_0 represent the correction of the principal distance and the principal point; l_x, l_y represent the residuals. Then, Eq. (5) can be written as:

$$\boldsymbol{v} = \boldsymbol{B}\hat{\boldsymbol{x}} - 1 \tag{6}$$

where $\boldsymbol{v} = [\boldsymbol{v}_x \quad \boldsymbol{v}_y]^{\mathrm{T}}$, $\boldsymbol{l} = [\boldsymbol{l}_x \quad \boldsymbol{l}_y]^{\mathrm{T}}$, $\hat{\boldsymbol{x}}$ denotes the correction vector of the camera calibration parameters (ΔX_S, ΔY_S, ΔZ_S, $\Delta\varphi$, $\Delta\omega$, $\Delta\kappa$, Δf, Δx_0, Δy_0), $\boldsymbol{B}$ denotes the coeffcient matrix, and $\boldsymbol{P}$ denotes the weight matrix.

When the extrinsic parameters of the stereo cameras are obtained in advance, the geometric relationship can be viewed as the constraint conditions:

$$\boldsymbol{E} = [\hat{\boldsymbol{R}}_{21} \quad \hat{\boldsymbol{R}}_{31} \quad \hat{\boldsymbol{R}}_{32} \quad (\boldsymbol{R}_i^l)^{\mathrm{T}}(\boldsymbol{T}_i^r - \boldsymbol{T}_i^l)]^{\mathrm{T}} - \boldsymbol{C}_1 = \boldsymbol{0} \tag{7}$$

where $\boldsymbol{T}_i^l$ and $\boldsymbol{R}_i^l$ represent the left camera pose in the rover frame, $\boldsymbol{T}_i^r$ and $\boldsymbol{R}_i^r$ represent the right camera pose in the rover frame, $\boldsymbol{C}_1$ repsents a constant vector, $\hat{\boldsymbol{R}}_{jk} = ((\boldsymbol{R}_i^l)^{\mathrm{T}}\boldsymbol{R}_i^r)_{jk}$ represents the j - throw and k - th column value of $\boldsymbol{R}$, and $\boldsymbol{E}$ represents the function. The linearization of Eq. (7) can be viewed as follows:

$$\boldsymbol{J}\hat{\boldsymbol{x}} + \boldsymbol{W} = \boldsymbol{0} \tag{8}$$

where $\boldsymbol{W} = -\boldsymbol{E}^0 - \boldsymbol{C}_1$, $\boldsymbol{J} = [\boldsymbol{J}_i \quad \underset{3\times 3}{\boldsymbol{0}}]^{\mathrm{T}}$. $\boldsymbol{E}^0$ represents the function value in the current iteration of the camera pose, $\boldsymbol{J}_i = \dfrac{\delta \boldsymbol{E}}{\delta \hat{\boldsymbol{x}}_i}$, and δ indicates a derivative. By combining Eqs (6) and (8), we can obtain the correction vector of the camera calibration parameters by an indirect adjustment method with restrictive conditions:

$$\hat{\boldsymbol{x}} = (\boldsymbol{N}_{BB}^{-1} - \boldsymbol{N}_{BB}^{-1}\boldsymbol{J}^{\mathrm{T}}\boldsymbol{N}_{JJ}^{-1}\boldsymbol{J}\boldsymbol{N}_{BB}^{-1})\boldsymbol{W}_x - \boldsymbol{N}_{BB}^{-1}\boldsymbol{J}^{\mathrm{T}}\boldsymbol{N}_{JJ}^{-1}\boldsymbol{W} \tag{9}$$

where $\boldsymbol{W}_x = \boldsymbol{B}^{\mathrm{T}}\boldsymbol{P}\boldsymbol{l}$, $\boldsymbol{N}_{BB} = \boldsymbol{B}^{\mathrm{T}}\boldsymbol{P}\boldsymbol{B}$, $\boldsymbol{N}_{JJ} = \boldsymbol{J}\boldsymbol{N}_{BB}^{-1}\boldsymbol{J}^{\mathrm{T}}$, the unit weight variance is $\sigma_0 = \sqrt{\boldsymbol{V}^{\mathrm{T}}\boldsymbol{P}\boldsymbol{V}/(n-u+s)}$, n, u, s is the number of measurements, camera calibration parameters, and constraint conditions, respectively. The error of the camera calibration parameter is alculated by the equations $\sigma_{\hat{x}\hat{x}} = \sigma_0\sqrt{\boldsymbol{Q}_{\hat{x}\hat{x}}}$ and $\boldsymbol{Q}_{\hat{x}\hat{x}} = \boldsymbol{N}_{BB}^{-1} - \boldsymbol{N}_{BB}^{-1}\boldsymbol{J}^{\mathrm{T}}\boldsymbol{N}_{JJ}^{-1}\boldsymbol{J}\boldsymbol{N}_{BB}^{-1}$, which can be viewed as the theoretical precision.

In practice, to improve the estimation precision of the camera calibration parameters, the mast should be moved with a large pitch angle, which will increase the number of control points in the image of the rover structure.

2.3 Precise camera pose estimation

The precise camera pose is the key to improving the calculation precision of the terrain occlusion based on the stereo image sequence. Because the navigation camera is used more frequently than the panoramic camera when the Yutu - 2 rover works on orbit, and the field of view angle is 47 degrees, the camera image can contain the sky, the structure of the rover and the solar wings. However, the field of view of the panoramic sequence image is such that it is difficult to include both the sky and the structure, so it is necessary to adjust the pitch angle of the panoramic camera. To limit the influence of the rotation error of the mast on the terrain occlusion angles, a block BA model with a sequence of stereo images, the control points on the structure of the rover and the length information of the solar cell is proposed.

In Fig. 2, the stereo cameras will rotate anti - clockwise from the rear of the rover at a certain angle interval (between 10 and 20 degrees) to acquire sequential stereo images. At location S_1, the left camera frame is set as the initial frame. In locations S_1; S_2, SIFT [13] is used to match the left images first, and a set of matching image points p_1^l, p_2^l is obtained. Then, a set of matching image points p_1^r, p_2^r in the right images is obtained by matching the stereo images with the correlation coefficient matching method. Finally, the correlation coefficient matching algorithm is used to match and check p_1^r, p_2^r. The mismatched points are then eliminated to obtain multi - degree overlapping image

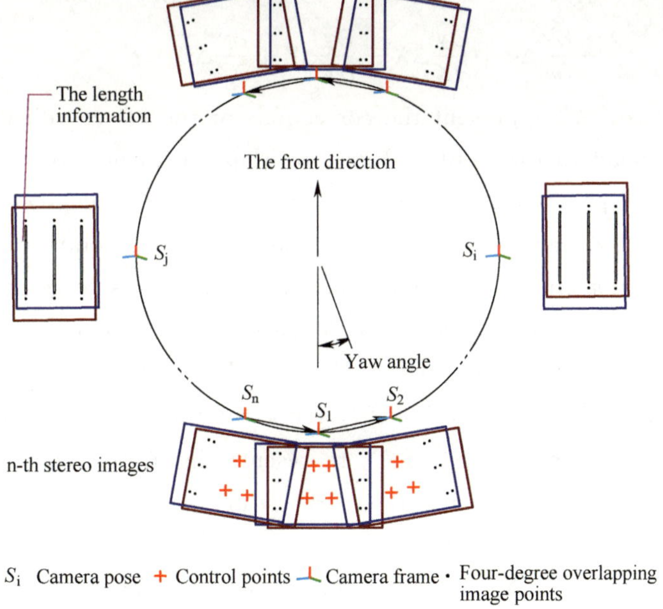

Fig. 2　Loop - closed block BA model

points. The stereo images at locations S_1, S_2 can be correlated by several uniformly distributed multi - degree overlapping image points, and a block BA model will be established to obtain the camera pose at location S_2 in the initial frame. In this way, the latter camera pose can be also acquired in the initial frame.

Due to the image point extraction precision and the small overlapping area of the stereo cameras at adjacent locations, the camera pose acquires a cumulative error, and the network of the block BA model is weak. In this case, the camera poses at different loca - tions will deviate from the actual values. In Fig. 2, the structure of the Yutu - 2 rover can provide several control points in the rover frame, which will be in the view of field of the cameras at locations S_n, S_1, S_2. In the left and right directions S_i, S_j of the rover, the solar wing cells can provide length information, which is the distance between the corners of the solar cells, and the corners are located in the multi - degree overlapping image points. By incorporating this control information into the block BA model, the camera frame in all locations can be incorporated from the initial frame to the rover frame, and the loop - unclosed BA model can be forcibly transformed into a loop - closed block BA model.

With the calibration parameters of the cameras as the known values and the multi - degree overlapping image points as input, Eq. (5) can be rewritten as follows:

$$\begin{cases} v_x = \dfrac{\partial x}{\partial X_S}\Delta X_S + \dfrac{\partial x}{\partial Y_S}\Delta Y_S + \dfrac{\partial x}{\partial Z_S}\Delta Z_S + \dfrac{\partial x}{\partial \varphi}\Delta\varphi + \dfrac{\partial x}{\partial \omega}\Delta\omega \\ \quad + \dfrac{\partial x}{\partial \kappa}\Delta\kappa + \dfrac{\partial x}{\partial X}\Delta X + \dfrac{\partial x}{\partial Y}\Delta Y + \dfrac{\partial x}{\partial Z}\Delta Z - l_x \\ v_y = \dfrac{\partial y}{\partial X_S}\Delta X_S + \dfrac{\partial y}{\partial Y_S}\Delta Y_S + \dfrac{\partial y}{\partial Z_S}\Delta Z_S + \dfrac{\partial y}{\partial \varphi}\Delta\varphi + \dfrac{\partial y}{\partial \omega}\Delta\omega \\ \quad + \dfrac{\partial y}{\partial \kappa}\Delta\kappa + \dfrac{\partial y}{\partial X}\Delta X + \dfrac{\partial y}{\partial Y}\Delta Y + \dfrac{\partial y}{\partial Z}\Delta Z - l_y \end{cases} \tag{10}$$

where ΔX, ΔY, ΔZ represent the corrections of the 3D coordinates of the feature points that correspond to the multi - degree overlapping image points. When the image points of the control points are selected from the set of multi - degree overlapping image points, ΔX, ΔY, ΔZ can be set to zero, and the image points construct the error equation (10). When n multi - degree overlapping image points correspond to the corners of the cells, the constraint condition of the distance can be viewed as:

$$(X_i - X_j)^2 + (Y_i - Y_j)^2 + (Z_i - Z_j)^2 = C_2 \tag{11}$$

where $(X, Y, Z)_i$ and $(X, Y, Z)_j$ represent the 3D coordinates of the feature points M_i and M_j, respectively, and C_2 represents the known constant, which is the length information associated with the corners of the cells. Equation (11) can be rewritten as:

$$F = (X_i - X_j)^2 + (Y_i - Y_j)^2 + (Z_i - Z_j)^2 - C_2 = 0 \tag{12}$$

The linearization of the function F can be viewed as:

$$\boldsymbol{H}_i \Delta\overline{\boldsymbol{X}}_i + \boldsymbol{H}_j \Delta\overline{\boldsymbol{X}}_j - \boldsymbol{F}^0 - \boldsymbol{C}_2 = \boldsymbol{0} \tag{13}$$

where $\boldsymbol{F}^0$ represents the function value in the current iteration, $\boldsymbol{H}_i = \frac{\delta \boldsymbol{F}}{\delta \boldsymbol{X}_i}$, $\boldsymbol{H}_j = \frac{\delta \boldsymbol{F}}{\delta \boldsymbol{X}_j}$.

To improve the calculation stability and precision of the loop - closed block BA model, a strict geometric relationship of the stereo camera should also be invoked, as shown in Eq. (7). Bycombining Eqs (10), (13) and (7), we can obtain the correction vector of the camera pose using the indirect adjustment method with restrictive conditions. The precision of $\boldsymbol{R}_{\text{S-R}}$ can also be obtained.

2.4 Total least squares solution of the camera self - calibration and the camera pose estimation problem

In the camera self - calibration and camera pose estimation procedure described above, the BA method with restrictive conditions is the basic solution based on the OLS method. In photogrammetry, BA by means of OLS is a repeated process of space resection and forward intersection that can refine a visual reconstruction to produce jointly optimal camera poses and a 3D structure by minimizing the robustified squared sum of all possible reprojection errors [26]. Several improved algorithms based on BA have been proposed to accomplish high - precision camera pose estimation, such as sparse BA [27], sparse sparse BA [28], g2o [29], relative BA [30], and incremental BA [31], which use Euclidean *XYZ* coordinates to describe the locations of natural features. Zhao et al. [32] indicated that parallaxBA outperforms BA, sparse BA, sparse sparse BA and g2o in terms of computing convergence, efficiency and accuracy. In Ref. [33], three problem - independent damping modifications of the Gauss - Newton method are presented: the line search method; the Levenberg - Marquardt algorithm (LM); and LM with Powell dogleg. However, these methods, which are based on OLS, have several shortcomings: they must satisfy the condition that during optimization, the coefficient matrix is assumed be error - free, with only the observation having an error. However, in reality, the 2D coordinates of the image points and the 3D coordinates of the control points should be viewed as the observations, indicating that the coefficient matrices in Eqs (5) and (10) also include random errors. Thus, Eqs (5) and (10) should be further extended to the errors in variables (EIV) model so all errors will be considered [34, 35], which is based on the TLS solution.

The TLS method is first presented by Adcock [36]. The TLS theory points out the criterion of minimizing the sum of squared residuals in between the coefficient matrix and the observation vector. At this basis, Deming and Stephan [37] proposed the TLS solution in the statistical sense. In Ref. [38], Golub and Van Loan proposed the famous singular value decomposition (SVD) method, which can obtain the total squares solution based on the EIV model fast. However, those studies only focus on the application to the approximation of curves. In Ref. [39], Yi and Jue first studied the TLS solution of space resection in the computer vision and photogrammetry area, which was based on the collinearity equations. Due to that the 3D coordinates of the ground point and the 2D image point all exist errors, the TLS method is used. TLS is one of several linear parameter estimation techniques that have been devised to compensate for data errors. Furthermore, TLS is the method of fitting that is appropriate when there are errors in both the observation vector and the coefficient matrix. So it can establish a more practical and suitable model. On the base of this model, the errors in the observation vector and the coefficient matrix can be corrected at the same time. In Ref. [40], the weighted TLS estimation technique is more realistic than OLS in the camera calibration problem.

When TLS is applied, Eqs (5) and (10) can be regarded as:

$$\bar{\boldsymbol{v}} = (\bar{\boldsymbol{B}} + \boldsymbol{v}_{\bar{B}})\bar{\boldsymbol{x}} - \bar{\boldsymbol{l}} \tag{14}$$

where $\bar{\boldsymbol{v}} = [\boldsymbol{v}_x \quad \boldsymbol{v}_y]^{\mathrm{T}}$, $\bar{\boldsymbol{l}} = [\boldsymbol{l}_x \quad \boldsymbol{l}_y]^{\mathrm{T}}$, $\bar{\boldsymbol{x}}$ denotes the correction vector of the unknown parameters, including the camera calibration parameters (ΔX_S, ΔY_S, ΔZ_S, $\Delta\varphi$, $\Delta\omega$, $\Delta\kappa$, Δf, Δx_0, Δy_0) in the camera self - calibration on orbit procedure (Section 2. 2) and the camera pose (ΔX_S, ΔY_S, ΔZ_S, $\Delta\varphi$, $\Delta\omega$, $\Delta\kappa$, ΔX, ΔY, ΔZ) in the camera pose estimation procedure (Section 2. 3), $\bar{\boldsymbol{B}}$ denotes the coefficient matrix, $\boldsymbol{v}_{\bar{B}}$ denotes the random error in $\bar{\boldsymbol{B}}$, $\bar{\boldsymbol{P}}$ denotes the weight matrix. The stochastic model for Eq. (14) is $\begin{bmatrix} \bar{\boldsymbol{v}} \\ \boldsymbol{V}_{\bar{B}} = \mathrm{vec}\, \boldsymbol{v}_{\bar{B}} \end{bmatrix} \sim N\left(\begin{bmatrix} 0 \\ 0 \end{bmatrix}, \sigma_0^2 \begin{bmatrix} \boldsymbol{Q} & \boldsymbol{0} \\ \boldsymbol{0} & \boldsymbol{I}_m \otimes \boldsymbol{Q} \end{bmatrix}\right)$ vec denotes the operator that stacks one column of a matrix beneath the previous one, n denotes the number of observations, m denotes the number of correction vectors of unknown parameters, $\otimes$ denotes the Kronecker - Zehfuss product of the matrices, which is defined as $\boldsymbol{G} \otimes \boldsymbol{H} = [\boldsymbol{G}_{ij} \cdot \boldsymbol{H}]$, $\boldsymbol{Q} = \bar{\boldsymbol{P}}^{-1}$, σ_0^2 denotes the unknown variance component, and $\boldsymbol{I}_m$ denotes the $m \times m$ identity matrix. In Section 2. 2, the restrictive conditions only include the geometric relationship, such as Eq. (7). Section 2. 3 includes the restrictive conditions of the geometric relationship and length information, such as Eqs (7) and (12). The linearization of these restrictive conditions can be viewed as:

$$\bar{\boldsymbol{J}}\bar{\boldsymbol{x}} + \bar{\boldsymbol{W}} = \boldsymbol{0} \tag{15}$$

Then, the combination of Eqs (14) and (15) can be written as the Lagrange target function Φ:

$$\Phi(\bar{\boldsymbol{v}}, \boldsymbol{V}_{\bar{B}}, \lambda) = \bar{\boldsymbol{v}}^{\mathrm{T}}\bar{\boldsymbol{P}}\bar{\boldsymbol{v}} + \boldsymbol{V}_{\bar{B}}^{\mathrm{T}}(\boldsymbol{I}_m \otimes \bar{\boldsymbol{P}})\boldsymbol{V}_{\bar{B}} +$$

$$2\gamma^{\mathrm{T}}(\bar{\boldsymbol{v}} - (\bar{\boldsymbol{B}} + \boldsymbol{v}_{\bar{B}})\bar{\boldsymbol{x}} + \bar{\boldsymbol{l}}) + 2\lambda^{\mathrm{T}}(\bar{\boldsymbol{J}}\bar{\boldsymbol{x}} + \bar{\boldsymbol{W}}) \tag{16}$$

where γ, λ denote the vector of unknown Lagrange multipliers. Hence, our method use a Lagrange target function to solve the self - calibration BA model and the loop - closed block BA model, which is different from the Refs [39, 40]. When the first - order con - dition equations are set to zero, the unknowns can be viewed as:

$$\frac{\delta\Phi}{2\delta\bar{\boldsymbol{v}}^{\mathrm{T}}} = \bar{\boldsymbol{v}}^{\mathrm{T}}\bar{\boldsymbol{P}} + \gamma^{\mathrm{T}} = \boldsymbol{0} \Rightarrow \bar{\boldsymbol{v}} = -\bar{\boldsymbol{P}}^{-1}\gamma \tag{17. a}$$

$$\frac{\delta\Phi}{2\delta\boldsymbol{V}_{\bar{B}}^{\mathrm{T}}}=\boldsymbol{V}_{\bar{B}}^{\mathrm{T}}(\boldsymbol{I}_m\otimes\bar{\boldsymbol{P}})-\gamma^{\mathrm{T}}(\bar{\boldsymbol{x}}^{\mathrm{T}}\otimes\boldsymbol{I}_n)=\boldsymbol{0}\Rightarrow\boldsymbol{v}_{\bar{B}}=\bar{\boldsymbol{P}}^{-1}\gamma\bar{\boldsymbol{x}}^{\mathrm{T}}$$

$$=-\bar{\boldsymbol{v}}\bar{\boldsymbol{x}}^{\mathrm{T}} \tag{17. b}$$

$$\frac{\delta\Phi}{2\delta\bar{\boldsymbol{x}}^{\mathrm{T}}}=-(\bar{\boldsymbol{B}}+\boldsymbol{v}_{\bar{B}})^{\mathrm{T}}\gamma+\bar{\boldsymbol{J}}^{\mathrm{T}}\lambda=\boldsymbol{0}\Rightarrow\bar{\boldsymbol{J}}^{\mathrm{T}}\lambda=(\bar{\boldsymbol{B}}+\boldsymbol{v}_{\bar{B}})^{\mathrm{T}}\gamma \tag{17. c}$$

In Eq. (17. b), $\boldsymbol{v}_{\bar{B}}\bar{\boldsymbol{x}}=\boldsymbol{I}_n\boldsymbol{v}_{\bar{B}}\bar{\boldsymbol{x}}=\mathrm{vec}(\boldsymbol{I}_n\boldsymbol{v}_{\bar{B}}\bar{\boldsymbol{x}})=(\bar{\boldsymbol{x}}^{\mathrm{T}}\otimes\boldsymbol{I}_n)\mathrm{vec}\boldsymbol{v}_{\bar{B}}=(\bar{\boldsymbol{x}}^{\mathrm{T}}\otimes\boldsymbol{I}_n)\boldsymbol{V}_{\bar{B}}$, such that

$$\frac{\delta\Phi}{2\delta\boldsymbol{V}_{\bar{B}}^{\mathrm{T}}}=\boldsymbol{V}_{\bar{B}}^{\mathrm{T}}(\boldsymbol{I}_m\otimes\bar{\boldsymbol{P}})-\gamma^{\mathrm{T}}(\bar{\boldsymbol{x}}^{\mathrm{T}}\otimes\boldsymbol{I}_n)=0$$

$$\Rightarrow\mathrm{vec}\,\boldsymbol{v}_{\bar{B}}=\boldsymbol{V}_{\bar{B}}=(\boldsymbol{I}_m\otimes\bar{\boldsymbol{P}})^{-1}(\boldsymbol{x}\otimes\boldsymbol{I}_n)\gamma=((\boldsymbol{I}_m\bar{\boldsymbol{x}})\otimes(\bar{\boldsymbol{P}}^{-1}\boldsymbol{I}_n))\gamma.$$
$$=(\bar{\boldsymbol{x}}\otimes\bar{\boldsymbol{P}}^{-1})\gamma=(\bar{\boldsymbol{x}}\otimes\bar{\boldsymbol{P}}^{-1})\mathrm{vec}\gamma=\mathrm{vec}(\bar{\boldsymbol{P}}^{-1}\gamma\bar{\boldsymbol{x}}^{\mathrm{T}})$$

The combination of Eqs. (14), (17. a) and (17. b) indicates the following:

$$\bar{\boldsymbol{v}}=-\bar{\boldsymbol{P}}^{-1}\gamma=(\bar{\boldsymbol{B}}+\boldsymbol{v}_{\bar{B}})\bar{\boldsymbol{x}}-\bar{\boldsymbol{l}}\Rightarrow\gamma=\dot{\boldsymbol{P}}^{-1}(-\bar{\boldsymbol{B}}\,\bar{\boldsymbol{x}}+\bar{\boldsymbol{l}}) \tag{18}$$

where $\dot{\boldsymbol{P}}=\bar{\boldsymbol{P}}^{-1}(1+\bar{\boldsymbol{x}}^{\mathrm{T}}\bar{\boldsymbol{x}})$.

The combination of Eqs (18) and (17. c) indicates the following:

$$\bar{\boldsymbol{J}}^{\mathrm{T}}\lambda=(\bar{\boldsymbol{B}}+\boldsymbol{v}_{\boldsymbol{B}})^{\mathrm{T}}\dot{\boldsymbol{P}}^{-1}(-\bar{\boldsymbol{B}}\,\bar{\boldsymbol{x}}+\bar{\boldsymbol{l}})\Rightarrow\bar{\boldsymbol{x}}$$
$$=\boldsymbol{M}((\bar{\boldsymbol{B}}+\boldsymbol{v}_{\bar{B}})^{\mathrm{T}}\dot{\boldsymbol{P}}^{-1}\bar{\boldsymbol{l}}-\bar{\boldsymbol{J}}^{\mathrm{T}}\lambda) \tag{19}$$

where $\boldsymbol{M}=((\bar{\boldsymbol{B}}+\boldsymbol{v}_{\bar{B}})^{\mathrm{T}}\dot{\boldsymbol{P}}^{-1}\bar{\boldsymbol{B}})^{-1}$。

Finally, the combination of Eqs (19) and (15) indicates the following:

$$\bar{\boldsymbol{J}}\,\boldsymbol{M}((\bar{\boldsymbol{B}}+\boldsymbol{v}_{\bar{B}})^{\mathrm{T}}\dot{\boldsymbol{P}}^{-1}\bar{\boldsymbol{l}}-\bar{\boldsymbol{J}}^{\mathrm{T}}\lambda)+\bar{\boldsymbol{W}}=\boldsymbol{0}\Rightarrow\lambda$$
$$=(\bar{\boldsymbol{J}}\,\boldsymbol{M}\bar{\boldsymbol{J}}^{\mathrm{T}})^{-1}(\bar{\boldsymbol{J}}\,\boldsymbol{M}(\bar{\boldsymbol{B}}+\boldsymbol{v}_{\boldsymbol{B}})^{\mathrm{T}}\dot{\boldsymbol{P}}^{-1}\bar{\boldsymbol{l}}+\bar{\boldsymbol{W}}) \tag{20}$$

Due to the combination of λ and γ, $\boldsymbol{v}_{\bar{B}}$, $\bar{\boldsymbol{v}}$ and $\bar{\boldsymbol{x}}$ should be obtained iteratively with the initial values. When the initial values of $\bar{\boldsymbol{x}}$ are given as $\boldsymbol{x}_0$, γ can be obtained from Eq. (18). Using Eqs (17. a) and (17. b), $\boldsymbol{v}_{\bar{B}}$, $\bar{\boldsymbol{v}}$ can also be obtained. In addition, λ can be obtained using Eq. (20). Finally, the new value of $\bar{\boldsymbol{x}}$ can be obtained using Eq. (19). During the iterative calculation, this procedure will be repeated until the number of convergences reaches a threshold of 0.003. In the initial procedure, due to the similar resolution, $\bar{\boldsymbol{P}}$ is set as the identity matrix. In the iterative procedure, $\bar{\boldsymbol{P}}$ will be recalculated by the varying weight method [41].

The TLS adjustment is biased, and the unit weight variance is $\sigma_0=\sqrt{(\bar{\boldsymbol{v}}^{\mathrm{T}}\bar{\boldsymbol{P}}\,\bar{\boldsymbol{v}}+\boldsymbol{V}_{\bar{B}}^{\mathrm{T}}(\boldsymbol{I}_m\otimes\bar{\boldsymbol{P}})\boldsymbol{V}_{\bar{B}})/(n-u+s)}$, where n represents the number of measurements, u represents the number of the unknown parameters, and s represents the number of constraint conditions with Eqs. (7) and (12). The error of the unknown parameters is approximated [42] as:

$$\sigma_{\bar{x}\bar{x}}\approx\sigma_0\times\sqrt{[(\boldsymbol{N}-\boldsymbol{v}_0\boldsymbol{I})+\bar{\boldsymbol{J}}^{\mathrm{T}}\bar{\boldsymbol{J}}]^{-1}(\boldsymbol{N}+\bar{\boldsymbol{J}}^{\mathrm{T}}\bar{\boldsymbol{J}})[(\boldsymbol{N}-\boldsymbol{v}_0\boldsymbol{I})+\bar{\boldsymbol{J}}^{\mathrm{T}}\bar{\boldsymbol{J}}]^{-1}} \tag{21}$$

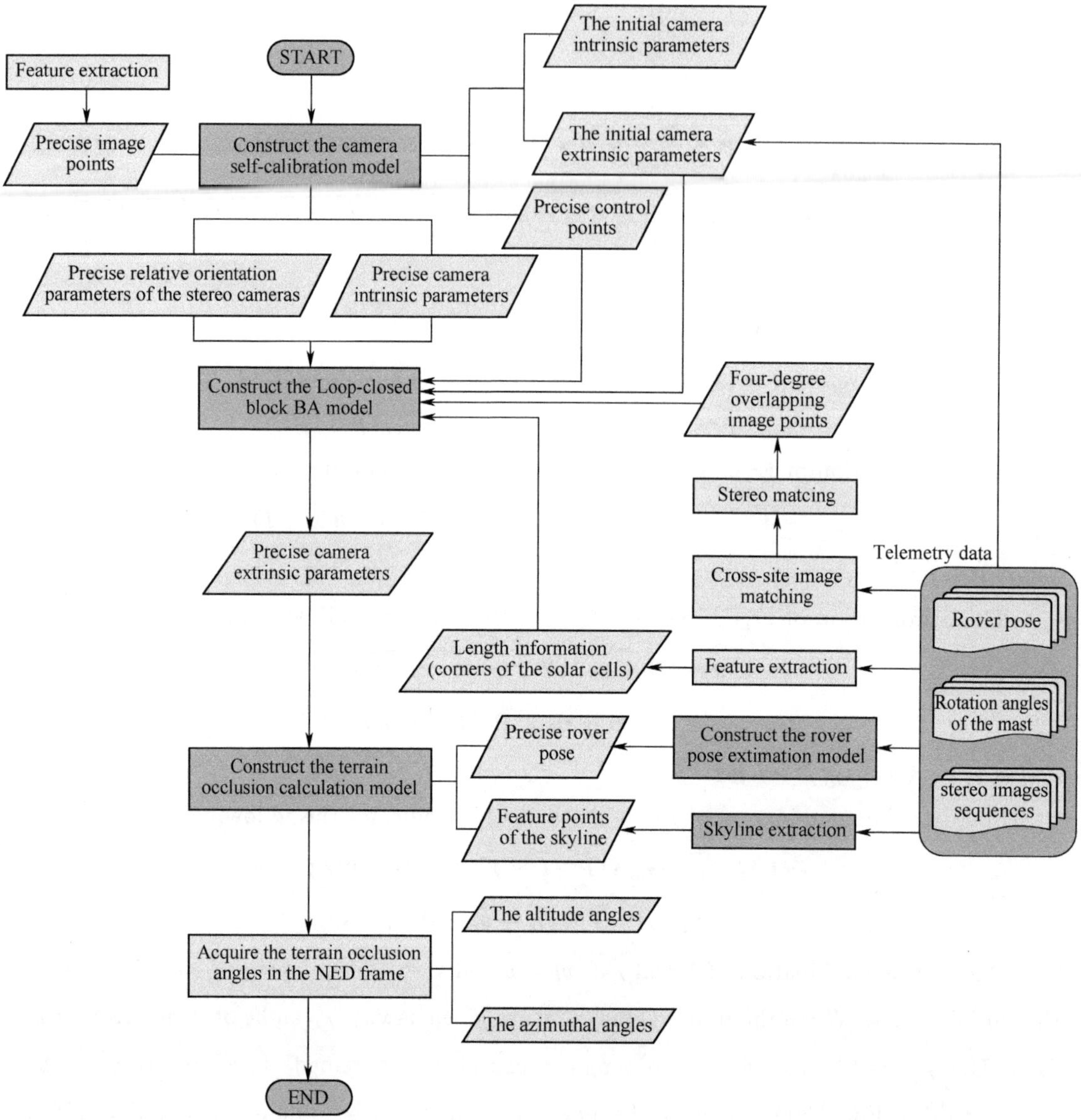

Fig. 3　The precise terrain occlusion prediction workflow for Yutu - 2 rover stereo cameras.

where $\boldsymbol{v}_0=\bar{\boldsymbol{v}}^{\mathrm{T}}\bar{\boldsymbol{P}}\bar{\boldsymbol{v}}+\boldsymbol{V}_{\bar{B}}^{\mathrm{T}}(\boldsymbol{I}_m\otimes\bar{\boldsymbol{P}})\boldsymbol{V}_{\bar{B}}$, $\boldsymbol{N}=\bar{\boldsymbol{B}}^{\mathrm{T}}\bar{\boldsymbol{P}}\bar{\boldsymbol{B}}$, and $\boldsymbol{I}$ denotes the identity matrix.

2.5　Work flow

The workflow is shown as Fig. 3 and comprises the following stages.

(i) The initial camera extrinsic parameters can be obtained through a transformation of the rover's coordinate system [4] and the telemetry data (including the rover pose and the rotation angles of the mast). Meanwhile, the initial camera intrinsic parameters can be provided by the BA method in ground experiments, as shown in Eq. (9). The pixel level

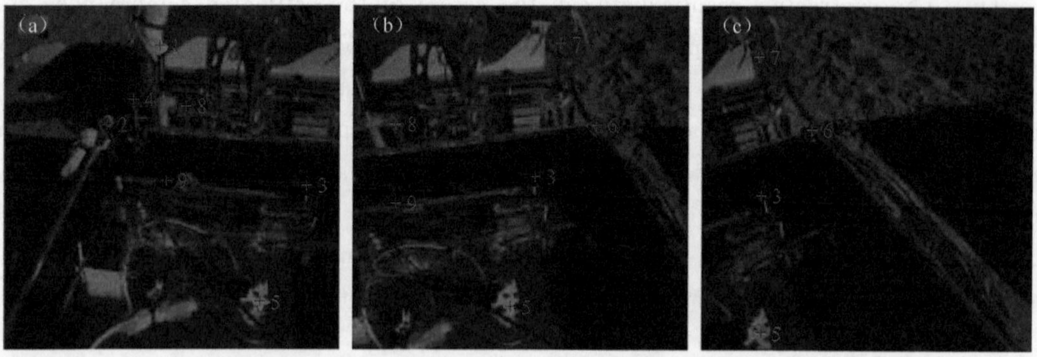

Fig. 4 Distribution of extracted image points in the left navigation camera at station A2.

image points are detected by classical Harris corner detector [43]. Based on the initial camera intrinsic and extrinsic parameters, the image points and the corresponding control points, the camera self-calibration model on orbit is constructed. At the same time, the geometric relationship should be viewed as the constraint conditions. Then, the TLS solution is given, as shown in Eq. (20). The camera calibration parameters (X_S, Y_S, Z_S, φ, ω, κ, f, x_0, y_0) can be obtained and are considered known. Then the relative orientation parameters of the stereo cameras can be obtained.

(ii) When the stereo image sequences in a station are acquired, SIFT is used to match the left images, and the cross-site matched image points are obtained. Based on those image points, the correlation coefficient matching method is applied in stereo matching procedure. Then the multidegree (e. g., four-degree) overlapping image points in the stereo images sequence can be obtained. Based on the camera intrinsic parameters and relative orientation parameters, the initial camera extrinsic parameters, the multi-degree overlapping image points, the corresponding control points and the length information (distance between the corners of the solar cells), the loop-closed block BA model is constructed, as shown in Eqs (10) and (13). Then, the TLS solution is given, as shown in Eqs (20) and (21). The following parameters can be obtained and are considered known: the camera pose (X_S, Y_S, Z_S, φ, ω, κ), the error of the camera pose.

Table 1 Rotation angles of the mast at station A2/degrees

	Fig. 4 (a)	Fig. 4 (b)	Fig. 4 (c)
Pitch angle	−39.9	−39.9	−39.9
Yaw angle	−169.9	169.8	149.8
Expansion angle	0.1	0.1	0.1

(iii) Based on the stereo images sequence, the feature points of the skyline are obtained by the proposed skyline autoextraction and manual procedure. The rover pose estimation results and the error of the rover pose can be obtained by WTLS [4].

(iv) The contrastive results between the errors of the estimated camera pose and rover pose, the measurement precision of telemetry rover pose and the rotation angles of the mast are available for the incremental analysis of actual state. Based on the feature points of the skyline, the actual camera pose and the rover pose, the occlusion calculation model is constructed, as shown in Eq. (3). Then, the altitude angles and azimuthal angles of the feature points of the skyline in the NED frame are obtained. Through contrasting the altitude angles between the skyline, the Sun and "QueQiao" satellite, we can fairly guide a decision to enter work state or a sleep state for the Yutu-2 rover.

3 Experimental validation and analysis

3.1 Self-calibration results of the Yutu-2 model rover cameras in the outdoor field

Before CE-4 mission started, the proposed algorithm has been tested with a model rover, which has same size and functions as Yutu-2. It is common knowledge that the desert in Dunhuang city is very similar to the Mars and Lunar surface, which is the test site of the China's first lunar rover. The 3D coordinates of all the feature points in the structure of the model rover are measured by two electronic stations, whose angle measurement precision is $2'$ and whose coordinate measurement precision is 0.12mm. At station A2, nine image points are extracted from the images at three navigation camera positions, which correspond to feature points. Thus, six images are invoked. The rotation angles of the mast are shown in Table1. At those positions, several feature points are reflected in two or three images, which form two-degree and three-degree overlapping image points with degrees of overlap of 60% and 20%, respectively, as shown in Fig. 4. The pixel level image points are detected by the Harris corner detector method [43] and the manual matching process. The precision of the extracted image points is nearly 1.0 pixels, and the weight matrix is set as an identity matrix.

In this section, the OLS method (including Eqs (6) and (8)) and the new TLS method proposed in Section 2.4 (including Eqs (14) and (15)) are used to solve the self-calibration BA model. The number of convergences is 5. There are 54 camera calibration parameters and 18 constraint conditions with Eq. (7). In both algorithms, the distortion parameters of the left navigation camera are k_1, k_2, p_1, $p_2 = -2.15\times10^{-8}$, 3.11×10^{-14}, 1.46×10^{-7} and -3.13×10^{-7}, respectively, and the distortion parameters of the right navigation camera are k_1, k_2, p_1, $p_2 = -2.22\times10^{-8}$, 3.09×10^{-14}, -5.32×10^{-9} and -2.85×10^{-7}, respectively. To investigate these two solutions based on different numbers of selected control points, several comparison tests should be performed. The unit weight variance σ_0 and the precision of the calibration parameters at station A2 (a) are shown in Tables 2 and 3, respectively.

Table 2 The precisions of the calibration parameters for the left navigation camera at station A2 (a)

Number		σ_0/pixels	σ_{X_s}/mm	σ_{Y_s}/mm	σ_{Z_s}/mm	σ_φ/degrees	σ_ω/degrees	σ_κ/degrees	σ_f/pixels	σ_{x0}/pixels	σ_{y0}/pixels
4	OLS	0.70	1.78	2.21	3.22	0.0983	0.0696	0.1211	1.31	3.3	3.02
	TLS	0.19	0.09	0.06	0.07	0.0023	0.0022	0.0101	0.05	0.07	0.06
6	OLS	0.47	1.93	2.49	2.84	0.1084	0.0757	0.1426	1.36	3.41	3.00
	TLS	0.14	0.07	0.05	0.06	0.0017	0.0016	0.0085	0.06	0.07	0.06
8	OLS	0.37	1.38	3.79	3.44	0.1278	0.1437	0.087	1.25	3.12	2.69
	TLS	0.12	0.05	0.04	0.05	0.0015	0.0014	0.0067	0.04	0.05	0.05
8+	OLS	0.41	3.48	6.67	4.25	0.2146	0.2684	0.0439	1.31	3.31	2.89
	TLS	0.06	0.03	0.02	0.02	0.0007	0.0007	0.0034	0.03	0.04	0.05

Table 3 The precisions of the calibration parameters for the right navigation camera at station A2 (a)

Number		σ_0/pixels	σ_{X_s}/mm	σ_{Y_s}/mm	σ_{Z_s}/mm	σ_φ/degrees	σ_ω/degrees	σ_κ/degrees	σ_f/pixels	σ_{x0}/pixels	σ_{y0}/pixels
4	OLS	0.7	1.94	2.27	3.61	0.1026	0.0797	0.1230	1.37	3.43	3.01
	TLS	0.19	0.1	0.06	0.08	0.0023	0.0022	0.0149	0.05	0.08	0.06
6	OLS	0.47	1.93	1.76	2.83	0.0819	0.0658	0.1047	1.35	3.41	3.01
	TLS	0.14	0.08	0.05	0.06	0.0017	0.0017	0.0122	0.06	0.07	0.06
8	OLS	0.37	1.35	3.41	3.71	0.1388	0.1295	0.0829	1.25	3.12	2.69
	TLS	0.12	0.06	0.04	0.05	0.0015	0.0015	0.0091	0.04	0.05	0.05
8+	OLS	0.41	4.09	4.01	4.2	0.2229	0.1689	0.0863	1.33	3.32	2.89
	TLS	0.06	0.03	0.02	0.02	0.0007	0.0007	0.0028	0.03	0.04	0.05

In this section, the OLS method (including Eqs (6) and (8)) and the new TLS method proposed in Section 2.4 (including Eqs (14) and (15)) are used to solve the self-calibration BA model. The number of convergences is 5. There are 54 camera calibration parameters and 18 constraint conditions with Eq. (7). In both algorithms, the distortion parameters of the left navigation camera are k_1, k_2, p_1, $p_2 = -2.15\times10^{-8}$, 3.11×10^{-14}, 1.46×10^{-7} and -3.13×10^{-7}, respectively, and the distortion parameters of the right navigation camera are -2.22×10^{-8}, 3.09×10^{-14}, -5.32×10^{-9} and -2.85×10^{-7}, respectively. To investigate these two solutions based on different numbers of selected control points, several comparison tests should be performed. The unit weight variancer σ_0 and the precision of the calibration parameters at station A2 (a) are shown in Tables 2 and 3, respectively.

In Tables 2 and 3, the first columns indicate the number of selected control points. In the case of 4 control points, the control points correspond to "2, 4, 8, 9". In the case of 6 control points, the control points correspond to "2, 4, 5, 6, 8, 9". In the case of 8 control points, the control points correspond to all nine points except "6." In the case of 8+ control points, the control points correspond to all nine points except "3". An analysis of the unit weight variances in Tables 2 and 3 shows that the precisions of the camera calibration parameters calculated by our method are higher than those of the

OLS solution. The comparison of the results with different control points, such as the cases with 4, 6 and 8 control points, shows that the use of more control points results in higher precision of the camera calibration parameters with our method and the OLS solution. In the cases of 8+ and 8 control points, the different numbers of selected control points lead to different precisions of the camera calibration parameters based on the OLS solution. With 8 control points, the unit weight variance is 0.37pixels. With 8+ control points, the unit weight variance is 0.41pixels in the OLS solution, which indicates that the different distribution of selected control points provides a different precision in the camera calibration parameter estimation. Because point "3" is a three degree overlapping image point, the OLS solution with image point "3" will have a higher precision in the camera calibration parameters at station A2 (a) than the solution without image point "3". With 8 control points, the unit weight variance is 0.12 pixels. With 8+ control points, the unit weight variance is 0.06 pixels in the TLS solution, which indicates that the effect of the distribution of control points is not obvious.

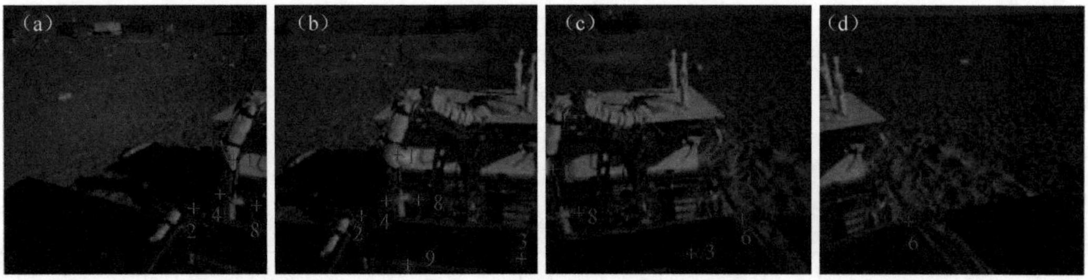

Fig. 5　Distribution of selected image points in the left navigation camera at station A2

To provide basic data for Section 2.3 and WTLS [4], the intrinsic parameters of the left navigation camera are f, x_0, y_0 = 1179.58pixels, −9.67pixels, and −13.14pixels, respectively, and the intrinsic parameters of the right navigation camera are 1183.53pixels, −10.44pixels, −32.62pixels, respectively, which are from the TLS solution with 8+ control points. The unit weight variance is 0.06pixels, as shown in the fifth row and third column of Tables 2 and 3. Then the geometric relationship of the stereo navigation camera can be obtained. The 3D coordinates X_1, Y_1, Z_1 of all nine selected image points can be recalculated using the intersection method [4]. Then, the precision can be obtained as $\Delta=\sqrt{(X_1-X_2)^2+(Y_1-Y_2)^2+(Z_1-Z_2)^2}$, where X_2, Y_2, Z_2 denote the measured 3D coordinates of the control points. The precision statistics are shown in Table 4.

In Table 4, "OLS" indicates that the calibration parameters of the stereo cameras from the OLS solution with 8 control points are used, and "TLS" indicates that the calibration parameters of the stereo cameras from the TLS solution with 8+ control points are used. The mean precisions of all image points are 1.42mm in OLS and 1.28mm in

TLS, which indicates that the 3D reconstruction precision can be improved by TLS in comparison to OLS.

Table 4 Precision of the 3D coordinates of the image points in the stereo camera calibration method/mm

Number	OLS	TLS
1	2. 24	2. 24
2	1. 19	1. 13
3	0. 83	0. 77
4	1. 77	0. 91
5	2. 17	2. 26
6	2. 11	2. 17
7	1. 23	1. 09
8	0. 66	0. 46
9	0. 58	0. 52

3. 2 Estimation results of the Yutu – 2 model rover camera pose in the outdoor field

At station A2, stereo image sequences are obtained when the pitch angle of the mast is − 19. 9 degrees. Eight image points are extracted from the navigation images in four positions, which correspond to the set of nine feature points in Fig. 4. The selected image points ("1, 2, 3, 4, 6, 7, 8, 9") are shown in Fig. 5, and the rotation angles of the mast are shown in Table 5. The pixel level image points are detected by the Harris corner detector method [43] and the manual matching process. The precision of the extracted image points is nearly 1. 0 pixels.

Through an analysis of the field of view of the navigation camera, the sky and the control points of the model rover structure can be simultaneously projected into the camera image when the pitch angle of the mast is from −24 degrees to −10 degrees. In practice, the pitch angle of the mast is − 20 degrees when the navigation cameras take photos towards the rear of the model rover. In other directions, the pitch angle of the mast is approximately 0 degrees, which makes the image contain more of the sky. However, it is necessary to ensure that the sequence of adjacent stereo images with different pitch angles of the mast contains nearly 20% overlapping areas. At station A2, the stereo navigation cameras take photos 18 times anti – clockwise from the rear. Six pairs of image points were automatically matched in adjacent camera positions for a total of 116 matching points, of which 108 were four – degree overlapping image points. The distance between the corners of the solar cells from left to right is chosen as the constraint condition for the image points, and there are 20 constraint conditions of the image points. The number of iterations is 5. The initial values of the camera external parameters are obtained by the coordinate transformation procedure between the model rover frame and the camera frame based on the rotation angles of the mast in Table 5. Then, the ordinary least squares method (abbreviated as OLS, including Eqs (10), (13) and (7)) and the total least squares method proposed in Section 2.4 (abbreviated as TLS, including Eqs (14) and (15)) are used to solve the loop - closed block BA model. There are 216 camera pose parameters and 108 constraint conditions with Eq. (7).

Different sequences of stereo images are used to compare the loop - closed block BA method to the loop - unclosed block BA method:

(a) The first case involves a sequence of stereo images in four locations S_1, S_2, …, S_i, with control points "1, 2, 4, 8" and 10 constraint conditions for the image points. There are 48 camera pose parameters and 24 constraint conditions with Eq. (7). In this case, the solutions with OLS and TLS are abbreviated as OLS - a and TLS - a, respectively.

(b) The second case involves a sequence of stereo images from the back locations S_1; S_2 to the front (eight locations) with control points "1, 2, 4, 8" and 10 constraint conditions of the image points. There are 96 camera pose parameters and 48 constraint conditions with Eq. (7). The solutions with OLS and TLS in this case are abbreviated as OLS - b and TLS - b, respectively.

(c) The third case involves a sequence of stereo images at thirteen locations S_1, S_2, …, S_j with control points "1, 2, 4, 8" and 20 constraint conditions of the image points. There are 156 camera pose parameters and 78 constraint conditions with Eq. (7). In this case, the solutions with OLS and TLS are abbreviated as OLS — c and TLS — c, respectively.

The unit weight variance σ_0 is shown in Table 6, and the precision $\sigma_{\bar{x}\bar{x}} = [\sigma_{X_S}\ \sigma_{Y_S}\ \sigma_{Z_S}\ \sigma_{\varphi}\ \sigma_{\omega}\ \sigma_{\kappa}]^T$ of the external parameters of the 36 camera positions is shown in Fig. 6. As shown in Table 6, for the case of the loop - closed block BA model, σ_0 is 1.20 pixels in OLS and 0.94 pixels in TLS, which indicates that the estimation precision of the camera pose from our method is higher than that in the OLS solution. TLS is also more precise than OLS for the case of the loop - unclosed block BA model.

As shown in Fig. 6, the precisions of the camera pose in the OLS - b and OLS - c solutions are much lower than the other solutions. The zoom view of the precisions in the OLS - a, OLS, TLS - a, TLS - b, TLS - c and TLS solutions are shown in Fig. 6 (a~f), respectively. When the OLS and TLS solutions are applied to the block BA model, several other conclusions can be obtained:

(1) In the loop - unclosed block BA model, the percentage of image points corresponding to the control points in all of the image points is higher, and the precision of the camera pose is higher. For example, in case "a", there are 28 image points, of which 4 correspond to control points. Thus, the percentage is 14.3%. In case "b", the image

points include 4 image points and 48 image points without 3D measurement coordinates; thus, the percentage is 7.7%. In case "c", the image points include 4 image points and 78 image points without 3D measurement coordinates; thus, the percentage is 4.9%. In the loop-closed block BA model, the image points include 8 control points and 108 image points without 3D measurement coordinates; thus, the percentage is 6.9%. In the TLS solutions, the precision of the camera pose estimation is highest in the case of TLS-a and lowest in the case of TLS-c. However, the OLS solutions in the loop-unclosed block BA model are not consistent with this trend. In cases "b" and "c", the precision of the camera position from the 16-th position to the 26-th position is less than 500mm in the OLS solution, which indicates that the uneven distribution of image points has a negative effect on the loop-unclosed block BA model. As shown in Table 6, TLS can effectively ensure the convergence and achieve high precision camera pose estimation. As shown in Fig. 6, the precision of the stereo camera pose with the TLS solutions is higher than that with the OLS solutions.

Table 5　Rotation angles of the mast at station A2/degrees

	Fig. 5 (a)	Fig. 5 (b)	Fig. 5 (c)	Fig. 5 (d)
Pitch angle	−19.9	−19.9	−19.9	−19.9
Yaw angle	150.1	169.9	169.8	149.8
Expansion angle	0.1	0.1	0.1	0.1

Table 6　Unit weight variance σ_0 of the block BA model at station A2/pixels

	Loop-unclosed			Loop-closed
	a	b	c	
OLS	0.64	7.32	1.54	1.20
TLS	0.63	1.02	1.09	0.94

(2) The precision of the camera altitude angles can be improved using the loop-closed block BA model over the telemetry precision of the rotation angles of the mast. In the loop-closed block BA model, TLS is more dependent on the even distribution of the multi-degree overlapping image points, as compared with OLS. The more uniform the multi-degree overlapping image points are distributed, the higher $\sigma_{\bar{X}\bar{X}}$ in OLS or TLS is. The telemetry precision of the rotation angle of the mast is 0.3 degrees (i.e., 0.0052 rad), and the zoom views of Fig. 6 (d~f) show that the precision of the camera altitude angles can be improved by using the loop-closed block BA model with the TLS solution when the yaw angles of the mast are from +170 degrees to +70 degrees and −170 degrees to −70 degrees. In other words, the rotation angles of the mast can be replaced by the calculated camera altitude angles. In other camera positions, the network stability of

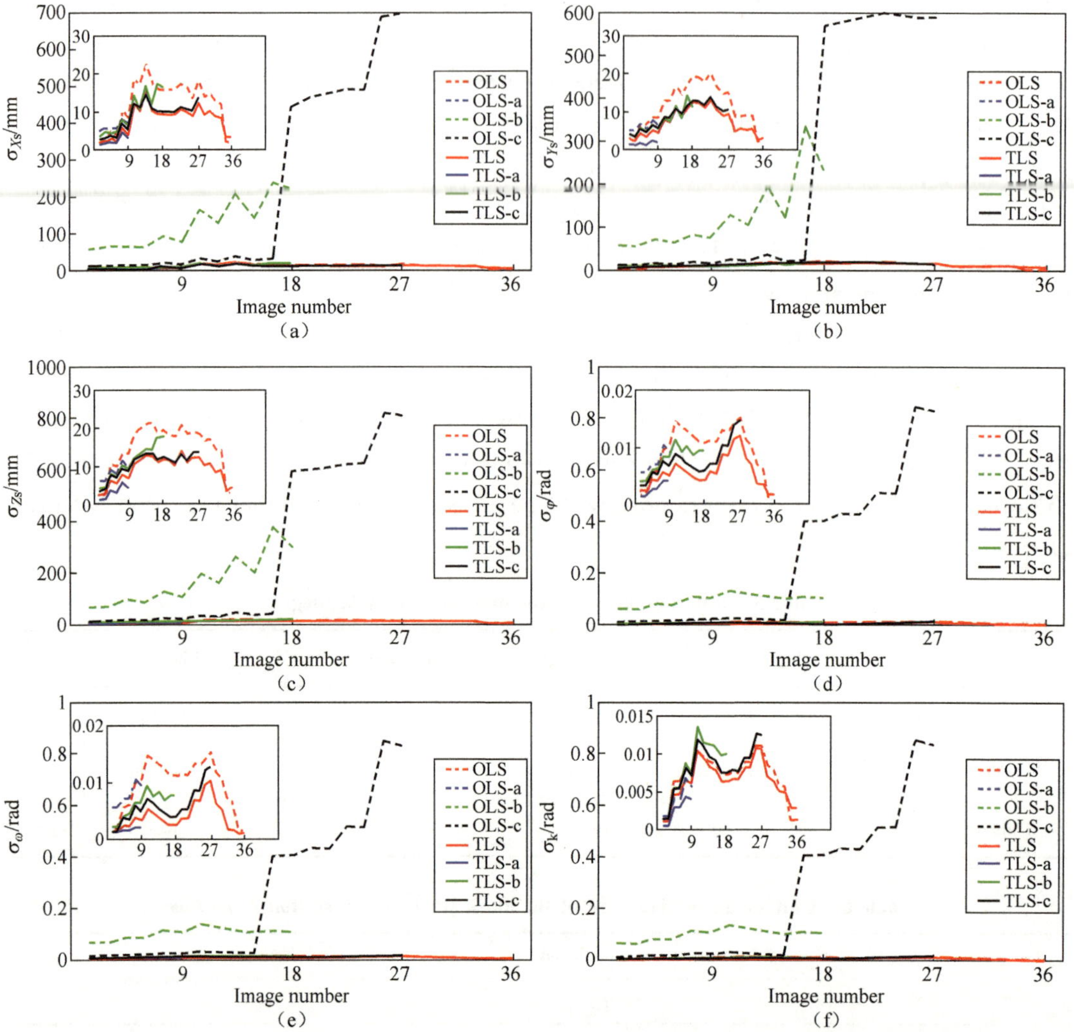

Fig. 6　Precision $\sigma_{\bar{x}\bar{x}}$ of the stereo navigation camera pose at station A2: (a～c) correspond to the camera positions X_S, Y_S, Z_S, respectively; (d～f) correspond to the camera altitude anglesu φ, ω; κ, respectively. The horizontal axis represents image number, whichis from 1 to 36

the loop－closed block BA model is poor due to the concentrated distribution of the multi－degree overlapping points, which indicates that the precision of the camera altitude angles is less than 0.0052 rad. The zoom views of Fig. 6 (a～c) show that the precision of the camera position parameters with the TLS solution is poor; it is less than 10 mm. Therefore, reducing the imaging interval of the stereo camera sequence to increase the proportion of overlapping areas and evenly distributing the multi－degree overlapping image points will improve the stability of the loop－closed BA model and the precision of the external camera parameters. In TLSa, the telemetry rotation angles of the mast can be replaced by the calculated camera altitude angles.

In OLS and TLS, the 3D reconstruction precision of all eight image points ("1, 2, 3, 4, 6, 7, 8, 9") are also counted with the measurement coordinates of the corresponding control points. The results are shown in Table 7. The mean precisions of all of the image points are 3. 25mm in OLS and 2. 79mm in TLS, indicating that the 3D reconstruction precision can be better improved by TLS than OLS.

Table 7　Precision of the 3D coordinates of the image points in the camera pose estimation method/mm

Number	OLS	TLS
1	4. 61	4. 44
2	2. 91	2. 44
3	3. 26	2. 82
4	3. 64	3. 10
6	2. 64	2. 12
7	3. 33	2. 71
8	3. 41	2. 94
9	2. 23	1. 80

In the procedure of calculating the terrain occlusion angles based on the stereo image sequence, it is necessary to determine the precision of the external camera angle at every position. When the precision of the external camera angles is higher than 0. 3 degrees, the proposed loop - closed block BA model can effectively improve the altitude angle precision between the camera frame and the rover frame. In that case, the proposed model can eliminate the influence of the measurement errors of the rotation angles of the mast on the calculation of the terrain occlusion angles.

3. 3　Visual terrain occlusion angle estimation results in the indoor field

The model rover operates in an indoor test field that is similar to the lunar surface environment. The telemetry pose of the model rover at station A is obtained by the guidance, navigation and control (GNC) system of the model rover, which is shown in Table 8.

In practice, the distance between the feature points of the skyline and the model rover is very large. Thus, the true coordinates of the feature points cannot be measured in the NED frame. In the indoor test field, the control points simulate the feature points of the skyline, and the 3D coordinates in the NED frame are obtained by two electronic stations. Equation (2) is established based on the premise that the feature points of the skyline are far away and cannot be used in the indoor test field. In this section, using $\boldsymbol{R}_{\text{NED-R}}$ that transforms between the model rover frame and the NED frame, the measured position of the model rover and $\boldsymbol{R}_{\text{S-R}}$ as inputs, the measured 3D coordinates of the control points in the NED frame are transformed into the camera frame X, Y, Z. The altitude angle of the control points is $\eta_S = \text{atan}(Y/\sqrt{Z^2+X^2})$, and the azimuthal angle is $\tau_S =$ atane(X/Z). In this procedure, the measured position of the model rover can be viewed as actual data, indicating that the measurement does not have errors. Thus, the procedure of calculating η_S; τ_S is the inverse process of Eq. (2), and the precision will be influenced by the error of $\boldsymbol{R}_{NED\text{-}R}$ and $\boldsymbol{R}_{S\text{-}R}$. In the camera frame, η_S, τ_S of the feature points corresponding to the control points can also be obtained by Eq. (1). The difference between the terrain occlusion angles (η_S; τ_S) in these two procedures is calculated. The larger the errors in $\boldsymbol{R}_{NED\text{-}R}$ and $\boldsymbol{R}_{S\text{-}R}$ are, the larger the difference is, and vice versa. $\boldsymbol{R}_{NED\text{-}R}$ and $\boldsymbol{R}_{S\text{-}R}$ can be obtained in two ways: one is directly from the telemetry pose of the model rover and the telemetry rotation angles of the mast (named "The direct way"), and the other is a combination of the procedure presented in Section 2.3 and WTLS [4] (named "OLS" and "TLS"). Therefore, the difference in the terrain occlusion angles is an effective way to investigate whether the proposed algorithm can replace the telemetry pose parameters of the model rover and the mast and improve the calculation precision of the terrain occlusion angles. The same intrinsic parameters of the stereo cam - eras are adopted in those two ways.

The ninth image taken by the left navigation cameraat stationA is shown in Fig. 7, in which eleven feature points are extracted using the Hough line detection method [44], and the image points are marked with red crosses. The extraction precision of the image points is 0.5 pixels. Eighteen pairs of stereo images are obtained by the navigation cameras at station A. The navigation camera of the model rover obtained 18 pairs of stereo images with an imaging interval of 20 degrees. $\boldsymbol{R}_{S\text{-}R}$ can be calculated using the loop - closed block BA method based on the OLS solution and the TLS solution. Then $\boldsymbol{R}_{NED\text{-}R}$ is calculated using WTLS [4].

In the loop - closed block BA method, 9 control points on the structure of the model rover are selected, and 12 pairs of image points are automatically matched in adjacent camera positions, for a total of 225 matching points, of which 216 are four - degree overlapping image points. The distance between the corners of the solar cells in the left to right direction of the model rover is chosen as the constraint condition of the image points, and there are 15 image points. There are 5 iterations, 216 camera pose parameters, and 108 constraint conditions with Eq. (7). Using the loop - closed block BA model based on the OLS solution, the unit weight variance is 0.95 pixels, and the camera pose precisions of the ninth camera position at station A are 0.350 degrees, 0.186 degrees and 0.202 degrees. Using the loop - closed block BA model based on the TLS solution, the unit weight variance is 0.86 pixels, and the camera pose precisions of the ninth camera position at station A are 0.262 degrees, 0.163 degrees and 0.143 degrees. It is necessary to set the model rover pose at the initial station as actual values, and 25 tie points, namely, four - degree overlapping points, are selected in the overlapping areas of the stereo image at adjacent stations. The unit weight variance is 0.22mm. The position precisions of the

Fig. 7　Distribution of the extracted image points in the ninth image taken by the left camera at station A

current rover pose are 0. 389mm, 0. 391mm and 0. 534mm, and the altitude angle precisions are 0. 940 degrees, 0. 499 degrees and 0. 447 degrees, which are higher than the telemetry angle precision of the model rover (1 degree; 0. 0175 rad). The results of the terrain occlusion angles by using "The direct way", "OLS" and "TLS" are shown in Table 9. In "The direct way", "OLS" and "TLS", the average differences of the altitude angles are −0. 23 degrees, −0. 18 degrees and −0. 12 degrees, respectively. At the same time, the average differences of the azimuthal angles are −0. 18 degrees, −0. 16 degrees and −0. 09 degrees, respectively. These results show that the absolute value of the difference in the occlusion terrain angles based on the TLS solution is smaller than that of the direct method with telemetry data and the OLS solution, which indicates that the calculation precision of the terrain occlusion angles using the proposed algorithm is higher. During the continuous movement of the model rover, there will be accumulated errors or even outliers in the rover telemetry pose information. The proposed algorithm can better obtain the terrain occlusion angles and ensure high precision.

Table 8　Telemetry pose of the model rover at station A

Telemetry location /mm			Telemetry altitude angles /degrees		
X	Y	Z	φ_R	ω_R	κ_R
956. 2	15898. 0	−371. 0	−0. 537	0. 802	115. 576

Table 9　Difference in the altitude angles and azimuthal angles of the control points in the camera frame at station A/degrees

Number	Altitude angle			Azimuthal angle		
	The direct way	OLS	TLS	The direct way	OLS	TLS
1	0.27	0.17	0.07	0.09	0.08	0.00
2	−0.02	−0.18	−0.06	−0.2	−0.16	−0.05
3	−0.17	−0.18	−0.13	−0.04	−0.26	−0.08
4	−0.10	−0.18	−0.19	−0.16	−0.16	−0.07
5	−0.32	−0.19	−0.15	−0.16	−0.16	−0.08
6	−0.21	−0.18	−0.11	−0.23	−0.22	−0.08
7	−0.26	−0.18	−0.11	−0.11	−0.10	−0.06
8	−0.23	−0.19	−0.14	−0.06	−0.05	−0.07
9	−0.41	−0.20	−0.14	−0.28	−0.20	−0.12
10	−0.27	−0.19	−0.12	−0.34	−0.19	−0.15
18	−0.22	−0.19	−0.09	−0.33	−0.18	−0.18

Without loss of generality, the proposed algorithm is tested by using the model rover's telemetry data in other stations.

3.4　Case 1: The station C

The telemetry pose of the model rover at station C is obtained by the GNC system of the model rover, which is shown in Table 10.

Table 10　Telemetry pose of the model rover at station C

Telemetry location /mm			Telemetry altitude angles / degrees		
X	Y	Z	φ_R	ω_R	κ_R
758.0	1259.3	−170.1	0.365	1.001	69.320

The ninth image at station C is shown in Fig. 8, in which seven feature points are extracted using the Hough line detection method [44], and the image points are marked with red crosses. The extraction precision of the image points is 0.5 pixels. Eighteen pairs of stereo images are obtained by the navigation cameras at station C. In the loop - closed block BA method, 9 control points on the structure of the model rover are selected, and 20 pairs of image points are automatically matched, for a total of 369 matching points, of which 360 are four - degree overlapping image points. There are 15 constraint conditions. There are 5 iterations, 216 camera pose parameters, and 108 constraint conditions with Eq. (7).

Using the loop - closed block BA model based on OLS, the unit weight variance is 0.75 pixels, and the camera pose precisions of the ninth camera position are 0.273

Fig. 8　Distribution of the extracted image points in the ninth image taken by the left navigation camera at station C

degrees, 0.145 degrees and 0.157 degrees. Using the loop - closed block BA model based on TLS, the unit weight variance is 0.67 pixels, and the camera pose precisions of the ninth camera position are 0.204 degrees, 0.127 degrees and 0.113 degrees. When WTLS [4] is applied, the altitude angle precisions of the model rover are 0.750 degrees, 0.439 degrees and 0.418 degrees, which are higher than the telemetry angle precision of the model rover (1 degree; 0.0175 rad). The results of the terrain occlusion angles by using "The direct way", "OLS" and "TLS" are shown in Table 11.

Table 11　Difference in the altitude angles and azimuthal angles of the control points in the camera frame at station C/degrees

Number	Altitude angle			Azimuthal angle		
	The direct way	OLS	TLS	The direct way	OLS	TLS
16	−0.15	−0.16	−0.09	−0.09	−0.06	−0.10
17	−0.26	−0.15	−0.06	−0.17	−0.13	−0.05
22	−0.12	−0.05	0.01	−0.14	−0.13	−0.05
23	−0.22	−0.13	−0.08	−0.17	−0.18	−0.09
24	−0.17	−0.13	−0.12	−0.19	−0.19	−0.09
30	−0.24	−0.18	−0.16	−0.19	−0.15	−0.19
40	−0.38	−0.21	−0.15	−0.22	−0.19	−0.13

In "The direct way", "OLS" and "TLS", the average differences of the altitude angles are −0.22 degrees, −0.15 degrees and −0.09 degrees, respectively. At the same time, the average differences of the azimuthal angles are −0.17 degrees, −0.15 degrees and −0.10 degrees, respectively. These results show that the proposed algorithm can better obtain the terrain occlusion angles with high precision.

3.5 Case 2: The station D

The telemetry pose of the model rover at station D is shown in Table 12. The ninth image at station D is shown in Fig. 9, in which eight feature points are also extracted using the Hough line detection method [44], and the image points are marked with red crosses. Eighteen pairs of stereo images are obtained by the navigation cameras at station D.

Table 12 Telemetry pose of the model rover at station D

Telemetry location/mm			Telemetry altitude angles/ degrees		
X	Y	Z	φ_R	ω_R	κ_R
5723.0	−14053.1	−396.0	1.324	0.584	85.875

Due to the field of view of the camera, the control points on the structure of the model rover can not be selected in the stereo images. When the loop - closed block BA method is applied, the camera pose at other location should be acquired in the initial frame (the first camera frame). Thus, the rotation matrix between the initial frame and the model rover frame should be obtained by the telemetry rotation angles of the mast mechanism. In that case, the loop - closed block BA method only limits the cumulative error of the camera pose at other locations.

In the loop - closed block BA method, 12 pairs of image points are automatically matched in adjacent camera positions, for a total of 216 four - degree overlapping image points. There are 15 constraint conditions of the image points. There are 5 iterations, 216 camera pose parameters, and 108 constraint conditions with Eq. (7). Using the loop - closed block BA model based on OLS, the unit weight vari ance is 1.14 pixels, and the camera pose precisions of the ninth camera position at station D are 0.431 degrees, 0.228 degrees and 0.252 degrees. Using the loop - closed block BA model based on TLS, the unit weight variance is 0.96 pixels, and the camera pose precisions of the ninth camera position are 0.357 degrees, 0.193 degrees and 0.241 degrees. When WTLS [4] is applied, the altitude angle precisions of the model rover are 0.819 degrees, 0.744 degrees and 0.907 degrees, which are higher than the telemetry angle precision of the model rover (1.0 degrees; 0.0175 rad). The results of the terrain occlusion angles by using "The direct way", "OLS" and "TLS" are shown in Table 13. In "The direct way", "OLS" and "TLS", the average differences of the altitude angles are − 0.38 degrees, − 0.35 degrees and −0.31 degrees, respectively. The average differences of the azimuthal angles are −0.32

degrees, −0.30 degrees and −0.26 degrees, respectively. These results show that the proposed algorithm can better obtain the terrain occlusion angles and ensure high precision.

Fig. 9　Distribution of the extracted image points in the ninth image taken by the left camera at station D

Table 13　Difference in the altitude angles and azimuthal angles of the control points in the camera frame at station D/degrees

Number	Altitude angle			Azimuthal angle		
	The direct way	OLS	TLS	The direct way	OLS	TLS
6	−0.41	−0.38	−0.34	−0.3	−0.27	−0.21
7	−0.38	−0.39	−0.23	−0.21	−0.19	−0.11
8	−0.37	−0.34	−0.29	−0.35	−0.33	−0.22
9	−0.36	−0.31	−0.35	−0.33	−0.32	−0.36
10	−0.40	−0.41	−0.34	−0.34	−0.32	−0.32
11	−0.36	−0.34	−0.29	−0.34	−0.39	−0.21
12	−0.35	−0.30	−0.31	−0.37	−0.27	−0.28
14	−0.11	−0.09	−0.12	−0.41	−0.40	−0.41

3.6　Visual terrain occlusion angle estimation results in the outdoor field

In order to fully and comprehensively analyse the visual terrain occlusion estimation precision based on the proposed algorithm, outdoor field experiments should be carried out. The telemetry pose of the model rover at station D1 is obtained by the GNC system of the model rover, which is shown in Table 14.

Table 14　Telemetry pose of the model rover at station D1

Telemetry location /mm			Telemetry altitude angles /degrees		
X	Y	Z	φ_R	ω_R	κ_R
38254.8	44343.4	1802.24	−1.637	1.086	−168.462

At station D1, the stereo navigation cameras take photos 18 times anti - clockwise from the rear when the pitch angle of the mast is −19.93 degrees. About twenty to forty pairs of image points are automatically matched in adjacent camera positions for a total of 378 matching points, of which 110 are six - degree overlapping image points and 258 are four - degree overlapping image points. The distance between the corners of the solar cells from left to right is chosen as the constraint condition for the image points, and there are 10 constraint conditions of the image points. Eight image points are extracted from the navigation images for four camera positions, which correspond to the set of nine feature points in Fig. 4. The selected image points ("1, 2, 3, 4, 6, 7, 8, 9") are shown in Fig. 5. The pixel level image points are detected by the Harris corner detector method [43] and the manual matching process. The precision of the extracted image points is nearly 1.0 pixels. The number of iterations is 5. When WTLS [4] is applied, the altitude angle precisions of the model rover are 0.5672 degrees, 0.8938 degrees and 0.6245 degrees, which are higher than the telemetry angle precision of the model rover (1.0 degrees; 0.0175 rad). Fig. 10 shows the feature points of the skyline extracted from navigation images at station D1.

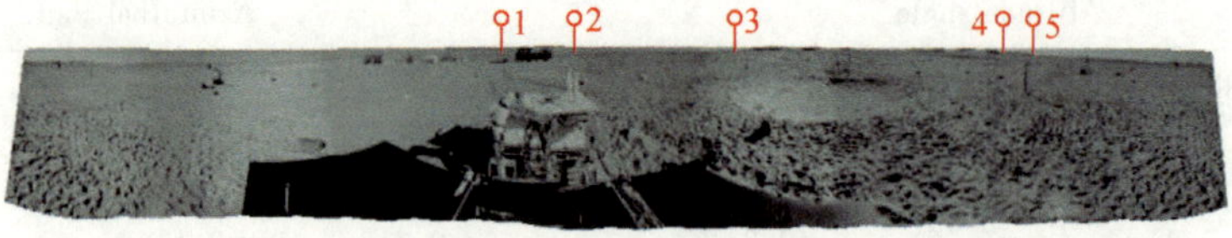

Fig. 10　Distribution of the extracted image points in the panoramic image taken by the left navigation camera at station D1.

Those feature points are located in the mountain peaks or lie on the border between the mountain and the ground. In actual measurements, the zenith angle and the azimuthal angle of the selected feature points are measured by two electronic stations, whose angle measurement precision is 2′. Thus, the measured altitude angle and the azimuthal angle can be viewed as true data.

In Fig. 10, the feature points "1", "2", "3" and "4, 5" are reflected in the image number "1, 2", "35, 36", "29, 30" and "21, 22", respectively. It is noted that the altitude angle is the complement of the zenith angle. In this section, the altitude angles and azimuthal angles of the feature points in the NED frame can be obtained by two ways: (1) the first way is directly from the telemetry pose of the model rover and the telemetry rotation angles of the mast (named "The direct way"); (2) the second way is a combination of the procedure presented in Section 2. 3 and WTLS [4] (named "OLS" and "TLS"). In OLS, an indirect adjustment solution with restrictive conditions is applied in the camera pose estimation procedure. In TLS, the total least squares method proposed in Section 2. 4 (including Eqs (14) and (15)) are used to solve the loopclosed block BA model. The same intrinsic parameters of the stereo cameras are adopted in those two ways. Different stereo images are used to compare the loop - closed block BA method to the loop - unclosed block BA method. In the loop - unclosed block BA model, a sequence of stereo images from the back locations S_1, S_2 to the front (nine locations) with control points "1, 2, 4, 8" and 10 constraint conditions of the image points are involved. There are 108 camera poses and 54 constraint conditions with Eq. (7). The solutions with OLS and TLS in this case are abbreviated as OLS - b and TLS - b, respectively. σ_0 is shown in Table 15, and $\sigma_{\bar{x}\bar{x}} = [\sigma_{Xs}\ \sigma_{Ys}\ \sigma_{Zs}\ \sigma_{\varphi}\ \sigma_{\omega}\ \sigma_{\kappa}]^T$ is shown in Fig. 11.

Table 15　Unit weight variance σ_0 of the block BA model at station D1/pixels

	Loop - unclosed	Loop - closed
OLS	4. 10	0. 67
TLS	0. 57	0. 52

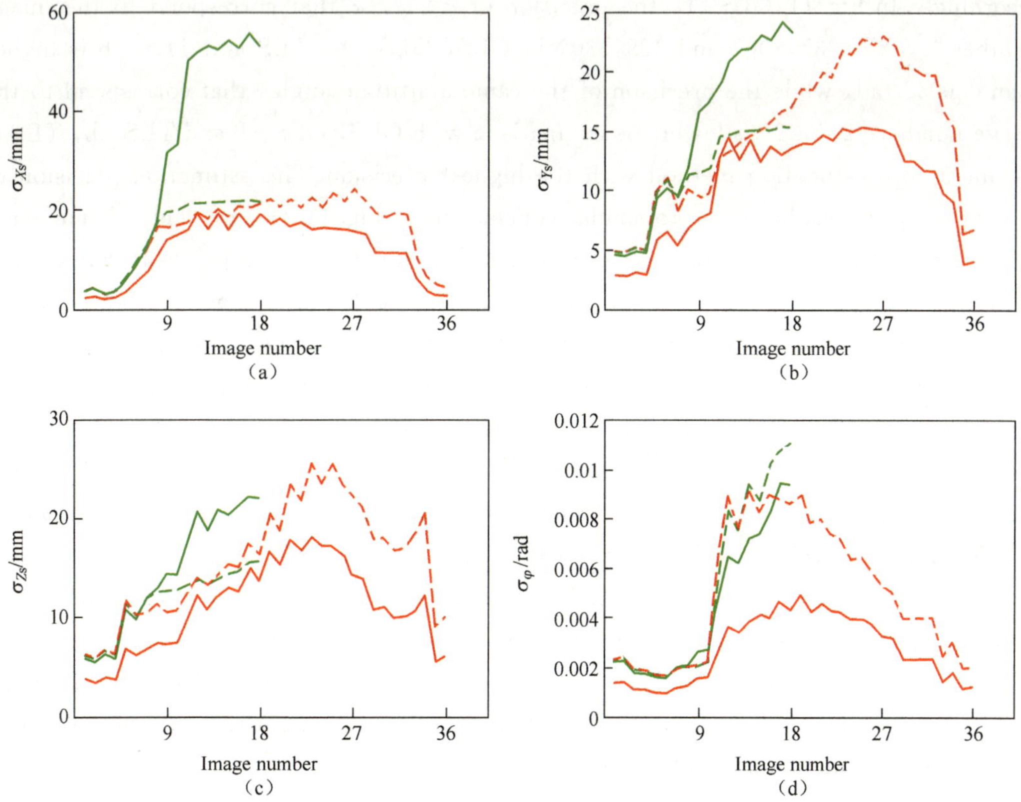

(a)　(b)　(c)　(d)

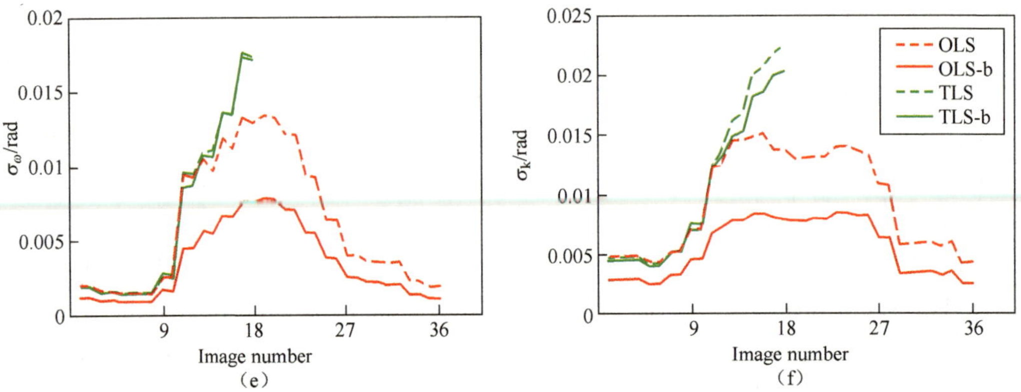

Fig. 11 Precision of the stereo navigation camera pose estimation at station D1: (a～c) correspond to the camera positions X_S, Y_S, Z_S, respectively; (d～f) correspond to the camera altitude angles φ, ω, κ, respectively. The horizontal axis represents image number, which is from 1 to 36

As shown in Table 15, for the case of the loop - closed block BA model, σ_0 is 0.67 pixels in OLS and 0.52 pixels in TLS, which indicates that the estimation precision of the camera pose from our method is higher than that in the OLS solution. σ_0 is 4.10 pixels in OLS - b and 0.57 pixels in TLS - b. Thus, the TLS solution is also more precise than the OLS solution in the loop - unclosed block BA model. As shown in Fig. 11, the estimation precision of the camera pose in OLS - b and TLS - b is much lower than OLS and TLS accordingly. In Fig. 11 (d～f), the precision of φ, ω, κ that correspond to the image number "1, 2", "35, 36" and "29, 30" in OLS, OLS - b, TLS and TLS - b is higher than 0.0052 rad, while the precision of the camera altitude angles that correspond to the image number "21, 22" is lower. As comparison with OLS, OLS - b and TLS - b, TLS is a camera pose estimation method with the highest precision. The estimation precision of the camera positions has come to similar conclusions in Fig. 11 (a～c). This phenomenon demonstrates that the camera pose estimation precision that corresponds to the images with the control points is higher, as comparison with the images without control points. When the proposed algorithm is applied in the terrain occlusion prediction problem, the difference in the altitude angles and the azimuthal angles of the feature points in the rover frame are listed in Table 16. The altitude angles and the azimuthal angles of the feature points "1, 2, 3" in TLS are closer to the true data compared with OLS and the direct way. Meanwhile, the altitude angles and the azimuthal angles of the feature points "4, 5" in TLS and OLS are further from the true data compared with "The direct way". In "The direct way", "OLS" and "TLS", the average differences of the altitude angles are −0.49 degrees, −0.43 degrees and −0.37 degrees, respectively. At the same time, the average differences of the azimuthal angles are −0.39 degrees, −0.37 degrees and −0.29 degrees, respectively. These results show that the proposed algorithm can better obtain the terrain occlusion angles and ensure high precision.

Table 16 Difference in the altitude angles and the azimuthal angles of the feature points in the rover frame /degrees

Number	Altitude angle			Azimuthal angle		
	The direct way	OLS	TLS	The direct way	OLS	TLS
1	−0.48	−0.30	−0.20	−0.32	−0.28	−0.15
2	−0.41	−0.33	−0.26	−0.35	−0.30	−0.18
3	−0.55	−0.42	−0.38	−0.36	−0.31	−0.23
4	−0.50	−0.56	−0.51	−0.46	−0.49	−0.46
5	−0.51	−0.55	−0.50	−0.45	−0.49	−0.43

3.7 Applications of visual terrain occlusion prediction in CE - 4 mission

In CE - 4 mission, the teleoperation system is responsible for terrain occlusion prediction, visual localization and mapping, path planning and other operations with respect to environmental perception, movement decision and science exploration. The navigation cameras play an important role in terrain occlusion prediction assisting to create instructions for deployment motion, antenna rotation and rover operation. On January 10th, 2019, two stereo image sequences are obtained at station A when the pitch angle of the mast is −2.9 degrees and −31.9 degrees. It is necessary to set the Yutu - 2 rover pose at the landing station as actual values, and 9 tie points are selected in the overlapping areas of the stereo images at station A and the landing station. When WTLS [4] is applied, the unit weight variance is 0.25mm. The estimation location of the Yutu - 2 rover at station A is −12832.14mm, −1688.85mm and 489.98mm, and the estimation altitude angles are −3.038 degrees, −3.546 degrees and 179.888 degrees. The altitude angle precisions are 1.114 degrees, 0.843 degrees and 0.637 degrees.

Fig. 12 shows the skylines extracted from images of the Yutu - 2 rover' stereo cameras using the proposed skyline extraction method, and the red curve is the skyline. In Fig. 12, the white line denotes the tie points between the stereo images in the different pitch angles of the mast, and the number of the tie points is 48. Manual selection should be also utilized when the image is potentially overexposed. In the loop - closed block BA method, control points on the structure of the Yutu - 2 rover are not selected due to the field of view of the camera. 8 pairs of image points are automatically matchedin adjacent camera positions (Fig. 12 (a, b)), for a total of 240 matching points, of which 192 are four - degree overlapping image points. The distance between the corners of the solar cells in the left to right direction of the Yutu - 2 rover (Fig. 12 (b)) is chosen as the constraint condition of the image points, and the number of the constraint condition is 10. There are 5 iterations, 32 camera poses, and 10 constraint conditions with Eq. (7). The precision of the extracted image points is nearly 1.0 pixels. Using the loop - closed block BA model based on the TLS solution, the unit weight variance is 0.75 pixels, and the calculated results are shown in Fig. 13.

Fig. 12　The extracted skyline in navigation images from the Yutu - 2 rover.
(a) The pitch angle of the mast is −2. 9 degrees. (b) The pitch angle of the mast is −31. 9 degrees. The yellow lime denotes the constraint condition of the image points. (For interpretation of the references to color in this figure legend, the reader is referred to the web version of this article.)

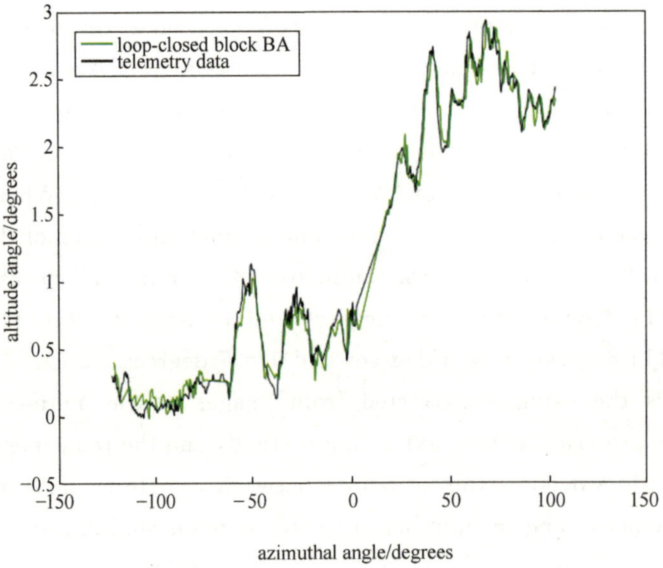

Fig. 13　The terrain occlusion calculation results

As shown in Fig. 13, the black curve denotes the terrain occlusion calculation result by using the telemetry data. The green curve represents the terrain occlusion calculation result by the loop - closed block BA method. The "telemetry data" method is applied in case that the rover pose and the rotation angles of the mast are all from the telemetry data. The mean absolute value of the difference between the altitude angles is 0. 075 degrees.

When the Yutu - 2 rover is on orbit, precise estimation of the camera pose and the rover pose using the proposed algorithm should be combined with the precision of the telemetry rotation angles of the mast and the telemetry angles of the Yutu - 2 rover. When the estimation precision of the proposed algorithm is higher than that of the telemetry data, the proposed algorithm is used to obtain high - precision terrain occlusion angles. Otherwise, the terrain occlusion angles are obtained directly from the telemetry rotation angles of the mast and the telemetry angles of the Yutu - 2 rover. Thus, the final result of the terrain occlusion angles of the skyline is shown in Fig. 14, which is represented by the green curve.

In Fig. 14, the red curve represents the azimuthal angle and the altitude angle of the Sun. For viewing convenience, the altitude angles of the Sun are all set as −42 degrees when the altitude angle is smaller than −42 degrees. The blue curve represents the azimuthal angle and the altitude angle of the "QueQiao" satellite. The green curve represents the terrain occlusion calculation result by the loop - closed block BA method. According to the intersection point A ofthe above red line and the green line, the Yutu - 2 rover should stop working before the moment of the sunset. From am 3: 30 on January 13th, 2019 to pm 8: 20 on January 29th, 2019, the Yutu - 2 rover is in the shadow region, which indicates that the Yutu - 2 rover enters a sleeping state. In analysis of the intersection point B of the above red line and the green line, the Yutu - 2 rover can enter work state from pm 8: 20 on January 29th, 2019. There are no intersection points between the blue line and the green line, which indicates that the "Queqiao" satellite is always in the view of the Yutu - 2 rover's antenna.

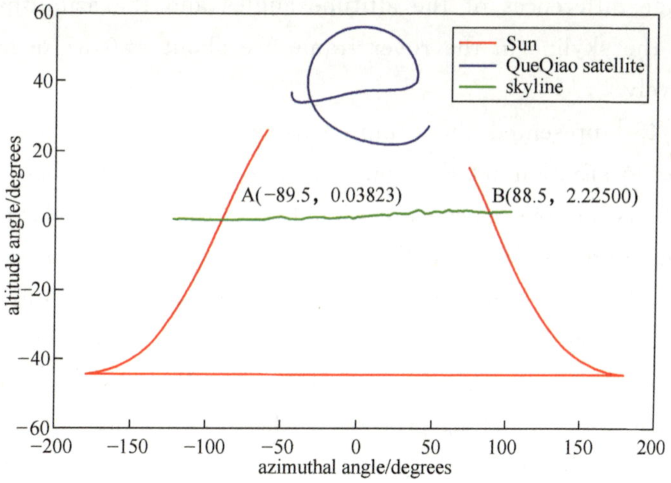

Fig. 14　The terrain occlusion prediction results

4　Conclusion

This paper studies the terrain occlusion prediction problem of China's CE - 4 mission and proposes a terrain occlusion angle prediction algorithm based on stereo image sequences of the Yutu - 2 rover' visual system, which is verified by ground test data and applied in the on orbit mission. The results show the following:

(1) The proposed algorithm can automatically recognize the skyline and calculate the terrain occlusion angles (azimuthal and altitude angles) with high precision, which solves the problem of skyline occlusion prediction in the simulated lunar environment. First, the camera self - calibration method based on control points is adopted using a strict geometric relationship of the stereo camera as the constraint condition. The calibration parameters of the stereo cameras are obtained using the TLS solution, which can perform a precise geometric calibration of the Yutu - 2 rover stereo cameras on orbit. In this part, the unit weight variance is 0.06 pixels. Then a loop - closed block BA model with a stereo image sequence is then proposed to obtain the camera pose in the rover frame. Multi - degree overlapping image points, the control points on the structure of the Yutu - 2 rover, the strict geometric relationship between the stereo cameras and the length information of the solar wing cells are used in this method. Then, the TLS solution is also used instead of the OLS solution to improve the precision of the estimated camera pose, which can effectively eliminate the measurement angular error of the mast mechanism in the Yutu - 2 rover (0.3 degrees). The unit weight variance in this method is 0.52 pixels. Finally, the proposed algorithm uses a rover pose estimation procedure based on tie points at adjacent stations [4], and the angular precision of the lunar rover is higher than 1.0 degrees in the NED frame. This method can further limit the accumulated measurement errors of the telemetry angles of the Yutu - 2 rover and improve the calculation precision of the terrain occlusion angles. The average differences of the altitude angles and the azimuthal angles of the feature points of the skyline in the rover frame are about −0.37 degrees and −0.29 degrees, respectively.

Liu et al. [45] presented the comprehensive results of landing site topographic mapping and rover localization in CE - 3 mission. The common point between the proposed algorithm and [45] is the rover pose estimation. Due to the equal localization precision and the higher computational efficiency, WTLS [4] is used in the rover pose estimation procedure instead of BA. In additional, [45] cannot provide the solution of the terrain occlusion prediction problem. Moreover, the camera self - calibration on orbit, camera pose estimation and skyline extraction are not studied in Ref. [45].

(2) Due to the stereo matching procedure, the proposed algorithm is closely related to the distribution of the multi - degree overlapping image points and the measured precision of the control points on the structure of the Yutu - 2 rover. In the loop - closed block BA model, multi - degree overlapping image points with a limited distribution in one image will substantially affect the estimation precision of the camera pose. In addition, the overlapping area of the adjacent stereo images is small, which will lead to an unstable calculation of the block BA model. The control points of the Yutu - 2 rover structure make the precision in estimating the camera poses in the first and last camera positions

extremely high; the farther in front of the Yutu - 2 rover the camera is, the lower the precision. Therefore, the distribution of the multi - degree overlapping image points, the overlapping area of adjacent stereo images and the number of control points are important factors that affect the loop - closed block BA model.

(3) The results of this paper have high precision, application stability and the proposed algorithm has been applied to the Yutu - 2 rover in a simulated ground test environment and on orbit missions.

Acknowledgements

YouQing Ma, Yang Jia and ShaoChuang Liu are supported by the National Natural Science Foundation of China (No. 41601494).

References

[1] W. Wu, C. Li, Z. Wei, H. Zhang, J. Liu, W. Wen, Y. Su, X. Ren, J. Yan, D. Yu, G. Dong, C. Wang, Z. Sun, E. Liu, J. Yang, Z. Ouyang, Lunar farside to be explored by Chang'e - 4, Nature Geoscience. 12 (2019) 222 - 223, https://doi. org/10. 1038/ s41561 - 019 - 0341 - 7.

[2] Y. Jia, Y. Zou, J. Ping, C. Xue, J. Yan, Y. Ning, The scientific objectives and payloads of Chang'E - 4 mission, Planetary and Space Science. 162 (2018) 207 - 215, https://doi. org/10. 1016/j. pss. 2018. 02. 011.

[3] J. Huang, Z. Xiao, J. Flahaut, M. Martinot, J. Head, X. Xiao, M. Xie, L. Xiao, Geological characteristics of Von Karman Crater, Northwestern South Pole - Aitken Basin: Chang'E - 4 Landing Site Region, Journal of Geophysical Research: Planets 123 (2018) 1684 - 1700, https://doi. org/10. 1029/2018JE005577.

[4] M. Y. Qing, L. S. Chuang, W. Bo, Z. Shuo, M. Li, P. Song, Weighted total least squares for the visual localization of a planetary rover, Photogrammetric Engineering & Remote Sensing 84 (2018) 605 - 618, https://doi. org/10. 14358/ PERS. 84. 10. 605.

[5] R. Giubilato, S. Chiodini, M. Pertile, S. Debei, An Evaluation of ROS - compatible Stereo Visual SLAM Methods on a nVidia Jetson TX2, Measurement. 140 (2019) 161 -170, https://doi. org/10. 1016/j. measurement. 2019. 03. 038.

[6] S . Chiodini, M. Pertile, R. Giubilato, F. Salvioli, M. Barrera, P. Franceschetti, S. Debei, Experimental Evaluation of a Camera Rig Extrinsic Calibration Method Based on Retro - reflective Markers Detection, Measurement. 140 (2019) 47 - 55, https://doi. org/10. 1016/j. measurement. 2019. 03. 036.

[7] F. Cozman, E. Krotkov, C. Guestrin, Outdoor visual position estimation for planetary rovers, Autonomous Robots. 9 (2000) 135 - 150, https://doi. org/ 10. 1023/ A: 1008966317408.

[8] A. Oiri, K. Nagatani, K. Yoshida, Global positioning for Planetary Rovers based on Panoramic Skyline Image, in: in: Proceedings of the 2010 JSME Conference on Robotics and Mechatronics, 2010, https://doi.org/10.1299/jsmermd.2010._1a2-b21_1.

[9] A. Nefian, X. Bouyssounousse, L. Edwards, M. Dille, T. Kim, E. Hand, T. Fong, Infrastructure Free Rover Localization, in: International Symposium on Artificial Intelligence, Robotics and Automation in Space (i-SAIRAS), 2014.

[10] X. Bouyssounouse, A. V. Nefian, A. homas, L. Edwards, M. Deans, T. Fong, Horizon based orientation estimation for planetary surface navigation, IEEE International Conference on Image Processing. IEEE (2016) 4368-4372, https://doi.org/10.1109/ICIP.2016.7533185.

[11] S. J. Dumble, Gibbens, P W, Efficient terrain-aided visual horizon based attitude estimation and localization, Journal of Intelligent and Robotic Systems. 78 (2015) 205-221, https://doi.org/10.1007/s10846-014-0043-8.

[12] T. Ahmad, G. Bebis, E. E. Regentova, A. Nefian, AMachine Learning Approach to Horizon Line Detection Using Local Features. Advances in Visual Computing, Springer, Berlin, Heidelberg, 2013, pp. 181-193.

[13] D. G. Lowe, Distinctive Image Features from Scale-Invariant Keypoints, Int J Comput Vision. 60(2004)91-110,https://doi.org/10.1023/b: visi.0000029664.99615.94.

[14] T. Ahmad, G. Bebis, E. Regentova, A. Nefian, T. Fong, in: An Experimental Evaluation of Different Features and Nodal Costs for Horizon Line Detection, Springer, Cham, 2014, https://doi.org/10.1007/978-3-319-14249-4_19.

[15] T. Ahmad, G. Bebis, M. Nicolescu, A. Nefian, An Edge-Less Approach to Horizon Line Detection, in: International Conference on Machine Learning and Applications, IEEE, 2015,pp. 1095-1102. https://doi.org/10.1109/ICMLA.2015.67.

[16] T. Ahmad, G. Bebis, E. Regentova, A. Nefian, T. Fong, Coupling dynamic programming with machine learning for horizon line detection, International Journal on Artificial Intelligence Tools. 24 (2015) 1540018, https://doi.org/ 10.1142/s0218213015400187.

[17] V. A. Grishin, Maslov, IA, Horizon Line Stability Observations over the Sea, The Journal of Navigation 71(2018)437-453,https://doi.org/10.1017/S0373463317000650.

[18] S. Fefilatyev, V. Smarodzinava, L. O. Hall, D. B. Goldgof, Horizon Detection Using Machine Learning Techniques, in: International Conference on Machine Learning and Applications. IEEE Computer Society, 2006, pp. 17-21, https://doi.org/10.1109/ICMLA.2006.25.

[19] T. G. McGee, R. Sengupta, K. Hedrick, Obstacle detection for small autonomous aircraft using sky segmentation, in: in: Proceedings of the 2005 IEEE International Conference on. IEEE, 2005, pp. 4679-4684, https://doi.org/ 10.1109/ROBOT.2005.1570842.

[20] S. Thurrowgood, D. Soccol, R. J. D. Moore, D. Bland, M. V. Srinivasan, A vision based system for altitude estimation of UAVs, in: Intelligent Robots and Systems, IEEE/RSJ International Conference on. IEEE, 2009, pp. 5725 - 5730, https://doi. org/10. 1109/IROS. 2009. 5354041.

[21] G. C. H. E. De. Croon, C. De. Wagter, B. D. W. Remes, R. Ruijsink, Sky segmentation approach to obstacle avoidance, in: Aerospace Conference, 2011, pp. 1 - 16. https://doi. org/10. 1109/AERO. 2011. 5747529.

[22] B. Grelsson, M. Felsberg, F. Isaksson, Highly accurate attitude estimation via horizon detection, Journal of Field Robotics 33 (2016) 967 - 993, https://doi. org/10. 1002/rob. 21639.

[23] J. Canny, A computational approach to edge detection, IEEE Transactions on Pattern Analysis & Machine Intelligence 6 (1986) 679 - 698, https://doi. org/10. 1109/TPAMI. 1986. 4767851.

[24] D. B. Gennery, Generalized camera calibration including fish - eye lenses, Int J Comput Vision. 68 (2006) 239 - 266, https://doi. org/10. 1007/s11263 - 006 - 5168 - 1.

[25] S. Zhang, S. Liu, Y. Ma, C. Qi, H. Ma, H. Yang, Self calibration of the stereo vision system of the Chang'e - 3 lunar rover based on the bundle block adjustment, ISPRS J Photogramm. 128 (2017) 287 - 297, https://doi. org/10. 1016/j. isprsjprs. 2017. 04. 004.

[26] Y. Jeong, D. Nistér, D. Steedly, R. Szeliski, I. S. Kweon, Pushing the envelope of modern methods for bundle adjustment. Computer Vision & Pattern Recognition, IEEE, 2010, https://doi. org/10. 1109/CVPR. 2010. 5539795.

[27] M. I. Lourakis, A. A. Argyros, SBA: A software package for generic sparse bundle adjustment, Acm T Math Software 36 (2009), https://doi. org/10. 1145/ 1486525. 1486527.

[28] K. Konolige, W. Garage, Sparse Sparse Bundle Adjustment, in: BMVC, 2010, pp. 102. 1 - 102. 11, https://doi. org/10. 5244/C. 24. 102.

[29] R. Kümmerle, G. Grisetti, H. Strasdat, K. Konolige, W. Burgard, g2o: A general framework for graph optimization, in: Robotics and Automation (ICRA), 2011 IEEE International Conference, 2011, pp. 3607 - 3613.

[30] J. L. Blanco, J. González - Jiménez, J. A. Fernández - Madrigal, Sparser relative bundle adjustment (srba): constant - time maintenance and local optimization of arbitrarily large maps, in: Robotics and Automation (ICRA), 2013 IEEE International Conference, 2013, pp. 70 - 77, https://doi. org/10. 1109/ICRA. 2013. 6630558.

[31] V. Indelman, A. Melim, F. Dellaert, Incremental light bundle adjustment for robotics navigation, in: Intelligent Robots and Systems (IROS), 2013, pp. 1952 - 1959, https://doi. org/10. 1109/IROS. 2013. 6696615.

[32] L. Zhao, S. Huang, Y. Sun, L. Yan, Dissanayake G, ParallaxBA: bundle adjustment

using parallax angle feature parametrization, Int J Robot Res. 34 (2015) 493 - 516, https://doi. org/10. 1177/0278364914551583.

[33] N. BöRlin, P. Grussenmeyer, Bundle adjustment with and without damping, Photogramm. Rec. 28 (2013) 396 - 415, https://doi. org/10. 1111/phor. 12037.

[34] X. Fang, Weighted total least squares: necessary and sufficient conditions, fixed and random parameters, J. Geodesy. 87 (2013) 733 - 749, https://doi. org/10. 1007/s00190 - 013 - 0643 - 2.

[35] V. Mahboub, On weighted total least - squares for geodetic transformations, J. Geodesy. 86 (2012) 359 - 367, https://doi. org/10. 1007/s00190 - 011 -0524 -5.

[36] R. J. Adcock, A problem in least squares, The Analyst. 5 (1878) 53 - 54, https: //doi. org/10. 2307/2635758.

[37] W. E. Deming, F. F. Stephan, On a least squares adjustment of a sampled frequency table when the expected marginal totals are known, The Annals of Mathematical Statistics 11 (1940) 427 - 444, https://doi. org/10. 1214/aoms/ 1177731829.

[38] G. H. Golub, C. F. Van Loan, An analysis of the total least squares problem, SIAM Journal on Numerical Analysis 17 (1980) 883 - 893, https://doi. org/10. 1137/ 0717073.

[39] C. Yi, L. Jue, Performing space resection using total least squares, The International Archives of the Photogrammetry, Remote Sensing and Spatial Information Sciences 2008 (37) (2003) 11.

[40] S. Doğan, Calibration of Digital Amateur Cameras for Optical 3D Measurements With Element - Wise Weighted Total Least Squares (EW - WTLS) and Classical Least Squares: A Comparison, Acta Montanistica Slovaca 18 (2013).

[41] Z. Wang, Principles of photogrammetry: (with remote sensing). Press of Wuhan Technical University of Surveying and Mapping 574 (1990) pages.

[42] B. Schaffrin, A note on constrained total least - squares estimation, Linear Alg. Appl. 417 (2006) 245 - 258, https://doi. org/10. 1016/j. laa. 2006. 03. 044.

[43] C. G. Harris, M. J. Stephens, A Combined Corner and Edge Detector. Proceedings 4th Alvey Vision Conference, 1988, pp. 147 - 151, https://doi. org/10. 5244/C. 2. 23.

[44] L. A. F. Fernandes, Oliveira, M M, Real - time line detection through an improved Hough transform voting scheme, Pattern Recognit. 41 (2008) 299 - 314, https://doi. org/10. 1016/j. patcog. 2007. 04. 003.

[45] Z. Liu, K. Di, M. Peng, W. Wan, B. Liu, L. Li, H. Chen, High precision landing site mapping and rover localization for Chang'e - 3 mission, Sci China - Phys Mech Astron 58 (2015) 1 - 11, https://doi. org/10. 1007/s11433 - 014 - 5612 - 0.

A Photogrammetric – Photometric Stereo Method for High – Resolution Lunar Topographic Mapping Using Yutu – 2 Rover Images

Man Peng[1], Kaichang Di[1,2], Yexin Wang[1], Wenhui Wan[1], Zhaoqin Liu[1], Jia Wang[3] and Lichun Li[3]

1. State Key Laboratory of Remote Sensing Science, Aerospace Information Research Institute, Chinese Academy of Sciences, Beijing 100101, China; pengman@radi.ac.cn (M. P.); wangyx716@radi.ac.cn (Y. W.); wanwh@radi.ac.cn (W. W.); liuzq@radi.ac.cn (Z. L.)
2. CAS Center for Excellence in Comparative Planetology, Hefei 230026, China
3. Beijing Aerospace Control Center, Beijing 100094, China; 15210106156@139.com (J. W.); lichunmail@163.com (L. L.)

Abstract: Topographic products are important for mission operations and scientific research in lunar exploration. In a lunar rover mission, high – resolution digital elevation models are typically generated at waypoints by photogrammetry methods based on rover stereo images acquired by stereo cameras. In case stereo images are not available, the stereo – photogrammetric method will not be applicable. Alternatively, photometric stereo method can recover topographic information with pixel – level resolution from three or more images, which are acquired by one camera under the same viewing geometry with different illumination conditions. In this research, we extend the concept of photometric stereo to photogrammetric – photometric stereo by incorporating collinearity equations into imaging irradiance model. The proposed photogrammetric – photometric stereo algorithm for surface construction involves three steps. First, the terrain normal vector in object space is derived from collinearity equations, and image irradiance equation for close – range topographic mapping is determined. Second, based on image irradiance equations of multiple images, the height gradients in image space can be solved. Finally, the height map is reconstructed through global least – squares surface reconstruction with spectral regularization. Experiments were carried out using simulated lunar rover images and actual lunar rover images acquired by Yutu – 2 rover of Chang'e – 4 mission. The results indicate that the proposed method achieves high – resolution and high – precision surface reconstruction, and outperforms the traditional photometric stereo methods. The proposed method is valuable for ground – based lunar surface reconstruction and can be applicable to surface reconstruction of Earth and other planets.

Keywords: photometric stereo; photogrammetric - photometric stereo; close - range topographic mapping; surface reconstruction; Yutu - 2 rover images; lunar exploration

1 Introduction

In lunar exploration, topographic mapping products, such as digital elevation model (DEM) and digital orthophoto map (DOM), are important for mission operations and scientific research. Orbital mapping provides large - area products for topographic analysis, geological characterization and other scientific investigations [1 - 3], and ground - based mapping using rover stereo imagery has been routinely applied in landing missions to support surface operations and in - situ scientific explorations [4 - 8] . Image matching is of particular importance in topographic mapping using rover stereo images [9]. However, sometimes the application of image matching is limited due to stereo imaging quality and insufficient image texture. Moreover, stereo images could not be acquired due to payload limit under some circumstances. For example, Chang' e - 4 lander carried a terrain camera, which is located on top of the lander [10] . It can rotate 360° horizontally to obtain a panoramic view of the lunar surface, but without stereo imaging capability. In these cases, ground - based high - precision topographic mapping becomes very challenging.

In the field of computer vision, shape from shading (SfS) technique has long been studied to estimate the shape of an object from one image on the basis of the surface photometry, the position of light, and the viewing directions [11, 12] . As an extension to the original single image SfS, the photometric stereo method can recover albedo and local surface shape at pixel - level with more than two images of the same scene taken under different illumination conditions [13] . Photometric stereo techniques have been developed for lunar and planetary surface reconstruction [14 - 18] to improve the quality of mapping products and generate pixel - level DEMs where image matching fails [14, 19] . It is noted that these lunar and planetary researches all use orbital images to generate mapping products, and both the viewing angle and illumination angle of the images are different.

Following the work of Woodham [13], the traditional photometric stereo method assumes orthographic projection and Lambert reflectance model. This simple reflectance model can be reasonable for certain types of materials, and orthographic projection is appropriate when objects are far away from the camera, e. g. , in the case of orbital images. Since the rover imaging are significantly off - nadir, they cannot be simplified as orthographic projection, perspective projection with collinearity equations should be more accurate. Meanwhile, various lunar reflectance models can be used for lunar rover mapping. Thus, it would be valuable to integrate photometric stereo with collinearity equations and lunar reflectance model for rover image - based surface reconstruction when stereo - photogrammetric method is not applicable.

This paper proposes a novel photogrammetric - photometric stereo (PPS) method based on the collinearity equation and reflectance model for ground - based mapping of the lunar surface. The PPS method is able to generate a height map from multiple rover images acquired under different illumination conditions and the same viewing condition. The paper is organized as follows: Section 2 summarizes related studies, and Section 3 describes the proposed PPS method in detail. The experimental results using simulated images and lunar surface images captured by Yutu - 2 navigation camera (Navcam) are presented in Section 4. Section 5 provides a brief discussion and the concluding remarks.

2 Related Work

Topographic reconstruction of the lunar surface is critical to engineering applications and scientific research. Traditional 3D reconstruction techniques based on photogrammetry have been widely applied in lunar topographical mapping using stereo orbital images [20] or stereo rover images [4, 5] . Some software packages have been developed for planetary image mapping, such as ASP [21], SOCET SET [22], dense matcher [23], and Planetary3D [24] .

In general, the key techniques of orbital image mapping include construction of geometric model, bundle adjustment, dense image matching, and point cloud generation. So far, most studies use a rigorous physical sensor model based on collinearity equations, e. g. , geometric parameters of a terrain camera (TC), lunar reconnaissance orbiter camera (LROC), narrow angle camera (NAC) images, Chang'e - 1 (CE - 1), Chang'e - 2 (CE - 2) images are refined to improve mapping accuracy based on a rigorous sensor model [25 - 29] . Recently, the rational function model (RFM) for geometric modelling has been developed [30 - 33] . After block adjustment based on RFM, the residual is less than 1 pixel, precise topographic models are generated from CE - 1, CE - 2, and NAC images [31, 33, 34] . Image matching is a crucial step in orbital photogrammetric processing. The scale invariant feature transform (SIFT) algorithm has been used for feature matching [35] . Based on matched feature points, epipolar constraints and spatial constraints are used to generate dense points. Delaunay triangulation is used to predict the approximate corresponding locations of other non - feature points [36, 37] . Dense matching algorithm, such as semi - global matching (SGM), has been adopted in orbital mapping [38, 39] . Recently, to deal with matching images with illumination differences, a photoclinometry assisted image matching (PAM) approach is developed in the image matching to create pixel - wise matches [40] .

Compared with the orbital data, ground images obtained by rover cameras can provide high accuracy in situ measurements at centimeter resolution. Stereo rover images have been routinely used to generate DEMs and DOMs for mission operations and scientific investigations in Chang'e-3 and Chang'e-4 rover missions [4, 5].

The early algorithms of SfS were initiated since 1970s [11, 12], and laid the foundation of photoclinometric surface reconstruction. Based on the image irradiance equation, with corresponding reflectance model, these algorithms estimate the surface slopes and heights of each pixel. Many SfS methods are built upon specific geometry conditions and limited properties of reflectance [13, 16, 17]. Although such methods can reconstruct reasonable 3D models, they are restricted to benchmark datasets and indoor images. Planetary images (e.g., Moon and Mars images) generally have subtle albedo variations and trivial reflectance diversities [41], these characteristics make them suitable for reconstructing DEM using SfS techniques. Actually, SfS has been further developed and applied in planetary mapping and relevant scientific research [42-44]. For instance, the SfS derived DEMs (together with stereo derived DEMs) were used for safety assessment of the Mars exploration rovers (MER) landing sites [45], the candidate Mars science laboratory (MSL) landing sites evaluation [46], and topographic analysis of Chang'e-4 landing site [47]. Moreover, SfS derived DEMs have also been used in spectral analysis of the lunar surface [48], lunar soil strength estimation based on Chang'e-3 images [49], concreted fragments analysis at Chang'e-4 landing site [50], and fine-scale analysis of craters and rocks at the Chang'e-4 landing site [40]. Photoclinometric surface reconstruction performs well in local scale reconstruction but has poor large scale accuracy. Hence, in recent years, methods have been developed to incorporate existing low-resolution 3D information (e.g., low-resolution DEMs from photogrammetry or laser altimetry) into photoclinometry [16, 17, 51, 52] and improved results have been achieved.

To overcome the uncertainty from the single-image SfS, photometric stereo method has been developed to estimate the local surface orientation by using several images of the same surface taken from the same viewpoint but under illuminations from different directions [13]. The method first recovers surface gradients from the overdetermined equations, and then integrates the gradient field to determine the 3D surface. As multiple images with different illuminations yield redundancy, the photometric stereo method can improve the accuracy of surface reconstruction. It has been applied to close-range and small-scale scene reconstruction [53]. In the field of planetary mapping, the multi-image shape from shading (MI-SfS) algorithm was developed [14]. The core concept of MI-SfS is similar to photometric stereo and it was able to produce high resolution lunar DEMs when image matching fails due to lack of distinctive texture. This method was further investigated in lunar mapping [19, 54], and pixel-level resolution mapping results at Chang'e-4 and Chang'e-5 landing sites are achieved [19].

In regard to photoclinometric surface reconstruction for rover images, several methods have been proposed to derive detailed topography. For Chang'e-3 rover images, a linear approximation photoclinometry algorithm is applied to generate a DTM in 3-

dimensional pixel scale, and the relationship between disparity and height is used to calculate the soil strength [49] . Wu et al. [40] presented an approach by integrating photogrammetry and photoclinometry for centimeter - resolution modelling from Yutu - 2 stereo Pancam images. Firstly, photogrammetry processing of stereo images is completed to generate a DEM and an orthoimage mosaic of the landing site [40] . Then the DEM is refined to higher resolution using photoclinometry. The same procedure is also applied to derive the detailed topography for target crater analysis [50] . In summary, this rover - based method mainly include stereo photogrammetry processing and photoclinometry refinements, stereo photogrammetry provides an initial DEM and phototoclinometry retrieves fine - grained details and complement the limitations of dense image matching. However, while the rover - based methods depend on the photogrammetry processing of stereo images, PPS approach integrates photometric stereo with collinearity equations, it can reconstruct topography directly from multiple monocular images. As multiple images with different illuminations provide redundancy, PPS can provide reliable reconstruction results. Furthermore, as the images are obtained by the rover cameras at the same location, it is not necessary for images co - registration.

It is well noted that the traditional photometric stereo algorithm assumes orthographic projection and Lambertian surface. The method has been extended from different aspects to deal with general materials, unknown lighting and reflection [55, 56], and perspective camera model [57] . Particularly, Tankus and Kiryati [57] introduced the image irradiance equations under the perspective projection model with identity rotation matrix, and developed the solution to a set of three image irradiance equations, which provided better accuracy than that under orthographic projection. The orthographic projection used in traditional photometric stereo is not appropriate for rover images. Firstly, the distance between rover camera and the lunar surface is close, secondly a pitch angle is usually set for the rover camera, the rotation matrix is not the identity matrix. Thus, the traditional photometric stereo method should be improved to accommodate rover images. The proposed PPS method integrates photogrammetry collinearity equations and lunar reflectance model, which extends the photometric stereo method for ground - based lunar topographic mapping.

3 Photogrammetric - Photometric Stereo Method

The framework of the PPS for lunar rover images is illustrated in Figure 1. First, the imaging irradiance equation of close - range topographic mapping is introduced based on collinearity equations and lunar reflectance model, and the relationship between surface height value and image gray value can be obtained. It can be applied in general cases of photogrammetry, when the rotation matrix is not the identity matrix. Second, the solution of photometric stereo for close - range topographic mapping is derived based on multiple perspective image irradiance equations, and height derivatives can be estimated by iterative calculation. Finally, the global least – squares surface reconstruction with spectral regularization is used to attain the height value from the estimated gradient.

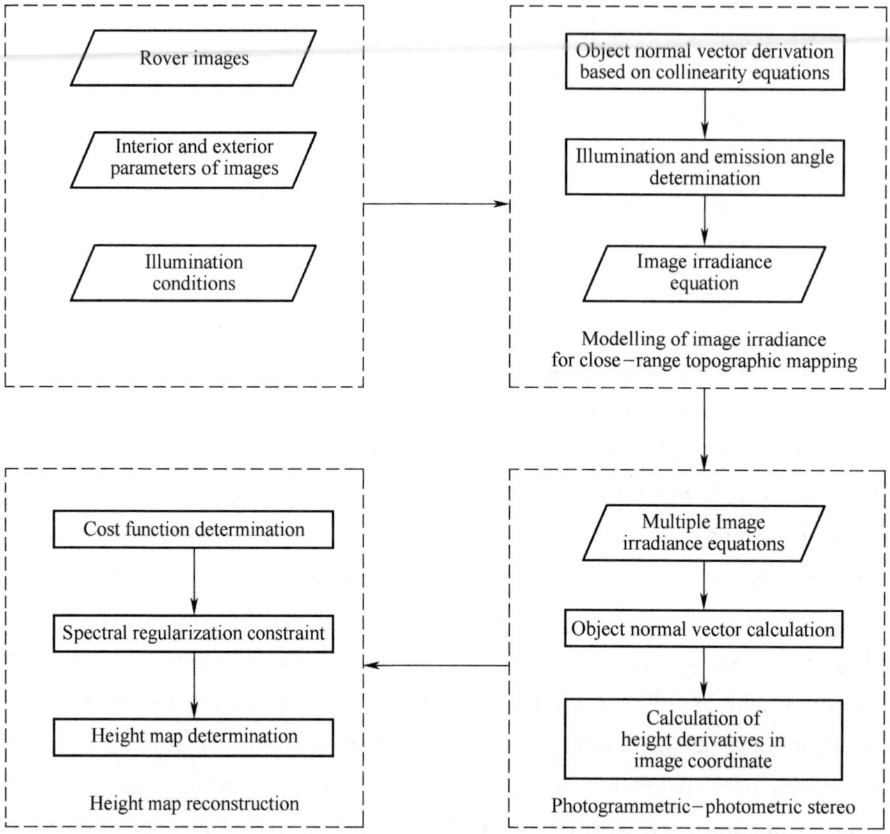

Figure 1 Framework of photogrammetric – photometric stereo (PPS) method

3.1 Modelling the Image Irradiance for Close – Range Topographic Mapping

The modelling of the image formation process defines the process between the sur – face shape and the projected image intensity, which is crucial for photometric stereo problem. The image intensity of a pixel depends mainly on the projection model, the surface reflectance property, surface orientations, and the direction of the illumination. Figure 2 shows relationships of the rover image formation, where L is the incident illumination direction vector, N is the normal vector of the surface, and E is the viewing direction, and the incidence angle i is formed between the vector of solar ray and the normal vector of the surface, the emission angle e is formed between the vector pointing to the sensor and the normal vector of the surface, and α is the phase angle between the solar ray and camera.

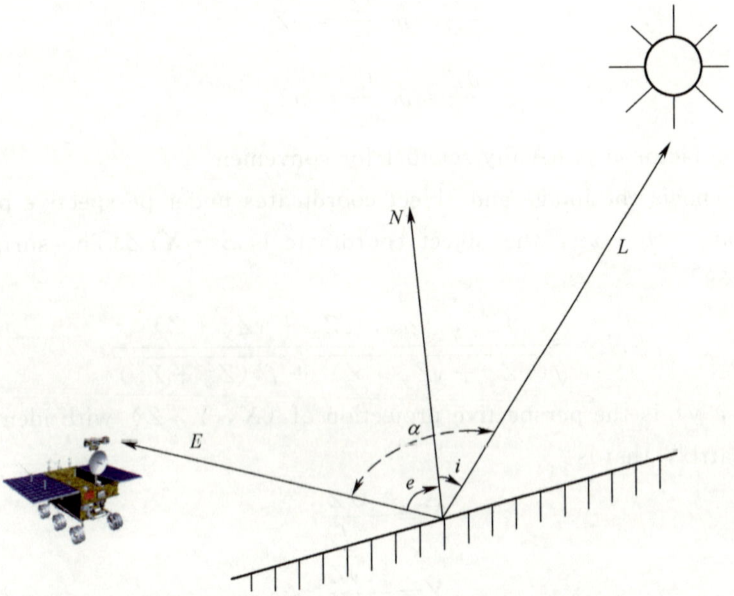

Figure 2　Illustration of rover image formation

3.1.1　Coordinates in Traditional Photometric Stereo vs. Coordinates in PPS

As Figure 3a shows, for most photometric stereo applications [55, 56], the origin of the object coordinate system is at the optical center S of the camera, f is the focal length, the axes of X, and Y are parallel to the x and y axes of the image plane, respectively. Furthermore, orthographic projection is assumed, so that the camera is considered to be far from the surface relative to the size of the scene. The light source is regarded as a point source at infinite distance and therefore constant illumination is assumed over the entire scene.

Assuming height Z is differentiable with respect to X and Y, the normalized surface normal $\hat{N}$ can be represented in terms of partial derivatives of Z with respect to X and Y [58]

$$\hat{N}=\frac{\left(-\frac{\partial Z}{\partial X},\ -\frac{\partial Z}{\partial Y},\ 1\right)}{\sqrt{\left(\frac{\partial Z}{\partial X}\right)^2+\left(\frac{\partial Z}{\partial Y}\right)^2+1}} \tag{1}$$

where $\left(\frac{\partial Z}{\partial X},\ \frac{\partial Z}{\partial Y}\right)$ represents the partial derivatives of Z in X and Y directions.

The orthographic projection is widely used in traditional photometric stereo for or-bital mapping. The partial derivatives of height Z w. r. t. x and y, which are described as $\frac{\partial Z}{\partial x}$ and $\frac{\partial Z}{\partial y}$, are always used to represent the surface normal, and as the resolution is the same for each pixel, there is

$$\frac{\partial Z}{\partial X}=m\frac{\partial Z}{\partial x}=mZ_x$$
$$\frac{\partial Z}{\partial Y}=m\frac{\partial Z}{\partial y}=mZ_y \tag{2}$$

The scaling factor m is usually set to 1 for convenience.

Figure 3b shows the image and object coordinates under perspective projection. The image coordinate is $o-xy$, the object coordinate is $S-XYZ$. The surface normal is derived as in [59]

$$\hat{N}=\frac{(fZ_x,\ fZ_y,\ xZ_x+yZ_y+Z)}{\sqrt{(xZ_x+yZ_y+Z)^2+f^2(Z_x^2+Z_y^2)}} \tag{3}$$

where (x, y) is the perspective projection of (X, Y, Z) with identity matrix as the rotation matrix, that is

$$X=-\frac{xZ}{f},$$
$$Y=-\frac{yZ}{f}, \tag{4}$$

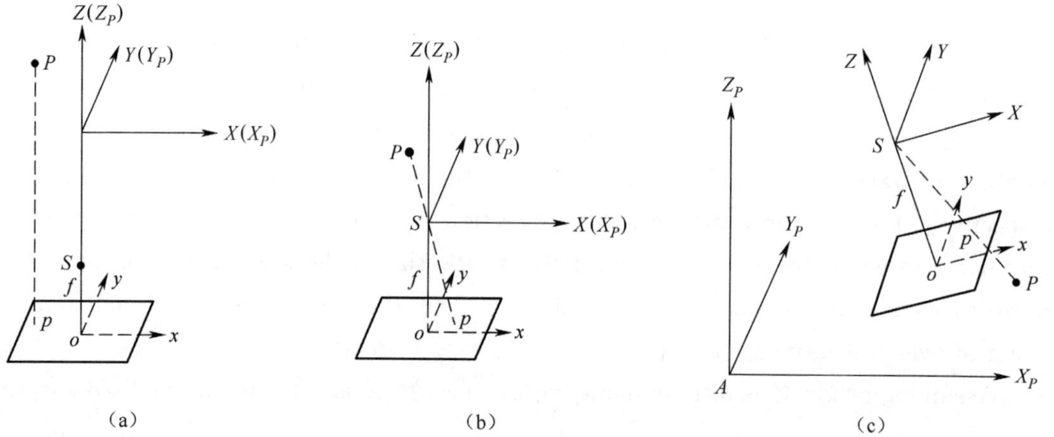

Figure 3　Schematic representation of different imaging conditions (a) image and object coordinates for photometric stereo under orthographic projection (PSOP), (b) image and object coordinates for photometric stereo under perspective projection with identity matrix (PSPP), (c) image and object coordinates for PPS

However, to facilitate topographic mapping, rover localization, and path planning, rotation angles of rover cameras are used [5]. The rotation angles are not small angles, and, thus, cannot be approximated as zero. Figure 3c shows the image and object coordinates for PPS. The image coordinate is $o-xy$, the object coordinate is $A-X_PY_PZ_P$, and it is well noted that the axes of X_P, Y_P in object space are not parallel to x, y axes of the image plane. For rover orthorectified images, the axes of X_P, Y_P in object space are parallel to x, y axes of the image plane, traditional photometric stereo can be used. However, new image irradiance equations should be developed for original rover images with perspective projection. In summary, the following aspects should be considered when developing a proper image irradiance equation for rover - based lunar surface mapping: (1) The pitch angle of the camera cannot be ignored, and the rotation matrix cannot be treated as identity matrix; (2) Objects are not often located very far from the camera as the camera is close to the ground; (3) The axes of X_P, Y_P are not parallel to x, y of the image plane. Thus the surface normal should be deduced based on the principles of photogrammetry and photometric stereo.

3.1.2 Modelling of Surface Normal Based on Collinearity Equations

To make the photometric stereo algorithm useful, a general imaging irradiance equation is introduced based on photogrammetric collinearity equations, illumination conditions and camera interior and exterior orientation parameters.

As the object coordinates X, Y are unknowns, the surface normal can be derived from the collinearity equations.

$$x - x_0 = -f \frac{a_1(X - X_s) + b_1(Y - Y_s) + c_1(Z - Z_s)}{a_3(X - X_s) + b_3(Y - Y_s) + c_3(Z - Z_s)}$$

$$y_0 - y = -f \frac{a_2(X - X_s) + b_2(Y - Y_s) + c_2(Z - Z_s)}{a_3(X - X_s) + b_3(Y - Y_s) + c_3(Z - Z_s)} \tag{5}$$

In the equations, (x, y) represents the coordinate of the image point in the image plane coordinate system; (X, Y, Z) is the object coordinate in the object coordinate system; (X_S, Y_S, Z_S) represents three translation components of the exterior orientation parameters of the camera; $(a_1, a_2, a_3, \ldots, c_3)$ stands for the elements of the rotation matrix; (x_0, y_0) is the principle point position, f represents the principal distance of the camera.

In order to simplify the derivation process, the following symbols are used

$$\begin{aligned} \widetilde{X} &= X - X_S \\ \widetilde{Y} &= Y - X_S \\ \widetilde{Z} &= Z - Z_S \\ \widetilde{x} &= x - x_0 \\ \widetilde{y} &= y_0 - y \end{aligned} \tag{6}$$

The derivatives can be calculated as

$$\begin{aligned} \frac{\partial \widetilde{Z}}{\partial \widetilde{X}} &= \frac{\partial \widetilde{Z}}{\partial x} \cdot \frac{\partial x}{\partial \widetilde{X}} + \frac{\partial \widetilde{Z}}{\partial y} \cdot \frac{\partial y}{\partial \widetilde{X}} \\ \frac{\partial \widetilde{Z}}{\partial \widetilde{Y}} &= \frac{\partial \widetilde{Z}}{\partial x} \cdot \frac{\partial x}{\partial \widetilde{Y}} + \frac{\partial \widetilde{Z}}{\partial y} \cdot \frac{\partial y}{\partial \widetilde{Y}} \end{aligned} \tag{7}$$

$$
\begin{aligned}
\frac{\partial x}{\partial \tilde{X}} &= -f\,\frac{\left[\left(a_1 + c_1\frac{\partial \tilde{Z}}{\partial \tilde{X}}\right)(a_3\tilde{X} + b_3\tilde{Y} + c_3\tilde{Z}) - \left(a_3 + c_3\frac{\partial \tilde{Z}}{\partial \tilde{X}}\right)(a_1\tilde{X} + b_1\tilde{Y} + c_1\tilde{Z})\right]}{(a_3\tilde{X} + b_3\tilde{Y} + c_3\tilde{Z})^2} \\
\frac{\partial y}{\partial \tilde{X}} &= f\,\frac{\left[\left(a_2 + c_2\frac{\partial \tilde{Z}}{\partial \tilde{X}}\right)(a_3\tilde{X} + b_3\tilde{Y} + c_3\tilde{Z}) - \left(a_3 + c_3\frac{\partial \tilde{Z}}{\partial \tilde{X}}\right)(a_2\tilde{X} + b_2\tilde{Y} + c_2\tilde{Z})\right]}{(a_3\tilde{X} + b_3\tilde{Y} + c_3\tilde{Z})^2} \\
\frac{\partial x}{\partial \tilde{Y}} &= -f\,\frac{\left[\left(b_1 + c_1\frac{\partial \tilde{Z}}{\partial \tilde{Y}}\right)(a_3\tilde{X} + b_3\tilde{Y} + c_3\tilde{Z}) - \left(b_3 + c_3\frac{\partial \tilde{Z}}{\partial \tilde{Y}}\right)(a_1\tilde{X} + b_1\tilde{Y} + c_1\tilde{Z})\right]}{(a_3\tilde{X} + b_3\tilde{Y} + c_3\tilde{Z})^2} \\
\frac{\partial y}{\partial \tilde{Y}} &= f\,\frac{\left[\left(b_2 + c_2\frac{\partial \tilde{Z}}{\partial \tilde{Y}}\right)(a_3\tilde{X} + b_3\tilde{Y} + c_3\tilde{Z}) - \left(b_3 + c_3\frac{\partial \tilde{Z}}{\partial \tilde{Y}}\right)(a_2\tilde{X} + b_2\tilde{Y} + c_2\tilde{Z})\right]}{(a_3\tilde{X} + b_3\tilde{Y} + c_3\tilde{Z})^2}
\end{aligned}
\tag{8}
$$

By substituting Equation (8) into Equation (7), the following equations are derived:

$$
\begin{aligned}
\frac{\partial \tilde{Z}}{\partial \tilde{X}}(a_3\tilde{X} + b_3\tilde{Y} + c_3\tilde{Z}) &= \frac{\partial \tilde{Z}}{\partial x}\left(-f\left(a_1 + c_1\frac{\partial \tilde{Z}}{\partial \tilde{X}}\right) - x\left(a_3 + c_3\frac{\partial \tilde{Z}}{\partial \tilde{X}}\right)\right) \\
&\quad + \frac{\partial \tilde{Z}}{\partial y}\left(f\left(a_2 + c_2\frac{\partial \tilde{Z}}{\partial \tilde{X}}\right) + y\left(a_3 + c_3\frac{\partial \tilde{Z}}{\partial \tilde{X}}\right)\right) \\
\frac{\partial \tilde{Z}}{\partial \tilde{Y}}(a_3\tilde{X} + b_3\tilde{Y} + c_3\tilde{Z}) &= \frac{\partial \tilde{Z}}{\partial x}\left(-f\left(b_1 + c_1\frac{\partial \tilde{Z}}{\partial \tilde{Y}}\right) - x\left(b_3 + c_3\frac{\partial \tilde{Z}}{\partial \tilde{X}}\right)\right) \\
&\quad + \frac{\partial \tilde{Z}}{\partial y}\left(f\left(b_2 + c_2\frac{\partial \tilde{Z}}{\partial \tilde{Y}}\right) + y\left(b_3 + c_3\frac{\partial \tilde{Z}}{\partial \tilde{Y}}\right)\right)
\end{aligned}
\tag{9}
$$

Since the ($\tilde{X}$, $\tilde{Y}$, $\tilde{Z}$) is translation of (X, Y, Z) by (X_s, Y_s, Z_s), according to the derivative rule, we have

$$
\begin{aligned}
\frac{\partial \tilde{Z}}{\partial \tilde{X}} &= \frac{\partial Z}{\partial X} \\
\frac{\partial \tilde{Z}}{\partial \tilde{Y}} &= \frac{\partial Z}{\partial Y},
\end{aligned}
\tag{10}
$$

Substituting Equation (10) into Equation (9), by grouping all terms in $\partial Z/\partial X$, $\partial Z/\partial Y$ on the left hand side of the equations, then Equation (9) is rewritten as follows:

$$
\begin{aligned}
\frac{\partial Z}{\partial X} &= \frac{(-a_1 f - a_3 x)\frac{\partial \tilde{Z}}{\partial x} + (a_2 f + a_3 y)\frac{\partial \tilde{Z}}{\partial y}}{(a_3\tilde{X} + b_3\tilde{Y} + c_3\tilde{Z}) + (c_1 f + c_3 x)\frac{\partial \tilde{Z}}{\partial x} - (c_2 f + c_3 y)\frac{\partial \tilde{Z}}{\partial y}} \\
\frac{\partial Z}{\partial Y} &= \frac{(-b_1 f - b_3 x)\frac{\partial \tilde{Z}}{\partial x} + (b_2 f + b_3 y)\frac{\partial \tilde{Z}}{\partial y}}{(a_3\tilde{X} + b_3\tilde{Y} + c_3\tilde{Z}) + (c_1 f + c_3 x)\frac{\partial \tilde{Z}}{\partial x} - (c_2 f + c_3 y)\frac{\partial \tilde{Z}}{\partial y}}
\end{aligned}
\tag{11}
$$

Dividing the right side of Equation (11) with $\tilde{Z}$ and replacing $p=\left(\frac{1}{\tilde{Z}}\right)\left(\frac{\partial \tilde{Z}}{\partial x}\right)=\left(\frac{\partial \ln \tilde{Z}}{\partial \tilde{Z}}\right)\left(\frac{\partial \tilde{Z}}{\partial x}\right)=\frac{\partial \ln \tilde{Z}}{\partial x}$ and $q=\left(\frac{1}{\tilde{Z}}\right)\left(\frac{\partial \tilde{Z}}{\partial y}\right)=\left(\frac{\partial \ln \tilde{Z}}{\partial \tilde{Z}}\right)\left(\frac{\partial \tilde{Z}}{\partial y}\right)=\frac{\partial \ln \tilde{Z}}{\partial y}$, which is similar with the definition in Tankus and Kiryati [57], finally the normal vector of real terrain can be represented with the height derivatives in image coordinates, focal length, and the rotation matrix elements.

$$\frac{\partial Z}{\partial X}=\frac{(-a_1 f-a_3 x)p+(a_2 f+a_3 y)q}{\left(a_3\frac{a_1\tilde{x}+a_2\tilde{y}-a_3 f}{c_1\tilde{x}+c_2\tilde{y}-c_3 f}+b_3\frac{b_1\tilde{x}+b_2\tilde{y}-b_3 f}{c_1\tilde{x}+c_2\tilde{y}-c_3 f}+c_3\right)+(c_1 f+c_3 x)p-(c_2 f+c_3 y)q}$$
$$\frac{\partial Z}{\partial Y}=\frac{(-b_1 f-b_3 x)p+(b_2 f+b_3 y)q}{\left(a_3\frac{a_1\tilde{x}+a_2\tilde{y}-a_3 f}{c_1\tilde{x}+c_2\tilde{y}-c_3 f}+b_3\frac{b_1\tilde{x}+b_2\tilde{y}-b_3 f}{c_1\tilde{x}+c_2\tilde{y}-c_3 f}+c_3\right)+(c_1 f+c_3 x)p-(c_2 f+c_3 y)q} \tag{12}$$

The parameters (x_0, y_0) and f can be obtained from calibrated results of the rover camera, $(a_1, a_2, a_3, \ldots, c_3)$ are determined using rover localization results [5]. Notice that when the rotation matrix is identity matrix, there is

$$\frac{\partial Z}{\partial X}=\frac{-f_p}{1+xp-yq}$$
$$\frac{\partial Z}{\partial Y}=\frac{fq}{1+xp-yq} \tag{13}$$

The unit surface normal is

$$\hat{N}=\frac{(-f_p, f_q, xp-yq+1)}{\sqrt{f^2(p^2+q^2)+(xp-yq+1)^2}} \tag{14}$$

which is the same as that of Tankus and Kiryati [57].

3.1.3 The Image Irradiance Equation

From Figure 2, the emitting vector between the image and the camera is E,

$$E=-(X-X_S, Y-Y_S, Z-Z_S) \tag{15}$$

After normalization, the vector is

$$\hat{E}=\frac{\left(-\frac{X-X_S}{Z-Z_S}, -\frac{Y-Y_S}{Z-Z_S}, 1\right)}{\sqrt{\left(\frac{X-X_S}{Z-Z_S}\right)^2+\left(\frac{Y-Y_S}{Z-Z_S}\right)^2+1}} \tag{16}$$

According to the collinearity equations

$$\begin{bmatrix} X-X_S \\ Y-Y_S \\ Z-Z_S \end{bmatrix}=\begin{bmatrix} a_1 & a_2 & a_3 \\ b_1 & b_2 & b_3 \\ c_1 & c_2 & c_3 \end{bmatrix}\begin{bmatrix} x-x_0 \\ y-y_0 \\ -f \end{bmatrix} \tag{17}$$

Substituting the Equation (17) into Equation (16), then the Equation (16) is rewritten as follows

$$\hat{E}=\frac{\left(-\frac{a_1\tilde{x}+a_2\tilde{y}-a_3f}{c_1\tilde{x}+c_2\tilde{y}-c_3f},\ -\frac{b_1\tilde{x}+b_2\tilde{y}-b_3f}{c_1\tilde{x}+c_2\tilde{y}-c_3f},\ 1\right)}{\sqrt{\left(\frac{a_1\tilde{x}+a_2\tilde{y}-a_3f}{c_1\tilde{x}+c_2\tilde{y}-c_3f}\right)^2+\left(\frac{b_1\tilde{x}+b_2\tilde{y}-b_3f}{c_1\tilde{x}+c_2\tilde{y}-c_3f}\right)^2+1}} \tag{18}$$

Defining the light source as a vector $L=(p_s, q_s, 1)$, (p_s, q_s) can be calculated from azimuth and solar elevation angle, so that the illumination angle and emission angle can be determined as

$$\cos i=\hat{N}\cdot L=\frac{-\frac{\partial Z}{\partial X}p_s-\frac{\partial Z}{\partial Y}q_s+1}{\sqrt{\left(\frac{\partial Z}{\partial X}\right)^2+\left(\frac{\partial Z}{\partial Y}\right)^2+1}\cdot\sqrt{p_s^2+q_s^2+1}} \tag{19}$$

$$\cos e=\hat{N}\cdot\hat{E}$$

$$=\frac{\frac{\partial Z}{\partial X}\left(\frac{a_1\tilde{x}+a_2\tilde{y}-a_3f}{c_1\tilde{x}+c_2\tilde{y}-c_3f}\right)+\frac{\partial Z}{\partial Y}\left(\frac{b_1\tilde{x}+b_2\tilde{y}-b_3f}{c_1\tilde{x}+c_2\tilde{y}-c_3f}\right)+1}{\sqrt{\left(\frac{\partial Z}{\partial X}\right)^2+\left(\frac{\partial Z}{\partial Y}\right)^2+1}\cdot\sqrt{\left(\frac{a_1\tilde{x}+a_2\tilde{y}-a_3f}{c_1\tilde{x}+c_2\tilde{y}-c_3f}\right)^2+\left(\frac{b_1\tilde{x}+b_2\tilde{y}-b_3f}{c_1\tilde{x}+c_2\tilde{y}-c_3f}\right)^2+1}}$$

Various models can be used to determine the reflectance depending on the applications and data, such as Lambert, Lommel - Seeliger model [60], lunar - Lambert model [61], and sophisticated Hapke model [60, 62]. Hapke model has been used in shape from shading when measurements of the surface properties are available. To combine photometric image information of high resolution and DEM data of comparably low resolution, Hapke IMSA and AMSA reflectance models, with two different formulations of the single - particle scattering function being used [16]. The same method is further applied to combine Global Lunar DTM 100m (GLD100) and radiance image data [48]. Compared with lunar - Lambert model, Lommel - Seeliger model is a simple analytical model. It can describe the photometric properties of zero phase angle, and has been widely applied in various lunar spectral datasets, such as UV/VIS of Clementine [63], lunar reconnaissance orbiter camera [64], moon mineralogy mapper [65], Chang'e - 1 interference imaging spectrometer (IIM) [66]. The lunar - Lambert model is a combination of the Lambertian model and the Lommel - Seeliger law. The Lambert model describes the reflectance of bright surfaces very well. In contrast to the Lambert model, the Lommel - Seeliger model describes dark surfaces better, and has been successfully used in planetary photoclinometry as part of the lunar - Lambert model [61]. Based on three LROC NAC images, Lommel - Seeliger model appears to be suitable for the purpose of the relief retrieval [67]. According to Lin et al. [68], the reflectance values are the same when phase angle difference is less than 5°. Considering the change of phase angles of rover images are small, and the ratio of the measurements from two images can cancel out effects of the intrinsic reflectivity of the surface materials (see Equation (22) below), Lommel - Seeliger model is chosen in this research. It can be approximated as follows

$$I(x, y)=p(g)\left[\frac{\cos i}{\cos i+\cos e}\right] \tag{20}$$

p (g) represents reflectance variation with the phase angle g.

After substituting the Equation (19) into Equation (20), image irradiation equation can be finally expressed as

$$I(x, y)=\rho\frac{\dfrac{\left(-\dfrac{\partial Z}{\partial X}p_s-\dfrac{\partial Z}{\partial Y}q_s+1\right)}{\sqrt{p_s^2+q_s^2+1}}}{\dfrac{\left(-\dfrac{\partial Z}{\partial X}p_s-\dfrac{\partial Z}{\partial Y}q_s+1\right)}{\sqrt{p_s^2+q_s^2+1}}+\dfrac{\dfrac{\partial Z}{\partial X}\left(\dfrac{a_1\tilde{x}+a_2\tilde{y}-a_3f}{c_1\tilde{x}+c_2\tilde{y}-c_3f}\right)+\dfrac{\partial Z}{\partial Y}\left(\dfrac{b_1\tilde{x}+b_2\bar{y}-b_3f}{c_1\tilde{x}+c_2\tilde{y}-c_3f}\right)+1}{\sqrt{\left(\dfrac{a_1\tilde{x}+a_2\tilde{y}-a_3f}{c_1\tilde{x}+c_2\tilde{y}-c_3f}\right)^2+\left(\dfrac{b_1\tilde{x}+b_2\tilde{y}-b_3f}{c_1\tilde{x}+c_2\tilde{y}-c_3f}\right)^2+1}}} \tag{21}$$

where ρ is the terrain - dependent albedo which differs by pixel.

3.2 Perspective Photometric Stereo for Close - Range Topographic Mapping

Denoting the images as $\{I(x, y)_k\}_{k=0}^{n-1}$, the corresponding illuminations $\{L_k\}_{k=0}^{n-1}$, and vector of emittance $\{V_k\}_{k=0}^{n-1}$, and assuming there are no shadow in the images, dividing the *ith* image by the *kth* image can eliminate the albedo variable, that is

$$\frac{I_i(x, y)}{I_k(x, y)}=\frac{(N_Xp_{s_i}+N_Yq_{s_i}+1)}{(N_Xp_{s_k}+N_Yq_{s_k}+1)}\frac{\|\vec{V}\|(N_Xp_{s_k}+N_Yq_{s_k}+1)+\|\vec{L}_k\|(N_XV_x+N_YV_y+1)}{\|\vec{V}\|(N_Xp_{s_i}+N_Yq_{s_i}+1)+\|\vec{L}_i\|(N_XV_x+N_YV_y+1)} \tag{22}$$

where i, k = 1, 2, 3, $I\neq k$, and $\|\vec{V}\|$ represents the denominator of emittance vector $\hat{E}$, $(p_{s_i}, q_{s_i}, 1)$, $(p_{s_k}, q_{s_k}, 1)$ are the incident sun vector of the *kth* and *ith* image, $\|\vec{L}_k\|$ is the mode of the incident sun vector of the *kth* image $\|\vec{L}_i\|$ is the mode of the incident sun vector of the *ith* image, and

$$N_X=-\frac{\partial Z}{\partial X}$$

$$N_Y=-\frac{\partial Z}{\partial Y}$$

$$V_x=\frac{a_1\tilde{x}+a_2\tilde{y}-a_3f}{c_1\tilde{x}+c_2\tilde{y}-c_3f}$$

$$V_y=\frac{b_1\tilde{x}+b_2\tilde{y}-b_3f}{c_1\tilde{x}+c_2\tilde{y}-c_3f} \tag{23}$$

After rearrangement, Equation (22) can be rewritten as

$$A_{ik}N_X^2 + B_{ik}N_Y^2 + C_{ik}N_XN_Y + D_{ik}N_X + E_{ik}N_Y + F_{ik} = 0 \tag{24}$$

where

$$A_{ik} = I_i(x, y)(\|\vec{V}\| p_{s_i}p_{s_k} + \|\vec{L}_i\| V_x p_{s_k}) - I_k(x, y)(\|\vec{V}\| p_{s_i}p_{s_k} + \|\vec{L}_k\| V_x p_{s_i})$$

$$B_{ik} = I_i(x, y)(\|\vec{V}\| q_{s_i}q_{s_k} + \|\vec{L}_i\| V_y q_{s_k}) - I_k(x, y)(\|\vec{V}\| q_{s_i}q_{s_k} + \|\vec{L}_k\| V_y q_{s_i})$$

$$C_{ik} = I_i(x, y)(\|\vec{V}\| p_{s_i}q_{s_k} + \|\vec{V}\| p_{s_k}q_{s_i} + \|\vec{L}_i\| V_x q_{s_k} + \|\vec{L}_i\| V_y p_{s_k})$$
$$- I_k(x, y)(\|\vec{V}\| p_{s_i}q_{s_k} + \|\vec{V}\| p_{s_k}q_{s_i} + \|\vec{L}_k\| V_x q_{s_i} + \|\vec{L}_i\| V_y p_{s_i})$$

$$D_{ik} = I_i(x, y)(\|\vec{V}\| p_{s_i} + \|\vec{V}\| p_{s_k} + \|\vec{L}_i\| V_x + \|\vec{L}_i\| p_{s_k})$$
$$- I_k(x, y)(\|\vec{V}\| p_{s_i} + \|\vec{V}\| p_{s_k} + \|\vec{L}_k\| V_x + \|\vec{L}_k\| p_{s_i})$$

$$E_{ik} = I_i(x, y)(\|\vec{V}\| q_{s_i} + \|\vec{V}\| q_{s_k} + \|\vec{L}_i\| V_y + \|\vec{L}_i\| q_{s_k})$$
$$- I_k(x, y)(\|\vec{V}\| q_{s_i} + \|\vec{V}\| q_{s_k} + \|\vec{L}_k\| V_y + \|\vec{L}_k\| q_{s_i})$$

$$F_{ik} = I_i(x, y)(\|\vec{V}\| + \|\vec{L}_i\|) - I_k(x, y)(\|\vec{V}\| + \|\vec{L}_k\|) \tag{25}$$

In the previous research of perspective photometric stereo with identity matrix [57], the Lambert model is generally adopted, and the image irradiance equations only include the illumination angle, so that after equation division the system is linear and the p, q can be solved directly. As our model combines collinearity equations and Lommel-Seeliger model, in order to solve the derivatives p, q, the unknowns N_X and N_Y must be determined firstly. Because the reflectance represents ratio of energy reflection, its value must be greater than zero, thus the estimated value of N_X, N_Y are constrained so that it will not exceed the feasible range.

$$\hat{N} \cdot L,\ \hat{N} \cdot E > 0 \tag{26}$$

As the system of Equation (24) is non-linear for two unknowns, there may exist multiple solutions for the system, reasonable initial values of N_X, N_Y are essential to improve the convergence speed and determine the results. Considering that the ground photogrammetry covers small areas and there are small changes in N_X, N_Y for the whole images, initial values of N_X, N_Y are assigned as a small value. Then Levenberg-Marquardt least-squares minimization algorithm [69] is adopted to solve the unknown parameters for each pixel.

After determining the surface normal (N_X, N_Y), p and q can be solved directly using Equation (12).

3.3 Height Estimation from Estimated Height Gradient

In order to solve the height value from the estimated height derivatives p and q, the global least-squares surface reconstruction with spectral regularization is used [70]. In this algorithm, the cost function is in the form of matrix-algebra, and the matrix Sylvester equation can be solved efficiently and directly.

Assuming Dx and Dy are difference matrices [71], the partial derivatives of the height Z in x - direction and y - direction can be calculated by

$$\frac{\partial Z}{\partial x} = ZD_x^T \qquad (27)$$
$$\frac{\partial Z}{\partial y} = D_y Z$$

so that on the base of calculated gradient field p and q, the cost function is formed by attaining the nearest gradient, represented as follows

$$E(Z) = \| ZD_x^T - p \|^2 + \| D_y Z - q \|^2 \qquad (28)$$

To attain the minimum of the cost function, the matrix equation can be deduced

$$D_y^T D_y Z + ZD_x^T D_x - D_y^T q - pD_x = 0 \qquad (29)$$

In order for surface regularization, the orthogonal basis function is introduced to improve the cost function, and the surface can be described by

$$Z = VMU^T \qquad (30)$$

where U and V are matrices consisting of set of basis functions, M is the coefficient matrix that represents the surface. After substituting Equation (30) into the Sylvester equation in Equation (29), the matrix equation is

$$D_y^T D_y VMU^T + VMU^T D_x^T D_x - D_y^T q - pD_x = 0 \qquad (31)$$

After pre - multiplying the equation by V^T and post - multiplying by U, the function becomes

$$V^T D_y^T D_y VM + MU^T D_x^T D_x U - V^T (D_y^T q + pD_x) U = 0 \qquad (32)$$

The normal equation is a rank deficient Sylvester equation which is solved by means of Householder reflections and the Bartels - Stewart algorithm [70]. After M is solved the height Z can be reconstructed.

As gradients p and q are the derivatives of the natural logarithm of the height, which is defined as $p = \frac{\frac{\partial \tilde{Z}}{\partial x}}{\tilde{Z}} = \frac{\partial \ln \tilde{Z}}{\partial x}$ and $q = \frac{\frac{\partial \tilde{Z}}{\partial y}}{\tilde{Z}} = \frac{\partial \ln \tilde{Z}}{\partial y}$, exponent of Z is finally calculated to obtain the relative height.

4 Experimental Analysis

4.1 Quantitative Evaluation Measures

As the reconstructed heights are integrated results, which are relative height, the height map cannot be compared with the reference height directly. The following three quantitative evaluation measures are used in the experiments: mean error between the normals, root mean square error between the normals, and normalized F - norm of height differences.

Mean error between the normals (MEANN)

MEANN is widely used in photometric stereo studies [71], the normal error represents the angle between different vectors. Let N_{calc} represents the calculated normal vector, and N_{ref} denotes the reference normal vector, for image with $M \times N$ pixels $MEANN$ can be expressed as

$$MEANN = \frac{\left| \sum_{i,j} \cos^{-1}(N_{calc} \cdot N_{ref}) \right|}{M \times N} \tag{33}$$

$MEANN$ is used to evaluate the accuracy of estimated normal vector. The lower the $MEANN$ is, the higher accuracy of the reconstructed height.

Normalized F-norm of height differences (NFD)

As the reconstructed height cannot be compared with the reference height directly, the reconstructed height matrix and reference height matrix are normalized as $\overline{H}_{calc}$, $\overline{H}_{ref}$, then the NFD is calculated as

$$NFD = \frac{\| \overline{H}_{calc} - \overline{H}_{ref} \|_F}{max(\| \overline{H}_{calc} \|_F, \| \overline{H}_{ref} \|_F)} \tag{34}$$

where $\| \overline{H}_{calc} - \overline{H}_{ref} \|_F$ is Frobenius norm (F-norm) of the difference matrix between $\overline{H}_{calc}$ and $\overline{H}_{ref}$, $\| \overline{H}_{calc} \|_F$ is Frobenius norm of $\overline{H}_{calc}$, $\| \overline{H}_{ref} \|_F$ is Frobenius norm of $\overline{H}_{ref}$. Frobenius norm of the difference matrix can be used as Euclidean distance of the two matrices. NFD is the normalized F-norm within the range of [0, 1] to depict the similarity between the two matrices. The smaller the NFD is, the higher accuracy the reconstructed result is.

4.2 Experimental Analysis for Simulated Data

The mapping accuracy of the PPS should be investigated. Due to the difficulty of obtaining ground-truth height information of images under different illumination conditions for rover cameras, simulated images are used to validate the effectiveness and the accuracy of the proposed method. The simulated images are generated from digital projection with a virtual camera by using the existing Chang'e-2 DEM and DOM of an area of lunar surface. The resolutions of DOM and DEM are 4.5m, and the vertical accuracy of the DEM is tens of centimeters. As the resolution of DEM generated by rover image is 0.02m [8], in order to obtain the corresponding simulated rover images, DEM and DOM have been considered as 0.02m with 1400×1000 pixels, as shown in Figure 4.

Given the interior and exterior orientation parameters, pixel size, image size, and illumination conditions for the virtual camera, the corresponding simulated images can be generated using back projection techniques based on the rigorous sensor model, i.e., the collinearity equations. The distortion parameters are assumed to be equal to zero, while the exterior orientation parameters of different images are defined differently. Height values from the DEM are used as ground truth. Based on the illumination conditions,

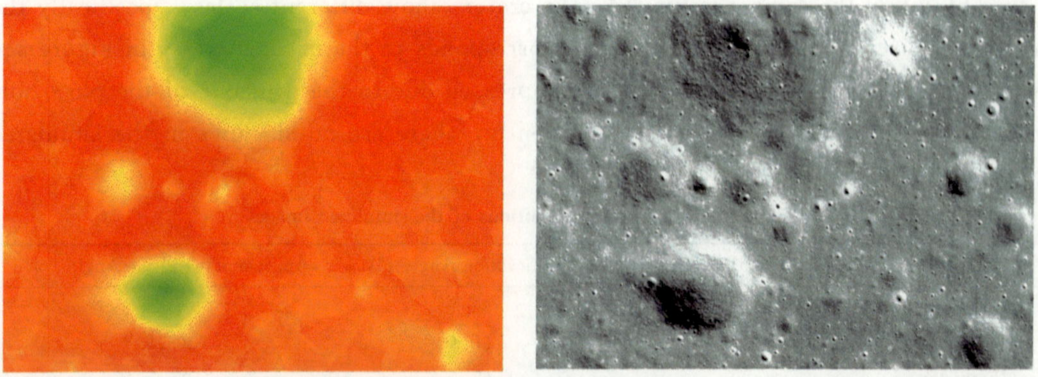

(a)　　　　　　(b)

Figure 4　(a) DEM and (b) DOM for rover image simulation

Chang'e - 2 DEM and DOM, ambient lighting, diffuse lighting, and specular lighting vectors are calculated, respectively, and Lommel - Seeliger model is used to derive the simulated images. Figure 5 shows three simulated images under different illumination

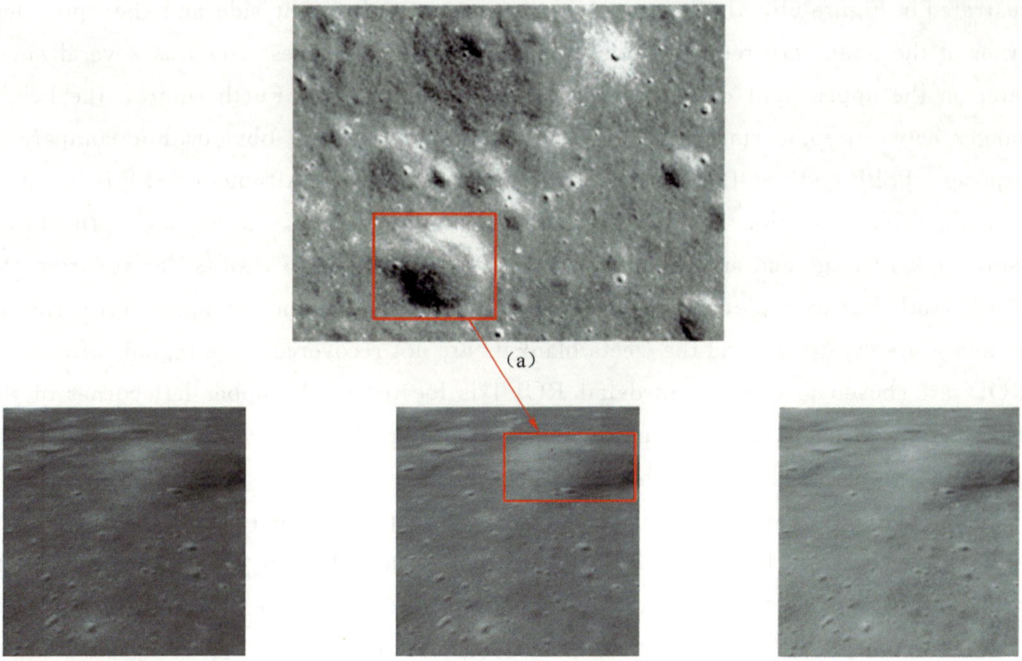

(a)

(b)　　　　　　(c)　　　　　　(d)

Figure 5　Simulated images under different lighting conditions (a) simulate image of solar azimuth angle 90°and elevation angle 55° (b) simulate image of solar azimuth angle 90°and elevation angle 60° (c) simulate image of solar azimuth angle 90°and elevation angle 65° (d)

conditions. Table 1 lists illumination conditions of the simulated images. There are tens of small craters in the image, and a larger crater is clear on the right side. As the field of view (FOV) of the camera is limited, the craters show different scales, specifically the crater in the red rectangular of Figure 5a corresponds to the marked crater in Figure 5c, other craters in the distance have been compressed due to the principle of the pin - hole model. In order to verify the method efficiently, there are no shadows in the simulated images.

Table 1　Illumination conditions of the simulated images

Image ID	Solar Azimuth Angle (°)	Solar Elevation Angle (°)	Approximated Phase Angle (°)
Figure 5b	90	55	107.5
Figure 5c	90	60	103.3
Figure 5d	90	65	99.0

Figure 6 shows height map of ground truth and estimated results using PPS, PSPP [57], and PSOP methods. After replacing the normalized surface normal (12) with Equations (14) and (1), respectively, the height gradient (p, q) can be calculated, PSPP and PSOP results are generated from the corresponding height gradient (p, q). It is noted that the height values have been normalized to the same scale [0, 1]. As illustrated in Figure 6b, the height of large craters at the right side and the upper left corner of the image are recovered distinctly. Small scale features, such as several small crater on the upper right corner of image, are also recovered. Furthermore, the height changes between some small craters and surrounding area are obvious. For comparison purposes, PSPP and PSOP results are shown in Figure 6c, d. Although PSPP is based on the perspective projection model with identity rotation matrix, the errors of the PSPP result are very large and similar with that of PSOP. These two results deviate from the ground truth. For example, the large craters at the right side and the upper left corner of the image are distorted, and the ejecta blankets are not recovered. Two regions of interest (ROI) are chosen to compare in detail. ROI 1 is located at the upper left corner of the image, it includes a large crater and several small craters. ROI 2 includes the largest crater in the image, the height values change greatly in this region.

The enlarged view of ROI 1 and the height maps of ground truth and the three methods are shown in Figure 7. The height distribution of PPS result is the same as the ground truth. As shown in Figure 7c, the details of the terrain are obtained, e. g., both the ejecta blankets of two big craters and the rim of the small crater has been correctly recovered using PPS method. As shown in Figure 7d, e, the resultant distributions of elevations are different from the ground truth, regions of two craters in Figure 7c are merged into a larger region, and the details of craters cannot be recognized. The differences are caused by the lack of interior and exterior elements. A profile along the diagonal direction in Figure 7a is extracted and shown in Figure 8, the left y - axis is the normalized height, the right y - axis represents the real height. The height of virtual camera is. −16.5m, the axis of elevation points down. As Figure 8 shows, the PPS

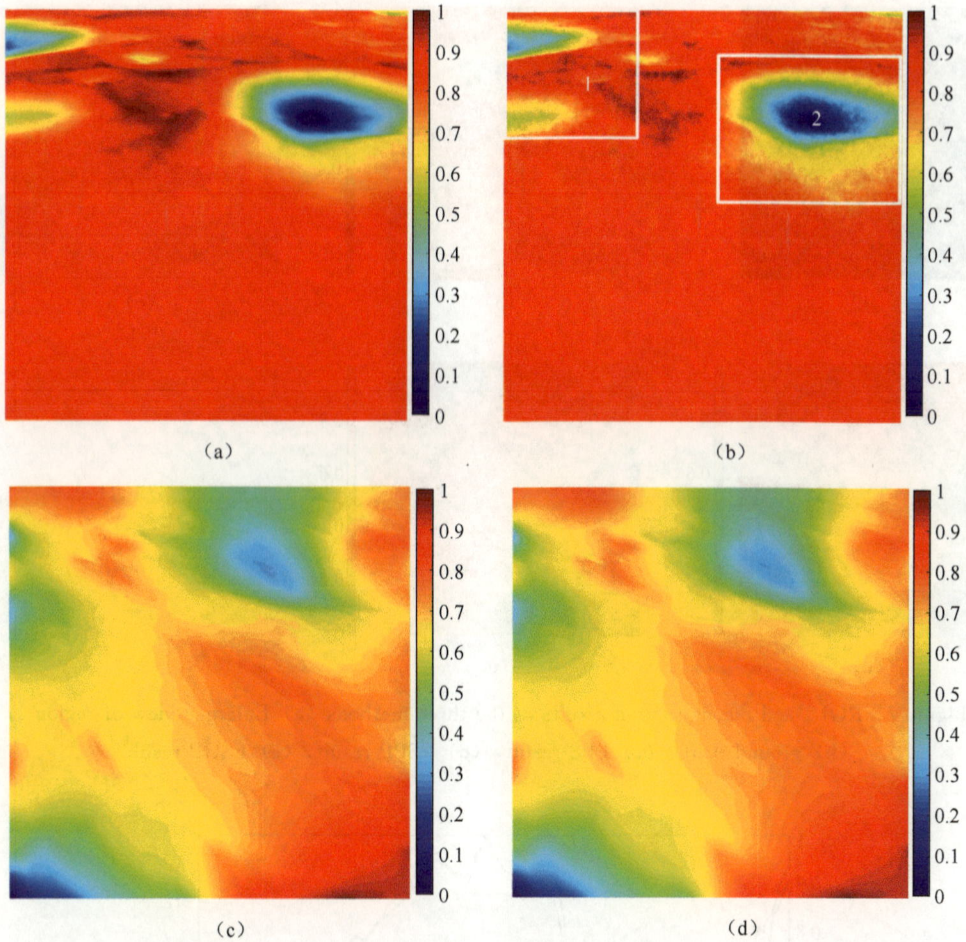

Figure 6　Height map of ground truth and results of three methods (a) Height map of ground truth, (b) PPS result, (c) PSPP result, (d) PSOP result

results of ROI 1 extract the height change of the craters, and the height profile of the PPS results is well consistent with the ground truth. Converting to real height difference, the reconstruction errors are at millimeter to centimeter level. It is noted that differences in PPS result at the beginning part is greater than the end part. The larger errors may be due to the elevation of the craters located at the top of the image change frequently, the small error of the height gradients in this area may affect the integrated results.

For ROI 2, Figure 9a shows a large crater where there is significant height difference between the rim and bottom, Figure 9c shows the details of the whole crater, including the bottom, wall, and ejecta blanket, has been well recovered using PPS method. Alt - hough there are several small craters located at the bottom side of the image, the height difference is small, the ground truth in Figure 9b does not show the corresponding ter - rain detail, and so does the PPS result. Figure 9d, e show a distorted part of a crater,

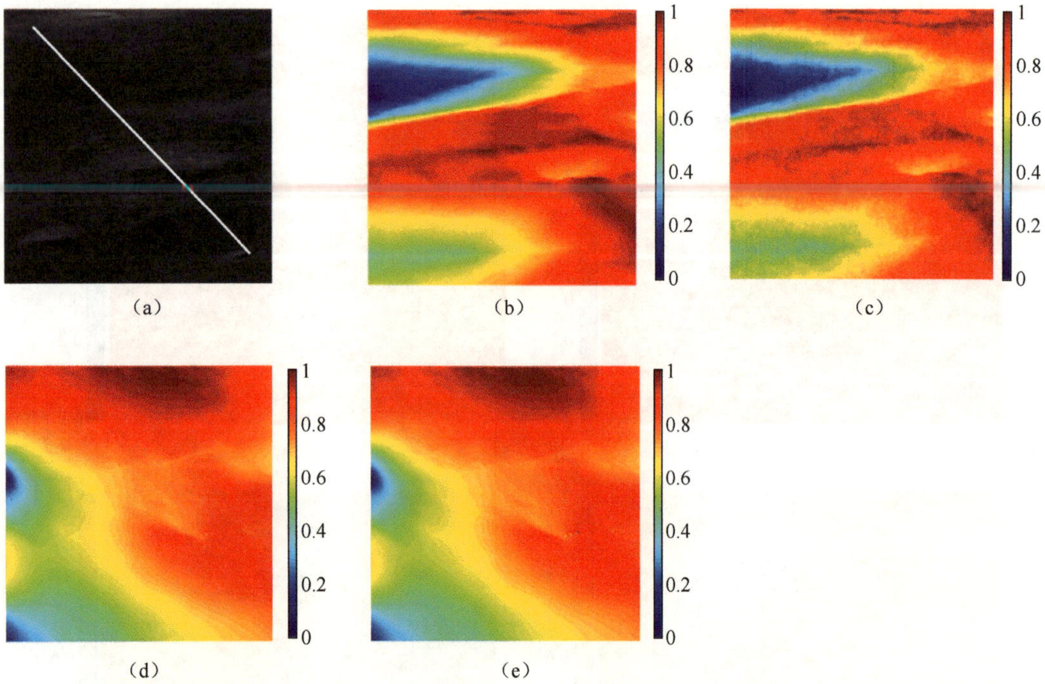

Figure 7　ROI 1 and reconstruction results of the three methods (a) Enlarged view of region 1, (b) ground truth, (c) PPS result, (d) PSPP result, (e) PSOP result

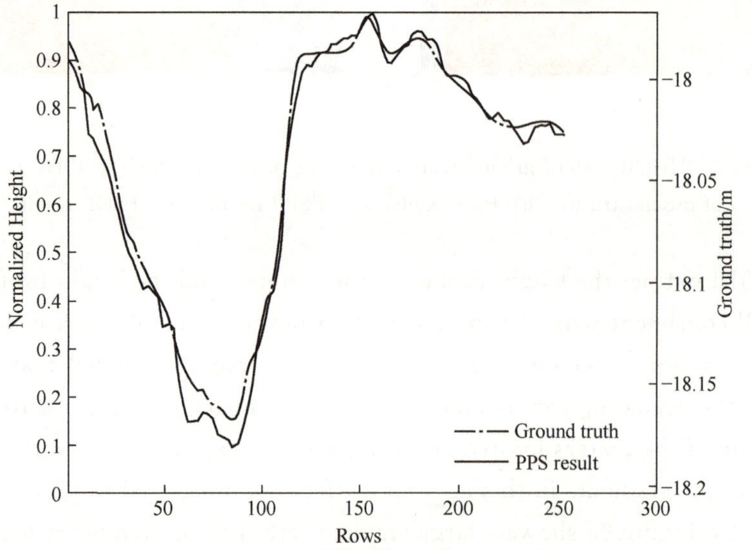

Figure 8　Height profile of ROI 1 in simulated imagery

which is different from the ground - truth shown in Figure 9b, and the ejecta blanket is not correctly recovered. This is caused by the ignorance of rotation matrix in the PSPP and PSOP methods. Figure 10 shows the height profiles along the white line in Figure 9a, which demonstrate a significant decrease along the wall of crater. According to the ground truth, the depth of the big crater is about 0. 2m. It is observed that the errors of PPS result are very small at millimeter to centimeter level. It seems the bottom of the crater shows relatively larger deviations, which needs further investigation.

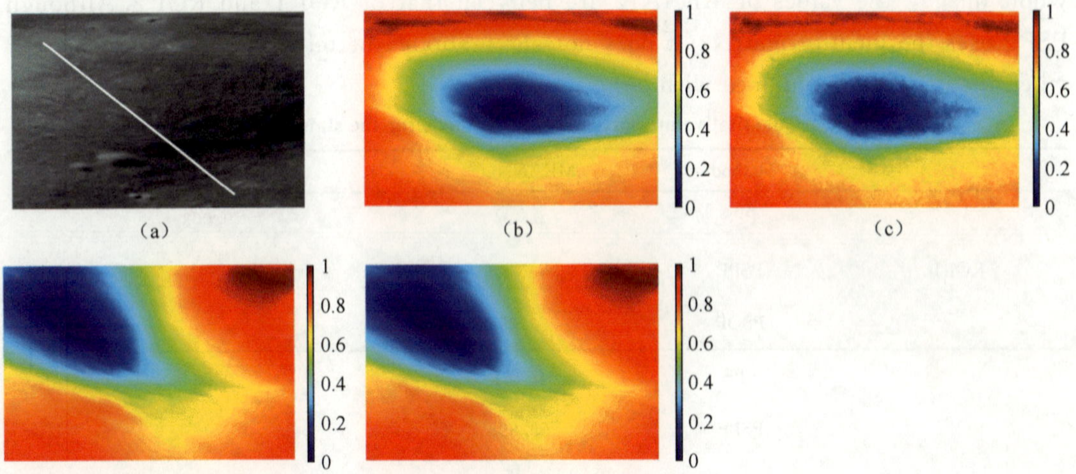

Figure 9　ROI 2 and reconstruction results of the three methods (a) Enlarged view of region 2, (b) ground truth (c) PPS result, (d) PSPP result, (e) PSOP result

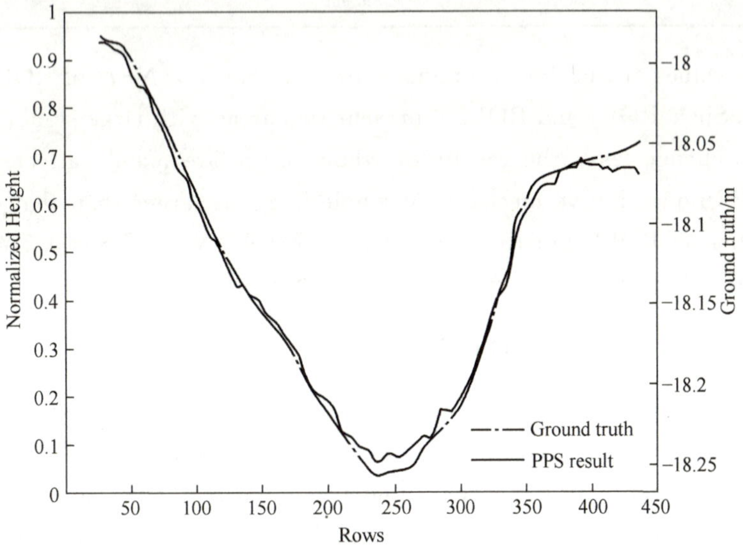

Figure 10　Height profile of ROI 2 in simulated imagery

To quantitatively evaluate the three methods for the simulated image, *MEANN* are obtained from these results. As the elevation coordinates of PSPP and PSOP are different from ground truth, the heights cannot be transferred and compared directly, *NFD* is only calculated for PPS. As the normal vectors of three methods correspond to different object coordinates, it is worth noting that the normal vectors of PSPP and PSOP were multiplied by rotation matrix before calculating the corresponding *MEANN*.

From the quantitative results of Table 2, both PSPP and PSOP methods show large errors, which are around 60°. The *MEANN* are similar for ROI 1 and ROI 2. For the whole image, the values of *MEANN* are between that of ROI 1 and ROI 2. Although PSPP uses the identity matrix, it does not show notable better performance than PSOP when the images are obtained with large pitch angles.

Table 2 Quantitative measures of three methods for the simulated images

	Methods	MEANN (°)	NFD
ROI 1	PPS	0.494	0.003
	PSPP	59.799	—
	PSOP	60.065	—
ROI 2	PPS	0.734	0.092
	PSPP	59.998	—
	PSOP	59.855	—
whole image	PPS	0.324	0.042
	PSPP	59.963	—
	PSOP	59.960	—

MEANN values of PPS for two regions are less than 1°, *NFD* of ROI 1 is less than that of ROI 2. Since ROI 1 and ROI 2 represent two areas with large slope, the elevation of other areas change less, the results of whole image are mainly affected by the two regions. The PPS quantitative measures of whole image is larger than that of ROI 1 and smaller than that of ROI 2. For the whole image, *MEANN* of PPS is 0.324°, and *NFD* is 0.042, which show good performance of the PPS method.

4.3 Experimental Analysis for Yutu-2 Data

Yutu-2 rover of Chang'e-4 mission carries three pairs of stereo cameras: Pancam, Navcam, and Hazcam [8]. The Navcam is mounted on the mast of the rover, and is mainly used for navigation of the rover. Table 3 lists the geometric parameters of Navcam stereo cameras. Normally, at each waypoint, the rover takes a sequence of Navcam images at every 20° in yaw direction at a fixed pitch angle [5]. As a special experiment, the rover obtained 3 pair of Navcam images at a fixed waypoint under different illuminations on 7 August 2019, within the 8th lunar day of the mission. Figure 11 shows the three left images used in our experiment. Their imaging conditions are summarized in Table 4. It can be observed that image tones and shadows change with the change of illumination conditions. The shadow regions are common on the rover images due to complex interactions of geometry and illumination. Figure 12 shows shadows inside craters or behind a boulder.

Table 3　Geometric parameters of Chang'e - 4 Navcam stereo cameras

Stereo baseline	0.27m
Focal length	1189 pixels
Pixel size	1024×1024
FOV	46.4°×46.4°

Table 4　Illumination conditions of the images used in the experiment

Image ID	Solar Azimuth Angle (°)	Solar Elevation Angle (°)	Approximated Phase Angle (°)
94741	−70.8	17	37.7
94914	−71.8	16.2	38.6
95633	−77.1	11.4	44.1

(a)　　(b)　　(c)

Figure 11　Navcam images of the left camera under different illumination conditions
(a) 94741, (b) 94914, (c) 95633

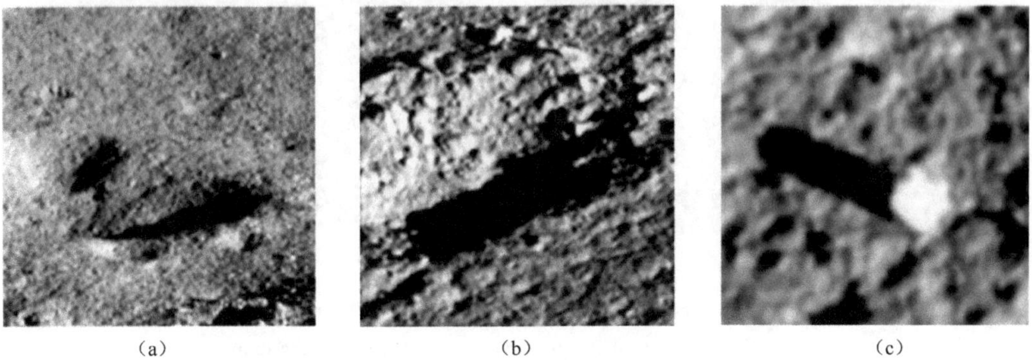

(a)　　(b)　　(c)

Figure 12　Examples of shadows (a) shadow inside two craters,
(b) shadow inside a crater, (c) shadow behind a boulder

In order to detect shadow pixels in the images, each image is segmented using 2D

Otsu thresholding method [72], which extends the Otsu method from one - dimensional to two - dimensional. The 2D Otsu algorithm is based on two - dimensional gray histogram. It utilizes the gray level of each pixel, as well as the average gray level of pixels in its neighborhood, and performs better than the traditional Otsu's method in case of low signal - to - noise ratio (SNR). Figure 13a - c show the shadow maps of the three images generated by the 2D OTSU method, the final shadow map is obtained after using union operation of the three shadow maps of (a), (b), and (c). Figure 13d shows the final shadow map used in the experiment. It can be seen that the lower parts of the images are full of shadows. This is due to the solar elevation angles of the three images are all small, 17° or less.

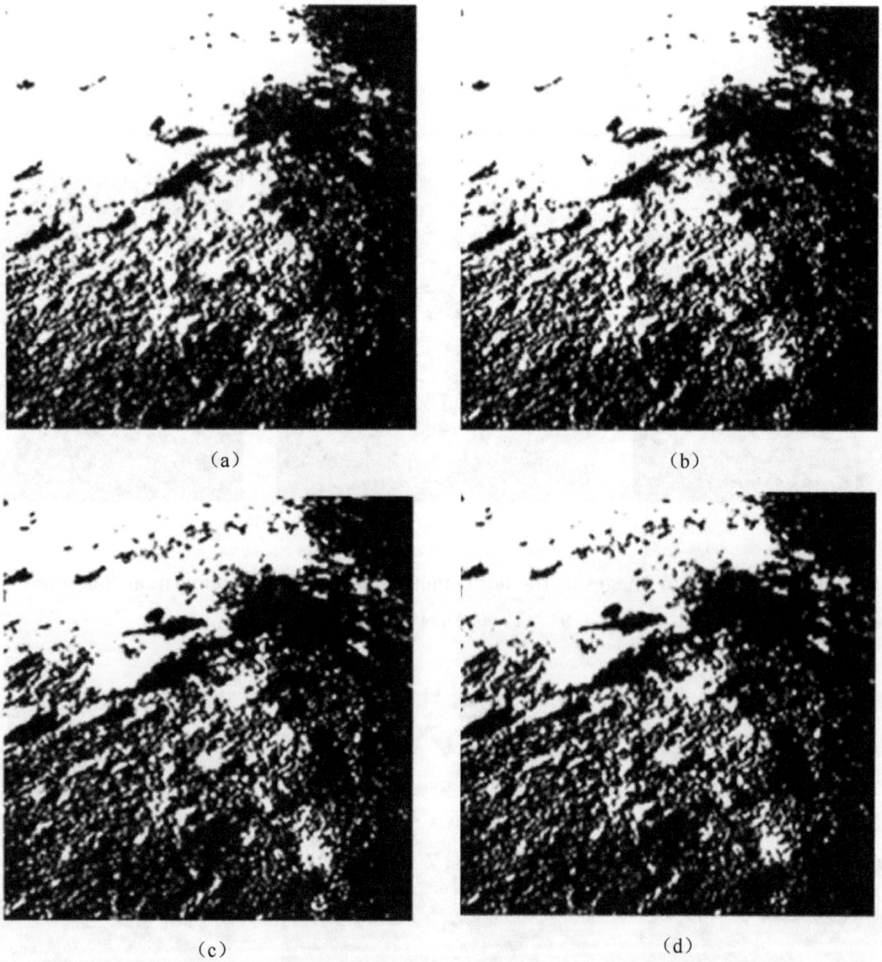

Figure 13　Shadow maps (a) Shadow map of image 94741, (b) Shadow map of image 94914, (c) Shadow map of image 95633, (d) Final shadow map

To quantitatively evaluate the accuracy of the PPS method, the semi - global match - ing (SGM) [38] is used on the first pair of stereo images to achieve height map as ground truth. Semi - global matching is performed to attain the disparity map, where the energy function is built on the determined disparity range of each pixel. The energy function used is as follows

$$E(D) = \sum_{p} C(p, D_p) + \sum_{q \in N_P} P_1 T[|D_p - D_q| = 1] + \sum_{q \in N_P} P_2 T[|D_p - D_q| > 1] \tag{35}$$

where $C(p, D_p)$ is the matching cost of pixel p with disparity D_p, P_1, P_2 represent constant penalty coefficients for all pixels q in the neighborhood of p.

Least squares matching is then performed to attain sub - pixel matching result. After that, 3D coordinates are calculated by space intersection using interior and exterior parameters, finally the height map can be generated.

As Figure 13d shows, although the lower part of image is full of shadows, there exist some continuous non - shaded areas on the upper part that can be used to reconstruct height map using PPS, PSPP and PSOP methods. Figure 14a shows the detail of the upper part image, Figure 14b shows the corresponding shadow map, two regions labelled as 1 and 2 are used to validate PPS method.

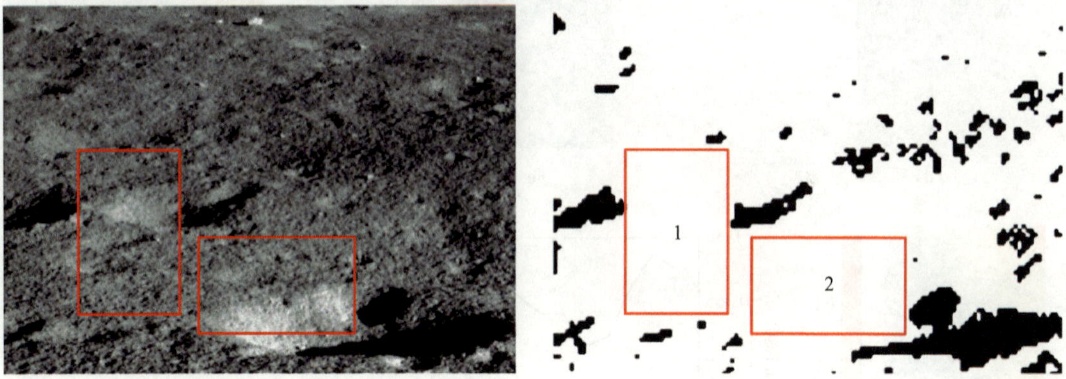

(a)　　　　　　　　(b)

Figure 14　(a) Two sub - regions of the image, (b) Shadow map

Figure 15 shows enlarged views of the area ROI 1 outlined in Figure 14b, the area includes lunar regolith and the rim of a small crater. SGM results and results of three methods have been normalised. As Figure 15b shows, the heights change continuously in a small range, and the rim of the small crater is recovered correctly. The height distribu - tion of PPS result in Figure 15c is similar with that of SGM result, where the upper part is higher than the lower part. Figure 15d, e show SGM and PPS result of ROI 1 with corresponding hillshade map. The rims of three small craters in Figure 15e are clear, and subtle terrain details are displayed. As the boulders are too small, they cannot be recognised in Figure 15d, e. Figure 15f, g represent PSPP and PSOP results. As PSPP and PSOP do not include the rotation matrix, the reconstructed results are inconsistent with the SGM result. The height distributions are similar where the left part is higher than

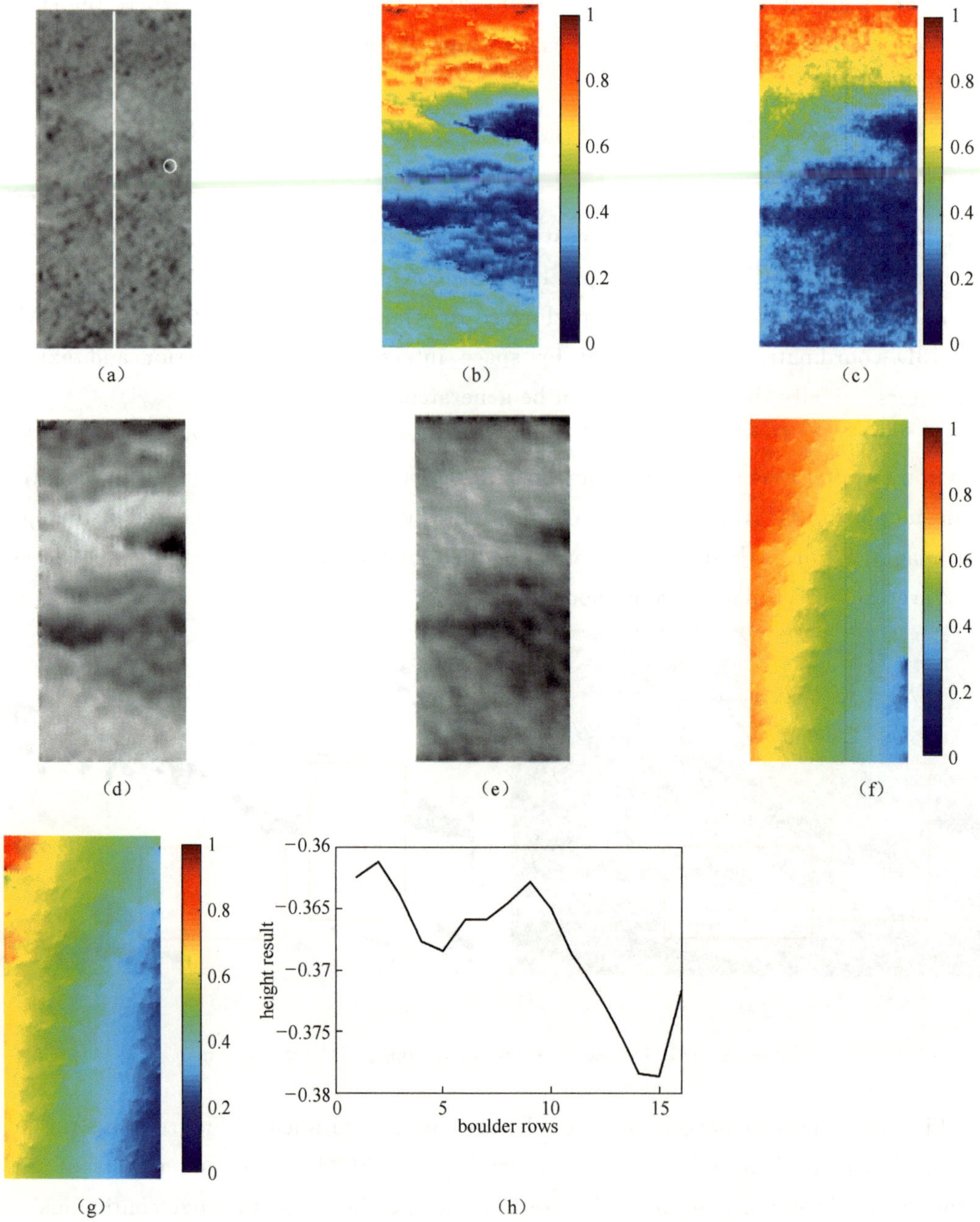

Figure 15　ROI 1 and reconstruction results of the three methods
(a) ROI 1 image, (b) SGM result, (c) PPS result, (d) shaded SGM result,
(e) shaded PPS result, (f) PSPP result, (g) PSOP result, (h) Profile of the boulder
(marked by the white circle in (a)) from PPS result

the right part. The elevation axis of Yutu - 2 points up, the origin of elevation axis located at the centre of rover body chassis. Figure 15h shows the profile of the boulder marked by the white circle in Figure 15a. Figure 16 shows SGM and PPS results of the profile along the white line in Figure 15a, the left y - axis is the normalized height, the right y - axis represents the real height. According to the SGM result, height range of the test area is about 0. 18m. Most of the normalised height of PPS result is larger than that of SGM result, the trend of the curves is consistent.

Enlarged view of ROI 2 is shown in Figure 17a - c show SGM result and normalised PPS result, respectively. Although there exist some isolated height values in the PPS result, the overall distribution of PPS result is close to that of SGM result, and the lower - right corner of ROI 2 in the crater has the lowest height values. Figure 17d, e show SGM and PPS results of ROI 2 with hillshading effects. Figure 17f, g show PSPP and PSOP results with large deviations, indicating that these two methods are not applicable. Being similar with Figure 15d, e, there is a small rotation angle between the PSPP result and the PSOP result. Figure 17h shows the profile of the boulder marked by the white circle in Figure 17a. The profiles of PPS result and SGM result along the white line are shown in Figure 18. As the white line crosses a part of the small crater, the height values decrease continuously, and the maximum height range is about 0. 2m. The PPS result is generally consistent with the SGM results with difference at centimeter level, demonstrating the effectiveness of the PPS method.

Table 5 summarizes the errors of two regions from different methods. The *MEANN* values of PSPP and PSOP are large, which are almost ten times the result of PPS. For the PPS method, *MEANN* and *NFD* of ROI 1 are smaller than those of ROI 2, indicating the method performs better for ROI 1 than for ROI 2, which is in good agreement with the profile maps. Due to the presence of image noises and other errors, the quantitative evaluation values are larger than that of the simulated images.

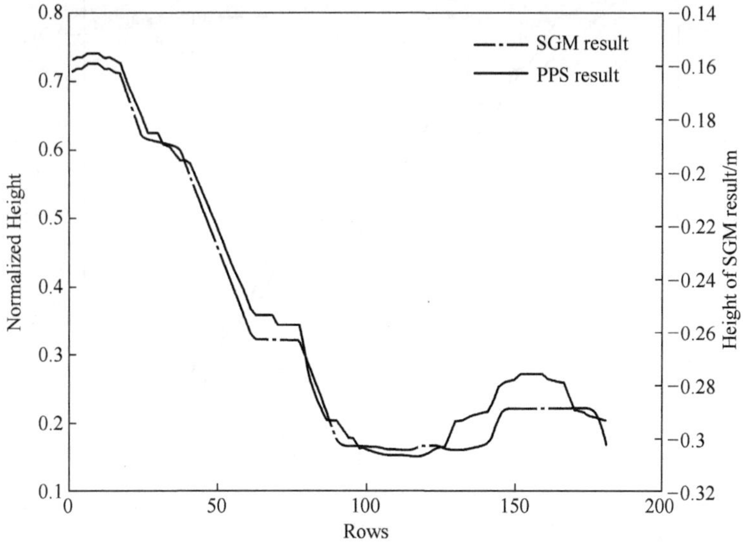

Figure 16　Height profile of ROI 1 in Yutu - 2 rover imagery

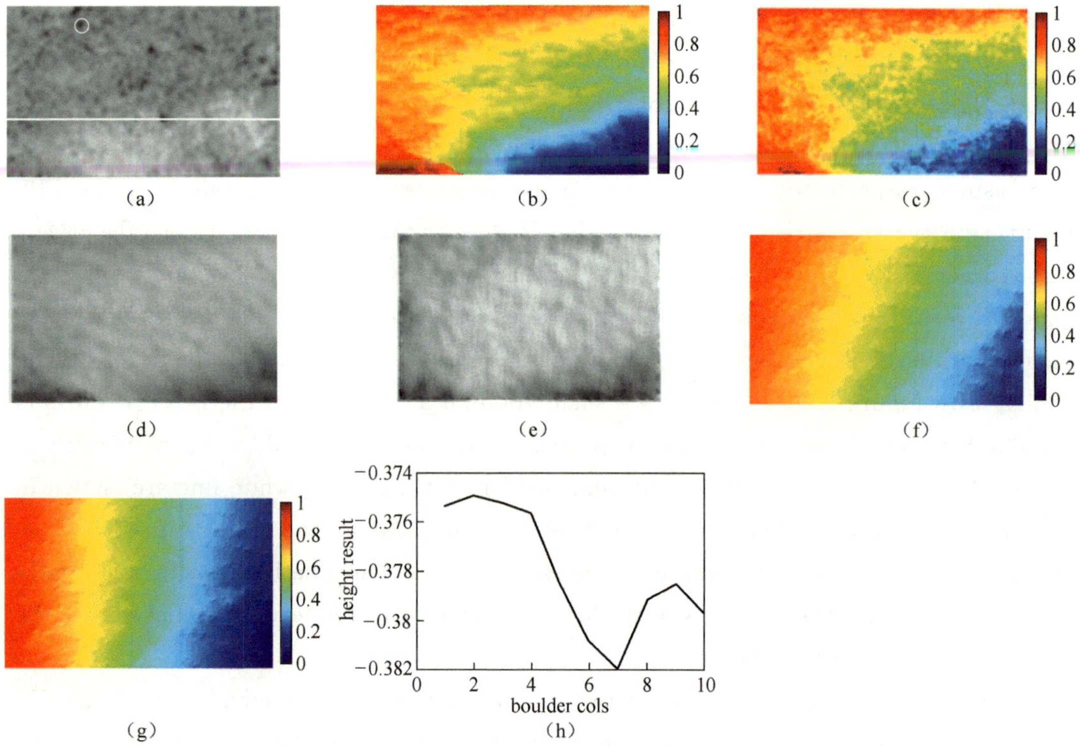

Figure 17　ROI 2 and reconstruction results of the three methods (a) ROI 2 image, (b) SGM result, (c) PPS result, (d) shaded SGM result, (e) shaded PPS result, (f) PSPP result, (g) PSOP result, (h) Profile of the boulder (marked by the white circle in (a)) from PPS result

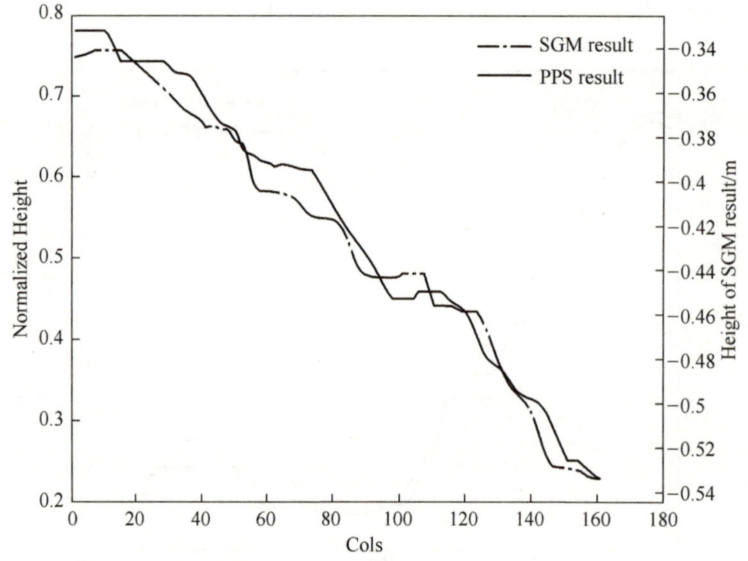

Figure 18　Height profile of ROI 2 in Yutu - 2 rover imagery

Table 5　Quantitative measures of three methods for Yutu 2 images

	Methods	MEANN (°)	NFD
ROI 1	PPS	8.274	0.336
	PSPP	93.792	—
	PSOP	102.746	—
ROI 2	PPS	9.459	0.386
	PSPP	86.786	—
	PSOP	101.843	—

As rover images are largely oblique, PPS results show higher accuracy than that of PSPP and PSOP. For general cases of PSOP, multiple images under the same viewing geometry with different illumination conditions are available, and the DEM is unknown. To further compare PPS with PSOP, three orthorectified images are generated from corresponding images with mean elevation value. Height map is generated using PSOP based on these orthorectified images. The PPS height result is transformed from image space into object space in order for comparison with the PSOP result. The steps of PPS DEM generation is as follows: (1) After SGM matching, the object coordinates of each point can be obtained; (2) The height value is replaced with PPS normalized result; (3) DEM with 0.005m resolution is generated using Kriging interpolation method.

Figure 19 shows orthorectified images of ROI 1. Figure 20 shows the normalized DEM from SGM (a), normalized DEM from PPS (b), and PSOP result based on orthorectified images. As Figure 20b shows, the height distribution is similar with that of Figure 19a, most of the rim of small crater is reconstructed. The upper part of Figure 20c is lower than that of Figure 20a, the rim of small crater is not reconstructed completely.

(a)　(b)　(c)

Figure 19　Orthorectified images of ROI 1 in Yutu-2 rover imagery

(a) ROI 1 of 94741, (b) ROI 1 of 94914, (c) ROI 1 of 95633

Following the same procedure, the orthorectified images of ROI 2 is shown in Figure

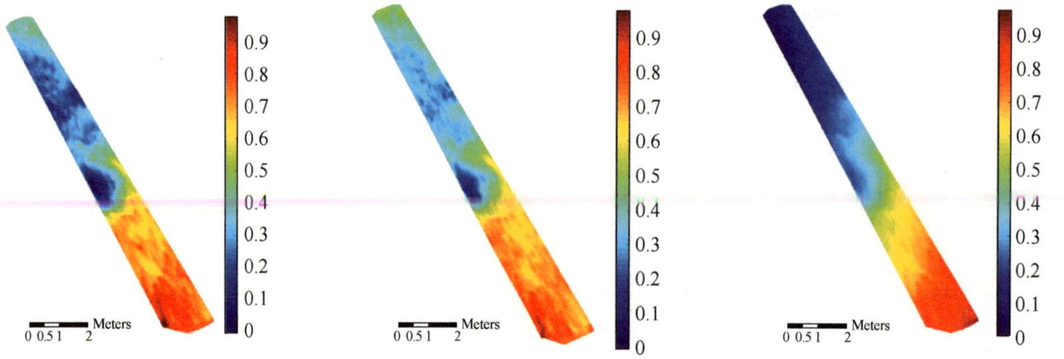

(a) (b) (c)

Figure 20 Reconstruction DEM results of ROI 1 by three methods

(a) SGM interpolation result, (b) PPS interpolation result, (c) PSOP result from orthorectified images

21. Figure 22 shows the normalized DEMs of three methods of ROI 2. The overall height distribution of Figure 22b is close to that of SGM interpolation result, the corerage of crater is also similar. As shown in Figure 22c, the left part of reconstructed result is lower than that of Figure 22a. Using the SGM result as a reference, the visual compqrison results in Figures 20 and 22 indicate that PPS result is generally better that that of PSOP with orthorectified images.

As the height values are all normalized, statistical measures including NIQE [73] and BRISQUE [74, 75] were adopted to evaluate three DEM results. NIQE compares image to a default model computed from images of natural scenes. A smaller score indicates better perceptual quality. The BRISQUE model is based on a pretrained imaging model, the lower values represent better quality. Table 6 summarizes quantitative measures of DEMs generated by three methods. For PPS result, the NIQE and BRISUE values are the smallest for ROI 1 and ROI 2. These statistical measures also demonstrate that the quality of the DEM derived from the PPS method is better than that derived from PSOP with orthorectified images. However, if the DEM of stereo images are available, the DEM result based on photoclinometry will show great improvement [40] .

(a) (b) (c)

Figure 21 Orthorectified images of ROI 2 in Yutu-2 rover imagery

(a) ROI 2 of 94741, (b) ROI 2 of 94914, (c) ROI 2 of 95633

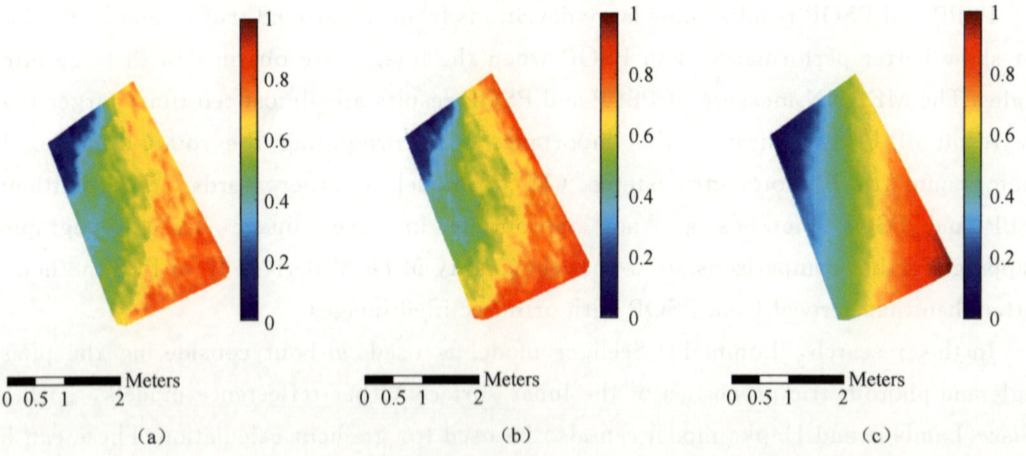

Figure 22　Reconstruction results of ROI 2 by three methods

(a) SGM interpolation result, (b) PPS interpolation result, (c) PSOP result for orthorectified images

Table 6　Quantitative measures of DEMs by three methods for Yutu - 2 images

	Methods	NIQE	BRISQUE
ROI 1	SGM	11.15	45.41
	PPS	10.81	45.26
	PSOP for orthorectified images	11.24	46.15
ROI 2	SGM	7.819	47.47
	PPS	7.540	47.37
	PSOP for orthorectified images	8.230	48.31

5　Conclusions and Discussion

This paper proposes a photogrammetric - photometric stereo (PPS) method to recover the height map from multiple rove images taken by a single camera under the same viewing geometry with changing illumination conditions. The new imaging irradiance model is developed based on collinearity equations and lunar reflectance model. The relationship between surface normal and elevation gradients derived in the method is the key to the final estimation of height map. The experimental results of simulated images and actual Yutu - 2 rover images show that the proposed PPS method algorithm can reconstruct pixel - level height that preserves terrain details, while the reconstructed height map is consistent with ground truth. The height reconstruction error varies between millimeter to centimeter level. Because of the Navcam images taken at the same waypoint with different illumination conditions is limited, and that the differences of azimuth angles and elevation angles among the experimental images are small, the terrain reconstruction errors of the actual rover images are generally larger that of the simulated images. This implies that the illumination differences of multiple images used in photometric stereo affect the resulting height accuracy.

PSPP and PSOP results show large deviations from the ground truth, and PSPP does not show better performance than PSOP when the images are obtained with large pitch angles. The MEANN measure of PSPP and PSOP results are almost ten times larger than the result of PPS, indicating the importance of incorporating the rotation matrix in photogrammetric – photometric stereo (PPS) model. In other words, the traditional PSPP and PSOP methods are not appropriate in rover image based topographic mapping. Further comparisons show that the quality of DEM derived from PPS method is better than that derived from PSOP with orthorectified images.

In this research, Lommel – Seeliger model is used without considering the phase angle and photometric properties of the lunar surface. Other reflectance models, such as lunar – Lambert and Hapke model can also be used for gradient calculation. These can be investigated and compared in future research.

The proposed approach enriches the photometric stereo theory, and extends the orthographic and perspective projection with identity matrix in classical photometric stereo to general cases in close – range topographic mapping. It can also be used for surface reconstruction of other planets with the corresponding reflectance model. In the near future, we will also validate the performance of the model using more ground – based images. Moreover, as the reconstructed height map are of relative height values, further study can be performed by integrating stereo matching and PPS method to improve the details and accuracy of rover image based topographic mapping.

Acknowledgments

The Chang'e – 4 mission was carried out by the Chinese Lunar Exploration Program. We thank the Lunar and Deep Space Exploration Science Applications Center of the National Astronomical Observatories for providing the Chang'e – 2 DOM and DEM. Conflicts of Interest: The authors declare no conflict of interest.

References

[1] Di, K.; Zhu, M.; Yue, Z.; Lin, Y.; Wan, W.; Liu, Z.; Gou, S.; Liu, B.; Pang, M.; Wang, Y.; et al. Topographic Evolution of Von Kármán Crater Revealed by the Lunar Rover Yutu. Geophys. Res. Lett. 2019, 46, 12764 – 12770.

[2] Qiao, L.; Ling, Z.; Fu, X.; Li, B. Geological characterization of the Chang'e – 4 landing area on the lunar farside. Icarus 2019, 333, 37 – 51.

[3] Yue, Z.; Di, K.; Liu, Z.; Michael, G.; Jia, M.; Xin, X.; Liu, B.; Peng, M.; Liu, J. Lunar regolith thickness deduced from concentric craters in the CE – 5 landing area. Icarus 2019, 329, 46 – 54.

[4] Liu, Z.; Di, K.; Peng, M.; Wan, W.; Liu, B.; Li, L.; Yu, T.; Wang,

B. ; Zhou, J. ; Chen, H. High precision landing site mapping and rover localization for Chang'e-3 mission. Sci. China Ser. G Phys. Mech. Astron. 2015, 58, 1-11.

[5] Liu, Z. ; Di, K. ; Li, J. ; Xie, J. ; Cui, X. ; Xi, L. ; Wan, W. ; Peng, M. ; Liu, B. ; Wang, Y. ; et al. Landing site topographic mapping and rover localization for Chang'e-4 mission. Sci. China Inf. Sci. 2020, 63, 1-12.

[6] Di, K. ; Liu, Z. ; Liu, B. ; Wan, W. ; Peng, M. ; Wang, Y. ; Gou, S. ; Yue, Z. ; Xin, X. ; Jia, M. ; et al. Chang'e-4 lander localization based on multi-source data. J. Remote Sens. 2019, 23, 177-184.

[7] Di, K. ; Liu, Z. ; Wan, W. ; Peng, M. ; Liu, B. ; Wang, Y. ; Gou, S. ; Yue, Z. Geospatial technologies for Chang'e-3 and Chang'e-4 lunar rover missions. Geo-Spat. Inf. Sci. 2020, 23, 87-97.

[8] Wang, Y. ; Wan, W. ; Gou, S. ; Peng, M. ; Liu, Z. ; Di, K. ; Li, L. ; Yu, T. ; Wang, J. ; Cheng, X. Vision-Based Decision Support for Rover Path Planning in the Chang'e-4 Mission. Remote Sens. 2020, 12, 624.

[9] Peng, M. ; Wan, W. ; Wu, K. ; Liu, Z. ; Li, L. ; Di, K. ; Li, L. ; Miao, Y. ; Zhang, L. Topographic mapping capability analysis of Chang'e-3 Navcam stereo images and three-dimensional terrain reconstruction for mission operations. J. Remote Sens. 2014, 18, 995-1002.

[10] Jia, Y. ; Zou, Y. ; Ping, J. ; Xue, C. ; Yan, J. ; Ning, Y. The scientific objectives and payloads of Chang'E-4 mission. Planet. Space Sci. 2018, 162, 207-215.

[11] Horn, B. K. Understanding image intensities. Artif. Intell. 1977, 8, 201-231.

[12] Horn, B. K. P. Height and gradient from shading. Int. J. Comput. Vis. 1990, 5, 37-75.

[13] Woodham, R. J. Photometric Method For Determining Surface Orientation from Multiple Images. Opt. Eng. 1980, 19, 191139.

[14] Lohse, V. ; Heipke, C. Multi-image shape-from-shading: Derivation of planetary digital terrain models using clementine images. Int. Arch. Photogramm. Rem. Sens. 2004, 35, 828-833.

[15] Lohse, V. ; Heipke, C. ; Kirk, R. L. Derivation of planetary topography using multi-image shape-from-shading. Planet. Space Sci. 2006, 54, 661-674.

[16] Grumpe, A. ; Belkhir, F. ; Wöhler, C. Construction of lunar DEMs based on reflectance modelling. Adv. Space Res. 2014, 53, 1735-1767.

[17] Wu, B. ; Liu, W. C. ; Grumpe, A. ; Wöhler, C. Construction of pixel-level resolution DEMs from monocular images by shape and albedo from shading constrained with low-resolution DEM. ISPRS J. Photogramm. Remote Sens. 2018, 140, 3-19.

[18] Alexandrov, O. ; Beyer, R. A. Multiview Shape-From-Shading for Planetary Images. Earth Space Sci. 2018, 5, 652-666.

[19] Liu, W. C.; Wu, B. An integrated photogrammetric and photoclinometric approach for illumination - invariant pixel - resolution 3D mapping of the lunar surface. ISPRS J. Photogramm. Remote Sens. 2020, 159, 153 - 168.

[20] Burns, K. N.; Speyerer, E. J.; Robinson, M. S.; Tran, T.; Rosiek, M. R.; Archinal, B. A.; Howington - Kraus, E. Digital elevation models and derived products from LROC NAC stereo ob - servations. ISPRS - International Archives of the Photogrammetry. Remote Sens. Spat. Inf. Sci. 2012, 39, 483 - 488.

[21] Shean, D.; Alexandrov, O.; Moratto, Z. M.; Smith, B. E.; Joughin, I.; Porter, C.; Morin, P. An automated, open - source pipeline for mass production of digital elevation models (DEMs) from very - high - resolution commercial stereo satellite imagery. ISPRS J. Photogramm. Remote Sens. 2016, 116, 101 - 117.

[22] DeVenecia, K.; Walker, A. S.; Zhang, B. New approaches to generating and processing high resolution elevation data with imagery. In Proceedings of the Photogrammetric Week, Heidelberg, Germany, 3 - 9 September 2007; pp. 1442 - 1448.

[23] Re, C.; Cremonese, G.; Dall' Asta, E.; Forlani, G.; Naletto, G.; Roncella, R. Performance evaluation of DTM area - based matching reconstruction of Moon and Mars. In Proceedings of the SPIE Remote Sensing, Edinburgh, UK, 24 - 27 September 2012; p. 8537.

[24] Hu, H.; Wu, B. Planetary3d: A photogrammetric tool for 3d topographic mapping of planetary bodies. ISPRS Ann. Photogramm. Remote Sens. Spat. Inf. Sci. 2019, Ⅳ-2/W5, 519 - 526.

[25] Haruyama, J.; Ohtake, M.; Matsunaga, T.; Morota, T.; Honda, C.; Yokota, Y.; Ogawa, Y. Selene (Kaguya) terrain camera observation results of nominal mission period. In Proceedings of the 40th Lunar and Planetary Science Conference, Woodlands, TX, USA, 23 - 27 March 2009.

[26] Tran, T.; Rosiek, M. R.; Beyer, R. A.; Mattson, S.; Howington - Kraus, E.; Robinson, M. S.; Archinal, B. A.; Edmunson, K.; Harbour, D.; Anderson, E. Generating digital terrain models using LROC NAC images. In Proceedings of the ASPRS/CaGIS 2010 Fall Specialty Conference, Orlando, FL, USA, 15 - 19 November 2010.

[27] Di, K.; Liu, Y.; Liu, B.; Peng, M.; Hu, W. A self calibration bundle adjustment method for photogrammetric processing of Chang'E - 2 stereo lunar imager. IEEE Trans. Geosci. Remote Sens. 2014, 52, 5432 - 5442.

[28] Wu, B.; Guo, J.; Zhang, Y.; King, B.; Li, Z.; Chen, Y. Integration of Chang'E - 1 Imagery and Laser Altimeter Data for Precision Lunar Topographic Modeling. IEEE Trans. Geosci. Remote Sens. 2011, 49, 4889 - 4903.

[29] Wu, B.; Hu, H.; Guo, J. Integration of Chang'E - 2 imagery and LRO laser altimeter data with a combined block adjustment for precision lunar topographic

modeling. Earth Planet. Sci. Lett. 2014，391，1 - 15.

[30] Liu，Y.；Di，K. Evaluation of Rational Function Model for Geometric Modeling of Chang ’ e - 1 CCD Images. ISPRS Int. Arch. Photogramm. Remote Sens. Spat. Inf. Sci. 2012，38，121 - 125.

[31] Liu，B.；Liu，Y.；Di，K.；Sun，X. Block adjustment of Chang’E - 1 images based on rational function model. In Proceedings of the Remote Sensing of the Environment：18th National Symposium on Remote Sensing of China，Wuhan，China，20 - 23 October 2012；Volume 9158，p. 91580G.

[32] Liu，B.；Xu，B.；Di，K.；Jia，M. A solution to low RFM fitting precision of planetary orbiter images caused by exposure time changing. In Proceedings of the International Archives of the Photogrammetry，Remote Sensing and Spatial Information Sciences，Prague，Czech Republic，12 - 19 July 2016；pp. 441 - 448.

[33] Hu，H.；Wu，B. Block adjustment and coupled epipolar rectification of LROC NAC images for precision lunar topographic mapping. Planet. Space Sci. 2018，160，26 - 38.

[34] Liu，B.；Jia，M.；Di，K.；Oberst，J.；Xu，B.；Wan，W. Geopositioning precision analysis of multiple image triangulation using LROC NAC lunar images. Planet. Space Sci. 2018，162，20 - 30.

[35] Wu，B.；Zeng，H.；Hu，H. Illumination invariant feature point matching for high - resolution planetary remote sensing images. Planet. Space Sci. 2018，152，45 - 54.

[36] Furukawa，Y.；Ponce，J. Accurate，Dense，and Robust Multi - View Stereopsis. IEEE Trans. Pattern Anal. Mach. Intell. 2007，32，1362 - 1376.

[37] Wu，B.；Zhang，Y.；Zhu，Q. Integrated point and edge matching on poor textural images constrained by self - adaptive triangulations. ISPRS J. Photogramm. Remote Sens. 2012，68，40 - 55.

[38] Hirschmuller，H. Accurate and Efficient Stereo Processing by Semi - Global Matching and Mutual Information. IEEE Trans. Pattern Anal. Mach. Intell. 2005，30，807 - 814.

[39] Hu，H.；Chen，C.；Wu，B.；Yang，X.；Zhu，Q.；Ding，Y. Texture - Aware Dense Image Matching using Ternary Census Transform. ISPRS Ann. Photogramm. Remote Sens. Spat. Inf. Sci. 2016，3，59 - 66.

[40] Wu，B.；Li，Y.；Liu，W. C.；Wang，Y.；Li，F.；Zhao，Y.；Zhang，H. Centimeter - resolution topographic modeling and fine - scale analysis of craters and rocks at the Chang ’ E - 4 landing site. Earth Planet. Sci. Lett. 2021，553，116666.

[41] Taguchi，M.；Funabashi，G.；Watanabe，S.；Fukunushi，H. Lunar albedo at hydrogen Lyman α by the NOZOMI/UVS. J. Earth Planets Space 2000，52，645 - 647.

[42] Kirk, R. L. A Fast Finite - element Algorithm for Two - dimensional Photoclinometry. Ph. D. Thesis, Division of Geological and Planetary Sciences, California Institute of Technology, Pasadena, CA, USA, 1987; pp. 165 - 258.

[43] Kirk, R. L. ; Barrett, J. M. ; Soderblom, L. A. Photoclinometry Made Simple. In Proceedings of the ISPRS Commission Ⅳ, Working Group Ⅳ/9 2003, "Advances in Planetary Mapping" Workshop, Houston, TX, USA, 22 March 2003.

[44] O'Hara, R. ; Barnes, D. A new shape from shading technique with application to Mars Express HRSC images. ISPRS J. Photogramm. Remote Sens. 2012, 67, 27 - 34.

[45] Kirk, R. L. ; Howington - Kraus, E. ; Redding, B. ; Galuszka, D. ; Hare, T. M. ; Archinal, B. A. ; Soderblom, L. A. ; Barret, J. A. High - resolution topomapping of candidate MER landing sites with Mars orbiter camera narrow - angle images. J. Geophys. Res. 2003, 108, 8088.

[46] Beyer, R. A. ; Kirk, R. L. Meter - scale slopes of candidate MSL landing sites from point photoclinometry. Space Sci. Rev. 2012, 170, 775 - 791.

[47] Wu, B. ; Li, F. ; Hu, H. ; Zhao, Y. ; Wang, Y. ; Xiao, P. ; Li, Y. ; Liu, W. C. ; Chen, L. ; Ge, X. ; et al. Topographic and Geomorphological Mapping and Analysis of the Chang' E - 4 Landing Site on the Far Side of the Moon. Photogramm. Eng. Remote Sens. 2020, 86, 247 - 258.

[48] Wöhler, C. ; Grumpe, A. ; Berezhnoy, A. ; Bhatt, M. U. ; Mall, U. Integrated topographic, photometric and spectral analysis of the lunar surface: Application to impact melt flows and ponds. Icarus 2014, 235, 86 - 122.

[49] Gao, Y. ; Spiteri, C. ; Li, C. ; Zheng, Y. C. Lunar soil strength estimation based on Chang'E - 3 images. Adv. Space Res. 2016, 58, 1893 - 1899.

[50] Ding, C. ; Xiao, Z. ; Wu, B. ; Li, Y. ; Prieur, N. C. ; Cai, Y. ; Su, Y. ; Cui, J. Fragments delivered by secondary craters at the Chang' e 4 landing site. Geophys. Res. Lett. 2020, 47, e2020GL087361.

[51] Jiang, C. ; Douté, S. ; Luo, B. ; Zhang, L. Fusion of photogrammetric and photoclinometric information for high - resolution DEMs from Mars in - orbit imagery. ISPRS J. Photogramm. Remote Sens. 2017, 130, 418 - 430.

[52] Tenthoff, M. ; Wohlfarth, K. ; Wöhler, C. High Resolution Digital Terrain Models of Mercury. Remote Sens. 2020, 12, 3989.

[53] Ackermann, J. ; Langguth, F. ; Fuhrmann, S. ; Goesele, M. Photometric stereo for outdoor webcams. In Proceedings of the IEEE Conference on Computer Vision and Pattern Recognition, Providence, RI, USA, 16 - 21 June 2012; pp. 262 - 269.

[54] Liu, W. C. ; Wu, B. ; Wöhler, C. Effects of illumination differences on photometric stereo shape - and - albedo - from - shading for precision lunar surface reconstruction. ISPRS J. Photogramm. Rem. Sens. 2018, 136, 58 - 72.

[55] Herbort, S. ; Wöhler, C. An introduction to image - based 3D surface

reconstruction and a survey of photometric stereo methods. 3D Res. 2011，2，1 - 17.

[56] Ackermann, J.；Goesele，M. A survey of photometric stereo techniques. Found. Trends Comput. Graph. Vis. 2015，9，149 - 254.

[57] Tankus, A.；Kiryati，N. Photometric stereo under perspective projection. In Proceedings of the Tenth IEEE International Conference on Computer Vision (ICCV' 05)，Beijing，China，17 - 21 October 2005；pp. 611 - 616.

[58] Argyriou，V.；Petrou，M. Photometric Stereo：An Overview. Adv. Imaging Electron Phys. 2009，156，1 - 54.

[59] Tankus, A.；Sochen，N.；Yeshurun，Y. Shape - from - shading under perspective projection. Int. J. Comput. Vis. 2005，63，21 - 43.

[60] Hapke，B. Theory of Reflectance and Emittance Spectroscopy；Cambridge University Press：Cambridge，UK，2012；ISBN 978 - 1 - 139-21750 - 7.

[61] McEwen，A. S. Photometric functions for photoclinometry and other applications. Icarus 1991，92，298 - 311.

[62] Hapke，B. Bidirectional reflectance spectroscopy：The coherent backscatter opposition effect and anisotropic scattering. Icarus 2002，157，523 - 534.

[63] Besse，S.；Sunshine，J.；Staid，M.；Boardman，J.；Pieters，C.；Guasqui，P.；Maralet，E.；McLaughlin，S.；Yokota，Y.；Li，J. - Y. A visible and near - infrared photometric correction for Moon Mineralogy Mapper (M3) . Icarus 2013，222，229 - 242.

[64] Hillier, J. K.；Buratti，B. J.；Hill，K. Multispectral photometry of the Moon and absolute calibration of the Clementine UV/Vis camera. Icarus 1999，141，205 - 225.

[65] Sato，H.；Denevi，B. W.；Robinson，M. S.；Boyd，A.；Hapke，B. W.；McEwen，A. S.；Speyerer，E. J. Photometric normalization of LROC WAC global color mosaic. Lunar Planet. Sci. 2001，XLII，636.

[66] Wu，Y.；Besse，S.；Li，J. Y.；Combe，J. P.；Wang，Z.；Zhou，X.；Wang，C. Photometric correction and in - flight calibration of Chang' E - 1 Interference Imaging Spectrometer (IIM) data. Icarus 2013，222，283 - 295.

[67] Bondarenko，N. V.；Dulova，I. A.；Kornienko，Y. V. Photometric Functions and the Improved Photoclinometry Method：Mature Lunar Mare Surfaces. In Proceedings of the Lunar and Planetary Science Conference，The Woodlands，TX，USA，16 - 20 March 2020；p. 1845.

[68] Lin，H.；Yang，Y.；Lin，Y.；Liu，Y.；Wei，Y.；Li，S.；Hu，S.；Yang，W.；Wan，W.；Xu，R.；et al. Photometric properties of lunar regolith revealed by the Yutu - 2 rover. Astron. Astrophys. 2020，638，A35.

[69] Gill, P. E.；Murray，W. Algorithms for the solution of the nonlinear least - squares problem. SIAM J. Numer. Anal. 1978，15，977 - 992.

[70] Harker, M.; O'Leary, P. Least squares surface reconstruction from gradients: Direct algebraic methods with spectral, Tikhonov, and constrained regularization. In Proceedings of the IEEE Computer Vision Pattern Recognition, Colorado Springs, CO, USA, 20-25 June 2011; pp. 2529-2536.

[71] Quéau, Y.; Lauze, F.; Durou, J. D. Solving uncalibrated photometric stereo using total variation. J. Math. Imaging Vis. 2015, 52, 87-107.

[72] Sha, C.; Hou, J.; Cui, H. A robust 2D Otsu's thresholding method in image segmentation. J. Vis. Commun. Image Represent. 2016, 41, 339-351.

[73] Mittal, A.; Soundararajan, R.; Bovik, A. C. Making a Completely Blind Image Quality Analyzer. IEEE Signal Process. Lett. 2013, 3, 209-212.

[74] Mittal, A.; Moorthy, A. K.; Bovik, A. C. No-Reference Image Quality Assessment in the Spatial Domain. IEEE Trans. Image Process. 2012, 21, 4695-4708. [PubMed]

[75] Mittal, A.; Moorthy, A. K.; Bovik, A. C. Referenceless Image Spatial Quality Evaluation Engine. In Proceedings of the 45th Asilomar Conference on Signals, Systems and Computers, Pacific Grove, CA, USA, 1-4 November 2011.

中国探月工程科学成果（第二辑）
航天科技图书出版基金资助出版

嫦娥四号科学研究创新成果
（下册）

刘继忠　严　俊
邹永廖　唐玉华　等　编著

中国宇航出版社
·北京·

目　录

---

① 为了保持与原刊载媒体一致，本书沿用了每篇论文发表时对该词的表达方式，全书未做统一，far - side 等词同理。

Stratigraphy versus artefacts in the Chang'e – 4 low – frequency radar

Elena Pettinelli[1], Sebastian E. Lauro[1], Elisabetta Mattei[1]
Barbara Cosciotti[1] and Francesco Soldovieri[2]

1. Dipartimento di Matematica e Fisica, Università degli Studi Roma Tre, Rome, Italy
2. Istituto per il Rilevamento Elettromagnetico dell'Ambiente, Consiglio Nazionale delle Ricerche, Naples, Italy.

The Chang'e – 4 (CE4) mission is performing the first ground penetrating radar (GPR) measurements on the far side of the Moon to study the subsurface structure and geological evolution of the Von Kármán crater. In their Article, Zhang et al. [1] interpret a series of reflectors in the low – frequency radar data in terms of a complex stratigraphy formed by different ejecta deposits coming from nearby craters. Here we show that these reflectors are instead artefacts generated by the system design and antenna configuration. This problem could also affect future data collected by the CE4 low – frequency system as well as by other missions that use a similar radar design, such as the one onboard the rover of the Martian Tianwen – 1 mission.

The CE4 dual – frequency GPR is mounted on the Yutu – 2 rover and is similar to that on the Chang'e – 3 (CE3) Yutu rover[2] deployed on the Moon's near side. On both rovers, the low – frequency unshielded monopole antennas are installed on the back of the rover's metallic body, in a configuration that can generate large electromagnetic disturbances capable of interfering with the radar echoes coming from the lunar subsurface[3]. This problem has been extensively discussed for the CE3 low – frequency GPR system, for which it has been proven that several interfaces previously interpreted as stratigraphic features were in fact signal artefacts[4]. On the basis of this evidence, the CE4 low – frequency data should be analysed with caution.

Zhang et al. [1] processed and analysed low – frequency data collected during the first three lunar days (about 163m of radar profile with a time window of 7,000ns) and interpreted such data according to the amplitude of the reflections, under the assumption that each reflector is real and caused by the dielectric contrast between adjacent lunar materials. The authors divided the radar cross – section into three vertical subsections (Fig. 2 in Zhang et al. [1]): a high reflection unit (0 – 2,000ns), a low reflection unit (2,000 – 3,700ns) and a relatively high reflection unit (3,700 – 7,000ns). These units are interpreted as ejecta from craters (~130m), mare basalt (~110m) and ejecta from the Leibnitz crater (⩾200m), respectively.

To verify the physical meaning of the radar signals presented in Zhang et al.[1], we compare low-frequency CE3 data collected at the Zi Wei crater landing site (near side) with CE4 data collected at the Von Kármán crater landing site (far side). The direct comparison between the two datasets is possible because the radar systems and rover configurations are virtually identical, and data are freely available on a public repository[5]. We perform the comparison using raw data acquired with the same system parameters, selecting a subset of contiguous radar traces (10m) of the first 60m of the radar sections collected at the two sites. Note that the data collected on the near side are limited[4], as after several tens of metres the Yutu rover got stuck in the lunar soil[4]. Nevertheless, the analysed subsets are well representative of the entire radar cross-section (Supplementary Fig. 1).

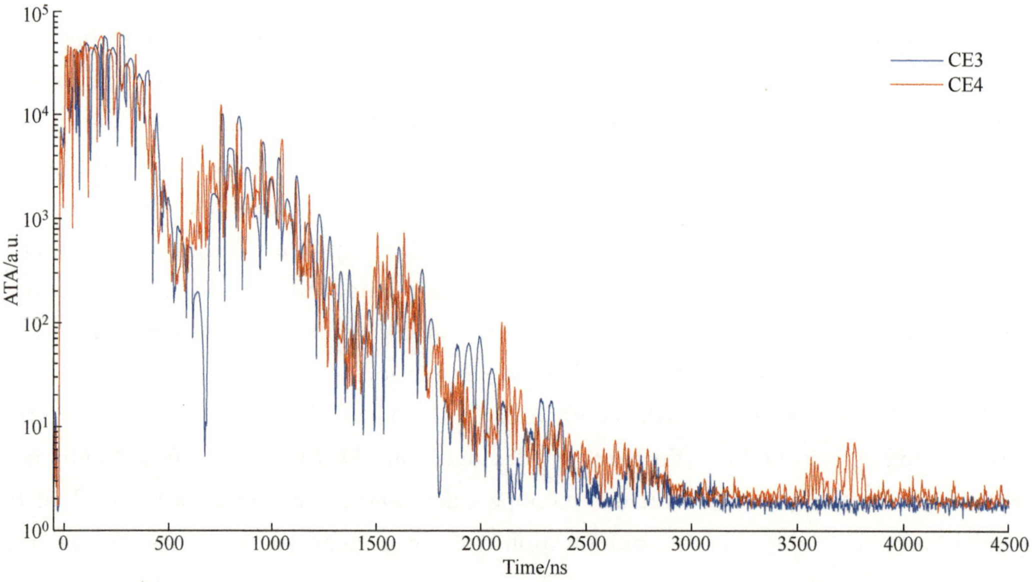

Fig. 1 ATA plot of the CE3 and CE4 data. The plots are generated by (1) applying a temporal filter to remove very-low-frequency components; (2) rectifying all traces of the radar cross-section; and (3) averaging such traces to a single trace. In this way, both noise and signals are preserved[6]. The minimum signal level near zero indicates the noise floor of the instrument, which occurs for both curves at about 3,000ns. The plot refers to 10m subsets of the CE3 and CE4 radar cross-sections and a time window of 7,000ns (as in Zhang et al.[1])

For each dataset, we compute the average time amplitude (ATA) plot[6] (Fig. 1), which is commonly used in GPR data analysis to: (1) determine the signal penetration (attenuation) from the overall decay of the ATA curve; and (2) detect the presence of time-invariant coherent noise due to the so-called system residual response, which indicates how fast the signal falloff is after the pulse has been transmitted[7]. Two major features are visible in the ATA plot: first, the two curves have an identical overall time decay and reach the noise floor at about 3,000ns; second, the first 2,000ns on the curves are very similar, with almost coincident peaks and troughs. After 2,000ns, the curves do not completely overlap, but still exhibit a very similar trend. The main peaks visible on the ATA plot (Fig. 1) correspond to the constant spatial bands detectable on CE3 and CE4 radar cross - sections (Fig. 2a, b); these perfectly horizontal bands are present on both sections at the same delay time, up to about 2,000ns. The only difference between the two radar sections is given by the presence, in the CE4 data, of some discontinuous features underlying the constant bands below 500ns.

The time - frequency analysis, performed by computing the spectrograms of the CE3 and CE4 data subsets in the same time window (4, 500ns) of the radar cross - sections, highlights the main spectral components of the signals in the frequency interval 0 - 100MHz (Fig. 2c, d). Both CE3 (Fig. 2c) and CE4 (Fig. 2d) spectrograms are dominated across the entire time window by a low - frequency (0 - 30MHz) signal component. Different from CE3, however, below 500ns the CE4 spectrogram also contains higher - frequency components (20 - 70MHz). The low - frequency components correspond to the wider pulses in the ATA plot, whereas the high - frequency components correspond to the sharp narrow peaks (Fig. 1 and Supplementary Fig. 1). Moreover, below 3, 000ns the spectral content in the CE3 data becomes constant, with an almost flat spectrum that is representative of white noise. Such noise is less evident on the CE4 10m spectrogram, but it is clearly recognizable in the spectrogram of the entire radar section (Supplementary Fig. 3).

The combined analysis of the ATA plot, the radar sections and the spectrograms indicate that the first 2, 000ns of the CE4 radar cross - section are dominated by time - invariant coherent noise, which appears in the section as a sequence of perfectly horizontal bands[7] (Fig. 2b and Supplementary Fig. 2). These bands are characterized by a low - frequency content (0 - 30MHz) (Fig. 2d) that is well below the nominal band of the antenna and thus generated by the system residual response, which is usually associated with the supporting structure and cabling of the antenna system[6]. None of these bands can be ascribed to real reflectors or dielectric interfaces between units, as suggested by Zhang et al.[1] (high reflection unit). Moreover, the overall decay trends on the ATA plot, being identical on CE3 and CE4, cannot be ascribed to the electromagnetic signal attenuation in the lunar soil, but rather it is the consequence of the slow falloff of the system residual response[6]. Below 2,000ns, such a decay trend does not appreciably change and both curves reach the noise floor at about 3, 000ns (Fig. 1). In the time interval 3,000 - 4,500ns, CE3 and CE4 spectrograms show a uniform spectral content dominated by the noise (Fig. 2c, d and Supplementary Fig. 3). The ATA plot (Fig. 1 and Supplementary Fig. 1) and the CE4 radar cross - section of the entire dataset (Supplementary Fig. 2) indicate that the signal rises again above the noise floor at about

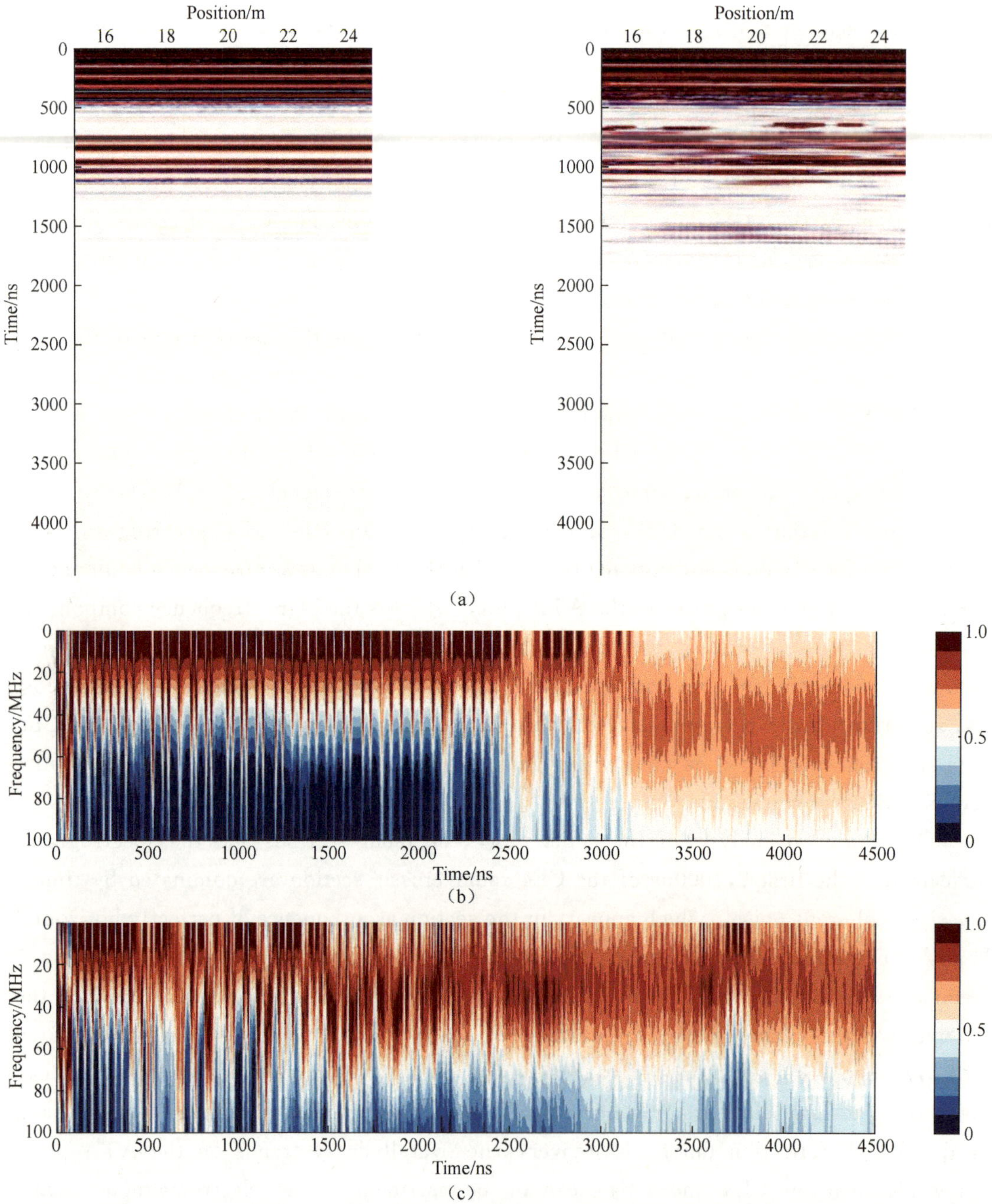

Fig. 2　Comparison of low - frequency CE3 and CE4 radar cross - sections and average spectrograms. a, b, A 10m subset of data collected by CE3 at the Zi Wei crater (a) and by CE4 at the Von Kármán crater (b) landing sites. The time window has been limited to 4, 500ns to better highlight the first 2,000ns of the radar sections. Standard seismic colour has been used: red, high signal amplitude; blue, low signal amplitude. c, d, CE3 (c) and CE4 (d) spectrograms for the 10m radar data subsets. A spectrogram is calculated for each radar trace by computing a normalized short - time Fourier transform with a 30ns time window, after which the spectrograms are averaged along the 10m section

3,700ns. However, the spectrograms of CE4 (Fig. 2d and Supplementary Fig. 3) around 3,700ns clearly show a frequency content typical of the system coherent noise (0 - 30MHz), suggesting that below 3,000ns no real reflections are present.

On the basis of our analysis, we conclude that the system noise has been misinterpreted as a real sequence of reflectors by Zhang et al. [1] and thus their stratigraphic interpretation of the low-frequency radar data collected at the Von Kármán crater landing site does not have any physical meaning. Our analysis, however, highlights the presence of a few weak radar signals between 500ns and 1,800ns, which could be ascribed to real lunar subsurface features. We also urge caution in analysing the data from the GPR in the upcoming Mars rover on the Tianwen-1 mission, as the low-frequency antennas and system configuration are similar to those mounted on the CE3 and CE4 Yutu rovers[8].

References

[1] Zhang, J. et al. Lunar regolith and substructure at Chang'E-4 landing site in South Pole-Aitken basin. Nat. Astron. https://doi.org/10.1038/s41550-020-1197-x (2021).

[2] Li, C. et al. The Moon's farside shallow subsurface structure unveiled by Chang'E-4 lunar penetrating radar. Sci. Adv. 6, eaay6898 (2020).

[3] Angelopoulos, M., Redman, D., Pollard, W. H., Haltigin, T. W. & Dietrich, P. Lunar ground penetrating radar: minimizing potential data artifacts caused by signal interaction with a rover body. Adv. Space Res. 54, 2059-2072 (2014).

[4] Li, C. et al. Pitfalls in GPR data interpretation: false reflectors detected in lunar radar cross sections by Chang'e-3. IEEE Trans. Geosci. Remote Sens. 56, 1325-1335 (2017).

[5] Data Publishing and Information Service System of China's Lunar Exploration Program (National Astronomical Observatories of China, 2020); http://moon.bao.ac.cn

[6] Butler, D. K. Near-Surface Geophysics (Society of Exploration Geophysicists, 2005).

[7] Diamanti, N. & Annan, A. P. Understanding the use of groundpenetrating radar for assessing clandestine tunnel detection. Lead. Edge 38, 453-459 (2019).

[8] Zhou, B. et al. The Mars rover subsurface penetrating radar onboard China's Mars 2020 mission. Earth Planet. Phys. 4, 345-354 (2020).

Geological characterization of the Chang'e – 4 landing area on the lunar farside

Le Qiao* , Zongcheng Ling, Xiaohui Fu, Bo Li

Shandong Provincial Key Laboratory of Optical Astronomy and Solar – Terrestrial Environment, Institute of Space Sciences, Shandong University, Weihai 264209, China

Abstract: On January 3, 2019, China's Chang'e – 4 probe, accomplishes the first – ever soft landing on Moon's farside in the Von Kármán crater floor mare volcanism plain (177.59° E, 45.46° S) within the South Pole – Aitken (SPA) basin. Its surface *in – situ* measurements would provide the first "ground truth" for lunar farside volcanism and SPA materials. We present a detailed geologic context, topographical, morphological, quantitative geochemical and mineralogical characterization of a 20×20km area centered at the Chang'e – 4 landing site, using the latest multi – source high – resolution orbital data sets. The landing area is very flat at a baseline of 30m, with 90% of surface having slope <5°, and deficient of surface boulder exposures. The landing area mare is superposed by abundant sub – kilometer, block – poor circular impact craters and secondary craters in chains and clusters. Mare basalts at the landing site is dominated by low – titanium basalts, and appear to be less mafic and iron – depleted than all typical kinds of retuned lunar basalt samples and may represent a new kind of mare basalt. The landing site mare is heavily sculptured by a set of linear parallel ridged and furrowed textures formed by ballistic ejecta from the Finsen crater, which are characterized by a more feldspathic composition, while with elevated orthopyroxene abundance, and may represent compositional properties of the primitive SPA basin floor materials. We propose the Finsen ejecta and mare basalt outcrops at the landing area as prioritized targets for Chang'e – 4 surface geological and spectroscopic explorations. Follow – on mission data interpret ations should consider the mixing effect of local mare basalts and distant feldspathic materials.

1 Introduction: China's Change'e – 4 mission to the lunar farside

Lunar farside, the hemisphere of the Moon that permanently faces away for the view of Earth due to tidal locking effect, has never seen by human before the Space Age and serves as a yearning destination for aeronautical exploration endeavors. In 1959, the Soviet Luna 3 mission first spotted the lunar farside and presented its fundamental difference from the nearside of the Moon. Since then, extensive remote sensing explorations over the past half century had revealed substantial knowledge of the lunar farisde, including Lunar Orbiter, Apollo (orbital observations), Clementine, Lunar Prospector, Kaguya, Chang'e - 1/2, Chandrayaan - 1, Lunar Reconnaissance Orbiter (LRO) and GRAIL. Numbers of lunar farside landed missions have ever been proposed, for instance, the farside Tsiolkovsky crater as the target for Apollo 17 crewed mission (Wilhelms, 1993), the MoonRise NASA mission concept (Jolliff et al., 2010) and the Farside Explorer proposed for ESA's Cosmic Vision program (Mimoun et al., 2012). However, due to many operational difficulties for lunar farside landing, especially communication relay and the highly rugged lunar farside surface, the lunar farside remains as an unexplored sector for soft - landing missions until today.

On January 3,2019, China's Chang'e - 4 (CE - 4) probe, carrying the Yutu - 2 rover, accomplished the first - ever soft landing on Moon's farside. The landing site (177.59°E, 45.46°S, Di, 2019) is located in the mare plain emplaced within the Von Kármán crater floor in the South Pole - Aitken (SPA) basin. The Chang'e - 4/Yutu - 2 will investigate the geological and compositional characteristics, and shallow subsurface structure of the landing area using its set of instruments including several cameras (namely Landing Camera, Terrain Camera and Panoramic Camera), Lunar Penetrating Radar (LPR) and Visible and Near - Infrared Imaging Spectrometer (VNIS) (Wu et al., 2017; Jia et al., 2018), with a much higher spatial resolution which cannot be achieved by previous and current remote sensing missions. Surface in - situ measurements implemented by the Chang'e - 4 mission will provide the first "ground truth" for lunar farside and the SPA basin.

Recently, Huang et al. (2018) presented a remarkable documentation on the detailed 3 - D geological characteristics of the Chang'e - 4 landing target region (45.0—46.0°S, 176.4—178.8°E, ~50×30km in size) using available remote sensing data, including surface morphology, topography, mineral composition, regolith and ejecta thicknesses, stratigraphy and chronology, which provide an instructive framework for Chang'e - 4 surface measurements. In this contribution, we build on the important initial survey by Huang et al. (2018) and present a dedicated geological characterization of the more local Chang'e - 4 landing area (a 20×20km area centered at the landing site) in a more insightful perspective using additional new multi - source high - resolution orbital data sets. We investigate (1) the regional geological context, (2) detailed and comprehensive topography, morphology and morphometry and (3) quantitative geochemistry and mineralogy of the 20×20km landing area. These analysis results provide additional timely and noteworthy references for Chang'e - 4 surface exploration planning and follow - on mission data interpretations.

2 Data and methodology

We first analyze the morphology and topography of the CE－4 landing area, including its geologic setting within the SPA basin, the Von Kármán crater and its vicinity, using the latest, moderate resolution images from the LRO Wide－Angle Camera (LROC WAC, 100m/pixel, Robinson et al., 2010) and Kaguya/SELENE－TC (Terrain Camera) + LRO－LOLA (Lunar Orbiter Laser Altimeter) merged topography (SLDEM2015; 512 pixels/degree spatial sampling and ～3－4m vertical altimetric accuracy; Barker et al., 2016). We then undertake a detailed characterization of the topography and morphology of the landing area using original LRO LOLA laser altimetric data (～10cm altimetric precision; Smith et al., 2010), high－resolution Kaguya TC stereogrammetry digital terrain model (DTM) (10m pixel size and ～3－4m altimetric accuracy; Haruyama, 2012; Barker et al., 2014), Kaguya TC (10m pixel size; Haruyama et al., 2008) and LROC Narrow－Angle Camera (LROC NAC, up to 0.60m/pixel, Robinson et al., 2010) images. Each individual LROC NAC frame is generated from raw NAC EDR (experiment data record) image through photometrically correction and map－projection using the USGS's Integrated Software for Imagers and Spectrometers (ISIS3; *e. g.*, Anderson et al., 2004) according to the terms of the LROC EDR Data Products Software Interface Specification (Bowman－Cisneros, 2010). We investigate the iron and titanium elemental contents of the studied landing area based on Kaguya Multiband Imager (MI) data (20m/pixel) using algorithms reported by Lemelin et al. (2015) and Otake et al. (2012), respectively. The errors for deriving the FeO and TiO_2 contents were estimated to be ～1 wt% and 0.43 wt%, respectively. At last, quantitative mineralogical analysis for major rock－forming minerals (olivine, orthopyroxene, clinopyroxene and plagioclase) is based on the global mineral abundance maps (60m/pixel) produced by Lemelin et al. (2019) from Kaguya MI spectrometer data through radiative transfer modeling calculations. The Kaguya MI－extracted mineral abundance has an $1\sigma=7$ wt% uncertainty from the Lunar Soil Characterization Consortium (LSCC) data (Taylor et al., 2001; Lemelin et al., 2019).

3 Results and interpretations

3.1 Geologic context of Chang'e－4 landing area

The Chang'e－4/Yutu－2 lander and rover combination successfully landed on the floor of the 186－km－diameter Von Kármán crater. The Von Kármán crater sits in the northwest central region of the SPA basin and is centered at 44.45°S, 176.25°E (LOLA

2011 control network). Its degraded topographic and morphologic characteristics had leaded earlier geologic mapping effort to designate it as a Nectarian - aged crater (Stuart - Alexander, 1978). Recent geologic mapping investigation reported an older, pre - Nectarian, ~3.97Ga age for this crater (Yingst et al., 2016). The landing site is ~350km northwest from the SPA ellipse center (53°S, 169°W) and ~700km from SPA basin outer ellipse (Fig. 1a; Garrick - Bethell and Zuber, 2009). As the largest (2400× 2050km, Garrick - Bethell and Zuber, 2009), deepest (> 8km; Petro and Pieters, 2004) and the oldest impact structure (more than four billion years old; e. g., Hiesinger et al., 2012; Garrick - Bethell and Miljković, 2018) recognized on the Moon, the SPA basin should have excavated the deepest materials of lunar interior (maximum excavation depth up to over 100km; e. g., Schultz and Crawford, 2011; Potter et al., 2012). Recent numerical simulations show the SPA basin - formation impact should have certainly exposed lunar mantle materials to the farside surface (Melosh et al., 2017), yet spectroscopic studies revealed very spare occurrence of olivine (Lucey, 2004; Yamamoto et al., 2010, 2012; Ohtake et al., 2014). Having landed within the central region of this unique terrain of Moon, the expected surface in - situ geological and geochemical investigations of Chang'e - 4 mission will provide an unprecedented opportunity to address a number of important questions regarding lunar basin formation and subsequent effects, earlier bombardment history of the Earth - Moon system (and even for the inner solar system), composition and physical characteristics of lunar deep interior (lower crust and possible lunar upper mantle) (NRC, 2007, 2011).

Regionally, the Von Kármán crater is located at the transition from the more rugged SPA basin rim structure to relatively flat (at the scale of hundreds of kilometers) basin floor (Fig. 1a). The Von Kármán crater is enclosed by several other large impact craters, namely Leibnitz in the north, Finsen in the northeast, Von Kármán M in the south and Oresme in the west (Fig. 1b). These impact craters are characterized by observable cross - cutting and superposition relationships, enabling us to establish the stratigraphic sequences of these craters. The Von Kármán M crater is clearly superposed and largely deconstructed (especially the northern rim; thus relatively older) by the Von Kármán crater. From the residual southern portion of the rim structure, the Von Kármán M crater has been estimated to have a diameter of ~245km (Fig. 1b; Head III et al., 2010). The ~240 - km - diamater Leibnitz crater lies in the immediate north exterior of Von Kármán crater. Emplacement of the south rim of Leibnitz crater has buried and deconstructed the north rim crest of Von Kármán crater. The Leibnitz impact event should have also spewed massive ejecta onto the northern floor of Von Kármán crater, resulting in its elevated topography and rugged and hummocky terrains (Fig. 1b). Recent geological study reported a Nectarian, ~3.88Ga age for the Leibnitz crater (Yingst et al., 2016). The Oresme crater, with a diameter of ~76km, seems to have comparable topographic degradation state with Von Kármán crater. Previous geologic mapping has attributed an ancient, pre - Nectarian age for this crater (Stuart - Alexander, 1978). The superposition relationship between the two craters cannot be unambiguously determined, while examination of new LROC WAC images show the development of the Oresme crater structure appears to has propagated into the exterior ejecta deposits of Von Kármán crater, suggesting Oresme probably postdate the Von Kármán crater (Fig. 1b). Another regional prominent feature is the 73 - km - diameter Finsen crater to the northeast of Von Kármán crater. The Finsen crater is characterized by well - preserved primitive large crater structures, including wall terrains, continuous ejecta blanket and central peak formation. Extensive bright ray patterns and radial secondary craters are clearly observed, in optical images, to be associated with Finsen crater, which serve as diagnostic evidences to attribute Finsen crater as a Copernianaged crater (Hawke et al., 2004). The mare deposits within Von Kármán and Leibnitz crater floors have been heavily superimposed by ejecta from the Finsen crater (Fig. 1b).

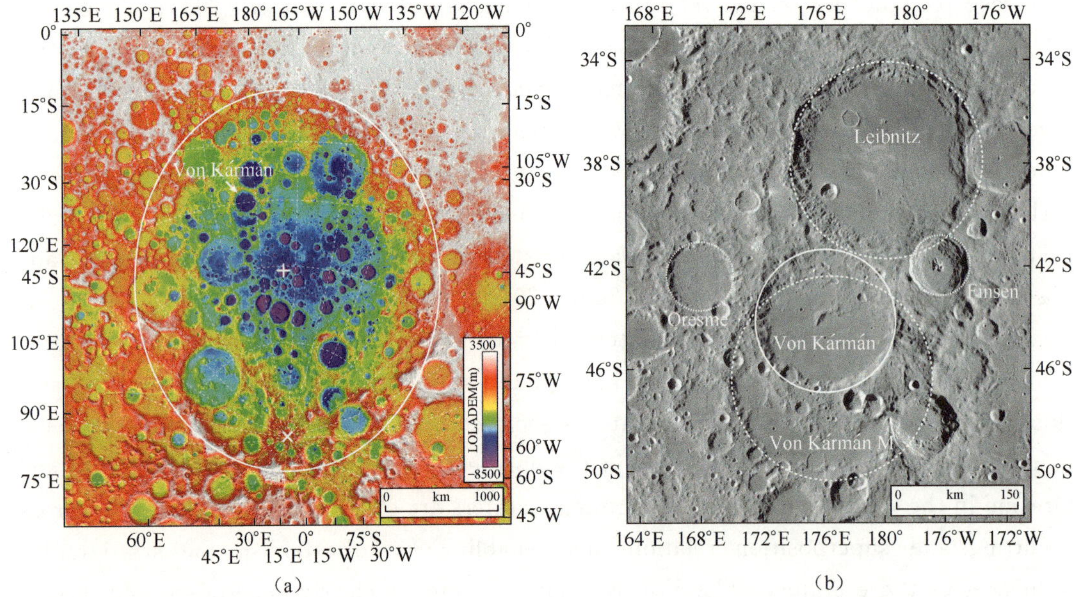

Fig. 1 (a) The topographic context of Von Kármán crater within the South Pole - Aitken (SPA) basin, LOLA DEM topography (128 pixel/degree) overlaid on LROC WAC mosaic. The white oval and cross mark the best - fit topographic outer ellipse and center of SPA basin, perspectivity (Garrick - Bethell and Zuber, 2009), the white multiplication sign marks the south pole, and the location of Von Kármán crater is pointed out. The projection is stereographic centered at 191°E, 53°W, and north is up. (b) The regional vicinity of Von Kármán crater, LROC WAC image mosaic. The bold white circle marks the rim locations of Von Kármán crater and the dotted white circles mark the rim locations of Von Kármán M, Leibnitz, Oresme and Finsen craters (Head III et al., 2010). The projection is lambert conformal projection, centered at 176°E, 44.5°S, and north is up

The Von Kármán crater itself is 186km in diameter, with a rim - tofloor depth of ~3.5km (Fig. 2), which shows the Von Kármán crater has a much smaller depth - to -

diameter value than young (post - Imbrian) complex lunar craters (Kalynn et al., 2013); we attributed this to the ancient age, degraded topographic characteristics and subsequent lava filling within crater floor. New LRO altimetry and image data show this crater probably developed a ringed peak - cluster structure (though, poorly reserved, pointed out by arrows in Fig. 2b). From the potential remnants of this ringed peak - cluster structure, we measure a ~40km ring diameter, which lies approximately on the trend line of ring diameter/rim - crest diameters for over 40 peak - ring and ringed peakcluster basins on the Moon (Baker et al., 2011). Peak — ring craters/basins are generally regarded as a morphologic transition from smaller complex craters to multi - ringed basins, while the specific formation mechanism is still unknown and includes interpretations of downward and radial movement of the "displaced zone" below the transient cavity (Melosh, 1989; Spudis, 1993), differential scaling between crater sizes and volume of impact melt (e. g., Grieve and Cintala, 1992), "nested melt - cavity" model (Head, 2010), emplacement of the outwardly collapsed crust basement above the collapsed transient cavity rim (Morgan et al., 2016), etc. The poor preservation state of the peak ring structure is probably a combined result from subsequent ejecta deposition (dominantly from Leibnitz crater), lava flooding and topographic degradation.

Topographic and compositional analyses can provide fundamental information of the geological characteristics and evolution history of the landing area. The Von Kármán crater interior floor is characterized by a wide range of topographic and morphological characteristics, and shows a distinct dichotomy between the northern and the southern portions (Fig. 2). The northern crater interior is dominated by ejecta deposits, with very rough and rugged surface textures (with slopes typically$>5°$; Fig. 2d) and a typical highland, feldspar - dominated composition. The southern portion of crater interior is largely filled with mare lava deposits, characterized by relatively flat surface (with slopes dominantly$<4°$; Fig. 2d) and an iron - rich basaltic composition. These mare plains occupy an area of $1.063 \times 10^4 km^2$. Elevation of the western mare is generally higher than the southern part, which show a topographic decline of ~350m across a distance of ~80km (Fig. 2b, c). Population study of superposed impact craters using Kaguya images reported a 3.35Ga model age for the Von Kármán mare flows (Haruyama et al., 2009). An updated crater population study using LROC WAC images yielded a relatively younger model age of 3.15Ga (Pasckert et al., 2018). In addition, Pasckert et al. (2018) also observed a second older model age of 3.75Ga, which might be related to an older buried flow. Two bulge/dome - like structures are identified in the western margin of the interior mare units (pointed out by white triangles in Fig. 2b). The northern bulge is circular in shape, 14×16km in diameter, and stands ~160m from the surrounding flat mare unit (Fig. 3c, d). The southern bulge is irregular in shape, and shows coalescing outline patterns (Pasckert et al., 2018), with a measured size of 40×20km; it stands nearly ~350m from the surrounding mare (Fig. 3a, b). No summit endogenetic pits have been observed to be associated with these bulges. In addition, these bulges are characterized by comparable elevations with the relatively higher portion of the western mare units (Fig. 2b). These morphologic and topographic observations suggest, these bulges, along with the western mare regions, may represent the latest phase of volcanic lava infilling activities within Von Kármán crater floor.

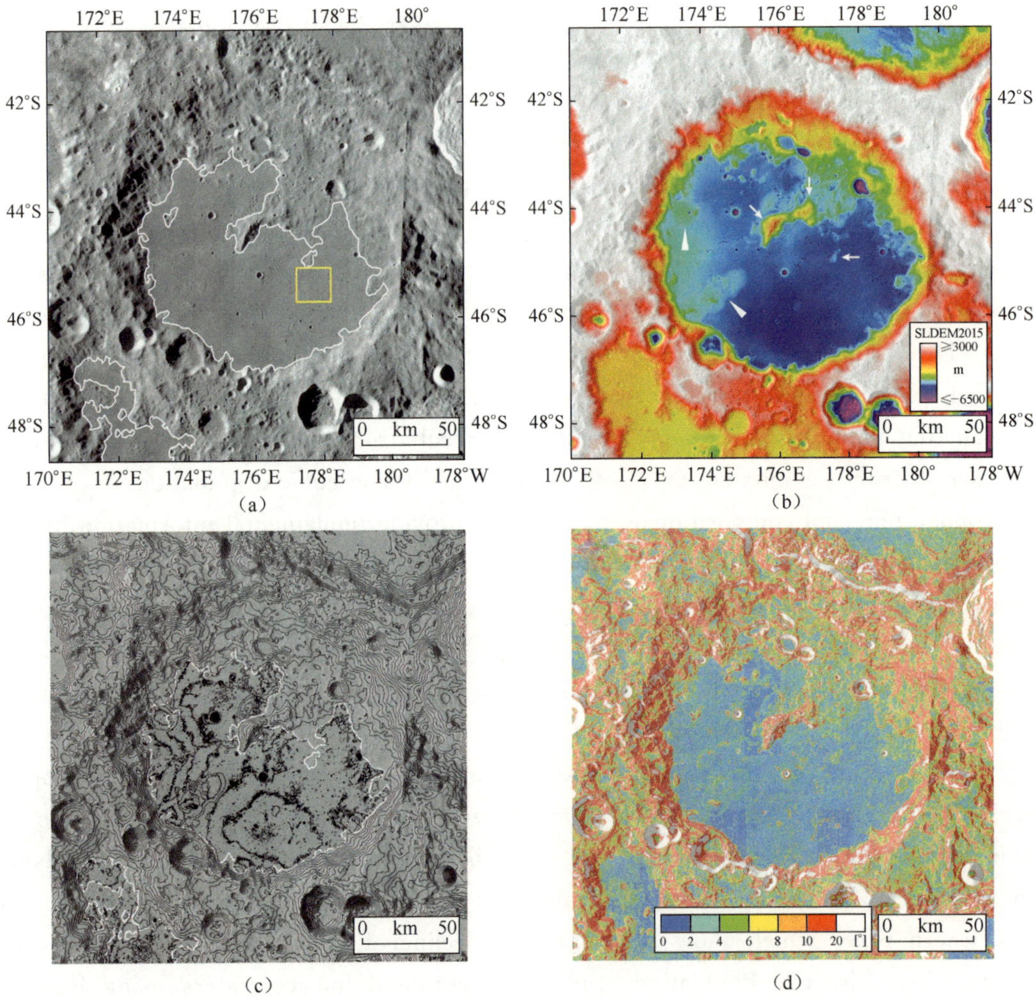

Fig. 2　Von Kármán crater: (a) LROC WAC image, the yellow square marks the studied 20×20km landing area (Section 3) and the white outlines delineate the extent of mare basalt deposits, (b) SLDEM2015 (Kaguya/SELENE - TC + LRO - LOLA merged DEM) topography overlaid on WAC image, the white triangles points to the two bulge/dome - like structures (Fig. 3) and the white arrows mark the possible ringed peak - cluster structure, (c) contour map overlaid on WAC image, contour interval is 50m for the mare region (outlined by white polygons) and 300m for non - mare region, (d) SLDEM2015 - derived slope map (baseline ~180m) overlaid on WAC image. All panels are projected into lambert conformal projection, centered at 176°E, 44.5°S, and north is up (For interpretation of the references to colour in this figure legend, the reader is referred to the web version of this article)

The floor terrains within Von Kármán crater are superposed by abundant impact craters with a wide range of diameters, and the northern ejecta deposits and the southern mare deposits show significant impact population discrepancy (Fig. 2a). Five craters > 5km in diameter are counted on the northern ejecta deposits, while all superposed craters on the southern mare is smaller than 5km; this suggest a lengthy period between the deposition of the two surfaces, probably on the scale of billions of years. On the mare deposits, several relatively large (diameters up to ~4.8km), fresh impact craters are identified, which are characterized by prominent ray patterns, ejecta blanket, and sharp rim structures (Fig. 2a).

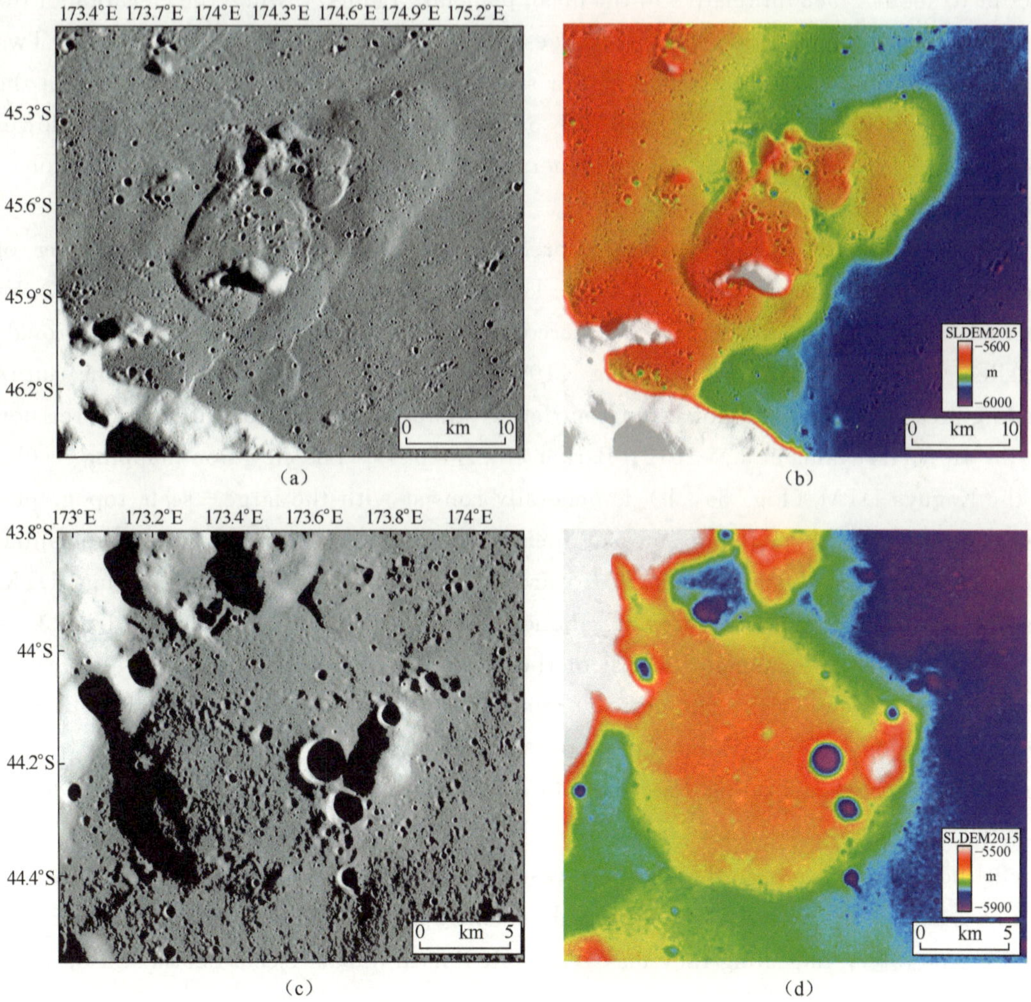

Fig. 3 The two dome - like structures (upper panels: the north one; lower panels: the south one) observed at the western margin of Von Kármán crater floor mare (Fig. 2b): (a, c) Kaguya TC image mosaic map and (b, d) SLDEM 2015 topography overlain on TC mosaic. All panels are projected into lambert conformal projection, centered at 176°E, 44.5°S, and north is up.

3.2 General topography of the landing area

General topography analysis of the landing area can provide important information for the Chang'e-4 mission planning and scientific investigations. We first employ the original LOLA laser range (and inferred elevation) measurement spots (also known as "shots") of lunar surface to characterize the general topography of the landing area, as LOLA shots provide the highest altimetric precision (~10cm; Smith et al., 2010). During years of operations, LOLA has accumulated over 70,000 measurements for the landing area (Fig. 4a). The total topographic relief of the 20×20km landing area is ~120m (from about −6000m to about −5880m relative to the mean planetary radius), while the elevation of the vast majority of the surface (99.3%) ranges from −5960m to −5900m (Fig. 4b). Two elliptic areas, one in the west edge with a size of about 9×6km and another one in the southwest corner with a size of about 7×3.5km, are characterized by relatively lower elevations. These two elliptic areas are generally ~10−20m lower than the rest portion of the studied landing area.

Despite the remarkable altimetric precision of LOLA shots, their sampling of topography of the landing area (a middle-latitude region) is still sparse (Fig. 4a). We alternatively refer to the Kaguya TC stereogrammetry DTM (Fig. 5), which provide excellent surface topography sampling (10m/pixel) and spatial coverage (the entire landing area and ~99% of global lunar surface), though less accurate altimetric precision (~3−4m; Haruyama et al., 2012; Barker et al., 2014). The elevation mapping results of the Kaguya DTM (Fig. 5a, b) is generally consist with the large-scale topography revealed by the LOLA shot measurements (Fig. 4a), while provide substantial topographic details of small geologic features. on the hight-resolution Kaguga DTM maβ, the most prominent features are abundant depressions with size ranging from 100s of meters to over one kilometer. Most of these relatively small depressions are generally circular in map view, which should be normal lunar impact craters. The several relatively large depressions (>~1km) are, however, dominated by irregular elongate and furrowed patterns, indicating they are more likely grooves sculptured by ejecta debris from distance basins/craters (Section 3.3; Huang et al., 2018). In addition, pervasive parallel ridges are observed in the south and east parts of the studied area, where Chang'e-4/Yutu-2 landed on. These ridges extend dominantly in the NE-SW direction, radial to the Finsen crater, indicating they may represent Finsen impact ejecta debris and deposits (Section 3.3; Huang et al., 2018).

Kaguya TC DTM-derived topographic slope map, with a baseline of 30m, shows the entire landing area is very flat, with 90% of surface having slopes<5° (Fig. 5c, d), which would definitely facilitate surface exploration efforts, including rover navigability and mobility, and surface operations and experiments. Local elevated slopes (>5°) mainly occurs at the inner walls of relatively larger (>~200m in diameter) and well-

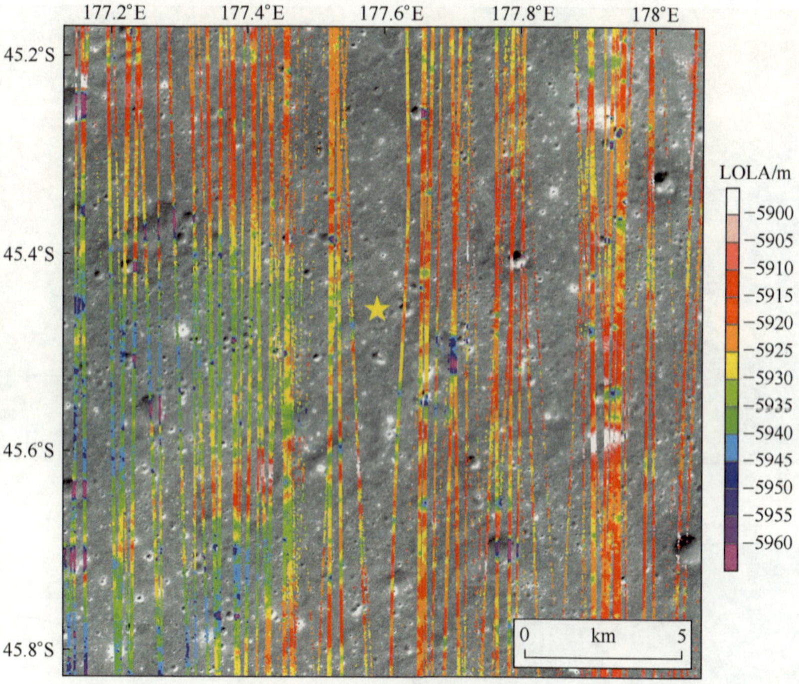

(a)

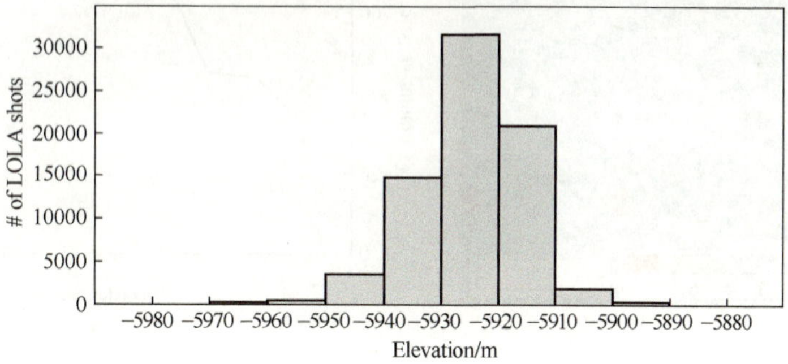

(b)

Fig. 4 (a) LOLA RDR (Reduced Data Record) shot elevation measurements of Chang'e - 4 landing area, overlain on Kaguya TC image mosaic, and (b) their histogram. The yellow star in panel A denotes the Chang'e - 4 landing site. All the maps for the studied landing area in this paper are projected into a lambert conformal projection, centered at 177.6°E, 44.5°S, and north is up (For interpretation of the references to colour in this figure legend, the reader is referred to the web version of this article)

preserved impact craters. NE - SW - trending ridges, while as prominent features on the topography maps (Fig. 5a, b), are indistinguishable on the slope maps, which may be ascribed to the less altimetric precision and longer baseline of the slope data.

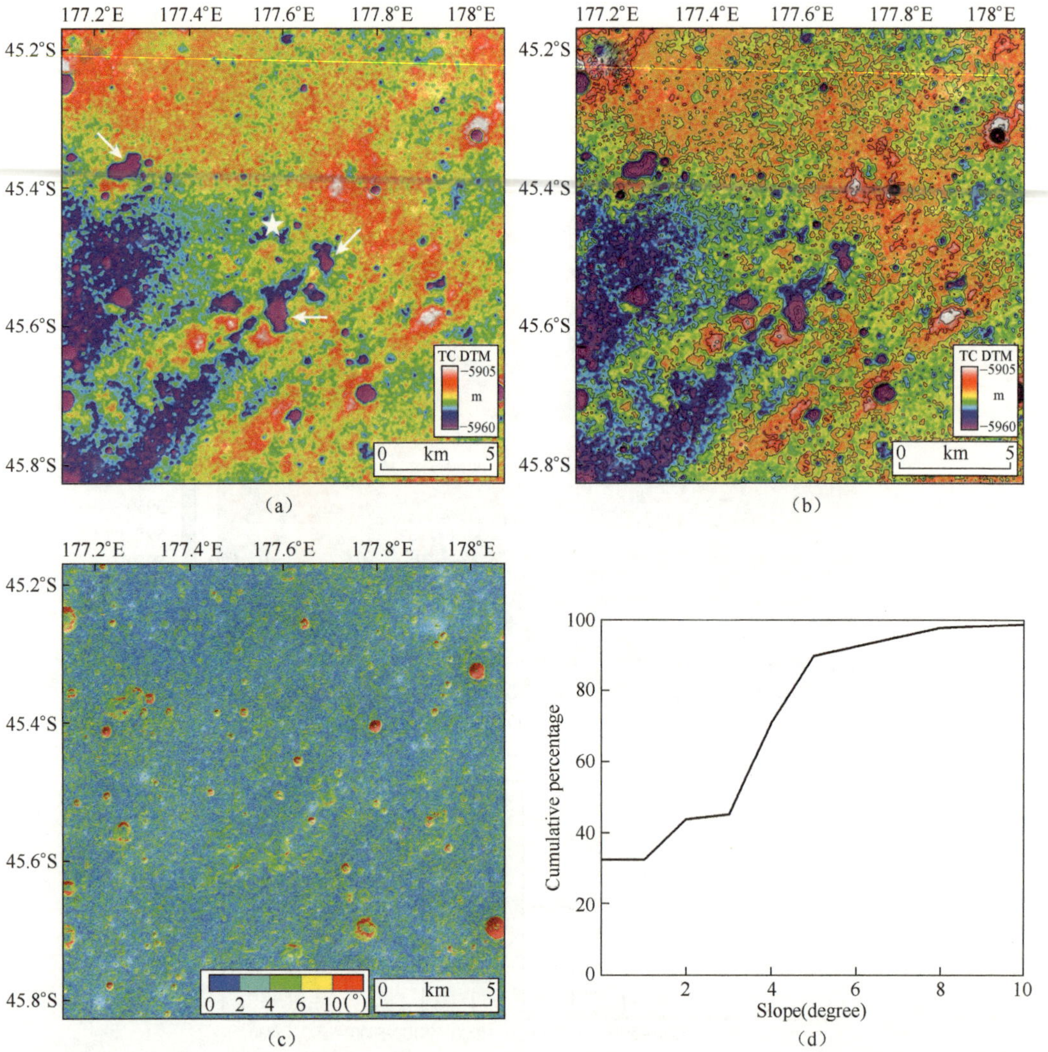

Fig. 5　Kaguya TC topography maps of the Chang'e - 4 landing area: (a) TC DTM topography (10m/pixel) overlain on TC image mosaic, (b) 10m - interval contours overlain on topography, (c) TC DTM - derived slope map overlain on TC image mosaic and (d) the cumulative percentage distribution of slopes with 1° interval. In panel A, the white star denotes the Chang'e - 4 landing site and the white arrows point to the relatively large, elongate depressions

3.3　Morphology of the landing area

Morphology analysis can identify various resultant deposits of many geological processes, thus providing us an insight into the regional geological evolution history of the landing area. In the studied landing area, the major landforms are very flat mare lava plain deposits (Figs. 5c and 6), typical for lava - filling impact basins/craters on the lunar nearside. Similar with mare plain elsewhere on the Moon, lava flow fronts and eruption vents/fractures are not observed within the landing area, suggesting the regional lava-filling are characterized by relatively low effusion rate, cooling-limited lava flows (Head and Wilson, 2017; Qiao et al., 2016a, 2016b), and/or the initially emplaced lava flow morphology has been largely degraded. In addition, other endmember volcanic features, like volcanic cones, pyroclastic deposits, mare domes, sinuous rilles, are not identified in the studied area, showing the regional volcanic eruptions are relatively less complicated than other volcanic concentrations on the Moon. Examination of submeter LROC NAC images (up to ~0.60m/pixel) show the landing area surface is highly deficient of surface boulder exposures, verified by surface images returned from the Chang'e-4/Yutu-2 (e. g., panoramic image obtained by Terrain Camera onboard the lander: http://www. cnsa. gov. cn/n6759533/c6805143/content. html). This contrasts with the Chang'e-3/Yutu landing area, where the regional mare surface is dotted with many small rock fragments and abundant boulders occurring within and surrounding several craters (Basilevsky et al., 2013; Qiao et al., 2016b, 2016c). The observed surface blockiness discrepancy can be attributed to the much older surface model age of the Chang'e-4 landing area mare (over 3 Ga; the Chang'e-3 mare can be as young as ~2 Ga (Morota et al., 2011)) and (the resultant) thicker surface regolith layer.

Circular craters (likely mainly primary impact craters) are the dominant morphologic features superposed on the landing area mare surface (Fig. 6). The largest circular impact crater occurs at the southeastern edge of the studied landing area (Fig. 6) and has a diameter of ~1.1km and a rim-to-floor depth of ~150m. These superposed craters are characterized by a wide range of preservation status, ranging from fresh carters with sharp rim structures (Fig. 7a) and highly degraded craters nearly beyond recognition (Fig. 7b). At the moderately-sloped (~5°) walls of these degraded crater, elephant hidelike textures are frequently observed (arrow pointed out in Fig. 7b); these are generally interpreted as indicator of regolith creep motion (e. g., Melosh, 2011). Circular impact craters are generally deficient of boulders in crater interior and ejecta blanket (Fig. 7b). A few very fresh craters have developed extensive bright ray pattern, ejecta and scattered boulders (Fig. 7a), which represent the least mature materials in the studied landing area. However, these fresh craters are still much less blocky than blocky craters in other landing sites, like the Cone crater (Apollo 14) and Zi Wei crater (Chang'e-3).

Another group of prominent landforms in the landing area is the suite of features related to secondary impact cratering, including elongate, chains or clusters of small craters radial to the source primary crater (Fig. 7c, d), and individual irregular craters with "herringbone" shapes (Fig. 6). These secondaries also show a wide range of preservation states; however, they generally lack ray patterns and blocky ejecta (even for those with the best preservation status), indicating these secondary craters were formed by much less energetic events than primary impact cratering. Their distinctive morphological patterns provide instructive insights for tracking their source region. Recent

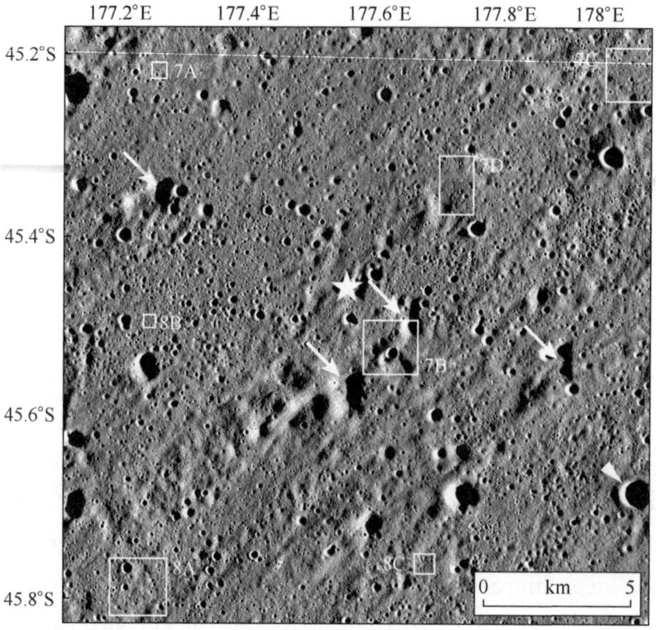

Fig. 6 Kaguya TC mosaic map (10m/pixel) of the studied 20×20km landing area. The white star denotes the Chang'e - 4 landing site, the white triangle points to the largest circular impact crater in the landing area, white arrows point to the relatively large, north - south trending secondary craters and the white boxes show the locations of Figs. 7 and 8

work by Huang et al. (2018) present remarkable morphological characterization and origin analyses of these secondary craters. In particular, they traced the relatively large, north - south trending secondaries (pointed out in Fig. 6) to the Antoniadi crater in the over 700km south. However, detailed mapping of Antoniadi secondaries showed Antoniadi was formed by an oblique impact with the projectile coming from the northwest, and the Von Kármán crater lies in distant up - range direction, where secondary craters should be much less abundant (Sruthi and Kumar, 2014). We suggest other relatively large and young craters along the north - south trend can be additional source crater candidates, including Boyle (57km), Abbe (64km), Abbe H (25km) and Minnaert (137km) (Fig. A1 in Appendix B).

A prominent set of linear parallel ridged (Huang et al., 2018) and furrowed textures is observed on high - resolution Kaguya (Fig. 6) and LRO images. This lineament set consists of two major components: broad parallel ridges ~0.3 to 1.2km wide and up to ~20—40km in length (extending beyond the studied landing area) and a series of smaller parallel grooves hundreds of meters wide and up to ~5km long. This distinctive texture is prevalently arranged in the NE - SW direction and radial to the Finsen crater, indicating they are genetically related with the Finsen impact. Similar radial textures, though much larger, has been observed and documented around a number of lunar impact basins including Imbrium and Orientale (Hartmann, 1963, 1964; Head, 1976; Schultz and

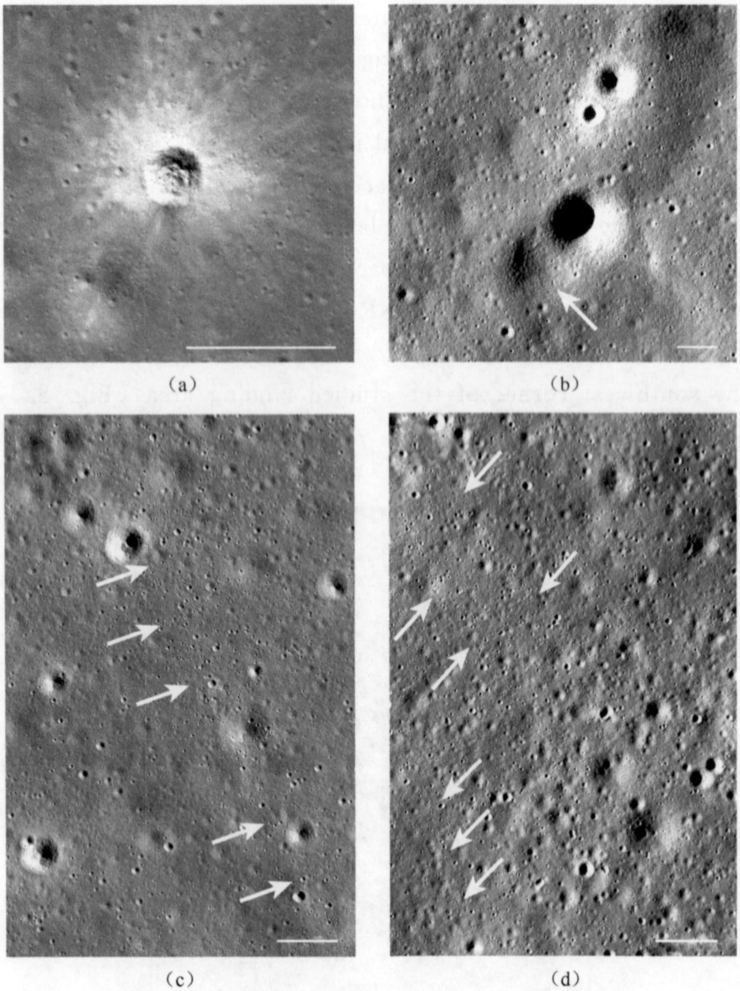

Fig. 7　(a) One of the regional freshest circular impact craters, ~80m in diameter, at the northeast corner of the landing area, LROC NAC frame M189435642L, 0.62m/pixel, 54.3° incidence angle. (b) Local surface ~1.5km southeast from the landing site, shows several relatively large craters (~100—400m diameter), with different preservation status, LROC NAC frames M134022629L/R, 1.27m/pixel, 77.6° incidence angle. The arrow points to elephant hide-like textures at the walls of a degraded crater. (c, d) Second crater chains and clusters (marked by aligned arrow sets) in the northeast of the landing area, LROC NAC frame M161142858L (1.16m pixel size and 71.6° incidence angle) for panel c, and M134022629R (1.27m pixel size and 77.6° incidence angle) for panel d. The scales are all 200m

Crawford, 2016). Given the relatively smaller size of the Finsen crater (compared with giant impact basins), these radial textures are more likely to be sedimentary in origin due to the erosional and depositional effects by ballistic ejecta from the Finsen craterforming impact (Head, 1976). Kaguya TC DTM topography (Fig. 5a) measurements shows the heights of these ridges can be up to ~6 m, consistent with estimated Finsen crater ejecta thickness at the studied landing area (Huang et al., 2018). These lineament features, especially the deposited ridges, should contain a considerable portion of distant foreign materials from the Finsen crater, and Chang'e - 4/Yutu - 2 just landed on this Finsen ejecta - bearing area. This may provide additional scientific chances for Chang'e - 4/Yutu - 2 surface in - situ geological and compositional measurements, given the landing area mare lava plain is relatively homogeneous in surface characteristics and materials of the SPA - basin floor materials have been buried; the latter host the compositional information and other important characteristics of SPA impact basin, which is one of the key scientific questions of lunar and planetary science (NRC, 2007, 2011). Parallel linear grooves in NW - SE trend, while much smaller (10 — 40m wide and 100 — 600m long) are also observed at the southwest corner of the studied landing area (Fig. 8a), indicating a genetic relation to craters other than Finsen, like the 82 - km - diameter Alder crater in the southeast vicinity.

(a)

(b)

(c)

Fig. 8 (a) NW - SE trending linear grooves in the southwest corner of the landing area, LROC NAC frame M134022629R, 1.27m/pixel, 77.6° incidence angle. Sinuous troughs at the (b) west margin (LROC NAC M161142858L) and (c) south margin (LROC NAC M134022629R) of the landing site. The scales are all 200m

We also find a number of small sinuous troughs, generally 20—40m wide and 200—350m long, in the lava plain of the studied landing area (Fig. 8b, c). Abundant similar sinuous ridges have also been documented in the vicinity of a fresh ～3.8 - km - diamter crater ～20km west of the studied landing area (Huang et al., 2018). While the spatial occurrence density of these sinuous troughs is much less sparse in the landing area, indicating a potential genetic relation to impact events, like impact - induced seismic disturbance of local surface and shallow subsurface substrates, which awaits on - going geophysical investigations by the Yutu - 2 ground penetrating radars.

3.4 Chemical composition of the landing area

Kaguya MI - derived elemental maps show the studied landing area is fairy homogeneous in surface iron and titanium contents, especially for titanium content (Fig. 9). The majority (> 80%) of the landing area has calculated FeO content between 11 and 13 wt% and > 95% of the regional surface has TiO_2 content between 1 and 2 wt%. This estimated titanium content is comparable with Apollo 15 pigeonite basalts (Papike et al., 1976) and reveals the landing area mare deposits are dominated by low - titanium basalts (Giguere et al., 2000).

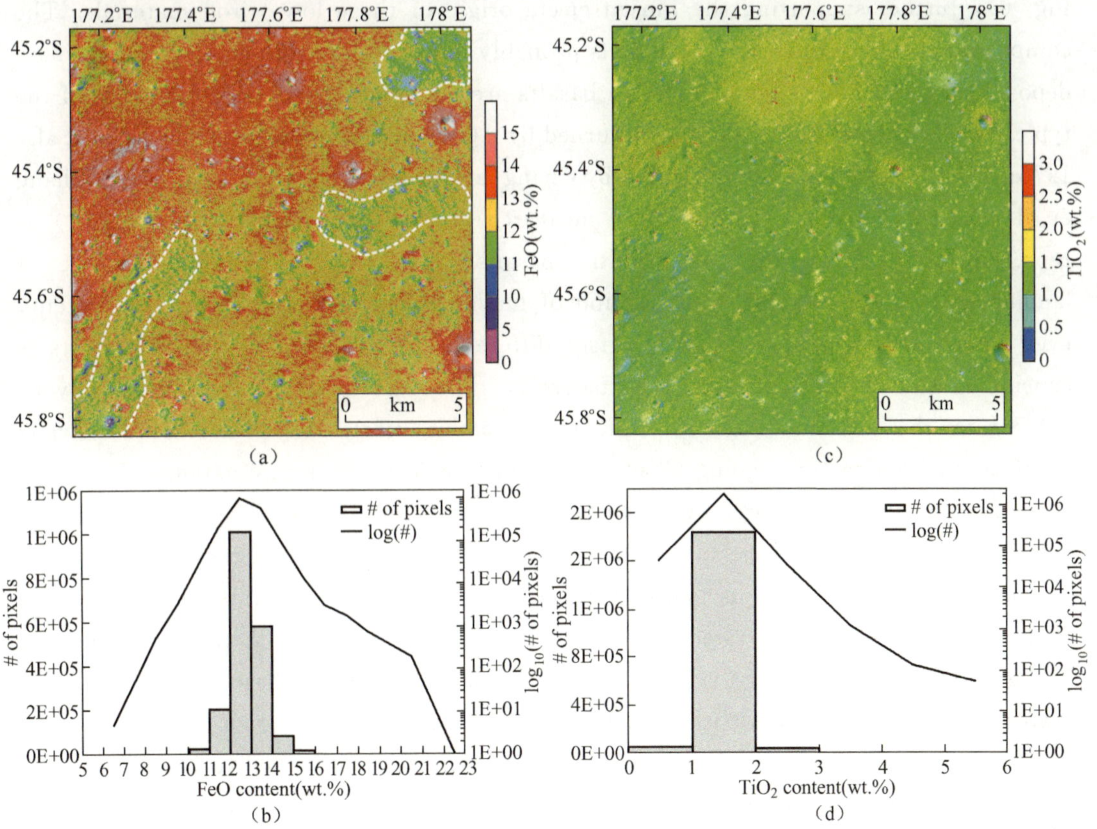

Fig. 9 FeO (a) and TiO_2 (c) abundance maps of the studied landing area (overlain on Kaguya TC image mosaic) and their histograms (b and d), calculated from Kaguya MI mosaic using the Lemelin et al. (2015) and Otake et al. (2012) algorithms, respectively. The dashed patches in panel A marks the strip - shape areas with decreased iron contents

High - resolution iron mapping by Kaguya MI spectrometer image also reveals considerable geochemical variations across the studied landing area (Fig. 9a, b) . Firstly, the studied landing area surface can be divided into two major units with slightly different iron contents: a northern belt region with relatively elevated FeO abundance (~13—14 wt%, mean value of 13.3 wt% with 1δ=0.7) and a broader region with slightly lower FeO abundance (~12—13 wt%, mean value of 12.7 wt% with 1δ=0.7). A null hypothesis test is performed to show the mean FeO abundance of the Finsen ejecta unit is statistically different from that of the north mare unit (Appendix A and Table A1). Cross - check with iron mapping at a much large extent covering the entire Von Kármán crater and its surroundings (Ling et al., 2019) shows the south low - iron unit, on which Chang'e - 4 landed, belongs to a much more extensive region radial to the Finsen crater, suggesting the lowiron signature may represent compositional characteristics of Finsen ejecta materials or a mixture of Finsen ejecta and local mare basalts. Geomorphological investigation on Kaguya TC images also show the south portion are superposed with many linear parallel ridged and furrowed textures radial to the Finsen crater (Section 3.3 and Fig. 6), further supporting the Finsen ejecta origin of these low - iron materials. The composition of the northern belt region is probably more close to that of landing area mare deposits, which shows the landing area basalts are more iron - depleted than most of the typical kinds of lunar basalt samples returned by Apollo and Luna missions (Papike et al., 1976; Fig. 11) and in - situ measured by Chang'e - 3 mission (Ling et al., 2015; Zhang et al., 2015; Fig. 11). If the MI - measured composition (mostly of regolith) is representative of that of the local primitive basalts, the mare basalts at the Chang'e - 4 landing area may represent a unique kind of lunar basalts with decreased iron content, even when the typical 2 wt% FeO offset difference between lunar soil and basalts is considered (Gillis et al., 2004). Alternatively, this observed low - iron signal may be subject to the contamination of distant feldspathic ejecta materials, which awaits clarification from the on - going Chang'e - 4 surface *in - situ* investigations. Secondly, tens of regionally larger craters (~330m - 1.1km diameter) are observed to have slightly iron - rich interiors and ejecta blankets (> 14 wt% FeO), suggesting a potential compositional inhomogeneous subsurface basaltic layer. These crater ejecta is also characterized by slightly elevated titanium content (~2—3 wt%) than the surrounding background mare plain (Fig. 9c). From the estimated depths of these craters (0.084× crater diameter; Melosh, 1989), the depth of the underlying iron - rich basalts, also the thickness upper limit of the surface lava deposits, can be constrained to be ~30—90m. Moreover, we also identified several strip - shape regions dotted by abundant small pits with decreased iron content (10—12 wt% FeO; Fig. 9a). We interpreted these low - iron characteristics in these strip - shape regions as ejectas from other distant craters, with lower iron content than Finsen crater ejecta.

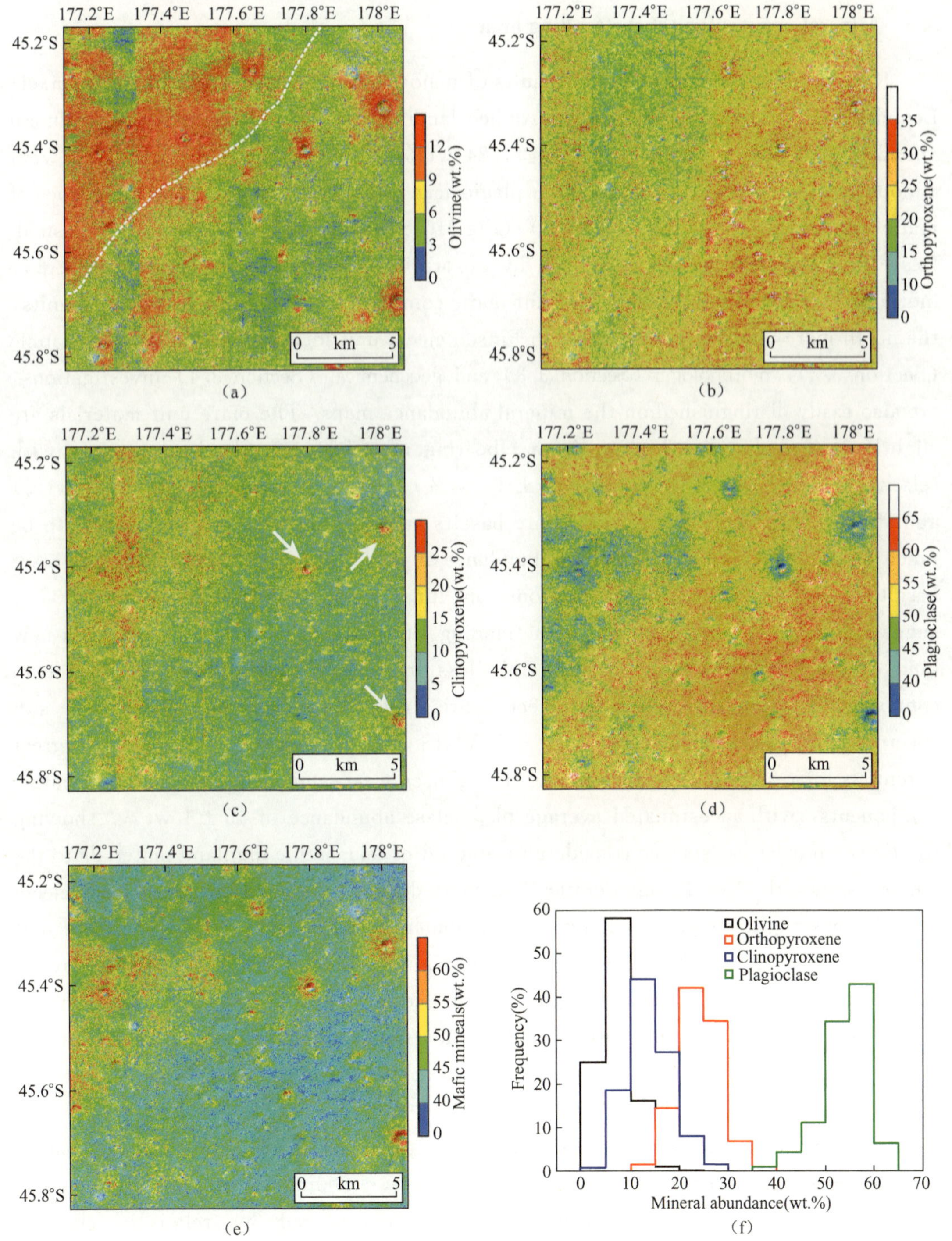

Fig. 10　Kaguya MI－retrieved abundance mapping results (Lemelin et al., 2019) of major rock－forming minerals of the studied landing area (overlain on Kaguya TC image mosaic): (a) olivine, (b) orthopyroxene, (c) clinopyroxene, (d) plagioclase, (e) mafic minerals, and (f) their histogram in percentage frequency in 5 wt% interval. The arrows in panel C point to the several craters with elevated clinopyroxene abundance. The dashed line in panel a delineates the boundary of the north mare plain unit and south Finsen ejecta unit (Section 3.4 and Fig. 9a).

3.5 Mineral composition of the landing area

Kaguya MI - retrieved mapping results of major rock - forming minerals (60m/pixel; Lemelin et al., 2019) reveal that the studied landing area has average estimated mineral abundances of 7 wt% olivine (1δ=3%), 24 wt% orthopyroxene (1δ=4%), 14 wt% clinopyroxene (1δ=4%), and 54 wt% plagioclase (1δ=5%), with a total abundance of mafic component of 46 wt% (1δ=5%) (Fig. 10). Kaguya MI spectral profiles of small fresh impact craters are characterized by absorptions at the 950nm band (Fig. A2), indicating orthopyroxene as the dominant mafic component. The two major surface units, the north mare plain unit and the south Finsen ejecta unit identified based on topographic (Section 3.2), morphologic (Section 3.3) and geochemical (Section 3.4) investigations, are also easily distinguished on the mineral abundance maps. The mare unit materials are slightly more mafic than Finsen ejecta - bearing materials near the landing site, with relatively higher average olivine (9 ± 3 wt%) and clinopyroxene (16 ± 4 wt%) abundances (Fig. 10a, c, e). The mare basalts at the studied landing area seems to be much less mafic (total mafic mineral abundance of 48±5 wt%) than returned lunar mare basaltic samples from several locations on lunar nearside (Lucey et al., 2006), suggesting mare basalts emplaced at the landing site on lunar farside may present a new kind of mare basalts. Alternatively, this less mafic composition may be subject to contamination from the feldspathic eject debris from Finsen crater, surface in - situ spectroscopic measurements by Yutu - 2 VNIS at the specimen - scale show great potentials for resolving this uncertainty. The Finsen ejecta unit contains more feldspathic components, with an estimated average plagioclase abundance of 45±4 wt%, showing the Finsen impact has spewed considerable amount of feldspathic highland materials to the landing site on the Von Kármán crater floor mare deposits. In addition, the Finsen ejecta unit also has slightly higher orthopyroxene abundance (25±4 wt%) than the mare unit (23±4 wt%) (A null hypothesis test is performed to show the difference between the mean orthopyroxene abundance of the Finsen ejecta unit and thenorth mare unit is statistically significant (Appendix A and Table A1)), consistent with the recent M^3 spectroscopic identification of low - calcium pyroxene - bearing materials ejected from the Finsen crater (Huang et al., 2018).

Regional large craters also exhibit mineral compositional anomalies, indicating vertical mineralogy variations. Several large craters (~600m to ~1.1km in diameter) superposed on south Finsen ejecta unit are characterized by relatively elevated clinopyroxene abundance, but comparable with that of the north mare unit (pointed out in Fig. 10c), suggesting these craters have penetrated the overlying Finsen ejecta and exposed the underlying more mafic basaltic materials, consistent with the identification of high - calcium pyroxene - bearing components through M^3 spectroscopic analysis (Huang et al., 2018). In addition, the iron - rich craters (Section 3.4) also exhibit higher olivine and total mafic component abundance, suggesting the underlying (up to ~30—90m deep) iron-rich basalts are also more olivine-rich.

4 Discussion and references for surface exploration planning and *in-situ* measurement interpretations

Our detailed geological characterization of the very local landing area presented above provide substantial context and references for Chang'e-4/Yutu-2 surface exploration mission planning and subsequent in-situ measurement interpretations. Mare basaltic deposits at the landing area appear to much less mafic than previously collected and in-situ measured lunar basalt samples (Papike et al., 1976; Lucey et al., 2006; Ling et al., 2015), with an estimated average FeO content of ~13 wt% and a total mafic mineral abundance of ~48 wt% (Sections 3.4, 3.5 and Fig. 11). In addition, a possible underlying basaltic layer at a depth up to ~30—90m, characterized by slightly higher iron and titanium contents, olivine and total mafic mineral abundances, is identified, but it is still much less mafic than typical returned lunar basalts from lunar nearside. These unique less mafic signatures suggest the Chang'e-4 landing area basalts may represent a new type (or more new types) of mare basalts (and/or be contaminated by low-iron feldspathic ejecta materials) and surface in-situ spectroscopic investigations by Yutu-2 VNIS will provide an extraordinary opportunity to characterize lunar farside basalts and volcanism at a much finer scale which has been hardly achieved by previous and current orbital investigations.

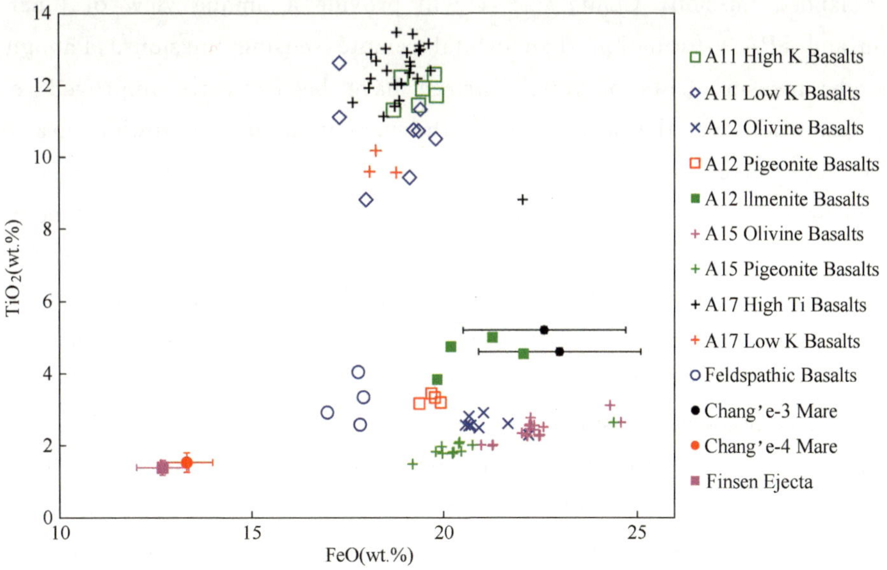

Fig. 11　Comparison of iron and titanium contents for mare basalts and Finsen crater ejecta materials at Chang'e-4 landing site (average values±1δ) with all typical kinds of lunar basalts (Papike et al., 1976) and Chang'e-3 surface *in-situ* measurements (Ling et al., 2015)

The landing area mare plain is heavily superimposed by ejecta deposits dominantly from the Copernian - aged 73 - km - diameter Finsen crater to the northeast and the Chang'e - 4 just landed on this Finsen ejecta - bearing unit. The Finsen ejecta - bearing unit is characterized by a more feldspathic composition (higher plagioclase abundance and lower iron, olivine and clinopyroxene contents), but elevated orthopyroxene abundance. Being located at the central portion of the enormous SPA basin, Finsen impact - exposed materials would inevitably bring compositional and geological information of parts of the primitive SPA basin floor materials. This provides a remarkable occasion for Chang'e - 4 surface geological and compositional measurements, given the SPA materials have been largely buried by subsequent mare lava flows at the landing area. Previous spectroscopic and geologic studies have presented substantial evidences that SPA materials contain more mafic components than the highland feldspathic crust, with the pervasive occurrences of low - Ca pyroxenes (e. g., Tompkins and Pieters, 1999; Pieters et al., 2001; Cahill et al., 2009; Hurwitz and Kring, 2015; Moriarty III and Pieters, 2018). However, the specific origin of these SPA materials and their implications for a number of key questions regarding lunar deep interior composition, multiring basin formation mechanism and earlier lunar evolution and bombardment history are highly debated and includes scenarios of differentiated impact melt sheet (e. g., Vaughan and Head, 2014; Hurwitz and Kring, 2015), exposures of lunar lower crust and/or upper mantle materials (e. g., Pieters et al., 2001; Ohtake et al., 2014) or Mg suite intrusions (e. g., Klima et al., 2011). Yutu - 2 surface in - situ spectroscopic measurements will provide the first "ground truth" for assessing these hypotheses.

As a landed mission, Chang' e - 4 will provide a unique view of lunar farside volcanism and SPA - formation than orbital remote sensing missions. Though orbital spectroscopic measurements of lunar surface have been greatly improved in spatial sampling resolution over the past decades, the current highest resolution measurements (down to ~20m pixel size provided by Kaguya MI) are still subject to spectral mixing effect. Our orbital data - based geological study also shows the Chang'e - 4 landing area is characterized by complicated mixing of local mare basalts, distant ejected feldspathic materials and other possible ejecta materials. Combined analyses with surface images obtained by a set of cameras and VNIS spectroscopic measurements at the specimen - scale would permit us to pinpoint the detailed origin and characteristics of specific targets, especially surface boulder exposures.

In summary, we suggest Finsen ejecta materials occurring at the parallel linear ridged and furrowed patterns (should be easily recognized in Chang' e - 4/Yutu - 2 surface images, e. g., Chang' e - 4 lander panoramic image: http://www. cnsa. gov. cn/n6759533/c6805143/content. html), and mare basalts exposed in the vicinity of craters of variable diameters as prioritized targets for surface in - situ geological and spectroscopic measurements, as their geological context and origin can be better understood and contain important information regarding to lunar farside volcanism and SPA basin formation. Surface small boulders exposed at the landing region surface are also important measuring objects, as they are subject to the least degree of mixing effect and preserve the primitive compositional information. Interpretations of Chang'e - 4/Yutu - 2 regolith measurements should consider the mixing effect of local mare basalts and distant feldspathic materials.

5 Conclusions

On the basis of detailed geological characterization of the Chang'e - 4 landing area presented above, we draw the following conclusions:

(1) The Chang'e - 4/Yutu - 2 landed on the floor of the pre - Nectarian aged Von Kármán crater, in the northwest central portion of the SPA basin of lunar farside, the largest and oldest identified impact structure on the Moon. The Von Kármán crater is enclosed by several regionally large craters, some of which postdate the Von Kármán crater and have spewed ejecta materials to Von Kármán crater floor.

(2) Locally, the Chang'e - 4 landing site is located on the mare basalt plain emplaced at the south floor of Von Kármán crater. The local landing area (20×20km) is very flat at the baseline of 30m, with 90% of surface having slope<5°, and deficient of surface boulder exposures, which should facilitate Yutu - 2 surface roving and experiments.

(3) The major morphologic features superposed on the landing area mare are sub - kilometer circular impact craters, which generally lack boulders in crater interior and surrounding ejecta. Secondary craters, occurring in chains or clusters, are also pervasive in the landing area.

(4) The mare basalts at the landing area is dominated by low - titanium basalts, and appear to be less mafic and more iron - depleted than most of the typical kinds of returned lunar basalt samples, suggesting the landing site mare basalt may represent a new type of mare basalts, or are subject to considerable level of contamination by distant low - iron materials. A possible underlying mare layer at a depth up to ~30—90m, with slightly higher abundances of iron element and mafic minerals, is identified. Surface in - situ spectroscopic measurements by Yutu - 2 VNIS will provide the first "ground truth" of lunar farside volcanism.

(5) The Chang'e - 4 landing site mare is heavily sculptured by a set of linear parallel ridged and furrowed textures formed by ballistic ejecta from the Copernian - aged Finsen crater. These Finsen ejecta materials have a more feldspathic composition, with elevated orthopyroxene abundance, and may represent compositional properties of the primitive SPA basin floor materials. Yutu - 2 surface in - situ spectroscopic measurements of these materials would provide key information of SPA basin formation and lunar deep interior.

(6) We propose the Finsen ejecta and mare basalt outcrops present at the landing area as prioritized targets for Chang'e - 4/Yutu - 2 surface geological and spectroscopic

explorations. Follow - on mission data interpretations should consider the mixing effect of local mare basalts and distant feldspathic materials.

Acknowledgments

We sincerely appreciate the Editor - in - Chief Rosaly Lopes for her very efficient and helpful editorial handling, and the reviewers for evaluating this work. This work is partly supported by the Open Fund of State Key Laboratory of Remote Sensing Science (Grant Nos. OFSLRSS201807 and OFSLRSS201802), National Natural Science Foundation of China (No. 41703063), China Postdoctoral Science Foundation (Nos. 2018T110682 and 2017M610421) and Natural Science Foundation of Shandong Province (JQ201511 and ZR2019MD015). All original LRO data can be found in the NASA's Planetary Data System (http://pds. nasa. gov/), the original Kaguya/ SELENE data are archived at SELENE Data Archive (https://darts. isas. jaxa. jp/planet/pdap/selene/) and the Kaguya Multiband Imager mineral maps are available from the United States Geological Survey at the Planetary Data System's Cartography and Imaging Sciences Node Annex (http://astrogeology. usgs. gov/pds/annex).

Appendix A. Null hypothesis test

Independent samples t - test is performed to assess whether the MI - retrieved mean FeO content and orthopyroxene abundance of the Finsen ejecta units at the Chang'e - 4 landing site are statistically different from those of the north mare unit (Sections 3. 4, 3. 5 and Figs. 9a, 10b in the main text). The null hypothesis is that the population means of the Finsen ejecta unit and the mare unit are equal. If the t value of our data is greater than t distribution critical value at a given confidence level (i. e., 95%), then the null hypothesis is rejected. The null hypothesis test results are presented in Table A1, which shows the calculated t values are all greater than the t distribution (two - tailed) critical values at 95% confidence level. Thus, the null hypothesis can be rejected and the FeO content and orthopyroxene abundance differences between the Finsen ejecta unit and north mare unit are statistically significant.

Table A1 Results of null hypothesis tests for assessing whether the Kaguya MI - retrieved mean FeO content and orthopyroxene abundance differences between Finsen ejecta and the north mare unit are statistically significant. N is the number of MI FeO or orthopyroxene abundance map pixels in each analyzed unit, df is degree of freedom, t is t value of our data, and t_c is t distribution (two - tailed) critical value at the 95% confidence level

Composition	Mare			Finsen ejecta			df	t	t_c
	N	Mean value	1σ	N	Mean value	1σ			
FeO (%)	553, 918	13.3	0.7	898, 368	12.7	0.7	1, 452, 284	541, 79	1.96
Orthopyroxene (%)	49, 284	23	4	80, 205	25	4	129, 487	120.47	1.96

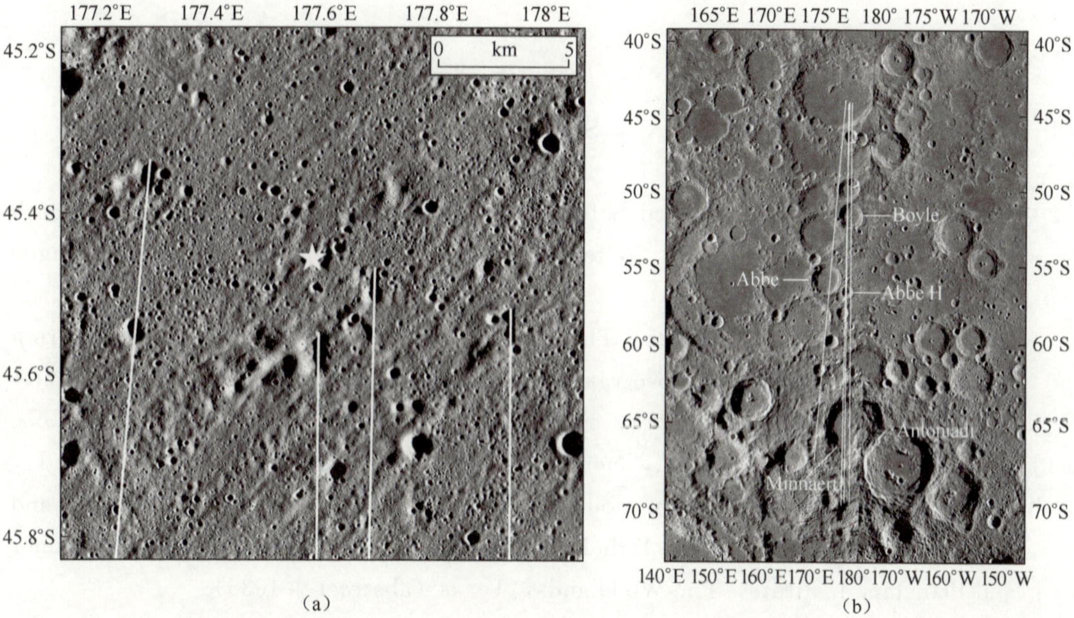

Fig. A1　(a) Local relatively large north－south trending secondary craters observed at the studied 20×20km landing area, the white star denotes the Chang'e－4 landing site and the white lines mark trend of these secondaries, which locate at the starting point of the lines. Kaguya TC image mosaic, projected into a lambert conformal projection, centered at the landing site (177.6°E, 44.5°S), and north is up.　(b) Great circles extended from trends of these secondaries. Several large craters along the trend are identified as additional source crater candidates other than Antoniadi, LROC WAC image mosaic in a same projection of panel A

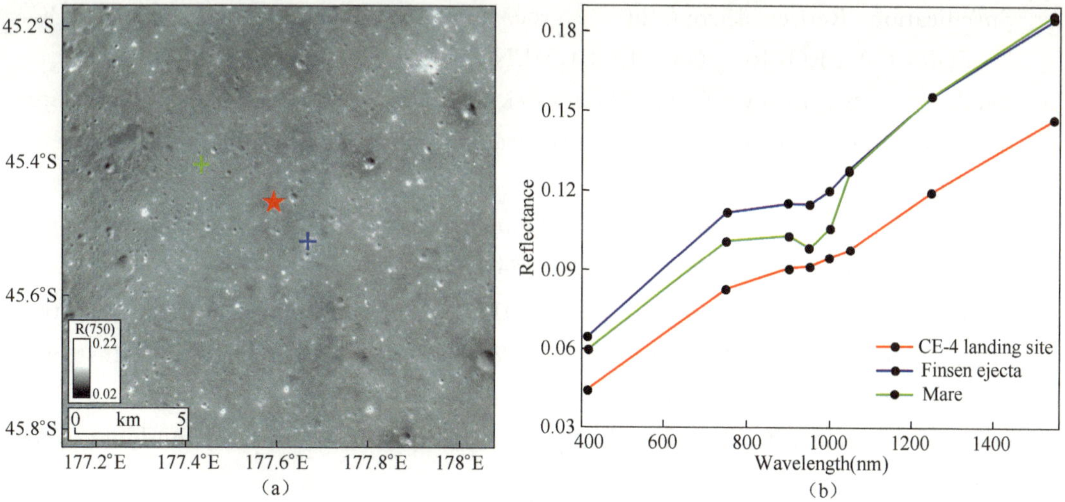

Fig. A2　(a) Kaguya MI 750nm reflectance (R (750)) map and (b) MI spectral profiles of the Chang'e－4 landing site (its location marked by the red star in panel a) and two small fresh craters in the south Finsen ejecta unit (blue cross in panel a) and north mare plain unit (green cross in panel a). The systematic differences between the near－infrared (NIR) band data and ultraviolet－visible (UVVIS) data have been corrected (Lemelin et al., 2015) and the spectral profiles in panel B are given as averages of 4×4 pixel windows to increase the signal to noise ratios

References

[1] Anderson, J. A., Sides, S. C., Soltesz, D. L., Sucharski, T. L., Becker, K. J., 2004.

[2] Modernization of the Integrated Software for Imagers and Spectrometers. In: 35th Lunar and Planetary Science Conference. Lunar and Planetary Institute, League City, Texas (abstract #2039).

[3] Baker, D. M. H., et al., 2011. The transition from complex crater to peak - ring basin on the Moon: new observations from the Lunar Orbiter Laser Altimeter (LOLA) instrument. Icarus 214, 377—393.

[4] Barker, M. K., Mazarico, E., Neumann, G. A., Smith, D. E., Zuber, M. T., 2014. Merging digital elevation models from the Lunar Orbiter Laser Altimeter and Kaguya Terrain Camera. In: 45th Lunar and Planetary Science Conference. Lunar and Planetary Institute, The Woodlands, Texas (abstract #1635).

[5] Barker, M. K., Mazarico, E., Neumann, G. A., Zuber, M. T., Haruyama, J., Smith, D. E., 2016.

[6] A new lunar digital elevation model from the Lunar Orbiter Laser Altimeter and SELENE Terrain Camera. Icarus 273, 346—355.

[7] Basilevsky, A. T., Head, J. W., Horz, F., 2013. Survival times of meter - sized boulders on the surface of the Moon. Planet. Space Sci. 89, 118—126.

[8] Bowman - Cisneros, E., 2010. LROC EDR/CDR Data Product Software Interface Specification. Retrieved from. http://lroc. sese. asu. edu/data/LRO - L - LROC - 2 - EDRV1. 0/LROLRC _ 0001/DOCUMENT/LROCSIS. PDF.

[9] Cahill,J. T. S., Lucey, P. G., Wieczorek, M. A., 2009. Compositional variations of the lunar crust: results from radiative transfer modeling of central peak spectra. J. Geophys. Res. Planets 114, E09001.

[10] Di, K. C., et al., 2019. Chang'e - 4 lander localization based on multi - source data. J. Remote. Sens. 23 (1), 177—184.

[11] Garrick - Bethell, I., Miljković, K., 2018. Age of the lunar South Pole - Aitken basin. 49th Lunar and Planetary Science Conference. Lunar and Planetary Institute, The Woodlands, Texas (abstract #2633).

[12] Garrick - Bethell, I., Zuber, M. T., 2009. Elliptical structure of the lunar South Pole - Aitken basin. Icarus 204, 399—408.

[13] Giguere, T. A., Taylor, G. J., Hawke, B. R., Lucey, P. G., 2000. The titanium contents of lunar mare basalts. Meteorit. Planet. Sci. 35, 193—200.

[14] Gillis, J. J., Jolliff, B. L., Korotev, R. L., 2004. Lunar surface geochemistry: global concentrations of Th, K, and FeO as derived from lunar prospector and Clementine data. Geochim. Cosmochim. Acta 68, 3791—3805.

[15] Grieve, R. A. F., Cintala, M. J., 1992. An analysis of differential impact melt - crater scaling and implications for the terrestrial impact record. Meteoritics 27, 526—538.

[16] Hartmann, W. K., 1963. Radial structures surrounding lunar basins, I: the imbrium system. Commun. Lunar Planet. Lab. 2, 1—16.

[17] Hartmann, W. K., 1964. Radial structures surrounding lunar basins, II: Orientale and other systems; Conclusions. Communications of the Lunar and Planetary Laboratory, vol. 2. pp. 175—191.

[18] Haruyama, J., et al., 2012. Lunar global digital terrain model dataset produced from SELENE (Kaguya) Terrain Camera stereo observations. In: 43rd Lunar and Planetary Science Conference. Lunar and Planetary Institute, The Woodlands, Texas (abstract #1200).

[19] Haruyama, J., et al., 2008. Global lunar - surface mapping experiment using the Lunar Imager/Spectrometer on SELENE. Earth Planets Space 60, 243—255.

[20] Haruyama, J., et al., 2009. Long lived - volcanism on the lunar farside revealed by SELENE Terrain Camera. Science 323, 905—908.

[21] Hawke, B. R., et al., 2004. The origin of lunar crater rays. Icarus 170, 1—16.

[22] Head, J. W., 1976. Evidence for the sedimentary origin of Imbrium sculpture and lunar basin radial texture. Moon Planets 15, 445—462.

[23] Head, J. W., 2010. Transition from complex craters to multi - ringed basins on terrestrial planetary bodies: scale - dependent role of the expanding melt cavity and progressive interaction with the displaced zone. Geophys. Res. Lett. 37.

[24] Head III, J. W., et al., 2010. Global distribution of large lunar craters: implications for resurfacing and impactor populations. Science 329, 1504—1507.

[25] Head, J. W., Wilson, L., 2017. Generation, ascent and eruption of magma on the Moon: new insights into source depths, magma supply, intrusions and effusive/explosive eruptions (part 2: predicted emplacement processes and observations). Icarus 283, 176—223.

[26] Hiesinger, H., et al., 2012. New crater size - frequency distribution measurements of the South Pole - Aitken basin. In: 43rd Lunar and Planetary Science Conference. Lunar and Planetary Institute, The Woodlands, Texas (abstract #2863).

[27] Huang, J., et al., 2018. Geological characteristics of Von Kármán crater, northwestern South Pole - Aitken basin: Chang'E - 4 landing site region. J. Geophys. Res. Planets 123, 1684—1700.

[28] Hurwitz, D., Kring, D. A., 2015. Potential sample sites for South Pole - Aitken basin impact melt within the Schrödinger basin. Earth Planet. Sci. Lett. 427, 31—36.

[29] Jia, Y., Zou, Y., Ping, J., Xue, C., Yan, J., Ning, Y., 2018. The scientific

objectives and payloads of Chang'E－4 mission. Planet. Space Sci. 162，207—215.
[30] Jolliff，B. L.，Shearer，C. K.，Papanastassiou，D. A.，Alkalai，L.，MoonRise，T.，2010.
[31] MoonRise：South Pole－Aitken basin sample return mission for solar system science. In：2010 Annual Meeting of the Lunar Exploration Analysis Group. Lunar and Planetary Institute，Washington，DC (abstract ＃3072).
[32] Kalynn，J.，Johnson，C. L.，Osinski，G. R.，Barnouin，O.，2013. Topographic characterization of lunar complex craters. Geophys. Res. Lett. 40，38—42.
[33] Klima，R. L.，et al.，2011. New insights into lunar petrology：distribution and composition of prominent low－Ca pyroxene exposures as observed by the Moon Mineralogy Mapper (M^3). J. Geophys. Res. Planets 116，E00G06.
[34] Lemelin，M.，Lucey，P. G.，Song，E.，Taylor，G. J.，2015. Lunar central peak mineralogy and iron content using the Kaguya Multiband Imager：reassessment of the compositional structure of the lunar crust. J. Geophys. Res. Planets 120，869—887.
[35] Lemelin，M.，Lucey，P. G.，Miljković，K.，Gaddis，L. R.，Hare，T.，Ohtake，M.，2019. The compositions of the lunar crust and upper mantle：spectral analysis of the inner rings of lunar impact basins. Planet. Space Sci. 165，230—243.
[36] Ling，Z.，et al.，2015. Correlated compositional and mineralogical investigations at the Chang'e－3 landing site. Nat. Commun. 6，8880.
[37] Ling，Z.，Jolliff，B. L.，Liu，C.，Qiao，L.，Cao，H.，Zhang，J.，Fu，X.，Li，B.，Liu，J.，2019. A close view of Chang'e－4 landing site and science questions to be answered by Yutu－2.
[38] In：50th Lunar and Planetary Science Conference. Lunar and Planetary Institute，The Woodlands，Texas (abstract ＃2330).
[39] Lucey，P. G.，2004. Mineral maps of the Moon. Geophys. Res. Lett. 31，L08701.
[40] Lucey，P.，et al.，2006. Understanding the lunar surface and space－Moon interactions. Rev. Mineral. Geochem. 60，83—219.
[41] Melosh，H. J.，1989. Impact Cratering：A Geologic Process. Oxford University Press，London.
[42] Melosh，H. J.，2011. Planetary Surface Processes. Cambridge University Press，Cambridge.
[43] Melosh，H. J.，et al.，2017. South Pole－Aitken basin ejecta reveal the Moon's upper mantle. Geology 45，1063—1066.
[44] Mimoun，D.，et al.，2012. Farside explorer：unique science from a mission to the farside of the moon. Exp. Astron. 33，529—585.
[45] Morgan，J. V.，et al.，2016. The formation of peak rings in large impact craters. Science 354，878—882.

[46] Moriarty III, D. P., Pieters, C. M., 2018. The character of South Pole - Aitken Basin: patterns of surface and subsurface composition. J. Geophys. Res. Planets 123, 729—747.

[47] Morota, T., et al., 2011. Timing and characteristics of the latest mare eruption on the Moon. Earth Planet. Sci. Lett. 302, 255—266.

[48] National Research Council, 2007. The Scientific Context for Exploration of the Moon. National Academy Press, Washington, D. C.

[49] National Research Council, 2011. Vision and Voyages for Planetary Science in the Decade 2013—2022. National Academy Press, Washington, D. C.

[50] Ohtake, M., et al., 2014. Geologic structure generated by large - impact basin formation observed at the South Pole - Aitken basin on the Moon. Geophys. Res. Lett. 41, 2738—2745.

[51] Otake, H., Ohtake, M., Hirata, N., 2012. Lunar iron and titanium abundance algorithms based on SELENE (Kaguya) Multiband Imager data. In: 43rd Lunar and Planetary Science Conference. Lunar and Planetary Institute, The Woodlands, Texas (abstract #1905).

[52] Papike, J. J., Hodges, F. N., Bence, A. E., Cameron, M., Rhodes, J. M., 1976. Mare basalts: crystal chemistry, mineralogy, and petrology. Rev. Geophys. 14, 475—540.

[53] Pasckert, J. H., Hiesinger, H., van der Bogert, C. H., 2018. Lunar farside volcanism in and around the South Pole - Aitken basin. Icarus 299, 538—562.

[54] Petro, N. E., Pieters, C. M., 2004. Surviving the heavy bombardment: ancient material at the surface of South Pole - Aitken Basin. J. Geophys. Res. Planets 109, E06004.

[55] Pieters, C. M., Head, J. W., Gaddis, L., Jolliff, B., Duke, M., 2001. Rock types of South Pole - Aitken basin and extent of basaltic volcanism. J. Geophys. Res. Planets 106, 28001—28022.

[56] Potter, R. W. K., Collins, G. S., Kiefer, W. S., McGovern, P. J., Kring, D. A., 2012. Constraining the size of the South Pole - Aitken basin impact. Icarus 220, 730—743.

[57] Qiao, L., Head, J. W., Wilson, L., Kreslavsky, M. A., Xiao, L., 2016a. Compound flow fields in southwest Mare Imbrium: Geomorphology, source regions, and implications for lunar basin filling. In: 47th Lunar and Planetary Science Conference. Lunar and Planetary Institute, The Woodlands, Texas (abstract #2038).

[58] Qiao, L., Liu, X., Zhao, J., Wei, Y., Xiao, L., 2016b. Geological investigations of Luna 17, Apollo 15 and Chang'E - 3 landing sites at Mare Imbrium of the Moon (in Chinese). SCI. SIN. Phys. Mech. Astronom. 46, 029603.

[59] Qiao, L., Xiao, Z., Zhao, J., Xiao, L., 2016c. Subsurface structures at the

Chang'e-3 landing site: interpretations from orbital and in-situ imagery data. J. Earth Sci. 27, 707—715.

[60] Robinson, M. S., et al., 2010. Lunar Reconnaissance Orbiter Camera (LROC) instrument overview. Space Sci. Rev. 150, 81—124.

[61] Schultz, P. H., Crawford, D. A., 2011. Origin of nearside structural and geochemical anomalies on the Moon. Geol. Soc. Am. Spec. Pap. 477, 141—159.

[62] Schultz, P. H., Crawford, D. A., 2016. Origin and implications of non-radial Imbrium Sculpture on the Moon. Nature 535, 391—394.

[63] Smith, D. E., et al., 2010. The Lunar Orbiter Laser Altimeter investigation on the Lunar Reconnaissance Orbiter mission. Space Sci. Rev. 150, 209—241.

[64] Spudis, P. D., 1993. The Geology of Multi-Ring Impact Basins. Cambridge University Press, Cambridge.

[65] Sruthi, U., Kumar, P. S., 2014. Volcanism on farside of the Moon: new evidence from Antoniadi in South Pole Aitken basin. Icarus 242, 249—268.

[66] Stuart-Alexander, D. E., 1978. Geologic Map of the Central Far Side of the Moon. United States Geological Survey Map 1—1047, Reston, VA.

[67] Taylor, L. A., Pieters, C. M., Keller, L. P., Morris, R. V., McKay, D. S., 2001. Lunar mare soils: space weathering and the major effects of surface-correlated nanophase Fe. J. Geophys. Res. Planets 106, 27985—27999.

[68] Tompkins, S., Pieters, C. M., 1999. Mineralogy of the lunar crust: results from Clementine. Meteorit. Planet. Sci. 34, 25—41.

[69] Vaughan, W. M., Head, J. W., 2014. Impact melt differentiation in the South Pole-Aitken basin: some observations and speculations. Planet. Space Sci. 91, 101—106.

[70] Wilhelms, D. E., 1993. To a Rocky Moon: A geologist's History of Lunar Exploration. University of Arizona Press, Tucson.

[71] Wu, W. R., et al., 2017. Design of Chang'E-4 lunar farside soft—landing mission (in Chinese). J. Deep Space Explor. 4 (2), 111—117.

[72] Yamamoto, S., et al., 2010. Possible mantle origin of olivine around lunar impact basins detected by SELENE. Nat. Geosci. 3, 533—536.

[73] Yamamoto, S., et al., 2012. Olivine-rich exposures in the South Pole-Aitken Basin. Icarus 218, 331—344.

[74] Yingst, R. A., Chuang, F. C., Berman, D. C., Mest, S. C., 2016. Geologic mapping of the Planck Quadrangle of the Moon (LQ-29). In: 47th Lunar and Planetary Science Conference. Lunar and Planetary Institute, The Woodlands, Texas (abstract #1188).

[75] Zhang, J., et al., 2015. Volcanic history of the Imbrium basin: a close-up view from the lunar rover Yutu. Proc. Natl. Acad. Sci. 112, 5342—5347.

Sub – surface stratification and dielectric permittivity distribution at the Chang'E – 4 landing site revealed by the lunar penetrating radar

Ruonan Chen[1], Yi Xu[1], Minggang Xie[2], Ling Zhang[1,3], Shengli Niu[1], and Roberto Bugiolacchi[1,4]

1. State Key Laboratory of Lunar and Planetary Sciences, Macau University of Science and Technology, Macau, PR China e – mail: yixu@must. edu. mo
2. College of Science, Guilin University of Technology, Guilin, PR China
3. College of Geo – exploration Science and Technology, Jilin University, Changchun, PR China
4. Earth Sciences, University College London, London, UK

Abstract: Context. In 2019, China's Chang'E – 4 (CE – 4) probe landed on the far side of the Moon: a first in lunar exploration. The Lunar Penetrating Radar (LPR) mounted on the Yutu – 2 rover allows the mapping of the near – surface structure and the dielectric permittivity of the landing area. The dielectric properties of the lunar soil affect the propagation of the LPR signals, which can be used to infer the depth of sub – surface boundaries and derive the composition of the component materials.

Aims. Our objectives are to estimate the fine – resolution spatial distribution of relative permittivity and to improve the interpretation of the geological processes combined with the radargram of the CE – 4 landing area.

Methods. We used a modified method that combines the F – K migration and the minimum entropy of the ground penetrating radar (GPR) signals to estimate the velocity and permittivity values; this has the advantage of obtaining the appropriate velocity and permittivity, even with the incomplete or unnoticeable hyperbolic curves in the radar image

Results. The sub – surface stratification of the CE – 4 landing area is seen in the first 31 lunar days of the LPR data. A fine – resolution dielectric permittivity profile ranging from ~2. 3 to ~6. 3 is obtained with our method, and the actual depths of the observed prominent sub – surface interfaces are determined, giving a maximum average depth of ~38m. The thickness of the regolith layer is in the range of ~5. 7 – 15. 6m, with an average of 11. 8m. The permittivity of the near – surface regolith (<30cm) is ~2. 78±

0.01, the bulk density is 1.57±0.01g cm^{-3}, which is close to the results of ~1.61g cm^{-3} at the Apollo 15 landing area. The permittivity map is consistent with the radargram; the regolith and the paleo-regolith layer have relatively low permittivity and low echo strengths, while the rock debris has high permittivity and shows strong echos in the radargram. Two buried craters of different diameters beneath the navigation sites 4-11 and 16-31 are revealed in the radar profile. The permittivity distribution map can show detailed variations of material properties both inside and outside craters.

Keywords: Moon - planets and satellites; surfaces - techniques; radar astronomy

1 Introduction

Most of the lunar surface is covered by a shallow layer of regolith that potentially carries the record of billions of years of surface modifications. The electromagnetic properties of the components in the lunar regolith control the propagation of the electromagnetic (EM) waves in the lunar substrate: a better understanding of these properties is key to a more reliable interpretation of remote sensing data, which in turn improves our understanding of the evolution of the Moon. The complex dielectric permittivity of the sub-surface of the Moon is determined largely by its bulk density and, chemically, by titanium abundances (Olhoeft & Strangway 1975; Strangway et al. 1977). Consequently, a correct estimation of the complex dielectric permittivity can be used to estimate these two properties.

Our understanding of the dielectric properties of the lunar soil comes mostly from remote sensing measurements and analysis of the samples returned by the Apollo missions. The relative permittivity (the real part of the complex dielectric permittivity) of the Surveyor regolith samples ranges from 2.4 to 3.7 and for the Luna samples from 1.7 to 4.4 (Carrier et al. 1991).

Higher frequency electromagnetic radar and radiometry measurements of the lunar surface have been made from Earth (Dewitt & Stodola 1949; Evans & Pettengill 1963; Keihm & Langseth 1975; Pettengill 1978; Marpaung & Lu 2016), a lunar orbit (Tyler & Howard 1973; Peeples et al. 1978; Sharpton & Head III 1982; Ono et al. 2009), and directly at the surface (Fang et al. 2014). Given the high resolution of the Ground Penetrating Radar (GPR) on board the lunar rover compared to the remote sensing methods, this represents a unique opportunity to derive a detailed sub-surface profile. On January 3 2019, China's Chang'E-4 (CE-4) probe touched down at 177.5991°E, 45.4446°S in the south of the Von Kármán crater (diameter = ~186km; central coordinates as 176.2°E, 44.4°S) located inside the ancient South Pole-Aitken (SPA) basin (Li et al. 2020). Subsequently, the Yutu-2 rover was assigned to a scientific exploration mission. The lunar penetrating radar (LPR) on board the rover provides an unprecedented opportunity to investigate substructures and dielectric properties of the lunar regolith on the far side of the Moon, which is inaccessible from Earth - based investigations.

The topography, morphology, and geochemistry of the region around the landing site have been studied in detail (Huang et al. 2018; Qiao et al. 2019; Guo et al. 2021); it is relatively flat with a minimal slope and populated by secondary craters, which form chains and clusters, as seen from highresolution Kaguya and Lunar Reconnaissance Orbiter (LRO) images (Huang et al. 2018; Qiao et al. 2019). In addition, clearly detectable texture and lineations arranged in the NE - SW direction suggest that they may be formed by ejecta from the Finsen crater (Guo et al. 2021). Kaguya MI - derived elemental maps show that the materials at the landing site are dominated by lowtitanium basalts, a composition akin to the Apollo 15 pigeonite basalts (Qiao et al. 2019; Guo et al. 2021).

Based on the estimation of the iron - titanium abundances at the landing area, the penetrating depth of the low - frequency channel could reach ~300 - 450m (Zhang et al. 2020, 2021a; Lai et al. 2020; Yuan et al. 2021), depending on the permittivity value used for time - depth conversion. Compared to the low - frequency channel, the CE - 4 LPR's high - frequency channel has a high range resolution, which is suited to deriving fine stratigraphic substructures in more detail at shallow depths (~45m) (Lai et al. 2019; Li et al. 2020; Zhang et al. 2021b; Zhou et al. 2021). The high - frequency results so far show the lunar regolith layer to be relatively thick, about ~12m, with the dominant source of materials probably consisting of Finsen ejecta (Gou et al. 2019; Lin et al. 2020).

Lai et al. (2019); Dong et al. (2020a, b); Li & Zhang (2021); Song et al. (2021); Giannakis et al. (2021); Wang et al. (2021) calculated the dielectric permittivity value or its depth profiles at the landing site to derive other attributes, including bulk density and mineralogical abundances. Most studies use the hyperbola fitting method due to its simplicity and computational efficiency (Dong et al. 2020a, b; Lai et al. 2019; Song et al. 2021). Giannakis et al. (2021) improved the method by considering the permittivity as varying with depth, while Li & Zhang (2021) used the migration method and separated the diffraction from reflections to estimate the permittivity.

Although the hyperbola fitting method is widely used, it has drawbacks such as a high dependence on the signal - to - noise ratio (S/N) of the LPR signals, and the distributions of scatters. It is difficult to identify the hyperbolic structure at a depth where the S/N is low (Daniels 2004); consequently, the inverted permittivity values can carry relatively large errors. The derivation of permittivity distribution along the horizontal direction is also limited by the absence of a hyperbola shaped signal. The permittivity map obtained by (Dong et al. 2020b; Li & Zhang 2021; Song et al. 2021) was produced by interpolation based on a couple of sparse points. The resolution is too coarse to reveal the finer details, which are necessary for a reliable mapping of the geological stratigraphy. The sectional permittivity map in Giannakis et al. (2021) shows additional sub - layers within approximately the first 10m of the regolith layer, but it does not provide the permittivity variation at deeper depths and across the horizontal plane.

To minimise the shortcomings of the traditional hyperbolafitting methods and obtain a high - resolution permittivity map, we applied a new method that combines the F - K migration and the minimum entropy of the GPR signals to estimate the appropriate electromagnetic velocity of lunar materials. Then, we calculated a relative permittivity distribution map along the exploration path of the Yutu - 2 rover with the highest resolution so far using the high frequency LPR data during the first 31 lunar days (Fig. 1). Furthermore, the permittivity map was used to derive the actual thickness of the sub - surface layers and the density variations of materials, and to improve the geologic understanding of the sub - surface stratigraphy.

2 Data

The GPR on board the Yutu - 2 rover is a dual - frequency, carrierfree, time - domain imaging radar. A pair of monopoles operating at 40 - 80MHz with central frequency of 60MHz are adopted into the low - frequency system with a vertical resolution of 3.75m in vacuum for the purpose of detecting the upper, shallow crust terrain. Another set of identical bow - tie antennas with central frequency of 450MHz and a bandwidth of 500MHz are utilised as the high - frequency system with a high range resolution of 0.3m in vacuum to detect the shallow substructure and the thickness of the lunar soil (Fang et al. 2014). When combined, these complementary data sets offer the opportunity to investigate the lunar substrata to build a reliable geological profile (Su et al. 2014). Radar data are collected and stored track by track. Each trace of data contains the relative amplitude intensity of the radar signal and its corresponding description information. Here, we employed the high - frequency channel (500MHz) data to study the shallow substructures and their dielectric properties. At the time of writing, the Yutu - 2 rover has already travelled ~690m during the first 31 lunar days, as shown in Fig. 1.

We first performed signal processing on the CH - 2 LPR data, mainly involving the removal of duplicated data, denoising, dielectric loss and geometric diffusion compensation, bandpass filtering, zero - time adjustment, and Kirchhoff migration. The redundant data were generated when the rover stopped to perform other experiments and were removed accordingly. The background noise was removed to eliminate ringing noises showing as horizontal and periodic events (Kim et al. 2007). Band - pass filtering was applied to reduce the lower and the higher frequency noises. Dielectric loss and geometric diffusion compensation (gain) was applied to compensate for the amplitudes of the LPR echoes at depth. The Kirchhoff migration method was applied to focus the received radar echoes into their correct locations. Because the receiving antenna is turned on 28.203 ns earlier than the transmitting antenna, the timing lag of each data track was compensated for.

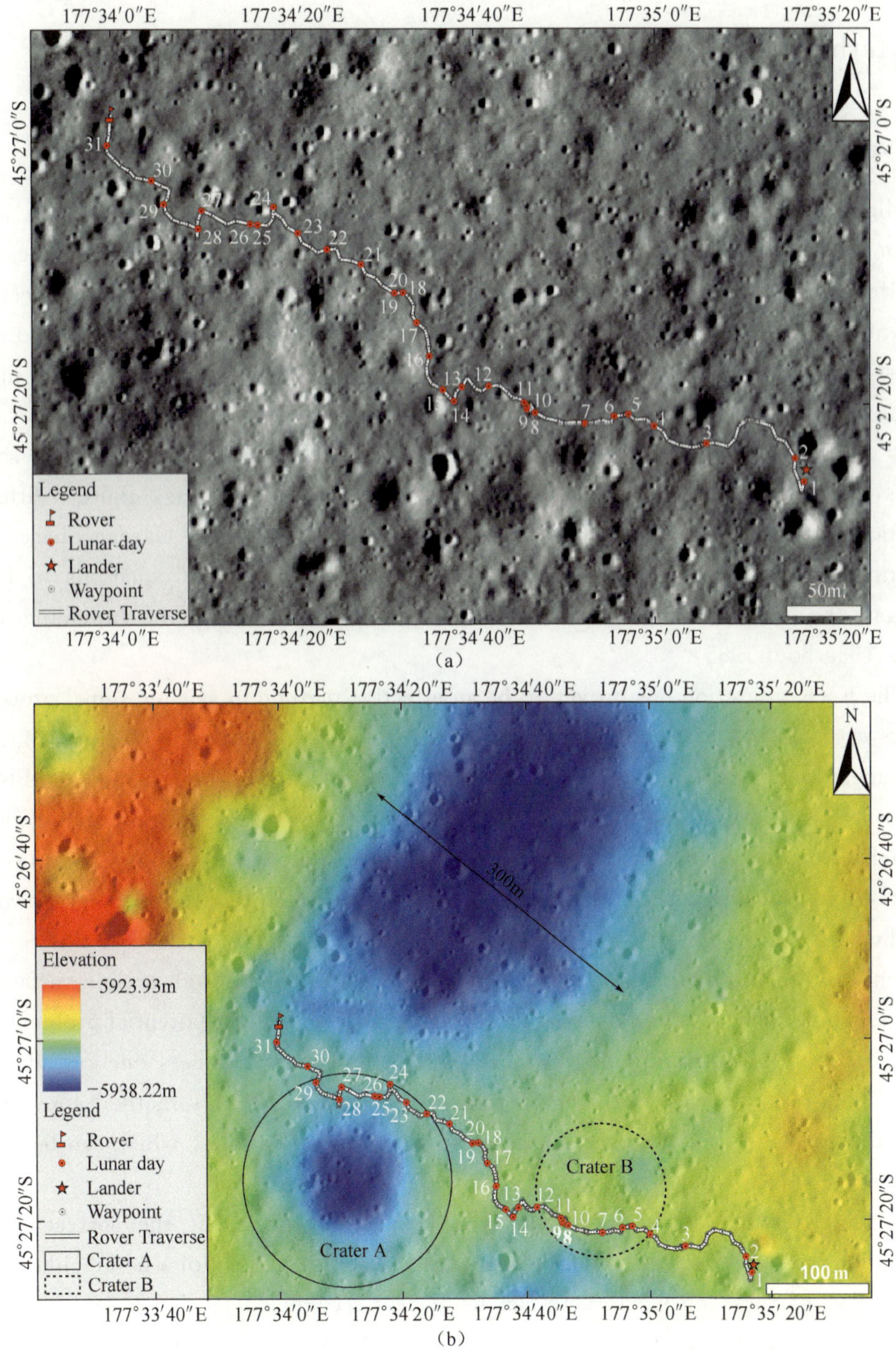

Fig. 1 Routing path covered by the Yutu－2 rover on the Moon.

(a) The base map is a Lunar Reconnaissance Orbiter (LRO) Narrow Angle Camera (NAC) image (File ID: M1314237625LR). (b) High－resolution NAC digital terrain model of CE－4 landing area. The red star marks the position of the CE－4 lander. The white line represents the traverse of the Yutu－2 rover, and the red points indicate the waypoints. The black circle outlines buried crater A, and the dashed circle shows buried crater B

3 Method

3.1 Velocity inversion method

Migration refers to the movement from the observed data to their true spatial positions. In a typical B - scan profile, any scatterer shows up as a hyperbola because of the different travel times of the EM wave propagating from the scatterer to the receiver at different locations, as the antenna is moving along the scanning direction. Therefore, the primary purpose for migration on GPR B - scan profiles is to produce a sharp and distinct peak at the location of the object and a little significant structure elsewhere. In fact, a library of migration algorithms (Yilmaz 2001) has been primarily developed from seismic images, including the Kirchhoff migration method (Schneider 1978), finite - difference migration (French 1974), Stolt F - K migration (Stolt 1978), and Gazdag phase - shift migration (Gazdag 1978). Those methods make use of the wave equations to back - propagate the surface data to the scattering sub - terrain based on the exploding source model (Claerbout 1985).

The F - K migration has proven to be an effective method in seismic signal processing (Peters et al. 1994), as well as in GPR signal processing (Özdemir et al. 2014). It is worth mentioning some assumptions for the F - K migration; for example, GPRs are located near to the ground, the signal being processed is free of noise and clutters, the material is homogeneous, and the ground surface is flat. However, those assumptions are probably unreasonable in the detection process of the GPR. Antennas are always located at some height (e. g. 0. 3m for the high frequency GPR of the Yutu - 2 rover). The resulting noise and clutter usually distort the shape of the hyperbola. The rough ground surface may introduce some random time - delays to GPR signals. Despite these potential problems, the F - K migration has proven to work well in GPR applications. Besides, one key advantage of F - K migration is that the Fourier transformation used in this method converts the differential operations in the time domain into algebraic operations, which can be quickly computed.

Velocity estimation is critical for the F - K migration. If the correct velocity distribution is used, the energy will be condensed into a few pixels for a point - like target. The location of these buried targets can be clearly recognised. A lower estimation will cause under - migration, and a higher estimation will cause over - migration.

Entropy is a measure of the information content of a signal (Pun 1980). Calculating the entropy of the GPR B - scan profile is an efficient way to estimate the velocity as the input of the F - K migration. In contrast to maximum entropy, which is associated with randomness or uncertainty, minimum entropy is an expression of simplicity or certainty. In terms of GPR B - scan profiles, the migrated images with the sharp and distinct peak at the location of the object have small entropies (Flores - Tapia & Pistorius 2010; Marpaung & Lu 2016). Therefore, minimising the entropy could be a way to find the accurate velocity. The minimum entropy deconvolution approach is used to minimise the inverse of the varimax norm to approximate the entropy of the seismic signals (Wiggins 1978; Wu & Barba 1998). This approach is widely used in obtaining a simple structured signal and has proven very successful in the deconvolution of seismic signals. Subsequently, Xu & Miller (2001) and Xu et al. (2003) combined minimum entropy image restoration with the F - K migration approach to accurately locate shallow sub - surface objects in GPR profiles. They determine the characteristics of the 'focus' using an entropy - like criterion, namely, the inverse of the varimax norm (Wiggins 1978) because of its ease of computation.

Our approach is as follows: for a measured GPR wave field P of size $M \times N$, we first divide it into a certain number of subwave fields P_i of size $M_i \times N_i$, where i is ith of sub - wave field, and the sub - wave fields P_i are a subset of the full radargram P. As the entropy value depends on the pixel intensity, we divide the whole image based on the electric field strength. The varimax norm R, defined as the sum of the normalized squares of the variances of the amplitude of signals, is an approximation of the inverse of entropy (Wiggins 1978); we use the varimax norm R as the criterion for judging the degree of focus in practical computation. There are two key considerations for selecting subwave field in the real detection of GPR data: The first is that one computational sub - wave field is generally chosen to be of a relatively uniform electric field strength, and the whole GPR profile is divided both in the distance (x) and time (t) directions. The second is for some special cases. There may be many scatterers under the lunar surface, such as small rocks or fractured breccia lens, underground slope terrain, or other geological units, causing noticeable hyperbolas and oblique echoes, and so on. In such cases, the computational sub - wave fields are selected as regions, including the complete hyperbolas and oblique echoes.

The procedure to obtain the velocity in each sub - wave field P_i is as follows: (1) set a reasonable material velocity set; (2) perform regular F - K migration at each velocity v_i on the preprocessed data; (3) calculate varimax norm of each migration data $R(P_i)$:

$$R(P_i) = \left[\frac{\sum_{m=1}^{M_i} \sum_{n=1}^{N_i} P_i^4(m, n)}{\left[\sum_{m=1}^{M_i} \sum_{n=1}^{N_i} P_i^2(m, n)\right]^2} \right]^{-1} \tag{1}$$

(4) search the appropriate root - mean - square (RMS) velocity v_{RMS} corresponding to the minimum value of $R(P_i)_{\min}$; (5) transform v_{RMS} to the local velocity v_{local} using the Dix equation (Dix 1955) by:

$$v_{\mathrm{local}} = \sqrt{\frac{v_{\mathrm{RMS2}}^2 t_2 - v_{\mathrm{RMS1}}^2 t_1}{t_2 - t_1}} \tag{2}$$

where $v_{RMS(n)}$ and $t_{(n)}$ are the RMS velocity and travel time of the nth sub - wave field.

After all sub - wave fields are processed following these steps, we obtain the whole local velocity distribution by interpolation.

The relative permittivity can then be calculated:

$$\epsilon_r = \left(\frac{c}{v_{\text{local}}}\right)^2 \tag{3}$$

where c is the velocity in vacuum. The depth is calculated with the local velocity and corresponding propagation time. The depth d_i of the current sub - field P_i is determined by the local velocity $v_{\text{local}i}$, time interval Δt_i, and the depth d_{i-1} of the previous sub - field P_{i-1}:

$$d_i = d_{i-1} + v_{\text{local}i} \Delta t_i \tag{4}$$

3.2 Method validation

We present an example of CH2 data with a classical hyperbolashaped structure to verify the performance of our method, as shown in Fig. 2. This section of the echo is located at a shallow depth of ~46 - 81ns in the vertical direction and ~332 - 343m in the horizontal direction. A hyperbola with multiple reflections can be seen in the black box in

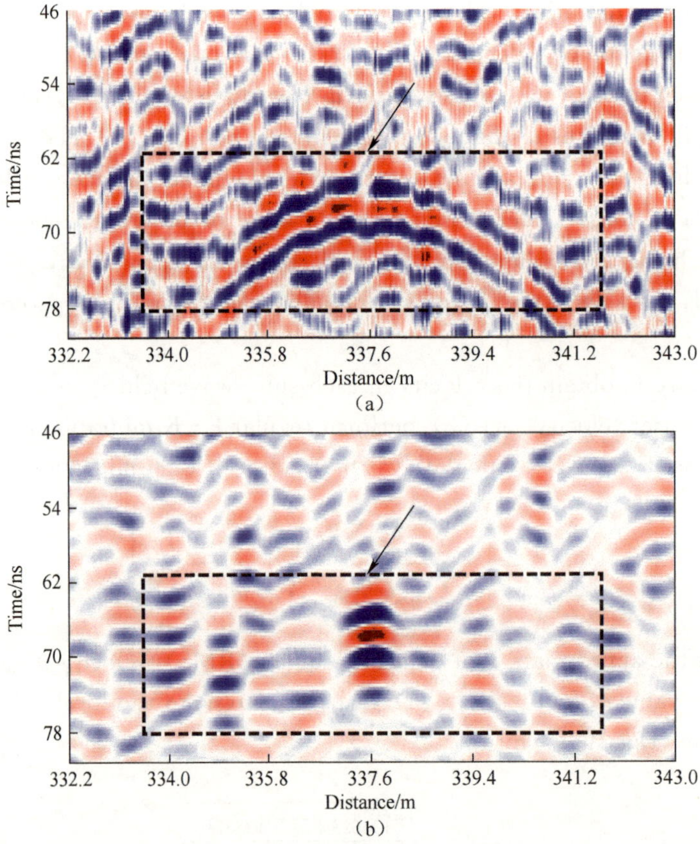

Fig. 2 Example with a hyperbola - shaped structure

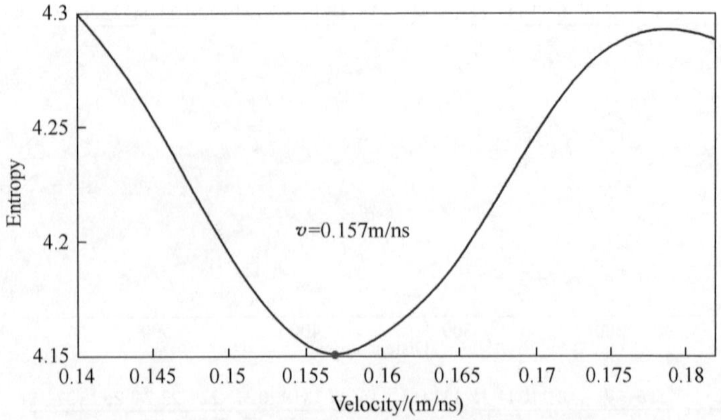

(c)

Fig. 2　Example with a hyperbola - shaped structure. (Continued)

(a) Classical hyperbola structure before migration. (b) Point - like structure after migration.
(c) Variation of entropy with electromagnetic wave velocity in the material

Fig. 2a, whose vertices are pointed by a black arrow, illustrating the data before migration. This hyperbola echo was probably produced by a discrete scatterer (Daniels 2004). The hyperbola condenses into a series of point - like echo after migration processing, which is presented in Fig. 2b, indicating that the velocity calculated by our method is realistic. Figure 2c shows the variation of entropy with the electromagnetic wave velocities in the material. The minimum entropy is around 4.15, and its corresponding velocity is $v = 0.157\text{m} \cdot \text{ns}^{-1}$; this velocity value will be used to calculate the local velocity and the actual depth.

Similarly, we use the same criterion to focus on other echoes with or without typical hyperbolic structures. Another example of waves without clear hyperbolic structures is shown in Fig. A. 2, which is also used to illustrate the effectiveness of our method.

4　Results

4.1　Stratigraphic structure

The reflection intensity of the high - frequency GPR data after data processing including Kirchhoff migration is presented in Fig. 3, where the black dashed lines in Fig. 3b show the stratification, and the numbers indicate the depth value corresponding to the position of the dashed lines, which are generated using the proposed method. Two reflectors with the average depth of ~13.9m and ~32.3m are horizontally continuous through the routing path, and they can divide the sub - surface structure of the landing site into three geologic sub - layers: ~0 - 13.9m (Layer I), ~13.9 - 32.3m (Layer II), and ~32.3- 38.1m (Layer III), marked on the left side of Fig. 3b.

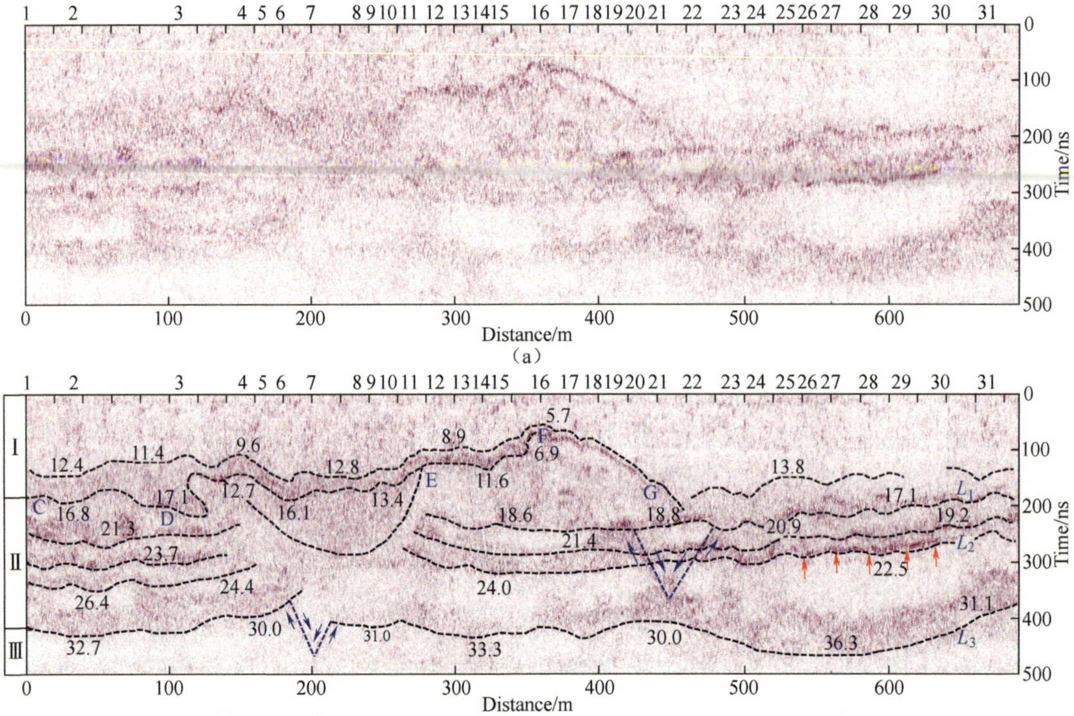

Fig. 3　Shallow sub - surface structure imaged by CH2 data during the first 31 lunar days. The colour bar represents the amplitude of the radar signals. The sites of 31 lunar days are indicated by the numbers at the *top of the panel*.

(a) High - frequency LPR profile. (b) Interpreted stratification structure. Black dashed lines mark the boundary of the reflectors. The blue dashed lines indicate the fault plane, and the blue arrows represent the relative movement direction of the current layer. Black numbers represent the depth in metres varying with positions. On the left side of the figure, we show the division of the radar sub - surface structures: I, II, and III

Reflections in the surface layer (the first layer within Layer I in Fig. 3) are relatively weak overall, with some slight fluctuations of stronger reflections. Surface materials have been pulverised into fine - grained particles by impacts (Melosh 2011) and space weathering processes (Pieters et al. 2000; Taylor et al. 2001), which may contain ejecta mostly from the neighbouring Finsen crater (Li et al. 2020; Huang et al. 2018; Hu et al. 2019; Gou et al. 2019). Moreover, some relatively strong signals with hyperbolic shape are probably produced by discrete scatterers (Daniels 2004). These scatterers were interpreted as ejecta rocks during the formation of Finsen crater, including an unknown fraction of local rocks. Another scenario is that these scatterers with relatively large permittivity in the regolith layer are probably broken pieces of glass - bearing breccia projectiles that were excavated from the consolidated bottom and walls of pre - existing small craters within the lunar regolith rather than from deeper hard igneous rocks (Lin et al. 2020). The thickness of the surface layer is ~5.7 - 15.6m with an average depth of 11.8m based on the permittivity value and the two - way time delay of the radar signals, our results are consistent with ~11.1m from Wang et al. (2021).

The echoes become weaker at the bottom (~9 - 13m) of this layer from the site that is 21st from the end, suggesting that the density of rocks and particle size in this local region are lower than those of the materials above. Between sites 16 and 22 in Fig. 3a, there is a well - defined and smooth down - slope reflection layer, which may represent a section of the crater wall of the buried impact crater. The Yutu - 2 rover travelled close to a surface depression located to the southwest of the travel route, which is shown in the surface elevation map in Fig. 1b, the elevation values corresponding to each point are displayed in Fig. A. 1. The reflector becomes flat between sites 22 and 31, where the Yutu - 2 rover moved along the edge of the circular depression. This large depression southwest of the travel path is defined as Crater A, which is depicted as a solid black circle in Fig. 1b.

The first sub - surface reflector represents the lower interface of the fine - grained particles, which is located in the ~8 - 22m range of the Finsen ejecta deposits, as estimated by ballistic sedimentation models (Huang et al. 2018; Xie et al. 2020; Xu et al. 2021). Beneath this fine - grained surface layer, there is a layer with stronger reflectance and upward slope from sites 1 to 16. The thickness of this coarse - grained layer varies widely from ~1.9 to ~7.3m at different horizontal positions. The baseline of this reflector is also undulating; it stays generally unchanged at depths of ~17m from C to D, fluctuates from ~12.8 - 16.1m from D to E, and then goes up to the shallowest depth of ~5.7m at F, and finally gradually falls down to the depth of ~13.8m from G to the end (Fig. 3b). Point F might represent the rim of a buried crater. The materials with higher reflectivity in this layer are believed to be mostly a mixture of coarse - grained deposits both of Finsen primary ejecta and the ejecta of crater A (Fig. 1b). Since the surface layer and the underlying layer may have the same material source, we classify them as Layer I.

Layer II comprises at least four ejecta overlays and contains multiple reflectors from depths of ~13 to ~36m. On the profile, these reflections' layers are continuous overall, and the average thicknesses of each layer are ~5.89, ~2.33, ~2.36, ~8.41m, and so on, respectively. Each deposited sub - layer is composed of two sub - stratigraphic regions. The upper part is relatively homogeneous and weak, because the surface material have undergone a longer period of weathering processes than the lower strata (Zhang et al. 2021b), while the lower part represents a larger reflectivity, since this part of material is distributed randomly and heterogeneously; it is thought to represent ejecta that are not completely weathered and that feature large rocks.

A small bowl - shaped crater below sites 4 - 11, named crater B, which was buried by the mixture of Finsen primary ejecta and local crater A ejecta deposits, disrupt the continuity of reflectors. The echoes inside the crater are chaotic and strong, while echoes outside are relatively weak. The crater A (~200m in diameter) shown in Fig. 1b also disturbs the structure of the layered ejecta as seen in the right part of Fig. 3. For example, the upper two ejecta layers are mixed, and they blend into one as the rover moved from site 21 to the end. Impact - induced features such as faults and dislocations also appear under the crater wall. A V - shaped structure beneath sites 6 - 9, denoted by blue dashed lines in Fig. 3b, might represent a fault or fracture caused by the shockwave of the impact event. Similarly, there is anotherV - shaped structure below navigation points 20 - 22. We provide a zoomed - in picture of these two regional structures, as shown in Figs. 4 and 5, respectively.

We interpreted Layer II as the multiple deposited ejecta materials mixed with excavated local materials. These ejecta materials may come from several impact craters inside and outside Von Kármán crater, such as the Late Imbrian Maksutov and Early Imbrian Orientale Basin, and so on, and several unnamed local craters (Zhang et al. 2021b; Xu et al. 2021). Since the age of the neighbouring Alder crater is greater than the surface mare units (Lu et al. 2021), it might be excluded from the source. Also, the Copernican Zhinyu crater next to the CE - 4 landing site (3. 75km in diameter) is too small to have formed a detectable ejecta layer at the CE - 4 landing site.

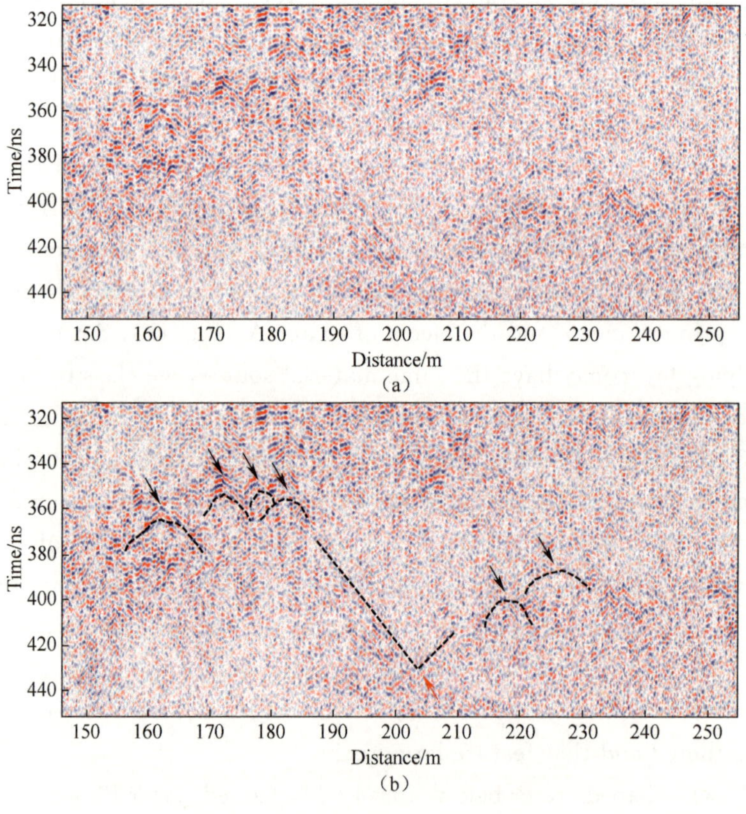

Fig. 4 'V' structure at $x \approx 200$m.

(a) Original radargram. (b) V structure and hyperbolas are marked by dashed lines and arrows. Red arrow points to the bottom of the V structure

Radar echoes in Layer III are weak and relatively homogeneous, and they are most

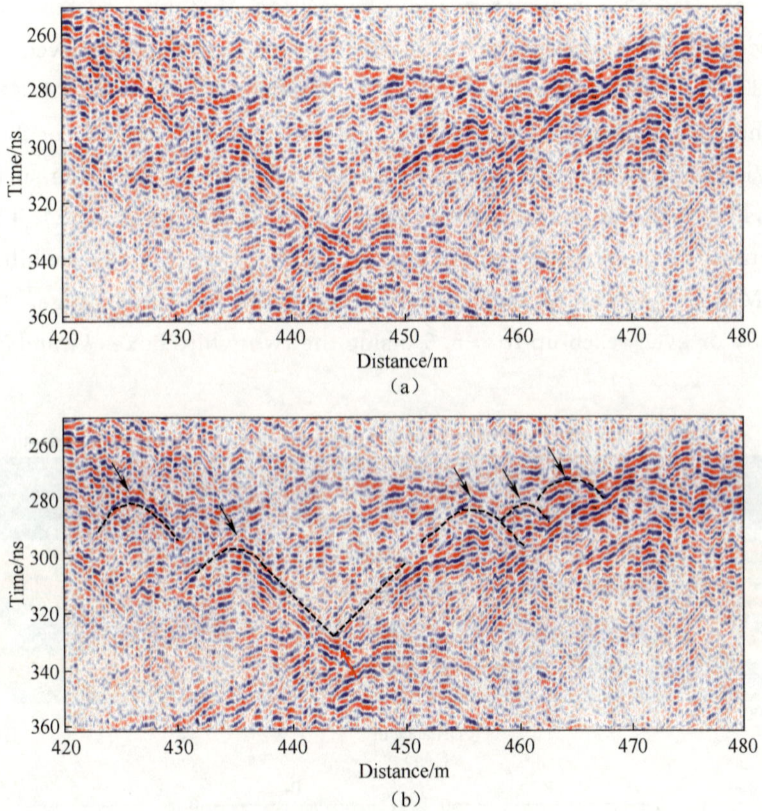

Fig. 5 'V' structure at $x\approx450$m.

(a) Original radargram. (b) V structure and hyperbolas are depicted by dashed lines and arrows. The red arrow points to the bottom of the V structure

likely caused by reflection or the scattering of grain particles, regarded as paleoregolith. The fine grain materials in this layer are probably subject to space weathering and frequent small impacts after mare filling events within Von Kármán. Below ~500ns, no prominent radar features are detectable.

4.2 Dielectric permittivity

The relative permittivity distribution estimated with the proposed method is illustrated in Fig. 6a, which is approximately in the 2.3 - 6.3 range. The resolution in the horizontal direction is in the range of ~10 - 50m, and the one in the vertical direction is from 45 to 60ns.

The relative permittivity of materials in the surface layer (the top layer within Layer I in Fig. 6) ranges from ~2.3 to ~3.7, indicating that the top surface materials have similar dielectric properties to the typical regolith (Carrier et al. 1991). The average permittivity in the shallow surface part (<30cm) is ~2.78 ± 0.01, and the bulk density is ~1.57 ± 0.01g · cm^{-3}, which is close to ~1.61g · cm^{-3} at the Apollo 15 landing area

(Mitchell et al. 1972), supporting the observation that the surface materials are comparable with Apollo 15 basalts according to the elemental maps derived from MI data (Qiao et al. 2019). Based on the FeO and TiO_2 abundances of 13.1 and 1.8wt% (Guo et al. 2021), the rock density of the surface materials is estimated to be 2.78g • cm^{-3} (Huang & Wieczorek 2012), inferring a porosity of about 43.5% at the surface, a value somewhat smaller than the typical porosity of 45% (Fa & Wieczorek 2012). The layer with strong reflectivity beneath the regolith layer from site 1 – 21 exhibits a larger permittivity. Most of the values are approximately in the 2.9 – 4.6 range, but some are larger than ~5, or even reach up to ~5.6 inside the two white boxes b_1 and b_2, as shown in Fig. 6a.

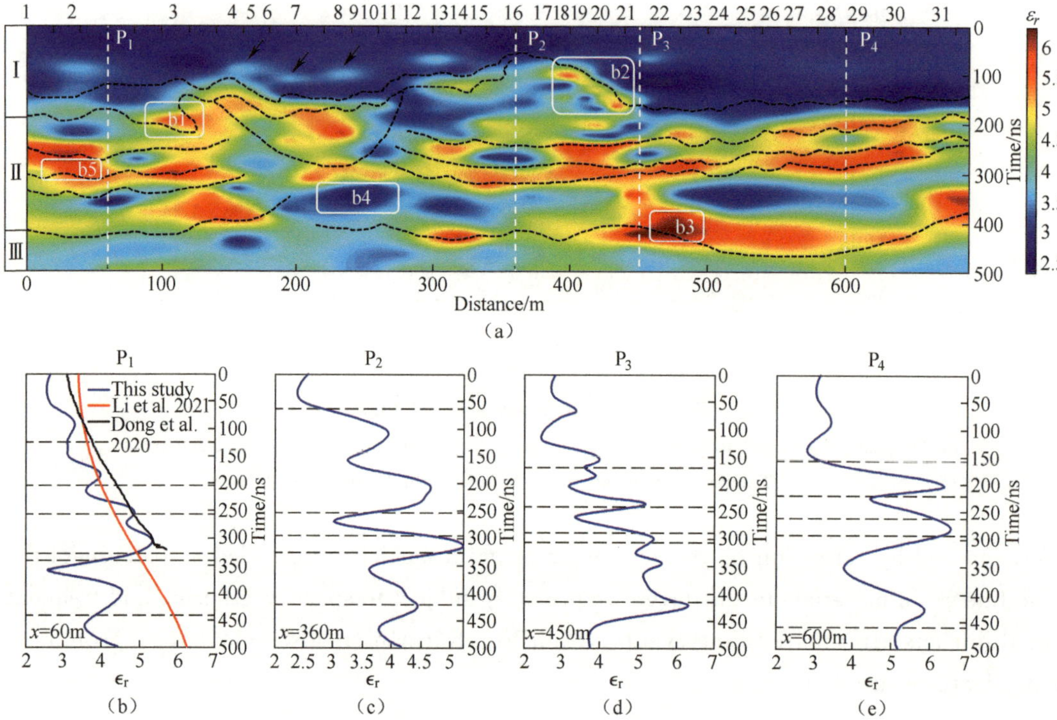

Fig. 6　Permittivity distribution determined by the CH2 data during the first 31 lunar days. The colour bar represents the relative permittivity values. The sites of 31 lunar days are indicated by the numbers at the top of the panel. (a) Permittivity distribution calculated by our method, where black dashed lines depict the boundary of the reflectors, and white dashed lines P_1 – P_4 show the positions of permittivity profiles of the four vertical distribution in *b* – *e*. *b* – *e* Detailed permittivity varying with depths at a single horizontal position, located at x = 60, 360, 450, and 600m, respectively. The grey horizontal dashed lines in *b* – *e* are the time values of the intersections between the four profiles (P_1 – P_4) and reflect or boundaries shown in a. The comparison of permittivity with Dong et al. (2020b) and Li & Zhang (2021) at x=60m, as shown in b.

Some irregular and abnormal scatterers are obviously observed at the regolith layer, the fine resolution permittivity map in Fig. 6a adds further detail compared to prior work.

For instance, there are three regions of notably large values reaching ~3.5 - 3.7 indicated by three black arrows under the sites from 4 to 11, probably due to the buried scatterers. Although these abnormally large values also appear in Song et al. (2021), our results are in closer agreement with the positions of the scatterers.

The permittivity from site 21 to 31 of Layer I is lower than that in the upper regolith part, supporting the notion that the materials within Crater A are mainly composed of small particles (Rogers & Head 1961; Ogolo et al. 2015).

In Layer II, the unevenly distributed permittivity values have a wide range of fluctuations, varying from ~2.6 to ~6.3 along the route of the rover. Overall, the left (from site 1 to 5) and right parts (from site 12 to 31) are obviously larger than the middle part, as shown in Fig. 6a. These materials with higher permittivity values, similarly to the ones in the white box b_3 in Fig. 6a, are local. The materials with small permittivity are considered to be paleoregolith or highly weathered ejecta with smaller particles, similarly to the ones in white box b_4. Recurring regularly, each ejecta unit in Layer II is composed of two sub - layers: the weathering part overlying the coarse materials. For example, the permittivity in box b_5 in Fig. 6a has a large value of ~5.8; this part of material is located in the lower part of the current deposits layer. A consistent value of ~5.6 in Dong et al. (2020b) is also shown in a similar region.

We collected four permittivity profiles varying with depths to highlight changes in the vertical direction, as shown in the white dashed lines P_1 - P_4 in Fig. 6a. The horizontal positions of the profiles are at x = 60m, x = 360m, x = 450m, and x = 600m, respectively. The vertical distributions, shown in Figs. 6b - e, display a general trend with permittivity increasing with the depth and fitting well compared to Dong et al. (2020a) and Li & Zhang (2021) as show in Fig. 6b. In addition, the fine resolution distribution map also reveals obvious jitters in the depth profile caused by the multi - layered sub - surface structure.

At the bottom, the region where the permittivity values become smaller and more uniform than the upper layer is split into Layer III. Most of the values range from ~2.9 to ~4.6, and large values only appeared at boundary positions connected to the upper layer, which is influenced by ejecta materials. Due to the compaction of the material, the density of material in Layer III becomes greater than that in the regolith layer, with a relatively larger permittivity.

Our methodology has the advantage that the appropriate velocity can be efficiently obtained even if the hyperbolic curves are incomplete or unnoticeable. In a sub - field image with a hyperbolic structure, its tailing will converge to a point through migration in terms of the appropriate velocity, and thus the image has the smallest entropy. Similarly, after migration with the appropriate velocity, profiles without noticeable hyperbolic structures have minimum entropy. Thus, the whole permittivity profile obtained by our method offers greater detail and is useful for the interpretation of the radar profile.

5 Discussion

With the estimated fine - resolution permittivity map and the migrated radargram, detailed structures and variations of material properties of the two buried craters can be unveiled. The bowl - shaped degraded crater B is below the navigation sites from 4 to 11, and its rough boundaries are traced by dashed lines, as shown in Fig. 7a. The reflectance inside the crater is chaotic and greater than that in the regolith layer, while the reflectance outside the crater, especially on the right side, is relatively weaker. There are also some obvious hyperbolic structures inside the radar feature, marked by short dashed lines.

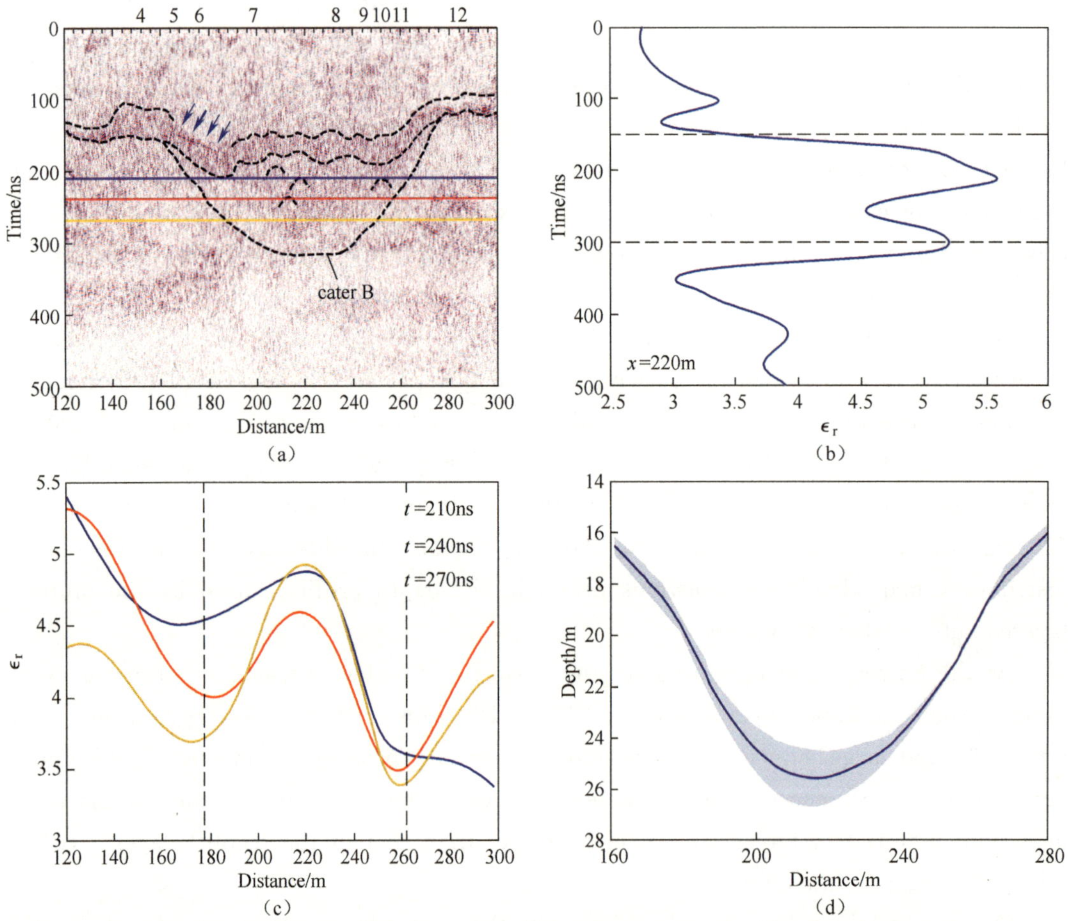

Fig. 7 Radargram, relative permittivity, and absolute depths of the crater B. (a) Radar profile of crater B. The colour bar represents the amplitude of the radar signals. Black dashed lines show the outlines of layered structures and crater B, and discrete scatterers. Blue arrows point to a sharp reflection. (b) Permittivity varied with depth at $x = 220$m. (c) Permittivity varied with horizontal distance at three time delays (blue, red and yellow line) in *a*; the time delays are at $t = 210$ns, $t = 240$ns, and $t = 270$ns, respectively. (d) Absolute depths of the outline of crater B and its error bar (blue shaded area)

The permittivity in crater B and surrounding materials is shown in Fig. 6a. It ranges from ~3.4 to ~5.8 inside the crater and becomes smaller on its two sides, reaching maximum values in the middle of the crater. The changing trend of permittivity values with depth at the horizontal position of $x=220$m is illustrated in Fig. 7b, and these values are extracted from the permittivity map. The permittivity remains small between ~2.7 and 3.4 at first, and then it increases rapidly starting from ~150ns and reaches the maximum value of ~5.6 at ~210ns; finally, it quickly drops to a lower value of ~3 - 3.9 from ~210ns to the end. Horizontally, the changing trend at different depths is shown in Fig. 7c. The two grey dashed lines are the horizontal positions of the intersections of the blue line and the crater wall in Fig. 7a. Correspondingly, inside the crater, the larger the permittivity values at the three depths (t = 210ns, t = 240ns, t = 270ns), the greater the reflectance (Fig. 7a). Based on the permittivity, the depth of the boundary of crater B is shown in Fig. 7d. The blue line is the average depth of the boundary, and the shaded area indicates the error bar. The diameter of the crater seen in radar profile is D~115±5m, and its bottom depth is d~25.75±0.95m.

The materials inside crater B with strong reflectance and large permittivity are likely to contain sections of breccia lenses, which were formed by the impact process. After the formation of its transient crater, the brecciated and fractured walls of the cavity and the elevated rim area become highly unstable and collapse inward, partially infilling the crater with lenses of brecciated materials (Heiken et al. 1991). The depth of the crater is deeper than the ~19.55±0.85m predicted by applying a standard depth - to - diameter ratio of 0.17 for fresh small craters (Stopar et al. 2017). The difference (~6.0m) between the bottom depth and the apparent depth is smaller than the theoretical range of thickness (~6.5- 9.8m) calculated using the same diameter, which is typically 1/2 to 1/3 of the apparet depth of a simple cvater (melosh 2011). Evidently, this part of the breccia lens represents only a sub - section of the entire structure. This is in accordance with the fact that the rover most probably did not cross its centre. A line of very sharp and obvious reflections is highlighted by a series of blue arrows in the left rim of this crater, as shown in Fig. 7a. The permittivity of this reflection is as high as 5.1, which may represent breccia or a melt formed by the impact.

The big depression in the right part of Fig. 3 (starting from the 16th day) is likely to represent an underground expression of crater profile A on the DEM map (Fig. 1b) (Lai et al. 2021). The downslope reflected layer from F to G in Fig. 3b may represent a section of the crater wall of the buried impact crater. Three continuous reflectors may depict the profile of this crater from G towards the end, which are denoted as L_1, L_2, and L_3 in Fig. 3b, respectively. The correlation coefficient between the distance from the rover to the centre of crater A and the elevation of layer L_1 is 0.68 between site 23 and 29, greater than layer L_2 and L_3 as shown in Fig. 3b, indicating that layer L_1 may represent the cross - section of the impact crater A (Fig. 1b) and no other large - scale impact ejecta was deposited around the area between the formation of Finsen and this impact crater. At the same time, part of the ejecta materials from crater A would be deposited in the area, shown on the left in Fig. 6. They would be mixed with Finsen ejecta, and even add to the materials inside the small crater B. Hence, the materials inside the small crater B (Fig. 7a) are probably a mixture of a breccia lens and the ejecta deposits from Finsen and crater A.

Not only were the relative shallow echoes affected by the impact, the upper two layers merge together into one from site 21 to 31, encompassing the deeper echoes. A continuous horizontal echo line with strong reflection at deeper depths is highlighted by red arrows in Fig. 3b, and it may be caused by the squeezing of the shock waves generated during the formation of this impact. The materials in this location with high permittivity value of $\sim 5.5 \pm 0.5$ may represent impact melts or a breccia lens produced by the impact. Along sites 30 - 31, the echoes grow slightly stronger as the rover moves farther away from the centre of crater A (Fig. 1b). These rising underground echoes may be affected by both this large crater and the irregular depression (width$\approx$300m) in the northwest direction of the path, as from Fig. 1b. Therefore, in addition to autochthonous ejecta, the modification of the surface and topography of the Chang'E - 4 landing area also includes this autochthonous ejecta, such as the deposits from this large crater or from the northwest depression.

6 Conclusions

The sub - surface stratification of the CE - 4 landing area is seen in the first 31 lunar days of LPR data. To aid in the interpretation of geological processes, we applied an effective velocity estimation method to obtain a fine resolution permittivity profile of the shallow sub - surface structure. The method has the advantage that the appropriate velocity can be efficiently obtained whether the waves are with or without significant diffraction structures; further, the high - resolution dielectric permittivity map and actual depths can be calculated by the velocities.

The shallowest surface layer is represented by the regolith layer, with an average thickness of ~11.8m and a dielectric constant between ~2.3 and 3.7. The top surface part (< 30cm) is ~2.78 on average, and the corresponding surface bulk density is $\sim 1.57 g \cdot cm^{-3}$, indicating that the surface material is consistent with the Apollo 15 landing site. There are multiple horizontally continuous sub - layers with different thicknesses in Layer II, each ejecta unit comprising two sub - layers: a weathered part overlying coarser materials. This layering phenomenon is also shown in our fine permittivity map.

Two buried craters of different sizes beneath sites 4 - 11 (crater B) and 21 - 31 (crater A) have come to light through the radargram, with their permittivity values firstly calculated in our permittivity map. The relative permittivity inside crater B is as large as 3.4 – 5.8. We interpret the data as showing the materials inside crater B to include not only the breccia lens as it formed, in addition to Finsen ejecta, but also ejecta from local craters, such as those from the larger crater A. A segment of the radar profiles (L_1) with an average depth of ~17m may represent the cross – section of the impact crater A. As for the echoes climbing slightly as the rover moved farther away from the centre of crater A, the modification of the surface and topography of the Chang'E – 4 landing area probably includes this autochthonous ejecta, such as the deposits from crater A or from the northwestern depression. The influence of the northwestern depression on the radar echoes is worth further study in future work.

Acknowledgements

The Chang'E – 4 data used in this work is processed and produced by the Ground Research and Application System (GRAS) of China's Lunar and Planetary Exploration Program can be downloaded from https://moon.bao.ac.cn/web/enmanager/home. This work is supported by the Science and Technology Development Fund of Macau (0089/2018/A3, 0042/2018/A2, 0049/2020/A1).

References

[1] Carrier, W. D., Olhoeft, G. R., & Mendell, W. 1991, Lunar Sourcebook: A User's Guide to the Moon (Cambridge: Cambridge University Press), 530.

[2] Claerbout, J. F. 1985, Imaging the Earth's Interior (Hoboken: Blackwell scientific publications Oxford), 1.

[3] Daniels, D. J. 2004, Ground Penetrating Radar (London, UK: The Institution of Engineering and Technology).

[4] Dewitt, J., & Stodola, E. 1949, Proceedings of the IRE, 37, 229.

[5] Dix, C. H. 1955, Geophysics, 20, 68.

[6] Dong, Z., Fang, G., Zhao, D., et al. 2020a, Geophys. Res. Lett., 47, e2020GL089264.

[7] Dong, Z., Feng, X., Zhou, H., et al. 2020b, Remote Sens., 12, 629.

[8] Evans, J. V., & Pettengill, G. H. 1963, J. Geophys. Res., 68, 423.

[9] Fa, W., & Wieczorek, M. A. 2012, Icarus, 218, 771.

[10] Fang, G. – Y., Zhou, B., Ji, Y. – C., et al. 2014, Res. Astron. Astrophys., 14, 1607.

[11] Flores – Tapia, D., & Pistorius, S. 2010, EURASIP J. Adv. Signal Process., 2010, 1.

[12] French, W. S. 1974, Geophysics, 39, 265.

[13] Gazdag, J. 1978, Geophysics, 43, 1342.

[14] Giannakis, I., Zhou, F., Warren, C., & Giannopoulos, A. 2021, Geophys. Res. Lett., 48, e2021GL092866.

[15] Gou, S., Di, K., Yue, Z., et al. 2019, Earth Planet. Sci. Lett., 528, 115829.
[16] Guo, D., Fa, W., Zeng, X., Du, J., & Liu, J. 2021, Icarus, 359, 114327.
[17] Heiken, G. H., Vaniman, D. T., & French, B. M. 1991, Lunar Sourcebook, a User's Guide to the Moon (Cambridge: Cambridge University Press).
[18] Hu, X., Ma, P., Yang, Y., et al. 2019, Geophys. Res. Lett., 46, 9439.
[19] Huang, Q., & Wieczorek, M. A. 2012, J. Geophys. Res. Planets, 117, E05003 Huang, J., Xiao, Z., Flahaut, J., et al. 2018, J. Geophys. Res. Planets, 123, 1684.
[20] Keihm, S. J., & Langseth, M. G. 1975, Icarus, 24, 211.
[21] Kim, J.-H., Cho, S.-J., & Yi, M.-J. 2007, Geosci. J., 11, 75.
[22] Lai, J., Xu, Y., Bugiolacchi, R., et al. 2020, Nat. Commun., 11, 1.
[23] Lai, J., Xu, Y., Bugiolacchi, R., et al. 2021, Geophys. Res. Lett., 48, e2021GL095133.
[24] Lai, J., Xu, Y., Zhang, X., et al. 2019, Geophys. Res. Lett., 46, 12783.
[25] Li, C., & Zhang, J. 2021, Remote Sens., 13, 4542.
[26] Li, C., Su, Y., Pettinelli, E., et al. 2020, Sci. Adv., 6, 6898.
[27] Lin, H., Lin, Y., Yang, W., et al. 2020, Geophys. Res. Lett., 47, e2020GL087949 Lu, Y., Wu, Y., Michael, G. G., et al. 2021, Icarus, 354, 114086.
[28] Marpaung, D. H. N., & Lu, Y. 2016, IEEE Geosci. Remote Sens. Lett., 13, 661.
[29] Melosh, H. J. 2011, Planetary Surface Processes, (Cambridge: Cambridge University Press), 13.
[30] Mitchell, J., Houston, W., Scott, R., et al. 1972, Lunar Planet. Sci. Conf. Proc., 3, 3235.
[31] Ogolo, N. A., Akinboro, O. G., Inam, J. E., Akpokere, F. E., & Onyekonwu, M. O. 2015, in SPE Nigeria annual international conference and exhibition, OnePetro.
[32] Olhoeft, G., & Strangway, D. 1975, Earth Planet. Sci. Lett., 24, 394.
[33] Ono, T., Kumamoto, A., Nakagawa, H., et al. 2009, Science, 323, 909.
[34] Özdemir, C., Demirci, Ş., Yiğit, E., & Yilmaz, B. 2014, Math. Problems Eng., 2014, 1.
[35] Peeples, W. J., Sill, W. R., May, T. W., et al. 1978, J. Geophys. Res. Solid Earth, 83, 3459.
[36] Peters, L., Daniels, J., & Young, J. 1994, Proc. IEEE, 82, 1802.
[37] Pettengill, G. H. 1978, Ann. Rev. Astron. Astrophys., 16, 265.
[38] Pieters, C. M., Taylor, L. A., Noble, S. K., et al. 2000, Meteor. Planet. Sci., 35, 1101.
[39] Pun, T. 1980, Signal Process., 2, 223.
[40] Qiao, L., Ling, Z., Fu, X., & Li, B. 2019, Icarus, 333, 37.

[41] Rogers, J. J. , & Head, W. B. 1961, J. Sediment. Res. , 31, 467.
[42] Schneider, W. A. 1978, Geophysics, 43, 49.
[43] Sharpton, V. L. , & Head III, J. W. 1982, J. Geophys. Res. Solid Earth, 87, 10983.
[44] Song, H. , Li, C. , Zhang, J. , et al. 2021, Remote Sens, 13.
[45] Stolt, R. H. 1978, Geophysics, 43, 23.
[46] Stopar, J. D. , Robinson, M. S. , Barnouin, O. S. , et al. 2017, Icarus, 298, 34.
[47] Strangway, D. W. , Olhoeft, G. R. , Massey, H. S. W. , et al. 1977, Philos. Trans. R. Soc. London. Ser. A, Math. Phys. Sci. , 285, 441.
[48] Su, Y. , Fang, G. - Y. , Feng, J. - Q. , et al. 2014, Res. Astron. Astrophys. , 14, 1623.
[49] Taylor, L. A. , Pieters, C. M. , Keller, L. P. , Morris, R. V. , & McKay, D. S. 2001, J. Geophys. Res. Planets, 106, 27985.
[50] Tyler, G. L. , & Howard, H. T. 1973, J. Geophys. Res. , 78, 4852.
[51] Wang, R. , Su, Y. , Ding, C. , et al. 2021, Remote Sens. , 13, 3679.
[52] Wiggins, R. A. 1978, Geoexploration, 16, 21.
[53] Wu, H. —S. , & Barba, J. 1998, IEEE Trans. Syst. Man, Cybernetics, Part B, 28, 227.
[54] Xie, M. , Liu, T. , & Xu, A. 2020, J. Geophys. Res. Planets, 125, e2019JE006113.
[55] Xu, X. , & Miller, E. L. 2001, SPIE, 4394, 742.
[56] Xu, X. , Miller, E. , & Rappaport, C. 2003, IEEE Trans. Geosci. Remote Sens. , 41, 1804.
[57] Xu, L. , Zhang, X. , Qiao, L. , & Jialong. 2021, AJ, 162, 29.
[58] Yilmaz, Ö. 2001, Seismic data analysis: Processing, inversion, and interpretation of seismic data (Society of exploration geophysicists) .
[59] Yuan, Y. , Zhu, P. , Xiao, L. , et al. 2021, Earth Planet. Sci. Lett. , 569, 117062.
[60] Zhang, L. , Li, J. , Zeng, Z. , et al. 2020, Geophys. Res. Lett. , 47, e2020GL088680.
[61] Zhang, J. , Zhou, B. , Lin, Y. , et al. 2021a, Nat. Astron. , 5, 25.
[62] Zhang, L. , Xu, Y. , Bugiolacchi, R. , et al. 2021b, Earth Planet. Sci. Lett. , 564, 116912.
[63] Zhou, H. , Feng, X. , Ding, C. , et al. 2021, IEEE Trans. Geosci. Remote Sens. , 1.

Appendix A: Processing flow of high - frequency data of LPR.

The main procedures of data processing of the high - frequency data are as follows.

The first step is to remove duplicated data. As the raw data of each lunar day are stored in PDS files in different segments separately, it is necessary to merge them to represent all 31 lunar day data as a B - scan profile. There are many redundant echoes reaching the LPR receivers at the same position when the rover stops moving, and needed to be removed.

The second step is Denoising. Ringing usually appears as horizontal and periodic events and is a common type of coherent noise in a GPR profile. In order to eliminate these ringing noises, we subtracted the horizontal average signal at each A - scan trace.

The third step is gain. The amplitude of electromagnetic waves will attenuate during the propagation in the materials, mainly due to the following reasons: dielectric loss and geometric diffusion. As the depth increases, the amplitude of the signals becomes smaller. Therefore, we need to highlight them by gain compensation to make them easier to analyse. It is worth noting that both signals and noise are amplified at the same time.

The fourth step is band - pass filtering. The bandwidth of transmitted signals of high frequency data is from 250MHz to 750MHz. However, the received signals have noise at other frequencies. Moreover, the actual bandwidth of the received signals is lower than the nominal parameters. Through spectrum analysis, we choose the band - pass filter with a frequency range of 150 - 650MHz.

The fifth step is time - lag adjustment. The receiver is operated 28. 203 ns earlier than the transmitter. Invalid data collected during this period should be removed.

The final step is Kirchhoff Migration. The migration method for imaging used in this paper is the Kirchhoff migration. We use the amplitudes of the electric field $E(z=0)$ of the received signals to calculate the depth of z_A in Eq. (A. 1):

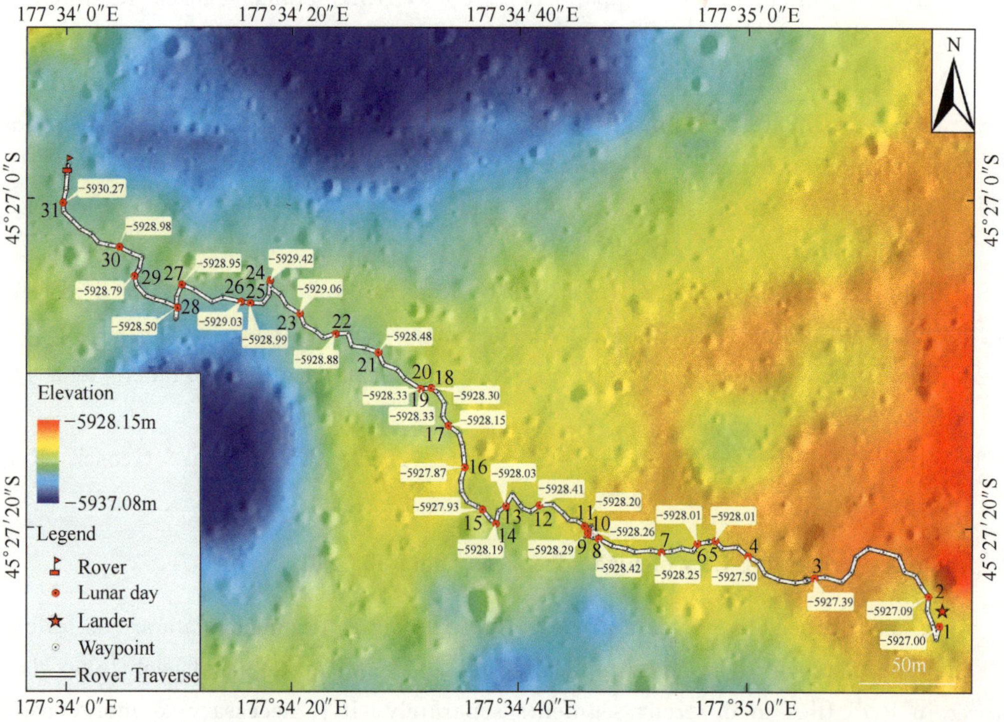

(a)

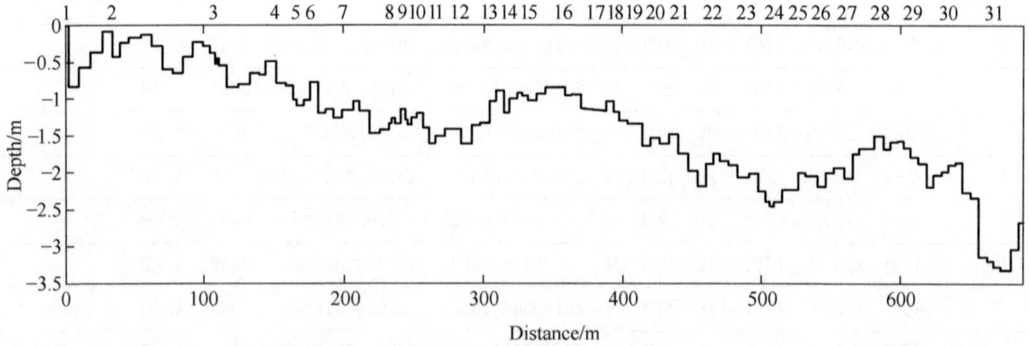

(b)

Fig. A. 1　Routing path of Yutu-2 rover and the surface elevation values for each site.
(a) Path of Yutu-2 with elevation values at each site. (b) Elevation values profile read from LPR data

$$E_A(x_A, z_A, t=0) = d_{\frac{1}{2}}(t) \otimes \int_{Lx} \frac{\cos\phi}{\sqrt{2\pi r}} E\left(x, z=0, t=\frac{r}{c}\right) dx \qquad \text{(A. 1)}$$

where ϕ is the view angle of the target, r is the distance from the antennas to the hyperbolic curves.

Table A. 1　　IDs of LPR CH-2 data

CE4 _ GRAS _ LPR －2B _ SCI _ N _ 20190104004000 _ 20190109213900 _ 0001 _ A. 2B
CE4 _ GRAS _ LPR －2B _ SCI _ N _ 20190109213901 _ 20190111150200 _ 0002 _ A. 2B
CE4 _ GRAS _ LPR －2B _ SCI _ N _ 20190111150201 _ 20190112102100 _ 0003 _ A. 2B
CE4 _ GRAS _ LPR －2B _ SCI _ N _ 20190131080301 _ 20190201054300 _ 0005 _ A. 2B
CE4 _ GRAS _ LPR －2B _ SCI _ N _ 20190201054301 _ 20190208101200 _ 0006 _ A. 2B
CE4 _ GRAS _ LPR －2B _ SCI _ N _ 20190208101201 _ 20190209053100 _ 0007 _ A. 2B
CE4 _ GRAS _ LPR －2B _ SCI _ N _ 20190209053101 _ 20190209130800 _ 0008 _ A. 2B
CE4 _ GRAS _ LPR －2B _ SCI _ N _ 20190209130801 _ 20190210033400 _ 0009 _ A. 2B
CE4 _ GRAS _ LPR －2B _ SCI _ N _ 20190210033401 _ 20190211100800 _ 0010 _ A. 2B
CE4 _ GRAS _ LPR －2B _ SCI _ N _ 20190301065101 _ 20190301155600 _ 0012 _ A. 2B
CE4 _ GRAS _ LPR －2B _ SCI _ N _ 20190301155601 _ 20190302021300 _ 0013 _ A. 2B
CE4 _ GRAS _ LPR －2B _ SCI _ N _ 20190302131101 _ 20190302160000 _ 0016 _ A. 2B
CE4 _ GRAS _ LPR －2B _ SCI _ N _ 20190303052101 _ 20190303100000 _ 0018 _ A. 2B
CE4 _ GRAS _ LPR －2B _ SCI _ N _ 20190310092701 _ 20190311064400 _ 0020 _ A. 2B
CE4 _ GRAS _ LPR －2B _ SCI _ N _ 20190311064401 _ 20190312052700 _ 0021 _ A. 2B
CE4 _ GRAS _ LPR －2B _ SCI _ N _ 20190312052701 _ 20190312110000 _ 0022 _ A. 2B
CE4 _ GRAS _ LPR －2B _ SCI _ N _ 20190329122801 _ 20190331080000 _ 0023 _ A. 2B
CE4 _ GRAS _ LPR －2B _ SCI _ N _ 20190409110001 _ 20190411050000 _ 0032 _ A. 2B
CE4 _ GRAS _ LPR －2B _ SCI _ N _ 20190428060001 _ 20190429153000 _ 0033 _ A. 2B
CE4 _ GRAS _ LPR －2B _ SCI _ N _ 20190430080001 _ 20190501120000 _ 0034 _ A. 2B

Continued Table A. 1

CE4 _ GRAS _ LPR －2B _ SCI _ N _ 20190508070001 _ 20190509073000 _ 0036 _ A. 2B
CE4 _ GRAS _ LPR －2B _ SCI _ N _ 20190509080001 _ 20190509110000 _ 0037 _ A. 2B
CE4 _ GRAS _ LPR －2B _ SCI _ N _ 20190528000001 _ 20190528080000 _ 0039 _ A. 2B
CE4 _ GRAS _ LPR －2B _ SCI _ N _ 20190529070001 _ 20190529100000 _ 0040 _ A. 2B
CE4 _ GRAS _ LPR －2B _ SCI _ N _ 20190529100001 _ 20190529120000 _ 0041 _ A. 2B
CE4 _ GRAS _ LPR －2B _ SCI _ N _ 20190530070001 _ 20190530160000 _ 0042 _ A. 2B
CE4 _ GRAS _ LPR －2B _ SCI _ N _ 20190607030001 _ 20190609140000 _ 0043 _ A. 2B
CE4 _ GRAS _ LPR －2B _ SCI _ N _ 20190627040001 _ 20190628070000 _ 0044 _ A. 2B
CE4 _ GRAS _ LPR －2B _ SCI _ N _ 20190706073001 _ 20190707040000 _ 0047 _ A. 2B
CE4 _ GRAS _ LPR －2B _ SCI _ N _ 20190707110001 _ 20190708050000 _ 0048 _ A. 2B
CE4 _ GRAS _ LPR －2B _ SCI _ N _ 20190726000001 _ 20190727080000 _ 0049 _ A. 2B
CE4 _ GRAS _ LPR －2B _ SCI _ N _ 20190728010001 _ 20190729020000 _ 0050 _ A. 2B
CE4 _ GRAS _ LPR －2B _ SCI _ N _ 20190805020001 _ 20190805050000 _ 0052 _ A. 2B
CE4 _ GRAS _ LPR －2B _ SCI _ N _ 20190805050001 _ 20190806120000 _ 0053 _ A. 2B
CE4 _ GRAS _ LPR －2B _ SCI _ N _ 20190825003001 _ 20190825053000 _ 0054 _ A. 2B
CE4 _ GRAS _ LPR －2B _ SCI _ N _ 20190827060001 _ 20190827100000 _ 0060 _ A. 2B
CE4 _ GRAS _ LPR －2B _ SCI _ N _ 20190903020001 _ 20190903123000 _ 0061 _ A. 2B
CE4 _ GRAS _ LPR －2B _ SCI _ N _ 20190904153001 _ 20190905083000 _ 0067 _ A. 2B
CE4 _ GRAS _ LPR －2B _ SCI _ N _ 20190923090001 _ 20190923110000 _ 0068 _ A. 2B
CE4 _ GRAS _ LPR －2B _ SCI _ N _ 20191023023001 _ 20191024073000 _ 0091 _ A. 2B
CE4 _ GRAS _ LPR －2B _ SCI _ N _ 20191025020001 _ 20191025073000 _ 0092 _ A. 2B
CE4 _ GRAS _ LPR －2B _ SCI _ N _ 20191101060001 _ 20191103033000 _ 0094 _ A. 2B
CE4 _ GRAS _ LPR －2B _ SCI _ N _ 20191121083001 _ 20191124031500 _ 0095 _ A. 2B
CE4 _ GRAS _ LPR －2B _ SCI _ N _ 20191201011501 _ 20191202054500 _ 0097 _ A. 2B
CE4 _ GRAS _ LPR －2B _ SCI _ N _ 20191221030501 _ 20191221123000 _ 0099 _ A. 2B
CE4 _ GRAS _ LPR －2B _ SCI _ N _ 20191222124001 _ 20191223083000 _ 0102 _ A. 2B
CE4 _ GRAS _ LPR －2B _ SCI _ N _ 20191230143501 _ 20191231062000 _ 0108 _ A. 2B
CE4 _ GRAS _ LPR －2B _ SCI _ N _ 20191231080001 _ 20191231120000 _ 0109 _ A. 2B
CE4 _ GRAS _ LPR －2B _ SCI _ N _ 20191231133001 _ 20200101041000 _ 0111 _ A. 2B
CE4 _ GRAS _ LPR －2B _ SCI _ N _ 20200120021001 _ 20200120084000 _ 0112 _ A. 2B
CE4 _ GRAS _ LPR －2B _ SCI _ N _ 20200121015001 _ 20200121142500 _ 0113 _ A. 2B
CE4 _ GRAS _ LPR －2B _ SCI _ N _ 20200218022501 _ 20200219023000 _ 0119 _ A. 2B
CE4 _ GRAS _ LPR －2B _ SCI _ N _ 20200219074001 _ 20200220102000 _ 0120 _ A. 2B
CE4 _ GRAS _ LPR －2B _ SCI _ N _ 20200228070001 _ 20200229091000 _ 0122 _ A. 2B
CE4 _ GRAS _ LPR －2B _ SCI _ N _ 20200318083501 _ 20200321035000 _ 0123 _ A. 2B
CE4 _ GRAS _ LPR －2B _ SCI _ N _ 20200328024001 _ 20200329102000 _ 0124 _ A. 2B

Continued Table A. 1

CE4 _ GRAS _ LPR －2B _ SCI _ N _ 20200417070501 _ 20200418110000 _ 0125 _ A. 2B
CE4 _ GRAS _ LPR －2B _ SCI _ N _ 20200419030501 _ 20200419125000 _ 0126 _ A. 2B
CE4 _ GRAS _ LPR －2B _ SCI _ N _ 20200427043501 _ 20200428095500 _ 0128 _ A. 2B
CE4 _ GRAS _ LPR －2B _ SCI _ N _ 20200615025001 _ 20200616030000 _ 0129 _ A. 2B
CE4 _ GRAS _ LPR －2B _ SCI _ N _ 20200624103001 _ 20200625135000 _ 0134 _ A. 2B
CE4 _ GRAS _ LPR －2B _ SCI _ N _ 20200715053001 _ 20200716022500 _ 0135 _ A. 2B
CE4 _ GRAS _ LPR －2B _ SCI _ N _ 20200716031001 _ 20200716100000 _ 0136 _ A. 2B
CE4 _ GRAS _ LPR －2B _ SCI _ N _ 20200717022501 _ 20200717061500 _ 0137 _ A. 2B
CE4 _ GRAS _ LPR －2B _ SCI _ N _ 20200724050501 _ 20200725062000 _ 0139 _ A. 2B
CE4 _ GRAS _ LPR －2B _ SCI _ N _ 20200814014001 _ 20200814075500 _ 0140 _ A. 2B
CE4 _ GRAS _ LPR －2B _ SCI _ N _ 20200815011501 _ 20200815044500 _ 0141 _ A. 2B
CE4 _ GRAS _ LPR －2B _ SCI _ N _ 20200822052501 _ 20200823065500 _ 0143 _ A. 2B
CE4 _ GRAS _ LPR －2B _ SCI _ N _ 20200912023501 _ 20200913083000 _ 0144 _ A. 2B
CE4 _ GRAS _ LPR －2B _ SCI _ N _ 20200914020501 _ 20200914044000 _ 0145 _ A. 2B
CE4 _ GRAS _ LPR －2B _ SCI _ N _ 20200922052501 _ 20200922130000 _ 0147 _ A. 2B
CE4 _ GRAS _ LPR －2B _ SCI _ N _ 20201011055001 _ 20201012105500 _ 0148 _ A. 2B
CE4 _ GRAS _ LPR －2B _ SCI _ N _ 20201013020001 _ 20201013054000 _ 0149 _ A. 2B
CE4 _ GRAS _ LPR －2B _ SCI _ N _ 20201020045501 _ 20201020062000 _ 0150 _ A. 2B
CE4 _ GRAS _ LPR －2B _ SCI _ N _ 20201020114501 _ 20201021115500 _ 0151 _ A. 2B
CE4 _ GRAS _ LPR －2B _ SCI _ N _ 20201110032001 _ 20201112041500 _ 0153 _ A. 2B
CE4 _ GRAS _ LPR －2B _ SCI _ N _ 20201119112501 _ 20201120143000 _ 0155 _ A. 2B
CE4 _ GRAS _ LPR －2B _ SCI _ N _ 20201219044001 _ 20201220084500 _ 0163 _ A. 2B
CE4 _ GRAS _ LPR －2B _ SCI _ N _ 20210108070001 _ 20210109061500 _ 0164 _ A. 2B
CE4 _ GRAS _ LPR －2B _ SCI _ N _ 20210109063001 _ 20210110070500 _ 0165 _ A. 2B
CE4 _ GRAS _ LPR －2B _ SCI _ N _ 20210117080001 _ 20210118070500 _ 0167 _ A. 2B
CE4 _ GRAS _ LPR －2B _ SCI _ N _ 20210118112501 _ 20210119100500 _ 0170 _ A. 2B
CE4 _ GRAS _ LPR －2B _ SCI _ N _ 20210206053501 _ 20210207092000 _ 0172 _ A. 2B
CE4 _ GRAS _ LPR －2B _ SCI _ N _ 20210208020001 _ 20210208070000 _ 0173 _ A. 2B
CE4 _ GRAS _ LPR －2B _ SCI _ N _ 20210217032001 _ 20210217125500 _ 0178 _ A. 2B
CE4 _ GRAS _ LPR －2B _ SCI _ N _ 20210308043501 _ 20210309093000 _ 0179 _ A. 2B
CE4 _ GRAS _ LPR －2B _ SCI _ N _ 20210310021501 _ 20210310041000 _ 0180 _ A. 2B
CE4 _ GRAS _ LPR －2B _ SCI _ N _ 20210318023001 _ 20210319111000 _ 0182 _ A. 2B
CE4 _ GRAS _ LPR －2B _ SCI _ N _ 20210406074001 _ 20210407051000 _ 0183 _ A. 2B
CE4 _ GRAS _ LPR －2B _ SCI _ N _ 20210407092001 _ 20210408124000 _ 0184 _ A. 2B
CE4 _ GRAS _ LPR －2B _ SCI _ N _ 20210416063001 _ 20210417105000 _ 0186 _ A. 2B
CE4 _ GRAS _ LPR －2B _ SCI _ N _ 20210604074001 _ 20210606083000 _ 0193 _ A. 2B
CE4 _ GRAS _ LPR －2B _ SCI _ N _ 20210614050001 _ 20210615071000 _ 0195 _ A. 2B

Continued Table A. 1

CE4 _ GRAS _ LPR －2B _ SCI _ N _ 20210704014001 _ 20210705071500 _ 0196 _ A. 2B
CE4 _ GRAS _ LPR －2B _ SCI _ N _ 20210706020001 _ 20210706052000 _ 0197 _ A. 2B
CE4 _ GRAS _ LPR －2B _ SCI _ N _ 20210713051001 _ 20210714100000 _ 0199 _ A. 2B
CE4 _ GRAS _ LPR －2B _ SCI _ N _ 20210714173001 _ 20210714180000 _ 0200 _ A. 2B

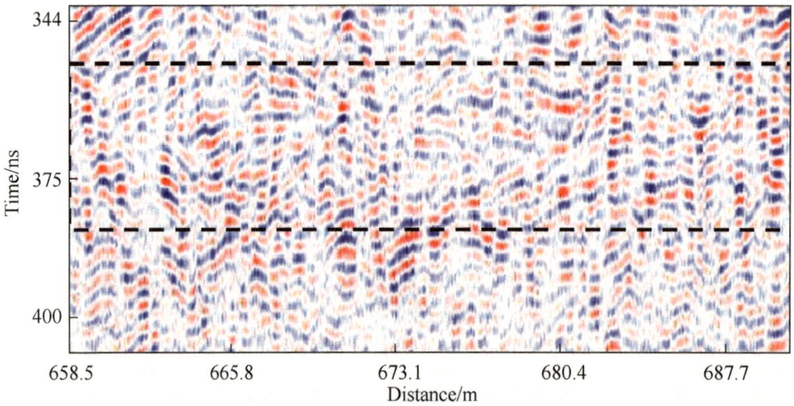

(a)

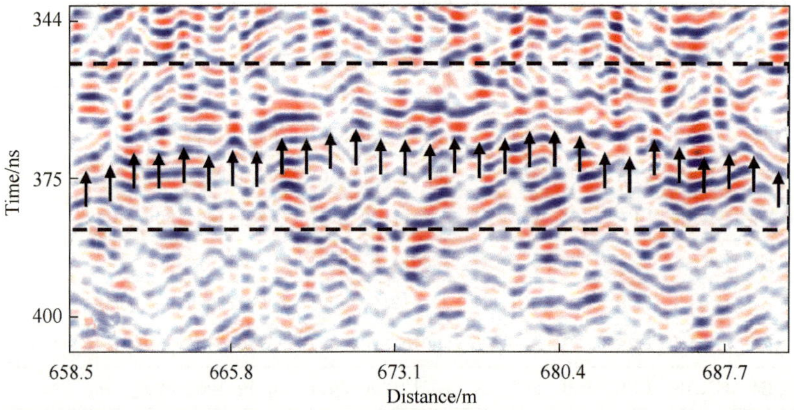

(b)

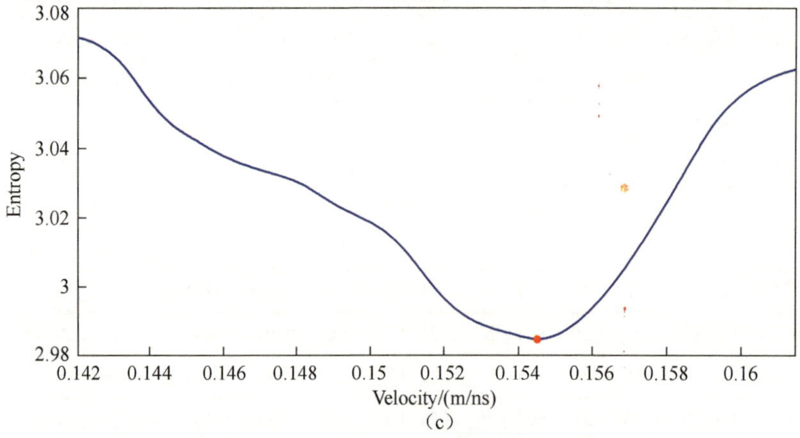

(c)

Fig. A. 2　Example of waves without significant diffraction.

(a) Area of fields with non－obvious diffraction but reflection. (b) Horizontal reflector after migration depicted by a set of black arrows in the black rectangle. (c) Variation of entropy with electromagnetic wave velocity in the material

The gardening process of lunar regolith by small impact craters: A case study in Chang'E – 4 landing area

Ke Shi[1,2], Zongyu Yue[1,4], Kaichang Di[1,4],
Jianzhong Liu[3,4], Zehua Dong[5]

1. State Key Laboratory of Remote Sensing Science, Aerospace Information Research Institute, Chinese Academy of Sciences, Beijing 100101, China
2. University of Chinese Academy of Sciences, Beijing 100049, China
3. Institute of Geochemistry Chinese Academy of Sciences, Guiyang 550002, China
4. CAS Center for Excellence in Comparative Planetology, Hefei 230026, China
5. School of Information and Electronics, Beijing Institute of Technology, Beijing 100081, China

Abstract: On January 3, 2019, Chang'E – 4 (CE – 4) probe successfully achieved soft landing on the farside of the Moon. A large number of small craters in the landing area can be found in the high – resolution remote sensing images. These small craters played a vital role in the evolution of lunar regolith because of the overturning and mixing of lunar surface material through excavating and ejecting during impact processes. However, the research of the gardening process by small craters and the evolution of farside lunar regolith are still insufficient. This research aims to make quantitative analysis to the evolution of the lunar regolith from regional small impacts and apply them in the CE – 4 landing area. Firstly, the processes of excavation and ejection of a single small impact on the lunar surface are analyzed, during which the formation and distribution of lunar regolith are numerically modeled. Then, based on the remote sensing observation and the analysis of regional impact flux, a group of impact events with resulting lunar regolith are modeled by Monte Carlo method. Thirdly, the above analysis is applied in the CE – 4 landing area, and the thickness, grain size distribution, and overturning times of the lunar regolith are calculated. Finally, the influencing factors are discussed and evolution process of lunar regolith is summarized. The results indicate that the formation of lunar regolith is controlled by larger impact events, while the surficial overturning and mixing of lunar regolith are mainly caused by smaller impact events. In addition, the lunar regolith is mostly formed in the earlier times because of the high impact flux at that time. For example, the absolute model age of the CE – 4 landing area is about 3.5Ga, while the lunar regolith had mostly been formed before 3.0 Ga and changed slowly afterwards. Our model beyond the observational capability of the Yutu - 2's lunar penetrating radar (LPR), and the results can contribute to better understanding of the regional lunar regolith formation and evolution.

Keywords: Lunar regolith; Gardening process; Small crater; Chang'E - 4; Lunar penetrating radar

1 Introduction

Lunar regolith is defined as the mixture of rock debris, mineral particles and a small amount of molten glass on the surface of the Moon (Shoemaker et al., 1967a; Shuster et al., 2010), which is generally believed that it is mainly the product of impact including meteorite impact, micro meteorite impact, and solar wind high - energy particle bombardment (Melosh, 1989). It's important to note that the lunar regolith usually refers to the lunar Surficial Regolith layer above the Upper Megaregolith layer produced by ejecta deposition or collapse modifications of large impact craters (Shoemaker et al., 1967b; Shoe - maker et al., 1968; Wilcox et al., 2005; Richardson and Abramov, 2020). The formation and evolution of the lunar regolith are closely related to impact. The depth and structure of lunar regolith are mainly determined by the meteorite impacts, which can excavate and splatter the material from the deep to the surface, and some original surficial material can be also buried in the process. In previous studies, the above overturning and mixing processes of surficial material caused by impact is defined as impact gardening (Gault et al., 1974; Arnold, 1975). Because there is no atmosphere on the Moon and few tectonic activities existed in last 3.0Ga (Wilhelms et al., 1987), the information of meteorite impact along with micro meteorite impact and solar wind particle bombardment has been mostly recorded in the lunar surface materials (Crawford et al., 2010; Shuster et al., 2010; Lin et al., 2020). Therefore, understanding of lunar impact gardening process is significant to the research of the lunar impact history, lunar surface rock fragmentation, lunar regolith formation and surficial structure, and even the content of some materials, such as the solar wind high - energy particles and water (Fruchter et al., 1978; Crider and Vondrak, 2003; Hurley et al., 2012; Costello et al., 2018).

Since the first high - resolution image of the Moon was sent back by Ranger - 7 in 1964 (https://www.jpl.nasa.gov/missions/ranger - 7), the lunar regolith has been concerned and its evolution has been substantially studied. Oberbeck et al. (1973) developed a Monte Carlo model to describe impact events and analyze the relationship between the growth of lunar regolith and impact events. Since then this method has been widely used in the research of lunar regolith (e.g., Quaide and Oberbeck, 1975; Hörz, 1977; Cashore and Woronow, 1985). Gault et al. (1974) proposed a groundbreaking model for the mixing of the weathering layer (previously called the Surficial Regolith layer), in which the impact flux is assumed to obey Poisson probability distribution. Based on this, Gault et al. (1974) analyzed the probability of a point at a certain depth affected by the impact event, and described the overturning rate of lunar regolith at different times and depths. Afterwards, Arnold (1975), Fischer and Pieters (1995), Li and Mustard (2005) carried out further research based on this model, and analyzed the impact gardening effect on the spatial distribution of lunar regolith in more detail. Recently, the lunar regolith evolution model was further developed along with the advancement of computing technology and research progresses of crater production and crater degradation. Richardson (2009) proposed a Cratered Terrain Evolution Model, which comprehensively studied the generation and disappearance of impact craters and ejecta. Huang et al. (2017) added a regolith transport model based on the CTEM model to study the impact transport and mixing effect of lunar mare/highland contact. Costello et al. (2018) updated the classical model proposed by Gault et al. (1974) with the novel research results of impact flux and crater scaling law. Liu et al. (2019) analyzed the long - term gardening process on impact melts in the case of three impact basins of Serenitatis, Crisium, and Imbrium. These studies have greatly enhanced our understanding of the impact gardening process and the lunar regolith evolution. However, because of the absence of the necessary data sets, the previous studies mainly focused on the spatial distribution of lunar regolith and lacked verification using in - situ high - resolution stratigraphic data. In this paper, based on the latest LROC NAC image and in - situ stratigraphic data from Chang'E - 4 (CE - 4) mission, we studied the impact gardening and evolution process of the lunar regolith. The results from our model are compared with the in - situ observations from Yutu - 2 rover, providing further information on the evolution process of the lunar regolith. Therefore, the analysis of the impact gardening and evolution process of lunar regolith in the CE - 4 landing area can contribute to the understanding of the evolution of lunar surface weathering layer and then provide some information for future sampling missions.

CE - 4 is the first soft landing mission on lunar farside and the probe successfully landed in the Von Kármán crater inside the South Pole - Aitken basin on January 3, 2019 (Di et al., 2019). The Yutu - 2 rover has worked for 27 lunar days and traveled 652.6m (till February 19, 2021). The Yutu - 2 rover is equipped with a lunar penetrating radar (LPR), which acquires information of surficial structures of lunar regolith (Jia et al., 2018). The LPR has two detection channels that can emit electromagnetic waves of two frequencies. The channel 1 has a low frequency of 60MHz with the maximum detection depth ~500m and a spatial resolution of ~10m, while the channel 2 has a high frequency of 500MHz with the maximum detection depth ~50m and a spatial resolution of ~ 30cm (Zhang et al., 2021). The reflection intensity of LPR signal is related to the state of underground materials and the detection result will help to reveal the information of subsurface stratigraphic structure (Zhang et al., 2021; Dong et al., 2021). These valuable data provide us with information of the lunar surface and shallow subsurface, which is crucial to the understanding of the impact mixing of lunar surface materials and the evolution process of lunar regolith (Hu et al., 2019; Lin et al., 2020). By combining the latest Yutu-2 LPR data and impact gardening technology, the main purpose of this research is to establish a model of the fine stratigraphic structure of the subsurface lunar regolith and analyze the evolution process of lunar regolith by small impact craters in the case of CE-4 landing area.

2 Lunar regolith evolution model with simple craters

Our evolution model with simple craters (EMSC) is based on remote sensing observation of the small craters, and combing the existing models for the lunar regolith formation and distribution during impact, the distribution of lunar regolith grain size, the lunar craters production and chronology, and the terrain degradation, to model the regional lunar regolith evolution.

2.1 Lunar regolith formation and distribution during impacts

The impact cratering process can be divided into contacting and compression stage, excavation stage, and modification stage (Melosh, 1989). When the asteroid contacts the surface of the planet, a strong shock wave is produced and both the target and projectile are heavily compressed. The shock wave moves rapidly outward in the target and backward to the tail of projectile, and a rarefaction wave is reflected back when the shock wave arrived the tail of the projectile. The rarefaction wave propagates much faster in the compressed material and unloads the high pressure caused by the shock wave. When the head of rarefaction wave catches the shock wave, the first stage of the impact process terminates. In the second stage, the irreversible plastic deformation during the first stage causes the excavation flow, which results in a transition crater and the inside material is radially ejected and an ejecta blanket is formed surrounding the transient crater. In the third stage, because of the gravity instability, some material on the rim and wall of the transient crater slides to the crater bottom and a breccia lens is produced in the crater center (Melosh, 1989, 2011; David, 2007; Osinski et al., 2011). The lunar regolith is mainly formed within the process and the typical distribution for a small crater is shown in Fig. 1.

According to observations and experimental studies, the maximum excavation depth of a small impact is equal to ~1/3 of the transitional crater diameter (Hirabayashi et al., 2018). The material above the maximum excavation depth is migrated outwardly and radially during the excavation stage, and it is finally distributed above the original surface. On the other hand, the material below the maximum excavation depth is compressed and moved downwardly and left inside the crater (Melosh, 1989). Most of ejecta fall back to the planetary surface because of the gravity of the planet, and they deposit surrounding the crater and its thickness gradually decreases with the increase of distance to the crater rim until discontinuous. The ejecta and the debris inside the crater are the main source of the lunar regolith. The ejecta thickness e_{out} of a small impact crater can be described as a function of the distance to the crater center and the crater diameter by a simple formula (Mcgetchin et al., 1973; Sharpton, 2014; Hirabayashi et al., 2018):

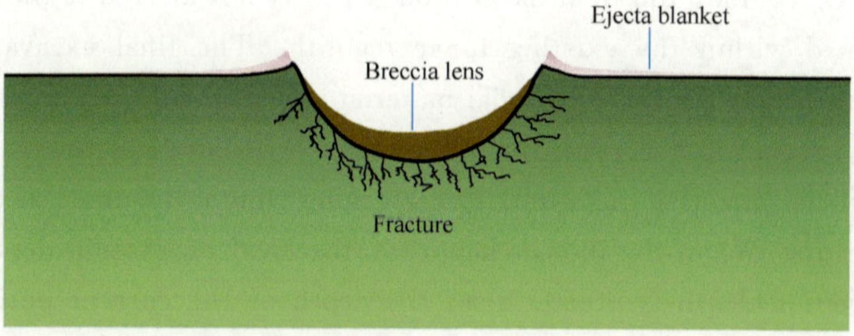

Fig. 1　Typical distribution of the lunar regolith for a small crater. It is notable that an ejecta blanket is formed surrounding the crater rim, and a breccia lens appears in the crater center

$$e_{out} = \sigma R^{1.01}\left(\frac{r}{R}\right)^{-\kappa}, \quad r > R \tag{1}$$

where R is the radius of the final crater in m, r is the distance to the center of the crater in m, and σ and κ are the empirical parameters. For small simple craters, σ and κ can be taken as 0.014 and 3, respectively (Mcgetchin et al., 1973; Sharpton, 2014; Hirabayashi et al., 2018).

After the impact excavation, some molten materials and debris on the rim and wall slips back into the crater and a breccia lens is formed, which is also considered as the lunar regolith. The thickness of this part of the regolith can be derived as the difference between the depth of the transitional crater and the final crater. For small impact crater, the typical thickness of the regolith within the breccia lens deposited inside the impact crater is given by Hirabayashi et al. (2018):

$$e_{in} = \delta R\left(1 - \left(\frac{r}{R}\right)^2\right), \quad r \leqslant R \tag{2}$$

where δ is the depth parameter and it is ~0.34 for most simple craters (Stopar et al., 2017; Hirabayashi et al., 2018). Similarly, the transitional crater can be also considered as a rotating paraboloid and its depth can be described as a function of the distance to the crater center (Melosh, 1989):

$$d_{exc} = d_{max} - \frac{d_{max}}{R_t^2} r^2, \quad r \leqslant R_t \tag{3}$$

where d_{exc} is the excavation depth at the distance r from the center of the crater, d_{max} is the maximum excavation depth of the impact crater and R_t is the radius of the transition crater. Melosh (1989) pointed out that the transition crater diameter is ~0.8 times of the final crater diameter for simple craters. Therefore, the parameters in the above equations can all be calculated if the radius of the final crater is known.

It is important to note that the excavation depth is derived relative to the original surface before impact, and then the final excavation depth of the impact is influenced by the preexisting terrain. For example, if one crater is formed inside an earlier crater, the final excavation depth will be increased provided the later impact penetrates the existing lunar regolith. Of course, the final excavation depth is not altered if the later impact is completely formed within the existing lunar regolith. The final excavation depth can represent the depth that the lunar surficial material is disturbed.

Another byproduct in analyzing the lunar regolith thickness is the overturning times of the lunar surficial material. The initial overturning time is assumed as zero before the impact event occurs. When the impact happens, the local excavation depth is calculated and if the excavation depth is greater than the depth of the current position, the local material would be overturned once. Similarly, the overturning time would be increased by one if the current depth is within the extent of the excavation depth of another impact. Thus, the overturning times of the material at different depths in the study area can be derived when all the impacts have occurred.

2.2 Regolith grain size

The impact of meteorite on the planet will produce a strong shock wave, and the material in the target area will be broken. The grain size of lunar regolith is also of significance. A lot of studies have shown that an impact will eject a large number of blocks, and the larger blocks are concentrated near the rim of the crater, while the smaller blocks can eject to a far distance, and the size of the largest fragment is exponentially related to the diameter of the impact crater (Moore, 1971; Bart and Melosh, 2010; Singer et al., 2020). In order to calculate the maximum size of the fragment, the total mass of the ejecta needs to be considered. Liu et al. (2019) pointed out that the volume of the excavated material is equivalent to a torus-like shape with thickness of 1/3 maximum excavation depth and width of transition crater diameter. Therefore, the total volume of excavated material in a single impact event can be calculated as follows (Liu et al., 2019):

$$V_{exc}=\frac{\pi R_t^3}{15} \tag{4}$$

Based on the spatial distribution investigations of the fragments on the lunar craters, earth craters, nuclear explosion experiments and laboratory experiments, Melosh (1989) summarized the maximum mass of fragment of the impact crater and pointed out that the mass of the largest fragment in the ejecta produced by small simple impact events $m_{f\max}$ can be related to the total mass of the ejecta M_e (Melosh, 1989) as follows:

$$m_{f\max}=0.8M_e^{0.8} \tag{5}$$

The total mass of the ejecta can be calculated via multiplying the total volume of the ejecta (Eq. (4)) by the density of the material, from which we can get the size of the largest fragment in the ejecta in an impact event:

$$size_{f\max} = \left(\frac{0.8\left(\frac{\pi \rho R_t^3}{15}\right)^{0.8}}{\rho} \right)^{1/3} \tag{6}$$

where ρ is the density of the material. Many researches have shown that the size of the ejecta decreases rapidly with the increased distance to the center of the crater, and similar to lunar regolith, the cumulative frequency distribution of the mass of the fragments showing an exponential relationship (Melosh, 1989; Krishna and Kumar, 2016). According to the observation of the impact crater and landslides on the Earth, we speculate that larger materials seem to have a larger friction coefficient and tend to stay on the crater wall, while small materials have a smaller friction coefficient and are easier to slide down to the crater bottom. It is undeniable that there may be some large fragments at the bottom of the crater. The changeable rule of the grain size inside the crater is still lack of research and the detailed law needs to be further testified. Further, by analogy with the laws of lunar regolith inside and outside the crater, it can be assumed that the size of the ejecta outside the crater decreases exponentially with the distance to the center of the crater, and the size of the ejecta inside the crater changes in a parabolic relationship with the distance:

$$size_{out} = \sigma' size_{f\max} r^{-b'} \tag{7}$$

$$size_{in} = \delta' size_{f\max} \left(1 - \left(\frac{r}{R}\right)^2\right) \tag{8}$$

where σ' and δ' are the integrity coefficients, both of which are related to the size of the material before impact. At first, the material integrity factor is 1, then the impact event occurred making the material broken and the strength decrease. Therefore, the coefficient decreases accordingly. When the material is almost completely broken ($size \leqslant 10^{-3}$ m), the coefficient is approximately zero. In the process of impact damaging, the shear force plays an important role and the research of rock mechanics indicates that under the mechanism of shear failure, the stress and strain show a linear relationship from the beginning of damage (integrity factor $=0$) to complete broken (integrity factor $=1$) (Stesky et al., 1974; Byerlee, 1978; Collins et al., 2004). Therefore, the coefficients can be calculated as $\sigma' = size_{bf}/size_0$, $\delta' = size_{bf}/size_0$. However, in the process of impact excavation, if the excavation depth of a certain position exceeds the historical accumulated lunar regolith thickness, it can excavate the initial material and for the impact proportion coefficient of penetration, the coefficient $\delta' = size_{org}/size_0$. Where $size_{bf}$ is grain size of the local material before each impact, $size_{org}$ is the initial material size of the study area and $size_0$ is the size of the bedrock, which can be taken as a large diameter because it should not influence the results. The exponent b' is related to the cumulative frequency distribution of fragment mass, which can be taken as 0.85 (Melosh, 1989).

It is worth noting that the above calculation for the lunar regolith grain size is applicable to primary crater characterized by the very high speed, while for secondary crater, the relatively low impact speed leads to less shock pressures on the target material during cratering process, and much of the ejecta grains will simply be expelled "as is", having been overturned and relocated, but without much modified in other properties. Therefore, the existence of the secondary crater will also make some large fragments appear on the Moon surface (Bart and Melosh, 2010), which is also consistent with the in-situ exploration data (Ding et al., 2020; Gou et al., 2020).

2.3 Impact events

For a specific area on the Moon at some moment, the happening of the impact events should be random, and the random events can be described by Monte Carlo simulation. Monte Carlo method was first introduced to simulate the production process of impact craters by Woronow (1977) and have been often used to simulate the production and evolution of impact craters on the Moon surface (Chapman and McKinnon, 1986; Richardson, 2009; Michael et al., 2018; Liu et al., 2019).

Now we consider a region with an area of $1km^2$ (or others in the same geological unit) and the lower left corner as the origin of the coordinates, and simulate the impact events in the region with Monte Carlo analysis. In the selected area, if we have identified N_c impact craters with diameter larger than D_m by remote sensing interpretation, we should simulate $N_c(D > D_m)$ impact events and analyze the lunar regolith evolution with them. Our simulation grid is 1000×1000, which means that the spatial resolution of the Monte Carlo simulation is 1m, and the boundary condition is set as open boundary. The impact crater size range generated in the simulation is the same as that observed by remote sensing (5 - 317.6m, see Section 3.1).

In the case of CE-4 landing area, the location and size of each impact crater can be obtained through remote sensing observation and interpretation. However, the time of each impact event is difficult to be accurately dated. According to the previous study, the cumulative frequency distribution of impact craters should conform to the power-law relationship which can be described by the lunar production function (Neukum, 1983; Costello et al., 2018). An often-quoted lunar production function is an eleventh-order polynomial advised by Neukum (Neukum, 1983; Neukum et al., 2001):

$$\log(N_{cum}(D, t)) = \sum_{i=0}^{11}(a_i(\log D)^i) \tag{9}$$

The function describes the cumulative number of impact craters $N_{cum}(D, t)$ with the diameter larger than D per km^2 until time t, where $a_1 \sim a_{11}$ are constants that can be found in Neukum (1983) and a_0 can be calculated provided $N(1, t)$ is known, which is derived according to the chronology function as follows:

$$N(1, t) = 5.44 \times 10^{-14}(e^{6.93t} - 1) + 8.38 \times 10^{-4}t \tag{10}$$

where t is the age in Ga. Therefore, $N(1, t)$ can be obtained provided the surface age t has been known through remote sensing observations. For example, the surface age of the CE-4 landing area is 3.5Ga (Gou et al., 2021) (See section 3.1), and $N(1, t)$ equals to 0.0048km^{-2}. As a result, all the parameters in Eq. (9) are known and the partial differential equations of N can also be derived. For a given time t, the partial derivative of N with respect to t indicates the variation of the cumulative cratering rate with the diameter at the given time (Fig. 2a). For a given diameter D, the partial derivative of N with respect to t indicates the variation of the cumulative cratering rate with the time at the given diameter (Fig. 2b). These partial derivatives are the basis to set the cumulative cratering rate at different time or different diameters in the Monte Carlo simulations. It is noteworthy that: (1) the number of small impact craters is much more than the large ones; (2) the early cratering rate is much higher than that of today; (3) large impact craters are mostly formed in earlier times.

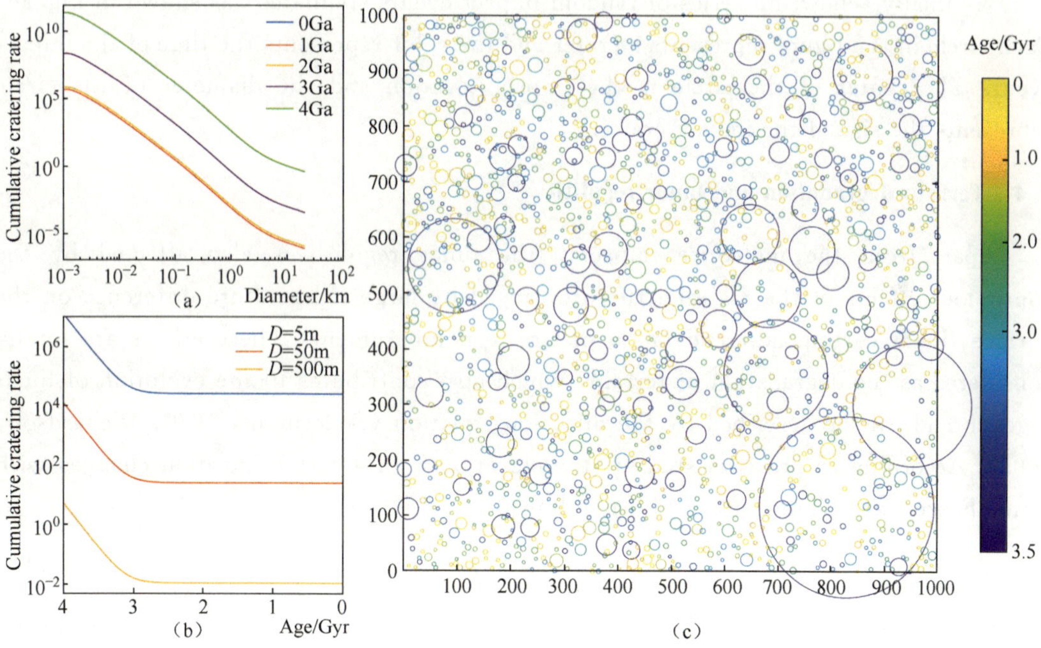

Fig. 2　Random impact events.

(a) The cumulative cratering rate at different time for different diameters, note that the curves of 0 Ga and 1 Ga almost coincide; (b) Cumulative impact rates for craters with different diameters on the Moon; (c) The impact events within 1km^2 around CE-4 landing site obtained by remote sensing interpretation and Monte Carlo simulation, within which each circle represents a crater produced by an impact event, and the color represents the time of the impact event

In the simulation of impact events each impact event is described by three properties: the position, the size and the emplacement order (or it can also be considered as the crater age). These three properties can all be generated by Monte Carlo simulation. In this research, we studied the evolution of lunar regolith in the case of the landing area of CE-4 specifically. We identified and counted the impact craters within 1km^2 in the landing site through remote sensing interpretation (see section 3.1). In order to facilitate comparison with the in - situ detection data (i. e. , the LPR data), the Monte Carlo simulation of impact craters production in our EMSC model is based on the results of remote sensing observation. Therefore, in the Monte Carlo simulation, only the emplacement order of craters is described by random numbers, while the location of the crater center and the size of the crater are determined from remote sensing interpretation. However, impact crater is the most common topographic feature on the lunar surface and its generation and evolution need to accord with the corresponding physical laws in nature. Although we created a series of random numbers to describe the impact order, the simulation of impact order is not a bunch of chaotic and irregular random numbers completely, the overall distribution of these random numbers conforms to the laws reflected by the impact crater production function and chronology function of Eq. (9), (10) and their partial derivatives, as described above. Based on this, we carried out Monte Carlo simulation of impact events in the study area.

We finally achieved a series of random impact events simulated, as shown in Fig. 2c. Each circle in the figure represents a crater and the color represents the time of the impact event. The center of the circle is the impact location and the diameter of the circle represents the size of the impact crater.

2.4 Terrain degradation and lunar regolith

Apart from the meteorite impacts, the lunar regolith is also affected by the continuous effects of the material rupture caused by large temperature difference on the surface, lunar earthquake, and micro meteorite bombardment. These effects are totally considered as the degradation of terrain, which also contributes to the evolution of lunar regolith and can be described by the diffusion equation (Soderblom, 1970; Richardson, 2009). According to previous studies, the elevation h of a certain position changes with time (Fassett and Thomson, 2014) as follows:

$$\frac{\partial h}{\partial t} = \kappa \nabla^2 h \tag{11}$$

where κ is the terrain diffusion rate. The research shows that the terrain diffusion rate of recent 3.0 Ga is basically constant, with an average value of $\kappa = 5.5\text{m}^2/Myr$ and for the 3.1 - 3.6 Ga, the terrain diffusion rate is about $6 \sim 20\text{m}^2/Myr$ (Fassett and Thomson, 2014). Thus, the terrain diffusion rate is calculated about $5.5 \sim 15.7\text{m}^2/Myr$ for the 3.0 - 3.5 Ga according to a polynomial fitting. Finally, the variable diffusion rate of $\kappa = 5.5 \sim 15.7\text{m}^2/Myr$ for 3.0 - 3.5 Ga and the average diffusion rate of $\kappa = 5.5\text{m}^2/Myr$ after 3.0 Ga is used in this research.

It should be noted that the terrain diffusion rate κ is a function of time, but for the research of terrain evolution, it is not necessary to know the specific value of diffusivity with time and the average terrain diffusion rate takes a more important role in characterizing the state of terrain evolution (Skianis et al. , 2008; Fassett and Thomson,

2014). Using the average diffusion rate, we can calculate the distribution of lunar regolith thickness and grain size at different historical times such as 3. 0Ga, 2. 0Ga and 1. 0Ga (See section 3. 4). This will not affect our final results even if the diffusion rate changes slightly with time.

In implementation of our model, terrain degradation and impact gardening are carried out at the same time, but they are independent of each other in the calculations. The terrain degradation is implemented uniformly in the study area, and the partial differential equation Eq. (11) is solved numerically by spatio - temporal discrete method. Taking the accumulated regolith thickness as the elevation of the terrain, the terrain degradation is calculated every 0. 01 Ga during running of the model. For each calculation the detailed process is to apply Eq. (11) in both x and y directions independently, and the second - order central difference scheme is used in solving the partial differential equation to update the spatial distribution of lunar regolith thickness.

3 Applications to the Chang'E - 4 landing area

3. 1 Remote sensing interpretation of impact craters

The landing area of CE - 4 is in the Von Kármán crater inside the South Pole - Aitken Basin on the farside of the Moon (Di et al. , 2019; Lin et al. , 2020). Recent researches show that the material up to the depth of dozens meters underneath CE - 4 landing area is mostly the ejecta from Finsen crater, and the geological age of Finsen crater is about 3. 5 Ga (Gou et al. , 2021). There are many craters with different diameters that can be identified in high - resolution LRO NAC images. In order to analyze the evolution process of lunar regolith material under the effect of impact gardening in detail, we try to find all craters, according to the comprehensive judgment of crater rim, shadow, terrain and other factors. With the add - in of "CraterTools" in ArcGIS (Kneissl et al. , 2011), we delineated 2049 craters with diameters greater than 5m in the 1km^2 area around the CE - 4 landing point in the LRO NAC image at a spatial resolution of 0. 89m/pixel (the data can be downloaded from Lunar Orbital Data Explorer https://ode. rsl. wustl. edu/Moon/indexProduct Search. aspx). The crater diameter ranges from 5. 0 to 317. 6m, and the mapped craters are shown in Fig. 3a.

Based on the results of remote sensing interpretation and the diameter of the impact crater within the study area, the cumulative number of impact craters is counted and the cumulative size frequency distribution (CSFD) is drawn in the double logarithmic coordinate system and the scatter diagram is shown in Fig. 3b. The standard lunar equilibrium function (Trask, 1966) is used to analyze the equilibrium diameter of impact craters in the study area, which is plotted as red dash line in Fig. 3b. The equilibrium line shows that the slope of the CSFD curve changes gradually at the "knee" position, indicating that the small impact craters in the region are equilibrium. Moreover, as mentioned above, the current geological unit of CE－4 landing area starts from the initial surface completely buried by the ejecta of Finsen impact crater. Therefore, in this research the geological age of the study area should be the same as Finsen crater. On the basis of the statistical dating results of the Finsen impact crater by Gou et al. (2021), the chronological curve (production line) of the landing area is drawn as the blue one in the Fig. 3b, of which two gray lines are the upper and lower limits of the dating error respectively. The geological age is reliable and correspond to the research results of others (Ivanov et al., 2018; Gou et al., 2021), which is obtained by the standard statistical dating analysis of the impact crater, using the production function (PF) and chronology function (CF) from Neukum (1983), and the counting area is a large flat zone at the bottom of the Finsen crater. According to the intersection relationship of CSFD plot, production plot and equilibrium plot, it can be found that the small impact crater in the area is equilibrium, and the equilibrium diameter is around 100 to 200m.

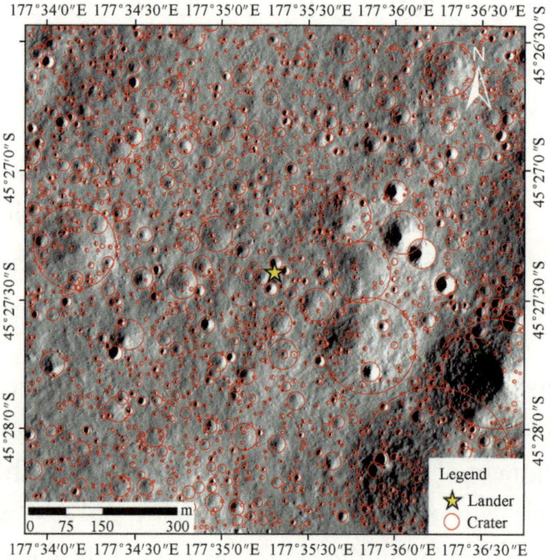

(a)

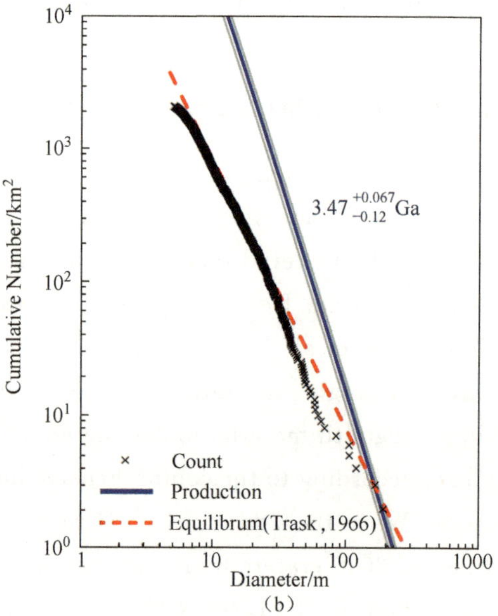

(b)

Fig. 3 Spatial distribution of impact craters within 1km² of CE－4 landing point and CSFD.

(a) Spatial distribution of impact craters. A total of 2049 impact craters were identified. The base image is an orthorectified LRO NAC and the landing point is in the center of the image as shown by the star. (b) CSFD of the impact craters, in which the red is equilibrium line, the blue is the production curve and the gray are the limits of the dating error, respectively. (For interpretation of the references to color in this figure legend, the reader is referred to the web version of this article.)

In addition, the impact events outside the research area may also have some effect on the regolith inside through their ejecta. Therefore, some impact events outside the area are also included our model. Finally, we identified a total of 6093 impact craters within 50 ×30km around the landing site base on mosaic image (Jia et al., 2021). These impact events will be involved in the subsequent calculation of lunar regolith thickness, excavation depth, overturning times and grain size.

It is noteworthy that in our simulation the locations of impact craters are the true locations from the observation results, and the order of the impact events is random. Meanwhile, the small craters themselves in the CE - 4 landing area are also randomly distributed (Jia et al., 2021), thus the impact craters produced by our Monte Carlo simulation based on remote sensing observation are still stochastic in nature.

3.2 Lunar regolith thickness, excavation depth, and overturning times in the CE - 4 landing area

It is necessary to create a thickness model of the lunar regolith when analyzing the impact gardening process. At first, the terrain was hypothesized as flat, and the thickness of the lunar regolith is 0 everywhere. When an impact event occurs, the lunar regolith thickness at a location will be derived according to Eq. (1) or (2). In subsequent impacts, the thickness of the lunar regolith will be calculated as follows. Firstly, the existing lunar regolith will be excavated out according to Eq. (3) if the position is inside the impact crater, that is, its thickness will be firstly reduced. Secondly, the thickness of the preexisted lunar regolith will be set to zero if the excavation depth is larger, and the thickness after the later impact will be derived according to Eq. (2) with the parameters of the newly formed crater. On the other hand, the lunar regolith will be accumulated if the excavation depth is less than the preexisted thickness of lunar regolith. As a result, the thickness of the lunar regolith will be evolved with the impact events one by one, and the spatial distribution of lunar regolith thickness caused by impact gardening in the study area can be obtained when all the impact events are executed.

The terrain degradation process is also included in analyzing the lunar regolith thickness, during which the time is discretized and the thickness of lunar regolith is updated every 0.01 Ga by solving the partial differential equation Eq. (11).

Fig. 4a shows the distribution of lunar regolith thickness in the study area. The vertical and horizontal coordinates are the distances to the lower left corner, and the color indicates the thickness of lunar regolith. In addition, four lines as shown in Fig. 4b～e are drawn to indicate the variation of the lunar regolith thickness in vertical and horizontal directions, respectively. The maximum thickness is ～10.30m, and in most places the thickness is less than 8m. The statistical results indicate that the average thickness is ～3.18m and in ～86.53% of the areas, the thickness is over 1m, ～47.39% over 2m and ～22.83% over 5m. It is clear that (1) The lunar regolith thickness is controlled by the large craters, and the maximum thickness appears in the center of the largest crater. (2) The lunar regolith is mainly composed of breccia lens inside the crater and the proportion of ejecta outside the crater is very small. (3) If one small crater appears inside the larger one, the lunar regolith is decreased instead of being increased. The reason is that in many cases small impacts cannot penetrate the preexisting lunar regolith and then eject them outwardly.

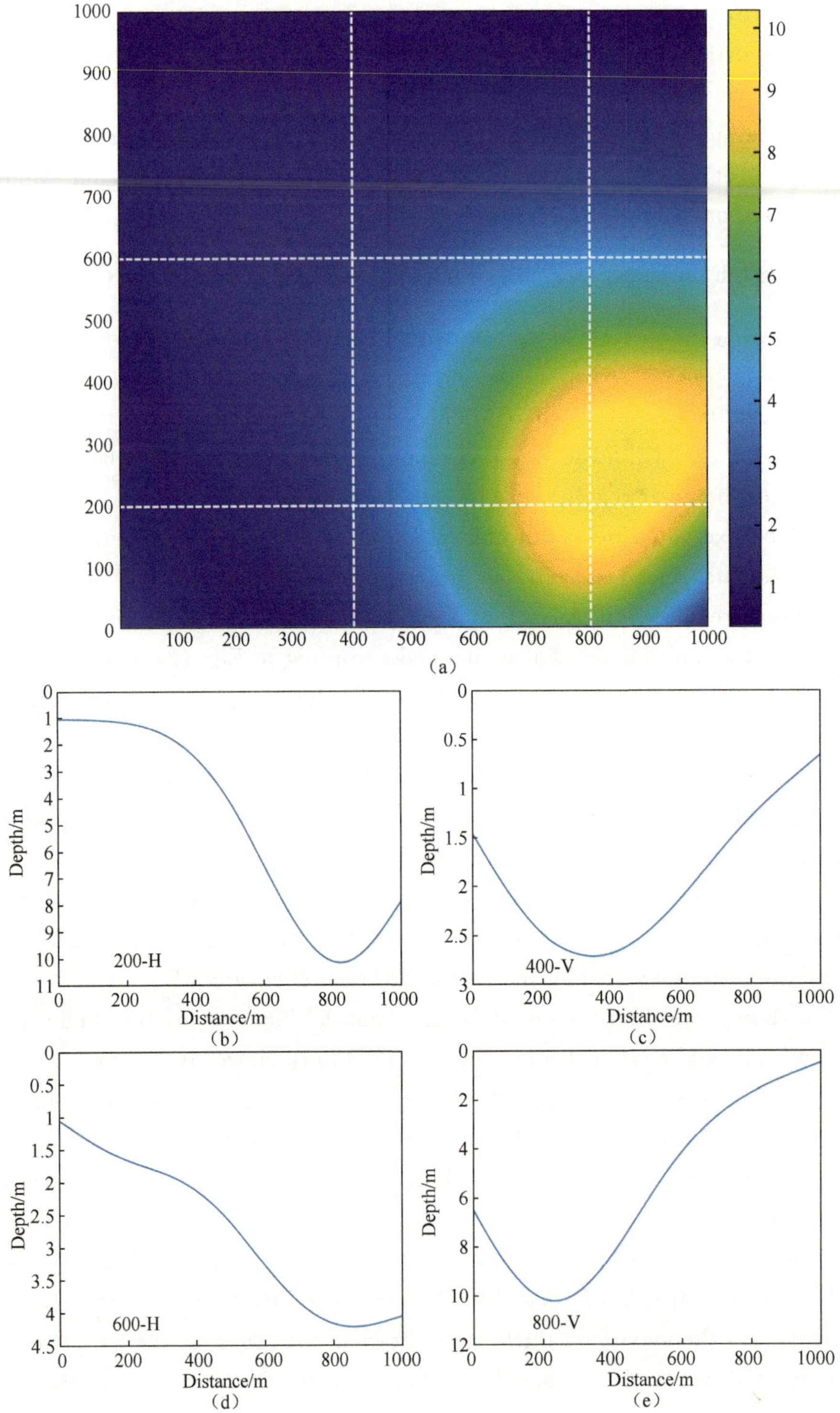

Fig. 4　Thickness of lunar regolith.

(a) Spatial distribution of lunar regolith thickness in the study area, and the horizontal and vertical coordinates indicate the distance to the lower left corner of the study area. The white dotted lines, corresponding to the horizontal 200m, the vertical 400m, the horizontal 600m and the vertical 800 m respectively are the locations of the profiles of lunar regolith thickness in (b)～(e). Note the vertical scales are heavily exaggerated in (b)～(e)

Fig. 5a shows the excavation depth in the study area, which is closely related to the lunar regolith thickness and can represent the maximum disturbing depth of the impact events. Similar to the lunar regolith thickness, the excavation depth is also controlled by the large craters. Different to the lunar thickness distribution, the excavation depth is not influenced by the small craters if they are inside the large ones. The maximum excavation depth is ~83.84m in the center of the largest crater in this area. It can be concluded that the material beneath this depth keeps the original structure. The average maximum excavation depth is ~4.16m and ~15.03% of the areas are deeper than this. In addition, the area percentage for the excavation over 1m, 5m, and 10m are ~20.58%, ~14.15%, and ~10.71%, respectively.

Fig. 5b~e show the overturning times at the depths of 10m, 5m, 1m, and 0.1m, respectively. The results are closely related the excavation depth in Fig. 5a. For example, only the center of the large craters is disturbed one time at the depth of 10m. At the depth of 5m, some areas inside the large craters are disturbed two times where the small impact craters are formed. At the depth of 1m, more area is overturned two times, and some area

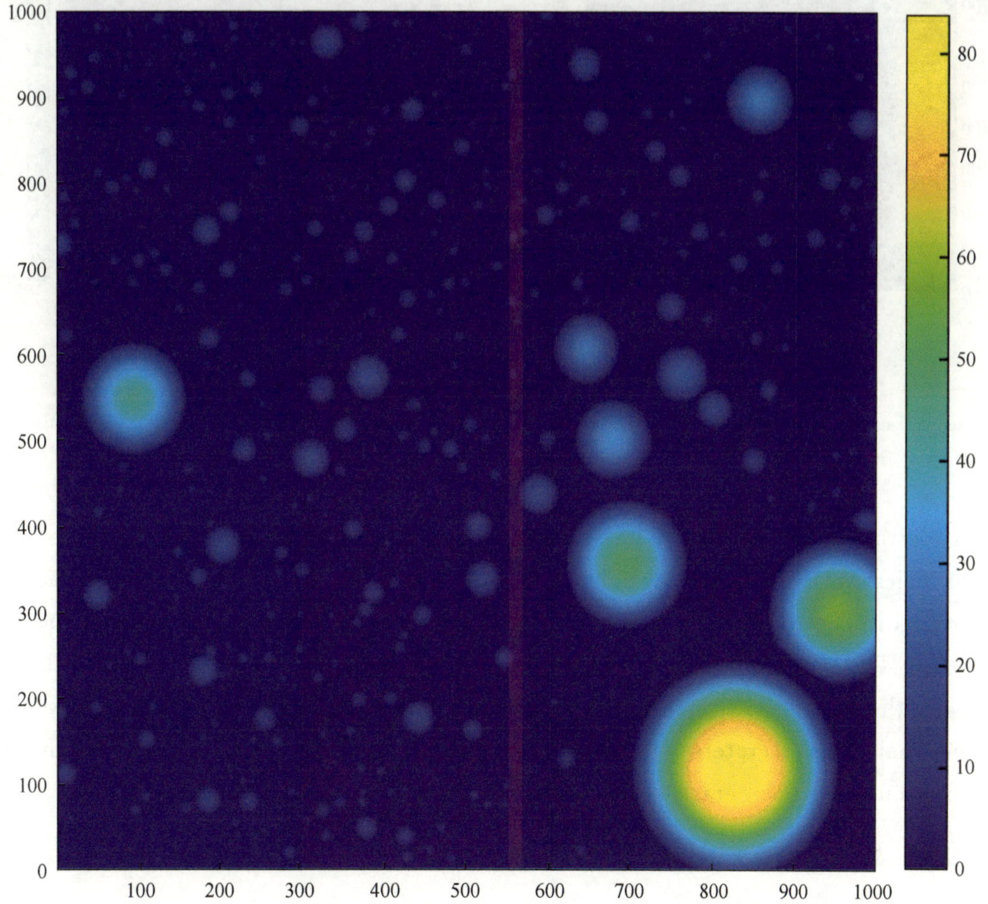

(a)

Fig. 5　Excavation depth and overturning times in the study area

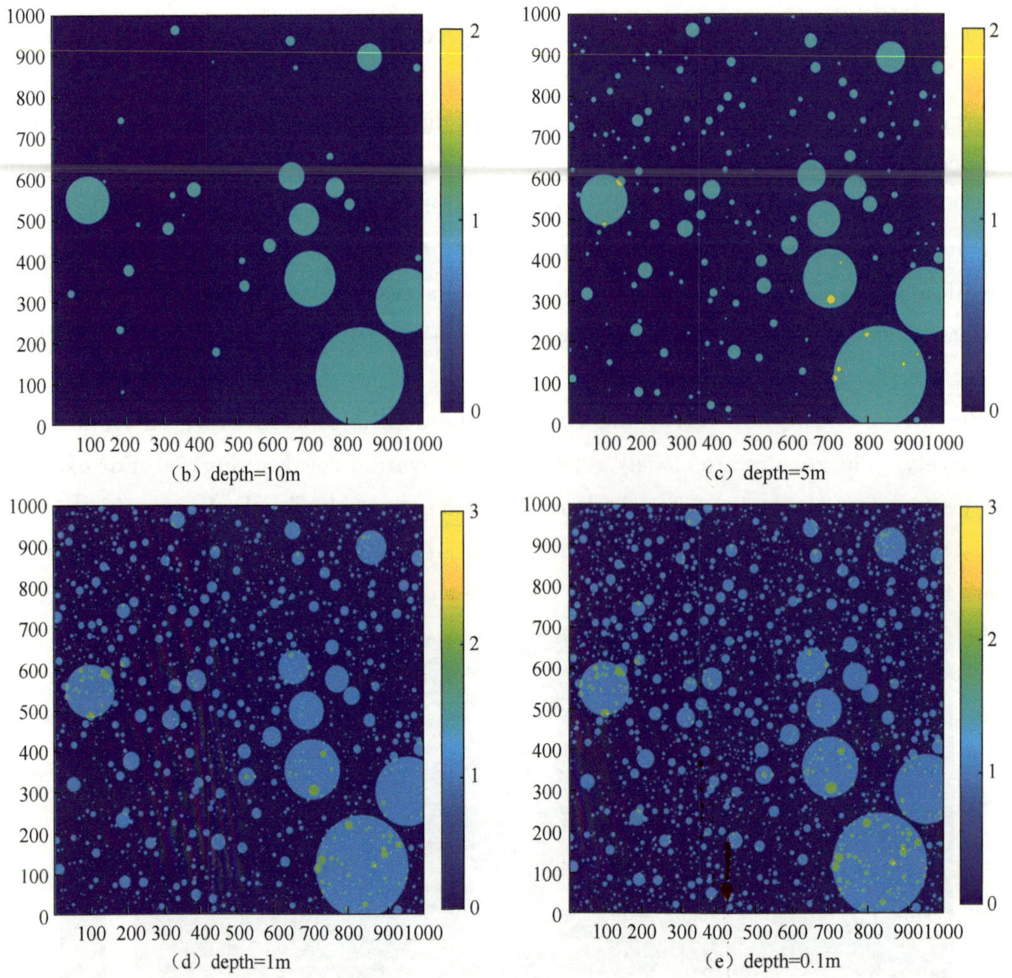

Fig. 5　Excavation depth and overturning times in the study area

(a) The excavation depth in the area. (b) The overturning times at the depth of 10m. (c) The overturning times at the depth of 5m. (d) The overturning times at the depth of 1m. (e) The overturning times at the depth of 0. 1m

is even overturned three times. The overturning times difference between the depths of 5m and 1m indicates that the shallow subsurface material is much easier to be overturned than the material in depth, and the reason is that much more small impacts occurred than the large ones. The situation is more evident in the surficial material at the depth of 0. 1m. However, the maximum overturning times is only three times at this depth, the reason is that the small impact craters with a diameter of less than 100m are in equilibrium, and many of them have been erased, which are not included in the model.

3. 3　Grain size of lunar regolith

The average grain size of the ejecta from the Finsen crater, which is the lunar Upper Megaregolith layer in the CE-4 landing area, is ~0. 06m calculated by using Eqs. (6)~

(7) and the thickness is ~10.47m calculated by using Eq. (1), This grain size is regarded as the initial materials size in the study area. The impact event is again assumed as occurred one by one, and size of the material caused by impact is recorded. When an impact event occurs, the maximum size of the fragment produced by the impact is first calculated by Eq. (6), and then the size of the fragments at a location will be derived according to Eq. (7) or (8). It is worth noting that when the material is broken to a certain extent, the later fragment will no longer follow the above law. Therefore, if the grain size is less than 10^{-3} m, further crushing will not be calculated. Finally, the distribution of the material size can be obtained.

Fig. 6a~d show the size profiles of lunar regolith along the directions of horizontal 200m, vertical 400m, horizontal 600m and vertical 800m, respectively. It is clear from Fig. 6 that (1) The size of lunar regolith gradually increases with the depth and the size of near surface material is very small while the size of deep material is large. (2) The size change of lunar regolith material is not smooth, but closely related to the impact event. A few large impact events produced large - scale blocks, while massive small impact events produced much finer grains underneath the shallow surface. (3) The grain size is not strictly increased from top to bottom, and occasionally there are smaller grains in the deep, which is because the later impact pushed the previously broken material to the deep.

3.4 Evolution of lunar regolith

The surface age of the CE - 4 landing area is 3.5Ga (Gou et al., 2021), and at that time the study area is covered by the ejecta of Finsen crater. For simplicity, the surface of the area is regarded as a flat and homog enous plane without crater. According to the lunar impact fluxes and impact history displayed in Fig. 2a and b, it can be calculated that 949 (46.33%) impact events occurred before 3.0Ga, 1333 (65.04%) impact events occurred before 2.0Ga, and 1691 (82.52%) impact events occurred before 1.0Ga. Therefore, we can analyze the thickness varia tion of the lunar regolith with the 949, 1333, and 1691 impact events, respectively. Fig. 7 shows the distribution of lunar regolith thickness at 3.0Ga, 2.0Ga and 1.0Ga. Fig. 7a~c are the distribution of lunar regolith thickness in the study area at different times. Similarly, the coordinates are the distance to the lower left corner, and the color indicates the thickness of lunar regolith. Meanwhile, three lines as shown in Fig. 7d~f are plotted to indicate the variation of the lunar regolith thickness in horizontal 200m at different times. Fig. 7g~i are three size profiles of lunar regolith sizes along the directions of horizontal 200m at the different times.

Considering the impact process is random, Fig. 7 may only represent one case of the evolution process of regional lunar regolith under the effect of impact gardening and the simulation results. However, the laws reflected in the figure are universal. The average value and variance of lunar regolith thickness during the evolution are shown in Table 1.

Furthermore, we calculated the average size of the regolith material along the

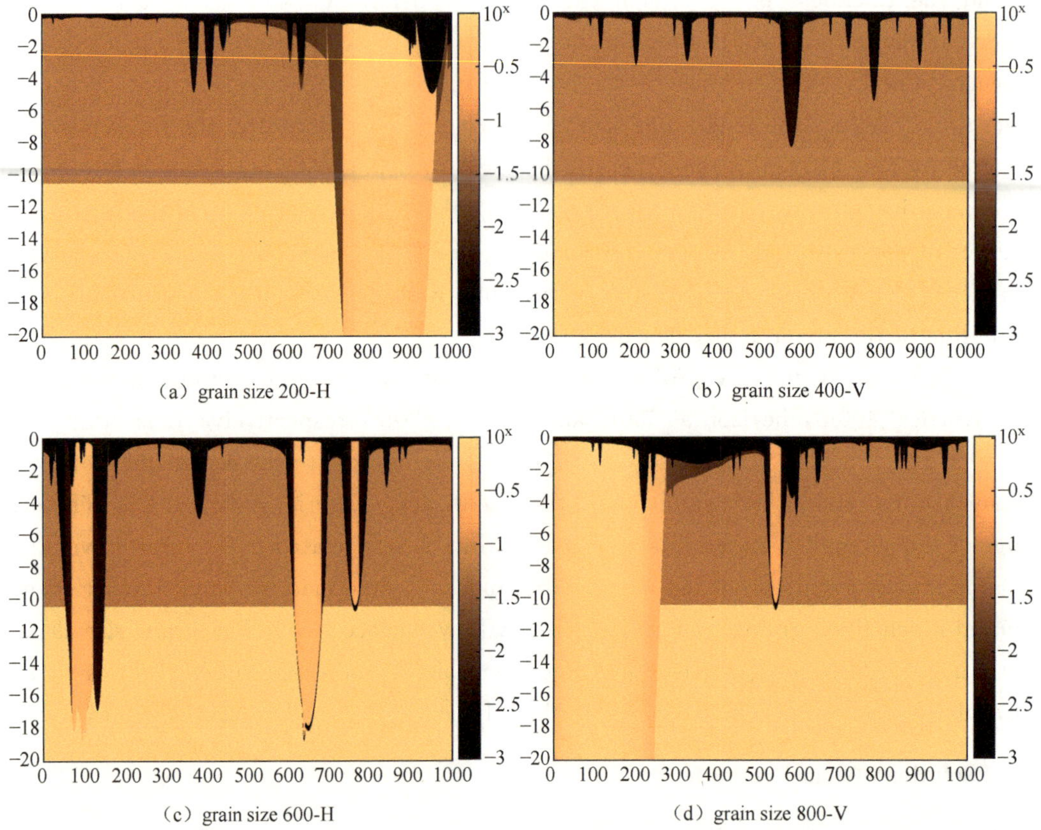

Fig. 6　Size profiles of lunar regolith along the directions of horizontal 200m (a), vertical 400m (b), horizontal 600m (c) and vertical 800m (d) (unit: meter). The color indicates the material size (unit: m), the darker the color, the smaller the size. Note the vertical scales are heavily exaggerated

directions of horizontal 200m at different times and depth (Fig. 8a). The result indicates that the average grain size rapidly in creases from the surface to the depth of 2m. Fig. 8b is the enlarged part of Fig. 8a up to the depth of 2.0m, indicating that the average grain size increases continuously.

Fig. 7 and Table 1 indicates that the overall structure of lunar regolith was basically formed at 3.0Ga, and changed little for a long time. This is consistent with the result in Fig. 8. It can be found from Fig. 8a that (1) The average size of the regolith material increases with the depth and remains essentially unchanged after 3.0Ga. (2) The lunar regolith was mainly controlled by large - scale impact events. After the formation of lunar regolith, it was continuously affected by the subsequent impacts, and was buried and pushed into the deep. In addition, it is worth noting that some larger impact events penetrated the Finsen ejecta and produced some larger materials. From Fig. 8b, it can be found that numerous small impact events shatter the near surface material constantly, consequently the average size of lunar regolith material are reduced generally and mixed homogeneously, forming the current lunar regolith. These fragments can be found in the data of Yutu - 2 rover on the surface of the Moon (Gou et al. , 2020).

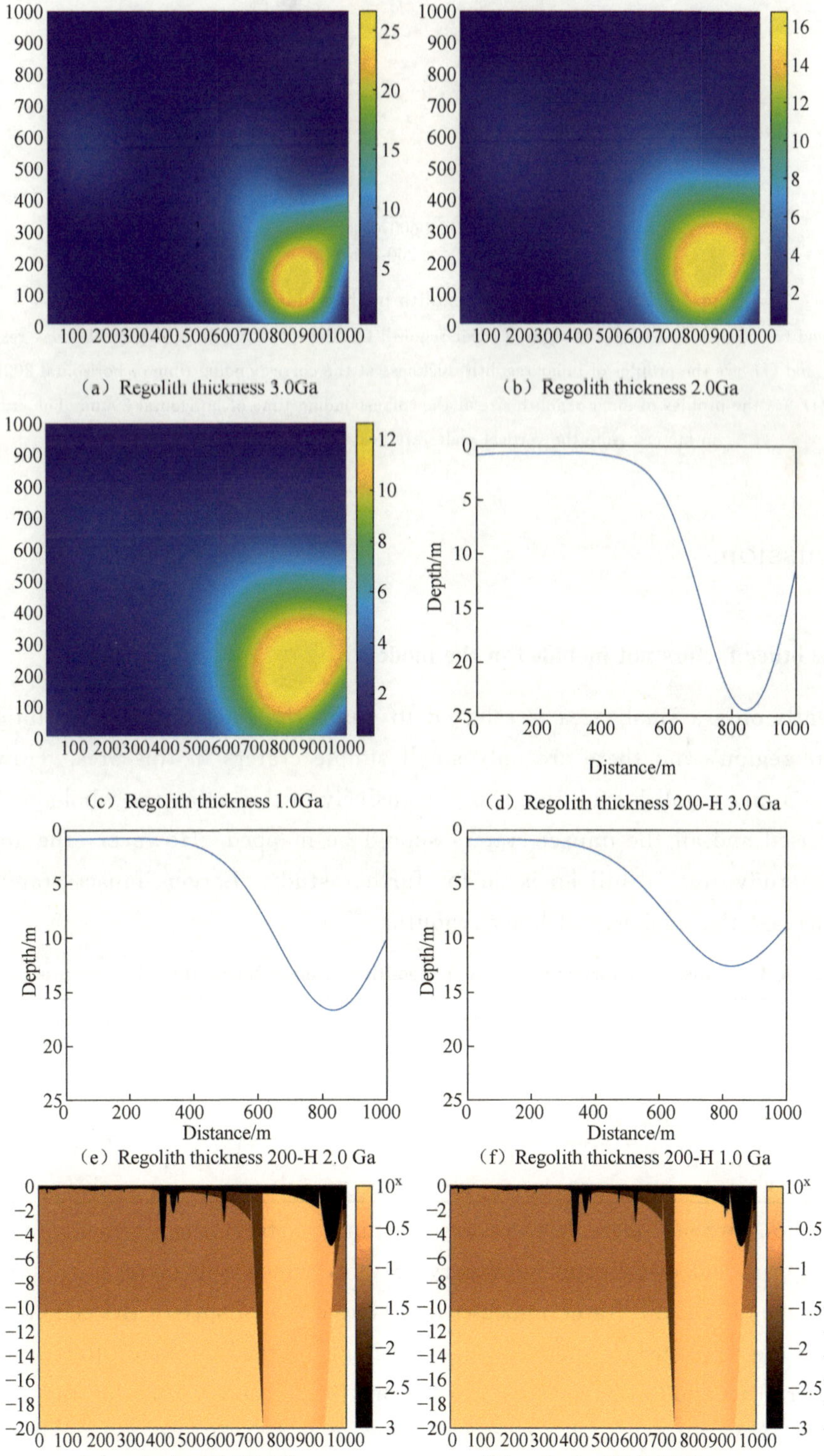

(a) Regolith thickness 3.0Ga　(b) Regolith thickness 2.0Ga

(c) Regolith thickness 1.0Ga　(d) Regolith thickness 200-H 3.0 Ga

(e) Regolith thickness 200-H 2.0 Ga　(f) Regolith thickness 200-H 1.0 Ga

(g) grain size 200-H 3.0 Ga　(h) grain size 200-H 2.0 Ga

Fig. 7　Evolution of lunar regolith in the study area

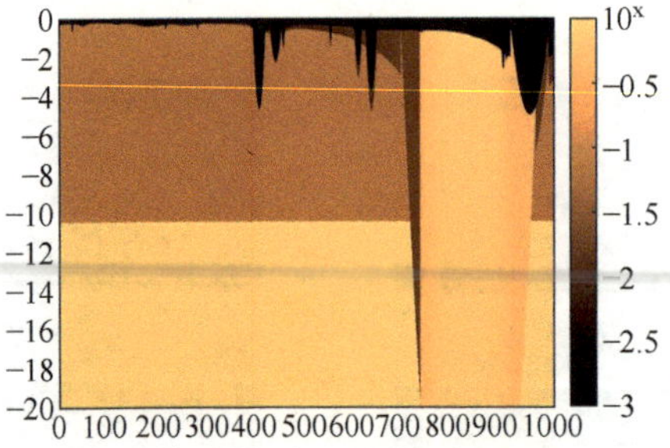

(i) grain size 200-H 1.0 Ga

Fig. 7　Evolution of lunar regolith in the study area (contcnude)

(a), (b) and (c) are the spatial distribution of lunar regolith thickness at 3.0Ga, 2.0Ga and 1.0Ga, respectively; (d), (e) and (f) are the profiles of lunar regolith thickness at the corresponding time of horizontal 200m; (g), (h) and (i) are the profiles of lunar regolith size at the corresponding time of horizontal 200m. The units are all in meters. Note the vertical scales are heavily exaggerated in (d)～(i)

4　Discussions

4.1　Some other factors not included in the model

Currently only a small area is selected to study the lunar regolith evolution in the landing site region, and there are only small simple craters in this area. However, to evaluate the lunar regolith evolution comprehensively in this area, the whole geologic unit should be used and all the impact craters should be mapped. However, the appropriate size of the study area is still an issue for further study. Larger impact craters nearby perhaps increase the thickness of lunar regolith.

Table 1　Statistical information of lunar regolith thickness during the evolution process

Time	Mean / m	Variance
3.0Ga	3.9311	29.8243
2.0Ga	3.5565	15.9595
1.0Ga	3.3344	10.3426
Nowadays	3.1781	7.2726

In addition, when a meteorite impacts the lunar surface, the ejecta also disturbs the surficial material and even forms secondary craters, which will increase the overturning times. For the secondary crater impacted at relative lower speed, the regolith material may simply be overturned and relocated without further broken, thus these small secondary craters should have little effect on the structure of the lunar regolith. Furthermore, it is difficult to distinguish the secondary craters from the small craters and there is no significant difference for the impact excavation and rolling mixing process between the secondary craters and the primary craters. Thus, we have not dealt with the secondary crater additionally during the calculation and all craters are processed in the same way.

Moreover, it is worth noting that the small craters in the study area are in a state of equilibrium and the counted 2049 craters (≥5m) represent only a small fraction of all the craters occurred. According to the estimation of the production function of lunar impact crater, there are about 21, 000 craters lager than 10m in diameter occurred in this surface. Therefore, our modeling is based on the premise that the underlying Surficial Regolith layer can be modeled using only the visible impact craters emplaced on the surface over the past 3.5Ga. However, if only the currently visible craters are modeled, the results will be underestimated. Our study shows that the equilibrium diameter of the impact crater in this region is around 100 to 200m. Wilcox et al. (2005) estimated the thickness of lunar regolith based on the equilibrium diameter of impact crater population, and pointed out that the regolith thickness in the area with equilibrium diameter of 100m is about 7 - 15m, which is larger than our calculation result. The possible reasons are that our study area is small, and there are fewer large impact craters that having a significant effect on the structure of lunar regolith. However, it is still an important issue in planetary geology research when the state of crater density equilibrium occurs (Richardson, 2009). And as mentioned in this paper, the thickness and structure of lunar regolith in the study area are mainly controlled by larger craters without density equilibrium, and a large number of small craters being saturated will only have an influence on the overturning times of the material and not alter the modeling and analysis of material evolution.

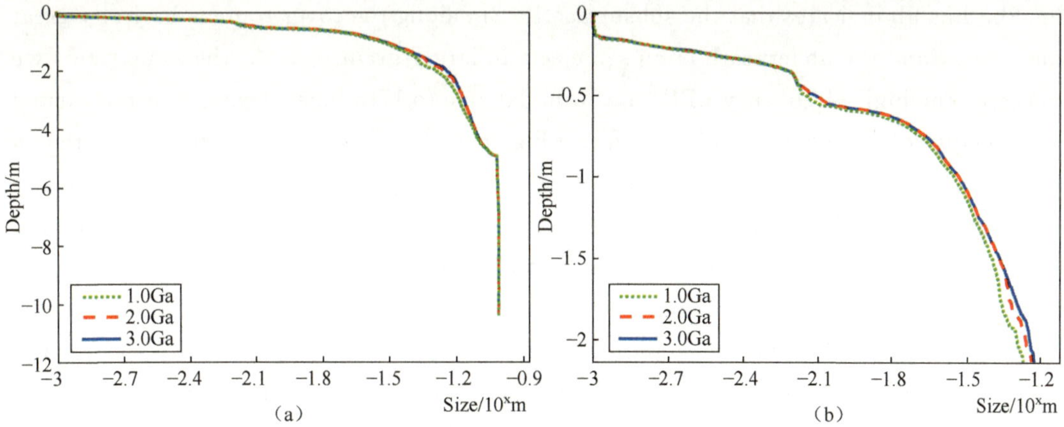

Fig. 8　The average size of the regolith material along the horizontal 200m at different times. Panel (b) is the partial enlarged figure of Panel (a) up to the depth of 2m

Finally, the target mechanic properties will be changed with the impacts, which will further influence the excavation depth and ejecta thickness in later impacts. Some detailed analyses of these factors are even ambiguous and they are out of the scope of the current research. An updated version of the model is expected to include some of these factors in the future.

4.2 Comparison with other researches and observations

Before the landing of CE-4, Huang et al. (2018) used the method of the concentric craters proposed by Quaide and Oberbeck (1968) to analyze the impact crater in the pre-landing area, and pointed out that the regolith thickness in the study area was 2.5 - 7.5m. Based on the investigation of the small crater at the bottom of the Von Kármán crater, Fu et al. (2020) reported that the thickness of the Finsen ejecta in the CE-4 landing area is ~18m. According to Eq. (1) and Eq. (7), we can calculate that the thickness of lunar regolith in the landing area from the Finsen ejecta is ~10.47m, and the grain size is ~0.06m. Our results are slightly different from the above results, and the reasons are that the calculation method is different. Their results are mostly based on the strength discontinuity in the target, while we are based on the impact gardening effect of small impact craters in the study area. Besides, as we have addressed, there may be some large craters far away outside the study area, and their ejecta can influence the lunar regolith in this area. In addition, our model further calculates the overturning times of the material and the changes of the grain size, which can provide more detailed information on the lunar regolith than the above studies and the in-situ detection data.

The LPR data of Yutu-2 provides an opportunity to reveal the regolith structure of the subsurface (Ding et al., 2020; Li et al., 2020). The route of Yutu-2 until 17th lunar days is shown in Fig. 9a (white line). We calculated the regolith material size along the Yutu-2 rover's route with the vertical resolution of 1cm and drew a profile, as shown in Fig. 9b. Fig. 9b indicates that the subsurface 1 - 2m along the rover route is basically great fine debris, and the underneath is Finsen ejecta in larger grain size. At the same time, we processed the high-frequency LPR data from day one to 17th lunar days and reconstructed the structure of the shallow sub-surface (Fig. 9c). The most obvious feature is that a discontinuity appears at about 10m underneath, which likely means that the thickness of the lunar Surficial Regolith layer of the CE-4 landing area is about 10m. The LPR detection result is thicker than the lunar regolith thickness calculated by our model and there are three possible reasons, as we mentioned above, which are summarized as follows: (1) The small impact craters in the study area are in equilibrium, and a large number of historical impact craters have been erased, which are not included in our model; (2) Our study area is small, and the number of large impact crater in this region is relatively small. (3) The regolith thickness in the study area may be affected by large impact events outside the area. Unfortunately, the spatial resolution of the high frequency channel of LPR is about 30cm and affected by noise and other factors such that it can't provide finer information in the depth about 2m, where our model can provide more detailed information of underground regolith material structure.

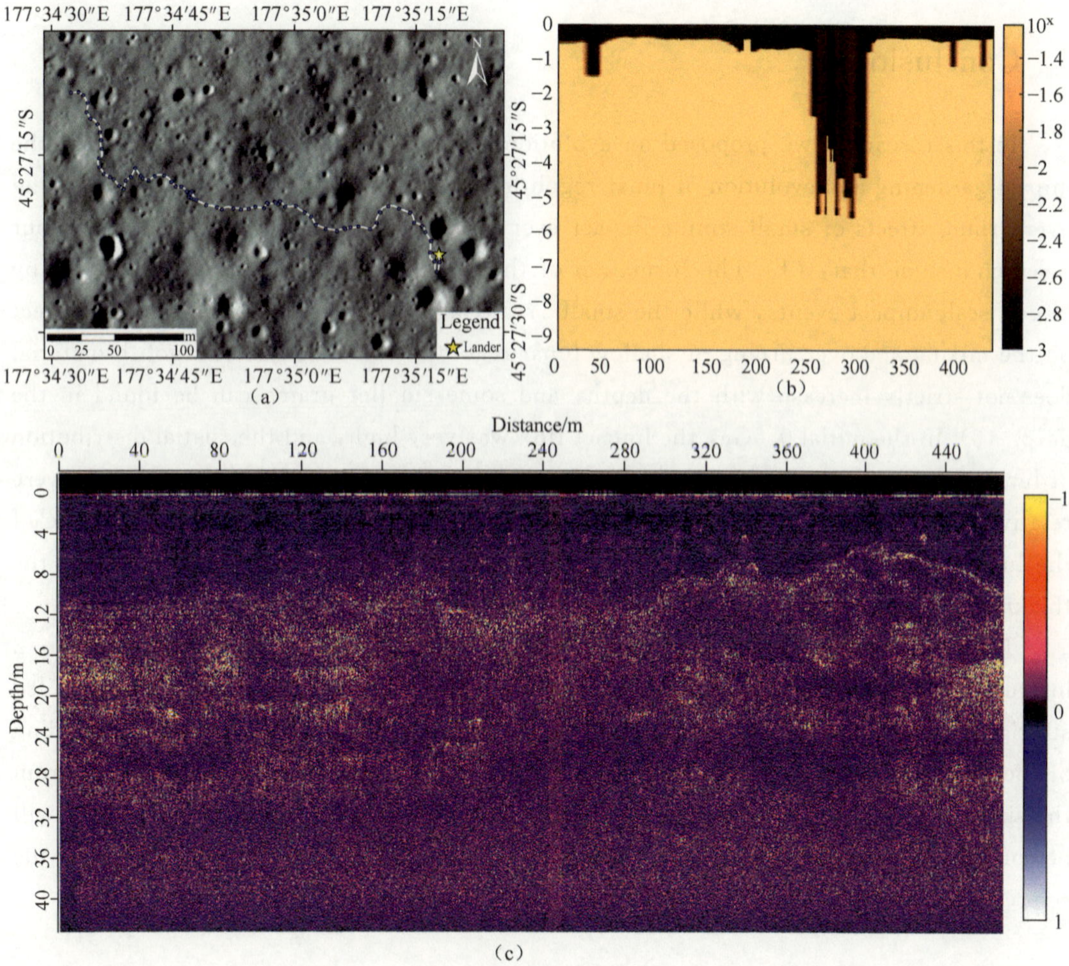

Fig. 9　The regolith along the Yutu - 2 route.

(a) The route of Yutu - 2 until 17th lunar days. (b) The size of the regolith material along the Yutu - 2 rover. (c) The subsurface structure from the lunar penetrating radar data, where the color represents the normalized radar amplitude

4.3　Evolution of lunar regolith in the Chang'E - 4 landing area

A possible scenario of the lunar regolith evolution is as follows. At 3.5Ga, the ejecta from the Finsen crater arrived here and formed the socalled megaregolith with the grain size ~0.06m. Because of the much higher impact flux and relatively large impact magnitude in the initial 0.5Ga, the main structure of the regolith was quickly determined. In the subsequent 3.0Ga, the impact flux was greatly reduced and the impact magnitude also became smaller, which mainly resulted in disturbing the shallow material and decreasing the grain size of regolith. To sum up, it can be found that the formation of lunar regolith is mainly determined by large - scale impact events, while the homogenous mixing and overturning of lunar regolith is mainly affected by small - scale impact events. In addition, except for the impact gardening effect, the terrain degradation also resulted in more uniform thickness of lunar regolith and smoother lunar surface.

5 Conclusion

In this research, we proposed an evolution model with simple craters to reveal the impact gardening and evolution of lunar regolith, and meanwhile analyzed the mixing and overturning effects of small simple impact events on lunar regolith. The findings of our research include that: (1) The formation of the lunar regolith is mainly determined by large - scale impact events, while the small - scale impact events have an important effect on the mixing and overturning of shallow lunar regolith. (2) The size of regolith material does not strictly increase with the depth, and some smaller grains can be found in the deep. (3) In the initial 0.5Ga, the impact flux was very high, and the spatial distribution of lunar regolith thickness was determined rapidly at that time. The impact events were relatively small in the subsequent 3.0Ga and mainly disturbed the shallow subsurface of the lunar regolith. Moreover, terrain degradation made the surface smoother and the thickness of lunar regolith more homogeneous.

The results of the model beyond the observational capability of the Yutu - 2 instrument package on the shallow subsurface structure and gives more detailed stratigraphic structure and evolution processes. This study can improve our understanding of the lunar regolith formation and evolution processes. In the future, the influence from the size of the study area, large complex craters, secondary craters and other factors will be considered. The current model can be used in future sample return missions to provide reference information about the variation of grain size with the depth.

Acknowledgements

The authors gratefully thank the LROC team for making the NAC images available. This work was supported by the B - type Strategic Priority Program of the Chinese Academy of Sciences (Grant No. XDB41000000), National Natural Science Foundation of China (Grant No. 41972321), and Key Research Program of Frontier Sciences, CAS (Grant No. QYZDY - SSW - DQC028).

References

[1] Arnold, J. R., 1975. A Monte Carlo model for the gardening of the lunar regolith. Moon 13, 159 - 172. https://doi.org/10.1007/BF00567513.

[2] Bart, G., Melosh, H. J., 2010. Distribution of boulders ejected from lunar craters. Icarus 209, 337 - 357. https://doi.org/10.1016/j.icarus.2010.05.023.

[3] Byerlee, J., 1978. Friction of rocks. PAGEOPH 116, 615 - 626. https://doi.org/10.1007/BF00876528.

[4] Cashore, J., Woronow, A., 1985. A new Monte Carlo model of lunar megaregolith development. J. Geophys. Res. Solid Earth 90, C811 - C815. https://doi.org/10.1029/JB090iS02p0C811.

[5] Chapman, C.R., McKinnon, W.B., 1986. Cratering of Planetary Satellites. University of Arizona Press, Satellites, Tucson.

[6] Collins, G.S., Melosh, H.J., Ivanov, B.A., 2004. Modeling damage and deformation in impact simulations. Meteorit. Planet. Sci. 39, 217 - 231. https://doi.org/10.1111/j.1945-5100.2004.tb00337.x.

[7] Costello, E.S., Ghent, R.R., Lucey, P.G., 2018. The mixing of lunar regolith: vital updates to a canonical model. Icarus 314, 327 - 344. https://doi.org/10.1016/j.icarus.2018.05.023.

[8] Crawford, I.A., Fagents, S.A., Joy, K.H., Rumpf, M.E., 2010. Lunar palaeoregolith deposits as recorders of the galactic environment of the solar system and implications for astrobiology. Earth Moon Planet. 107, 75 - 85. https://doi.org/ 10.1007/s11038-010-9358-z.

[9] Crider, D., Vondrak, R., 2003. Space weathering of ice layers in lunar cold traps. Adv. Space Res. 31, 2293 - 2298. https://doi.org/10.1016/S0273 - 1177 (03) 00530-1.

[10] David, A.K., 2007. Guidebook to the Geology of Barringer Meteorite Crater, Arizona (a. k. a. Meteor Crater). Houston: Lunar and Planetary Institute. LPI Contribution No. 1355.

[11] Di, K., Liu, Z., Liu, B., Wan, W., Peng, M., Wang, Y., Gou, S., Yue, Z., Xin, X., Jia, M., Niu, S., 2019. Chang'e-4 lander localization based on multi-source data. J. Remote Sens. 23, 177 - 184. https://doi.org/10.11834/jrs.20199015.

[12] Ding, C., Xiao, Z., Wu, B., Li, Y., Prieur, N., Cai, Y., Su, Y., Cui, J., 2020. Fragments delivered by secondary craters at the Chang'e - 4 landing site. Geophys. Res. Lett. 47 https://doi.org/10.1029/2020GL087361 e2020GL087361.

[13] Dong, Z., Fang, G., Zhou, B., Di, Z., Gao, Y., Ji, Y., 2021. Properties of lunar Regolith on the Moon's Farside unveiled by Chang'E-4 lunar penetrating radar. J. Geophys. Res. Planets 126. https://doi.org/10.1029/2020JE006564 e2020JE006564.

[14] Fassett, C.I., Thomson, B.J., 2014. Crater degradation on the lunar maria: topographic diffusion and the rate of erosion on the moon. J. Geophys. Res. Planets 119, 2255 - 2271. https://doi.org/10.1002/2014JE004698.

[15] Fischer, E.M., Pieters, C.M., 1995. Lunar surface aluminum and iron concentration from Galileo solid state imaging data, and the mixing of mare and highland materials. J. Geophys. Res. Planets 100, 23279 - 23290. https://doi.org/10.1029/95JE02359.

[16] Fruchter, J., Rancitelli, L., Evans, J., Perkins, R., 1978. Lunar surface

processes and cosmic ray histories over the past several million years. In: 9th Lunar and Planetary Science Conference. The Houston, Texas.

[17] Fu, X., Qiao, L., Zhang, J., Ling, Z., Li, B., 2020. The subsurface structure and stratigraphy of the Chang'E-4 landing site: orbital evidence from small craters on the Von Kármán crater floor. Res. Astron. Astrophys. 20, 59-70. https://doi.org/ 10.1088/1674-4527/20/1/8.

[18] Gault, D., Hörz, F., Brownlee, D., Hartung, J., 1974. Mixing of the lunar regolith. In: 5th Lunar and Planetary Science Conference. The Houston, Texas.

[19] Gou, S., Yue, Z., Di, K., Wang, J., Wan, W., Liu, Z., Liu, B., Peng, M., Wan, W., He, Z., Xu, R., 2020. Impact melt breccia and surrounding regolith measured by Chang'e-4 rover. Earth Planet. Sci. Lett. 544, 116378 https://doi.org/10.1016/j.epsl.2020.116378.

[20] Gou, S., Yue, Z., Di, K., Cai, Z., Liu, Z., Niu, S., 2021. Absolute model age of lunar Finsen crater and geologic implications. Icarus 354, 114046. https://doi.org/10.1016/j.icarus.2020.114046.

[21] Hirabayashi, M., Howl, B. A., Fassett, C. I., Soderblom, J. M., Minton, D. A., Melosh, H. J., 2018. The role of breccia lenses in regolith generation from the formation of small, simple craters: application to the Apollo 15 landing site. J. Geophys. Res. Planets 123, 527-543. https://doi.org/10.1002/2017JE005377.

[22] Hörz, F., 1977. Impact cratering and regolith dynamics. Phys. Chem. Earth 10, 3-15. https://doi.org/10.1016/0079-1946(77)90003-9.

[23] Hu, X., Ma, P., Yang, Y., Zhu, M., Jiang, T., Lucey, P. G., Sun, L. Z., Zhang, H., Li, C., Xu, R., He, Z., Lin, H., Huang, C., Sun, Y., 2019. Mineral abundances inferred from in situ reflectance measurements of Chang'E-4 landing site in south pole-Aitken Basin. Geophys. Res. Lett. 46, 9439-9447. https://doi.org/10.1029/2019GL084531.

[24] Huang, Y., Minton, D. A., Hirabayashi, M., Elliott, J. R., Richardson, J. E., Fassett, C. I., Zellner, N. E. B., 2017. Heterogeneous impact transport on the Moon. J. Geophys. Res. Planets 122, 1158-1180. https://doi.org/10.1002/2016JE005160.

[25] Huang, J., Xiao, Z., Flahaut, J., Martinot, M., Head, J., Xiao, X., Xie, M., Xiao, L., 2018. Geological characteristics of Von Kármán crater, northwestern south pole-Aitken Basin: Chang'E-4 landing site region. J. Geophys. Res. Planets 123, 1684-1700. https://doi.org/10.1029/2018JE005577.

[26] Hurley, D. M., Lawrence, D. J., Bussey, D. B. J., Vondrak, R. R., Elphic, R. C., Gladstone, G. R., 2012. Two-dimensional distribution of volatiles in the lunar regolith from space weathering simulations. Geophys. Res. Lett. 39, L09203.

[27] Ivanov, M. A., Hiesinger, H., van der Bogert, C. H., Orgel, C., Pasckert, J. H., Head, J. W., 2018. Geologic history of the northern portion of the south

pole - Aitken Basin on the moon. J. Geophys. Res. Planets 123, 2585 - 2612. https://doi.org/10.1029/ 2018JE005590.

[28] Jia, Y., Zou, Y., Ping, J., Xue, C., Yan, J., Ning, Y., 2018. The scientific objectives and payloads of Chang'E - 4 mission. Planet. Space Sci. 162, 207 - 215. https://doi.org/ 10.1016/j.pss.2018.02.011.

[29] Jia, M., Di, K., Yue, Z., Liu, B., Wan, W., Niu, S., Liu, J., Cheng, W., Lin, Y., 2021. Multiscale morphologic investigation of craters in the Chang'e - 4 landing area. Icarus 355, 114164. https://doi.org/10.1016/j.icarus.2020.114164.

[30] Kneissl, T., Gasselt, S.V., Neukum, G., 2011. Map - projection - independent crater size - frequency determination in GIS environments - new software tool for ArcGIS. Planet. Space Sci. 59, 1243 - 1254. https://doi.org/10.1016/j.pss.2010.03.015.

[31] Krishna, N., Kumar, P.S., 2016. Impact spallation processes on the moon: a case study from the size and shape analysis of ejecta boulders and secondary craters of Censorinus crater. Icarus 264, 274 - 299. https://doi.org/10.1016/j.icarus.2015.09.033.

[32] Li, L., Mustard, J.F., 2005. On lateral mixing efficiency of lunar regolith. J. Geophys. Res. Planets 110. https://doi.org/10.1029/2004JE002295.

[33] Li, C., Su, Y., Pettinelli, E., Xing, S., Ding, C., Liu, J., Ren, X., Lauro, S., Soldovieri, F., Zeng, X., Gao, X., Chen, W., Dai, S., Liu, D., Zhang, G., Zuo, W., Wen, W., Zhang, Z., Zhang, X., Zhang, H., 2020. The Moon's farside shallow subsurface structure unveiled by Chang'E - 4 lunar penetrating radar. Sci. Adv. 6 https://doi.org/10.1126/ sciadv.aay6898 eaay6898.

[34] Lin, H., Lin, Y., Yang, W., He, Z., Hu, S., Wei, Y., Xu, R., Zhang, J., Liu, X., Yang, J., Xing, Y., Yu, C., Zou, Y., 2020. New insight into lunar regolith - forming processes by the lunar rover Yutu - 2. Geophys. Res. Lett. 47 https://doi.org/10.1029/ 2020GL087949 e2020GL087949.

[35] Liu, T., Michael, G., Engelmann, J., Wünnemann, K., Oberst, J., 2019. Regolith mixing by impacts: lateral diffusion of basin melt. Icarus 321, 691 - 704. https://doi.org/ 10.1016/j.icarus.2018.12.026.

[36] Mcgetchin, T.R., Settle, M., Head, J.W., 1973. Radial thickness variation in impact crater ejecta: implications for lunar basin deposits. Earth Planet. Sci. Lett. 20, 226 - 236. https://doi.org/10.1016/0012 - 821X (73) 90162 - 3.

[37] Melosh, H.J., 1989. Impact Cratering: A Geologic Process. Oxford University Press, New York.

[38] Melosh, H.J., 2011. Planetary Surface Processes. Cambridge University Press, Cambridge.

[39] Michael, G., Basilevsky, A., Neukum, G., 2018. On the history of the early meteoritic bombardment of the moon: was there a terminal lunar cataclysm? Icarus 302, 80 - 103. https://doi.org/10.1016/j.icarus.2017.10.046.

[40] Moore, H. J. , 1971. Large blocks around lunar craters. In: Analysis of Apollo 10 Photography and Visual Observations. NASA, Washington, DC, pp. 26 - 27. NASA spec. Publ. SP - 232.

[41] Neukum, G. , 1983. Meteoritenbombardement und Datierung Planetarer Oberflachen. Munich. In: Habilitation Dissertation for Faculty Membership. Ludwig - Maximilians - Univ, German (English translation, 1984: Meteorite bombardment and dating of planetary surfaces).

[42] Neukum, G. , Ivanov, B. A. , Hartmann, W. K. , 2001. Cratering records in the inner solar system in relation to the lunar reference system. Space Sci. Rev. 96, 55 - 86. https:// doi. org/10. 1023/A: 1011989004263.

[43] Oberbeck, V. R. , Quaide, W. L. , Mahan, M. , Paulson, J. , 1973. Monte Carlo calculations of lunar Regolith thickness distributions. Icarus 19, 87 - 107. https:// doi. org/10. 1016/ 0019 - 1035 (73) 90141 - 3.

[44] Osinski, G. R. , Tornabene, L. L. , Grieve, R. A. F. , 2011. Impact ejecta emplacement on terrestrial planets. Earth Planet. Sci. Lett. 310, 167 - 181. https://doi. org/10. 1016/j. epsl. 2011. 08. 012.

[45] Quaide, W. L. , Oberbeck, V. R. , 1968. Thickness determinations of the lunar surface layer from lunar impact craters. J. Geophys. Res. 73, 5247 - 5270. https://doi. org/ 10. 1029/JB073i016p05247.

[46] Quaide, W. L. , Oberbeck, V. , 1975. Development of the mare regolith: some model considerations. Moon 13, 27 - 55. https://doi. org/10. 1007/BF00567506.

[47] Richardson, J. E. , 2009. Cratering saturation and equilibrium: a new model looks at an old problem. Icarus 204, 697 - 715. https://doi. org/10. 1016/j. icarus. 2009. 07. 029.

[48] Richardson, J. E. , Abramov, O. , 2020. Modeling the formation of the lunar upper megaregolith layer. Planet. Sci. J. 1, 2.

[49] Sharpton, V. L. , 2014. Outcrops on lunar crater rims: implications for rim construction mechanisms, ejecta volumes and excavation depths. J. Geophys. Res. Planets 119, 154 -168. https://doi. org/10. 1002/2013JE004523.

[50] Shoemaker, E. M. , Batson, R. M. , Holt, H. E. , Morris, E. C. , Rennilson, J. J. , Whitaker, E. A. , 1967a. Surveyor V: television pictures. Science 158, 642 - 652. https://doi. org/ 10. 1126/science. 158. 3801. 642.

[51] Shoemaker, E. M. , Batson, R. M. , Holt, H. E. , Morris, E. C. , Rennilson, J. J. , Whitaker, E. A. , 1967b. Television observations from surveyor III. In: Surveyor III Mission Report, Part II: Scientific Results, pp. 9 - 67. NASA JPL Technical Report, 32 - 1177.

[52] Shoemaker, E. M. , Batson, R. M. , Holt, H. E. , Morris, E. C. , Rennilson, J. J. , Whitaker, E. A. , 1968. Television observations from surveyor VII. In: Surveyor VII: A Preliminary Report, NASA Special Paper 173. Government Printing Office, Washington, D. C. , U. S, pp. 13 - 81.

[53] Shuster, D. L. , Balco, G. , William, S. C. , Fernandes, V. A. , Garrick - Bethell, L. , Weiss, B. P. , 2010. A record of impacts preserved in the lunar regolith. Earth Planet. Sci. Lett. 290, 155 - 165. https://doi. org/10. 1016/j. epsl. 2009. 12. 016.

[54] Singer, K. N. , Jolliff, B. L. , Mckinnon, W. B. , 2020. Lunar secondary craters and estimated ejecta block sizes reveal a scale - dependent fragmentation trend. J. Geophys. Res. Planets 125. https://doi. org/10. 1029/2019JE006313 e2019JE006313.

[55] Skianis, G. A. , Vaiopoulos, D. , Evelpidou, N. , 2008. Solution of the linear diffusion equation for modelling erosion processes with a time varying diffusion coefficient. Earth Surf. Process. Landf. 33, 1491 - 1501. https://doi. org/10. 1002/esp. 1619.

[56] Soderblom, L. A. , 1970. A model for small - impact erosion applied to the lunar surface. J. Geophys. Res. 75, 2655 - 2661. https://doi. org/10. 1029/JB075i014p02655.

[57] Stesky, R. M. , Brace, W. F. , Riley, D. K. , Bobin, P. Y. , 1974. Frictionin faulted rock at high temperature and pressure. Tectonophysics 23, 177 - 203. https://doi. org/10. 1016/ 0040 - 1951 (74) 90119 - X.

[58] Stopar, J. D. , Robinson, M. S. , Barnouin, O. S. , McEwen, A. S. , Speyerer, E. J. , Henriksen, M. R. , Sutton, S. S. , 2017. Relative depths of simple craters and the nature of the lunar regolith. Icarus 298, 34 - 48. https://doi. org/10. 1016/j. icarus. 2017. 05. 022.

[59] Trask, N. J. , 1966. Size and spatial distribution of craters estimated from ranger photographs. In: Propul. Lab. Tech. Rep. 32 - 800, Pasadena, Cal, 252 S.

[60] Wilcox, B. B. , Robinson, M. S. , Thomas, P. C. , Hawke, B. R. , 2005. Constraints on the depth and variability of the lunar regolith. Meteorit. Planet. Sci. 40, 695 - 710. https://doi. org/10. 1111/j. 1945 - 5100. 2005. tb00974. x.

[61] Wilhelms, D. E. , McCauley, J. F. , Trask, N. J. , 1987. The geologic history of the Moon. U. S. Geological Survey Professional Paper 1348, USA.

[62] Woronow, A. , 1977. Crater saturation and equilibrium: a Monte Carlo simulation. J. Geophys. Res. 82, 2447 - 2456. https://doi. org/10. 1029/JB082i017p02447.

[63] Zhang, J. , Zhou, B. , Lin, Y. , Zhu, M. , Song, H. , Dong, Z. , Gao, K. , Di, K. , Yang, W. , Lin, H. , Yang, J. , Liu, E. , Wang, L. , Lin, Y. , Li, C. , Yue, Z. , Yao, Z. , Ouyang, Z. , 2021. Lunar regolith and substructure at Chang'E - 4 landing site in South Pole - Aitken basin. Nat. Astron. 5, 25 - 30. https://doi. org/10. 1038/s41550 - 020 - 1197 - x.

Rock Location and Property Analysis of Lunar Regolith at Chang'E - 4 Landing Site Based on Local Correlation and Semblance Analysis

Hanjie Song[1], Chao Li[2,3], Jinhai Zhang[2,3], Xing Wu[1], Yang Liu [1] and Yongliao Zou[1]

1. State Key Laboratory of Space Weather, National Space Science Center, Chinese Academy of Sciences, Beijing 100190, China; songhanjie@nssc. ac. cn (H. S.); wuxing@nssc. ac. cn (X. W.); zouyongliao @nssc. ac. cn (Y. Z.)

2. Key Laboratory of Earth and Planetary Physics, Institute of Geology and Geophysics, Chinese Academy of Sciences, Beijing 100029, China; superlee@mail. iggcas. ac. cn (C. L.); zjh@mail. iggcas. ac. cn (J. Z.)

3. Innovation Academy for Earth Science, Chinese Academy of Sciences, Beijing 100029, China

Abstract: The Lunar Penetrating Radar (LPR) onboard the Yutu - 2 rover from China's Chang'E - 4 (CE - 4) mission is used to probe the subsurface structure and the near - surface stratigraphic structure of the lunar regolith on the farside of the Moon. Structural analysis of regolith could provide abundant information on the formation and evolution of the Moon, in which the rock location and property analysis are the key procedures during the interpretation of LPR data. The subsurface velocity of electromagnetic waves is a vital parameter for stratigraphic division, rock location estimates, and calculating the rock properties in the interpretation of LPR data. In this paper, we propose a procedure that combines the regolith rock extraction technique based on local correlation between the two sets of LPR high - frequency channel data and the common offset semblance analysis to determine the velocity from LPR diffraction hyperbola. We consider the heterogeneity of the regolith and derive the relative permittivity distribution based on the rock extraction and semblance analysis. The numerical simulation results show that the procedure is able to obtain the high - precision position and properties of the rock. Furthermore, we apply this procedure to CE - 4 LPR data and obtain preferable estimations of the rock locations and the properties of the lunar subsurface regolith.

Keywords: lunar penetrating radar; local correlation; semblance analysis; rock location; property analysis; regolith

1 Introduction

The surface of the Moon is covered by regolith, which records at least a 4 - billion - year history of meteoroid impacts and implantation of solar wind. The thickness of the regolith varies from about 2m beneath the lunar maria to more than 10m beneath the lunar highland, respectively, which is correlated with the age of the lunar surface [1]. The estimation of regolith thickness usually uses indirect measurements including seismic experiments [2], microwave remote sensing [3], and impact crater morphology and frequency distribution of the crater diameters [4]. Unlike the Apollo Lunar Sounder Experiment [5] and Lunar Radar Sounder [6], Chang'E - 3 conducted the first in situ Lunar Penetrating Radar (LPR) exploration with a wide frequency band and high spatial resolution in the Imbrium basin, showing the detailed structure of lunar regolith [7, 8]. In addition, Chang'E - 4 (CE - 4) is conducting the first in situ exploration on the farside of the Moon in the Von Kármán crater at the South Pole - Aitken (SPA) basin [9, 10].

The LPR aboard on the Yutu - 2 Rover of CE - 4 is a nanosecond imaging radar which is carrier - free and operates in the time domain [11], and has two channels (CH - 1 and CH - 2) with center frequencies at 60 and 500MHz, respectively [10]. CH - 1 is used to map the structure of the shallow lunar crust with meter - level resolution and the CH - 2 is used to detect the structure of regolith with a depth resolution of 0.3m, which contains one transmitting antenna and two receiving antennas of different offsets - namely, CH - 2A and CH - 2B [12]. The LPR measurements suggest an underestimation of the global lunar regolith thickness by other methods and reveal a vast volume from the last volcanic eruption [8]. Yutu - 2 has obtained a 425m LPR profile along the rover tracks in the first 16 lunar days, displaying clear and complex subsurface structures of the landing area, which would help us to reveal the fine structure of lunar regolith.

The subsurface velocity of electromagnetic waves is a vital parameter for stratigraphic division, rock location estimates, and calculating rock properties including relative permittivity, density, and content of FeO and TiO_2 during the interpretation of LPR data. Meanwhile, the location of rocks in regolith can cause diffractions in LPR data, and the vertex position of the diffractions can help estimate the possible location of rocks. This is different from the main reflections from the layers, which are useful in stratigraphic analysis. Analysis of the rock location and the properties of the lunar regolith can help to reveal the formation and evolution history of the landing site. Previous studies on geological stratification and parameter inversion of the lunar regolith were mainly based on the CH - 2B data only [3, 8, 12]. For example, Feng et al. proposed a hyperbola fitting method for radar velocity analysis in the CH - 2B radar - gram [12]. Dong et al. calculated the parameters of the regolith by relative reflection amplitudes [13]. Lai et al.

acquired the radar velocity based on the two - way delay method [14]. Hu et al. proposed an adaptive rock extraction method based on local similarity constraints to achieve the rock location and quantitative analysis for regolith [15]. Dong et al. analyzed the regolith properties from the LPR data of Chang'E - 4 based on the 3D velocity spectrum [16]. However, a complexsub surface structure and interference of noise always result in incomplete, interlaced, and amplitude - varying hyperbolas which cause the inaccurate velocity analysis of field LPR data [16], and the previous studies cannot reach a satisfied accuracy. In addition, the determinationof the rock location or the vertex position of the diffractions is the key procedure to obtain a high - precision diffraction velocity analysis spectrum. Although Zhang et al. used two sets of CH - 2A and CH - 2B data to estimate the electrical parameters and the iron - titanium content of regolith [17], the subsurface velocity estimation is not as accurate as expected due to the limitation of their method.

In this paper, two sets of CH - 2A and CH - 2B data are used to estimate velocity and the relative permittivity; thus, the other properties of lunar regolith including density and content of FeO and TiO_2 can be calculated based on the relative permittivity. Firstly, we consider the measurement similarity [18, 19] between the two sets of CH - 2A and CH - 2B data based on the local correlation, and develop an extraction method to determine the vertex position of the diffractions and estimate the rock location. Secondly, we apply the normalized velocity spectrum based on the semblance analysis to estimate the subsurface velocity of the lunar regolith [20]. Finally, we utilize the interpolation method to obtain the velocity of the subsurface structure and derive the distribution of the relative permittivity.

2 Materials and Methods

2.1 LPR Data Processing

The CE - 4 LPR is a dual - frequency ground penetrating radar (GPR) system, operating at 60MHz (low - frequency) with a frequency band of 40~80MHz and 500MHz (highfrequency) with frequency band of 250~750MHz [11, 21]. The CH - 2 radar data were collected during the first 16 lunar days along the Yutu - 20s traverse of about 425m, as shown in Figure 1, which is consistent with the results acquired by Lin et al. [22].

To reveal the near - surface structure of the regolith, we processed and interpreted the high - frequency LPR data (CH - 2A and CH - 2B) by excluding the effects of the electromagnetic coupling with the rover's metallic body. The CH - 2 antenna is mounted at the bottom of the lunar rover about 0.3m away from the ground, and the space between two adjacent antenna elements (one transmitting antenna and two receiving antennae) is about 0.16m [23]. The receiving antenna for CH - 2A data is closer to the transmitting antenna than that of CH - 2B, and thus has a lower signal - to - noise ratio.

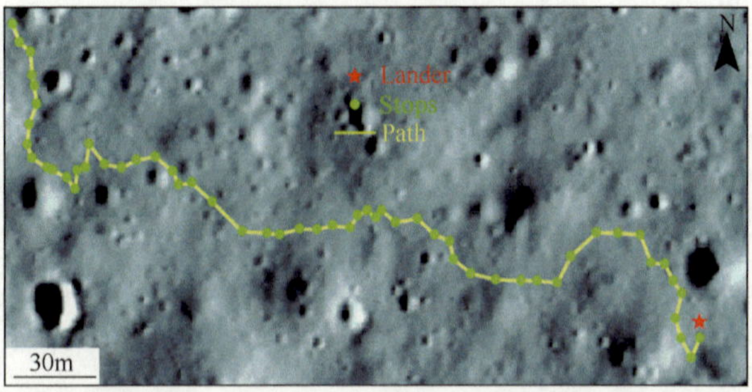

Figure 1　The location of Chang'E－4 landing site and the traverse of Yutu－2 rover for the first 16 lunar days. The lunar background was obtained using the imagery by the Lunar Reconnaissance Orbiter Camera

Therefore, we developed a series of procedures to process the CH－2A and CH－2B data as follows: (1) data deleting; (2) time delay removal; (3) band－pass filter application; (4) removal of DC bias; (5) background removal; (6) mean spatial filter application. Then, we could obtain the high－resolution CH－2A and CH－2B data with 1958 samples with 0.3125ns time intervals and 11, 661 traces of 0.0365m spatial intervals, as shown in Figure 2.

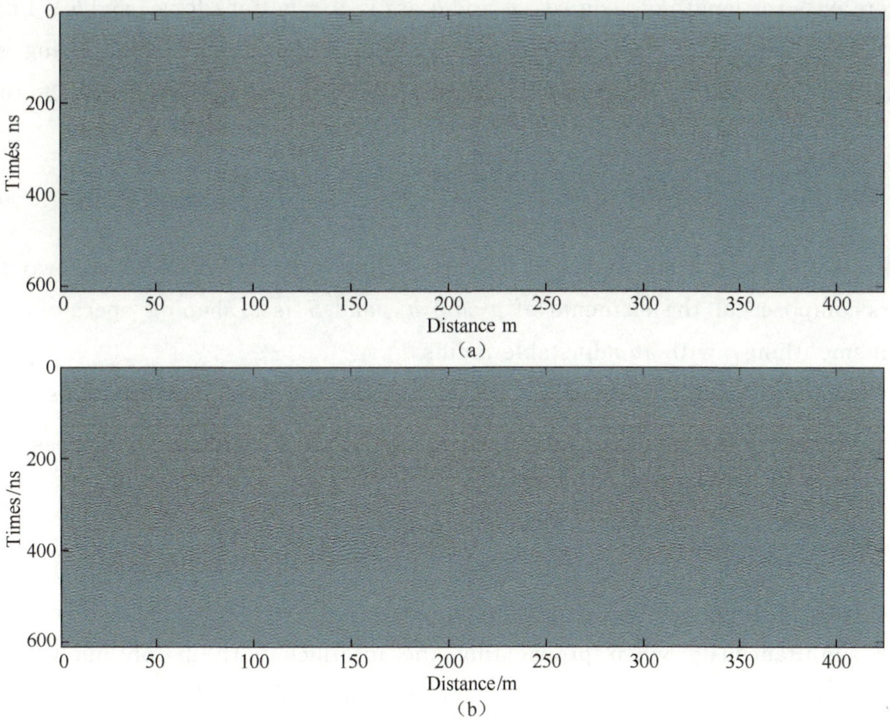

Figure 2　High－frequency LPR profile along the track of rover Yutu－2 after the series of data processing. (a) The CH－2A data; (b) the CH－2B data

2.2 Local Correlation.

The processed LPR CH - 2A and CH - 2B data are always affected by noise significantly, and the strong coherence of the noise may cause local maximums [15]. Here, we introduce local correlation [18] to determine the rock position in order to take advantage of both LPR CH - 2A and CH - 2B data simultaneously.

The global correlation coefficient between two discrete signals, a_i and b_i can be defined as [18]:

$$\gamma = \frac{\sum_{i=1}^{N} a_i b_i}{\sqrt{\sum_{i=1}^{N} a_i^2 \sum_{i=1}^{N} b_i^2}}, \quad (1)$$

which can be represented as two least - squares inverse based on a linear algebra notation:

$$\gamma = \sqrt{\gamma_1 \gamma_2}, \quad (2)$$

$$\gamma_1 = \arg\min_{\gamma_1} \| \boldsymbol{b} - \gamma_1 \boldsymbol{a} \|^2 = (\boldsymbol{a}^T \boldsymbol{a})^{-1} (\boldsymbol{a}^T \boldsymbol{b}) \quad (3)$$

$$\gamma_2 = \arg\min_{\gamma_2} \| \boldsymbol{a} - \gamma_2 \boldsymbol{b} \|^2 = (\boldsymbol{b}^T \boldsymbol{b})^{-1} (\boldsymbol{b}^T \boldsymbol{a}) \quad (4)$$

where N is the length of a signal, $\boldsymbol{a}$ and $\boldsymbol{b}$ are vector notions for a_i and b_i. Localizing Equations (3) and (4) were used to add regularization to inversion. Using shaping regularization, the scalars can turn into vectors [18] —i. e., scalars γ_1 and γ_2 turn into vectors $\boldsymbol{c}_1$ *and* $\boldsymbol{c}_2$, defined as:

$$\boldsymbol{c}_1 = [\lambda_1^2 \boldsymbol{I} + \boldsymbol{S}(\boldsymbol{A}^T \boldsymbol{A} - \lambda_1^2 \boldsymbol{I})]^{-1} \boldsymbol{S} \boldsymbol{A}^T \boldsymbol{b} \quad (5)$$

$$\boldsymbol{c}_2 = [\lambda_2^2 \boldsymbol{I} + \boldsymbol{S}(\boldsymbol{B}^T \boldsymbol{B} - \lambda_1^2 \boldsymbol{I})]^{-1} \boldsymbol{S} \boldsymbol{B}^T \boldsymbol{a} \quad (6)$$

where $\lambda_1 = \| \boldsymbol{a}^T \boldsymbol{a} \|$ and $\lambda_2 = \| \boldsymbol{b}^T \boldsymbol{b} \|$ are scaling controls, $\boldsymbol{A}$ and $\boldsymbol{B}$ are two diagonal operators composed of the elements of $\boldsymbol{a}$ and $\boldsymbol{b}$, and $\boldsymbol{S}$ is a shaping operator, such as Gaussian smoothing, with an adjustable radius.

The componentwise of vectors $\boldsymbol{c}_1$ and $\boldsymbol{c}_2$ defines the local correlation as $\boldsymbol{c}$, whose elements are given by $c_i = \sqrt{c_{1,i}^* c_{2,i}}$ $(1 \leqslant i \leqslant N)$. The local correlation is a measure of the similarity between two signals [15, 19].

2.3 Semblance Analysis

The high - frequency electromagnetic wave will be reflected, diffracted, and refracted simultaneously when propagating the interface with discontinuous dielectric properties, such as boulders, voids, and soil inhomogeneities [24]. The LPR CH - 2A and CH - 2B data represent the common offset profiles, and the responses analyzed in semblance analysis are diffraction hyperbola instead of reflection hyperbola in the common midpoint data. The normal moveout equation was reconfigured for diffraction trajectories in the common offset profile, thus the traveltime was approximated as:

$$t(x)=\sqrt{t(x_0)^2+\frac{4(x-x_0)^2}{v_{st}^2}} \tag{7}$$

where x is the position along the common offset profile, x_0 is surface position vertically above the diffracting target, $t(x_0)$ is the two - way traveltime to the target when the antenna is positioned at x_0 [25], and v_{st} is the stacking velocity that is similar to the root - mean - square velocity. Velocity analysis is a reliable way to estimate stacking velocity and the accuracy depends on the signal - to - noise ratio of data, which influences the quality of picking. The semblance analysis was performed along the rover's traverse, which corresponds to the diffraction hyperbola in the LPR profile.

The calculated stacking velocity by semblance analysis could be transferred to interval velocity through Dix inversion [26] as:

$$v_{interval,n}=\sqrt{\frac{v_{rms,n}^2 t_n - v_{rms,n-1}^2 t_{n-1}}{t_n - t_{n-1}}} \tag{8}$$

where $v_{interval,n}$ is the interval velocity of n - th layer and v_{rms} is the root - mean - square velocity.

Semblance analysis provides a measure of the coherency of energy along trial trajectories defined by the substitution of trial pairs of v_{st} and $t(x_0)$, which was conducted over an aperture of L traces and $2M+1$ samples as:

$$S=\frac{1}{L}\frac{\sum_{j=i-M}^{i+M}\left(\sum_{k=0}^{L-1}Q_{j,k}\right)^2}{\sum_{j=i-M}^{i+M}\left(\sum_{k=0}^{L-1}Q_{j,k}^2\right)}, \tag{9}$$

where L is 11, 661 in LPR data, i and j are time sample indices, k is a trace number and $Q_{j,k}$ is the amplitude of the j - th sample of the k - th trace [27]. The selection of M would not influence the peak location of contours for semblance analysis. A proper selection of M would gain a better tradeoff between resolution (traveltime and velocity) and noise compression. Our sensitivity analysis indicates that a value of $M=3$ would give a reasonable resolution and less noise level, so here we set $M=3$.

2.4 Property Calculation

The relative permittivity without considering the electric conductivity and magnetic permeability can be approximated by the propagating velocity as [8]:

$$\varepsilon=\left(\frac{c}{v}\right)^2 \tag{10}$$

where c is the speed of light in vacuum, which is 0.3m/ns. The relation between the relative permittivity and density of lunar regolith [28] is:

$$\rho=\log_{1.919}(\varepsilon) \tag{11}$$

where ρ is the bulk density with the unit of (g/cm^3). Laboratory measurements show that the loss tangent represents the ratio of the imaginary part of the relative permittivity to its real part, which depends on bulk density. The ($FeO+TiO_2$) abundance of lunar regolith could be estimated based on the loss tangent and bulk density, which is beyond the scope of this study.

3 Results

3.1 Simulation Data Results

We verify the effectiveness of the proposed method in a synthetic model using the finite difference time domain (FDTD) method [29], as shown in Figure 3a. The synthetic model consists of a heterogeneous background and several anomalous rocks with random relative permittivities less than 5.0 [24]. The dominant frequency of a Ricker wavelet is 500MHz, the time windows and the sampling interval are 120 and 0.02ns, respectively, and the trace interval is 0.005m.

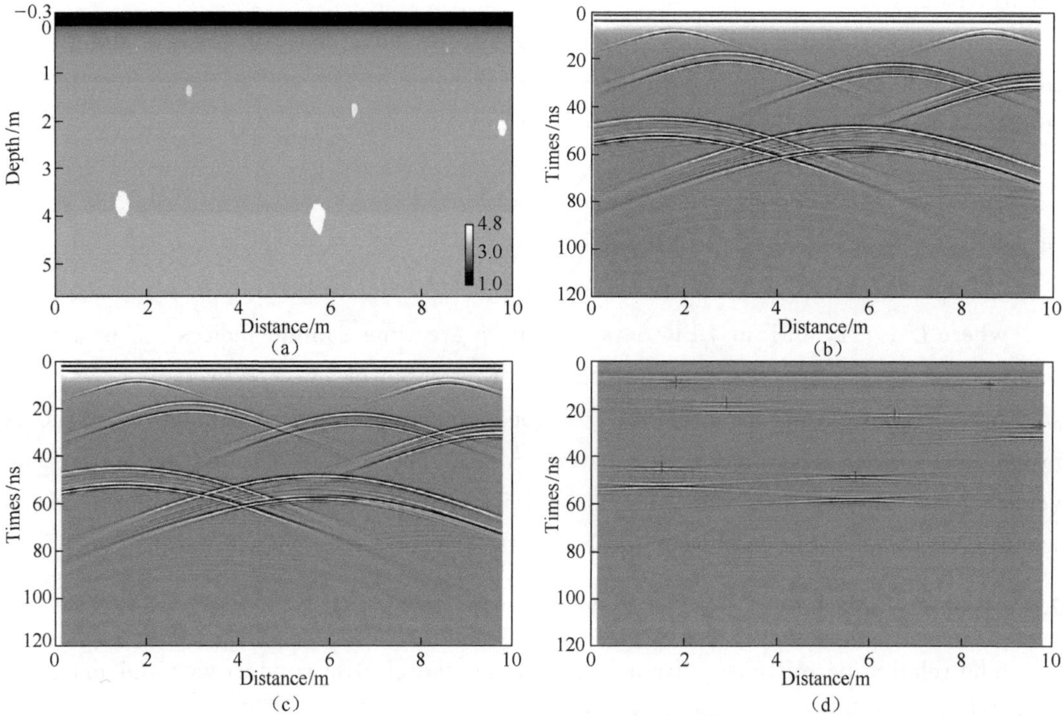

Figure 3 (a) The synthetic regolith model; (b) forward result with offset of 0.16m; (c) forward result with offset of 0.32m; (d) the local correlation defined in Section 2.2 for the two filtered data. The red-cross indicates the vertex position of the diffractions and the white color indicates the maximum value of local correlation.

The forward simulation results of CH-2A with the offset of 0.16m and CH-2B with the offset of 0.32m are illustrated in Figure 3b, c, respectively. The double diffraction hyperbolas were generated due to the upper and bottom surfaces of the rocks [16]. We focused on the upper diffraction hyperbola to reduce the error of velocity estimation. Furthermore, we demonstrate the results by applying the local correlation based on the frequency wavenumber (FK) fan filtered data in Figure 3d, where the red - cross indicates the vertex position of the diffractions. Dong et al. determined the vertex positions of the diffractions by directly applying a 3D velocity spectrum for the whole data [16], which would cause a relatively large error because the maximum value position of semblance analysis would be significantly influenced by the noise. The vertex position of the diffraction determinations based on local correlation would have higher accuracy than the direct 3D velocity spectrum. Accurate determination of vertex positions of the diffractions is the basis of high - precision semblance analysis for the velocity.

Based on the picking positions in Figure 3d, we could scan the stacking velocity within the local scope of the simulation data. We display the contours from the semblance analysis for the velocity ranges from 0. 1 to 0. 3m/ns within a 4ns time window in Figure 4. The vertex positions of the diffractions are shown in the upper right corner of the graph. The stacking velocity and the calculated depth of each rock are listed in Table 1.

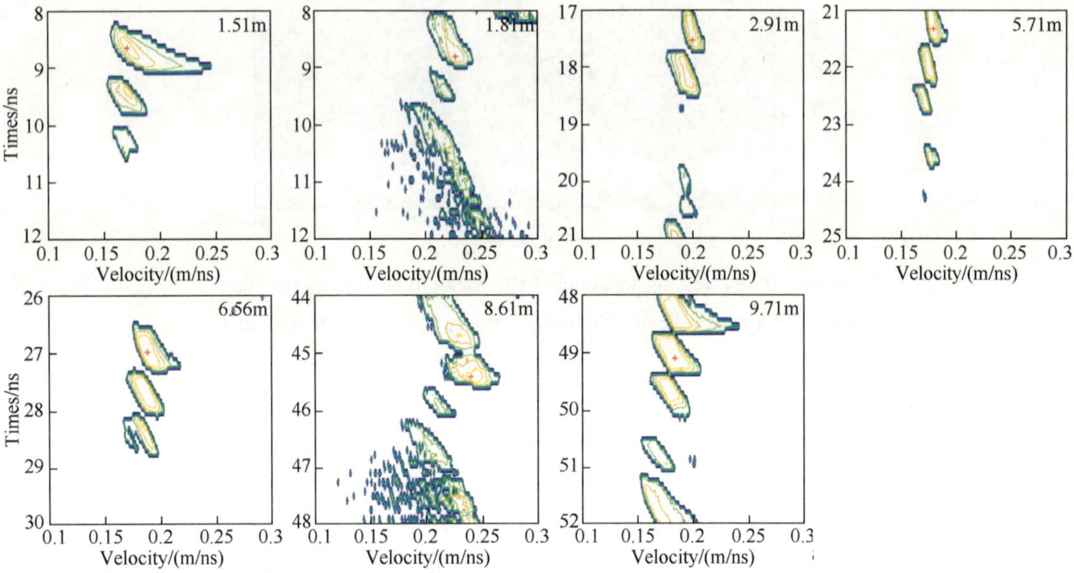

Figure 4　Semblance analysis contours of selected diffraction hyperbolas for the simulation model, where the normalized stacked amplitude threshold of 0. 6 was used.
The red - cross indicates the maximum value 1 of the semblance analysis

Table 1　The estimated results of velocity and depth for the simulation model

Rocks	Distance (m)	Relative Permittivity	Time (ns)	Stacking Velocity (m/ns)	Depth (m)
1	1. 51	4. 5	44. 64	0. 171	3. 517
2	1. 81	3. 5	8. 80	0. 227	0. 699

（续）

Rocks	Distance (m)	Relative Permittivity	Time (ns)	Stacking Velocity (m/ns)	Depth (m)
3	2.91	4.0	17.52	0.200	1.452
4	5.71	4.1	48.32	0.179	4.025
5	6.56	4.5	22.00	0.188	1.768
6	8.61	4.5	9.44	0.240	0.833
7	9.71	4.8	27.12	0.184	2.195

The scanning positions are well consistent with the actual positions of rocks except the two rocks in the shallow layer (Figure 5). The inconsistency may be due to the high velocity in the shallow layers which results in flat diffraction hyperbola, thus the accuracy of semblance analysis for velocity is insufficient. Thus, the simulation shows that our proposed procedures based on local correlation and semblance analysis is effective to obtain the subsurface parameters.

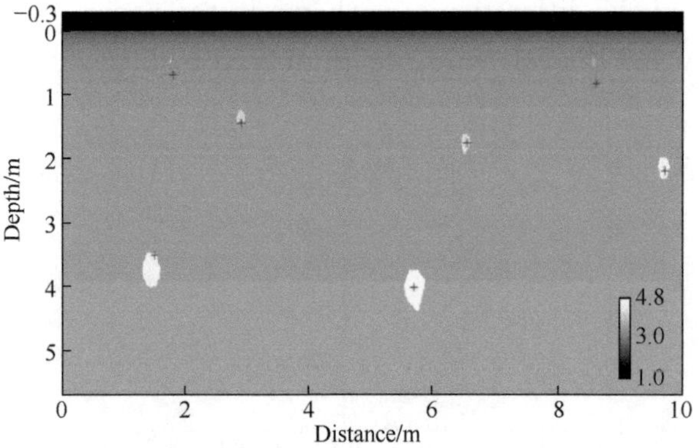

Figure 5　The scanning location of the rocks in the synthetic model. The red - cross indicates the scanning location of the diffractions

3.2　LPR Data Results

The amplitude of echoes in the first ~150ns (length $N = 501$ with 0.3125ns time interval), with a depth of about 10m for the LPR CH - 2A and CH - 2B data, is displayed in Figure 6a, b, respectively. The local correlation for the filtered CH - 2A and CH - 2B data is shown in Figure 6c, where the red - cross indicates the vertex position of the diffractions after selection.

The semblance analysis contour of diffraction hyperbolas for the LPR CH - 2B data was based on the rock location extraction by applying the local correlation, and the results are shown in Figure 7. The red - cross in the contour maps represents the maximum value of the semblance analysis. The stacking velocity for the rocks was obtained by picking the maximum value of the semblance analysis, and the calculated depths are listed in Table 2.

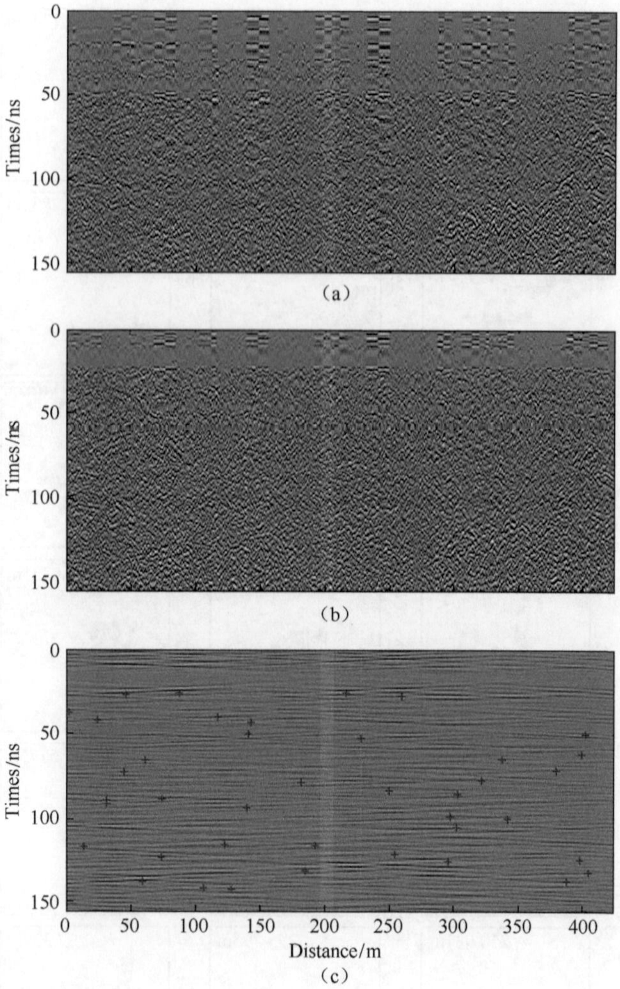

Figure 6　The profile of lunar penetrating radar (LPR) data within the region of 0～150 ns. (a) The CH－2A data; (b) the CH－2B data; (c) the local correlation defined in Section 2.2 for the filtered CH－2A and CH－2B data. The red－cross indicates the vertex position of the diffractions and the white color indicates the maximum value of local correlation

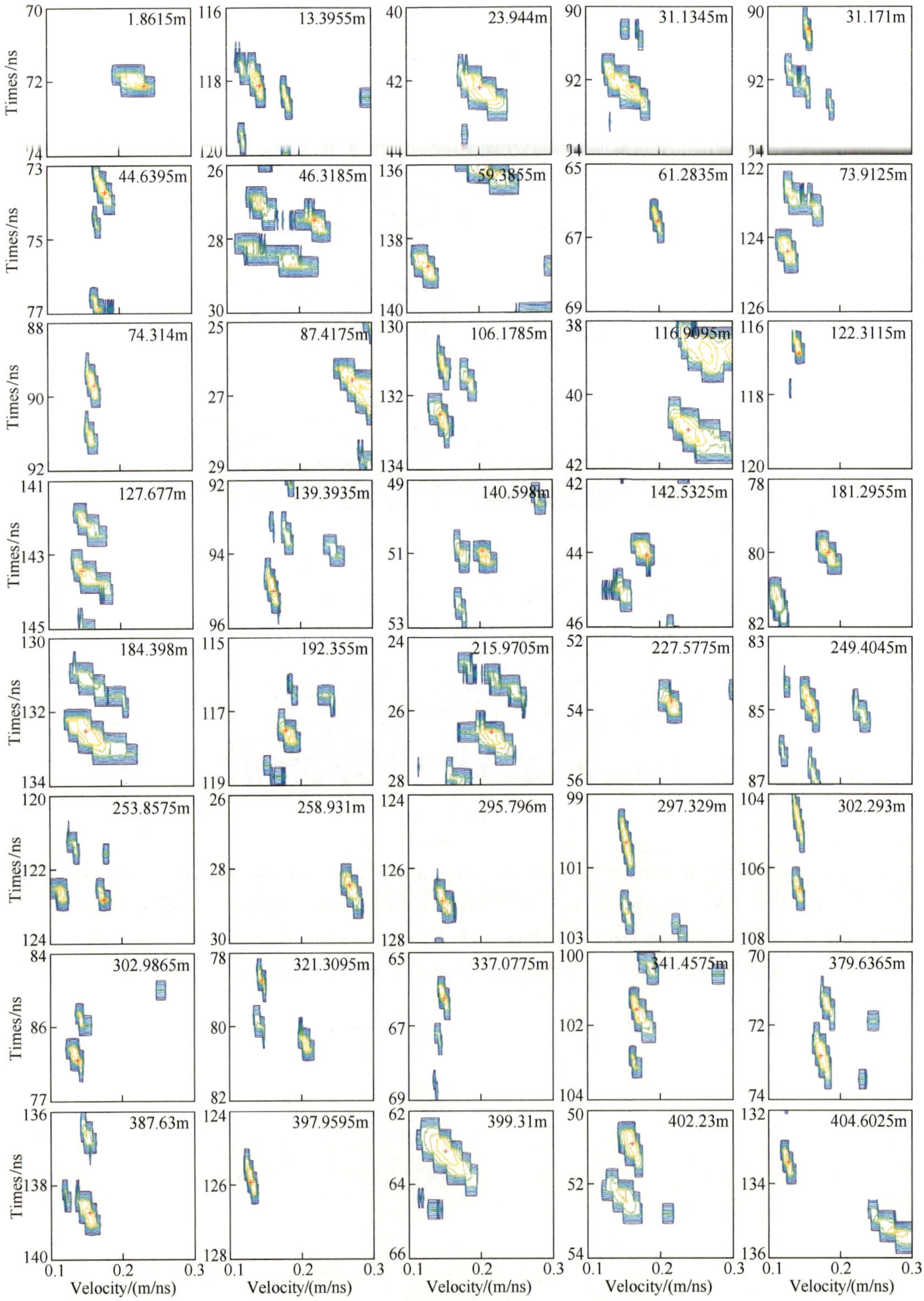

Figure 7 Semblance analysis contour of diffraction hyperbolas after screening the LPR CH－2B data, where the normalized stacked amplitude threshold of 0.6 was used. The red－cross indicates the maximum value of 1 for the semblance analysis

Table 2　The estimated results of velocity and depth for the LPR data

Rocks	Distance (m)	Time (ns)	Stacking Velocity (m/ns)	Depth (m)
1	1.8615	38.1250	0.234	4.461
2	13.3955	118.1250	0.146	8.623
3	23.9440	42.8125	0.206	4.410
4	31.1345	92.5000	0.180	8.325
5	31.1710	90.6250	0.156	7.069
6	44.6395	73.7500	0.179	6.601
7	46.3185	27.5000	0.221	3.039
8	59.3855	138.7500	0.129	8.949
9	61.2835	66.5625	0.197	6.556
10	73.9125	124.375	0.127	7.898
11	74.3140	89.6875	0.163	7.310
12	87.4175	26.5625	0.273	3.626
13	106.1785	142.5000	0.145	10.331
14	116.9095	40.9375	0.238	4.872
15	122.3115	116.875	0.142	8.298
16	127.6770	143.4375	0.146	10.471
17	139.3935	95.0000	0.162	7.695
18	140.5980	50.9375	0.203	5.170
19	142.5325	44.0625	0.181	3.988
20	181.2955	80.0000	0.183	7.320
21	184.3980	132.5000	0.150	9.938
22	192.355	117.5000	0.178	10.458
23	215.9705	26.5625	0.215	2.855
24	227.5775	53.7500	0.214	5.751
25	249.4045	85.0000	0.161	6.843
26	253.8575	122.8125	0.174	10.685
27	258.9310	28.7500	0.276	3.968
28	295.7960	126.8750	0.123	7.803
29	297.3290	100.3125	0.152	7.624
30	302.2930	106.5625	0.142	7.566
31	302.9865	86.8750	0.137	5.951
32	321.3095	78.7500	0.145	5.709
33	337.0775	66.2500	0.148	4.903
34	341.4575	101.5625	0.164	8.328
35	379.6365	72.8125	0.172	6.262
36	387.6300	138.7500	0.153	10.614
37	397.9595	125.9375	0.129	8.123
38	399.3100	63.1250	0.150	4.734
39	402.2300	50.9375	0.160	4.075
40	404.6025	133.4375	0.127	8.473

Meanwhile, we compared the diffraction hyperbola for No. 30 and No. 34 rocks with the CH - 2B data. As shown in Figure 8, the white dotted lines represent the diffraction hyperbola at 302. 2930m and 106. 5625 ns for the stacking velocity of 0. 142m/ns (Figure 8a) and at 341. 3845m and 101. 5625 ns for the stacking velocity of 0. 165m/ns (Figure 8b), respectively. The trend of the hyperbola shows great consistency with the CH - 2B data, which indicates the semblance analysis procedure can obtain the stacking velocity with high precision.

We applied the spline image interpolation method based on the 40 selected rocks to obtain the stacking velocity of the substructures. The stacking velocity and the rootmean - square velocity approach to the same when the source - receiver offset approaches zero for the isotropic layered model. The interval velocity can be calculated using the Dix formula as shown in Figure 9a. Subsequently, the estimated relative permittivity of the subsurface structure along the Yutu - 2 rover's traverse, as shown in Figure 9b, could be obtained based on the interval velocity distribution. The subsurface velocity and the relative permittivity are important parameters for calculating other physical properties. Based on the relative permittivity of the subsurface structure without considering the echo power attenuation, we could acquire the subsurface properties of lunar regolith, which are important for geological interpretation.

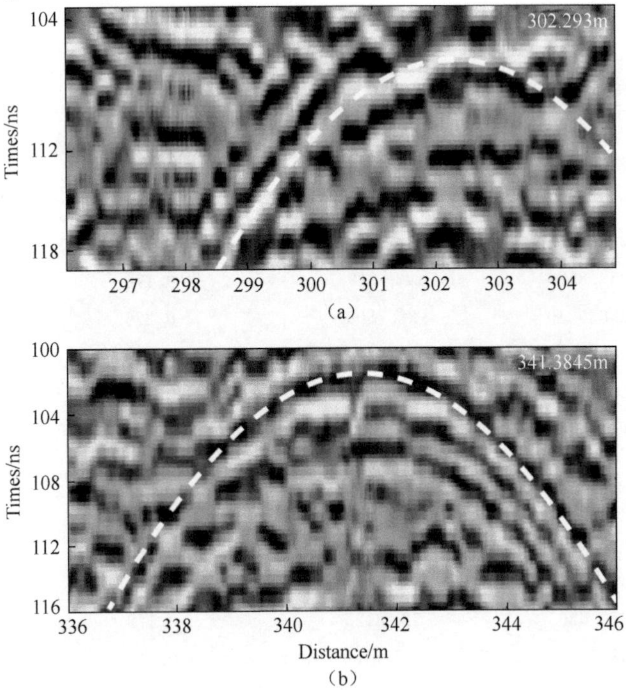

Figure 8 The comparison between the diffraction hyperbola and the CH - 2B data within the local region. (a) The white dotted lines represent the diffraction hyperbola at 302. 2930m and 106. 5625 ns for the stacking velocity of 0. 142m/ns; (b) the white dotted lines represent the diffraction hyperbola at 341. 3845m and 101. 5625ns for the stacking velocity of 0. 165m/ns

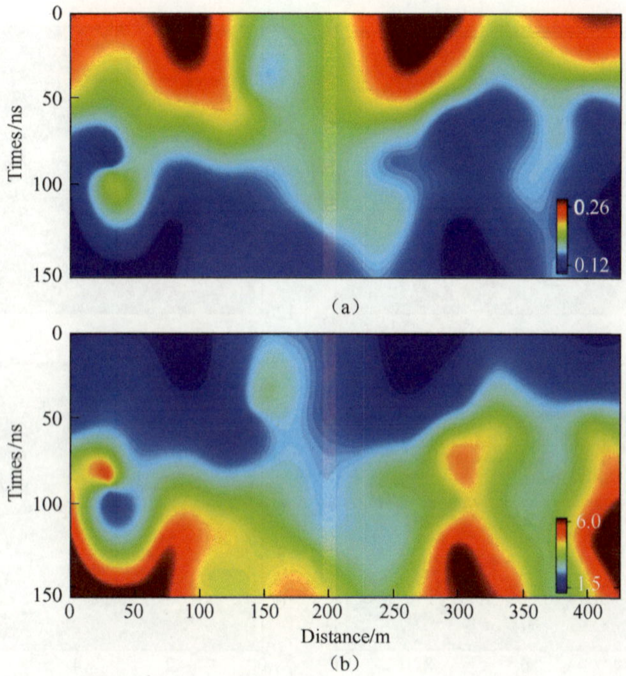

Figure 9　The estimated subsurface velocity and relative permittivity distribution along the Yutu－2 rover route. (a) The subsurface velocity distribution; (b) the subsurface relative permittivity distribution

4　Discussion

The noise interference and incomplete hyperbola will produce errors in semblance analysis. Additionally, the region selection of the data will influence the determination of the vertex position of the diffractions, which will affect the accuracy of semblance analysis. The determination of rock locations (i. e., the vertex position of the diffractions) is the most important procedure to obtain reliable semblance analysis results. Our analysis indicates that the semblance analysis in the shallow layers of the simulation model always has a larger bias than that in the deep layers. One possible reason is that the flat diffraction hyperbola would influence the accuracy of semblance analysis. Thus, a high－precision semblance analysis method such as the weighted AB semblance [30] or new migration methods [31] should be developed to improve the accuracy in the future studies. Additionally, LPR data processing should consider the local attributes to separate the data with higher accuracies.

We display the forward result with small Gaussian noise (mean is 0 and variance is 0.005) added and the corresponding local correlation results in Figure 10 to evaluate the validity of the combined method with noise. The results show that small noise has little influence on the local correlation results, and the combined method is valid to derive accurate velocity, permittivity, and bulk density distribution.

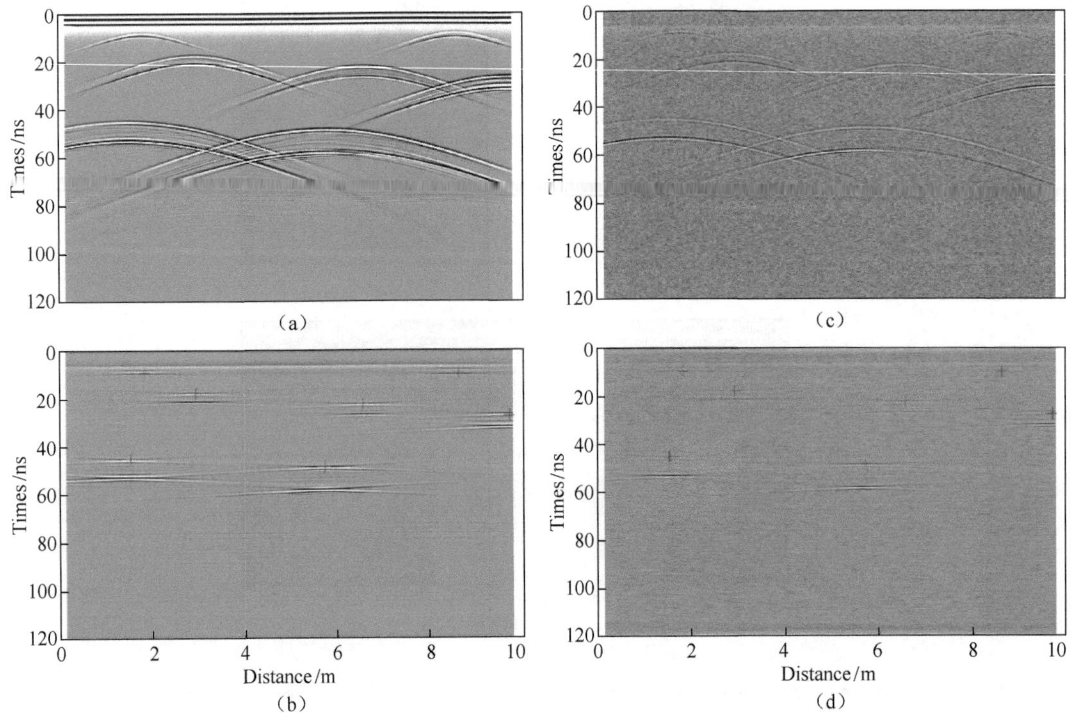

Figure 10 (a) Forward result; (b) the local correlation results; (c) forward result with Gaussian noise added; (d) the local correlation results with Gaussian noise added. The red - cross indicates the vertex position of the diffractions and the white color indicates the maximum value of local correlation

The local correlation and semblance analysis were combined to estimate the velocity structure, which can provide a good estimation of abnormal rock location and relative permittivity. The future multiple - input multiple - output radar onboard the lander from China's Chang'E - 5 (CE - 5) mission can help the detection of the regolith structure and the position of the underlying abnormal rocks in the landing site. The combined method by using local correlation and semblance analysis could be helpful for further processing on the CE - 5 radar data. Furthermore, the method can be used in the site selection of the International Lunar Research Station (ILRS) [32] and similar missions in the future.

5 Conclusions

We derived the similarity between filtered LPR CH - 2A and CH - 2B data based on the local correlation, which would help to reveal the rock location distribution in the lunar regolith. Subsequently, we applied the semblance analysis method for the diffraction hyperbola to obtain the subsurface velocity structure and the relative permittivity distribution. The other properties of lunar regolith including bulk density and FeO and TiO_2 abundance can be calculated based on the relative permittivity. A procedure that combined the local correlation and semblance analysis has an advantage over traditional methods [12 - 16] in terms of accuracy and robustness in the interpretation of the high - frequency LPR data.

The velocity structure and the properties of the subsurface were derived from diffraction hyperbolas in the high - frequency LPR data, which could provide a good structural constraint for understanding the formation and evolution of the lunar regolith at the landing site. The combined procedure is an excellent choice for estimating the subsurface velocity and the other properties of the lunar regolith. Compared with the methods by analyzing Apollo's samples [1] and by the microwave remote sensing techniques [3] to derive the physical parameters of lunar regolith, the application of LPR has great advantages both in terms of the range and the accuracy. Therefore, it is very important to develop parameter estimation methods based on the LPR data, especially its high - frequency data. The procedure combined with the local correlation and semblance analysis, proposed in this paper, is of great significance in the interpretation of LPR high - frequency data.

Acknowledgments

We thank Lin H. L. for his assistance with the manuscript. We would like to thank the editor and reviewers for their reviews that improved the content of this paper. The data reported in this work are archived at http://moon. bao. ac. cn/searchOrder _ pdsData. search.

References

[1] McKay, D. S.; Heiken, G.; Basu, A.; Blanford, G.; Simon, S.; Reedy, R.; French, B. M.; Papike, J. The lunar regolith. In Lunar Source - Book: A User's Guide to the Moon; Heiken, G., Ed.; Cambridge University Press: New York, NY, USA, 1991; Volume 7, pp. 285 - 357.

[2] Nakamura, Y.; Latham, G. V.; Dorman, H. J. Apollo Lunar Seismic Experiment - Final summary. J. Geophys. Res. Space Phys. 1982, 87, A117 - A123.

[3] Fa, W.; Jin, Y. A primary analysis of microwave brightness temperature of lunar surface from Chang - E 1 multi - channel radiometer observation and inversion of regolith layer thickness. Icarus 2010, 207, 605 - 615.

[4] Neish, C. D.; Barnes, J. W.; Sotin, C.; MacKenzie, S.; Soderblom, J. M.; Le Mouélic, S.; Kirk, R. L.; Stiles, B. W.; Malaska, M. J.; Le Gall, A.; et al. Spectral properties of titan's impact craters imply chemical weathering of its surface. Geophys. Res. Lett. 2015, 42, 3746 - 3754.

[5] Porcello, L.; Jordan, R.; Zelenka, J.; Adams, G.; Phillips, R.; Brown, W.;

Ward, S.; Jackson, P. The Apollo lunar sounder radar system. Proc. IEEE 1974, 62, 769 - 783.

[6] Ono, T.; Kumamoto, A.; Kasahara, Y.; Yamaguchi, Y.; Yamaji, A.; Kobayashi, T.; Oshigami, S.; Nakagawa, H.; Goto, Y.; Hashimoto, K.; et al. The Lunar Radar Sounder (LRS) Onboard the KAGUYA (SELENE) Spacecraft. Space Sci. Rev. 2010, 154, 145 - 192.

[7] Xiao, L.; Zhu, P.; Fang, G.; Xiao, Z.; Zou, Y.; Zhao, J.; Zhao, N.; Yuan, Y.; Qiao, L.; Zhang, X.; et al. A young multilayered terrane of the northern Mare Imbrium revealed by Chang'E - 3 mission. Science 2015, 347, 1226 - 1229.

[8] Zhang, J.; Yang, W.; Hu, S.; Lin, Y.; Fang, G.; Li, C.; Peng, W.; Zhu, S.; He, Z.; Zhou, B.; et al. Volcanic history of the Imbrium basin: A close - up view from the lunar rover Yutu. Proc. Natl. Acad. Sci. USA 2015, 112, 5342 - 5347.

[9] Li, C.; Su, Y.; Pettinelli, E.; Xing, S.; Ding, C.; Liu, J.; Ren, X.; Lauro, S. E.; Soldovieri, F.; Zeng, X.; et al. The Moon's farside shallowsub surface structure unveiled by Chang'E - 4 Lunar Penetrating Radar. Sci. Adv. 2020, 6, eaay6898.

[10] Zhang, J.; Zhou, B.; Lin, Y.; Zhu, M.; Song, H.; Dong, Z.; Gao, Y.; Di, K.; Yang, W.; Lin, H.; et al. Lunar regolith and substructure at Chang'E - 4 landing site in South Pole - Aitken basin. Nat. Astron. 2020, 1 - 6.

[11] Fang, G.; Zhou, B.; Ji, Y.; Zhang, Q.; Shen, S.; Li, Y.; Guan, H.; Tang, C.; Gao, Y.; Lu, W.; et al. Lunar Penetrating Radar onboard the Chang'e - 3 mission. Res. Astron. Astrophys. 2014, 14, 1607 - 1622.

[12] Feng, J.; Su, Y.; Ding, C.; Xing, S.; Dai, S.; Zou, Y. Dielectric properties estimation of the lunar regolith at CE - 3 landing site using lunar penetrating radar data. Icarus 2017, 284, 424 - 430.

[13] Dong, Z.; Fang, G.; Ji, Y.; Gao, Y.; Wu, C.; Zhang, X. Parameters and structure of lunar regolith in Chang'E - 3 landing area from lunar penetrating radar (LPR) data. Icarus 2017, 282, 40 - 46.

[14] Lai, J.; Xu, Y.; Zhang, X.; Tang, Z. Structural analysis of lunar subsurface with Chang'E - 3 lunar penetrating radar. Planet. Space Sci. 2016, 120, 96—102.

[15] Hu, B.; Wang, D.; Zhang, L.; Zeng, Z. Rock Location and Quantitative Analysis of Regolith at the Chang'e - 3 Landing Site Based on Local Similarity Constraint. Remote Sens. 2019, 11, 530.

[16] Dong, Z.; Feng, X.; Zhou, H.; Liu, C.; Zeng, Z.; Li, J.; Liang, W. Properties Analysis of Lunar Regolith at Chang'E - 4 Landing Site Based on 3D Velocity Spectrum of Lunar Penetrating Radar. Remote Sens. 2020, 12, 629.

[17] Zhang, L.; Zeng, Z.; Li, J.; Huo, Z.; Wang, K.; Zhang, J.; Huang, L. Parameter Estimation of Lunar Regolith from Lunar Penetrating Radar Data.

Sensors 2018，18，2907.

[18] Fomel，S. Local seismic attributes. Geophysics 2007，72，29－33.

[19] Liu，G.；Fomel，S.；Jin，L.；Chen，X. Stacking seismic data using local correlation. Geophysics 2009，74，V43—V48.

[20] Fisher，E.；McMechan，G. A.；Annan，A. P. Acquisition and processing of wide－aperture ground－penetrating radar data. Geophysics 1992，57，495－504.

[21] Su，Y.；Fang，G.；Feng，J.；Xing，S.；Ji，Y.；Zhou，B.；Gao，Y.；Li，H.；Dai，S.；Xiao，Y.；et al. Data processing and initial results of Chang'e－3 lunar penetrating radar. Res. Astron. Astrophys. 2014，14，1623－1632.

[22] Lin，H.；Lin，Y.；Yang，W.；He，Z.；Hu，S.；Wei，Y.；Xu，R.；Zhang，J.；Liu，X.；Yang，J.；et al. New Insight Into Lunar Regolith－Forming Processes by the Lunar Rover Yutu－2. Geophys. Res. Lett. 2020，47，e2020GL087949.

[23] Angelopoulos，M.；Redman，D.；Pollard，W. H.；Haltigin，T. W.；Dietrich，P. Lunar ground penetrating radar：Minimizing po－tential data artifacts caused by signal interaction with a rover body. Adv. Space Res. 2014，54，2059－2072.

[24] Lv，W.；Li，C.；Song，H.；Zhang，J.；Lin，Y. Comparative analysis of reflection characteristics of lunar penetrating radar data using nu－merical simulations. Icarus 2020，350，113896.

[25] Ristic，A. V.；PetrovaČki，D.；Govedarica，M. A new method to simultane－ously estimate the radius of a cylindrical object and the wave propagation velocity from GPR data. Comput. Geosci. 2009，*35*，1620－1630.

[26] Dix，C. H. Seismic velocities from surface measurements. Geophysics 1955，20，68－86.

[27] Neidell，N. S.；Taner，M. T. Sembalnce and other coherency measures for multichannel data. Geophysics 1971，34，859－891.

[28] Carrier，W. D.；Olhoeft，G. R.；Mendell，W. Physical properties of the lunar surface. In Lunar Source－Book：A User's Guide to the Moon；Heiken，G.，Ed.；Cambridge University Press：New York，NY，USA，1991；Volume 9，pp. 475－594.

[29] Irving，J.；Knight，R. Numerical modeling of ground－penetrating radar in 2－D using MATLAB. Comput. Geosci. 2006，32，1247－1258.

[30] Ebrahimi，S.；Kahoo，A. R.；Chen，Y.；Porsani，M. A high－resolution weighted AB semblance for dealing with ampli－tude－variationwith－offset phenomenon. Geophysics 2017，82，V85－V93.

[31] Sun，H.；Gao，C.；Zhang，Z.；Liao，X.；Wang，X.；Yang，J. High－Resolution Anisotropic Prestack Kirchhoff Dynamic Focused Beam Migration. IEEE Sens. J. 2019，20，11753－11760.

[32] Pei，Z.；Liu，J.；Wang，Q.；Kang，Y.；Zou，Y.；Zhang，H.；Zhang，Y.；He，H.；Wang，Q.；Yang，R. Overview of lunar exploration and International Lunar Research Station. Chin. Sci. Bull. 2020，65，2577－2586.（In Chinese）

Physical and Mechanical Characteristics of Lunar Soil at the Chang'E－4 Landing Site

Zhencheng Tang[1,2], Jianjun Liu[1,2], Xing Wang[1,2], Xin Ren[1], Wangli Chen[1], Wei Yan[1], Xiaoxia Zhang[1], Xu Tan[1], Xingguo Zeng[1], Dawei Liu[1], Hongbo Zhang[1,2], Weibin Wen1, Wei Zuo[1,2], Yan Su[1,2], Jianfeng Yang[3], and Chunlai Li[1,2]

1. Key Laboratory of Lunar and Deep Space Exploration, National Astronomical Observatories, Chinese Academy of Sciences, Beijing, China
2. School of Astronomy and Space Science, University of Chinese Academy of Sciences, Beijing, China
3. Xi'an Institute of Optics and Precision Mechanics, Chinese Academy of Sciences, Xi'an, China

Abstract: Chang'E－4, with the Yutu－2 rover, is the first lunar probe to successfully land and conduct a tour on the far side of the Moon from early 2019. We analyze the physical and mechanical characteristics of lunar soil through the in situ terrain data collected by the panoramic camera onboard the Yutu－2 rover. With the slip ratio and wheel sinkage obtained by the derived Digital Orthophoto Map (DOM) and Digital Elevation Model (DEM), the mechanical parameters of lunar soil are derived from the slip－sinkage model. These mechanical parameters and wheel size of the rover are used to obtain the pressure－sinkage curves, which can estimate the lunar soil strength. The experimental results indicate that the soil strength at the Chang'E－4 landing site is much higher than that at the Chang'E－3 landing site. The discrepancies in lunar soil strength between the two landing sites may be related to the local surface topography and degree of space weathering.

Plain Language Summary: The knowledge of the physical and mechanical characteristics of lunar soil is of fundamental importance because it is the basis for mineral resource exploration and engineering activity aimed at the construction of lunar bases. The Chang'E－4 landing site is within the mare floor of the Von Kármán crater inside the South Pole－Aitken (SPA) basin. The ejecta from the nearbyimpact craters has covered the Chang'E－4 landing area. Besides, the lunar soil of the Chang'E－4 landing area has been relatively mature. Here, a method of analyzing the characteristics of lunar soil is introduced to provide an opportunity to understand the physical and mechanical properties of the mature lunar soil at the Chang'E－4 landing site. With the terrain data collected by the panoramic camera onboard the Yutu - 2 rover, we can get the relationship curve of the interaction between the rover wheels and lunar soil, such as the pressure - sinkage curve. The pressure - sinkage curves can intuitively reflect the discrepancies of lunar soil strength at the Chang'E - 4 and Chang'E - 3 landing sites, which may bear a big relationship to the local surface morphology and degree of space weathering.

1 Introduction

China's Chang'E - 4 (CE - 4) lunar probe, carrying the Yutu - 2 rover, became the first space mission to touch down on the far side of the Moon on 3 January 2019. The landing site (177.5991°E, 45.4446°S) is within the mare floor of the Von Kármán crater inside the South Pole - Aitken (SPA) basin (Liu et al., 2019). Figure 1 shows the landing position of CE - 4 and the exploration route of the Yutu - 2 rover in the first two lunar days. The latest research suggests that the CE - 4 landing site has experienced complex geological processes and is distinct from the Apollo, Luna, and Chang'E - 3 (CE - 3) landing sites in chemical compositions (Ling et al., 2019). The ejecta from the nearby impact craters has covered the CE - 4 landing area, which may originate from the lunar mantle (Li et al., 2019). Additionally, the lunar soil of the CE - 4 landing area has been relatively mature (Gou et al., 2020). CE - 4 provides a unique opportunity to assess the soil properties on the far side of the Moon. By comparing the data collected during the previous Apollo, Luna, and CE - 3 missions, the similarities and differences of lunar soil properties between those sites on the nearside and the CE - 4 landing site on the farside of the Moon can be determined.

The lunar surface material (regolith or "lunar soil") is a complex mixture of five basic particle types: crystal line rock fragments, mineral fragments, breccias, agglutinates, and glasses. The relative proportion of each particle type varies from place to place and is dependent on the mineralogy of the source rocks and the geologic processes that the rocks have undergone (Heiken et al., 1991). The processes of the continuous impacts of meteorites, the bombardment of interstellar energetic particles, and the huge diurnal temperature variation have produced the lunar soil (Jolliff et al., 2006; Li et al., 2015). The knowledge of the physical and mechanical properties of the soil is of fundamental importance because it is the basis for engineering activity aimed at the construction of lunar bases and for mineral resource exploration (Gromov, 1998). And this knowledge is also important to design terramechanics models for interactions with rovers, landers, scoops, and other vehicles with the lunar surface (Slyuta, 2014). Current research about the physical and mechanical properties of lunar soil is mainly based on the in situ measurements and the successfully returned samples of Apollo and Luna missions (Carrier et al., 1991; Mitchell et al., 1972). In addition to the above, the study on the wheel - soil interaction also provides a useful method to figure out the

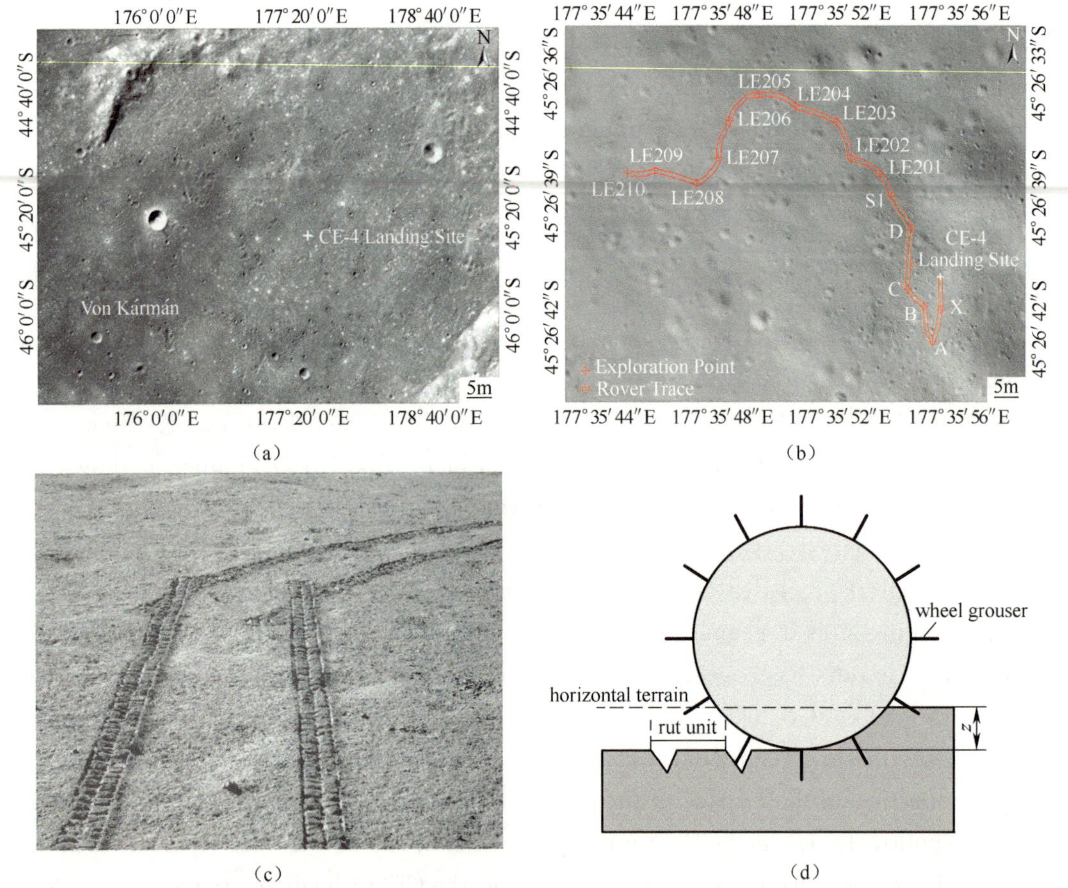

Figure 1　CE－4 landing area and the track of the Yutu－2 rover. (a) CE－4 landing site; (b) the exploration route of the Yutu－2 rover in the first two lunar days; (c) panoramic image of the track; (d) diagram of the contact between rover wheel and lunar soil.

properties of lunar soil.

Many studies have been conducted on the factors that influence the wheel－soil interaction by the ground experiments and numerical analysis. For the calculation of the slip ratio, Gustafsson (1997) and Ward and Iagnemma (2008) used measuring instruments to estimate the slip ratio, but a variety of sensors were required to participate. Nister et al. (2004) and Ding et al. (2009) estimated the slip ratio from track images, but they did not conduct further research on the formation mechanism of tracks. Li et al. (2016) estimated the slip ratio based on the frequency domain characteristics of track images. This method was robust to complex illumination, and the slip ratio could be estimated only by track images collected by the camera. Besides, the mechanical models of the wheel－soil interaction can be derived from the results of classical terramechanics, such as the model reflecting the slip－sinkage effect. When the rover slips, there is a certain functional relationship between the wheel sinkage and the slip ratio. Vasil'ev et al. (1969) established a linear relationship between sinkage and slip ratio. However, according to the theory of the terramechanics, there is a nonlinear relationship between the wheels and the soil (Bekker, 1969), so Jin et al. (2013) introduced the exponential term of slip ratio to establish the dynamic sinkage model. Lyasko (2010) proposed an analytical model of wheel dynamic sinkage based on slip - sinkage effect, and the dynamic sinkage was related to the static sinkage and slip ratio. Moreover, the measurement can be made by analyzing the imagery data collected during the tour of the rover. The wheel tracks can be identified in the imagery data. They provide the information about the physical and mechanical characteristics of lunar soil.

CE - 4 is the backup of its predecessor, CE - 3. Both rovers of the CE - 4 and CE - 3 have the same weight (135kg) and wheel size (the wheel diameter is 30cm and the wheel width is 15cm). Unlike several Luna and Apollo landing spacecraft, the lander and rover of CE - 4 and CE - 3 were not equipped with any dedicated instrument for measuring soil physical properties. The area around the CE - 4 landing site has been imaged by a pair of panoramic cameras (PCAM) onboard the Yutu - 2 rover. During the operation on the lunar surface, these cameras have collected high spatial resolution images of the lunar terrain, which can be used to reconstruct the topography of the lunar surface. The physical and mechanical characteristics of lunar soil can be figured out based on these in situ terrain data.

An alternative application of the terrain data produced by PCAM images to measure the wheel sinkage and slip ratio is introduced in this paper. Using the slip - sinkage model, the mechanical parameters of lunar soil can be calculated, such as cohesive modulus, frictional modulus, and sinkage exponent of the soil. In terms of these mechanical parameters and the wheel size of the lunar rover, the pressure - sinkage curve is plotted to directly display the mechanical properties of lunar soil. In section 2, we describe the method of calculating the slip ratio and sinkage based on the terrain data produced by PCAM images, and analysis method of mechanical characteristics of lunar soil. The results are presented in section 3. Section 4 includes a discussion of the results and the comparison with other landing sites. Section 5 is the summary and implication of this study.

2 Data Processing and Methods

2.1 Data Processing

The wheels of the Yutu - 2 rover were constantly in contact with soil and left tracks on the lunar surface, as shown in Figure 1c. The imagery data acquired at the exploration sites of LE202, LE204, LE206, LE208, and LE210 in the second lunar day of the Yutu - 2 rover were selected. The positions of these exploration sites are shown in Figure 1b. The Digital Elevation Model (DEM) and Digital Orthophoto Map (DOM) were generated from the panoramic camera images. The slip ratio was estimated from the track image with DOM. The sinkage was calculated from the elevation values extracted from the corresponding DEM before and after the occurrence of the track.

The technical processes of producing DEM and DOM are as follows (Wang et al., 2015). First, the raw image pairs are preprocessed by radiometric correction, geometry correction, and color restoration. These input images that represent the lunar surface well are selected for further research. The interior orientation parameters are necessary inputs and are obtained from the ground calibration experiment. With the exterior orientation parameters, the positions and attitudes of image pairs are determined. Second, identical points on the left and the right image are matched by combining Scale Invariant Feature Transform (SIFT) operator and Harris operator. Third, the coarse points are eliminated by applying Random Sample Consensus (RANSAC) method. And the matching accuracy of the image pairs is better than 0.2 pixels with the above matching methods. Fourth, measurement adjustments of regional networks are performed to optimize the original external orientation elements of the imagery data. With these optimized external orientation elements, the three - dimensional coordinates of dense identical points are calculated using forward intersection, and a triangulated irregular network (TIN) can be constructed with the scatter points. Finally, the DEM is generated by Kriging interpolation with the TIN, and the root - mean - square (RMS) error of the reprojection error (internal relative error of DEM) is better than 0.2 pixels after the measurement adjustment (Liu et al., 2019). And the DOM for a single site is generated after digital differential rectification. Considering the need for positioning and navigation, as well as processing speed, we select 0.01m as the average resolution of DEM and DOM. The internal relative error of the single - site DEM is less than 0.6mm within 10m and a more detailed description about the accuracy of the topography data is shown in Text S1 in the supporting information.

2.2 Calculation of Slip Ratio

The slip ratio represents the deviation degree between the actual linear velocity and the theoretical circular velocity of the wheel, which is defined in Equation 1. By analyzing the interaction between the wheel and soil and the formation mechanism of the wheel track, Li et al. (2016) established the Equation 2 to estimate the slip ratio based on the characteristics of the track image. By comparing Equations 1 and 2, the slip ratio can be also described as the deviation degree between the practical distance and the theoretically calculated distance. In our work, the track data with easily visible wheel imprints was selected from the DOM in order to calculate the slip ratio. The distance of pixels of the sample row of ruts represents the actual distance the rover traveled. The commanded distance can be calculated from the total number of wheel grousers in the selected track. And slip ratio s of a section of the wheel ruts is obtained by Equation 3.

$$s = 1 - \frac{\nu}{r\omega} \tag{1}$$

where ν is linear velocity of the wheel, ω is circular velocity of the wheel, and r is wheel radius.

$$s = 1 - \frac{u n_L N}{2\pi r n_f} \tag{2}$$

where u is the actual distance represented by one pixel, n_L is number of wheel grousers, N is sampling points of a section of the track, and n_f is number of rut unit in the sample row of the track.

$$s = 1 - \frac{L_p}{(n_t / n_L) \times C_w} \tag{3}$$

where L_p is the actual distance the rover traveled, n_t is the total number of wheel grousers of a track, n_t/n_L is the total number of wheel revolutions, and C_w is perimeter of a wheel. And the number of wheel grousers of Yutu-2 rover is 12, which is shown in Figure 1d.

2.3 Calculation of Wheel Sinkage

The wheel sinkage was produced as the rover traveling. In classical terramechanics, wheel sinkage is regarded as the distance between the horizontal flat terrain and the lowest point of the wheel sinking in the soil (Wong, 2008). The ruts left by wheels have experienced three times of wheel rolling, but the increase of rut depth is minimal to negligible after the second time of wheel rolling (Holm, 1969). The wheels of the Yutu-2 rover have the structure of wheel grousers. Wheel sinkage is measured as the depth from the undisturbed lunar soil and depression caused by the flat part of the wheels between grousers (Gao et al., 2017). The contact between rover wheel and lunar soil is shown in Figure 1d.

The elevation values before and after the occurrence of wheel ruts are extracted from the corresponding DEMs and the elevation difference between which is the initial sinkage of wheels (see Figure S1). However, systematic errors are unavoidable since the image data of the same area before and after the occurrence of wheel ruts are taken at different sites. Therefore, to obtain a more accurate elevation difference, the systematic error needs to be subtracted in the rut sampling line. Given that the nonrut areas are not disturbed by the rover, the elevation values in the same area at two corresponding sites should maintain the same value. The variations in the elevation values of the same areas without tracks from two corresponding sites thus represent the systematic errors. In this work, the corresponding positions were manually matchedusing the DOMs, the systematic errors were then calculated from DEMs in the same areas without ruts. The absolute value of the difference between the initial sinkage and the systematic error value represents the real wheel sinkage z of the rover. The elevation accuracy of the sinkage can reach the submillimeter level (see Text S1).

2.4 Analytical Method of Mechanical Characteristics of Lunar Soil

2.4.1 Sinkage Model

As the rover moved forward, the lunar soil was sheared by the wheels, resulting in the slipping and sinking of the wheels. Therefore, the wheel sinkage is composed of static deformation sinkage and dynamic slip sinkage. The static wheel sinkage z_0 can be calculated by Equation 4 (Bekker, 1969).

$$z_0 = \left[\frac{3W}{b\left(\frac{k_c}{b}+k_\phi\right)\sqrt{D}\,(3-n)}\right]^{\frac{2}{2n+1}} \tag{4}$$

where W is vertical wheel load, b is wheel width, D is wheel diameter, k_c is cohesive modulus, k_ϕ is frictional modulus, and n is sinkage exponent of soil.

On the basis of the static sinkage z_0, the slip - sinkage effect as a dynamic response should be taken into account, when calculating the dynamic sinkage of the rover wheel. In the microgravity environment of the lunar surface, the rover wheels will slip in the loose soil. The wheel sinkage is related to the slip ratio when the rover slips. The relationship between wheel sinkage and slip ratio can be described by Lyasko′s model, which is obtained by theoretical analysis based on Equation 4 and has been verified by experiments of terrestrial dry sand (Lyasko, 2010). Mitchell et al. (1972) found that the mechanical properties of lunar soil were very similar to that of terrestrial dry sand. Therefore, Lyasko′s model is selected for the prediction of dynamic sinkage, which is shown in Equation 5 (Lyasko, 2010).

$$z = \frac{1+m_1 s}{1-m_2 s} z_0 \tag{5}$$

where the Lyasko′s model suggests $m_1 = 1$ and $m_2 = 0.5$.

2.4.2 Pressure - Sinkage Model

The penetrometers were employed to measure soil strength at different locations and depths during the Apollo 15 and Luna 17 mission (Carrier et al., 1991). These measurements can produce a relation curve between pressure p and sinkage z. The Bekker model is commonly used to estimate the behaviors of pressure and sinkage, which is shown in Equation 6 (Bekker, 1969). Bekker model had been well verified in the researches of trafficability parameters of Apollo lunar rovers and could be applied to the wheel - soil interaction (Mitchell et al., 1974).

$$p = kz^n \tag{6}$$

where k is stiffness modulus ($k = \frac{k_c}{b} + k_\phi$).

However, for a light - loaded vehicle like Yutu - 2 rover, the amount of wheel sinkage is related to the wheel size. An improved pressure - sinkage model had modified the Bekker model by considering the effect of the wheel size and was suitable for the rover with wheel diameter less than 50cm (Meirion & Spenko, 2011). The improved pressure - sinkage model is described as Equation 7.

$$p = kD^{m} z_{n} \tag{7}$$

where m is diameter exponent.

The m values were derived from the tests of terrestrial dry sand. Although lunar soil is different from that of terrestrial dry sand in composition, the mechanical characteristics are quite similar (Mitchell et al., 1972). Therefore, for dry sand applying to the lunar landing sites, $m = 0.39$ is suggested in Equation 7 (Meirion & Spenko, 2011).

To illustrate the proposed method of analysis of soil strength, we have analyzed the track images of CE - 4. Furthermore, the proposed method was also tested on the track images of CE - 3 and the results were compared with Gao et al. (2016). n was set to 0.8 in the work of Gao et al. (2016), but the mechanical parameter derived from the ground tests could not reflect the mechanical properties of lunar soil well. In our work, n is derived from the in situ terrain data produced by the panoramic camera image taken from Yutu - 2 rover, which reflects the mechanical properties of lunar soil more intuitively.

3 Results

3.1 Mechanical Parameters of Lunar Soil at the CE - 4 Landing Site

The relationship about the slip ratio s, the sinkage z, and the mechanical parameters (k_c, k_ϕ, n) of lunar soil is established by Equations 4 and 5. With this relationship and the dataset of slip ratio and sinkage obtained by data processing (see Figure S2 and Table S1), the mechanical parameters of lunar soil were solved out by the nonlinear least square fitting method. Here, the initial values of three mechanical parameters (k_c, k_ϕ, n) were set to 0.140, 0.820, and 1.000, respectively (Heiken et al., 1991). The experimental values of these mechanical parameters are derived from the five selected exploration sites, which are shown in Table 1.

Table 1　The Experimental Values of Mechanical Parameters

Mechanical parameters	Test number				
	LE202	LE204	LE206	LE208	LE210
Cohesive modulus, k_c (N/cm^2)	0: $220^{+0.017}_{-0.011}$	$0.295^{+0.009}_{-0.007}$	$0.146^{+0.006}_{-0.007}$	$0.317^{+0.008}_{-+0.011}$	$0.169^{+0.007}_{-0.016}$
Frictional modulus, k_ϕ (N/cm^2)	$0.912^{+0.012}_{-0.019}$	$0.891^{+0.022}_{-0.024}$	$0.904^{+0.021}_{-0.025}$	$0.949^{+0.010}_{-0.023}$	$0.887^{+0.018}_{-0.017}$
Sinkage exponent, n	$0.787^{+0.017}_{-0.009}$	$0.747^{+0.012}_{-0.022}$	$0.732^{+0.011}_{-0.015}$	$0.710^{+0.018}_{-0.008}$	$0.756^{+0.012}_{-0.021}$

Note. The uncertainty of mechanical parameters is calculated after introducing the elevation accuracy of the sinkage (see Table S1).

3.2 Pressure - Sinkage Curves at Different Landing Sites

The values of n and m are fixed in Equation 7. The pressure - sinkage relation is derived from the stiffness modulus k, and the wheel diameter D. For comparison, the median values of n and k are selected to calculate the pressure p by Equation 7. The corresponding $kD^{0.39}$ and n are listed in Table 2 which illustrates the overall pressure - sinkage relations. The pressure p becomes larger with the increase of the overall value of $kD^{0.39}$ and n. The pressure - sinkage curves based on Equation 7 are presented in Figure 2.

Table 2 Soil Strength Analysis Based on Images and Improved Pressure - Sinkage Model

Landing site	Vehicle	Vehicle mass per wheel, M (kg)	Wheel diameter, D (cm)	Sinkage, z (cm)	Median $kD^{0.39}$ ($N/cm^{2.8}$)	Median sinkage exponent, n
1# Luna 17	Lunokhod - 1	94.5	51	1.07—2.54	0.5	0.8
1# Apollo 15	LRV	177	82	0.37—1.99	2.45	0.8
1# CE—3	Yutu	22.5	30	0.56—1.15	0.55	0.8
2# CE—3	Yutu	22.5	30	0.25—0.93	1.19	0.616
2# CE—4	Yutu—2	22.5	30	0.23—0.78	3.44	0.747

Note. 1# refers to the method adopted by Gao et al. (2016). The parameters of the soil strength at the Lunar 17, Apollo 15, and CE - 3 landing sites are from Gao et al. (2016). 2# refers to the method proposed in our work.

4 Discussions

The pressure - sinkage curves reflect the mechanical properties of lunar soil and can be used to compare lunar soil strength at different landing sites. Figure 2 shows the results of pressure - sinkage curves derived from the median value of mechanical parameters of the lunar soil at the CE - 4 and CE - 3 landing sites with the proposed method, and the comparative results obtained by Gao et al. (2016) at the CE - 3, Luna - 17, and Apollo - 15 landing sites. In general, lunar soil strength at the CE - 4 landing site nearby is greater than that at the CE - 3 landing site.

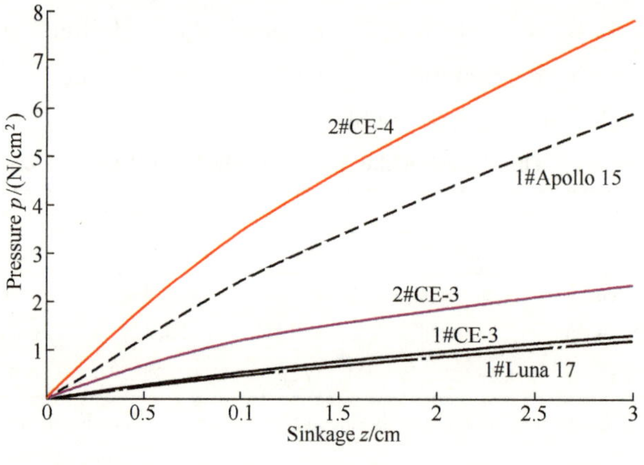

Figure 2 The pressure - sinkage curves for CE - 4, CE - 3, Luna 17, and Apollo 15. 1# refers to the method adopted by Gao et al. (2016). The pressure - sinkage curves at the Lunar 17, Apollo 15, and CE - 3 landing sites are from Gao et al. (2016). 2# refers to the method proposed in our work.

In Figure 2, the similarities of the pressure - sinkage curves imply that the soil strength at the CE - 4 landing site nearby is close to that at the Apollo 15 landing site, and CE - 3 and Luna 17 landing sites are similar in soil strength. These discrepancies of lunar soil strength may be related to the local surface morphology and degree of space weathering of each exploration area. The typical landscapes in four exploration areas of Luna 17, Apollo 15, CE - 3, and CE - 4 are presented in Figure S3. The entire route of the Yutu rover onboard CE - 3 did not leave the ejecta blanket of Crater Zi Wei (Xiao et al., 2015). And the Lunokhod - 1 rover onboard Luna 17 moved around the western rim of the Crater Borya. Basilevsky et al. (2015) suggested that the geological features of Yutu and Lunokhod - 1 exploration sites could be very similar. Both exploration sites are within the slopes near the rims of the young craters, and strongly affected by the impact ejecta. However, the lunar surface of CE - 4 and Apollo 15 exploration sites are relatively smooth and have been no modification of similar - sized impact craters to Crater Zi Wei and Crater Borya. Previous in situ measurements of the physical properties of lunar soil found that the soil near the rims of craters tended to be softer than those in the intercrater areas (Heiken et al., 1991). Our result is consistent with those obtained by in situ measurements. Besides, a large number of excavated rocks covered both exploration routes of Yutu and Lunokhod - 1 rovers. These rock fragments were not completely weathered and crushed, which suggested that the exposure ages of two exploration areas were relatively younger and the soils were relatively immature. Compared with the CE - 3 and Luna 17 landing areas, less exposed rocks and more degraded impact craters were observed within routes of Yutu - 2 and Apollo 15 rovers, which indicates that the exposure time in CE - 4 and Apollo 15 landing areas are much longer. Therefore, the soil in CE - 4 and Apollo 15 landing areas are more mature. The Is/FeO maturity index of the CE - 4 landing site (82 ± 15) is higher than that of the CE - 3 landing site (~53), which indicates that the lunar soil of the CE - 4 landing site is more mature than that of the CE - 3 landing site (Gou et al., 2020; Wang et al., 2017). Agglutinates are the major product of space weathering, the proportion of which in lunar soil increases with the soil maturity. Moreover, it was found that the maturity (Is/FeO) was positively correlated with the percentage of agglutinates, and lunar soil with higher maturity may have a higher percentage of agglutinates (Liu et al., 2020; Morris, 1976). Therefore, the percentage of agglutinates in lunar soil of the CE - 4 landing site is higher than that of the CE - 3 landing site. The presence of the agglutinates significantly influences the mechanical properties of the soil. The highly irregular agglutinates may improve the shear strength (Gustafson et al., 2008). The entire soil strength also varies with the proportion of the agglutinates (Carrier et al., 1972). And lunar soil strength of the CE - 4 landing site is higher than that of the CE - 3 landing site. Consequently, the soil strength at the CE - 4 and Apollo 15 landing sites could be greater than that at the CE - 3 and Luna 17 landing sites due to longer exposure time and a greater degree of space weathering.

5 Conclusions

The paper presents a method for quantitative analysis of lunar soil strength combined with slip - sinkage effect and pressure - sinkage relation. The Yutu - 2 rover track images of the CE - 4 mission are applied in this method to perform lunar soil analysis. The wheel slip ratio is estimated by the DOMs with the track, and the wheel sinkage is obtained through the DEMs before and after the occurrence of the track. And the multiple mechanical parameters of lunar soil are acquired by the slip - sinkage model. With these mechanical parameters and the wheel size of lunar rover, the pressure - sinkage curves are accurately obtained to display the soil strength at different landing sites. The similarity and difference of the soil strength at different landing sites may be related to the local surface morphology and degree of space weathering. The mechanical properties of lunar soil may be affected by the change of maturity. The proposed method only uses imagery data containing wheel tracks without in situ soil penetrometer data. The findings provide a reference for the further study of mechanical characteristics of lunar soil and give support for the implementation of relevant scientific experiments and engineering tasks in the later Chang'E program.

However, the proposed method of estimating the slip ratio based on the track images is merely suitable for the ideal conditions without sideslip. Further research will be carried out on the feature extraction method of track image if the rover is in the state of sideslip and steering, which can cope with different working conditions and complex environments on the lunar surface.

References

Basilevsky, A. T., Abdrakhimov, A. M., Head, J. W., Pieters, C. M., Wu, Y. Z., & Xiao, L. (2015). Geologic characteristics of the Luna 17/Lunokhod 1 and Chang'E - 3/Yutu landing sites, Northwest Mare Imbrium of the Moon. *Planetary and Space Science*, *117*, 385—400. https://doi. org/10. 1016/j. pss. 2015. 08. 006

Bekker, M. G. (1969). *Introduction to terrain - vehicle systems*. Michigan: University of Michigan Press.

Carrier, W. D., Bromwell, L. G., & Martin, R. T. (1972). Strength and compressibility of returned lunar soil. Paper presented at the 3rd Lunarand Planetary Science Conference on Lunar and Planetary Science Institute, Houston, TX.

Carrier, W. D., Olhoeft, G. R., & Mendell, W. (1991). Physical properties of the lunar surface. In G. H. Heiken D. T. Vaniman & B. M. French (Eds.), *Lunar sourcebook: A user's guide to the Moon* (pp. 475 — 594). Cambridge: Cambridge University Press.

Ding, L. A., Gao, H. B., Deng, Z. Q., Yoshida, K., & Nagatani, K. (2009).

Slip ratio for lugged wheel of planetary rover in deformable soil: Definition and estimation. Paper presented at IEEE/RSJ International Conference on Intelligent Robots and Systems on IEEE, New York.

Gao, H. B., Lv, F. T., Yuan, B. F., Li, N., Ding, L., Li, N. X., et al. (2017). Sinkage definition and visual detection for planetary rovers wheels on rough terrain based on wheel - soil interaction boundary. *Robotics and Autonomous Systems*, *98*, 222—240. https://doi. org/10. 1016/j. robot. 2017. 09. 011

Gao, Y., Spiteri, C., Li, C. L., & Zheng, Y. C. (2016). Lunar soil strength estimation based on Chang'E - 3 images. Advances in Space Research, 58 (9), 1893—1899. https://doi. org/10. 1016/j. asr. 2016. 07. 017

Gou, S., Yue, Z. Y., Di, K. C., Wan, W. H., Liu, Z. Q., Liu, B., et al. (2020). In situ spectral measurements of space weathering by Chang'e - 4rover. *Earth and Planetary Science Letters*, *535*, 116117.

https://doi. org/10. 1016/j. epsl. 2020. 116117Gromov, V. (1998). Physical and mechanical properties of lunar and planetary soils. *Earth*, *Moon*, *and Planets*, *80* (1—3), 51—72. https://doi. org/10. 1023/a: 1006353510204

Gustafson, R. J., White, B. C., & Gustafson, M. A. (2008). Development of a high fidelity lunar soil simulant. Paper presented at the AIP Conference Proceedings on Space Technology and Applications International Forum - STAIF 2008, Springer - Verlag GmbH.

Gustafsson, F. (1997). Slip - based tire road friction estimation. *Automatica*, *33* (6), 1087—1099. https://doi. org/10. 1016/s0005—1098 (97) 00003—4

Heiken, G. H., Vaniman, D. T., & French, B. M. (1991). Lunar sourcebook: A user's guide to the Moon. Cambridge: Cambridge University Press.

Holm, I. C. (1969). Multi - pass behaviour of pneumatic tires. *Journal of Terramechanics*, *6* (3), 47—71. https://doi. org/10. 1016/0022—4898 (69) 90128—1

Jin, D., Li, J., Dang, Z., Chen, B., & Zou, M. (2013). Study on model for sinkage of lunar rover wheel under slip. [滑转条件下月球车轮沉陷模型研究]. *Acta Aeronautica et Astronautica Sinica*, *34* (5), 1215—1221. (in Chinese). https://doi. org/10. 7527/S1000—6893. 2013. 0207

Jolliff, B. L., Wieczorek, M. A., Shearer, C. K., & Neal, C. R. (2006). *New views of the Moon*: *Reviews in mineralogy and geochemistry* (Vol. 60). Chantilly, Virginia: The Mineralogical Society of America.

Li, C. L., Liu, D. W., Liu, B., Ren, X., Liu, J. J., He, Z. P., et al. (2019). Chang' E - 4 initial spectroscopic identification of lunar far - side mantlederived materials. *Nature*, *569* (7756), 378—382. https://doi. org/10. 1038/s41586—019—1189—0

Li, C. L., Liu, J. J., Ren, X., Zuo, W., Tan, X., Wen, W. B., et al. (2015).

The Chang' e 3 mission overview. *Space Science Reviews*, *190* (1－4), 85－101. https://doi. org/10. 1007/s11214－014－0134－7

Li, N., Gao, H., Lv, F., Ding, L., Liu, Z., Yu, H., & Zongquan, D. (2016). Wheel trace imprint image frequency domain analysis and roverwheel slip ratio estimation. [车辙图像频域分析及星球车车轮滑转率估计方法]. Journal of Astronautics, 37 (11), 1356－1364. (in Chinese). https://doi. org/10. 3873/j. issn. 1000－1328. 2016. 11. 010

Ling, Z. C., Qiao, L., Liu, C. Q., Cao, H. J., Bi, X. Y., Lu, X. J., et al. (2019). Composition, mineralogy and chronology of mare basalts andnon－mare materials in Von Kármán crater: Landing site of the Chang'E－4 mission. *Planetary and Space Science*, *179*, 104741. https://doi. org/10. 1016/j. pss. 2019. 104741

Liu, D. W., Zhang, Y. Z., Zhang, G. L., Liu, B., & Li, C. L. (2020). An empirical model to estimate abundance of nanophase metallic iron ($npFe^0$) in lunar soils. *Remote Sensing*, *12* (6), 1047. https://doi. org/10. 3390/rs12061047

Liu, J. J., Ren, X., Yan, W., Li, C. L., Zhang, H., Jia, Y., et al. (2019). Descent trajectory reconstruction and landing site positioning of Chang'E－4 on the lunar farside. *Nature Communications*, *10* (1), 1－10. https://doi. org/10. 1038/s41467－019－12278－3

Lyasko, M. (2010). Slip sinkage effect in soil－vehicle mechanics. *Journal of Terramechanics*, *47* (1), 21－31. https://doi. org/10. 1016/j. jterra. 2009. 08. 005

Meirion, G. G., & Spenko, M. (2011). A modified pressure－sinkage model for small, rigid wheels on deformable terrains. *Journal of Terramechanics*, *48* (2), 149－155. https://doi. org/10. 1016/j. jterra. 2011. 01. 001

Mitchell, J. K., Houston, W. N., Carrier, W. D., & Costes, N. C. (1974). Apollo soil mechanics experiment S－200. Archives of Dermatology and Syphilology, 53 (1), 73－73. https://doi. org/10. 1097/01634989－199312000－00006

Mitchell, J. K., Scott, R. F., Houston, W. N., Costes, N. C., Carrier, W. D., & Bromwell, L. G. (1972). Mechanical properties of lunar soil: Density, cohesion, and angle of internal friction. Paper presented at the 3rd Lunar and Planetary Science Conference on Lunar and Planetary Science Institute, Houston, TX.

Morris, R. V. (1976). Surface exposure indices of lunar soils: A comparative FMR study. Paper presented at the 7th Lunar and Planetary Science Conference on Lunar and Planetary Science Institute, Houston, TX.

Nister, D., Naroditsky, O., & Bergen, J. (2004). Visual odometry. Paper presented at the IEEE Computer Society Conference on Computer Vision and Pattern Recognition, Princeton, USA.

Slyuta, E. N. (2014). Physical and mechanical properties of the lunar soil (a review). *Solar System Research*, *48* (5), 330－353. https://doi. org/10. 1134/S0038094614050050

Vasil'ev, A. V., Dokychaeva, E. N., & Utkin-Lubovtsov, O. L. (1969). Effect of tracked tractor design parameters on tractive performance. Moscow: Mashinostroenie.

Wang, W. R., Ren, X., Wang, F. F., Liu, J. J., & Li, C. L. (2015). Terrain reconstruction from Chang'E-3 PCAM images. *Research in* Astronomy and Astrophysics, 15 (7), 1057—1067. https://doi.org/10.1088/1674—4527/15/7/013

Wang, Z. C., Wu, Y. Z., Blewett, D. T., Cloutis, E. A., Zheng, Y. C., & Chen, J. (2017). Submicroscopic metallic iron in lunar soils estimated from the in situ spectra of the Chang'E-3 mission. *Geophysical Research Letters*, *44*, 3485—3492. https://doi.org/10.1002/2017GL072652 Ward, C. C., & Iagnemma, K. (2008). A dynamic-model-based wheel slip detector for mobile robots on outdoor terrain. *IEEE Transactions on Robotics*, *24* (4), 821 — 831. https://doi.org/10.1109/tro.2008.924945

Wong, J. Y. (2008). *Theory of ground vehicles*. New York: John Wiley & Sons.

Xiao, L., Zhu, P., Fang, G., Xiao, Z., Zou, Y., Zhao, J., et al. (2015). A young multilayered terrane of the northern Mare Imbrium revealed by Chang'E-3 mission. Science, 347 (6227), 1226 — 1229. https://doi.org/10.1126/science.1259866

A Novel Approach for Permittivity Estimation of Lunar Regolith Using the Lunar Penetrating Radar Onboard Chang'E - 4 Rover

Ruigang Wang[1,2], Yan Su[1,2], Chunyu Ding[3], Shun Dai[1], Chendi Liu[1,2], Zongyu Zhang[1,2], Tiansheng Hong[1,2], Qing Zhang[1,2] and Chunlai Li[1,2]

1. Key Laboratory of Lunar and Deep Space Exploration, National Astronomical Observatories, Chinese Academy of Sciences, Beijing 100101, China; rgwang@nao.cas.cn (R. W.); dais@nao.cas.cn (S. D.); cdliu@nao.ac.cn (C. L.); zhangzy@nao.cas.cn (Z. Z.); hongts@nao.cas.cn (T. H.); zhangq@nao.cas.cn (Q. Z.); licl@nao.cas.cn (C. L.)
2. School of Astronomy and Space Science, University of Chinese Academy of Sciences, Beijing 100049, China
3. Department of Earth and Space Sciences, Southern University of Science and Technology, Shenzhen 518000, China; dingcy@sustech.edu.cn

Abstract: Accurate relative permittivity is essential to the further analysis of lunar regolith. The traditional hyperbola fitting method for the relative permittivity estimation using the lunar penetrating radar generally ignored the effect of the position and geometry of antennas. This paper proposed a new approach considering the antenna mounting height and spacing in more detail. The proposed method is verified by numerical simulations of the regolith models. Hence the relative permittivity of the lunar regolith is calculated using the latest high - frequency radar image obtained by the Yutu - 2 rover within the first 24 lunar days. The simulation results show that the relative permittivity is underestimated when derived by the traditional method, especially at the shallow depth. The proposed method has improved the accuracy of the estimated lunar regolith relative permittivity at a depth of 0 - 3m, 3 - 6m, and 6 - 10m by 35%, 14%, and 9%, respectively. The thickness of the lunar regolith at the Chang'E 4 landing site is reappraised to be 11.1m, which improved by ~8% compared with previous studies.

Keywords: moon; Chang' E 4 mission; relative permittivity; lunar regolith; lunar penetrating radar

1 Introduction

The airless Moon is the closest extraterrestrial object to the Earth. It is also the most frequently explored planetary body with 129 exploration missions completed so far [1]. On 3 January 2019, China's Chang'E - 4 spacecraft successfully landed in the Von Karmen crater within the South Pole - Aitken basin (SPA) on the far side of the Moon [2]. It is the first in situ exploration for human beings on the far side of the Moon, which is conducive to unveiling the mystery of the lunar subsurface structures [3].

The lunar regolith is the transitional zone between the solid Moon and the free space, which contains essential information about the geological evolution of the Moon [4]. The study of lunar regolith is critical to better understand the origin and evolution of the lunar surface activities. Relative permittivity describes the ability of dielectric materials to store and release energy. It is an important electromagnetic property, which is closely related to electromagnetic (EM) wave velocity, bulk density, and loss tangent of lunar materials [4]. Therefore, the accurate measurement and estimation of the relative permittivity are very important for lunar radar observations [5]. There are three methods to estimate the relative permittivity of the lunar materials: laboratory sample measurement, remote sensing, and in situ detection [6 - 11]. As to sample measurement, it is the most accurate and direct method compared with the other two methods. Laboratory measurements of Apollo samples show that the relative permittivity of lunar regolith has a broad range from 2.3 to 6.5 [4], with a typical value of 2.7 [12, 13]. However, only the dielectric properties of the sampling position around the Apollo landing sites can be tested, and the number of sample collection sites is limited. The laboratory environment is quite different from that of the Moon. The moisture and temperature in the laboratory environment probably affect the accuracy of the relative permittivity test [6]. For radar remote sensing measurement of the dielectric properties of the lunar surface [7, 14], the relative permittivity can be obtained by combining the time delay and layer thickness measured by other methods like laser altimetry [15 - 17]. The advantage of orbiting and ground - based radar can be used to detect the dielectric properties over large areas. However, the spatial resolution of this method is not as good as that of in situ detection. For example, the ALSE (Apollo lunar sounder experiment) onboard Apollo 17 had a resolution of 300m in free space, while the Chang'E - 3 (CE - 3) radar has a range resolution of meters level [18, 19].

The lunar penetrating radar (LPR) onboard the CE - 3 mission made the first in situ radar survey of the subsurface structures of the Moon [19]. The Chang'E - 4 (CE - 4) spacecraft is a backup of the CE - 3, so that the scientific instruments are almost the same [20]. The LPR consists of two channels, namely low - and high - frequency channels [21]. The low - frequency channel operates at 60MHz, with a bandwidth of 40MHz to 80MHz, and a range resolution of meters level. The high - frequency channel has a center frequency of 500MHz, and its bandwidth ranges from 250MHz to 750MHz [21]. The range resolution of the high - frequency channel is better than 0.3m [19]. The relative permittivity can be derived by several different methods using the LPR, e.g., the method of the surface reflection [11], the dual - antenna inversion [22], and the hyperbola fitting [10, 23 - 26]. The surface reflection method is affected by the direct coupling wave, ground reflection, the lunar surface roughness, which can only estimate the relative permittivity of the shallowest layer of the lunar regolith. The surface relative permittivity is estimated to be ~2.9 and ~2.01 at CE - 3 and Chang'- 4 landing sites, respectively [11, 27]. Ding et al. [28] assumed a dichotomy boundary presence at a rocky hill region detected by the Yutu - 1 rover. The relative permittivity is derived by comparing the subsurface radar reflectors with the actual interpolation depth, and the result is ~9 [28]. Zhang et al. [22] used the LPR dual high - frequency data to inverse the relative permittivity of lunar regolith at the CE - 3 landing site. Dong et al. [29] estimated the wave velocity and the relative permittivity of the CE - 4 landing site by the 3D velocity spectrum method.

The hyperbola fitting method has been widely applied to relative permittivity estimation both in Ground Penetrating Radar (GPR) and the LPR field. The average relative permittivity of the lunar regolith at the CE - 3 landing site is estimated to be ~3.2 based on the high - frequency LPR radar image [10]. Three layers at the CE - 3 landing site can be recognized by the relative permittivity distribution [23, 24], and the error analysis between different hyperbolic shape recognitions is discussed [26]. The relative permittivity of lunar materials is deduced to be 3.5 on average within the depth range of ~12m at the CE - 4 landing site [2]. However, in previous works on LPR [2, 10, 22 - 25], the traditional hyperbola fitting method assumes that antennas are close to the surface. It is not the case for the LPR antenna system, which has an antenna height of 0.3 m, and spacing of 0.16m [21]. The uncertainty of relative permittivity estimation caused by the antenna position and geometry has not been assessed and evaluated by the previous works.

In this paper, a new EM wave spreading model and the relative permittivity calculation algorithm are proposed, considering the influence of the antenna height and spacing. To verify the performance of the new approach, we used the finite - difference time - domain (FDTD) method to simulate the EM response of various models, including homogeneous models and stochastic models. We analyzed the effects of antenna height and spacing. Meanwhile, we calculated and compared the relative permittivity of the lunar regolith by both traditional and new methods.

2 Geological Background, Data Collections, and Methodology

2.1 Geological Context of the CE - 4 Landing Site

The Chinese CE - 4 spacecraft has successfully landed on the floor of the Von Karman

Crater in the South Pole - Aitken basin (SPA; Figure 1a). The SPA is the largest basin discovered in the solar system, and the age is estimated to be 4.2Ga [30]. The materials from the upper layer of the lunar mantle to the depth of ~100km might be excavated to the surface, including olivine and low - calcium pyroxene in the SPA [1, 31]. The Von Karman crater was formed in the SPA, and the age is dated to ~3.6Ga [32]. The CE - 4 landing site is located on the ejecta ray of the Finsen crater, which contributes an ejecta deposit of ~7m calculated by the model proposed by Pike et al. [33]. The thickness of the lunar regolith is estimated to be ~12m by the LPR, which is thicker than that of the CE - 3 landing site [2, 34 - 36]. The loss tangent of lunar materials is estimated to be ~0.005, which is close to that of typical lunar regolith [2, 35]. It shows that the EM attenuation rate at CE - 4 is less than that of regolith at the CE - 3 landing site [37]. The origin of those regolith developed on the ejecta materials source from nearby craters (e.g., Finsen, Alder, Von Karman L, and L' [2]).

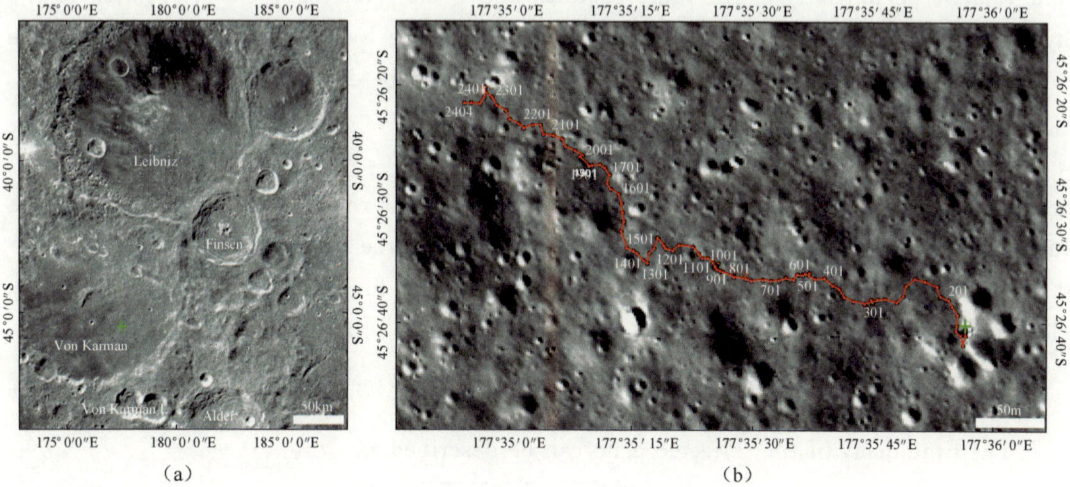

Figure 1　The landing site of the CE - 4 and the traveling route of the Yutu - 2 rover within the first 24 lunar days. (a) Geological background of the CE - 4 landing area. The base map is obtained by Chang'E 2 camera. The image IDs include CE2_GRAS_DOM_07m_J240_38S175E_A, CE2_GRAS_DOM_07m_J201_38S175W_A, CE2_GRAS_DOM_07m_K101_45S175W_A, CE2_GRAS_DOM_07m_K136_45S175E_A. (b) The traveling path of Yutu - 2 rover (red line). The base map is obtained by an LRO camera. The image ID is LRO_M1303619844. The green crosses both in (a, b) indicate the CE - 4 lander. The red dots mark the navigation points of the rover

2.2　LPR Data Collections and Processing

The Yutu - 2 rover has walked for 589.6m during the first 24 lunar days (Figure 1b). The LPR obtains 266,073 and 48,282 traces of high - frequency data (antenna 2B) and low - frequency data, respectively. In LPR data collection, 1 - bit quantization and non - uniform sampling methods were adopted [21]. A variable gain method is used to compress the strong signals and amplify the weak signals. Besides, to improve the signal-to-noise ratio, each trace of LPR data results from the accumulation of multiple measurements. Thus, it is necessary to pre-process the raw data to recover the actual LPR signal [19].

The flow of data processing of the LPR starts from level 2B data. The processes in this study mainly include trace editing, band-pass filtering, background removal, time delay adjustment, and spherical and exponential compensation (SEC) gain setting [2].

2.3 Traditional Hyperbolic Fitting Method

The hyperbola fitting method has been widely used in the GPR field [38]. It is based on the assumption that the depth of reflectors is much greater than the height and spacing of the GPR antenna so that the geometry and position of the antenna can be ignored. The buried reflector within the lunar regolith forms a hyperbola curve in the radar image [38]. The relative permittivity of the lunar regolith above the reflector is closely related to the hyperbolic curve pattern, which can be obtained by the hyperbolic fitting method. The velocity of the LPR radar pulse in the lunar regolith can be described as

$$v=\frac{c}{\sqrt{\varepsilon_r\mu_r}} \tag{1}$$

where μ_r is the relative permeability of the medium; ε_r is the relative permittivity of the lunar materials; c is the speed of the EM wave in the free space (here we suppose $c=0.3$m/ns). Generally, the relative permeability of the lunar regolith is approximately close to 1 [6], so that Equation (1) can be rewritten as follows

$$\varepsilon_r=\left(\frac{c}{v}\right)^2 \tag{2}$$

The time delay of the reflected echo can be described as

$$t=\frac{2\sqrt{H^2+(x-x_0)^2}}{v} \tag{3}$$

where $(x, 0)$, $(x_0, -H)$ are the position of the Yutu-2 rover and the reflector (as shown in Figures 2 and 3). The variables t and x can be obtained from the hyperbolic curve in the radar image. H, v, and x_0 can be derived by the hyperbola fitting. Once the v is derived, the relative permittivity of the lunar material can be simply calculated by the Equation (2).

2.4 New Method

The hyperbola fitting method has good performance in the GPR field. However, this is not the case for LPR as shown in Figures 2 and 3. The layout of the LPR antennas is quite special. The transmitting and receiving antennas are allocated at the bottom of the Yutu-2 rover, and it has to be lifted to avoid obstacles while the LPR travels on the lunar surface. The antenna height and spacing are ~0.3m and 0.16m, respectively [21]. A

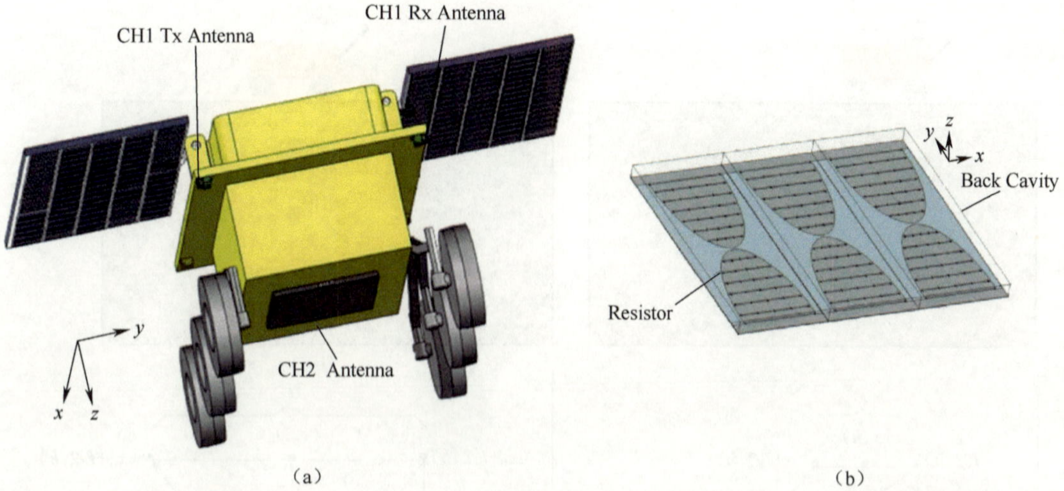

Figure 2　The Yutu2 Rover and the structure of the CH2 Antenna [21]. (a) Schematic diagram of Yutu - 2 Rover and the layout of the LPR antenna; (b) structure of the high - frequency antenna

new method considering antenna height and spacing is required.

According to Snell's law, the following equations can be derived.

$$\frac{\sin\theta_{d1}}{\sin\theta_{d2}}=\frac{\sin\theta_{u1}}{\sin\theta_{u2}}=\sqrt{\varepsilon} \tag{4}$$

where, θ_{u1}, θ_{u2}, θ_{d1} and θ_{d2} are the incident angle and the refraction angle of the up - going wave and the downward - going wave, respectively. According to the geometric relationship, as shown in Figure 3a, c, the following equations can be derived.

$$\sin\theta_{d1}=\frac{\left|x-\frac{L}{2}-x_1\right|}{\sqrt{h^2+\left(x-\frac{L}{2}-x_1\right)^2}} \tag{5}$$

$$\sin\theta_{d2}=\frac{|x_0-x_1|}{\sqrt{H^2+(x_0-x_1)^2}} \tag{6}$$

$$\sin\theta_{u1}=\frac{|x_0-x_2|}{\sqrt{H^2+(x_0-x_2)^2}} \tag{7}$$

$$\sin\theta_{u2}=\frac{\left|x+\frac{L}{2}-x_2\right|}{\sqrt{h^2+\left(x+\frac{L}{2}-x_2\right)^2}} \tag{8}$$

where h and L are the antenna height and antenna spacing, respectively. The position of the echo reflector, the transmitting and receiving antenna are $(x_0, -H)$, $(x_0-L/2, h)$, $(x_0+L/2, h)$, respectively. The position of the incident and emission points are

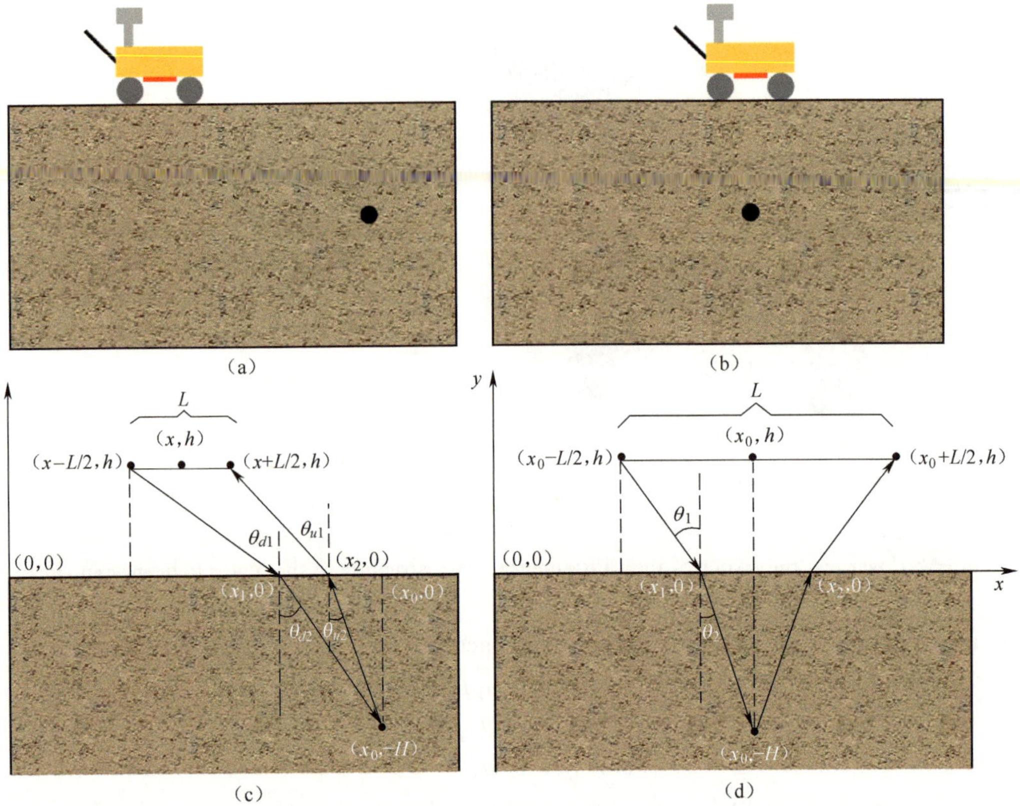

Figure 3　Schematic diagram of the propagation routine of electromagnetic waves. (a, b) Schematic diagram of Yutu - 2 rover, regolith, and rocks. The rock is horizontally far away from the rover in (a) and is horizontally in the middle of the rover in (b - d) illustrates the signal propagation routine of (a) and (b), respectively. The antenna height and spacing are L and h, respectively. The red points in (c, d) are the transmitting antenna and the receiving antenna, the blue point is the middle of the antenna, representing the position of the rover

$(x_1, 0)$ and $(x_2, 0)$, respectively. Assuming the propagation distances in the free space and lunar material are l_1 and l_2, respectively. Hence, the two - way travel time t can be re - described as

$$t = \frac{l_1}{c} + \frac{l_2}{v} \tag{9}$$

where l_1 and l_2 can be expressed as

$$l_1 = \sqrt{h^2 + \left(x - \frac{L}{2} - x_1\right)^2} + \sqrt{h^2 + \left(x + \frac{L}{2} - x_2\right)^2} \tag{10}$$

$$l_2 = \sqrt{H^2 + (x_0 - x_1)^2} + \sqrt{H^2 + (x_0 - x_2)^2} \tag{11}$$

Suppose the reflector is horizontally in the middle of the transmitting antenna and the receiving antenna as shown in Figure 3b, d. The incident angle and the refraction angles are θ_1 and θ_2, respectively. The propagation distances of the radar pulse in free space and lunar material are l_{01} and l_{02}, respectively. t_0 is the time delay of the reflected echo. According to the geometric relationship, the following equations can be derived

$$h \cdot \tan\theta_1 + H \cdot \tan\theta_2 = \frac{L}{2} \tag{12}$$

$$t_0 = \frac{l_{01}}{c} + \frac{l_{02}}{v} \tag{13}$$

$$l_{01} = \frac{2h}{\cos\theta_1} \tag{14}$$

$$l_{02} = \frac{2H}{\cos\theta_2} \tag{15}$$

Suppose $\theta = \theta_2$, Equations (12) — (15) can be combined as

$$\frac{h\sqrt{\varepsilon}\sin\theta}{\sqrt{1-\varepsilon\sin^2\theta}} + H\tan\theta = \frac{L}{2} \tag{16}$$

$$\frac{h}{\sqrt{1-\varepsilon\sin^2\theta}} + H\frac{\sqrt{\varepsilon}}{\cos\theta} = \frac{t_0}{2}c \tag{17}$$

Combining the Equations (4) — (8), (16) — (17), we can obtain a group of equations as Equation (18). In Equation (18), x_1, x_2, ε, H, θ, t_0, and x_0 are unknown variables, x, t, h, L, and c are the known variables. The number of equations is smaller than those of unknown variables. To calculate these equations, we have to obtain at least two unknown variables by other methods to constrain the solution of the above equations.

The hyperbolic echo pattern of the buried object has a symmetrical feature with the peak. Thus, x_0 can be obtained by the hyperbolic fitting method, and t_0 can be obtained by finding the peak of the hyperbola from the radar image. Therefore, the unknown variables that remained are x_1, x_2, ε, H, and θ. Equation (18) can be used to calculate these five unknown variables so that relative permittivity can be obtained.

Specifically, as for a certain point (x, t) on the hyperbola curve, once the peak of the corresponding hyperbola curve (x_0, t_0) is determined, one relative permittivity ε can be obtained by combining (x, t), (x_0, t_0), and Equation (18). In other words, one relative permittivity can be derived by one point on the hyperbola curve. For a hyperbola curve on the radar image, the relative permittivity of the new method is the average of all calculated relative permittivities.

$$\begin{cases} \dfrac{\left| x - \dfrac{L}{2} - x_1 \right|}{\sqrt{h^2 + \left(x - \dfrac{L}{2} - x_1 \right)^2}} - \sqrt{\varepsilon} \dfrac{|x_0 - x_1|}{\sqrt{H^2 + (x_0 - x_1)^2}} = 0 \\ \dfrac{\left| x + \dfrac{L}{2} - x_2 \right|}{\sqrt{h^2 + \left(x + \dfrac{L}{2} - x_2 \right)^2}} - \sqrt{\varepsilon} \dfrac{|x_0 - x_2|}{\sqrt{H^2 + (x_0 - x_2)^2}} = 0 \\ \dfrac{\sqrt{h^2 + \left(x - \dfrac{L}{2} - x_1 \right)^2} + \sqrt{h^2 + \left(x + \dfrac{L}{2} - x_2 \right)^2}}{c} + \\ \dfrac{\sqrt{H^2 + (x_0 - x_1)^2} + \sqrt{H^2 + (x_0 - x_2)^2}}{\dfrac{c}{\sqrt{\varepsilon}}} = t \\ \dfrac{h\sqrt{\varepsilon}\sin\theta}{\sqrt{1 - \varepsilon\sin^2\theta}} + H\tan\theta = \dfrac{L}{2} \\ \dfrac{h}{\sqrt{1 - \varepsilon\sin^2\theta}} + H\dfrac{\sqrt{\varepsilon}}{\cos\theta} = \dfrac{t_0}{2}c \end{cases} \tag{18}$$

2.5 Regolith Modeling and Numerical Simulation

2.5.1 Simplified Modeling

Homogeneous models are the most ideal model, of which the relative permittivity is set to be the same both in the horizontal and the vertical direction (as shown in Figure 4a). Besides, previous studies on the CE - 4 landing site and the Apollo sample measurement have indicated that the lunar regolith is not vertically homogeneous, and the relative permittivity increases with the depth. Therefore, we also established the regolith model with the relative permittivity increasing from 2 to 4 (as shown in Figure 4b).

2.5.2 Stochastic Modeling

The interior structure of the lunar regolith is rather complicated, which includes numerous buried rock fragments of different sizes [4, 39, 40]. The relative permittivity distribution of lunar regolith is not homogeneous, which exists stochastic disturbances in general [4]. In some places, the disturbance is uniform both vertically and horizontally, while in some other places, obvious horizontal disturbance can be seen [9]. Thus, we established two different stochastic models with fewer and more horizontal disturbances, respectively, as shown in Figure 4c, d. Here, we applied the stochastic modeling method to simulate the lunar regolith, which is closer to the real situation according to the statistic [22, 41]. Previous works have applied this method both in the GPR and LPR numerical simulations [41 - 43]. In addition to the simulation method, the effectiveness of the new method can also be verified by laboratory experiments.

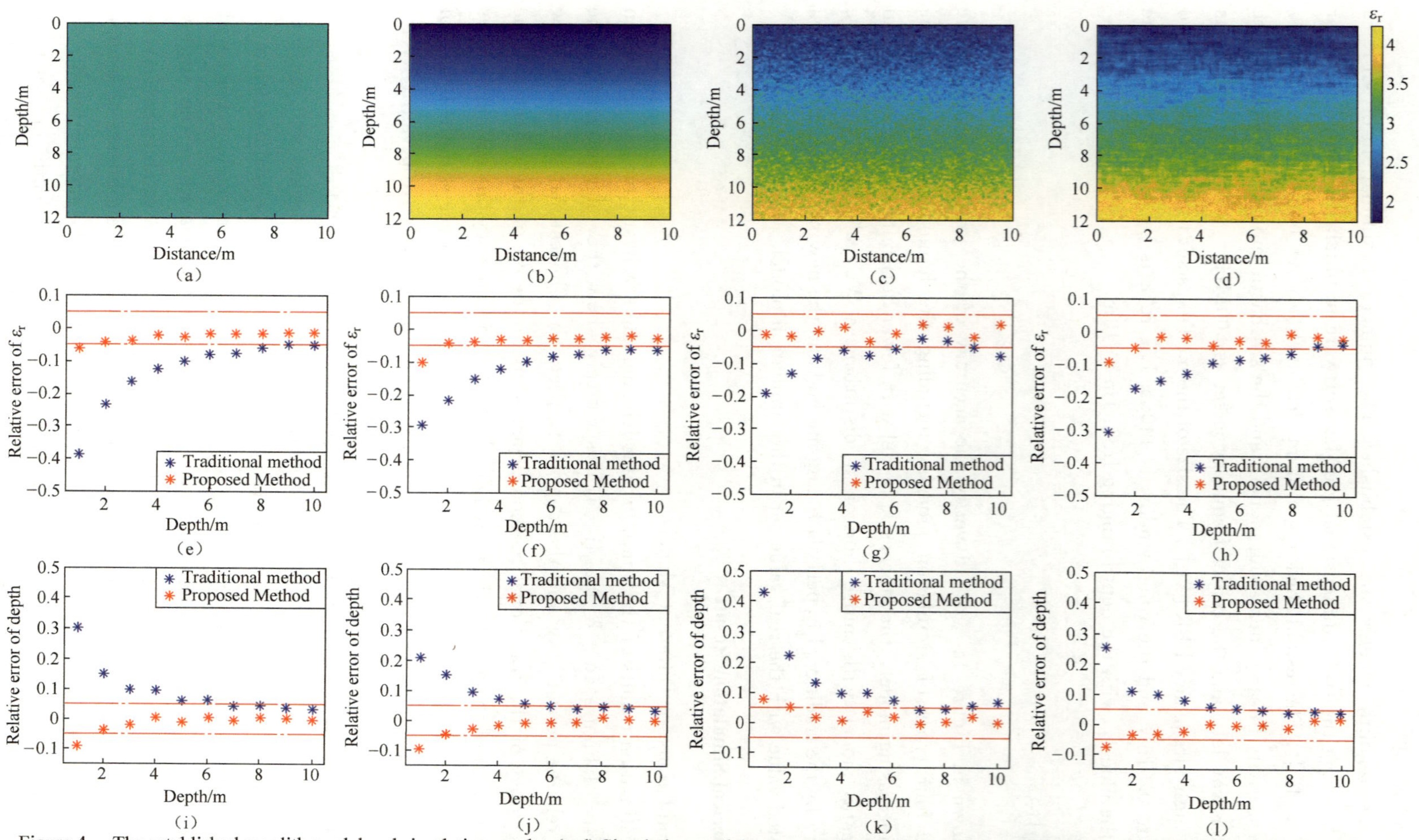

Figure 4　The established regolith model and simulation results. (a-d) Simulation models, representing the homogeneous model, the vertically increasing model, the stochastic model with fewer horizontal disturbances and the stochastic model with more horizontal disturbances; (e-h) indicate the calculated relative permittivity by the four models, respectively; (i-l) represent the derived depth of reflectors in the four models, respectively. The red dash line indicates the error of ±5%. Besides the relative permittivity, the rock depth is another important parameter for geological interpretation, it can be calculated by the two methods and can be used to compare the two methods. Therefore, we showed the estimated rock depth along with the dielectric constant

2.5.3　FDTD Simulation

Once the regolith models are established. The numerical simulation of LPR is performed by using the two-dimensional gprMax software to obtain the simulated radar images [44, 45]. The established regolith models include the homogeneous model, relative permittivity linearly increasing model, and the stochastic model. The input source is the Ricker wavelet, and the center frequency is set as 500MHz, which is the same as that of the high-frequency LPR [21]. The size of the models is set to 12×10m, the grid size is set to 0.01m, and the time window is 140ns. The rock is set horizontally in the middle of the model, with the depth ranging from 1m to 10m.

3　Results

As shown in Figure 4a-d, to verify the accuracy of the proposed method, we established four types of simulation models, including the homogeneous model, the vertically increasing model, the stochastic model with fewer horizontal disturbances, and the stochastic model with more horizontal disturbances. We calculated the relative permittivity by the traditional hyperbola fitting method and the new method and compared the estimated value with the real value set in the simulation model.

3.1　Numerical Simulation Result

3.1.1　Simulations for Different Models

The estimation results of the simplified models and the stochastic models are shown in Figure 4e-l. Figure 4e-h and 4i-l, respectively show the relative error of the estimated relative permittivity and the depth, respectively. The estimated results show that the accuracy of the traditional method depends on the depth of the reflector (blue dots). The result of the new method (red dots) has less dependency on depth and shows good robustness compared with the hyperbola fitting method. Compared with the results of the simplified models, there exists disturbance in the estimated relative permittivity for the stochastic models.

3.1.2　The Influence of Antenna Height and Spacing on Both Methods

The traditional hyperbola fitting method ignores the antenna height and spacing, which causes errors for dielectric constant estimation, especially at the shallow depth. As shown in Figure 4, the estimation accuracy of the hyperbola fitting method increases with the decrease in depth. In particular, the error is intensely upheaved at the shallow depth. To further study the influence of antenna height, we established several homogeneous models with different antenna heights and spacings. The influence of antenna height on the accuracy of both methods is illustrated in Figure 5. The antenna heights are set to be 0.2m, 0.25m, 0.3m, and 0.35m, respectively. The rock depths are set to be 1m, 2m, 3m, 5m, 7m, and 10m. The results show that the accuracy of the hyperbola fitting method has a significant dependency on the depth of the reflector, especially at the shallow depth. The new method has less dependency on the depth of the reflector compared with the traditional method (Figure 5e – h). Most of the relative errors are within 5%, and all of the relative errors are within 10% for the new method. Although subtle discrepancy occurs at a depth less than 2m for the new method, the results of the new method are barely influenced by the antenna height (red results of Figure 5).

Furthermore, we also considered the influence of antenna spacing on both methods. The various antenna height is set to be 0.1m, 0.16m, 0.22m, and 0.28m, respectively. The rock depths are set to be 1m, 2m, 3m, 5m, 7m, and 10m. The calculated results are shown in Figure 6a – h. Both the relative error of estimated relative permittivity and depth are analyzed. The results show that the accuracy caused by the antenna spacing has less influence on both methods compared with that of antenna height. The subtle discrepancy can be observed for the hyperbola fitting method at the shallow depth. In conclusion, we can infer that the proposed method is much more robust to calculate the relative permittivity compared with the traditional method.

3.2 The high – Frequency LPR Radar Image within the First 24 Lunar Days

Eighty – three hyperbolic echo patterns are recognized in the high – frequency LPR radar image within the first 24 lunar days (red curves in Figure 7). The radar survey distance within the first 24 lunar days is ~589.6m, and its routine is shown in Figure 1b.

As shown in Figure 7, only obvious hyperbola – like curves are selected in the radar image, and some detailed hyperbola curves are plotted in Figure 8. We applied both the hyperbola fitting method and our proposed method to calculate the relative permittivity. The estimated results are plotted in Figure 9a. The difference of the calculated relative permittivity by the two methods increases with the decrease in depth (Figure 9a), which is consistent with those of the simulation results (Figure 4). The traditional method underestimated the relative permittivity, especially at the shallow depth. Based on the relative permittivity calculated by the proposed method, we obtained an empirical relationship between the relative permittivity and two – way travel time (red line in Figure 9a), expressed as Equation (19).

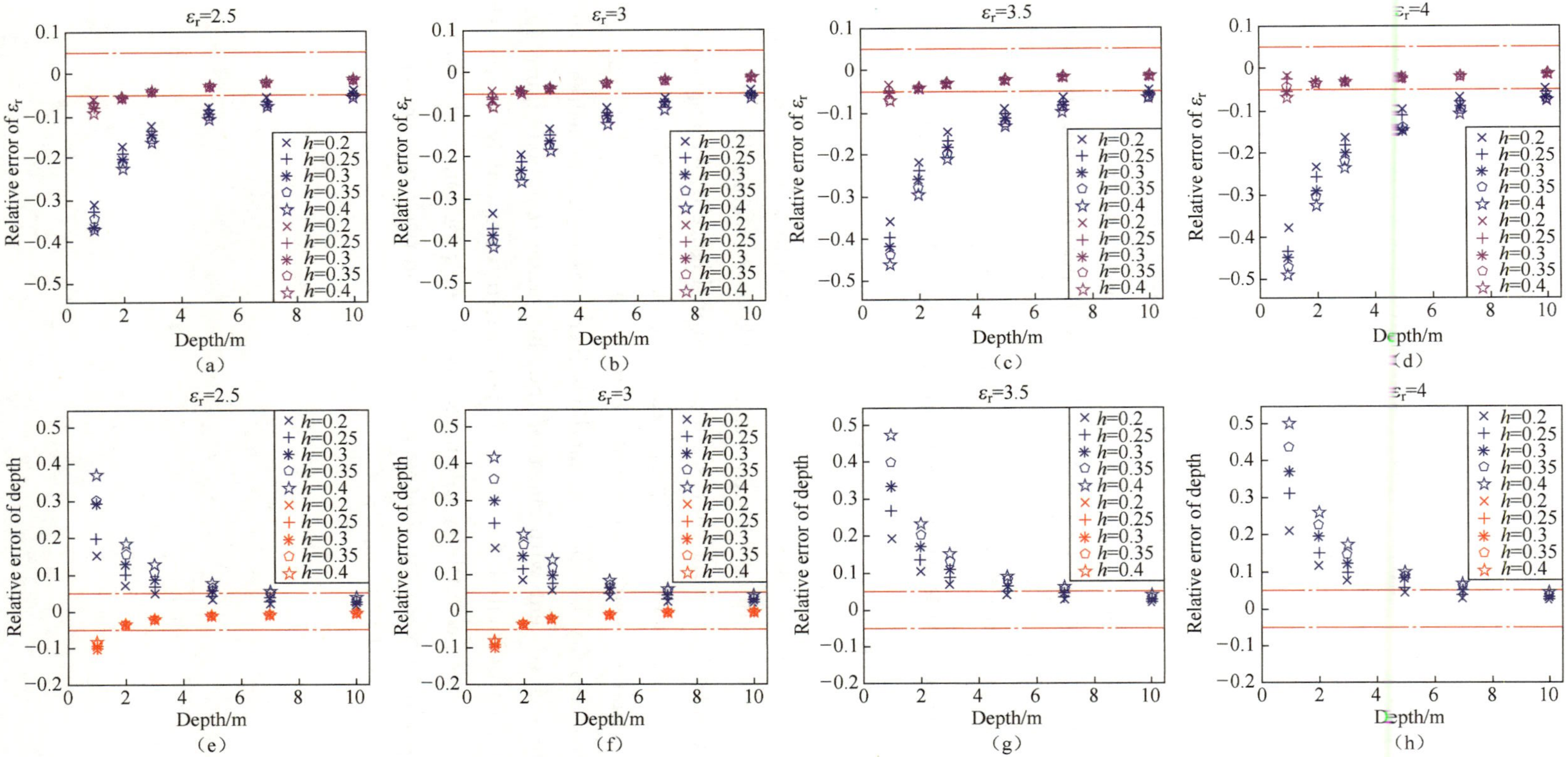

Figure 5 The influence of antenna height on the hyperbola fitting method and the proposed method. Blue and red markers indicate the results of the hyperbola fitting method and the proposed method, respectively. (a-d) Represent the result of calculated relative permittivity for the simulation models with the relative permittivity of 2.5, 3, 3.5, and 4, respectively. (e-h) Represent the results of the calculated depth for the simulation models with the relative permittivity of 2.5, 3, 3.5, and 4, respectively. The two red dash lines indicate the ±5% error

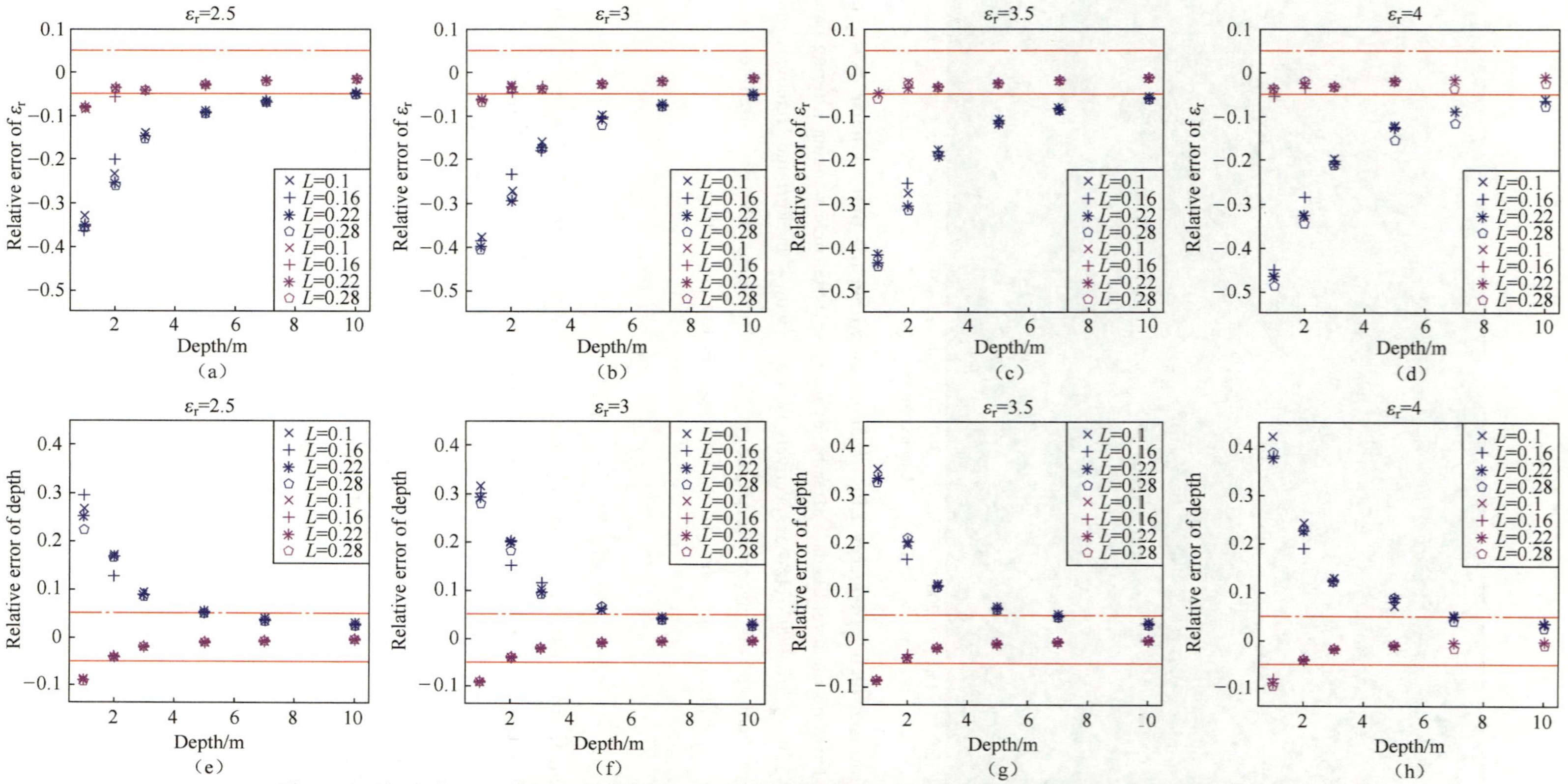

Figure 6　The influence of antenna spacing on both methods. Blue and red markers indicate the results of the hyperbola fitting method and the proposed method, respectively. (a-d) Represent the result of the simulation models with the relative permittivity of 2.5, 3, 3.5, and 4, respectively. (e-h) Represent the result of the simulation models with the relative permittivity of 2.5, 3, 3.5, and 4, respectively. The two red dash lines indicate the ±5% error

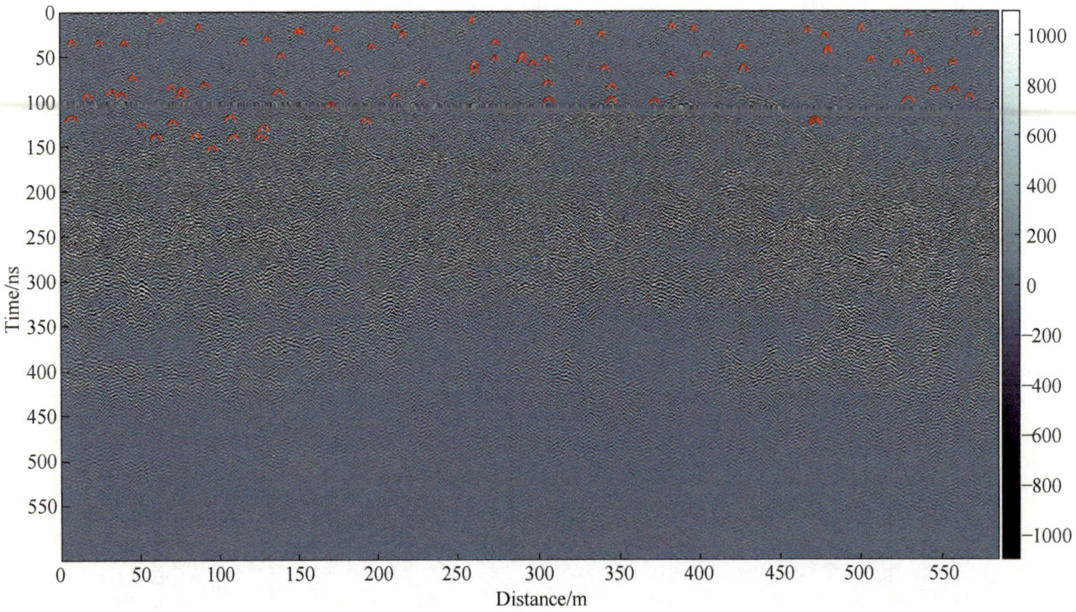

Figure 7　The processed high - frequency LPR radar image within the first 24 lunar days. The radar image is obtained based on the level 2B data after bandpass filter (the corresponding filtering parameters are set to 100, 250, 750, 900MHz, respectively), de - wow, background removal, and SEC gain. The red curves indicate the hyperbola picked in the radar image. The data used for imaging is available on the website http://moon. bao. ac. cn/(accessed on 10March 2021)

$$\varepsilon = \frac{4.9t + 152.9}{t + 67.2} \tag{19}$$

Based on the fitting result, we transformed the two - way travel time versus depth for accurate estimation of the lunar regolith thickness at the CE - 4 landing site. As shown in Figure 9b, the blue and red lines represent the depth derived by the relative permittivities that are calculated by the traditional hyperbola fitting method and the proposed method, respectively. The calculated regolith depths are 12m and 11.1m, respectively. Therefore, the estimated thickness of the lunar regolith at the CE - 4 landing site is improved ~8% compared with that calculated by Li et al. [2].

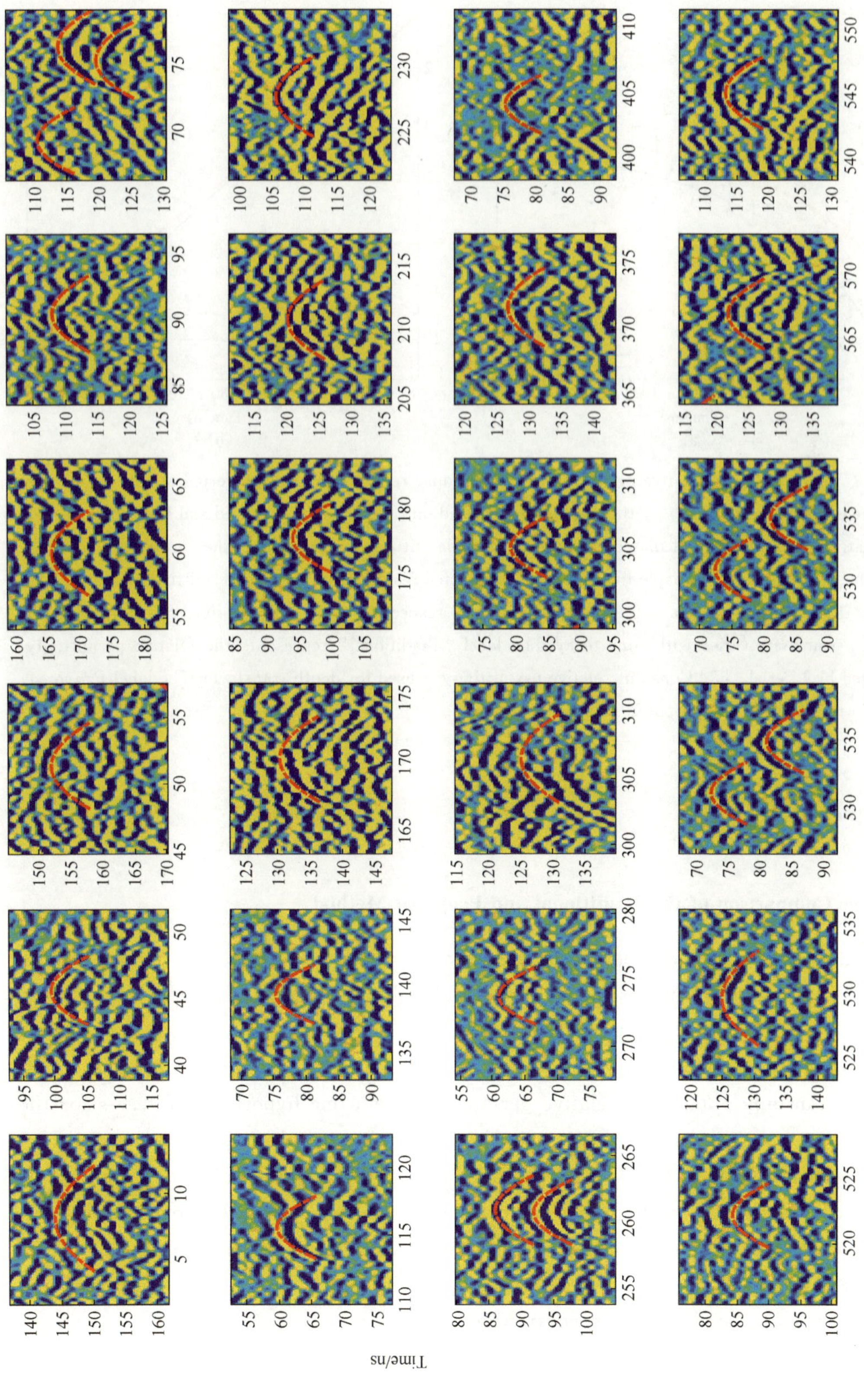

Figure 8　Several detailed structures of the hyperbola curve observed on the radar image. The red dash curve anchor the profile of the detected hyperbolic echo pattern. Each small region of the radar image is extracted from Figure 7

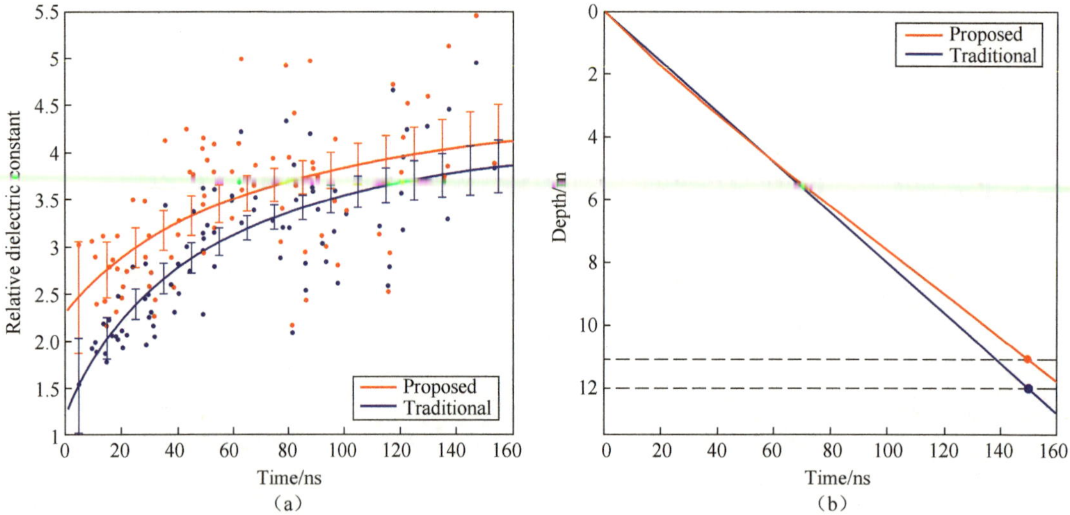

Figure 9　The calculated relative permittivities of the lunar regolith and the converted depth. (a). The calculated relative permittivity by the two methods based on the LPR data. The red and blue dot indicates the relative permittivity calculated by the hyperbola fitting method and the proposed method, respectively. The green and maple line represents the fitted result of the relative permittivity estimated by the hyperbola fitting method and the proposed method, respectively. (b) The derived depth by different relative permittivity varies with time delay. The label "Traditional" represents the relative permittivity calculated by Li et al. [2] Constant relative permittivity is used for depth transform. The label *Proposed* indicates the result of the proposed method

4　Discussion

4.1　The Comparison of the Traditional and Proposed Method

Both methods use the hyperbola curves to calculate the relative permittivity. The hyperbola fitting method assumes that the antenna height and spacing are zero, and only one equivalent relative permittivity can be obtained for all selected points from a hyperbola curve, which reflects the equivalent dielectric distribution above the reflector. In other words, it can only obtain one relative permittivity from one hyperbola curve. As for the proposed method, the relative permittivity is not derived from the hyperbola curve pattern or curve fitting. Instead, once the position of the peak point of the hyperbola is determined, each point on the hyperbola curve can calculate one relative permittivity according to the geometric relationship. In general, a relative permittivity can be obtained at each picked point on a specific hyperbola curve, the result of the new method is the average of all calculated relative permittivities.

4.2 The Influence of Antenna Height and Antenna Spacing

The hyperbola curve fitting method is based on a simplified model and ignores the antenna height and spacing. The estimation error at the shallow depth caused by the simplified model is more obvious than that at the deep depth. This is mainly because antenna height and spacing are comparable to the reflector depth when the reflector is at the shallow depth, in which case the antenna height and spacing should not be ignored. However, at the deep depth, the depth of the reflector is much smaller than the antenna height and spacing, the influence of the antenna height and spacing on the simplified model of the hyperbola fitting method becomes subtle. As for the proposed method, although subtle discrepancy can be observed at the shallow depth, the estimation accuracy has less dependence on antenna height and spacing.

5 Conclusions

The hyperbola fitting method is based on the simplified model, which ignores the antenna height and spacing. Simulation results show that the influence of antenna height on the traditional method increases with the decrease in depth. The previous works using the hyperbola fitting method have underestimated the relative permittivity of the lunar regolith, especially at the shallow depth. The proposed method has good performance at both shallow and deep depth, and the accuracy is less dependent on the depth of reflectors. We selected eighty - three obvious hyperbola curves in the LPR radar image and calculated the relative permittivity both by the traditional and proposed methods. The result shows that the proposed method improved the calculated relative permittivity at a depth of 0 - 3m, 3 - 6m, and 6 - 10m by 35%, 14%, and 9%, respectively. Finally, the estimated thickness of the lunar regolith at the CE - 4 landing site is ~11.1m by the proposed method, which improved by ~ 8% compared with that of the traditional method.

Acknowledgments

The authors thank the team members of the Ground Research and Application System of National Astronomical Observatories.

References

[1] Ip, W. H.; Yan, J.; Li, C.; Ouyang, Z. Preface: The Chang'e - 3 lander and rover mission to the Moon. Res. Astron. Astrophys. 2014, 14, 1511 - 1513.

[2] Li, C.; Liu, D.; Liu, B.; Ren, X.; Liu, J.; He, Z.; Zuo, W.; Zeng, X.;

Xu, R.; Tan, X.; et al. Chang'E-4 initial spectroscopic identification of lunar far-side mantle-derived materials. Nature 2019, 569, 378.

[3] Li, C.; Su, Y.; Pettinelli, E.; Xing, S.; Ding, C.; Liu, J.; Ren, X.; Lauro, S. E.; Soldovieri, F.; Zeng, X.; et al. The Moon's farside shallow subsurface structure unveiled by Chang'E-4 lunar penetrating radar. Sci. Adv. 2020, 6, 6898.

[4] Carrier, W. D.; Olhoeft, G. R.; Mendell, W. Physical properties of the lunar surface. In Lunar Source Book; Heiken, G. D., Vaniman, D. T., French, B. M., Eds.; Cambridge University Press: New York, NY, USA, 1991; pp. 475-594.

[5] Campbell, B. A.; Hawke, B. R.; Thompson, T. W. Regolith composition and structure in the lunar maria: Results of long-wavelength radar studies. J. Geophys. Res. 1997, 102, 19307-19320.

[6] Olhoeft, G. R.; Strangeway, G. D. Dielectric properties of the first 100meters of the Moon. Earth Planet. Sci. Lett. 1975, 24, 394-404.

[7] Ono, T.; Kumamoto, A.; Nakagawa, H.; Yamaguchi, Y.; Oshigami, S.; Yamaji, A.; Kobayashi, T.; Kasahara, Y.; Oya, H. Lunar radar sounder observations of subsurface layers under the nearside maria of the Moon. Science 2009, 323, 909-912.

[8] Phillips, R. J.; Adams, G. F.; Brown, W. E.; Eggleton, R. E.; Jackson, P.; Jordan, R.; Linlor, W. J.; Peeples, W. J.; Porcello, L. J.; Ryu, J. Apollo Lunar Sounder Experiment; Technical Report; NASA: Washington, DC, USA, 1973.

[9] Ding, C.; Li, C.; Xiao, Z.; Su, Y.; Xing, S.; Wang, Y.; Feng, J.; Dai, S.; Xiao, Y.; Yao, M. Layering Structures in the Porous Material Beneath the Chang'e-3 Landing Site. Earth Space Sci. 2020, 7, e2019EA000862.

[10] Fa, W.; Zhu, M.; Liu, T.; Jeffrey, B. P. Regolith stratigraphy at the Chang'E-3 landing site as seen by lunar penetrating radar. Geophys. Res. Lett. 2015, 42, 10179-10187.

[11] Dong, Z.; Fang, G.; Ji, Y.; Gao, Y.; Wu, C.; Zhang, X. Parameters and structure of lunar regolith in Chang'E-3 landing area from lunar penetrating radar (LPR) data. Icarus 2017, 282, 40-46.

[12] McKay, D.; Heiken, G.; Basu, A.; Blanford, G.; Simon, S.; Reedy, R.; French, B.; Papike, J. The lunar regolith. In Lunar Source-Book; Heiken, G. D., Vaniman, D. T., French, B. M., Eds.; Cambridge University Press: New York, NY, USA, 1991; pp. 285-356.

[13] Hagfor, T. Backscattering from an undulating surface with application to radar returns from the Moon. Geophys. Res 1964, 97, 13319-13346.

[14] Fa, W.; Wieczorek, M. A. Regolith thickness over the lunar nearside: Results from earth-based 70-cm Arecibo radar observations. Icarus 2012, 218, 771-787.

[15] Watters, T. R.; Campbell, B.; Carter, L.; Leuschen, C.; Plaut, J.; Picardi, G.; Orosei, R.; Safaeinili, A.; Clifford, S.; Farrell, W.; et al. Radar sounding of the Medusae fossae formation mars: Equatorial ice or dry, low-density deposits? Science 2007, 318, 1125-1128.

[16] Orosei, P.; Rossi, A.; Cantini, F.; Graziella. Radar sounding of Lucus Planum, Mars, by MARSIS. Geophys. Res. 2017, 122, 1405-1418.

[17] Carter, L. M.; Campbell, B. A.; Holt, J. W.; Phillips, R.; Putzing, N.; Mattei, S.; Seu, R.; Okubo, C.; Egan, A. Dielectric properties of lava flows west of Ascraeus Mons, Mars. Geophys. Res. Lett. 2009, 36, L23204.

[18] Porcello, L. J.; Zelenka, J. S.; Adams, G. F.; Jackson, P. L.; Jordan, R. L.; Phillips, R. J.; Brown, W. E.; Ward, S. H. The Apollo lunar sounder radar system. Proc. IEEE 2017, 62, 769-788.

[19] Su, Y.; Fang, G.; Feng, J.; Xing, S.; Ji, Y.; Zhou, B.; Gao, Y.; Li, H.; Dai, S.; Xiao, Y.; et al. Data processing and initial results of Chang'e-3 lunar penetrating radar. Res. Astron. Astrophys. 2014, 14, 1623-1632.

[20] Jia, Y.; Zou, Y.; Ping, J.; Xue, C.; Yan, J.; Ning, Y. The scientific objectives and payloads of Chang'E-4 mission. Planet. Space Sci. 2018, 162, 207-215.

[21] Fang, G.; Zhou, B.; Ji, Y.; Zhang, Q.; Shen, S.; Li, Y.; Guan, H.; Tang, C.; Gao, Y.; Ye, S.; et al. Lunar Penetrating Radar onboard the Chang'e-3 mission. Res. Astron. Astrophys. 2014, 14, 1607-1622.

[22] Zhang, L.; Zeng, Z.; Li, J.; Lin, J.; Hu, Y.; Wang, X.; Sun, X. Simulation of the Lunar Regolith and Lunar-Penetrating Radar Data Processing. IEEE J. Sel. Top. Appl. Earth Obs. Remote Sens. 2018, 11, 2.

[23] Lai, J.; Xu, Y.; Zhang, X.; Tang, Z. Structural analysis of lunar subsurface with ChangE-3 lunar penetrating radar. Planet. Space Sci. 2016, 120, 96-102.

[24] Feng, J.; Su, Y.; Ding, C.; Xing, S.; Dai, S.; Zou, Y. Dielectric properties estimation of the lunar regolith at CE-3 landing site using lunar penetrating radar data. Icarus Int. J. Sol. Syst. Stud. 2017, 284, 424-430.

[25] Fa, W. Bulk Density of the Lunar Regolith at the Chang'E-3 Landing Site as Estimated from Lunar Penetrating Radar. Earth Space Sci. 2020, 7, e2019EA000801.

[26] Ding, C.; Xiao, Z.; Su, Y.; Cui, J. Hyperbolic reflectors determined from peak echoes of ground penetrating radar. Icarus 2021, 358, 114280.

[27] Dong, Z.; Fang, G.; Zhou, B.; Di, Z.; Gao, Y.; Ji, Y. Properties of Lunar Regolith on the Moon's Farside unveiled by Chang'E Lunar Penetrating Radar. J. Geophys. Res. Planets 2021, 126, e2020JE006564.

[28] Ding, C.; Cai, Y.; Xiao, Z.; Su, Y. A rocky hill on the continuous ejecta of Ziwei crater revealed by the Chang'e-3 mission. Earth Planet. Phys. 2020, 4, 1-

6.

[29] Dong, Z.; Feng, X.; Zhou, H.; Liu, C.; Zeng, Z.; Li, J.; Liang, W. Properties Analysis of Lunar Regolith at Chang'E-4 Landing Site Based on 3D Velocity Spectrum of Lunar Penetrating Radar. Remote Sens. 2020, 12, 629.

[30] Hiesinger, H.; Bogert, C.; Pasckert, J. H.; Schmedemann, N.; Robinson, M. S.; Jolliff, B. L.; Petro, N. New Crater Size - Frequency Distribution Measurements of the South Pole - Aitken Basin. In Proceedings of the Lunar and Planetary Science Conference, The Woodlands, TX, USA, 19-23 March 2012.

[31] Melosh, H. J.; Kendall, J.; Horgan, B.; Johnson, C. B.; Bowling, T. South Pole - Aitken basin ejecta reveal the Moon' s upper mantle. Geology 2017, 45, 1063-1066.

[32] Huang, J.; Xiao, Z.; Flahaut, J.; Martinot, M.; Head, J.; Xiao, X.; Xie, M.; Xiao, L. Geological Characteristics of Von Kármán Crater, Northwestern South Pole - Aitken Basin: Chang' E - 4 Landing Site Region. J. Geophys. Res. Planets 2018, 123, 1684-1700.

[33] Pike, R. J. Depth/diameter relations of fresh lunar craters: Revision from spacecraft data. Geophys. Res. Lett. 1974, 1, 291-294.

[34] Xiao, L.; Zhu, P.; Fang, G.; Xiao, Z.; Zou, Y.; Zhao, J.; Zhao, N.; Yuan, Y.; Qiao, L.; Zhang, X.; et al. A young multilayered terrane of the northern Mare Imbrium revealed by Chang' E - 3 mission. Science 2015, 347, 1226-1229.

[35] Lai, J.; Xu, Y.; Zhang, X.; Yan, L.; Qi, Y.; Xu, M.; Zhou, B.; Dong, Z.; Di, Z. Comparison of Dielectric Properties and Structure of Lunar Regolith at Chang'e-3 and Chang' e - 4 Landing Sites Revealed by Ground - Penetrating Radar. Geophys. Res. Lett. 2019, 46, 12783-12793.

[36] Xiao, Z.; Ding, C.; Xie, M.; Cai, Y.; Cui, J.; Zhang, K.; Wang, J. Ejecta from the Orientale basin at the Chang' Elanding site. Geophys. Res. Lett. 2021, 48, e2020GL090935.

[37] Ding, C.; Xiao, Z.; Wu, B.; Li, Y.; Prieur, N.; Cai, Y.; Su, Y.; Cui, J. Fragments Delivered by Secondary Craters at the Chang' E - 4 Landing Site. Geophys. Res. Lett. 2020, 47, e2020GL087361.

[38] Daniels, D. Ground Penetrating Radar; The Institution of Electrical Engineers: London, UK, 2004.

[39] Wilcox, B. B.; Robinson, M. S.; Thomas, P. C.; Hawke, B. R. Constraints on the depth and variability of the lunar regolith. Meteorit. Planet. Sci. 2005, 40, 695-710.

[40] Dai, S.; Su, Y.; Xiao, Y.; Feng, J.; Xing, S.; Ding, C. Echo simulation of lunar penetrating radar: Based on a model of inhomogeneous multilayer lunar regolith structure. Res. Astron. Astrophys. 2014, 12, 1642-1653.

[41] Ding, C.; Su, Y.; Xing, S.; Dai, S.; Xiao, Y.; Feng, J.; Liu, J.; Li, C. Numerical Simulations of the Lunar Penetrating Radar and Investigations of the Geological Structures of the Lunar Regolith Layer at the CE - 3 Landing Site. Int. J. Antennas Propag. 2017, 2017, 3013249.

[42] Hu, Y. S.; Zeng, Z. F.; Li, J.; Liu, F. Simulation and processing of LPR onboard the rover of Chang'E - 3 mission: Based on multilayer lunar regolith structure stochastic media model. In Proceedings of the International Conference on Ground Penetrating Radar, Hong Kong, China, 13 - 16 June 2016.

[43] Jiang, Z.; Zeng, Z.; Liu, F.; Li, W. Simulation and analysis of GPR signal based on stochastic media model with an ellipsoidal autocorrelation function. J. Appl. Geophys. 2013, 99, 91 - 97.

[44] Giannopoulos, A. Modelling ground penetrating radar by GprMax. Constr. Build. Mater 2005, 19, 755 - 762.

[45] Warren, C.; Giannopoulos, A.; Giannakis, I. GprMax: Open source software to simulate electromagnetic wave propagation for ground penetrating radar. Comput. Phys. Commun. 2016, 209, 163 - 170.

Dusty Mafic Rocks Along the Path of Chang'E - 4 Rover: Initial Analysis of the Image Cubes of the Onboard Visible and Near - Infrared Imaging Spectrometer

Xing Wang[1,2], Jianjun Liu[1,2], Dawei Liu[1], Hai Huang[1,2], Lin Guo[1,2], Qing Zhang[1,2], Yuan Chen[1,2], Bin Liu[1], Wangli Chen[1], Xin Ren[1], and Chunlai Li[1,2]

1. Key Laboratory of Lunar and Deep Space Exploration, National Astronomical Observatories, Chinese Academyof Sciences, Beijing, China

2. School of Astronomy and Space Science, University of Chinese Academyof Sciences, Beijing, China

Abstract: The Yutu - 2 rover of Chang'E - 4 encountered four special rocks along its path. The onboard Visible and Near - infrared Imaging Spectrometer (VNIS) collected their spectral data at close range. In this work, we focus on the visible/near - infrared image cubes captured by the VNIS. Simple classification was performed using the K - means algorithm based on the 750/900nm spectral ratios to highlight the surface material variations. In the case of narrow measurements of phase angle, the classification results were not significantly affected by viewing geometry. Varying degrees of dust coverage on the rock surfaces were revealed, which may be attributed to ejecta deposition and dust electrostatic levitation. The spectral shapes of the less dust - covered surfaces indicated that these rocks are dominated by mafic materials, suggesting they may not originate from the lunar upper crust. The influence of dust coverage on spectral measurement of rocks was also confirmed, which significantly attenuates the absorption depths.

Plain Language Summary: The Chang'E - 4 (CE - 4) mission is the first probe to successfully land on the lunar farside. The rover of CE - 4 mission, Yutu - 2, has come across four curious rocks during its exploration. To reveal their compositions, Yutu - 2 rover conducted close investigations and captured their spectral images using the onboard Visible and Near - infrared Imaging Spectrometer instrument (VNIS). We analyzed these data and found an interesting phenomenon that these rock surfaces are partly covered by lunar dust. Dust coverage obscures the information from the rocks themselves, making it difficult to decipher the rock compositions if not carefully handled. Fortunately, benefiting from the spatial insight provided by the VNIS, we analyzed the areas with less dust coverage on rock surfaces and determined that these rocks are likely dominated by mafic materials, suggesting that they may not originate from the lunar upper crust.

Key Points:

• The spectral variations of the rock surfaces are revealed using the visible/ near - infrared image cubes collected by the Chang'E - 4 (CE - 4) rover

• The four rocks encountered by the CE - 4 rover may be all dominated by mafic materials but with different degrees of dust coverage

• The absorption depths of the rock surface spectra in the field of view are significantly attenuated as a result of dust coverage

1 Introduction

China's Chang'E - 4 (CE - 4), which isthe first probe to land on the lunar farside, has been working stably for over 2 years since its touching down on 3 January 2019. The Yutu - 2 rover of CE - 4 has traveled over 700m across the floor of Von Kármán crater until the end of its 29th lunar day (Figure S1 in Supporting Information S1) . In order to analyze the surface materials along the path, a Visible and Near - infrared Imaging Spectrometer (VNIS) is mounted on the front of Yutu - 2 rover. A number of surface soil spectra were acquired by the VNIS and their photometry and spectroscopy properties have already been investigated in many studies (Chen et al. , 2020; Gou et al. , 2019; Gou et al. , 2021; Gou, Di et al. , 2020; Gou, Yue, Di, Wan et al. , 2020; Gou, Yue, Di, Wang et al. , 2020; Huang et al. , 2020; Li, Liu et al. , 2019; Lin, He et al. , 2020; Lin, Yang et al. , 2020; Lin, Y. Lin et al. , 2020; Ma et al. , 2020; Qi et al. , 2020; Yang et al. , 2020) . Additionally, VNIS also captured four unique rocks along the Yutu - 2's path in its third, thirteenth, 26th and 27th lunar day, respectively (we refer to them as R1 to R4 in the following sections and the specific VNIS measurement numbers are shown in the corresponding parentheses) . This is the first time close - up spectral images have been collected of rocks exposed at the lunar surface and the undisturbed states of these rocks were recorded.

Lunar dust can be defined as the $<20\mu m$ fraction of the soil (Liu & Taylor, 2011) . Dust activity is one of several considerable surface processes on the Moon (Szalay et al. , 2018) . Two and a half years after Surveyor 3 landed on the lunar surface, the Apollo 12 mission, which touched down nearby, examined the equipment of Surveyor 3 and demonstrated the dust accumulation and adhesion. Given that the rocks were exposed for much longer periods, dust would inevitably be deposited on their surfaces. However, dust coverage could obscure the optical information of the rock, significantly affecting the geological interpretation, especially on the in - situ measurements such as CE - 3 and CE -

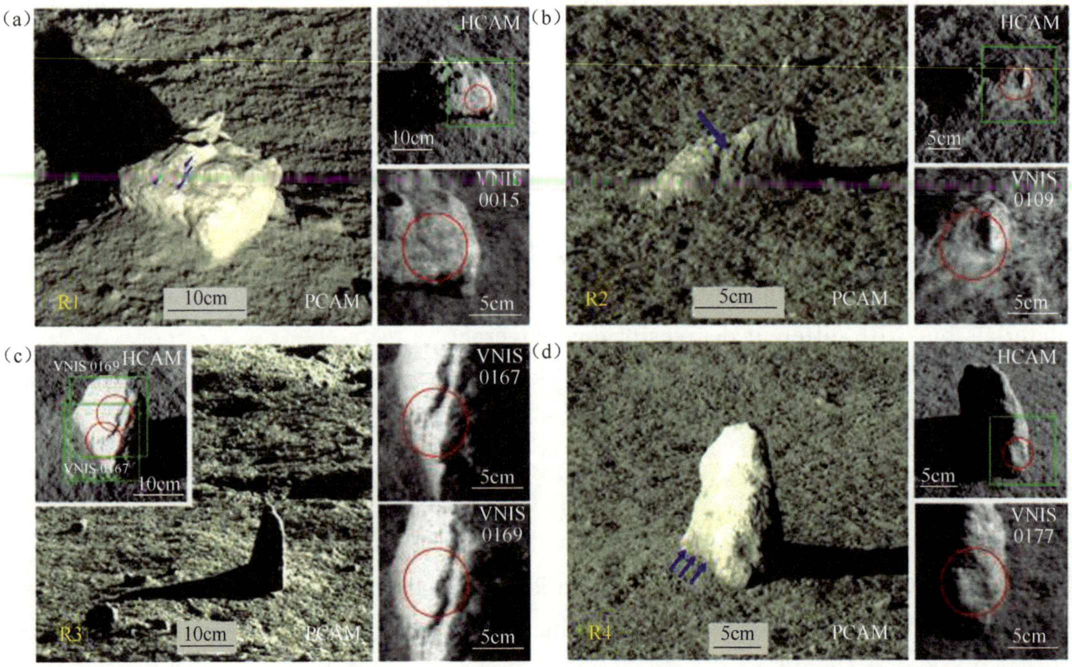

Figure 1　The images of the four rocks encountered by Yutu - 2 along its path. (a - d) are the relevant images of R1 - R4, respectively. The image data for each rock include the Panoramic Camera (PCAM) color mode image [the approximate center wavelengths are: R - 640nm, G1 - 540nm, G2 - 540nm, B - 470nm (Jin et al., 2015; Ren et al., 2014)], the Hazard Avoidance Camera (HCAM) image and the Visible and Near - infrared Imaging Spectrometer (VNIS) reflectance image at 750nm. The green box in each HCAM image roughly encompasses the detection region of the corresponding VNIS image. The red circles in the HCAM and VNIS images are the regions of the field of views of VNIS. Two VNIS measurements of R3 were conducted at different distances

4. It is thus necessary to analyze the dust coverage on the rock surfaces before figuring out their compositions through the optical data. Photometric measurements were used to identify the dust coverage on the rock surfaces in the previous study (Yan et al., 2019). In terms of the prominent spectral differences of rock and soil, spectral measurements may also help (Figure S2 in Supporting Information S1).

Table1　K - Means Classification Rules Based on the 750/900nm Spectral Ratios

Classification	Range of 750/900nm spectral ratio
Class1	>1.183
Class2	1.055 - 1.183
Class3	<1.055

Most previous studies on CE - 4 VNIS data were based on the continuous spectrum derived from connecting the average spectrum of visible/near infrared (VIS/NIR) image cube in the field of view (FOV) with the short wave infrared (SWIR) one - pixel spectrum. In this work, we focus more on the VIS/NIR image cube which provides great insight into both spatial and spectral dimensions. According to the spectral variations on the rock surface, we give a comprehensive analysis of the dust coverage and the possible rock composition. The effect of dust coverage on in - situ spectral detection and future work on spectral modeling are further discussed.

2 Instrument and Data Sets

The VNIS is a special acousto - optic tunable filter spectroscopy with two detection channels. A complementary metal - oxide - semiconductor (CMOS) produces a 256 by 256 pixels image in VIS/NIR range (450 - 950nm) with the spatial and spectral resolutions of 0.6 mm/pixel and 2.4 - 6.5nm, respectively. A SWIR (900 - 2,395nm) spectrometer obtain a one - pixel spectrum with the spectral resolution of 3.6 - 9.6nm. The FOVs of the CMOS imager and SWIR spectrometer are 8.5° and 3.58°, respectively (Li, Wang et al., 2019).

In this work, the reflectance is defined as radiance factor (RADF; Hapke, 2012). We converted the VNIS Level 2B radiance data to RADF using the solar irradiance method. RADF is given by

$$\mathrm{RADF}(\lambda) = \frac{\pi I(\lambda)}{\int J(\lambda) F(\lambda) d\lambda}$$

where $I(\lambda)$ is the VNIS radiance measured at wavelength λ, $J(\lambda)$ is the solar irradiance at 1 AU (Gueymard, 2004) and $F(\lambda)$ the spectral response function of the VNIS detector. We also obtained the continuous spectra in the range of 450 - 2,395nm using the same method in Li, Liu, et al. (2019) (described in Text S1 in Supporting Information S1). The continuous spectrum of each measurement is shown in Figure S3 in Supporting Information S1.

3 Methods

3.1 Image Cubes and Spectra Analysis

Given that the signal at the end of the instrument detection range may be unstable, we only employed the VNIS image cube data in the range of 450 - 900nm. The rock section in each image cube was extracted and analyzed, and the pixels with RADF $<$ 0.035 at 750nm were regarded as shadows and excluded. We first smoothed the VNIS image cubes in the spectral dimension using the Savitzky - Golay algorithm (Savitzky & Golay, 1964) with 15 points and first polynomial order. To highlight the absorption variations on the rock surface, we then calculated the 750/900nm spectral ratio, which is sensitive to the strength of the ferrous absorption around 1μm. The 750/900nm spectral ratios of the four rocks were treated as one sample set and classified using the K - means clustering algorithm (MacQueen, 1967) . The classification rules were obtained in Table 1. In addition, we also selected an area of the soil surface near each rock for comparison.

The band depths of the continuous spectra shown in Figure S3 in Supporting Information S1 were also estimated. The continuums of the spectra were removed using the convex hull method (Clark & Roush, 1984) . Then, based on the continuum - removed spectra, the absorption bands near 1 and 2μm were fitted by the fourth and second order polynomials, respectively. The band depths were calculated by 1 minus the minimum value of the fitting polynomials. The reason why we adopted different fitting polynomials is that the signal - to - noise ratios of the VNIS data are different near 1 and 2μm.

3.2 Calculation of Viewing Geometry

The viewing geometry on rock surface could be variable due to the irregular shape. For this reason, the photometric effects caused by the changes of viewing geometry should be taken into account. In order to estimate the viewing geometry on rock surface, the Hazard Avoidance Camera (HCAM) stereo image pairs were employed to recover the topographic information (the images of HCAM and corresponding VNIS were collected at the same location) . The three - dimensional coordinate and surface normal of a certain point on rock surface can be obtained after the point pairs were matched (in local east, north, up Cartesian coordinate system) . The vector between the point on rock surface and the position of VNIS is defined as emission light vector. The incident light vector can be easily calculated by the known solar incident angle and azimuth angle recorded in the header file of VNIS measurement. In this way, the incident angle, emission angle and phase angle of any pixel on rock surface can be obtained by simple vector algebra operation.

4 Results and Discussions

4.1 Effects of Viewing Geometry

Intriguing variations of absorption depth around 1μm on the four rock surfaces are shown in their 750/900nm spectral ratios (Figure S4 in Supporting Information S1) and classification results (Figure 2) . In general, these spectral variations may depend not only on changes of composition and mineralogy but also on changes of viewing geometry. In particular, the observational angle of each pixel could vary when the in - situ target is rock. To better understand how any photometric effects might affect the classification results, we calculated the geometric information from the HCAM stereo images (taking

R1 and R3 as examples, shown in Figure S5 in Supporting Information S1) . Since these rocks are relatively small, the phase angles of the pixels on the rock surfaces do not change much. The relationships between 750/900nm spectral ratio, 750nm reflectance and viewing geometry (incidence angle, emission angle and phase angle) in each pixel are shown in Figure 3 and Figure S6 in Supporting Information S1, respectively. It was found that the 750/900nm spectral ratios of both rocks did not exhibit noticeable trend with the viewing geometry, but the 750nm reflectance did. The results here indicated that in the case of such small change of phase angle, viewing geometry may still have appreciable effects on the absolute brightness of the rock surfaces, but less on their absorption depth. In terms of the classification rules, different classes can still be distinguished under different observational angles. It is thus reasonable to believe the changes of viewing geometry may not have significant impacts on the classification results, at least for the relatively narrow phase angle ranges provided by these data sets.

4.2 Property Variations on the Rock Surface

The above analysis demonstrates that the property changes could be mainly responsible for the spectral variations observed on each rock surface. Factors such as mineral distribution, space weathering and dust coverage can all lead to these sorts of variations. Based on the classification results and the corresponding spectral shapes shown in Figure 2, we attempted to recognize what possible materials each class represents on the rock surface.

The most significant difference among Class 1 - 3 is the absorption depth around 1μm. The absorption feature of Class 3 is the weakest. When we compared the average spectrum of Class 3 on each rock with that of the selected nearby soil surface, we found they are quite similar in spectral shape. Previous spectral investigations of Martian rocks confirmed that the spectral characteristics of dust - covered rocks would be somewhat similar to that of soil or regolith (e. g. , Johnson et al. , 2004) . It should be consistent in lunar spectroscopy, which implies Class 3 possibly represent the rock surface completely covered by dust. Compared with Class 3, Class 1 and Class 2 have relatively strong absorption features, suggesting that they may contain the spectral information from the rock itself. The spectral difference between Class 1 and 2 probably indicate the difference in dust coverage. It should be noted that, in lunar airless environment, different degrees of space weathering may also contribute to the spectral difference between Class 1 and 2. The spectral effects of dust coverage and space weathering on rock surface could be similar that both would attenuate the absorption characteristics (Noble et al. , 2012; Pieters et al. , 1993; Pieters & Noble, 2016) . Some external disturbances, such as micrometeorite impacts that may be observed on the surface of R1 (as blue arrows indicate in Figure 1a), would produce such heterogeneous space weathering on rock surface, and these sorts of processes may also remove the surface dust. Thus, the effects of dust coverage and space weathering could be coupled and difficult to fully distinguish when conducting spectral measurement of lunar rock surface. Even so, Class 1 should represent the relatively fresh surfaces on the rocks with minimum dust coverage and/or smallest degree of space weathering, while Class 2 could be the space weathered rock surface without completely covered by dust.

In addition, since Class 1 and 2 may contain the spectral information about the rocks, the mineral composition of these rocks could be determined accordingly. For the mafic rock - forming minerals, spectra of pyroxene have diagnostic absorption bands around 1 and 2μm at visible and near - infrared wavelengths, while the spectra of olivine only present a broad complex absorption near 1μm. Their absorption characteristics are caused by the crystal field transitions of Fe^{2+} (Burns, 1993). Pure plagioclase is featureless at visible and near - infrared wavelengths, but would show a weak absorption band around 1.25μm if containing Fe^{2+} (Cheek et al., 2013). All the four rocks detected by Yutu - 2 show prominent absorption features at 1μm which can be seen from Class 1 and 2, indicating these rocks are more likely dominated by mafic materials, instead of plagioclase - rich materials from the lunar upper crust.

4.3 Dusty Rocks Along the Probing Path

Detail analysis of the dust coverage on the rocks is as follow. R1 is the first rock explored by Yutu - 2. Many reports about its composition and origin were proposed but are still in dispute (Chen et al., 2020; Gou et al., 2019; Gou, Di et al., 2020; Huang et al., 2020; Lin, He et al., 2020; Ma et al., 2020). It is a cube - shaped rock with ~8cm height and a distinct cavity appears on its top (Figure 1a). When comparing the classification results with the rock image in Figure 1a, the spectra in the depressed areas on the rock was found to be classified into Class 3, which is consistent with the tendency of these areas to accumulate dust, while Class 1 and Class 2 correspond to the raised areas or the small pits possibly created by micrometeorite impacts (blue arrows in Figure 1a). R2 is a small elongated rock lying next to several impact craters and its overall height is less than 5cm. There are very few pixels on R2's surface that were classified into Class 1 and 2. As shown in the Panoramic Camera (PCAM) and HCAM image, the middle part of the rock may have been mostly buried by nearby soil (blue arrows in Figure 1b). The spectral classification gives a consistent result that the area showing relatively strong absorption features only occurs on its upper surface. Besides, most observed parts in the FOV are the buried rock surface, which agrees with that the spectral measurement of R2 (VNIS 0109) is quite similar to that of nearby pure lunar soil (Figure S7 in Supporting Information S1). R3 and R4 are similar in shape, with similar heights of ~14cm. Both of them are isolated rocks standing upright on the ground. The spectral measurement of R3 were conducted on a highly inclined facet of the rock and the classification results indicated the inclined facet could also be covered by dust. For R4, dust coverage is

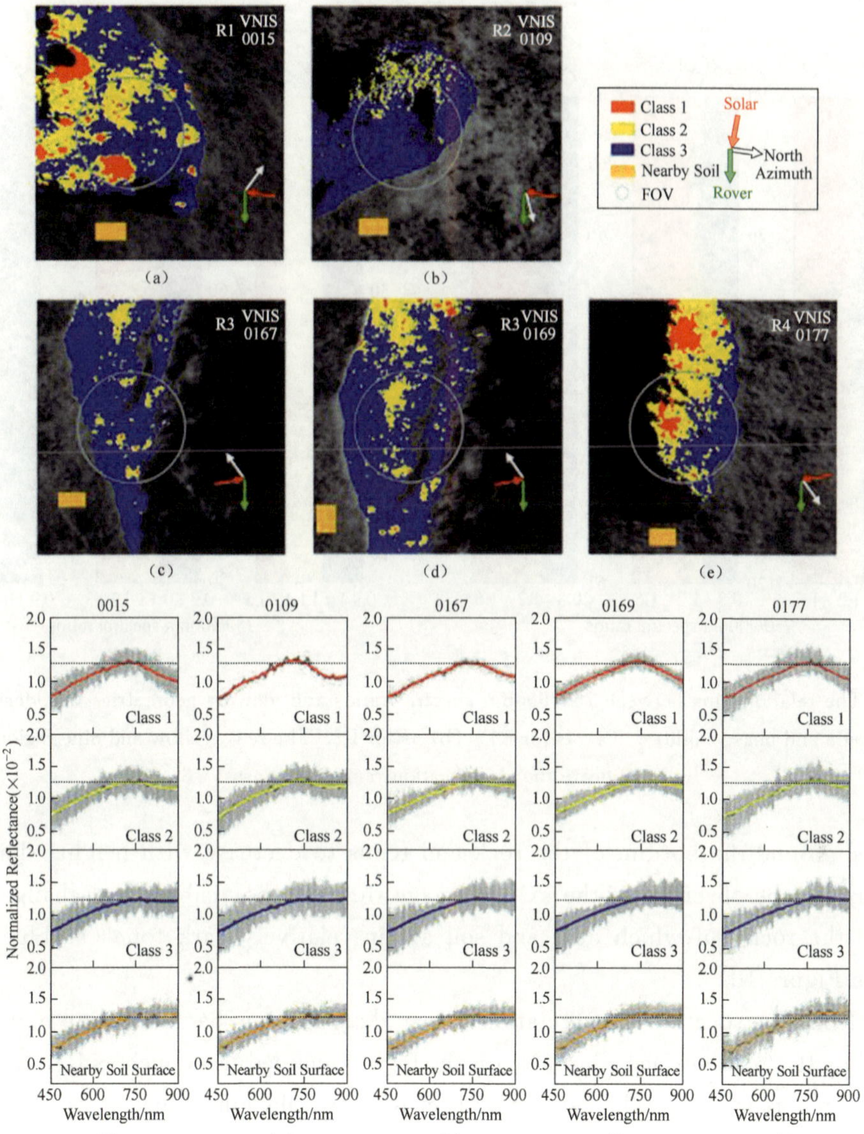

Figure 2　The classification results of the four rocks and the corresponding average spectra. (a) - (e) shows the classification of each rock surface, and selected nearby soil surfaces are marked in orange. The field of views (FOVs) of the Visible and Near - infrared Imaging Spectrometer (VNIS) are indicated in gray circles. The viewing geometry of each measurement is shown by arrows in relative azimuths. (f - j) plot the average spectra of Class 1 - 3 and nearby soil surfaces corresponding to (a - e), respectively. Note that all the spectra are normalized by the equation $r_{\text{norm}} = r_i / \sum_{i=1}^{i=n} r_i$, where i is the band number (Lin, Y. Lin et al., 2020). The background gray point sets are the groups of the original spectra from individual pixels in each class. The gray solid lines are the average spectra and the colored lines are the corresponding smoothed spectra. For clarity, a gray dashed line is marked in each plot, which is a horizontal line across the normalized smoothed reflectance value at 750nm

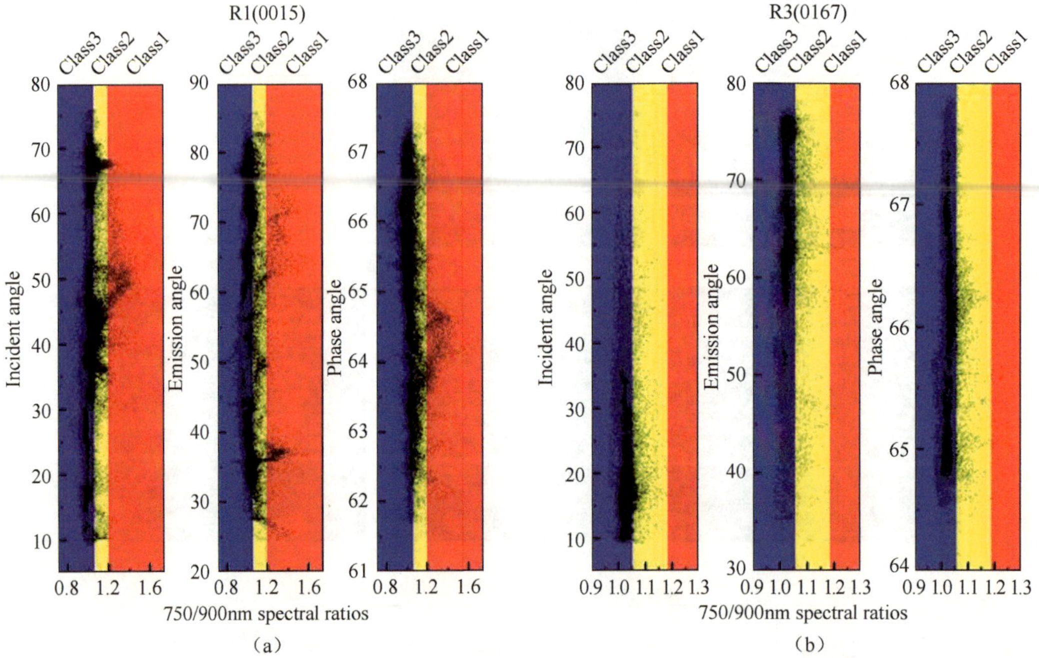

Figure 3　The relationships between 750/900nm spectral ratios and viewing geometries (incidence angle, emission angle and phase angle). (a) Is for R1, (b) is for R3. The red, yellow and blue regions in each frame indicate the classification ranges of Class 1 – 3

distributed around the bottom of the rock and tends to decrease with height. The variations may also be observed in the PCAM image through the visible tonal changes at the bottom of the rock, in which dust and soil are in relatively dark tone (as blue arrows indicate in Figure 1d).

Based on the spectral interpretation, dust deposits can be found not only on the depressed or flat rock surface, but also on the highly inclined facet such as R3. Some dust deposition and transport mechanisms may lead to these phenomenons. It has been found that the Moon is surrounded by a permanent, asymmetric dust cloud produced by the continual meteoroid bombardments to the lunar surface (Horányi et al., 2015). The bombardments can generate dust rains and then fall back down onto the lunar surface (Szalay & Horányi, 2016), including the upward rock surface. But near the lunar surface, electrostatically lofted processes also contribute to the evolution and transport of dust (e. g., Szalay et al., 2018). The levitated charged dust could be easily attached to the rock surface. The inclined facet of the rock maybe coated by dust in this process. This dust deposition mechanism was also found on the surface of CE – 3 detected rocks and the reported upper line of dust deposition can be up to about 28cm above the ground (Yan et al., 2019). These long – term dust activities could make dust coverage common on the lunar rocks. However, later disturbances such as micrometeorite impacts may alter the distribution of dust on the rock surfaces (e. g., the rocks in the dashed cyan box of Figure

S8 in Supporting Information S1) .

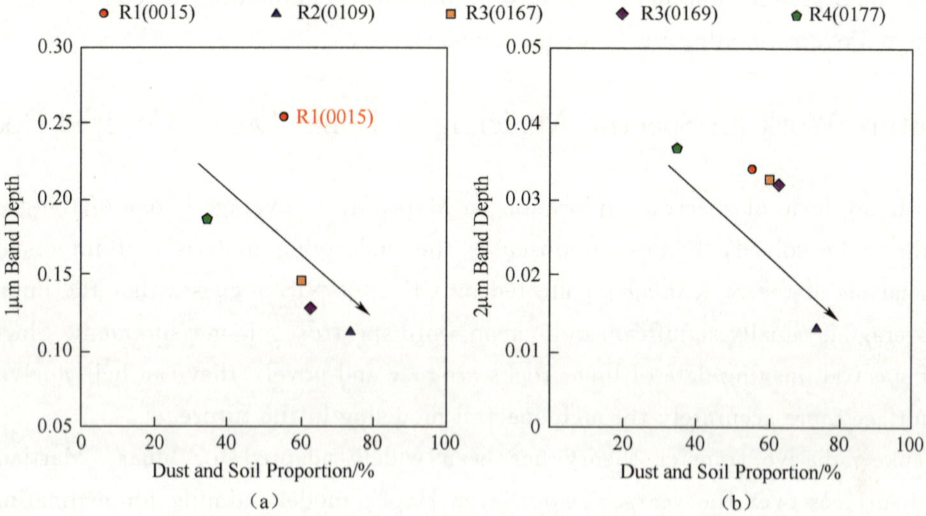

Figure 4　The influence of the dust and soil observed in the field of view (FOV) on the continuous spectra of the rocks. The 1 and 2μm absorption depths of the continuous spectra obviously decrease with the increase of the proportion of dust or soil observed in the FOV. Note that the point 0015in (a) may be an outlier, because the 1μm absorption depth of Visible and Near - infrared Imaging Spectrometer 0015 measurement is probably deepened resulting from the anomalous signal (Figure S10 in Supporting Information S1)

4.4　Implication for the VNIS Data Application

The VNIS onboard Yutu - 2 recorded detailed spatial and spectral information. However, more careful analysis is required for these in - situ data because the detected targets often exhibit more complex textural and composition characteristics than the orbital detection. For example, shadows have a pronounced effect on the in - situ measured albedo (Figure S9 in Supporting Information S1; Qi et al., 2020) .

More notably, dust coverage also significantly affects the absorption features of the detected spectra. Based on the image cubes collected by the VNIS, the proportion of each material distributed in the FOV was determined by pixel counting. For these four rocks, in addition to Class 1 - 3, shadow and soil surface are also involved. The statistics are shown in Table S2 in Supporting Information S1. Note that the effect of soil surface on the average spectrum in the FOV could be the same as that of dust covered on rock surface because of their similar spectral shapes. Figure 4 presents the influence of the dust and soil observed in the FOV on the in - situ measured spectra. Obviously, the 1 and 2μm absorption depths of the continuous spectra decrease with the increasing proportion of dust or soil observed in the FOV. The observed dust or soil in the FOV will definitely attenuate the absorption depth, even though the targets are rocks. As a result, a direct quantitative investigation on the rock composition without considering the effect of dust coverage could produce unreliable results, perhaps underestimating the contents of mafic material and overestimating the contents of plagioclase.

5 Futuro Work on Spoctral Modeling of Lunar Dust – Covered Rocks

With any form of spectral exploration on Mars, dust coverage is one of the greatest problems to be solved, because it obscures the underlying materials of interest. Our initial analysis of the rock images collected by CE – 4 VNIS suggests that the impact of dust coverage is equally significant in lunar in – situ spectroscopic measurement. Since this type of spectral imaging data of lunar rocks are new and novel, they can help analyze the lunar surface more accurately through spectral modeling in the future.

Hapke radiative transfer theory has been widely adapted for lunar, Martian and asteroid surfaces over the years. Two – layer Hapke model, aiming for estimating the bidirectional reflectance of a two – layer medium, was proven to be effective in solving such dust – covered problems, which was once used to model the spectra of laboratory dust – coated rocks and radiometric calibration targets on Martian rover (Johnson & Grundy, 2001; Johnson et al., 2003, 2004, 2006). The model is presented in Second 10.6.2 of Hapke (2012) and also briefly described in supplementary materials for readers' convenience. Only the single – scattering albedo of each layer ω_U, ω_L, the phase function of each layer P_U (g), P_L (g) and the optical thickness τ need to be solved. In the case of spectral modeling of dust – covered lunar rocks, it is reasonable to assume the single – scattering albedo of the dust – covered layer ω_U is the same as lunar soil due to their similar spectral properties, which could be easily acquired from the nearby soil surface. Isotropic or single – term HG phase function are suggested (Johnson et al., 2004). The optical thickness τ can be defined as $\tau = -3tln(p)/2D$, where t, p and D are the coating thickness, porosity and grain size, respectively (Johnson et al., 2003). It should be noted that the opposition effect is assumed to be ignored for simplicity, the model may thus not work well for small phase angle. Besides, the single – scattering albedo of the lower layer ω_L, which is the rock surface in this case, could be helpful to investigate the space weathering on lunar rock surface. However, additional laboratory works are needed to study this process and how to distinguish its spectral effect from dust coverage. In a word, the model gives a great insight to separate the information of un – known underlying material from dust coverage and would benefit from multiple viewing geometry observations if future mission could be conducted.

6 Conclusions

The Chang'E – 4 rover, Yutu – 2, captured the spectral data of four rocks along its exploration path. In this work, we present a new perspective to study the material distributions on the rock surface using the image cubes collected by the VNIS. Simple classification based on the 750/900nm spectral ratio was conducted to identify the compo-sition on the rock surfaces, the results of which were found to be not significantly affected by viewing geometry in the relatively narrow phase angle ranges provided by the data sets. Benefitting from spatial and spectral information provided by the image cubes, it was found that the surfaces of the four detected rocks were covered by varing degrees of dust, while the spectral shapes of the less dust - covered surfaces indicated that these rocks were possibly dominated by mafic materials.

Dust coverage would affect the spectral measurements of the rocks in significantly attenuating the absorption depth of the continuous spectra. Therefore, the issue about dust coverage should be appreciated in future lunar in - situ missions. Since optical detection cannot penetrate the dust coverage, some dust removal tools and methods should be taken into consideration.

Acknowledgments

The authors would like to thank Jeffery R. Johnson and Briony Horgan for theirthoughtful and constructive reviews thatsignificantly improved our manuscript. They also thank the team members of the Ground Research and ApplicationSystem, who have contributed to data receiving and release. This work is supported by National Natural Science Foundation of China (Grant No. 11941002and11803056) and the Strategic Priority Research Program of Chinese Academy of Sciences (XDB—41000000).

References

[1] Burns, R. G. (1993) . Mineralogical applications of crystal field theory (2nd ed. , pp. 551) . Cambridge University Press.

[2] Cheek, L. C. , Donaldson Hanna, K. L. , Pieters, C. M. , Head, J. W. , & Whitten, J. L. (2013) . The distribution and purity of anorthosite across the Orientale basin: New perspectives from Moon Mineralogy Mapper data. Journal of Geophysical Research: Planets, 118 (9), 1805 - 1820. https://doi.org/10.1002/jgre.20126.

[3] Chen, J. , Ling, Z. , Qiao, L. , He, Z. , Xu, R. , Sun, L. , et al. (2020). Mineralogy of Chang'e - 4 landing site: Preliminary results of visible and near - in - frared imaging spectrometer. Science China Information Sciences, 63 (4), 1 - 12. https://doi.org/10.1007/s11432 - 019 - 2768 - 1.

[4] Clark, R. N. , & Roush, T. L. (1984) . Reflectance spectroscopy: Quantitative analysis techniques for remote sensing applications. Journal of Geophysical Research:

Solid Earth, 89 (B7), 6329 - 6340. https: //doi. org/10. 1029/JB089iB07p06329.

[5] Gou, S. , Di, K. , Yue, Z. , Liu, Z. , He, Z. , Xu, R. , et al. (2019) . Lunar deep materials observed by Chang'e - 4 rover. Earth and Planetary Science Letters, 528, 115829. https: //doi. org/10. 1016/j. epsl. 2019. 115829.

[6] Gou, S. , Di, K. , Yue, Z. , Liu, Z. , He, Z. , Xu, R. , et al. (2020). Forsteritic olivine and magnesium - rich orthopyroxene materials measured by Chang'e - 4 rover. Icarus, 345, 113776. https: //doi. org/10. 1016/j. icarus. 2020. 113776.

[7] Gou, S. , Yue, Z. , Di, K. , Bugiolacchi, R. , Wan, W. , Yang, M. , & Ye, L. (2021) . Geologically old but freshly exposed rock fragments encountered by Yutu - 2 rover. Journal of Geophysical Research: Planets, 126 (3), e2020JE006565. https: //doi. org/10. 1029/2020JE006565.

[8] Gou, S. , Yue, Z. , Di, K. , Wan, W. , Liu, Z. , Liu, B. , et al. (2020) . In situ spectral measurements of space weathering by Chang'e - 4 rover. Earth and Planetary Science Letters, 535, 116117. https: //doi. org/10. 1016/j. epsl. 2020. 116117.

[9] Gou, S. , Yue, Z. , Di, K. , Wang, J. , Wan, W. , Liu, Z. , et al. (2020). Impact melt breccia and surrounding regolith measured by Chang'e - 4 rover. Earth and Planetary Science Letters, 544, 116378. https: //doi. org/10. 1016/j. epsl. 2020. 116378.

[10] Gueymard, C. A. (2004) . The sun's total and spectral irradiance for solar energy applications and solar radiation models. Solar Energy, 76 (4), 423 - 453. https: //doi. org/10. 1016/j. solener. 2003. 08. 039.

[11] Hapke, B. (2012) . Theory of reflectance and emittance spectroscopy. Cambridge University Press.

[12] Horányi, M. , Szalay, J. R. , Kempf, S. , Schmidt, J. , Grün, E. , Srama, R. , & Sternovsky, Z. (2015) . A permanent, asymmetric dust cloud around the Moon. Nature, 522, 324 - 326. https: //doi. org/10. 1038/nature14479.

[13] Huang, J. , Xiao, Z. , Xiao, L. , Horgan, B. , Hu, X. , Lucey, P. , et al. (2020). Diverse rock types detected in the lunar South Pole - Aitken Basin by the Chang' E - 4 lunar mission. Geology, 48 (7), 723 - 727. https: //doi. org/10. 1130/G47280. 1.

[14] Jin, W. , Zhang, H. , Yuan, Y. , Yang, Y. , Shkuratov, Y. G. , Lucey, P. G. , & Wang, Z. (2015) . In situ optical measurements of Chang'E - 3 landing site in mare imbrium: 2. Photometric properties of the regolith. Geophysical Research Letters, 42 (20), 8312 - 8319. https: //doi. org/10. 1002/2015GL065789.

[15] Johnson, J. R. , & Grundy, W. M. (2001) . Visible/near - infrared spectra and two - layer modeling of palagonite - coated basalts. Geophysical Research Letters, 28 (10), 2101 - 2104. https: //doi. org/10. 1029/2000GL012669.

[16] Johnson, J. R., Grundy, W. M., & Lemmon, M. T. (2003). Dust deposition at the Mars Pathfinder landing site: Observations and modeling of visible/near - infrared spectra. Icarus, 163 (2), 330 - 346. https://doi.org/10.1016/S0019 - 1035 (03) 00084 - 8.

[17] Johnson, J. R., Grundy, W. M., & Shepard, M. K. (2004). Visible/near - infrared spectrogoniometric observations and modeling of dust - coated rocks. Icarus, 171 (2), 546 - 556. https://doi.org/10.1016/j.icarus.2004.05.013.

[18] Johnson, J. R., Sohl - Dickstein, J., Grundy, W. M., Arvidson, R. E., Bell, J., Ⅲ, Christensen, P., & Shepard, M. K. (2006). Radiative transfer modeling of dust - coated Pancam calibration target materials: Laboratory visible/near - infrared spectrogoniometry. Journal of Geophysical Research: Planets, 111 (E12). https://doi.org/10.1029/2005JE002658.

[19] Li, C., Liu, D., Liu, B., Ren, X., Liu, J., He, Z., et al. (2019). Chang'E - 4 initial spectroscopic identification of lunar far - side mantle - derived mate - rials. Nature, 569 (7756), 378 - 382. https://doi.org/10.1038/s41586 - 019 - 1189 - 0.

[20] Li, C., Wang, Z., Xu, R., Lv, G., Yuan, L., He, Z., & Wang, J. (2019). The scientific information model of Chang'e - 4 visible and near - IR imaging spectrometer (VNIS) and in - flight verification. Sensors, 19 (12), 2806. https://doi.org/10.3390/s19122806.

[21] Lin, H., He, Z., Yang, W., Lin, Y., Xu, R., Zhang, C., et al. (2020). Olivine - norite rock detected by the lunar rover Yutu - 2 likely crystallized from the SPA - impact melt pool. National Science Review, 7 (5), 913 - 920. https://doi.org/10.1093/nsr/nwz183.

[22] Lin, H., Lin, Y., Yang, W., He, Z., Hu, S., Wei, Y., et al. (2020). New insight into lunar regolith - forming processes by the lunar rover Yutu - 2. Geophysical Research Letters, 47 (14), e2020GL087949. https://doi.org/10.1029/2020GL087949.

[23] Lin, H., Yang, Y., Lin, Y., Liu, Y., Wei, Y., Li, S., et al. (2020). Photometric properties of lunar regolith revealed by the Yutu - 2 rover. Astronomy & Astrophysics, 638, A35. https://doi.org/10.1051/0004 - 6361/2020378596.8.

[24] Liu, Y., & Taylor, L. A. (2011). Characterization of lunar dust and a synopsis of available lunar simulants. Planetary and Space Science, 59 (14), 1769 - 1783. https://doi.org/10.1016/j.pss.2010.11.007.

[25] Ma, P., Sun, Y., Zhu, M. H., Yang, Y., Hu, X., Jiang, T., et al. (2020). A plagioclase - rich rock measured by Yutu - 2 Rover in Von Kármán crater on the far side of the Moon. Icarus, 350, 113901. https://doi.org/10.1016/j.icarus.2020.113901.

[26] MacQueen, J. (1967). Some methods for classification and analysis of multivariate

observations. In Proceedings of the fifth Berkeley symposium on mathematical statistics and probability (pp. 281 – 297) .

[27] Noble, S. K. , Keller, L. P. , Christoffersen, R. , & Rahman, Z. (2012). Space weathering of lunar rocks. In 43th lunar and planetary science conference.

[28] Pieters, C. M. , Fischer, E. M. , Rode, O. , & Basu, A. (1993) . Optical effects of space weathering: The role of the finest fraction. Journal of Geophysical Research: Planets, 98 (E11), 20817 – 20824. https://doi.org/10.1029/93JE02467.

[29] Pieters, C. M. , & Noble, S. K. (2016) . Space weathering on airless bodies. Journal of Geophysical Research: Planets, 121 (10), 1865 – 1884. https://doi.org/10.1002/2016JE005128.

[30] Qi, X. , Ling, Z. , Zhang, J. , Chen, J. , Cao, H. , Liu, C. , et al. (2020). Photometric normalization of chang'e – 4 visible and near – infrared imaging Spectrometer datasets: A combined study of in – situ and laboratory spectral measurements. Remote Sensing, 12 (19), 3211. https://doi.org/10.3390/rs12193211.

[31] Ren, X. , Li, C. L. , Liu, J. J. , Wang, F. F. , Yang, J. F. , Liu, E. H. , & Zhao, R. J. (2014) . A method and results of color calibration for the Chang'e – 3 terrain camera and panoramic camera. Research in Astronomy and Astrophysics, 14 (12), 1557. https://doi.org/10.1088/1674 – 4527/14/12/005.

[32] Savitzky, A. , & Golay, M. J. E. (1964) . Smoothing and differentiation of data by simplified least squares procedures. Analytical Chemistry, 36 (8), 1627 – 1639. https://doi.org/10.1021/ac60214a047.

[33] Szalay, J. R. , & Horányi, M. (2016) . Lunar meteoritic gardening rate derived from in situ LADEE/LDEX measurements. Geophysical Research Letters, 43, 4893 –4898. https://doi.org/10.1002/2016GL069148.

[34] Szalay, J. R. , Poppe, A. R. , Agarwal, J. , Britt, D. , Belskaya, I. , Horányi, M. , et al. (2018) . Dust phenomena relating to airless bodies. Space Science Reviews, 214 (5) . https://doi.org/10.1007/s11214 – 018 – 0527 – 0.

[35] Yan, Q. , Zhang, X. , Xie, L. , Guo, D. , Li, Y. , Xu, Y. , & Xiao, L. (2019). Weak dust activity near a geologically young surface revealed by Chang'E – 3 mission. Geophysical Research Letters, 46 (16), 9405 – 9413. https://doi.org/10.1029/2019GL083611.

[36] Yang, Y. , Lin, H. , Liu, Y. , Lin, Y. , Wei, Y. , Hu, S. , et al. (2020). The effects of viewing geometry on the spectral analysis of lunar regolith as inferred by in situ spectrophotometric measurements of Chang'E – 4. Geophysical Research Letters, 47 (8), e2020GL087080. https://doi.org/10.1029/2020GL087080.

Vision - Based Decision Support for Rover Path Planning in the Chang'e - 4 Mission

Yexin Wang[1], Wenhui Wan[1], Sheng Gou[1], Man Peng[1], Zhaoqin Liu[1], Kaichang Di[1], Lichun Li[2], Tianyi Yu[2], JiaWang[2] and Xiao Cheng[2]

1. State Key Laboratory of Remote Sensing Science, Aerospace Information Research Institute, Chinese Academy of Sciences, Beijing 100101, China; wangyx716@radi. ac. cn (Y. W.); wanwh@radi. ac. cn (W. W.); gousheng@radi. ac. cn (S. G.); pengman@radi. ac. cn (M. P.); dikc@radi. ac. cn (K. D.)

2. Beijing Aerospace Control Center, Beijing 100094, China; lichunmail@163. com (L. L.); i_am_yty@sina. com (T. Y.); 15210106156@139. com (J. W.); vlcx26@163. com (X. C.)

Abstract: In planetary rover missions, rover path planning is critical to ensure the safety and efficiency of the rover traverse and in situ explorations. In the Chang'e - 4 (CE - 4) mission, we have proposed and developed vision - based decision support methods comprising obstacle map generation and path evaluation for the rover. At each waypoint along the rover traverse, digital elevation model (DEM) is automatically generated, which is then used for obstacle map generation and path searching. For path evaluation, the searched path and the predicted wheel trajectories are projected onto the original images captured by different cameras to coincide with the real observation scenario. The proposed methods have been applied in the CE - 4 mission to support teleoperation of the rover, and examples of multiple applications used in the mission are presented in this paper. Under the support of the proposed techniques, by the end of the 14th lunar day (3 February 2020), the rover has already travelled 367. 25 meters on the far side of the Moon.

Keywords: lunar exploration; Chang'e - 4 mission; Yutu - 2 rover; obstacle map; path planning

1 Introduction

The Chang'e - 4 (CE - 4) probe, including a lander and a rover, successfully lifted off on 8 December 2018 from Xichang Satellite Launch Center, and soft landed in the Von

Kármán crater inside the South Pole – Aitken basin at 10：26 a. m. on 3 January 2019 [1]. The success of the mission marked CE – 4 as the first probe landing on the far side of the Moon [2]. After separation from the lander on the same day, the rover Yutu – 2 (Jade Rabbit – 2) started surface traversing and in situ exploration, under teleoperation from the Earth control center.

To ensure safety and eciency of surface exploration, it is critical to plan a secure and accurate path for the rover to travel and approach science targets. There are three levels of path planning in the CE – 4 mission: overall mission planning, period planning, and unit planning [3, 4]. Overall mission planning employs the orbiter images to plan rough directions and paths for the rover to approach the science targets. Period planning designs a route that spans tens of meters, consisting of several waypoints and unit planning paths between them. Unit path planning is fundamental and provides the final path for the rover to traverse. The length of a unit planning path between adjacent waypoints is usually less than ten meters. In this paper, we mainly discuss vision – based support for unit path planning.

Generally, the unit path – planning process includes several stages: traversability map, or cost map generation [3, 5, 6], automatic path searching, path evaluation, and command sequence generation. The rover then executes the path according to the uplinked command sequence. As the rover carries several types of stereo cameras, vision – based support is used to guide the planning process. To support the traversability map establishment and all the subsequent processes, stereo camera images have been routinely used to generate a digital elevation model (DEM) and digital orthophoto map (DOM) of the terrain, which are further used to recognize obstacles and hazards to derive an obstacle map [4].

In the Chang'e – 3 (CE – 3) mission, the path evaluation work was mainly achieved by expert check with path visualizing on the corresponding DOM. That is to say that the path planning and evaluation are both in overhead view (DEM and DOM). Due to the mounting positions of the cameras and the imaging angles, the resolution of the DEM decreases as the optical path length increases, though the DEM can be interpolated with the same grid spacing. Therefore, the areas far away from the rover on the DEM and DOM may have more distortion than the closer areas because of 3D measurement error and interpolation, etc. [7]. The distortions can make it difficult for experts to interpret the terrain, inspect, and evaluate the searched path.

In the CE – 4 mission, we have proposed and developed vision – based decision support methods to generate an obstacle map and evaluate the searched path in both overhead view (DEM and DOM) and perspective view (DEM and original images). In the former view, the path – searching results are projected on the DOM in perspective view to coincide with the actual observation angle. In the latter, the path – searching results are mapped and back – projected onto different original images: images captured by navigation camera (Navcam) and their stitched mosaic, and images captured by hazard avoidance camera (Hazcam). The latter strategy takes full advantage of the original images resolutions, and therefore provides a more detailed texture than that of DEM. Based on the back - projection method, multiple applications can be carried out to support decision - making during the mission operation, such as searched path evaluation, prediction of imaging area, etc.

2 Data

Same as in the CE - 3 mission, the rover of CE - 4 mission carries three pairs of stereo cameras: panoramic camera (Pancam), Navcam, and Hazcam [8]. The Pancam and Navcam are mounted on the top mast of the lunar rover, which can be rotated in both yaw and pitch directions. The Hazcam is fixed at the front panel of the rover. Among them, the Navcam and Hazcam are engineering payloads and parts of the guidance, navigation, and control (GNC) system, which is an important engineering system aiming at supporting engineering applications such as navigation, positioning, path planning, and obstacle avoidance, etc. [9].

Vision - based decision support methods use images captured by Navcam and Hazcam of the GNC system, hence the parameters of these two cameras are shown in Table 1. Normally at each waypoint, the rover takes a sequence of Navcam images at every 20 degree in yaw direction at a fixed pitch angle. The overlap angle between adjacent images is about 26. 6 degrees [8]. The Hazcam uses fisheye lenses to provide a wide field of view (FOV) of the area starting from the front wheels of the rover. Before the probe launch, the intrinsic and structure parameters of both Navcam and Hazcam have been precisely calibrated [10 - 12].

Table 1　Basic parameters of Navcam and Hazcam [9]

Payload	FOV (°)	Focal Length (mm)	Baseline (mm)	Image Size (pixel×pixel)	Working Mode
Navcam	46. 6	17. 7	270	1024×1024	Fixed roll angle and changeable pitch and yaw angles, imaging along yaw angle at an interval of 20 degrees. Power on every waypoint.
Hazcam	120	7. 3	100	1024×1024	Fixed on the front panel of the rover. Power on at certain waypoints.

3 Methodology

The vision - based path - planning decision support consists of two major steps: DEM and obstacle map generation before automatic path searching, and evaluation of the path afterwards. Figure 1 shows the overall flowchart of vision - based decision support for rover path planning in CE - 4 mission, with our work introduced in this paper highlighted. In this section, techniques for DEM and obstacle map generation, and path planning are briefly introduced. Then evaluation of the searched path using projection strategy is elaborated in detail. In addition, the error analysis of the proposed strategy is presented at the end of this section.

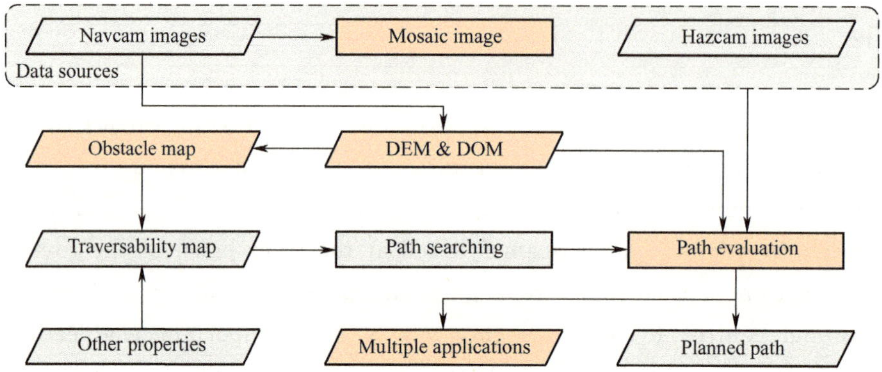

Figure 1 Overall flowchart of vision - based decision support for CE - 4 rover path planning

3.1 DEM and Obstacle Map Generation

DEM is the fundamental data for obstacle analysis and path planning. A DEM and the corresponding DOM are automatically generated using Navcam stereo images at each waypoint immediately after the images are downlinked. The DEM usually covers a range of 16m from the rover, with a grid spacing of 0.02m.

First, feature points matching on neighboring stereo images are performed using their epipolar images generated based on intrinsic and extrinsic parameters of the Navcam. Bundle adjustment is then applied to all the images to refine their extrinsic parameters. After building a triangular irregular network (TIN) based on the matched feature points, for each pixel on the left image of every pair, dense matching is implemented with control and prediction from the TIN in order to obtain a matched point on the right image. To achieve a more precise 3D terrain map, least square matching method is then used to obtain sub - pixel accuracy. After that, 3D coordinates are calculated using the refined image extrinsic parameters and accurate matching results. Finally, DEM is accomplished by these dense 3D points using Kriging interpolation. For more detailed information, please refer to our previous work [7].

Based on the generated DEM, terrain features are analyzed by performing window - based calculations using several statistical metrics: slope, roughness, and elevation change. The slope reflects the degree of surface steepness. Roughness reflects the extent of terrain fluctuation in the window area. The elevation change is the difference between the minimum and maximum heights inside the window. By setting reasonable thresholds according to the rover mechanism such as climbing capability, obstacle crossing capability, etc., the analyzed three features are integrated to generate the obstacle map [13, 14]. The obstacle map, combined with light and communication conditions, are integrated into a traversability map for automatic path searching.

3.2 Path Searching

For unit path planning, on the basis of the established traversability map, the path from the current waypoint to the next is searched by path - searching methods such as A* [3, 15], or SD* lite [16]. Please note that the mechanical capabilities of the rover, such as climbing ability (<30 degree), obstacle crossing capability (<200mm) [17], maximum speed 200m/h [18], are also taken into account during path searching. The automatically searched path is described by coordinates of start, turning, and end points of the path, as well as the curvature of each segment between these points. Please note that the coordinates of the points represent the rover center positions.

3.3 Path Evaluation

3.3.1 Traditional Path Evaluation in Overhead View

Before uplinking the command sequence to the rover, the automatically searched path should be carefully evaluated and checked by experts. In CE - 3 mission, together with the searched path that represents the rover center, wheels trace calculated according to the relationship between the rover center and the wheels, are visualized on the DOM of the current waypoint and checked in overhead view. In CE - 4 mission, we retain this visualization method as one of the evaluation options. However, due to reasons described in the previous section, the texture of the DOM gets worse as the optical path length grows. Therefore, new path evaluation method has been developed and applied in teleoperation of Yutu - 2 rover.

3.3.2 Path Projection into Perspective Views

In CE - 4 mission, new approaches were applied to support searched path evaluations, which include: (A) Project the searched path on 3D perspective view of DOM; (B) project the searched path back on the original Hazcam images, Navcam images, and their mosaic image. The details are as follows.

1. Project on DOM in 3D perspective view

Assign the DEM as elevations of both the DOM surface and the searched path so that they can be visualized in 3D perspective view. For better visualization effect, the elevations are exaggerated with a factor of 2. In this way of visualization, the searched path goes up and down vividly along with the terrain reliefs.

2. Project back on Hazcam and Navcam images

As described in the previous section, the texture of the DOM deforms worse as the optical path length grows. According to the generation process of the DEM and DOM, two – three pixels on the Navcam images make one pixel on the DEM and the DOM. As the optical path length grows, the number increases to five – six pixels make one when far away from the rover.

Therefore, except projecting the path on DOM that the texture problem still remains, we project the searched path and other auxiliary information back onto the original images taken by different cameras. In these perspective views, the resolutions of the original images can be made full use of and without texture deformation or distortion. Three different kinds of images have been used for path projection, including original Navcam image, Navcam image mosaic, and original Hazcam image.

Figure 2 is the flowchart of the back – projection processes on three different kinds of images. Generally, first the searched path is sampled into 2D points according to the path – searching result, and their corresponding elevation values are obtained by interpolation from the DEM generated from the Navcam images. Then the mapping relationships between the three and two – dimensional information are established based on the camera models of both Navcam and Hazcam. Finally, the 3D information of the searched path is mapped and back – projected on multi – source 2D images to support expert decisions. Processes of back projecting on different images are not the same, therefore the rest of the sub – section will introduce each procedure separately.

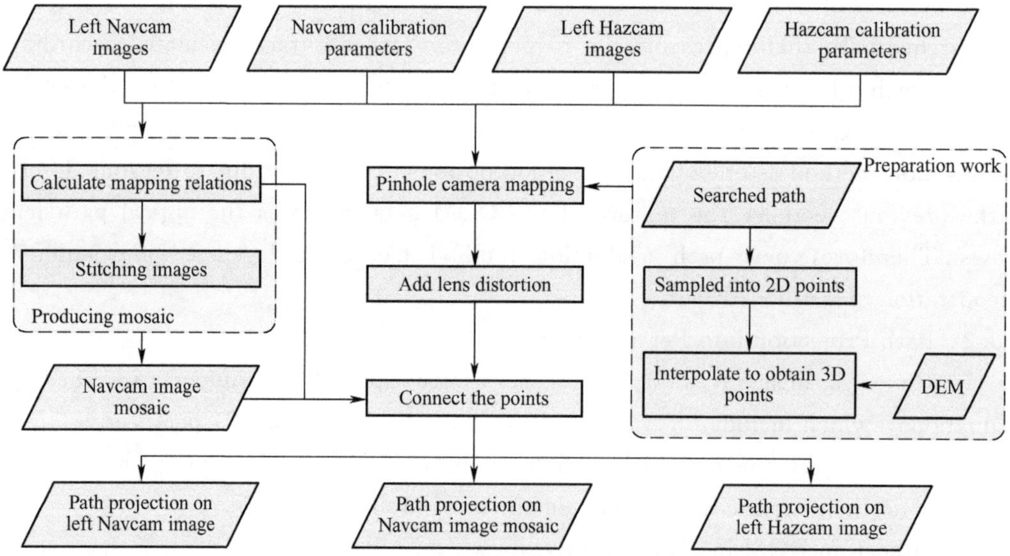

Figure 2　Flowchart of the back – projection processes on three different kinds of images

(1) Original left Navcam image

According to the searched path description information, the path is sampled into discrete two – dimensional points from the start to the end points with proper intervals. By interpolation from the corresponding DEM, their corresponding elevation values can be obtained.

The back - projection model of the Navcam is divided into two parts: pinhole camera model and lens distortion model. First, transform the 3D coordinates of the sampled path points from the work coordinate system to the image coordinate system based on the pinhole camera model and the calibrated Navcam parameters. The resultant coordinates are distortion free. The final image coordinates of the sampled points are obtained by adding distortions to the undistorted ones according to the distortion model of the Navcam. The distortion model of the Navcam in Yutu - 2 rover considers both radial and tangential distortions. The complete distortion model is as follows:

$$x = x_d + (x_d - x_0)(k_1 r + k_2 r^2 + k_3 r^3 + \alpha) + \beta(y_d - y_0) + p_1[r + 2(x_d - x_0)^2] + 2p_2(x_d - x_0)(y_d - y_0)$$

$$y = y_d + (y_d - y_0)(k_1 r + k_2 r^2 + k_3 r^3) + p_2[r + 2(y_d - y_0)^2] + 2p_1(x_d - x_0)(y_d - y_0) \quad (1)$$

where k_1, k_2, k_3 are the radial distortion coefficients; p_1, p_2 are the tangential distortion coefficients, α and β are the orthogonal correction coefficients in the x and y directions, respectively. In Equation (1), (x, y) are the ideal undistorted image coordinates of a point p, $r = (x_d - x_0)^2 + (y_d - y_0)^2$, where (x_d, y_d) are image coordinates of point with distortions in the coordinate system with origin at distortion center point (x_0, y_0). As the distortion model is established from distortion coordinates to distortion free ones, the distorted coordinates are solved iteratively in practice.

According to the above procedure, each sampled point is mapped from the 3D space to the 2D image. The projected path is finally acquired by linking all the projected points on the original Navcam image.

(2) Mosaic image stitched by a sequence of original left Navcam images

As the FOV of the Navcam is limited, only one Navcam image is not enough to show the entire projected path in most cases. In practice, the Navcam would continuously rotate in the yaw direction to take 6 - 7 pairs or more as a routine. Therefore, projecting the searched path on apanoramic image stitched by the sequence of images taken by one of the Navcam can give a wider FOV and more intuitional visualization effect.

All the images captured by the left Navcam are used to extract and match feature points using SIFT algorithm [19], and then RANSAC is applied to eliminate the outliers [20]. To obtain a high - precision stitched image, bundle adjustment is applied to refine the orientation information of each image using corresponding points of each pair of adjacent images. The mapping relations of the images calculated using the refined extrinsic parameters are used to stitch the images [21]. Finally, the image mosaic is obtained by gain compensation and multi - band blending after stitching [22]. Figure 3 shows a sequence of images captured by the left Navcam at waypoint LE00302 and the resultant image mosaic obtained by the method described above.

While in the process of stitching the sequence of images, the refined orientation

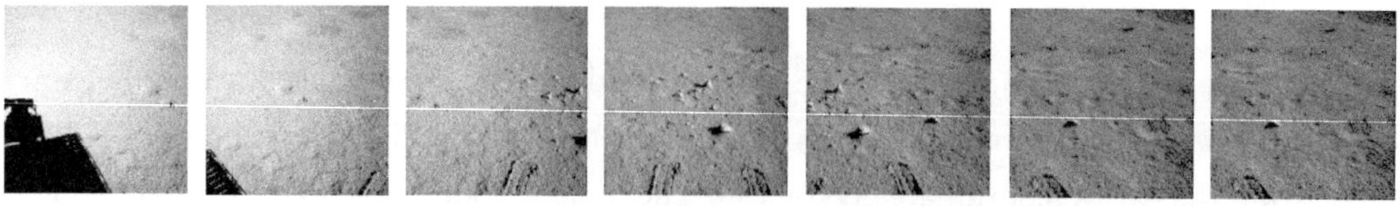

(a)

(b)

Figure 3　Original Navcam images and mosaic at waypoint LE00302. (a) A sequence of original images captured by the left Navcam; (b) the resultant image mosaic

information is stored. For each sampled point on the searched path, the stored orientation information helps to decide on which image the point would be back-projected by judging whether the projected coordinates exceed the boundary of each image. Since the adjacent images have overlap areas, there are always 2 - 3 projections on different images calculated by this judgment criterion. As the lens distortion increases as the distance from the point to the image center increase, the point closest to the image center will be picked as the final projected point, and its coordinates on the mosaic image are calculated according to the stored orientation information.

(3) Original left Hazcam image

As the Navcam normally cannot image the area range from the front wheel of the rover to about 1.2 - 1.6m away from the rover, part of the searched path cannot be seen and projected onto the Navcam (mosaic) image. The Hazcam, fixed on the front panel of the rover, can effectively complement the closer area where Navcam cannot image. Therefore, projecting the planned path onto the Hazcam image can reflects the terrain influence of the planned path at areas that cannot be imaged by Navcam.

The projection models of the Hazcam include a pinhole imaging model and a distortion model, wherein the imaging model is similar to Navcam model, but with different calibrated parameters values. Since Hazcam uses fisheye lenses, the distortion model is different from Navcam. The distortion model of the Hazcam is divided into two parts. The first part is as same as Equation (1). According to Equation (1), the obtained image coordinates are marked as intermediate results (x', y'), and then the final rectified coordinates are obtained by:

$$x = x' \tan(t)/t$$
$$y = y' \tan(t)/t \quad (2)$$

In Equation (2),

$$t = \frac{\sqrt{x'^2 + y'^2}}{f}$$

where f is the focal length of the Hazcam.

In CE - 4 mission operations, the Navcam captures images at each waypoint. Therefore, at each waypoint, the searched path is back - projected onto the Navcam image mosaic as the main decision support information. Sometimes the projection on single Navcam image is applied for particular occasions. When Hazcam images are available, the planned path is also back - projected onto the Hazcam image to provide complementary support information. Moreover, by providing proper 3D information, the back - projection approaches are not limited to support path evaluation. They can be extended in more prediction and evaluation applications, as illustrated in the following sections.

3.3.3 Error Analysis

The back - projection procedure involves two error propagation processes: (1) the error of the DEM generated by the Navcam stereo images where the 3D coordinates of the sampled path points are obtained; (2) projection of the 3D points onto different type of images.

(1) The first error propagation process and the accuracy evaluations can be found in our previous work [7, 23]. According to the calibration parameters of Yutu - 2 rover and the error propagation principle, the error in the depth direction for Navcam is less than 0.11m at distance of 10m, and for Hazcam is less than 0.2m at distance of 5m.

(2) As the projection processes from 3D to 2D Navcam images and 2D Hazcam images are similar, they have the same error propagation process, yet with different parameters values. Denote the 3D coordinates in the work coordinate system as (X_c, Y_c, Z_c), the obtained image coordinates through pinhole model are (x_p, y_p), which has:

$$\begin{cases} x_p = \dfrac{X_c}{Z_c} f \\ y_p = \dfrac{Y_c}{Z_c} f \end{cases} \quad (3)$$

According to the law of error propagation [24], the errors of the corresponding image coordinates are calculated by the following equations:

$$\begin{cases} \sigma_{x_p} = \sqrt{\left(\dfrac{f}{Z_c}\right)^2 {\sigma_{X_c}}^2 + \left(\dfrac{x}{Z_c}\right)^2 {\sigma_{Z_c}}^2} \\ \sigma_{y_p} = \sqrt{\left(\dfrac{f}{Z_c}\right)^2 {\sigma_{Y_c}}^2 + \left(\dfrac{y}{Z_c}\right)^2 {\sigma_{Z_c}}^2} \end{cases}, \quad (4)$$

where σ_{X_c}, σ_{Y_c}, and σ_{Z_c} are measurement errors calculated in the first error propagation process, (x, y) represent the image coordinates which come from the image in the first error propagation process. In addition, (σ_{x_p}, σ_{y_p}) are the projection errors after propagation effect.

4 Results

4.1 Obstacle Map Generation and Support of Path Searching

After the rover reaches a new waypoint, Navcam is powered on to capture a sequence of stereo images. Then DEM and DOM with 0.02m resolution are automatically generated as the basis for obstacle analysis and path searching. Figure 4 shows an example of DEM and DOM at waypoint LE00203, which were automatically generated from 6 pairs of Navcam images captured at a fixed pitch angle. The mapped area covers from about 1.6m in front of the rover to about 16m.

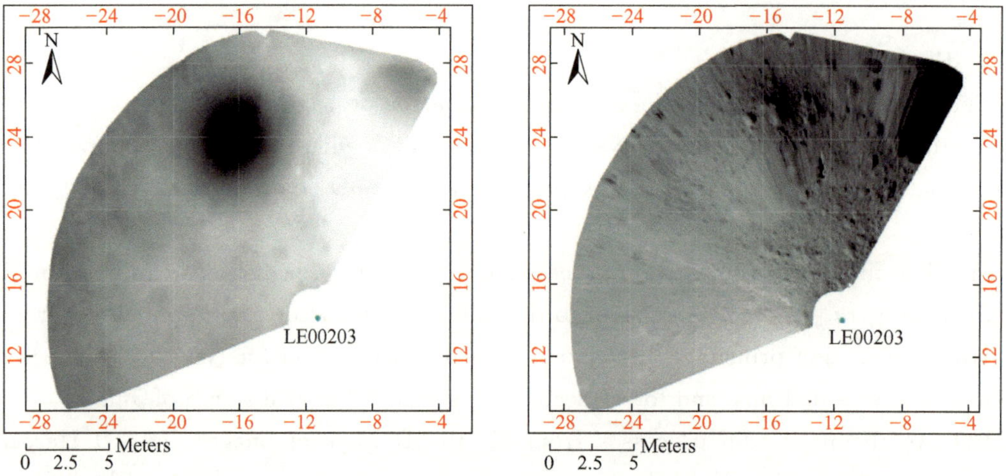

(a) (b)

Figure 4 Automatically generated DEM (a) and DOM (b) from Navcam stereo images at waypoint LE00203

The obstacle map calculated based on the generated DEM at this waypoint is shown in Figure 5, where the red and green areas represent obstacles and safe areas, respectively. As can be seen from the figure, a crater with diameter about 8.7m is in front of the rover. Based on this information and other conditions described in Section 3.2, the path from waypoint LE00203 to LE00204 is searched, which is shown on the obstacle map in Figure 5.

4.2 Evaluation of the Path

After path searching, vision - based evaluation is applied. This section introduces the process of evaluating searched paths from waypoint LE00207 to LE00208 in CE - 4 mission.

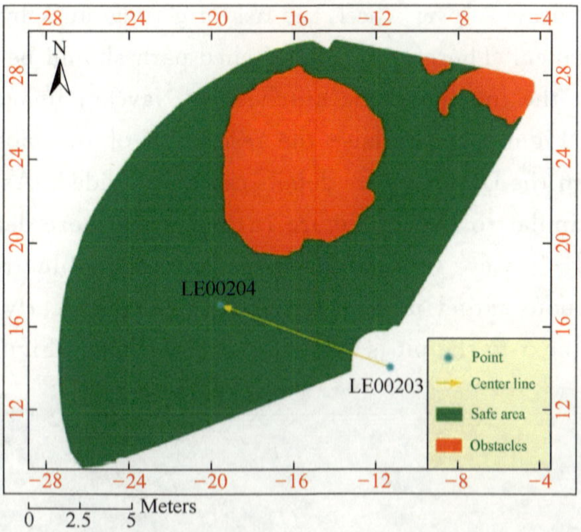

Figure 5　Obstacle map at waypoint LE00203 and a searched path calculated by its support

In Figure 6, the searched path from the current waypoint LE00207 to the next waypoint LE00208 is overlain on DOM. As described before, the traditional visualization of the searched path for decision support is to draw the center line, wheels trace, and course lines on the DOM, as shown in Figure 6.

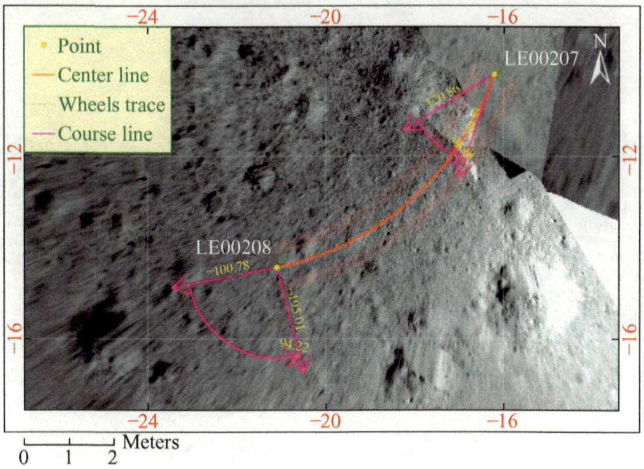

Figure 6　DOM at waypoint LE00207 and a searched path in overhead view

Then more evaluation work was carried out, and the results are shown in Figure 7. First, the searched path was projected on the DOM in 3D perspective view, which is shown in Figure 7a. The elevations are exaggerated twice so that the terrain relief is easier to identify. As can be seen from the zoomed - in view in Figure 7c, the left and right wheels would pass over two pits (where arrows 2 and 3 point to) just before approaching the destination LE00208. Due to limitation of the DOM resolution and texture deformation, only large craters can be identified, the small rocks and pits are impossible to be recognized. In Figure 7c, arrow 1 points to something that looks like a rock or a pit.

Due to the design of Yutu - 2 rover wheel, no risk of getting stuck by a rock in the sieved - like wheel should be taken. Therefore, the searched path should be checked using back - projection strategy on the image mosaic stitched by Navcam images which has subtler texture, as shown in Figure 7b. Because the resolution of the mosaic is at least two - three times higher than the DOM, more details can be provided. As shown in Figure 7b, the viewing angle is similar to that of Figure 7a, yet with more detailed texture. In the corresponding zoomed - in view (shown in Figure 7d) of the blue rectangle annotated in Figure 7b, the ambiguous target pointed by arrow 1 can be clearly identified as a small pit, and a small rock next to the pit is also clearer than that in Figure 7c.

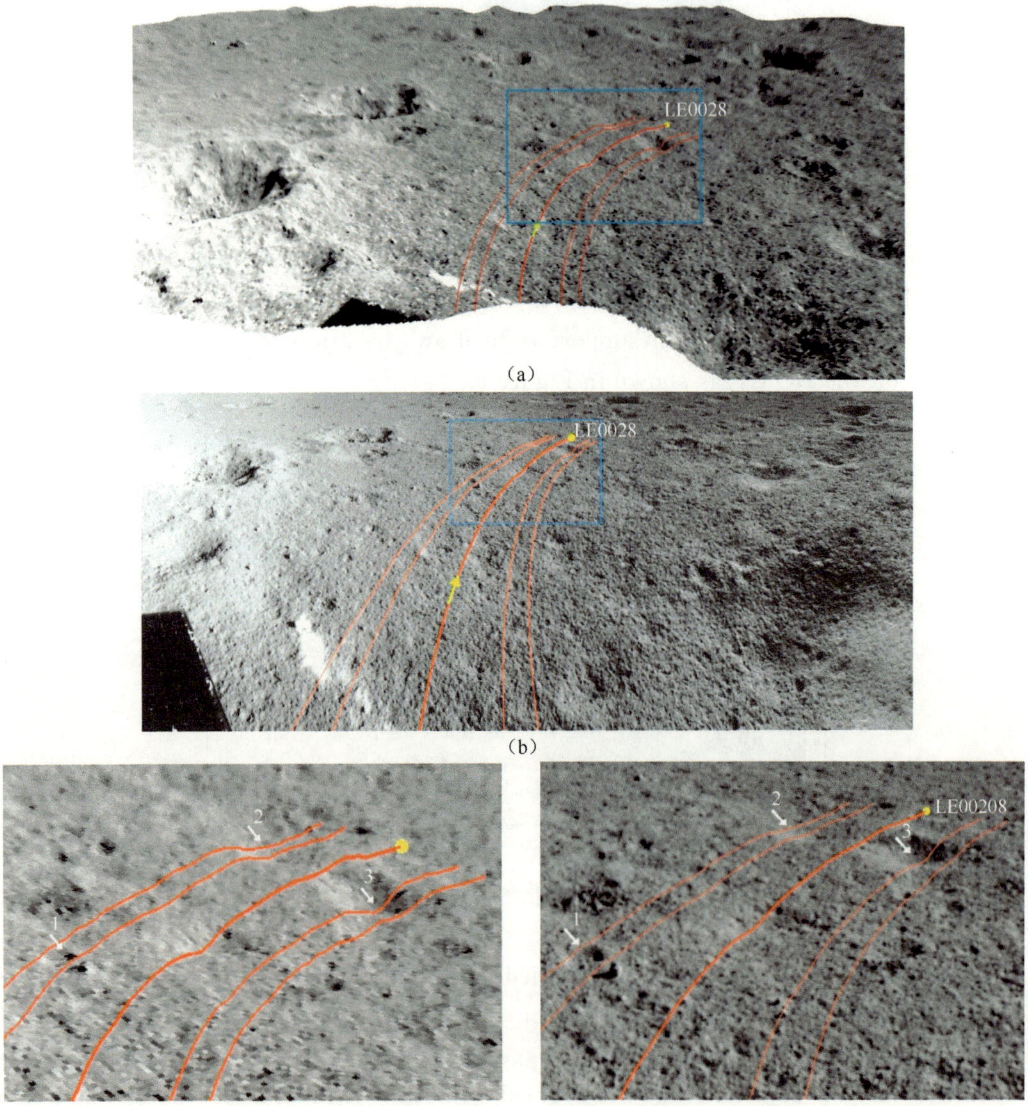

Figure 7 Comparisons of searched path projections on DOM in 3D perspective view and on Navcam image mosaic at waypoint LE00207: (a) The projected path on DOM in 3D perspective view; (b) projection results of the searched path on the Navcam image mosaic; (c) a zoomed - in view of the blue rectangle on (a); (d) a zoomed - in view of the blue rectangle on (b)

From the above evaluation and analysis, the searched path was not optimal and has room for improvement. Therefore, the first searched path, denoted as the Path 1, was then fine - tuned to obtain Path 2. The refined path is shown in Figure 8a. The comparison of the back - projection on Navcam mosaic image of the two paths are shown in Figure 8b. Path 2 drifts left from the first one, which makes the wheels avoiding the two pits before the end point. Moreover, the small pit and the rock near the left wheel trace are also be kept far away. From the fluctuation extent of the projected wheel trace, Path 2 is more appropriate than Path 1.

As introduced above, the Navcam cannot image the foot of the rover, therefore the projection of the path on that part is missing. For evaluation of this part, the two paths were projected on the left Hazcam image, as shown in Figure 9. As can be seen from the figure, in the 2nd path, the right wheels would pass the edge of a pit (pointed by the black arrow in Figure 9). From the fluctuation extent of the projected wheels trajectories, the influence on the wheels can be ignored. Finally, Path 2 was executed to reach waypoint LE00208.

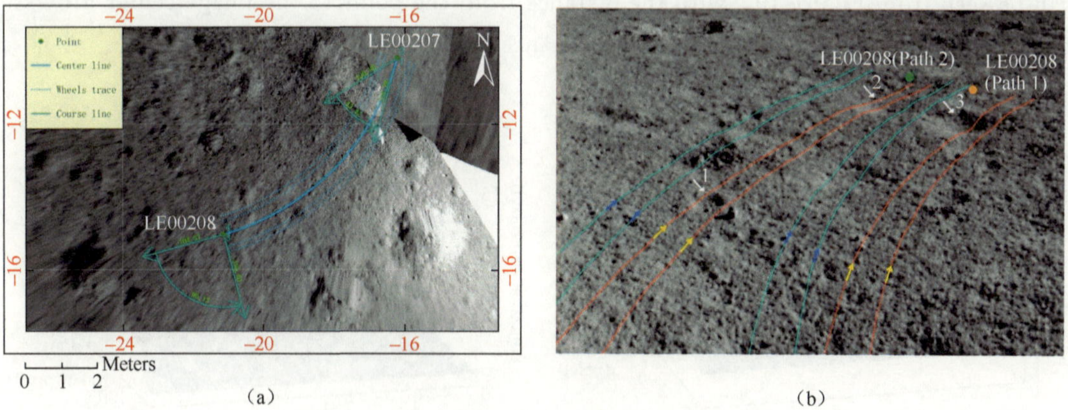

(a)　　(b)

Figure 8　The 2nd planned path and comparison of the two paths: (a) The 2nd planned path at waypoint LE00207; (b) a zoomed - in view of comparisons of two paths at waypoint LE00207 projected on the Navcam image mosaic

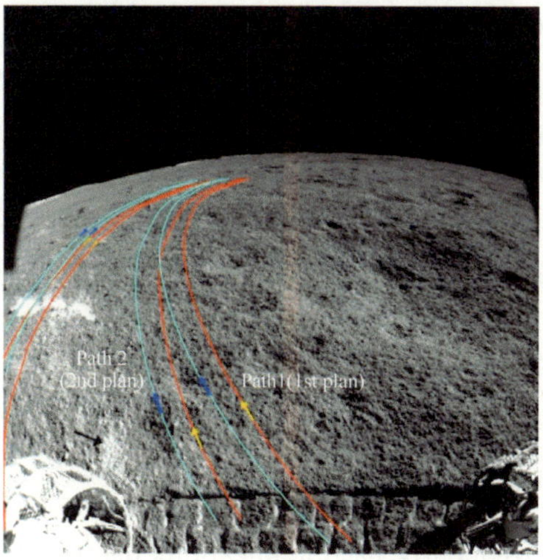

Figure 9　Comparisons of two planned paths at waypoint LE00207 projected on the left Hazcam image

4. 3　Error Analysis Results

According to the error propagation analysis in the previous section, projection error analysis results are shown in Figure 10. For Navcan, since the searched path starts from the current waypoint and generally has a length less than 10m, error from 0 to 10meters is analyszed. For Hazcam, as the effective working distance is shorter, we analyze the error from 0 to 5m. The overall trend in both analysis results shown in Figure 10 indicates that the error grows as the distance from the rover and the drift from the image center grows. From [7, 23], the range measurement error is inversely proportional to the square of the distance. Therefore, in Equation (4) combining two error propagation effects, the projection error is approximately linear inversely proportional to the distance. In addition, in the first error propagation process, the image point position affects the error, where the points farther from the image center get larger error. This effect also propagates in the second process reflected in Equation (4). From the above analyses, as can be seen from Figure 10, with the distance from the point to the image center grows, the error increases, and with the image distance grows the error increases.

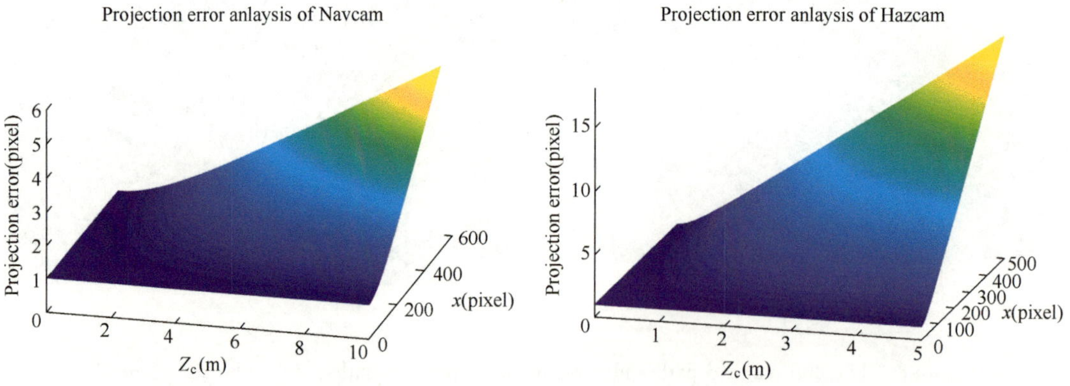

(a)　　(b)

Figure 10　Projection error analyses results for both Navcam (a) and Hazcam (b)

From the above error analyses, it can be seen that under the assurance of high calibration accuracy, the error of the planned path projection is within the acceptable range, and can correctly reflect the searched path. It should be noted that since there is no additional point position error from Navcam images to their stitched image mosaic, the error of projection on the Navcam image mosaic is the same as that on the Navcam image.

5　Discussion

5. 1　Support Other Applications

Inspired by the most frequently used applications described in the above section, by providing proper information, the projection method can be extended and used in more applications. Here introduces two of them implemented in CE - 4 mission.

1. Wheels trace prediction of rover spin

From Figure 8a it can be seen that sometimes the rover needs to make turns at the start or (and) the end points. According to the design of the rover mobility sub - system [17] which contains the locations of the front, middle, and rear wheels, combining with their transformation relationships to the rover center and the cameras, the wheel trajectories can be determined by them and projected together with the searched paths on the Navcam and Hazcam images.

As shown in Figure 8a, the rover would make an anti - clockwise turn of 43.42° at the start point, and an anti - clockwise turn of 90.13° at the end point. Based on the path projection method, the predicted wheel traces of rover spin at the end point of Path 2 are calculated and illustrated in Figure 11. From the figure, it can be seen that the middle wheels are relatively stable, the right front wheels are affected by the terrain in the middle of the turn, and the left rear wheel would pass over the shallow pit. By the guide of these information, the depth of the pit and slopes of the areas are carefully checked and appeared safe according to the rover capabilities.

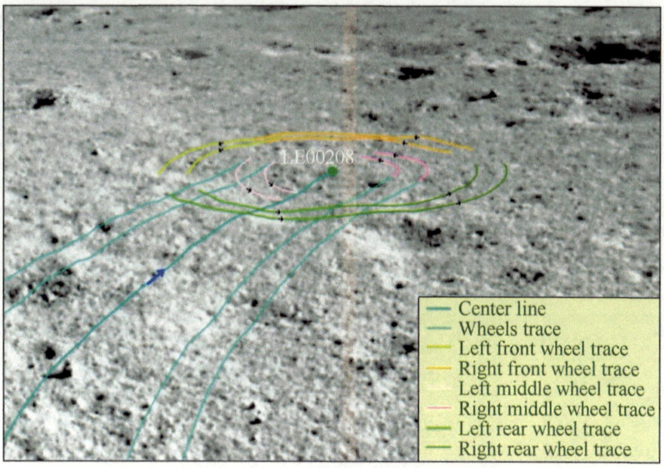

Figure 11　Projection application of wheel trace prediction of rover spin on the Navcam image mosaic

2. Imaging area prediction of other sensors

The projection approach was also extended to implement imaging area prediction for other payloads. One of the typical applications is predicting science target imaging area of Visible and Near - Infrared Imaging Spectrometer (VNIS) onboard the Yutu - 2 rover. The VNIS, consisting of a visible and near - infrared (VNIR) imaging spectrometer and a shortwave infrared (SWIR) spectrometer, aims at investigating the mineralogical compositions of lunar surface materials [25]. The VNIS is fixed on the rover front panel, the FOVs of VNIR and SWIR are 8.5° and 3.6°, respectively [26, 27]. Due to the fixed mounting and small FOVs, approaching and pointing to the science target entirely depend on the movement of the rover [28]. Moreover, sometimes an interesting science target (e. g., a rock) may be very small. Therefore, precise imaging area predictions are crucial to plan and refine the rover path to make sure that the target is within the FOV of the VNIS after the movement.

According to the mounting and calibration parameters of Navcam, Hazcam, and VNIS, the rigid transformation relations, including rotation and translation from one to another, can be calculated. Therefore, the FOV areas of both VNIR and SWIR can be projected on the Navcam and Hazcam images to provide guidance for the rover movement. Figure 12 shows the process of approaching a target at the beginning of the 3rd lunar day since Yutu - 2 started its work. The rover was awaked at waypoint LE00301 where the rock can be observed by Navcam. By a short movement to arrive LE00302, the imaging area of VNIS on Navcam image captured at this waypoint was predicted as shown in Figure 12a, in which the deformed rectangle represents the FOV of VNIR and the deformed circle represents the FOV of SWIR. It can be seen that the SWIR imaging area was only partial on the target. Therefore, by the support of these information, the rover adjusted its position and gesture, and moved to waypoint LE00303. The imaging area of VNIS was predicted again on Hazcam image as this time the target is closer (shown in Figure 12b). Under the guidance of the prediction, the rover completed the target approaching task. The acquired VNIR image is shown in Figure 12c with FOV of SWIR marked as a red circle. The actual captured image indicates that the science target approaching task guided by imaging area prediction was implemented successfully. The VNIR and SWIR data of the rock and other science targets have contributed significantly to mineralogical study, surface composition research and other fields on the lunar far side [29].

(a)　(b)　(c)

Figure 12　VNIS imaging area prediction: (a) prediction result on a Navcam image at waypoint LE00302; (b) prediction result on a Navcam image at waypoint LE00303; (c) the image obtained by VNIS at waypoint LE00303 (the circle represents the FOV of SWIR)

5. 2　Error Source Analysis

From the path evaluation applications introduced in Section 4. 2, it can be found that the projection errors (i. e., deviations from the true projection points) of the searched path at the farthest area from the rover are less than the amount that analyzed in Section 4. 3. This is because the overall terrain has no severe elevation steps, which leads to high matching accuracy of the corresponding points. Moreover, camera parameters with high calibration accuracy assure the precise measurement of the DEM.

However, in the extended application that predicts the imaging area of VNIS, we found the predicted area is a bit different from the obtained image, as shown in Figure 12b, c. The deviation may be caused by one or more of the following reasons: (1) Calibration errors. Although the parameters of the Navcam and Hazcam are calibrated with high precision, the transformation error i. e., from the VNIS coordinate system to the work coordinate system and to the Navcam coordinate system, may combine with the slight camera calibration errors and cause the deviation. (2) Rover localization errors. The position and orientation of the rover are calculated by visual localization method, whose accuracy relies on feature matching and sensor parameters.

According to the above preliminary analysis of the causes, error compensation work can be carried out from the following aspects in the future. (1) At certain waypoints, the cameras can acquire images of the real wheel trajectories of the last few passing waypoints. The searched path can be projected on this kind of images and overlaid with these real wheel trajectories for comparisons. (2) Collect more VNIS imaging area predictions results and real VNIS images for comparisons. By comparison data of the above two aspects, together with more complicated error distribution and propagation simulations, more clues of error correction or compensation will be found out.

6　Conclusions

The paper presents vision - based decision support methods for the generation of obstacle map and evaluation of path - planning results for Yutu - 2 rover in the CE - 4 mission. At each waypoint, 0. 02m resolution DEM and the corresponding DOM are automatically generated by Navcam stereo images, and obstacle map is established based on terrain analysis to support path searching. Based on the camera models and calibrated parameters, the paths and wheel traces are mapped from the 3D coordinates gained from DEM to the 2D Navcam images, Navcam image mosaic, and Hazcam images. The methods provide intuitional visualization of the path projection results, which conforms to the actual observation angle and makes full use of the image resolution of Navcam and Hazcam images.

The vision - based decision support methods and its applications in multiple aspects have been directly used in the CE - 4 mission to support teleoperation of the rover. In addition, we presented some typical and extended applications in this paper. While Yutu - 2 continues exploring the far side of the Moon, the proposed methods are routinely applied to ensure safe traversing and data acquisition. Under the support of the proosed techniques, by the end of the 14th lunar day (3 February 2020), the rover has already travelled 367.25 meters on the far side of the Moon [30].

Acknowledgments

The authors thank the Lunar and Deep Space Exploration Science Applications Center of the National Astronomical Observatories for providing the VNIS data.

References

[1] Di, K.; Liu, Z.; Liu, B.; Wan, W.; Peng, M.; Wang, Y.; Gou, S.; Yue, Z.; Xin, X.; Jia, M.; et al. Chang'e - 4 lander localization based on multi - source data. J. Remote Sens. 2019, 23, 177 - 184.

[2] Ye, P.; Sun, Z.; Zhang, H.; Zhang, L.; Wu, X.; Li, F. Mission design of Chang'e - 4 probe system. Sci. Sin. Technol. 2019, 49, 124 - 137. [CrossRef]

[3] Yu, T.; Fei, J.; Li, L.; Cheng, X. Study on path planning method of lunar rover. J. Deep Space Explor. 2019, 6, 384 - 390.

[4] Liu, Z.; Di, K.; Peng, M.; Wan, W.; Liu, B.; Li, L.; Yu, T.; Wang, B.; Zhou, J.; Chen, H. High precision landing site mapping and rover localization for Chang'e - 3 mission. Sci. China - Phys. Mech. Astron. 2015, 58, 1 - 11.

[5] Helmick, D. M.; Angelova, A.; Livianu, M.; Matthies, L. H. Terrain Adaptive Navigation for Mars Rovers. In Proceedings of the 2007 IEEE Aerospace Conference, Big Sky, MT, USA, 3 - 10 March 2007; pp. 1 - 11.

[6] Carsten, J.; Rankin, A.; Ferguson, D.; Stentz, A. Global Path Planning on Board the Mars Exploration Rovers. In Proceedings of the 2007 IEEE Aerospace Conference, Big Sky, MT, USA, 3 - 10 March 2007; p. 1125.

[7] Peng, M.; Wan, W.; Wu, K.; Liu, Z.; Li, L.; Di, K.; Li, L.; Miao, Y.; Zhan, L. Topographic mapping capbility analysis of Chang'e - 3 Navcam stereo images and three - dimensional terrain reconstruction for mission operations. J. Remote Sens. 2014, 18, 995 - 1002.

[8] Liu, Z.; Wan, W.; Peng, M.; Zhao, Q.; Xu, B.; Liu, B.; Liu, Y.; Di, K.; Li, L.; Yu, T.; et al. Remote sensing mapping and localizatin techniques for teleoperation of Chang'e - 3 rover. J. Remote Sens. 2014, 18, 971 - 980.

[9] Chen, J.; Xing, Y.; Teng, B.; Mao, X.; Liu, X.; Jia, Y.; Zhang, J.; Wang, L. Guidance, Navigation and Control Technologies of Chang'E - 3 Lunar Rover. Sci.

China Technol. Sci. 2014, 44, 461 - 469.

[10] Wang, B.; MAO, X.; Tang, G. A Calibration Method for Fish - Eye Camera of Lunar Rover. J. Astronaut. 2011, 32, 933 - 939.

[11] Zhang, S.; Liu, S.; Ma, Y.; Qi, C.; Yang, H. M. Self calibration of the stereo vision system of the Chang'e - 3 lunar roverbased on the bundle block adjustment. Isprs J. Photogramm. Remote Sens. 2017, 128, 287 - 297.

[12] Zhang, S.; Xu, Y.; Liu, S.; Yan, D. Calibration of Chang'e - 3 Lunar Rover Stereo - camera System Based on Control Field. Geomat. Inf. Sci. Wuhan Univ. 2015, 40, 1509 - 1513.

[13] Zhou, J.; Xie, Y.; Zhang, Q.; Wu, K.; Xia, Y.; Zhang, Z. Research on mission planning in teleoperation of lunar rovers. Sci. China Inf. Sci. 2014, 44, 441 -451.

[14] Wang, Y.; Peng, M.; Di, K.; Wan, W.; Liu, Z.; Yue, Z.; Xing, Y.; Mao, X.; Teng, B. Vision Based Obstacle Detection Using Rover Stereo Images. In Proceedings of the International Archives of the Photogrammetry, Remote Sensing and Spatial Information Sciences ISPRS Geospatial Week, Enschede, The Netherlands, 10—14 June 2019; pp. 1471 - 1477.

[15] Wu, W.; Zhou, J.; Wang, B.; Liu, C. Key technologies in the teleoperation of Chang'E - 3 Jade Rabbit rover. Sci. China Inf. Sci. 2014, 44, 425 - 440.

[16] Li, Q.; Jia, Y.; Peng, S.; Han, L. Top Design and Implementation of the Lunar Rover Mission Planning. J. Deep Explor. 2017, 4, 58 - 65.

[17] Zhang, Y.; Xiao, J.; Zhang, X.; Liu, D.; Zou, H. Design and implementation of Chang'E - 3 rover location system. Sci. China Technol. Sci. 2014, 44, 483 - 491.

[18] Shen, Z.; Zhang, W.; Jia, Y.; Sun, Z. System Design and Technical Characteristics Analysis of Chang'e - 3 Lunar Rover. Spacecr. Eng. 2015, 24, 8 - 13.

[19] Lowe, D. G. Distinctive Image Features from Scale - Invariant Keypoints. Int. J. Comput. Vis. 2004, 60, 91 - 110.

[20] Fischler, M. A.; Bolles, R. C. Random sample consensus: A paradigm for model fitting with applications to image analysis and automated cartography. Commun. ACM 1981, 24, 381 - 395.

[21] Liang, J.; Liu, Z.; Di, K. Construction of a 3D Measurable Virtual Reality Enviroment for Planetary Exploration Applications. Spacecr. Eng. 2012, 21, 28 - 34.

[22] Brown, M.; Lowe, D. G. Automatic Panoramic Image Stitching Using Invariant Features. Int. J. Comput. Vis. 2007, 74, 59 - 73.

[23] Di, K.; Li, R. Topographic Mapping Capability Analysis of Mars Exploration Rover 2003 Mission Imagery. In Proceedings of the 5th International Symposium on

Mobile Mapping Technology (MMT 2007), Padua, Italy, 28 - 31 May 2007.

[24] Wang, Z. Principles of Photogrammetry (with Remote Sensing); Publishing House of Surveying and Mapping: Beijing, China, 1990; pp. 312 - 313.

[25] Jia, Y.; Zou, Y.; Ping, J.; Xue, C.; Yan, J.; Ning, Y. The scientific objectives and payloads of Chang'E - 4 mission. Planet. Space Sci. 2018, 162, 207 - 215.

[26] He, Z. P.; Wang, B. Y.; Lv, G.; Li, C. L.; Yuan, L. Y.; Xu, R.; Chen, K.; Wang, J. Y. Visible and near - infrared imaging spectrometer and its preliminary results from the Chang'E 3 project. Rev. Sci. Instrum. 2014, 85, 083104.

[27] Li, C.; Wang, Z.; Xu, R.; Lv, G.; Yuan, L.; He, Z.; Wang, J. The Scientific Information Model of Chang'E - 4 Visible and Near - IR Imaging Spectrometer (VNIS) and In - Flight Verification. Sensors 2019, 19, 2806.

[28] Liu, Z.; Di, K.; Li, J.; Xie, J.; Cui, X.; Xi, L.; Wan, W.; Peng, M.; Liu, B.; Wang, Y.; et al. Landing Site Topographic Mapping and Rover Localization for Chang'E - 4 Mission. Sci. China - Phys. Mech. & Astron.. accepted.

[29] Li, C.; Liu, D.; Liu, B.; Ren, X.; Liu, J.; He, Z.; Zuo, W.; Zeng, X.; Xu, R.; Tan, X.; et al. Chang'E - 4 initial spectroscopic identification of lunar far -side mantle - Deriv. materials. Nature 2019, 569, 378 - 382.

[30] China's Lunar Rover Travels over 367 Meters on Moon's Far Side. Available online: https://www.chinadaily.com.cn/a/202002/03/WS5e37ce2aa310128217274769.html (accessed on 3 February 2020).

Topographic and Geomorphological Mapping and Analysis of the Chang'E – 4 Landing Site on the Far Side of the Moon

Bo Wu, Fei Li, Han Hu, Yang Zhao, Yiran Wang, Peipei Xiao, Yuan Li, Wai Chung Liu, Long Chen, Xuming Ge, Mei Yang, Yingqiao Xu, Qing Ye, Xueying Wu, and He Zhang

Bo Wu, Han Hu, Yiran Wang, Peipei Xiao, Yuan Li, Wai Chung Liu, Long Chen, and Xuming Ge are with the Department of Land Surveying and Geo – Informatics, The Hong Kong Polytechnic University, Hung Hom, Kowloon, Hong Kong (bo.wu@polyu.edu.hk).

Fei Li, Yang Zhao, Mei Yang, Yingqiao Xu, Qing Ye, Xueying Wu, and He Zhang are with the China Academy of Space Technology, 104 Youyi Street, Haidian, Beijing, P. R. China (kuokuolee@163.com).

Abstract: The Chinese lunar probe Chang'E – 4 successfully landed in the Von Kármán crater on the far side of the Moon. This paper presents the topographic and geomorphological mapping and their joint analysis for selecting the Chang'E – 4 landing site in the Von Kármán crater. A digital topographic model (DTM) of the Von Kármán crater, with a spatial resolution of 30m, was generated through the integrated processing of Chang'E – 2 images (7 m/pixel) and Lunar Reconnaissance Orbiter (LRO) Laser Altimeter (LOLA) data. Slope maps were derived from the DTM. Terrain occlusions to both the Sun and the relay satellite were studied. Craters with diameters ≥ 70m were detected to generate a crater density map. Rocks with diameters ≥ 2 m were also extracted to generate a rock abundance map using an LRO narrow angle camera (NAC) image mosaic. The joint topographic and geomorphological analysis identified three subregions for landing. One of them, recommended as the highest – priority landing site, was the one in which Chang'E – 4 eventually landed. After the successful landing of Chang'E – 4, we immediately determined the precise location of the lander by the integrated processing of orbiter, descent and ground images. We also conducted a detailed analysis around the landing location. The results revealed that the Chang'E – 4 lander has excellent visibility to the Sun and relay satellite; the lander is on a slope of about 4.5° towards the southwest, and the rock abundance around the landing location is almost 0. The developed methods and results can benefit future soft-landing missions to the Moon and other celestial bodies.

1 Introduction

On 3 January 2019, the Chinese lunar probe Chang'E-4, carrying the Jade Rabbit-2 lunar rover, successfully landed in the Von Kármán crater in the northwestern South Pole-Aitken (SPA) basin on the far side of the Moon (Figure 1). Chang'E-4 was the first spacecraft to make a soft landing on the lunar far side. The Chang'E-4 lander and the Jade Rabbit-2 rover are now exploring the surface and subsurface of the Von Kármán crater's mare-covered floor with their onboard scientific instruments. Their tasks include investigating the compositions of mare basalt and the subsurface structure of the regolith of this far side region (Wu et al. 2017), and finding clues that could provide insight into the early geologic history of the Moon (Wilhelms, John, and Trask 1987; Huang et al. 2018; Li et al. 2019; Di et al. 2019).

Chang'E-4 is part of the second phase of the Chinese Lunar Exploration Program (Wu et al. 2017; Jia et al. 2018). This probe is a former backup of the Chang'E-3 probe (Jia et al. 2018), which made a successful lunar landing in the Sinus Iridum region in the Mare Imbrium on 14 December 2013 (Wu et al. 2014a). Unlike the landing site of the Chang'E-3, the targeted landing region of Chang'E-4 is on the far side of the Moon; therefore, a relay satellite was sent to the Earth-Moon Lagrangian Point 2 (L2) halo orbit in May 2018, which serves as a communication relay station between Earth and the lunar far side (Jia et al. 2018).

The selection of the Chang'E-4 landing site includes several steps. In the first step, two craters located inside the SPA basin, including the Von Kármán crater and the nearby Chretien crater, were selected. This is because a similar latitude to the Chang'E-3 landing site was preferred (i. e., at about 45°S in latitude) and due to the temperature and communication constraints. At the latitude of about 45°S inside the SPA basin, the Von Kármán crater and Chretien crater are superior to others in terms of crater size and roughness of the crater floor. In the second step, two candidate landing regions located inside the Von Kármán crater and Chretien crater, each covering an area of about 50×30 square kilometers, were selected for detailed analysis in terms of surface elevations, terrain slopes, and terrain occlusions to sun illumination and telecommunication. The candidate landing region located in the southeast floor of the Von Kármán crater, outlined in Figure 1b by the green box, was chosen based on its superiority in all these aspects.

To ensure a safe and successful landing, a careful analysis of the topography and geomorphology (De Rosa et al. 2012; Wu et al. 2014a) of the targeted landing region for further selecting a suitable landing site is critical. Topographic features, such as surface slopes and occlusions, along with geomorphological features such as the distributions of craters and boulders/rocks, were mapped and analyzed to identify the best suitable site for

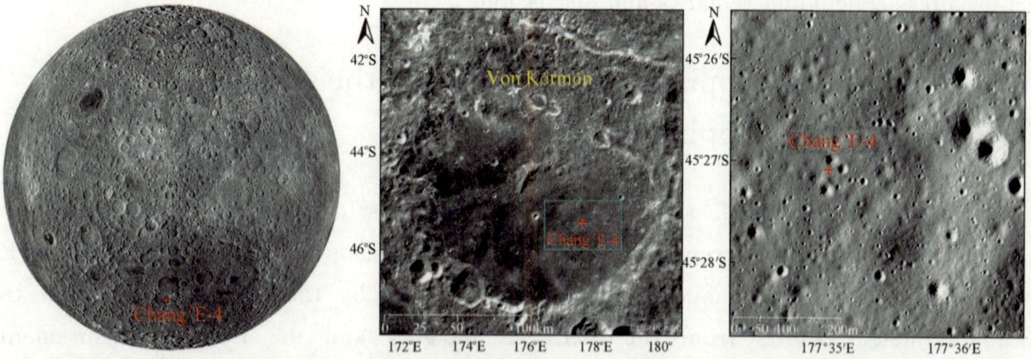

(a)　　　　　　　　(b)　　　　　　　　(c)

Figure 1　The Chang'E-4 landing site. (a) The landing site shown in a global view of the far side of the Moon; (b) an enlarged view of the landing site inside the Von Kármán crater shown in a Chang'E-2 image, with the green box indicating the targeted landing region; and (c) a further enlarged view of the landing site shown in a LRO NAC image

landing. In this research, we created digital topographic models (DTMs) of the Von Kármán crater at a spatial resolution of 30m through the integrated processing of Chang'E-2 images and Lunar Reconnaissance Orbiter (LRO) Laser Altimeter (LOLA) data (Wu, Hu, Guo 2014b). Based on the DTMs, we scrutinized the surface slopes and terrain occlusions to the solar illumination and the relay satellite. Based on the Chang' E-2 images and the high-resolution LRO narrow angle camera (NAC) images, we analyzed the crater distribution and rock abundance inside the targeted landing region. The topographic and geomorphological features were then analyzed jointly to identify three subregions for landing. Chang' E-4 eventually landed in the subregion of the highest priority on 3 January 2019. After the landing, we precisely located the lander by using the images collected both in orbit and during the descent phase, together with a ground panorama collected by the lander's overhead camera upon landing. A detailed terrain occlusion analysis around the landing location was performed, and a 1.5m/pixel DTM was generated using the shape-from-shading (SfS) approach (Wu et al. 2018a). This made it possible to conduct detailed topographic analyses to support the surface operations of the lander and rover.

This paper focuses on the targeted landing region inside the Von Kármán crater. It is organized as follows. The next section describes the methods and results regarding topographic mapping, with the topic of geomorphological mapping reserved for the section "Geomorphologic Mapping and Analysis of the Landing Region". The joint topographic and geomorphological analysis with the purpose of identifying subregions for landing are presented in the subsequent section. The section "Landing Site Localization and Analysis" includes further results regarding the precision localization of the lander and a detailed analysis of the surface slopes and occlusions around the landing site. Finally, the last section offers concluding remarks and discussion.

2 Topographic Mapping and Analysis of the Landing Region Topographic Mapping

High - resolution DTMs are the vital data resources for topographic analysis of the targeted landing region. Currently, the Selenological and Engineering Explorer (SELENE) and LRO digital elevation model (SLDEM) generated by the co - registration of the photogrammetric results from the SELENE imagery and the LOLA measurements (Barker et al. 2016), provides full coverage of the targeted landing region at a spatial resolution of about 60 m. The SLDEM offers a typical vertical accuracy of 3 to 4 m. The SLDEM has favorable geometric accuracies but sparse spatial resolution. In this research, we used the Chang' E - 2 images (7m/pixel) covering the Von Kármán crater to generate a DTM with a spatial resolution of 30 m by the integrated processing with the LOLA measurements (Wu, Hu, Guo 2014b).

The Chang' E - 2 camera consists of two linear push broom sensors with a convergence angle of about 25° that are used to formulate an along - track stereo configuration of the images (Wu, Hu, Guo 2014b). There is an overlap of about 50% between the images in the adjacent tracks. We developed a combined block adjustment approach for the integrated processing of the multiple - track Chang'E - 2 images and the LOLA measurements to generate high - precision DTMs. By doing so, inconsistencies can be reduced between the multipletrack Chang' E2 images, and between the three - dimensional (3D) measurements from the Chang' E - 2 images and the LOLA measurements. This approach was used for the integrated processing of the Chang' E2 images and LOLA measurements covering the Von Kármán crater and thus improved the orientation parameters of Chang'E - 2 images. Dense image matching was carried out on the stereo Chang' E - 2 images to generate dense matches, using a texture - aware semiglobal matching algorithm (Hu et al. 2016). 3D point clouds were then generated from the matching results based on the improved image orientation parameters, and finally, a DTM of the region with a spatial resolution of 30m was interpolated from the 3D point clouds. The generated DTM was compared with the SLDEM covering the same region. The results indicated that the former has a degree of geometric accuracy similar to the latter but offers better spatial resolution (Hu and Wu 2019).

Figure 2a shows a 3D view of the generated DTM covering the Von Kármán crater. The right column shows an enlarged view of the targeted landing region as marked with a red box in the left column. Based on the generated DTM, the Von Kármán crater measures about 180km in diameter, with a rimto - floor depth of about 5km. The crater floor has a diameter of about 140km, and there is a central peak of about 1.55km in height in the middle part of the crater. Within the targeted landing region, the terrain surface has a minimum height of − 6076m and a maximum height of − 5776m, respectively, with respect to a reference sphere of radius 1737. 4 km. Figure 2b shows the generated ortho − image mosaics. Based on the generated DTM, the Chang'E − 2 images were ortho − rectified and mosaicked to generate an orthographic image mosaic covering the Von Kármán crater, as shown in the left column of Figure 2b, which has the same resolution of 7 m/pixel as the original Chang'E − 2 images. Inside the targeted landing region, LRO NAC images were collected to generate an image mosaic with a higher spatial resolution (1. 5 m/pixel), as shown in the right column of Figure 2b.

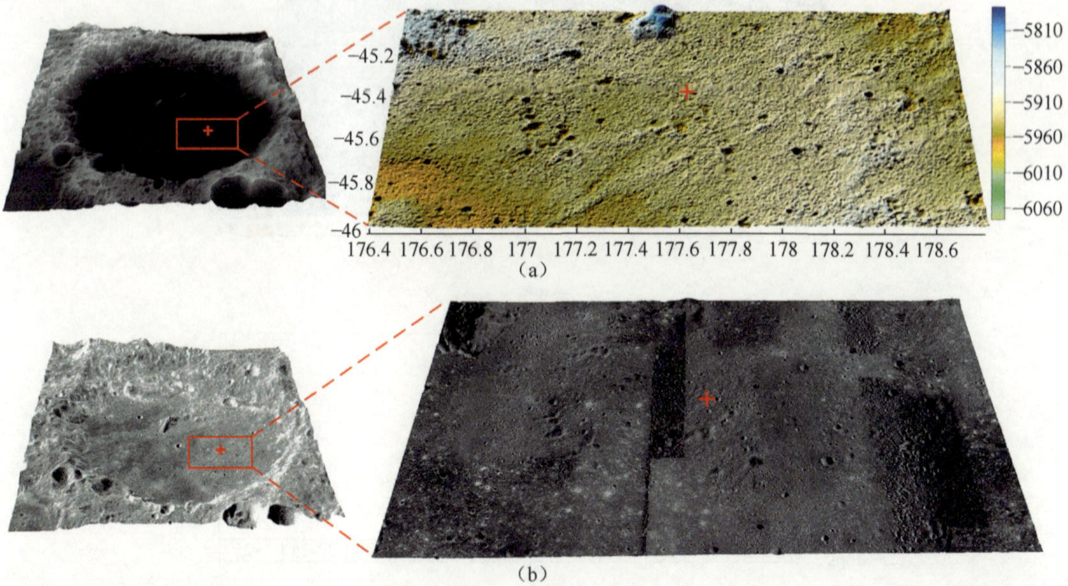

Figure 2 Topographic mapping results of the Von Kármán crater. (a) 3D view of the DTM (30m/pixel), (b) 3D view of the orthographic image mosaics projected on the DTM. The left column shows the Chang'E − 2 image mosaic (7m/pixel), and the right column shows the LRO NAC image mosaic (1. 5m/pixel) in the targeted landing region. The red cross indicates the landing location

Slope Analysis

Surface slope is an important engineering constraint in landing site evaluation to ensure the stability of the lander during touchdown, which is the magnitude of the derivatives of the gridded topographic model (i. e. , the DTM) at a specific baseline. With two grid cells required for slope calculation along one direction, the corresponding baseline for slope analysis is two times the spatial resolution of the DTM, which is 60m for the DTM generated from the Chang'E − 2 images. For any point in the DTM, slopes along the east − west and north − south directions are calculated based on the distance and elevation difference of the two neighboring grid cells, and the square root of the sum of their squares is calculated as the maximum slope at that point. The slope calculation method is consistent with those provided in popular software systems, such as

ArcGIS. After the slopes of all the points in the DTM have been calculated, a slope map can be generated. Figure 3 shows the slope map and statistics of the targeted landing region at a baseline of 60m. It can be seen that most of the region (99.7%) has slopes of less than 15°, and 96.9% of the region has slopes of less than 8°. Only 3.1% of the region has slopes larger than 8°, and those areas are mainly distributed inside some craters.

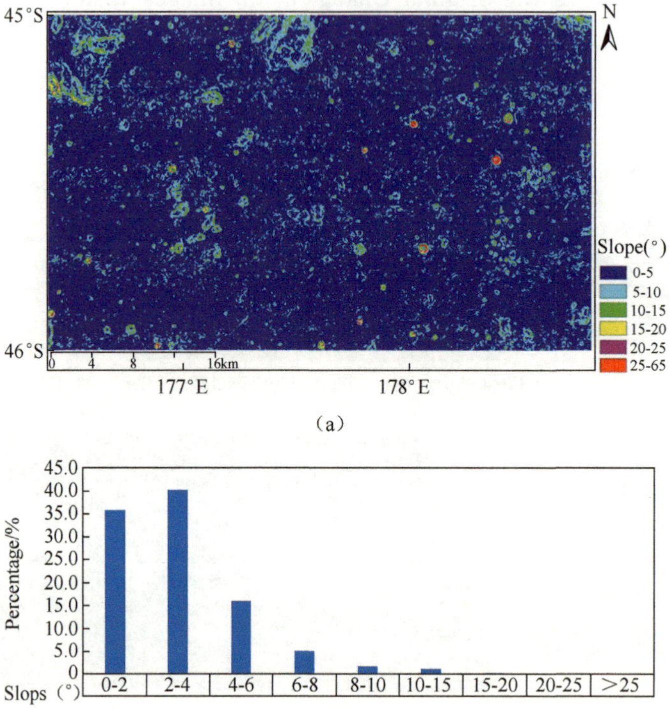

Figure 3 Slope map and statistics of the targeted landing region. (a) The slope map at a baseline of 60m and (b) histogram of the slope statistics

A baseline of 8m is further imposed for slope analysis, considering the lander footprint size of about 4.5m plus a buffer for attitude adjustment of the lander before touchdown. However, slope analysis at an 8 - m baseline requires a higher resolution DTM. There is only one stereo pair of LRO NAC images covering a small portion of the middlenorth part of the landing region that could be used to generate a DTM of 4 m resolution. No high - resolution DTMs are available for other areas inside the landing region. In our previous work (Wang and Wu 2017), we presented a correlation analysis of slopes derived from the same terrain surface with different baselines. The results indicated that it is possible to estimate slopes at short baselines from slopes calculated from relatively longer baselines. The same method was used here for the estimation of slopes at different baselines. The 4 - m resolution DTM from the stereo NAC images and the corresponding 30 - m resolution DTM from the Chang'E - 2 images, both covering the same local area inside the landing region, were used for correlation analysis of slopes at baselines of 8m and 60m. A slope amplification function with multipliers in the range of [1. 1, 1. 6] was obtained from the correlation analysis in this local area, which was then used for estimation of slopes at 8 - m baseline in other areas inside the landing region, based on the assumption that the influences of different baselines on slopes should be similar for this local area and its nearby region. Therefore, we also compute an aggressive (the lower amplification bound of 1. 1) and a conservative (the upper amplification bound of 1. 6) estimation of the slopes at the 8m baseline based on the original slopes at 60m baseline. In the case of conservative estimations, about 3. 4% of the areas have slopes larger than 8°, and this value only rises to about 5%, even in the aggressive case. Similarly, the areas with larger slopes are located inside the craters.

Occlusion Analysis

As the targeted landing region is on the floor of the Von Kármán crater and the height difference from the floor to the crater rim is about 5km, it is crucial to analyze the possible terrain occlusions to solar illumination and communication signals from the relay satellite. The lander and rover rely on solar energy to charge the batteries of the onboard instruments. Also, communications and data transfer between the Earth control center and the lander require that the latter stays in the sight of the relay satellite. Therefore, areas with maximum visibility of the Sun and the relay satellite, that is, areas that feature the least amount of occlusion from the terrain, are preferred when evaluating landing site suitability.

In the analysis of the occlusion conditions caused by the terrain, a search radius of 200km surrounding the targeted landing region is considered, to enclose the entire rim of the Von Kármán crater and some high mountains outside the crater that may cause occlusions. Due to the long distances involved when calculating terrain occlusion angles, the curvature of the lunar surface cannot be ignored. An approach based on the geodesic line (Karney 2013) was developed to analyze the altitude angles of terrain occlusions, as illustrated in Figure 4. For each target point A (λ_1, φ_1) at longitude λ_1 and latitude φ_1 in the landing region, a geodesic line l is determined by a specific azimuth angle α. Given a specific distance s_{12}, the geographic coordinates of the search point B (λ_2, φ_2) can be determined based on the geodesic line (Karney 2013). The height of point B can be sampled from the DTM. Then, the 3D coordinates of B can be projected to a local Cartesian coordinate system of the target point A, in which the altitude and distance of B, with respect to the target point A, can be determined, and an altitude angle θ can be calculated to represent the terrain occlusion of point B to A. The process just described is repeated for all of the azimuth angles and the entire search radius of 200km. The intervals for the azimuth angle α and geodesic distance s_{12} are set to be 1° and the spatial resolution of the DTM, respectively, considering the accuracy in capturing the topographic features and the processing time. Next, the maximum occlusion angle for point A along each azimuth direction can be determined, and the maximum occlusion angle among all the azimuth directions from 0° to 360° can be obtained. This procedure is further repeated for each grid point in the targeted landing region to generate the terrain occlusion map shown in Figure 5.

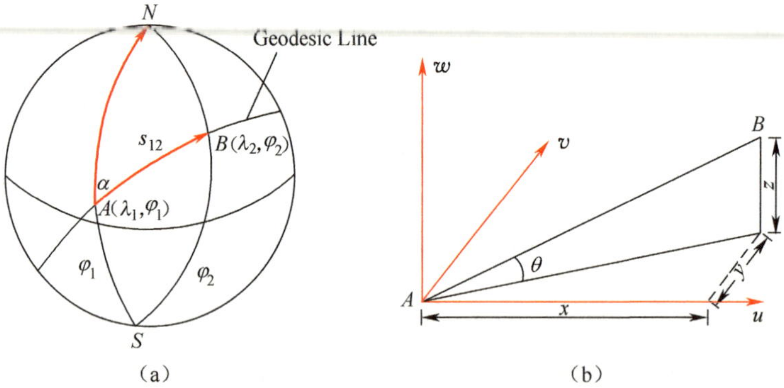

Figure 4　Computation of terrain occlusions based on the geodesic line. (a) The terrain occlusion is searched along the geodesic line for a specific target point A and sampled by a uniform interval. (b) The searched point B is projected to the local Cartesian coordinate system of target point A and the altitude angle of terrain occlusion is calculated

Figure 5 shows the terrain occlusion maps of the targeted landing region. Figure 5a shows the maximum altitude angle of terrain occlusion for each point inside the region, and Figure 5b shows the corresponding azimuth of the maximum altitude angle for each point. For most of the flat areas inside the landing region, the maximum terrain altitude angle is less than 10°. Also, most of the maximum altitude angles are caused by the rims of the Von Kármán crater in the southeast and east parts, as indicated by the azimuth map of terrain occlusion in Figure 5b. In the targeted landing region located in the southeast part of the crater floor, as shown in Figure 1, the southeast and east parts of the crater rim create larger terrain occlusions because they are closer to the landing region. In contrast, terrain occlusions in the northwest corner of Figure 5b show that the terrain occlusions in this part come from the northwest direction, either from the rim of the Von Kármán crater far to the northwest at a far distance or from the nearby central peak of Von Kármán crater, also located in the northwest of the landing region. The large terrain occlusions indicated in red or yellow in Figure 5a are inside the craters and along the slopes of the hills.

For a period of six months after the landing of Chang'E-4, from 3 January to 3 July 2019 (the designed working period of Chang'E-4), the solar azimuth and altitude of each point inside the landing region can be obtained from ephemeris information and compared with the altitude angle of the terrain occlusion along the same azimuth. The solar occlusion conditions are derived for each point by summarizing the possible occlusions of

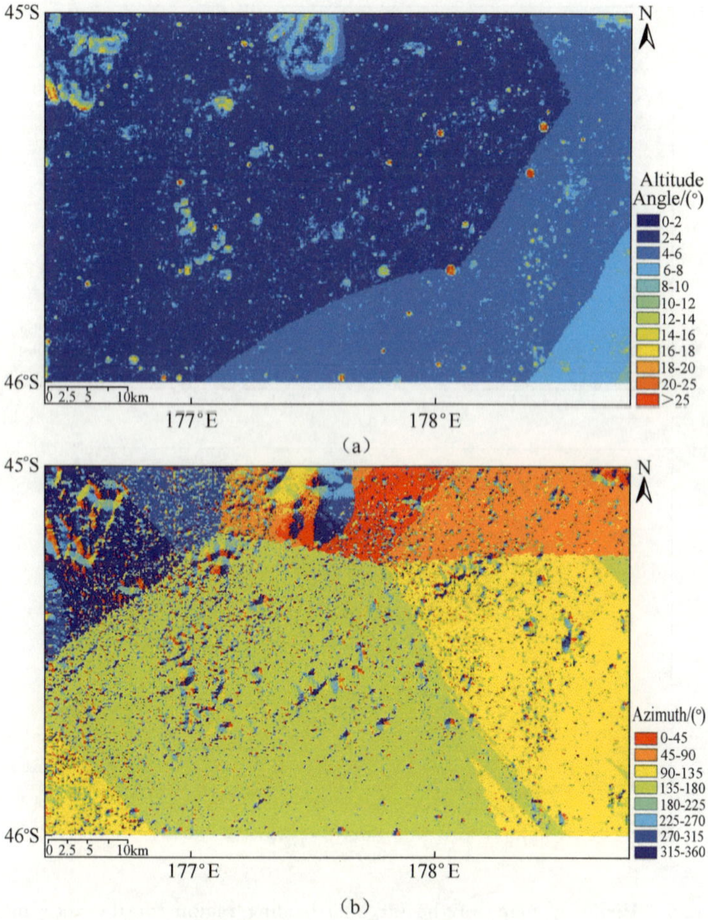

Figure 5　Terrain occlusion maps of the targeted landing region. (a) The maximum altitude angles of terrain occlusions for each point in all azimuth directions, (b) the corresponding azimuth angles for the maximum occlusion of each point

all the solar azimuth directions. Figure 6a presents the resulting percentage of visibility of the Sun during the six - month period over the entire region, with red representing occlusions to the Sun. It should be noted that the lander and rover hibernate during the lunar night and they will automatically wake up when the solar altitude angle is greater than 10°; therefore, only the time slots with solar altitude angles greater than 10° (43.7% of the designed working period of six months) were considered in the above occlusion analysis. It can be seen from Figure 6a that a majority (99.8%) of the landing region has 100% visibility of the Sun during the six - month period (excluding the time when the solar altitude angles are ≤ 10°); however, in some areas, especially inside small craters and near hills, the solar visibility is as low as 86%.

Figure 6b presents the terrain occlusions to the relay satellite. As mentioned earlier, a relay satellite was sent to the Earth - Moon L2 point and stayed there in a halo orbit to

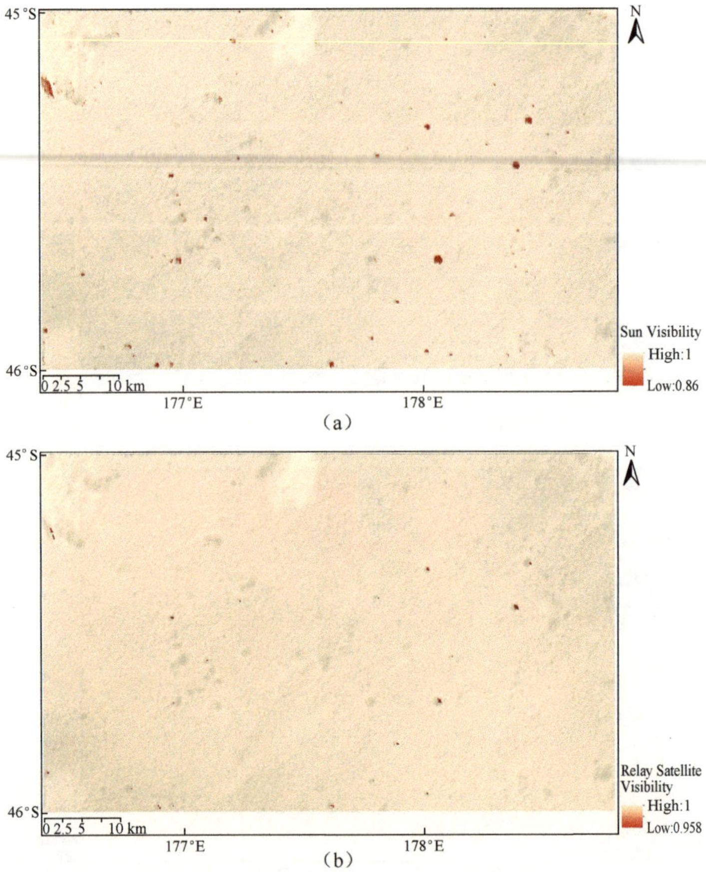

Figure 6　Visibility maps of the targeted landing region for the six - month period after the landing of Chang'E - 4 from 3 January to 3 July 2019.
(a) Visibility map to the Sun and (b) visibility map to the relay satellite

provide continuous communication between the Earth control center and the Chang'E - 4 lander on the far side of the Moon. For the same period of six months after the landing of Chang'E - 4, the azimuth and altitude of the relay satellite with respect to each point inside the landing region were estimated from the orbit design of the relay satellite. An occlusion map to the relay satellite for the entire landing region was generated following the same process as for occlusion to the Sun, as shown in Figure 6b. As before, the time slots with solar altitude angles less than 10°, when the lander and rover hibernate, were ignored in the occlusion analysis. From Figure 6b, the relay satellite is 100% visible over the majority (99.98%) of the landing region during the six - month period of interest. Nevertheless, there are several places, shown in red, where the relay satellite is visible only for 95.8% of the time. Fortunately, these places are rare and are distributed mainly inside a few craters or around the foot of the hill in the northwest of the landing region. The occlusions to the relay satellite are less severe compared with the occlusions to the Sun, as can be noted from Figure 6. This is reasonable considering that the altitudes of the relay satellite, with respect to the targeted landing region, are consistently larger than ～25° by the design of the halo orbit.

3 Geomorphologic Mapping and Analysis of the Landing Region Crater Distribution

Craters are common geomorphologic features on the Moon.

Craters are considered possible hazards in landing site evaluation due to their internal slopes and protuberant rims. Therefore, a lower crater density is preferred in selecting the landing site. Also, Chang'E - 4 had to avoid landing in any small craters (a few meters to dozens of meters in diameter) to allow the maximum maneuvering capability of the rover (Brady et al. 2009; Wu et al. 2014a).

We developed an active machine - learning algorithm (Wang and Wu 2019) for automatic crater detection in the targeted landing region. The 7 m/pixel Chang'E - 2 image mosaic was used for crater detection inside the landing region. We aimed to identify all craters larger than 70m in diameter, which is equivalent to 10 pixels on the image, as the performance of our algorithm decreases for craters smaller than 10 pixels in diameter. It should be noted that the LRO NAC image mosaic was not used for crater detection here. Even though it offers higher resolution, LRO NAC imagery has varying illumination conditions and unbalanced image contrasts, as shown in the right column of Figure 2b, which will influence the crater detection and result in biases in crater density analysis across the landing region. By contrast, the Chang'E - 2 image mosaic has consistent illumination and contrast conditions, as can be seen in the left column of Figure 2b. After the automatic crater detection on the Chang'E - 2 image mosaic using the algorithm (Wang and Wu 2019), a manual checking process was further carried out to ensure the reliability of the detected craters, by digitizing missing craters and removing artifacts with the assistance of a grid with a cell size of 70m × 70m overlaid on the image mosaic. Ultimately, 9391 craters with diameters ranging from 70m to 1405 m were detected in the landing region. Figure 7a shows the distribution of those detected craters.

To understand the crater distribution in the targeted landing region, a crater density map depicting crater numbers in a unit area was generated, as shown in Figure 7b. The map was obtained by using a moving circular window with an area of $10km^2$ to count the numbers of craters. Smaller density represents a sparser distribution of craters and thus a relatively safer situation for landing. Figure 7b shows that the eastern part of the targeted landing region has relatively higher crater density compared with the central and western parts. The average crater density in the targeted landing region is about 60 craters over a circular window of $10km^2$ area.

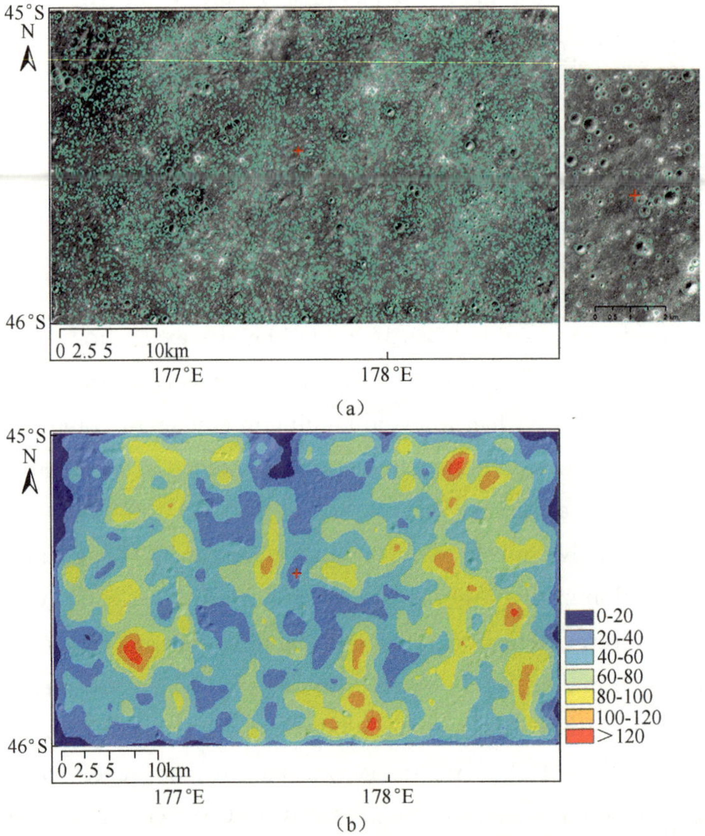

Figure 7　(a) Distribution of craters (≥ 70m in diameter) in the targeted landing region. The red cross indicates the landing location, and the inset figure on the right shows an enlarged view of the detected craters around the landing location. (b) The crater density map in terms of crater numbers calculated using a moving circular window

Rock Abundance

Rocks are another type of major feature on the lunar surface that can be hazardous for the lander and rover. Rock abundance, defined as the cumulative fractional area covered by rocks with respect to their diameters, is another crucial factor in evaluating the suitability of finding a landing site for any landing mission (Golombek and Rapp 1997; Golombek et al. 2003; De Rosa et al. 2012; Wu et al. 2018b). For example, one criterion for selecting the landing site of National Aeronautics and Space Administration's (NASA's) In Sight mission on Mars was that the rock abundance had to be less than 10% (Golombek et al. 2017); and for the ExoMars mission of the European Space Agency, the rock abundance must be less than 7% (Pajola et al. 2017). In this research, a rock abundance constraint of 7% was considered in landing site evaluation given the ability of the lander and rover to tolerate rocks during their maneuvering.

We developed a rock detection algorithm (Li and Wu 2018) to extract rocks from the LRO NAC image mosaic. The algorithm detects rocks based on their particular brightness distributions and relief properties presented in the image. The details pertaining to this rock detection algorithm can be found in our previous work. Using the algorithm, a total of 21 004 rocks were identified from the LRO NAC image mosaic (1.5m/pixel) covering the targeted landing region. The minimum size of rocks that were detected is 2m (the longest diameter of the rock boundary), and the maximum size is 17.24m. The detected rocks were manually checked by two independent operators to further ensure the reliability of rock detection. The distribution of the final detected rocks is shown in Figure 8a.

Figure 8a shows the distribution of rocks larger than 2m within the targeted landing region. However, for landing site assessment, we will need to understand the overall distribution of rocks of all sizes. A rock abundance model developed by Li and Wu (2018) is used here, which allows us to infer the overall distribution of rocks of all sizes from rocks above a particular size. The rock abundance model has the following exponential format:

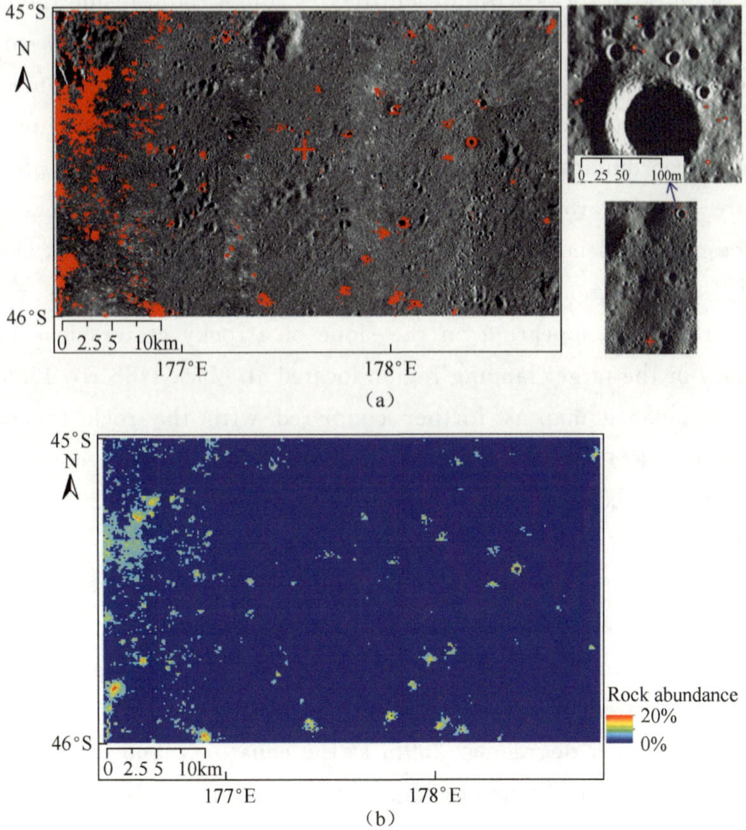

Figure 8　Rock abundances in the targeted landing region. (a) Distribution of rocks larger than 2m in diameter overlaid on the LRO NAC image mosaic. The red cross indicates the landing location, and the inset figures on the right show enlarged views of the detected rocks. (b) The rock abundance map showing the cumulative fractional area of rocks in the region

$$F_k(D)=ke^{-\left(0.5648+\frac{0.01285}{k}\right)D} \tag{1}$$

where D is the rock diameter, k is rock abundance, and $F_k=$ (D) is the fractional area covered by rocks with diameters larger than D. Based on the rock detection result, the rock abundances were estimated using the above rock abundance model to generate a rock abundance map of the targeted landing region, as shown in Figure 8b. It should be noted that the rock abundance model can only be used to estimate the overall distribution of rocks in an area from the size-frequency distribution of the detected rocks. If no rock is detected from the image in a local area, this either indicates that the area is indeed rock-free, or there might be smaller rocks, but they are ignored. The latter case requires higher-resolution image data for checking the existence of smaller rocks. Nevertheless, the rock abundance map generated from the method above can still be used to analyze the relative rock abundances of different local areas in the targeted landing region.

The rock abundance map shown in Figure 8b was generated by calculating the fractional area of detected rocks within each tile of 250m×250m inside the landing region and by further extrapolating the rock abundance values for each tile based on Equation 1. The tile size of 250m was selected considering its statistical significance (Li and Wu 2018) and meaningfulness for comparing with LRO Diviner Radiometer data, as described below. According to the rock abundance map and the rock distribution map, as shown in Figure 8, most rocks are located around rocky impact craters and slopes of hills. Within the targeted landing region, 99.9% of the region has a rock abundance of < 7%, and 97.1% of the region has a rock abundance of < 1%. The highest rock abundance in the region is 18.21%, appearing on the slope of a rocky crater close to the south-western boundary of the target landing region located at (176.4488°E, 45.8222°S).

Our rock abundance map is further compared with the rock concentration map obtained from the LRO Diviner Radiometer data. Based on the contrasting thermal conductivities between rocks and lunar regolith, Bandfield et al. (2011) derived a rock concentration map (http://ode.rsl.wustl.edu/moon/index.aspx) using the LRO Diviner Radiometer data. The same reference system (Mean Earth/Polar Axis) and the same map projection (Mercator) were used for the Diviner map and the LRO NAC image mosaic, and we found that the two maps matched well in general by referring to the locations and shapes of some big craters commonly visible on them. The Diviner map has a spatial resolution of 128 pixels per degree (~250m at the equator), and each pixel presents an aerial fraction of rocks with diameters larger than ~1m. To compare with the Diviner map, the areal fraction of rocks ≥ 1m was extrapolated from our detected rocks (≥ 2m in diameter), which were then binned into pixels of the same scale of the Diviner bins, using the rock abundance model as shown in Equation 1. The differences between our results and the Diviner map in the landing region are shown in Figure 9a. The average absolute difference between our map and the Diviner map is 0.25%. The greatest positive difference is 12.8%, and the greatest negative difference is −2.2%, as shown in Figure

9b. The first row in Figure 9b shows the case in which the greatest positive difference appears. The fraction area of rocks (≥1m) derived from our rock detection result is as high as 15.1%, while the Diviner result is only 2.3%, leading to a positive difference of 12.8%. Our higher value is more reasonable than the Diviner result considering that the areal fraction of the detected rocks (≥2m) already reaches 8.2%. The reason for this kind of discrepancy might be that craters with various slopes may show a range of temperatures with the Diviner measurements, which may consequently lead to uncertainties and errors (Li and Wu 2018; Bandfield et al. 2011). The second row in Figure 9b shows the greatest negative difference of −2.2%. This type of discrepancy can be attributed to the unfavorable illumination conditions in which rocks cannot be identified in the dark shadows. Fortunately, such shadowed areas cover less than 1% of the landing region.

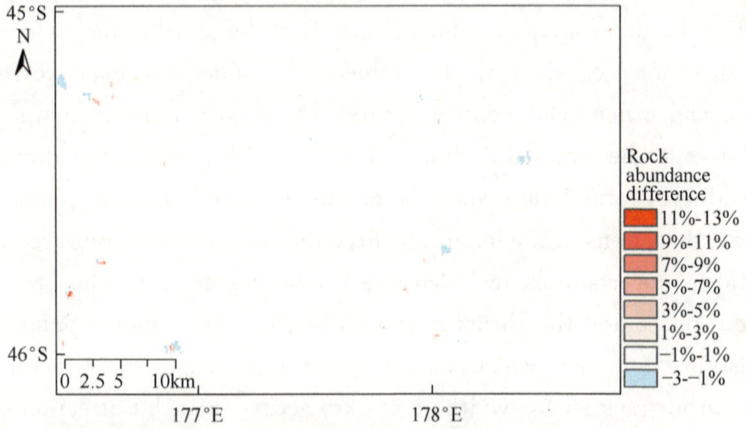

(a)

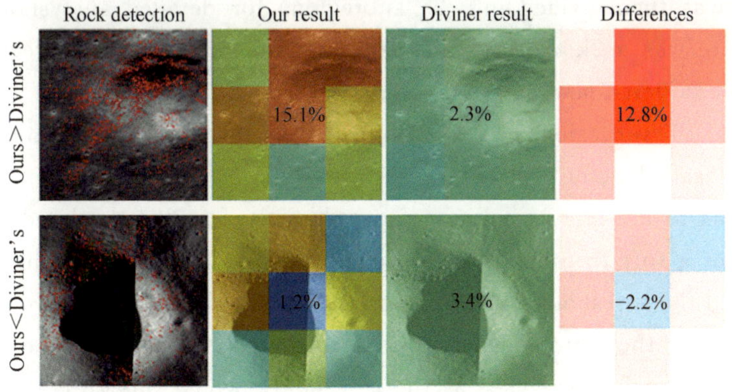

(b)

Figure 9　Comparison between the rock abundances from our results and the LRO Diviner Radiometer data in the targeted landing region. (a) Difference map of rock abundances (ours minus the Diviner map) and (b) enlarged views of examples for detailed comparison

4 Joint Topographic and Geomorphological Analysis for Landing Site Selection

The previously described topographic factors, including surface slopes and occlusions, and the geomorphologic factors, including crater distribution and rock abundance, are critical considerations when selecting suitable landing sites for Chang'E-4 to avoid hazards during touchdown and ensure favorable maneuverability of the robotic rover. Therefore, we carried out a joint topographic and geomorphologic analysis to identify subregions inside the targeted landing region that could be favorable for a safe landing.

The Chang'E-4 lander had an orbital inclination angle ranging from 80° to 100° from the east by orbit design dependent on the launching window. There were various uncertainties in orbit determination which might affect the actual landing (Li et al. 2017). In the longitude direction, considering the influence of midcourse orbit correction, near-Moon braking, and other orbit control errors, there were certain errors in the orbital inclination relative to the nominal design, which would be propagated along the longitude direction. In addition, simulation analysis of the powered descent phase also revealed kilometer-scale deviations in the longitude direction caused by the powered descent. In the latitude direction, uncertainties included the prediction error of the start point of the powered descent phase and the thrust deviation of the 7500N motor onboard the lander, which were also in kilometer scale. Considering the above engineering factors, a rectangle centered at the orbit track with a width of ±5km across the orbit direction and a length of ±15km along the orbit direction was defined as a subregion for landing. The targeted landing region was then divided into 25 subregions for detailed analysis, as shown in Figure 10a. There was a 2km offset between any two neighboring subregions, and the areas beyond the targeted landing region were ignored in the analysis.

For all the 25 subregions, we examined five specific criteria to evaluate their suitability for a safe landing, including surface slopes, terrain occlusions to the Sun, terrain occlusions to the relay satellite, crater densities, and rock abundances. For each criterion, we set a soft threshold and scored the subregion based on the percentage of areas that exceed the threshold. For the slope threshold, 8° was applied on the slopes at the 8m baseline. For the crater density, the average crater density of the landing region plus its standard deviation was used as the threshold, which was 78 craters (≥ 70m in diameter) per $10km^2$. For the rock abundance, 7% was used as the threshold. For the occlusions to both the Sun and relay satellite, we directly counted the percentages of areas with any occlusions during the six-month period after landing with respect to the total area of the subregion. It should be noted that some of these threshold parameters (e.g., slopes and rock abundance) were considered mainly based on the engineering constraints

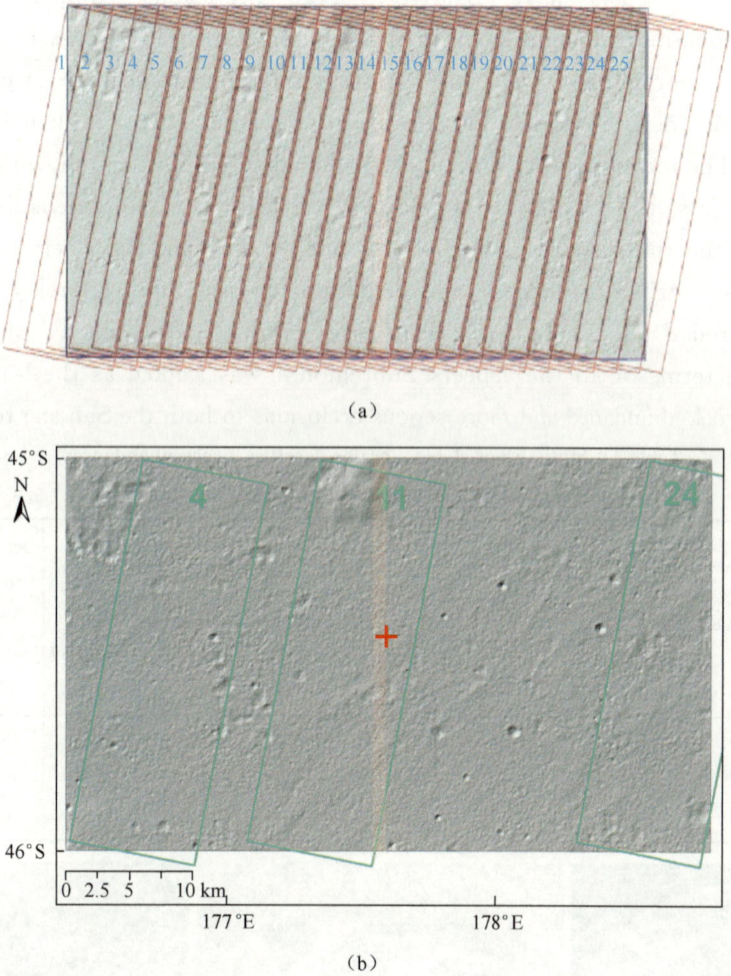

Figure 10　Joint analysis for selecting the subregion for landing. (a) 25 subregions covering the targeted landing region; (b) the three short - listed subregions, with 11 being the highest priority, where Chang'E - 4 actually landed. The red cross indicates the exact landing location of Chang'E - 4. The underlying image is a shaded relief of the DTM of the targeted landing region

described previously. Some threshold parameters (e. g., the crater density) might not directly relate to an engineering constraint, but they worked fine as we were interested in the subregions of relatively favorable conditions.

We carried out the above analysis for all 25 subregions. The subregions (1 - 8) in the western part of the targeted landing region have relatively large slopes and rock abundances. Their crater densities are all below the threshold. The subregions (9 - 15) in the middle part have fewer slopes and rock abundances. Their crater densities are gradually increased. The subregions (16 - 25) in the eastern part have relatively large crater densities and slopes. Their rock abundances are moderate. For the occlusions to the Sun and the relay satellite, the first two subregions in the west and several subregions in the east (e.g., 19 – 23) have relatively large occlusions compared with other subregions. Comparing the subregions in the western, central, and eastern parts of the landing region, subregions 4, 11, and 24 show overall superiorities as compared with others in their neighborhoods. Therefore, they were short – listed for further consideration. The locations of the three short – listed subregions are shown in Figure 10b.

Table 1 lists the statistics of slopes, occlusions, crater densities, and rock abundances of the three short – listed subregions 4, 11, and 24. Their average slopes, crater densities, rock abundances, and occlusions to the Sun and relay satellite were further compared. From Table 1, it can be seen that subregion 11 shows the best performance in terms of all the aspects. Subregion 4 was ranked as the last, as it has a relatively high rock abundance and more serious occlusions to both the Sun and relay satellite.

Table 1　Statistics of slopes, crater densities, rock abundances, and occlusions of the three short – listed subregions

Subregion	Slope		Crater density		Rock abundance		Occlusion to the Sun (%)	Occlusion to the relay satellite (%)	Priority as landing site
	Average (°)	% of areas ≥ 8°	Average (no. craters per10km^2)	% of areas ≥ 78 craters per 10km^2	Average (%)	% of areas ≥ 7%			
4	2.39	2.37	58	0.00	0.04	0.031	0.44	0.02	3
11	2.32	1.96	46	0.00	0.00	0.000	0.32	0.00	1
24	2.65	3.04	81	0.38	0.00	0.011	0.41	0.00	2

(a)　(b)　(c)

Figure 11　Localization of the Chang'E – 4 lander. (a) Three small craters marked as red dots on the ground panorama image. (b) The corresponding positions of the craters marked on the descent image, which was geo – referenced with the LRO NAC image. The red cross indicates the lander location obtained by optimal cartographic triangulation. (c) The LRO NAC image (M1303619844LR) taken on 1 February 2019, showing the real locations of the lander and rover as indicated by the white arrows

Subregion 11 was finally selected as the landing site with the highest priority, and Chang'E – 4 actually landed in this subregion, with the exact location marked by a red cross in Figure 10b.

5　Landing Site Localization and Analysis Localization of the Chang’E－4 Lander

On 3 January 2019, Chang’E－4 successfully landed in the Von Kármán crater. Identifying the exact location of the lander within the landing region and with respect to other surface features as soon as possible after landing was critical for planning science and engineering activities in the initial stages of surface exploration. The identification of other nearby features was also needed in traverse planning for the rover over the course of the mission. This was accomplished within two hours of landing before the rover drove off the lander. The related procedures involved Doppler positioning of the orbit track, reconstructing the entry, descent, and landing (EDL) process using returned descent images, and locating common features in the orbiter, descent, and ground images.

The navigation team determined the location of the lander at 45.5°S, 177.6°E, by Doppler positioning and orbit determination, which offered an initial location of the lander. A series of descent images collected by a descent camera mounted on the lander during the EDL process were used to reconstruct the lander’s descent track, and the endpoint of the descent track was already close to the lander location. Finally, a ground panorama image, taken by the terrain camera mounted on top of the lander after landing, was used for the precise localization of the lander. As shown in Figure 11a, three meter-sized craters were manually identified on the ground panorama, and their centers were marked with red dots. Their corresponding positions on the descent image were then manually identified and marked with red dots as shown in Figure 11b. The descent image was geo-referenced to the LRO NAC image (M134022629L), as shown in the background in Figure 11b. The LRO NAC image was ortho-rectified using the Integrated Software for Imagers and Spectrometers software, which was already in the Mean Earth/Polar Axis reference system with planetocentric coordinates. The latitude and longitude coordinates of the lander were then calculated based on the optimal cartographic triangulation (Li et al. 2004) using these three terrain features, which were in the same cartographic system of the used LRO NAC image. The elevation of the lander was measured from the DTM of the same location. The determined landing location of Chang’E－4 was at 45.4561°S, 177.5892°E, and at －5926m elevation. The elevation was with respect to a reference sphere of radius 1737.4km.

One week after the landing (on 11 January 2019), the LROC team at Arizona State University also identified the location of the Chang’E－4 lander at 45.457°S, 177.589°E, plus or minus 20 meters (http://lroc.sese.asu.edu/posts/1087), based on the co-registration of the Chang’E－4 descent frames and the NAC image (M1298916428LR). Liu et al. (2019) also located the Chang’E－4 landing site as 45.4446°S, 177.5991°E with an elevation of −5935 m using the digital orthophoto map and DTM of Chang'E−2 as geo−reference data, whose relative position to the surrounding terrain features is consistent with our results.

About one month later, after the successful landing of Chang'E−4, NASA's LRO passed by the Chang'E−4 landing region on 1 February 2019. The LRO NAC collected an image (M1303619844LR) of the landing region with a spatial resolution of 0.85 m/pixel, as shown in Figure 11c, on which the Chang'E−4 lander (the white dot with a dark shadow) and the Jade Rabbit−2 rover (the dark dot) can be clearly seen, as indicated by the white arrows. Comparing Figure 11b and 11c, it can be seen that the lander's location, as determined by our approach, is generally consistent with its real location, as imaged by the LRO NAC.

Analysis Around the Landing Site

After determining the lander's location, the analysis of its surrounding environment is vital for planning the subsequent science and engineering activities. We first examined the detailed terrain occlusions at the landing site. The approach described in the section "Occlusion Analysis" was adopted to analyze the occlusions at the landing site induced by the terrain, and the results are shown in Figure 12. Specifically, 360 geodesic lines were cast from the landing location at an azimuth interval of 1° as shown in Figure 12a. The position with the largest altitude angle induced by the nearby terrain was denoted by the endpoint of each geodesic line. Places with no geodesic lines visible are due to the terrain occlusions that happened in close ranges. Similarly, the relatively large occlusions were caused by the east−west rim of the Von Kármán crater. Overall, the terrain occlusion angles are generally less than 3° at the landing location. Figure 12b illustrates the relationships among the terrain occlusions, solar altitude angles, and altitude angles of the relay satellite at the landing location for the six−month period after landing, from 3 January to 3 July 2019. It can be seen that the solar altitude angles and the altitude angles of the relay satellite are both much higher than the terrain occlusion angles at the landing location, which indicates its excellent visibility to the Sun and relay satellite.

For a detailed analysis of surface slopes and geometric information of small surface features around the landing location, higher resolution DTMs are required. However, at the time when Chang'E−4 landed, no high−resolution stereo pair of LRO NAC images had been collected around the landing site. Therefore, generating high−resolution DTMs from stereo images using the traditional photogrammetric approach was not possible. We developed an innovative SfS method for generating pixel−wise lunar DTMs from a single image with the constraint of a lower−resolution DTM (Wu et al. 2018a). Given a single image with a known light source, the SfS method is able to refine a lower−resolution DTM covering the same image area to higher resolution, the same as the image resolution, by using shading information to optimize regularization with respect to shape reconstruction (Kirk et al. 2003). Details about this SfS method can be found in our

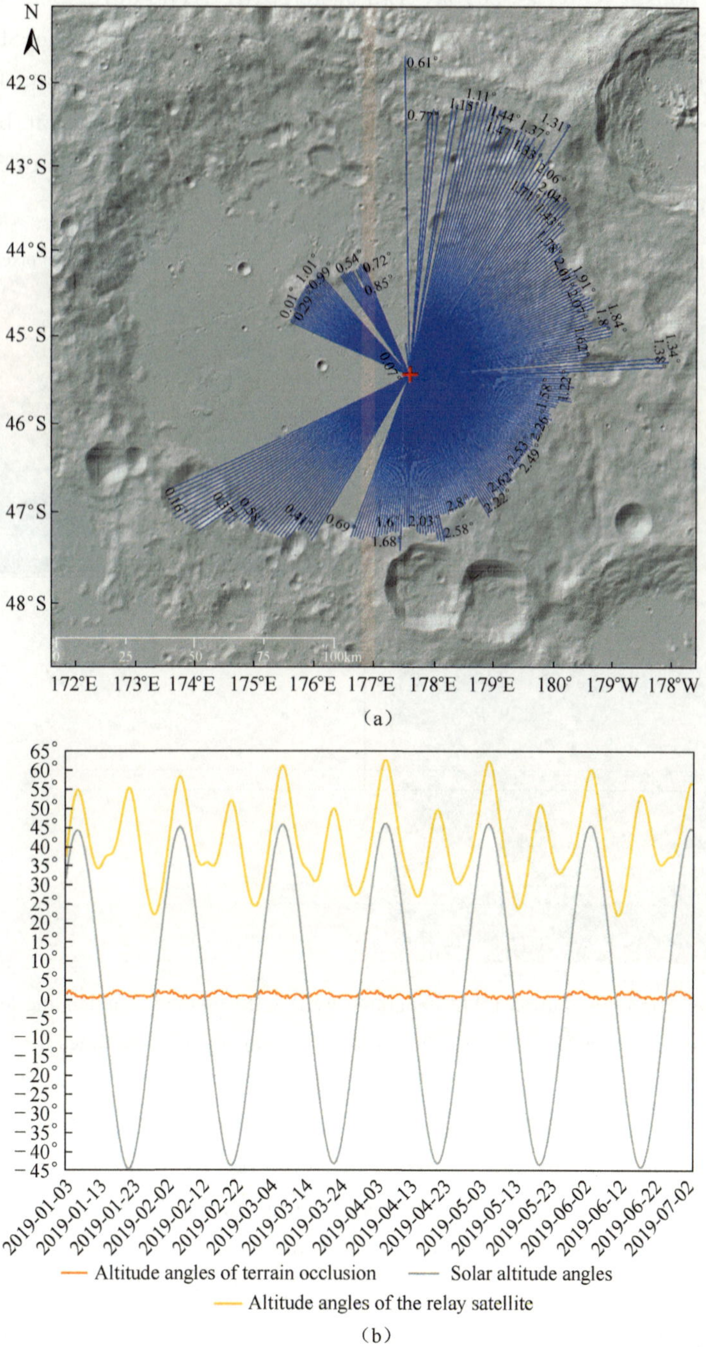

Figure 12　Occlusion analysis at the landing site. (a) The geodesic lines cast from the landing location (the red cross) at an interval of 1° and (b) the relationships among the terrain occlusions, solar altitude angles, and altitude angles of the relay satellite at the landing location for the sixmonth period after landing

previous publications (Wu et al. 2018a; Liu et al. 2018). The developed SfS method was applied to the 1.5m/pixel LRO NAC image (M134022629L), and the SLDEM, with a resolution of 60m used as the constraint, from which a high-resolution DTM (1.5 m/pixel) was generated, as shown in Figure 13. The differences in elevation between the SfS DTM and a photogrammetric DTM generated later using a stereo pair of NAC images is less than 5% (Liu and Wu 2020).

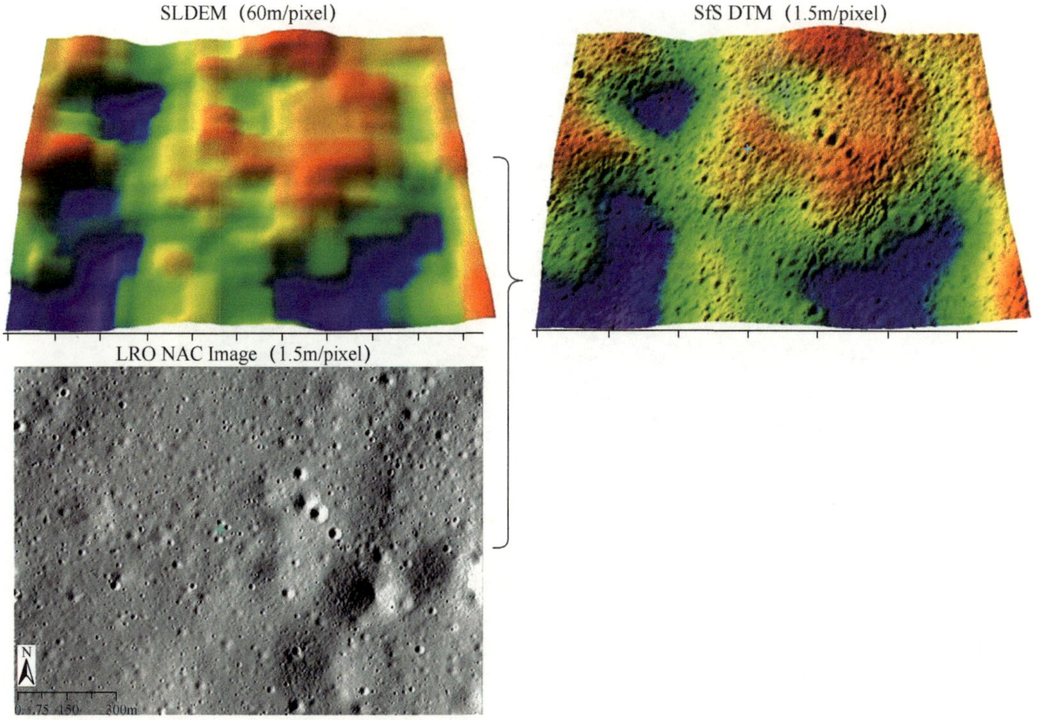

Figure 13 High-resolution DTM generated by the SfS method. The landing location is marked with a cyan cross. The elevations of the DTMs are exaggerated three times for better visualization

The upper row of Figure 13 shows a side-by-side comparison of the low-resolution SLDEM and the generated high-resolution SfS DTM, the latter showing overall consistent geometry with the former but providing much better details of the terrain features. The landing location of Chang'E-4 is marked in the SfS DTM and LRO NAC image with cyan crosses. It can be seen that the small craters around the landing location are clearly visible on the SfS DTM, but they are entirely missing on the SLDEM.

Based on the 1.5m/pixel SfS DTM, a slope map with a baseline of 3m was generated as shown in Figure 14a, from which it can be seen that the eastern part of the landing site has relatively larger slopes and the western part is relatively flat. For any point in the DTM, an aspect is calculated, which is the angle from the direction of the maximum slope at that point to the north direction. An aspect map was then generated to show the orientations of the slopes, as shown in Figure 14b. From the slope and aspect maps, the slope of the lander location is measured to be about 4.5°, and its aspect is toward southwest with an angle of 205° from the north.

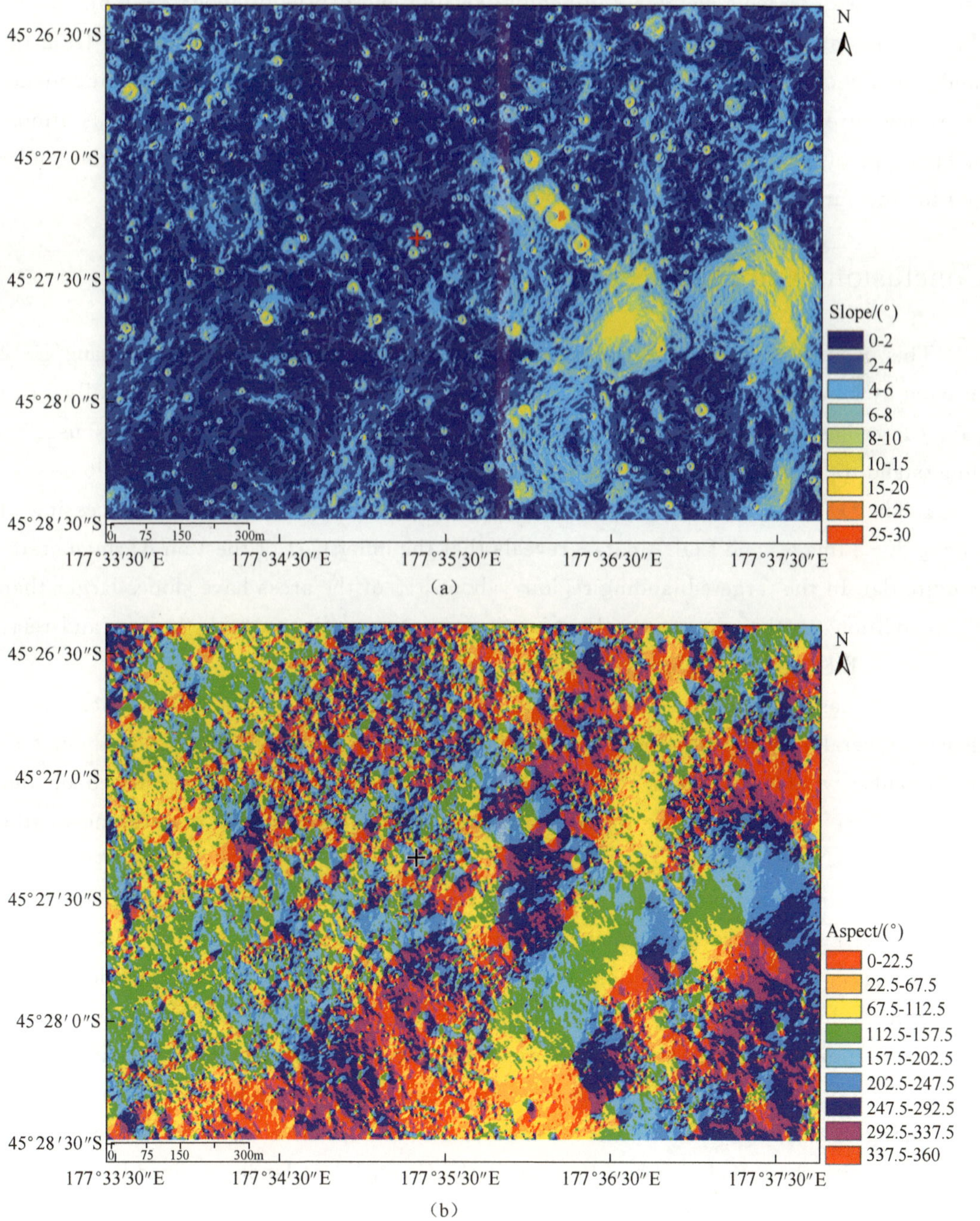

Figure 14 Detailed slope analysis around the landing site. (a) The slope map calculated with a baseline of 3m and (b) the corresponding aspect map. The lander's location is marked on the maps with red and black crosses, respectively

The high - resolution DTM also enables the measurement of the detailed geometric information of small terrain features (e. g. , diameters and depths of small craters) around the landing location. For instance, the three small craters surrounding the lander (see in Figure 13) have diameters of about 20 - 30m and depths of about 2 - 2. 6m. These three craters all have a depth - diameter ratio of about 0. 1. The shortest distance from the lander to any of these three craters is about 13m. From the aforementioned rock abundance map, measurements show that the rock abundance around the landing location is almost 0. This type of detailed information is vital for the operation of instruments on the lander and for the surface operations of the rover.

Conclusions and Discussion

This paper presents the efforts concerning the landing site selection for the Chang'E - 4 mission by jointly analyzing topographic and geomorphological features. Additionally, this paper describes the localization of the landing site and the analytical methods used to understand its surrounding area. The investigation leads to the following conclusions:

1. The DTM, with a resolution of 30m generated from the integrated processing of Chang'E - 2 images and LOLA data, reveals that the inner part of the Von Kármán crater is quite flat. In the targeted landing region, about 5% of the areas have slopes larger than 8°. In addition, within the region there is satisfactory visibility to both the Sun and relay satellite, which are not occluded by the terrain surface in most areas.

2. The eastern part of the targeted landing region has relatively higher crater density. Overall, on average the crater density is about 60 craters (≥ 70m in diameter) in a circular window of $10km^2$. Most of the landing region is rock - free based on our investigation, and most of the rocks are located around large impact craters and on the slopes of hills.

3. The joint analysis of surface slopes, occlusions to the Sun and relay satellite, crater densities, and rock abundances identified three subregions inside the targeted landing region. The one located close to the center of the region was recommended as the site with the highest priority, and the Chang'E - 4 eventually landed at this site.

4. The integrated processing of orbiter, descent, and ground images identified the precise location of the Chang'E - 4 lander within two hours after its successful landing, positioning it at 45. 4561°S, 177. 5892°E, and at - 5926m in elevation. This location is consistent with the actual observed location of the Chang'E - 4 lander as fixed by the LRO NAC image one month later.

5. Detailed occlusion analysis around the landing location reveals that the lander has excellent visibility to the Sun and relay satellite. A detailed slope analysis around the landing location using an SfS DTM with a high resolution of 1. 5m/pixel indicates that the lander is on a slope of about 4. 5° toward the southwest. The rock abundance around the

landing location is almost 0.

The analytical methods described in this paper have demonstrated their effectiveness in real operations of the Chang'E－4 mission. The developed methods and results can benefit future missions of soft landings on the Moon and other celestial bodies.

Acknowledgments

This work was funded by China Academy of Space Technology（Project No：17CPIT/HK0102）and a grant from the Research Grants Council of Hong Kong（Research Impact Fund－Project No：R5043－19）. The work was also supported by a grant from the National Natural Science Foundation of China（Project No：41671426）. The authors would like to thank everyone who worked on the archives of the imagery and datasets used in this research for making them publicly available.

References

[1] Bandfield，J. L.，R. R. Ghent，A. R. Vasavada，D. A. Paige，S. J. Lawrence and M. S. Robinson. 2011. Lunar surface rock abundance and regolith fines temperatures derived from LRO Diviner Radiometer data. Journal of Geophysical Research：Planets 116（E12）.

[2] Barker，M. K.，E. Mazarico，G. A. Neumann，M. T. Zuber，J. Haruyama and D. E. Smith. 2016. A new lunar digital elevation model from the Lunar Orbiter Laser Altimeter and SELENE Terrain Camera. Icarus 273：346－355.

[3] Brady，T.，E. Robertson，C. Epp，S. Paschall and D. Zimpfer. 2009. Hazard detection methods for lunar landing. Pages 1－8 in Proceedings of the 2009 IEEE Aerospace Conference.

[4] De Rosa，D.，B. Bussey，J. T. Cahill，T. Lutz，I. A. Crawford，T. Hackwill，S. van Gasselt，G. Neukum，L. Witte and A. McGovern. 2012. Characterization of potential landing sites for the European Space Agency's Lunar Lander project. Planetary and Space Science 74（1）：224－246.

[5] Di，K.，M. Zhu，Z. Yue，Y. Lin，W. Wan，Z. Liu，S. Gou，B. Liu，M. Peng，Y. Wang，S. Niu，J. Zhang，J. Li，J. Xie，L. Xi，J. Yang and B. Xue. 2019. Topographic evolution of Von Kármán crater revealed by the lunar rover Yutu－2. Geophysical Research Letters 46（22）：12764－12770.

[6] Golombek，M. and D. Rapp. 1997. Size frequency distributions of rocks on Mars and Earth analog sites：Implications for future landed missions. Journal of Geophysical Research：Planets 102（E2）：4117－4129.

[7] Golombek，M. P.，A. Haldemann，N. Forsberg－Taylor，E. N. Di Maggio，R. D. Schroeder，B. M. Jakosky，M. Mellon and J. Matijevic. 2003. Rock size－

frequency distributions on Mars and implications for Mars Exploration Rover landing safety and operations. Journal of Geophysical Research 108 (E12): 8086.

[8] Golombek, M., D. Kipp, N. Warner, I. J. Daubar, R. Fergason, R. L. Kirk and B. A. Campbell. 2017. Selection of the InSight landing site. Space Science Reviews 211 (1 - 4): 5 - 95.

[9] Hu, H. and B. Wu. 2019. Planetary3D: A photogrammetric tool for 3D topographic mapping of planetary bodies. ISPRS Annals of the Photogrammetry, Remote Sensing and Spatial Information Sciences Ⅳ - 2/W5: 519 - 526.

[10] Hu, H., C. Chen, B. Wu, X. Yang, Q. Zhu and Y. Ding. 2016. Textureaware dense image matching using ternary census transform. ISPRS Annals of the Photogrammetry, Remote Sensing and Spatial Information Sciences III - 3: 59 - 66.

[11] Huang, J., Z. Xiao, J. Flahaut, M. Martinot, J. Head, X. Xiao, M. Xie and L. Xiao. 2018. Geological characteristics of Von Kármán crater, northwestern south Pole Aitken basin: Chang' E - 4 landing site region. Journal of Geophysical Research: Planets (123): 1684 - 1700.

[12] Jia, Y., Y. Zou, J. Ping, C. Xue, J. Yan and Y. Ning. 2018. The scientific objectives and payloads of Chang'E - 4 mission. Planetary and Space Science 162: 207 - 215.

[13] Karney, C. F. 2013. Algorithms for geodesics. Journal of Geodesy 87 (1): 43 - 55.

[14] Kirk, R. L., E. Howington - Kraus, B. Redding, D. Galuszka, T. M. Hare, B. A. Archinal, L. A. Soderblom and J. M. Barrett. 2003. High - resolution topomapping of candidate MER landing sites with Mars Orbiter Camera narrow - angle images. Journal of Geophysical Research: Planets 108 (E12): 8088.

[15] Li, C., D. Liu, B. Lu, X. Ren, J. Liu, Z. He, W. Zuo, X. Zeng, R. Xu, X. Tan, X. Zhang, W. Chen, R. Shu, W. Wen, Y. Su, H. Zhang and Z. Ouyang. 2019. Chang'E - 4 initial spectroscopic identification of lunar far - side mantle - derived materials. Nature 569: 378 - 382.

[16] Li, F., H. Zhang, X. Wu, J. Ma and L. Lu. 2017. Influence analysis of terrain of the far side of the Moon on soft - landing. Journal of Deep Space Exploration 4 (2): 143 - 149.

[17] Li, R., K. Di, L. H. Matthies, W. M. Folkner, R. E. Arvidson and B. A. Archinal. 2004. Rover localization and landing - site mapping technology for the 2003 Mars exploration rover mission. Photogrammetric Engineering and Remote Sensing 70 (1): 77 - 90.

[18] Li, Y. and B. Wu. 2018. Analysis of rock abundance on lunar surface from orbital and descent images using automatic rock detection. Journal of Geophysical Research: Planets 123 (5): 1061 - 1088.

[19] Liu, W. C., B. Wu and C. Wöhler. 2018. Effects of illumination differences on photometric stereo shape - and - albedo - from - shading for precision lunar surface

reconstruction. ISPRS Journal of Photogrammetry and Remote Sensing 136：58 - 72.

[20] Liu，W. C. and B. Wu. 2020. An integrated photogrammetric and photoclinometric approach for illumination - invariant pixelresolution 3D mapping of the lunar surface. ISPRS Journal of Photogrammetry and Remote Sensing 159：153 - 168.

[21] Liu，J.，X. Ren，W. Yan，C. Li，H. Zhang，Y. Jia，X. Zeng，W. Chen，X. Gao，D. Liu，X. Tan，X. Zhang，T. Ni，H. Zhang，W. Zuo，Y. Su and W. Wen. 2019. Descent trajectory reconstruction and landing site positioning of Chang'E - 4 on the lunar farside. Nature Communications 10：4229.

[22] Pajola，M.，S. Rossato，E. Baratti，R. Pozzobon，C. Quantin，J. Carter and P. Thollot. 2017. Boulder abundances and size - frequency distributions on Oxia Planum - Mars：Scientific implications for the 2020 ESA ExoMars rover. Icarus 296：73 - 90.

[23] Wang，Y. and B. Wu. 2017. Improved large - scale slope analysis on Mars based on correlation of slopes derived with different baselines. International Archives of the Photogrammetry，Remote Sensing and Spatial Information Sciences 42.

[24] Wang，Y. and B. Wu. 2019. Active machine learning approach for crater detection from planetary imagery and digital elevation models. IEEE Transactions on Geoscience and Remote Sensing 57（8）：5777 - 5789.

[25] Wilhelms，D. E.，F. John and N. J. Trask. 1987. The Geologic History of the Moon. Washington，D. C.：U. S. Geological Survey.

[26] Wu，B.，F. Li，L. Ye，S. Qiao，J. Huang，X. Wu and H. Zhang. 2014a. Topographic modeling and analysis of the landing site of Chang'E - 3 on the Moon. Earth and Planetary Science Letters 405：257 - 273.

[27] Wu，B.，H. Hu and J. Guo. 2014b. Integration of Chang'E - 2 imagery and LRO laser altimeter data with a combined block adjustment for precision lunar topographic modeling. Earth and Planetary Science Letters 391：1 - 15.

[28] Wu，B.，W. C. Liu，A. Grumpe and C. Wöhler. 2018a. Construction of pixel - level resolution DEMs from monocular images by shape and albedo from shading constrained with low - resolution DEM. ISPRS Journal of Photogrammetry and Remote Sensing 140：3 - 19.

[29] Wu，B.，J. Huang，Y. Li，Y. Wang and J. Peng. 2018b. Rock abundance and crater density in the candidate Chang'E - 5 landing region on the Moon. Journal of Geophysical Research - Planets 123（12）：3256 - 3272.

[30] Wu，W. R.，Q. Wang，Y. H. Tang，J. Z. Liu and L. L. Lu. 2017. Design of Chang'E - 4 lunar farside soft - landing mission. Journal of Deep Space Exploration 4（2）：111 - 117.

Centimeter – resolution topographic modeling and fine – scale analysis of craters and rocks at the Chang'E – 4 landing site

BoWu[1], YuanLi[1], Wai ChungLiu[1], YiranWang[1], FeiLi[2], YangZhao[2], HeZhang[2]

1. Department of Land Surveying and Geo – Informatics, The Hong Kong Polytechnic University, Hung Hom, Kowloon, Hong Kong, PR China

2. China Academy of Space Technology, 104 Youyi Street, Haidian, Beijing, PR China

Abstract: Chang'E – 4 was the first lunar mission to successfully land on the far side of the Moon. The stereo images collected by the panoramic camera (Pancam) mounted on the Jade Rabbit – 2 rover have provided the most detailed in situ observations of the landing site along the rover traverse. This paper presents our efforts in centimeter – resolution topographic modeling and fine – scale analysis of craters and rocks at the Chang'E – 4 landing site using the Pancam images. We first used an integrated photogrammetric and photoclinometric approach to generate a digital elevation model (DEM) and an orthoimage mosaic of the landing site and a 3D model of the Chang'E – 4 lander itself, all with spatial resolutions of 1cm/pixel. Craters with diameters of decimeters to meters and rocks with sizes of decimeters were then extracted and analyzed. The results indicate that Chang'E – 4 landed on a terrain slope of 4° toward 72. 3° southwest, with one foot sunk in the regolith 1. 5cm deeper than the other foot on the front side of the lander. Fine – scale analysis of the small craters at the Chang'E – 4 landing site showed overall a similar power – law trend as large craters of hundreds of meters, but there is a clear gap between their size – frequency distributions, indicating the accumulated impact flux appears to be different for these two crater populations. The average depth – to – diameter ratio of the small craters around the landing site was 0. 061, obviously less than those of craters on the kilometer scale. Fine – scale analysis of rocks revealed that the Chang'E – 4 landing site has a rock abundance of 0. 21%, which agrees well with previous results based on orbital data. Joint analysis of craters and rocks indicates that more rocks are located inside craters with greater depth – to – diameter ratios on such a fine scale. The 3D topographic modeling approach and fine – scale analysis of craters and rocks presented in this paper provide new insights into lunar mapping and geological study.

Article history: Received 26 June 2020 Received in revised form 30 September 2020 Accepted 4 November 2020 Available online 16 November 2020 Editor: W. B. McKinnon
Keywords: Chang'E - 4 Pancam photogrammetry photoclinometry craters rocks

1　Introduction

On January 3, 2019, the Chinese Chang'E - 4 probe successfully landed in the Von Kármán crater in the northwestern South Pole - Aitken (SPA) basin on the far side of the Moon (Liu et al., 2019; Di et al., 2019; Wu et al., 2020). Since then, the Chang'E - 4 lander and Jade Rabbit - 2 rover have been exploring the lunar surface and subsurface of the landing region (Li et al., 2019; Gou et al., 2020; Ding et al., 2020).

Topographic modeling and analysis are important in evaluation of the landing site to support surface operations and for scientific studies. For instance, analysis of surface slopes and roughness based on topographic data is crucial for landing site evaluation (De Rosa et al., 2012) and rover path planning (Ma et al., 2020). Detailed topographic data of fresh craters and fragments in and around the craters are helpful to identify the origin of the fragments and the provenances of surface materials (Ding et al., 2020). Before the landing of Chang'E - 4, the topography and geomorphology of the candidate landing region (45.0°—46.0°S, 176.4°—178.8°E) within the Von Kármán crater floor were carefully studied and analyzed using orbital data. Wu et al. (2020) generated a digital el - evation model (DEM) of the candidate landing region with a spatial resolution of 30m/pixel using images (7m/pixel) from Chang'E - 2 and data from the Lunar Reconnaissance Orbiter (LRO) Laser Al - timeter (LOLA; Smith et al., 2010), and an orthoimage mosaic (1m/pixel) using images from the LRO Narrow - Angle Camera (NAC). Based on the DEM and the orthoimage mosaic, they analyzed in detail the surface slopes, terrain occlusions, and distributions of craters and rocks within the candidate landing region to select the most suitable landing site for Chang'E - 4. Huang et al. (2018) also investigated the topographic and geological characteristics of the candidate landing region for Chang'E - 4 using orbital data. The or bital data used in previous works revealed the topographic and geomorphological characteristics on a large scale, but their relatively low resolution compared to ground - based imagery limited their use for the study of fine topographic and geomorphological details.

In contrast with the orbital data, ground images collected by cameras onboard the rover can facilitate in situ measurements and analysis of the lunar surface on the millimeter to centimeter scales. This is very important for surface operations such as rover ma - neuvers and for scientific explorations at the landing site (Di et al., 2019; Liu et al., 2020). For instance, Ma et al. (2020) used the images collected by the navigation camera (Navcam) and the panoramic camera (Pancam) on board the Jade Rabbit - 2 rover to estimate the rover's precise location using photogrammetry technologies. Liu et al. (2015) combined the rover images collected at the Chang'E - 3 landing site with orthoimages generated from orbital and descent images to generate a local DEM with a resolution of 0.02m/pixel, which allowed analysis of detailed topographic characteristics to support the rover's teleoperation tasks. This technology was further applied at the Chang'E - 4 landing site for the localization and navigation of the Jade Rabbit - 2 rover (Liu et al., 2020).

However, the DEMs generated from the rover images using the traditional technology of photogrammetry may suffer from problems in image matching and stereo coverage. Photogrammetry requires stereo images that cover the same area for 3D modeling, whereas stereo coverage is not always guaranteed even with a stereo camera system due to possible occlusions of the terrain surface. Dense image matching on stereo images is also a challenging task because lunar surface images usually lack favorable textural conditions. As a result, the DEMs generated from rover images by photogrammetry usually have serious interpolation defects, resulting in less effective spatial resolution.

Photoclinometry is a technique in which 3D information of surfaces is retrieved based on the photometric content of images with known orientations (Horn, 1990). Unlike photogrammetry, which depends on stereo images and the quality of image matching, photoclinometry allows the recovery of pixel - wise topographic details from a single image. However, because photoclinometry strongly relies on the local topography, subtle errors may accumulate and propagate with distance, leading to unfavorable geometric accuracy at large scales, at which photogrammetry excels. Therefore, the integration of photogrammetry and photoclinometry facilitates the generation of pixel - wise DEMs with accuracy comparable to photogrammetric products. Photogrammetry and photoclinometry have been successfully integrated for topographic modeling of the lunar surface from orbital images and have shown performance superior to either technology alone (Grumpe et al., 2014; Wu, Liu et al., 2018; Liu and Wu, 2020). For rover images collected on the ground, it is natural to imagine that the integration strategy still works. Compared with orbital images, however, rover images pose more challenges, such as the significant differences between images collected at different stations and the large variations in illumination, imaging perspective geometry, and spatial resolution.

Analyses of craters and rocks have also been focused on relatively large scales on the lunar surface. For example, the spatial distribution and density of craters used for landing site evaluation are on the order of dozens of meters (De Rosa et al., 2012; Wu et al., 2014; 2020). The size - frequency distribution of craters used to infer surface age lies mainly on the kilometer scale (Neukum et al., 2001). Rocks have been investigated during landing site evaluation as potential hazards for the lander and rover. Bandfield et al. (2011) generated a rock concentration map of the lunar surface on a nearly global scale using measurements from the LRO Diviner radiometer with a spatial resolution of 128 pixels/degree. De Rosa et al. (2012) used LRO NAC images with resolutions be - tween 0. 6 and 1. 0m/pixel and identified hundreds of rocks several meters in size in the vicinity of the chosen landing sites of the European Space Agency's Lunar Lander mission. Li and Wu (2018) studied the size - frequency distributions (SFDs) of rocks in several regions (the Chang'E - 3, Luna 17, and Luna 23 landing sites) of the lunar surface using high - resolution orbital and descent images and proposed a rock abundance model with an exponential form. The associations between craters and rocks have also been investigated. Basilevsky et al. (2013) and Watkins et al. (2019) studied the SFDs of boulders around small young impact craters to infer the relationship between the boulder populations and crater ages. Ghent et al. (2014) used Diviner rock abundance data to constrain the rate of breakdown of ejecta blocks on craters with diameters of 19 - 97km. However, fine - scale analysis of craters (decimeters to meters in size) and rocks (decimeters in size) has not been sufficient in the past.

Section 2 of this paper presents an integrated photogrammetric and photoclinometric approach for centimeter - resolution 3D modeling from rover images, from which the topographic details and pose information of the Chang'E - 4 lander are precisely measured. The generated centimeter - resolution 3D models in the vicinity of the Chang'E - 4 landing site also enable fine - scale analysis of craters and rocks in the region, as presented in Section 3. Section 3 also presents a comparison of fine - scale analysis of craters and rocks with those at large scales in previous works. Section 4 summarizes our findings and presents our concluding remarks.

2 Integration of photogrammetry and photoclinometry for centimeter - resolution topographic modeling from rover images

An approach that integrates photogrammetry and photoclinometry was developed for high - resolution (1cm) topographic modeling of the Chang'E - 4 landing site using stereo Pancam images (with the best spatial resolution of about 3mm/pixel) acquired by the Jade Rabbit - 2 rover. The approach begins with the photogrammetric processing of the stereo Pancam images, including tie point matching, bundle adjustment, co - registration of image blocks, dense image matching, and 3D point cloud generation. The output of the photogrammetric process is a DEM and an orthoimage mosaic of the landing site and a 3D model of the lander itself. The photogrammetric DEM is then refined to higher resolution based on photoclinometry using Pancam images. The photoclino - metric refinement is constrained by the overall geometry derived from photogrammetry to ensure a high level of geometric accuracy. The framework of the approach is illustrated in Fig. 1.

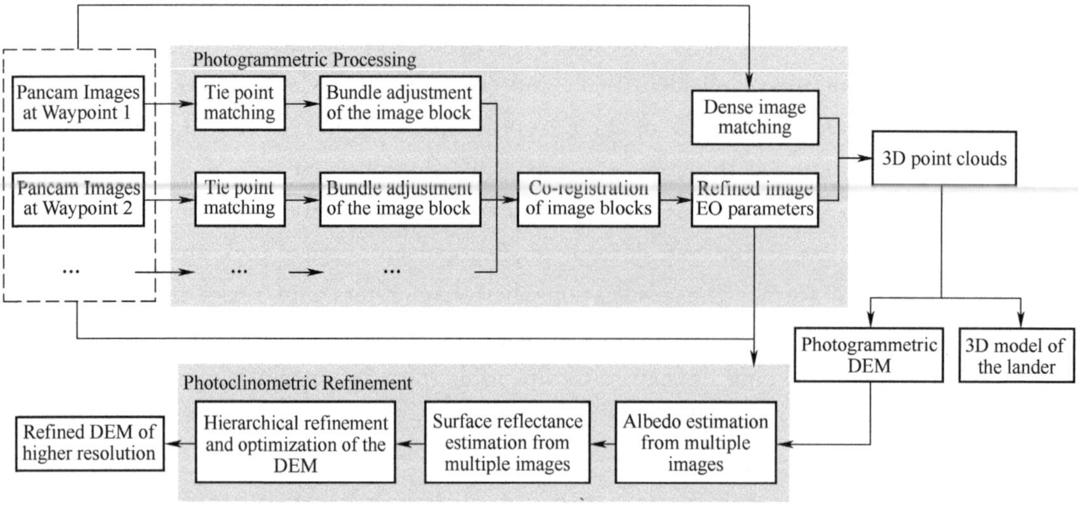

Fig. 1 Framework of the integrated photogrammetric and photoclinometric approach

2.1 Photogrammetric processing of rover images

The Pancam onboard the Jade Rabbit-2 rover has two identical cameras separated by a rigid base with a length of 270 mm. Each camera has an image size of 2352 pixels ×1728 pixels and a focal length of 50 mm. The mast of the Pancam allows the cameras to rotate a full 360°to obtain a panoramic view (Jia et al., 2018). The photogrammetric processing of stereo Pancam images involves the interior orientation (IO) and exterior orientation (EO) parameters of the images. The IO parameters include the focal length of the camera, the location of the principal point, and possible lens distortions. They are calibrated in the laboratory before the launch, as detailed in Table S1. The EO parameters, including the positions and orientations of the camera during image collection, are measured by onboard sensors in a local coordinate system (LCS) defined in the rover's body-fixed frame.

The IO and EO parameters together enable calculation of the 3D object-space coordinates of a pair of matched points in the stereo images using well-known co-linearity equations (Wang, 1998). However, those parameters might include errors; therefore, a bundle adjustment process (Wu, 2017) is conducted to improve the image orientation parameters and the consistency within an image block. In this study, bundle adjustment is performed for each image block collected by the rover at each stop (waypoint) separately. This is because it is very difficult to obtain tie points between images collected at different waypoints, due to the large differences in illumination directions and camera perspectives, so that to connect different image blocks. Within each image block, pairs of tie points on the images are automatically extracted and matched using the scale-invariant feature transform (SIFT) algorithm (Lowe, 2004). They are then used in the bundle adjustment of the image block to refine the image orientation parameters and to achieve the internal consistence of the 3D modeling within the block.

After bundle adjustment of each image block at different waypoints, the blocks are co - registered to the united LCS defined in the lander's body - fixed frame based on a 3D conformal transformation using a few manually selected tie points between the image blocks. At least three tie points are necessary to calculate the transformation parameters from each image block to the united LCS based on the least - squares solution (Agarwal and Mierle, 2012).

After the transformation of individual image blocks to the united LCS, the EO parameters of the images are transferred to the united LCS using the corresponding transformation parameters determined in the previous step. Dense image matching (Wu et al., 2011) is then carried out to obtain dense matched points, from which 3D point clouds can be generated by photogrammetric space intersection using the image EO parameters. The 3D point clouds are used to generate 3D models of the region, including a DEM of the terrain surface and a 3D model of the lander. An orthoimage mosaic covering the same region is also generated based on the Pancam images and the DEM.

2.2 Photoclinometric refinement of the photogrammetric DEM

Previous photogrammetric processing of rover images generates an initial DEM with relatively sparse spatial resolution, and it will be further refined to a higher spatial resolution using photoclinometry (McEwen, 1991). Photoclinometry allows retrieval of fine - grained topographic details of the scene and complements the limitations of dense image matching in photogrammetry, such as those caused by poor image textures and a lack of overlapping. The photoclinometric refinement approach is extended from our previous study (Wu, Liu et al., 2018; Liu and Wu, 2020) to process multiple images collected by the rover on the lunar surface.

For the images on the lunar surface, the relationship between the image intensity values and the surface can be modeled as in Equation (1):

$$I = AG(Z) \tag{1}$$

where the image intensity value I is the product of the surface albedo A and the geometric reflectance of the surface $G(Z)$, with Z represents the elevations in a DEM. Instead of using surface normals or slopes, the elevations are used in the equation, as our purpose is to directly retrieve the elevations of a refined DEM from photoclinometry using an initial DEM as a constraint. $G(Z)$ can be modeled using the Lunar - Lambert model (McEwen, 1991), which describes how the lunar surface reflects the incoming illumination as in Equation (2):

$$G(Z) = (1 - \lambda(g))\mu_0(Z) + 2\lambda(g)\frac{\mu_0(Z)}{\mu_0(Z) + \mu(Z)} \tag{2}$$

where $\mu_0(Z)$ refers to the cosine of the incidence angle, and $\mu(Z)$ refers to the cosine of the viewing angle. λ(g) is a weighting function that is dependent on the phase angle g (McEwen, 1991; Lohse and Heipke, 2004). It can be modeled by a 3rd - order polynomial function of g as described in Lohse et al. (2006). $G(Z)$ is dependent on the illumination and viewing conditions, so a point on a DEM can correspond to various reflectance values when multiple images are used. In contrast, the surface albedo A is expected to remain invariant, so a point on a DEM should correspond to only one albedo value.

To facilitate photoclinometric refinement, the expected $G(Z)$ value for each node of the DEM must be determined from the images. First, each DEM node is projected to each image based on the EO parameters obtained from the previous photogram - metric process, and the corresponding image intensity values are assigned to the node accordingly. Based on Equation (1), the surface albedo for each node is then inferred from its intensity value and the reflectance value derived from the initial DEM under the same illumination condition of the image. Finally, the reflectance values $G(Z)$ corresponding to the refined DEM can be calculated using the image intensities and the inferred surface albedo, which are further used to determine the elevation values of the refined DEM. Computation of the surface albedo is hence critical to this refinement process. It is derived from multiple images using Equation (3):

$$\begin{cases} A_{x,y} = E[A_{i \in n}]_{x,y} \\ A_{i \in n} = \dfrac{I_{i \in n}}{G_{i \in n}(Z_{DEM})} \end{cases} \tag{3}$$

where A_i is the albedo computed, based on the i th image inten - sity I_i and its corresponding reflectance value $G_i(Z_{DEM})$ derived from the initial photogrammetric DEM under the same illumination conditions as the ith image. The albedo of a DEM node $A_{x,y}$ can then be determined by evaluating all albedo candidates from the n images. Because the computation of A_i is sensitive to factors such as shadows, occlusions, and the fine - scale quality of the DEM, it is possible to encounter individual A_i with extreme or invalid values (e. g., A_i is negative or zero). Therefore, the median operator $E[*]$ is selected to determine the albedo $A_{x,y}$ to ensure robust estimation.

An albedo map can be generated after computing the albedo for each DEM node using Equation (3). The resulting albedo map is then smoothed using a Gaussian filter to eliminate spike noises. Considering the relatively smooth terrain and homogeneous sur - face albedo in local regions, this albedo map is valid for the next step of photoclinometric refinement.

For photoclinometric refinement, a higher - resolution DEM is first interpolated from the initial photogrammetric DEM. The geometric reflectance values for each DEM node can be estimated by rearranging Equation (1) to solve for $G(Z)$ as:

$$\overline{G}_{i \in n} = \frac{I_{i \in n}}{A} \tag{4}$$

where $\overline{G}_{i\in n}$ is the estimated reflectance with the illumination and viewing conditions of the *i th image, and* A is the corresponding albedo of the DEM node. The elevation values of the interpolated higher - resolution DEM are then optimized by minimizing the differences in the reflectance values between the DEM and the estimated values from multiple images using the following Equation:

$$arg \min_{Z_{DEM}} e_{phcl}(\sigma) = \sum_{i\in n} g(\overline{G}_{i\in n} - G_i(Z_{DEM}), \sigma)^2 \quad (5)$$

where $\overline{G}_{i\in n}$ is the estimated reflectance from the ith image using Equation (4). $G_i(Z_{DEM})$ is the reflectance computed from the current DEM with the same illumination and viewing conditions of the ith image using Equation (2). e_{phcl} is the error function that measures the difference between the estimated reflectance values and that of the current DEM. Equation (5) is constructed for each DEM node, and a relaxation strategy is used in the refinement process, which is iterative in nature. A hierarchical coarse - to - fine architecture is implemented to evaluate the error function e_{phcl} at multiple scales, which is realized by setting a predefined scale parameter of σ in Equation (5).

2.3 Generated topographic models of the Chang'E - 4 landing site and evaluation

In this study, 71 pairs of Pancam images collected by the Jade Rabbit - 2 rover at three waypoints close to the landing location were used for high - resolution topographic modeling and analysis. At waypoint 1 (WP1), the rover collected seven stereo pairs of Pan - cam images that covered the back side of the lander. The rover then moved to the other side of the lander at WP2 and collected six stereo pairs of images that covered the front side of the lander. The rover further traveled to WP3 and collected 58 stereo pairs of images that covered the vicinity of the landing location (about 50 ×50m). Fig. 2 shows the Pancam images and the locations of the waypoints and the Chang'E - 4 lander (Wu et al., 2020).

The spatial resolutions of the Pancam images decrease as the range increases. The highest spatial resolution is about 3 mm/pixel at approximately 3 m from the camera, and the lowest spatial resolution is about 5 cm/pixel at approximately 30 m from the camera. The previously described photogrammetric processing and photoclinometric refinement methods were applied to the Pancam images, and DEMs of the landing site were generated with a uni - fied spatial resolution of 1 cm/pixel. Fig. 3 shows a side - by - side comparison of the DEMs generated by photogrammetry and after photoclinometric refinement, respectively.

The DEM covers an area of about 50 ×50m, and the Chang'E - 4 lander is located in the southeast part of the area. The dark regions in the DEM were not modeled, either due to occlusion or due to their distances from the camera. Although the same spatial res - olution of 1 cm/pixel was applied for both DEMs, the photogrammetric DEM shows a

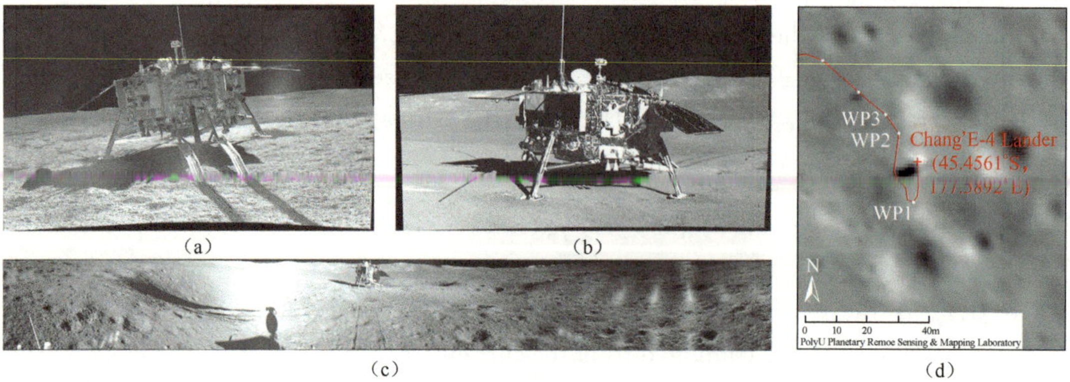

Fig. 2　Mosaics of the Pancam images (left images of the stereo pairs used for mosaicking) collected at the waypoints (WP). (a) An image mosaic at WP1; (b) An image mosaic at WP2; (c) A panoramic image mosaic at WP3; (d) Locations of the WPs and the Chang'E－4 lander (the base map is an LROC NAC image)

much lower effective resolution, especially for regions relatively far from the camera (e.g., the upper part of the crater in the northeast) and regions without sufficient coverage of stereo images. The former difficulty is due to the substantial decrease in the density of the photogrammetric point clouds generated by dense image matching at far ranges, which leads to severe interpolation artifacts in the photogrammetric DEM. Photogram－metry also has no way to generate 3D results for regions without sufficient coverage of stereo images, and serious interpolation artefacts appear in these regions, as shown in Fig. 3 (c). In contrast, the integrated photoclinometric approach incorporates the photometric information into the 3D reconstruction process based on the photogrammetric DEM, resulting in better effective resolution, as seen in Fig. 3 (d).

Fig. 3 (c) and (d) show a clear improvement in the DEM from the integrated photoclinometric approach relative to the photogrammetric DEM. It is apparent from Fig. 3 (c) that the rover tracks and the terrain details between the tracks cannot be recov－ered by photogrammetry due to the absence of dense matching points and possible occlusions in one of the stereo images. In the refined DEM, the rover tracks and terrain details are successfully recovered because photoclinometry can capture subtle topographic details and can work with only a single image.

To evaluate the accuracy of the generated DEM from the Pancam images, a DEM with a spatial resolution of 1.5 m/pixel generated from stereo LRO NAC images at the Chang'E－4 landing site using a photogrammetric approach followed by a photoclinometric refinement (Liu and Wu, 2020) was used for comparison, as shown in Fig. 4 (a). The LROC team also generated a DEM with a resolution of 5 m/pixel of the Chang'E－4 landing area using a stereo pair of NAC images, which was not used here due to its lower

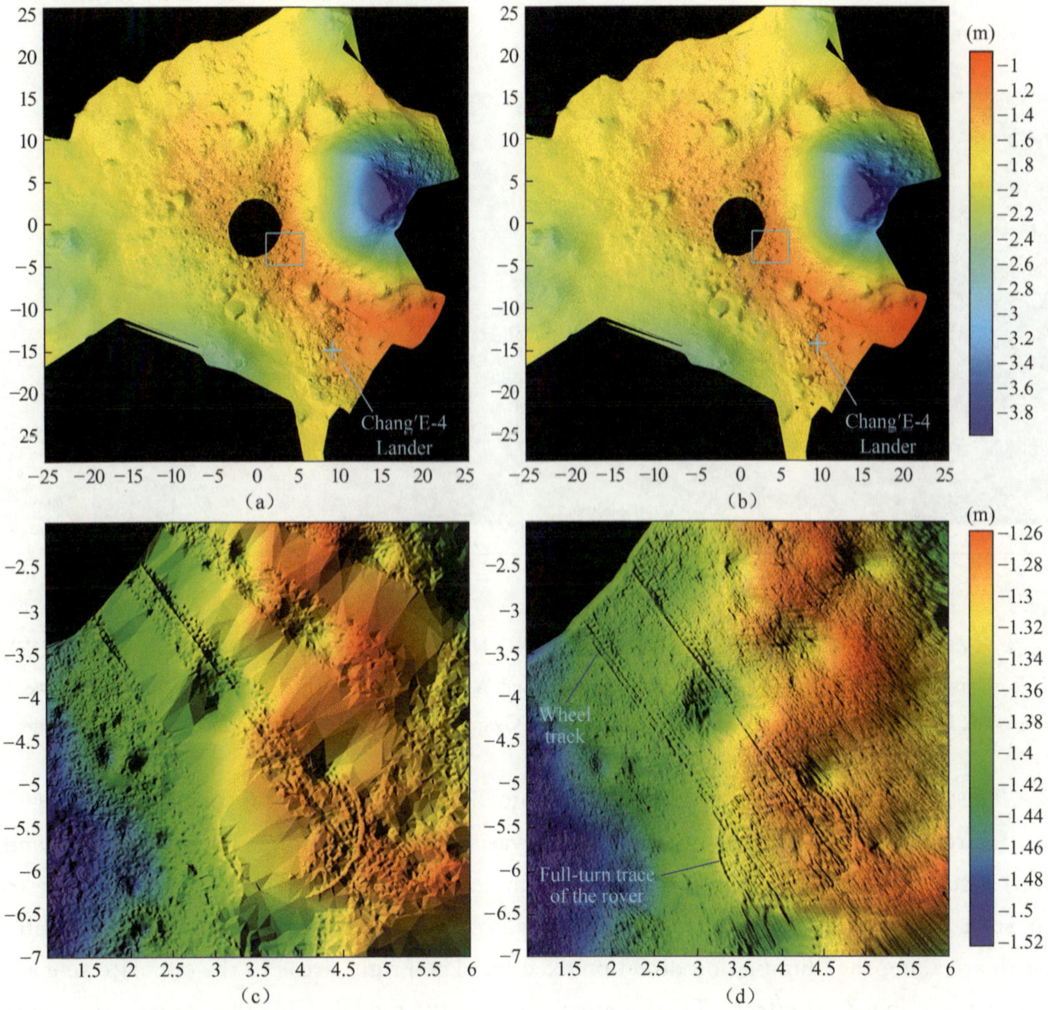

Fig. 3　Color - shaded views of the generated DEMs. (a) The photogrammetric DEM (1 cm/pixel); (b) Refined DEM by photoclinometry (1 cm/pixel); (c) and (d) are the enlarged views of the local region marked by rectangles in (a) and (b). (For interpretation of the colors in the figure (s), the reader is referred to the web version of this article)

spatial resolution. The differences in elevation between the NAC DEM and the DEM from the Pancam images were calculated for a difference map, as shown in Fig. 4 (b), in which red indicates that the Pancam DEM is lower than the NAC DEM and blue indicates that it is higher. According to the difference map, the elevations of the Pancam DEM are generally consistent with those of the NAC DEM, with 68% of the area showing elevation differences of less than ±0.5 m and 50% showing differences of less than ±0.3 m. The largest positive difference (red) of 1.2m appears at the lander location because the lander is removed from the Pancam DEM but preserved in the NAC DEM. The negative elevation difference (blue) increases in further regions, possibly due to the reduced measurement accuracy of elevations far from the Pancam images and due to the uncertainties in the NAC DEM itself.

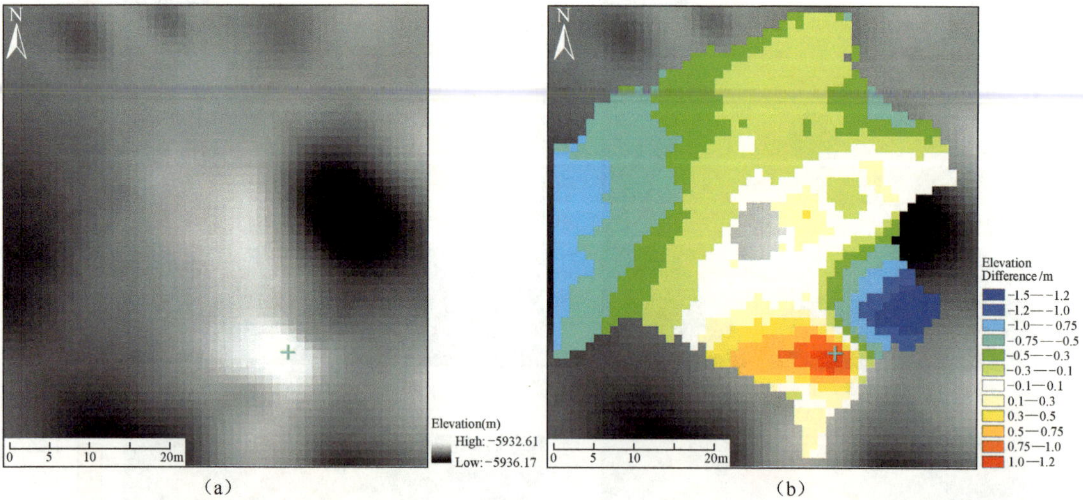

Fig. 4 Comparison between DEMs generated from LRO NAC images and Pancam images. (a) The DEM (1.5 m/pixel) generated from LRO NAC images (Liu and Wu, 2020); (b) Elevation difference map between the NAC DEM and the Pancam DEM (1 cm/pixel). Cyan cross indicates location of Chang'E-4 lander

The generated albedo map from the photoclinometric process is presented in Fig. 5 (a) for reference, in which moderate albedo variations can be noticed around craters and rover tracks. It should be noted that the bright rays visible on the albedo map are due to the solar glare (see the inset Pancam image in Fig. 5a) reflected by the lander itself, which are along the illumination direction exactly. The bright areas in the opposite side are due to the opposition effect. They were not perfectly modeled by the Lunar-Lambert model. But fortunately, they didn't apparently affect the surface reconstruction because of the constraint from the photogrammetric DEM.

An orthoimage mosaic with a spatial resolution of 1cm/pixel was generated from the Pancam images of the same area of the DEM, as shown in Fig. 5 (b). The rover tracks can be clearly seen on the orthoimage mosaic, which allows an evaluation of the planimetric accuracy of the 3D mapping results. Forty width measurements of the wheel tracks were obtained from various distances from the rover locations (WP3 and WP1) and compared with the actual track width (15 cm) (Ding et al., 2020). Table S2 summarizes the comparison results. Overall, the measured track widths agree well with the actual value, with a root mean square error (RMSE) of 1.1cm from a distance range of 4 to 7.8 m from the camera.

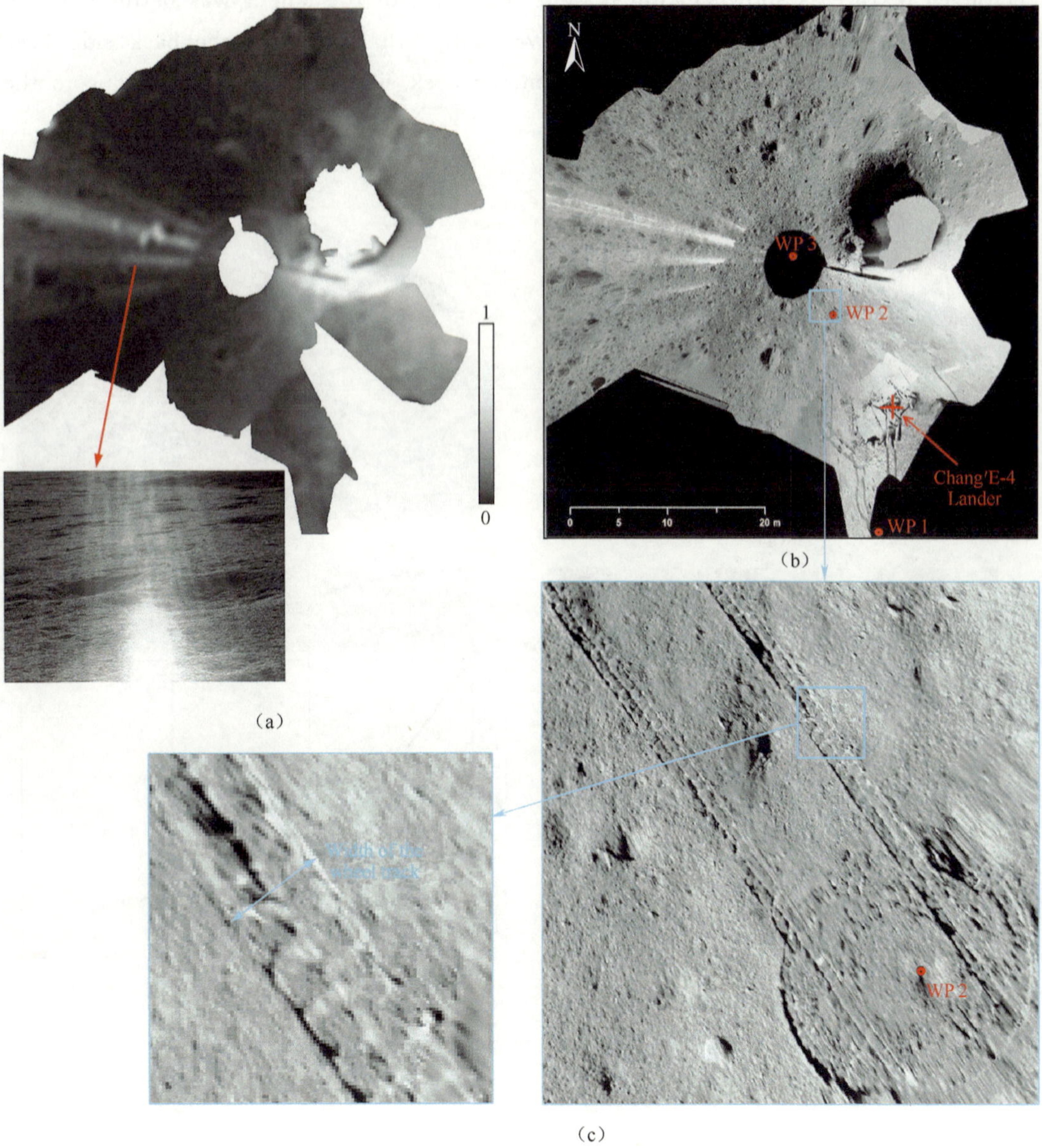

Fig. 5 (a) The generated albedo map from photoclinometric processing; (b) The generated orthoimage mosaic (1cm/pixel); (c) Enlarged views of the local area on the orthoimage mosaic, where the rover wheel tracks are clearly visible. The width of the wheel tracks is about 15cm and the separation of the two wheel tracks is about 70cm

2.4 3D model of the Chang'E－4 lander and measurement of its orientation and tilt

The previous photogrammetric process was used to generate a 3D model of the Chang'E－4 lander and extract it as a separate 3D model for convenience of analysis. The grayscale Pancam images were converted to RGB mode with artificial colors using a Bayer color filter array (Jia et al., 2018), so that the 3D model of the lander was mapped with a colorized texture as shown in Fig. 6. The front side of the lander was better modeled because the image quality (see Fig. 2b) was better than that for the back side (see Fig. 2a). Part of the rover descent ramp on the back side was not modeled due to the shadow of the lander itself.

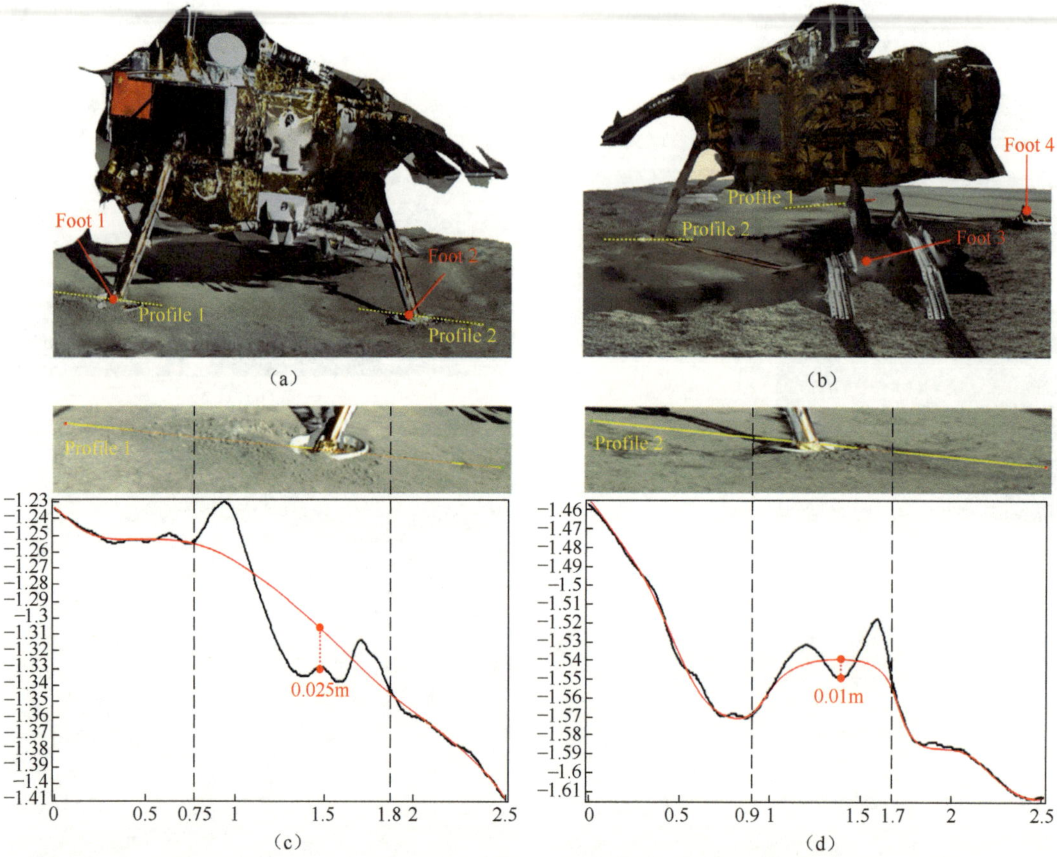

Fig. 6　3D model of Chang'E-4 lander and terrain profiles under the two feet. (a) Front side of the lander; (b) Back side of the lander; (c) Profile 1 under Foot 1; (d) Profile 2 under Foot 2

Using the 3D model of the lander, 3D profiles of the terrain under the lander's two feet were measured, as shown in Fig. 6 (c) and (d). When the lander touched down on the lunar surface, surface regolith was pushed away, forming small mounds around the lander's feet. Based on the terrain profiles in the local regions surrounding the feet, the original terrain profiles (before touchdown) were estimated using Bézier curves (red curves). The elevation differences between the original and actual terrain profiles after touchdown were approximately 0.025m and 0.01m for Foot 1 and Foot 2, respectively. The elevation at Foot 1 is about −1.33m, and the elevation at Foot 2 is about −1.55m, which may in-dicate that the Chang'E-4 lander landed on a slope and that Foot 1 touched down on the surface before Foot 2 and with a stronger impact force into the regolith.

To further analyze the orientation and tilt information of the lander itself, the 3D coordinates of the four feet of the Chang'E-4 lander were measured in the LCS defined in the rover body-fixed frame at WP3 and further fitted by a plane as shown in Figure S1. According to the normal vector (−0.067, −0.021, 0.997) of the fitted plane, the Chang'E-4 lander has a tilt angle of 4°, toward 72.3°southwest, after landing on the lunar surface. This is consistent with our previous finding based on a 1m-resolution DEM generated from the LRO NAC images (Wu et al., 2020).

3 Fine-scale analysis of craters and rocks at the Chang'E-4 landing site

The DEM and orthoimage mosaic with a 1cm/pixel resolution at the Chang'E-4 landing site generated from the previous steps enable a fine-scale analysis of the craters and rocks at the landing site at the centimeter level. The following section presents the fine-scale analysis results, comparisons with corresponding analysis on a large scale, and implications for lunar geology.

3.1 Crater distribution

Craters are common features on the Moon, and their distribution patterns may reveal the geological history and impact process of the lunar surface. Based on the 1cm/pixel orthoimage mosaic, automatic crater detection was performed using an active machine-learning algorithm (Wang and Wu, 2019), followed by careful manual checking to remove possible incorrect detections and digitize missing craters. Finally, 1023 craters with diameters greater than 10cm (equivalent to 10 pixels on the orthoimage) were detected at the Chang'E-4 landing site. Their diameters ranged from 10cm to 28.6m. Fig. 7 (a) shows the distribution of the detected craters. The enlarged view in the right of Fig. 7 (a) shows examples of small craters with diameters of decimeters or meters.

It is interesting to see the size-frequency distribution of craters of such small sizes because they have seldom been reported. The size-frequency distribution of the above-detected craters is shown in Fig. 7 (c) as purple squares. For a comparison study, data for the larger craters in a larger area covering the entire Von Kármán crater floor obtained from a separate study (Wang and Wu, 2020) were used here for analysis, as shown in Fig. 7 (b). These craters are larger than 350m in diameter. Their size-frequency distribution is also included in Fig. 7 (c) as black squares, which indicate an estimated surface age of about 3.65 Ga of the Von Kármán crater mare floor. Although craters with diameters smaller than 10m are not meaningful for estimation of the surface age (Neukum et al., 2001), the size-frequency distribution of very small craters (purple squares) shows overall a similar power-law trend as the large craters (black squares). However, a clear gap can be seen between the purple squares and the solid black line (i. e., the modeled curve from the black squares), which indicates that using the standard crater production function (Neukum et al., 2001), the accumulated impact flux appears to be different for the two populations. This gap may indicate different surface ages of the Von Kármán crater floor on a large scale and the Chang'E－4 landing site on a fine scale, possibly due to resurfacing events around the landing site.

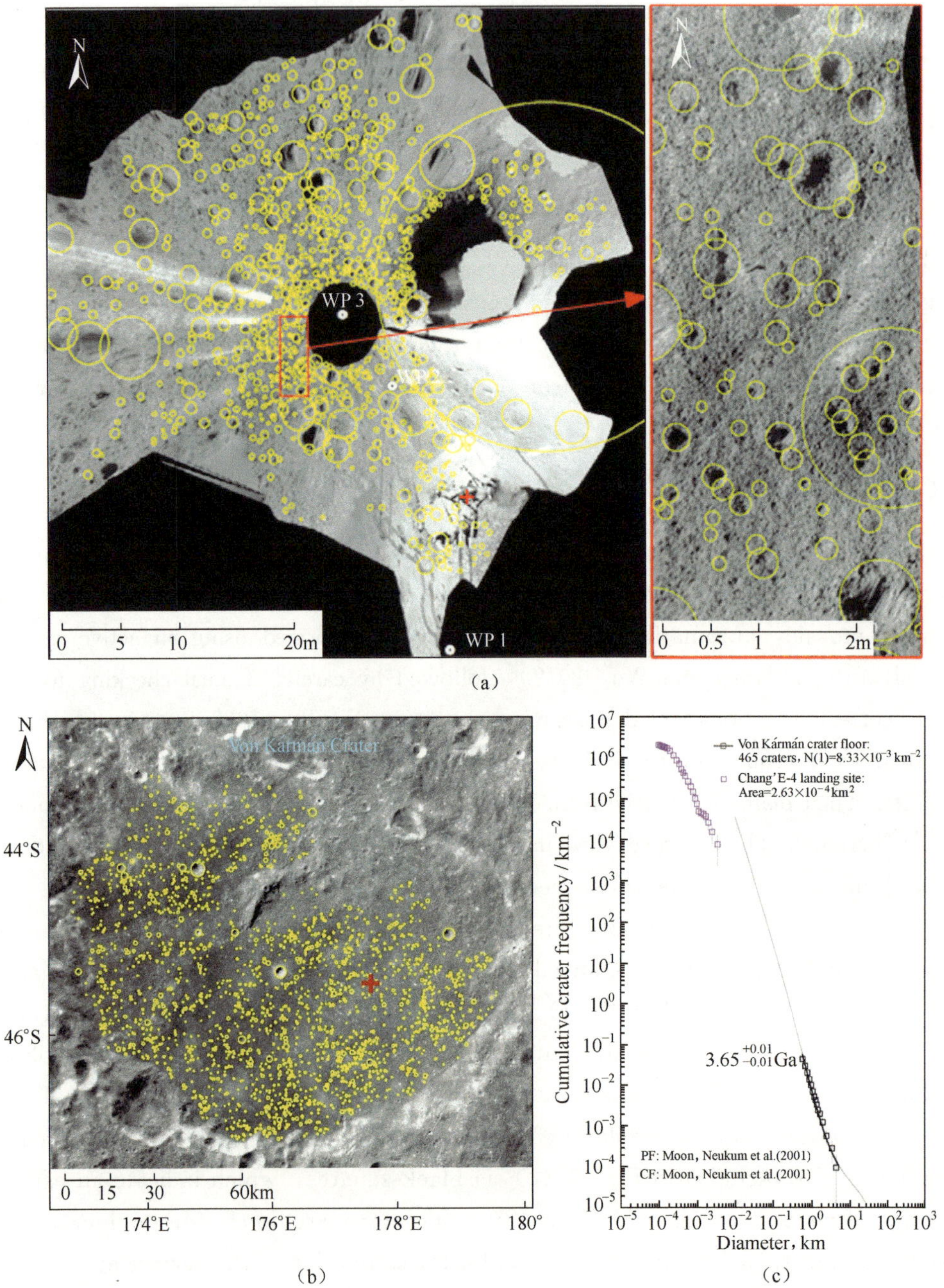

Fig. 7 (a) Distribution of craters (≥10cm in diameter) at the Chang'E－4 landing site; (b) Distribution of craters (≥350m) on the Von Kármán crater floor; (c) Size－frequency distributions of the two crater populations. Red crosses in (a) and (b) indicate the lander location

It is interesting to see the depth - to - diameter ratio of craters of such small sizes. Analysis of the crater depth - to - diameter ratio is useful to formulate models of crater degradation and modification (Pike and Spudis, 1987) and to provide implications for changes in crater morphology (Wünnemann and Ivanov, 2003). Previous studies of the crater depth - to - diameter ratio mainly concerned craters on the kilometer scale, whereas the DEM with a 1cm/pixel resolution generated in this study enables the measurement of the depth information of craters with sizes of decimeters.

We carefully measured the depth information of 310 craters (>0.1m in diameter) with 3D morphological information well pre - served in the DEM. The craters were mainly distributed in the regions near each of the rover locations (WP1, WP2, and WP3), which offers sufficient coverage and textural details to derive 3D information, as shown in Fig. 8 (a). For comparison, craters on the order of kilometers in the SPA basin were selected from a global crater catalogue in a separate study (Wang and Wu, 2020), of which the global crater catalogue by Robbins (2018) was used as a reference in the checking and verification of the craters, and the depth information of each crater was made available as derived from a quasi - global DEM of a spatial resolution of 60m (Barker et al., 2016), by using Gaussian fitting (Wang and Wu, 2020) of the elevation profiles from the DEM. In total, 128, 498 craters (≥1km in diameter) in the SPA basin were selected, as shown by yellow circles in Fig. 8 (b). Fig. 8 (c) is a box - and - whisker plot of the depth - to - diameter ratios for craters at the Chang'E - 4 landing site and in the SPA basin within various diameter ranges. In each diameter range, the whiskers represent the minimum and maximum depth - to - diameter ratios of all craters within this range, and the upper quartile, median, and lower quartile are marked by the box. The left part of Fig. 8 (c) shows the depth - to - diameter ratios of the small craters at the Chang'E - 4 landing site (range: 0.02 to 0.12; average depth - to - diameter ratio: 0.061). The right part of Fig. 8 (c) shows the depth - to - diameter ratios of craters on the kilometer scale in the SPA basin, ranging from 0.3 to nearly 0.

The depth - to - diameter ratios of craters vary for different types of craters (Kalynn et al., 2013). Previous research has suggested that secondary craters are generally relatively shallow with depth - to - diameter ratios of about 0.1 or less (Pike, 1971; McEwen and Bierhaus, 2006). Referring to the average depth - to - diameter ratio of 0.061 for the small craters at the Chang'E - 4 landing site (red line in Fig. 8c), craters with diameters from 1km to about 30km in the SPA basin have obviously larger depth - to - diameter ratios. This indicates that the crater morphology appears to be different for these two crater populations, which may be related to the strength of the target material (Senft and Stewart, 2007) or impactor speed (Prieur et al., 2018). This may also indicate that the small craters at the Chang'E - 4 landing site could either be secondary craters or largely degraded primaries.

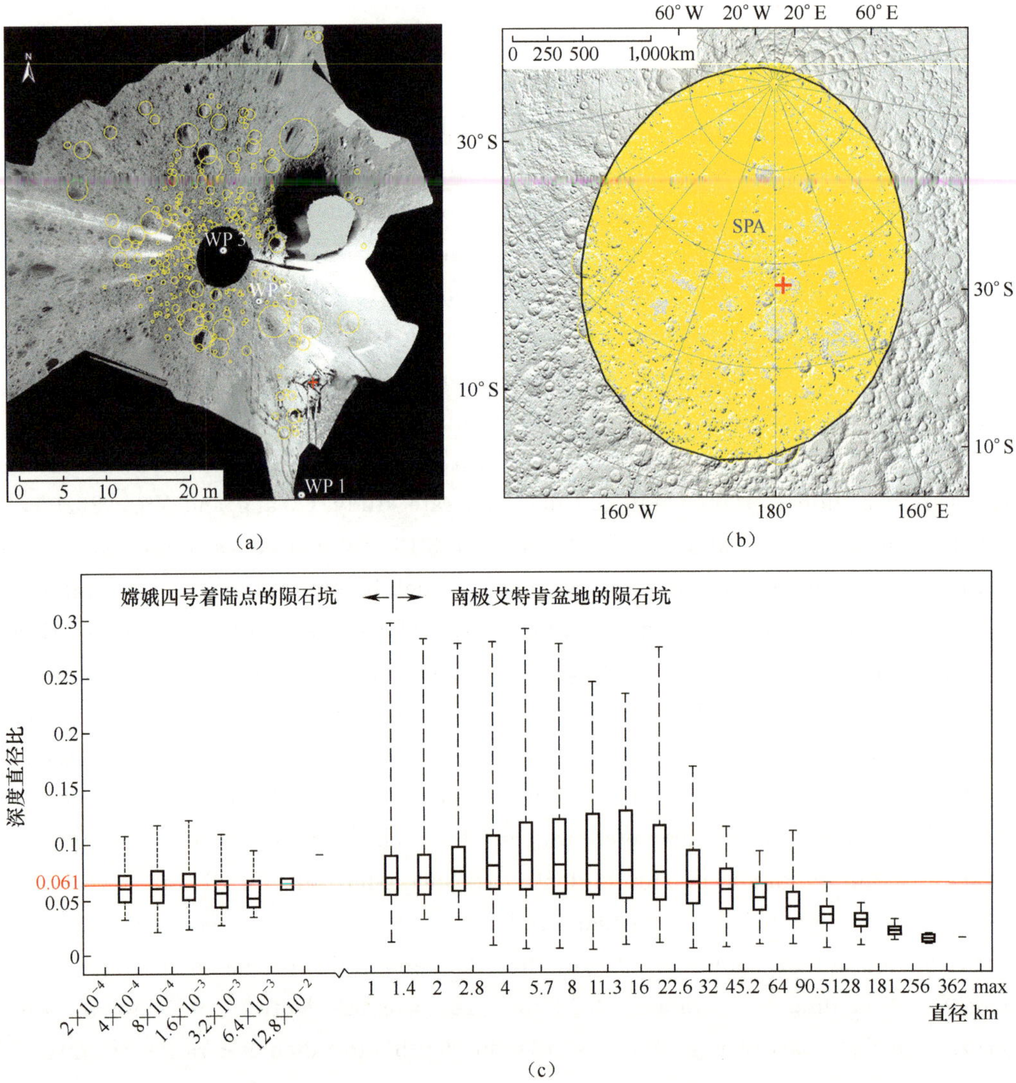

Fig. 8　Comparison of crater depth - to - diameter ratios between (a) small craters (>0. 1m) at Chang'E - 4 landing site; and (b) large craters (>1km) in the SPA basin; (c) Box - and - whisker plot of crater depth - to - diameter ratios. Red crosses in (a) and (b) indicate the lander location

3. 2　Rock abundance

Rock abundance, defined as the cumulative fractional area of rocks versus their diameters (Golombek and Rapp, 1997; Bandfield et al., 2011; Li and Wu, 2018), is not only an essential criterion for the selection of a safe landing site but also an important means for rock - related scientific studies (Ding et al., 2020). Because rocks are relatively smaller features than craters, rover images collected from the ground are favorable for investigation of the distributions and physical properties of rocks.

To analyze the rock distribution and abundance at the Chang'E - 4 landing site, an interactive method similar to that proposed by Di et al. (2016) was used to extract rocks from the Pancam images. A total of 547 rocks with a minimum diameter of 12cm and a maximum diameter of 0.45m were extracted, as shown in Fig. 9 (a). In general, most small rocks with diameters of less than 50cm were concentrated around the center of the study area because the spatial resolutions of the Pancam images decrease from near (3mm/pixel) to far (5cm/pixel), as mentioned previously, so small rocks were unrecognizable at far ranges. As shown in Fig. 9 (a), a rock with the minimum size of 12cm was located near the center of the area, occupying about 4 pixels in the Pancam image. A rock with the maximum size of 0.45m was found near the border of the study area, which is represented by about 10 pixels in the Pancam image. Fig. 9 (b) shows the size-frequency distribution of the extracted rocks (blue triangles) at the Chang'E-4 landing site.

Li and Wu (2018) presented a rock abundance model based on a size-frequency distribution analysis of rocks identified on high-resolution orbital, descent, and ground images that cover various terrain types on the lunar surface. It has an exponential format as follows:

$$F_k(D) = k \cdot e^{-(0.5648+0.01258/k)\cdot D} \tag{6}$$

where D is the rock diameter, k indicates the rock abundance, and F_k (D) is the fractional area covered by rocks with diameters greater than D. It should be noted that the power-law function was not considered in this research, as it is more suitable for fitting the distributions of relatively larger rocks (e. g., boulders $>$ 25cm in diameter); however, the power-law function may overestimate the number of rocks of smaller diameters (Golombek and Rapp, 1997). Also, an exponential form of the rock abundance model can be better explained by the fracture and fragmentation theory (Gilvarry, 1961; Golombek and Rapp, 1997).

Fig. 9 (b) shows that the actual size-frequency distribution of the extracted rocks can mostly fitted by the rock abundance curve based on Equation (6), with a rock abundance value k=0.21%, thus indicating that the Chang'E-4 landing site has a very small value of rock abundance (0.21%). As mentioned above, the spatial resolution of the Pancam images varies from about 3mm/pixel to 5cm/pixel from near to far ranges, so rocks smaller than 8cm in diameter (rocks that occupy at least two pixels at the far ranges) are underestimated at the far ranges. This means that the size-frequency curve for rocks less than 8cm in diameter is under the rock abundance curve, as indicated by the first dashed line in Fig. 9 (b). For rocks larger than 0.3m in diameter, the two curves are not consistent because only four rocks in the study area are larger than 0.3m, and the inadequacy of rocks within this range introduces randomness into the statistics.

In general, the rock abundance result at the Chang'E-4 landing site obtained from this study shows good agreement with the estimated rock abundance values of relatively larger rocks in previous studies (Wu et al., 2020; Bandfield et al., 2011). Wu et al.

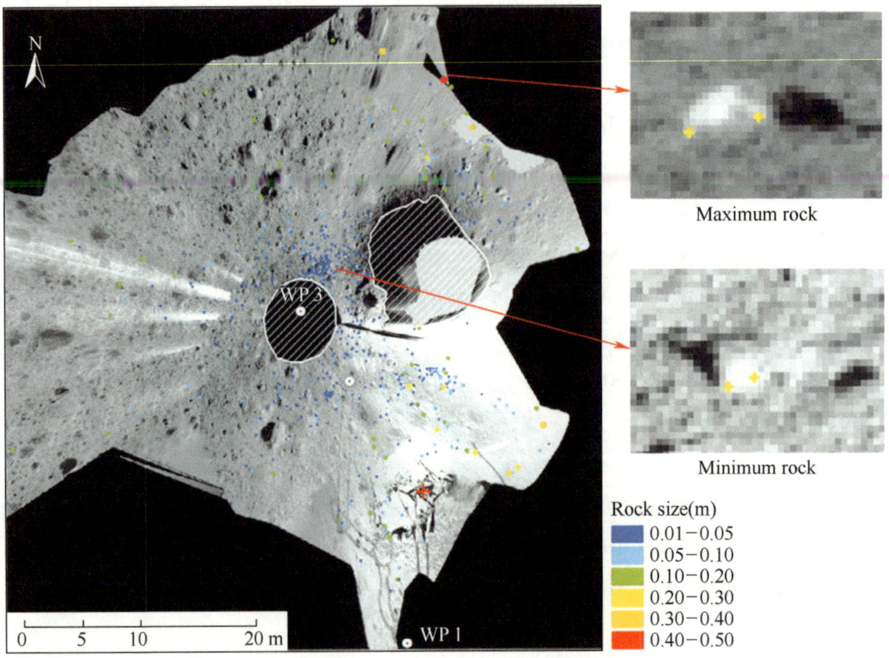

(a)

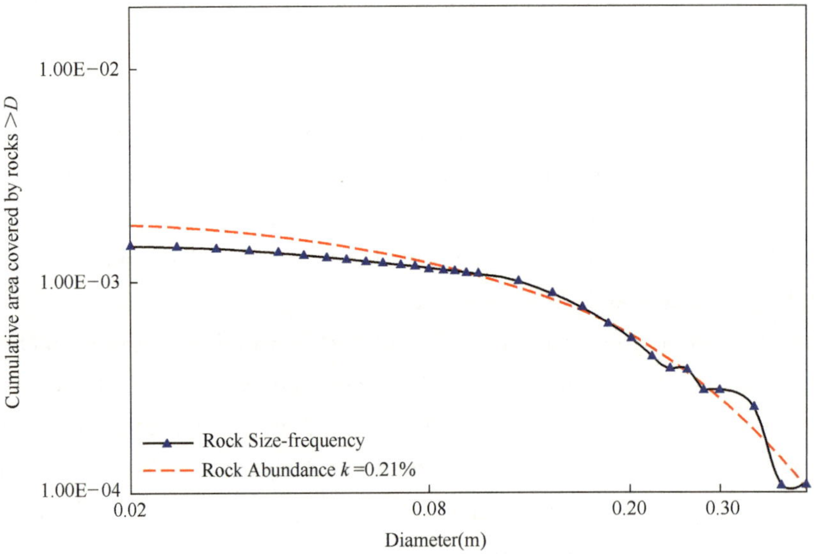

(b)

Fig. 9 (a) Rock distribution at Chang'E-4 landing site. Rocks of various sizes are represented by different colors. Yellow crosses in inset images to the right indicate start and end points for measurement of rock size shown on Pancam images. Red cross in left figure indicates lander location, and shaded parts represent regions where rocks cannot be recognized due to shadows or lack of image coverage. (b) Rock size-frequency distribution and rock abundance curve

(2020) used high - resolution LRO NAC images (1m/pixel) to investigate the rock abundance in the targeted landing region of Chang'E - 4 on the Von Kármán crater floor (a local region of about 50 × 30 square kilometers), and the results indicated that the cumulative areal fraction of rocks larger than 2m was almost 0 around the location of the Chang'E - 4 lander. Bandfield et al. (2011) generated a global rock concentration map of the lunar surface based on the LRO Diviner Radiometer data, and the corresponding rock abundance for rocks larger than 1m was 0.2% at the Chang'E - 4 landing site as interpolated from the rock concentration map.

3.3 Relationships among slopes, craters, and rocks

Maps of extracted craters and rocks at the Chang'E - 4 landing site were co - registered for joint analysis, as shown in Fig. 10. To facilitate the analysis, the slopes of the study area were also calculated based on the DEM with a resolution of 1cm/pixel. The slopes were classified into categories with an interval of 5°for slopes less than 25°, also shown in Fig. 10. Fig. 10 suggests an apparent relationship between the slopes and craters in which most slopes of 5°to 10°or greater (grayscale) appear inside the craters and slopes of less than 5° (yellow) mostly appear outside the craters.

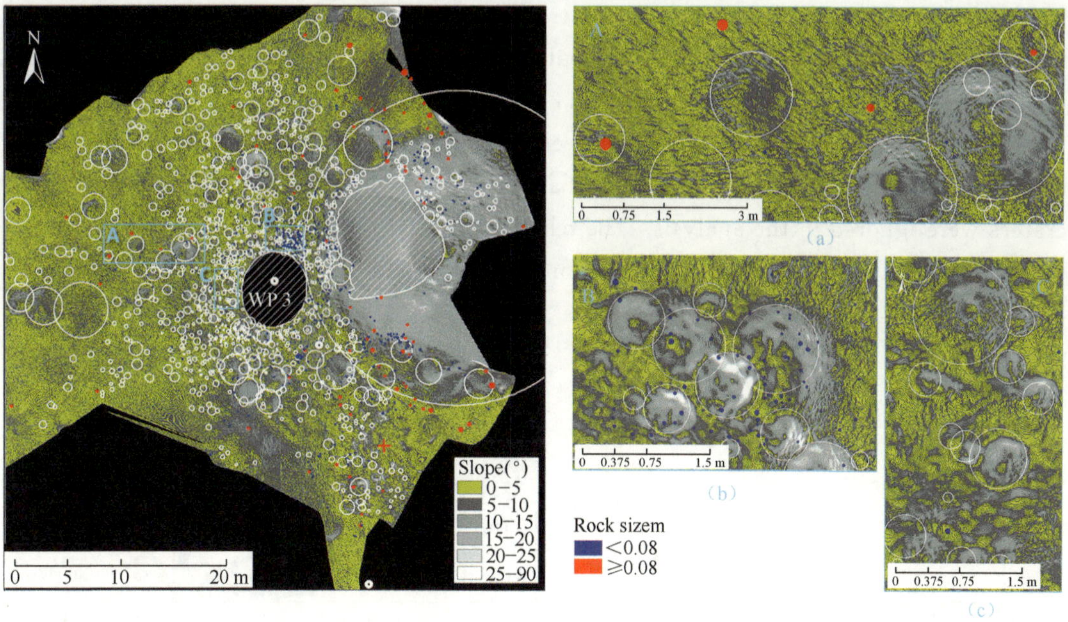

Fig. 10　Joint analysis of slopes, craters, and rocks at Chang'E - 4 landing site. Inset figures on right show enlarged views of areas marked by cyan boxes on left. Red cross denotes lander location

The rocks were grouped into two categories: relatively large rocks (≥0.08m) that we believe have been completely extracted, as shown in red in Fig. 10, and smaller rocks (<0.08m), shown in blue, that we believe have been underestimated due to the de - creasing spatial resolution of the Pancam images. Fig. 10 shows that most of the relatively large rocks were distributed inside or near the largest crater (28.6m in diameter) in the east part of the area. Several isolated large rocks were also found 20 to 30m from the largest crater, as shown in the enlarged view of area A in Fig. 10. Most of the small rocks (<0.08m) were concentrated inside some small craters (≤1m in diameter), as shown by the examples in Fig. 10. Because 93% of craters in the study area were smaller than 1m in diameter, it can be assumed that most of the large rocks (≥0.08m in diameter) were the ejecta of the largest crater in this area, and ejection can reach at least 30m from the rim of the largest crater.

Fig. 10 shows enlarged views of the local areas of B and C. Because they were near the camera (<8 m to WP3), we believe that the craters and rocks in these two areas have been reliably extracted. The rock distributions in these two areas, however, showed different patterns even though the craters in both areas had similar sizes (approximately 10cm to 1m in diameter). The craters in area B featured more rocks and larger slopes than those in area C, lead-ing to the assumption that craters with larger depth-to-diameter ratios were likely to have more rocks inside or near them.

To verify the above assumption, 117 craters (diameter range: 10cm to 1.8m) within 8m of WP3 and rocks located inside these craters were selected for further analysis. The crater depth-to-diameter ratios and the fractional areas covered by rocks are plotted in Fig. 11 (a), which shows an apparent trend in which the latter increased as the former increased. However, some craters (red dots in Fig. 11a), including some with depth-to-diameter ratios of approximately 0.1, had few or no rocks inside them, possibly because the Pancam images had low solar elevation angles and the craters were concaved into the terrain, resulting in shadows that prevent rock extraction. Therefore, these craters were ignored in the analysis. The relationship between the depth-to-diameter ratios of the remaining craters and the fractional area covered by rocks can be presented by a linear function, as shown in Fig. 11 (a).

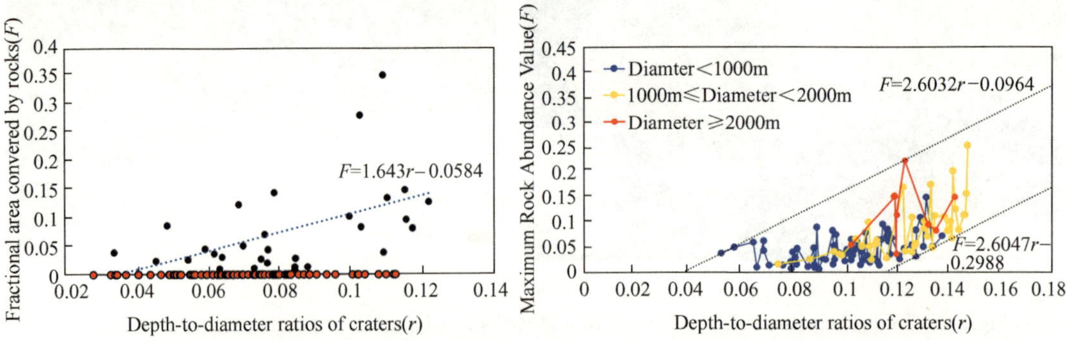

(a)　　　　(b)

Fig. 11　Relationship between rocks inside craters and depth-to-diameter ratios of craters. (a) Fractional area covered by rocks inside craters vs. depth-to-diameter ratios of craters (<1.8m in diameter) on fine scale at the Chang'E-4 landing site. The red dots represent craters with few or no rocks detected inside them due to problems such as shadows, which were ignored in the least-squares fitting; (b) Maximum rock abundance vs. depth-to-diameter ratio of craters (>200m in diameter) on large scale in the Oceanus Procellarum (Li and Wu, 2018)

Li and Wu (2018) studied the relationship between the crater depth - to - diameter ratios and rock abundances for craters larger than 200m in diameter in the Oceanus Procellarum on the near side of the Moon. They found linear relationships similar to those shown in Fig. 11 (b). The results on both fine and large scales in - dicate that craters with depth - to - diameter ratios of approximately 0. 03 or less were likely to have no rocks inside them. This may suggest that as lunar craters evolve (Head, 1975; Pike, 1971), lunar regolith may gradually cover the craters and the rocks inside the craters may degrade, resulting in lower depth - to - diameter ratios and rock abundances.

It should be noted that there are other factors that may influence whether or not a crater produces rocks and the distribution of rocks around craters. Wilcox et al. (2005) emphasized that knowledge of regolith thickness is important for the interpretation of materials on the lunar surface, as a crater must excavate beneath the regolith and impact bedrock in order to produce rocks. Bart and Melosh (2010) presented that larger rocks lie preferentially closer to the crater rim, and small rocks occur at all distances, dependent on the ejection velocities. They also stated that the number of rocks inside (rather than outside) a crater reflects the terrain type, with the more fractured terrain producing more internal rock, presumably as slump blocks in the modification stage of crater formation. These factors need to be considered as well in order to have a better understanding of the relationship between the distribution of rocks and the morphologies and ages of craters.

4 Conclusions and discussion

In this paper, we present an integrated photogrammetric and photoclinometric approach for the generation of 3D models with a resolution of 1cm/pixel for the Chang'E - 4 landing site using rover images collected on the ground. We report the results of fine - scale analysis of the craters and rocks at the Chang'E - 4 landing site based on the generated high - resolution topographic models. The developments and results lead to the following conclusions:

1) The DEM generated from an integrated photogrammetric and photoclinometric approach offers better effective spatial resolution that cannot be achieved solely with photogrammetry and offers favorable geometric accuracy.

2) The generated 3D model of the lander reveals that Chang'E - 4 landed on a terrain slope of 4°toward 72. 3°southwest, with one foot sunk into the regolith 1. 5cm deeper than the other foot on the front side of the lander.

3) Fine - scale analysis of craters with diameters of decimeters to meters shows a power - law trend similar to that of large craters hundreds of meters in size. However, there is a clear gap between their size - frequency distributions, indicating the accumulated impact flux appears to be different for these two crater populations. The average depth - to - diameter ratio of the small craters at the Chang'E - 4 landing site was 0. 061, which is

obviously lower than the depth - to - diameter ratios of craters with diameters between 1km and about 30km in the SPA basin, indicating the small craters at the Chang'E - 4 landing site were either secondary craters or largely degraded primaries.

4) Fine - scale analysis of rocks revealed that the size - frequency distribution of rocks at the Chang'E - 4 landing site is well rep - resented by the rock abundance model of an exponential format (Li and Wu, 2018), with a rock abundance of 0.21%, which agrees well with previous studies based on large - scale analysis using LRO NAC images (Wu et al., 2020) and LRO Diviner Radiometer data (Bandfield et al., 2011).

5) Joint analysis of craters and rocks indicated that most of the relatively large rocks (≥80cm in diameter) were the ejecta of the largest crater in the east part of the study area. The fractional areas covered by rocks inside craters increased as the crater depth - to - diameter ratios increased, and craters with depth - to - diameter ratios smaller than 0.03 were likely to have no rocks inside. This observation is generally consistent with the analysis results of craters larger than 200m in diameter in the Oceanus Procellarum (Li and Wu, 2018).

It should be noted that the generated 3D models and orthoimage mosaic from the rover images may still have defects due to the challenging illumination conditions and occlusions. It would be highly desirable to integrate rover images at multiple waypoints to provide better coverage. Further in - depth scientific interpretation of the observed results regarding the fine - scale analysis of craters and rocks would also be desirable. Nevertheless, the integrated photogrammetric and photoclinometric approach for refined 3D topographic modeling from rover images has significance in planetary mapping and can be useful in other similar applications. The fine - scale analysis of craters and rocks presented in this study and their comparative analysis on a large scale provide new in - sights into lunar surface geology.

Acknowledgements

This work was funded by the China Academy of Space Technology (Project No: 17CPIT/HK0102) and a grant from the Research Grants Council of Hong Kong (Research Impact Fund Project No: R5043 - 19). The work was also supported by a grant from the National Natural Science Foundation of China (Project No: 41671426). The Chang'E - 4 Pancam images are available at the Data Release and Information Service System of China's Lunar Exploration Program (http://moon.bao.ac.cn/searchOrder_dataSearchData.search). The authors would like to thank all those who worked on the archive of the datasets to make them publicly available.

References

[1] Agarwal, S., Mierle, K., 2012. Ceres Solver: Tutorial & Reference, vol. 2. Google Inc., p. 72.

[2] Bandfield, J. L., Ghent, R. R., Vasavada, A. R., Paige, D. A., Lawrence, S. J., Robinson, M. S., 2011. Lunar surface rock abundance and regolith fines temperatures derived from LRO Diviner Radiometer data. J. Geophys. Res., Planets 116.

[3] Barker, M., Mazarico, E., Neumann, G., Zuber, M., Haruyama, J., Smith, D., 2016. A new lunar digital elevation model from the Lunar Orbiter Laser Altimeter and SELENE Terrain Camera. Icarus 273, 346 - 355.

[4] Bart, G. D., Melosh, H. J., 2010. Distributions of boulders ejected from lunar craters. Icarus 209 (2), 337 - 357.

[5] Basilevsky, A. T., Head, J. W., Horz, F., 2013. Survival times of meter - sized boulders on the surface of the Moon. Planet. Space Sci. 89, 118 - 126.

[6] De Rosa, D., Bussey, B., Cahill, J. T., Lutz, T., Crawford, I. A., Hackwill, T., van Gasselt, S., Neukum, G., Witte, L., McGovern, A., Grindrod, P. M., Carpenter, J. D., 2012. Characterisation of potential landing sites for the European Space Agency's Lunar Lander project. Planet. Space Sci. 74 (1), 224 - 246.

[7] Di, K., Zhu, M. H., Yue, Z., Lin, Y., Wan, W., Liu, Z., Gou, S., Liu, B., Peng, M., Wang, Y., 2019. Topographic evolution of Von Kármán Crater revealed by the lunar rover Yutu - 2. Geophys. Res. Lett. 46, 12764 - 12770.

[8] Di, K., Xu, B., Peng, M., Yue, Z., Liu, Z., Wan, W., Li, L., Zhou, J., 2016. Rock size - frequency distribution analysis at the Chang'E - 3 landing site. Planet. Space Sci. 120, 103 - 112.

[9] Ding, C., Xiao, Z., Wu, B., Li, Y., Prieur, N. C., Cai, Y., Su, Y., Cui, J., 2020. Fragments delivered by secondary craters at the Chang'e - 4 landing site. Geophys. Res. Lett. 47 (7).

[10] Ghent, R. R., Hayne, P. O., Bandfield, J. L., Campbell, B. A., Allen, C. C., Carter, L. M., Paige, D. A., 2014. Constraints on the recent rate of lunar ejecta breakdown and implications for crater ages. Geology 42 (12), 1059—1062.

[11] Gilvarry, J., 1961. Fracture of brittle solids. Distribution function for fragment size in single fracture (theoretical). J. Appl. Phys. 32 (3), 391—399.

[12] Golombek, M., Rapp, D., 1997. Size - frequency distributions of rocks on Mars and Earth analog sites: implications for future landed missions. J. Geophys. Res., Planets 102, 4117 - 4129.

[13] Gou, S., Yue, Z., Di, K., Wan, W., Liu, Z., Liu, B., Peng, M., Wang,

Y., He, Z., Xu, R., 2020. In situ spectral measurements of space weathering by Chang'e - 4 rover. Earth Planet. Sci. Lett. 535, 116117.

[14] Grumpe, A., Belkhir, F., Wöhler, C., 2014. Construction of lunar DEMs based on re - flectance modelling. Adv. Space Res. 53, 1735 - 1767.

[15] Head, H. W., 1975. Processes of lunar crater degradation - changes in style with geologic time. Moon 12 (3), 299 - 329.

[16] Horn, B. K. P., 1990. Height and gradient from shading. Int. J. Comput. Vis. 5 (1), 37 - 75.

[17] Huang, J., Xiao, Z., Flahaut, J., Martinot, M., Head, J., Xiao, X., Xie, M., Xiao, L., 2018. Geological characteristics of Von Kármán crater, northwestern South Pole - Aitken Basin: Chang'E - 4 landing site region. J. Geophys. Res., Planets 123, 1684—1700.

[18] Jia, Y., Zou, Y., Ping, J., Xue, C., Yan, J., Ning, Y., 2018. The scientific objectives and payloads of Chang'E - 4 mission. Planet. Space Sci. 162, 207 - 215.

[19] Kalynn, J., Johnson, C. L., Osinski, G. R., Barnouin, O., 2013. Topographic characterization of lunar complex craters. Geophys. Res. Lett. 40 (1), 38 - 42.

[20] Li, C., Liu, D., Liu, B., Ren, X., Liu, J., He, Z., Zuo, W., Zeng, X., Xu, R., Tan, X., 2019. Chang'E - 4 initial spectroscopic identification of lunar far - side mantle - derived materials. Nature 569, 378 - 382.

[21] Li, Y., Wu, B., 2018. Analysis of rock abundance on lunar surface from orbital and descent images using automatic rock detection. J. Geophys. Res., Planets 1235 (5), 1061—1088.

[22] Liu, J., Ren, X., Yan, W., Li, C., Zhang, H., Jia, Y., Zeng, X., Chen, W., Gao, X., Liu, D., 2019. Descent trajectory reconstruction and landing site positioning of Chang'E - 4 on the lunar farside. Nat. Commun. 10, 1 - 10.

[23] Liu, W. C., Wu, B., 2020. An integrated photogrammetric and photoclinometric ap - proach for illumination - invariant pixel - resolution 3D mapping of the lunar surface. ISPRS J. Photogramm. Remote Sens. 159, 153 - 168.

[24] Liu, Z., Di, K., Li, J., Xie, J., Cui, X., Xi, L., Wan, W., Peng, M., Liu, B., Wang, Y., 2020. Landing site topographic mapping and rover localization for Chang'e - 4 mission. Sci. China Inf. Sci. 63, 140901.

[25] Liu, Z., Di, K., Peng, M., Wan, W., Liu, B., Li, L., Yu, T., Wang, B., Zhou, J., Chen, H., 2015. High precision landing site mapping and rover localization for Chang'e - 3 mission. Sci. China, Phys. Mech. Astron. 58, 1 - 11.

[26] Lohse, V., Heipke, C., 2004. Multi - image shape - from - shading: derivation of planetary digital terrain models using Clementine images. Int. Arch. Photogramm. Remote Sens. 35, 828 - 833.

[27] Lohse, V., Heipke, C., Kirk, R. L., 2006. Derivation of planetary topography using multi - image shape - from - shading. Planet. Space Sci. 54, 661 - 674.

[28] Lowe, D. G., 2004. Distinctive image features from scale - invariant keypoints. Int. J. Comput. Vis. 60, 91 - 110.

[29] Ma, Y., Liu, S., Sima, B., Wen, B., Peng, S., Jia, Y., 2020. A precise visual localisation method for the Chinese Chang'e - 4 Yutu - 2 rover. Photogramm. Rec. 35, 10 - 39.

[30] McEwen, A. S., 1991. Photometric functions for photoclinometry and other applica - tions. Icarus 92, 298 - 311.

[31] McEwen, A. S., Bierhaus, E. B., 2006. The importance of secondary cratering to age constraints on planetary surfaces. Annu. Rev. Earth Planet. Sci. 34, 535 -567.

[32] Neukum, G., Ivanov, B. A., Hartmann, W. K., 2001. Cratering records in the inner solar system in relation to the lunar reference system. Space Sci. Rev. 96 (1 - 4), 55 - 86.

[33] Prieur, N. C., Rolf, T., Wünnemann, K., Werner, S. C., 2018. Formation of simple impact craters in layered targets: implications for lunar crater morphology and regolith thickness. J. Geophys. Res., Planets 123 (6), 1555—1578.

[34] Pike, R., 1971. Genetic implications of the shapes of Martin and lunar craters. Icarus 15 (3), 384 - 395.

[35] Pike, R. J., Spudis, P. D., 1987. Basin - ring spacing on the Moon, Mercury, and Mars. Earth Moon Planets 39, 129—194.

[36] Robbins, S. J., 2018. A new global database of lunar impact craters >1—2km: crater locations and sizes, comparisons with published databases, and global analysis. J. Geophys. Res., Planets 124 (4), 871 - 892.

[37] Senft, L. E., Stewart, S. T., 2007. Modeling impact cratering in layered surfaces. J. Geophys. Res., Planets 112, E11.

[38] Smith, D. E., Zuber, M. T., Neumann, G. A., Lemoine, F. G., Mazarico, E., Torrence, M. H., McGarry, J. F., Rowlands, D. D., Head, J. W., Duxbury, T. H., 2010. Initial observations from the lunar orbiter laser altimeter (LOLA). Geophys. Res. Lett. 37.

[39] Wang, Z., 1998. The research program of the full digitized and automatic mapping system. J. Wuhan Techn. Univ. Surv. Map. 23, 4.

[40] Wang, Y., Wu, B., 2019. Active machine learning approach for crater detection from planetary imagery and digital elevation models. IEEE Trans. Geosci. Remote Sens. 57, 5777 - 5789.

[41] Wang, Y., Wu, B., 2020. A new global catalogue of lunar craters (⩾1km) with 3D information and preliminary results of global analysis. Int. Arch. Photogramm. Remote Sens. Spat. Inf. Sci. 43, 1171 - 1176.

[42] Watkins, R. N., Jolliff, B. L., Mistick, K., Fogerty, C., Lawrence, S. J., Singer, K. N., Ghent, R. R., 2019. Boulder distributions around young, small

lunar impact craters and implications for regolith production rates and landing site safety. J. Geophys. Res., Planets 124, 2754 – 2771.

[43] Wilcox, B. B., Robinson, M. S., Thomas, P. C., Hawke, B. R., 2005. Constraints on the depth and variability of the lunar regolith. Meteorit. Planet. Sci. 40 (5), 695 –710.

[44] Wu, B., Li, F., Hu, H., Zhao, Y., Wang, Y., Xiao, P., Li, Y., Liu, W. C., Chen, L., Ge, X., 2020. Topographic and geomorphological mapping and analysis of the Chang'E – 4 landing site on the far side of the Moon. Photogramm. Eng. Remote Sens. 86, 247 – 258.

[45] Wu, B., Liu, W. C., Grumpe, A., Wöhler, C., 2018. Construction of pixel – level resolution DEMs from monocular images by shape and albedo from shading constrained with low – resolution DEM. ISPRS J. Photogramm. Remote Sens. 140, 3 – 19.

[46] Wu, B., 2017. Photogrammetry – 3D from imagery. In: Richardson, D., Castree, N., Goodchild, M. F., Kobayashi, A., Liu, W., Marston, R. A. (Eds.), The International Encyclopedia of Geography. John Wiley & Sons, Ltd., New York, pp. 1– 13.

[47] Wu, B., Li, F., Ye, L., Qiao, S., Huang, J., Wu, X., Zhang, H., 2014. Topographic modeling and analysis of the landing site of Chang'E – 3 on the Moon. Earth Planet. Sci. Lett. 405, 257 – 273.

[48] Wu, B., Zhang, Y., Zhu, Q., 2011. A triangulation – based hierarchical image match – ing method for wide – baseline images. Photogramm. Eng. Remote Sens. 77 (7), 695 – 708.

[49] Wünnemann, K., Ivanov, B., 2003. Numerical modelling of the impact crater depth – diameter dependence in an acoustically fluidized target. Planet. Space Sci. 51, 831 – 845.

Ejecta From the Orientale Basin at the Chang'E - 4 Landing Site

Zhiyong Xiao[1,2], Chunyu Ding[1,3], Minggang Xie[4], Yuzhen Cai[1], Jun Cui[1,2,5], Ke Zhang[6], and Juntao Wang[6]

1. Planetary Environmental and Astrobiological Research Laboratory, School of Atmospheric Sciences, Sun Yatsen University, Zhuhai, China

2. Center for Excellence in Comparative Planetology, Chinese Academy of Science, Hefei, China

3. Department of Earth and Space Sciences, Southern University of Science and Technology, Shenzhen, China

4. College of Science, Guilin University of Technology, Guilin, China

5. Key Laboratory of Lunar and Deep Space Exploration, National Astronomical Observatories, Chinese Academy of Sciences, Beijing, China

6. Institute of Geochemistry, Chinese Academy of Sciences, Guiyang, China

Abstract: Radargrams obtained by the Yutu - 2 rover reveal that the post - mare deposits at the Chang'E - 4 landing site are ~45m thick, consistent with estimations based on orbital observations. Besides obvious ejecta from Finsen, polarized interpretations exist on whether or not the Alder crater is another contributor to the post - mare deposits, although predicted thicknesses of ejecta from all potential source craters are not adequately large compared to observations. We recognize that the rover has been exploring along a shared crater wall of two secondaries from Finsen. Mechanics of secondary impacts suggests that the discontinuous layers of coarse and fine materials observed in the radargram are older than Finsen. Cross - cutting relationships and crater density comparisons show that Alder is older than the mare basalts, and the Orientale basin was the major source for the post - mare materials. Lower - crust materials excavated by Orientale constituted a substantial portion of the surface regolith detected by the rover.

Plain Language Summary: Understanding surface processes at the lunar farside and revealing potential compositions related with SPA are two scientific targets of the Chang'E - 4 mission. Orbital observations and radargrams obtained by the rover show that the mare basalts at the landing area are covered by > 30m thick deposits. Finsen has delivered bright rays to this area, but whether or not Alder is another source has been debated. Considering all potential sources that even include Alder, predicted thickness of post -

mare materials is substantially less than observations. We recognize that Yutu - 2 is exploring along the shared wall of two secondaries from Finsen, and subsurface materials are older than Finsen. Orientale's secondaries are traceable on top of Alder. The crater density on Orientale's ejecta deposits is comparable with that of the mare surface at the landing site, and morphological comparison of secondaries indicates that substantial ejecta had been deposited in Von Kármán before the volcanism in Leibnitz was ceased. The limited time window suggests that Orientale's ejecta were the only suitable source for the post - mare materials, forming the discontinuous layers of coarse and fine materials revealed in radargrams. Surface regolith examined by Yutu - 2 should contain lower - crust materials that were excavated by Orientale.

Key Points:

- The Yutu - 2 rover has been exploring along the rim of Finsen's secondaries, and pre - Finsen materials dominate the subsurface
- Crater density comparisons reveal that Orientale is a major source of the post - mare ejecta, explaining the radargram obtained
- Surface regolith detected by Yutu - 2 is dominated by lower - crust materials ejected by Orientale, not related with SPA

1 Introduction

As the first successful surface exploration mission at the lunar farside, the Chang'E - 4 lander and the Yutu - 2 rover (Wu et al., 2019) have been exploring the lunar South Pole - Aitken basin (i. e., SPA). Together with the other sophisticated scientific payloads (Jia et al., 2018), the lunar penetrating radar (i. e., LPR) and the visible - near - infrared spectrometer (i. e., VNIS) onboard the rover are returning invaluable data on subsurface structures and surface compositions, aiming at understanding both surface processes at the lunar farside and possible compositions related with SPA. Interpretations of planetary remote sensing data usually face nonuniqueness (Baker, 2015). Likewise, data obtained by both the LPR and VNIS have already yielded diverged geological interpretations. For example, based on VNIS reflectance spectra for surface regolith, it is debated whether or not substantial olivine could be indicated (Gou et al., 2019, 2020; Huang et al., 2020; Li et al., 2019; Lin et al., 2019); LPR radargrams revealed layered structures in the subsurface, and inconsistent interpretations exist about their origins (Lai et al., 2020, 2019; Li et al., 2020; Zhang et al., 2020). Regional context is the framework for the interpretation of remote sensing data, and the emerged debates are largely caused by incomplete understanding of the geological history of the landing area.

The landing site is located on a mare surface, which has been covered by younger impact ejecta from various sources. Geological mapping based on crosscutting relationships (Fortezzo et al., 2020; Yingst et al., 2017) and age determination based on crater counts (Haruyama et al., 2009; Huang et al., 2018; Ling et al., 2019; Pasckert et al., 2018) both suggested a lower - Imbrian age for the mare basalts, formed around 3.6 — 3.7 billion years ago. Bright impact rays and secondaries formed by the Eratosthenian - aged Finsen crater are pronounced in the landing area (Huang et al., 2018), and abundant secondaries younger than Finsen are also visible (Figure 1). Based on small craters that have and have not excavated mare basalts, the post - mare debris is estimated to be mostly >34m thick (Huang et al., 2018; Li et al., 2020; Ling et al., 2019). This estimation is consistent with the depth distribution of echo patterns in LPR radargrams, as the first sharp transition from weak to strong echoes, which was interpreted as the boundary between the younger debris cover and deeper competent mare basalts, is located at ~45m (Lai et al., 2020; Zhang et al., 2020).

Insufficient local context has led to disagreements about the possible provenance of post - mare deposits. A persistent debate is the relative age between the Alder crater and the mare unit in the landing area (Figure 1a). Geological mapping (Fortezzo et al., 2020; Yingst et al., 2017) and recent crater statistics (Lu et al., 2020) suggested that Alder is older. Based on a belt of raised topography on the mare unit (Figure 1b), one interpretation argued that Alder was younger (Di et al., 2019), although topographic bumps on lunar mare that are not related with impact ejecta are common. On the other hand, referring to reported stratigraphic ages of impact craters, a ballistic sedimentation model that has considered both primary ejecta deposition and excavation of local materials by secondary impacts (Xie, Liu et al., 2020) was employed to estimate the thickness of ejecta delivered by various post - mare craters around the landing site (Huang et al., 2018; Lai et al., 2020). For craters that are stratigraphically younger than the mare basalts, only a few were capable to deliver nonnegligible ejecta (>0.2m) to the landing area, considering that ejecta thickness decreases exponentially away from the parent crater (Huang et al., 2018; Lai et al., 2020). The estimated total thickness of primary ejecta from all known source craters is less than 10m (Huang et al., 2018). Even assuming Alder had delivered impact ejecta to the mare surface (~8m thick; Lai et al., 2020), the estimated total thickness of primary ejecta plus local excavation is much less than the observed thickness of post - mare deposits. Nevertheless, Alder has been frequently cited as a post - mare crater to explain the thick debris cover (e. g., Zhang et al., 2020).

Sorting out the post - mare stratigraphy is critically important to interpret the scientific data returned. Reflectance spectra obtained from orbit showed that at pixel scales of ~100 — 200 m, surface regolith in the landing area is generally non - basaltic in composition (Huang et al., 2018; Ling et al., 2019). It has been a consensus that ejecta from Finsen are the major source of the surface regolith (Li et al., 2019; Gou et al., 2020; Huang et al., 2020; Lin et al., 2019). However, ballistic sedimentation models

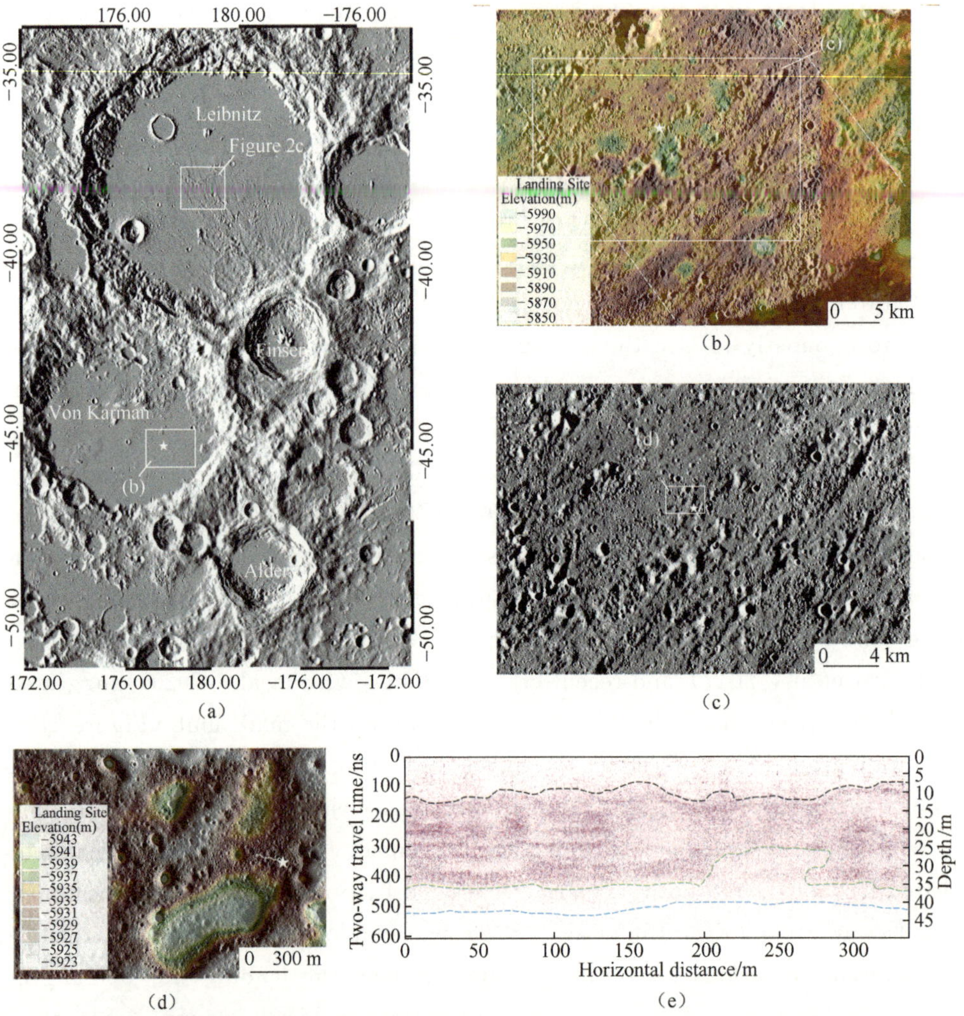

Figure 1　Regional and local context of the Chang'E－4 landing site (white stars). (a) The landing site is located on a mare unit in the floor of Von Kármán. The canvas uses the shade relief map based on the LOLA global DTM. (b) Lineations sourced from Finsen dominate the regional topography. The two white dashed lines mark the high topography area on the mare reported by Di et al. (2019). The base image is a Kaguya TC mosaic that is overline by a Selene－LOLA DTM (SLDEM; Barker et al., 2016). (c) The landing site is located within a chain of secondaries, which is ～140km to the center of Finsen (Kaguya TC). (d) The route of Yutu－2 in the first 12 lunar days (white curve) was located on a shared wall of two secondaries from Finsen. The base image is LROC NAC _ DTM _ CHANGE4 _ E458S1775 overlying LROC NAC M178833263L. (e) Radargrams obtained by the high－frequency LPR in the first 12 lunar days. The warm and cold colors are relative strengths of echoes. The black curve denotes the boundary between surface fine materials and deeper coarse materials, the green curve denotes the bottom of the coarse materials, and echo strengths are undifferentiated from noises at the blue curve. The strength and spatial distribution of echoes are referred for the subjective assignment of the boundaries

predicted a dominance of mare basalts in the debris cover that was excavated by distal ejecta (Lai et al., 2020). Essentially, lunar regolith is a weathering product (McKay et al., 1991). LPR radargrams (Figure 1e) showed that the top ～10－13 m thick materials have similar dielectric properties with typical lunar regolith (Ding et al., 2020; Lai et al., 2019; Li et al., 2020). This thickness is much larger than that on similar - aged lunar mare, requiring unusually fast growth for the entire regolith column at the landing site (Lai et al., 2019). For comparison, the predicted thickness of Finsen's ejecta and their local excavation was <8m at the landing site (Huang et al., 2018; Lai et al., 2020), which is contradicted with the top ~10 - 13m thick materials being entirely developed in Finsen's ejecta (Lai et al., 2019). One hydrocode simulation argued that Finsen's ejecta were >30m thick at the landing site (Di et al., 2019), which was not supported by observations based on reflectance spectra (Li et al., 2020; Qiao et al., 2019) or predictions based on ballistic sedimentation models (Huang et al., 2018; Lai et al., 2020). On the other hand, discontinuous layering of weak and strong echoes are observed in the LPR radargrams (Figure 1e), with the consensus being emplaced as distal ejecta (Lai et al., 2020, 2019; Li et al., 2020; Zhang et al., 2020). However, emplacement of distal ejecta does not occur via simple carpeting. Observations of various populations of secondary craters suggested that excavation of local materials and simultaneous resurfacing by ejecta delivered from uprange secondaries are vitally important (Xiao, 2016, 2018). Therefore, the formation of the layered echoes in the radargram (Figure 1e) needs an explanation in the frame of ejecta deposition.

Here we provide a new perspective on the geologic context of the Chang'E - 4 landing site. We notice that while Finsen's ejecta indeed have dominated the landing area, locally, both the landing site and rover route are located on top of a shared crater wall of two secondaries formed by Finsen. Mechanics of secondary impacts suggests that the subsurface materials are older than Finsen (Section 3. 1). Analyses of LPR radargram confirm that competent mare basalts occur at depths larger than ~45m (Section 3. 2). Based on crater density comparisons, we independently verify that Alder is older than the mare basalts and further show that the Orientale basin is an unnoticed major source (Section 3. 3). The emplacement mechanism of basin ejecta is investigated to explain the alternatively strong and weak echoes in the radargram (Section 4. 1). With a reconstructed local stratigraphy, interpretation of compositional data obtained by the rover is reorientated (Section 4. 2).

2 Data and Methods

The morphology and geometry of the landing area are studied at different scales using data obtained by the Narrow Angle Camera (NAC) and Wide Angle Camera (WAC) of the Lunar Reconnaissance Orbiter Camera (LROC; Robinson et al., 2010), the Terrain Camera (TC; Haruyama et al., 2008) onboard the Kaguya mission, and the Lunar Orbiter Laser Altimeter (LOLA; Smith et al., 2010). The IDs and sources of imagery and topography data used in this study are summarized in Table S1.

Crater densities are used to represent relative ages to identify additional sources of post-mare deposits besides Finsen. The cease of mare volcanism in Von Kármán occurred >1 billion years before the formation of the Finsen crater, and numerous impacts have occurred on the Moon during this time (Huang et al., 2018; Lu et al., 2020). Previous studies screened possible source craters based on their stratigraphic ages and distances to the landing area, and major ejecta suppliers such as the Leibnitz crater and the Ingenii and Imbrium basins were safely ruled out (Huang et al., 2018; Ling et al., 2019; Lai et al., 2020). Except for the debated Alder crater, no other pre-Finsen craters were identified that might have contributed nonnegligible ejecta to the post-mare deposits (Huang et al., 2018; Lai et al., 2020). Meanwhile, discrepancies between stratigraphic ages and absolute model ages derived from crater counts have been widely noticed for given craters (Xie, Xiao et al., 2020). Instead of globally cataloging absolute model ages for impact craters using crater statistics, this study uses the candidate early Imbrian source craters that were reported to be slightly older than the mare basalts at the landing area (Lai et al., 2020). For the mare surface at the landing site, the spatial densities of craters larger than 1 and 2km in diameters, $N(1)$ and $N(2)$ respectively, are referred from Huang et al. (2018), Pasckert et al. (2018), and Ling et al. (2019). The Orientale Basin has caused voluminous melting and delivered widespread light plains across the Moon (Meyer et al., 2016), and their $N(1)$ values were referred as the crater density of Orientale. Lu et al. (2020) reported model ages of ~3.5 and 3.2–3.3 Ga for Alder and the landing mare, respectively. The results are substantially less than both their stratigraphic ages (Yingst et al., 2017) and results from other crater statistics (Huang et al., 2018; Ling et al., 2019; Pasckert et al., 2018). To provide an independent evaluation, several areas on both the continuous ejecta deposits and floor of the Alder crater were selected as counting areas (Figure S1). For the completeness of crater collection and avoiding possible contamination by secondaries, only craters larger than 10 pixels of the base images were used (Wang et al., 2020). Table S2 in the appendices enumerates the crater densities compared. All the $N(1)$ and $N(2)$ values compared are derived directly from crater counts instead of being fitted from smaller craters using a production function.

Radargrams obtained by the high-frequency lunar penetrating radar (Fang et al., 2014) were used to study the internal structures of the post-mare deposits. The rover has traveled over 500m by September 2020. For the purpose of this study, data obtained in the first 12 lunar days (until January 1, 2020) are adequate (~340m). Processing of the LPR data includes the following steps: removal of redundant data, dewow filtering, amplitude compensation, background subtraction, and band-pass filtering. To facilitate geological interpretation of the radargram, migration transformation was performed to highlight echo contrasts (Figure 1e). In the supplementary information, Text S1 shows the detailed processing procedure, and Figures S2–S7 are the intermediate results.

3 Results

3.1 New Perspective on the Local Context

The topography of the landing area is mainly shaped by the northeast - to - southwest trending lineations (Figure 1), which are remanent topography of chains of secondaries formed by Finsen (Huang et al., 2018). The landing site is located within such a chain (Figures 1b and 1c), which exhibits similar morphology and strike with the other secondaries of Finsen (Figure 1c). Low - sun images with sub - meter pixel scales show that both the landing site and rover path are located on an east - to - west trending ridge (Figure 1d). The ridge is the shared wall of two heavily degraded secondaries in the chain (Figure 1c). The two secondaries have rim - to - rim lengths of >1km and widths of ~700m, and the landing site is ~6 and 20m above the floors of the two secondaries, respectively (Figure 1d). Topography degradation has substantially reduced the original rim heights, broadening the rim width, and the current depth - to - diameter ratios are < 0.02 and local slopes are <2° (Figure S8).

Materials beneath the rover route are older than Finsen. Crater walls are structurally deformed target materials that are subsequently exposed by the collapse of transient cavities, and excavated target materials are deposited on crater rims (Melosh, 1989). Therefore, walls of secondaries consist of pre - impact target, and Finsen's ejecta that have formed the secondaries shown in Figure 1d were distributed outside of the chain. Irregularly shaped secondaries in chains and clusters are most likely formed by a cluster of primary ejecta, instead of a single fragment (Schultz & Gault, 1985). Depending on the spatial dispersion of primary ejecta, secondaries in a chain exhibit various morphology and geometry (Xiao, 2016), so the initial geometry of the two secondaries shown in Figure 1d was unpredictable. On the other hand, for a chain of secondaries, materials exhibiting flow characteristics are commonly visible across secondaries in the downrange direction, which are ejecta launched by secondary impacts in the uprange direction of the chain (Luchhitta, 1977; Xiao, 2016). Figure 2a shows the morphology of such flows across secondaries that are ~140km to the center the Jackson crater, which has a similar rim - to - rim diameter (D = ~ 71km) with Finsen. Self - destruction among secondaries by immediately followed sweeping flows occurs in a spatially discontinuous manner (Xiao, 2016), and layering of flow deposits is usually observed (Figure 2a). Cratering mechanics suggests that most ejecta launched by subsonic secondary impacts are local target materials, instead of shattered primary ejecta that formed the secondaries (Schultz & Gault, 1985). As the original geometry of secondaries and local thickness of subsequent flows cannot be constrained, initial stratigraphic column of the two secondaries cannot be quantitatively constrained. Nevertheless, the post - degradation

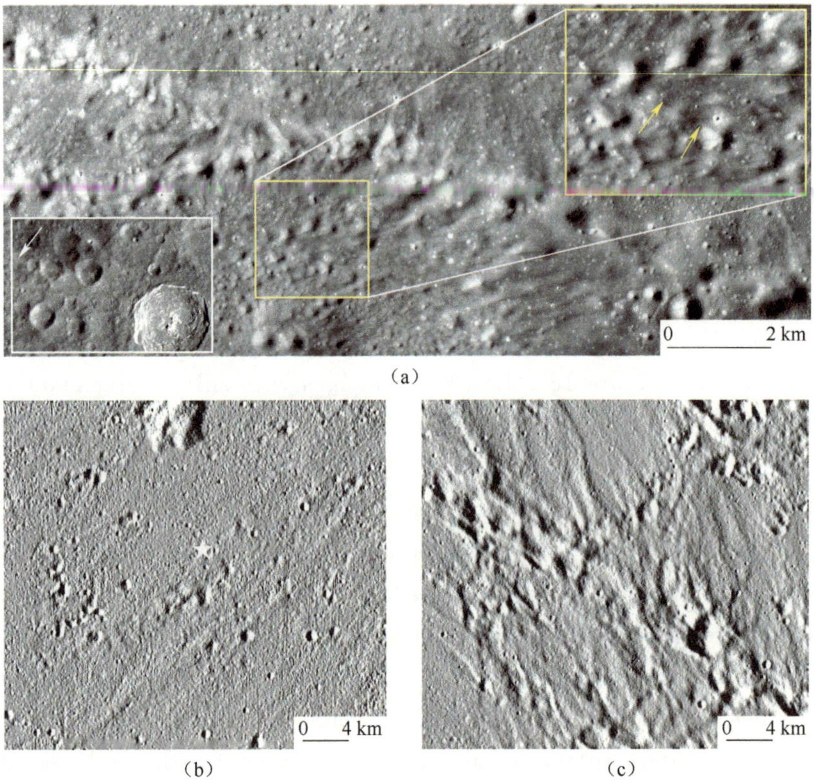

Figure 2　Formation and degradation of lunar secondaries. (a) Secondaries of the Jackson crater are resurfaced by flow - like materials that are ejecta delivered by secondary impacts at the uprange portion of this chain. Herringbone patterns are constructed by the flows, and yellow arrows in the inset point to layers in the flow. Jackson has a similar diameter with Finsen, and both the sizes and ballistic ranges of secondaries shown here are comparable with those at the Chang'E - 4 landing site (Figure 1d). The white arrow in the lower - left inset shows the location of the main figure. (b) Finsen's secondaries on the mare unit in Von Kármán are much more degraded than those on the slightly younger mare unit in Leibnitz (c). The white star denotes the landing site. The base image of (a) is from Kaguya TC, and that of (b) and (c) is shade relief map constructed from SLDEM

walls of secondaries, at least for the remnant height of ~20m (Figure 1d), are dominated by pre - Finsen materials.

3.2 Relative Permittivity of Pre - finsen Materials

The high - frequency LPR is capable to resolve materials down to ~520nm before echo strengths are undifferentiated from noises (Figure S9). In radargrams obtained by the low - frequency LPR, the debris cover on top of the integrated mare basalts occurs as a weak and more or less a homogenous unit (~600ns), and deeper materials exhibit much stronger echoes (Lai et al., 2020; Zhang et al., 2020). The high - frequency LPR features higher spatial resolutions (Fang et al., 2014), and three major units with different relative echo strengths are recognized in the debris cover (Figure 1e). The first unit, including the surface regolith, exhibits relatively homogenous and weak echoes (Figure 1e), indicating a similar bulk permittivity with that of typical regolith (Ding et al., 2020; Lai et al., 2020, 2019; Li et al., 2020). The second unit is composed of discontinuous and uneven stacks of strong and weak echoes (Li et al., 2020), and the contacts are transitional instead of abrupt (Figure 1e). Relative permittivity of lunar materials is mainly affected by the bulk density (Carrier et al., 1991), so compared to the first unit, the second unit is mainly composed of relative dense materials, with pockets of more porous materials are interbedded. Deeper echoes are similarly weak as the first unit, and a slightly strong horizon is visible at ~510ns before noises dominating the signal (Figure 1e).

Geometry of hyperbolic echo patterns in the radargram is employed to estimate the relative permittivity of subsurface materials (see supplementary text for method; Daniels, 1989), so the two-way travel times can be translated to depths (Figure 1e). Twelve reflectors that have integrated hyperbolic shapes are recognized in the first 20-230 ns of the radargram, and the Figure S10 shows their morphology. The reflectors are caused by materials with a higher relative permittivity than the surrounding, which are normally formed by competent boulders. The derived relative permittivities are ~2.6-4.3, with an average value is ~3.4 (Table S3). Therefore, materials within the equivalent depths of 230 ns (~18 m) have similar bulk densities with typical lunar regolith (Figure S11; Carrier et al., 1991), concurring the overall weak echoes detected by the low-frequency LPR (Lai et al., 2020; Zhang et al., 2020). Although materials deeper than ~10m exhibit relatively strong echoes (Figure 1e), the entire second unit follows an overall similar echo pattern, suggesting that materials at ~10—37m are coarse debris instead of competent rocks. In conclusion, we support that integrated mare basalts that were not substantially disturbed by ejecta deposition occur at depths >~45m (Zhang et al., 2020), and the debris cover is composed by layers of coarse and fine particles.

3.3 Orientale as a Dominate Source of Pre-Finsen Ejecta

The ~45m thick post-mare deposits are dominated by distal ejecta from pre-Finsen craters and excavated local basalts are minor, since the surface materials are generally non-basaltic in composition (Huang et al., 2018). Finsen's secondaries at the landing area (Figure 2b) are more degraded than those with similar ballistic ranges on the floor of Leibnitz (Figure 2c). This sharp contrast indicates that before Finsen was formed, the mare unit in Von Kármán had been covered by thicker porous materials than that in Leibnitz, so both the initiation of ejecta flows by secondary impacts and post-impact degradation by mass wasting were enhanced in Von Kármán. As predicted by lunar crater chronology, the mare volcanism in Leibnitz occurred only ~380 million years later than that in Von Kármán (Pasckert et al., 2018). Therefore, before the cease of mare vocalism in Leibnitz, a limited number of craters were able to deliver a significant amount of ejecta to the landing area.

Unlike the obvious secondaries formed by Finsen (Figure 1), older craters have not left typomorphic topography evidence at the landing site. Crater populations collected on both the floor and different ejecta facies of Alder uniformly exhibit a much larger density in terms of both $N(1)$ and $N(2)$ values than the mare basalts in Von Kármán (Table S2). This result verifies both geological maps that assigned a Pre-Imbrian age for Alder (Fortezzo et al., 2020) and the other crater density comparisons (Lu et al., 2020). On the other hand, crater densities on Orientale's light plains (Meyer et al., 2016) are comparable with that on the mare unit in Von Kármán (Huang et al., 2018; Ling et al., 2019; Pasckert et al., 2018), which are further verified by their overlapped model ages (~3.7Ga) as predicted using the same crater chronology system (Ling et al., 2019; Whitten et al., 2011). In conclusion, Alder is much older than the mare basalts in Von Kármán, and the Orientale basin is a suitable candidate that has delivered ejecta to the landing site.

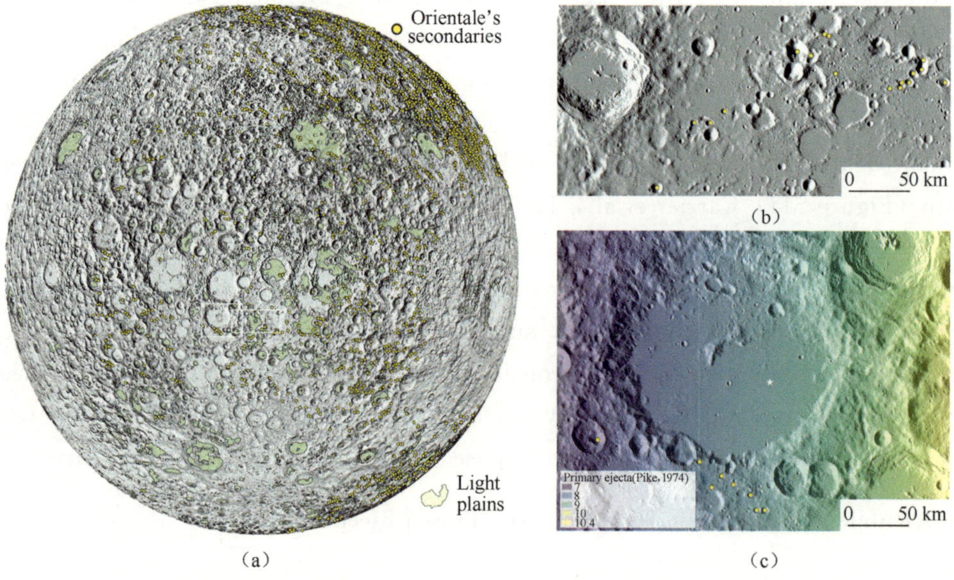

Figure 3 Existence of Orientale deposits at the Chang'E-4 landing site. (a) Spatial distributions of Orientale's secondaries (yellow dots; Guo et al., 2018) and global distribution of light plains (Meyer et al., 2018) at the lunar farside. Not all the light plains were formed by Orientale. The canvas is the shade relief map constructed based on the LOLA global DTM, and the Orientale basin is located at the right edge. (b) Orientale's secondaries (yellow dots) are traceable on top of Finsen's ejecta. (c) Orientale's ejecta at the landing area (white star) are ~8m thick as predicted by Pike (1974), but Orientale's secondaries nearby are as large as ~7km (yellow dots), representing up to hundreds of meters of primary ejecta. Panels (b) and (c) are based on SLDEM shade relief maps, and the location is marked in panel (a)

As a first-order estimation, ballistic sedimentation model (Pike, 1974; Xie, Liu et al., 2020) predicted that the Orientale basin has delivered >8m thick primary ejecta to the landing site (Figure 3), and their local excavated materials are ~118m thick (Lai et al., 2020). Together with the crater density comparison (Table S2), this estimation suggests that ejecta delivered by the Orientale basin cannot be ignored as a major component of the post-mare deposits. However, the actual thickness of primary ejecta and their local excavation can deviate substantially from predictions, as ballistic sedimentation models are based on simplified assumptions, such as azimuthally symmetric ejecta deposition, similar cratering scaling laws for secondary and primary impacts, and ignorance of downrange flows triggered by secondaries. Orientale's secondaries as large as ~10km are visible around the Von Kármán crater, and some are superposed on the ejecta deposits of Alder (Figure 3), further proving that Alder is older than Orientale. Primary ejecta that formed such secondaries had equivalent dimensions of >500m (Guo et al., 2018), which are comparable with the thickness of light plains formed by Orientale (Meyer et al., 2016). The co-occurrence of light plains and secondaries formed by Orientale (Figure 3a) suggests that with identical ballistic ranges, deposition of distal ejecta can be substantially different, although the detailed mechanism awaits further investigations. Compared to the other potential lower-Imbrian aged source craters (Lai et al., 2020), Orientale is the only candidate that is capable of accounting for a substantial, if not most of the post-mare deposits at the landing site.

4 Discussion

4.1 Sandwich Structures in Orientale's Distal Ejecta

Orientale being the dominate source of the post-mare deposits is consistent with a thicker debris cover on the mare basalts in Von Kármán before volcanism was stopped in Leibnitz (Section 3.3). This interpretation further corroborates observations by the low-frequency LPR that the top ~45m materials are composed by more porous materials than competent basalts (Zhang et al., 2020), since Orientale's distal ejecta should be porous breccias (Cai et al., 2020). The interpretation also explains the much larger regolith thickness at the landing site than same-aged lunar mare (Lai et al., 2019), as the starting materials for regolith development were already porous breccia instead of competent rocks. Besides these concordances, the actual thickness of both Orientale's ejecta and their locally excavated materials is sensu stricto not precisely known, since both ejecta thickness and their emplacement mechanism are spatially heterogeneous (Figure 3c).

The second unit shown in Figure 1e was interpreted as layers of ejecta delivered by different but unknown source craters older than Finsen (Li et al., 2020; Lai et al.,

2020; Zhang et al., 2020). We provide an alternative interpretation that links the origin of the second unit to the emplacement of Orientale's ejecta (Figure 3e). Ejecta emplacement is a highly dynamical process. The different landing times of individual ejecta fragments in a swarm and their residual momentum would cause horizontal shearing within the deposits (Figure 2a; Hörz et al., 1983). Layers of fluidized materials have also been observed feeding the light plains formed by Orientale (Meyer et al., 2016). Seismic reflections in the upper 220 meters of the Cayley Plains at the Apollo 16 landing site also feature alternative weak and strong echoes, which were formed by Imbrium's ejecta (Cooper et al., 1974). Therefore, the stacks of relative dense materials and lens of more porous materials in the second unit of the radargram (Figure 3e) may be breccias that have different bulk densities emplaced by Orientale's distal ejecta. This interpretation is also applicable for both the first and third units (Figure 1e), if assuming that the entire ~45 m post-mare deposits were emplaced by Orientale.

4.2 Ancient Highland Crust Detected by the Yutu-2 Rover

The new perspective that Orientale's ejecta being the dominate source of the post-mare deposits completely changes provenance interpretations for the surface regolith at the landing site. VNIS cannot unambiguously differentiate plagioclase, because plagioclase exhibits rather weak absorption features in VNIS wavelengths of 450-2,395nm, which are easily masked by a minor presence of other mafic minerals (Cheek & Pieters, 2014). Using different algorithm parameters to estimate mineral components based on VNIS data, the interpreted mineralogy was sharply different, such as the possible percentage of olivine (Li et al., 2019; Gou et al., 2019; Lin et al., 2019). For the deconvolution of VNIS data, one critical parameter in the center of discussion is whether or not plagioclase and other phases were adequately considered as end members (Huang et al., 2020). Current interpretations favor that both the surface regolith and some rock fragments measured by VNIS are noritic that is rich in plagioclase and pyroxene (Huang et al., 2020). We argued that Orientale's ejecta is the major source of the surface regolith, which should be ancient highland crustal materials, since Orientale did not excavate mantle materials as predicted by numerical modeling (Johnson et al., 2016; Potter et al., 2013). The noritic composition of surface regolith further indicates that Orientale's ejecta at the landing site were mainly from the lower crust of the Moon, largely unrelated with SPA.

5 Conclusion

There are many interpretations of the surface compositional and subsurface structures at the Chang'E-4 landing site, but it has been a consensus that Finsen's ejecta are the dominate source for the surface regolith, while older craters that might have contributed ejecta to the post - mare deposits are not known. We reveal that both the landing site and the rover route are located on a shared wall of two heavily degraded secondaries from Finsen (Figure 1d), so that local materials are older. Different preservation states of Finsen's secondaries suggest that during the short hiatus of mare vocalism in Von Kármán and Leibnitz, substantial ejecta were deposited at the landing area. Crater density comparisons and overlapping relationships preclude Alder and verify Orientale as the dominate source. This recognition provides a sententious explanation for the missed source craters, origin of layered interior structures in radargrams, cause of the thicker - than - normal local regolith, and reason of different preservation states of Finsen's secondaries. We appeal for a re - evaluation on interpretations of compositional data obtained by the rover, since the subsurface materials may be lower - crust materials excavated by the Orientale basin.

Acknowledgment

The Chang'E - 4 mission was carried out by the Chinese Lunar Exploration Program, and the scientific data are provided by China National Space Administration. The authors are supported by the B - type Strategic Priority Program of the Chinese Academy of Sciences [grant number XDB41000000], the National Natural Science Foundation of China [grant number 41773063, 42004099, 41941003, 41773065, and 41490634], the Pre - research Project on Civil Aerospace Technologies of China National Space Administration [grant number D020101 and D020202], the Science and Technology Development Fund of Macau [grant number 0042/2018/A2], the Chinese Academy of Sciences Interdisciplinary Innovation Team, and the Key Program of Frontier Science of Chinese Academy of Sciences (Grant QYZDY - SSW - DQC028).

References

[1] Baker, V. R. (2015). Planetary geomorphology: Some historical/analytical perspectives. Geomorphology, 250, 8 - 17.

[2] Barker, M. K., Mazarico, E., Neumann, G. A., Zuber, M. T., Haruyama, J., & Smith, D. E. (2016). A new lunar digital elevation model from the lunar orbiter laser altimeter and SELENE terrain camera. *Icarus*, 273, 346 — 355. https://doi. org/10. 1016/j. icarus. 2015. 07. 039

[3] Cai, Y., Xiao, Z., Ding, C., & Cui, J. (2020). Fine debris flows formed by the Orientale Basin. Earth and Planetary Physics, 4, 212—222.

[4] Carrier, W. D., Olhoeft, G. R., & Mendell, W. (1991). Physical properties of the lunar surface. In G. H. Heiken, D. T. Vaniman, & B. M. French (Eds.), Lunar sourcebook: A user's guide to the Moon (pp. 285 — 356). Cambridge:

Cambridge University Press.
[5] Cheek, L. C., & Pieters, C. M. (2014). Reflectance spectroscopy of plagioclase - dominated mineral mixtures: implications for characterizing lunar anorthosites remotely. American Mineralogist, 99 (10), 1871 - 1892.
[6] Cooper, M. R., Kovach, R. L., & Watkins, J. S. (1974). Lunar near - surface structure. Reviews of Geophysics and Space Physics, 12, 291 - 308.
[7] Daniels, J. J. (1989). Fundamentals of ground penetrating radar. Proceedings of the Symposium on the Application of Geophysics to Engineering an Environmental Problems (pp. 62 - 142). SAGEEP 89, Golden, Colorado.
[8] Ding, C., Xiao, Z., Wu, B., Li, Y., Prieur, N. C., Cai, Y., et al. (2020). Fragments delivered by secondary craters at the Chang'e 4 landing site. Geophysical Research Letters, 47, e2020GL087361. https://doi. org/10. 1029/2020GL087361.
[9] Di, K., Zhu, M. —H., Yue, Z., Lin, Y., Wan, W., Liu, Z., et al. (2019). Topographic evolution of Von Kármán crater revealed by the lunar rover Yutu - 2. Geophysical Research Letters, 46, 12764 — 12770. https://doi. org/ 10. 1029/2019GL085252.
[10] Fang, G., Zhou, B., Ji, Y., Zhang, Q., Shen, S., Li, Y., et al. (2014). Lunar penetrating radar onboard the Chang'e - 3 mission. Research in Astronomy and Astrophysics, 14 (12), 1607 - 1622. https://doi. org/10. 1088/1674—4527/ 14/12/009.
[11] Fortezzo, C. M., Spudis, P. D., & Harrel, S. L. (2020). Release of the Digital Unified Global Geologic Map of the Moon At 1:5,000,000 - Scale. Paper presented at 51st Lunar and Planetary Science Conference, Lunar and Planetary Institute, Houston, TX. https://www. hou. usra. edu/meetings/lpsc2020/pdf/2760. pdf.
[12] Gou, S., Di, K., Yue, Z., Liu, Z., He, Z., Xu, R., et al. (2019). Lunar deep materials observed by Chang'e - 4 rover. Earth and Planetary Science Letters, 528, 115829. https://doi. org/10. 1016/j. epsl. 2019. 115829.
[13] Gou, S., Di, K., Yue, Z., Liu, Z., He, Z., Xu, R., et al. (2020). Forsteritic olivine and magnesium - rich orthopyroxene materials measured by Chang' e - 4 rover. Icarus, 345, 113776. https://doi. org/10. 1016/j. icarus. 2020. 113776.
[14] Guo, D., Liu, J., Head, J. W., III, & Kreslavsky, M. A. (2018). Lunar Orientale impact basin secondary craters: Spatial distribution, size - frequency distribution, and estimation of fragment size. Journal of Geophysical Research: Plan, 123, 1344 - 1367. https://doi. org/10. 1029/2017JE005446.
[15] Haruyama, J., Matsunaga, T., Ohtake, M., Morota, T., Honda, C., Yokota, Y., et al. (2008). Global lunar surface mapping experiment using the Lunar Imager/Spectrometer on SELENE. Earth Planetary and Space, 60, 243 -256.

[16] Haruyama, J., Ohtake, M., Matsunaga, T., Morota, T., Honda, C., Yokota, Y., et al. (2009). Long-lived volcanism on the lunar farside revealed by SELENE Terrain Camera. Science, 323 (5916), 905 - 908. https://doi. org/10. 1126/science. 1163382.

[17] Hörz, F., Ostertag, R., & Rainey, D. A. (1983). Bunte Breccia of the Ries continuous deposits of large impact craters. Reviews of Geophysics and Space Physics, *21*, 1667-1725.

[18] Huang, J., Xiao, Z., Flahaut, J., Martinot, M., Head, J., Xiao, X., et al. (2018). Geological characteristics of Von Kármán crater, northwestern south pole-Aitken Basin: Chang'E-4 landing site region. Journal of Geophysical Research: Plan, 123, 1684-1700. https://doi. org/10. 1029/2018JE005577.

[19] Huang, J., Xiao, Z., Xiao, L., Horgan, B., Hu, X., Lucey, P., et al. (2020). Diverse rock types detected in the lunar South Pole-Aitken Basin by the Chang' E - 4 lunar mission. Geology, 48 (7), 723 - 727. https://doi. org/10. 1130/G47280. 1.

[20] Jia, Y., Zou, Y., Ping, J., Xue, C., Yan, J., & Ning, Y. (2018). The scientific objectives and payloads of Chang' E - 4 mission. Planetary and Space Science, 162, 207-215.

[21] Johnson, B. C., Blair, D. M., Collins, G. S., Melosh, H. J., Freed, A. M., Taylor, G. J., et al. (2016). Formation of the Orientale lunar multiring basin. Science, 354, 441-444.

[22] Lai, J., Xu, Y., Bugiolacchi, R., Meng, X., Xiao, L., Xie, M., Liu, B., et al. (2020). First look by the Yutu-2 rover at the deep subsurface structure at the lunar farside. Nature Communications, 11, 3426. https://doi. org/10. 1038/s41467—020—17262—w.

[23] Lai, J., Xu, Y., Zhang, X., Xiao, L., Yan, Q., Meng, X., et al. (2019). Comparison of dielectric properties and structure of lunar regolith at Chang'e-3 and Chang' e - 4 landing sites revealed by ground penetrating radar. Geophysical Research Letters, 46, 12783—12793. https://doi. org/10. 1029/2019GL084458.

[24] Li, C., Liu, D., Liu, B., Ren, X., Liu, J., He, Z., et al. (2019). Chang'e-4 initial spectroscopic identification of lunar far-side mantle-derived materials. Nature, 569(7756), 378—382. https://doi. org/10. 1038/s41586-019-1189-0.

[25] Li, C., Su, Y., Pettinelli, E., Xing, S., Ding, C., Liu, J., et al. (2020). The Moon's farside shallow subsurface structure unveiled by Chang'e-4 lunar penetrating radar. Science Advances, 6 (9), eaay6898. https://doi. org/10. 1126/sciadv. aay6898

[26] Ling, Z., Qiao, L., Liu, C., Cao, H., Bi, X., Lu, X., et al. (2019). Composition, mineralogy and chronology of mare basalts and non-mare materials in Von Kármán crater: Landing site of the Chang'e-4 mission. Planetary and

Space Science, *179* (104), 741. https://doi. org/10. 1016/j. pss. 2019. 104741.

[27] Lin, H., He, Z., Yang, W., Lin, Y., Xu, R., Zhang, C., et al. (2019). Olivine - norite rock detected by the lunar rover Yutu - 2 likely crystallized from the SPA impact melt pool. National Science Review, *7*, 913 - 920. https://doi. org/10. 1093/nsr/nwz183.

[28] Luchhitta, B. K. (1977). Crater clusters and light mantle at the Apollo 17 site: A result of secondary impact from Tycho. Icarus, 30, 80 - 96.

[29] Lu, Y., Wu, Y., Michael, G. G., Ma, J., Cai, W., & Qin, N. (2020). Chronological sequence of Chang'E - 4 landing zone within Von Kármán crater. Icarus, 354, 114086. https://doi. org/10. 1016/j. icarus. 2020. 114086

[30] McKay, D. S., Heiken, G., Basu, A., Blanford, G., Simon, S., Reedy, R., et al. (1991). The lunar regolith. In G. H. Heiken, D. T. Vaniman, & B. M. French (Eds.), Lunar sourcebook: A user's guide to the Moon (pp. 285 - 356). Cambridge: Cambridge University Press.

[31] Melosh, H. J. (1989). Impact Cratering: A Geologic Process. New York, NY: Oxford Univ. Press.

[32] Meyer, H. M., Denevi, B. W., Boyd, A. K., & Robinson, M. S. (2016). The distribution and origin of lunar light plains around Orientale basin. Icarus, 273, 135—145. https://doi. org/10. 1016/j. icarus. 2016. 02. 014.

[33] Meyer, H. M., Robinson, M. S., Denevi, B. W. & Boyd, A. K. (2018) A new global map of light plains from the Lunar Reconnaissance Orbiter. Paper presented at the 49th Lunar and Planetary Science Conference, Lunar and Planetary Institute, Houston, TX. https://www. hou. usra. edu/meetings/lpsc2018/pdf/1474. pdf.

[34] Pasckert, J. H., Hiesinger, H., & van der Bogert, C. H. (2018). Lunar farside volcanism in and around the South Pole - Aitken basin. Icarus, 299, 538—562. https://doi. org/10. 1016/j. icarus. 2017. 07. 023.

[35] Pike, R. J. (1974). Ejecta from large craters on the Moon: Comments on the geometric model of McGetchin et al. Earth and Planetary Science Letters, 23 (3), 265—271. https://doi. org/10. 1016/0012 - 821X.

[36] Potter, R. W. K., Kring, D. A., Collins, G. S., Kiefer, W. S., & McGovern, P. J. (2013). Numerical modeling of the formation and structure of the Orientale impact basin. Journal of Geophysical Research: Planets, 118, 963 - 979. https://doi. org/10. 1002/jgre. 20080.

[37] Qiao, L., Ling, Z., Fu, X., & Li, B.(2019). Geological characterization of the Chang'e - 4 landing area on the lunar farside. Icarus, 333, 37 - 51.

[38] Robinson, M. S., Brylow, S. M., Tschimmel, M., Humm, D., Lawrence, S. J., Thomas, P. C., et al. (2010). Lunar Reconnaissance Orbiter Camera (LROC) instrument overview. Space Science Reviews, 150, 81 - 124.

[39] Schultz, P., & Gault, D. E. (1985). Clustered impacts: Experiments and implications. Journal of Geophysical Research, 90 (B5), 3701－3732. https://doi. org/10. 1029/JB090iB05p03701.

[40] Smith, D. E., Zuber, M. T., Jackson, G. B., Cavanaugh, J. F., Neumann, G. A., et al. (2010). The Lunar Orbiter Laser Altimeter investigation on the Lunar Reconnaissance Orbiter Mission. Space Science Reviews, 150, 209 - 241.

[41] Wang, Y., Xie, M., Xiao, Z., & Cui, J. (2020). The minimum confidence limit for diameters in crater counts. Icarus, 341, 113645. https://doi. org/10. 1016/j. icarus. 2020. 113645.

[42] Whitten, J., Head, J. W., Staid, M., Pieters, C. M., Mustard, J., Clark, R., et al. (2011). Lunar mare deposits associated with the Orientale impact basin: New insights into mineralogy, history, mode of emplacement, and relation to Orientale Basin evolution from Moon Mineralogy Mapper (M3) data from Chandrayaan - 1. Journal of Geophysical Research: Planets, 116, E00G09. https://doi. org/10. 1029/2010JE003736.

[43] Wu, W., Li, C., Zuo, W., Zhang, H., Liu, J., Wen, W., et al. (2019). Lunar farside to be explored by Chang'e - 4. Nature Geoscience, 12 (4), 222－223. https://doi. org/10. 1038/s41561 - 019 - 0341 - 7.

[44] Xiao, Z. (2016). Size - frequency distribution of different secondary crater populations: 1. Equilibrium caused by secondary impacts. Journal of Geophysical Research, Planets, 121, 2404－2425. https://doi. org/10. 1002/2016JE005139.

[45] Xiao, Z. (2018). On the importance of self - secondaries. Geoscience Letters, 5, 17.

[46] Xie, M., Liu, T., & Xu, A. (2020). Ballistic sedimentation of impact crater ejecta: Implications for the provenance of lunar samples and the resurfacing effect of ejecta on the lunar surface. Journal of Geophysical Research: Planets, 125, e2019JE006113. https://doi. org/10. 1029/2019JE006113.

[47] Xie, M., Xiao, Z., Zhang, X., & Xu, A. (2020). The provenance of regolith at the Chang'e - 5 candidate landing region. Journal of Geophysical Research: Plan, 125, e2019JE006112. https://doi. org/10. 1029/2019JE006112.

[48] Yingst, R., Chuang, F., Berman, D., & Mest, S. (2017). Geologic Mapping of the Planck Quadrangle of the Moon (LQ－29). Paper presented at the 48th Lunar and Planetary Science Conference, Lunar and Planetary Institute, Houston, TX. https://www. hou. usra. edu/meetings/lpsc2017/pdf/1680. pdf.

[49] Zhang, L., Li, J., Zeng, Z., Xu, Y., Liu, C., & Chen, S. (2020). Stratigraphy of the Von Kármán Crater Based on Chang'E - 4 Lunar Penetrating Radar Data. Geophysical Research Letters, 47, e2020GL088680. https://doi. org/10. 1029/2020GL088680.

Time – Frequency Attribute Analysis of Channel 1 Data of Lunar Penetrating Radar

Chenyang Xu[1,2], Gongbo Zhang[1,2], Jianmin Zhang[1,2] and Zhuo Jia[1,2]

1. College of Geo – exploration Science and Technology, Jilin University, Changchun 130026, China; cyxu15@mails. jlu. edu. cn (C. X.); gongbozhang@outlook. com (G. Z.)
2. Ministry of Land and Resources Key Laboratory of Applied Geophysics, Jilin University, Changchun 130026, China

Abstract: The Lunar Penetrating Radar (LPR) carried by the Chang'E – 3 (CE – 3) and Chang'E – 4 (CE – 4) mission plays a very important role in lunar exploration. The dual – frequency radar on the rover (DFR) provides a meaningful opportunity to detect the underground structure of the CE – 3 landing site. The low – frequency channel (channel 1) maps the underground structure to a depth of several hundred meters, while the high – frequency channel (channel 2) can observe the stratigraphic structure of gravel near the surface. As the low – frequency radar image is troubled by unknown noise, time – frequency analysis of a single trace is applied. Then, a method named complete ensemble empirical mode decomposition (CEEMD) is conducted to decompose the channel 1 data, and the Hilbert transform gives us the chance for further data analysis. Finally, combined with regional geology, previous studies, and channel 2 data, a usability analysis of LPR channel 1 data provides a reference for the availability of the CE – 4 LPR data.

Keywords: lunar penetrating radar; data analysis; complete ensemble empirical mode decomposition (CEEMD); Hilbert transform

1 Introduction

Exploration of the lunar subsurface structure is a powerful force for the analysis of the stratigraphic thickness, time, and contact relationships, which exert a vital role not only on theorizing the evolution of the subsurface structure, but also on estimating the total resources of the lunar subsurface.

In order to detect the shallow geological structure of the moon, spaceborne radar and synthetic aperture radar (SAR) are mainly used. As early as 1972, Apollo 17 carried out on – board radar detection [1]. It operated at 5, 15, and 150MHz. In addition, it mainly detected the underground geological structure of the moon. Since 20 November 2007, the

Lunar - Radar Sounder (LRS) on Japan's Kaguya probes (SELENE) has been surveying the moon underground. The LRS uses the high - frequency (HF) band (5MHz) to enable the acquisition of subsurface data at depths of several kilometers [2]. The small SAR was launched on the moon in September 2007 and it carried out the Indian Chandrayaan mission. It mapped the scattering characteristics of the moon's pole and determined the existence and extent of polar ice [3, 4]. The Mini - RF instrument on the Lunar Reconnaissance Orbiter (LRO) is a SAR, and it works with bistatic radar from the Arecibo Observatory to conduct data analysis of the region. The results show that there may be near - surface sediments of water ice [4, 5].

Unlike the spaceborne radar described above, the lunar radar on Chang'E - 3 (CE - 3) is a field instrument for underground structure detection [6]. The dual - frequency radar (DFR) offers a great opportunity to map underground structures to depths of a few hundred meters from the CH - 1 and its near - surface stratigraphic structure re - aggregates from the CH - 2 [7] of Mare Imbrium.

The Yutu rover (YR) landed at a height of 340. 4875°E, 44. 1189°N in a new area that had never been surveyed in the largest basin, the Mare Imbrium. Ancient basalts and young basalts are distributed here, and the last volcanic activity also stopped here [8]. The CE - 3 Lunar Penetrating Radar (LPR) data provide a new perspective for analyzing subsurface structures. In the beginning, the LPR data primary results were put forward by the National Astronomical Observatory of the Chinese Academy of Sciences (NAOC) [9]. The preliminary analysis of LPR, especially CH - 1, showed that there are nine underground layers from the surface to a depth of about 360m [6]. The existence of three basaltic areas detected by the LPR is the same as the distribution of the three major basaltic areas in Mare Imbrium [10 - 14]. By processing the original CH - 2 data, the near -surface structure was inferred. The parameters of cemented carbide were calculated by Dong et al. [15] and Zhang et al. [16]. Wang et al. [17] and Zhang et al. [18] developed a new method for processing CH - 2 data. Hu et al. proposed a rock extraction method by using local similarity constraints to realize rock location and quantitative analysis [19].

The basis for interpretation when using CH - 1 data is its credibility. However, Li et al. demonstrated some false reflectors in the CH - 1 data [20]. In order to analyze and decompose the CH - 1 data, an analysis method that can be used in nonlinear data is necessary.

In recent years, empirical mode decomposition (EMD) [21] has developed into a data analysis method. EMD has demonstrated appropriate versatility in the application of signals for extracting from random data with a complex noise environment. Although EMD has proven to be useful, it still has some unresolved diffculties. The main disadvantage of EMD is the constant appearance of pattern blending. Signals of different scales can be combined into an intrinsic mode function (IMF).

In order to conquer the frequent occurrence of pattern mixing, a new method of noise assisted data analysis (NADA) has been proposed. Its name is ensemble EMD (EEMD) [22]. The core idea of this method is to average the results of EMD and apply it to the original signal. The EEMD method can solve the pattern blending, but it creates a new pattern.

In the use of complete ensemble empirical mode decomposition (CEEMD) [23], all the stages after decomposition will add some specific noise. Next, a unique residue is calculated to obtain each pattern. The final decomposition is finished, and the error is negligible. Compared to EMD and EEMD, CEEMD provides a better mode spectral separation and fewer screening iterations while also reducing computational costs.

As mentioned by Li et al. [20], the signal - to - noise ratio (SNR) of the CH - 1 data is extremely low and there are many false reflectors in the CH - 1 data. It is important to be able to extract useful information from the CH - 1 data. In order to analyze and decompose the CH - 1 data, an analysis method, which combines CEEMD and time - frequency analysis, is developed. This method can not only analyze the LPR CH - 1 data of CE - 3, but also provide an analysis method for the subsequent Chang'E - 4 (CE - 4) radar data.

In this paper, time - frequency analysis is applied to the troubled data. Then, CEEMD is conducted to decompose the channel 1 data, and the Hilbert transform gives us the chance for further data analysis. Furthermore, combined with geological information, the previous summary, and the channel 2 data, usability analysis of the LPR channel 1 data has been conducted, which can provide a reference for the usability of the CE - 4 LPR data.

2 Materials and Methods

2.1 Data Explanation and Preprocessing

The YR launched by CE - 3 was the first to land on the moon after the Soviet Union in 1976. Specifically, the YR used it to detect the surface of the landing site on the north side of the mare and several instruments in the ground. Its trajectory stretched to 114.8m near the sinus (Figure 1). In this installment, the data description and pretreatment of the LPR CH - 1 data will be introduced.

The DFR on the YR offers a meaningful chance to map the underground structure about several hundred meters deep from the CH - 1 (60MHz) and its stratigraphic structure near the surface from the CH - 2 (500MHz). For the underground structure, the CH - 1 data are selected. According to the recording parameters, real conditions, and data condition, a set of LPR data pretreatment procedures (Table 1) was projected. The data can be preprocessed to access the CH - 1 image (Figure 2).

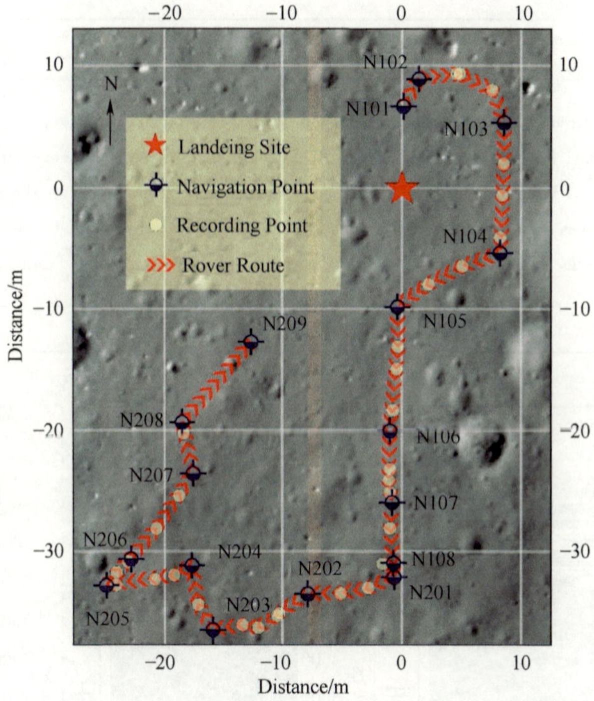

Figure 1　Walking track of Yutu. This picture has come from the descent camera on the Chang'E－3 (CE－3) lander

Table 1　Data preprocessing instructions

	Processing	Interpretative Statement
i	Data reading	The data is decomposed into 9 parts. According to their storage format (*. psd, which is the standard storage format for aerospace and aerospace), data and location information will be read one by one.
ii	Data registration	The data is split into 9 parts and ought to be stitched together.
iii	Time lag adjustment	The arrival time of the radar echo is delayed by 69. 664ns, which corresponds to the start time of the recorded data.
iv	Deep useless data deleting	As the SNRs of Part 1 and Part 2 are low and the rest of the data are not collected, the last 10,000 ns of data without the study value needs to be deleted.
v	Low SNR traces deleting	Some parts with low SNR should be removed.

Continued Table

	Processing	Interpretative Statement
vi	Traces selecting	Rover patrols are uneven because the rover may stop at a waypoint to collect other scientific data, for example, APXS data, VNIS data, and so forth. However, LPR never stops collecting data, which results in multiple acquisitions of multiple traces at the same place. The repeating traces are stacked and averaged. Two sets of data are obtained. One is all data without a copy track, and the other is a repeating track that has been stacked and averaged at the waypoint.
vii	Location	Location information is added to the image.

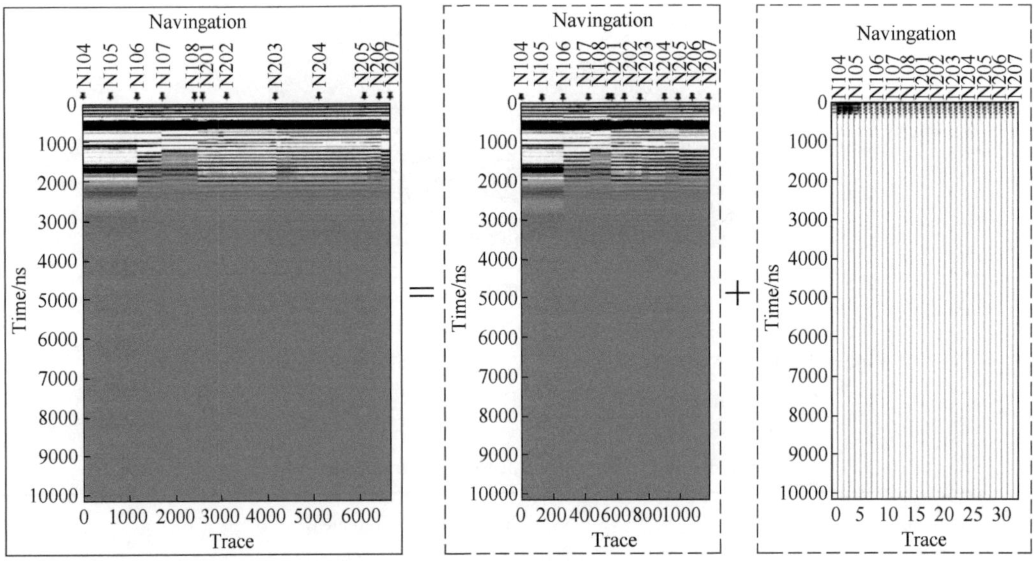

Figure 2　Lunar Penetrating Radar (LPR) CH-1 data from N104 to N207 after preprocessing.

(a) All traces from N104 to N207. (b) The traces after deleting the repeated traces.

(c) The stagnation traces at each recording and navigation point from N104 to N207

There are many kinds of space electromagnetic waves on the moon; the terrain of the moon is very complex under the adverse circumstances, so the signal-to-noise ratio (SNR) of LPR data is low. The repeated gathers of each stop point are superimposed, the purpose of which is to improve the SNR of the single trace at the waypoint [24], and provide more accurate and reasonable one-dimensional data for the subsequent analysis of the frequency-domain (FD) and the time-frequency domain (TFD).

In this paper, several analysis and processing methods are utilized, for example, the Fourier transform, short-time Fourier transform, CEEMD, and Hilbert transform. The Fourier transform (FT) and short-time Fourier transform (STFT) are commonly used as data analysis methods, the principle of which will not be repeated here. This section mainly introduces the CEEMD and Hilbert transform.

2.2 CEEMD

CEEMD comes from EMD and EEMD. The EMD method resolves a given signal into a series of IMFs. These IMFs are components that satisfy certain conditions. The reconstructed original signal is [22]

$$x=\sum_{i=1}^{n} c_i + r_n \tag{1}$$

where c_i is the i th IMF and r_n is the residue of data x after n IMFs are extracted. The decomposition flowchart of EMD is shown in Figure 3. The termination conditions of the EMD in the figure are: (1) The number of extreme points is the same as or, at most, one of the zero crossings; (2) the average value of the envelope defined by the local extremum at any point is zero.

As mentioned above, the main disadvantage of EMD is the frequent occurrence of mode overlap, which is defined as a single IMF composed of signals of different scales, or a signal of likeness scale in varying IMF components. In order to overcome the frequent appearance of mode overlap, several new noise - assisted data analysis methods are proposed, such as variational mode decomposition (VMD) [25] and ensemble EMD (EEMD) [22]. The EEMD method can be expressed as follows [23]:

1. Generate

$$x^i[n]=x[n]+w^i[n] \tag{2}$$

where $w^i[n](i=1,\ \cdots,\ I)$ are different implementations of white Gaussian noise,

2. Each $x^i[n](i=1,\ \cdots,\ I)$ is fully resolved by EMD to get their modes $IMF_k^i[n]$, where $k=1$, K represents the modes,

3. Set up $\overline{IMF_k}$ to be the k - th pattern of $x[n]$, obtained as the corresponding average IMF_k^i [23]:

$$\overline{IMF_k}[n]=\frac{1}{I}\sum\nolimits_{i=1}^{I} IMF_k^i[n] \tag{3}$$

The resulting decomposition of EEMD solves the problem of mode mixing; however, it introduces new ones. In the technique of CEEMD, the decomposition mode will be recorded as $\widetilde{IMF_k}$ and be proposed to count a unique first residue as

$$r_1[n]=x[n]-\widetilde{IMF_1}[n] \tag{4}$$

$\widetilde{IMF_1}[n]$ is obtained in the same way as EEMD. Then, calculate the first EMD mode over aggregate $r_1[n]$ plus various realizations of a given noise, obtaining $\widetilde{IMF_2}$ through averaging. The next residue can be defined as $r_2[n]=r_1[n]-\widetilde{IMF_2}[n]$. The rest of the mode continues working until the requirements are met.

First, we define an operator $E_j(\cdot)$, which, given a signal, produces the j - th mode from EMD [23]. w^i is the white noise with $N(0,\ 1)$. If $x[n]$ is the target datastore, the

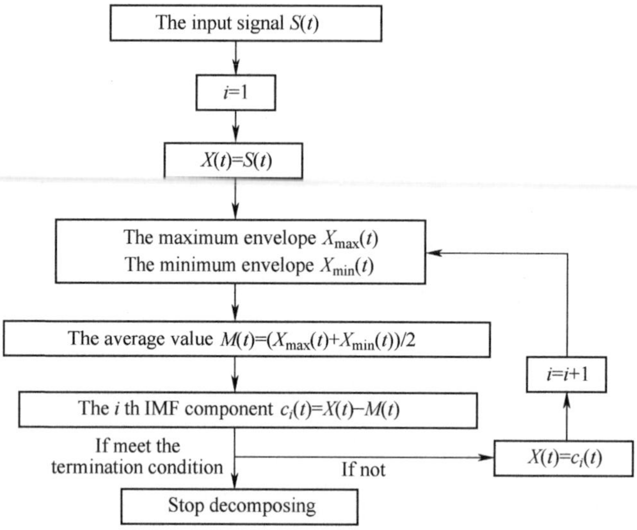

Figure 3　The decomposition flowchart of empirical mode decomposition (EMD)

method can be described by the following algorithm [23]:

1. Disintegrate by EMD I realizations $x[n]+\varepsilon_0 w^i[n]$ to obtain their first modes and calculate

$$\widetilde{IMF}_1[n]=\frac{1}{I}\sum_{i=1}^{I} IMF_1^i[n]=\overline{IMF}_1[n] \tag{5}$$

2. In the first phase ($k=1$), obtain the first residue as in Equation (4).

3. Decompose realizations $r_1[n]+\varepsilon_1 E_1(w^i[n])$, $i=1$, …, I, until their first EMD mode and define the second mode [23]:

$$\widetilde{IMF}_2[n]=\frac{1}{I}\sum_{i=1}^{I} E_1(r_1[n]+\varepsilon_1 E_1(w^i[n])) \tag{6}$$

4. For $k=2$, …, K, calculate the k - th residue [23]:

$$r_k[n]=r_{(k-1)}[n]-\widetilde{IMF}_k[n] \tag{7}$$

5. Decompose realizations $r_k[n]+\varepsilon_k E_k(w^i[n])$, $i=1$, …, I, until their first EMD mode and define the $(k+1)$ - th mode as [23]

$$\widetilde{IMF}_{(k+1)}[n]=\frac{1}{I}\sum_{i=1}^{I} E_1(r_k[n]+\varepsilon_k E_k(w^i[n])) \tag{8}$$

6. Go to step 4 to continue to get the next k.

Repeat steps 4 - 6. The direct residual part cannot be further decomposed, and the final residue satisfies [23]

$$R[n]=x[n]-\sum_{k=1}^{K} \widetilde{IMF}_k \tag{9}$$

K is the total number of modes. Therefore, the given signal $x[n]$ can be written as

$$x[n]=\sum_{k=1}^{K} \widetilde{IMF}_k+R[n] \tag{10}$$

Equation (10) indicates that the data decomposition has been completed and the original data can be reconstructed.

2.3 Hilbert Transform

After introducing the algorithm of CEEMD, we describe the principle of the Hilbert transform.

For any time series $X(t)$, we can get its Hilbert transform result:

$$Y(t)=\frac{1}{\pi}\int_{-\infty}^{\infty}\frac{X(\tau)}{t-\tau}d\tau \tag{11}$$

The Hilbert inverse transformation is

$$X(t)=H^{-1}[Y(t)]=-\frac{1}{\pi t}*Y(t)=-\frac{1}{\pi}\int_{-\infty}^{\infty}\frac{Y(t-\tau)}{\tau}d\tau=\frac{1}{\pi}\int_{-\infty}^{\infty}\frac{Y(t+\tau)}{\tau}d\tau \tag{12}$$

For a real signal $X(t)$, there are many kinds of complex methods, and it is the simplest and most effective to obtain the analytical signal by the Hilbert transform method. Assuming that the Hilbert transform of $X(t)$ is $Y(t)$, the expression of the parsed signal is

$$Z(t)=X(t)+iY(t)=a(t)e^{i\theta(t)} \tag{13}$$

where $a(t)$ and $\theta(t)$ are the instantaneous amplitude and phase of the signal, respectively, and they can be written in the following form:

$$a(t)=\sqrt{X^2(t)+Y^2(t)} \tag{14}$$

$$\theta(t)=\tan^{-1}\frac{X(t)}{Y(t)} \tag{15}$$

The signal's instantaneous frequency can be obtained by deriving the instantaneous phase:

$$\omega(t)=\frac{d\theta(t)}{dt} \tag{16}$$

Unlike the Fourier transform, a and ω in the equation are not constant, but a function of time, so the Hilbert spectrum can characterize the temporal variation of a data sequence; the instantaneous frequency is relative to the analytical signal, which is the analytical signal phase.

The frequency obtained by the signal FT reflects the distribution characteristics of the signal with time, and the frequency of the signal with time cannot be obtained. The instantaneous frequency reflects the frequency of the signal at a certain time, which can better describe the characteristics of the signal. This is also the main difference between them. The Hilbert transform can only deal with the stationary signal of a single frequency, which has great limitations for non - stationary signals with multiple frequencies at a certain time. Therefore, the Hilbert transform must meet some constraints in the application to obtain a physical instantaneous frequency.

3 Results

3.1 Analysis in Frequency Domain and the Time - Frequency Domain

The FT and the STFT are performed to obtain the frequency spectrum (Figure 4) and the time - frequency spectrum (Figure 5) (et al., N104, N106, N202, and N206, for examples), and, thus, the time - frequency spectra will expose the issue and the kind of processing that should be done.

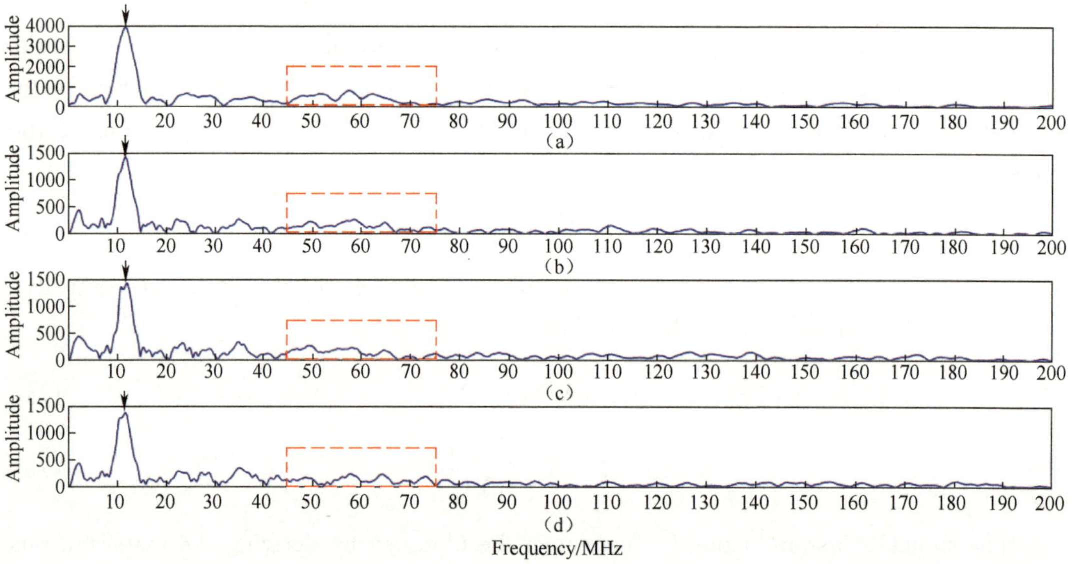

Figure 4 Frequency spectrum of traces.
(a) N104. (b) N106. (c) N202. (d) N206.

It can be seen from the frequency spectra of the sample traces (Figure 4) that in the frequency domain, the maximum energy of all sample gathers is around 12MHz (as indicated by the arrow), whereas at around 60MHz (the center frequency of CH - 1), the energy is not prominent.

Take a look at the time - frequency spectra of the sample traces:

(1) As the arrows in the N104 time - frequency spectrum of STFT show, there are 150, 100, and 60MHz electromagnetic energies in the data throughout the entire time delay. Obviously, the energies at 150 and 100MHz are not the transmitting signals. The electromagnetic energies at those frequencies may have come from the lunar space or the

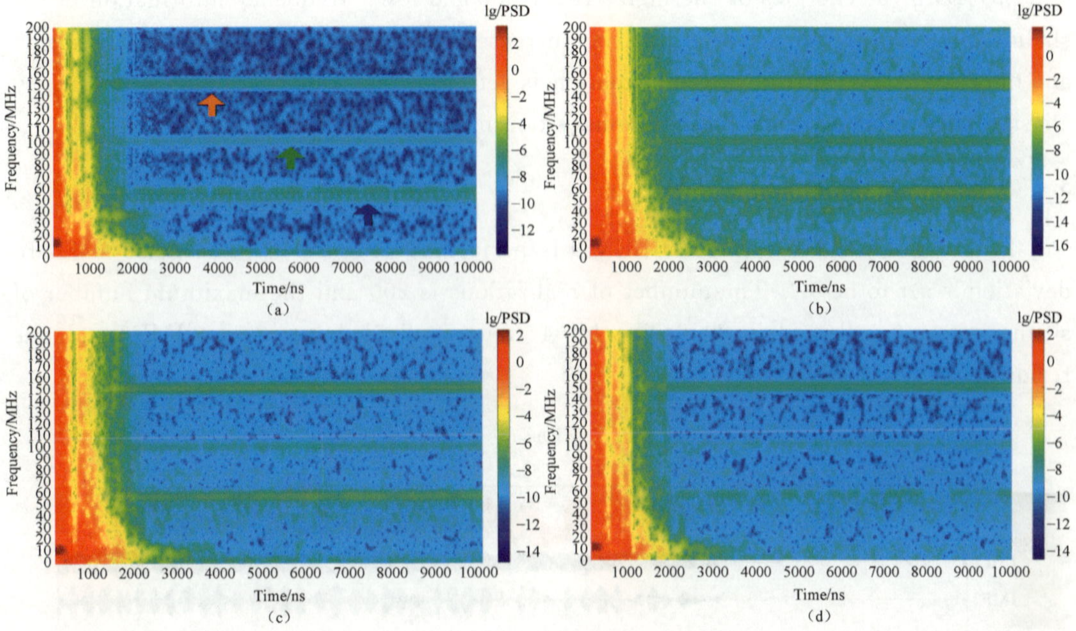

Figure 5 Time - frequency spectrum of
(a) N104, (b) N106, (c) N202, and (d) N206 traces.

instrument (needs further research).

(2) Compared to the shallow signal, the energy of the deep electromagnetic signal is very weak. Gain should be processed to highlight the deep information (Figure 6).

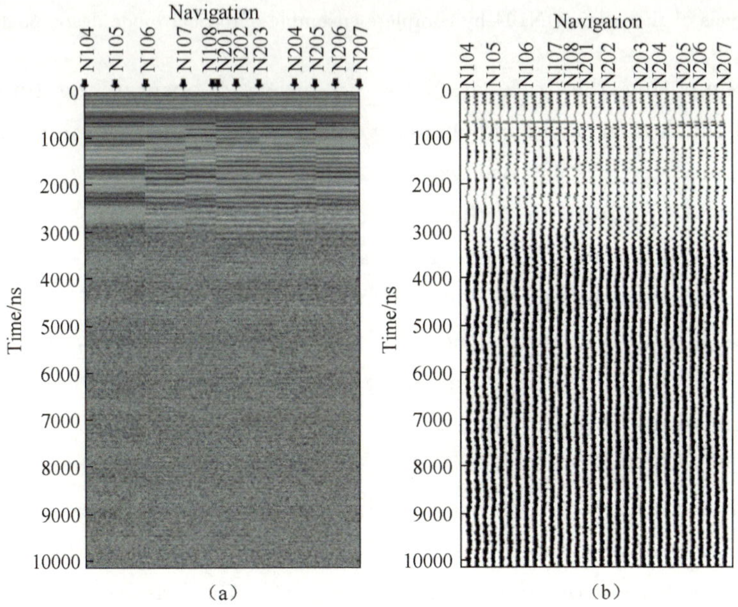

Figure 6 (a) LPR CH - 1 overall data and (b) stagnation points' traces after gain

(3) Both the energies of the high - frequency and low - frequency information in the shallow part are strong. As the time delay increases, high - frequency energy is absorbed, and only some low - frequency information is left. For this kind of nonlinear frequency aliasing phenomenon, a nonlinear decomposition method must be processed.

3.2 CEEMD of the Trace at N104

In terms of the N104 trace, CEEMD is performed (Figure 7). The noise standard deviation is set to be 0.2. The number of realizations is 200 and the maximum number of sifting iterations allowed is 2000. The N104 trace is divided into seven IMFs and one residual.

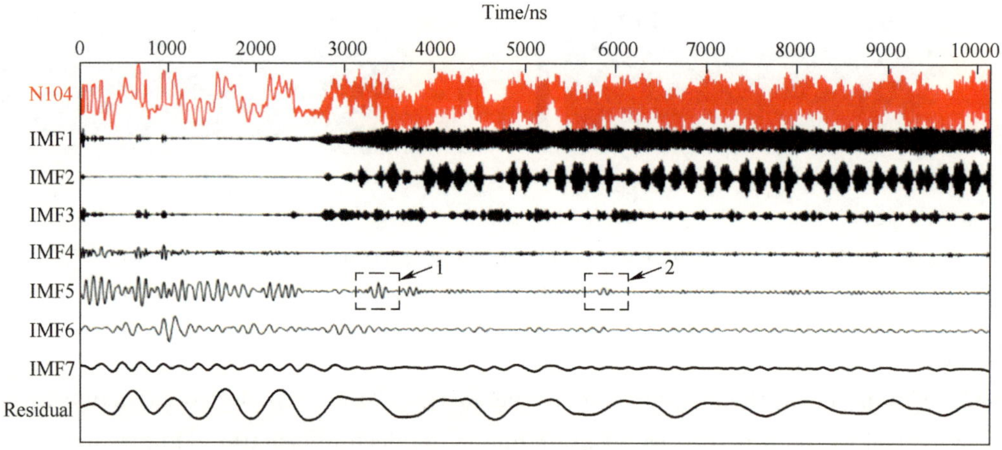

Figure 7　Analysis of the trace at N104 by complete ensemble empirical mode decomposition (CEEMD)

After decomposing the trace at N104, in order to observe the time - frequency information of the data in each component, we performed a short - time Fourier transform for each component (Figure 8). Analyze each component one by one:

(1) In the STFT spectrum of the IMF1 component, energy is concentrated mainly at 150MHz, and a small amount of energy is at 100 and 60MHz; this component can be considered a high - frequency noise component.

(2) In the STFT spectrum of the IMF2 and IMF3 components, the main energy is collected near 60MHz, and some of the energy scattered around it is considered to be the component containing the useful signal.

(3) The main energy in the STFT spectrum of the IMF4 component is collected near 30MHz, which is considered to be a lower - frequency noise component.

(4) Among the IMF components, IMF5 is worthy of attention. Two obvious energy information are observed from the component curve (in the IMF5 curve in Figure 7) and its STFT spectrum (Figure 8e), which are around 3500 and 5800 ns, respectively. The reflection of these two positions was used for reflection in the articles by Xiao et al. [6]

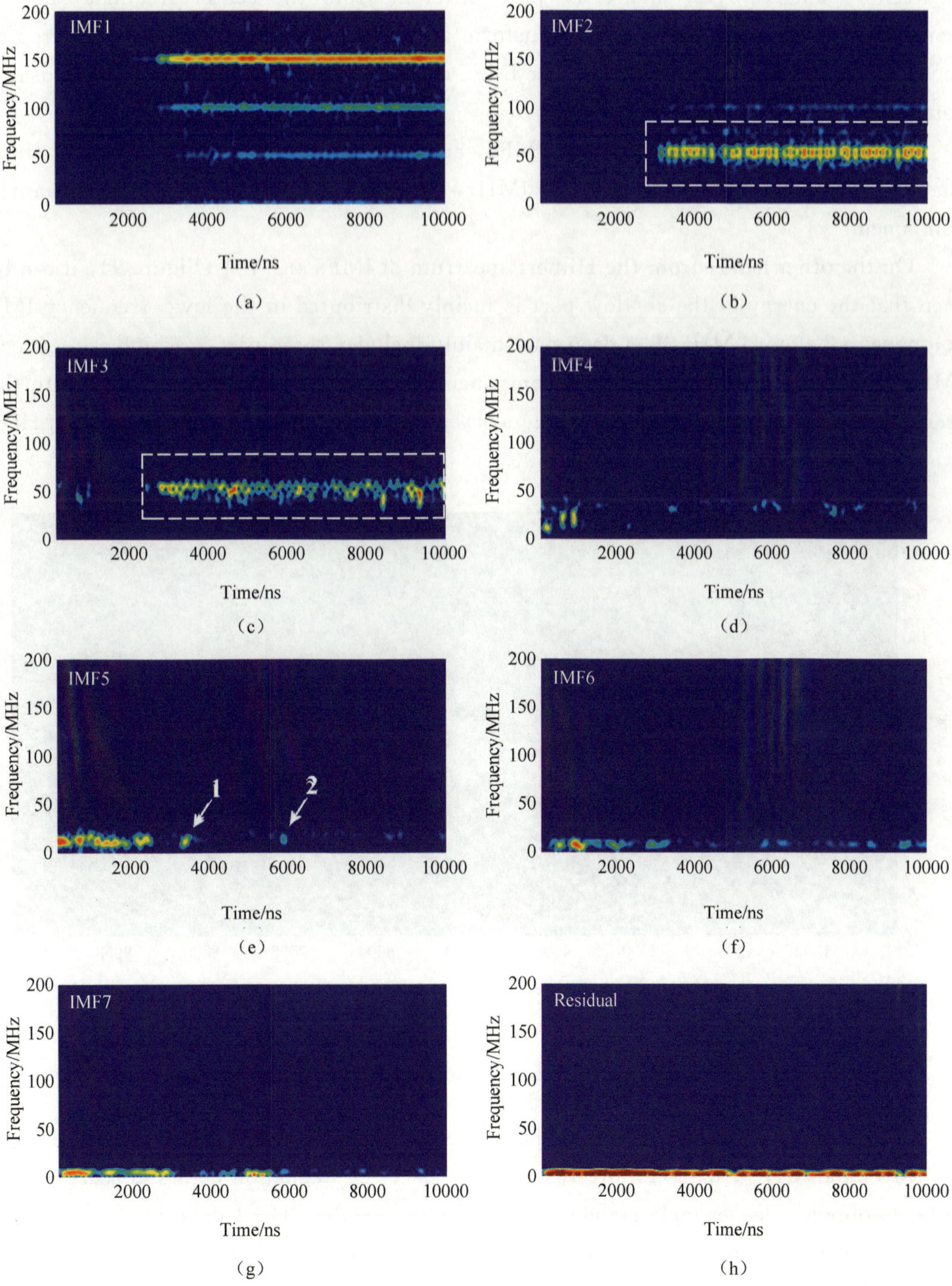

Figure 8　Time - frequency spectrum of each intrinsic mode function (IMF) decomposed from the trace at N104.

(a) IMF1. (b) IMF2. (c) IMF3. (d) IMF4. (e) IMF5. (f) IMF6. (g) IMF7. (h) Residual.

and Zhang et al. [10], but it can be seen from the STFT spectrum that the frequency is between 10 and 15MHz, and it is quite different from the center frequency of the transmitter. Li et al. also used the S transform and LPR ground data to verify the trap of these two pitfalls [20]. In summary, the IMF5 component is considered to be the unusable component.

(5) The STFT spectra of IMF5, IMF6, and IMF7 and the residual component show that the energy is collected below 10MHz, which is a low - frequency unwanted component.

On the other hand, from the Hilbert spectrum of IMFs at N104 (Figure 9), it can be seen that the energy in the shallow part is mainly distributed in the low - frequency IMF components below 20MHz. The deep part mainly includes three parts, a high - frequency IMF component of 150MHz, an IMF component of 60MHz, which is consistent with the transmission frequency, and a low - frequency pitfall component of 20MHz or less. Here again, an aliasing component around 3500 and 5800ns can be found.

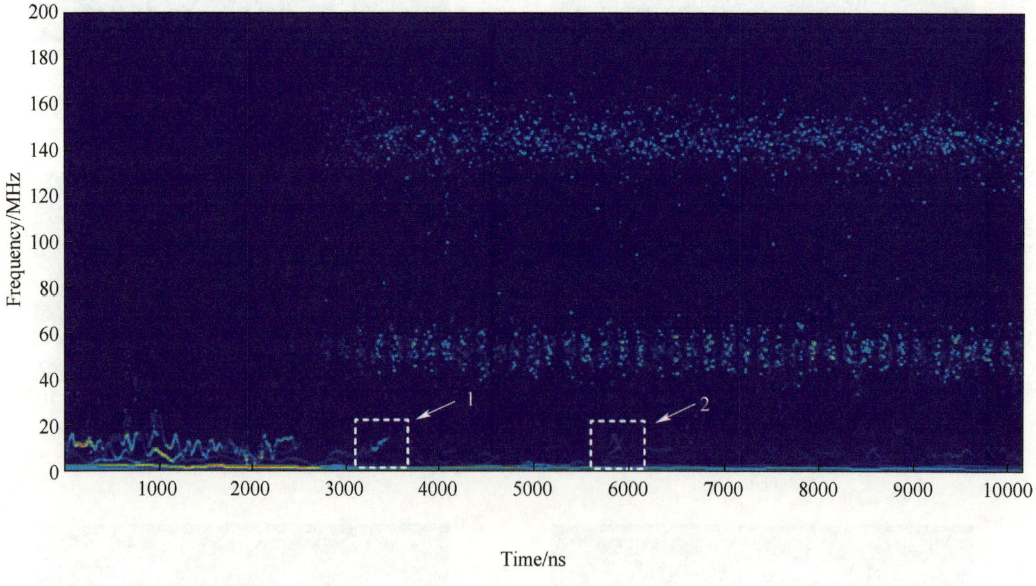

Figure 9 Hilbert spectrum of IMFs from N104

After decomposing and analyzing the single trace at N104 as an example, we perform CEEMD on all CH - 1 data (Figure 10) and the traces of stagnation points (Figure 11) in order to obtain a deeper understanding and utilization of the CH - 1 data.

After comprehensively observing the CEEMD components of the overall CH - 1 data and the CEEMD components of the stagnation points' traces, it can be found that in IMF5 (shown by the arrows in Figures 10 and 11), two pitfalls at 3500 and 5800 ns can be found (IMF5 in the stagnation points' traces is more obvious than IMF5 in the overall CH - 1 data).

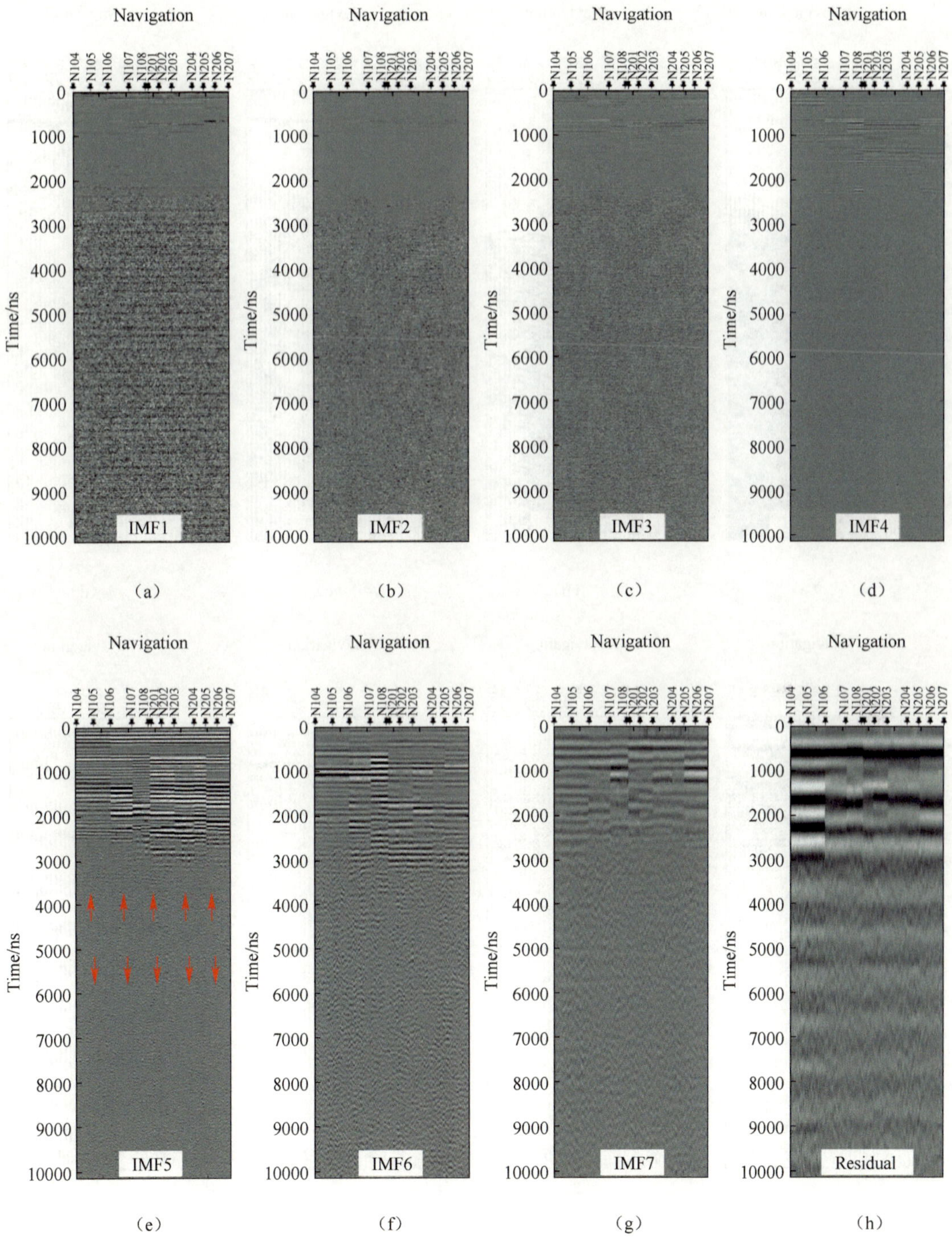

Figure 10 IMFs of LPR CH - 1 overall data.

(a) IMF1. (b) IMF2. (c) IMF3. (d) IMF4. (e) IMF5. (f) IMF6. (g) IMF7. (h) Residual.

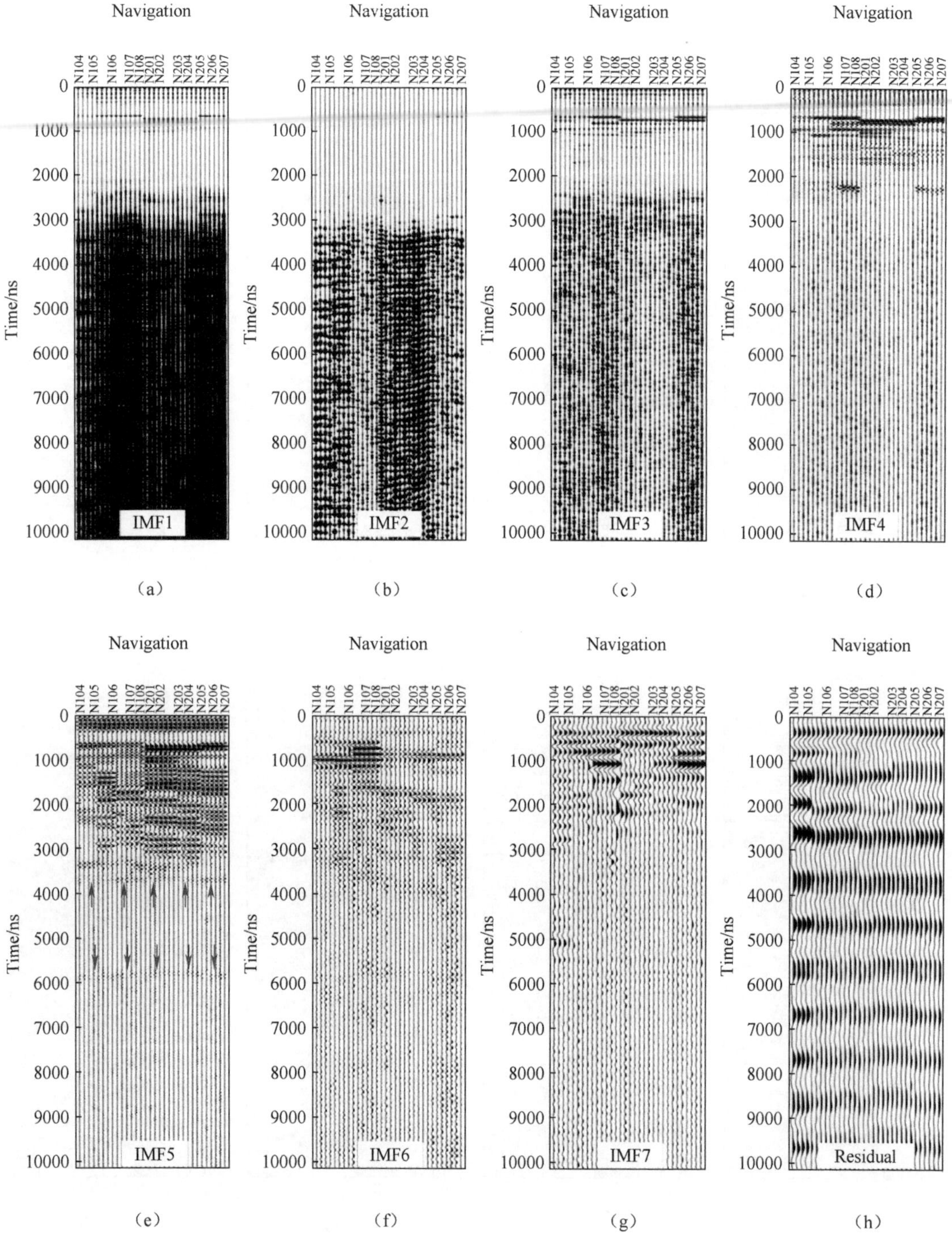

Figure 11　IMFs of stagnation points' traces in Figure 7.

(a) IMF1. (b) IMF2. (c) IMF3. (d) IMF4. (e) IMF5. (f) IMF6. (g) IMF7. (h) Residual.

3.3 Data Interpretation

As can be known from the above analysis, only the frequencies of the IMF2 and IMF3 components are in accordance with the radar's transmission frequency and are available. These two components are superimposed for further analytical interpretation. Figure 12a shows the data after IMF3 is superimposed with IMF4, and Figure 12b plots all the traces. From these two figures, especially in Figure 12b, it is found that the SNR of the data below 3000 ns is very low, making it difficult to pick up effective and convincing events. Therefore, we will focus on the shallow part.

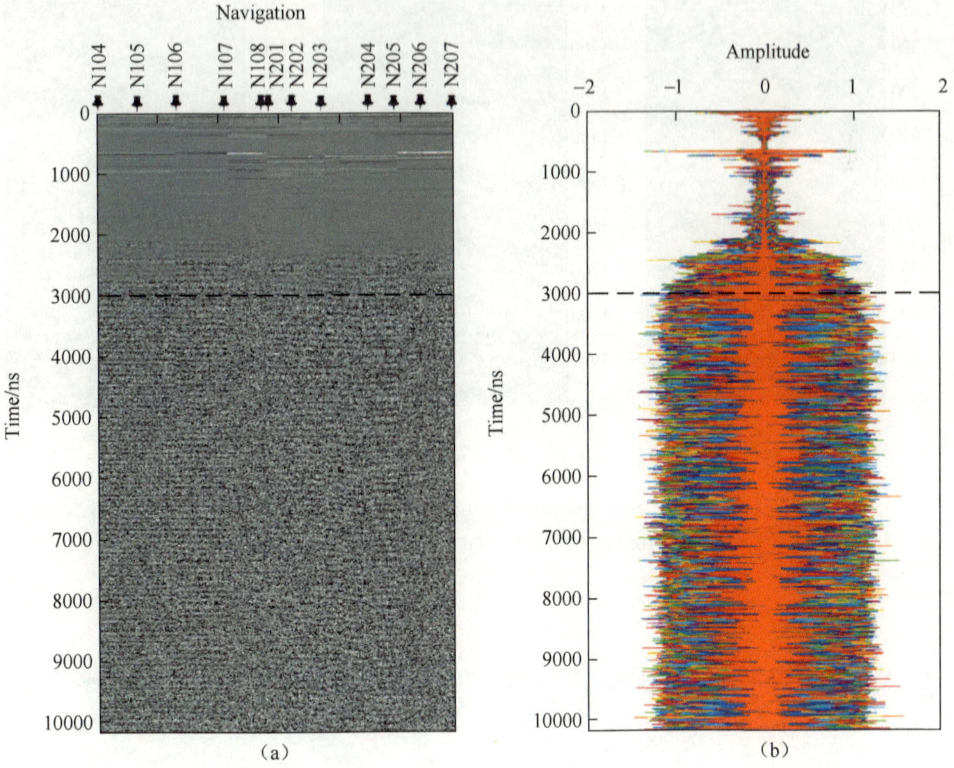

Figure 12　(a) Superimposed component of IMF3 and IMF4 and (b) all the traces

Figure 13 shows the information within the shallow part of 3000ns, but the events below 300 ns are not continuous. It is necessary to be very careful to interpret such information. This paper aims to decompose and analyze the CH－1 data, and the deep information no longer needs deeper explanation. Instead, information within the first 200 ns corresponding to the CH－2 data is selected for analysis. According to the hierarchical information of Xiao et al. [6], Zhang et al. [10], Fa et al. [11], and Zhang et al. [14], there are three well-recognized interfaces (Figure 14), namely:

1. The interface between the ejected materials of the Chang'E－3 crater and the lunar regolith;

2. The interface between the lunar regolith and paleoregolith;

3. The bedrock interface.

Figure 15 shows the information within 200 ns of the CH－1 data. There are three distinct events at the same position corresponding to the CH－2 data. It proves the rationality of stratification at the shallow part.

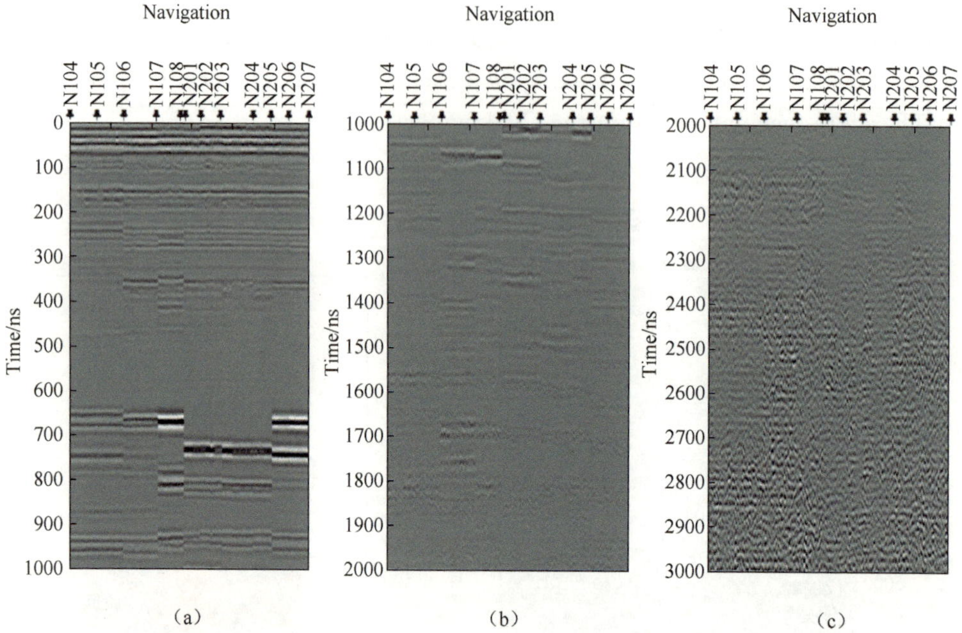

Figure 13　Superimposed component of IMF3 and IMF4.

(a) 0－1000, (b) 1000－2000, and (c) 2000－3000ns.

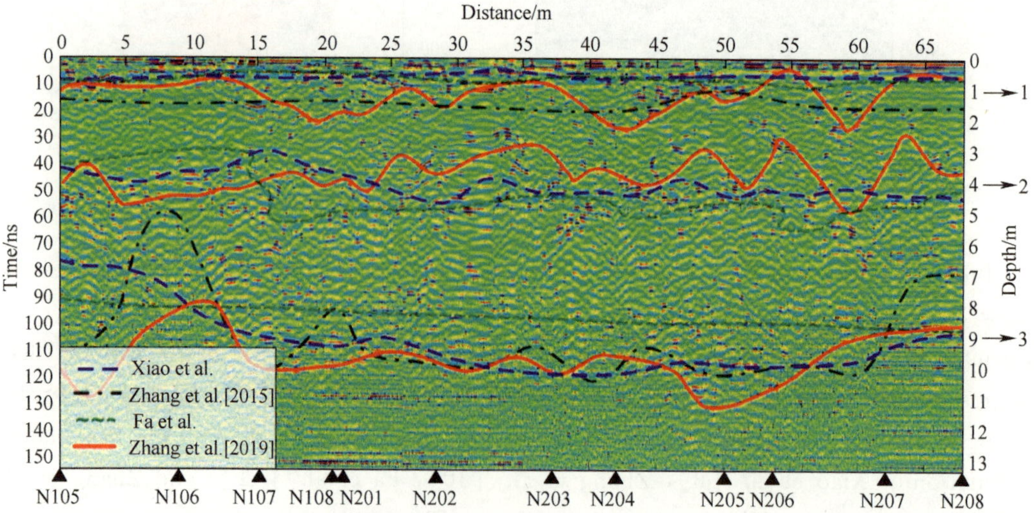

Figure 14　Stratification on CE－3 landing site by CH－2 data

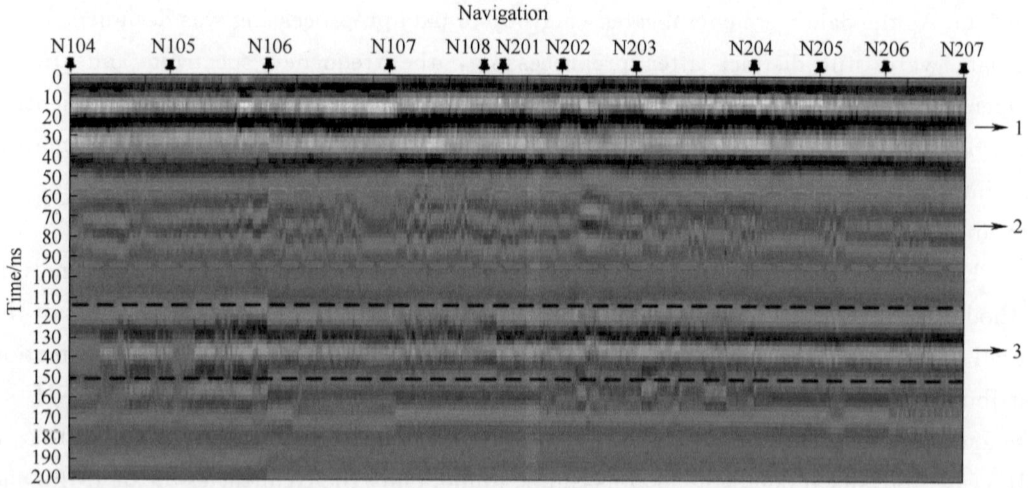

Figure 15　Stratification on CE - 3 landing site by CH - 1 data

4　Discussion

In the use of CEEMD to process radar data, Li et al. used the CEEMD decomposition method not only to perform time - frequency analysis but also to perform denoising and target recognition on GPR data [26, 27]. They mainly used CEEMD to decompose GPR data, used the Hilbert transform to analyze the data, and then performed denoising accordingly. The method in this paper not only uses CEEMD to decompose the data, but also uses the STFT and Hilbert transform to analyze the single component and the overall data comprehensively, which improves the application of CEEMD in ground - penetrating - radar processing.

On the other hand, regarding the use of CH - 1 data from the lunar radar, although Li et al. [20] believed that Xiao et al. [6] and Zhang et al. [10] had objections to the use of CH - 1 data, and used the ground experimental data and S transform to support their view, they did not perform in - depth decomposition and analysis of the overall CH - 1 data. In this paper, the non - linear decomposition method is used to deeply analyze the false reflectors and available information of CH - 1 data.

The most important significance of this paper is that the LPR data of CE - 4 is being collected on the moon, and the CH - 1 data collected by CE - 4 can be used by the CEEMD combined with the short - time Fourier transform and the Hilbert in the case of an extremely low SNR. The idea of special transformation is analyzed to improve the credibility of the CE - 4 data.

5　Conclusions

The LPR equipped on the YR measured the lunar geological structure in Northern

Imbrium. As the data were complicated, a group of data preprocessing was designed. When the data were still distinct after preprocessing, the frequency spectrum and time - frequency spectrum of LPR CH - 1 data were carried out for analysis. Subsequently, CEEMD helped to analyze the data, and the IMFs indicated the information distribution. Finally, combining with the previous studies and LPR CH - 2 data, we proved the credibility of the CH - 1 data at the shallow part.

CEEMD has been put forward as an adaptive time - frequency LPR data analysis method. It has been verified as quite multifunctional in extracting signals. The IMFs of the overall CH - 1 data and the stagnation points' traces describe the frequency information distribution. After comprehensively observing the CEEMD components of the overall CH - 1 data and the CEEMD components of the stagnation points' traces, it can be found that in IMF5, two pitfalls at 3500 and 5800 ns can be found. Only the frequencies of the IMF2 and IMF3 components are in accordance with the radar's transmission frequency and are available. These two components are superimposed for further analytical interpretation. Superimposed components provide us with the chance to analyze geological information at the shallow part.

This paper aims to decompose and analyze the CH - 1 data, and the deep information no longer needs a deeper explanation. Instead, information within the first 200 ns corresponding to the CH - 2 data is selected for analysis. According to the hierarchical information of former researchers, there are three well - recognized interfaces. As the CE - 4 LPR data are going to be published soon, CEEMD, which can prove the availability of the data, provides solid help and support for the articles published on the CE - 4 radar data.

Those results obtained in this paper provide valuable reference for understanding LPR data and valuable information for future lunar exploration missions.

Abbreviations

LPR	Lunar Penetrating Radar
CE - 3	Chang'E - 3
CE - 4	Chang'E - 4
CH - 1	The low - frequency channel
CH - 2	The high - frequency channel
HF	high - frequency
CEEMD	Complete Ensemble Empirical Mode Decomposition
DFR	dual - frequency radar on the rover
SAR	synthetic aperture radar
LRS	Lunar - Radar Sounder
SELENE	Japan's Kaguya probes

LRO	Lunar - Reconnaissance - Orbiter
YR	Yutu rover
NAOC	National Astronomical Observatory of the Chinese Academy of Sciences
EMD	Empirical - Mode - Decomposition
IMF	intrinsic mode function
EEMF	Ensemble Empirical - Mode - Decomposition
NADA	new method of noise assisted data analysis
SNR	Signal - to - noise ratio
FD	frequency - domain
TFD	time - frequency domain
FT	Fourier transform
STFT	short - time Fourier transform
VMD	Variational Mode Decomposition

References

[1] Phillips, R. J.; Adams, G. F.; Brown, W. E., Jr.; Eggleton, R. E.; Zelenka, J. S. Apollo Lunar Sounder Experiment. Apollo 17 Prelim. Sci. Rep. 1973, 330, 22.

[2] Ono, T.; Kumamoto, A.; Nakagawa, H.; Yamaguchi, Y.; Oshigami, S.; Yamaji, A.; Kobayashi, T.; Kasahara, Y.; Oya, H. Lunar radar sounder observations of subsurface layers under the nearside maria of the moon. Science 2009, 323, 909 - 912.

[3] Spudis, P. D.; Bussey, B.; Lichtenberg, C.; Marinelli, B.; Nozette, S. Mini - sar: An imaging radar for the chandrayaan - 1 mission to the moon. Curr. Sci. 2005, 96, 533 -539.

[4] Calla, O. P. N.; Mathur, S.; Jangid, M.; Gadri, K. L. Circular polarization characteristics of south polar lunar craters using chandrayaan - 1 mini - sar and lro mini - rf. Earth Moon Planets 2015, 115, 83 - 100.

[5] Patterson, G. W.; Stickle, A. M.; Turner, F. S.; Jensen, J. R.; Jakowatz, C. V. Bistatic radar observations of the moon using mini - rf on lro and the arecibo observatory. Icarus 2016, 283, 2 - 19.

[6] Xiao, L.; Zhu, P.; Fang, G.; Xiao, Z.; Zou, Y.; Zhao, J.; Zhao, N.; Yuan, Y.; Qiao, L.; Zhang, X.; et al. A young multilayered terrane of the northern mare imbrium revealed by chang'e - 3 mission. Science 2015, 347, 1226 - 1229.

[7] Fang, G. - Y.; Zhou, B.; Ji, Y. - C.; Zhang, Q. - Y.; Shen, S. - X.; Li, Y. - X.; Guan, H. F.; Tang, C. J.; Gao, Y. Z.; Lu, W.; et al. Lunar penetrating radar onboard the chang'e - 3 mission. Res. Astron. Astrophys. 2014, 14, 1607 - 1622.

[8] Wilhelms, D. E.; Mccauley, J. F.; Trask, N. J. The Geologic History of the Moon; United States Government Publishing Office: Washington, DC, USA, 1987.

[9] Su, Y. ; Fang, G. -Y. ; Feng, J. -Q. ; Xing, S. -G. ; Ji, Y. -C. ; Zhou, B. ; Gao, Y. Z. ; Li, H. ; Dai, S. ; Xiao, Y. ; et al. Data processing and initial results of Chang'e-3 lunar penetrating radar. Res. Astron. Astrophys. 2014, 14, 1623-1632.

[10] Zhang, J. ; Yang, W. ; Hu, S. ; Lin, Y. ; Fang, G. ; Li, C. ; Peng, W. ; Zhu, S. ; He, Z. ; Zhou, B. Volcanic history of the imbrium basin: A close-up view from the lunar rover yutu. Proc. Natl. Acad. Sci. USA 2015, 112, 5342-5347.

[11] Fa, W. ; Zhu, M. H. ; Liu, T. ; Plescia, J. B. Regolith stratigraphy at the chang'e-3 landing site as seen by lunar penetrating radar. Geophys. Res. Lett. 2015, 42, 179-187.

[12] Lai, J. L. ; Xu, Y. ; Zhang, X. P. ; Tang, Z. S. Structural analysis of lunar subsurface with Chang'E-3 lunar penetrating radar. Planet. Space Sci. 2016, 120, 96-102.

[13] Zhang, L. ; Zeng, Z. ; Li, J. ; Lin, J. ; Hu, Y. ; Wang, X. ; Sun, X. Simulation of the lunar regolith and lunar-penetrating radar data processing. IEEE J. Sel. Top. Appl. Earth Obs. Remote. Sens. 2018, 11, 655-663.

[14] Zhang, L. ; Zeng, Z. F. ; Li, J. ; Huang, L. ; Huo, Z. J. ; Wang, K. ; Zhang, J. M. A Story of Regolith Told by Lunar Penetrating Radar. Icarus 2019, 321, 148-160.

[15] Dong, Z. ; Fang, G. ; Ji, Y. ; Gao, Y. ; Zhang, X. Parameters and structure of lunar regolith in chang'e-3 landing area from lunar penetrating radar (lpr) data. Icarus 2016, 282, 40-46.

[16] Zhang, L. ; Zeng, Z. ; Li, J. ; Huang, L. ; Huo, Z. ; Wang, K. ; Zhang, J. Parameter Estimation of Lunar Regolith from Lunar Penetrating Radar Data. Sensors 2018, 18, 2907.

[17] Wang, K. ; Zeng, Z. ; Zhang, L. ; Xia, S. ; Li, J. A Compressive Sensing-Based Approach to Reconstructing Regolith Structure from Lunar Penetrating Radar Data at the Chang'E-3 Landing Site. Remote Sens. 2018, 10, 1925.

[18] Zhang, J. ; Zeng, Z. ; Zhang, L. ; Lu, Q. ; Wang, K. Application of Mathematical Morphological Filtering to Improve the Resolution of Chang'e-3 Lunar Penetrating Radar Data. Remote Sens. 2019, 11, 524.

[19] Hu, B. ; Wang, D. ; Zhang, L. ; Zeng, Z. Rock Location and Quantitative Analysis for Regolith on Chang'e 3 Landing Site Based on Local Similarity Constraint. Remote Sens. 2019, 11, 530.

[20] Li, C. ; Xing, S. ; Lauro, S. E. ; Su, Y. ; Pettinelli, E. Pitfalls in gpr data interpretation: False reflectors detected in lunar radar cross sections by chang'e-3. IEEE Trans. Geosci. Remote Sens. 2017, 56, 1325-1335.

[21] Huang, N. E. ; Shen, Z. ; Long, S. R. ; Wu, M. C. ; Shih, H. H. ; Zheng, Q. ; Yen, N. C. ; Tung, C. C. ; Liu, H. H. The empirical mode decomposition and the hilbert spectrum for nonlinear and non-stationary time series analysis. Proc. R. Soc.

London Ser. A 1998, 454, 903 - 995.

[22] Wu, Z. ; Huang, N. E. Ensemble empirical mode decomposition: A noise - assisted data analysis method. Adv. Adapt. Data Anal. 2009, 1, 1 - 41.

[23] Torres, M. E. ; Colominas, M. A. ; Schlotthauer, G. ; Flandrin, P. A complete ensemble empirical mode decomposition with adaptive noise. In Proceedings of the IEEE International Conference on Acoustics, Speech and Signal, Prague, Czech Republic, 22 - 27 May 2011; pp. 22 - 27.

[24] Jia, Z. ; Liu, S. ; Zhang, L. ; Hu, B. ; Zhang, J. Weak Signal Extraction from Lunar Penetrating Radar Channel 1 Data Based on Local Correlation. Electronics 2019, 8, 573.

[25] Cheng, S. ; Liu, S. ; Guo, J. ; Luo, K. ; Zhang, L. ; Tang, X. Data Processing and Interpretation of Antarctic Ice - Penetrating Radar Based on Variational Mode Decomposition. Remote Sens. 2019, 11, 1253.

[26] Li, J. ; Liu, C. ; Zeng, Z. ; Chen, L. GPR Signal Denoising and Target Extraction with the CEEMD Method. IEEE Geosci. Remote Sens. Lett. 2015, 12, 1615 - 1619.

[27] Li, J. ; Zeng, Z. ; Liu, C. ; Huai, N. ; Wang, K. A study on Lunar regolith quantitative random model and lunar penetrating radar parameter inversion. IEEE Geosci. Remote Sens. Lett. 2017, 14, 1953 - 1957.

Evaluating the Thickness and Stratigraphy of Ejecta Materials at the Chang'e – 4 Landing Site

Luyuan Xu[1,2], Xunyu Zhang[1,2], Le Qiao[3], and Jialong Lai[1,2,4]

1. State Key Laboratory of Lunar and Planetary Sciences, Macau University of Science and Technology, Macau, People's Republic of China; xyzhang@must.edu.mo
2. CNSA Macau Center for Space Exploration and Science, Macau, People's Republic of China
3. Shandong Key Laboratory of Optical Astronomy and Solar – Terrestrial Environment, School of Space Science and Physics, Institute of Space Sciences, Shandong University, Weihai, Shandong 264209, People's Republic of China
4. School of Science, Jiangxi University of Science and Technology, Ganzhou, People's Republic of China

Abstract: The landing site of China's Chang'e – 4 (CE – 4) probe is located on the mare basalts on the floor of the Von Kármán crater on the lunar far side. The Von Kármán crater is inside the ancient and highly cratered South Pole – Aitken basin, which has experienced complex emplacement sequences of both near and distant ejecta materials. These issues complicated the interpretation of the CE – 4 surface in situ measurements of the visible and near – infrared spectrometer and the lunar penetrating radar (LPR) onboard Yutu – 2 rover. To evaluate the sources and amounts of all principle foreign materials at the CE – 4 landing site, we thoroughly examine the ejecta delivered by craterforming events that occurred later than the formation of the mare basalts at the CE – 4 landing site. We found a total of 16 craters that may have delivered ejecta thicker than 10cm level superposed on the mare basalts at the CE – 4 landing site. Crater Finsen, Von Kármán L, Von Kármán L′, and Maksutov are the top four major foreign material sources, and each of them contributed ejecta thicker than 1m. Our surveys confirm that the ejecta from Finsen crater are the most dominant foreign materials in the uppermost few meters at the CE – 4 landing site and the total impact ejecta deposited upon the mare basalts at the landing site is estimated to be thinner than 30m. We found that the estimations from Pike's model are the most consistent with the Yutu – 2 LPR observations.

Unified Astronomy Thesaurus concepts: The Moon (1692); Lunar craters (949).

1 Introduction

On 2019 January 3, China's Chang'e – 4 (CE – 4) lunar probe successfully landed in the Von Kármán crater on the lunar far side and deployed the Yutu－2 rover. The landing site（coordinated at 45. 4446°S， 177. 5991°E； Li et al. 2019） was later officially named as Statio Tianhe by the International Astronomical Union（IAU）. Von Kármán is an impact crater with a diameter of ～186km and is located inside the heavily cratered South Pole－Aitken（SPA） basin（Figure 1）， the largest and oldest known impact structure on the Moon（Spudis et al. 1994； Garrick－Bethell & Zuber 2009）. Although Von Kármán is an ancient crater with a pre－Nectarian age（Losiak et al. 2009； Yingst et al. 2017）， the crater floor was resurfaced by younger multiple－phase Imbrian－aged mare basalts（Yingst & Head 1997； Yingst et al. 2017； Pasckert et al. 2018）. After those basalts were deposited， the impact ejecta of many postdated craters on the entire Moon successively superposed on these mare basalt units（termed "post－mare craters" in this study）， creating the current geologic settings of the CE－4's landing site.

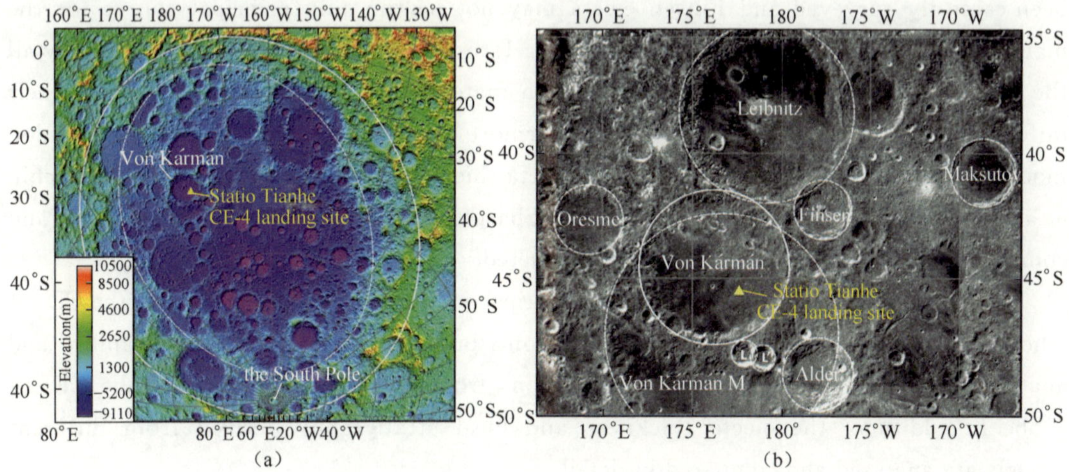

Figure 1　The location and surrounding context of Statio Tianhe， the Chang'e－4（CE－4） landing site. （a） The topography of the CE－4 landing site within the Von Kármán crater shown on a Lunar Orbiter Laser Altimeter（LOLA） color－shaded relief map with a resolution of 128 pixels/degree. The white ovals mark the best－fit topographic inner ring and outer ellipse of the SPA basin according to Garrick－Bethell & Zuber（2009）. The projection is orthographically centered on the SPA basin（53°S， 191°E）. （b） The craters around Von Kármán crater shown on a CE－2 Digital Orthophoto Map（DOM） with a resolution of 7 m/pixel（IDs： J101， J102， J139， J140， J201， J202， J239， J240， K101， K102， K135， K136， K201， K202， K235， and K236）. The solid and dashed circles show the size of the labeled craters according to Robbins（2019）. Labels L and L′ represent the Von Kármán L and Von Kármán L′ craters， respectively. This map is shown in the Mercator projection centered on the Von Kármán crater（44. 45°S， 176. 25°E）.

Understanding the origin and amounts of foreign materials at the CE－4 landing site is essential for analyzing the local stratigraphy and geologic history. Before the landing of CE－4， Huang et al.（2018） reported its landing site is mainly superposed by the ejecta from at least four nearby（all <140km in distance from the landing site） craters： Finsen， Von

Kármán L, Von Kármán L′, and Antoniadi. After the landing, the visible and near - infrared spectrometer (VNIS) onboard the Yutu - 2 rover found that the mafic materials around the landing site are dominated by orthopyroxene, different from the clinopyroxenedominated mare basalts. Some researchers (Huang et al. 2018; Hu et al. 2019; Li et al. 2019) suggested that these materials are likely originated from the nearby Finsen crater, a ~73 - km - diameter crater of the Eratosthenian period (Wilhelms & Byrne 2009; Fortezzo et al. 2020), since the landing site is crossed by ejecta rays from Finsen and the Finsen ejecta is also dominated by orthopyroxene. In addition, the lunar penetrating radar (LPR) onboard the Yutu - 2 rover also suggests that the uppermost stratigraphy at the CE - 4 landing site is a fine regolith layer underlain by several impact ejecta deposit layers, including that of Finsen crater (Lai et al. 2019; Li et al. 2020; Zhang et al. 2021a).

However, the foreign materials delivered from relatively distant sources have not yet been carefully surveyed and Finsen ejecta may not be the only principal source for the materials detected by the Yutu - 2 rover. Therefore, it is important to carefully survey all the possible principal sources of these foreign materials on the Moon and evaluate their influence on the CE - 4 landing site. Furthermore, various empirical ejecta distribution models have been widely used to estimate ejecta thickness in different studies (McGetchin et al. 1973; Pike 1974; Housen et al. 1983; Sharpton 2014), while few assessments and comparisons of these models have been conducted.

In this study, we examine all possible principal sources of foreign materials at the CE - 4 landing site, quantitatively evaluate their contributions to the local stratigraphy, and analyze their effects on the in situ observations from the VNIS and LPR of the CE - 4 probe. In addition, the ejecta thickness and local stratigraphy derived from different models are analyzed and compared in detail.

2 The Identification of Post - mare Craters for the CE - 4 Landing Site

Prior crater counting analyses for different areas determined the ages of the Von Kármán mare units as 3.75—3.15 Ga (Haruyama et al. 2009; Huang et al. 2018; Pasckert et al. 2018; Ling et al. 2019), roughly corresponding to the Late Imbrian Epoch (3.80—3.20 Ga; Stöffler & Ryder 2001). Therefore, we focus on craters contemporaneous with and younger than the Late Imbrian Epoch (i. e., "post - mare craters") throughout the surface of the Moon in this study.

The database of Robbins (2019) is used to obtain the sizes and locations of lunar craters. The stratigraphic ages of lunar craters and basins mainly refer to the most recent geologic map of Fortezzo et al. (2020), which is the first unified global lunar geologic map at a 1: 5,000,000 - scale. Cross - cutting and superposition relationships are also considered if the stratigraphy of the crater is not identified in the geologic map of Fortezzo et al. (2020). The Lunar Reconnaissance Orbiter Camera (LROC) Wide Angle Camera (WAC) mosaic image is used to evaluate the geologic context of these craters.

3 Models and Modifications

3.1 The Thickness of the Primary Ejecta

Multiple models were proposed to estimate the primary ejecta thickness δ (r) from an impact crater at a spherical distance r from the crater center:

1. Pike's model (Pike 1974):

$$\delta = 0.033R_t(r/R_t)^{-3.0} \tag{1}$$

2. McGetchin's model (McGetchin et al. 1973):

$$\delta = 0.14R_t^{0.74}(r/R_t)^{-3.0}\text{(in m)} \tag{2}$$

3. Housen's model (Housen et al. 1983):

$$\left(\frac{\delta}{R_{\text{at}}}\right) = \frac{A_1(e_r-2)}{2\pi}(\sin 2\theta)^{e_r-2}r^{*-e_r} \times \left[1+\frac{4e_r-5}{3}\left(\frac{r^*}{\sin 2\theta}\right)^{-(e_r-2)/2}\frac{A_2}{r^*}\right]\text{(in the gravity regime)} \tag{3}$$

and (4) Sharpton's model (Sharpton 2014):

$$\delta = \begin{cases} 0.014R^{1.01}(r/R)^{-3} & \text{for simple crater} \\ 3.95R^{0.399}(r/R)^{-3} & \text{for complex crater} \end{cases} \text{(in m)} \tag{4}$$

where R_t is the transient rim radius of the crater, R_{at} is the apparent transient radius at the surface, $r^* = \frac{r}{R_{\text{at}}}$ is the nondimensional distance from the crater center, and R is the final rim radius of the crater. The transient radius at the surface R_{at} is approximately 83% of the transient rim radius R_t (Baldwin 1963; Pike 1974). For Housen's model (Equation (3)), θ is the ejecta launch angle, $A_1 = 0.2$ is constant, and A_2 and e_r are target property-related parameters (Housen et al. 1983). By adopting the typical impact angle of $\theta = 45°$ and widely used target parameters ($A_2 = 0.82$, $e_r = 2.61$) for Ottawa sand, Equation (3) can be simplified as (Xu & Xie 2020):

$$\delta = 0.0100 \times R_t\left(\frac{r}{R_t}\right)^{-2.6145} + 0.011724 \times R_t\left(\frac{r}{R_t}\right)^{-3.9217} \tag{5}$$

noting that the ejecta thickness δ estimated from Housen's model is highly determined by the specific ejection angle and target property, which will be further discussed in Section 4.2.

In addition, if the distance between the center of the impact crater and the ejecta delivery area is considerably large, the distances of the ejecta after ballistic trajectories are not greatcircle distances and the ejecta cannot be treated simply as being deposited on a flat surface. Therefore, two kinds of correction should be taken into consideration: (1) distance correction and (2) deposition correction. We use the improved ballistic sedimentation model of Haskin et al. (2003) and Xie et al. (2020a) in this section. This method assumes that the ejecta launched from each impact crater is symmetric and the ejecta emplacements are not affected by the pre - existing topography, as with previous models (Section 3.1).

3.1.1 The Distance Correction

For simplicity, we assume that primary ejecta from an impact crater with a final rim radius of R is launched from the transient cavity radius at the surface R_{at}, with a velocity of v (R_S) and an ejection angle of θ measured from the horizon, and falls on a square of interest (SOI) with a length of $L_{SOI}=\frac{1}{4}R_{at}$ after a ballistic trajectory, as shown in Figure 2.

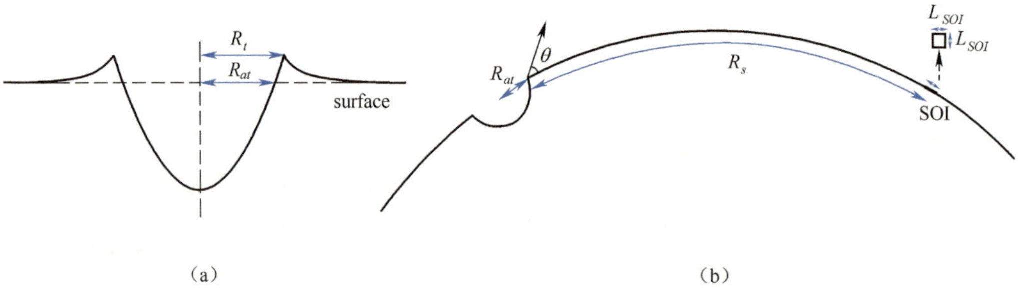

Figure 2 (a) Schematic showing the rim radius, R_t, and the apparent radius, R_{at} (i.e., the radius at the surface), of a transient cavity. (b) Side view showing a primary transient crater, the launch angle θ and the travel distance R_S of the ejecta, and the location and size, LSOI, of the SOI

We consider $R_t=0.84R$ for simple craters (Melosh 1989) and $R_t=R_{SC}^{0.15}R^{0.85}$ for complex craters (Croft 1985), where $R_{SC}=9.5$km is the transition radius from simple to complex craters (Pike 1980).

In addition, considering that R_{at} is generally substantially smaller than the average radius $R_0=1737.1$km of the Moon, the arc distance from the center of the impact crater to the location of R_{at} can be treated as approximately equal to R_{at}. Therefore, the spherical distance from the launch point to the ejecta delivery area SOI, is:

$$R_S \approx d_{sphere} - R_{at} \tag{6}$$

where d_{sphere} is the spherical distance from the center of the crater to the SOI region.

The velocity of the primary ejecta (Melosh 1989, p.87; Haskin et al. 2003, Equation (9)) is:

$$v(R_s)=\frac{\sqrt{R_0 g \tan X}}{\sqrt{\tan X \cos^2\theta+\sin\theta\cos\theta}} \tag{7}$$

where g is the surface gravity of the Moon and $X=R_S/(2R_0)$.

The actual distance of the ejecta is:

$$r(R_s)=R_{at}+\frac{v^2(R_s)\sin(2\theta)}{g}. \tag{8}$$

Assuming the typical ejection angle of $\theta=45°$ and substituting Equation (7) into Equation (8), Equation (8) can be further simplified to:

$$r(R_s)=R_{at}+\frac{2R_0\tan X}{\tan X+1} \tag{9}$$

noting that the revised distance, $r(R_s)$, is irrespective of the value of surface gravity, g.

3.1.2　The Deposition Correction

For the region of a SOI with a radical length of $L_{SOI}=\frac{1}{4}R_{at}$, as shown in Figure 2 (b), the inner and outer boundaries from the center of the primary crater are $r_{inner}=r(d_{sphere}-L_{SOI}/2)$ and $r_{outer}=r(d_{sphere}+L_{SOI}/2)$, respectively. Considering that ejecta thickness decreases as a power function with distance, the average thickness of the emplaced ejecta within the SOI is approximately equal to the thickness at the geometry mean distance at $\bar{r}=\sqrt{r_{inner}r_{outer}}$.

Combining the above terms with Equations (1) − (4), the primary ejecta thickness at the SOI region, δ_{SOI}, can be derived as:

$$\delta_{SOI}=\frac{S_{ring-flat}}{S_{ring}}\delta(\bar{r}) \tag{10}$$

where $S_{ring}=2\pi R_0^2\ (\cos((d_{sphere}-L_{SOI}/2)\ R_0)-\cos((d_{sphere}+L_{SOI}/2)\ /R_0))$ is the area of a ring for a spherical planetary body and $S_{ring\_flat}=\pi\ (r_{outer}^2-r_{inner}^2)$ is the area of a ring emplaced on a flat surface.

3.2　The Depth of Local Materials

The local ejecta stratigraphy is determined not only by the amount of primary ejecta but also by the degree of mixing of the primary ejecta and the local materials excavated by the ejecta. Taking the mixing model of Petro & Pieters (2004, 2006, 2008), which is modified from the initial model of Oberbeck et al. (1975), the ratio, μ', of the thickness of the local materials to the primary ejecta can be estimated by:

$$\mu'=\begin{cases}\mu, & \mu\leqslant 5\\ \mu/2+2.5, & \mu>5\end{cases}$$

$$\text{and } \mu=0.0183r^{0.87}\ (\text{in km}) \tag{11}$$

4 Evaluating the Amounts of Ejecta Materials On the Mare Basalts at the Chang'e – 4 Landing Site

For our estimations, the lower threshold of the ejecta thickness is based on the vertical resolution of the Yutu – 2 LPR, which is no less than 15cm for the high – frequency channel and greater than 1m for the low frequency channel (Lai et al. 2019, 2020). Therefore, we take 10cm as the threshold and consider that only the ejecta estimated to be thicker than 10cm level are potentially distinguished from LPR observations. However, we should note that the modeled thickness of distal ejecta beyond continuous ejecta blankets may be of high uncertainty, so we only use this threshold as a rough lower limit. All four models, Pike (1974), McGetchin et al. (1973), Housen et al. (1983), and Sharpton (2014), are used for thickness estimations in this study, and we only survey the craters that delivered primary ejecta thicker than the threshold based on the maximum values among these four models.

4.1 Possible Sources of Foreign Materials at the CE – 4 Landing Site

4.1.1 Thickness of Primary Ejecta from Large Post – mare Craters ($D \geqslant 100$km)

Since the larger impact cratering events can produce larger craters and deliver more ejecta to other locations (assuming the other factors are the same), we first surveyed the thickness of large craters' ($D \geqslant 100$km) primary ejecta present at the CE – 4 landing site in this subsection.

According to the geologic ages summarized by the geologic map of Fortezzo et al. (2020), there are a total of 14 impact craters larger than 100km throughout the surface of the Moon within or younger than the Late Imbrian Epoch (Figure 3 (a) and Table 1). Among them, six craters (Hausen, Humboldt, Tsiolkovskiy, Antoniadi, Aitken, and Fizeau) contributed ejecta >10cm in thickness to the CE – 4 landing site accordingto Pike's model (the model that predicts the maximum ejecta thickness, as referred in the sections below).

According to Pike's model (Table 1), most of the ejecta from the large post – mare craters ($D \geqslant 100$km) to the CE – 4 landing site is from Antoniadi crater of the Late Imbrian Epoch, with an ejecta thickness of 57cm. The Hausen crater is the only Eratosthenian – aged crater that contributed considerable amounts of ejecta, with a thickness of 13cm. All large postmare craters ($D \geqslant 100$km) on the Moon considered in this study delivered a total ejecta thickness of 1.5m at the CE – 4 landing site.

4.1.2 Thickness of Ejecta from Small Post – mare Craters ($D<100$km)

For any lunar post – mare crater ($D<100$km) beyond a spherical distance of 950km

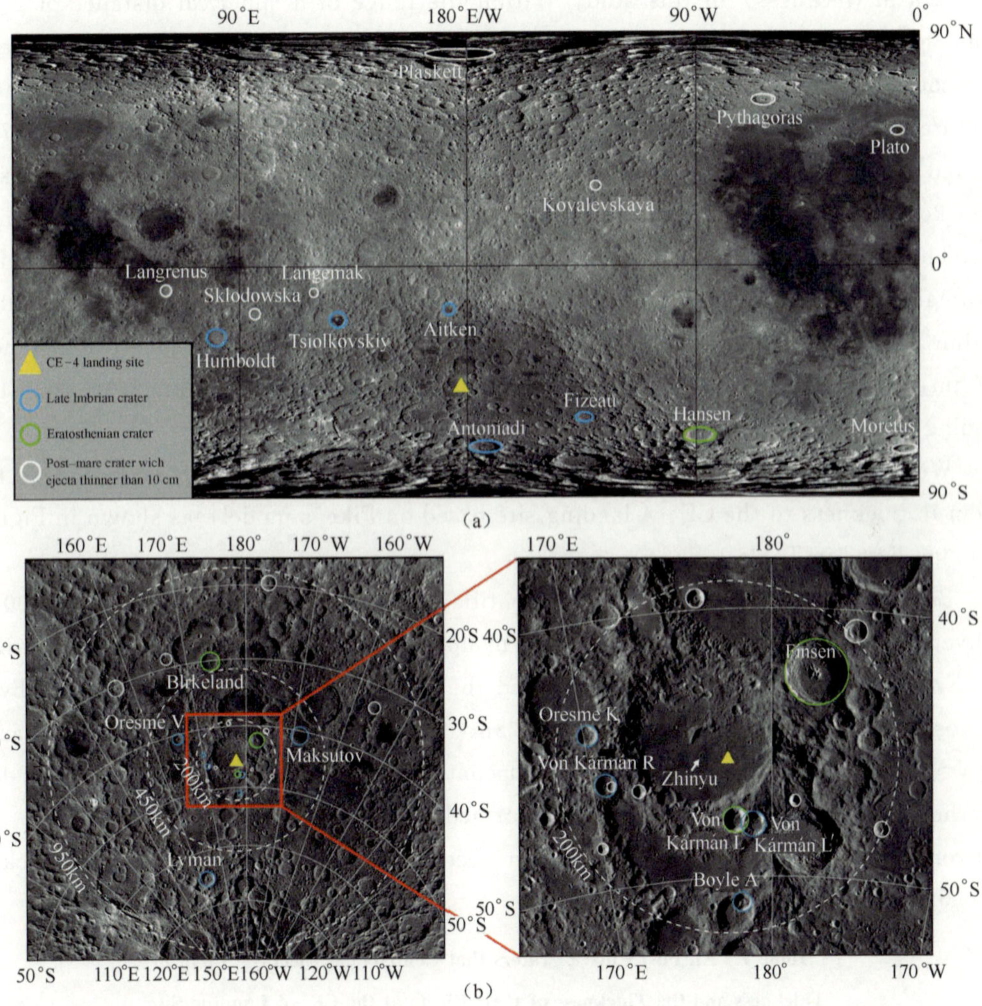

Figure 3　Locations of all post - mare craters identified in this study: (a) craters with $D \geqslant 100$km (simple cylindrical projection) and (b) craters with $D < 100$km (stereographic projection center at the CE - 4 landing site) . The base image is a LROC WAC mosaic image. The dashed white circles in (b) mark the examined boundaries of spherical distances of 950, 450, and 200km from the CE - 4 landing site. The colored craters are the post - mare craters that delivered ejecta >10cm in thickness to the mare units at the CE - 4 landing site (Late Imbrian craters are in cyan and Eratosthenian craters are in green), and the white craters are the post - mare craters that delivered ejecta <10cm in thickness. The yellow triangle marks the location of the CE - 4 landing site

from the CE - 4 landing site, the maximum thickness of its ejecta on the mare basalts at the CE - 4 landing site is ~9cm (i. e. , within the threshold of 10cm in this study; based on Pike's model) . Therefore, we can narrow the examination range in this study. Similarly, for any impact crater smaller than 50km in diameter and beyond a spherical distance of 450km from the CE - 4 landing site, and for any impact crater smaller than 25km in diameter and beyond a spherical distance of 200km, the upper limits of ejecta thickness were estimated to be ~8cm and ~9cm, respectively, which are considered neglectable

(<10cm in thickness) in this study. Within the range of a spherical distance of 200km from the CE-4 landing site, there are no smaller post-mare craters (D<10km) nearby that could deliver ejecta >10cm in thickness based on our observations and calculations (Figure 3 (b)). For example, the Copernican crater Zhinyu (with a diameter of 3.75km located ~30km from the landing site, Figure 3 (b)) delivered ejecta with a thickness less than 2cm (from the four models in Section 3) superposed on the CE-4 landing site. Therefore, we exclude the contributions from craters with D<10km in this study. In summary, we examine post-mare craters in the following order: (1) 50km≤D<100km within a spherical distance of 950km, (2) 25km≤D<50km within a spherical distance of 450km, and (3) 10km≤D<25km within a spherical distance of 200km from the CE-4 landing site in this section.

We find a total of 10 small post-mare craters (D<100km) that contributed ejecta >10cm in thickness to the CE-4 landing site based on Pike's model, as shown in Figure 3 (b) and listed in Table 1.

Based on Pike's model, all these identified small post-mare craters (D<100km) delivered a total ejecta thickness of ~17.1m upon the CE-4 landing site. Among them, the Eratosthenian crater Finsen contributed the most ejecta (~9.4m), followed by the Eratosthenian crater Von Kármán L (~3.2m) and the Late Imbrian crater Von Kármán L′ (~2.1m). Taking the most dominant contributor (Finsen crater) as the boundary, all these small craters (D < 100km) formed earlier than Finsen crater (i.e., they delivered ejecta deposited below the Finsen ejecta) and contributed a total ejecta thickness of ~5.5m (Table 1).

Table 1　All Post-mare Craters that Contributed Ejecta >10cm in Thickness and the Thickness of their Ejecta at the CE-4 Landing Site[a]

Crater	Stratigraphy (Fortezzo et al. 2020)[b]	Diameter (km)	Longitude (°)	Latitude (°)	Spherical Distance (km)	Ejecta Thickness (m)			
						Pike's model	McGetchin's model	Housen's Model	Shapton's model
Humboldt	Ic2	204.25	80.96	−27.18	2283.3	0.11	0.03	0.11	0.05
Tsiolkovskiy	Ic2	183.71	129.05	−20.38	1421.4	0.24	0.06	0.22	0.12
Antoniadi	Ic2	137.89	−172.99	−69.40	739.9	0.57	0.15	0.48	0.28
Aitken	Ic2	132.98	173.05	−16.50	885.6	0.30	0.08	0.27	0.15
Fizeau	Ic2	106.79	−134.08	−58.21	958.8	0.11	0.03	0.11	0.06
Lyman	Ic2	84.60	162.47	−64.96	642.7	0.17	0.05	0.15	0.08
Maksutov	Ic2	84.53	191.36	−40.79	335.4	**1.13**	**0.32**	**0.86**	**0.55**
OresmeV	Ic2	53.98	165.41	−40.70	305.4	0.33	0.10	0.27	0.16
Von Kármán L′c	Ic2	29.04	178.93	−48.00	82.0	**2.06**	**0.73**	**1.38**	**1.00**
Von Kármán R	Ic2	25.44	170.76	−46.04	145.6	0.23	0.09	0.19	0.11
Oresme K	Ic2	23.52	170.21	−44.09	164.0	0.13	0.05	0.11	0.06
Boyle A	Ic2	22.97	178.14	−51.00	168.3	0.11	0.04	0.09	0.05

Continued Table1

Crater	Stratigraphy (Fortezzo et al. 2020)[b]	Diameter (km)	Longitude (°)	Latitude (°)	Spherical Distance (km)	Ejecta Thickness (m)			
						Pike's model	McGetchin's model	Housen's Model	Shapton's model
Total ejecta thickness of Late Imbrian craters						5.5	1.7	4.2	2.7
Hausen	Ec	165.01	−88.57	−65.20	1550.8	0.13	0.03	0.13	0.06
Birkeland	Ec	81.65	173.99	−30.16	471.6	0.37	0.10	0.31	0.18
Finsen	Ec	73.10	182.31	−42.32	140.4	**9.39**	**2.73**	**5.94**	**4.57**
Von Kármán L	Ec	29.65	177.90	−47.83	72.2	**3.22**	**1.14**	**2.09**	**1.57**
Total ejecta thickness of Eratosthenian craters						13.1	4.0	8.5	6.4
Total ejecta thickness for all large post-mare craters ($D \geqslant 100$km)						1.5	0.4	1.3	0.7
Total ejecta thickness for all small post-mare craters ($D < 100$km)						17.1	5.4	11.4	8.3
Total ejecta thickness for all post-mare craters						18.6	5.7	12.7	9.0

Notes.

[a] The crater sizes and locations are according to the crater database of Robbins (2019) and the values marked in bold font show the four thickest ejecta.

[b] Ic2 indicates Late Imbrian craters, and Ec indicates Eratosthenian craters.

[c] We provisionally use the name of Von Kármán L′ according to Huang et al. (2018), which is not an IAU-approved name.

4.2 Comparisons of Primary Ejecta Thicknesses between the Four Ejecta Models

All 16 post-mare craters, including the large ($D \geqslant 100$km) and small ($D < 100$km) craters identified in this study, delivered total primary thicknesses of 18.6, 5.7, 12.7, and 9.0m at the CE-4 landing site based on the ejecta models of Pike (1974), McGetchin et al. (1973), Housen et al. (1983), and Sharpton (2014), respectively (Table 1).

Pike's model gives the largest thickness estimates among the four models as shown in Table 1. For the same identified craters, primary ejecta thicknesses derived from McGetchin's model and Shapton's model are only ~1/3 and ~1/2 of those from Pike's model, respectively. This indicates that the ejecta thicknesses estimated from these three models are roughly proportional to the same exponential term of $(r/R_t)^{-3.0}$ (Equations (1), (2), and (4)).

Pike's and McGetchin's models are derived from the same observation data sets (McGetchin et al. 1973; Pike 1974), and their discrepancy is due to the fact that Pike's model excluded two obviously unreliable data points (i.e., Copernicus crater and Imbrium basin). In this respect, Pike's model may provide more reliable estimations than McGetchin's model, which is also supported by recent studies (e.g., Xie & Zhu 2016; Xu & Xie 2020).

Both Sharpton's and Pike's model comes from actual ejecta measurements, their discrepancy is possibly due to two aspects: (1) Sharpton's model underestimates the ejecta thickness by a factor of ~2 or Pike's model overestimates the ejecta thickness by a factor of ~2 (Sharpton's model itself has a measurement uncertainty of ~±30%;

Sharpton 2014); (2) the inversion relationship from the final radius to the transient radius derived by Croft (1985) is unsuitable for the craters identified in this study, resulting in larger transient radiuses for Pike's model (and also for MacGetchin and Housen's models) and thus thicker ejecta.

The discrepancy between Housen's model and another three models is more complicated. The ejecta thickness estimated from Housen's model is highly determined by the specific ejection angle (θ) and target property (A_2 and e_r) but, in reality, those parameters may be varied for each specific impact event and ejecta deposition process. In this study, we use the typical ejection angle of $\theta = 45°$ and widely used target parameters ($A_2 = 0.82$, $e_r = 2.61$) for Ottawa sand and, if the estimated thickness and the actual thickness are inconsistent, the inconsistency is more likely due to that the choice of those parameters may be not suitable for the specific case at the CE-4 landing site, rather than the model validation itself. For example, if the ejection angle varies within $\pm 20\%$ from the typical value (i. e., θ varies from $36° - 54°$), the ejecta thickness will decrease $<3\%$ compared with the current result for the case of Finsen's ejecta to the CE-4 landing site. Similarly, if the target parameters A_2 or e_r vary within $\pm 20\%$ from the typical value, the ejecta thickness will vary within $\pm 3\%$ and decrease $<64\%$ compared with the current result, respectively. Therefore, we must note that the ejecta thicknesses derived from Housen's model are insensitive to the minor variations from the typical values of parameters θ and A_2 used in this study, but sensitive to the variations of target parameter e_r.

In addition, ejecta thicknesses estimated from Housen's model have different trends for near and distant craters. For most distant craters (beyond $\sim 5\ R_t$; Table 1), Housen's model gives similar ejecta estimates compared to those of Pike, while for a few nearer craters (Finsen, Von Kármán L, Von Kármán L′, and Maksutov; all within $\sim 5\ R_t$), their ejecta thicknesses derived from Housen's model are ~2/3 of those from Pike's model. This is because that for the nearer craters, both the terms $(r/R_t)^{-2.6145}$ and $(r/R_t)^{-3.9217}$ in Equation (3) play important roles in the thickness estimations, while for distance craters, only the low-order term $(r/R_t)^{-2.6145}$ plays an important role.

4.3 The Principal Contributors of Foreign Materials at the CE-4 Landing Site

The first five lunar days of Yutu-2 LPR observations show that the uppermost stratigraphy upon the local basalts at the CE-4 landing site have four obvious reflectors (Lai et al. 2019) that may correspond to various ejecta layers from four main impact events. Therefore, we find the craters delivering the top four thick ejecta to the CE-4 landing site as the principal contributors: (1) Finsen crater of the Eratosthenian Period (~9.4m; based on Pike's model; similarly hereinafter), (2) Von Kármán L crater of the Eratosthenian Period (~3.2m), (3) Von Kármán L′ crater of the Late Imbrian Epoch (~2.1m), and (4) Maksutov crater of the Late Imbrian Epoch (~1.1m). This differs slightly from the results of Huang et al. (2018) in that the ejecta from Maksutov crater outweighing Antoniadi crater (also a Late Imbrian crater contributing an ejecta 57cm thick) plays a more important role than previously thought (Huang et al. 2018).

Because the ratio, μ', is only determined by the distance, r, between the center of the impact crater and the ejecta destination, the ratio of the thickness of local materials to the primary ejecta is irrelevant to the choice of ejecta models. Therefore, Finsen crater caused the thickest local materials and thus the thickest mixing zones (i. e., ejecta deposits, which are the combination of primary ejecta and local materials) at the CE-4 landing site, followed by Von Kármán L and Maksutov craters, regardless of the ejecta models, as shown in Table 2.

5 Discussion

5.1 The Emplacement Sequence of these Principal Ejecta at the CE-4 Landing Site

For simplicity, we only discuss the emplacement sequence of the top four principal source craters (Table 2). Although, according to Fortezzo et al. (2020), the two Eratosthenian craters Finsen and Von Kármán L formed later than the two Late Imbrian craters Von Kármán L′ and Maksutov, the relative age between the two craters in the same period is unknown and it is impractical to directly compare the absolute model ages of these four craters by crater counting because the countable areas of Von Kármán L and Von Kármán L′ craters are too small.

Table 2　The Thickness of Local Materials and Ejecta Deposits of All Post-mare Craters Contributing Ejecta >10cm in Thickness to the CE-4 Landing Site[a]

Crater	Ratio of Primary Ejecta[b]	Pike's Model (m)		McGetchin's Model (m)		Housen's Model (m)		Shapton's Model (m)	
		Local Materials	Ejecta Deposits[c]	Local Materials	Ejecta Deposits	Local Materials	Ejecta Deposits	Local Materials	Ejecta Deposits
Humboldt	0.09	1.12	1.23	0.26	0.29	1.13	1.24	0.54	0.60
Tsiolkovskiy	0.12	1.82	2.07	0.43	0.49	1.67	1.90	0.89	1.00
Antoniadi	0.16	3.08	3.65	0.78	0.92	2.56	3.04	1.50	1.77
Aitken	0.15	1.77	2.07	0.45	0.53	1.56	1.83	0.86	1.00
Fizeau	0.14	0.70	0.81	0.19	0.22	0.67	0.78	0.34	0.39
Lyman	0.17	0.83	1.00	0.24	0.28	0.77	0.92	0.41	0.49
Maksutov	**0.26**	**3.27**	**4.40**	**0.92**	**1.24**	**2.48**	**3.34**	**1.59**	**2.14**
Oresme V	0.27	0.87	1.20	0.27	0.37	0.72	1.00	0.42	0.58
VonKármán L′	**0.54**	**1.74**	**3.80**	**0.62**	**1.36**	**1.17**	**2.54**	**0.85**	**1.85**
VonKármán R	0.42	0.33	0.56	0.12	0.21	0.27	0.46	0.16	0.27
Oresme K	0.39	0.19	0.32	0.07	0.12	0.17	0.28	0.09	0.16
Boyle A	0.39	0.17	0.28	0.06	0.10	0.15	0.24	0.08	0.13
Hausen	0.11	1.06	1.20	0.26	0.29	1.03	1.16	0.52	0.58

Continued Table2

Crater	Ratio of Primary Ejecta[b]	Pike's Model (m)		McGetchin's Model (m)		Housen's Model (m)		Shapton's Model (m)	
		Local Materials	Ejecta Deposits[c]	Local Materials	Ejecta Deposits	Local Materials	Ejecta Deposits	Local Materials	Ejecta Deposits
Birkeland	0. 21	1. 40	1. 78	0. 40	0. 51	1. 20	1. 51	0. 69	0. 87
Finsen	**0. 43**	**12. 69**	**22. 08**	**3. 69**	**6. 43**	**8. 02**	**13. 96**	**6. 17**	**10. 74**
Von Kármán L	**0. 57**	**2. 44**	**5. 66**	**0. 87**	**2. 01**	**1. 58**	**3. 67**	**1. 19**	**2. 76**

Notes

[a] The crater sizes and locations are according to the crater database of Robbins (2019) and the values marked in bold font show top four primary ejecta in Table 1.

[b] The ratios of primary ejecta to ejecta deposits are computed by $1/(1+\mu')$.

[c] Ejecta deposits are the sum of primary ejecta and local materials.

The Diviner rock abundance (RA; Bandfield et al. 2011) of crater ejecta shows an inverse correlation with crater age (Ghent et al. 2014). Because the Diviner - derived RA around Finsen crater is higher than that around Von Kármán L crater (Figure 4), we consider Finsen crater to be younger than Von Kármán L crater. However, since the RA around crater Maksutov and Von Kármán L′ are low and difficult to compare, it is impractical to deduce the relative ages of these two craters by using this method. Instead, we refer to the geologic map of Stuart - Alexander (1978) and obtain that Maksutov crater is older than Von Kármán L′ crater.

Given the above discussions, the most likely emplacement sequence of these four craters' ejecta at the CE - 4 landing site (from oldest to youngest) is Maksutov, Von Kármán L′, Von Kármán L, and Finsen (scenarios of Figures 5 and 6).

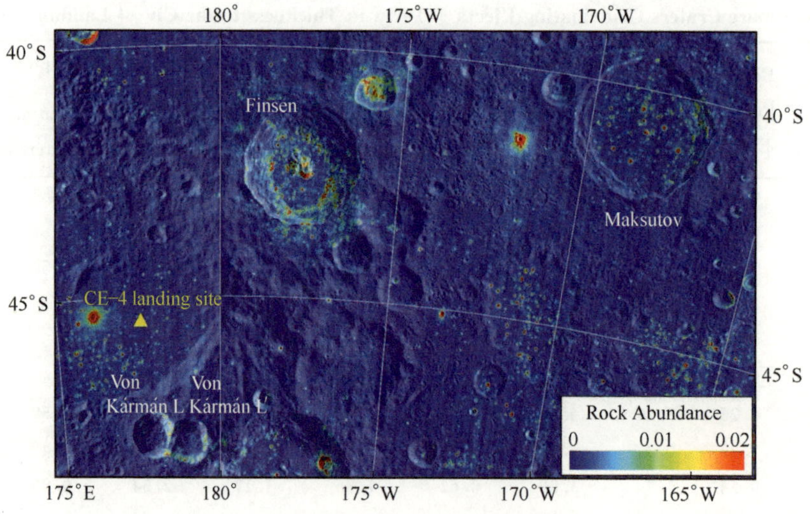

Figure 4　Diviner - derived rock abundance (Bandfield et al. 2011) around Finsen, Von Kármán L, Von Kármán L′, and Maksutov craters. This map is shown in the Mercator projection centered on Von Kármán crater

5. 2　The Stratigraphic Structure at the CE－4 Landing Site

If there is no local material excavation or mixing during each ejecta emplacement and no regolith production processes afterwards, these foreign materials from the four principal craters ought to be deposited as discrete layers in sequence, as shown in Figure 5. In these scenarios, the Finsen ejecta should be the topmost layer among the four craters' ejecta with depths of ～9. 4m, 2. 7m, 5. 9m, and 4. 6m based on the ejecta models of Pike (1974), McGetchin et al. (1973), Housen et al. (1983), and Sharpton (2014), respectively.

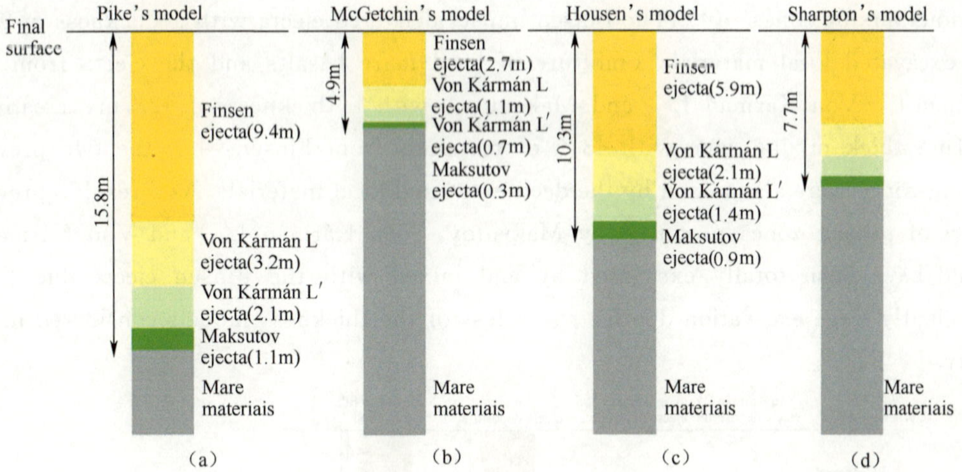

Figure 5　The emplacement stratigraphy at the CE－4 landing site without the consideration of local mixing and regolith production. (a) — (d) Ejecta thickness estimated from the four models with a sequence of older (bottom) to younger (top) of ejecta: from Maksutov, Von Kármán L′, Von Kármán L, and Finsen craters

However, the final stratigraphic structure is not only determined by the primary ballistically deposited ejecta but also by the excavated local materials mixed with the ejecta. Therefore, the final stratigraphic layers reflect the mixture of the primary ejecta and the excavated local materials (i. e. , ejecta deposits) .

The endmember scenario is that the foreign materials mixed homogenously with the local materials (i. e. , a mixing zone homogenously combined of primary ejecta and local materials) . If so, the stratigraphic sequence should be much different from Figure 5. As time passed, each crater's ejecta would have emplaced atop the previously deposited ejecta sequentially, excavated previous local materials, and mixed with them. The stratigraphy after homogenous mixture processes is shown in Figure 6. Taking Figure 6 (a) as an example: (1) first, the Maksutov impact delivered ejecta with a thickness of 1. 1m and excavated local materials (local mare basalts) with a thickness of 3. 3m, creating a 4. 4m－thick mixing zone with 26% of materials from Maksutov crater and 74% from local mare basalts. (2) Then, the Von Kármán L′ impact delivered ejecta with a thickness of 2. 1m and excavated local materials (mixture of local mare basalts and Maksutov ejecta) with a thickness of 1.7m, creating a 3.8 m-thick mixing zone with 54% of materials from Von Kármán L′ crater and 46% of materials from the previous mixture (i.e., Maksutov ejecta and local mare basalts). Because the excavating depth of the Von Kármán L′ ejecta (1.7m) is shallower than the previously created mixing zone (4.4m), a 2.7m-thick pre-existing mixing zone remains. (3) Similarly, the Von Kármán L impact delivered ejecta with a thickness of 3.2m and excavated local materials (mixture of the local mare basalts and the ejecta from Von Kármán L′ and Maksutov) with a thickness of 2.4m, creating a 5.7m-thick mixing zone with 57% of materials from Von Kármán L and leaving two previous mixing zones. (4) The Finsen impact delivered ejecta with a thickness of 9.4m and excavated local materials (mixture of local mare basalts and the ejecta from Von Kármán L, Von Kármán L′, and Maksutov) with a thickness of 12.7m, creating a 22.1m-thick mixing zone with 43% of materials from Finsen, but the two previous mixing zones were also mixed by the deep excavated local materials. As a result, previous layers of mixing zones produced by Maksutov, Von Kármán L′, and Von Kármán L would have been totally excavated by and mixed with the Finsen ejecta due to its sufficiently large excavation depth, regardless of the thickness models considered in this study.

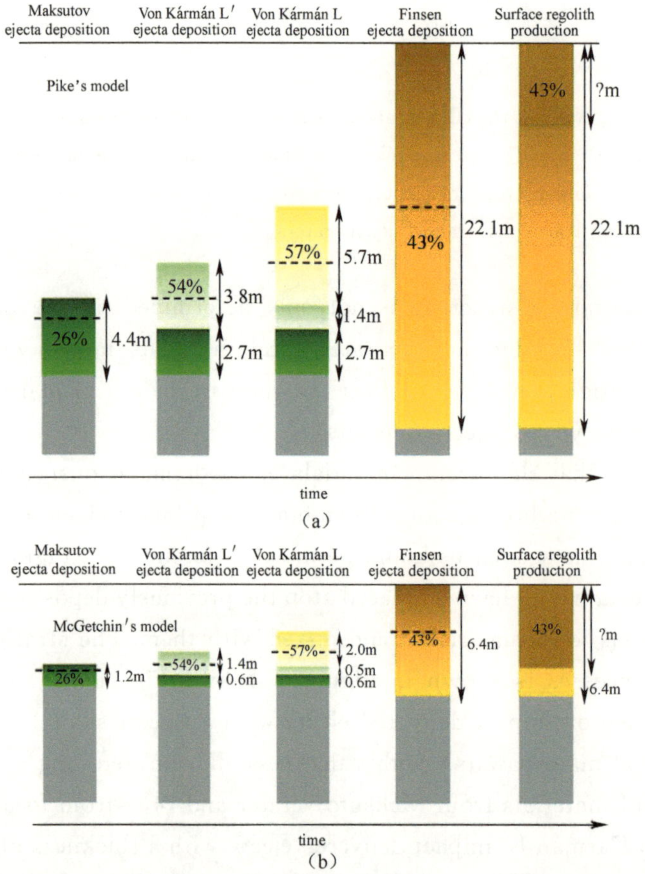

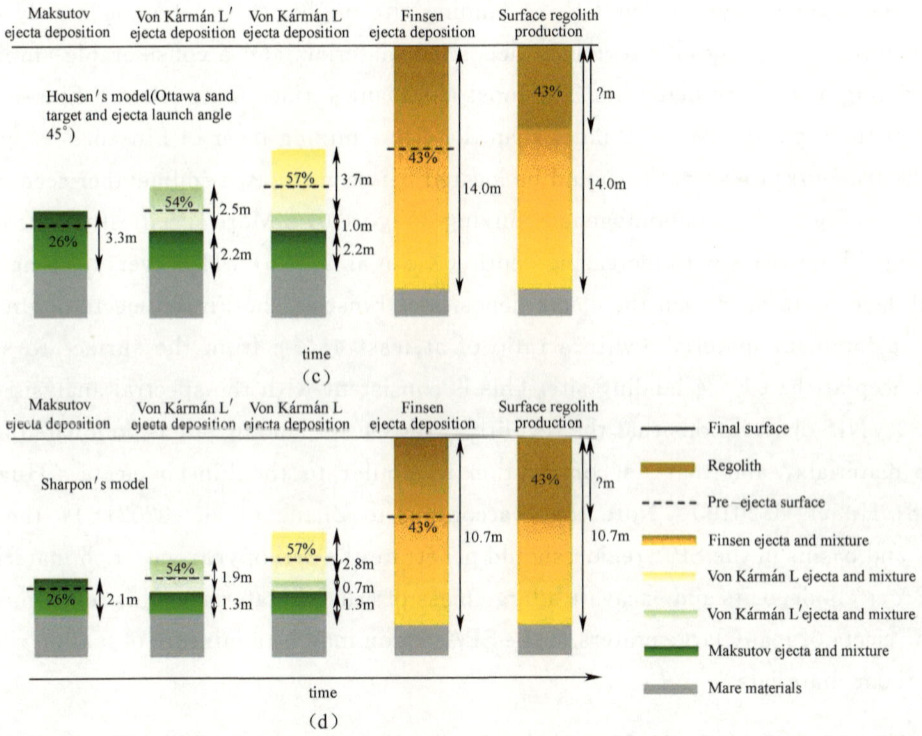

Figure 6　Illustration of the homogeneous mixing scenario showing the evolution of the stratigraphy with time at the CE - 4 landing site. The mixture processes are illustrated under different emplacement orders((a) — (d) for the ejecta sequence of Maksutov, Von Kármán L′, Von Kármán L, and Finsen, from older to younger) and thickness models (a for Pike's mode, b for McGetchin's model, c for Housen's model, and d for Sharpton's model) . The percentage values are the percentage of the foreign materials at the topmost layer and the regolith thickness can be determined by other methods

The mixture process produced a total thickness of 22.1m (Figure 6 (a)), 6.4 m (Figure 6 (b)), 14.0 m (Figure 6 (c)), and 10.7m (Figure 6 (d)) based on the ejecta model of Pike (1974), McGetchin et al. (1973), Housen et al. (1983), and Sharpton (2014), respectively, including the primary ejecta and local materials of each ejecta - delivering event. Taking the ejecta deposits of all post - mare craters (Tables 1 and 2) into consideration, the impact ejecta upon the mare basalts at the CE - 4 landing site ought to be no thicker than 30m based on the maximum value (Pike' s model) .

After the emplacement of Finsen ejecta, regolith production processes would further excavate and mix the near surface materials, making a fine - grained regolith layer with a proportion of Finsen ejecta remaining if the regolith is thinner than the topmost mixing zone (Figure 6) . The actual regolith thickness can be determined from other methods, such as LPR (e. g. , Lai et al. 2019; Li et al. 2019; Zhang et al. 2021a) or crater morphology (e. g. , Quaide & Oberbeck 1968; Bart et al. 2011; Fa et al. 2014) . As a result, the topmost layer at the CE - 4 landing site ought to be a homogenously mixed regolith layer containing all previously deposited materials and a considerable amount of pre - existing local mare materials. The most abundant surface materials are Finsen ejecta (43%) if the regolith layer is thinner than the ejecta mixing layer of Finsen.

The true mixture scenario should be located between the two endmember scenarios of no mixing (Figure 5) and homogenous mixing (Figure 6) . More specifically, the degree of mixing will decrease with increasing depth (Xie et al. 2020b) . However, as long as the regolith layer is thinner than the ejecta deposits of Finsen, the Finsen ejecta ought to be the most dominant material (with a ratio of at least 43%) from the surface to several meters deep at the CE - 4 landing site. This is consistent with the spectral analysis of the Yutu - 2 VNIS observations that the regolith at the CE - 4 landing site is dominated by non - mare materials, and that its composition is similar to the Finsen ejecta (Huang et al. 2018; Hu et al. 2019) . Note that, according to Zhang et al. (2021b), the large craters and basins in the SPA region should penetrate the orthopyroxene - rich materials of the SPA ejecta deposits and excavate a large mass of crustal materials. This result indicates that the ejecta of many large craters in the SPA region may be a mixture of mafic - rich and mafic - poor materials.

5.3 Comparison between the Model - derived Stratigraphy and the LPR Observations

The first five lunar days of the LPR observations of Yutu - 2 show three distinct subsurface layers ~40m deep, possibly corresponding to a fine - grained regolith layer, a layer of ejecta deposits, and a layer of possible mare basalts (hr1, hr2, and hr3 in Figure 7; Lai et al. 2019) . On the basis of the CE - 4 in situ radar measurements, only the stratigraphy estimated from Pike's model (22.1m in total thickness) are consistent with the LPR observations (Figure 7), while the ejecta deposit thicknesses estimated from the other three models (6.4m for McGetchin's model, 14.0m for Housen's model, and 10.7m for Sharpton's model) are far thinner than the sum thickness of hr1 and hr2 of the LPR observations. This is also consistent with several previous studies (e. g. , Krüger et al. 2014; Xie & Zhu 2016; Xu & Xie 2020) that showed that Pike's model gives better constraints on the distribution of primary ejecta than other models. In particular, since the impact ejecta deposits on mare basalts at the CE - 4 landing site are no thicker than 30m, we confirmed that the hr3 layer (~26—40 m deep) should not be ejecta deposits but mare basalts.

Therefore, our estimates based on Pike's model show that the hr1, hr2, and hr3 layers reflected in radar measurements can correspond to a fine - grained regolith layer (0—~11m deep) dominated by Finsen ejecta, a combination layer (~11—26 m deep) of coarse ejecta deposits (i. e. , the mixture zone of Finsen's primary ejecta, pre - Finsen primary ejecta, and local basalt materials), and a local mare basalt layer (~26—40m deep; Figure 7), respectively. Since the true mixture scenario should be located between the no mixing (Figure 5 (a)) and homogenous mixing scenarios (Figure 6 (a)), the Finsen ejecta would not completely mix with ejecta from previous craters and the layers dominated by previous ejecta may still exist. The second layer (hr2 in Figure 7) observed in the first five lunar days of the LPR observations can be further divided into three sublayers of ~11−17m, ~17−20m, and ~20−26m deep (hr2−1, hr2−2, and hr2−3 in Figure 7 (c)), which also correspond well to the mixing zones dominated by Von Kármán L ejecta, Von Kármán L′ ejecta, and Maksutov ejecta, respectively. Each sublayer is mixed with Finsen ejecta and prematerials (Figure 7).

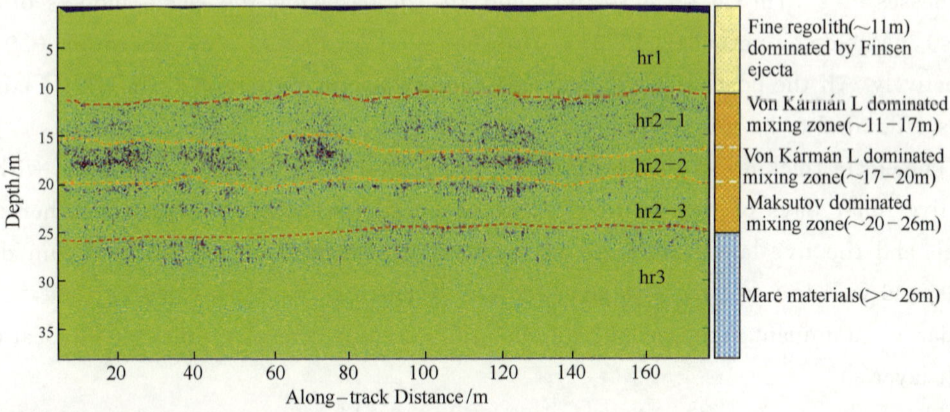

Figure 7　The first five lunar days of the LPR radargram of Yutu-2 and the resultant geologic interpretations. The LPR data were processed with repetitive data removal, including filtering, background removal, amplitude compensation, migration, and topographic correction. The topographic correction is referenced to the elevation of the landing site. For further details, refer to Lai et al. (2019, 2020). "hr" stands for horizontal reflector, hr1 for fine-grained regolith dominated by Finsen ejecta, hr2−1 for a mixture zone dominated by Von Kármán L ejecta, hr2−2 for a mixture zone dominated by Von Kármán L′ ejecta, hr2−3 for a mixture zone dominated by Maksutov ejecta, and hr3 for local mare basalts. All the hr1, hr2−1, hr2−2, and hr2−3 layers also contain Finsen ejecta, pre-existing ejecta, and local mare materials

6　Conclusions

In this study, we examined all possible sources (post-mare craters) of foreign materials on the mare basalts at the CE-4 landing site and found 16 craters potentially delivering primary ejecta >10cm level in thickness based on Pike's model. The total ejecta thicknesses of these 16 craters are 18.6m, 5.7m, 12.7m, and 9.0m based on the models of Pike (1974), McGetchin et al. (1973), Housen et al. (1983), and Sharpton (2014), respectively.

Four craters are principal foreign material sources to the CE-4 landing site based on Pike's model: Finsen (9.4m thick), Von Kármán L (3.2m thick), Von Kármán L′

(2.1m thick), and Maksutov (1.1m thick). Their most likely emplacement sequence is: Maksutov, Von Kármán L′, Von Kármán L, and Finsen from older to younger. Finsen ejecta should be the most dominant materials from the surface to several meters deep at the CE-4 landing site with a ratio of at least 43% regardless of the model choice, as long as the regolith layer is thinner than the ejecta deposits of Finsen, which is consistent with the VNIS observations of Yutu-2 (Huang et al. 2018; Hu et al. 2019).

Taking the local material excavations and mixtures into consideration, these four principal source craters produce ejecta deposits at the CE-4 landing site with total thicknesses of 22.1m, 6.4m, 14.0m, and 10.7m based on the ejecta models of Pike (1974), McGetchin et al. (1973), Housen et al. (1983), and Sharpton (2014), respectively. All the post-mare impact ejecta upon local mare basalts at the CE-4 landing site is estimated to be thinner than 30m.

The ejecta thicknesses estimated from Pike's model are the most consistent with the first five lunar days of the Yutu-2 LPR observations compared with the other three models and the five layers reflected in the LPR observations (from 0 to ~40m deep) correspond well to a fine regolith layer, a Von Kármán L-dominant mixing zone, a Von Kármán L′-dominant mixing zone, a Maksutov-dominant mixing zone, and a local mare basalt layer.

The authors thank Dr. Minggang Xie for valuable discussions when writing the manuscript. This work was supported by the pre-research project on Civil Aerospace Technologies No. D020202 funded by China National Space Administration (CNSA), the Science and Technology Development Fund, Macau SAR, the National Key R&D Program of China (No. 2020YFE0202100), the pre-research project on Civil AerospaceTechnologies No. D020204 funded by CNSA, the Science and technology project of Jiangxi education department (Grant GJJ200821), and the Scientific Research Starting Foundation for scholars from Jiangxi University of Science and Technology (Grant jxxjbs18017). All LROC and LOLA data can be downloaded from the USGS Astrogeology Science Center (https://astrogeology.usgs.gov/maps/) and CE-2 Digital Orthophoto Map (DOM) 7m data are available at the Data Release and Information Service System of China's Lunar Exploration Program (http://moon.bao.ac.cn).

References

[1] Baldwin, R.B. 1963, The Measure of the Moon (Chicago, IL: Univ. Chicago Press).

[2] Bandfield, J.L., Ghent, R.R., Vasavada, A.R., et al. 2011, JGRE, 116, E12.

[3] Bart, G.D., Nickerson, R.D., Lawder, M.T., et al. 2011, Icar, 215, 485.

[4] Croft, S.K. 1985, JGR, 90, C828.

[5] Fa, W., Liu, T., Zhu, M.H., et al. 2014, JGRE, 119, 1914.

[6] Fortezzo, C. M., Spudis, P. D., & Harrel, S. L. 2020, LPI, 2326, 2760.

[7] Garrick - Bethell, I., & Zuber, M. T. 2009, Icar, 204, 399.

[8] Ghent, R. R., Hayne, P. O., Bandfield, J. L., et al. 2014, Geo, 42, 1059.

[9] Haruyama, J., Ohtake, M., Matsunaga, T., et al. 2009, Sci, 323, 905.

[10] Haskin, L. A., Moss, B. E., & McKinnon, W. B. 2003, M&PS, 38, 13.

[11] Housen, K. R., Schmidt, R. M., & Holsapple, K. A. 1983, JGRB, 88, 2485.

[12] Hu, X., Ma, P., Yang, Y., et al. 2019, GeoRL, 46, 9439.

[13] Huang, J., Xiao, Z., Flahaut, J., et al. 2018, JGRE, 123, 1684.

[14] Krüger, T., Sturm, S., & Kenkmann, T. 2014, Proc. LPSC, 45, 1834.

[15] Lai, J., Xu, Y., Bugiolacchi, R., et al. 2020, NatCo, 11, 1.

[16] Lai, J., Xu, Y., Zhang, X., et al. 2019, GeoRL, 46, 12783.

[17] Li, C., Liu, D., Liu, B., et al. 2019, Natur, 569, 378.

[18] Li, C., Su, Y., Pettinelli, E., et al. 2020, SciA, 6, eaay6898.

[19] Ling, Z., Qiao, L., Liu, C., et al. 2019, P&SS, 179, 104741.

[20] Losiak, A., Wilhelms, D. E., Byrne, C. J., et al. 2009, LPI, 40, 1532.

[21] McGetchin, T. R., Settle, M., & Head, J. W. 1973, E&PSL, 20, 226.

[22] Melosh, H. J. 1989, Impact Cratering: A Geologic Process (Oxford: Oxford Univ. Press), 245.

[23] Oberbeck, V. R., Hörz, F., Morrison, R. H., et al. 1975, Moon, 12, 19.

[24] Pasckert, J. H., Hiesinger, H., & van der Bogert, C. H. 2018, Icar, 299, 538.

[25] Petro, N. E., & Pieters, C. M. 2004, JGRE, 109, E6.

[26] Petro, N. E., & Pieters, C. M. 2006, JGRE, 111, E9.

[27] Petro, N. E., & Pieters, C. M. 2008, M&PS, 43, 1517.

[28] Pike, R. J. 1974, E&PSL, 23, 265.

[29] Pike, R. J. 1980, Proc. LPSC, 11, 2159.

[30] Quaide, W. L., & Oberbeck, V. R. 1968, JGR, 73, 5247.

[31] Robbins, S. J. 2019, JGRE, 124, 871.

[32] Sharpton, V. L. 2014, JGRE, 119, 154.

[33] Spudis, P. D., Gillis, J. J., & Reisse, R. A. 1994, Sci, 266, 1848.

[34] Stöffler, D., & Ryder, G. 2001, Chronology and Evolution of Mars (Dordrecht: Springer), 9.

[35] Stuart - Alexander, D. E. 1978, US Geological Survey, Map I - 1047.

[36] Wilhelms, D. E., & Byrne, C. J. 2009, Index of Strata, http://www.imageagain.com/Strata/.

[37] Xie, M., Liu, T., & Xu, A. 2020a, JGRE, 125, e06113.

[38] Xie, M., Xiao, Z., Zhang, X., et al. 2020b, JGRE, 125, e06112.

[39] Xie, M., & Zhu, M. - H. 2016, E&PSL, 440, 71.

[40] Xu, L., & Xie, M. 2020, JGRE, 125, e06506.
[41] Yingst, R. A., Chuang, F. C., Berman, D. C., et al. 2017, LPI, 1964, 1680.
[42] Yingst, R. A., & Head, J. W., III 1997, JGRE, 102, 10909.
[43] Zhang, J., Zhou, B., Lin, Y., et al. 2021a, NatAs, 5, 25.
[44] Zhang, X., Xie, M., & Xiao, Z. 2021b, Icar, 358, 114214.

Intermittent volcanic activity detected in the Von Kármán crater on the farside of the Moon

Yuefeng Yuan[1], Peimin Zhu[1], Long Xiao[2], Jun Huang[2], Edward J. Garnero[3], Jian Deng[1], Fenghua Wang[4], Yuqi Qian[2], Na Zhao[5], Wengang Wang[1], Weiwei Li[1]

1. Hubei Subsurface Multi-scale Imaging Key Laboratory, Institute of Geophysics and Geomatics, China University of Geosciences, Wuhan 430074, China

2. State Key Laboratory of Geological Processes and Mineral Resources, Planetary Science Institute, China University of Geosciences, Wuhan 430074, China

3. School of Earth and Space Exploration, Arizona State University, Tempe, AZ 85287, USA

4. Institute of Rock and Soil Mechanics, Chinese Academic of Sciences, Wuhan 430071, China

5. College of Finance & Information, Ningbo University of Finance and Economics, Ningbo 315175, China

Abstract: Sensitive to permittivity differences, lunar penetrating radar (LPR) can detect stratification due to material property differences in a medium. We examined the subsurface structures of Von Kármán crater within the South Pole-Aitken (SPA) basin on the farside of the Moon using LPR data obtained by the Chang'E-4 mission. Various distinct strata with depth are clearly recognized, consistent with multiple periods of intermittent lava flows, paleo-regolith, and crater ejecta. Numerous concave and arc shape structures observed at various depths are consistent with buried craters and broken rocks, indicating that the ancient lunar surfaces which are currently buried by up to 30m deep in Von Kármán, as well as the current surface, suffered numerous impacting events. An imaged subsurface thick paleo-regolith infers a quiescent period with low volcanic activity within the Von Kármán crater. Following the quiescent period (shallower in the profile), a relatively young period of volcanism in the Von Kármán crater is inferred. The mapped stratigraphy and chronology of these events present a self-consistent history of the Von Kármán crater within the SPA, which suggests connections to events on the lunar nearside.

Article history: Received 10 June 2020 Received in revised form 7 June 2021 Accepted 13 June 2021 Available online 30 June 2021 Editor: W. B. McKinnon

Keywords: Chang'e-4 lunar penetrating radar Von Kármán crater the lunar farside

1 Introduction

Unraveling lunar crater subsurface structure is essential for understanding their formation as well as early lunar history. The South Pole - Aitken (SPA) basin, the Moon's largest and oldest impact structure, has a well - preserved impact history, and thus contains essential information for understanding early lunar history of the farside of the Moon (Wilhelms et al., 1987, Head et al., 1993). To investigate this region, China's Chang'E - 4 (CE - 4) robotic lunar mission (Di et al., 2019) was carried out in the Von Kármán crater inside the SPA (Fig. 1). CE - 4 included the Yutu - 2 rover to image subsurface structures beneath the mare of the lunar farside using lunar penetrating radar (LPR), which proved effective for exploring the nearside lunar shallow crust in the previous Chang'E - 3 mission (Xiao et al., 2015, Yuan et al., 2017, 2020).

The Von Kármán crater formed in the pre - Nectarian period (Losiak et al., 2009, Yingst et al., 2017) from 4. 56 to 3. 92Ga (Wilhelms et al., 1987), and is located in the northwestern SPA basin. The reported olivine - rich materials (Li et al., 2020) in the landing region have not been detected so far within the Von Kármán (Huang et al., 2020). The floor of Von Kármán crater experienced several stages of basaltic flooding during the Imbrian period (the 650 Myr that followed the Nectarian) (Wilhelms et al., 1979, Yingst and Head, 1997, Yingst et al., 2017, Pasckert et al., 2018), as well as ejecta delivered from nearby impacts, including that from the creation of Finsen, Alder, Von Kármán L, and the similar - sized Von Kármán L' craters (Huang et al., 2018). However, maria age dating and depositing process analyses in Von Kármán crater differ. For example, Haruyama et al. (2009) deduced a model age for maria in Von Kármán of 3. 35Ga, while the mare units dated by Pasckert et al. (2018) are 3. 15 Ga and 3. 75 Ga which might be related to younger and older buried flows, respectively. Huang et al. (2018) selected three single geological units in Von Kármán and dated their ages between 3. 5Ga and 3. 6Ga. LPR in the Von Kármán crater can thus directly image the subsurface to better constrain stratigraphy and chronology implied from surface images. Recent LPR studies (e. g. Lai et al., 2020, Zhang et al., 2020a, b) showed several strata interfaces beneath the Yutu - 2 surveying path, but the origins of these strata, especially the shallow materials, are not well constrained.

LPR data collected over nine lunar - days are used in this study, and the rover path was traced by photographs from the navigation camera. Yutu - 2 was designed to navigate flat surfaces and to explore sites near crater rims as well as diverse types of exposed rocks (Huang et al., 2020). Numerous craters of various diameters were detected along the rover - path from the images of the navigation camera (Fig. 1 and Fig. S1). The distinct imprint left by the wheel of Yutu - 2 rover indicates the surface of explored area consists of a fine - grained regolith.

Fig. 1　Chang'E 4 landing site and rover path. (a) Lunar Reconnaissance Orbiter Camera (LROC) Wide - Angle Camera (WAC) mosaic of Von Kármán crater. The yellow star is the landing site of CE - 4 (45. 46°S, 177. 59°E) (Di et al., 2019) . (b) Yutu - 2 rover path and lunar penetrating radar sites. The Yutu - 2 rover path (yellow line) with LPR acquisition locations (blue dots) numbered in order of collection, in the Von Kármán crater. Inset images were taken by the Yutu - 2 navigation camera, and display exposed rocks (e. g., locations 23, 53), and local craters (e. g., locations 9, 53, and 60). The distinct Yutu - 2 rover wheel tracks (locations 23 and 60) indicate a fine grain regolith. Larger blue text indicates locations of the nine lunar nights. Background image is a LROC Narrow - Angle Camera (NAC) high resolution image (captured from LROC QuickMap, https://quickmap . lroc . asu . edu)

2 Data and methods

2.1 Lunar penetrating radar

The LPR carried by Yutu-2 rover consists of two channels, Channel-1 (CH1) and Channel-2 (CH2), with different center frequencies (60MHz and 500MHz, respectively) and bandwidths (40MHz and 450MHz, respectively). Their center frequencies correspond to wavelengths of 5m and 0.6m in vacuum, respectively. CH1 aims to resolve structures up to hundreds of meters deep with vertical resolution of a few meters, and CH2 focuses on the shallow regolith at the top of lunar surface (uppermost 10's of m) with vertical resolution of 0.15-0.3m. CH2 has two identical receivers: antenna A and B (CH2A and CH2B, respectively). In this paper, CH2A is not used due to the low quality of data returned from that channel.

The LPR sends high-frequency electromagnetic pulses to penetrate the lunar subsurface. For a given position of the LPR on the lunar surface, a radar trace consists of the echoes reflected from subsurface interfaces with contrasting dielectric properties. Thus, all LPR traces along the rover track can be combined together to construct a radar profile, i.e., amplitudes of reflected events versus time, or equivalently, depth below all the positions. The CH1 and CH2B LPR traces are 20480ns and 640ns in duration, respectively, with a sampling interval of 2.5ns and 0.3125ns, respectively.

2.2 LPR data processing

As the system of navigation and work mode of LPR aboard the Yutu-2 rover is the same as that of Chang'E-3 (CE-3) mission (e.g. Xiao et al., 2015, Yuan et al., 2017, Lai et al., 2019, Li et al., 2020, some of the same inherent problems in the CE-3 LPR data are present in the CE-4 dataset. These include repeated records collected at the same position, unequal spacing between the trace collection locations caused by speed variation of Yutu-2, and interference signals probably generated by electromagnetic coupling with the rover's metallic body (Angelopoulos et al., 2014, Li et al., 2018). We consider and mitigate all of these issues in our data processing to improve the quality of radar profiles constructed from the CH1 and CH2B data (Figs. S2—S8).

Firstly, the LPR waveform variation between adjacent radar traces was measured, resulting in 3368 and 7463 valid traces being selected from the raw LPR data from CH1 and CH2B, respectively, after delay time removing. Radar wave propagation in the subsurface experiences amplitude loss from geometric (spherical wavefront) spreading, as well as absorption (e.g., Xiao et al., 2015). A spherical spreading correction and absorption-attenuation compensation correction were applied to the radar profiles based on the wave propagation distance which is estimated from a dielectric model (Fig. 2).

This model was derived from the surface electric experiment at the Apollo - 17 site which has a dielectric discontinuity at a depth of ~43m (Simmons et al., 1973), similar to a shallow interface in our radar profile. We later constrain the dielectric model by wavefield migration tests, and adopt the value 3 as the average relative dielectric constant for the shallow LPR profile.

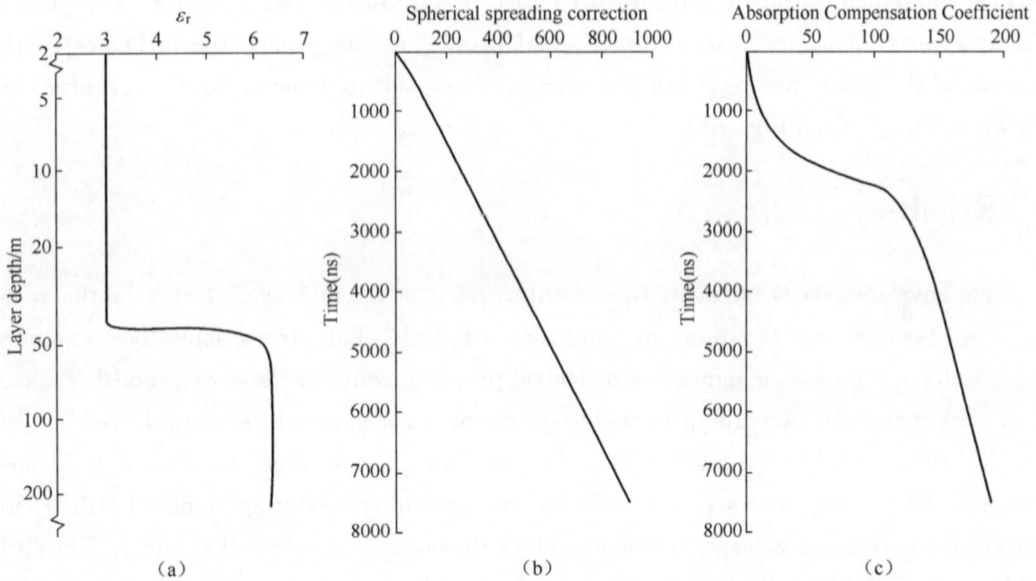

Fig. 2　Parameters for amplitude compensation. (a) The relative dielectric constant adopted in this study. (b) The time dependency of the spherical spreading correction. (c) The time dependency of coefficient

(Porcello et al., 1974, Wang et al., 2010) for absorption attenuation compensation

In the initial profile, uniform horizontal stripes with strong energy signal were present in the upper part of the profile (Fig. S2 and S3), which is likely due to the strong energy of the direct radar wave from the source antenna to the receiver antenna. As those strong amplitude stripes are mainly horizontal, the wave - number filter was designed to remove zero wave number (i. e., horizonal) interferences. A consequence of this filter is that horizontal interfaces along the rover path would not be readily detectable. To reveal distinct deeper structures, noise and inherent interference signals which are dominated in the 5 - 20MHz frequency range in CH1 must be removed (as demonstrated with the LPR data in the CE - 3 mission by Li et al., 2018). We have thus filtered all LPR traces of the CH1 profiles in the frequency domain with 30 - 60MHz bandpass filter. A 200 - 500MHz bandpass filter was selected for CH2B after testing. Additionally, a Gaussian smoothing image filter, was applied to sharpen reflections from deep strata.

We further apply the method of Stolt's F - K migration (Stolt, 1978) for CH2B profile to migrate the diffracted waves, whereas migration was not applied to CH1 data as the deeper structures become flatter so that few diffracted waves were recorded. Un -

equal trace collection location spacing caused by the variation of Yutu-2's speed was also considered by means of migrating separately between the trace collection locations (Fig. S7b). Addition-ally, the surface relief along the Yutu-2 path can considerably influence mapped interface morphology at depth, especially for shallow structures. Therefore, we correct for the topography on the basis of the relative elevation which is derived from the vertical component of position recorded by LPR (Fig. S7 and S8a). As the radar data is simply approximated to a two-dimensional profile in this paper, the influence of the attitude of the radar antenna is not considered. The resulting lunar radar depth profiles are shown in Fig. 3 (and Fig. S8).

3 Results

We have observed, presented, and interpreted different wave-forms in the radar profiles based on the variations in amplitude strength. The strong laterally-extended radar reflection interfaces indicate a dielectric property contrast between regolith/rock or rock/rock materials. According to this pattern, over dozen interfaces are inferred in LPR profiles (Fig. 3). Terrestrial studies (e. g. Russell and Stasiuk, 1997, Rust and Russell, 2000) demonstrate the efficacy of ground penetrating radar (GPR) for detecting and imaging volcanic sediments (lava flows, pyroclastic rocks, etc.). The LPR profiles presented here thus can similarly map deposits of volcanic origin. Basalt deposits on the lunar surface within the Von Kármán crater exposed in the Zhinyu crater wall (roughly 20m of stratified layers at the top of the crater, Fig. 4) provide a direct evidence for intermittent superimposed lava flows. This is in good accordance with LPR reflections (Fig. 3), and thus is consistent with an origin of lava flows, pyroclastic rocks, and ejected materials or regolith.

3.1 Profile of LPR CH2B

3.1.1 Shallow layered structures

In the LPR profile of CH2B (Fig. 3b), relatively strong echoes are apparent from the surface (***S***) down to a distinct interface (***A***) that averages ~2.5m depth, suggesting this surficial layer between ***S*** and ***A*** (*hereafter denoted* ***S*** - ***A***) is surface regolith which has suffered prolonged impacting. Thus, buried craters along ***A*** were likely filled with ejecta from adjacent craters. Large craters, such as Von Kármán L and L' (Fig. 1a) may have also contributed to the growth of ***S*** - ***A*** (each with predicted ejecta blanket thickness less than ~1m).

Below ***S*** - ***A***, reflections in stratum ***A*** - ***B*** are relative weak with slight fluctuations of increased reflectance. Buried craters are absent, suggesting contemporaneously deposited materials from the same source. Crater excavation analyses support this, concluding the top materials of the eastern floor of Von Kármán crater contain ejecta from the neighboring Finsen crater (Huang et al., 2018). The depth of ***B*** varies from ~9 to 16m, consistent with Finsen ejecta thickness estimations at the CE－4 landing site.

Reflectance in stratum ***B*** －***B'*** is stronger than in ***A*** －***B***, with strong and sometimes relatively sharp fluctuations. Diffracted reflections likely caused by broken rocks and/or impact craters are present. Additionally, ***B*** － ***B'*** contains large (5 － 10m) thickness variations, inferring impact－born ***B*** horizon topography. Reflectance in this stratum is similar to that in surface regolith (***S*** －***A***), and thus may similarly be local regolith, but subsequently interrupted by Finsen ejecta.

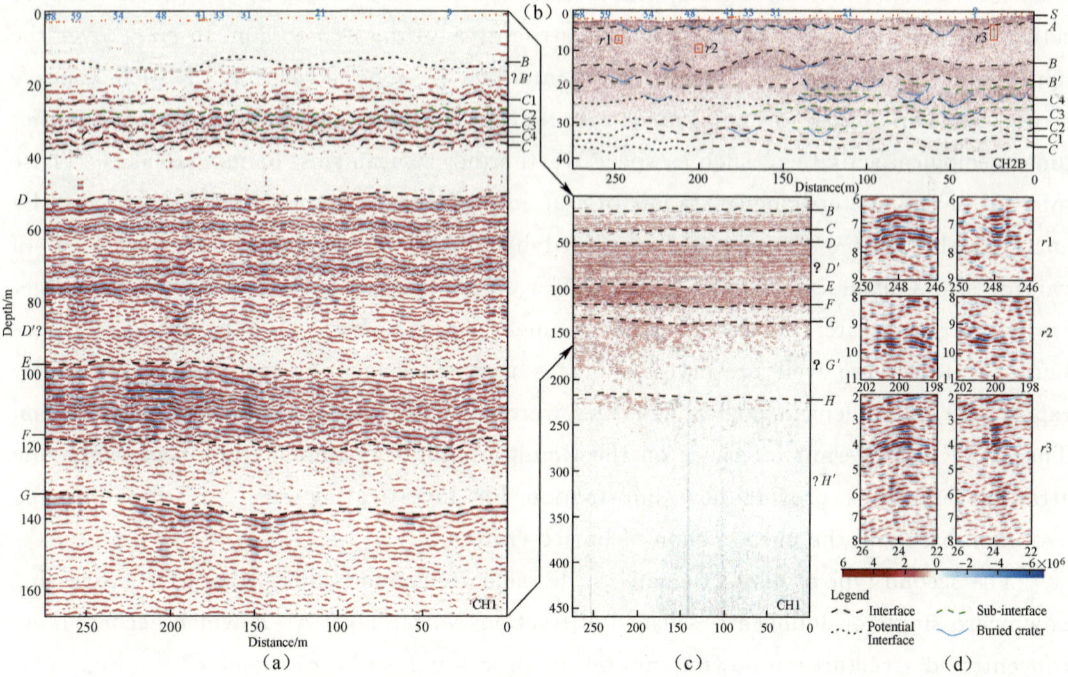

Fig. 3　LPR profiles and subsurface reflectance. (a) The layered subsurface structure along the Yutu－2 path imaged by LPR CH1 data, and (b) CH2B data. The rover progressed from right－to－left in the images, with sites of the 9 lunar nights indicated by the blue numbers at the top of panels (a) and (b) (which correspond to Fig. 1b). The intensity of the red and blue shading, shown as the color scale bar at the right bottom, represents reflectance amplitude strength. Black long－dashed lines outline interfaces which can be tracked continuously across the CH1 and CH2B profiles; green dashed lines demark interfaces that are present in just part of the Yutu－2 path; black short－dashed lines highlight interfaces having high uncertainty. Blue lines indicate clustered reflections of filled concave structures in (b) and may correspond to buried impact craters. (c) An expansion of panel (a) that includes greater depths. (d) Three boxed regions from panel (b) which are zoomed in (***r*1**, ***r*2**, and ***r*3**), *exhibiting arc－shaped reflections before migration (left column) and more concentrated structures after migration (right column). All axes are distances in meters*

Beneath ***B*** －***B'***, multiple distinct reflection interfaces (***C*1**, ***C*2**, ***C*3**, ***C*4** and ***C***) bound ~2－3m thick layers, consistent with intermittent mare flow deposition. Buried impact craters indicate a paleo Von Kármán crater floor subsequently covered by adjacent crater (and within the Von Kármán crater) ejecta deposits. The family of ***C*** horizon reflections (***C1***, ***C2***, etc.) are reduced in amplitude with depth due to depth - sensitively limitations for CH2B which typically loses resolution below 20m depth, though some impact craters are visible (Fig. 3b).

3.1.2 Small structures

Two distinctive types of small - scale reflections are apparent. The first is the arc - shaped reflections with strong amplitude variations caused by syncline or concave - shaped reflectors. Fig. 3b shows various concave structures (from ~5 to 20m in size) over the upper 35m in depth from surface ***S*** to interface ***C***. The formation of such concave structures is attributed to small impact events from falling projectiles, as opposed to other lunar geological activities, such as space weathering, volcanism, or moonquakes. These inferred buried craters suggest a history of impact events and deposition hiatuses. By analogy, the surface environment traversed by Yutu - 2 contains dozens of craters of varying size (diameters from a few to 10's of m, Fig. S1), as well as much larger craters in the greater nearby SPA environment (e. g., Fig. 1). The fact that these buried craters were well preserved indicates that subsequent deposition was relatively rapid, and subsequent geological activities were not to such an extent as to disturb them. These LPR data permit inference on the timing of the depositional rate of an overlaying stratum, and thus provide new information for investigating stratum formation and evolution based on the phenomenon of buried craters.

The second type of distinct small - scale radar reflection is more complex, possessing downward single or double arc - shaped diffractions within layer ***A*** - ***B*** which become more concentrated structures following migration processing (***r1***, ***r2***, and ***r3*** in Fig. 3d). Distinct from the concave crater shapes, these structures may correspond to large ejected rocks. Explaining the observed diffractions with buried rocks 2 - 15m in size, is supported by synthetic tests of LPR diffractions from buried rock (Zhao et al., 2018) (Fig. S9), and provides new evidence for regolith formation by frequent impacts.

3.2 Profile of LPR CH1

In the LPR profiles of CH1, deeper interfaces (***D*** - ***H***) (Fig. 3a and 3c) indicate earlier periods of geological processes contributing to deep mare basement formation within the Von Kármán crater. The CH1 band is designed to explore deep structures with low vertical resolution, thus interfaces in the uppermost 17m (where ***A***, *B* and ***B'*** are detected by CH2B with higher vertical resolution, Fig. 3b) are not resolved, consistent with a relative fine grain size in this depth shell. At greater depths (~17 - 37m depth range), strong continuous reflections are observed in the CH1 profile (Fig. 3a), corresponding to strata ***C1***, ***C2***, ***C3***, ***C4*** and ***C*** in the CH2B profile (Fig. 3b).

Stratum ***C*** –***D*** lacks significant reflectance (Fig. 3a), similar to that above interface ***B***, and is thus similarly interpreted as finegrained paleo – regolith from a hiatus in rapid deposition (of, for example, lava flows). This finding is similar to a quiescent period inferred from a homogeneously weak reflectance layer in LPR data from the Chang'E – 3 (CE – 3) mission (Yuan et al., 2017) on the nearside of the Moon. Additional weak amplitudes (thus relative fine – grained materials) are apparent in strata ***D***′– ***E***, ***F*** –***G*** (Fig. 3a) and ***G***′– ***H*** (Fig. 3c), though the interfaces ***D***′ and ***G***′ are poorly constrained due to weaker reflectance at these greater depths. Nearly horizontal alternating strong – and – weak reflections in stratum ***D*** –***E*** (Fig. 3a) are interpreted as layered lava flows, similar to those in ***B*** –***C***.

Strong reflections in ***E*** – ***F*** (Fig. 3a), however, show interlacing horizontal reflectors, suggesting continuous and relatively rapid material deposition. These reflection patterns are similar to GPR profiles from terrestrial pyroclastic flow deposits (Russell and Stasiuk, 1997), suggesting that some evolutionary processes on the lunar surface may be similar to those on Earth. We can speculate that pyroclastic tephra and tuff are less easily stratified in the lunar vacuum environment, so that sub – paralleled reflections in ***E*** – ***F***, and the clear and continuous interface ***E*** are consistentwith a subsequent lava flow. Additional evidence from further work will be required to validate this hypothesis.

Deeper interfaces ***G*** and ***H*** feature relative strong reflections at average depths 136 and 222m, respectively. Associated strata ***G*** –***G***′ and ***H*** –***H***′ possess strong amplitudes and fluctuations similar to ***E*** –***F*** (albeit lower amplitude), but with relatively weak lower interfaces due to limited transmitted energy from these greater depths due to wave attenuation. These reflections can be interpretedas interfaces of early periods of Imbrian lava materials.

3.3 Ejecta of Finsen and Alder crater

The ejecta from large craters adjacent to Von Kármán crater (e. g., Finsen and Alder craters) is expected to be deposited as strata with a thickness which can be estimated by ejecta thickness models (e. g., Sharpton, 2014). Foreign ejecta thickness T can be generally described as $T = aR_t^b\ (r/R_t)^c$, where r is the distance to the crater center, and R_t is the radius of the transient crater, all in meters, and parameters a, b, and c are constants used to match observed thicknesses (corresponding parameters are shown in Table S1). For large, complex craters, R_t should be estimated from measurable crater properties using a basic geometric model (e. g., Thomson et al., 2009). However, due to the complex original surface located on the rim of large craters, this study simply calculates R_t from $R_t = 0.84R_{SC}^{0.15}R^{0.85}$ (Croft, 1985, Xie et al., 2020) for the complex crater, where $R_{SC} = 9.5$km is adopted for the transition diameter of simple and complex craters (Pike, 1980), and R is the final rim – to – rim radius.

Moreover, the thickness for locally mixed materials is addedby μT , where ratio $\mu = 0.0183r^{0.87}$ (for $\mu \leqslant 5$) (Oberbeck, 1975, Petro and Pieters, 2006). Then we calculate Finsen crater (73km in diameter) and Alder crater (82km in diameter) ejecta thicknesses at the CE-4 landing site using different foreign ejecta models (McGetchin et al., 1973, Pike, 1974, Housen et al., 1983, Petro and Pieters, 2008, Sharpton, 2014) together with the correction estimate model of locally mixed materials (Fig. 5). Their crater center to landing site distances is 140.7km and 133.4km, respectively.

Fig. 4 Exposed lava flow deposition on the rim of Zhinyu crater. The thicker white dashed line delineates an obvious continuous interface with coarse materials. Thinner dashed lines indicate interfaces interpreted as past lava flows. The Zhinyu crater is visible in the center of the Von Kármán crater in Fig. 1, with a diameter of ~3.8km, which is ~30.8km away from CE-4 landing site. Roughly 20m of stratified layers are seen on the crater wall near the rim/surface. (Images were captured from LROC QuickMap, https://quickmap.lroc.asu.edu.)

Finsen crater ejecta includes widespread linear features containing high albedo or surface terrain pointing to Finsen (Huang et al., 2018). The fine-grained Finsen ejecta deposited on the surface is consistent with fine, homogeneous material composing stratum ***A*-*B***. In contrast, very few radial ejecta features can be traced back to the neighboring Alder crater, suggesting Alder ejecta deposits, which should similarly be fine-grained, have since been buried.

After removing the contribution of crater Von Kármán L and L' (~2m in total, Huang et al., 2018), the thickness of Finsen ejecta (inferred to be stratum ***S*-*B*** in Fig. 2b) around the CE-4 landing site is ~10m on average (blue asterisks in Fig. 5 (a)), which is consistent with the estimation of the Pike model (Pike, 1974) after the local-mixing correction (T=11.0m). Moreover, stratum ***F*-*G*** with weak radar reflections in the LPR profile averages 17.7m thick (blue asterisks in Fig. 5 (b)), which is in good agreement with the estimated thickness of Alder ejecta at the CE-4 landing site using the Pike model (T = 18.6m). The slight differences between observed and estimated thicknesses can be attributed to erosion and compaction of the upper lava flow (as well as ejecta model and radar depth prediction uncertainties). This good accordance of observed and estimated thickness suggests that stratum ***F*** − ***G*** was probably formed by ejecta from the Alder crater. Stratum ***G***′ − ***H*** is excluded because of its unconstrained top interface.

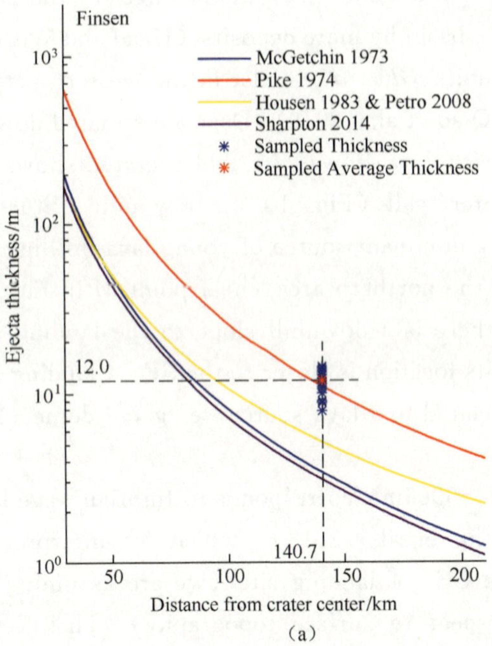

Fig. 5　Thickness predictions of ejecta deposits. The ejecta thickness of (a) Finsen crater and (b) Alder crater, estimated as a function of distance based on different foreign ejecta models (McGetchin et al., 1973, Pike, 1974, Housen et al., 1983, Petro and Pieters, 2008, Sharpton, 2014) after applying a correction for locally mixed materials. The horizontal and vertical gray dashed lines represent distance from the crater center to the CE − 4 landing site, and the crater ejecta thickness at CE − 4, respectively. The blue asterisks are the observed stratum thicknesses determined from the LPR profile by every 10min rover distance, and the red asterisk is their average

4　Discussion

4.1　Stratum origins

Fig. 6 displays the Von Kármán crater floor topography, which averages roughly 6000m below the lunar average, but has significant topographical variations. In order to investigate the origin of these strata, this paper traced topographical variations, flow thicknesses, and flow dips across the CE − 4 traverse.

A first − order feature near the CE − 4 site is downhill slope to the southwest (along cross − section NN' towards N) and uphill slope to the southeast (along cross − section MM' towards M') that terminates near a dome − like structure (*d*3) near the Von

Karman crater wall (near M’). Dome $d3$ contains arc - shaped grooves, and drops ~80m over ~10km, then an additional 40m over ~35km to - wards the CE - 4 site. The domes are likely expressions of subsurface magmatic intrusion (i. e., laccoliths), possibly with surface extrusion; the location of the dome ($d3$) at the edges of the basin is consistent with earlier flexural loading of the basin by mare deposits (Head and Wilson, 1992, Hiesinger, 2006). Other dome - like units ($d1$, $d2$) exist in the western part of Von Kármán crater (Pasckert et al., 2018, Qiao et al., 2019). Deposits situated down - gradient from domes would be consistent with lava flows. As Alder crater shows no evidence for breaching the Von Kármán crater wall (Fig. 1) or flow unit (Stuart - Alexander, 1978), dome $d3$ may have been a dominant source of young lava infilling (i. e., stratum $\boldsymbol{B}'-\boldsymbol{C}$). Compared to dome $d3$, the northern area (near point M in Fig. 6) is an alternate source of this young lava as there is a downhill slope to the landing site along cross - section MM’, and furthermore its location is closer to the CE - 4 landing site (though lacking evidence for any structures related to a lava source, e. g., a dome - like unit).

The deepest Von Kármán crater region (−5965m) corresponds to Imbrian - age lava flow material (Pasckert et al., 2018, Huang et al., 2018), which we interpret as corresponding to stratum $\boldsymbol{C}-\boldsymbol{D}'$ beneath the CE - 4 landing site (we are assuming all dielectric interfaces are flat - lying with respect to surface topography). The deeper imaged strata, $\boldsymbol{H}-\boldsymbol{H}'$, $\boldsymbol{E}-\boldsymbol{F}$ and $\boldsymbol{D}-\boldsymbol{D}'$, show decreasing values (~3, 2, and 0m) of average elevation changes across the Yutu - 2 path, which intend that the topographical changes of strata be - come smaller (more and more flat) from deep to shallow. As these stata dip towards the east, we suggest that their materials originated from the west, e. g., from the largest dome - like unit $d1$ in Von Kármán crater (Fig. 6).

4.2 Intermittent lava flows

The low and relatively homogeneous reflectance $\boldsymbol{C}-\boldsymbol{D}$ stratum (Fig. 3a), interpreted as fine - grained regolith from extended qui - escence in section 3. 2, is referred to as **Pr3** (for paleo - regolith). Additional potential paleo - regolith layers are observed at 89 and 203m depth (Fig. 3a and 3c), hereafter, **Pr2** and **Pr1**, respectively (but are poorly constrained due to ambiguous top interfaces). It is noteworthy that the thickness of **Pr3** is ~14m, significantly exceeding the average thickness of 5m on lunar maria and 10m on highland surfaces (Taylor, 1982, McKay et al., 1991). As finest - grain lunar materials are due to continuous space weathering and relatively frequent small impacts, very few large - scale resurfacing events (e. g., lava flows) may have occurred at beneath CE - 4 during the formation of Pr3. On the other hand, previous work inferred a major peak of volcanic activity between 3. 6Ga and 3. 2Ga within the SPA basin and the crater count in maria Von Kármánis is 3. 15Ga age (Pasckert et al., 2018), suggesting that this thick regolith might have formed after 3. 15Ga, in the early Eratosthenian period. This implies

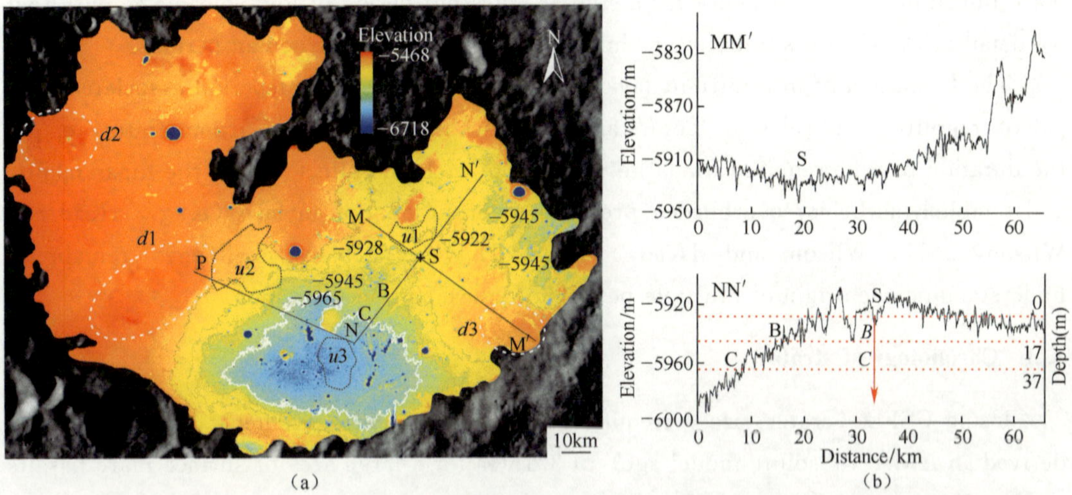

Fig. 6　Topographic profiles near CE－4 landing site. (a) SLDEM2015 topography (Barker et al., 2016) in the Von Kármán crater on a base map image from the LROC Wide－Angle Camera (WAC). The color scale denotes deeper (blue) and shallower (orange, red) elevations. *d*1, *d*2, *d*3 are dome－like units (white dashed ellipses). The ages of mare units *u*1, *u*2 and *u*3 (black dotted regions) were dated in our previous study (Huang et al., 2018). Variable－sized small, dark blue regions are topographical expressions of present－day surface craters. Constant elevation contour lines (white and gray) and associated elevations are shown, with points S, B, and C correspond to the intersection of interfaces S, B, C (discussed in the text and Fig. 3) with the surface topography. Three cross－section lines are shown, MM', NN', and NP. (b) and (c), Topographic profiles across CE－4 landing site (S) corresponding to MM' and NN' cross－section lines. Horizontal dashed red lines in (c) correspond to the topographical elevations of the CE－4 location (point S), points B and C in (a). We are assuming all dielectric interfaces are flat－lying with respect to surface topography

gentle deposition processes during this period, which is in agreement with a quiescent period in the nearside northern Mare Imbrium, from a boundary between the Imbrian and Eratosthenian periods (Yuan et al., 2017). This quiescent period thus infers a low activity of mare volcanism during the early Eratosthenian period.

Following the inferred quiescence period, the ***B*－*C*** stratum deposited above the ***C*－*D*** stratum may correspond to the second peak of volcanic activities (hereafter, **Em**, for Eratosthenian mare unit) on the lunar farside between 2.2－2.5Ga (Haruyama et al., 2009, Pasckert et al., 2018). Then the observations of this mare－like stratum suggest a new, younger period of volcanic activity within Von Kármán crater.

The LPR profile presented here suggests the scale of farside lava flows is similar to those on the nearside (Xiao et al., 2015, Yuan et al., 2017). For example, evidence for relatively thin (2－10m) lava flows as evidenced from nearside pit and crater walls (Denevi et al., 2012, Robinson et al., 2012), is similar to our lava flow thicknesses (e.g., the layers in strata ***B*－*C*** and ***D*－*D'*** are ～2－5m thick), which match well with Zhinyu crater wall layer thicknesses (Fig. 4). Also, the nearside CE－3 landing site (in mare Imbrium) shows evidence for a similar quiescent period of low volcanic activity that we detail at the CE-4 site between the Imbrian and the Eratosthenian periods.

The formation of intermittent lava flows and a long quiescent period associated with paleo-regolith could relate to factors and processes inside the Von Kármán crater, e. g., the duration of time required for large volumes of melt to collect within the lunar mantle and establish sufficient overburden pressure to erupt on the surface (e. g., Head and Wilson, 2017, Wilson and Head, 2017). However, many challenges remain in understanding the origin of multiple periods of volcanism on the Moon.

4.3 Chronology of strata

Using CSFD (crater size-frequency distribution) measurements, previous work derived an AMA (absolute model age) of 3.15Ga for a large area of surface mare basalts in Von Kármán crater floor (Pasckert et al., 2018), which corresponds to the area surrounding dome-like units $d1$ and $d2$ and lower elevation areas. We propose this surface layer of mare basalts exposed in the lower elevation areas (i. e. the southern, blue regions in Fig. 6) corresponds to $\boldsymbol{D}-\boldsymbol{D}'$ (denoted as **Im4**). Crater degradation and burial from younger ero-sional and depositional events will result in younger than actual CSFD-estimated ages. We thus suggest material in the upper part of **Im4** should be older than 3.15Ga. Seen from the nearly horizontal strong stripe reflections in $\boldsymbol{D}-\boldsymbol{D}'$, there may be as many as seven lava flow sub-periods with relatively even eruption intervals. Additionally, a buried lava flow with an AMA of 3.75Ga, corresponding to the oldest lava flow in our profile $\boldsymbol{H}-\boldsymbol{H}'$ (**Im1**), was inferred to be an underlying mare deposit in the Von Kármán crater (Pasckert et al., 2018).

4.4 Stratigraphic interpretation of explored region

Three small mare units $u1$, $u2$ and $u3$ (near locations S, P and N, respectively, in Fig. 6) were dated with AMAs of 3.6Ga, 3.5Ga and 3.6Ga, respectively (Huang et al., 2018). The age of 3.6Ga in area $u1$ near point S may be due to old lava flow materials from the north. As area $u3$ near point N has elevations of at least 80m below surrounding areas (Fig. 6), and the thickness of lava flow deposits generally decreases fromtheir source (presumed to be point P, which is the largest dome-like unit on the western Von Kármán crater floor), we suggest the second oldest period of lava flow deposit $\boldsymbol{G}-\boldsymbol{G}'$ (**Im2**) has the age of ~3.6Ga. A younger period of volcanic activity may have produced stratum $\boldsymbol{E}-\boldsymbol{F}$ (**Im3**) at ~3.5Ga (derived from $u2$), where pyroclastic materials were deposited first, then overlaid by a subsequent flow from point P. The Alder ejecta is an episode between the **Im2** and **Im3** flows between 3.15-3.75Ga-which represents a peak period of volcanic activity within the Von Kármán crater inside SPA basin.

Finally, we construct the possible stratigraphy and evolutionary history of the Von Kármán crater floor based on the LPR strata combined with consideration of the surface geology around CE－4 landing site，as well as in other parts of the Von Kármán crater (Fig. 7). Five periods of large mare basalt flows from regional lavainfilling events were inferred (**Em**，**Im1**，**Im2**，**Im3** and **Im4**). The oldest flow (Im1) might have been erupted from a western volcanic vent on the Von Kármán crater floor，forming the deepest imaged Imbrian－aged stratum. After the formation of a paleo－regolith **Pr1**，the second lava flow period occurred from a western source (**Im2**). The Alder crater impact occurred southeast of Von Kármán crater，and deposited a fine ejecta stratum above **Im2**. The third period of volcanism produced a large amount of pyroclastic materials which overlaid the Alder ejecta，followed by a lava flow that covered the surface，forming **Im3**.

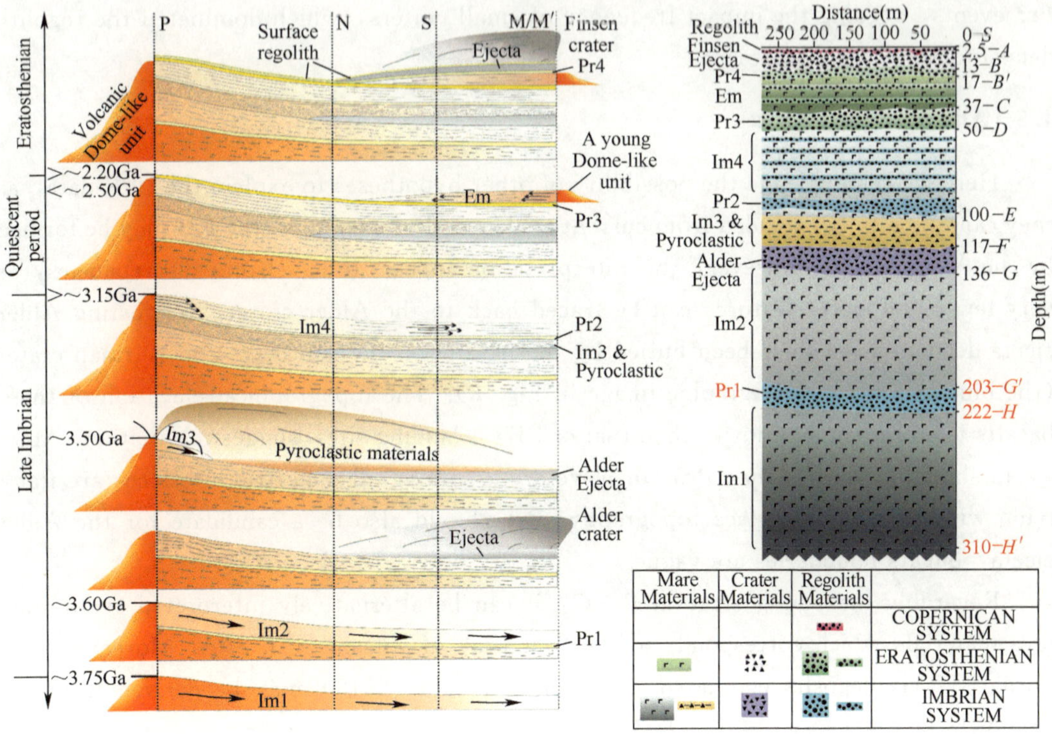

(a)　　　　　　　　　　　　(b)

Fig. 7　Possible geological evolution diagram in Von Kármán crater inferred from this study. (a) Geological evolution diagram in Von Kármán crater along the composite Fig. 6 cross－section P－N－S－M (or M’) from 3.75Ga (bottom) to earlier than 2.2Ga (top)，displaying an interpretation for temporal evolution of strata creation. Lunar geologic era names are printed on left，along with ages，and depiction of lava flow dome－like unit evolution (orange “hills”，e. g.，d1in Fig. 6). Strata include Imbrium lava flows (**Im1 － 4**)，paleoregolith layers (**Pr1 － 4**)，an Eratosthenian mare unit (**Em**)，as well as pyroclastic and ejecta deposits. Arrows indicate inferred mare flow directions. Ages and layer thicknesses are not to scale. The surface regolith was formed after deposition of Finsen crater ejecta. (b) Present day strata interpretation of LPR profiles beneath the Yutu－2 path. Left side symbols correspond to strata interpretations in (a)，and right side presents the depths and strata interface letters (from Fig. 3). Text and numbers in red have large uncertainties

Subsequent intermittent lava flows with quiescence intervals resulted in **Im4**. Paleo - regolith **Pr2** formed between **Im3** and **Im4**. A quiescent period, with little vol - canism and no large impact events, occurred in the Eratosthenian period after the deposition of **Im4**, forming the thick paleo - regolith **Pr3**. This was followed by mare basalts (deposited as unit **Em**) from the eastern Von Kármán crater floor. A thin paleo - regolith **Pr4** subsequently formed on top of **Em**, which was later buried by the ejecta of the Finsen crater from the northeast of Von Kármán crater. Finally, a distinct thin surface regolith formed in our explored area. The southwestern maria regions in Von Kármán crater are expected to have thicker regolith due to the long, continuous formation time. Additionally, the thin surface regolith and **Pr4** are thinner than the early **Pr1**, **Pr2** and **Pr3** events, arguing the impact frequency of small craters, which dominated the regolith deposition rate, has decreased.

4.5 Alternative hypotheses

Here we acknowledge the possibility of other hypotheses to explain the LPR data, as they cannot be fully excluded from current evidence. For example, the **Pr3** may be formed by the Alder ejecta. However, this interpretation appears to have some discrepancies, as very few radial ejecta features can be traced back to the Alder crater, suggesting Alder ejecta deposits have since been buried by the basalts on the southern Von Kármán crater (the blue area in the topographic image in Fig. 6). The topographical elevation of these basalts is morethan 60m lower than that of CE - 4 landing site, suggesting that any Alder ejecta should be deeper than **Pr3** when we are assuming all dielectric interfaces are flat - lying with respect to surface topography. **Pr1** should also be a candidate for the Alder ejecta, but its boundaries are vague.

Regarding the young stratum ***B*** - ***C***, it can be alternatively interpreted as Imbrain lava materials which corresponds to the first peak of volcanism in SPA. Moreover, the buried 3.75Ga deposits estimated by Pasckert et al. (2018) may be derived from the shallower layers (e. g., **Im2**, **Im3** or even **Im4**), and then other **Im** units and Alder ejecta would be accordingly older. A challenge with this hypothesis is explaining the homogeneous stratum ***C*** - ***D***, where a long hiatus of volcanism in the Von Kármán crater during the Imbrian period is inferred, because lava flows were typically more active in the Imbrian period compared to those in the Eratosthe - nian period (Haruyama et al., 2009, Pasckert et al., 2018). Concerning the uncertainty of the AMAs calculated from different exposed regions in Von Kármán crater, the exact ages of these strata are only plausible results referred to previous studies (e. g. Huang et al., 2018, Pasckert et al., 2018).

Yutu - 2 is still operating on the far side of the Moon, thus additional data will be available to better constrain these geological interpretations.

5 Conclusions

There is good agreement between surface geological features and the structures detected in lunar LPR profiles, supporting the reliability of imaged subsurface structures. Numerous concave and arc shape structures observed at various depths are consistent with buried craters and rocks, indicating that the paleo floor of Von Kármán, as well as the present - day surface, experienced numerous impacting events. The observed strata thicknesses match well with predictions of ejecta thickness from Finsen and Alder craters.

Multiple periods of volcanism are revealed by LPR profiles, which likely originated from dome - like units on the maria floor of Von Kármán. A subsequent younger lava flow unit following the quiescent period provides evidence for young volcanism within Von Kármán. Finally, the most possible stratigraphy and chronology in maria Von Kármán crater within the SPA are constructed, representing an example of evolutionary processes on the farside of the Moon. It is a remarkable fact that some alternative hypotheses cannot be ignored, so more evidence is stillneeded to constrain and improve our model.

Acknowledgements

The authors would like to thank the editor W. McKinnon, reviewer B. Thomson and an anonymous reviewer for constructive reviews and comments that helped improve the manuscript. This work was supported by National Natural Science Foundation of China [41904119, 41830214], the Fundamental Research Funds for the Central Universities, China University of Geosciences (Wuhan) [CUG2106109], China Postdoctoral Science Foundation [2018M642950], the Pre - research Project on Civil Aerospace Tech - nologies [D020101] of CNSA, Zhejiang Provincial Natural Science Foundation of China [LQ18A030001], and Ningbo Natural Science Foundation [2019A610090]. Data presented in this paper are hosted at http://moon . bao . ac . cn/index _ en . jsp.

References

[1] Angelopoulos, M., Redman, D., Pollard, W. H., et al., 2014. Lunar ground penetrating radar: minimizing potential data artifacts caused by signal interaction with a rover body. Adv. Space Res. 54, 2059 - 2072. https://doi. org/10. 1016/j. asr. 2013. 09. 014.

[2] Barker, M. K., Mazarico, E., Neumann, G. A., Zuber, M. T., Haruyama, J., Smith, D. E., 2016. A new lunar digital elevation model from the Lunar Orbiter Laser Altimeter and SELENE Terrain Camera. Icarus 273, 346 - 355. https://doi. org/10. 1016/j. icarus. 2015. 07. 039.

[3] Croft, S. K., 1985. The scaling of complex craters. J. Geophys. Res. 90 (S02), C828 - C842. https://doi. org/10. 1029/JB090iS02p0C828.

[4] Denevi, B. W., et al., 2012. Physical constraints on impact melt properties from Lunar Reconnaissance Orbiter Camera images. Icarus 219, 665 - 675. https://doi. org/10. 1016/j. icarus. 2012. 03. 020.

[5] Di, K. C., et al., 2019. Chang'e - 4 lander localization based on multi - source data. J. Remote. Sens. 23, 177 - 184. https://doi. org/10. 11834/jrs. 20199015.

[6] Haruyama, J., et al., 2009. Long - lived volcanism on the lunar farside revealed by SELENE Terrain Camera. Science 323, 905 - 908. https://doi. org/10. 1126/science. 1163382.

[7] Head, J. W., Wilson, L., 1992. Lunar mare volcanism: stratigraphy, eruption conditions, and the evolution of secondary crusts. Geochim. Cosmochim. Acta 56, 2155 - 2175. https://doi. org/10. 1016/0016 - 7037 (92) 90183 - J.

[8] Head, J. W., Wilson, L., 2017. Generation, ascent and eruption of magma on the Moon: new insights into source depths, magma supply, intrusions and effusive/explosive eruptions (part 2: predicted emplacement processes and observations). Icarus 283, 176 - 223. https://doi. org/10. 1016/j. icarus. 2016. 05. 031.

[9] Head, J. W., et al., 1993. Lunar impact basins: new data for the western limb and far side (Orientale and South Pole - Aitken basins) from the first Galileo flyby. J. Geophys. Res. 98, 17, 149 - 17, 181. https://doi. org/10. 1029/93JE01278.

[10] Hiesinger, H., 2006. New views of lunar geoscience: an introduction and overview. Rev. Mineral. Geochem. 60, 1 - 81. https://doi. org/10. 2138/rmg. 2006. 60. 1.

[11] Housen, K. R., Schmidt, R. M., Holsapple, K. A., 1983. Crater ejecta scaling laws: fundamental forms based on dimensional analysis. J. Geophys. Res., Planets 88, 2485. https://doi. org/10. 1029/JB088iB03p02485.

[12] Huang, J., Xiao, Z., Flahaut, J., et al., 2018. Geological characteristics of Von Kármán crater, northwestern South Pole - Aitken Basin: Chang'E - 4 landing site region. J. Geophys. Res., Planets123, 1684 - 1700. https://doi. org/10. 1029/2018JE005577.

[13] Huang, J., Xiao, Z., Xiao, L., et al., 2020. Diverse rock types detected in the lunar South Pole - Aitken Basin by the Chang'E - 4 lunar mission. Geology 48, 723 - 727. https://doi. org/10. 1130/G47280. 1.

[14] Lai, J., Xu, Y., Zhang, X., Xiao, L., Yan, Q., Meng, X., et al., 2019. Comparison of dielectric properties and structure of lunar regolith at Chang'e - 3 and Chang'e - 4 landing sites revealed by ground - penetrating radar. Geophys. Res. Lett. 46, 12783 - 12793. https://doi. org/10. 1029/2019GL084458.

[15] Lai, J., Xu, Y., Bugiolacchi, R., et al., 2020. First look by the Yutu - 2 rover at the deep subsurface structure at the lunar farside. Nat. Commun. 11, 3426.

https://doi. org/10. 1038/s41467 - 020 - 17262 - w.

［16］ Li，C.，et al.，2018. Pitfalls in GPR data interpretation：false reflectors detected in lunar radar cross sections by Chang'e - 3. IEEE Trans. Geosci. Remote Sens. 56，1325 - 1335. https://doi. org/10. 1109/TGRS. 2017. 2761881.

［17］ Li，C.，Su，Y.，Pettinelli，E.，Xing，S.，Ding，C.，et al.，2020. The Moon's farside shallow subsurface structure unveiled by Chang'E - 4 Lunar Penetrating Radar. Sci. Adv. 6，eaay6898. https://doi. org/10. 1126/sciadv. aay6898.

［18］ Losiak，A.，et al.，2009. A new lunar impact crater database. In：Lunar and Planetary Science Conference.

［19］ McGetchin，T. R.，Settle，M.，Head，J. W.，1973. Ejecta from large craters on the Moon：discussion. Earth Planet. Sci. Lett. 20，226. https://doi. org/10. 1016/0012 - 821X（74）90115 - 0.

［20］ McKay，D. S.，Heiken，G.，Basu，A.，et al.，1991. The Lunar Regolith. Lunar Sourcebook，vol. 7. Cambridge University Press，Chap，pp. 285 - 356.

［21］ Oberbeck，V. R.，1975. The role of ballistic erosion and sedimentation in lunar stratig - raphy. Rev. Geophys. 13，337 - 362. https://doi. org/10. 1029/RG013i002p00337.

［21］ Pasckert，J. H.，Hiesinger，H.，van der Bogert，C. H.，2018. Lunar farside volcanism in and around the South Pole - Aitken basin. Icarus 299，538 - 562. https://doi. org/10. 1016/j. icarus. 2017. 07. 023.

［22］ Petro，N. E.，Pieters，C. M.，2006. Modeling the provenance of the Apollo 16 regolith. J. Geophys. Res. 111，E09005. https://doi. org/10. 1029/2005JE002559.

［23］ Petro，N. E.，Pieters，C. M.，2008. The lunar - wide effects of basin ejecta distribution on the early megaregolith. Meteorit. Planet. Sci. 43，1517 - 1529. https://doi. org/10. 1111/j. 1945 - 5100. 2008. tb01025. x.

［24］ Pike，R. J.，1974. Ejecta from large craters on the Moon：comments on the geometric model of McGetchin et al. Earth Planet. Sci. Lett. 23，265 - 274. https://doi. org/10. 1016/0012 - 821X（74）90114 - 9.

［25］ Pike，R. J.，1980. Control of crater morphology by gravity and target type - Mars，Earth，Moon. In：Lunar and Planetary Science Conference Proceedings.

［26］ Porcello，L. J.，et al.，1974. The Apollo lunar sounder radar system. Proc. IEEE 62，769 - 788. https://doi. org/10. 1109/PROC. 1974. 9517.

［27］ Qiao，L.，Ling，Z.，Fu，X.，Li，B.，2019. Geological characterization of the Chang'e - 4 landing area on the lunar farside. Icarus 333，37 - 51. https://doi. org/10. 1016/j. icarus. 2019. 05. 029.

［28］ Robinson，M. S.，et al.，2012. Confirmation of sublunarean voids and thin layering in mare deposits. Planet. Space Sci. 69，18 - 27. https://doi. org/10. 1016/j. pss. 2012. 05. 008.

[29] Russell, J. K., Stasiuk, M. V., 1997. Characterization of volcanic deposits with ground - penetrating radar. Bull. Volcanol. 58, 515 - 527. https://doi. org/10. 1007/s004450050159.

[30] Rust, A. C., Russell, J. K., 2000. Detection of welding in pyroclastic flows with ground penetrating radar: insights from field and forward modeling data. J. Volcanol. Geotherm. Res. 95, 23 - 34. https://doi. org/10. 1016/S0377 - 0273 (99) 00119 - 5.

[31] Sharpton, V. L., 2014. Outcrops on lunar crater rims: implications for rim construc - tion mechanisms, ejecta volumes and excavation depths. J. Geophys. Res., Planets 119, 154 - 168. https://doi. org/10. 1002/2013JE004523.

[32] Simmons, G., Strangway, D., Annan, P., 1973. Surface electrical properties experi - ment. Apollo 17 Preliminary Science Report. NASA 15, 330.

[33] Stolt, R. H., 1978. Migration by Fourier transform. Geophysics63, 23 - 48. https://doi. org/10. 1190/1. 1440826.

[34] Stuart - Alexander, D. E., 1978. Geologic map of the central far side of the Moon. USGS, map I - 1047, 1: 5M scale.

[35] Taylor, S. R., 1982. Lunar and terrestrial crusts: a contrastin origin and evolution. Phys. Earth Planet. Inter. 29, 233 - 241. https://doi. org/10. 1016/0031 - 9201 (82) 90014 - 0.

[36] Thomson, B. J., Grosfils, E. B., Bussey, D. B. J., Spudis, P. D., 2009. A new technique for estimating the thickness of mare basalts in Imbrium Basin. Geophys. Res. Lett. 36, L12201. https://doi. org/10. 1029/2009GL037600.

[37] Wang, Z., Li, Y., Jiang, J., Li, D., 2010. Lunar surface dielectric constant, regolith thickness, and ^{3}He abundance distributions retrieved from the microwave brightness temperatures of CE - 1 Lunar Microwave Sounder. Sci. China Earth Sci. 53, 1365 - 1378.

[38] Wilhelms, D. E., John, F., Trask, N. J., 1987. The geologic history of the Moon. U. S. Geological Survey Professional Paper 1348.

[39] Wilhelms, D. E., Howard, K. A., Wilshire, H. G., 1979. Geologic map of the south side of the Moon.

[40] Wilson, L., Head, J. W., 2017. Generation, ascent and eruption of magma on the Moon: new insights into source depths, magma supply, intrusions and effusive/explosive eruptions (part 1: theory). Icarus 283, 146 - 175. https://doi. org/10. 1016/j. icarus. 2015. 12. 039.

[41] Xiao, L., Zhu, P., Fang, G., Xiao, Z., Zou, Y., Zhao, J., Zhao, N., Yuan, Y., et al., 2015. A young multilayered terrane of the northern Mare Imbrium revealed by Chang'E - 3 mission. Science 347, 1226 - 1229. https://doi. org/10. 1126/science. 1259866.

[42] Xie, M., Liu, T., Xu, A., 2020. Ballistic sedimentation of impact crater ejecta:

implication for resurfacing and provenance of lunar samples. J. Geophys. Res., Planets125, e2019JE006113. https://doi. org/10. 1029/2019JE006113.

[43] Yingst, R. A., Chuang, F., Berman, D., Mest, S., 2017. Geologic mapping of the Planck quadrangle of the Moon (LQ - 29). In: Lunar and Planetary Science Conference. 48 Abstract #1680.

[44] Yingst, R. A., Head, J. W., 1997. Volumes of lunar lava ponds in South Pole - Aitken and Orientale basins: implications for eruption conditions, transport mechanisms, and magma source regions. J. Geophys. Res. 102, 10909 - 10931. https://doi. org/10. 1029/97JE00717.

[45] Yuan, Y., Wang, F., Zhu, P., Xiao, L., Zhao, N., 2020. New constraints on the young lava flow profile in thc northern Mare Imbrium. Geophys. Res. Lett. 47, e2020GL088938. https://doi. org/10. 1029/2020GL088938.

[46] Yuan, Y., Zhu, P., et al., 2017. The 3 - D geological model around Chang'E - 3 landing site based on lunar penetrating radar Channel 1 data. Geophys. Res. Lett. 44, 6553 - 6561. https://doi. org/10. 1002/2017GL073589.

[47] Zhang, J., Zhou, B., Lin, Y., et al., 2020a. Lunar regolith and substructure at Chang'E - 4 landing site in South Pole - Aitken basin. Nat. Astron. https://doi. org/10. 1038/s41550 - 020 - 1197 - x.

[48] Zhang, L., Li, J., Zeng, Z., Xu, Y., Liu, C., Chen, S., 2020b. Stratigraphy of the Von Kármán crater based on Chang'E - 4 Lunar Penetrating Radar data. Geophys. Res. Lett. 47, e2020GL088680. https://doi. org/10. 1029/2020GL088680.

[49] Zhao, N., Zhu, P., Yuan, Y., et al., 2018. The shallow subsurface structures of Chang'E - 3 landing site based on the wavefield characteristics of LPR Channel - 2B data. Adv. Space Res. 62, 884 - 889. https://doi. org/10. 1016/j. asr. 2018. 05. 023.

Chronology of the Basalt Units Surrounding Chang'e - 4 Landing Area

Zongyu Yue[1,2], Ke Shi[1,3], Gregory Michael[4], Kaichang Di[1,2], Sheng Gou[1,5], Jianzhong Liu[2,6] and Shengli Niu[5]

1. State Key Laboratory of Remote Sensing Science, Aerospace Information Research Institute, Chinese Academy of Sciences, Beijing 100101, China; yuezy@radi.ac.cn (Z. Y.); shike191@mails.ucas.ac.cn (K. S.); dikc@radi.ac.cn (K. D.)
2. Center for Excellence in Comparative Planetology, Chinese Academy of Sciences, Hefei 230026, China; liujianzhong@mail.gyig.ac.cn
3. University of Chinese Academy of Sciences, Beijing 100049, China
4. Institute of Geological Sciences, Freie Unversitaet Berlin, Malteser Strasse 74 - 100, Haus D, 12249 Berlin, Germany; gregory.michael@fu-berlin.de
5. State Key Laboratory of Lunar and Planetary Sciences, Macau University of Science and Technology, Macau 999078, China; 1909853ese30001@student.must.edu.mo
6. Institute of Geochemistry, Chinese Academy of Sciences, Guiyang 550081, China

Abstract: The Chang'e - 4 (CE - 4) lunar probe, the first soft landing spacecraft on the far side of the Moon, successfully landed in the Von Kármán crater on 3 January 2019. Geological studies of the landing area have been conducted and more intensive studies will be carried out with the in situ measured data. The chronological study of the maria basalt surrounding the CE - 4 landing area is significant to the related studies. Currently, the crater size - frequency distribution (CSFD) technique is the most popular method to derive absolute model ages (AMAs) of geological units where no returned sample is available, and it has been widely used in dating maria basalt on the lunar surface. In this research, we first make a mosaic with multi - orbital Chang'e - 2 (CE - 2) images as a base map. Coupled with the elevation data and FeO content, nine representative areas of basalt units surrounding the CE - 4 landing area are outlined and their AMAs are derived. The dating results of the nine basalt units indicate that the basalts erupted from 3.42 to 2.28 Ga ago in this area, a period much longer than derived by previous studies. The derived chronology of the above basalt units establishes a foundation for geological analysis of the returned CE - 4 data.

Keywords: lunar chronology; basalt units; CE - 4 landing area; absolute model age; crater sizefrequency distribution (CSFD)

1 Introduction

Chang'e - 4 (CE - 4) was the second soft landing mission in the China's Lunar Exploration Program, and the probe successfully landed on the far side of the Moon in the Von Kármán crater inside the South Pole - Aitken (SPA) basin on 3 January 2019 [1, 2]. The Yutu - 2 rover was then successfully released from the lander on the same day, and it is still working as of now. The SPA basin is an elliptical impact basin with axes measuring 2400×2050km [3], and it is considered as one of the three major lunar terranes according to its unique geochemistry and petrologic history [4]. Using three - dimensional numerical impact simulations, ref. [5] demonstrates that the SPA must have sampled the Moon's upper mantle to depths of 100km. Therefore, a number of fundamental lunar science questions, such as the composition of the lunar upper mantle [6] and the reason that far side maria differ from their near side counterparts [1], are expected to be addressed with the returned CE - 4 data. As a result, many studies related to the CE - 4 landing area have been recently carried out (e. g. , [7 - 13]), and more works are in progress with the returned CE - 4 data. To support reliable analysis and interpretation of the scientific data from the CE - 4 mission, it is necessary to create a reliable chronology of the geological units surrounding the CE - 4 landing area.

Chronology here refers to the occurrence order of a series of geological events, and the occurrence of the geological events can be expressed with the absolute age or the relative age. Figure 1 shows the context of the CE - 4 landing area. The local relative chronology surrounding the CE - 4 landing area may be determined according to the superpositional relationships of different geological units. However, determining their absolute model ages (AMAs) is always preferable and particularly important when it is difficult to derive the relative ages between two non - adjacent geological units [14]. The principle of determining the AMA of lunar geological units was elaborated in [15] and was the subject of many papers (e. g. , [14 - 16]). The rationale of the approach is to fit the observed crater sizefrequency distribution (CSFD) of a counting area that represents the geological unit to a known crater production function (PF) (e. g. , [17]), which is further used to derive the AMA along with a chronology function (CF) calibrated to radiometric dating from lunar samples [16]. Currently, the method has been widely used in dating lunar mare basalts, e. g. , [17 - 19]. With this method, the AMA of the mare unit on the floor of Von Kármán crater has been derived as 3. 35Ga [20], 3. 6Ga [7], and 3. 15Ga [21]. However, the mare surface on the floor of Von Kármán crater was interpreted to be emplaced by one episode of basaltic volcanism in these studies, which can only give a rough estimation of the average AMA of the maria basalt. As to the mare deposits in Von Kármán M crater [20], an AMA of 3. 38Ga was derived, while [21] mapped two mare basalt units in Von Kármán M crater and derived AMAs of 3. 38Ga and 3. 52Ga, respectively. In addition, the AMA of the basalt unit in Leibnitz crater was derived as 3.44Ga [20] and 3.37Ga [21]. The mare basalt in Leibnitz X crater (the crater diameter is about 19.71km), which is inside Leibnitz crater, was dated as 3.68Ga [21]. The different dating results of the above studies to the same maria unit may arise from the different data sources and counting areas. Therefore, a strict basalt unit mapping and AMA dating is urgent for the analysis of data acquired by the CE-4 rover. In addition, because there are many ejecta in this area, CE-4 rover would probably encounter the fragments from the maria basalts nearby. Thus, the AMAs of them are also necessary to understand the whole history of the basalt formation in this area.

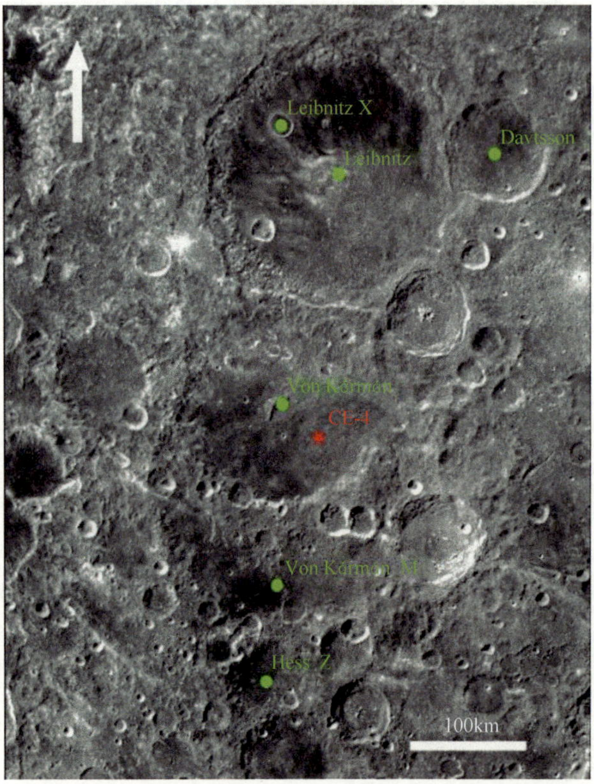

Figure 1　Geographical context of the CE4 landing area. The central location is (177.07°E, 43.19°S). The basemap is Chang'e-2 (CE-2) Digital Orthophoto Map (DOM) mosaic. The crater names are from the International Astronomical Union data base (https://planetarynames.wr.usgs.gov/, accessed on 15 November 2021). North is up

In this research, nine basalt geological units were mapped in the CE-4 landing area according to the difference of brightness, FeO content, and the topography. Superposed craters in the counting areas used for deriving the AMAs of the above geological units were mapped in mosaics generated from the Chang'e-2 (CE-2) digital orthophoto map (DOM) [22]. The AMAs of all the nine measurement areas were obtained with the CSFD method.

2 Materials and Methods

2.1 Data

Craters for dating AMAs were mapped in a DOM mosaic generated from CE-2 orbital images. CE-2 mission was China's second lunar spacecraft and it was launched on 1 October 2010. The onboard stereo camera obtained high-resolution images, and the returned data were processed with radiometric, geometric, and photometric calibration [22]. A DOM was finally generated with a resolution of ~7m/pixel and divided into multiple regions. The data are now publicly available through the Data Release and Information Service System of China's Lunar Exploration Program (http://moon.bao.ac.cn, accessed on 15 November 2021). Twelve DOM files were downloaded and merged to produce a DOM mosaic in ArcGIS®.

The merged digital elevation model from the Lunar Orbiter Laser Altimeter and SELENE Terrain Camera (designated as "SLDEM") [23] was used for topographic analysis of the geological context surrounding the CE-4 landing site. The product has a typical vertical accuracy of ~3-4m and a resolution of ~42m/pixel around the CE-4 landing area [23]. To reveal the topographic detail of the region, the SLDEM was color coded and was registered.

The FeO content product derived from the Kaguya Multiband Imager data [24] is also used to evaluate the boundaries of maria basalt units in this research. The product is released with a spatial resolution of ~60m/pixel between 50°N and 50°S and nine maria basalts units studied in this research are included.

2.2 Methods

The process of determining the AMA of the geological unit includes mapping the surface unit and identifying the superposed craters. The AMAs can be derived by comparing the CSFD of the counted craters with the CF [15]. Here, we present the principles of mapping the measurement areas and the crater counting method. We used Craterstats (http://www.geoinf.fu-berlin.de/mex4/software/craterstats2/, accessed on 15 November 2021) to derive the AMA of each geological unit [25].

2.2.1 Maria Basalt Measurement Areas Mapping

It only makes sense to assess the AMA of a surface with the same geological history, i.e., a geologically homogeneous region that had undergone the same geological processes [25]. Significant interruptions to the homogeneity should be considered as another geological unit. For the maria basalt in this study, this means only one episode of basaltic volcanism should be emplaced in each geological unit.

In maria units mapping, the brightness differences, spectral ratios, FeO and TiO_2

contents, and morphology have usually been used to ensure homogeneity of the investigated basalts, e.g., [17 – 19, 26 – 28]. In this study, we combined the derived FeO content, brightness difference from CE – 2 DOM mosaic, and topographical difference from SLDEM to map the maria basalt units. The mapped measurement areas are the intersection of these above products if they are inconsistent. In addition, areas with identifiable secondary craters should be excluded in the geological mapping [25]. Finally, nine maria basalt units were outlined see below.

2.2.2 Crater Counting

The craters superimposed on the measurement areas have been mapped on the CE – 2 DOM mosaics with the ArcGIS add – in "CraterTools", which measures crater diameter correctly regardless of map projection [29]. CraterTools can work well with three reference points on the crater rims, even for craters where a part of the rim is obscured or even absent. When manually digitizing a crater, the three points on the crater rim should be as evenly distributed as possible, which can give the minimum error [30]. The secondary craters should be excluded in the dating process [31, 32], and they are usually recognized in chains or clusters with irregular shape. This judgement is inevitably subjective, and we have conducted an analysis of spatial randomness and clustering of the mapped craters superimposed on each geological unit following the method of mean closest two neighbors' distance (MC2ND) [32].

3 Results

3.1 Maria Basalt Units in Von Kármán Crater

The floor of Von Kármán crater (Figure 2) was mapped as mare basalts by [33], and the crater floor has significantly higher FeO contents than the surrounding terrain (Figure 2c). However, the crater floor is extensively overprinted by ejecta from nearby craters [34], making the FeO content much lower in the northern and eastern part of the crater floor. In addition, the fresh Zhinyu crater obviously excavated mare basalt deposits from the subsurface resulted in much higher FeO content (Figure 2c). Thus, we selected a small area labeled as 'a' in Figure 2 to date the maria basalt in the Von Kármán crater floor (Figure 2a, d). Figure 2B shows two local upland structures in the SLDEM, and they are more clearly shown in Figure 2e, f, respectively. In this research, they are mapped as units 'b' and 'c' for subsequent surface dating.

Our CSFD measurements of the mare units inside Von Kármán crater show AMAs of 3.42Ga, 3.34Ga, and 3.34Ga for the basalt units 'a', 'b', and 'c', respectively (Figure 3). The analysis results of spatial randomness and clustering of the mapped craters are on the top of each group of CSFD fitting curve. It is clear that, for the crater diameters used

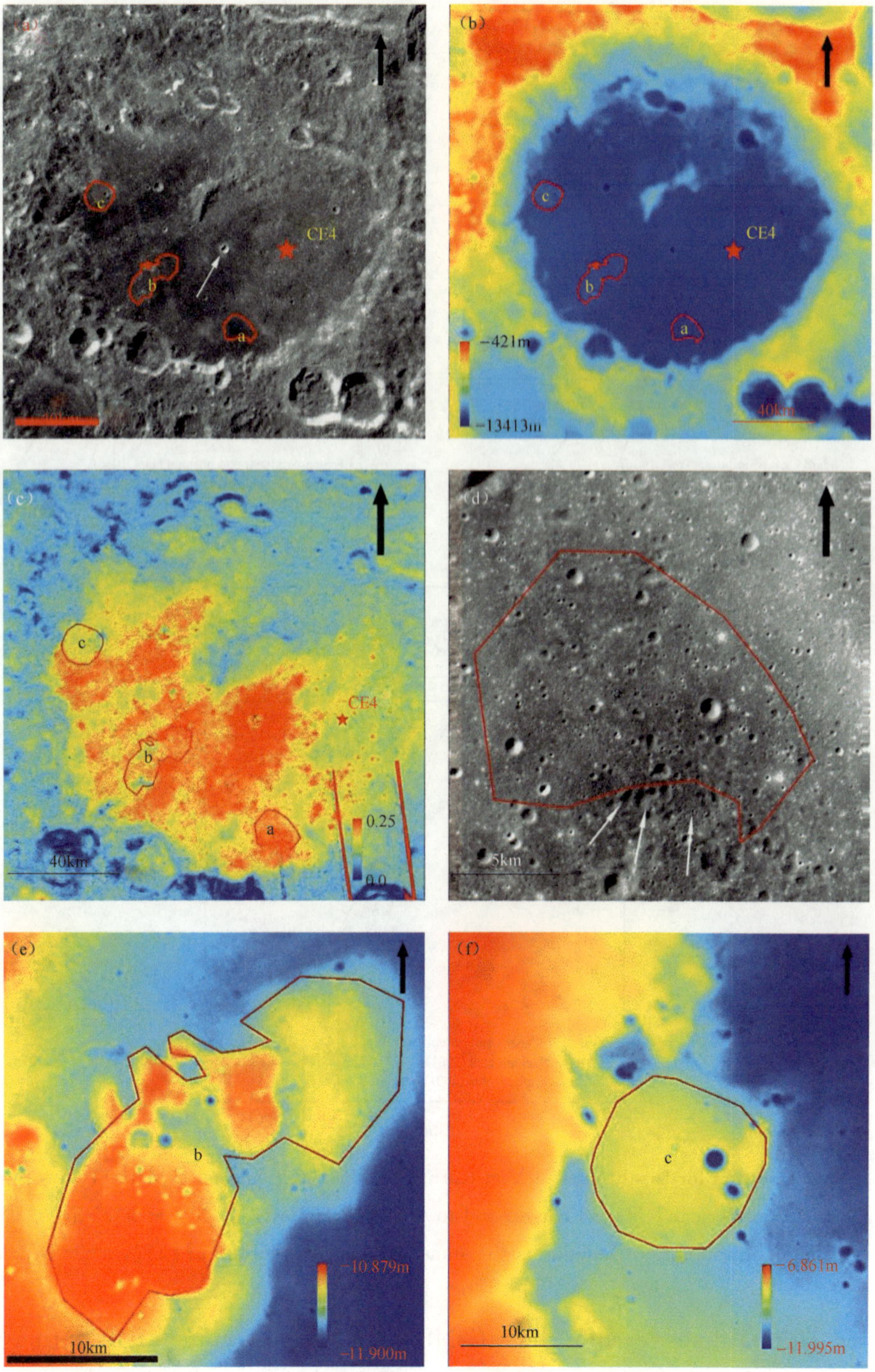

Figure 2　Three basalt measurement areas mapped by red polygons in Von Kármán crater.

(a) CE－2 DOM mosaic (central location: 175.6°E, 44.64°S) of the Von Kármán crater with CE－4 landing site labeled. The fresh crater pointed by the white arrow is Zhinyu crater. (b) Local topography shown in SLDEM; the elevation ranges from －13, 413m to －421m as coded from blue to red in this figure. (c) FeO content map [24] in Von Kármán crater; the values range from 0 to 24.73% as coded from blue to red in this figure. (d) Basalt unit 'a' shown in CE－2 DOM mosaic, with the white arrows indicating the chained secondaries. The area is 138 km^2. (e) Basalt unit 'b' is located in a topographical elevated terrain, and the elevation ranges from －11, 990m to －10, 839m as coded from blue to red in this figure. The area is 267km^2. (f) Basalt unit 'c' is located in a topographical elevated terrain, and the elevation ranges from －11, 995m to －6861m as coded from blue to red in this figure. The red polygon in each sub－figure is the outlined representative crater count area. The area is 151km^2. North is up in all sub－figures

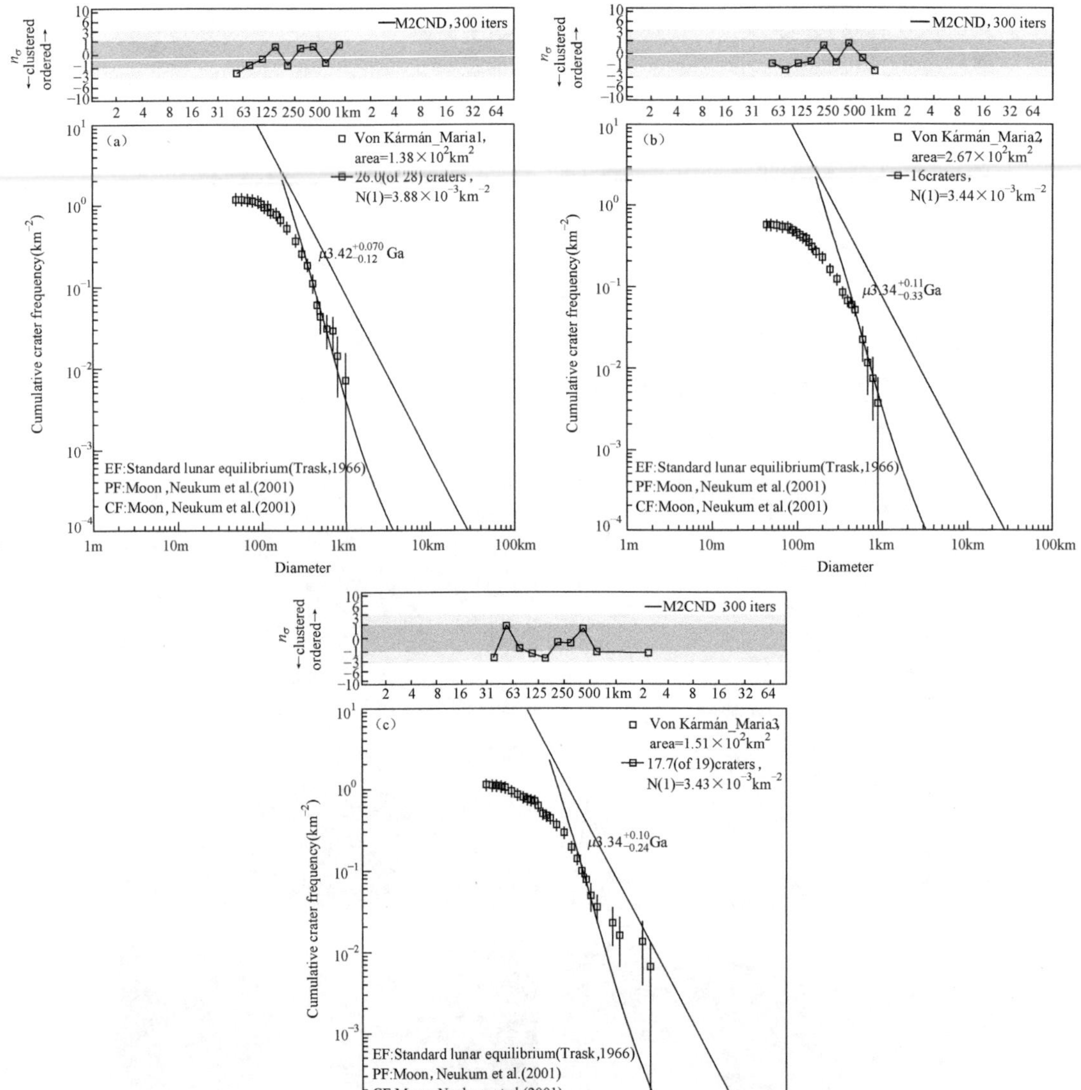

Figure 3　Dating results for basalt units in Von Kármán crater. Von Karman _ Maria1, Von Karman _ Maria2, and Von Karman _ Maria3 correspond to the basalt units ‘a’, ‘b’, and ‘c’ in Figure 2, respectively. The equilibrium function is the standard lunar equilibrium of [35], while both the production and chronology functions are from [36].

for surface dating, almost each of the MC2ND of the mapped craters falls within $\pm 3\sigma$ of the expectation value derived from Monte Carlo simulation, indicating that the spatial distribution of the mapped craters is random. Figure 3 also includes the areas of the counting regions, with the values ranging from 138km^2 to 267km^2.

3.2　Maria Basalt Units in Von Kármán M Crater

Most of the floor of Von Kármán M crater is superposed by Von Kármán crater, and the exposed southern half of the crater is filled with mare basalts [33]. The eastern part of the floor is covered by the ejecta from Alder crater (Figure 1), and there are several massifs protruding from the basalt units (Figure 4a). Similar to [21], two basalt units are mapped in this study (Figure 4), in which the raised massifs are excluded. The basalt units are relatively flat, and the topography gradually rises up from southeast to northwest (Figure 4b). The FeO contents of the basalt units are obviously much higher than surrounding terrains (Figure 4c).

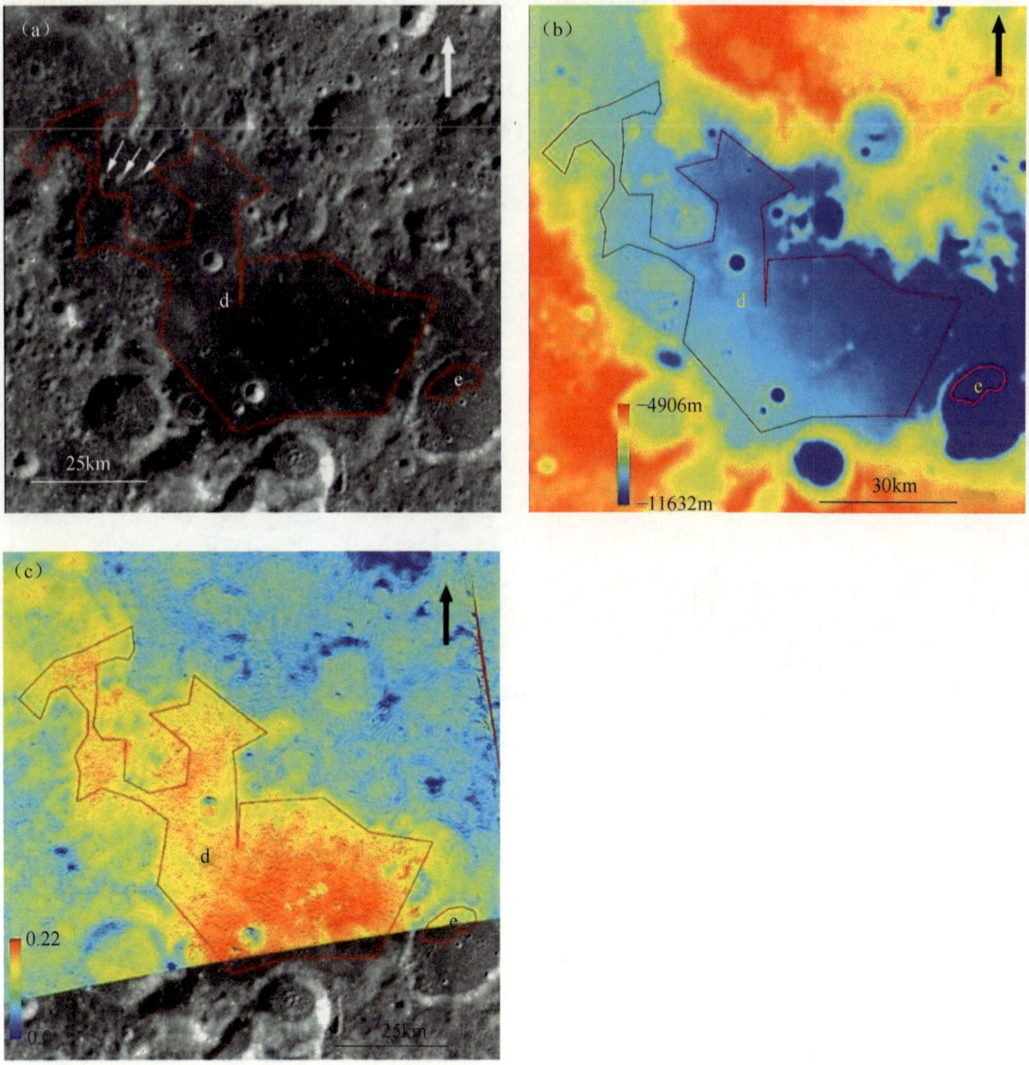

Figure 4　Two basalt measurement areas d ($2320km^2$) and e ($59km^2$) mapped by red polygons in Von Kármán M crater. (a) CE-2 DOM mosaic (central location: 173.6°E, 48.76°S) of the Von Kármán M crater. The raised massifs are indicated by the white arrows. (b) Local topography shown in SLDEM, and the elevation ranges from -11,632m to -4906m as coded from blue to red in this figure. (c) FeO content map [24] in Von Kármán M crater, and the maximum value is up to 22.28% as coded from blue to red in this figure. The red polygon in each sub-figure is the outlined representative crater count area. North is up in all sub-figures

The CSFD measurements of the mare units in Von Kármán M crater show AMAs of 2.36 Ga and 2.28 Ga for the basalt units 'd' and 'e', respectively (Figure 5). For the basalt unit 'd' (counting area is 2320km^2), the minimum result of the random analysis is about -6σ of the expectation value derived from Monte Carlo simulation at the diameter of ~144m, and the random analysis results for D>=~200m are within $\pm 3\sigma$ of the expectation value. The minimum diameter for fitting the isochron line is at ~400m in this study, which thus does not affect the reliability of the dating results. For the basalt unit 'e' (counting areas is 59.3km^2), the analysis results of spatial randomness and clustering of the mapped craters fall within $\pm 3\sigma$ of the expectation value derived from Monte Carlo simulation for all the diameters. The AMA difference (0.08Ga) between the two basalt units is small, indicating the basalt in these two areas are probably from one episode.

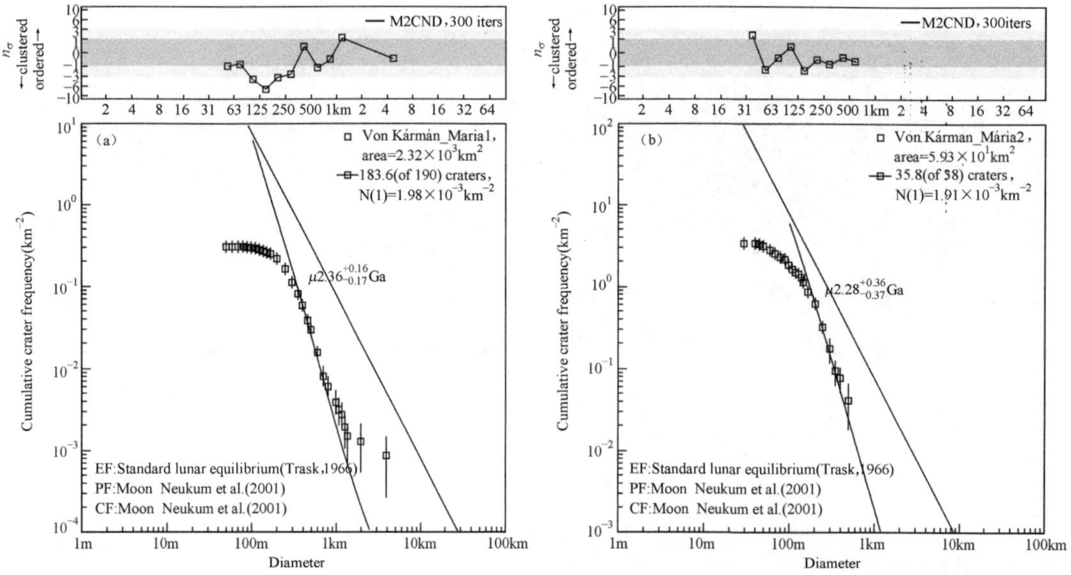

Figure 5　Dating results for basalt units in Von Kármán M crater. Von KarmanM _ Maria1 and VonKarmanM _ Maria2 correspond to the basalt units 'd' and 'e' in Figure 4. The equilibrium function is the standard lunar equilibrium of [35], while both the production and chronology functions are from [36]

3.3　Maria Basalt Unit in Hess Z Crater

Figure 6 shows the mapped basalt unit (counting area is 530km^2) and the derived AMA in Hess Z crater. The southeastern part of the Hess Z crater rim was destroyed by the ejecta from the Hess crater nearby, which was superposed by the basalt herein. There are several clustered secondaries in the eastern floor of the Hess Z crater, which are deliberately avoided when map the countering area. The mapped basalt unit is relatively flat, although the topography is gradually lowered from east to west.

The CSFD measurement of the mare unit in Davisson crater shows an AMA of 3.27Ga (Figure 8). For the craters used for dating, the MC2ND of the mapped craters

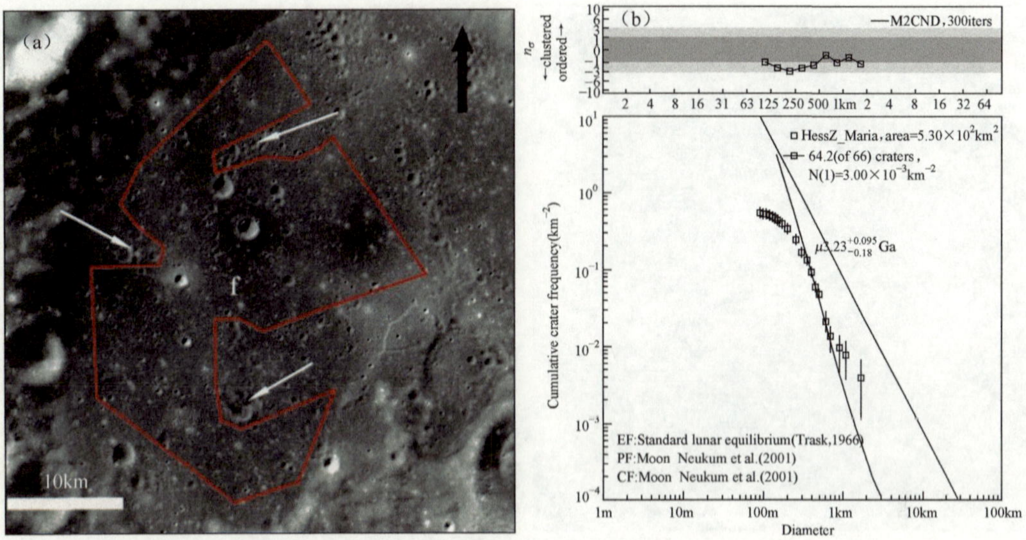

Figure 6 (a) CE-2 DOM mosaic of the mapped basalt unit in Hess Z crater ($530km^2$) (central location: 173.49°E, 51.94°S); North is up. (b) derived AMA. The equilibrium function is the standard lunar equilibrium of [35], while both the production and chronology functions are from [36]. The red polygon in the left sub-figure is the outlined representative crater count area, and the arrows indicate the excluded secondary craters

falls within $\pm 3\sigma$ of the expectation value derived from Monte Carlo simulation, indicating mapped craters in the counting area are random in spatial distribution.

The CSFD measurement of the mare unit in Hess Z crater shows an AMA of 3.23Ga (Figure 6). The mapped craters are randomly distributed in the counting area, as indicated by the spatial randomness and clustering analysis result (Figure 6).

3.4 Maria Basalt Unit in Davisson Crater

Most of the floor of Davisson crater is covered by basalt, as indicated by the high FeO content (Figure 7c). A central peak protrudes above the floor and several chained ejecta are distributed in the southeastern of the crater floor. They are deliberately excluded when mapping the countingarea (Figure 7a). The mapped basalt unit is relatively flat, and it is lowest in the southern part of the floor (Figure 7b).

3.5 Maria Basalt Units in Leibnitz and Leibnitz X Crater

The southeast part of the Leibnitz crater is superposed by the ejecta from Finsen crater (Figure 1), and a large amount of linear ejecta are distributed in the floor of Leibnitz crater (Figure 9a). Thus, we carefully selected a small area in the north part of the crater floor as the counting area, which is covered by basalts as proven by the much higher FeO content than surrounding ejecta (Figure 9c). It is interesting to note that the north part is higher than the south part in the floor of Leibnitz crater (Figure 9b),

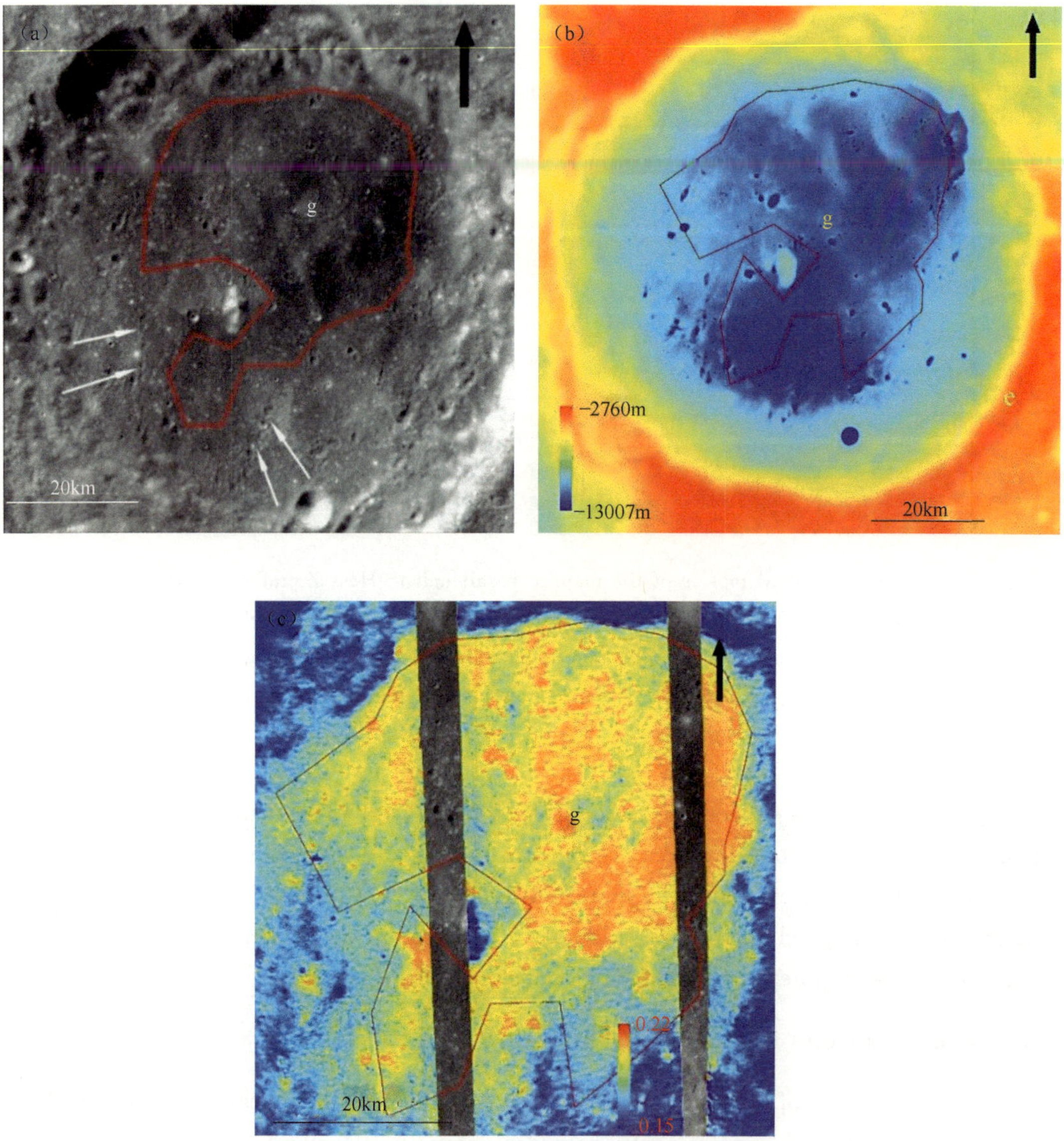

Figure 7 (a) CE-2 DOM mosaic (central location: 174.88° W, 37.52° S) of the mapped basalt unit (1,280km²) in Davisson crater. The central peak and clustered secondaries (arrows) are excluded when map the counting area. (b) Local topography shown in SLDEM, and the elevation ranges from −13,007m to −2,760m as coded from blue to red in this figure. (c) FeO content map [24] of the counting area, and the values range from 1.94% to 21.94% as coded from blue to red in this figure. The red polygon in each sub-figure is the outlined representative crater count area. North is up in all sub-figures

although the ejecta is from south.

Leibnitz X crater is in the northwest of Leibnitz crater floor, and the floor of Leibnitz X crater can be divided into two parts according to the brightness difference (Figure 9a). We selected the east part of the floor as the counting area, which also has a higher FeO content than the west part (Figure 9c). The selected counting area seems unaffected by the

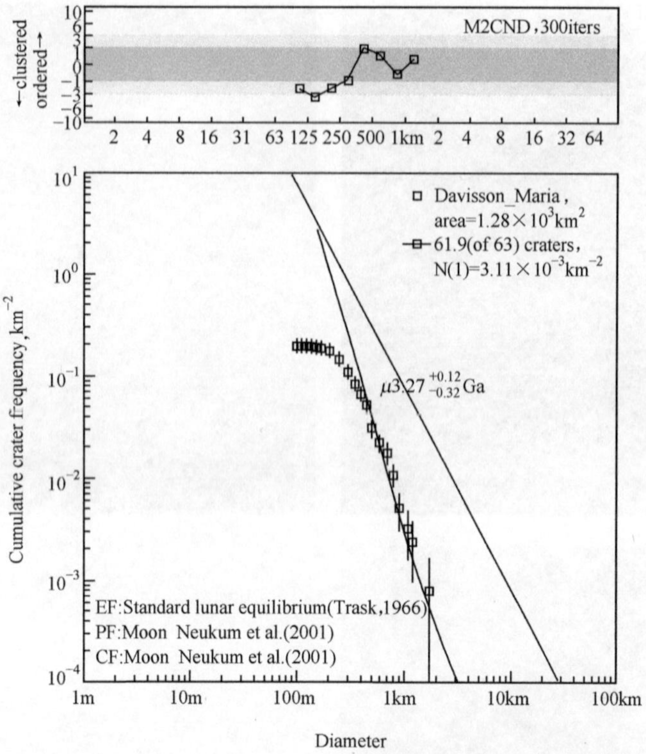

Figure 8　Dating results for basalt unit mapped in Figure 7. The equilibrium function is the standard lunar equilibrium of [35], while both the production and chronology functions are from [36]

linear ejecta from the Finsen crater. It is more interesting to note that the whole Leibnitz X crater, including its floor, is much higher than the surrounding terrains (Figure 9b). In addition, the FeO content of the maria in the counting area for Leibnitz X crater (Figure 9c) is lower than that for Leibnitz crater, indicating they are probably from different sources.

The CSFD measurements of the mare units in Leibnitz and Leibnitz X craters show AMAs of 2.61Ga and 3.05Ga (Figure 10). For the basalt unit in Leibnitz crater (i. e., unit 'h' in Figure 9), the random analysis is about -6σ of the expectation value at the diameter of 144m, and the random analysis results for D $>=$ ~400m are within $\pm 3\sigma$ of the expectation value. Therefore, this again does not affect the reliability of our dating results, as the minimum diameter for fitting the isochron line is at ~400m. For all the counted craters in the mare unit in Leibnitz X crater (i. e., unit 'i' in Figure 9), the spatial randomness and clustering analysis result of the mapped craters falls within $\pm 3\sigma$ of the expectation value derived from Monte Carlo simulation. It is obvious that the basalt unit in Leibnitz X crater is older than that in Leibnitz crater. Given this, coupled with the difference in FeO content, it is reasonable to conclude that they formed in two episodes of basalt formation.

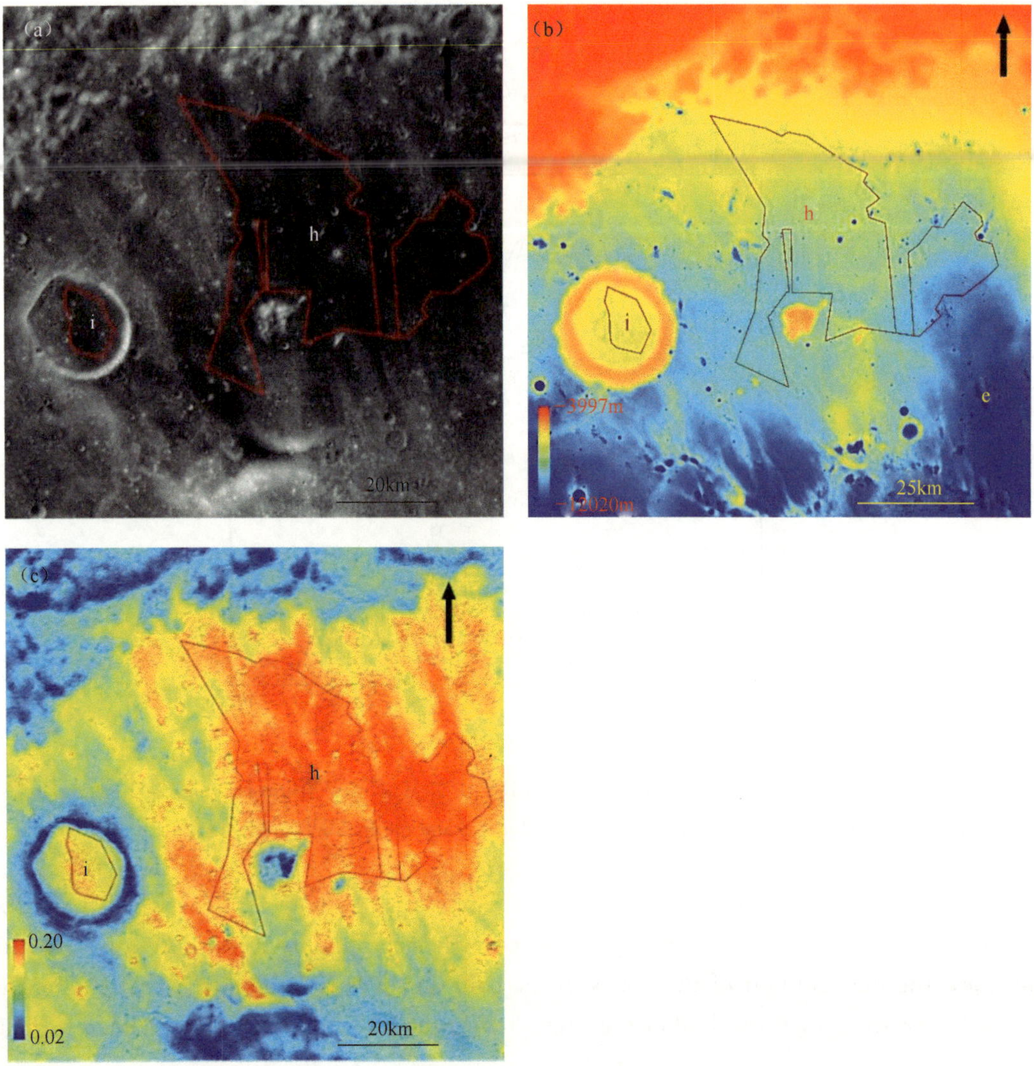

Figure 9　(a) CE - 2 DOM mosaic (central location: 178.91°E, 36.23°S) of the mapped basalt units 'h' ($1450km^2$) and 'i' ($127km^2$) in Leibnitz and Leibnitz X craters. The central peak and clustered secondaries are excluded in the counting area. (b) Local topography shown in SLDEM, and the elevation ranges from −12,020m to −3997m as coded from blue to red in this figure. (c) FeO content map [24] of the counting areas, and the values range from 2.27% to 19.69% as coded from blue to red in this figure. The red polygons in each sub - figure are the outlined representative crater count areas. North is up in all sub - figures

3.6　Overview for AMAs of Basalt Units Surrounding CE - 4 Landing Site

Table 1 lists all the derived AMAs for the basalt units mapped in this study, and the N (1), i.e., the number of craters with a diameter larger than or equal to 1km per km^2, derived by fitting a polynomial proposed in [36] to observed CSFDs, is also included for convenient comparison with other studies. The youngest basalt unit (i.e., unit 'e') is in

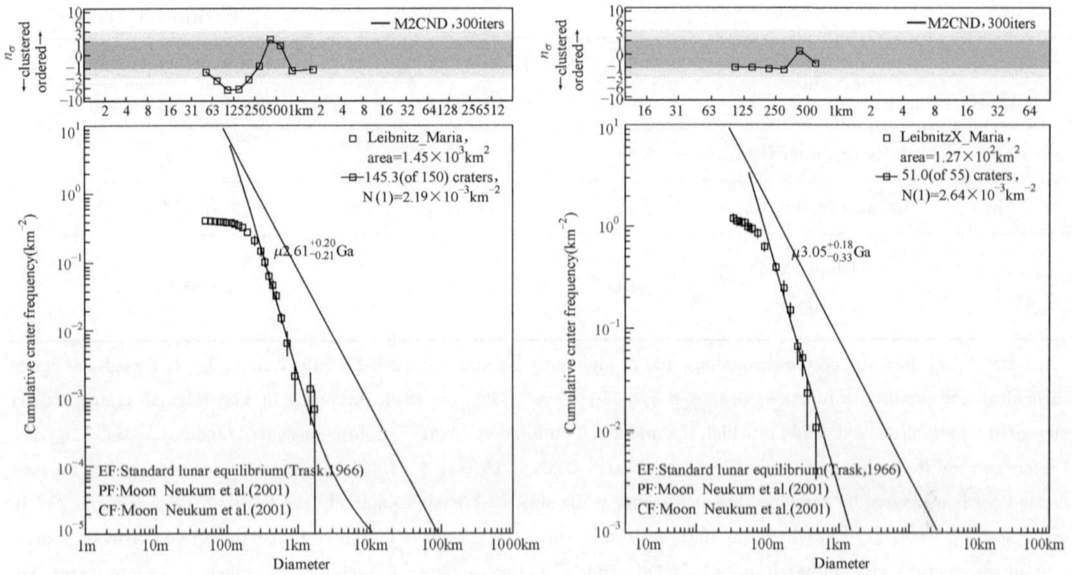

Figure 10　Dating results for the basalt units mapped in Figure 9. Leibnitz _ Maria and LeibnitzX _ Maria correspond to the basalt units 'h' and 'i' in Figure 9. The equilibrium function is the standard lunar equilibrium of [35], while both the production and chronology functions are from [36]

the Von Kármán M crater, and unit 'd' in Von Kármán M crater is the second youngest among the studied basalt units. This perhaps certificates that the basalt inside Von Kármán M crater formed later than that in other places in the study region. Basalt unit 'h' in Leibnitz crater floor is the third youngest among the basalt units, and they are the only three basalt units younger than 3.0Ga in this study. The oldest basalt unit 'a' is in the Von Kármán crater floor, which has an AMA of 3.42Ga, and the three basalt units in the Von Kármán crater floor are the three oldest in this study. The similarity of the derived AMAs of the three basalt units inside Von Kármán crater indicates that the basalt should be first emplaced in the Von Kármán crater floor in this area. The second oldest basalt unit is in Davisson crater floor, and the above four basalt units belong to the Late Imbrian while others belong to Eratosthenian, according to [26].

Table 1　Results of AMAs for the groups of basalt units in this study

Basalt Unit	Location	N (1) (×10^{-3}km^{-2})	AMAs (Ga)	
			This Study	Previous Studies[1]
a	Von Kármán crater floor	3.98	3.42	3.35*; E−I†; 3.6‡; 3.15#;
b		3.44	3.34	
c		3.43	3.34	
d	Von Kármán M crater floor	1.98	2.36	3.38##; E−I†; 3.38**
e		1.91	2.28	3.52##; E−I†; 3.38**
f	Hess Z crater floor	3.00	3.23	E−I†; 3.51##

Continued Table 1

Basalt Unit	Location	N (1) ($\times 10^{-3}km^{-2}$)	AMAs (Ga)	
			This Study	Previous Studies[1]
g	Davisson crater floor	3.11	3.27	—
h	Leibnitz crater floor	2.19	2.61	3.37###;
i	Leibnitz X crater floor[1]	2.64	3.05	3.68##;

[1] Ref. [20] used the crater chronology and production functions proposed by [15], ref. [7, 21] used the crater chronology and production functions proposed by [36]; * ref. [20] in which the maria in Von Kármán crater floor is measured together; ** ref. [20] in which the maria in Von Kármán M crater floor is measured together; + ref. [26] E-I refers to near the Imbrian-Eratosthenian boundary, i. e., 3.26 Ga; ‡ ref. [7] in which the maria in Von Kármán crater floor is measured by combing three subareas, while subarea 2 is approximately the same to the basalt unit 'a' in this research; # ref. [21] in which the maria in Von Kármán crater floor is measured by combing two subareas, which contain the counting areas of basalt units 'a', 'b', and 'c' in this research; ## ref. [21] in which the counting areas are roughly equivalent to that in this research; ### ref. [21] in which the counting areas are much larger than that in our research.

4 Discussion

4.1 Comparison with Previous Studies

Table 1 also lists the dating results in previous studies. For the maria in Von Kármán crater, the dated AMAs range from 3.34 to 3.42 Ga for the three counting areas mapped in our research. The results are similar to the previous results (3.26Ga for [26]; 3.35Ga for [20]; 3.6Ga for [7]; 3.15Ga for [21]). Thus, we have much confidence to analyze the data returned by CE-4 rover, which is now traversing to this area. For the results of maria in Hess Z crater, there is also not too much difference (0.28Ga) [21, 26]. However, there is slight difference of 0.63Ga in the case of Leibnitz X crater, and large differences of ~1.0Ga [20, 21, 26] in the cases of Von Kármán M crater and Leibnitz crater. For the AMA difference between this research and that of [21], given similar counting areas are mapped, the possible reason might be different imagery data used. In our study, the CE-2 DOM mosaic at a resolution of 7m/pixel was used, while the wide angle camera (WAC) images (~100m/pixel) from the Lunar Reconnaissance Orbiter (LRO) were used in [21]. In fact, ref. [21] also found that their dating AMAs based on the WAC images are indeed much older than those based on narrow angle camera (NAC) images (~1m/pixel) from the LRO for the maria basalts in Antoniadi crater. For the difference of the derived AMAs between this research and that of [20], given similar image resolution used in the two studies, the most probable reason is the different locations and sizes of the counting areas. We cannot find the counting area of [20].

However, ref. [20] usually mapped the basalt in the floor of Von Kármán M crater and Leibnitz crater as one unit. If the basalt was formed through different episodes, this is unsuitable. In addition, we obtained the AMA of the maria in the floor of Davisson crater for the first time.

In early geological mapping [26], most mare deposits within SPA were mapped as late Imbrian. Ref. [20] first showed that some mare deposits in the lunar farside may have much younger ages up to Eratosthenian. Ref. [37] also proposed that the lunar farside was volcanically active for nearly as long as the lunar nearside. Our dating results surrounding the CE - 4 landing area support this conclusion and indicate that small - scale lunar farside volcanism might have continued longer than previously thought.

4.2 Implications for CE - 4 Mission

The dating results for the maria basalts inside Von Kármán crater are significant to the interpretation of CE - 4 data. It is interesting to note that AMAs of the three basalt units are very similar, indicating that the basalts in Von Kármán crater were formed within a short period. CE - 4 landed on linear impact ejecta (Figures 1 and 2), and Finsen crater was the major source [7, 9 - 11, 34]. The thickness of the ejecta might be several tens of meters in the landing area [10, 11, 34, 38]. However, the ejecta here appear linear and are not continuous [34], and CE - 4 rover may probably happen to encounter local mare basalt when it traverses westward, especially around new craters (e. g. , Zhinyu crater) which can excavate subsurface material [39]. If CE - 4 rover will traverse the exposed mare basalt in future, the dating results of the mare basalt inside the Von Kármán crater can provide information for analyzing the lunar regolith evolution and crater degradation, combing the spectral data and lunar ground penetrating radar data. For example, according to the data collected by the ground - penetrating radar onboard the CE -4 rover, several mare basalt layers were interpreted by [10, 11]. The ejecta from Zhinyu crater also reveal at least three episodes of the basalt eruption [39], and the dating results can help to provide chronological information.

The dating results for other mare basalt is also useful in relevant studies. For example, through spatial density investigation and randomness analysis, ref. [40] pointed out that secondary craters cover at least one third of the 2. 4°×1° area surrounding CE -4 landing site. Therefore, it is very likely that CE - 4 rover would encounter the fragments from the mare basalts inside Von Kármán crater dated in the manuscript, and the results in the research can provide geological context to these fragments. For example, the compositions of the fragments can be analyzed by the CE - 4 rover, and their sources can thus be determined by comparing the compositions with the dated mare basalts, whose formation times have been constrained in this study.

5 Conclusions

In this research, we dated the AMAs of nine basalt measurement areas surrounding the CE-4 landing area using the crater counting method, including the mare deposits in Von Kármán, Von Kármán M, Hess Z, Davisson, Leibnitz, and Leibnitz X craters. Our results indicate that the youngest mare deposit is in the floor of Von Kármán M crater with AMAs of 2.28Ga and 2.36Ga. This research provides both systematic information on the history of the basalt in the region and constraints for the interpretation of the in situ measured data from the CE-4 rover. The results point to a much longer-lived volcanism in the region than was previously thought (e.g., ref. [26] proposed that most mare deposits in the SPA basin were formed during the late Imbrian epoch, i.e., 3.85-3.8Ga.), which is also beneficial to the understanding of the evolution of lunar volcanism globally. We suggest that CE-4 rover may return data on local genetic basalts while it keeps moving westward, which will contribute to our understanding of basalt formation in the lunar farside.

Acknowledgments

The authors gratefully acknowledge China's Lunar Exploration Program at National Astronomical Observatories of Chinese Academy of Sciences for providing CE-2 DOM data (http://moon.bao.ac.cn, accessed on 15 November 2021).

References

[1] Wu, W.; Li, C.; Zuo, W.; Zhang, H.; Liu, J.; Wen, W.; Su, Y.; Ren, X.; Yan, J.; Yu, D.; et al. Lunar farside to be explored by Chang'e-4. Nat. Geosci. 2019, 12, 222-223.

[2] Di, K.; Liu, Z.; Liu, B.; Wan, W.; Peng, M.; Wang, Y.; Gou, S.; Yue, Z.; Xin, X.; Jia, M.; et al. Chang'e-4 lander localization based on multi-source data. J. Remote Sens. 2019, 23, 177-184.

[3] Garrick-Bethell, I.; Zuber, M.T. Elliptical structure of the lunar South Pole-Aitken basin. Icarus 2009, 204, 399-408.

[4] Jolliff, B.L.; Gillis, J.J.; Haskin, L.A.; Korotev, R.L.; Wieczorek, M.A. Major lunar crustal terranes: Surface expressions and crust-mantle origins. J. Geophys. Res. Planets 2000, 105, 4197-4216.

[5] Melosh, H.J.; Kendall, J.; Horgan, B.; Johnson, B.C.; Bowling, T.; Lucey, P.G.; Taylor, G.J. South Pole-Aitken basin ejecta reveal the Moon's upper mantle. Geology 2017, 45, 1063-1066.

[6] Moriarty, D. P. , III; Pieters, C. M. The Character of South Pole - Aitken Basin: Patterns of Surface and Subsurface Composition. J. Geophys. Res. Planets 2018, 123, 729 - 747.

[7] Huang, J. ; Xiao, Z. ; Flahaut, J. ; Martinot, M. ; Head, J. ; Xiao, X. ; Xie, M. ; Xiao, L. Geological Characteristics of Von Kármán Crater, Northwestern South Pole-Aitken Basin: Chang'e - 4 Landing Site Region. J. Geophys. Res. Planets 2018, 123, 1684 - 1700.

[8] Li, C. ; Liu, D. ; Liu, B. ; Ren, X. ; Liu, J. ; He, Z. ; Zuo, W. ; Zeng, X. ; Xu, R. ; Tan, X. ; et al. Chang'E - 4 initial spectroscopic identification of lunar far - side mantle - derived materials. Nature 2019, 569, 378 - 382.

[9] Gou, S. ; Di, K. ; Yue, Z. ; Liu, Z. ; He, Z. ; Xu, R. ; Lin, H. ; Liu, B. ; Peng, M. ; Wan, W. ; et al. Lunar deep materials observed by Chang'e - 4 rover. Earth Planet. Sci. Lett. 2019, 528, 115829.

[10] Zhang, J. ; Zhou, B. ; Lin, Y. ; Zhu, M. - H. ; Song, H. ; Dong, Z. ; Gao, Y. ; Di, K. ; Yang, W. ; Lin, H. ; et al. Lunar regolith and substructure at Chang'E - 4 landing site in South Pole - Aitken basin. Nat. Astron. 2021, 5, 25 - 30.

[11] Lai, J. ; Xu, Y. ; Bugiolacchi, R. ; Meng, X. ; Xiao, L. ; Xie, M. ; Liu, B. ; Di, K. ; Zhang, X. ; Zhou, B. ; et al. First look by the Yutu - 2 rover at the deep subsurface structure at the lunar farside. Nat. Commun. 2020, 11, 3426.

[12] Zhang, L. ; Li, J. ; Zeng, Z. ; Xu, Y. ; Liu, C. ; Chen, S. Stratigraphy of the Von Kármán Crater Based on Chang'E - 4 Lunar Penetrating Radar Data. Geophys. Res. Lett. 2020, 47, e2020GL088680.

[13] Lin, H. ; Xu, R. ; Yang, W. ; Lin, Y. ; Wei, Y. ; Hu, S. ; He, Z. ; Qiao, L. ; Wan, W. In Situ Photometric Experiment of Lunar Regolith With Visible and Near - Infrared Imaging Spectrometer On Board the Yutu - 2 Lunar Rover. J. Geophys. Res. Planets 2020, 125, e2019JE006076.

[14] Stöffler, D. ; Ryder, G. ; Ivanov, B. A. ; Artemieva, N. A. ; Cintala, M. J. ; Grieve, R. A. F. Cratering History and Lunar Chronology. Rev. Mineral. Geochem. 2006, 60, 519 - 596.

[15] Neukum, G. Meteoriten Bombardement und Datierung Planetarer Oberflächen. Habilitation Dissertation, University of Munich, Munich, Germany, 1983.

[16] Stöffler, D. ; Ryder, G. Stratigraphy and Isotope Ages of Lunar Geologic Units: Chronological Standard for the Inner Solar System. Space Sci. Rev. 2001, 96, 9 - 54.

[17] Hiesinger, H. ; Jaumann, R. ; Neukum, G. ; Head, J. W. , III. Ages of mare basalts on the lunar nearside. J. Geophys. Res. Planets 2000, 105, 29239 - 29275.

[18] Hiesinger, H. ; Head, J. ; Wolf, U. ; Jaumann, R. ; Neukum, G. Ages and stratigraphy of mare basalts in Oceanus Procellarum, Mare Nubium, Mare Cognitum, and Mare Insularum. J. Geophys. Res. Planets 2003, 108, 5065.

[19] Hiesinger, H.; Head, J. W., III; Wolf, U.; Jaumann, R.; Neukum, G. Ages and stratigraphy of lunar mare basalts in Mare Frigoris and other nearside maria based on crater size - frequency distribution measurements. J. Geophys. Res. Planets 2010, 115, E03003.

[20] Haruyama, J.; Ohtake, M.; Matsunaga, T.; Morota, T.; Honda, C.; Yokota, Y.; Abe, M.; Ogawa, Y.; Miyamoto, H.; Iwasaki, A.; et al. Long - Lived Volcanism on the Lunar Farside Revealed by SELENE Terrain Camera. Science 2009, 323, 905 - 908.

[21] Pasckert, J. H.; Hiesinger, H.; van der Bogert, C. H. Lunar farside volcanism in and around the South Pole - Aitken basin. Icarus 2018, 299, 538 - 562.

[22] Zuo, W.; Li, C.; Zhang, Z. Scientific data and their release of Chang'E - 1 and Chang'E - 2. Chin. J. Geochem. 2014, 33, 24 - 44.

[23] Barker, M. K.; Mazarico, E.; Neumann, G. A.; Zuber, M. T.; Haruyama, J.; Smith, D. E. A new lunar digital elevation model from the Lunar Orbiter Laser Altimeter and SELENE Terrain Camera. Icarus 2016, 273, 346 - 355.

[24] Lemelin, M.; Lucey, P. G.; Gaddis, L. R.; Hare, T.; Ohtake, M. Global map products from the Kaguya Multiband Imager at 512 ppd: Minerals, FeO and OMAT. In Proceedings of the 47th Lunar and Planetary Science Conference, The Woodlands, TX, USA, 21 - 25 March 2016; p. 2994.

[25] Michael, G. G.; Neukum, G. Planetary surface dating from crater size - frequency distribution measurements: Partial resurfacing events and statistical age uncertainty. Earth Planet. Sci. Lett. 2010, 294, 223 - 229.

[26] Wilhelms, D. E.; McCauley, J. F.; Trask, N. J. The Geologic History of the Moon; U. S. Geological Survey Professional Paper 1348; U. S. Geological Survey: Denver, CO, USA, 1987.

[27] Nelson, D.; Koeber, S.; Daud, K.; Robinson, M.; Watters, T.; Banks, M.; Williams, N. Mapping Lunar Maria Extents and Lobate Scarps Using LROC Image Products. In Proceedings of the 45th Lunar and Planetary Science Conference, The Woodlands, TX, USA, 17 - 21 March 2014; p. 2861.

[28] Ivanov, M. A.; Hiesinger, H.; van der Bogert, C. H.; Orgel, C.; Pasckert, J. H.; Head, J. W. Geologic History of the Northern Portion of the South Pole - Aitken Basin on the Moon. J. Geophys. Res. Planets 2018, 123, 2585 - 2612.

[29] Kneissl, T.; van Gasselt, S.; Neukum, G. Map - projection - independent crater size -frequency determination in GIS environments - New software tool for ArcGIS. Planet. Space Sci. 2011, 59, 1243 - 1254.

[30] Yue, Z.; Di, K.; Liu, Z.; Michael, G.; Jia, M.; Xin, X.; Liu, B.; Peng, M.; Liu, J. Lunar regolith thickness deduced from concentric craters in the CE - 5 landing area. Icarus 2019, 329, 46 - 54.

[31] McEwen, A. S.; Bierhaus, E. B. The Importance of Secondary Cratering to Age

Constraints on Planetary Surfaces. Annu. Rev. Earth Planet. Sci. 2006, 34, 535 - 567.

[32] Michael, G. G. ; Platz, T. ; Kneissl, T. ; Schmedemann, N. Planetary surface dating from crater size - frequency distribution measurements: Spatial randomness and clustering. Icarus 2012, 218, 169 - 177.

[33] Stuart - Alexander, D. E. Geologic Map of the Central Far Side of the Moon; U. S. Geological Survey IMAP 1047; U. S. Geological Survey: Denver, CO, USA, 1978.

[34] Di, K. ; Zhu, M. - H. ; Yue, Z. ; Lin, Y. ; Wan, W. ; Liu, Z. ; Gou, S. ; Liu, B. ; Peng, M. ; Wang, Y. ; et al. Topographic evolution of Von Kármán crater revealed by the lunar rover Yutu - 2. Geophys. Res. Lett. 2019, 46, 12764 - 12770.

[35] Trask, N. J. Size and spatial distribution of craters estimated from the ranger photographs. In Ranger VIII and IX, Part IIExperimenters' Analyses and Interpretations; Technical Report No. 32 - 800; Jet Propulsion Laboratory, California Institute of Technology: Pasadena, CA, USA, 1966; pp. 252 - 263.

[36] Neukum, G. ; Ivanov, B. A. ; Hartmann, W. K. Cratering records in the inner solar system in relation to the lunar reference system. Space Sci. Rev. 2001, 96, 55 - 86.

[37] Pasckert, J. H. ; Hiesinger, H. ; van der Bogert, C. H. Small - scale lunar farside volcanism. Icarus 2015, 257, 336 - 354.

[38] Fu, X. - H. ; Qiao, L. ; Zhang, J. ; Ling, Z. - C. ; Li, B. The subsurface structure and stratigraphy of the Chang'E - 4 landing site: Orbital evidence from small craters on the Von Kármán crater floor. Res. Astron. Astrophys. 2020, 20, 8.

[39] Gou, S. ; Yue, Z. ; Di, K. ; Bugiolacchi, R. ; Zhu, M. - H. ; Pinet, P. ; Cai, Z. Mare basalt flooding events surrounding Chang'e - 4 landing site as revealed by Zhinyu crater ejecta. Icarus 2021, 360, 114370.

[40] Jia, M. ; Di, K. ; Yue, Z. ; Liu, B. ; Wan, W. ; Niu, S. ; Liu, J. ; Cheng, W. ; Lin, Y. Multi - scale morphologic investigation of craters in the Chang'e - 4 landing area. Icarus 2021, 355, 114164.

Lunar regolith and substructure at Chang'E – 4 landing site in South Pole – Aitken basin

Jinhai Zhang[1,10], Bin Zhou[2,10], Yangting Lin[1], Meng – Hua Zhu[3,10], Hanjie Song[4], Zehua Dong[2], Yunze Gao[2], Kaichang Di[5], Wei Yang[1], Hongyu Lin[6], Jianfeng Yang[7], Enhai Liu[8], Lei Wang[4], Yi Lin[1], Chao Li[1], Zongyu Yue[5], Zhenxing Yao[1] and Ziyuan Ouyang[9]

1. Key Laboratory of Earth and Planetary Physics, Institute of Geology and Geophysics, Chinese Academy of Sciences, Beijing, China
2. Key Laboratory of Electromagnetic Radiation and Detection Technology, Aerospace Information Research Institute, Chinese Academy of Sciences, Beijing, China
3. State Key Laboratory of Lunar and Planetary Sciences, Macau University of Science and Technology, Macau, China
4. National Space Science Center, Chinese Academy of Sciences, Beijing, China
5. State Key Laboratory of Remote Sensing Science, Aerospace Information Research Institute, Chinese Academy of Sciences, Beijing, China
6. Beijing Institute of Space Mechanics and Electricity, Beijing, China
7. Xian Institute of Optics and Precision Mechanics, Chinese Academy of Sciences, Xi'an, China
8. Instituteof Optics and Electronics, Chinese Academy of Sciences, Chengdu, China
9. Institute of Geochemistry, Chinese Academy of Sciences, Guiyang, China
10. These authors contributed equally: Jinhai Zhang, Bin Zhou, Meng – Hua Zhu

The South Pole – Aitken (SPA) basin is the oldest and largest impact structure on the Moon, and it gives particular insight on the lunar interior composition[1–3]. However, the surface of the SPA basin has been substantially modified by consequent impacts and basalt flooding. The exploration of the surficial material and the substructure of the SPA basin is one of the main scientific goals of the Chinese spacecraft Chang'E – 4 that landed in the Von Kármán crater inside the SPA basin[4,5]. Here we report the lunar penetrating radar profiles along the track of the lunar rover Yutu – 2, which show a three – unit substructure at the landing site. The top unit consists of the ~12 – m – thick lunar regolith and ~120m multilayered ejecta that were delivered from several adjacent craters. The middle unit is the mare basalts filling the Von Kármán crater. The lowest unit is another ejecta layer with a thickness of ≥200m, likely from the Leibnitz crater. These discoveries fully support the local stratigraphy and geological explanation presented pre viously[6]. Our results reveal that the surface materials at the Chang'E – 4 landing site are unambiguously dominated by the ejecta from the Finsen crater with a minor contribution from other neighbouring craters. The regolith measured by Yutu – 2 is representative of the initial lunar deep interior materials, rather than the later erupted basalts.

The subsurface structure of the Moon preserves clues on the lunar early history but it is still poorly understood. The Apollo 17 and Kaguya spacecraft conducted radar observation to detect the global portraits of the subsurface[7–9]; however, the orbital radar measurements have relatively low resolution, and the local structures, especially the shallow fine structures, are not well known. Most studies have focused on the thickness of the so – called lunar regolith, a fine – grained layer globally covering the surface of the Moon[10,11]. The lunar regolith thickness has been investigated by in situ drilling[12], microwave remote sensing[13,14], and topography and spectral statistics of small crater ejecta[15–17], with a range from ~2m in the young lunar maria to more than 10m on the old highlands[12]. The Chang'E – 3 mission conducted the first in situ lunar penetrating radar (LPR) detection in the Imbrium basin on the nearside of the Moon in 2013, which showed the detailed substructures of lunar regolith, multilayered ejecta and breccia of underlying bedrock, substantially improving our understanding of the Moon's surface evolution and volcanic eruption history at the landing site[18,19]. However, the subsurface structures of the Moon's farside are still unclear so far, in particular for the floor of the South Pole – Aitken (SPA) basin, the formation of which is thought to have excavated out deep materials down to the lower crust or upper mantle of the Moon.

The Von Kármán crater is filled by thick mare basalts and surrounded by several large craters (for example, the Finsen, Alder, Von Kármán L and L′, and Leibnitz craters; see Extended Data Fig. 1). The LPR onboard lunar rover Yutu – 2 emitted two – frequency electromagnetic waves (that is, 60MHz and 500MHz) towards the lunar subsurface. The low – frequency channel (60MHz, channel 1) has a spatial resolution of ~10m and the maximum detection depth is ~500m; the high – frequency channel (500MHz, channel 2) has a spatial resolution of ~30cm, but with a shallow detection depth of ~50m. The amplitude of the reflections is proportional to the contrast of dielectric constants of the lunar regolith or rocks, which are dominated by the porosities and compositions[20–22]. Strong reflections occur at the boundaries between the regolith (fine – grained soil dominated), the ejecta layer (rock clast dominated), and the brecciated zone of bedrock and bedrocks, and within the ejecta layer and the brecciated zone of bedrock (where the rock clasts are larger than the spatial resolution of the LPR; Methods). In contrast, no or weak reflections occur within the fine – grained regolith (except for a few of scattered rock clasts) or homogeneous basalt layers.

Within the first three lunar days, the Yutu – 2 rover conducted a 163m LPR profile along its track (Fig. 1 and Extended Data Figs. 1 – 3). The LPR profiles (Figs. 2 and 3) were obtained after a series of data – processing procedures, such as decoding, denoising and removing duplicative traces (Methods and Extended Data Figs. 4 - 6). The high - frequency LPR profile was further processed using the depth migration method (Methods), to pursue the true depths and correct shapes of the reflectors[18]. For the low - frequency LPR channel, the reflection depth was determined directly from the travel time of the signal, using the relative dielectric model at the landing site[20-22] (see the details in Methods).

Figure 2 shows the low - frequency LPR profile, in which the subsurface was mainly divided into three (high reflection, low reflection and a relatively high reflection) units according to the amplitude of the reflections. The relatively high - reflection unit at the bottom is ≥200m thick. According to the stratigraphy at the landing site[6,23,24] (Extended Data Fig. 2), this unit is likely from the Leibnitz crater. The numerical simulation indicates that the impact forming a Leibnitz - sized crater (~245km in diameter) can deposit ejecta with a thickness of ~200m at the Chang'E - 4 landing site (Methods), in agreement with the thickness observed by the LPR. Therefore, this relatively high - reflection unit at the bottom is interpreted as the ejecta layer from the Leibnitz crater. The low - reflection unit, with a thickness of ~110m in the middle, shows obviously weak reflections compared with the upper and lower units. These weak reflections are possibly from the mare basalts, because the Von Kármán crater was flooded by one or several eruptions of mare basalts that cover the ejecta of the Leibnitz crater[23,24]. The thickness of the mare basalts in the Von Kármán crater was estimated to be ~120m from the crater shape method[25], which is well consistent with the thickness of this low - reflection unit. Thus, the low - reflection unit is deciphered as mare basalts from multiple eruptions with brief interruptions.

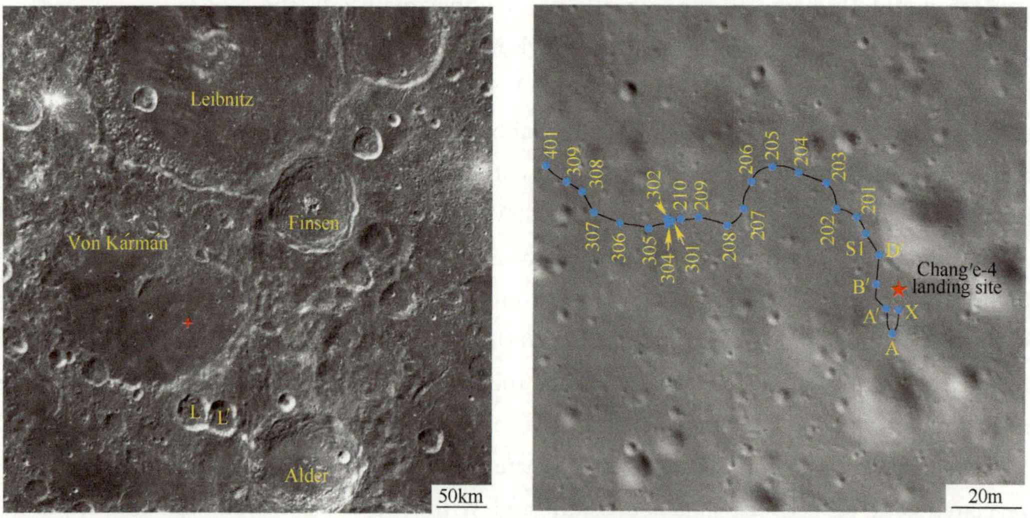

(a)　　　　(b)

Fig. 1 The location of the Chang'E - 4 landing site and the track of the Yutu - 2 rover. a, The location of the Chang'E - 4 landing site at the farside of the Moon (red cross). The main craters in the area are labelled: Finsen, Alder, Leibnitz, and Von Kármán L and L' craters. The background map was obtained by Chang'E - 2 (http://moon. bao. aç. cn). b, The track of the Yutu - 2 rover. The yellow numbers and characters are the unique name of each exploration point. The image was obtained by the descending camera under the Chang'E - 4 lander. More landing site information is shown in Extended Data Fig. 1

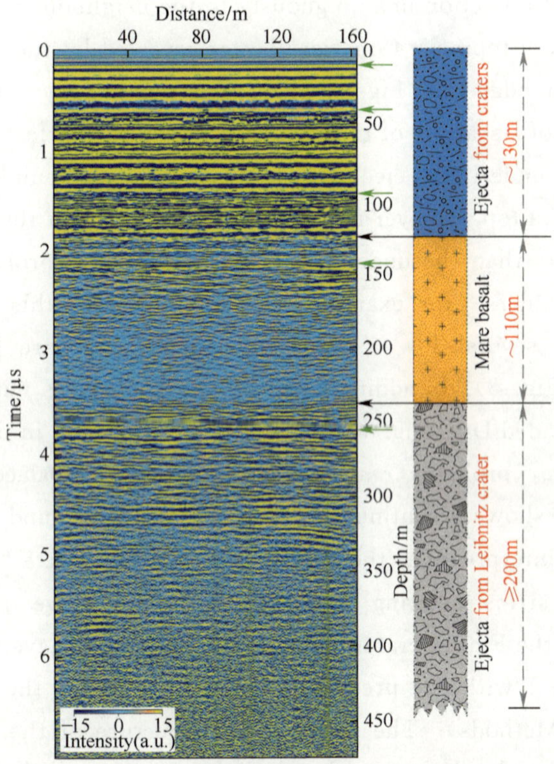

Fig. 2　Low - frequency LPR profile along the track of the Yutu - 2 rover. Left: the LPR profile obtained along the trace from the exploration point X to 401 in Fig. 1b. Right: the geological interpretation and depth estimation of each layer. The strength of the LPR reflections are denoted by the relative amplitude. The aqua colour in the LPR profile represents no or weak reflection, and the strong colour contrasts (for example, light yellow versus dark blue) denote strong reflections. In the right panel, the red number denotes the thickness of each unit and the black solid arrows show the boundaries of the three main units. The depth coordinates are converted by the lunar regolith model (Methods and Extended Data Fig. 7). The green arrows in the left panel represent the possible boundaries between sublayers with different reflection signal features. a. u., arbitrary units

The LPR profiles reveal a high - reflection unit, with a total thickness of ~120m, above the mare basalt unit (Figs. 2 and 3). This high - reflection unit appears complicated and shows obvious variation in the reflection signal features with depth (Fig. 2). Several structures with strong reflectors can be identified in the LPR profiles, likely from the crater ejecta and brecciated bedrocks. The crater ejecta was estimated to be ~70m thick, based on the different altitudes between the landing site and the mare basalt[26]. The heterogeneity of this high - reflection unit is consistent with the regional geological context[23,24] (Extended Data Figs. 1 and 2) that several large impact craters are located around the landing site and their ejecta are deposited at the landing site. However, as the low - frequency LPR profile has a relatively low spatial resolution, we cannot connect each individual reflection unambiguously to the neighbouring crater. Instead, the detailed structures of the topmost 45m were better revealed by the high - frequency LPR profile (Fig. 3 and Extended Data Figs. 8—10). The regolith layer (layer A) shows few reflectors, consisting of a scarcity of rocks (>30cm) on the surface (Extended Data Fig. 3). In contrast, the underlying ejecta layer (layer B) is complicated, composed of multiple sublayers. The top sublayer is continuous and ~5m in thickness, which shows relatively fewer scatters than the underlying ones and hence is probably indicative of the small size of rock boulders (for example, ~30cm). Under this sublayer, there is a ~17m sublayer with several low - reflection laminae that are discontinuous in the horizontal direction (Fig. 3). In addition, several concave areas with low reflections are also recognized (Extended Data Fig. 8b), between 15 and 30m in depth and between 20 and 120 m in distance, probably resulting from the buried palaeocraters. Our high - frequency LPR profile shows a continuous reflection pattern around the depth of ~24m, which was previously interpreted as the bottom of the Finsen ejecta[27]. As the Chang'E - 4 mission landed on the strips radiating from the Finsen crater, we attribute both layers A and B to the ejecta of the Finsen crater. The total thickness of layers A and B is ~34m, which is well consistent with the predicted ejecta thickness of the Finsen crater at the landing site (~30m; Methods). The multilayered structure of the Finsen ejecta can be attributed to asteroid bombardments with a declining intensity with time, which smashed more rock boulders at shallower depths; consequently, deeper layers can preserve larger boulders due to less exposure to meteorite impacts.

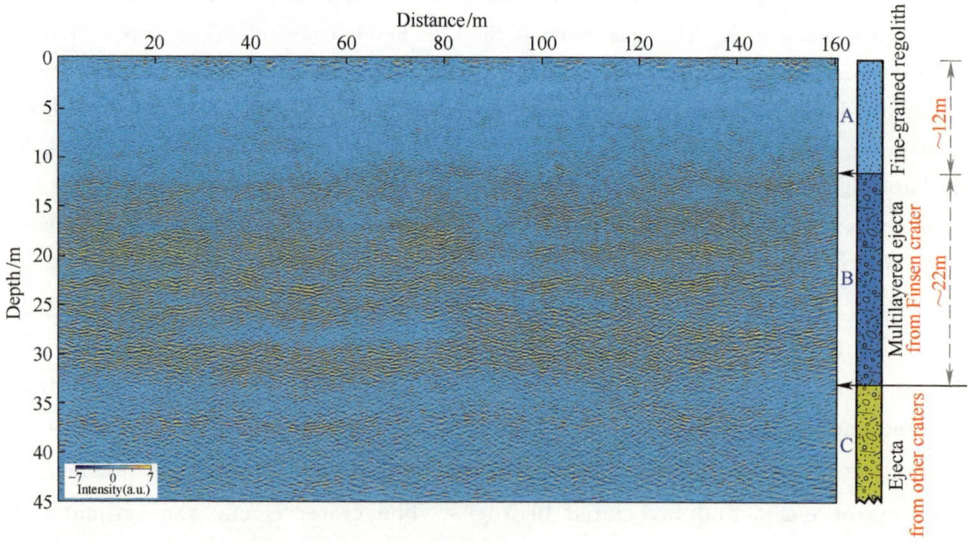

Fig. 3　Imaging results of the high - frequency LPR using the depth migration. The function of the depth migration on recovering the true depth and shapes of the reflectors is shown in Extended Data Figs. 8 and 9. The colours, characters and numbers are the same as in Fig. 2. The fine - grained regolith (layer A) and multilayered ejecta (layer B), with a total thickness of ~34km, are thought to be ejecta of the Finsen crater, where the upper layer (layer A) was subsequently bombarded into fine - grained regolith

The LPR profiles reveal that the heavy asteroid bombardments, thick ejecta deposits from adjacent craters and multi - episodic volcanic eruptions[28] took place within the SPA basin, by which the exposed materials of the lunar deep interior and the basin - forming impact records have been severely modified. These derived subsurface structures are well consistent with expectations of the shallow structure of the lunar upper crust[29], supporting the idea that large - scale cratering plays a substantial role in the formation of the subsurface structure of the Moon[7]. These discoveries shed light on the complex asteroid bombardment and volcanic eruption history of the SPA basin. The subsurface structures are indicative for landing - site selection in future explorations, especially for sample - return missions.

Methods

Brief introduction of the LPR. Te LPR on the Yutu - 2 rover of the Chang'E - 4 mission is the same as that on the Yutu rover of the Chang'E - 3 mission at Imbrium basin in 2013[30-32]. Te LPR can image the interior of the lunar regolith and can detect the geological structure of the subsurface by analysing the electromagnetic waves refected from the subsurface afer stimulating impulsive sources. Te LPR was designed to detect the subsurface structure, based on the radar signals refected at interfaces of objects with different dielectric constants, which are determined mainly by composition (for example, FeO and TiO_2 contents) and porosity. Hence, the LPR could detect the boundaries between the porous lunar regolith, ejecta, basalt lava layers and bedrock. In addition, the lunar regolith and ejecta layers commonly contain large rock boulders (larger than the spatial resolution of the radar wave), and the signals refected at the interfaces of these large boulders can also be detected. In contrast, homogeneous regions within the fine - grained regolith and the interior of basalt layers show weak (or even no) radar reflection.

Two broadband monopoles are chosen as the channel 1 antennas (one transmitting antenna and one receiving antenna), which are respectively mounted on two bottom sides of the lunar rover's top board, spaced about 800mm apart. The antenna is 1,150mm in length and 12mm in diameter. The dominant frequency for the source of channel 1 is 60MHz. A set of bow - tie antennas are chosen as the channel 2 transceiver antenna, which are mounted at the bottom of the lunar rover, about 30cm away from the ground. Each antenna element is 336mm in length and 120mm in width, and the space between the antenna elements is about 160mm. Channel 2 works with a dominant frequency of 500MHz, which penetrates shallower than channel 1 but provides a better spatial resolution. The depth resolutions of channels 1 and 2 are < 10m and < 30cm, respectively, based on the ground experiments[31]. Extended Data Fig. 1 shows the landing site of Chang'E - 4 and the track of lunar rover Yutu - 2.

Processing of the LPR data. An LPR profile with a total length of ~163m was carried out in the first three lunar days by the Yutu - 2 rover (Extended Data Fig. 1d). Extended Data Fig. 4a shows 1,973 traces obtained by channel 1. It is difficult to identify the deep reflections as the original data are too noisy. We reduced the random noise by applying the sixth - order Butterworth band - pass filter between 10 and 80MHz on each trace and a two - dimensional median filter[18]. The results are shown in Extended Data Fig. 4b, where the deepest reflections that can be identified from the profile are at ~6.5μs. According to Extended Data Fig. 5, we can determine that the time delay before the first arrival for channel 1 is about 57.5ns, which has been removed in Extended Data Fig. 4b. The singular value decomposition[33] of the low - frequency LPR profile (Extended Data Fig. 6) further separates the reflections and diffractions. We divide the singular values into three segments according to their energy characteristics, 1 - 10, 11 - 200 and 201 — 1,973 (Supplementary Fig. 5), and the results correspond to the low —, bandand high - pass eigenimages (Extended Data Fig. 6b - d), which mainly present the reflections, diffractions and random noises, respectively.

It is noticed that the profile processed with the low - pass eigenimage filtering is consistent with Fig. 2, where we can again identify the two boundaries at ~130 and ~240m, respectively. The LPR signal patterns change across both boundaries. The reflection signals gradually vanish at ~440m (Extended Data Fig. 6b). The diffractions mainly appear in the band - pass eigenimage (Extended Data Fig. 6c) but with some energy weakening at the strong reflection zones shown in Extended Data Fig. 6b, due to energy leakage of high - frequency components during the singular value decomposition[33]; thus, we could not identify more layers from the band - pass eigenimage according to the energy pattern of diffractions. Nevertheless, the results of eigenimage filtering still confirm that the low - frequency LPR profile consists of reflections and diffractions from the underground structures, without apparent systematic artefacts over the lunar surface. Furthermore, the results of eigenimage filtering provide us with more information to confirm the rationality of the three main units divided by the two notable geological boundaries at ~130 and ~240m shown in Fig. 2.

Channel 2 has two receiving antennas (2A and 2B) and channel 2B has recorded 4, 416 high - quality traces. For detailed information on processing the data (including removing the direct currents, noise attenuation and weak signal enhancement), please refer to Zhang et al.[18]. We applied the sixth - order Butterworth filter with a band - pass between 100 and 800MHz.

The exact coordinates of each trace are not available; thus, we assume that the rover is moving at a constant speed between two adjacent stops, and, therefore, we take a uniform spatial interval of 0.0365m between two adjacent traces. The error is fairly small as the coordinates at some control points are well estimated (see blue dots in Extended Data Fig. 1d). The route of the Yutu - 2 rover is generally horizontal, and the maximum vertical deviation from the horizontal baseline is no more than 1m; thus, we disregard the topography of the patrol area. The effect associated with the topography can be reconsidered in the future after the coordinates of all traces are comprehensively estimated from other methods[34].

Verifying the function of depth migration. Seismic exploration has developed many advanced techniques to detect terrestrial reservoirs buried under complex deep substructures, such as signal processing and seismic migration that are essential for extracting weak signals and high - accuracy imaging of substructures[35,36]. These methods have been applied to process the LPR data obtained by Chang'E - 3[18].

In general, each underground scattering object would be mapped into a hyperbola in the profile due to the wave propagation effects; thus, it is difficult to identify substructures directly from the ground - based profile. Therefore, the depth migration is crucial for imaging complex structures using the reflected seismic waves[35-38] and LPR[18]. It can reduce the propagation effects and recover both the shapes and the true depth of scattering objects. The migrated profile would be much closer to the actual position, compared with the original profile, which is important for interpreting the structures and the process of the lunar subsurface.

Zhang et al.[18] examine the detectability of several typical structures: syncline, anticline, dipping structures, circular and rectangular blocks, buried in a linear increasing gradient model of relative dielectric constant. They show that it is difficult to identify complex structures directly from the synthetic profile, as a lot of crossed hyperbolas can be observed due to wave propagation effects; in contrast, most gently dipping structures are well recovered by depth migration using the one - way wave equation method[37-39].

In this study, we can see that the depth migration, with the one - dimensional model shown in Extended Data Fig. 7, can move the hyperbolas back towards the true positions of the scattering points or reflecting interfaces, as shown in Extended Data Fig. 8. Extended Data Fig. 9 further shows some local details in the rectangles in Extended Data Fig. 8. Obviously, the migration results show much clearer scattering objects and more continuous reflectors than the original LPR profile, and most scattering objects are well imaged. Therefore, the depth migration is necessary for high - resolution imaging of the complex structures and objects buried in the lunar regolith.

Extended Data Fig. 7 shows a reasonable background model (or macro model) at the landing site, based on some prior information from lunar geological studies[12,24] and the sublayer depth estimated by the depth migration (Extended Data Fig. 8). Although a laterally homogeneous macro model was used for the depth migration, the imaging results still show many clear structures, especially for the continuous structures and scattering objects. A finer model including lateral variation would be helpful for better accuracy in relocating deep reflectors.

As pointed out by Zhang et al.[18], the narrow receiver array of the LPR (two receivers are 16 and 32cm away from the source, respectively) leads to the difficulty in receiving the reflected waves from nearly vertical structures under the rover; thus, we would not image steep dips, as we have too weak or even no signal reflected from them. Therefore, we can faithfully identify only the upper and bottom surfaces of the objects from the migration results, as we could not observe their vertical boundaries. Nevertheless, the migration results still provide us with solid support in interpreting the lunar subsurface, as the artefacts are greatly reduced by focusing the scattering waves and by relocating the reflectors much closer to their true positions.

The relative dielectric model at the landing site. For the depth migration, we need a velocity model, at least a one-dimensional model (that is, the gradient model). The velocity of the electromagnetic waves propagating in homogeneous media can be obtained by

$$v = \frac{c}{\sqrt{\varepsilon_r}}$$

where c is the speed of light in vacuum and ε_r is the relative dielectric permittivity. The relative dielectric permittivity of the lunar regolith is nearly independent of frequency above 1MHz and temperature[12], and it is dominantly controlled by the bulk density approximately by

$$\varepsilon_r' = 1.92^{\rho}$$

where ρ is the bulk density in g cm^{-3} which varies with depth z (in centimetres) in a hyperbolic form[12]

$$\rho = 1.92\frac{z + 12.2}{z + 18}$$

From the depth migration results (Fig. 3 and Extended Data Fig. 8b) and the low-frequency profile (Fig. 2 and Extended Data Fig. 6), we can identify several sublayers; thus, we can build up the bulk relative dielectric model of the lunar regolith and the subsurface at the Chang'E-4 landing site as follows

$$\varepsilon_r = \begin{cases} 1.92^{\rho}, & 0 < z \leqslant h \\ 4.5 \text{ for ejecta} & h < z \\ 6.0 \text{ for basalt} & h < z \end{cases}$$

where $h = 12$m is the depth of the lunar regolith layer (Fig. 3).

We determined the model by guessing an initial model according to the geological interpretation (Figs. 2 and 3) where the ejecta or brecciated top of mare basalt have a fixed relative dielectric permittivity of 4.5 and the mare basalt has a relative dielectric permittivity of 6 (refs.[12,21,22]); then, we converted the time coordinate into the depth coordinate using a fine depth interval (0.05m) according to $v = c/\sqrt{\varepsilon_r}$; finally, we updated the model to make sure that the depth given in the model is consistent with the depth of each geological layer in the LPR profile (Figs. 2 and 3). Our final model after several iterations is shown in Extended Data Fig. 7.

Distinguishing different layers in the LPR profiles. We distinguish different layers according two assumptions: (1) the amplitude of reflections is proportional to the contrast of dielectric parameters above and below the reflector; (2) a continuous region of reflections with a similar pattern should be interpreted as a geologic layer formed under similar conditions. We can observe two notable geological boundaries at the depths of ~130 and ~240m (Fig. 2), which have the strongest pattern mutations in the low - frequency LPR profile; in addition, we can distinguish more layers with apparent pattern mutations in the depth migration results of the high - frequency LPR profile (Fig. 3). Lv et al. have performed comparative analyses of LPR data using numerical simulations and provided some criteria for distinguishing different layers[40], which provide more experiments and details on distinguishing different layers.

The consistency of the low - frequency channel and the high - frequency channel. Extended Data Fig. 10 shows the comparison between the low - frequency channel and the high - frequency channel. Obviously, the amplitude clipped parts in the low - frequency channel (mainly within 0.31μs, as shown in Extended Data Fig. 5) can be well constrained by the high - frequency channel within 0.62μs.

The uncertainty of geological interpretation. The depth coordinates are converted by the relative dielectric model (Extended Data Fig. 7), which may vary if we use a completely different model according to some new constraints. The model here is constructed after the Apollo samples[12], which is the best model we can provide. The uncertainty of the depth estimation is small at shallow depths, for example, being 11cm at 70cm, and it increases with the depth[18].

Numerical simulation for the impact cratering process and ejecta thickness distribution. The ejecta thickness can be described as a power function of the range x (measured from the crater centre), $T=T_{tr}(x/R)^{-b}$, where T is the estimated thickness at distance x; T_{tr} is the estimated ejecta thickness at the rim of the transient crater; R is the transient crater radius; and b is the exponent parameter within a range from 2.3 to 3.3. In this equation, both T_{tr} and b are independent unknowns. Although these two parameters can be estimated according to the experiences derived from the laboratory experiments, a small variation of the selected exponent b can substantially affect the estimation of ejecta thickness at a location far away from the crater centre. The numerical simulation allows systematic parametric studies on the effect of target properties such as porosity and strength on ejecta distribution, and therefore could provide a more precise estimate of the ejecta thickness.

In this work, we use the multi - material, multi - rheology two - dimensional iSALE (Dellen version) shock physics code[41,42] to model the impact cratering process and ejecta distribution of 200 - km - diameter Leibnitz crater and 72 - km - diameter Finsen crater (Supplementary Fig. 1). The iSALE is based on the SALE (Simplified Arbitrary Lagrangian Eulerian) hydrocode[43] and has been used to simulate both small - scale laboratory experiments[44] and large - scale lunar basin formation (for example, refs.[45-49]).

In our simulations, we use the analytic equation of state for dunite[50,51] to represent the composition of the impactor and the Tillotson equation of state for gabbroic anorthosite to represent the composition of the target. However, for the 200 - km - diameter Leibnitz crater, its formation may have penetrated through the crust layer and excavated the deep materials. Therefore, we assume that the target of the Leibnitz crater is a crust with a thickness of ~20km overlaying the dunite mantle material. This is reasonable because the average crustal thickness within the SPA basin is ~20km (ref.[52]). Material strength and damage is accounted for based on the model of Collins et al.[41] and Ivanov et al.[53]. A temporary weakening model, acoustic fluidization[54,55], is also considered to facilitate collapse in the impact cratering process. The detailed parameters for the impact crater simulations are listed in Supplementary Table 1.

To generate a crater comparable in size to the Leibnitz crater, the impactor diameter was varied between 5 and 24km. The computational domain covers an area of 300km in the lateral and vertical directions with a cell size of 1km × 1km in the high - resolution zone. For the Finsen crater, we varied the impactor diameter from 5 to 24km. The computational domain covers an area of 100km in the lateral and vertical directions with a cell size of 0.06km × 0.06km in the high - resolution zone. For both impact cratering simulations, the resolution of 20 cells per projectile radius is suitable for the calculation of ejecta properties[48,56]. We varied the impact velocity between 10 and 20kms^{-1} for both simulations to cover a suitable velocity for asteroid hitting the Moon[57,58]. These impact velocities could also represent moderately oblique incidence angles because the vertical component of the impact velocity mainly controls the diameter of the craters. In all simulations, we assigned a lunar surface gravitational acceleration of 1.62ms^{-2}. Owing to the axisymmetric nature of the two - dimensional model, all simulations considered only vertical impacts. Though vertical impacts are highly unlikely, they provide a reasonable proxy for the most common angle of impact (45°) on planetary surfaces. In this work, we run 40 impact models in total (20 for the Leibnitz crater and 20 for the Finsen crater). All simulations were stopped at 2h (modelling time) after impact.

We track the ejecta trajectories and physical states by using Lagrangian tracers in iSALE, which were initially placed in each computational cell and represent the matter originally in that cell throughout the simulation. We consider the tracer as ejecta that have ballistic trajectories, cross the pre - impact target surface and deposit beyond the transient crater rim. We record its launch angle and launch velocity by interpolating the nodal velocities when its trajectories interact with the pre - impact surface. Using these parameters, the parabolic trajectory of each tracer and final position were calculated[48,56]. Note, ejecta with velocities below 2.4kms^{-1} (Moon's escape velocity) were thought to move along parabolic trajectories and could eventually fall back to the lunar surface. The surrounding surface of the crater was subdivided into discrete rings. The ejecta thickness was calculated from the number of tracers that landed at a given distance from the point of impact. This approach for the calculation of ejecta thickness has been validated against laboratory impact experiments into sand[44]. The entrainment of local material into the ejecta blanket upon landing is not taken into account. At larger distance, this process may be non-negligible[56,59]. It is noted that we do not simulate the formation of the ejecta plume in our models; vapourized material (material with a density $< 300 kgm^{-3}$) is removed from the computational mesh to expedite simulation time. This assumes that the expanding vapour plume and small ejected fragments do not have any effect on crater formation and that the drag of the vapour on the ejected particles is negligible, which is a reasonable assumption for ejecta forming layers several hundreds of metres thick. Artemieva et al.[60] showed that only a relatively thin layer of fine-grained ejecta and dust particles interacting with the vapour plume are finally deposited on top of the ballistic ejecta.

Supplementary Fig. 2 shows the snapshots of the impact cratering process for an impactor with a diameter of 6.2km and an impact velocity of $15 kms^{-1}$. The impact produces a complex crater with a diameter of ~78 km, analogous to the size of the Finsen crater. The produced centre peak is ~2km below the pre-impact surface, consistent with that of the Finsen crater. We calculate the average profile of topography for the Finsen crater (Supplementary Fig. 2d) from the Lunar Orbiter Laser Altimeter (LOLA) observations in a crater-centred projection over the white region of Supplementary Fig. 1. The profiles of elevation derived from the impact simulation and observations match well, suggesting that our impact model could reasonably reproduce the formation of the Finsen crater. Supplementary Fig. 3 shows the ejecta thickness as a function of radial distance from the rim of the Finsen crater to 200km from the crater centre. The ejecta deposit near the crater rim is ~1.9km and decreases with increasing radial distance from the crater centre. The decrease of ejecta thickness with distance can be described by a power law with an exponent of −3.2, which is consistent with laboratory experiments[44,61] and ejecta thickness observations for the lunar crater[29]. The Chang'E-4 landing site is ~138km from the centre of the Finsen crater. According to our simulation, ~32-m-thick ejecta from the Finsen crater deposited at the Chang'E-4 landing site, which is similar to the thickness derived from the LPR (Fig. 3).

Supplementary Fig. 4a shows the structure of the simulated crater by an impactor with a diameter of 14km and an impact velocity of $16 kms^{-1}$. The structure is ~230km in diameter, which is similar to the size of the Leibnitz crater. As the Leibnitz crater was filled by mare basalts and its morphologies have been destroyed by subsequent impacts (Supplementary Fig. 1), we cannot calculate its topographic profiles from the LOLA observations as we did for the Finsen crater. Here we just assume that the impact model could reasonably reproduce the similar-size structure of the Leibnitz crater and estimate its ejecta thickness along the radial distance from the crater centre. Supplementary Fig. 4b shows the ejecta thickness variation of the Leibnitz-sized crater along the radial distance beyond the crater. The Chang'E-4 landing site is ~230km from the centre of the

Leibnitz crater. According to our estimation, ejecta with a thickness of ~200m from the Leibnitz crater deposited at the Chang'E－4 landing site.

References

[1] Ohtake, M. et al. Geologic structure generated by large－impact basin formation observed at the South Pole－Aitken basin on the Moon. Geophys. Res. Lett. 41, 2738－2745 (2014).

[2] Melosh, H. J. et al. South Pole－Aitken basin ejecta reveal the Moon's upper mantle. Geology 45, 1063－1066 (2017).

[3] Moriarty, D. P. III & Pieters, C. M. Te character of South Pole－Aitken basin: patterns of surface and subsurface composition. J. Geophys. Res. Planets 123, 729－747 (2018).

[4] Li, C. et al. Chang'e－4 initial spectroscopic identification of lunar far－side mantle－derived material. Nature 569, 378－382 (2019).

[5] Wu, W. et al. Lunar farside to be explored by Chang'e－4. Nat. Geosci. 12, 222－223 (2019).

[6] Stuart－Alexander, D. E. Geologic Map of the Central Far Side of the Moon Investigations Map Series I－1047 (USGS, 1978).

[7] Peeples, W. J., Sill, W. R. & May, T. W. Orbital radar evidence for lunar subsurface layering in Maria Serenitatis and Crisium. J. Geophys. Res. Solid Earth 83, 3459－3468 (1978).

[8] Sharpton, V. L. & Head, J. W. Stratigraphy and structural evolution of southern Mare Serenitatis: a reinterpretation based on Apollo Lunar Sounder Experiment data. J. Geophys. Res. Solid Earth 87, 10983－10998 (1982).

[9] Ono, T. et al. Lunar radar sounder observations of subsurface layers under the nearside maria of the Moon. Science 323, 909－912 (2009).

[10] Shkuratov, Y. G., Kaidash, V. G., Kreslavsky, M. A. & Opanasenko, N. V. Absolute calibration of the Clementine UVVIS data: comparison with ground－based observation of the moon. Sol. Syst. Res. 35, 29－34 (2001).

[11] Petro, N. E. & Pieters, C. M. Te lunar－wide effects of basin ejecta distribution on the early megaregolith. Meteorit. Planet. Sci. 43, 1517－1529 (2008).

[12] Heiken, G. H., Vaniman, D. T. & French, B. M. Lunar Sourcebook: A User's Guide to the Moon (Cambridge Univ. Press, 1991).

[13] Montopoli, M., Tognolatti, P., Marzano, F. S., Pierdicca, M. & Perrotta, G. Remote sensing of the Moon sub－surface from a spaceborne microwave radiometer aboard the European Student Moon Orbiter (ESMO). In Proc. IEEE International Geoscience and Remote Sensing Symposium (IGARSS) 4451－4454 (2007).

[14] Fa, W. Z. & Jin, Y. Q. A primary analysis of microwave brightness temperature

of lunar surface from Chang'e－1 multi－channel radiometer observation and inversion of regolith layer thickness. Icarus 207，605－615 (2010).

[15] Hawke，B. R. & Head，J. W. Impact melt on lunar crater rims. In Impact and Explosion Cratering (eds Roddy，D. J. et al.) 815－841 (Pergamon Press，1977).

[16] Wilcox，B. B.，Lucey，P. G. & Gillis，J. J. Mapping iron in the lunar mare：an improved approach. J. Geophys. Res. Planets 110，E11001 (2005).

[17] Neish，C. D. et al. Spectral properties of Titan's impact craters imply chemical weathering of its surface. Geophys. Res. Lett. 42，3746－3754 (2015).

[18] Zhang，J. H. et al. Volcanic history of the Imbrium basin：a close－up view from the lunar rover Yutu. Proc. Natl Acad. Sci. USA 112，5342－5347 (2015).

[19] Xiao，L. et al. A young multilayered terrane of the northern Mare Imbrium revealed by Chang'e－3 mission. Science 347，1226－1229 (2015).

[20] Choy，T. C. Effective Medium Theory：Principles and Applications (Oxford Univ. Press，2015)

[21] Huang，Q. & Wieczorek，M. A. Density and porosity of the lunar crust from gravity and topography. J. Geophys. Res. Planets 117，E05003 (2012).

[22] Olhoef，G. R. & Strangway，D. W. Dielectical properties of the first 100 meters of the Moon. Earth Planet. Sci. Lett. 24，394－404 (1975).

[23] Ivanov，M. A. et al. Geologic history of the northern portion of the South Pole－Aitken basin on the Moon. J. Geophys. Res. Planets123，2585－2612 (2018).

[24] Huang，J. et al. Geological characteristics of Von Kármán crater，northwestern South Pole－Aitken basin：Chang'e－4 landing site region. J. Geophys. Res. Planets 123，1684－1700 (2018).

[25] Du，J. et al. Tickness of lunare mare basalts：new results based on modeling the degradation of partially buried craters. J. Geophys. Res. Planets 124，2430－2459 (2019).

[26] Di，K. et al. Topographic evolution of Von Kármán crater revealed by the lunar rover Yutu－2. Geophys. Res. Lett. 46，12764－12770 (2019).

[27] Li，C. et al. Te Moon's farside shallow subsurface structure unveiled by Chang'e－4 lunar penetrating radar. Sci. Adv. **6**，eaay6898 (2020).

[28] Moriarty，D. P. III & Pieters，C. M. Te nature and origin of Mafic Mound in the South Pole－Aitken basin. Geophys. Res. Lett. 42，7907－7915 (2015).

[29] McGetchin，T. R.，Settle，M. & Head，J. W. Radial thickness variation in impact crater ejecta：implications for lunar basin deposits. Earth Planet. Sci. Lett. 20，226－236 (1973).

[30] Fang，G. Y. et al. Lunar penetrating radar onboard the Chang'e－3 mission. Res. Astron. Astrophys. 14，1607－1622 (2014).

[31] Su，Y. et al. Data processing and initial results of Chang'e－3 lunar penetrating radar. Res. Astron. Astrophys. 14，1623－1632 (2014).

[32] Shen, S. et al. Design of dual - channel nonuniform digital sampling receiver for lunar penetrating radar in Chang'e - 3 rover. IEEE Aerosp. Electron. Syst. Mag. 31, 22 - 31 (2016).

[33] Cagnoli, B. & Ulrych, T. J. Singular value decomposition and wavy reflections in ground - penetrating radar images of base surge deposits. J. Appl. Geophys. 48, 175 - 182 (2001).

[34] Di, K. C. et al. Chang'e - 4 lander localization based on multi - source data. J. Remote Sens. 23, 177 - 180 (2019).

[35] Claerbout, J. F. Imaging the Earth's Interior (Blackwell Scientific, 1985).

[36] Yilmaz, ö. Seismic Data Processing (Society of Exploration Geophysicists, 1987).

[37] Etgen, J., Gray, S. H. & Zhang, Y. An overview of depth imaging in exploration geophysics. Geophysics 74, WCA5 - WCA17 (2009).

[38] Stoffa, P. L. et al. Split - step Fourier migration. Geophysics 55, 410 - 421 (1990).

[39] Wu, R. S. Wide - angle elastic wave one - way propagation in heterogeneous media and an elastic wave complex - screen method. J. Geophys. Res. 99, 751 - 766 (1994).

[40] Lv, W. et al. Comparative analysis of reflection characteristics of lunar penetrating radar data using numerical simulations. Icarus 350, 113896 (2020).

[41] Collins, G. S., Melosh, H. J. & Ivanov, B. A. Modeling damage and deformation in impact simulations. Meteorit. Planet. Sci. 39, 217 - 231 (2004).

[42] Wünnemann, K., Collins, G. S. & Melosh, H. J. A strain - based porosity model for use in hydrocode simulations of impacts and implications for transient - crater growth in porous targets. Icarus 180, 514 - 527 (2006).

[43] Amsden, A. A., Ruppel, H. M. & Hirt, C. W. SALE: A Simplified ALE Computer Program for Fluid Flow at all Speeds LANL Report LA - 8095 (LANL, 1980).

[44] Wünnemann, K., Zhu, M. H. & Stöffler, D. Impacts into quartz sand: crater formation, shock metamorphism, and ejecta distribution in laboratory experiments and numerical models. Meteorit. Planet. Sci. 51, 1762 - 1794 (2016).

[45] Melosh, H. J. et al. The origin of lunar mascon basins. Science 340, 1552 - 1555 (2013).

[46] Potter, R. W. K. et al. Constraining the size of the South Pole - Aitken basin impact. Icarus 220, 730 - 743 (2012).

[47] Potter, R. W. K. et al. Numerical modeling of the formation and structure of the Orientale impact basin. J. Geophys. Res. 118, 963 - 979 (2013).

[48] Zhu, M. - H., Wünnemann, K. & Potter, R. W. K. Numerical modeling of the ejecta distribution and formation of the Orientale basin on the Moon. J. Geophys. Res. 120, 118 - 2, 134 (2015).

[49] Zhu, M. H. et al. Are the Moon's nearside - farside asymmetries the result of a giant impact? J. Geophys. Res. 124, 2117 - 2140 (2019).

[50] Ahrens, T. J. & O'Keefe, J. D. Shock melting and vaporization of lunar rocks and minerals. In Impact and Explosion Cratering (eds Roddy, D. J. et al.) 639 - 656 (Pergamon Press, 1977).

[51] Benz, W., Cameron, A. G. W. & Melosh, H. J. The origin of the Moon and the single - impact hypothesis III. Icarus 81, 113 - 131 (1989).

[52] Wieczorek, M. A. et al. The crust of the Moon as seen by GRAIL. Science 339, 671 - 675 (2013).

[53] Ivanov, B. A., Melosh, H. J. & Pierazzo E. in Large Meteorite Impacts and Planetary Evolution IV (eds Gibson, R. L. & Reimold, W. U.) 29 - 49 (Geological Society of America, 2010).

[54] Melosh, H. J. Acoustic fluidization: a new geologic process? J. Geophys. Res. 84, 7513 - 7520 (1979).

[55] Melosh, H. J. & Ivanov, B. A. Impact crater collapse. Annu. Rev. Earth Planet. Sci. 27, 385 - 415 (1999).

[56] Zhu, M. H., Wünnemann, K. & Artemieva, N. Effects of Moon's thermal state on the impact basin ejecta distribution. Geophys. Res. Lett. 44, 11292 - 11300 (2017).

[57] Bottke, W. F. et al. An Archaean heavy bombardment from a destabilized extension of the asteroid belt. Nature 485, 78 - 81 (2012).

[58] Le Feuvre, M. & Wieczorek, M. A. Nonuniform cratering of the Moon and a revised crater chronology of the inner Solar System. Icarus 214, 1 - 20 (2011).

[59] Hörz, F., Ostertag, R. & Rainey, D. A. Bunte Breccia of the Ries: continuous deposits of large impact crater. Rev. Geophys. Space Phys. 21, 1667 - 1725 (1983).

[60] Artemieva, N. A. et al. Ries crater and suevite revisited - observation and modeling. Part II: modeling. Meteorit. Planet. Sci. 48, 590 - 627 (2013).

[61] Stöffler, D. et al. Experimental hypervelocity impact into quartz sand: distribution and shock metamorphism of ejecta. J. Geophys. Res. 80, 4062 - 4077 (1975).

[62] Fortezzo, C. M. & Hare, T. M. Completed digital renovation of the 1 : 5,000,000 lunar geologic map series. Lunar Planet. Sci. Conf. 44, 2623 (2013).

Acknowledgements

The Chang'E - 4 mission was carried out by the Chinese Lunar Exploration Program, and the data were provided by the China National Space Administration. We thank D. Moriarty for helpful suggestions, which greatly improved the clarity and readability of the manuscript. This work was supported by Key Research Program of Frontier Sciences, Chinese Academy of Sciences (QYZDJ - SSW - DQC001) and the National Nature Science

Foundation of China (41490631 and 41941002). M. - H. Z. was supported by the Science and Technology Development Fund, Macau (0079/2018/A2) and the pre - research project on Civil Aerospace Technologies No. D020202 funded by China National Space Administration (CNSA). M. - H. Z. acknowledges the iSALE developers. H. S. was supported by the Beijing Municipal Science and Technology Commission (Z181100002918003) and the pre - research project on Civil Aerospace Technologies No. D020201 funded by China National Space Administration (CNSA). We thank the Supercomputing Laboratory of IGGCAS for providing computing resources.

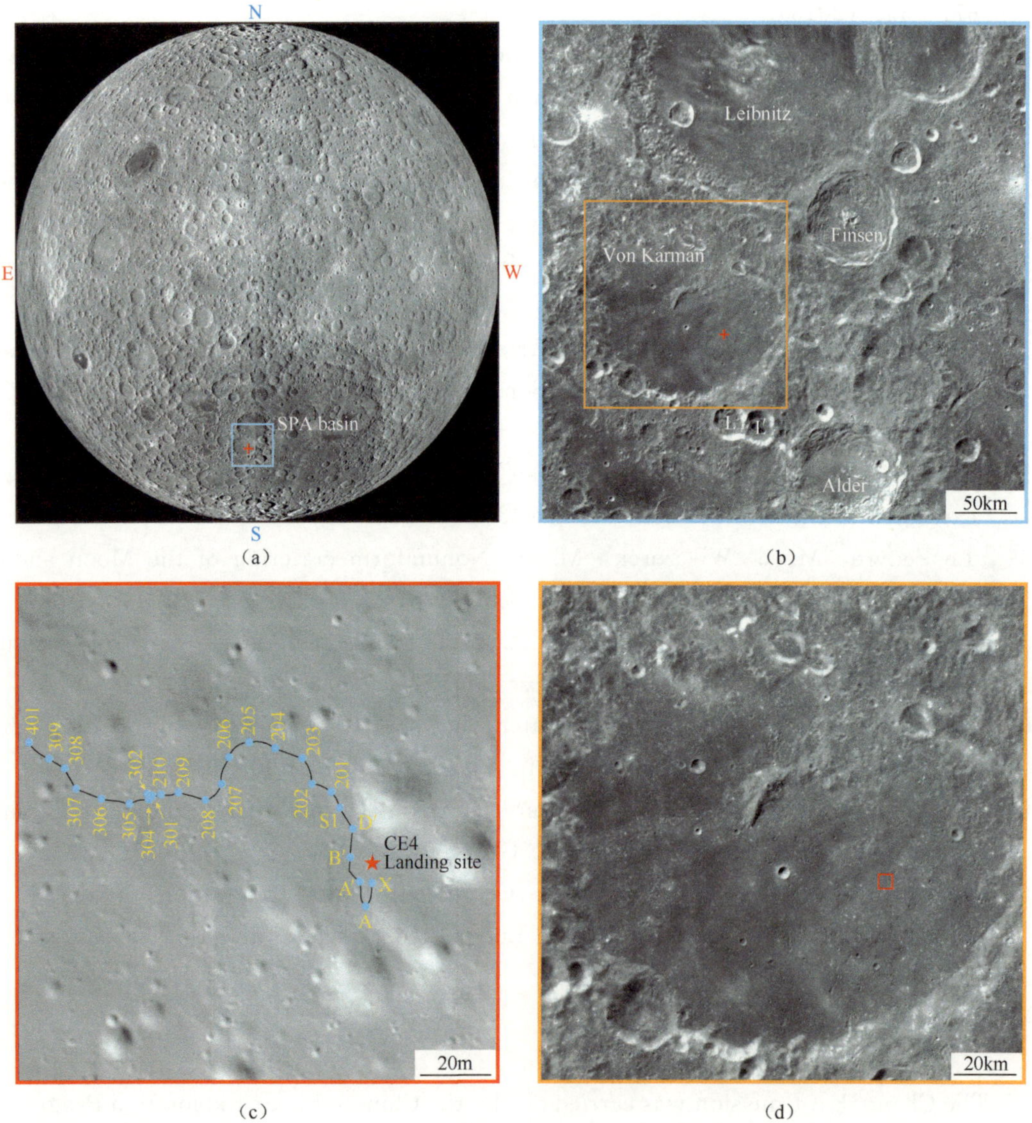

Extended Data Fig. 1　The location of Chang'E - 4 landing site and the track of Yutu - 2 rover. a, The farside of the Moon; b, Local details in the rectangle in a; c, Local details in the rectangle in b; d, Local details in the rectangle in c. a is obtained by the Lunar Resonant Orbiter (http: //lroc. sese. asu. edu/posts/298); b and c are obtained by Chang'E - 2 (http: //moon. bao. ac. cn); d is obtained by the descending camera under the Chang'E - 4 lander

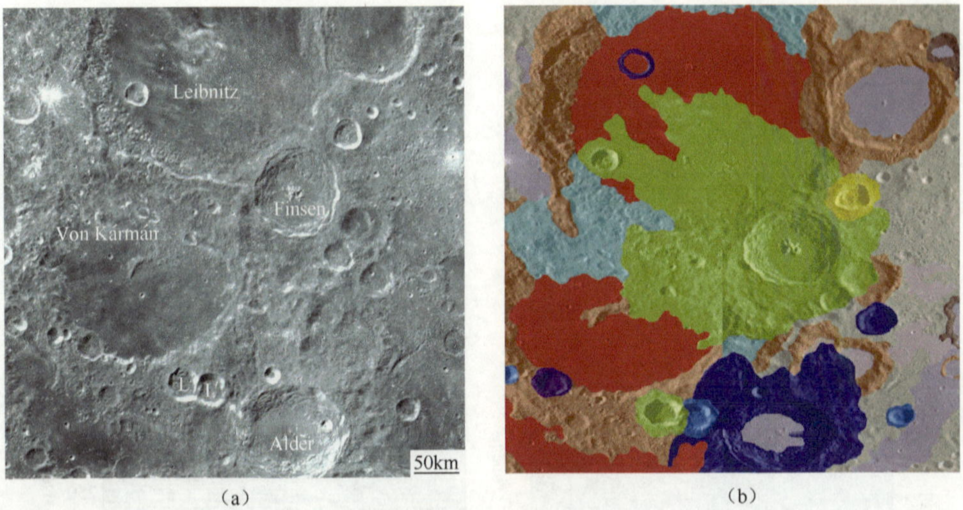

Extended Data Fig. 2　High resolution image of the Chang'E－4 landing site and the regional geology. a, The Chang'E－4 landing site; b, The regional geology that modified from https://quickmap. lroc. asu. edu[62]

Extended Data Fig. 3　The rareness of rocks on the lunar surface at the Chang'E－4 landing site. The landscape images were taken with the panorama camera on the rover Yutu－2 (a) and by the terrain camera on the Chang'E－4 lander (b). The distance between two rows of wheels is ～80cm

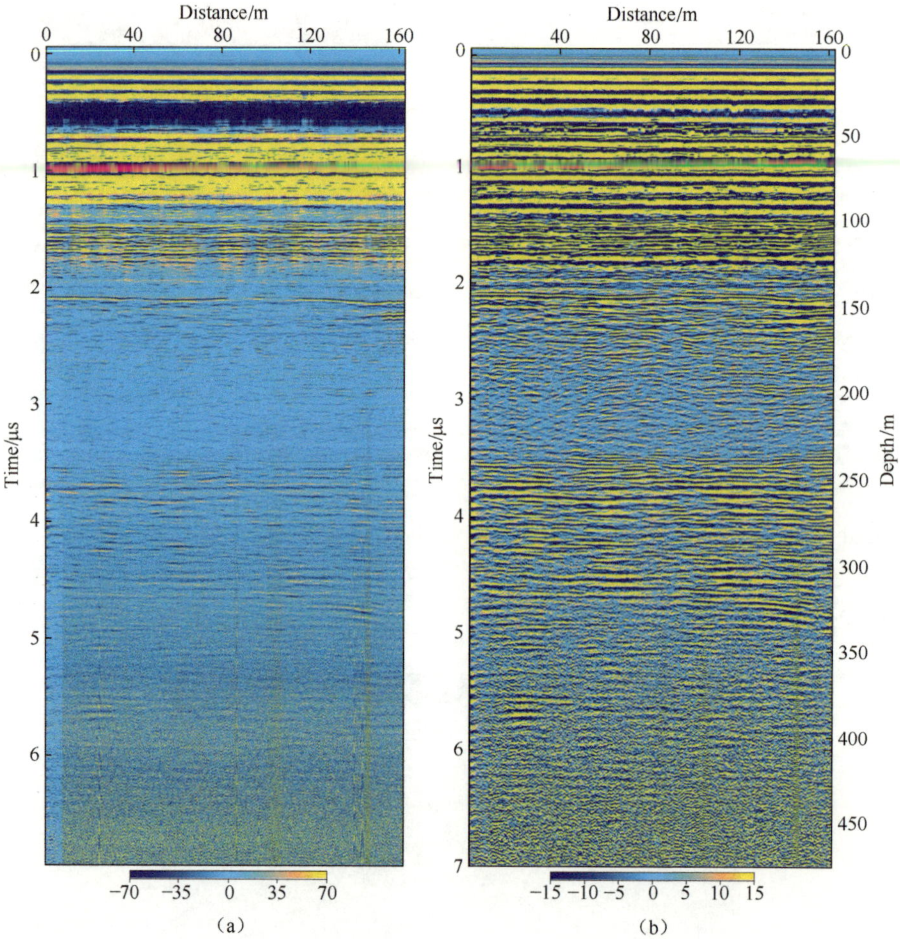

Extended Data Fig. 4　The LPR data of Channel - 1. **a**, The raw data profile; b, The profile after using a bandpass filter between 10 and 80MHz for each trace and a 2D median filter using parameters of (3, 10). The depth is converted from the travel time using a relative dielectric model shown in Extended Data Fig. 7. The processed data shown in **b** is exactly the same as that shown in Fig. 2, which is shown here to examine the processing results compared with the original data shown in **a**

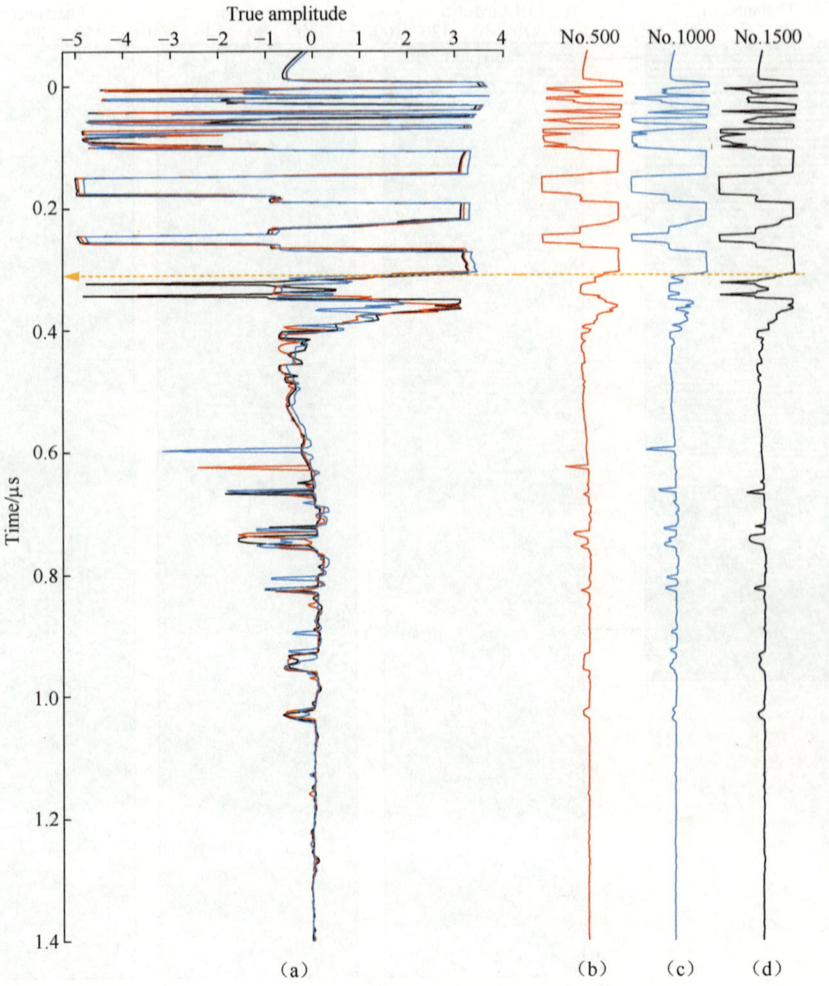

Extended Data Fig. 5　Local zoom in of original waveforms of low - frequency LPR Channel from 0 to 1. 4μs. a, The stack of three waveforms at the trace no. 500, 1000, and 1500. b, c, d, The independent waveforms at trace no. 500, 1000, and 1500, respectively. Note that only the waveforms within 0. 31μs (indicated by the orange arrow) have been apparently clipped and the rest are reliable

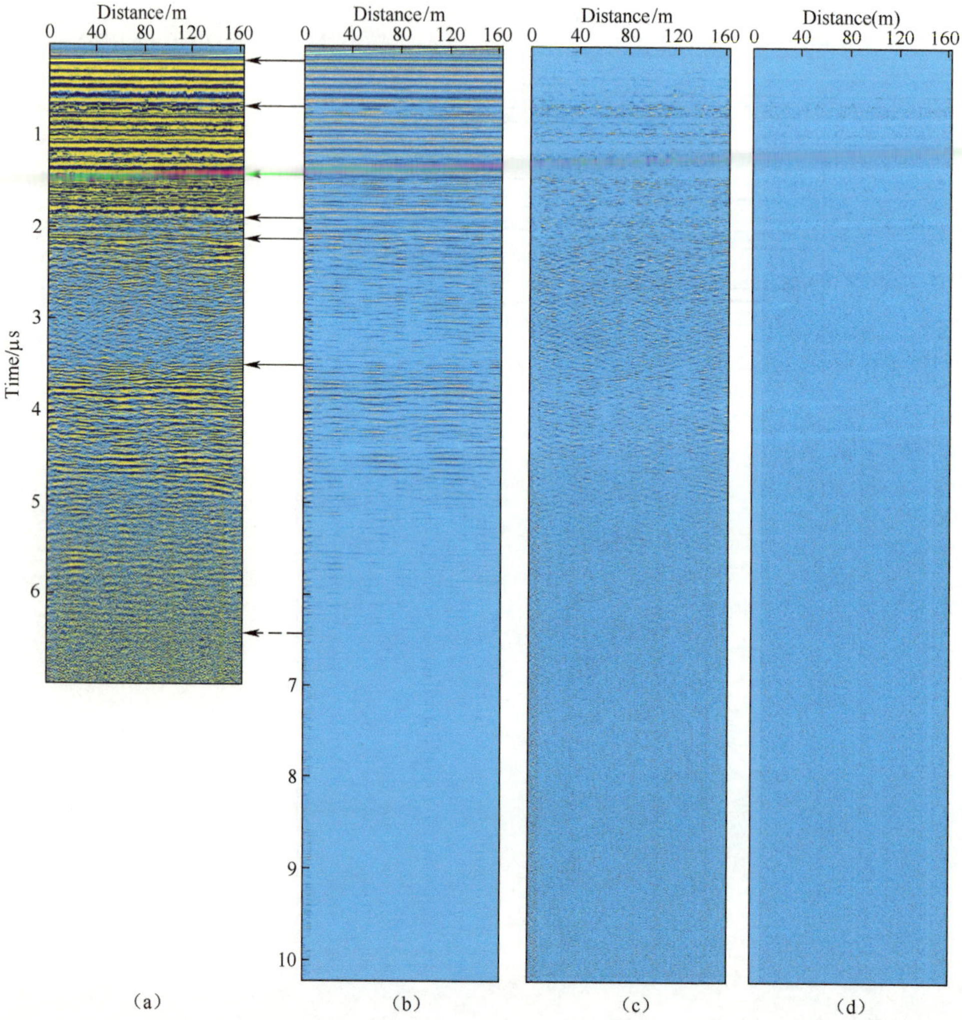

Extended Data Fig. 6　Singular value decomposition of low-frequency channel. a, The same profile as shown in Fig. 2. b, The low-pass eigenimage[33] (only keeping eigenvalues from 1 to 10); c, the band-pass eigenimage (only keeping eigenvalues from 11 to 200); d, the high-pass eigenimage (only keeping eigenvalues from 201 to 1973). b～d share the same color scale but is slightly less clipped than a for the convenience of pattern comparison. The singular values are shown in Supplementary Fig. 5

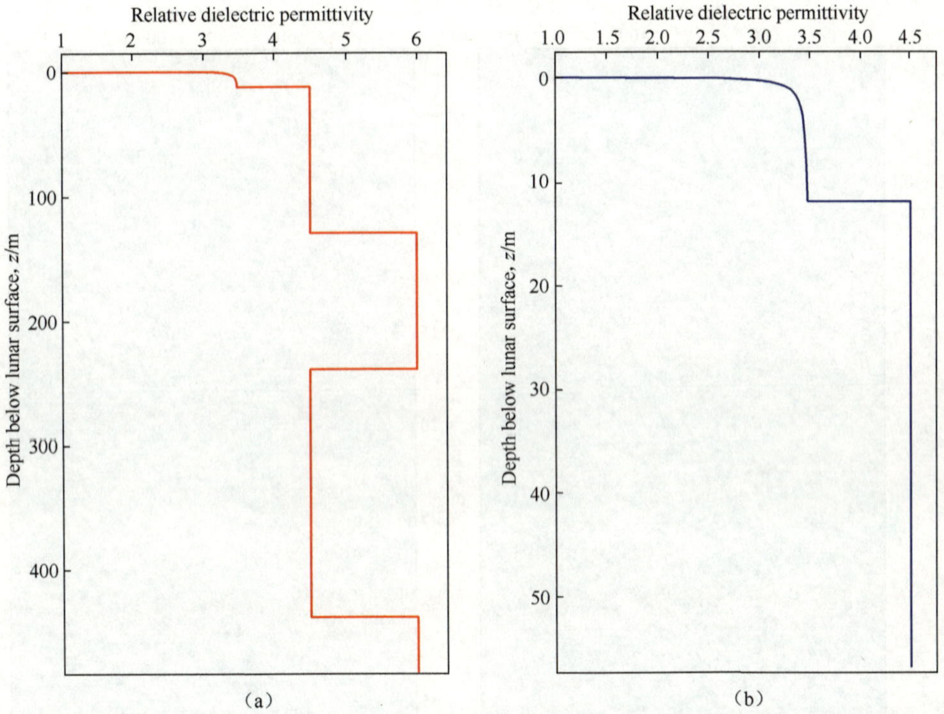

Extended Data Fig. 7　The lunar regolith model of the relative dielectric constant. a, The whole model; b, Local details of a within 60m. This model is constructed after the Apollo samples[12], which can be used to convert the travel time to depth in Fig. 2 and Extended Data Fig. 4

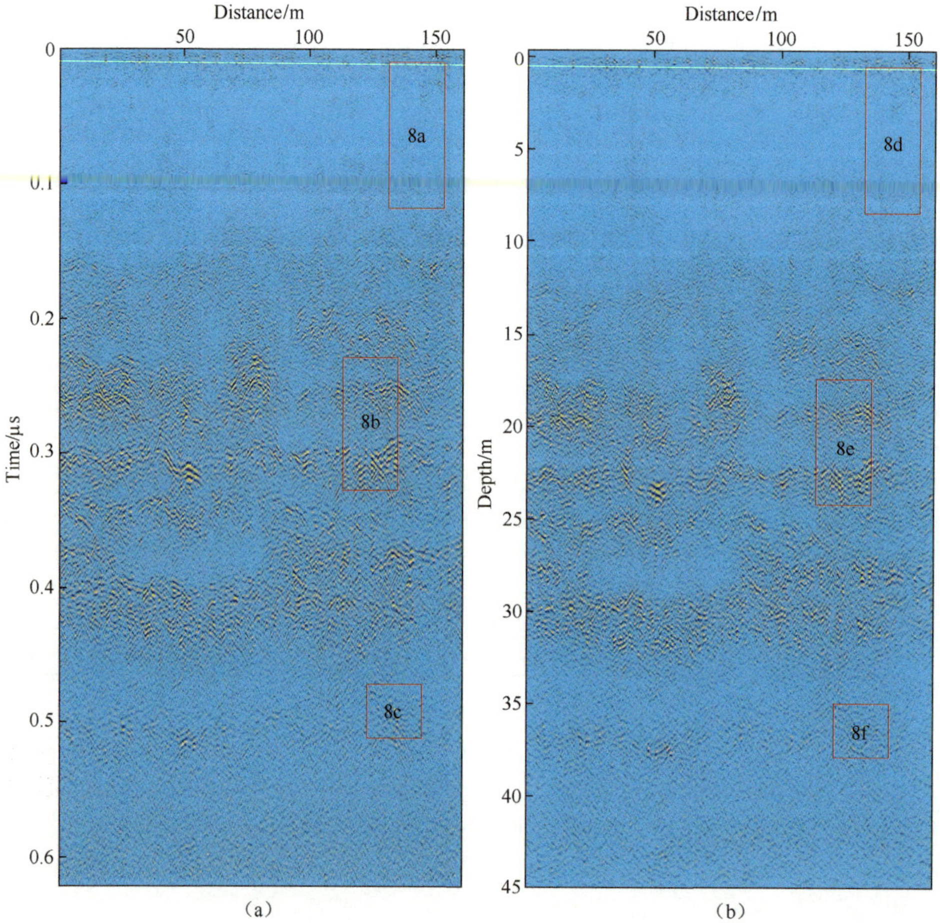

Extended Data Fig. 8　Comparison between the original LPR profile and the migration result. a, The original LPR profile; b, The migration result. The data are obtained by the high-frequency Channel-2B. The local details in the rectangles are shown in Extended Data Fig. 9

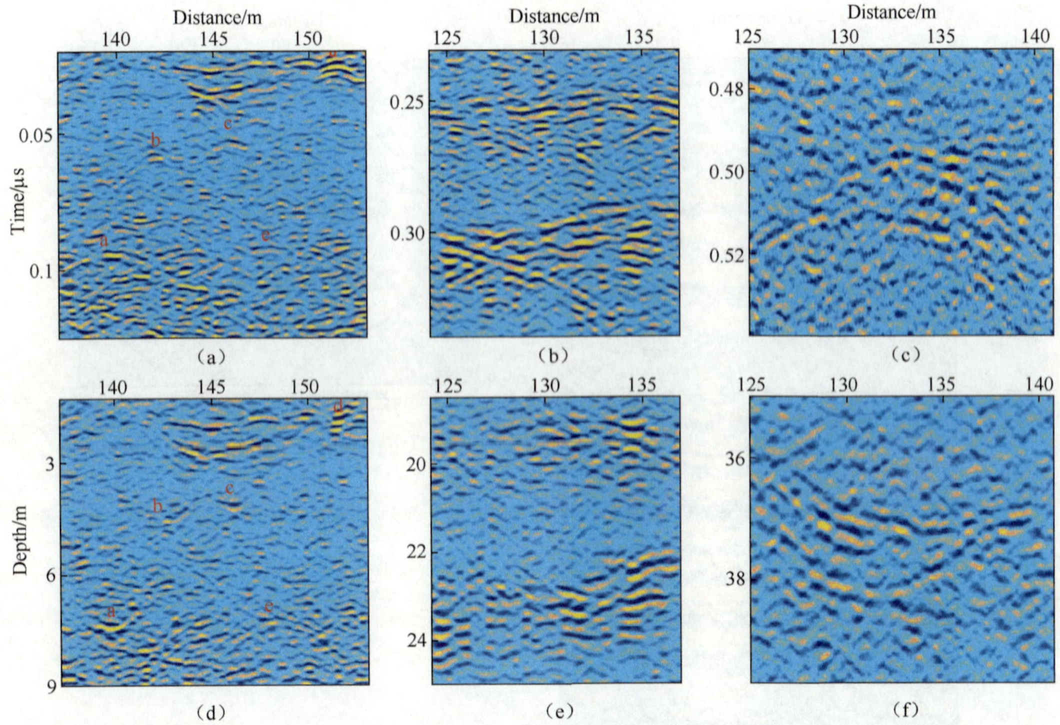

Extended Data Fig. 9　Local details in the rectangles in Extended Data Fig. 8.
a–**c**, The original LPR profiles; **d**–**f**,
The corresponding migration results of **a**–**c**, respectively

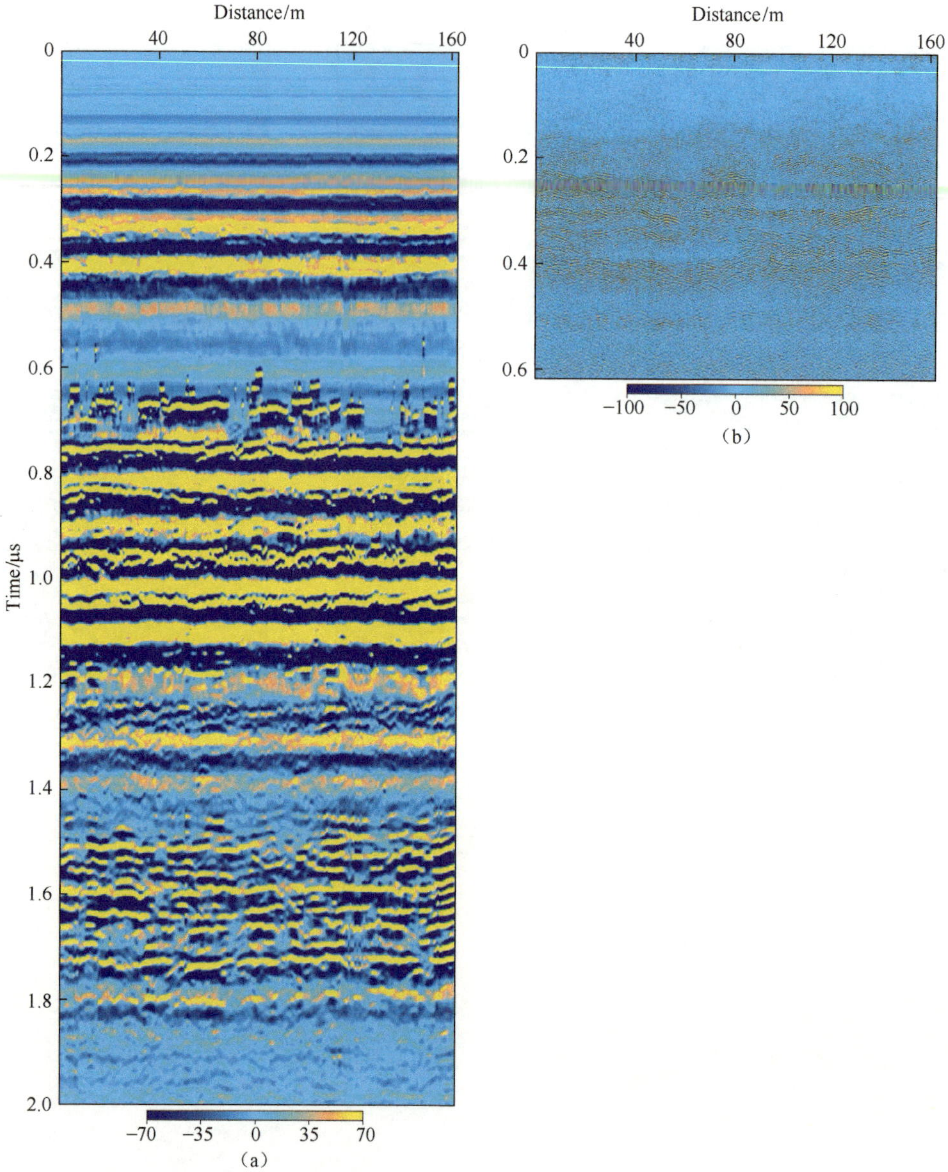

Extended Data Fig. 10　Comparison between the low - frequency and high - frequency LPR channels. a, Low - frequency channel; b, High - frequency channel. The high - frequency channel can well constrain the subsurface structures within 0.62μs

Reply to: Stratigraphy versus artefacts in the Chang'e - 4 low - frequency radar

Jinhai Zhang[1], Bin Zhou[2] and Yangting Lin[1]

1. Key Laboratory of Earth and Planetary Physics, Institute of Geology and Geophysics, Chinese Academy of Sciences, Beijing, China
2. Key Laboratory of Electromagnetic Radiation and Detection Technology, Aerospace Information Research Institute, Chinese Academy of Sciences, Beijing, China

Pettinelli et al.[1] argue that almost all of the low - frequency data collected by the dual - frequency penetrating radar (LPR) onboard the Chang'e - 4 (CE4) rover and presented in our Letter[2] are system noise and have no physical meaning. Their conclusions are mainly based on two arguments: (1) the similarity of the average time amplitude (ATA) curves between the LPR on Chang'e - 3 (CE3) and CE4 (Fig. 1 in Pettinelli et al.[1]), despite them being deployed in two different hemispheres; and (2) the similar horizontal lines (probably time - invariant coherent system noise) in the low - frequency LPR sections from CE3 and CE4 (fig. 2 in Pettinelli et al.[1]).

We agree that the ATA curves for raw low - frequency data from CE4 and CE3 have similar features, since they were dominated by the strong electromagnetic disturbances and noise that drowned out the relatively weak echoes. The similarity in the major features of the ATA plots computed from the short track (say 10m) between CE3 and CE4 demonstrates that both LPR payloads are reproducible and almost identical. We computed the ATA curves for CE3 and CE4 using all CE3 data (up to 114m) and days 1 - 4 of CE4 (Fig. 1). Our plot confirms their similar overall time - decay trends and the presence of some strong and coincident peaks and troughs between CE3 and CE4. Yet differences in the ATA curves after 1.1μs of travel time are readily discernible (Fig. 1a). The system disturbance and noise are reproducible and recognizable; hence they can be largely reduced by subtracting the averages traced from each curve. The background - subtracted ATA curves are plotted in Fig. 1b, and the differences between CE3 and CE4 become significant. Furthermore, we compared the ATA plots between two fragments of the Yutu - 2 rover track (days 1 - 4 and 12 - 16, Fig. 1c). Except for the remaining system noise, which appears as several strong and coincident peaks within the 0.6 - 1.1μs travel time, clear echoes from the travel time later than 1.1μs can be recognized in the ATA plots (Fig. 1).

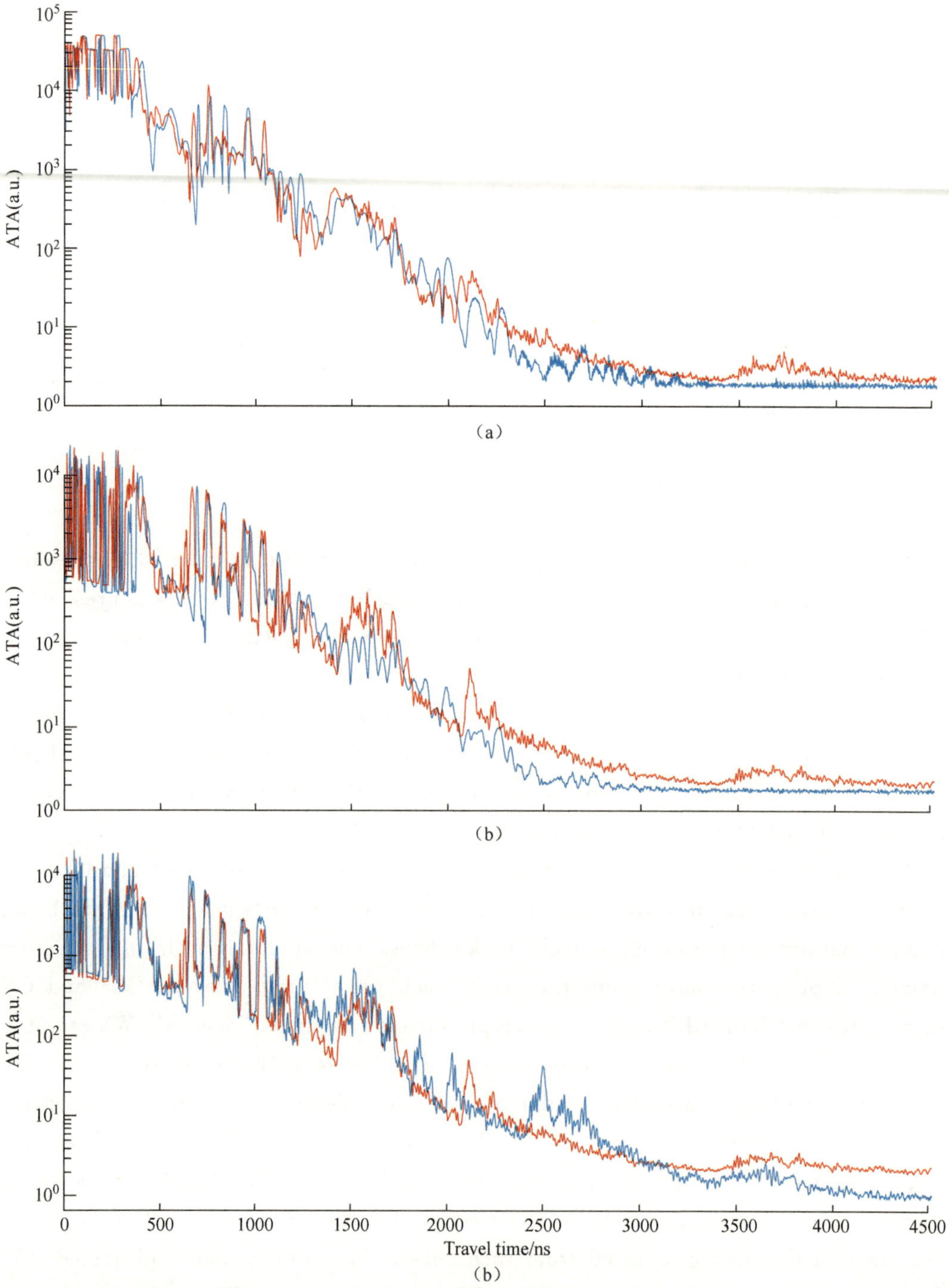

Fig. 1　Comparison of the ATA plots between the low - frequency LPR data. a, The ATA curves were computed from the raw low - frequency LPR data acquired by CE3 along its ~114 - m - long track (blue line) and by CE4 over lunar days 1 - 4 (red line). b, The same data as in a but with the background subtracted. c, The ATA curves of two subsets of CE4 data acquired over lunar days 1 - 4 (red line) and 12 - 16 (blue line). The noise floor of CE4 LPR data (>4μs) for days 12 - 16 is lower by a factor of ~2 than that for days 1 - 4 due to the adjustment of two parameters of the LPR (that is, the time window and data accumulation times). This adjustment has the effect of reducing the white noise but does not shift the peak positions in the ATA curves

Pettinelli et al.[1] compared the low – frequency LPR sections between CE4 and CE3, using a subset of data collected within a 10m track for each. Several strong straight lines can be recognized in both LPR sections, and they appear at the same time. We agree that these coincident straight lines were dominated by the system background as depicted in the ATA plots discussed above. To reduce this strong coincident background, we process CE4 LPR data for the lunar days 1 – 16 by bandpass filtering within 10 – 80MHz or 30 – 80MHz, respectively (Figs. 2 and 3). The processed data clearly show the signals after 1.1μs. Although the straight lines (probably the system background) before 1.1μs remain, they are readily recognizable and were not explained as reflections of subsurface structures. The system background can be further reduced by subtracting the average traces followed by the bandpass filtering (Supplementary Fig. 1c). However, the combination of both processes could not significantly enhance the signals, especially for those after 1.1μs, in comparison with the bandpass filtering alone (Supplementary Fig. 1b, d). This additional process is therefore not necessary.

After reducing the system background by bandpass filtering, the low – frequency LPR sections of CE4 are evidently different from those of CE3 (Fig. 2). Although several strong lines within the time interval of 0.6 – 1.1μs remain, they are not as strictly straight as those before 0.6μs. This feature is more significant for data processed with a bandpass filter of 30 – 80MHz (Supplementary Fig. 1d). In addition, the reflections between these strong lines are also different from the invariant coherent noise (Fig. 2). It is evident that the narrow and strong reflections within 1.4 – 1.8μs and 2.0 – 2.2μs have local slopes between traces 1, 600 and 2, 600 (Figs. 2 and 3), indicative of subsurface boundaries

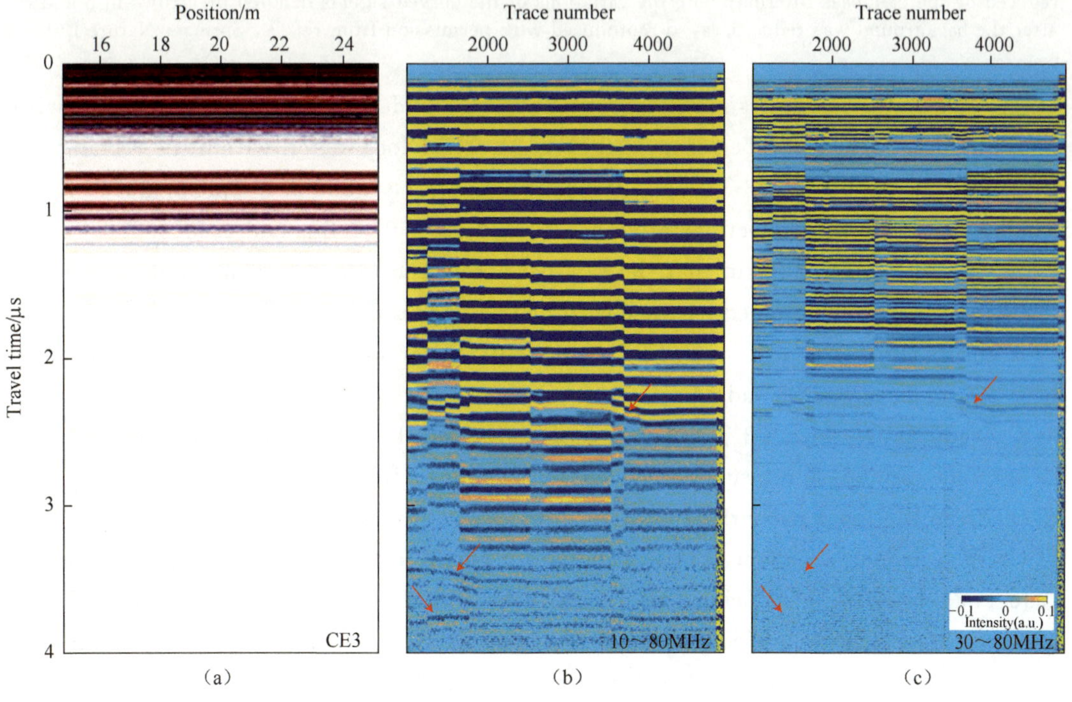

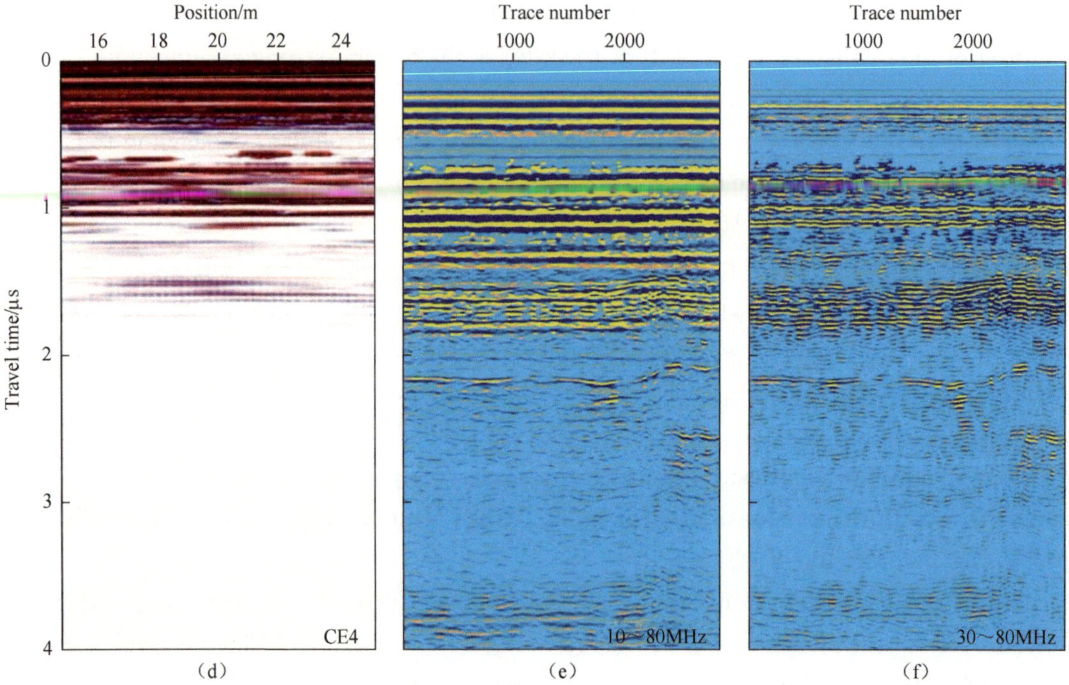

Fig. 2 Comparison of the low-frequency LPR sections of CE3 and CE4 within 0-4μs travel time. a-c, CE3 data. d-f, CE4 data. a and d are from Pettinelli et al.[1], with a travel distance of 10m. b, c, Our results for CE3, whose travel distance is about 114m with a 10-80MHz bandpass filter applied (b) and a 30-80MHz bandpass filter applied (c). e, f, Our results for CE4, whose travel distance is about 424m (the whole dataset for lunar days 1-16) with a 10-80MHz bandpass filter applied (e) and a 30-80MHz bandpass filter applied (f). Note that the spatial interval after trace number 2,159 is doubled due to using a doubled sampling time duration. The system background in our results has been greatly reduced by the bandpass filtering; note the variations of the curved signals denoted by arrows in b and c after the background was reduced. a, d reproduced with permission from ref.[1], Springer Nature Ltd

rather than the invariant system noise; in addition, a wide zone (3.6-5.0μs) filled with weak but clear signals can be recognized (Fig. 3). More local detail within the rectangular windows (1.3-2.6 and 4.3-5.6μs, respectively) can be found in Supplementary Fig. 2, which exhibits evident reflections around 1.4-1.8μs, 2.0-2.2μs and 4.4-5.0μs.

In summary, we confirm that the raw data from the low-frequency radars onboard CE3 and CE4 were dominated by the electromagnetic disturbances and noise, leading their ATA curves and the 10-m-long LPR sections to appear similar. However, we state that these strong disturbances and noise can be largely reduced by subtracting the data averaged over several lunar days (3 in the case of ref.[2] and 16 here) from each trace. The background can be significantly reduced by bandpass filtering, further enhancing the echoes from the lunar subsurface stratigraphy, especially for deep reflections after 1.1μs. Of course, proper selection of the bandpass filter is critical to retaining weak low-frequency echoes while reducing background noise, as shown in Fig. 2b, c, e, f. These new data show that most of the residual signals observed by the low-frequency channel of CE4 are real reflections from subsurface structures rather than artefacts.

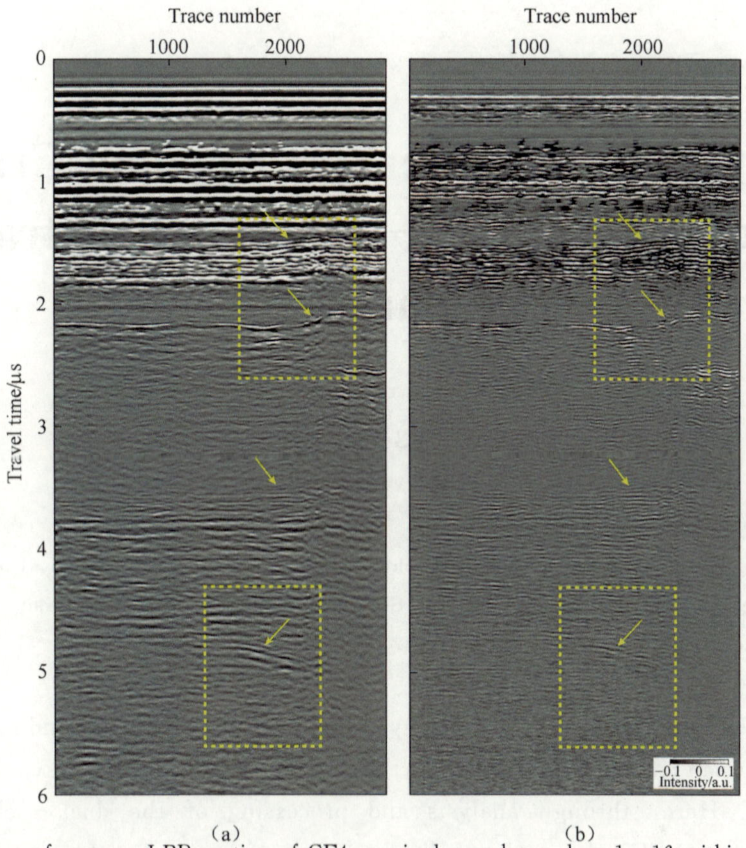

Fig. 3　The low - frequency LPR section of CE4 acquired over lunar days 1 - 16 within 0 - 6μs travel time. a, b, The same datasets shown in Fig. 2e (a) and Fig. 2f (b) but with a longer travel time. The yellow arrows mark deep reflections that have significant travel time variations at different time windows. More local details within the yellow dashed rectangles are shown in Supplementary Fig. 2

Acknowledgements

This work was supported by Key Research Program of Frontier Sciences, Chinese Academy of Sciences (grant number QYZDJ - SSW - DQC001 to Y. L.) and the National Nature Science Foundation of China (grant numbers 41490631 to Y. L. and 41941002 to J. Z.). We thank the Supercomputing Laboratory of IGGCAS for providing computing resources.

References

[1] Pettinelli, E., Lauro, S. E., Mattei, E., Cosciotti, B. & Soldovieri, F. Stratigraphy versus artefacts in the Chang'e - 4 low - frequency radar. Nat. Astron. https://doi. org/10. 1038/s41550 - 021 - 01432 - x (2021).

[2] Zhang, J. et al. Lunar regolith and substructure at Chang'E - 4 landing site in South Pole - Aitken basin. Nat. Astron. https://doi. org/10. 1038/s41550 - 020 - 1197 - x (2021).

Stratigraphy of the Von Kármán Crater Based on Chang'E-4 Lunar Penetrating Radar Data

Ling Zhang[1], Jing Li[1], Zhaofa Zeng[1],
Yi Xu[2], Cai Liu[1], and Shengbo Chen[1]

1. College of Geo-exploration Science and Technology, Jilin University, Changchum, China
2. Space Science Institute, Lunar and Planetary Science Laboratory, Macau University of Science and Technology, Macau, China

Abstract: The Chang'E-4 (CE-4) achieved the first successful soft landing on the floor of the Von Kármán crater located in the northeast of the South Pole-Aitken (SPA) basin on the Moon. Here, through analysis and processing of the dual-channel lunar penetrating radar (LPR) data, the shallow regolith structure and deep geological strata below the landing site are exposed. The high-frequency data (CH-2, 500MHz) show that the Von Kármán crater underwent remodeling because of ejecta from multiple craters within a thickness of 40m. Under the surface fine-grained regolith, superpositions among these crater materials and mare basalt units establish shallow regolith stratigraphic sequences. The low-frequency data (CH-1, 60MHz) show a deep stratigraphic structure to the depth of 360m that indicates the previous geological evolution history of the Von Kármán crater. The LPR results provide an important scientific basis for understanding the geological history of the far side of the Moon.

Plain Language Summary: The Chang'E-4 (CE-4) landed in the Von Kármán crater, which is located on the far side of the Moon. The lunar penetrating radar (LPR) loaded on the Yutu-2 rover can be used to determine the shallow lunar regolith structure and deep geological strata of the Von Kármán crater. In this article, first, different processing procedures focused on the two channels of LPR data (CH-1 and CH-2) are designed, and clear radar images that can be used to explain the underground structure of the Von Kármán crater are obtained. The radar images are then combined with previous geological information to reveal the subsurface geological structure of the Von Kármán crater. The Von Kármán crater underwent various geological processes, including basaltic magmatism, ejecta coverage, and the interaction of volcanic eruptive clastics and lava.

The regolith layer on the surface has a thickness of more than 10m, and below the regolith layer is a multilayer formed by the ejecta material of surrounding impacts, which is as thick as more than 30m. Underneath are three basalt magma covers.

Key Points:

• The CH - 2 LPR data reveal that the landing site underwent remodeling because of ejecta from multiple craters within a thickness of 40 m

• The CH - 1 LPR data provide the deep stratigraphic structure within 360m, including more than three basalt magma activities

• The LPR data demonstrate that the Von Kármán crater was landscaped by various geological processes

1 Introduction

On 3 January 2019, the Chang'E - 4 (CE - 4) landed at 177.6°E, 45.5°S, on the South Pole - Aitken (SPA) basin, achieving the second Chinese lunar landing and the first human soft landing on the far side of the Moon (Liu et al., 2019). The CE - 4 landed in the SPA basin and set down on the floor of the Von Kármán crater, located in the northeast of the SPA basin and representing a typical landform type in the SPA basin. The SPA basin is the largest impact structure on the Moon (diameter = 2, 600km) and one of the largest impact structures in the solar system (Stuart - Alexander, 1978). The elevation difference from its depths to the tops of the highest surrounding peaks is over 15km (Stuart - Alexander, 1978), almost twice the height of Mount Everest on Earth. The SPA is impressive for several reasons (Melosh et al., 2017). First, significant impact events can remove surficial materials from local areas and bring material from beneath the impact craters to or closer to the surface. The SPA is the deepest impact basin on the Moon (>8km). The deepest lunar crustal materials should be exposed on the surface. Then, the lower crust of the Moon can be revealed in areas within the SPA, which is something that cannot be achieved anywhere else on the Moon.

The Von Kármán crater (186km diameter) shows distinctive geologic characteristics and has a complex geological evolution history. It flooded after its formation by lava flows in the Late Imbrian Period. The basalt plains were subsequently reshaped by materials ejected from the surrounding young craters, such as the Finsen crater and Alder crater (Huang et al., 2018; Moriarty & Pieters, 2018).

The main scientific objective of this mission is to determine the mineral composition of the surface materials and perform geological characterization of the landing site by two principal instruments, lunar penetrating radar (LPR) and a visible and near - infrared imaging spectrometer (VNIS), on the Yutu - 2 rover (Li et al., 2019). The subsurface structure of the CE - 4 landing site contains a direct record of the sequence of major geological events and the geological history of this area. The LPR aboard the Yutu - 2 rover was intended to estimate the subsurface structure and the depth of the subsurface layers. It includes two - channel antennas with different penetration depths: a low - frequency channel (CH - 1, 60MHz) and a high - frequency channel (CH - 2,500MHz). The CH - 1 can penetrate the subsurface structure to a depth of several hundred meters, and the CH - 2 acknowledges the detection of the shallow lunar regolith structure (Fang et al., 2014; Jia et al., 2018).

The previous Chang'E - 3 (CE - 3) two - channel LPR data provided impressive results for the shallow regolith structure of Mare Imbrium. In particular, the CE - 3 LPR CH - 1 result indicated that there are more than nine subsurface layers from the surface to a depth of approximately 360m (Xiao et al., 2015). It detected three underlying basalt units at depths of 195,215, and 345m. The CH - 2 (500MHz) LPR data revealed several major stratigraphic zones from the surface to a depth of approximately 20m along the survey line (Fa et al., 2015; Zhang et al., 2018, 2019). In addition, the radar measurements indicate an underestimation of the global lunar regolith thickness by other methods and reveal a vast volume from the last volcanic eruption (Yuan et al., 2017; Zhang et al., 2015). As the backup of CE - 3, CE - 4 LPR provides the opportunity to investigate the subsurface stratigraphic and shallow regolith structure at the Von Kármán crater. This new information offers significant scientific evidence that advances knowledge of the far side of the Moon.

Previous studies of lunar surface mineralogy and morphology analysis demonstrate that the Finsen crater is the primary source of surface ejecta at the CE - 4 landing site. The spectral features of the CE - 4 landing site are considerably different from spectral features typical of mare basalts, which indicates that the CE - 4 landing site was modified by the emplacement of ejecta from the Finsen crater. The ejecta of the Finsen crater can represent deep - seated materials emplaced at the time that the SPA basin was formed (Li et al., 2019). However, what is the thickness of the Finsen ejecta? Where is the Alder ejecta, below or above the mare basalt unit? How can we explain and verify the geological evolution of the Von Kármán crater? The LPR data provide quantitative evidence to answer the above scientific questions.

The lunar surface characteristics and material composition of the Von Kármán crater were speculated based on the lunar reconnaissance orbiter camera (LROC) images, in situ spectral observations, topography data, and other means (Di et al., 2019; Fu et al., 2020). These results demonstrated that the regolith is considered to constitute the uppermost layer in the Von Kármán crater, with a thickness of approximately 2. 5 to 7. 5 m. Under the regolith is the Finsen low - calcium - pyroxene (HCP) - bearing ejecta layer. The local infilled higher - calcium - pyroxene - bearing mare basalt beneath the HCP - bearing ejecta materials is greater than 310m. However, the above conclusion is based on speculation and statistical data. These inferences lack convincing evidence. The LPR is a direct way to reconstruct the subsurface structure and material composition. The previous work used the CH－2 LPR data of the first two lunar days (total distance is approximately 110m) to indicate that the shallow subsurface can be divided into three units. Unit 1 (up to 12m) consists of lunar regolith, Unit 2 (depth range, 12 to 24m) consists of coarser materials with embedded rocks, and Unit 3 (depth range, 24 to 40m) contains alternating layers of coarse and fine materials (Chisenga et al., 2019; Lai et al., 2019; C. Li et al., 2020).

In this paper, we use all 12 lunar days of LPR data, for which the total profile distance is more than 300m, to explain the shallow lunar structure and the deep geological strata in the landing area. To guarantee data accuracy, we present a complete LPR data preprocessing strategy. We estimate the radar velocity in regolith according to the physical properties of the Apollo 16 samples and build a homogeneous velocity model (Heiken et al., 1991). Then, we propose the LPR least squares Kirchhoff migration (LSM) to improve image resolution (J. Li et al., 2020; Nemeth et al., 1999). This study provides reliable LPR data applications for CE－4 exploration, such as stratigraphic sequences and significant geological events in the SPA. The LPR data can be used to reconstruct and estimate the stratigraphic for different geological units at the CE－4 landing site.

2　The CE－4 Landing Site

The Von Kármán crater (diameter $D = 186$km; central coordinates 44.4°S, 176.2°E; Figure 1b) is in the center of the SPA (Figure 1a). The Von Kármán crater has experienced a series of complicated geological processes since the crater was formed in the pre－Nectarian period (Fu et al., 2020; Huang et al., 2018; Losiak et al., 2009; Moriarty & Pieters, 2018; Paskert et al., 2018; Wilhelms et al., 1979; Yingst & Head, 1997; Yingst et al., 2017). The Von Kármán crater runs from the pre－Nectarian SPA basin to the sectarian Von Kármán crater (Losiak et al., 2009; Yingst et al., 2017). It is younger than the Von Kármán M basin and older than the Leibnitz crater. Mare basalts flooded parts of the Von Kármán crater floor during the Imbrian period (Wilhelms et al., 1979; Paskert et al., 2018; Yingst & Head, 1997), but a portion of the central peak remained exposed near the center of the crater. It has higher FeO and TiO^2 contents and lower elevations than other parts of the basin.

The impact structures in the regional vicinity of the Von Kármán crater span the vast expanse of geological time. The main geological activities of the Von Kármán crater can be traced in Figure 1c. The ejecta from the Leibnitz crater reshaped the northern area of the Von Kármán crater before the arrival of the Imbrian period (Losiak et al., 2009; Yingst et al., 2017). During the Imbrian period, mare basalts flooded the crater floor, and the Alder crater (82km in diameter) was formed, which affected the geological structure of the Von Kármán crater (Huang et al., 2018). In the long period that followed, the Von Kármán crater may have experienced multiple periods of basalt coverage and a series of geological processes (Fu et al., 2020; Huang et al., 2018). At the end of the Imbrian

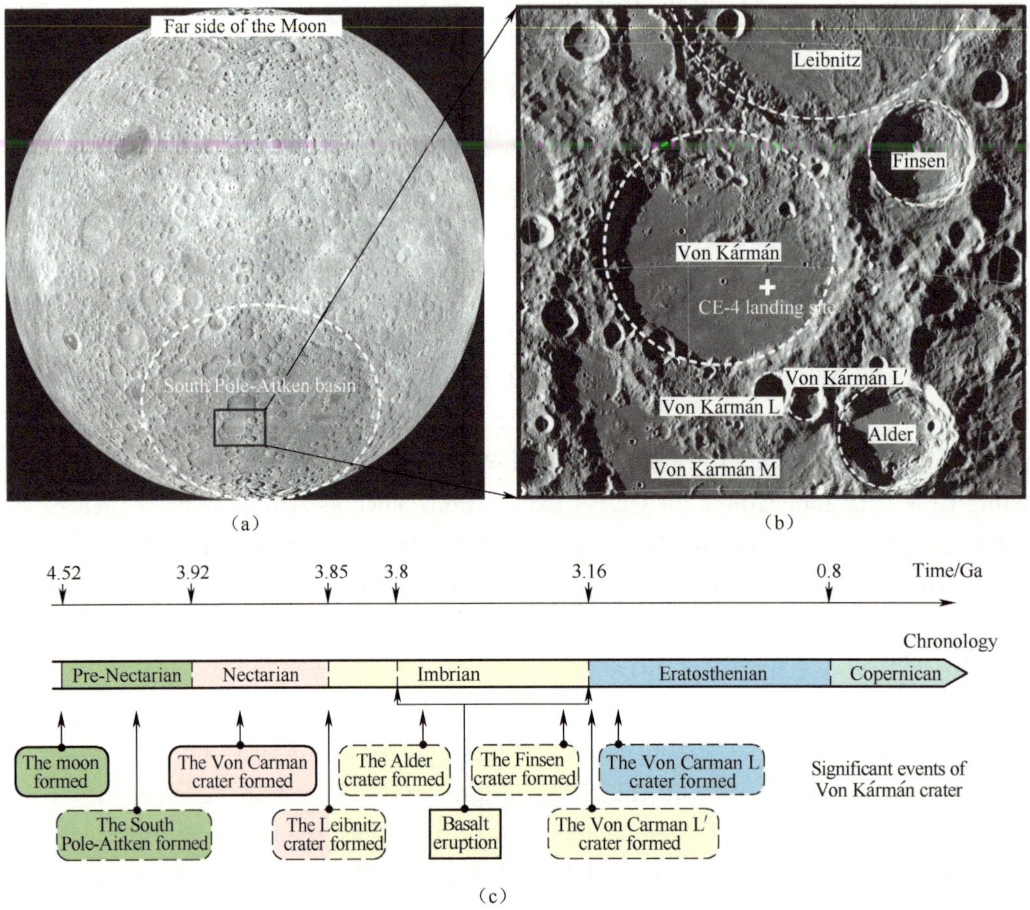

Figure 1　(a) The South Pole - Aitken basin. The map derived from an image from the lunar reconnaissance orbiter's wide - angle camera. (b) The Chang'E - 4 landed in the Von Kármán crater. The map derived from an image from the lunar reconnaissance orbiter's narrow - angle camera. (c) Geological evolution history of the Von Kármán crater

period and the beginning of the Eratosthenian period, a Finsen crater (73km in diameter) with a significant influence on the surface shaping of the Von Kármán crater was formed, along with two smaller craters (Von Kármán L′ and Von Kármán L) with less influence on the surface shaping (Paskert et al., 2018; Wilhelms et al., 1979; Yingst & Head, 1997). In the subsequent long Eratosthenian and Copernican periods, apart from the impacts of small meteorites and solar wind, the Von Kármán crater did not experience more landscaping (McKay et al., 1991). Superpositioning of these crater materials and mare basalt units establishes local stratigraphic sequences. We try to combine the LPR processed results to reveal the sequence of the major geological events and the geological reconstruction history in the Von Kármán crater (Figure 1c).

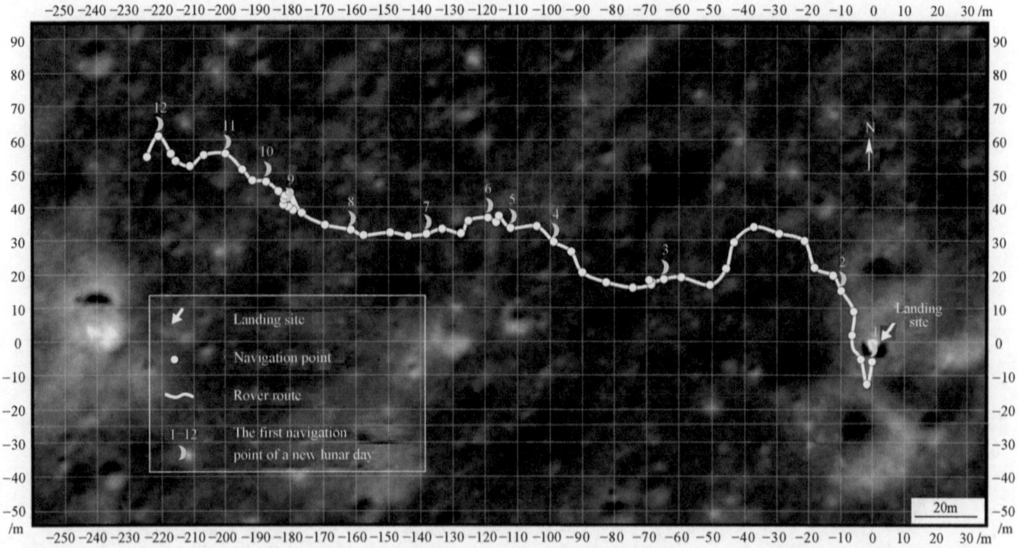

Figure 2 Yutu - 2's path on the far side of the Moon during the 12 lunar days. (The context image was taken by the lunar reconnaissance orbiter camera on 3 June 2019. ID: M1314237625LR)

3 Data Processing

Compared with the LPR data of CE - 3, the LPR of CE - 4 achieved a breakthrough in the detection area. The far side of the Moon is more challenging to detect, and the type of geological structure changes. The signalto - noise ratio (SNR) of CE - 4 LPR data was higher than that of CE - 3 LPR data. In addition, the performance of the Yutu - 2 rover and LPR system was dramatically improved. To date, the Yutu - 2 Rover has worked more than 14 lunar days and profiled more than 400m of horizontal coverage. During the whole 12 lunar days, the Yutu - 2 rover equipped with LPR traveled almost 320m (Figure 2).

According to the characteristics of the CE - 4 LPR data and the previous experience of CE - 3 LPR data processing (Li et al., 2017; Zhang et al., 2018), we designed processing procedures for the dual - channel LPR data. A detailed description of the entire process is provided in the supporting information (Tables S1 and S2 and Figures S1 - S19). Previous work divided the regolith into three layers with different grain sizes based on the reflected energy distribution of the CH - 2 data from the first and second lunar days (C. Li et al., 2020). Based on the 12 lunar day CH - 2 data (Figure 3a), we formed a more detailed division of horizons (Figure 3b). Due to the differences in the abundance of ilmenite and rocks in the regolith, the penetration depth of the CE - 4 LPR data is approximately 3 times deeper than that of CE - 3 (Lai et al., 2019). The sufficient penetrating depth is greater than 40m.

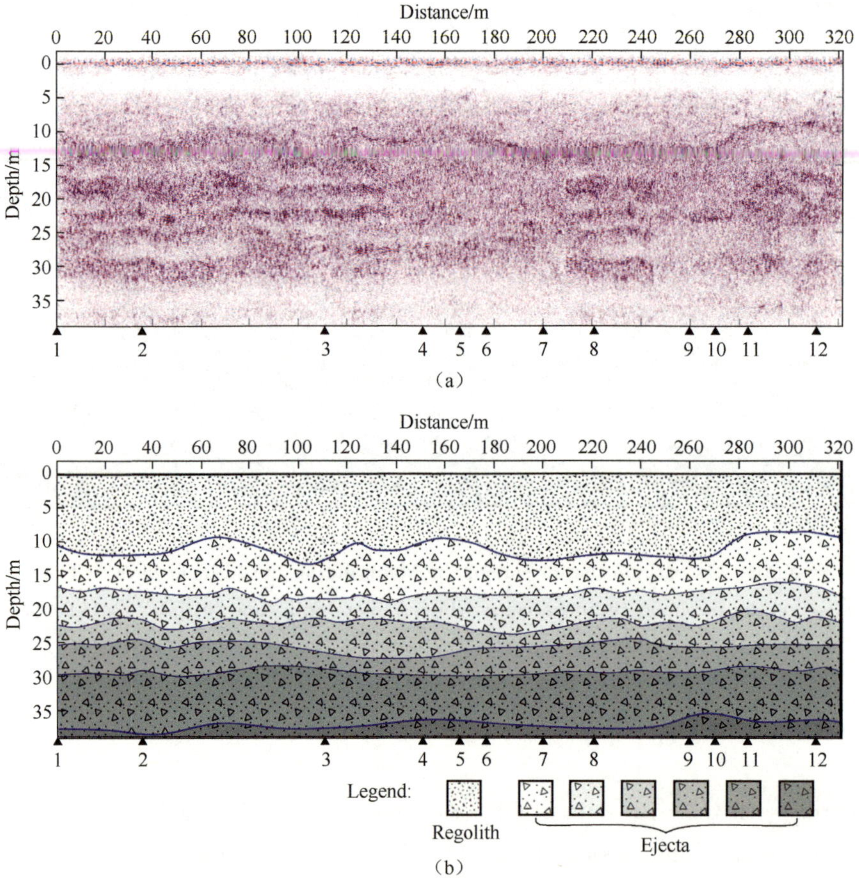

Figure 3　Results and interpretation of the CH－2 data

(a) The initial results of the CH－2 data. (b) Interpretations of the CH－2 data.

Due to pitfalls in the CH－1 LPR data of CE－3 (Li et al., 2018), we need to be extra careful in the CH－1 data processing of CE－4 (Figure 4a, more details in Table S2 and Figures S11－S19). Heuristic stratification is performed on the layer identification (Figure 4b).

4　Results

Based on the dual－channel LPR data of CE－4 and the geological events of the Von Kármán crater, the geological structure of the landing site was analyzed. Figure 4b shows the proposed stratigraphy of the landing area. Table 1 explains the composition, type, and thickness of each unit.

The impact ejecta from meteorites constitutes the uppermost layer of the reconstructed stratigraphic column. They formed a fine－grained layer termed "regolith." The CH－2 data (Figure 3b) shows that the thickness of this fine grain regolith layer

(*Layer i* in Figure 4b and Table 1) is approximately 12m. There is a significant lateral depth change in the grain regolith layer, which is also confirmed in Ding's work (Ding et al., 2020). Below the fine - grained regolith layer, an interactive layer with visible stratification is formed (Figure 3b). This layer is speculated to be alternating layers of ejecta and local materials (*Layer ii* in Figure 4b and Table 1). The ejecta is from different craters formed at the end of the Imbrian period and the beginning of the Eratosthenian period, such as the Finsen, Von Kármán L′, and Von Kármán L.

The CH - 1 LPR data reveal the structure from 50 to 360m (Figure 4a). The central part of the Von Kármán crater is most likely basalt, which mixes the alternating layers of ejecta and local materials. Between two basaltic units (*iii* and *v* in Figure 4b and Table 1), there is an interpreted paleoregolith layer of a 20m thickness (iv in Figure 4b and Table 1). The LPR result is reflected in the paleoregolith layer sandwiched among lava flows, which is supported by a simple radar reflection/transmission model (Ishiyama & Kumamoto, 2019). The radar performance characteristics (Figure 4a) of *vii* and *ix* in Figure 4b are different from those of Units *iii* and v. Layer *vii* is an interlayer that is formed by several lava flows. Layer *ix* probably contains many pyroclastic rocks. The layer depth and the thickness of the ejecta at the CE - 4 landing site indicate that Layer *viii* is probably formed by the ejecta from the Alder crater.

From the dual - channel LPR results and interpretations of the proposed stratigraphy at the CE - 4 landing site, two possible facts are known:

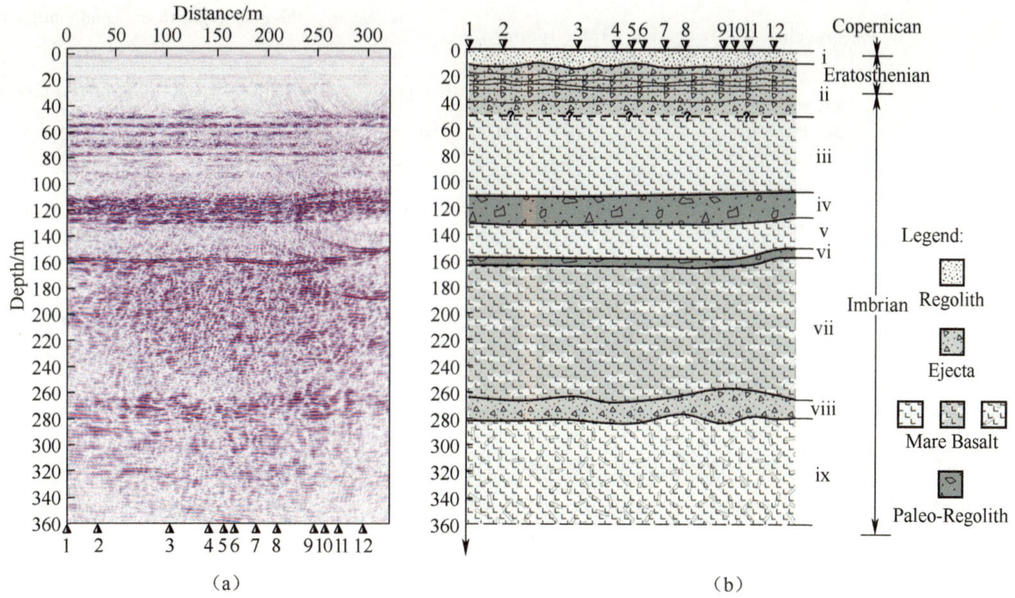

Figure 4　Result and interpretation of the CH - 1 data. (a) The initial results of CH - 1 data. (b) Proposed stratigraphy of the CE - 4 landing site

Table 1　Geological Units of the CE－4 Landing Site

Unit symbol	Unit name	Unit thickness and position	Unit description
i	Regolith	～12m (0—12m)	Different－sized asteroidal materials collide with the surface in the Eratosthenian and Copernican period. As time passes, these collisions and the surface materials create a fine－grained layer termed "regolith."
ii	Alternating layers of ejecta and local materials	～38m (12～50m)	Some craters are formed near the Von Kármán crater at the end of the Imbrian period and the beginning of the Eratosthenian period, such as Finsen, Von Kármán L′, and Von Kármán L. Alternating layers: The ejecta from these different craters interacts with local materials.
iii	Basalt	60m (50－110m)	During the Imbrian period, mare basalts flood the Von Kármán crater floor. This layer is the youngest basalt layer, enriched with HCP－bearing mare basalt.
iv	Paleoregolith	20m (110－130m)	During the time between the two basalt covers, asteroidal materials cultivate the local materials and form the paleoregolith, the same mechanism as the regolith.
v	Basalt	28m (130－158m)	This layer is enriched with mare basalt and similar to Layer *iii*.
vi	Paleoregolith	5m (158－163 m)	This layer is the paleoregolith unit and similar to Layer *iv*.
vii	Basalt	200m (163－263m)	This layer is a mare basalt unit different from Units *iii* and *v*. It is probably an interlayer with several lava flows.
viii	Ejecta from Alder crater	17 m (263－280m)	According to the depth and the thickness, this layer is probably formed by the ejecta from the Alder crater.
ix	Basalt	＞80m (280 to ＞360m)	This layer is the mare basalt unit. Different from Units *iii*, *v*, and *vii*, this layer may contain many pyroclastic rocks.

1. At the end of the Imbrian period and the beginning of the Eratosthenian period, the Von Kármán crater experienced more complex geological activities than other craters in the same period. We know this to be true because the ejecta from these different craters interacts with local materials, such as the Finsen, the Von Kármán L′, and the Von Kármán L craters.

2. In the Imbrian period, under the background of global magma flooding, the Von Kármán crater experienced no less than four periods and three types of basalt magma

coverage. The earliest magma contains many pyroclastic rocks. The magmatic activity in the middle period may not have been so rapid. Under moderate movement, lava flows formed multilayered basaltic lavas. The following magmatic activity was relatively pure and straightforward. In addition, possibly because the energy inside the Moon weakened, the amount of magma in the last two periods decreased, forming two thinner basalt layers. This information suggests that the source materials and types of basalt magma are different during different stages of the Imbrian period in the Von Kármán crater region and even in the SPA Basin.

5 Conclusions

In this paper, we utilize 12 lunar days of LPR data to reconstruct the general sequence of geological events in the CE - 4 landing area. LPR CH - 2 reveals the fine subsurface structure of the regolith layer within a thickness of 40m. The CE - 4 landing site suggests the superposition of multiple ejecta layers, such as the mixed zone between the Finsen ejecta and mare basalt, the thicknesses of the Finsen ejecta, and the basaltic regolith. Under the fine - grained regolith, we interpret that the interaction of these foreign ejecta materials with local materials formed a complex near - surface structure. The results of LPR CH - 1 refine the interpretation of the established stratigraphic column at the CE - 4 landing site and offer an explanation of the geological activity events of the Von Kármán crater. The Von Kármán crater was landscaped by various geological processes, including basaltic magmatism, ejecta coverage, and the interaction of volcanic eruptive clastics and lava. This is in agreement with known geological activity events. These observations will likely lead to discoveries and constraints on the stratigraphic structure and geologic evolution of the Moon and provide an important scientific basis for understanding the geological history of the far side structure of the Moon.

Acknowledgments

We thank the China National Space Administration for the open data of the Chinese Academy of Sciences, Granb XDB 41000000, the National Natural Science Sponsorship Program by the China Association for Science and Technology (2019QNRC001), and the Natural Science Foundation of JilinProvince (20200201216JC) .

References

[1] Chisenga, C., Yan, J., Zhao, J., Deng, Q., & Barriot, J. P. (2019). Density structure of the Von Kármán crater in the northwestern South Pole - Aitken basin: Initial subsurface interpretation of the Chang'E - 4 landing site region. Sensors, 19

(20), 4445. https://doi. org/10. 3390/s19204445

[2] Di, K., Zhu, M., Yue, Z., Lin, Y., Wan, W., Liu, Z., et al. (2019). Topographic evolution of Von Kármán crater revealed by the lunar rover Yutu - 2. Geophysical Research Letters, 46, 12, 764 - 12, 770. https://doi. org/10. 1029/2019GL085252

[3] Ding, C., Xiao, Z., Wu, B., Li, Y., Prieur, N. C., Cai, Y., et al. (2020). Fragments delivered by secondary craters at the Chang'E 4 landing site. Geophysical Research Letters, 47 (7), E2020GL087361. https://doi. org/10. 1029/2020GL087361

[4] Fa, W., Zhu, M., Liu, T., & Plescia, J. B. (2015). Regolith stratigraphy at the Chang'E - 3 landing site as seen by lunar penetrating radar. Geophysical Research Letters, 42 (23), 179 - 185. https://doi. org/10. 1002/2015GL066537

[5] Fang, G., Zhou, B., Ji, Y., Zhang, Q., Shen, S., Li, Y., et al. (2014). Lunar penetrating radar onboard the Chang'E - 3 mission. Research in Astronomy and Astrophysics, 14 (12), 1607 - 1622. https://doi. org/10. 1088/1674 - 4527/14/12/009

[6] Fu, X. H., Qiao, L., Zhang, J., Ling, Z., & Li, B. (2020). The subsurface structure and stratigraphy of the Chang'E - 4 landing site: Orbital evidence from small craters on the Von Kármán crater floor. Research in Astronomy and Astrophysics, 20 (1), 59 - 70. https://doi. org/10. 1088/1674 - 4527/20/1/8

[7] Heiken, G. H., Vaniman, D. T., & French, B. (1991). Physical properties of the lunar surface, in Lunar sourcebook: A user guide to the Moon (pp. 530 - 558). Cambridge: U. K. Cambridge Univ. Press.

[8] Huang, J., Xiao, Z., Flahaut, J., Martinot, M., Head, J., Xiao, X., et al. (2018). Geological characteristics of Von Kármán crater, northwestern South Pole - Aitken basin: Chang'E - 4 landing site region. Journal of Geophysical Research: Planets, 123, 1684 - 1700. https://doi. org/10. 1029/2018JE005577

[9] Ishiyama, K., & Kumamoto, A. (2019). Volcanic history in the Smythii basin based on SELENE radar observation. Scientific Reports, 9 (1), 14502. https://doi. org/10. 1038/s41598 - 019 - 50296 - 9

[10] Jia, Y., Zou, Y., Ping, J., Xue, C., Yan, J., & Ning, Y. (2018). The scientific objectives and payloads of Chang'E - 4 mission. Planetary & Space Science, 162, 207 - 215. https://doi. org/10. 1016/j. pss. 2018. 02. 011

[11] Lai, J., Xu, Y., Zhang, X., Xiao, L., Yan, Q., Meng, X., et al. (2019). Comparison of dielectric properties and structure of lunar regolith at Chang'E - 3 and Chang'E - 4 landing sites revealed by ground penetrating radar. Geophysical Research Letters, 46, 12, 783 - 12, 793. https://doi. org/10. 1029/2019GL084458

[12] Li, C., Liu, D., Liu, B., Ren, X., Liu, J., He, Z., et al. (2019). Chang'E - 4 initial spectroscopic identification of lunar far side mantle - derived materials.

Nature, 569 (7756), 378 - 382. https://doi. org/10. 1038/s41586 - 019 - 1189 - 0

[13] Li, C., Su, Y., Pettinelli, E., Xing, S., Ding, C., Liu, J., et al. (2020). The Moon's farside shallow subsurface structure unveiled by Chang'E - 4 Lunar Penetrating Radar. Science Advances, 6, (9), eaay6898. https://doi. org/10. 1126/sciadv. aay6898

[14] Li, C., Xing, S., Lauro, S. E., Su, Y., Dai, S., Feng, J., et al. (2018). Pitfalls in GPR data interpretation: False reflectors detected in lunar radar cross sections by Chang'E - 3. IEEE Transactions on Geoscience and Remote Sensing, 56 (3), 1325 - 1335. https://doi. org/10. 1109/tgrs. 2017. 2761881

[15] Li, J., Chen, Y., & Zeng, Z. F. (2020). Estimated lunar regolith structure based on the least - squares Kirchhoff migration of CE - 3 Lunar Penetrating Radar data. In IEEE Geoscience and Remote Sensing Letters, X (pp. 1 - 5). https://doi. org/10. 1109/LGRS. 2020. 2986033

[16] Li, J., Zeng, Z. F., & Liu, C. (2017). A study on lunar regolith quantitative random model and lunar penetrating radar parameter inversion. IEEE Geoscience and Remote Sensing Letters, 14 (11), 1953 - 1957. https://doi. org/10. 1109/LGRS. 2017. 2743618

[17] Liu, J., Ren, X., Yan, W., Li, C., Zhang, H., Jia, Y., et al. (2019). Descent trajectory reconstruction and landing site positioning of Chang'E - 4 on the lunar far side. Nature Communications, 10 (1), 4229. https://doi. org/10. 1038/s41467 - 019 - 12278 - 3

[18] Losiak, A., Wilhelms, D. E., Byrne, C. J., Kevin, T., Weider, S. Z., Kohout, T., et al. (2009). A new lunar impact crater database. Houston, TX: Paper presented at 40th Lunar and Planetary Science Conference.

[19] McKay, D. S., Heiken, G., Basu, A., Blanford, G., Simon, S., Reedy, R., et al. (1991). The lunar regolith. In G. H. Heiken, D. T. Vaniman, & B. M. French (Eds.), Lunar sourcebook: A user's guide to the Moon (pp. 285 - 356). Cambridge: Cambridge University Press.

[20] Melosh, H. J., Kendall, J., Horgan, B., Johnson, B. C., Bowling, T., Lucey, P. G., & Taylor, G. J. (2017). South pole - Aitken basin ejecta reveals the Moon's upper mantle. Geology, 45 (12), 1063 - 1066. https://doi. org/10. 1130/G39375. 1

[21] Moriarty, D. P. III, & Pieters, C. M. (2018). The character of south pole Aitken basin: Patterns of surface and subsurface composition. Journal of Geophysical Research: Planets, 123, 729 - 747. https://doi. org/10. 1002/2017JE005364

[22] Nemeth, T., Wu, C., & Schuster, G. T. (1999). Least - squares migration of incomplete reflection data. Geophysics, 64 (1), 208 - 221. https://doi. org/10. 1190/1. 1444517

[23] Paskert, J. H., Hiesinger, H., & van der Bogert, C. H.(2018). Lunar far side volcanism in and around the south pole - Aitken basin. Icarus, 299, 538 - 562. https://doi. org/10. 1016/j. icarus. 2017. 07. 023

[24] Stuart - Alexander, D. E.(1978). Geologic map of the central far side of the Moon. Washington, DC: U. S. Geological Survey. https://doi. org/10. 3133/i1047

[25] Wilhelms, D. E., Howard, K. A., & Wilshire, H. G. (1979). Geologic map of the south side of the Moon. Washington, DC: U. S. Geological Survey. https://doi. org/10. 3133/i1162

[26] Xiao, L., Zhu, P., Fang, G., Xiao, Z., Zou, Y., Zhao, J., et al. (2015). A young multilayered terrane of the northern Mare Imbrium revealed by Chang'E - 3 mission. Science, 347 (6227), 1226 - 1229. https://doi. org/10. 1126/science. 1259866

[27] Yingst, R. A., Chuang, F. C., Berman, D. C., & Mest, S. C. (2017). In Geologic mapping of the Planck quadrangle of the Moon (LQ - 29), 48th Lunar and Planetary Science Conference. The Woodlands, TX.

[28] Yingst, R. A., & Head, J. W. (1997). Volumes of lunar lava ponds in South Pole - Aitken and Orientale basins: Implications for eruption conditions, transport mechanisms, and magma source regions. Journal of Geophysical Research, 102 (E5), 10909 - 10931. https://doi. org/10. 1029/97JE00717

[29] Yuan, Y., Zhu, P., Zhao, N., Xiao, L., & Qiao, L. (2017). The 3D geological model around Chang'E - 3 landing site based on lunar penetrating radar channel - 1 data. Geophysical Research Letters, 44, 6553 - 6561. https://doi. org/10. 1002/2017GL073589

[30] Zhang, J., Yang, W., Hu, S., Lin, Y., Fang, G., Li, C., et al. (2015). Volcanic history of the Imbrium basin: A close - up view from the lunar rover Yutu. Proceedings of the National Academy of Sciences of the United States of America, 112 (17), 5342 - 5347. https://doi. org/10. 1073/pnas. 1503082112

[31] Zhang, L., Zeng, Z., Li, J., Huang, L., Huo, Z., Zhang, J., & Huai, N. (2019). A story of regolith told by lunar penetrating radar. Icarus, 321, 148 - 160. https://doi. org/10. 1016/j. icarus. 2018. 11. 006

[32] Zhang, L., Zeng, Z., Li, J., Lin, J., Hu, Y., Wang, X., & Sun, X. (2018). Simulation of the lunar regolith and lunar - penetrating radar data processing. IEEE Journal of Selected Topics in Applied Earth Observations and Remote Sensing, 11 (2), 655 - 663. https://doi. org/10. 1109/JSTARS. 2017. 2786476

Rock abundance and evolution of the shallow stratum on Chang'e - 4 landing site unveiled by lunar penetrating radar data

Ling Zhang[1,2], Yi Xu[1], Roberto Bugiolacchi[1,3], Bin Hu[2], Cai Liu[2], Jialong Lai[1,4], Zhaofa Zeng[2], Zhijun Huo[2]

1. State Key Laboratory of Lunar and Planetary Sciences, Macau University of Science and Technology, Macau, China
2. College of Geo - exploration Science and Technology, Jilin University, Changchun 130026, China
3. Earth Sciences, University College London, London, UK
4. School of Science, Jiangxi University of Science and Technology, Ganzhou, China

Abstract: The Yutu - 2 rover, part of the Chang'e - 4 (CE - 4) mission to the lunar farside, has been exploring the Von Kármán crater (VK) within the South Pole - Aitken (SPA) gathering data on its geology. The Lunar Penetrating Radar's (LPR) 500MHz antenna reveals the subsurface structure up to 40m in depth, which helps to construct models on the evolution of the shallow stratum. This work presents the analysis and interpretation of the data collected during the first twenty lunar days, from January 2019 to July 2020. Below an 11±4 - meter surface regolith layer, the radar image reveals well - defined reflectors that are interpreted as discrete depositional layers. These marked density boundaries could indicate a geologically brief time interval between each arrival, which prevented the materials to undergo a deeper homogenization through space weathering. To investigate further, we generated a subsurface rock distribution map using the correlation of dual - channel antennas data from the same channel. Clusters of rocks located at the bottom of each ejecta stratum are revealed, probably sorted by size due to the statistically less frequent occurrence of larger impact events admixing the deeper - seated materials. The bowl - shaped outline discovered in both rock and LPR energy distribution maps is interpreted to represent a 150±30m wide and 10±5m deep paleo - crater that was exca - vated within three thin ejecta layers. The CE - 4 LPR reveals a complex geological history within the Von Kármán crater with no fewer than four impact events delivering meters thick ejecta at the CE - 4 landing area following the end of the infilling phase.

Keywords: lunar penetrating radar Chang'e - 4 mission Von Kármán crater regdith ejecta shallow stratigrophic structure

1 Introduction

On 3 January 2019 the Chang'e - 4 (CE - 4) mission comprising of a lander and a rover (Yutu - 2) accomplished the first soft landing on the lunar farside and it remains in operation at the time of publication. It landed on the South Pole - Aitken (SPA) Basin and set down on the floor of the Von Kármán crater [Liu et al., 2019] with the aim at probing the surface and subsurface characteristics of this region previously explored only from orbiting crafts.

The impact that formed the SPA Basin, the largest known crater in the Solar System, could have potentially blasted into the lunar lower crust and exhumed upper mantle materials [Stuart - Alexander, 1978], thus offering a rare opportunity for sampling the inner structure of the Moon. The Von Kármán crater, with a 186km diameter and ~3km depth, is located near the center of the basin (central position coordinates 44. 45°S 176. 25°E (Figure S1) and it is superimposed on the much older Von Kármán M crater (pre - Nectarian). North of the Von Kármán crater is the Leibniz crater (245km), while in the northeast is the smaller Finsen crater (72km) [Wilhelms, 1987]. The Von Kármán crater has experienced a convoluted geological history starting from its for mation in the Pre - Nectarian period [Huang et al., 2018]. Key events included the arrival of ejecta from the Leibnitz impact which reshaped the northern morphology of the crater [Carrier et al., 1991; Yingst and Head, 1997]. Volcanic events during the subsequent Imbrian period produced the infill of the crater basin by basalts [Carrier et al., 1991; Fu et al., 2020; Huang et al., 2018; Lai et al., 2020; Paskert et al., 2018; Yingst and Head, 1997]. Between the end of the Late Imbrian and the beginning of the Early Eratosthenian periods, the Finsen crater, and to a lesser extent Von Kármán L' and Von Kármán L, significantly reshaped the morphology of the Von Kármán crater. No other major geological events occurred there - after, apart from surface reworking by relatively small impacts and the continuous gardening of the lunar regolith [Lin et al., 2020]. Although remote sensing has improved our current understanding of the region, the subsurface stratigraphy had not been accessible until now, if not by indirect inference [Carrier et al., 1991; Fu et al., 2020; Huang et al., 2018; Paskert et al., 2018; Yingst and Head, 1997].

The LPR equipped on the Yutu - 2 rover in CE - 4 mission was designed to probe the subsurface geological structure and the thickness of the lunar regolith at high resolution [Fang et al., 2014; Jia et al., 2018]. To reach its scientific targets, two different channels were implemented in the design of the LPR system. The first channel is a low -

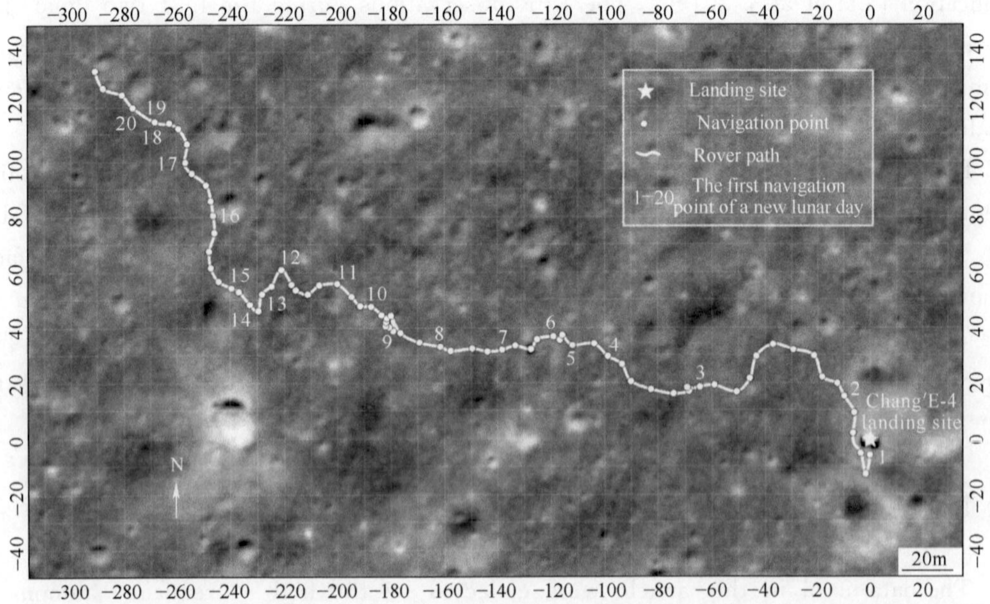

Fig. 1　The path of Yutu – 2 rover during the first twenty lunar days from January 2019 to July 2020. Background NAC image M1314237625LR [Robinson et al., 2010]

frequency antenna with a center frequency of 60MHz and bandwidth of 41MHz, which is mainly used to detect the geological structure of the upper, shallow crust. Its vertical resolution is 1 – 2m in the mare basalt and the penetration depth is no less than 360m [Zhang et al., 2020a]. The second channel is a high – frequency antenna with a center frequency of 500MHz and bandwidth of 500MHz, which is used to detect the thickness and structure of the lunar regolith layer [Fang et al., 2014; Xiao et al., 2015]. Its vertical resolution is 0. 3m in vacuum and can reach the depth of no less than 40m [Li et al., 2020]. More details of the LPR instrument and data specifications are provided at 'Appendix A' in the Supporting Materials.

An identical LPR instrument, part of the CE – 3 mission that targeted Mare Imbrium, successfully detected nine underground layers around the landing site [Dong et al., 2016; Fa et al., 2015; Feng et al., 2016; Hu et al., 2019; Lai et al., 2015; Xiao et al., 2015; Zhang et al., 2015, 2019, 2020b]. The energy amplitude, the hyper – bolic morphology formed by the diffraction of the boulders, and the reflection time differences contained in the radar data were used to calculate the parameters of the regolith on CE – 3 landing site, including the dielectric constant, density, electrical conduc – tivity, and estimate the $FeO + TiO_2$ fraction [Dong et al., 2016; Feng et al., 2016; Zhang et al., 2018]. Compared to the CE – 3 mission, the LPR of CE – 4 has been able to probe deeper into the surface substrate due to the lower titanium content of the VK's crater basalts, which acts as a barrier to the electromagnetic energy in the radar range. It has also been operating for a much longer time and still gathering data at the time of publication [Lai et al. , 2019] . The LPR observations during the first two lunar days reveal that the uppermost shallow layer is primarily made of low loss, highly porous, granular materials with embedded boulders of different sizes and that the underlying mare basalts are not detected at depths of less than 40m [Li et al., 2020].

In this study, we present the high - frequency LPR data collected during the first twenty lunar days, from January 2019 to July 2020. The energy distribution of the data and the resulting radar images reveal an extensive stratification of the subsurface, although with variations and local discontinuities. The rock distribution derived from the data also points to changes in rock size and internal layered structure of breccia. This paper adds to our understanding of the structure and evolution of the shallow subsurface structure on the farside of the Moon.

2 Data

The data used in this article were collected by the high - frequency antenna of CE - 4's LPR during the first twenty lunar days from January 2019 to July 2020 (IDs in Table S1). In this time, the Yutu - 2 rover equipped with the LPR had travelled for more than 450 meters (Fig. 1).

The scientific objective of the high - frequency channel is to conduct high resolution structural imaging within a depth of tens of meters in the Von Kármán crater. To produce an informative radar image, we built on the experience of processing similar data from the CE - 3 LPR data [Zhang et al., 2019; Zhao et al., 2014], and developed a set of new processing procedures for the CE - 4 LPR data:

1. Data format conversion - We selected the second channel data collected by antenna B. CE - 4 LPR data complies with the standard storage format in aviation and spaceflight field (. PDS). We converted the raw data into a binary format for processing, analysis, and imaging.

2. Ordering of the stitched data - The data of different lunar days are stored in different multiple files. After reading the data files one by one, they need to be stitched together to form a complete radar profile.

3. Deletion of duplicate redundant traces - Yutu - 2 rover sometimes stops at specific locations to collect VNIS data. However, the LPR antennas do not stop collecting radar data. It leads to repeated acquisitions in the same place, forming redundant and repeated traces. This step deletes the duplicate traces.

4. Time lag adjustment - When the high - frequency LPR channel is operative, the receiving antenna is turned on 28. 203ns earlier than the transmitting antenna. Therefore, this information is redundant and must be deleted.

5. Removal of the background noise - The strong reflection off the surface and the lateral noise of the instrument affect the quality of the data. This step removes the

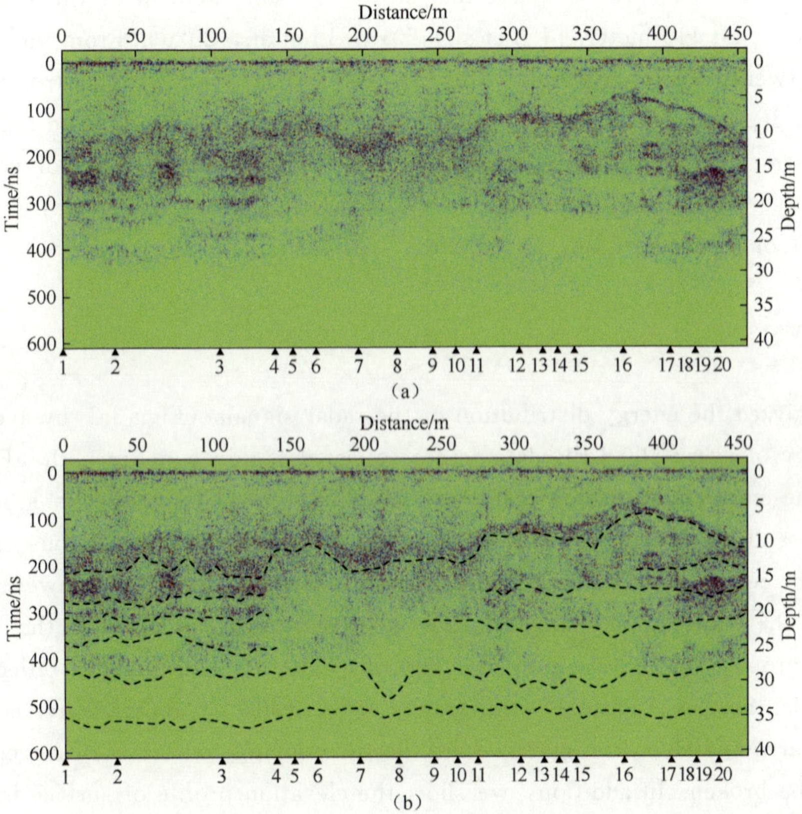

Fig. 2　(a) High - frequency channel LPR image during the first 20 lunar days from January 2019 to July 2020. (b) Inconsistency of stratification in the high - frequency radar image

horizontal background signals by subtracting the averaging trace.

6. Bandpass filtering to remove high and low frequency noise - The bandwidth of the high - frequency channel is 250MHz - 750MHz and the bandpass filtering of the same bandwidth was set to remove the signal noise outside this range.

7. Boosting weaker signals - Since the signal of the emitted electromagnetic wave becomes weaker from propagation with increasing depth, the energy of the reflected signal also weakens in strength. Consequently, a gain stage is applied as a boost.

8. Appling a migration technique - The migration method can improve the clarity of image and the time - depth conversion ($\varepsilon = 5$) to obtain the radar image used for interpretation.

Fig. 2a shows the high - frequency LPR image after processing. Further details of the LPR image in proper scale can be seen in Figure S2 and Data. mp4. 20 - lunar - day high - frequency radar image gives a perspective of the structure on the lunar farside up to a depth of about 40 meters. Compared with CE - 3 LPR data (Figure S3), the CE - 4 LPR achieves a higher signal - to - noise ratio and a deeper detection depth.

Continuous interfaces at depths of ~10m, ~27m (Fig. 2, Figure S2 and Data.

mp4) and ～34m (Figure S2 and Data. mp4) are consistent with the reported LPR results for the first 110 meters [Lai et al., 2019; Li et al., 2020]. From the radar image of the full twenty lunar days data, the first reflector feature with a strong continuity appears at a depth of ～10m: this is interpreted as a layer of fine - grained regolith (Fig. 2b). This stratum overlies a more cohesive layer with admixed rocks of various sizes (probably brec - cia), with some regions showing either strong (X - axle: 0 - 130m and 300 - 450m) or poor (X - axle: 130 - 300m) stratification.

3 Analysis

We analyzed the energy distribution of the radar signals (Fig. 3a) by averaging the signal energy of every 100 LPR data traces collected within a ～5m path (Figure S4). Based on the energy distribution map, several strong subsurface reflections between the depth of 10 and 25 m could be identified, which appear mostly continuous except for a section at the center of the image. Here, a bowl - shaped pattern of the power spectrum is observed in the same region characterized by poor stratification. Based on the energy map of LPR we propose that a buried impact crater disrupted the continuity of reflectors in the stratum (Fig. 3b). The ambient stratigraphy affected by the impact crater is illustrated with white arrows shown in Fig. 3b where the surrounding continuity of the ejecta layers appears to be broken. In addition, we show the elevation profile of surface is correlated with that of the first subsurface reflector at depth of ～11m in the supplemen - tary Fig. S5, indicating the surface topography is affected by the underlying presumed impact crater although the surface has been heavily modified by later impacts, gravity and space weathering effects. The diameter of the crater is 150±30m and its depth 10±5m.

After landing, the Yutu - 2 rover explored the lunar surface with an onboard panorama camera (PCAM), LPR, and with a visible and near - infrared imaging spectrometer (VNIS) targeting rock fragments lying as case studies. Overall, the region around the Chang'E - 4 landing site is characterized by the absence of boulders, based on both orbit and in - situ observations; however, several fresh craters encountered by the Yutu - 2 rover have blocky rims and interiors in sharp contrast with the much more common heavily degraded crater population in the area. The exposed fragments are less than 10cm long and identified as compacted regolith clumps formed by secondary impacts and impact meltconglutinated regolith breccia projectiles, not subsurface rock breccia [Lin et al., 2020; Ding et al., 2020]. The LPR can estimate rock distribution in the regolith using the correlation of two antenna data of the CE - 4 high frequency channel [Hu et al., 2019]. The process of estimating rock abundances is shown in 'Appendix C' in the Supporting Materials. Fig. 4 represents the rock distribution map in the landing area.

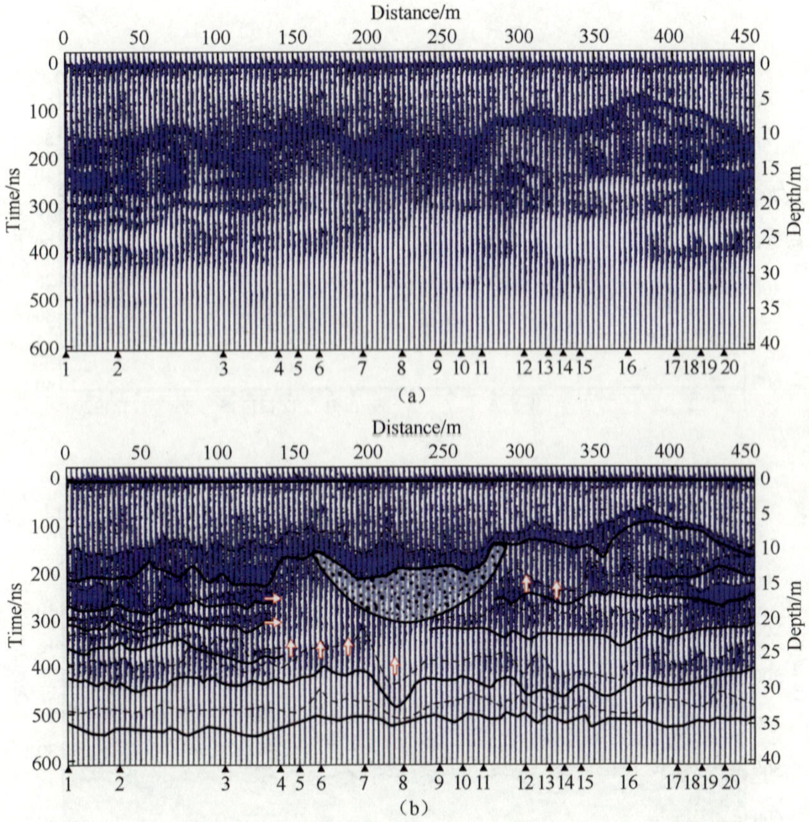

Fig. 3　(a) Energy distribution traces of the high - frequency LPR data.
(b) Stratification in LPR energy traces

4　Result and discussion

Based on the current understanding of the evolution of the Von Kármán crater supplemented by the new data and interpretation, we propose a four - layer model of the subsurface stratigraphy (Fig. 5):

(Ⅰ) A surface layer of comminuted particles (regolith). From the geologically quiet Eratosthenian through the Copernican periods the surface materials have been modified both mechanically, by micro - impacts [e. g., Szalay and Horányi, 2016], and geochemically, by various other space weathering processes [Pieters et al., 2000; Taylor et al., 2001]. In particular, the components of the uppermost regolith layer are likely to contain an important fraction of ejecta materials from the Finsen crater, based on the chronology of the surrounding craters derived from recent studies, and on the thickness of the crater ejecta and composition of the local regolith [Ding et al., 2020; Gou et al., 2019, 2020; Huang et al., 2018; Lai et al., 2019; Lu et al., 2021].

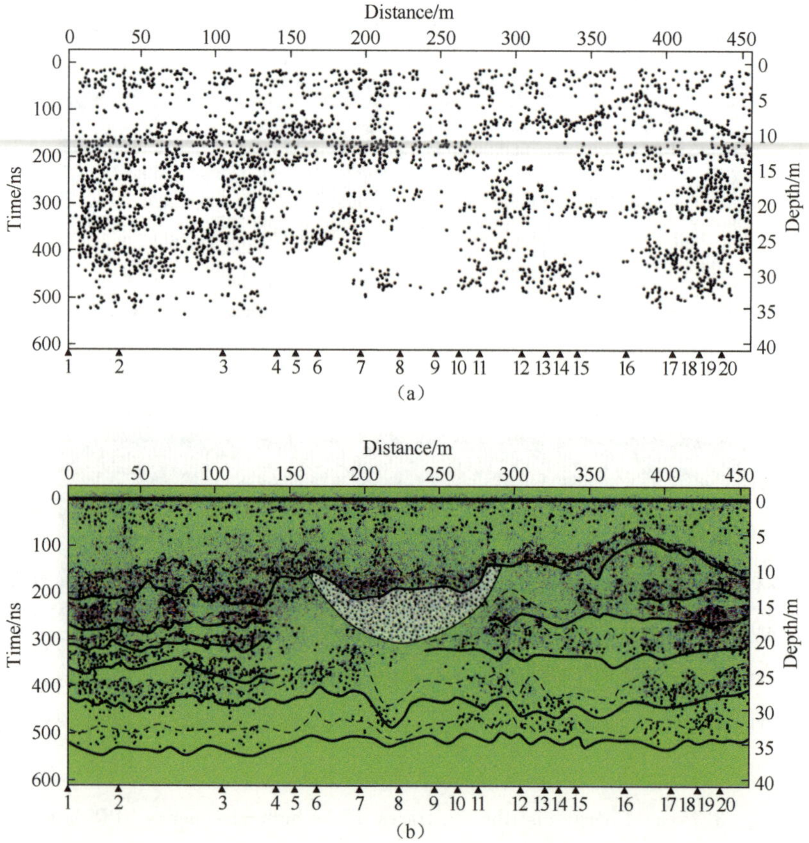

Fig. 4 (a) Rock distribution from LPR data. (b) Rock distribution and stratification in LPR data

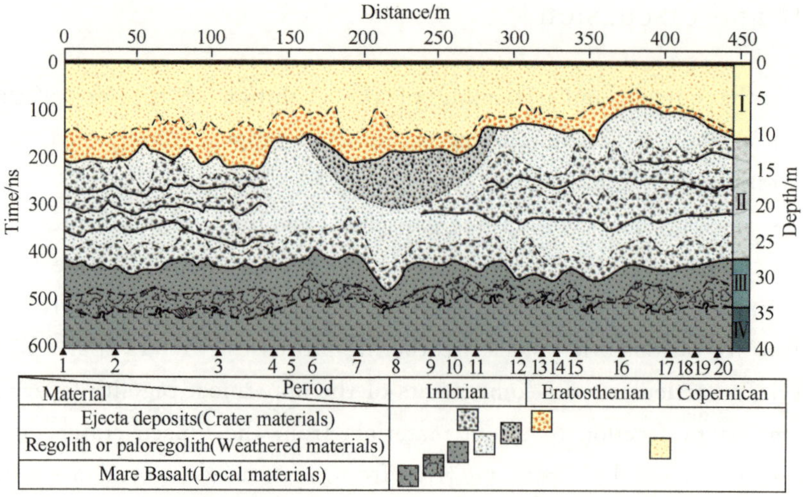

Fig. 5 The stratigraphic division of CE - 4 high - frequency radar data

Table1　Thickness of subsurface layers based on the rock distribution map

Layer		Thickness (m)	Thickness of weathered skin (m)
LayerI：Finsen Ejecta		11±4	8±4
LayerII	Ejecta 1	4±2	2±2
	Ejecta 2	3±1	2±1
	Ejecta 3	4±2	2±2
	Ejecta 4	6±3	3±3
LayerIII：	paleo - regolith layer	7±3	4±2

（Ⅱ）Ejecta layers from several major impact craters. Towards the end of the Imbrian period and the beginning of the Eratosthenian period，ejecta from craters both inside and outside the Von Kármán crater（craters Von Kármán L，Von Kármán L'，Orem，etc. as in Figure S1）built layered deposits over the basalt bedrock. This section comprises of at least four（major）ejecta overlays. During the interval between each ejecta arrival，exposed surface materials are subjected to mechanical weathering，resulting in the lowering of the average rocks' size - frequency distribution. By this mechanism，each stage of ejecta deposition would have resulted in two sub - stratigraphic regions，with the upper materials comprising mostly of finer particles populated by few scattered rock blocks，while the lower materials with a more heterogeneous size distribution. The weathering period of each deposition appears to have been geologically brief，thus preventing complete homogenization. The crater size - frequency distribution measurements of the Finsen crater and the mare basalts within the VK crater suggest that the time interval between the last basalt filling events and the Finsen impact could have been as short as 100 - 200Ma，including the other four main ejecta depositions mentioned earlier [Lu et al.，2021]. The thickness of each ejecta layer and its weathered skin are summarized in Table 1. Based on mathematical models of ejecta thickness [McGetchin et al.，1973] and [Pike，1974]，the meters - thick ejecta layers，as from Fig. 5 and listed in Table 1，can be used to constrain both the distance and the size of the source impacts. The estimated thickness of ejecta deposits from Von Kármán L and Von Kármán L' is ~ 2m and 2m，respectively [Huang et al.，2018；Lai et al.，2020]. Based on the modeling results，Imbrium impact events such as Maksutov delivered meter - thick ejecta deposits at the CE - 4 site [Lai et al.，2020]. However，the lateral variations in the 20 - day radar image are not conclusive based on current data：as the exploration mission from the LPR continues，it will potentially provide further insights into the potential source of the ejecta.

（Ⅲ）A small number of observed reflectors in the third continuous interface. They could represent a paleo - regolith layer that had time to develop on the surface of the (then) exposed basalts，which was subsequently buried by impact ejecta. The regolith layer can also be split into two sub - stratigraphic structures each characterized by different grain size and rock blocks distribution.

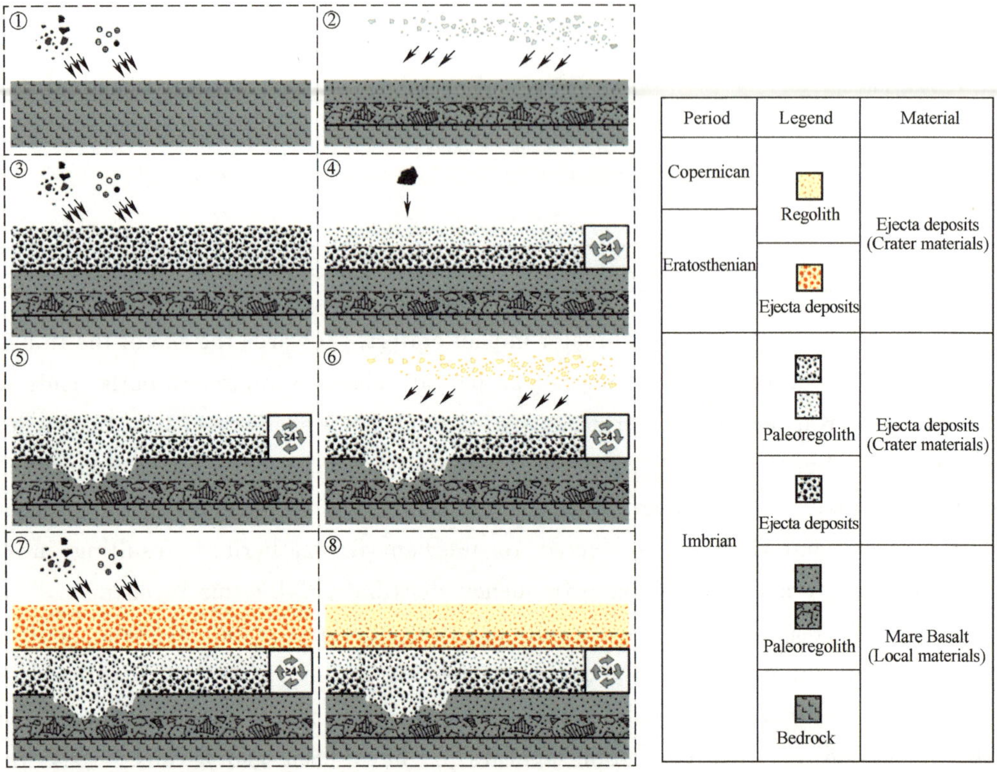

Fig. 6　Evolution of the shallow subsurface structure at the CE - 4 landing site

(Ⅳ) No obvious reflections are detected from the deepest signals, suggesting materials at this depth to be relatively homogenous. Therefore, this layer could represent the basalt bedrock as also indicated by crater geomorphological evidence [Huang et al., 2018; Fu et al., 2020]. Another possibility is that rock breccia/fractured basalts are indeed present at such depths, but they reside beyond the radar detection limit and clear reflections are not generated.

Based on our interpretation of the data, we propose the following evolutionary scenario of the shallow subsurface structure of the Von Kármán crater (Fig. 6 and Evolution. mp4):

(1) During the Imbrian period, the crater interior was the theatre of several extrusive events that formed the thick basalt bedrock. Today this is mantled by a veneer of both autochthonous and allochthonous unconsolidated materials (breccias, ejecta mate - rials and regolith).

(2) Towards the end stage of the Imbrian period, the exposed mare basalts endured both the disaggregating and constructive (i. e., agglutinates) effects of space weathering culminating in a paleo - regolith layer characterized by two sub - stratigraphic structures with different grain sizes and rock blocks distributions (from① to ② in Fig. 6).

(3) Towards the end of the Imbrian period and up to the beginning of the Eratosthenian, ejecta from different craters both inside and outside of Von Kármán crater were deposited in the CE - 4 landing site area. This region was at the center of at least four crater ejecta blankets depositions. In the time periods between each arrival, the exposed surface materials were exposed to weathering. Consequently, each stage of ejecta produced two sub - stratigraphic structures: the upper materials dominated by fine particle sizes with fewer rock blocks, and the lower materials that are coarser with a higher density of rock blocks (from② to ④ in Fig. 6).

(4) After receiving more than four crater ejecta blankets depositions, at one point, a relatively large meteorite struck the surface, which caused the obliteration of the layered structure (from④ to ⑤ in Fig. 6).

(5) In the early Eratosthenian, ejecta from the Finsen impact reached the CE - 4 landing site. No other significant ejecta event followed after. From this period on into the Copernican, the uppermost materials weathered into the present regolith (from⑥ to ⑧ in Fig. 6).

The supplemental evolution video (Evolution. mp4) shows the evolutionary scenario of CE - 4 surveying area (Fig. 6): Four main ejecta delivery events occurred atop the paleoregolith stratum that had developed on the mare surface; one relatively large impact event disrupted the layered subsurface structure at the point of contact. Later, the crater was buried by a substantial amount of ejecta material from the Finsen impact. Eventually, a ~ 11m thick regolith layer built up due to space weathering and micrometeorite bombardments.

5 Conclusion

In this study, we have processed the high - frequency LPR data collected by the Chang'e - 4 radar onboard the Yutu - 2 rover during the first twenty lunar days of exploration. The subsurface structure and stratigraphy of the Von Kármán crater within a depth of nearly 40 meters in the landing area was derived and used to build a complex geological scenario.

The constructed LPR image reveals localized and, in places, continuous, strong variations in reflection indicating physical discontinuities most probably relating to changes in the materials' density. The energy distribution from the data helps to further investigate the variations in the stratification pattern. There is evidence based on the shape of the discontinuity that a larger meteorite impacted the surface at some point in time, causing the obliteration of the target's layered structure.

The rock distribution derived from the LPR data also indicates a clear stratification structure formed by multiple ejecta which is preserved below the regolith layer. The thickness information of each ejecta layer and its inferred weathering degree can help to constrain the chronology of nearby impact craters and improve our models of ejecta distribution.

Acknowledgements

All of the authors appreciate the Chang'e－4 team for making such a great successful mission and providing scientific data. This work is supported by the Science and Technology Development Fund of Macau (0089/2018/A3, 121/2017/A3, 2018/2018/A2, 0049/2020/A1), the Preresearch Project on Civil Aerospace Technologies of China National Space Administration (D020101), the Postdoctoral Innovation Talent Support Program of China (BX20200148), China Postdoctoral Science Foundation funded project (2020M681041) and the National Natural Science Foundation of China (41874134, 41574109).

References

[1] Carrier, W., et al., 1991. Lunar Sourcebook: A User's Guide to the Moon. Cambridge Univ. Press, New York.

[2] Ding, C., et al., 2020. Fragments delivered by secondary craters at the Chang'E－4 landing site. Geophys. Res. Lett. 47 (7). https://doi. org /10 . 1029 /2020GL087361.

[3] Dong, Z., et al., 2016. Parameters and structure of lunar regolith in Chang'E－3 land－ing area from lunar penetrating radar (LPR) data. Icarus 282, 40－46. https://doi . org /10. 1016/j. icarus. 2016 . 09 . 010.

[4] Fa, W., et al., 2015. Regolith stratigraphy at the Chang'E－3 landing site as seen by lunar penetrating radar. Geophys. Res. Lett. 42 (23), 10179－10187. https://doi . org /10. 1002/2015GL066537.

[5] Fang, G., et al., 2014. Lunar penetrating radar onboard the Chang'e－3 mission. Res. Astron. Astrophys. 14 (12), 1607－1622. https://doi. org /10. 1088/1674－4527/14 /12 /009.

[6] Feng, J., et al., 2016. Dielectric properties estimation of the lunar regolith at CE－3 landing site using lunar penetrating radar data. Icarus 284, 424－430. https://doi. org /10. 1016 /j. icarus. 2016. 12. 005.

[7] Fu, X., et al., 2020. The subsurface structure and stratigraphy of the Chang'E－4 land－ing site: orbital evidence from small craters on the Von Kármán crater floor. Res. Astron. Astrophys. 20 (1). https://doi . org /10 . 1088 /1674—4527 /20 / 1 /8 . Gou.

[8] Gou, S., et al., 2019. Lunar deep materials observed by Chang'e－4 rover. Earth Planet. Sci. Lett. 528. https://doi . org /10 . 1016 /j . epsl . 2019 . 115829.

[9] Gou, S., et al., 2020. In situ spectral measurements of space weathering by Chang'e－4 rover. Earth Planet. Sci. Lett. 535. https://doi . org /10 . 1016 /j . epsl . 2020 . 116117.

[10] Hu, B., et al., 2019. Rock location and quantitative analysis for regolith on

Chang'e－3 landing site based on local similarity constraint. Remote Sens. (Basel) 11 (5), 530. https: //doi . org /10 . 3390 /rs11050530.

[11] Huang, J., et al., 2018. Geological characteristics of Von Kármán crater, northwestern South Pole Aitken basin: Chang'E 4 landing site region. J. Geophys. Res., Plan－ets 123, 1684－1700. https: //doi . org /10 . 1029 /2018JE005577.

[12] Jia, Y., et al., 2018. The scientific objectives and payloads of Chang'E－4 mission. Planet. Space Sci. 162, 207－215. https: //doi . org /10 . 1016 /j . pss. 2018. 02. 011.

[13] Lai, J., et al., 2015. Structural analysis of lunar subsurface with Chang'E－3 lunar pen－etrating radar. Planet. Space Sci. 120, 96－102. https: //doi . org / 10 . 1016 /j . pss . 2015 . 10 . 014.

[14] Lai, J., et al., 2019. Comparison of dielectric properties and structure of lunar regolith at Chang'e－3 and Chang'e－4 landing sites revealed by ground－penetrating radar. Geophys. Res. Lett. 46 (22), 12783－12793. https: //doi . org /10 . 1029 /2019GL084458.

[15] Lai, J., et al., 2020. First look by the Yutu－2 rover at the deep subsurface structure at the lunar farside. Nat. Commun. 11, 3426. https: //doi . org /10 . 1038 /s41467－020－17262－w.

[16] Li, C., et al., 2020. The Moon's farside shallow subsurface structure unveiled by Chang'E－4 lunar penetrating radar. Sci. Adv. 6 (9). https: //doi . org /10 . 1126 /sciadv. aay6898.

[17] Liu, J., et al., 2019. Descent trajectory reconstruction and landing site positioning of Chang'E－4 on the lunar farside. Nat. Commun. 10 (1). https: //doi . org / 10 . 1038 /s41467－019－12278－3.

[18] Lin, H., et al., 2020. New insight into lunar regolith－forming processes by the lunar rover Yutu－2. Geophys. Res. Lett. 47 (14), e2020GL087949. https: // doi . org /10 . 1029 /2020GL087949.

[19] Lu, et al., 2021. Chronological sequence of Chang'E－4 landing zone within Von Kármán crater. Icarus 354 (15), 114086. https: //doi . org /10 . 1016 /j . icarus. 2020 . 114086.

[20] McGetchin, T. R., et al., 1973. Radial thickness variation in impact crater ejecta: implications for lunar basin deposits. Earth Planet. Sci. Lett. 20 (2), 226－236. https: //doi . org /10 . 1016 /0012－821x (73) 90162－3.

[21] Paskert, J., et al., 2018. Lunar farside volcanism in and around the South Pole－Aitken basin. Icarus 299, 538－562. https: //doi . org /10 . 1016 /j . icarus . 2017. 07. 023.

[22] Pieters, C. M., et al., 2000. Space weathering on airless bodies: resolving a mystery with lunar samples. Meteorit. Planet. Sci. 35 (5), 1101－1107. https: //doi . org /10 . 1111 /j . 1945－5100 . 2000 . tb01496 . x.

[23] Pike, R. J., 1974. Ejecta from large craters on the Moon: comments on the

geometric model of McGetchin et al. Earth Planet. Sci. Lett. 23 (3), 265 - 271. https: //doi . org /10 . 1016 /0012 - 821X (74) 90114 - 9.

[24] Robinson, M. S., et al., 2010. Lunar reconnaissance orbiter camera (LROC) instru - ment overview. Space Sci. Rev. 150, 81 - 124. https: //doi . org /10 . 1007 /s11214 - 010 - 9634 - 2.

[25] Stuart - Alexander, D. E., 1978. Geologic Map of the Central Farside of the Moon. U. S. Geological Survey, Washington, DC.

[26] Szalay, J. R., Horányi, M., 2016. Lunar meteoritic gardening rate derived from in situ LADEE/LDEX measurements. Geophys. Res. Lett. 43, 4893 - 4898. https: //doi . org /10 . 1002 /2016GL069148.

[27] Taylor, L. A., et al., 2001. Lunar mare soils: space weathering and the major effects of surface - correlated nanophase Fe. J. Geophys. Res. 106, 27, 985 - 28, 000. https: //doi . org /10 . 1029 /2000JE001402.

[28] Wilhelms, D. E., 1987. The Geologic History of the Moon. U. S. Geological Survey, Washington, DC.

[29] Xiao, L., et al., 2015. A young multilayered terrane of the northern Mare Imbrium revealed by Chang'E - 3 mission. Science 347 (6227), 1226 - 1234. https: //doi . org /10 . 1126 /science . 1259866.

[30] Yingst, R. A., Head, J. W., 1997. Volumes of lunar lava ponds in South Pole - Aitken and Orientale basins: implications for eruption conditions, transport mechanisms, and magma source regions. J. Geophys. Res., Planets102, 10909 - 10931. https: //doi . org /10 . 1029 /97JE00717.

[31] Zhang, J., et al., 2015. Volcanic history of the Imbrium basin: a close - up view from the lunar rover Yutu. Proc. Natl. Acad. Sci. USA 112 (17), 5342 - 5348. https: //doi . org /10 . 1073 /pnas . 1503082112.

[32] Zhang, L., et al., 2018. Parameter estimation of lunar regolith from lunar penetrating radar data. Sensors 18 (9). https: //doi . org /10 . 3390 /s18092907.

[33] Zhang, L., et al., 2019. A story of regolith told by lunar penetrating radar. Icarus 321, 148 - 160. https: //doi . org /10 . 1016 /j . icarus . 2018 . 11. 006.

[34] Zhang, L., et al., 2020a. Stratigraphy of the Von Kármán crater based on Chang' E - 4 lunar penetrating radar data. Geophys. Res. Lett. 47 (15), e2020GL088680. https: //doi . org /10 . 1029 /2020GL088680.

[35] Zhang, L., et al., 2020b. The subsurface structure on the CE - 3 landing site: LPR CH - 1 data processing by shearlet transform. Pure Appl. Geophys. 177, 3459 - 3474. https: //doi . org /10 . 1007 /s00024 - 020 - 02417 - z.

[36] Zhao, N., et al., 2014. The preliminary processing and analysis of LPR channel - 2B data from Chang'E - 3. Sci. China, Phys. Mech. Astron. 57 (12), 2346 - 2353. https: //doi . org /10 . 1007 /s11433 - 014 - 5590 - 2.

Self - Organization Characteristics of Lunar Regolith Inferred by Yutu - 2 Lunar Penetrating Radar

Xiang Zhang[1,2,3], Wenmin Lv[1,2,3], Lei Zhang[1,2], Jinhai Zhang[1,2], Yangting Lin[1,2] and Zhenxing Yao[1,2]

1. Key Laboratory of Earth and Planetary Physics, Institute of Geology and Geophysics, Chinese Academy of Sciences, Beijing 100029, China; zhangxiang191 @ mails. ucas. ac. cn (X. Z.); lvwenmin18 @ mails. ucas. ac. cn (W. L.); zhangl @ mail. iggcas. ac. cn (L. Z.); linyt @ mail. iggcas. ac. cn (Y. L.); yaozx@mail. iggcas. ac. cn (Z. Y.)
2. Innovation Academy of Earth Science, Chinese Academy of Sciences, Beijing 100029, China
3. College of Earth and Planetary Sciences, University of Chinese Academy of Sciences, Beijing 100029, China

Abstract: Most previous studies tend to simplify the lunar regolith as a homogeneous medium. However, the lunar regolith is not completely homogeneous, because there are weak reflections from the lunar regolith layer. In this study, we examined the weak heterogeneity of the lunar regolith layer using a self - organization model by matching the reflection pattern of both the lunar regolith layer and the top of the ejecta layer. After a series of numerical experiments, synthetic results show great consistency with the observed Chang'E - 4 lunar penetrating radar data and provide some constraints on the range of controlling parameters of the exponential self - organization model. The root mean square permittivity perturbation is estimated to be about 3% and the correlation distance is about 5 - 10cm. Additionally, the upper layer of ejecta has about 1 - 2 rocks per square meter, and the rock diameter is about 20 - 30cm. These parameters are helpful for further study of structural characteristics and the evolution process of the lunar regolith. The relatively small correlation distance and root mean square perturbation in the regolith indicate that the regolith is mature. The weak reflections within the regolith are more likely to be due to structural changes rather than material composition changes.

Keywords: lunar penetrating radar; numerical simulation; lunar regolith; self - organization model

1　Introduction

The structure of the lunar shallow surface is complex due to long - term impacting, sputtering and stacking processes; in addition, the Moon has no strong cementation due to the absence of interstitial water. Thus, the heterogeneity on the Moon can be well retained and is much stronger than that on the Earth. The remote - sensing spectrum and radar detection results show that there is a thick regolith layer on the top of the Von Kármán crater [1 - 3]. This provides an excellent opportunity for detection of the structural characteristics and the stacking mode of the lunar regolith, which are critical for understanding the formation process of the lunar regolith. However, in the past, the lunar regolith has usually been assumed to be a uniform layered medium [4 - 8]. A series of recent studies have shown that the lunar regolith is essentially inhomogeneous [9 - 16] but the specific structural characteristics of the lunar regolith are still unclear.

Radar is widely used in the exploration of the Moon and Mars, and has become a necessary tool to detect the underground structure of extraterrestrial bodies [17 - 23]. The exploration of the Moon, Mars, and other planets with penetrating radars may benefit from the long experience in using the same approaches (the same type of radars and models to interpret the measurements) in studies carried out on Earth. The Yutu - 2 rover traveled a long distance (>500m) with a fine spatial sampling interval (~3 - 5cm) on the far side of the Moon. Such an observation provides a significant opportunity to explore the local fine structure under the rover's path. The lunar regolith at the landing site has a thickness of up to ~12m [24 - 26], which is much thicker than that detected on the near side of the Moon by the Chang'E - 3 lunar penetrating radar (LPR) [27, 28].

Surprisingly, even though the thickness of the lunar regolith is high (>10m), the Chang'E - 4 LPR profile still shows some weak reflections in the lunar regolith, and their amplitudes are far smaller than those of the reflections from the underlying ejecta. This indicates that the lunar regolith is not homogeneous but weakly heterogeneous. The weak heterogeneity of the lunar regolith does not seriously influence the imaging and inversion [24, 29, 30] because it is too weak compared with the reflections from the ejecta. However, further study of these weak heterogeneous characteristics would be helpful in revealing the process of lunar regolith formation, particularly to the most recent stage of local small impacts [31]. Of course, long - distance heavy bombardments may also convey an amount of small - particle ejecta and contribute to such an ultra - thick layer of fine regolith on the far side of the Moon [32].

Most traditional numerical simulations of LPR data assume layered models [4 - 8] and cannot well represent the heterogeneity of the lunar regolith. Recently, several works have considered the heterogeneity [9 - 15] but do not take typical models into account (e. g., cracks and ejecta). Lv et al. [16] constructed four typical models and performed numerical simulations using the time-domain finite-difference method, which can better present the wave propagation in the lunar regolith with strong heterogeneity. Some works noted the existence of weak heterogeneity of the lunar regolith [11, 33-35]. However, to date, no numerical simulation has been conducted specifically on the weak heterogeneity of the lunar regolith, probably due to three aspects: 1. lack of methods describing heterogeneous media; 2. technical difficulty in constructing random models; 3. excessive computational cost of numerical simulations [16] compared with traditional methods. In this study, we assumed that the weak heterogeneity of the lunar regolith obeys the self-organization model [36], which is basically a stochastic approach for dealing with scattering waves through randomly inhomogeneous structures and is well known in the seismology field; then, we used the finite-difference time-domain method [16] to carry out a series of numerical simulations using various models; next, we analyzed the influence of different parameters on the weak reflection characteristics; finally, we constrained the possible range of each controlling parameter through grid searching techniques by matching the numerical simulation results with the Yutu-2 LPR observations, in terms of waveforms, envelopes, and reflection patterns.

2 Modeling and Simulation

The weak heterogeneity of the lunar regolith can be described by a stochastic model [36]. We keep a small part of the underlying ejecta by adding some sputtering blocks to the stochastic model [16]. This can provide a reliable reference for checking the amplitude variations from weak reflections of the lunar regolith to strong reflections of ejecta.

First, we generate a 2D dielectric permittivity background according to the typical 1D lunar regolith model [37-39] (Figure 1) $\varepsilon_r = 1.919^{\rho}$ and $\rho = 1.92(z+12.2)/(z+18)$, where ε_r is the relative dielectric permittivity, ρ (g/cm^3) is the density, and z (cm) is the depth. The velocity of radar wave propagation within the lunar regolith can be converted from the dielectric permittivity as $v = c/\sqrt{\varepsilon_r}$, where v is the velocity of the radar wave and c is the speed of light in vacuum [40]. Then, we add some random perturbations and some ejecta to the background model to generate the model.

The 2D model has a width of 15m and a depth of 16m (Figure 2). From top to bottom, it includes a 0.3m thick vacuum medium (with a relative dielectric permittivity of 1) between the rover bottom and the lunar surface, a 12m thick random medium representing the lunar regolith, and a ~4m thick random medium with some rocky blocks (i.e., ejecta layer). The numerical simulation method and parameter settings are consistent with those of Lv et al. [16]. The finite-difference method [41, 42] is used to discretization the spatial partial derivatives of Maxwell's curl equation [43, 44], and only the 2D transverse electric pattern is considered. The radar frequency is 400MHz to avoid visible numerical dispersion. We can analyze the reflection characteristics of the lunar regolith by either a snapshot of the wavefield at a given time or a waveform at a given position.

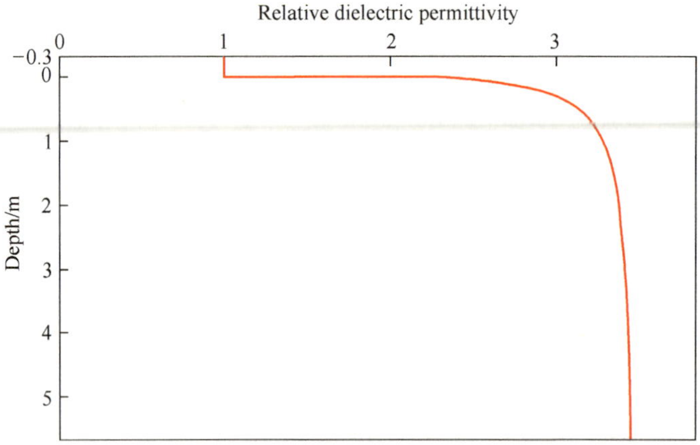

Figure 1　The 1D background model of relative dielectric permittivity of the lunar regolith. There is a 0.3m vacuum layer between the radar antennas and the lunar surface

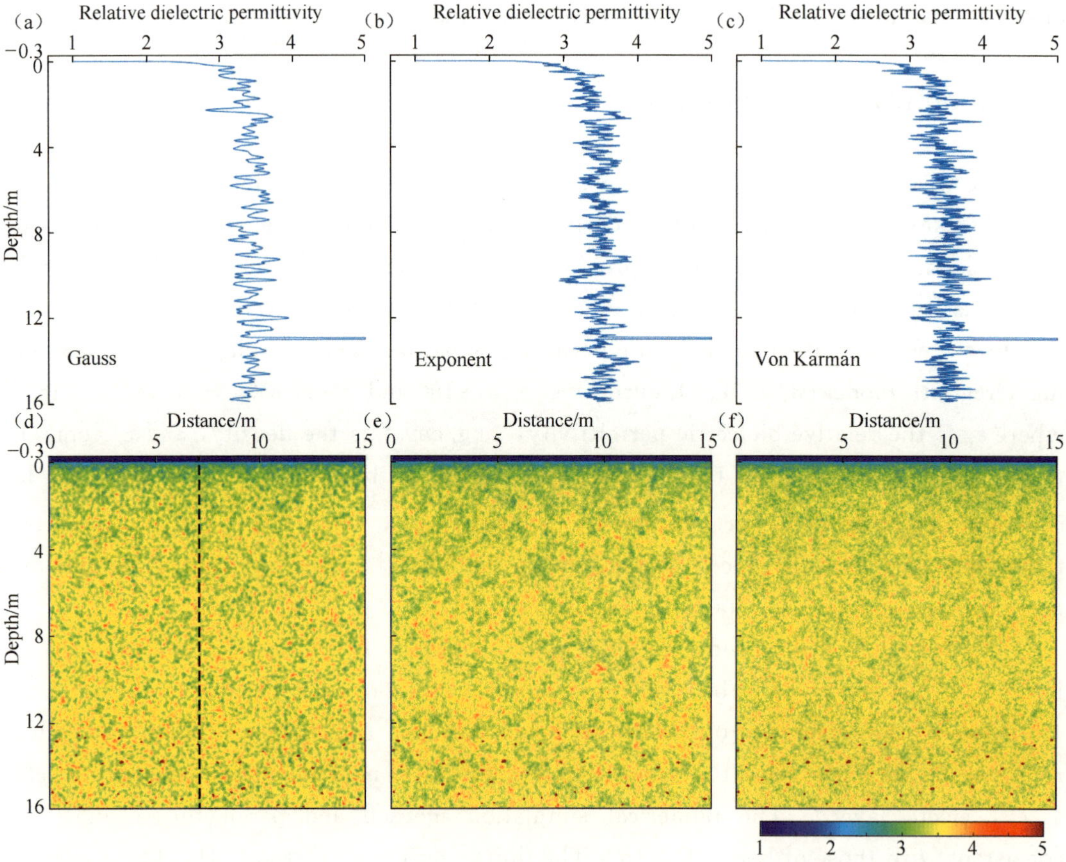

Figure 2　The Gaussian, exponential, and Von Kármán (zero - order) types of random media. (a - c) Fluctuation of the 1D probe along the center of the 2D model. (d - f) 2D relative dielectric permittivity models for three types of random media. The color bar represents the relative dielectric permittivity

2.1 Self - Organization Model

The self - organization model is an important approach for describing stochastic media. The controlling parameters generally include the mean value, variance (or standard deviation), and autocorrelation function (ACF). The ACF and its corresponding power spectral density function (PSDF) are the two most important parameters to reflect the distribution characteristics of random media. The most widely used autocorrelation functions include Gaussian, exponential, and Von Kármán types [45 - 48]. In the stochastic process theory, the PSDF is the Fourier transformation of ACF [49]. To characterize the random media, we define the ACF of the fractional permittivity fluctuation as:

$$R(\boldsymbol{x}) \equiv \langle \xi(\boldsymbol{y})\xi(\boldsymbol{y}+\boldsymbol{x}) \rangle \tag{1}$$

where non - dimensional quantity $\xi(\boldsymbol{x}) \equiv \delta\varepsilon_r(\boldsymbol{x})/\varepsilon_r^0$ is the fractional. uctuation of the permittivity, ε_r^0 is the mean permittivity, and $\delta\varepsilon_r$ is the perturbation of permittivity. This provides a statistical measure of the spatial scale and the magnitude of medium inhomogeneity.

The most tractable ACF is Gaussian ACF (Figure 2a, d), as given by:

$$R(\boldsymbol{x}) = R(r) = \varepsilon^2 e^{-r^2/a^2} \tag{2}$$

where a is the correlation distance. For the 3D case, the PSDF is:

$$P(\boldsymbol{m}) = P(m) = \varepsilon^2 \sqrt{\pi^3}\, a^3 e^{-m^2a^2/4} \tag{3}$$

where $\boldsymbol{m}$ is the wavenumber vector and $m = |\boldsymbol{m}|$. The Gaussian ACF is used to describe the medium with poor short - wavelength components because the PSDF goes rapidly to zero for a large $\boldsymbol{m}$. Many models are based on the Gaussian ACF due to its solid mathematical background [50, 51].

Another commonly used model is the exponential ACF (Figure 2b, e):

$$R(\boldsymbol{x}) = R(r) = \varepsilon^2 e^{-r/a} \tag{4}$$

For the 3D case, the PSDF is

$$P(\mathbf{m}) = P(m) = \frac{8\pi\varepsilon^2 a^3}{(1+a^2m^2)^2} \qquad \propto (am)^{-4} \text{ for } am \gg 1 \tag{5}$$

There is an extension of the exponential ACF, called Von Kármán ACF (Figure 2c, f):

$$R(\boldsymbol{x}) = R(r) = \frac{\varepsilon^2 2^{1-\kappa}}{\Gamma(\kappa)}\left(\frac{r}{a}\right)^{\kappa} K_{\kappa}\left(\frac{r}{a}\right) \qquad \text{for } \kappa = 0 \sim 1 \tag{6}$$

where $\Gamma(\kappa)$ is the gamma function and K_κ is the κ - th order modified Bessel function of the second kind. For the 3D case, the PSDF is:

$$P(\mathbf{m}) = P(m) = \frac{8\pi^{3/2}\Gamma(\kappa+3/2)\varepsilon^2 a^3}{\Gamma(\kappa)(1+a^2m^2)^{\kappa+3/2}} \qquad \propto (am)^{-2\kappa-3} \quad \text{for } am \gg 1 \tag{7}$$

Compared with the Gaussian ACF, the Von Kármán ACF (including exponential ACF) is more suitable for describing the random permittivity inhomogeneity of the lunar regolith due to the power law property of the Von Kármán ACF [36].

We assume that the randomness is spatially stationary, and we individually add Gaussian, exponential, and Von Kármán (zero - order) random fractional fluctuations to the background model. The ejecta are then added at the bottom of the lunar regolith (Figure 2).

Figure 3 shows the numerical simulation results of these three models. Because of the poor short - wavelength component of Gaussian ACF random media, their reflection characteristics show more long - wavelength structures. Compared with the Gaussian media, the Von Kármán media (including the exponential type) show more short - wavelength structures. Due to the intense sputtering and stacking process, the heterogeneity of the lunar regolith should be more consistent with the Von Kármán (including the exponential type) media [35, 36, 52]. However, we cannot well distinguish its difference from the exponential ACF due to the limited radar wave resolution, even though the zero - order Von Kármán model has greater heterogeneity at small scales. Therefore, we choose the exponential ACF to construct random media in the following experiments. This can simplify the model and facilitate the discussion of the results.

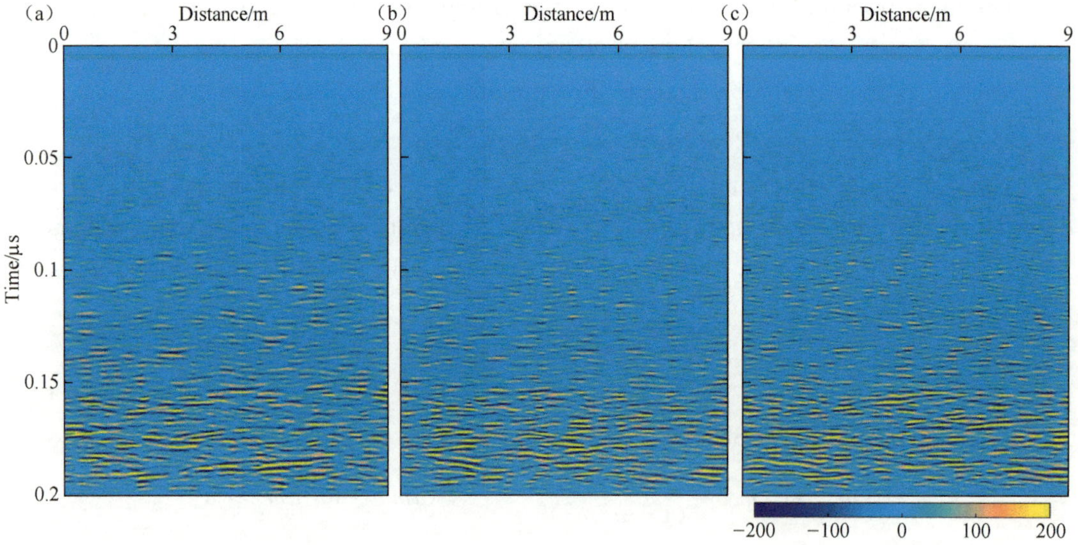

Figure 3 The simulated LPR profiles of the three models shown in Figure 2.
(a) Gaussian ACF random media; (b) exponential ACF random media; (c) Von Kármán ACF media.
The color bar represents intensity (arbitrary units)

2.2 The Model of Lunar Regolith

The correlation distance and root mean square (RMS) perturbation are two key parameters to describe the self - organization random media; thus, we concentrate on these two parameters in the numerical simulations. These two parameters have a relatively clear physical meaning: the correlation distance can well constrain the variation range of regolith permittivity and the RMS perturbation can well present the extent of regolith permittivity variation. Different groups of correlation distance and RMS perturbation limit the physical properties and structural characteristics of the lunar regolith.

We established different models of 5 - 20cm correlation distance and 1 - 7% RMS perturbation. Figure 4 shows three exponential ACF random media with different correlation distances, namely, 5, 10, and 20cm, respectively. Figure 5 shows three exponential ACF random media with different RMS perturbations of 3, 5, and 7%, respectively. Sputtering stones were also added below the 12m depth of the lunar regolith.

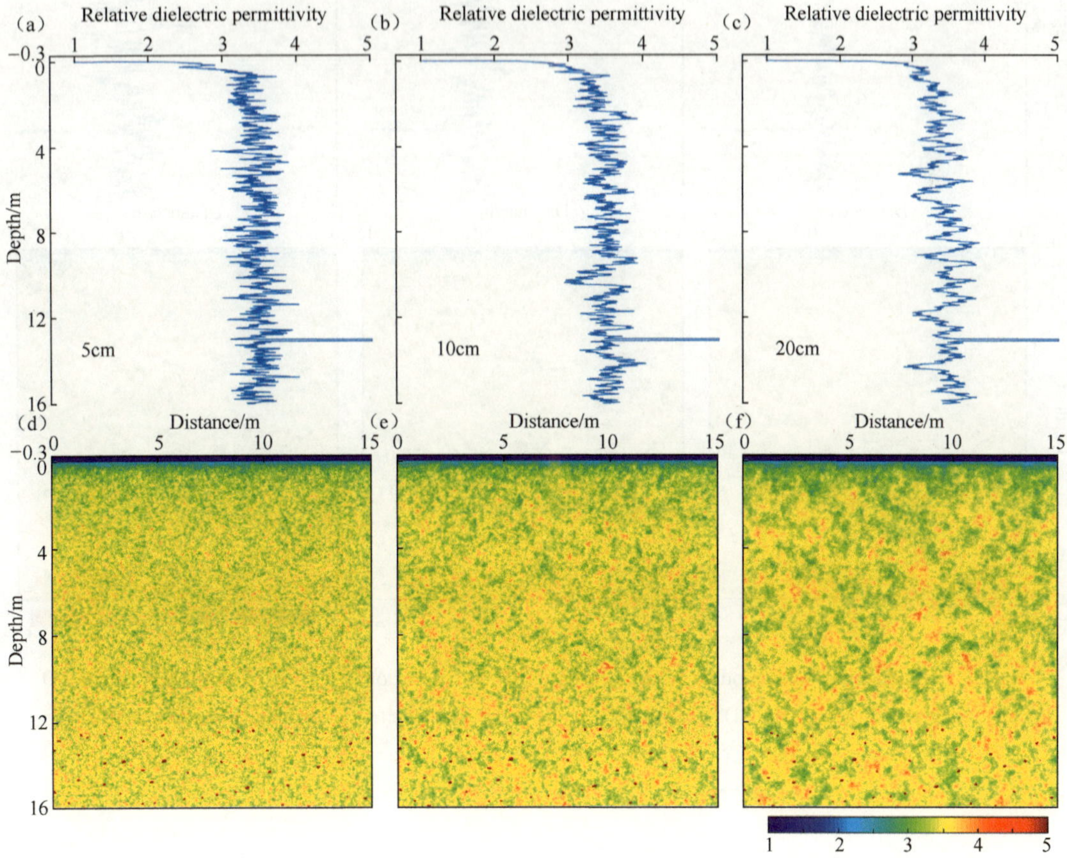

Figure 4　Exponential ACF models with different correlation distances. (a - c) Fluctuation of the 1D probe along the center of 2D model. (d - f) 2D relative dielectric permittivity models with three correlation distances. The correlation distances were (a, d) 5cm, (b, e) 10cm, and (c, f) 20cm, respectively. The RMS perturbation is 5%

For the sake of simplicity, we only select the critical groups of controlling parameters by plotting them together, as shown in Figure 6, although we actually performed a vast number of numerical simulations. Figure 6 well presents the sensitivity of each controlling parameter. Obviously, a large correlation distance corresponds to weak reflections; in addition, a large RMS perturbation corresponds to strong reflections. Therefore, we can fit the Yutu - 2 LPR data by searching for different correlation distances and RMS perturbations to obtain the subsurface features of the lunar regolith at the Chang'E - 4 landing site.

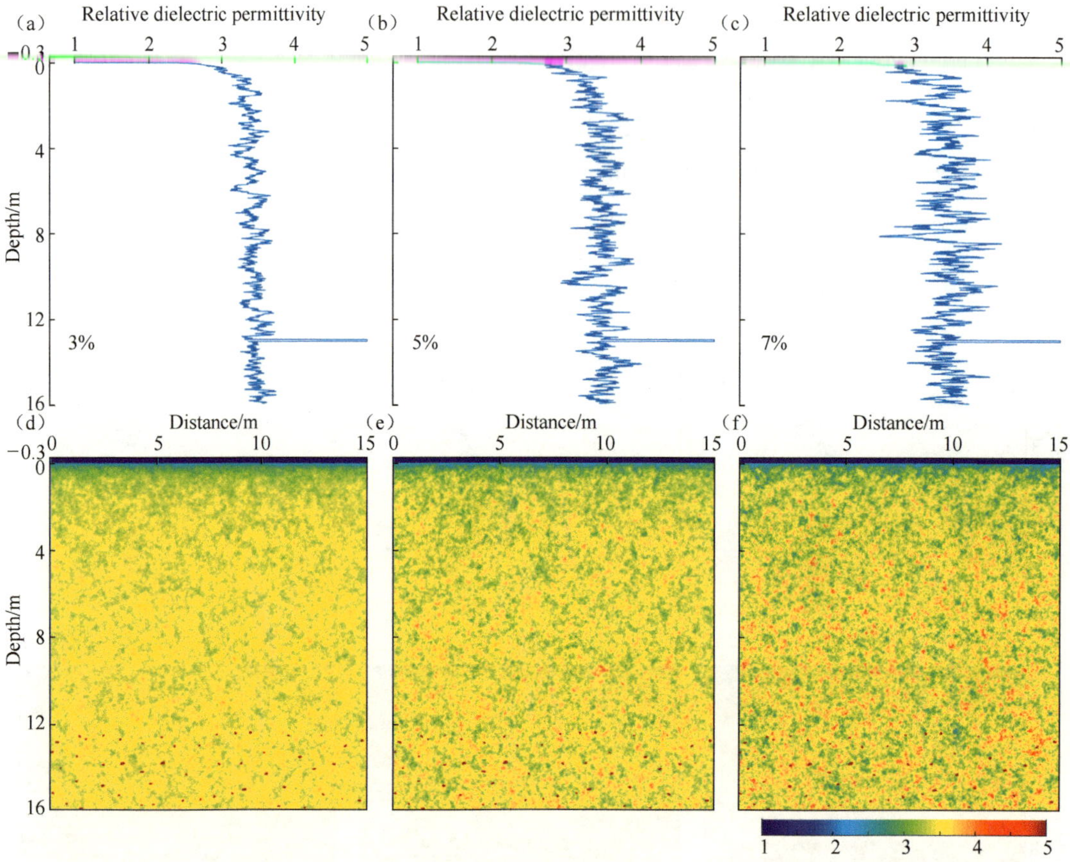

Figure 5　Exponential ACF models with different RMS perturbations. (a - c) Fluctuation of the 1D probe along the center of the 2D model. (d - f) 2D relative dielectric permittivity models with three perturbations. The RMS perturbation were (a, d) 3%, (b, e) 5%, and (c, f) 7%, respectively. The correlation distance is 10cm

2.3　The Diameter and Stacking Density of Rocky Blocks

A 4m thick ejecta layer at the bottom was used to qualitatively identify the rock size and spatial density in the shallow surface of ejecta. The physical properties of the lunar regolith were further determined by the relative amplitude of weak reflections within the lunar regolith compared with the strong reflections from the ejecta layer. We set up three kinds of models: without rocky blocks (Figure 7a), with several ~20cm rocks (Figure 7b), and with a ~1m rock (Figure 7c). The simulated LPR profiles of these three models are shown in Figure 8.

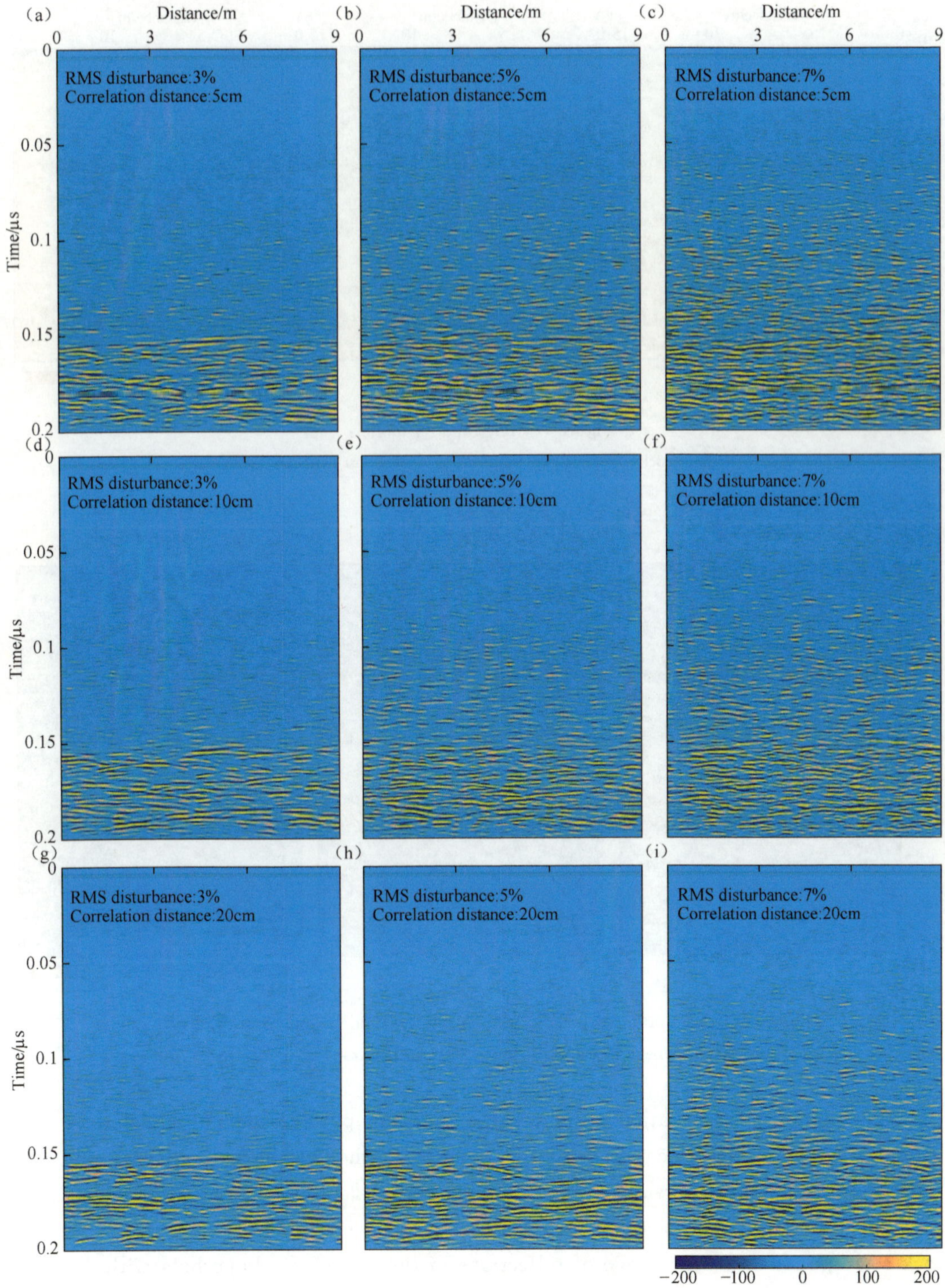

Figure 6　Numerical simulation results using different parameters. The three columns from left to right are with RMS perturbation of 3%, 5%, and 7%, respectively. The three rows from top to bottom are with correlation distance of 5, 10, and 20cm, respectively

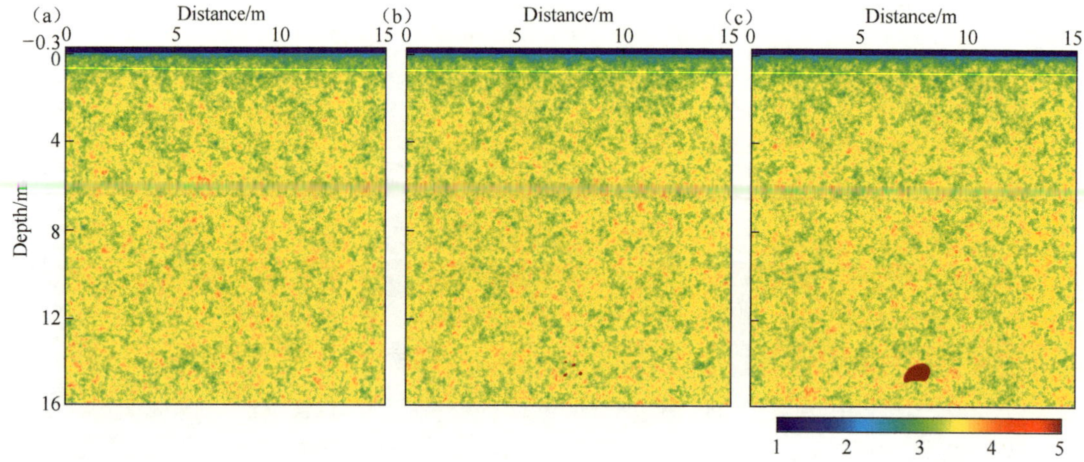

Figure 7 The random model:

(a) without rocky blocks; (b) with～20cm rocks; (c) with a～1 m rock.

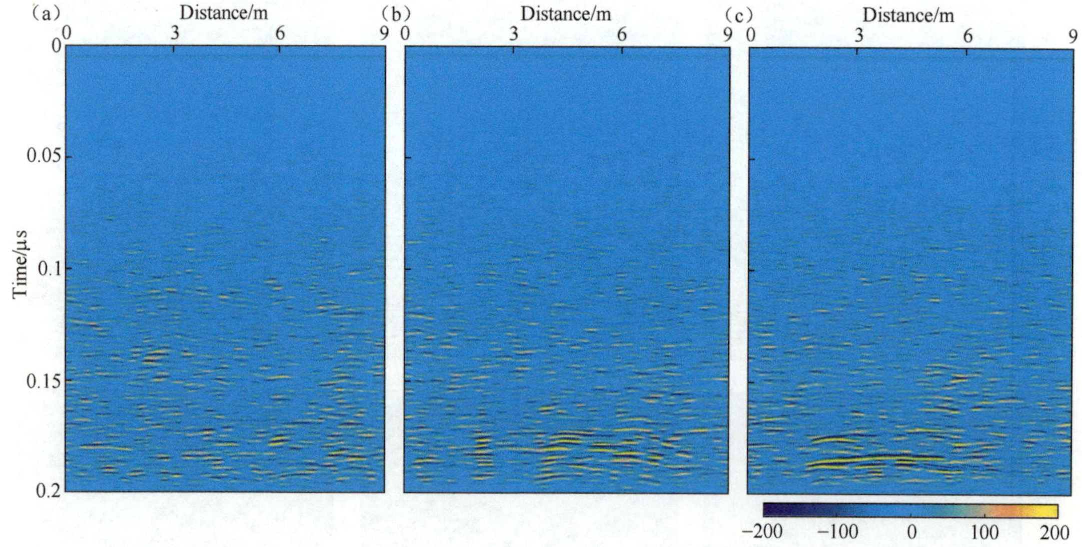

Figure 8 The simulated LPR profiles using the models shown in Figure 7.

(a) Without rocky blocks; (b) with～20cm rocks; (c) with a～1m rock.

The numerical simulation results show that the purely random model does not indicate an obvious interface between the lunar regolith and the ejecta layer, and the reflection patterns in different time windows are similar (Figure 8a). After adding the rocky blocks at the bottom, an obvious interface appears (Figure 8b). In contrast, a large rock corresponds to a wide range of reflections in the shape of a hyperbola with a slight curvature (Figure 8c). Therefore, a single big rock (e. g. , ≥1m) or without rocky blocks does not match the LPR observations. There must be a stack of rocks at the bottom of the model. We further used different stacking densities (i. e. , low, moderate, and high) of small rocks and constrained their diameters (Figure 9). Our results show that the rock diameters should range from ～20 to ～30cm, and the stacking density of rocks should be about 1－2 within each square meter, as shown in Figures 9 and 10.

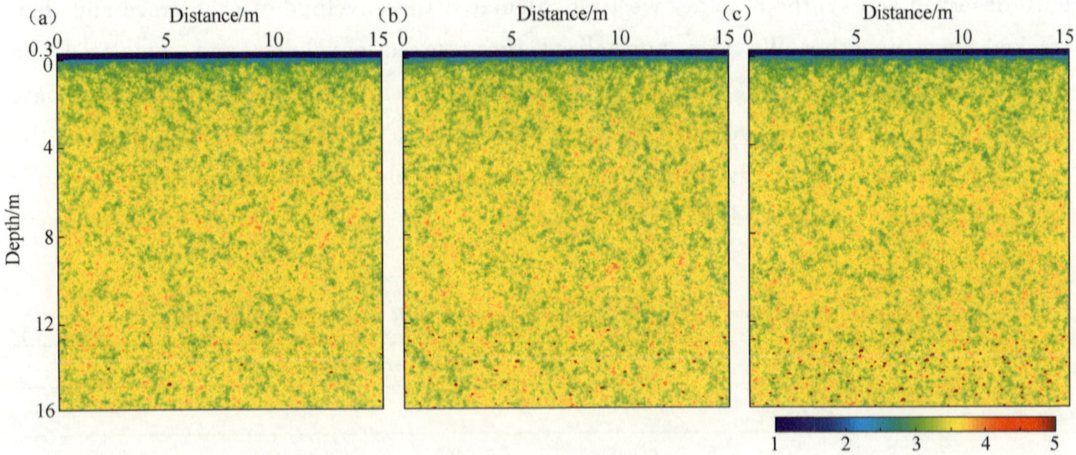

Figure 9　The ejecta model with small

(a) medium; (b) and large; (c) spatial densities of rocks.

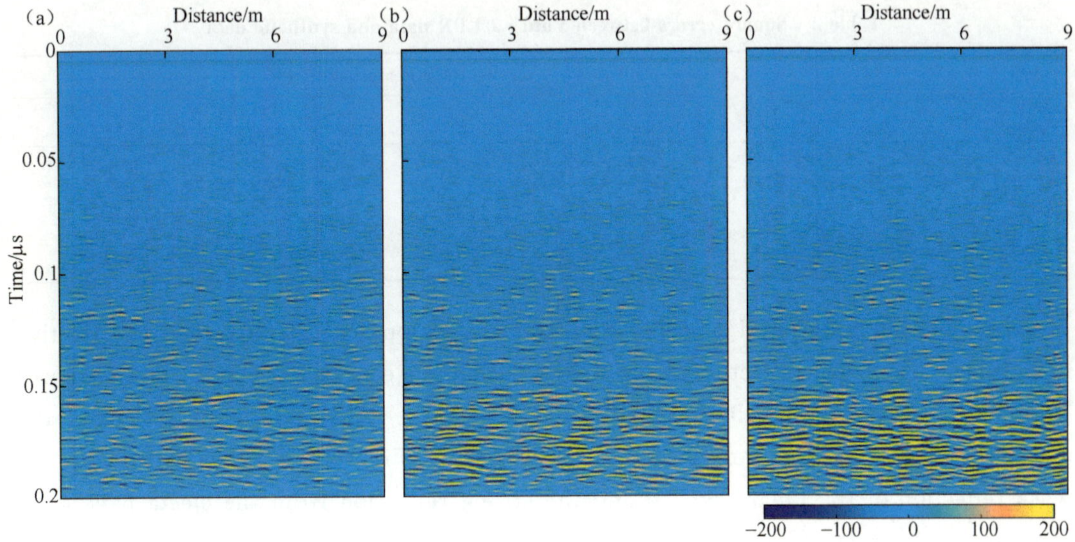

Figure 10　The simulated LPR profiles using the models shown in Figure 9.

(a) With small spatial density of rocks; (b) with medium spatial density of rocks; (c) with large spatial density of rocks.

3　Results

To evaluate the structural characteristics of the regolith near the Chang'E－4 landing site, we quantitatively compared the Yutu－2 radar data with the simulated data. We adopted the envelope－stacking method [53], which discards phase information and only considers energy variations. This method has been successfully applied to seismograms for characterizing small - scale spatial heterogeneity of the mantle [54, 55]. We took a small segment of the Yutu - 2 LPR profile processed by Zhang et al. [24] as a reference. For both observed and synthetic data, we first calculated the envelope of each trace and then stacked them together. We also normalized the stacked envelopes and calculated the correlation coefficients as well as square errors between the observed and synthetic data (Tables 1 and 2). We only selected the data from 0.06 to 0.2μs (~5 - 16m) because the reflection patterns and energy of the first few meters could not be well considered by our numerical simulation method. More explanations on this can be found in the Discussion section.

Table 1　Correlation coefficients between Yutu - 2 LPR data and synthetic data

Correlation Distance (cm)	RMS Perturbation (%)			
	1	3	5	7
5	0.9469	0.9609	0.9590	0.9147
10	0.9446	0.9617	0.9527	0.9309
20	0.9471	0.9394	0.9140	0.9336

Table 2　Square errors between Yutu - 2 LPR data and synthetic data

Correlation Distance (cm)	RMS Perturbation (%)			
	1	3	5	7
5	7.08	3.23	4.21	8.82
10	8.34	3.25	4.60	8.51
20	8.04	7.78	13.22	6.19

As shown in Figures 11 and 12, the stacked envelopes can well reflect the relative magnitude variation of reflections from the regolith and ejecta layers. As shown in Figures 13 and 14, when the correlation coefficient is large or the square error is small, the stacked envelope also fits well with the observed data (Figure 11). Obviously, both the weak reflection of the lunar regolith and the strong reflection from the ejecta have good consistency in terms of radar profiles (Figure 15) and stacked envelopes (Figure 11). Therefore, we can determine the best group of the two controlling parameters: the correlation distance of the self - organization model is about 5 - 10cm and the RMS perturbation is about 3%. In contrast, the other RMS perturbations (e.g., 1%, 5%, and 7%) show significant deviations from the stacked envelopes of the observed data. This indicates that the proposed method is sensitive to constraining the range of RMS perturbation. In addition, with the increasing value of RMS perturbation from 1% to 7%, the correlation coefficient changes accordingly and has a consistently varying trend. This indicates that the proposed method is feasible and robust for the evaluation of the RMS perturbation.

The correlation distance (about 5 – 10cm) is reasonable because it is smaller than the spatial resolution (about 30cm) of the high – frequency (500MHz) channel of the LPR [56]; otherwise, we would observe a clear diffraction or reflection if the correlation distance was comparable to or bigger than 30cm. The RMS perturbation of 3% is also reasonable for weak reflections of the lunar regolith, considering that a much higher perturbation or contrast would cause more evident reflections. Such a small RMS perturbation means that there is no significant variation of regolith materials; instead, local weak variations of structures due to an uneven deposit or gardening would be expected. The diffractions would be evident if the scattering body (with a much higher permittivity compared with that of the regolith) is close to the ground surface [16]; however, we did not observe clear diffractions within the whole regolith layer from 0 to 12 m, even around the upper surface. This indicates that the permittivity variation within the regolith should be relatively gentle. Therefore, we conclude that the RMS perturbation of 3% and the correlation distance (about 5 – 10cm) well reflect the physical property of the regolith variations.

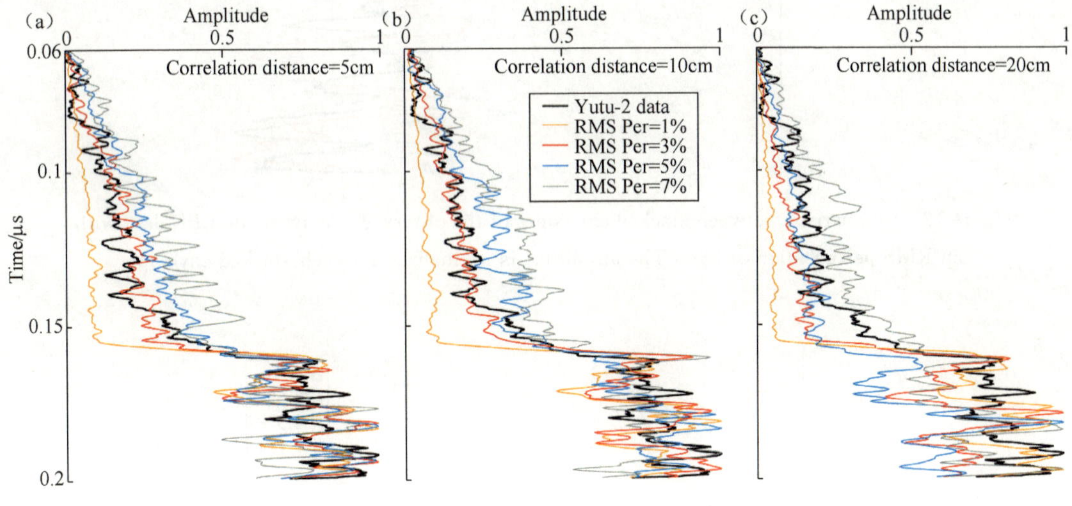

Figure 11　Comparison between stacked envelopes of the observed and synthetic LPR data using different correlation distances and RMS perturbations. (a) Correlation distance is 5cm. (b) Correlation distance is 10cm. (c) Correlation distance is 20cm. The amplitude is normalized for each stacked envelope

Figure 12 shows that a correlation distance of 5 – 10cm is the best estimation, because that of 20cm has a lower correlation coefficient and larger square errors compared with the observed LPR data. These results also indicate that the proposed method is not as sensitive to constraining the correlation distance within a range of 5 to 10cm, at least compared with its high sensitivity to constraining the RMS perturbation. Considering the relatively small correlation distance and RMS perturbation in the regolith, we argue that this is more likely to be due to structural changes rather than material composition changes.

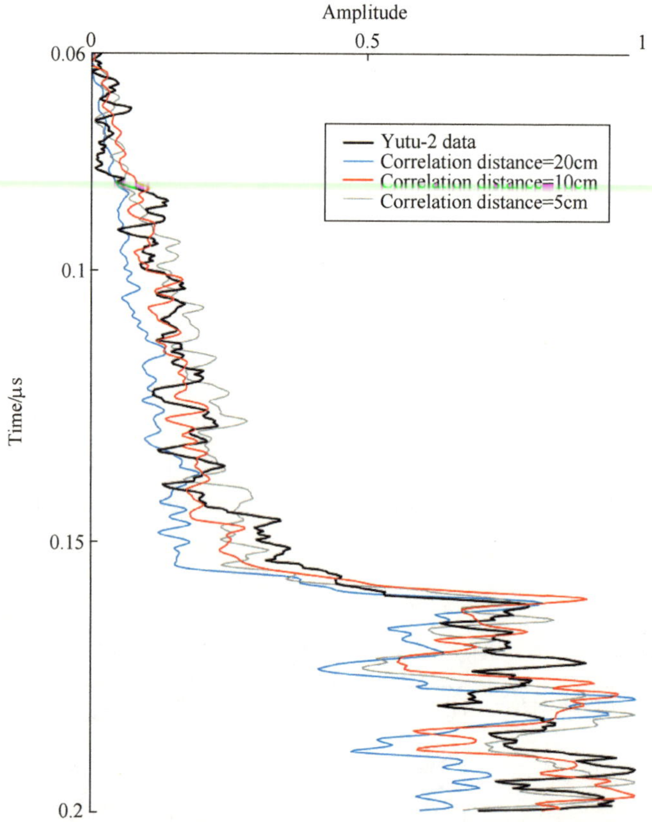

Figure 12　Comparison between stacked envelopes of the observed and synthetic LPR data with an RMS perturbation of 3%. The amplitude is normalized for each stacked envelope

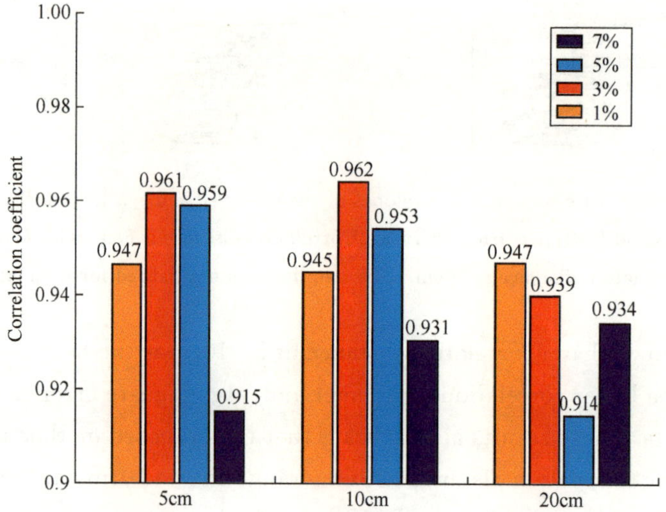

Figure 13　Comparison of correlation coefficients between stacked envelopes of the observed and synthetic LPR data. These correlation coefficients are listed in Table 1

The size (or diameter) of rocks is estimated to be about 20 - 30cm, which is reasonable because it is comparable with the spatial resolution (about 30cm) of the LPR payload [56]. The stacking density of rocks is only about 1 - 2 per square meter, which is consistent with the expectation of the shallow structure of the ejecta [57]. Note that the total depth of this layer is up to 4 m (Figure 15). Such a thick layer with low spatial density of rocks may indicate that it experienced a completely different gardening history, because it is significantly different from the upper fine regolith layer and the typical rocky ejecta layer (at~20m depth). It may be the transitional zone from ejecta to the regolith.

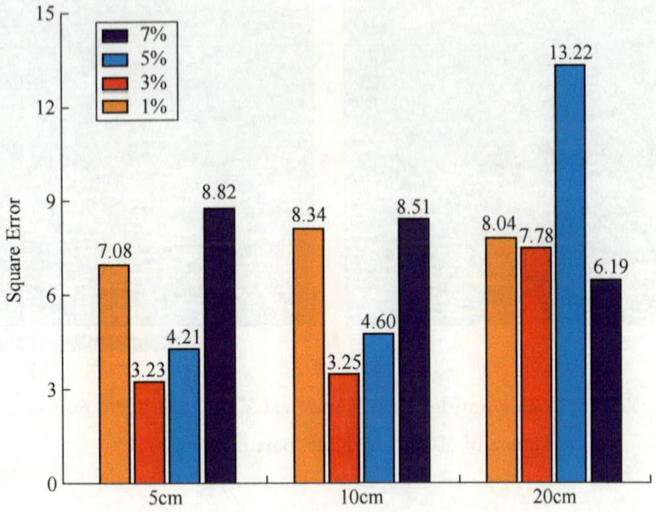

Figure 14　Comparison of square errors between stacked envelopes of the Yutu - 2 LPR data and the synthetic data. These square errors are listed in Table 2

The LPR signals detected by the Chang'E - 4 mission in the Von Kármán crater mainly appear below 12m, and there are almost no strong reflected waves in the shallow area. Our results show that the weak heterogeneity of lunar regolith is only about 3%, and its correlation distance is no more than 10cm. Such a low degree of heterogeneity indicates that the lunar soil is mature in the Von Kármán crater. Consequently, it allows radar wave energy to penetrate more deeply, up to 45m. In contrast, the lunar regolith is less mature in the Imbrium basin [27] compared with that in the Von Kármán crater. This kind of regolith would have a strong scattering effect, particularly close to the upper surface [16]. As a result, it would prevent radar waves from penetrating more deeply; thus, the LPR signals detected by the Chang'E - 3 mission in the Imbrium basin are mainly concentrated within 5m, with almost no clear deep reflections. This indicates that the regolith maturity may be reflected by the radar wave energy and the maximum penetrating depth. Therefore, in addition to the difference in chemical components, the maturity or heterogeneity of the regolith layer also contribute to the LPR reflections.

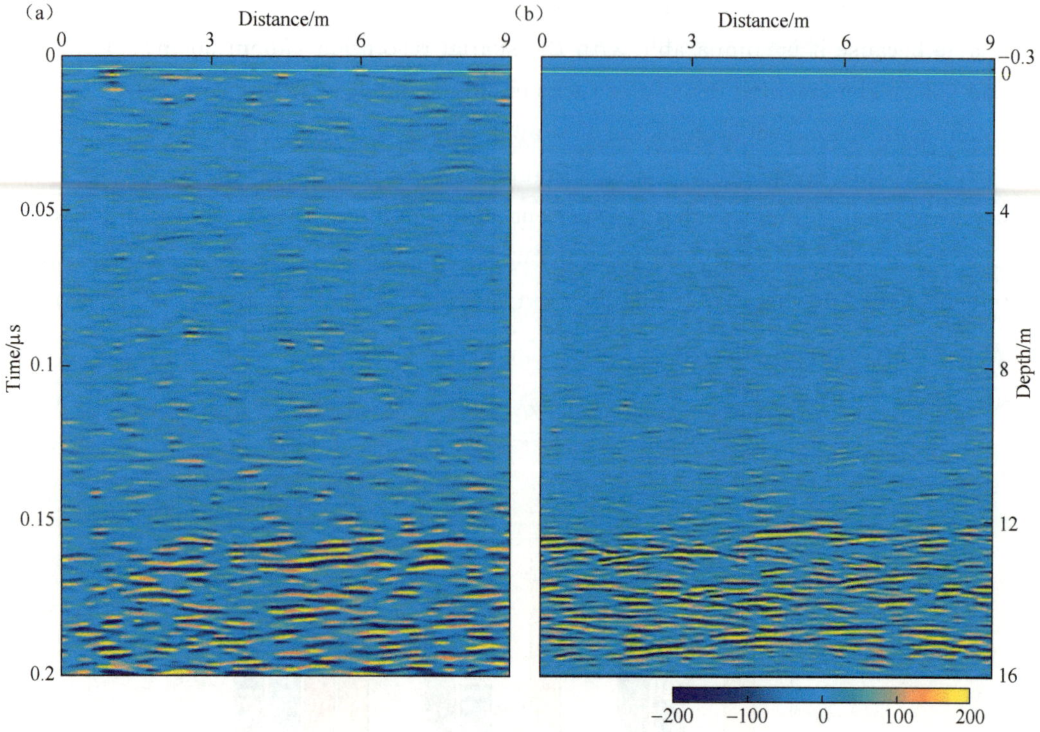

Figure 15　(a) Yutu - 2 radar profile. (b) Numerical simulation LPR profile with correlation distance of 10cm and RMS perturbation of 3%

4　Discussion

The observed data have relatively strong reflections within the depth of ~1 - 2m on the top of the lunar regolith, but we could not explain them at the current stage because their simulation must consider the antenna effect and 3D models. Our code is currently able to process 2D random media but cannot well consider the antenna effect; in contrast, gprMax [58] can consider the antenna effect but it has significant difficulty in constructing random models [16]. Nevertheless, both the stacked envelope and reflection pattern are well matched between the synthetic and observed LPR data from ~5 to 16m, as shown in Figures 11 and 15. This verifies the feasibility of the proposed method.

As shown in Figures 11 and 12, the stacked envelopes of the synthetic data can well match those of the Yutu - 2 radar data before 0.16μs (corresponding to the bottom of the regolith layer), compared with those after 0.17μs (corresponding to the top of ejecta layer). The amplitudes of reflections from rocks after 0.17μs are also influenced by the rocks' position and depth; however, it is not necessary to better match the stacked envelopes of this part because we concentrate on the weak heterogeneity of the regolith layer, and the ejecta layer is only introduced to qualitatively evaluate the relative amplitude variations from the regolith layer to the ejecta layer. The local peak amplitudes of the stacked envelopes after 0. 17μs are generally comparable to each other; therefore, they can well present the general energy of reflections within the ejecta layer and may be a good reference to evaluate the general energy of reflections within the regolith layer.

In this paper, we only present 12 models with three correlation distances (i. e. , 5, 10, and 20cm) and four RMS perturbations (i. e. , 1, 3, 5, and 7%). However, we actually tested a large number of various models to estimate the possible ranges of all parameters. Similarly, significant efforts were also made for estimating the rock diameter and spatial density. We used the 2D finite - difference time - domain method to simulate the LPR data [16, 26]; unfortunately, even for a 2D model, and using an appropriate setting for the finite - difference method to avoid evident artifacts [16], processing would take about one day (~24h) when running on a supercomputing cluster with 40 cores; that is, 3D numerical simulations are almost unaffordable at present, even though the radar waves are actually propagating through 3D random media.

The self - organization characteristics of the lunar regolith are only inferred by the Yutu - 2 LPR data using numerical simulations, which may be corroborated using a drilling core. However, all existing drilling cores on the Moon are located on the near side, and the maximum length is no more than 3m [18, 59, 60]. This indicates that our model cannot be verified from the perspective of geology until deeper drilling can be carried out on the far side of the Moon (to at least 5m).

5 Conclusions

We used three self - organization random models with different autocorrelation functions to simulate the lunar regolith on the far side of the Moon. We also added randomly distributed rocks under the 12m thick regolith to simulate the ejecta. Through a series of comparative analyses of the simulated and observed LPR data, we found that the structure of the lunar regolith in the landing area of Chang'E - 4 can be expressed by an exponential autocorrelation random model. Numerical simulations were proven to be capable of quantitatively estimating statistical parameters of the regolith. The correlation distance was estimated to be about 5 - 10cm and the RMS perturbation was about 3%. The ejecta under the lunar regolith is relatively sparse, only containing about 1 - 2 rocks per square meter; in addition, the diameter of the rocks is about 20 - 30cm, which is comparable to but smaller than the spatial resolution of the high - frequency LPR. These results well constrain the ranges of the controlling parameters of self - organization models and provide a qualitative evaluation of the spatial density of the rocks on the top of ejecta layer. The relatively small correlation distance and RMS perturbation in the regolith indicate that the regolith is mature. The weak reflections within the regolith at the Chang'E - 4 landing site are more likely to be due to local weak changes of structures rather than material composition changes.

References

[1] Di, K.; Zhu, M.; Yue, Z.; Lin, Y.; Wan, W.; Liu, Z.; Gou, S.; Liu, B.; Peng, M.; Wang, Y.; et al. Topographic Evolution of von Kármán Crater Revealed by the Lunar Rover Yutu-2. Geophys. Res. Lett. 2019, 46, 12764-12770.

[2] Lin, H.; Lin, Y.; Yang, W.; He, Z.; Hu, S.; Wei, Y.; Xu, R.; Zhang, J.; Liu, X.; Yang, J.; et al. New Insight into Lunar Regolith-Forming Processes by the Lunar Rover Yutu-2. Geophys. Res. Lett. 2020, 47, e2020GL087949.

[3] Gou, S.; Yue, Z.; Di, K.; Wang, J.; Wan, W.; Liu, Z.; Liu, B.; Peng, M.; Wang, Y.; He, Z.; et al. Impact Melt Breccia and Surrounding Regolith Measured by Chang'e-4 Rover. Earth Planet. Sci. Lett. 2020, 544, 116378.

[4] Fa, W. Simulation for Ground Penetrating Radar (GPR) Study of the Subsurface Structure of the Moon. J. Appl. Geophys. 2013, 99, 98-108.

[5] Dai, S.; Su, Y.; Xiao, Y.; Feng, J.; Xing, S.; Ding, C. Echo Simulation of Lunar Penetrating Radar: Based on a Model of Inhomogeneous Multilayer Lunar Regolith Structure. Res. Astron. Astrophys. 2014, 14, 1642-1653.

[6] Su, Y.; Fang, G.; Feng, J.; Xing, S.; Ji, Y.; Zhou, B.; Gao, Y.; Li, H.; Dai, S.; Xiao, Y.; et al. Data Processing and Initial Results of Chang'e-3 Lunar Penetrating Radar. Res. Astron. Astrophys. 2014, 14, 1623-1632.

[7] Zhao, N.; Zhu, P.; Yuan, Y.; Zhang, B.; Deng, J. The Shallow Subsurface Structures of Chang'e-3 Landing Site Based on the Wavefield Characteristics of LPR Channel-2B Data. Adv. Space Res. 2018, 62, 884-889.

[8] Zhang, L.; Li, J.; Zeng, Z.; Xu, Y.; Liu, C.; Chen, S. Stratigraphy of the Von Kármán Crater Based on Chang'e-4 Lunar Penetrating Radar Data. Geophys. Res. Lett. 2020, 47, e2020GL088680.

[9] Hu, Y.; Zeng, Z.; Li, J.; Liu, F. Simulation and Processing of LPR Onboard the Rover of Chang'e-3 Mission: Based on Multilayer Lunar Regolith Structure Stochastic Media Model. In Proceedings of the International Conference on Ground Penetrating Radar, Hong Kong, China, 13-16 June 2016.

[10] Ding, C.; Su, Y.; Xing, S.; Dai, S.; Xiao, Y.; Feng, J.; Liu, D.; Li, C. Numerical Simulations of the Lunar Penetrating Radar and Investigations of the Geological Structures of the Lunar Regolith Layer at the Chang'e 3 Landing Site. Int. J. Antenn. Propag. 2017, 2017, 1-11.

[11] Feng, J.; Su, Y.; Ding, C.; Xing, S.; Dai, S.; Zou, Y. Dielectric Properties Estimation of the Lunar Regolith at Ce-3 Landing Site Using Lunar Penetrating Radar Data. Icarus 2017, 284, 424-430.

[12] Lai, J. L.; Xu, Y.; Zhang, X.; Tang, Z. Lunar Regolith Stratigraphy Analysis

Based on the Simulation of Lunar Penetrating Radar Signals. Adv. Space Res. 2017, 60, 2099 -2107.

[13] Li, J. ; Zeng, Z. ; Liu, C. ; Huai, N. ; Wang, K. A Study on Lunar Regolith Quantitative Random Model and Lunar Penetrating Radar Parameter Inversion. IEEE. Geosci. Remote Sens. 2017, 14, 1953 - 1957.

[14] Zhang, L. ; Zeng, Z. ; Li, J. ; Huang, L. ; Huo, Z. ; Wang, K. ; Zhang, J. Parameter Estimation of Lunar Regolith from Lunar Penetrating Radar Data. Sensors 2018, 18, 2907.

[15] Zhang, L. ; Zeng, Z. ; Li, J. ; Huang, L. ; Huo, Z. ; Zhang, J. ; Huai, N. A Story of Regolith Told by Lunar Penetrating Radar. Icarus 2019, 321, 148 - 160.

[16] Lv, W. ; Li, C. ; Song, H. ; Zhang, J. ; Lin, Y. Comparative Analysis of Reflection Characteristics of Lunar Penetrating Radar Data Using Numerical Simulations. Icarus 2020, 350, 113896.

[17] Boisson, J. ; Heggy, E. ; Clifford, S. M. ; Yoshikawa, K. ; Anglade, A. ; Lognonne, P. Radar Sounding of Temperate Permafrost in Alaska: Analogy to the Martian Midlatitude to High - Latitude Ice - Rich Terrains. J. Geophys. Res. Planets 2011, 116.

[18] Yang, W. ; Lin, Y. New Lunar Samples Returned by Chang'e - 5: Opportunities for New Discoveries and International Collaboration. Innovation 2021, 2, 100070.

[19] Fa, W. ; Wieczorek, M. A. Regolith Thickness over the Lunar Nearside: Results from Earth - Based 70 - CM Arecibo Radar Observations. Icarus 2012, 218, 771 - 787.

[20] Lai, J. ; Xu, Y. ; Bugiolacchi, R. ; Meng, X. ; Xiao, L. ; Xie, M. ; Liu, B. ; Di, K. ; Zhang, X. ; Zhou, B. ; et al. First Look by the Yutu - 2 Rover at the Deep Subsurface Structure at the Lunar Farside. Nat. Commun. 2020, 11, 3426.

[21] Tian, H. ; Zhang, T. ; Jia, Y. ; Peng, S. ; Yan, C. Zhurong: Features and Mission of China's First Mars Rover. Innovation 2021, 2, 100121.

[22] Jia, Y. ; Zou, Y. ; Ping, J. ; Xue, C. ; Yan, J. ; Ning, Y. The Scientific Objectives and Payloads of Chang'e - 4 Mission. Planet. Space Sci. 2018, 162, 207 - 215.

[23] Li, C. ; Zuo, W. ; Wen, W. ; Zeng, X. ; Gao, X. ; Liu, Y. ; Fu, Q. ; Zhang, Z. ; Su, Y. ; Ren, X. ; et al. Overview of the Chang'e - 4 Mission: Opening the Frontier of Scientific Exploration of the Lunar Far Side. Space Sci. Rev. 2021, 217, 35.

[24] Zhang, J. ; Zhou, B. ; Lin, Y. ; Zhu, M. ; Song, H. ; Dong, Z. ; Gao, Y. ; Di, K. ; Yang, W. ; Lin, H. ; et al. Lunar Regolith and Substructure at Chang'e - 4 Landing Site in South Pole - Aitken Basin. Nat. Astron. 2021, 5, 25 - 30.

[25] Li, C. ; Su, Y. ; Pettinelli, E. ; Xing, S. ; Ding, C. ; Liu, J. ; Ren, X. ; Lauro, S. ; Soldovieri, F. ; Zeng, X. ; et al. The Moon's Farside Shallow Subsurface Structure Unveiled by Chang'E - 4 Lunar Penetrating Radar. Sci. Adv. 2020, 6,

eaay6898.

[26] Zhang, L.; Xu, Y.; Bugiolacchi, R.; Hu, B.; Liu, C.; Lai, J.; Zeng, Z.; Huo, Z. Rock Abundance and Evolution of the Shallow Stratum on Chang'e-4 Landing Site Unveiled by Lunar Penetrating Radar Data. Earth Planet. Sci. Lett. 2021, 564, 116912.

[27] Zhang, J.; Yang, W.; Hu, S.; Lin, Y.; Fang, G.; Li, C.; Peng, W.; Zhu, S.; He, Z.; Zhou, B.; et al. Volcanic History of the Imbrium Basin: A Close-up View from the Lunar Rover Yutu. Proc. Natl. Acad. Sci. USA 2015, 112, 5342-5347.

[28] Chen, S.; Meng, Z.; Cui, T.; Lian, Y.; Wang, J.; Zhang, X. Geologic Investigation and Mapping of the Sinus Iridum Quadrangle from Clementine, Selene, and Chang'e-1 Data. Sci. China. Phys. Mech. 2010, 53, 2179-2187.

[29] Song, H.; Li, C.; Zhang, J.; Wu, X.; Liu, Y.; Zou, Y. Rock Location and Property Analysis of Lunar Regolith at Chang'e-4 Landing Site Based on Local Correlation and Semblance Analysis. Remote Sens. 2020, 13, 48.

[30] Li, C.; Zhang, J. Velocity Analysis Using Separated Diffractions for Lunar Penetrating Radar Obtained by Yutu-2 Rover. Remote Sens. 2021, 13, 1387.

[31] Cai, Y.; Fa, W. Meter-scale Topographic Roughness of the Moon: The effect of Small Impact Craters. J. Geophys. Res. Planets 2020, 125, e2020JE006429.

[32] Gou, S.; Yue, Z.; Di, K.; Cai, Z.; Liu, Z.; Niu, S. Absolute Model Age of Lunar Finsen Crater and Geologic Implications. Icarus 2021, 354, 114046.

[33] Hu, B.; Wang, D.; Zhang, L.; Zeng, Z. Rock Location and Quantitative Analysis of Regolith at the Chang'e 3 Landing Site Based on Local Similarity Constraint. Remote Sens. 2019, 11, 530.

[34] Fa, W.; Zhu, M.; Liu, T.; Plescia, J. B. Regolith Stratigraphy at the Chang'e-3 Landing Site as Seen by Lunar Penetrating Radar. Geophys. Res. Lett. 2015, 42, 10179-10187.

[35] Onodera, K.; Kawamura, T.; Tanaka, S.; Ishihara, Y.; Maeda, T. Numerical Simulation of Lunar Seismic Wave Propagation: Investigation of Subsurface Scattering Properties near Apollo 12 Landing Site. J. Geophys. Res. Planets 2021, 126, e2020JE006406.

[36] Sato, H.; Fehler, M. C.; Maeda, T. Seismic Wave Propagation and Scattering in the Heterogeneous Earth, 2nd ed.; Springer: Berlin/Heidelberg, Germany, 2012.

[37] Olhoeft, G. R.; Strangway, D. W. Dielectric Properties of the First 100m of the Moon. Earth Planet. Sci. Lett. 1975, 24, 394-404.

[38] Heiken, G.; Vaniman, D.; French, B. M. Lunar Sourcebook, a User's Guide to the Moon; CUP Archive: Cambridge, UK, 1991.

[39] Dong, Z.; Fang, G.; Zhao, D.; Zhou, B.; Gao, Y.; Ji, Y. Dielectric Properties of Lunar Subsurface Materials. Geophys. Res. Lett. 2020, 47, e2020GL089264.

[40] Conyers, L. B. ; Goodman, D. Ground - Penetrating Radar: An Introduction for Archaeologists; AltaMira Press: Walnut Creek, CA, USA, 1997.

[41] Yee, K. S. Numerical Solution of Initial Boundary Value Problems Involving Maxwell's Equations in Isotropic Media. IEEE. Trans. Antenn. Propag. 1966, 14, 302 - 307.

[42] Taflove, A. ; Hagness, S. Computational Electrodynamics: The Finite - Difference Time - Domain Method; Artech House, Inc. : Norwood, MA, USA, 2005; pp. 629 - 670.

[43] Feng, D. ; Dai, Q. ; Ji - Shan, H. ; Gang, H. Finite Difference Time Domain Method of GPR Forward Simulation. Prog. Phys. Geog. 2006, 2, 630 - 636.

[44] Irving, J. ; Knight, R. Numerical Modeling of Ground - Penetrating Radar in 2 - D Using Matlab. Comput. Geosci. 2006, 32, 1247 - 1258.

[45] Levander, A. R. ; Holliger, K. Small - Scale Heterogeneity and Large - Scale Velocity Structure of the Continental Crust. J. Geophys. Res. 1992, 97, 8797 - 8804.

[46] Ikelle, L. T. ; Yung, S. K. ; Daube, F. 2 - D Random Media with Ellipsoidal Autocorrelation Functions. Geophysics 1993, 58, 1359 - 1372.

[47] Witte, O. ; Roth, M. ; Müller, G. Ray Tracing in Random Media. Geophys. J. Int. 1996, 124, 159 - 169.

[48] Frankel, A. ; Clayton, R. W. Finite Difference Simulations of Seismic Scattering: Implications for the Propagation of Short - Period Seismic Waves in the Crust and Models of Crustal Heterogeneity. J. Geophys. Res. 1986, 91, 6465 - 6489.

[49] Goff, J. A. ; Jordan, T. H. Stochastic Modeling of Seafloor Morphology: Inversion of Sea Beam Data for Second - Order Statistics. J. Geophys. Res. 1988, 93, 13589 - 13608.

[50] Holliger, K. Upper - Crustal Seismic Velocity Heterogeneity as Derived from a Variety of P - Wave Sonic Logs. Geophys. J. Int. 1996, 125, 813 - 829.

[51] Wu, R. ; Xu, Z. ; Li, X. Heterogeneity Spectrum and Scale - Anisotropy in the Upper Crust Revealed by the German Continental Deep - Drilling (KTB) Holes. Geophys. Res. Lett. 1994, 21, 911 - 914.

[52] Weber, R. C. ; Knapmeyer, M. ; Panning, M. ; Schmerr, N. ; Tong, V. C. H. ; Garcia, R. A. Modeling Approaches in Planetary Seismology. In Extraterrestrial Seismology; Tong, V. C. H. , Garcia, R. A. , Eds. ; Cambridge University Press: Cambridge, UK, 2015; pp. 140 - 156.

[53] Shearer, P. M. ; Earle, P. S. Observing and Modeling Elastic Scattering in The Deep Earth. Adv. Geophys. 2008, 50, 167 - 193.

[54] Mancinelli, N. J. ; Shearer, P. M. Reconciling Discrepancies Among Estimates of Small - scale Mantle Heterogeneity from PKP Precursors. Geophys. J. Int. 2013, 195, 1721 -1729.

[55] Shearer, P. M. ; Earle, P. S. The Global Short - period Wavefield Modelled with a Monte Carlo Seismic Phonon Method. Geophys. J. Int. 2004, 158, 1103 - 1117.

[56] Fang, G. ; Zhou, B. ; Ji, Y. ; Zhang, Q. ; Shen, S. ; Li, Y. ; Guan, H. ; Tang, C. ; Gao, Y. ; Lu, W. ; et al. Lunar Penetrating Radar Onboard the Chang'e - 3 Mission. Res. Astron. Astrophys. 2014, 14, 1607 - 1622.

[57] Mcgetchin, T. R. ; Settle, M. ; Head, J. W. Radial Thickness Variation in Impact Crater Ejecta: Implications for Lunar Basin Deposits. Earth Planet. Sci. Lett. 1972, 20, 226 - 236.

[58] Giannopoulos, A. Modelling Ground Penetrating Radar by gprMax. Constr. Build. Mater. 2005, 19, 755 - 762.

[59] Jolliff, B. L. ; Wieczorek, M. A. ; Shearer, C. K. ; Neal, C. R. New Views of the Moon; Mineralogical Society of America: Chantilly, VA, USA, 2006.

[60] Deng, X. J. ; Zheng, Y. H. ; Jin, S. Y. ; Yao, M. ; Zhao, Z. H. ; Li, H. L. ; Jiang, S. Q. ; Wang, G. X. Design and Implementation of Sampling, Encapsulating, and Sealing System of Chang'e - 5. Sci. Sin. Technol. 2021, 51, 753 - 762.

Heterogeneous Weathering Process of Lunar Regolith Revealed by Polarimetric Attributes Analysis of Chang'E-4 Lunar Penetrating Radar Data Acquired During the Yutu-2 Turnings

Haoqiu Zhou[1], Xuan Feng[1], Chunyu Ding[2],
Zejun Dong[1], Cai Liu[1], and Wenjing Liang[1]

1. College of Geo-exploration Science and Technology, Jilin University, Changchun, China
2. Institute of Moon-base Exploration and Observation, Institute of Advance Study, Shenzhen University, Shenzhen, China

Abstract: This paper provides analyses to quantify the maturity of regolith within 24m depth using a novel polarimetric attributes analysis of lunar penetrating radar (LPR) data. The results demonstrate that LPR signals were mainly from subsurface rock fragments of different sizes and were rarely from strata interfaces. The averages and ranges of maturities for the near-surface (0～2m), the fine-grained regolith (2～12m), and the coarse-grained ejecta (12～24m) are 68.98% (0%～99.82%), 82.61% (41.52% ～ 96.56%), and 72.33% (5.91% ～ 96.66%), respectively. Local regions with anomalously low maturities on the lunar surface may be due to the dense materials formed by local impacts or the rock fragments from distal impacts; a newly discovered sandwich structure with low maturity at a depth between 12 and 18m is formed by the heterogeneous weathering in a paleo-crater. Based on these new insights, we infer a heterogeneous weathering process of the regolith.

Plain Language Summary: The entire lunar surface is considered to be covered by a layer of regolith. The study of regolith weathering degree almost focused on the near-surface because the traditional techniques cannot detect the regolith at large depths. The Chang'E-4 lunar penetrating radar (LPR) can detect the regolith structure within dozens of meters depth. However, previous publication of LPR results only use the macro layering model to interpret the regolith structure. This study innovatively extracts new properties from the LPR data acquired while the rover was turning and estimated the quantitative maturity of regolith within ~24m depth at lunar farside. We found that the LPR can mainly detect the subsurface rock fragments that survived weathering and rarely the interfaces of strata because the materials of different strata are mixed at interfaces and make it gradational. Our results also reveal the spatial difference of weathering in the regolith. We investigate the formations and material compositions of several interesting regions with unusual weathering degrees. Based on these new insights, we establish a model to illustrate the weathering process of the regolith at the CE－4 landing site.

Key words:

• We quantify the maturity of the lunar regolith within 24m depth by amalyzing the tunar penetrating radar (LRR) data acquired during the rover turnings

• LPR Signals were mainty from subsurface rock fragments of different sizesand were rarely from strata interfaces

• Inferring the heterogeneous weathering processby comperehensively investigating the regions with anomalous in the regolith

1 Introduction

The entire lunar surface is considered to be covered by a layer of regolith which is formed by meteorite impacts and gardening and consists of unconsolidated rock and mineral fragments, breccias, glasses, and agglutinates (Gou et al., 2019; McKay et al., 1991). Most of the knowledge of regolith such as optical maturity and surface roughness was acquired by remote sensing and mainly characterize the regolith properties at the near－surface. The regolith over the full depth records the lunar geological evolution. It is generally believed that regolith at larger depths contains more and larger rock fragments than shallower regolith (Bart & Melosh, .2010; McKay et. al., 1991). However, the regolith at larger depths is rarely studied as it is beyond the detection of remote sensing techniques.

The Chinese Chang' E－4 probe landed on the floor of the Von Kármán crater in the South Pole－Aitken basin (Liu et al., 2019) and the lunar penetrating radar (LPR) (Fang et al., 2014; Li et al., 2020; Su et al., 2014) carried by the Yutu－2 rover are now mapping the fine structure of subsurface at lunar farside. The Von Kármán crater was covered by several periods of basalt flows and the ejecta from the adjacent craters during the Imbrian and Eratoshenian periods (Head III et al., 2012; Huang et al., 2018; Stuart－Alexander, 1978), which have resulted in complex subsurface compositions and structures.

The previous LPR sounding results have revealed the macro layering structure of lunar regolith. The top unit 1 is a fine－grained regolith with a depth of ~0－12m and is interpreted to be the ejecta from the nearby Finsen crater (Gou et al., 2021; Hu et al., 2019; Li et al., 2019); the middle unit 2 with a depth of 12－24m corresponds to strong reflections and it is considered to be the superposition of multiple ejecta sources from nearby craters; the bottom unit 3 has a depth of 24 - 40m and contains alternating layers of fine - grained materials and rocks (Lai et al., 2019; Li et al., 2020; Zhang, Zhou et al., 2021). However, these impressive works only provide the macro layering structure of regolith and its maturity remains unknown.

In other studies, the LPR data are applied to the analysis of subsurface properties and rock quantity in the regolith (Dong et al., 2020; Lai et al., 2021a, 2021b; Zhang, Xu et al., 2021; Zhou, Feng, Ding, et al., 2021). However, the rock localization using LPR data mainly depends on the reflection signals from the subsurface and the results in previous works cannot determine whether these signals are from horizontally distributed rocks or continuous geological interfaces (Zhou, Feng, Ding, et al., 2021) because the regolith maturity over the full depth has not been obtained.

The LPR system onboard the Yutu - 2 rover adopted the time - trigger mode, which means the LPR transmitted and received signals by a certain time interval. However, the rover did not move at a constant speed (Fang et al., 2014; Feng et al., 2017; Su et al., 2014). The obtained LPR data are therefore related to three different situations, namely: the data acquired when the rover was moving, stopped, and was turning. In previous works, only the first type of data was selected to analyze the regolith (Li et al., 2020; Zhou, Feng, Dong, et al., 2021). However, the LPR data acquired during the rover turnings is also a type of data that worth analysis. The rings distributed along the Yutu - 2 path (See Figures 1c - 1e) indicate that the LPR acquired data with different orientations at the same positions and these data can be used to obtain polarimetric attributes and better characterize the subsurface properties.

In this paper, we extract and analyze the polarization information from LPR data acquired when the rover was turning during the 15 lunar days. The results provide unprecedented quantitative maturity information and newly discovered fine structures of regolith within ~24m depth at lunar farside. The results reveal the heterogeneity of space weathering on the Moon and produce a fine evolution process of the regolith at Chang'E - 4 landing site.

2 Methods

2.1 Analysis of the LPR Data Acquired During the Rover Turnings

When the rover was turning, the collected LPR data is a function of the rotation angles. The differences in the LPR data of different measuring angles are related to the Sinclair matrix of the subsurface (Cloude & Pottier, 1996). We select the rover heading direction when the rover started to turn as the x - axis, and the direction perpendicular to the x - axis as the y - axis as Figure. 1e shows. Supposing that the turning angle of the rover/antennas is θ, the acquired turning data is given by (Dong et. al., 2021, 2022; Sinclair, 1950; van Zyl et al., 1987):

$$M(t, s, \theta) = E_r^T S E_t = \begin{bmatrix} \sin\theta \\ \cos\theta \end{bmatrix}^T \begin{bmatrix} S_{hh}(t, s) & S_{hv}(t, s) \\ S_{vh}(t, s) & S_{vv}(t, s) \end{bmatrix} \begin{bmatrix} \sin\theta \\ \cos\theta \end{bmatrix} \quad (1)$$

where E_t and E_r are the polarization modes of the transmitting and receiving antennas and are the same for the LPR system. S represents the Sinclair matrix of the subsurface. t and s denote the travel time of an LPR signal and the turning position along the Yutu－2 route, respectively. (1) can be further simplified as follows:

$$M(t, s, \theta) = \frac{1}{2}[\gamma(t, s)\sin 2\theta - \beta(t, s)\cos 2\theta + \alpha(t, s)] \quad (2)$$

where $\alpha = S_{hh} + S_{vv}$, $\beta = S_{hh} - S_{vv}$, and $\gamma = 2S_{hv}$ are the Pauli components to be obtained (Cloude & Pottier, 1996). The acquired rover turning LPR data is composed of sine and cosine functions and a constant threshold. Subsequently, the least－squares fitting is performed as follows:

$$\arg\min L(\alpha, \beta, \gamma) = \arg\min \| D - M \|^2 \quad (3)$$

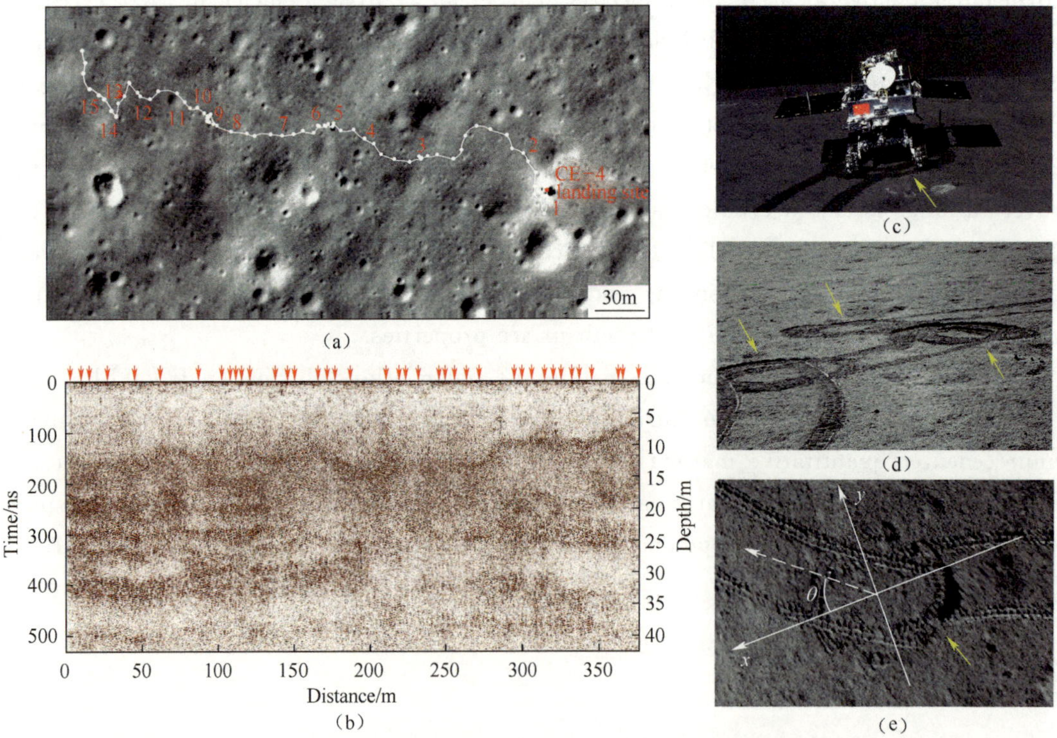

Figure 1　Yutu－2 route and Lunar Penetrating Radar data. (a) The Yutu－2 route within the first 15 lunar days. (b) The acquired CH－2B LPR data. The red arrows indicate the positions to extract rover turning data. (c－e) Show the details of the rover route, the yellow arrows indicate the turning positions. A coordinate is established in (e) to analyze the acquired data when the rover was turning. The rover heading direction when the rover started to turn is selected as the x－axis

where D represents the acquired rover turning LPR data. By computing the minimum value of L, the polarization information S_{hh}, S_{vv}, and S_{hv} can be extracted.

2.2 Freeman Decomposition

Freeman decomposition (FD) is a type of classic polarimetric analysis technique, which is a physical model - based, three - component decomposition method (Freeman & Durden, 1998). The key idea of FD method is that for each scattering point, its scattering mechanism is contributed by three basic scattering components, which refer to a surface - like scattering (p_s), a double - bounce scattering (p_d), and a volume scattering (p_v) (Feng et al., 2017; Freeman & Durden, 1998; Zhou, Feng, Dong, Liu, Liang, & An, 2022, Zhou, Feng, Dong, Liu, & Liang, 2022). By decomposing the polarimetric scattering covariance matrix, the power of the three components can be obtained (See Supporting Information. S1). In this study, the p_s and p_v components are adopted to derive the regolith maturity.

2.3 Quantitation of Regolith Maturity

The Freeman theory indicates that the surface - like scattering (p_s) characterizes the power of surface - like scatterings, such as the signals from the surfaces of strata and relatively coarse - grained boulders. Relatively fine - grained rocks can not generate high strength p_s. If quantities of boulders or rock fragments concentrate in a certain region, the electromagnetic (EM) waves will scatter many times in this region, which will lead to the high strength of volume scattering (p_v) component (Freeman & Durden, 1998). The amplitudes of p_s and p_v components are related to the differences in relative permittivity, the grain size, and the clutter degree of rocks within the regolith (See Supporting Information S1). The grain size and the clutter degree of rocks is a measure of regolith maturity (McKay et al., 1991), whereas relative permittivity is related to the FeO and TiO_2 content (Dong et al., 2017, 2020; Feng et al., 2017; Lai et al., 2019). Therefore, we can adopt the relevance of the two components to characterize the regolith maturity as follows:

$$RM(t, x) = \left[1 - \sqrt{\frac{p_s(t, x)p_v(t, x)}{\max[p_s(t, x)p_v(t, x)]}}\right] \times 100\% \qquad (4)$$

where RM can be used to represent the regolith maturity. The verification can be found in Supporting Information. S1. The assumption of Equation. 4 is that the ejecta layer with a depth of 12 - 24m at the CE - 4 landing site contains totally unweathered materials. Otherwise, the RM represents the relative maturity of the investigated zone.

3 Results and Discussions

3.1 Quantitative Regolith Maturity

The LPR is equipped with two types of channels (CH - 1 and CH - 2), with center frequencies of 60 and 500MHz and bandwidths of about 40 and 450MHz, respectively.

CH - 2 channel contains one transmitting and two receiving bow - tie antennas, known as CH - 2A and CH - 2B, with different offsets of 0.16 and 0.32m (Fang et al., 2014). The CH - 1 antenna is located at the tail of the rover, whereas the CH - 2 antenna is at the bottom of the rover (Fang et al., 2014; Su et al., 2014). To obtain accurate polarization information, we should select the LPR data which are acquired at the same position for each turning, which means the center of transmitting and receiving antennas should be close to the center of the rover. Therefore, CH - 1 data is excluded and the CH - 2B data within 24m depth with a higher signal - to - noise ratio is adopted in this study.

Figure 2a is the instantaneous amplitude profile of migrated CH - 2B LPR image within 24m depth. Figures 2b and 2c present the surface - like and volume scattering components (p_s and p_v) which correspond to the models in Figures 2d - 2g, respectively. p_s mainly characterizes the strata interfaces and the surface of boulders whereas p_v mainly demonstrates the rock fragments of different grain sizes. The results demonstrate that the regolith is mainly composed of rock fragments that survived weathering whereas continuous strata interfaces are rare in the regolith. A reasonable interpretation is that the materials of different strata are mixed at interfaces which makes the boundary gradational and invisible (Giannakis et al., 2021). The received LPR signals are mainly from the dense fragments located at the bottoms of strata and survived the weathering (Zhang, Xu et. al., 2021).

Subsequently, the p_s and p_v components can be applied to estimating the regolith maturity along the Yutu - 2 route in Figure. 3a. The subsurface can be divided into three units by depths. The near - surface region with a depth of 0 - 2m has variable maturities ranging from 0% to 99.82% with an average of 68.98%; the maturities of the fine - grained regolith (2 - 12m) vary from 41.52% to 96.56% with an average of 82.61%; coarse - grained ejecta layer (12 - 24m) has a variable maturity of 5.91%- 96.66% with an average of 72.33%.

Although the 0 - 10m is convinced to be the fine - grained Finsen ejecta, the 0 - 2m region presents distinct lower maturities than the 2 - 10m region. A possible reason is that the surface of the CE - 4 landing area may consist of lithic material from secondary cratering whereas the 2 - 10m region below consists mainly of native materials which experienced longer time weathering (Ding et al., 2021). Besides, both maturities of near - surface and coarse ejecta regions have significant spatial differences whereas that of the fine - grained regolith layer is generally of higher degrees and lower variations.

3.2 Heterogeneous Lunar Surface Along the Yutu - 2 Route

The surface - like scattering component dominates at the top of Figure 2b but it is not uniform along the Yutu - 2 route. Figure 3b presents average curves of the two scattering components and average regolith maturity within 0 - 2m. The red arrows indicate the regions where surface - like scatterings are greater than the volume scatterings. Regions 2 - 4 show the high strengths of both two scatterings, demonstrating their high relative permittivities and roughness of the lunar surface. Therefore, the regolith in these regions shows lower maturities. These regions are likely to contain dense material on the rough surface (Ding et al., 2021) and can be described by the hybrid model composed of Figures 2d and 2e. Regions 1, 5, and 6 exhibit moderate strengths of the surface - like scattering and low strengths of volume scattering, which indicate the moderate relative permittivities and smoother surface and correspond to moderate regolith maturities. These regions are likely to contain a smooth surface with small fragments or without fragments and can be described by the hybrid model composed of Figures 2d and 2g or the single model in Figure 2d.

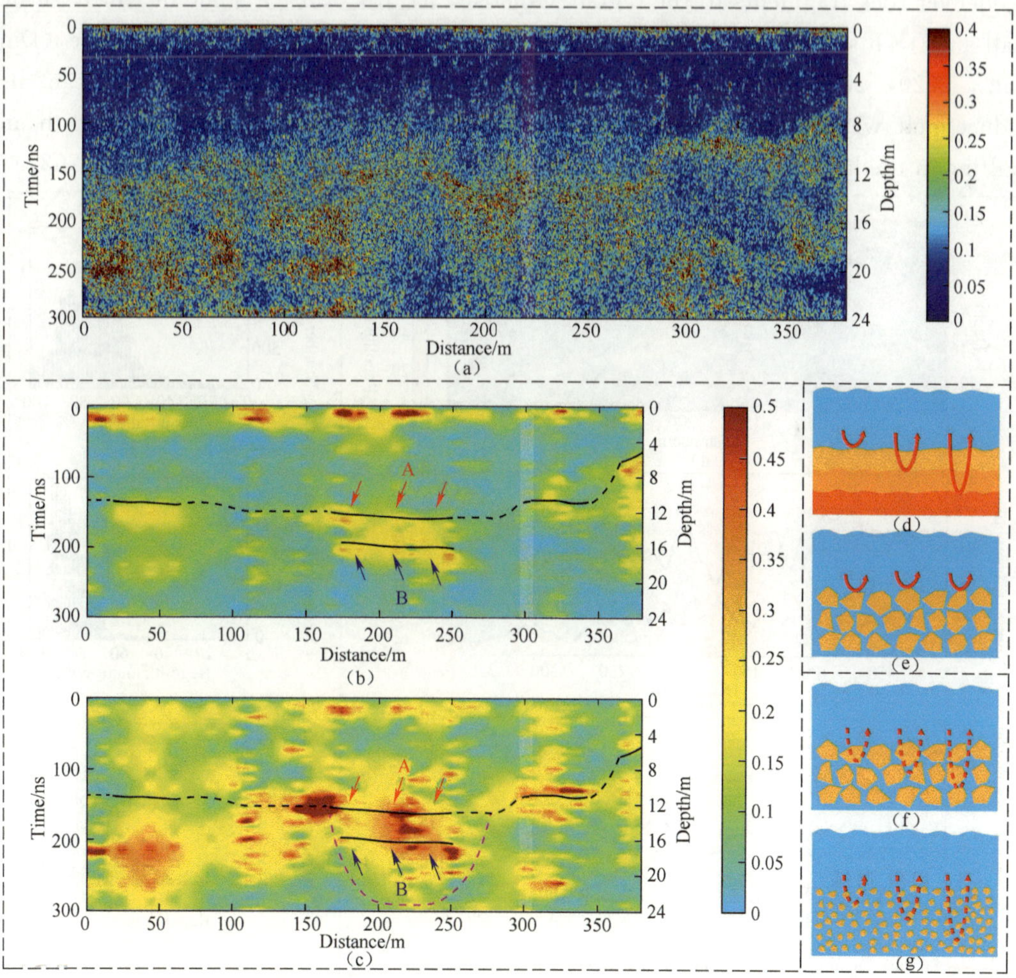

Figure 2　The Lunar Penetrating Radar polarimetric attributes processing results. (a) The instantaneous amplitude profile of migrated LPR image. (b) The surface - like scattering component (p_s) corresponding to the signals from continuous rough interfaces in (d) or the surfaces of boulders in (e). (c) The volume scattering component (p_v) corresponding to the signals from different sizes of boulders of strata interiors in (f and g). The curved arrows in (d - g) represent the schematic propagating paths of electromagnetic waves transmitted by LPR. A and B indicate the boundaries of the newly discovered sandwich structure in the paleo - crater

Besides, the blue arrows indicate the regions where volume scatterings dominate and are with low strengths of both the two scattering components. Therefore, regions 7 - 11 demonstrate a lunar surface with low relative permittivities, moderate roughness, and high maturities, also corresponding to the hybrid model composed of Figures 2d and 2g.

The results reveal the heterogeneous lunar surface along the Yutu - 2 route and this heterogeneity is mainly due to the uneven space weathering. The regions with high maturities are likely to experience more impacting events whereas the regions with low maturities or high dielectric constants may compose of fresh ejecta from local impact or dense materials like melt - bearing impact breccias. Besides, the distal impact events can also deliver rock fragments to the region (Melosh, H. J. , 1989), some of which formed small and fresh secondary craters and increased the roughness of the lunar surface (Ding et al. , 2020, 2021; Gou et al. , 2019; Lin et al. , 2020) . These events can also produce melt welded impact breccia in the near - surface of fine - grained regolith and resulting in the horizontal heterogeneity of the lunar surface (Ding et al. , 2020, 2021) .

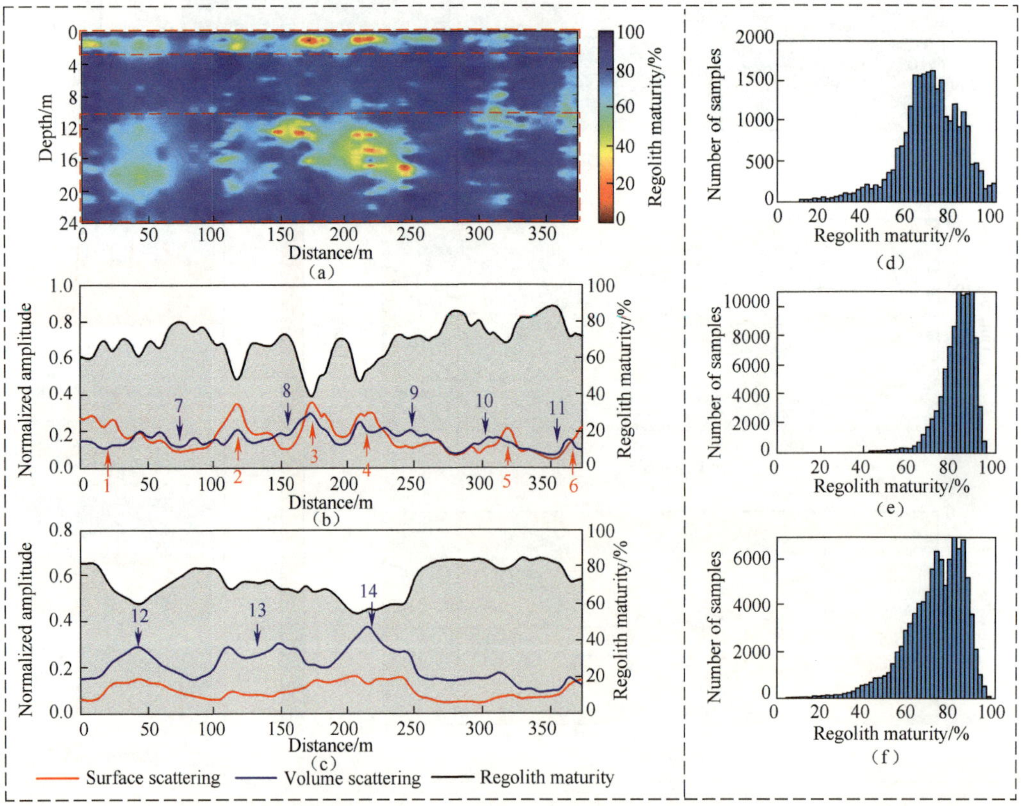

Figure 3　The regolith maturity and analyses. (a) The regolith maturity within 0 - 24m depth. The black curve in (b and c) represent the average maturity of 0 - 2m and 12 - 24m depths along the Yutu - 2 route, respectively. The red and blue curves in (b and c) denote the average p_s and p_v within the same depth ranges. The red numbers 1 - 6 indicate the regions with unusual strong surface - like scattering and the blue numbers 7 - 14 mark the regions with anomalously strong volume scattering. (d - f) Are the distribution of regolith maturity of three - layered regions of different depths. (d) The near - surface region. (0 - 2m) (e) the fine - grained regolith layer (2 - 12m) . (f) The coarse - grained ejecta layer (12 - 24m)

3.3　Newly Discovered Sandwich Structure in the Ejecta Layer

Figure 3c presents average curves of the two scattering components and regolith maturity within 12 – 24m. In this depth range, volume scattering dominates along the Yutu – 2 route and three anomalies (12 – 14) are observed. Regions 12 and 13 are two separated areas of concentrated rocks; different strengths of surface – like scatterings indicate smaller rock sizes and higher regolith maturity in region 13. Besides, region 14 has been interpreted as a buried crater with large quantities of rocks (Lai, Cui, et al., 2021; Zhang, Xu et al., 2021; Zhou, Feng, Ding, et al., 2021), presenting the minimum of regolith maturity along the Yutu – 2 route. The regolith below 12m depth also shows horizontal heterogeneity. Therefore, we use the discontinuous black curves in Figures 2b and 2c to illustrate the bottom of the fine – grained regolith. The dashed lines denote the regions with weaker surface – like scatterings and higher regolith maturities.

It is interesting that we newly observe another continuous interface B within the buried crater (the blue arrows in Figures 2b and 2c). The newly discovered interface is of higher strengths compared to boundary A between the buried crater and the fine – grained regolith, which indicates that the materials on the two sides of interface B have a higher difference in dielectric constants than those of interface A. In other words, the compositions of the materials between A and B are more close to those of the fine – grained regolith. The fine – grained regolith is convinced to be the ejecta from the nearby Finsen crater, so the materials between A and B are also the Finsen ejecta but with lower maturity. These materials are likely to have survived the gardening on the lunar surface, resulting in their unique properties compared to both the materials of the underlying and overlying regolith. Besides, B is the bottom of the buried crater when it was covered by the Finsen ejecta. Compared to the excavated depth of the buried crater observed on radargram, it is likely to have experienced degradation before the coverage of Finsen ejecta.

3.4　Indications for Regolith Formation

Based on the regolith maturity and the analysis before, a hypothesis of the weathering process for the regolith within ～24m depth at the CE – 4 landing site is possibly established as follows (Figure 4):

1. Stage (a and b) postdated the superposition of multiple coarse ejecta from nearby caters. During this period, the lunar surface was likely to experience multiple and important impacts events of different extents that reshaped the structure of the regolith. These impacting events resulted in the undulation and discontinuity of the lunar surface. Today, we can only recognize the largest crater from the LPR radargram located at ～170 – 290m on the Yutu – 2 route (Zhang, Xu et al. 2021; Zhou, Feng, Ding, et. al., 2021). However, the heterogeneous rock sizes and maturities of the rest regions

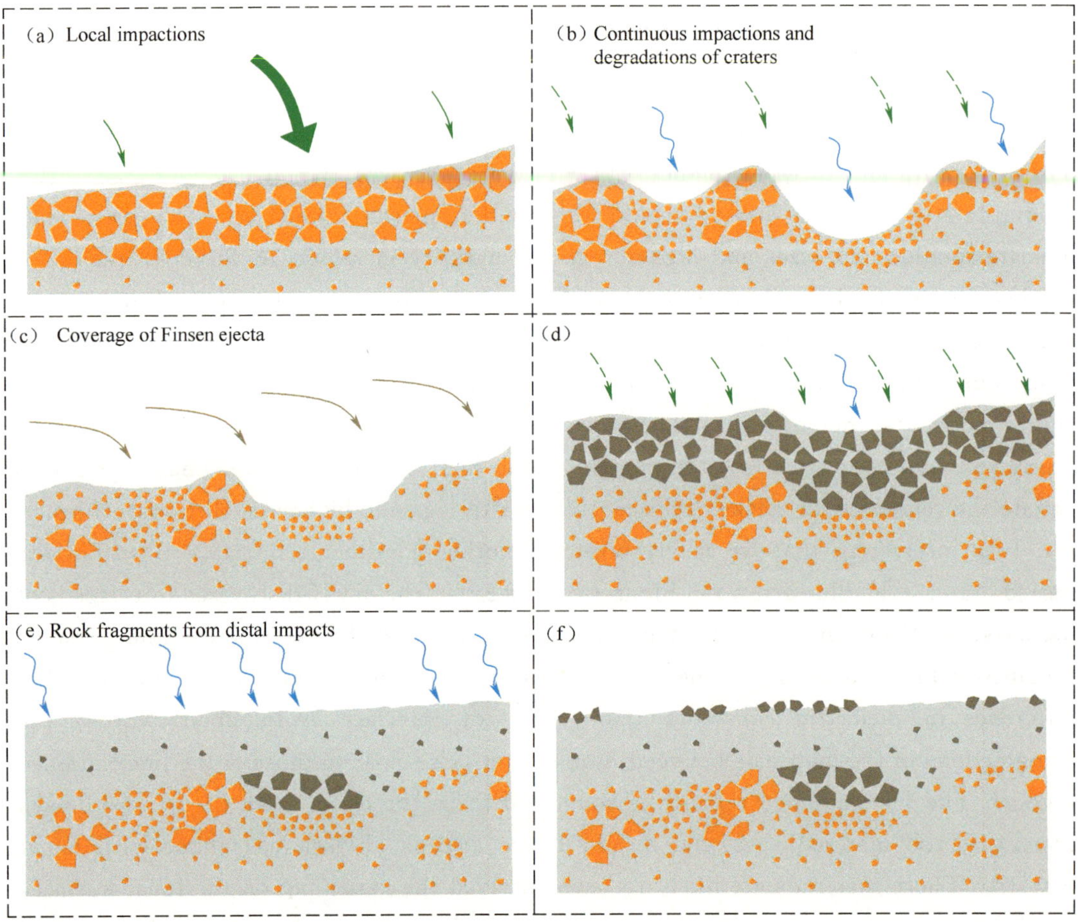

Figure 4　Fine Evolution of the Regolith at the CE – 4 landing site. (a) Impact events of different extents reshaped the structure of coarse – grained regolith. (b) Erosion and small – scale impactions constantly modified the lunar surface. (c) Coverage of Finsen ejecta. (d) Formation of fine – grained regolith. (e) Delivery of fragments by distal impact events. (f) Current regolith structure. The grains of the rocks represent the maturities of regolith. The orange – colored anomalies represent the superposition of multiple ejecta from nearby caters and the gray – colored anomalies denote the Finsen ejecta. The green arrows and the dashed arrows represent the single and multiple impact events, respectively; whereas the undulated arrows and the gray arrows indicate the accumulations of multi – source materials and Finsen ejecta, respectively

(0 – 170m and 290 – 370m) probably indicate other craters that have degraded.

2. Stage (b and c): erosion and small – scale impactions constantly modified the lunar surface and the coarse ejecta layer was probably grained during this stage. The heterogeneous weathering resulted in the asymmetry of the large crater and the ejecta in its right sliding wing are of higher maturity. The formation of two separated rock – concentrated areas at the left of the large crater was completed. Besides, debris materials accumulated and deposited in the craters; small craters were highly degraded and became invisible.

3. Stage (c and d): the coarse ejecta from the eastern Finsen crater buried the zone (Li et al., 2020; Zhang, Zhou et al., 2021). Only the larger and not fully degraded crater was visible on the topography of the lunar surface.

4. Stage (d and e): the fine - grained regolith of unconsolidated materials accumulated gradually from the continuous impaction and gardening of the Finsen ejecta materials. These modifications probably lasted for a long time and simultaneously flattened the lunar surface. Most coarse ejecta materials filled in the degraded crater survive the gardening on the lunar surface resulting in its unique properties compared to both the materials of the underlying and overlying regolith.

5. Stage (e and f): the distal impact events delivered rock fragments to the surface of the landing zone, some of which formed small and fresh secondary craters (Ding et al., 2020, 2021). These events also produce melt welded impact breccia in the near - surface of fine - grained regolith and resulting in the horizontal heterogeneity of the lunar surface.

4 Conclusions

This study provides quantitative maturity and fine structure of regolith at lunar farside using polarimetric attributes analysis of CE - 4 LPR data. The polarization information is extracted from LPR data acquired during the rover turnings. Local regions with anomalously low maturities are observed on the lunar surface which may be due to the dense materials formed by local impact or rock fragments delivered from distal impacts. A sandwich structure with low maturity is newly discovered within the ejecta layer which is formed by the heterogeneous weathering in a paleo - crater. We also found that LPR signals were mainly from subsurface rock fragments of different sizes and were rarely from strata interfaces. Based on these new insights, we infer a heterogeneous weathering process of the regolith.

The polarimetric information is important and useful to characterize the subsurface and is applied to planetary exploration for the first time in this paper. Both the Chinese Tianwen - 1 Mars mission and the ESA's ExoMars 2022 mission are equipped with polarimetric rover radar. Our results are pivotal and instructive for the interpretation of polarimetric data from future Mars rover radar.

Acknowledments

We thank the China National Space Administration for providing the scientific data used in this study. This work was supported by the National Key Research and Development Program of China under Grant 2021 YFC1523401 and the National Natural Science Foundation of China under Grant 41974129.

References

[1] Bart, G. D. , & Melosh, H. J. (2010) . Distributions of boulders ejected from lunar craters Icarus, 209 (2), 337 - 357.

[2] Cloude, S. R. , & Pottier, E. (1996) . A review of target decomposition theorems in radar polarimetry. IEEE Transactions on Geoscience and Remote Sensing, 34 (2), 498 - 518.

[3] Ding, C. , Xiao, Z. , Wu, B. , Li, Y. , Prieur, N. C. , Cai, Y. , et. al. (2020). Fragments delivered by secondary craters at the Chang'e 4 landing site.

[4] Geophysical Research Letters, 47, e2020GL087361. https: //doi. org/10. 1029/2020gl087361

[5] Ding, C. , Xiao, Z. , Wu, B. , Li, Z. , Su, Y. , Zhou, B. , et. al. (2021). Rock fragments in shallow lunar regolith: Constraints by the lunar penetrating radar onboard the Chang' E - 4 mission. Journal of Geophysical Research: Planets, 126, e2021JE006917. https: //doi. org/10. 1029/2021je006917

[6] Dong, Z. , Fang, G. , Ji, Y. , Gao, Y. , Wu, C. , & Zhang, X. (2017). Parameters and structure of lunar regolith in Chang'E - 3 landing area from lunar penetrating radar (LPR) data. Icarus, 282, 40 - 46. https: //doi. org/10. 1016/j. icarus. 2016. 09. 010

[7] Dong, Z. , Feng, X. , Zhou, H. , Liu, C. , Lu, Q. , & Liang, W. (2022). Assessing the effects of induced field rotation on water ice detection of Tianwen - 1 full - polarimetric Mars rover penetrating radar. IEEE Transactions on Geoscience and Remote Sensing, 60, 1 - 13. https: //doi. org/10. 1109/tgrs. 2021. 3138463

[8] Dong, Z. , Feng, X. , Zhou, H. , Liu, C. , & Sato, M. (2021) . Effects of induced field rotation from rough surface on H - alpha decomposition of full - polarimetric GPR. IEEE Transactions on Geoscience and Remote Sensing, 59 (11), 9192 - 9208.

[9] Dong, Z. , Feng, X. , Zhou, H. , Liu, C. , Zeng, Z. , Li, J. , & Liang, W. (2020) . Properties analysis of lunar regolith at Chang' E - 4 landing site based on 3D velocity spectrum of lunar penetrating radar. Remote Sensing, 12, 629. https: //doi. org/10. 3390/rs12040629

[10] Fang, G. , Zhou, B. , Ji, Y. , Zhang, Q. , Shen, S. , Li, Y. , et. al. (2014). Lunar penetrating radar onboard the Chang'E - 3 mission. Research in Astronomy and Astrophysics, 14 (12), 1607 - 1622.

[11] Feng, J. , Su, Y. , Ding, C. , Xing, S. , Dai, S. , & Zou, Y. (2017). Dielectric properties estimation of the lunar regolith at CE - 3 landing site using lunar penetrating radar data. Icarus, 284, 424 - 430. https: //doi. org/10. 1016/j. icarus. 2016. 12. 005

[12] Freeman, A., & Durden, S. L. (1998). A three - component scattering model for polarimetric SAR Data. IEEE Transactions on Geoscience and Remote Sensing, 36 (3), 963 - 973.

[13] Giannakis, I., Zhou, F., Warren, C., & Giannopoulos, A. (2021). Inferring the shallow layered structure at the Chang' E - 4 landing site: A novel interpretation approach using lunar penetrating radar. Geophysical Research Letters, 48 (16), e2021GL092866. https://doi.org/10.1029/2021gl092866

[14] Gou, S., Di, K., Yue, Z., Liu, Z., He, Z., Xu, R., et. al. (2019). Lunar deep materials observed by Chang'e - 4 rover. Earth and Planetary Science Letters, 528 (15), 115829. https://doi.org/10.1016/j.epsl.2019.115829

[15] Gou, S., Yue, Z., Di, K., Cai, Z., Liu, Z., & Niu, S. (2021). Absolute model age of lunar Finsen crater and geologic implications. Icarus, 354 (15), 115829. https://doi.org/10.1016/j.icarus.2020.114046

[16] Head, J. W., III, Fassett, C. I., Kadish, S. J., Smith, D. E., Zuber, M. T., Neumann, G. A., & Mazarico, E. (2012). Global distribution of large lunar craters: Implications for resurfacing and impactor populations. Science, 329 (5998), 1504 - 1507.

[17] Hu, X., Ma, P., Yang, Y., Zhu, M., Jiang, T., Lucey, P. G., et. al. (2019). Mineral abundances inferred from in situ reflectance measurements of Chang' E - 4 landing site in South Pole - Aitken basin. Geophysical Research Letters, 46, 9439 - 9447. https://doi.org/10.1029/2019gl084531

[18] Huang, J., Xiao, Z., Flahaut, J., Martinot, M., Head, J., Xiao, X., et. al. (2018). Geological characteristics of Von Kármán crater, northwestern South Pole Aitken basin: Chang' E - 4 landing site region. Journal of Geophysical Research: Planets, 123, 1684 - 1700. https://doi.org/10.1029/2018je005577

[19] Lai, J., Cui, F., Xu, Y., Liu, C., & Zhang, L. (2021). Dielectric properties of lunar materials at the Chang'e - 4 landing site. Remote Sensing, 13, 4056. https://doi.org/10.3390/rs13204056

[20] Lai, J., Xu, Y., Bugiolacchi, R., Wong, H. K., Xu, L., Zhang, X., et. al. (2021). A complex paleo - surface revealed by the Yutu - 2 rover at the lunar farside. Geophysical Research Letters, 48, e2021GL095133. https://doi.org/10.1029/2021gl095133

[21] Lai, J., Xu, Y., Zhang, X., Xiao, L., Yan, Q., Meng, X., et. al. (2019). Comparison of dielectric properties and structure of lunar regolith at Chang'e - 4 landing site in South Pole - Aitken basin. Geophysical Research Letters, 46, 12783 -12793. https://doi.org/10.1029/2019gl084458

[22] Li, C., Liu, D., Liu, B., Ren, X., Liu, J., He, Z., et. al. (2019). Chang' E - 4 initial spectroscopic identification of lunar far side mantle - derived materials. Nature, 569 (7756), 378 - 382.

[23] Li, C., Su, Y., Pettinelli, E., Xing, S., Ding, C., Liu, J., et. al. (2020). The Moon's farside shallow subsurface structure unveiled by Chang'E-4 lunar penetrating radar. Science Advances, 6 (9), eaay6898. https://doi.org/10.1126/sciadv.aay6898

[24] Lin, H., Lin, Y., Yang, W., He, Z., Hu, S., Wei, Y., et. al. (2020). New insight into lunar regolith forming processes by the lunar rover Yutu-2. Geophysical Research Letters, 47, e2020GL087949. https://doi.org/10.1029/2020gl087949

[25] Liu, J., Ren, X., Yan, W., Li, C., Zhang, H., Jia, Y., et. al. (2019). Descent trajectory reconstruction and landing site positioning of Chang' E-4 on the lunar farside. Nature Communications, 10, 4229. https://doi.org/10.1038/s41467-019-12278-3

[26] McKay, D.S., Heiken, G., Basu, A., Blanford, G., Simon, S., Reedy, R., et. al. (1991). The lunar regolith. In G. H. Heiken, D. T. Vaniman, & B. M. French (Eds.), Lunar sourcebook: A user's guide to the Moon (pp. 285-356). Cambridge University Press.

[27] Melosh, H.J. (1989). Impact cratering: A geologic process. Oxford Univ. Press.

[28] Sinclair, G. (1950). The transmission and reception of elliptically polarized waves. Proceedings of the IRE, 38, 148-151. https://doi.org/10.1109/jrproc.1950.230106

[29] Stuart-Alexander, D. E. (1978). Geologic map of the central far side of the Moon I-1047. US Geological Survey.

[30] Su, Y., Fang, G., Feng, J., Xing, S., Ji, Y., Zhou, B., et. al. (2014). Data processing and initial results of Chang'E-3 lunar penetrating radar. Research in Astronomy and Astrophysics, 14 (12), 1623-1632.

[31] van Zyl, J.J., Zebker, H.A., & Elachi, C. (1987). Imaging radar polarization signatures: Theory and observation. Radio Science, 22 (4), 529-543.

[32] Zhang, J., Zhou, B., Lin, Y., Zhu, M., Song, H., Dong, Z., et. al. (2021). Lunar regolith and substructure at Chang' E-4 landing site in South Pole-Aitken basin. Nature Astronomy, 5, 25-30. https://doi.org/10.1038/s41550-020-1197-x

[33] Zhang, L., Xu, Y., Bugiolacchi, R., Hu, B., Liu, C., Lai, J., et. al. (2021). Rock abundance and evolution of the shallow stratum on Chang'e-4 landing site unveiled by lunar penetrating radar data. Earth and Planetary Science Letters, 564, 116912. https://doi.org/10.1016/j.epsl.2021.116912

[34] Zhou, H., Feng, X., Ding, C., Dong, Z., Liu, C., Zhang, Y., & Meng, Z. (2021). Yutu-2 radar sounding evidence of a buried crater at Chang' E-4 landing site. IEEE Transactions on Geoscience and Remote Sensing, 60, 1-19. https://doi.org/10.1109/tgrs.2021.3090528

[35] Zhou, H., Feng, X., Dong, Z., Liu, C., & Liang, W. (2021). Application of denoising CNN for noise suppression and weak signal extraction of lunar penetrating radar data. Remote Sensing, 13 (4), 779. https://doi.org/10.3390/rs13040779

[36] Zhou, H., Feng, X., Dong, Z., Liu, C., & Liang, W. (2022). Multi-parameter adaptive target classification using full-polarimetric GPR: A novel approach to landmine detection. IEEE Journal of Selected Topics in Applied Earth Observations and Remote Sensing, 15, 2592-2606. https://doi.org/10.1109/jstars.2022.3159305

[37] Zhou, H., Feng, X., Dong, Z., Liu, C., Liang, W., & An, Y. (2022). Local Freeman decomposition for robust imaging and identification of subsurface anomalies using misaligned full-polarimetric GPR data. Remote Sensing, 14 (3), 804. https://doi.org/10.3390/rs14030804

Yutu - 2 Radar Sounding Evidence of a Buried Crater at Chang'E - 4 Landing Site

Haoqiu Zhou, *Student Member*, *IEEE*, Xuan Feng, *Senior Member*, *IEEE*, Chunyu Ding, Zejun Dong, Cai Liu, Yan Zhang, and Zhiguo Meng, *Member*, *IEEE*

Haoqiu Zhou, Zejun Dong, Cai Liu, and Zhiguo Meng are with the College of Geo - Exploration Science and Technology, Jilin University, Changchun 130026, China (e - mail: zhouhq20@mails. jlu. edu. cn; dzj19@mails. jlu. edu. cn; liucai@jlu. edu. cn; mengzg@jlu. edu. cn).

Xuan Feng is with the College of Geo - Exploration Science and Technology, Jilin University, Changchun 130026, China, and also with the Key Laboratory of Applied Geophysics, Ministry of Natural Resources, Jilin University, Changchun 130026, China (e - mail: fengxuan@jlu. edu. cn).

Chunyu Ding is with the Department of Earth and Space Science, Southern University of Science and Technology, Shenzhen 518055, China (e - mail: dingcy@sustech. edu. cn).

Yan Zhang is with the College of Earth Science, Jilin University, Changchun 130061, China (e - mail: yan_zhang@jlu. edu. cn).

Abstract: Buried craters within tens of meters of lunar regolith are rarely studied but are significant for understanding the evolution of surface processes on the Moon. Here, we first report the evidence of an intact buried crater within the layered strata at Chang'E - 4 (CE - 4) landing site revealed by the lunar penetrating radar (LPR). The time - frequency comparative analysis method based on the variational mode decomposition (VMD) and the rock quantitative analysis method based on the local unit correlation (LUC) are proposed and applied to the processing and analysis of LPR data within 15 lunar days. The results presented by the two methods provide evidence of a buried crater at the CE - 4 landing site and simultaneously reveal the rock - concentrated structure within the buried crater. According to the results, it is considered that the filling materials within the buried crater have survived the impaction and gardening during the formation of the overlying fine - grained regolith. Recent works have proposed that the near - surface material at the CE - 4 landing site is mainly the lunar mantle materials excavated from the nearby Finsen crater. Therefore, the buried crater probably preserves the initial lunar mantle materials.

Index Terms: Chang'E - 4 (CE - 4), impacting crater, local unit correlation (LUC), lunar penetrating radar (LPR), rock localization and quantitative analysis, time -

frequency comparative analysis.

I　INTRODUCTION

ON JANUARY 3, 2019, the Chinese Chang'E - 4 (CE - 4) probe achieved the first soft landing on the far side of the Moon. The CE - 4 landing site (177.59 E, 45.46 S) is on the floor of the Von Kármán crater in the South Pole - Aitken (SPA) basin [1]. The SPA basin, with a diameter of approximately 2500km and a depth of 15 km, is the largest impact structure on the Moon [2] - [4], which can provide clues to the early evolution of the Moon, basin - scale impact melt deposits, nature of large impact basins, and their formation and modification processes [5].

The Von Kármán crater is located in the SPA basin and has a diameter of ~186km. According to the previous regional surveys, it has experienced a complex geological evolution [6], [7]. After the occurrence of two magmatism events in the Von Kármán crater during the Imbrian period, the Alder and Finsen craters ejected materials into it during the Eratoshenian period [8] - [10]. Apart from the Finsen crater, the Von Kármán L and Von Kármán L' craters also contributed to the ejecta in the Von Kármán crater [5]. Recent researches indicate that the near - surface materials measured by the CE - 4 rover were mainly the ejecta from the Finsen crater, with possible contributions from the Alder crater [11] - [20]. The thickness of the ejecta from these two craters is, respectively, predicted to be ~30 and ~35m [11], which, however, was not supported by observations based on reflectance spectra or predictions based on ballistic sedimentation models [21].

The lunar penetrating radar (LPR) carried by the Yutu - 2 rover provides a unique opportunity to map the lunar shallow structure with a high precision resolution on the far side of the Moon [22] - [24]. The LPR is equipped with two types of channels (CH - 1 and CH - 2), with center frequencies of 60 and 500MHz and bandwidths of 40 and 450MHz, respectively [25], [26]. Besides, the pulselength of CH - 1 and CH - 2 are approximately 50 and 5 ns [25], respectively. CH - 1 channel is a pair of dipole antennas deployed on the tail of the rover, and CH - 2 channel contains one transmitting and two receiving bow - tie antennas, known as CH - 2A and CH - 2B, with different offsets of 0.16 and 0.32 m [25]. The CH - 2 antennas are ~0.3m above the ground [25]. The scientific objective of CH - 1 is to detect the lunar crust, while the goal of CH - 2 is to map the fine structure of the near - surface layers [25] - [28]. During the Chang'E - 3 (CE - 3) mission, the LPR, which had the same frequencies and bandwidths as in CE - 4, helped to reveal the complex geological processes in the Imbrium basin on the near side of the Moon. More than nine subsurface layers were identified by CH - 1 radargram, and those layers record the history of the volcano activities in the different periods at the CE - 3 landing site [29], [30]. CH - 2 radargram has mapped the shallow regolith with two and/or three sublayers at a depth of ~10m, and those structures are orderly interpreted as the fine regolith, ejecta from Ziwei crater, and paleoregolith, respectively [29], [30].

Compared with the CE - 3 mission, the CE - 4 radargram shows deeper layered reflectors due to the radar pulse propagation into the low attenuation materials [31], [32]. In the previous studies, the CH - 2 LPR data within the first two or three lunar days revealed a three - unit regolith structure. Unit 1 (the fine - grained regolith) has a depth of 0 - 12 m developed from the ejecta material of the Finsen crater; unit 2 has a depth of 12 - 24 m and contains a substantial number of randomly distributed rocks; unit 3 has a depth of 24 - 40m and contains alternating layers of fine - grained materials and rocks [19], [24], [31], [32]. In this article, based on the proposed new techniques, we process and analyze the LPR data within 15 lunar days and reveal a buried crater at the CE - 4 landing site.

The filed LPR signals are always the amplitude - modulated and frequency - modulated signals. Previous studies [28], [33], [34] have proposed the empirical mode decomposition (EMD) [35] method to analyze the time - frequency attributes of LPR data. However, the EMD method, known by its name, is an empirical method without strict mathematical derivation, which leads to its modal aliasing, and poor denoising ability [36]. Thus, the variational mode decomposition (VMD) method [36] was developed to process the nonstationary signals. This new method has a strict mathematical foundation leading to its stronger signal processing ability beyond the traditional EMD method [36]. As an advanced technique, the VMD method has been applied to ground penetrating radar (GPR) signal processing in recent years [37] - [39].

The subsurface rock localization and quantitative analysis is an important scientific aim of LPR exploration which will help to better understand the subsurface structure and the evolution of lunar regolith. In the previous studies, the local similarity [40] - [42] has been applied to achieving this aim [43], [44]. The local similarity uses both CH - 2A and CH - 2B LPR data with different offsets which aims to suppress random noise interference and highlight similar signals [43]. However, this method can be only applied to each single trace signal and cannot analyze the horizontal feature of radargram. Here, we proposed the local unit correlation (LUC) method to obtain the distribution of subsurface rocks. This method is more simple but effective and can analyze the 2 - D feature of the LPR data.

Overall, in this article, we performed the time - frequency comparative analysis based on the VMD method and the rock quantitative analysis based on the LUC method to process and analyze the LPR data within 15 lunar days. The results provide evidence of a buried crater within the layered strata at the CE - 4 landing site and simultaneously reveal the rock - concentrated structure within the buried crater. Finally, we propose an interpretation of the results and summarize an evolution process of the lunar regolith at the CE - 4 landing site.

Ⅱ　METHODOLOGY

A　VMD Method

The VMD method, similar to the EMD method, aims to decompose a signal into several intrinsic mode functions (IMFs). An IMF is a harmonic signal with time-varying amplitude and frequency. The real expression of the kth IMF [36] is given by

$$u_k(t)=A_k(t)\cos(\varphi_k(t)) \tag{1}$$

where $A_k(t)$ and $\varphi_k(t)$ represent the amplitude and the phase.

The complex analytic signal of $u_k(t)$ can be computed using Hilbert transform

$$f_k(t)=A_k(t)e^{i_k(t)}=\left(\delta(t)+\frac{i}{\pi t}\right)*u_k(t) \tag{2}$$

where $\delta(t)$ is the Dirac function.

It is assumed that the IMFs are the signals with finite bandwidth around the respective center frequency $d(\varphi_k(t))/dt$. Because $u_k(t)$ and $\varphi_k(t)$ are unknown so a variable frequency $\omega_k(t)$ is supposed and mixed into the signal to move the spectrum to the baseband, which can be represented as follows:

$$f_k(t)e^{-i\omega_k t}=A_k(t)e^{i\varphi_{ik}(t)}e^{-i\omega_k t}=\left[\left(\delta(t)+\frac{i}{\pi t}\right)*u_k(t)\right]e^{-i\omega_k t} \tag{3}$$

Subsequently, the matching between ω_k and $d(\varphi_k(t))/dt$ is the main problem which is to minimize the sum of the modal function bandwidths and can be represented using a variational model [36]

$$\min_{\{u_k\},\{\omega_k\}}\left\{\sum_k\left\|\partial_t\left[\left(\delta(t)+\frac{j}{\pi t}\right)*u_k(t)\right]e^{-i\omega_k t}\right\|_2^2\right\}$$
$$s.t.\ \sum_k u_k=f \tag{4}$$

where $u_k(t)$ represents the k th IMF with a center frequency ω_k. The computation seeks K IMFs under the constraint that the sum of IMFs is equal to the initial signal f, which is equivalent to minimize the sum of the bands between ω_k and $d(\varphi_k(t))/dt$.

In order to obtain the optimal solution, the VMD method utilizes a penalty factor β and a Lagrangian multiplier λ to transform the variational mode into an augmented Lagrange function [36] as follows:

$$\begin{aligned}&Lg(\{u_k\},\{\omega_k\},\lambda)\\&=\beta\sum_k\left\|\partial_t\left[\left(\delta(t)+\frac{j}{\pi t}\right)*u_k(t)\right]e^{-j\omega_k t}\right\|_2^2\\&\quad+\left\|f-\sum_k u_k(t)\right\|_2^2+\left\langle\lambda,f-\sum_k u_k(t)\right\rangle.\end{aligned} \tag{5}$$

Subsequently, the alternate direction method of multipliers (ADMM) is adopted to solve the objective function in (5). Assuming that the signal composes of k IMFs, u_k, ω_k, and λ_k can be updated using the following formulas [36]:

$$\begin{cases} \hat{u}_k^{n+1}(\omega) = \dfrac{\hat{f}(\omega) - \sum_{i=1}^{k-1} \hat{u}_k^{n+1}(\omega) - \sum_{i=k+1}^{K} \hat{u}_i^n(\omega) + \dfrac{\hat{\lambda}^n(\omega)}{2}}{1 + 2\delta(\omega - \omega_k^n)^2} \\ \omega_k^{n+1} = \dfrac{\int_0^\infty \omega \mid \hat{u}_k^{n+1}(\omega) \mid^2 d\omega}{\int_0^\infty \mid \hat{u}_k^{n+1}(\omega) \mid^2 d\omega} \\ \hat{\lambda}^{n+1}(\omega) = \hat{\lambda}^n(\omega) + \tau\left(\hat{f}(\omega) - \sum_{i=1}^{K} \hat{u}_i^{n+1}(\omega)\right) \end{cases} \tag{6}$$

where $\hat{f}(\omega)$, $\hat{u}_k(\omega)$, and $\hat{\lambda}(\omega)$ are the Fourier transform of $f(t)$, $u_k(t)$, and $\lambda(t)$. The parameter n represents iteration epoch, and τ represents the noise margin parameter. The initial parameters are set to zero. The iteration will continue until [36]

$$\frac{\sum_k \left\| u_k^{n+1}(t) - u_k^n(t) \right\|_2^2}{\left\| u_k^{n+1}(t) - u_k^n(t) \right\|_2^2} \leqslant \varepsilon \tag{7}$$

where ε represents the preset error. Subsequently, a residual $r(t)$ can be derived using the initial signal $f(t)$ and the IMFs [36]

$$r(t) = f(t) - \sum_{i=1}^{K} u_i(t) \tag{8}$$

The number of IMFs K is an important parameter to be specified, which should be determined based on the shape of the spectrum, the purpose of analysis, and a large number of tests. Modal aliasing and false components will occur when K is too small and too large, respectively. In this study, the LPR signals are decomposed into several IMFs. And the comparative analysis is proposed to analyze the IMFs of different center frequencies.

B LUC Method

The CH-2 channel contains one transmitting and two receiving bow-tie antennas, known as CH-2A and CH-2B, with different offsets of 0.16 and 0.32m [25]. In this article, both the CH-2A and CH-2B data of different offsets are adopted to obtain the spatial distribution of subsurface rocks. In the previous studies, the local similarity [40]-[42] has been applied to achieving this aim [43], [44], which uses both CH-2A and CH-2B LPR data with different offsets and aims to suppress random noise interference and highlight similar signals [43]. Although the local similarity can use a regularized mathematical inversion method to eliminate the sliding window in the local correlation calculation, this method can be only applied to each single trace signal and cannot analyze the horizontal feature of the radargram. Therefore, in this article, the rock localization results are obtained by computing the LUC between the CH - 2A and CH - 2B data. The LUC method derives the maximum correlation of each pair of image units from the two radargrams. The image units are the 2 - D regions, whose sizes are settable. The LUC is given by

$$L_{f_1, f_2}(i, j) = \max_{t_0, x_0}\left[\sum_{-\infty}^{+\infty}\sum_{-\infty}^{+\infty} f_1^{i,j}(t, x) f_2^{i,j}(t+t_0, x+x_0)\right] \quad (9)$$

where

$$f_{1,2}^{i,j}(t, x) = f_{1,2}\left(i-\frac{w_1}{2}: i+\frac{\omega_1}{2}, j-\frac{w_2}{2}: j+\frac{\omega_2}{2}\right) \quad (10)$$

$f_1(t, x)$ and $f_2(t, x)$ represent the image units from CH - 2A and CH - 2B radargram. w_1 and w_2 denote the length and width of the unit, respectively. Equations (9) and (10) represent that for the radargram pixel at position (i, j), two $w_1 \times w_2$ units of data centered on position (i, j) from two channels are selected to calculate a 2 - D cross correlation; subsequently, the maximum value of the 2 - D cross - correlation result is assigned to the LUC result at position (i, j). A soft threshold function is subsequently utilized to attenuate the noise interference, which is given by [43]

$$\hat{L} = \begin{cases} L-\zeta, & L > \xi \\ 0, & L \leqslant \xi \end{cases} \quad (11)$$

where ζ represents the settable threshold value.

Ⅲ　NUMERICAL VALIDATION OF LUC METHOD

A　LUC Test of a Simple Model

A simple model in Fig. 1 (a) is applied to the test of the LUC method. Three targets are immersed in a uniform medium at the same depth. The permittivity of the background medium is set to 4 which is similar to that of lunar regolith; the permittivity of the buried rocks is set to 7. 5, which is similar to that of buried debris [12], [28]. In order to exclude the effects of attenuation on the amplitude of signals, the electrical conductivity is set to 0. The thickness of the top vacuum layer increases linearly. The depth of the slanted interface is from 0m at left to 0. 4m at right. The transmitting and receiving antennas are moving at the depth of 0m with a movement of 0. 04m for each time, which is similar to the field LPR acquisition; the antenna intervals of the two radargrams are 0. 16 and 0. 32m, respectively, corresponding to the intervals of CH - 2A and CH - 2B antennas [25]. In this model, the rootmean - square velocity (v_{rms}) presented by the three targets is different. Twice simulations with different antenna intervals are performed using the 500 - MHz wavelet and the radargrams after deleting the reflections of the slanted interface are shown in Fig. 1 (b) and (c).

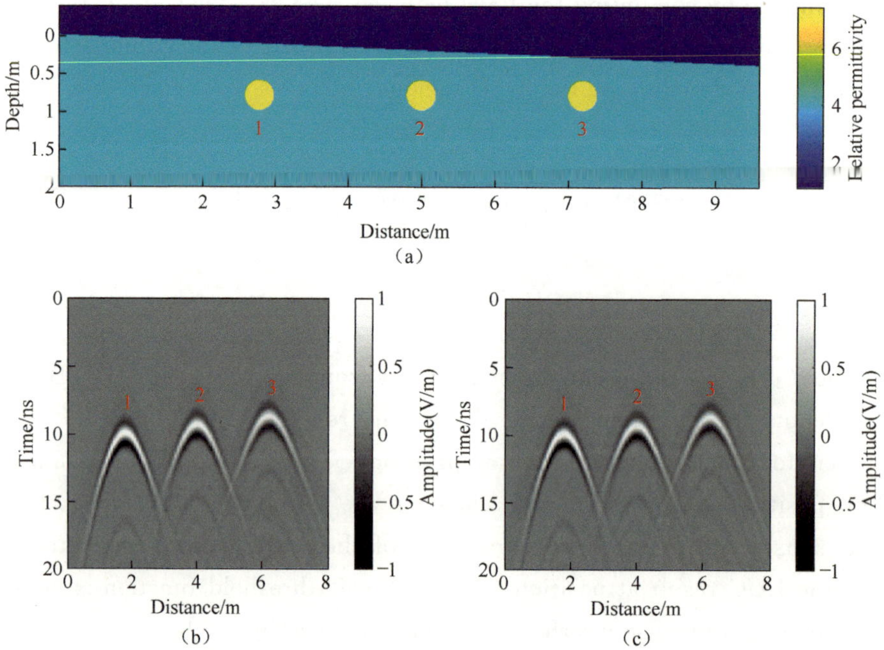

Fig. 1　Simple model and its simulated results. (a) Dielectric constant model. (b) and (c) Simulated results with antenna intervals are 0.16 and 0.32m, respectively

Subsequently, the Gaussian noise is added into the radargrams using the following formula:

$$S_n = \frac{S}{\max(|S|)} + \theta \cdot \mathrm{Randn}(0,\ 1) \tag{12}$$

where S_n and S represent the noise added signal and the original signal, respectively. Randn (0, 1) denotes the Gaussian noise with the mean and variance are 0 and 1, respectively. θ represents the standard deviation which is given by

$$\theta = \frac{\max(|S|)}{10^{\frac{SNR}{20}}} \tag{13}$$

where SNR is the signal-to-noise ratio (SNR) of the noise. In this test, the SNR is set to be 5. The noise added data are, respectively, shown in Fig. 2 (a) and (b) and the signals from the targets are immersed in the noise. Fig. 2 (c) and (d) shows the radargrams after Kirchhoff migrations with a velocity of 0.185m/ns. Due to the difference of v_{rms}, the qualities of the migration results are different. The applied velocities are suitable to target 2 but high for target 1 and low for target 3. The velocity model is unknown for the field LPR data and we cannot obtain a perfect migration result without distortion, so the simple model can also test the effects of the improperly migrated results with distortions on the rock localization of the LUC method.

The LUC result with vertical and horizontal window sizes 50 and 10 is shown in Fig. 2 (e), which can clearly present the positions of the three targets with much less noise

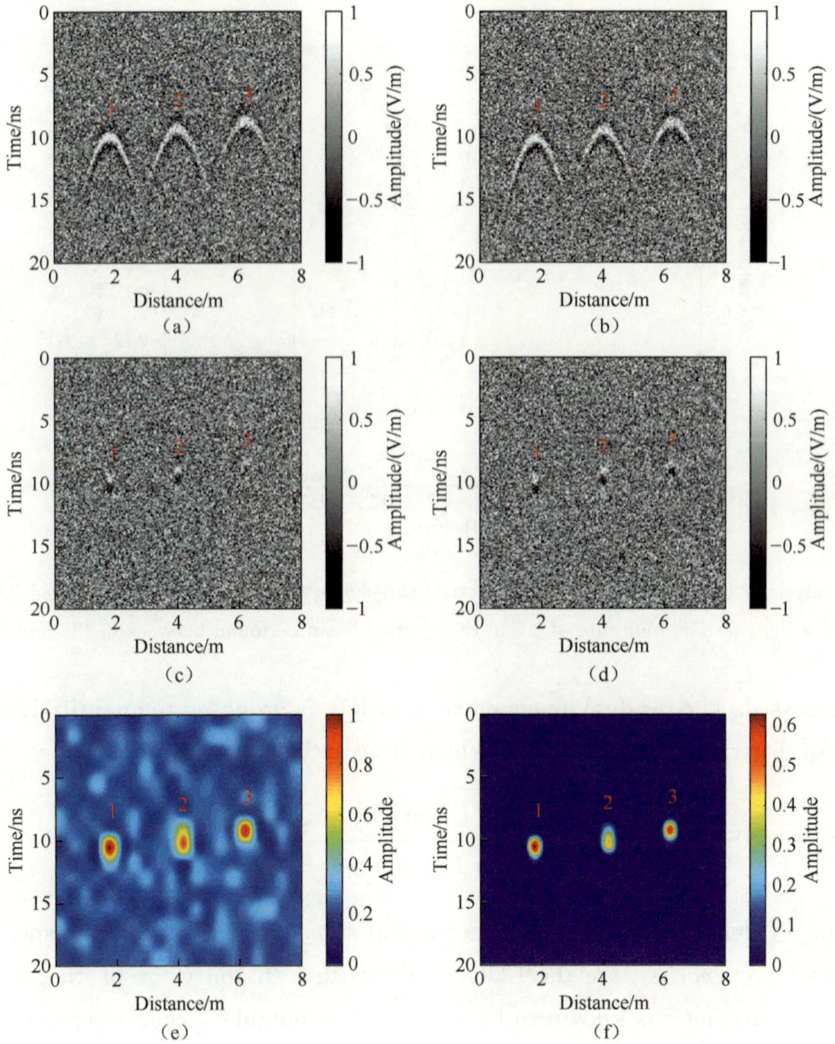

Fig. 2　(a) and (b) Noise added data. (c) and (d) Migration results of (a) and (b) with a velocity of 0.185 m/ns, respectively. (e) and (f) LUC results before and after the soft threshold function processing (ζ=0.38)

interference. It is because the noises in the two radargrams do not correlate but the signals from the same target are similar. Fig. 2 (f) shows the LUC result after the soft threshold function processing in (11) with ζ= 0.38. The procedure further suppresses the residual interference of noise. The positions of the targets can be extracted by computing the local maximums. Overall, the results indicate that the noise interference and the distortions due to the migration with inaccurate velocities have little effect on the LUC results.

B　Soft Threshold Selection for LUC

In this section, the soft threshold selection for LUC method for the radar profiles in

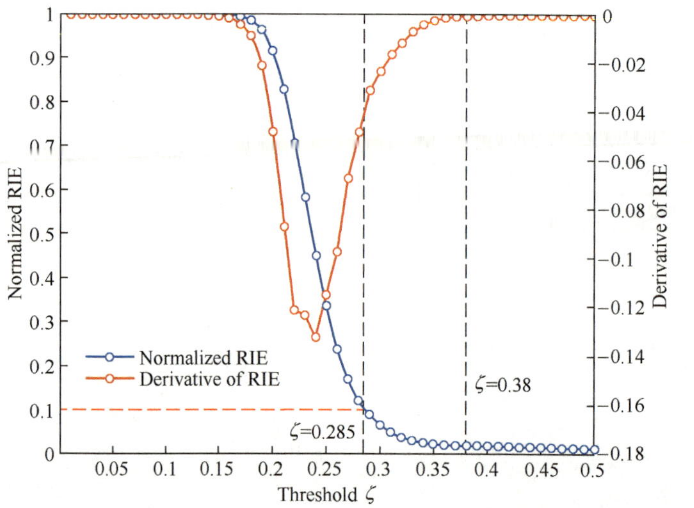

Fig. 3　Analysis of LUC results with different soft thresholds ζ. The red line (normalized RIE=0. 1) provides the possible minimum of ζ and the optimal ζ can be found between 0. 285 and 0. 38

Fig. 1 is discussed. The residual image entropy (RIE) is proposed to quantitatively analyze the effects of different soft threshold ξ which is given by

$$\mathrm{RIE}_i = \frac{\left[\sum_{m=1}^{M} \sum_{n=1}^{N} \left| L_i(m, n) - L_{i-1}(m, n) \right|^2\right]^2}{\sum_{m=1}^{M} \sum_{n=1}^{N} \left| L_i(m, n) - L_{i-1}(m, n) \right|^4} \tag{14}$$

where M and N denote the vertical and horizontal sizes of a radar profile, respectively. $L_i(m, n)$ and $L_{i-1}(m, n)$ are the LUC results with i th and $(i-1)$ th ζ at (m, n) position. This parameter is known to be an easily computable, accurate approximation to entropy [45] – [48]. The larger RIE value indicates the greater interference of noises.

Different soft thresholds ζ from 0 to 0. 5 are adopted to the LUC result in Fig. 2 (e) and the normalized RIE curve and its derivative curve are shown in Fig. 3. The RIE will decrease with the increase of soft threshold ξ because more noise interference is suppressed by greater ζ. However, when ζ reaches 0. 38, the variation of image entropy slows down and both the RIE curve and its derivative curve are close to zero. It is because the noise interference has been completely deleted.

To illustrate, the LUC results with ζ=0. 3, 0. 34, 0. 37, and 0. 38 are presented in Fig. 4. The red points in Fig. 4 indicate the computed positions of local maximums. The number of red points can indicate the interference level of noise, which decreases with the increase of ζ. When ζ reaches 0. 38, the local maximums only appear at the positions of the three targets, which correspond to the results in Fig. 3. However, if the ζ is set to a high level, the signals of the three targets will also be suppressed. Therefore, the ξ in Fig. 2 (f) is set to 0. 38.

Besides, this is a simple test for LUC; for the field LPR data, the radargram is always complex and the RIE curve may not as ideal as Fig. 3 shows so the ξ cannot be determined only by RIE. In the field cases, the RIE is an ancillary method that can provide the possible maximum of ξ, and the main way to determine ζ is to compare the LUC results of different ζ. Taking Fig. 2 (e) and (f), for example, the ζ can be determined if it can mute most of the clutters and noise in the background regions on the radargrams and can clearly present the signals of targets. Therefore, we define a minimum to provide a range of possible thresholds to help the determination process. The ζ when the normalized RIE decreases to 0.1, which indicates that, compared to the case of $\zeta=0$, the most of noises have been removed. As Fig. 3 shows, the red line (normalized RIE = 0.1) provides the possible minimum of ζ and the optimal ζ can be found between 0.285 and 0.38.

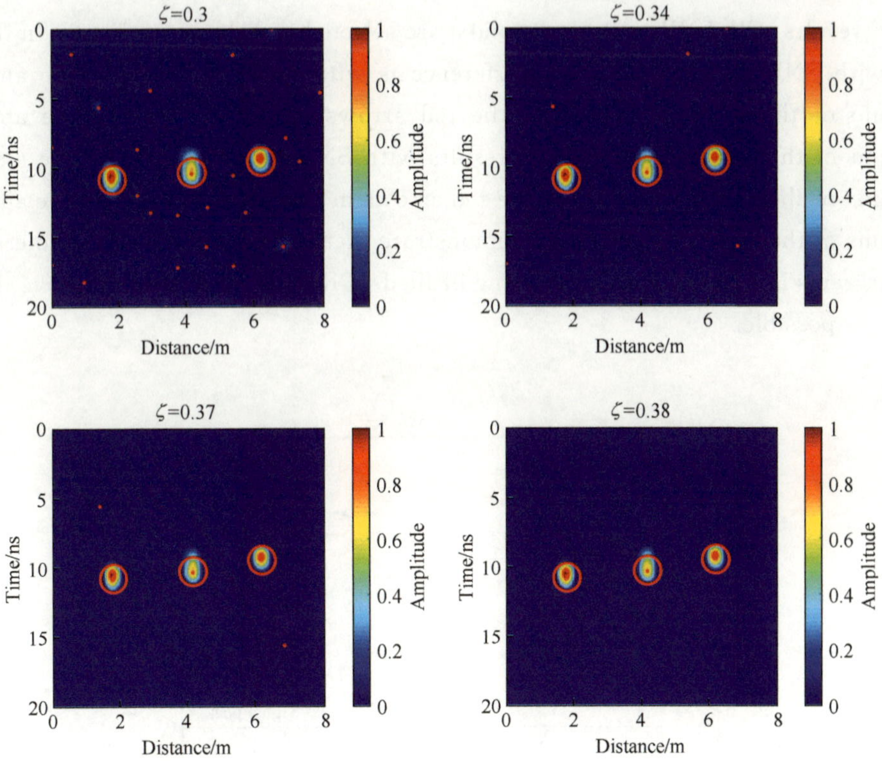

Fig. 4　LUC results with soft thresholds ζ of 0.3, 0.34, 0.37, and 0.38. The red points in the figure indicate the computed positions of local maximums

C　Noise Tolerance of LUC

In this section, the noise tolerance of the LUC method for the radar profiles in Fig. 1

is discussed. Different noises are added into the simulation results with SNR from 0 to 20. It can be speculated that when the noise reaches a high level, the target signals will be totally immersed in the noise and cannot be identified. To visualize the relative change of the target signals and the noise, a peak deviation parameter is adopted, which is given by

$$PD = \max(L_S) - \max(L_N - L_S) \tag{15}$$

where L_N represents the LUC results of the radargrams with different levels of noise and L_S represents the LUC results of the radargrams with no noise. The PD indicates the deviation between the peaks of L_S and $(L_N - L_S)$; when PD=0, the SNR indicates the noise tolerance of the radargrams for LUC.

The PD curve and the image entropy RIE curve with SNR from 0 to 20 are shown in Fig. 5. The RIE will decrease with the increase of SNR but it cannot indicate the noise tolerance. The peak deviation curve will increase from below zero to above zero with the increase of SNR; when SNR $<$ 1, PD $<$ 0, and when SNR$\geqslant$1, PD $>$ 0. To illustrate, the LUC results with SNR = 0, 1, 2, and 3 are selected and shown in Fig. 6. In the LUC results with SNR = 0, the noise interference is with high amplitudes. The amplitude maximums of the noise interference (the red arrows in Fig. 6) exceed the amplitude maximums of the targets. In the LUC results with SNR = 1, 2, and 3, the amplitude maximums of the noise interference (the red arrows in Fig. 6) are less than the amplitude maximums of the targets. The results demonstrate that SNR = 1 is the noise tolerance for the radargrams in Fig. 1. In the processing of filed LPR data, we should improve the SNR as much as possible.

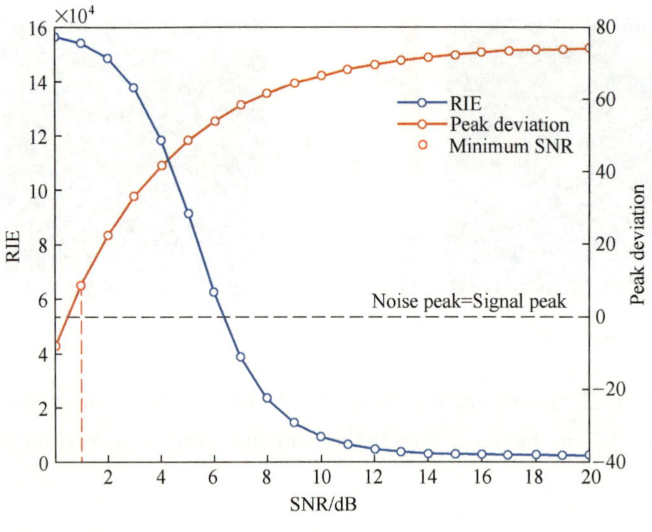

Fig. 5　RIE and PD of LUC results with different levels of noise

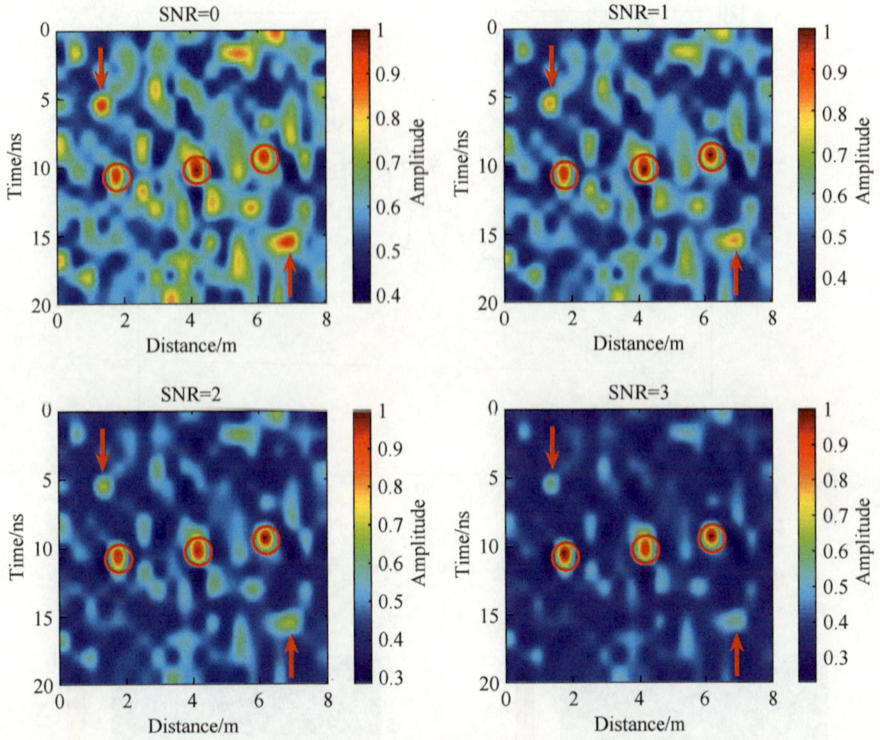

Fig. 6　LUC results with different levels of noise

D　Window Sizes Selections of LUC

In this section, the effects of the vertical and horizontal window sizes (w_1 and w_2) on the LUC result are discussed. The w_1 and w_2 increase from 1 to 100. When $w_2=10$, the relation between w_1 and the maximums of LUC results is presented by the blue curve in Fig. 7; when $w_1=50$, the relation between w_2 and the maximums of LUC results is presented by the orange curve in Fig. 7. The results indicate that the amplitudes of LUC results will increase with the increase of both two window sizes.

Fig. 8 presents four LUC results with vertical window sizes $w_1=1$, 50, 100, and 200 ($w_2=10$). When $w_1=1$, the window size is smaller than the pulselength so the LUC result shows some vertical discontinuities; when the window size exceeds the pulselength, i. e., $w_1=$ 50, 100, and 200, the horizontal resolution will decrease with the increase of w_1.

Fig. 9 presents four LUC results with horizontal window sizes $w_2=$ 1, 10, 30, and 50 ($w_1=$ 50). This case involves no horizontal discontinuity when $w_2=1$; the horizontal resolution will decrease with the increase of w_2; when $w_2=50$, distortions appear in the result, which is because the horizontal window simultaneously covers the signals of two

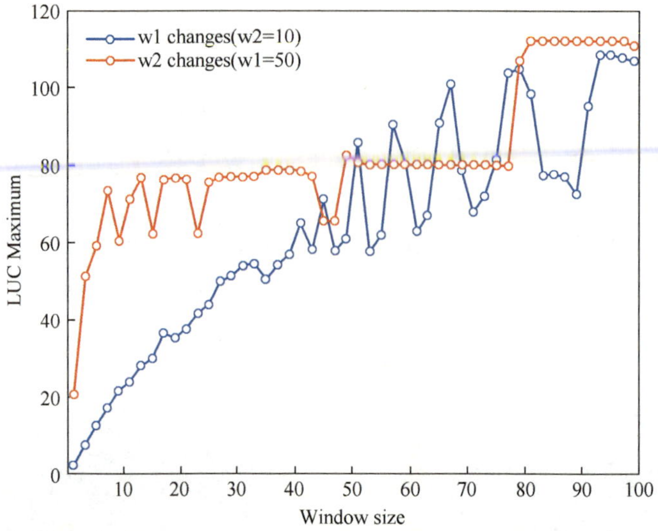

Fig. 7　Maximums of LUC results with different window sizes

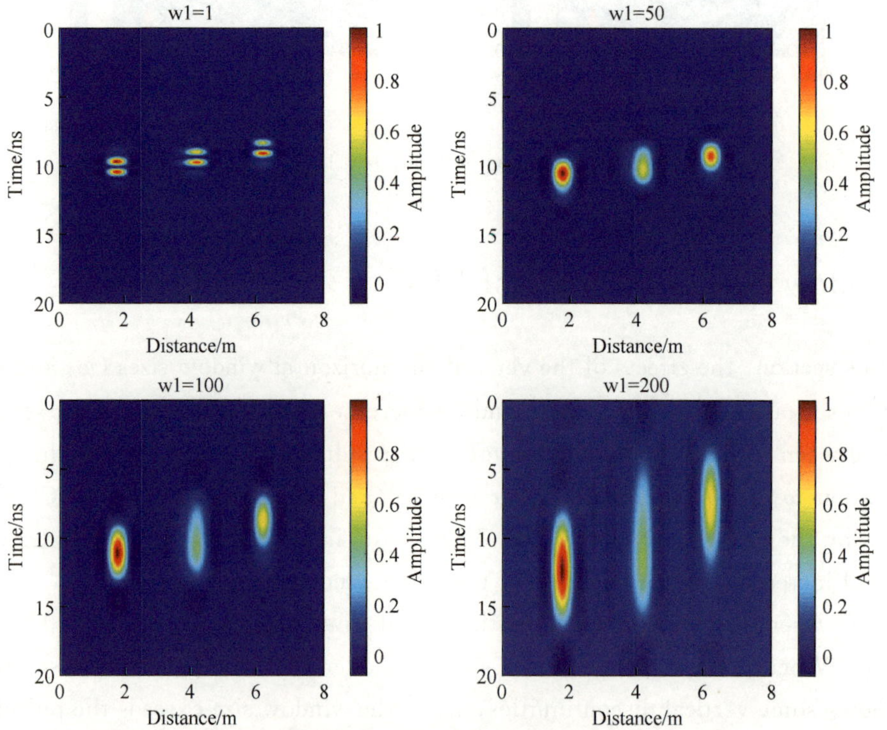

Fig. 8　LUC results with different vertical window sizes w_1 ($w_2 = 10$)

targets.

Subsequently, the combined effects of noises and windows w_1 and w_2 on LUC are discussed. Fig. 10 presents the relations of the maximums of LUC results for signals and noises with w_1 and w_2, respectively. Fig. 10 (a) and (b) shows the cases when SNR =

5, whereas Fig. 10 (c) and (d) shows the cases when SNR = 1. The results indicate that, when SNR = 5, moderately enlarging the window sizes can increase the deviations between signals and noises, thus suppressing the noise. However, when SNR = 1, extremely increasing w_1 and w_2 will instead strengthen the effect of noise. Overall, we can conclude the following.

1) w_1 should be larger than the length of radar pulse.

2) Smaller w_1 and w_2 will result in higher resolutions.

3) If the noises are of relatively low levels, the noise interference can be suppressed by moderately increasing w_1 and w_2.

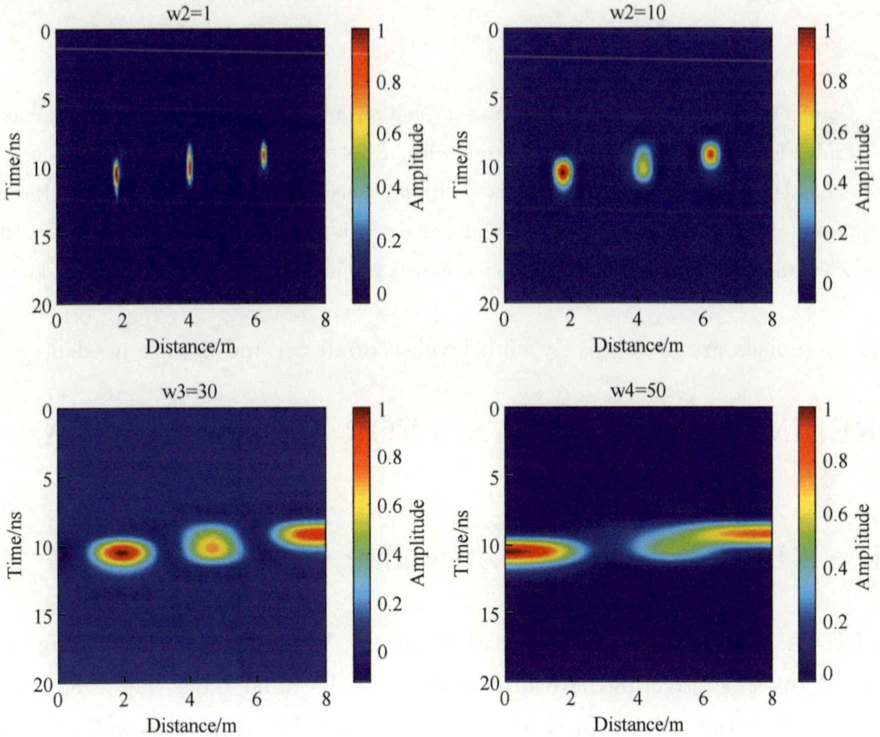

Fig. 9　LUC results with different horizontal window sizes w_2 ($w_1 = 50$)

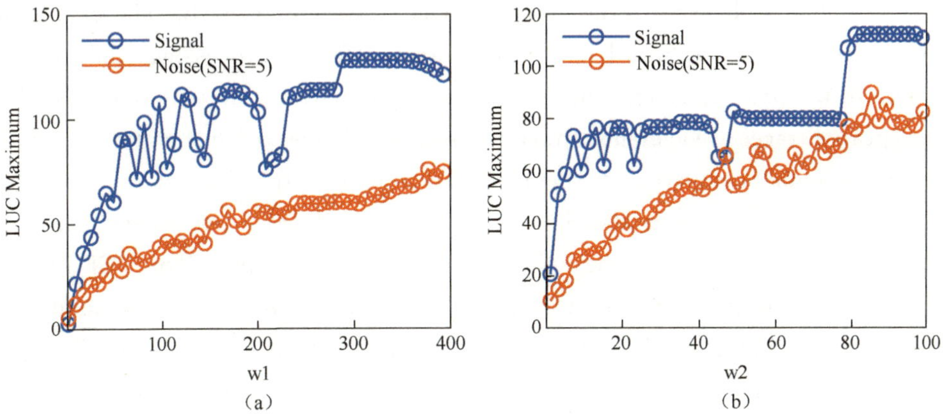

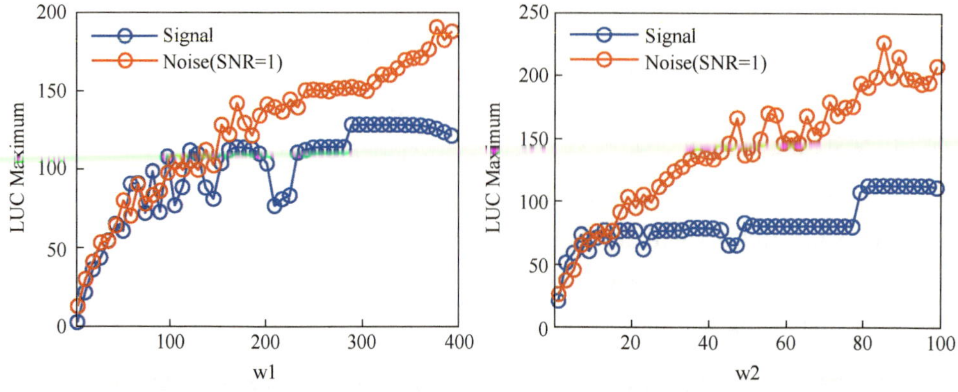

(c)　　　　(d)

Fig. 10　Combined effects of noise and windows w_1 and w_2 on LUC. (a) Relations of the maximums of LUC results for signals and noises with w_1 when SNR = 5, w_2 = 10. (b) Relations of the maximums of LUC results for signals and noises with w_2 when SNR = 5, w_1 = 50. (c) Relations of the maximums of LUC results for signals and noises with w_1 when SNR = 1, w_2 = 10. (d) Relations ZXof the maximums of LUC results for signals and noises with w_2 when SNR = 1, w_1 = 50

4) If the noises are of relatively high levels, small w_1 and w_2 are needed.

Ⅳ　PRELIMINARY ANALYSIS FOR THE LPR DATA

A　CH-2 LPR Data at CE-4 Landing Site

The CE-4 landing site was on the floor of the Von Kármán crater [Fig. 11 (a)]. The Yutu-2 rover's traveling path during the first 15 lunar days was ~380 m in total [Fig. 11 (b)]. The red point represents the lander and the white dots indicate navigation points. The white curve illustrates the Yutu-2 rover traveling route, and red numbers represent the different lunar days of Yutu-2 rover measurement on the Moon.

Fig. 12 (a) and (b) shows the processed CH-2A and CH-2B. The details of the data processing are in Table Ⅰ, which contains the preprocessing and processing procedures. In the preprocessing, the vertical misalignment of different traces is adjusted and the repeated traces are averaged first; subsequently, due to the start time delay of the transmitting antenna compared to the receiving antenna, the data of 0 - 28.203ns is muted; the data after 560 ns is also deleted as the data is beyond the penetrating depth of LPR [31]. In the processing, the spherical exponential compensation (SEC) is first conducted to make deep information more visible; subsequently, the background removal and bandpass filtering are adopted to suppress the clutters and noises; finally, the Kirchhoff migration is used to suppress the diffractions and reconstruct the subsurface structure.

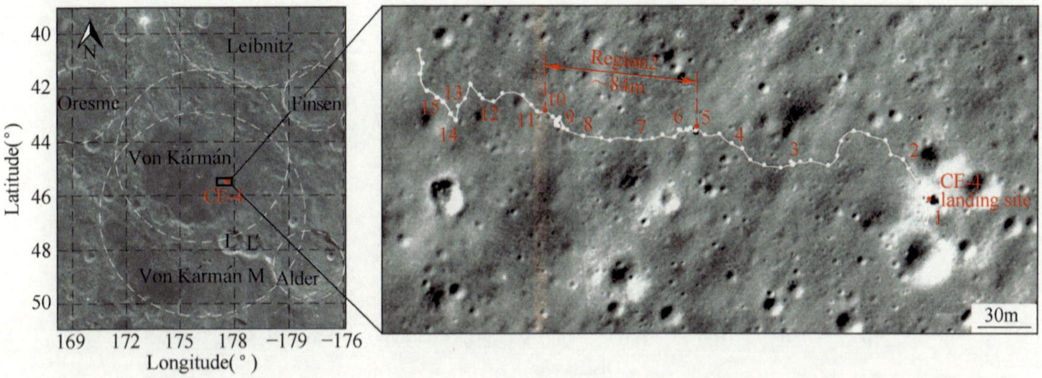

(a)　　　　(b)

Fig. 11　Geological content of the CE－4 landing site and processed LPR CH－2B radargram. (a) CE－4 landing site. The background map is from the digital orthophoto model (DOM) data of CE－1 with CE1_GRAS_DOM_120m_001_GLOBAL_A. (b) Yutu－2's path on the far side of the Moon during the first 15 lunar days. The Lunar Reconnaissance Orbiter Camera Narrow Angle Camera (LROC NAC) image is with NAC_DTM_CHANGE4_M1303619844_140CM. The red point represents the lander and the white dots indicate navigation points. The white curve illustrates the Yutu－2 rover traveling route, and red numbers represent the different lunar days of Yutu－2 rover measurement on the Moon. The length of R2 on radargram is ～130m but its spatial length on lunar surface is estimated to be ～84m

It can be found that the quality of CH－2B is better than that of CH－2A. The deep information below ～300ns in CH－2B radargram is more visible than that in CH－2A radargram. In this study, the CH－2B with the better quality is selected for further VMD processing.

B　VMD Processing Results

Fig. 12 (c) presents the average frequency amplitude spectrum of the CH－2B LPR data, which shows an impressive double－peak structure beyond the rather than normal GPR spectrum on Earth. The frequency absence between the two peaks may be related to the special attenuation of lunar regolith. The spectrum can be roughly divided into four regions which are two main peak frequency bands, a frequency band below the left peak and a frequency band beyond the right peak. Therefore, the VMD processing adopts four IMFs. As the IMFs in Fig. 13 and their average frequency amplitude spectra in Fig. 14 show, the IMF2 and IMF3 are the two main peak frequency components; IMF1 contains the ultralow frequency component and IMF4 contains much high－frequency (HF) noise beyond the frequency band of LPR which is to be deleted. We informally defined the two peaks as the low－frequency (LF) and the HF part of CH－2B. We subsequently recompose the four IMFs into the LF data and the HF data, where LF ＝ IMF1＋IMF2 and HF ＝ IMF3. The average frequency amplitude spectra of the LF and HF data are presented in Fig. 15 (a). The radargram of the LF and HF data are shown in Fig. 15 (b) and (c), respectively.

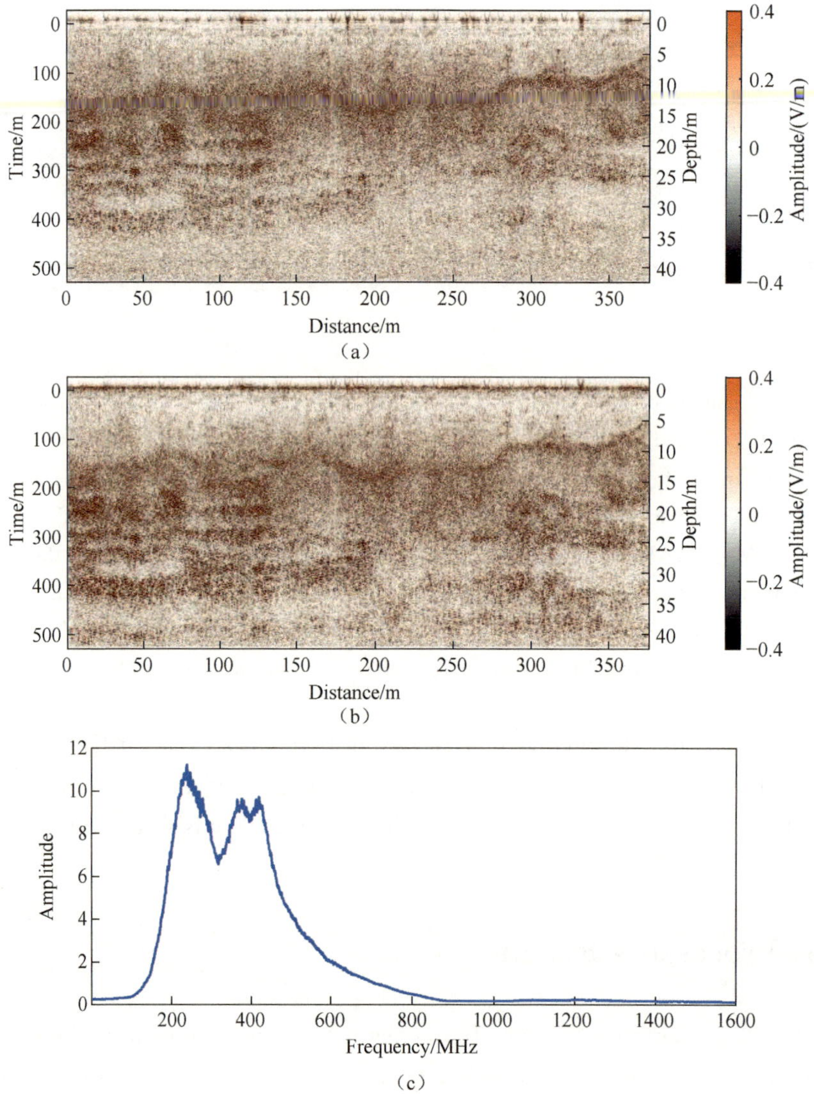

Fig. 12　(a) CH－2A LPR data. (b) CH－2B LPR data. The processed CH－2B radargram within the first 15 lunar days. The left vertical axis is the two－way traveling time of radar echoes. The right vertical axis is the inversed depth from two－way traveling time with velocity assumed to 0.16m/ns [32]. The horizontal direction shows the survey distance (m) of the LPR. (c) Average frequency amplitude spectra for all traces of the CH－2B data in (b)

TABLE Ⅰ　DETAILS OF DATA PROCESSING

Processing	Details
Traces amending	Adjusting the vertical misalignment of traces based on the phase of a strong reflection event

Continued Table

Processing	Details
Traces selecting	The rover might stop at some points on the way to collect other scientific data. But LPR never stop acquisition, which resulted in repeated acquisition at a same position. We average the repeated traces
Time lag adjustment	There is a 28. 203 ns delay for the start time of the transmitting antenna compared to the receiving antenna
Useless data deleting	Signals after 560 ns are removed since these signals are not reliable
Attenuation compensation	Conducting the spherical exponential compensation (SEC) to make deep information more visible
Background removal	Removing the average data along the rover path
Band - pass filtering	Adopting band - pass filtering to suppress the lowfrequency and high - frequency noise
Migration	Using the Kirchhoff migration to suppress the diffracted wave

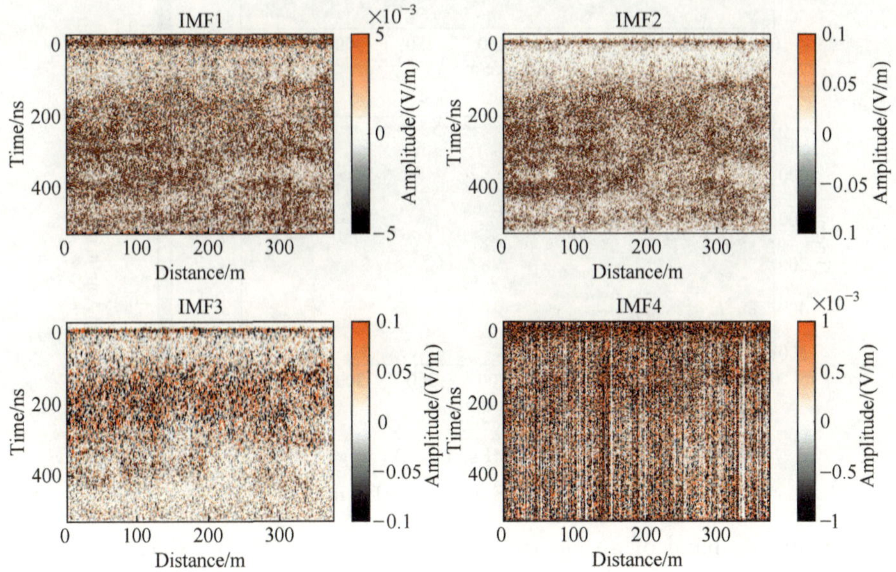

Fig. 13　Profiles of the four IMFs of CH - 2B LPR data generated by the VMD method

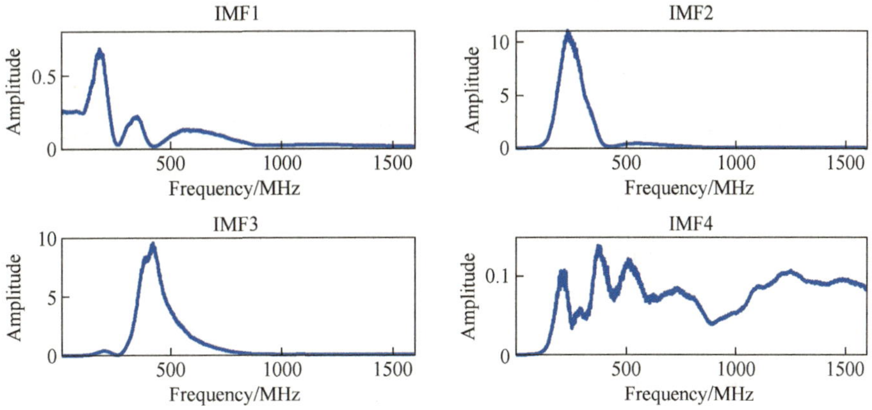

Fig. 14　Average frequency amplitude spectra for all traces of the IMFs in Fig. 13

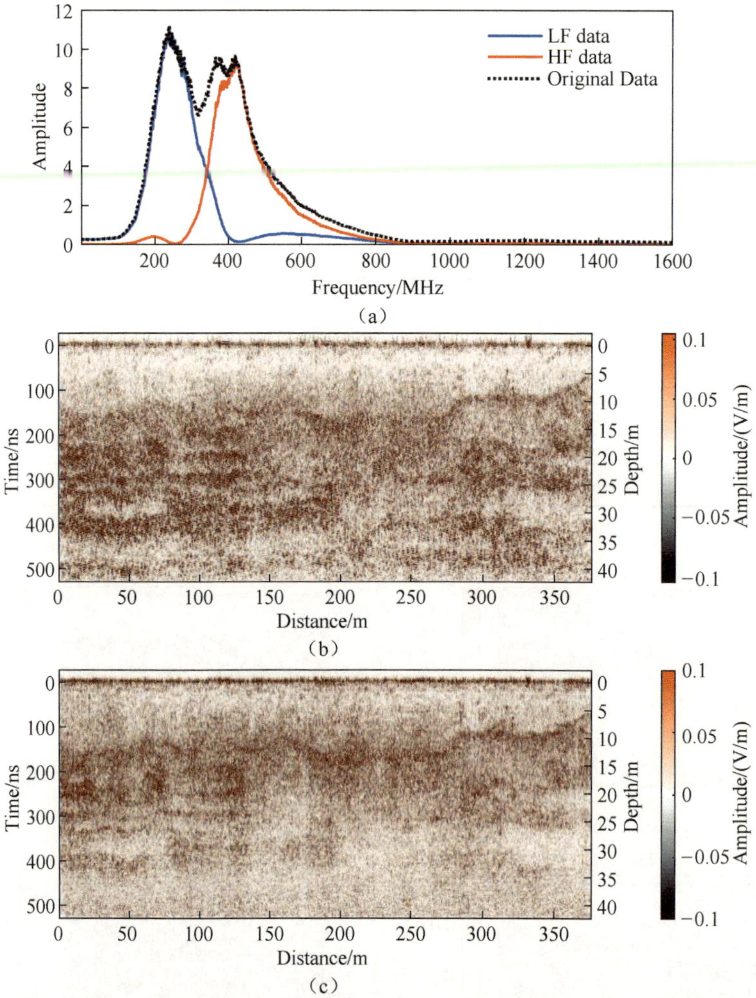

Fig. 15　Recomposition results of IMFs. (a) Average frequency amplitude spectra of recomposed LF and HF data. (b) LF data. (c) HF data.

As the frequency band of an electromagnetic wave is closely related to the detection resolution of LPR, we speculate that the LF data and HF data may contain the signals from the subsurface objects of different sizes. In order to verify the conjecture, we will perform a series of synthetic tests in Section Ⅴ.

Ⅴ　SPECIFIC NUMERICAL TEST OF VMD AND LUC METHODS

A　VMD Test of a Gravel Model

To figure out the relation between the rock size and the LF and HF data of LPR, we construct a gravel model in Fig. 16. The region with a distance <3m contains relatively coarse - grained rocks and the region with distance >3m contains relatively fine - grained rocks. The grain sizes of the two types of rock are approximately 0.5 and 0.1m, respectively. The settings of the electrical parameters are similar to the model in Section Ⅲ-A. The antenna interval is set to 0.32m, which is the same as that of the CH-2B channel.

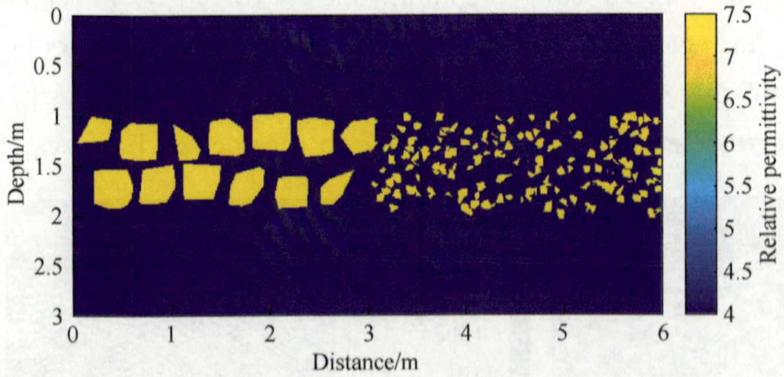

Fig. 16　Gravel permittivity model for VMD processing test. The region with distance <3m contains the coarse-grained rocks and the region with distance >3m contains the fine-grained rocks. The grain sizes of the two types of rock are approximately 0.5 and 0.1m, respectively. In order to exclude the effects of attenuation, the electrical conductivity is set to 0

To simultaneously test the capability of the VMD method and analyze the relationship between the rock size and the LF and HF data, two special wavelets are synthesized as the sources for simulations. The two wavelets and their frequency spectra are shown in Fig. 17 (a) - (d), respectively. We informally name the two wavelets as W1 and W2. The center frequencies and frequency bands are 230 and 60-420MHz for W1 and 420 and 260-700 MHz for W2, which are with similar center frequencies and frequency bands to that of the LF and HF data of LPR in Fig. 15 (a).

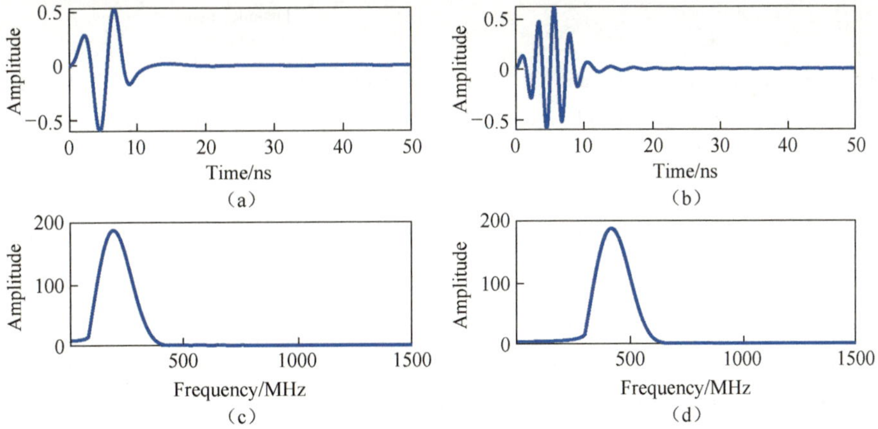

Fig. 17　Wavelets W1 and W2 used in the simulations. (a) and (b) Wavelets in time domain. (c) and (d) Frequency spectra of (a) and (b), respectively

Fig. 18 (a) and (b) presents the simulation results using the two types of wavelets in Fig. 17 (a) and (b). Their migration results and Hilbert amplitudes are shown in Fig.

18 (c) - (f), respectively. The migration velocities are set to 0.15m/ns, which correspond to the background permittivity in Fig. 16. All the velocities mentioned in this article refer to the propagating velocities of electromagnetic waves in different mediums. The red lines in Fig. 18 (e) and (f) divide the profiles into two regions, i. e., the regions of 0 - 3 and 3 - 6 m. In Fig. 18 (e), the amplitudes in the 0 - 3 - m region are greater than those in the 3 - 6 - m region whereas the opposite phenomenon appears in Fig. 18 (f). The results demonstrate that the wavelet W1 is more sensitive to the rocks with sizes of ~0.5m whereas the wavelet W2 is more sensitive to the rocks with sizes of ~0.1m.

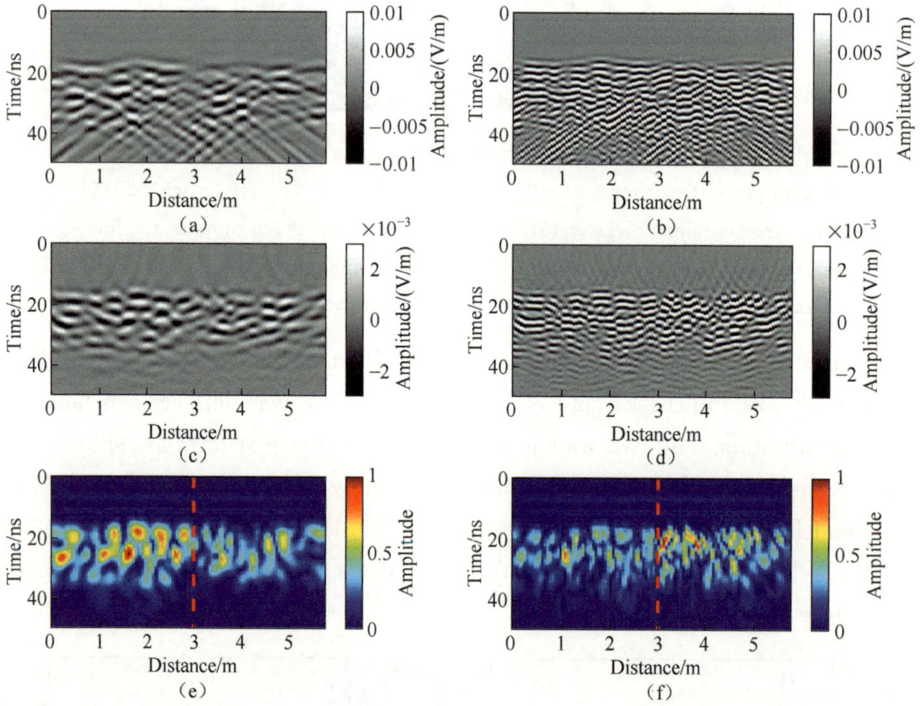

Fig. 18 Simulation results. (a) and (b) Simulated radargrams using the model in Fig. 16 and the wavelets in Fig. 17 (a) and (b), respectively. (c) and (d) Kirchhoff migration results of (a) and (b) with a velocity of 0.15m/ns. (e) and (f) Hilbert amplitudes of (c) and (d)

The results are reasonable. The center frequency of wavelet W1 and the background velocity of the model are 230MHz and 0.15m/ns, respectively. The wavelength corresponding to the center frequency is approximately 0.65m, whereas the rocks with sizes of 0.1m are below the minimum resolution of W1 if considering the 1/4 times of the wavelength as the minimum resolution. However, the W1 is with a frequency band of approximately 60 - 420MHz and the components with relatively higher frequencies but lower amplitudes can detect the rocks with sizes of 0.1 m, so the result in Fig. 18 (e) shows that the amplitudes in 0 - 3 - m region are greater than those in 3 - 6 - m region.

For the wavelet W2, both two types of rocks are in the resolution range of detection. However, because the W2 can easily detect the rocks with sizes of 0.1m, the reflected and scattered signals from the large amount of debris in the 3 - 6 - m region are superimposed, which results in the greater amplitudes in this region in Fig. 18 (f).

Subsequently, the two data in Fig. 18 (c) and (d) are composed into the synthetic data in Fig. 19 (a) and the VMD is conducted for the decomposition test. Similar to the processing in Section Ⅳ - B, the data is decomposed into four IMFs; the recomposed LF and HF data are presented in Fig. 19 (c) and (d) and the spectra before and after VMD are compared in Fig. 19 (b). Although the radar profiles of the results show some distortions and the spectra of LF and HF data before and after VMD have some differences (e. g., the yellow and blue curves in Fig. 19 (b) do not exactly coincide), the profiles and the spectra are similar in general which can verify the capability of VMD method.

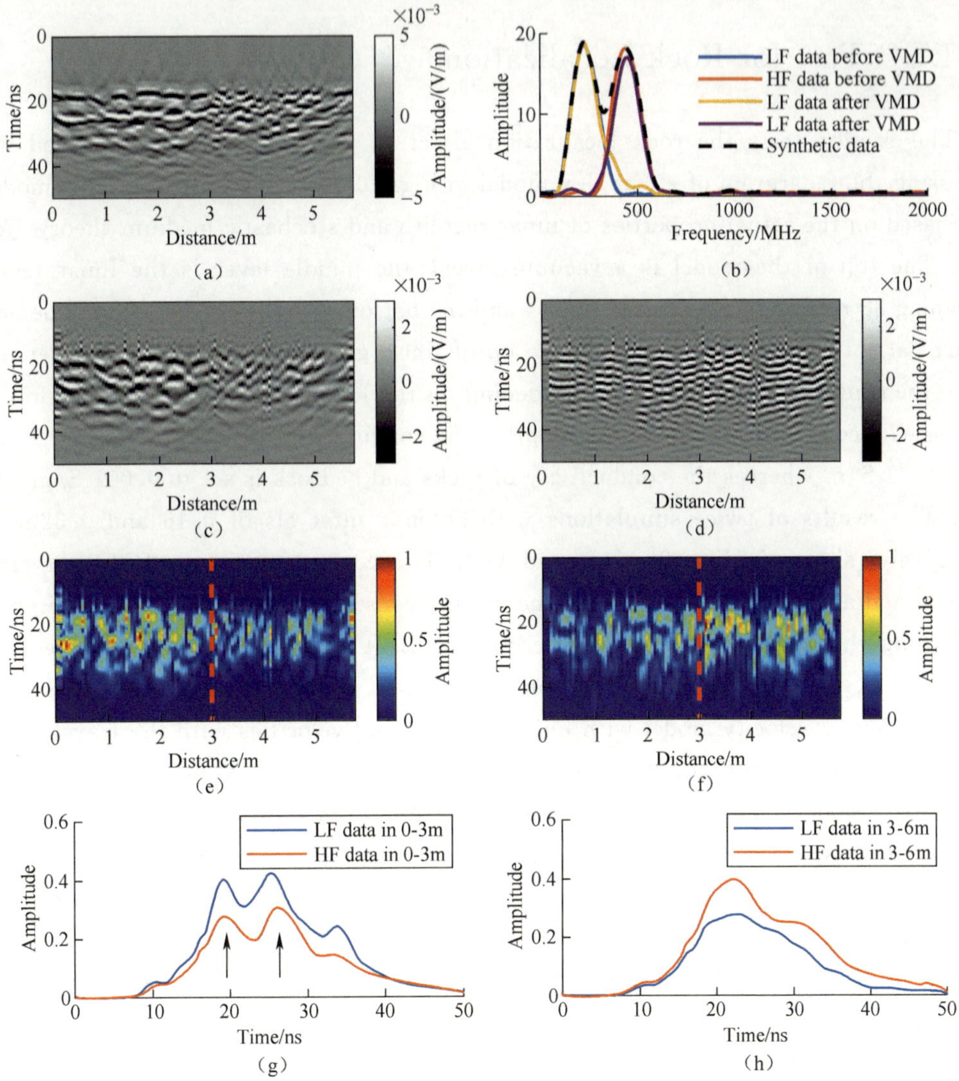

Fig. 19　VMD test results. (a) Synthetic data made up of Fig. 18 (c) and (d). (b) Spectra comparison before and after VMD. (c) and (d) LF and HF data after VMD processing on (a). (e) and (f) Hilbert amplitudes of (c) and (d). (g) and (h) Average amplitudes for the traces in the two regions (0 - 3 and 3 - 6m) in (e) and (f), respectively

Furthermore, the Hilbert amplitudes of the LF and HF data after VMD are computed and presented in Fig. 19 (e) and (f). The results are more cluttered but similar to the results in Fig. 18 (e) and (f); the opposite horizontal variations of the amplitudes for LF and HF data can also be observed, which further verifies the capability of VMD. Fig. 19 (g) and (h) shows the average amplitudes for the traces in the two regions (0 - 3 and 3 - 6m) in Fig. 19 (e) and (f), respectively. It is clear that the average amplitude of LF data is greater than that of HF data in the 0 - 3 - m region whereas the average amplitude of HF data is greater than that of LF data in the 3 - 6 - m region. Besides, the average amplitude in the 0 - 3 - m region contains two peaks [the black arrows in Fig. 19 (g)] which indicate the locations of the two layers made up of coarse - grained rocks in the model.

B　LUC Test for Rock Localization

This section tests the rock localization effect of the proposed LUC method when processing the radargram of a complex model in Fig. 20 (a) [28], [49]. This model is built based on the actual properties of lunar regolith and stochastic medium theory [28], [50]. The top of the model is a vacuum layer, the middle layer is the lunar regolith containing 38 randomly distributed rocks, and the bottom layer is the bedrock. The near - surface part of regolith is generally made up of fine - grained materials with fewer rocks due to the continuous impaction and gardening on the lunar surface [19]. Therefore, the grain sizes ascend with the increasing depth. The conductivity of background of 0 - 5m is set to 10^{-7} S/m whereas the conductivity of rocks and bedrock is set to 0.001 S/m [28], [49]. The results of twice simulations with antenna intervals of 0.16 and 0.32m are, respectively, shown in Fig. 20 (b) and (c). The center frequency of the wavelet is 500MHz and antennas are 0.3m above the ground consistent with the LPR measurement.

During the processing of the experimental LPR data acquired on the field, the accurate velocity model used for migration is unknown. Therefore, to get more close to the field situation, a velocity model with linearly decreasing velocities with the traveling time is applied to the migration of the simulated data instead of the real model. The values of the velocity model vary from 0.18 m/ns at the top to 0.16 m/ns at the bottom. The radargrams of two offsets after migrations are shown in Fig. 21 (a) and (b). It is evident that the migrated results are not perfect with residual diffracted waves but more consistent with field cases.

Fig. 21 (c) shows the original result of LUC between Fig. 21 (a) and (b). Fig. 21 (d) presents the LUC result with soft threshold function $\zeta = 0.1$. By extracting the local maximum in Fig. 21 (d), we obtain the result in Fig. 22 (a). The right vertical axis of Fig. 22 (a) is the inverted depth from two - way traveling time with the velocity model used in migrations. Subsequently, the comparison between Fig. 22 (a) and the rock locations in the model of Fig. 20 (a) is performed. The result in Fig. 22 (b) indicates

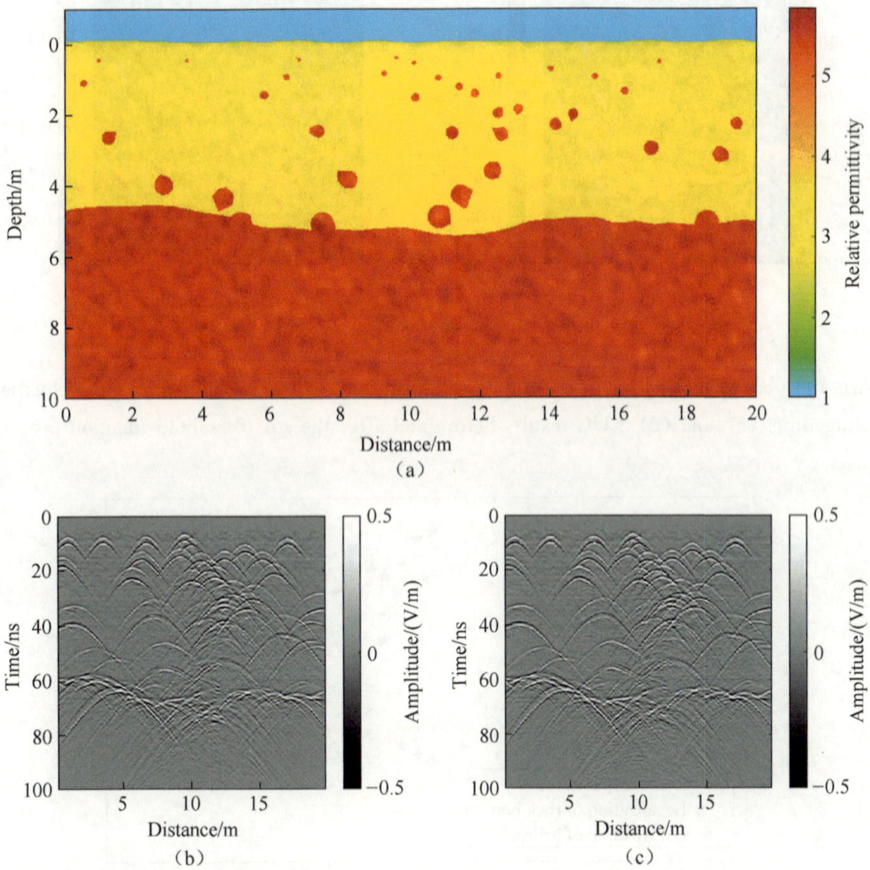

Fig. 20　Complex lunar regolith model and its simulated results.
(a) Dielectric constant model [28], [45]. (b) and (c) Simulated results with antenna intervals are 0.16 and 0.32m, respectively

that the 38 rocks are detected. The detected points are not perfectly matched with the rocks as the depths converted by the velocity model are not accurate; the blue points indicate the false alarm points which are near the matched points. These pairs of points are resulted from the signals of the top and bottom interfaces of the same rocks.

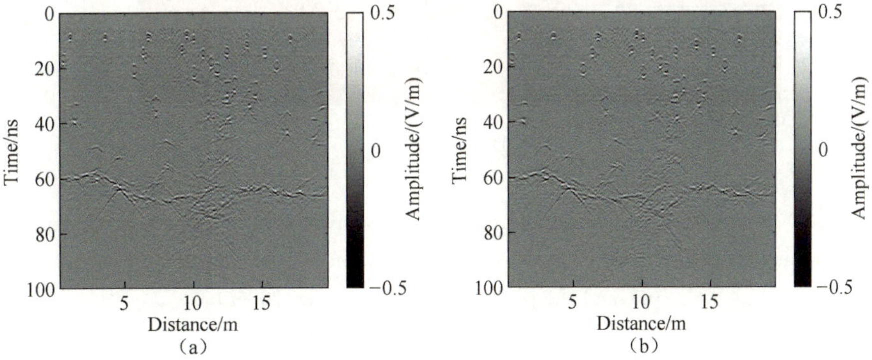

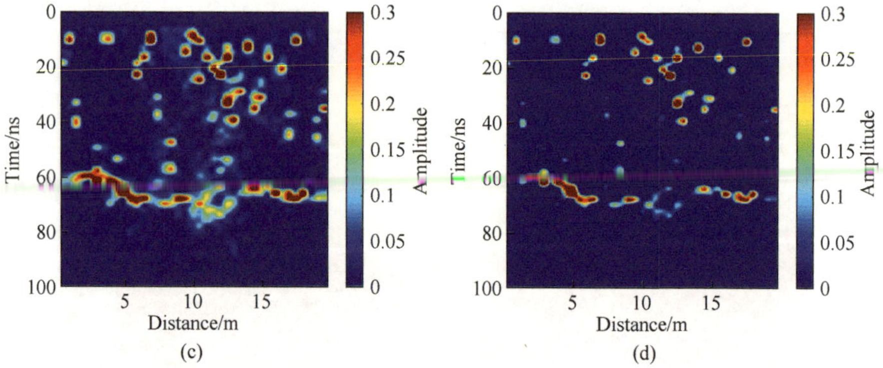

Fig. 21 (a) and (b) Migration results of Fig. 20 (b) and (c) with linearly decreasing velocities with the traveling time. (c) and (d) LUC results before and after the soft threshold function processing

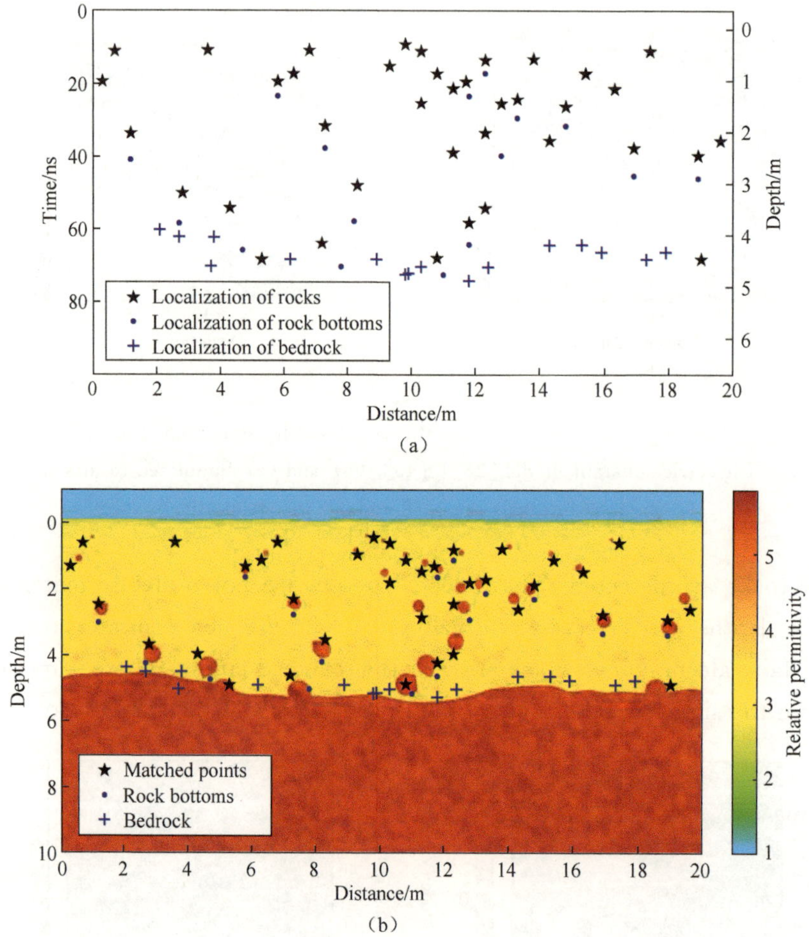

Fig. 22 (a) Positions of local maximums. The right vertical axis of is the inverted depth from two—way traveling time with velocity model used in migrations. (b) Comparison between the points in (a) and the positions of rocks in Fig. 20 (a)

Besides, the LUC will also detect the signals from the bedrock, which results in the 16 false alarm points in the example. During the LUC processing of field LPR data, we will not know a priori which signals originate from stratigraphic interfaces and which originate from discrete rocks, so the effects of bedrock must be included in the analysis. Assuming that the stratigraphic interfaces uniformly distribute in the horizontal direction similar to the model in Fig. 20 (a), it is likely that the effects of bedrock are also uniform in the horizontal direction. To illustrate, we, respectively, counted the number of points in Fig. 22 (a) per 2m in the horizontal direction; the two histograms/curves in Fig. 23 presents the results with and without the points of bedrock, respectively. Both two curves demonstrate that 10—14m is a rock-rich region which corresponds to the model in Fig. 20 (a). Furthermore, the Pearson correlation coefficient is adopted to analyze the relevance between the two curves, which is given by

$$\rho_{X,Y}=\frac{E((X-E(X))\cdot(Y-E(Y)))}{\sqrt{E(X^2)-E^2(X)}\cdot\sqrt{E(Y^2)-E^2(Y)}} \tag{16}$$

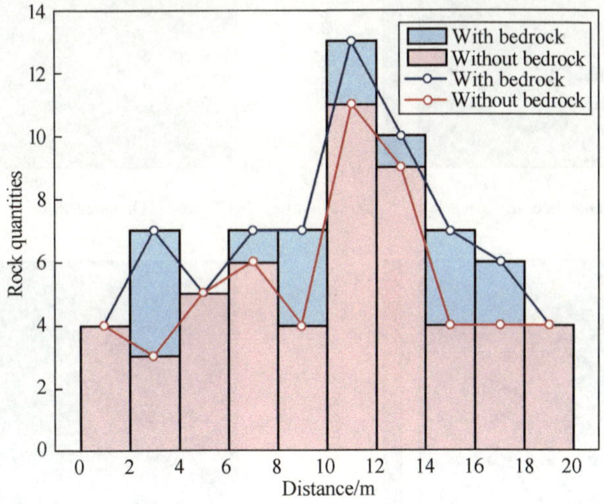

Fig. 23　Rock distribution statistics in horizontal direction. The two histograms/curves represent the case with and without the effects of bedrock. The Pearson correlation coefficient of the two curves is computed to be 0. 8582

where X and Y are the data of two curves and $E(\cdot)$ represents the expectation of the data. The ρ between the two curves is computed to be 0. 8582, which is between 0. 8 and 1 and indicates the two curves are extremely relevant [50]. Therefore, if we study the horizontal variation of rock quantities, the effects of the stratigraphic interfaces are uniform in the horizontal direction and will have little effect on the results of rock distribution analysis.

Subsequently, the test with noise added data is performed. The noises with SNR = 10 are added into the radargrams in Fig. 20 (b) and (c). Fig. 24 (a) and (b) presents the noise added radargrams whereas Fig. 24 (c) and (d) shows the migration results.

The LUC result of the two noise added data is shown in Fig. 24 (e), which contains much more noise compared to that in Fig. 21 (c). Fig. 24 (f) presents the LUC result with soft threshold function $\zeta=0.25$.

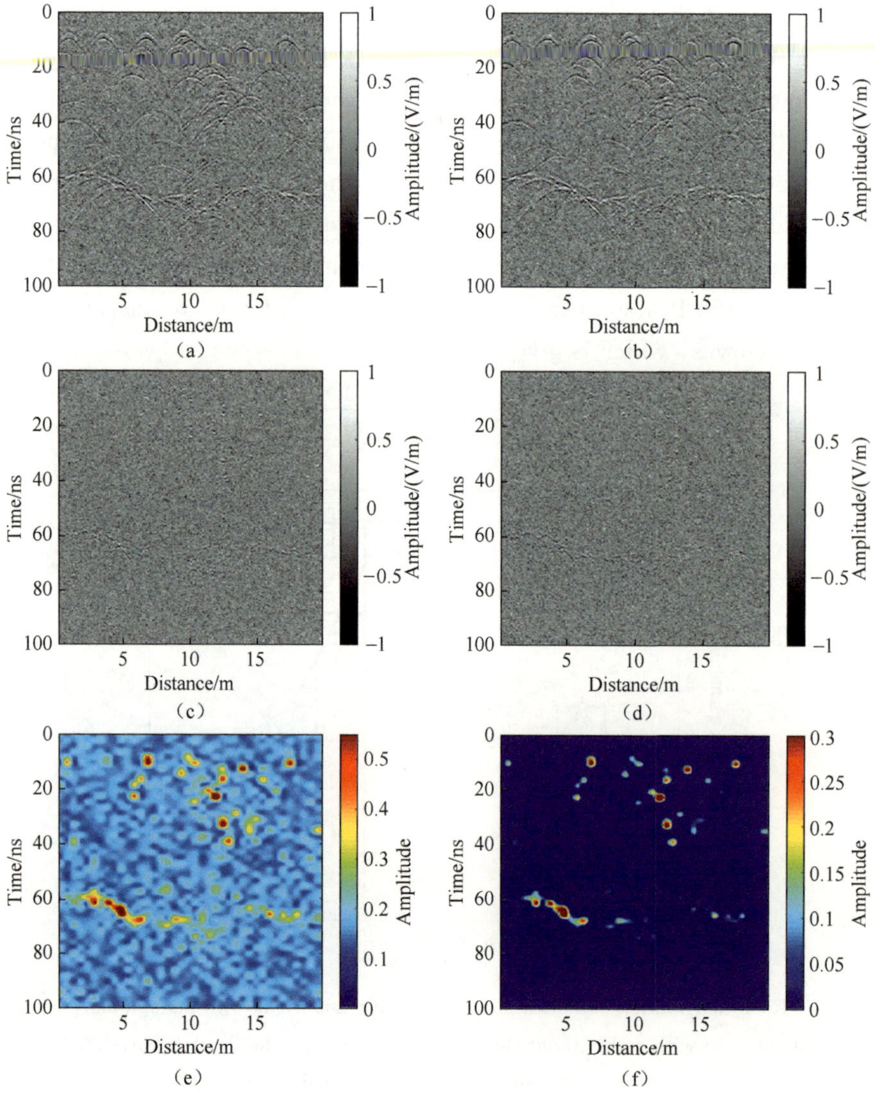

Fig. 24 (a) and (b) Noise added data of Fig. 20 (b) and (c). (c) and (d) Migration results of (a) and (b) with linearly decreasing velocities with the traveling time. (e) and (f) LUC results before and after the soft threshold function processing ($\zeta=0.25$)

The rock localization result with noise interference is shown in Fig. 25. Compared to Fig. 22, seven rocks are undetected in the result of noise added case. The possible reason is that the amplitude of the noise is far beyond the amplitude of signals from these rocks. The undetected rate is computed to be 18.42%.

Besides, in the noise added case, the effects of bedrock also exist; the same analysis method in Fig. 23 is also adopted and the result is shown in Fig. 26. The result can also clearly indicate the rock – rich region within 10 – 14 m. The Pearson correlation coefficient between the two curves is computed to be 0. 9317, which is also between 0. 8 and 1 and indicates that the two curves are extremely relevant.

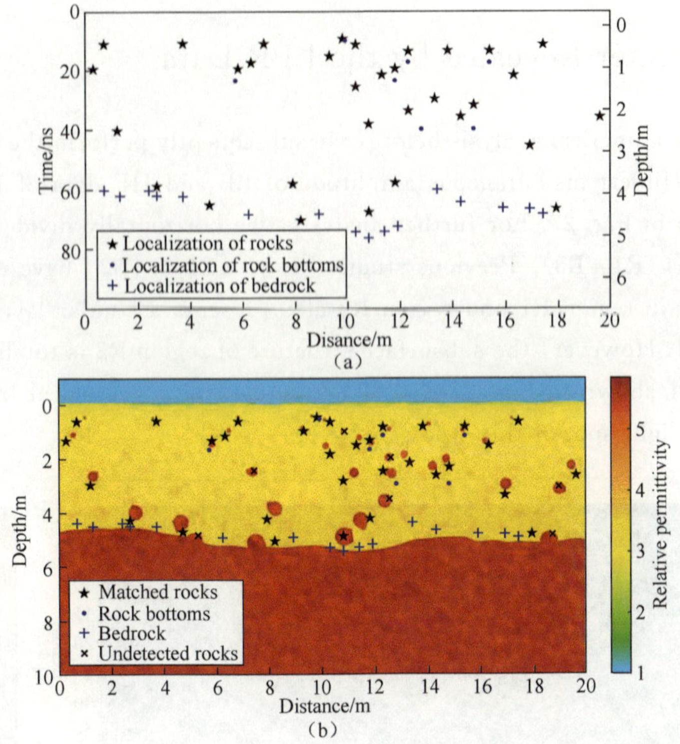

Fig. 25　(a) Positions of local maximums obtained using the noise added data. The right vertical axis is the inverted depth from two – way traveling time with velocity model used in migrations. (b) Comparison between the points in (a) and the positions of rocks in Fig. 20 (a)

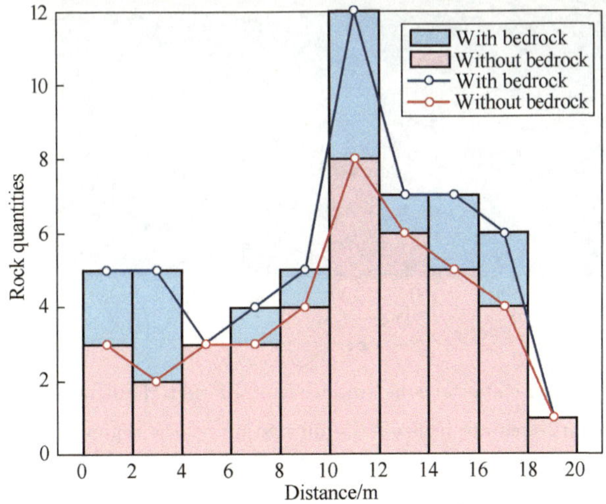

Fig. 26　Rock distribution statistics in horizontal direction. The two histograms/curves represent the case with and without the effects of bedrock. The Pearson correlation coefficient of the two curves is computed to be 0. 9317

Ⅵ LPR DATA INTERPRETATIONS

A Buried Crater Revealed by the LPR Data

Based on the numerical analysis before, we subsequently perform the interpretation of LPR data. The Hilbert instantaneous amplitude of LF and HF data of LPR is obtained which are shown in Fig. 27. For further analysis, we horizontally divided the radargram into three regions (R1 - R3). Previous studies [19], [24], [32] have revealed that the layered structure in region R1 and region R3 also presents a similar layered structure to that of region R1. However, the subsurface structure of region R2 is totally different from that of region R1 shown in Fig. 27 (a). Therefore, a new geological interpretation for region R2 is the light spot of this study.

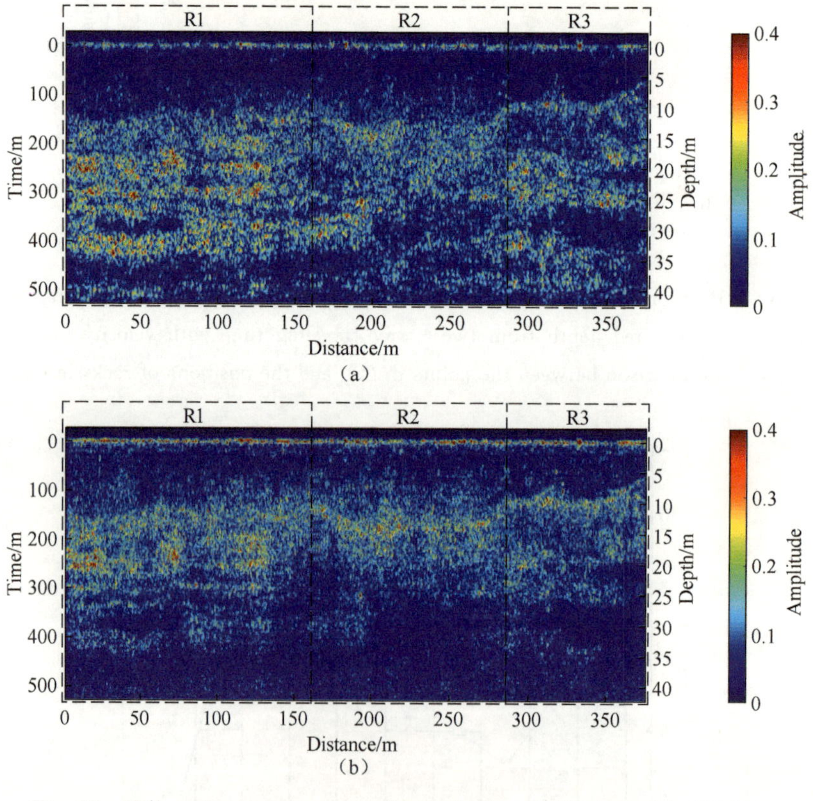

Fig. 27　Hilbert instantaneous amplitude of LF and HF data. (a) LF data. (b) HF data. The radargrams are horizontally divided into three regions by the dotted rectangles

Fig. 28 illustrates the proposed subsurface structure contains seven discrete layers (named E1 - E7) and a buried crater (named C1). E1 is the fine - grained regolith of ~12m, and E2 - E7 are the superposition of multiple ejecta sources from nearby caters

[19], [24], [31], [32]. Notably, the interior layered structure of region R2 is absent compared to those in region R1 and R3, and the radar profile shows strong evidence of a buried crater with clear tow side - wings and bowl - like bottom shown in Fig. 28 (a). The possible buried crater is informally denoted as C1 in this study and its location lies to the west of the CE - 4 lander. The formation of C1 discontinued the multilayered structure observed in the previous studies [19], [24], [31], [32], which also demonstrate that the multistrata and its overlying fine - grained regolith are of different sources.

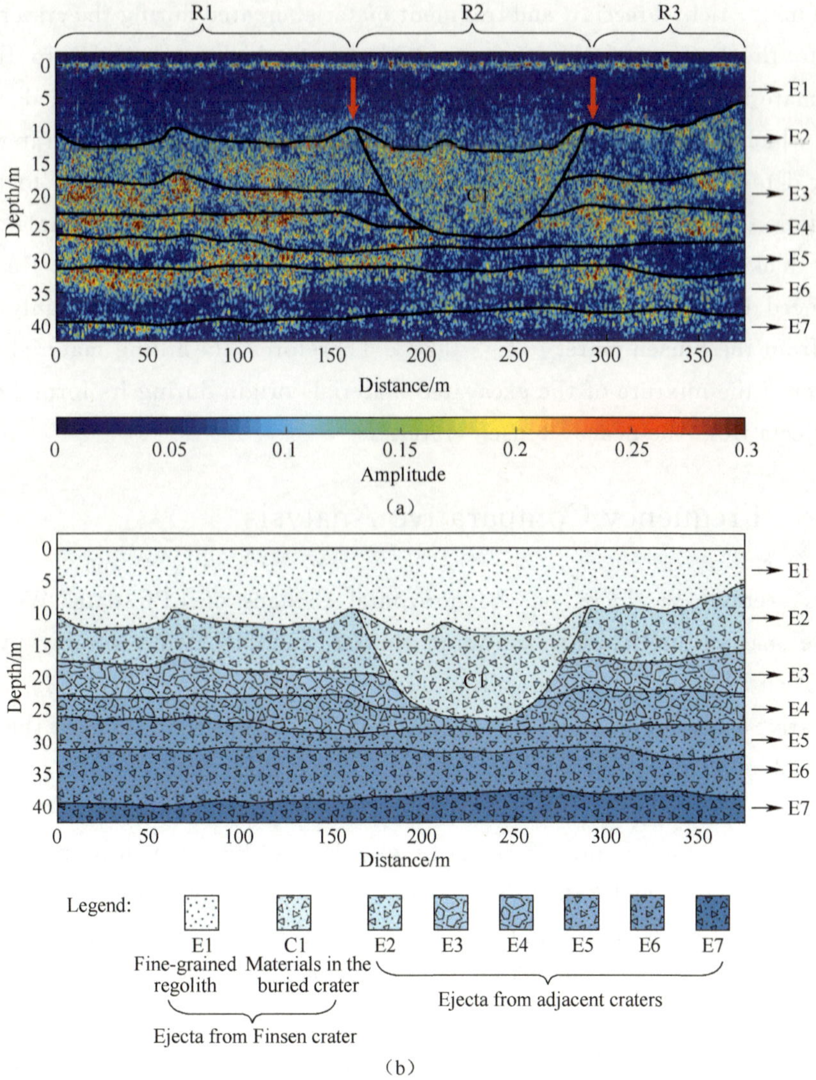

Fig. 28　Subsurface structure observed by the CH - 2 LPR and diagram of each stratigraphic division. (a) Layered subsurface structure observed by the CH - 2 LPR and position of the buried crater. Black solid lines indicate the interfaces of each stratum, and two red arrows marked the side - wings of the buried crater. The base in (a) is the Hilbert imaging result of the LF part of CH - 2B data. (b) Diagram of new geological interpretation of the CH - 2 radargram. The different geological symbols are denoted as C1 and E1 - E7, respectively

The distance between the two red arrows in Fig. 28 is measured to be ~130m, and the crater bottom is at the depth of ~26m. However, the positions of the two red arrows on the lunar surface, which is shown in Fig. 11 (b), are approximately at the navigation points 5 and 10 and the spatial length of the buried crater is ~ 84m. Assuming the maximum excavation depth of the simple crater is 1/13 that of the final crater diameter [51], it can be inferred that the buried crater C1 has excavated materials from a depth of ~6.5m, and the penetrating depth of the crater is treble time than that of excavation depth. The melt-rich, breccia, and fragment material created during the crater formation backfilled to the bottom of the crater, which has a thickness similar to that of the excavated materials. The bowl shape represents the position of the buried crater, the bottom of which has a depth of ~ 16 m relative to the edge of the crater, but the penetrating depth of the buried crater has a depth of ~19.4m according to the impact crater formation theory [51]. Therefore, the buried crater is possible a simple crater with a degraded thickness of ~3.4m and the rover path is close to its diameter. Recent works have convinced that the near-surface material at CE-4 landing is mainly the ejecta excavated from the Finsen crater [11]-[20]. Therefore, the filling material within the buried crater is the mixture of the excavated materials origin during its formation and the overlying ejecta from the nearby Finsen crater.

B Time-Frequency Comparative Analysis

In this section, we focus on the horizontal changes of LPR data. We propose a comparative analysis method for analyzing the LF and HF data. Notably, the central frequency of an LPR signal will be shifted due to the attenuation [52], [53]. Therefore, the comparative analysis can only be performed for the LF and HF signals at the same time and it is suitable for analyzing the horizontal changes of LPR data.

The radargram of Fig. 27 (a) and (b) are averaged with regions R1, R2, and R3 and shown in Fig. 29. In the LF and HF data of region R1 and R3, two clear and similar peaks are present between 18 and 26m [Fig. 29 (a), (c), (d), and (f)]. These two peaks correspond to the signals of E3 and E4. However, there is no evident layered reflector found [Fig. 29 (b) and (e)] in the region R2, which can be interpreted as the impacting meteorite penetrated the E2 and E3 during the formation of the C1. The bottom of the final crater is located in the E4. Compared with the signals of E2 and E3 in Fig. 29 (a) and (c), the signals in the region of C1 in Fig. 29 (b) lack the peaks and are more cluttered. Based on the numerical test in Section Ⅴ-A, the absence of the two peaks in R2 of LF data indicates the absence of stratigraphic layers. Therefore, it is proper that the proposed buried crater resulted in the horizontal changes and strata absence of subsurface structure.

Comparing the results of HF data in Fig. 29 (d)-(f), apart from the strata

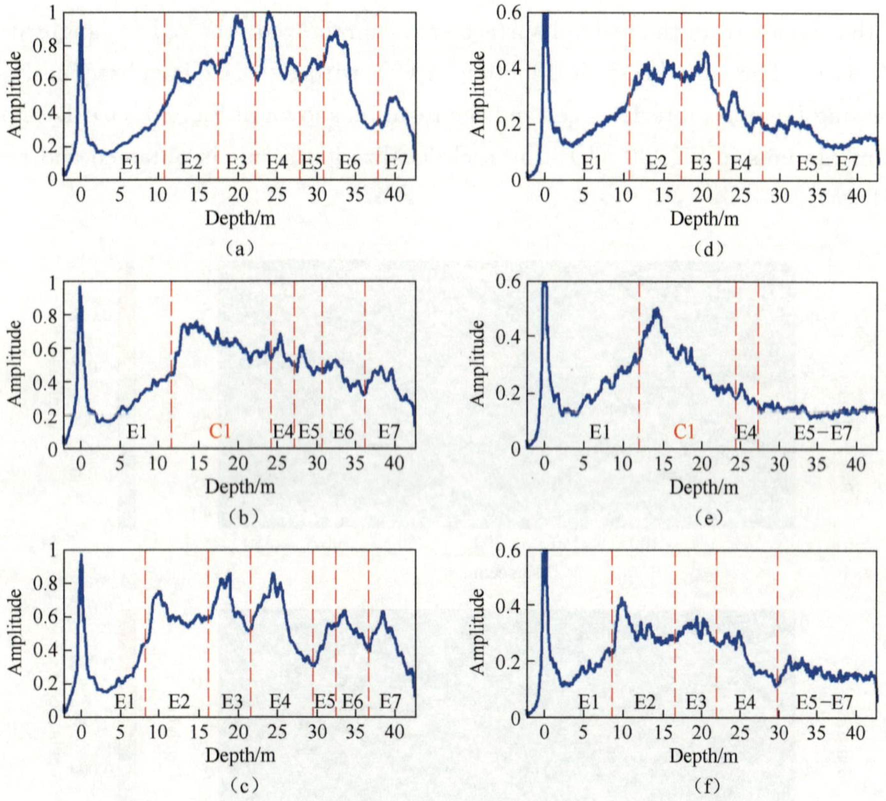

Fig. 29　Average amplitudes of traces of the three regions in Fig. 27 (a) and (b). (a) R1 in LF data. (b) R2 in LF data. (c) R3 in LF data. (d) R1 in HF data. (e) R2 in HF data. (f) R3 in HF data

absence, the average energy of the reflected signal from the shallow of C1 in R2 is greater than those in E2 of R1 and R3 [Fig. 29 (d) - (f)] which is also observed in the radargram in Fig. 27 (b). Based on the numerical test in Section Ⅴ-A, the shallow of the buried crater must be filled with more randomly distributed rocks with sizes of approximately 0.1m comparing to R1 and R3. In other words, the result demonstrates that the filling material in C1 is different from the material of E2 in R1 and R3 and may contain large quantities of randomly distributed rocks.

C　Rock Quantitative Analysis

To verify the results in the acknowledgment section, the rock quantitative analysis is performed. Fig. 30 (a) presents the results of LUC based on the CH-2A and CH-2B data in Fig. 12 (a) and (b). Due to the attenuation and different qualities of the two data, the amplitudes of the CH-2A and CH-2B data below 300ns are imbalanced, which will attenuate the energy of LUC results in this region, and some signals from rocks may be muted in the soft threshold processing. Therefore, the rock localization results below 25 m depth are not credible and are waived [the red rectangle at the bottom of Fig. 30 (a)]. Besides, the signals from the ground surface are also removed [the red rectangle at the top of Fig. 30 (a)]. The soft threshold function in (9) with $\xi=0.02$ is subsequently applied to suppressing the noise interference, and the result is shown in Fig. 30 (b). By extracting the local maximums in Fig. 30 (b), the rock localization result is obtained and presented in Fig. 31 (a).

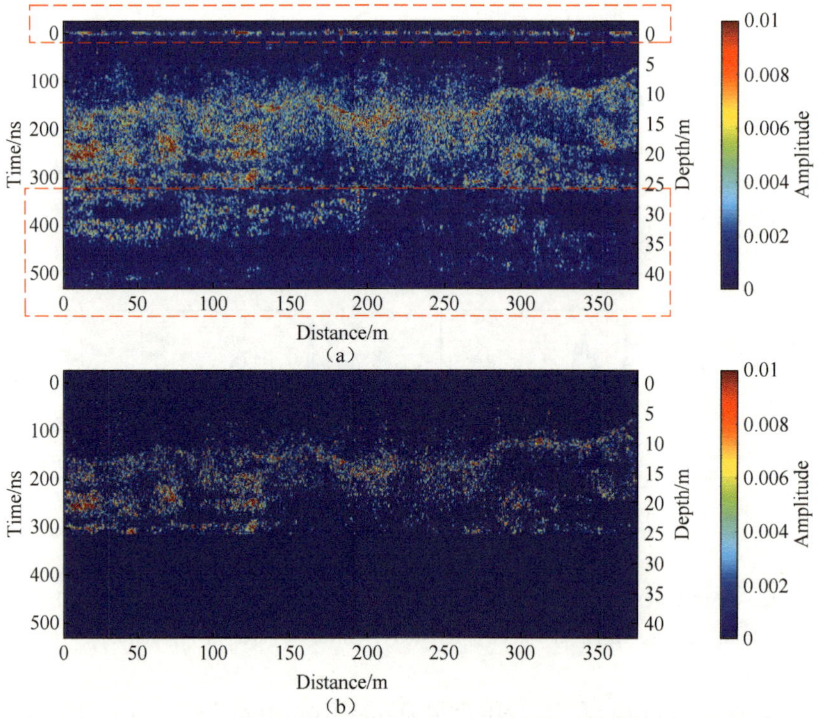

Fig. 30　LUC results based on the CH-2A and CH-2B data. (a) Initial result of LUC. The red rectangles represent the reflections from the ground surface and the structure deeper than 25m depth, which need to be muted. (b) LUC result after the processing of the soft threshold function with $\xi=0.02$ and deletion of the data in the rectangle of (a)

Further statistical analysis is performed for the localization result. Fig. 31 (b) presents the local rock statistic result and each point in Fig. 31 (b) represents the rock quantity in a region of 5-m length and 1-m depth. The rock distribution in the buried crater shows an evident bowl-like shape [the red curve in Fig. 31 (b)] and the shallow of the buried crater is filled with a larger quantity of rocks compared to the R1 and R3 in the same depth (10-15m). Overall, the result of rock quantitative analysis is coincident with that of the time-frequency comparative analysis before.

Fig. 31 (c) - (e) presents the averaged statistical results of the three divided regions R1-R3, respectively. Similar to the results in Section Ⅵ-B, the rock distributions of R1 and R3 present evident layered structure [E2 and E3 in Fig. 31 (c) and (e)] which, however, is absent in R2 and replaced by a bowl-like distribution in Fig. 31 (b), which also verifies the buried crater. More importantly, the shallow part of the buried crater is concentrated and filled with a large number of rocks. There may be two different cases.

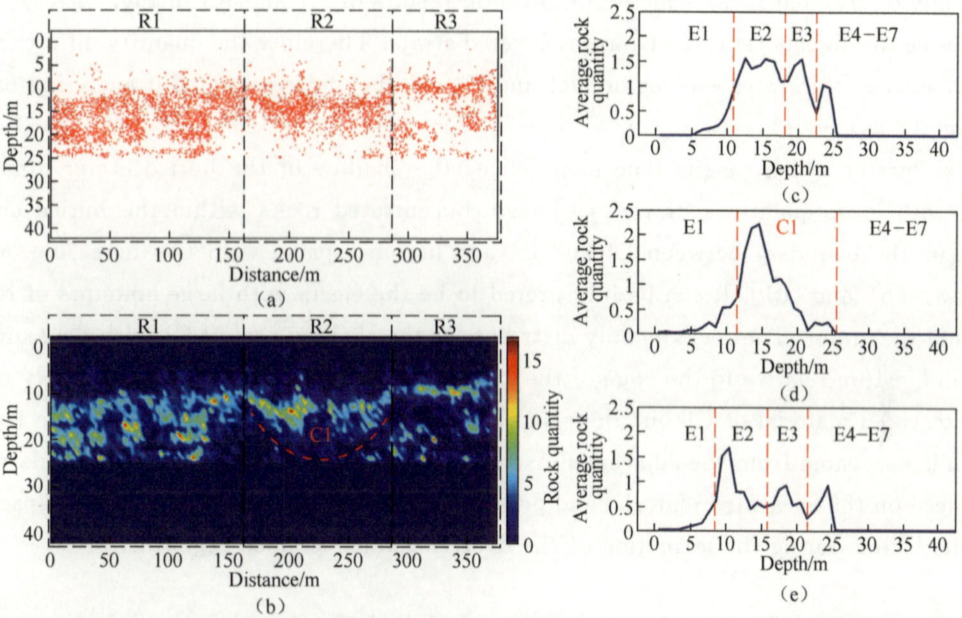

Fig. 31 Results of rock localization and quantitative analysis. (a) Result of rock localization. Each red point represents a rock position. (b) Local rock statistic result and each point in the figure represents the rock quantity in a region of 5 - m length and 1 - m depth. (c) Average rock statistical result in R1. (d) Average rock statistical result in R2. (e) Average rock statistical result in R3

Case 1: The buried rocks are the only possible reflecting or scattering materials in the subsurface in this region and the signals detected by the LUC method are all from the randomly distributed rocks. In this case, the rock quantity at ~13m depth in the crater even exceeds the maximum in the rock - rich region R1 [E2 and E3 in Fig. 31 (c)].

Case 2: The subsurface in this region contains layered strata. According to the analysis in the last section, the HF signals from E2 and E3 in R1 and R3 are weaker than the LF signals which indicates the possible layered strata may be in these regions. Meanwhile, the greater amplitudes of HF data in the crater compared to the other two regions in the same depth verify the rock concentrated structure within the crater. Because the LUC method will also detect the interfaces, the localization results of R1 and R3 in Fig. 31 (a) may mainly be the signals reflected from the layered strata rather than the randomly distributed rocks. There are two subcases.

Subcase 1: There is a distinct interface between the crater and the overlying stratum, i. e., the interface of E1 and E2 extends to the top of the buried crater. In this case, the signals in the crater detected by the LUC method will also contain the signals from this interface. However, according to the numerical test in Section Ⅴ - B, the effects of the interface are uniform in the horizontal direction. Therefore, the conclusion of this subcase is similar to Case 1 of no strata.

Subcase 2: There is no distinct interface between the crater and the overlying stratum. In this case, the signals in the crater detected by the LUC method will all be from the randomly distributed rocks and the localization results of R1 and R3 in Fig. 31 (a) may mainly be the signals reflected from the layered strata. Therefore the quantity difference of rocks between the crater and regions R1 and R3 may be underestimated if analyzing based on Fig. 31 (c) - (e).

No matter which case is true in practice, the shallow of the buried crater must be filled with large quantities of rocks. These concentrated rocks within the buried crater make up the boundary between E1 and C1 and become sparse with the increasing depth [Fig. 31 (b) and (d)]. It can be interpreted to be the ejecta with large amounts of rocks infilling the crater. The rocks mainly distribute at the shallow part of C1 with a maximum depth of ~10m relative to the edge of the crater. Therefore, the rocks are not likely to be the excavated material of C1 but the same ejecta with the overlying near - surface layer, which is excavated from the adjacent Finsen crater. These coarse - grained materials were emplaced on the excavated material and preserved by the crater and survived the impaction and gardening during the formation of the overlying fine - grained regolith.

D　Numerical Validation Using a Buried Crater Model

To further verify the results, a buried crater model is estimated which is shown in Fig. 32. Similar to the model in Fig. 20 (a), the conductivity of background and rocks is set to 10^{-7} and 0.001S/m, respectively. According to the previous analysis in Sections Ⅴ- A, Ⅵ- B, and Ⅵ- C, the LF and HF data with different bands can identify the rocks with a size of ~0.1m and the boulders with a size of ~0.5m. In fact, based on this theory, as long as the sizes of boulders are greater than ~0.5m, the identifications are all applicable. Besides, the field LPR data shows that the materials within the buried crater contain larger quantities of rocks with sizes of ~0.1m than the surrounding regions. Therefore, in this model, the buried crater is designed to be filled with large quantities of rocks with a size of ~0.1m; the crater destroyed the original strategies so the boulders with a size of ~0.5m distributed around the crater; due to the squeezing effects during the formation of the crater, the boulders are not perfectly distributed horizontally.

First, the test of LF and HF data analysis is conducted. The two wavelets of different frequency bands in Fig. 17 are adopted in the simulations and the results of LF and HF data are shown in Fig. 33 (a) and (b), respectively. The antennas are 0.3m above the ground and the antenna and acquisition intervals are 0.32 and 0.04 m, respectively; these simulation parameters are all consistent with the field CH - 2B LPR measurement. Fig. 33 (c) and (d) presents the Kirchhoff migration results of Fig. 33 (a) and (b) using a linear varying velocity model from 0.185m/ns at the top to the 0.13m/ns at the bottom. Fig. 33 (e) and (f) shows the Hilbert amplitudes of Fig. 33 (c) and (d). The red lines in the figures divide the radargrams into three regions, i. e., regions 0—2.5, 2.5—7.5, and 7.5—10m, and region 2.5—7.5 m is the region of the buried crater. The results demonstrate that the LF data is sensitive to the interface between the crater and the overlying stratum. The signal amplitudes of the boulders around the crater are greater than the relatively fine-grained rocks with a size of ~0.1m within the crater (40-60 ns). A few signals at ~50ns in LF data may be from the bottom of the crater. Besides, the HF data shows the opposite result comparing to the LF data. The signal amplitudes of the rocks in the crater are much greater than those of the crater surface and the boulders around the crater.

Subsequently, the LUC test is also performed for the buried crater model. Similar to the field CH-2A and CH-2B measurements, the simulations adopted a 500 MHz wavelet and two different antenna intervals of 0.16 and 0.32m. The antennas are 0.3m above the ground and the acquisition interval is also 0.04m. The simulation results are presented in Fig. 34 (a) and (b). Fig. 34 (c) and (d) presents the Kirchhoff migration results of Fig. 34 (a) and (b) using a linear varying velocity model from 0.185m/ns at the top to the 0.13m/ns at the bottom. The bottom of the crater is invisible due to the sloping walls and that the scattered wave from the randomly distributed rocks overlapped, which is consistent with the field LPR data. Fig. 34 (e) and (f) presents the LUC result before and after the soft threshold function processing ($\zeta = 0.09$). By extracting the local maximum in Fig. 34 (f), we obtain the result in Fig. 35 (a). The right vertical axis of Fig. 35 (a) is the inverted depth from two-way traveling time with the velocity model used in migrations. The effects of the interface between the crater and its overlying stratum are included in Fig. 35 (a). Subsequently, we, respectively, counted the number of points in Fig. 35 (a) per meter in the horizontal direction and the results are shown in Fig. 35 (b). The rock quantity of the true model is also shown in Fig. 35 (b) and the Pearson correlation coefficient of the two curves is computed to be 0.9249. The results indicate that even considering the effects of the stratum interface, the LUC can also reveal the rock concentrated structure in the buried crater.

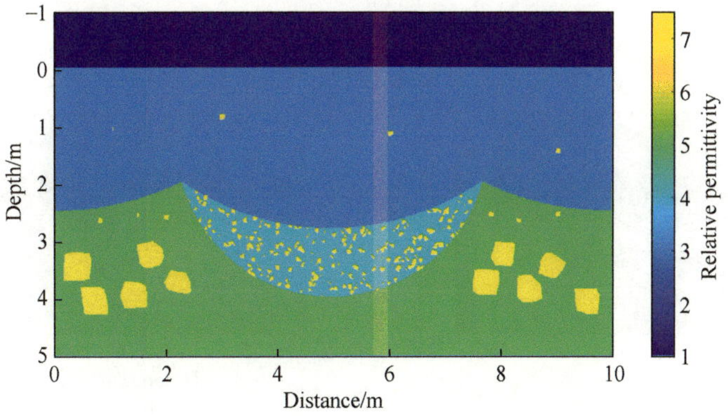

Fig. 32　Permittivity model of a buried crater

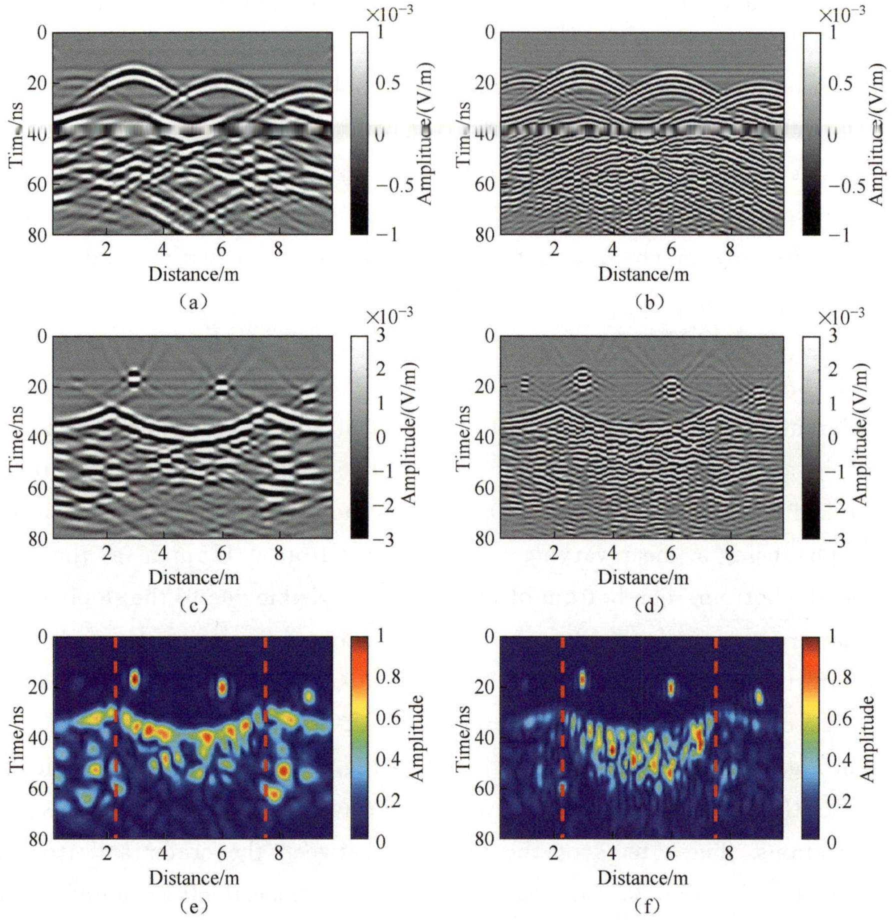

Fig. 33　Test of LF and HF data analysis. (a) and (b) Simulated LF and HF data using the wavelets of different frequency bands in Fig. 17. (c) and (d) Kirchhoff migration results of (a) and (b) using a linear varying velocity model from 0.185m/ns at the top to the 0.13 m/ns at the bottom. (e) and (f) Hilbert amplitudes of (c) and (d). The region between the two red lines is the position of the buried crater

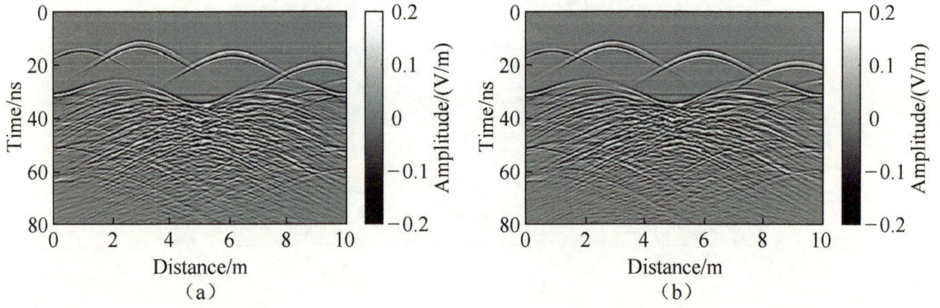

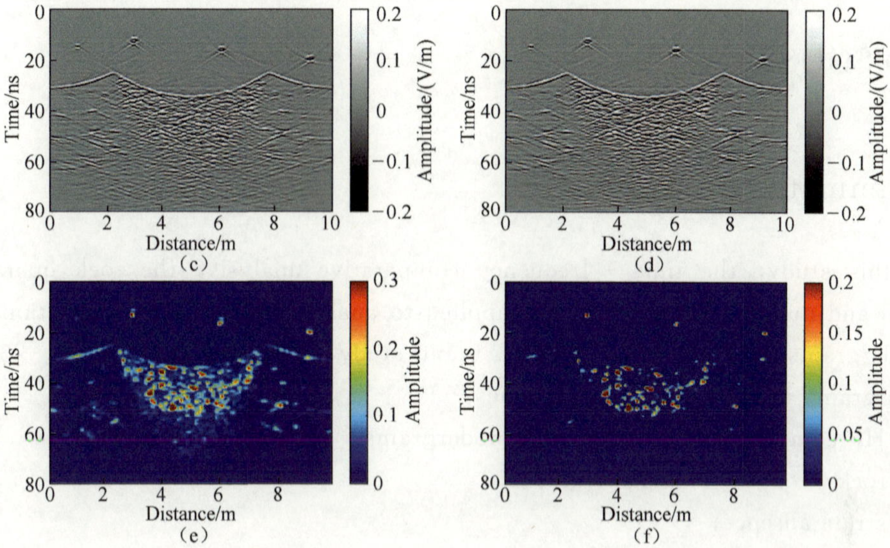

Fig. 34　LUC test for the buried crater model. (a) and (b) Simulated 500−MHz CH - 2A and CH - 2B data with antenna intervals of 0. 16 and 0. 32m, respectively. (c) and (d) Kirchhoff migration results of (a) and (b) using a linear varying velocity model from 0. 185 m/ns at the top to the 0. 13m/ns at the bottom. (e) and (f) LUC result before and after the soft threshold function processing (ζ=0. 09)

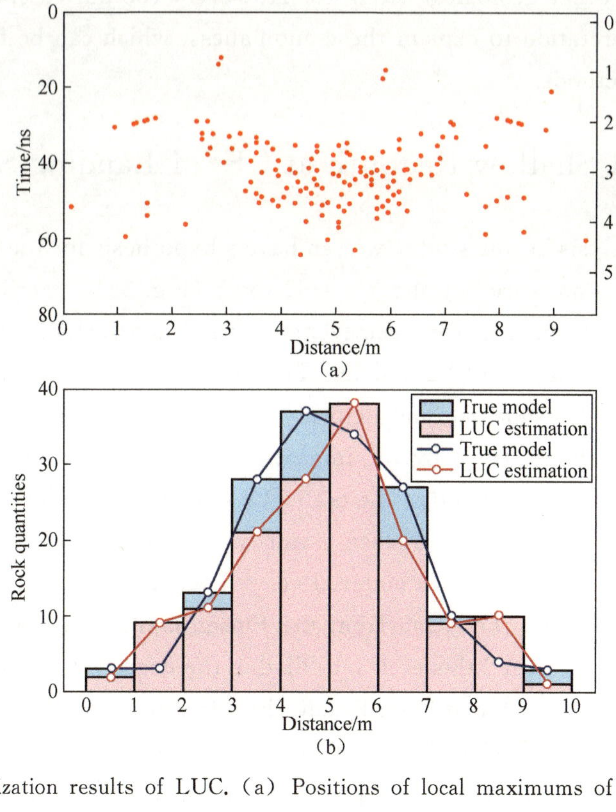

Fig. 35　Rock localization results of LUC. (a) Positions of local maximums of Fig. 34 (f). The right vertical axis is the inverted depth from two - way traveling time with velocity model used in migrations. (b) Rock distribution statistics in horizontal direction. The Pearson correlation coefficient of the two curves is computed to be 0. 9249

Ⅶ DISCUSSION

A Summative Analysis

In this study, the time - frequency comparative analysis, the rock quantitative analysis, and numerical simulation are applied to analyze the horizontal variation of the LPR radargrams. For the focus region R2 in our study, i. e., the region ~160 - 280 m on the radargram, there are several anomalies:

1) HF component dominates in the radargrams;

2) rock - concentrated structure;

3) strata absence;

4) clear features of a crater: tow side - wings and bowl - like bottom.

A buried crater is proposed to interpret these anomalies. Besides, there may be some other possibilities, e. g., the bowl - shaped reflection could also be caused by the topography of the deeper layers; the rock - concentrated structure and the strata absence may be generated by other geological activities. However, the buried crater is a tentative and reasonable interpretation to explain these anomalies, which can be further confirmed by other detection methods.

B Evolution of Shallow Regolith at CE - 4 Landing Site

Based on the analysis in this study, we can have a hypothesis for the evolution process of the shallow regolith measured by the Yutu - 2 rover (Fig. 36).

After the E2 was formed, an impact event occurred and formed the crater C1. The meteorite penetrated the layers of E2 and E3. The bottom of the final crater is located in the E4. The impact event squeezed and deformed the regolith [Fig. 36 (a) and (b)]. Simultaneously, the debris formed by the impact event was ejected and emplaced on the E2 - E4 [Fig. 36 (b)]. Subsequently, the crater C1 and the overlaying debris underwent weathering and gardening before the Finsen crater was formed in the East. Recent works have convinced that the near - surface material (e. g., regolith and rock fragment) at the CE - 4 landing site is mainly the ejecta from the Finsen crater [11] - [13]. The ejecta from the eastern impact covered this area and filled in the degraded crater C1 and form the initial layer of E1 [Fig. 36 (b) and (c)]. Due to the different distances from Finsen on the rover path, the thickness of E1 decreases from east to west (Fig. 28). Finally, a finegrained regolith, which is mainly composed of unconsolidated materials, accumulated gradually from the continuous impaction and gardening of the lunar surface materials [15]. Simultaneously, the material with large amounts of rocks in the crater C1 survived the

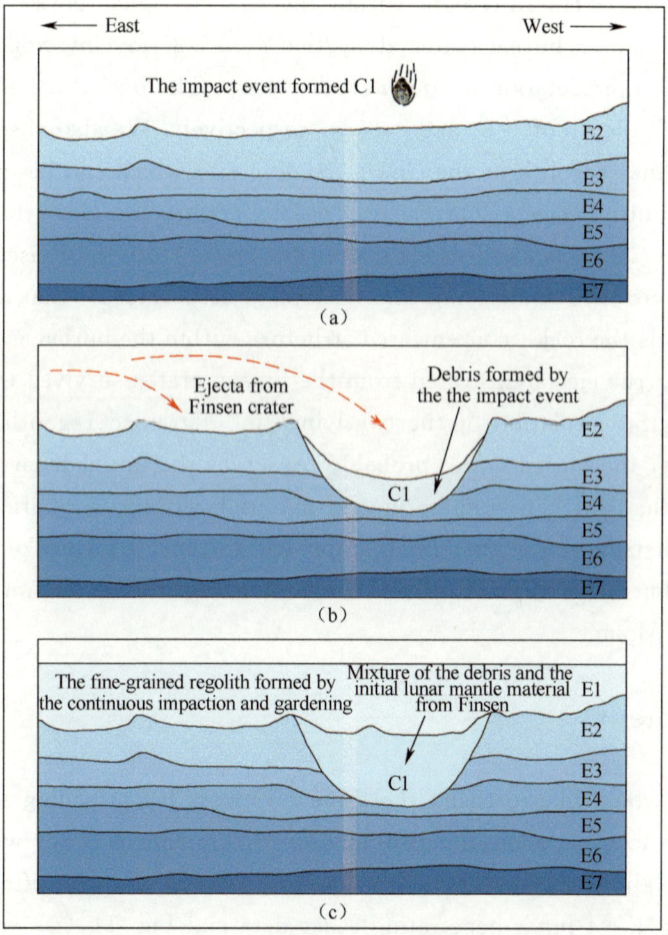

Fig. 36 Evolution process of the shallow regolith measured by Yutu - 2 rover. (a) Impact event that formed C1. (b) Finsen crater formed in the East and its ejecta was emplaced on the C1. (c) Continuous impaction and gardening formed the fine - grained regolith

impact events and gardening which shows a boulder - concentrated structure. Recent works have convinced that the near - surface material excavated from the Finsen crater at the CE - 4 landing site probably contains the lunar mantle materials [13] - [20]. Therefore, the buried crater preserves the possible initial lunar mantle material. Overall, the buried crater provides a significant chance to study the ejecta from the Finsen crater even the initial lunar deep materials.

Ⅷ CONCLUSION

This study proposes the time - frequency comparative analysis method based on the VMD and the rock quantitative analysis method based on the LUC. The model tests verify the capabilities of the proposed methods. Subsequently, the methods are applied to the analysis of the 500 - MHz LPR data within the first 15 lunar days. According to the results, we first report a buried crater along the Yutu - 2 traveling route and propose a detailed geological interpretation for our new discovery. The buried crater has a minimum diameter and height depth of ~84 and ~26m, respectively. We also observed the layered structure in the lunar regolith at the CE - 4 landing site which can be interpreted as the superposition of multiple ejecta sources from nearby craters. Besides, the ~12 - m fine - grained regolith at CE - 4 landing site is mainly the ejecta from the Finsen crater and is of different sources to the underlying multistrata. Furthermore, the rock quantitative analysis also reveals the rock - concentrated structure within the buried crater which can be interpreted as that the ejecta excavated from the Finsen crater survived the impaction and gardening during the formation of the overlying fine - grained regolith. Combining the geological context, the buried crater probably preserves the initial lunar mantle material. We recommend that further studies on the material within the buried crater can be conducted. The distribution of the buried crater within tens of meters of lunar regolith is rarely studied, but it is significant for understanding the evolution of the surface processing of the Moon.

Acknowledgment

The authors would like to thank the three reviewers for providing so many valuable advice which help to greatly improve the quality of this article. They would also like to thank the China National Space Administration for providing the scientific data used in this study. The Chang'E - 4 lunar penetrating radar data and Fig. 11 (a) are available from the Data Publishing and Information Service System of China Lunar Exploration Program. It can be downloaded from the following URL: http://moon. bao. ac. cn. The base image in Fig. 11 (b) was taken by the Lunar Reconnaissance Orbiter's Narrow - Angle Camera on June 3, 2019, which is available online from http: //pds. lroc. asu. edu/ data/LRO - L - LROC - 5 - RDR - V1. 0/LROLRC _ 2001/DATA/SDP/NAC _ DTM/ CHANGE4/.

References

[1] J. Liu, et al., "Descent trajectory reconstruction and landing site positioning of Chang'E - 4 on the lunar farside," Nature Commun., vol. 10, no. 1, p. 4229, Sep. 2019, doi: 10. 1038/s41467 - 019 - 12278 - 3.

[2] I. Garrick - Bethell and M. T. Zuber, "Elliptical structure of the lunar South Pole - Aitken basin," *Icarus*, vol. 204, no. 2, pp. 399 - 408, Dec. 2009, doi: 10. 1016/j. icarus. 2009. 05. 032.

[3] J. Haruyama, et al., "Long - lived volcanism on the lunar farside revealed by

SELENE terrain camera," Science, vol. 323, no. 5916, pp. 905 - 908, Feb. 2009. [Online]. Available: https: //science. sciencemag. org/content/323/5916/905

[4] J. H. Pasckert, H. Hiesinger, and C. H. van der Bogert, "Lunar farside volcanism in and around the South Pole - Aitken basin," Icarus, vol. 299, no. 1, pp. 538 - 562, Jan. 2018, doi: 10. 1016/j. icarus. 2017. 07. 023.

[5] J. Huang, et al., "Geological characteristics of Von Kármán crater, northwestern South Pole - Aitken Basin: Chang'E - 4 landing site region," J. Geophys. Res., Planets, vol. 123, no. 7, pp. 1684—1700, Jul. 2018, doi: 10. 1029/2018JE005577.

[6] J. W. Head et al., "Global distribution of large lunar craters: Implications for resurfacing and impactor populations," Science, vol. 329, no. 5998, pp. 1504 - 1507, Sep. 2010. [Online]. Available: https: //science. sciencemag. org/content/329/5998/1504

[7] D. E. Stuart - Alexander, Geologic Map of the Central Far Side of the Moon (US Geological Survey). 1978, doi: 10. 3133/i1047.

[8] Z. Ling et al., "Composition, mineralogy and chronology of mare basalts and non - mare materials in Von Kármán crater: Landing site of the Chang'E - 4 mission," Planet. Space Sci., vol. 179, Dec. 2019, Art. no. 104741, doi: 10. 1016/j. pss. 2019. 104741.

[9] L. Qiao, Z. Ling, X. Fu, and B. Li, "Geological characterization of the Chang'E - 4 landing area on the lunar farside," Icarus, vol. 333, pp. 37 - 51, Nov. 2019, doi: 10. 1016/j. icarus. 2019. 05. 029.

[10] J. Lai et al., "First look by the Yutu - 2 rover at the deep subsurface structure at the lunar farside," Nature Commun., vol. 11, no. 1, p. 3426, Jul. 2020, doi: 10. 1038/s41467 - 020 - 17262 - w.

[11] K. Di et al., "Topographic evolution of Von Kármán crater revealed by the lunar rover Yutu - 2," Geophys. Res. Lett., vol. 46, no. 22, pp. 12, 764 - 12, 770, Nov. 2019, doi: 10. 1029/2019GL085252.

[12] C. Ding et al., "Fragments delivered by secondary craters at the Chang'E - 4 landing site," Geophys. Res. Lett., vol. 47, no. 7, Apr. 2020, Art. no. e2020GL087361, doi: 10. 1029/2020GL087361.

[13] S. Gou et al., "Impact melt breccia and surrounding regolith measured by Chang'E - 4 rover," Earth Planet. Sci. Lett., vol. 544, Aug. 2020, Art. no. 116378, doi: 10. 1016/j. epsl. 2020. 116378.

[14] S. Gou et al., "Lunar deep materials observed by Chang'E - 4 rover," Earth Planet. Sci. Lett., vol. 528, Dec. 2019, Art. no. 115829, doi: 10. 1016/j. epsl. 2019. 115829.

[15] S. Gou, Z. Yue, K. Di, Z. Cai, Z. Liu, and S. Niu, "Absolute model age of lunar Finsen crater and geologic implications," Icarus, vol. 354, no. 15, Jan. 2021, Art. no. 115829, doi: 10. 1016/j. icarus. 2020. 114046.

[16] X. Hu et al., "Mineral abundances inferred from in situ reflectance measurements of Chang'E - 4 landing site in South Pole - Aitken basin," Geophys. Res. Lett., vol. 46, no. 16, pp. 9439 - 9447, Aug. 2019, doi: 10.1029/2019GL084531.

[17] H. Lin et al., "New insight into lunar regolith - forming processes by the lunar rover Yutu 2," Geophys. Res. Lett., vol. 47, no. 14, Jul. 2020, Art. no. e2020GL087949, doi: 10.1029/2020GL087949.

[18] P. Ma et al., "A plagioclase - rich rock measured by Yutu - 2 rover in von Kármán crater on the far side of the Moon," Icarus, vol. 350, Nov. 2020, Art. no. 113901, doi: 10.1016/j. icarus. 2020.113901.

[19] J. Zhang et al., "Lunar regolith and substructure at Chang'E - 4 landing site in South Pole - Aitken basin," Nature Astron., vol. 5, no. 1, pp. 25 - 30, Sep. 2020, doi: 10.1038/s41550 - 020 - 1197 - x.

[20] C. Li et al., "Chang'E - 4 initial spectroscopic identification of lunar farside mantle - derived materials," Nature, vol. 569, no. 7756, pp. 378 - 382, May 2019, doi: 10.1038/s41586 - 019 - 1189 - 0.

[21] Z. Xiao et al., "Ejecta from the orientale basin at the Chang'E - 4 landing site," *Geophys. Res. Lett.*, vol. 48, no. 3, Feb. 2021, Art. no. e2020GL090935, doi: 10.1029/2020GL090935.

[22] Z. Dong et al., "Properties analysis of lunar regolith at Chang'E - 4 landing site based on 3D velocity spectrum of lunar penetrating radar," Remote Sens., vol. 12, no. 4, p. 629, Feb. 2020, doi: 10.3390/rs12040629.

[23] W. Lv, C. Li, H. Song, J. Zhang, and Y. Lin, "Comparative analysis of reflection characteristics of lunar penetrating radar data using numerical simulations," Icarus, vol. 350, Nov. 2020, Art. no. 113896, doi: 10.1016/J. ICARUS. 2020.113896.

[24] L. Zhang, J. Li, Z. Zeng, Y. Xu, C. Liu, and S. Chen, "Stratigraphy of the von Kármán crater based on Chang'E - 4 lunar penetrating radar data," Geophys. Res. Lett., vol. 47, no. 15, Jul. 2020, Art. no. e2020GL088680, doi: 10.1029/2020GL088680.

[25] G. - Y. Fang et al., "Lunar penetrating radar onboard the Chang'E - 3 mission," Res. Astron. Astrophys., vol. 14, no. 12, pp. 1607 - 1622, Dec. 2014, doi: 10.1088/1674 - 4527/14/12/009.

[26] Y. Su et al., "Data processing and initial results of Chang'E - 3 lunar penetrating radar," Res. Astron. Astrophys., vol. 14, no. 12, pp. 1623 - 1632, Dec. 2014, doi: 10.1088/1674 - 4527/14/12/010.

[27] W. Fa, M. Zhu, T. Liu, and J. B. Plescia, "Regolith stratigraphy at the Chang'E - 3 landing site as seen by lunar penetrating radar," Geophys. Res. Lett., vol. 42, no. 23, pp. 10, 179 - 10, 187, Dec. 2015, doi: 10.1002/2015GL066537.

[28] L. Zhang et al., "Simulation of the lunar regolith and lunarpenetrating radar data processing," IEEE J. Sel. Topics Appl. Earth Observ. Remote Sens., vol. 11, no. 2,

pp. 655 – 663, Feb. 2018, doi: 10. 1109/JSTARS. 2017. 2786476.

[29] L. Xiao et al., "A young multilayered terrane of the northern mare imbrium revealed by Chang'E – 3 mission," Science, vol. 347, no. 6227, pp. 1226 – 1229, Mar. 2015, doi: 10. 1126/science. 1259866.

[30] J. Zhang et al., "Volcanic history of the Imbrium basin: A close – up view from the lunar rover Yutu," Proc. Nat. Acad. Sci. USA, vol. 112, no. 17, pp. 5342 – 5347, Apr. 2015, doi: 10. 1073/pnas. 1503082112.

[31] J. Lai et al., "Comparison of dielectric properties and structure of lunar regolith at Chang'E – 3 and Chang'E – 4 landing sites revealed by ground – penetrating radar," Geophys. Res. Lett., vol. 46, no. 22, pp. 12, 783 – 12, 793, Oct. 2019, doi: 10. 1029/2019GL084458.

[32] C. Li et al., "The Moon's farside shallow subsurface structure unveiled by Chang'E – 4 lunar penetrating radar," Sci. Adv., vol. 6, no. 9, Feb. 2020, Art. no. eaay6898, doi: 10. 1126/sciadv. aay6898.

[33] L. Zhang, Z. Zeng, J. Li, J. Lin, and J. Zhang, "Lunar penetrating radar data processing and analysis based on CEEMD," in Proc. 17th Int. Conf. Ground Penetrating Radar (GPR), Rapperswil, Switzerland, 2018, pp. 635 – 638.

[34] J. Zhang, Z. Zeng, L. Zhang, Q. Lu, and K. Wang, "Application of mathematical morphological filtering to improve the resolution of Chang'E – 3 lunar penetrating radar data," Remote Sens., vol. 11, no. 5, p. 524, Mar. 2019, doi: 10. 3390/rs11050524.

[35] N. E. Huang et al., "The empirical mode decomposition and the Hilbert spectrum for nonlinear and non – stationary time series analysis," Proc. Roy. Soc. London A, Math., Phys. Eng. Sci., vol. 454, no. 1971, pp. 903 – 995, Mar. 1998.

[36] K. Dragomiretskiy and D. Zosso, "Variational mode decomposition," IEEE Trans. Signal Process., vol. 62, no. 3, pp. 531 – 544, Feb. 2014, doi: 10. 1109/TSP. 2013. 2288675.

[37] X. Zhang, E. Nilot, X. Feng, Q. Ren, and Z. Zhang, "IMF – slices for GPR data processing using variational mode decomposition method," Remote Sens., vol. 10, no. 3, p. 476, Mar. 2018, doi: 10. 3390/rs10030476.

[38] S. Cheng, S. Liu, J. Guo, K. Luo, L. Zhang, and X. Tang, "Data processing and interpretation of antarctic ice – penetrating radar based on variational mode decomposition," Remote Sens., vol. 11, no. 10, p. 1253, May 2019, doi: 10. 3390/rs11101253.

[39] Z. Dong et al., "Date processing for lunar penetrating radar of Chang'E – 4 based on variational mode decomposition," in Proc. 18th Int. Conf. Ground Penetrating Radar (GPR), Golden, CO, USA, 2020, pp. 424 – 427.

[40] S. Fomel, "Applications of plane – wave destruction filters," Geophysics, vol. 67, no. 6, pp. 1946 – 1960, Nov. 2002, doi: 10. 1190/1. 1527095.

[41] S. Fomel, "Shaping regularization in geophysical - estimation problems," Geophysics, vol. 72, no. 2, pp. R29 - R36, Mar. 2007, doi: 10. 1190/1. 2433716.

[42] S. Fomel, "Local seismic attributes," Geophysics, vol. 72, no. 3, pp. A29 - A33, May 2007, doi: 10. 1190/1. 2437573.

[43] B. Hu, D. Wang, L. Zhang, and Z. Zeng, "Rock location and quantitative analysis of regolith at the Chang'E 3 landing site based on local similarity constraint," Remote Sens., vol. 11, p. 530, Mar. 2019, doi: 10. 3390/rs11050530.

[44] H. Song, C. Li, J. Zhang, X. Wu, Y. Liu, and Y. Zou, "Rock location and property analysis of lunar regolith at Chang'E - 4 landing site based on local correlation and semblance analysis," Remote Sens., vol. 13, no. 1, p. 48, Dec. 2020, doi: 10. 3390/rs13010048.

[45] R. A. Wiggins, "Minimum entropy deconvolution," Geoexploration, vol. 16, nos. 1 - 2, pp. 21 - 35, Apr. 1978, doi: 10. 1016/0016 - 7142 (78) 90005 - 4.

[46] H. -S. Wu and J. Barba, "Minimum entropy restoration of star field images," IEEE Trans. Syst., Man Cybern. B, Cybern., vol. 28, no. 2, pp. 227 - 231, Apr. 1998, doi: 10. 1109/3477. 662762.

[47] X. Xu, E. L. Miller, and C. M. Rappaport, "Minimum entropy regularization in frequency - wavenumber migration to localize subsurface objects," IEEE Trans. Geosci. Remote Sens., vol. 41, no. 8, pp. 1804 - 1812, Aug. 2003, doi: 10. 1109/TGRS. 2003. 813497.

[48] H. Zhou, X. Feng, Z. Dong, C. Liu, and W. Liang, "Application of denoising CNN for noise suppression and weak signal extraction of lunar penetrating radar data," Remote Sens., vol. 13, no. 4, p. 779, Feb. 2021, doi: 10. 3390/rs13040779.

[49] L. Zhang et al., "A story of regolith told by lunar penetrating radar," Icarus, vol. 321, no. 15, pp. 148 - 160, Mar. 2019, doi: 10. 1016/j. icarus. 2018. 11. 006.

[50] Z. Dong, X. Feng, H. Zhou, C. Liu, and M. Sato, "Effects of induced field rotation from rough surface on H - Alpha decomposition of fullpolarimetric GPR," IEEE Trans. Geosci. Remote Sens., early access, Feb. 2, 2021, doi: 10. 1109/TGRS. 2021. 3052547.

[51] H. J. Melosh, Impact Cratering: A Geologic Process. New York, NY, USA: Oxford Univ. Press, 1989.

[52] C. Ding, Z. Xiao, Y. Su, J. Zhao, and J. Cui, "Compositional variations along the route of Chang'E - 3 yutu rover revealed by the lunar penetrating radar," Prog. Earth Planet. Sci., vol. 7, no. 1, pp. 1 - 11, Jul. 2020, doi: 10. 1186/s40645 - 020 - 00340 - 4.

[53] S. E. Lauro et al., "Electromagnetic signal penetration in a planetary soil simulant: Estimated attenuation rates using GPR and TDR in volcanic deposits on Mount Etna," J. Geophys. Res., Planets, vol. 122, no. 7, pp. 1392 - 1404, Jul. 2017, doi: 10. 1002/2016JE005192.

Application of Denoising CNN for Noise Suppression and Weak Signal Extraction of Lunar Penetrating Radar Data

Haoqiu Zhou[1,2], Xuan Feng[1,2,3], Zejun Dong[1,2],
Cai Liu[1,3] and Wenjing Liang[1,2]

1. College of Geo - Exploration Science and Technology, Jilin University, Changchun 130026, China; zhouhq20@mails.jlu.edu.cn (H. Z.); dzj19@mails.jlu.edu.cn (Z. D.); liucai@jlu.edu.cn (C. L.); liangwj@jlu.edu.cn (W. L.)
2. Science and Technology on Near - Surface Detection Laboratory, Wuxi 214035, China
3. Key Laboratory of Geophysical Exploration Equipment, Ministry of Education, Jilin University, Changchun 130026, China

Abstract: As one of the main payloads mounted on the Yutu - 2 rover of Chang'E - 4 probe, lunar penetrating radar (LPR) aims to map the subsurface structure in the Von Kármán crater. The field LPR data are generally masked by clutters and noises of large quantities. To solve the noise interference, dozens of filtering methods have been applied to LPR data. However, these methods have their limitations, so noise suppression is still a tough issue worth studying. In this article, the denoising convolutional neural network (CNN) framework is applied to the noise suppression and weak signal extraction of 500MHz LPR data. The results verify that the low - frequency clutters embedded in the LPR data mainly came from the instrument system of the Yutu rover. Besides, compared with the classic band - pass filter and the mean filter, the CNN filter has better performance when dealing with noise interference and weak signal extraction; compared with Kirchhoff migration, it can provide original high - quality radargram with diffraction information. Based on the high - quality radargram provided by the CNN filter, the subsurface sandwich structure is revealed and the weak signals from three sub - layers within the paleo - regolith are extracted.

Keywords: Chang'E - 4; lunar penetrating radar (LPR); convolutional neural network (CNN); noise suppression; weak signal extraction

1 Introduction

After China's Chang'E - 3 (CE - 3) mission was completed, the Chang'E - 4 (CE -4)

probe was launched on 8 December 2018, and achieved the first - ever soft landing on the farside of the moon on January 3, 2019. The landing site of CE - 4 is on the floor of Von Kármán crater in the South Pole - Aitken (SPA) basin [1]. With a diameter of approximately 2500km and the deepest depth of 13km, the SPA basin is the largest impact structure on the moon and can provide important clues to the internal structure of the moon [2 - 7]. Recent works indicate that the near - surface materials measured by the CE - 4 rover were mainly the ejecta from the Finsen crater, with possible contributions from the Alder, the Von Kármán L, and the Von Kármán L' craters [7 - 17]. It is also convincing that the ejecta from the Finsen crater probably contains the lunar mantle materials [18 - 21].

The Yutu - 2 rover, carried by CE - 4, is mounted with a lunar penetrating radar (LPR) which is used to detect the lunar subsurface structure [22]. Similar to CE - 3 [23, 24], the LPR possesses two channels (CH - 1 and CH - 2) with center frequencies of 60 and 500MHz, respectively [23 - 29]. The CH - 2 consists of one transmitting antenna and two receiving antennas (also known as CH - 2A and CH - 2B) of different offsets [30 - 32]. The mission of CH - 2 is to map the fine structure of the lunar regolith [23 - 32].

The CH - 2 LPR amplitude - modulated and frequency - modulated signals are always masked by clutters and noises of large quantity [33]. The clutters with relatively low frequency may generate from the coupling of antennas and the instrument system, as there are many kinds of payloads mounted on the rover [22]. The processing of LPR data is challenging especially in noise suppression. In previous researches, many filtering methods have been applied to LPR data, such as band - pass filtering [34], bi - dimensional empirical mode decomposition (BDEMD) [35], complete ensemble empirical mode decomposition (CEEMD) [36], mathematical morphology filtering [37], and variational mode decomposition (VMD) [38]. However, these methods have their limitations and cannot provide a satisfactory result. More importantly, due to the attenuation of electromagnetic waves, the signals of great depth are quite weak and immersed within the noises. The noises and weak signals are of similar amplitudes and frequencies and cannot be easily classified using traditional denoising methods. Therefore, it is of vital importance to develop a more effective filtering method for LPR data processing. Convolutional neural network (CNN) [39 - 41] is a type of deep learning network architecture that is widely applied to various fields. In 2017, Zhang et al. proposed a CNN model for image denoising [42, 43]. The denoising CNN model innovatively removed the pooling layers from the initial CNN model, as pooling layers may erase the characteristics of high - frequency noise.

In this article, the denoising CNN framework is applied to CE - 4 LPR data filtering. The low - frequency clutters used in the CNN training are extracted from CE - 3 LPR data using the empirical mode decomposition (EMD) [44]. Therefore, the achievement of noise suppression can also verify that the low - frequency clutters embedded in the CE - 4 are the same as that of CE - 3 and are generated from the instrument system of the rover. Based on the well trained CNN filter, the low - frequency clutters and high - frequency noises are both suppressed in the field LPR profile and the weak signals in the deep are extracted. To illustrate the superiority of the CNN filter on noise suppression, a classic band - pass filter and a mean filter are respectively conducted as comparisons. For discussing the results of weak signal extraction, the combined methods of the band - pass filter and the mean filter as well as the combined methods of the band - pass filter and the Kirchhoff migration are respectively conducted as comparisons. Finally, based on the high - quality radargram provided by the CNN filter, the subsurface sandwich structure is revealed and the weak signals from three sub - layers within the paleo - regolith are extracted.

2 Materials and Methods

2.1 The Initial CH - 2B LPR Data

The walking route for the Yutu - 2 rover on the moon for the first 12 lunar days is shown in Figure 1a [11]. The curve is ~320 meters in length. A series of processing steps, including the traces amending, the traces selecting, the time lag adjustment, the attenuation compensation, and the background removal [10, 11, 25 - 31] (The details are in Table A1) are conducted to obtain the initial CH - 2B profile. The original profile and its average frequency amplitude spectrum are respectively shown in Figure 1b, c. Considering that the center frequency of the CH - 2B LPR signal is 500MHz with a bandwidth of 450MHz [23, 24], the radargram in Figure 1b is masked by low - frequency clutters and high - frequency noises of large quantity. Simultaneously, as the signals from the deep are very weak and their amplitudes are similar to the noise, the signals below 400ns are more severely affected by the noise interference and the reflections from the strata are invisible or hard to recognize. Therefore, the key point of this article is to suppress the noises and extract the deep weak signals.

2.2 Methodology of Denoising CNN

The key idea of deep learning is to deepen the depth of the network for enhancing the accuracy of nonlinear matching. The CNN is a method of deep learning. It integrates the input data into a hierarchical stack of operations such as convolution operations, convergent operations, and nonlinear activation function mapping, and extracts features from the original data [39 - 41]. This process is called "feed - forward operation." The last layer of the CNN compares the operation result with the target task, converts it into the objective function, obtains the error between the calculated value and the real value, and then uses the error back - propagation (BP) algorithm to propagate the error back. The error is assigned to the neurons of each layer, and then the parameters of each layer can be

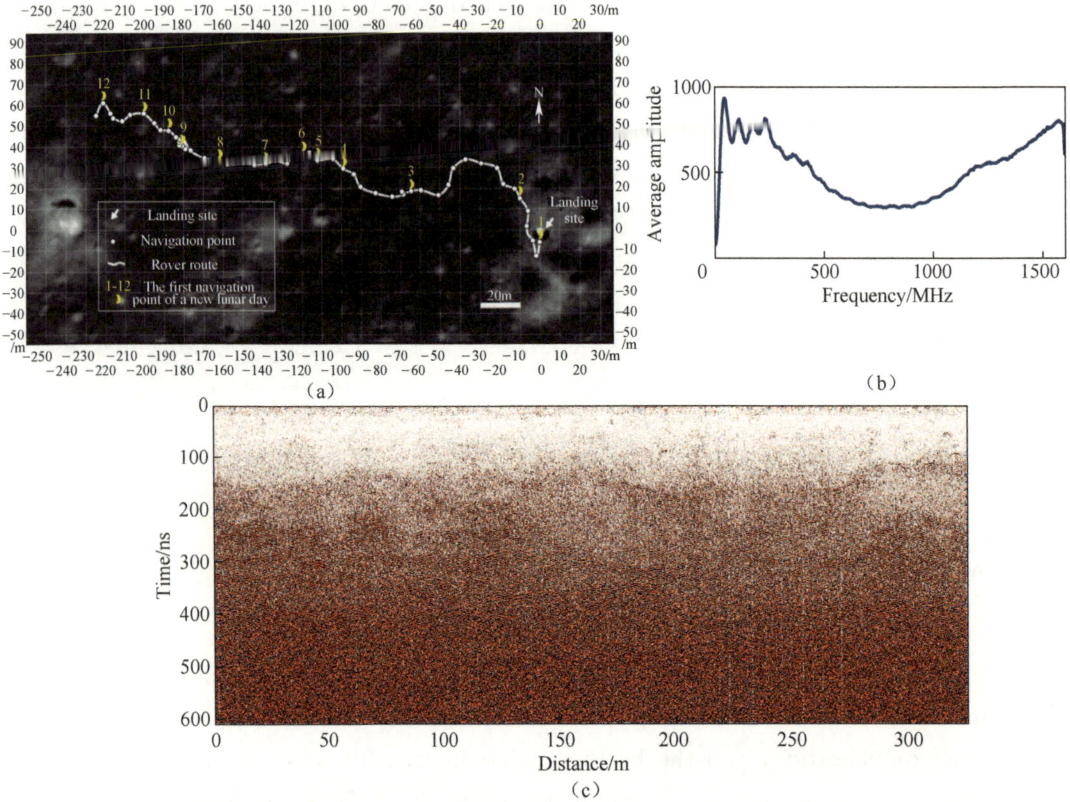

Figure 1 (a) The traveling route of Yutu - 2 rover within the first 12 lunar days [11]. (b) The initial CH - 2B LPR radargram acquired along the path in (a) without filtering. (c) The average frequency amplitude spectrum of the radargram in (b), which is masked by low - frequency clutter and high - frequency noise of large quantities

updated. This process is called "feed - back operation." The iterations will continue until the objective function converges, which means the network is well trained. When new data are input, they can be predicted by the network's feed - forward operation.

Denoising CNN is a CNN framework for image denoising. It innovatively removed the pooling layer from the initial CNN model. The framework of denoising CNN is shown in Figure 2 [42, 43]. In this model, the batch normalization (BN) and the rectified linear unit (Relu) are applied. These two training techniques are used to avoid the vanishing gradient problem [42, 43]. The mathematical form of BN is given by [42, 43]:

$$P_{out} = \frac{P_{in} - \mathrm{mean}(P_{in})}{\sqrt{\mathrm{var}(P_{in})}} \cdot \gamma + \beta \tag{1}$$

where P_{in} and P_{out} represent the input and output of a layer. Mean (P_{in}) and var (P_{in}) are the average value and standard deviation of P_{in}; γ and β are the learnable parameters.

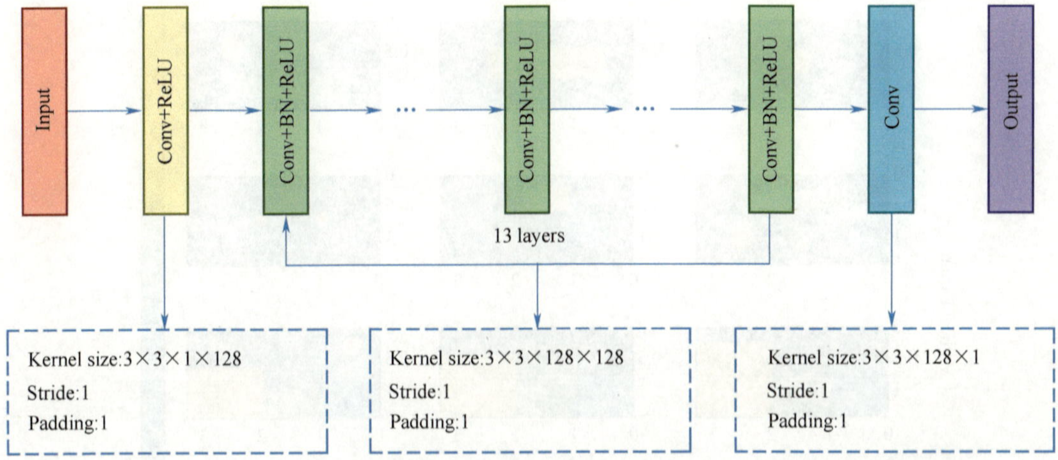

Figure 2　The framework of denoising convolutional neural network (CNN) [42, 43]

The Relu is a type of nonlinear activation function widely applied in deep learning [42, 43]. It always follows the convolutional layer, and its mathematical expression is shown as follow:

$$P_{out} = \max(0,\ P_{in} \cdot w + b) \tag{2}$$

where w and b are the convolutional kernel and the threshold, respectively.

3　Results

3.1　CNN Training

The key of CNN training is to use the simulated GPR data with low - frequency clutters and high - frequency noises as the training data. The GPR simulations for different types of relative permittivity models are conducted using the finite - difference time - domain (FDTD) method. The models as well as their simulated data are shown in Figure 3a, b. The models contain different target combinations with different relative permittivity.

As we have discussed in Section 2.1, the LPR data are masked by low - frequency clutters and high - frequency noises of large quantities. The low - frequency clutter is considered to generate from the coupling of antennas and the instrument system, so it is relatively stable. As CE - 4 is the backup of CE - 3, we decide to extract 200 low - frequency clutters from CH - 2B data of CE - 3 using empirical mode decomposition (EMD) [44]. If the filtering result is satisfactory, the speculation for the source of low - frequency clutters can also be proved. The extracted clutters as well as their average frequency amplitude spectrum are respectively shown in Figure 4a, b. Besides, the high - frequency noise can be easily simulated by Gaussian noise. Subsequently, the low -

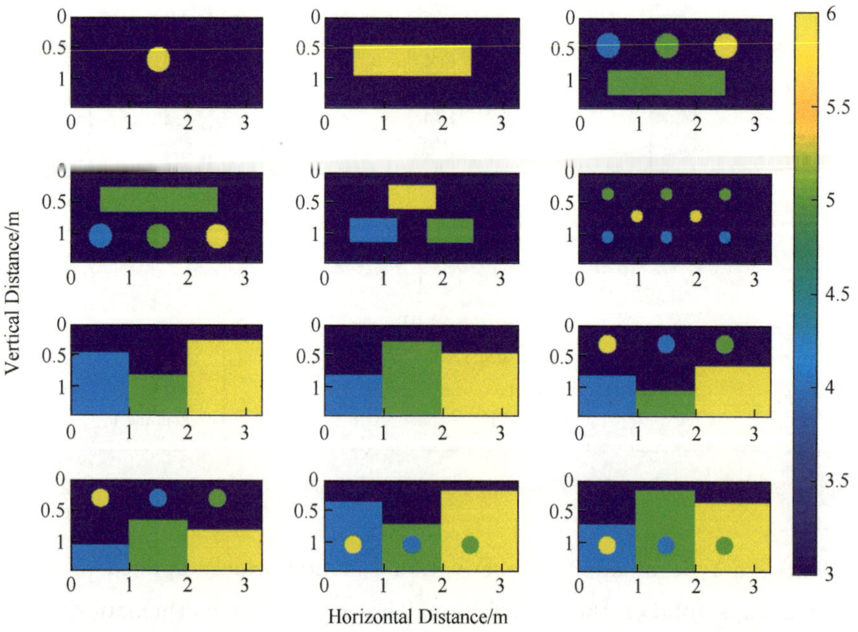

(a)

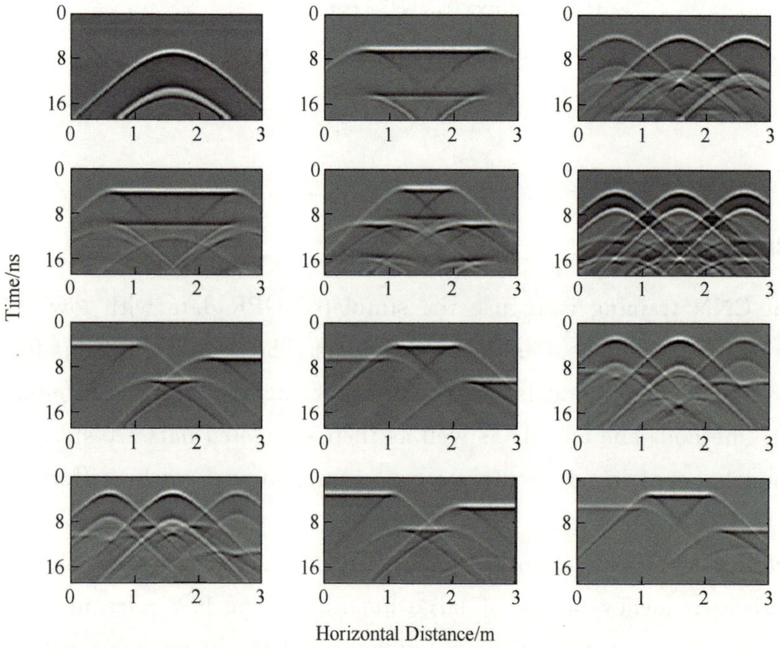

(b)

Figure 3　Training models and simulated data. (a) Training models. (b) Simulated data.

frequency clutters and high - frequency noises are embedded into the simulated GPR data using the formula below:

$$Image = Label + \alpha_1 LC + \alpha_2 HN \tag{3}$$

where *Label* represents the simulated GPR data; *LC* and *HN* are low - frequency clutters and high - frequency noises, respectively; *Image* denotes the data masked by *LC* and *HN*. The α_1 and α_2 are the intensity coefficients. The greater α_1 and α_2 correspond to the lower signal - to - noise ratio (SNR).

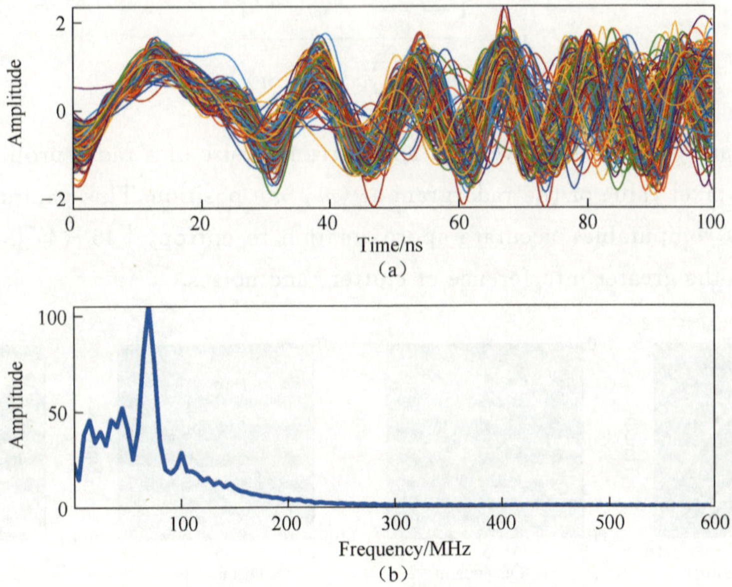

Figure 4　The 200 extracted low - frequency clutters from Chang'E - 3 LPR data and its average frequency amplitude spectrum. (a) The extracted low - frequency clutters. (b) Average frequency amplitude spectrum of the signals in (a).

Compared with the random noises, the clutters are more stable and, more importantly, time - related. Therefore, before adding clutters, the lengths of training radar profiles are re - sized into the length of clutters, which can make the training data remaining the features of clutters. Besides, the additions of *LC* are conducted for a single trace of the simulated data at a time. For each trace, we randomly draw one of the 200 clutters and add it to the signal.

Here, a test is conducted to determine the ranges of these two parameters in the CNN training process. The thresholds should contain most possible cases of different noise levels and the maximums of these two parameters should guarantee that the signals are almost immersed in clutters or noises. Besides, wider ranges of α_1 and α_2 will result in generating more training images which will decelerate the convergence of CNN training and make it more difficult and time - consuming. Therefore, we need to select the smallest ranges of α_1 and α_2 while covering most of the possible noise cases.

The tests with $\alpha_1 = 0, 1, 1.5$, and 2 and $\alpha_2 = 0, 1, 3$, and 5 are shown in Figures 5 and 6, which are based on the simulated data of the first radargram in Figure 3b. It is evident that the signals from the target gradually become invisible with the increase in α_1 and α_2.

To provide quantitative results, the image entropy is conducted for different images with different α_1 and α_2. The image entropy is given by [45 - 47]:

$$R=\frac{\left[\sum_{m=1}^{M}\sum_{n=1}^{N}a^2(m,n)\right]^2}{\sum_{m=1}^{M}\sum_{n=1}^{N}a^4(m,n)} \tag{4}$$

where M and N denote the vertical and horizontal size of a radar profile; $a(m, n)$ represents the pixel value of the radargram at (m, n) position. This parameter is known to be an easily computable, accurate approximation to entropy [45 - 47]. The larger R value indicates the greater interference of clutters and noises.

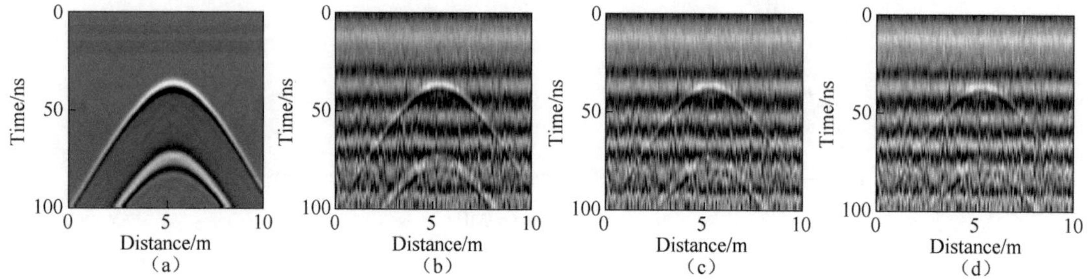

Figure 5　The comparison of radar profiles with different parameter α_1.

(a) The normalized radargram with $\alpha_1=0$. (b) $\alpha_1=1$. (c) $\alpha_1=1.5$. (d) $\alpha_1=2$.

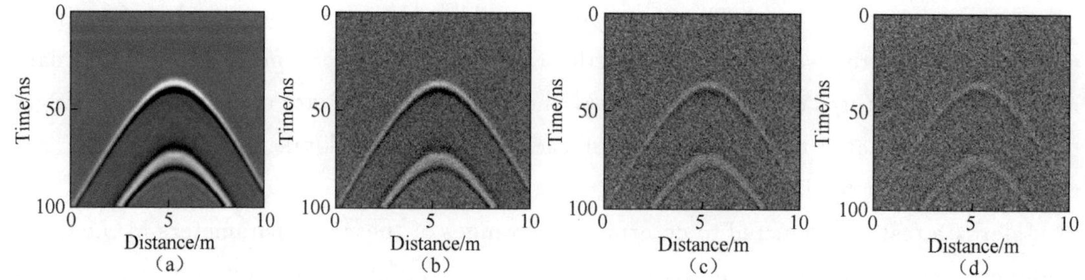

Figure 6　The comparison of radar profiles with different parameter α_2.

(a) The normalized radargram with $\alpha_2=0$. (b) $\alpha_2=1$. (c) $\alpha_2=3$. (d) $\alpha_2=5$.

Figure 7a shows the image entropy of the image with different α_1 and α_2; the blue and orange curves represent the clutters added and random noise added cases, respectively. Figure 7b presents the derivatives of the curves in Figure 7a. Both curves of image entropy rise with the increase in α_1 and α_2; when $\alpha_1 = 2$and $\alpha_2 = 5$, the curves gradually become flat corresponding to the zero values of derivatives in Figure 7b. In this case, the clutters and noises become the main ingredients of the images. It is also feasible to keep increasing the α_1 and α_2, but the images will become complete clutters and noises without signals. These useless images will not strengthen the training result but make the training more time - consuming. Therefore, $\alpha_1 \in [0, 2]$, $\alpha_2 \in [0, 5]$ are reasonable thresholds that are the smallest and can cover most of the possible noise cases.

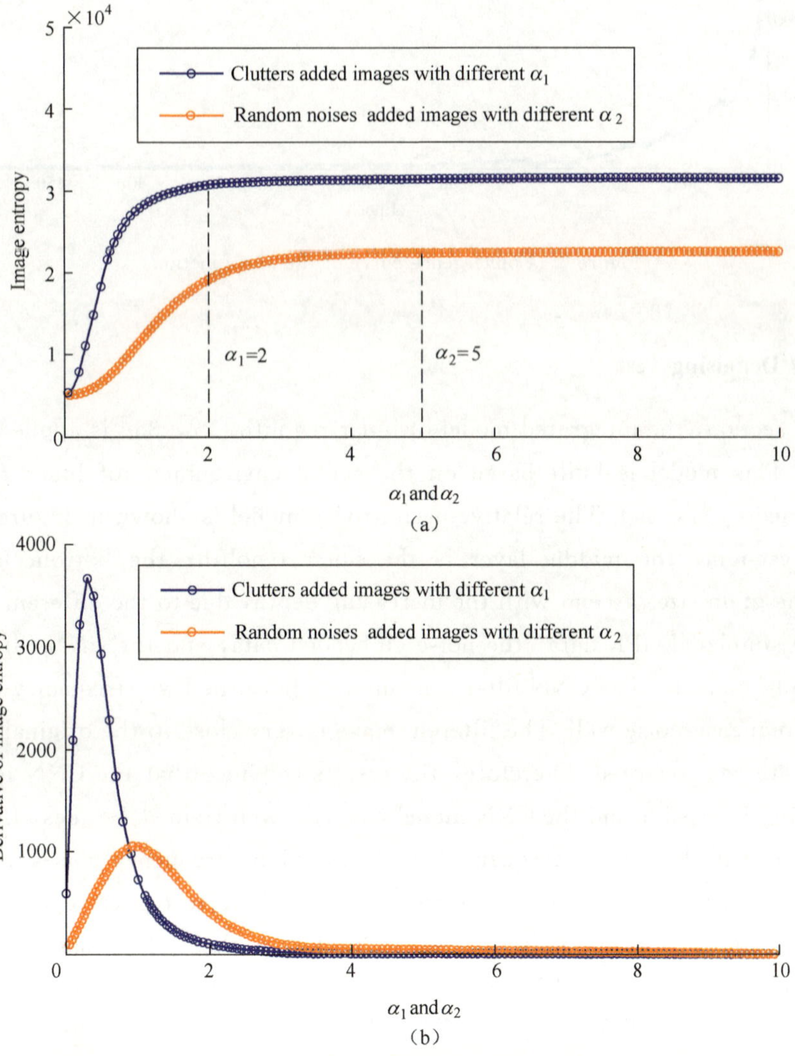

Figure 7　The image entropy analysis of α_1 and α_2 tests.

(a) Images entropy of the radar profiles with different α_1 and α_2. (b) The derivatives of the curves in (a).

To expand the training data set and enhance the performance of CNN, the widely used data augmentation method [42, 43] is subsequently applied. The augmentation contains horizontal and vertical flip. Finally, the augmented data are applied to the CNN training. The convergence curve of the loss function is shown in Figure 8. The objective function converges at approximately 200 iterations.

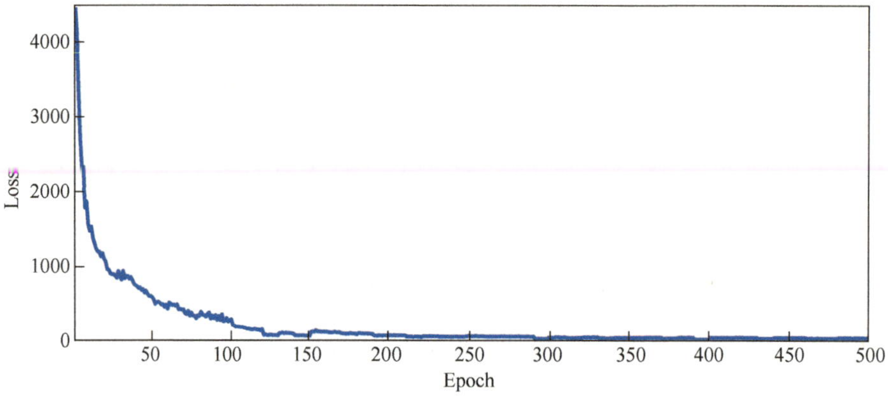

Figure 8　Convergence curve of the loss function

3. 2　Model Denoising Test

In this section, the integrated model of lunar regolith [27, 35] is applied to test the CNN filter. This model is built based on the actual environment of lunar regolith and stochastic media [27, 35]. The relative permittivity model is shown in Figure 9. The top layer is a vacuum; the middle layer is the lunar regolith; the bottom layer is the bedrock. The grain sizes ascend with the increasing depth, due to the different weathering degree. The simulated GPR data, the noise embedded data, and the CNN filtering result are shown in Figure 10. The CNN filter can suppress both the low - frequency clutters and the high - frequency noise well. The filtered image is very close to the original image, the hyperbolas are well restored. Therefore, the results convince that the CNN for the GPR data denoising is feasible and the CNN model has been well trained. Besides, the signals in the radargram with both low - frequency clutter and high - frequency noises in Figure 10c are almost totally immersed in the interference and are hard to recognize. In the test, the amplitudes of the clutters and noises are set to be close to those of the signals, the case of which is similar to the field LPR data with the great depths. However, the CNN filter

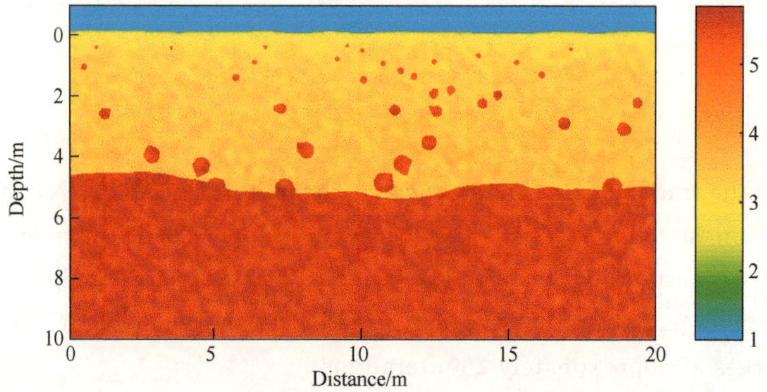

Figure 9　The integrated relative permittivity model of lunar regolith [27, 35]

shows its stable ability to suppress the interference and restore the signals.

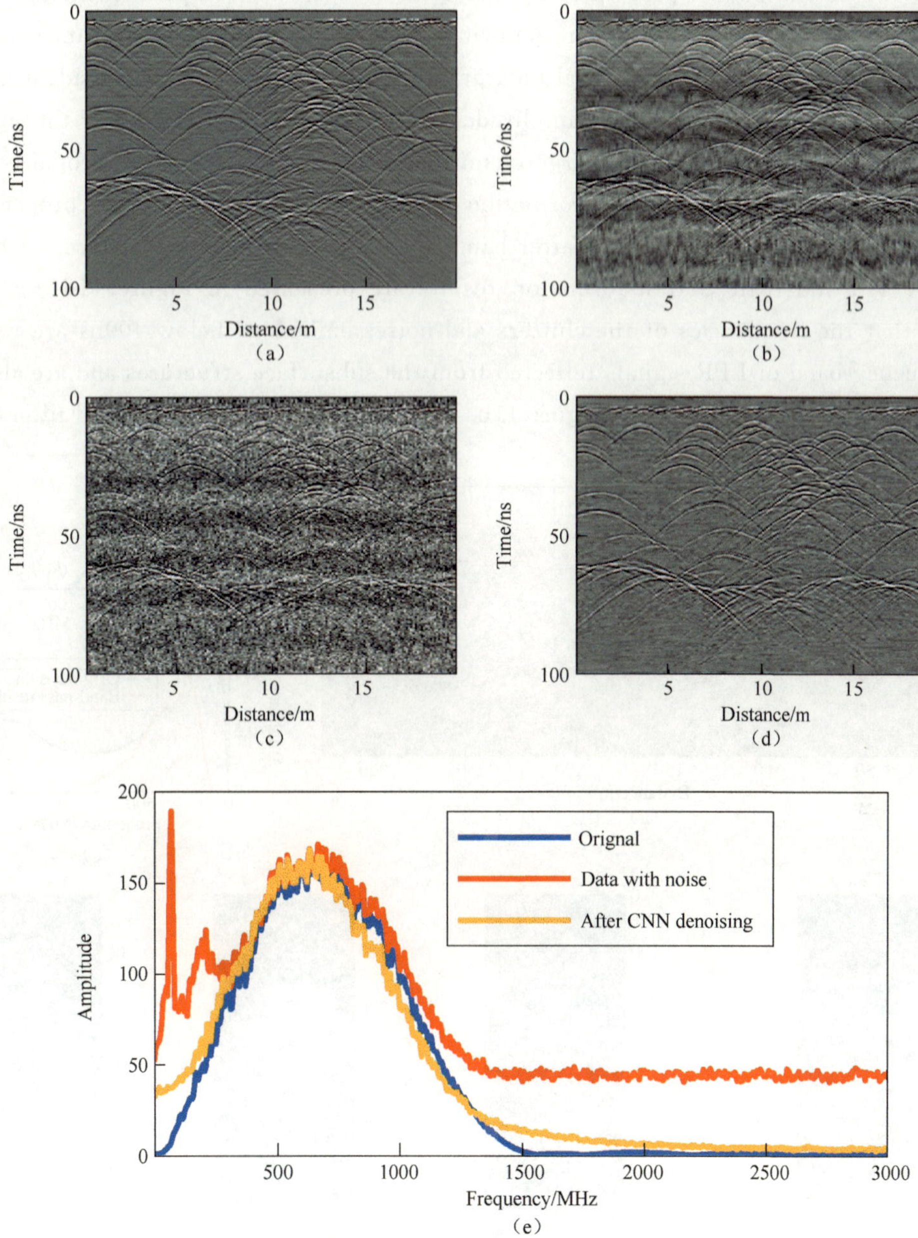

Figure 10　Model denoising test. (a) Simulated data; (b) data with low - frequency clutters; (c) data with low - frequency clutters and high - frequency noises; (d) denoising result; (e) comparison of average frequency amplitude spectrum

3.3　Denoising and Weak Signal Extraction of LPR Data

In this section, the well - trained denoising CNN model is applied to the denoising and weak signal extraction of LPR data. To well illustrate the superiority of the denoising CNN method, the classic band - pass filter is first conducted as the comparison method. The results of the band - pass filtering are shown in Figure 11a - c. The radargram contains much less noise compared to the initial radargram in Figure 1c which is also evident in the comparison of the average frequency amplitude spectra in Figure 11c. However, the signals below 400ns are still masked with large quantities of noise and cannot be recognized. To explain the phenomenon, the S transformation [48] is conducted to analyze the property of the noise embedded in the LPR data after band - pass filtering. Four signals are extracted in Figure 11a and their S transformation results are presented in Figure 11d - g. It is obvious that the frequencies of the clutters and noises embedded below 400ns are within the frequency band of LPR signals reflected from the subsurface structures and are also in the frequency band of the filter in Figure 11b. Therefore, the frequency - based filter such

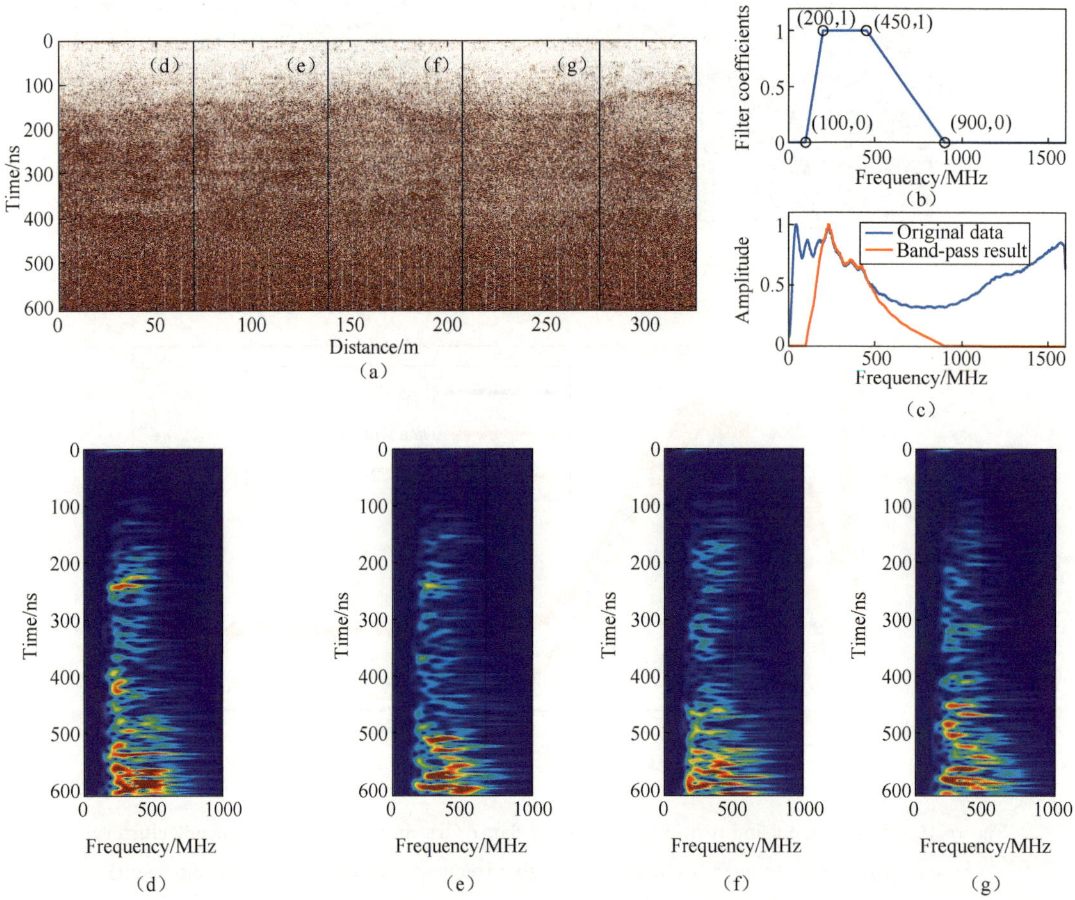

Figure 11　Band - pass filtering test. (a) The radargram after band - pass filtering. The signals marked by black lines are the signals used for the S transformation analysis. (b) Parameters of the band - pass filter. (c) The average frequency amplitude spectrum of the LPR data before and after band - pass filtering. (d - g) Present the S transformation results of the marked signals in (a)

as the band - pass filter cannot suppress the clutter and noise below 400ns.

This phenomenon can be interpreted to be the result of the magnification of the ringing effect and noise. Due to the attenuation of electromagnetic waves [29], the signals from the deep are weak and their amplitudes are close to the ringing signals. In order to make the deep signals more visible, the attenuation compensation [10, 11, 25 - 31] is necessary, resulting in the magnification of ringing signals and noise. This interference is of the same frequency band as the useful LPR signals and is difficult to suppress.

Subsequently, the CNN filter is applied to the LPR data. As a frequency - based filter such as the band - pass filter cannot suppress the clutter and noise below 400 ns, we simultaneously conducted the classical mean filter as a comparison method with the CNN filter. The mean filter is a typical linear filter, which involves a template to the target pixel on the image that includes the neighboring pixels around it (the 3 × 3 sized pixel region around the target pixel constitutes a filtering template). The filtering process removes the target pixel itself and replaces the original pixel value with the average of all the pixels in the template.

Figure 12a, b presents the results of the mean filter and CNN filter. Besides, Figure 12c, d shows the radargram after both band - pass and mean filtering and the radargram after both band - pass and Kirchhoff migration, respectively. For analyzing the processing results, Figure 12e compares the average frequency amplitude spectra of the results.

The result after only mean filtering cannot suppress the low - frequency clutter and high - frequency noise, which is also evident in the spectra comparison. As the purple curve in Figure 12e shows, the mean filter can suppress large quantities of high - frequency noises but the result is much worse than the CNN and band - pass filtering. Besides, the mean filtering has almost no attenuation ability for low frequency clutters.

The CNN filltering can suppress the clutters and noises well, especially for the signals below 400ns (Figure 12b). An evident interface at ～500 ns and another unclear interface at ～600ns can be found, which are pointed out by the double - way black arrows (Figure 12b). Furthermore, we test a combined method of conducting both the band - pass and the mean filters (Figure 12c). The CNN and the combined methods are similar in the spectra results, but the image quality of the combined method result falls flat. The imaging result of the combined method is still masked with many clutters and noises and only the unclear interface at ～500ns can be observed whereas the signals at ～600ns are immersed in the noise. Some strips of interference (the blue arrows in Figure 12c), which can also be observed in the original radargram (Figure 1a) or the mean filtering result (Figure 12a), are not completely removed. However, these interferences are well attenuated in the CNN filtering results. Therefore, the result of CNN filtering can present a radargram with higher quality compared with the combined method of band - pass and mean filltering.

Besides, the migration processing after band - pass filtering, which is commonly adopted in the previous studies [9 - 11], can present a similar result (Figure 12b) to that

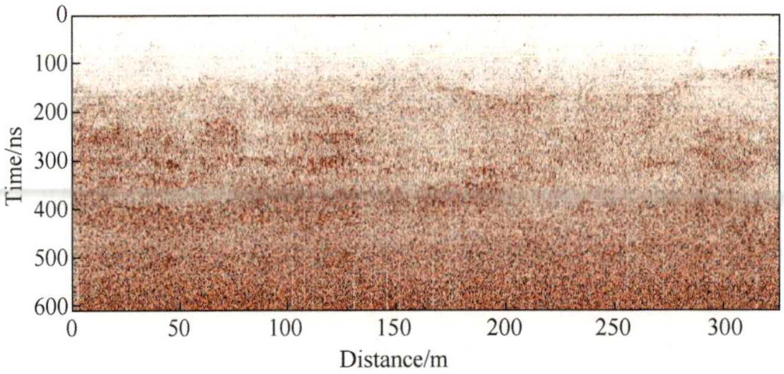

(a)

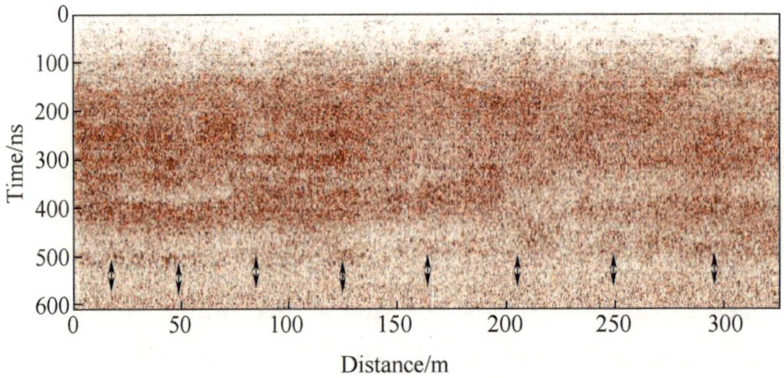

(b)

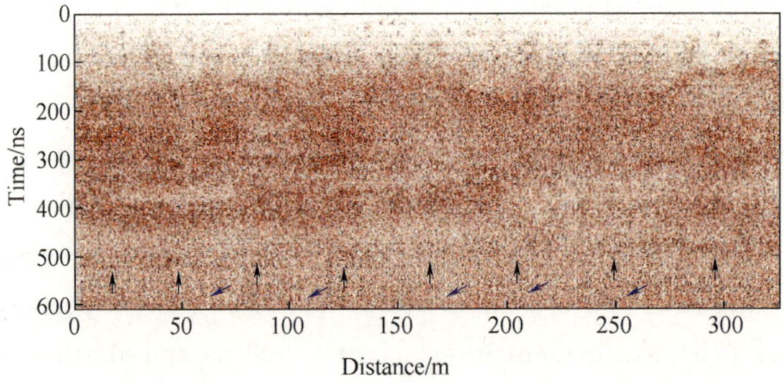

(c)

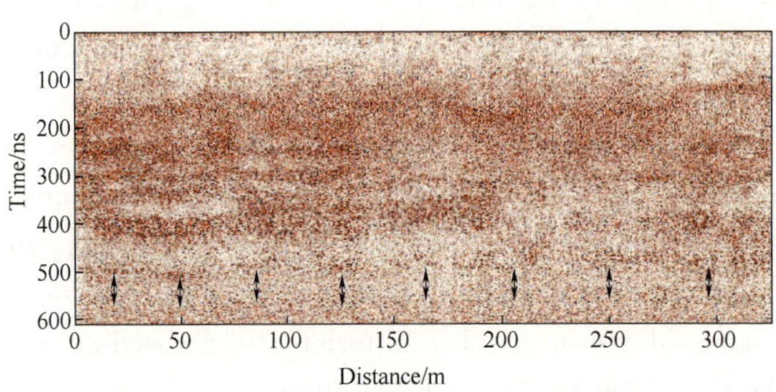

(d)

Figure 12　Result and comparative analysis of the applied filtering methods

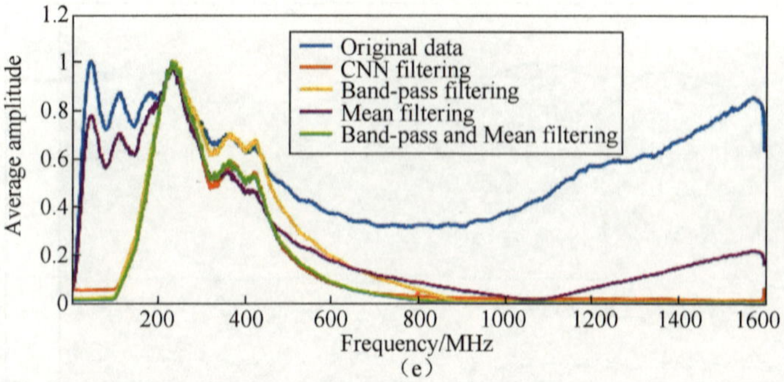

(e)

Figure 12　Result and comparative analysis of the applied filtering methods. (continue)

(a) The radargram after the mean filtering; (b) the radargram after the CNN filtering; (c) the radargram after the band - pass filtering and the mean filtering; (d) the radargram after the band - pass filtering and Kirchhoff migration, the velocity used in migration is 0.14m/ns [9]; (e) the comparison of the average frequency amplitude spectra. The black arrows in (b - d) indicate the extracted weak reflected signals. The blue arrows in (c) point out the strips of interference

of the CNN filtering. However, the result of migration has attenuated the diffracted waves and is much more suitable for stratigraphic divisions [9 - 11] whereas the CNN filter can obtain an original radargram with the image of high quality and information of diffracted waves. The diffracted wave contains important information for property analysis of LPR data [30, 32, 34, 37, 38, 49 - 52] and the high - quality image can guarantee the accuracy of the analysis. Overall, the result of migration can provide clues to stratigraphic divisions and the result of the CNN filter can provide high - quality radargram for property analysis.

3.4　The Subsurface Sandwich Structure at CE - 4 Landing Site

In Section 3.3, the superiority of the CNN method has been verified. Based on the extracted signals from the radargram, a subsurface structure can be summarized as Figure 13 shows. The radargram shows a fine - coarse - fine sandwich structure [9 - 11, 53].

The top 0 - 10m is the fine - grained regolith, which is mainly composed of unconsolidated materials accumulating gradually from the continuous impaction and gardening of the lunar surface materials [12]. The second layer with a depth of 10 - 35m is the coarsegrained material which can be interpreted as the superposition of multiple ejecta sources from nearby craters [9 - 11]. The signals from this layer contain many reflections from subsurface boulders which have survived the impaction and gardening during the formation of the overlying fine - grained regolith.

The analysis for the signals >35m is the focus of our studies. The third layer also contains fine - grained materials similar to the surface layer which can be interpreted to be the paleo - regolith. Besides, based on the CNN filtering result, three sublayers can be observed within this layer. The interface at ~40m is evident and continuous and another

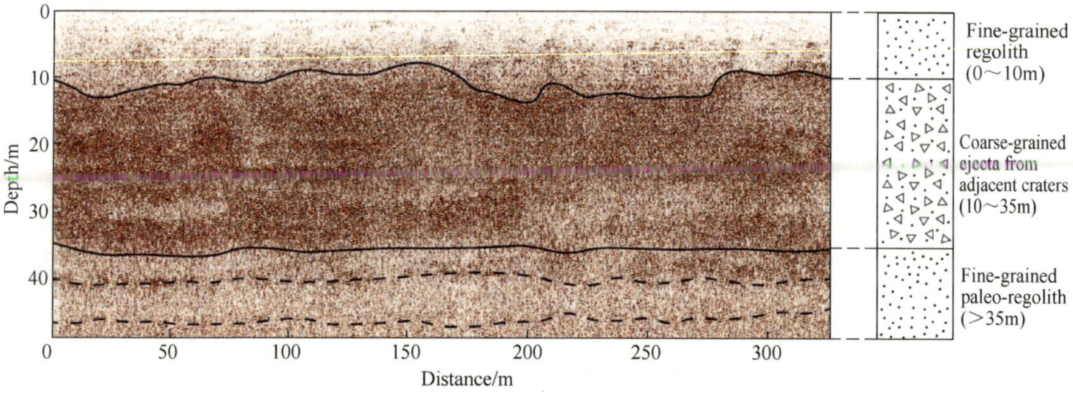

Figure 13　Stratigraphic divisions result and interpretation. The base of the figure is the radargram after CNN filtering. The vertical axis is the transformed depth from two - way traveling time with velocity assumed to 0. 16m/ns [11]. The radargram shows a fine - coarse - fine sandwich structure. The black curves indicate the interfaces of the sandwich structure. The dotted curves represent the interfaces of the sub - layers within the paleo - regolith

interface at ～48m is unclear but also continuous. Although the depths are beyond the estimated penetrating depth [50], the CNN filter can also make the weak signal visible. The top sublayer has a few reflections. This layer can be interpreted as a transition zone and may contain the mixture of the paleo - regolith material and the overlying coarse ejecta. The mixed material was formed during the placement of the overlying material. Furthermore, the continuous reflected signals within the fine - grained material at ～40m and ～48m are not likely to be the reflections from the boulders within the layers but the signals from horizontal interfaces of two layers with different dielectric properties. Therefore, the paleo - regolith contains three kinds of materials with different dielectric properties and sources. The materials were probably excavated from more distant and different craters relative to the local impactions. However, it is still difficult to find multiple sources of different dielectric properties within such a small region. Considering the geologic content and the depths of the layers, a possible interpretation is that these materials may have different contents of FeO and TiO_2, which are related to the magmatic activities during the Imbrian period [7 - 10, 14 - 17]. The material may be the ejecta containing the basalt material excavated from different nearby craters. Due to the different degrees of mixing with the local materials, the ejecta contains different contents of FeO and TiO_2, which results in the difference in dielectric properties. Besides, the formation time intervals of the three sub - layers are relatively longer than that of the first two layers (0 - 10m and 10 - 35m). Each sub - layer had been through long - time gardening before its overlying sub - layer was formed; this process of lunar surface homogenized the material and gradually formed the layered fine - grained paleo - regolith.

4 Discussions

4.1 Detailed and Quantitative Comparisons of Filtering Methods

In this section, we conducted detailed and quantitative comparisons of the methods mentioned in Section 3.3. The comparisons mainly focus on the denoising results of the LPR radargram below 35m. Figure 14a - e presents the radar profiles below 450 ns (>~35 m) with no filtering, mean filtering, band - pass filtering, band - pass and mean filtering, and CNN filtering, respectively. The mean filtering and band - pass filtering have almost no effect on the suppression of noise and clutter interference. The blue arrows point out the strips of interference on the radargrams; this interference will increase the difficulties of stratigraphic interpretation. The combined method of band - pass and mean filtering and the CNN method are the two methods that can present the clear interface at 500 ns. However, the radargram provided by the combined method is not clean compared to the result of CNN filtering and some strips of interference are still visible.

Subsequently, three regions (the black rectangles in Figure 14d, e) on the results of the combined method and CNN method are extracted and presented in Figure 15. Figure 15a - c corresponds to the regions 1 - 3 in Figure 14d, e, respectively. The red up - toward and down - toward arrows indicate the signals at 500 and 600ns, respectively. The results presented by the CNN filter are cleaner than that of the combined method. The events in radargram after CNN processing are more visible and continuous and can be well recognized. The signals on the results of the combined method, especially pointed by the down - toward arrows, are still masked by strong noise. Besides, the strips of interference pointed by the blue arrows are still evident on the results of the combined method but well suppressed on the CNN filtering results. Comparing the images in the rectangles, the CNN method reduces the effects of the distortion and well reconstructs and smooths the events.

Furthermore, an objective quantitative comparison is conducted based on the image entropy given by (4) in Section 3.1. The results are shown in Table 1. The image entropy of CNN filtering results are all lower than those of band - pass and mean filtering combined method. According to the analysis in Figures 5 - 7, Section 3.1, the low image entropy corresponds to low noise and clutter interference. Therefore, these results are consistent with the image analysis before and objectively verify the results.

Table 1　Image entropy comparison of the filtering results

Images	CNN Filtering	Band - Pass and Mean Filtering
Figure 14d, e	1.9075×10^6	2.1871×10^6
Figure 15a	5.7698×10^4	7.3160×10^4
Figure 15b	6.3342×10^4	7.4902×10^4
Figure 15c	5.5177×10^4	7.0545×10^4

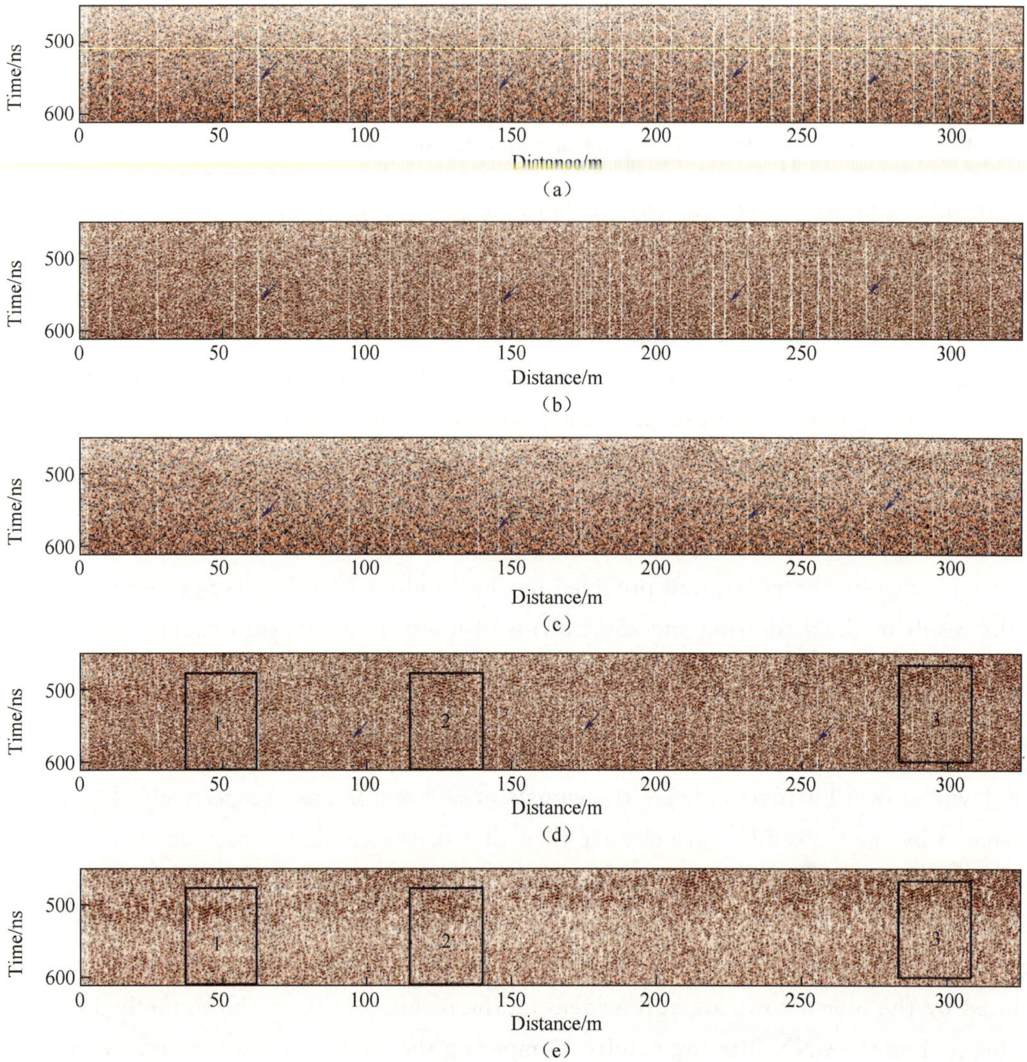

Figure 14　Comparisons of radar profiles below 450ns (depth below ～35m). (a) Original radargram with no filtering processing. (b) Radargram after mean filtering. (c) Radargram after band－pass filtering. (d) Radargram after the combined method of band－pass and mean filtering. (e) Radargram after CNN filtering. The blue arrows indicate the strips of interference. The three rectangular regions 1－3 are magnified and presented in Figure 15

4.2　Why the CNN Filter Performs Better?

Finally, we will discuss why the CNN filter performs better than the combined method of band－pass and mean filter. First, the spectra of the original data and the data after processing of the two methods are extracted from Figure 12e and presented in Figure 16. According to the spectra, the result of CNN filtering seems worse than the band－pass and mean filter; the curves of CNN result in the low － frequency and high － frequency

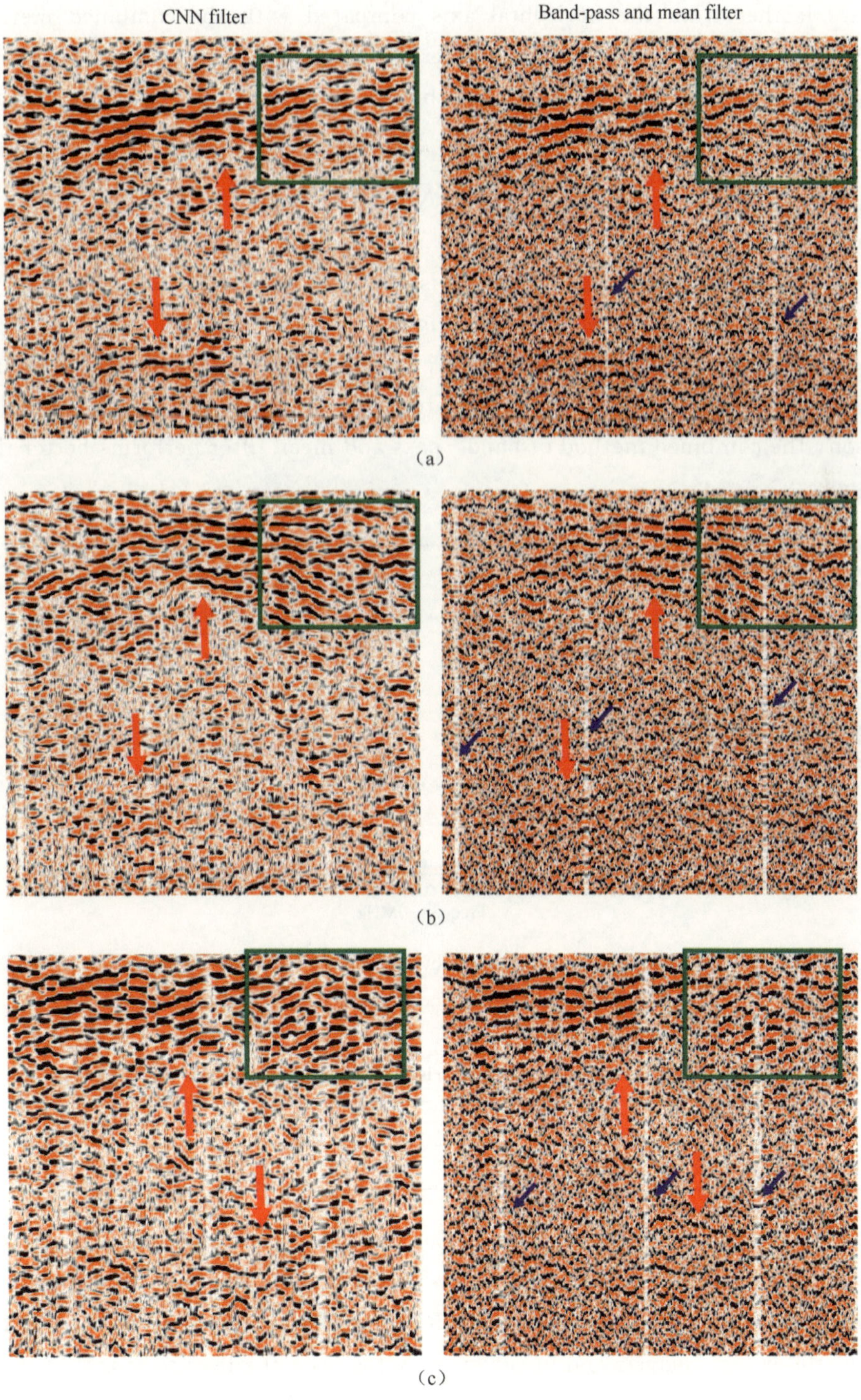

Figure 15　Detailed comparisons of the magnified images in the three rectangular regions of Figure 14d，e. (a) Region 1. (b) Region 2. (c) Region 3. The red up - toward and down - toward arrows indicate the signals at 500 and 600ns，respectively. The blue arrows point out the strips of interference

region are farther from the horizontal axis compared with the combined method. We further calculate the mean integral relative errors (MIRE) between the original signals and the denoised results. The MIRE is given by:

$$E = \frac{\int_f (a_O(f, x) - a_D(f, x))df}{N_x} \tag{5}$$

where a_O and a_D denote the original data and the denoised result in the $f-x$ domain, respectively. f represents the frequency. N_x is the trace number. The Evalue can present the difference between the spectra of the original data and the denoised result. In most cases, it is of a better denoising effect with a greater E value. The comparison of the E value is shown in Table 2. The results indicate that, according to the spectra and MIRE comparison, the combined method of band - pass and mean filter performs better than the CNN fillter.

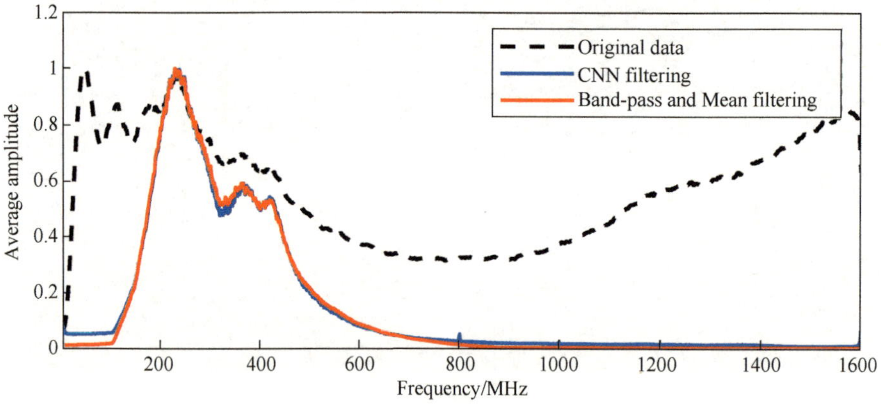

Figure 16　Detailed spectra comparison between the CNN filtering result and the combined band - pass and mean filtering result

Table 2　Mean integral relative error comparison of the filtering results in Figure 12d, e

CNN Filtering	Band - Pass and Mean Filtering
1.9278×10^3	2.4477×10^3

However, according to the image analysis and image entropy comparison, we have concluded that the CNN filter performs better. The results are contradictory but reasonable. The spectra analysis and the MIRE describe the features of the single trace signals but the image analysis and image entropy focus on the feature of the 2D images.

Essentially, a band - pass filter is a 1D filter that focuses on the spectra of single traces and cannot handle the matching of trace signals in the horizontal directions. Therefore, the band - pass filter cannot suppress the horizontal distortion such as the strips of interference in the radargram. By combining the mean filter, the result becomes better. However, the mean filter has limitations and cannot fully suppress the noise and clutter. According to the result in Figure 14b, the mean filter is sensitive to noise and should be performed after other filters like the band - pass filter. Besides, Figure 12e indicates that the mean filter has almost no attenuation ability for low frequency clutters which is because the effect of a mean filter depends on the size of a filtering template. Similarly, for the horizontal distortions with a low wave number such as the strips of interference, the mean filter will perform poorly.

The key of the CNN framework is the convolution layer which can be expressed as follows:

$$P_{in}(x, y) * w(x, y) = \sum_{x_0=a_1}^{a_2} \sum_{y_0=b_1}^{b_2} P_{in}(x_0, y_0) w(x - x_0, y - y_0) \quad (6)$$

where $P_{in}(x, y)$ and $w(x, y)$ are the input image and convolutional weight, respectively. a_1, a_2 and b_1, b_2 represent the horizontal and vertical ranges of the calculation region.

A CNN filter is essentially a 2D filter. The convolutional processing can extract the features of radargram and is performed for the whole 2D image in each layer. It can not only suppress the noise but also handle the matching of trace signals in the horizontal directions, which makes the events more evident, smoothing, and continuous in Figure 15. These advantages of CNN are the basis of the further geologic interpretations.

5 Conclusions

In this article, the denoising CNN framework is applied to the denoising of 500MHz LPR data. The low - frequency clutter extracted from CE - 3 LPR data using the EMD method is applied to the training of CNN. The results verify that the low - frequency clutters embed - ded in the LPR data mainly came from the instrument system of the Yutu rover. Besides, the frequency - based filter such as the classic band - pass filter cannot suppress the interference due to the magnification of ringing clutters as the signals and clutters have the same frequency bands. Furthermore, compared with the combined methods of the band - pass filter and the mean filter, the CNN filter has better performance when dealing with the noise interference and weak signal extraction; compared with the combined methods of the band - pass filter and the Kirchhoff migration, it can provide original high - quality radargram with diffraction information. Based on the high - quality radargram provided by the CNN filter, the subsurface sandwich structure is revealed and the weak signals from three sub - layers within the paleo - regolith are extracted. The materials in the three sub - layers probably contain different contents of FeO and TiO_2, which are related to the magmatic activities during the Imbrian period. The materials may be the ejecta containing the basalt material from nearby craters.

Acknowledgments

We thank the China National Space Administration for providing the scientific data used in this study. The scientific data are available at Data Publishing and Information Service System of China's Lunar Exploration Program (http://moon. bao. ac. cn (accessed on 19 January 2021)).

Appendix A

The appendix contains a Table of data processing procedures. The procedures are conducted to obtain the initial radargram in Figure 1c.

Table A1 Data processing procedures before filtering

Processing	Details
Traces amending	Adjusting the longitudinal displacement of traces, based on the phase of a strong reflection event.
Traces selecting	The rover might stop at some points on the way to collect other scientific data. But LPR never stops measurement, which resulted in the repeated acquisition of multiple traces at the same location. We average the repeated traces.
Time lag adjustment	There is a 28. 203 ns delay for the start time of the transmitting antenna compared to the receiving antenna.
Attenuation compensation	Conducting SEC makes deep information more visible.
Background removal	Removing the average background data.

References

[1] Liu, J.; Ren, X.; Yan, W.; Li, C.; Zhang, H.; Jia, Y.; Zeng, X.; Chen, W.; Gao, X.; Liu, D.; et al. Descent trajectory reconstruction and landing site positioning of Chang'E-4 on the lunar farside. Nat. Commun. 2019, 10, 4229.

[2] Haruyama, J.; Ohtake, M.; Matsunaga, T.; Morota, T.; Honda, C.; Yokota, Y.; Abe, M.; Ogawa, Y.; Miyamoto, H.; Iwasaki, A.; et al. Long-lived volcanism on the lunar farside revealed by SELENE Terrain Camera. Science 2009, 323, 905-908.

[3] Stuart-Alexander, D. E. Geologic Map of the Central Far Side of the Moon I-1047. In US Geological Survey; 1978. Available online: https://pubs. er. usgs. gov/publication/i1047 (accessed on 4 December 2019).

[4] Head, J. W., III; Fassett, C. I.; Kadish, S. J.; Smith, D. E.; Zuber, M. T.;

Neumann, G. A.; Mazarico, E. Global distribution of large lunar craters: Implications for resurfacing and impactor populations. Science 2010, 329, 1504-1507.

[5] Garrick-Bethell, I.; Zuber, M. T. Elliptical structure of the lunar South Pole-Aitken basin. Icarus 2009, 204, 399-408.

[6] Pascket, J. H.; Hiesinger, H.; van der Bogert, C. H. Lunar far side volcanism in and around the south pole Aitken basin. Icarus 2018, 299, 538-562.

[7] Huang, J.; Xiao, Z.; Flahaut, J.; Martinot, M.; Head, J.; Xiao, X.; Xie, M.; Xiao, L. Geological characteristics of Von Kármán crater, northwestern South Pole Aitken basin: Chang'E-4 landing site region. J. Geophys. Res. Planets 2018, 123, 1684-1700.

[8] Lai, J.; Xu, Y.; Bugiolacchi, R.; Meng, X.; Xiao, L.; Xie, M.; Liu, B.; Di, K.; Zhang, X.; Zhou, B.; et al. First look by the Yutu-2 rover at the deep subsurface structure at the lunar farside. Nat. Commun. 2020, 11, 3426.

[9] Zhang, J.; Zhou, B.; Lin, Y.; Zhu, M.; Song, H.; Dong, Z.; Gao, Y.; Di, K.; Yang, W.; Lin, H.; et al. Lunar regolith and substructure at Chang'E-4 landing site in South Pole-Aitken basin. Nat. Astron. 2020.

[10] Zhang, L.; Li, J.; Zeng, Z.; Xu, Y.; Liu, C.; Chen, S. Stratigraphy of the Von Kármán Crater based on Chang'E-4 lunar penetrating radar data. Geophys. Res. Lett. 2020, 47, e2020GL088680.

[11] Li, C.; Su, Y.; Pettinelli, E.; Xing, S.; Ding, C.; Liu, J.; Ren, X.; Lauro, S. E.; Soldovieri, F.; Zeng, X.; et al. The Moon's farside shallow subsurface structure unveiled by Chang'E-4 lunar penetrating radar. Sci. Adv. 2020, 6, eaay6898.

[12] Gou, S.; Yue, Z.; Di, K.; Cai, Z.; Liu, Z.; Niu, S. Absolute model age of lunar Finsen crater and geologic implications. Icarus 2021, 354, 115829.

[13] Ding, C.; Xiao, Z.; Wu, B.; Li, Y.; Prieur, N. C.; Cai, Y.; Su, Y.; Cui, J. Fragments delivered by secondary craters at the Chang'E-4 landing site. Geophys. Res. Lett. 2020, 47, e2020GL087361.

[14] Di, K.; Zhu, M.; Yue, Z.; Lin, Y.; Wan, W.; Liu, Z.; Gou, S.; Liu, B.; Peng, M.; Wang, Y. Topographic evolution of Von Kármán crater revealed by the lunar rover Yutu-2. Geophys. Res. Lett. 2019, 46, 12764-12770.

[15] Hu, X.; Ma, P.; Yang, Y.; Zhu, M.; Jiang, T.; Lucey, P. G.; Sun, L.; Zhang, H.; Li, C.; Xu, R.; et al. Mineral abundances inferred from in situ reflectance measurements of Chang'E-4 landing site in South Pole-Aitken basin. Geophys. Res. Lett. 2019, 46, 9439-9447.

[16] Qiao, L.; Ling, Z.; Fu, X.; Li, B. Geological characterization of the Chang'e-4 landing area on the lunar farside. Icarus 2019, 333, 37-51.

[17] Ling, Z.; Qiao, L.; Liu, C.; Cao, H.; Bi, X.; Lu, X.; Zhang, J.; Fu, X.; Li,

B.; Liu, J. Composition, mineralogy and chronology of mare basalts and non-mare materials in Von Kármán crater: Landing site of the Chang'E-4 mission. Planet. Space Sci. 2019, 179.

[18] Li, C.; Liu, D.; Liu, B.; Ren, X.; Liu, J.; He, Z.; Zuo, W.; Zeng, X.; Xu, R.; Tan, X.; et al. Chang'E-4 initial spectroscopic identification of lunar far side mantle-derived materials. Nature 2019, 569, 378-382.

[19] Gou, S.; Di, K.; Yue, Z.; Liu, Z.; He, Z.; Xu, R.; Lin, H.; Liu, B.; Peng, M.; Wan, W.; et al. Lunar deep materials observed by Chang'e-4 rover. Earth Planet. Sci. Lett. 2019, 528, 115829.

[20] Gou, S.; Yue, Z.; Di, K.; Wang, J.; Wan, W.; Liu, Z.; Liu, B.; Peng, M.; Wang, Y.; He, Z.; et al. Impact melt breccia and surrounding regolith measured by Chang'e-4 rover. Earth Planet. Sci. Lett. 2020, 544, 116378.

[21] Ma, P.; Sun, Y.; Zhu, M.; Yang, Y.; Hu, X.; Jiang, T.; Zhang, H.; Lucey, P. G.; Xu, R.; Li, C.; et al. A plagioclase-rich rock measured by Yutu-2 rover in Von Kármán crater on the far side of the Moon. Icarus 2020, 350, 113901.

[22] Jia, Y.; Zou, Y.; Ping, J.; Xue, C.; Yan, J.; Ning, Y. The Scientific Objectives and Payloads of Chang'E-4 Mission. Planet. Space Sci. 2018, 38, 118-130.

[23] Fang, G.; Zhou, B.; Ji, Y.; Zhang, Q.; Shen, S.; Li, Y.; Guan, H.; Tang, C.; Gao, Y.; Lu, W.; et al. Lunar Penetrating Radar onboard the Chang'e-3 mission. Res. Astron. Astrophys. 2014, 14, 1607-1622.

[24] Su, Y.; Fang, G.; Feng, J.; Xing, S.; Ji, Y.; Zhou, B.; Gao, Y.; Li, H.; Dai, S.; Xiao, Y.; et al. Data processing and initial results of Chang'e-3 lunar penetrating radar. Res. Astron. Astrophys. 2014, 14, 1623-1632.

[25] Fa, W.; Zhu, M.; Liu, T.; Plescia, J. B. Regolith stratigraphy at the Chang'E-3 landing site as seen by lunar penetrating radar. Geophys. Res. Lett. 2015, 42, 10179-10187.

[26] Zhang, J.; Yang, W.; Hu, S.; Lin, Y.; Fang, G.; Li, C.; Peng, W.; Zhu, S.; He, Z.; Zhou, B.; et al. Volcanic history of the Imbrium basin: A close-up view from the lunar rover Yutu. Proc. Natl. Acad. Sci. USA 2015, 112, 5342-5347.

[27] Zhang, L.; Zeng, Z.; Li, J.; Huang, L.; Huo, Z.; Zhang, J.; Huai, N. A Story of Regolith Told by Lunar Penetrating Radar. Icarus 2019, 321, 148-160.

[28] Xiao, L.; Zhu, P.; Fang, G.; Xiao, Z.; Zhu, Y.; Zhao, J.; Zhao, N.; Yuan, Y.; Qiao, L.; Zhang, X.; et al. A young multilayering terrane of the northern Mare Imbrium revealed by Chang'E-3 mission. Science 2015, 347, 1226-1229.

[29] Ding, C.; Xiao, Z.; Su, Y.; Zhao, J.; Cui, J. Compositional variations along the route of Chang'e-3 Yutu rover revealed by the lunar penetrating radar. Prog. Earth Planet. Sci. 2020, 7.

[30] Zhang, L.; Zeng, Z.; Li, J.; Huang, L.; Huo, Z.; Wang, K.; Zhang, J. Parameter Estimation of Lunar Regolith from Lunar Penetrating Radar

Data. Sensors 2018, 18, 2907.

[31] Hu, B. ; Wang, D. ; Zhang, L. ; Zeng, Z. Rock Location and Quantitative Analysis of Regolith at the Chang'e 3 Landing Site Based on Local Similarity Constraint. Remote Sens. 2019, 11, 530.

[32] Song, H. ; Li, C. ; Zhang, J. ; Wu, X. ; Liu, Y. ; Zou, Y. Rock Location and Property Analysis of Lunar Regolith at Chang'E-4 Landing Site Based on Local Correlation and Semblance Analysis. Remote Sens. 2021, 13, 48.

[33] Zhang, L. ; Hu, B. ; Jia, Z. ; Xu, Y. ; Huang, L. The Subsurface Structure on the CE-3 Landing Site: LPR CH-1 Data Processing by Shearlet Transform. Pure Appl. Geophys. 2020, 177, 3459-3474.

[34] Feng, J. ; Su, Y. ; Ding, C. ; Xing, S. ; Dai, S. ; Zou, Y. Dielectric properties estimation of the lunar regolith at CE-3 landing site using lunar penetrating radar data. Icarus 2017, 284, 424-430.

[35] Zhang, L. ; Zeng, Z. ; Li, J. ; Lin, J. ; Hu, Y. ; Wang, X. ; Sun, X. Simulation of the lunar regolith and lunar penetrating radar data processing. IEEE J. Sel. Top. Appl. Earth Observ. Remote Sens. 2018, 11, 655-663.

[36] Zhang, L. ; Zeng, Z. ; Li, J. ; Lin, J. ; Zhang, J. Lunar Penetrating Radar Data Processing and Analysis Based on CEEMD. In Proceedings of the 17th International Conference on Ground Penetrating Radar (GPR), Rapperswil, Switzerland, 18-21 June 2018; pp. 635-638.

[37] Zhang, J. ; Zeng, Z. ; Zhang, L. ; Lu, Q. ; Wang, K. Application of Mathematical Morphological Filtering to Improve the Resolution of Chang'E-3 Lunar Penetrating Radar Data. Remote Sens. 2019, 11, 524.

[38] Dong, Z. ; Feng, X. ; Zhou, H. ; Lu, Q. ; Liu, C. ; Zeng, Z. ; Li, J. ; Liang, W. Date Processing for Lunar Penetrating Radar of Chang'E-4 based on Variational Mode Decomposition. In Proceedings of the 18th International Conference on Ground Penetrating Radar (GPR), Golden, CO, USA, 14-19 June 2020; pp. 424-427.

[39] Hubel, D. H. ; Wiesel, T. N. Receptive fields and functional architecture of monkey striate cortex. J. Physiol. 1968, 195, 1.

[40] LeCun, Y. ; Bottou, L. ; Bengio, Y. ; Haffner, P. Gradient-Based Learning Applied to Document Recognition. Proceed. IEEE 1998, 86, 11.

[41] Krizhevsky, A. ; Sutskever, I. ; Hinton, G. E. ImageNet classification with deep convolutional neural networks. Adv. Neural inf. Process. Syst. 2012, 1097-1105.

[42] Zhang, K. ; Zuo, W. ; Chen, Y. ; Meng, D. ; Zhang, L. Beyond a Gaussian Denoiser: Residual Learning of Deep CNN for Image Denoising. IEEE Trans. Image Process. 2017, 26, 7.

[43] Zhang, K. ; Zuo, W. ; Zhang, L. FFDNet: Toward a Fast and Flexible Solution for CNN-Based Image Denoising. IEEE Trans. Image Process. 2018, 27, 9.

[44] Huang, N. E. ; Shen, Z. ; Long, S. ; Wu, M. ; Shih, H. ; Zheng, Q. ; Yen, N. ;

Tung, C. ; Liu, H. The empirical mode decomposition and the Hilbert spectrum for nonlinear and non - stationary time series analysis. Proc. R. Soc. Lond. Seri. AMath. Phys. Eng. Sci. 1998, 454, 903 - 995.

[45] Wiggins, R. Minimum entropy deconvolution. Geoexploration 1978, 16, 21 - 35.

[46] Wu, H. ; Barba, J. Minimum entropy restoration of star field images. IEEE Trans. Syst. Man Cybernet. 1998, 28, 227 - 231.

[47] Xu, X. ; Miller, E. L. ; Rappaport, C. R. Minimum Entropy Regularization in Frequency - Wavenumber Migration to Localize Subsurface Objects. IEEE Trans. Geosci. RemoteSens. 2003, 41, 8.

[48] Stockwell, R. G. ; Man - sinha, L. ; Lowe, R. P. Localization of the Complex Spectrum: The S Transform. IEEE Trans. Signal Process. 1996, 44, 998 - 1001.

[49] Dong, Z. ; Fang, G. ; Zhao, D. ; Zhou, B. ; Gao, Y. ; Ji, Y. Dielectric Properties of Lunar Subsurface Materials. Geophys. Res. Lett. 2020 , 47.

[50] Lai, J. ; Xu, Y. ; Zhang, X. ; Xiao, L. ; Yan, Q. ; Meng, X. ; Zhou, B. ; Dong, Z. ; Zhao, D. Comparison of dielectric properties and structure of lunar regolith at Chang'e - 3 and Chang'e - 4 landing sites revealed by ground - penetrating radar. Geophys. Res. Lett. 2019, 46, 12783 - 12793.

[51] Lai, J. ; Xu, Y. ; Zhang, X. ; Tang, Z. Structural analysis of lunar subsurface with Chang'E - 3 lunar penetrating radar. Planet. Space Sci. 2016, 120, 96 - 102.

[52] Dong, Z. ; Feng, X. ; Zhou, H. ; Liu, C. ; Zeng, Z. ; Li, J. ; Liang, W. Properties Analysis of Lunar Regolith at Chang'E - 4 Landing Site Based on 3D Velocity Spectrum of Lunar Penetrating Radar. Remote Sens. 2020, 12, 629.

[53] Xiao, Z. ; Ding, C. ; Xie, M. ; Cai, Y. ; Cui, J. ; Zhang, K. ; Wang, J. Ejecta from the Orientale basin at the Chang' E - 4 landing site. Geophys. Res. Lett. 2021, e2020GL090935.

嫦娥4号着陆区冯·卡门撞击坑溅射物的识别及年代分析

李圣卉[1,2]，刘建忠[1,3]，刘敬稳[1]，王俊涛[1]，张敬宜[1,2]，傅泽华[4]，张珂[1,2]

LI ShengHui[1,2]，LIU JianZhong[1,3]，LIU JingWen[1]，WANG JunTao[1]，ZHANG JingYi[1,2]，FU ZeHua[4] and ZHANG Ke[1,2]

1. 中国科学院地球化学研究所月球与行星科学研究中心，贵阳 550081
2. 中国科学院大学，北京 100049
3. 中国科学院比较行星学卓越创新中心，合肥 230026
4. 澳门科技大学，澳门 999078
1. Center for Lunar and Planetary Sciences，Institute of Geochemistry，Chinese Academy of Sciences，Guiyang 550081，China
2. University of Chinese Academy of Sciences，Beijing 100049，China
3. Center for Excellence in Comparative Planetology，Chinese Academy of Sciences，Hefei 230026，China
4. Macau University of Science and Technology，Macau 999078，China

摘要： 撞击坑是月球、火星和水星等固态行星表面重要的地貌和构造，其形成时产生的溅射物包含了大量地质信息，是认识撞击坑形成过程和区域地质演化的重要研究对象。本文以冯·卡门撞击坑为研究对象，结合月球勘测轨道飞行器（LRO）影像数据、激光高度计（LOLA）地形数据、Clementine 多光谱数据等对冯·卡门撞击坑的溅射物进行综合解析。冯·卡门撞击坑的溅射物后期改造比较严重，只有南部区域保存相对完整。选取连续溅射物中受其他撞击坑溅射物影响较小的区域，并以宽角相机（WAC）数据和高分辨率地形相机（TC）数据为基础，利用撞击坑提取工具 CraterTools 对均一的区域进行撞击坑统计分析，进而利用月球产率和年代学函数推导出绝对模型年龄，获得了冯·卡门撞击坑的年龄为 4.15Ga。本文主要强调了多源数据在识别溅射物时的重要性，以及年代分析时排除周围溅射物影响的必要性。

关键词： 冯·卡门撞击坑；多源数据；溅射物；年代分析

Abstract： Impact craters are a type of important geological landforms on the surfaces of solid planets such as the Moon，Mars and Mercury. The impact ejecta provides valuable geological information for understanding the impacting process and regional geological evolution of the planet. In this study，the ejecta of Von Kármán crater was investigated by combining Lunar Reconnaissance Orbiter（LRO）image data，Lunar Orbiter Laser Altimeter（LOLA）topographic data and Clementine multi - spectral data. Our results indicate that most of the ejecta within Von Kármán crater have been seriously alternated. The continuous ejecta that minor affected by other impact craters were selected and analyzed by using the Crater - Tools, based on Wide Angle Camera (WAC) and Terrain Camera (TC) image data. Then, the dating analysis was carried out, yielded an age of 4.15Ga (i.e., pre - Nectarian) for Von Kármán crater. Our studies demonstrate that multi - source data is useful in the identification of ejecta and the necessity of excluding the influence of surrounding ejecta in the dating analysis.

Keywords: Von Kármán crater; Multi - source data; Ejecta; Chronological analysis

撞击坑是研究月球必不可少的一类地质单元，其几乎覆盖了整个月球（Osinski *et al.*，2011）。撞击成坑作为一个重要而独特的地质过程，对月球表面的地质特征产生了很大的影响，其产生的溅射物的影响范围更为广泛（Alvarez *et al.*，1980；Collins *et al.*，2012；Öhman and Kring，2012）。根据其形态特征，撞击坑的内部从外向内依次为坑壁、坑底、中央峰（撞击坑类型不同而有所差别），而撞击坑外部则普遍分布着大量的撞击溅射物，包括连续溅射物和不连续溅射物。其中，连续溅射物是撞击坑形成过程中被挖掘的物质向四周抛射形成，又被称为坑缘，通常分布于撞击坑的最高点至撞击坑外围平坦区域之间的地带；不连续溅射物则是指溅射沉积及次生产物（Osinski *et al.*，2011；Collins *et al.*，2012；Glass and Simonson，2013；丁孝忠等，2014；郭弟均等，2016）。溅射物不仅能为认识撞击坑内部物质成分提供信息，还是研究地层序列、撞击坑形成过程和区域地质演化的重要对象。

冯·卡门撞击坑位于月球背面南极-艾肯（South Pole - Aitken，SPA）盆地的西北部（图 1），其中心坐标为：44.45°S、176.25°E（Qiao *et al.*，2019），直径约为 186km（Huang *et al.*，2018；Ling *et al.*，2019），内部被玄武岩填充（Stuart - Alexander，1978；Pasckert *et al.*，2018），是嫦娥 4 号的着陆区域（吴伟仁等，2017），也是目前月球背面唯一具有就位探测数据的区域。在冯·卡门撞击坑及其周边区域，地形复杂，撞击坑数量多且相互叠加、相切，没有明确的界线，各个撞击坑的溅射物相互混合，交错分布，难以区分（Qiao *et al.*，2019）。

在以往的地质图编研中，对冯·卡门撞击坑的溅射物进行了提取，如 1∶500 万的月球背面中心地质图中，以 Apollo 数据为基础，提取的冯·卡门撞击坑的溅射物主要集中于西南部（Stuart - Alexander，1978）；在 1∶250 万的 Planck 区域地质图中，利用影像数据和高程数据识别出的冯·卡门撞击坑的溅射物在四周都有分布（Yingst *et al.*，2017）。但是识别溅射物时利用的数据较为单一，且提取的溅射物以坑缘为主，还包括一部分坑壁物质，并未将溅射物进行连续与不连续的划分。

撞击坑年龄的判定方法，从早期的撞击坑叠置法，到后期的撞击坑大小-频率统计以及如今的绝对模型年龄，经历了不断的演化和提升。但对冯·卡门撞击坑的年龄一直存在争议，主要存在两种观点：一是冯·卡门撞击坑的年龄为酒海纪。在早期的地质图绘制中，人们利用撞击坑相互之间的叠置关系，初步判定其为雨海纪和酒海纪年龄之间（Stuart - Alexander，1978），之后的学者根据判定的月海的年龄支持冯·卡门撞击坑是酒海纪撞击坑（Pasckert *et al.*，2018；Ling *et al.*，2019）。另一种则认为冯·卡门撞击坑

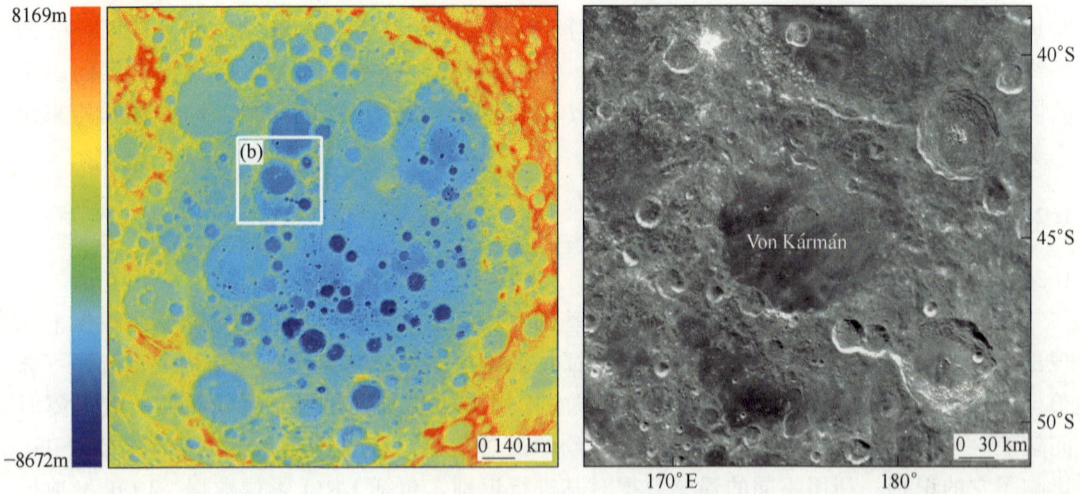

(a)　　　　(b)

图 1　南极-艾肯盆地的数字高程模型（a）及嫦娥 4 号着陆区冯·卡门撞击坑（b）

Fig. 1　Digital elevation model of the South Pole - Aitken Basin (a) and the Von Kármán crater (b) where the Chinese Chang'e - 4 landed

是前酒海纪撞击坑，在《月球的地质历史》一书中基于早期数据，利用大小-频率统计将年龄定为前酒海纪（Wilhelms *et al*.，1987）；在之后的 Planck 区域地质图中，利用高清影像数据并结合撞击坑大小-频率分布统计定年法对冯·卡门撞击坑进行定年，获得年龄为 3.97Ga，即前酒海纪（Yingst *et al*.，2017）；目前，此观点被大多数人所支持（Huang *et al*.，2018）。虽然前人对冯·卡门撞击坑已开展了一定的分析研究，但由于对该区域进行撞击坑统计定年所使用的数据分辨率有限，并且选择的撞击坑的大小及分布区域也存在一定的影响，因此对该区域进行定年的精确度有待进一步提高。

本文将以冯·卡门撞击坑为研究对象，结合影像数据、地形数据、光谱数据对冯·卡门撞击坑的溅射物进行识别，充分发挥各类数据的优势，进行综合解析，提取连续溅射物以及不连续溅射物；然后根据其周边撞击坑的溅射物的影响程度对冯·卡门撞击坑连续溅射物进行单元划分，以 WAC 和 TC 数据为基础提取撞击坑进行定年分析，获得更加准确的冯·卡门撞击坑的年龄。可为溅射物的模拟提供参考数据，为深入认识月球撞击过程和地层划分等提供约束条件，为此区域的地质演化过程乃至整个 SPA 盆地的地质演化过程提供参考。

1　数据和方法

1.1　数据

本文使用的数据包括：(1) 日本月亮女神地形相机（TC）数据（Kato *et al*.，2010），分辨率为 10m/pixel，数据来源于 SELENE Data Archive；(2) 美国月球勘测轨道飞行器（LRO）的宽角相机数据（WAC），分辨率为 100m/pixel（Robinson *et al.*，2010）；(3) LRO 的激光高度计获得的月球 DEM（Digital Elevation Model）数据，分辨率约为 59m/pixel（Smith *et al.*，2010）；(4) Clementine 的 UVVIS 多光谱数据的多个波段，分辨率为 100m/pixel（Nozette *et al.*，1994）。其中（2）、（3）、（4）数据均来源于美国 PDS 行星数据节点。

1.2 方法

1.2.1 溅射物的识别方法

遥感探测数据是进行月球地质解译的基础数据，不同的地质结构在遥感探测数据上呈现出不同的形态、纹理和反照率等信息特征，借助这些特征信息对地质结构进行解译（郭弟均等，2016）。溅射物是撞击坑形成过程中挖掘出的物质，相对于背景物质，其出露时间较短，年龄较新，呈现出明显不同于背景的遥感信息，但是由于后期的破坏较为严重，影响了它的识别。利用不同的遥感数据对其进行识别，包括 LRO 影像数据、LOLA 地形数据、Clementine 多光谱数据和计算获得的光学成熟度（optical maturity parameter，OMAT）（Pieters *et al.*，1994；Lucey *et al.*，2000）数据，同时进行多数据融合识别，最后选择最优的结果。

光学成熟度的计算方法为（Lucey *et al.*，2000）：

$$\mathrm{OMAT}=\sqrt{(R_{750}-x_0)^2+\left(\frac{R_{950}}{R_{750}}-y_0\right)^2} \tag{1}$$

R_{750} 表示波长为 750nm 的光在一个象元或区域中的反射率，R_{950} 表示波长为 950nm 的光在一个象元或区域中的反射率，x_0 和 y_0 在 Clementine 多光谱数据中的值分别为 0.08 和 1.19。OMAT 值越大表示该象元或区域越不成熟，形成时间越晚。

1.2.2 定年方法

常用的定年方法包括相对定年和绝对定年两种方法。其中，相对定年主要有地层叠置法、撞击坑形态法、撞击坑大小-频率法、月壤成熟度法（郭弟均等，2014）；绝对定年则分为样品同位素测年法和绝对模型年龄法。本文主要使用地层叠置法、撞击坑大小-频率法和绝对模型年龄法，具体介绍如下：

地层叠置法：应用 Nicolaus Steno（1669）提出的地层三定律，即叠加定律（Harris，1975）、原始水平律和原始连续律（Harris，1979）进行相对定年分析，获得撞击坑的相对年龄。

撞击坑大小-频率分布统计法（crater size - frequency distributions，CSFDs）：此方法基于两个假设，一是在地理位置上，撞击坑的形成是随机分布的，具有随机性；二是在时间尺度上，撞击坑被破坏的速度要远远小于其形成的速度。在此基础上，同一个天体表面，原始撞击坑密度越大的区域，年龄越老。此种方法具有累积分布和相对分布两种表达形式（Crater Analysis Techniques Working Group，1979；Neukum *et al.*，2001）。

绝对模型年龄法（absolute model ages，AMAs）：结合样品年龄和采样区撞击坑分布获得产率函数，再利用年代学函数，获得绝对模型年龄。利用 CSFDs 确定绝对模型年龄包括两个基本步骤：(1) 识别均一地质单元，并测量其面积；(2) 精确识别并量测该区域内主撞击坑的数目与直径（Hartmann，1970；Neukum *et al.*，1975，2001；Neukum，

1983；Hiesinger *et al*. ，2012）。

利用撞击坑提取工具 CraterTools（Kneissl *et al*. ，2011）对均一地质区域进行 CSFDs 测量。将获得的撞击坑的数量和直径输出到 Craterstats II（Michael and Neukum，2010；Michael *et al*. ，2012，2016；Michael，2013），再利用（Neukum *et al*.，2001）月球产率和年代学函数推导出 AMAs，结果以累积分布图的形式呈现。

2　结果

2.1　冯·卡门撞击坑的溅射物

图 2 分别呈现了冯·卡门撞击坑的溅射物在 WAC 影像、DEM 影像、Clementine UVVIS 750nm 反射率影像和光学成熟度影像的界线特征。从 WAC 影像（图 2a）可以清楚地观察到撞击溅射物的纹理，尤其是冯·卡门撞击坑的东部区域，但由于冯·卡门撞击坑周围地形复杂，撞击坑之间相互叠加，严重影响了撞击溅射物界线的判别。在 DEM 数据中（图 2b），冯·卡门撞击坑的溅射物的界线在南部可以辨认，其他区域无明显边界，无法详细区分。在 Clementine UVVIS 750nm 反射率影像（图 2c）中，溅射物具有较高的反射率，比背景更加明亮，形成了一定的差别，但是溅射物的界线仍旧较为模糊。在 OMAT 影像（图 2d）中溅射物与背景形成强烈对比，在撞击坑周边的光学成熟度具有向外延展的趋势，可根据这种趋势识别溅射物。

为了更加准确地识别出冯·卡门撞击坑的连续溅射物和不连续溅射物，本文将结合不同类型的数据来识别其溅射物。针对连续溅射物，本文利用由 DEM 影像提取的坡度图（图 3a）和 OMAT 影像图（图 2d）初步识别冯·卡门撞击坑的连续溅射物，并与周边撞击坑溅射物进行区分，进而通过 WAC 影像的纹理、结构特征提取边界部分（图 3b），最终结合两种影像的优势更好地识别连续溅射物。而针对不连续溅射物，首先将 750nm 反射率影像和光学成熟度影像进行叠加（图 3c），使不连续溅射物的分布特征更加明显，从而识别不连续溅射物的整体轮廓；再分类显示 OMAT 影像（图 3d），使不连续溅射物的边界信息更加清晰易分辨，进而识别不连续溅射物的细节部分。

利用上述方法对冯·卡门撞击坑的解译结果如图 4a 所示，从外向内依次为不连续溅射物、连续溅射物和坑内物质（坑壁、坑底、中央峰），其中坑内物质主要用来判定连续溅射物的内部界线，为将冯·卡门撞击坑的溅射物进行完整表达，提取的溅射物存在部分推测。冯·卡门撞击坑的连续溅射物以东北-西南方向线为界，其西北侧连续溅射物保留的较窄，而东南侧则较宽；不连续溅射物集中在南部区域，即冯·卡门 M 撞击坑内的玄武岩上。

为了能够更加准确地分析冯·卡门撞击坑溅射物的分布并获得其连续溅射物的均一地质单元，本文将冯·卡门撞击坑周边的撞击坑的溅射物也进行了识别（图 4b），其中 Leibnitz、Alder 和 Finsen 撞击坑的不连续溅射物可识别，其他撞击坑的均未保留。从东南到西部的一系列小型撞击坑规模虽小，但是由于其分布在冯·卡门撞击坑的连续溅射物上，也对冯·卡门撞击坑溅射物产生了一定的影响，这些撞击坑内部结构简单，连续溅射物保存也都相对完整，只有东南方向的两个撞击坑冯·卡门 L 和冯·卡门 L'（Huang *et*

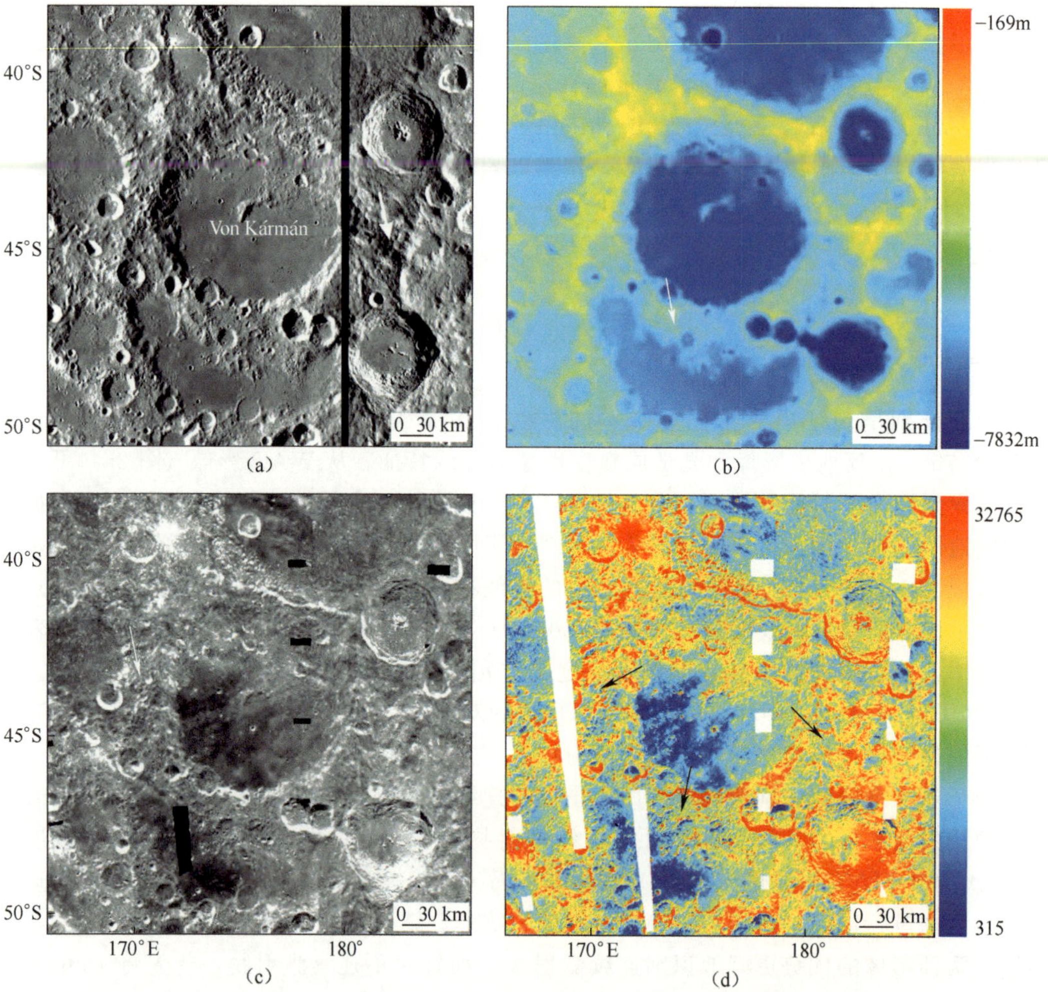

图 2　冯·卡门撞击坑的溅射物在不同数据中的展现

(a) WAC 影像中，白色箭头处显示清晰的溅射物纹理；(b) DEM 影像中，白色箭头处显示溅射物的界线；(c) UVVIS 750nm 反射率影像中，溅射物具有较高的反射率；(d) OMAT 影像中，黑色箭头标出撞击坑周边具有向外延展的趋势

Fig. 2　The occurrences of different types of data in the ejecta from the Von Kármán crater

(a) LROC WAC image mosaic of the Von Kármán area: the white arrow shows the ejecta texture; (b) DEM of the same area: the white arrow shows the boundary of the ejecta; (c) UVVIS 750nm reflectance image of the same area: the ejecta has a higher reflectivity; (d) The OMAT image: the black arrows show the surrounding of the impact crater has a tendency to extend outward

al. , 2018）存在部分叠加，溅射物混合严重，难以区分；位于西部的 Cajori 和 Oresme 撞击坑破坏严重，溅射物保留非常有限，其中 Oresme 撞击坑距离冯·卡门撞击坑比较近，其连续溅射物部分覆盖在冯·卡门撞击坑溅射物上，Cajori 撞击坑距离冯·卡门撞击坑较远，其连续溅射物对冯·卡门撞击坑的影响不显著；位于北部的 Leibnitz 撞击坑明显覆盖了部分冯·卡门撞击坑区域，其溅射物甚至覆盖了部分冯·卡门撞击坑的内部，对冯·卡门撞击坑影响较大；东北部的 Finsen 撞击坑保留完整，其溅射物具有明显的地形隆起和放射性纹理，次级坑长链分布广泛，这些溅射物不同程度地覆盖了周围的撞击坑；

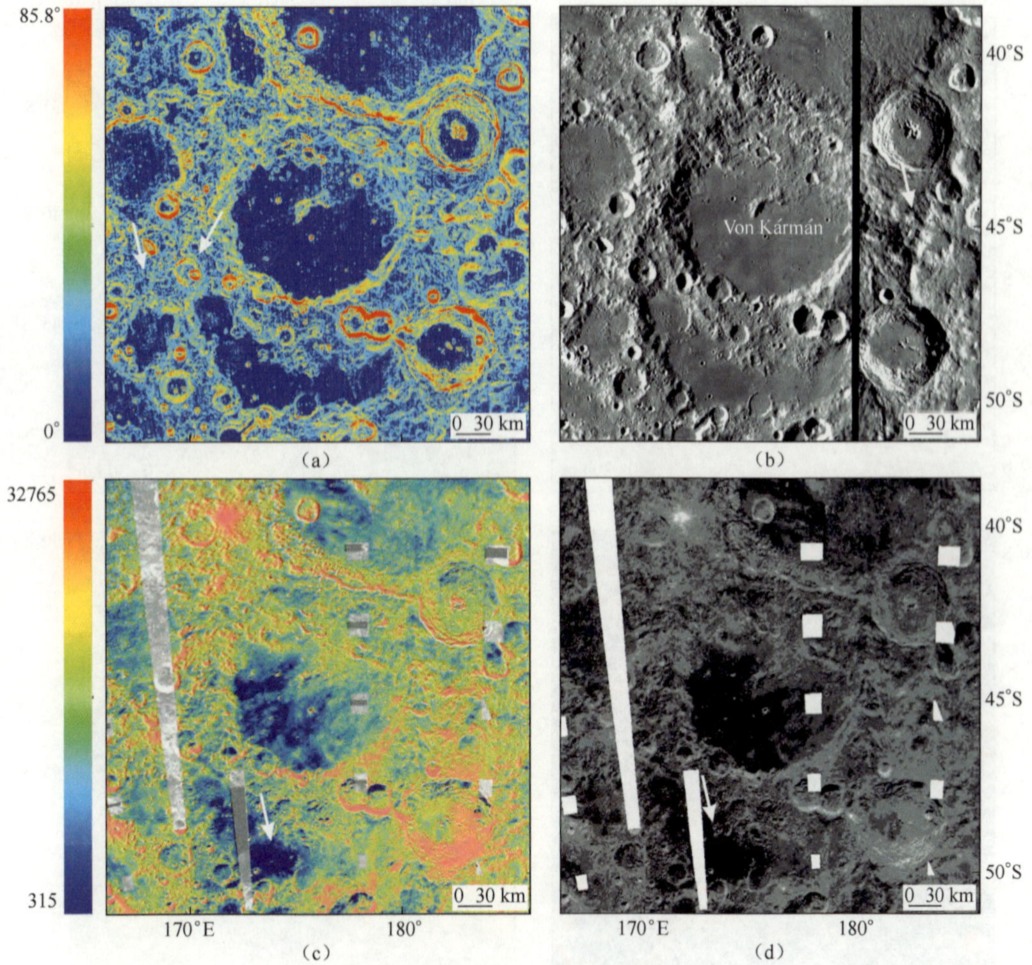

图 3　溅射物在处理后数据中的展现

(a) 坡度图中，不同撞击坑的溅射物呈现不同的坡度走向（白色箭头）；(b) WAC 影像图呈现纹理特征；
(c) 750nm 反射率影像与 OMAT 影像叠加图，白色箭头指出不连续溅射物边界；
(d) OMAT 影像分类图，白色箭头指出溅射物边界结构化

Fig. 3　The processed data of the ejecta from the Von Kármán crater

(a) slope image of the Von Kármán area, the ejecta from different craters shows different slope trends (white arrow); (b) LROC WAC image mosaic shows the texture of ejecta; (c) superposition image of 750nm reflectivity image and OMAT image, the white arrow shows the boundary of discontinuous ejecta; (d) classification of OMAT image, the white arrow shows the ejecta boundary structuring

位于东南部的 Alder 撞击坑保留较为完整，其溅射物在四周都有分布，较易识别，其西北部受小型撞击坑影响较大。

基于对周围撞击坑溅射物的提取结果，发现冯·卡门撞击坑的连续溅射物以南部保存最为完整，几乎没有受到后期大型撞击的影响；东部保存虽相对完整，但是受到 Finsen 和 Alder 撞击坑的影响，尤其是 Finsen 撞击坑溅射物的影响，混合较为严重；而在东北部和北部区域，连续溅射物基本没有保留，都被后期形成的 Leibnitz 和 Finsen 撞击坑所覆

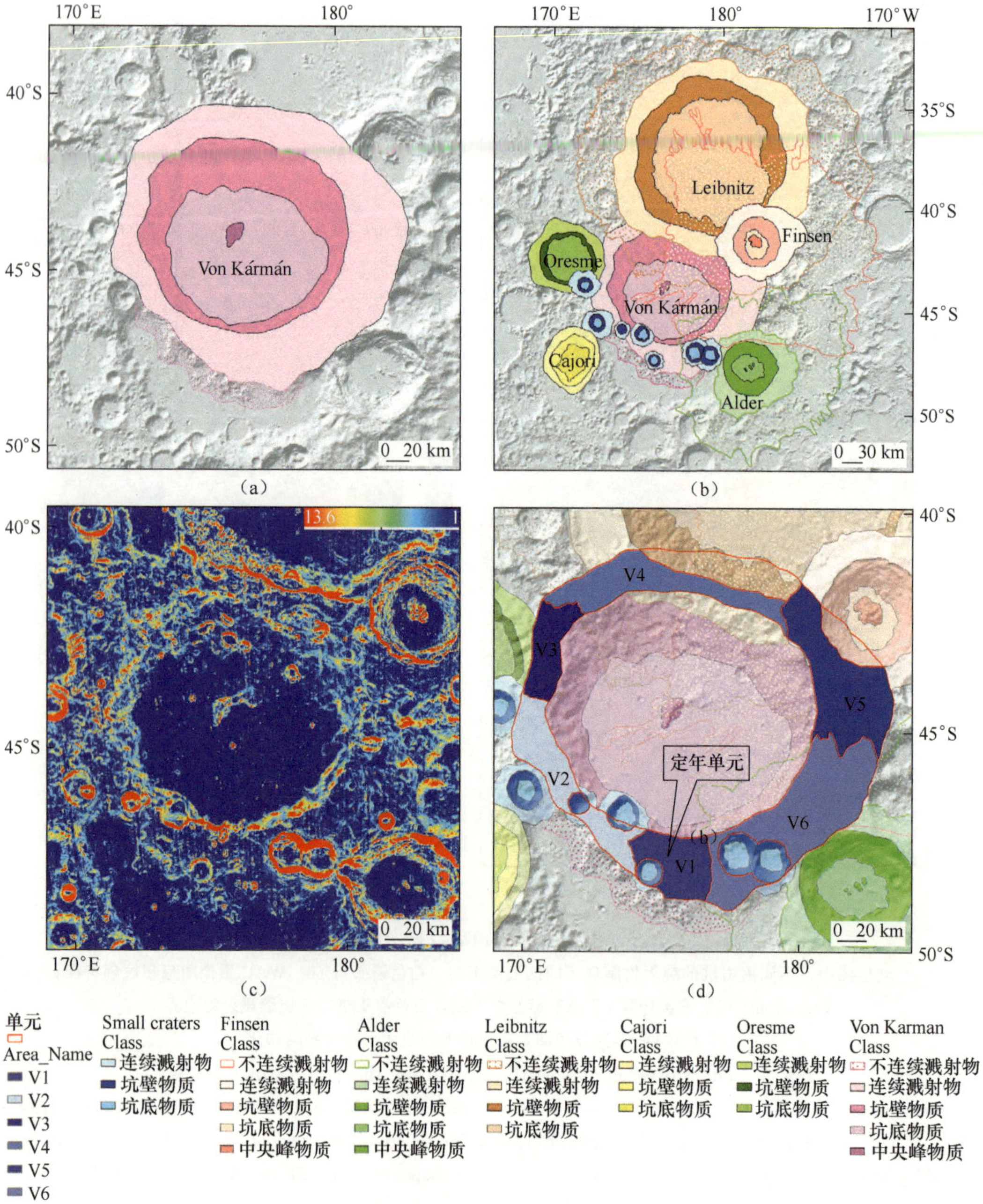

图 4　冯·卡门撞击坑溅射物分布图 (a)、撞击坑周围溅射物分布图 (b)、地形粗糙度图 (c) 及定年的单元图 (d)

Fig. 4　Ejecta from the Von Kármán crater (a) and its surroundings (b), roughness image of landforms (c) and dating units (d)

盖；在西部连续溅射物有部分残留，但它不仅受到大型撞击坑 Cajori 和 Oresme 撞击坑的影响，还被后期许多小型撞击坑覆盖，溅射物较为复杂。

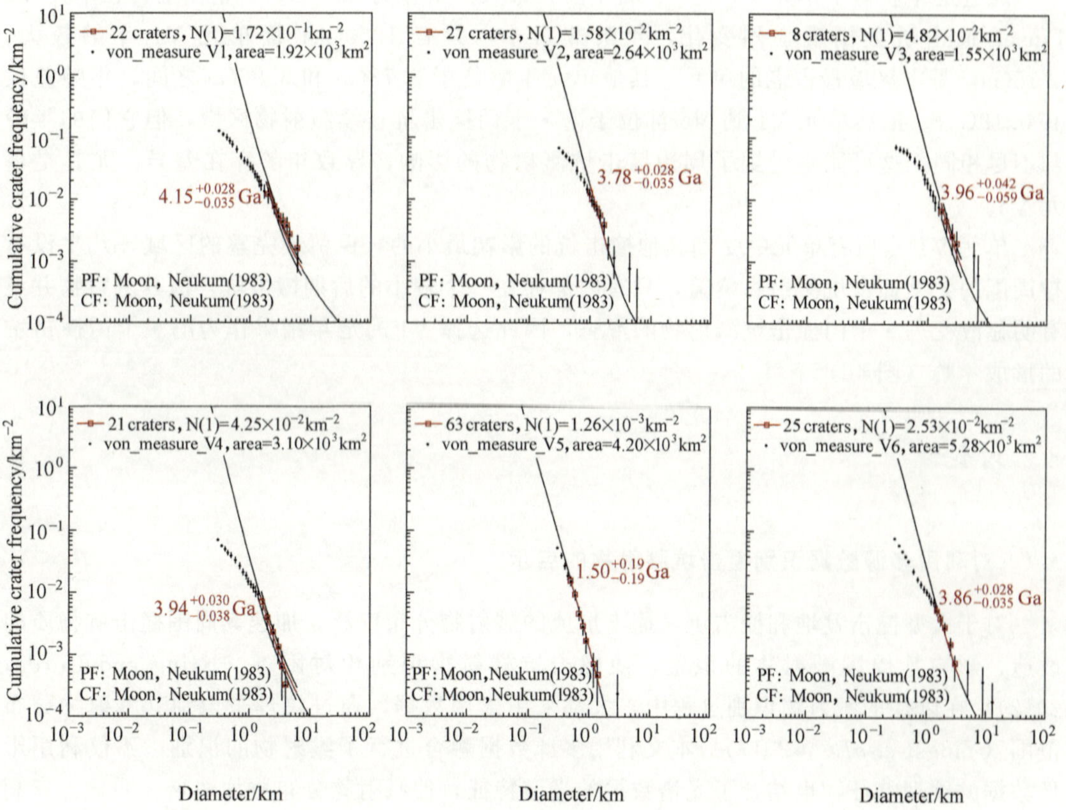

图 5　冯·卡门撞击坑单元 1-单元 6 年龄图

单元 V1 的年龄为 4.15Ga，年龄最大，是本区域最古老的单元；单元 V5 的年龄为 1.50Ga，年龄最小；单元 V2、V3、V4、V6 的年龄分别为 3.78Ga、3.96Ga、3.94Ga、3.86Ga，最大年龄差仅有 0.18Gyr

Fig. 5　The ages of the 6 units in the Von Kármán crater

Unit V1 has an age of 4.15Ga, the biggest and the oldest unit in the region; Unit V5 has an age of 1.50Ga and is the youngest; The ages of units 2, 3, 4, and 6 are 3.78Ga, 3.96Ga, 3.94Ga, and 3.86Ga, respectively, and the maximum age difference is only 0.18Gyr

2.2　冯·卡门撞击坑区域的年龄

根据冯·卡门撞击坑及周边撞击坑的连续溅射物和不连续溅射物的分布和影响特征并结合地形粗糙度图（图 4c），可将冯·卡门撞击坑的溅射物分成 6 个单元（图 4d）。V1：冯·卡门撞击坑南部几乎没有受到周围大型撞击坑溅射物影响的粗糙度较低的平坦区域；V2：西南部主要被后期小型撞击坑影响的粗糙度较高的区域；V3：Oresme 撞击坑溅射物影响的区域；V4：Leibnitz 撞击坑溅射物影响的区域；V5：以 Finsen 撞击坑溅射物影响为主的区域；V6：以 Alder 撞击坑溅射物影响为主的区域。

利用 CraterTools 工具提取撞击坑时，将 WAC 影像和 TC 影像相结合，以保证提取的撞击坑的准确性。将 6 个单元的撞击坑均进行统计并排除次级坑，基于大小-频率分布统计定年法进行相对定年分析，最后利用 CraterstatsⅡ获得各个地质单元的绝对模型年龄（图 5）。

通过定年获得了单元 V1 到 V6 的年龄，最大年龄差为 2.65Gyr。整体上，在单元 V5 Finsen 撞击坑处出现强烈变化，年龄非常小，只有 1.50Ga；而单元 V1 年龄最大，4.15Ga，是本区域最古老的单元。其他单元年龄位于 3.78Ga 和 3.96Ga 之间，年龄差仅有 0.18Gyr。虽然单元 V1 到 V6 都位于冯·卡门撞击坑连续溅射物区域，但它们的年龄却不尽相同。这可能是受到了周边撞击坑溅射物的影响，导致年龄存在差异，尤其是单元 V5。

单元 V1 是所有单元中受到其他撞击坑的影响最小的，保留最完整的区域，几乎没有被周围的大型撞击坑溅射物覆盖，只有内部存在一个较小的后期撞击坑，但其溅射物并没有明显改变冯·卡门撞击坑溅射物的形态，因此选择 V1 的定年结果作为冯·卡门撞击坑的形成年龄（图 4d)。

3 讨论

3.1 对利用多源数据识别撞击坑溅射物的启示

对于大型撞击盆地和撞击坑，撞击形成的溅射物分布广泛，加速其周围撞击坑物质的改造，影响其周围溅射物的识别，使撞击坑溅射物识别相对困难（Kring and Durda，2012)。在以往的溅射物识别过程中，大都采用影像数据，通过纹理来识别其溅射物分布范围（Yingst *et al.*，2017)。本文利用多源数据融合进行了溅射物的识别，不仅利用影像数据的纹理特征，也结合了光谱数据的光谱特征，使溅射物分布更加清晰、明显，有利于溅射物的识别与判断。同时，将溅射物划分为连续溅射物和不连续溅射物（郭弟均等，2016)，使识别的溅射物更加准确。

坡度图显示出明显的坡度对比，突出了溅射物与背景的差异性，相邻撞击坑间则可通过坡度的趋势进行初步的划分，这简化了溅射物的提取工作，再结合 WAC 影像图进行细节提取，提高了工作效率。OMAT 图像充分利用了月壤成熟度，增强了影像中元素的对比度，以彩色显示，使不同物质间差别增大，与 750nm 反射率影像融合使不连续溅射物纹理清晰，形态可见，两个数据间相互补充，表达效果明显提高。OMAT 影像分类图利用月壤成熟度值将影像进行了分类，使数据简单化，溅射物的边界结构化，有明确的分界线，识别起来更加简单可操作。

冯·卡门撞击坑虽然受后期撞击坑溅射物影响较大，但仍有部分保留，且其明显叠加在冯·卡门 M 撞击坑上，它可能挖掘出了相对较老的前酒海纪冯·卡门 M 的物质（Pasckert *et al.*，2018)，它与周边撞击坑影响都小的区域可能是 SPA 盆地物质，对于 SPA 盆地的研究具有重要的意义。根据叠覆关系和定年结果发现，撞击坑年龄越老，其溅射物保留得越少，尤其是不连续溅射物，撞击坑年龄越小，其溅射物保留得越完好，尤其是爱拉托逊纪的撞击坑，其溅射物纹理清晰易识别。

3.2 冯·卡门撞击坑的形成年龄

尽管前人已对冯·卡门撞击坑年龄分析开展了多项工作，但目前关于其形成年代尚未获得统一的认识，有酒海纪（Stuart－Alexander，1978；Pasckert *et al.*，2018；Ling *et*

图 6　WAC 和 TC 影像的比较

(a) 选择的区域的位置；(b) WAC 影像图；(c) TC 影像图

Fig. 6　The comparison between the WAC and TC images

(a) the position of the area; (b) the WAC image; (c) the TC image

al．，2019）与前酒海纪之争（Wilhelms *et al*．，1987；Losiak *et al*．，2009；Yingst *et al*．，2017；Huang *et al*．，2018）。之所以产生这样不一致的结果，主要是受到后期周围撞击坑溅射物的影响，未能准确提取定年撞击坑。

以往都是利用 WAC 影像进行撞击坑的提取（Yingst *et al*．，2017；Pasckert *et al*．，2018），它覆盖的范围广，各个区域都能够显示。但 WAC 影像分辨率有限，其部分撞击坑边界显示不清晰，会影响撞击坑的精确识别（图 6b）。而 TC 影像具有一定的倾角，导致撞击坑内部及坡度较大的区域出现盲区，给撞击坑的识别带来了困难；但是正因为其存在倾角，其显示的撞击坑边界更加明显，易于识别，且其分辨率较高，提取的边界更加准确（图 6c）。因此本文结合 WAC 和 TC 影像对撞击坑进行提取，既保证了撞击坑数量的全面，又提高了其识别的精度，为精确定年奠定了基础。

获得的年龄结果中，从单元 V1 到 V6 年龄起伏较大，其累积频率与年代曲线部分重合，存在较大偏移，撞击坑直径越小，其曲线坡度越缓；而相对频率分布都较分散，没有明显的聚集（图 7）。利用年代曲线将撞击坑大小-频率分布划分为几部分，使获得的年龄结果更加可信。可以发现，在撞击坑直径为 300500m 时（绿色），获得的绝对模型年龄都较小，其与年代曲线的重合度太低，出现了“翻转”（Kneissl *et al*．，2016）；当撞击坑直径较大时（红色，每个单元依据匹配度而选择不同的直径），获得的绝对模型年龄都较大，但是许多单元都只有几个撞击坑，虽然获得了年龄，但其可信度太低，偶然性太强；位于曲线中间的部分（黄色和蓝色），获得的绝对模型年龄多数在 3.6～4.0Ga 之间，只有单元 V5 较小，且此区域的撞击坑大小-频率分布要优于前两者。

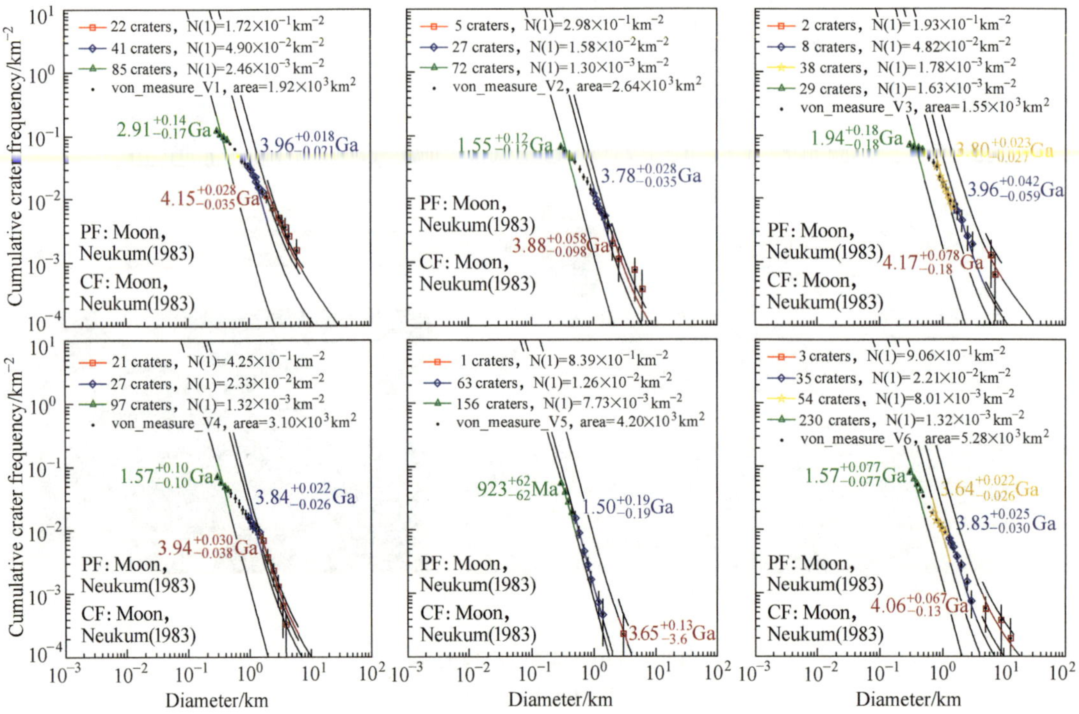

图 7　各单元 CSFDs 和模型年龄的累积图

Fig. 7　The CSFDs and derived model ages of the units in cumulative

选择的 300500m 撞击坑曲线明显偏离，这是因为此研究区域的小型撞击坑受到微陨石轰击、太阳风、宇宙射线和风化作用的影响并被后期溅射物覆盖，仅有部分仍然保留。单元 V5 中比较匹配没有明显偏离，是因为这里地层较新，小型撞击坑保留的相对较多。每个单元曲线划分的几个不同区域，年龄有差别，这是由于每一次溅射物的重新覆盖都会对表面年龄产生影响，故代表了此区域经历了几次地质演化。受到后期撞击坑溅射物影响越大，年龄则越偏小，大型撞击坑溅射物对撞击坑年龄判定的影响大于小型撞击坑，因此选择 V1 作为冯・卡门撞击坑的年龄，可信度较高。

单元 V1 选择了直径大于 2km 的撞击坑（匹配度高），共 22 个，获得的绝对模型年龄为 4.15Ga，在前酒海纪范围内，即冯・卡门撞击坑的年龄为前酒海纪，在其后期可能受到来自其正北面的 Leibnitz 撞击坑和西南方向的 Cajori 等撞击坑溅射物的影响，曲线坡度变缓，获得另一绝对模型年龄 3.96Ga。

为了进一步对比，本文对整个研究区域进行定年，获得冯・卡门撞击坑的年龄曲线（图 8），其位于中间的部分（蓝）匹配度要优于大型撞击坑区域（红），倾向于其年龄为 3.89Ga，即酒海纪。其大型撞击坑区域与单元 V1 相比，其年龄较小，撞击坑数量少，匹配度低，因此可信度较低，但是它的存在证明了冯・卡门撞击坑应有更古老的年龄。

经过对比发现单元 V1 的年龄更加真实可信，且通过溅射物进行单元划分可以有效排除后期撞击坑溅射物的影响，使获得的撞击坑年龄更接近其真实年龄值，其可信度明显提高，这将为后期进一步的工作提供坚实的基础。

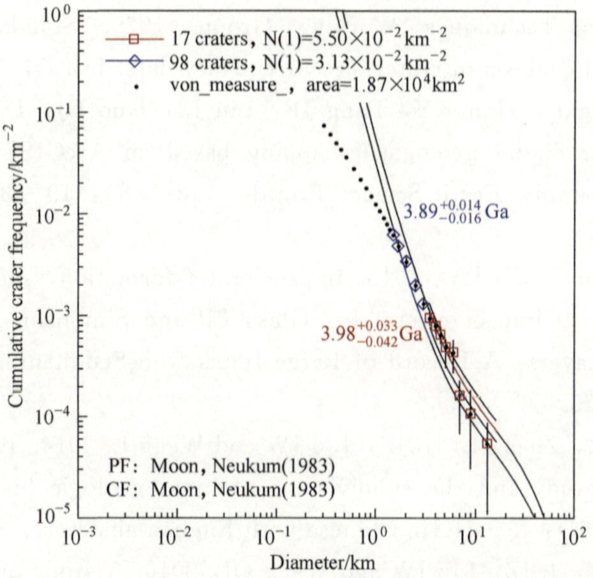

图 8　整个研究区域的年龄

Fig. 8　The ages of the whole study area

4　结论

本文利用多源数据对撞击坑溅射物进行了识别和定年分析，主要获得以下结论：

(1) 利用影像数据的纹理特征，并结合光谱数据的成分差异性，可以提高撞击溅射物识别的准确性；

(2) 冯·卡门撞击坑的连续溅射物和不连续溅射物都受到后期撞击影响，破坏较为严重，基于对周边撞击坑溅射物的提取结果，发现冯·卡门撞击坑的连续溅射物以南部保存最为完整；

(3) 本文在精确识别撞击坑边界的基础上，选择溅射物保留相对完整的区域作为定年对象，获得了冯·卡门撞击坑的年龄为 4.15Ga，即前酒海纪。

致谢　本文所使用的 TC 数据来源于 SELENE Data Archive；LRO 的 WAC 数据、DEM 数据和 Clementine 的 UVVIS 多光谱数据均来源于美国 PDS 行星数据节点，在此致以真诚的感谢。感谢中国科学院地质与地球物理研究所岳宗玉老师及另外一位审稿人的宝贵意见。另外，还要感谢中国科学院地球化学研究所各位老师和同学们的帮助。

References

[1] Alvarez LW, Alvarez W, Asaro F and Michel HV. 1980. Extraterrestrial cause for the Cretaceous - Tertiary extinction. Science, 208 (4448): 1095 - 1108.

[2] Collins GS, Melosh HJ and Osinski GR. 2012. The impact - cratering process.

Elements, 8 (1): 25 - 30.

[3] Crater Analysis Techniques Working Group. 1979. Standard techniques for presentation and analysis of crater size - frequency data. Icarus, 37 (2): 467 - 474.

[4] Ding XZ, Wang L, Han KY, Pang JF, Liu JZ, Guo DJ, Ding WC and Ju YJ. 2014. The lunar digital geological mapping based on ArcGIS: Taking the Arctic region as an example. Earth Science Frontiers, 21 (6): 19 - 30 (in Chinese with English abstract).

[5] Glass BP and Simonson BM. 2013. Impact crater formation, shock metamorphism, and distribution of impact ejecta. In: Glass BP and Simonson BM (eds.). Distal Impact Ejecta Layers: A Record of Large Impacts in Sedimentary Deposits. Berlin: Springer, 15 - 75.

[6] Guo DJ, Liu JZ, Zhang L, Ji JZ, Liu JW and Wang L. 2014. The methods of lunar geochronology study and the subdivisions of lunar geologic history. Earth Science Frontiers, 21 (6): 45 - 61 (in Chinese with English abstract).

[7] Guo DJ, Liu JZ, Ji JZ, Liu JW and Wang QL. 2016. A study of lunar typical crater ejecta and its implications for lunar geologic mapping. Acta Petrologica Sinica, 32 (1): 53 - 63 (in Chinese with English abstract).

[8] Harris EC. 1975. The stratigraphic sequence: A question of time. World Archaeology, 7 (1): 109 - 121.

[9] Harris EC. 1979. The laws of archaeological stratigraphy. World Archaeology, 11 (1): 111 - 117.

[10] Hartmann WK. 1970. Lunar cratering chronology. Icarus, 13 (2): 299 - 301.

[11] Hiesinger H, Van Der Bogert CH, Pasckert JH, Schmedemann N, Robinson MS, Jolliff B and Petro N. 2012. New crater sizefrequency distribution measurements of the South Pole - Aitken Basin. In: 43rd Lunar and Planetary Science Conference. Houston: Lunar and Planetary Institute.

[12] Huang J, Xiao ZY, Flahaut J, Martinot M, Head J, Xiao X, Xie MG and Xiao L. 2018. Geological characteristics of Von Kármán crater, northwestern South Pole - Aitken Basin: Chang'E - 4 landing site region. Journal of Geophysical Research: Planets, 123 (7): 1684 - 1700.

[13] Kato M, Sasaki S, Takizawa Y and The Kaguya Project Team. 2010. The Kaguya mission overview. Space Science Reviews, 154 (1 - 4): 3 - 19.

[14] Kneissl T, Van Gasselt S and Neukum G. 2011. Map - projection - independent crater size - frequency determination in GIS environments: New software tool for ArcGIS. Planetary and Space Science, 59 (11 - 12): 1243 - 1254.

[15] Kneissl T, Michael GG and Schmedemann N. 2016. Treatment of nonsparse cratering in planetary surface dating. Icarus, 277: 187 - 195.

[16] Kring DA and Durda D. 2012. A Global Lunar Landing Site Study to Provide the Scientific Context for Exploration of the Moon. Houston: Lunar and Planetary

Institute.

[17] Ling ZC, Qiao L, Liu CQ, Cao HJ, Bi XY, Lu XJ, Zhang J, Fu XH, Li B and Liu JZ. 2019. Composition, mineralogy and chronology of mare basalts and non-mare materials in Von Kármán crater: Landing site of the Chang'E-4 mission. Planetary and Space Science, 179: 104741.

[18] Losiak A, Wilhelms DE, Byrne CJ, Thaisen KG, Weider SZ, Kohout T, O'Sullivan K and Kring DA. 2009. A new lunar impact crater database. In: 40th Lunar and Planetary Science Conference. Houston: Lunar and Planetary Institute.

[19] Lucey PG, Blewett DT, Taylor GJ and Hawke BR. 2000. Imaging of lunar surface maturity. Journal of Geophysical Research: Planets, 105 (E8): 20377-20386.

[20] Michael GG and Neukum G. 2010. Planetary surface dating from crater size-frequency distribution measurements: Partial resurfacing events and statistical age uncertainty. Earth and Planetary Science Letters, 294 (3-4): 223-229.

[21] Michael GG, Platz T, Kneissl T and Schmedemann N. 2012. Planetary surface dating from crater size-frequency distribution measurements: Spatial randomness and clustering. Icarus, 218 (1): 169-177.

[22] Michael GG. 2013. Planetary surface dating from crater size-frequency distribution measurements: Multiple resurfacing episodes and differential isochron fitting. Icarus, 226 (1): 885-890.

[23] Michael GG, Kneissl T and Neesemann A. 2016. Planetary surface dating from crater size-frequency distribution measurements: Poisson timing analysis. Icarus, 277: 279-285.

[24] Neukum G, König B, Fechtig H and Storzer D. 1975. Cratering in the Earth-Moon system: Consequences for age determination by crater counting. In: 6th Lunar and Planetary Science Conference Proceedings. New York: Pergamon Press, 2597-2620.

[25] Neukum G. 1983. Meteorite bombardment and dating of planetary surfaces. Translation of Meteoriten bombardement und Datierung planetarer Oberflächen. Dissertation on Attaining Venia Legendi (Tenure) in the Geophysics Department in the Faculty of Geological Sciences. Munich: Ludwig-Maximilians University, 1-186.

[26] Neukum G, Ivanov BA and Hartmann WK. 2001. Cratering records in the inner solar system in relation to the lunar reference system. Space Science Reviews, 96 (1-4): 55-86.

[27] Nozette S, Rustan P, Pleasance LP, Kordas JF, Lewis IT, Park HS, Priest RE, Horan DM, Regeon P, Lichtenberg CL, Shoemaker EM, Eliason EM, McEwen AS, Robinson MS, Spudis PD, Acton CH, Buratti BJ, Duxbury TC, Baker DN, Jakosky BM, Blamont JE, Corson MP, Resnick JH, Rollins CJ, Davies ME, Lucey PG, Malaret E, Massie MA, Pieters CM, Reisse RA, Smith DE, Sorenson TC, Breugge RWV and Zuber MT. 1994. The clementine mission to the moon:

Scientific overview. Science, 266 (5192): 1835 - 1839.

[28] Öhman T and Kring DA. 2012. Photogeologic analysis of impact melt - rich lithologies in Kepler crater that could be sampled by future missions. Journal of Geophysical Research: Planets, 117 (E12): E00H08.

[29] Osinski GR, Tornabene LL and Grieve RAF. 2011. Impact ejecta emplacement on terrestrial planets. Earth and Planetary Science Letters, 310 (3 - 4): 167 - 181.

[30] Pasckert JH, Hiesinger H and Van Der Bogert CH. 2018. Lunar far side volcanism in and around the South Pole - Aitken Basin. Icarus, 299: 538 - 562.

[31] Pieters CM, Staid MI, Fischer EM, Tompkins S and He G. 1994. A sharper view of impact craters from clementine data. Science, 266 (5192): 1844 - 1848.

[32] Qiao L, Ling ZC, Fu XH and Li B. 2019. Geological characterization of the Chang'e - 4 landing area on the lunar farside. Icarus, 333: 37 - 51.

[33] Robinson MS, Brylow SM, Tschimmel M, Humm D, Lawrence SJ, Thomas PC, Denevi BW, Bowman - Cisneros E, Zerr J, Ravine MA, Caplinger MA, Ghaemi FT, Schaffner JA, Malin MC, Mahanti P, Bartels A, Anderson J, Tran TN, Eliason EM, McEwen AS, Turtle E, Jolliff BL and Hiesinger H. 2010. Lunar reconnaissance orbiter camera (LROC) instrument overview. Space Science Reviews, 150 (1 - 4): 81 -124.

[34] Smith DE, Zuber MT, Neumann GA, Lemoine FG, Mazarico E, Torrence MH, McGarry JF, Rowlands DD, Head III JW, Duxbury TH, Aharonson O, Lucey PG, Robinson MS, Barnouin OS, Cavanaugh JF, Sun XL, Liiva P, Mao DD, Smith JC and Bartels AE. 2010. Initial observations from the Lunar Orbiter Laser Altimeter (LOLA). Geophysical Research Letters, 37 (18): L18204.

[35] Stuart - Alexander DE. 1978. Geologic Map of the Central Far Side of the Moon. Washington: US Geological Survey, Report nr 1047.

[36] Wilhelms DE, Mccauley JF and Trask NJ. 1987. The geologic history of the moon. No. 1348.

[37] Wu WR, Wang Q, Tang YH, Yu GB, Liu JZ, Zhang W, Ning YM and Lu LL. 2017. Design of Chang'E - 4 lunar farside soft - landing mission. Journal of Deep Space Exploration, 4 (2): 111 - 117 (in Chinese with English abstract).

[38] Yingst RA, Chuang FC, Berman DC and Mest SC. 2017. Geologic mapping of the Planck quadrangle of the Moon (LQ - 29). In: 48th Lunar and Planetary Science Conference. Woodlands: Lunar and Planetary Institute, 1680.

参考文献

[1] 丁孝忠，王梁，韩坤英，庞健峰，刘建忠，郭弟均，丁伟翠，剧远景. 2014. 基于ArcGIS的月球数字地质填图：以月球北极地区为例. 地学前缘，21 (6)：19 - 30.

[2] 郭弟均，刘建忠，张莉，籍进柱，刘敬稳，王梁. 2014. 月球地质年代学研究方法及月面历史划分. 地学前缘，21 (6)：45 - 61.

[3] 郭弟均，刘建忠，籍进柱，刘敬稳，王庆龙. 2016. 月球典型撞击坑溅射物研究及对月球地质编图的意义. 岩石学报，32 (1)：53 - 63.

[4] 吴伟仁，王琼，唐玉华，于国斌，刘继忠，张玮，宁远明，卢亮亮. 2017. “嫦娥 4 号”月球背面软着陆任务设计. 深空探测学报，4 (2)：111 - 117.

Apollo 11 和嫦娥四号着陆区撞击坑退化对比分析

史珂[1,2]，岳宗玉[1,4*]，邸凯昌[1,4]，刘建忠[3,4]，牛胜利[5]

1. 中国科学院 空天信息创新研究院，遥感科学国家重点实验室，北京 100101
2. 中国科学院大学，北京 100049
3. 中国科学院 地球化学研究所，贵阳 550002
4. 中国科学院 比较行星学卓越创新中心，合肥 230026
5. 澳门科技大学 月球与行星科学国家重点实验室，澳门

摘要： Apollo 11 和嫦娥四号（Chang'E－4）是人类探月历史上的里程碑，它们的着陆区分别位于月球正面和背面。对两个着陆区内不同退化程度撞击坑的统计和对比分析有助于揭示研究区域的地质年龄和演化历史，对月球地质研究有着重要的意义。本文使用 LRO NAC 影像和 DTM 产品对两个着陆区附近 $1km^2$ 范围内撞击坑的退化进行分析，通过目视解译识别撞击坑并根据形貌将其分为不同的类别，然后对各类撞击坑进行统计，最后根据统计结果分析区域的地质年龄和撞击坑退化速度。结果表明 Apollo 11 和 Chang'E－4 着陆区附近直径在 5～300m 的撞击坑累计大小频率分布规律基本相同，撞击坑累积数量随直径的减小呈指数关系增加；撞击坑最初退化速度较快，随着退化程度的增加，退化速度急剧降低；两个区域的地质年龄相近，撞击历史相似。

关键词： Apollo 11；嫦娥四号；着陆区；撞击坑

Comparative Analysis of the Degradation of Impact Craters Between the Apollo 11 and Chang'E－4 Landing Areas

SHI Ke[1,2]，YUE Zong－yu[1,4]，DI Kai－chang[1,4]，LIU Jian－zhong[3,4]，NIU Sheng－li[5]

1. State Key Laboratory of Remote Sensing Science，Aerospace Information Research Institute，Chinese

Academy of Sciences, Beijing 100101, China
2. University of Chinese Academy of Sciences, Beijing 100049, China
3. Institute of Geochemistry, Chinese Academy of Sciences, Guiyang 550002, China
4. CAS Center for Excellence in Comparative Planetology, Hefei 230026, China
5. State Key Laboratory of Lunar and Planetary Sciences, Macau University of Science and Technology, Macau, China

Abstract: The Apollo 11 and Chang'E－4 missions are milestones in the history of lunar exploration. Their landing areas are located in the near and far sides of the Moon, respectively. The statistical and comparative analyses of the degradation states of the impact craters can be helpful to reveal the geological age and evolution history of the study areas in the Moon, and are of great significance to the study of lunar geology. In this paper, by using the LRO NAC images and DTM products, the degradations of impact craters within two 1 km^2 areas surrounding the two landing points are analyzed. Firstly, the impact craters are identified by using visual interpretation, and then they are classified into five categories according to their morphologies. Subsequently, the statistics of various impact crater are made. Finally, the geological ages of the study areas and the degradation speeds of impact craters are calculated based on the above results of statistical analyses. The results show that the cumulative size－frequency distributions of impact craters with diameters between 5 and 300 meters near the Apollo 11 and Chang'E－4 landing sites are generally similar, and the cumulative number of impact craters is increased exponentially with the decrease of their diameters. The degradation speeds of impact craters are relatively fast at first, and then they are decreased rapidly with the increase of degradation degrees. The geological ages and impact histories of impact craters in the two regions are generally similar.

Keywords: Apollo 11; Chang'E－4; landing area; impact crater

0　引言

撞击坑是月球表面最显著的地貌特征，在月球地质学的研究中具有重要地位（欧阳自远等，2002；欧阳自远，2005）。月球的遥感观测起源于 20 世纪 60 年代，借助影像数据撞击坑的系统研究也随即展开。例如，Baldwin（1965）和 Trask（1967）等开展了撞击坑的统计和形貌研究，Kreiter（1960）和 Neukum 等（1975）进行了撞击坑年代学的研究，Pike（1967）、Soderblom 和 Lebofsky（1972）等揭示了月球撞击历史、月球地表演化等信息。但是，早期的影像整体分辨率较低，因此大多数研究只限于对大型撞击坑的分析。近年来，月球勘测轨道器（LRO，Lunar Reconnaissance Orbiter）等探测任务获取了高分辨率的月表影像，为我们在小区域内开展撞击坑的相关研究提供了重要数据来源。1976 年 Apollo 11 任务首次将人类送上月球正面，2019 年 Chang'E－4 任务在月球背面首次实现了软着陆，这两次任务在月球探测史上都具有里程碑意义。有资料表明，Apollo 11 和 Chang'E－4 的着陆区均位于月海玄武岩区，其地形相对平坦，地质特征比较相近（Di et al. ，2019；Iqbal et al. ，2019）。因此，利用高分辨率数据对这两个着陆区的撞击坑进行详细的对比研究，可为分析月球正面和背面的演化异同提供重要信息，同时也可以为后期的探测任务提供参考。

撞击坑从形成伊始就开始逐渐退化，撞击坑退化特征的研究是撞击坑形貌学的深化，对分析所在区域的撞击历史及其演化历史都具有重要价值。撞击坑的退化过程和微陨石撞击、太阳风等月球内外部环境密切相关，撞击坑的退化会改变撞击坑的形态，退化消失会使撞击坑数量减少，从而影响区域撞击坑的形貌特征和统计特性（李坤，2013）。撞击坑退化的研究起源于 20 世纪 60 年代，Marcus（1967）和 Soderblom（1970）等在撞击坑形貌学的研究基础上开展了撞击坑的退化分析；之后 Mcgill（1974）、Head（1975）和 Basilevsky（1976）等对月球表面的撞击坑按照退化程度的不同进行了分类；Craddock 和 Howard（2000）及 Richardson（2009）等借助计算机技术对撞击坑的形态变化、退化速度等进行了定量化研究；近年来，借助高分辨率的影像，李坤（2013）、Fassett 和 Thomson（2014）、Xiao 和 Werner（2015）、Xie 等（2017，2019，2020）对撞击坑的退化规律、区域撞击历史等进行了详细的研究，提升了人们对撞击坑退化的认知。对区域撞击坑退化的研究能够确定区域的相对年龄，进而可以推测区域地表的撞击历史，这对月球科学研究有着重要意义。

本文依据 LRO NAC 影像识别研究区域撞击坑，并结合高程数据进行撞击坑的量测和统计，依据量测和统计结果进行撞击坑的分类和退化分析，最后根据统计分析结果对比分析 Apollo11 和 Chang'E－4 两个着陆区的撞击坑的退化情况，分析它们的地质年龄和撞击历史。

1　研究区与数据

1.1　研究区域概况

Apollo 11 探测器在 1969 年 7 月 20 日着陆于月球正面的静海西南边缘，首次将人类送上月球，着陆点的坐标为（0° 40′27″ N，23° 28′23″ E）（Iqbal et al.，2019）。静海是一个大型的撞击盆地，边缘很不规则，面积约为 42 万 km^2，其内部被后期的熔岩流覆盖（Rajmon and Spudis，2004）。2019 年 1 月 3 日，Chang'E－4 探测器成功软着陆于月球背面的南极-艾特肯盆地里的冯 · 卡门撞击坑内，着陆点坐标为（45°27′25″ S，177°35′17″ E），随后两器分离释放玉兔二号月球车（邸凯昌等，2019）。南极-艾特肯盆地是月球上最大的撞击盆地，直径约 2 500 km，冯 · 卡门撞击坑是其内部的一个撞击坑，其坑底被后期喷出的黑色玄武岩充填，较为平坦（Di et al.，2019）。依据着陆区的遥感影像和高程数据分别制作 Apollo 11 和 Chang'E－4 着陆点附近 1 km^2 的三维地形（图 1）。

从图 1 中可以发现两个着陆区地形总体上起伏均不大，比较平坦，相比之下 Apollo 11 着陆点周围有几个比较大的撞击坑深度稍微大一些，而 Chang'E－4 着陆点周围地形更加平坦，撞击坑的深度都不大。

1.2　数据源

LRO 是美国国家航空航天局（NASA）“新太空探索计划”的首个任务，于 2009 年发射升空，已环绕月球拍摄了大量的高分辨率影像，其空间分辨率最优为 0.5 m，高分辨率的影像是月球小撞击坑退化研究的首选数据源。Apollo 11 和 Chang'E－4 着陆区撞击坑退化的研究使用的是 LRO 的窄角相机 NAC 影像和利用 NAC 立体影像制作的数字地面模型 DTM 及正射影像产品，数据可从 Lunar Orbital Data Explorer 网站（https://ode.rsl.wustl.edu/moon/indexProductSearch.aspx）下载。我们使用的 Apollo 11 着陆区正射影像的空间分辨率为 0.5 m，DTM 的空间分辨率为 2 m；Chang'E-4 着陆区正射影像的空间分辨率为 1 m，DTM 的空间分辨率为 5 m。

(a)　　(b)

(a) Apollo 11 着陆点（着陆点位于图示中心，中心处正上方为北方向）。底图为 LRO NAC 正射影像，编号 NAC_DTM_APOLLO11_M150361817_50CM，高程数据为 NAC DTM，编号 NAC_DTM_APOLLO11；
(b) Chang'E-4 着陆点（着陆点位于图示中心，中心处正上方为北方向）。底图为经过正射矫正的 LRO NAC 影像，编号 M1303619844LR，高程数据为 NAC DTM，编号 NAC_DTM_CHANGE4

图 1　着陆点附近 1 km^2 的三维地形

Fig. 1　The 3D topographies of two 1 km^2 areas surrounding the landing sites

2　方法

本文基于 NAC 正射影像和 DTM 产品通过人工的目视解译识别和圈定着陆点周围的撞击坑，接着根据撞击坑的形貌按照不同的退化程度进行分类，然后对撞击坑进行统计分析，确定着陆区域的地质年龄以及不同退化程度撞击坑的年龄，最后根据撞击坑的深度直径比值的变化计算撞击坑的退化速度，并根据统计的结果对研究区域的地质年龄和撞击历史进行综合分析，总体技术路线如图 2 所示。

2.1　撞击坑识别与区域定年

撞击坑识别是指从影像中提取和圈定出撞击坑，识别的方法主要有传统的人工目视解译与计算机自动识别两大类（李坤，2013）。其中目视解译方法需要研究人员在影像上手动标绘撞击坑，但是大区域的撞击坑识别工作量比较大，效率比较低。计算机自动识别是依据影像特征和高程信息，利用合理的算法借助计算机进行自动识别。主要方法有边缘检测、面向对象、霍夫变换等；借助计算机强大的运算能力，这些方法对大区域撞击坑的识别效率很高，但是存在误判、漏判等问题，尤其是对比较小的撞击坑识别精度不是很高(刘宇轩等，2012)。因为本文的研究区域在着陆点附近比较小的范围内，区域内撞击坑尺寸普遍较小，所以我们采用传统的目视解译方法来识别撞击坑。将 NAC 正射影像和 DTM 精确配准，裁剪出着陆点周围 1km^2 的影像，使用 ArcGIS 中的二次开发工具

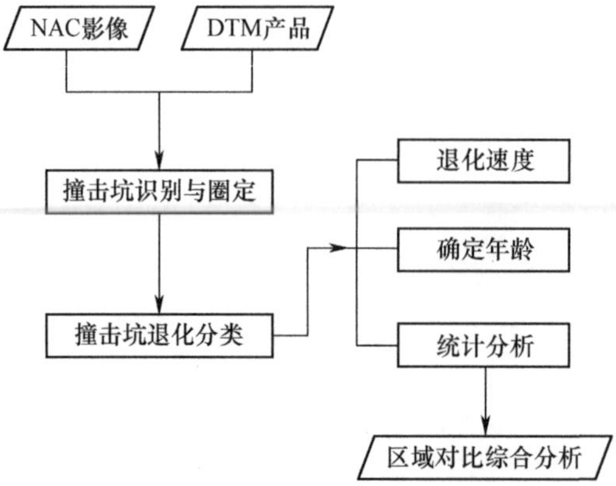

图 2　技术路线图

Fig. 2　The chart of technical procedure

“CraterTools ”（Kneissl et al.，2011），通过识别撞击坑边缘均匀分布的三个点圈定撞击坑，之后再使用“CraterSurvey”（Liu et al.，2018）工具计算撞击坑的边缘点、深度等信息。Wang 等（2020）的研究表明在影像上识别和统计出的撞击坑直径大于 10 个像素时比较可靠，所以我们对识别的撞击坑进行筛选，仅将直径大于 5 m 的撞击坑保留下来。

依据撞击坑的大小频率分布可以确定研究区域的地质年龄。Neukum（1983），Stiffler 和 Ryder（2001）的研究指出月球上同一地区在不同时期、相同直径范围的撞击坑大小频率关系是相似的，所以可以根据月球表面撞击坑的生产函数来确定研究区域的地质年龄。结合 Apollo 任务和 Luna 任务采集到的月球样品的同位素绝对定年，Neukum 等（2001）给出了一个修正的月球撞击坑定年方程，研究区域地质单元的年龄 T（Ga）可以表示为区域中每 km^2 内直径大于等于 1km 的撞击坑个数 N（1）的函数：

$$N(1)=5.44\times10^{-14}(e^{6.93T}-1)+8.38\times10^{-4}T \tag{1}$$

依据该关系由研究区域内撞击坑的统计分析就可以确定出该区域的地质年龄。

2.2　撞击坑退化分类与年龄确定

月球表面的撞击坑从形成开始就不断地受到月震、火山、高能粒子、陨石撞击等的影响，其形貌特征一直在变化，从最初的具有新鲜明亮辐射纹、陡峭坑壁的形态，到后来撞击坑边缘模糊直至最后完全消失，这一过程称为撞击坑的退化（Baldwin，1949）。根据 Mcgill 和 Wise（1972 ）、Basilevsky（1976）等的研究，按照撞击坑不同的退化程度将撞击坑分成 A、AB、B、BC 和 C 五类，其中 A 类为最新鲜的撞击坑，C 类为退化最严重的撞击坑。据此，我们依据撞击坑的形态、阴影等特征，对研究区内不同退化程度的撞击坑进行判别：A 类撞击坑边缘清晰完整尖锐，四周分布有明显的辐射纹或碎石块、溅射物，或某个方向明亮的溅射物分布较远，超过 2 倍的半径，撞击坑深度较大，阴影非常明显；AB 类撞击坑边缘清晰完整较为尖锐，四周有明显的溅射物分布，或某个方向溅射物分布不超过 2 倍的半径，撞击坑深度较大，阴影比较明显，四周看不到明亮的辐射纹或碎石块分布；B 类撞击坑边缘清晰完整较为尖锐，四周看不到明显的溅射物分布，撞击坑度深不大，阴影明显；BC 类撞击坑边缘不完整，某些部分比较模糊，四周看不到溅射物分布，撞击坑深度较小，有阴影但不明显；C 类撞击坑边缘很模糊不完整，退化严重，四周看不到溅射物分布，撞击坑深度很小，基本没有阴影。图 3 展示了不同退化类型撞击坑的示例。

自然状态下撞击坑从形成到完全消失，会经历从 A→C 的连续退化过程，这期间经历的时间为撞击坑的总生命周期。Basilevsky（1976）的研究表明，撞击坑刚刚形成时形态为 A，之后大约经过总生命周期的 3%形态会从 AB 转变到 B，然后大约经过总生命周期的 20%形态会从 B 转变到 BC，之后大约经过总生命周期的 50%形态会从 BC 转变到 C。此外 Basilevsky 和 Head（2012）还给出了一个简单的撞击坑总生命周期估计公式，直径小于 160m 撞击坑的总生命周期计算式为公式（2），直径大于 160m 的撞击坑总生命周期则用公式（3）计算：

$$E = 2.5D \tag{2}$$

$$E = 8D - 900 \tag{3}$$

式中，D 为撞击坑的直径（m）；E 为撞击坑的总生命周期（Ma）。根据上述的分析及公式（2）和（3），可以计算出不同退化程度的撞击坑的年龄，结果见第 3.2 节。

2.3　撞击坑退化速度计算

由于不同大小的撞击坑的退化速度不相同，且撞击坑尺寸越小退化速度越快（Head III et al. ，2010），所以对撞击坑退化速度的衡量应采用一个无量纲量。我们通过撞击坑的深度直径比值随时间的变化来衡量撞击坑的退化速度，具体可以通过深径比对时间的偏导数来计算，如式（4）所示（Basilevsky et al. ，2014），撞击坑形成之初深度直径比值较大，之后撞击坑退化深径比逐渐减小，到最后撞击坑基本消失时深径比趋于 0。

$$-\frac{\partial(d/D)}{\partial t} = -R\left(\frac{\partial(d/D)}{\partial N}\right) \tag{4}$$

式中，d/D 表示深度直径比；R 为撞击坑的生产率；N 为深径比大于 d/D 的撞击坑个数。对公式（4）积分得到公式（5）：

$$d/D = A - BN + (A_0 - A)e^{-CN} \tag{5}$$

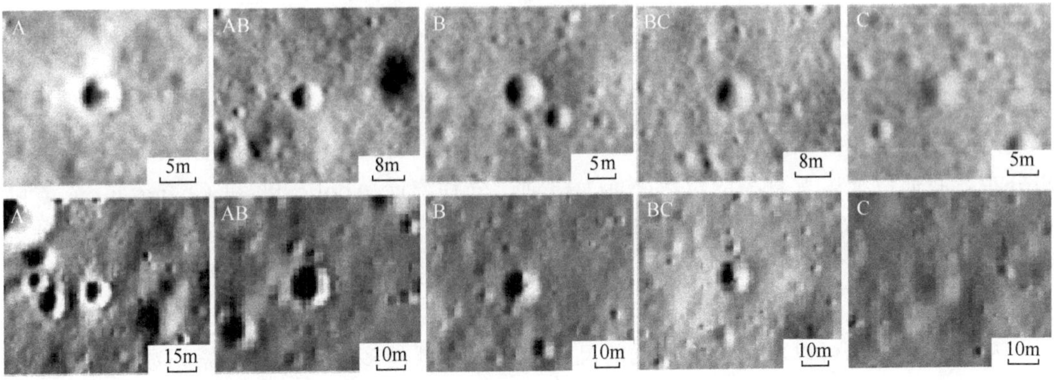

第一行为 Apollo 11 着陆区的撞击坑，第二行为 Chang'E－4 着陆区的撞击坑。
撞击坑的类别标注在图中左上角，向上为北向

图 3　不同退化类型撞击坑的示例

Fig. 3　Examples of impact craters of different degradation categories

式中，A_0 为撞击坑最初的深度直径比；A、B 和 C 是常数，可以根据研究区域内撞击坑的观测值，通过最小二乘拟合求得。

3　结果与讨论

3.1　撞击坑的提取与退化分类

按照前文给出的方法，在经过处理的正射影像上通过目视解译识别和圈定着陆点附近 $1km^2$ 内的撞击坑。Apollo 11 着陆点附近共提取直径 5m 以上的撞击坑 1359 个，最大的撞击坑直径为 295.68m，按照撞击坑的直径将其分为不同的区段，统计每个区段内撞击坑的个数，如图 4a 所示，同时统计撞击坑的总体累计大小频率分布信息，在对数坐标系下绘制分布图，如图 5a 所示；与此类似，在 Chang'E－4 着陆点附近共提取直径 5m 以上的撞击坑 1329 个，最大的撞击坑直径为 317.57m，根据撞击坑的直径分为不同的区段，绘制每个区段的撞击坑个数统计图，如图 4b 所示，同时在对数坐标系下绘制着陆点附近撞击坑累计大小频率分布图，如图 5b 所示。

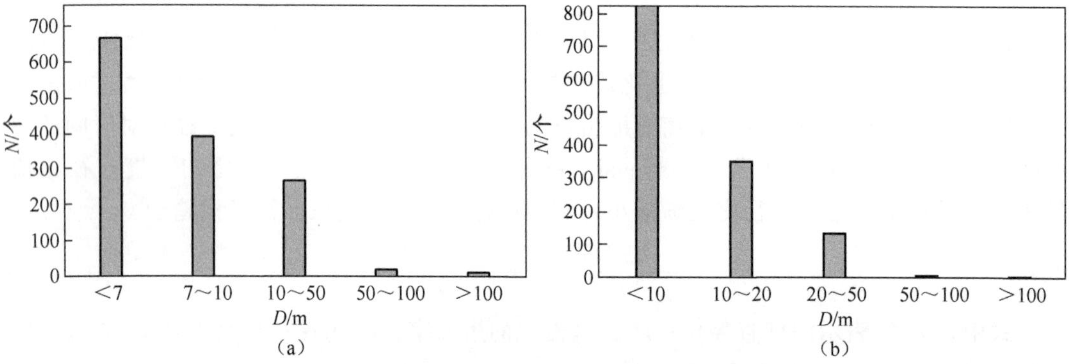

(a) Apollo 11 着陆区的统计结果；(b) Chang'E－4 着陆区的统计结果

图 4　不同尺寸的撞击坑个数统计图

Fig. 4　Statistics of the numbers of impact craters of different sizes

依据上述不同退化程度的撞击坑的影像特征，通过目视判别将撞击坑从最新鲜到几乎完全退化分为 A、AB、B、BC 和 C 五类 。为了清楚地看到不同退化程度的撞击坑大小频率分布，我们以撞击坑直径的对数坐标为横轴，分别统计 Apollo 11 和 Chang'E－4 着陆点附近每一类撞击坑的累计个数并绘制累计大小频率分布图（图 6），同时根据统计结果分别计算 Apollo 11 和 Chang'E－4 着陆点附近不同退化程度的撞击坑所占的比例，绘制饼状统计图（图 7）。

从图 4 和图 5 可以看出，撞击坑的尺寸越小数量就越多，撞击坑的累积数量随着撞击坑直径以类似指数的关系增加。图 5 中 Chang'E－4 着陆点附近的撞击坑直径接近 5m 时累计大小频率分布曲线斜率的绝对值略微有减小的趋势，这和我们使用的正射影像分辨率有关，因为在 Apollo 11 着陆区影像的分辨率为 0.5m，优于在 Chang'E－4 着陆区的影像分辨率 1m，所以在 Chang'E－4 着陆区直径 5m 的撞击坑已经少于 10 个像素，可能存在一些漏判导致识别出来的小撞击坑数量减少，从而使得曲线斜率绝对值减小。通过在双对数坐标中趋势拟合可以发现不完全统计的拐点为 6.6m。这种现象在图 6 中同样可以看出，图 6 中的 C 类，BC 类，B 类撞击坑的累计分布曲线和总体的撞击坑累计分布曲线的变化趋势基本相同，而其他类别的撞击坑因为数量比较少，撞击坑累计大小频率分布随尺寸的指数变化趋势不是很明显。

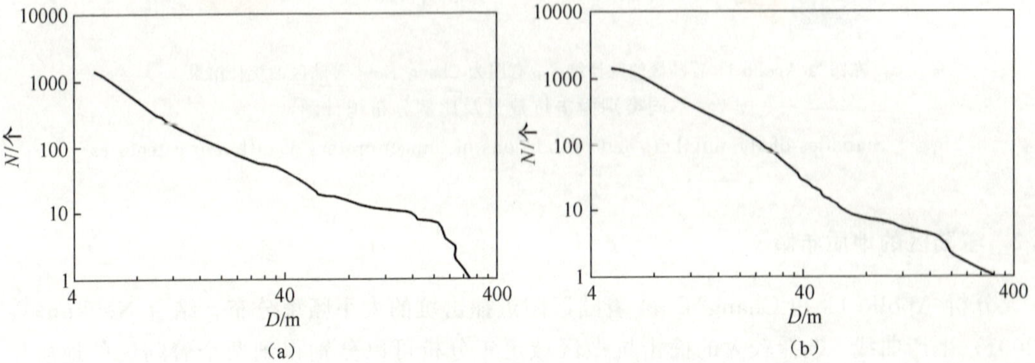

（a）Apollo 11 着陆区的统计结果；（b）Chang'E－4 着陆区的统计结果

图 5　撞击坑累计大小频率分布统计图

Fig. 5　Statistics of the cumulative size－frequency distributions of impact craters

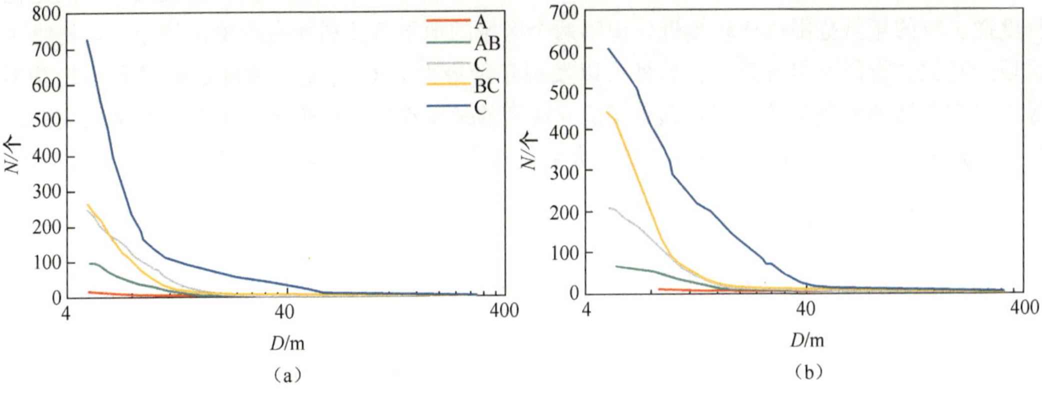

（a）Apollo 11 着陆区的统计结果；（b）Chang'E－4 着陆区的统计结果

图 6　各类撞击坑累计大小频率分布统计图

Fig. 6　Statistics of the cumulative size－frequency distributions of various impact craters

图 7 给出了不同类别撞击坑的数量分布统计，从中可以发现越是新鲜的撞击坑数量则越少。进一步对比 Apollo 11 和 Chang'E－4 着陆区的撞击坑统计信息可以发现，Apollo 11 着陆点附近新鲜的撞击坑和退化严重的撞击坑的数量和比例均比 Chang'E－4 着陆区要多，说明 Apollo 11 着陆区的地质年龄可能比 Chang'E－4 着陆区要大一些（详见 3.2 节）。

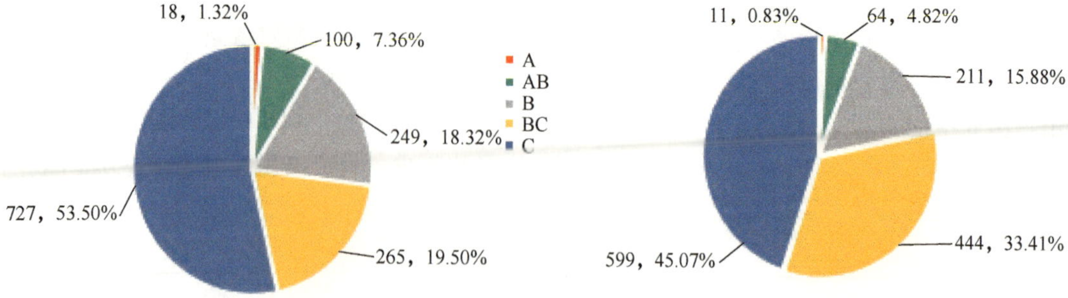

左图为 Apollo 11 着陆区的统计结果；右图为 Chang'E-4 着陆区的统计结果

图 7 不同类别撞击坑数量及比例分布统计图

Fig. 7 Statistics of the numbers and proportions of impact craters of different categories

3.2 着陆区的地质年龄

分析 Apollo 11 和 Chang'E-4 着陆区附近撞击坑的大小频率分布，结合 Neukum 等（2001）生产曲线，选择较大的撞击坑做区域定年分析可以分别得到两个着陆区的地质年龄，如图 8 所示。结果表明两区域地质年龄比较接近，Apollo 11 着陆区地质年龄 $3.70^{+0.052}_{-0.082}$ Ga，与同位素绝对定年的结果 3.53～3.72Ga 相一致（Iqbal et al.，2019），Chang'E-4 着陆区地质年龄 $3.56^{+0.110}_{-0.610}$ Ga，与前人的研究结果 3.7Ga 很接近（Gou et al.，2020）。同时根据前文所述，利用公式（2）和（3）由撞击坑的直径计算出不同退化程度撞击坑的年龄分布范围，将其分别列在表格中如表 1 所示。

从上述结果可以看出，Apollo 11 和 Chang'E-4 着陆区的地质年龄比较接近，不同退化程度撞击坑的年龄范围也比较接近，说明两个区域的地质条件和撞击历史比较相似，使得两区域的对比分析研究很有意义。另外可以发现年龄最大的 C 类撞击坑确定的最大年龄并不等于研究区域的地质年龄，而是远小于研究区域的地质年龄。这个年龄上限主要是较大的撞击坑通过公式（3）计算出来的，说明有大量的撞击坑已经完全退化而无法辨认了。

3.3 撞击坑的退化速度与深径比

使用 ArcGIS 的二次开发工具“CraterSurvey”在 DTM 产品基础上，对前面识别出的每一个撞击坑统计 8 个方向的深度、直径信息，计算并统计每个撞击坑的深度直径比值，如果深径比小于 0.01 则认为撞击坑完全退化。分别统计 Apollo 11 和 Chang'E-4 着陆区附近深径比（d/D）大于 0.01 的撞击坑累计个数，结果如图 9 所示。由图 9 可以看出，两个着陆区撞击坑深径比的累积分布曲线形状相似，随着深径比的增大撞击坑个数均急剧减少。但是经过对比能够发现 Chang'E-4 着陆区深径比大于 0.01 的撞击坑数量明显偏少，累计分布曲线不光滑。接下来分别针对 Apollo 11 和 Chang'E-4 着陆区计算撞击坑的退化速度，依据统计信息绘制撞击坑的深径比 d/D 和累计个数 N 的散点图，设定公式（5）为目标函数进行最小二乘拟合计算，得到拟合曲线如图 10 所示。其中拟合参数 A_0 表示撞击坑形成时的初始深径比，设为 0.18，计算得到 Apollo 11 着陆区的拟合参数 $A=0.03523$，$B=9.373\times10^{-5}$，$C=0.06852$；Chang'E-4 着陆区拟合参数 $A=$

0.03639，$B=4.990\times10^{-4}$，$C=0.6451$。

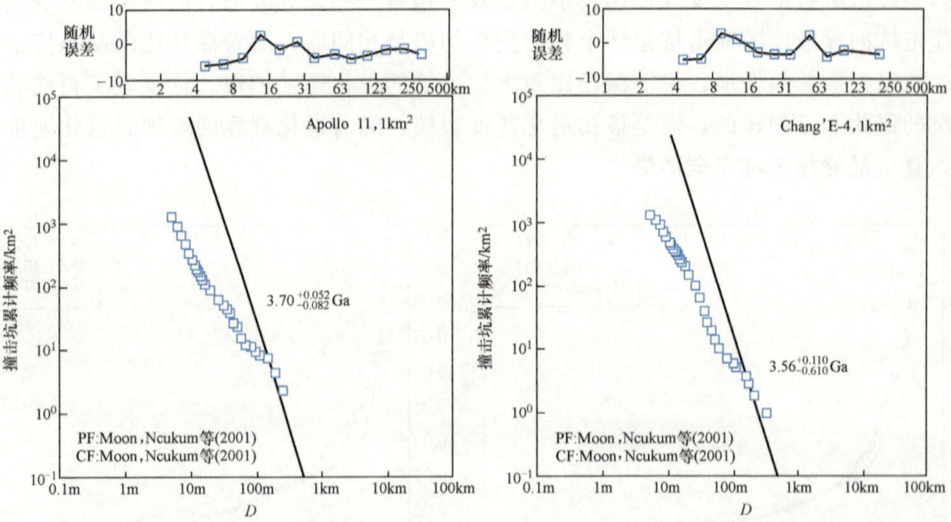

图 8　Apollo 11 着陆区（a）和 Chang'E-4 着陆区（b）的地质年龄

Fig. 8　Geological ages of the (a) Apollo 11 landing area and (b) Chang'E-4 landing area

表 1　Apollo 11 和 Chang'E-4 着陆区不同类型撞击坑的年龄范围

Table 1　Age variations of different categories of impact craters in the landing areas of the Apollo 11 and Chang'E-4

Apollo 11 着陆区				
撞击坑类型	A、AB	B	BC	C
年龄/Ma	0～3.63	3.35～22.39	14.12～293.09	146.09～1122.56
Chang'E-4 着陆区				
撞击坑类型	A、AB	B	BC	C
年龄 / Ma	0～4.37	3.63～24.21	12.50～375.40	149.59～1640.18

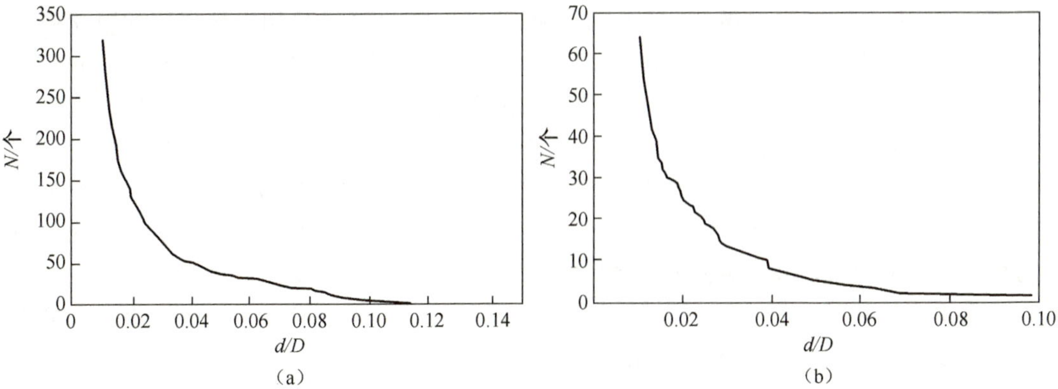

(a) Apollo 11 着陆区；(b) Chang'E-4 着陆区

图 9　撞击坑深径比累计频率分布图

Fig. 9　The cumulative frequency distribution diagrams of the crater depths versus diameter ratios of impact craters

图 10 显示，Apollo 11 着陆区撞击坑退化速度曲线拟合效果较好，而 Chang'E－4 着陆区因为撞击坑数量较少及深径比较小拟合效果稍差一些。观察拟合曲线可以发现两个着陆区撞击坑的深径比随撞击坑累计个数的变化趋势是相同的，随着深径比的减小撞击坑的累计个数呈指数关系增加，说明深径比越大撞击坑的退化速度越快。这反映了自然状态下撞击坑的退化不是匀速的，而是最初退化速度很快，随着退化程度的增加，退化速度迅速减慢，直至最终撞击坑完全消失。

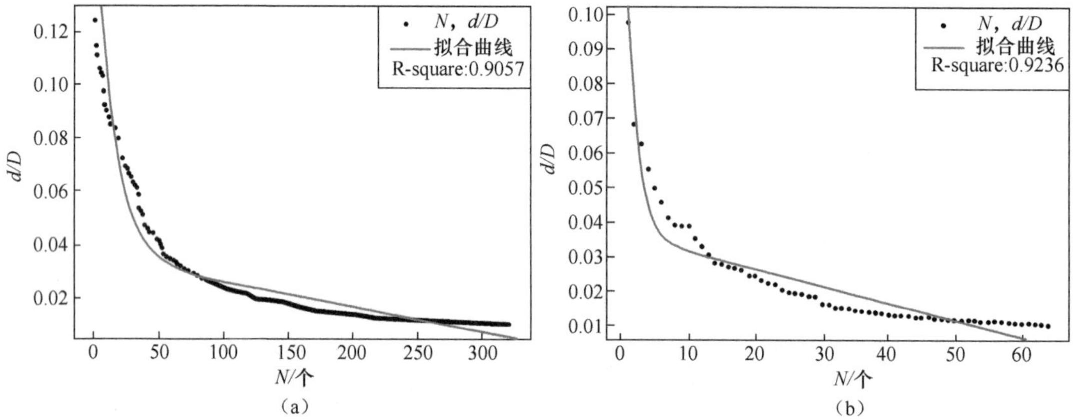

(a) Apollo 11 着陆区；(b) Chang'E－4 着陆区

图 10 撞击坑退化速度曲线拟合

Fig. 10 The fitting curves of degradation velocities of impact craters

4 结论

撞击坑是类地行星表面最常见和最为重要的地质构造之一，在行星科学的研究中有着至关重要的作用。Apollo 11 和 Chang'E－4 是人类探月历史上的里程碑，它们的着陆区是重要的研究区域，本文主要对 Apollo 11 和 Chang'E－4 着陆区附近撞击坑的退化做对比分析，结果表明两个着陆区附近每个不同退化类型的以及总体的撞击坑累计大小频率分布曲线的变化趋势都基本相同，撞击坑的个数随着直径的减小呈指数关系增加，直径在 10m 以下的小尺寸撞击坑的数量非常多，直径在 100m 以上的大尺寸撞击坑数量则很少；不同退化类型的撞击坑数量分布并不均匀，越新鲜的撞击坑个数则越少，最新鲜的 A 类撞击坑占比很小，而退化严重的 C 类撞击坑占比很大；撞击坑从形成伊始逐渐退化，但是退化速度并不均匀，最初撞击坑的深径比 d/D 较大退化速度较快，随着退化程度的增大，深径比 d/D 呈指数关系迅速衰减退化速度逐渐减慢。两个着陆区的地质年龄和撞击历史比较接近，Apollo 11 着陆区相对更古老一些。

Apollo 11 和 Chang'E－4 的着陆区分别位于月球正面和背面，具有重要的研究价值，研究区域内撞击坑的统计分析对推断区域的地质年龄和演化历史有着重要的意义，然而撞击坑的形成与演化是一个非常复杂的过程，人们对撞击坑相关科学问题的认知仍然十分有限。受到影像分辨率的限制，我们的统计分析结果存在一些不足之处，有些退化严重的撞击坑可能没有识别出来，Chang'E－4 着陆点附近的撞击坑深径比普遍偏小，在未来借助高分辨率的影像，特别是更加精确的高程模型可以做更加精细的对比分析。

参考文献

[1] Baldwin R B. 1949. The face of the moon. Chicago：The University of Chicago Press.

[2] Baldwin R B. 1965. The crater diameter－depth relationship from Ranger VII photographs. Astronomical Journal，70（8）：545－547.

[3] Basilevsky A T，Head J W. 2012. Age of Giordano Bruno crater as deduced from the morphology of its secondaries at the Luna 24 landing site. Planetary and Space Science，73（1）：302－309.

[4] Basilevsky A T，Kreslavsky M A，Karachevtseva I P ，Gusakova E N. 2014. Morphometry of small impact craters in the Lunokhod－1 and Lunokhod－2 study areas. Planetary and Space Science，92：77－87 Basilevsky A T. 1976. On the rate of evolution of small lunar craters.

[5] In：Proceedings of the 7th Lunar Science Conference ，Houston，TX. New York：Pergamon Press，1005－1020.

[6] Craddock R A，Howard A D. 2000. Simulated degradation of lunar impact craters and a new method for age dating farside mare deposits. Journal of Geophysical Research，105（E8）：20387－20410.

[7] Di K C，Zhu M H，Yue Z Y，Lin Y T，Wan W H，Liu Z Q，Gou S，Liu B，Peng M，Wang Y X，Niu S L，Zhang J H，Li J，Xie J F，Xi L H，Yang J F，Xue B. 2019. Topographic evolution of von kármán crater revealed by the lunar rover Yutu－2. Geophysical Research Letters，46（22）：12764－12770.

[8] Fassett C I，Thomson B J. 2014. Crater degradation on the lunar maria：Topographic diffusion and the rate of erosion on the Moon. Journal of Geophysical Research：Planets，119（10）：2255－2271.

[9] Gou S，Yue Z Y，Di K C，Wan W H，Liu Z Q，Liu B，Peng M，Wang Y X，He Z P，Xu R. 2020. In situ spectral measurements of space weathering by Chang'E－4 rover. Earth and Planetary Science Letters，535：116117.

[10] Head III J W，Fassett C I，Kadish S J，Smith D E，Zuber M T，Neumann G A，Mazarico E. 2010. Global distribution of large lunar craters：Implications for resurfacing and impactor populations. Sci－ence，329（5998）：1504－1507.

[11] Head J W. 1975. Processes of lunar crater degradation：Changes in style with geologic time. The Moon，12（3）：299－329.

[12] Iqbal W，Hiesinger H，van der Bogert C H. 2019. Geological mapping and chronology of lunar landing sites：Apollo 11. Icarus，333：528－547.

[13] Kneissl T，van Gasselt S，Neukum G. 2011. Map－projection－independent crater size－frequency determination in GIS environments－New software tool for ArcGIS. Planetary and Space Science ，59（11－12）：1243－1254.

[14] Kreiter T J. 1960. Dating lunar surface features by using crater frequencies. Publications of the Astronomical Society of the Pacific , 72 (428): 393 - 398.

[15] Liu Z Q, Yue Z Y, Michael G, Gou S, Di K C, Sun S J, Liu J Z. 2018. A global database and statistical analyses of (4) Vesta craters. Icarus, 311: 242 - 257.

[16] Marcus A H. 1967. A Stochastic Model of the formation and survival of lunar craters: VI. Initial depth, distribution of depths, and lunar history. Icarus, 6 (1 - 3): 56 - 74.

[17] Mcgill G E, Wise D U. 1972. Regional variations in degradation and density of Martian craters. Journal of Geophysical Research , 77 (14): 2433 - 2441.

[18] Mcgill G E. 1974. Morphology of lunar craters: A test of lunar erosional models. Icarus, 21 (4): 437 - 447.

[19] Neukum G, Ivanov B A, Hartmann W K. 2001. Cratering records in the inner solar system in relation to the lunar reference system. Space Science Reviews, 96 (1 - 4): 55 - 86.

[20] Neukum G, Kinig B, Arkani - Hamed J. 1975. A study of lunar impact crater size - distributions. The Moon, 12 (2): 201 - 229.

[21] Neukum G. 1983. Meteoriten bombardement und datierung planetarer oberflächen. Germany: University of München.

[22] Pike R J. 1967. Schroeter′s rule and the modification of lunar crater impact morphology. Journal of Geophysical Research, 72 (8): 2099 - 2106.

[23] Rajmon D, Spudis P. 2004. Distribution and stratigraphy of basaltic units in Maria Tranquillitatis and Fecunditatis: A Clementine perspective. Meteoritics & Planetary Science, 39 (10): 1699 - 1720.

[24] Richardson J E. 2009. Cratering saturation and equilibrium: A new model looks at an old problem. Icarus, 204 (2): 697 - 715.

[25] Soderblom L A, Lebofsky L A. 1972. Technique for rapid determination of relative ages of lunar areas from orbital photography. Journal of Geophysical Research, 77 (2): 279 - 296.

[26] Soderblom L A. 1970. A model for small - impact erosion applied to the lunar surface. Journal of Geophysical Research , 75 (14): 2655 - 2661.

[27] Stöffler D, Ryder G. 2001. Stratigraphy and isotope ages of lunar geologic units: Chronological standard for the inner solar system. Space Science Reviews, 96 (1 - 4): 9 - 54.

[28] Trask N J. 1967. Distribution of Lunar craters according to morphology from Ranger VIII and IX photographs. Icarus, 6 (1 - 3): 270 - 276.

[29] Wang Y C, Xie M G, Xiao Z Y, Cui J. 2020. The minimum confidence limit for diameters in crater counts. Icarus, 341: 113645.

[30] Xiao Z Y, Werner S C. 2015. Size - frequency distribution of crater populations in equilibrium on the Moon. Journal of Geophysical Research: Planets, 120 (12):

2277 - 2292.

[31] Xie M G, Liu T T, Xu A A. 2020. Ballistic Sedimentation of impact crater ejecta: Implications for the provenance of lunar samples and the resurfacing effect of ejecta on the lunar surface. Journal of Geophysical Research: Planets, 125 (5): e2019JE006113.

[32] Xie M G, Xiao Z Y, Xu A A. 2019. Tim - dependent production functions of lunar simple craters on layered targets with consideration of topographic degradation. Geophysical Research Letters, 46 (20): 10987 - 10996.

[33] Xie M G, Zhu M H, Xiao Z Y, Wu Y Z, Xu A A. 2017. Effect of topography degradation on crater size - frequency distributions: Implications for populations of small craters and age dating. Geophysical Research Letters, 44 (20): 10171 -10179.

[34] 邸凯昌，刘召芹，刘斌，万文辉，彭嫚，王晔昕，芶盛，岳宗玉，辛鑫，贾萌娜，牛胜利 . 2019. 多源数据的嫦娥四号着陆点定位 . 遥感学报，23 (1)：177 - 184.

[35] 李坤 . 2013. 月球小型撞击坑形貌特征及退化规律研究 . 硕士学位论文 . 武汉：武汉大学.

[36] 刘宇轩，刘建军，牟伶俐，李春来 . 2012. 撞击坑识别方法综述 . 天文研究与技术，9 (2)：203 - 212.

[37] 欧阳自远，邹永廖，李春来，刘建忠 . 2002. 月球某些资源的开发利用前景 . 地球科学——中国地质大学学报，27 (5)：498 - 503.

[38] 欧阳自远 . 2005. 月球科学概论 . 北京：中国宇航出版社.

第5章　月球背面环境新数据新认识

月球没有全球性的偶极磁场和大气层，太阳风和宇宙线几乎可以无阻碍地撞击月表，在月壤和月球稀薄外逸层的演化过程中发挥着重要作用。嫦娥四号着陆器和巡视器分别搭载了月表中子与辐射剂量探测仪（Lunar Neutron Detector，LND）、中性原子分析仪（Advanced Small Analyzer for Neutrals，ASAN），为人们研究月表空间环境、太阳风和宇宙线与月球及月壤的相互作用提供了良好的机遇。

另外，在嫦娥四号着陆器上的两根月球车导轨的末端底部，安装有4个与月壤直接接触的温度计，首次对月球背面的表层月壤进行了温度就位测量。

5.1　月背粒子辐射环境特征

宇宙线通常由太阳高能粒子（SEPs）、银河宇宙线（GCRs）和异常宇宙线（ACRs）组成。SEPs起源于太阳，通常发生在太阳活跃期，并伴随着太阳耀斑和日冕物质抛射事件，粒子的能量从几个MeV到GeV量级变化。GCRs被认为起源于银河系内的超新星爆发，但是不能排除超高能量粒子来源于银河系外的可能性。ACRs则被认为起源于日球层外的局地星际介质中的中性原子，这些原子通常具有比较高的第一电离势，包括质子（H）、氦（He）、氮（N）、氧（O）、氖（Ne）和氩（Ar）。在太阳平静期内，宇宙线由GCRs主导。GCRs的主要成分包括：H（约87%）、He（约12%）、重核（约1%）。此外，GCRs是持续不断的，SEPs则是零星发生的，通常发生在太阳活动高年。

月球表面和地球表面的辐射环境是截然不同的。月表因接近真空和缺乏全球性磁场的保护，时刻遭受着来自宇宙线的轰击，因此月表的辐射环境由高能离子主导；相反，地表因受到稠密大气层和地磁场的保护，高能离子很难到达地表，因此地表的辐射主要由伽马射线（能量极高的光子）和低能阿尔法粒子（氦核）贡献。月球上粒子辐射环境较为复杂，除了银河宇宙线、太阳高能粒子以外，还有银河宇宙线或太阳粒子事件与月壤产生的反照粒子，以及天然放射性核素放出的射线。宇宙线及其与月表相互作用产生的次级粒子会对宇航员和精密仪器的安全构成严重威胁，比如，宇航员暴露在过量的宇宙线中，他们患上白内障、癌症以及中枢神经退化性疾病的风险会增加；仪器暴露在过量的宇宙线中，高能带电粒子可能会使仪器内部一些电导率较低的绝缘材料被击穿而丧失绝缘功能，从而影响到仪器的正常使用。近年来，国内外对月球粒子辐射环境的探测主要是在绕月轨道上开展的辐射剂量以及能谱的测量，观测高度都高于50km，在嫦娥四号之前，人类还未对月球表面的辐射剂量进行过主动测量。

嫦娥四号着陆器搭载的月表中子与辐射剂量探测仪测量月球表面的辐射剂量率、通量，以及月球表面快中子的能谱。LND 的主要目的是获取月球表面的剂量率随时间的变化，以及线性能量转移（LET）谱，为人类登月做准备，同时还可开展日球层物理的研究。

LND 由 10 个分区的 Si 固体探测器（SSD）构成一个粒子望远镜，用于测量带电粒子（0.5MeV 到数 MeV 的电子、8～35MeV 的质子和 17～75MeV/nuc 的重核）。它具有一种特殊的三明治几何结构，稀有材料或硅材料夹在两个固体探测器 Si 之间，用于观测热中子和快中子（伽马射线）。快中子对地下水很敏感，对理解月球表面物质的混合过程很重要。

根据 LND 在 2019 年 1 月 3 日～1 月 12 日、1 月 31 日～2 月 10 日两个月昼获取的观测数据（图 5-1），月表的粒子辐射剂量率为 13.2μGy/h（Si），其中中性粒子（中子和伽马射线）的辐射剂量率为 3.1μGy/h（Si），约占总比例的 23%。品质因子为 4.32，剂量当量约为 60μSv/h，是火星表面和空间站内部的 2～3 倍，地球单次航班遭受剂量的 5～10 倍，地球表面（北京）的 300 倍。

比较太阳平静期（2019.01—2020.02）LND 和近地航天器（ACE、SOHO 和 STEREO-A）的观测数据，在约 10～100MeV/nuc 能量范围内 LND 的宇宙线能谱与近地航天器得到的能谱是一致的（见图 5-2）。具体地，LND 与近地航天器对 H 和 CNO 的观测结果在误差范围内是一致的，通量比的平均值分别为 1.05±0.15 和 1.08±0.16；而 He 和重离子，近地航天器与 LND 的观测结果在 1.7σ 内相当，通量比的平均值分别为 1.30±0.18 和 1.24±0.21，表明对于这些低能宇宙线的能谱，月表环境对它们的影响是可以忽略的；考虑到更高能的宇宙线更不易受到月表环境的影响，因此可以将该结论进一步推广为月表环境对这些宇宙线能谱的影响是可以忽略的。这些宇宙线能谱最直接的作用是能够直接为月背的宇宙线及与其相关的研究提供输入数据，进而促进对月表的质子、中子和伽马发射能谱的了解。在结果中，LND 测量的宇宙线能谱，特别是质子能谱，与广泛使用的 CRÈME 模型给出的预测值之间存在显著差异：对于 CRÈME96 和 CRÈME2009 模型，LND 与它们之间的通量比值的平均值分别达到 1.69±0.17 和 2.25±0.23，为 CRÈME 模型的改进提供很强的约束，宇宙线能谱也为其他相关的理论模型提供检验和约束。

提取 ^{3}He/^{4}He 的通量比发现，在约 12MeV/nuc 有显著增强，LND 测得的 ^{3}He/^{4}He 通量比值与 GALPROP 模型给的预测值之间存在很显著的差异，为 GALPROP 模型以及其他相关的理论模型提供了很强的约束。LND 的观测还证实了宇宙线通量的晨昏对称性，这将为未来的载人登月任务着陆时间和宇航员月表出舱活动时间的选择提供重要指引。

根据 LND 两年内（2019 年 1 月—2020 年 12 月）的观测数据，月表粒子辐射在硅中的平均总吸收剂量率为（12.66±0.45）μGy/h，中性粒子吸收剂量率为（2.67±0.16）μGy/h。辐射剂量率在 2 年内随时间出现缓慢的下降，LET 谱的变化则很小。另外，2020 年 12 月的太阳活动末期，由于银河宇宙线福布斯下降，辐射剂量率降低了 10%左右（图 5-3）。

LND 不仅是辐射剂量仪器，还是一个太阳高能粒子探测器，可监控太阳爆发活动及其对日地空间环境的影响。突发的太阳爆发（如耀斑、太阳日冕物质抛射）产生的太阳高

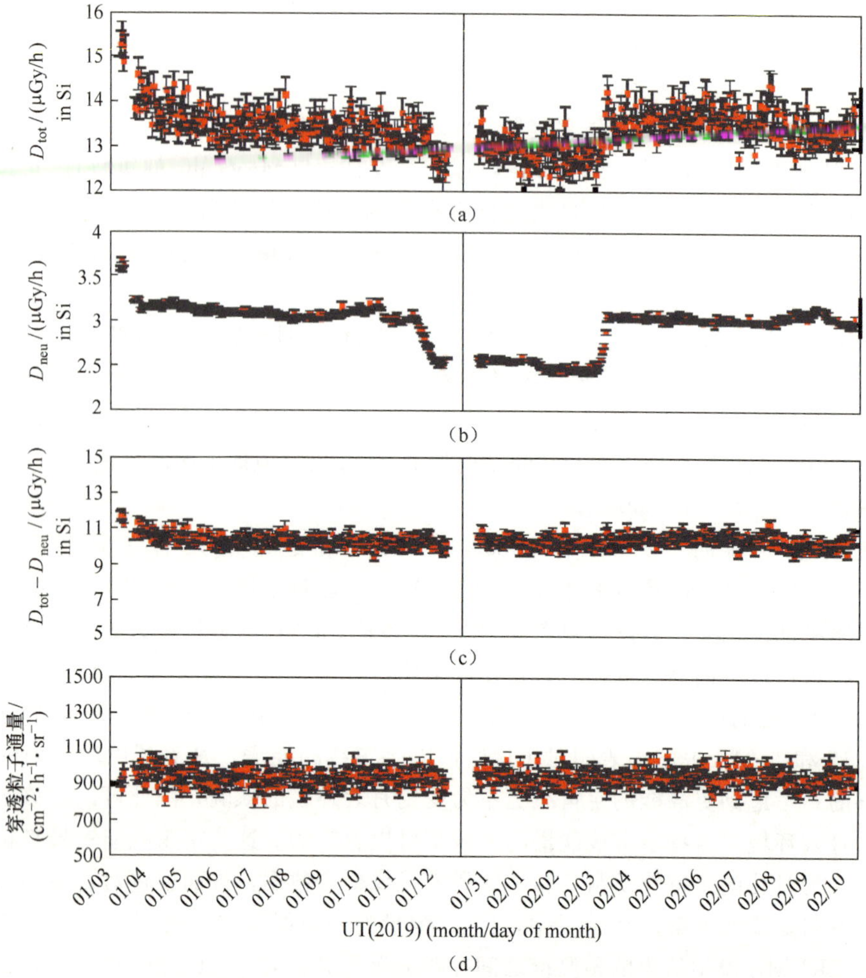

图 5-1　嫦娥四号着陆后前两个月昼 LND 测量的月球辐射环境随时间的变化

左侧显示的是第一个月昼的结果，右侧显示的是第二个月昼的结果

(a) LND 中硅探测器 B 测量的总吸收剂量率；(b) LND 中硅探测器 C1 测量的中性粒子吸收剂量率；(c) 带电粒子的总吸收剂量率；(d) 穿透粒子通量随时间变化

能粒子（SEP）可以直接到达月球表面，对仪器和工作人员造成辐射伤害，因而 LND 对太阳高能粒子的监测对探月工程有直接的意义。2019 年 5 月 6 日，LND 在月球背面探测到了第一个 SEP 事件，这个事件峰值流量非常小，持续时间也很短，是一个典型的脉冲型太阳高能粒子事件（impulsive SEPs）。根据不同能量段的粒子流量曲线，计算出了电子和质子开始到达月球的时间以及开始时间和对应的速度（$1/\beta=c/v$）（见图 5-4），可以发现，无论是电子还是质子，都表现出很好的速度弥散分布，即速度快的粒子先到达，速度慢的粒子后到达，通过对数据点进行线性拟合，可以推出粒子从加速区释放的时间以及粒子到达月球前在行星际空间传输的距离，结果显示电子释放比质子要早 1 个小时（图 5-4）。进一步分析这个 SEP 事件的质子能谱，结合 SOHO 卫星和 LND 的数据，发现能量拐点大概为 2.5MeV。

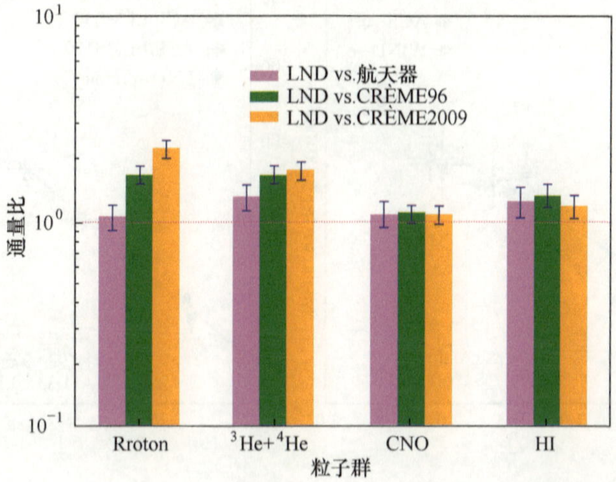

图 5-2　嫦娥四号和近地航天器对宇宙线各组成分的

通量测量结果的比值的平均值（紫色），及嫦娥四号对宇宙线各组成分的通量测量结果和

CRÈME 模型给出的预测值的比值的平均值（绿色对应 CRÈME96，橙色对应 CRÈME2009），

其中，1.0 的比率用红色虚线表示，作为比较的参照

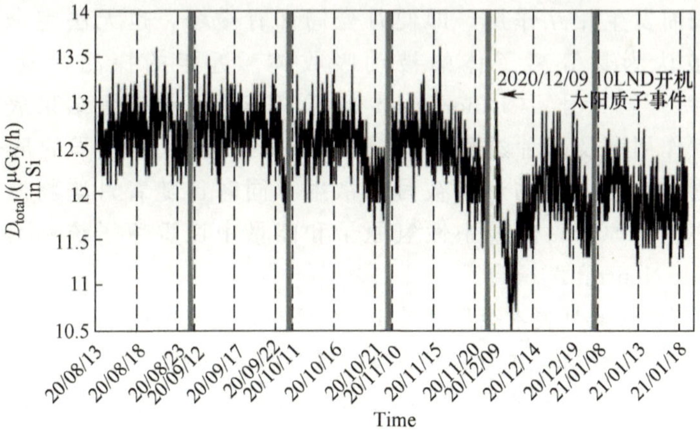

图 5-3　太阳质子爆发期间辐射剂量的变化情况

一般认为脉冲型 SEP 会处在磁场连接性比较好的位置，因为加速的粒子可以很快且有效地传播至观测点，形成一个脉冲型的时间曲线。但这个事件却完全不符合这个经典的范畴，这个 SEP 事件起源于一个 M1.0 级耀斑，结合模型发现太阳源区距离地球的磁力线足点非常远。分析其机理，其中一个原因可能跟日冕物质抛射驱动的激波有关，激波可以加速粒子，然后激波在向外传播的过程中和地球的磁力线链接了起来；另外一种解释可能是剧烈变化的磁场为粒子的传输临时搭建了一个通道，和大结构的太阳势场完全不同。但目前尚未有任何直接的观测能证实以上两种解释。

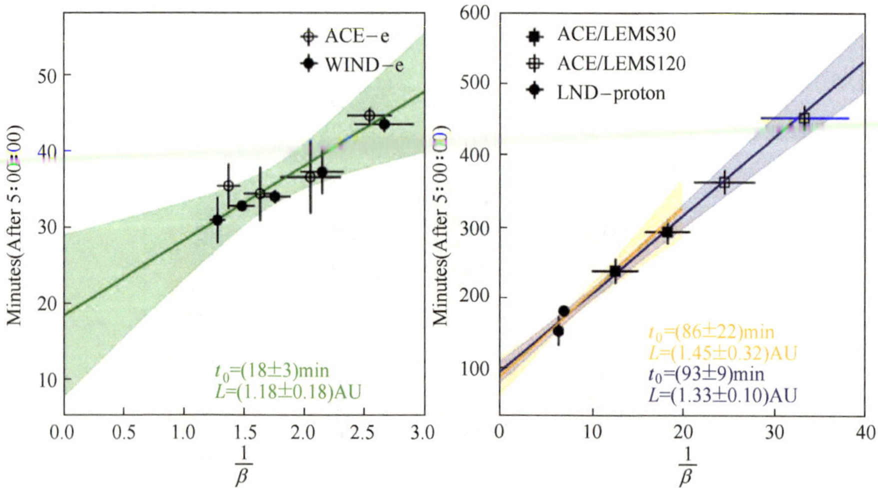

图 5 - 4　粒子事件开始的时间和粒子能量的关系

5.2　太阳风与月面相互作用

月球既没有大气层，也没有全球性的磁场，超声速的太阳风能够直接到达月球表面，并与月球表面发生相互作用。即便月壳局部有磁场，也无法完全阻止这种情况的发生，月球被认为是等离子体的被动吸收体。近年的探测活动［如月亮女神(Kaguya)、嫦娥一号/二号、Chandrayaan - 1 和 IBEX 等］观测结果表明，太阳风与月球表面的相互作用涉及多种过程，0.1%～1%的太阳风以质子的形式、高达 21%的太阳风以能量中性氢原子的形式被月面散射，同时伴随着月壤物质被太阳风离子溅射出的能量中性重原子，能量中性氢原子和能量中性重原子统称为能量中性原子(ENA，Energetic Neutral Atom)。

中性原子探测仪获取月表米级尺度的 ENA 分布，研究太阳风与月壤相互作用的小尺度特征，揭示其中的微观物理过程。ASAN 能量范围为 10eV 至 10keV，平均能量分辨率 $\Delta E/E$ 约为 30%（实际的分辨率取决于测量粒子的能量，氢原子的能量分辨率在 16%～130%之间变化）。基于飞行时间质谱分析技术，中性原子探测仪可以实现成分分辨，质量分辨率（$m/\Delta m$）约为 2，能够区分氢、氦和氧组（≥16amu）。仪器的时间分辨率为 4.6s。中性原子探测仪视场为 37°×30°（垂直方向为 37°，水平方向为 30°，半高宽为 28°×14°），视轴斜向下指向月球表面，与水平方向夹角为 30°，所产生的梯形视场边界围成的面积约为 1.57m²。

ASAN 的观测结果表明，当太阳风离子能量高于 $0.1E_{sw}$ 时，氢 ENA 能谱与 Chandrayaan - 1 和 IBEX 的大尺度观测数据一致性较好；而在低于 $0.1E_{sw}$ 时，ASAN 测得的氢 ENA 通量要高于 Chandrayaan - 1 和 IBEX 的结果（图 5 - 5）。从能谱结构可以确定与 ENA 生成过程有关的两个分量，高于 $0.1E_{sw}$ 的分量主要是太阳风离子被反向散射后生成的氢 ENA，$0.1E_{sw}$ 能量以下的分量则主要是月表物质溅射的 ENA。ENA 通量随能量下降而迅速上升的原因，可能是 ENA 溅射产率的变化以及月壤中氢含量的变化。

相对于 Chandrayaan－1 和 IBEX 在轨道上对 ENA 的大尺度探测，ASAN 的探测视场仅覆盖了月表 1.57m^2的面积，视场范围内局地月壤特性（如疏松程度、颗粒大小以及成分等）也可能影响能谱，尤其是 ENA 能量较低时。此外 ENA 能谱也可能会随着太阳天顶角（SZA）和观测极角的变化而变化，因此，特定的观测角度也可能会导致特殊的能谱结构。

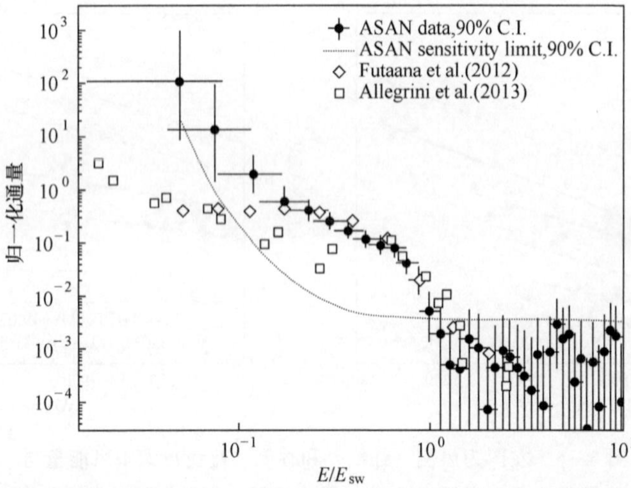

图 5－5　黑色实心圆：2019/5/1 06：18—10：45 UTC ASAN 观测到的 ENA 能谱，相对于太阳风的能量 E_{sw} 做了归一化。垂直误差棒对应 90％的置信区间（C. I.），水平误差棒对应能谱响应的半高宽。点虚线为 90％置信区间，在这个区间之下 ASAN 有效信号将无法与噪声区分。空心菱形：Chandrayaan－1 的氢 ENA 能谱，同样经太阳风离子能量归一化。空心正方形：IBEX 的氢 ENA 能谱，同样进行过归一化

进一步分析 2019 年 1 月 11 日至 2020 年 10 月 12 日的 ENA 能谱，发现在大多数月昼，晨侧的 ENA 微分通量高于昏侧。结合 ARTEMIS 卫星的同期观测数据，分析发现晨侧和昏侧不同能量范围的 ENA 微分通量与太阳风状态参数如通量、密度及动压呈正相关关系。能量范围在 100～600eV 时相关系数较高，峰值约在 140～333eV；在能量较低的氢原子（＜100eV）中，微磁层的影响不明显，而在 100eV 至 600eV 的能量范围内，微磁层影响明显。这种差异表明存在两种似乎起源不同的氢原子，一种能量低于 100eV，而另一种能量大于 100eV，后者可能与太阳风质子轰击月壤直接相关，能量小于 100eV 的 ENA 通量可能会被来自月球表面和仪器内表面溅射的 ENA 污染。

通过统计晨侧和昏侧的太阳风能量与 ENA 截止能量及温度的关系（图 5－6）发现，在相同能量的太阳风轰击下，昏侧的 ENA 的截止能量和温度低于晨侧，说明在昏侧太阳风可能受月球微磁层影响，被磁异常上方的静电场所减速，使能谱变窄、温度降低。利用晨侧与昏侧的 ENA 截止能量之差，可估算太阳风在昏侧的减速，进而计算静电势，静电势和太阳风减速率都随太阳风能量的变化而变化。静电势范围为 50～260V（95％置信区间为 30～350V）。太阳风的减速率为 12％～18％（95％置信区间为 7.5％～30％），

Kaguya 观测发现南极艾特肯盆地附近磁异常内部存在＋150V 的电势，此时初始太阳风能量约为 630eV。嫦娥四号 ASAN 观测到的太阳风减速也可能是由南极艾特肯盆地附近磁异常引起的。幸运的是，我们发现两个分别在晨侧和昏侧的事件，即初始太阳风能量（约 680eV）类似于上面提到的 Kaguya 的观测（约 630eV）。晨侧（680eV）和昏侧（550eV）的 ENA 截止能量差约为 100eV，与之前在绕月轨道的观测结果一致。与 Chandrayaan－1 和 Kaguya 卫星的观测结果一致。

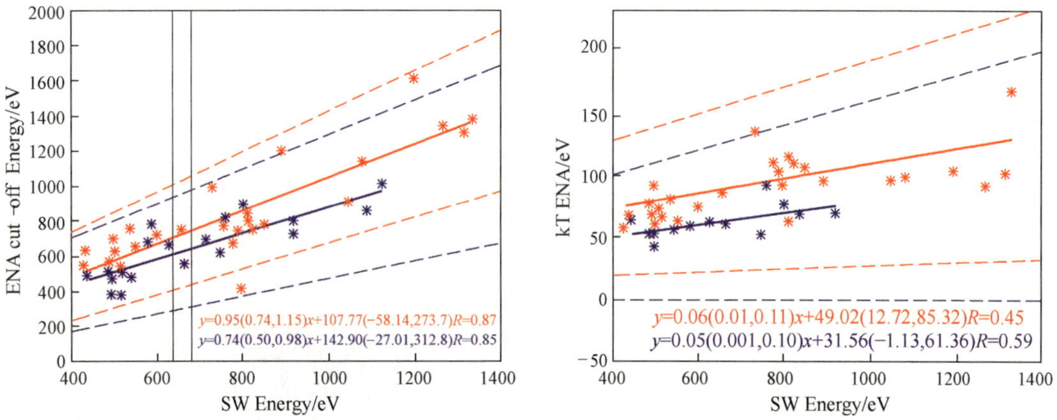

图 5－6　左图为晨侧（红线）和昏侧（蓝线）太阳风能量与 ENA 截止能量的关系。虚线表示 95％置信区间。矩形显示了两个分别在晨昏侧观测的太阳风能量（约 680eV）相似的事件。右图为晨侧（红线）和昏侧（蓝线）的太阳风能量与 ENA 绝对温度的关系。虚线是 95％置信区间

5.3　月球微磁层机理研究

南极艾特肯盆地是月球上最大最古老的撞击盆地，它有很特殊的表面特性，例如低反照率并且在盆地北向边缘还有一些磁异常区域。从 Chandrayaan－1 在南极艾特肯盆地的 ENA 观测中，并没有发现 ENA 散射率与月表高程、反照率以及铁、钍含量之间的关系。嫦娥四号中性原子分析仪由于其特殊的观测几何，在研究 ENA 能谱与月表物质的依赖性方面具有一定优势，但目前 ENA 能谱对表面性质的依赖性尚不明确。

月球广泛分布的局部剩磁，某些地方的强度可达上百 nT，它们可以阻碍太阳风的运动，局部形成磁层结构，称为微磁层。然而，以往所有的观测证据均来自距离月面几十千米甚至更远的航天器，Chadrayaan－1 只发现磁异常区 ENA 通量减弱，推测可能与微磁层有关。微磁层的形成机制以及内部结构目前仍然不清楚，在月面是什么样更是无从知晓。

嫦娥四号的着陆点位于月球最大磁异常区——雨海对趾区的东部边缘。当嫦娥四号位于磁异常下游时（昏侧），中性原子探测仪测量的粒子通量整体低于其位于磁异常上游时（晨侧）的测量值。结合全球 Hall MHD 模拟，证实了月球微磁层是造成通量差异的原因（图 5－7）。进一步分析发现，微磁层对太阳风的作用主要是偏转和减

速，而不能完全遮挡太阳风，其平均遮挡效率约为 50%。此外，微磁层只有在离子惯性长度足够小（小于 120km）时才可能形成，且其遮挡效率与太阳风动压反相关。ASAN 的观测结果，有助于人们认识月球微磁层的形成条件和内部结构，同时也为进一步探测微磁层提供了线索。

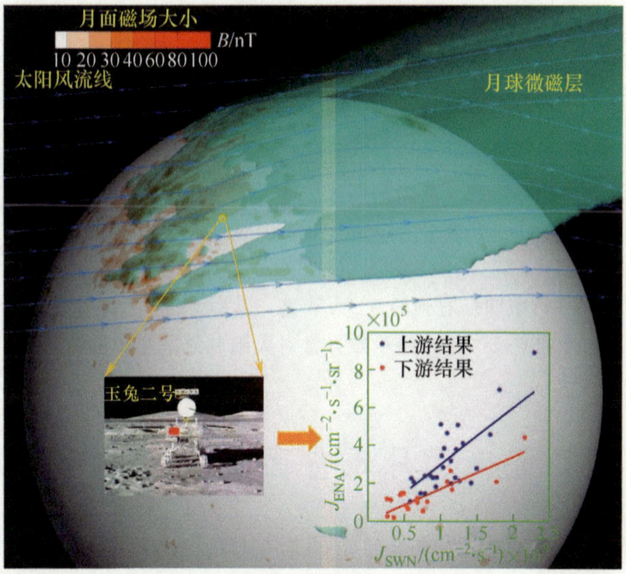

图 5-7　嫦娥四号在月面观测到 ENA 通量晨昏不对称，经 MHD 仿真验证，是微磁层对太阳风的遮挡造成的

利用嫦娥四号月面 ENA 测量数据以及 ARTEMIS 两颗在轨卫星数据实现了太阳风与月球磁异常相互作用的首次天地联合观测。研究人员通过分析数据找到一次难得的事件，事件中太阳风离子惯性长度只有 50km 左右（低于 98%以上的太阳风），理论上最有利于形成微磁层，从而为验证微磁层中是否存在密度空腔提供了良好机会。事件中 ARTEMIS 卫星在磁异常下游看到了明显的激波结构，这与离子惯性长度小有利于形成微磁层的预期一致。然而，通过分析嫦娥四号 ENA 数据发现虽然有 2/3 的太阳风被磁异常挡住了，但仍然有 1/3 的太阳风穿过了磁异常并到达月面。此外，在穿越磁异常的过程中，太阳风的能量从 492eV 降到了 325eV（图 5-8）。这些结果表明，即使在最有利于形成微磁层的情况下，磁异常仍然不能完全屏蔽太阳风（穿透率 1/3），并且太阳风的速度也没有降到 0。因此，所谓的微磁层中心密度空腔区是不存在的，磁异常对太阳风的主要作用是偏转和减速，从而造成月面法向太阳风通量降低，在一定程度上保护了月面。这些研究结果表明，所谓的月球微磁层不同于一般意义上的磁层，在它的中心磁场区并没有密度空腔，而只有太阳风偏转带来的边界层。另外，在磁异常上游也没有分离的弓激波，而只有下游的尾激波。这些大大提高了人们对于太阳风与月球磁异常相互作用的认识，为研究空间物理中太阳风与其他小尺度磁场的相互作用提供了参考。

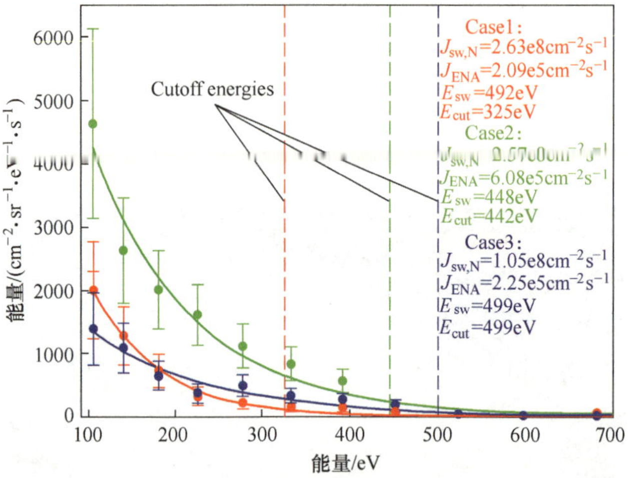

图 5-8　嫦娥四号看到的 ENA 能谱，其中 Case1 位于微磁层内，Case2 和 Case3 用于展示磁异常对太阳风的减速和偏转作用

5.4　就位温度测量揭示月球背面月壤热物理性质

在嫦娥四号着陆器上的两根月球车导轨的末端底部，安装有 4 个与月壤直接接触的温度计（图 5-9 中的 T1～T4），每 900s 测量一次月壤的温度，测量精度为 0.3K，首次对嫦娥四号着陆点的月球背面温度测量数据进行了分析，计算了着陆点月壤的热导率。研究结果首次揭示了月球背面月壤的热物理性质，表明月壤具有非常好的隔热性，为全月月表热物理性质的遥感分析提供了新的“地面真值”。

图 5-9　嫦娥四号着陆器温度计布局，彩色圆点分别指示四个温度计（T1～T4）的位置

分析温度测量的结果表明，白天月壤的温度受太阳辐射和周围环境热辐射的影响，温度在日出后快速上升，正午附近时达到最大值，在傍晚附近急剧下降。夜间月表温度的变

化主要受月壤的热物理性质影响，温度缓慢下降（图5-10a）。在月球当地时间正午附近时，四个温度计的测量温度突然降低，而后急剧上升（图5-10b），此变化与着陆器的阴影运动相关。

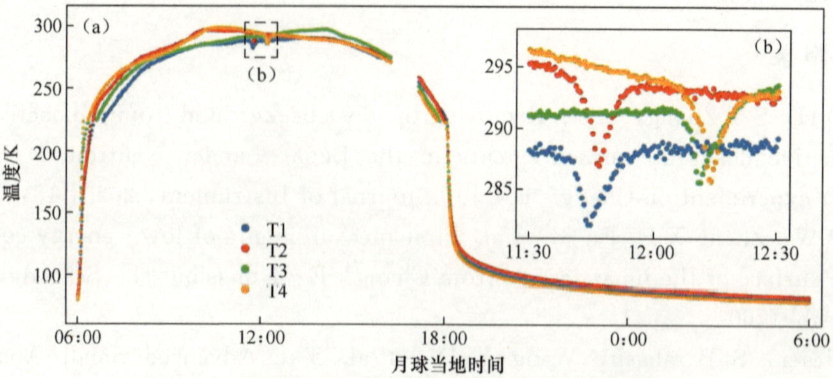

图5-10 嫦娥四号着陆点着陆后第三个月的月壤温度。

横坐标为24小时制的月球当地时间，纵坐标是测量的月壤温度

(a) 彩色圆点分别表示四个温度计测量的温度；(b) 正午附近时温度的变化特征

对着陆点的表面月壤温度和热物理性质的详细分析结果显示着陆点的月壤颗粒平均粒径约为15μm，月壤的温度变化在0.4m深以下可以忽略不计（图5-11a）。热导率中的热传导分量在月表约为 1.53×10^{-3} W/(m·K)，在月壤内1m深处约为 8.48×10^{-3} W/(m·K)（图5-11c）。

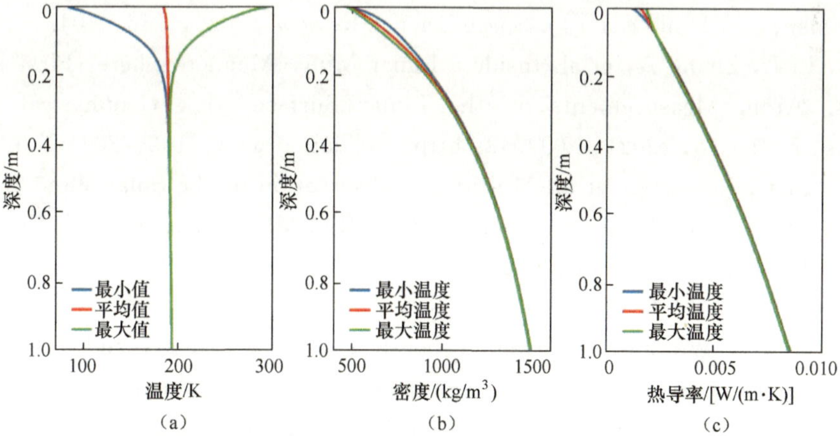

图5-11 嫦娥四号着陆点月壤粒径为15μm的温度、密度和热导率剖面图

(a) 从月表到1m深度的最小、平均和最大温度的剖面图，(b) 和 (c) 从月表到1m深度的密度剖面图和热导率中的热传导分量剖面图，对应 (a) 中的最小、平均和最大温度

5.5 小结

LND和ASAN目前仍在月球背面正常工作，源源不断地向地面输送观测数据。长期的观测不仅可获取更多事件的观测资料，同时还可得到宇宙线和太阳风长期变化带来的月

表环境的变化。如 LND 可提供太阳周内的月表粒子辐射环境数据；而随着玉兔二号的月面巡视，ASAN 观测将实现更大的空间覆盖，探寻轨道器遥感与月球车原位观测之间的关系。嫦娥四号中性原子和粒子辐射研究团组将持续关注 LND 和 ASAN 新的观测数据，并得到新的研究成果。

主要成果附录

[1] Hou D H, S Y Zhang, et al. Removing the dose background from radioactive sources from active dose rate measurements in the Lunar Lander Neutron & Dosimetry (LND) experiment on Chang' E 4 [J]. Journal of Instrument, 2020, 15: 1.

[2] Luo P W, Zhang X P, Fu S, et al. First measurements of low - energy cosmic rays on the surface of the lunar farside from Chang' E - 4 mission [J]. Sci Adv, 2022, 8 (2): eabk1760.

[3] M Wieser, S Barabash, Wang X D, et al. The Advanced Small Analyzer for Neutrals (ASAN) on the Chang' E - 4 Rover Yutu - 2 [J]. Space Science Reviews, 2020, 216: 73. https: //doi. org/10. 1007/s11214 - 020 - 00691 - w.

[4] S Zhang, R F Wimmer - Schweingruber, J Yu, et al. First measurements of the radiation dose on the lunar surface [J]. Sci Adv, 2020, (6, eaaz1334).

[5] WANG H Z, XIAO C, SHI Q Q, et al. Energetic Neutral Atom Distribution on the Lunar Surface and Its Relationship with Solar Wind Conditions [J]. The Astrophysical Journal Letters, 2021, 922: L41. https: //doi. org/10. 3847/2041 - 8213/ac34f3.

[6] Wimmer Robert F, et al. The Lunar Lander Neutron and Dosimetry (LND) Experiment on Chang'E 4 [J]. Space Science Review, 2020, 216: 104.

[7] Xie L, Li L, Zhang A, et al. Inside a Lunar Mini - Magnetosphere: First Energetic Neutral Atom Measurements on the Lunar Surface [J]. Geophysical Research Letters, 2021, 48, e2021GL093943. https: //doi. org/10. 1029/2021GL093943

[8] Xie L, Li L, Zhang A, et al. Multipoint Observation of the Solar Wind Interaction with Strong Lunar Magnetic Anomalies by ARTEMIS Spacecraft and Chang'E - 4 Rover [J]. The Astrophysical Journal Letters, 2022, 937: L5 (5pp). https: //doi. org/10. 3847/2041 - 8213/ac903f.

[9] Xu Z, Guo J, Wimmer - Schweingruber R F, et al. Primary and albedo protons detected by the Lunar Lander Neutron and Dosimetry experiment on the lunar farside [J]. Front Astron Space Sci, 2022, 9: 974946. doi: 10. 3389/fspas. 2022. 974946.

[10] Xu Z G, Guo J N, Robert F Wimmer - Schweingruber, et al. First Solar Energetic Particles Measured on the Lunar Far - side [J]. The Astrophysical Journal Letters, 2020, 902: L30.

[11] Zhang A B, et al. Emission of energetic neutral atoms measured on the lunar surface by Chang' E - 4 [J] . Planetary and Space Science, 2020, 189.

[12] 谢良海，张爱兵，李磊，等. 嫦娥四号能量中性原子观测揭示太阳风与月面相互作用新特征 [J]. 空间科学学报，2022，42 (1)：11 - 24.

[13] 张珅毅，等 . LND 两年月表辐射剂量测量结果 [J] . 空间科学学报，2021，41：3.

Removing the dose background from radioactive sources from active dose rate measurements in the Lunar Lander Neutron & Dosimetry (LND) experiment on Chang'E 4

D. Hou[1,2,3], S. Zhang[1,2,3], J. Yu[4], R. F. Wimmer - Schweingruber[1,4], S. Burmeister[4], H. Lohf[4], B. Yuan[1,3], G. Shen[1,3], C. Wang[1,2], X. Hou[5] and B. Ren[5]

1. National Space Science Center, Chinese Academy of Sciences, Beijing, China
2. University of Chinese Academy of Science, Beijing, China
3. Beijing Key Laboratory of Space Environment Exploration, Beijing, China
4. Christian Albrechts University, Kiel, Germany
5. NO, 18th Research Institute, China Electronics Technology Group Corporation, Tianjin, China

Abstract: The Lunar Lander Neutron & Dosimetry (LND) experiment is part of the scientific payload of the Chinese Chang'E 4 spacecraft which landed on the Moon on January 3, 2019. The LND measures the radiation environment on the surface of the moon in preparation of future manned missions to the Moon. There are, however, also four radioactive sources on the lander, a radioisotope thermoelectric generator (RTG) and three radioisotope heater units (RHUs) which provide heat and power for the instruments on the Chang'E 4 lander. The radiation emitted by these radioactive sources leads to a non - negligible background which interferes with the measurements of LND.

The aim of this paper is to describe the method that how to remove the background from these radioactive sources on the Chang'E 4 lander. We measured the effect of the RTG and RHUs on LND in a laboratory on Earth which is a very different environment from that on the Moon. We discuss how to take these major differences into account using a combination of scaling laws and Monte - Carlo simulations.

Keywords: Models and simulations; Radiation calculations; Particle detectors; Space instrumentation

1　Introduction

The Lunar Lander Neutron & Dosimetry (LND) experiment (as shown in Fig. 1)[1] on China's Chang'E 4 spacecraft which landed on the Moon on January 3, 2019. LND was designed and selected to measure both the charged and neutral particle radiation environment on the lunar surface in preparation of human exploration. In order to survive the very cold and long lunar nights, the Chang'E 4 lander spacecraft relies on a combination of one radioisotope thermoelectric generator (RTG) and three radioisotope heater units (RHU1 - 3). These contribute a significant background to the LND measurements of absorbed dose rate which needs to be accounted for in order to provide an accurate assessment of the lunar radiation environment. LND was calibrated in a heavily shielded nuclear facility in Tianjin, China, and is now measuring the radiation environment on the far side of the Moon. Thus, the calibration environment was characterized by a full containment of the experiment, while half of its current measurement environment on the Moon points towards open space. This may be an extreme example for the differences between calibration of a nuclear detector and its final use, but nevertheless, it serves as an instructive example on how to account for such differences.

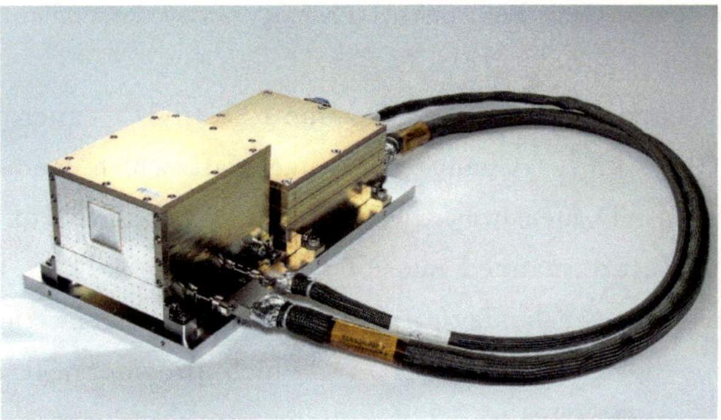

Fig. 1　The Lunar Lander Neutrons &Dosimetry (LND)

The active component of the four radioactive sources (one RTG, three RHUs) is plutonium oxide ($^{238}PuO_2$), which provides heat and power through the α decay of ^{238}Pu. The radioactive sources continuously produce neutrons and gamma - rays[2]. These particles are highly penetrating and can pass through the various shielding and spacecraft structures and cause a substantial background which is also measured by LND. The backgrounds to LND's measurements contributed by each radioactive source were measured with LND as well as with a Bonner sphere and a gamma dosimeter in the summer of 2018, i. e., shortly before launch (December 7, 2018). These measurements provide an accurate calibration of the RHU/RTG induced background, including spectra of the energy depositions in individual detectors, as well as LND's scientific data[1]. The RTG and RHU1 have the same power (120W) and RHU2 and RHU3 are both identical, smaller heater units of 5W each.

The half - life of ^{238}Pu for radioactive α - decay is 87.7 years[3]. Within the decay chain, neutrons are produced by (α, n) reactions. The (α, n) reactions with low Z isotopes (especially $^{17}O/^{18}O$, as this reaction is energetically not allowed for ^{16}O.) in RTG/RHUs is the primary source of the neutron background. Another neutron source of the RTG/RHUs is the spontaneous fission of ^{238}Pu with a half time of 5×10^{10} years[4]. Because this is much longer than the α - decay half - life, we have neglected it in the further considerations. The reactions producing the gamma - ray background from the RTG/RHUs are similar to the those for the neutron background mentioned above and include (1) spontaneous fission of ^{238}Pu; (2) induced fission; (3) (α, nγ) reactions with light elements present in the RTG/RHUs, especially $^{17}O/^{18}O$. In addition to these reactions, radioactive decay always produces gamma rays to allow the nucleus to de - excite [5]. The exact amount of the RTG and RHU's plutonium, as well as detailed information about its interior position and the shielding provided by the RTG and RHUs are, however, not known to us. Therefore, we did not model the (unknown) internal structure of the RTG/RHUs, but used published measurements of the gammaray spectra of the RTG[10] used by the ESA/NASA Ulysses space mission[11]. We scaled it to the measurements acquired with LND at the Tianjin facility. For the neutrons, we used the spectrum published by Taherzadeh[4], who measured it for plutonium dioxide fuel, and scaled it to the LND data. Because the half - life of ^{238}Pu is much longer than the mission lifetime of Chang'E 4, we may consider the background radiation caused by the RTG/RHUs to be comparatively stable and we can estimate the background radiation of LND on the lunar surface using ground test results from Earth after the corrections discussed in the following sections.

2　Radioactive source background test: measurements

LND's sensor head consists of ten silicon solid - state detectors (A - J) [1], each 500 microns thick. The second detector, B, is used for measuring the total absorbed dose rate on the lunar surface. The detectors B, C, and D are packed as close together as possible as shown in Fig. 2. We define the inner segment of C as C_1 and the outer one as C_2. The neutral particles are determined by measuring the energy deposition in C_1 in anti - coincidence with detectors B and D as well as C_2. So the absorbed dose rate of neutral particles selected by $C_1\overline{BDC2}$ can be called neutral absorbed dose rate.

During the calibration measurements, all unnecessary equipment was removed from the test room to simplify the setup as much as possible. LND was placed in the correct positions and orientations with respect to the RTG/RHUs and, when appropriate,

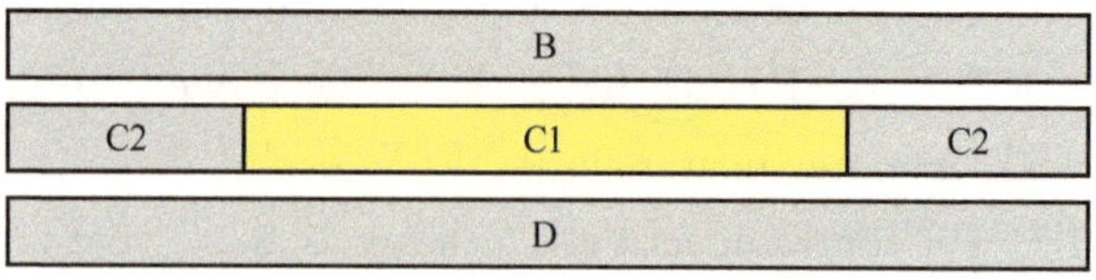

Fig. 2 LND measures neutral particles with the inner segment of detector C, C1, in anti-coincidence with all surrounding detector segments (B, D, and C2)

aluminum shielding was added to simulate the shielding between RTG/RHUs and LND in the Chang'E 4 lander. The omni-present background from cosmic rays and natural activity was accounted for by subtracting data from background runs without the RTG/RHUs. For these background runs LND was kept in the same position and orientation as for the runs with the respective RTG/RHUs. Fig. 3 shows the spectra acquired with the B detector (magenta solid line) and the C1 detector segment (blue solid line) with the RTG present. The corresponding background measurements are shown as dashed lines. The signal is clearly substantially larger than the natural background present in the calibration facility, which illustrates the importance of accounting for the influence of the RTG and RHUs in the measurements on the Moon. A photograph of the experimental setup for this case is shown in Fig. 4. This procedure was repeated for the three RHUs.

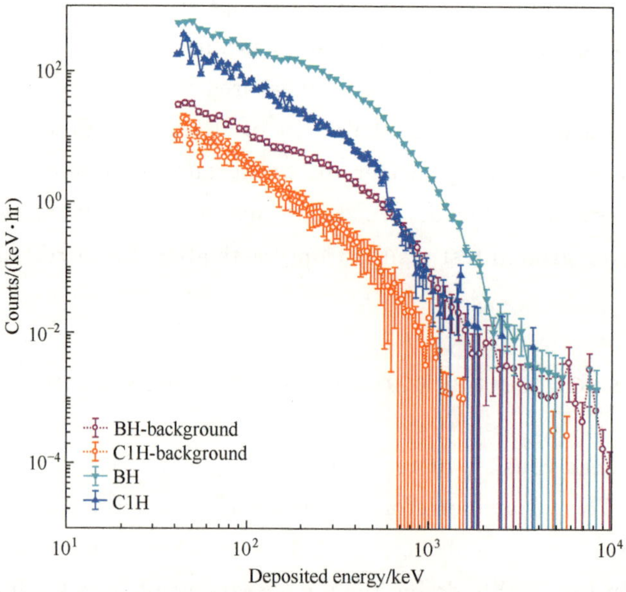

Fig. 3 The deposited energy spectrum in detector B and C1 with the Chang'E 4 RTG present (solid lines, filled symbols) and the corresponding background measurements (dashed lines, empty symbols). The Green solid line shows the spectrum of energy deposited in the B detector, while the blue solid line shows the spectrum of energy deposited by neutral particles in detector C1, i. e., using the anti-coincidence discussed in the text. The dashed curves show the corresponding background measurements

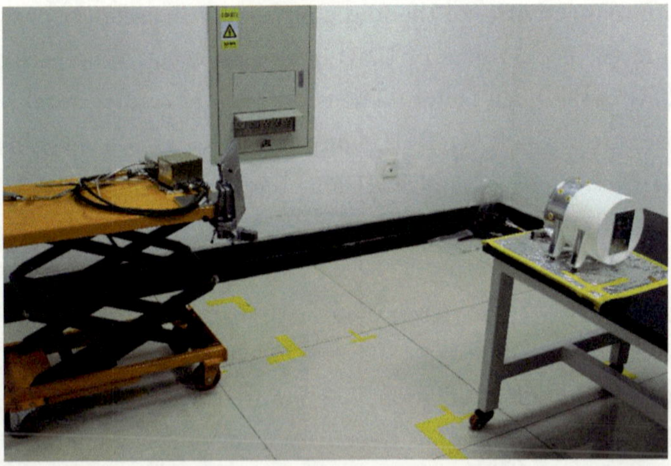

Fig. 4　Test setup for the RTG in the Tianjin facility. LND is on the right of the yellow vehicle, while the RTG is set on the black laboratory table. The Al plate also on the right corner of the yellow vehicle, is mounted between the RTG and LND to mimic the shielding by the Chang'E 4 lander vehicle

Absorbed dose is the energy deposited in a medium by ionizing radiation per unit mass. Thus, the absorbed dose[6] to Silicon (unit: μGy) is given by the equation

$$\text{Absorbed Dose} = 1.602 * 10^{-7}(\mu\text{Gy}) * \Delta(\text{MeV})/\Delta M(\text{g}) \quad (2.1)$$

where ΔE is the total deposited energy in keV, and ΔM is the mass of detector in grams.

Tab. 1 lists the total absorbed dose rates measured by the LND's B detector with the RTG/RHUs present, the corresponding background absorbed dose rate, as well as the differences, i. e., the absorbed dose rates due to the neutrons and gamma rays emitted by the RTG/RHUs. Similarly, the results of neutral radiation dose rates detected by C1 are shown in Tab. 2.

Tab. 1　Total absorbed dose rate for RTG/RHUs measured in B. Dr=absorbed dose rate

Radioactive source	$Dr_B/(\mu Gy/h)$	$Dr_{bkgnd}/(\mu Gy/h)$	$Dr_B - Dr_{bkgnd}$
RTG	2.21±0.08	0.11±0.03	2.10±0.09
RHU1	0.74±0.05	0.11±0.03	0.63±0.06
RHU2	2.22±0.08	0.11±0.02	2.11±0.08
RHU3	1.45±0.06	0.11±0.02	1.34±0.06
total	6.62±0.11	0.44±0.05	6.18±0.15

Tab. 2　Total (neutral) absorbed dose rate for RTG/RHUs measured in C1

Radioactive source	$Dr_n/(\mu Gy/h)$	$Dr_{h,bkgnd}/(\mu Gy/h)$	$Dr_n - Dr_{nbkgnd}$
RTG	0.72±0.07	0.03±0.02	0.69±0.07
RHU1	0.24±0.04	0.03±0.02	0.21±0.04
RHU2	0.76±0.07	0.03±0.01	0.73±0.07
RHU3	0.53±0.06	0.03±0.01	0.50±0.06
total	2.25±0.12	0.13±0.03	2.13±0.12

Due to the different activities of the four sources and the relative position of the LND, their contribution to the background of LND is also different. The background caused by the RHU2 and RTG is the main factor. The experimental results in table 1 show that the total absorbed dose rate of the RTG/RHUs on LND is 6. 18±0. 15μGy/h, and the neutral absorbed dose rate from the RTG and RHUs at the Tianjin facility is 2. 13±0. 12μGy/h. In the following we will derive an estimate for this background on the Moon.

3　Simulation of the laboratory setup

In the Tianjin calibration, radiation was emitted in all directions, not necessarily isotropically, because of the internal structures and self - shielding of the RTG and RHUs. Neutrons and gamma rays which have scattered of the walls, ceiling, and floor contribute to the measurements made by LND. On the lunar surface, however, there is no scattering off the walls and ceiling, but only off the lunar surface on which Chang'E 4 landed. Therefore, we need to determine the fraction of the particles detected by LND which were not scattered (i. e., came directly from the source) or were only reflected off the floor, as opposed to all particles which could reach LND. The simulation of this Tianjin calibration test is described below, some sample particle trajectories are indicated in Fig. 5.

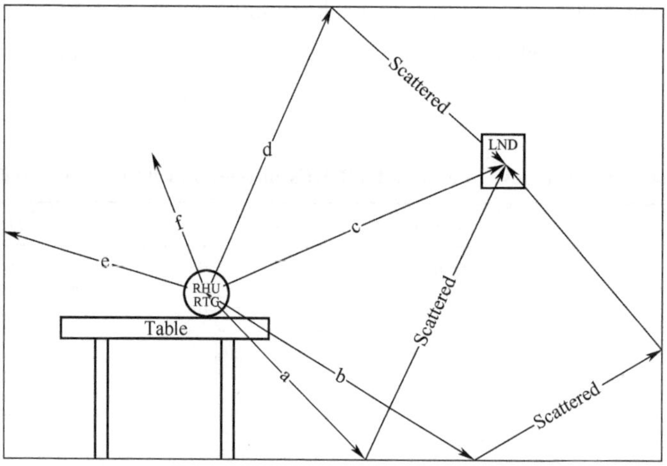

Fig. 5　Illustration of the scattering by floor, walls, and ceiling of the test facility. This scattering results in an enhanced background compared to the situation on the Moon. Trajectories 'a' stands for the particles that only scattered by floor before entering LND; trajectories 'c' stands for the particles enter LND directly without scattering; trajectories 'b', 'd' stand for the particles that scattered by walls or ceiling before entering LND; trajectories 'e' stands for particles 'absorbed' by floor, walls or ceiling; 'f' stands for the particles that scattered by atmosphere

The simulation model was set up as shown in Fig. 6. We used a custom GEANT4 setup with QGSP _ BERT _ HP, which at lower energies uses the High Precision neutron models and cross section data based on the ENDF/B - VII[8,9]. We tracked every particle so as to know the path by which it reached LND, i. e., whether they hit LND directly or were scattered of the floor, walls, or ceiling. This was done for all four sources.

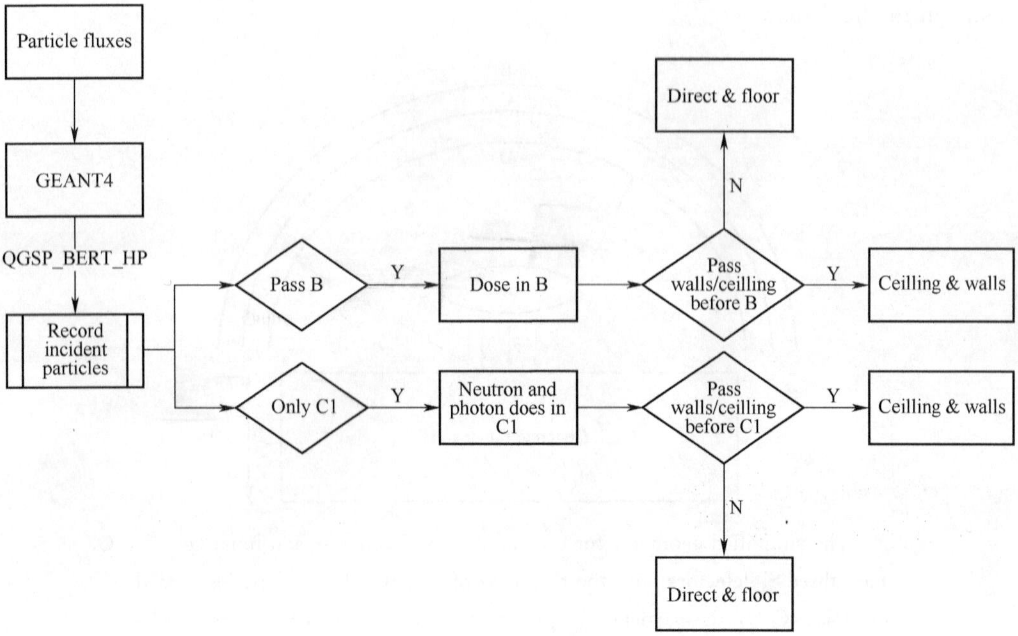

Fig. 6　Logic schematic diagram of simulation

We recorded all particles which hit LND and were detected in detector C1 only (i. e., a neutral particle) or in detector B. For particles detected in B, we added the energy deposition to the energy already accumulated in B (to derive the total absorbed dose in B). For particles detected in C1 only, we accumulated the "neutral dose" as well. Subsequently, we filtered for those particles which hit B (C1) directly (i. e., without scattering off the walls, ceiling, or floor) and those which reached B (C1) via walls, ceiling, and floor, keeping track of which was their first scattering "surface". If their first scattering was on the ceiling or walls, their contribution was accepted for the simulation of the laboratory situation, but not for the simulation of the Moon. If their first and only scattering "surface" was the floor, or they reached LND without scattering, they were accepted for both cases. All other cases were only accepted for the simulation of the laboratory situation. This allows us to discriminate between the situation for the Moon and the laboratory using the same (time consuming) simulation. In other words, we only had to simulate the (more complicated) laboratory setup, but not the Moon, which saved considerable computation time. In fact, the composition of the lunar soil and concrete is not exactly the same (especially for water content), so the scattering effect on the neutrons and gamma rays is also different. The Monte Carlo program was used to simulate the lunar soil[12] and concrete[7,8], using the very much simplified geometry as shown in Fig. 7. We use the neutron (gamma) spectrum of PuO_2 for concrete and lunar soil simulation. We define the dose ratio = dose (lunar) /dose (concrete), so the dose ratio in detector B is DR_B, while the neutral dose ratio in detector C is DR_C. The results are shown in the Tab. 3.

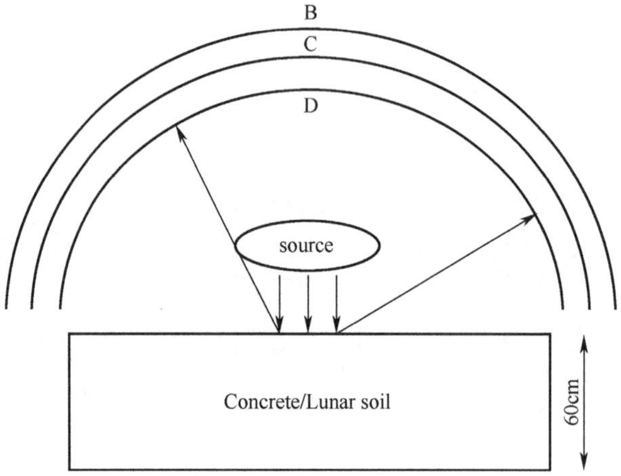

Fig. 7 The simplified geometry for the simulation of concrete and lunar soil. B, C, D are three Si detectors with the thickness of 0.5mm. If the particles pass the detector B/C/D, the particles are considered to be scattered by concrete/Moon

Tab. 3 The simulation results for neutron and gamma scattered by lunar soil or concrete. Dose ratio (DR) = Dose (lunar) /Dose (concrete)

Particle type	DR_B	DR_C
Neutron	59%±7.6%	100%±16%
gamma	88%±11%	92%±10%

4 Results

In this section we compare the simulation and the measurements results of Tianjin calibration test. For the rest of this paper we introduce the following notation $A_{b,e}$ (s), where the 'A' stands for the dose rate (measurements (M) or simulations (S)); the 'b' stands for the type of particles (neutrons or gamma rays); the 'e' stands for the environment (lab, floor, ceiling, etc.); and 's' is the type of source (RTG, RHU1 - 3). For example, some specific definitions of RHU2 and RHU3 are shown in Tab. 4.

We simulated neutron and gamma radiation emitted by the RTG and three RHUs. Consider, for example, the gamma - ray spectrum of RHU2. From our simulations we can determine the energy deposited in detector B, and convert it to the (simulated) dose (rate) measured by B by dividing by its mass (and exposure duration). This can be done for all particles (i. e., the laboratory condition) or only those particles which hit B directly or only via the floor of the laboratory (as on the Moon). Because we simulate the radiation dose measured with the LND B detector, we can easily filter out all gamma rays which were scattered by the walls and ceiling of the test room. The radiation dose produced by the particles which entered detector B directly or only passing through the floor before entering detector B is $S_{g,df}$ (RHU2). Similarly, we can simulate the radiation dose which produced by same number (N) of neutrons, and their direct contribution to the radiation dose, $S_{nd(RHU2)}$, or the one via the floor, $S_{nf(RHU2)}$.

Tab. 4　Some specific definitions of RHU2 and RHU3. In the definition, 'M' stands for measurements, 'S' stands for simulation; 'n' stands for neutron, 'g' stands for gamma; 'ng' stands for neutron *and* gamma, 'df' stands for the particles that entered detector directly (d) or only passing through the floor (f) before entering detector, 'cw' stands for the particles that pass through ceiling or walls before entering detector, 't' stands for 'total', i. e., all particles that entered detector

Source	Measure	Simulate (direct/floor)	Simulate (ceiling/walls)	Simulate (total)
RHU2	$M_{ng,lab}$ (RHU2)	$S_{g,df}$ (RHU2)	$S_{g,cw}$ (RHU2)	$S_{g,t}$ (RHU2)
		$S_{n,df}$ (RHU2)	$S_{n,cw}$ (RHU2)	$S_{n,t}$ (RHU2)
RHU3	$M_{ng,lab}$ (RHU3)	$S_{g,df}$ (RHU3)	$S_{g,cw}$ (RHU3)	$S_{g,t}$ (RHU3)
		$S_{n,df}$ (RHU3)	$S_{n,cw}$ (RHU3)	$S_{n,t}$ (RHU3)

For each of the four sources (RTG, RHU1, RHU2, RHU3) we have the measured dose rates in B and C1. In the following, we will attempt to derive a weighting factor x (y) for gamma rays (neutrons) which is equivalent to the probability of measuring a gamma ray (neutron) from the RTG/RHU, as shown in eq. (4. 1) and eq. (4. 2). It turns out that we only need to determine the ratio m = x/y, as we shall see shortly in eq. (4. 3) and eq. (4. 4).

$$x = N_{gamma}/N_t \tag{4. 1}$$

$$y = N_{neutron}/N_t \tag{4. 2}$$

where N_{gamma} is the number of gamma from RTG/RHU in one hour, and $N_{neutron}$ is the number of neutron from RTG/RHU in one hour, and $N_t = N_{neutron} + N_{gamma}$. So '$S_{g,t}$ (RHU2) × (x×N_t) /N' is the dose rate due to gamma rays emitted by RHU2 in the simulation.

RHU2 and RHU3 are identical for our purposes (and have the same heating power of 5W), and so we can use the following formulas (4. 3) and (4. 4) to calculate the ratio of weighting factors for neutrons and gamma rays, m=x/y,

$$S_{g,\ t}(RHU2) * (x * N_t)/N + S_{n,\ t}(RHU2) * (y * N_t)/N = M_{ng,\ lab}(RHU2) \tag{4. 3}$$

$$S_{g,\ t}(RHU3) * (x * N_t)/N + S_{n,\ t}(RHU3) * (y * N_t)/N = M_{ng,\ lab}(RHU3) \tag{4. 4}$$

From our simulations we can derive the ratio of doses from gamma rays (neutrons) which hit an LND detector directly or via a single scattering with the dose rate from all gamma rays (neutrons) which interact with that detector. This probability, P_{RHU2} is given by eq. (4. 5) below.

From table 1, we can take the background corrected value for $M_{ng,lab}$ (RHU2) (2. 11±0. 08μGy) and multiply it by P_{RHU2} (97%) to obtain the expected value for the radiation background due to RHU2 on the Moon. These corrected values are given in table 4, the total background due to the combination of RTG and RHUs is 5. 20±0. 56μGy/h.

$$P_{RHU2} = (m * S_{g,df}(RHU2) + S_{n,df}(RHU2))/(m * S_{g,t}(RHU2) + S_{n,t}(RHU2)) \quad (4.5)$$

$$D_{Moon2} = M_{ng,lab}(RHU2) * P_{RHU2} \quad (4.6)$$

Tab. 5　Result for radiation dose rates in B for direct/floor from RTG/RHU

Radioactive source	P_{RX}	Lunar background/(μGy/h)
RTG	74%	1. 56±0. 29
RHU1	51%	0. 32±0. 11
RHU2	97%	2. 05±0. 35
RHU3	95%	1. 27±0. 3
total	/	5. 20±0. 56

As discussed for the example of RHU2, the probabilities for direct or single-bounce contributions to the dose rate in B, P_{RX}, are shown in the second column of the Tab. 6. The third column is the thus corrected contribution of the RTG or RHUs to the measured dose rate on the Moon. Obviously, although the radioactive sources RTG and RHU1 have the largest heating power (120W), the probability to detect their radiation in LND is quite different, which is due to the different distances and shielding between them and the LND. Moreover, because LND is much farther away from RHU1 than from the RTG, the contribution of radiation scattered by the walls and ceiling is higher than for the RTG and our simple model may not be very accurate for this large difference in distances. Because the contribution from RHU1 is also small compared to the others, we include this uncertainty in our error estimate. But for RHU2 and RHU3, the radiation dose rate measured in Tianjin is dominated by the particles which reach LND without scattering off the walls or ceiling, because of the closer distance between the two radioactive sources and LND.

The radiation dose rates from the RTG/RHUs measured in C1 show similar results as the radiation dose rates for B analyzed above. We use the same method to calculate the neutral channel (C1) efficiency of background radiation dose rates from the RTG/RHUs. The results are shown in Tab. 6. The total neutral radiation dose rate is 1. 69 ± 0. 46μGy/h.

Tab. 6　Result of neutral radiation dose rates direct/floor from RTG/RHUs

Radioactive source	P_{RX}	Moon background/(μGy/h)
RTG	64%	0.44±0.37
RHU1	30%	0.06±0.10
RHU2	97%	0.71±0.19
RHU3	96%	0.48±0.17
total	/	1.69±0.46

5　Error estimation

While an exact quantification of the errors associated with the many steps of the procedure described in this paper is probably not possible, we attempt to estimate them as well as we can. There are a number of sources of errors, some physical, some statistical, and some from modeling, which we summarized in the following. The room in which the RTG/RHU calibrations was simplified for our simulations, we did not simulate every corner and door. We account for this uncertainty with an overall 3% error. Similarly, the spectrum of the RTG and RHUs is unknown to us but we use the one for ESA's Ulysses mission[11] and Taherzadeh measured[4]. We account for this uncertainty by an overall 5% error. Finally, GEANT4 itself is not free of uncertainties and we include another 5% error to account for this. Adding quadratically, we arrive at an error estimate of 7.7%. The uncertainties given in Tab. 1, 2, 5 and 6 were calculated using standard error propagation for equations (4.3), (4.4), and (4.5) to arrive at a final error of 0.56μGy/h for detector B and 0.46μGy/h for detector C1.

6　Summary and conclusions

In this paper we have used the LND instrument aboard Chang'E 4 to describe how the background from RTG and RHUs in measured dose rates can be corrected based on simulations and calibrations. The method involves measuring and modeling the background from each RTG/RHU in a calibration facility. The situation in free space (or in our case on the Moon) can not be calibrated without breaking strict regulations in radiation protection legislation, but it can be simulated. We have shown that carefully keeping track of the simulated radiation allows us to reuse the same simulation as performed for the calibration facility and simply ignore those particles that reached the detector via interactions with the walls or ceiling. This method allowed us to provide an estimate of the radiation background due to the RTG and RHUs, Previous measurements by Chandrayaan-1 and LRO of the dose rate in lunar orbit values range from 9 to 13μGy/h for the total dose rate from charged and neutral particles. We found a value of 5.2±0.56μGy/h for the total dose rate due to the RTG and RHUs, while their contribution to dose rate from neutral particles (neutrons and gamma rays) was 1.69 ± 0.46μGy/h. This is a substantial contribution and the background corrections discussed in this paper are indeed important.

Acknowledgments

This paper is supported by the Beijing Municipal Science and Technology Commission, Grant No. Z181100002918003, and National Natural Science Foundation of China Grant 11603028, and National Natural Science Foundation of China Grant: 41941001. The Lunar Lander Neutron and Dosimetry (LND) instrument is supported by the German Space Agency, DLR, and its Space Administration under grant 50 JR 1604 to the Christian-Albrechts-University (CAU) Kiel. We would like to thank the following facilities for supporting the calibration of LND: CERN/CERF in Geneva, Switzerland; NIRS, HIMAC in Chiba, Japan; ATI in Vienna, Austria; SFE at NSSC China, We thank the Lunar Exploration and Space Engineering Center as well as the China National Space Administrator (CNSA) and DLR, Germany for their unwavering support of Chang'E4.

References

[1] R. F. Wimmer-Schweingruber et al., The Lunar Lander Neutron and Dosimetry (LND) Experiment on Chang'E 4, manuscript to be submitted to Space Sci. Rev. (2019).

[2] V. P. Chechev et al., The Evaluation of 238Pu, 240Pu, *and 242Pu* Decay Data, in AIP Conference Proceedings. Vol. 769: International Conference on Nuclear Data for Science & Technology, American Institute of Physics, College Park U. S. A. (2005).

[3] NuDat2.5., http://www.nndcbnl.gov/nudat2/ (2011).

[4] M. Taherzadeh and P. J. Gingo, Neutron Radiation Characteristics of Plutonium Dioxide Fuel, Nucl. Technol. 15 (1972) 396.

[5] R. A. Neff, M. E. Anderson, A. R. Campbell and F. X. Haas, Some Neutron and Gamma Radiation Characteristics of Plutonium Cermet Fuel for Isotopic Power Sources, in Proceedings of the National Symposium on Natural and Manmade Radiation in Space, Washington U. S. A. (1972), pg. 853.

[6] C. Zeitlin et al., Comparisons of high-linear energy transfer spectra on the ISS and in deep space, Space Weather 17 (2019) 396.

[7] R. Khabaz, Analysis of neutron scattering components inside a room with concrete walls, Appl. Radiat. Isot. 95 (2015) 1.

[8] Geant4 collaboration, Geant4's Guide for Application Developers, Version Geant4. 10.0 (2013).

[9] G. F. Cao et al., Test of hadronic interaction models in GEANT4 at low energy using the BESIII data, Nucl. Instrum. Meth. A 606 (2009) 700.

[10] E. A. Hawley, Radiation Measurements on the Qualification RTG, Technical report (1984).

[11] K.-P. Wenzel, R. G. Marsden, D. E. Page and E. J. Smith, The Ulysses Mission, Astron. Astrophys. Suppl. 92 (1992) 207.

[12] R. Battiston, M. T. Brunetti, F. Cervelli, C. Fidani and M. Menichelli, A Moon-borne electromagnetic calorimeter, Astrophys. Space Sci. 323 (2009) 357.

First measurements of low – energy cosmic rays on the surface of the lunar farside from Chang'E – 4 mission

Pengwei Luo[1,2], Xiaoping Zhang[1,2], Shuai Fu[1,2], Yong Li[1,2], Cunhui Li[3], Jinbin Cao[4]

1. State Key Laboratory of Lunar and Planetary Sciences, Macau University of Science and Technology, Macau, China
2. CNSA Macau Center for Space Exploration and Science, Macau, China
3. Science and Technology on Vacuum Technology and Physics Laboratory, Lanzhou Institute of Physics, Lanzhou, China
4. School of Space and Environment, Beihang University, Beijing, China

Abstract: Human activities on the lunar surface are severely constrained by the space radiation dominated by cosmic rays (CRs). Here, we report the first measurements of the low – energy (about 10 to 100MeV/nuc) CR spectra on the lunar surface from China's Chang'E – 4 (CE – 4) mission around the solar minimum 24/25. The results show that for the proton, helium, CNO, and heavy – ion groups, the ratios (ratio errors) of the CE – 4 fluxes to those from the near – earth spacecraft are 1.05 (0.15), 1.30 (0.18), 1.08 (0.16), and 1.24 (0.21), respectively, and to those predicted by the models [CRÈME96 and CRÈME2009] are instead [1.69 (0.17), 2.25 (0.23)], [1.66 (0.17), 1.76 (0.18)], [1.08 (0.11), 1.07 (0.11)], and [1.33 (0.18), 1.17 (0.15)]. Moreover, a notable enhancement of $^3He/^4He$ ratio is observed at ~12MeV/nuc, and the CR dawn – dusk symmetry is confirmed. These results provide valuable insights into the CRs on the lunar farside surface and will benefit future lunar exploration.

Introduction

The Moon is a barren dusty body with an atmosphere about 13 orders of magnitude thinner than on Earth and without a shielding global magnetic field: the result in a drastically different radiation environment from our planet. In general, the radiations mainly associated with cosmic rays (CRs) can easily reach the lunar surface, especially for the high linear energy transfer charged particles from the CR hadronic components with energies at a few megaelectron volts and above. CRs typically consist of solar energetic particles (SEPs), galactic CRs (GCRs), and anomalous CRs (ACRs). SEPs, originating from the Sun, are frequently associated with intense solar activities like solar flares and coronal mass ejections (CMEs), and they are believed to be accelerated to the energies at megaelectron volt or above (and perhaps up to gigaelectron volt) via CME - driven shock waves (1, 2). GCRs are deemed to come from the explosions of supernovae within our own Milky Way Galaxy (3), but an extra - galactic origin for ultrahigh - energy particles cannot be ruled out (4). The supernova remnants are considered to be powerful accelerators of CR particles up to the energies $\sim 10^{17}$ eV, as reported in the literature (5 - 7). The ACR component was first discovered by Garcia - Munoz et al. (8) when an anomalous enhancement was observed in the low - energy end of the He spectrum during solar quiet times in 1972. To explain the origin of ACRs, a favored interpretation was soon proposed by Fisk et al. (*9*). Specifically, neutral atoms of the local interstellar medium drift into the inner heliosphere, where they are singly ionized by solar ultraviolet radiation and/or charge - exchanged with solar wind ions, and then are picked up by the solar wind and transported to the outer heliosphere where they are accelerated to hundreds of megaelectron volts, presumably at the solar wind termination shock (10, 11). ACR components have been confirmed in several species of elements with high first ionization potentials (FIPs) like H, He, N, O, Ne, and Ar in the outer heliosphere, while in the inner heliosphere all of them except H have been observed (12 - 14). Most elements with FIPs lower than that of H have been deflected by the interplanetary magnetic field as they approach the heliosphere because they have been ionized in the local interstellar medium (11, 14). GCRs mainly consist of protons ($\sim 87\%$), helium ($\sim 12\%$), and heavier nuclei ($\sim 1\%$) (15), dominating the radiation environment in deep space during solar quiet periods (16). GCRs are continuous, while SEPs are sporadic and can occur at any point of a solar cycle, although most likely at the solar maximum (17). Further details about GCRs and ACRs can be found in our recent work (18).

Gamma rays and low - energy alpha particles make a predominantly contribution to the Earth's radiation dose, while ions of high atomic number and energy are the major contributors of deep space radiation, such as on the lunar surface (19). Astronauts and precision instruments on deep - space missions are exposed to substantial amounts of radiation, arising from high linear energy transfer ions and secondary particles (proton, neutron, gamma ray, electron, positron, heavy ions, and many other elementary particles) produced by interactions with the elements in the lunar regolith (20, 21). Arguably, one of the most serious challenges in extraterrestrial travel derive from safety concerns relating to exposure to space radiation, which effective protection measures could escalate costs beyond reach (19). Studies show that long - term exposure to high - energy radiation may carry a heavy health risk, for instance, increasing the risk of developing cataracts (22), cancer (23, 24), and degeneration of the central nervous system (*25*). Furthermore, the charged highenergy particles may cause the dielectric breakdown of the materials with sufficiently low conductivity within an instrument (26). Therefore, it is essential to produce an accurate estimation of the space radiation environment in consideration of the safety of both astronauts and instruments as well as the cost control. Overestimation will lead to a notable increase in manufacturing costs due to severe mass constraints in spaceflight, whereas underestimation will pose a serious threat to the safety of both astronauts and instruments.

An accurate evaluation of the radiation environment on the lunar surface, such as the determination of radiation dose rate along with the type and fluxes of radiation particles, is an important prerequisite for planning an extended human presence on the lunar surface. Before the launch of the Chang'E - 4 (CE - 4) mission, the assessment of the lunar radiation environment was mainly based on the data obtained by the Radiation Monitor (RADOM) and the Cosmic Ray Telescope for the Effects of Radiation Experiment (CRaTER) onboard the Chandrayaan - 1 and the Lunar Reconnaissance Orbiter (LRO) spacecraft, respectively. The Chandrayaan - 1/RADOM, a miniature dosimeter - spectrometer, was aimed at monitoring the local radiation environment both en route and in lunar orbit, including measurements of the total absorbed dose rate, the flux of surrounding energetic particles, and the spectrum of the deposited energy from high - energy particles (27). The LRO/CRaTER is designed to characterize the radiation environment to serve NASA's future lunar return missions. The LRO/CRaTER linear energy transfer measurements are used to constrain the models of the biological effects of ionizing radiation combined with the models of radiation transport through materials in the lunar environment, and they also provide valuable information about the radiation effects on electronic systems (28).

The lunar radiation environment is far more complex than previously thought, with many factors likely to affect it, such as the lunar exosphere, dust grains, and lunar magnetic anomalies. The lunar exosphere is rather tenuous at around 10^{-9} mbar: This was first detected by in situ measurements of the Apollo missions and found to consist of atoms and light molecular species, such as Ar, He, Ne, Na, K, and H (29). Dust grains above the surface are regarded to come from ejecta produced by the continual bombardment of the Moon by interplanetary micrometeoroids and small grains lifted by plasma - induced near - surface electric fields (30). The existence of lunar magnetic anomalies, i. e., regions of local magnetization, has been confirmed (31 - 33). Some magnetic anomalies have surface fields of up to hundreds of nanoteslas. The landing site of the CE - 4 (177. 6°E, 45. 4°S) (34) is located near the edge of the largest lunar magnetic anomaly and called the Imbrium antipode, centered at (162°E, 33°S) with a radius larger than 600km (32); however, the influence of the major antipodal impact on the charged particles downstream remains unclear. Furthermore, the radiation environment near the lunar surface may also be affected by many other factors, such as the solar wind reflected particles, the local terrain obscuring, and the secondary particles produced by the interactions between CRs and the nuclei of lunar regolith.

As reported by Mazur et al. (35), a clear altitude dependence of the dose rate was found by the LRO/CRaTER. The Chandrayaan - 1/RADOM also observed the variation of the particle flux at lunar altitudes in the range between 92 and 118km (27). On 3 January 2019, China successfully sent a lander (CE - 4) and a rover (Yutu - 2) to the Von Kármán crater in the South Pole Aitken basin, the largest known impact crater on the Moon and one of the largest in the solar system (36). They are both still active at the time of writing. One of the payloads onboard the lander is the Lunar Lander Neutrons and Dosimetry (LND) experiment. To have an insight into the radiation environment, Zhang et al. (37) investigated the radiation dose rate and the linear energy transfer spectrum on the surface of the lunar farside based on the data collected by the CE - 4/LND during the first two lunar days. It was found that the average absorbed dose rate from charged particles is consistent with the LRO/CRaTER measurements when the LRO flew over the location of the CE - 4. They also propose that the contribution of neutral particles (neutrons and gamma rays, mainly produced by the interactions between CRs and the lunar regolith) to the absorbed dose rate is nonnegligible with a fraction up to $(23 \pm 17)\%$ around solar minimum. To evaluate the influence of the abovementioned factors (the lunar exosphere, the dust grains, and the lunar magnetic anomalies, etc.) on the CRs, an effective method is to compare the CR spectral difference between the CE - 4 measurements and those made from space.

In addition, chemistry modifications and space weathering of physical matter by the radiation interactions of energetic particles at megaelectron volt are ubiquitous on the Moon due to the strong penetration ability of these particles (26). As an almost airless body, the radiation interactions of the Moon can deepen our understanding of radiation - induced chemistry and space weathering. For example, the radiation interactions at the Moon can be applied to detect the near - surface deposits of water ice via the absorption of low - energy neutrons produced by CR interactions [see (38) and references therein]. CRs are an important source of chemical modification and space weathering, and thus, the CR spectrum on the lunar surface plays a key role in the studies of the evolution of the lunar surface. However, our data of the CR spectrum on the lunar surface are limited and those derived from CR particle flux models are generally taken as an approximate substitution. Therefore, the CR spectrum on the lunar surface can provide these CR - related studies with actual measurements and tools to refine related models (39 - 43). Moreover, the differential flux ratio of ^3He to ^4He, which is a powerful tool to constrain the CR propagation models, was measured in various experiments as reported by Wu and Chen (44) and the literature therein. These experimental results were in the energy range between ~100MeV/nuc and several GeV/nuc, with fewer below ~100MeV/nuc.

Benefiting from the CE - 4/LND CR measurements, we aim to (i) report the low - energy CR spectrum on the surface of the lunar farside during the solar quiet period

(January 2019 to February 2020), (ii) inspect the influence of lunar surroundings on the CR spectrum on the lunar surface, (iii) calculate the differential flux ratio of ^{3}He to ^{4}He at low energies (about 10 to 100MeV/nuc), and (iv) verify the CR dawn - dusk symmetry.

RESULTS CR energy spectrum

The low - energy CR spectra measured by the CE - 4/LND during the weakest solar minimum 24/25 (18, 45, 46) are shown in Fig. 1. These ions are separated into four groups according to their respective atomic number (Z): protons ($Z=1$), He ($Z=2$), CNO group ($Z=6$ to 8), and heavy - ion (HI) group ($Z\geqslant 10$, including Ne, Na, Mg, Al, Si, S, Ar, Ca, Fe, and Ni). Besides the CE - 4/LND measurements, the CR spectra both observed by the near - earth spacecraft [Advanced Composition Explorer (ACE), Solar and Heliospheric Observatory (SOHO), and Solar Terrestrial Relations Observatory Ahead (STEREO - A)] and modeled by the Cosmic Ray Effects on Micro - Electronics (CRÈME) models (39, 40) are also plotted in the same period. In Tab. 1, we list the time interval of the CE - 4/LND that is studied here. Of note here is that the Moon was outside of the Earth's magnetotail during the CE - 4/LND working periods. To make a more intuitive and quantitative comparison, the differential flux ratios of the CE - 4/LND measurements to the spacecraft observations and the numerical modeling results are calculated and displayed in Fig. 2, and the weighted average flux ratios of each particle group over different kinetic energies are shown in Fig. 3. From Fig. 3, we can see that the CE - 4/LND data are comparable to those observations from the near - earth spacecraft within 1.7 SD (1.7σ) for all particle groups, while the flux ratios of the CE - 4/LND measurements to the CRÈME predictions reach up to a level of 1.7 with more than 4.0 σ for the proton. A notable discrepancy in the predictions between the CRÈME96 (39) and CRÈME2009 (40) models at energies below ~20MeV/nuc for the proton and He and below ~5MeV/nuc for the CNO and HI groups, respectively, is also worth noting, ascribing to the differences in their respective embedded dependency models (Fig. 1). To be specific, the CRÈME96 model may suffer partial contamination from the SEPs at low energies because of its use of the Nymmik model (47), while the revised Nymmik model (48) is applied in the CREME2009 code and the influence from SEPs is fully removed.

From the proton spectrum shown in Fig. 1A, it is found that the observations from the CE - 4/LND (red upward triangles) are comparable with those from the SOHO/EPHIN (Electron Proton Helium Instrument; blue squares) and the STEREO - A/LET (Low - Energy Telescope; violet diamonds) with an average flux ratio of (1.05±0.15) (violet bar in the first column of Fig. 3), but the comparisons of these observed results with those predicted by the models (aquamarine solid and lime dashed lines for the CRÈME96 and CRÈME2009 models, respectively) exhibit obvious deviations. Specifically, the observed fluxes are smaller than the predictions from the CRÈME96

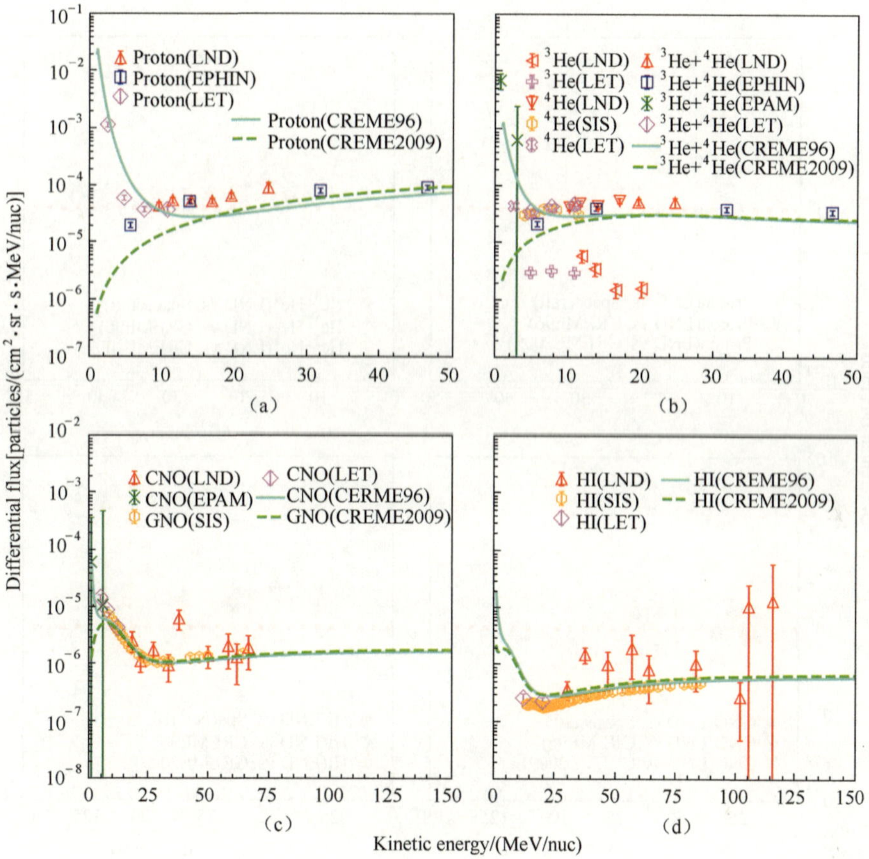

Fig. 1　Comparisons of the CR fluxes between the CE－4/LND measurements and the in situ measurements made at 1 AU and the predictions of the CRÈME models. (a) For proton. (b) for He. (c) for CNO group. (d) for HI group (Ne, Na, Mg, Al, Si, S, Ar, Ca, Fe, and Ni). Note that EPHIN, SIS and EPAM, and LET are the four instruments aboard the SOHO, ACE, and STEREO－A spacecraft, respectively. See the text for details

Tab.1 Time interval used in this study from the CE－4/LND

No.	Start time	End time
1	2019－01－03　06:00:00	2019－01－12　11:00:00
2	2019－01－30　17:00:00	2019－02－10　07:00:00
3	2019－03－30　12:00:00	2019－04－10　08:00:00
4	2019－07－27　02:00:00	2019－08－06　09:00:00
5	2019－08－25　03:00:00	2019－09－05　09:00:00
6	2019－09－24　02:00:00	2019－10－03　04:00:00
7	2019－10－23　02:00:00	2019－11－03　09:00:00
8	2019－11－22　04:00:00	2019－12－03　07:00:00
9	2019－12－21　02:00:00	2020－01－01　10:00:00
10	2020－01－20　02:00:00	2020－01－31　07:00:00
11	2020－02－18　02:00:00	2020－03－01　05:00:00

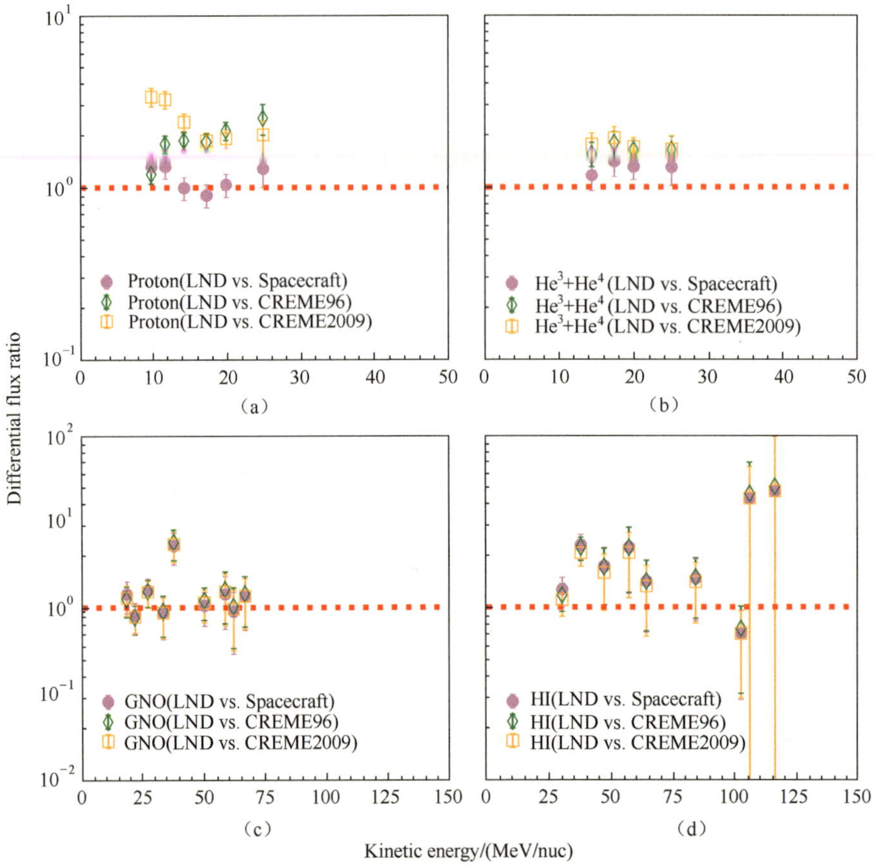

Fig. 2 CR flux ratios of the CE-4/LND measurements to the in situ measurements made at 1 AU and to the CRÈME modeling results. (a) For proton. (b) for He. (c) for CNO group. (d) for HI group (Ne, Na, Mg, Al, Si, S, Ar, Ca, Fe, and Ni). The ratio of 1.0 is indicated by a red dotted line for reference

model at energies below ~10MeV/nuc, while for energies above ~10MeV/nuc the opposite is true by a respective factor of (0.69±0.17) and (1.25±0.23) (green and orange bars in the first column of Fig. 3 for the CRÈME96 and CRÈME2009 models, respectively). There is also an obvious difference in the predictions between the CRÈME96 and CRÈME2009 models in the energy range between ~20 and 50MeV/nuc, where the CRÈME2009 model gives a relatively larger prediction than the CRÈME96 model. The proton flux ratios of the CE-4/LND measurements to the near-earth spacecraft observations are shown in Fig. 2a, along with are those of the CE-4/LND measurements to the CRÈME modeling results. For the flux ratios of the CE-4/LND measurements to the spacecraft observations, the values of these data points (filled violet circles) are comparable to 1.0 within 1.6σ with slight fluctuations. As for the flux ratios of the CE-4/LND measurements to the CRÈME modeling results (green diamonds and orange squares for the CRÈME96 and CRÈME2009 models, respectively), they vary in the ranges of (1. 20±0. 15 to 2. 54±0. 51) and (1. 88±0. 22 to 3. 38±0. 42) for CRÈME96 and CRÈME2009 at energies ranging in (9. 77 and 24. 88) MeV/nuc, respectively. In addition, a prominent enhancement centered at ~20MeV/nuc, which corresponds to the contribution of the ACR component, was observed in the CR proton spectrum by the Voyager spacecraft around the solar minimum spanning from the years 1993 to 1999 [at 63. 7 and 49. 6 astronomical units (AU) for Voyager 1 and Voyager 2, respectively] (14), but no such enhancement is observed in our measurements, which confirms that these low-energy (~10 to ~50MeV) protons are of galactic origin during solar quiet times in the inner heliosphere.

The energy spectrum of He is shown in Fig. 1b. It is evident that the agreement of energy spectrum above ~5MeV/nuc between the CE-4/LND (red downward and red upward triangles for ^{4}He and ^{3}He + ^{4}He, respectively) and the spacecraft [orange hexagons, blue squares, and violet diamonds for ^{4}He, ^{3}He + ^{4}He, and ^{3}He + ^{4}He from the ACE/SIS (Solar Isotope Spectrometer), the SOHO/EPHIN, and the STERED-A/LET, respectively] relativly good, while the comparisons of the ^{3}He + ^{4}He predictions from the CRÈME models (aquamarine solid and lime dashed lines for the CRÈME96 and CRÈME2009 models, respectively) with those observations from the CE-4/LND and the spacecraft exhibit obvious deviations in general. We find that the ^{3}He + ^{4}He observations from the CE-4/LND are consistent with those from the spacecraft (SOHO/EPHIN, blue squares in Fig. 1b) within 1. 7σ, as shown by the violet bar in the second column of Fig. 3, and are larger than those predictions made by the CRÈME models as well by a respective average factor of (0. 66±0. 17) and (0. 76±0. 18) (green and orange bars in the second column of Fig. 3 for the CRÈME96 and CRÈME2009 models, respectively). The predictions from the CRÈME96 and CRÈME2009 models above ~20MeV overlap almost completely. It seems that the predicted curve from the CRÈME96 model can reflect the variation trend of the combined data collected by the ACE/EPAM (Electron, Proton, and Alpha Particle Monitor; green thin-cross symbols for ^{3}He + ^{4}He) and the STEREO-A/LET (violet thick-cross symbols for ^{4}He) at energies below ~5MeV/nuc with limited accuracy. Also, we can see an enhancement centered at ~20MeV/nuc (see lime dashed line) in the energy range between ~10 and ~50MeV/nuc, which can be ascribed to the contribution of the ACR component (49). In addition, the ^{3}He + ^{4}He flux ratios of the CE-4/LND measurements to the spacecraft observations and the CRÈME modeling results are shown in Fig. 2b. For the flux ratio of the CE-4/LND measurement to the spacecraft observation (SOHO/EPHIN, blue squares in Fig. 1B), the values of these data points (filled violet circles), varying in the range of (1. 17±0. 22 to 1. 42±0. 27) at energies from 14. 18 to 24. 81MeV/nuc, are consistent with 1. 0 within 1. 6σ. Concerning the flux ratios of the CE-4/LND measurements to the CRÈME modeling results (green diamonds and orange squares for the CRÈME96 and CRÈME2009 models, respectively), they vary in the ranges of (1. 57

± 0.25 to 1.82 ± 0.29) and (1.66 ± 0.33 to 1.94 ± 0.31) for CRÈME96 and CRÈME2009, respectively. Also, the CE-4/LND can distinguish He isotopic components, ^3He and ^4He, which conduces to the extraction of the flux ratio of ^3He to ^4He, as will be shown in the next subsection.

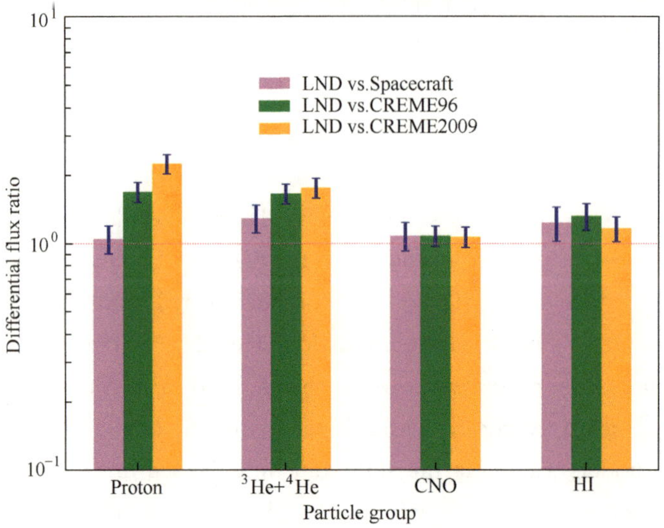

Fig. 3　Similar to Fig. 2, but are averaged over the energies of each group. See the text for details

Figure 1C presents the energy spectrum for the CNO group. The predictions from the CRÈME models (aquamarine solid and lime dashed lines for the CRÈME96 and CRÈME2009 models, respectively) match the observations from the ACE/SIS (orange hexagons) relatively closely, but are slightly lower than the STEREO-A/LET (violet diamonds) measurements. Meanwhile, the observations from the CE-4/LND (red upward triangles) are consistent with those from the ACE/SIS and comparable to those predictions made by the CRÈME models as well as with fluctuations. More specifically, the CNO observations from the CE-4/LND are consistent with those predictions from the CRÈME96 and CRÈME2009 models within 1.0σ with a respective ratio of (1.08 ± 0.11) and (1.07 ± 0.11) (green and orange bars in the third column of Fig. 3 for the CRÈME96 and CRÈME2009 models, respectively) and are also comparable to those from the ACE/SIS with a ratio of (1.08 ± 0.16) within 1.0σ (violet bar in the third column of Fig. 3). Furthermore, the CNO flux ratios of the CE-4/LND measurement to the ACE/SIS observations are shown in Fig. 2C, together with those of the CE-4/LND measurements to the predictions made by the CRÈME models. As for the flux ratios of the CE-4/LND measurements to the ACE/SIS observations, these data points (filled violet circles) fluctuate slightly around the line with a value of 1.0. For the flux ratios of the CE-4/LND measurements to the CRÈME modeling results (green diamonds and orange squares for the CRÈME96 and CRÈME2009 models, respectively), they overlap almost completely and vary in the ranges of (0.77 ± 0.28 to 5.82 ± 2.25) and (0.78 ± 0.29 to 5.49 ± 2.13)

for CRÈME96 and CRÈME2009 in the energy range between 18. 45 and 66. 75MeV/nuc, respectively. For some energy points beyond the ACE/SIS measurements, the differential flux values for ratio calculations are taken from those predictions made by the CRÈME2009 model as substitutions because CRÈME2009 can reproduce the ACE/SIS observations comparatively well at the energies nearby. Besides, we can see a bump centered at ~10MeV/nuc due to the ACR component as well (49), as shown by the CRÈME2009 model (lime dashed line). It is also essential to point out that the CNO spectra of the ACE/SIS and the STEREO - A/LET are separately extracted from their respective spectra for elements C, N, and O. Here, the spectrum with data points detected directly by the ACE/SIS [STEREO - A/LET] is specified as the original spectrum. For each component element, the data points detected directly by the ACE/SIS [STEREO - A/LET] in the original spectrum are relatively sparse. To obtain an extensive spectrum with denser data points, the interpolation with the Bezier Curves method (50) is applied to the original spectrum of each component element. However, only the nominal values of differential fluxes for these interpolated data points can be acquired based on this method. To further estimate the absolute error of the differential flux for a specific interpolated data point in the extensive spectrum, the average relative error of differential fluxes of its two nearest data points on both sides in the original spectrum is taken as its relative error, and then its corresponding absolute error can be obtained simply. As for the spectrum of the CNO group, the kinetic energy value is adopted directly from those of the energy channels of its component elements, and the nominal value of differential flux for the corresponding energy is calculated by summing up that of each component element. Concerning the absolute error of the differential flux, it is the summation of those from its component elements in quadrature. On the basis of the strategy mentioned above, data points of the CNO group from the ACE/SIS [STEREO - A/LET] can be acquired, as shown in Fig. 1c, marked as CNO (SIS) [CNO (LET)] in the legend.

The energy spectrum for the HI group is presented in Fig. 1d, which is obtained by following the same methodology applied to the CNO group. The observations from the ACE/SIS are in comparatively good agreement with the predictions from the CRÈME models at energies above ~40MeV/nuc, while divergency is seen between the predictions and those from the ACE/SIS and the STEREO - A/LET at energies below ~40MeV/nuc. Moreover, the data points measured by the CE - 4/LND fluctuate around those made by the near - earth spacecraft due to the relatively low statistics and complex constituents of the HI group. The HI observations made by the CE - 4/LND are consistent with those from the ACE/SIS within 1.2σ with a ratio of (1.24 ± 0.21) (violet bar in the fourth column of Fig. 3) and are also comparable to the predictions from the CRÈME96 and CRÈME2009 models within 1.9σ with a respective ratio of (1.33 ± 0.18) and (1.17 ± 0.15) (orange and green bars in the fourth column of Fig. 3). Besides, the HI flux ratios of the CE－4/LND measurements to the ACE/SIS observations as well as those of the CE－4/LND measurements to the CRÈME modeling results are shown in Fig. 2d. Concerning the flux ratios of the CE－4/LND measurements to the ACE/SIS observations, these data points (filled violet circles) fluctuate intensely around the line with a value of 1.0 at energies between 30.45 and 110.03MeV/nuc. The predictions of CRÈME96 are larger than those made by CRÈME2009 by a factor of (0.15±0.23) for the flux ratios of the CE－4/LND measurements to the CRÈME modeling results (green diamonds and orange squares for the CRÈME96 and CRÈME2009 models, respectively).

We can conclude that the CR fluxes on the lunar surface measured by the CE－4/LND instrument at energies ～10 to ～100MeV/nuc are consistent with those observed near the Earth. Specifically, the CE－4/LND measurements are comparable to the spacecraft observations within 1.0σ for the proton and the CNO group. The ratios for He and the HI group are (1.30±0.18) and (1.24±0.21), consistent with 1 within 1.7σ and 1.2σ, respectively. Because the higher－energy particles are less sensitive to the lunar physical environment, we can infer that the higher－energy CRs on the lunar surface are likely to be comparable to those near the Earth. As mentioned earlier, the largest magnetic anomaly Imbrium antipode is located northwest of the CE－4/LND. Therefore, the influence of lunar magnetic anomalies, as well as other lunar environment factors on the CR spectrum (>～10MeV/nuc), is negligible. Of note here is that systematic errors for the CE－4/LND and the near－earth spacecraft have been considered in this paper. Specifically, a relative systematic error of 10% is separately conservatively assigned to the CE－4/LND and the ACE/SIS, SOHO/EPHIN, and STEREO－A/LET spacecraft, which mainly arises from various factors when determining the geometric factor, such as the geometry configuration and the detection efficiency.

In addition, it is essential to point out that the weighted averages of the flux ratios over different energies for each particle group shown in Fig. 3 are derived from the data of Fig. 2 with a standard weighted least－square procedure method, of which the detailed description can be found in the literature [(51), pp. 15－18].

Differential flux ratio of ^{3}He to ^{4}He

The differential flux ratios of ^{3}He to ^{4}He above ～100MeV/nuc have been reported in previous experiments, such as the balloon－borne experiments [IMAX92 (52, 53), BESS (54), and BESS－Polar II (55)] and the space instruments [AMS－01 (56) and PAMELA (57)]. Here, we show the results based on the combined data collected by the CE－4/LND and the STEREO－A/LET along with the predictions from the Galactic Propagation (GALPROP) model (41) within the energy range between ～5 and ～20MeV/nuc. The predictions from the GALPROP model in the left column (Fig. 4, a and c) and the right column (Fig. 4, b and d) are based on a conventional reacceleration model (58, 59) and a plain diffusion model (59), respectively, of which the isotopic abundances have been tuned to reproduce those from the ACE observations (60). The subfigures in the upper row (Fig. 4, a and b) show the comparisons of differential fluxes between the spacecraft observations and the GALPROP model's predictions for ^{3}He and ^{4}He with solar modulation potentials (Φ) ranging from 100 to 1000 MV; the corresponding value ranges of differential fluxes for ^{3}He and ^{4}He are indicated by shaded pink and shaded sky - blue areas limited by two thin dotted and two thin dashed lines, respectively. The predicted differential fluxes of ^{3}He and ^{4}He with two different modulation potentials, Φ=314 and 490MV, are indicated by thick dotted pink and thick dashed sky - blue lines, respectively. In the GALPROP model, Φ is an important parameter in determining CR fluxes, and we have chosen two values (314 and 490MV) to represent the modulation level during the period 2019 - 2020. Φ = 314MV is directly obtained from http://cosmicrays.oulu.fi/phi/Phi_mon.txt, and Φ = 490MV is calculated with the method from Usoskin et al. (61) using the average Oulu neutron monitor count rates during the working periods of the CE - 4/LND. The comparisons of flux ratios of ^{3}He to ^{4}He between the observations from the CE - 4/LND and those from the STEREO - A/LET and the predictions from the GALPROP model are shown in Fig. 4 (C and D). The filled red circles [filled violet pentagons] with error bars denote the observation from the CE - 4/LND [STEREO - A/LET] and the shaded red [shaded violet] areas correspond to the 1 SD band, while the dotted and dashed blue lines denote the predictions from the GALPROP model with Φ= 314 and 490MV, respectively. The shaded orange areas show the predicted value range of flux ratio of ^{3}He to ^{4}He with modulation potentials Φ in the range between 100 and 1000MV from the GALPROP model. It is worth noting that the lower limit of kinetic energy for the GALPROP model is 10MeV/nuc; thus, there is no prediction at energies below 10MeV/nuc.

The observed flux ratios of ^{3}He to ^{4}He in the energy range between ~5 and ~20MeV/nuc are within (12.43±2.78)%. Also, we can see a prominent enhancement at ~12MeV/nuc. The values of the observations first decrease gradually from (9.38±0.64)% to (6.52±0.45)%, with increasing kinetic energy in the range between 4.90 and 10.95MeV/nuc, and then increase sharply to reach a peak (12.43±2.78)% at 11.99MeV/nuc. With the continuous increase in the kinetic energy, the values of the observations decrease gradually again to the minimum value of (3.16±0.84)% at 16.66MeV/nuc and then remain relatively stable. The GALPROP model cannot give a good prediction to the observation from the CE - 4/LND and the third data point from the STEREO - A/LET. The values of the observations vary dramatically, while those of the GALPROP modeling results remain stable with the increase of kinetic energy. The anomaly may ascribe to the contribution of the ACR component, which can be further confirmed by the poor consistency of differential fluxes between the observations and the predictions in the energy range between ~10 and ~50MeV/nuc for He shown in Fig. 4 (A and B). As mentioned earlier, there is an enhancement in the energy spectrum due to the ACR component in the energy range between ~10 and ~50MeV/nuc for He shown in Fig. 1b. The predicted differential fluxes

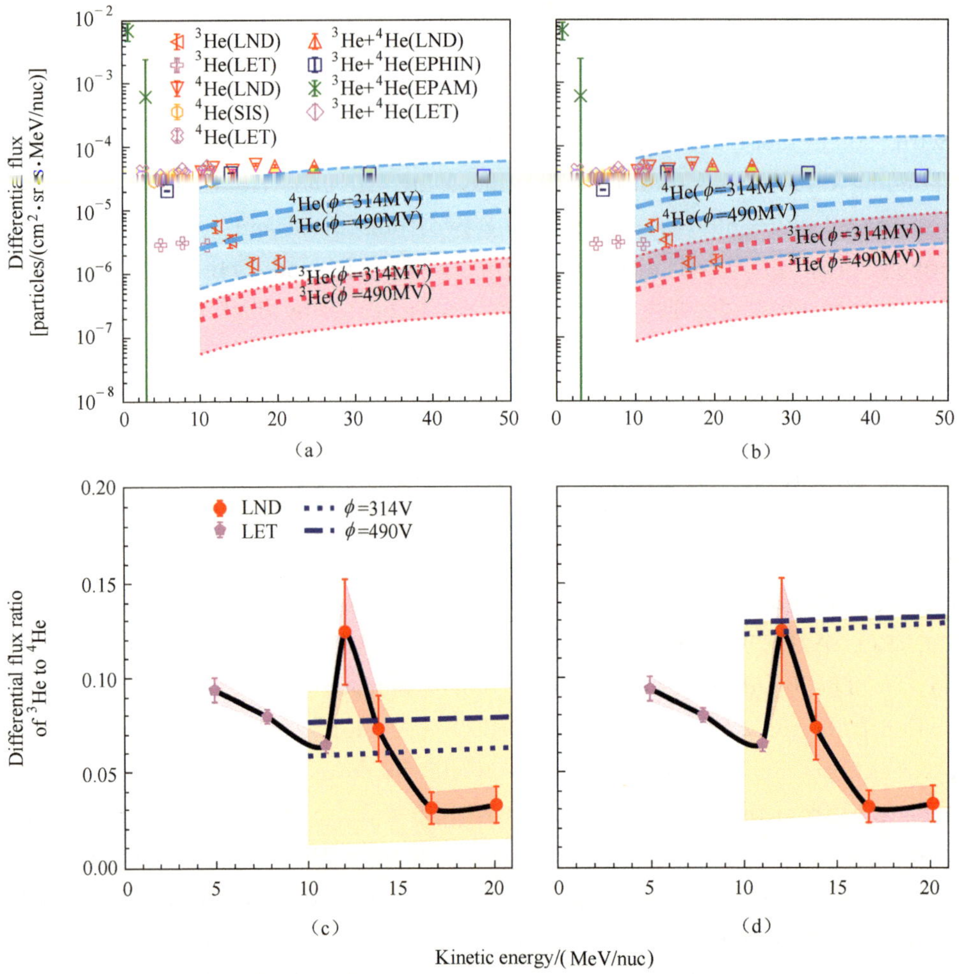

Fig. 4　The CR energy spectra of ^{3}He and ^{4}He and the ratios of ^{3}He to ^{4}He. (A and B) The thick dashed sky-blue [dotted pink] lines denote the predictions from the GALPROP model with the specified modulation potentials for ^{4}He [^{3}He], and the shaded sky-blue [pink] areas limited by two thin dashed sky-blue [dotted pink] lines correspond to the value range of the GALPROP modeling results with modulation potentials $\Phi = 100$ to 1000 MV for ^{4}He [^{3}He]. (C and D) The filled red circles with error bars denote the observation from the CE-4/LND and the shaded red areas show the 1 SD band, while the filled violet pentagon with error bars and the shaded violet areas denote the observation from the STEREO-A/LET. The dashed and dotted blue lines denote the GALPROP modeling results with $\Phi = 490$ and 314 MV, respectively, and the shaded orange areas denote the value range of the predictions from the GALPROP model with $\Phi = 100$ to 1000 MV. Note that the predictions in (a) and (c) are from the GALPROP conventional reacceleration model, and in (b) and (d) are from the GALPROP plain diffusion model

of ^{3}He and ^{4}He based on the conventional reacceleration model with $\Phi = 314$ and 490MV shown in Fig. 4a are both lower than the observations, while those based on the plain diffusion model perform a little better shown in Fig. 4b. However, the difference between the observations and the predictions in the flux ratio of ^3He to ^4He is still large. Of note is that the weighted average (51) of the differential fluxes of ^4He from the CE - 4/LND has been used to calculate the flux ratios of ^3He to ^4He in consideration of its relatively uniform distribution in the energy range of ^3He.

Dawn - dusk symmetry of the CR spectrum

The differential flux ratios between the dawn and dusk sides of these particle groups are shown in Fig. 5. The flux ratios from the spacecraft lie on the line with a value of 1.0 within the error bar for all particle groups, which indicates that the CR flux in deep space is isotropic as expected. The flux ratios from the CE - 4/LND have similar distributions, although fluctuations in the flux ratios for the CNO and HI groups are observed, which could be attributed to the relative low statistics of these two particle groups. Thus, both the low - energy CR spectra in the lunar local morning and the lunar local afternoon are consistent. The dawn - dusk symmetry in the low - energy CR spectrum during solar quiet times around the solar minimum is confirmed, which could also be extended to the higher

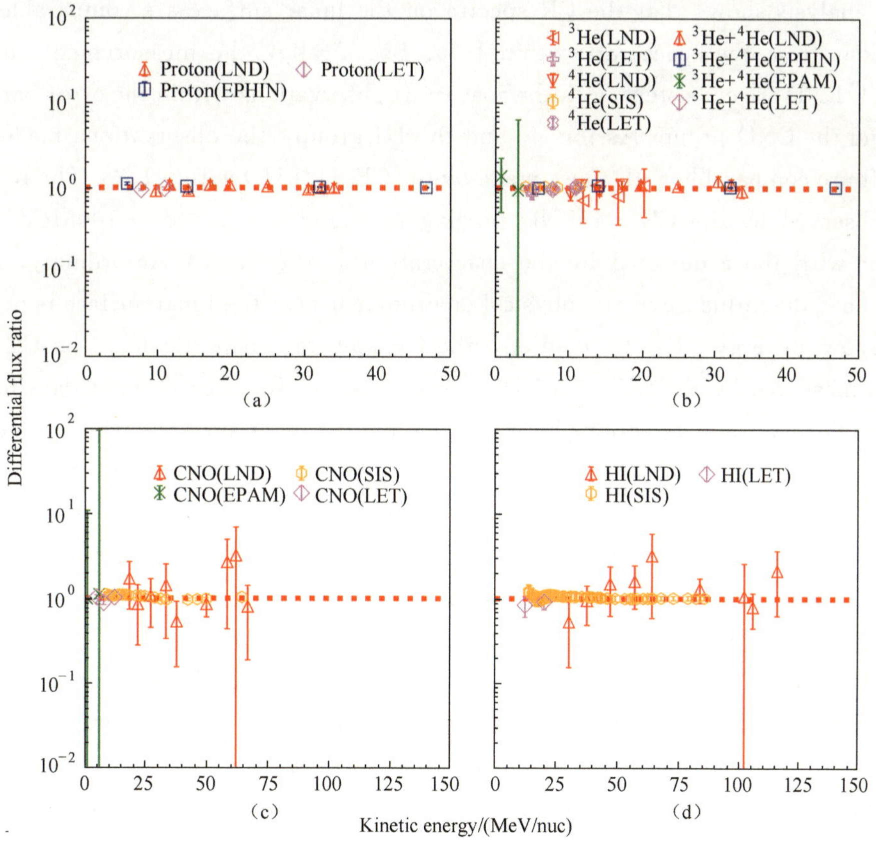

Fig. 5　The ratios of the observed CR fluxes in the lunar local morning to those in the lunar local afternoon. (a) For proton. (b) for He. (c) for CNO group. (d) for HI group (Ne, Na, Mg, Al, Si, S, Ar, Ca, Fe, and Ni). The ratio of 1.0 is indicated by a red dotted line for reference

energy range of the CR spectrum since the solar modulation weakens with the increase of the energies of particles.

Discussion

Crewed missions to the Moon are essential for both scientific and commercial endeavors. The safety of astronauts and the reliability of sensitive instruments are severely compromised by the harsh radiation environment at the lunar surface: However, our understanding of its nature is far from complete.

In this study, on the basis of the CE - 4/LND measurement around the extraordinarily quiet solar cycle 24/25 minimum, the low - energy CR spectrum on the surface of the lunar farside is reported and it is further compared with measurements from several spacecraft near the Earth, such as the ACE, the SOHO, and the STERE - A. The differential flux ratio of ^{3}He to ^{4}He is extracted, and the dawn - dusk symmetry of the CR spectrum on the lunar surface is considered too.

Our analysis shows that the CR spectra on the lunar surface are comparable to the observations from the spacecraft within 1.7σ. Specifically, the measurements made by the CE - 4/LND are consistent with the spacecraft observations within the error bar for the proton and the CNO group. As for He and the HI group, the observations made by the spacecraft are comparable with those made by the CE - 4/LND within 1.7σ. Thus, the CR spectra observed by the CE - 4/LND ranging at energies $\sim$10 to $\sim$100MeV/nuc are consistent with those detected by the spacecraft near the Earth: Accordingly, we can conclude that the influence of the physical environment near the lunar surface is negligible for these low - energy CRs. In addition, the CR spectra, in particular, the CR proton spectrum measured by the CE - 4/LND, can provide in situ precise input data for CR - related studies like the interactions between CRs and the lunar surface, which can enhance our understanding on the proton, neutron, and gamma emission spectroscopy from the lunar surface. Further, they can also be used to test theoretical models such as CRÈME, which diverge the CE - 4/LND measurements remarkably for the proton. Meanwhile, the flux ratios of ^{3}He to ^{4}He from the CE - 4/LND diverge considerably from the predictions of the GALPROP model: This can be used to improve the model and other related theoretical models. We confirm the dawn - dusk symmetry of the CR spectrum on the lunar surface, which will guide the selections of landing times for future crewed lunar missions and extravehicular activities on the lunar surface.

Although these findings can deepen our understanding of the radiation environment on the lunar surface, they are still far from exhaustive: The energy range of the CR spectrum measured by the CE - 4/LND is limited, and the precisions of some measurements need to be improved. At present, the CE - 4/LND is still in operation on the lunar farside and continuing harvesting data. This will lead to the production of more precise spectra of the various elements of interest.

Materials and Methods CR observations

In this work, we have used CR measurements from two sources: One is from the LND instrument aboard the CE－4 mission, and the other is from three near－earth spacecraft in space (ACE, SOHO, and STEREO－A). The ACE and SOHO spacecraft orbit the Sun－Earth first Lagrangian (L1) point in a halo orbit, as illustrated in Fig. 6. The STEREO－A is a space－based observatory orbiting the Sun just inside 1 AU. These data were collected from 3 January 2019 to 1 March 2020 (see Tab. 1 for details) during the solar minimum 24/25. Note that the working periods of these instruments are not identical due to the different cadences of their data products used in this work. However, the influence of these differences is essentially negligible considering the relatively long total working times. The 5－min level－2, 256－s level－2, 10－min level－1, and 1－min level－3 average flux data [in units of particles/ (cm^2 • sr • s • MeV/nuc)] of the ACE/EPAM, the ACE/SIS, the STEREO－A/LET, and the SOHO/EPHIN are applied, respectively. A brief introduction to these instruments is as follows.

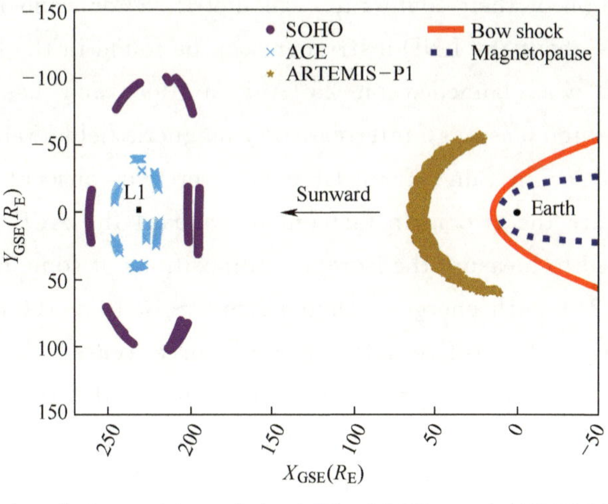

Fig. 6　Illustration of the relative positions of the ACE, SOHO, and ARTEMIS－P1 spacecraft in the X－Y GSE plane during the working periods of the CE－4/LND. The ARTEMIS mission consists of two identical probes, P1 and P2, in orbit around the Moon (74), whose trajectories can approximate the position of the Moon. The bow shock model and the magnetopause model are obtained from Fairfield (75). R_E is the radius of the Earth

The CE－4 mission, which is the first mission to the surface of the lunar farside, is composed of a lander (CE－4), a rover (Yutu－2), and a relay satellite [Queqiao relay satellite, which was successfully launched on 21 May 2018 and is currently orbiting the Earth－Moon second Lagrangian (L2) point]. Four scientific payloads are aboard the lander and the rover, respectively. More details of the CE－4 mission can be found in the literature (62). The LND instrument aboard the lander is developed to determine the dose rate and the linear energy transfer spectrum as well as the particle flux on the surface of the lunar farside, which contributes to the studies related to the lunar surface environment and heliospheric science (36). The results of the time series - based dose rate and the linear energy transfer spectrum have been reported on the basis of the data collected by the LND during the first two lunar days (37), and on 6 May 2019, the LND instrument measured its first SEP event on the lunar farside (63). The LND instrument, mounted in the payload compartment of the lander, is composed of an electron box (EB) and a sensor head (SH). The EB is connected with the instrument control unit as the electrical and data interface of the LND with the lander. The SH is mounted close to the upper deck of the lander with a resealable door to ensure an unobstructed field of view into the sky. The door keeps open during working hours in the lunar daytime but closed duringthe cold lunar night to protect the payloads inside. The SH has a telescope configuration, a stack of ten 500 - m - thick dual - segment Si solid - state detectors labeled A through J as shown in figure 4 of (36), to ensure the detection of charged - particle radiations and the identification of its composition. The differential flux of a specific type of charged particles in a certain kinetic energy range can be obtained on the basis of the information of their count rate, geometric factor, and kinetic energy range interval. More details about the LND instrument can be found in the literature (36, 64). The ACE spacecraft was launched on 25 August 1997 and has been continuously monitoring the solar wind plasmas, interplanetary magnetic fields, and energetic particles (both SEPs and CRs) at the Sun - Earth L1 point over two consecutive solar cycles. The SIS and the EPAM are the two main instruments aboard the ACE spacecraft. The SIS instrument is designed to measure the isotopic compositions of some energetic nuclei from He to Ni ($Z=2$ to 28) with energies ranging from ~10 to ~100MeV/nuc (65). The EPAM instrument consists of five telescopes of three types: two Low Energy Foil Spectrometers (LEFS) designed for the measurement of the fluxes and directions of electrons above 30keV, two Low Energy Magnetic Spectrometers (LEMS) for the fluxes and directions of ions, and one Composition Aperture (CA) for the compositions of ions (66). For more information about the ACE mission, the reader can refer to the literature (67).

The SOHO spacecraft aims to study the Sun from its deep core to the outer corona and the solar wind (68). There are 12 payloads in total aboard the SOHO spacecraft, including the Comprehensive Suprathermal and Energetic Particle Analyzer (COSTEP). The COSTEP is designed to measure the energetic particles emitted by the Sun, and it consists of two instruments: the EPHIN and the Low Energy Ion and Electron Instrument (LION). Resembling the ACE/SIS, the SOHO/EPHIN measures the fluxes of energetic particles (electron, proton, and helium) from the solar corona during large solar events, while particles from GCRs and ACRs are measured during the solar quiet times. Further details of the SOHO mission can be found in the literature (68).

The STEREO mission, consisting of two nearly identical spacecraft, STEREO - A and STEREO Behind (STEREO - B), is in orbit around the Sun at ~1 AU (69, 70). The STEREO - A is moving ahead of the Earth, and the STEREO - B is moving behind. The STEREO, carrying four instrument packages, was launched on 25 October 2006. The LET is one of four sensors making up the SEP instrument of the In situ Measurements of Particles and CME Transients (IMPACT) investigation. The LET can measure the elemental compositions, energy spectra, angular distributions, and arrival times of ions ($Z = 1$ to 28) over the energy range between ~3 and ~30MeV/nuc and identify the rare isotope ^{3}He and trans - iron nuclei with Z=30 to 83 (70). Of note here is that communications with the STEREO - B have been lost since 1 October 2014 due to multiple hardware anomalies affecting control of the spacecraft orientation. More details about the STEREO and the LET can be found in (69, 70), respectively.

CR models

For comparison with spacecraft measurements, two widely used CR numerical models are used here: the CRÈME (39, 40, 71, 72) and the GALPROP (41).

The CRÈME model (https://creme. isde. vanderbilt. edu/), first proposed by the Naval Research Laboratory, is a widely used semi - empirical model that can provide a numerical evaluation of the radiation effects on the electronic systems in spacecraft orbiting near the Earth (39, 40). CRÈME96 is an upgraded version released in 1996, which can offer a more accurate description of the solar cycle variation in the GCR flux by relating the solar cycle variation to the historical sunspot number (39). The contribution to the CR spectrum from the ACR component is also included in this version. In addition, an announcement stating that the CRÈME96 model is valid only for the years between 1950 and 1997 was made on the CRÈME website in 2010. Concurrently, CRÈME2009, a revised version, was released. However, CREME96 has been extensively used after 1997 (73). In this work, both results from the CRÈME96 and CRÈME2009 models were adopted. CRÈME2009 is the latest improved version, of which the method to calculate the Wolf number has been optimized compared to that of CRÈME96 (73). CRÈME2009 is more advantageous for the prediction of the modulation level up to 1 year beforehand by using the sunspot number as a representation of solar modulation compared with other models (40).

The GALPROP model (https://galprop. stanford. edu/webrun/), a numerical code for the calculations of the propagation of relativistic charged particles and the diffuse emissions produced during their propagation, is widely used in high - energy astrophysics and its related areas. The GALPROP code has been a sophisticated, efficient, and configurable tool for high - energy astrophysics through the evolution of several decades since the 1990s. The code is devoted to incorporating as many realistic astrophysical inputs as possible, together with the latest theoretical developments, which aims to calculate the propagation of CR nuclei, antiprotons, electrons, and positrons, as well as the diffuse gamma rays and synchrotron emission in the same framework (41). In this work, the GALPROP webRun v54, the latest stable version, was used.

Calculation method

For a specified kind of particles detected in the channel at energies between E_i and E_f, its differential flux F, in units of particles/(cm^2 · sr · s · MeV/nuc), can be calculated by

$$F = C(GF \cdot \Delta t \cdot \Delta E) \tag{1}$$

where GF is the geometric factor for the detection of these particles in units of cm^2 · sr; Δt is the measurement live time of a good data record, which can be substituted by the cadence of data product approximately under conditions of low and moderate count rates; ΔE, in units of MeV/nuc, is the interval of the energy channel $E_f - E_i$; and C is the total number of signal counts accumulated in Δt. The corresponding error of F, δF, can be given as

$$\delta F = F \cdot \sqrt{(\delta C/C)^2 + (\delta GF/GF)^2} \tag{2}$$

where δC and δGF are the errors of C and GF, respectively.

The average value of differential flux in a specified period can be obtained by

$$\overline{F} = \sum_{i=1}^{N}(F_i \cdot \Delta t_i) / \sum_{i=1}^{N}\Delta t_i \tag{3}$$

where N is the total number of good data records in this specified period, and F_i and Δt_i are the differential flux and the measurement live time of the i th good data record, respectively. The corresponding error can be written as

$$\delta\overline{F} = \sqrt{\sum_{i=1}^{N}(\delta F_i \cdot \Delta t_i)^2} / \sum_{i=1}^{N}\Delta t_i \tag{4}$$

Using these equations, the average values of differential flux in this specified period can be calculated on the basis of the flux data collected by the spacecraft and the CE4/LND. Of note here is that the GF of the CE-4/LND has been updated (see the Supplementary Materials for details).

In addition, the daily average differential flux $\overline{F}_{pre}$ predicted by the CRÈME models can be obtained by

$$\overline{F}_{pre} = \left(\sum_{j=1}^{M}F_j\right)/M \tag{5}$$

where M is the total number of days in this specified period and F_j is the average differential flux of the j th day. No error is provided in the predicted results from the CRÈME models.

Acknowledgments

We are grateful for the support from the team members of the Ground Application and Research System (GRAS), who contributed to data receiving and preprocessing. We also thank R. Bugiolacchi for helping polish the manuscript.

References

[1] G. Li, G. P. Zank, W. K. M. Rice, Acceleration and transport of energetic particles at CME-driven shocks. Adv. Space Res. 32, 2597-2602 (2003).

[2] S. Fu, Y. Jiang, V. Airapetian, J. Hu, G. Li, G. Zank, Effect of star rotation rate on the characteristics of energetic particle events. Astrophys. J. Lett. 878, 36 (2019).

[3] V. G. Sinitsyna, V. Y. Sinitsyna, Y. I. Stozhkov, Red dwarf stars as a new source type of galactic cosmic rays. Astron. Nachr. 342, 342-346 (2021).

[4] C. P. Auger, Observation of a large-scale anisotropy in the arrival directions of cosmic rays above 8×10^{18} eV. Science 357, 1266-1270 (2017).

[5] E. G. Berezhko, G. F. Krymskiĭ, Acceleration of cosmic rays by shock waves. Soviet Phys. Uspekhi 31, 27-51 (1988).

[6] S. P. Reynolds, Supernova remnants at high energy. Annu. Rev. Astron. Astrophys. 46, 89-126 (2008).

[7] E. Amato, P. Blasi, Cosmic ray transport in the Galaxy: A review. Adv. Space Res. 62, 2731-2749 (2018).

[8] M. Garcia-Munoz, G. M. Mason, J. A. Simpson, A new test for solar modulation theory: The 1972 May-July low-energy galactic cosmic-ray proton and helium spectra. Astrophys. J. 182, L81-L84 (1973).

[9] L. A. Fisk, B. Koslovsky, R. Ramaty, An interpretation of the observed oxygen and nitrogen enhancements in low-energy cosmic rays. Astrophys. J. 190, L35-L37 (1974).

[10] M. E. Pesses, J. R. Jokipii, D. Eichler, Cosmic ray drift, shock wave acceleration, and the anomalous component of cosmic rays. Astrophys. J. 246, L85-L88 (1981).

[11] A. C. Cummings, E. C. Stone, Composition of anomalous cosmic rays. Space Sci. Rev. 130, 389-399 (2007).

[12] D. Hovestadt, O. Vollmer, G. Gloeckler, C. Y. Fan, Differential energy spectra of low-energy (<8.5 MeV per nucleon) heavy cosmic rays during solar quiet times. Phys. Rev. Lett. 31, 650-653 (1973).

[13] F. B. McDonald, B. J. Teegarden, J. H. Trainor, W. R. Webber, The anomalous abundance of cosmic-ray nitrogen and oxygen nuclei at low energies. Astrophys. J. 187, L105 (1974).

[14] A. C. Cummings, E. C. Stone, C. D. Steenberg, Composition of anomalous cosmic rays and other heliospheric ions. Astrophys. J. 578, 194-210 (2002).

[15] J. A. Simpson, Elemental and isotopic composition of the galactic cosmic-rays.

Annu. Rev. Nucl. Part S 33, 323 - 382 (1983).

[16] I. C. Task Group on Radiation Protection in Space On behalf of the ICRP, G. Dietze, D. T. Bartlett, D. A. Cool, F. A. Cucinotta, X. Jia, I. R. McAulay, M. Pelliccioni, V. Petrov, G. Reitz, T. Sato, ICRP PUBLICATION 123: Assessment of radiation exposure of astronauts in space. Ann. ICRP 42, 1 - 339 (2013).

[17] C. Zeitlin, A. W. Case, H. E. Spence, N. A. Schwadron, M. Golightly, J. K. Wilson, J. C. Kasper, J. B. Blake, M. D. Looper, J. E. Mazur, L. W. Townsend, Y. Iwata, Measurements of galactic cosmic ray shielding with the CRaTER instrument. Space Weather 11, 284 - 296 (2013).

[18] S. Fu, L. Zhao, X. Zhang, P. Luo, Y. Li, Comparison of anomalous and galactic cosmic - ray oxygen at 1 au during 1997 - 2020. Astrophys. J. Lett. 920, L12 (2021).

[19] M. Durante, F. A. Cucinotta, Physical basis of radiation protection in space travel. Rev. Mod. Phys. 83, 1245 - 1281 (2011).

[20] J. D. Wrbanek, S. Y. Wrbanek, "Space radiation and impact on instrumentation technologies"(NASA Technical Note NASA/TP - 2020 - 220002, 2020); https: // ntrs. nasa. gov/api/citations/20200001895/downloads/20200001895. pdf.

[21] Y. Li, X. Zhang, W. Dong, Z. Ren, T. Dong, A. Xu, Simulation of the production rates of cosmogenic nuclides on the Moon based on Geant4. J. Geophys. Res. Space Physics 122, 1473 - 1486 (2017).

[22] J. Garrett, S. Valluri, M. S. Mendonca, R. M. Bigsby, J. Lopez, A. Caperell - Grant, J. Nees, J. R. Dynlacht, The protective effect of estrogen against radiation cataractogenesis is dependent upon the type of radiation. Radiat. Res. 194, 557 - 565 (2020).

[23] F. A. Cucinotta, M. Durante, Cancer risk from exposure to galactic cosmic rays: Implications for space exploration by human beings. Lancet Oncol. 7, 431 - 435 (2006).

[24] J. Chen, S. Yun, T. Dong, Z. Ren, X. Zhang, Investigate the radiation - induced damage on an atomistic DNA model by using Geant4 - DNA toolkit. Nucl. Instrum. Methods Phys. Res. Sect. B 494, 59 - 67 (2021).

[25] F. A. Cucinotta, M. Alp, F. M. Sulzman, M. Wang, Space radiation risks to the central nervous system. Life Sci. Space Res. 2, 69, 54 (2014).

[26] N. A. Schwadron, J. F. Cooper, M. Desai, C. Downs, M. Gorby, A. P. Jordan, C. J. Joyce, K. Kozarev, J. A. Linker, Z. Mikíc, P. Riley, H. E. Spence, T. Török, L. W. Townsend, J. K. Wilson, C. Zeitlin, Particle radiation sources, propagation and interactions in deep space, at Earth, the Moon, Mars, and beyond: Examples of radiation interactions and effects. Space Sci. Rev. 212, 1069 - 1106 (2017).

[27] T. P. Dachev, B. T. Tomov, Y. N. Matviichuk, P. S. Dimitrov, S. V. Vadawale, J. N. Goswami, G. de Angelis, V. Girish, An overview of RADOM results for Earth and Moon radiation environment on Chandrayaan – 1 satellite. Adv. Space Res. 48, 779 – 791 (2011).

[28] H. E. Spence, A. W. Case, M. J. Golightly, T. Heine, B. A. Larsen, J. B. Blake, P. Caranza, W. R. Crain, J. George, M. Lalic, A. Lin, M. D. Looper, J. E. Mazur, D. Salvaggio, J. C. Kasper, T. J. Stubbs, M. Doucette, P. Ford, R. Foster, R. Goeke, D. Gordon, B. Klatt, J. O' Connor, M. Smith, T. Onsager, C. Zeitlin, L. W. Townsend, Y. Charara, CRaTER: The cosmic Ray telescope for the effects of radiation experiment on the lunar eeconnaissance orbiter mission. Space Sci. Rev. 150, 243 – 284 (2010).

[29] P. B. Crandall, J. J. Gillis – Davis, R. I. Kaiser, Untangling the origin of molecular hydrogen in the lunar exosphere. Astrophys. J. 887, (2019).

[30] M. Horányi, Z. Sternovsky, M. Lankton, C. Dumont, S. Gagnard, D. Gathright, E. Grün, D. Hansen, D. James, S. Kempf, B. Lamprecht, R. Srama, J. R. Szalay, G. Wright, The lunar dust experiment (LDEX) onboard the lunar atmosphere and dust environment explorer (LADEE) mission. Space Sci. Rev. 185, 93 – 113 (2014).

[31] L. L. Hood, A. Zakharian, J. Halekas, D. L. Mitchell, R. P. Lin, M. H. Acuna, A. B. Binder, Initial mapping and interpretation of lunar crustal magnetic anomalies using Lunar Prospector magnetometer data. J. Geophys. Res. Planets 106, 27825 – 27839 (2001).

[32] D. L. Mitchell, J. S. Halekas, R. P. Lin, S. Frey, L. L. Hood, M. H. Acuña, A. Binder, Global mapping of lunar crustal magnetic fields by Lunar Prospector. Icarus 194, 401 – 409 (2008).

[33] H. Tsunakawa, F. Takahashi, H. Shimizu, H. Shibuya, M. Matsushima, Surface vector mapping of magnetic anomalies over the Moon using Kaguya and Lunar Prospector observations. J. Geophys. Res. Planets 120, 1160 – 1185 (2015).

[34] J. Liu, X. Ren, W. Yan, C. Li, H. Zhang, Y. Jia, X. Zeng, W. Chen, X. Gao, D. Liu, X. Tan, X. Zhang, T. Ni, H. Zhang, W. Zuo, Y. Su, W. Wen, Descent trajectory reconstruction and landing site positioning of Chang'E – 4 on the lunar farside. Nat. Commun. 10, 4229 (2019).

[35] J. E. Mazur, W. R. Crain, M. D. Looper, D. J. Mabry, J. B. Blake, A. W. Case, M. J. Golightly, J. C. Kasper, H. E. Spence, New measurements of total ionizing dose in the lunar environment. Space Weather 9, S07002 (2011).

[36] R. F. Wimmer – Schweingruber, J. Yu, S. I. Böttcher, S. Zhang, S. Burmeister, H. Lohf, J. Guo, Z. Xu, B. Schuster, L. Seimetz, J. L. Freiherrvon Forstner, A. Ravanbakhsh, V. Knierim, S. Kolbe, H. Woyciechowski, S. R. Kulkarni, B. Yuan, G. Shen, C. Wang, Z. Chang, T. Berger, C. E. Hellweg, D. Matthiä, D.

Hou, A. Knappmann, C. Büschel, X. Hou, B. Ren, Q. Fu, The Lunar Lander Neutron and Dosimetry (LND) experiment on Chang'E 4. Space Sci. Rev. 216, 40 (2020).

[37] S. Zhang, R. F. Wimmer - Schweingruber, J. Yu, C. Wang, Q. Fu, Y. Zou, Y. Sun, C. Wang, D. Hou, S. I. Böttcher, S. Burmeister, L. Seimetz, B. Schuster, V. Knierim, G. Shen, B. Yuan, H. Lohf, J. Guo, Z. Xu, J. L. Freiherrvon Forstner, S. R. Kulkarni, H. Xu, C. Xue, J. Li, Z. Zhang, H. Zhang, T. Berger, D. Matthiä, C. E. Hellweg, X. Hou, J. Cao, Z. Chang, B. Zhang, Y. Chen, H. Geng, Z. Quan, First measurements of the radiation dose on the lunar surface. Sci. Adv. 6, eaaz1334 (2020).

[38] R. C. Elphic, V. R. Eke, L. F. A. Teodoro, D. J. Lawrence, D. B. J. Bussey, Models of the distribution and abundance of hydrogen at the lunar south pole. Geophys. Res. Lett. 34, L13204 (2007).

[39] A. J. Tylka, J. H. Adams, P. R. Boberg, B. Brownstein, W. F. Dietrich, E. O. Flueckiger, E. L. Petersen, M. A. Shea, D. F. Smart, E. C. Smith, CREME96: A revision of the cosmic ray effects on micro - electronics code. IEEE Trans. Nucl. Sci. 44, 2150 - 2160 (1997).

[40] J. H. Adams Jr., A. F. Barghouty, M. H. Mendenhall, R. A. Reed, B. D. Sierawski, J. W. Watts Jr., R. A. Weller, CRÈME: The 2011 revision of the cosmic ray effects on micro - electronics code. IEEE Trans. Nucl. Sci. 59, 3141 - 3147 (2012).

[41] A. E. Vladimirov, S. W. Digel, G. Jóhannesson, P. F. Michelson, I. V. Moskalenko, P. L. Nolan, E. Orlando, T. A. Porter, A. W. Strong, GALPROP WebRun: An internet - based service for calculating galactic cosmic ray propagation and associated photon emissions. Comput. Phys. Commun. 182, 1156 - 1161 (2011).

[42] J. Chen, S. Yun, T. Dong, Z. Ren, X. Zhang, Validation of Geant4 physics models for nuclear beams in extended media. Nucl. Instrum. Methods Phys. Res., Sect. B 434, 113 - 119 (2018).

[43] S. Fu, L. Zhao, G. P. Zank, M. Wang, Y. Jiang, An ACE/CRIS - observation - based galactic cosmic rays heavy nuclei spectra model II. Sci. China Phys. Mech. Astron. 63, 219511 (2020).

[44] J. Wu, H. Chen, Revisit cosmic ray propagation by using 1H, 2H, 3He and 4He. Phys. Lett. B789, 292 - 299 (2019).

[45] R. Hajra, Weakest solar cycle of the space age: A study on solar wind - magnetosphere energy coupling and geomagnetic activity. Solar Phys. 296, 33 (2021).

[46] S. Fu, X. Zhang, L. Zhao, Y. Li, Variations of the galactic cosmic rays in the recent solar cycles. Astrophys. J. Suppl. Ser. 254, 37 (2021).

[47] R. A. Nymmik, M. I. Panasyuk, T. I. Pervaja, A. A. Suslov, A model of galactic cosmic ray fluxes. Int. J. Radiat. Appl. Instrum. Part D Nucl. Tracks Radiat. Meas. 20, 427-429 (1992).

[48] ISO TS 15390, Space environment (natural and artificial) —Galactic cosmic ray model. ISO 15390: 2004 (E) (2004).

[49] R. A. Mewaldt, A. C. Cummings, E. C. Stone, Anomalous cosmic rays: Interstellar interlopers in the heliosphere and magnetosphere. Eos Trans. Am. Geophys. Union 75, 185-193 (1994).

[50] A. Riskus, G. Liutkus, An improved algorithm for the approximation of a cubic bezier curve and its application for approximating quadratic bezier curve. Inf. Technol. Control 42, 303-308 (2013).

[51] P. A. Zyla, R. M. Barnett, J. Beringer, O. Dahl, D. A. Dwyer, D. E. Groom, C. -J. Lin K. S. Lugovsky, E. Pianori, D. J. Robinson, C. G. Wohl, W. -M. Yao, K. Agashe, G. Aielli, B. C. Allanach, C. Amsler, M. Antonelli, E. C. Aschenauer, D. M. Asner, H. Baer, S. Banerjee, L. Baudis, C. W. Bauer, J. J. Beatty, V. I. Belousov, S. Bethke, A. Bettini, O. Biebel, K. M. Black, E. Blucher, O. Buchmuller, V. Burkert, M. A. Bychkov, R. N. Cahn, M. Carena, A. Ceccucci, A. Cerri, D. Chakraborty, R. Sekhar Chivukula, G. Cowan, G. D' Ambrosio, T. Damour, D. de Florian, A de Gouvêa, T. De Grand, P. de Jong, G. Dissertori, B. A. Dobrescu, M. D' Onofrio, M. Doser, M. Drees, H. K. Dreiner, P. Eerola, U. Egede, S. Eidelman, J. Ellis, J. Erler, V. V. Ezhela, W. Fetscher, B. D. Fields, B. Foster, A. Freitas, H. Gallagher, L. Garren, H. -J. Gerber, G. Gerbier, T. Gershon, Y. Gershtein, T. Gherghetta, A. A. Godizov, M. C. Gonzalez-Garcia, M. Goodman, C. Grab, A. V. Gritsan, C. Grojean, M. Grünewald, A. Gurtu, T. Gutsche, H. E. Haber, C. Hanhart, S. Hashimoto, Y. Hayato, A. Hebecker, S. Heinemeyer, B. Heltsley, J. J. Hernández-Rey, K. Hikasa, J. Hisano, A. Höcker, J. Holder, A. Holtkamp, J. Huston, T. Hyodo, K. F. Johnson, M. Kado, M. Karliner, U. F. Katz, M. Kenzie, V. A. Khoze, S. R. Klein, E. Klempt, R. V. Kowalewski, F. Krauss, M. Kreps, B. Krusche, Y. Kwon, O. Lahav, J. Laiho, L. P. Lellouch, J. Lesgourgues, A. R. Liddle, Z. Ligeti, C. Lippmann, T. M. Liss, L. Littenberg, C. Lourengo, S. B. Lugovsky, A. Lusiani, Y. Makida, F. Maltoni, T. Mannel, A. V. Manohar, W. J. Marciano, A. Masoni, J. Matthews, U. -G. Meißner, M. Mikhasenko, D. J. Miller, D. Milstead, R. E. Mitchell, K. Mönig, P. Molaro, F. Moortgat, M. Moskovic, K. Nakamura, M. Narain, P. Nason, S. Navas, M. Neubert, P. Nevski, Y. Nir, K. A. Olive, C. Patrignani, J. A. Peacock, S. T. Petcov, V. A. Petrov, A. Pich, A. Piepke, A. Pomarol, S. Profumo, A. Quadt, K. Rabbertz, J. Rademacker, G. Raffelt, H. Ramani, M. Ramsey-Musolf, B. N. Ratcliff, P. Richardson, A. Ringwald, S. Roesler, S. Rolli, A. Romaniouk, L. J. Rosenberg,

J. L. Rosner, G. Rybka, M. Ryskin, R A Ryutin, Y. Sakai, G. P. Salam, S. Sarkar, F. Sauli, O. Schneider, K. Scholberg, A. J. Schwartz, J. Schwiening, D. Scott, V. Sharma, S. R. Sharpe, T. Shutt, M. Silari, T. Sjöstrand, P. Skands, T. Skwarnicki, G. F. Smoot, A. Soffer, M. S. Sozzi, S. Spanier, C. Spiering, A. Stahl, S L Stone, Y. Sumino, T. Sumiyoshi, M. J. Syphers, F. Takahashi, M. Tanabashi, J. Tanaka, M. Taševský, K. Terashi, J. Terning, U. Thoma, R. S. Thorne, L. Tiator, M. Titov, N. P. Tkachenko, D. R. Tovey, K. Trabelsi, P. Urquijo, G. Valencia, R. Van de Water, N. Varelas, G. Venanzoni, L. Verde, M. G Vincter, P. Vogel, W. Vogelsang, A. Vogt, V. Vorobyev, S. P. Wakely, W. Walkowiak, C. W. Walter, D. Wands, M. O. Wascko, D. H. Weinberg, E. J. Weinberg, M. White, L. R. Wiencke, S. Willocq, C. L. Woody, R. L. Workman, M. Yokoyama, R. Yoshida, G. Zanderighi, G. P. Zeller, O. V. Zenin, R. - Y. Zhu, S. - L. Zhu, F. Zimmermann, J. Anderson, T. Basaglia, V. S. Lugovsky, P. Schaffner, W. Zheng, Review of particle physics. Prog. Theor. Exp. Phys. 2020, 083C01 (2020).

[52] G. A. De Nolfo, A measurement of cosmic ray deuterium from 0.5 - 2.9 GeV/nucleon. AIP Conf. Proc. 528, 425 - 428 (2000).

[53] W. Menn, M. Hof, O. Reimer, M. Simon, A. J. Davis, A. W. Labrador, R. A. Mewaldt, S. M. Schindler, L. M. Barbier, E. R. Christian, K. E. Krombel, J. F. Krizmanic, J. W. Mitchell, J. F. Ormes, R. E. Streitmatter, R. L. Golden, S. J. Stochaj, W. R. Webber, I. L. Rasmussen, The absolute flux of protons and helium at the top of the atmosphere using IMAX. Astrophys. J. 533, 281 - 297 (2000).

[54] J. Z. Wang, E. S. Seo, K. Anraku, M. Fujikawa, M. Imori, T. Maeno, N. Matsui, H. Matsunaga, M. Motoki, S. Orito, T. Saeki, T. Sanuki, I. Ueda, K. Yoshimura, Y. Makida, J. Suzuki, K. Tanaka, A. Yamamoto, T. Yoshida, T. Mitsui, H. Matsumoto, M. Nozaki, M. Sasaki, J. Mitchell, A. Moiseev, J. Ormes, R. Streitmatter, J. Nishimura, Y. Yajima, T. Yamagami, Measurement of cosmic - ray hydrogen and helium and their isotopic composition with the BESS experiment. Astrophys. J. 564, 244 - 259 (2002).

[55] Nicolas Picot - Clemente, K Abe, H Fuke, S Haino, T Hams, M Hasegawa, A Horikoshi, A Itazaki, K C Kim, T Kumazawa, A Kusumoto, M H Lee, Y Makida, S Matsuda, Y Matsukawa, K Matsumoto, J W Mitchell, A A Moiseev, J Nishimura, M Nozaki, R Orito, J F Ormes, K Sakai, M Sasaki, E S Seo, Y Shikaze, R Shinoda, R E Streitmatter, J Suzuki, Y Takasugi, K Takeuchi, K Tanaka, N Thakur, T Yamagami, A Yamamoto, T Yoshida, K Yoshimura, Precise measurements of hydrogen and helium isotopes with BESS - Polar II, in 35th International Cosmic Ray Conference, Volume 301 - 35th International Cosmic Ray Conference (ICRC2017) - Session Cosmic - Ray Direct. CRD - direct measurements (Proceedings of Science, 2017).

[56] M. Aguilar, J. Alcaraz, J. Allaby, B. Alpat, G. Ambrosi, H. Anderhub, L. Ao, A. Arefiev, L. Arruda, P. Azzarello, M. Basile, F. Barao, G. Barreira, A. Bartoloni, R. Battiston, R. Becker, U. Becker, L. Bellagamba, J. Berdugo, P. Berges, B. Bertucci, A. Biland, V. Bindi, G. Boella, M. Boschini, M. Bourquin, G. Bruni, M. Buénerd, J. D. Burger, W. J. Burger, X. D. Cai, P. Cannarsa, M. Capell, D. Casadei, J. Casaus, G. Castellini, I. Cernuda, Y. H. Chang, H. F. Chen, H. S. Chen, Z. G. Chen, N. A. Chernoplekov, T. H. Chiueh, Y. Y. Choi, F. Cindolo, V. Commichau, A. Contin, E. Cortina - Gil, D. Crespo, M. Cristinziani, T. S. Dai, C. dela Guia, C. Delgado, S. di Falco, L. Djambazov, I. D' Antone, Z. R. Dong, M. Duranti, J. Engelberg, F. J. Eppling, T. Eronen, P. Extermann, J. Favier, E. Fiandrini, P. H. Fisher, G. Flügge, N. Fouque, Y. Galaktionov, M. Gervasi, F. Giovacchini, P. Giusti, D. Grandi, O. Grimm, W. Q. Gu, S. Haino, K. Hangarter, A. Hasan, V. Hermel, H. Hofer, W. Hungerford, M. Ionica, M. Jongmanns, K. Karlamaa, W. Karpinski, G. Kenney, D. H. Kim, G. N. Kim, K. S. Kim, T. Kirn, A. Klimentov, R. Kossakowski, A. Kounine, V. Koutsenko, M. Kraeber, G. Laborie, T. Laitinen, G. Lamanna, G. Laurenti, A. Lebedev, C. Lechanoine - Leluc, M. W. Lee, S. C. Lee, G. Levi, C. H. Lin, H. T. Liu, G. Lu, Y. S. Lu, K. Lübelsmeyer, D. Luckey, W. Lustermann, C. Maña, A. Margotti, F. Mayet, R. R. McNeil, M. Menichelli, A. Mihul, A. Mujunen, S. Natale, A. Oliva, F. Palmonari, M. Paniccia, H. B. Park, W. H. Park, M. Pauluzzi, F. Pauss, R. Pereira, E. Perrin, A. Pevsner, F. Pilo, M. Pimenta, V. Plyaskin, V. Pojidaev, M. Pohl, N. Produit, L. Quadrani, P. G. Rancoita, D. Rapin, D. Ren, Z. Ren, M. Ribordy, E. Riihonen, J. Ritakari, S. Ro, U. Roeser, R. Sagdeev, D. Santos, G. Sartorelli, P. Saouter, C. Sbarra, S. Schael, A. S. von Dratzig, G. Schwering, E. S. Seo, J. W. Shin, E. Shoumilov, V. Shoutko, T. Siedenburg, R. Siedling, D. Son, T. Song, F. R. Spada, F. Spinella, M. Steuer, G. S. Sun, H. Suter, X. W. Tang, S. C. C. Ting, S. M. Ting, N. Tomassetti, M. Tornikoski, J. Torsti, J. Trümper, J. Ulbricht, S. Urpo, E. Valtonen, J. Vandenhirtz, E. Velikhov, B. Verlaat, I. Vetlitsky, F. Vezzu, J. P. Vialle, G. Viertel, D. Vité, H. von Gunten, S. W. Wicki, W. Wallraff, J. Z. Wang, K. Wiik, C. Williams, S. X. Wu, P. C. Xia, S. Xu, Z. Z. Xu, J. L. Yan, L. G. Yan, C. G. Yang, J. Yang, M. Yang, S. W. Ye, H. Y. Zhang, Z. P. Zhang, D. X. Zhao, F. Zhou, Y. Zhou, G. Y. Zhu, W. Z. Zhu, H. L. Zhuang, A. Zichichi, B. Zimmermann, P. Zuccon, Isotopic composition of light nuclei in cosmic rays: Results from AMS - 01. Astrophys. J. 736, 105 (2011).

[57] O. Adriani, G. C. Barbarino, G. A. Bazilevskaya, R. Bellotti, M. Boezio, E. A. Bogomolov, M. Bongi, V. Bonvicini, S. Bottai, A. Bruno, F. Cafagna, D. Campana, P. Carlson, M. Casolino, G. Castellini, C. D. Donato, C. D. Santis,

N. D. Simone, V. D. Felice, V. Formato, A. M. Galper, A. V. Karelin, S. V. Koldashov, S. Koldobskiy, S. Y. Krutkov, A. N. Kvashnin, A. Leonov, V. Malakhov, L. Marcelli, M. Martucci, A. G. Mayorov, W. Menn, M. Mergè, V. V. Mikhailov, E. Mocchiutti, A. Monaco, N. Mori, R. Munini, G. Osteria, F. Palma, B. Panico, P. Papini, M. Pearce, P. Picozza, M. Ricci, S. B. Ricciarini, R. Sarkar, V. Scotti, M. Simon, R. Sparvoli, P. Spillantini, Y. I. Stozhkov, A. Vacchi, E. Vannuccini, G. Vasilyev, S. A. Voronov, Y. T. Yurkin, G. Zampa, N. Zampa, Measurements of cosmic - ray hydrogen and helium isotopes with thepamelaexperiment. Astrophys. J. 818, 68 (2016).

[58] A. W. Strong, I. V. Moskalenko, Models for galactic cosmic - ray propagation. Adv. Space Res. 27, 717 - 726 (2001).

[59] V. S. Ptuskin, I. V. Moskalenko, F. C. Jones, A. W. Strong, V. N. Zirakashvili, Dissipation of magnetohydrodynamic waves on energetic particles: Impact on interstellar turbulence and cosmic - ray transport. Astrophys. J. 642, 902 - 916 (2006).

[60] M. E. Wiedenbeck, N. E. Yanasak, A. C. Cummings, A. J. Davis, J. S. George, R. A. Leske, R. A. Mewaldt, E. C. Stone, P. L. Hink, M. H. Israel, M. Lijowski, E. R. Christian, T. T. von Rosenvinge, The origin of primary cosmic rays: Constraints from ACE elemental and isotopic composition observations. Space Sci. Rev. 99, 15 - 26 (2001).

[61] I. G. Usoskin, K. Alanko, K. Mursula, G. A. Kovaltsov, Heliospheric modulation strength during the neutron monitor era. Solar Phys. 207, 389 - 399 (2002).

[62] C. Li, W. Zuo, W. Wen, X. Zeng, X. Gao, Y. Liu, Q. Fu, Z. Zhang, Y. Su, X. Ren, F. Wang, J. Liu, W. Yan, X. Tan, D. Liu, B. Liu, H. Zhang, Z. Ouyang, Overview of the Chang'E - 4 mission: Opening the frontier of scientific exploration of the lunar far side. Space Sci. Rev. 217, 35 (2021).

[63] Z. Xu, J. Guo, R. F. Wimmer - Schweingruber, J. L. Freiherr von Forstner, Y. Wang, N. Dresing, H. Lohf, S. Zhang, B. Heber, M. Yang, First solar energetic particles measured on the lunar far - side. Astrophys. J. 902, L30 (2020).

[64] D. Hou, S. Zhang, J. Yu, R. F. Wimmer - Schweingruber, S. Burmeister, H. Lohf, B. Yuan, G. Shen, C. Wang, X. Hou, B. Ren, Removing the dose background from radioactive sources from active dose rate measurements in the Lunar Lander Neutron & Dosimetry (LND) experiment on Chang'E 4. J. Instrum. 15, P01032 (2020).

[65] E. C. Stone, C. M. S. Cohen, W. R. Cook, A. C. Cummings, B. Gauld, B. Kecman, R. A. Leske, R. A. Mewaldt, M. R. Thayer, B. L. Dougherty, R. L. Grumm, B. D. Milliken, R. G. Radocinski, M. E. Wiedenbeck, E. R. Christian,

S. Shuman, T. T. von Rosenvinge, The solar isotope spectrometer for the advanced composition explorer. Space Sci. Rev. 86, 357 - 408 (1998).

[66] R. E. Gold, S. M. Krimigis, S. E. Hawkins, III, D. K. Haggerty, D. A. Lohr, E. Fiore, T. P. Armstrong, G. Holland, L. J. Lanzerotti, Electron, proton, and alpha monitor on the advanced composition explorer spacecraft. Space Sci. Rev. 86, 541 - 562 (1998).

[67] E. C. Stone, A. M. Frandsen, R. A. Mewaldt, E. R. Christian, D. Margolies, J. F. Ormes, F. Snow, The Advanced Composition Explorer. Space Sci. Rev. 86, 1 - 22 (1998).

[68] V. Domingo, B. Fleck, A. I. Poland, The SOHO mission: An overview. Solar Phys. 162, 1 - 37 (1995).

[69] M. L. Kaiser, The STEREO mission: An overview. Adv. Space Res. 36, 1483 - 1488 (2005).

[70] R. A. Mewaldt, C. M. S. Cohen, W. R. Cook, A. C. Cummings, A. J. Davis, S. Geier, B. Kecman, J. Klemic, A. W. Labrador, R. A. Leske, H. Miyasaka, V. Nguyen, R. C. Ogliore, E. C. Stone, R. G. Radocinski, M. E. Wiedenbeck, J. Hawk, S. Shuman, T. T. von Rosenvinge, K. Wortman, The Low - Energy Telescope (LET) and SEP Central Electronics for the STEREO Mission. Space Sci. Rev. 136, 285 - 362 (2008).

[71] R. A. Weller, M. H. Mendenhall, R. A. Reed, R. D. Schrimpf, K. M. Warren, B. D. Sierawski, L. W. Massengill, Monte Carlo simulation of single event effects. IEEE Trans. Nucl. Sci. 57, 1726 - 1746 (2010).

[72] M. H. Mendenhall, R. A. Weller, A probability - conserving cross - section biasing mechanism for variance reduction in Monte Carlo particle transport calculations. Nucl. Instrum. Methods Phys. Res. Sect. A 667, 38 - 43 (2012).

[73] A. I. Mrigakshi, D. Matthiä, T. Berger, G. Reitz, R. F. Wimmer - Schweingruber, Assessment of galactic cosmic ray models. J. Geophys. Res. Space Phys. 117, A08109 (2012).

[74] V. Angelopoulos, The ARTEMIS Mission. Space Sci. Rev. 165, 3 - 25 (2011).

[75] D. H. Fairfield, Average and unusual locations of the Earth' s magnetopause and bow shock. J. Geophys. Res. 76, 6700 - 6716 (1971).

[76] S. Agostinelli, J. Allison, K. Amako, J. Apostolakis, H. Araujo, P. Arce, M. Asai, D. Axen, S. Banerjee, G. Barrand, F. Behner, L. Bellagamba, J. Boudreau, L. Broglia, A. Brunengo, H. Burkhardt, S. Chauvie, J. Chuma, R. Chytracek, G. Cooperman, G. Cosmo, P. Degtyarenko, A. Dell′ Acqua, G. Depaola, D. Dietrich, R. Enami, A. Feliciello, C. Ferguson, H. Fesefeldt, G. Folger, F. Foppiano, A. Forti, S. Garelli, S. Giani, R. Giannitrapani, D. Gibin, J. J. Gómez Cadenas, I. González, G. Gracia Abril, G. Greeniaus, W. Greiner, V. Grichine, A. Grossheim, S. Guatelli, P. Gumplinger, R. Hamatsu, K.

Hashimoto, H. Hasui, A. Heikkinen, A. Howard, V. Ivanchenko, A. Johnson, F. W. Jones, J. Kallenbach, N. Kanaya, M. Kawabata, Y. Kawabata, M. Kawaguti, S. Kelner, P. Kent, A. Kimura, T. Kodama, R. Kokoulin, M. Kossov, H. Kurashige, E. Lamanna, T. Lampén, V. Lara, V. Lefebure, F. Lei, M. Liendl, W. Lockman, F. Longo, S. Magni, M. Maire, E. Medernach, K. Minamimoto, P. Mora de Freitas, Y. Morita, K. Murakami, M. Nagamatu, R. Nartallo, P. Nieminen, T. Nishimura, K. Ohtsubo, M. Okamura, S. O′Neale, Y. Oohata, K. Paech, J. Perl, A. Pfeiffer, M. G. Pia, F. Ranjard, A. Rybin, S. Sadilov, E. di Salvo, G. Santin, T. Sasaki, N. Savvas, Y. Sawada, S. Scherer, S. Sei, V. Sirotenko, D. Smith, N. Starkov, H. Stoecker, J. Sulkimo, M. Takahata, S. Tanaka, E. Tcherniaev, E. Safai Tehrani, M. Tropeano, P. Truscott, H. Uno, L. Urban, P. Urban, M. Verderi, A. Walkden, W. Wander, H. Weber, J. P. Wellisch, T. Wenaus, D. C. Williams, D. Wright, T. Yamada, H. Yoshida, D. Zschiesche, Geant4—A simulation toolkit. Nucl. Instrum. Methods Phys. Res. Sect. A 506, 250 - 303 (2003).

[77] J. Allison, K. Amako, J. Apostolakis, H. Araujo, P. Arce Dubois, M. Asai, G. Barrand, R. Capra, S. Chauvie, R. Chytracek, G. A. P. Cirrone, G. Cooperman, G. Cosmo, G. Cuttone, G. G. Daquino, M. Donszelmann, M. Dressel, G. Folger, F. Foppiano, J. Generowicz, V. Grichine, S. Guatelli, P. Gumplinger, A. Heikkinen, I. Hrivnacova, A. Howard, S. Incerti, V. Ivanchenko, T. Johnson, F. Jones, T. Koi, R. Kokoulin, M. Kossov, H. Kurashige, V. Lara, S. Larsson, F. Lei, O. Link, F. Longo, M. Maire, A. Mantero, B. Mascialino, I. McLaren, P. Mendez Lorenzo, K. Minamimoto, K. Murakami, P. Nieminen, L. Pandola, S. Parlati, L. Peralta, J. Perl, A. Pfeiffer, M. G. Pia, A. Ribon, P. Rodrigues, G. Russo, S. Sadilov, G. Santin, T. Sasaki, D. Smith, N. Starkov, S. Tanaka, E. Tcherniaev, B. Tome, A. Trindade, P. Truscott, L. Urban, M. Verderi, A. Walkden, J. P. Wellisch, D. C. Williams, D. Wright, H. Yoshida, Geant4 developments and applications. IEEE Trans. Nucl. Sci. 53, 270 - 278 (2006).

[78] J. Allison, K. Amako, J. Apostolakis, P. Arce, M. Asai, T. Aso, E. Bagli, A. Bagulya, S. Banerjee, G. Barrand, B. R. Beck, A. G. Bogdanov, D. Brandt, J. M. C. Brown, H. Burkhardt, P. Canal, D. Cano - Ott, S. Chauvie, K. Cho, G. A. P. Cirrone, G. Cooperman, M. A. Cortés - Giraldo, G. Cosmo, G. Cuttone, G. Depaola, L. Desorgher, X. Dong, A. Dotti, V. D. Elvira, G. Folger, Z. Francis, A. Galoyan, L. Garnier, M. Gayer, K. L. Genser, V. M. Grichine, S. Guatelli, P. Guèye, P. Gumplinger, A. S. Howard, I. Hřivnáčová, S. Hwang, S. Incerti, A. Ivanchenko, V. N. Ivanchenko, F. W. Jones, S. Y. Jun, P. Kaitaniemi, N. Karakatsanis, M. Karamitros, M. Kelsey, A. Kimura, T. Koi, H. Kurashige, A. Lechner, S. B. Lee, F. Longo, M. Maire, D. Mancusi, A. Mantero, E. Mendoza, B. Morgan, K. Murakami, T. Nikitina, L. Pandola, P.

Paprocki, J. Perl, I. Petrović, M. G. Pia, W. Pokorski, J. M. Quesada, M. Raine, M. A. Reis, A. Ribon, A. Ristić Fira, F. Romano, G. Russo, G. Santin, T. Sasaki, D. Sawkey, J. I. Shin, I. I. Strakovsky, A. Taborda, S. Tanaka, B. Tomé, T. Toshito, H. N. Tran, P. R. Truscott, L. Urban, V. Uzhinsky, J. M. Verbeke, M. Verderi, B. L. Wendt, H. Wenzel, D. H. Wright, D. M. Wright, T. Yamashita, J. Yarba, H. Yoshida, Recent developments in Geant4. Nucl. Instrum. Methods Phys. Res. Sect. A 835, 186 - 225 (2016).

[79] G. Santin, Normalisation modelling sources. Geant4 (2007).

The Advanced Small Analyzer for Neutrals (ASAN) on the Chang'E – 4 Rover Yutu – 2

M. Wieser[1], S. Barabash[1], X. – D. Wang[1]
A. Grigoriev[2], A. Zhang[3,4], C. Wang[3,4], W. Wang[3]

1. Swedish Institute of Space Physics, Box 812, 98128 Kiruna, Sweden
2. Institut de Recherche en Astrophysique et Planétologie, 9, avenue du Colonel Roche, BP 44346 – 31028 Toulouse Cedex 4, France
3. National Space Science Center (NSSC), Chinese Academy of Sciences (CAS), No. 1 Nanertiao, Zhongguancun, Haidian district, Beijing, 100190, China
4. University of Chinese Academy of Sciences, No. 19 (A) Yuquan Road, Shijingshan district, Beijing, 100049, China

Abstract: The Advanced Small Analyzer for Neutrals (ASAN) is a compact mass resolving energetic neutral atom analyzer flown on the Yutu – 2 rover of the Chinese Chang'E – 4 mission to the Moon. ASAN measures energetic neutral atoms in the energy range from 10eV to 10 keV with typically 30% energy resolution. The field – of – view is a single angular pixel of 37° × 30° pointing at lunar surface. Energetic neutral atoms enter the instrument through a charge particle deflection system and are then ionized on a diamond – like carbon conversion surface. Positive ions from the conversion surface are energy analyzed using an electrostatic analyzer and their velocity is subsequently determined in a time – of – flight section using ceramic channel electron multipliers as detectors. The obtained mass resolution $m/\Delta m$ is about 2, comfortably separating neutral hydrogen, helium, and the oxygen – group (≥16 amu). A full energy spectrum for energetic neutral atoms is obtained in 3 seconds. Switching the charged particle deflection system off, ASAN also directly measures positive ions with similar performance as for energetic neutral atoms. Including electronics the instrument weighs 970 g and consumes 3.4W.

Keywords: Mass resolving energetic neutral atom analyzer; Lunar surface; Solar wind – regolith interaction; Chang'E – 4; Yutu – 2

1 Introduction

The surface of the Moon is covered by regolith, a layer of fine powdered rock (e. g. Clark et al. 2002). Exposure of the lunar crust material to micrometeoroids impacts, solar wind particles, cosmic ray particles, and solar photons, space weathering, is responsible for forming and modifying the regolith. This space weathering turns the rock of the upper crust into a porous layer of glassy, irregularly shaped, sharp, abrasive, chemically reduced, μm - sized grains.

As the Moon only has a very tenuous atmosphere (or exosphere), the supersonic solar wind precipitates directly onto the surface where permitted to do so by lunar local magnetic fields (Lue et al. 2011). The lunar regolith was thought to be a passive absorber of the solar wind (Feldman et al. 2000; Crider and Vondrak 2002), but observations from the recent missions, such as Kaguya, Chang'E - 1 and - 2, Chandrayaan - 1, ARTEMIS, and IBEX have shown that this view is too simplistic and that the interaction of the solar wind with the lunar surface is rather dynamic with a variety of unexpected processes involved. Through sputtering, backscattering or electron emission, the solar wind interaction with the regolith creates a zoo of particles that are emitted from the surface: sputtered atoms (Hoffman and Hodges 1975), electrons (Willis et al. 1973; Halekas et al. 2009), ions (Saito et al. 2008; Holmström et al. 2010), and energetic neutral atoms (ENA) (Wieser et al. 2009, 2010; McComas et al. 2009). These particles are then a source for the lunar exosphere and important component of the near - Moon plasma environment. Precipitating solar wind protons result in chemical reactions with the regolith, producing e. g. hydroxyl (OH) groups or even water molecules embedded in the regolith matrix (Pieters et al. 2009). The water molecules can be then released to space, migrate, and be trapped in cold polar region creating ice deposits. Therefore, the solar wind is considered as one of possible sources of water on the Moon. This type of interaction is not limited to the Moon, but happens as well on other atmosphere - less bodies that are exposed to the solar wind, e. g. Mercury or rocky asteroids and contributes to the ice production there. Opposed to the reaction products embedded in the regolith matrix, the emitted particles, e. g., backscattered or sputtered atoms, can be measured remotely from larger distances, e. g. from orbit, to give insights into the global scale of this interaction process and thus the solar wind implantation rates.

Remote sensing using neutrals however has its limits, it relies on obtaining reliable ground truth measurements and the microphysics of the particle interaction of the surface is difficult to assess from tens or hundreds of kilometres away. The particle populations measured from orbit may have been modified before reaching the observer making interpretation a challenge. Since it is difficult or impossible to reproduce the lunar surface conditions in the laboratory, instrumentation placed on lunar landers is the only option to obtain such measurements.

In 2015, an opportunity opened with a small instrument package onboard of the Yutu - 2 rover of the Chinese Chang'E - 4 mission to the Moon. The Advanced Small Analyzer for Neutrals (ASAN) is the 8th member of the *SWIM family* (Wieser and Barabash 2016), measuring energy and mass distributions of energetic neutral atoms emitted from the surface. In collaboration with the National Space Science Center (NSSC), the ASAN instrument was adapted for operation on Yutu - 2. ASAN is accommodated inside of the thermally insulated payload bay of Yutu - 2 with the entrance system protruding outwards. Located about 0.5m above the surface, ASAN covers with its field - of - view an ellipsoidal area of about $0.96m^2$, located about 0.4 - 2.3m in front of the rover, consisting of - in the ideal case - undisturbed regolith. This unique observation geometry is only available from a rover. The rover's mobility allows investigating locations different in topology and mineralogy, solar wind illumination conditions, lunar local times, and scattering angles and separate contribution of all these factors.

2 Lunar Science and Measurement Requirements

The Advanced Small Analyzer for Neutrals (ASAN) was designed to contribute to the following high - level science questions:

A) *What is the microphysics of the solar wind - surface interaction processes?* With microphysics we mean physical processes of particle interaction with solid material of complex composition and highly irregular structure on the scale down to 10s μm (e. g. Heiken et al. 1974; Heiken 1975, Zbik et al. 2010). The microphysics of the solar wind - surface interaction process that results in the emission energetic neutral atoms and ions is poorly understood. This is mainly because so far observations were all done from orbital distances (e. g. Wieser et al. 2009, 2010; McComas et al. 2009). From such a distant vantage point any particle measurement shows a convolution of the surface properties over large areas, the actual surface interaction and the distribution of the impinging solar wind. To untangle these components a ground truth measurement on the lunar surface is needed. Understanding the microphysics of the backscattering process may e. g. shed light on the formation of hydroxyl or water molecules in the upper most regolith layer. Also, understanding of the microphysics of the interaction is vital for correct interpretation of the ENA measurements from orbit.

B) *How is the lunar exosphere formed and how is it maintained?* Both the ions coming directly from the solar wind as well as energized lunar ions precipitating on the Moon's surface, resulting in sputtering. Atoms sputtered from the regolith are one of the sources for the tenuous and dynamic lunar exosphere (Stern 1999). Since the energies of the sputtered atoms extend to few 10s eV, they are characterized as ENAs. Measurements of these ENA are crucial for understanding the contribution of sputtering to the formation of the Moon's or similar other solar system bodies' exosphere. In-situ surface measurements are the most efficient technique to investigate, due to the difficulty of reproducing the lunar surface conditions in the laboratory.

These science questions translated to measurement requirements as shown in Tab. 1. Determining the surface conditions and the parameters of the precipitating solar wind at the landing site are key to address the these science questions. For surface morphology and conditions we will rely on images from the Panoramic Camera (PCAM) and on data from the Visible and Near-Infrared Imaging Spectrometer (VNIS) instruments onboard of the rover. There is unfortunately no solar wind monitor on the rover or the lander, instead solar wind parameters are obtained from data measured by the WIND and ACE spacecrafts, forward propagated to the lunar surface. The effect of local lunar magnetic anomalies on the conditions at the landing site will be estimated using hybrid modeling of the plasma in the near lunar environment.

3 Instrument Description

3.1 Principle of Operation

The ASAN instrument is a single pixel energetic neutral atom and ion instrument with coarse mass resolution. To detect energetic neutral atoms, ASAN uses surface ionization to convert impinging energetic neutral atoms to positive ions. The mass and energy of the particles is then determined using a combination of an electrostatic analyzer and a surface interaction based time-of-flight section. When detecting energetic neutral atoms, a charged particle deflection system at the entrance rejects ions. To detect positive ions, this system is switched off and ions are directly guided into the electrostatic analyzer for further analysis. Figure 1 shows schematically the ion optical elements and the path of particles in the sensor.

Tab. 1　ASAN requirement matrix

	Measurement requirement	Justification	ASAN sensor performance
Particles	ENAs, positive ions	Solar wind-surface interaction products	ENAs, positive ions
Energy range	10eV-4keV (ENAs) 10eV/q-4keV/q (ions)	Solar wind proton energy defined the upper limit, sputtered particles define the lower limit	10eV-10keV (ENAs) 10eV/q-10keV/q (ions)
Mass resolution	ENAs: H, "heavy" Ions: H+, He+, O+	Hydrogen and heavier abundant species should be separated	Mass groups centered at 1, 2, 4, (8), 16, 28 amu (ENAs) or amu/q (ions)

Continued Table

	Measurement requirement	Justification	ASAN sensor performance
Energy resolution	ENAs：100% Ions：10%（FWHM）	ENA："smooth" spectra，no beams or spectral lines Ions：Driven by ENA mode design	ENAs：16%-140% Ions：8.6%（FWHM）
Time resolution	1min	Temporal variation of the solar wind	3s
UV suppression	$>10^6$	Ability to point at a sun illuminated lunar surface	（$>10^8$ for coincidence signals，by analysis）

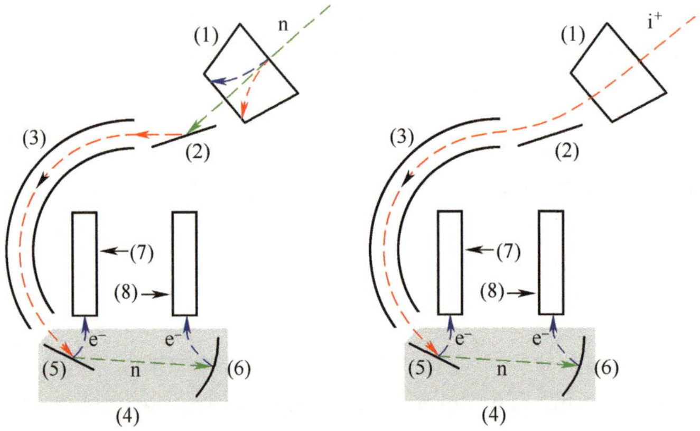

Fig. 1　Ion optical elements. Left：Energetic neutral atom detection mode. Right：Ion detection mode. （1） ion deflection system，（2） charge conversion surface，（3） Electrostatic analyzer，（4） Time-of-flight cell consisting of：（5） start surface，（6） stop surface，（7） start channel electron multiplier，and （8） stop channel electron multiplier

Energetic neutral atoms with energy E and mass m enter the instrument through the deflection system and hit the charge conversion surface that is biased to a positive potential V_{CS}. During the interaction with the conversion surface the particles will lose a fraction η of their energy and some will be converted to single charged positive ions. These ions are then pre-accelerated by the conversion surface potential into the electrostatic analyzer for energy filtering. A stepping high voltage V_{ESA} is applied to the inner plate of the analyzer to select the passing center energy $E_{ESA}=k \cdot V_{ESA}$， with k the analyzer constant. From the exit of the analyzer， ions are post accelerated by V_{PACC} onto the start surface of the time-of-flight section. At the start surface the ion looses a fraction η_S of its energy and propagates through the time-of-flight section with energy E_{TOF}. The ion also is most likely converted back to neutral atom at the start surface. The velocity of the atom is determined by measuring the time-of-flight t between the detection of secondary electrons emitted from the start and the stop surfaces that are separated by the distance s.

For energetic neutral atoms, the energy E_{ENA} and the mass m_{ENA} is determined as follows:

$$E_{ESA}=E_{ENA}(1-\eta_{CS})+V_{CS}\cdot q \tag{1}$$

$$E_{ESA}=k\cdot V_{ESA} \tag{2}$$

$$E_{TOF}=(E_{ESA}+V_{PACC}\cdot q)(1-\eta_S) \tag{3}$$

$$m_{ENA}=2E_{TOF}t^2/s^2 \tag{4}$$

$$m_{ENA}=2(k\cdot V_{ESA}+V_{PACC}\cdot q)(1-\eta_S)t^2/s^2=(\alpha t)^2 \tag{5}$$

where α is an calibration dependent factor.

Solving for the incident neutral atom energy:

$$E_{ENA}=(k\cdot V_{ESA}-V_{CS}\cdot q)/(1-\eta_{CS}) \tag{6}$$

with both $\eta_S<1$ and $\eta_{CS}<1$.

If the ion rejection system is switched off, the energy E/q and the mass m/q of an ion is determined as follows:

$$E_{ESA}/q=E_{ION}/q=k\cdot V_{ESA} \tag{7}$$

$$E_{TOF}/q=(E_{ESA}/q+V_{PACC})(1-\eta_S) \tag{8}$$

$$m_{ION}/q=2E_{TOF}/q\cdot t^2/s^2 \tag{9}$$

$$m_{ION}/q=2(k\cdot V_{ESA}/q+V_{PACC})(1-\eta_S)t^2/s^2=(\beta t)^2 \tag{10}$$

with β a factor similarly to α obtained during instrument calibration.

Solving for the incident ion energy per charge:

$$E_{ION}/q=k\cdot V_{ESA} \tag{11}$$

The approximate expressions for mass are implemented in the sensor's onboard software using tabulated values for α and β depending on the actual energy setting of the electrostatic analyzer and the pre - and post - acceleration potentials.

3.2 The Instrument

3.2.1 Functional Blocks

The ASAN instrument consists of a sensor head and an electronic box integrated into a single unit (Fig. 2). ASAN is bolted through four long, thermally insulating legs onto an inner wall of the payload bay of the Yutu - 2 rover. ASAN shares much of the electrical architecture and the mechanical form factor of the *SWIM family* sensors (Wieser and Barabash 2016).

The sensor head contains the one - time door, all of the ion optical elements and the channel electron multiplier detectors. The electronic box houses three stacked electronics boards; a power subsystem board, a digital processing board including front - end electronics, time to - digital converter and the telemetry interfaces, and a high - voltage board. The interaction between the different functional blocks is shown in Fig. 3 and described in more detail below.

3.2.2 Entrance Subsystem

A one - time spring - loaded door protects the entrance aperture of the instrument.

Fig. 2　ASAN flight model: the sensor head with the entrance system is pointing upwards in front and the electronics box located behind

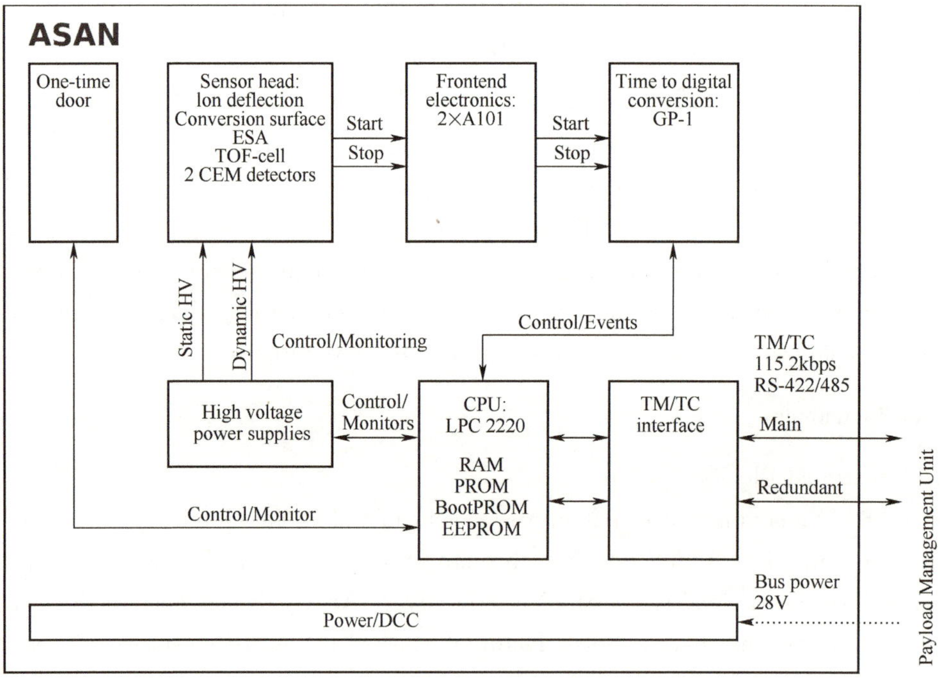

Fig. 3　ASAN functional blocks

The cover reduces the risk of particle, mostly dust, contamination during landing and egress of the rover from the lander. The cover is held closed by an UHMwPE (Ultra High Molecular weight Polyethylene) string guided over two redundant high temperature resistors. To open the door, the resistors are powered and thus heated up to 200℃; this melts the string and releases the spring - loaded door. A micro - switch is used to detect that the door was released and that the resistors can be unpowered. In case of failure of the micro - switch, the resistors are automatically unpowered after 30 seconds. Once opened, the door cannot be closed again. The entrance subsystem also contains the charged particle deflection system. When configured for the detection of energetic neutral atoms, an electric field perpendicular to the nominal particle trajectory deflects ions away from the aperture leading to the charge conversion surface. Ions up to 12keV/q are deflected this way. The deflection voltage is set to 0V when the instrument is configured to detect ions. Electrical fields from the deflection system are screened by a 96% transparent wire harp placed at the exit of the deflection system to prevent disturbance of the electrical field configuration at the conversion surface.

When the deflection system is switched on, deflected ions may hit one of the two deflection system electrodes where they are most likely neutralized and possibly forward scattered towards the charge conversion surface. Such energetic neutral atoms would be difficult to distinguish from the neutral foreground signal and need to be minimized. This is achieved by a combination of rough surfaces that maximize the isotropic scattering and geometric constraints minimizing the view factor of the conversion surface from possible scatter surfaces. These measures suppress ions by an estimated factor of >100 for incident 3keV protons. Given that the neutral particle flux for the observation geometry used is typically an order of magnitude larger than the ion flux, this is considered sufficient.

Furthermore, the possible presence of such ions up to 10keV/q can be identified using the ion mode of the instrument. The opposite case in ion mode where ions are generated from incident neutrals on the conversion surface is more problematic and discussed in Sect. 3. 4. 5.

Both the deflection system and the door are mounted in an assembly that is thermally insulated from the rest of the instrument. This minimizes thermal inputs through the entrance system during daytime when the entrance system is illuminated by the Sun and reduces thermal losses during nighttime. The entrance system also defines the angular size of the field - of - view and the full width half maximum (FWHM) of the angular response.

3. 2. 3　ENA to Ion Conversion System

Energetic neutral atoms need to be converted to ions prior further energy and mass analysis. ASAN uses surface interaction to convert energetic neutral atoms to positive ions. The same technique was successfully used in the Analyzer of Space Plasmas and Energetic neutral Atoms (ASPERA - 3 and 4) onboard Mars Express and Venus Express (Barabash et al. 2006, 2007), in the Chandrayaan Energetic Neutrals Analyzer flown on the Chandrayaan - 1 mission to the Moon (Barabash et al. 2009) and is also used in the Energetic Neutral Analyzer instrument that is en route to Mercury on Bepi Colombo (Kazama et al. 2007). In the energy range of interest for ASAN, surface conversion is the only technique available given the requirement for energetic neutral atom measurements down to 10eV. In ASAN, neutral atoms are reflected at grazing incidence from a diamond like carbon (DLC) coated glass disk. Diamond like carbon surfaces show excellent charge conversion yields to positive ions (Wieser et al. 2005) even at energies below 100eV. Positive ions generated this way are mostly singly charged. Newly converted positive ions also exhibit angular scattering on the charge conversion surface. To improve the collection efficiency of these positive ions, the charge conversion surface is positively biased up to 500V. Ions are thus accelerated into the electrostatic analyzer. When configured for ion detection, the charge conversion surface is passive and only serves as electrostatic lens electrode to guide ions directly into the electrostatic analyzer. The ionization efficiency on the charge conversion surface depends strongly on the energy and the species of the incident energetic neutral atom. This is the main reason for the drop in geometric factor for energetic neutral atoms below 100eV (see e. g. Fig. 13). ASAN uses the positive ion fraction from the conversion surface mainly for heritage reasons (Barabash et al. 2006, 2007, 2009; Kazama et al. 2007). ASAN also inherits the ion - optics optimized for positive ions from the SWIM - family of instruments (Wieser and Barabash 2016). The use of the negative ion fraction would have advantages, as the charge conversion efficiency for some species is up to an order of magnitude better (Wieser et al. 2005), especially at energies below 100eV. However, using the negative ion fraction would make an electron suppression system in front of the electrostatic analyzer necessary to prevent saturation of the time - of - flight section with electrons. Given the limited development resources that were available for ASAN, such a modification remains to be implemented in future members of this instrument family.

3.2.4 Energy Analysis System

Positive ions are energy analyzed in a 127° cylindrical electrostatic analyzer. To simplify the design, only the inner analyzer plate is on a variable potential, the outer plate is kept on ground potential all the time. The small analyzer gap of only 2mm results in a rather small energy pass band for ions of $\Delta E_{ION} \approx 8.6\% \ E_{ION}$. Ions converted from energetic neutral atoms are pre - accelerated between the conversion surface and the electrostatic analyzer, resulting in a much wider relative energy pass band width at low energies for energetic neutral atoms. The energy resolution for energetic neutral atoms is also affected by the energy straggling on the conversion surface. The resulting energy resolution ranges from $\Delta E_{ENA} \approx 16\%$ to $140\% \ E_{ENA}$, depending on energy and species (see Sect. 3. 4. 3).

3.2.5 Time - of - Flight System

After energy analysis, ions are post accelerated into the surface interaction based time - of - flight cell. The time - of - flight start signal is generated by scattering the ions on a 9mm × 25mm polished W (100) single crystal at grazing incidence. A secondary electron is likely emitted in the process, collected and detected by a ceramic channel electron multiplier (CCEM) type KBL 408 from Sjuts Optotechnik, Göttingen, Germany. To increase the secondary electron yield on the W (100) surface, the ions incoming to the cell are accelerated with a 400V potential. Particles travel on to an Al_2O_3 coated stop surface where they hit at near normal incidence. Again, a secondary electron is emitted, collected and detected by another CCEM as stop signal for the time - of - flight measurement. The time - of - flight cell, the start and stop detectors as well as the required electrical components near the detectors are mechanically tightly integrated with the rest of the ion optics (Fig. 4). Fig. 5 shows simulated nominal particle trajectories through all ion optical elements for both ions and energetic neutrals detection.

Due to its compact nature, the resulting mass resolution of the time - of - flight system is limited. The two dominant factors that broaden a peak in the time - of - flight spectrum are the uncertainty of the time - of - flight path length due to the unknown hit position of a particle on the start surface and the only statistically known energy loss during this interaction. These two effects contribute between 6ns and 40ns to the peak width. The energy spread from the electrostatic analyzer, variations in the secondary electron travel time from start or stop surfaces to the corresponding CEMs, the variation of the propagation delay in the CEMs itself and the variation in propagation delay in the preamplifiers due to variations in input signal amplitude contribute combined with about 9 ns. All factors combined, the resulting mass resolution m/Δm of the system is about 2.

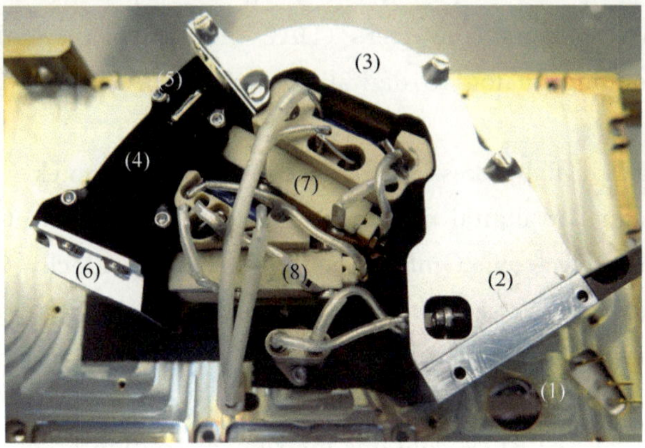

Fig. 4　ASAN sensor head with cover and charged particle deflection system removed. Numbers correspond to numbers used in Fig. 3: (1) charged particle deflection system (not mounted in picture), (2) charge conversion assembly hidden under cover, (3) electrostatic analyzer hidden under cover, (4) time - of - flight cell, (5) start surface, (6) stop surface, (7) start channel electron multiplier, and (8) stop channel electron multiplier both embedded in insulating PEEK blocks

3. 2. 6　Ultraviolet Suppression

All surfaces of ion optical elements in the entrance system, charge conversion system, electrostatic analyzer and the time - of - flight system are coated with electrically conductive Acktar 'Fractal black' (Acktar Ltd., Kiryat - Gat, Israel) to suppress ultraviolet (UV) photons and scattered particles. The UV response was only validated by analysis, based on heritage from previous similarly built instruments: The black coating has a hemispherical reflectance of about 2% in the ultraviolet range. Including geometrical constraints, each reflection from a blackened surface provides an ultraviolet flux reduction of approximately 10^3. With the instrument mounted on the rover and the sun illuminating the instrument aperture, three or more reflections on a blackened surface are needed to reach the time - of - flight system. For photons scattered from the lunar surface two or more reflections from blackened inside surfaces are needed but the lunar ultraviolet albedo is only a few percent. The time - of - flight system helps to further suppress signals from UV photons resulting in a UV suppression of $>10^8$ for coincidence events. In - flight data does indeed not show significant signs of background that could be attributed to UV photons.

3.2.7 Electronics

Electronics are distributed onto three different boards that are stacked below the sensor head inside the electronics box: a high voltage board, a front - end and digital processing board, and a power converter board.

The high voltage board contains two voltage multiplier stages, providing the main high voltages of +5kV and -5kV respectively. Voltages are regulated down to the desired output voltages using IRF - built and qualified opto - couplers. The high voltage board also contains digital to analogue converters (DAC) to control the high voltage outputs as well as the corresponding analog to digital converters (ADC) to generate monitoring values for housekeeping.

The front - end and digital processing board accommodates Amptek A101 preamplifiers to convert the CCEM signal to digital pulses. Subsequently, a TDC - GP1 Time to Digital Converter (acam - messelectronic GmbH, Stutensee - Blankenloch, Germany) is used to measure the time between start and stop pulses. A LPC2220 microprocessor controls the data acquisition and the telemetry interface. During startup, a hardcoded boot - loader is used to copy the onboard software from a 64kByte PROM into RAM for execution. An additional 4 kByte serial EEPROM contains instrument configuration and look - up tables. The EEPROM contents can be patched via tele - command, if required. The telemetry interface consists of two redundant asynchronous 112.5 kbps bidirectional RS422/485 links. Switching from the main to the redundant telemetry interface is automatic; the sensor will answer on the interface that last received a valid command.

ASAN internal low voltages are generated from the 28V spacecraft power bus using two insulating power converters providing the instrument internal voltages +3.3V, +5V, +12V and -12V. The inrush current during power - on is limited using a soft - start mechanism to conform to spacecraft requirements.

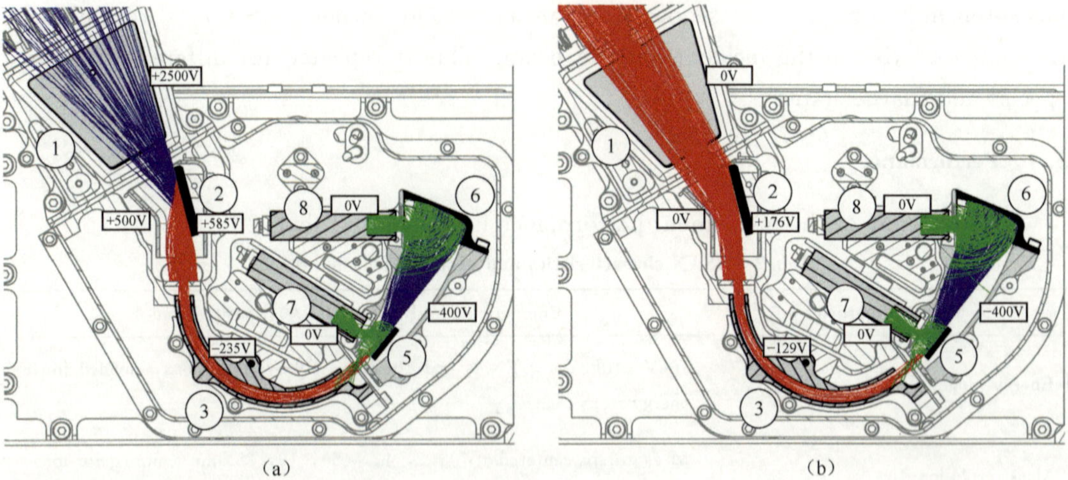

Fig. 5　Simulated trajectories of particles relevant for detection with the instrument set to neutral mode a) and ion mode b). Trajectories are shown for energetic neutral atoms (blue), ions (red) and electrons (green). In panel a), 500eV energetic neutral atoms enter from the top left, are charge converted to positive ions, energy analyzed and finally detected in the time - of - flight cell. In panel b), 500eV/q ions enter from the top left and are redirected towards the electrostatic analyzer, energy analyzed and finally detected in the time - of - flight cell. Numbers in circles identify the same elements as in Figs. 1 and 4. Electrode potentials are shown in rectangular boxes

3.3　Calibration

The ASAN instrument was calibrated at the calibration facility of the Swedish Institute of Space physics in Kiruna, Sweden. A basic description of the system can be found in Wuest et al. (2007), although the system has been significantly upgraded since. The facility is capable to provide an ion beam in the energy range from 30eV/q to 50keV/q. Mass separation is done using a Wien filter in the beam line. The ion beam intensity is monitored using a removable Faraday cup. The base pressure in the main chamber is typically 10^{-7} mbar. A four - axis turntable allows both translation and rotation movements of the instrument under test relative to the incident particle beam. The control software of the turntable permits to run complex translation and rotation patterns without user invention.

For the calibration of ASAN, ion beams consisting of H^+, H_2^+, N^+ and O^+ with energies of 100, 200, 500, 1300 and 3000eV/q were used. Energetic neutral atoms were produced by increasing the pressure in the last part of the beam line to 2×10^{-6} mbar resulting in about 1% of the ions in the beam being neutralized due to collisions with the residual gas. The remaining ions in the beam were electrostatically removed prior entering the main chamber. The rather low intensity of the resulting energetic neutral beam needed to be compensated by longer integration times on the instrument. Calibration measurements consist of a sequence of measurements where ASAN is rotated and translated relative to the incident particle beam. This is repeated for different energies, species and charge states.

3.4 Performance

An overview of the instrument performance is shown in Tab. 2.

Tab. 2 ASAN characteristics and performance summary

Measured particles	Energetic neutral atoms (ENA), positive ions
Energy range	10eV - 10keV for ENAs and 10eV/q - 10keV/q for ions, divided in 48 energy steps each
Mass resolution Energy resolution E/ΔE	Mass groups centered at: 1, 2, 4, (8), 16, 28 amu (amu/q for ions) 16% - 140% for ENAs (typically 20% for H, 50% for O) 8.6% for ions
Field - of - view	ENA: 37° × 30° Ions: 12.3°×11.2°
Geometric factor w/o detection probability for hydrogen @ 1.3keV	ENA: 2.4 e - 5cm^2 sreV/eV Ions: 6.1e - 6cm^2 sreV/eV Notes: Ion optics are optimized for ENA measurements. Values for other species and energies, see Fig. 9.
Detection probability for hydrogen @ 1.3keV Time resolution	Both ions and ENAs: 0.45 (start), 0.05 (stop), 0.02 (coincidence). Values for other species and energies, see Tab. 5 31.25 ms per energy step, 3s for combined ENA and ion energy spectrum
Mass incl. harness	970g
Power	28 V bus 3.4W (7W cover opening)
Dimensions	170.5mm×120.5mm×105mm
Average telemetry	1000bps (CCSDS packet stream)
Operation temperature	−20 to +65℃
Storage temperature	−50 to +70℃
Actuators	One - time operable door

Tab. 3 Angular response

Particle type	Field - of - view (el°×az°)	FWHM (el° × az°)
ENAs	(37±5)×(30±5)	(28±4)×(14±1)
Ions	(12.3±0.9)×(11.2±0.5)	(7.0±1.0)×(4.9±0.5)

Tab. 4　Approximate mass groups

Mass group centered at	Representative species	Mass bin range
m/q=1	H，H^+	6 - 11
m/q=2	H_2，H_2^+，He^{++}	12 - 16
m/q=4	He，He^+	17 - 22
m/q=8	n/a	23 - 30
m/q=16	O，O^+	31 - 50
m/q≈28	Si - group Si^+， - group	51 - 63

3.4.1　Angular Response

Both field - of - view and angular pixel size are different for ions and neutrals，but energy and species independent. The bore - sight direction is independent of energy and species，and with good approximation also independent of charge state. Fig. 6 shows the instrument accommodation with the bore - sight direction and both azimuth and elevation angles indicated. The shape of the angular response is well approximated by a Gaussian distribution centered round the bore - sight direction. Tab. 3 summarizes the angular response.

3.4.2　Mass Resolution

Uncorrelated event rates from the start and stop detector as well as coincidence rates from the time - to - digital converter are recorded in dedicated counters. In case of a coincidence event，the measured time - of - flight t is converted onboard to an energy independent mass bin number m_b by multiplying the time - of - flight t with an energy step dependent tabulated constant α_i：

$$m_b = \min(\text{floor}(t \cdot \alpha_i), 63) \tag{12}$$

with i the energy step number，the floor () function returning the largest integer smaller than the argument and the min () function returning the smaller of its two arguments. For onboard calculations it is assumed for simplicity that there is no energy lost during the interaction with the start surface. The obtained mass bin number m_b in the range from 0 to 63 inclusive and ideally proportional to the square - root of the particle mass (Eqns. 5 and 10). Mass bin 63 is special in the sense that it collects all events that would otherwise result in a calculated mass bin number larger than 63. For first - order calculations a lookup table (Tab. 4) can be used to determine the species. The Si - group also contains other elements than Si with $m \approx 28$amu (e. g. Al or Mg)，but species with $m \geqslant 56$ amu (e. g. Fe) are unlikely to contribute.

Better mass separation is obtained by interpreting the mass spectrum as vector **M** with 64 elements corresponding to the 64 mass bins and then decomposing it into a linear combination of several calibrated response vectors $\mathbf{M_m}$，where the index **m** corresponds to the species：

$$\mathbf{M} = \text{sum}(k_m \cdot \mathbf{M_m}) \tag{13}$$

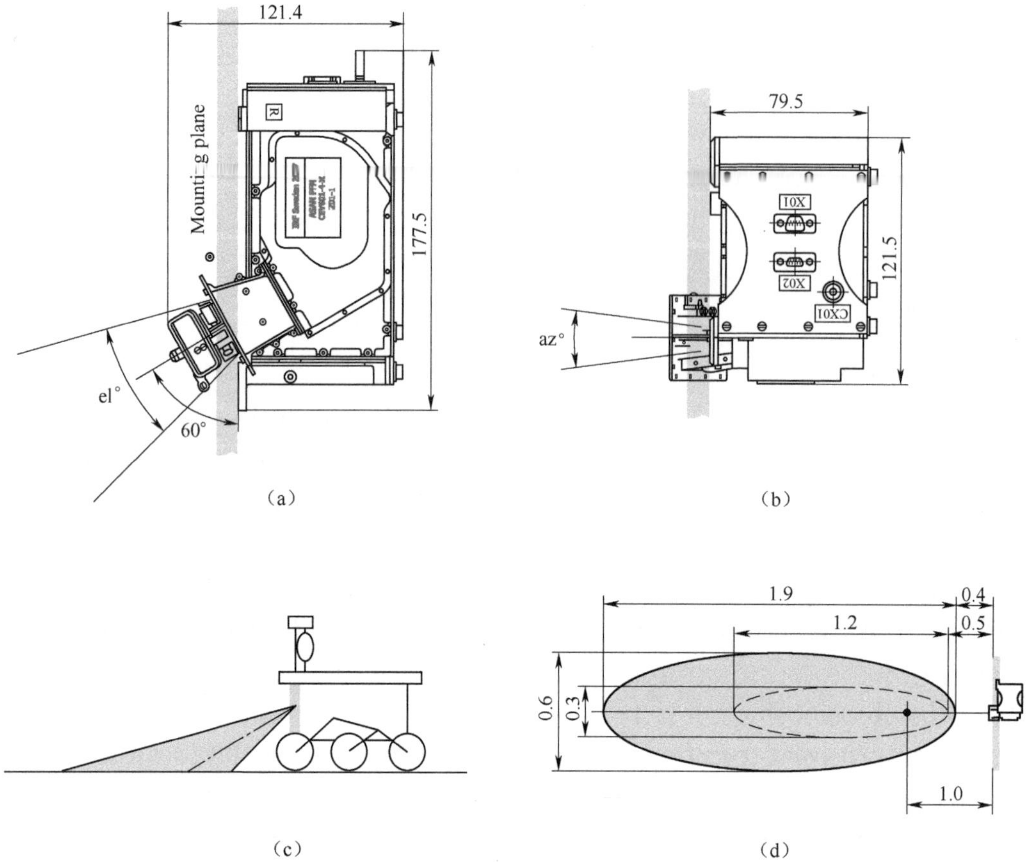

Fig. 6 (a) and (b) ASAN field of view. Instrument dimensions are in millimeters. (c) Accommodation on the Yutu - 2 rover. (d) Top view of the approximate ENA field - of - view projected onto a flat lunar surface (solid line) with full width half maximum contour (dashed line) and the boresight direction indicated (black dot). Dimensions are in m rounded to 0.1m. The grey vertical bar on the right hand side of the plot represents the mounting plane as shown in panel (b). The area of the outer ellipse without rounding is $0.96m^2$, the area enclosed by the full width half maximum contour $0.24m^2$

The response vectors $\mathbf{M_m}$ describe the probability of a particle of mass m being detected in a specific mass bin (Fig. 7). The scaling factor k_m represents the contribution of each species to the spectrum **M**. The response vectors were determined using measured time - of - flight spectra for $m=1$, 2, 14 and 28 amu and subsequent fitting of smooth empirical functions to the observed peak shapes, resulting in the vectors $\mathbf{M_1}$, $\mathbf{M_2}$, $\mathbf{M_{14}}$ and $\mathbf{M_{28}}$. These directly fitted vectors were then interpolated to obtain $\mathbf{M_4}$, and $\mathbf{M_{16}}$. $\mathbf{M_8}$ was not determined because there is no species with 8 amu per charge expected.

The decomposition of **M** can e. g. be done using a least square fitting procedure. Care must be taken that the response vectors used in such a procedure are sufficiently linear independent to obtain a converging solution, e. g. [$\mathbf{M_1}$, $\mathbf{M_2}$, $\mathbf{M_4}$, $\mathbf{M_{16}}$, $\mathbf{M_{28}}$] is a usable set, whereas [$\mathbf{M_1}$, $\mathbf{M_{14}}$, $\mathbf{M_{16}}$] is not.

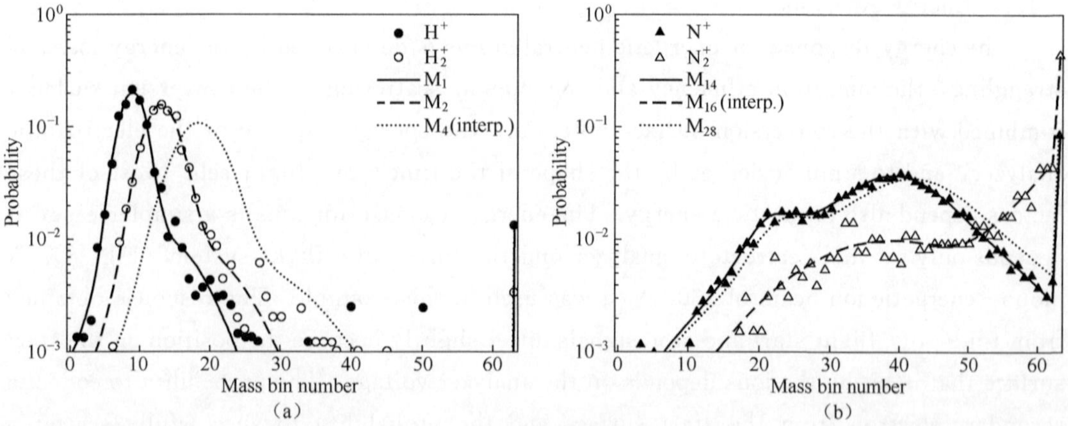

Fig. 7　Response vectors M_m for different ion species entering the time - of - flight cell. a) $\mathbf{M_1}$ and $\mathbf{M_2}$ are smooth empirical functions fitted to mass spectra of H^+ (filled circles) and H_2^+ (open circles) measured during instrument calibration. $\mathbf{M_4}$ is interpolated from $\mathbf{M_2}$ by scaling the time - of - flight data. $\mathbf{M_8}$ was not determined as there is no likely ion associated with a mass of 8 amu. b) $\mathbf{M_{14}}$ and $\mathbf{M_{28}}$ are fitted to measured spectra of N^+ (filled triangles) and N_2^+ (open triangles). $\mathbf{M_{16}}$ (for O^+) is interpolated from $\mathbf{M_{14}}$. The difference between $\mathbf{M_{14}}$ and $\mathbf{M_{16}}$ is small however. While $\mathbf{M_{28}}$ was determined using N_2^+, it is thought to be representative for the Si^+ group too. The peak in $\mathbf{M_{14}}$ around bin 20 and part of the leading shoulder between bins 20 and 50 in $\mathbf{M_{28}}$ are due to recoils from the start surface

The relation between mass bin number and particle mass (per charge) is the same for ions and energetic neutral atoms entering the instrument as energetic neutral atoms are converted to single charged positive ions at the charge conversion surface before they are energy analyzed and enter the time - of - flight cell.

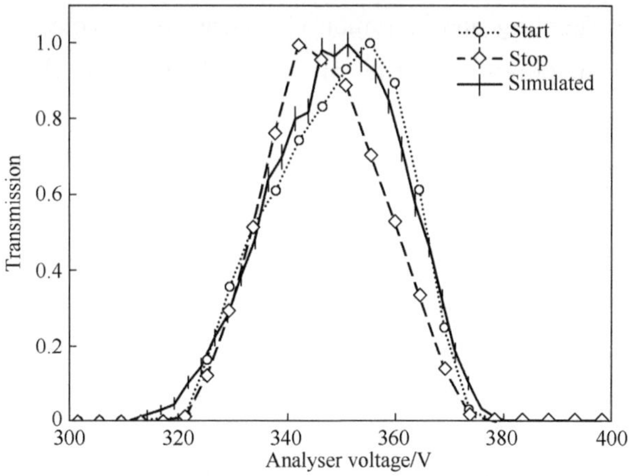

Fig. 8　Measured energy response for ions to a fixed mono - energetic 1300eV N^+ beam. The response obtained using start - (circles) and stop - counters (diamonds) is shown. The corresponding energy resolution $\Delta E/E$ using start counters is 8. 6% with ΔE the full width half maximum. For comparison, the solid line shows a simulated response for the start signal for the same conditions as the measured data

3.4.3 Energy Analysis

The energy response for energetic neutral atoms is determined by the energy loss and straggling, the ionization efficiency and the angular scattering at the conversion surface, combined with the conversion surface potential, the energy response of the electrostatic analyzer, and to a minor degree by the shape of the time - of - flight cell. Most of these factors depend also on particle energy. The energy response for ions is a simpler case; it depends only on the electrostatic analyzer and the time - of - flight system (Fig. 8). A mono - energetic ion beam of 1300eV/q was used in this example. The response obtained from time - of - flight start and stop signals differ slightly because the position on the start surface that is hit by the ions depends on the analyzer voltage. The probability to collect a secondary electron from the start surface and the probability to successfully scatter a particle towards the stop surface both depend on the hit position on the start surface - but not in the same way. This results in a small difference of the peak position for start and stop signals as shown in Fig. 8. The corresponding analyzer constant $k = E/V_{ESA}$ of the analyzer is 3.7 with E the ion energy and V_{ESA} the analyzer voltage. The value increases slightly to 3.93 below 500eV/q, due to fringe field effects between the exit of the analyzer and the entrance of the time - of - flight cell.

The similar response to energetic natural atoms is more difficult to obtain because of the much lower neutral beam intensities in the calibration system. The non - linear response in energy of the conversion surface does not allow directly determining the energy response of an individual energy bin using a fixed incident neutral beam energy. Several measurements with fixed incident neutral beam energy were combined into an empirical model to obtain the energy response of each individual energy bin to energetic neutral atoms over a wide energy range. The empirical model contains energy and mass dependent energy loss, energy straggling and ionization efficiency on the conversion surface, the pre - acceleration caused by the potential of the conversion surface and the electrostatic analyzer. The small asymmetries introduced by the time - of - flight section (Fig. 8) are not included in the model as they are too small to have a notable effect when measuring neutral atoms.

An example of one such measurement together with the fit from the empirical model is shown in Fig. 9. A larger number of such measurements were used to build and tune the empirical model.

There are two different energy tables stored in the instrument, one for ions and one for energetic neutral atoms. Due to the mass dependent energy loss for neutrals on the charge conversion surface and the pre - acceleration of the neutrals converted to ions, the energy step centers and energy resolution are different for ions, neutral hydrogen and neutral oxygen (the latter serving as a reference case for "heavy" neutrals). The two onboard tables are built such that a near dense energy coverage is achieved for ions and for energetic neutral hydrogen.

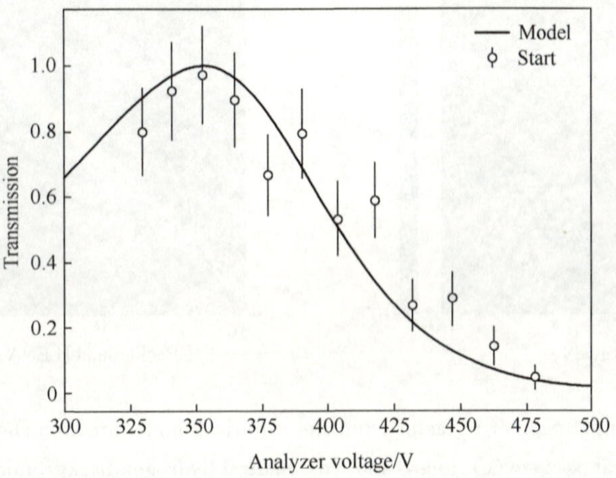

Fig. 9　Energy response of the complete sensor to fixed mono - energetic 1300eV N^0 beam when varying the voltage on the electrostatic analyzer. Time - of - flight start signals (circles) and the empirical model used for interpolation to other energies (solid line) is shown. Error bars shown are Poisson errors

Due to different energy loss and energy straggling for energetic neutral oxygen on the conversion surface, each energy bin corresponds to two different energy intervals, one for energetic neutral hydrogen and one for energetic neutral oxygen. Fig. 10 shows the energy coverage for neutral hydrogen atoms, neutral oxygen atoms and ions, obtained from the empirical model fitted to the actual measurements, for all energy bins. Figure 11 shows the corresponding energy resolution for different species obtained from the same model.

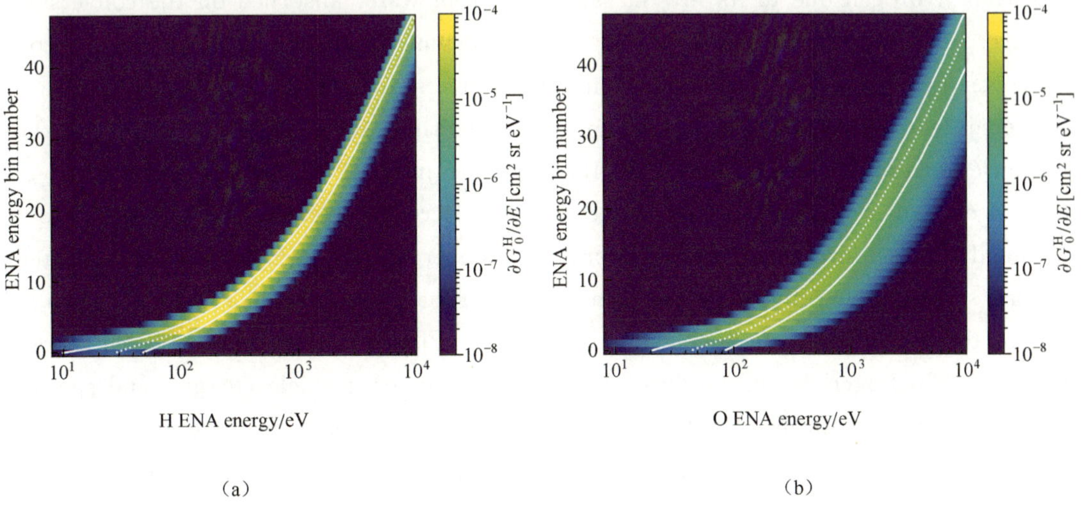

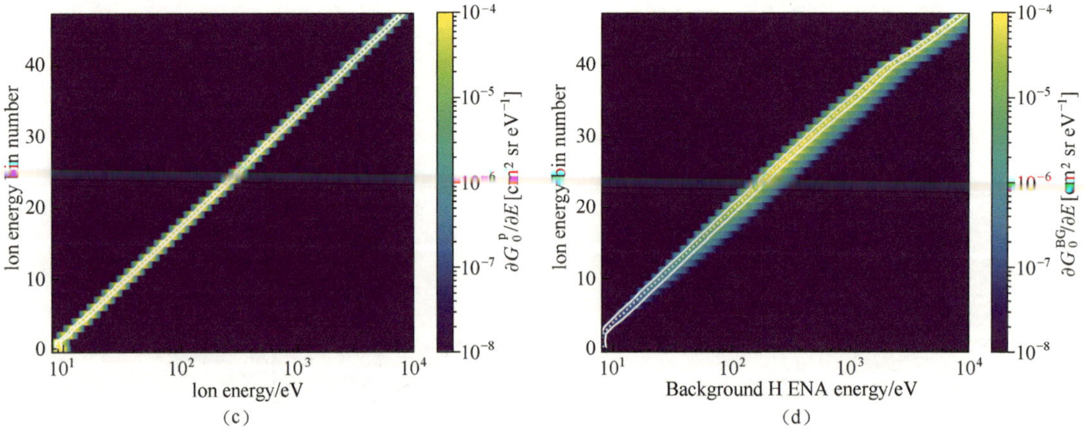

Fig. 10 Energy differential geometric factor from the empirical model for each energy bin for (a) neutral hydrogen (b) and neutral oxygen (c) ions, and (d) neutral hydrogen background in ion mode (see Sect. 3. 4. 4). The dotted white line indicates the maximum in each energy bin, the solid white lines correspond to the 50% levels

3.4.4 Secondary Particles from the Conversion Surface

Energetic neutral particles hitting the charge conversion surface are nominally forward scattered. If the energy of the neutral particle is sufficiently high, secondary particles from the charge conversion surface itself may be emitted through single collisional sputtering (recoil) or electronic sputtering. Such particles can, provided having a positive charge state, subsequently be detected and be mistaken for incident neutral particles. The energy of a recoil particle depends on the ratio of incident particle mass to recoil particle mass and the total scattering angle (Niehus et al. 1993). For the expected energy spectrum and mass composition (Wieser et al. 2009), the most important recoil species is hydrogen due to its low mass, e. g. from water absorbed on the conversion surface. For the geometry present in the ASAN instrument, the energy spectrum of recoil hydrogen created by an incident hydrogen atom is almost indistinguishable from a forward scattered hydrogen atom. The process may be responsible for a slight asymmetry in the response function as shown in (Fig. 10a). If the incident particle is heavier, e. g. neutral oxygen or helium, a proton recoil may be created that can be separated from nominally ionized foreground hydrogen only by its energy. The recoil energy is estimated using a single elastic collision between an incident primary particle and a hydrogen atom at rest on the conversion surface (Niehus et al. 1993). Figure 12 shows the modeled energy bins where hydrogen recoils produced by different primary particle energies and species would register. The probability for getting a proton recoil depends on the amount of water absorbed on the conversion surface at the time of measurement - something that is poorly constraint. A rough estimate for the amount of recoils from an instrument internal surface can be obtained using calibration data from the time - of - flight cell: For 1700eV N^+

hitting the start surface, about 6% of the particles generating a time - of - flight event were hydrogen recoils (Fig. 7).

While the amount of hydrogen recoils from heavier incident species is expected to be low, careful comparison of the obtained mass spectra from different observation times will be needed to constrain the amount of recoils present. Secondary particles originating from electronic sputtering have low energies (<100eV) and consist mainly of heavy elements from the conversion surface. For these to be efficiently produced, heavy high - energy (>1keV) incident neutral atoms are needed. While the electronic sputtering products will pollute the lowest energy bins, their presence can be inferred from the presence of the parent neutrals at higher energies.

3.4.5 Geometric Factor and Efficiencies

The differential number flux $J^s(E_i)$ [$sr^{-1}cm^{-2}eV^{-1}s^{-1}$] seen by each energy step i centered at an energy E_i[eV], and species s [one out of neutral hydrogen (H), neutral oxygen (O) or protons and other ions (p)] is given by

$$J^s(E_{i,s}) = C^s(E_i)/[G^s(E_i) \cdot E_i \cdot \Delta t] \tag{14}$$

with $C^s(E_i)$ [#] the number of counts accumulated during the measurement interval Δt [s], and $G^s(E_i)$ is the geometric factor [cm^2 sreV/eV] including the detection probabilities valid for species s and energy step i. The measurement interval Δt is always 31.25ms long. The dead time including the high voltage settling time between different measurement intervals is negligible.

The geometric factor G is separated in two factors ε and G_0:

$$G^s(E_i, s) = \varepsilon * G_0^s(E_i) \tag{15}$$

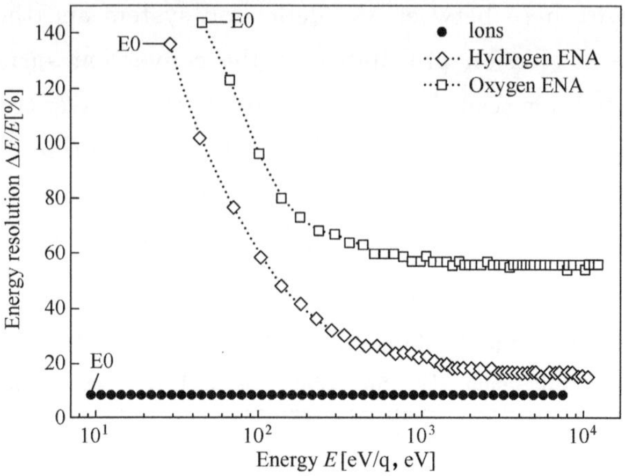

Fig. 11　Fitted energy resolution depending on energy for all 48 energy bins for ions (filled circles), neutral hydrogen (open diamonds) and neutral oxygen (open squares). Data points are placed at the peak energy of each energy bin. The peak energy is species dependent. The left most data point in each curve corresponds to energy step E0. The values for E0 to E2 for neutral atoms should be used with caution due to uncertainties when determining the ionization efficiency at the conversion surface. Full - width - half - maximum values are used for ΔE. The dotted lines are shown to guide the eye

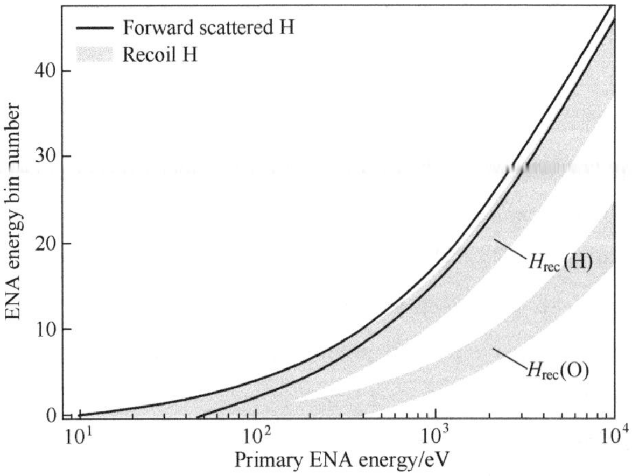

Fig. 12　Predicted energy bin numbers where hydrogen recoils from the conversion surface appear in the energy mass matrix (grey areas) depending on primary species and energy. Hydrogen recoil positions from incident hydrogen, H_{rec} (H), and incident oxygen, H_{rec} (O), are shown. Nominally forward scattered hydrogen registers between the two black lines

with ε the detection probability for start, stop or coincidence events, and G_0^s the geometric factor without detection probability (sometimes also called 'geometric factor without efficiency') depending on energy E_i and species s.

G_0^s (E_i) contains all elements affecting particles prior entering the time-of-flight cell: it includes, besides the effective aperture area and the angular field of view, the transparency of the wire harp between the deflection system and the conversion surface assembly, the charge conversion probability at the conversion surface, the probability that a newly generated ion can be collected from the conversion surface, and the transmission function of the electrostatic analyzer. The detection probability ε depends on the detector bias voltage, the aging of the detectors and the secondary electron emission probability at the start and stop surfaces in the time-of-flight cell. There are two detectors in the instrument giving three different detection event types: start, stop and coincidence.

Tab. 5　Detection probabilities (a) for 1300eV/q ions or 1keV neutrals from calibration data, (b) for <300eV neutrals from in-flight data

Mass [amu/q]	ε_{START}	ε_{STOP}
(a)		
1	0.45+/−0.03	0.040+/−0.002
2	0.45+/−0.04	0.046+/−0.003
3	0.41+/−0.03	0.048+/−0.003

Continued Table

Mass [amu/q]	ε_{START}	ε_{STOP}
12	0.28+/−0.02	0.022+/−0.002
14	0.26+/−0.02	0.023+/−0.002
28	0.21+/−0.02	0.021+/−0.002
(b)		
1	0.59+/−0.07	0.09+/−0.02
≥28	0.30+/−0.05	0.11 +/−0.03

Each event type has a different detection probability ε. Values for ε can be obtained using the start −, stop − and coincidence event counters C_{START}, C_{STOP}, and C_{COINC} (Funsten et al. 2005):

$$\varepsilon_{START} = C_{COINC} / C_{STOP} \tag{16}$$

$$\varepsilon_{STOP} = C_{COINC} / C_{START} \tag{17}$$

$$\varepsilon_{COINC} = \varepsilon_{START} \cdot \varepsilon_{STOP} \tag{18}$$

Depending on which counter, (i. e., start, stop, or coincidence) is used in Eq. (14) to calculate differential flux, the corresponding ε value must be used to calculate G in Eq. (15).

As ε is determined by particles in the time − of − flight section, the values obtained are valid for both ions and energetic neutral atoms entering the instrument. Values for ε_{START} and ε_{STOP} were obtained during calibration using a 1300eV beam of single charged ions. The probabilities depend on mass per charge of the particles entering the time − of − flight cell (Tab. 5a). In − flight data confirmes that ε_{START} and ε_{STOP} are energy dependent (Tab. 5b). The significant larger value of ε_{STOP} can be explained by the smaller angular scattering on the start surface at lower energies. As particles are post − accelerated by 400 V before they interact with the surfaces in the time − of − flight system, the values shown in Tab. 5b are expected to be valid down to the lowest energy settings of the instrument. Nevertheless, for best accuracy the values should always be validated using in − flight data.

During calibration, G_0 has been determined for energetic neutral hydrogen (H) and for protons (p). These geometric factors were then interpolated to the full energy range of the instrument. The interpolated values for protons are valid for any ion mass, but a possible background from large neutral atom flux needs to be taken into account. An empirical model using independently measured charge state fractions (Wieser et al. 2002, 2005) was used to predict the geometric factor for neutral oxygen based on the data from energetic neutral hydrogen. Other neutral species are less well constrained. Figures 13a and 13b show values interpolated over the full energy range for energetic neutral atoms and ions respectively. Note that the energy steps E0 and E1 are poorly constrained by the model.

In ion mode, ASAN is also sensitive to energetic neutral atoms. This is because energetic neutral atoms cannot be hindered from reaching the conversion surface in ion mode where they may convert to positive ions and are subsequently indistinguishable from original ions. This becomes important when the neutral particle flux is of a similar order of magnitude or even larger than the ion flux. In the lunar environment, mainly neutral hydrogen contributes to this effect. This makes it possible to correct an ion spectrum under the condition that an energetic neutral hydrogen spectrum is available for similar measurement conditions. In the following this background due to energetic neutral hydrogen is treated like a separate species '*BG*' with its own energy table E_{BG} and geometric factor G^{BG}.

As the bias potential of the conversion surface where these additional ions are generated is not zero, the energy of the energetic neutral atoms E_i^{BG} contributing to a specific energy bin i is different from the center energy E_i^p in ion mode. To obtain the background-corrected number of ion counts $C^{p*}(E_i)$, background counts $C^{BG}(E_i^{BG})$ must be subtracted from the number of counts $C^p(E_i)$ obtained in ion mode during energy step i. Note that the same energy step number i is used for E_i^p and E_i^{BG}, though the two energy values are not identical:

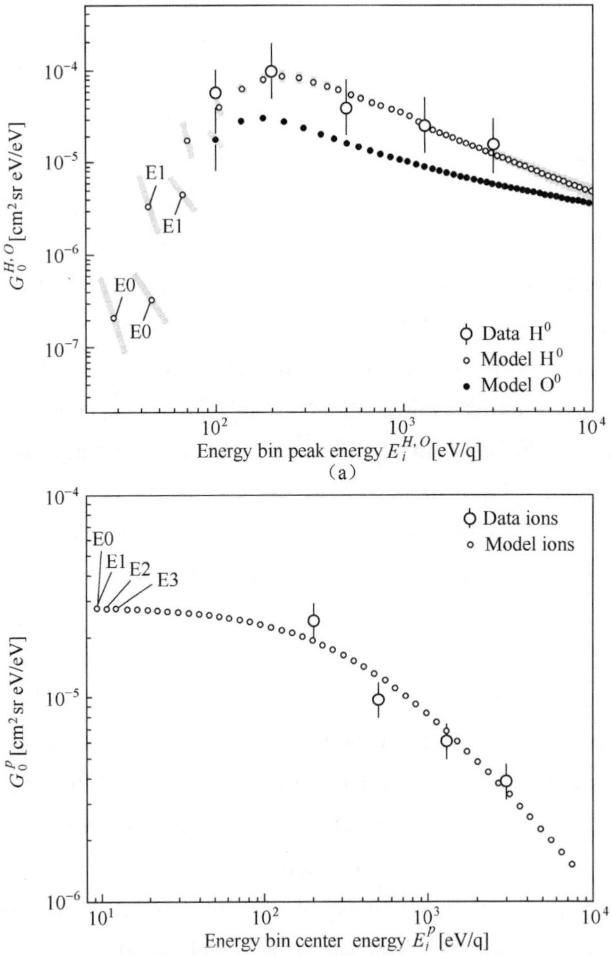

(a)

(b)

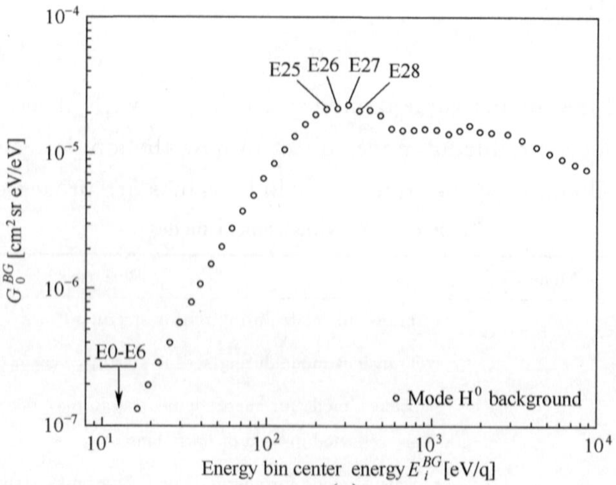

(c)

Fig. 13 (a) Geometric factor without detection probability interpolated for neutral hydrogen G_0^H and neutral oxygen G_0^O. An ion - optical model of the sensor is used to obtain a fit to the available data points. The fitted data points refer to the peak energy for energetic neutral hydrogen E_i^H and oxygen E_i^O in each energy step i. The gray bars illustrate model dependent uncertainties originating from uncertainties when determining the ionization efficiency at the conversion surface at low energies. (**b**) Geometric factor without detection probability for positive ions of any mass. An ion - optical model of the sensor is used to obtain a fit to the available data points. The fitted data points refer to the center energy E_i^p of each energy step i. (c) Geometric factor without detection probability G_0^{BG} (E_i^{BG}) for background energetic neutral hydrogen in ion mode. The data points refer to the center energy E_i^{BG} of energetic neutral hydrogen contributing to each energy step i in ion mode. The contribution of the first 7 energy steps is negligible and outside the y - scale of this plot.

$$C^{p*}(E_i) = C^p(E_i) - C^{BG}(E_i^{BG}) \tag{19}$$

with the background contribution C^{BG} (E_i^{BG}) obtained from an independent measurement of the differential energetic neutral atom flux J^H in neutral mode:

$$C^{BG}(E_i^{BG}) = J^H(E_i^{BG}) \cdot G^H(E_i^{BG}) \cdot E_i^{BG} \cdot \Delta t \tag{20}$$

As the energy levels E_i^{BG} do not match the energy levels E_i^H used in neutral mode, the differential energetic neutral flux J^H (E_i^{BG}) must be interpolated from the energy levels E_i^H used in the neutral mode measurement to the energy levels E_i^{BG}. The energy levels do not match due to different bias potentials on the conversion surface and different energy stepping used in neutral and ion mode respectively.

A background corrected differential ion flux J^{p*} is then obtained by

$$J^{p*}(E_i) = C^{p*}(E_i) / [G^p(E_i) \cdot E_i \cdot \Delta t] \tag{21}$$

Fig. 13c shows the values of G_0^{BG} (E_i^{BG}) versus energy. G^{BG} (E_i^{BG}) is obtained by multiplying these with a detection probability ε.

3.5　Instrument Operation

3.5.1　Instrument Modes

There are five different instrument modes (Tab. 6), with three of them producing science data. The science producing modes differ in how the ion deflection system and the energy tables are used and how the time-of-flight values are processed.

Table 6　ASAN instrument modes

Mode name	Mode ID	Purpose
Safe	0	Transient mode during sensor startup
Normal	1	Transient mode during sensor startup, sensor cover opening
TOF	2	Science mode for energetic neutral atoms, raw time—of—flight events are reported instead of mass bins
EXM	3	Science mode for energetic neutral atoms, standard science mode
ION	4	Science mode for background ions

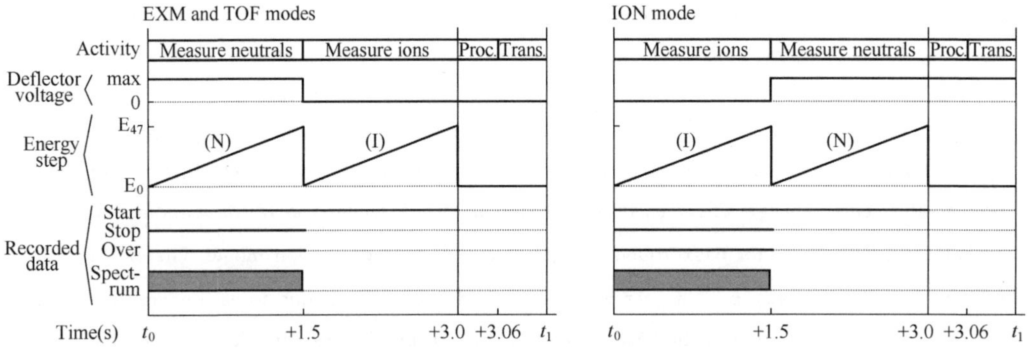

Fig. 14　Cyclogram for the three science modes. t_0: start of cycle, t_1 start of next cycle. Most data is recorded during the first energy sweeps: start-, stop- and over-counters ('Start', 'Stop' and 'Over') and a mas spectrum ('Spectrum') in EXM and ION modes or a list of time-of-flight events in TOF mode for each energy step E_0 to E_{47}. In the second energy sweep only start-counters are recorded. The two energy sweeps use different lookup tables for measurements of neutrals (N) and ions (I). The voltage applied to the deflector system is zero during ion measurements and non-zero during measurement of neutrals. Internal post processing takes typically 60ms ('Proc.'), followed by typically 0s to 3s waiting time for data transmission ('Trans')

3.5.2　Science Mode Measurement Cycle

A full measurement cycle of ASAN in a science mode consists of a pair of energy sweeps of 48 steps with a stepping time of 31.25ms per step. Internal lookup tables are used to set high voltages depending on energy step for the ion deflection system, the charge conversion system and the electrostatic analyzer. The lookup tables also provide the value α needed to normalize the time-of-flight (Eq. (12)). During the first energy sweep, time-of-flight start-, time-of-flight stop-, and over-event counters along with a mass spectrum or time - of - flight event list are recorded; during the second energy sweep only time - of - flight start counters are recorded. Over - events are valid time - of - flight events where the time - of - flight exceeded 1.6μs but was still shorter than 7.6us, the time - out value for the time - to - digital converter. Over - events cannot be converted to mass bins because they would result in an overflow in the calculation. A full measurement cycle takes 3.06s including data processing (Fig. 14). Before the next measurement cycle can be started, sufficient data buffer space must be available (see Sect. 3.5.3). Depending on how often data is requested by the spacecraft, this may lengthen the total measurement cycle time to 5 - 6s.

The data structures generated in TOF, EXM or ION modes (Tab. 7) are identical in size but differ slightly in the organization of segment #5. In TOF mode, an event counter followed by the time - of - flight of up to 63 individual particles is reported instead of the 64 mass bins. If there are less than 63 time - of - flight measurements in an energy bin, the unused measurements are filled with zeros to get a fixed structure size. If there were more than 63 time - of - flight events, the time - of - flight of the excess events are not reported. The event counter however always contains the actual number of time - of - flight events, independent of the number that could be reported. The factor 48 in the number of elements in segments #1 to #5 corresponds to the number of energy steps.

Table 7　ASAN science data structure for TOF, ION and EXM modes (excluding CCSDS header)

Segment	Content	Uncompressed structure size (bits)
#0	Status information (instrument mode, binning, etc.)	26×8
#1	48 start counters from first energy sweep	48×16
#2	48 stop counters from first energy sweep	48×16
#3	48 over counters from first energy sweep	48×16
#4	48 start counters from second energy sweep	48×16
#5	TOF mode:	48× (1+63) ×16
	48 × (1 event counter + 63 time—of—flight events)	
	ION and EXM mode:	48×64×16
	48 × 64 mass bins	

In all three science modes, segments #1 to #5 contain only unsigned 16 - bit counters or unsigned 16 - bit time - of - flight values, which are before transmission to spacecraft linear - to - logarithmic encoded to 8 - bit to save telemetry bandwidth. The encoding scheme used is shown in Tab. 8. The scheme compresses unsigned 16 bit integers to unsigned 8 bit integer values using three different domains: Values up to 63 are directly encoded, values between 64 and 255 retain the first 6 significant bits and values 256 and larger retain the first 5 significant bits. The lossless encoding of the smallest values guarantees that small signals are not affected and can be recovered during data analysis through integration over time. For values larger than 63, the encoding scheme is lossy due to the fact that non - significant bits are simply truncated. With optimal decoding, the reconstructed values have an average error of about ±1% between 64 and 255 and about ±2.5% above. After linear - to - logarithmic encoding, lossless Rice - Golomb compression (Golomb 1966; Rice et al. 1993) is applied to segments #1 to #5 to further reduce the data volume.

Table 8 Linear to logarithmic encoding table. Values are represented as binary numbers with the most significant bit on the left. Letters a to e in the uncompressed value represent bits that are copied verbatim to the encoded value, the letter x in the uncompressed value represents bits that are discarded in the encoded value

16 bit uncompressed integer	8 bit encoded integer
0000 0000 000a bcde	000a bcde
0000 0000 001a bcde	001a bcde
0000 0000 01ab cdex	010a bcde
0000 0000 1abc dexx	011a bcde
0000 0001 abcd xxxx	1000 abcd
0000 001a bcdx xxxx	1001 abcd
0000 01ab cdxx xxxx	1010 abcd
0000 1abc dxxx xxxx	1011 abcd
0001 abcd xxxx xxxx	1100 abcd
001a bcdx xxxx xxxx	1101 abcd
01ab cdxx xxxx xxxx	1110 abcd
1abc dxxx xxxx xxxx	1111 abcd

3.5.3 Telemetry Handling

The communication between the Yutu - 2 Payload Management Unit (PMU) and ASAN is done in a master - slave fashion: the PMU is periodically requesting data packets from ASAN by polling.

Housekeeping data is emitted by ASAN in the 30 byte long fixed size PMU housekeeping packet format and send on request from the PMU. As the bandwidth available in the housekeeping channels is limited, more comprehensive "private" housekeeping packets are inserted into the science data stream.

Science data is requested periodically by the PMU through polling. However, internally, ASAN generates asynchronously data packets based on a subset of the CCSDS (Consultative Committee for Space Data Systems) standard (https://public.ccsds.org). This is heritage from the PRisma Ion Mass Analyzer (PRIMA) instrument (Wieser and Barabash 2016). The CCSDS formatted science packets vary greatly in size and are generated independently of the PMU science data request. To map these CCSDS packets to PMU science data requests, the CCSDS science packet flow is buffered on ASAN and interpreted as byte stream. A PMU formatted science data packet with its own header is generated, as soon as a minimum number of bytes are available in the buffer and a science packet request arrives. This results in PMU science packet frames not being aligned with CCSDS packet boundaries. On ground, this fragmentation process is undone and the original CCSDS formatted science data is recovered. As a consequence, ASAN may temporarily hold the start of a new measurement cycle until there is sufficient free space in the internal packet buffer. This in turn will occasionally result in an uneven distribution of sampling start times.

3. 5. 4　Standard Operation Sequences

For operation on the lunar surface, the number of commands necessary to run ASAN was minimized. Two standard command sequences were developed, corresponding to different mission phases:

1. A commissioning sequence without high voltages but including the opening of the sensor door. This sequence is run at the very beginning of the surface operations to maximize the time available for improved sensor outgassing.

2. A standard operation sequence with 30 minutes of science operations. This sequence contains 25 minutes EXM - mode followed by 5 minutes ION - mode and is intended to be repeated as many times as possible at different locations, rover orientations and lunar local times. The total duration of the sequence is adjustable by expanding or shrinking the time spent in EXM - and ION - mode proportionally.

3. 6　Sample Data

Chang - E'4 landed on the Moon on January 3rd, 2019, in the Von Kármán Crater on the lunar far - side. Following the Yutu - 2 rover deployment, the ASAN instrument commissioning was completed on January 11th, 2019 after which regular science operations started. ASAN has typically two operation windows during each lunar day, one during the local morning and the other during the local evening. As of January 29th, 2020, 41 individual science operation sequences were run - resulting in more than 49h of science data. A non - massresolved quick - look energy spectrum recorded at the end of the first lunar day, on January 31th, 2019, is shown in Fig. 15. Further analysis of these data show that, except at the very lowest energies, almost the entire signal is due to energetic neutral hydrogen, justifying the use of the geometric factor for hydrogen. This quick - look data has not been corrected for the effect of possible hydrogen recoils or electronically sputtered hydrogen from the conversion surface that could lead to an artificial flux enhancement at very low energies.

4　Summary

The Advanced Small Analyzer for Neutrals is a compact energetic neutral atom instrument for the Yutu - 2 rover of the Chang'E - 4 mission. It is designed to detect energetic neutral atoms emitted from the lunar surface in the energy range from 10eV to 10keV with coarse mass resolution. An additional ion mode is used to estimate backgrounds. The instrument includes a power system, data processing, a high-voltage power system and the sensor head in one compact unit. ASAN has undergone successful testing and calibration and was shown to fulfill all performance requirements. The instrument was successfully deployed onto the Moon and in-flight performance after more than one year on the lunar surface is nominal, with no signs of degradation.

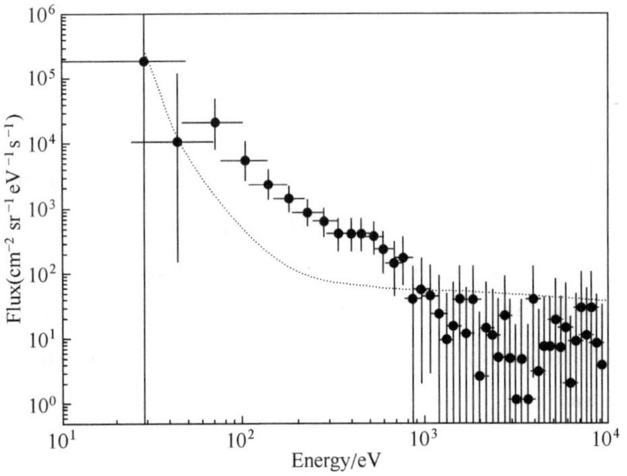

Fig. 15　Energetic neutral atom spectrum obtained on 2019-01-31 between 19:11 and 19:37 UTC, with background subtracted (black circles). Horizontal error bars correspond to the full width half maximum of the corresponding energy bin; vertical bars are 90% confidence intervals (Feldman and Cousins 1998) for the flux after subtracting backgrounds. The dotted line is a 90% confidence sensitivity limit, below which the signal cannot be distinguished from the no-signal case

Acknowledgements

The Advanced Small Analyzer for Neutrals (ASAN) instrument was developed by the Swedish Institute of Space Physics (IRF) in Kiruna, Sweden, and the National Space Science Center (NSSC) in Beijing, China, as a joint project. It is supported by the Swedish National Space Agency (SNSA) under grants 95/11 and 122/18, the National Natural Science Foundation of China (NSFC) under grant 41941001, and the Beijing Municipal Science and Technology Commission under grant 181100002918003. The data describing the geometric factor of the instrument is freely available at: https://data.irf.se/wieser2020ssr/. Reuse of this data should be credited to the ASAN instrument team at the Swedish Institute of Space Physics (IRF) in Kiruna, Sweden, and the National Space Science Center (NSSC) in Beijing, China. ASAN science data are provided through the China National Space Administration and the National Astronomical

Observatories of China, Chinese Academy of Sciences.

References

[1] S. Barabash, R. Lundin, H. Andersson, K. Brinkfeldt, A. Grigoriev, H. Gunell, M. Holmström, M. Yamauchi, K. Asamura, P. Bochsler, P. Wurz, R. Cerulli - Irelli, A. Mura, A. Milillo, M. Maggi, S. Orsini, A. Coates, D. Linder, D. Kataria, C. Curtis, K. Hsieh, B. Sandel, R. Frahm, J. Sharber, J. Winningham, M. Grande, E. Kallio, H. Koskinen, P. Riihelä, W. Schmidt, T. Säles, J. Kozyra, N. Krupp, J. Woch, S. Livi, J. Luhmann, S. McKenna - Lawlor, E. Roelof, D. Williams, J. - A. Sauvaud, A. Fedorov, J. - J. Thocaven, The analyzer of space plasmas and energetic atoms (ASPERA - 3) for the Mars express mission. Space Sci. Rev. 126 (1 - 4), 113 - 164 (2006). https://doi.org/10.1007/s11214-006-9124-8.

[2] S. Barabash, J. - A. Sauvaud, H. Gunell, H. Andersson, A. Grigoriev, K. Brinkfeldt, M. Holmström, R. Lundin, M. Yamauchi, K. Asamura, P. Bochsler, W. Baumjohann, T. Zhang, A. Coates, D. Linder, D. Kataria, C. Curtis, H. Hsieh, B. Sandel, A. Fedorov, C. Mazelle, J. - J. Thocaven, M. Grande, H. E. Koskinen, E. Kallio, R. Riihela, J. Kozyra, N. Krupp, J. Woch, J. Luhmann, S. McKenna - Lawlor, S. Orsini, R. Cerulli - Irelli, M. Mura, M. Milillo, M. Maggi, E. Roelof, P. Brandt, C. Russell, K. Szego, J. Winningham, R. Frahm, J. Scherrer, J. Sharber, P. Wurz, P. Bochsler, The Analyser of Space Plasmas and Energetic Atoms (ASPERA - 4) for the Venus Express mission. Planet. Space Sci. 55 (12), 1772 - 1792 (2007). https://doi.org/10.1016/j.pss.2007.01.014.

[3] S. Barabash, A. Bhardwaj, M. Wieser, R. Sridharan, T. Kurian, S. Varier, E. Vijayakumar, V. Abhirami, K. V. Raghavendra, S. V. Mohankumar, M. B. Dhanya, S. Thampi, A. Kazushi, H. Andersson, F. Yoshifumi, M. Holmström, R. Lundin, J. Svensson, S. Karlsson, R. D. Piazza, P. Wurz, Investigation of the solar wind - Moon interaction onboard Chandrayaan - 1 mission with the SARA experiment. Curr. Sci. 96 (4), 526 - 532 (2009).

[4] B. E. Clark, B. Hapke, C. Pieters, D. Britt, Asteroid space weathering and regolith evolution, in *Asteroids III*, ed. by W. F. Bottke Jr., A. Cellino, P. Paolicchi, R. P. Binzel (University of Arizona Press, Tucson, 2002), pp. 585 -599.

[5] D. H. Crider, R. R. Vondrak, Hydrogen migration to the lunar poles by solar wind bombardment of the Moon. Adv. Space Res. 30 (8), 1869 - 1874 (2002).

[6] G. J. Feldman, R. D. Cousins, Unified approach to the classical statistical analysis of small signals. Phys. Rev. D 57, 3873 - 3889 (1998).

[7] W. C. Feldman, D. J. Lawrence, R. C. Elphic, B. L. Barraclough, S. Maurice, I. Genetay, A. B. Binder, Polar hydrogen deposits on the Moon. J. Geophys. Res. 105 (E2), 4175 - 4196 (2000).

[8] H. O. Funsten, R. W. Harper, D. J. McComas, Absolute detection efficiency of space - based ion mass spectrometers and neutral atom imagers. Rev. Sci. Instrum. 76 (5), 053301 (2005).

[9] S. W. Golomb, Run - length encodings. IEEE Trans. Inf. Theory 12 (3), 399 (1966).

[10] J. S. Halekas, G. T. Delory, R. P. Lin, T. J. Stubbs, W. M. Farrell, Lunar prospector measurements of secondary electron emission from lunar regolith. Planet. Space Sci. 57 (1), 78 - 82 (2009).

[11] G. Heiken, Petrology of lunar soils. Rev. Geophys. 13 (4), 567 - 587 (1975).

[12] G. H. Heiken, D. S. McKay, R. W. Brown, Lunar deposits of possible pyroclastic origin. Geochim. Cosmochim. Acta 38 (11), 1703 - 1718 (1974).

[13] J. H. Hoffman, R. R. Hodges, Molecular gas species in the lunar atmosphere. Earth Moon Planets 14 (1), 159 - 167 (1975).

[14] M. Holmström, M. Wieser, S. Barabash, Y. Futaana, A. Bhardwaj, Dynamics of solar wind protons reflected by the moon. J. Geophys. Res. 115 (A6) 06 (2010).

[15] Y. Kazama, S. Barabash, M. Wieser, K. Asamura, P. Wurz, Development of an LENA instrument for planetary missions by numerical simulations. Planet. Space Sci. 55 (11), 1518 - 1529 (2007).

[16] C. Lue, Y. Futaana, S. Barabash, M. Wieser, M. Holmström, A. Bhardwaj, M. B. Dhanya, P. Wurz, Strong influence of lunar crustal fields on the solar wind flow. Geophys. Res. Lett. 38 (3), 02 (2011).

[17] D. J. McComas, F. Allegrini, P. Bochsler, P. Frisch, H. O. Funsten, M. Gruntman, P. H. Janzen, H. Kucharek, D. J. McComas, E. Möbius, D. B. Reisenfeld, N. A. Schwadron, Lunar backscatter and neutralization of the solar wind: First observations of neutral atoms from the Moon. Geophys. Res. Lett. 36, 12 (2009).

[18] H. Niehus, W. Heiland, E. Taglauer, Low energy ion scattering at surfaces. Surf. Sci. Rep. 17, 213 - 303 (1993).

[19] C. M. Pieters, J. N. Goswami, R. N. Clark, M. Annadurai, J. Boardman, B. Buratti, J. P. Combe, M. D. Dyar, R. Green, J. W. Head, C. Hibbitts, M. Hicks, P. Isaacson, R. Klima, G. Kramer, S. Kumar, E. Livo, S. Lundeen, E. Malaret, T. McCord, J. Mustard, J. Nettles, N. Petro, C. Runyon, M. Staid, J. Sunshine, L. A. Taylor, S. Tompkins, P. Varanasi, Character and spatial distribution of OH/H_2O on the surface of the Moon seen by M3 on Chandrayaan - 1. Science 326 (5952), 568 - 572 (2009).

[20] R. Rice, P. - S. Yeh, W. Miller, Algorithms for high - speed universal noiseless coding in 9th Computing in Aerospace Conference, (American Institute of Aeronautics and Astronautics, 1993) https://doi. org/10. 2514/6. 1993 - 4541.

[21] Y. Saito, S. Yokota, T. Tanaka, K. Asamura, M. N. Nishino, M. Fujimoto, H. Tsunakawa, H. Shibuya, M. Matsushima, H. Shimizu, F. Takahashi, T. Mukai, T. Terasawa, Solar wind proton reflection at the lunar surface: low energy ion measurement by map - pace onboard Selene (Kaguya). Geophys. Res. Lett. 35, 12 (2008).

[22] S. A. Stern, The lunar atmosphere: history, status, current problems, and context. Rev. Geophys. 37 (4), 453 - 491 (1999). https: //doi. org/10. 1029/1999RG900005.

[23] M. Wieser, S. Barabash, A family for miniature, easily reconfigurable particle sensors for space plasma measurements. J. Geophys. Res. Space Phys. 121 (12), 11588 - 11604 (2016).

[24] M. Wieser, P. Wurz, K. Bruning, W. Heiland, Scattering of atoms and molecules off a magnesium oxide surface. Nucl. Instrum. Methods B 192, 370 - 380 (2002).

[25] M. Wieser, P. Wurz, P. Bochsler, E. Moebius, J. Quinn, S. Fuselier, A. Ghielmetti, J. DeFazio, T. Stephen, R. Nemanich, NICE: an instrument for direct mass spectrometric measurement of interstellar neutral gas. Meas. Sci. Technol. 16, 1667 - 1676 (2005).

[26] M. Wieser, S. Barabash, Y. Futaana, M. Holmström, A. Bhardwaj, R. Sridharan, M. Dhanya, P. Wurz, A. Schaufelberger, K. Asamura, Extremely high reflection of solar wind protons as neutral hydrogen atoms from regolith in space. Planet. Space Sci. 57 (14 - 15), 2132 - 2134 (2009).

[27] M. Wieser, S. Barabash, Y. Futaana, M. Holmström, A. Bhardwaj, R. Sridharan, M. B. Dhanya, A. Schaufelberger, P. Wurz, K. Asamura, First observation of a mini - magnetosphere above a lunar magnetic anomaly using energetic neutral atoms. Geophys. Res. Lett. 37 (5), 03 (2010).

[28] R. F. Willis, M. Anderegg, B. Feuerbacher, B. Fitton, Photoemission and secondary electron emission from lunar surface material, in *Photon and Particle Interactions with Surfaces in Space*, ed. by R. J. L. Grard (Springer, Dordrecht, 1973), pp. 389 - 401.

[29] M. Wuest, D. S. Evans, R. von Steiger (eds.). Calibration of Particle Instruments in Space Physics. SR - 007 (International Space Science Institute, 2007).

[30] M. S. Zbik, R. L. Frost, Y. - F. Song, Y. - M. Chen, High porosity chained aggregates from the topsoil of the lunar regolith dust, in *Proceedings of the 9th Australian Space Conference* (Australian Academy of Science, 2010), pp. 55 - 65.

First measurements of the radiation dose on the lunar surface

Shenyi Zhang[1,2,3,4], Robert F. Wimmer - Schweingruber[1,5], Jia Yu[5], Chi Wang[1], Qiang Fu[6,7], Yongliao Zou[1], Yueqiang Sun[1,2,4], Chunqin Wang[1,2,4], Donghui Hou[1,2,3,4], Stephan I. Böttcher[5], Sönke Burmeister[5], Lars Seimetz[5], Björn Schuster[5], Violetta Knierim[5], Guohong Shen[1,2,4], Bin Yuan[1,2,4], Henning Lohf[5], Jingnan Guo[5,8,9], Zigong Xu[5], Johan L. Freiherr von Forstner[5], Shrinivasrao R. Kulkarni[5], Haitao Xu[1], Changbin Xue[1], Jun Li[1], Zhe Zhang[10], He Zhang[11], Thomas Berger[12], Daniel Matthiä[12], Christine E. Hellweg[12], Xufeng Hou[13], Jinbin Cao[14], Zhen Chang[1,2,4], Binquan Zhang[1,2,4], Yuesong Chen[1], Hao Geng[1], Zida Quan[1,2,4]

1. National Space Science Center, Chinese Academy of Sciences, Beijing, PR China
2. Beijing Key Laboratory of Space Environment Exploration, Beijing, PR China
3. University of Chinese Academy of Science, Beijing, PR China
4. Key Laboratory of Environmental Space Situation Awareness Technology, Beijing, PR China
5. Institute of Experimental and Applied Physics, Christian - Albrechts - University, Kiel, Germany
6. National Astronomical Observatories, Chinese Academy of Sciences, Beijing, PR China
7. Key Laboratory of Lunar and Deep Space Exploration, Chinese Academy of Sciences, Beijing, PR China
8. University of Science and Technology of China, Hefei, PR China
9. CAS Center for Excellence in Comparative Planetology, Hefei, PR China
10. Lunar Exploration and Space Engineering Center, Beijing, PR China
11. China Academy of Space Technology, Beijing, PR China
12. German Aerospace Center (DLR), Institute of Aerospace Medicine, Cologne, Germany
13. 18th Research Institute, China Electronics Technology Group Corporation, Tianjin, PR China
14. BeiHang University, Beijing, PR China

Abstract: Human exploration of the Moon is associated with substantial risks to astronauts from space radiation. On the surface of the Moon, this consists of the chronic exposure to galactic cosmic rays and sporadic solar particle events. The interaction of this radiation field with the lunar soil leads to a third component that consists of neutral particles, i. e., neutrons and gamma radiation. The Lunar Lander Neutrons and

Dosimetry experiment aboard China's Chang'E 4 lander has made the first ever measurements of the radiation exposure to both charged and neutral particles on the lunar surface. We measured an average total absorbed dose rate in silicon of 13.2 ± 1 μGy/hour and a neutral particle dose rate of 3.1 ± 0.5μGy/hour.

Introduction

The Moon is the next stepping stone for human space exploration, and several nations have announced plans for its exploration by humans. Space radiation exposure is one of the major risks for astronauts' health (1 - 3) as the chronic exposure to galactic cosmic rays (GCRs) may have late health effects such as induction of cataract (4), cancer (5 - 7), or degenerative diseases of the central nervous system (8) or other organ systems (9, 10). Moreover, exposure to large solar particle events (SPEs) in a situation with insufficient shielding may cause severe acute effects (11). The exposure to GCR is inevitable but generally contributes a low dose rate compared to the sporadic, unpredictable, but sometimes very intense SPEs in which solar energetic particles are accelerated close to the Sun by solar flares and coronal mass ejections. The nucleonic component of GCR consists mainly of protons (~87%), helium (~12%), and heavier nuclei (~1%) (12). These nuclei have very high energy and are therefore highly penetrating. Because of their single charge, protons are only weakly ionizing, and helium nuclei are four times more ionizing per nucleus. The remaining 1% of nuclei are high (H) atomic number (Z) and energy (E) elements (HZE) that contribute to radiation damage disproportionally according to the square of their nuclear charge, Z, resulting in very dense ionization along their tracks. Because of nuclear fragmentation and other complex interactions with matter, their exact effects on humans are uncertain but may be considerable (13, 14).

It appears that there have been no active (i. e., time resolved) measurements of the radiation dose rate on the surface of the Moon until the Chinese Chang'E 4 mission landed in the von Karman crater on the far side of the Moon on 3 January 2019 at 02:26 UTC. During the Apollo missions, astronauts carried dosimeters with them (15) to the Moon, but time-resolved radiation data from the surface of the Moon were never reported (16). Here, we report radiation dose rate measurements with previously unseen accuracy from the surface of the Moon.

For the assessment of the radiation exposure, the relevant quantities have to be measured by the detector systems: The absorbed dose, D, is the ratio of the energy (E; usually measured in keV) deposited in a detector and the mass, m, of the detector and is expressed in units of Gray (Gy = J/kg). Division by the accumulation time results in the measured dose rate (expressed in Gy/hour). Using a combination of two detectors in coincidence, one measures the distribution of energies deposited in a detector to obtain the linear energy transfer (LET) spectrum [usually in units of keV per micrometer (keV/μm)]. This spectrum is integrated with so-called quality factors, Q, used as biological weights to obtain the dose equivalent, H, which is expressed in units of Sievert (Sv = J/kg). The exact procedures are defined by the International Commission on Radiation Protection (17). Because the human body is not made of silicon, and to make dose, dose rate, and LET measurements more easily comparable to others, one normally converts the values measured in Si to the corresponding quantities in water using a constant dose conversion factor of 1.30 (18).

The Lunar Lander Neutrons and Dosimetry (LND) experiment is described in more detail in the literature (19), but we summarize the pertinent information here for convenience. The LND is mounted in the payload compartment of the Chang' E 4 lander. The red arrow in Fig. 1 points at the reclosable door that protects LND from the cold lunar nights but is open during lunar daytime. The LND consists of a stack of 10 dual-segment silicon solid-state detectors (SSDs), A to J, as shown in the main part of Fig. 2. Total absorbed dose and dose rate are measured in detector B, and the absorbed dose (rate) from neutral particles is measured in the inner segment of the C detector, C1, with the closely spaced detectors B and D as well as the outer segment of C, C2, serving as anticoincidence to discriminate against charged particles. The LET is then determined as discussed above from the dE/dx measured using three different combinations of detector pairs with different counting rates and average path lengths. Penetrating particles are measured by requiring signals in all 10 detectors.

Fig. 1　View of the Chang' E 4 lander with the location of the LND sensor head indicated by the red arrow. LND is mounted in the Chang' E 4 payload compartment; the lid at the tip of the red arrow is closed at night to protect LND from the cold lunar night. Photo credit: Chinese National Space Agency (CNSA) and National Astronomical Observatories of China (NAOC)

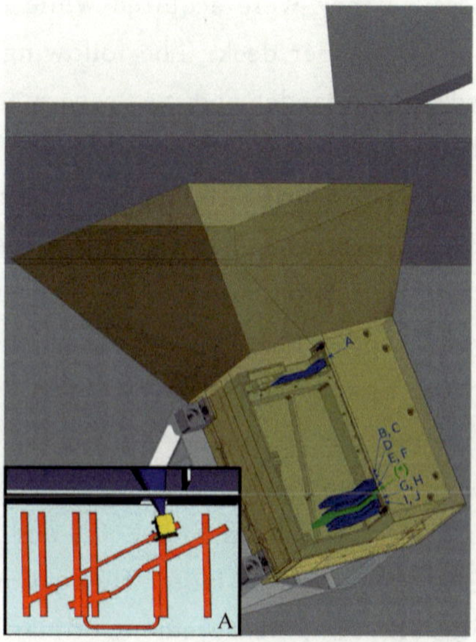

Fig. 2　Schematic view of the LND sensor head and its accommodation in the Chang'E 4 payload compartment. The LND detector system consists of 10 dual-segment silicon SSDs (A to J) shown and labeled in blue. They are arranged such as to form a particle telescope that views the sky through an opening of the payload compartment. The structure shown in green absorbs thermal neutrons and is irrelevant for this paper. This opening in the payload compartment (indicated by gray walls) is closed during the lunar night and reopened in the lunar morning. Multilayer insulation is shown in gold and insulates the LND sensor head, which is mounted to the side panel of the payload compartment with an Al bracket also shown in gray. Inset A shows LND' s location on the payload panel (in pale blue) together with its NH_3 thermal control system (TCS) indicated in red

Materials and Methods

Fig. 3 shows time-resolved measurements acquired by LND from 3 January to 12 January and from 31 January to 10 February 2019, i. e., when the lander was not hibernating during the intensely cold lunar night. To survive this extreme thermal environment, the Chang' E 4 lander contains a radioisotope thermoelectric generator (RTG) and three radioisotope heater units (RHUs); in addition, the Yutu-2 rover is also equipped with an RTG. We have measured the contributions of the lander RTG and RHUs in August 2018 before the launch of the spacecraft and have thus determined their contribution to be 5.2 ± 0.6μGy/hour to the total dose rate and 1.7 ±0.4μGy/hour to the neutral dose rate (20). These values have been subtracted from the measurements shown in Fig. 3, which shows (from top to bottom) (A) the evolution of total dose rate with time and the same for neutral dose rate (B) and charged-particle dose rate (C), as well as the flux of penetrating particles (D). The first four data points in Fig. 3A lie above the remaining data points because they were acquired while the Yutu - 2 rover with its RTG was still piggybacked on the lander deck. The following data points show the dose rate after the rover had rolled off the lander and was separated from the lander and LND by at least 7m. The last 11 hours in the first lunar day show a further decrease in dose rate that also extends to 3 February 2019. During this period of time, the liquid NH_3 thermal control system (TCS; shown in red in inset A to Fig. 2) of the lander had been activated, provided additional shielding, and modulated high - energy neutrons (21). The contribution of (predominantly) high - energy neutrons is shown in Fig. 3B and clearly exhibits a drop at the end of the first lunar day, which also extends throughout the time period when the TCS was active. As expected, the dose rate from charged particles (shown in Fig. 3C) shows no cotemporal decrease, nor does the flux of penetrating particles (Fig. 3D). The dose rates shown in Fig. 3 (A and C), as well as the flux of penetrating particles (Fig. 3D), show much larger fluctuations than the dose rate from neutral particles shown in Fig. 3B. This measurement primarily records the recoil energy transferred by neutrons to the Si nuclei of the C1 detector segment and the energy deposited by low - energy ($E_\gamma < 1$MeV) γ rays. Because there is no directional information, the geometric factor for this measurement is much larger than that for penetrating particles. The fluctuations seen in Fig. 3 (A, C, and D) are thus due to statistical fluctuations and to fluctuations in the number of helium nuclei contributing to the total dose rate and even rarer contributions from heavy ions. The lander was again put into hibernation on 10 February 2019.

The LND measures time - resolved LET spectra that are remarkably invariant. Fig. 4 shows such spectra acquired during the first lunar day. The black curve (black circles) shows the LET spectrum for the time period when the rover was still piggybacked on the lander, the red curve (and red squares) shows the same quantity while the lid that protects the LND sensor head from the cold night was still closed, and the purple upward triangles show data with the lid opened (as in Fig. 1). Last, the blue downward triangles show the LET spectrum acquired while the TCS was active. It is nearly impossible to distinguish any differences in these spectra. LET spectra are only measured for charged particles, so their invariance is consistent with the finding from Fig. 3 (C and D) that the charged particle dose rate and flux hardly varied at all during the LND measurements. Thus, the clear differences in measured dose rates seen in the course of the first two lunar days must be due to changes in the dose rate from neutral particles, which is also borne out by Fig. 3B and underlines the importance of also measuring the dose rate from neutral particles.

Results

Ignoring the data acquired while Yutu - 2 was still piggybacked on the lunar lander module and while the TCS was active [11 January 2019 (14: 41) to 3 February 2019 (07:

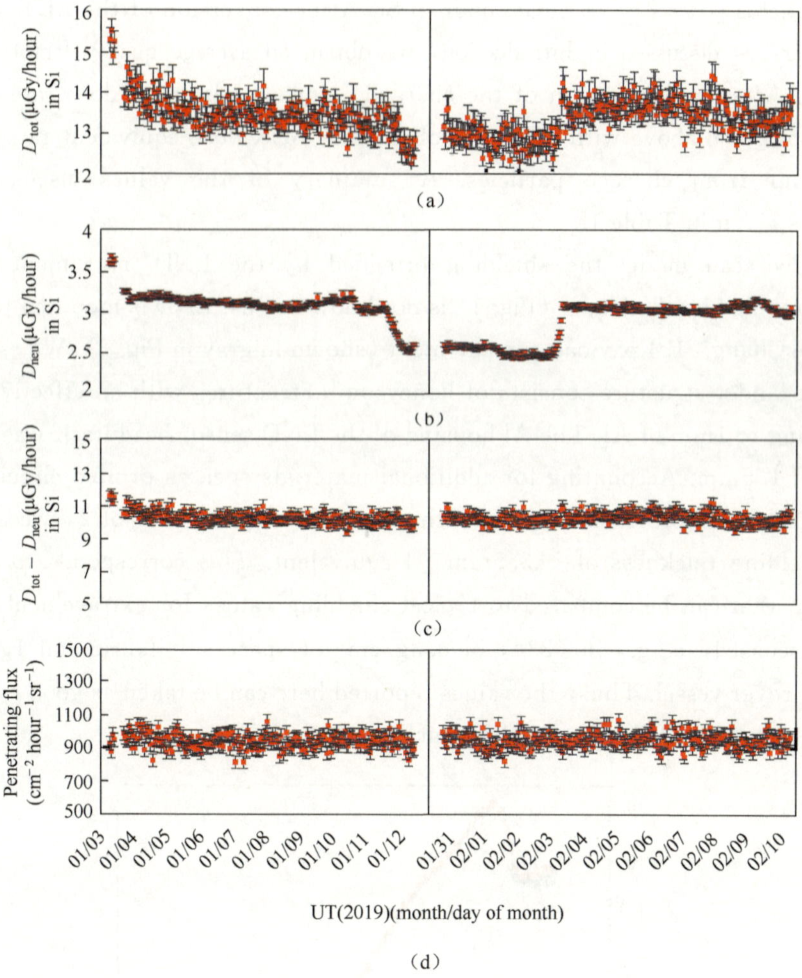

Fig. 3　Temporal evolution of the radiation environment on the Moon as measured by LND on Chang'E 4 during the first and second lunar day after Chang' E4 landed. The left - hand panels show data for the first lunar day, and the right - hand panels show data for the second lunar day. (a) Total absorbed dose rate measured with the LND B detector. (b) Neutral particle dose rate recorded in the LND C1 silicon detector. (c) Total absorbed dose rate from charged particles only [i. e., (a and b)]. The known background from the RTG and RHUs (20) has been subtracted from the values reported in (a) and (b). (d) Temporal evolution of the flux of penetrating particles. UT, universal time

53)], we measured an average total dose rate in silicon of 13. 2 ± 1 and 3. 1 ± 0. 5μGy/hour in silicon for neutral particles. The background due to the RTG and RHUs was measured in August 2018 (20), and the values reported above have been corrected accordingly; the corresponding errors are reported separately in Tab. 1. Thus, we find that neutral particles contributed a nonnegligible fraction of 23 ± 8% to the total dose (22).

Subtracting the neutral contribution, we find that the average absorbed dose rate due to charged particles is 10.2 ± 1.1μGy/hour in Si. After conversion of the LET spectrum to LET in water, as discussed in Introduction, we obtain an average quality factor of $\langle Q \rangle$ = 4.3 ± 0.7. After multiplication of the charged - particle absorbed dose rate (in water) measurement given above with $\langle Q \rangle$, we obtain the GCR dose equivalent rate of 57.1 ± 10.6μSv/hour from charged particles. A summary of the values discussed in this paragraph is given in Table 1.

While we can model the shielding provided by the LND instrument itself, the shielding provided by the lander (Fig. 1) is not known to us. LND is mounted to the inside panel of the Chang' E 4 payload compartment (shown in gray in Fig. 2). We estimate that most of the lander structure consists of honeycomb structures with an effective thickness corresponding to 1mm of Al. The Al housing of the LND sensor head itself was milled to a thickness of 1.5mm. Accounting for additional materials such as printed circuit boards in the sensor head and projection effects with a factor of square root of 2, we arrive at an average shielding thickness of ~3.5mm Al equivalent. This corresponds to a shielding of ~1g/cm^2 that can be compared to typical shielding values for extravehicular activities (EVAs) discussed, e.g., in (23), of 0.3g/cm^2 of space suit fabric and 1g/cm^2 for a pressurized rover vessel. Thus, the values reported here can be taken as good estimates for the dose rate during EVAs on the lunar surface.

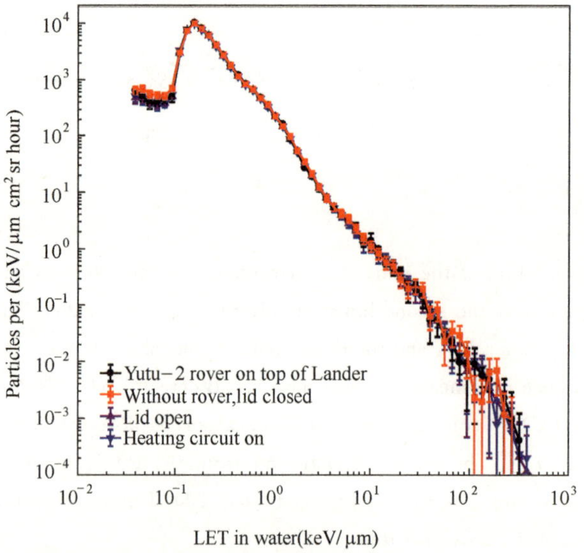

Fig. 4 Linear Energy Transfer (LET) spectra (converted to LET in water) measured during four different time periods show only small variations. The data shown as black circles were acquired before lander and rover separation; data shown in red after the rover left the lander but with the lid still closed; data in purple: normal operation of LND with the lid open; data in blue: with the Chang'E 4 heater circuit activated toward the end of the lunar day

Tab. 1　Summary of measurements of the radiation dose rate measured in μGy/hour on the lunar surface. The errors of the background dose rate from the RTG/RHUs (20) are considered systematic errors and have been added quadratically when reporting the final values in the rightmost column

Dose rate/(μGy/hour)	Measured	Background	Final in Si
Total	18.4±0.4	5.2±0.6	13.2±0.7
Neutral	4.7±0.1	1.7±0.5	3.1±0.5
Charged	13.7±0.4	3.5±0.8	10.2±0.9

Interpretation

To put the values reported here into context, we briefly summarize measurements by the Cosmic Ray Telescope for the Effects of Radiation (CRaTER) on Lunar Reconnaissance Orbiter (LRO) (24). Incidentally, LRO flew over the location of Chang' E 4 at 01: 30 on 2 February 2019. At this time, both LND and CRaTER saw identical heliospheric conditions, and CRaTER measured a dose rate of 13.29μGy/hour (reported by the CRaTER team as converted to water and to the lunar surface) with its D1 and D2 detectors (25) while LND measured a dose rate of 10.2 ± 1.1μGy/hour in silicon (13.2 ± 1.4μGy/hour converted to water) for charged particles (26). CRaTER uses a factor of 1.33 to convert dose rate from Si to water (27); therefore, it is more convenient to compare the dose rate measured in Si by LND (10.2 ± 1.1 Gy/hour) and CRaTER (10.0μGy/hour). These two values are equal within uncertainties. Thus, the differences in shielding by the instruments themselves and the two spacecraft (28) have no noticeable effect on their measured dose rates.

Discussion

LND measured an average dose equivalent of 1369μSv/day on the surface of the Moon. For the same time period, the dose equivalent onboard the International Space Station (ISS) as measured with the DOSIS 3D DOSTEL instruments (29) was 731μSv/day with contributions only from GCR of 523μSv/day. The additional ~208μSv/day is due to protons while crossing the South Atlantic Anomaly. Therefore, the daily GCR dose equivalent on the surface of the Moon is around a factor of 2.6 higher than the dose inside the ISS. Because the Sun is currently still in an extended activity minimum (30), the dose rate from GCR reported here may be considered as an upper limit for human exploration of the Moon during conditions of low solar activity. Settlements on the Moon will provide additional shielding because they will be buried beneath layers of lunar regolith. While this would decrease the dose rate from charged particles, the absolute contribution from neutrons is expected to increase for shielding constructed from in situ resources, as borne out by measurements with the Apollo 17 Lunar Neutron Probe Experiment. These showed that the flux of thermal and epithermal neutrons increases significantly up to a depth of approximately 150g/cm^2 (31).

LND measured the radiation environment on the surface of the Moon at this precision for the first time. In addition, due to the fact that we are now approaching solar minimum conditions, the contributions from GCR can be seen as upper estimations for the GCR dose. In the time period reported here, no SPE was observed from the surface of the Moon. Such events can increase the dose by orders of magnitude behind only thin shielding (32).

Acknowledgments

We thank the following facilities and personnel for supporting the calibration of LND: CERN/CERF in Geneva, Switzerland; NIRS, HIMAC in Chiba, Japan; ATI in Vienna, Austria; SFE at NSSC China. We thank the Lunar Exploration and Space Engineering Center and O. Angerer from DLR, Germany, for their unwavering support of Chang'E 4 and the China National Space Administration (CNSA) for data acquisition and dissemination.

References

[1] Task Group on Radiation Protection in Space; ICRP Committee 2, G. Dietze, D. T. Bartlett, D. A. Cool, F. A. Cucinotta, X. Jia, I. R. McAulay, M. Pelliccioni, V. Petrov, G. Reitz, T. Sato, ICRP, 123. Assessment of radiation exposure of astronauts in space. ICRP Publication 123. Ann. ICRP 42, 1 - 339 (2013).

[2] NASA, Space Radiation Cancer Risk Projections and Uncertainties - 2012 Rep. NASA - STD - 3001, VOLUME 1, (2013).

[3] U. Straube, T. Berger, G. Reitz, R. Facius, C. Fuglesang, T. Reiter, V. Damann, M. Tognini, Operational radiation protection for astronauts and cosmonauts and correlated activities of ESA Medical Operations. Acta Astronaut. 66, 963 - 973 (2010).

[4] F. A. Cucinotta, F. K. Manuel, J. Jones, G. Iszard, J. Murrey, B. Djojonegro, M. Wear, Space radiation and cataracts in astronauts, Radiat. Res. 156 (5 Pt. 1), 460 - 466 (2001).

[5] P. Y. Chang, F. A. Cucinotta, K. A. Bjornstad, J. Bakke, C. J. Rosen, N. Du, D. G. Fairchild, E. Cacao, E. A. Blakely, Harderian gland tumorigenesis: Low - dose and LET response. Radiat. Res. 185, 449 - 460 (2016).

[6] D. M. Sridharan, A. Asaithamby, S. M. Bailey, S. V. Costes, P. W. Doetsch, W. S.

Dynan, A. Kronenberg, K. N. Rithidech, J. Saha, A. M. Snijders, E. Werner, C. Wiese, F. A. Cucinotta, J. M. Pluth, Understanding cancer development processes after HZE - particle exposure: Roles of ROS, DNA damage repair and inflammation. Radiat. Res. 183, 1 - 26 (2015).

[7] F. A. Cucinotta, Review of NASA approach to space radiation risk assessments for Mars exploration. Health Phys. 108, 131 - 142 (2015).

[8] R. Jandial, R. Hoshide, J. D. Waters, C. L. Limoli, Space - brain: The negative effects of space exposure on the central nervous system. Surg. Neurol. Int. 9, 9 (2018).

[9] R. L. Hughson, A. Helm, M. Durante, Heart in space: Effect of the extraterrestrial environment on the cardiovascular system. Nat. Rev. Cardiol. 15, 167 - 180 (2018).

[10] G. A. Nelson, Space radiation and human exposures, a primer. Radiat. Res. 185, 349 - 358 (2016).

[11] L. W. Townsend, J. H. Adams, S. R. Blattnig, M. S. Clowdsley, D. J. Fry, I. Jun, C. D. McLeod, J. I. Minow, D. F. Moore, J. W. Norbury, R. B. Norman, D. V. Reames, N. A. Schwadron, E. J. Semones, R. C. Singleterry, T. C. Slaba, C. M. Werneth, M. A. Xapsos, Solar particle event storm shelter requirements for missions beyond low Earth orbit. Life Sci. Space Res. 17, 32 - 39 (2018).

[12] J. A. Simpson, Elemental and isotopic composition of the galactic cosmic rays. Annu. Rev. Nucl. Part. Sci. 33, 323 - 382 (1983).

[13] National Research Council 1996. Radiation Hazards to Crews of Interplanetary Missions: Biological Issues and Research Strategies. Washington, DC: The National Academies Press. https://doi. org/10. 17226/5540.

[14] F. A. Cucinotta, M. Durante, Cancer risk from exposure to galactic cosmic rays: Implications for space exploration by human beings. Lancet Oncol. 7, 431 - 435 (2006).

[15] R. A. English, R. E. Bensotz, J. Vernon Builey, C. M. Barnes, Apollo Experience Report - Protection against radiation, NASA Technical Note NASA TN D - 7080 Retrieved from: https://ntrs. nasa. gov/archive/nasa/casi. ntrs. nasa. gov/19730010172. pdf.

[16] The Apollo Lunar - Surface Experiments Package [ALSEP; (33)] provided time - resolved measurements of suprathermal ions (34) and charged particles (35) from Earth's magnetosphere and sheath, but this is unimportant for dosimetric purposes as the measured particles have insufficient energy to penetrate even the outermost layers of a space suit.

[17] ICRP, 1990 Recommendations of the International Commission on Radiological Protection, ICRP Publication 60. Ann. ICRP 21, 1 - 3 (1991).

[18] C. Zeitlin, L. Narici, R. R. Rios, A. Rizzo, N. Stoffle, D. M. Hassler, B. Ehresmann, R. F. Wimmer - Schweingruber, J. Guo, N. A. Schwadron, H. E.

Spence, Comparisons of high - linear energy transfer spectra on the ISS and in deep space. Space Weather 17, 396 - 418 (2019).

[19] R. F. Wimmer - Schweingruber, J. Yu, S. I. Böttcher, S. Zhang, S. Burmeister, H. Lohf, J. Guo, Z. Xu, B. Schuster, L. Seimetz, Johan L. Freiherr von Forstner, A. Ravanbakhsh, V. Knierim, S. Kolbe, H. Woyciechowsky, S. R. Kulkarni, B. Yuan, G. Shen, C. Wang, Z. Chang, T. Berger, C. E. Hellweg, D. Matthiä, D. Hou, A. Knappmann, C. Büschel, X. Hou, B. Ren, Q. Fu, The Lunar Lander Neutron and Dosimetry (LND) Experiment on Chang'E 4. Space Sci. Rev. 216, 104 (2020). https: //doi. org/10. 1007/s11214 - 020 - 00725 - 3.

[20] D. Hou, S. Zhang, J. Yu, R. F. Wimmer - Schweingruber, S. Burmeister, H. Lohf, B. Yuan, G. Shen, C. Wang, X. Hou, B. Ren, Removing the Dose Background from Radioactive Sources from Active Dose Rate Measurements in the Lunar Lander Neutron & Dosimetry (LND) Experiment on Chang'E 4, J. Instrumentation, 15 (1), P01023 - P01032 (2020), https: //doi. org/10. 1088/1748 - 0221/15/01/ P01032. The background measurements were performed separately for the RTG and all RHUs. This does not account for the possibility of neutron multiplication (36) in which neutrons from one RHU lead to nuclear reactions in another RHU and thus enhance the neutral radiation.

[21] The Chang'E 4 payload thermal control system (TCS) circulates ammonia (NH_3) from the warm regions in the lander to the walls of the payload compartment where it provides heat during the late lunar evening, night, and early morning. During lunar daytime, the TCS that serves the payload compartment is empty. Six heat pipes service the panel to which LND is mounted. They are indicated in red in inset A of Fig. 2, where LND is shown in yellow.

[22] The remarkably high dose rate from neutral particles reported here may be at least partially explained by neutron multiplication (36) and the production of secondary particles by the interaction of the GCR with the structure of the Chang' E 4 lander and LND.

[23] J. W. Wilson, B. M. Anderson, F. A. Cucinotta, J. Ware, C. J. Zeitlin, Spacesuit Radiation Shield Design Methods, SAE Report 2006 - 01 - 2110, (2006).

[24] H. E. Spence and Crater Science Team, An Overview of Results from the Lunar Reconnaissance Orbiter (LRO) Cosmic Ray Telescope for the Effects of Radiation (CRaTER), in Annual Meeting of the Lunar Exploration Analysis Group, LPI Contributions, 1595, 66 (2010).

[25] The effect of the varying distance on the effective shielding by the Moon is included in the dose rates published by the CRaTER team, as discussed on their website (http: //prediccs. sr. unh. edu/craterweb/algorithms. html).

[26] LND measures total dose and dose from neutral particles separately. The chargedparticle dose is calculated as the difference.

[27] N. A. Schwadron, T. Baker, B. Blake, A. W. Case, J. F. Cooper, M. Golightly, A. Jordan, C. Joyce, J. Kasper, K. Kozarev, J. Mislinski, J. Mazur, A. Posner, O. Rother, S. Smith, H. E. Spence, L. W. Townsend, J. Wilson, C. Zeitlin, Lunar radiation environment and space weathering from the Cosmic Ray Telescope for the Effects of Radiation (CRaTER). J. Geophys. Res. 117, E00H13 (2012).

[28] LND's B detector sees varying amounts of shielding because of its location inside LND and the Chang'E 4 lander; the minimum shielding is 25μm of Al, 50μm of Kapton, and 0.5mm of Si. The CRaTER D1 and D2 detectors are also shielded by varying material thicknesses from the instrument and LRO spacecraft (19). If used in coincidence with other detectors, CRaTER's 148 - μm D1 detector is shielded by a 0.82 - mm - thick Al entrance window, and its 1 - mm - thick D2 detector is thus shielded by 0.82mm of Al and 148μm of Si.

[29] T. Berger, S. Burmeister, D. Matthiä, B. Przybyla, G. Reitz, P. Bilski, M. Hajek, L. Sihver, J. Szabo, I. Ambrozova, F. Vanhavere, R. Gaza, E. Semones, E. G. Yukihara, E. R. Benton, Y. Uchihori, S. Kodaira, H. Kitamura, M. Boehme, DOSIS & DOSIS 3D: Radiation measurements with the DOSTEL instruments onboard the Columbus Laboratory of the ISS in the years 2009 - 2016. J. Space Weather Space Clim. 7, A08 (2017).

[30] B. Kakad, A. Kakad, D. S. Ramesh, G. S. Lakhina, Diminishing activity of recent solar cycles (22 - 24) and their impact on geospace. J. Space Weather Space Clim. 9, A1 (2019).

[31] D. S. Woolum, D. S. Burnett, M. Furst, J. R. Weiss, Measurement of the lunar neutron density profile. Moon 12, 231 - 250 (1975).

[32] N. A. Schwadron, F. Rahmanifard, J. Wilson, A. P. Jordan, H. E. Spence, C. J. Joyce, J. B. Blake, A. W. Case, W. de Wet, W. M. Farrell, J. C. Kasper, M. D. Looper, N. Lugaz, L. Mays, J. E. Mazur, J. Niehof, N. Petro, C. W. Smith, L. W. Townsend, R. Winslow, C. Zeitlin, Update on the worsening particle radiation environment observed by CRaTER and implications for future human deep - space exploration. Space Weather 16, 289 - 303 (2018).

[33] Apollo 14 Preliminary Science Report, NASA, Special Publication SP - 272 (1971).

[34] H. K. Hills, J. W. Freeman Jr., Suprathermal Ion Detector Experiment (Lunar Ionosphere Detector), Charged - Particle Lunar Environment Experiment, in Apollo 14 Preliminary Science Report, NASA, Special Publication SP - 272 (1971), pp. 175 -183.

[35] B. J. O'Brien, D. L. Reasoner, Charged - Particle Lunar Environment Experiment, in Apollo 14 Preliminary Science Report, NASA, Special Publication SP - 272 (1971), pp. 193 - 213.

[36] M. Taherzadeh, Neutron Radiation Characteristics of Plutonium Dioxide Fuel, Technical Report 32 - 1555, Revision 1, Jet Propulsion Laboratory, California Institute of Technology, Pasadena, California, December 1, (1972).

Energetic Neutral Atom Distribution on the Lunar Surface and Its Relationship with Solar Wind Conditions

H. Z. Wang[1,2], C. Xiao[1,2], Q. Q. Shi[1,2], R. L. Guo[1,2], C. Yue[3], L. H. Xie[2], J. Zhang[1], A. B. Zhang[4,5], M. Wieser[6], Y. Saito[7], M. N. Nishino[5], M. Nowada[1], Q. G. Zong[3], A. W. Degeling[1], A. M. Tian[1], S. Y. Fu[3], H. Zhang[8], J. Chen[1], T. X. Zhang[8], J. Liu[2], C. Y. Han[2], W. S. Shang[1], and S. C. Bai[1]

1. Laboratory of Optical Astronomy and Solar - Terrestrial Environment, School of Space Science and Physics, Institute of Space Sciences, Shandong University, Weihai, Shandong, 264209, People's Republic of China; sqq@sdu. edu. cn, zhang _ jiang@sdu. edu. cn
2. State Key Laboratory of Space Weather, National Space Science Center (NSSC), Chinese Academy of Sciences, Beijing, People's Republic of China
3. School of Earth and Space Sciences, Peking University, Beijing, People's Republic of China
4. Beijing Key Laboratory of Space Environment Exploration, National Space Science Center, Chinese Academy of Sciences, Beijing, People's Republic of China
5. University of Chinese Academy of Sciences, Beijing, People's Republic of China
6. Swedish Institute of Space Physics, Kiruna, Sweden
7. Institute of Space and Astronautical Science, Japan Aerospace Exploration Agency, 3 - 1 - 1 Yoshinodai, Chuo - ku, Sagamihara, Kanagawa 252 - 5210, Japan
8. Key Laboratory of Earth and Planetary Physics, Institute of Geology and Geophysics, Chinese Academy of Sciences, Beijing, People's Republic of China

Abstract: The Advanced Small Analyzer for Neutrals (ASAN) on board the Chang'E - 4 Yutu - 2 rover first detected energetic neutral atoms (ENAs) originating from the lunar surface at various lunar local times on the lunar farside. In this work, we examine the ENA energy spectra, obtained in the first 23 lunar days from 2019 January 11 to 2020 October 12, and find a higher ENA differential flux on the lunar dawnside than on the duskside. Combined with Acceleration, Reconnection, Turbulence and Electrodynamics of the Moon's Interaction with the Sun (ARTEMIS) data, we analyze the correlation between the ENA differential flux and solar wind parameters, such as flux, density, dynamic pressure, and velocity, for each ASAN energy channel on the dawnside and duskside. The results show that ENA differential flux is positively correlated with solar wind flux, density, and dynamic pressure and relatively lower on the duskside than on the dawnside. To determine the relationship between solar wind energy and ENA energy, we analyze the correlation between solar wind energy and ENA cutoff energy and temperature on the dawnside and duskside. The results show that the ENA cutoff energy and temperature are lower on the duskside than on the dawnside at the same solar wind energy. The difference between the ENA - solar wind observation on the dawnside and duskside is possibly caused by solar wind deflection and deceleration on the duskside, which can be attributed to the interaction between solar wind and the lunar magnetic anomalies located nearby in the northwestern direction of the Chang'E - 4 landing site.

Unified Astronomy Thesaurus concepts: Solar wind (1534); Lunar surface (974); Space plasmas (1544); The Moon (1692)

1 Introduction

The lunar surface is exposed to solar wind protons due to a lack of a global magnetic field or dense atmosphere. The interactions between the solar wind and lunar surface have been investigated for several decades. Although most of the solar wind protons are neutralized when impacting on the lunar surface (Crider & Vondrak 2002), the in situ observations from Chang'E, Kaguya, Chandrayaan - 1, and Interstellar Boundary Explorer (IBEX) missions reveal that 0.1%-1% of solar wind protons are backscattered from the lunar surface (Saito et al. 2008; Wang et al. 2010), 10%~20% are converted to energetic neutral atoms (ENAs; McComas et al. 2009; Wieser et al. 2009), and the remaining part are implanted into the lunar regolith to produce OH/H_2O, as previously found for nonmagnetized regions of the surface (Li & Garrick - Bethell 2019; Wang et al. 2021). The reflection ratio would be much higher over largescale regions of magnetic anomalies (average ~10%, locally ~50%; Saito et al. 2010; Lue et al. 2011). Among these processes, measurement of ENAs is considered to be one of the best ways to provide useful information on the interactions between the solar wind protons and the lunar surface, such as the surface composition and the location and strength of magnetic anomalies (localized regions of strong crustal magnetization with surface magnetic field strengths on the order of tens to hundreds of nT; Futaana et al. 2006).

The observations of ENAs have been made under different plasma conditions (solar wind, the Earth's magnetosheath, and the Earth's plasma sheet) and lunar regions (e.g., magnetic anomalies) from ENA instruments on board the IBEX and Chandrayaan - 1 missions (Futaana et al. 2006). During about three - quarters of the lunar orbit, the Moon is immersed in the solar wind that provides a predominant proton flux at around 1 keV and a typical speed of about 400km s^{-1}. However, when the Moon enters into the Earth's magnetosheath and plasma sheet, the plasma density, temperature, and Mach numbers of the space environment around the Moon differ significantly from those in the solar wind. When the Moon is in the undisturbed solar wind, the intensity of the reflected neutral atom flux changes corresponding with the solar wind incident angle and the intensity of the solar wind proton flux (Wieser et al. 2009). The scattering function of backscattered ENAs was derived using the Chandrayaan-1 Energetic Neutral Atom (CENA) data, and it is also used to map a global solar wind reflection ratio from the lunar surface (Schaufelberger et al. 2011; Vorburger et al. 2013). A reduced ENA flux is observed above the magnetic anomalies, which indicates a local ion shielding effect (Wieser et al. 2010; Vorburger et al. 2012). The shielding efficiencies depend on the magnetic field strength, the geometric structure of the magnetic anomalies, and the solar wind dynamic pressure (Wieser et al. 2010; Vorburger et al. 2012). Allegrini et al. (2013) compared the ENA energy spectra in the Earth's magnetosheath observed by IBEX and found that the ENA intensities are generally higher in the magnetosheath. However, Lue et al. (2016) found that the ENA scattering process remains unchanged in Earth's magnetosheath compared with the undisturbed solar wind condition from the Chandrayaan-1 ENA measurements. They also observed the shielding effect of magnetic anomalies in the magnetosheath. When the Moon enters into the Earth's magnetotail, the solar wind is shielded by Earth's magnetic field; however, the energy of protons in the plasma sheet is higher, even though the proton flux is reduced compared with the solar wind proton flux. Harada et al. (2014) found that the plasma sheet protons can also be scattered as ENAs from the lunar surface, and magnetic shielding of the lunar surface is less effective than in the solar wind due to the broad velocity distributions of the plasma sheet protons. The solar wind conditions vary dynamically, with significant variations in parameters such as the solar wind velocity, proton density, and proton flux. Futaana et al. (2012) found a good correlation between the solar wind velocity and the temperature of the ENAs. Funsten et al. (2013) found that the ENA reflection coefficient exhibits a strong dependence on the incident solar wind speed.

These orbital observations promote our understanding of the interactions between the plasma and the lunar surface. However, studies of ENAs based on in situ observations on the lunar surface had not been conducted until the Chang'E-4 mission.

The Advanced Small Analyzer for Neutrals (ASAN) on board the Chang'E-4 Yutu-2 rover achieved the first in situ observation of ENAs on the lunar surface. Chang'E-4 landed on the lunar farside surface on 2019 January 3 at 177°.6E, 45°.5S within the South Pole-Aitken basin (Di et al. 2019; Liu et al. 2019; Wu et al. 2019; Liu et al. 2020; Wieser et al. 2020), as shown in Fig. 1 (a). The South Pole-Aitken basin is the largest confirmed impact basin of the Moon; it has unusual surface properties, such as a low albedo, and coincides with a cluster of strong magnetic anomalies located on the northern rim of the basin. Vorburger et al. (2015) studied ENA observations in the South Pole-

Aitken basin and tried to find a correlation between the ENA map and local surface features. However, they only found ENA depletion at the magnetic anomalies in the South Pole - Aitken basin and no correlation with elevation, albedo, or compositions (iron and thorium) using the total ENA flux data. The dependence of the ENA spectra on surface properties remains unclear because the observation's geometry area was large. Zhang et al. (2020) presented the first result of ASAN and found that the reflection ratio derived from ASAN data has a good agreement with previous works from the Chandrayaan - 1 and IBEX missions. However, the influence of the variation of the ENA spectra on the lunar surface with solar wind parameters remains unclear. In the present work, we study the ENA distribution and the relationship between the ENA differential flux and solar wind parameters from Acceleration, Reconnection, Turbulence and Electrodynamics of the Moon's Interaction with the Sun (ARTEMIS) observations, such as flux, density, velocity, and dynamic pressure.

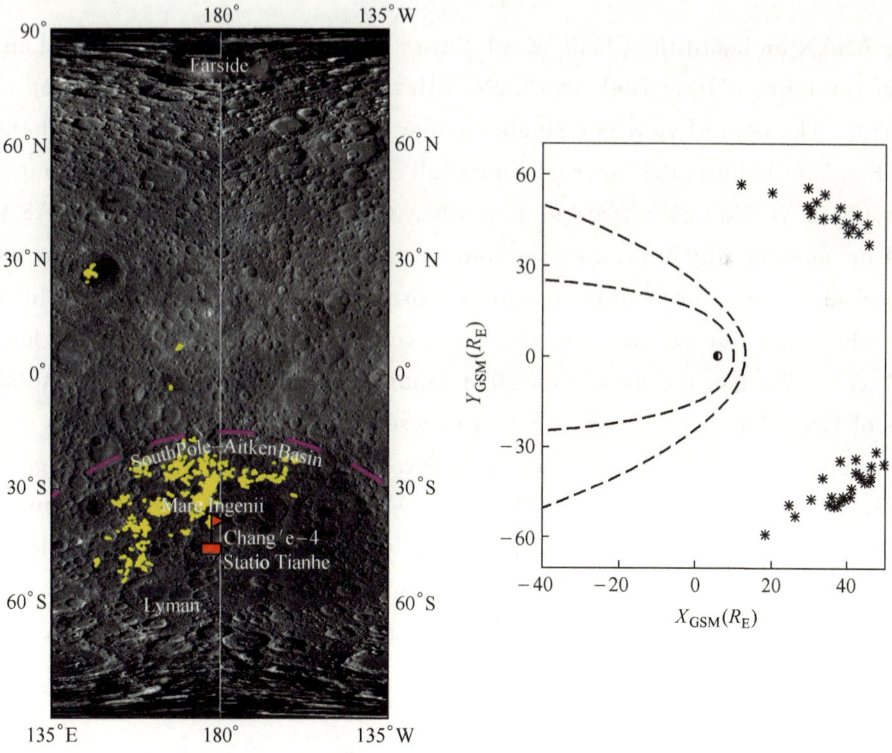

(a)　　　　　　　　　　　　(b)

Fig. 1　(a) Chang'E - 4 landing site in the selenography coordinate system (overlaid on the Lunar Reconnaissance Orbiter Wide Angle Camera base map). The yellow area indicates the magnetic anomaly region. The magenta dashed line indicates the boundary of the South Pole - Aitken basin. (b) Overview of the ARTEMIS positions in the GSM coordinate system during the operation of ASAN, along with typical boundaries for the terrestrial bow shock (Fairfield 1971) and magnetopause (Shue et al. 1997)

This paper consists of five sections. The introduction and background of this study are described in Section 1. Section 2 provides a brief description of the ARTEMIS and Chang'E-4 ASAN data used in this study. In Section 3, we present the main results obtained through this study: ENA observations over 17 lunar days, the statistical correlation between lunar ENA differential flux and solar wind parameters, and the relationship between the solar wind energy and ENA cutoff energy on the dawnside and duskside. In Section 4, we discuss significant features of the statistical results and physics obtained from our statistical study. Finally, the major conclusions of this paper are provided in Section 5.

2 Data and Methods

2.1 ASAN Instrument

The ASAN on board the Chang'E-4 Yutu-2 rover is an energetic neutral analyzer, whose energy range is from 10eV to 10keV with typically 30% energy resolution (Wieser et al. 2020). The field of view is a single angular pixel of 37° × 30° pointing at the lunar surface. Fig. 1 (b) shows the lunar positions (all in the solar wind) in the Geocentric Solar Magnetospheric (GSM) coordinate system when the ASAN is operated. The ASAN was powered on aperiodically for tens of minutes, on average, in two operation windows during each lunar day, one during the local morning from 07:00 to 09:00 lunar local time and the other during the local afternoon from 14:00 to 16:00 lunar local time (Fig. A1 (a)). We use the data from 2019 January 11 to 2020 April 28, a total of 54 intervals of ENA data over 23 lunar days. In these operations, 31 observations are in the local morning, and 23 observations are in the local afternoon. The observation geometries (solar azimuth angle, solar zenith angle, solar elevation angle, solar elevation angle of rover, solar azimuth angle of rover) during each operation, along with the lunar local time, are shown in Fig. A1 (b)-(f). The definitions of these angles are shown in the Appendix. In this study, we only used the hydrogen ENA data as shown in Fig. A1 (g).

2.2 ARTEMIS Data

The ARTEMIS probes were originally part of the five Time History of Events and Macroscale Interactions during Substorms (THEMIS) spacecraft in the near-Earth magnetosphere. In early 2010, they gradually started to move outward to their final orbits at the lunar distances of 60 Earth radii (RE). The ARTEMIS mission consists of a pair of identically instrumented spacecraft that have been in orbit around the Moon since mid-2011 (Angelopoulos 2011; Sibeck et al. 2011). Both probes measure in situ low-energy ion (up to 25keV) and electron (up to 30keV) distributions with an electrostatic analyzer (ESA; McFadden et al. 2008). In this study, we used the ion measurements from the two ARTEMIS probes (ARTEMIS - P1 and ARTEMIS - P2) to study the variation of the lunar plasma environment during the observation times of the Chang'E-4 ASAN instrument. The mean values of the solar wind velocity, density, pressure, and ion flux obtained from ARTEMIS during each ASAN operation event are shown in Fig. A1 (h) - (k). The solar wind energy was calculated by $E_{sw} = 0.5 * (1.6 * 10^{-27}) * (V_{sw} * 10^3)^2 / (1.6 * 10^{19})$.

3 Observations

An overview of the ENA energy spectra distribution during the first 23 lunar days is shown in Fig. 2. As we mentioned in previous sections, the operation time of ASAN is aperiodic, which means that ENAs are not always measured on both dawnside and duskside in each lunar day (e. g., not on the 1st, 9th, and 15th lunar days). However, based on the 17 lunar days with both dawnside and duskside measurements, we find that the ENA differential flux is higher on the dawnside, except on the 12th lunar day. The solar wind parameters and the ENA integral flux on the dawnside and duskside in each lunar day are shown in Figure A2. From previous studies, we know that the intensity of the ENA flux varies with the solar wind incident angle and the intensity of the solar wind proton flux (Wieser et al. 2009). Here we find that the ENA fluxes obtained on the duskside are generally lower than those on the dawnside, even though the intensity of the solar wind flux is higher on the duskside with similar solar elevation angles, such as on the 7th, 13th, 16th, 20th, 21st, and 22nd lunar days. Comparing the 12th and 13th lunar days in Fig. 2 and A2, we can find that the ENA differential flux is higher on the duskside than on the dawnside if both the solar wind flux and dynamic pressure are higher on the duskside. When the ASAN operates on the lunar duskside, the solar wind protons are coming from the northwestern direction, passing through the magnetic anomalies in the South Pole - Aitken basin (yellow points in Fig. 1 (a)). During the ASAN operation on the dawnside, the solar wind protons are coming from the northeastern direction, in which there are few magnetic anomalies.

Statistics of the correlation between the ENA differential flux and the solar wind integrated flux at each ASAN energy channel on both the dawnside and duskside are shown in Fig. 3. The data were removed when the ENA differential flux was less than the data sensitivity limit line. First, from the statistical result between the ENA differential flux and the solar wind flux, we can find that the ENA differential flux has a positive correlation with the solar wind flux on both the dawnside and duskside, which is consistent with previous studies (Wieser et al. 2009). The linear fitting results of Fig. 3 are shown in Tab. A1 ($df_{ENA} = c * f_{SW} + k$), where c represents the reflection coefficient. A positive k could indicate that there is another source. There are also few negative k values in the table, as the data set might not be large enough. The correlation coefficient is

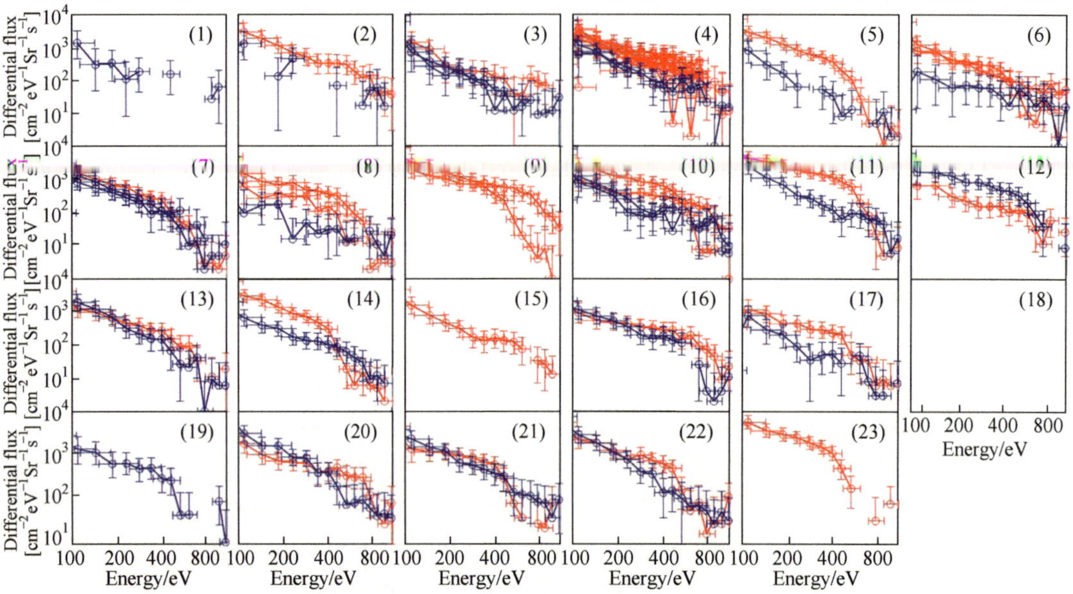

Fig. 2　Summary of ASAN ENA spectral distribution over 23 lunar days from 2019/01 to 2020/10. The red and blue lines show the ENA spectral distribution in the lunar dawnside and duskside, respectively. The horizontal error bars indicate the lower and upper limits of the energy bin. The vertical error bars indicate the lower and upper limits of the differential flux. Only data over the sensitivity limit are shown here. The ASAN was powered on aperiodically so that ENAs are not always (e. g., the 1st, 9th, 15th, 19th, and 23rd lunar days) measured at both dawnside and duskside in each lunar day

higher at ENA energies between 105 and 523eV, up to 0. 7 between 140 and 333eV. The correlation coefficients between the integrated solar wind ion flux and the differential fluxes of ENAs in the E_{ENA}/E_{SW} bin (normalize the measured ENA energy by the solar wind energy) are shown in Fig. A3 (we remove the data when the ENA energy is less than 100eV). We can find that the coefficient on the duskside is relatively larger than those on the dawnside within 0. 2 – 0. 9 E_{ENA}/E_{SW} bin, which indicates that the production of ENA in this energy range on the duskside is more directly influenced by the solar wind than that on the dawnside. And c is largest when the ratio of $E_{ENA}/E_{SW} \sim 0.1 - 0.2$ (Tab. A2), which indicates that the protons lose approximately 90% of their initial energy when turned into ENAs during their interaction with the lunar surface. The correlations between the ENA differential flux and the solar wind proton density and dynamic pressure are shown in Fig. 4. The ENA differential flux also has a positive correlation with the solar wind density and dynamic pressures in each energy channel. And the correlation coefficient between the ENA differential flux and density is higher on the dawnside. However, the correlation coefficient between the ENA differential flux and dynamic pressure is higher on the duskside. The linear fitting results of Figure 4 are shown in Tab. A3. The statistics of the relationship between the ENA differential flux and the solar wind velocity are also analyzed with the same method; however, no significant correlation was found (not shown). There is a positive correlation between the solar zenith angle of the rover and the ENA flux/Nsw (ENA flux was normalized by solar wind density), and the correlation coefficient is about 0.4 (Fig. A4). The relationship between the solar azimuth angle of the rover and the ENA differential flux remains unclear because the data are not completely available (not shown).

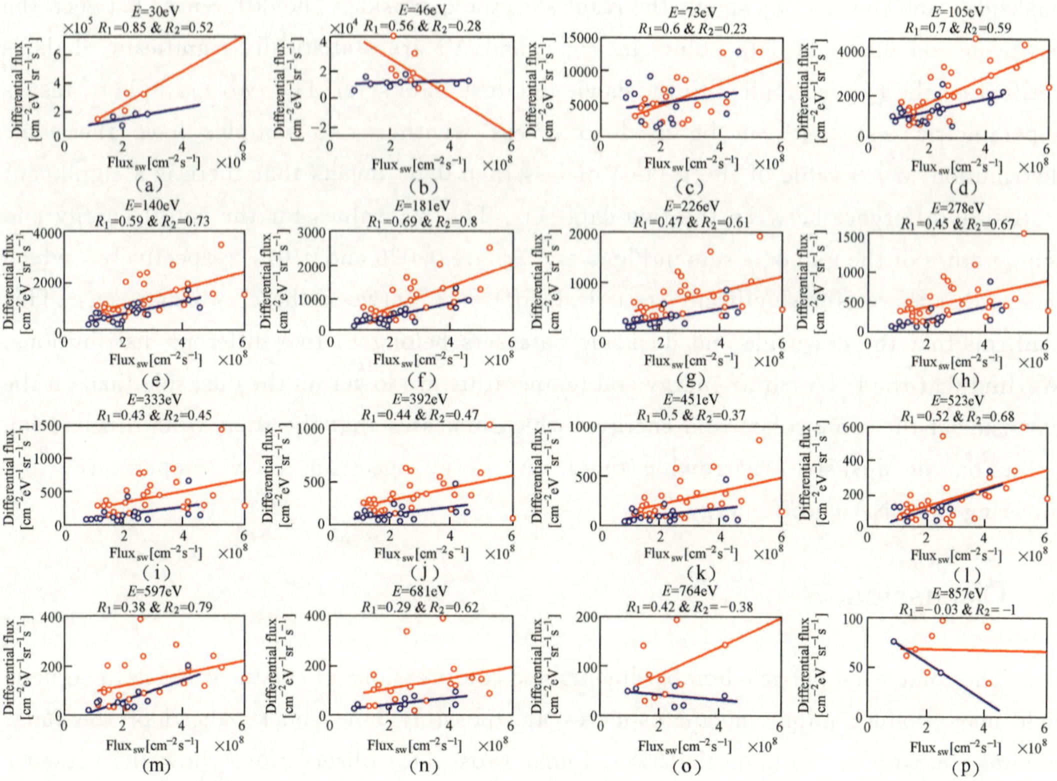

Fig. 3　Relationship between the ENA differential flux and solar wind flux at individual energies of ASAN from 30 to 857eV. The red circles indicate the data on the lunar dawnside, and the blue circles indicate the data on the lunar duskside. Here R1 and R2 represent the correlation coefficients on the dawnside and duskside, respectively. The linear fitting results of the integrated solar wind flux and the ENA differential flux in each ASAN energy bin are shown in Tab. A1, $df_{ENA} = c * f_{SW} + k$

To quantify the relationship between solar wind energy and ENA energy, we show the relationship between solar wind kinetic energy (determined by the bulk solar wind velocity) and ENA cutoff energy on the dawnside and duskside (Fig. 5). The relationship between the ENA temperature (calculation method of Yue et al. 2016 shown in the Appendix) and solar wind kinetic energy on the dawnside and duskside is shown in Fig. A5. When calculating the temperature in Fig. A5, we excluded several events that might be without a magnetic shielding effect and outside of the lunar mini-magnetosphere or deflected solar wind particles affected by the neighboring magnetic anomalies (Xie et al. 2021). These deflected particles increase the penetration rate to a certain extent that might have an effect on temperature. We show the 95% confidence interval (dashed lines) as error bars of linear fit parameters in Fig. 5 and A5. The red lines are relatively higher than the corresponding blue lines in Fig. 5 and A5. The calculated ENA temperature is of the same magnitude as the result using the bestfitting parameters for the Maxwell - Boltzmann distribution in Futaana et al. (2012). The blue line shows the linear fitting results on the duskside, and the red line shows the results on the dawnside. The differences between the dawnside and duskside data points in Fig. 5 and A5 are statistically significant. This is verified by the t - test applied to the dawn and dusk data sets. The two - sample t - test is a parametric test that can be used to assess whether two samples have the same distribution; a p - value of the t - test of less than 0. 05 means that there is a significant statistical difference between the two data sets. The p - values for the cutoff energy and temperature of the two data sets in Fig. 5 and A5 are 0. 026 and 0. 01, respectively, which means that there is a significant statistical difference between these two data sets. This confirms that the dawnside and duskside data sets belong to two different distributions. We find that the ENA cutoff energy and temperature are lower on the duskside than on the dawnside at the same solar wind energy, which indicates that the solar wind deceleration occurs on the duskside, narrowing the ENA energy spectrum (low temperature) and lowering the ENA cutoff energy.

4 Discussion

The solar wind particle bombardments and compressions of the lunar crustal magnetic field may produce mini - magnetospheres and possibly bow shocks ahead of the lunar minimagnetospheres (Lin et al. 1998). Lunar Prospector observations show the presence of an electron cavity at low altitudes over the Crisium, which is the first evidence of the presence of a mini - magnetosphere (Halekas et al. 2008). More recently, the reduction of the ENA flux and reflection of the solar wind protons within the lunar magnetic anomalies further suggest that mini - magnetospheres effectively shield the solar wind plasma (Wieser et al. 2010; Lue et al. 2011; Vorburger et al. 2012; Futaana et al. 2013). The solar wind proton reflection efficiency, largely correlated with crustal magnetic field strength, is 5%-12%, on average, and can be up to 50% of the solar wind flux (Saito et al. 2008; Lue et al. 2011; Poppe et al. 2017). The reduction of the ENA energy exceeding 150eV within the magnetic anomalies arises from the shielding effect due to solar wind deflection resulting from the minimagnetosphere (Wieser et al. 2010). This is the first observation of a mini - magnetosphere above a lunar magnetic anomaly using ENA data and shows that the ENA differential flux can be influenced by a mini - magnetosphere.

As shown in Fig. 2 and A2, the ENA differential flux is still lower on the duskside, even when the solar wind flux is higher on the duskside. The dawn - dusk asymmetries of the ENA differential flux can be attributed to the shielding effects of the lunar mini -

magnetosphere on the duskside, which is supported by 3D magnetohydrodynamic modeling (Xie et al. 2015, 2021). During dawnside measurements, the solar wind comes from northeast of the rover. In this case, the solar wind can bombard the lunar regolith directly, and the backscattered ENA will be detected by the ASAN. However, when the ASAN operates at the lunar duskside, the solar wind from northwest of the rover is most likely to be influenced by the lunar mini - magnetosphere produced by the magnetic anomalies on the northwestern rim of the South Pole - Aitken basin.

In Figures 3 and 4, the measured ENA flux is mainly dependent on solar wind flux, density, and dynamic pressure, which is consistent with the previous observations by Chandrayaan - 1 and IBEX (Wieser et al. 2009). The correlation coefficient is higher in energy channels from ~100 to~600eV, and the peak is from about 140 to 333eV. Wieser et al. (2010) found that the mini - magnetosphere is hardly visible in lower - energy hydrogen atoms (<100eV), whereas it is pronounced in the energy range from 150 to 600eV. They concluded that this discrepancy shows the presence of two populations of hydrogen atoms that seem to have different origins. One population has lower energies, below 100eV, while the other exhibits energies larger than 150eV. The latter population is probably directly related to the impinging solar wind protons because the mini - magnetosphere is visible in images produced from these hydrogen atoms (Wieser et al. 2010). The hydrogen ENA energy spectrum measured by the ASAN agrees well with those measured by Chandrayaan - 1 and IBEX above 100eV, whereas the ASAN data show a higher flux below 100eV compared with the Chandrayaan - 1 and IBEX data (Zhang et al. 2020). It is also proposed that the component above 100eV mostly consists of solar wind protons directly backscattered as hydrogen ENAs, while below this energy, ENAs sputtered from the surface material become important. The proton ENA flux at energies less than 100eV may be contaminated by the sputtered ENAs from the lunar surface and the internal surface of the instrument (Wieser et al. 2020; Zhang et al. 2020). The calculated correlation coefficient between the solar wind proton flux observed from ARTEMIS and ENA flux confirms that the ENA energy ranges from 100 to 600eV originates from solar wind protons, which supports the previous result of Wieser et al. (2010). The origin of fluxes with energies below 100eV is puzzling and will be investigated in future works. We also examined the relationship between the scattering ratio and the solar wind parameters but found no obvious correlation.

Recent observations by the Kaguya mission at 30km altitude showed that some magnetic anomalies decelerate and deflect the solar wind ions, which indicates that the large - scale electrostatic potentials form above the lunar magnetic anomalies (Saito et al. 2012). They analyzed a magnetic anomaly in the South Pole - Aitken basin and found that both protons and alpha particles are decelerated by ~150eV/q (electron volts per unit charge). The creation of ambipolar electric fields over lunar magnetic anomalies has also been reproduced based on particle - in - cell simulations of solar wind interactions with

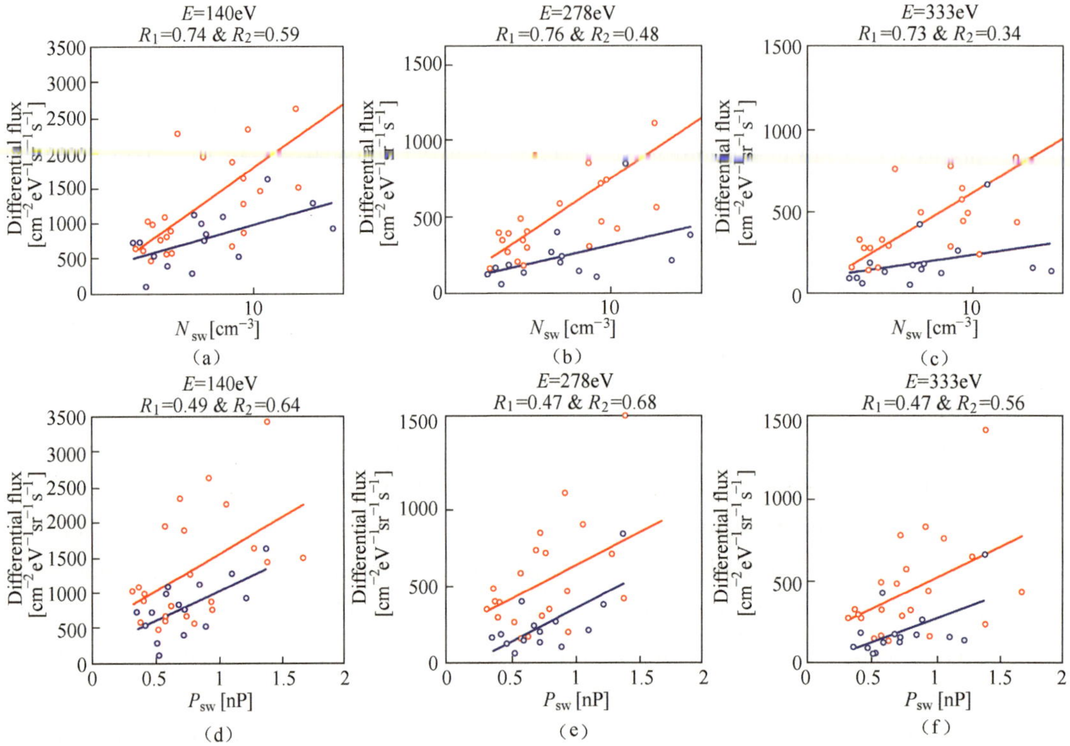

Fig. 4　Relationship between the ENA differential flux and solar wind density (a) – (c) and dynamic pressure (d) – (f) at individual ASAN energies. The red circles indicate the data on the lunar dawnside, and the blue circles indicate the data on the lunar duskside. Here R1 and R2 represent the correlation coefficients on the dawnside and duskside, respectively. The linear fitting results of the solar wind density and pressure and the ENA differential flux in each ASAN energy bin are shown in Tab. A2

lunar magnetic anomalies (Poppe et al. 2012). Recently, Futaana et al. (2013) found an electrostatic potential larger than +135eV/q inside the Gerasimovic anomaly using the data from the neutral atom energy spectrometer SARA/CENA on board Chandrayaan-1, which is in agreement with the result from the Kaguya observations. This interaction process has also been revealed by numerical simulations using magnetohydrodynamic, hybrid, and kinetic/particle-in-cell numerical simulations. These simulations indicate that the large-scale ambipolar and/or Hall electrostatic fields may be the primary mechanism of the interaction between the plasma and lunar crustal magnetic fields. In Fig. 5, we find that the ENA cutoff energy and temperature are lower on the duskside than on the dawnside at the same solar wind energy. The solar wind observed by ARTEMIS cannot be influenced by this electrostatic potential above the lunar magnetic anomalies because the interaction region is most likely located at low altitudes, e. g., ~30km altitude (Kurata et al. 2005), which is lower than ARTEMIS's orbital altitude. On the duskside, the solar wind might be influenced by the lunar mini-magnetosphere so that the solar wind would be decelerated by the electrostatic potentials above the magnetic anomalies. Therefore, to generate ENAs with the same energy, solar wind protons on the duskside with higher energies are required before their interaction with the surface due to deceleration inside the mini - magnetosphere. The solar wind deceleration on the duskside can be estimated from the difference between the ENA cutoff energy on the dawnside and that on the duskside, and then the electrostatic potential is calculated. The electrostatic potential was estimated by the ENA cutoff energy difference between the dawnside and duskside fitting results (solid red and blue lines in Fig. 5). The deceleration rate of the solar wind energy is derived by dividing the ENA cutoff energy difference by the initial solar wind energy, as shown in Figure A6. The 95% confidence interval is also displayed. We can find that the electrostatic potential and deceleration rate both vary with solar wind energy. The potential range is 50 - 260V (30 - 350V for the 95% confidence interval). The deceleration rate of the solar wind energy is 12% - 18% (7.5% - 30% for the 95% confidence interval). We compared our results with previous works. Electrostatic potential larger than +135V inside the Gerasimovic anomaly is confirmed using the ESW - TENA model by Futaana et al. (2012); the differences in the solar wind energy between the outside (525eV) and inside (390eV) of the anomaly is 135eV. The presence of a +150V electric potential inside the South Pole - Aitken magnetic anomaly was found by Kaguya observations (Fig. 5 in Saito et al. 2012), in which case, the initial solar wind energy was about 630eV (Fig. 5 in Saito et al. 2012). The deceleration of the solar wind observed by the Chang'E - 4 ASAN is also possibly caused by South Pole - Aitken magnetic anomaly. Fortunately, we find two events with an initial solar wind energy (about 680eV) similar to the Kaguya observation mentioned above (about 630eV) on both the dawnside and duskside (marked with the rectangle in Fig. 5). The difference in the ENA cutoff energy between the dawnside (680eV) and duskside (550eV) is about 130eV, which is in agreement with the previous results.

This study confirms that the ENA differential flux at energies between 100 and 600eV is probably directly related to the impinging solar wind protons, because the lunar minimagnetosphere is shielding the impinging solar wind on the duskside and results in a lower hydrogen atom flux. Therefore, the correlation coefficient between the solar wind proton flux and the ENA flux is highest in this energy range. Using backscattered ENA data and ARTEMIS data, we also give information on the deceleration of the solar wind protons. The large electric potential produced by charge separation also influences the near - surface environment above the magnetic anomalies, such as lunar swirls and water production (Kramer et al. 2011; Li & Garrick - Bethell 2019). As a new technique, the ENA observation plays an important role in the study of airless bodies' interaction with the plasma environment. More work needs to be done on investigating how lunar materials influence the interaction between the plasma and the surface and its effects on the lunar exosphere.

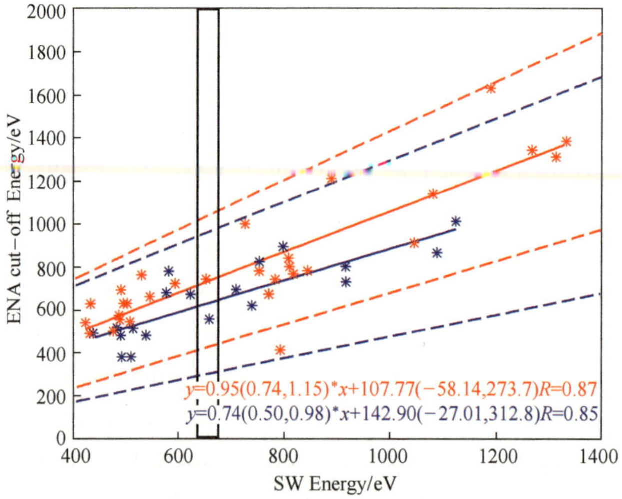

Fig. 5　Relationships between solar wind energy and ENA cutoff energy on the dawnside (red line) and duskside (blue line). The dashed lines indicate 95% confidence intervals. The rectangle indicates two events in which the solar wind energy was similar (about 680eV) on both the dawnside and duskside

5　Conclusion and Perspective

Based on the ASAN data obtained from the Chang'E－4 mission, we study the ENA energy spectral distribution on the lunar farside surface and its relationship with solar wind conditions. Major concluding remarks are as follows.

(1) The relationships between the ENA differential flux and solar wind parameters, including solar wind flux, number density, dynamic pressure, and the velocity in individual energy channels, are studied on both the dawnside and duskside. The results show that the measured ENA flux strongly depends on the solar wind density and dynamic pressure, being consistent with the previous observations by Chandrayaan－1 and IBEX. It indicates that a portion of the solar wind ions are backscattered from the surface as ENAs. The correlation coefficients between the ENA differential flux and solar wind integrated flux and dynamic pressure are higher on the duskside than on the dawnside. However, a correlation between the measured ENA flux and solar wind velocity cannot be excluded at this stage.

(2) The relationship between the solar wind energy and ENA energy on the dawnside and duskside shows that the production of ENAs with the same energy requires a higher solar wind energy on the duskside, indicating the solar wind may be decelerated by ambipolar and Hall electrostatic fields within a lunar mini－magnetosphere.

We thank all the members of the Chang'E－4 ASAN and ARTEMIS ESA instrument teams. Funding: This research was supported by the National Natural Science Foundation of China (grants 41941001, 41974189, 41731068, 41773065, 41961130382, 41941001 41974191, and 42074194), the Key Research Program of Frontier Sciences, CAS, grant No. QYZDY-SSW-DQC028, the Royal Society NAF \ R1 \ 191047, the Specialized Research Fund for State Key Laboratories, the Strategic Priority Research Program of the Chinese Academy of Sciences (grant No. XDB 41000000), China Space Agency project D020303, the open fund of the Key Laboratory of Lunar and Deep Space Exploration, the Chinese Academy of Sciences, research grant National Key R&D Program of China 2020YFE0202100, and the National Key Research and Development Project (2019YFE0123300). This project has received funding from the European Union's Horizon 2020 research and innovation program under grant agreement No. 871149. The Advanced Small Analyzer for Neutrals (ASAN) instrument was developed by the Swedish Institute of Space Physics (IRF) in Kiruna, Sweden, and the National Space Science Center (NSSC) in Beijing, China, as a joint project. It is supported by the Swedish National Space Agency (SNSA) under grants 95/11 and 122/18, the National Natural Science Foundation of China (NSFC) grant 41941001, and the Beijing Municipal Science and Technology Commission under grant 181100002918003. We thank the data processing team of the National Astronomical Observations (CAS). Competing interests: Authors declare no competing interests. Data and materials availability: The ARTEMIS data are available at http://artemis.ssl.berkeley.edu.

Appendix

A.1 The Definition of Angles

The solar azimuth angle is the angle between the projection of the solar direction in the tangent plane at the rover location and the north direction. The value is positive for the clockwise direction, and the range is from −180° to 180°.

The solar azimuth angle is the angle between the solar direction and the normal of the lunar surface at the rover location, and the range is from 0° to 90°. The solar elevation angle is the complement angle of the solar azimuth angle.

The rover coordinate system is defined as follows. The origin is the center of the baseplate of the rover structure. The X-axis is in the baseplate, and $+X$ points in the direction of the mast installation (the forward direction). The Z-axis is along the normal of the baseplate, and $+Z$ points to the lunar surface. The Y-, Z-, and X-axes form a right-hand coordinate system.

The solar azimuth angle of the rover is the angle between the projection of the solar direction in the $X-Y$ plane of the rover coordinate system and the rover $+X$ direction. The value is positive for the clockwise direction, and the range is from −180° to 180°.

The solar elevation angle of the rover is the complement angle between the solar direction and the $-Z$ direction of the rover coordinate system, and the range is from 0° to 90°.

A. 2　Calculation Method of ENA Temperature

Considering relativistic effects, the particle phase - space density (f) is calculated using the differential energy flux (j) as

$$f = \frac{\gamma^3 m_0 j}{v^2}, \text{ or } f[\mathrm{s^3/m^6}]$$
$$= 6.25 \times 10^{16} \gamma^3 \frac{m_0[\mathrm{kg}] j [\mathrm{cm^{-2} MeV^{-1} s^{-1} sr^{-1}}]}{v^2[\mathrm{m/s}]}, \tag{A1}$$

where γ is the relativistic factor, m_0 is the particle's rest mass, and v is the velocity.

Tab. A1

The Linear Fitting Results of the Integrated Solar Wind Flux and the Differential Flux in Each ASAN Energy Bin, $df_{ENA} = c * f_{SW} + k$

E/eV	dawn _ c	dawn _ k	dawn _ r²	dusk _ c	dusk _ k	dusk _ r²
30	1.22E-03	-1.44E+04	0.73	1.29E-04	1.15E+05	0.04
46	−1.31E-04	5.31E+04	0.32	−8.38E-06	1.73E+04	0.00
73	1.66E-05	1.32E+03	0.36	5.98E-06	3.90E+03	0.05
105	5.85E-06	4.20E+02	0.49	2.91E-06	6.36E+02	0.34
140	3.17E-06	3.97E+02	0.34	2.73E-06	9.51E+01	0.55
181	3.20E-06	3.24E+01	0.48	2.09E-06	1.12E+00	0.62
226	1.37E-06	2.51E+02	0.22	1.02E-06	5.89E+01	0.36
278	1.04E-06	2.12E+02	0.20	1.08E-06	−1.24E+01	0.48
333	8.43E-07	1.60E+02	0.18	6.24E-07	1.39E+01	0.22
392	7.02E-07	1.38E+02	0.19	4.59E-07	2.27E+01	0.24
451	6.60E-07	6.64E+01	0.25	3.43E-07	2.76E+01	0.16
523	5.31E-07	1.73E+01	0.27	5.94E-07	−2.07E+01	0.48
597	2.36E-07	6.98E+01	0.15	4.14E-07	−2.53E+01	0.64
681	2.09E-07	6.41E+01	0.08	1.24E-07	1.74E+01	0.39
764	2.78E-07	2.54E+01	0.18	−2.72E-08	4.38E+01	0.04
857	−5.46E-09	6.88E+01	0.00	−1.85E-07	9.30E+01	0.98

Tab. A2

The Linear Fitting Results of the Integrated Solar Wind Flux and the Differential Flux in Each Ratio of E_{ENA}/E_{SW}, $df_{ENA} = c * f_{SW} + k$

E_{ENA}/E_{SW}	dawn _ c	dawn _ k	dawn _ r²	dusk _ c	dusk _ k	dusk _ r²
0.1-0.2	1.15E-05	−14.57	0.55	7.15E-06	274.61	0.53
0.2-0.3	4.65E-06	550.53	0.07	6.26E-06	36.65	0.39

（续）

E_{ENA}/E_{SW}	dawn _ c	dawn _ k	dawn _ r^2	dusk _ c	dusk _ k	dusk _ r^2
0.3－0.4	3.45E－06	278.35	0.09	4.07E－06	－51.68	0.52
0.4－0.5	2.48E－06	234.17	0.09	2.17E－06	3.89	0.38
0.5－0.6	2.21E－06	171.93	0.11	1.63E－06	3.11	0.48
0.6－0.7	1.58E－06	147.30	0.09	9.75E－07	20.68	0.25
0.7－0.8	1.44E－06	85.77	0.10	6.40E－07	31.85	0.35
0.8－0.9	1.25E－06	35.85	0.16	6.01E－07	15.73	0.27

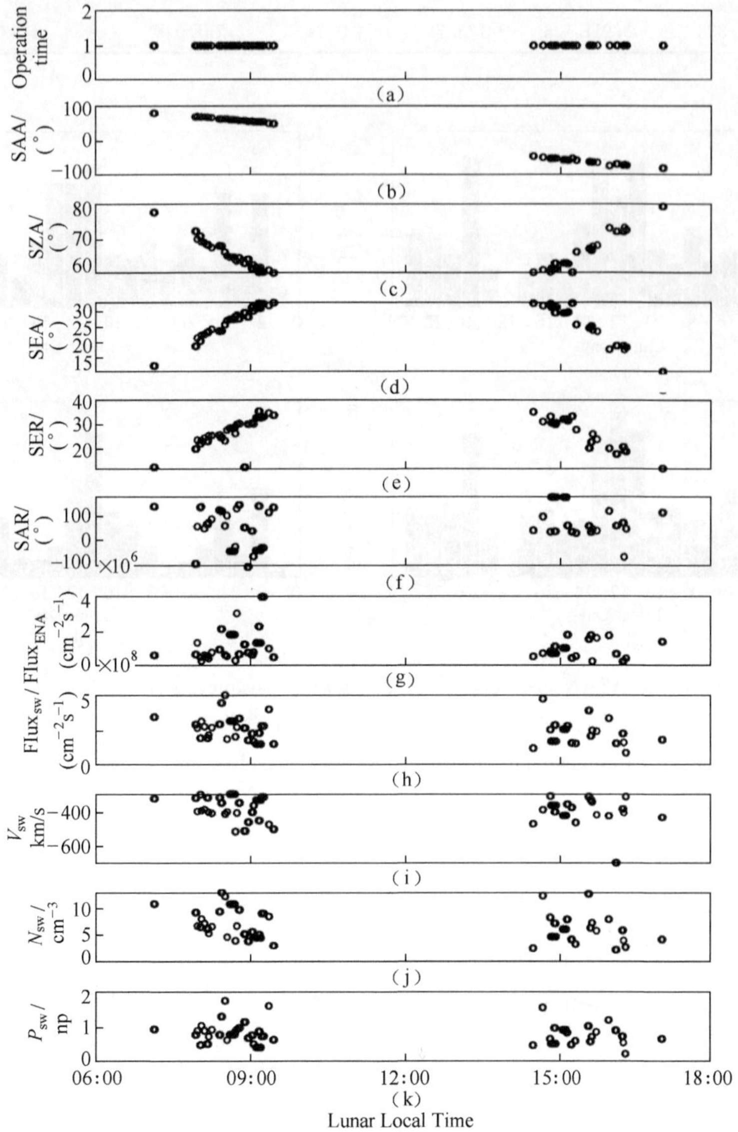

Fig. A1　Overview of ASAN geometries and ARTEMIS data. (a) ASAN operation times. (b) solar azimuth angle. (c) solar zenith angle. (d) solar elevation angle. (e) solar elevation angle of rover. (f) solar azimuth angle of rover. (g) ENA flux. (h) solar wind flux. (i) solar wind velocity. (j) solar wind density. (k) solar wind dynamic pressure as a function of the lunar local time

Tab. A3

The Linear Fitting Results of the Integrated Solar Wind Density and Pressure and the Differential Flux in Each ASAN Energy Bin

	E/eV	dawn _ c	dawn _ k	dawn _ r^2	dusk _ c	dusk _ k	dusk _ r^2
	140	1.65E+02	90.38	0.60	4.93E+01	396.71	0.24
Density	278	7.14E+01	35.12	0.57	1.75E+01	81.18	0.12
	333	6.06E+01	−0.66	0.54	7.52E+00	111.52	0.03
	140	1.15E+03	364.86	0.28	8.12E+02	177.03	0.38
Pressure	278	4.31E+02	204.21	0.22	4.32E+02	−103.58	0.43
	333	3.91E+02	123.76	0.24	2.75E+02	−33.36	0.26

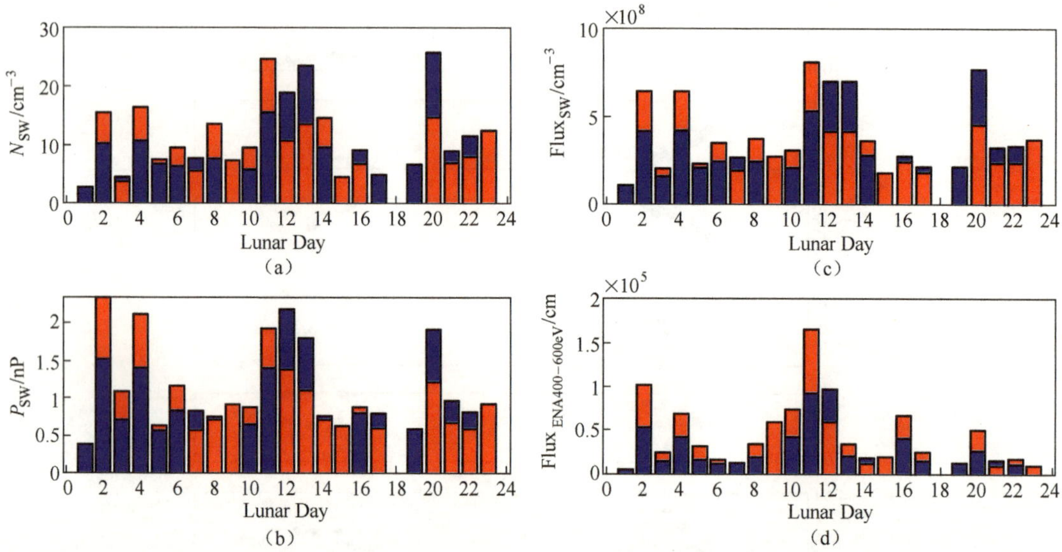

Fig. A2　Overview of the ASAN geometries and solar wind parameters over 23 lunar days. (a)-(b) Solar wind density and pressure. (c)-(d) Solar wind flux and ENA flux with 200-600eV. The red bars indicate the mean data on the dawnside of the Moon, and the blue bars indicate the mean data on the duskside of the Moon

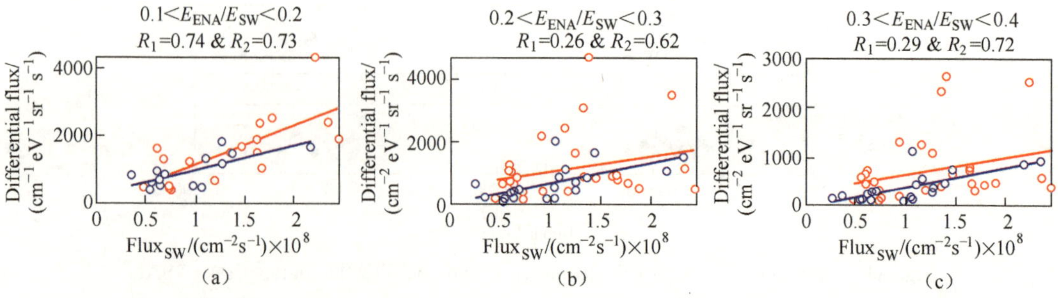

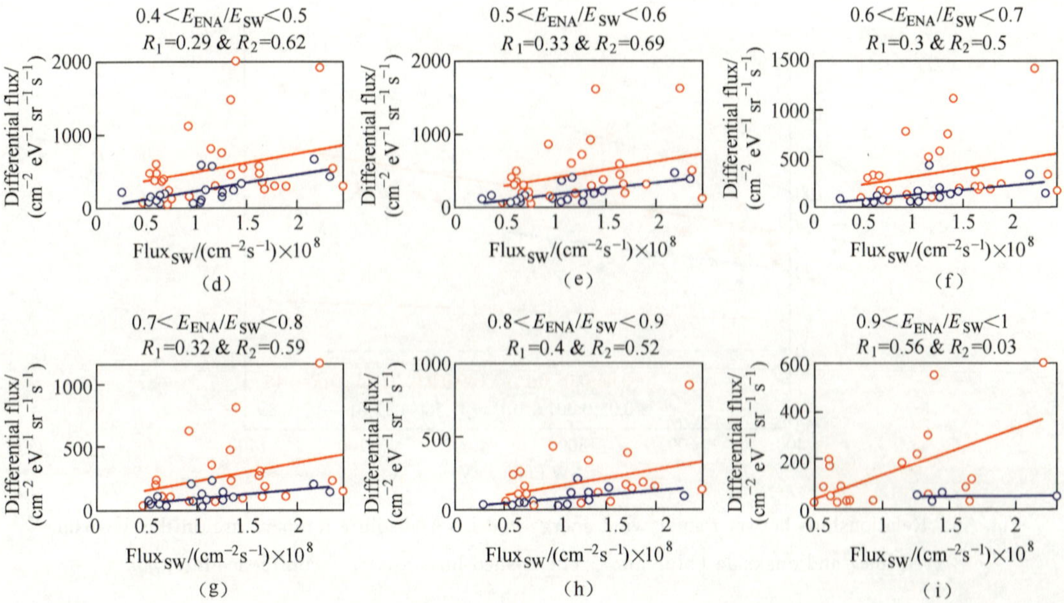

Fig. A3　Correlation coefficients between the integrated solar wind ion flux and the differential fluxes of ENAs in the E _ ENA/E _ SW bin (normalize the measured ENA energy by the solar wind energy). The red and blue lines show the results on the dawnside and duskside, respectively. Here R_1 and R_2 represent the correlation coefficients on the dawnside and duskside, respectively

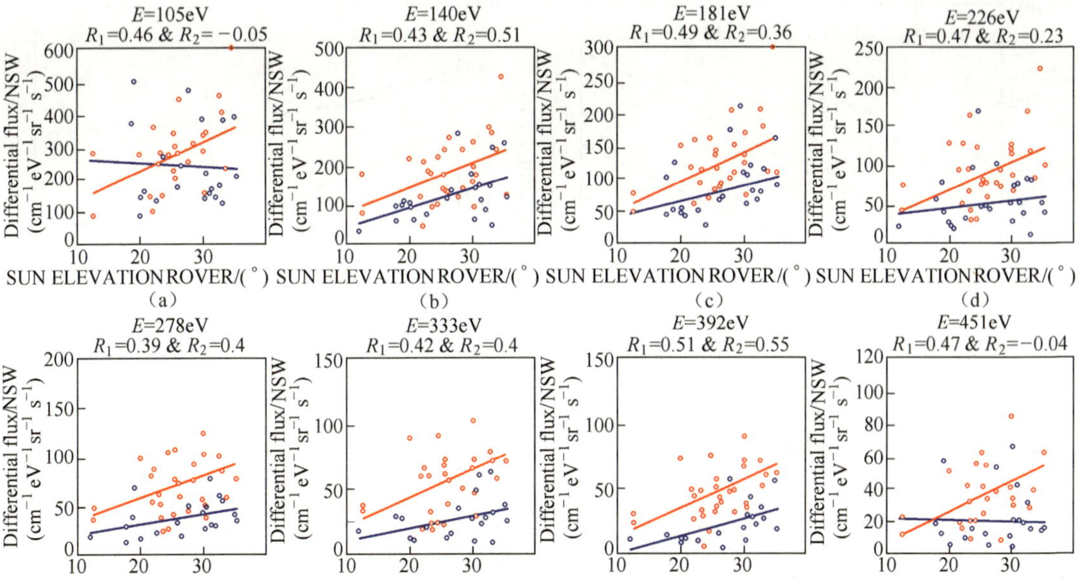

Fig. A4　Relationship between the ENA differential flux/Nsw and the solar elevation angle of the rover at an individual energy of ASAN from 105 to 451eV. The red circles indicate the data on the lunar dawnside, and the blue circles indicate the data on the lunar duskside. Here R_1 and R_2 represent the correlation coefficients on the dawnside and duskside, respectively

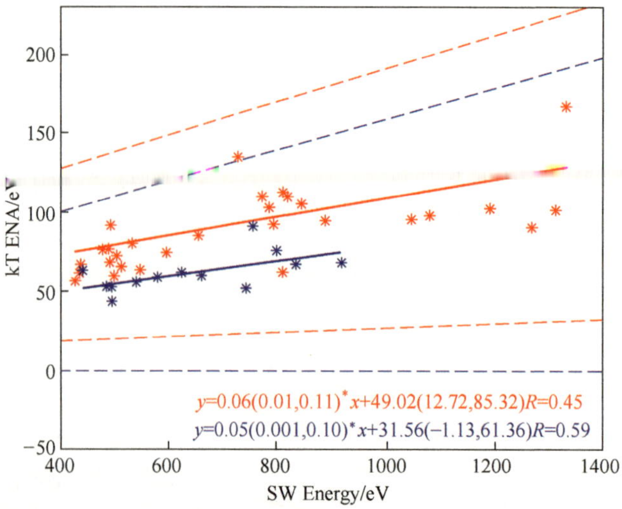

Fig. A5 Relationships between solar wind energy and ENA absolute temperature on the dawnside (red line) and duskside (blue line). The dashed lines are 95% confidence intervals

The γ and v are calculated using the relations

$$E_k=(\gamma-1)m_0c^2 \text{ and } \frac{v^2}{c^2}=1-\frac{1}{\left(\frac{E_k}{m_0c^2}+1\right)^2} \tag{A2}$$

where c is the light speed and E_k is the particle's kinetic energy.

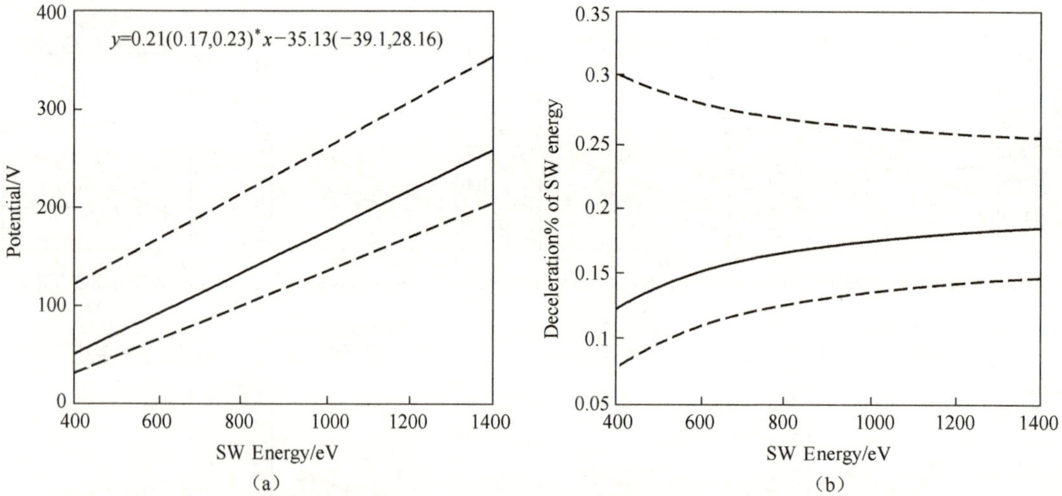

Fig. A6 The possible solar wind electrostatic potential and deceleration percentage of the solar wind are derived from Fig. 5. The dashed lines are 95% confidence intervals

The integrated particle density is calculated as

$$n=\int_{v_{\min}}^{v_{\max}}\int_0^{\pi}2\pi\sin(\alpha)v^2 f\,\mathrm{d}\alpha\,\mathrm{d}v \tag{A3}$$

where α is the pitch angle.

The parallel and perpendicular pressures are calculated as

$$P_{\parallel}=\int_{v_{\min}}^{v_{\max}}\int_{0}^{\pi}2\pi\sin(\alpha)\cos^{2}(\alpha)m_{0}v^{4}f\,\mathrm{d}\alpha\,\mathrm{d}v \tag{A4}$$

and

$$P_{\perp}=\int_{v_{\min}}^{v_{\max}}\int_{0}^{\pi}\pi\sin^{3}(\alpha)m_{0}v^{4}f\,\mathrm{d}\alpha\,\mathrm{d}v \tag{A5}$$

The temperature is calculated as

$$T=P/\mathrm{nK} \tag{A6}$$

References

[1] Allegrini, F., Dayeh, M. A., Desai, M. I., et al. 2013, P&SS, 85, 232
[2] Angelopoulos, V. 2011, SSRv, 165, 3
[3] Crider, D. H., & Vondrak, R. R. 2002, AdSpR, 30, 1869
[4] Di, K., Liu, Z., Liu, B., et al. 2019, JRemS, 23, 177
[5] Fairfield, D. H. 1971, JGR, 76, 6700
[6] Funsten, H. O., Allegrini, F., Bochsler, P. A., et al. 2013, JGRE, 118, 292
[7] Futaana, Y., Barabash, S., Holmström, M., & Bhardwaj, A. 2006, P&SS, 54, 132
[8] Futaana, Y., Barabash, S., Wieser, M., et al. 2012, JGR, 117, E05005
[9] Futaana, Y., Barabash, S., Wieser, M., et al. 2013, GeoRL, 40, 262
[10] Halekas, J. S., Delory, G. T., Brain, D. A., Lin, R. P., & Mitchell, D. L. 2008, P&SS, 56, 941
[11] Harada, Y., Futaana, Y., Barabash, S., et al. 2014, JGRA, 119, 3573
[12] Kramer, G. Y., Besse, S., Dhingra, D., et al. 2011, JGR, 116, E00G18
[13] Kurata, M., Tsunakawa, H., Saito, Y., et al. 2005, GeoRL, 32, L24205
[14] Li, S., & Garrick-Bethell, I. 2019, GeoRL, 46, 14318
[15] Lin, R. P., Mitchell, D. L., Curtis, D. W., et al. 1998, Sci, 281, 1480
[16] Liu, J., Ren, X., Yan, W., et al. 2019, NatCo, 10, 4229
[17] Liu, Z., Di, K., Li, J., et al. 2020, Sci. China Inf.. Sci, 63, 140901
[18] Lue, C., Futaana, Y., Barabash, S., et al. 2011, GeoRL, 38, L03202
[19] Lue, C., Futaana, Y., Barabash, S., et al. 2016, JGRA, 121, 432
[20] McComas, D. J., Allegrini, F., Bochsler, P., et al. 2009, GeoRL, 36, L12104
[21] McFadden, J. P., Carlson, C. W., Larson, D., et al. 2008, SSRv, 141, 277
[22] Poppe, A. R., Halekas, J. S., Delory, G. T., & Farrell, W. M. 2012, JGR, 117, A09105
[23] Poppe, A. R., Halekas, J. S., Lue, C., & Fatemi, S. 2017, JGRE, 122, 771
[24] Saito, M. H., Miyashita, Y., Fujimoto, M., et al. 2008, GeoRL, 35, L07103
[25] Saito, Y., Yokota, S., Asamura, K., et al. 2010, SSRv, 154, 265
[26] Saito, Y., Nishino, M. N., Fujimoto, M., et al. 2012, EP&S, 64, 83

[27] Schaufelberger, A., Wurz, P., Barabash, S., et al. 2011, GeoRL, 38, L22202
[28] Sibeck, D. G., Angelopoulos, V., Brain, D. A., et al. 2011, SSRv, 165, 59
[29] Shue, J. -H., Chao, J. K., Fu, H. C., et al. 1997, JGR, 102, 9497
[30] Vorburger, A., Wurz, P., Barabash, S., et al. 2012, JGRA, 117, A07208
[31] Vorburger, A., Wurz, P., Barabash, S., et al. 2013, JGRA, 118, 3937
[32] Vorburger, A., Wurz, P., Barabash, S., et al. 2015, P&SS, 115, 57
[33] Wang, H. Z., Zhang, J., Shi, Q. Q., et al. 2021, ApJL, 907, L32
[34] Wang, X. -D., Bian, W., Wang, J. -S., et al. 2010, GeoRL, 37, L07203
[35] Wieser, M., Barabash, S., Futaana, Y., et al. 2009, P&SS, 57, 2132
[36] Wieser, M., Barabash, S., Futaana, Y., et al. 2010, GeoRL, 37, L05103
[37] Wieser, M., Barabash, S., Wang, X. -D., et al. 2020, SSRv, 216, 73
[38] Wu, W., Li, C., Zuo, W., et al. 2019, NatGe, 12, 222
[39] Xie, L., Li, L., Zhang, A., et al. 2021, GeoRL, 48, e93943
[40] Xie, L., Li, L., Zhang, Y., et al. 2015, JGRA, 120, 6559
[41] Yue, C., An, X., Bortnik, J., et al. 2016, GeoRL, 43, 7804
[42] Zhang, A., Wieser, M., Wang, C., et al. 2020, P&SS, 189, 104970

The Lunar Lander Neutron and Dosimetry (LND) Experiment on Chang'E 4

Robert F. Wimmer - Schweingruber[1,2], Jia Yu[1], Stephan I. Böttcher[1], Shenyi Zhang[2,3,4], Sönke Burmeister[1], Henning Lohf[1], Jingnan Guo[1,5,6], Zigong Xu[1], Björn Schuster[1], Lars Seimetz[1], Johan L. Freiherr von Forstner[1], Ali Ravanbakhsh[1], Violetta Knierim[1], Stefan Kolbe[1], Hauke Woyciechowski[1], Shrinivasrao R. Kulkarni[1], Bin Yuan[2], Guohong Shen[2], Chunqing Wang[2], Zheng Chang[2], Thomas Berger[7], Christine E. Hellweg[7], Daniel Matthiä[7], Donghui Hou[2,3], Alke Knappmann[1], Charlotte Büschel[1], Xufeng Hou[8], Baoguo Ren[8], Qiang Fu[9]

1. Institute of Experimental and Applied Physics, Kiel University, 24118 Kiel, Germany
2. National Space Science Center, Chinese Academy of Sciences, Beijing, China
3. University of Chinese Academy of Sciences, Beijing, China
4. Beijing Key Laboratory of Space Environment Exploration, Beijing, China
5. CAS Key Laboratory of Geospace Environment, Dept. of Geophys. & Planet Sci., University of Science and Technology of China, Hefei 230026, China
6. CAS Center for Excellence in Comparative Planetology, Hefei 230026, China
7. Institute of Aerospace Medicine, German Aerospace Center, Linder Höhe, 51147 Cologne, Germany
8. China Electronics Technology Group Corporation, NO, 18th Research Institute, Tianjing, China
9. National Astronomical Observatories, Chinese Academy of Sciences, Beijing, China

Abstract: Chang'E 4 is the first mission to the far side of the Moon and consists of a lander, a rover, and a relay spacecraft. Lander and rover were launched at 18: 23 UTC on December 7, 2018 and landed in the von Kármán crater at 02: 26 UTC on January 3, 2019. Here we describe the Lunar Lander Neutron & Dosimetry experiment (LND) which is part of the Chang'E 4 Lander scientific payload. Its chief scientific goal is to obtain first active dosimetric measurements on the surface of the Moon. LND also provides observations of fast neutrons which are a result of the interaction of high - energy particle radiation with the lunar regolith and of their thermalized counterpart, thermal neutrons, which are a sensitive indicator of subsurface water content.

Keywords: Space radiation; Moon; Dosimetry; Neutrons; Exploration

List of Acronyms

ADC	Analog to Digital Converter/Conversion
ATI	ATomInstitut，(Technical University Vienna)
CRaTER	Cosmic Ray Telescope for the Effects of Radiation
dps	data product scheduler
EB	Electronics Box
EM	Engineering Model
ESA	European Space Agency
EVA	Extravehicular Activity
FM	Flight Model
FOV	Field Of View
FPGA	Field - Programmable Gate Array
FRED	Flight Radiation Environment Detector
FS	Flight Spare
FWHM	Full Width at Half Maximum
GCR	Galactic Cosmic Ray
GEANT4	GEometry ANd Tracking 4
ICU	Instrument Control Unit
IRAS	Iionizing RAdiation Sensor
LET	Linear Energy Transfer
LND	Lunar Lander Neutrons and Dosimetry Experiment
LVPS	Low - Voltage Power Supply
MLI	Multi - Layer Insulation
MSL	Mars Science Laboratory
PCB	Printed Circuit Board
PHA	Pulse - Height Analysis
PIPS	Passivated Implanted Planar Silicon
PTB	Physikalisch Technische Bundesanstalt
RAD	Radiation Assessment Detector on MSL
RHU	Radioactive Heater Unit
RTG	Radioisotope Thermal Generator
SEP	System Engineering Plan
SH	Sensor Head
SPA	South Pole Aitken (basin)
SPI	Serial Peripheral Interface
SSD	Solid State Detector
TID	Total Ionizing Dose

1 Introduction

China is currently implementing a series of lunar missions which shall culminate in a crewed station on the lunar surface. The most recent major step in this endeavouring is the challenging Chang'E 4 mission to the far side of the Moon, to the South - Pole Aitken (SPA) basin (Xu et al. 2018). Chang'E 4 landed inside the von Kármán crater at 177.5991°E, 45.4446° S in selenographic coordinates at an elevation of - 5,935m (Liu et al. 2019).

The SPA basin is the largest known impact crater on the Moon and one of the largest in the solar system. It is clear that a record of such a large impact is key to understanding large impacts in the solar system and thus constrains the impact history in the solar sys - tem (Potter et al. 2012). In addition, it gives insights into the evolution of such large - scale impact basins (Head et al. 1993), and reflects the composition of the lower crust and upper mantle of the Moon. Inside the SPA basin lies the relatively smooth von Kármán impact crater which is the landing site for Chinese lunar mission Chang'E 4 (Ye et al. 2017; Wu et al. 2018). The coordinates of the landing site and its geological context of the von Kármán crater and SPA basin is discussed in Huang et al. (2018).

In preparation of human exploration of the Moon and solar system, the Lunar Lander Neutrons and Dosimetry (LND) experiment aboard Chang'E 4 is performing the first active dosimetric measurements on the lunar surface. Radiation is one of the main concerns in human space flight and is a potentially limiting factor for crewed long - term missions such as a mission to the Moon or to Mars (Cucinotta and Chappell 2011). The complex radiation field of lunar missions is summarized briefly in the following two paragraphs. It is expected to exhibit large temporal variations in the intensity as well as the composition of the radiation and is quite different from that measured on the surface of Earth (Gordon et al. 2004; Haino et al. 2004; Neher 1971; Kremer et al. 1999). Therefore, active dosimetry, which provides time - resolved measurements of the radiation exposure, is a crucial requirement for human spaceflight in general and equally for human exploration of the Moon.

Apart from sporadic solar energetic particle events the radiation environment on the lunar surface is dominated by galactic cosmic radiation and secondary albedo particles created in interactions of the primary particles with the lunar regolith. Primary GCR nuclei from hydrogen to iron are considered relevant for human exposure in space, the relevant energy range spanning from approximately 100MeV/n up to 100GeV/n or more for shielded environments. The estimated human exposure for lightly shielded environments on the lunar surface from GCR is on the order of 200μGy/d and 600 to 800μSv/d (Adamcyk et al. 2011) during solar minimum conditions. Measured dose rates, however, are expected to differ from these values due to experimental conditions, detector sensitivities, the lack of self-shielding provided by the human body, etc. A large fraction of the biologically relevant dose equivalent is due to heavy ions. Also, albedo neutrons are more relevant for the dose equivalent than the dose due to the fact that they can deposit a large fraction of their energy very localized through fragmentation processes of target nuclei (in other words, neutrons have a very large quality factor).

Looper et al. (2013) provide estimates for lunar albedo particle spectra resulting from GCR at a 50km orbit. The results show signiflcant contributions from protons, neutrons, gammas, electrons and positrons similar to the estimates by Matthiä and Berger (2017) for the Martian surface. The flux of albedo protons is estimated to exceed the primary GCR flux at energies below approximately 50MeV. The dosimetric relevance of the albedo particles depends on the experimental conditions, for instance the solar activity and the local regolith composition, and, especially for neutrons, on the detector material. Looper et al. (2013) estimated the contribution of albedo particles to the measured dose rate in silicon in their detector conflguration to only a few percent. Out of geometrical considerations, however, the albedo contribution to LND on the lunar surface is expected to be larger. Additionally, the sensitivity of silicon detectors to neutrons is small which means that the expected dose rate in tissue from the prevailing albedo neutron field is larger than the measured dose in a silicon detector. We have estimated in preliminary model calculations for LND conditions the total albedo contribution to the dose rate to approximately 20% in silicon and 25% in tissue.

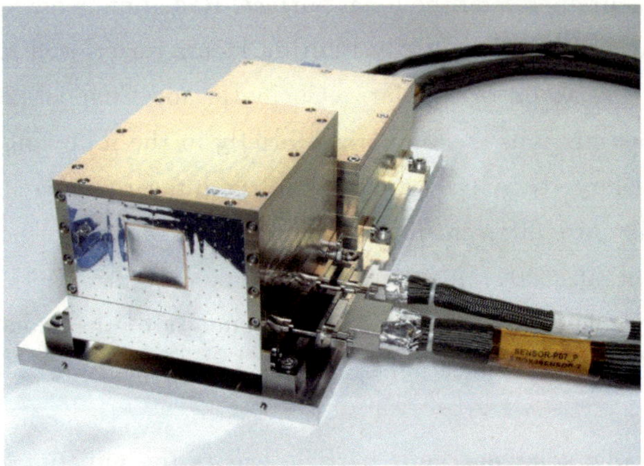

Fig. 1　Photograph of LND which consists of a sensor head (SH, front) and electronics box (EB, behind), as well as the power and data harnesses which connect the SH with the EB. The harness to the instrument control unit (ICU) is not shown

The LND experiment consists of segmented silicon Solid State Detectors (SSDs) which form a particle telescope that measures the charged particle radiation. A new geometrical arrangement and combination with conversion foils allows LND to also measure the electrically neutral component, neutrons and γ-rays. These contribute a non

– negligible fraction to the radiation dose (Reitz et al. 2012). The largest part of the absorbed radiation dose comes from electrically charged galactic cosmic rays (GCR), sporadic solar particle events, orinside the Earth's magnetosphere – from so – called "trapped" particles in the radiation belts. On the Moon, the latter is unimportant, but is replaced by an additional, secondary radiation source which is due to the interaction of the GCR with the lunar soil (regolith). Apart from charged albedo particles (Wilson et al. 2012), this secondary radiation contains neutral γ – and neutron radiation. The latter are highly relevant in a radio – biological context (Ottolenghi et al. 2013).

Apart from providing information about the temporal variation of the dose rate, LND also determines the average quality factor $\langle Q \rangle$ by measuring the Linear Energy Transfer (LET) spectrum of the radiation field. For calculating $\langle Q \rangle$ from the measured LET spec – trum, the LET dependence of Q as described in ICRP60 and ICRP103 (Streffer 2007) is used. This $\langle Q \rangle$ can be folded with the measured energy dose D to determine the dose equivalent H. Radiation measurement with planar Si – SSDs telescopes is a well – established method in dosimetry because it is robust and easy to handle (see e. g. Reitz et al. 2005; Labrenz et al. 2015; Berger et al. 2017). As just described, it also provides all required information for dosimetric measurements. Its disadvantage is that it can not directly measure the tissue – equivalent dose rate and that its count rate in a non – isotropic radiation field depends on the orientation of the detector. The multiple segments of LND allow us to compensate for the latter disadvantage by using the ratios of coincidence count rates from the various segments, very similar to what has recently been done for the radiation field on the Martian surface (Wimmer – Schweingruber et al. 2015).

Measurements of the radiation field on the lunar surface are crucial to prepare future human exploration of the Moon. LND determines the required dosimetric quantities as well as provides measurements of the lunar neutron and γ – radiation environment. In this paper we describe the LND, starting with its scientific objectives and measurement requirements (Sect. 2), an in – depth discussion of the LND design (Sect. 3), LND – internal data processing (Sect. 4. 1) and data products (Sect. 4). Fig. 1 shows the LND sensor head (SH) in the front and the LND electronics box (EB) in the rear.

2　Science Objectives

Despite the aim of landing humans on the Moon in the not too distant future, radiation measurements in the vicinity of the Moon are remarkably scarce. Fairly recent measurements in lunar orbit were provided by the Radiation Dose Monitor (RADOM) on board Chandrayaan – 1 (Dachev et al. 2011). The spacecraft reached its operational 100km circular orbit on November 12, 2008. Measurements showed a dose rate of 0. 227mGy per day averaged over 3545 hours of measurement time (20/11/2008 to 18/5/2009). During the last three months of the mission (20/05/2009 to 28/08/2009) the spacecraft reached a

200km orbit. The dose rate increased to 0.257mGy per day owing to the reduction of the lunar shadow effect for cosmic rays and to the increase of the cosmic ray fiux related to the reduced solar activity. The minimum of solar activity was reached shortly after the end of the mission in December 2009 and it was not expected that significantly larger radiation exposures due to galactic cosmic rays than the 0.257mGy per day would occur in these specific orbits (Reitz et al. 2012). This expectation was borne out by measurements in interplanetary space (Mewaldt et al. 2010). Newer measurements have been provided by the Cosmic Ray Telescope for the Effects of Radiation (CRaTER) instrument (Spence et al. 2010) on board the Lunar Reconnaissance Orbiter (LRO). CRaTER measured a radiation exposure of about 0.22 - 0.27mGy per day in its 50km orbit (Spence 2010). Chandrayaan - 1 and LRO are the only recent missions which have reported data about the radiation environment of the Moon apart from the Apollo 17 *subsurface* measurements of low - energy neutrons by Woolum et al. (1975).

In comparison with these meagre orbital data, there is a real dearth of data on the lunar surface. The current knowledge about the radiation environment on the surface of the Moon is based exclusively on calculations using radiation transport models (e. g. Reitz et al. 2012) with input parameters from models for the galactic cosmic ray spectra and for solar particle events. While such models are based on well - established physical principles, uncertainties in the inputs and interpretations can result in significant differences in the predicted total absorbed dose rate or total dose equivalent rate (Matthiä et al. 2017) and come with considerable uncertainties (Mrigakshi et al. 2012). Measurements of the lunar neutron density at depths of 20 - 400g/cm^2 within the lunar subsurface were performed during the Apollo 17mission (Woolum et al. 1975). These measurements were performed to determine the depth profile of neutrons in the lunar regolith. This is an important factor if one aims to determine time scales of the lunar surface mixing processes (Eugster et al. 1970). These authors stressed the importance of measuring the spectrum of neutrons at the lunar surface, but to our knowledge, this has never been done.

The high biological effectiveness of neutrons has been studied by multiple authors (e. g. Ottolenghi et al. 2013; Streffer 2007; Baiocco et al. 2016). This is due to the fact that neutrons can easily interact with nuclei in biological tissue and produce secondary charged particles. These low - energy short - range particles interact with the tissue and exhibit a high ionization density and therefore high potential to damage living cells by the well - known mechanisms such as direct DNA damage or production of free radicals. Thus, determination of the neutron spectrum at the lunar surface is a crucial measurement in preparation of future human exploration of the Moon.

Thus, LND has the following two major science objectives:

1. Dosimetry for human exploration of the Moon: Determination of time series of dose rate and of LET spectra in the complex radiation fleld of the lunar surface: Radiation is one of the largest long - term risks in manned spaceflight. LND will measure the dose for a

manned lunar mission as well as the temporal variation of the dose rate. For the interpretation of these dosimetric data, the quality factor, Q, needs to be known. This is derived from the LET spectra measured by LND as described in Sect. 1. A novel contribution of LND is to also provide measurements of the neutral (neutron and γ - ray) radiation which is an important contribution to the radiation dose. Measuring the neutron spectrum at the surface also provides important information to understand lunar surface mixing processes.

2. Contribution to heliospheric science: Determination of particle fluxes and their temporal variations on the far side of the Moon: There are already a number of spacecraft close to Earth (especially at Lagrange point L1) that measure the heliospheric particle radiation environment in the vicinity of Earth. Such a heliospheric "cluster" of spacecraft is augmented by LND because it provides an additional, and displaced measurement location. This new data can contribute to the understanding of particle propagation and transport in the heliosphere through measurements of the onset times of particle events and the timing of changes in the energetic particle flux. This is an active fleld of research in heliophysics and LND can contribute to it by providing such data. LND provides data at high cadence (up to one minute) to allow precise determination of the onset times. Solar particle events have different spectra than the GCR that do not extend to as high energies, but can nevertheless be a danger for astronauts on the surface of or orbiting the Moon if their high - energy flux is sufficiently high (e. g. Guo et al. 2018).

In addition to the two aforementioned major science objectives, LND also has a technological demonstration objective. The extremely large cross section of gadolinium (Gd) to capture thermal neutrons and subsequently emit a conversion electron has been exploited for thermal neutron measurements on Earth, but to our knowledge never before in space (see Sect. 3 for an indepth discussion). Thermal and epithermal neutrons serve as sensitive probes for the shallow subsurface water (proton) (Feldman et al. 1998a, 1998b) and FeO content (Elphic et al. 1998) and their quantitative measurement is therefore of great interest for planetary exploration. So - called fast neutrons with kinetic energies of several MeV are created by spallation reactions when galactic cosmic rays (GCRs) interact with the lunar soil. These neutrons scatter off and interact with the nuclei in the lunar soil, thereby losing energy, initially by elastic and inelastic collisions. Because protons and neutrons have nearly identical mass, a large proton content in the soil will lead to a reduction of the flux of fast neutrons and an increase in the flux of epithermal (0. 3eV to 500keV, Elphic et al. 1998) and thermal (0. 001 to 0. 3eV) neutrons. The ensuing change in the spectrum of neutrons can thus be an indicator of subsurface water. There are, however, also considerable ambiguities involved in the measurement of neutrons *only*. At epi - thermal and thermal neutron energies, some elements such as F, Fe, and Ti have large scattering cross sections, and can also influence the neutron energy spectrum if they are abundant enough (Lingenfelter et al. 1961; Feldman et al. 2000; Gasnault et al.

2000). Thus, knowledge of the flux of thermal neutrons can provide *additional* information for planetary exploration. LND's technological objective is thus to demonstrate the technological readiness level of using a Gd - based thermal - neutron detection technique for space - based exploration missions. The very large cross section of Gd for thermal neutrons has the potential to reduce the size of detectors for thermal neutrons. However, because of the long and cold lunar nights, Chang'E 4 has to rely on Radioisotope Thermoelectric Generators (RTGs) and several Radioactive Heater Units (RHUs) on the lander and rover, and so the scientific success of this method can not be ascertained at this point. The following two science questions illustrate the potential of this measurement technique.

1. Determine the subsurface water content in the South - Pole Aitken Basin: Subsurface water on the Moon is a precious resource which would be needed for long - term human presence of humans on the Moon. It is thus critical to understand the amount of water on the Moon. Because the South - Pole Aitken Basin has some permanently shadowed regions, it is believed that it may harbor signiflcant amounts of water (Feldman et al. 2001). LND can detect thermal neutrons and is thus sensitive to the subsurface water content. The detector arrangement described in Sect. 3 could determine the flux of thermal neutrons averaged over a small area beneath the lander and a larger area surrounding the lander. It could thus provide information about small - scale variations in the subsurface proton content of the lunar soils at the landing site. See Sect. 3 for a more thorough discussion. At the time of writing this paper, the influence of the RTG and RHUs on this measurement is still being assessed.

2. Determine the FeO content in the South - Pole Aitken Basin: The South - Pole Aitken Basin is the second largest known impact crater in the solar system. It is therefore also a "deep hole" in the crust of the Moon and allows us to peer into mantle material of the Moon. Measurements of Lunar Prospector and Clementine show different results for the abundance of heavy elements (which are dominated by FeO) (Lawrence et al. 2002) which is puzzling, but also of high importance to understanding the origin of the Moon and thus the solar system.

To achieve the science objectives listed above, LND is designed to perform several measurements of the lunar radiation environment, which are discussed in more detail in Sect. 4 and summarized here:

1. Time series of the charged and neutral particle dose rate in Si at a cadence of up to 1 minute.

2. Coarse LET - spectra at a cadence of up to 1 minute.

3. Coarse charged particle spectra at a cadence of up to 1 minute.

4. Coarse energy deposit spectrum of neutral particles at a cadence of up to 1 minute.

5. Count rates of thermal neutrons at a cadence of 10 minutes.

6. High - resolution measurements of all these quantities as well as heavy ions at a

cadence of 1 hour.

3 Instrument Description

LND is mounted inside the $-Y$ payload compartment of the Chang'E 4 lander and consists of a Sensor Head (SH) and an Electronics Box (EB) (see Fig. 1). The LND EB connects to the Instrument Control Unit (ICU, not shown) which serves as the electrical and data interface of LND with the lander. The sensor head is mounted on a bracket close to the upper deck of the lander which has a re-closeable opening which allows an unobstructed field of view (FOV) into the sky. This opening is closed during lunar nights to conserve heat inside the payload compartment. The LND FOV points approximately 13° from the direction of the normal to the lander deck or along ($\mathbf{n}$ = (0.9591, 0.0644, 0.2772) in (sky, west, south) coordinates) to avoid an obstruction by a spacecraft structure. Because the GCR radiation field is isotropic, this has no effect on LND's measurements.[1] The small detector signals are pre-amplified, shaped, and converted to digital signals inside the SH. They are driven to the EB via SPI (Serial Peripheral Interface) where further analysis of the signals is performed. SH and EB are connected by two harnesses, one for power supply, heater, thermistor and bias voltage of detectors, and the other one for ADCs' control signals and data readout (see Fig. 1). The design with a separate sensor head and electronics box was driven by the requirement to keep the sensor head as cool as possible in the challenging thermal environment of the Chang'E 4 lander.

Fig. 2 shows a functional block diagram of LND. The left-hand side of it (divided by the vertical dashed line) shows the functions performed in the LND SH, the right-hand side the LND EB functions. The LND detector system, preampliflers, shapers and analog to digital converters (ADCs) are all contained in the LND SH. Twenty charge sensitive pre-ampliflers are integrated onto a printed-circuit board (PCB) which amplifles trigger signals generated by the LND detector system. The pre-ampliflers are followed by 30 pairs of shaper and ADC (analog to digital conversion) circuits which shape and then sample the amplifled signals before they are sent to the LND EB via the harness. The digitized values from the SH are pulse-height analyzed (PHA) by the analog FPGA (Field-Programmable Gate Array) in the EB. The analog FPGA also controls the ADCs, performs digital flltering, the frontend trigger processing, and housekeeping data acquisition. The data from the analog FPGA are sent to the digital FPGA for additional processing and storage before being sent to the ICU. The additional processing in the FPGAs is discussed in Sect. 4.1. The LND EB also contains the LVPS board, which supplies latch-up protected power for LND as well as the bias voltage for the LND SSDs.

1 For the measurement of thermal neutrons this small tilt results in a small correction due to the projection effect (cos (13°)).

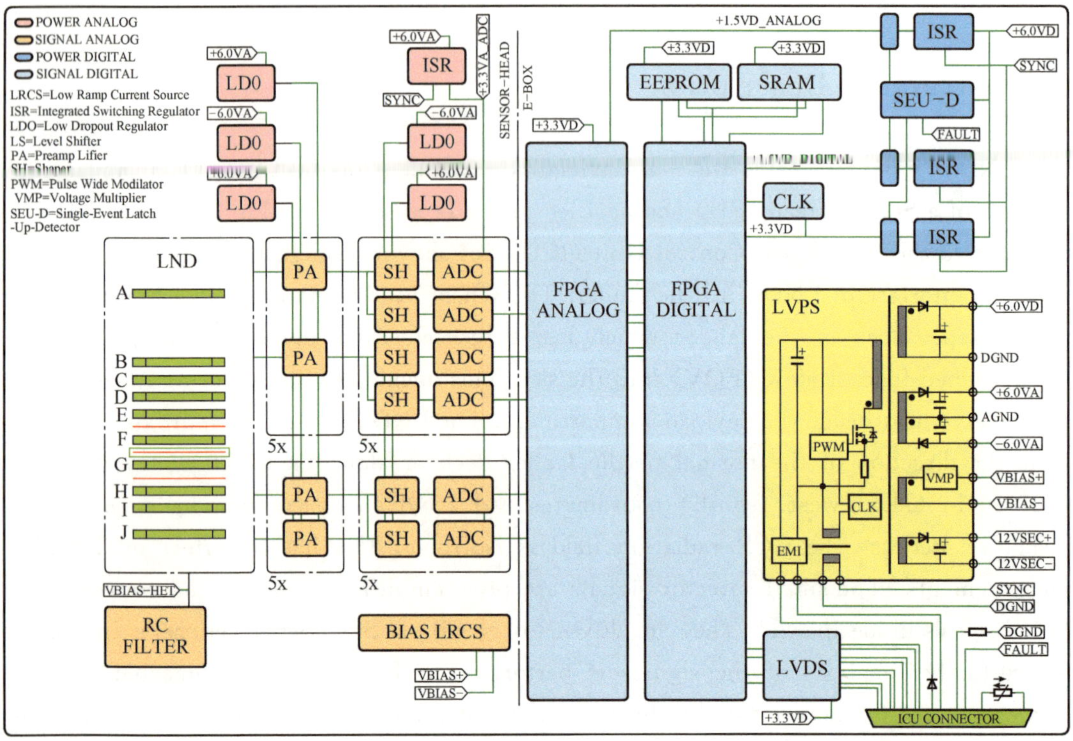

Fig. 2　Functional block diagram of LND. All 10 detectors (indicated by A - J inside the dash - dotted rect - angle marked LND) and their front - end electronics (pre - ampliflers (PA), shapers (SH), and analog - to - digital converters (ADCs)) are accommodated in the sensor head (left half of the schematic). All other electronics such as the analog and digital fleld programmable gate arrays (FPGAs) low - voltage power supply (LVPS) and are housed in the electronics box (E - Box, to the right of the dash - dotted vertical line)

3.1　Mechanical and Thermal Design Considerations

As already mentioned, LND is mounted in the $-Y$ compartment of Chang'E 4 Lander and is tilted by 13 degrees, as shown in Fig. 3. The sensor head sits on a bracket in order to achieve a unobstructed FOV through the top window of $-Y$ compartment. Additionally, the position of the sensor head is very close to the window not only for the purpose of unobstructed FOV but also to radiate as much heat to the sky as possible via the second surface mirror tape with a reflectivity to absorptivity ratio of ~8 which is applied to the front panel of the LND SH. This keeps the sensor head cool and reduces the importance of leakage current as a noise source. The size of the top window is 16cm × 23cm and is open during lunar day time and closed during the cold lunar night. The LND SH is further insulated from the warm payload compartment by a "skirt" of MLI (10 layers, emissivity 0.035) which is also shown in Fig. 3. This connects from the top of the window and completely surrounds the LND SH, effectively thermally decoupling LND SH from the payload compartment.

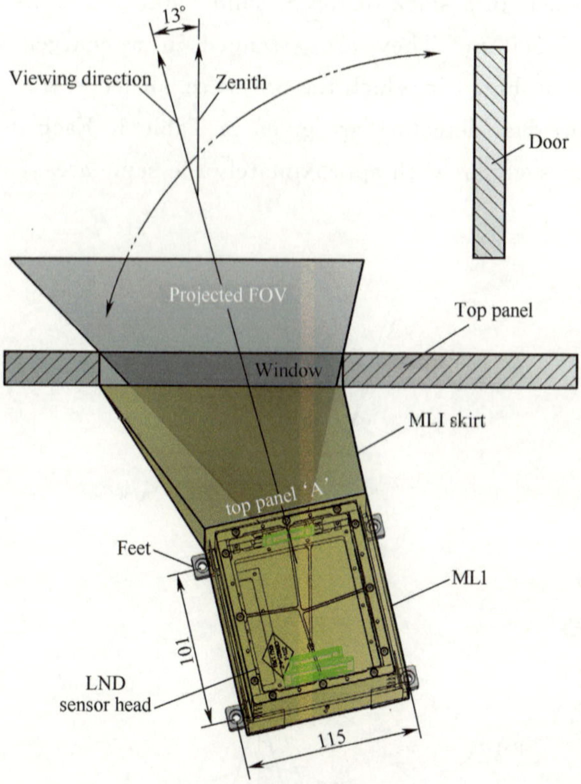

Fig. 3　Schematic view of the LND sensor head in the payload compartment of the Chang' E 4 lander. LND is tilted by 13 degrees with respect to the zenith direction, as indicated. There is no tilt in the other direction. The door is closed during lunar nights and open while LND is measuring. Distances between mounting feet are given in mm, the LND detectors are indicated in green and shown for reference. The LND sensor head is covered by MLI except for its zenith - pointing face, which acts as a radiator. The MLI - skirt insulates the rest of the payload compartment from the cold sky

The temperature at the interface of the LND bracket and its mounting panel of the lander was modeled to vary from −20℃ to +55℃ in the operational case and −60℃ to +70℃ in the non - operational case. The interface temperature of the LND electronics box and the lander is expected to lie between −20℃ and +55℃ in the operational case and −70℃ to +70℃ in the non - operational case. During the first lunar day, temperatures varied between −7. 97℃ and 48. 10℃, with the highest temperatures occurring on the sensor head analog board, and the lowest in electronics box.

3. 2　The LND Sensor Head

This section gives an overview of the LND sensor head, more detailed descriptions of the different arrangements are given farther down in this paper, as are several of the flgures which are referred to here for the first time.

The LND SH consists of a stack of ten Si solid - state detectors (SSDs) each with a nominal thickness[2] of 500μm. They are arranged in a charged - particle telescope conflguration as shown in Fig. 4 in which the SSDs are labeled A through J. The measured thicknesses of the individual detectors are given in Table 1. Each detector is segmented into an inner and outer segment with approximately the same areas, as shown in Fig. 5.

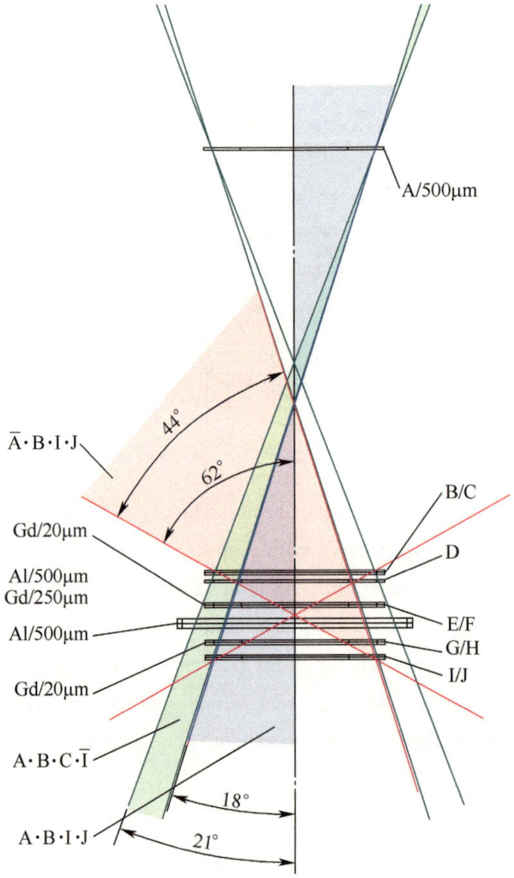

Fig. 4　The LND telescope consists of 10 segmented 500μm - thick Si solid - state detectors (SSDs), A - J. Detectors B/C, E/F, G/H, and I/J are mounted in a sandwich conflguration, the inner segments of the A and B detectors, A1 & B1, span the telescope opening. See text for a more detailed discussion. Angles are measured from the middle to middle of the detector edges, as discussed in Sect. 3. 3. 1

[2] Their measured thicknesses are given in Tab. 1.

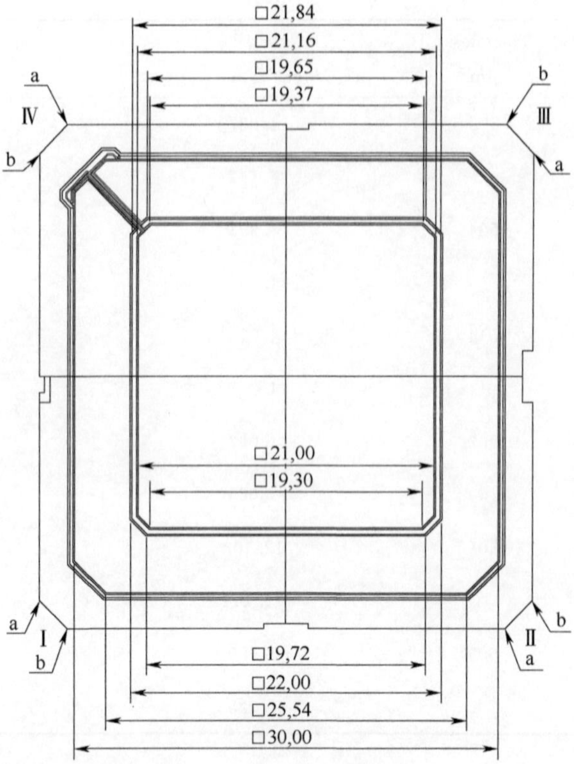

Fig. 5　LND detector layout shows the sensitive areas of LND's detectors. The inner segments are referred to as "1", the outer segments as "2". Dimensions are in mm. The area of the inner segment is 440mm². The outer segment and cross-talk ring are connected and their area adds up to 435.6mm², uncertainties in the measurements in this Figure are less than 0.01mm

Table 1　LND detector segments, measured detector thicknesses and gains

Detector/segment	Thickness/ μm	Gain 1/ (keV/mV)	Gain 2/ (keV/mV)
A1	514	12.425	186.476
A2		12.530	183.837
B1	509	12.447	179.555
B2		12.800	199.022
C1	505	12.613	159.538
C2		12.400	178.805
D1	508	12.322	182.632
D2		12.480	183.416

Continued Tab. 1

Detector/segment	Thickness/ μm	Gain 1/ (keV/mV)	Gain 2/ (keV/mV)
E1	502	4.037	—
E2		4.142	—
F1	504	3.993	—
F2		4.051	—
G1	514	4.073	—
G2		4.070	—
H1	502	4.062	—
H2		4.065	—
I1	505	12.403	163.653
I2		12.144	158.402
J1	509	3.990	—
J2		4.025	—

This telescope conflguration is used to detect charged particle radiation and identify its composition. Fast neutrons and γ - radiation are detected in the inner segment of the C detector in anti - coincidence with all other detector segments. A 20μm thin Gd foil is sandwiched between detectors E & F and G & H and is used to detect thermal neutrons. A 250μm thick Gd foils is sandwiched between two 500μm Al plates (shown in green) to shield the upward (downward) thermal neutron flux from the E & F (G & H) detectors and thus allow directional information, as discussed below in Sect. 3. 4.

Figure 5 shows the detector segments and detailed dimensions.

3.3 Detectors, Segmentation, and Gain Stages

As already discussed, the detectors on LND are all segmented, some are read out with two energy gains. Table 1 gives the conversion factors from ADC value to energy (in keV) for LND's detector segments. For instance, detector segment A1 is read out at Low gain (A1L) and High gain (A1H), while the two segments of detector E, E1 and E2, are both read out with one gain. This information is only needed when analyzing PHA data. The conversion has already been applied to the published LND data.

The geometry factors of LND are deflned by coincident measurements in A1 and B1 up to a depth of detector C. For signals in detectors D - J, a coincident measurement in A1 and D1 is required, as well as a larger energy deposition in D1 than in D2. This results in a geometry factor of 0.58cm^2sr for A1 & B1 and 0.55cm^2sr for A1 & D1 with uncertainties less than 1%. The geometric factors of LND were calculated using GEANT4 (Agostinelli et al. 2003) for an isotropic flux of particles, i. e., from 4π. 4π geometry factors are given here because there is also a (small) "return" flux of secondary particles from the Moon.[3] The low - energy part of this return flux is also measured by LND as discussed in Sect. 4. 6.

For dosimetric quantities shielding by structures in the lander, etc. is not that important because astronauts will always be shielded to some extent. In the lightest possible shielding during an extravehicular activity (EVA) astronauts will at least be shielded by their spacesuits which typically provide a shielding of 0. 3g/cm^2 by their fabric, 1g/cm^2 by the helmet, and 5g/cm^2 by the backpack (Wilson et al. 2006). This shielding is thus also fairly inhomogeneous, similar to what we estimate for the Chang'E 4 lander. LND is accommodated in the very lightly shielded payload compartment which is attached to one of the sides of the lander. While we do not know the exact equivalent thickness of the walls of the payload compartment, we can safely assume that they were constructed from honeycomb structures to save mass. If we assume that it is constructed from two 0. 7mm thick Al sheets and a much thinner internal honeycomb structure, we obtain a shielding of about 0. 4g/cm^2, comparable to a spacesuit. Of course, the lander provides more shielding in the direction of the lander, but this covers a considerably smaller solid angle than the lightly shielded payload compartment. Nevertheless, we also estimate the shielding provided by the lander in this direction. Let us assume that the lander wet mass was approximately 1200kg and has side lengths of 3m and a height (excluding landing gear) of 1. 5m, i. e., a volume of $3 \times 3 \times 1.5 = 18$m^3. The Apollo landers had a dry mass of approximately one third of the wet mass (including propellant), so we assume a dry mass of about 400kg for the Chang'E 4 lander. Thus its average density after landing is a little over 20kg/m^3 and it provides a shielding of approximately $3 \times 400/18$kg/m^2 or 7g/cm^2 from this side. This is comparable to the shielding provided by the astronaut's backpacks. Thus, being inside the lightly shielded payload compartment of the Chang'E 4 lander, LND provides a reasonable estimate for the exposure of astronauts to the radiation fleld on the lunar surface. These considerations also allow us to use a larger geometric factor (22. 23cm^2 sr) by using the combination of the B (B1 & B2) and I (I1 & I2) detectors which provides high counting statistics by integrating over a larger solid angle. Again, the geometry factor given here is calculated for a 4π radiation field.

3. 3. 1　Charged - Particle Telescope

A schematic view of the LND SH charged - particle telescope is shown in Fig. 4. It can be divided into an upper and lower half. The upper four detectors (A - D) form a detector arrangement similar to the Flight Radiation Environment Detector (FRED) (Möller et al.

---

[3] The CRaTER team report a global average ratio of lunar protons to galactic cosmic ray protons of 0. 38± 0. 02 for the CRaTER D4 + D6 detections with incident energies between 60MeV and 150MeV (Wilson et al. 2012).

2011, 2013a, 2013b) which was based on the Ionizing RAdiation Sensor (IRAS[4]) which was developed as part of ESA's Exomars program. The lower half of LND (detectors E - J) consists of a closely packed stack of six more Passivated Implanted Planar Silicon (PIPS) detectors and an Al - Gd - Al absorber for thermal neutrons. Two pairs (E & F and G & H) clamp a 20μm thin Gd foil, while I & J complete the charged - particle telescope.

The full opening angle of 29.4° of the LND FOV (Field of View) is determined by the inner segments of detectors A and B (A1 and B1) and the outer segment of detector J (J2) as sketched in Fig. 4. Note that this angle is not shown in Fig. 4. The opening angles (measured from the middle to the middle of the detector edges) of detector combinations relevant for measuring the dosimetric quantities are given in Fig. 4. As stated in Sect. 3.3, the relevant geometry factors are 0.58cm^2 sr for A1 & B1 and 0.55cm^2 sr for A1 & D1 with uncertainties less than 1%.

This arrangement allows to determine which ions stop in the various detectors and gives a good estimate of the penetrating particle flux. The resulting rather longish telescope ensures that the full energy deposit of a particle which triggers B1 and A1 is measured for stopping ions or that it is identifled as a penetrating ion. The primary energy ranges for ions stopping in B - I are given in Table 6. For dosimetric measurements discussed in Sect. 4.4 a higher count rate is advantageous and both inner and outer segments of the relevant detectors are used. Because electrons scatter much more than ions, this telescope conflguration is less sensitive to their primary energy, although some information is still retained as can be seen in Fig. 14 and discussed there.

The telescope aperture is covered by a layer of thin aluminized Kapton foil (50.8μm Kapton and 25.4μm of Al). The size of this window corresponds to the opening of the telescope spanned by the A1 · B1 detector segments projected onto the front housing surface of the LND SH. The energy which particles loose in this protective foil is small compared to the energy loss in the front detector, A, and has been accounted for in the values of energy ranges given in Table 6.

LND can measure the composition of the charged particles using the combination of the energy deposited in the front detector (A) and the total deposited energy in the (B - I) detector stack. This allows LND to measure the composition of the stopping particles, which is important for dosimetric purposes but also for heliospheric science. LND provides energy spectra up to 30MeV/nuc, i. e., above the often - seen "knee" in the interplanetary energy spectrum of SEPs. This "knee" is an important indicator of the acceleration process

[4] ESA's original Exomars mission included the Humboldt lander with its Pasteur payload. The Pasteur payload included the Ionizing Radiation Assssment Sensor (IRAS, Wimmer - Schweingruber et al. 2008), and parts of its design are used in LND. After assembly of the IRAS prototype, it was reused as FRED in sevaral highaltitude balloon flights (Möller et al. 2013a, 2013b) and was also qualified for use in civil aircraft (Möller 2013). FRED used an IRAS - telescope of our 300μm Si - SSDs and was calibrated with neutrons at PTB and iThemba (Möller 2013; Möller et al. 2013b).

(Reames 2014). Of course the majority of all particles will penetrate LND, nevertheless, their energy deposit remains proportional to the square of their nuclear charge and, therefore, LND measures and discriminates all major minimally ionizing particle species. Knowledge of the particle composition is important for dosimetry and helps to better understand the LET spectrum (see Sect. 4. 4).

3. 4 Neutral Particle Detection

LND's top telescope consists of four segmented Si - SSDs (A - D in Fig. 4). The three lower detectors (B, C, and D) are packed as close together as possible using a special process developed by Canberra (now Mirion technologies), the provider of the SSDs. The inner segment of the C detector, C1, is surrounded by the B and D detectors as well as the outer segment of the C detector, C2. Thus, any signal measured in C1 in anti - coincidence with B, D, and C2, will be due to a neutral particle (neutron or gamma ray). This configuration allows us to measure neutrons in the energy range $1 \lesssim E_n \lesssim 20$MeV with a maximum energy deposition of ~13% of their primary energy (Fig. 7). Of course, the anti - coincidence is not perfect, but a GEANT4 simulation showed an efficiency of $99.807\% \pm 10^{-3}\%$.

Figure 7 shows the geometry factor for the detection of neutrons as a function of energy. In it, incident energy is shown along the x - axis, and the energy deposited in the detector is shown along the y - axis. The geometric factor for the measurement is color coded as shown in the color bar on the right and was simulated using GEANT4 version 10 - 1 - patch - 2[5] (Agostinelli et al. 2003; Allison et al. 2006, 2016). The dashed red line marks the previously discussed elastic scattering of neutrons with Si - nuclei, it marks the 13. 3% maximum energy transfer. Close inspection shows that the recoil nuclei deposit somewhat less energy than the allowed maximum which is due to the so - called plasma - effect (Tove and Seibt 1967; Möller 2013). Because the knock - on Si nuclei are highly ionizing and have only a very short range in Si, the electric field used to separate electron - hole pairs can not fully penetrate into the region of ionization and a fraction of the electrons and holes can recombine before they are separated. This effect results in a small correction which is indicated by the blue dotted line and is not included in the GEANT4 physics lists. Figure 9 shows the expected measured energy spectrum in C1 only (i. e. in anti - coincidence with the all other detectors and detector segments).

Figure 6 shows LND's response to γ - rays. It shows that γ - rays can be measured up to approximately 1MeV. Above that energy, the geometry factor rapidly decreases. The diagonal line (highlighted with black dots) at the lowest energies is due to the photo - electric effect in which a γ - ray transfers energy to a photo - electron. This effect dominates LNDs γ - ray response at low energies because the recoil electrons do not have

---

[5] Using the physics lists QGSP _ BERT _ HP, G4EmLivermorePhysics, and G4RadioactiveDecayPhysics.

enough energy to exit the LND detector and thus deposit all their energy in the 500μm detector. To the right of the photo-electric effect, one sees the increasing importance of Compton scattering (highlighted with blue dots) which dominates for energies between ~100keV and a few MeV. At higher incident photon energies, the photo effect (shown with red dots) begins to contribute. LNDs detectors are virtually transparent for higher energies of γ radiation. The horizontal lines which can be seen at low energy depositions are due to γ-rays emitted by various materials inside the LND SH, lead and tin in the electronics, and gadolinium from the Gd foils and absorber discussed in the following paragraphs. These lines are highlighted by colored dots and called out in the legend panel. The enhanced response at 17MeV $< E_\gamma <$ 23MeV (surrounded by a dashed black line) is due to photo-neutron reactions (Anderson et al. 1969).

As has become obvious in the previous discussion, LND can not truly discriminate between neutrons and γ-rays. Nevertheless, the different instrument response functions for γ-rays and neutrons (Figs. 6 and 7) result in different sensitivities to the two dominant neutral particle species. Energy depositions below 1MeV are dominated by γ s whereas at higher deposited energies, neutrons gain in importance.

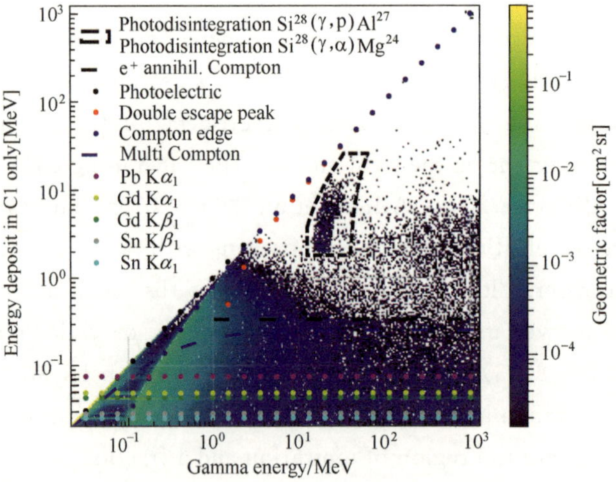

Fig. 6　Geometry factor for the detection of gamma rays. Colored dotted lines indicate various interaction processes discussed in the text

3.5　Thermal Neutrons

We now move on to discuss the lower six detectors (E-J) of LND which are mounted in two different sandwich conflgurations. In one, detectors E & F and G & H clamp a very thin ($\approx$ 20μm) Gd foil, as discussed in the caption of Fig. 8. The (natural) Gd-foil has a very large cross section[6] (48'800 barn) for thermal ($v_{th} \sim$ 2200m/s) neutron

[6] Leinweber et al. (2006) and Abdushukurov (2010) give different values for the cross sections, here we cite those of Leinweber et al. (2006).

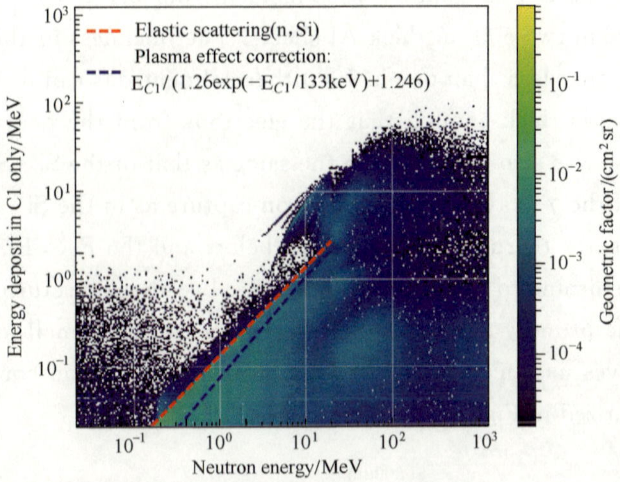

Fig. 7 Geometry factor for the detection of neutrons. The red line shows the maximum recoil energy transfer. The blue dotted line shows the plasma effect

capture (Leinweber et al. 2006; Abdushukurov 2010) and therefore is used to detect thermal neutrons. The cross section is dominated by those of the ^{155}Gd (60’700 barns) and ^{157}Gd (254’000 barns) isotopes with natural abundances of 14.8% and 15.7% (Leinweber et al. 2006). The ($1/e$) mean free path of thermal neutrons in natural Gd is easily calculated to be on the order of 7μm. After a thermal neutron has been captured in a ^{155}Gd or ^{157}Gd nucleus, the excited ^{156}Gd or ^{158}Gd nucleus decays via internal conversion emitting a single conversion electron (in~80% or~60% of all cases) or γ-decay. Most of the conversion electrons have energies between~80 and~300keV (Fogelberg and Bäcklin 1972). The electrons have a range of several microns in the Gd foil and therefore a signiflcant fraction of them can escape the foil and hit the neighboring sandwich detector. Thus, the ≈ 20μm Gd foil is thicker than the mean free path of thermal neutrons and than the range of the conversion electrons. Therefore, we expect less counts of conversion electrons in the detector which is shielded from the source by the thin Gd foil than in the detector which is "upwind" of the Gd foil. Using the other LND detectors as an anti-coincidence, this provides a clean measurement of thermal neutrons. The γ-ray from neutron capture can also be detected in anti-coincidence with surrounding detectors, providing an independent measurement of the thermal neutrons. The lowermost detector sandwich (I & J) is a copy of the LND B & C-sandwich and serves as the flnal detector in the stack. Detector J serves as a discriminator for particles stopping in the particle telescope spanned by detectors A-I.

Thermal neutrons emitted from the soil are very slow and a large fraction of them will return to the Moon because their speeds are less than the escape speed from the Moon, 2.4km/s. To discriminate such ballistic thermal neutrons (which are sensitive to the subsurface proton (water) content) returning to the Moon from neutrons coming from beneath the lander, the E & F sandwich is shielded from below by a ~250μm thick Gd foil which is encased in two 500μm thick Al sheets. The thickness of this Gd absorber was optimized to ensure that less than one in 1000 thermal neutrons can penetrate through the Gd. The Al encasing is thick enough that the electrons from the neutron capture cannot escape. Its thickness was also chosen to be the same as that of the Si-SSDs to also absorb a similar fraction of the γ-rays from the neutron capture as in the Si-SSDs. The G & H sandwich then measures thermal neutrons from below and the E & F sandwich measures thermal neutrons returning to the Moon from above. Because the return flux averages over a large area and the primary flux from below averages over a small area, the difference between the two gives us a measure of the local subsurface proton content. The detector geometry is summarized in Fig. 8.

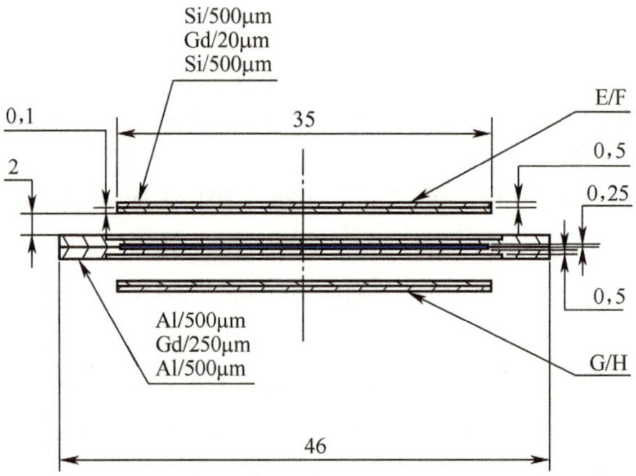

Fig. 8　The Gd-sandwich detector concept consists of two pairs of SSD detectors indicated by E/F and G/H which clamp a 20μm thin Gd foil. That foil is too thin to be visible in this drawing but captures thermal neutrons and emits conversion electrons as discussed in the text. The 250μm thick Gd-foil shown in blue is encased in Al and effectively shields the upper Gd-sandwich from thermal neutrons from the soil and the bottom Gd-sandwich from thermal neutrons from above. The Al encasing has the same thickness as the Si-SSDs. The side lengths of all (quadratic) Gd-foils is 34mm, i. e., sufficiently larger than that of the active areas of the SSDs (30mm, cf. Fig. 5), the side lengths of the SSD-Si wafer is 35mm and indicated in the drawing. The other distances are defined by the overall telescope design shown in Fig. 4

As already discussed in Sect. 2, LND's capability to detect thermal neutrons is primarily a technology demonstration. The concept of using neutron spectroscopy to determine lunar composition dates back to the pre-Apollo era (Lingenfelter et al. 1961), a comprehensive introduction to the method is given in Feldman et al. (2000) and Gasnault et al. (2000). Among the major elements present of the Moon (see, e. g., Chap. 8 in Heiken et al. 1991 for the distinction of elements). Fe and Ti are especially important in this context. The Fe abundance in lunar soils varies considerably, as does that of Ti, but somewhat less. The flux of fast neutrons is proportional to the abundance of these two major elements in the soil (Gasnault et al. 2000), whereas it is nearly independent of the abundance of the lighter major elements. Part of the reason for this is that Fe and Ti contain more neutrons than protons whereas the lighter spallation products tend to have equal numbers of neutrons and protons. Thus Fe and Ti produce a larger number of fast neutrons than the other major elements. After they are produced, these neutrons are moderated to lower energies as they scatter elastically and inelastically off nuclei in the lunar soil. Because of their nearly identical masses, fast neutron measurements are strongly affected by the presence of protons (hydrogen) (Maurice et al. 2000). The LND C1 detector measures fast neutrons (see Sect. 3.4 and Fig. 9) with energies above ~10MeV. These neutrons have to originate from the Moon (with some of them from the lander) because their speed greatly exceeds the escape speed from the Moon.

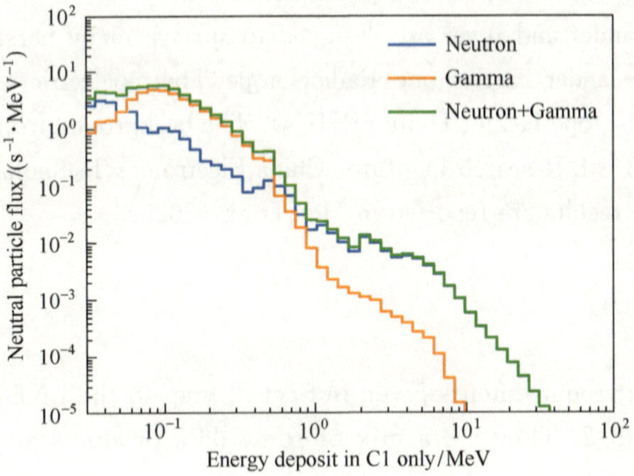

Fig. 9　Modeled spectrum of energy deposits in C1 in anti-coincidence with all other detector channels. Energy deposits above 1MeV are dominated by neutrons

The fast neutrons loose energy through collisions with nuclei in the soil, this process leads to a cascade down to thermal energies. Because neutrons and protons (in water) have approximately the same mass, neutrons are moderated to thermal energies more efficiently in minerals with high proton (water) content. Thus, the ratio of epithermal to thermal neutron flux has been shown to be a strong and quantitative indicator of sub-surface water (Feldman et al. 2000). LND has no detector for epi-thermal neutrons, but only detectors for fast and thermal neutrons. On the other hand, its neutron detectors alone (i. e., C1 with its anti-coincidence, as well as E/F and G/H) are very small and light compared to, e. g., Lunar Prospector's gamma-ray and neutron spectrometers (Feldman et al. 1999). Thus, the technology demonstration aim of LND is to determine whether a solid-state, Si-based neutron instrument could provide information about the water (proton) content of lunar soils. Of course, such a demonstration would be easier on an orbiting platform because one could compare this method with known and established abundances of water and other relevant elements such as Fe and Ti. We are currently modeling the response of LND to subsurface composition with GEANT4. On an orbiter, a number light-weight Si-Gd-Si sandwich detectors could point in different directions, which would result in different detection efficiencies for thermal neutrons, but equal ones for fast neutrons. This would provide additional constraints on the flux of thermal neutrons and we expect that this would improve the signal-to-noise ratio of LND's measurements as information is built up over several orbits (Wimmer-Schweingruber et al. 2008).

Model spectra of neutral particles are shown in Fig. 9 and exhibit relatively low count rates. Integrated over all energies, we expect a count rate of about 1.3 counts per second in the absence of the RTGs/RHUs. At low energies, this count rate is dominated by gammas. As one can easily see in Fig. 9, neutrons start to dominate the count rate above energy depositions of about 700keV, at energies above ~1MeV, we expect only about 0.006 gamma, but 0.06 neutron counts per second.[7]

The Chang'E 4 lander and rover are designed to survive many harsh lunar day and night cycles. Therefore, the lander carries one Radioisotope Thermoelectric Generator (RTG) for power and three Radioisotope Heater Units (RHUs). The background from the RTG and RHUs was measured at the 18-th Research Institute, China Electronics Technology Group Corporation in Tianjin, China. The results are reported in Hou et al. (2020).

4 Data Products

The measurement requirements given in Sect. 2 map to the LND data products which are summarized in Tab. 2. They are a mix of some data products at high time resolution (one minute), medium time resolution (10 minutes), and low time resolution data (one hour). The different cadences of LND's data products allow the investigation of rapid variations such as are, e.g., expected during the onset of a solar particle event. At the same time, no telemetry is wasted with sending *all* information at this high cadence. This chapter gives a detailed description of LND's data processing and data products.

4.1 Data Processing Inside LND

LND data processing is summarized in Fig. 10 which we discuss from left to right here. When a particle hits one or more of LND's detector segments, the L1 trigger checks which of LND's 30 detector (gain) channels registered an energy deposition larger than their trigger thresholds. These are given in the box beneath the left-most entry of Fig. 10. This information as well as the measured values are passed on to the L2 trigger which performs a number of tasks. It checks whether the event has triggered detectors which are relevant for the LET data product (different detectors can contribute). If so, the counter for valid LET events (L2 [0]) is incremented by one. Similarly, if an event is classifled

[7] Building up statistics over time, one may attempt to invert the neutron and gamma spectra similarly to what was done for the Radiation Assessment Detector (RAD) on NASA's Mars Science Laboratory (MSL) (Köhler et al. 2011), albeit with considerably lower counting statistics and fidelity of the result.

as fulfllling the conditions to contribute to the determination of TID (Total Ionizing Dose), the counter for valid TID events (L2 [1]) is incremented by one. This is repeated for neutral particles detected in C1 (L2 [2]) and detectors E - F (L2 [3] - L2 [6]). Such events are valid if energy was deposited in only one of these detectors, i. e., in anti-coincidence with all other detectors. Finally, if an event is identifled as being due to a heavy ion, the L2 [7] counter is incremented by one. An event which does not satisfy these trigger conditions is dropped. All processing up to this point is performed in the analog FPGA (see Fig. 2).

Tab. 2　List of LND science data products which are included in telemetry at 1, 10, and 60 - minute time resolution as indicated by the headings

1 - minute data products		
Data product	# counters	Comments
Protons	14	2 counters for B, one each in C, D, E, F, G, H, and I, 5 for penetrating. See Table 6 for energy ranges.
Electrons[a]	7	4 counters for electrons stopping in B, C, D, 3 counters for electrons stopping in E and F, one for G, H, and I combined. See Table 7 for energy ranges.
Neutrals	4	4 counters for $20 < E < 200$, $200 < E < 10^3$, $10^3 < E < 10^4$, $10^4 < E < 10^5$ keV
Dosimetry[b]	8	8 counters for $20<E<95$, $95<E<320$, $320<E<1076$keV, and $1.076<E<3.62$, $3.62<E<12.2$, $12.2<E<41$, $41<E<138$, $138<E<390$MeV
LET spectrum[b]	8	same bins as dosimetry
10 - minute data products		
Data product	# counters	Comments
Thermal neutrons	8	1 each for E1, E2, F1, F2, G1, G2, H1, H2, covering $\sim 55 < E < \sim 75$keV
Electrons[c]	16	16 counters for electrons stopping in E, F, G, H (4 counters for each detector). See Table 7 for energy ranges.
^{3}He	5	3 counters in B, one each in C and D (stopping). See Table 6 for energy ranges.
^{4}He	10	3 counters in B, one box in C, D, E, F, G, H, and I. See Table 6 for energy ranges.
CNO	9	2 counters in B, one box in C, D, E, F, G, H, and I. See Table 6 for energy ranges.
heavy ions	9	2 counters in B, one box in C, D, E, F, G, H, and I. See Table 6 for energy ranges.
penetating	7	^{3}He, ^{4}He, and heavy ions. See Table 6 for energy ranges.
60 - minute data products		
Data product	# entries	Comments
"Xmas plot"	150. 64	Full information
PHA words	n/a	16 PHA buffers of 1 kByte each per hour for diagnostic and calibration purposes.

[a]Deflned in Table 7

[b]Bin edges will be adjusted during the course of the mission

[c]The counters are deflned in Table 7

After this processing by the L2 trigger, a valid event is passed on to the L3 trigger which is implemented in the digital FPGA. This trigger further classifies events and increments the appropriate counters. We being this discussion with the L3 [1] and L3 [2] triggers because they are the simplest ones. An event which was identified by the L2 trigger as an event seen only in C1 (in anti - coincidence with all other detector segments) is further analyzed and the spectrum of energy depositions in the C1 detector is incremented accordingly. This is indicated by "+1 fast n histogram" in Fig. 10. The same is done for events seen in detectors E - H only (L3 [2] trigger: "+1 thermal n histogram"). Because these events only triggered single detectors, their processing ends here. For all other events, a more complicated processing chain is applied. An event which satisfies the L3 [0] condition to contribute to TID has triggered LND's B detector. Depending on the energy deposited in detector B, the corresponding histogram counter is incremented by one. While TID is only a number, LND can not measure it as such. The spectrum of energy depositions thus accumulated needs to be summed on Earth to derive TID, see Sect. 4. 4. This event is further processed by checking whether it also triggered detector I, as depicted in Fig. 10. Events which are classified as L3 [3] events have triggered the B detector and at least one other detector, they are processed in the same manner. The flow chart in the right of Fig. 10describes the further processing chain.

The histograms which are thus accumulated by LND can be one - dimensional or two - dimensional. TID and some LET spectra are accumulated in one - dimensional histograms whereas events that triggered several detectors augment two - dimensional histograms. We refer to the collection of these histograms as the Xmas plot (see Sect. 4. 2 and Fig. 11). The quantities used to calculate the entry "pixel" in the Xmas plot are calculated according to the quantities summarized in Tab. 5, the corresponding energy ranges are given in Tab. 6. The individual entry types are discussed in the following subsections.

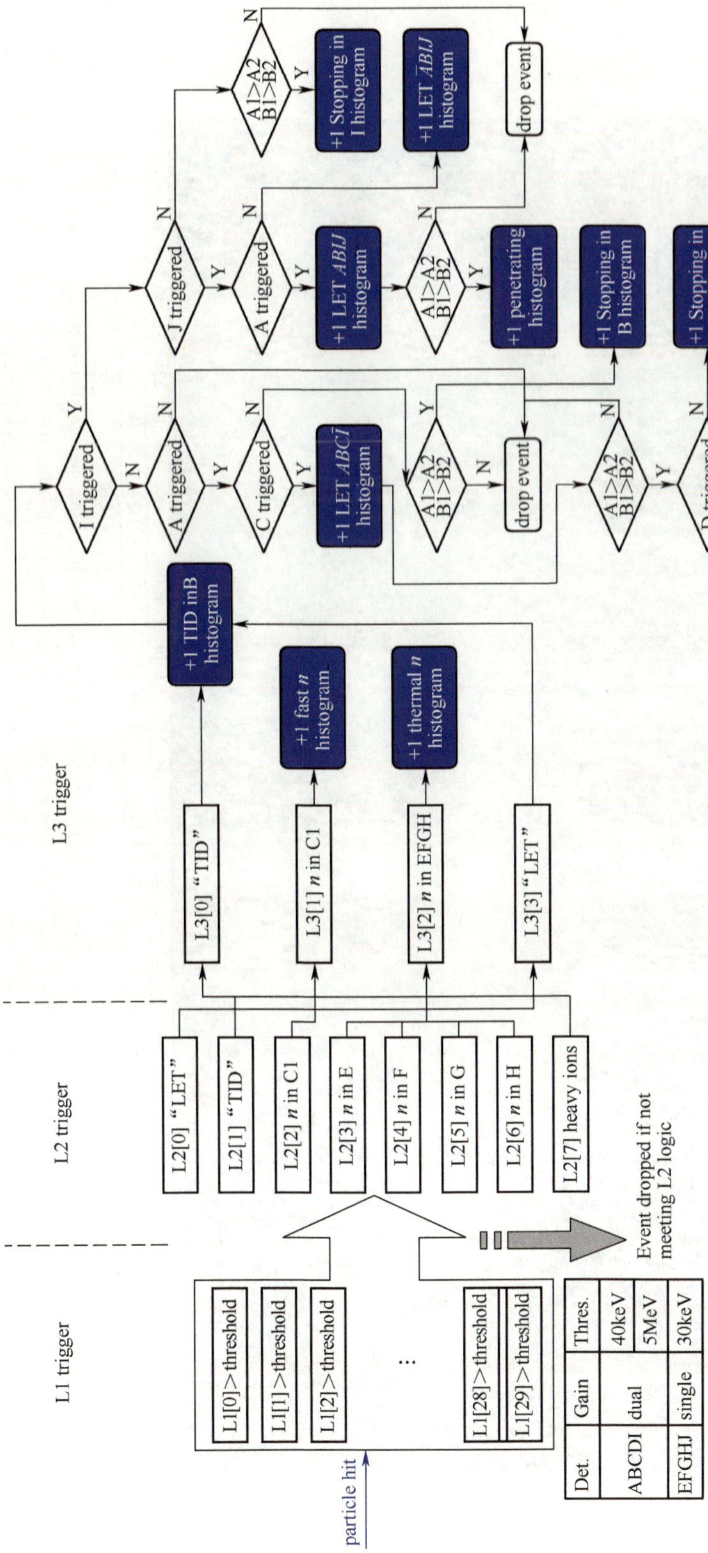

Fig. 10　LND's trigger logic is divided into three levels (L1, L2, and L3). Processing is distributed into the analog FPGA (L1 & L2) and digital FPGA (L3)

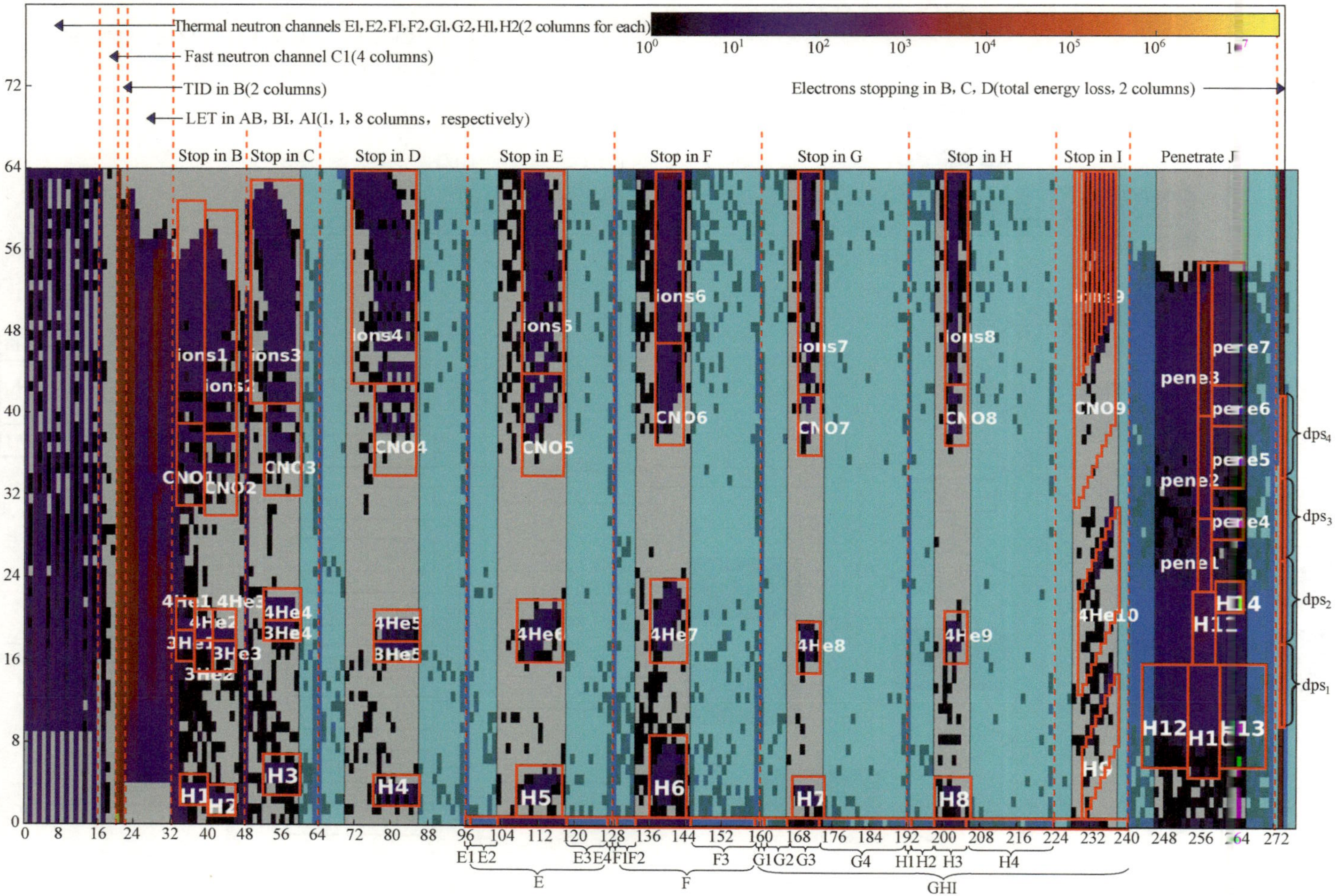

Fig. 11　LND "Xmas plot". It is divided into several vertical regions which contain information about different particle populations. The principal divisions are indicated in this figure which is based on GEANT4 simulations. Heavy ions are over－represented to make them visible. The regions indicated with red boundaries in this Figure will be adapted to reflect updated calibration values. They are labeled according to the entries given in Tables 3, 6, and 7. The regions shaded in cyan are not transmitted back to Earth. "dps" stands for "data product scheduler" and is an electron data product, discussed in Sect. 4.7

4.2　Fig. 11 – the "Xmas Plot"

Fig. 11 shows a "Xmas plot[8]" which is LND's principal data product, all others are derived from it. It contains all data of the particles measured by LND. The Xmas plot is actually an image of LND's accumulator memory, the different regions are explained in more detail in Tab. 3 and are summarized below. Each pixel of the Xmas plot is a counter which is incremented as discussed below. LND memory contains two Xmas plots, an active and an inactive one. Upon measurement of a particle, the active one is updated as described below. This accumulation is ended after 3600 seconds when the Xmas plot is turned to inactive mode. At this moment, the counters in the previously inactive Xmas plot are reset to zero and this memory region (Xmas plot) is turned to active mode. The previously active Xmas plot is now inactive and is read out and sent to telemetry over the course of the next hour. Thus the two Xmas plots alternate between active and inactive mode and provide continuous data coverage. Regions shaded in cyan in Fig. 11 are not put into telemetry because they don't contain physical data and are dominated by noise.

The Xmas plot is divided up column wise into different regions which are defined in Tab. 3. Each column has 64 rows. Each point in this 274 × 64 matrix serves as a counter which is incremented when the LND L3 trigger identifies this to be the correct entry. The integer counters are 4 Bytes, 6 of the 32 bits serve as Hamming - encoding, single - event upset correction.

Tab. 3　Column ranges for the different regions in the Xmas plot. Each column contains 64 rows or entries. Columns which are not mentioned in this table are not transmitted back to Earth and are shaded in Fig. 11

Xmas plot region	Columns		
	Start	End	
Netutral particles	0	19	
detected in E1	0	1	$E1 \cdot \overline{E2} \cdot \overline{A} \cdot \overline{B} \cdot \overline{C} \cdot \overline{D} \cdot \overline{F} \cdot \overline{G} \cdot \overline{H} \cdot \overline{I} \cdot \overline{J}$
detected in E2	2	3	$E2 \cdot \overline{E1} \cdot \overline{A} \cdot \overline{B} \cdot \overline{C} \cdot \overline{D} \cdot \overline{F} \cdot \overline{G} \cdot \overline{H} \cdot \overline{I} \cdot \overline{J}$
detected in F1	4	5	$F1 \cdot \overline{F2} \cdot \overline{A} \cdot \overline{B} \cdot \overline{C} \cdot \overline{D} \cdot \overline{E} \cdot \overline{G} \cdot \overline{H} \cdot \overline{I} \cdot \overline{J}$
detected in F2	6	7	$F2 \cdot \overline{F1} \cdot \overline{A} \cdot \overline{B} \cdot \overline{C} \cdot \overline{D} \cdot \overline{E} \cdot \overline{G} \cdot \overline{H} \cdot \overline{I} \cdot \overline{J}$
detected in G1	8	9	$G1 \cdot \overline{G2} \cdot \overline{A} \cdot \overline{B} \cdot \overline{C} \cdot \overline{D} \cdot \overline{E} \cdot \overline{F} \cdot \overline{H} \cdot \overline{I} \cdot \overline{J}$
detected in G2	10	11	$G2 \cdot \overline{G1} \cdot \overline{A} \cdot \overline{B} \cdot \overline{C} \cdot \overline{D} \cdot \overline{E} \cdot \overline{F} \cdot \overline{H} \cdot \overline{I} \cdot \overline{J}$
detected in H1	12	13	$H1 \cdot \overline{H2} \cdot \overline{A} \cdot \overline{B} \cdot \overline{C} \cdot \overline{D} \cdot \overline{E} \cdot \overline{F} \cdot \overline{G} \cdot \overline{I} \cdot \overline{J}$
detected in H2	14	15	$H2 \cdot \overline{H1} \cdot \overline{A} \cdot \overline{B} \cdot \overline{C} \cdot \overline{D} \cdot \overline{E} \cdot \overline{F} \cdot \overline{G} \cdot \overline{I} \cdot \overline{J}$
detected in C1	16	19	$C1 \cdot \overline{C2} \cdot \overline{A} \cdot \overline{B} \cdot \overline{D} \cdot \overline{E} \cdot \overline{F} \cdot \overline{G} \cdot \overline{H} \cdot \overline{I} \cdot \overline{J}$
Dosimetry	20	31	

[8] We call it the "Xmas plot" because we sent this data summary plot to colleagues around the world as our Xmas card in 2017. We had to choose a name for this plot.

Continued Table 3

Xmas plot region	Columns		
	Start	End	
TID in B	20	21	B(if energy deposit in B is bigger than 40keV)
LET $ABC\overline{I}$	[illegible]	[illegible]	$A \cdot B \cdot \overline{C} \cdot \overline{I}$
LET $\overline{A}BIJ$	23	23	$\overline{A} \cdot B \cdot I \cdot J$
LET $ABIJ$	24	31	$A \cdot B \cdot I \cdot J$
Charged particles	32	273	
stopping in B	32	47	$A \cdot B \cdot \overline{C} \cdot \overline{I} \cdot (E_{A1} > E_{A2}) \cdot (E_{B1} > E_{B2})$
stopping in C	48	59	$A \cdot B \cdot C \cdot \overline{D} \cdot \overline{I} \cdot (E_{A1} > E_{A2}) \cdot (E_{B1} > E_{B2})$
stopping in D	70	85	$A \cdot B \cdot C \cdot D \cdot \overline{E} \cdot \overline{I} \cdot (E_{A1} > E_{A2}) \cdot (E_{B1} > E_{B2}) \cdot (E_{D1} > E_{D2})$
stopping in E	103	117	$A \cdot B \cdot C \cdot D \cdot E \cdot \overline{F} \cdot \overline{I} \cdot (E_{A1} > E_{A2}) \cdot (E_{B1} > E_{B2}) \cdot (E_{D1} > E_{D2})$
stopping in F	133	144	$A \cdot B \cdot C \cdot D \cdot E \cdot F \cdot \overline{G} \cdot \overline{I} \cdot (E_{A1} > E_{A2}) \cdot (E_{B1} > E_{B2}) \cdot (E_{D1} > E_{D2})$
stopping in G	166	173	$A \cdot B \cdot C \cdot D \cdot E \cdot F \cdot G \cdot \overline{H} \cdot \overline{I} \cdot (E_{A1} > E_{A2}) \cdot (E_{B1} > E_{B2}) \cdot (E_{D1} > E_{D2})$
stopping in H	198	205	$A \cdot B \cdot C \cdot D \cdot E \cdot F \cdot G \cdot H \cdot \overline{I} \cdot (E_{A1} > E_{A2}) \cdot (E_{B1} > E_{B2}) \cdot (E_{D1} > E_{D2})$
stopping in I	228	237	$A \cdot B \cdot C \cdot D \cdot I \cdot \overline{J} \cdot (E_{A1} > E_{A2}) \cdot (E_{B1} > E_{B2}) \cdot (E_{D1} > E_{D2})$
Penetrating	246	265	$A \cdot B \cdot I \cdot J \cdot (E_{A1} > E_{A2}) \cdot (E_{B1} > E_{B2})$
Electrons	273	273	stopping in B if $16 \cdot \log_2((E_A + E_{B1}) \cdot E_A / (8000 \cdot 4000)) < 0$
			stopping in C if $16 \cdot \log_2((E_A + E_{B1} + E_C) \cdot E_A / (10000 \cdot 4000)) < -16$
			stopping in D if $16 \cdot \log_2((E_A + E_B + E_C + E_{D1}) \cdot E_A / (1200 \cdot 6000)) < 0$

We now proceed to give the explanation of the Xmas plot starting from the left and moving through it to the right. The left most region of the Xmas plot contains the accumulators for neutral particles (Sect. 4. 3), it is followed (from left to right) by two regions for the dosimetric quantities total ionizing dose (TID) and linear energy transfers (LETs) (Sect. 4. 4). To the right of these are memory regions for the accumulation of particles stopping in LNDs detectors B–I (Sect. 4. 5). Penetrating particles are mapped to the region indicated by "Penetrate J" (Sect. 4. 6). The single column on the very right contains electron data, as does the narrow strip at the bottom of the plot (Sect. 4. 7). The summed counts in the regions indicated by red counters and white names are read out and telemetered at a cadence of 1 or 10 minutes, while the entire Xmas plot with the full information is put into telemetry every hour. The 10 and 1 minute data are explained in more detail in Sects. 4. 9 and 4. 10. Figure 11 shows an annotated Xmas plot, the annotations are referred to in the following paragraphs.

4. 3　Neutral Particles

Neutral particle measurements are based on measurements in single - detectors or detector segments in anti - coincidence with all other detectors/segments. The energy deposits measured in individual detectors are histogrammed and telemetered down as part of the Xmas plot, some lower-resolution data products are sent at higher time resolution, as summarized in Tab. 3. Because the Chang'E 4 lander has an RTG and three RHUs, small energy deposits in the single detector are dominated by the background from the lander's RTG and RHUs. This needs to be accounted for when interpreting LND's measurements and using them to predict human exposure on the surface of the Moon.

4.3.1 Thermal Neutrons

The left-most 16 columns contain the histograms of the energy deposits in detectors E, F, G, and H. Each detector has two segments (1: central, 2: outer segment), so there are 8 histograms in total. The energy deposit is accumulated in 128 channels, it is wrapped around once to fit into the format of the Xmas plot, thus each histogram covers two columns in the Xmas plot. An example will help understand the structure:

The histogram of energies measured in E1, E_{E_1}, is shown in the two left-most columns of Fig. 11 (see also Tab. 3). The 128 bins are accumulated based on the following formula:

$$\text{bin}\# = 16 \cdot \log_2 (E_{E_1}/20) \tag{1}$$

where the energy values E_{E_1} and 20 are in keV. Such histograms are shown in Fig. 12 and are shown in a color scale in the Xmas plot. Thus, to extract the E1 histogram one only needs to extract the two left-most columns of the Xmas plot. This is fully analogous for all thermal neutron detectors. The calibration data for thermal neutrons shown in Fig. 12 shows peaks at 43,71, 131, and 173keV which are due to X-ray emission from the Gd foil and internal conversion (IC) electrons from the neutron capture in the Gd foil. Counts in the two smaller peaks at 131 and 173keV are much rarer than those at 71keV and are consequently only included in the Xmas-plot, but not in the 10-minute data products.

The number of counts due to thermal neutrons is determined by taking the differences E-F and H-G. Because of the very short mean free path of thermal neutrons in Gd of ~7μm (see Sect. 3.5) and the limited range of the conversion electrons we only expect a small fraction of thermal neutrons to be measured in F and in G, thus these two detectors can serve as background measurements but at the cost of a somewhat lower efficiency for detecting thermal neutrons. The number of thermal neutron counts is then divided by the detection efficiency for thermal neutrons, $\eta_{tn} \approx 1/50$, which was established during the calibration at ATI in Vienna in September 2017. Together with the area of LND's detectors (given in Fig. 5) LND can thus determine the flux of thermal neutrons on the surface of the Moon.

4.3.2 Fast Neutrons (and Gamma Radiation)

The inner segment of the C detector, C1, is also read out in anti-coincidence with all other LND detectors. It thus responds nearly exclusively to neutral particles which do not interact with the surrounding detectors and the outer segment of C, C2. Its energy histogram is stored in 256 bins and is wrapped around three times in the Xmas plot, see Fig. 11. The histogram bins are calculated according to Eq. (2)

$$\text{bin}\# = 16 \cdot \log_2 (E_{C1}/20) \tag{2}$$

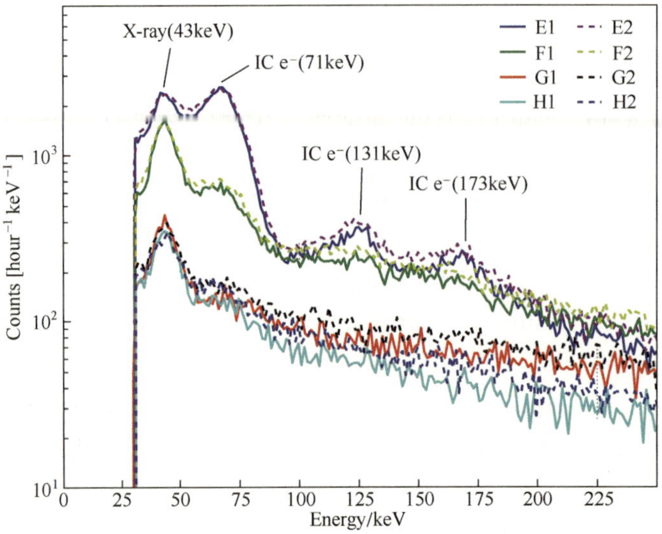

Fig. 12　Thermal neutron measurements acquired with the LND flight spare unit at the Atominstitut (ATI) in Vienna in September 2017 during one hour. Solid and dashed lines show measurements in the inner and outer segments (1 and 2) of LND's detectors E, F, G, and H. The beam hit LND from the front, so detectors G and H do not see thermal neutrons. Signatures include X - rays at 43keV and internal conversion electrons at 71, 131, and 173keV

where E_{C_1} and 20 are in keV, i. e., exactly as in Eq. (1). Thus the differential count rate of neutral particles, **z**, is easily computed.

Figure 6 shows the geometry factors G_γ for γ - rays, Fig. 7 those for neutrons in C1. The vector of differential count rates in the energy bins of C1, **z**, is then given by

$$\mathbf{z} = G_\gamma \cdot \mathbf{J}_\gamma + G_n \cdot \mathbf{J}_n \tag{3}$$

where $\mathbf{J}_\gamma$ and $\mathbf{J}_n$ are the vectors containing the differential fluxes of γ - rays and neutrons, respectively. Inspection of Figs. 6 and 7 shows that an energy deposition in C1 above ~1MeV is very unlikely to be due to γ - rays, but much more probable for neutrons with energies $E_n >$ 10MeV, as is discussed in more detail in Sect. 3. 4. For energy depositions below ~1MeV, LND can not discriminate between neutrons and γ - rays. Inferences about the neutron and γ spectra will have to rely on an extrapolation of the higher - energy neutron spectrum which should be based on modeling results. The process to obtain fast neutron spectra is based on similar work already performed for MSL/RAD (Köhler et al. 2014; Guo et al. 2017).

4.4　Dosimetry: Total Ionizing Dose, TID, and Linear Energy Transfer, LET

4.4.1　Total Ionizing Dose

While Total Ionizing Dose (TID) is only one number, it can not be computed by LND and therefore needs to be calculated on ground. To measure TID the energy spectrum of all particles measured in detector B (i. e., in *both* the inner *and* the outer segments of B, B1 and B2) is accumulated in 128 bins according to Eq. (4).

$$\text{bin}\# = 8 \cdot \log_2 (E_B/20) \tag{4}$$

where E_B and 20 are measured in keV. On ground, TID is then computed using Eq. (5),

$$\text{TID} = \sum_{i=8}^{127} \sqrt{E_i \cdot E_{i+1}} \cdot \text{counts}_i \tag{5}$$

where i is the bin number and

$$E_i = 20 \cdot 2^{i/8}\ [\text{keV}] \tag{6}$$

Note that the average energy of the bin is expressed here as the harmonic mean. LND uses logarithmically spaced bins. It could also be expressed by the usual arithmetic mean, or also by the logarithmic mean which is larger than the harmonic mean and smaller than the arithmetic mean. Which of these expressions is used for the average energy of a bin is ambiguous. The choice should depend on the observed spectrum of energy deposits in B. Values reported in Zhang et al. (2020) are calculated according to Eqs. (5) and (6).

In order to derive the final, physical value of TID, the value calculated using Eq. (5) needs to be converted into Joules and divided by the detector's mass, $m = 1.04$g

$$\text{TID}_{\text{final}} = 1.602 \cdot 10^{-13} \cdot \text{TID}/m \tag{7}$$

where TID from Eq. (5) is expressed in keV and m in grams, and $\text{TID}_{\text{final}}$ in Gy (J/kg). A coarse value for TID is also given in the LND 1 - minute data, see Sect. 4. 10. An example of a TID spectrum is shown in Fig. 13a and was acquired by LND in February 2018 during predelivery tests. The threshold of ~40keV is clearly seen, as are the cosmic muons between 100 and 200keV. The low energy deposits below that of cosmic muons are due to X - rays in the laboratory which are created by the interaction of high - energy GCR products with the walls, floor, and ceiling of the laboratory, as well as the small background activity.

4.4.2 Linear Energy Transfer Spectra

The Xmas plot contains three LET spectra defined by the detector combinations given in the three first lines of Table 4. Their locations in the Xmas plot are given in Tab. 3. The first three lines in Table 4 define the relevant quantities measured by LND. The combinations in lines 4 and 5 define two LET spectra which are more relevant from a radiation protection point of view. Line 3 gives the narrowest view cone and highest LET resolution, while line 4 gives a slightly wider view cone with somewhat higher count rate. The combination given in line 5, finally, gives a very broad opening angle which is very similar to that of the Dostel instrument which has been operating on the International Space Station (ISS) for more than a decade (Berger et al. 2016, 2017). A low - resolution version (only 8 bins) of the LET spectrum 2 is sent to telemetry every minute, thus providing high time resolution LET measurements. An example for an LET spectrum is shown in Fig. 13b and was acquired in February 2018 during pre - delivery test. It is based on high - resolution data from the Xmas plot.

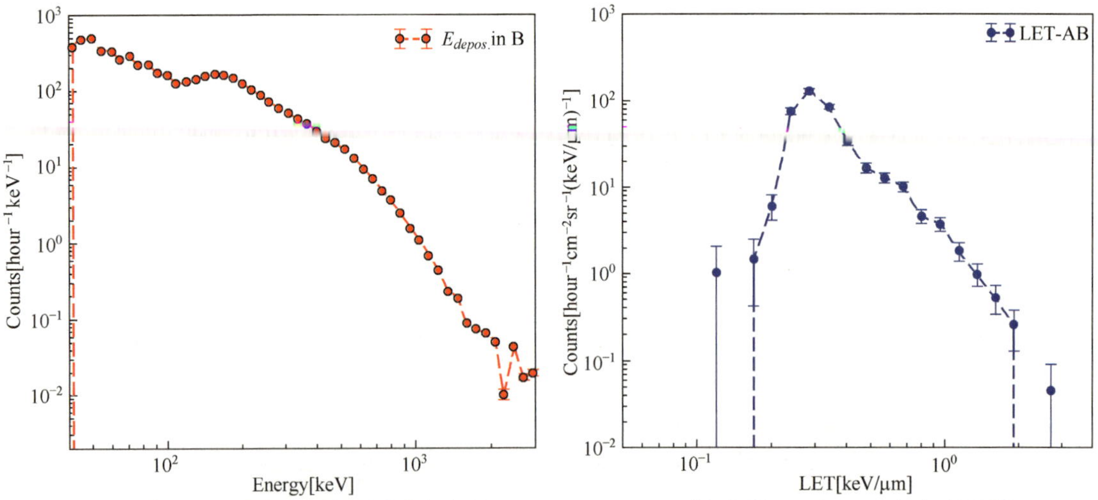

(a) Spectrum of energy depositions in the B detector when LND was standing upright, pointing at the zenith in our laboratory in Kiel. The threshold of ~40 keV is clearly seen, as are the cosmic muons between 100 and 200 keV. The red dashed curve was acquired using LNDs high-speed streaming mode, the red dots are data as stored in the "X-mas card".

(b) LET spectrum acquired with LND standing upright in our laboratory, pointing at the zenith. The energy deposit of minimally-ionizing muons is clearly seen around 0.3 keV/micron.Particles with lower primary energy contribute to higher LETs. The data shown correspond to the LET spectrum described in line 5 of Tab. 4.

Fig. 13　TID (left) and LET (right) acquired by LND prior to re - delivery to NSSC in February 2018. Higher energy deposits than those visible are expected for heavy ions (GCRs), but these are only very rarely seen on Earth at sea level (i. e., in our labs in Kiel)

Tab. 4　Properties of the various LET spectra. The first three lines are the measurements performed by LND while the following two lines are more relevant from a radiation protection point of view, as discussed in the text. Det.: detector coincidence used to deflne the LET ⟨L⟩: mean path length through detector, $\hat{L}$: median of path length, var (L): variance of pat length, g: 4π geometric factor calculated with GEANT4. 2 ∪ 3 and 1 ∪ 3 in rows 4 and 5 indicate the logical union of conditions 2 and 3 (($\overline{A}\cdot B\cdot I\cdot J$) ∪ ($A\cdot B\cdot I\cdot J$)) and of 1 and 3 (($A\cdot B\cdot C\cdot \overline{I}$) ∪ ($A\cdot B\cdot I\cdot J$))

	Det.	# bins	⟨L⟩ [μm]	$\hat{L}$ [μm]	var (L) [μm]	g [mm²sr]
1	$A\cdot B\cdot C\cdot \overline{I}$	64	526	527	367	64
2	$\overline{A}\cdot B\cdot I\cdot J$	64	619	593	15055	2019
3	$A\cdot B\cdot I\cdot J$	8×64	514	515	962	158
4	$2\cup 3\approx B\cdot J$	64	611	584	14774	2177
5	$1\cup 3\approx A\cdot B$	64	517	517	823	222

The energy for the LET is measured in the B detector and is converted to bin number as given in Eq. (8),

$$\text{bin}\# = 4\cdot \log_2 (E_B/20) \tag{8}$$

where energies (E_B and 20) are again measured in keV. Because a high LET can be due to either heavy ions or to stopping or nearly stopping ions, LND gives one LET spectrum which allows to discriminate between these two types of energy deposits in a 2-dimensional LET histogram. It is defined by

$$\text{bin}\#_x = 2 \cdot \log_2(E_I/E_A) + 4,\quad \text{bin}\#_y = 4 \cdot \log_2(E_B/20) \tag{9}$$

where energies are again measured in keV. As can be seen in Eq. (9), it uses the energy deposits in the A and I detectors in addition to the energy deposit in B, so it uses the detector combination with the smallest geometry factor and smallest path length dispersion, and thus highest resolution. The bins in the x direction are defined completely analogously to the x axis for the penetrating charged particles which are described in Sect. 4. 6.

To determine the LET spectra, one needs to divide by the average path lengths, $\langle L \rangle$, given in Table 4. The x-axis of the LET spectrum is thus given by

$$x_i = E_i/\langle L \rangle = 20 \cdot 2^{i/4}/\langle L \rangle[\text{keV}/\mu\text{m}],\quad y_i = \text{counts}_i \tag{10}$$

4. 5 Charged Particles - Ions

LND identifies charged particles by measuring the energy deposition in different combinations of its SSDs. The mean energy loss per unit path length of a charged projectile particle with nuclear charge Z_1 and velocity v in a detector is given by the well-known Bethe-Bloch equation (Bethe 1930; Bloch 1933),

$$\frac{\mathrm{d}E}{\mathrm{d}x} = -\frac{Z_1^2 e^4 n_e}{4\pi \cdot \varepsilon_0^2 \beta^2 c^2 m_e} \cdot \left[\ln\left(\frac{2m_e\beta^2c^2}{\langle E_B \rangle}\right) - \ln(1-\beta^2) - \beta^2\right] \tag{11}$$

where $\beta = v/c$, c the speed of light, E_B is the ionization energy of the target (173eV for Si), and n_e is the electron density in the target ($6.7\times10^{29}\ \text{m}^{-3}$ for Si), m_e the electron mass, and ε_0 the permittivity of free space. The expression in the square brackets simplifies to a constant plus a slowly changing ln (β^2) in the energy range covered by LND. Thus, we may approximate the energy loss in the front most detector, A, as

$$E_A \propto \frac{Z_1^2 m_1}{E_{\text{tot}}} \tag{12}$$

where E_{tot} is the measured initial energy of the stopping projectile particle, and E_A is the energy deposited in the front detector (A). The product of the total energy and the energy deposited in A (Eq. (12)) is then given by

$$Y \doteq E_{\text{tot}} \cdot E_A \propto E_{\text{tot}} \cdot \frac{Z_1^2 m_1}{E_{\text{tot}}} \propto Z_1^2 m_1 \approx \text{const.} \tag{13}$$

This quantity, Y, only depends on the particle species, i. e., its nuclear charge, Z_1, and mass, m. Except for some very weak dependencies on ln (β^2) all other quantities have can-celed out. Thus this quantity only depends on the particle's properties which allows us to discriminate even some isotopes of the various elements. Y as defined in Eq. (13) is the quantity plotted as the y-axis of the Xmas plot for particles which stop in detectors B, C, and D. Stopping particles up to a penetration depth of detector D are identified by the condition that there is no energy deposited in detector E. Because the following detectors, F, G, H, I, and J all have non-sensitive material in front of them, and detectors E, F, G, and H have only single high-gain pre-amplifiers,[9] the analogous quantity can not be used in the same manner. For them, the quantities given in Table 5 are plotted along the y-axis of the Xmas plot.

Table 5　Quantities used as the x and y-axis for the stopping charged-particle regions of the Xmas plot. Det: Detector in which the particle stops. The last line gives the quantities used for penetrating particles (column J in the Xmas plot)

Det.	x-axis
stopping in B	$8 \cdot \log_2((E_A + E_B)/E_A)$
stopping in C	$8 \cdot \log_2((E_A + E_B + E_C)/E_A) - 8$
stopping in D	$16 \cdot \log_2((E_A + E_B + E_C + E_{D1})/E_A) - 24$
stopping in E	$16 \cdot \log_2((E_A + E_B + E_C + E_{D1})/E_A) - 24$
stopping in F	$16 \cdot \log_2((E_A + E_B + E_C + E_{D1})/E_A) - 24$
stopping in G	$16 \cdot \log_2((E_A + E_B + E_C + E_{D1})/E_A) - 24$
stopping in H	$16 \cdot \log_2((E_A + E_B + E_C + E_{D1})/E_A) - 24$
stopping in I	$2 \cdot \log_2(E_I/E_A) + 8$
Det.	y-axis
stopping in B	$4 \cdot log_2((\mathrm{E_A} + \mathrm{E_B}) \cdot \mathrm{E_A}/(8000 \cdot 4000))$
stopping in C	$4 \cdot \log_2((E_A + E_B + E_C) \cdot E_A/(10000 \cdot 4000)) + 4$
stopping in D	$4 \cdot \log_2((E_A + E_B + E_C + E_{D1}) \cdot E_A/(1200 \cdot 6000)) - 8$
stopping in E	$4 \cdot \log_2((E_A + E_B + E_C + E_{D1}) \cdot E_A)/(1200 \cdot 6000)) + 2$
stopping in F	$4 \cdot \log_2((E_A + E_B + E_C + E_{D1}) \cdot E_A)/(1200 \cdot 6000)) + 7$
stopping in G	$4 \cdot \log_2((E_A + E_B + E_C + E_{D1}) \cdot E_A)/(1200 \cdot 6000)) + 7$
stopping in H	$4 \cdot \log_2((E_A + E_B + E_C + E_{D1}) \cdot E_A)/(1200 \cdot 6000)) + 8$
stopping in I	$8 \cdot \log_2((E_A + E_B + E_C + E_D + E_I)/4000)$
Penetrating Particles	
x-axis	$4 \cdot \log_2(E_i/E_A) + 16$
y-axis	$4 \cdot \log_2((E_{B1} + E_C + E_D)/100)$

The quantity plotted along the x-axis of the Xmas plot is given by the ratio of the total deposited energy, E_{tot}, and the energy measured in the front detector, E_A

$$X \doteq E_{tot}/E_A \tag{14}$$

[9] Because they are primarily used to detect thermal neutrons.

Using the approximation of Eq. (12) this ratio is proportional to $E_{tot}^2/(Z_1^2 \cdot m)$, i. e., is sensitive to the total (primary) energy of the stopping particle.

For stopping particles, the quantities for the x and y-axis are summarized in Tab. 5. As indicated by red counters in Fig. 11 the combination (X, Y) identifies individual elements or element groups for all particles stopping in detectors B through I. The grey-shaded strip marked by "Stop in B" has two such counters for protons and three each for both He isotopes. Particles with higher total energy tend to lie at larger values of $X \propto E_{tot}^2$. The neighboring strip marked by "Stop in C" shows the same kind of data for particles stopping in C, etc.

Energetic electrons need to be treated separately because of their high penetration power. A similar identification scheme is used for them, but their resulting memory addresses ((X, Y) coordinates) in the Xmas plot are given in Tab. 7, as explained in Sect. 4. 7.

4. 5. 1　Energy Ranges for Stopping Ions

As particles enter the front foil of LND and penetrate the A detector they loose energy. For a particle to be registered, it must trigger the B detector which gives the minimum energy it must have in order to be measured by LND. If it has more energy, it may penetrate the B detector and stop in C, etc. The stopping energies are given in Tab. 6. They are based on GEANT4 simulations of the LND sensor and include all relevant layers such as the Gdabsorber, its surrounding Al-housing, etc. The simulations were performed for an isotropic source on top of the LND entrance foil.

4. 6　Penetrating Charged Particles

The spectrum of energetic particles, especially the GCR, has a substantial flux of particles with energies which exceed the stopping power of LND. They are measured in the penetrating channel of LND which is defined by particles which penetrate all detectors, and stop in or penetrate the J detector. LND can not discriminate particles stopping in the last detector, J, from particles that penetrate it because J serves as the anti-coincidence detector for particles stopping in LND. Most of these penetrating particles are close to minimally ionizing, i. e., they have a nearly constant energy loss across LND. Barely penetrating particles have primary energies which result in smaller energy losses in the front detectors A, B, C, D, than in the rear detectors I and J. To discriminate barely stopping particles from penetrating particles, the Xmas plot is redefined as a 32×64 matrix with its X and Y axis coordinates computed according to the rows marked "Penetrating Particles" in Tab. 5. This is the definition of the rightmost panel of the Xmas plot marked as "Penetrate J". One easily recognizes the fully relativistic particles as a nearly straight line from the bottom to the top and the nearly stopping particles which turn to the right, i. e., towards larger values of $4 \log_2 (E_I/E_A)$. Particles which veer to the left move upwards through LND, i. e., away from the Moon. They are part of the secondary particle population which is created by the interaction of (mainly) GCRs with the lunar soil. Penetrating electrons lie at the bottom of this panel and are clearly separated from barely penetrating protons, but not from fully relativistic protons which is as expected.

4.7 Charged Particles - Electrons

The calculation of the (X, Y) addresses for electrons in the Xmas plot results in vanishing Y values and so most electron data products are plotted along the X axis ($Y = 0$) of Fig. 11.[10] Because electrons are abundant, there are also one - and ten - minute electron data products. These are indicated in Fig. 11 by E, F, GHI, and E1,. . . E4, ... H1, ... H4. The primary energy ranges corresponding to these one - and ten - minute data products are given in Tab. 7. Electrons stopping in the B, C, and D detectors are mapped to the two right - most columns of the Xmas plot, columns 273 and 274. Those addresses are calculated as $16 \cdot \log_2(T)$, where T is the total deposited energy (measured in keV). These predominantly low - energy electrons have four one - minute data products indicated by dps1 - dps4 in Fig. 11, their ranges of *deposited* energy is also given in Tab. 7. The name of these data products, dps, stands for "data product scheduler", one of the processes running in the digital FPGA.

The increasing penetration depth of electrons with increasing primary energy can be used for an initial, coarse estimate of the electron spectrum. Fig. 14 shows how this works. In the left - hand panel, Fig. 14a, the simulated counts are shown for an isotropic, E^{-1} incident electron spectrum. One easily sees that increasing primary energy of electrons results in increasing penetration depth into the detector stack. electrons need to have a kinetic energy of at last 300keV to trigger the B detector. The results for the F detector (and to a lesser degree the H detector) appear to "stick out" of the trend for the other detectors. Its geometry factor is considerably larger than that for D or E. This is also expected, because the trigger condition is $B \cdot C \cdot D \cdot E \cdot F \cdot \overline{G}$ and there is the Al - Gd - Al sandwich between the F and G detectors which stops some particles before they can reach G and trigger this anti - coincidence. The blue curve shows results for penetrating electrons with primary energies exceeding 3MeV. The right - hand panel (Fig. 14b) shows the response of the dps1 - dps4 bins of the Xmas plot to the same electron spectrum. It is dominated by electrons in the primary energy range $0.5 < E < 2\text{MeV}$. These four data products are part of the LND 1 - minute data, as are those marked E, F, and GHI in the Xmas plot.

---

[10] The transformation that defines the Xmas plot results in near - horizontal lines for ions, but for electrons it does not work as well. Nevertheless, electrons end up with small values for the Y coordinate and remain well separated from the ions. We treat these as vanishing Y - values in the Xmas plot.

Tab. 6　List of primary energy ranges that protons, ^3He, ^4He, CNO, and heavy ions need to have to stop in the LND detectors. "B dps_i", "pene i" and H10 – H14 refer to the multiple red counters for protons (ions) in the Xmas plot, Fig. 11, which are discussed in Sects. 4.9 and 4.10. Example: A proton must have a primary energy $15.9 < E < 18.4$MeV so that it will stop in D. The lower half of the table shows the name of the corresponding energy bins in the upper half of this table. The positions of these energy bins in the X – mas plot are indicated with white labels in Fig. 6. The lowest part of the table gives the minimum energy needed to stop in or penetrate J

Stop in	Proton [MeV]	^3He [MeV/nuc]	^4He [MeV/nuc]	CNO [MeV/nuc]	Heavy ions [MeV/nuc]
B dps_1	9.0 – 10.6	10.6 – 11.4	8.9 – 9.5	16.6 – 20.5	24.6 – 37.7
B dps_2	10.7 – 12.7	11.4 – 12.6	9.6 – 10.9	20.5 – 23.5	31.0 – 46.4
B dps_3	—	12.9 – 14.8	10.9 – 12.6	—	—
C	12.8 – 15.7	15.0 – 18.5	12.8 – 15.7	25.5 – 29.3	38.2 – 58.5
D	15.9 – 18.4	18.7 – 21.7	15.9 – 18.5	31.0 – 36.4	47.5 – 68.7
E	18.6 – 21.0	21.7 – 24.6[a]	18.5 – 21.0[a]	37.3 – 38.6	52.1 – 78.9
F	21.2 – 29.2	24.9 – 34.4	21.0 – 29.3	42.3 – 59.1	67.1 – 105.2
G	29.6 – 31.3	34.8 – 36.4	29.3 – 31.4	54.5 – 62.6	91.6 – 122.2
H	31.5 – 33.0	36.9 – 38.6	31.4 – 32.8	57.8 – 66.4	79.8 – 131.0
I	33.4 – 34.5	39.0 – 40.4	32.8 – 34.4	61.2 – 72.8	101.6 – 132.5
Stop in	Proton	^3He	^4He	CNO	Heavy ions
B dps_1	H1	3He1	4He1	CNO1	ions1
B dps_2	H2	3He2	4He2	CNO2	ions2
B dps_3	—	3He3	4He3	—	—
C	H3	3He4	4He4	CNO3	ions3
D	H4	3He5	4He5	CNO4	ions4
E	H5	4He6[a]	4He6[a]	CNO5	ions5
F	H6	4He7	4He7	CNO6	ions6
G	H7	4He8	4He8	CNO7	ions7
H	H8	4He9	4He9	CNO8	ions8
I	H9	4He10	4He10	CNO9	ions9
Penetrating					
pene1		56.5 – 182.9	48.6 – 180.8		
pene2				≥88.5[b]	≥150.4[b]
pene3				95.9 – 147.0	≥111.4[b]
pene4		40.7 – 56.5	34.8 – 45.9		
pene5				84.5 – 91.6	
pene6				70.3 – 93.8	78.9 – 106.4
pene7				75.3 – 78.9	105.2 – 170.7
H10	≥156.2				
H11	42.3 – 139.2	≥198.3	≥198.3		
H12					
H13					
H14	34.9 – 40.8				

[a] ^3He and ^4He share the dps boxes from detectors E to I

[b] 10% of maximum used instead of 50% maximum

Tab. 7　Deposited and primary energy ranges (E_{dep} and E_{prim}) for electrons measured in LND detectors and/or dps counters/counters. Note that these show very large variability as electrons scatter a lot. The column half max gives the FWHM range around the modeled maximum, the following column gives the range between ± 20% of the maximum. The positions of these energy bins in the Xmas plot are indicated with black labels in Fig. 11. For instance, Xmas - E3 refers to the electron region marked by E3 in the Xmas plot (rows 118 - 126, column 0). The last column gives the time resolution at which these data products are put into telemetry

Electron channel	Bin edges Low	Bin edges High	E_{dep} [MeV]		Time res. [minutes]
Column 273	Row #s				
e^-dps_1	10	17	0.5 - 0.707		1
e^-dps_2	18	25	0.707 - 1		1
e^-dps_3	26	33	1 - 1.414		1
e^-dps_4	34	41	1.414 - 2		1
Row 0	Column #s		E_{prim} [MeV] half max	E_{prim} [MeV] 20%-20%	
Xmas - E	97	126	1.3 - 2.7	1.2 - 4.3	1
Xmas - F	129	158	1.9 - 5.2	1.6 - 8.5	1
Xmas - GUHUI	160	239	3.0 - 9.0	2.7 - 16.3	1
Xmas - E1	96	96	1.5 - 4.2	1.3 - 9.8	10
Xmas - E2	97	102	1.3 - 2.8	1.2 - 5.4	10
Xmas - E3	118	126	1.5 - 2.8	1.3 - 4.3	10
Xmas - E4	127	127	2.0 - 3.4	1.8 - 4.9	10
Xmas - F1	128	128	2.2 - 6.7	1.8 - 12.3	10
Xmas - F2	129	132	2.0 - 5.3	1.7 - 9.3	10
Xmas - F3	145	158	2.1 - 5.7	1.8 - 8.9	10
Xmas - F4	159	159	3.0 - 6.2	2.4 - 8.5	10
Xmas - G1	160	160	3.1 - 7.2	2.7 - 12.9	10
Xmas - G2	161	165	2.8 - 6.5	2.6 - 10.7	10
Xmas - G3	166	172	2.8 - 6.1	2.6 - 10.2	10
Xmas - G4	173	191	2.9 - 7.5	2.7 - 9.8	10
Xmas - H1	192	192	3.6 - 16.1	2.9 - 29.7	10
Xmas - H2	193	197	3.3 - 12.5	2.8 - 24.7	10
Xmas - H3	198	204	3.1 - 12.2	2.8 - 20.5	10
Xmas - H4	205	223	3.3 - 11.6	2.9 - 22.5	10

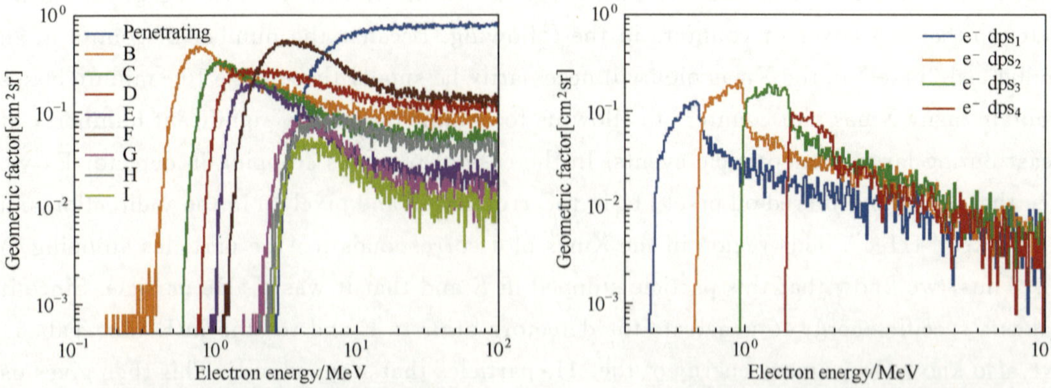

(a) These simulated count spectra show that electrons need at least 300 keV to be identified in the B, while the probability to be seen in this detector peaks at approximately 0.7 MeV. This peak depends on the incident spectrum, of course. "Deeper" detectors have higher detection thresholds, allowing for a coarse identification of the spectral shape of electrons.

(b) These simulated count spectra show that electrons need at least 0.5 MeV to be identified in the e^- dps_1 data product, while the probability to be seen in this data product peaks at approximately 0.7 MeV. This dps "threshold" increases with increasing primary energy which allows us to give a very coarse estimate of the electron spectrum, see also Fig. 14a.

Fig. 14　Electron penetration depths inside LND give information about their primary energy. The mapping is not 1∶1 though, and an inversion process as described by Eq. (15) will be needed to obtain higher fidelity electron spectra

Note that determining the "true" electron spectrum will require an inversion approach which accounts for the LND instrument response function, $K_{e^-}(E_{\text{det}}, E_{\text{prim}})$, which gives the geometry factors for an electron of primary energy E_{prim} to be measured at an energy E_{det}. This can be achieved using constrained optimization or other optimization techniques (Böhm et al. 2007; Kharytonov et al. 2009) which solve the Fredholm integral given in Eq. (15).

$$\mathbf{z}(E_{\text{det}}) = \int_0^\infty K_{e^-}(E_{\text{det}}, E_{\text{prim}}) f(E_{\text{prim}}) \mathrm{d}E_{\text{prim}} \tag{15}$$

where $f(E_{\text{prim}})$ is the primary spectrum and $\mathbf{z}(E_{\text{det}})$ are the measured counts in the energy bins E_{prim}.

4.8　1 - Hour Data Products

The Xmas plot (Fig. 11) is LND's principal data product and is accumulated over the course of an hour. As explained in Sect. 4.2, there are two Xmas plots, one active (i. e., being accumulated) and one inactive (i. e., being sent to telemetry). It contains the high-resolution data needed to discriminate between different particle species, especially the rare heavy ions. The data products contained in the Xmas plot have been discussed in Sects. 4.2 - 4.7.

4.9　10 - Minute Data Products

To allow for higher time resolution, LND reads out certain regions in the active Xmas plot more often than once an hour. These regions are shown as red boxes in Fig. 11 and we refer to them as boxes or counters in the following. Because the number of counts in an individual "pixel" in the Xmas plot will necessarily be small, these large 10 – minute boxes contain many Xmas plot counters to allow us to obtain statistically significant numbers, at least during large solar particle events. In the case of particles stopping in detector E, we see that a box is centered on pixel 112 in the x direction and pixel 20 in the y direction and is marked "4He6". This region in the Xmas plot corresponds to ^4He particles stopping in E. Thus, we know that this particle stopped in E and that it was a ^4He particle. Since it takes a specific energy to penetrate the detectors prior to E and to stop in E (see Tab. 6) we also know the primary energy of the ^4He particles that stopped in E. This then gives us a point in the differential flux of ^4He particles.

The red box which we have placed around the ^4He particles stopping in E also serves as a counter in LND. The LND processing scheduler ensures that the sum of all the counters in the individual pixels in this box is determined every 10 minutes. This sum is then put into the telemetry. Note that this counter now contains the number of counts accumulated in this box since the Xmas plot was last initialized. The number of counts in this box are thus cumulative for every 10 minute interval in an hour. The total number of counts in the past 10 minutes is thus determined by computing (on ground) the difference between the latest counter and the preceding one. These differences have already been computed in the published data. These 10 – minute "counters" are 24 – bit integer counters, no compression is performed on the counters. The data products available at 10 – minute time resolution are listed in Tab. 2.

The energy ranges for the various counter boxes are given in Tab. 6. Their locations can be seen in Fig. 11 and their meanings are obvious from it and Tab. 6.

4.10 1 – Minute Data Products

LND also provides data at a time resolution of 1 minute. The logics are exactly the same as for the 10 – minute data, counters are accumulated in boxes and read out once every minute. They are cumulative, i. e., one needs to take the difference between the current value and the preceding one to obtain the count rate. The data products available at 1 – minute time resolution are listed in Table 2. The energy ranges for the various ions are also given in Table 6, the positions of these one – minute counter boxes are shown in Fig. 11.

4.11 Dead – Time Corrections

The accumulation of a Xmas plot happens during a full hour, but is stopped for 500μs every second. Thus, the accumulation time of the Xmas plot is not 3600 seconds, but 3598.2seconds. This is a small correction, but mentioned here for completeness sake. Similarly, the *first* dps_1 data product in every hour is only accumulated for 59 seconds, all others for 60 seconds per minute, and, the *first* dps_{10} data product in one hour is only accumulated for 599 seconds, all others for 600 seconds every 10 minutes. Corrections for these small discrepancies are applied to the published data.

Tab. 8 Accumulation times and dead times to be considered when producing LND data products. The "first" are underlined because this applies only to the first dps packet per hour. Dead time corrections are discussed in Sect. 4. 11. All counters associated with a data product given in Fig. 10 contribute to its dead time and thus their sum must be multiplied by 7μs to obtain the dead time

Type of data product	Accumulation time/s
Xmas plot	3598. 2
first dps_1 per hour	59
other dps_1	60
first dps_{10} per hour	599
other dps_{10}	600
Type of measurement	Dead time [μs]
TID, LET, charged particles	(L2 [0] +L2 [1] +L2 [7]) · 7
Neutrons in C1	L2 [2] · 7
Thermal neutrons in E, F, G, H	(L2 [3] +L2 [4] +L2 [5] +L2 [6]) · 7

In addition to the correct accumulation times of the different data products, some dead time corrections need to be applied to LND's data. An approximation of the dead time that occurs while LND reads out its detectors can be calculated by multiplying the number of detector readouts by 7μs. The number of readouts can be taken from the L2 trigger counts, but it needs to be considered which L2 triggers contribute to which data product. This is easiest to understand by inspecting Fig. 10. The simplest situation is when LND detects a neutron in detector C1, as this affects only one single detector and trigger, L2 [2]. There is also only one corresponding L3 trigger, L3 [1]. The dead time is then calculated by multiplying the number of L2 [2] triggers by a dead time of 7μs. A more complicated dead time correction is that for TID because it gets contributions from L2 [0] (LET) via the L3 [3] (LET) trigger, but also the L2 [1] (TID) and L2 [7] (heavy ions) via the L3 [0] trigger path to LND's TID data product (shown in blue in Fig. 10). Thus, the dead time associated with TID is the sum of the counts in L2 [0], L2 [1], and L2 [7] multiplied by 7μs. Finally, the dead time correction for the thermal neutron channels in E, F, G and H is done using the sum of the trigger counts in L2 [3] through L2 [6]. The eight L2 counters are in telemetry at 1 minute resolution, so this correction can be made at LND's highest measuring cadence. For the data products sent at lower cadence, the L2 counters are summed up over the respective time intervals to calculate the corresponding dead time.

LND's accumulation times and dead times are summarized in Table 8. They need to be considered by correcting the nominal accumulation time, d*t* (3600, 600 or 60 seconds, respectively), to

$$dt_{true} = \text{accumulation time} - \text{dead time} \tag{16}$$

where the accumulation time and dead time are taken from Table 8. Count rates are thus corrected for dead - time effects using the following expression,

$$\text{true count rate} = \text{measured count rate} \cdot \frac{dt}{dt_{true}} \tag{17}$$

where dt is the nominal, uncorrected accumulation time for a data product. Dead time corrections have been applied to the published data.

4.12 Calibration of LND

The LND data products are provided in calibrated values, i. e., after all conversions from instrumental measurement units to the reported physical units have been applied. The calibration of LND was performed using the flight model (FM), the flight spare (FS) unit, and also the engineering model (EM). Because LND provides very diverse measurements and data products, its calibration was an extended effort which included a number of facilities in Europe and China. These are summarized in Tab. 9 and will be published elsewhere in the literature. LND itself was not calibrated with fast neutrons, but uses the calibration of the Flight Radiation Environment Detector (FRED, Möller 2013) which also served as a prototype for the fast neutron detection capabilities of LND. FRED was calibrated with neutrons at the Physikalisch - Technische Bundesanstalt (PTB) in Braunschweig (5MeV $< E_n <$ 19MeV) and with higher energy neutrons at iThemba in South Africa (Möller 2013). FRED contains near - identical detectors B, C, D in the same stack configuration as LND. The main difference lies in their thicknesses (300μm in FRED and 500μm in LND). These are accounted for by modeling the slightly different responses with GEANT4.

Tab. 9　Overview of calibrations performed with LND.
FM: flight model, FS: flight spare, EM: engineering model

Calibration		LND model	Time frame
Facility	Purpose		
CAU Kiel	Bi - 207 gammas	FM, FS, EM	Summer 2017
CERN/CERF	dose	FM	2017 - 06 - 17 - 2017 - 06 - 18
HIMAC	ions: H, He, C, O, Ar	FM	2017 - 06 - 30 - 2017 - 07 - 04
ATI	thermal neutrons	FM, FS	2017 - 09 - 25 - 2017 - 09 - 28
18th Institute	RTG & RHU background	FS	2018 - 07 - 09 - 2018 - 07 - 12
NSSC	electrons	EM	2018 - 08 - 17 - 2018 - 08 - 20

Acknowledgements

The Lunar Lander Neutron and Dosimetry (LND) instrument is supported by the

German Aerospace Center, DLR Space Administration, Bonn, under grant 50 JR 1604 to the Christian – Albrechts – University (CAU) Kiel and supported by Beijing Municipal Science and Technology Commission, Grant No. Z181100002918003 to the National Space Science Center (NSSC) and the National Natural Science Foundation of China Grant: 41941001. J. G. is supported by the Strategic Priority Program of the Chinese Academy of Sciences (Grant No. XDB41000000) and the CNSA pre – research Project on Civil Aerospace Technologies (Grant No. D020104). We would like to thank the following facilities for allowing us to calibrate LND: CERN/CERF in Geneva, Switzerland; HIMAC in Chiba, Japan; ATI in Vienna, Austria; SEF in NSSC, Beijing, China. We thank Oliver Angerer at DLR for his unwavering support of LND. We also thank the LESEC/CAST their joint effort for LND and the anonymous reviewer for the very helpful comments.

References

[1] D. Abdushukurov, Gadolinium Foils as Converters of Thermal Neutrons in Detectors of Nuclear Radiation. Physics Research and Technology (2010). Nova Science Publishers, Incorporated. http: //books. google. de/books? id=SqmkSgAACAAJ.

[2] A. M. Adamcyk, M. S. Clowdsley, G. D. Qualls, S. B. Blattnig, K. T. Lee, D. J. Fry, N. N. Stoffle, L. C. Simonsen, T. C. Slaba, S. A. Walker, in IEEE Aerospace Conference 2011, March 05, 2011 – March 12, 2011, Big Sky, MT, United States (2011). NASA Johnson Space Center, Houston, TX, United States, documents number JSC – CN – 21080.

[3] S. Agostinelli, J. Allison, K. Amako, J. Apostolakis, H. Araujo, P. Arce, M. Asai, D. Axen, S. Banerjee, G. Barrand, F. Behner, L. Bellagamba, J. Boudreau, L. Broglia, A. Brunengo, H. Burkhardt, S. Chauvie, J. Chuma, R. Chytracek, G. Cooperman, G. Cosmo, P. Degtyarenko, A. dell'Acqua, G. Depaola, D. Diet – rich, R. Enami, A. Feliciello, C. Ferguson, H. Fesefeldt, G. Folger, F. Foppiano, A. Forti, S. Garelli, S. Giani, R. Giannitrapani, D. Gibin, J. J. Gómez Cadenas, I. González, G. Gracia Abril, G. Greeniaus, W. Greiner, V. Grichine, A. Grossheim, S. Guatelli, P. Gumplinger, R. Hamatsu, K. Hashimoto, H. Hasui, A. Heikkinen, A. Howard, V. Ivanchenko, A. Johnson, F. W. Jones, J. Kallenbach, N. Kanaya, M. Kawabata, Y. Kawabata, M. Kawaguti, S. Kelner, P. Kent, A. Kimura, T. Kodama, R. Kokoulin, M. Kossov, H. Kurashige, E. Lamanna, T. Lampén, V. Lara, V. Lefebure, F. Lei, M. Liendl, W. Lockman, F. Longo, S. Magni, M. Maire, E. Medernach, K. Minamimoto, P. Mora de Freitas, Y. Morita, K. Murakami, M. Nagamatu, R. Nartallo, P. Nieminen, T. Nishimura, K. Ohtsubo, M. Okamura, S. O'Neale, Y. Oohata, K. Paech, J. Perl, A. Pfeiffer, M. G. Pia, F. Ranjard, A. Rybin, S. Sadilov, E. di Salvo, G. Santin, T. Sasaki, N. Savvas, Y. Sawada, S. Scherer, S. Sei, V. Sirotenko, D. Smith, N. Starkov, H. Stoecker, J.

Sulkimo, M. Takahata, S. Tanaka, E. Tcherniaev, E. Safai Tehrani, M. Tropeano, P. Truscott, H. Uno, L. Urban, P. Urban, M. Verderi, A. Walkden, W. Wander, H. Weber, J. P. Wellisch, T. Wenaus, D. C. Williams, D. Wright, T. Yamada, H. Yoshida, D. Zschiesche (Geant4 Collaboration), Nucl. Instrum. Methods Phys. Res., Sect. A 506, 250 (2003).

[4] J. Allison, K. Amako, J. Apostolakis, H. Araujo, P. A. Dubois, M. Asai, G. Barrand, R. Capra, S. Chauvie, R. Chytracek, G. A. P. Cirrone, G. Cooperman, G. Cosmo, G. Cuttone, G. G. Daquino, M. Donszelmann, M. Dressel, G. Folger, F. Foppiano, J. Generowicz, V. Grichine, S. Guatelli, P. Gumplinger, A. Heikkinen, I. Hrivnacova, A. Howard, S. Incerti, V. Ivanchenko, T. Johnson, F. Jones, T. Koi, R. Kokoulin, M. Kossov, H. Kurashige, V. Lara, S. Larsson, F. Lei, O. Link, F. Longo, M. Maire, A. Mantero, B. Mascialino, I. McLaren, P. M. Lorenzo, K. Minamimoto, K. Murakami, P. Nieminen, L. Pandola, S. Parlati, L. Peralta, J. Perl, A. Pfeiffer, M. G. Pia, A. Ribon, P. Rodrigues, G. Russo, S. Sadilov, G. Santin, T. Sasaki, D. Smith, N. Starkov, S. Tanaka, E. Tcherniaev, B. Tome, A. Trindade, P. Truscott, L. Urban, M. Verderi, A. Walkden, J. P. Wellisch, D. C. Williams, D. Wright, H. Yoshida, IEEE Trans. Nucl. Sci. 53 (1), 270 (2006). https://doi.org/10.1109/TNS.2006.869826.

[5] J. Allison, K. Amako, J. Apostolakis, P. Arce, M. Asai, T. Aso, E. Bagli, A. Bagulya, S. Banerjee, G. Barrand, B. Beck, A. Bogdanov, D. Brandt, J. Brown, H. Burkhardt, P. Canal, D. Cano - Ott, S. Chauvie, K. Cho, G. Cirrone, G. Cooperman, M. Cortés-Giraldo, G. Cosmo, G. Cuttone, G. Depaola, L. Desorgher, X. Dong, A. Dotti, V. Elvira, G. Folger, Z. Francis, A. Galoyan, L. Garnier, M. Gayer, K. Genser, V. Grichine, S. Guatelli, P. Guèye, P. Gumplinger, A. Howard, I. Hřivnáčová, S. Hwang, S. Incerti, A. Ivanchenko, V. Ivanchenko, F. Jones, S. Jun, P. Kaitaniemi, N. Karakatsanis, M. Karamitros, M. Kelsey, A. Kimura, T. Koi, H. Kurashige, A. Lechner, S. Lee, F. Longo, M. Maire, D. Mancusi, A. Mantero, E. Mendoza, B. Morgan, K. Murakami, T. Nikitina, L. Pandola, P. Paprocki, J. Perl, I. Petrović, M. Pia, W. Pokorski, J. Quesada, M. Raine, M. Reis, A. Ribon, A. R. Fira, F. Romano, G. Russo, G. Santin, T. Sasaki, D. Sawkey, J. Shin, I. Strakovsky, A. Taborda, S. Tanaka, B. Tomé, T. Toshito, H. Tran, P. Truscott, L. Urban, V. Uzhinsky, J. Verbeke, M. Verderi, B. Wendt, H. Wenzel, D. Wright, D. Wright, T. Yamashita, J. Yarba, H. Yoshida, Nucl. Instrum. Methods Phys. Res., Sect. A, Accel. Spectrom. Detect. Assoc. Equip. 835, 186 (2016). https://doi.org/10.1016/j.nima.2016.06.125.

[6] D. W. Anderson, A. J. Bureau, B. C. Cook, T. J. Englert, Phys. Rev. 183, 978 (1969). https://doi.org/10.1103/PhysRev.183.978.

[7] G. Baiocco, S. Barbieri, G. Babini, J. Morini, D. Alloni, W. Friedland, P. Kundrát, E. Schmitt, M. Puchalska, L. Sihver, Sci. Rep. 6, 34033 (2016). https://doi.org/

10. 1038/srep34033.

[8] T. Berger, B. Przybyla, D. Matthiä, G. Reitz, S. Burmeister, J. Labrenz, P. Bilski, T. Horwacik, A. Twardak, M. Hajek, M. Fugger, C. Hofstätter, L. Sihver, J. K. Palfalvi, J. Szabo, A. Stradi, I. Ambrozova, J. Kubancak, K. P. Brabcova, F. Vanhavere, V. Cauwels, O. Van Hoey, W. Schoonjans, A. Parisi, R. Gaza, E. Semones, E. G. Yukihara, E. R. Benton, B. A. Doull, Y. Uchihori, S. Kodaira, H. Kitamura, M. Boehme, J. Space Weather Space Clim. 6, A39 (2016). https://doi. org/10. 1051/swsc/2016034.

[9] T. Berger, S. Burmeister, D. Matthiä, B. Przybyla, G. Reitz, P. Bilski, M. Hajek, L. Sihver, J. Szabo, I. Ambrozova, F. Vanhavere, R. Gaza, E. Semones, E. G. Yukihara, E. R. Benton, Y. Uchihori, S. Kodaira, H. Kitamura, M. Boehme, J. Space Weather Space Clim. 7, A8 (2017). https://doi. org/10. 1051/swsc/2017005.

[10] H. A. Bethe, Ann. Phys. 5, 325 (1930).

[11] F. Bloch, Ann. Phys. 16, 285 (1933).

[12] E. Böhm, A. Kharytonov, R. F. Wimmer - Schweingruber, Astron. Astrophys. 473, 673 (2007). https://doi. org/10. 1051/0004 - 6361: 20077216.

[13] F. A. Cucinotta, L. J. Chappell, Radiat. Res. 176, 102 (2011). https://doi. org/10. 1667/RR2540. 1.

[14] T. P. Dachev, B. T. Tomov, Y. N. Matviichuk, P. S. Dimitrov, S. V. Vadawale, J. N. Goswami, G. De Angelis, V. Girish, Adv. Space Res. 48, 779 (2011). https://doi. org/10. 1016/j. asr. 2011. 05. 009.

[15] R. C. Elphic, D. J. Lawrence, W. C. Feldman, B. L. Barraclough, S. Maurice, A. B. Binder, P. G. Lucey, Science 281, 1493 (1998). https://doi. org/10. 1126/science. 281. 5382. 1493.

[16] O. Eugster, F. Tera, D. S. Burnett, G. J. Wasserburg, Earth Planet. Sci. Lett. 8, 20 (1970). https://doi. org/10. 1016/0012 - 821X (70) 90094 - 4.

[17] W. C. Feldman, B. L. Barraclough, S. Maurice, R. C. Elphic, D. J. Lawrence, D. R. Thomsen, A. B. Binder, Science 281, 1489 (1998a). https://doi. org/10. 1126/science. 281. 5382. 1489.

[18] W. C. Feldman, S. Maurice, A. B. Binder, B. L. Barraclough, R. C. Elphic, D. J. Lawrence, Science 281, 1496 (1998b). https://doi. org/10. 1126/science. 281. 5382. 1496.

[19] W. C. Feldman, B. L. Barraclough, K. R. Fuller, D. J. Lawrence, S. Maurice, M. C. Miller, T. H. Prettyman, A. B. Binder, Nucl. Instrum. Methods Phys. Res., Sect. A 422 (1 - 3), 562 (1999). https://doi. org/10. 1016/S0168 - 9002 (98) 00934 - 6.

[20] W. C. Feldman, D. J. Lawrence, R. C. Elphic, D. T. Vaniman, D. R. Thomsen, B. L. Barraclough, S. Maurice, A. B. Binder, J. Geophys. Res. 105 (E8), 20347

(2000). https: //doi. org/10. 1029/1999JE001183.

[21] W. C. Feldman, S. Maurice, D. J. Lawrence, R. C. Little, S. L. Lawson, O. Gasnault, R. C. Wiens, B. L. Barraclough, R. C. Elphic, T. H. Prettyman, J. T. Steinberg, A. B. Binder, J. Geophys. Res. 106, 23231 (2001). https: //doi. org/10. 1029/2000JE001444.

[22] B. Fogelberg, A. Bäcklin, Conversion electron study of the reaction ^{157}Gd $(n, \gamma)^{158}$ Gd. Tech. rep. The Swedish Research Councils's Laboratory (1972).

[23] O. Gasnault, C. d'Uston, W. C. Feldman, S. Maurice, J. Geophys. Res. 105 (E2), 4263 (2000). https: //doi. org/10. 1029/1999JE001124.

[24] M. Gordon, P. Goldhagen, K. Rodbell, T. Zabel, H. Tang, J. Clem, P. Bailey, IEEE Trans. Nucl. Sci. 51 (6), 3427 (2004).

[25] J. Guo, C. Zeitlin, R. Wimmer - Schweingruber, D. M. Hassler, J. Köhler, B. Ehresmann, S. Böttcher, E. Böhm, D. E. Brinza, Life Sci. Space Res. 14, 12 (2017). https: //doi. org/10. 1016/j. lssr. 2017. 06. 001.

[26] J. Guo, C. Zeitlin, R. F. Wimmer - Schweingruber, T. McDole, P. Kühl, J. C. Appel, D. Matthiä, J. Krauss, J. Köhler, Astron. J. 155 (1), 49 (2018).

[27] S. Haino, T. Sanuki, K. Abe, K. Anraku, Y. Asaoka, H. Fuke, M. Imori, A. Itasaki, T. Maeno, Y. Makida et al., Phys. Lett. B 594 (1 - 2), 35 (2004).

[28] J. W. Head, S. Murchie, J. F. Mustard, C. M. Pieters, G. Neukum, A. McEwen, R. Greeley, E. Nagel, M. J. S. Belton, J. Geophys. Res. 98, 17149 (1993). https: //doi. org/10. 1029/93JE01278.

[29] G. H. Heiken, D. T. Vaniman, B. M. French, Lunar Sourcebook, a User's Guide to the Moon (1991).

[30] D. Hou, S. Zhang, J. Yu, R. Wimmer - Schweingruber, S. Burmeister, H. Lohf, B. Yuan, G. Shen, C. Wang, X. Hou, B. Ren, J. Instrum. 15 (01), P01032 (2020). https: //doi. org/10. 1088/1748 - 0221/15/01/p01032.

[31] J. Huang, Z. Xiao, J. Flahaut, M. Martinot, J. Head, X. Xiao, M. Xie, L. Xiao, J. Geophys. Res., Planets 123 (7), 1684 (2018). https: //doi. org/10. 1029/2018JE005577.

[32] A. Kharytonov, E. Böhm, R. F. Wimmer - Schweingruber, Astron. Astrophys. 495, 663 (2009). https: //doi. org/10. 1051/0004 - 6361: 200810854.

[33] J. Köhler, B. Ehresmann, C. Martin, E. Böhm, A. Kharytonov, O. Kortmann, C. Zeitlin, D. M. Hassler, R. F. Wimmer - Schweingruber, Nucl. Instrum. Methods Phys. Res., Sect. B 269, 2641 (2011). https: //doi. org/10. 1016/j. nimb. 2011. 07. 021.

[34] J. Köhler, C. Zeitlin, B. Ehresmann, R. F. Wimmer - Schweingruber, D. M. Hassler, G. Reitz, D. E. Brinza, G. Weigle, J. Appel, S. Böttcher, E. Böhm, S. Burmeister, J. Guo, C. Martin, A. Posner, S. Rafkin, O. Kortmann, J. Geophys. Res., Planets (2014). https: //doi. org/10. 1002/2013JE004539.

[35] J. Kremer, M. Boezio, M. e. Ambriola, G. Barbiellini, S. Bartalucci, R. Bellotti,

D. Bergström, U. Bravar, F. Cafagna, P. Carlson et al., Phys. Rev. Lett. 83 (21), 4241 (1999).

[36] J. Labrenz, S. Burmeister, T. Berger, B. Heber, G. Reitz, J. Space Weather Space Clim. 5 (27), A38 (2015). https://doi.org/10.1051/swsc/2015039.

[37] D. J. Lawrence, W. C. Feldman, R. C. Elphic, R. C. Little, T. H. Prettyman, S. Maurice, P. G. Lucey, A. B. Binder, J. Geophys. Res., Planets 107, 5130 (2002). https://doi.org/10.1029/2001JE001530.

[38] G. Leinweber, D. P. Barry, M. J. Trbovich, J. A. Burke, N. J. Drindak, H. D. Knox, R. V. Ballad, R. C. Block, Y. Danon, L. I. Severnyak, Nucl. Sci. Eng. 154 (3), 261 (2006). https://doi.org/10.13182/NSE05-64.

[39] R. E. Lingenfelter, E. H. Canfleld, W. N. Hess, J. Geophys. Res. 66 (9), 2665 (1961). https://doi.org/10.1029/JZ066i009p02665.

[40] J. Liu, X. Ren, W. Yan, C. Li, H. Zhang, Y. Jia, X. Zeng, W. Chen, X. Gao, D. Liu, X. Tan, X. Zhang, T. Ni, H. Zhang, Z. Wei, Y. Su, W. Wen, Nat. Commun. 10, 4229 (2019). https://doi.org/10.1038/s41467-019-12278-3.

[41] M. D. Looper, J. E. Mazur, J. B. Blake, H. E. Spence, N. A. Schwadron, M. J. Golightly, A. W. Case, J. C. Kasper, L. W. Townsend, Space Weather 11 (4), 142 (2013). https://doi.org/10.1002/swe.20034.

[42] D. Matthiä, T. Berger, Life Sci. Space Res. 14, 57 (2017). https://doi.org/10.1016/j.lssr.2017.03.005. Radiation on the Martian Surface: Model Comparisons with Data from the Radiation Assessment Detector on the Mars Science Laboratory (MSL/RAD): Results from the 1st Mars Space Radiation Modeling Workshop.

[43] D. Matthiä, D. M. Hassler, W. de Wet, B. Ehresmann, A. Firan, J. Flores-McLaughlin, J. Guo, L. H. Heilbronn, K. Lee, H. Ratliff, R. R. Rios, T. C. Slaba, M. Smith, N. N. Stoffle, L. W. Townsend, T. Berger, G. Reitz, R. F. Wimmer-Schweingruber, C. Zeitlin, Life Sci. Space Res. 14, 18 (2017). https://doi.org/10.1016/j.lssr.2017.06.003.

[44] S. Maurice, W. C. Feldman, D. J. Lawrence, R. C. Elphic, O. Gasnault, C. d'Uston, I. Genetay, P. G. Lucey, J. Geophys. Res. 105 (E8), 20365 (2000). https://doi.org/10.1029/1999JE001151.

[45] R. A. Mewaldt, A. J. Davis, K. A. Lave, R. A. Leske, E. C. Stone, M. E. Wiedenbeck, W. R. Binns, E. R. Christian, A. C. Cummings, G. A. de Nolfo, M. H. Israel, A. W. Labrador, T. T. von Rosenvinge, Astrophys. J. Lett. 723, L1 (2010). https://doi.org/10.1088/2041-8205/723/1/L1.

[46] T. Möller, Measurement of the radiation exposure for high altitude flights in the polar region. Dissertation. Christian-Albrechts-Universität zu Kiel, Kiel, Germany (2013).

[47] T. Möller, B. Ehresmann, J. Labrenz, L. Panitzsch, in 20th Symposium on European Rocket and Balloon Programmes and Related Research, ed. by L.

Ouwehand. ESA Special Publication, vol. 700 (2011), pp. 641 - 644.

[48] T. Möller, B. Ehresmann, J. Labrenz, L. Panitzsch, in 21st ESA Symposium on European Rocket and Balloon Programmes and Related Research. ESA Special Publications (2013a), https://doi.org/10.13140/2.1.4096.2565, 978-92-9092-285-8

[49] T. Möller, T. Berger, S. Böttcher, S. Burmeister, B. Ehresmann, B. Heber, J. Labrenz, L. Panitzsch, R. Wimmer-Schweingruber, in Proc. Int. Cosmic Ray Conf. (2013b). http://www.cbpf.br/~icrc2013/papers/icrc2013-0422.pdf.

[50] A. I. Mrigakshi, D. Matthiä, T. Berger, G. Reitz, R. F. Wimmer-Schweingruber, J. Geophys. Res. Space Phys. 117, A08109 (2012). https://doi.org/10.1029/2012JA017611.

[51] H. Neher, J. Geophys. Res. 76 (7), 1637 (1971).

[52] A. Ottolenghi, V. Smyth, K. Trott, Radiat. Meas. 57, 68 (2013).

[53] R. W. K. Potter, G. S. Collins, W. S. Kiefer, P. J. McGovern, D. A. Kring, Icarus 220, 730 (2012). https://doi.org/10.1016/j.icarus.2012.05.032.

[54] D. V. Reames, Sol. Phys. 289, 977 (2014). https://doi.org/10.1007/s11207-013-0350-4.

[55] G. Reitz, R. Beaujean, E. Benton, S. Burmeister, T. Dachev, S. Deme, M. Luszik-Bhadra, P. Olko, Radiat. Prot. Dosim. 116, 374 (2005). https://doi.org/10.1093/rpd/nci262.

[56] G. Reitz, T. Berger, D. Matthiae, Planet. Space Sci. 74, 78 (2012). https://doi.org/10.1016/j.pss.2012.07.014

[57] H. E. Spence (Crater Science Team), in Annual Meeting of the Lunar Exploration Analysis Group. LPI Contributions, vol. 1595 (2010), p. 66.

[58] H. E. Spence, A. W. Case, M. J. Golightly, T. Heine, B. A. Larsen, J. B. Blake, P. Caranza, W. R. Crain, J. George, M. Lalic, A. Lin, M. D. Looper, J. E. Mazur, D. Salvaggio, J. C. Kasper, T. J. Stubbs, M. Doucette, P. Ford, R. Foster, R. Goeke, D. Gordon, B. Klatt, J. O'Connor, M. Smith, T. Onsager, C. Zeitlin, L. W. Townsend, Y. Charara, Space Sci. Rev. 150, 243 (2010). https://doi.org/10.1007/s11214-009-9584-8.

[59] C. Streffer, Radiat. Prot. Dosim. 127 (1-4), 2 (2007). https://doi.org/10.1093/rpd/ncm246.

[60] P. Tove, W. Seibt, Nucl. Instrum. Methods 51 (2), 261 (1967). https://doi.org/10.1016/0029-554X (67) 90012-2.

[61] J. W. Wilson, B. M. Anderson, F. A. Cucinotta, J. Ware, C. Zeitlin, in SAE Technical Paper (SAE International, 2006). https://doi.org/10.4271/2006-01-2110.

[62] J. K. Wilson, H. E. Spence, J. Kasper, M. Golightly, J. Bern Blake, J. E. Mazur, L. W. Townsend, A. W. Case, M. Dixon Looper, C. Zeitlin, N. A. Schwadron, J.

Geophys. Res., Planets 117,E00H23(2012). https://doi. org/10. 1029/2011JE003921.

[63] R. F. Wimmer - Schweingruber, J. Koehler, E. C. Gonzales, B. Heber, S. Boettcher, S. Burmeister, L. Seimetz, B. Schuster, in AGU Fall Meeting Abstracts, vol. 2008 (2008), pp. P53C - 1458.

[64] R. F. Wimmer - Schweingruber, J. Köhler, D. M. Hassler, J. Guo, J. Appel, C. Zeitlin, E. Böhm, B. Ehresmann, H. Lohf, S. I. Böttcher, S. Burmeister, C. Martin, A. Kharytonov, D. E. Brinza, A. Posner, G. Reitz, D. Matthiä, S. Rafkin, G. Weigle, F. Cucinotta, Geophysical Research Letters (2015). https://doi. org/10. 1002/2015GL066664.

[65] D. S. Woolum, D. S. Burnett, M. Furst, J. R. Weiss, Moon 12, 231 (1975). https://doi. org/10. 1007/BF00577879.

[66] W. Wu, Y. Tang, L. Zhang, D. Qiao, Sci. China Inf. Sci. 61, 040305 (2018). https://doi. org/10. 1007/s11432 - 017 - 9202 - 1.

[67] L. Xu, Y. L. Zou, J. Wu, in Lunar and Planetary Science Conference. Lunar and Planetary Inst. Technical Report, vol. 49 (2018), p. 1856.

[68] P. Ye, Z. Sun, H. Zhang, F. Li, Sci. China, Technol. Sci. 60, 658 (2017). https://doi. org/10. 1007/s11431 - 016 - 9034 - 6.

[69] S. Zhang, R. F. Wimmer - Schweingruber, J. Yu, C. Wang, Q. Fu, Z. Yxhas, Y. Sun, W. C, D. Hou, B. S I, S. Burmeister, L. Seimetz, B. Schuster, G. Shen, B. Yuan, H. Lohf, J. Guo, Z. Xu, J. von Forstner, S. R. Kulkarni, H. Xu, C. Xue, J. Li, Z. Zhang, H. Zhang, T. Berger, C. Hellweg, D. Matthiä, X. Hou, B. Ren, Z. Chang, B. Zhang, Y. Chen, H. Geng, Z. Quan, Sci. Adv. (2020, in press). https://doi. org/10. 1126/sciadv. aaz1334.

Inside a Lunar Mini – Magnetosphere: First Energetic Neutral Atom Measurements on the Lunar Surface

Lianghai Xie[1,2,3], Lei Li[1,2,3], Aibing Zhang[2,4], Yiteng Zhang[1,2,3], Jinbin Cao[5], Martin Wieser[6], Xiaoping Zhang[7], Jingdong Wang[1,2,3], Bin Zhou[1,2,3], Yongyong Feng[1,2,3], Linggao Kong[2], Yong Li[7], and Fang Wang[8]

1. State Key Laboratory of Space Weather, National Space Science Center, Chinese Academy of Sciences, Beijing, China
2. National Space Science Center, Chinese Academy of Sciences, Beijing, China
3. Joint Research and Development Center of Chinese Science Academy and Shen County, Shandong, China
4. University of Chinese Academy of Sciences, Beijing, China
5. School of Space and Environment, Beihang University, Beijing, China
6. Swedish Institute of Space Physics, Kiruna, Sweden
7. State Key Laboratory of Lunar and Planetary Sciences, Macau University of Science and Technology, Macau, China
8. National Astronomical Observatories of China, Chinese Academy of Sciences, Beijing, China

Abstract: Due to lack of in situ observations, people still know little about the formation conditions and the internal structure of lunar mini – magnetosphere. Here we present the first energetic neutral atom (ENA) measurement on lunar surface by the Chang'E – 4 mission. It is found that when downstream from the Imbrium antipode magnetic anomalies, the measured ENA fluxes are generally decreased, suggesting a mini – magnetosphere that can shield the lunar surface from the solar wind. Further studies show that the shielding effect can only be found when the ion inertia length is small (<120km) and some solar wind ions can penetrate into the mini – magnetosphere with a penetrating efficiency positively correlated with the solar wind dynamic pressure. When the ion inertia length is large, no shielding effect can be found. With the help of Hall MHD simulations, we confirm the existence of a mini – magnetosphere and the shielding effect anti – correlated with ion inertia length.

Plain Language Summary: Different from other airless bodies, the Moon possesses a large number of crustal magnetic fields, known as magnetic anomalies. The interaction between the solar wind and the magnetic anomalies may form a small – scale magnetosphere, called lunar mini - magnetosphere (LMM). So far, there have been only a few observations of LMM from orbit, and the truth on the ground is still unknown. Here we present the first observations of the LMM on the lunar surface by Chang'E - 4, which find some specific characteristics associated with LMM: (a) The LMM is usually found in a dense solar wind with a small ion inertia length; (b) the LMM cannot fully shield the lunar surface from the solar wind ions; (c) more solar wind ions can penetrate the LMM when the solar wind dynamic pressure is higher. These findings can motivate future studies about LMM and also be helpful to understand the formation of mini - magnetospheres elsewhere.

1 Introduction

Without a global magnetic field or a significant atmosphere, the Moon was previously thought to have a simple interaction with the solar wind, in which the impinging solar wind ions could be completely absorbed by the lunar surface on the dayside, leaving a lunar wake on the nightside (Colburn et al., 1971; Ness et al., 1968). However, later observations showed that the Moon possesses a large number of local magnetic anomalies, which were first observed by the Apollo missions (Dyal et al., 1970, 1974), and then by the Lunar Prospector (LP) and the Kaguya missions (Hood et al., 2001; Mitchell et al., 2008; Tsunakawa et al., 2015). Some magnetic anomalies have surface fields of hundreds of nT, which are strong enough to deflect the solar wind ions and form a magnetosphere - like structure called the lunar mini - magnetosphere (LMM). The LMM is defined as a small - scale magnetic structure caused by the solar wind interaction with the local magnetic anomalies, which can shield the lunar surface from the solar wind. Strictly speaking, both a shock and a plasma density cavity are needed to identify a LMM. Previously, the bow shocks associated with the magnetic anomalies were first observed by LP (Halekas et al., 2006; Lin et al., 1998) and then confirmed by ARTEMIS (Halekas et al., 2014). However, the expected plasma density cavity has only been seen by LP on two consecutive orbits under some special conditions, including low altitude, high solar zenith angle (SZA), and short ion inertia length (Halekas et al., 2008). It seems that the density cavity is not easy to form in LMM, possibly caused by ion penetration on this small scale. On the other hand, the density cavity may only form at a very low altitude near the lunar surface, which is too low to be seen by an orbiter. Recently, both Kaguya and Chandrayaan - 1 observed a large number of solar wind protons reflected from the area with strong magnetic anomalies (Lue et al., 2011; Saito et al., 2010). Meanwhile, Chandrayaan - 1 found that less solar wind ions could reach the lunar surface to be backscattered as energetic neutral atoms (ENAs) near the same area, leaving a void in the ENA image (Vorburger et al., 2012, 2015; Wieser et al., 2010). These observations suggest that at least some lunar magnetic anomalies can generate a LMM to shield the surface from the solar wind.

One question is under what conditions the lunar magnetic anomalies can form a mini - magnetosphere. According to the hybrid simulations of the solar wind interaction with dipoles with a range of strengths (Omidi et al., 2002), the field strengths of the lunar magnetic anomalies are not strong enough to generate a shock - like interaction, but can only bring some whistler or magnetosonic waves. However, later studies showed that the non - dipolar nature of the lunar crustal fields makes them have a larger lateral extent than a dipole, which could enhance the efficiency to stand off the solar wind and reduce the field strength needed to form a mini - magnetosphere (Harnett & Winglee, 2003). Such a multi - dipole effect was also found in the recent Hall MHD simulations done by Xie et al. (2015), in which the lunar crustal field was modeled by five dipoles with a surface strength of 480 nT and a mini - magnetosphere could form even near the subsolar region. Furthermore, Xie et al. (2015) found that the formation of LMM favors higher SZA and smaller ion inertia length, and the simulation results are consistent with the observations by both LP and Chang'E - 2. Additionally, the reflected solar wind ions can help to slow down and compress the solar wind flow, and also play a role in the formation of LMM (Fatemi et al., 2014; Halekas et al., 2014).

The other question: What are the differences between the LMM and the Earth's magnetosphere? The typical size of the LMM is about 100km, which is far smaller than that of the Earth's magnetosphere and comparable to the inertia length of the solar wind protons. On such a scale, the magnetic field lines are frozen in the electron fluid, instead of in the ion fluid due to the Hall effect, and the ions can penetrate across the field lines at a distance depending on the ion gyroradius. An anti - moonward electric field will be established due to the decoupling of ions from electrons, as observed by Kaguya (Saito et al., 2012), which can accelerate electrons but decelerate ions, and help to form a mini - magnetosphere. Nevertheless, the ion penetration can cause a decrease in the jump of velocity across the magnetic obstacle, which may be unfavorable for the formation of a shock. Additionally, the ion penetration can reduce magnetic shielding and compensate plasma depletion inside the mini - magnetosphere. For this reason, most observations of the LMM only find shock - like magnetic enhancements, but no density cavity (Halekas et al., 2006). So far, people still know little about the internal structure of the LMM, which has only been observed several times with the ENA measurements. Wieser et al. (2010) found that the backscattered ENA flux could be reduced by about 50% due to the magnetic shielding associated with the LMM. Furthermore, Vorburger et al. (2012) found that the shielding efficiency favors lower solar wind dynamic pressure. However, all these measurements are obtained from orbit, and the truth on the ground is still unknown. In this study, we present the first ENA observations relevant to the LMM on the lunar surface by Chang'E - 4 (CE - 4) mission. Moreover, the formation condition of the LMM as well as its shielding effect will be discussed.

2　Observations

CE－4 is the first mission landing on the far side of the Moon. As shown in Fig. 1a, the landing site (177.6°E, 45.4°S) (Liu et al., 2019) is near the southeast edge of the largest magnetic anomalies called Imbrium antipode, which is centered at 162°E, 33°S with a radius of more than 600km (Mitchell et al., 2008). Previously, the LP spacecraft found evidence of a mini－magnetosphere near the Imbrium antipode at SZA about 60° and with a small ion inertia length of about 55km (Lin et al., 1998). Additionally, the ENA meas－urements from Chandrayaan－1 indicated that the shielding efficiency of the mini－magnetosphere near the Imbrium antipode could be in a range of 27%-74% (Vorburger et al., 2012). For the CE－4 mission, it would be possible to see the tail of the mini－magnetosphere if the solar wind flows from the west (Fig. 1a). The Advanced Small Analyzer for Neutrals (ASAN) onboard the rover of CE－4 is an ENA analyzer, which can measure the ENAs

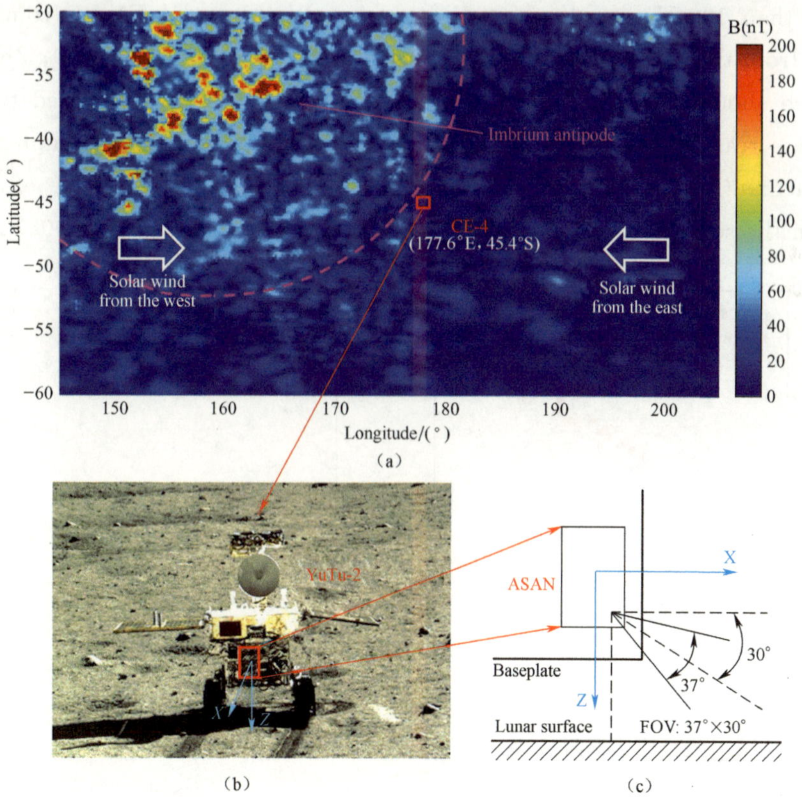

Fig. 1　(a) The map of crustal magnetic fields around the CE－4's landing site. The magnetic fields are plotted according to the 450° spherical harmonic model of the lunar crustal magnetic fields (Tsunakawa et al., 2015). The magenta circle shows the region of the Imbrium antipode and the small red rectangle marks the position of CE－4 rover. (b and c) show the accommodation geometry of Advanced Small Analyzer for Neutrals (ASAN) onboard the rover (YuTu－2), in which the rover normally moves in the $+X$ direction, the Z axis points to the lunar surface, and ASAN is mounted on the $+X$ panel of the rover with a field of view of 37° × 30°

emitted from the lunar surface in an energy range from 10eV to 10keV typically with a 30% energy resolution (Wieser et al., 2020). The ASAN is mounted on the $+X$ panel of the rover (Fig. 1b) with a field of view of $37° \times 30°$, and its boresight points forward ($+X$ direction) at an angle of 30° from the lunar surface (Fig. 1c).

Here, we analyse 46 intervals of ENA data from ASAN during the period between January 11, 2019 and April 28, 2020. Twenty - seven of them are measured in the morning (7: 00 - 9: 30 local time), and the left are measured in the afternoon (14: 30 - 17: 00 local time), as shown in Fig. 2a. Normally, ASAN is downstream from the Imbrium antipode for the afternoon measurements, when the solar wind flows from the west and we can expect a mini - magnetosphere. On the contrary, ASAN is upstream from the Imbrium antipode for the morning measurements, when the solar wind flows from the east. Previously, Zhang et al. (2020) reported some preliminary results of ASAN measurements, in which significant hydrogen (H) ENA fluxes were observed with a cutoff energy determined by the solar wind energy. However, the H ENA flux at the energy less than 100eV may be contaminated by the sputtered ENAs from the lunar surface and from the internal surface of the instrument (Wieser et al., 2020; Zhang et al., 2020), which should be carefully treated when studying the ENAs related to the solar wind. For this reason, here we only consider the H ENAs with energy higher than 100 eV, and calculate the differential ENA flux (J_{ENA}) by integrating the effective flux (higher than the sensitivity limit of ASAN) over an energy range above 100eV.

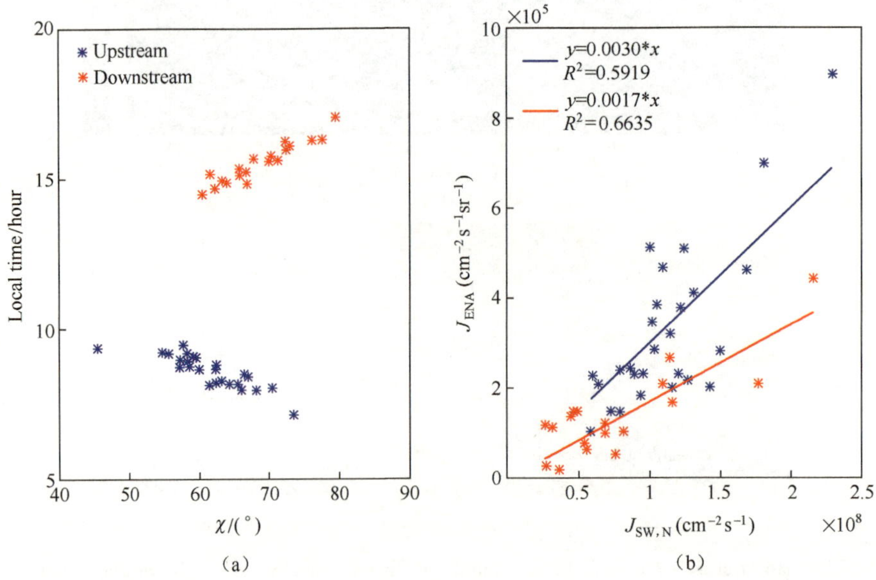

Fig. 2　(a) The positions of the Advanced Small Analyzer for Neutrals measurements, where χ is the incident angle of solar wind, defined as the angle between the solar wind velocity and the normal of local surface. (b) the dependence of the differential energetic neutral atom (ENA) flux J_{ENA} on the normal component of the solar wind flux $J_{SW,N}$; the blue and the red asterisks show the measurements obtained upstream and downstream from the magnetic anomalies, respectively; the blue and the red lines indicate the fitting lines for the upstream measurements and the downstream measurements, respectively

Strictly speaking, the J_{ENA} measured by ASAN should be determined by the solar wind flux, the orientation of the rover, and the local topography. Previously, the directional scattering function of the backscattered ENA from the lunar surface was obtained by the Chandrayaan - 1 ENA observations, which is a function of the SZA and the scattering angles in both the azimuthal and the polar directions (Schaufelberger et al., 2011; Vorburger et al., 2013). However, as shown in Figure S1, the ASAN data only cover a limited range of the azimuth angle, from which no clear trend can be identified and we cannot obtain the scattering function with these limited data. In addition, we have checked the slope angles measured in the rover coordinate system, and found that the slope angles are usually smaller than 5° and the average value is about 1.7° (Fig. S1), suggesting that the rover is moving on a relatively flat surface. There might be some influences from some local craters or stones on the ASAN measurements. However, such an influence seems random and cannot bring an overall variation of the angular distribution. Consequently, here we focus on discussing the dependence on the solar wind flux and neglect the effects of other factors. According to the Chandrayaan - 1 observations (Vorburger et al., 2013), the backscattered ENA flux at a given SZA is proportional to the solar wind flux and the cosine of the SZA. However, it is found that there is still a difference between the SZA and the incident angle of the solar wind χ (defined as the angle between the solar wind velocity and the normal of local surface). Moreover, the dependence of J_{ENA} on the $J_{\mathrm{SW}} \cdot \cos\chi$ is stronger than on $J_{\mathrm{SW}} \cdot \cos \mathrm{SZA}$, where J_{SW} is the solar wind flux. As a result, here we use $J_{\mathrm{SW}} \cdot \cos\chi$ to represent the normal component of J_{SW}, denoted as $J_{\mathrm{SW,N}}$. With measurements from ARTEMIS mission (Angelopoulos, 2011), we obtain the averaged number density and the velocity of the solar wind for each ASAN measurement, and then we can calculate the $J_{\mathrm{SW,N}}$. We use upstream and downstream measurements to denote the ENA flux measured upstream and downstream from the Imbrium magnetic anomalies, respectively. As shown in Figure 2b, the J_{ENA} for both the upstream measurements and the downstream measurements almost linearly increase with $J_{\mathrm{SW,N}}$ (some deviations may be caused by the difference in directional scattering), which are consistent with the Chandrayaan - 1 observations. In addition, J_{ENA} for the downstream measurements is generally smaller than that for the upstream measurements (Fig. 2b), suggesting magnetic shielding associated with the LMM. However, it seems that there are some solar wind protons can penetrate across the mini - magnetosphere to reach the lunar surface, resulting in the positive correlation of J_{ENA} with $J_{\mathrm{SW,N}}$ for the downstream measurements. It is important to know how many solar wind ions can be shielded by LMM. Due to lack of the knowledge about the directional scattering, it is not easy to make a quantitative calculation of the shielding efficiency. We select four intervals of the downstream measurements and two intervals of the upstream measurements. As shown in Table S1, the downstream intervals (Dn1 and Dn2, or Dn3 and Dn4) have similar solar wind incident angle χ and sun azimuth angle ϕ to the upstream intervals (Up1 or Up2), but the relative magnitude of the ENA fluxes ($J_{ENA}/J_{SW,N}$) for downstream intervals are usually smaller than those for the upstream intervals. Such a difference should be mainly caused by the shielding effect of the magnetic anomaly, since the observation geometries are similar. Then we use the ratio of $J_{ENA}/J_{SW,N}$ for the downstream intervals to $J_{ENA}/J_{SW,N}$ for the upstream intervals to calculate the penetrating efficiency of the solar wind (η_{SW}, defined as the ratio of the solar wind flux inside the mini - magnetosphere to the undisturbed solar wind flux). As shown in Table S1, the η_{SW} for the intervals of Dn1, Dn2, and Dn3 are 0.43, 0.52, and 0.36, respectively. The corresponding shielding efficiencies are 0.57, 0.48, and 0.64, respectively, which are consistent with the Chandrayaan - 1 observations (the shielding efficiency of the Imbrium antipode ranges from 0.27 to 0.74, Vorburger et al., 2012). However, the η_{SW} for the interval Dn4 is as large as 1.6, implying that the solar wind is not shielded by the LMM. Instead, there may be some extra solar wind ions deflected from the neighboring magnetic anomalies. A similar enhanced flux region next to the reduced flux region was also observed by Chandrayaan - 1 (Futaana et al., 2013; Wieser et al., 2010). The reason might be that the ion inertia length (L_{pi}) for Dn4 is very large (162km), which can bring significant ion penetration and then cancel out the formation of a LMM. These results suggest that not all the downstream measurements are made within a LMM.

The penetrating effect associated with L_{pi} can also be found in Figure 3a, where $J_{ENA}/J_{SW,N}$ is generally smaller than the upstream measurements when the L_{pi} is small (the red asterisks), and comparable to the upstream measurements when the L_{pi} is large (the green asterisks). In addition, higher $P_{d,N}$ may bring larger $J_{ENA}/J_{SW,N}$ according to the Chandrayaan - 1 observations (Vorburger et al., 2012), where $P_{d,N} = \rho V_{SW}^2 \cos^2 \chi$ is the normal component of the solar wind dynamic pressure, and ρ and V_{SW} are the density and the velocity of the solar wind, respectively. As shown in Figure 3b, only the red asterisks show a positive correlation with $P_{d,N}$, which is a characteristic of LMM that is compressed by the solar wind. As for the green asterisks, $J_{ENA}/J_{SW,N}$ seems to be anti - correlated with $P_{d,N}$, with a magnitude comparable to or larger than the upstream measurements. These features imply that there should be no LMM or magnetic shielding for these green asterisks. For these reasons, we divide the downstream measurements into two groups: one is called the "inside" measurements that refer to the downstream measurements within LMM (shown as the red asterisks); the other is called the "outside" measurements that refer to the downstream measurements without magnetic shielding (shown as the green asterisks).

After normalizing the ENA energy with E_{SW} (the kinetic energy of the solar wind) and the ENA fluxes with $10^{-5} J_{SW,N}$, we obtain the normalized energy spectra for all the upstream, inside, and outside measurements. As shown in Figure 3c, the normalized ENA fluxes for the inside measurements are generally lower than those for the upstream

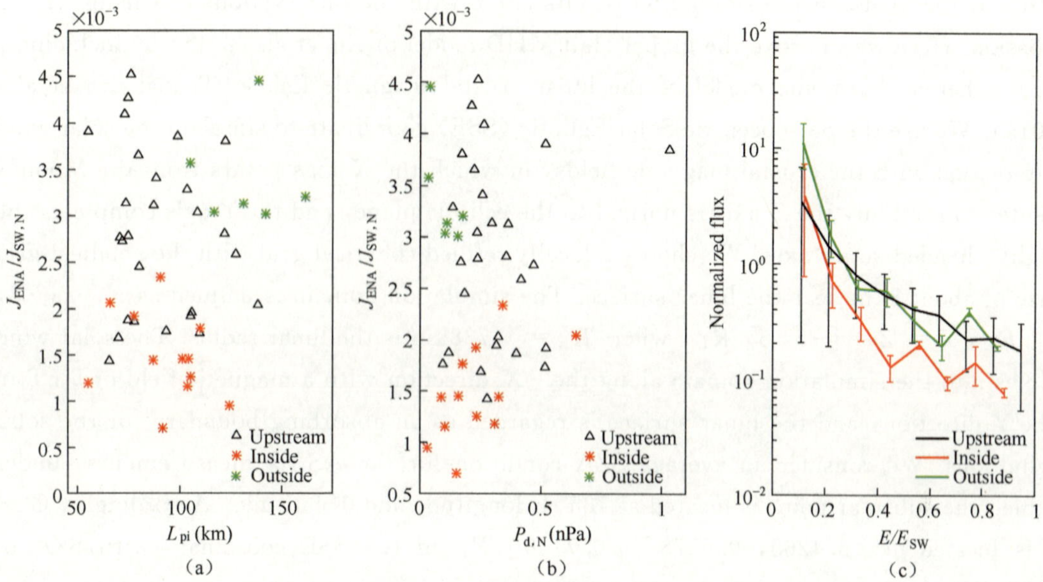

Fig. 3 (a and b) show the dependences of the normalized energetic neutral atom (ENA) flux ($J_{ENA}/J_{SW,N}$) on the ion inertia length (L_{pi}) and the normal component of the solar wind dynamic pressure ($P_{d,N}$), respectively, in which the gray triangles indicate the measurements obtained upstream from the magnetic anomalies, and the red and green asterisks indicate the measurements obtained downstream from the magnetic anomalies but inside and outside the LMM, respectively. (c) the normalized energy spectra, in which the ENA energies (E) are normalized by the kinetic energy of the solar wind (E_{SW}) and the differential fluxes are normalized by $10^{-5}\,J_{SW,N}$; the gray, red, and green circles represent the average normalized flux with an increment of about 0.1 E_{SW} for the upstream, inside, and outside measurements, respectively

measurements. Furthermore, such a difference is more significant at higher energy, implying that the solar wind ions have been decelerated when penetrating into the mini-magnetosphere, which reflects the shielding effect of the inside measurements. Besides, we find that the ENA fluxes for the outside measurements are approximately equal to those for the upstream measurements in the high energy range. When lower than 0.2 E_{SW}, the outside measurements even have higher ENA fluxes than the upstream measurements, suggesting some extra solar wind ions that have been deflected and decelerated by the neighboring magnetic anomalies. This may be the reason for the relatively large $J_{ENA}/J_{SW,N}$ of the green asterisks shown in Fig. 3a and 3b.

3　Comparison With Simulations

Previously, Xie et al. (2015) studied the LMM with 3D Hall MHD simulations, and found that the ratio of the pressure balance distance to the ion inertia length is important for the formation of the mini-magnetosphere. Moreover, ion penetration due to the Hall effect is revealed and the simulation results agree with the observations of Chang' E - 2 mission. Here we improve the global Hall MHD model of Xie et al. (2015) by including a 450° spherical harmonic model of the lunar crustal magnetic fields (Tsunakawa et al., 2015). We use the Selenocentric Solar Ecliptic (SSE) coordinate to simulate the solar wind interaction with the crustal magnetic fields, in which the X axis points from the Moon's center to the Sun, the Z axis is normal to the ecliptic plane, and the Y axis completes the right - handed set of axes. We choose a locally refined spherical grid with the smallest grid size of about 8km near the lunar surface. The simulation domain is defined as $X = [-8, 2]\ R_L$, $Y = Z = [-5, 5]\ R_L$, where $R_L = 1{,}738$km is the lunar radius. The solar wind flows into the simulation domain along the $-X$ direction with a magnetic field of 5 nT in the Y direction, and the lunar surface is regarded as an absorbing boundary for the solar wind ions. We consider an average SZA condition for the ASAN measurements, under which the subsolar point is located at 125° E longitude and 0° latitude. Accordingly, CE - 4 is located at [0.4265, 0.5578, - 0.7120] R_L in the SSE coordinate with SZA of about 65°.

Four cases have been studied, with different solar wind conditions and simulation results as shown in Figure 4. Case 1 represents the typical solar wind conditions for the ASAN measurements, where a mini - magnetosphere is formed with a long tail hanging over CE - 4's landing site (Fig. 4a). Meanwhile, the solar wind is decelerated and deflected by the magnetic anomalies with $\eta_{SW}=0.41$ (Fig. 4b), which is comparable to the typical η_{SW} shown in Tab. S1. By comparing Case 2 with Case 1, we find that lower $P_{d,N}$ can bring smaller η_{SW} when the ion inertia length is the same (Fig. 4c). By comparing Case 4 with Case 3, we find that large ion inertia length can cancel out the formation of the mini - magnetosphere even though $P_{d,N}$ is the same (Fig. 4d and 4e). These results are consistent with the observations shown in Figures 3a and 3b. Additionally, we find that the solar wind speed at the location of CE - 4 is reduced by about 16% for Case 1, 36% for Case 2, and 35% for Case 3. Meanwhile, the solar wind velocity is deflected by an angle of about 11° for Case 1, 32° for Case 2, and 30° for Case 3, respectively. These results suggest that the solar wind with a lower dynamic pressure can be more easily deflected by the magnetic anomalies, which is also in accordance with the ASAN observations. However, we fail to see a density void in the mini - magnetosphere. It implies that the mini - magnetosphere is not easy to form a plasma void due to ion penetration, and the decrease in the solar wind flux on the lunar surface is mainly caused by the deflection and deceleration of the solar wind near the surface. It should be noted that here we do not intend to make a quantitative comparison between the numerical simulation and the ASAN observation, considering the limitations and uncertainties both in the model and the observation, for example, the near surface accuracy and spatial resolution of the lunar crustal magnetic field model, and the error in ENA flux based on ASAN measurement. As a result, we only give a qualitative comparison about the effects of different parameters, such as the ion inertia length and the solar wind dynamic pressure, on the formation of LMM.

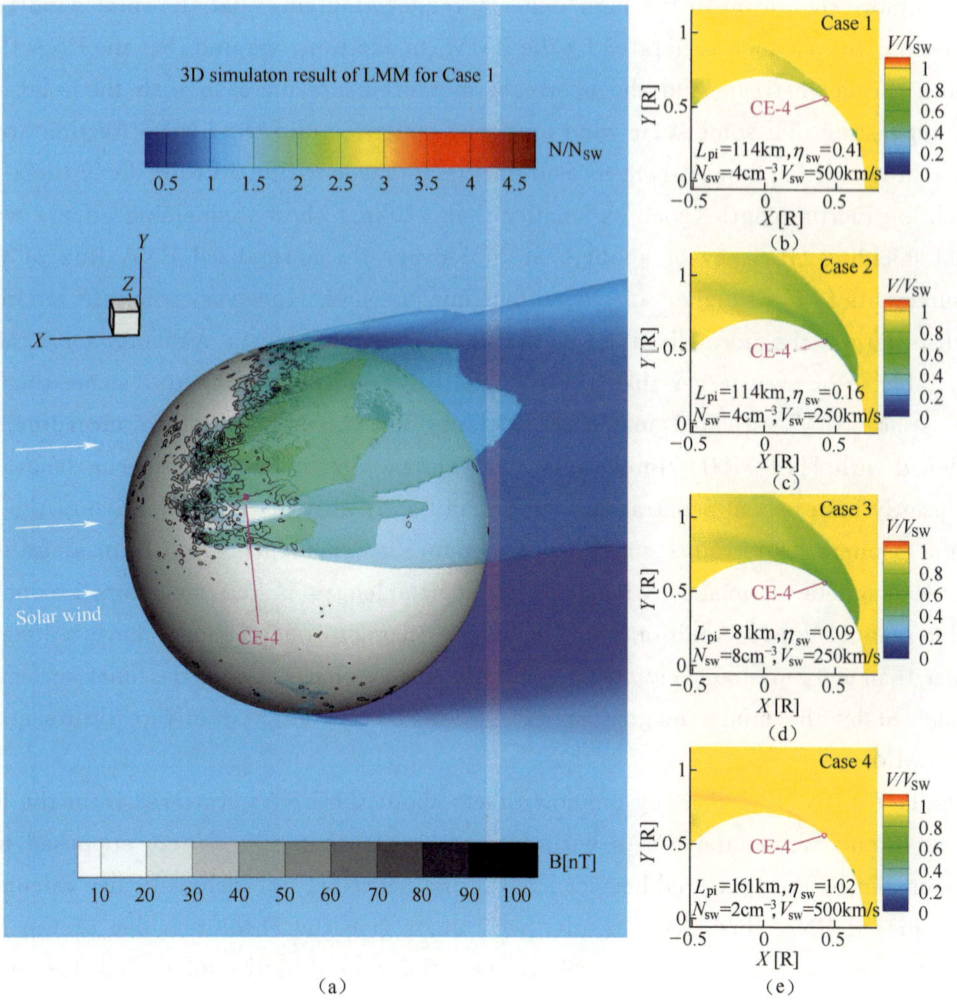

Fig. 4　(a) 3D view of the simulation result for Case 1, where the white arrows indicate the direction of the solar wind, the central ball represents the lunar body with gray contours to show the magnitudes of the crustal magnetic fields, the 3D colored contours show the number densities normalized by the solar wind number density (N_{SW}), and the magenta circle indicates the location of CE - 4; (b - e) are the results in the XY plane at $Z = -0.712R_L$ for Cases 1, 2, 3, and 4, respectively, in which the white disk indicates the lunar body, overlapped by the simulation conditions, the magenta circle marks the location of CE - 4, and the colored contours indicate the speed normalized by the solar wind speed (V_{SW})

4　Summary and Discussion

We analyze the ENA data measured by the CE - 4 rover, near the edge of the Imbrium antipode magnetic anomalies. The ENA fluxes obtained downstream from the magnetic anomaly are generally smaller than those obtained upstream from the magnetic anomalies, which is more significant in the high - energy range, implying that the solar wind might have been deflected and decelerated by the LMM. In addition, we find that the ENA fluxes for both the downstream and the upstream measurements increase with the solar wind flux, suggesting that some solar wind ions may penetrate into the LMM. Further studies show that not all the downstream measurements are made within a LMM. Only those with a small ion inertia length (such as smaller than 120km) show a shielding effect, with a typical shielding efficiency of about 0.5. Moreover, the normalized ENA flux of these measurements favors a higher solar wind dynamic pressure. However, when the ion inertia length is large, the normalized ENA fluxes of the downstream measurements can be comparable to or even larger than those of the upstream measurements, suggesting that there is no LMM - related magnetic shielding for these downstream measurements. Combined with Hall MHD simulations, we confirm that a large ion inertia length can significantly increase ion penetration, which can finally obstruct the formation of the mini - magnetosphere. Our results suggest that a mini - magnetosphere can form at Imbrium antipode magnetic anomalies when the ion inertia length is relatively small (such as smaller than 120km). Additionally, the mini - magnetosphere should have a long tail (longer than 600 km) to envelop CE - 4, implying that a large area of the lunar surface can be shielded by the mini - magnetosphere. These results might motivate future studies about LMM.

So far, there are still many unknown facts about the ENAs scattered from the lunar surface. In this work, the ENAs with energies lower than 100 eV are supposed to be contaminated by the sputtered heavier ENAs, and hence are not included when calculating the integrated ENA flux. As a result, we cannot give the absolute value of the ENA reflection ratio, but just discuss its relative variation caused by the solar wind flux and the magnetic anomalies. In addition, we have only analyzed 46 intervals of data. Moreover, these data only cover a part of the scattering angle of the backscattered ENAs. Consequently, we cannot give the angular scattering function at present. Nevertheless, both the energy spectrum and the angular scattering function of the backscattered ENAs are important to understand the interaction between the solar wind and the lunar surface. In the future, with more ENA data from ASAN, we hope to establish the angular scattering function of the backscattered ENAs near the surface and compare them with the Chandrayaan - 1 observations from orbit.

Acknowledgments

This work is supported by the National Natural Science Foundation of China (NSFC) grant 41941001 and the National Key R&D Program of China (grant no.

2020YFE0202100). The Advanced Small Analyzer for Neutrals (ASAN) instrument was developed by the Swedish Institute of Space Physics (IRF) in Kiruna, Sweden, and the National Space Science Center (NSSC) in Beijing, China as a joint project. It is supported by the Swedish National Space Agency, SNSA, under grants 95/11 and 122/18. This work is also supported by the Science and Technology Development Fund (FDCT) of Macau (008/2017/AFJ) and by the Youth Innovation Promotion Association of the Chinese Academy of Sciences.

References

[1] Angelopoulos, V. (2011). The ARTEMIS mission. *Space Science Reviews*, 165, 3 - 25. https://doi.org/10.1007/s11214-010-9687-2.

[2] Colburn, D. S., Mihalov, J. D., & Sonett, C. P. (1971). Magnetic observations of the lunar cavity. *Journal of Geophysical Research*, 76, 2940 - 2957. https://doi.org/10.1029/ja076i013p02940.

[3] Dyal, P., Parkin, C. W., & Daily, W. D. (1974). Magnetism and the interior of the Moon. *Reviews of Geophysics*, 12 (4), 568 - 591. https://doi.org/10.1029/rg012i004p00568.

[4] Dyal, P., Parkin, C. W., & Sonett, C. P. (1970). Apollo 12 magnetometer: Measurement of a steady magnetic field on the surface of the moon. *Science*, 169, 762 - 764. https://doi.org/10.1126/science.169.3947.762.

[5] Fatemi, S., Holmström, M., Futaana, Y., Lue, C., Collier, M. R., Barabash, S., & Stenberg, G. (2014). Effects of protons reflected by lunar crustal magnetic fields on the global lunar plasma environment. *Journal of Geophysical Research: Space Physics*, 119, 6095 - 6105. https://doi.org/10.1002/2014ja019900.

[6] Futaana, Y., Barabash, S., Wieser, M., Lue, C., Wurz, P., Vorburger, A., et al. (2013). Remote energetic neutral atom imaging of electric potential over a lunar magnetic anomaly. *Geophysical Research Letters*, 40, 262 - 266. https://doi.org/10.1002/grl.50135.

[7] Halekas, J. S., Brain, D. A., Mitchell, D. L., Lin, R. P., & Harrison, L. (2006). On the occurrence of magnetic enhancements caused by solar wind interaction with lunar crustal fields. *Geophysical Research Letters*, 33, L08106. https://doi.org/10.1029/2006gl025931.

[8] Halekas, J. S., Delory, G. T., Brain, D. A., Lin, R. P., & Mitchell, D. L. (2008). Density cavity observed over a strong lunar crustal magnetic anomaly in the solar wind: A mini-magnetosphere? *Planetary and Space Science*, 56 (7), 941 - 946. https://doi.org/10.1016/j.pss.2008.01.008.

[9] Halekas, J. S., Poppe, A. R., McFadden, J. P., Angelopoulos, V., Glassmeier, K.-H., & Brain, D. A. (2014). Evidence for small-scale collision-less shocks at the moon from artemis. *Geophysical Research Letters*, 41, 7436 - 7443. https://doi.

org/10. 1002/2014gl061973.

[10] Harnett, E. M., & Winglee, R. M. (2003). 2. 5 – D fluid simulations of the solar wind interacting with multiple dipoles on the surface of the Moon. *Journal of Geophysical Research: Space Physics*, 108 (A2). https://doi. org/10. 1029/2002ja009617.

[11] Hood, L. L., Zakharian, A., Halekas, J., Mitchell, D. L., Lin, R. P., Acuña, M. H., & Binder, A. B. (2001). Initial mapping and interpretation of lunar crustal magnetic anomalies using Lunar Prospector magnetometer data. Journal of Geophysical Research: Planets, 106 (E11), 27825 – 27839. https://doi. org/10. 1029/2000je001366.

[12] Lin, R. P., Mitchell, D. L., Curtis, D. W., Anderson, K. A., Carlson, C. W., McFadden, J., et al. (1998). Lunar surface magnetic fields and their interaction with the solar wind: Results from Lunar Prospector. Science, 281 (5382), 1480 – 1484. https://doi. org/10. 1126/science. 281. 5382. 1480.

[13] Liu, J., Ren, X., Yan, W., Li, C., Zhang, H., Jia, Y., et al. (2019). Descent trajectory reconstruction and landing site positioning of Chang'E – 4 on the lunar farside. Nature Communications, 10 (1), 1 – 10. https://doi. org/10. 1038/s41467 – 019 – 12278 – 3.

[14] Lue, C., Futaana, Y., Barabash, S., Wieser, M., Holmström, M., Bhardwaj, A., et al. (2011). Strong influence of lunar crustal fields on the solar wind flow. Geophysical Research Letters, 38, L03202. https://doi. org/10. 1029/2010gl046215.

[15] Mitchell, D. L., Halekas, J. S., Lin, R. P., Frey, S., Hood, L. L., Acuña, M. H., & Binder, A. (2008). Global mapping of lunar crustal magnetic fields by Lunar Prospector. Icarus, 194 (2), 401 – 409. https://doi. org/10. 1016/j. icarus. 2007. 10. 027.

[16] Ness, N. F., Behannon, K. W., Taylor, H. E., & Whang, Y. C. (1968). Perturbations of the int erplanetary magnetic field by the lunar wake. Journal of Geophysical Research, 73, 3421 – 3440. https://doi. org/10. 1029/ja073i011p03421.

[17] Omidi, N., Blanco – Cano, X., Russell, C. T., Karimabadi, H., & Acuna, M. (2002). Hybrid simulations of solar wind interaction with magnetized asteroids: General characteristics. Journal of Geophysical Research: Space Physics, 107 (A12), SSH – 12 – 1. https://doi. org/10. 1029/2002ja009441.

[18] Saito, Y., Nishino, M. N., Fujimoto, M., Yamamoto, T., Yokota, S., Tsunakawa, H., et al. (2012). Simultaneous observation of the electron acceleration and ion deceleration over lunar magnetic anomalies. Earth Planets and Space, 64 (2), 83 – 92. https://doi. org/10. 5047/eps. 2011. 07. 011.

[19] Saito, Y., Yokota, S., Asamura, K., Tanaka, T., Nishino, M., Yamamoto, T., et al. (2010). In – flight Performance and Initial Results of Plasma Energy Angle and Composition Experiment (PACE) on SELENE (Kaguya). Space Science Reviews,

154 (1 - 4), 265 - 303. https://doi. org/10. 1007/s11214 - 010 - 9647 - x.

[20] Schaufelberger, A., Wurz, P., Barabash, S., Wieser, M., Futaana, Y., Holmström, M., et al. (2011). Scattering function for energetic neutral hydrogen atoms off the lunar surface. Geophysical Research Letters, 38, L22202. https://doi. org/10. 1029/2011gl049362.

[21] Tsunakawa, H., Takahashi, F., Shimizu, H., Shibuya, H., & Matsushima, M. (2015). Surface vector mapping of magnetic anomalies over the Moon using Kaguya and Lunar Prospector observations. Journal of Geophysical Research: Planets, 120, 1160 - 1185. https://doi. org/10. 1002/2014je004785.

[22] Vorburger, A., Wurz, P., Barabash, S., Wieser, M., Futaana, Y., Bhardwaj, A., & Asamura, K. (2015). Imaging the South Pole - Aitken basin in backscattered neutral hydrogen atoms. Planetary and Space Science, 115, 57 - 63. https://doi. org/10. 1016/j. pss. 2015. 02. 007.

[23] Vorburger, A., Wurz, P., Barabash, S., Wieser, M., Futaana, Y., Holmström, M., et al. (2012). Energetic neutral atom observations of magnetic anomalies on the lunar surface. Journal of Geophysical Research: Space Physics, 117 (A7), 1 - 9. https://doi. org/10. 1029/2012ja017553.

[24] Vorburger, A., Wurz, P., Barabash, S., Wieser, M., Futaana, Y., Lue, C., et al. (2013). Energetic neutral atom imaging of the lunar surface. Journal of Geophysical Research: Space Physics, 118 (7), 3937 - 3945. https://doi. org/10. 1002/jgra. 50337.

[25] Wieser, M., Barabash, S., Futaana, Y., Holmström, M., Bhardwaj, A., Sridharan, R., et al. (2010). First observation of a mini - magnetosphere above a lunar magnetic anomaly using energetic neutral atoms. Geophysical Research Letters, 37, L05103. https://doi. org/10. 1029/2009gl041721.

[26] Wieser, M., Barabash, S., Wang, X. - D., Grigoriev, A., Zhang, A., Wang, C., & Wang, W. (2020). The Advanced Small Analyzer for Neutrals (ASAN) on the Chang'E -4 Rover Yutu - 2. Space Science Reviews, 216 (4). https://doi. org/10. 1007/s11214 - 020 - 00691 - w.

[27] Xie, L., Li, L., Zhang, Y., Feng, Y., Wang, X., Zhang, A., & Kong, L. (2015). Three - dimensional Hall MHD simulation of lunar minimagnetosphere: General characteristics and comparison with Chang'E - 2 observations. Journal of Geophysical Research: Space Physics, 120 (8), 6559 - 6568. https://doi. org/10. 1002/2015ja021647.

[28] Zhang, A., Wieser, M., Wang, C., Barabash, S., Wang, W., Wang, X., et al. (2020). Emission of energetic neutral atoms measured on the lunar surface by Chang'E - 4. Planetary and Space Science, 189 (15). https://doi. org/10. 1016/j. pss. 2020. 104970.

Multipoint Observation of the Solar Wind Interaction with Strong Lunar Magnetic Anomalies by ARTEMIS Spacecraft and Chang'E－4 Rover

Lianghai Xie[1,2,3], Lei Li[1,2,3], Aibing Zhang[2], Huizi Wang[4], Quanqi Shi[4], Jindong Wang[1,2,3], Yiteng Zhang[1,2,3], Bin Zhou[1,2,3], Yongyong Feng[1,2,3], Jinbin Cao[5], Dafei Li[6], and Shaojin Han[6]

1. State Key Laboratory of Space Weather, National Space Science Center, Chinese Academy of Sciences, Beijing, 100190, People's Republic of China; xielianghai@nssc. ac. cn
2. National Space Science Center, Chinese Academy of Sciences, Beijing, 100190, People's Republic of China
3. Joint Research and Development Center of Chinese Science Academy and Shen county, Shandong, 252400, People's Republic of China
4. Laboratory of Optical Astronomy and Solar－Terrestrial Environment, School of Space Science and Physics, Institute of Space Sciences, Shandong University, Weihai, Shandong, 264209, People's Republic of China
5. School of Space and Environment, Beihang University, Beijing, 100191, People's Republic of China
6. Beijing Aerospace Control Center, Beijing, 100048, People's Republic of China

Abstract: A shock or a mini－magnetosphere was once thought to be formed by the solar wind interaction with strong lunar magnetic anomalies. However, the full structure of a mini－magnetosphere has never been verified and whether a mini－magnetosphere can be completely formed remains a controversy. In this work, we present a unique multipoint observation of such an interaction by the ARTEMIS spacecraft and the Chang'E－4 rover. Both solar wind deceleration and penetration are observed by the Chang'E－4 rover on the lunar surface near the magnetic anomaly. Meanwhile, a shock is observed by the ARTEMIS spacecraft downstream from the magnetic anomaly. It is suggested that the magnetic anomaly cannot stand off the solar wind, and there is no shock but just a boundary layer near the magnetic anomaly. Accordingly, a mini－magnetosphere is not completely formed and the downstream shock observed the ARTEMIS spacecraft just corresponds to a trailing shock.

Keywords: Solar wind; Lunar science; Space plasmas

1 Introduction

Differing from the Earth, the Moon has neither a global magnetic field nor a significant atmosphere. Particles from the interplanetary space, such as solar wind ions, galactic cosmic rays, and micrometeoroids, can directly bombard the lunar surface, resulting in a relatively harsher space environment. However, it has been found that the Moon possesses a large number of local crustal magnetic fields, known as magnetic anomalies (Hood et al. 2001; Mitchell et al. 2008; Tsunakawa et al. 2015), which can deflect the solar wind and form a smallscale structure with lower solar wind flux. Previously, both the Lunar Prospector (LP) and the ARTEMIS spacecraft observed shock-like structures near some strong magnetic anomalies (Lin et al. 1998; Halekas et al. 2006, 2014). Meanwhile, a reduced solar wind flux was found in the central magnetic anomaly region by the energetic neutral atom (ENA) observations of the Chandrayaan-1 spacecraft (Wieser et al. 2010; Vorburger et al. 2012), suggesting the magnetic shielding of the lunar surface from the solar wind. As a result, it seems that these strong magnetic anomalies can stand off the solar wind and form a local protecting structure, which is similar to the Earth's magnetosphere but with a smaller size, and hence it is known as a lunar mini-magnetosphere (LMM).

The question is how a magnetic field on the scale of the ion gyroradius (or ion inertial length) can form a shock. Previously, lunar magnetic anomalies were thought to be not strong enough to generate a shock but only some whistler or magnetosonic waves, according to hybrid simulations (Omidi et al. 2002), while Hall MHD simulations showed that the nondipolar nature of lunar crustal fields could increase the lateral extent of the field and help to form a shock (Harnett & Winglee 2003; Xie et al. 2015). Additionally, there are some solar wind protons (~10%) reflected from the strong magnetic anomalies (Saito et al. 2010; Lue et al. 2011). These reflected protons can also help to slow down the solar wind and form a shock-like structure (Fatemi et al. 2014; Halekas et al. 2014). Another question is whether an LMM can completely be formed. Up to now, people have just found enhancements in the magnetic field and the plasma density near some strong magnetic anomalies, but without a density cavity. These enhancements may correspond to either the sheath of a LMM or the nose of a magnetosonic wake (Omidi et al. 2002). A density cavity was once seen by LP under some special solar wind conditions (Halekas et al. 2008), including a high solar wind density ($15.8\,cm^{-3}$), a low solar wind velocity ($30\,km\ s^{-1}$), a low solar wind temperature (1.11eV), a high solar zenith angle (SZA; ~83°), and a low altitude (~30km). Nevertheless, such a density cavity lied far downstream (~10° in longitude) from the magnetic anomaly, which may only represent the tail of an LMM or a magnetosonic wake. Besides, the reduced solar wind flux indicated by the ENA measurements may be just caused by a deflected solar wind velocity rather than a density cavity. Consequently, it is still uncertain whether there is a density cavity in the central magnetic anomaly region. If there is no central density cavity, the "mini - magnetosphere" may be not completely formed, and the term "mini - magnetosphere" seems inappropriate.

So far, there is no multipoint observation of the solar wind interaction with lunar magnetic anomalies, and hence the full structure of the interaction is still unknown. In addition, both the simulations and the observations have suggested that the shock - like structure favors a smaller ion inertial length and a stronger magnetic field. Here we present a unique multipoint observation of the solar wind interaction with a group of strong lunar magnetic anomalies, in which the Chang'E - 4 (CE - 4) rover is located on the lunar surface near the magnetic anomalies and the ARTEMIS spacecraft are in orbit. Moreover, the multipoint observation happens under a special solar wind condition with a very small ion inertial length. As a result, this work provides a good chance to check whether a bow shock as well as a complete mini - magnetosphere can be caused by the solar wind interaction with lunar magnetic anomalies.

2 Observations

The ARTEMIS mission is the extension of the THEMIS mission, which has two identical probes, P1 and P2 (Angelopoulos 2011). From 2011 April on, both probes operate in a high - eccentricity orbit of ~100km × 19, 000km, with an orbit period of ~26hr. When one probe is in the lunar wake, the other one is usually in the solar wind. In this way, ARTEMIS mission can measure the plasma in both the solar wind and the lunar wakes, with the fluxgate magnetometer (Auster et al. 2008) and the electrostatic analyzer (McFadden et al. 2008) instruments. the CE - 4 mission is the first mission soft - landed on the lunar far side. The Advanced Small Analyzer for Neutrals (ASAN) onboard the rover of the CE - 4 mission is an ENA analyzer, which can measure the solar wind ions that are backscattered from the lunar surface as ENAs (Wieser et al. 2020; Zhang et al. 2020). The incident solar wind flux on the lunar surface can be inferred with the measured ENA flux (Xie et al. 2021). During 01:00 - 08:00 UT on 2019 December 31, ARTEMIS P1 was in the undisturbed solar wind upstream from the lunar wake and observed a very special solar wind, which had a high number density (~ $20cm^{-3}$), a low velocity (~$300km\ s^{-1}$), and a low temperature (~7eV). These solar wind parameters are similar to those for the density cavity event observed by LP (Halekas et al. 2008). Meanwhile, ARTEMIS P2 flew across the lunar wake and observed a shock around 02:15 UT before entering the wake, with an altitude of about 800km. The shock is downstream from a group of strong magnetic anomalies (Imbrium, Serenitatis, and Crisium antipodes; see Figure 1). During the same period of interest, the CE - 4 rover was located on the lunar surface near the Imbrium antipode anomaly. An ENA spectrum was measured by ASAN of CE - 4 during 05:10 - 06:06 UT on 2019 December 31, which was about 3 hr after the shock seen by ARTEMIS P2, but still in the period of 01:00 - 08:00 UT, when ARTEMIS P1 saw the special solar wind. Previously, we have found that the ion inertial length and the normal component of the solar wind dynamic pressure are two key parameters to determine the interaction between the solar wind and the lunar magnetic anomaly (Xie et al. 2021). Here we find that these two parameters only change a little (less than 2%) between the ARTEMIS P2 and the CE - 4 measurements. In addition, similar shocks were also observed by ARTEMIS P2 in the previous 2 days and the following 2 days, though the shock intensities were relatively weaker due to the different solar wind conditions, implying that a long - period shock could be caused by the solar wind interaction with these strong magnetic anomalies. As the time difference between the CE - 4 and the ARTEMIS P2 measurements is about 3 hr, and the solar wind conditions for these two measurements are similar, we think that CE - 4 and ARTEMIS P2 can approximately provide a two - point observation of the same interaction. The observing geometries of ARTEMIS and CE - 4 during this period are shown in Figure 1, where the Selenocentric Solar Ecliptic (SSE) coordinate is used, whose X - axis points from the

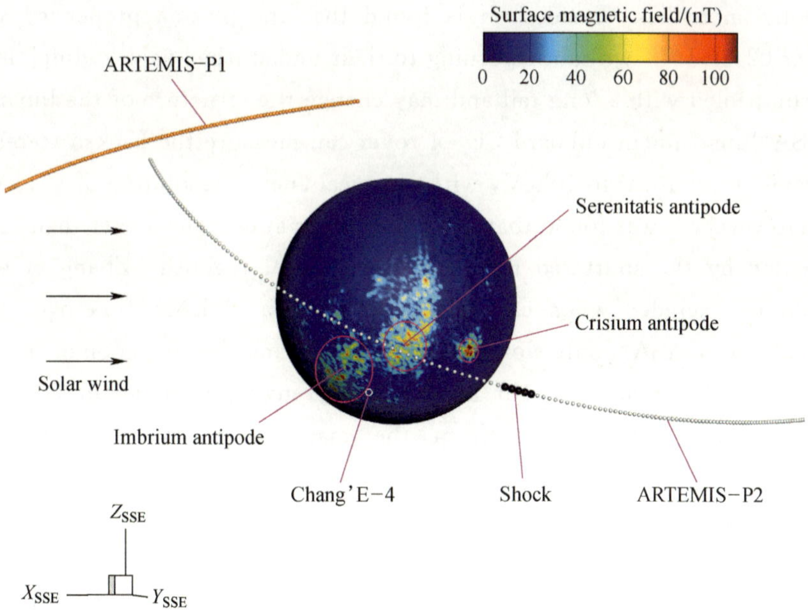

Fig. 1　Observing geometries of ARTEMIS and CE - 4 during 01:00 - 08:00 UT on 31 December 2019. The central sphere represents the lunar body with color contours to show the crustal magnetic field on the lunar surface. The white circle on the lunar surface indicates the position of CE - 4. The magenta circles indicate the magnetic anomalies at the Imbrium antipode, Serenitatis antipode, and Crisium antipode, respectively. The yellow and gray dots show the trajectories of ARTEMIS P1 and P1, respectively. The solid black dots indicate the time around 02:15 UT when ARMTEMIS P2 encounters the shock. The geometries are shown in the SSE coordinate, in which the solar wind flows in the opposite direction to the X - axis and the Z - axis is normal to the ecliptic plane.

Moon's center to the Sun, the Z-axis is normal to the ecliptic plane, and the Y-axis completes the right-handed set of axes.

The shock observed by ARTEMIS P2 is shown in Figure 2, where we can see obvious changes in the plasma properties around 02:15 UT, including an ion deceleration (Figure 2 (a)), an electron heating (Figure 2 (b)), and some magnetic field fluctuations (Figure 2 (c)). Furthermore, both the number density and the magnetic field are significantly enhanced by factors of 2.1 (Figure 2 (d)) and 2.3 (Figure 2 (g)), respectively. Meanwhile, the average electron temperature increases from 6.9eV to 8.6eV (Figure 2 (e)), and there is also a deflection in the solar wind velocity (mainly in the Y direction), apart from the deceleration in the X direction (Figure 2 (f)). The total change in the velocity is ~46km s^{-1}, which is larger than both the Alfvén velocity of ~29km s^{-1} and the magnetosonic velocity of ~42km s^{-1} of the upstream solar wind, implying a discontinuity in the flow. In addition, the magnetic field lines have been rotated across the discontinuity (Figure 2 (g)). All of these features suggest a strong shock observed by ARTEMIS P2. Known from Figure 1, the shock is downstream from a group of strong magnetic anomalies, suggesting that the shock is caused by the solar wind interaction with these magnetic anomalies. Besides, it is found that the plasma properties are further changed after 02:30 UT, without returning to their undisturbed values, implying that the shock is accompanied with a long tail and may change the structure of the lunar wake.

The ASAN instrument onboard CE-4 rover can measure the backscattered ENAs in an energy range from 10eV to 10keV, with a typical energy resolution of 30% (Wieser et al. 2020). However, it was found that the ENA fluxes at energies lower than 100eV could be contaminated by the sputtered ENAs from the local regolith (Zhang et al. 2020), which should be excluded when calculating the integrated ENA flux over the energy (J_{ENA}). In addition, ASAN only works in the morning and the afternoon of the local lunar day, with a typical period of ~1hr and an SZA ranging from about 50° to 80°. In particular, ASAN is downstream from the magnetic anomaly for the afternoon measurements but upstream from the magnetic anomaly for the morning measurements. Previously, Zhang et al. (2020) found that the ENA spectra measured by ASAN showed a cutoff energy (E_{cut}) that was approximately equal to the solar wind energy (E_{sw}), at which the ENA flux can drop to the instrument sensitivity level. Additionally, Xie et al. (2021) found that the J_{ENA} measured by CE-4 could almost linearly increase with the normal component of the solar wind flux on the lunar surface ($J_{sw,N}$), but the J_{ENA} measured downstream from the magnetic anomaly were generally lower, suggesting a magnetic shielding caused by the magnetic anomaly. Wang et al. (2021) found that that the E_{cut} downstream from the magnetic anomaly was lower than those upstream from the magnetic anomaly, implying solar wind deceleration by the magnetic obstacle.

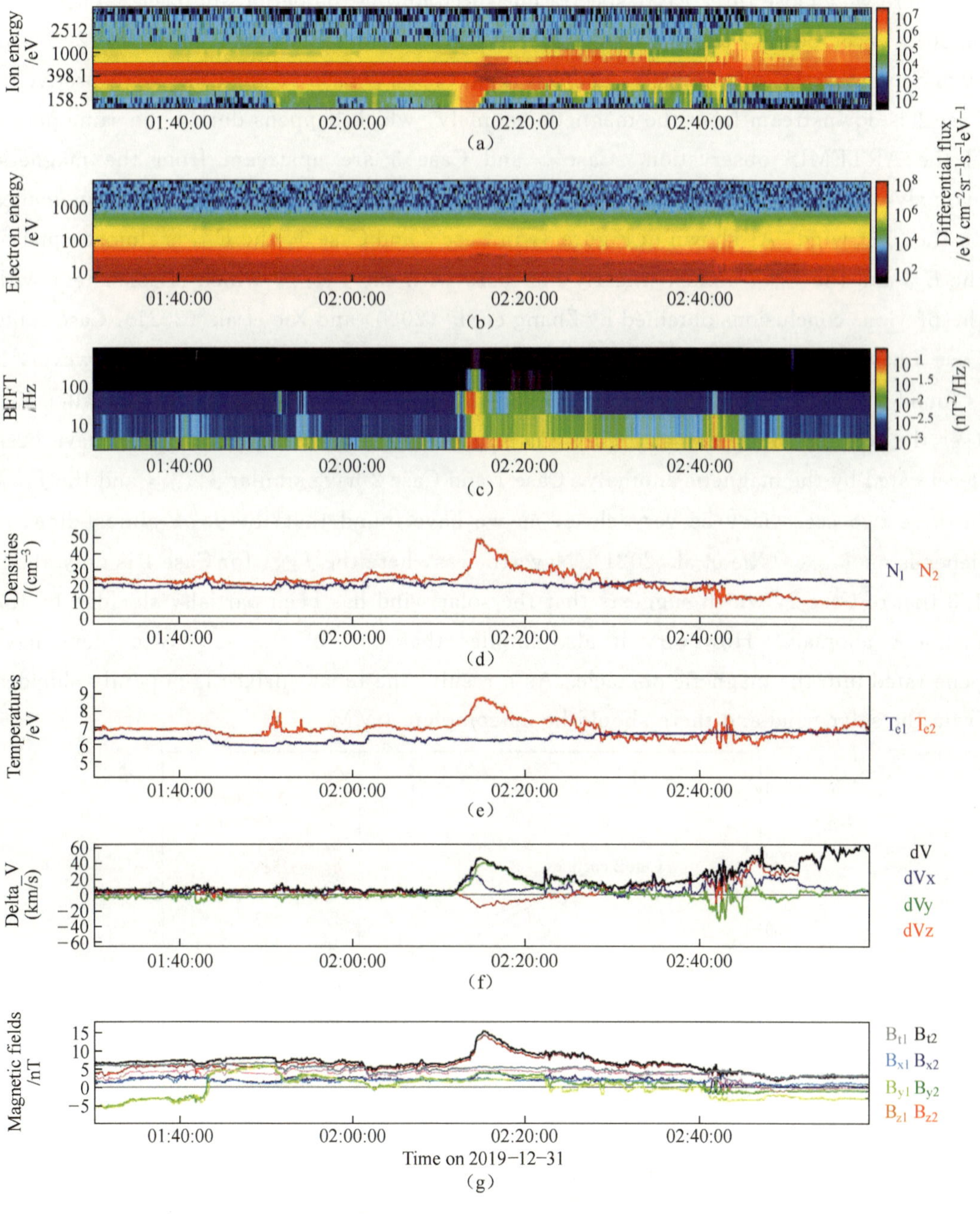

Figure 2　Shock observed by ARTEMIS.

(a)-(c) show the ion energy spectrum, the electron energy spectrum, and the fast Fourier transform of magnetic field fluctuations measured by ARTEMIS P2; (d)-(g) show the number densities and the electron temperatures, the velocity differences, and the magnetic fields, respectively, in which subscripts of 1 and 2 indicate the measurements of P1 and P2, respectively, and the lines in different colors represent the different components in SSE coordinate. The black line in (f) shows the magnitude of velocity change between the upstream P1 measurements and the downstream P2 measurements. The gray and the black lines in (g) show the magnitudes of magnetic field measured by P1 and P2, respectively

Here we choose three ENA spectra measured during the periods of 05:10 – 06:06 UT on 2019 December 31, 09:43 – 10:37 UT on 2020 October 12, and 07:11 – 08:12 UT on 2020 September 13, respectively, and call them Case 1, Case 2, and Case 3, respectively. Case 1 is downstream from the magnetic anomaly, which happens during the same period of the ARTEMIS observation. Case 2 and Case 3 are upstream from the magnetic anomaly, which are used to compare with Case 1 to show the effect of magnetic anomalies on the solar wind. As shown in Figure 3, in Case 2 and Case 3, the E_{cut} is almost equal to the E_{sw} and the J_{ENA} almost linearly increases with the $J_{sw,N}$, which is consistent with the previous conclusions obtained by Zhang et al. (2020) and Xie et al. (2021). Case 1 and Case 3 have similar E_{sw}, and we may expect similar E_{cut} in these two cases. However, it is found that the ENAs in Case 1 are more gathered at lower energies with a smaller E_{cut} (see the red and blue lines in Figure 3), suggesting that the solar wind ions have been decelerated by the magnetic anomaly. Case 1 and Case 2 have similar $J_{sw,N}$, and the J_{ENA} in these two cases may be very close, as we have found that the J_{ENA} almost linearly depends on $J_{sw,N}$ (Xie et al. 2021). Nevertheless, here the J_{ENA} for Case 1 is only about 1/3 that of Case 2, which suggests that the solar wind has been partially shielded by the magnetic anomaly. However, it also implies that part of the solar wind ions have penetrated into the magnetic obstacle. As a result, the lunar surface is not fully shielded from the solar wind and there should be no complete LMM.

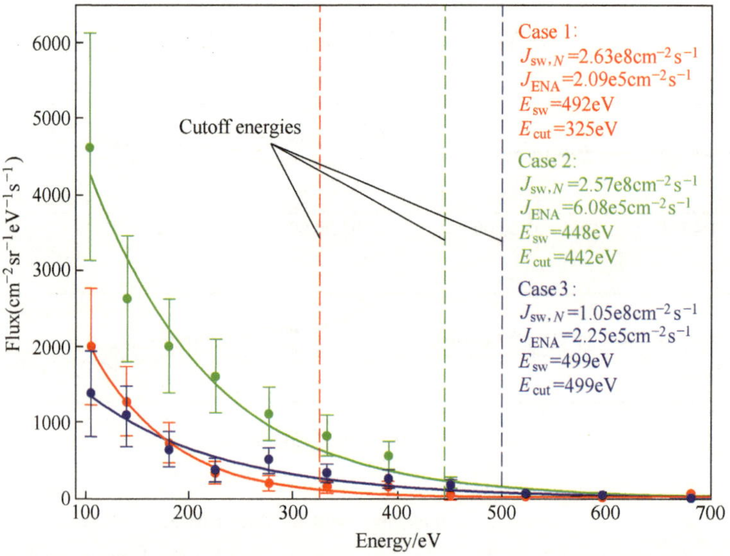

Figure 3　ENA spectra observed by CE – 4. The red, green, and blue dots show the measured spectra of Case 1, Case 2, and Case 3, respectively, in which Case 1 is downstream from the magnetic anomaly and correlated with the LMM observed by ARTEMIS, while Case 2 and Case 3 are upstream from the magnetic anomaly with similar solar wind conditions to those of Case 1. The solid lines are the fitting results of the three spectra with exponential functions. The dashed lines indicate the cutoff energies, at which the ENA flux is equal to 5.4% of its initial flux at 105eV. Texts on the right show the parameters of the three cases, in which $J_{sw,N}$ is the normal component of the solar wind flux on the lunar surface; J_{ENA} is the integrated ENA flux over the energy; E_{sw} is the solar wind energy; and E_{cut} is the cutoff energy

3　Discussions and Implications

Based on the multipoint observations shown above, we obtain a new physical picture for the solar wind interaction with lunar magnetic anomalies. As shown in the Figure 4, a lunar magnetic anomaly has many substructures, and the interaction between the solar wind and the magnetic anomaly can be regarded as the sum of the solar wind interactions with all of these substructures. For an individual substructure, the magnetic field is not strong enough to stop the solar wind, but only deflect the solar wind. The deflection is not obvious near the nose of the interaction region and a large number of solar wind particles can penetrate across the magnetic field and impact the lunar surface. Accordingly, there is no shock but only a magnetosonic wake in this region, where the solar wind can be slightly decelerated and compressed. When moving downstream, the solar wind can be further deflected by the following substructures, and the magnetosonic wake is getting more and more significant. Moreover, the magnetosonic wakes caused by different substructures may overlap each other, which jointly form a boundary layer near the lunar surface. Besides, some solar wind ions can be reflected by the magnetic anomaly, which can help decelerate and compress the incoming solar wind. Finally, the solar wind can flow horizontally or even outwardly from the local surface, and a trailing shock appears downstream from the magnetic anomaly. It suggests that a bow shock as well as a complete mini - magnetosphere are not formed here, even if the interaction is expected to be in an obvious fluid manner.

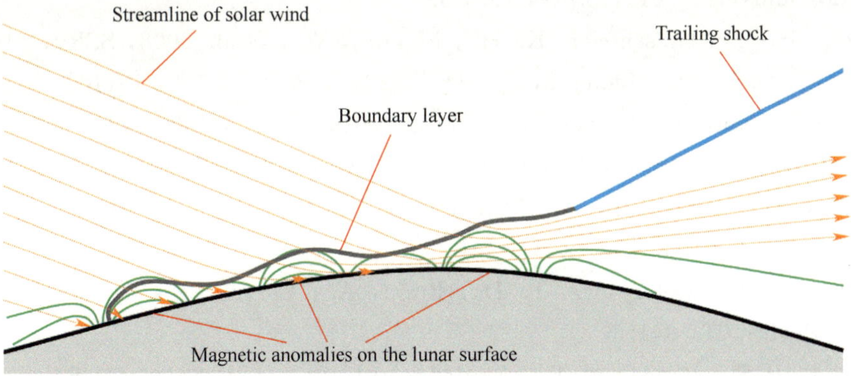

Figure 4　Schematic diagram of the solar wind interaction with lunar magnetic anomalies. The green lines indicate the magnetic anomalies on the lunar surface. The dotted red lines indicate the streamlines of the solar wind. The solid gray line indicates the boundary layer near the magnetic anomalies. The solid cyan line shows the trailing shock downstream from the magnetic anomalies

Our results imply that a lunar mini - magnetosphere may be never completely formed. As a result, the term of "min - imagnetosphere" should be not appropriate, which needs to be carefully used or redefined. In addition, if the lunar surface is not fully shielded by the magnetic anomaly, all applications associated with the magnetic shielding, such as the formation of a lunar swirl and the protection of a lunar station, need to be reassessed. It should be noted that a small - scale density cavity ($<$10km) may exist in the central magnetic anomaly region with a very strong magnetic field ($\sim$500nT), according to the particle - in - cell simulations done by Deca et al. (2014, 2015). Such a small - scale density cavity is very hard to be observed but may be important for the formation of lunar swirls, which needs to be further investigated in the future.

This work is supported by the National Key R&D Program of China (grant No. 2020YFE0202100). This work is also supported by the National Natural Science Foundation of China (NSFC) grants 42174216 and 41941001. L. Xie is supported by the Youth Innovation Promotion Association of the Chinese Academy of Sciences. ARTEMIS data are publicly available at http://artemis.ssl.berkeley.edu and NASA's CDAWeb. The ASAN data of Chang'E - 4 mission is available through the Data Release and Information Service System of China's Lunar Exploration Program (https://moon.bao.ac.cn/searchOrder_dataSearchData.search) and the ASAN data used in this paper can be found at NSSDC Space Science Article Data Repository (https://sadr-en.nssdc.ac.cn/detail?
dataSetId=852493495695310848&version=V1&dataSetType=personal&tag=1).

References

[1] Angelopoulos, V. 2011, SSRv, 165, 3.
[2] Auster, H. U., Glassmeier, K. H., Magnes, W., et al. 2008, SSRv, 141, 235.
[3] Deca, J., Divin, A., Lapenta, G., et al. 2014, PhRvL, 112, 151102.
[4] Deca, J., Divin, A., Lembège, B., et al. 2015, JGRA, 120, 6443.
[5] Fatemi, S., Holmstrom, M., Futaana, Y., et al. 2014, JGRA, 119, 6095.
[6] Halekas, J. S., Brain, D. A., Mitchell, D. L., Lin, R. P., & Harrison, L. 2006, GeoRL, 33, L08106.
[7] Halekas, J. S., Delory, G. T. D., Brain, A., Lin, R. P., & Mitchell, D. L. 2008, P&SS, 56, 941.
[8] Halekas, J. S., Poppe, A. R., McFadden, J. P., et al. 2014, GeoRL, 41, 7436.
[9] Harnett, E. M., & Winglee, R. 2003, JGRA, 108, 1088.
[10] Hood, L. L., Zakharian, A., Halekas, J., et al. 2001, JGR, 106, 27825.
[11] Lin, R. P., Mitchell, D. L., Curtis, D. W., et al. 1998, Sci, 281, 1480.
[12] Lue, C., Futaana, Y., Barabash, S., et al. 2011, GeoRL, 38, L03202.
[13] McFadden, J. P., Carlson, C. W., Larson, D., et al. 2008, SSRv, 141, 277.
[14] Mitchell, D. L., Halekas, J. S., Lin, R. P., et al. 2008, Icar, 194, 401.
[15] Omidi, N., Blanco - Cano, X., Russell, C. T., Karimabadi, H., & Acuna, M.

2002, JGRA, 107, 1487.

[16] Saito, Y. , Yokota, S. , Asamura, K. , et al. 2010, SSRv, 64, 4.

[17] Tsunakawa, H. , Takahashi, F. , Shimizu, H. , Shibuya, H. , & Matsushima, M. 2015, JGRE, 120, 1160.

[18] Vorburger, A. , Wurz, P. , Barabash, S. , et al. 2012, JGRA, 117, A07208.

[19] Wang, H. Z. , Xiao, C. , Shi, Q. Q. , et al. 2021, ApJL, 922, L41.

[20] Wieser, M. , Barabash, S. , Futaana, Y. , et al. 2010, GeoRL, 37, L05103.

[21] Wieser, M. , Barabash, S. , Wang, X. D. , et al. 2020, SSRv, 216, 73.

[22] Xie, L. , Li, L. , Zhang, A. , et al. 2021, GeoRL, 48, e2021GL093943.

[23] Xie, L. , Li, L. , Zhang, Y. , et al. 2015, JGRA, 120, 6559.

[24] Zhang, A. , Wieser, M. , Wang, C. , et al. 2020, P&SS, 189, 104970.

Primary and albedo protons detected by the Lunar Lander Neutron and Dosimetry experiment on the lunar farside

Zigong Xu[1], Jingnan Guo[2,3],
Robert F. Wimmer-Schweingruber[1,4], Mikhail I. Dobynde[2],
Patrick Kühl[1], Salman Khaksarighiri[1] and Shenyi Zhang[4]

1. Institute of Experimental and Applied Physics, Kiel University, Kiel, Germany
2. School of Earth and Space Sciences, University of Science and Technology of China, Hefei, China
3. CAS Center for Excellence in Comparative Planetology, Hefei, China
4. National Space Science Center, Beijing, China

Abstract: The Lunar Lander Neutron and Dosimetry (LND) Experiment aboard the Chang'E-4 Lander on the lunar far-side measures energetic charged and neutral particles and monitors the corresponding radiation levels. During solar quiet times, galactic cosmic rays (GCRs) are the dominating component of charged particles on the lunar surface. Moreover, the interaction of GCRs with the lunar regolith also results in upward-directed albedo protons which are measured by the LND. In this work, we used calibrated LND data to study the GCR primary and albedo protons. We calculate the averaged GCR proton spectrum in the range of 9-368MeV and the averaged albedo proton flux between 64.7 and 76.7MeV from June 2019 (the seventh lunar day after Chang'E-4's landing) to July 2020 (the 20th lunar day). We compare the primary proton measurements of LND with the Electron Proton Helium INstrument (EPHIN) on SOHO. The comparison shows a reasonable agreement of the GCR proton spectra among different instruments and illustrates the capability of LND. Likewise, the albedo proton measurements of LND are also comparable with measurements by the Cosmic Ray Telescope for the Effects of Radiation (CRaTER) during solar minimum. Our measurements confirm predictions from the Radiation Environment and Dose at the Moon (REDMoon) model. Finally, we provide the ratio of albedo protons to primary protons for measurements in the energy range of 64.7-76.7MeV which confirm simulations over a broader energy range.

Keywords: LND; Moon; lunar radiation environment; galactic cosmic rays; lunar albedo protons; data calibration; instrumentation

1 Introduction

The charged particle radiation environment on the lunar surface consists of Galactic Cosmic Rays (GCR), a small contribution from Anomalous Cosmic Rays (ACR), and a highly variable, sporadic contribution from Solar Energetic Particles (SEPs). In addition, secondary albedo particles are created primarily by the GCR interaction with the lunar regolith (Treiman, 1953; Dorman, 2004; Wilson et al. , 2012). In this work we focus on measurements by the Lunar Lander Neutron and Dosimetry experiment (LND, Wimmer - Schweingruber et al. , 2020) of GCR primary and secondary (albedo) protons and compare them with state - of - art model predictions (Dobynde and Guo, 2021). The space radiation measured by LND is a key concern for human space flight and may pose limits on long - term crewed missions to the Moon or Mars (Cucinotta and Chappell, 2011).

The interaction of high - energy particles with the lunar regolith results in the production of secondary particles. Some of these particles can escape from the soil and can be measured as albedo particles. Obviously, they will also contribute to the radiation exposure of astronauts on the lunar surface, but they also provide information about conditions beneath the lunar surface.

Spence et al. (2013) (Mitrofanov et al. , 2016). Schwadron et al. (2016) (Spence et al. , 2010, CRaTER). Wilson et al. (2012).

Wilson et al. (2012) first distinguished protons measured by the Cosmic Ray Telescope for the Effects of Radiation (Spence et al. , 2010, CRaTER) instrument from different directions and constructed the first albedo proton yield map between 60 and 150MeV. They found that the ratio of albedo to primary protons is uniformly distributed over the lunar surface within ~10% uncertainty and that the average ratio of upward to downward particle during the minimum between solar cycles 23 and 24 (in2009 - 2011) was about 0.38 ± 0.02. Simulations by Spence et al. (2013) show that such upward albedo particles can contribute a significant amount (~8. 62%) to the radiation dose with albedo protons accounting for 3. 1% of the total dose rate. Moreover, albedo neutrons have been used to detect and determine the subsurface hydrogen content (Mitrofanov et al. , 2016). Likewise, Schwadron et al. (2016) used the CRaTER instrument to detect albedo protons and provided evidence for the existence of hydrated material in the lunar regolith at the polar regions, based on small variations of the proton yield. A follow - up study by Schwadron et al. (2017) analyzed the difference of the lunar albedo protons yield for the lunar sunrise and sunset terminators.

(Looper et al. , 2013) presented the CRaTER measurements of radiation environment near the lunar surface including GCRs and albedo particles. They also utilized a GEANT4 simulation to model the energy distribution of albedo particles, and the response of different detectors to various particle species that reach the instrument on the 50km height lunar orbit.

More recently, Dobynde and Guo (2021) developed the Radiation Environment and Dose at the Moon (REDMoon) model that describes the detailed radiation environment both on the lunar surface and in the lunar soil. In this model, the lunar body is described by concentric spherical layers of different densities and soil composition is based on Apollo 17 drill core results (McKinney et al. , 2006). The model is based on the GEANT4 Monte-Carlo particle transport code (Agostinelli et al. , 2003) and the FTFP_BERT_HP physics list was used to calculate the GCR-induced radiation environment on and below the lunar surface. Primary GCRs have an isotropic direction (in the upper half-sphere) and the output particle types and energy at each soil depth or zenith angle are recorded. Results from different solar modulation conditions are also derived.

On 3 January 2019, China successfully landed the Chang'E-4 spacecraft on the far side of the Moon inside the von Kármán Crater. As one of its scientific payloads, Lunar Lander Neutron and Dosimetry experiment (LND, Wimmer-Schweingruber et al. , 2020) aboard the Chang'E-4 Lander is a small instrument with the primary objective of monitoring the radiation level on the lunar surface. LND consists of ten 500μm silicon detectors (labeled A-J) that are assembled as shown in Figure 1. The primary data products of LND include dynamic Total Ionizing Dose rate (TID) and Linear Energy Transfer (LET) spectra. The radiation dose rate is the radiation energy deposited by incoming radiation per unit time and unit mass of the absorber and is often measured in silicon detectors. LET is the energy that an ionizing particle transfers to the material per unit path length. Knowledge of both of these quantities is crucial in preparing for human spaceflight and human exploration of the Moon. Apart from these dosimetric quantities, LND also measures primary charged-particle energy spectra. Using the energy loss in its individual detectors, LND measures the primary energy of particles in the energy range between ~10 MeV/nuc and a few hundred MeV/nuc and distinguishes different particle species, including electrons, protons, ^{4}He, its isotope ^{3}He, and heavy ions such as carbon, oxygen, nitrogen, and iron. The post-launch performance of LND and its capability of measuring protons have been partially verified through analysis of a weak and impulsive SEP event in May 2019 (Xu et al. , 2020), with proton energy up to ~20MeV. Moreover, LND measures upward-directed albedo particles using the same method as was successfully applied using data from the Radiation Assessment Detector (RAD, Hassler et al. , 2012) on Mars. Appel et al. (2018) determined the flux of albedo particles in the energy range 100-200MeV on the Martian surface using a two-dimensional count density histogram. Here we use a similar method to distinguish upward proton fluxes from downward ones.

Zhang et al. (2020) displayed the first measurement of TID and LET spectra during the first two lunar days after LND was switched on. During the solar minimum of Solar Cycle (SC) 24/25, the average total absorbed dose rate on the lunar surface reached 13. 2

±1 μGy/h. This value is consistent with the radiation dose that is measured by CRaTER in orbit after the conversion to the surface in January 2019. A comparison with the radiation level on the Mars surface shows that the dose rate on the lunar surface is about 15% higher than that on Mars when GCR intensity reached a maximum during the deep solar minimum. Furthermore, the lunar surface dose rate is about two times higher than that measured in the International Space Station (ISS) by the 3D-DOSTEL instrument (Berger et al., 2020).

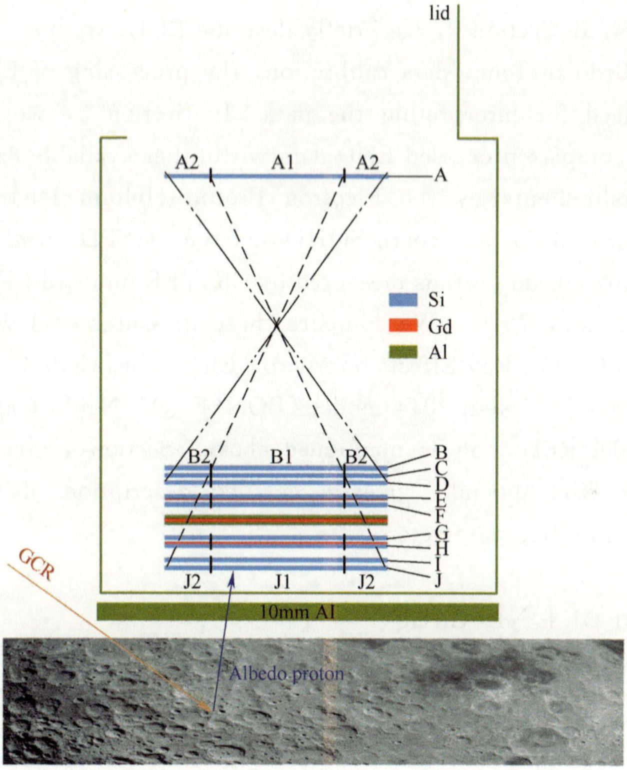

Fig. 1　Schematic view of the LND sensor head located above the lunar regolith with a 10-mm Al equivalent shielding (brown) between LND and the lunar soil. The LND sensor head consists of 10 segmented 500μm-thick Si detectors A-J. The Gd (red) and Al (brown) absorbers are designed to detect uncharged particles. The 10-mm Al sheet is added to mimic the shielding by the material between the sensor head and the lunar soil. On the lunar surface, the interaction of high-energy GCR particles (mainly protons) with the soil generates secondary, upward-moving protons. A more detailed description of LND is given in Wimmer-Schweingruber et al. (2020)

More than 3 years into the Chang'E 4 mission, most of LND's detectors are still in good condition and operating well despite the grueling temperature differences between lunar day and night. Nevertheless, since a mishap on the dawn of the third lunar day some segments of four of the 10 detectors in LND have begun to suffer from increased noise levels. We believe that a premature opening of LND's lid led to a severe drop in the temperature of LND's sensor head. The mounting of LND's detectors led to unintended thermo-mechanical stress on the front detector (A) and the three detectors at the end of the particle telescope (H, I, and J). An adjustment of the thresholds of the affected detector segments has led to some changes in the LND data products compared to those presented in Wimmer-Schweingruber et al. (2020). The changes are described in the appendix.

In this work, we provide a detailed calibration of the GCR proton spectra measured by LND and present the first measurements of the albedo protons by LND. The paper is organized as follows: In Section 2, we briefly describe LND, its measurement principles for primary and albedo protons, data calibration, the processing of LND data, and the simulation setup used for interpreting the data. In Section 3, we present the LND measurements and compare processed LND data with other available data sets, i.e., the primary GCR measurements by the Electron Proton Helium INstrument (EPHIN, Müller-Mellin et al., 1995) on-board SOHO and the (HEPD, Ambrosi et al., 2018; Bartocci et al., 2020) albedo protons measured by CRaTER onboard LRO (Wilson et al., 2012; Schwadron et al., 2016). We compare these measurements with the numerical cosmic ray model of Cosmic Ray Effects on Micro-Electronics Code (Tylka et al., 1997, CREME96), Badhwar-O'Neill 2014 model (BON14, O' Neill et al., 2015) and the lunar radiation model REDMoon as mentioned above. Section 4 gives a summary and discussion. Finally, the appendix gives a detailed description of the current LND configuration and calibration parameters.

2 Preparation of LND data

2.1 Instrument and the simulation set-up

A schematic structure of the LND sensor head is given in Figure 1. It consists of ten silicon detectors, each with a nominal thickness of 500μm. Detectors are arranged in a charged-particle telescope configuration and labeled A through J from top to bottom. Each detector is segmented into an inner and an outer segment (labeled 1 and 2, respectively), each with approximately the same area. LND uses coincidence measurements to detect the charged particles which requires particles to pass through the uppermost detector A and at least trigger the next detector, B. In the middle of the sensor head, Al and Gd absorbers (shown in olive and red in Figure 1) are used to detect the upward and downward flux of thermal neutrons (Gd has a very large crosssection for thermal neutrons). These absorbers increase the scattering of particles inside the sensor head and absorb part of the particle energy without providing a measurement thereof.

Above detector A, the front window is covered by a lid that opens when LND operates, allowing particles to arrive at the detector without any obstacles. During lunar nights, the lander closes the lid to keep the sensor head and detectors warm enough for them to survive the cold lunar night. The payload compartment of the Chang'E - 4 lander which contains other scientific instruments is located beneath the LND sensor head and thus provides extra shielding for upward albedo particles before they reach LND. However, the details of that shielding are unknown and an estimation of 10 - mm Al shielding equivalent was provided by the Chang'E - 4 team. We thus add a 10 - mm Al sheet between LND and the lunar soil in our model for estimating the albedo contribution to LND as shown in Fig. 1. Consequently, the shielding material raises the lowest energy that albedo protons are required to have in order to arrive at the LND detector. A more detailed description of LND is given in Wimmer - Schweingruber et al. (2020).

We employed the GEANT4 toolkit (Agostinelli et al., 2003), version 10. 4. 1, using the QGSP _ BERT physics list to simulate the detector response to different particles from different directions at the relevant energies.

To simulate the particles from above, we placed a squareshaped planar source right above the front detector A with a size that is larger than the size of detector A and fully covers the field of view (FOV) of LND, i. e., the inner dashed line of the combination of A1 and B1 in Fig. 1. Similarly, a square particle source is placed below the 10 mm - Al sheet when simulating upward particles. The sources are isotropic and the energy spectrum is a power law with index - 1, i. e., logarithmically flat, to allow easy scaling to any other spectral index[1], as discussed, e. g., in Guo et al. (2019), but is a standard process. The spectrum is in differential flux units. We then obtain the averaged geometric factors from such a simulation and use them to convert detected counts to flux.

2. 2　Xmas plot

The Xmas plot[2] is the primary data product of LND that is generated onboard and represents the memory space of the LND flight model. All LND measurements are stored in a 274×64 matrix though not all elements of this matrix are transmitted to Earth. The memory space refreshes every hour; hence the time cadence of the Xmas plot is 1h. The Xmas plot can be divided up column - wise into the following regions (from columns 0 - 274): thermal neutrons, fast neutrons, TID, LET, and charged particles. More details of the data products can be found in Wimmer - Schweingruber et al. (2020, Sec. 4). In this section, we focus on charged particles and explain the generation of the corresponding two -dimensional density histogram in the Xmas plot which is important to understand the calibration of LND data. Part of the charged particles measurement is displayed in panels (a) and (b) of Fig. 2.

---

1 The "logarithmically flat" spectrum appears "flat" or horizontal when plotted in a log - log plot. Thus, bins are evenly spaced in a logarithmic scale and have the same number of counts per bin

2 We call this plot "Xmas plot" because our group in Kiel used a similar plot as our 2017 Xmas card

The charged particles in the Xmas plot can be further divided up into two parts, stopping and penetrating particles. The stopping particles have energies between ~8 and ~35MeV/nuc. LND thus measures SEPs, ACRs, and lower energy GCRs in this energy range. When arriving at LND on the lunar surface, particles within this energy range can first trigger and penetrate detector A, then pass the following detectors, depositing (part of) their kinetic energy. Depending on their primary energy, some of these particles will stop in one of the detectors between B and I (as indicated by B, C, etc. along the top of Fig. 2a) and hence deposit all of their energy. The higher their primary energy, the deeper they get in the detector stack. The upper limit is ~35MeV/nuc for protons and helium nuclei stopping in detector I, but different for heavier ions (Wimmer - Schweingruber et al., 2020). A small uncertainty in their primary energy remains due to noise in the individual detectors. Particles with energies above 35 MeV/ nuc will penetrate detector I and stop in detector J or penetrate it, i. e., all detectors from A to J are triggered. Penetrating particles may come from above or below. Panels (a) and (b) of Fig. 2 show those parts of the Xmas plot that belong to stopping and penetrating particles. A precise description of the quantities used to populate the Xmas plot are given in Table 5 of Wimmer - Schweingruber et al. (2020), but a short summary is given here for convenience. For particles stopping between detectors B and H, the values along the x axis are a monotonous function of the ratio $E_{\text{total}}/E_{\text{A}}$. The y axis is ordered according to the product $E_{\text{total}} \cdot E_{\text{A}}$. The ratio and product are mapped to row and column values internally in the instrument. Thus, the column regions marked by B, C, D are similar to traditional $\mathrm{d}E/\mathrm{d}x$ - E plots, but rotated and compressed into $\mathrm{d}E/\mathrm{d}x \cdot E_{\text{total}}$ vs. $\mathrm{d}E/\mathrm{d}x/E_{\text{total}}$ space. In this space the "1/E" energy loss appears as a horizontal line and the penetrating particles do not show up because they are detected in the next detector. The horizontal (energy) axis is compressed by the division by E_{A}. Particles stopping in detectors E, F, etc. are treated somewhat differently, see Wimmer - Schweingruber et al. (2020) for more details. The quantities used for penetrating particles and particles stopping in detector I are similar. The x axis is ordered by E_I/E_A and the y axis is ordered by the sum of energy depositions in detectors B, C, and D (see Section 2. 4. 1 for the exact definition of penetrating particles). Stopping particles populate columns 32 - 240, while penetrating particles are between 246 and 266. The data in the empty regions in Fig. 2A are not transmitted back to Earth.

The proton channels that are used in this work are outlined by the black boxes in panels (a) and (b) of Fig. 2 and further indicated by H1 - H14. We call them DPS boxes, where DPS stands for "data product scheduler". Count numbers inside DPS boxes are read out from memory every minute. Therefore the cadence of primary proton data is 1min. These high - timeresolution proton data are important during the onsets of SEP events (Xu et al., 2020). The pixels with high count rates around row 20 in Fig. 2a are ^{4}He counts. Heavier ions would lie at even higher rows.

One of the advantages of the Xmas plot data product is that one can define one's own data product. For this work, we defined a mask for upward - moving protons. It is marked by the blue box in panel (b) of Figure 2. The positions of those pixels were determined manually based on simulation results. Because such particles are rare, the 1 - h time resolution of the Xmas plot is more than sufficient. 2. In the following sections, we explain this data product in depth.

The Xmas plot in Fig. 2 was accumulated between the seventh and the 20th lunar day (from 27 June 2019 to 26 July 2020)① . This was a period in the deep solar minimum between solar cycles 24 and 25. The minimum solar modulation and solar activity during that period led to the highest GCR flux since space age (Fu et al., 2021). It offers the opportunity to validate LND's performance using GCR data without the interruption by SEPs. As discussed above and in the appendix, the thresholds of several detectors had to be increased, the period we use for this work begins after all these changes have been made.

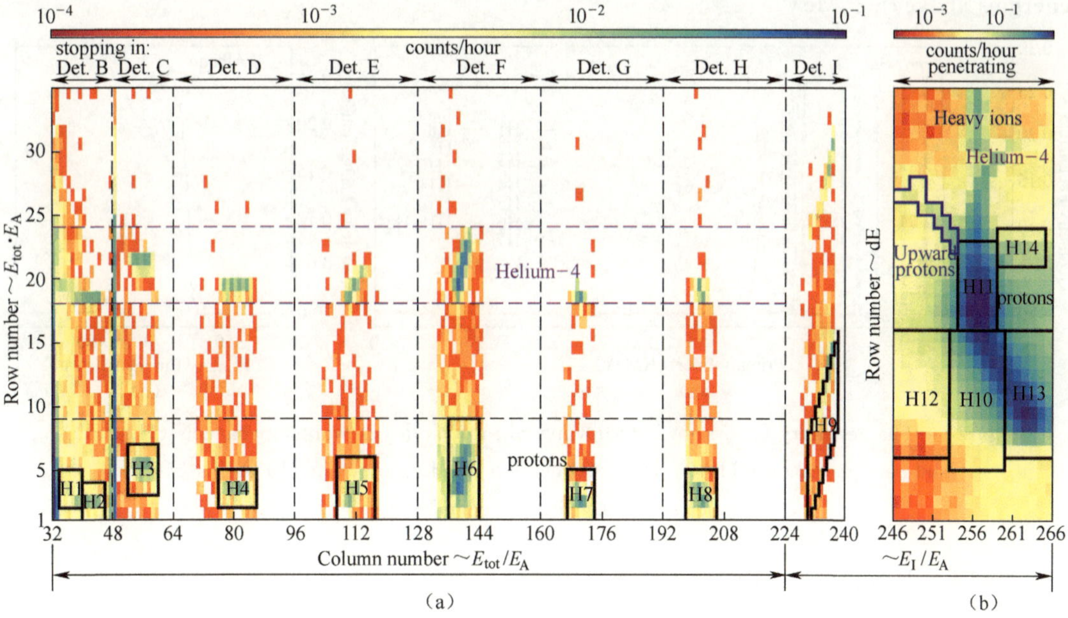

Fig. 2　The (a) panel displays the "Xmas plot" of part of the charged particles stopping between detectors B and I. The empty area represents the part of the LND memory that is not transmitted to the ground. Panel (b) is the Xmas plot between columns 246 and 266. The primary proton channels and albedo proton channel are defined accordingly in those two panels, marked by black and blue boxes. The quantities used to calculate the entry "pixel" in the Xmas plot are calculated according to the quantities summarized in Tab. Five in Wimmer - Schweingruber et al. (2020).

① One lunar day is approximately 28 Earth days long. The periods when LND was on are listed in https: //www. ieap. uni - kiel. de/et/change4/data _ by _ lunar _ day/info

2.3 Detection of downward - moving protons

2.3.1 Stopping and penetrating protons

Stopping and penetrating protons populate DPS boxes H1 - H14, their geometry factors are shown in panel (a) of Figure 3. The DPS boxes H1 - H9 count the protons that stop in detectors B to I①. DPS boxes H10 - H14 record penetrating protons with energies above ~35MeV. The energy bins in H10, H11, and H14 are used in this paper for the first time, they extend the measurement capability of LND beyond the energy of stopping particles. Unlike those for stopping particles, the geometry factors of penetrating particles partially overlap, as can be seen in panel (a) of Figure 3. Moreover, the measurement of penetrating channels has a larger uncertainty due to uncertainties in the determination of the background as will be discussed in Section 2.3.3. Currently, data from the penetrating channels are not yet publicly available because they still require a better calibration. With the onset of solar activity, we expect this to be achieved with some large SEP events with energies above ~35MeV.

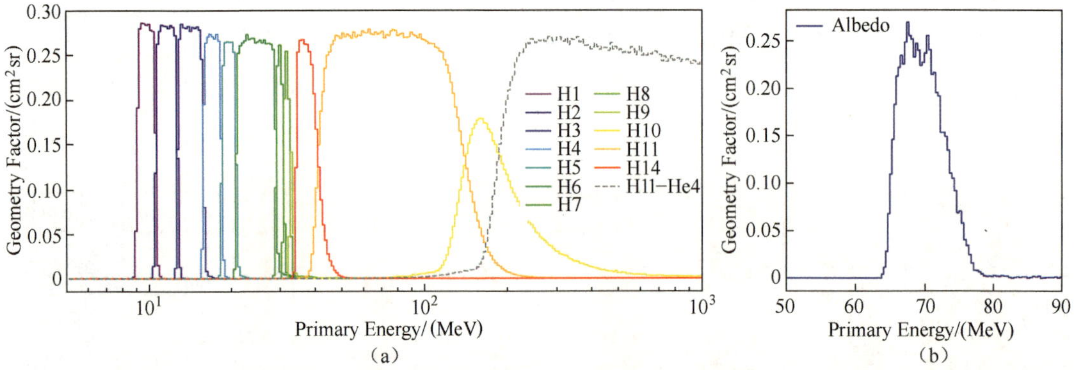

Figure 3　The geometry factors for both downward protons in (a) stopping channels (H1 - H9) and penetrating ones (H10, H11, H14), and albedo protons in (b) are based on an input spectrum with power index $\gamma = -1$

2.3.2 Calculation of fluxes

The differential flux, dF is calculated according to Eq. 1 which is valid for all particle species, including albedo protons,

$$\mathrm{d}F = \frac{N}{\mathrm{d}E \cdot \overline{G} \cdot \mathrm{d}t} \tag{1}$$

① Note that channel H9 has a relatively narrow energy range (bin width) and a smaller geometry factor than the other channels. The reason for this is an error in the definition of the corresponding DPS box (between columns 224 and 240) which does not contain the bulk of the particle population stopping in detector I. As shown in panel (a) of Figure 2, nearly half of the pixels with higher density lie outside of the black box, leading to the reduced geometry factor. This defect was not resolved in time for the flight model and hence the updated geometry factors given in the appendix should be used to make use of this channel where N is the number of counts in the DPS boxes, dE is the energy bin width determined from the response functions shown in panel (a) of Figure 3, dt is the accumulation time of the whole measurement, and includes possible dead time corrections (Wimmer - Schweingruber et al., 2020), and $\overline{G}$ is the weighted geometry factor of each energy bin. The method that was used to calculate the averaged geometry factors is given in the appendix and their values are given in Table S3. The geometry factors of primary protons and albedo protons as a function of energy are given in panels (a) and (b) of Figure 3.

We note that, despite the intention of the concept of geometry factors, the averaged geometry factors derived *via* Eq. S1 in the supplementary material do depend on the input spectra. As explained in appendix, the contribution of highenergy particles results in non - negligible contributions in the count rates for low - energy channels. Thus, the spectral shape beyond ~ 100MeV affects the count rate in the penetrating channels. The stopping channels are less affected, as discussed in the appendix.

In Table S3 we provide the averaged geometry factors derived for two different input spectra; one is a power - law spectrum with index - 1 (column 3), where the particles are uniformly distributed in logarithmic energy space. The other one is the numerical CREME96 GCR model spectrum (Tylka et al., 1997) for the 2019 solar minimum (column 4).

2.3.3　Correction of the contribution from ^4He and heavy ions to the proton flux

Close inspection of panel (a) of Figure 2 shows that ^4He and heavy ions contaminate the proton DPS boxes H1 - H9. While ^4He is well separated from protons (its counts can be seen to populate the Xmas plot around row number 20), the Xmas plot entries between ^4He and protons are clearly populated by a substantial background. This background which mainly affects the low - energy proton DPS boxes is primarily due to high - energy He and heavier ions which create secondary particles after they interacted with detector A. The secondary particles trigger B after which A is read out. Because the geometry factor for interaction with A is large and the flux of high - energy (GCR) ions is also high, this process happens sufficiently often to result in this background.

The penetrating channels shown in panel (b) of Figure 2 are affected by the noisy detector segments A2, I1&I2, and J1. The effect of this malfunction has been accounted for by increasing the thresholds of these detectors and disabling A2 in the LND level - 3 data processing logics (see Wimmer - Schweingruber et al., 2020, for a discussion). This results in the asymmetry shape seen in panel (b). In the supplementary material, we give the detailed explanations of this asymmetric structure. Moreover, minimally - ionizing protons are affected more than ^4He because of their smaller energy deposition in the detectors. Thus the detection efficiency for minimally - ionizing protons (measured in H10) is reduced because of the increased thresholds in I and J. The minimally - ionizing protons populate primarily H10, whereas minimally ionizing ^4He populates primarily

H11. Thus, as can be seen in panel (a) of Figure 3 the protons measured in H11 are strongly affected by minimally - ionizing ^{4}He (dashed line in Figure 3).

In order to quantify these different sources of contamination and reduced detection efficiencies, we ran extensive GEANT4 simulations of LND. We used the simulation setup introduced in sec. 2.1 to model elements from hydrogen to iron, thus including all relevant GCR species. Instead of calculating the geometry factors for detecting these elements in their instrument channels (i. e., in the appropriate boxes in the Xmas plot), we derived the geometry factors for measuring them in the proton channels (H1 - H14) as a function of energy. These simulation data were then analyzed using the current LND configuration (incl. thresholds). Thus we could estimate the contamination of the H1 - H14 proton channels by high - energy (GCR) helium and - to a lesser extent - heavy ions. Of course, ^{4}He is rarer than hydrogen, but the contribution of GCR 4He to the stopping proton channels (H1 - H7) is on the order of a few percent, but larger for H8 and H9 (cf. Tab. S3 of supplementary material). Similarly, minimally - ionizing ^{4}He affects the penetrating, but not minimally - ionizing protons in H11, only 36% of all particles in H11 are protons (see respective entry in Tab. S3 of supplementary material).

To check the influence of different GCR models on the correction factors, we compare the results of the CREME 96 (Tylka et al., 1997) and BON14 GCR models (O' Neill et al., 2015). The correction factors are given in columns 5 and 6 of Tab. S3 of supplementary material. Both models have comparable values, except for the factor of channel H1, which measures protons with energy of about 10MeV. CREME96 suggests that about 95% of particles measured in H1 are protons, while this percentage for the BON14 is about 87%. This discrepancy is likely due to the different spectral shapes for protons below 10MeV. CREME96 includes low - energy (solar) particles while BON14 only includes GCR protons (and heavier ions). Thus, the ratio of "true" protons in H1 to "background protons" is larger for CREME96 than for BON14.

We note that the contamination from GCR ions discussed above is negligible during SEP events because of the higher fluxes at low energies (softness of the spectra) and the higher proton abundance compared to heavier ions. However, the same contamination processes will affect the measurements, but to a different extent that depends on the exact spectra of the solar particles. Because the GCR - induced background is always present, we can subtract the background which was measured before the SEP event from the measurements taken during the event.

2.4 Detection of albedo protons

2.4.1 Measurement principle and response function

As alluded to in Section 2.2 both downward and upward pointing particles contribute to the "penetrating" channels in the Xmas plot between columns 246 and 266 (panel (b) of Figure 2). The extent of the y - axis shown covers penetrating protons and ^{4}He nuclei, heavier nuclei do not contribute to the region shown here.

The x and y axes in panel (b) of Figure 2 are defined as $4 \cdot \log_2 (E_I/E_A) + 16$ and $4 \cdot \log_2 ((E_{B1} + E_C + E_D)/100)$ respectively, where E_I, E_A, E_C, E_D represent the total energy deposited in both the inner *and* outer segments of the corresponding detectors, and E_{B1} is the energy deposition in the inner segment of detector B. The row number is determined by the total energy deposited by particles in detectors B1, C, and D. The Bethe - Bloch equation (Bethe, 1930; Bloch, 1933), $dE \propto \frac{Z^2}{ME}$, gives the energy deposited by particles in the silicon detectors; high - energy particles tend to deposit less energy than lower - energy particles (E less than ~1GeV) and heavy ions deposit more energy than light elements due to their higher nuclear charge, Z. Therefore, in the Xmas plot, the row number of protons with high primary energy is smaller than the low energy proton, and the protons are at the bottom of the 2D histogram. ^{4}He and heavy ions are located above them.

The column number of a particle in the Xmas plot depends on the E_I to E_A ratio. Suppose particles in particular the minimum ionizing particles, deposit similar energy in the front detector A and the bottom detector I, i. e. $E_I \sim E_A$, particles will be placed in the middle column of the penetrating histogram, i. e. , in column 256. When the particle deposit more energy in detector I than detector A ($E_I > E_A$), the quantities for the x - axis are larger than 256, and those particles will be added to the pixels in the right half of panel (b). On the other hand, if a particle deposits less in detector I than in detector A, it will be placed in the left half of panel (b). In principle, the former corresponds to particles moving downward, since particles first trigger detector A with larger incident energy and trigger I in the end with smaller incident energy after they lose their energy in the previous detectors, and the latter represents particles moving upward. Here we only focus on particles with energies of up to a few tens to a few hundred MeV/nuc and for which their average energy loss in silicon detectors decreases with the increased primary energy.

From simulations we thus determine the region marked in blue in panel (b) of upward - moving protons and calculate the corresponding geometry factor as shown in the panel (b) of Figure 3. The energy range of this albedo proton channel is (64. 7 - 76. 7 MeV), with an average geometry factor of 0. 18cm^{2}sr for an $dJ/dE \propto E^{-1}$ spectrum. The energy range is defined by the 10%- of - maximum criterion, and the uncertainty in these limits was estimated at the 5 and 10% of maximum and 90 - 95% of maximum levels. Given the steep flanks of this energy channel, these uncertainties are realistic.

Fortunately, our simulations show that the noise issue and the changes in LND configuration do not affect LND's response to albedo protons. They are too much away from the affected region in the Xmas plot, as can be seen in panel (b) of Figure 2.

2. 4. 2 Background subtraction

The count rates of albedo protons between 64. 7 and 76. 7MeV in the Xmas plots are affected by a background which is caused by high - energy protons from both directions and needs to be subtracted. To better visualize this background, we plot the distributions of the count rates with statistical uncertainties as error bar along the Y - axis from columns 246 to 253 in the different panels of Figure 4. The blue dots are the averaged counts per hour in each pixel, and the orange dots are the count rates of albedo protons in the relevant pixels.

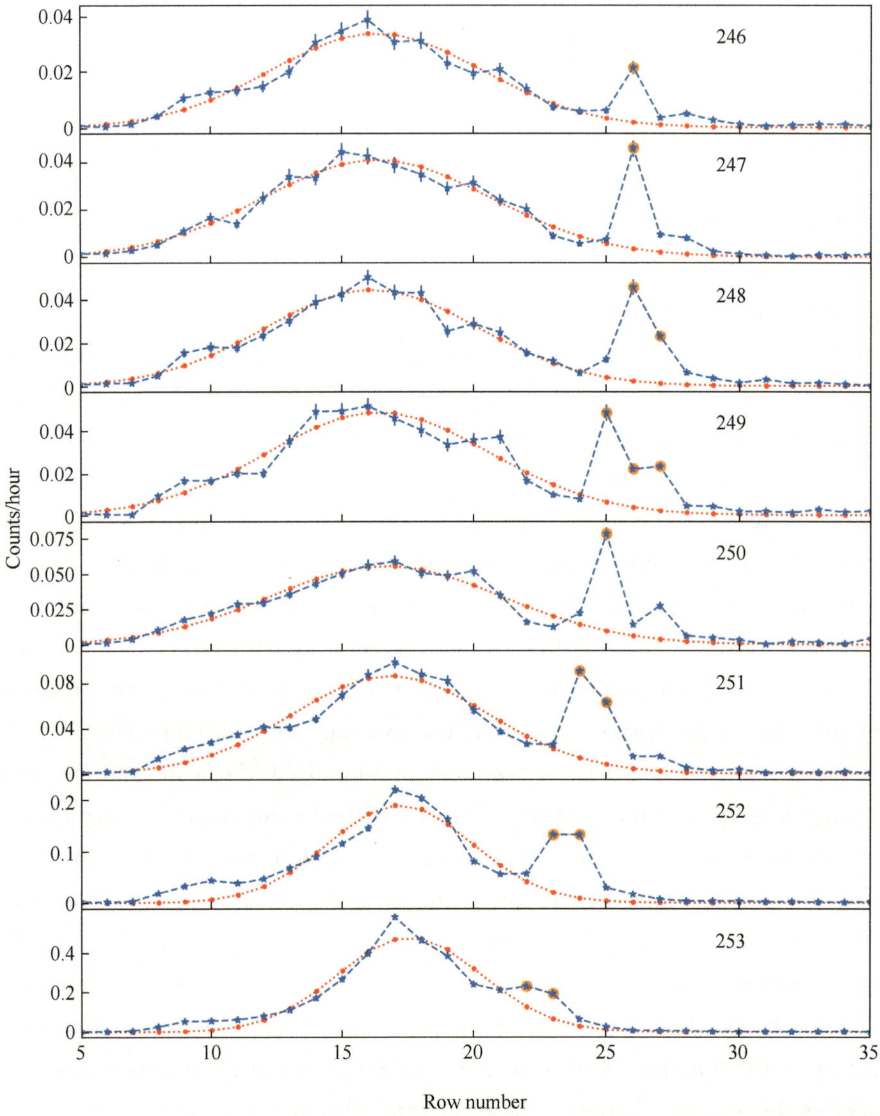

Figure 4　The distribution of count rates along the Y - axis between columns 246 and 253 "Xmas plot". The blue dashed lines are the measurements and the red dashed lines are the fitted Gaussian distributions. The error bars are statistical uncertainties. The difference between the orange dots and the background gives the albedo proton - contributed count rates

We model the background along the y - axis with a Gaussian distribution. The fitted results are plotted as red dashed lines in Figure 4. Each panel is fitted separately with different parameters. Finally, the backgrounds of the albedo protons are estimated based on the fitted model and subtracted from the albedo proton data points (shown as orange data points) to determine the flux of albedo protons.

3 Measurements and comparison with model

Following the calibration processes mentioned before, we derived the averaged primary proton spectrum between 8 and 368 MeV and the averaged albedo proton flux between 64.7 and 76.7MeV for the lunar days between June 2019 and July 2020, i. e., when LND was on. No solar energetic protons were observed by LND on the lunar surface.

3.1 GCR primary proton spectrum

The obtained LND GCR proton differential flux is plotted as empty blue circles in Figure 5. The error bars in the x and y directions correspond to the widths of the energy bins and the flux uncertainty, respectively (The numerical values are also given in the last column of table S3 of appendix). The nine lowest energy channels (H1 - H9) are the averages of the 1 - min data products of stopping protons in the energy range between 9 and 35MeV. The rightmost three proton channels (from left to right, H14, H11, H10) show the penetrating proton flux with energy above 35 MeV. The detailed information of the bins is given in Table S3. We only show the statistical uncertainties of the differential flux. Since each channel accumulated a large number of particles during this 1 - year measurement reported here, the statistical errors of each channel are small, especially for the penetrating channels. Hence the error bars of those three channels are almost invisible. Moreover, as we explained in Section 2.3.3, the contamination caused by GCR ions has been subtracted.

Besides the measurements of LND, we also plot SOHO/EPHIN (grey square). SOHO/EPHIN provides measurements of protons with energies between 4.3 and 53 MeV. In this work, we use the EPHIN daily averaged level 3 proton fluxes derived from the PHA data according to the methods developed by Kühl et al. (2020). Those data have 16 energy bins provided by the SOHO/EPHIN team. The periods of EPHIN data are the same as those during which LND operated, i. e., to the times when Chang'E 4 was on during local lunar daytime. One sees that the proton spectrum drops in the energy channels above 50MeV. As explained by the EPHIN team, this change is unexpected and might not reflect the actual proton spectrum in space①. Bartocci et al. (2020, Table 1).

① P. Kühl, private communication, 2022

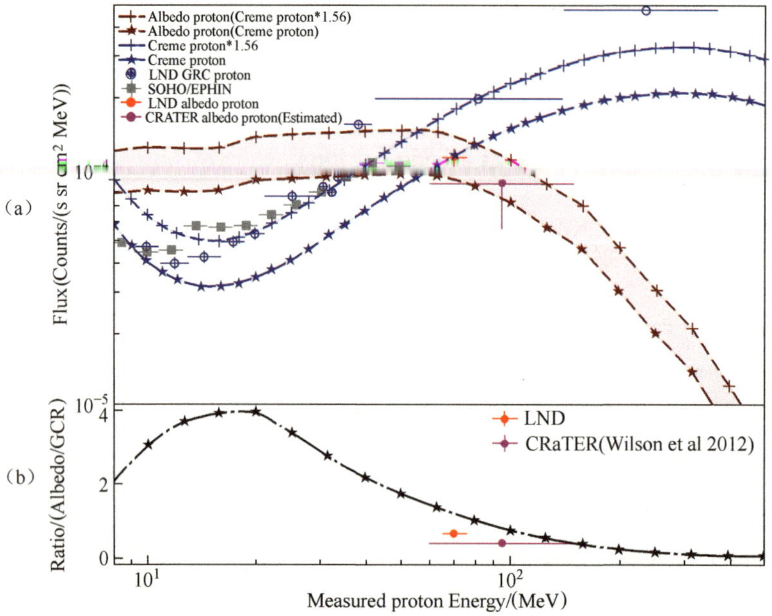

Figure 5　(a) The GCR primary proton spectra including both measurements (LND in empty blue circles, and SOHO/EPHIN in grey triangles) and modeled results (blue dashed lines). Albedo proton spectra are drawn in reddish colors. The simulation results from REDMoon are within the transparent red area. The LND measurement of albedo protons is the red circle. The error bar is the systematic uncertainty as defined in Section 2.4. The estimated albedo proton flux that we derived using the ratio of albedo to primary proton measured by LRO/CRaTER is the magenta data point. (b) The ratio of albedo to primary flux versus proton energy. The red and magenta points are the ratio of albedo to primary proton flux derived from LND data in this study and that from LRO/CRaTER data (Wilson et al., 2012), respectively. The black line shows the prediction from the REDMoon model. The detailed explanations are given in the main text, the values for the LND data pints are given in Tab. S3 of supplementary material.

As shown in Figure 5, the GCR proton spectrum measured by LND agrees well with that measured by SOHO/EPHIN in the energy range below 40MeV.

Table 1　The ratio of lunar albedo protons to primary protons

LND (64.7 - 76.7MeV)	CRaTER (60 - 150MeV)	simulation (65 - 76MeV)	simulation (60 - 150MeV)
2019 - 2020	2008 - 2009	2019	2019
64%±7	38%±2	80%	33%

Bartocci et al. (2020), Figure 5.

Moreover, we also compare the LND measurement of GCR protons with the CREME96 and BON14 cosmic ray models. CREME96 is a widely - used program that can simulate the ionizing - radiation environment in near - Earth space and predict the cosmic ray flux, based on the semi - empirical model of Nymmik et al. (1992). Their solar quiet model includes GCR, ACR, and a low - energy component below ~10MeV that originates from the Sun and interplanetary space. Likewise, BON14 is based on GCR measurements from particle detectors and can predict the GCR variations related to solar modulation which is an input parameter of the model. In Figure 5, we only plot the CREME96 cosmic ray spectra of 2019 as dashed blue lines since the BON14 predicts a very similar proton flux during solar minimum. The lower curve is the original spectrum generated by the CREME websitefor this period. Obviously, the GCR model underestimates the proton flux at a few tens of MeV during solar minimum 24/25, and the measurement is about 48 ±2% higher than the model spectra. This percentage is the averaged ratio between the measurement and the mode. The comparison with BON14 leads to the same conclusion. Therefore, we multiplied the CREME spectrum by a factor of 1.56 for the limited energy range which fits the LND measurement best. This implies that the actual solar modulation during the solar minimum 24/25 may be weaker than the modulation used in numerical models. A similar disagreement between models and experimental data has been reported by Mrigakshi et al. (2012) during the solar minimum 23/24 and further discussed by Matthiä et al. (2013). They discuss that the averaged sunspot number is not an accurate predictor of GCR fluxes during solar minimum and show substantial differences between the GCR ISO model (ISO, 2004) (which is a later version of the Nymmik et al. (1992) model) and the BON model. In fact, solar modulation has been reported to be reduced in 2019 - 2020 and GCR intensities at one au during this period reached the highest record in the space age (Fu et al., 2021).

3.2 Albedo protons

As discussed in Section 2.4, LND can resolve albedo protons and determine their flux between 64.7 and 76.7MeV. This is shown as the red circle in Figure 5 for the same time period as the GCR spectra. The background has been subtracted and the error of the flux is the resulting systematic uncertainty which is larger than the statistical error. The averaged flux at ~ 70.5MeV is about $1.12 \pm 0.09 \times 10^{-4}$ $(cm^2\ s\ sr\ MeV)^{-1}$ which is remarkably close to the primary GCR proton flux measured by LND in this energy range. Albedo protons are the product of high - energy GCRs which interact with the lunar soil. A small fraction of the reaction products can escape from the soil and thus forms the albedo population. This interaction is modeled by the REDMoon code (Dobynde and Guo, 2021) which then predicts the lunar proton albedo as shown by the red lines in Figure 5 and discussed in the following paragraph.

Here the REDMoon model takes the CREME GCR model as input to calculate the angle - and energy - resolved surface particle flux on the Moon. The result for the time period reported here is given in Figure 6 which shows the isotropically distributed downward flux (zenith angle 90 - 180 deg) at the lunar surface in the upper half of the figure. The lower half shows the angleand energy - resolved upward flux of (albedo) protons, the black rectangle marks the energy range which LND measures. The albedo protons are primarily due to GCR protons, but GCR helium and heavier ions contribute a small fraction. The REDMoon results shown as dashed red lines in Figure 5 were calculated with the unscaled CREME model as well as a CREME input spectrum scaled by a constant factor of 1.56, as discussed above. The region between these two model predictions is shaded in pink and represents the upward proton flux that is expected to be measured by LND. The decrease of the albedo protons flux at energy higher than ~100MeV reflects that the generation efficiency of secondary particles drops with energy.

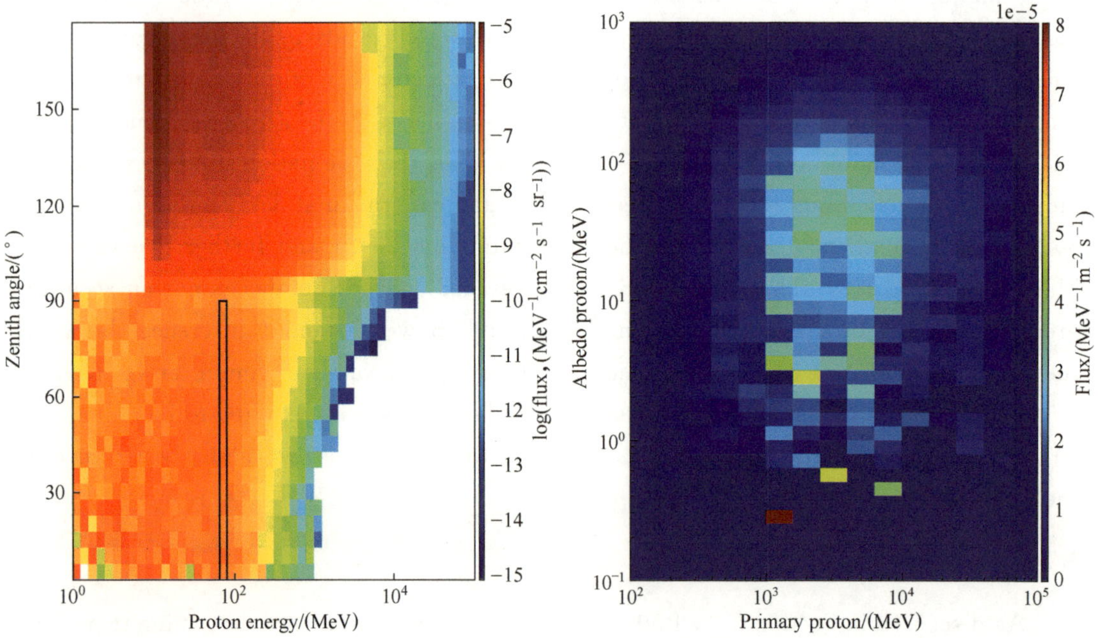

Figure 6　(a) The angular distribution of protons on the lunar surface including both albedo protons (0 - 90°) and GCR primary protons (90 - 180°). The plot is modified from panel (a) of Figure 4 in Dobynde and Guo (2021). (b): Albedo proton flux as a function of primary protons shows a maximum around albedo energies of 80MeV. These albedo particles are produced by primary protons with much higher energies. Both albedo particle distributions are derived by REDMoon model.

The lower panel of Figure 5 shows the ratio of upward (albedo) to downward (GCR) proton flux as derived by taking the ratio of REDMoon and CREME simulation results. Note that the ratio is the same for the scaled and non - scaled inputs of CREME spectra and that it is computed at the same energy of downward and upward propagating protons. The ratio peaks around 20MeV, where the albedo protons are more than two times higher than the primary GCR protons in that energy range. Above it, the ratio drops as the GCR flux increases and the albedo flux decreases. The high ratio (>1) is the consequence of the high - energy GCR making the main contribution to the low - energy albedo protons.

We determined the ratio of albedo protons to the primary protons in the energy range between 64.7 and 76.7MeV, as measured by LND. The average primary proton flux is estimated from the scaled GCR spectrum that fits the LND measurements. We found that the ratio is about 0.64±0.07 which means that the albedo protons are a significant contribution to the particle flux on the lunar surface in this energy range. Thus, about 39±6% of all protons around 70MeV are secondary albedo particles emitted from the lunar regolith. The second data point shown in magenta dots is the ratio reported by the CRaTER team (Wilson et al., 2012). The REDMoon simulation results agree well with the both the LND and CRATER measurements in their energy ranges.

Wilson et al. (2012) The CRaTER data point was only reported by Wilson et al. (2012) as a ratio of the number of upward (albedo) protons to that of downward protons measured within the same FOV and the same energy range for both particle populations. They found a value of 0.38±0.02 for this ratio which is different from the one reported by LND in this work. It is important to note that the CRaTER measurements are sensitive to a different energy range than LND, i.e., between 60 and 150MeV (with an average energy higher than the LND-seen albedo protons) and were obtained between 2009 and 2011. This time period also corresponds to a solar minimum and we don't expect the ratio of upward to downward protons to depend strongly on the detailed differences in the two solar minima between solar cycles 23/24 and 24/25.

Primary (GCR) protons interact with the lunar surface and can produce the albedo population reported here. Schwadron et al. (2017) discuss the various generation mechanisms in more detail. The overall shape of the ratio of albedo to primary proton can be understood without sophisticated modeling: Albedo protons necessarily must have lost a significant part of their energy in the uppermost layers of the lunar regolith. Because the primary (GCR) proton flux increases with energy in and beyond the energy range covered by LND, the ratio of albedo/primary proton flux must decrease with energy. This simple explanation is supported by REDMoon model calculations and were also obtained by Looper et al. (2013) in their simulation of the radiation environment on the lunar orbiter of 50km height. This also explains why the LND observations of the albedo to primary ratio is nearly two times higher than that of LRO/CRaTER. In Table 1, we list the ratios of different cases, including the ratios from LND in 2019, CRaTER in 2009, and the ratio from the simulation averaged over different energy. We always use the GCR spectrum of 2019 as the input spectrum in the simulation. Ignoring the time difference, we can directly compare the simulation with measurements in the same energy range. The modeled albedo to primary proton ratio averaged between in the LND energy range (65-76MeV) is 0.8, which is slightly higher than LND's measurement, but the albedo flux falls within the pink shaded band in the upper panel. For CRaTER one needs to average the model over a wider energy range which is given by CRaTER's energy range. Thus folding the model function with the CRaTER energy range, we obtain a value of 0.33, which is comparable with the CRaTER's result.

Wilson et al. (2012) only report the albedo to primary proton ratio, but not the upward proton flux. This can easily be estimated by multiplying the scaled CREME model flux in the appropriate energy range with the reported albedo ratio and is shown as the magenta point in the upper panel of Figure 5. The albedo proton flux that is calculated using the original (unscaled) GCR spectrum can be taken as a lower limit. LND measurements, CRaTER measurements, and the REDMoon simulation are consistent within the reported uncertainties.

4 Summary, discussion and conclusion

We have performed and reported a careful re - analysis of the LND calibration. LND data in the energy range between 9 and 368MeV agrees well with data from the SOHO/EPHIN instrument. LND data up to 35MeV are for well - resolved protons stopping in LND's detector stack; the remaining three data points are for penetrating protons.

Furthermore, we compared LND's proton spectrum with the CREME96 GCR model. We find that the CREME96 predictions underestimate the proton flux during reported time period in the solar minimum 24/25. The average ratio between LND measurements and the CREME96 model predictions is $\sim 1.48 \pm 0.02$. This may be due to the much weaker solar modulation during the deep solar minimum 24/25. In fact, GCR intensities at one au during this time period reached the highest recorded fluxes in the space age (Fu et al., 2021).

With its stack of 10 silicon detectors, LND is capable of measuring both the downward as well as the upward proton flux. Such albedo protons are the result of the interaction of high - energy GCR particles with the lunar regolith. The bulk of the albedo particles is created by the high - energy portion of the primary GCR protons, as shown in the right hand panel of Figure 6. This part of the GCR spectrum is less affected by solar modulation than the low - energy part. Therefore, when considering the effect of solar modulation on the ratio of upward to downward particles one needs to consider that this ratio is calculated at the same primary energy of the upward and downward - pointing particles. Because the low - energy part of the GCR spectrum is strongly affected by solar modulation, the up/down ratio is primarily affected by the low - energy downward flux and much less so by the upward flux which is primarily determined by the high - energy downward - pointing flux, as reported by Wilson et al. (2012). This effect is especially important during solar particle events. Comparing the measured upward to downward differential fluxes we find a ratio of 0.64 ± 0.07 in the energy range between 64.7 - 76.7MeV. This ratio is shown as the red data point in the lower panel of Figure 5. Because of the additional shielding from beneath provided by the Chang'E 4 lander, the energy ranges resolved by LND differ for the upward and downward particle populations. Furthermore, we scaled the upward to downward flux ratio reported by the CRaTER team using data prior to solar minimum 23/24 to the time period investigated here. We used the scaled CREME96 downward flux to determine the upward flux expected in the CRaTER energy range. The resulting upward flux (shown as the magenta data point in the lower panel of Figure 5) agrees well with LND's measurement and the REDMoon predictions, despite being valid for a different energy range (60 - 150MeV).

As already stated, the REDMoon simulation tool provides the upward (albedo) differential proton flux. We compare the REDMoon model results with the downward flux from the scaled CREME96 model to derive the ratio of upward to downward differential flux shows as the dash - dotted line in the lower panel of Figure 5. At energies below ~ 50MeV, the ratio is above one and it has a peak value above two near 20MeV. With increasing energy, the ratio decreases as the generation of albedo protons at this energy decreases. We observe that the modeled energy dependence of the albedo proton flux is confirmed by the combination of the LND and (scaled) CRaTER data. Luo et al. (2022).

LND is the first high - energy charged - particles telescope to be working on the lunar surface. We have presented measurements of low - energy cosmic ray protons in conjunction with albedo protons and that the latter contribute significantly to the particle flux on the lunar surface. We also show that the albedo proton flux is energy dependent, and that both primary and albedo proton fluxes are in good agreement with measurements from LRO/CRaTER. Obviously, due to their high flux and relatively low energy, albedo protons are an important contribution to the radiation to which astronauts would be exposed to on the surface of the Moon. Albedo protons are energetic enough to penetrate a space suit and have sufficiently low energy do stop in the body of an astronaut, thus depositing all their energy in the astronaut's tissue. The Sun is now becoming more active and future data from large solar particle events will be critical to understand their effect on the lunar radiation environment and its effect on human exploration of the Moon.

Acknowledgments

We thank the two reviewers for their thorough reviews which helped to improve this paper and the many individuals who were involved in conceiving, designing, building, testing, and launching LND and Chang'E 4.

References

[1] Agostinelli, S., Allison, J., Amako, K., Apostolakis, J., Araujo, H., Arce, P., et al. (2003). GEANT4 - a simulation toolkit. Nucl. Instrum. Methods Phys. Res. Sect. A Accel. Spectrom. Detect. Assoc. Equip. 506, 250 - 303. doi: 10.1016/S0168 - 9002 (03) 01368 - 8.

[2] Ambrosi, G., Bartocci, S., Basara, L., Battiston, R., Burger, W. J., Carfora, L., et al. (2018). The HEPD particle detector of the CSES satellite mission for

investigating seismo - associated perturbations of the Van Allen belts. Sci. China Technol. Sci. 61, 643 - 652. doi: 10.1007/s11431 - 018 - 9234 - 9.

[3] Appel, J. K., Köehler, J., Guo, J., Ehresmann, B., Zeitlin, C., Matthiä, D., et al. (2018). Detecting upward directed charged particle fluxes in the Mars science laboratory radiation assessment detector. Earth Space Sci. 5, 2 - 18. doi: 10.1002/2016EA000240.

[4] Bartocci, S., Battiston, R., Burger, W. J., Campana, D., Carfora, L., Castellini, G., et al. (2020). Galactic cosmic - ray hydrogen spectra in the 40 - 250 mev range measured by the high - energy particle detector (hepd) on board the cses - 01 satellite between 2018 and 2020. Astrophys. J. 901, 8. doi: 10.3847/1538 - 4357/abad3e.

[5] Berger, T., Matthiä, D., Burmeister, S., Zeitlin, C., Rios, R., Stoffle, N., et al. (2020). Long term variations of galactic cosmic radiation on board the International Space Station, on the Moon and on the surface of Mars. J. Space Weather Space Clim. 10, 34. doi: 10.1051/swsc/2020028.

[6] Bethe, H. A. (1930). Zur Theorie des Durchgangs schneller Korpuskularstrahlen durch Materie. Ann. Phys. 5, 325 - 400. doi: 10.1002/andp.19303970303.

[7] Bloch, F. (1933). Zur Bremsung rasch bewegter Teilchen beim Durchgang durch Materie. Ann. Phys. 16, 285 - 320. doi: 10.1002/andp.19334080303.

[8] Cucinotta, F. A., and Chappell, L. J. (2011). Updates to astronaut radiation limits: Radiation risks for never - smokers. Radiat. Res. 176, 102 - 114. doi: 10.1667/RR2540.1.

[9] Dobynde, M. I., and Guo, J. (2021). Radiation environment at the surface and subsurface of the moon: Model development and validation. JGR. Planets 126, e06930. doi: 10.1029/2021JE006930.

[10] Dorman, L. I. (2004). Cosmic rays in the earth's atmosphere and underground. Cosmic Rays Earth's Atmos. Undergr. 303. doi: 10.1007/978 - 1 - 4020 - 2113 - 8.

[11] Fu, S., Zhang, X., Zhao, L., and Li, Y. (2021). Variations of the galactic cosmic rays in the recent solar cycles. Astrophys. J. Suppl. Ser. 254, 37. doi: 10.3847/1538 - 4365/abf936.

[12] Hassler, D. M., Zeitlin, C., Wimmer - Schweingruber, R. F., Böttcher, S., Martin, C., Andrews, J., et al. (2012). The radiation assessment detector (RAD) investigation. Space Sci. Rev. 170, 503 - 558. doi: 10.1007/s11214 - 012 - 9913 - 1.

[13] ISO, I. (2004). Space environment (natural and artifical) - galactic cosmic ray model. Geneva, Switzerland: International Organization for Standardization, 15390.

[14] Kühl, P., Heber, B., Gómez - Herrero, R., Malandraki, O., Posner, A., and Sierks, H. (2020). The electron proton helium INstrument as an example for a space weather radiation instrument. J. Space Weather Space Clim. 10, 53. doi: 10.1051/swsc/2020056.

[15] Looper, M. D. , Mazur, J. E. , Blake, J. B. , Spence, H. E. , Schwadron, N. A. , Golightly, M. J. , et al. (2013). The radiation environment near the lunar surface: Crater observations and geant4 simulations. Space weather. 11, 142 - 152. doi: 10.1002/swe.20034.

[16] Luo, P. , Zhang, X. , Fu, S. , Li, Y. , Li, C. , and Cao, J. (2022). First measurements of low - energy cosmic rays on the surface of the lunar farside from chang'e - 4 mission. Sci. Adv. 8, eabk1760. doi: 10.1126/sciadv.abk1760.

[17] Matthiä, D. , Berger, T. , Mrigakshi, A. I. , and Reitz, G. (2013). A ready - to - use galactic cosmic ray model. Adv. Space Res. 51, 329 - 338. doi: 10.1016/j.asr.2012.09.022.

[18] McKinney, G. W. , Lawrence, D. J. , Prettyman, T. H. , Elphic, R. C. , Feldman, W. C. , and Hagerty, J. J. (2006). MCNPX benchmark for cosmic ray interactions with the Moon. J. Geophys. Res. 111, E06004. doi: 10.1029/2005JE002551.

[19] Mitrofanov, I. G. , Sanin, A. B. , and Litvak, M. L. (2016). Water in the Moon's polar areas: Results of LEND neutron telescope mapping. Dokl. Phys. 61, 98 - 101. doi: 10.1134/S1028335816020117.

[20] Mrigakshi, A. I. , Matthiä, D. , Berger, T. , Reitz, G. , and Wimmer - Schweingruber, R. F. (2012). Assessment of galactic cosmic ray models. J. Geophys. Res. 117, A08109. doi: 10.1029/2012JA017611.

[21] Müller - Mellin, R. , Kunow, H. , Fleißner, V. , Pehlke, E. , Rode, E. , Röschmann, N. , et al. (1995). Costep - comprehensive suprathermal and energetic particle analyser. Sol. Phys. 162, 483 - 504. doi: 10.1007/bf00733437.

[22] Nymmik, R. , Panasyuk, M. , Pervaja, T. , and Suslov, A. (1992). A model of galactic cosmic ray fluxes. Int. J. Radiat. Appl. Instrum. Part D. Nucl. Tracks Radiat. Meas. 20, 427 - 429. Special Section Galactic Cosmic Radiation: Constraints on Space Exploration. doi: 10.1016/1359 - 0189 (92) 90028 - T.

[23] O'Neill, P. , Golge, S. , and Slaba, T. (2015). Badhwar - O'Neill 2014 galactic cosmic ray flux model description. Tech. Rep.

[24] Schwadron, N. A. , Wilson, J. K. , Jordan, A. P. , Looper, M. D. , Zeitlin, C. , Townsend, L. W. , et al. (2017). Using proton radiation from the moon to search for diurnal variation of regolith hydrogenation. Planet. Space Sci. 162, 113 - 132. doi: 10.1016/j.pss.2017.09.012.

[25] Schwadron, N. A. , Wilson, J. K. , Looper, M. D. , Jordan, A. P. , Spence, H. E. , Blake, J. B. , et al. (2016). Signatures of volatiles in the lunar proton albedo. Icarus 273, 25 - 35. doi: 10.1016/j.icarus.2015.12.003.

[26] Spence, H. E. , Case, A. W. , Golightly, M. J. , Heine, T. , Larsen, B. A. , Blake, J. B. , et al. (2010). Crater: The cosmic ray telescope for the effects of radiation experiment on the lunar reconnaissance orbiter mission. Space Sci. Rev.

150, 243 - 284. doi: 10.1007/s11214 - 009 - 9584 - 8.

[27] Spence, H. E., Golightly, M. J., Joyce, C. J., Looper, M. D., Schwadron, N. A., Smith, S. S., et al. (2013). Relative contributions of galactic cosmic rays and lunar proton "albedo" to dose and dose rates near the moon. Space weather. 11, 643 - 650. doi: 10.1002/2013SW000995.

[28] Treiman, S. B. (1953). The cosmic - ray albedo. Phys. Rev. 91, 957 - 959. doi: 10.1103/PhysRev.91.957.

[29] Tylka, A. J., Adams, J. H., Boberg, R., Brownstein, B., Dietrich, W. F., Flueckiger, E. O., et al. (1997). CREME96: A revision of the cosmic ray effects on microelectronics code. IEEE Trans. Nucl. Sci. 44, 2150 - 2160. doi: 10.1109/23.659030.

[30] Wilson, J. K., Spence, H. E., Kasper, J., Golightly, M., Blake, J. B., Mazur, J. E., et al. (2012). The first cosmic ray albedo proton map of the moon. J. Geophys. Res. 117, 1 - 7. doi: 10.1029/2011JE003921.

[31] Wimmer - Schweingruber, R. F., Yu, J., Böttcher, S. I., Zhang, S., Burmeister, S., Lohf, H., et al. (2020). The lunar lander neutron and Dosimetry (LND) experiment on chang'E 4. Space Sci. Rev. 216, 104. doi: 10.1007/s11214 - 020 - 0725 - 3.

[32] Xu, Z., Guo, J., Wimmer - Schweingruber, R. F., Freiherr von Forstner, J. L., Wang, Y., Dresing, N., et al. (2020). First solar energetic particles measured on the lunar far - side. Astrophys. J. Lett. 902, L30. doi: 10.3847/2041 - 8213/abbccc.

[33] Zhang, S., Wimmer - Schweingruber, R. F., Yu, J., Wang, C., Fu, Q., Zou, Y., et al. (2020). First measurements of the radiation dose on the lunar surface. Sci. Adv. 6, eaaz1334. doi: 10.1126/sciadv.aaz1334.

First Solar Energetic Particles Measured on the Lunar Far - side

Zigong Xu[1], Jingnan Guo[2,3,1], Robert F. Wimmer - Schweingruber[1,4], Johan L. Freiherr von Forstner[1], Yuming Wang[2,3], Nina Dresing[1], Henning Lohf[1], Shenyi Zhang[4,5,6], Bernd Heber[1], and Mei Yang[7]

1. Institute of Experimental and Applied Physics, Kiel University, D - 24118 Kiel, Germany; xu@physik. uni - kiel. de

2. CAS Key Laboratory of Geospace Environment, University of Science and Technology of China, Hefei 230026, People's Republic of China; jnguo@ustc. edu. cn

3. CAS Center for Excellence in Comparative Planetology, Hefei 230026, People's Republic of China

4. National Space Science Center, Chinese Academy of Sciences, Beijing, People's Republic of China

5. Beijing Key Laboratory of Space Environment Exploration, Beijing, People's Republic of China

6. University of Chinese Academy of Science, Beijing, People's Republic of China

7. Beijing Institute of Spacecraft System Engineering, Beijing, People's Republic of China

Abstract: On 2019 May 6 the Lunar Lander Neutron & Dosimetry (LND) Experiment on board the Chang'E - 4 lander on the far - side of the Moon detected its first solar energetic particle (SEP) event with proton energies up to 21MeV. Combined proton energy spectra are studied based on the LND, SOHO/EPHIN, and ACE/EPAM measurements, which show that LND could provide a complementary data set from a special location on the Moon, contributing to our existing observations and understanding of space environment. We applied velocity dispersion analysis to the impulsive electron event and weak proton enhancement and show that electrons are released only 22 minutes after the flare onset and ~15 minutes after the type II radio burst, while protons are released more than one hour after the electron release. The beam - like in situ electrons and clear velocity dispersion indicate a good magnetic connection between the source and Earth. This is remarkable because stereoscopic remote - sensing observations from Earth and STEREO - A suggest that the SEPs are associated with an active region nearly 113° away from the magnetic footpoint of Earth. This suggests that these SEPs did not propagate along the nominal Parker spral normally assumed for ballistic mapping and that the release and propagation mechanism of electrons and protons are likely to differ significantly for this event.

Unified Astronomy Thesaurus concepts: Solar energetic particles (1491), Interplanetary physics (827), Lunar surface (974), Solar active regions (1974), Solar coronal mass ejection shocks (1997), Solar physics (1476), Space weather (2037), Solar active region magnetic fields (1975), Solar instruments (1499)

1 Introduction

The Chang'E - 4 mission, which consists of a lander, a rover, and a relay satellite, is the first mission to land on the far side of the Moon. It landed in the von Kármán Crater on 2019 January 3,02: 26 UTC. The Lunar Lander Neutron and Dosimetry experiment (LND) (Wimmer - Schweingruber et al. 2020) on board the lander of Chang'E - 4 is designed to take active dosimetry measurements on the surface of the Moon as its chief scientific goal. Apart from the primary objective of LND, which is to measure the radiation level on the lunar far - side preparing for astronaut missions (Zhang et al. 2020), the charged particle telescope also provides high - quality data of energetic particles and contributes to heliophysics. For example, LND provides proton and Helium - 4 spectra between 9 and 35MeVnuc^{-1}.

During the first year of the mission (2019), solar activity was minimal and LND detected only two small solar energetic particle (SEP) events on 2019 May 4 and 6, of which the second had sufficient counting statistics for protons between 9.0 and 21.0MeV to allow a meaningful analysis. This event was related to an active region located at E50, nearly 113° away from the Earth's nominal coronal magnetic footpoint, where an M1.0 class flare erupted before the SEPs onset and a narrow and slow coronal mass ejection (CME) appeared later. Combining remote - sensing observations of the solar source with in situ particle measurements from multiple spacecraft, we analyze and discuss the possible particle release and transport processes.

2 Observations

2.1 In Situ Measurements

Chang'E 4 landed on the far side of the Moon and can only operate during its local daytime because it is too cold at night. Therefore, LND only provides measurements ahead of the Earth's bow shock. It measures the energy that a particle deposits in its 10 detectors. LND can stop and thus identify the species and energy of charged particles up to 30MeV nuc^{-1}. In order to better understand the temporal variation of protons and electrons at other energies during the event, we also include observations from the Electron Proton Helium Instrument (EPHIN, Müller - Mellin et al. 1995) on board the Solar and Heliospheric Observatory (SOHO), the Electron Proton Alpha Monitor (EPAM, Gold et al. 1998) on board the Advanced Composition Explorer (ACE), and the 3D Plasma and Energetic Particle Investigation (3DP, Lin et al. 1995) on board Wind. Those data are presented in the top panel of Figure 1 and are discussed from top to bottom in the following paragraph.

After the flare eruption at 04:56 on May 6 (Tab. 1, more details in Section 2.4), which is marked as a red vertical dashed line in Fig. 1, LND detected the arrival of SEPs as shown in panel (a) of Fig. 1. The proton channel 9.0 - 21.0MeV shown here is the combination of the first five channels of LND's 1minute proton data, which are provided in the Appendix. Due to the low intensity of the event and poor counting statistics in the data, the flux is averaged over 30 minutes for this figure. The analysis described below will be performed using the highest possible time resolution. In panel (b) we display the 4.3 - 7.8MeV proton intensity profile measured by SOHO/EPHIN. However, the EPHIN measurements have data gaps due to limited telemetry during a SOHO roll maneuver on May 6. Thus, LND becomes the only instrument that observed the complete duration of this event at these energies.

Tab. 1 Time Line of the SEP Event on 2019 May 6

Time (UT)	Event	Characteristics	Refer to
2019.05.06			
04:56:00	Eruption of Flare and start of SXR	AR 12740@50°E; M1.0 D_E=113°; D_{STA}=7°	
04:57:23	Type III radio burst	SSRT, Wind/WAVES, STA/SWAVES	
05:00:22	EUV wave propagating toward west	AIA 193Å; $V\approx$500km s^{-1}	Figure 4
05:03:30	Type II radio burst	@230 MHz - 90MHz, Last for 6 minutes	
05:18:24	Electron release	VDA, IMF~1.18 au, beam - like	Figure 2
05:28:00	CME first appears at SOHO/LASCO C2	376km s^{-1a}; 326km s^{-1b}	Figure 4
05:30:59	310keV electron onset	Wind	Figure 2
05:31:00	CME at STA COR2	AW~20°, deflect to west	Figure 4
06:27:00	10.7 - 12.7 MeV proton release	LND; TSA	Figure 2
06:29:42	Proton release	VDA, IMF ~1.33 au	Figure 2
07:32:00	10.7 - 12.7 MeV proton onset	LND, Poisson - CUSUM (Lucas 1985)	Appendix

Notes. The 8.3 minute light travel time has been subtracted from all the times in this table. D_E is the longitudinal distance between active region and magnetic footpoint of the Earth; D_{STA} is the distance to the STA footpoint; SSRT= Siberian Solar Radio Telescope; AW=angular width; GOES=Geostationary Operational Environmental Satellite; VDA =velocity dispersion analysis; IMF=interplanetary magnetic field; TSA=Time shifted analysis.

[a] Velocity from Cactus catalog.

[b] Velocity from GCS fitting.

Both y - axes in panels (a) and (b) are plotted in linear scale in order to better show the SEP enhancement. Panel (c) shows the ion flux averaged over each 30minutes for two energy channels between 1 and 4.75MeV detected by the LEMS30 (Low - Energy Magnetic Spectrometer) detector, one of the telescopes of ACE/EPAM. The intensity profile shows clear velocity dispersion of ions that are mostly attributed to protons.

Finally the electron profile observed by Wind between 40 and 180keV is plotted in panel (d), which clearly shows electrons starting to arrive at around 5: 30 on May 6, slightly earlier than the energetic protons. Different energy channels show different onset times with higher energy electrons arriving earlier. This velocity dispersion feature is studied using a velocity dispersion analysis (VDA) method as discussed below. The second increase later that day is associated with another solar eruption, which is not considered in this study.

The remaining panels of Fig. 1 show the local solar wind plasma data from 1 minute Near-Earth Heliosphere Data (OMNI), including the magnetic field, solar wind speed, and proton density. The near-Earth space is rather calm around the time of the SEP event of interest, with no indication of transient interplanetary coronal mass ejections (ICMEs), shocks or stream interaction regions (SIR), which could otherwise contribute to possible local acceleration of particles around the onset of the SEP event. A preceding SIR on May 3/4 is indicated by the increase of magnetic fields, enhanced proton density and the slow rise of solar wind speed and was followed by a high-speed stream on May 4/5. We do not expect this preceding structure to have any influence on the SEP event reported here.

As shown in the bottom left panel of Fig. 1, at the time of the SEP event under study, the longitudinal separation between the active region and the magnetic footpoint of STA using ballistic back mapping is only 7°, which suggests that STA would be a perfect observer for these energetic particles. However, STA was already experiencing an ongoing SEP event at the time of the flare and no impulsive contribution is seen at STA. Thus all we can state is that the impulsive event reported here was too small to be seen above the background of the preceding event at STA.

2.2 Determination of Onset and Release Times

The small intensity of the SEP event along with the limited geometry factor of LND require careful analysis of the data and limit the accuracy of the determination of the event onset times. Therefore, we apply the Poisson-CUSUM method (Lucas 1985) and follow the procedure in Huttunen-Heikinmaa et al. (2005) to derive the onset time of each energy channel using LND's highest time resolution of 1minute. In the Appendix, we present the Poisson-CUSUM analysis on the different LND energy channels in more detail.

The onset times of the 9.0-10.6 and 10.7-12.7MeV channels are 08: 00±7 and 07: 32 ± 21minutes. The calculation of the uncertainties is explained in the Appendix. Assuming protons traveling scatter-free along a 1.2 au interplanetary magnetic field (IMF) line, which is calculated from the solar wind speed averaged over 8 hr before the SEP event, we derive that the 10.7 - 12.7 MeV protons need about 63minutes to

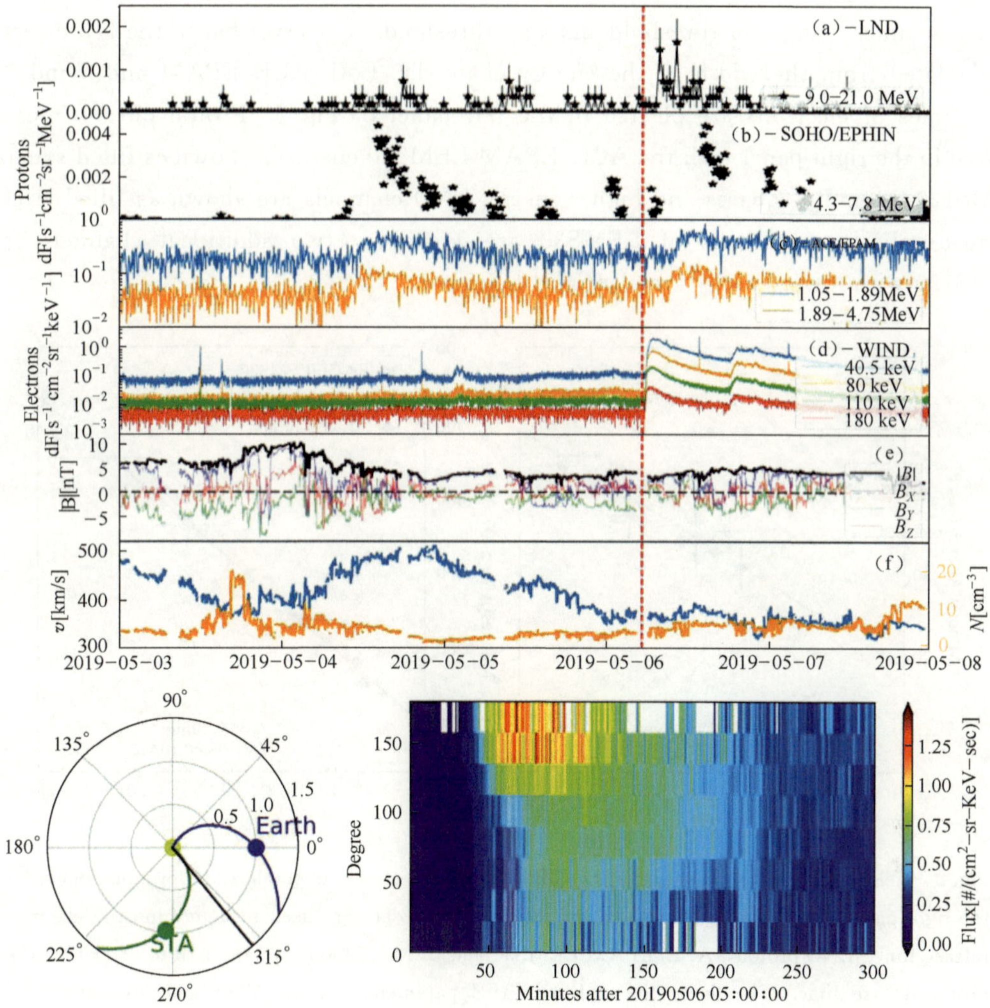

Fig. 1　In situ measurements near Earth, including (a) proton intensity at 9. 0 - 21. 0 MeV by LND, (b) proton intensity at 4. 3 - 7. 8MeV by SOHO/EPHIN, (c) ion intensity in 1. 05 - 4. 75MeV by LEMS120 of ACE/EPAM, (d) electron intensity in 40. 5keV - 180keV by Wind/3DP, (e) magnetic field magnitude and three components in the GSE coordinates, (f) solar wind velocity and plasma density from OMNI. The red line indicates the eruption time of solar flare on May 6. The radial direction of the active region as well as the nominal Parker spirals for Earth and STA are plotted in the HEEQ coordinate in the bottom left panel. The pitch angle distribution of electrons measured by Wind after 05: 00: 00 on May 6 is given in the bottom right panel. Electrons are flowing outwards toward the observer at 180 degrees

arrive at Earth and the release time is around 06: 29±21 minutes. The onset times for ACE/EPAM and Wind/3DP are determined as the time when the flux exceeds 3σ of the background signal. The latter is defined as the flux an hour before the SEP onset.

In Fig. 2, the onset times t_{onset} of different channels are plotted versus $1/\beta$ that is c/v, where c is the speed of light and v is the speed of particles with different energies. The uncertainties of the onset times at ACE and Wind are computed as the difference between the onset times using a 1σ threshold and a 3σ threshold. The error bar in the $1/\beta$ direction is calculated from the width of the energy channels. Both ACE/EPAM and Wind/3DP onset times of electrons are plotted in the left panel of Fig. 2. Proton onset times are plotted in the right panel with the ACE/EPAM LEMS30 channels shown as filled squares, LEMS120 as empty squares and high - energy LND channels are shown as filled circles. Because only two channels of LEMS30 are available, two additional channels from LEMS120 with lower energy are also used here.

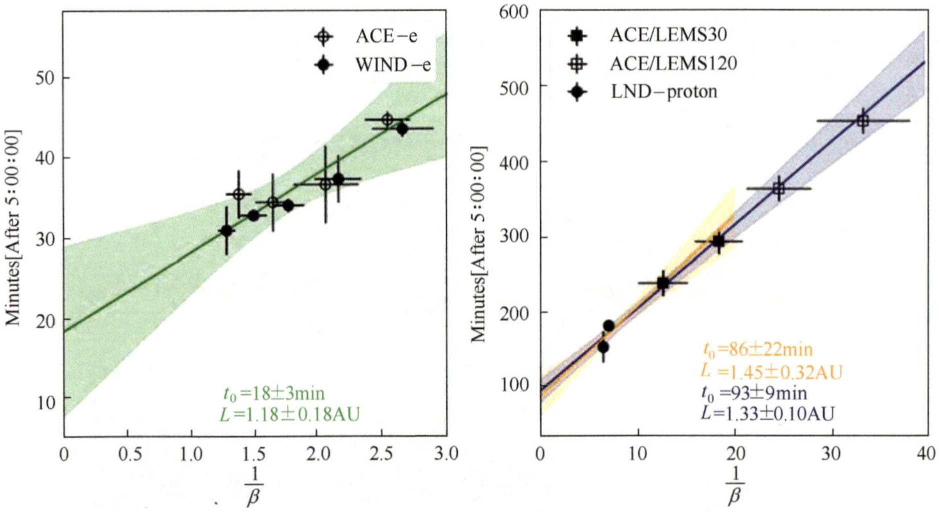

Fig. 2 Velocity dispersion analysis of the SEPs (electrons shown in the left and protons shown in the right panel) on 2019 May 6. Wind and ACE electron data are used to determine the electron release time. LND proton, ACE/EPAM LEMS30 and LEMS120 ion data are used to determine the proton release time. The linear fits and the fitted parameters of the VDA analysis with 95% confidence interval are marked in the plots. More details can be found in the text of Section 2. 2

We apply the VDA method using the measured onset times in different energy channels to determine the particle release time, t_0 and the length of the IMF spiral, L, along which particles propagated. We fit the measured onset times t_{onset} and inverse velocities, $1/v$, with the function $t_{\text{onset}}=t_0+L/v$ using orthogonal distance regression to account for uncertainties in both the x and y directions. The underlying assumption is that the particles that arrive earliest have undergone little scattering. The linear VDA fits are plotted as solid lines in the left and right panels of Fig. 2 for electrons and protons, respectively, and 95% confidence intervals are given as shaded regions. The fitting result based on protons from the LND and LEMS30 measurements is given in orange, while the result using all data is given in blue. The two fit results are consistent. The results for t_0 and L are also given in Tab. 1.

The results of the fits suggest that electrons are released 22±3 minutes after the flare eruption and that protons are released about 75 ± 12 minutes after the electron release. The IMF length inferred from electrons is $L_e = 1.18 \pm 0.18$ au, which is consistent with the Parker spiral derived from the in situ solar wind speed, $L = 1.2$au. For protons we obtain a length $L_p = 1.34 \pm 0.10$ au. L_e and L_p are consistent with each other and weighting them with their inverse errors we find an average $L = 1.28 \pm 0.15$ au.

The VDA method assumes that all particles stream along the same IMF, which may become invalid in a realistic condition. In order to assess the reliability of this assumption for this SEP event, we also checked the pitch - angle measurement by Wind. At the beginning of the SEP event, electrons show a clear anisotropy (enhanced intensity at ~180° pitch angle, bottom right panels of Fig. 1) suggesting that electrons first arrived along the IMF (Tab. 1). Consequently, these particles experienced little scattering, in agreement with the requirement of the VDA method. Unfortunately, proton observations with much smaller statistics make it difficult to identify a significant anisotropy of protons from Wind.

2.3 Energy Spectra

Combining the proton measurement by SOHO/EPHIN as shown in panel (b) of Fig. 1 and the ions registered by ACE/EPAM in panel (c), which are dominated by lower energy protons and also those of LND, we plot the proton spectra integrated between 06: 00 and 17: 00 on 2019 May 6 in the left panel of Fig. 3. The areas shaded in blue, green, and pink are the energy coverage of LND (9 - 35MeV), SOHO/EPHIN (4.3 - 7.8MeV), and ACE/EPAM (0.310 - 4.75MeV), respectively. The background spectra are also plotted and they are averaged between April 29 and May 3 (LND started working on April 29 for its fifth lunar day measurement). The energetic proton spectra during the event are well above the background as shown in the left panel of Figure 3, despite the large uncertainty in the LND data due to the low number of counts. The right panel of the figure shows the SEP spectra where the background spectrum has been subtracted. Because of the EPHIN data gaps during the impulsive phase of the SEP event as shown in Fig. 1, the flux of the SEP event measured by EPHIN is likely to be larger than the derived one, as pointed out by the upward - pointing green arrow in the right panel.

Despite this event being of a more impulsive nature, we use a classic double power - law spectrum to fit the data. As can be seen in Fig. 3, there is a break in the spectral slopes around 2.5MeV, a fit using a double power law gives power - law indices of -0.93 ± 0.18 and -3.14 ± 0.38 for the spectra below and above this break energy, respectively. Interestingly, the break energy falls right between the values for large gradual events (Mewaldt et al. 2012) and those determined for very small events by Joyce et al. (2020) using data from the Parker Solar Probe (Fox et al. 2016).

2.4 Remote - sensing Observations

On 2019 May 6, an impulsive M1.0 flare erupted from active region (AR) 12470

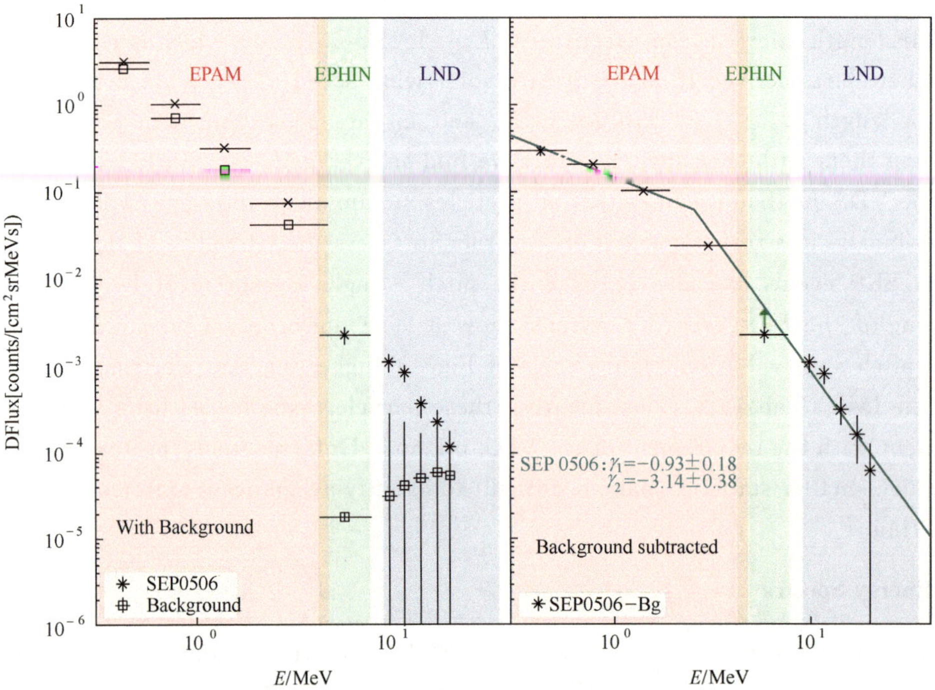

Fig. 3　Proton spectra of the SEPs on 2019 May 6 and the background spectra which are averaged between April 29 and May 3. The left panel shows the spectra including the background and the right panel shows the SEP spectra with background subtracted. Different shaded areas indicate the energy ranges covered by the different instruments from ~300keV to ~30MeV. More details can be found in the text of Section 2. 3

located at N08E50. The soft X－ray (SXR) flare had an onset at 04：56：00[8] and lasted for ~8 minutes as detected by the solar X－ray Imager on the GOES satellite. Since this is the only visible eruptive source on the Sun seen from Earth before the onset of the SEPs, we believe this is the solar counterpart of the SEPs measured in situ near the Earth and Moon. At nearly the same time as the SXR emission, a broadband type III radio burst starting from 240MHz was observed by not only the ground radio observatory Solar Radio Telescope (SSRT[9]) on Earth but also by the WAVES instrument on board the Wind spacecraft at Earth L1 point as well as by WAVES on board STEREO－A (STA). A type II radio burst between 230 and 90MHz indicating the existence of a coronal shock was reported by the SSRT during 05：03：30－05：09：30. The timeline of this event is given in Table 1.

At 05：00：20, an asymmetric EUV wave ($500km \cdot s^{-1}$) started propagating toward

8 The time for remote－sensing measurements in this study has subtracted the~8. 3 minute light travel time over 1 au distance.

9 http：//www. e－callisto. org/

the northwestern hemisphere from the source as observed in the 193 Å band of the Atmospheric Imaging Assembly (AIA) on board the Solar Dynamics Observatory. The outer edges of the wave front at different times, i. e., at 05: 00, 05: 02, 05: 07, 05: 12, 05: 17, and 05: 22 on May 6, are marked as colored dashed lines in the top panel of Fig. 4, which is explained in more detail below. After 05: 22, the wave continued expanding across the solar surface as a rather faint structure that does not contain a clear wave front.

About half an hour after the flare eruption, at 05: 28, a CME first appeared in the field of view of the Large Angle and Spectrometric Coronagraph Experiment (LASCO) C2 of SOHO with a projected (plane of the sky) speed of 376km s^{-1} and an angular width of 20°, as reported in the CACTUS CME catalog.[10] Simultaneously, STA COR2 also captured the same structure. We applied the Graduated Cylindrical Shell (GCS) model (Thernisien 2011) to the coronagraph observations from two directions to obtain the velocity and propagation direction of the CME. In the bottom panel of Fig. 4, we give the CME observation and outline the fitted CME structure in green at 05: 46 on May 6. The fitting results show that the linear speed of the CME front was about 326km · s^{-1} and the CME was deflected by about 10 degrees away from the location of the flare to the west, consistent with the direction of the EUV wave propagation.

In the top panel of Figure 4, we show a magnetogram (in gray shades) on which the results of a potential field sourcesurface (PFSS) model for this Carrington rotation 2217 have been overlaid.[11] Open magnetic field lines (here with positive polarity shown in green and negative polarity shown in red) that connect the photosphere and the PFSS source surface for the solar wind are generally considered to be the main channels for SEP propagation in the solar corona. The polarity inversion line is shown in blue. The abovementioned active region AR 12470 is in the right part of the plot and marked by a white arrow. The radial projection points of STA and Earth on the solar surface on May 6 are plotted as open circles in black and blue, respectively. The magnetic footpoints based on the ballistic back mapping of the solar wind propagating at an average solar wind speed (about 360km s^{-1} as observed in situ at Earth) are marked as filled circles. We note that the magnetic footpoint of STA is only 7° away from the location of AR 12470, which suggests STA is well connected to this active region. On the other hand, the longitudinal separation between the flare and Earth's footpoint is about 113° as displayed in Fig. 4. Also, the wave does not appear to persist to the Earth's footpoint.

---

10 http: //sidc. oma. be/cactus/

11 Source: Global Oscillation Network Group (GONG, https: //gong. nso. edu).

3 Summary and Discussion

On 2019 May 6, an SEP event was observed by LND on the far - side surface of the Moon. While it had only a low intensity, this is nevertheless the first SEP event with enough counting statistics that is detected by LND. This event was also detected by SOHO/EPHIN, which unfortunately only registered its decay phase, and also by Wind and ACE. The only possible solar source is a flare and its accompanying CME at AR 12470 located on the east hemisphere (E50°). The type II radio burst indicates the existence of a shock in the lower corona and the CME speed was fitted as 326km • s^{-1} by the GCS model.

The time profiles of electrons and protons clearly show velocity dispersion. According to the VDA analysis, which assumes that all particles propagate along the same IMF line arriving at Earth and based on the combined data of LND, Wind, and ACE, electrons were released about 22±3 minutes after the flare and type III radio burst and about 8 minutes after the high frequency type II radio burst, while protons were released at least 70 minutes later. The in situ velocity dispersion and anisotropy of electrons suggest that a direct magnetic connection from the source to Earth was established for these electrons. However, the separation between the flare location and the magnetic footpoint of Earth derived from the standard ballistic mapping is as large as 113°. This wide separation is remarkable. In the classical scenario of particle transport during impulsive events (Reames 1999) accelerated particles stream along an open IMF connecting the source and the observer with high - energy particles arriving earlier than lower energy ones. The in situ electron observation of this event at Earth (impulsive, velocity dispersion, and beam - like distribution) suggests a good connection to the flare, while the remotesensing observation contradicts this explanation.

A possible explanation could be that particles are accelerated by the coronal shock, which deviated westwards and reached the magnetic footpoint of Earth. For example, Rouillard et al. (2012) have found an association between EUV waves that track the lateral shock expansion (Veronig et al. 2010) and particle release in wide - spread SEPs. As shown in the top panel of Fig. 4, the EUV waves marked as colored dashed lines are propagating westward. The GCS fitting of the CME also indicates a nonradial and westward - deflected propagation toward the location of the Earth footpoint. However, the deflection of the CME is only about 10° and the EUV wave became rather faint after 05: 25 suggesting that the shock was unlikely to reach Earth's footpoint around the particle release time. Therefore, it is difficult to conclude that a deflected shock should be responsible for the initial particle acceleration and release processes.

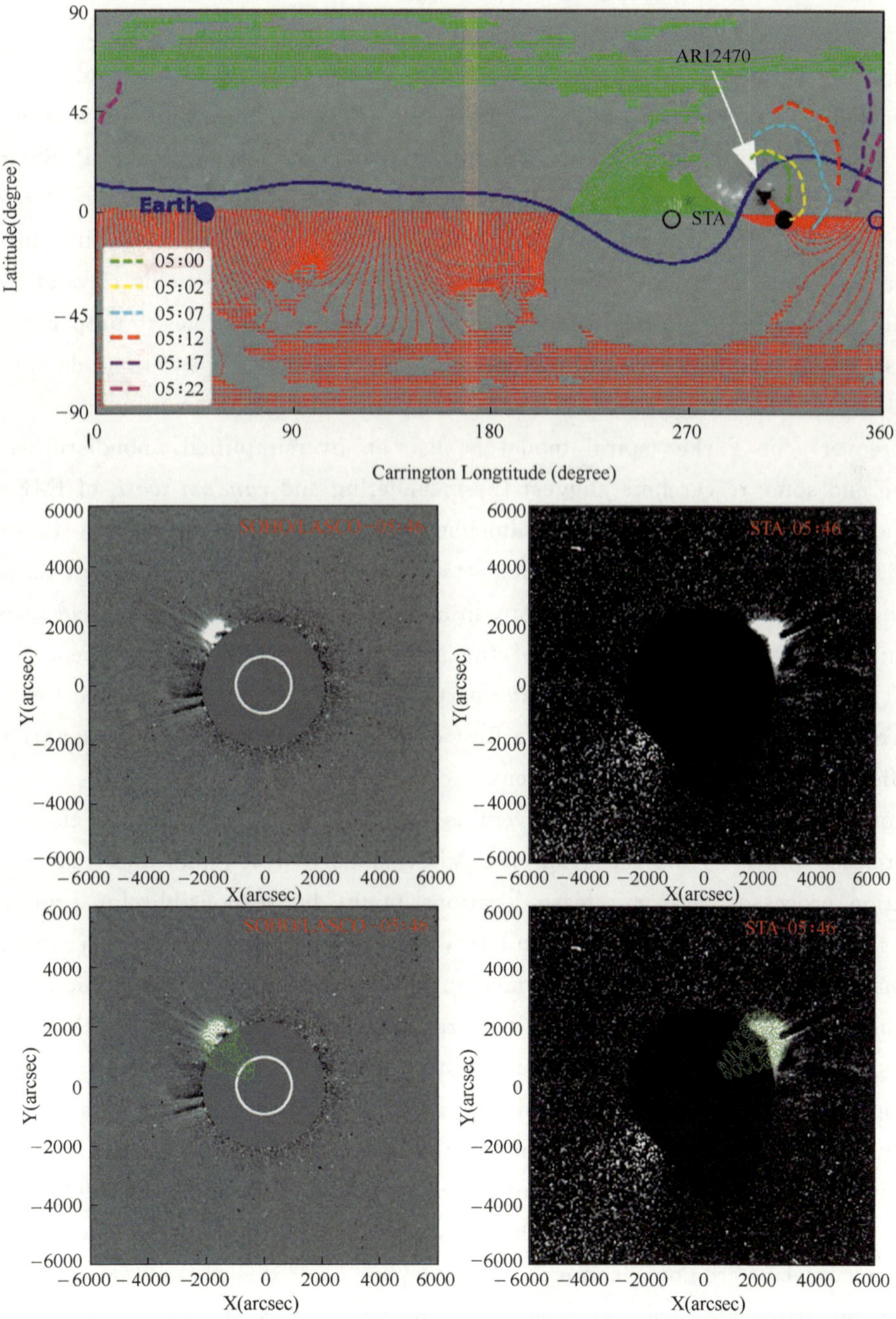

Fig. 4　Top: the synoptic ecliptic - plane field plot for Carrington rotation 2217. The projected locations of STEREO - A (STA) and Earth are added as open circles and their magnetic footpoints mapped back to 2.5 solar radii based on the ballistic model are shown as filled circles. The outer edges of an EUV wave front at different times, i.e., at 05: 00, 05: 02, 05: 07, 05: 12, 05: 17, and 05: 22 on May 6, are marked as colored dashed lines. Bottom: CME observation of SOHO/LASCO (left) and STA/Cor2 (right) at 05: 46 on 2019 May 6. The green mesh outlines the GCS reconstruction of the CME geometry. A more in - depth description of this figure can be found in Section 2.4

Klassen et al. (2018) proposed another scenario in which electrons accelerated by a flare can reach a distant magnetic footpoint (90° in their example found with STA) through an irregular magnetic field at the solar source. The PFSS extrapolation of the magnetic field (top panel of Fig. 4) before this eruption does not show any direct connection between the flare and the location of Earth's footpoint. The PFSS model, however, assumes a current-free field by definition, and is an idealized consideration of the solar corona, so the real magnetic configuration might be drastically different. Furthermore, solar eruptions often rearrange the solar magnetic fields through, e.g., magnetic reconnection, and might have created a path for the particles to propagate over a large distance in heliolongitude. Unfortunately, the available observations do not show any evidence of a rearrangement of the solar magnetic field due to eruptions.

Moreover, the Parker spiral model is also an oversimplified, nondisturbed IMF condition and some researchers suggest the meandering and *random walk* of IMF would affect the particle propagation in the heliosphere (e.g., Mazur et al. 2000; Laitinen et al. 2016). In addition, the cross-field transport due to pitch-angle scattering and diffusion also causes particles to propagate in longitude and be observed on field lines with footpoints far away from the center of the solar eruption (e.g., Wibberenz & Cane 2006). This means that one could observe particle events on poorly connected field lines. However, the beam-like nature of the first electrons to arrive does not support such a cross-field transport model for this event.

Another observation about this event is the long delay between the electron and proton release times as derived from the VDA model, which indicates that different acceleration processes and/or release locations might be responsible for protons and electrons. As protons are released much later, they may be more likely accelerated by the shock that, however, cannot be confirmed to have reached Earth's magnetic footpoint. This explanation would also be difficult to reconcile with the common path lengths of electrons and protons, L_e and L_p. In their study of the delay between the electron and proton onset times at STA, STB, and L1 for wide-spread events, Richardson et al. (2014) found that this time delay and the longitudinal separation between the flare location and backmapped spacecraft footpoint were correlated. In Tab. 1 we report a delay between the electron and protons release times that may contribute to the Richardson et al. (2014) results. Such delays are not uncommon.

To summarize, we have presented various observations related to the first SEP event ever detected on the Lunar far-side surface. (1) The energy spectra of LND are consistent with observations from other spacecraft, though this is a weak event requiring large background subtractions. (2) The proton onset time of LND is also consistent with observations from other spacecraft, suggesting that the instrument response appears to be consistent with expectations. (3) The observations show clear velocity dispersion. (4) The SEP event was associated with a widely separated (~113°) flare at E50° relative to

Earth.

The Lunar Lander Neutron and Dosimetry (LND) instrument is supported by the German Space Agency, DLR, and its Space Administration under grant 50 JR 1604 to the Christian - Albrechts - University (CAU) Kiel and supported by Beijing Municipal Science and Technology Commission, grant No. Z181100002918003 and National Natural Science Foundation of China Grant: 41941001 to the National Space Science Center (NSSC). The scientific data are provided by China National Space Administration. J. G. and Y. W. are supported by the Strategic Priority Program of the Chinese Academy of Sciences (grant Nos. XDB41000000 and XDA15017300), and the CNSA preresearch Project on Civil Aerospace Technologies (grant No. D020104). N. D. was supported under grant 50OC1302 by the Federal Ministry of Economics and Technology on the basis of a decision by the German Bundestag. We also would like to express our very great appreciation to Patrick Kühl, Lars Berger, Andreas Klassen, and an anonymous referee for their valuable suggestions on this work.

Appendix

LND Measurement and Poisson - CUSUM Analysis

The charged particle telescope of LND measures protons in 1 minute time resolution between 9.0 and 35MeV and the explicit energy bins are given in Tab. 6 of Wimmer - Schweingruber et al. (2020). Here, we present the 1 minute proton count rates at 9.0 - 10.6 MeV and 10.7 - 12.7 MeV during May 4 and 8 in Fig. 5. The blue curves are the LND measurement in 1 minute time resolution. Most of the time, zero count is registered by LND, even during the SEP event. One reason is the small intensity of the SEP event and the other is the small geometry factor of LND. Both of them cause the poor statistics.

In order to calculate the event onset time, the Poisson - CUSUM method is applied and the results are plotted in an orange line for each channel. Cumulative sum (CUSUM) control schemes are widely used in industrial applications because they are designed to give an indication of when there is a change in a process (Page 1954). In our case, the change is when the SEPs rise above the background. The traditional CUSUM schemes are applied to a normally distributed quantity. When the variable has a Poisson distribution, a Poisson - CUSUM should be used. In the measurement of LND, the number of counts in a fixed interval, for example, 1 minute, obey a Poisson distribution. Hence, the Poisson - CUSUM will be used to determined the onset time of the SEP event. By the definition of control schemes (Lucas 1985), the difference between the observed value Y_i and a reference value k are accumulated as a systematic change, which is:

$$S_i = \max(0,\ Y_i - k + S_{i-1}) \tag{A1}$$

The start value is $S_0 = 0$ in standard CUSUM.

The reference value k is determined below:

$$k = \frac{\mu_d - \mu_a}{\ln(\mu_d) - \ln(\mu_a)} \tag{A2}$$

where μ_a is the mean number of counts estimated for each channel during the pre-event background and μ_d is selected by a twoσ-shift criterion (Huttunen-Heikinmaa et al. 2005):

$$\mu_d = \mu_a + 2\sigma_a \qquad \text{(A3)}$$

where σ_a is the standard deviation of the pre-event background. In order to have a nonzero background, the pre-event background is integrated from April 29 to May 2.

When the systematic change S_i exceeds the decision value h, then the onset of the SEP event is determined. Here we give a small decision value $h=1$ according to the table in Lucas (1985). In order to reduce the false alarms due to the small h, we apply the following criterion: once an S_i exceeds the h, then the following 60 data points are checked as well. If all of them are larger than h, then the first signal is defined as the onset time of the SEP event.

We then apply the Poisson-CUSUM method for the one-minute data and the systematic changes are plotted as orange lines in Fig. 5. The red dashed vertical line marks the eruption time of the flare at 04:56 on May 6. The increases of the orange lines during the SEP are very clear and the onset time can be easily determined. The onset time of the 9.0-10.6 MeV and 10.7-12.7 MeV channels are at 08:00 and 07:32, respectively.

We calculated the uncertainty of the Poisson-CUSUM method by determining the average time differences between the Poisson-CUSUM onset time and the times of the closest two neighboring nonzero data points.

The Poisson-CUSUM profiles, especially the one in 9.0-10.6 MeV, indicate the existence of energetic particles registered by LND during May 4 and 6, which are not explicitly shown in Fig. 1 (a). The corresponding lower energy protons are also detected by SOHO/EPHIN and ACE/EPAM. Those particles may have the same solar origin as that of the gradual SEP event detected by STA starting on May 4 and lasting until May 8, i.e., the ongoing event that we mentioned in Section 2.1 during May 6.

References

[1] Fox, N. J., Velli, M. C., Bale, S. D., et al. 2016, SSRv, 204, 7.

[2] Gold, R. E., Krimigis, S. M., Hawkins, S. E., III, et al. 1998, SSRv, 86, 541.

[3] Huttunen-Heikinmaa, K., Valtonen, E., & Laitinen, T. 2005, A&A, 442, 673.

[4] Joyce, C. J., McComas, D. J., Christian, E. R., et al. 2020, ApJS, 246, 41.

[5] Klassen, A., Dresing, N., Gómez-Herrero, R., Heber, B., & Veronig, A. 2018, A&A, 614, A61.

[6] Laitinen, T., Kopp, A., Effenberger, F., Dalla, S., & Marsh, M. S. 2016, A&A, 591, A18.

[7] Lin，R.，Anderson，K.，Ashford，S.，et al. 1995，SSRv，71，125.
[8] Lucas，J. M. 1985，Technometrics，27，129.
[9] Mazur，J. E.，Mason，G. M.，Dwyer，J. R.，et al. 2000，ApJL，532，L79.
[10] Mewaldt，R.，Looper，M.，Cohen，C.，et al. 2012，SSRv，171，97.
[11] Müller-Mellin，R.，Kunow，H.，Fleißner，V.，et al. 1995，SoPh，162，483.
[12] Page，E. S. 1954，Biometrika，41，100.
[13] Reames，D. V. 1999，SSRv，90，413.
[14] Richardson，I. G.，von Rosenvinge，T. T.，Cane，H. V.，et al. 2014，SoPh，289，3059.
[15] Rouillard，A. P.，Sheeley，N. R.，Tylka，A.，et al. 2012，ApJ，752，44.
[16] Thernisien，A. 2011，ApJS，194，33.
[17] Veronig，A. M.，Muhr，N.，Kienreich，I. W.，Temmer，M.，& Vršnak，B. 2010，ApJL，716，L57.
[18] Wibberenz，G.，& Cane，H. V. 2006，ApJ，650，1199.
[19] Wimmer-Schweingruber，R. F.，Yu，J.，Böttcher，S. I.，et al. 2020，SSRv，216，104.
[20] Zhang，S.，Wimmer-Schweingruber，R. F.，Yu，J.，et al. 2020，SciA，6，eaaz1334.

Emission of energetic neutral atoms measured on the lunar surface by Chang'E-4

Aibing Zhang[1,2,3,4], Martin Wieser[5], Chi Wang[1,4,6], Stas Barabash[5], Wenjing Wang[1,2,3], Xiaodong Wang[5], Yongliao Zou[1], Lei Li[1,6], Jinbin Cao[7], Leif Kalla[5], Lei Dai[1,6], Johan Svensson[5], Linggao Kong[1,2,3], Magnus Oja[5], Bin Liu[1,2,3], Vesa Alatalo[5], Yiteng Zhang[1,6], Juha Talonen[5], Yueqiang Sun[1,2,3], Magnus Emanuelsson[5], Changbin Xue[1], Lei Wang[1], Fang Wang[8], Wenlong Liu[7]

1. National Space Science Center, Chinese Academy of Sciences, No. 1 Nanertiao, Zhongguancun, Haidian district, Beijing, 100190, China
2. Beijing Key Laboratory of Space Environment Exploration, No. 1 Nanertiao, Zhongguancun, Haidian district, Beijing, 100190, China
3. Key Laboratory of Science and Technology on Space Environment Situational Awareness, Chinese Academy of Sciences, No. 1 Nanertiao, Zhongguancun, Haidian district, Beijing, 100190, China
4. University of Chinese Academy of Sciences, No. 19 (A) Yuquan Road, Shijingshan district, Beijing, 100049, China
5. Swedish Institute of Space Physics, Box 812, SE-98128, Kiruna, Sweden
6. State Key Laboratory of Space Weather, Chinese Academy of Sciences, No. 1 Nanertiao, Zhongguancun, Haidian district, Beijing, 100190, China
7. Beihang University, No. 37 Xueyuan Road, Haidian District, Beijing, 100083, China
8. National Astronomical Observatories of China, Chinese Academy of Sciences, 20A Datun Road, Chaoyang District, Beijing, 100012, China

Abstract: The Advanced Small Analyzer for Neutrals (ASAN) is an international payload onboard the Chinese Chang'E-4 rover. It performed for the first time measurements of energetic neutral atoms (ENAs) on the lunar surface. We show a typical mass separated ENA energy spectrum measured by the ASAN. After normalizing to the impinging solar wind proton energy, good agreement to previous measurements from Chandrayaan-1 and IBEX is found. The hydrogen ENA albedo is estimated to be 32% for energy above 30eV, comparable to the estimates obtained from Chandrayaan-1 and IBEX. Particle fluxes at lower energies are generally higher than those observed by Chandrayaan-1 and IBEX, and possible reasons are discussed.

Keywords: ASAN; Chang'E-4; ENA; Energy spectrum; Albedo

1 Introduction

The Moon has neither an atmosphere nor a global magnetic field. Supersonic solar wind ions are able to reach the lunar surface directly and interact with it where they are not prevented from doing so by local crustal magnetizations. The Moon was previously thought to be a passive absorber of plasma (Feldman et al., 2000; Crider and Vondrak, 2002). However, observations from recent missions, such as Kaguya, Chang'E-1 and -2, Chandrayaan-1, ARTEMIS, and IBEX, have shown that the interaction of solar wind with the lunar surface is quite dynamic with a variety of processes involved, such as backscattering of 0.1-1% as protons (Saito et al., 2008; Lue et al., 2011; Wang et al., 2010) and up to 21% as hydrogen energetic neutral atoms (ENAs) (McComas et al., 2009; Wieser et al., 2009, Vorburger et al., 2013) and the occurrence of sputtered particle populations (Vorburger et al., 2014). Reported energy spectra for energetic neutral hydrogen show generally a cutoff corresponding to the solar wind proton energy (Wieser et al., 2009; Futaana et al., 2012; Rodríguez et al., 2012), while the lowest reported energies (<100eV) are more variable.

The Chandrayaan Energetic Neutral Atom (CENA) sensor of the Sub-keV Atom Reflecting Analyzer (SARA) (Barabash et al., 2009) investigated the dependence of the energetic neutral hydrogen albedo and the energy spectrum on various external parameters. The backscatter rate for energetic neutral hydrogen mostly depends on the solar zenith angle (Wieser et al., 2009). The angular scattering function for energetic neutral hydrogen derived from Chandrayaan-1 observations shows more scattering in the sunward direction (backscattering) than the forward scattering as the solar zenith angle increases (Vorburger et al., 2013). Using this scattering function, the global mean for energetic neutral hydrogen albedo of 16 5% (Vorburger et al., 2013) was obtained, while the value estimated from IBEX observations is 11 6% (Saul et al., 2013). On a local scale, the flux of emitted energetic neutral hydrogen varies in correlation to local crustal magnetic anomalies that modulate the intensity and energy spectrum of the solar wind ions impinging on the surface (Vorburger et al., 2013; Wieser et al., 2010).

The observations of Chandrayaan-1 were obtained as an average over large areas of several hundreds of km^2, and the observational geometry of IBEX is even larger, i. e., a quadrant of the Moon. These observations completely blurred out typical topography elements such as craters with varying solar wind illumination. A ground truth observation was still missing. Even a single point measurement directly measured on the lunar surface would calibrate previous remote sensing data and enable studying small-scale processes on the meter scale that could reveal more of the microphysics of the interaction of the solar wind with the lunar regolith. The Advanced Small Analyzer for Neutrals (ASAN) onboard the Chang'E-4 mission was designed to perform such a measurement directly on

the lunar surface.

2 Instrumentation

Chang'E-4 is the fourth mission of the Chinese lunar program and the first spacecraft that soft-landed on the lunar far side. Chang'E-4 carries a rover to explore the environment of the landing site. The Lunar Exploration and Space Engineering Center (LESEC) of the China National Space Administration (CNSA) announced the opportunity for international cooperation in October 2014. The ASAN was selected in 2016 as a payload of the Chang'E-4 rover to measure the ENA emissions from the lunar surface in situ. The ASAN was jointly developed by the Swedish Institute of Space Physics (IRF) and the National Space Science Center (NSSC), Chinese Academy of Sciences (CAS).

The ASAN measures the flux of ENAs emitted from the lunar surface in an energy range from 10 eV to 10keV with an average energy resolution $\Delta E/E$ of ~30% for ENAs. The actual resolution is energy dependent, and it varies between 16% and 130% for H. The mass resolution is sufficient to separate energetic neutral hydrogen from heavier species. The ASAN is a single-pixel instrument with a field of view of 37°×30° (28°×14° at full width at half maximum), with 37° in the vertical direction and 30° in the horizontal direction. The bore sight points toward the lunar surface with 30° tilting from the lunar horizon to a point located in front of the rover. The resulting trapezoidal outer boundary of the field of view footprint has an area of approximately 1.57 m^2. Measurements from the ASAN are designed to determine the distribution function of ENAs emitted from the lunar surface and the dependence of this distribution function on lunar topography and local time. The results can be used to investigate the interaction between solar wind and the lunar surface at a small scale, the influence of sputtering on the lunar surface on the formation, and the maintenance of the lunar exosphere.

The instrument is mounted on the +X panel of the Chang'E-4 rover (Fig. 1). The electronic box and sensor head are accommodated inside the thermally insulated payload bay with the entrance system protruding out through a small opening. A one-time openable protective cover was used to prevent lunar dust from entering during landing. The cover was permanently opened after the deployment of the rover to the surface. Combined with the mobility of the rover, this accommodation allows investigating undisturbed regolith away from the blast zone of the lander at different solar wind illumination angles.

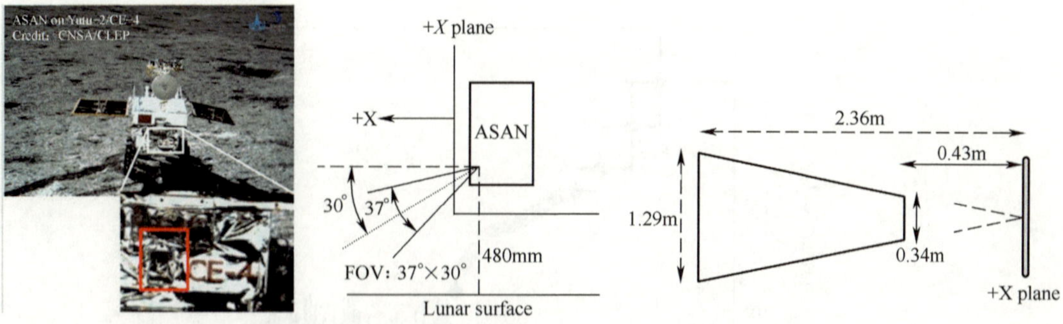

Fig. 1　Left panel: The ASAN (red square) on the rover, middle panel: Accommodation geometry (the rover moves nominally in the $+X$ direction, the dotted line shows the bore sight of FOV, 37° of FOV in the vertical direction is shown, and 30° bore sight tilting from the lunar horizon is also shown), right panel: Field of view footprint of the surface area from which the ASAN collects particles. (For interpretation of the references to colour in this figure legend, the reader is referred to the Web version of this article.)

3　Observations

3.1　Energy spectra

Chang'E - 4 landed at the preselected landing area at 177. 6° east longitude and 45. 5° south latitude on the far side of the Moon. The landing site, which is now called "Statio Tianhe", is located in the South Pole - Aitken (SPA) Basin, the largest, deepest and oldest crater. Fig. 1 (left) shows that the landing area is relatively flat with some small craters.

In the first six lunar days from January to June of 2019, measurements were carried out by ASAN in 21 intervals. Fig. 2 shows a typical average ENA energy spectrum for 2019/5/1 06: 18 -10: 45 UTC where the solar wind speed was 340km/s. Two domains are identifiable in the energetic neutral hydrogen spectrum, a directly backscattered component above 100eV which is almost entirely of solar wind origin, and a component below 100eV mostly due to surface sputtering. The only significant flux of heavier energetic neutral atoms (mass larger than 4 amu) is confined to energies below 100eV. This heavy component is highly variable from measurement interval to measurement interval. The ENA flux with masses in the range (1 amu $<$ m$\leqslant$4amu) is more difficult to separate due to the complexity of separating background signals. There is a possible weak heavy component above 5keV that may be due to backscattered heavy solar wind ions, e. g. multiply - charged oxygen, and this needs further investigations (Fig. 2 bottom).

The spectrum of backscattered hydrogen has a distinct knee at solar wind proton energy (Fig. 3).

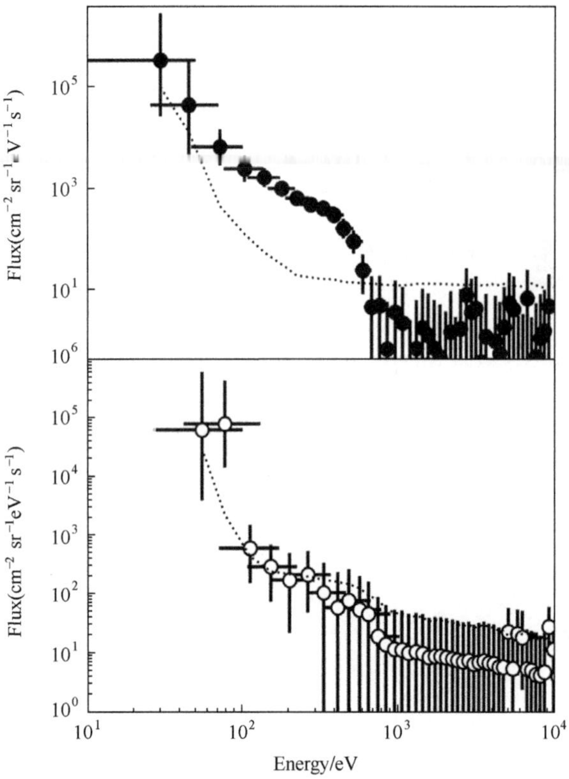

Fig. 2 Average ENA energy spectrum from 2019/5/1 06: 18 – 10: 45 UTC. Hydrogen ENA flux (top panel) and the ENA fluxes for masses larger than 4 amu (bottom panel) are shown. Fluxes below the sensitivity limit (dotted line in both panels) cannot be distinguished from zero flux

3.2 Data processing

The data are processed in several steps. First, the raw counts are integrated over the duration of a measurement session. In a next step, the background caused by the radioisotope heat unit (RHU) is subtracted. Then possible sources of uncertainty are evaluated. Finally, the foreground rate is converted to differential flux and albedo can be calculated.

3.2.1 Integration of raw counts

Because the count rate is low, raw counts during one measurement session are integrated to improve statistics. The duration of a measurement session depends on the operation plan of the rover. For the data on 2019/5/1, the session was from 06: 18 UTC to 10: 45 UTC.

3.2.2 Background subtraction

The RHU mounted at the back panel of the rover, i. e., on the opposite side of where

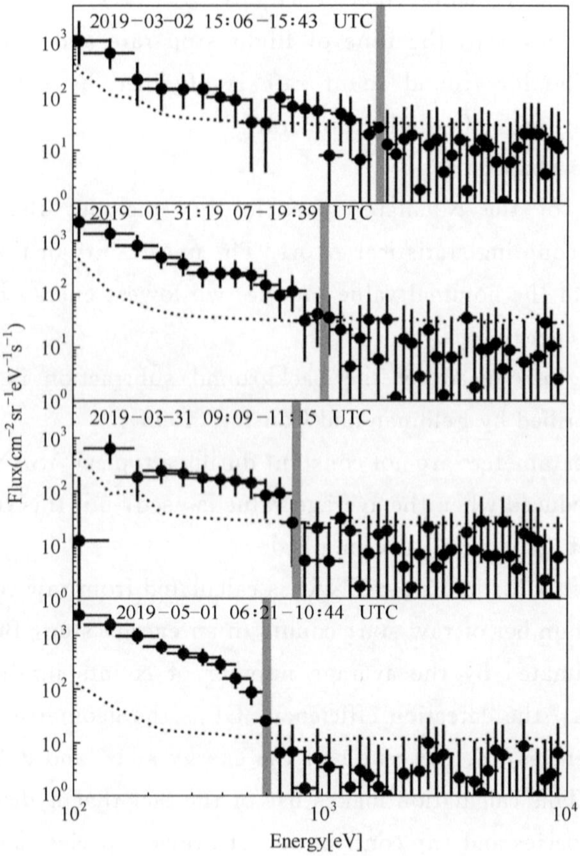

Fig. 3　Hydrogen ENA energy spectra under four typical solar wind speed conditions (the vertical bars show the mean energy of the solar wind), from top panel to bottom panel: 575km/s, 443km/s, 390km/s and 340km/s. The ENA cutoff energy follows the solar wind proton energy. The dashed line is a 90% confidence limit below which the flux cannot be distinguished from zero

the ASAN is mounted, creates a notable background due to gamma radiation originating from the RHU. Two different ways of background subtraction are used depending on the type of signal to be processed and which method is statistically better.

The typical hydrogen ENA spectrum has a cut - off at solar wind energy. Combined with the fact that the background signal introduced by RHU is independent of the energy step selected by the ASAN, the signal in the ENA energy bins from 4. 5keV to 8. 7keV is mainly due to the RHU. For the dominant hydrogen ENA signal, backgrounds are estimated by taking average count rates from these energy bins. The foreground count rate is obtained by subtracting the average background rate from raw start rate.

For the weaker ENA signal with masses larger than 4 amu, time - offlight data are directly evaluated. The background signal introduced by RHU appears in the time - of - flight spectrum as a peak mostly below 1 amu. An upper limit for the background rate

(r_{BG}) is estimated by the accidental coincidence rate $r_{BG}=r_{STA}*r_{STO}*t_{TOF}$, with r_{STA} the time-offlight start rate, r_{STO} the time of flight stop rate and t_{TOF} the time of flight window of 1.6μs. The foreground count rate is obtained by subtracting the average background rate from time-of-flight count rate.

3.2.3 Uncertainty evaluation

The uncertainty of the calculated particle flux includes the uncertainty of the geometric factor and counting statistical errors. The uncertainty of the geometric factor is +200% and −70% of the nominal value for the two lowest energy bins and +40% and −30% for the other energy bins.

The uncertainty introduced by the background subtraction is included using the unified approach described by Feldman and Cousins (1998).

The solar wind parameters are not constant during a typical ASAN operation interval, so some error is introduced when the average value is used, but this contribution is small.

3.2.4 Calculation of differential flux an albedo

The differential flux for hydrogen ENAs is calculated from raw counts using equation (1), where C is the number of raw start counts in an energy step, BG is the background from the RHU, estimated by the average number of counts in the energy bins from 4.5keV to 8.7keV, ε_H the detection efficiency, GF_{H0} the geometric factor for hydrogen ENAs without efficiencies, E_0 the center of the energy step, and dt the integration time for one energy step. This calculation makes use of the fact that hydrogen ENAs is by far the major observed species and any contribution of heavier species can be neglected.

$$f_H=\frac{C-BG}{\varepsilon_H\cdot GF_{H0}\cdot E_0\cdot dt}\tag{1}$$

For species with masses larger than 4 amu, coincidence data needs to be evaluated. The differential flux of heavy ENAs is calculated using equation (2), where C is the number of coincidence events recorded in the mass range of interest, r_{BG} is the background rate estimated by the number of accidental coincidences in the time-of-flight system, ε_X the coincidence detection probability, GF_{X0} the geometric factor for heavy ENAs without efficiencies, E_0 the center of the energy step, and dt the integration time for one energy step. It is important to note that because r_{BG} must be considered an upper limit, the calculated differential flux of heavy ENAs may be underestimated to some amount.

$$f_X=\frac{C-r_{BG}\cdot dt}{\varepsilon_X\cdot GF_{X0}\cdot E_0\cdot dt}\tag{2}$$

The differential flux observed from the lunar surface depends on the solar zenith angle and the rover orientation relative to the Sun. As the rover does not move during ASAN measurements, only flux reflected from the surface into a specific direction is observed. The scattering function from Vorburger et al., (2013) is used to infer the total flux emitted from the surface based on the available observation. Together with propagated solar wind parameters, an energetic neutral hydrogen albedo is calculated.

3.3 ENA albedo

An ENA albedo of the surface can be calculated by assuming an angular scattering function. We define the solar zenith angle (SZA), the scattering azimuth angle (ϕ) and the scattering polar angle (θ) in the same way as Vorburger et al., (2013). The ASAN energy spectra are first integrated over energy to obtain the directional flux j_{ENA} (SZA_0, ϕ_0, θ_0) (#/(cm^2 sr s)), and the subscript 0 means the angles at which the ASAN carried out the observation. The trapezoid method is used during integration for simplicity. Equation (3) gives the calculation of j_{ENA} (SZA_0, ϕ_0, θ_0), in which j_{Ei} (#/(cm^2 sr s eV)) is the differential flux at energy step E_i (eV).

$$j_{ENA}(SZA_0, \varphi_0, \theta_0) = \sum \frac{(j_{E_{i+1}} + j_{E_i}) \cdot (E_{i+1} - E_i)}{2} \tag{3}$$

Based on the ENA scattering function given by Vorburger et al., (2013), the value of the scattering function $f_s(SZA_0; \varphi_0; \theta_0)$ can be obtained. The ENA flux J_{ENA} (#/(cm^2 s)) after energy integration and angular integration is:

$$J_{ENA} = \frac{j_{ENA}(SZA_0, \varphi_0, \theta_0)}{f_s(SZA_0, \varphi_0, \theta_0)} \int f_s(SZA, \varphi, \theta) \cdot \sin\theta d\varphi d\theta \tag{4}$$

Due to the normalization of the scattering function, $\int f_s(SZA, \varphi, \theta) . \sin\theta d\varphi d\theta$ is equal to the cosine of SZA, so:

$$J_{ENA}(SZA) = \frac{j_{ENA}(SZA_0, \varphi_0, \theta_0)}{f_s(SZA_0, \varphi_0, \theta_0)} \cdot \cos SZA \tag{5}$$

With the solar wind flux J_{SW} (#/ (cm^2 s)), the reflection ratio $R_\perp$ for perpendicular incidence is:

$$R_\perp = \frac{J_{ENA}(SZA)}{J_{SW} \cdot \cos SZA} \tag{6}$$

Combining equations (5) and (6), we obtain:

$$R_\perp = \frac{j_{ENA}(SZA_0, \varphi_0, \theta_0)}{f_s(SZA_0, \varphi_0, \theta_0) \cdot J_{SW}} \tag{7}$$

The error source of $R_\perp$ includes the error of the directional ENA flux j_{ENA} and the error of the solar wind flux J_{SW}. The error of the directional ENA flux j_{ENA} comes from the error of the geometric factor of each energy step, and the error of the solar wind flux comes from the variable solar wind during the ASAN observations.

Using ASAN data of 2019 - 05 - 01 and with the corresponding solar zenith angle SZA = 58° and two scattering angles ϕ = - 31° and θ = 60° due to the actual rover orientation, we obtain a hydrogen ENA albedo of $R_\perp$ = 0.32 (0.19, 0.58) with a low energy cutoff of 30eV and a hydrogen ENA albedo of $R_\perp$ = 0.10 (0.07, 0.17) with a low energy cutoff of 46eV. The numbers in parentheses indicate the 25th and 75th percentiles.

4 Discussion

The landing site of Chang'E－4 is located at 177.6°E and 45.5°S. According to the crustal magnetic field model established based on data from Lunar Prospector (Michael et al., 2010), the magnetic field on the lunar surface can be estimated. The magnetic field at the landing site is predicted to have a magnitude of 6.2 nT. This rather small magnetic field is expected to have only a small effect on the incident solar wind ions. It must be noted that the Lunar Prospector model is based on measurements at typically 30km altitude extrapolated down to the surface. Localized stronger magnetization on a scale of km or less may exist but would not be resolved in the model, and it could alter the spectrum and the bulk direction of the impinging solar wind. The solar wind could be deflected, decelerated or heated (Wieser et al., 2010; Futaana et al., 2013), generally resulting in a softer ENA spectrum.

The hydrogen ENA energy spectrum measured by the ASAN agrees well with data measured by Chandrayaan－1 and IBEX above 10% of the solar wind energy E_{sw} (Fig. 4). Two components related to the generation process can be identified. The component above 0.1 E_{sw} mostly consists of solar wind protons directly backscattered as hydrogen ENAs, while below this energy, ENAs sputtered from the surface material become important. Sputtered ENAs have energies mainly in the range from 0.1eV to 100eV (Vorburger et al., 2014). Generally, the ASAN data show a higher flux below 0.1 E_{sw} compared with the Chandrayan－1 and IBEX data. The amount of this excess is highly variable. The variation may be due to variations in the sputtering yield and available hydrogen in the regolith. The larger flux at low energies observed by the ASAN could also arise from the fact that the ENA energy spectra from both Chandrayaan－1 and IBEX are averages over large areas (a quarter of the Moon for IBEX and several hundreds of km^2 for Chandrayaan－1) in orbit, so any local characteristics in the area of observation are blurred out. The ASAN field of view covers a surface area of only 1.57m^2. Local regolith features (e. g., porosity, grain size and composition) of the viewing area may affect the energy spectrum, especially at lower energies. The ASAN energy spectrum shown in Figs. 2 and 4 is obtained at a solar zenith angle SZA of 58° and scattering polar angle θ of 60°. The energy spectrum from Chandrayaan－1 is averaged over an SZA range from 0° to 90° and $\theta = 0°$ (i. e., nadir direction). The spectrum reported by IBEX is a weighted average over typically a quarter of the lunar disc covering wide ranges of SZA from 0° to 90° and θ from 0° to 90°. The ENA energy spectrum may vary with SZA and θ, so the particular viewing geometry of the ASAN may lead to a special energy spectrum when integrating over all emission angles. Furthermore, uncertainties in the geometric factor extrapolation downward from the lowest calibrated energy of 100eV could account for some of the difference at lower energies.

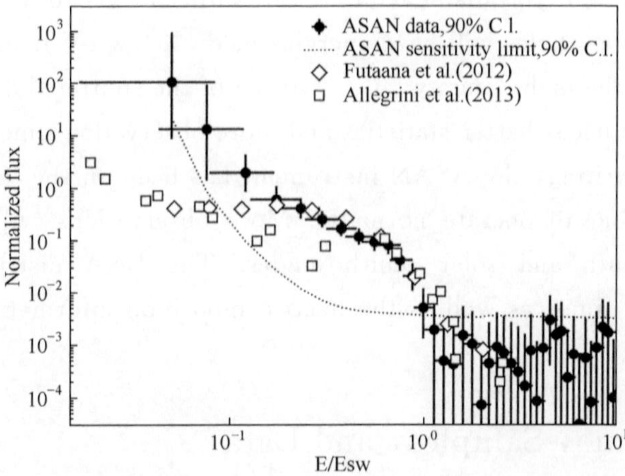

Fig. 4　Solid circles: Average hydrogen ENA energy spectrum observed by the ASAN on 2019/5/1 06: 18 - 10: 45 UTC, normalized to the solar wind energy Esw. Vertical error bars correspond to 90% confidence limits (C. I.) (Feldman and Cousins, 1998), horizontal error bars correspond to the full width at half maximum of the energy response. The dotted line represents a 90% confidence limit, below which the ASAN signal cannot be distinguished from the case where no signal is present. Open diamonds: Hydrogen ENA energy spectrum observed by SARA on Chandrayaan - 1 from (Futaana et al., 2012), again normalized to the solar wind energy. Open squares: Sample hydrogen ENA spectrum measured by IBEX (Allegrini et al., 2013), normalized in the same way

The hydrogen ENA albedo for energies above 30eV of 0. 32 with the 25th and 75th percentiles at 0. 19 and 0. 58, respectively, is in agreement with the values reported by Chandrayaan - 1 (Futaana et al., 2012), where an albedo of 0. 19 with the 25th and 75th percentiles at 0. 16 and 0. 21, respectively, was given, and IBEX, where a value of 0. 11±0. 06 (one sigma) was reported (Saul et al., 2013). The larger hydrogen ENA albedo values obtained from the ASAN data when integrating down to lower energies are likely due to the inclusion of the sputtered particle fraction. Another source of uncertainty is the local angular scattering function, which was assumed to be identical to the average function obtained by Vorburger et al., (2013).

The landing area is at the edge of a large - scale magnetic anomaly on the Aitken basin. This could result in a precipitating solar wind flux that is different from that estimated from time - delayed ACE data. For example, depending on the solar wind conditions, the landing site may end up in an enhanced flux area (Wieser et al., 2010), with higher precipitating solar wind flux resulting in an overestimation of the calculated hydrogen ENA albedo. Lacking a direct solar wind measurement on the lunar surface, modeling of the near Moon plasma environment is needed to resolve this case.

Similar to the estimation of the hydrogen albedo, a calculation similar as in Section 3. 1 can be used to estimate the sputtering yield of heavy atoms by the solar wind on the lunar surface. The currently available data are not sufficient yet to do this calculation, as most sputtered heavy particles will have energies below a few eV that cannot be detected by the ASAN. Only the high-energy tail (>10eV) of the sputtered distribution is visible and its interpretation needs better statistics and more observation time.

At the time of writing, the ASAN instrument has been on the lunar surface for one year and is continuing to operate normally. More observations will be conducted for different solar azimuth and solar zenith angles. The ENA distribution with lunar topography and local time, as well as the mass composition information, will be studied further.

Acknowledgments, Samples, and Data

The science data and auxiliary data are obtained by support from the China National Space Administration, Beijing Aerospace Control Center and China Academy of Space Technology. This work was supported by the National Natural Science Foundation of China (NSFC) [grant number 41941001] and Beijing Municipal Science and Technology Commission [grant number 181100002918003]. The ASAN has been supported by the Swedish National Space Agency (SNSA) [contract 95/11]. The solar wind data are from the Space Weather Prediction Center, NOAA (https://www.swpc.noaa.gov/products/ace-real-time-solar-wind/). The magnetic field data are from the Lunar Prospector mission. The data presented in Figs. 2 to 4 in this work can be downloaded online (https://data.irf.se/data/aibing2020pss/).

References

[1] Allegrini, F., Dayeh, M. A., Desai, M. I., Funsten, H. O., Fuselier, S. A., Janzen, P. H., McComas, D. J., Möbius, E., Reisenfeld, D. B., Rodríguez, M. D. F., Schwadron, N., Wurz, P., 2013. Lunar energetic atom (ENA) spectra measured by the interstellar boundary explorer (IBEX). Planet. Space Sci. 85, 232-242.

[2] Barabash, S., Bhardwaj, A., Wieser, M., Futaana, Y., Varier, S., Dhanya, D., Peter, Wurz, Asamura, K., 2009. Investigation of the solar wind-Moon interaction onboard Chandrayaan-1 mission with the SARA experiment. Curr. Sci. 96 (4), 526-532.

[3] Crider, D. H., Vondrak, R. R., 2002. Hydrogen migration to the lunar poles by solar wind bombardment of the Moon. Adv. Space Res. 30 (8), 1869-1874.

[4] Feldman, G. J., Cousins, R. D., 1998. Unified approach to the classical statistical analysis of small signals. Phys. Rev. D 57 (7), 3873-3889.

[5] Feldman, W. C., Lawrence, D. J., Elphic, R. C., Barraclough, B. L., 2000. Polar

hydrogen deposits on the Moon. J. Geophys. Res. 105 (E2), 4175－4195.

[6] Futaana, Y., Barabash, S., Wieser, M., Holmström, M., Lue, C., Wurz, P., Schaufelberger, A., Bhardwaj, A., Dhanya, M. B., Asamura, K., 2012. Empirical energy spectra of neutralized solar wind protons from the lunar regolith. J. Geophys. Res. 117, E05005.

[7] Futaana, Y., Barabash, S., Wieser, M., Lue, C., Wurz, P., Vorburger, A., Bhardwaj, A., Asamura, K., 2013. Remote energetic neutral atom imaging of electric potential over a lunar magnetic anomaly. Geophys. Res. Lett. 40, 262－266.

[8] Lue, C., Futaana, Y., Barabash, S., Wieser, M., Holmström, M., Bhardwaj, A., Dhanya, M. B., Wurz, P., 2011. Strong influence of lunar crustal fields on the solar wind flow. Geophys. Res. Lett. 38, L03202.

[9] McComas, D. J., Allegrini, F., Bochsler, P., Frisch, P., Funsten, H. O., Gruntman, M., Janzen, P. H., Kucharek, H., Mobius, E., Reisenfeld, D. B., Schwadron, N. A., 2009. Lunar backscatter and neutralization of the solar wind: first observations of neutral atoms from the Moon. Geophys. Res. Lett. 36, L12104.

[10] Michael, E. P., Nicholas, J. B., 2010. Global spherical harmonic models of the internal magnetic field of the Moon based on sequential and coestimation approaches. J. Geophys. Res. 115, E12007.

[11] Rodríguez, M. D. F., Saul, L., Wurz, P., Fuselier, S. A., Funsten, H. O., McComas, D. J., Möbius, E., 2012. IBEX－Lo observations of energetic neutral hydrogen atoms originating from the lunar surface. Planet. Space Sci. 60, 297－303.

[12] Saito, Y., Yokota, S., Tanaka, T., Asamura, K., Nishino, M. N., Fujimoto, M., Tsunakawa, H., Shibuya, H., Matsushima, M., Shimizu, H., Takahashi, F., Mukai, T., Terasawa, T., 2008. Solar wind proton reflection at the lunar surface: low energy ion measurement by MAP－PACE onboard SELENE (KAGUYA). Geophys. Res. Lett. 35, L24205.

[13] Saul, L., Wurz, P., Vorburger, A., Rodríguez, M. D. F., Fuselier, S. A., McComas, D. J., Möbius, E., Barabash, S., Funsten, H., Janzen, P., 2013. Solar wind reflection from the lunar surface: the view from far and near. Planet. Space Sci. 84, 1－4.

[14] Vorburger, A., Wurz, P., Barabash, S., Wieser, M., Futaana, Y., Lue, C., Holmström, M., Bhardwaj, A., Dhanya, M. B., Asamura, K., 2013. Energetic neutral atom imaging of the lunar surface. J. Geophys. Res. Space Phys. 118, 3937－3945.

[15] Vorburger, A., Wurz, P., Barabash, S., Wieser, M., Futaana, Y., Holmström, M., Bhardwaj, A., Asamura, K., 2014. First direct observation of sputtered lunar oxygen. J. Geophys. Res. Space Phys. 119, 709－722.

[16] Wang, X. D., Bian, W., Wang, J. S., Liu, J. J., Zou, Y. L., Zhang, H. B., , LüC, Liu, J. Z., Zuo, W., Su, Y., Wen, W. B., Wang, M., Ouyang, Z. Y., Li,

C. L., 2010. Acceleration of scattered solar wind protons at the polar terminator of the Moon: results from Chang'E-1/ SWIDs. Geophys. Res. Lett. 37, L07203.

[17] Wieser, M., Barabash, S., Futaana, Y., Holmström, M., Bhardwaj, A., Sridharan, R., Dhanya, M. B., Wurz, P., Schaufelberger, A., Asamura, K., 2009. Extremely high reflection of solar wind protons as neutral hydrogen atoms from regolith in space. Planet. Space Sci. 57, 2132-2134.

[18] Wieser, M., Barabash, S., Futaana, Y., Holmström, M., Bhardwaj, A., Sridharan, R., Dhanya, M. B., Schaufelberger, A., Wurz, P., Asamura, K., 2010. First observation of a mini-magnetosphere above a lunar magnetic anomaly using energetic neutral atoms. Geophys. Res. Lett. 37, L05103.

嫦娥四号能量中性原子观测揭示太阳风与月面相互作用新特征

谢良海[1,2]，张爱兵[2,3]，李磊[1,2]，王慧姿[4]，史全岐[4]
张江[4]，王文静[2]，WEISER Martin[5]，张艺腾[1,2]
孔令高[2]，冯永勇[1,2]，周斌[1,2]，王劲东[1,2]

1. 中国科学院国家空间科学中心 空间天气学国家重点实验室，北京 100190
2. 中国科学院国家空间科学中心，北京 100190
3. 中国科学院大学，北京 100049
4. 山东大学空间科学与物理学院，威海 264209
5. 瑞典空间物理研究所，基律纳 SE-981

摘要： 与地球不同，月球暴露在太阳风中。太阳风注入月面，与月壤相互作用，部分太阳风质子以能量中性原子（Energetic Neutral Atom，ENA）的形式被月表散射。另外，月球局部地区的磁异常能阻挡太阳风到达月面，并形成微磁层，成为月面天然的保护屏障。然而以往相关的观测数据都来自轨道器，月面的真实情况无从知晓。嫦娥四号任务搭载的中性原子探测仪首次在月面就位测量 ENA，为研究月面和太阳风相互作用提供了新的视角。本文综述了嫦娥四号的 ENA 探测，重点介绍了一些不同于以往遥感观测的新现象，包括月面 ENA 反射率较高，ENA 通量向低能段聚集，以及除了氢 ENA 还有其他重成分 ENA 等。分析上游太阳风观测数据发现，月面对太阳风的作用主要体现在 105～523eV 能量段，且在磁异常下游时 ENA 通量整体偏低。利用全球 Hall MHD 数值模拟，证明了微磁层是造成 ENA 通量降低的原因。同时，还发现月球微磁层的形成与太阳风动压以及离子惯性长度有关，微磁层内的静电场使得太阳风减速和偏转，对应的电势差为 50～260V。

关键词： 太阳风-月面相互作用；能量中性原子；微磁层；嫦娥四号

Chang'E - 4 Energetic Neutral Atom Observation Reveals New Features about the Solar Wind - Moon Interaction

XIE Lianghai[1,2], ZHANG Aibing[2,3], LI Lei[1,2], WANG Huizi[4], SHI Quanqi[4], ZHANG Jiang[4], WANG Wenjing[2], WEISER Martin[5], ZHANG Yiteng[1,2], KONG Linggao[2], FENG Yongyong[1,2], ZHOU Bin[1,2], WANG Jindong[1,2]

1. State Key Laboratory of Space Weather, National Space Science Center, Chinese Academy of Sciences, Beijing 100190
2. National Space Science Center, Chinese Academy of Sciences, Beijing 100190
3. University of Chinese Academy of Sciences, Beijing 100049
4. School of Space Science and Physics, Institute of Space Sciences, Shandong University, Weihai 264209
5. Swedish Institute of Space Physics, Kiruna SE - 981

Abstract: Different from the Earth, the solar wind can directly impact the lunar surface, and partly be scattered as Energetic Neutral Atoms (ENAs). However, the lunar magnetic crustal fields in some regions, called magnetic anomalies, can deflect the solar wind to form a mini - magnetosphere, shielding the surface. All previous understandings about these processes are obtained from orbit, and the truth on the lunar surface is still unknown. The Advanced Small Analyzer for Neutrals (ASAN) onboard Chang' E - 4 mission can detect the reflected Energetic Neutral Atoms (ENAs) from the lunar surface, which will provide new perspectives to study the solar wind interaction with the Moon. Here is a review on the recent works with the ENA data from ASAN, focusing on introducing some new discoveries by ASAN, such as a higher ENA reflection ratio, more ENAs gathered at lower energies, and some heavier ENAs other than the H ENA. Compare with the upstream solar wind data, it is found that the ENAs in the energy range of 105 ~ 523eV are closely related with the solar wind. Moreover, the ENA fluxes downstream from the magnetic anomalies are generally smaller. Combined with the global Hall MHD simulation, reduction in the ENA flux is confirmed to be caused by a mini - magnetosphere. Meanwhile, it is found the formation of the mini - magnetosphere is determined by the solar wind dynamic pressure and the ion inertia length, and the mini -

magnetosphere brings a deceleration to the solar wind, by a differential electrostatic potential of 50～260V.

Key words: Interaction between solar wind and lunar surface; Energetic Neutral Atom (ENA); Mini - magnetosphere; Chang'E - 4

0　引言

由于月球缺少全球性磁场及浓密大气层的保护，太阳风粒子可以毫无障碍地轰击月表。这些粒子的注入一方面会引起月表物质各种物理和化学属性的变化，如通过还原反应产生月球水[1-3]和纳米铁[4,5]，带来空间风化效应；另一方面通过反射、溅射等方式补充月球稀薄的气体外逸层[6,7]。因此，研究太阳风与月面相互作用对于理解月面的光谱特征以及月球稀薄大气的成分构成具有重要意义。关于太阳风月面反射，人们比较关心的是其反射率、能量分布和角分布等，Kaguya 卫星首先观测到约有 0.1%～1% 的太阳风质子会被月表反射回来，能量损失约为 30%[8]。另外，Chandrayaan - 1 和 IBEX 任务的中性原子探测仪发现部分太阳风质子获得一个电子后，以能量中性原子（Energetic Neutral Atom，ENA）的形式反射回来，平均反射率为 10%～ 20%，损失能量约为 50%[9]。Chandrayaan - 1 观测结果显示，ENA 的角分布并不是理论预期的前向散射，而是以后向散射为主，这可能与月表的微观结构有关[10]。另外，ENA 能量近似呈 Maxwell - Boltzmann 分布，其温度正比于入射太阳风速度，但其反射率似乎与入射太阳风速度无关，这与理论预期不符[11]。Funsten 等[12]利用 IBEX 数据分析发现，ENA 反射率在入射能量较低时与太阳风能量无关，但在入射能量较高时（>250 eV）反射率随入射太阳风能量增大而减小，与理论预期一致。但是，这些结果都是来自卫星观测，月面的情况还有待月面就位观测数据的进一步验证。此外，以往这些结果都只看到了反射的 H 原子，而对于可被溅射出来的 O、Si、Mg、Ca、Al 和 Fe 等重元素目前还没有观测。因此，太阳风溅射对月球气体外逸层的具体贡献目前仍不清楚。

另外，月球虽然没有像地球一样全球性的偶极场，但有广泛分布的月壳剩磁，称为磁异常[13,14]。这些磁异常主要分布在月球背面，水平方向的特征尺度约为 100km，月面的磁场强度为几十到几百 nT 不等。通常情况下，太阳风可以直接轰击月面并被月面吸收或散射。但在一些强磁异常区，太阳风会被磁场偏转和反射，从而大大降低到达月面的太阳风通量，形成一个类似于地球磁层一样的局部空腔结构。由于该结构的尺度小，通常称之为微磁层。月球微磁层是太阳系中最小的磁层结构。认识微磁层，一方面有助于理解太阳风与小尺度磁场的相互作用；另一方面，微磁层作为月面天然的屏障，是研究表面物质成分、月壤风化等与太阳风关系的最佳场所。

月球微磁层的早期观测证据来自月球勘探者号（Lunar Prospector，LP）卫星。利用这些数据，Lin 等[15]首先发现在雨海对跖区（Imbrium antipode）磁异常区存在类似于地球弓激波和磁鞘区的压缩及磁场低频扰动，之后 Halekas 等[16]发现这种激波结构更多地发生在离子回旋半径较小的时候。最近，Halekas 等[17]利用 ARTEMIS 卫星数据进一步证实了与月球微磁层相关的激波存在。然而，一个完整的磁层结构还应包含激波之下的等离子体空腔。这种空腔结构目前仅被 LP 卫星观测到一次，并伴随一些特殊的观测条件，例如低高度、高太阳天顶角以及小的离子惯性长度[18]。这表明月球微磁层由于尺度小，离子可能是非磁化的，部分离子可以穿透进去，从而不容易形成等离子空腔；另一种可能是空腔的高度非常低，无法被绕月飞行的卫星看到，这时可以借助 ENA 遥感观测推断是否有微磁层。Kaguya 和 Chandrayaan－1 在一些强的磁异常区都观测到大量反射的太阳风质子[19,20]。同时，这些区域的 ENA 发射率大大降低，在 ENA 图像上留下一个局部的空洞[21,22]。这些结果表明，太阳风在强磁异常区受到了磁场的有效阻挡，到达月面的太阳风通量大幅度减少，从而间接证实了微磁层及相应等离子空腔存在。然而，目前仍然没有月面微磁层的直接观测数据。

嫦娥四号是人类首个着陆在月球背面的探测器，其上搭载的中性原子探测仪可以就位测量被月面散射的太阳风 ENA 粒子。通过分析嫦娥四号的 ENA 数据，可以获得更加准确的太阳风在月面的散射信息，有助于更好地理解太阳风与月面的相互作用。此外，嫦娥四号着陆点（177.6°E，45.4°S）刚好位于月球最强磁异常区-雨海对跖区（Imbrium antipode）的东部边缘，当太阳风从西边吹过来时此处可能形成微磁层，从而为嫦娥四号就位观测月球微磁层提供了良好契机。

1　中性原子探测仪

嫦娥四号上搭载的中性原子探测仪（Advanced Small Analyzer for Neutrals，ASAN）由瑞典空间物理研究所（Swedish Institute of Space Physics，IRF）和中国科学院国家空间科学中心联合研制。该仪器搭载在嫦娥四号巡视器上，对月表能量中性原子和正离子的能量、通量和成分进行探测。中性原子探测仪传感器组成如图 1 所示[23]。中性原子探测仪有两种工作模式：中性原子探测模式和离子探测模式，可分时探测中性原子和正离子，其中中性原子探测模式为其主要探测模式。在中性原子探测模式下，能量中性原子经过离子偏转系统后进入仪器内部，被电荷转换面电离为正离子，正离子通过静电分析器完成能量分析，之后由飞行时间单元（离子打在起始面上产生二次电子，并被电子倍增器放大产生起始信号；离子与起始面作用后生成的中性原子打在终止面上，产生二次电子并被电子倍增器放大产生终止信号）确定其速度，通过能量和速度可以得到中性原子的质量（成分）信息。在离子探测模式下，离子偏转系统关闭，可实现正离子探测。中性原子探测仪正样如图 2 所示，前部为传感器，入口斜向上，后部为电子学箱，仪器总质量（包括电子学）为 970g，功耗 3.4W，使用瑞典空间物理所的定标设备完成了定标。

中性原子探测仪的能量范围为 10eV 至 10keV，分成 48 个能道，平均能量分辨率 $\Delta E/E$ 约为 30%（实际的分辨率取决于测量粒子的能量，氢原子的能量分辨率在 16%～130%之间变化）。基于飞行时间质谱分析技术，中性原子探测仪可以实现成分分辨，质量分辨率（$m/\Delta m$）约为 2，能够区分氢、氦和氧组（≥16 amu）。仪器的时间分辨率为 4.6 s。中性原子探测仪视场为 37°×30°（垂直方向为 37°，水平方向为 30°，半高宽为 28°×14°），视轴斜向下指向月表，与水平方向夹角为 30°，所产生的梯形视场边界围成的面积约为 1.57m^2[24]。中性原子探测仪安装在嫦娥四号巡视器的 $+x$ 舱板上（见图 3），仪器隔热安装在有效载荷舱内，传感器的入口系统通过 $+x$ 舱板的开孔伸出舱外。传感器入口采用一次性开启的保护盖防止着陆时月尘进入，巡视器到达月表后，保护盖永久打开。结合巡视器的机动性，中性原子探测仪可以在不同太阳风照射角度及巡视路径上不同地点处开展探测。

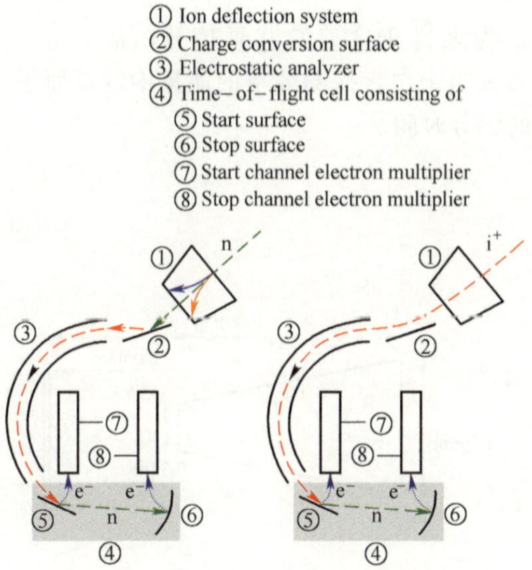

图 1　中性原子探测仪传感器

Fig. 1　Ion optical elements of the sensor

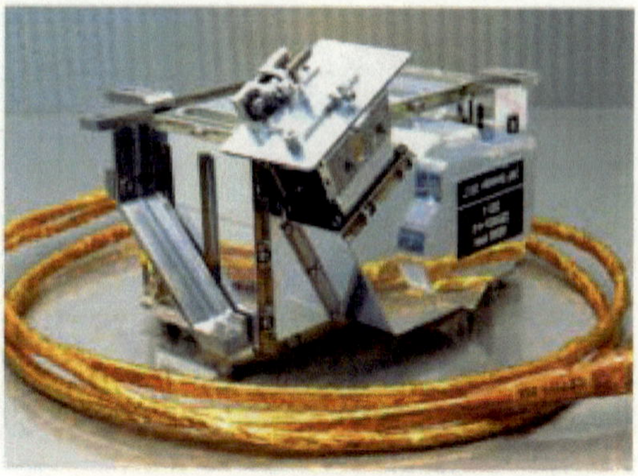

图 2　中性原子探测器正样

Fig. 2　ASAN flight model

2　ENA 数据处理及能谱结果

2019 年 1—6 月（6 个月昼）中性原子探测仪开机 21 次进行探测，单次科学探测的持续时间约 3h。数据分以下几个步骤处理。首先，由于计数率较低，将单次科学探测的原始计数进行积分以改善数据统计的质量，然后扣除由放射性同位素热源（RHU）造成的背景噪声，接着评估造成不确定性的可能因素。最后，将粒子计数转换为微分通量。

安装在巡视器后面板上的放射性同位素热源（位于中性原子探测仪安装位置的另一侧）会发出伽马射线，产生明显的背景干扰信号。背景扣除可使用两种不同的方法，选用哪种方法取决于要处理的信号类型，以及综合考量在统计学上哪一种更好。典型的能量中性氢原子能谱在太阳风能量处有一个截止值，一般约为 1keV。鉴于来自放射性同位素热源的背景信号与仪器的能道无关，而 4.5～8.7keV 之间能道的信号主要由放射性同位素热源产生，通过这些能道的平均计数率可估算背景信号，从 START 原始计数率中扣除平均背景计数率即可获得能量中性氢原子有效计数率。对于质量大于 4 amu 的较弱能量中性原子信号，需要估算飞行时间。放射性同位素热源产生的背景信号在飞行时间谱中有一个峰值，通常低于 1 amu。通过事件符合性可估算背景计数率（r_{BG}）的上限，即 $r_{BG}=r_{STA}\,r_{STO}\,t_{TOF}$。其中：$r_{STA}$为飞行时间起始信号计数率；$r_{STO}$为飞行时间终止信号计数率；tTOF 为飞行时间窗口长度，为 1.6μs。有效计数率通过从飞行时间计数率中扣除平均背景计数率 r_{BG}获得。

迄今为止，观测到的主要成分为能量中性氢原子。因此假设重粒子成分可忽略，能量中性氢原子的微分通量为

$$f_{\mathrm{H}}=\frac{C-s_{\mathrm{BG}}}{\varepsilon_{\mathrm{H}} G_{\mathrm{H0}} E_{0}\, \mathrm{d}t} \tag{1}$$

其中，C 为每个能道的 START 原始计数，s_{BG} 为来自 RHU 的背景信号，从 4.5～8.7keV 能道的平均计数估计，ε_{H} 为探测效率，G_{H0} 为未包含探测效率的能量中性氢原子的几何因子，E_0 为能道中心值，$\mathrm{d}t$ 为每个能道的积分时间。

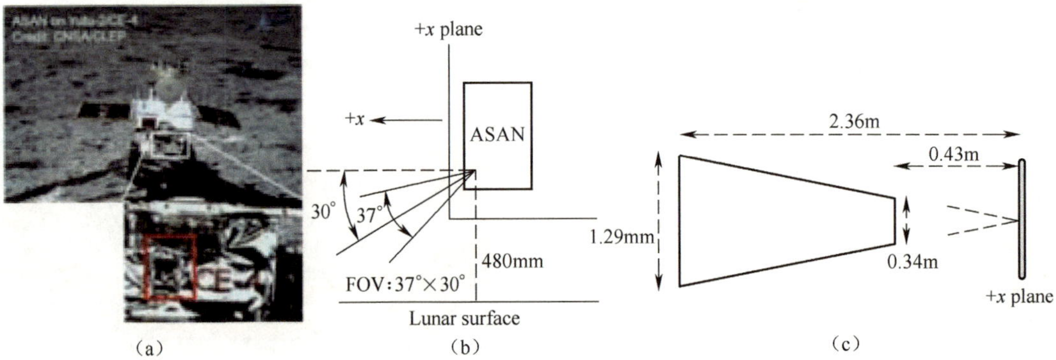

图 3　(a) 巡视器上的中性原子探测仪（红色正方形），(b) 仪器安装方位（巡视器前进方向为 $+x$ 方向，虚线显示的是视场的视轴），(c) 中性原子探测仪收集粒子的表面区域的视场范围

Fig. 3　(a) ASAN (red square) on the rover, (b) accommodation geometry (the rover moves nominally in the $+x$ direction, and the dotted line shows the bore sight of FOV), (c) field of view footprint of the surface area from which ASAN collects particles

质量大于 4 amu 的重粒子成分的微分通量使用

$$f_{x}=\frac{C-r_{\mathrm{BG}}\, \mathrm{d}t}{\varepsilon_{X} G_{X0} E_{0}\, \mathrm{d}t} \tag{2}$$

进行计算。其中 C 为在感兴趣的质量范围内的计数值，r_{BG} 为背景计数率，由飞行时间系统中事件符合性计数来估计，ε_{X} 为符合探测效率，G_{X0} 为未考虑探测效率的能量中性重粒子几何因子，E_0 为能道的中心，$\mathrm{d}t$ 为一个能道的积分时间。值得注意的是，因为 r_{BG} 为背景计数率的上限，所以可能在一定程度上低估能量中性重粒子的微分通量。

粒子通量的不确定度包括几何因子的不确定度和计数统计误差。对于两个最低的能道，几何因子的不确定度为标称值的＋200％和－70％，其他能道为＋40％和－30％。

图 4 显示了 2019 年 5 月 1 日 06：18－10：45 UTC 期间探测到的能量中性氢原子的平均能谱，当时太阳风速为 340km · s^{-1}。图 4 (a) 给出了能量中性氢原子通量，图 4 (b) 给出了质量大于 4 amu 的能量中性原子通量。当通量低于灵敏度极限（见图 4 中虚线）时，无法将其与零通量区分开。能量中性氢原子能谱中有两个域，能量大于 100eV 的通量几乎完全来自月面散射的太阳风，能量低于 100eV 的则可能来自表面溅射。较重的能量中性原子（质量大于 4 amu）集中在能量小于 100eV 范围内。每次测量的重原子成分的结果变化很大，可能是 5keV 以上的重原子，也可能是后向散射的太阳风重离子，

例如高价氧离子，这有待进一步研究（见图 4b）。由于背景噪声的影响，质量在 1 amu < $m \leqslant 4$ amu 的能量中性原子更加难以区分。

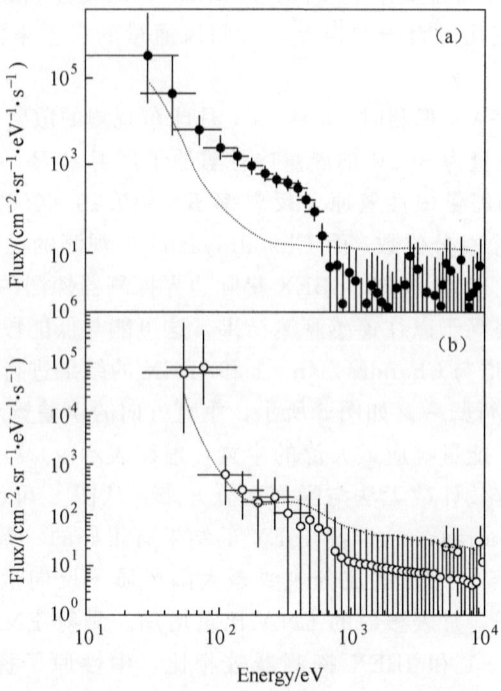

图 4　2019 年 5 月 1 日 06：18—10：45 UTC 期间测得的能量中性粒子平均能谱

Fig. 4　Average ENA energy spectrum from 06：18 to 10：45 UTC on May 1，2019

ASAN 测量到的 ENA 微分通量（j_{ENA}）与太阳天顶角（SZA，定义符号 α）及月球车相对于太阳的方向有关。由于月球车在 ASAN 测量期间不移动，因此只能观测来自特定方向的表面散射通量。如果月面的散射函数与轨道遥感得到的散射函数[25]相同，则基于 ASAN 观测数据，使用散射函数计算月表反射的 ENA 总通量；结合太阳风参数，可以计算氢 ENA 的反射率。首先对 ASAN 能谱进行能量积分，得到方向通量，近似写为

$$j_{\mathrm{ENA}}(\alpha_0,\ \phi_0,\ \theta_0)=\sum_{E}\frac{(j_{E_{i+1}}+j_{E_i})(E_{i+1}-E_i)}{2} \tag{3}$$

其中，下标 0 表示中性原子探测仪观测的角度，ϕ 为散射方位角、θ 为散射极角（定义见文献［24］），j_{E_i}（$\mathrm{cm^{-2}\cdot sr^{-1}\cdot s^{-1}\cdot eV^{-1}}$）为能道 E_i（eV）的微分通量。采用文献［25］的方法，可以得到 ASAN 视场方向上的散射函数 $f_s(\alpha_0,\ \phi_0,\ \theta_0)$，再结合太阳风通量 J_{sw}，可以计算垂直方向上的反射率

$$R_{\perp}=\frac{j_{\mathrm{ENA}}(\alpha_0,\ \phi_0,\ \theta_0)}{f_s(\alpha_0,\ \phi_0,\ \theta_0)\cdot J_{\mathrm{sw}}} \tag{4}$$

$R_{\perp}$的误差源包括 ENA 方向通量 j_{ENA}的误差和太阳风通量 J_{SW}的误差。ENA 方向通量 j_{ENA}的误差来源于各能道几何因子的误差，太阳风通量的误差来源于中性原子探测仪观测期间太阳风的变化。

2019 年 5 月 1 日 ASAN 观测时，$\alpha=58°$，月球车观测的散射方位角 $\phi=-31°$和散射极角 $\theta=60°$，低端截止能量为 30eV 的能量中性氢原子反射率 $R_{\perp}=0.32$（0.19，0.58），低端截止能量为 46eV 的能量中性氢原子反射率 $R_{\perp}=0.10$（0.07，0.17），括号中的数字表示第 25 个和第 75 个百分位数。而 Chandrayaan－1 观测的反射率为 0.19，第 25 和第 75 百分位数分别为 0.16 和 0.21。IBEX 星际边界探测器探测的反照率为 0.11 ± 0.06（1σ）。ASAN 测量结果略高于以往遥感观测结果，这可能与低能段的溅射有关。

为了将 ASAN 的能谱与 Chandrayaan－1 和 IBEX 的结果进行比较，利用太阳风能量 E_{sw}对中性原子的能量进行归一。如图 5 所示，垂直方向的误差棒对应于 90% 置信区间，水平方向的误差棒对应于能量响应最大值的半宽。虚线表示 90% 置信区间，低于该置信区间，中性原子探测仪有效计数无法与噪声区分开来。从图 5 可以看出，ASAN 能谱在大于 $0.1E_{\mathrm{sw}}$部分与 Chandrayaan－1 及 IBEX 的结果有很好的一致性。与 ENA 生成过程相关的有两种成分。高于 $0.1E_{\mathrm{sw}}$的成分主要是太阳风质子反向散射为能量中性氢原子，而低于该能量的中性原子，月表溅射的 ENA 比重增加。溅射 ENA 的能量主要在 0.1～100eV。与 Chandrayaan－1 和 IBEX 探测器数相比，中性原子探测仪数据显示在小于 $0.1E_{\mathrm{sw}}$以下有较大的通量。低端通量在不同次探测中变化很大。这种现象可能是由于溅射产额以及月壤中可用氢的变化造成的。Chandrayaan－1 和 IBEX 探测器的 ENA 能谱是在轨道上大面积区域的平均值。其中，IBEX 是月球的 1/4，Chandrayaan－1 是几百平方千米。因此，观测区域的任何局部特征都是模糊的。而中性原子探测仪的视场范围只有 $1.57m^2$，局部风化层特征例如孔隙率、粒度和成分也可能影响能谱，尤其是在能量较低

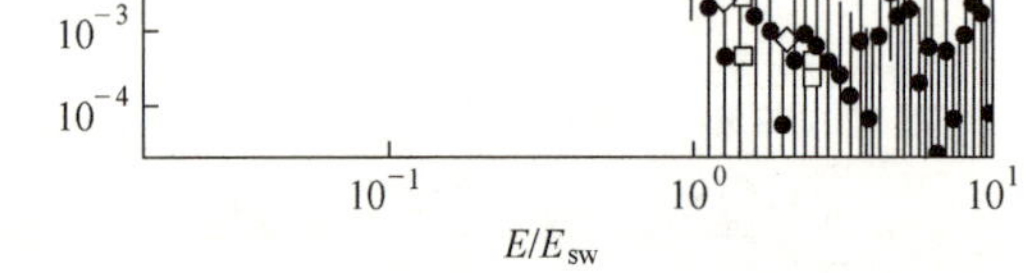

图 5　2019 年 5 月 1 日 06：18—10：45 UTC 观测的能量中性氢原子平均能谱

Fig. 5　Average hydrogen ENA energy spectra at 06：18—10：45 UTC on May 1，2019

时。图 4 和图 5 展示了太阳天顶角 $\alpha=58°$ 和散射极角 $\theta=60°$ 情况下的 ENA 能谱。Chandrayaan－1 的能谱是 α 范围为 0°～90°和 $\theta=0°$（即星下点方向）下的平均值。IBEX 的能谱是一个典型的 1/4 月球表面的加权平均值，覆盖范围广，α 范围为 0°～ 90°，θ 为 0°～90°。ENA 能谱可能随 α 和 θ 而变化。因此当在所有角度上积分时，中性原子探测仪的特定观测几何可能导致特殊的能谱。此外，最低定标能量为 100eV，外推得到的更低能量的几何因子存在不确性，这也是在低能量下不同探测器结果存在差异的另一个原因。

3　月球微磁层

ASAN 结果低能段通量较高的另一个原因可能是受到周围磁场的减速作用。如图 6 所示，虽然嫦娥四号着陆点（177.6°E，45.4°S）处的磁场比较弱（约为 6 nT），但在该着陆点西北边是月球最强的磁异常区，即雨海对跖区（Imbrium antipode）。该磁异常区的中心位于 162°E，33°S，半径达 600km[13]，嫦娥四号刚好位于该磁异常区的东南边缘。以往研究结果表明，该磁异常区可能形成微磁层[15]，并且可以屏蔽掉 27%～74% 的太阳风[22]。由图 6 可知，当太阳风西边吹过来时，嫦娥四号位于磁异常的下游，这时如果上游的磁异常形成微磁层，嫦娥四号可能被包裹在微磁层的磁尾里。由于微磁层可以使太阳风发生偏转及减速，从而会影响月面测到的 ENA 能谱。接下来利用 ASAN 观测的 ENA 数据，分析微磁层对 ASAN 观测结果的影响，并对月球微磁层的形成条件及其屏蔽效率展开讨论[26]。

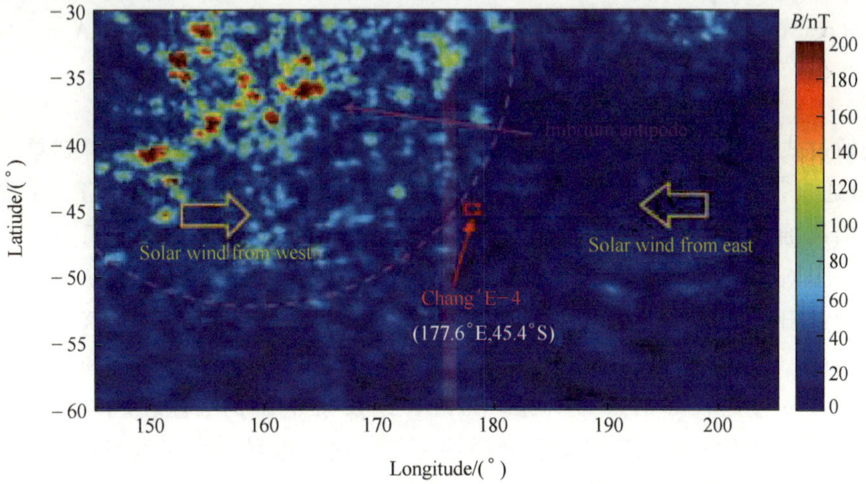

图 6　嫦娥四号着陆点周围的磁场分布

Fig. 6　Lunar crustal magnetic field strength distribution near the Chang'E－4 landing site

选取 2019 年 1 月 11 日至 2020 年 4 月 28 日期间中性原子探测仪的 46 个 ENA 能谱数据。如图 7（a）所示，其中蓝色、红色星号分别表示上游和下游的测量值，蓝色和红色线分别表示上游测量和下游测量的拟合线。中性原子探测仪开机时间集中在月球的上午和下午两个时间段，其中有 26 次测量发生在上午，对应的时间为 07：00－09：30 LT，其他测量则发生在下午，对应时间为 14：30－ 17：00 LT。结合图 6 发现，上午时段嫦娥四号是在雨海对跖区磁异常区的上游，而下午时段嫦娥四号是在磁异常区的下游，因此下游测量可能看到微磁层的尾巴。由上节讨论，ASAN 测到的 H－ENA 可能受到月表溅射的能量中性重粒子的污染，这些污染主要集中在低于 100eV 能量段[23,24]。因此，这里选取能量高于 100eV 的 ENA 通量研究磁异常对太阳风的影响。

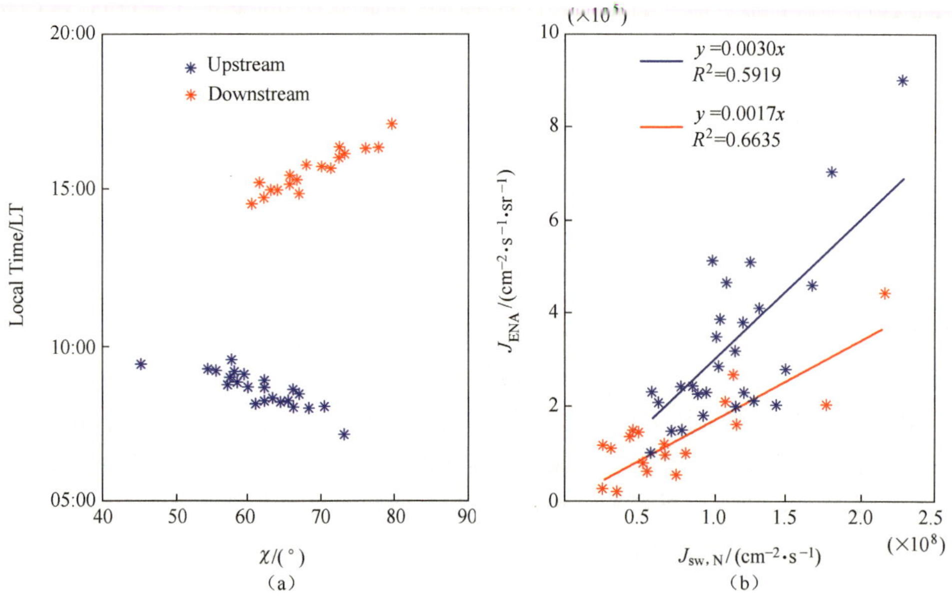

图 7　ENA 观测对应的月球地方时及 ENA 微分通量 J_{ENA} 与太阳风通量法向分量 $J_{sw;N}$ 之间的关系

Fig. 7　Local time for ASAN ENA measurement, dependence of the differential ENA flux J_{ENA} on the normal component of the solar wind flux $J_{sw,N}$

首先通过式（3）对 100eV 以上能量范围的 ENA 能谱进行积分，得到单次测量的 ENA 微分通量，即 J_{ENA}。根据 Chandrayaan－1 观测结果，月面散射的 ENA 微分通量决定于太阳风通量、太阳天顶角以及散射角等参数。通过对散射角求积分，可以得到 $J_{ENA}/J_{sw}\cos(\alpha)$，其中 J_{sw} 和 α 分别表示太阳风通量和太阳天顶角［25］。可以发现 J_{sw} 实际上是和太阳风法向通量成正比，即 J_{ENA} $J_{sw,N}$，其中 $J_{sw;N}=J_{sw}\cos\chi$ 是太阳风在月表的法向通量，χ 为太阳风与月表法向的夹角。如图 7（b）所示，上游和下游观测到的 J_{ENA} 均与 $J_{sw,N}$ 有比较好的线性关系。另外，还发现下游的 J_{ENA} 整体上小于上游的 J_{ENA}，这意味着这些下游观测可能和微磁层有关。这时磁异常位于嫦娥四号上游，如果形成了微磁层，嫦娥四号会被包裹在微磁层的磁尾里。由于微磁层对太阳风的屏蔽作用，造成到达月面的太阳风通量降低，相应月面反射的 EAN 通量也会降低，最终造成如图 7（b）所示的下游 ENA 通量整体偏低。

另外，下游测量到的 J_{ENA} 虽然较小但仍然正比于 $J_{sw,N}$，这表明微磁层可能只是部分屏蔽掉太阳风，一些太阳风质子可以穿透微磁层并到达月表，从而 J_{ENA} 与 $J_{sw,N}$ 仍然正相关。因此，有必要进一步估计有多少太阳风粒子能够穿透微磁层，即粒子的穿透效率 η_{sw}。由图 7（b）中的拟合曲线可知，没有微磁层时 J_{ENA} 与 $J_{sw,N}$ 的比值约为 0.003。如果忽略不同散射角分布函数带来的影响，可以将 0.003 看成是正常情况下 J_{ENA} 与 $J_{sw,N}$

的平均比值，其表征的是 ASAN 视场内 ENA 的平均反射率。将位于磁异常下游时测到的 $J_{ENA}/J_{sw,N}$ 除以 0.003，就得到了微磁层的穿透效率。

如图 8（a）所示，当离子惯性长度足够大（>120km）时，一些 η_{sw}（图 8 中绿色星号）可能接近或甚至大于 1.0，这意味着在这些时刻并没有形成微磁层，可能的原因是大部分离子都穿透进去了从而破坏了微磁层结构。因此，将这些数据归为外部测量（Outside measurements），其余的则为内部测量（Inside measurements），只有内部测量的 η_{sw} 才能表征微磁层的穿透效率。通过这种方式，最终得到微磁层的穿透效率为 0.23～0.78（如图 8a 中的红色星号），平均值为 0.57。此前，Chandrayaan - 1 在雨海对跖区磁异常区附近观测到的磁场屏蔽效率为 0.27～0.74[22]，对应的 η_{sw} 为 0.26～0.73，与这里的结果基本一致。

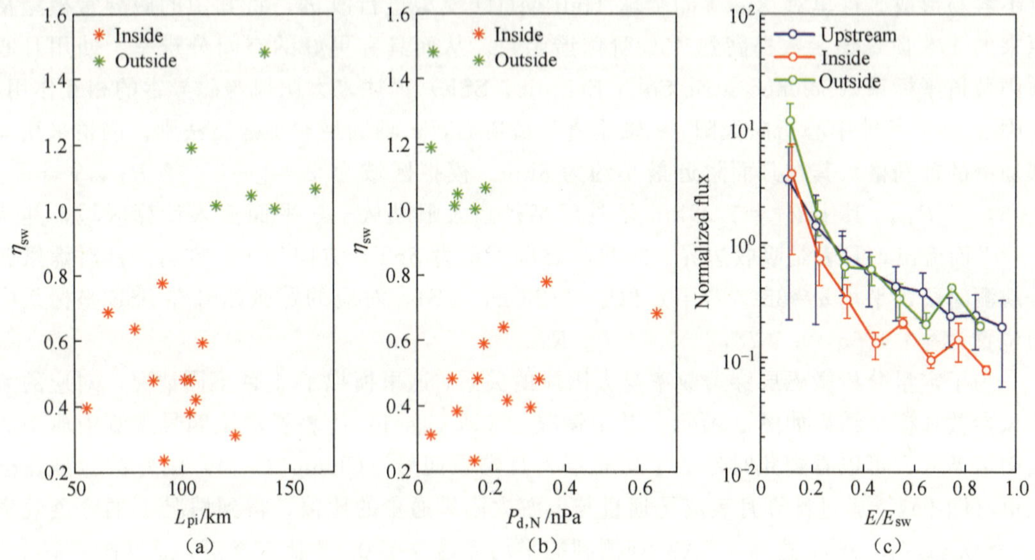

图 8　（a）和（b）分别为穿透效率 η_{sw}、离子惯性长度 L_{pi} 和太阳风动压法向分量 $P_{d,N}$ 的依赖关系，（c）归一化的能谱，不同点间的间隔约为 $0.1E_{sw}$

Fig. 8　Dependences of penetrating efficiency
(a) and the ion inertia length L_{pi}
(b) on the normal component of the solar wind dynamic pressure $P_{d,N}$, respectively
(c) normalized energy spectra with an increment of about $0.1\,E_{sw}$ for the upstream, inside, and outside measurements, respectively

此外，Chandrayaan - 1 结果显示 η_{sw} 与太阳风动压正相关。考虑到太阳风动压对微磁层的作用主要在垂直月面方向，这里检查 η_{sw} 与太阳风动压方向分量 $P_{d,N}$ 的关系，有

$$P_{d,N}=\rho v_{sw}^2\cos^2\chi \tag{5}$$

其中，ρ 和 v_{sw} 分别为太阳风的密度和速度。如图 8（b）所示，对于内部测量，η_{sw} 随 $P_{d,N}$ 增加而增加，这与 Chandrayaan - 1 的观测结果一致。对于外部测量，η_{sw} 通常大于 1.0，最大值接近 1.5，这意味着有额外的太阳风打到月面，可能来源是太阳风由旁边磁场偏转而来。另外，此时 η_{sw} 随 $P_{d,N}$ 增加而减小，这表明太阳风动压越低越容易被磁场

偏转。Chandrayaan-1 也对类似的通量增强区进行了观测，其一般发生在紧邻微磁层的周围，产生原因为太阳风被微磁层偏转，并在其外围形成一个堆积区[21]。此外，通过比较归一化的能谱（ENA 能量由太阳风动能 E_{sw}归一化，微分通量由 10^{-5} $J_{sw,N}$归一化），发现微磁层内的 ENA 通量整体小于没有微磁层时（上游时）的 ENA 通量（见图 8c）。并且，这种差异在高能段更明显，这表明离子在穿透微磁层时已经减速，从而造成高能段通量向低能段转移。对于外部测量，其与上游观测结果在高能段相差不大，但是其低能段（低于 0.2 E_{SW}）的 ENA 通量比上游结果还高。这部分多出来的低能粒子应该是由上游或旁边磁异常对太阳风的偏转和减速带来的。

此前，Xie 等[27]利用三维 Hall MHD 模型研究了月球微磁层，发现压力平衡距离与离子惯性长度之比对于微磁层的形成非常重要。当比值较小时，离子会穿透得更深，微磁层不容易形成。这里对 Xie 等的全球 Hall MHD 模型进行改进，将其中的磁异常模型从原来的 178 阶球谐函数提高到 450 阶球谐函数，从而具有更好的空间分辨率。使用月心太阳黄道坐标系（Selenocentric Solar Ecliptic，SSE）描述来太阳风与磁异常的相互作用，其中 x 轴从月球中心指向太阳，z 轴垂直于黄道面，y 轴满足右手螺旋法则。网格采用局部加密的球网格，其在月面附近最小约为 8km。模拟区域为 $x=[-8,\ 2]R_L$，$y=z=[-5,\ 5]R_L$，其中 $R_L=1738$km 是月球半径。太阳风从 $+x$ 平面流入计算区域，并从 $-x$ 平面流出，月表是吸收边界。行星际磁场固定为 5nT，方向为 $+y$ 方向。针对嫦娥四号观测时段，平均 $\alpha=65°$，日下点位于（125°E，0°S），对应的嫦娥四号在 SSE 坐标系中的位置为 [0.4265，0.5578，−0.7120] R_L。

为了定量分析微磁层穿透效率与太阳风的关系，这里模拟了 4 种不同情况，对应的太阳风参数和模拟结果如图 9 所示。其中情况 1（即 Case1）代表了嫦娥四号观测中典型的太阳风条件，可以看到此时形成了微磁层，且嫦娥四号（Chang'E-4）刚好被包裹在磁尾里（图 9a）。通过计算月表离子通量与上游太阳风通量的比值，得到情况 1 的穿透效率 η_{sw}为 0.41（图 9b），这与 ASAN 观测到的平均穿透效率 0.57 比较接近。通过比较情况 2 和情况 1，发现当离子惯性长度相同时，较低的 $P_{d,N}$可以带来较小的 η_{sw}（图 9c）。通过比较情况 4 和情况 3，发现即使 $P_{d,N}$相同，离子惯性长度较大时可能不会形成微磁层（图 9e 中 η_{sw}大于 1）。这些结果与图 8（a）和 8（b）所示情况一致。此外，模拟中还发现太阳风的减速和偏转，情况 1、2 和 3 中月面太阳风速度分别降低了 16%、36% 和 35%，同时速度分别偏转了约 11°、32°和 30°。这些结果表明，具有较低动压的太阳风更容易被磁异常偏转，这也与 ASAN 观测结果一致。然而，这里的模拟并没有在微磁层中看到密度空腔。这意味着由于离子穿透效应的存在，月球微磁层内部不容易形成等离子体空腔，月面太阳风通量的降低主要是由太阳风的偏转和减速造成的。

4 ENA 能谱与太阳风的关系

ENA 由太阳风粒子在月表反射产生，其能谱决定于入射太阳风参数及月面微观结构。Futaana 等[11]利用 Chandrayaan-1 数据发现 ENA 的温度与太阳风的速度正相关。前两节介绍 ASAN 发现 ENA 的截止能量与太阳风能量基本一致[24]，同时 ENA 能谱还受磁异常或者微磁层的影响[26]。Wieser 等[21]发现能量较低（<100eV）的氢原子受微磁层的

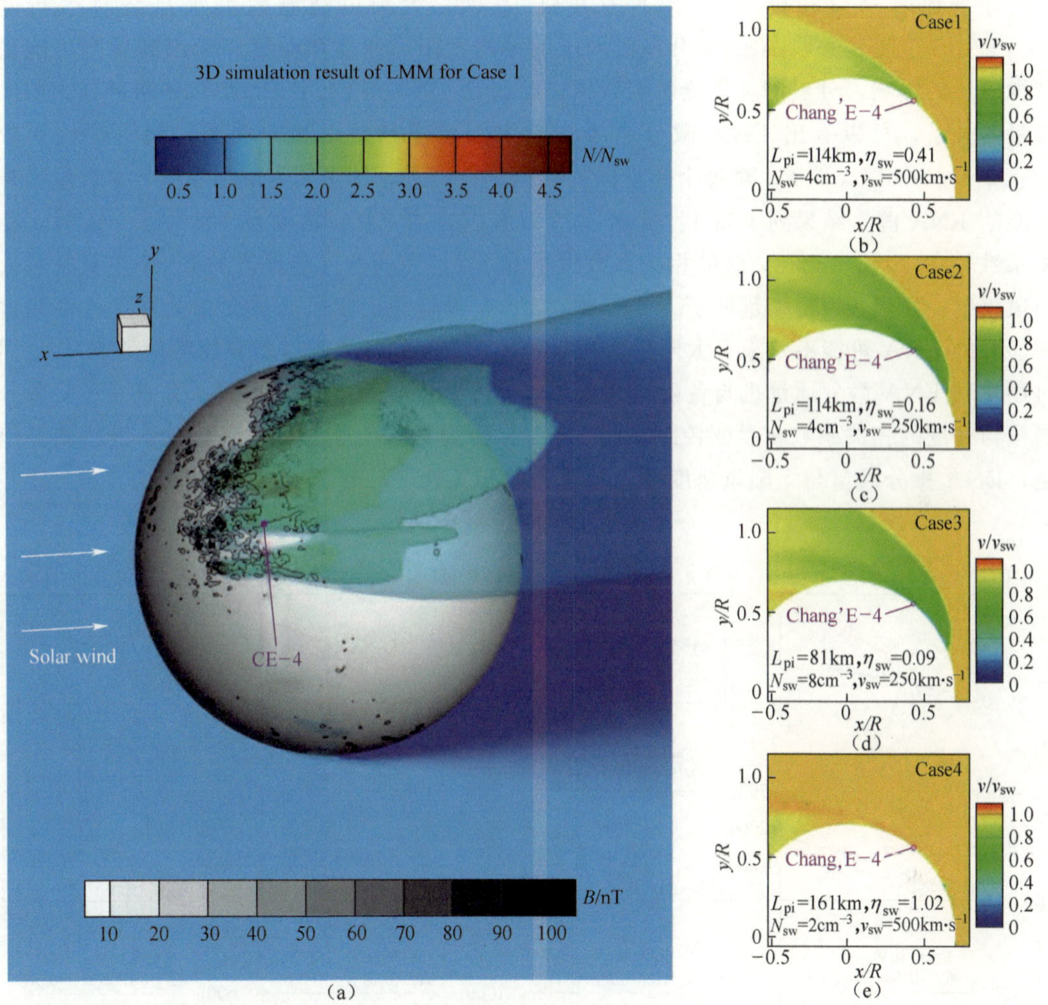

图 9　(a) 情况 1 的三维模拟结果，其中月球上的灰度图表示月球磁场大小，彩色透视图表示密度等值图，其大小利用太阳风密度 N_{sw} 进行了归一化，紫色圆圈为嫦娥四号（Chang'E-4）位置；(b)(c)(d)(e) 分别为情况 1、2、3 和 4 在 $z=-0.712R_L$ 的 xy 平面上的速度模拟结果

Fig. 9　(a) 3D view of the simulation result for Case 1, where the central ball represents the lunar body with gray contours to show the magnitudes of the crustal magnetic fields, and the 3D colored contours show the number densities normalized by the solar wind number density N_{sw}. The magenta circle indicates the location of Chang'E-4. (b)(c)(d)(e) are the results in the xy plane at $z=-0.712\ R_L$ for Cases 1, 2, 3, and 4, respectively

影响不明显，而对于 150～600eV 的 ENA，微磁层的影响比较明显。这种差异表明可能存在两种起源的氢原子，一种能量低于 100eV，而另一种能量大于 150eV。能量大于 150eV 的氢原子可能与太阳风质子轰击月壤直接相关，因为在这些氢原子产生的图像中可以看到微磁层[21]。另外，微磁层中存在离子和电子分离带来的静电场，该静电场造成太阳风在到达月面前减速，每单位电荷能量降低 135～150eV[28,29]。

这里使用 2019 年 1 月 11 日至 2020 年 10 月 12 日 23 个月昼的 54 次观测数据，其中在月球当地时早晨开机 31 次，下午开机 19 次。太阳风观测数据来自绕月飞行的 ARTEMIS 卫星，并取 ASAN 开机时间段的平均太阳风速度和密度。考虑到昏侧可能受到微磁层的影响，分晨侧和昏侧两组数据进行讨论。图 10 给出了 ENA 微分通量与太阳风通量的关系，可以看出 ENA 微分通量均与太阳风通量呈正相关关系。利用线性函数 $df_{ENA}=c\cdot df_{SW}+b$，对每个能量段的数据进行拟合，其中 c 为反射系数，b 表示另外可能的 ENA 源，结果列于表 1。从表 1 可以看出，当 ENA 能量为 105～523eV 时，相关系数较高，140～333eV 时的相关系数可达 0.7，这说明这一能量段的 ENA 主要来自太阳风反射。而当 ENA 能量低于 105eV 时，ENA 通量与太阳风通量无相关性，这说明这一能段的 ENA 可能不是来自太阳风，而是其他源，例如月表或仪器表面溅射的 ENA。同时还发现 ENA 微分通量也与各能道的太阳风密度和动压呈正相关关系，并且 ENA 微分通量与密度的相关系数在晨侧较高，而 ENA 微分通量与动压的相关系数在昏侧较高。但是，ENA 微分通量与太阳风速度并没有明显的相关性[30]。

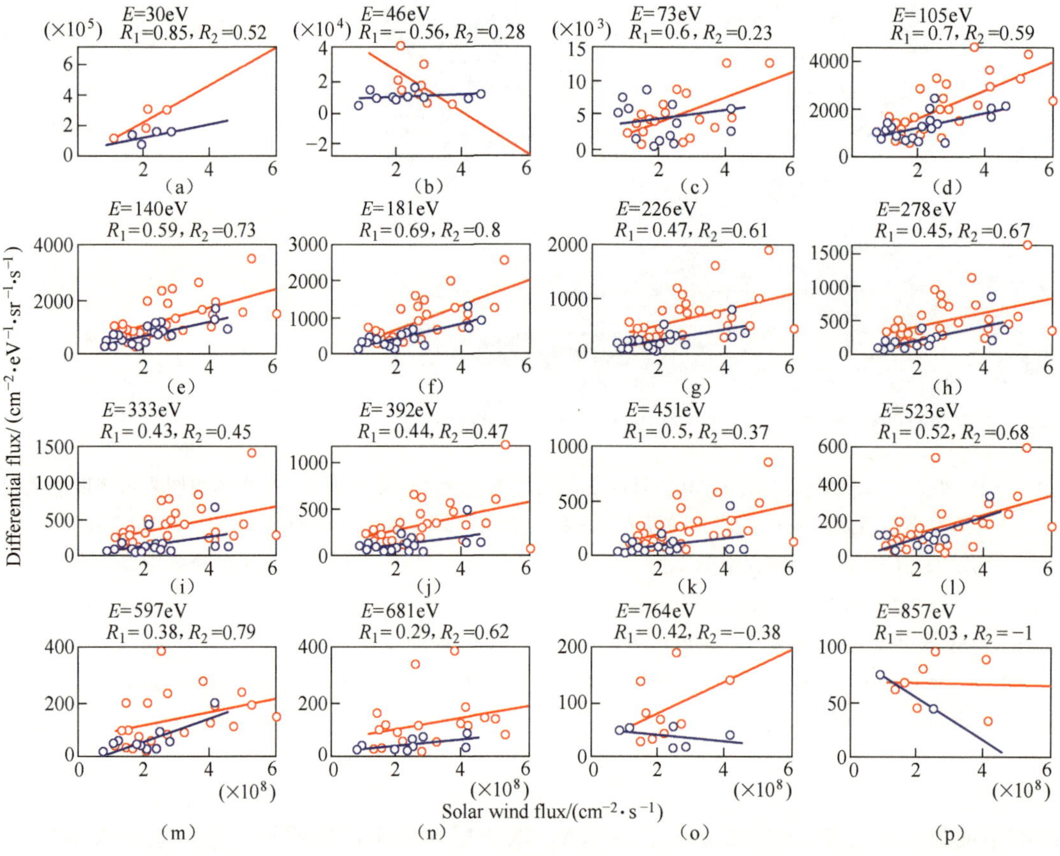

图 10　ENA 微分通量与太阳风通量的关系。红色圆圈表示在月球晨侧的数据，蓝色圆圈表示在月球昏侧的数据，R_1 和 R_2 分别表示晨侧和昏侧的相关系数

Fig. 10　Relationship between the ENA differential flux and solar wind flux. The red circles indicate the data on the lunar dawnside, and the blue circles indicate the data on the lunar duskside. R_1 and R_2 represent the correlation coefficients on the dawnside and duskside, respectively

表 1　各 ASAN 能量段内 ENA 微分通量与太阳风通量的线性拟合结果

Table 1　Linear fitting results of the integrated solar wind flux and the differential flux in each ASAN energy bin

能量/eV	晨侧 c	晨侧 k	晨侧 r^2	昏侧 c	昏侧 k	昏侧 r^2
30	1.22×10^{-3}	-1.44×10^{4}	0.73	1.29×10^{-4}	1.15×10^{5}	0.04
46	-1.31×10^{-4}	5.31×104	0.32	-8.38×10^{-6}	1.73×10^{4}	0.00
73	1.66×10^{-5}	1.32×10^{3}	0.36	5.98×10^{-6}	3.90×10^{3}	0.05
105	5.85×10^{-6}	4.20×10^{2}	0.49	2.91×10^{-6}	6.36×10^{2}	0.34
140	3.17×10^{-6}	3.97×10^{2}	0.34	2.73×10^{-6}	95.1	0.55
181	3.20×10^{-6}	32.4	0.48	2.09×10^{-6}	1.12	0.62
226	1.37×10^{-6}	2.51×10^{2}	0.22	1.02×10^{-6}	58.9	0.36
278	1.04×10^{-6}	2.12×10^{2}	0.20	1.08×10^{-6}	−12.4	0.48
333	8.43×10^{-7}	1.60×10^{2}	0.18	6.24×10^{-7}	13.9	0.22
392	7.02×10^{-7}	1.38×10^{2}	0.19	4.59×10^{-7}	22.7	0.24
451	6.60×10^{-7}	66.4	0.25	3.43×10^{-7}	27.6	0.16
523	5.31×10^{-7}	17.3	0.27	5.94×10^{-7}	−20.7	0.48
597	2.36×10^{-7}	69.8	0.15	4.14×10^{-7}	−25.3	0.64
681	2.09×10^{-7}	64.1	0.08	1.24×10^{-7}	17.4	0.39
764	2.78×10^{-7}	25.4	0.18	-2.72×10^{-8}	43.8	0.04
857	-5.46×10^{-9}	68.8	0.00	-1.85×10^{-7}	93.0	0.98

图 11（a）给出了晨昏两侧太阳风动能（由太阳风体积速度得出）与 ENA 截止能量之间的关系，图 11（b）为晨昏两侧 ENA 温度与太阳风动能的关系。在计算 ENA 温度时，这里排除了几个可能没有磁屏蔽效应或受邻近磁异常偏转影响的事件（即图 8 中绿色的点）。这些偏转的粒子在一定程度上增加了穿透速度，可能对温度有影响。在图 11（a）、（b）中，蓝线表示昏侧线性拟合结果，红线表示晨侧线性拟合结果。这里将 95% 置信区间（虚线）作为线性拟合参数的误差。计算得到的 ENA 温度与文献［11］中 Maxwell - Boltzmann 分布的最佳拟合参数的结果具有相同的量级。在图 11（a）、（b）中，红线均相对高于相应的蓝线，晨昏数据点之间的差异均具有统计学意义。对晨侧和昏侧数据集应用 t 检验，截止能量和温度的 p 值分别为 0.026 和 0.01，这说明晨昏数据集之间存在显著的统计学差异，从而证实了晨侧和昏侧数据集属于两个不同的分布。结果表明，在相同的太阳风能量下，昏侧 ENA 截止能量和 ENA 温度均低于晨侧 ENA，即在昏侧太阳风减速 ENA 能谱变窄。因此，ENA 截止能量降低，温度较低。

在昏侧，太阳风可能受到月球微磁层的影响，被磁异常上方的静电场所减速。因此，产生同样能量的 ENA。由于微磁层的减速作用，昏侧的太阳风质子在与月球表面相互作用之前，需要能量更高的太阳风质子。利用晨侧与昏侧的 ENA 截止能量之差，即将晨昏两侧的 ENA 截止能量的拟合结果相减（图 11 中的红色实线和蓝色实线），可估算太阳风在昏侧的减速，进而计算昏侧月表相对于太阳风的静电势。太阳风减速率可利用 ENA 截止能量差除以初始太阳风能量获得。

Kaguya[28]观测发现南极艾特肯盆地附近磁异常内部存在＋150V 的电势，此时初始太阳风能量约为 630eV。图 11 中的黑色方框标注了两个分别在晨侧和昏侧初始太阳风能量（约 680eV）与之相近的事件，晨侧（680eV）和昏侧（550eV）的 ENA 截止能量差约为 130eV，即存在＋130V 的电势差。由图 12 可以看出，静电势和太阳风减速率都随太阳风能量而变化，静电势为 50～360V（95% 置信区间为 30～350V），太阳风的减速率为 12%～18%（95%置信区间为 7.5%～30%）。这里的结果与前人基本一致[28,29]。

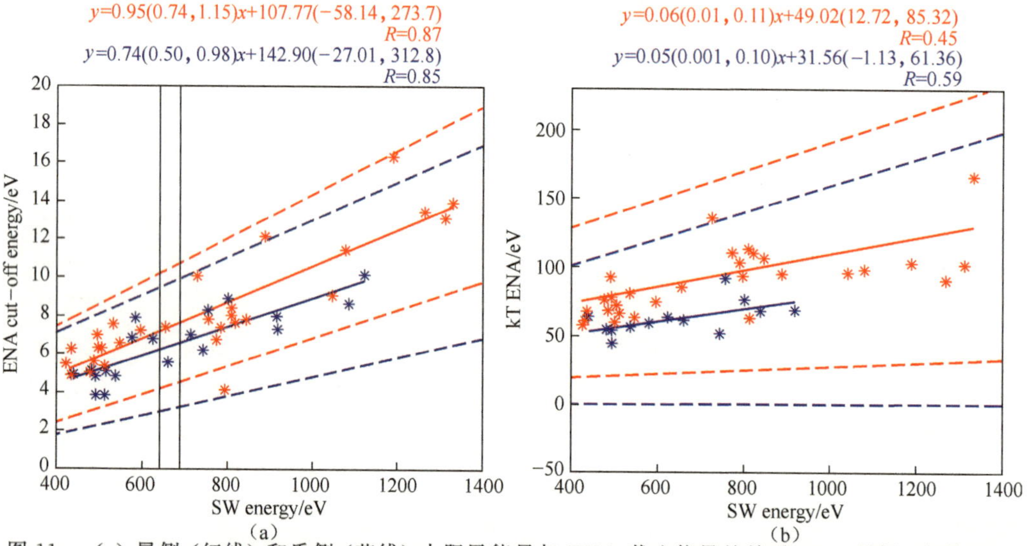

图 11　(a) 晨侧（红线）和昏侧（蓝线）太阳风能量与 ENA 截止能量的关系，(b) 晨侧（红线）和昏侧（蓝线）的太阳风能量与 ENA 温度的关系。虚线为 95% 置信区间

Fig. 11　Relationships between solar wind energy and ENA cutoff energy on the dawnside (red line) and duskside (blue line). The dashed lines indicate 95% confidence intervals

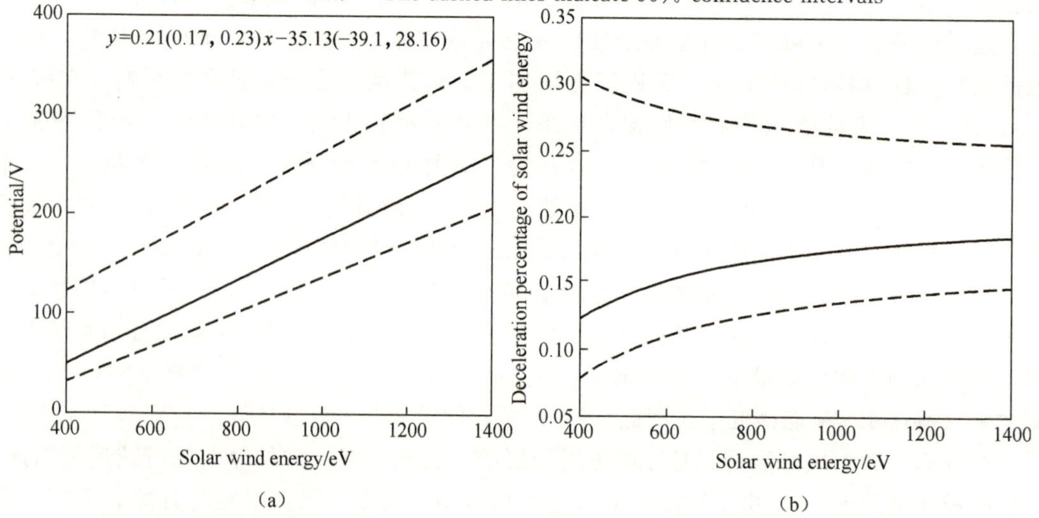

图 12　月表相对于太阳风的静电势 (a) 及太阳风减速百分比与太阳风能量的关系 (b)。虚线为 95% 置信区间

Fig. 12　(a) Lunar surface electrostatic potential relative to the solar wind, (b) relation between the solar wind deceleration percentage and the solar wind energy. The dashed lines are 95% confidence intervals

5　结论与展望

嫦娥四号搭载了国际上首个原位观测月面能量中性原子的探测仪。本文介绍了中性原子探测仪的工作原理，重点对月面 ENA 观测的最新研究成果进行了梳理和总结。主要结论如下。

（1）相比于以往遥测观测，月面就位测量的 ENA 反射率较高（约为 0.32），通量向低能段（低于 100eV）聚集，且首次测量到质量大于 4 amu 的重 ENA。

（2）太阳风对 ENA 的作用主要集中在 105～523eV 能量区间，ENA 的截止能量和温度均正比于太阳风能量，昏侧 ENA 的通量、截止能量和温度均低于晨侧 ENA。

（3）结合数值模拟，证明了微磁层是造成 ENA 晨昏差异的原因，该微磁层位于雨海对跖磁异常区下游，空间尺度达 600km，其可使太阳风通量减少约 43%，速度减小 12%～18%，对应的静电势为 50～260V。

（4）月球微磁层仅在离子回旋半径较小（<120km）时才可形成，部分太阳风质子仍可穿透微磁层，穿透效率在 0.23～0.78 之间，且正比于太阳风动压。

这些结果提高了对太阳风与月面相互作用的认识，但同时又带来了一些新的问题。首先，低能段 ENA 的具体来源仍然不清楚，氢 ENA 可能的来源包括太阳风月面溅射以及仪器内部表面的溅射。如果是前者，则可能与月壤多孔结构引起的多次散射有关，那将有助于进一步认识太阳风与月面的相互作用过程。其次，ASAN 测量到了较重的能量中性原子，为研究太阳风月面溅射对月球外逸层的贡献提供了实测数据，但是目前测量结果的背景噪声比较大，这给数据处理带来困难。再次，目前观测数据仍然有限，不能覆盖所有的散射角，因而 ENA 月面散射的角分布函数仍然不知道，这会对 ENA 反射率以及微磁层遮挡效率的计算带来影响。截至撰写本文时，ASAN 探测器仍然在月面工作，并且仪器状态良好。未来借助更多测量数据，将有望得到更为准确的 ENA 散射函数、反射率等信息，同时也为研究月球微磁层提供更多实测数据。另外，通过对重成分 ENA 探测数据的处理和分析，有望对太阳风月面物质溅射及其对月球稀薄大气的贡献进行定量评估。

参考文献

［1］PIETERS C M，GOSWAMI J N，CLARK R N，et al. Character and spatial distribution of OH/H_2O on the surface of the Moon seen by M^3 on chandrayaan - 1［J］. Sci - ence，2009，326（5952）：568 - 572.

［2］LI Shuai，GARRICK - BETHELL I. Surface water at lunar magnetic anomalies［J］. Geophysical Research Letters，2019，46（24）：14318 - 14327.

［3］WANG Huizi，ZHANG Jiang，SHI Quanqi，et al. Earth wind as a possible exogenous source of lunar surface hydration［J］. The Astrophysical Journal Letters，2021，907（2）：L32.

［4］NOBLE S K，PIETERS C M，KELLER L P. An experimental approach to understanding the optical effects of space weathering［J］. Icarus，2007，192（2）：

629 - 642.

[5] WU Yanxue, LI Xiongyao, YAO Wenqing, et al. Impact characteristics of different rocks in a pulsed laser irradiation experiment: simulation of micrometeorite bombardment on the Moon [J] . Journal of Geophysical Research: Planets, 2017, 122 (10): 1956 - 1967.

[6] WURZ P, ROHNER U, WHITBY J A, et al. The lunar exosphere: the sputtering contribution [J] . Icarus, 2007, 191 (2): 486 - 496.

[7] SARANTOS M, KILLEN R M, GLENAR D A, et al. Metallic species, oxygen and silicon in the lunar exosphere: upper limits and prospects for LADEE measurements [J] . Journal of Geophysical Research: Space Physics, 2012, 117 (A3): A03103.

[8] SAITO Y, YOKOTA S, TANAKA T, et al. Solar wind proton reflection at the lunar surface: low energy ion measurement by MAP - PACE onboard SELENE (KAGUYA) [J] . Geophysical Research Letters, 2008, 35 (24): L24205.

[9] WIESER M, BARABASH S, FUTAANA Y, et al. Extremely high reflection of solar wind protons as neutral hydrogen atoms from regolith in space [J]. Planetary and Space Science, 2009, 57 (14/15): 2132 - 2134.

[10] SCHAUFELBERGER A, WURZ P, BARABASH S, et al. Scattering function for energetic neutral hydrogen atoms off the lunar surface [J] . Geophysical Research Letters, 2011, 38 (22): L22202.

[11] FUTAANA Y, BARABASH S, WIESER M, et al. Empirical energy spectra of neutralized solar wind protons from the lunar regolith [J] . Journal of Geophysical Research: Planets, 2012, 117 (E5): E05005.

[12] FUNSTEN H O, ALLEGRINI F, BOCHSLER P A, et al. Reflection of solar wind hydrogen from the lunar surface [J] . Journal of Geophysical Research: Planets, 2013, 118 (2): 292 - 305.

[13] MITCHELL D L, HALEKAS J S, LIN R P, et al. Global mapping of lunar crustal magnetic fields by Lunar Prospector [J] . Icarus, 2008, 194 (2): 401 - 409.

[14] TSUNAKAWA H, TAKAHASHI F, SHIMIZU H, et al. Surface vector mapping of magnetic anomalies over the Moon using Kaguya and Lunar Prospector observations [J] . Journal of Geophysical Research: Planets, 2015, 120 (6): 1160 - 1185.

[15] LIN R P, MITCHELL D L, CURTIS D W, et al. Lunar surface magnetic fields and their interaction with the solar wind: results from Lunar Prospector [J] . Science, 1998, 281 (5382): 1480 - 1484.

[16] HALEKAS J S, BRAIN D A, MITCHELL D L, et al. On the occurrence of magnetic enhancements caused by solar wind interaction with lunar crustal fields [J]. Geophysical Research Letters, 2006, 33 (8): L08106.

[17] HALEKAS J S, POPPE A R, MCFADDEN J P, et al. Evidence for small - scale collisionless shocks at the moon from ARTEMIS [J] . Geophysical Research Letters, 2014, 41 (21): 7436 - 7443.

[18] HALEKAS J S, DELORY G T, BRAIN D A, et al. Density cavity observed over a strong lunar crustal magnetic anomaly in the solar wind: a mini - magnetosphere? [J] . Planetary and Space Science, 2008, 56 (7): 941 - 946.

[19] LUE C, FUTAANA Y, BARABASH S, et al. Strong influence of lunar crustal fields on the solar wind flow [J] . Geo - physical Research Letters, 2011, 38 (3): L03202.

[20] SAITO Y, YOKOTA Y, ASAMURA S, et al. In - flight performance and initial results of Plasma Energy Angle and Composition Experiment (PACE) on SELENE (Kaguya) [J] . Space Science Reviews, 2010, 154 (1/2/3/4): 265 - 303.

[21] WIESER M, BARABASH S, FUTAANA Y, et al. First observation of a mini - magnetosphere above a lunar magnetic anomaly using energetic neutral atoms [J]. Geophysical Research Letters, 2010, 37 (5): L05103.

[22] VORBURGER A, WURZ P, BARABASH S, et al. Energetic neutral atom observations of magnetic anomalies on the lunar surface [J] . Journal of Geophysical Research: Space Physics, 2012, 117 (A7): A07208.

[23] WIESER M, BARABASH S, WANG X D, et al. The advanced small analyzer for neutrals (ASAN) on the Chang ' E - 4 Rover Yutu - 2 [J] . Space Science Reviews, 2020, 216 (4): 73.

[24] ZHANG Aibing, WIESER M, WANG Chi, et al. Emission of energetic neutral atoms measured on the lunar surface by Chang'E - 4 [J] . Planetary and Space Science, 2020, 189: 104970.

[25] VORBURGER A, WURZ P, BARABASH S, et al. Energetic neutral atom imaging of the lunar surface [J] . Journal of Geophysical Research: Space Physics, 2013, 118 (7): 3937 - 3945.

[26] XIE Lianghai, LI Lei, ZHANG Aibing, et al. Inside a lunar mini - magnetosphere: first energetic neutral atom measurements on the lunar surface [J] . Geophysical Research Letters, 2021, 48 (14): e2021GL093943.

[27] XIE Lianghai, LI Lei, ZHANG Yiteng, et al. Three - dimensional Hall MHD simulation of lunar minimagnetosphere: general characteristics and comparison with Chang'E - 2 observations [J] . Journal of Geophysical Research: Space Physics, 2015, 120 (8): 6559 - 6568.

[28] SAITO Y, NISHINO M N, FUJIMOTO M, et al. Simultaneous observation of the electron acceleration and ion deceleration over lunar magnetic anomalies [J]. Earth, Planets and Space, 2012, 64 (2): 4.

[29] FUTAANA Y, BARABASH S, WIESER M, et al. Remote energetic neutral atom imaging of electric potential over a lunar magnetic anomaly [J] . Geophysical Research Letters, 2013, 40 (2): 262 - 266.

[30] WANG Huizi, XIAO Chao, SHI Quanqi, et al. Energetic neutral atom distribution on the lunar surface and its relationship with solar wind conditions [J] . The Astrophysical Journal Letters, 2021, 922 (2): L41.

LND两年月表辐射剂量测量结果

张珅毅[1,2,3,4]，侯东辉[1,2,3,4]，WIMMER - SCHWEINGRUBER R F[1,5]，孙越强[1,2,3]，王春琴[1,2,3]，常峥[1,2,3]，徐子贡[5]，沈国红[1,2,3]，袁斌[1,2,3]，薛长斌[1]

1. 中国科学院国家空间科学中心，北京 100190
2. 天基空间环境探测北京市重点实验室，北京 100190
3. 中国科学院空间态势感知重点实验室，北京 100190
4. 中国科学院大学，北京 100049
5. 德国基尔大学实验应用物理研究所，基尔 24118

摘要： 月球表面的辐射剂量是影响航天员安全和月表驻留时间的重要参数，通过对月表的粒子辐射测量可以为航天员的辐射安全防护提供重要依据。利用嫦娥四号着陆器上搭载的月表中子与辐射剂量探测仪二年的观测数据得到：月表粒子辐射在硅中的平均总吸收剂量率为 12.66±0.45μGy·h^{-1}，中性粒子吸收剂量率为 2.67±0.16μGy·h^{-1}，辐射剂量率随时间出现缓慢的下降，LET 谱的变化则很小。同时观测到了 2020 年 12 月太阳活动末期由于银河宇宙线福布斯下降导致的辐射剂量率降低。

关键词： 嫦娥四号；月表；辐射剂量；航天员

Radiation Dose of LND on the Lunar Surface in Two Years

ZHANG Shenyi[1,2,3,4], HOU Donghui[1,2,3,4], WINMMER - SCHWEINGRUBER R F[1,5], SUN Yueqiang[1,2,3], WANG Chunqin[1,2,3], CHANG Zheng[1,2,3], XU Zigong[5], SHEN Guohong[1,2,3], YUAN Bin[1,2,3], XUE Changbin[1]

1. National Space Science Center, Chinese Academy of Sciences, Beijing 100190
2. Beijing Key Laboratory of Space Environment Exploration, Beijing 100190

3. Key Laboratory of Environmental Space Situation Awareness Technology，Beijing 100190
4. University of Chinese Academy of Sciences，Beijing 100049
5. Institute of Experimental and Applied Physics，Christian - Albrechts - University，Kiel 24118

Abstract：The radiation dose on the lunar surface is an important parameter affecting the safety of astronauts and the residence time of the lunar surface. The measurement of the particle radiation on the lunar surface can provide an important basis for the radiation safety protection of astronauts. Based on the two - year observation data of lunar neutron and radiation dose detector on Chang'E - 4 lander，the average total absorbed dose rate of lunar particle radiation in silicon is 12.66±0.45μGy·h^{-1}，and the absorbed dose rate of neutral particle is 2.67±0.16μGy·h^{-1}. The radiation dose rate decreases slowly with time，but the change of LET spectrum is very small. The decrease of radiation dose rate due to the decrease of fubusch of GCR at the end of solar activity in December 2020 was observed.

Keywords：Chang' E - 4；Lunar Surface；Radiation dose；Astronant

0　引言

随着中国嫦娥四号/五号任务的圆满成功，月球科研站及月球基地等月球探测步伐也正在逐步开展，最终实现人类月球深空的长期驻留。空间辐射暴露是影响航天员健康的主要风险之一[1-3]，因为长期暴露在银河宇宙射线中可能引起航天员一系列缓发的健康问题，例如白内障、癌症以及中枢神经系统或其他器官系统[4,5]的退化性疾病。

月球表面的粒子辐射来源主要有三种：一是长期存在银河宇宙线（GCR）[6]，GCR 受到行星际磁场的调制发生缓慢的变化；二是偶发的太阳高能粒子事件，太阳发生耀斑及日冕物质抛射时会发出大量的高能粒子，在较短的时间内对月球表面产生强烈的辐射[7]；三是 SEPs 和 GCR 通过与月壤的作用产生的二次粒子，主要是中子和 γ 射线[8]。

嫦娥四号着陆器[9]在 2019 年 1 月 3 日 02：26UT 登陆月球背面的冯卡门陨石坑。在此之前，尚未对月球表面的辐射剂量率进行过主动测量（即时间分辨）。在阿波罗任务期间，航天员到达月球时虽然随身携带了剂量仪，但是在月球表面的时间段内的辐射数据从未进行过报告[10]。

为了评估辐射暴露，必须利用探测器系统测量相关量。吸收剂量 D 是沉积在探测器中的能量 E（单位：keV）与探测器质量（m）的比值，常用的单位为格瑞（Gy=J·kg^{-1}）将吸收剂量除以累积的时间可以得到剂量率（单位：Gy·h^{-1}）。利用两个探测器符合的逻辑工作方式以及其中一个探测器中的沉积能量分布，可以获得线性能量转移谱 LET（单位：keV·μm^{-1}）。由于人体不是由硅构成的，为了便于比较不同测量的剂量、剂量率以及 LET 谱，通常使用 1.30[11]的恒定剂量转换系数，将硅中的测量值转换为在水中相应的物理量。

嫦娥四号任务在 2019 年 1 月登陆月球背面后，目前已持续在轨工作超过 25 个月昼，月表中子与辐射剂量探测仪（Lunar Lander Neutrons and Dosimetry，LND)[12,13]对月表的粒子辐射环境和辐射剂量进行测量，包括质子、电子、重离子能谱，带电粒子和中性粒子的辐射剂量以及 LET 谱[14]。

LND 安装在嫦娥四号着陆器的有效载荷舱内，由 10 片双灵敏区硅固体探测器（SSD）组成，这 10 片硅探测器命名为 A－J。总吸收剂量和剂量率通过探测器 B 测量，中性粒子的吸收剂量和剂量率通过探测器 C 的内灵敏区 C1 测量，紧密靠近的探测器 B 和 D 以及探测器 C 的外灵敏区 C 2 作为反符合用于区别带电粒子，LET 谱由 dE/dx 确定。

1 数据结果

图 1 给出了 LND 探头在嫦娥四号着陆器的位置，图 1 中的红色箭头指向为 LND 的探测指向，为了保护 LND，舱门在寒冷的月夜关闭，在月昼时打开。图 2 显示了 LND 在 2019 年 5 月至 2021 年 1 月两年时间共 21 个月昼期间（着陆器未休眠期间）获得的辐射剂量测量结果。图 2（a）为辐射总剂量率，包括带电粒子和中性粒子（中子、γ 射线）的贡献；图 2（b）为中性粒子的辐射剂量率；图 2（c）为带电粒子的辐射剂量率；图 2（d）为穿透高能带电粒子的通量（质子＞34.8MeV，He＞40MeV），也代表了银河宇宙线的通量；图 2（e）为地面测量到的奥卢中子结果，其也代表了宇宙线的变化情况。通过图 2 可以看到 21 个月昼内月球表面的平均辐射剂量率为 12.66±0.45μGy·h^{-1}（Si），等效到人体中的辐射剂量率为 16.46μGy·h^{-1}（H_2O），其中，中性粒子的辐射剂量率为 2.67± 0.16μGy·h^{-1}（Si），约占总辐射剂量的 21%。从图 2 可以看到辐射总剂量率和中性粒子辐射剂量率的变化与穿透粒子的通量有明显的正相关，与空间物理学认知相符。

图 1 LND 探头在嫦娥四号着陆器的位置，红色箭头为视场方向

Fig. 1 LND sensor head (whose field of view is indicated by red arrow) is installed in the Chang'E－4 lander payload compartment

从 2020 年 10 月的第 21 个月昼开始辐射剂量率出现了缓慢的下降（见图 2 阴影部分），这是由于行星际磁场的变化造成了银河宇宙线的通量降低导致的，最直接的证据是 LND 测量到的同时间段的穿透粒子数量和奥卢中子数量也出现了降低。

图 3 显示了 LND 在第 7，9，13，16，20 月昼期间获得的 LET 谱结果。可以看到在近两年的时间内月表粒子辐射 LET 谱的变化很小，这是因为银河宇宙线的变化很小，更

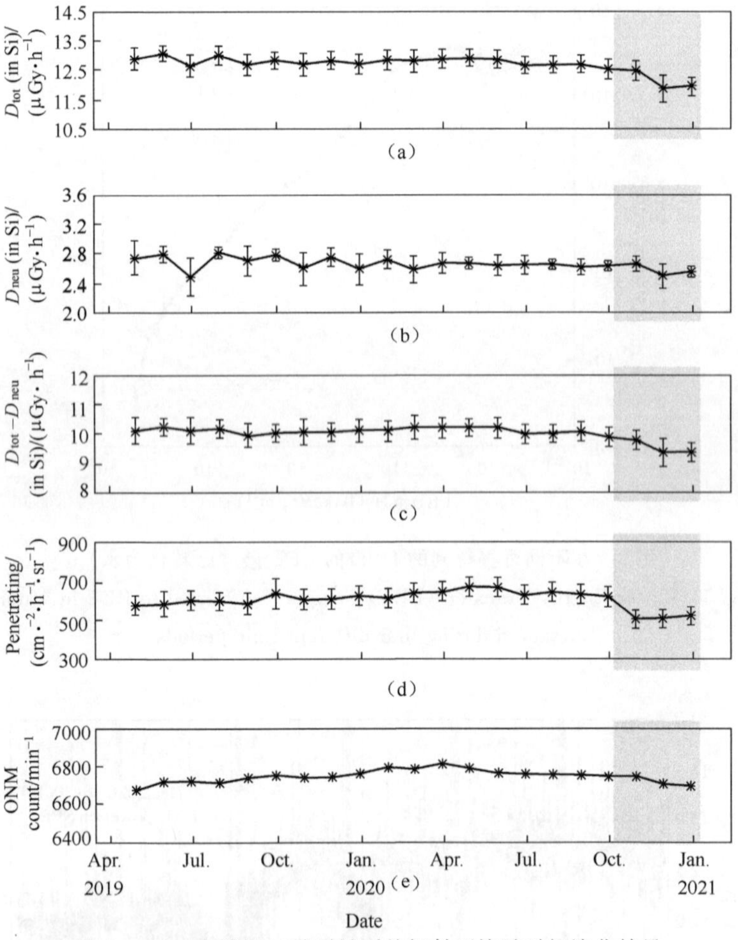

图 2　LND 在嫦娥四号测量到的辐射环境随时间演化结果

Fig. 2　Temporal evolution of the radiation environment on the Moon as measured by LND on Chang'E－4

大的变化需要更长期的观测。

LND 在轨期间遇到了两次太阳粒子爆发事件，分别为 2019 年 5 月 6 日[15]和 2020 年 12 月 1 日，2019 年 5 月的太阳质子事件中 LND 的 9～21MeV 高能质子通量出现了少量增长，但 LND 的辐射剂量率并没有出现明显的变化。这是因为太阳质子事件爆发等级不高，能够加速质子的通量和能量也较低，被仪器挡光层和第一片探测器 A 阻挡或者降低了能量，导致了其在 B 探测器的剂量测量中所占比例较小。2020 年 12 月 1 日的太阳活动中，磁场出现了明显的变化，keV 量级的粒子增长大约持续了 10 天，而 MeV 量级的高能粒子并没有出现明显变化，LND 在 2020 年 12 月 9 日开机，处于太阳活动的末期，从图 4 可以看到此时测量到的辐射剂量出现了一个明显的下降，这是因为太阳活动造成了行星际磁场的扰动，该扰动引起了银河宇宙线的福布斯下降导致辐射剂量降低，该福布斯下降使辐射剂量比正常水平少下降了约 10％。

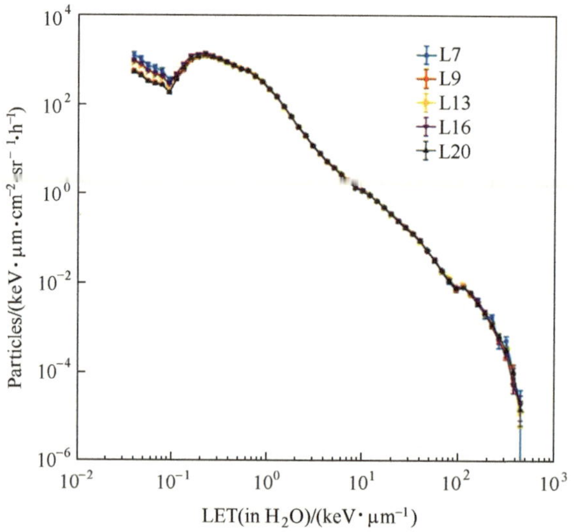

图 3　在不同月昼得到的 LND 的 LET 谱（已转化为水）

Fig. 3　Linear Energy Transfer (LET) spectra (converted to LET in water) measured during four different time periods

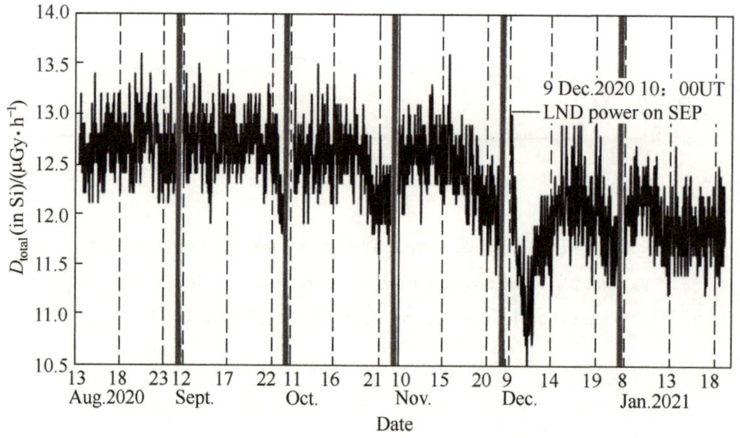

图 4　太阳质子爆发期间辐射剂量的变化情况

Fig. 4　Dose rate measured in period of SEP

2　分析讨论

将 LND 的测量结果与月船一号（Chandrayann－1）上的粒子载荷 RADOM[16] 和月球轨道器（LRO）上的载荷 CRaTER[17] 的测量结果进行比较，结果列于表 1。

由于 Chandrayann－1，LRO 和嫦娥四号进行的测量是在太阳平静期进行的，可以假设剂量率的差异主要是由轨道高度的差异造成的，即月球提供的屏蔽不同。然而这种简单的解释并不合理，通过将文中的结果与没有被月球遮挡的部分（立体角）进行比较可以看出，屏蔽与吸收剂量并不存在相应的比例关系。

表 1　不同时间段内不同轨道测量到的月球辐射剂量率

Table 1　Dose rate measured in different orbits of the Moon at different time

卫星	测量时间段	轨道高度 /km	辐射剂量 /（μGy・h⁻¹）	未被月球遮挡 /（%）	参考文献
Chandrayann-1	3 to 8 Nov. 2008	20000～30000	12.76	100	[18]
Chandrayann-1	20 May to 28 Aug. 2009	200	10.7	72	[18]
Chandrayann-1	20 Nov. 2008 to 18 May 2009	100	9.76	66	[18]
LRO	17 Sept. 2009 to 31 May 2010	50	9.17～11.25	62	[19]
Chang'E-4	1 May 2019 to 19 Jan. 2020	0	12.66	50	—

RADOM 在 2008 年离月球很远的地方测得的剂量率与 2019 年在月球表面测得的剂量率不可比。原因主要有两点：第一，其处在不同的太阳周期。例如，从 2015 年开始，美国好奇号上的辐射评估探测器 RAD 在火星观测到银河宇宙射线的剂量率增加了 50%，第二，LND 测量的能量范围与 RADOM 和 CRaTER 不同，LND 也能够测量更低能量的粒子，这些低能粒子比能量更高、穿透力更强的粒子更多，对剂量率的贡献也更大。

LND 与 CRaTER 测量能量范围不同对剂量结果的影响可以通过具体实例看到，通过对比同一时间段、同一位置（2019 年 2 月 2 日 01：30UT）LND 和 CRaTER 测量值可以看出 LND 和 CRaTER 覆盖不同能量测量范围对剂量的影响（见图 5）。此时，两台仪器都处于相同的日球层条件，但在各自测量上下能量截止阈值上存在差异。

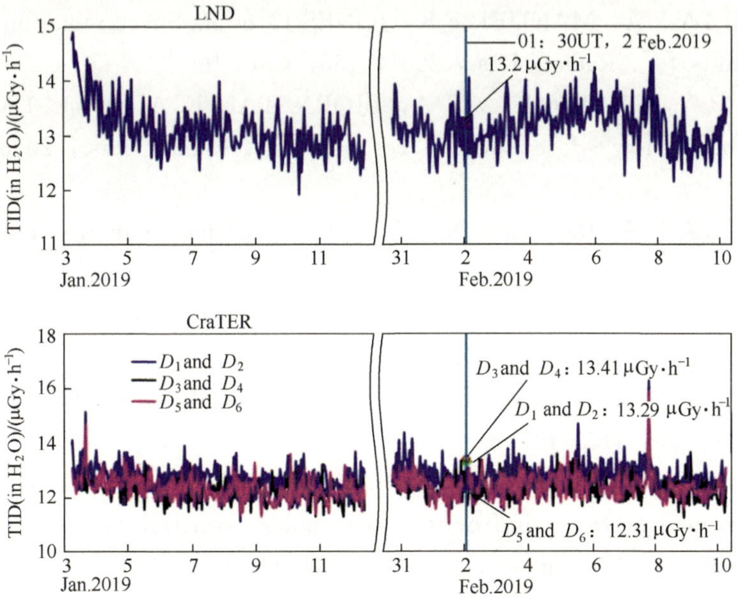

图 5　LND 和 CRaTER 的剂量率测量。CRaTER 测量到的剂量率已对月球提供的几何屏蔽进行了修正。垂直线标记 LRO 位于嫦娥四号正上方的时间点，指示剂量率是此时的测量值

Fig. 5　Dose rate measurements by LND and CraTER. Dose rates measured by CRaTER are corrected for the geometrical shielding provided by the Moon. The vertical line marks the time point when LRO was located right above Chang'E-4, and the dose rate values indicated in the panels are the values measured at this time

图5给出了LND和CRaTER在2019年初LND活动时间段的测量，垂直线表示LRO和嫦娥四号在月球经纬度大致相同的时间段，并在图中给出了测量剂量率值。结果表明，D1和D2的剂量率为13.2μGy·h^{-1}，LND剂量率为13.29μGy·h^{-1}，LND测量的剂量率与CRaTER的结果十分类似。需要注意的是，月球的屏蔽效应已经在公布的CRaTER剂量率数据中得到修正，但是如果没有详细考虑两种仪器的响应范围及其屏蔽差异，仍然不能比较CRaTER和LND的剂量率。虽然LND的B探测器由于其位于LND和嫦娥四号着陆器内受到不同程度的屏蔽，但CRaTER的D1和D2探测器也被来自仪器和LRO卫星的不同材料厚度屏蔽。CRaTER探测器D1（148μm）探测器由0.82mm厚的铝入口窗口屏蔽，其1mm厚的D2探测器被0.82mmAl和148μm的硅屏蔽。LND的B探测器在最小屏蔽方向上被25μm的Al及50μm的kapton和0.5mm的硅屏蔽。

本文的测量结果涵盖了LND在月表两年的辐射剂量结果，这些结果与未来人类月球任务的安排和规划直接相关。

参考文献

[1] NASA. Space Radiation Cancer Risk Projections and Uncertainties - 2012 [R]. Houston：NASA，2013.

[2] STRAUBE U，BERGER T，REITZ G，et al. Operational radiation protection for astronausts and cosmonauts and correlated activities of ESA Medical Operations [J]. Acta Astronaut.，2010，66：963 - 973.

[3] CUCINOTTA F A，MANUEL F K，JONES J，et al. Space radiation and cataracts in astronauts [J]. Radiat. Res.，2001，156：460 - 466.

[4] CHANG P Y，CUCINOTTA F A，BJORNSTAD K A，et al. Harderian gland tumorigenesis：low - dose and LET response [J]. Radiat. Res.，2016，185：449 - 460.

[5] CUCINOTTA F A. Review of NASA approach to space radiation risk assessments for Marsexploration [J]. Health Phys.，2015，108：131 - 142.

[6] SIMPSON J A. Elemental and isotopic composition of the galactic cosmic rays [J]. Ann Rev. Nucl. Part. Sci.，1983，33：323 - 382.

[7] REAMES D V. Particle acceleration at the sun and in the heliosphere [J]. Space Sci. Rev.，1990，90：413 - 491.

[8] BOYNTON，DROECE G F，MITROFANOV I G. High spatial resolution studies of epithermal neutron emission from the lunar poles：constraints on hydrogen mobility [J]. J. Geophys. Res. Planet.，2015，117 (12)：1991 - 2012.

[9] JIA Yingzhuo，ZOU Yongliao，XUE Changbin，et al. Scientific objectives and payloads of Chang'E - 4 missionormalsize [J]. Chin. J. Space Sci，2018，38 (1)：118 - 130（贾瑛卓，邹永廖，薛长斌，等. 嫦娥四号任务科学目标和有效载荷配置[J]. 空间科学学报，2018，38 (1)：118 - 130.

[10] ENGLISH R A，BENSOTZ R E，VERNON BUILEY J，et al. Apollo Experience Report -

Protection against radiation, NASA Technical Note NASA TN D-7080 [OL]. [1973-03-01]. https: //ntrs. nassa. gov/archive/nasa/casi. ntrs. nasa. gov/19730010172. pdf.

[11] ZEITLIN C, NARICI L, RIOS R R, et al. Comparisons of high - linear energy transfer spectra on the ISS and in deep space [J]. Space Weather, 2019, 17: 396-418.

[12] WIMMER-SCHWEINGRUBER R F, YU J, BÖTTCHER SI, et al. The Lunar Lander Neutron and Dosimetry (LND) experiment on Chang'E-4 [J]. Space Sci. Rev., 2020, 216: 104.

[13] HOU D, ZHANG S, YU J, et al. Removing the dose background from radio active sources from active dose rate measurements in the Lunar Lander Neutron & Dosimetry (LND) experiment on Chang'E-4 [J]. J. Instrum., 2020, 15 (1): P01023-P01032.

[14] ZHANG S, WIMMER - SCHWEINGRUBER R F, YU J, et al. First measurements of the radiation dose on the lunar surface [J]. Sci. Adv., 2020, 6. DOI: 10. 1126/sciadv. aaz1334.

[15] XU Zigong, GUO Jingnan, ROBERT F, et al. First solar energetic particles measured on the lunar far-side [J]. Astrophys. J. Lett., 2020, 902: 30.

[16] BHANDARI N. Chandrayaan-1: science goals [J]. J. Earth Syst. Sci., 2005, 114: 701-709.

[17] CHIN G, BRYLOW S, FOOTE M, et al. Lunar recon naissance orbiter overview: the instrument suite and mission [J]. Space Sci. Rev., 2007, 129: 391-491.

[18] DACHEV T P, TOMOV B T, YU N, et al. An overview of RADOM results for earth and moon radiation environment on Chandrayann-1 satellite [J]. Adv. Space Res., 2011, 48 (5): 779-791.

[19] SPENCE H E. Crater science team: an overview of results from the Lunar Reconnaissance Orbiter (LRO) Cosmic Ray Telescope for the effects of Radiation (CRaTER) [C] //Annual Meeting of the Lunar Exploration Analysis Group. Washington: LPI, 2010.

第6章　未来展望

嫦娥四号任务的成功，开创了人类探月史在月球背面上的先河，但这并不意味着我们已经解决了月球背面的科学问题，或者说已经彻底解决了着陆月球背面的所有技术难题，人类对月球背面探测的步伐不但不会停止，而且会加快。同样，虽然人类开展了130多次的月球探测活动、经历了60多年科学的沉淀和技术的发展，这并不表示不需要继续对它进行探测了。相反，月球本身还有很多很多的科学谜团需要我们去探索，月球独特的空间位置资源需要我们去挖掘、去利用，月球特殊而极端的环境资源也需要我们去克服、去反利用，更重要的是，月球丰富的物质资源更是呼吁着人类去开发、去利用。

6.1　月球本身存在诸多科学问题需要人类去揭示

在认识月球漫长的历史长河中，从手段上可划分为三大、三小的时期或阶段。三大时期包括以肉眼观测为主的手段、以望远镜为主的观测手段和以近距离探测及样品实验室分析为主的研究手段的三大认知期。而真正发挥对月球最详细、最完整、最精准、最深刻认知的几乎都来自最后一个时期。这一时期又可划分为三个阶段，即20世纪50年代末至70年代中期以美国与苏联之间太空争霸引发的第一个探月活动高潮期、从70年代中期至90年代初以第一个高潮期所获得的海量科学数据和返回月球样品研究为主的月球科学认知升华期、从90年代初至今的多航天大国参与的新一轮探月高潮期。不同时期、不同阶段对月球的认知程度都实现了革命性的发展，特别是近20年来，随着科学探测技术和实验分析技术的进步，今天对月球的认知相比于20世纪阿波罗时代同样发生了翻天覆地的变化。

虽然从20世纪50年代末至今的60多年里，人类开展了130多次探月活动，实现了月面着陆巡视就位探测（图6-1）和11次采样返回，大大推进了对月球认识的深度，构建了系列的有关月球形成和演化相关的理论体系，但这并不代表我们对月球的认识全了、透了。事实上，对月球的研究还存在包括月球的二分性、岩浆洋、月幔物质翻转、发电机、内部结构、火山作用与热演化历史等诸多的科学难题。

1）月球岩浆洋问题。月球岩浆洋是直接影响其早期演化的核心要素及其基本走向，是月球科学研究中最核心的关键问题之一。目前，对月球岩浆洋存疑的问题主要有：一是持续时间的问题。研究表明，在没有额外热源供应的情况下，月球岩浆洋的持续时间不会超过一千万年。作为岩浆洋演化晚期阶段的产物，斜长石的结晶时间跨度应该在一千万年之内，然而斜长石的定年结果表明其跨度可以达到2.7亿年。虽然有专家认为斜长石定年技术的不确定性较大以及撞击熔融可能重置年龄，但是该矛盾到目前为止依然没有得到合理的解释。二是月球斜长石微量元素的问题。斜长岩如果是结晶晚期通过结晶上浮形成，

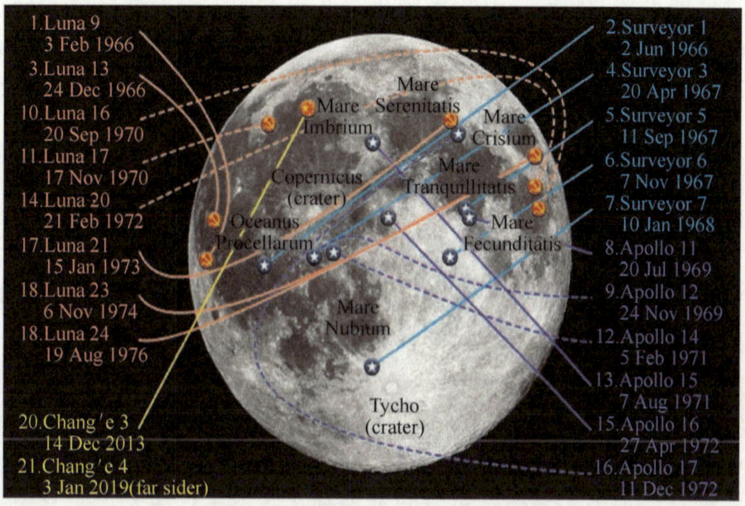

图 6-1　成功软着陆月球任务

那么结晶区间的斜长石中微量元素（或其比值）不会发生显著的变化。然而，返回样品斜长岩中诸多微量元素特征却无法通过斜长石分离结晶理论来解释。三是晚期残余岩浆中 Mg 的含量问题。按照现有理论模型，结晶晚期残余岩浆中的 Mg 含量已经非常低，但实际研究发现，返月样品中斜长岩同时结晶的基性矿物中 Mg 含量非常高。此外，还有一些问题尚未得到较好的解决，如斜长岩和镁质岩套的年龄重叠、斜长岩上浮过程中是否发生重新溶解、玄武岩普遍存在的 Eu 异常与经典模型预测的深部月幔不存在 Eu 异常的矛盾等等。未来，随着探测与研究的进一步发展，更多的样品（包含背面样品）的获取，凸显了岩浆洋冷却分异动力学过程模拟、斜长岩成因、温度梯度对斜长岩成分改造等关键要素有待解决。

2）月球发电机问题。现今的月球没有全球性偶极磁场，但存在岩石圈微弱磁场。阿波罗返回月球岩石样品剩磁研究表明，在 42 亿年前可能存在金属内核和磁场发电机，但是发电机的后期历史尚不清楚。最近的模拟研究认为即使月球核幔边界的辐射热流不足以支持月球的发电机，月球内核的凝固过程和其导致的在月球外核的热化学对流可以在 35 亿年前延长月球发电机的寿命；月球发电机是自然消退还是发生突变导致全球磁场的消失一直是研究月球磁场起源的焦点和难点。此外，最近 20 多年的探测与研究表明，月球上存在一些磁异常区，与地貌关系密切，且最强磁异常区聚集于撞击盆地的对跖点，这些极为异常的现象目前学术界仍未解决。最近有学者认为，直接观测月球极区表面磁场可以提供极区磁场分布的更准确信息，从而改变对月球磁异常区起源和月球发电机演化的认识。总之，月球发电机是否曾经存在、如何运行和停止是月球科学中最主要的科学问题之一。

3）月球内部圈层结构问题。月球内部圈层结构与月球早期形成和演化过程密切关联，对于未来破解月球科学重大问题，特别是其在整个月球演化进程中重大地质事件在时空上的分布特征及其形成机制至关重要。月震是约束月核、月幔和月壳厚度的最主要手段，能够为月球内部提供关键基础参数。虽然阿波罗任务布设了 4 台月震台网并在近 8 年的执行期内记录到了 13000 多次月震，提供了到目前为止有关月震和月球内部结构的基本认知，然而，由于当时仪器水平的限制，其探测结果存在很大的不确定性，学术分歧严重，制约了人们对月球内部结构的认识。目前人们对月球壳幔边界、核幔边界、月核大小及物质性质与状态等的认识仍极为有限且存在巨大争议，大大制约了我们对月球深部物质和演化历史的认识。

4）火山作用问题。火山作用是研究月球内地质营力作用的重要手段，也是研究月球热历史演化的核心要素，但目前许多涉及火山作用的问题仍然未得到解决。首先是月球火山活动的起始与结束时间的问题。月球火山活动是月球内部物质成分、热状态及其演化在月表的直接表现，准确限定其起始及结束时间等年代学特征，将对深化认识月球地质及热演化历史和模型提供关键证据。尽管早期返月样品分析结果显示，月球火山活动的活跃时间为 43 亿年至 31 亿年，后续对月球陨石样品的分析进一步将月球火山活动的时限范围延续到 29 亿年左右，但嫦娥五号返回样品分析结果将火山活动延续到 20 亿年左右。过去 20 多年来，基于全月影像数据及撞击坑统计定年方法，也发现了大量更加年轻的火山活动的证据，甚至把月球火山活动的结束时间推迟到了距今 1 亿年左右。其次是月球火山活动喷发规模、样式及随时间演化的问题。月球火山活动喷发规模及其演化直接反映了月球内部热状态及物质成分变化，也直接影响了火山活动产物的形貌特征。基于返回样品的岩石学分析及遥感数据形貌学计算，估计典型月球火山活动的喷发速率约为 10^5～10^6 m^3/s。基于对全月玄武岩单元年代学及厚度计算结果，目前对月球火山活动的喷发规模随时间的演化规律有了初步的认知，发现月球火山活动喷发规模在晚雨海纪（3.2 Ga～3.8 Ga）达到最大，其后迅速衰减，进入哥白尼纪时，几乎停止了喷发活动。但由于月球早期喷发的产物基本都被盆地溅射物覆盖了，且基于遥感数据对玄武岩厚度的估计结果不确定性较大，导致目前对月球火山活动的历史岩浆喷发通量变化仍约束较弱。除此之外，目前仍不清楚月球火山作用随时间变化的方式，是幕式喷发、逐渐减小还是持续喷发、逐渐减小。此外，特殊火山地貌成因及年龄尚存在较多争议。利用月表高分辨率影像数据，在月球正面月海地区发现了 70 余处不规则月海斑块（IMPs）。IMPs 内部表现出异常复杂的地形地貌特征，包括地势较高、“水疱”状的山丘单元，地势较低的圆丘状单元及块状单元，但到目前为止，IMPs 的具体成因是月球科学领域悬而未决的一大难题。近年来，在月球上发现了若干处的非月海火山穹隆建造，可能代表了月球火山活动的全新端元，但目前对其形成机制尚存争议。

5）月球水和挥发分问题。对月球极区水冰和挥发分的探测和研究是近年来月球科学的研究热点，更是未来月球探测计划和月球基地建立选址的重要战略资源。目前的遥感探测研究显示，月表的大部分水分布在月球的两极。月球永久阴影区的撞击实验也指示着南极艾特肯盆地的一些永久阴影区域和深陷处可能有水冰的存在。这些永久阴影区域处于不受阳光直接照射的区域，温度很低，有利于水冰的储存。除了水冰之外，南极艾特肯盆地的永久阴影区域还可能含有其他挥发性物质，如甲烷、氨和一氧化碳等。这些挥发性物质可能与水冰混合存在，形成一种冰态的混合物。但是，月球极区水冰探测结果主要基于雷达、中子、光谱等轨道遥感手段间接获得。因此，要获得月球极区水冰存在的直接证据，需要前往月球极区或永久阴影坑内通过着陆就位采样探测，勘明月球极区是否存在水冰，勘探水冰的储量和赋存形态。同样，基于遥感数据发现极区水冰分布的绝大部分区域与极移发生以前就存在的冷阱区域吻合，暗示着目前观测到的水冰可能比较古老。月球水冰被认为可能赋存于永久阴影区内的月壤表面以及极区的地下冷阱中，其形态可分为表面暴露水冰和地下埋藏水冰两类。最近几年的研究还发现，月球上的水和挥发分有从赤道向两极迁移的现象，但这一现象和理论观点目前还需通过进一步探测和研究来证明。总之，未来的月球水和挥发分的研究方向应沿着两大主线开展：一是对水和挥发分，特别是氦-3 等资源的开发利用前景的探测和研究；二是研究水和挥发分在月球早期演化过程中的制约作用。而后者一直是月球科学研究中的关键核心问题和热点难题。此外，最近的研究表明，月壤中的水和挥发分有向两极迁移的迹象，但其迁移机理尚不清楚，这是近年来研究的热点问题之一。

6）月球地质构造架构及其演化机制问题。月球地质构造架构及其演化机制是月球演化理论体系中的重要内涵，是揭示月球存在二元架构、三大地体等的关键要素。现今月球的地质构造架构是其长期演变至今在形貌构造上的具体体现。月球构造架构不但反映了月球早期地质构造运动、火山作用等内地质营力的结果，更是内地质作用和外力作用（如撞击、太空风化等）的综合产物。以往的探测任务主要着陆在中低纬度地区，因此就位探测数据和返回的样品具有很大的局限性。之前获取的就位探测数据和样品多为玄武岩和雨海盆地溅射物，还有部分高地物质。因此根据样品和就位数据获得的认识，主要是关于月球岩浆洋时期、雨海纪时期的地质演化过程的认知。我国嫦娥四号在南极区域着陆是一次重要的突破，首次获得了南极区域（SPA 盆地内部）的就位探测数据，为 SPA 盆地形成过程和月球深部物质组成提供了重要线索，但是仍然没有从根本上解决这些问题的争议。嫦娥五号获取的 2.0 Ga 的年轻玄武岩，将月球的火山活动历史延长了近 10 亿年，但是目前对 10 亿年以来的构造热运动的研究仍然不足。总之，月壳早期深部构造和晚期新构造活动的研究比较欠缺，导致在月球构造演化“两端”的研究明显不足，解决的关键主要有月壳早期应力和热状态、深部构造三维空间展布特征；全球整体性的月球晚期新构造活动及其分布规律等。

7）月球撞击作用问题。外来天体撞击是月球形成和演化过程中的重要地质营力，也是月球科学研究的重要方向，其撞击的效应问题一直是学术界的研究热点和重点之一。在太阳系形成的首个十亿年，内太阳系的撞击频率更高，形成了月球和类地行星上大部分直径大于 200 km 的撞击盆地，并改变了其内能释放的路径和过程。持续撞击作用形成的撞击坑为使用撞击坑的空间密度估算天体表面的模式年龄提供了载体，科学家从月面采集的样品为标定月球的撞击通量提供了同位素年龄的定标点。进而，月球的撞击历史得以缩放至其他类地行星，成为行星地质领域的重要定年方法。目前，月球南极艾特肯盆地的形成时间、月球关键撞击盆地的形成时间及后续重型撞击存在形式、月球撞击盆地的形成过程及其对内部结构的改造、雨海纪和埃拉托讯纪的撞击通量标定等是未来研究月球撞击作用的关键科学问题和重要方向。

8）月壤形成与时空演化问题。月壤蕴含了月球内外动力、太空风化、月表复杂环境、物质迁移与交换等诸多作用的信息，一直是月球科学研究的重要组成部分，是月球轨道遥感、着陆巡视以及就位资源利用等工程任务的最直接的探测对象。月壤主要由月表岩石/矿物碎屑和撞击熔融形成的胶结物组成，其形成过程主要受火山喷发、小天体撞击、陨石/微陨石轰击、太阳风/太阳耀斑/银河宇宙射线等粒子辐射以及温差热疲劳作用的综合影响，其记录了大量月球内外动力地质作用的信息，是人类认识月球形成与演化过程以及月球空间环境演变历史的主要窗口。月壤的形成与时空演化动力学过程是月球科学研究的重要内容，同时对于支撑月球工程探测，开展月球就位资源开发与利用也具有重要意义。但是，目前对月壤的认识仍存在诸多存疑问，如月壤的形成与时空演化过程有何差异、月壤的成熟度评价体系如何有效统一、月球表面是否存在广泛的氧化作用、永久阴影区月壤是如何形成与演化的等等，这些问题仍有待进一步探测和研究。

9）月球太空风化问题。在月壤形成过程中发生的一系列物理、化学和光谱特征的变化，统称为太空风化作用。由于没有大气层和偶极性磁场的保护，月球表面持续遭受微陨石撞击、太阳风及银河宇宙射线的直接辐射，这些过程极大地改造了月球表面物质的微观形貌、晶体结构和化学成分，同时影响了月面物质的光谱特征，进而影响了月球遥感和就位探测的分析结果。随着暴露时间的增加，新鲜月壤逐渐成熟并显示出独特的可见-近红外反射光谱变化规律（“变黑”、“变红”、吸收特征弱化）。深入研究月球上太空风化效应与机理，是认识月球表面物质演化和空间环境变化过程的关键。因此，未来的重点关注问题应该聚焦于月壤成熟度的精准表征、月球漩涡及其物质组织特性关系、月球极区环境下月壤的成熟和形成过程等关键科学问题。

10）月表辐射环境问题。由于缺少全球性内禀磁场及大气保护，月球表面可直接与来自周围的各种空间粒子相互作用，这些空间粒子包括银河宇宙射线、太阳风、太阳能量粒子、地球风、微流星体等。月面在这些空间粒子长期轰击作用下，一方面会发生各种物理变化，如晶格缺陷、碎片化和玻璃化等，另一方面还会发生各种化学变化，如产生纳米铁、OH/H_2O 和赤铁矿等。此外，月表在空间风化过程中也会释放各种粒子，包括反射、溅射、解吸附、热释放和扩散等带来的各种气体分子，这些气体分子最终构成月球稀薄的气体外逸层。而且，月球虽然没有全球性的内禀磁场，但是有广泛分布的局部月壳剩磁(称作磁异常)，一些强的磁异常可以阻挡和偏转太阳风，从而降低到达月表的太阳风通量，形成一个类似于地球磁层一样的局部保护结构，称作微磁层。了解这些原生的辐射和次生的辐射环境及其变化特征本身就是月球科学的重要组成内容，同时也是未来月球探测器、宇航员需要特别关注的工程技术难题。

6.2 月球资源开发利用的前景

月球独特的空间位置资源、环境资源和物质资源是探索更深远宇宙奥秘、更深刻认识与保护地球、更好实现人们对美好生活追求愿望的理想场地和战略要地。

相对于宇宙的浩瀚和其他天体的深远，月球是距离地球最近的天体，仅有约 38 万千米远，而且其无大气、只有地球六分之一的重力场，决定了它独特空间位置资源的巨大利用潜力，因此，人类探索更遥远、更浩瀚的宇宙需要它，从月球上发射更遥远的深空探测器能为人类节省太多太多的资源，而美国的月球门户及其 Artemis 计划就是一个鲜活的例证。

月球无大气、无磁场、弱重力和巨大的温差，特别是月球背面干净的电磁环境，决定了它环境的严酷性和独特性。无大气，太阳风、地球风、宇宙射线可长驱直入到月面，一方面使它成为开展月面基础科学试验研究、天文观测等的理想场所；另一方面，又是考验高温差、高辐射等严酷环境下在月球上开展各类科研作业的各类机器人和宇航员“抗压”

能力，进而提升技术水平和保障条件的最佳试验平台。

随着地球不可再生资源的不断匮乏和人民不断提升对生活品质的美好追求，向地外空间要资源、要发展是人类文明进程的必然选择，而月球丰富的氦-3等稀有气体、海量的金属和非金属矿产资源以及极区水冰等战略资源，如同月球“馈赠”我们人类的福利一样，随着人类航天技术的发展和科学认识的不断升华，开发利用月球必然是我们的首选目标，地-月大经济圈也必将是我们人类规模化走出地球、享受太空旅游的第一站。

6.3　国际月球探测未来发展格局

月球是地球的唯一天然卫星，是离地球最近的天体，也是人类走进深空的第一站。月球，因其独有的位置资源、环境资源和物质资源，已成为各国深空科学探测活动的最重要的规划。

20 世纪 50 年代末，随着苏联成功发射第一个月球探测器，标志着以月球为主目标天体的太空时代正式到来。这一事件随即引发了近 20 年的美、苏两国竞赛格局，形成了人类首次探月高潮。频繁的探测活动为世界留下了大量弥足珍贵的空间科学、空间技术、空间应用遗产。随后的探月活动进入了近 20 年的“寂静期”和月球科学认知的大发展期。人类在这一时期极大地丰富了对包括月球在内的太阳系及其各类天体的认识。进入 20 世纪 90 年代后，随着科学认知的不断升华和航天技术的迅猛发展，特别是月球极区永久阴影区存在水冰的间接证据的发现，国际上再次掀起了新的探月高潮。美国在这一时期引领了整个探月活动，并激发了包括中国在内的多国探月热情。经过一段时间酝酿，欧空局、日本、印度和我国等纷纷在进入新千年后启动了各自的探月计划。目前，真正有影响力并具备独立开展探月活动的国家和地区主要集中在美国、中国、欧空局、印度、俄罗斯和日本。

中国月球探测工程有序推进和一路凯歌，特别是嫦娥四号任务实现人类首个在月球背面的着陆巡视探测、嫦娥五号月球样品研究结果将月球岩浆活动年龄“延寿”近 10 亿年撼动了月球演化理论、嫦娥六号首次从月球背面取回样品等一系列进展，在某种意义上促进了各航天大国特别是美国的探月方略及其发展进程。2017 年，时任美国总统的特朗普签署了《太空政策 1 号指令》，要求美国航天局整合资源，加快美国探月步伐。2019 年，美国“阿尔忒弥斯计划”正式出台，宣布其载人登月时间提前至 2024 年，该计划意在月球南极构建“阿尔忒弥斯大本营”。目前国际上已有 44 个国家和组织签署了与阿尔忒弥斯计划的合作协议。

欧洲空间局在 2015 年提出了国际共建“月球村”的设想，并于 2022 年 7 月发布了新版空间探索路线图——《新世界 2030＋战略路线图》，但因受其特殊组织架构等因素的影响，欧空局的主要航天成员国都签署了《阿尔忒弥斯协议》，欧空局将与美国在阿尔忒弥斯任务中开展合作。

俄罗斯有完整的深空探测规划，希望通过其月球 25 号至 28 号的实施，提振民族自信心和凝聚力。遗憾的是，月球 25 号着陆失败，对俄罗斯航天界的打击非常大，其后续的月球 26 号、月球 27 号和月球 28 号任务能否如期执行尚难意料。

日本的航天实力很强，其深空探测以发展技术为主旨目标，“小而精”为其最大特点，围绕一个独特的创新目标进行突破，成绩也很斐然。对外合作上，日本与美国、欧空局、印度等开展了广泛的合作。日本的探月活动也启动很早，20 世纪 90 年代就发射了首个飞越月球的“羽衣号”，但其真正意义上的探月任务是于 2007 年初发射的月神 1 号。从其发展规划看，日本未来的探月活动是以参与阿尔忒弥斯计划为发展方向。

印度作为国际航天大国和深空探测领域“新秀”，其探月活动与我国探月工程几乎同步开展。2008 年，印度成功发射月船一号。2019 年，月船二号着陆失败。2023 年 8 月 23 日，月船三号成功着陆月球南极，大大增强了其民族自信力，也必将加快其探月计划的发展进程。

总之，未来的探月活动主要是以美国主导的阿尔忒弥斯计划和中国倡导的国际月球科研站为主。美国、中国、俄罗斯、印度和日本将引领未来较长时间里深空探测工程任务，欧洲空间局各航天成员国将深度参与，实施科学研究、技术试验和资源利用将是各国开展探月的主发展目标，而月球南极将成为未来探月活动主要的集聚地。

6.4 继续书写好中国探月故事是我们不懈的追求

“绕落回”三步走的大圆满、六战六捷的辉煌战果、月背就位探测先河的实现、新发现与系列原创成果的涌现，无不展示了我们嫦娥人的智慧，更是凸显了中国探月一张蓝图绘到底的决心和信心。

早在实施嫦娥四号和五号任务期间，中国在 2015 年就倡导共建共用共享“国际月球科研站”的发展理念，通过几年的宣介、推动以及嫦娥四号开展的实质性合作所取得的很好的成效和国际社会的响应，于 2021 年 6 月正式对外公布了共建国际月球科研站的发展路线图。

未来，我们探月的主任务将是在月球上构建服务于全人类科学探索、技术试验和资源利用的科研站，实现全月面可达和开展各类科研作业、资源开发并建立可在月面发射到更遥远深空开展科学活动的中转站。

未来，我们将利用月球科研站，进一步开展对月球更全、更深、更细的科学探测和研究，揭示月球重大地质事件的时空分布特征与形成机制、月球发动机、热历史等关键科学难题，实现“透明月球”并提升对月球的全新认知。

未来，我们将以月球为观测基地，利用其无大气、低背景辐射、遮挡地球干扰等绝佳物质资源，建设全波段、大规模、可升级、长寿命、长期连续运行的月基巡天观测平台，开展天文巡天观测研究，探寻宇宙黑暗时代和黎明之谜，以及解答“人类在宇宙中是否孤独”等重大科学命题。

未来，我们将利用月球低重力、无磁场、强辐射等独特环境资源，开展生命科学、基础物理学、材料科学等基础学科的实验研究，期望在极端环境下解开基础学科的一些科学疑团。

未来，我们还将利用月球科研站这一独特的平台，开展对太阳、地球、月球三位一体的“太阳三高、月球三源、地球三层”在同一平台、同一时段的大视场长周期的观测研究，实现日-地-月空间物理规律的新认识。

未来，我们还将依托科研站，开展月球能源、金属和非金属矿产、水等物质资源的开发利用，构建地-月大经济圈，实现人们对未来美好生活追求的愿望。

后　语

千山我独行、万籁懂曲章。2019 年 1 月 3 日，嫦娥四号成功实现了第一个造访月球背面的壮举，彰显了嫦娥人勇闯无人区的勇气和底气。那么，嫦娥四号探测区到底有何“曲章”？其蕴意又如何？从月球深部物质与“天外来客”的发现，到表面光谱的剖析和物质成分特性的解读；从形貌构造的精细刻画与撞击坑的相对定年，到地质演化历史的再造与撞击效应的厘定；从粒子辐射剂量的测量与中性原子特性的解译，到太空风化作用效应的演绎与月球微磁层成因理论的构建，无不绽放出中国科学家的智慧，更是印证了习近平总书记所说的“探月工程成果凝结着我国几代航天人的智慧和心血，从一个侧面展示了我们这些年在科技自立自强上取得的显著成就，充分展现了中国人的志气、骨气和底气”。

“任务带学科，学科促任务”，中国航天先驱赵九章先生如是说。如果说，嫦娥一号、二号卫星绕月科学探测所取得的科学认知，相比于美国、苏联乃至欧空局、日本的绕月探测所取得的科学成果，还微不足道，那是因为其时中国行星科学队伍还很弱小；如果说，嫦娥三号着陆在虹湾区，震惊国际航天界，但相比于 20 世纪 60—70 年代的美国、苏联着陆探测，我们的科学进展仍很不足；嫦娥四号不但实现了工程技术上的敢为人先，而且在科学无人区上取得了可喜进展，既是中国智慧对月球与行星科学发展的贡献，更昭示了嫦娥工程推动着中国月球与行星科学队伍不断成长逐步成熟的历史必然。

“嫦娥既是中国的，又属于全人类。”中国探月工程一直秉承着“外层空间是人类的共同疆域，空间探索是人类的共同事业。探月工程始终秉持平等互利、和平利用、合作共赢的原则。嫦娥四号任务中，德国中性原子探测仪、瑞典中子与辐射剂量探测仪、荷兰低频射电频谱仪等国际科学载荷的成功搭载，把中国嫦娥工程国际合作推向了高潮，而与合作国科学家共同研究发现的“月球背面粒子辐射是火星表面和空间站内部 2～3 倍、地球表面 300 倍等”的系列科研成果，是合作则共赢的鲜活例证。

路且长，唯奋斗方可圆梦。虽然人类对月球已经开展了 130 多次的探测活动，获取了对月球丰富的科学认知；虽然嫦娥四号开创了人类探月史在月球背面上着陆巡视探测的先河，也获得了对月球背面的一些新认识，但这还远远不够。月球，作为地球唯一的天然卫星，依然存在许许多多我们未知的科学谜团，需要我们人类共同去探索研究，而月球独特的空间位置、环境以及丰富的各类资源，又是天然的开展巡天观地、揭示宇宙奥秘和更好提升我们人类生活质量的理想场所和资源宝库，更是作为中转站飞往更遥远星际探测的不二选择。而正在实施的并作为中国倡导的国际月球科研站基本型的嫦娥七号和嫦娥八号任务，以及载人月球探测工程、国际月球科研站，就是我们“蓝图绘就、奋斗圆梦”的具体行动举措。

本书收集了过去几年来嫦娥四号所取得的主要科研成果。一方面是为了便于广大科技工作者查询了解最近的研究进展，另一方面是为了让受众更好地熟知当代中国嫦娥奔月的故事，更为重要的是希望有更多的青年学子树立热爱科学、潜心科研、奉献中国航天事业与行星科学的精神。